DERIVATIVES AND INTEGRALS

Basic Differentiation Rules

1. $\frac{d}{dx}[cu] = cu'$
2. $\frac{d}{dx}[u \pm v] = u' \pm v'$
3. $\frac{d}{dx}[uv] = uv' + vu'$
4. $\frac{d}{dx}\left[\frac{u}{v}\right] = \frac{vu' - uv'}{v^2}$
5. $\frac{d}{dx}[c] = 0$
6. $\frac{d}{dx}[u^n] = nu^{n-1}u'$
7. $\frac{d}{dx}[x] = 1$
8. $\frac{d}{dx}[|u|] = \frac{u}{|u|}(u'), \quad u \neq 0$
9. $\frac{d}{dx}[\ln u] = \frac{u'}{u}$
10. $\frac{d}{dx}[e^u] = e^u u'$
11. $\frac{d}{dx}[\log_a u] = \frac{u'}{(\ln a)u}$
12. [illegible] $)a^u u'$
13. $\frac{d}{dx}[\sin u] = (\cos u)u'$
14. $\frac{d}{dx}[\cos u] = -(\mathrm{s}$ [illegible]
15. [illegible] $c^2\, u)u'$
16. $\frac{d}{dx}[\cot u] = -(\csc^2 u)u'$
17. $\frac{d}{dx}[\sec u] = (\sec u$ [illegible]
18. [illegible] $\mathrm{sc}\, u \cot u)u'$
19. $\frac{d}{dx}[\arcsin u] = \frac{u'}{\sqrt{1 - u^2}}$
20. $\frac{d}{dx}[\arccos u] = \frac{-}{\sqrt{1}}$ [illegible]
21. $\frac{d}{dx}[\arctan u] = \frac{u'}{1 + u^2}$
22. $\frac{d}{dx}[\operatorname{arccot} u] = \frac{-u'}{1 + u^2}$
23. $\frac{d}{dx}[\operatorname{arcsec} u] = \frac{u'}{|u|\sqrt{u^2 - 1}}$
24. $\frac{d}{dx}[\operatorname{arccsc} u] = \frac{-u'}{|u|\sqrt{u^2 - 1}}$
25. $\frac{d}{dx}[\sinh u] = (\cosh u)u'$
26. $\frac{d}{dx}[\cosh u] = (\sinh u)u'$
27. $\frac{d}{dx}[\tanh u] = (\operatorname{sech}^2 u)u'$
28. $\frac{d}{dx}[\coth u] = -(\operatorname{csch}^2 u)u'$
29. $\frac{d}{dx}[\operatorname{sech} u] = -(\operatorname{sech} u \tanh u)u'$
30. $\frac{d}{dx}[\operatorname{csch} u] = -(\operatorname{csch} u \coth u)u'$
31. $\frac{d}{dx}[\sinh^{-1} u] = \frac{u'}{\sqrt{u^2 + 1}}$
32. $\frac{d}{dx}[\cosh^{-1} u] = \frac{u'}{\sqrt{u^2 - 1}}$
33. $\frac{d}{dx}[\tanh^{-1} u] = \frac{u'}{1 - u^2}$
34. $\frac{d}{dx}[\coth^{-1} u] = \frac{u'}{1 - u^2}$
35. $\frac{d}{dx}[\operatorname{sech}^{-1} u] = \frac{-u'}{u\sqrt{1 - u^2}}$
36. $\frac{d}{dx}[\operatorname{csch}^{-1} u] = \frac{-u'}{|u|\sqrt{1 + u^2}}$

Basic Integration Formulas

1. $\int kf(u)\, du = k\int f(u)\, du$
2. $\int [f(u) \pm g(u)]\, du = \int f(u)\, du \pm \int g(u)\, du$
3. $\int du = u + C$
4. $\int a^u\, du = \left(\frac{1}{\ln a}\right)a^u + C$
5. $\int e^u\, du = e^u + C$
6. $\int \sin u\, du = -\cos u + C$
7. $\int \cos u\, du = \sin u + C$
8. $\int \tan u\, du = -\ln|\cos u| + C$
9. $\int \cot u\, du = \ln|\sin u| + C$
10. $\int \sec u\, du = \ln|\sec u + \tan u| + C$
11. $\int \csc u\, du = -\ln|\csc u + \cot u| + C$
12. $\int \sec^2 u\, du = \tan u + C$
13. $\int \csc^2 u\, du = -\cot u + C$
14. $\int \sec u \tan u\, du = \sec u + C$
15. $\int \csc u \cot u\, du = -\csc u + C$
16. $\int \frac{du}{\sqrt{a^2 - u^2}} = \arcsin \frac{u}{a} + C$
17. $\int \frac{du}{a^2 + u^2} = \frac{1}{a} \arctan \frac{u}{a} + C$
18. $\int \frac{du}{u\sqrt{u^2 - a^2}} = \frac{1}{a} \operatorname{arcsec} \frac{|u|}{a} + C$

Definition of the Six Trigonometric Functions

Right triangle definitions, where $0 < \theta < \pi/2$.

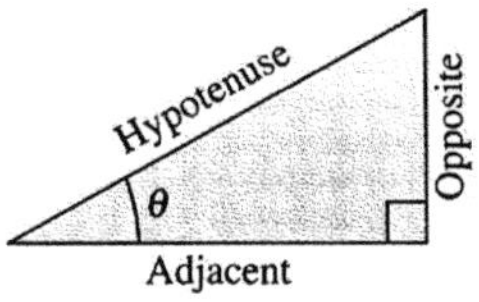

$$\sin\theta = \frac{\text{opp}}{\text{hyp}} \quad \csc\theta = \frac{\text{hyp}}{\text{opp}}$$

$$\cos\theta = \frac{\text{adj}}{\text{hyp}} \quad \sec\theta = \frac{\text{hyp}}{\text{adj}}$$

$$\tan\theta = \frac{\text{opp}}{\text{adj}} \quad \cot\theta = \frac{\text{adj}}{\text{opp}}$$

Circular function definitions, where θ is any angle.

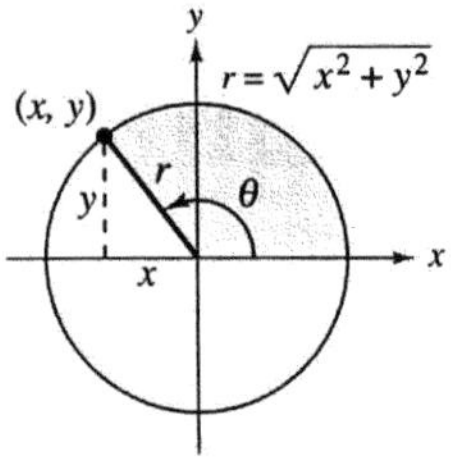

$$\sin\theta = \frac{y}{r} \quad \csc\theta = \frac{r}{y}$$

$$\cos\theta = \frac{x}{r} \quad \sec\theta = \frac{r}{x}$$

$$\tan\theta = \frac{y}{x} \quad \cot\theta = \frac{x}{y}$$

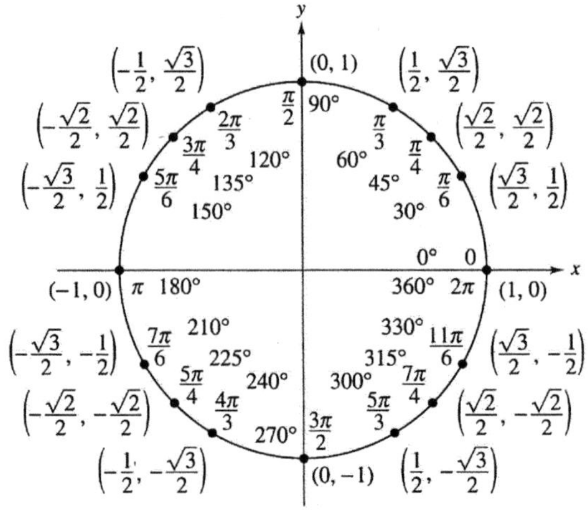

Reciprocal Identities

$$\sin x = \frac{1}{\csc x} \quad \sec x = \frac{1}{\cos x} \quad \tan x = \frac{1}{\cot x}$$

$$\csc x = \frac{1}{\sin x} \quad \cos x = \frac{1}{\sec x} \quad \cot x = \frac{1}{\tan x}$$

Tangent and Cotangent Identities

$$\tan x = \frac{\sin x}{\cos x} \quad \cot x = \frac{\cos x}{\sin x}$$

Pythagorean Identities

$$\sin^2 x + \cos^2 x = 1$$

$$1 + \tan^2 x = \sec^2 x \qquad 1 + \cot^2 x = \csc^2 x$$

Cofunction Identities

$$\sin\left(\frac{\pi}{2} - x\right) = \cos x \quad \cos\left(\frac{\pi}{2} - x\right) = \sin x$$

$$\csc\left(\frac{\pi}{2} - x\right) = \sec x \quad \tan\left(\frac{\pi}{2} - x\right) = \cot x$$

$$\sec\left(\frac{\pi}{2} - x\right) = \csc x \quad \cot\left(\frac{\pi}{2} - x\right) = \tan x$$

Reduction Formulas

$$\sin(-x) = -\sin x \quad \cos(-x) = \cos x$$

$$\csc(-x) = -\csc x \quad \tan(-x) = -\tan x$$

$$\sec(-x) = \sec x \quad \cot(-x) = -\cot x$$

Sum and Difference Formulas

$$\sin(u \pm v) = \sin u \cos v \pm \cos u \sin v$$

$$\cos(u \pm v) = \cos u \cos v \mp \sin u \sin v$$

$$\tan(u \pm v) = \frac{\tan u \pm \tan v}{1 \mp \tan u \tan v}$$

Double-Angle Formulas

$$\sin 2u = 2 \sin u \cos u$$

$$\cos 2u = \cos^2 u - \sin^2 u = 2\cos^2 u - 1 = 1 - 2\sin^2 u$$

$$\tan 2u = \frac{2 \tan u}{1 - \tan^2 u}$$

Power-Reducing Formulas

$$\sin^2 u = \frac{1 - \cos 2u}{2}$$

$$\cos^2 u = \frac{1 + \cos 2u}{2}$$

$$\tan^2 u = \frac{1 - \cos 2u}{1 + \cos 2u}$$

Sum-to-Product Formulas

$$\sin u + \sin v = 2 \sin\left(\frac{u + v}{2}\right) \cos\left(\frac{u - v}{2}\right)$$

$$\sin u - \sin v = 2 \cos\left(\frac{u + v}{2}\right) \sin\left(\frac{u - v}{2}\right)$$

$$\cos u + \cos v = 2 \cos\left(\frac{u + v}{2}\right) \cos\left(\frac{u - v}{2}\right)$$

$$\cos u - \cos v = -2 \sin\left(\frac{u + v}{2}\right) \sin\left(\frac{u - v}{2}\right)$$

Product-to-Sum Formulas

$$\sin u \sin v = \frac{1}{2}[\cos(u - v) - \cos(u + v)]$$

$$\cos u \cos v = \frac{1}{2}[\cos(u - v) + \cos(u + v)]$$

$$\sin u \cos v = \frac{1}{2}[\sin(u + v) + \sin(u - v)]$$

$$\cos u \sin v = \frac{1}{2}[\sin(u + v) - \sin(u - v)]$$

CALCULUS

with Analytic Geometry

EIGHTH EDITION

RON LARSON • ROBERT P. HOSTETLER
The Pennsylvania State University
The Behrend College

BRUCE H. EDWARDS
University of Florida

HOUGHTON MIFFLIN COMPANY
Boston New York

Vice President and Publisher: Jack Shira
Associate Sponsoring Editor: Cathy Cantin
Development Manager: Maureen Ross
Senior Development Manager: Claire Boivin
Editorial Assistant: Elizabeth Kassab
Supervising Editor: Karen Carter
Senior Project Editor: Patty Bergin
Editorial Assistant: Allison Seymour/Julia Keller
Production Technology Supervisor: Gary Crespo
Executive Marketing Manager: Michael Busnach
Senior Marketing Manager: Danielle Potvin
Marketing Coordinator: Nicole Mollica
Senior Manufacturing Manager: Priscilla Bailey

We have included examples and exercises that use real-life data as well as technology output from a variety of software. This would not have been possible without the help of many people and organizations. Our wholehearted thanks goes to all for their time and effort.

Custom Publishing Editor: Todd Corbin
Custom Publishing Production Manager: Kathleen McCourt
Project Coordinator: Anisha Sandhu

Cover Design: Galen B. Murphy
Cover Art: Photodisc

This book contains select works from existing Houghton Mifflin Company resources and was produced by Houghton Mifflin Custom Publishing for collegiate use. As such, those adopting and/or contributing to this work are responsible for editorial content, accuracy, continuity and completeness.

Printed in the United States of America.

ISBN: 0-618-63969-1
N-04847

4 5 6 7 8 9 – ACME – 07 06 05

Houghton Mifflin
Custom Publishing

222 Berkeley Street • Boston, MA 02116

Address all correspondence and order information to the above address.

Contents

Chapter 11 Vectors and the Geometry of Space 761

Chapter 12 Vector-Valued Functions 831

Chapter 13 Functions of Several Variables 883

Additional Appendices

A Word from the Authors

Welcome to *Calculus with Analytic Geometry*, Eighth Edition. Much has changed since we wrote the first edition, which we began writing in 1973—over 30 years ago. With each edition, we have listened to you, our users, and incorporated many of your suggestions for improvement.

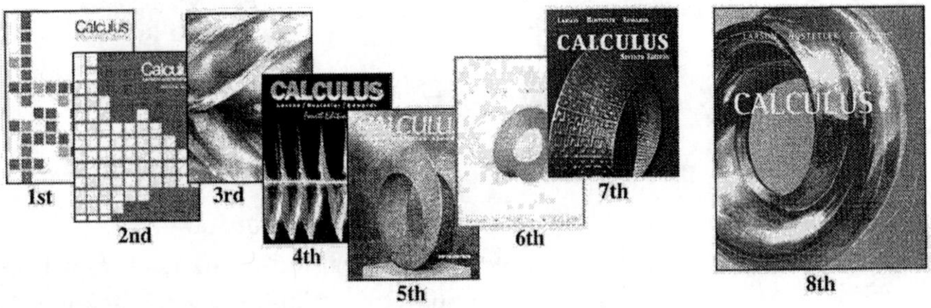

A Text Formed by Its Users

Through your support and suggestions, the text has evolved over eight editions to include these extensive enhancements:

- Comprehensive exercise sets containing a wide variety of problems such as skill-building exercises, applications, explorations, writing exercises, critical thinking exercises, and theoretical problems
- Abundant real-life applications that accurately represent the diverse uses of calculus
- Many open-ended activities and investigations
- Clear, uncluttered text presentation with full annotations and labels and a carefully planned page layout
- Comprehensive, four-color art program
- Comprehensive and mathematically rigorous text (The third semester of the Eighth Edition, in particular, is quite different when compared with the third semester of the First Edition.)
- Technology used throughout as both a problem-solving tool and an investigative tool
- A comprehensive program of additional resources available in print, on CD-ROM, and online
- With eight different volumes of the text available, you can choose the sequence, amount of content, and teaching approach that is best for you and your students (see pages xii–xiii)
- References to the history of calculus and to the mathematicians who developed it, including over 50 biographical sketches available on the HM mathSpace® Student CD-ROM
- References to over 50 articles from mathematical journals are available at *www.MathArticles.com*

What's New and Different in the Eighth Edition

In the Eighth Edition, we continue to offer instructors and students a text that is pedagogically sound, mathematically precise, and still comprehensible. There are many changes in the mathematics, prose, art, and design; the more significant changes are noted here.

- ***New Chapter Openers*** Each Chapter Opener has two parts: a description of the concepts that are covered in the chapter and a thought-provoking question about a real-life application from the chapter.
- ***New Introduction to Differential Equations*** The topic of differential equations is now introduced in Chapter 6 in the first semester of calculus, to better prepare students for their courses in disciplines such as engineering, physics, and chemistry. The chapter contains four sections: 6.1 *Slope Fields and Euler's Method*, 6.2 *Differential Equations: Growth and Decay*, 6.3 *Separation of Variables and the Logistic Equation*, and 6.4 *First-Order Linear Differential Equations*.
- **Revised Exercise Sets** The exercise sets have been carefully and extensively examined to ensure they are rigorous and cover all topics suggested by our users. Many new skill-building and challenging exercises have been added.
- **Updated Data** All data in the examples and exercise sets have been updated.
- **EduSpace®** *Eduspace®* combines numerous dynamic resources with online homework and testing materials to create a comprehensive online learning system. Students benefit from having immediate access to algorithmic tutorial practice, videos, and resources such as a color graphing calculator. Instructors benefit from time-saving grading resources, as well as dynamic instructional tools such as animations, explorations, and Computer Algebra System Labs.
- ***Student Study Guide*** The worked-out solutions to the odd-numbered text exercises are now provided in a printed supplement, on a CD-ROM, in *Eduspace®*, and at *www.CalcChat.com.*

Although we carefully and thoroughly revised the text by enhancing the usefulness of some features and topics and by adding others, we did not change many of the things that our colleagues and the over two million students who have used this book have told us work for them. *Calculus*, Eighth Edition, offers comprehensive coverage of the material required by students in a three-semester or four-quarter calculus course, including carefully stated theories and proofs.

We hope you will enjoy the Eighth Edition. We welcome any comments, as well as suggestions for continued improvement.

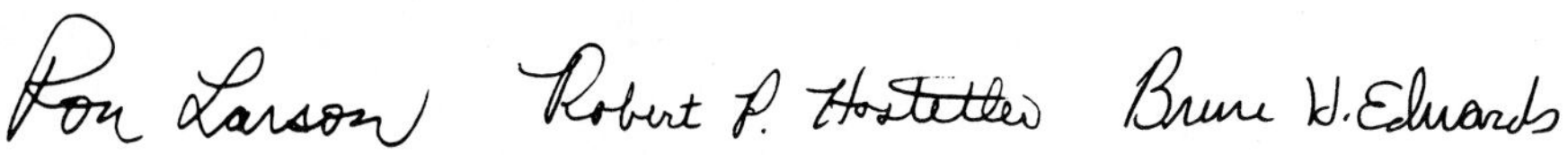

Integrated Learning System for Calculus

Over 25 Years of Success, Leadership, and Innovation

The best-selling program by Larson, Hostetler, and Edwards continues to offer instructors and students more flexible teaching and learning options for the calculus course.

Calculus Textbook Options

CALCULUS

The traditional calculus course is available in a variety of textbook configurations to address the different ways instructors teach—and students take—their classes.

Calculus I, II, and III

Calculus I and II

Calculus III

Calculus I

Calculus II

Calculus III

Also available for the *Calculus*, Eighth Edition, program by Larson, Hostetler, and Edwards

- Eduspace® online learning system
- HM mathSpace® Student CD-ROMs
- Instructional DVDs and videotapes

For more information on these—and more—electronic course materials, please turn to pages xv-xvii.

Also available from the Larson, Hostetler, and Edwards* Calculus *Author Team

CALCULUS WITH PRECALCULUS

To give more students access to calculus by easing the transition from precalculus, the following textbook sequence is available.

Precalculus and Calculus I

Calculus II

Calculus III

CALCULUS WITH EARLY TRANSCENDENTAL FUNCTIONS

For instructors who prefer to introduce the transcendental functions early in the calculus course, the following textbook sequence is available.

Calculus I, II, and III

Calculus I and II

Calculus III

CALCULUS WITH LATE TRIGONOMETRY

For instructors who introduce the trigonometric functions in the second semester, the following textbook is available.

Calculus I, II, and III

Integrated Learning System for Calculus

Comprehensive Calculus Resources

The Integrated Learning System for *Calculus*, Eighth Edition, addresses the changing needs of today's instructors and students. Recognizing that the calculus course is presented in a variety of teaching and learning environments, we offer extensive resources that support the textbook program in print, CD-ROM, and online formats.

- Online homework practice
- Testing
- Tutoring
- Graded homework
- Classroom management
- Online course
- Interactive resources

ONE
INTEGRATED LEARNING SYSTEM

The teaching and learning resources you need in the format you prefer

The Integrated Learning System for *Calculus*, Eighth Edition, offers dynamic teaching tools for instructors and interactive learning resources for students in the following flexible course delivery formats.

- Eduspace® online learning system
- HM mathSpace® Student CD-ROM
- Instructional DVDs and Videotapes
- HM ClassPrep with HM Testing CD-ROM
- Companion Textbook Websites
- Printed Resources

Enhanced! Eduspace® Online Calculus

Eduspace®, powered by Blackboard®, is ready to use and easy to integrate into the calculus course. It provides comprehensive homework exercises, tutorials, and testing keyed to the textbook by section.

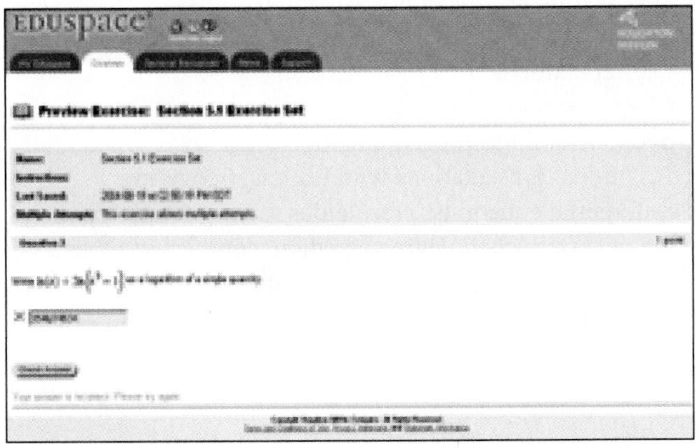

Features

- Algorithmically generated tutorial questions for unlimited practice
- Comprehensive exercise sets for graded homework
- Point-of-use links to additional tools, animations, simulations, videos, and exercises
- SMARTHINKING™ live, online tutoring for students
- Color graphing calculator
- Ample prerequisite skills review with customized student self-study plan
- Chapter tests
- Link to CalcChat
- Electronic version of all textbook exercises
- Links to detailed, stepped-out solutions to odd-numbered textbook exercises

Enhanced! HM mathSpace® CD-ROM

For the student, HM mathSpace® Student CD-ROM offers a wealth of learning resources keyed to the textbook by section.

Features

- Algorithmically generated tutorial questions for unlimited practice of prerequisite skills
- Point-of-use links to additional tools, animations, and simulations
- Link to CalcChat
- Color graphing calculator
- Chapter tests

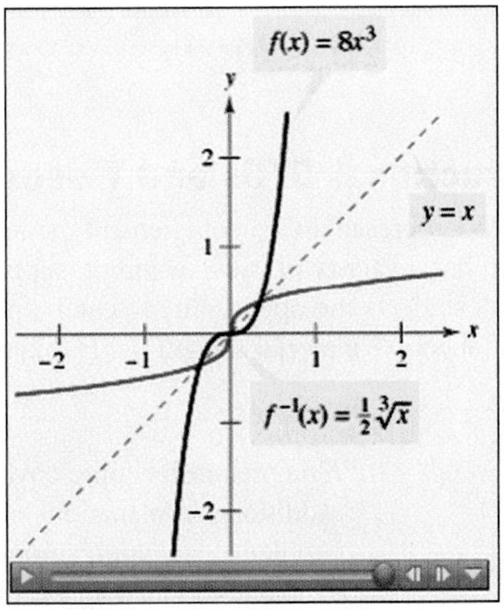

For additional information about the Larson, Hostetler, and Edwards Calculus program, go to *college.hmco.com/info/larsoncalculus*.

New! HM ClassPrep with HM Testing CD-ROM

This valuable CD-ROM contains an array of useful instructor resources keyed to the textbook.

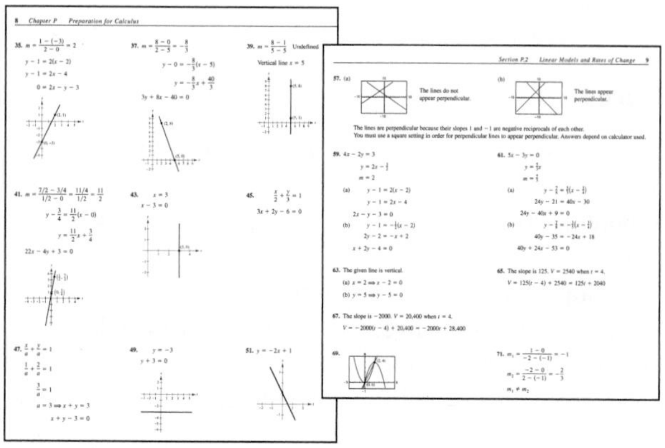

Features

- HM Testing test generator
- Solutions to all textbook exercises
- Digital textbook art
- Chapter summaries
- Section-level teaching strategies
- Additional presentations with exercises covering differential equations, precalculus review, rotation and the general second degree equation, complex numbers, and business and economic applications
- Graphing calculator programs

New! HM Testing

For the instructor, HM Testing is a robust test-generating system.

Features

- Comprehensive set of algorithmic test items
- Can produce chapter tests, cumulative tests, and final exams
- Online testing
- Gradebook function

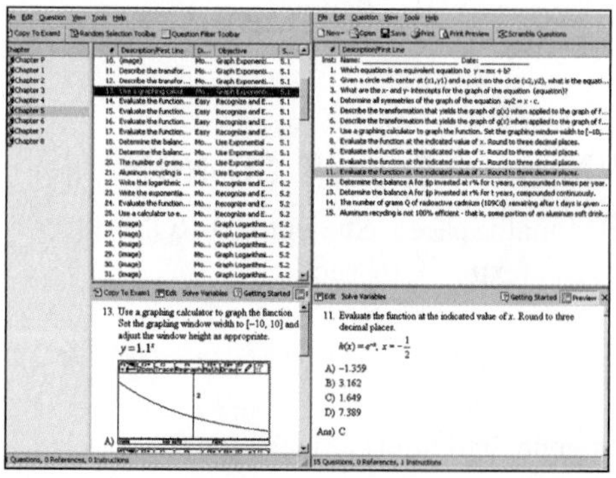

Enhanced! Instructional DVDs and Videos

These comprehensive video presentations complement the textbook topic coverage. This video presentation has a variety of uses, including supplementing an online or hybrid course, giving students the opportunity to catch up if they miss a class, and providing substantial course material for self-study and review.

Features

- Comprehensive topic coverage from Calculus I, II, and III
- Additional explanations of calculus concepts, sample problems, and applications
- Available in video and DVD format

Enhanced! Companion Textbook Website

The free Houghton Mifflin website at *math.college.hmco.com* contains an abundance of instructor and student resources.

Features

- Downloadable graphing calculator programs
- Textbook Appendices C–G, containing additional presentations with exercises covering differential equations, precalculus review, rotation and the general second-degree equation, complex numbers, and business and economic applications
- Algebra Review Summary
- Calculus Labs
- 3-D rotatable graphs

Printed Resources

For the convenience of both instructors and students, many of our teaching and learning resources are available as printed supplements, but are also available in electronic format.

For instructors

Complete Solutions Guide by Bruce Edwards

This instructor's resource contains worked-out solutions to all textbook exercises. It is available in three volumes: Volume I covers Chapters P–6, Volume II covers Chapters 7–11, and Volume III covers Chapters 11–15.

Instructor's Resource Manual by Ann Rutledge Kraus

This instructor's resource contains an abundance of resources keyed to the textbook by chapter and section, including chapter summaries, teaching strategies, multiple versions of chapter tests, final exams, and gateway tests, and suggested solutions to the Chapter Openers, Explorations, Section Projects, and Technology features in the text.

Test Item File

The *Test Item File* contains a sample question for every algorithm in HM Testing.

For students

Study and Solutions Guide by Bruce Edwards

This student resource contains detailed, worked-out solutions to all odd-numbered textbook exercises. It is available in two volumes: Volume I covers Chapters P–11 and Volume II covers Chapters 11–15.

For additional information about the Larson, Hostetler, and Edwards Calculus program, go to *college.hmco.com/info/larsoncalculus*.

Features

Chapter Openers

Each chapter opens with a real-life application of the concepts presented in the chapter, illustrated by a photograph. Open-ended and thought-provoking questions about the application encourage the student to consider how calculus concepts relate to real-life situations. A brief summary with a graphical component highlights the primary mathematical concepts presented in the chapter, and explains why they are important.

5

Logarithmic, Exponential, and Other Transcendental Functions

$\int_1^1 \frac{1}{t}\,dt = \ln 1 = 0$

$\int_1^{3/2} \frac{1}{t}\,dt = \ln \frac{3}{2} \approx 0.41$

$\int_1^2 \frac{1}{t}\,dt = \ln 2 \approx 0.69$

$\int_1^3 \frac{1}{t}\,dt = \ln 3 \approx 1.10$

In Chapter 5, you will see how the function $f(x) = 1/x$ can be used to define the natural logarithmic function. To do this, consider the definite integral

$$\int_1^x \frac{1}{t}\,dt.$$

When $x < 1$, the value of this definite integral is negative. When $x = 1$, the value is 0. When $x > 1$, the value is positive.

A geyser is a hot spring that erupts periodically when groundwater in a confined space boils and produces steam. The steam forces overlying water up and out through an opening on Earth's surface. The temperature at which water boils is affected by pressure. Do you think an *increase* or a *decrease* in pressure causes water to boil at a lower temperature? Why?

Brian Maslyar/Index Stock

321

Section Openers

Every section begins with an outline of the key concepts covered in the section. This serves as a class planning resource for the instructor and a study and review guide for the student.

Explorations

For selected topics, Explorations offer the opportunity to discover calculus concepts before they are formally introduced in the text, thus enhancing student understanding. This optional feature can be omitted at the discretion of the instructor with no loss of continuity in the coverage of the material.

Historical Notes

Integrated throughout the text, Historical Notes help students grasp the basic mathematical foundations of calculus.

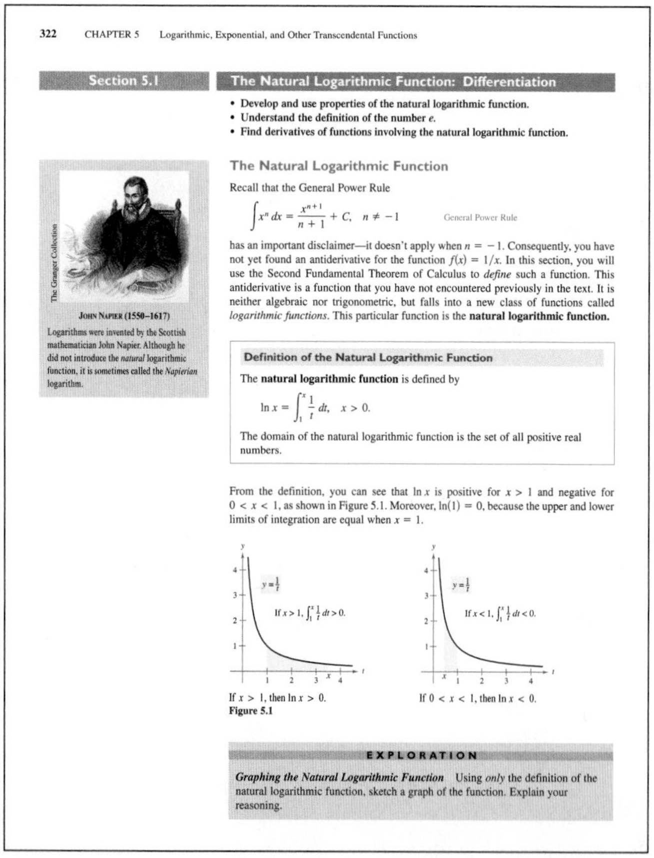

322 CHAPTER 5 Logarithmic, Exponential, and Other Transcendental Functions

Section 5.1 **The Natural Logarithmic Function: Differentiation**

- **Develop and use properties of the natural logarithmic function.**
- **Understand the definition of the number *e*.**
- **Find derivatives of functions involving the natural logarithmic function.**

The Granger Collection

JOHN NAPIER (1550–1617)

Logarithms were invented by the Scottish mathematician John Napier. Although he did not introduce the *natural* logarithmic function, it is sometimes called the *Napierian* logarithm.

The Natural Logarithmic Function

Recall that the General Power Rule

$$\int x^n\,dx = \frac{x^{n+1}}{n+1} + C, \quad n \neq -1 \qquad \text{General Power Rule}$$

has an important disclaimer—it doesn't apply when $n = -1$. Consequently, you have not yet found an antiderivative for the function $f(x) = 1/x$. In this section, you will use the Second Fundamental Theorem of Calculus to *define* such a function. This antiderivative is a function that you have not encountered previously in the text. It is neither algebraic nor trigonometric, but falls into a new class of functions called *logarithmic functions*. This particular function is the **natural logarithmic function.**

Definition of the Natural Logarithmic Function

The **natural logarithmic function** is defined by

$$\ln x = \int_1^x \frac{1}{t}\,dt, \quad x > 0.$$

The domain of the natural logarithmic function is the set of all positive real numbers.

From the definition, you can see that $\ln x$ is positive for $x > 1$ and negative for $0 < x < 1$, as shown in Figure 5.1. Moreover, $\ln(1) = 0$, because the upper and lower limits of integration are equal when $x = 1$.

$y = \frac{1}{t}$ If $x > 1$, $\int_1^x \frac{1}{t}\,dt > 0$.

If $x > 1$, then $\ln x > 0$.

$y = \frac{1}{t}$ If $x < 1$, $\int_1^x \frac{1}{t}\,dt < 0$.

If $0 < x < 1$, then $\ln x < 0$.

Figure 5.1

EXPLORATION

Graphing the Natural Logarithmic Function Using *only* the definition of the natural logarithmic function, sketch a graph of the function. Explain your reasoning.

342 CHAPTER 5 Logarithmic, Exponential, and Other Transcendental Functions

EXAMPLE 1 Verifying Inverse Functions

Show that the functions are inverse functions of each other.

$$f(x) = 2x^3 - 1 \quad \text{and} \quad g(x) = \sqrt[3]{\frac{x+1}{2}}$$

Solution Because the domains and ranges of both f and g consist of all real numbers, you can conclude that both composite functions exist for all x. The composition of f with g is given by

$$\begin{aligned} f(g(x)) &= 2\left(\sqrt[3]{\frac{x+1}{2}}\right)^3 - 1 \\ &= 2\left(\frac{x+1}{2}\right) - 1 \\ &= x + 1 - 1 \\ &= x. \end{aligned}$$

The composition of g with f is given by

$$\begin{aligned} g(f(x)) &= \sqrt[3]{\frac{(2x^3 - 1) + 1}{2}} \\ &= \sqrt[3]{\frac{2x^3}{2}} \\ &= \sqrt[3]{x^3} \\ &= x. \end{aligned}$$

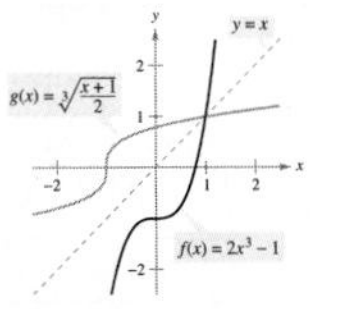

f and g are inverse functions of each other.
Figure 5.11

Because $f(g(x)) = x$ and $g(f(x)) = x$, you can conclude that f and g are inverse functions of each other (see Figure 5.11).

STUDY TIP In Example 1, try comparing the functions f and g verbally.

For f: First cube x, then multiply by 2, then subtract 1.

For g: First add 1, then divide by 2, then take the cube root.

Do you see the "undoing pattern"?

In Figure 5.11, the graphs of f and $g = f^{-1}$ appear to be mirror images of each other with respect to the line $y = x$. The graph of f^{-1} is a **reflection** of the graph of f in the line $y = x$. This idea is generalized in the following theorem.

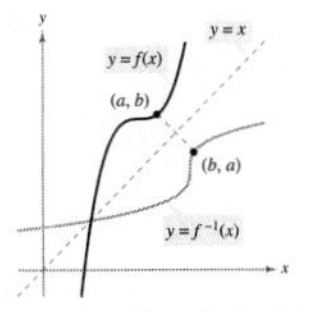

The graph of f^{-1} is a reflection of the graph of f in the line $y = x$.
Figure 5.12

THEOREM 5.6 Reflective Property of Inverse Functions

The graph of f contains the point (a, b) if and only if the graph of f^{-1} contains the point (b, a).

Proof If (a, b) is on the graph of f, then $f(a) = b$ and you can write

$$f^{-1}(b) = f^{-1}(f(a)) = a.$$

So, (b, a) is on the graph of f^{-1}, as shown in Figure 5.12. A similar argument will prove the theorem in the other direction.

Theorems

All Theorems and Definitions are highlighted for emphasis and easy reference. Proofs are shown for selected theorems to enhance student understanding.

Study Tip

Located at point of use throughout the text, Study Tips advise students on how to avoid common errors, address special cases, and expand upon theoretical concepts.

Graphics

Numerous graphics throughout the text enhance student understanding of complex calculus concepts (especially in three-dimensional representations), as well as real-life applications.

Example

To enhance the usefulness of the text as a study and learning tool, the Eighth Edition contains numerous Examples. The detailed, worked-out Solutions (many with side comments to clarify the steps or the method) are presented graphically, analytically, and/or numerically to provide students with opportunities for practice and further insight into calculus concepts. Many Examples incorporate real-data analysis.

Open Exploration

Eduspace® contains Open Explorations, which investigate selected Examples using computer algebra systems (*Maple*, *Mathematica*, *Derive*, and *Mathcad*). The icon identifies these Examples.

Notes

Instructional Notes accompany many of the Theorems, Definitions, and Examples to offer additional insights or describe generalizations.

SECTION 5.4 Exponential Functions: Differentiation and Integration 353

EXAMPLE 5 The Standard Normal Probability Density Function

Show that the *standard normal probability density function*

$$f(x) = \frac{1}{\sqrt{2\pi}} e^{-x^2/2}$$

has points of inflection when $x = \pm 1$.

Solution To locate possible points of inflection, find the x-values for which the second derivative is 0.

$$\begin{aligned} f(x) &= \frac{1}{\sqrt{2\pi}} e^{-x^2/2} && \text{Write original function.} \\ f'(x) &= \frac{1}{\sqrt{2\pi}} (-x) e^{-x^2/2} && \text{First derivative} \\ f''(x) &= \frac{1}{\sqrt{2\pi}} [(-x)(-x)e^{-x^2/2} + (-1)e^{-x^2/2}] && \text{Product Rule} \\ &= \frac{1}{\sqrt{2\pi}} (e^{-x^2/2})(x^2 - 1) && \text{Second derivative} \end{aligned}$$

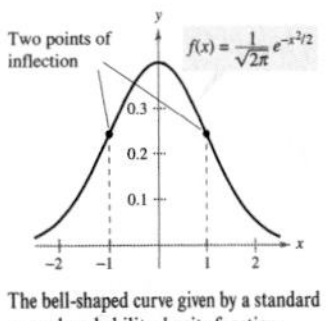

The bell-shaped curve given by a standard normal probability density function
Figure 5.22

So, $f''(x) = 0$ when $x = \pm 1$, and you can apply the techniques of Chapter 3 to conclude that these values yield the two points of inflection shown in Figure 5.22.

NOTE The general form of a normal probability density function (whose mean is 0) is given by

$$f(x) = \frac{1}{\sigma\sqrt{2\pi}} e^{-x^2/2\sigma^2}$$

where σ is the standard deviation (σ is the lowercase Greek letter sigma). This "bell-shaped curve" has points of inflection when $x = \pm\sigma$.

EXAMPLE 6 Shares Traded

The number y of shares traded (in millions) on the New York Stock Exchange from 1990 through 2002 can be modeled by

$$y = 36{,}663e^{0.1902t}$$

where t represents the year, with $t = 0$ corresponding to 1990. At what rate was the number of shares traded changing in 1998? *(Source: New York Stock Exchange, Inc.)*

Solution The derivative of the given model is

$$\begin{aligned} y' &= (0.1902)(36{,}663)e^{0.1902t} \\ &\approx 6973e^{0.1902t}. \end{aligned}$$

By evaluating the derivative when $t = 8$, you can conclude that the rate of change in 1998 was about

31,933 million shares per year.

The graph of this model is shown in Figure 5.23.

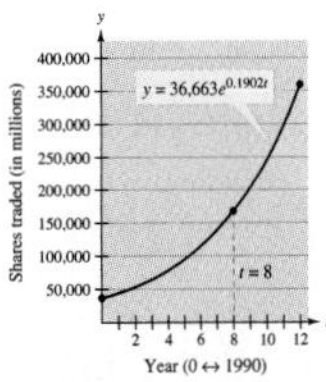

Figure 5.23

Exercises

The core of every calculus text, Exercises provide opportunities for exploration, practice, and comprehension. The Eighth Edition contains over 10,000 Section and Chapter Review Exercises, carefully graded in each set from skill-building to challenging. The extensive range of problem types includes true/false, writing, conceptual, real-data modeling, and graphical analysis.

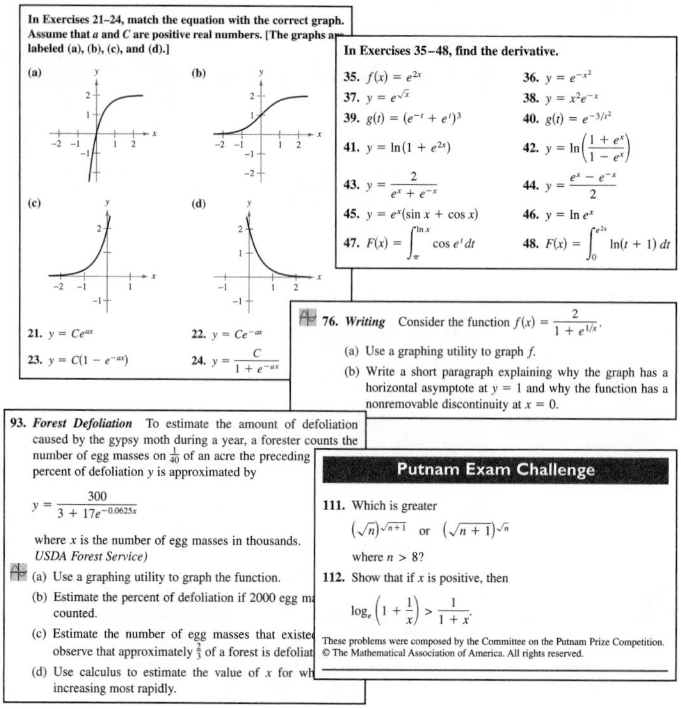
In Exercises 21–24, match the equation with the correct graph. Assume that a and C are positive real numbers. [The graphs are labeled (a), (b), (c), and (d).]

21. $y = Ce^{ax}$ 22. $y = Ce^{-ax}$

23. $y = C(1 - e^{-ax})$ 24. $y = \dfrac{C}{1 + e^{-ax}}$

In Exercises 35–48, find the derivative.

35. $f(x) = e^{2x}$ 36. $y = e^{-x^2}$

37. $y = e^{\sqrt{x}}$ 38. $y = x^2e^{-x}$

39. $g(t) = (e^{-t} + e^t)^3$ 40. $g(t) = e^{-3/t^2}$

41. $y = \ln(1 + e^{2x})$ 42. $y = \ln\left(\dfrac{1 + e^x}{1 - e^x}\right)$

43. $y = \dfrac{2}{e^x + e^{-x}}$ 44. $y = \dfrac{e^x - e^{-x}}{2}$

45. $y = e^x(\sin x + \cos x)$ 46. $y = \ln e^x$

47. $F(x) = \displaystyle\int_{\pi}^{\ln x} \cos e^t\,dt$ 48. $F(x) = \displaystyle\int_0^{e^{2x}} \ln(t + 1)\,dt$

76. *Writing* Consider the function $f(x) = \dfrac{2}{1 + e^{1/x}}$.

(a) Use a graphing utility to graph f.

(b) Write a short paragraph explaining why the graph has a horizontal asymptote at $y = 1$ and why the function has a nonremovable discontinuity at $x = 0$.

93. *Forest Defoliation* To estimate the amount of defoliation caused by the gypsy moth during a year, a forester counts the number of egg masses on $\frac{1}{40}$ of an acre the preceding ... percent of defoliation y is approximated by

$$y = \frac{300}{3 + 17e^{-0.0625x}}$$

where x is the number of egg masses in thousands. ... USDA Forest Service)

(a) Use a graphing utility to graph the function.

(b) Estimate the percent of defoliation if 2000 egg m... counted.

(c) Estimate the number of egg masses that existe... observe that approximately $\frac{2}{3}$ of a forest is defoliat...

(d) Use calculus to estimate the value of x for wh... increasing most rapidly.

Putnam Exam Challenge

111. Which is greater

$$(\sqrt{n})^{\sqrt{n+1}} \quad \text{or} \quad (\sqrt{n+1})^{\sqrt{n}}$$

where $n > 8$?

112. Show that if x is positive, then

$$\log_e\left(1 + \frac{1}{x}\right) > \frac{1}{1 + x}.$$

These problems were composed by the Committee on the Putnam Prize Competition. © The Mathematical Association of America. All rights reserved.

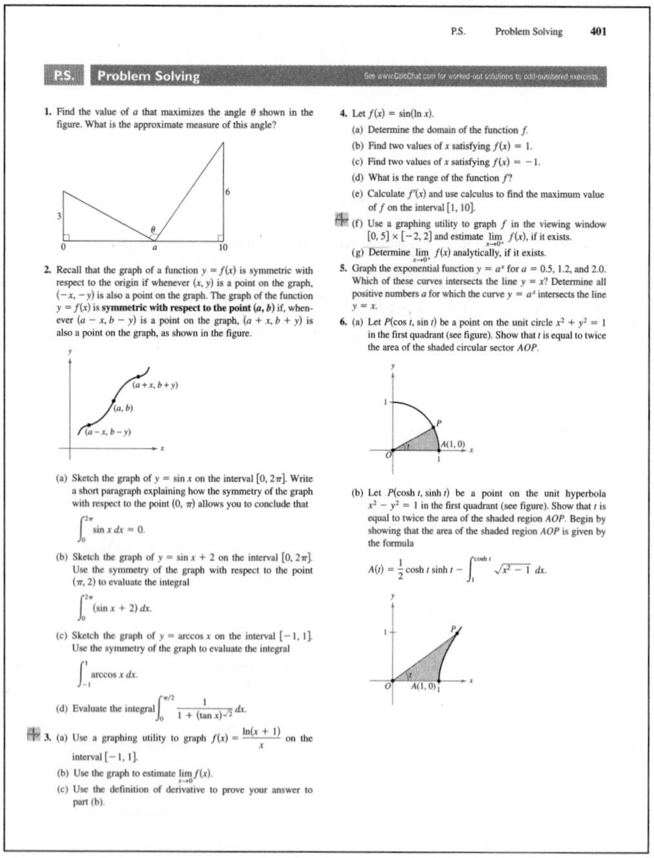
P.S. Problem Solving 401

P.S. Problem Solving See www.CalcChat.com for worked-out solutions to odd-numbered exercises.

1. Find the value of a that maximizes the angle θ shown in the figure. What is the approximate measure of this angle?

2. Recall that the graph of a function $y = f(x)$ is symmetric with respect to the origin if whenever (x, y) is a point on the graph, $(-x, -y)$ is also a point on the graph. The graph of the function $y = f(x)$ is **symmetric with respect to the point (a, b)** if, whenever $(a - x, b - y)$ is a point on the graph, $(a + x, b + y)$ is also a point on the graph, as shown in the figure.

(a) Sketch the graph of $y = \sin x$ on the interval $[0, 2\pi]$. Write a short paragraph explaining how the symmetry of the graph with respect to the point $(0, \pi)$ allows you to conclude that

$$\int_0^{2\pi} \sin x\,dx = 0.$$

(b) Sketch the graph of $y = \sin x + 2$ on the interval $[0, 2\pi]$. Use the symmetry of the graph with respect to the point $(\pi, 2)$ to evaluate the integral

$$\int_0^{2\pi} (\sin x + 2)\,dx.$$

(c) Sketch the graph of $y = \arccos x$ on the interval $[-1, 1]$. Use the symmetry of the graph to evaluate the integral

$$\int_{-1}^{1} \arccos x\,dx.$$

(d) Evaluate the integral $\displaystyle\int_0^{\pi/2} \frac{1}{1 + (\tan x)^{\sqrt{2}}}\,dx.$

3. (a) Use a graphing utility to graph $f(x) = \dfrac{\ln(x + 1)}{x}$ on the interval $[-1, 1]$.

(b) Use the graph to estimate $\lim_{x \to 0} f(x)$.

(c) Use the definition of derivative to prove your answer to part (b).

4. Let $f(x) = \sin(\ln x)$.

(a) Determine the domain of the function f.

(b) Find two values of x satisfying $f(x) = 1$.

(c) Find two values of x satisfying $f(x) = -1$.

(d) What is the range of the function f?

(e) Calculate $f'(x)$ and use calculus to find the maximum value of f on the interval $[1, 10]$.

(f) Use a graphing utility to graph f in the viewing window $[0, 5] \times [-2, 2]$ and estimate $\lim_{x \to 0^+} f(x)$, if it exists.

(g) Determine $\lim_{x \to 0^+} f(x)$ analytically, if it exists.

5. Graph the exponential function $y = a^x$ for $a = 0.5, 1.2,$ and 2.0. Which of these curves intersects the line $y = x$? Determine all positive numbers a for which the curve $y = a^x$ intersects the line $y = x$.

6. (a) Let $P(\cos t, \sin t)$ be a point on the unit circle $x^2 + y^2 = 1$ in the first quadrant (see figure). Show that t is equal to twice the area of the shaded circular sector AOP.

(b) Let $P(\cosh t, \sinh t)$ be a point on the unit hyperbola $x^2 - y^2 = 1$ in the first quadrant (see figure). Show that t is equal to twice the area of the shaded region AOP. Begin by showing that the area of the shaded region AOP is given by the formula

$$A(t) = \frac{1}{2}\cosh t \sinh t - \int_1^{\cosh t} \sqrt{x^2 - 1}\,dx.$$

P.S. Problem Solving

Each chapter concludes with a set of thought-provoking and challenging exercises that provide opportunities for the student to explore the concepts in the chapter further.

Technology

Throughout the text, the use of a graphing utility or computer algebra system is suggested as appropriate for problem-solving as well as exploration and discovery. For example, students may choose to use a graphing utility to execute complicated computations, to visualize theoretical concepts, to discover alternative approaches, or to verify the results of other solution methods. However, students are not required to have access to a graphing utility to use this text effectively. In addition to describing the benefits of using technology to learn calculus, the text also addresses its possible misuse or misinterpretation.

Additional Features

Additional teaching and learning resources are integrated throughout the textbook, including Section Projects, journal references, and Writing About Concepts Exercises.

Acknowledgments

We would like to thank the many people who have helped us at various stages of this project over the last 30 years. Their encouragement, criticisms, and suggestions have been invaluable to us.

For the Eighth Edition

James Pommersheim
Reed College

Kevin J. Leith
Albuquerque Community College

Andrew J. Guzo
Chatham High School, NJ

Mary J. Quadrini
East Greenwich High School, RI

Jim Burton
Vernon Verona Sherrill High School, NY

Guillermo Barberena III
South Hills High School, TX

Susan A. Natale
The Ursuline School, NY

Patrick Ward
Illinois Central College

Donna J. Gorton
Butler County Community College

Diane Zych
Erie Community College

Guy Hogan
Norfolk State University

Michael Frantz
University of La Verne

Darren Narayan
Rochester Institute of Technology

Stanley J. Brzezicki
Iroquois High School, PA

Leland E. Rogers
Pepperdine University

Paul Seeburger
Monroe Community College

Ashok Kumar
Valdosta State University

Alexander Arhangelskii
Ohio University

James Braselton
Georgia Southern University

Harvey Braverman
Middlesex County College

Jianzhong Su
University of Texas at Arlington

P.S. Crooke
Vanderbilt University

Stan Adamski
Owens Community College

Edith A. Silver
Mercer County Community College

Seth G. Armstrong
Southern Utah University

Desmond Stephens
Florida A&M University

Terence H. Perciante
Wheaton College

Linda A. Bolte
Eastern Washington University

Sudhir Goel
Valdosta State University

Donna Flint
South Dakota State University

For the Eighth Edition Technology Program

John Gosselin
University of Georgia

Oiyin Pauline Chow
Harrisburg Area Community College

Murray Eisenberg
University of Massachusetts at Amherst

Douglas B. Meade
University of South Carolina

Tim Chappell
Penn Valley Community College

Howard Speier
Chandler-Gilbert Community College

Marcelle Bessman
Jacksonville University

Jim Ball
Indiana State University

Julie M. Clark
Hollins University

Teri Murphy
University of Oklahoma

Jim Dotzler
Nassau Community College

Arek Goetz
San Francisco State University

Shahryar Heydari
Piedmont College

Reviewers of Previous Editions

Dennis Alber, *Palm Beach Junior College*; James Angelos, *Central Michigan University*; Raymond Badalian, *Los Angeles City College*; Kerry D. Bailey, *Laramie County Community College*; Harry L. Baldwin, Jr., *San Diego State City College*; Homer F. Bechtell, *University of New Hampshire*; Keith Bergeron, *United States Air Force Academy*; Norman Birenes, *University of Regina*; Brian Blank, *Washington State University*; Andrew A. Bulleri, *Howard Community College*; Christopher Butler, *Case Western Reserve University*; Dane R. Camp, *New Trier High School, IL*; Paula Castagna, *Fresno City College*; Jack Ceder, *University of California-Santa Barbara*; Charles L. Cope, *Morehouse College*; Barbara Cortzen, *DePaul University*; Jorge Cossio, *Miami-Dade Community College*; Jack Courtney, *Michigan State University*; James Daniels, *Palomar College*; Kathy Davis, *University of Texas*; Paul W. Davis, *Worcester Polytechnic Institute*; Luz M. DeAlba, *Drake University*; Nicolae Dinculeanu, *University of Florida*; Rosario Diprizio, *Oakton Community College*; Garret J. Etgen, *University of Houston*; Russell Euler, *Northwest Missouri State University*; Phillip A. Ferguson, *Fresno City College*; Li Fong, *Johnson County Community College*; Michael Frantz, *University of La Verne*; William R. Fuller, *Purdue University*; Dewey Furness, *Ricks College*; Javier Garza, *Tarleton State University*; K. Elayn Gay, *University of New Orleans*; Thomas M. Green, *Contra Costa College*; Ali Hajjafar, *University of Akron*; Ruth A. Hartman, *Black Hawk College*; Irvin Roy Hentzel, *Iowa State University*; Kathy Hoke, *University of Richmond*; Howard E. Holcomb, *Monroe Community College*; Eric R. Immel, *Georgia Institute of Technology*; Arnold J. Insel, *Illinois State University*; Elgin Honston, *Iowa State University*; Hikeaki Kaneko, *Old Dominion University*; Toni Kasper, *Borough of Manhattan Community College*; William J. Keane, *Boston College*; Timothy J. Kearns, *Boston College*; Ronnie Khuri, *University of Florida*; Frank T. Kocher, Jr., *Pennsylvania State University*; Robert Kowalczyk, *University of Massachusetts-Dartmouth*; Joseph F. Krebs, *Boston College*; David C. Lantz, *Colgate University*; Norbert Lerner, *State University of New York at Cortland*; Maita Levine, *University of Cincinnati*; Murray Lieb, *New Jersey Institute of Technology*; Beth Long, *Pellissippi State Technical College*; Ransom Van B. Lynch, *Phillips Exeter Academy*; Bennet Manvel, *Colorado State University*; Mauricio Marroquin, *Los Angeles Valley College*; Robert L. Maynard, *Tidewater Community College*; Robert McMaster, *John Abbott College*; Gordon Melrose, *Old Dominion University*; Darrell Minor, *Columbus*

State Community College; Maurice Monahan, *South Dakota State University*; Michael Montaño, *Riverside Community College*; Philip Montgomery, *University of Kansas*; David C. Morency, *University of Vermont*; Gerald Mueller, *Columbus State Community College*; Duff A. Muir, *United States Air Force Academy*; Charlotte J. Newsom, *Tidewater Community College*; Terry J. Newton, *United States Air Force Academy*; Donna E. Nordstrom, *Pasadena City College*; Larry Norris, *North Carolina State University*; Robert A. Nowlan, *Southern Connecticut State University*; Luis Ortiz-Franco, *Chapman University*; Barbara L. Osofsky, *Rutgers University*; Judith A. Palagallo, *University of Akron*; Eleanor Palais, *Belmont High School*, MA; Wayne J. Peeples, *University of Texas*; Jorge A. Perez, *LaGuardia Community College*; Darrell J. Peterson, *Santa Monica College*; Donald Poulson, *Mesa Community College*; Lila Roberts, *Georgia Southern University*; Jean L. Rubin, *Purdue University*; John Santomas, *Villanova University*; Barry Sarnacki, *United States Air Force Academy*; N. James Schoonmaker, *University of Vermont*; George W. Schultz, *St. Petersburg Junior College*; Richard E. Shermoen, *Washburn University*; Thomas W. Shilgalis, *Illinois State University*; J. Philip Smith, *Southern Connecticut State University*; Lynn Smith, *Gloucester County College*; Frank Soler, *De Anza College*; Enid Steinbart, *University of New Orleans*; Michael Steuer, *Nassau Community College*; Mark Stevenson, *Oakland Community College*; Anthony Thomas, *University of Wisconsin-Platteville*; Lawrence A. Trivieri, *Mohawk Valley Community College*; John Tweed, *Old Dominion University*; Carol Urban, *College of DuPage*; Marjorie Valentine, *North Side ISD, San Antonio*; Robert J. Vojack, *Ridgewood High School, NJ*; Bert K. Waits, *The Ohio State University*; Florence A. Warfel, *University of Pittsburgh*; John R. Watret, *Embry-Riddle Aeronautical University*; Carroll G. Wells, *Western Kentucky University*; Charles Wheeler, *Montgomery College*; Jay Wiestling, *Palomar College*; Paul D. Zahn, *Borough of Manhattan Community College*; August J. Zarcone, *College of DuPage*

We would like to thank the staff at Larson Texts, Inc., who assisted in preparing the manuscript, rendering the art package, and typesetting and proofreading the pages and supplements.

A special note of thanks goes to the instructors who responded to our survey and to the over 2 million students who have used earlier editions of the text.

On a personal level, we are grateful to our wives, Deanna Gilbert Larson, Eloise Hostetler, and Consuelo Edwards, for their love, patience, and support. Also, a special note of thanks goes out to R. Scott O'Neil.

If you have suggestions for improving this text, please feel free to write to us. Over the years we have received many useful comments from both instructors and students, and we value these very much.

Ron Larson

Robert P. Hostetler

Bruce H. Edwards

P Preparation for Calculus

Two types of race cars designed and built by NASCAR teams are short track cars and super-speedway (long track) cars. The super-speedway race cars are subjected to extensive testing in wind tunnels like the one shown in the photo. Short track race cars and super-speedway race cars are designed either to allow for as much downforce as possible or to reduce the amount of drag on the race car. Which design do you think is used for each type of race car? Why?

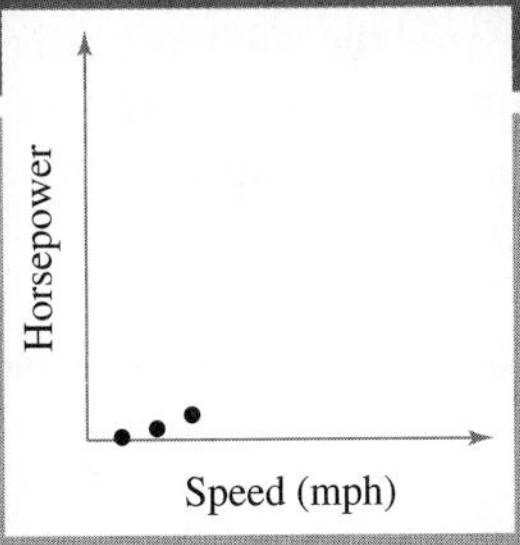

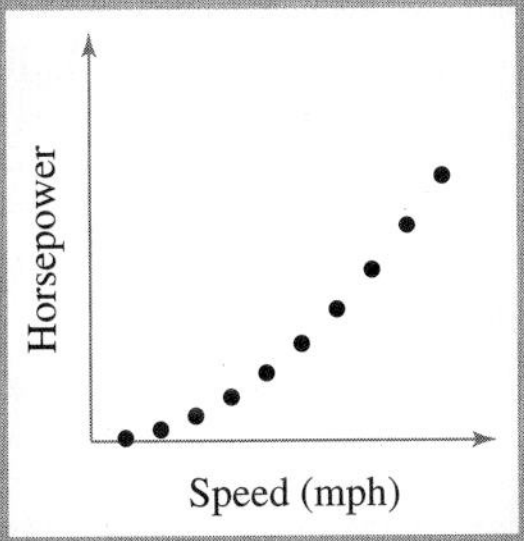

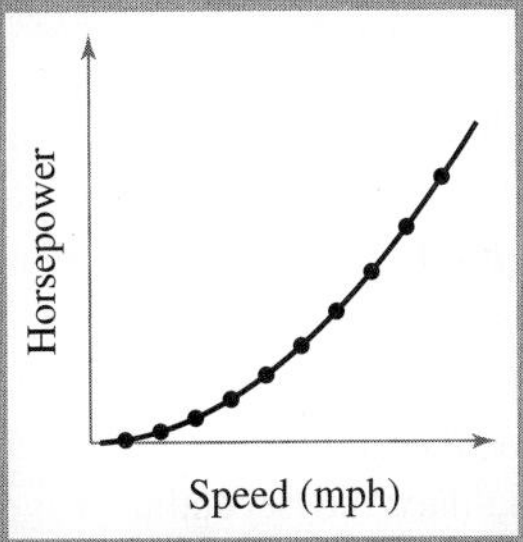

Mathematical models are commonly used to describe data sets. These models can be represented by many different types of functions, such as linear, quadratic, cubic, rational, and trigonometric functions. In Chapter P, you will review how to find, graph, and compare mathematical models for different data sets.

Carol Anne Petrachenko/Corbis

Section P.1 Graphs and Models

- **Sketch the graph of an equation.**
- **Find the intercepts of a graph.**
- **Test a graph for symmetry with respect to an axis and the origin.**
- **Find the points of intersection of two graphs.**
- **Interpret mathematical models for real-life data.**

René Descartes (1596–1650)
Descartes made many contributions to philosophy, science, and mathematics. The idea of representing points in the plane by pairs of real numbers and representing curves in the plane by equations was described by Descartes in his book *La Géométrie*, published in 1637.

The Graph of an Equation

In 1637 the French mathematician René Descartes revolutionized the study of mathematics by joining its two major fields—algebra and geometry. With Descartes's coordinate plane, geometric concepts could be formulated analytically and algebraic concepts could be viewed graphically. The power of this approach is such that within a century, much of calculus had been developed.

The same approach can be followed in your study of calculus. That is, by viewing calculus from multiple perspectives—*graphically*, *analytically*, and *numerically*—you will increase your understanding of core concepts.

Consider the equation $3x + y = 7$. The point $(2, 1)$ is a **solution point** of the equation because the equation is satisfied (is true) when 2 is substituted for x and 1 is substituted for y. This equation has many other solutions, such as $(1, 4)$ and $(0, 7)$. To find other solutions systematically, solve the original equation for y.

$$y = 7 - 3x \qquad \text{Analytic approach}$$

Then construct a **table of values** by substituting several values of x.

x	0	1	2	3	4
y	7	4	1	-2	-5

Numerical approach

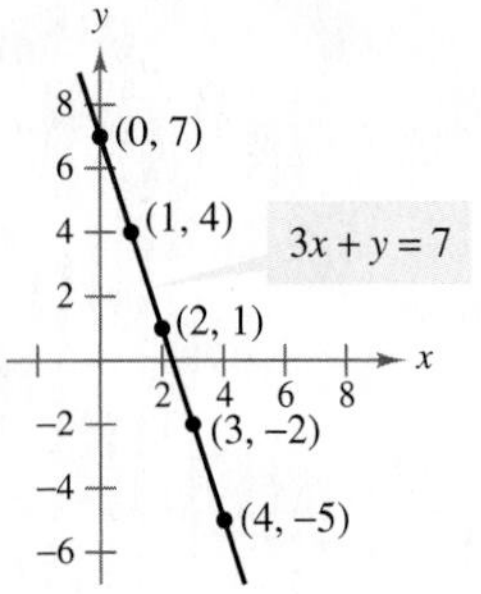

Graphical approach: $3x + y = 7$
Figure P.1

From the table, you can see that $(0, 7)$, $(1, 4)$, $(2, 1)$, $(3, -2)$, and $(4, -5)$ are solutions of the original equation $3x + y = 7$. Like many equations, this equation has an infinite number of solutions. The set of all solution points is the **graph** of the equation, as shown in Figure P.1.

NOTE Even though we refer to the sketch shown in Figure P.1 as the graph of $3x + y = 7$, it really represents only a *portion* of the graph. The entire graph would extend beyond the page.

In this course, you will study many sketching techniques. The simplest is point plotting—that is, you plot points until the basic shape of the graph seems apparent.

EXAMPLE 1 Sketching a Graph by Point Plotting

Sketch the graph of $y = x^2 - 2$.

Solution First construct a table of values. Then plot the points shown in the table.

x	-2	-1	0	1	2	3
y	2	-1	-2	-1	2	7

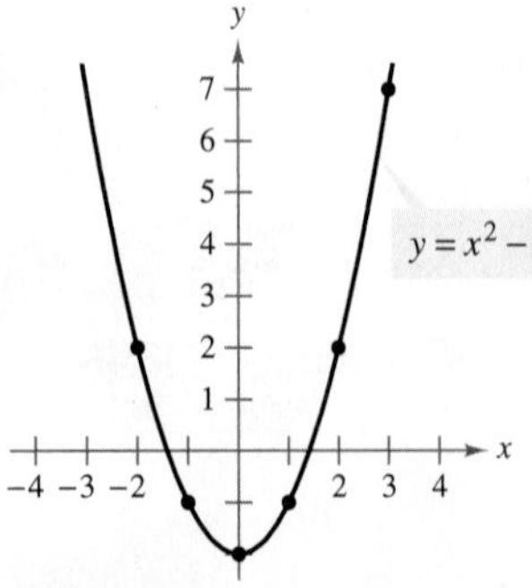

The parabola $y = x^2 - 2$
Figure P.2

Finally, connect the points with a *smooth curve*, as shown in Figure P.2. This graph is a **parabola.** It is one of the conics you will study in Chapter 10.

One disadvantage of point plotting is that to get a good idea about the shape of a graph, you may need to plot many points. With only a few points, you could badly misrepresent the graph. For instance, suppose that to sketch the graph of

$$y = \frac{1}{30}x(39 - 10x^2 + x^4)$$

you plotted only five points: $(-3, -3)$, $(-1, -1)$, $(0, 0)$, $(1, 1)$, and $(3, 3)$, as shown in Figure P.3(a). From these five points, you might conclude that the graph is a line. This, however, is not correct. By plotting several more points, you can see that the graph is more complicated, as shown in Figure P.3(b).

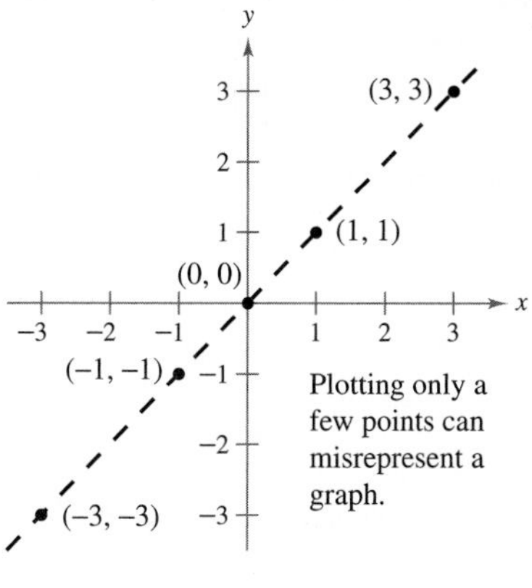

(a)

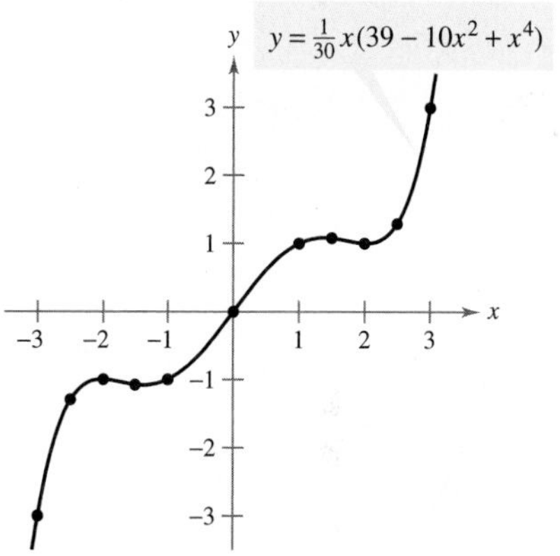

(b)

Figure P.3

EXPLORATION

Comparing Graphical and Analytic Approaches Use a graphing utility to graph each equation. In each case, find a viewing window that shows the important characteristics of the graph.

a. $y = x^3 - 3x^2 + 2x + 5$

b. $y = x^3 - 3x^2 + 2x + 25$

c. $y = -x^3 - 3x^2 + 20x + 5$

d. $y = 3x^3 - 40x^2 + 50x - 45$

e. $y = -(x + 12)^3$

f. $y = (x - 2)(x - 4)(x - 6)$

A purely graphical approach to this problem would involve a simple "guess, check, and revise" strategy. What types of things do you think an analytic approach might involve? For instance, does the graph have symmetry? Does the graph have turns? If so, where are they?

As you proceed through Chapters 1, 2, and 3 of this text, you will study many new analytic tools that will help you analyze graphs of equations such as these.

TECHNOLOGY Technology has made sketching of graphs easier. Even with technology, however, it is possible to misrepresent a graph badly. For instance, each of the graphing utility screens in Figure P.4 shows a portion of the graph of

$$y = x^3 - x^2 - 25.$$

From the screen on the left, you might assume that the graph is a line. From the screen on the right, however, you can see that the graph is not a line. So, whether you are sketching a graph by hand or using a graphing utility, you must realize that different "viewing windows" can produce very different views of a graph. In choosing a viewing window, your goal is to show a view of the graph that fits well in the context of the problem.

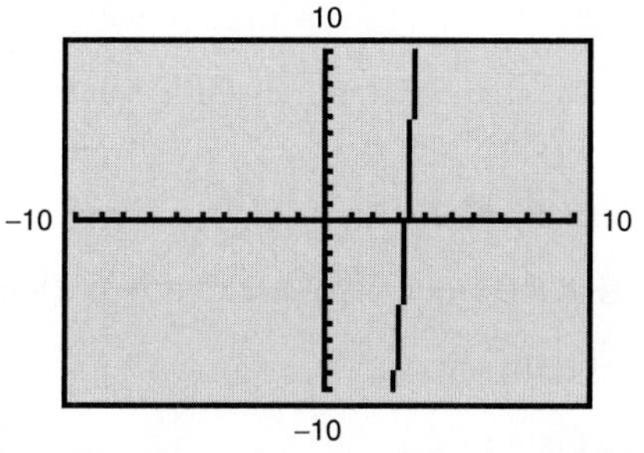

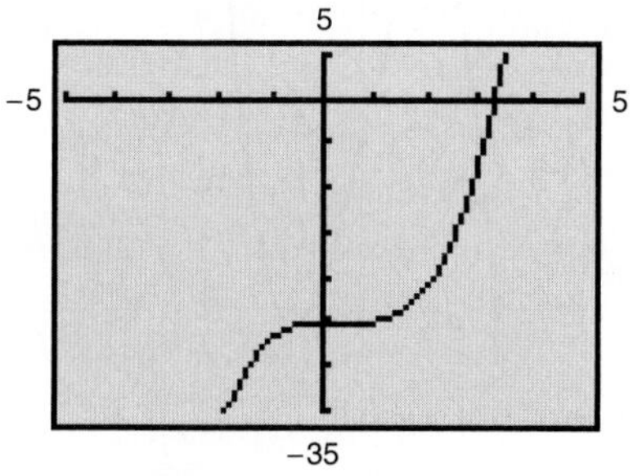

Graphing utility screens of $y = x^3 - x^2 - 25$
Figure P.4

NOTE In this text, the term *graphing utility* means either a graphing calculator or computer graphing software such as *Maple*, *Mathematica*, *Derive*, *Mathcad*, or the *TI-89*.

Intercepts of a Graph

Two types of solution points that are especially useful in graphing an equation are those having zero as their x- or y-coordinate. Such points are called **intercepts** because they are the points at which the graph intersects the x- or y-axis. The point $(a, 0)$ is an ***x*-intercept** of the graph of an equation if it is a solution point of the equation. To find the x-intercepts of a graph, let y be zero and solve the equation for x. The point $(0, b)$ is a ***y*-intercept** of the graph of an equation if it is a solution point of the equation. To find the y-intercepts of a graph, let x be zero and solve the equation for y.

NOTE Some texts denote the x-intercept as the x-coordinate of the point $(a, 0)$ rather than the point itself. Unless it is necessary to make a distinction, we will use the term *intercept* to mean either the point or the coordinate.

It is possible for a graph to have no intercepts, or it might have several. For instance, consider the four graphs shown in Figure P.5.

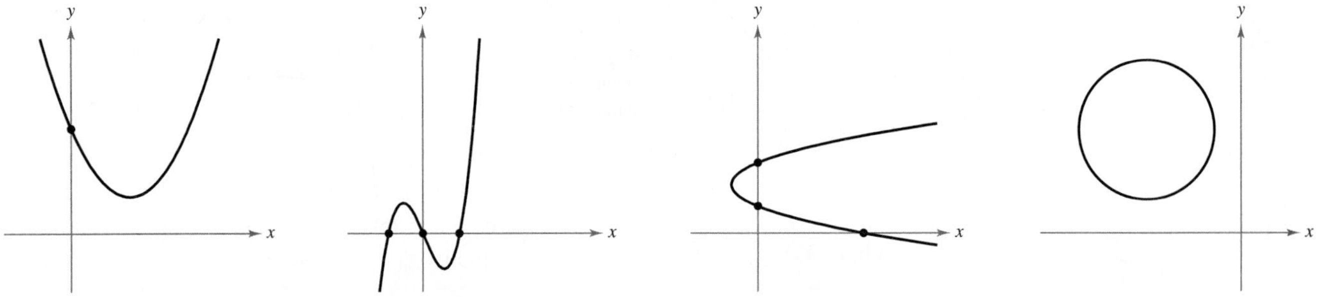

No x-intercepts
One y-intercept

Three x-intercepts
One y-intercept

One x-intercept
Two y-intercepts

No intercepts

Figure P.5

EXAMPLE 2 Finding x- and y-intercepts

Find the x- and y-intercepts of the graph of $y = x^3 - 4x$.

Solution To find the x-intercepts, let y be zero and solve for x.

$$x^3 - 4x = 0 \quad \text{Let } y \text{ be zero.}$$

$$x(x - 2)(x + 2) = 0 \quad \text{Factor.}$$

$$x = 0, 2, \text{ or } -2 \quad \text{Solve for } x.$$

Because this equation has three solutions, you can conclude that the graph has three x-intercepts:

$(0, 0), (2, 0),$ and $(-2, 0)$. *x-intercepts*

To find the y-intercepts, let x be zero. Doing this produces $y = 0$. So, the y-intercept is

$(0, 0)$. *y-intercept*

(See Figure P.6.)

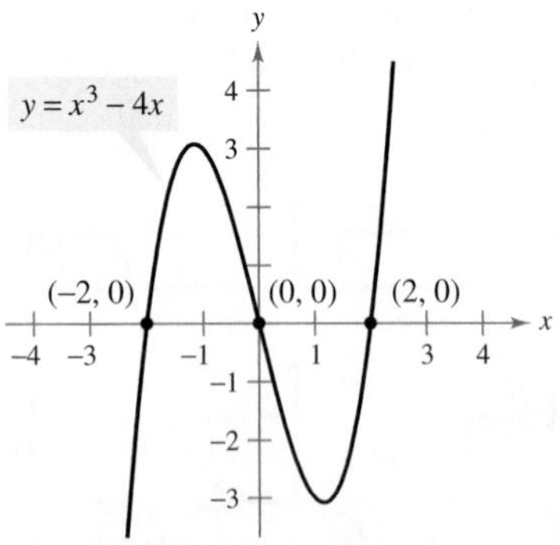

Intercepts of a graph
Figure P.6

TECHNOLOGY Example 2 uses an analytic approach to finding intercepts. When an analytic approach is not possible, you can use a graphical approach by finding the points at which the graph intersects the axes. Use a graphing utility to approximate the intercepts.

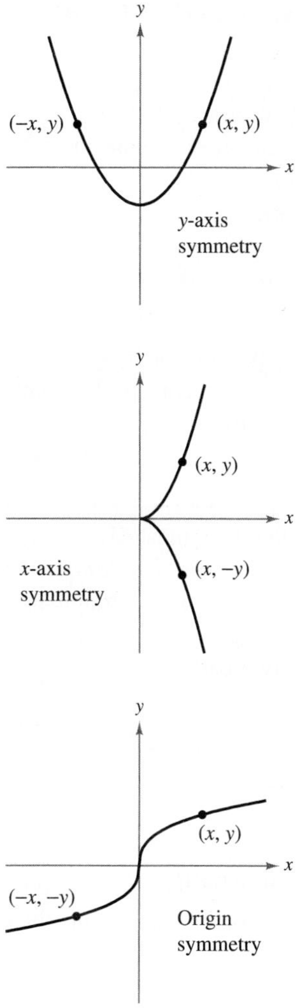

Figure P.7

Symmetry of a Graph

Knowing the symmetry of a graph *before* attempting to sketch it is useful because you need only half as many points to sketch the graph. The following three types of symmetry can be used to help sketch the graphs of equations (see Figure P.7).

1. A graph is **symmetric with respect to the *y*-axis** if, whenever (x, y) is a point on the graph, $(-x, y)$ is also a point on the graph. This means that the portion of the graph to the left of the y-axis is a mirror image of the portion to the right of the y-axis.
2. A graph is **symmetric with respect to the *x*-axis** if, whenever (x, y) is a point on the graph, $(x, -y)$ is also a point on the graph. This means that the portion of the graph above the x-axis is a mirror image of the portion below the x-axis.
3. A graph is **symmetric with respect to the origin** if, whenever (x, y) is a point on the graph, $(-x, -y)$ is also a point on the graph. This means that the graph is unchanged by a rotation of 180° about the origin.

Tests for Symmetry

1. The graph of an equation in x and y is symmetric with respect to the y-axis if replacing x by $-x$ yields an equivalent equation.
2. The graph of an equation in x and y is symmetric with respect to the x-axis if replacing y by $-y$ yields an equivalent equation.
3. The graph of an equation in x and y is symmetric with respect to the origin if replacing x by $-x$ and y by $-y$ yields an equivalent equation.

The graph of a polynomial has symmetry with respect to the y-axis if each term has an even exponent (or is a constant). For instance, the graph of

$$y = 2x^4 - x^2 + 2 \qquad \text{y-axis symmetry}$$

has symmetry with respect to the y-axis. Similarly, the graph of a polynomial has symmetry with respect to the origin if each term has an odd exponent, as illustrated in Example 3.

EXAMPLE 3 Testing for Origin Symmetry

Show that the graph of

$$y = 2x^3 - x$$

is symmetric with respect to the origin.

Solution

$$\begin{aligned} y &= 2x^3 - x && \text{Write original equation.} \\ -y &= 2(-x)^3 - (-x) && \text{Replace x by $-x$ and y by $-y$.} \\ -y &= -2x^3 + x && \text{Simplify.} \\ y &= 2x^3 - x && \text{Equivalent equation} \end{aligned}$$

Because the replacements yield an equivalent equation, you can conclude that the graph of $y = 2x^3 - x$ is symmetric with respect to the origin, as shown in Figure P.8.

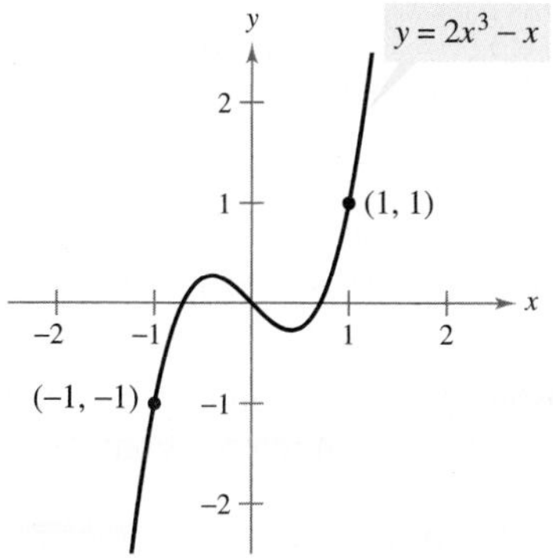

Origin symmetry
Figure P.8

EXAMPLE 4 Using Intercepts and Symmetry to Sketch a Graph

Sketch the graph of $x - y^2 = 1$.

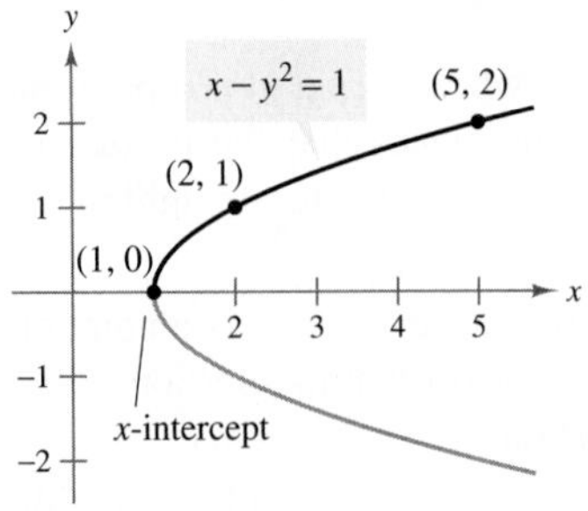

Figure P.9

Solution The graph is symmetric with respect to the x-axis because replacing y by $-y$ yields an equivalent equation.

$x - y^2 = 1$	Write original equation.
$x - (-y)^2 = 1$	Replace y by $-y$.
$x - y^2 = 1$	Equivalent equation

This means that the portion of the graph below the x-axis is a mirror image of the portion above the x-axis. To sketch the graph, first plot the x-intercept and the points above the x-axis. Then reflect in the x-axis to obtain the entire graph, as shown in Figure P.9.

TECHNOLOGY Graphing utilities are designed so that they most easily graph equations in which y is a function of x (see Section P.3 for a definition of **function**). To graph other types of equations, you need to split the graph into two or more parts *or* you need to use a different graphing mode. For instance, to graph the equation in Example 4, you can split it into two parts.

$y_1 = \sqrt{x - 1}$	Top portion of graph
$y_2 = -\sqrt{x - 1}$	Bottom portion of graph

Points of Intersection

A **point of intersection** of the graphs of two equations is a point that satisfies both equations. You can find the points of intersection of two graphs by solving their equations simultaneously.

EXAMPLE 5 Finding Points of Intersection

Find all points of intersection of the graphs of $x^2 - y = 3$ and $x - y = 1$.

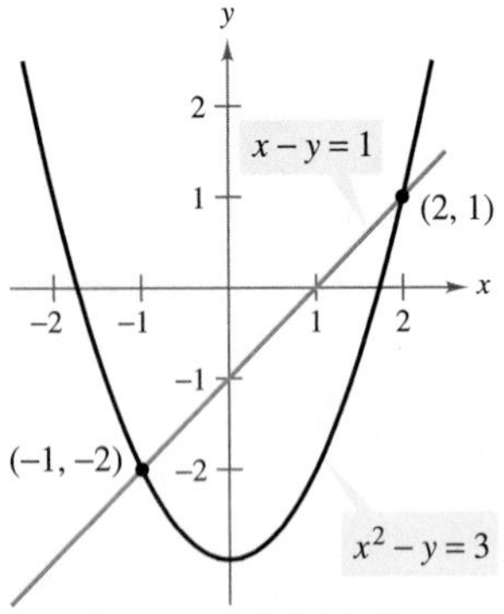

Two points of intersection
Figure P.10

STUDY TIP You can check the points of intersection from Example 5 by substituting into *both* of the original equations or by using the *intersect* feature of a graphing utility.

Solution Begin by sketching the graphs of both equations on the *same* rectangular coordinate system, as shown in Figure P.10. Having done this, it appears that the graphs have two points of intersection. You can find these two points, as follows.

$y = x^2 - 3$	Solve first equation for y.
$y = x - 1$	Solve second equation for y.
$x^2 - 3 = x - 1$	Equate y-values.
$x^2 - x - 2 = 0$	Write in general form.
$(x - 2)(x + 1) = 0$	Factor.
$x = 2$ or -1	Solve for x.

The corresponding values of y are obtained by substituting $x = 2$ and $x = -1$ into either of the original equations. Doing this produces two points of intersection:

$(2, 1)$ and $(-1, -2)$. Points of intersection

indicates that in the HM mathSpace® *CD-ROM and the online* Eduspace® *system for this text, you will find an Open Exploration, which further explores this example using the computer algebra systems* Maple, Mathcad, Mathematica, *and* Derive.

Mathematical Models

Real-life applications of mathematics often use equations as **mathematical models.** In developing a mathematical model to represent actual data, you should strive for two (often conflicting) goals: accuracy and simplicity. That is, you want the model to be simple enough to be workable, yet accurate enough to produce meaningful results. Section P.4 explores these goals more completely.

EXAMPLE 6 Comparing Two Mathematical Models

Gavriel Jecan/Corbis

The Mauna Loa Observatory in Hawaii has been measuring the increasing concentration of carbon dioxide in Earth's atmosphere since 1958.

The Mauna Loa Observatory in Hawaii records the carbon dioxide concentration y (in parts per million) in Earth's atmosphere. The January readings for various years are shown in Figure P.11. In the July 1990 issue of *Scientific American*, these data were used to predict the carbon dioxide level in Earth's atmosphere in the year 2035, using the quadratic model

$$y = 316.2 + 0.70t + 0.018t^2 \qquad \text{Quadratic model for 1960–1990 data}$$

where $t = 0$ represents 1960, as shown in Figure P.11(a).

The data shown in Figure P.11(b) represent the years 1980 through 2002 and can be modeled by

$$y = 306.3 + 1.56t \qquad \text{Linear model for 1980–2002 data}$$

where $t = 0$ represents 1960. What was the prediction given in the *Scientific American* article in 1990? Given the new data for 1990 through 2002, does this prediction for the year 2035 seem accurate?

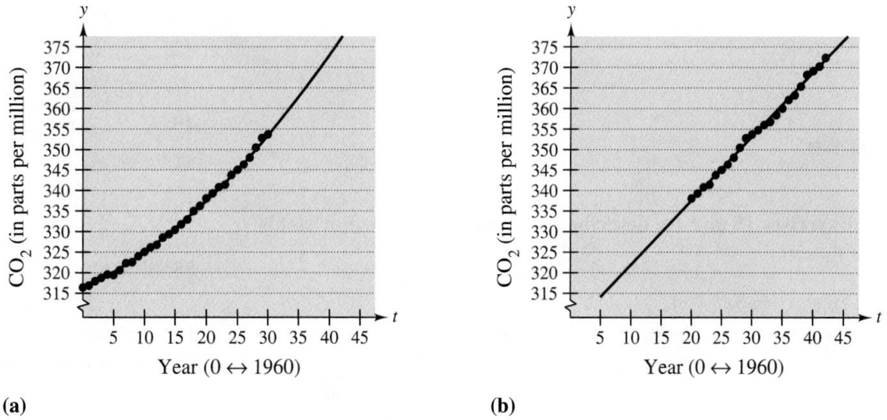

Figure P.11

Solution To answer the first question, substitute $t = 75$ (for 2035) into the quadratic model.

$$y = 316.2 + 0.70(75) + 0.018(75)^2 = 469.95 \qquad \text{Quadratic model}$$

So, the prediction in the *Scientific American* article was that the carbon dioxide concentration in Earth's atmosphere would reach about 470 parts per million in the year 2035. Using the linear model for the 1980–2002 data, the prediction for the year 2035 is

$$y = 306.3 + 1.56(75) = 423.3. \qquad \text{Linear model}$$

So, based on the linear model for 1980–2002, it appears that the 1990 prediction was too high.

NOTE The models in Example 6 were developed using a procedure called *least squares regression* (see Section 13.9). The quadratic and linear models have a correlation given by $r^2 = 0.997$ and $r^2 = 0.996$, respectively. The closer r^2 is to 1, the "better" the model.

Exercises for Section P.1

See www.CalcChat.com for worked-out solutions to odd-numbered exercises.

In Exercises 1–4, match the equation with its graph. [Graphs are labeled (a), (b), (c), and (d).]

(a)

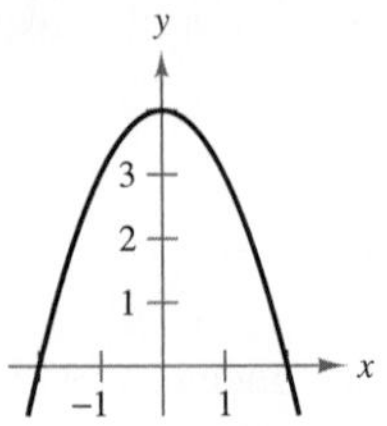

(b)

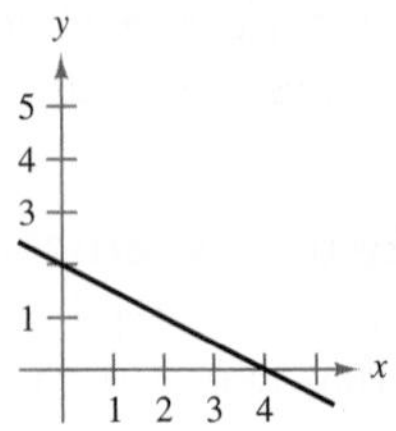

(c)

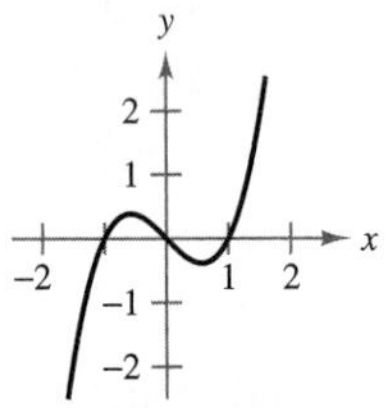

(d)

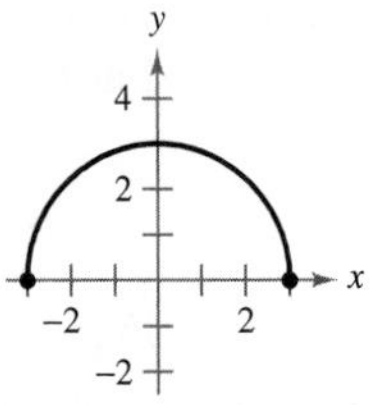

1. $y = -\frac{1}{2}x + 2$
2. $y = \sqrt{9 - x^2}$
3. $y = 4 - x^2$
4. $y = x^3 - x$

In Exercises 5–14, sketch the graph of the equation by point plotting.

5. $y = \frac{3}{2}x + 1$
6. $y = 6 - 2x$
7. $y = 4 - x^2$
8. $y = (x - 3)^2$
9. $y = |x + 2|$
10. $y = |x| - 1$
11. $y = \sqrt{x} - 4$
12. $y = \sqrt{x + 2}$
13. $y = \dfrac{2}{x}$
14. $y = \dfrac{1}{x - 1}$

In Exercises 15 and 16, describe the viewing window that yields the figure.

15. $y = x^3 - 3x^2 + 4$

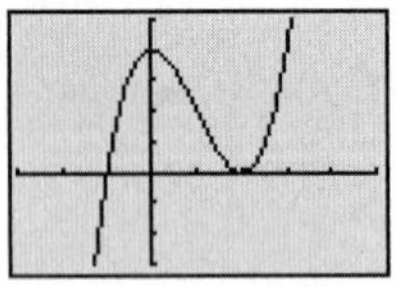

16. $y = |x| + |x - 10|$

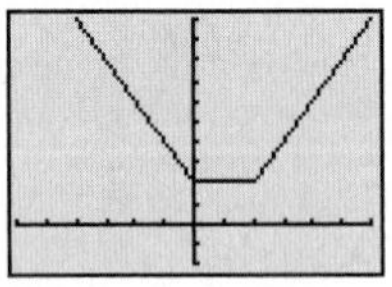

In Exercises 17 and 18, use a graphing utility to graph the equation. Move the cursor along the curve to approximate the unknown coordinate of each solution point accurate to two decimal places.

17. $y = \sqrt{5 - x}$ (a) $(2, y)$ (b) $(x, 3)$
18. $y = x^5 - 5x$ (a) $(-0.5, y)$ (b) $(x, -4)$

In Exercises 19–26, find any intercepts.

19. $y = x^2 + x - 2$
20. $y^2 = x^3 - 4x$
21. $y = x^2\sqrt{25 - x^2}$
22. $y = (x - 1)\sqrt{x^2 + 1}$
23. $y = \dfrac{3(2 - \sqrt{x})}{x}$
24. $y = \dfrac{x^2 + 3x}{(3x + 1)^2}$
25. $x^2y - x^2 + 4y = 0$
26. $y = 2x - \sqrt{x^2 + 1}$

In Exercises 27–38, test for symmetry with respect to each axis and to the origin.

27. $y = x^2 - 2$
28. $y = x^2 - x$
29. $y^2 = x^3 - 4x$
30. $y = x^3 + x$
31. $xy = 4$
32. $xy^2 = -10$
33. $y = 4 - \sqrt{x + 3}$
34. $xy - \sqrt{4 - x^2} = 0$
35. $y = \dfrac{x}{x^2 + 1}$
36. $y = \dfrac{x^2}{x^2 + 1}$
37. $y = |x^3 + x|$
38. $|y| - x = 3$

In Exercises 39–56, sketch the graph of the equation. Identify any intercepts and test for symmetry.

39. $y = -3x + 2$
40. $y = -\frac{1}{2}x + 2$
41. $y = \frac{1}{2}x - 4$
42. $y = \frac{2}{3}x + 1$
43. $y = 1 - x^2$
44. $y = x^2 + 3$
45. $y = (x + 3)^2$
46. $y = 2x^2 + x$
47. $y = x^3 + 2$
48. $y = x^3 - 4x$
49. $y = x\sqrt{x + 2}$
50. $y = \sqrt{9 - x^2}$
51. $x = y^3$
52. $x = y^2 - 4$
53. $y = \dfrac{1}{x}$
54. $y = \dfrac{10}{x^2 + 1}$
55. $y = 6 - |x|$
56. $y = |6 - x|$

In Exercises 57–60, use a graphing utility to graph the equation. Identify any intercepts and test for symmetry.

57. $y^2 - x = 9$
58. $x^2 + 4y^2 = 4$
59. $x + 3y^2 = 6$
60. $3x - 4y^2 = 8$

In Exercises 61–68, find the points of intersection of the graphs of the equations.

61. $x + y = 2$
$2x - y = 1$
62. $2x - 3y = 13$
$5x + 3y = 1$
63. $x^2 + y = 6$
$x + y = 4$
64. $x = 3 - y^2$
$y = x - 1$

The symbol indicates an exercise in which you are instructed to use graphing technology or a symbolic computer algebra system. The solutions of other exercises may also be facilitated by use of appropriate technology.

65. $x^2 + y^2 = 5$
$x - y = 1$

66. $x^2 + y^2 = 25$
$2x + y = 10$

67. $y = x^3$
$y = x$

68. $y = x^3 - 4x$
$y = -(x + 2)$

In Exercises 69–72, use a graphing utility to find the points of intersection of the graphs. Check your results analytically.

69. $y = x^3 - 2x^2 + x - 1$
$y = -x^2 + 3x - 1$

70. $y = x^4 - 2x^2 + 1$
$y = 1 - x^2$

71. $y = \sqrt{x + 6}$
$y = \sqrt{-x^2 - 4x}$

72. $y = -|2x - 3| + 6$
$y = 6 - x$

73. ***Modeling Data*** The table shows the Consumer Price Index (CPI) for selected years. *(Source: Bureau of Labor Statistics)*

Year	1970	1975	1980	1985	1990	1995	2000
CPI	38.8	53.8	82.4	107.6	130.7	152.4	172.2

(a) Use the regression capabilities of a graphing utility to find a mathematical model of the form $y = at^2 + bt + c$ for the data. In the model, y represents the CPI and t represents the year, with $t = 0$ corresponding to 1970.

(b) Use a graphing utility to plot the data and graph the model. Compare the data with the model.

(c) Use the model to predict the CPI for the year 2010.

74. ***Modeling Data*** The table shows the average numbers of acres per farm in the United States for selected years. *(Source: U.S. Department of Agriculture)*

Year	1950	1960	1970	1980	1990	2000
Acreage	213	297	374	426	460	434

(a) Use the regression capabilities of a graphing utility to find a mathematical model of the form $y = at^2 + bt + c$ for the data. In the model, y represents the average acreage and t represents the year, with $t = 0$ corresponding to 1950.

(b) Use a graphing utility to plot the data and graph the model. Compare the data with the model.

(c) Use the model to predict the average number of acres per farm in the United States in the year 2010.

75. ***Break-Even Point*** Find the sales necessary to break even ($R = C$) if the cost C of producing x units is

$C = 5.5\sqrt{x} + 10{,}000$ Cost equation

and the revenue R for selling x units is

$R = 3.29x.$ Revenue equation

76. ***Copper Wire*** The resistance y in ohms of 1000 feet of solid copper wire at 77°F can be approximated by the model

$$y = \frac{10{,}770}{x^2} - 0.37, \quad 5 \le x \le 100$$

where x is the diameter of the wire in mils (0.001 in.). Use a graphing utility to graph the model. If the diameter of the wire is doubled, the resistance is changed by about what factor?

Writing About Concepts

In Exercises 77 and 78, write an equation whose graph has the indicated property. (There may be more than one correct answer.)

77. The graph has intercepts at $x = -2$, $x = 4$, and $x = 6$.

78. The graph has intercepts at $x = -\frac{5}{2}$, $x = 2$, and $x = \frac{3}{2}$.

79. Each table shows solution points for one of the following equations.

(i) $y = kx + 5$ (ii) $y = x^2 + k$

(iii) $y = kx^{3/2}$ (iv) $xy = k$

Match each equation with the correct table and find k. Explain your reasoning.

(a)

x	1	4	9
y	3	24	81

(b)

x	1	4	9
y	7	13	23

(c)

x	1	4	9
y	36	9	4

(d)

x	1	4	9
y	−9	6	71

80. (a) Prove that if a graph is symmetric with respect to the x-axis and to the y-axis, then it is symmetric with respect to the origin. Give an example to show that the converse is not true.

(b) Prove that if a graph is symmetric with respect to one axis and to the origin, then it is symmetric with respect to the other axis.

True or False? **In Exercises 81–84, determine whether the statement is true or false. If it is false, explain why or give an example that shows it is false.**

81. If $(1, -2)$ is a point on a graph that is symmetric with respect to the x-axis, then $(-1, -2)$ is also a point on the graph.

82. If $(1, -2)$ is a point on a graph that is symmetric with respect to the y-axis, then $(-1, -2)$ is also a point on the graph.

83. If $b^2 - 4ac > 0$ and $a \neq 0$, then the graph of $y = ax^2 + bx + c$ has two x-intercepts.

84. If $b^2 - 4ac = 0$ and $a \neq 0$, then the graph of $y = ax^2 + bx + c$ has only one x-intercept.

In Exercises 85 and 86, find an equation of the graph that consists of all points (x, y) having the given distance from the origin. (For a review of the Distance Formula, see Appendix D.)

85. The distance from the origin is twice the distance from $(0, 3)$.

86. The distance from the origin is $K\,(K \neq 1)$ times the distance from $(2, 0)$.

Section P.2 Linear Models and Rates of Change

- **Find the slope of a line passing through two points.**
- **Write the equation of a line with a given point and slope.**
- **Interpret slope as a ratio or as a rate in a real-life application.**
- **Sketch the graph of a linear equation in slope-intercept form.**
- **Write equations of lines that are parallel or perpendicular to a given line.**

The Slope of a Line

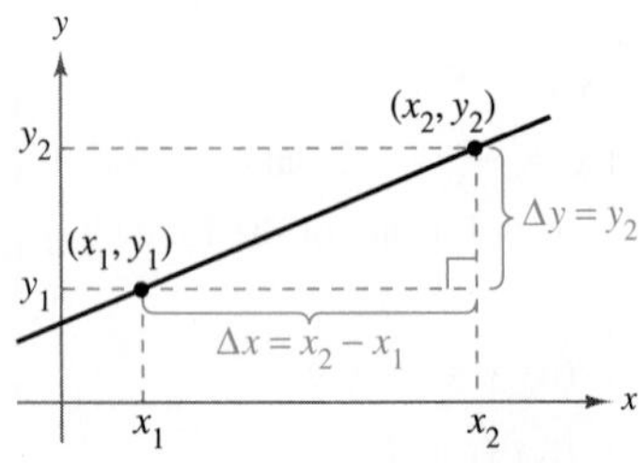

$\Delta y = y_2 - y_1 =$ change in y
$\Delta x = x_2 - x_1 =$ change in x
Figure P.12

The **slope** of a nonvertical line is a measure of the number of units the line rises (or falls) vertically for each unit of horizontal change from left to right. Consider the two points (x_1, y_1) and (x_2, y_2) on the line in Figure P.12. As you move from left to right along this line, a vertical change of

$$\Delta y = y_2 - y_1 \qquad \text{Change in } y$$

units corresponds to a horizontal change of

$$\Delta x = x_2 - x_1 \qquad \text{Change in } x$$

units. (Δ is the Greek uppercase letter *delta*, and the symbols Δy and Δx are read "delta y" and "delta x.")

Definition of the Slope of a Line

The **slope** m of the nonvertical line passing through (x_1, y_1) and (x_2, y_2) is

$$m = \frac{\Delta y}{\Delta x} = \frac{y_2 - y_1}{x_2 - x_1}, \qquad x_1 \neq x_2.$$

Slope is not defined for vertical lines.

NOTE When using the formula for slope, note that

$$\frac{y_2 - y_1}{x_2 - x_1} = \frac{-(y_1 - y_2)}{-(x_1 - x_2)} = \frac{y_1 - y_2}{x_1 - x_2}.$$

So, it does not matter in which order you subtract *as long as* you are consistent and both "subtracted coordinates" come from the same point.

Figure P.13 shows four lines: one has a positive slope, one has a slope of zero, one has a negative slope, and one has an "undefined" slope. In general, the greater the absolute value of the slope of a line, the steeper the line is. For instance, in Figure P.13, the line with a slope of -5 is steeper than the line with a slope of $\frac{1}{5}$.

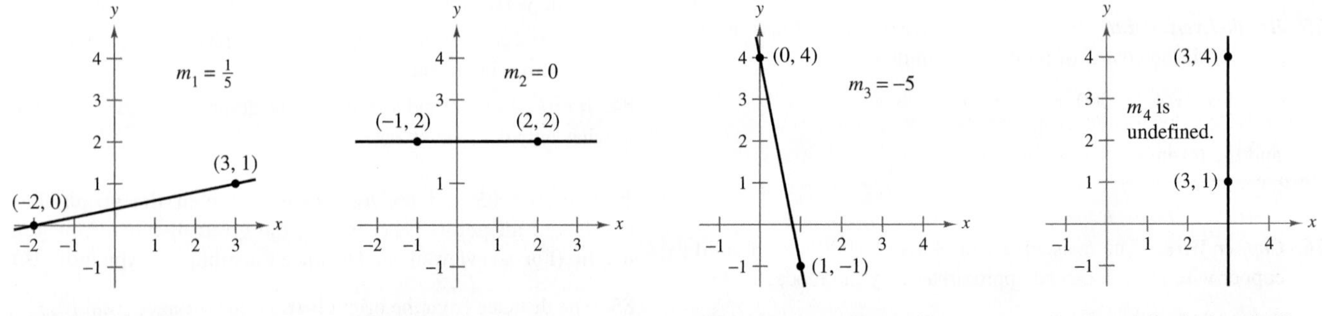

If m is positive, then the line rises from left to right.
If m is zero, then the line is horizontal.
If m is negative, then the line falls from left to right.
If m is undefined, then the line is vertical.
Figure P.13

EXPLORATION

Investigating Equations of Lines Use a graphing utility to graph each of the linear equations. Which point is common to all seven lines? Which value in the equation determines the slope of each line?

a. $y - 4 = -2(x + 1)$

b. $y - 4 = -1(x + 1)$

c. $y - 4 = -\frac{1}{2}(x + 1)$

d. $y - 4 = 0(x + 1)$

e. $y - 4 = \frac{1}{2}(x + 1)$

f. $y - 4 = 1(x + 1)$

g. $y - 4 = 2(x + 1)$

Use your results to write an equation of a line passing through $(-1, 4)$ with a slope of m.

Equations of Lines

Any two points on a nonvertical line can be used to calculate its slope. This can be verified from the similar triangles shown in Figure P.14. (Recall that the ratios of corresponding sides of similar triangles are equal.)

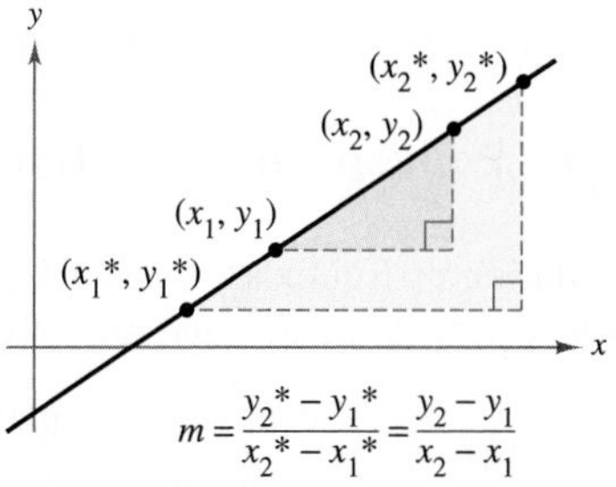

Any two points on a nonvertical line can be used to determine its slope.
Figure P.14

You can write an equation of a nonvertical line if you know the slope of the line and the coordinates of one point on the line. Suppose the slope is m and the point is (x_1, y_1). If (x, y) is any other point on the line, then

$$\frac{y - y_1}{x - x_1} = m.$$

This equation, involving the two variables x and y, can be rewritten in the form $y - y_1 = m(x - x_1)$, which is called the **point-slope equation of a line.**

Point-Slope Equation of a Line

An equation of the line with slope m passing through the point (x_1, y_1) is given by

$$y - y_1 = m(x - x_1).$$

EXAMPLE 1 Finding an Equation of a Line

Find an equation of the line that has a slope of 3 and passes through the point $(1, -2)$.

Solution

$y - y_1 = m(x - x_1)$	Point-slope form
$y - (-2) = 3(x - 1)$	Substitute -2 for y_1, 1 for x_1, and 3 for m.
$y + 2 = 3x - 3$	Simplify.
$y = 3x - 5$	Solve for y.

(See Figure P.15.)

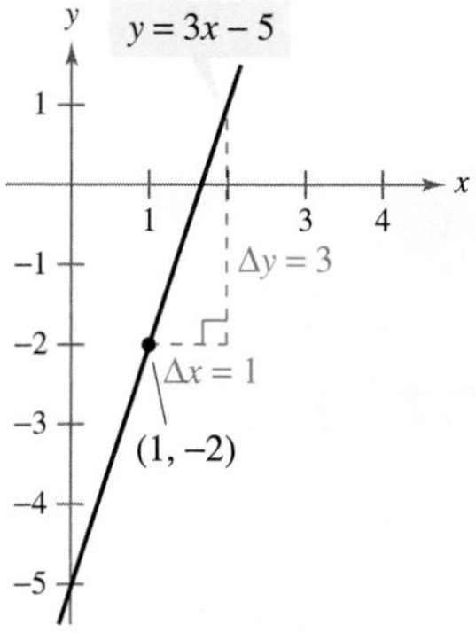

The line with a slope of 3 passing through the point $(1, -2)$
Figure P.15

NOTE Remember that only nonvertical lines have a slope. Consequently, vertical lines cannot be written in point-slope form. For instance, the equation of the vertical line passing through the point $(1, -2)$ is $x = 1$.

Ratios and Rates of Change

The slope of a line can be interpreted as either a *ratio* or a *rate*. If the x- and y-axes have the same unit of measure, the slope has no units and is a **ratio.** If the x- and y-axes have different units of measure, the slope is a rate or **rate of change.** In your study of calculus, you will encounter applications involving both interpretations of slope.

EXAMPLE 2 Population Growth and Engineering Design

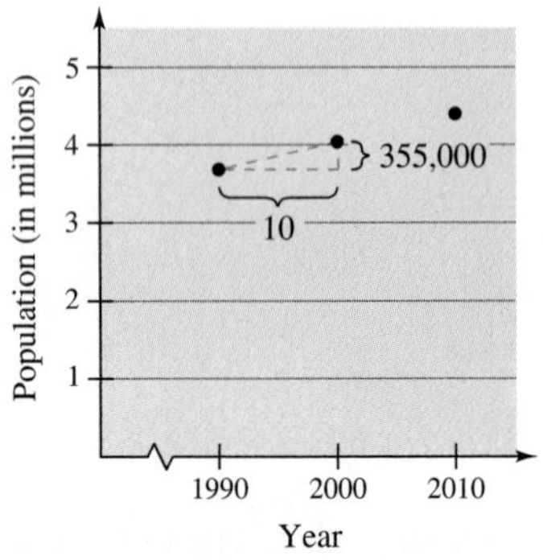

Population of Kentucky in census years
Figure P.16

a. The population of Kentucky was 3,687,000 in 1990 and 4,042,000 in 2000. Over this 10-year period, the average rate of change of the population was

$$\begin{aligned}\text{Rate of change} &= \frac{\text{change in population}}{\text{change in years}}\\ &= \frac{4{,}042{,}000 - 3{,}687{,}000}{2000 - 1990}\\ &= 35{,}500 \text{ people per year.}\end{aligned}$$

If Kentucky's population continues to increase at this same rate for the next 10 years, it will have a 2010 population of 4,397,000 (see Figure P.16). *(Source: U.S. Census Bureau)*

b. In tournament water-ski jumping, the ramp rises to a height of 6 feet on a raft that is 21 feet long, as shown in Figure P.17. The slope of the ski ramp is the ratio of its height (the rise) to the length of its base (the run).

$$\begin{aligned}\text{Slope of ramp} &= \frac{\text{rise}}{\text{run}} && \text{Rise is vertical change, run is horizontal change.}\\ &= \frac{6 \text{ feet}}{21 \text{ feet}}\\ &= \frac{2}{7}\end{aligned}$$

In this case, note that the slope is a ratio and has no units.

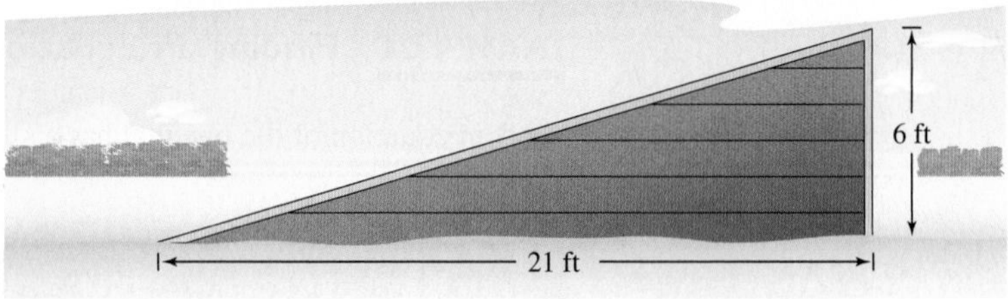

Dimensions of a water-ski ramp
Figure P.17

The rate of change found in Example 2(a) is an **average rate of change.** An average rate of change is always calculated over an interval. In this case, the interval is [1990, 2000]. In Chapter 2 you will study another type of rate of change called an *instantaneous rate of change*.

Graphing Linear Models

Many problems in analytic geometry can be classified in two basic categories: (1) Given a graph, what is its equation? and (2) Given an equation, what is its graph? The point-slope equation of a line can be used to solve problems in the first category. However, this form is not especially useful for solving problems in the second category. The form that is better suited to sketching the graph of a line is the **slope-intercept** form of the equation of a line.

> **The Slope-Intercept Equation of a Line**
>
> The graph of the linear equation
>
> $$y = mx + b$$
>
> is a line having a *slope* of m and a *y-intercept* at $(0, b)$.

EXAMPLE 3 Sketching Lines in the Plane

Sketch the graph of each equation.

a. $y = 2x + 1$ **b.** $y = 2$ **c.** $3y + x - 6 = 0$

Solution

a. Because $b = 1$, the y-intercept is $(0, 1)$. Because the slope is $m = 2$, you know that the line rises two units for each unit it moves to the right, as shown in Figure P.18(a).

b. Because $b = 2$, the y-intercept is $(0, 2)$. Because the slope is $m = 0$, you know that the line is horizontal, as shown in Figure P.18(b).

c. Begin by writing the equation in slope-intercept form.

$$3y + x - 6 = 0 \quad \text{Write original equation.}$$

$$3y = -x + 6 \quad \text{Isolate } y\text{-term on the left.}$$

$$y = -\frac{1}{3}x + 2 \quad \text{Slope-intercept form}$$

In this form, you can see that the y-intercept is $(0, 2)$ and the slope is $m = -\frac{1}{3}$. This means that the line falls one unit for every three units it moves to the right, as shown in Figure P.18(c).

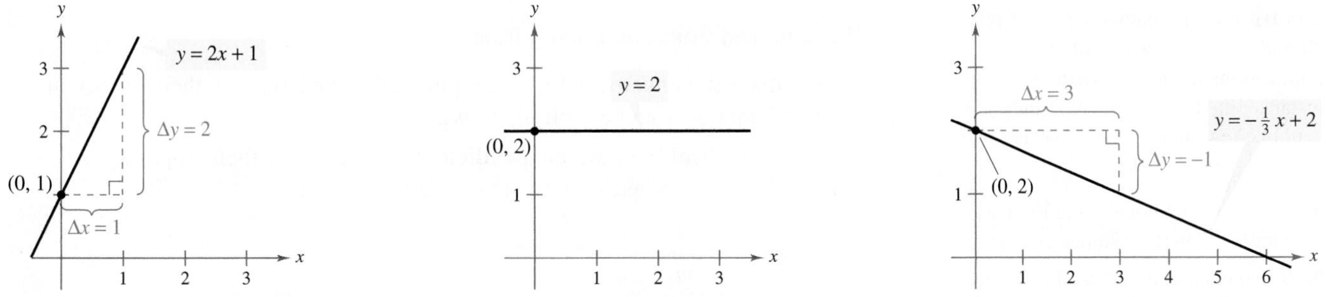

(a) $m = 2$; line rises (b) $m = 0$; line is horizontal (c) $m = -\frac{1}{3}$; line falls

Figure P.18

Because the slope of a vertical line is not defined, its equation cannot be written in the slope-intercept form. However, the equation of any line can be written in the **general form**

$$Ax + By + C = 0$$ General form of the equation of a line

where A and B are not *both* zero. For instance, the vertical line given by $x = a$ can be represented by the general form $x - a = 0$.

Summary of Equations of Lines

1. General form: $Ax + By + C = 0, \quad (A, B \neq 0)$
2. Vertical line: $x = a$
3. Horizontal line: $y = b$
4. Point-slope form: $y - y_1 = m(x - x_1)$
5. Slope-intercept form: $y = mx + b$

Parallel and Perpendicular Lines

The slope of a line is a convenient tool for determining whether two lines are parallel or perpendicular, as shown in Figure P.19. Specifically, nonvertical lines with the same slope are parallel and nonvertical lines whose slopes are negative reciprocals are perpendicular.

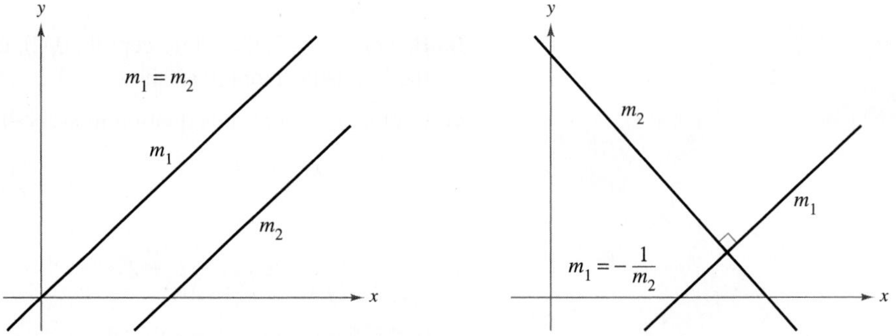

Parallel lines

Perpendicular lines

Figure P.19

STUDY TIP In mathematics, the phrase "if and only if" is a way of stating two implications in one statement. For instance, the first statement at the right could be rewritten as the following two implications.

a. If two distinct nonvertical lines are parallel, then their slopes are equal.

b. If two distinct nonvertical lines have equal slopes, then they are parallel.

Parallel and Perpendicular Lines

1. Two distinct nonvertical lines are **parallel** if and only if their slopes are equal—that is, if and only if $m_1 = m_2$.
2. Two nonvertical lines are **perpendicular** if and only if their slopes are negative reciprocals of each other—that is, if and only if

$$m_1 = -\frac{1}{m_2}.$$

EXAMPLE 4 Finding Parallel and Perpendicular Lines

Find the general forms of the equations of the lines that pass through the point $(2, -1)$ and are

a. parallel to the line $2x - 3y = 5$ **b.** perpendicular to the line $2x - 3y = 5$.

(See Figure P.20.)

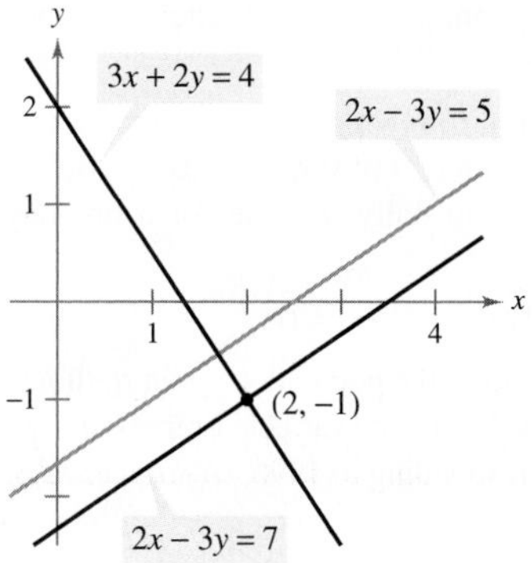

Lines parallel and perpendicular to $2x - 3y = 5$
Figure P.20

Solution By writing the linear equation $2x - 3y = 5$ in slope-intercept form, $y = \frac{2}{3}x - \frac{5}{3}$, you can see that the given line has a slope of $m = \frac{2}{3}$.

a. The line through $(2, -1)$ that is parallel to the given line also has a slope of $\frac{2}{3}$.

$$\begin{aligned} y - y_1 &= m(x - x_1) && \text{Point-slope form} \\ y - (-1) &= \tfrac{2}{3}(x - 2) && \text{Substitute.} \\ 3(y + 1) &= 2(x - 2) && \text{Simplify.} \\ 2x - 3y - 7 &= 0 && \text{General form} \end{aligned}$$

Note the similarity to the original equation.

b. Using the negative reciprocal of the slope of the given line, you can determine that the slope of a line perpendicular to the given line is $-\frac{3}{2}$. So, the line through the point $(2, -1)$ that is perpendicular to the given line has the following equation.

$$\begin{aligned} y - y_1 &= m(x - x_1) && \text{Point-slope form} \\ y - (-1) &= -\tfrac{3}{2}(x - 2) && \text{Substitute.} \\ 2(y + 1) &= -3(x - 2) && \text{Simplify.} \\ 3x + 2y - 4 &= 0 && \text{General form} \end{aligned}$$

TECHNOLOGY PITFALL The slope of a line will appear distorted if you use different tick-mark spacing on the x- and y-axes. For instance, the graphing calculator screens in Figures P.21(a) and P.21(b) both show the lines given by $y = 2x$ and $y = -\frac{1}{2}x + 3$. Because these lines have slopes that are negative reciprocals, they must be perpendicular. In Figure P.21(a), however, the lines don't appear to be perpendicular because the tick-mark spacing on the x-axis is not the same as that on the y-axis. In Figure P.21(b), the lines appear perpendicular because the tick-mark spacing on the x-axis is the same as on the y-axis. This type of viewing window is said to have a *square setting*.

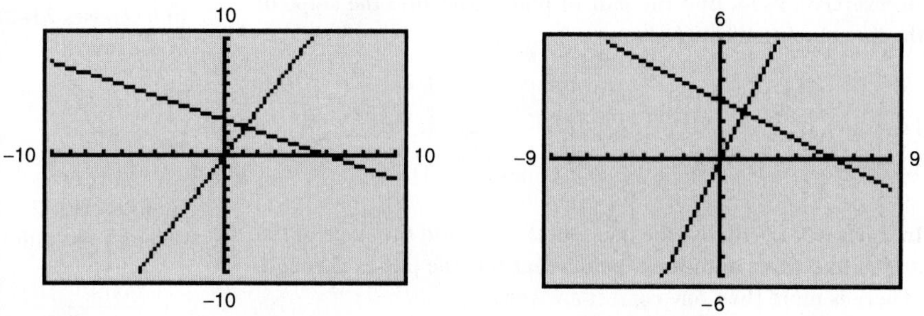

(a) Tick-mark spacing on the x-axis is not the same as tick-mark spacing on the y-axis.

(b) Tick-mark spacing on the x-axis is the same as tick-mark spacing on the y-axis.

Figure P.21

Exercises for Section P.2

See www.CalcChat.com for worked-out solutions to odd-numbered exercises.

In Exercises 1–6, estimate the slope of the line from its graph. To print an enlarged copy of the graph, go to the website *www.mathgraphs.com*.

1.
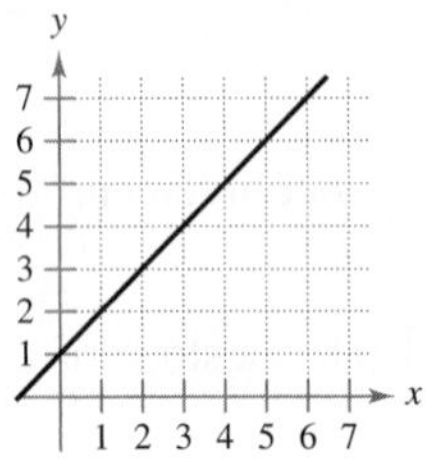

2.
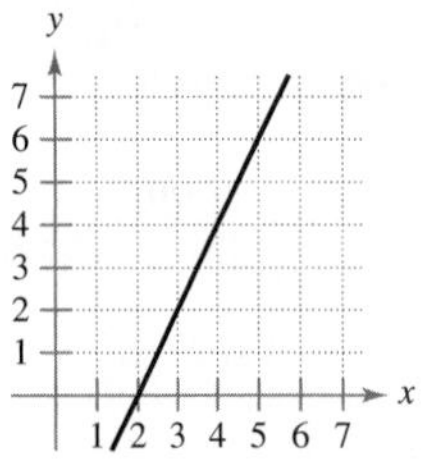

3.
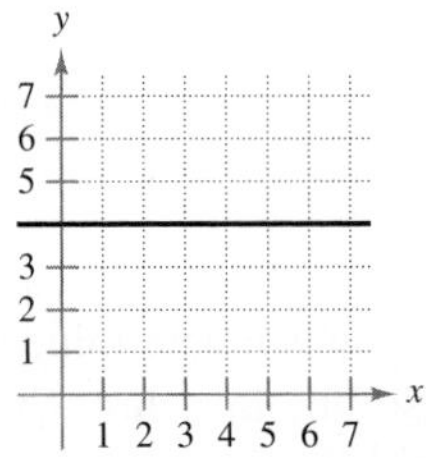

4.
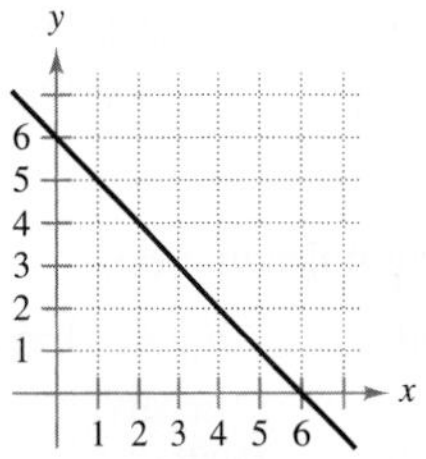

5.
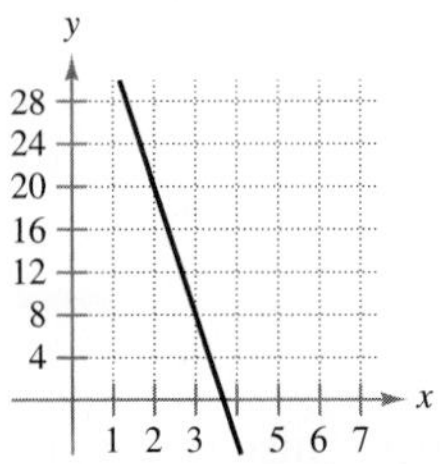

6.
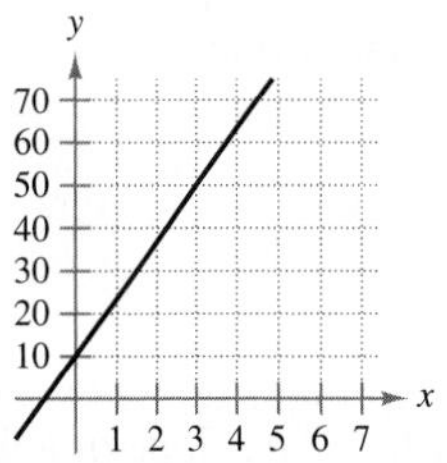

In Exercises 7 and 8, sketch the lines through the point with the indicated slopes. Make the sketches on the same set of coordinate axes.

Point	*Slopes*			
7. (2, 3)	(a) 1	(b) -2	(c) $-\frac{3}{2}$	(d) Undefined
8. $(-4, 1)$	(a) 3	(b) -3	(c) $\frac{1}{3}$	(d) 0

In Exercises 9–14, plot the pair of points and find the slope of the line passing through them.

9. $(3, -4), (5, 2)$

10. $(1, 2), (-2, 4)$

11. $(2, 1), (2, 5)$

12. $(3, -2), (4, -2)$

13. $\left(-\frac{1}{2}, \frac{2}{3}\right), \left(-\frac{3}{4}, \frac{1}{6}\right)$

14. $\left(\frac{7}{8}, \frac{3}{4}\right), \left(\frac{5}{4}, -\frac{1}{4}\right)$

In Exercises 15–18, use the point on the line and the slope of the line to find three additional points that the line passes through. (There is more than one correct answer.)

Point	*Slope*	*Point*	*Slope*
15. (2, 1)	$m = 0$	**16.** $(-3, 4)$	m undefined
17. (1, 7)	$m = -3$	**18.** $(-2, -2)$	$m = 2$

19. ***Conveyor Design*** A moving conveyor is built to rise 1 meter for each 3 meters of horizontal change.

(a) Find the slope of the conveyor.

(b) Suppose the conveyor runs between two floors in a factory. Find the length of the conveyor if the vertical distance between floors is 10 feet.

20. ***Rate of Change*** Each of the following is the slope of a line representing daily revenue y in terms of time x in days. Use the slope to interpret any change in daily revenue for a one-day increase in time.

(a) $m = 400$ (b) $m = 100$ (c) $m = 0$

21. ***Modeling Data*** The table shows the populations y (in millions) of the United States for 1996–2001. The variable t represents the time in years, with $t = 6$ corresponding to 1996. *(Source: U.S. Bureau of the Census)*

t	6	7	8	9	10	11
y	269.7	272.9	276.1	279.3	282.3	285.0

(a) Plot the data by hand and connect adjacent points with a line segment.

(b) Use the slope of each line segment to determine the year when the population increased least rapidly.

22. ***Modeling Data*** The table shows the rate r (in miles per hour) that a vehicle is traveling after t seconds.

t	5	10	15	20	25	30
r	57	74	85	84	61	43

(a) Plot the data by hand and connect adjacent points with a line segment.

(b) Use the slope of each line segment to determine the interval when the vehicle's rate changed most rapidly. How did the rate change?

In Exercises 23–26, find the slope and the y-intercept (if possible) of the line.

23. $x + 5y = 20$

24. $6x - 5y = 15$

25. $x = 4$

26. $y = -1$

In Exercises 27–32, find an equation of the line that passes through the point and has the indicated slope. Sketch the line.

Point	*Slope*	*Point*	*Slope*
27. (0, 3)	$m = \frac{3}{4}$	**28.** $(-1, 2)$	m undefined
29. (0, 0)	$m = \frac{2}{3}$	**30.** (0, 4)	$m = 0$
31. $(3, -2)$	$m = 3$	**32.** $(-2, 4)$	$m = -\frac{3}{5}$

In Exercises 33–42, find an equation of the line that passes through the points, and sketch the line.

33. $(0, 0), (2, 6)$ **34.** $(0, 0), (-1, 3)$

35. $(2, 1), (0, -3)$ **36.** $(-3, -4), (1, 4)$

37. $(2, 8), (5, 0)$ **38.** $(-3, 6), (1, 2)$

39. $(5, 1), (5, 8)$ **40.** $(1, -2), (3, -2)$

41. $\left(\frac{1}{2}, \frac{7}{2}\right), \left(0, \frac{3}{4}\right)$ **42.** $\left(\frac{7}{8}, \frac{3}{4}\right), \left(\frac{5}{4}, -\frac{1}{4}\right)$

43. Find an equation of the vertical line with x-intercept at 3.

44. Show that the line with intercepts $(a, 0)$ and $(0, b)$ has the following equation.

$$\frac{x}{a} + \frac{y}{b} = 1, \quad a \neq 0, b \neq 0$$

In Exercises 45–48, use the result of Exercise 44 to write an equation of the line.

45. x-intercept: $(2, 0)$
y-intercept: $(0, 3)$

46. x-intercept: $\left(-\frac{2}{3}, 0\right)$
y-intercept: $(0, -2)$

47. Point on line: $(1, 2)$
x-intercept: $(a, 0)$
y-intercept: $(0, a)$
$(a \neq 0)$

48. Point on line: $(-3, 4)$
x-intercept: $(a, 0)$
y-intercept: $(0, a)$
$(a \neq 0)$

In Exercises 49–56, sketch a graph of the equation.

49. $y = -3$ **50.** $x = 4$

51. $y = -2x + 1$ **52.** $y = \frac{1}{3}x - 1$

53. $y - 2 = \frac{3}{2}(x - 1)$ **54.** $y - 1 = 3(x + 4)$

55. $2x - y - 3 = 0$ **56.** $x + 2y + 6 = 0$

Square Setting **In Exercises 57 and 58, use a graphing utility to graph both lines in each viewing window. Compare the graphs. Do the lines appear perpendicular? Are the lines perpendicular? Explain.**

57. $y = x + 6, \; y = -x + 2$

(a)
Xmin = -10
Xmax = 10
Xscl = 1
Ymin = -10
Ymax = 10
Yscl = 1

(b)
Xmin = -15
Xmax = 15
Xscl = 1
Ymin = -10
Ymax = 10
Yscl = 1

58. $y = 2x - 3, \; y = -\frac{1}{2}x + 1$

(a)
Xmin = -5
Xmax = 5
Xscl = 1
Ymin = -5
Ymax = 5
Yscl = 1

(b)
Xmin = -6
Xmax = 6
Xscl = 1
Ymin = -4
Ymax = 4
Yscl = 1

In Exercises 59–64, write an equation of the line through the point (a) parallel to the given line and (b) perpendicular to the given line.

	Point	*Line*		*Point*	*Line*
59.	$(2, 1)$	$4x - 2y = 3$	**60.**	$(-3, 2)$	$x + y = 7$
61.	$\left(\frac{3}{4}, \frac{7}{8}\right)$	$5x - 3y = 0$	**62.**	$(-6, 4)$	$3x + 4y = 7$
63.	$(2, 5)$	$x = 4$	**64.**	$(-1, 0)$	$y = -3$

Rate of Change **In Exercises 65–68, you are given the dollar value of a product in 2004 *and* the rate at which the value of the product is expected to change during the next 5 years. Write a linear equation that gives the dollar value V of the product in terms of the year t. (Let $t = 0$ represent 2000.)**

	2004 Value	*Rate*
65.	\$2540	\$125 increase per year
66.	\$156	\$4.50 increase per year
67.	\$20,400	\$2000 decrease per year
68.	\$245,000	\$5600 decrease per year

In Exercises 69 and 70, use a graphing utility to graph the parabolas and find their points of intersection. Find an equation of the line through the points of intersection and graph the line in the same viewing window.

69. $y = x^2$
$y = 4x - x^2$

70. $y = x^2 - 4x + 3$
$y = -x^2 + 2x + 3$

In Exercises 71 and 72, determine whether the points are collinear. (Three points are *collinear* if they lie on the same line.)

71. $(-2, 1), (-1, 0), (2, -2)$ **72.** $(0, 4), (7, -6), (-5, 11)$

Writing About Concepts

In Exercises 73–75, find the coordinates of the point of intersection of the given segments. Explain your reasoning.

73.

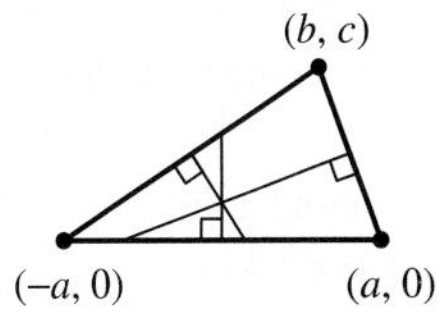

Perpendicular bisectors

74.

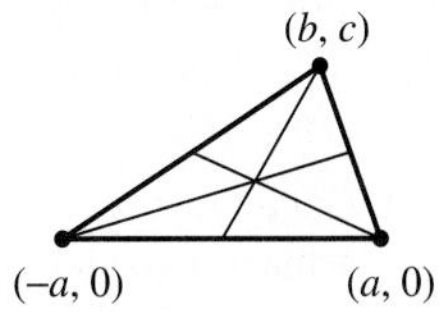

Medians

75.

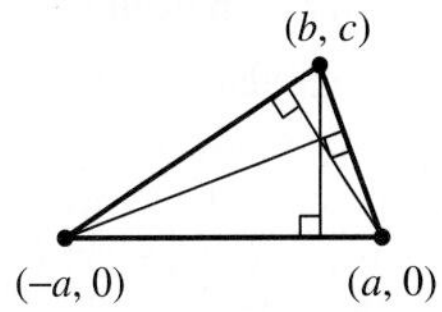

Altitudes

76. Show that the points of intersection in Exercises 73, 74, and 75 are collinear.

77. ***Temperature Conversion*** Find a linear equation that expresses the relationship between the temperature in degrees Celsius C and degrees Fahrenheit F. Use the fact that water freezes at 0°C (32°F) and boils at 100°C (212°F). Use the equation to convert 72°F to degrees Celsius.

78. ***Reimbursed Expenses*** A company reimburses its sales representatives \$150 per day for lodging and meals plus 34¢ per mile driven. Write a linear equation giving the daily cost C to the company in terms of x, the number of miles driven. How much does it cost the company if a sales representative drives 137 miles on a given day?

79. ***Career Choice*** An employee has two options for positions in a large corporation. One position pays \$12.50 per hour *plus* an additional unit rate of \$0.75 per unit produced. The other pays \$9.20 per hour *plus* a unit rate of \$1.30.

(a) Find linear equations for the hourly wages W in terms of x, the number of units produced per hour, for each option.

(b) Use a graphing utility to graph the linear equations and find the point of intersection.

(c) Interpret the meaning of the point of intersection of the graphs in part (b). How would you use this information to select the correct option if the goal were to obtain the highest hourly wage?

80. ***Straight-Line Depreciation*** A small business purchases a piece of equipment for \$875. After 5 years the equipment will be outdated, having no value.

(a) Write a linear equation giving the value y of the equipment in terms of the time x, $0 \le x \le 5$.

(b) Find the value of the equipment when $x = 2$.

(c) Estimate (to two-decimal-place accuracy) the time when the value of the equipment is \$200.

81. ***Apartment Rental*** A real estate office handles an apartment complex with 50 units. When the rent is \$580 per month, all 50 units are occupied. However, when the rent is \$625, the average number of occupied units drops to 47. Assume that the relationship between the monthly rent p and the demand x is linear. (*Note:* The term *demand* refers to the number of occupied units.)

(a) Write a linear equation giving the demand x in terms of the rent p.

(b) *Linear extrapolation* Use a graphing utility to graph the demand equation and use the *trace* feature to predict the number of units occupied if the rent is raised to \$655.

(c) *Linear interpolation* Predict the number of units occupied if the rent is lowered to \$595. Verify graphically.

82. ***Modeling Data*** An instructor gives regular 20-point quizzes and 100-point exams in a mathematics course. Average scores for six students, given as ordered pairs (x, y) where x is the average quiz score and y is the average test score, are (18, 87), (10, 55), (19, 96), (16, 79), (13, 76), and (15, 82).

(a) Use the regression capabilities of a graphing utility to find the least squares regression line for the data.

(b) Use a graphing utility to plot the points and graph the regression line in the same viewing window.

(c) Use the regression line to predict the average exam score for a student with an average quiz score of 17.

(d) Interpret the meaning of the slope of the regression line.

(e) The instructor adds 4 points to the average test score of everyone in the class. Describe the changes in the positions of the plotted points and the change in the equation of the line.

83. ***Tangent Line*** Find an equation of the line tangent to the circle $x^2 + y^2 = 169$ at the point (5, 12).

84. ***Tangent Line*** Find an equation of the line tangent to the circle $(x - 1)^2 + (y - 1)^2 = 25$ at the point $(4, -3)$.

Distance **In Exercises 85–90, find the distance between the point and line, or between the lines, using the formula for the distance between the point (x_1, y_1) and the line $Ax + By + C = 0$.**

$$\textbf{Distance} = \frac{|Ax_1 + By_1 + C|}{\sqrt{A^2 + B^2}}$$

85. Point: (0, 0)
Line: $4x + 3y = 10$

86. Point: (2, 3)
Line: $4x + 3y = 10$

87. Point: $(-2, 1)$
Line: $x - y - 2 = 0$

88. Point: (6, 2)
Line: $x = -1$

89. Line: $x + y = 1$
Line: $x + y = 5$

90. Line: $3x - 4y = 1$
Line: $3x - 4y = 10$

91. Show that the distance between the point (x_1, y_1) and the line $Ax + By + C = 0$ is

$$\text{Distance} = \frac{|Ax_1 + By_1 + C|}{\sqrt{A^2 + B^2}}.$$

92. Write the distance d between the point (3, 1) and the line $y = mx + 4$ in terms of m. Use a graphing utility to graph the equation. When is the distance 0? Explain the result geometrically.

93. Prove that the diagonals of a rhombus intersect at right angles. (A rhombus is a quadrilateral with sides of equal lengths.)

94. Prove that the figure formed by connecting consecutive midpoints of the sides of any quadrilateral is a parallelogram.

95. Prove that if the points (x_1, y_1) and (x_2, y_2) lie on the same line as (x_1^*, y_1^*) and (x_2^*, y_2^*), then

$$\frac{y_2^* - y_1^*}{x_2^* - x_1^*} = \frac{y_2 - y_1}{x_2 - x_1}.$$

Assume $x_1 \ne x_2$ and $x_1^* \ne x_2^*$.

96. Prove that if the slopes of two nonvertical lines are negative reciprocals of each other, then the lines are perpendicular.

True or False? **In Exercises 97 and 98, determine whether the statement is true or false. If it is false, explain why or give an example that shows it is false.**

97. The lines represented by $ax + by = c_1$ and $bx - ay = c_2$ are perpendicular. Assume $a \ne 0$ and $b \ne 0$.

98. It is possible for two lines with positive slopes to be perpendicular to each other.

Section P.3 Functions and Their Graphs

- **Use function notation to represent and evaluate a function.**
- **Find the domain and range of a function.**
- **Sketch the graph of a function.**
- **Identify different types of transformations of functions.**
- **Classify functions and recognize combinations of functions.**

Functions and Function Notation

A **relation** between two sets X and Y is a set of ordered pairs, each of the form (x, y), where x is a member of X and y is a member of Y. A **function** from X to Y is a relation between X and Y that has the property that any two ordered pairs with the same x-value also have the same y-value. The variable x is the **independent variable,** and the variable y is the **dependent variable.**

Many real-life situations can be modeled by functions. For instance, the area A of a circle is a function of the circle's radius r.

$$A = \pi r^2 \qquad \text{A is a function of r.}$$

In this case r is the independent variable and A is the dependent variable.

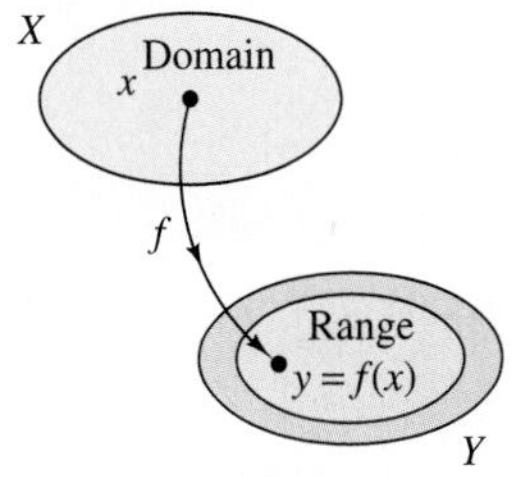

A real-valued function f of a real variable
Figure P.22

Definition of a Real-Valued Function of a Real Variable

Let X and Y be sets of real numbers. A **real-valued function f of a real variable x** from X to Y is a correspondence that assigns to each number x in X exactly one number y in Y.

The **domain** of f is the set X. The number y is the **image** of x under f and is denoted by $f(x)$, which is called the **value of f at x.** The **range** of f is a subset of Y and consists of all images of numbers in X (see Figure P.22).

Functions can be specified in a variety of ways. In this text, however, we will concentrate primarily on functions that are given by equations involving the dependent and independent variables. For instance, the equation

$$x^2 + 2y = 1 \qquad \text{Equation in implicit form}$$

defines y, the dependent variable, as a function of x, the independent variable. To **evaluate** this function (that is, to find the y-value that corresponds to a given x-value), it is convenient to isolate y on the left side of the equation.

$$y = \frac{1}{2}(1 - x^2) \qquad \text{Equation in explicit form}$$

Using f as the name of the function, you can write this equation as

$$f(x) = \frac{1}{2}(1 - x^2). \qquad \text{Function notation}$$

The original equation, $x^2 + 2y = 1$, **implicitly** defines y as a function of x. When you solve the equation for y, you are writing the equation in **explicit** form.

Function notation has the advantage of clearly identifying the dependent variable as $f(x)$ while at the same time telling you that x is the independent variable and that the function itself is "f." The symbol $f(x)$ is read "f of x." Function notation allows you to be less wordy. Instead of asking "What is the value of y that corresponds to $x = 3$?" you can ask "What is $f(3)$?"

FUNCTION NOTATION

The word *function* was first used by Gottfried Wilhelm Leibniz in 1694 as a term to denote any quantity connected with a curve, such as the coordinates of a point on a curve or the slope of a curve. Forty years later, Leonhard Euler used the word "function" to describe any expression made up of a variable and some constants. He introduced the notation $y = f(x)$.

In an equation that defines a function, the role of the variable x is simply that of a placeholder. For instance, the function given by

$$f(x) = 2x^2 - 4x + 1$$

can be described by the form

$$f(\quad) = 2(\quad)^2 - 4(\quad) + 1$$

where parentheses are used instead of x. To evaluate $f(-2)$, simply place -2 in each set of parentheses.

$$\begin{aligned} f(-2) &= 2(-2)^2 - 4(-2) + 1 && \text{Substitute } -2 \text{ for } x. \\ &= 2(4) + 8 + 1 && \text{Simplify.} \\ &= 17 && \text{Simplify.} \end{aligned}$$

NOTE Although f is often used as a convenient function name and x as the independent variable, you can use other symbols. For instance, the following equations all define the same function.

$$\begin{aligned} f(x) &= x^2 - 4x + 7 && \text{Function name is } f\text{, independent variable is } x. \\ f(t) &= t^2 - 4t + 7 && \text{Function name is } f\text{, independent variable is } t. \\ g(s) &= s^2 - 4s + 7 && \text{Function name is } g\text{, independent variable is } s. \end{aligned}$$

EXAMPLE 1 Evaluating a Function

For the function f defined by $f(x) = x^2 + 7$, evaluate each expression.

a. $f(3a)$ **b.** $f(b - 1)$ **c.** $\dfrac{f(x + \Delta x) - f(x)}{\Delta x}, \quad \Delta x \neq 0$

Solution

a. $$\begin{aligned} f(3a) &= (3a)^2 + 7 && \text{Substitute } 3a \text{ for } x. \\ &= 9a^2 + 7 && \text{Simplify.} \end{aligned}$$

b. $$\begin{aligned} f(b - 1) &= (b - 1)^2 + 7 && \text{Substitute } b - 1 \text{ for } x. \\ &= b^2 - 2b + 1 + 7 && \text{Expand binomial.} \\ &= b^2 - 2b + 8 && \text{Simplify.} \end{aligned}$$

c. $$\begin{aligned} \frac{f(x + \Delta x) - f(x)}{\Delta x} &= \frac{[(x + \Delta x)^2 + 7] - (x^2 + 7)}{\Delta x} \\ &= \frac{x^2 + 2x\Delta x + (\Delta x)^2 + 7 - x^2 - 7}{\Delta x} \\ &= \frac{2x\Delta x + (\Delta x)^2}{\Delta x} \\ &= \frac{\cancel{\Delta x}(2x + \Delta x)}{\cancel{\Delta x}} \\ &= 2x + \Delta x, \quad \Delta x \neq 0 \end{aligned}$$

STUDY TIP In calculus, it is important to communicate clearly the domain of a function or expression. For instance, in Example 1(c) the two expressions

$$\frac{f(x + \Delta x) - f(x)}{\Delta x} \quad \text{and} \quad 2x + \Delta x, \quad \Delta x \neq 0$$

are equivalent because $\Delta x = 0$ is excluded from the domain of each expression. Without a stated domain restriction, the two expressions would not be equivalent.

NOTE The expression in Example 1(c) is called a *difference quotient* and has a special significance in calculus. You will learn more about this in Chapter 2.

The Domain and Range of a Function

The domain of a function can be described explicitly, or it may be described *implicitly* by an equation used to define the function. The implied domain is the set of all real numbers for which the equation is defined, whereas an explicitly defined domain is one that is given along with the function. For example, the function given by

$$f(x) = \frac{1}{x^2 - 4}, \qquad 4 \le x \le 5$$

has an explicitly defined domain given by $\{x\colon\ 4 \le x \le 5\}$. On the other hand, the function given by

$$g(x) = \frac{1}{x^2 - 4}$$

has an implied domain that is the set $\{x\colon\ x \ne \pm 2\}$.

EXAMPLE 2 Finding the Domain and Range of a Function

a. The domain of the function

$$f(x) = \sqrt{x - 1}$$

is the set of all x-values for which $x - 1 \ge 0$, which is the interval $[1, \infty)$. To find the range observe that $f(x) = \sqrt{x - 1}$ is never negative. So, the range is the interval $[0, \infty)$, as indicated in Figure P.23(a).

b. The domain of the tangent function, as shown in Figure P.23(b),

$$f(x) = \tan x$$

is the set of all x-values such that

$$x \ne \frac{\pi}{2} + n\pi, \qquad n \text{ is an integer.} \qquad \text{Domain of tangent function}$$

The range of this function is the set of all real numbers. For a review of the characteristics of this and other trigonometric functions, see Appendix D.

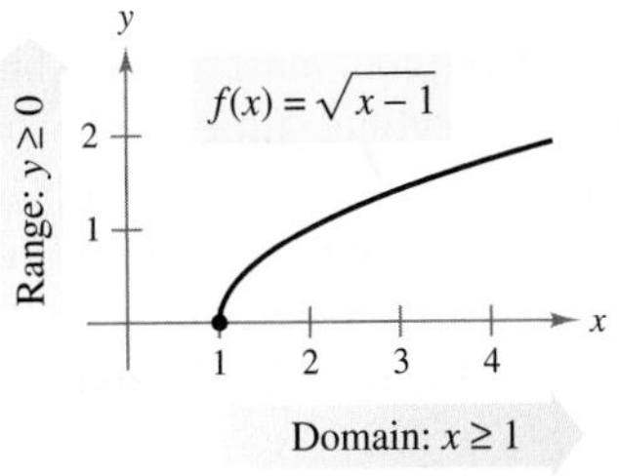

(a) The domain of f is $[1, \infty)$ and the range is $[0, \infty)$.

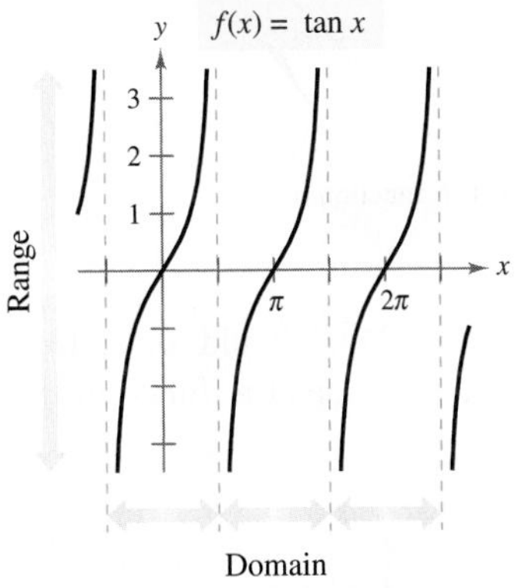

(b) The domain of f is all x-values such that $x \ne \frac{\pi}{2} + n\pi$ and the range is $(-\infty, \infty)$.

Figure P.23

EXAMPLE 3 A Function Defined by More than One Equation

Determine the domain and range of the function.

$$f(x) = \begin{cases} 1 - x, & \text{if } x < 1 \\ \sqrt{x - 1}, & \text{if } x \ge 1 \end{cases}$$

Solution Because f is defined for $x < 1$ and $x \ge 1$, the domain is the entire set of real numbers. On the portion of the domain for which $x \ge 1$, the function behaves as in Example 2(a). For $x < 1$, the values of $1 - x$ are positive. So, the range of the function is the interval $[0, \infty)$. (See Figure P.24.)

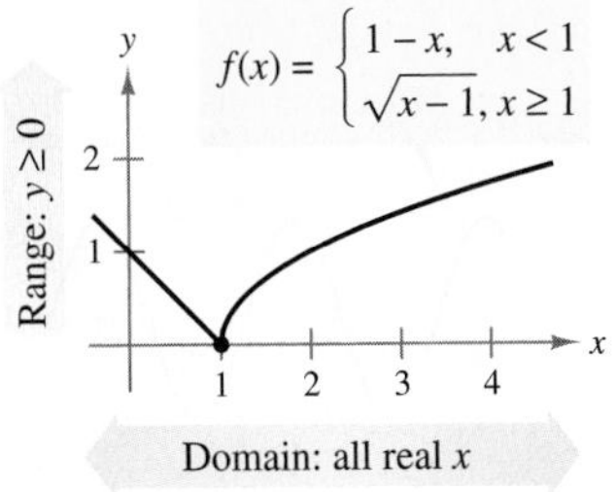

The domain of f is $(-\infty, \infty)$ and the range is $[0, \infty)$.

Figure P.24

A function from X to Y is **one-to-one** if to each y-value in the range there corresponds exactly one x-value in the domain. For instance, the function given in Example 2(a) is one-to-one, whereas the functions given in Examples 2(b) and 3 are not one-to-one. A function from X to Y is **onto** if its range consists of all of Y.

The Graph of a Function

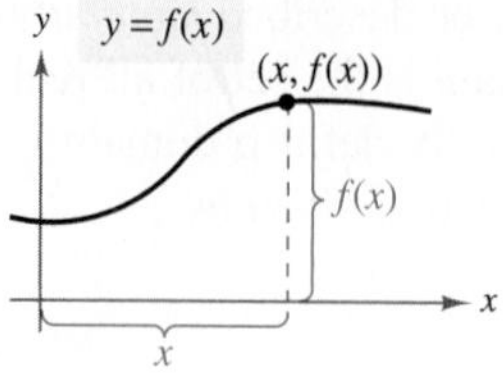

The graph of a function
Figure P.25

The graph of the function $y = f(x)$ consists of all points $(x, f(x))$, where x is in the domain of f. In Figure P.25, note that

$$x = \text{the directed distance from the } y\text{-axis}$$
$$f(x) = \text{the directed distance from the } x\text{-axis.}$$

A vertical line can intersect the graph of a function of x at most *once*. This observation provides a convenient visual test, called the **Vertical Line Test,** for functions of x. That is, a graph in the coordinate plane is the graph of a function of x if and only if no vertical line intersects the graph at more than one point. For example, in Figure P.26(a), you can see that the graph does not define y as a function of x because a vertical line intersects the graph twice, whereas in Figures P.26(b) and (c), the graphs do define y as a function of x.

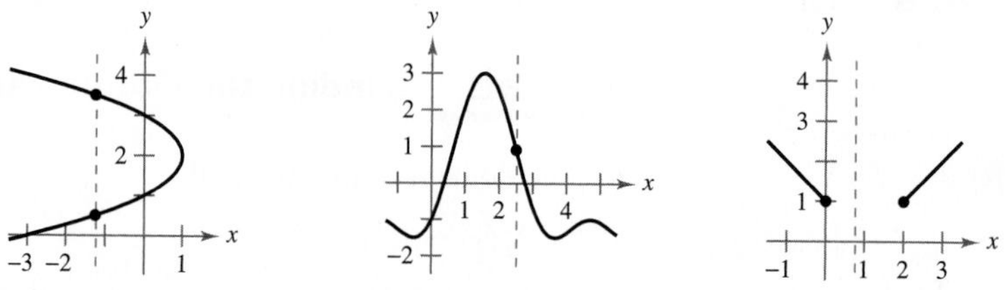

(a) Not a function of x **(b)** A function of x **(c)** A function of x
Figure P.26

Figure P.27 shows the graphs of eight basic functions. You should be able to recognize these graphs. (Graphs of the other four basic trigonometric functions are shown in Appendix D.)

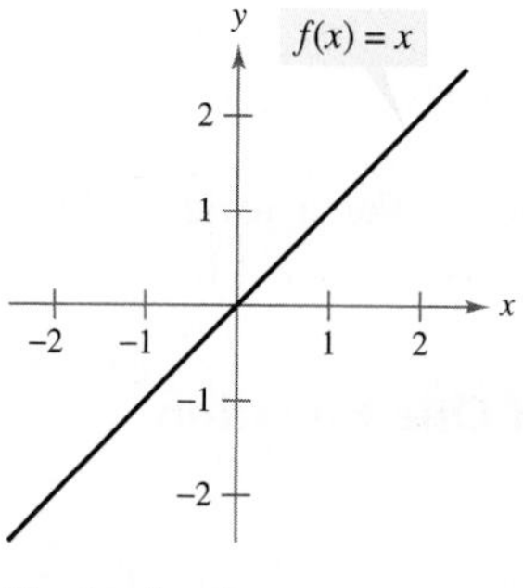

Identity function

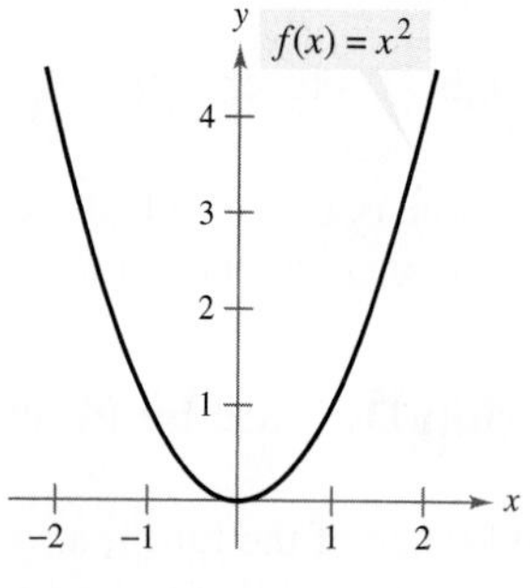

Squaring function

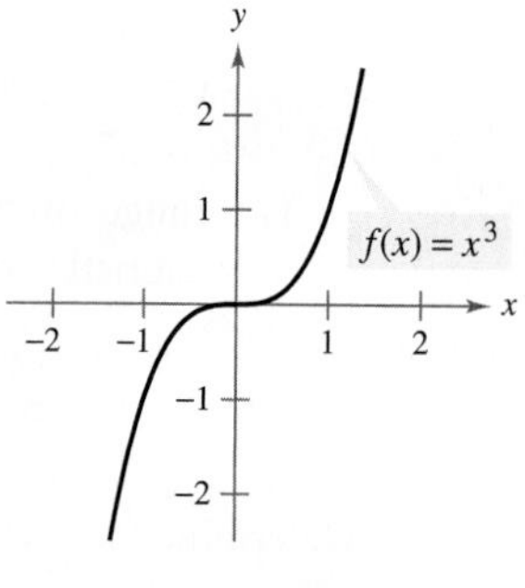

Cubing function

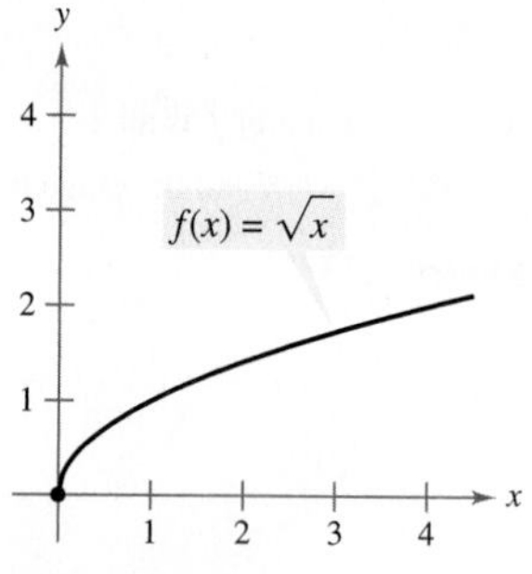

Square root function

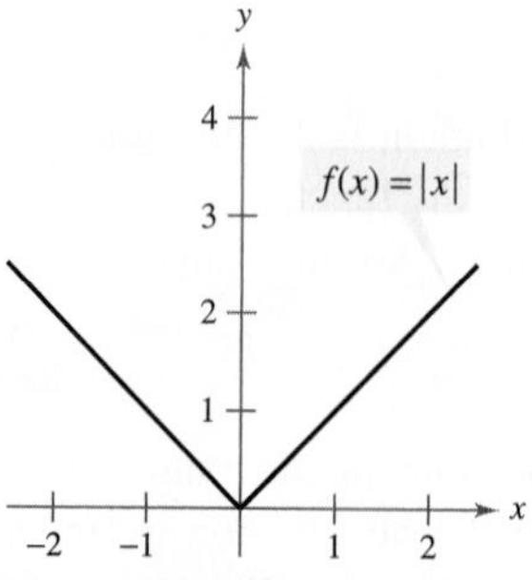

Absolute value function

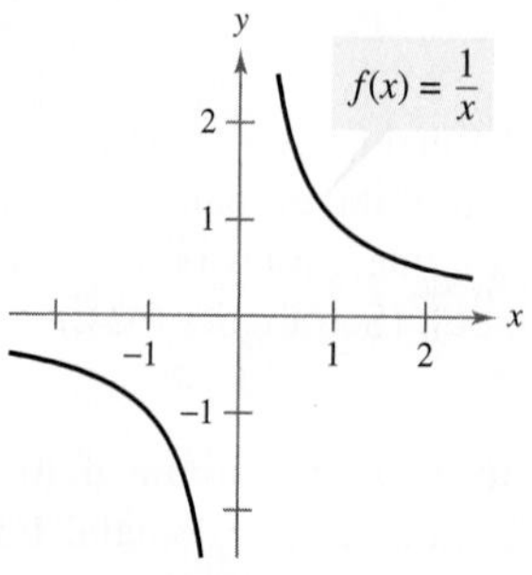

Rational function

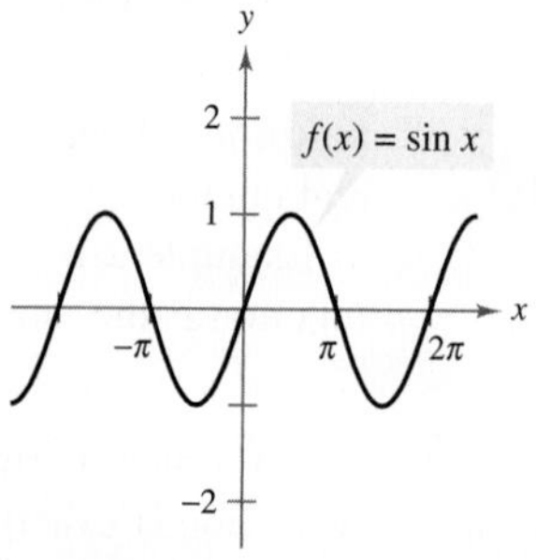

Sine function

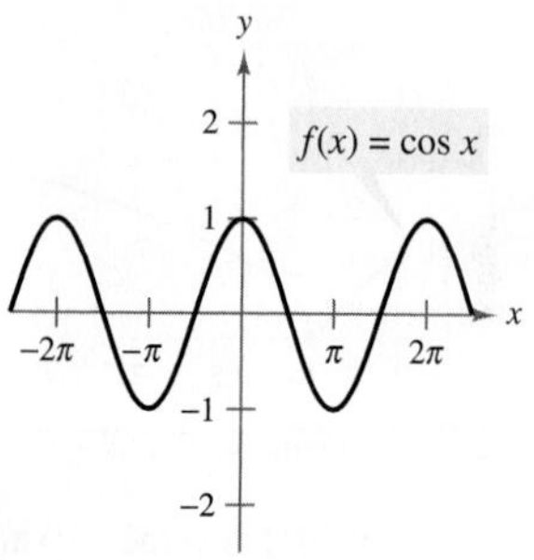

Cosine function

The graphs of eight basic functions
Figure P.27

EXPLORATION

Writing Equations for Functions
Each of the graphing utility screens below shows the graph of one of the eight basic functions shown on page 22. Each screen also shows a transformation of the graph. Describe the transformation. Then use your description to write an equation for the transformation.

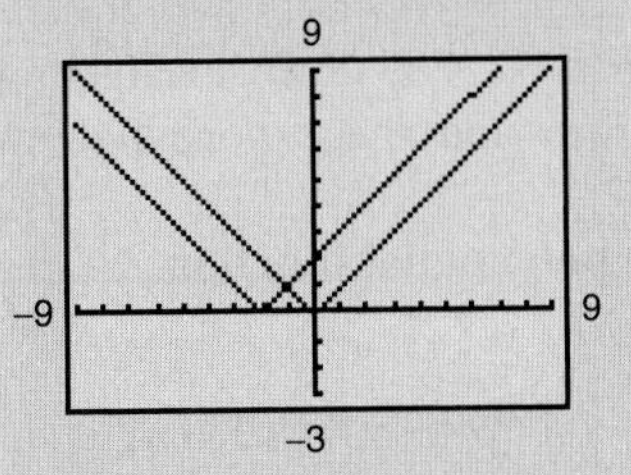

a.

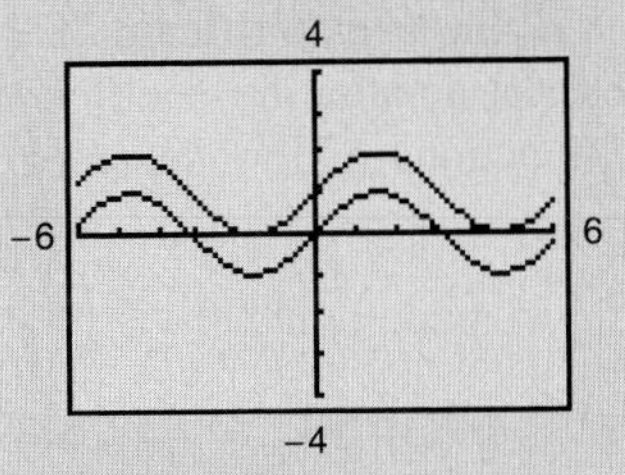

b.

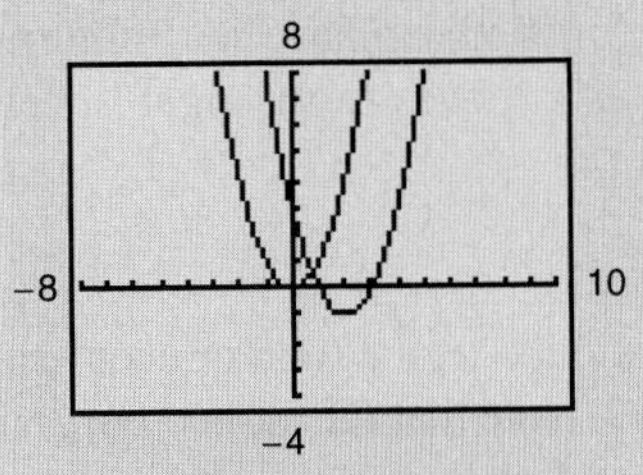

c.

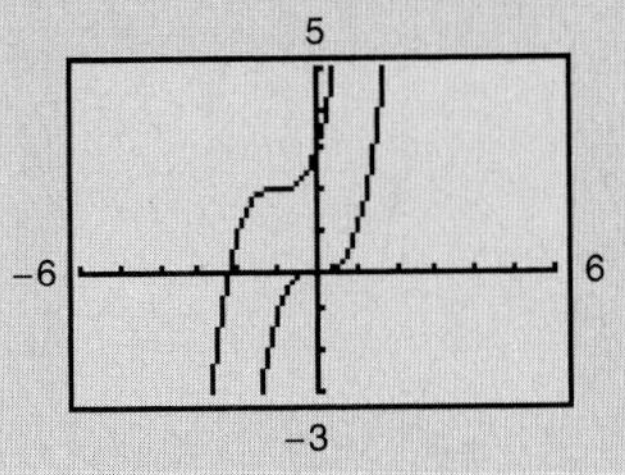

d.

Transformations of Functions

Some families of graphs have the same basic shape. For example, compare the graph of $y = x^2$ with the graphs of the four other quadratic functions shown in Figure P.28.

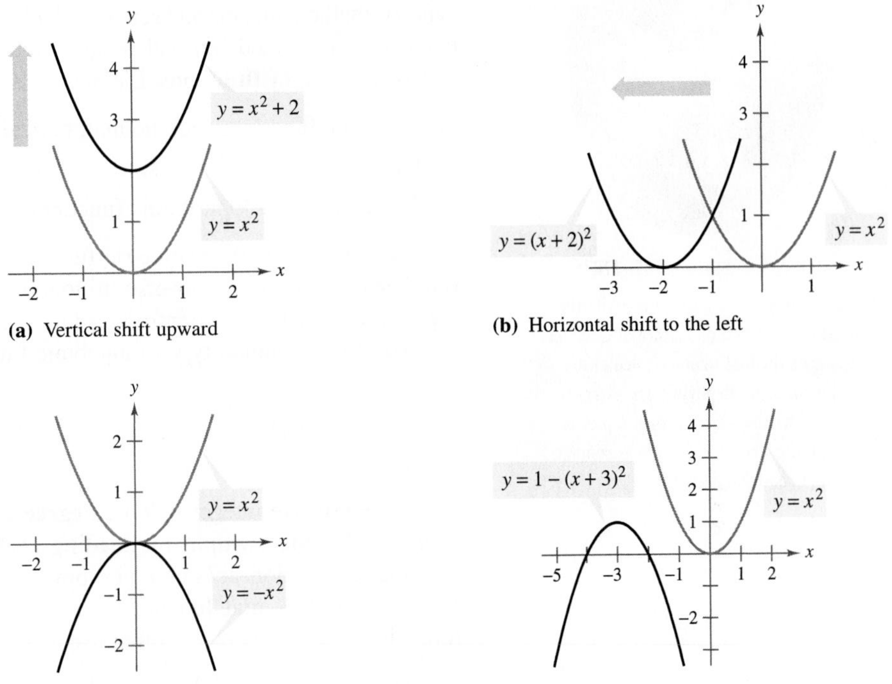

(a) Vertical shift upward
(b) Horizontal shift to the left
(c) Reflection
(d) Shift left, reflect, and shift upward

Figure P.28

Each of the graphs in Figure P.28 is a **transformation** of the graph of $y = x^2$. The three basic types of transformations illustrated by these graphs are vertical shifts, horizontal shifts, and reflections. Function notation lends itself well to describing transformations of graphs in the plane. For instance, if $f(x) = x^2$ is considered to be the original function in Figure P.28, the transformations shown can be represented by the following equations.

$y = f(x) + 2$	Vertical shift up 2 units
$y = f(x + 2)$	Horizontal shift to the left 2 units
$y = -f(x)$	Reflection about the x-axis
$y = -f(x + 3) + 1$	Shift left 3 units, reflect about x-axis, and shift up 1 unit

Basic Types of Transformations ($c > 0$)

Original graph:	$y = f(x)$
Horizontal shift c units to the **right**:	$y = f(x - c)$
Horizontal shift c units to the **left**:	$y = f(x + c)$
Vertical shift c units **downward**:	$y = f(x) - c$
Vertical shift c units **upward**:	$y = f(x) + c$
Reflection (about the x-axis):	$y = -f(x)$
Reflection (about the y-axis):	$y = f(-x)$
Reflection (about the origin):	$y = -f(-x)$

LEONHARD EULER (1707–1783)

In addition to making major contributions to almost every branch of mathematics, Euler was one of the first to apply calculus to real-life problems in physics. His extensive published writings include such topics as shipbuilding, acoustics, optics, astronomy, mechanics, and magnetism.

Classifications and Combinations of Functions

The modern notion of a function is derived from the efforts of many seventeenth- and eighteenth-century mathematicians. Of particular note was Leonhard Euler, to whom we are indebted for the function notation $y = f(x)$. By the end of the eighteenth century, mathematicians and scientists had concluded that many real-world phenomena could be represented by mathematical models taken from a collection of functions called **elementary functions.** Elementary functions fall into three categories.

1. Algebraic functions (polynomial, radical, rational)
2. Trigonometric functions (sine, cosine, tangent, and so on)
3. Exponential and logarithmic functions

You can review the trigonometric functions in Appendix D. The other nonalgebraic functions, such as the inverse trigonometric functions and the exponential and logarithmic functions, are introduced in Chapter 5.

The most common type of algebraic function is a **polynomial function**

$$f(x) = a_n x^n + a_{n-1}x^{n-1} + \cdot \cdot \cdot + a_2x^2 + a_1x + a_0, \qquad a_n \neq 0$$

where the positive integer n is the **degree** of the polynomial function. The constants a_i are **coefficients,** with a_n the **leading coefficient** and a_0 the **constant term** of the polynomial function. It is common practice to use subscript notation for coefficients of general polynomial functions, but for polynomial functions of low degree, the following simpler forms are often used.

Zeroth degree: $f(x) = a$ — Constant function

First degree: $f(x) = ax + b$ — Linear function

Second degree: $f(x) = ax^2 + bx + c$ — Quadratic function

Third degree: $f(x) = ax^3 + bx^2 + cx + d$ — Cubic function

FOR FURTHER INFORMATION For more on the history of the concept of a function, see the article "Evolution of the Function Concept: A Brief Survey" by Israel Kleiner in *The College Mathematics Journal.* To view this article, go to the website *www.matharticles.com.*

Although the graph of a nonconstant polynomial function can have several turns, eventually the graph will rise or fall without bound as x moves to the right or left. Whether the graph of

$$f(x) = a_n x^n + a_{n-1}x^{n-1} + \cdot \cdot \cdot + a_2x^2 + a_1x + a_0$$

eventually rises or falls can be determined by the function's degree (odd or even) and by the leading coefficient a_n, as indicated in Figure P.29. Note that the dashed portions of the graphs indicate that the **Leading Coefficient Test** determines *only* the right and left behavior of the graph.

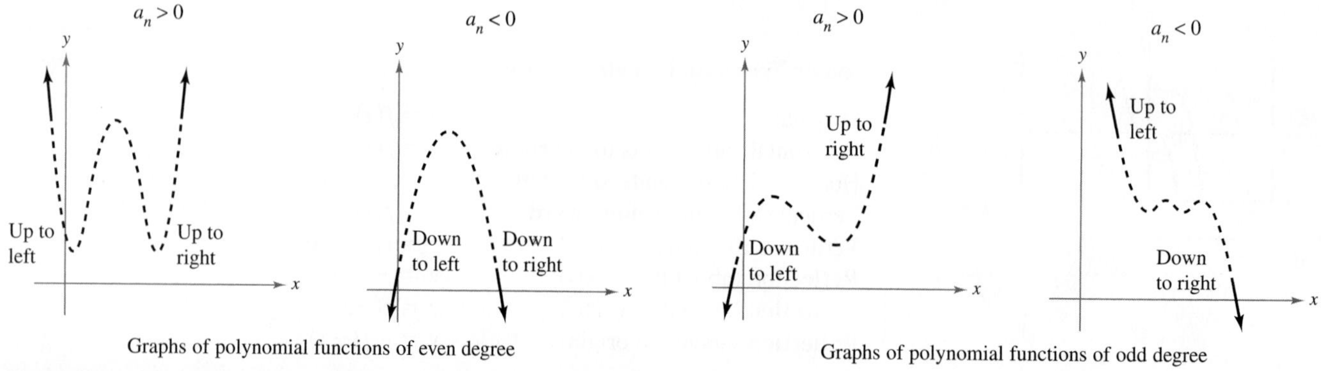

The Leading Coefficient Test for polynomial functions
Figure P.29

Just as a rational number can be written as the quotient of two integers, a **rational function** can be written as the quotient of two polynomials. Specifically, a function f is rational if it has the form

$$f(x) = \frac{p(x)}{q(x)}, \quad q(x) \neq 0$$

where $p(x)$ and $q(x)$ are polynomials.

Polynomial functions and rational functions are examples of **algebraic functions.** An algebraic function of x is one that can be expressed as a finite number of sums, differences, multiples, quotients, and radicals involving x^n. For example, $f(x) = \sqrt{x+1}$ is algebraic. Functions that are not algebraic are **transcendental.** For instance, the trigonometric functions are transcendental.

Two functions can be combined in various ways to create new functions. For example, given $f(x) = 2x - 3$ and $g(x) = x^2 + 1$, you can form the functions shown.

$$(f+g)(x) = f(x) + g(x) = (2x-3) + (x^2+1) \qquad \text{Sum}$$
$$(f-g)(x) = f(x) - g(x) = (2x-3) - (x^2+1) \qquad \text{Difference}$$
$$(fg)(x) = f(x)g(x) = (2x-3)(x^2+1) \qquad \text{Product}$$
$$(f/g)(x) = \frac{f(x)}{g(x)} = \frac{2x-3}{x^2+1} \qquad \text{Quotient}$$

You can combine two functions in yet another way, called **composition.** The resulting function is called a **composite function.**

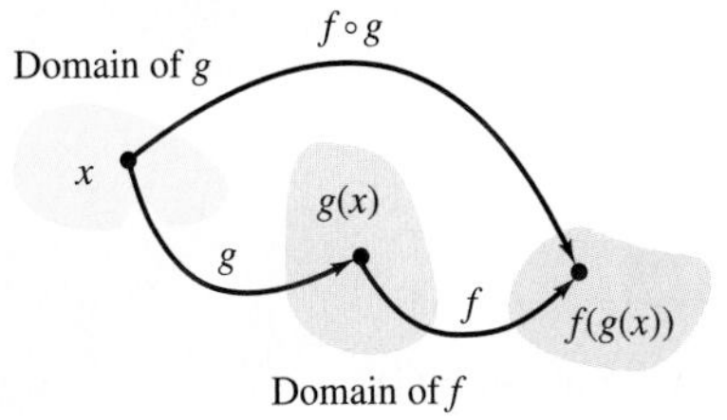

The domain of the composite function $f \circ g$
Figure P.30

Definition of Composite Function

Let f and g be functions. The function given by $(f \circ g)(x) = f(g(x))$ is called the **composite** of f with g. The domain of $f \circ g$ is the set of all x in the domain of g such that $g(x)$ is in the domain of f (see Figure P.30).

The composite of f with g may not be equal to the composite of g with f.

EXAMPLE 4 Finding Composite Functions

Given $f(x) = 2x - 3$ and $g(x) = \cos x$, find each composite function.

a. $f \circ g$ **b.** $g \circ f$

Solution

a.
$$\begin{aligned}(f \circ g)(x) &= f(g(x)) && \text{Definition of } f \circ g\\ &= f(\cos x) && \text{Substitute } \cos x \text{ for } g(x).\\ &= 2(\cos x) - 3 && \text{Definition of } f(x)\\ &= 2\cos x - 3 && \text{Simplify.}\end{aligned}$$

b.
$$\begin{aligned}(g \circ f)(x) &= g(f(x)) && \text{Definition of } g \circ f\\ &= g(2x-3) && \text{Substitute } 2x-3 \text{ for } f(x).\\ &= \cos(2x-3) && \text{Definition of } g(x)\end{aligned}$$

Note that $(f \circ g)(x) \neq (g \circ f)(x)$.

EXPLORATION

Use a graphing utility to graph each function. Determine whether the function is *even, odd,* or *neither.*

$f(x) = x^2 - x^4$

$g(x) = 2x^3 + 1$

$h(x) = x^5 - 2x^3 + x$

$j(x) = 2 - x^6 - x^8$

$k(x) = x^5 - 2x^4 + x - 2$

$p(x) = x^9 + 3x^5 - x^3 + x$

Describe a way to identify a function as odd or even by inspecting the equation.

In Section P.1, an x-intercept of a graph was defined to be a point $(a, 0)$ at which the graph crosses the x-axis. If the graph represents a function f, the number a is a **zero** of f. In other words, *the zeros of a function f are the solutions of the equation $f(x) = 0$.* For example, the function $f(x) = x - 4$ has a zero at $x = 4$ because $f(4) = 0$.

In Section P.1 you also studied different types of symmetry. In the terminology of functions, a function is **even** if its graph is symmetric with respect to the y-axis, and is **odd** if its graph is symmetric with respect to the origin. The symmetry tests in Section P.1 yield the following test for even and odd functions.

Test for Even and Odd Functions

The function $y = f(x)$ is **even** if $f(-x) = f(x)$.

The function $y = f(x)$ is **odd** if $f(-x) = -f(x)$.

NOTE Except for the constant function $f(x) = 0$, the graph of a function of x cannot have symmetry with respect to the x-axis because it then would fail the Vertical Line Test for the graph of the function.

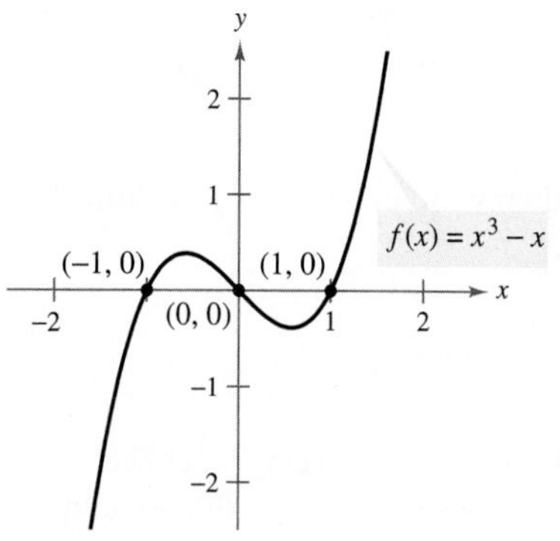

(a) Odd function

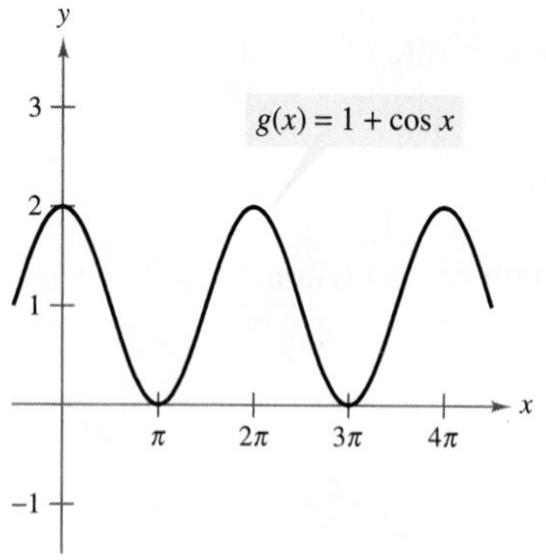

(b) Even function

Figure P.31

EXAMPLE 5 Even and Odd Functions and Zeros of Functions

Determine whether each function is even, odd, or neither. Then find the zeros of the function.

a. $f(x) = x^3 - x$ **b.** $g(x) = 1 + \cos x$

Solution

a. This function is odd because

$$f(-x) = (-x)^3 - (-x) = -x^3 + x = -(x^3 - x) = -f(x).$$

The zeros of f are found as shown.

$$x^3 - x = 0 \qquad \text{Let } f(x) = 0.$$

$$x(x^2 - 1) = x(x - 1)(x + 1) = 0 \qquad \text{Factor.}$$

$$x = 0, 1, -1 \qquad \text{Zeros of } f$$

See Figure P.31(a).

b. This function is even because

$$g(-x) = 1 + \cos(-x) = 1 + \cos x = g(x). \qquad \cos(-x) = \cos(x)$$

The zeros of g are found as shown.

$$1 + \cos x = 0 \qquad \text{Let } g(x) = 0.$$

$$\cos x = -1 \qquad \text{Subtract 1 from each side.}$$

$$x = (2n + 1)\pi,\ n \text{ is an integer.} \qquad \text{Zeros of } g$$

See Figure P.31(b).

NOTE Each of the functions in Example 5 is either even or odd. However, some functions, such as $f(x) = x^2 + x + 1$, are neither even nor odd.

Exercises for Section P.3

See www.CalcChat.com for worked-out solutions to odd-numbered exercises.

In Exercises 1 and 2, use the graphs of f and g to answer the following.

(a) Identify the domains and ranges of f and g.

(b) Identify $f(-2)$ and $g(3)$.

(c) For what value(s) of x is $f(x) = g(x)$?

(d) Estimate the solution(s) of $f(x) = 2$.

(e) Estimate the solutions of $g(x) = 0$.

1.

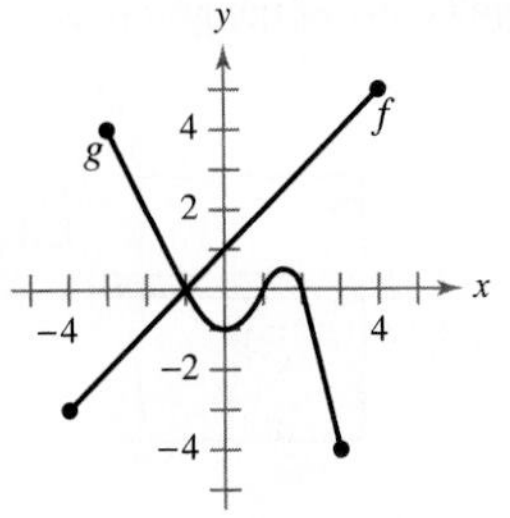

2.

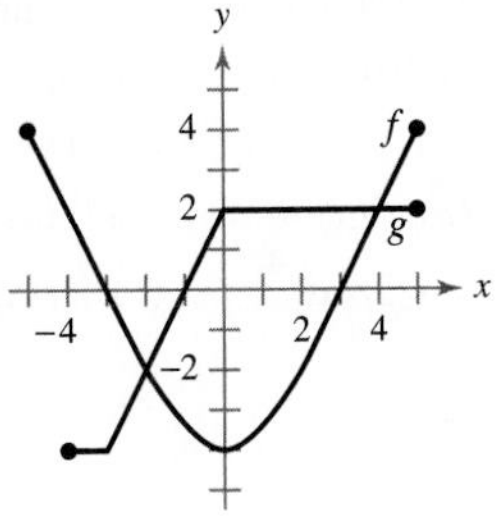

In Exercises 3–12, evaluate (if possible) the function at the given value(s) of the independent variable. Simplify the results.

3. $f(x) = 2x - 3$
(a) $f(0)$
(b) $f(-3)$
(c) $f(b)$
(d) $f(x - 1)$

4. $f(x) = \sqrt{x + 3}$
(a) $f(-2)$
(b) $f(6)$
(c) $f(-5)$
(d) $f(x + \Delta x)$

5. $g(x) = 3 - x^2$
(a) $g(0)$
(b) $g(\sqrt{3})$
(c) $g(-2)$
(d) $g(t - 1)$

6. $g(x) = x^2(x - 4)$
(a) $g(4)$
(b) $g(\frac{3}{2})$
(c) $g(c)$
(d) $g(t + 4)$

7. $f(x) = \cos 2x$
(a) $f(0)$
(b) $f(-\pi/4)$
(c) $f(\pi/3)$

8. $f(x) = \sin x$
(a) $f(\pi)$
(b) $f(5\pi/4)$
(c) $f(2\pi/3)$

9. $f(x) = x^3$

$$\frac{f(x + \Delta x) - f(x)}{\Delta x}$$

10. $f(x) = 3x - 1$

$$\frac{f(x) - f(1)}{x - 1}$$

11. $f(x) = \dfrac{1}{\sqrt{x - 1}}$

$$\frac{f(x) - f(2)}{x - 2}$$

12. $f(x) = x^3 - x$

$$\frac{f(x) - f(1)}{x - 1}$$

In Exercises 13–18, find the domain and range of the function.

13. $h(x) = -\sqrt{x + 3}$

14. $g(x) = x^2 - 5$

15. $f(t) = \sec \dfrac{\pi t}{4}$

16. $h(t) = \cot t$

17. $f(x) = \dfrac{1}{x}$

18. $g(x) = \dfrac{2}{x - 1}$

In Exercises 19-24, find the domain of the function.

19. $f(x) = \sqrt{x} + \sqrt{1 - x}$

20. $f(x) = \sqrt{x^2 - 3x + 2}$

21. $g(x) = \dfrac{2}{1 - \cos x}$

22. $h(x) = \dfrac{1}{\sin x - \frac{1}{2}}$

23. $f(x) = \dfrac{1}{|x + 3|}$

24. $g(x) = \dfrac{1}{|x^2 - 4|}$

In Exercises 25–28, evaluate the function as indicated. Determine its domain and range.

25. $f(x) = \begin{cases} 2x + 1, & x < 0 \\ 2x + 2, & x \geq 0 \end{cases}$

(a) $f(-1)$ (b) $f(0)$ (c) $f(2)$ (d) $f(t^2 + 1)$

26. $f(x) = \begin{cases} x^2 + 2, & x \leq 1 \\ 2x^2 + 2, & x > 1 \end{cases}$

(a) $f(-2)$ (b) $f(0)$ (c) $f(1)$ (d) $f(s^2 + 2)$

27. $f(x) = \begin{cases} |x| + 1, & x < 1 \\ -x + 1, & x \geq 1 \end{cases}$

(a) $f(-3)$ (b) $f(1)$ (c) $f(3)$ (d) $f(b^2 + 1)$

28. $f(x) = \begin{cases} \sqrt{x + 4}, & x \leq 5 \\ (x - 5)^2, & x > 5 \end{cases}$

(a) $f(-3)$ (b) $f(0)$ (c) $f(5)$ (d) $f(10)$

In Exercises 29–36, sketch a graph of the function and find its domain and range. Use a graphing utility to verify your graph.

29. $f(x) = 4 - x$

30. $g(x) = \dfrac{4}{x}$

31. $h(x) = \sqrt{x - 1}$

32. $f(x) = \frac{1}{2}x^3 + 2$

33. $f(x) = \sqrt{9 - x^2}$

34. $f(x) = x + \sqrt{4 - x^2}$

35. $g(t) = 2 \sin \pi t$

36. $h(\theta) = -5 \cos \dfrac{\theta}{2}$

Writing About Concepts

37. The graph of the distance that a student drives in a 10-minute trip to school is shown in the figure. Give a verbal description of characteristics of the student's drive to school.

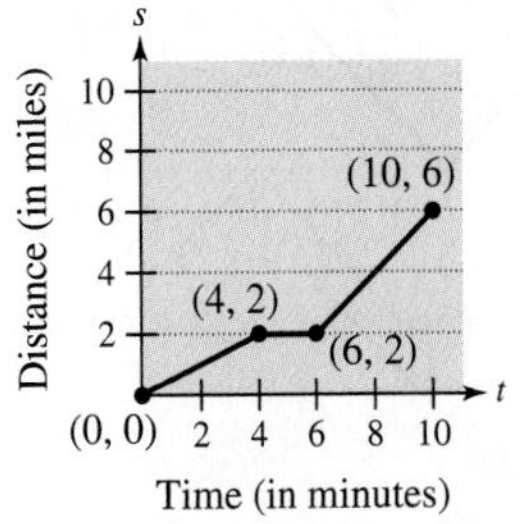

Writing About Concepts *(continued)*

38. A student who commutes 27 miles to attend college remembers, after driving a few minutes, that a term paper that is due has been forgotten. Driving faster than usual, the student returns home, picks up the paper, and once again starts toward school. Sketch a possible graph of the student's distance from home as a function of time.

In Exercises 39–42, use the Vertical Line Test to determine whether y is a function of x. To print an enlarged copy of the graph, go to the website *www.mathgraphs.com*.

39. $x - y^2 = 0$

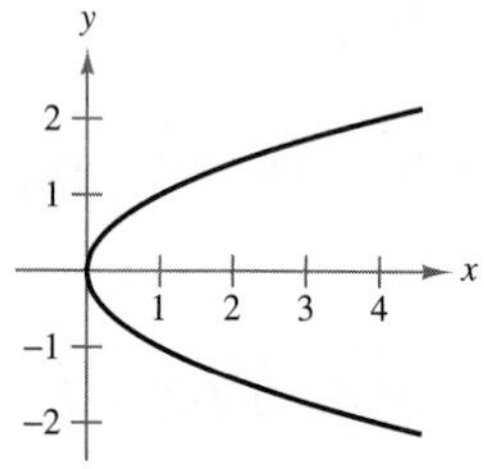

40. $\sqrt{x^2 - 4} - y = 0$

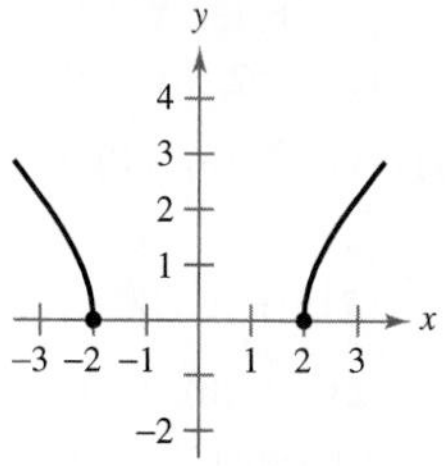

41. $y = \begin{cases} x + 1, & x \le 0 \\ -x + 2, & x > 0 \end{cases}$

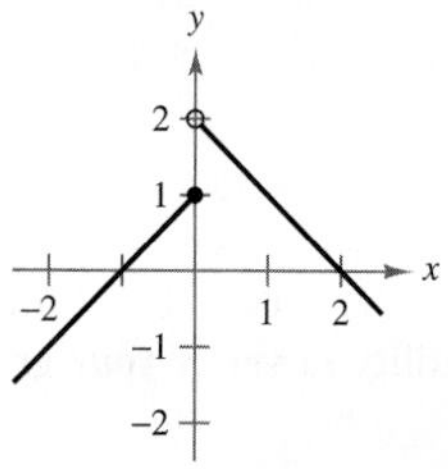

42. $x^2 + y^2 = 4$

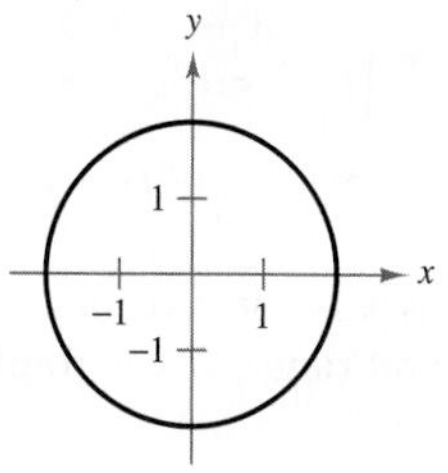

In Exercises 43–46, determine whether y is a function of x.

43. $x^2 + y^2 = 4$

44. $x^2 + y = 4$

45. $y^2 = x^2 - 1$

46. $x^2y - x^2 + 4y = 0$

In Exercises 47–52, use the graph of $y = f(x)$ to match the function with its graph.

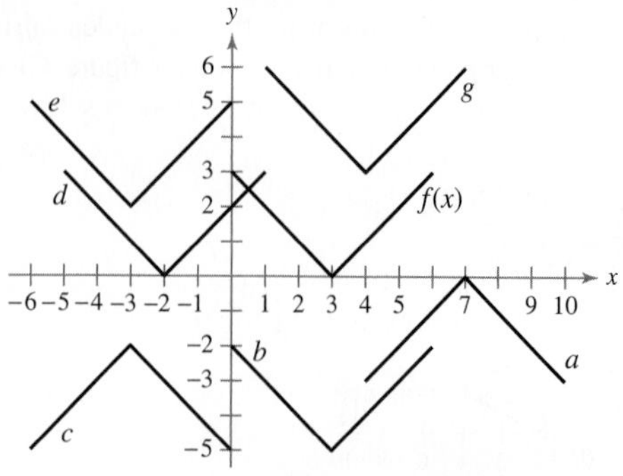

47. $y = f(x + 5)$

48. $y = f(x) - 5$

49. $y = -f(-x) - 2$

50. $y = -f(x - 4)$

51. $y = f(x + 6) + 2$

52. $y = f(x - 1) + 3$

53. Use the graph of f shown in the figure to sketch the graph of each function. To print an enlarged copy of the graph, go to the website *www.mathgraphs.com*.

(a) $f(x + 3)$ (b) $f(x - 1)$

(c) $f(x) + 2$ (d) $f(x) - 4$

(e) $3f(x)$ (f) $\frac{1}{4}f(x)$

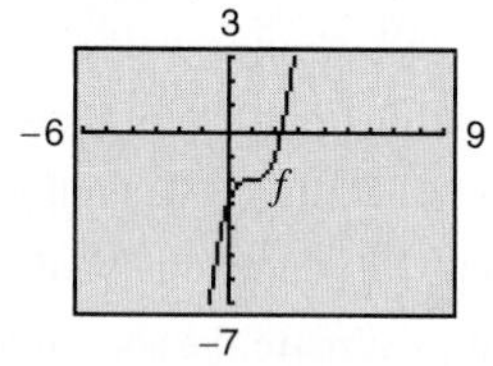

54. Use the graph of f shown in the figure to sketch the graph of each function. To print an enlarged copy of the graph, go to the website *www.mathgraphs.com*.

(a) $f(x - 4)$ (b) $f(x + 2)$

(c) $f(x) + 4$ (d) $f(x) - 1$

(e) $2f(x)$ (f) $\frac{1}{2}f(x)$

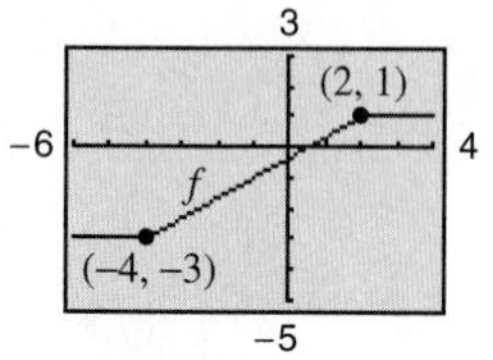

55. Use the graph of $f(x) = \sqrt{x}$ to sketch the graph of each function. In each case, describe the transformation.

(a) $y = \sqrt{x} + 2$ (b) $y = -\sqrt{x}$ (c) $y = \sqrt{x - 2}$

56. Specify a sequence of transformations that will yield each graph of h from the graph of the function $f(x) = \sin x$.

(a) $h(x) = \sin\left(x + \frac{\pi}{2}\right) + 1$ (b) $h(x) = -\sin(x - 1)$

57. Given $f(x) = \sqrt{x}$ and $g(x) = x^2 - 1$, evaluate each expression.

(a) $f(g(1))$ (b) $g(f(1))$ (c) $g(f(0))$

(d) $f(g(-4))$ (e) $f(g(x))$ (f) $g(f(x))$

58. Given $f(x) = \sin x$ and $g(x) = \pi x$, evaluate each expression.

(a) $f(g(2))$ (b) $f\left(g\left(\frac{1}{2}\right)\right)$ (c) $g(f(0))$

(d) $g\left(f\left(\frac{\pi}{4}\right)\right)$ (e) $f(g(x))$ (f) $g(f(x))$

In Exercises 59–62, find the composite functions $(f \circ g)$ and $(g \circ f)$. What is the domain of each composite function? Are the two composite functions equal?

59. $f(x) = x^2$
$g(x) = \sqrt{x}$

60. $f(x) = x^2 - 1$
$g(x) = \cos x$

61. $f(x) = \dfrac{3}{x}$
$g(x) = x^2 - 1$

62. $f(x) = \dfrac{1}{x}$
$g(x) = \sqrt{x + 2}$

63. Use the graphs of f and g to evaluate each expression. If the result is undefined, explain why.

(a) $(f \circ g)(3)$ (b) $g(f(2))$

(c) $g(f(5))$ (d) $(f \circ g)(-3)$

(e) $(g \circ f)(-1)$ (f) $f(g(-1))$

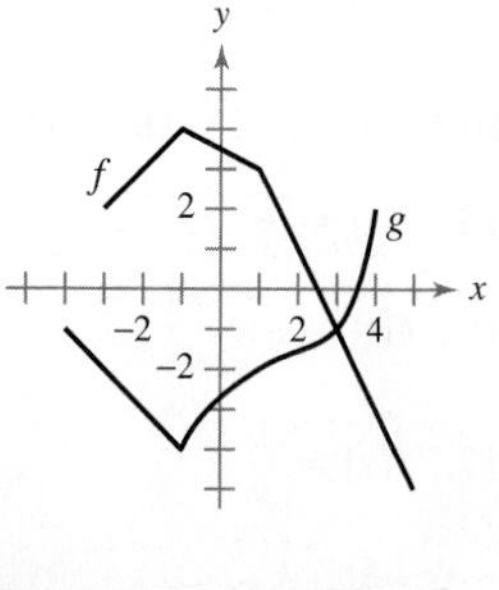

64. ***Ripples*** A pebble is dropped into a calm pond, causing ripples in the form of concentric circles. The radius (in feet) of the outer ripple is given by $r(t) = 0.6t$, where t is the time in seconds after the pebble strikes the water. The area of the circle is given by the function $A(r) = \pi r^2$. Find and interpret $(A \circ r)(t)$.

Think About It **In Exercises 65 and 66, $F(x) = f \circ g \circ h$. Identify functions for f, g, and h. (There are many correct answers.)**

65. $F(x) = \sqrt{2x - 2}$

66. $F(x) = -4\sin(1 - x)$

In Exercises 67–70, determine whether the function is even, odd, or neither. Use a graphing utility to verify your result.

67. $f(x) = x^2(4 - x^2)$

68. $f(x) = \sqrt[3]{x}$

69. $f(x) = x \cos x$

70. $f(x) = \sin^2 x$

Think About It **In Exercises 71 and 72, find the coordinates of a second point on the graph of a function f if the given point is on the graph and the function is (a) even and (b) odd.**

71. $\left(-\frac{3}{2}, 4\right)$

72. $(4, 9)$

73. The graphs of f, g, and h are shown in the figure. Decide whether each function is even, odd, or neither.

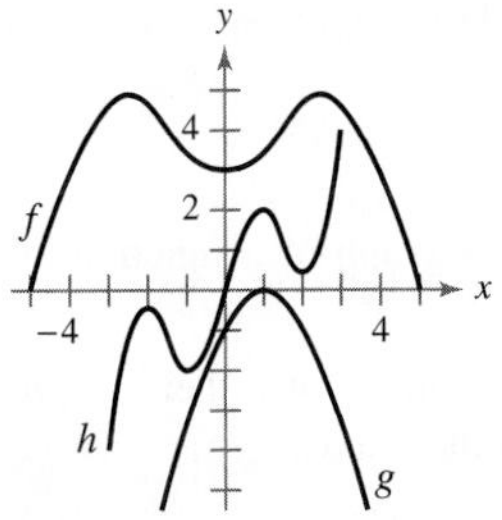

Figure for 73

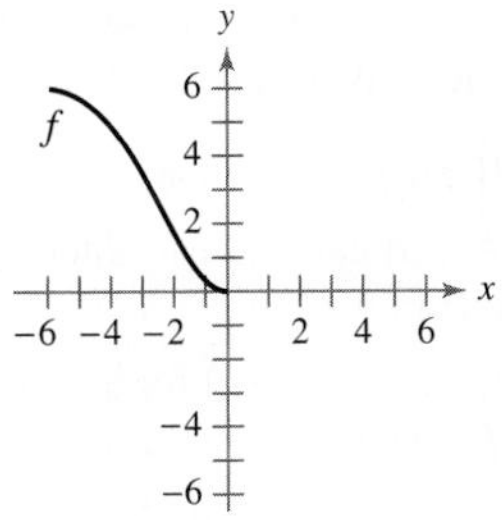

Figure for 74

74. The domain of the function f shown in the figure is $-6 \le x \le 6$.

(a) Complete the graph of f given that f is even.

(b) Complete the graph of f given that f is odd.

Writing Functions **In Exercises 75–78, write an equation for a function that has the given graph.**

75. Line segment connecting $(-4, 3)$ and $(0, -5)$

76. Line segment connecting $(1, 2)$ and $(5, 5)$

77. The bottom half of the parabola $x + y^2 = 0$

78. The bottom half of the circle $x^2 + y^2 = 4$

Modeling Data **In Exercises 79–82, match the data with a function from the following list.**

(i) $f(x) = cx$

(ii) $g(x) = cx^2$

(iii) $h(x) = c\sqrt{|x|}$

(iv) $r(x) = c/x$

Determine the value of the constant c for each function such that the function fits the data shown in the table.

79.

x	-4	-1	0	1	4
y	-32	-2	0	-2	-32

80.

x	-4	-1	0	1	4
y	-1	$-\frac{1}{4}$	0	$\frac{1}{4}$	1

81.

x	-4	-1	0	1	4
y	-8	-32	Undef.	32	8

82.

x	-4	-1	0	1	4
y	6	3	0	3	6

83. ***Graphical Reasoning*** An electronically controlled thermostat is programmed to lower the temperature during the night automatically (see figure). The temperature T in degrees Celsius is given in terms of t, the time in hours on a 24-hour clock.

(a) Approximate $T(4)$ and $T(15)$.

(b) The thermostat is reprogrammed to produce a temperature $H(t) = T(t - 1)$. How does this change the temperature? Explain.

(c) The thermostat is reprogrammed to produce a temperature $H(t) = T(t) - 1$. How does this change the temperature? Explain.

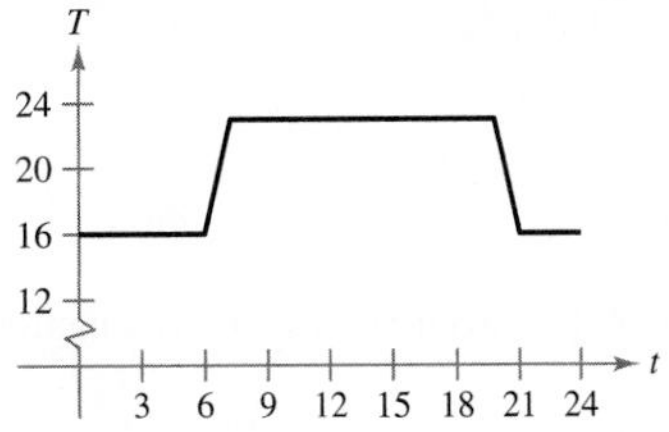

84. Water runs into a vase of height 30 centimeters at a constant rate. The vase is full after 5 seconds. Use this information and the shape of the vase shown to answer the questions if d is the depth of the water in centimeters and t is the time in seconds (see figure).

(a) Explain why d is a function of t.

(b) Determine the domain and range of the function.

(c) Sketch a possible graph of the function.

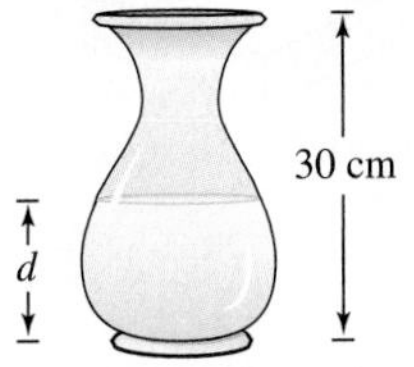

85. ***Modeling Data*** The table shows the average numbers of acres per farm in the United States for selected years. *(Source: U.S. Department of Agriculture)*

Year	1950	1960	1970	1980	1990	2000
Acreage	213	297	374	426	460	434

(a) Plot the data where A is the acreage and t is the time in years, with $t = 0$ corresponding to 1950. Sketch a freehand curve that approximates the data.

(b) Use the curve in part (a) to approximate $A(15)$.

86. ***Automobile Aerodynamics*** The horsepower H required to overcome wind drag on a certain automobile is approximated by

$$H(x) = 0.002x^2 + 0.005x - 0.029, \qquad 10 \le x \le 100$$

where x is the speed of the car in miles per hour.

(a) Use a graphing utility to graph H.

(b) Rewrite the power function so that x represents the speed in kilometers per hour. [Find $H(x/1.6)$.]

87. ***Think About It*** Write the function

$$f(x) = |x| + |x - 2|$$

without using absolute value signs. (For a review of absolute value, see Appendix D.)

88. ***Writing*** Use a graphing utility to graph the polynomial functions $p_1(x) = x^3 - x + 1$ and $p_2(x) = x^3 - x$. How many zeros does each function have? Is there a cubic polynomial that has no zeros? Explain.

89. Prove that the function is odd.

$$f(x) = a_{2n+1}x^{2n+1} + \cdot \cdot \cdot + a_3x^3 + a_1x$$

90. Prove that the function is even.

$$f(x) = a_{2n}x^{2n} + a_{2n-2}x^{2n-2} + \cdot \cdot \cdot + a_2x^2 + a_0$$

91. Prove that the product of two even (or two odd) functions is even.

92. Prove that the product of an odd function and an even function is odd.

93. ***Volume*** An open box of maximum volume is to be made from a square piece of material 24 centimeters on a side by cutting equal squares from the corners and turning up the sides (see figure).

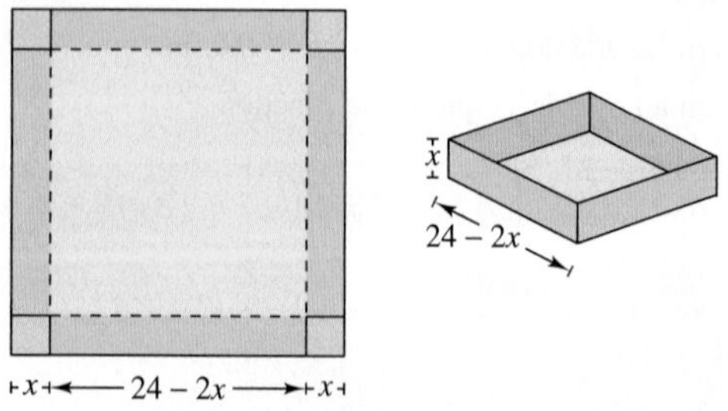

(a) Write the volume V as a function of x, the length of the corner squares. What is the domain of the function?

(b) Use a graphing utility to graph the volume function and approximate the dimensions of the box that yield a maximum volume.

(c) Use the *table* feature of a graphing utility to verify your answer in part (b). (The first two rows of the table are shown.)

Height, x	Length and Width	Volume, V
1	$24 - 2(1)$	$1[24 - 2(1)]^2 = 484$
2	$24 - 2(2)$	$2[24 - 2(2)]^2 = 800$

94. ***Length*** A right triangle is formed in the first quadrant by the x- and y-axes and a line through the point $(3, 2)$ (see figure). Write the length L of the hypotenuse as a function of x.

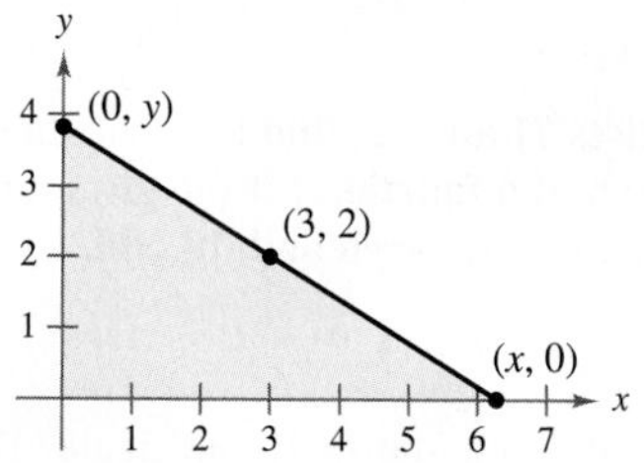

True or False? **In Exercises 95–98, determine whether the statement is true or false. If it is false, explain why or give an example that shows it is false.**

95. If $f(a) = f(b)$, then $a = b$.

96. A vertical line can intersect the graph of a function at most once.

97. If $f(x) = f(-x)$ for all x in the domain of f, then the graph of f is symmetric with respect to the y-axis.

98. If f is a function, then $f(ax) = af(x)$.

Putnam Exam Challenge

99. Let R be the region consisting of the points (x, y) of the Cartesian plane satisfying both $|x| - |y| \le 1$ and $|y| \le 1$. Sketch the region R and find its area.

100. Consider a polynomial $f(x)$ with real coefficients having the property $f(g(x)) = g(f(x))$ for every polynomial $g(x)$ with real coefficients. Determine and prove the nature of $f(x)$.

These problems were composed by the Committee on the Putnam Prize Competition.

Section P.4 Fitting Models to Data

- **Fit a linear model to a real-life data set.**
- **Fit a quadratic model to a real-life data set.**
- **Fit a trigonometric model to a real-life data set.**

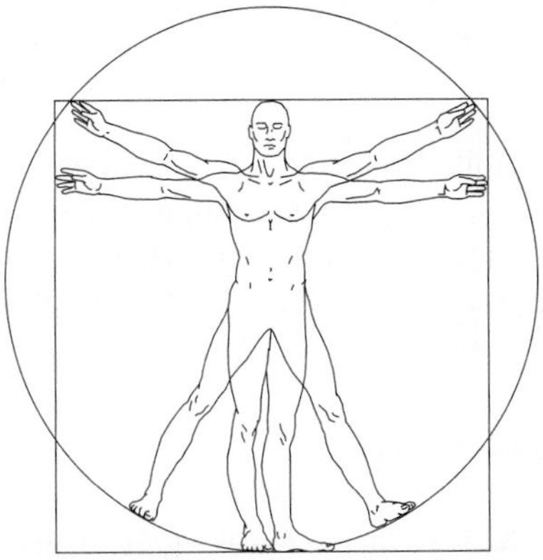

A computer graphics drawing based on the pen and ink drawing of Leonardo da Vinci's famous study of human proportions, called *Vitruvian Man*

Fitting a Linear Model to Data

A basic premise of science is that much of the physical world can be described mathematically and that many physical phenomena are predictable. This scientific outlook was part of the scientific revolution that took place in Europe during the late 1500s. Two early publications connected with this revolution were *On the Revolutions of the Heavenly Spheres* by the Polish astronomer Nicolaus Copernicus and *On the Structure of the Human Body* by the Belgian anatomist Andreas Vesalius. Each of these books was published in 1543 and each broke with prior tradition by suggesting the use of a scientific method rather than unquestioned reliance on authority.

One basic technique of modern science is gathering data and then describing the data with a mathematical model. For instance, the data given in Example 1 are inspired by Leonardo da Vinci's famous drawing that indicates that a person's height and arm span are equal.

EXAMPLE 1 Fitting a Linear Model to Data

A class of 28 people collected the following data, which represent their heights x and arm spans y (rounded to the nearest inch).

(60, 61), (65, 65), (68, 67), (72, 73), (61, 62), (63, 63), (70, 71),
(75, 74), (71, 72), (62, 60), (65, 65), (66, 68), (62, 62), (72, 73),
(70, 70), (69, 68), (69, 70), (60, 61), (63, 63), (64, 64), (71, 71),
(68, 67), (69, 70), (70, 72), (65, 65), (64, 63), (71, 70), (67, 67)

Find a linear model to represent these data.

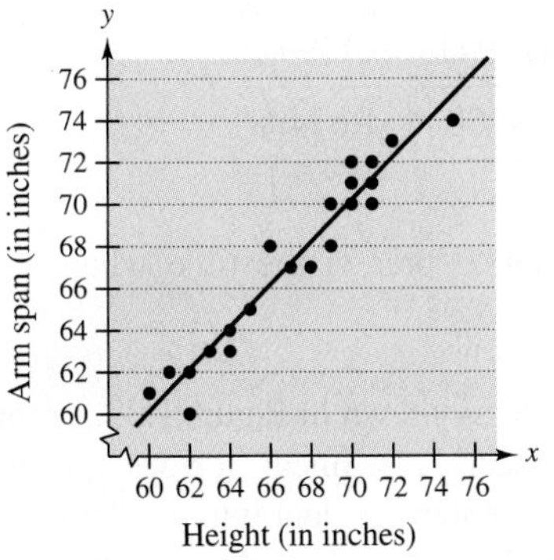

Linear model and data
Figure P.32

Solution There are different ways to model these data with an equation. The simplest would be to observe that x and y are about the same and list the model as simply $y = x$. A more careful analysis would be to use a procedure from statistics called linear regression. (You will study this procedure in Section 13.9.) The least squares regression line for these data is

$$y = 1.006x - 0.23. \qquad \text{Least squares regression line}$$

The graph of the model and the data are shown in Figure P.32. From this model, you can see that a person's arm span tends to be about the same as his or her height.

TECHNOLOGY Many scientific and graphing calculators have built-in least squares regression programs. Typically, you enter the data into the calculator and then run the linear regression program. The program usually displays the slope and y-intercept of the best-fitting line and the *correlation coefficient r*. The correlation coefficient gives a measure of how well the model fits the data. The closer $|r|$ is to 1, the better the model fits the data. For instance, the correlation coefficient for the model in Example 1 is $r \approx 0.97$, which indicates that the model is a good fit for the data. If the r-value is positive, the variables have a positive correlation, as in Example 1. If the r-value is negative, the variables have a negative correlation.

Fitting a Quadratic Model to Data

A function that gives the height s of a falling object in terms of the time t is called a position function. If air resistance is not considered, the position of a falling object can be modeled by

$$s(t) = \frac{1}{2}gt^2 + v_0t + s_0$$

where g is the acceleration due to gravity, v_0 is the initial velocity, and s_0 is the initial height. The value of g depends on where the object is dropped. On earth, g is approximately -32 feet per second per second, or -9.8 meters per second per second.

To discover the value of g experimentally, you could record the heights of a falling object at several increments, as shown in Example 2.

EXAMPLE 2 Fitting a Quadratic Model to Data

A basketball is dropped from a height of about $5\frac{1}{4}$ feet. The height of the basketball is recorded 23 times at intervals of about 0.02 second.* The results are shown in the table.

Time	0.0	0.02	0.04	0.06	0.08	0.099996
Height	5.23594	5.20353	5.16031	5.0991	5.02707	4.95146

Time	0.119996	0.139992	0.159988	0.179988	0.199984	0.219984
Height	4.85062	4.74979	4.63096	4.50132	4.35728	4.19523

Time	0.23998	0.25993	0.27998	0.299976	0.319972	0.339961
Height	4.02958	3.84593	3.65507	3.44981	3.23375	3.01048

Time	0.359961	0.379951	0.399941	0.419941	0.439941
Height	2.76921	2.52074	2.25786	1.98058	1.63488

Find a model to fit these data. Then use the model to predict the time when the basketball will hit the ground.

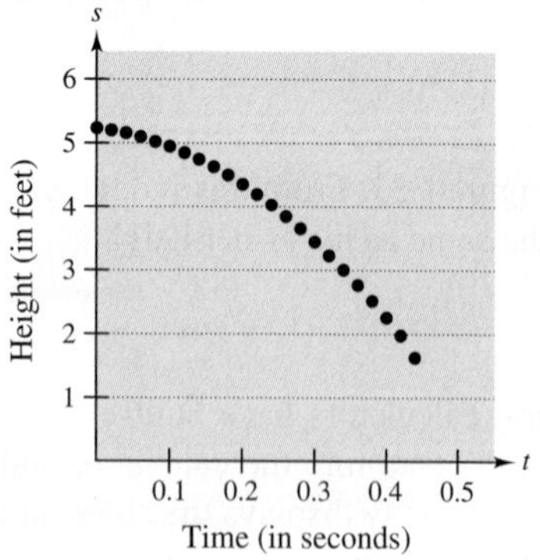

Scatter plot of data
Figure P.33

Solution Begin by drawing a scatter plot of the data, as shown in Figure P.33. From the scatter plot, you can see that the data do not appear to be linear. It does appear, however, that they might be quadratic. To check this, enter the data into a calculator or computer that has a quadratic regression program. You should obtain the model

$$s = -15.45t^2 - 1.30t + 5.234.$$ Least squares regression quadratic

Using this model, you can predict the time when the basketball hits the ground by substituting 0 for s and solving the resulting equation for t.

$$0 = -15.45t^2 - 1.30t + 5.234$$ Let $s = 0$.

$$t = \frac{1.30 \pm \sqrt{(-1.30)^2 - 4(-15.45)(5.234)}}{2(-15.45)}$$ Quadratic Formula

$$t \approx 0.54$$ Choose positive solution.

The solution is about 0.54 second. In other words, the basketball will continue to fall for about 0.1 second more before hitting the ground.

* *Data were collected with a Texas Instruments CBL (Calculator-Based Laboratory) System.*

Fitting a Trigonometric Model to Data

The plane of Earth's orbit about the sun and its axis of rotation are not perpendicular. Instead, Earth's axis is tilted with respect to its orbit. The result is that the amount of daylight received by locations on Earth varies with the time of year. That is, it varies with the position of Earth in its orbit.

What is mathematical modeling? This is one of the questions that is asked in the book *Guide to Mathematical Modelling*. Here is part of the answer.*

1. Mathematical modeling consists of applying your mathematical skills to obtain useful answers to real problems.
2. Learning to apply mathematical skills is very different from learning mathematics itself.
3. Models are used in a very wide range of applications, some of which do not appear initially to be mathematical in nature.
4. Models often allow quick and cheap evaluation of alternatives, leading to optimal solutions that are not otherwise obvious.
5. There are no precise rules in mathematical modeling and no "correct" answers.
6. Modeling can be learned only by *doing*.

EXAMPLE 3 Fitting a Trigonometric Model to Data

The number of hours of daylight on Earth depends on the latitude and the time of year. Here are the numbers of minutes of daylight at a location of 20° N latitude on the longest and shortest days of the year: June 21, 801 minutes; December 22, 655 minutes. Use these data to write a model for the amount of daylight d (in minutes) on each day of the year at a location of 20° N latitude. How could you check the accuracy of your model?

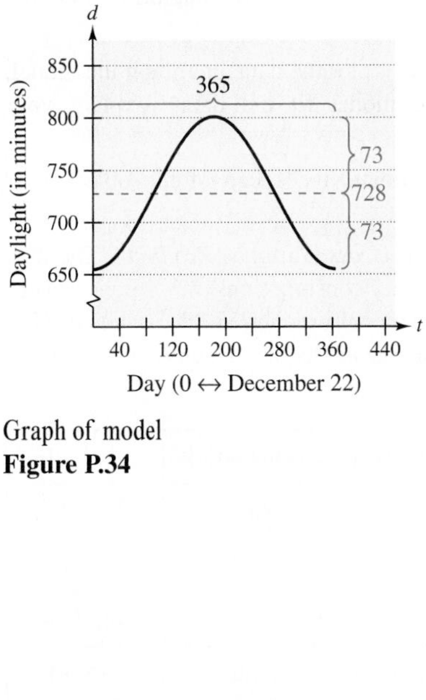

Graph of model
Figure P.34

Solution Here is one way to create a model. You can hypothesize that the model is a sine function whose period is 365 days. Using the given data, you can conclude that the amplitude of the graph is $(801 - 655)/2$, or 73. So, one possible model is

$$d = 728 - 73 \sin\left(\frac{2\pi t}{365} + \frac{\pi}{2}\right).$$

In this model, t represents the number of each day of the year, with December 22 represented by $t = 0$. A graph of this model is shown in Figure P.34. To check the accuracy of this model, we used a weather almanac to find the numbers of minutes of daylight on different days of the year at the location of 20° N latitude.

NOTE For more review of trigonometric functions, see Appendix D.

Date	*Value of t*	*Actual Daylight*	*Daylight Given by Model*
Dec 22	0	655 min	655 min
Jan 1	10	657 min	656 min
Feb 1	41	676 min	672 min
Mar 1	69	705 min	701 min
Apr 1	100	740 min	739 min
May 1	130	772 min	773 min
Jun 1	161	796 min	796 min
Jun 21	181	801 min	801 min
Jul 1	191	799 min	800 min
Aug 1	222	782 min	785 min
Sep 1	253	752 min	754 min
Oct 1	283	718 min	716 min
Nov 1	314	685 min	681 min
Dec 1	344	661 min	660 min

You can see that the model is fairly accurate.

* *Text from Dilwyn Edwards and Mike Hamson,* Guide to Mathematical Modelling *(Boca Raton: CRC Press, 1990). Used by permission of the authors.*

Exercises for Section P.4

See www.CalcChat.com for worked-out solutions to odd-numbered exercises.

In Exercises 1–4, a scatter plot of data is given. Determine whether the data can be modeled by a linear function, a quadratic function, or a trigonometric function, or that there appears to be no relationship between x and y. To print an enlarged copy of the graph, go to the website *www.mathgraphs.com*.

1.

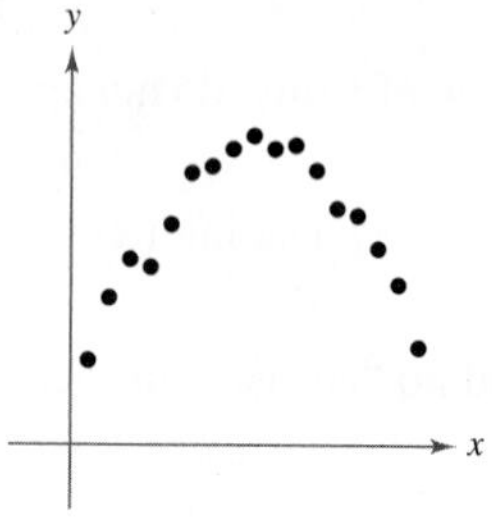

2.

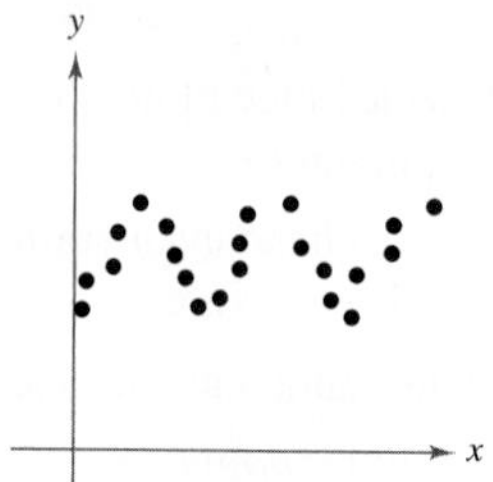

3.

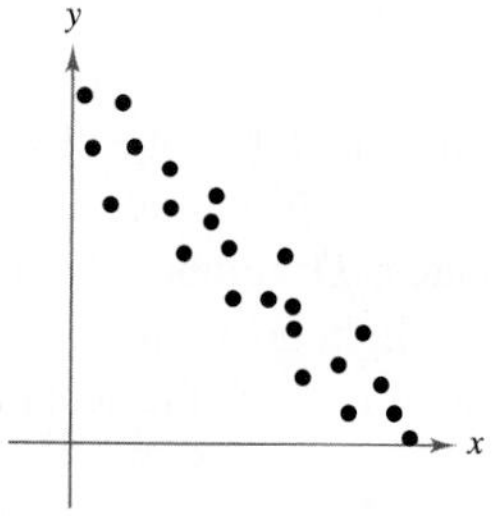

4.

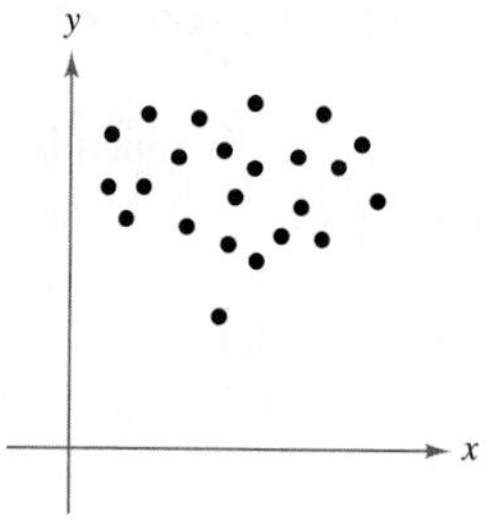

5. ***Carcinogens*** Each ordered pair gives the exposure index x of a carcinogenic substance and the cancer mortality y per 100,000 people in the population.

(3.50, 150.1), (3.58, 133.1), (4.42, 132.9),
(2.26, 116.7), (2.63, 140.7), (4.85, 165.5),
(12.65, 210.7), (7.42, 181.0), (9.35, 213.4)

(a) Plot the data. From the graph, do the data appear to be approximately linear?

(b) Visually find a linear model for the data. Graph the model.

(c) Use the model to approximate y if $x = 3$.

6. ***Quiz Scores*** The ordered pairs represent the scores on two consecutive 15-point quizzes for a class of 18 students.

(7, 13), (9, 7), (14, 14), (15, 15), (10, 15), (9, 7),
(14, 11), (14, 15), (8, 10), (15, 9), (10, 11), (9, 10),
(11, 14), (7, 14), (11, 10), (14, 11), (10, 15), (9, 6)

(a) Plot the data. From the graph, does the relationship between consecutive scores appear to be approximately linear?

(b) If the data appear to be approximately linear, find a linear model for the data. If not, give some possible explanations.

7. ***Hooke's Law*** Hooke's Law states that the force F required to compress or stretch a spring (within its elastic limits) is proportional to the distance d that the spring is compressed or stretched from its original length. That is, $F = kd$, where k is a measure of the stiffness of the spring and is called the *spring constant*. The table shows the elongation d in centimeters of a spring when a force of F newtons is applied.

F	20	40	60	80	100
d	1.4	2.5	4.0	5.3	6.6

(a) Use the regression capabilities of a graphing utility to find a linear model for the data.

(b) Use a graphing utility to plot the data and graph the model. How well does the model fit the data? Explain your reasoning.

(c) Use the model to estimate the elongation of the spring when a force of 55 newtons is applied.

8. ***Falling Object*** In an experiment, students measured the speed s (in meters per second) of a falling object t seconds after it was released. The results are shown in the table.

t	0	1	2	3	4
s	0	11.0	19.4	29.2	39.4

(a) Use the regression capabilities of a graphing utility to find a linear model for the data.

(b) Use a graphing utility to plot the data and graph the model. How well does the model fit the data? Explain your reasoning.

(c) Use the model to estimate the speed of the object after 2.5 seconds.

9. ***Energy Consumption and Gross National Product*** The data show the per capita electricity consumptions (in millions of Btu) and the per capita gross national products (in thousands of U.S. dollars) for several countries in 2000. *(Source: U.S. Census Bureau)*

Argentina	(73, 12.05)
Chile	(68, 9.1)
Greece	(126, 16.86)
Hungary	(105, 11.99)
Mexico	(63, 8.79)
Portugal	(108, 16.99)
Spain	(137, 19.26)
United Kingdom	(166, 23.55)

Bangladesh	(4, 1.59)
Egypt	(32, 3.67)
Hong Kong	(118, 25.59)
India	(13, 2.34)
Poland	(95, 9)
South Korea	(167, 17.3)
Turkey	(47, 7.03)
Venezuela	(113, 5.74)

(a) Use the regression capabilities of a graphing utility to find a linear model for the data. What is the correlation coefficient?

(b) Use a graphing utility to plot the data and graph the model.

(c) Interpret the graph in part (b). Use the graph to identify the three countries that differ most from the linear model.

(d) Delete the data for the three countries identified in part (c). Fit a linear model to the remaining data and give the correlation coefficient.

10. ***Brinell Hardness*** The data in the table show the Brinell hardness H of 0.35 carbon steel when hardened and tempered at temperature t (degrees Fahrenheit). *(Source: Standard Handbook for Mechanical Engineers)*

t	200	400	600	800	1000	1200
H	534	495	415	352	269	217

(a) Use the regression capabilities of a graphing utility to find a linear model for the data.

(b) Use a graphing utility to plot the data and graph the model. How well does the model fit the data? Explain your reasoning.

(c) Use the model to estimate the hardness when t is 500°F.

11. ***Automobile Costs*** The data in the table show the variable costs for operating an automobile in the United States for several recent years. The functions y_1, y_2, and y_3 represent the costs in cents per mile for gas and oil, maintenance, and tires, respectively. *(Source: American Automobile Manufacturers Association)*

Year	y_1	y_2	y_3
0	5.40	2.10	0.90
1	6.70	2.20	0.90
2	6.00	2.20	0.90
3	6.00	2.40	0.90
4	5.60	2.50	1.10
5	6.00	2.60	1.40
6	5.90	2.80	1.40
7	6.60	2.80	1.40

(a) Use the regression capabilities of a graphing utility to find a cubic model for y_1 and linear models for y_2 and y_3.

(b) Use a graphing utility to graph y_1, y_2, y_3, and $y_1 + y_2 + y_3$ in the same viewing window. Use the model to estimate the total variable cost per mile in year 12.

12. ***Beam Strength*** Students in a lab measured the breaking strength S (in pounds) of wood 2 inches thick, x inches high, and 12 inches long. The results are shown in the table.

x	4	6	8	10	12
S	2370	5460	10,310	16,250	23,860

(a) Use the regression capabilities of a graphing utility to fit a quadratic model to the data.

(b) Use a graphing utility to plot the data and graph the model.

(c) Use the model to approximate the breaking strength when $x = 2$.

13. ***Health Maintenance Organizations*** The bar graph shows the numbers of people N (in millions) receiving care in HMOs for the years 1990 through 2002. *(Source: Centers for Disease Control)*

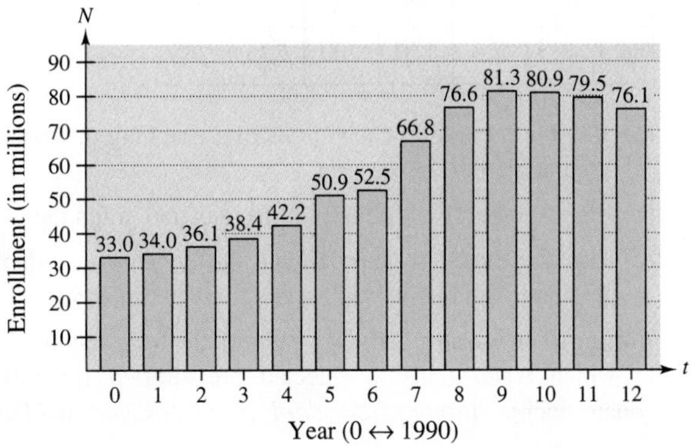

(a) Let t be the time in years, with $t = 0$ corresponding to 1990. Use the regression capabilities of a graphing utility to find linear and cubic models for the data.

(b) Use a graphing utility to graph the data and the linear and cubic models.

(c) Use the graphs in part (b) to determine which is the better model.

(d) Use a graphing utility to find and graph a quadratic model for the data.

(e) Use the linear and cubic models to estimate the number of people receiving care in HMOs in the year 2004.

(f) Use a graphing utility to find other models for the data. Which models do you think best represent the data? Explain.

14. ***Car Performance*** The time t (in seconds) required to attain a speed of s miles per hour from a standing start for a Dodge Avenger is shown in the table. *(Source: Road & Track)*

s	30	40	50	60	70	80	90
t	3.4	5.0	7.0	9.3	12.0	15.8	20.0

(a) Use the regression capabilities of a graphing utility to find a quadratic model for the data.

(b) Use a graphing utility to plot the data and graph the model.

(c) Use the graph in part (b) to state why the model is not appropriate for determining the times required to attain speeds less than 20 miles per hour.

(d) Because the test began from a standing start, add the point (0, 0) to the data. Fit a quadratic model to the revised data and graph the new model.

(e) Does the quadratic model more accurately model the behavior of the car for low speeds? Explain.

15. ***Car Performance*** A V8 car engine is coupled to a dynamometer and the horsepower y is measured at different engine speeds x (in thousands of revolutions per minute). The results are shown in the table.

x	1	2	3	4	5	6
y	40	85	140	200	225	245

(a) Use the regression capabilities of a graphing utility to find a cubic model for the data.

(b) Use a graphing utility to plot the data and graph the model.

(c) Use the model to approximate the horsepower when the engine is running at 4500 revolutions per minute.

16. ***Boiling Temperature*** The table shows the temperatures T (°F) at which water boils at selected pressures p (pounds per square inch). *(Source: Standard Handbook for Mechanical Engineers)*

p	5	10	14.696 (1 atmosphere)	20
T	162.24°	193.21°	212.00°	227.96°

p	30	40	60	80	100
T	250.33°	267.25°	292.71°	312.03°	327.81°

(a) Use the regression capabilities of a graphing utility to find a cubic model for the data.

(b) Use a graphing utility to plot the data and graph the model.

(c) Use the graph to estimate the pressure required for the boiling point of water to exceed 300°F.

(d) Explain why the model would not be correct for pressures exceeding 100 pounds per square inch.

17. ***Harmonic Motion*** The motion of an oscillating weight suspended by a spring was measured by a motion detector. The data collected and the approximate maximum (positive and negative) displacements from equilibrium are shown in the figure. The displacement y is measured in centimeters and the time t is measured in seconds.

(a) Is y a function of t? Explain.

(b) Approximate the amplitude and period of the oscillations.

(c) Find a model for the data.

(d) Use a graphing utility to graph the model in part (c). Compare the result with the data in the figure.

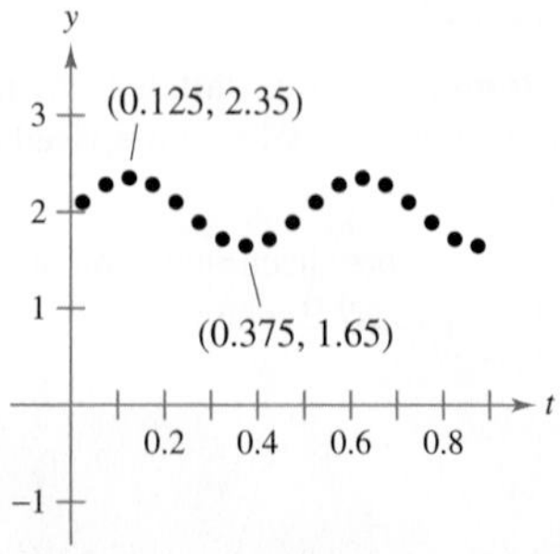

18. ***Temperature*** The table shows the normal daily high temperatures for Honolulu H and Chicago C (in degrees Fahrenheit) for month t, with $t = 1$ corresponding to January. *(Source: NOAA)*

t	1	2	3	4	5	6
H	80.1	80.5	81.6	82.8	84.7	86.5
C	29.0	33.5	45.8	58.6	70.1	79.6

t	7	8	9	10	11	12
H	87.5	88.7	88.5	86.9	84.1	81.2
C	83.7	81.8	74.8	63.3	48.4	34.0

(a) A model for Honolulu is

$$H(t) = 84.40 + 4.28 \sin\left(\frac{\pi t}{6} + 3.86\right).$$

Find a model for Chicago.

(b) Use a graphing utility to graph the data and the model for the temperatures in Honolulu. How well does the model fit?

(c) Use a graphing utility to graph the data and the model for the temperatures in Chicago. How well does the model fit?

(d) Use the models to estimate the average annual temperature in each city. What term of the model did you use? Explain.

(e) What is the period of each model? Is it what you expected? Explain.

(f) Which city has a greater variability of temperatures throughout the year? Which factor of the models determines this variability? Explain.

Writing About Concepts

19. Search for real-life data in a newspaper or magazine. Fit the data to a model. What does your model imply about the data?

20. Describe a possible real-life situation for each data set. Then describe how a model could be used in the real-life setting.

(a)
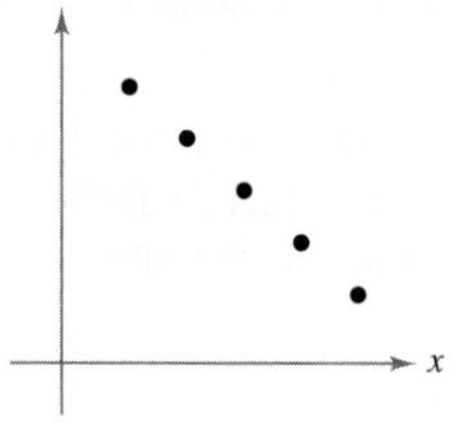

(b)
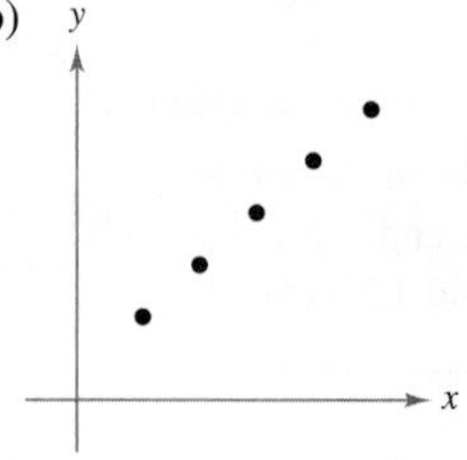

(c)
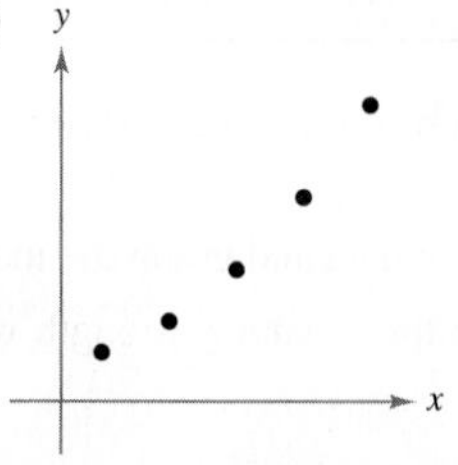

(d)
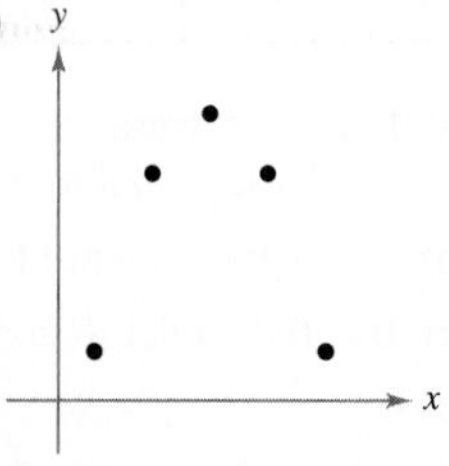

Review Exercises for Chapter P

See www.CalcChat.com for worked-out solutions to odd-numbered exercises.

In Exercises 1–4, find the intercepts (if any).

1. $y = 2x - 3$

2. $y = (x - 1)(x - 3)$

3. $y = \dfrac{x - 1}{x - 2}$

4. $xy = 4$

In Exercises 5 and 6, check for symmetry with respect to both axes and to the origin.

5. $x^2y - x^2 + 4y = 0$

6. $y = x^4 - x^2 + 3$

In Exercises 7–14, sketch the graph of the equation.

7. $y = \frac{1}{2}(-x + 3)$

8. $4x - 2y = 6$

9. $-\frac{1}{3}x + \frac{5}{6}y = 1$

10. $0.02x + 0.15y = 0.25$

11. $y = 7 - 6x - x^2$

12. $y = 6x - x^2$

13. $y = \sqrt{5 - x}$

14. $y = |x - 4| - 4$

In Exercises 15 and 16, describe the viewing window of a graphing utility that yields the figure.

15. $y = 4x^2 - 25$

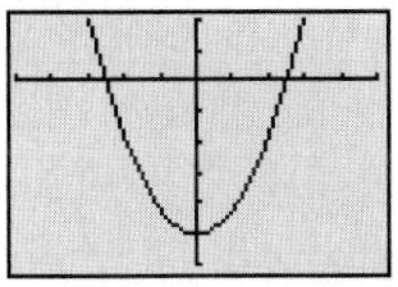

16. $y = 8\sqrt[3]{x - 6}$

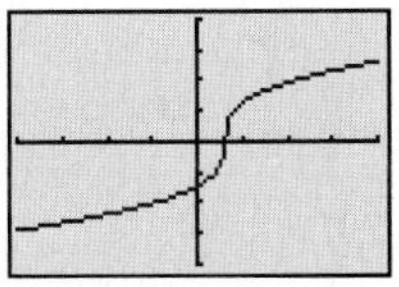

In Exercises 17 and 18, use a graphing utility to find the point(s) of intersection of the graphs of the equations.

17. $3x - 4y = 8$
$x + y = 5$

18. $x - y + 1 = 0$
$y - x^2 = 7$

19. ***Think About It*** Write an equation whose graph has intercepts at $x = -2$ and $x = 2$ and is symmetric with respect to the origin.

20. ***Think About It*** For what value of k does the graph of $y = kx^3$ pass through the point?

(a) $(1, 4)$ (b) $(-2, 1)$ (c) $(0, 0)$ (d) $(-1, -1)$

In Exercises 21 and 22, plot the points and find the slope of the line passing through the points.

21. $\left(\frac{3}{2}, 1\right), \left(5, \frac{5}{2}\right)$

22. $(7, -1), (7, 12)$

In Exercises 23 and 24, use the concept of slope to find t such that the three points are collinear.

23. $(-2, 5), (0, t), (1, 1)$

24. $(-3, 3), (t, -1), (8, 6)$

In Exercises 25–28, find an equation of the line that passes through the point with the indicated slope. Sketch the line.

25. $(0, -5), \quad m = \frac{3}{2}$

26. $(-2, 6), \quad m = 0$

27. $(-3, 0), \quad m = -\frac{2}{3}$

28. $(5, 4), \quad m$ is undefined.

29. Find equations of the lines passing through $(-2, 4)$ and having the following characteristics.

(a) Slope of $\frac{7}{16}$

(b) Parallel to the line $5x - 3y = 3$

(c) Passing through the origin

(d) Parallel to the y-axis

30. Find equations of the lines passing through $(1, 3)$ and having the following characteristics.

(a) Slope of $-\frac{2}{3}$

(b) Perpendicular to the line $x + y = 0$

(c) Passing through the point $(2, 4)$

(d) Parallel to the x-axis

31. ***Rate of Change*** The purchase price of a new machine is \$12,500, and its value will decrease by \$850 per year. Use this information to write a linear equation that gives the value V of the machine t years after it is purchased. Find its value at the end of 3 years.

32. ***Break-Even Analysis*** A contractor purchases a piece of equipment for \$36,500 that costs an average of \$9.25 per hour for fuel and maintenance. The equipment operator is paid \$13.50 per hour, and customers are charged \$30 per hour.

(a) Write an equation for the cost C of operating this equipment for t hours.

(b) Write an equation for the revenue R derived from t hours of use.

(c) Find the break-even point for this equipment by finding the time at which $R = C$.

In Exercises 33–36, sketch the graph of the equation and use the Vertical Line Test to determine whether the equation expresses y as a function of x.

33. $x - y^2 = 0$

34. $x^2 - y = 0$

35. $y = x^2 - 2x$

36. $x = 9 - y^2$

37. Evaluate (if possible) the function $f(x) = 1/x$ at the specified values of the independent variable, and simplify the results.

(a) $f(0)$ (b) $\dfrac{f(1 + \Delta x) - f(1)}{\Delta x}$

38. Evaluate (if possible) the function at each value of the independent variable.

$$f(x) = \begin{cases} x^2 + 2, & x < 0 \\ |x - 2|, & x \geq 0 \end{cases}$$

(a) $f(-4)$ (b) $f(0)$ (c) $f(1)$

39. Find the domain and range of each function.

(a) $y = \sqrt{36 - x^2}$ (b) $y = \dfrac{7}{2x - 10}$ (c) $y = \begin{cases} x^2, & x < 0 \\ 2 - x, & x \geq 0 \end{cases}$

40. Given $f(x) = 1 - x^2$ and $g(x) = 2x + 1$, evaluate each expression.

(a) $f(x) - g(x)$ (b) $f(x)g(x)$ (c) $g(f(x))$

41. Sketch (on the same set of coordinate axes) graphs of f for $c = -2, 0,$ and 2.

(a) $f(x) = x^3 + c$ (b) $f(x) = (x - c)^3$

(c) $f(x) = (x - 2)^3 + c$ (d) $f(x) = cx^3$

42. Use a graphing utility to graph $f(x) = x^3 - 3x^2$. Use the graph to write a formula for the function g shown in the figure. To print an enlarged copy of the graph, go to the website *www.mathgraphs.com.*

(a)

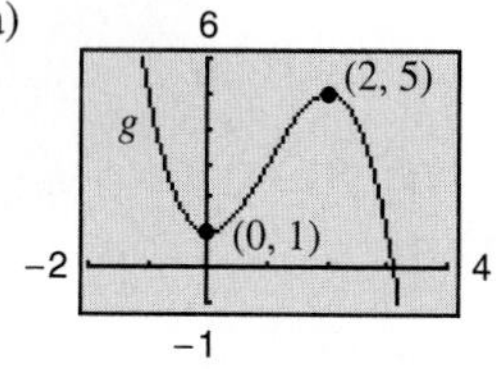

(b)

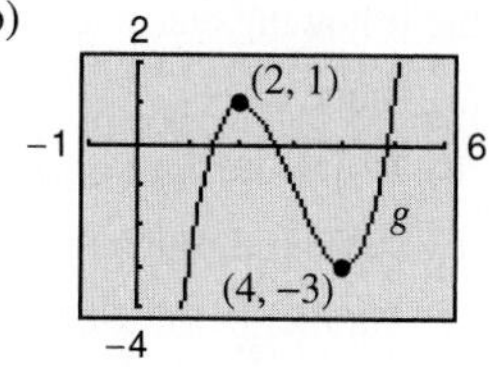

43. *Conjecture*

(a) Use a graphing utility to graph the functions f, g, and h in the same viewing window. Write a description of any similarities and differences you observe among the graphs.

Odd powers: $f(x) = x$, $g(x) = x^3$, $h(x) = x^5$

Even powers: $f(x) = x^2$, $g(x) = x^4$, $h(x) = x^6$

(b) Use the result in part (a) to make a conjecture about the graphs of the functions $y = x^7$ and $y = x^8$. Use a graphing utility to verify your conjecture.

44. *Think About It* Use the result of Exercise 43 to guess the shapes of the graphs of the functions f, g, and h. Then use a graphing utility to graph each function and compare the result with your guess.

(a) $f(x) = x^2(x - 6)^2$ (b) $g(x) = x^3(x - 6)^2$

(c) $h(x) = x^3(x - 6)^3$

45. *Area* A wire 24 inches long is to be cut into four pieces to form a rectangle whose shortest side has a length of x.

(a) Write the area A of the rectangle as a function of x.

(b) Determine the domain of the function and use a graphing utility to graph the function over that domain.

(c) Use the graph of the function to approximate the maximum area of the rectangle. Make a conjecture about the dimensions that yield a maximum area.

46. *Writing* The following graphs give the profits P for two small companies over a period p of 2 years. Create a story to describe the behavior of each profit function for some hypothetical product the company produces.

(a)

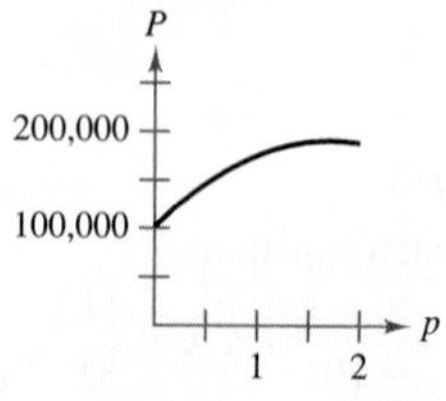

(b)

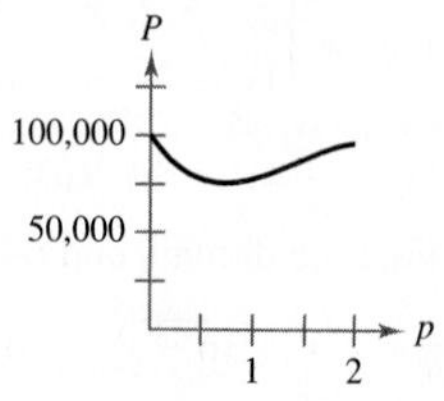

47. *Think About It* What is the minimum degree of the polynomial function whose graph approximates the given graph? What sign must the leading coefficient have?

(a)

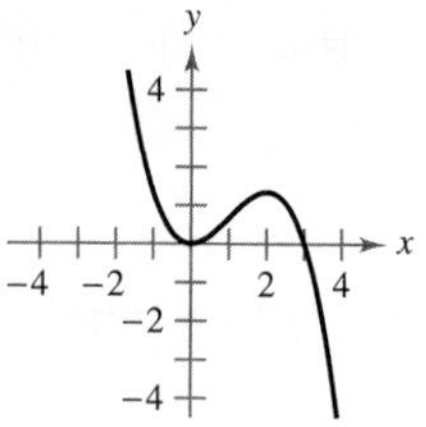

(b)

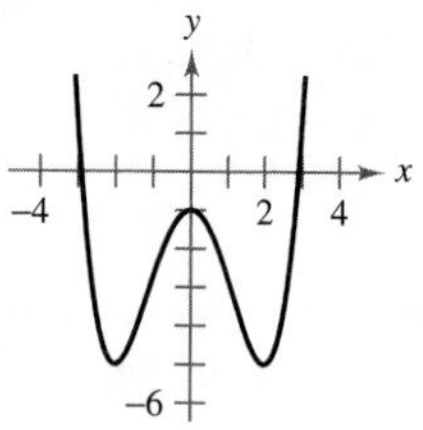

(c)

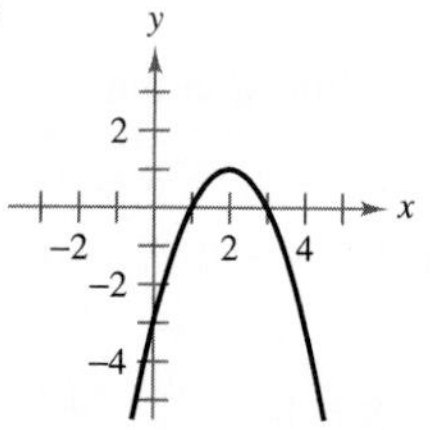

(d)

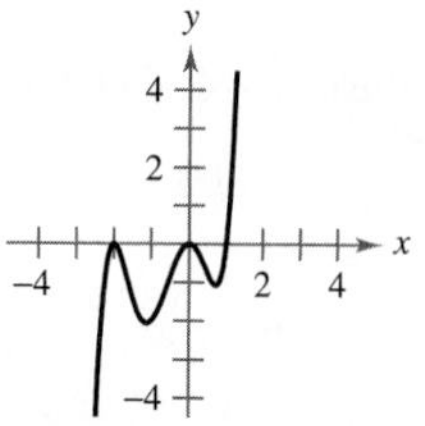

48. *Stress Test* A machine part was tested by bending it x centimeters 10 times per minute until the time y (in hours) of failure. The results are recorded in the table.

x	3	6	9	12	15	18	21	24	27	30
y	61	56	53	55	48	35	36	33	44	23

(a) Use the regression capabilities of a graphing utility to find a linear model for the data.

(b) Use a graphing utility to plot the data and graph the model.

(c) Use the graph to determine whether there may have been an error made in conducting one of the tests or in recording the results. If so, eliminate the erroneous point and find the model for the remaining data.

49. *Harmonic Motion* The motion of an oscillating weight suspended by a spring was measured by a motion detector. The data collected and the approximate maximum (positive and negative) displacements from equilibrium are shown in the figure. The displacement y is measured in feet and the time t is measured in seconds.

(a) Is y a function of t? Explain.

(b) Approximate the amplitude and period of the oscillations.

(c) Find a model for the data.

(d) Use a graphing utility to graph the model in part (c). Compare the result with the data in the figure.

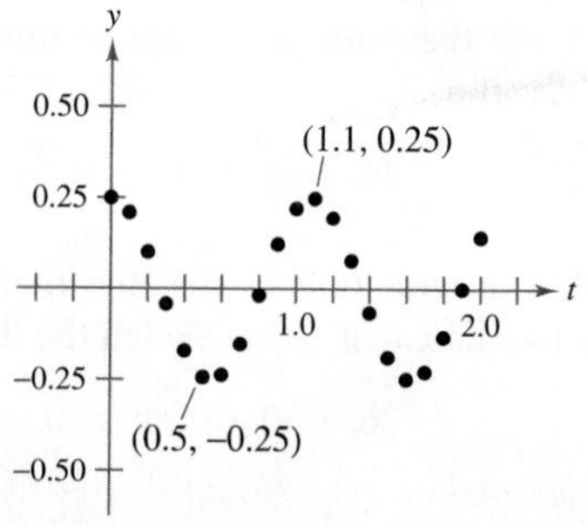

P.S. Problem Solving

See www.CalcChat.com for worked-out solutions to odd-numbered exercises.

1. Consider the circle $x^2 + y^2 - 6x - 8y = 0$, as shown in the figure.

(a) Find the center and radius of the circle.

(b) Find an equation of the tangent line to the circle at the point $(0, 0)$.

(c) Find an equation of the tangent line to the circle at the point $(6, 0)$.

(d) Where do the two tangent lines intersect?

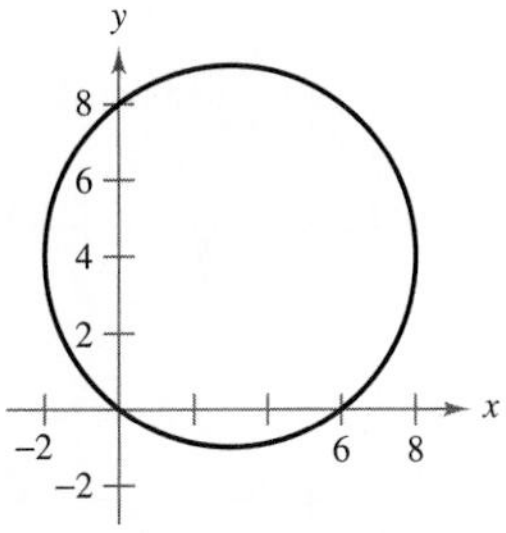

Figure for 1

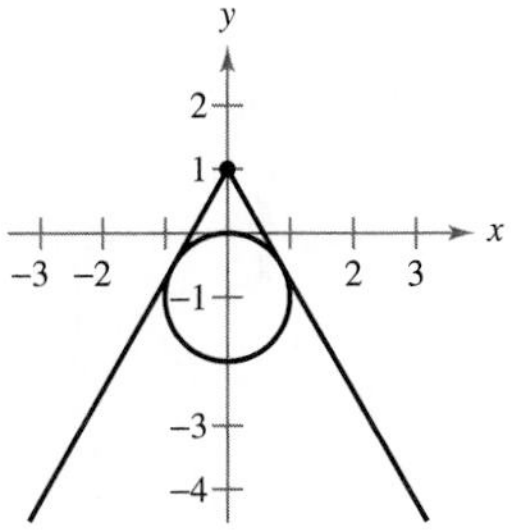

Figure for 2

2. There are two tangent lines from the point $(0, 1)$ to the circle $x^2 + (y + 1)^2 = 1$ (see figure). Find equations of these two lines by using the fact that each tangent line intersects the circle in *exactly* one point.

3. The Heaviside function $H(x)$ is widely used in engineering applications.

$$H(x) = \begin{cases} 1, & x \geq 0 \\ 0, & x < 0 \end{cases}$$

Sketch the graph of the Heaviside function and the graphs of the following functions by hand.

(a) $H(x) - 2$ (b) $H(x - 2)$ (c) $-H(x)$

(d) $H(-x)$ (e) $\frac{1}{2}H(x)$ (f) $-H(x - 2) + 2$

Institute of Electrical Engineers, London

OLIVER HEAVISIDE (1850–1925)

Heaviside was a British mathematician and physicist who contributed to the field of applied mathematics, especially applications of mathematics to electrical engineering. The *Heaviside function* is a classic type of "on-off" function that has applications to electricity and computer science.

4. Consider the graph of the function f shown below. Use this graph to sketch the graphs of the following functions. To print an enlarged copy of the graph, go to the website *www.mathgraphs.com.*

(a) $f(x + 1)$ (b) $f(x) + 1$ (c) $2f(x)$ (d) $f(-x)$

(e) $-f(x)$ (f) $|f(x)|$ (g) $f(|x|)$

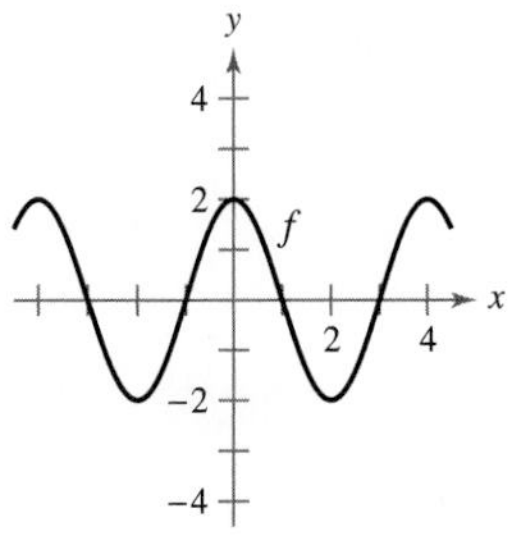

5. A rancher plans to fence a rectangular pasture adjacent to a river. The rancher has 100 meters of fence, and no fencing is needed along the river (see figure).

(a) Write the area A of the pasture as a function of x, the length of the side parallel to the river. What is the domain of A?

(b) Graph the area function $A(x)$ and estimate the dimensions that yield the maximum amount of area for the pasture.

(c) Find the dimensions that yield the maximum amount of area for the pasture by completing the square.

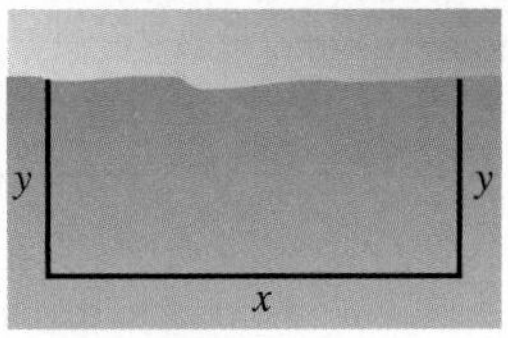

Figure for 5

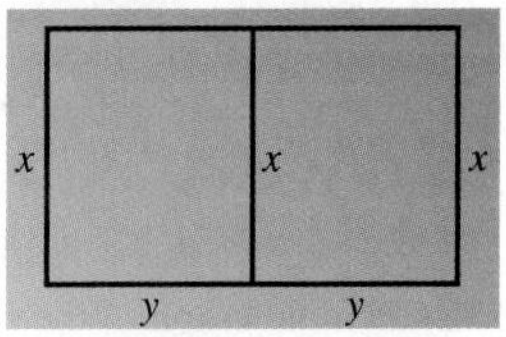

Figure for 6

6. A rancher has 300 feet of fence to enclose two adjacent pastures.

(a) Write the total area A of the two pastures as a function of x (see figure). What is the domain of A?

(b) Graph the area function and estimate the dimensions that yield the maximum amount of area for the pastures.

(c) Find the dimensions that yield the maximum amount of area for the pastures by completing the square.

7. You are in a boat 2 miles from the nearest point on the coast. You are to go to a point Q located 3 miles down the coast and 1 mile inland (see figure). You can row at 2 miles per hour and walk at 4 miles per hour. Write the total time T of the trip as a function of x.

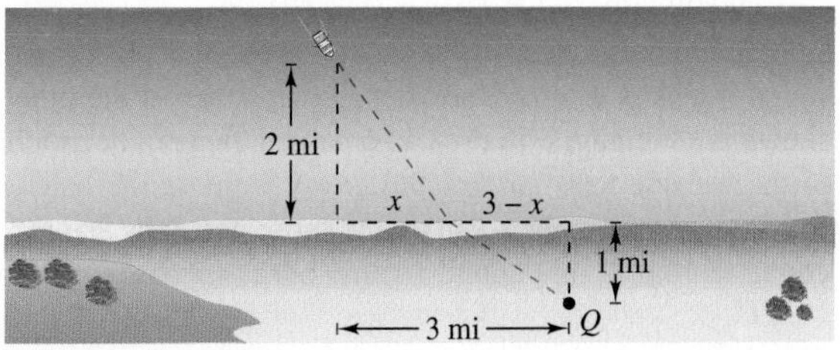

8. You drive to the beach at a rate of 120 kilometers per hour. On the return trip, you drive at a rate of 60 kilometers per hour. What is your average speed for the entire trip? Explain your reasoning.

9. One of the fundamental themes of calculus is to find the slope of the tangent line to a curve at a point. To see how this can be done, consider the point (2, 4) on the graph of $f(x) = x^2$.

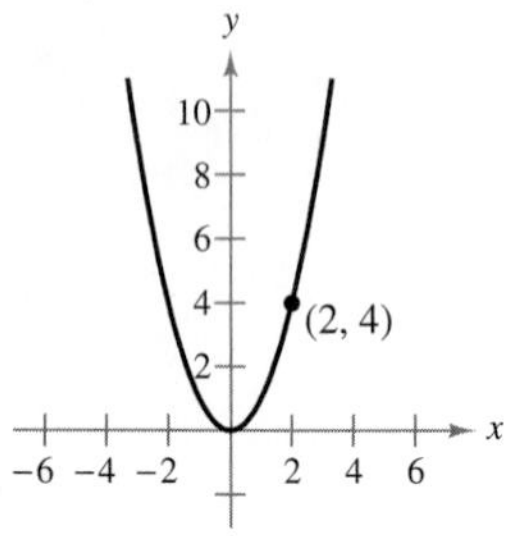

(a) Find the slope of the line joining (2, 4) and (3, 9). Is the slope of the tangent line at (2, 4) greater than or less than this number?

(b) Find the slope of the line joining (2, 4) and (1, 1). Is the slope of the tangent line at (2, 4) greater than or less than this number?

(c) Find the slope of the line joining (2, 4) and (2.1, 4.41). Is the slope of the tangent line at (2, 4) greater than or less than this number?

(d) Find the slope of the line joining (2, 4) and $(2 + h, f(2 + h))$ in terms of the nonzero number h. Verify that $h = 1, -1$, and 0.1 yield the solutions to parts (a)–(c) above.

(e) What is the slope of the tangent line at (2, 4)? Explain how you arrived at your answer.

10. Sketch the graph of the function $f(x) = \sqrt{x}$ and label the point (4, 2) on the graph.

(a) Find the slope of the line joining (4, 2) and (9, 3). Is the slope of the tangent line at (4, 2) greater than or less than this number?

(b) Find the slope of the line joining (4, 2) and (1, 1). Is the slope of the tangent line at (4, 2) greater than or less than this number?

(c) Find the slope of the line joining (4, 2) and (4.41, 2.1). Is the slope of the tangent line at (4, 2) greater than or less than this number?

(d) Find the slope of the line joining (4, 2) and $(4 + h, f(4 + h))$ in terms of the nonzero number h.

(e) What is the slope of the tangent line at the point (4, 2)? Explain how you arrived at your answer.

11. A large room contains two speakers that are 3 meters apart. The sound intensity I of one speaker is twice that of the other, as shown in the figure. (To print an enlarged copy of the graph, go to the website *www.mathgraphs.com.*) Suppose the listener is free to move about the room to find those positions that receive equal amounts of sound from both speakers. Such a location satisfies two conditions: (1) the sound intensity at the listener's position is directly proportional to the sound level of a source, and (2) the sound intensity is inversely proportional to the square of the distance from the source.

(a) Find the points on the x-axis that receive equal amounts of sound from both speakers.

(b) Find and graph the equation of all locations (x, y) where one could stand and receive equal amounts of sound from both speakers.

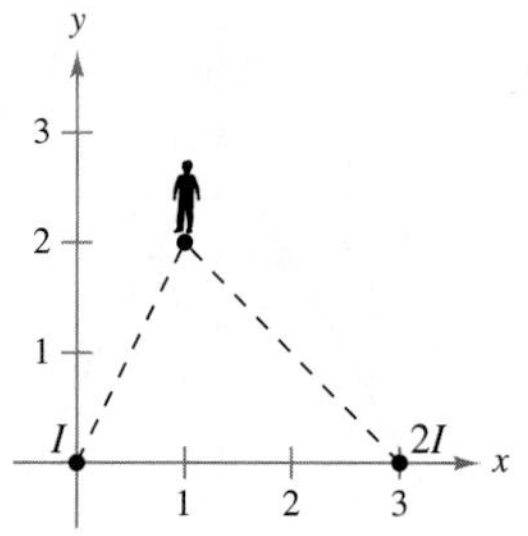

Figure for 11

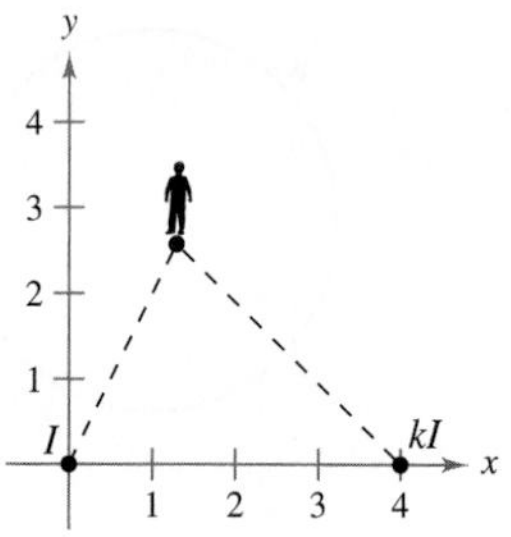

Figure for 12

12. Suppose the speakers in Exercise 11 are 4 meters apart and the sound intensity of one speaker is k times that of the other, as shown in the figure. To print an enlarged copy of the graph, go to the website *www.mathgraphs.com.*

(a) Find the equation of all locations (x, y) where one could stand and receive equal amounts of sound from both speakers.

(b) Graph the equation for the case $k = 3$.

(c) Describe the set of locations of equal sound as k becomes very large.

13. Let d_1 and d_2 be the distances from the point (x, y) to the points $(-1, 0)$ and $(1, 0)$, respectively, as shown in the figure. Show that the equation of the graph of all points (x, y) satisfying $d_1 d_2 = 1$ is $(x^2 + y^2)^2 = 2(x^2 - y^2)$. This curve is called a **lemniscate.** Graph the lemniscate and identify three points on the graph.

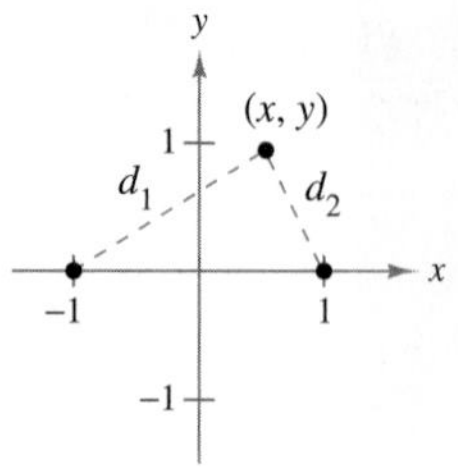

14. Let $f(x) = \dfrac{1}{1 - x}$.

(a) What are the domain and range of f?

(b) Find the composition $f(f(x))$. What is the domain of this function?

(c) Find $f(f(f(x)))$. What is the domain of this function?

(d) Graph $f(f(f(x)))$. Is the graph a line? Why or why not?

Limits and Their Properties

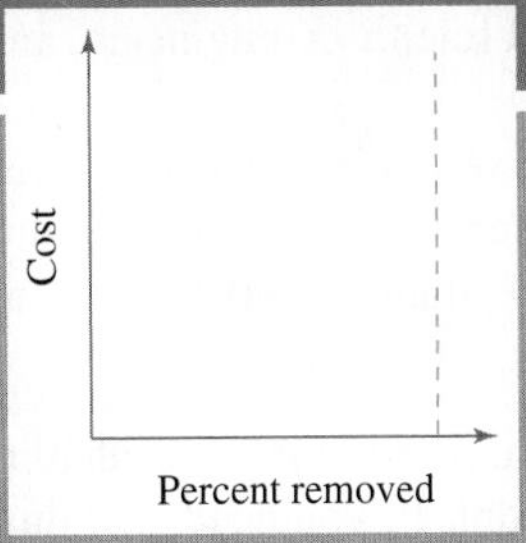

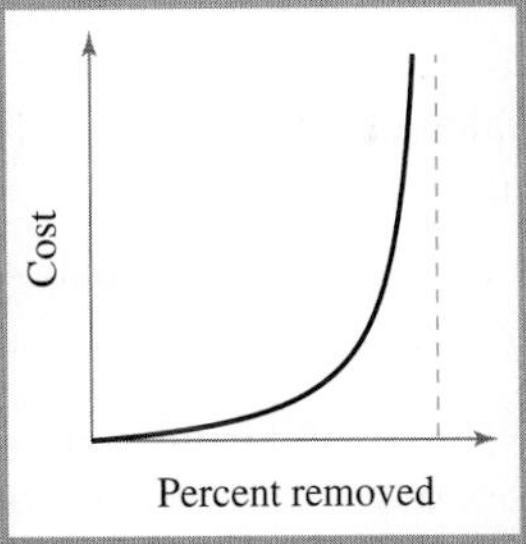

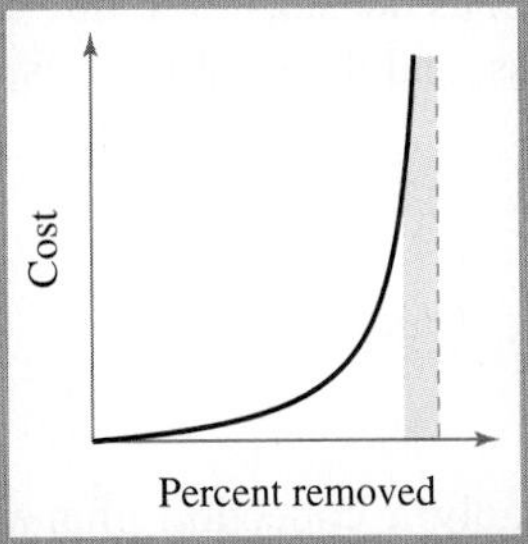

The limit process is a fundamental concept of calculus. One technique you can use to estimate a limit is to graph the function and then determine the behavior of the graph as the independent variable approaches a specific value. In Chapter 1, you will learn how to find limits of functions analytically, graphically, and numerically.

Utility companies use a platinum catalyst pollution scrubber to remove pollutants from smokestack emissions. Which do you think is more costly for a utility company to accomplish—removing the first 90 percent of the pollutants, or removing the last 10 percent? Why?

Jeremy Walker/Getty Images

Section 1.1 A Preview of Calculus

- **Understand what calculus is and how it compares with precalculus.**
- **Understand that the tangent line problem is basic to calculus.**
- **Understand that the area problem is also basic to calculus.**

STUDY TIP As you progress through this course, remember that learning calculus is just one of your goals. Your most important goal is to learn how to use calculus to model and solve real-life problems. Here are a few problem-solving strategies that may help you.

- Be sure you understand the question. What is given? What are you asked to find?
- Outline a plan. There are many approaches you could use: look for a pattern, solve a simpler problem, work backwards, draw a diagram, use technology, or any of many other approaches.
- Complete your plan. Be sure to answer the question. Verbalize your answer. For example, rather than writing the answer as $x = 4.6$, it would be better to write the answer as "The area of the region is 4.6 square meters."
- Look back at your work. Does your answer make sense? Is there a way you can check the reasonableness of your answer?

What Is Calculus?

Calculus is the mathematics of change—velocities and accelerations. Calculus is also the mathematics of tangent lines, slopes, areas, volumes, arc lengths, centroids, curvatures, and a variety of other concepts that have enabled scientists, engineers, and economists to model real-life situations.

Although precalculus mathematics also deals with velocities, accelerations, tangent lines, slopes, and so on, there is a fundamental difference between precalculus mathematics and calculus. Precalculus mathematics is more static, whereas calculus is more dynamic. Here are some examples.

- An object traveling at a constant velocity can be analyzed with precalculus mathematics. To analyze the velocity of an accelerating object, you need calculus.
- The slope of a line can be analyzed with precalculus mathematics. To analyze the slope of a curve, you need calculus.
- A tangent line to a circle can be analyzed with precalculus mathematics. To analyze a tangent line to a general graph, you need calculus.
- The area of a rectangle can be analyzed with precalculus mathematics. To analyze the area under a general curve, you need calculus.

Each of these situations involves the same general strategy—the reformulation of precalculus mathematics through the use of a limit process. So, one way to answer the question "What is calculus?" is to say that calculus is a "limit machine" that involves three stages. The first stage is precalculus mathematics, such as the slope of a line or the area of a rectangle. The second stage is the limit process, and the third stage is a new calculus formulation, such as a derivative or integral.

Precalculus mathematics ⇒ Limit process ⇒ Calculus

Some students try to learn calculus as if it were simply a collection of new formulas. This is unfortunate. If you reduce calculus to the memorization of differentiation and integration formulas, you will miss a great deal of understanding, self-confidence, and satisfaction.

On the following two pages some familiar precalculus concepts coupled with their calculus counterparts are listed. Throughout the text, your goal should be to learn how precalculus formulas and techniques are used as building blocks to produce the more general calculus formulas and techniques. Don't worry if you are unfamiliar with some of the "old formulas" listed on the following two pages—you will be reviewing all of them.

As you proceed through this text, come back to this discussion repeatedly. Try to keep track of where you are relative to the three stages involved in the study of calculus. For example, the first three chapters break down as shown.

Chapter P: Preparation for Calculus	Precalculus
Chapter 1: Limits and Their Properties	Limit process
Chapter 2: Differentiation	Calculus

The Mistress Fellows, Girton College, Cambridge

Grace Chisholm Young (1868–1944)

Grace Chisholm Young received her degree in mathematics from Girton College in Cambridge, England. Her early work was published under the name of William Young, her husband. Between 1914 and 1916, Grace Young published work on the foundations of calculus that won her the Gamble Prize from Girton College.

Without Calculus	**With Differential Calculus**
Value of $f(x)$ when $x = c$ 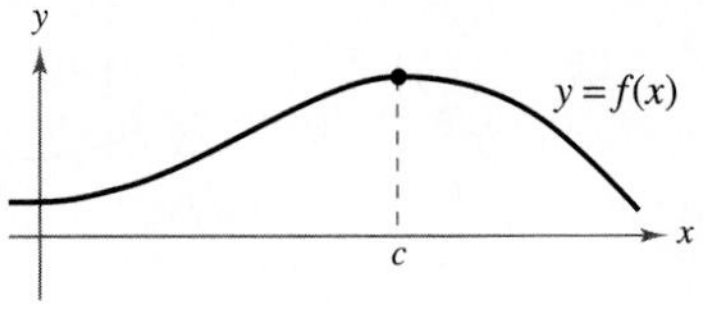	Limit of $f(x)$ as x approaches c 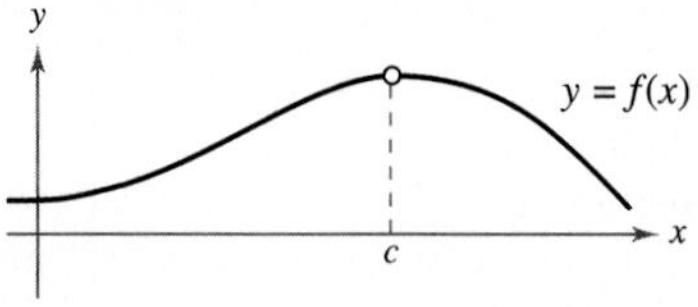
Slope of a line 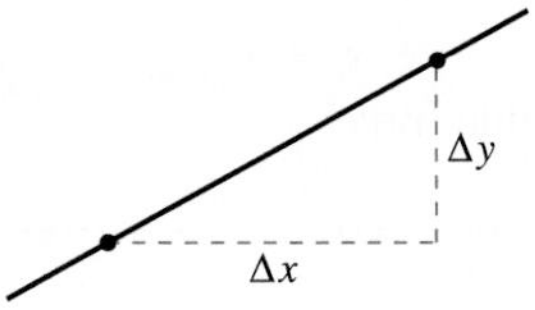	Slope of a curve 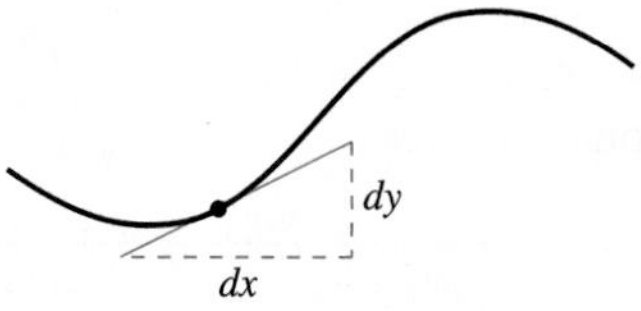
Secant line to a curve 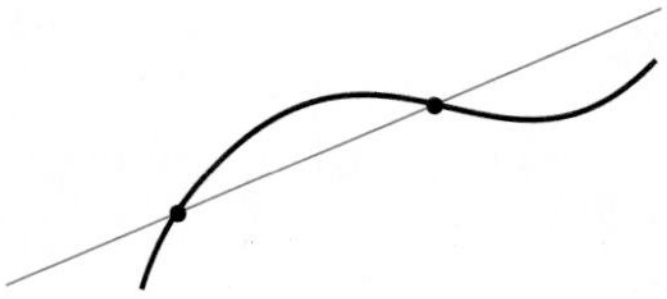	Tangent line to a curve 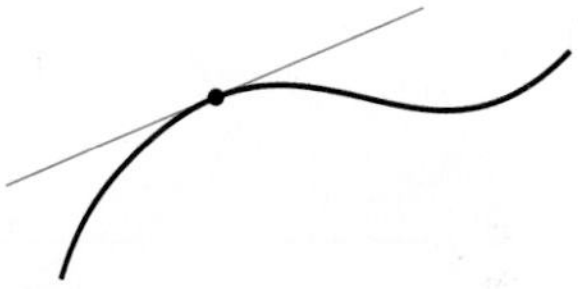
Average rate of change between $t = a$ and $t = b$ 	Instantaneous rate of change at $t = c$ 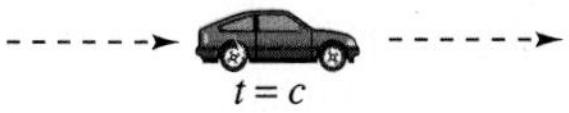
Curvature of a circle 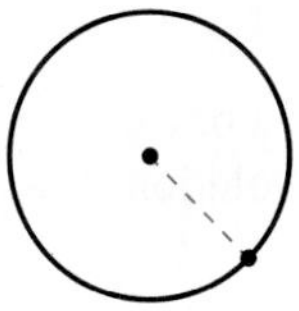	Curvature of a curve 
Height of a curve when $x = c$ 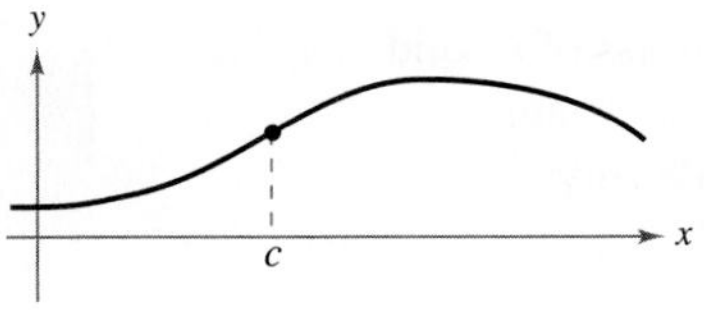	Maximum height of a curve on an interval 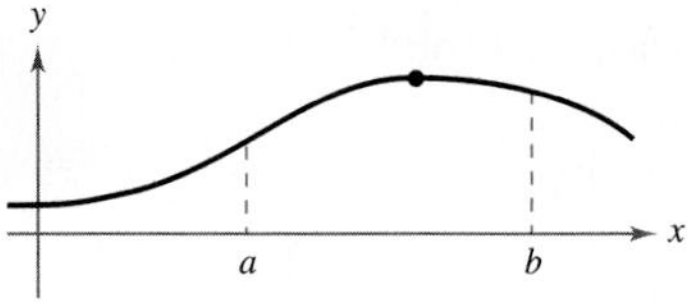
Tangent plane to a sphere 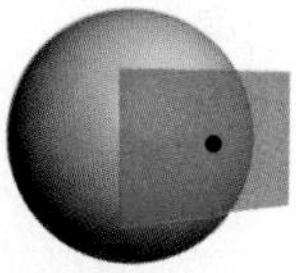	Tangent plane to a surface 
Direction of motion along a line 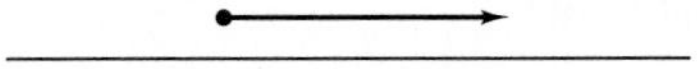	Direction of motion along a curve 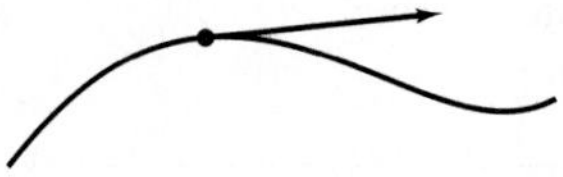

Without Calculus		With Integral Calculus	
Area of a rectangle		Area under a curve	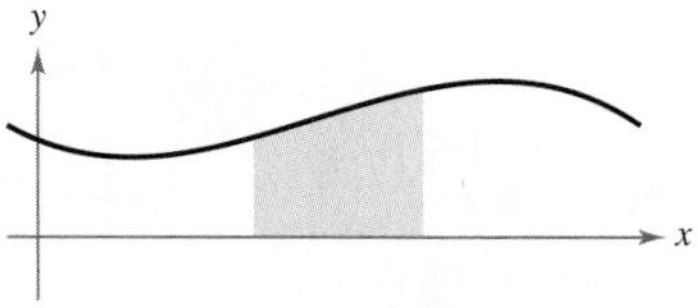
Work done by a constant force	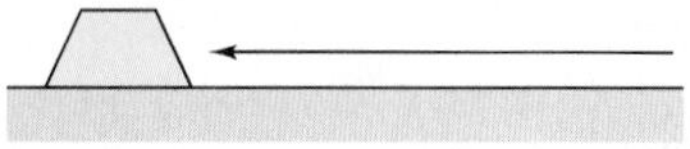	Work done by a variable force	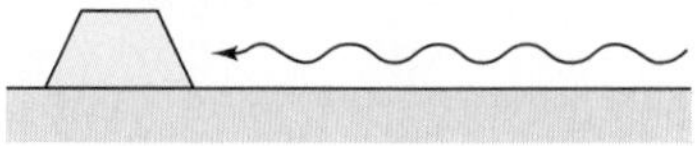
Center of a rectangle	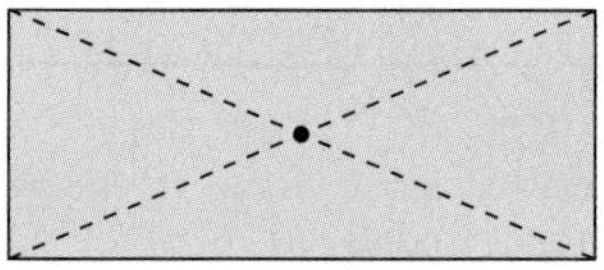	Centroid of a region	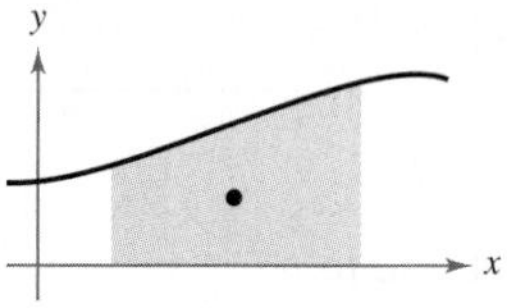
Length of a line segment	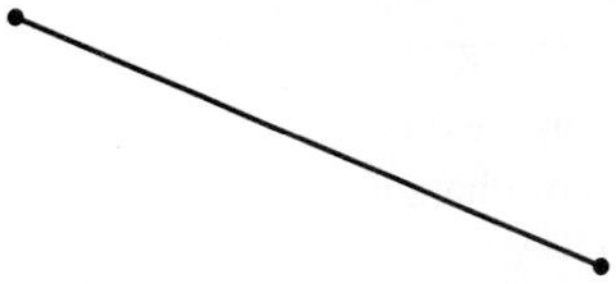	Length of an arc	
Surface area of a cylinder		Surface area of a solid of revolution	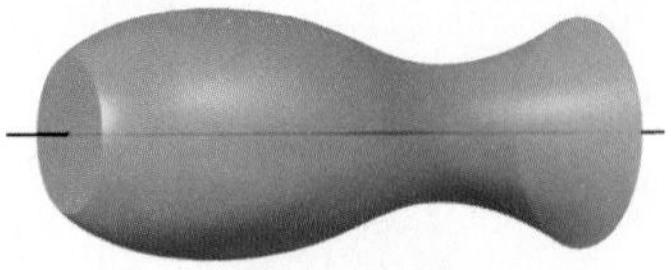
Mass of a solid of constant density		Mass of a solid of variable density	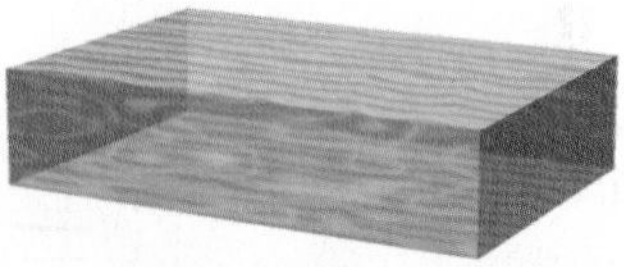
Volume of a rectangular solid	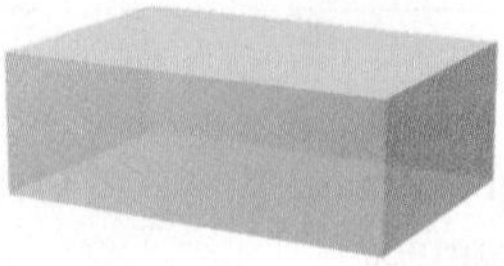	Volume of a region under a surface	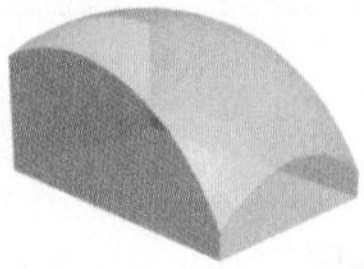
Sum of a finite number of terms	$a_1 + a_2 + \cdots + a_n = S$	Sum of an infinite number of terms	$a_1 + a_2 + a_3 + \cdots = S$

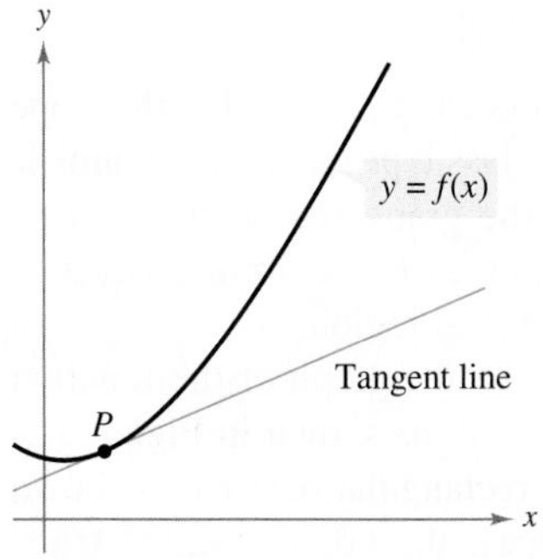

The tangent line to the graph of f at P
Figure 1.1

The Tangent Line Problem

The notion of a limit is fundamental to the study of calculus. The following brief descriptions of two classic problems in calculus—*the tangent line problem and the area problem*—should give you some idea of the way limits are used in calculus.

In the tangent line problem, you are given a function f and a point P on its graph and are asked to find an equation of the tangent line to the graph at point P, as shown in Figure 1.1.

Except for cases involving a vertical tangent line, the problem of finding the **tangent line** at a point P is equivalent to finding the *slope* of the tangent line at P. You can approximate this slope by using a line through the point of tangency and a second point on the curve, as shown in Figure 1.2(a). Such a line is called a **secant line.** If $P(c, f(c))$ is the point of tangency and

$$Q(c + \Delta x, f(c + \Delta x))$$

is a second point on the graph of f, the slope of the secant line through these two points is given by

$$m_{sec} = \frac{f(c + \Delta x) - f(c)}{c + \Delta x - c} = \frac{f(c + \Delta x) - f(c)}{\Delta x}.$$

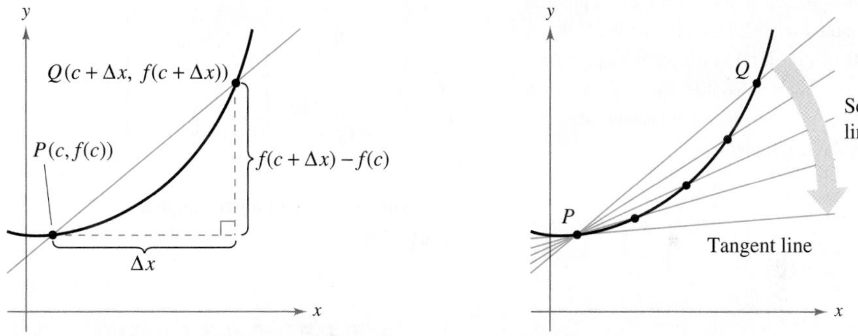

(a) The secant line through $(c, f(c))$ and $(c + \Delta x, f(c + \Delta x))$

(b) As Q approaches P, the secant lines approach the tangent line.

Figure 1.2

As point Q approaches point P, the slope of the secant line approaches the slope of the tangent line, as shown in Figure 1.2(b). When such a "limiting position" exists, the slope of the tangent line is said to be the **limit** of the slope of the secant line. (Much more will be said about this important problem in Chapter 2.)

EXPLORATION

The following points lie on the graph of $f(x) = x^2$.

$Q_1(1.5, f(1.5))$, $Q_2(1.1, f(1.1))$, $Q_3(1.01, f(1.01))$,
$Q_4(1.001, f(1.001))$, $Q_5(1.0001, f(1.0001))$

Each successive point gets closer to the point $P(1, 1)$. Find the slope of the secant line through Q_1 and P, Q_2 and P, and so on. Graph these secant lines on a graphing utility. Then use your results to estimate the slope of the tangent line to the graph of f at the point P.

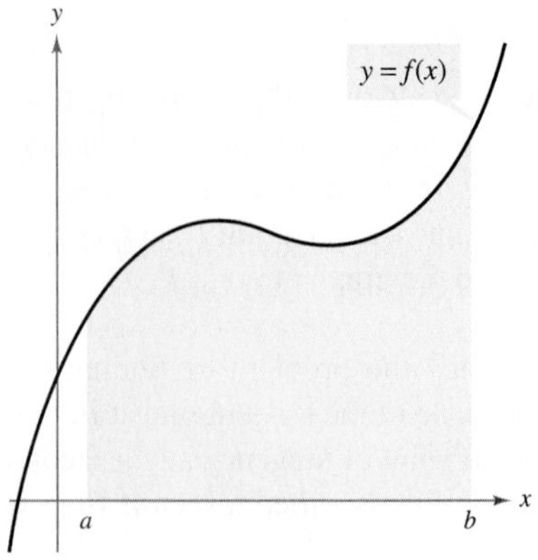

Area under a curve
Figure 1.3

The Area Problem

In the tangent line problem, you saw how the limit process can be applied to the slope of a line to find the slope of a general curve. A second classic problem in calculus is finding the area of a plane region that is bounded by the graphs of functions. This problem can also be solved with a limit process. In this case, the limit process is applied to the area of a rectangle to find the area of a general region.

As a simple example, consider the region bounded by the graph of the function $y = f(x)$, the x-axis, and the vertical lines $x = a$ and $x = b$, as shown in Figure 1.3. You can approximate the area of the region with several rectangular regions, as shown in Figure 1.4. As you increase the number of rectangles, the approximation tends to become better and better because the amount of area missed by the rectangles decreases. Your goal is to determine the limit of the sum of the areas of the rectangles as the number of rectangles increases without bound.

Historical Note

In one of the most astounding events ever to occur in mathematics, it was discovered that the tangent line problem and the area problem are closely related. This discovery led to the birth of calculus. You will learn about the relationship between these two problems when you study the Fundamental Theorem of Calculus in Chapter 4.

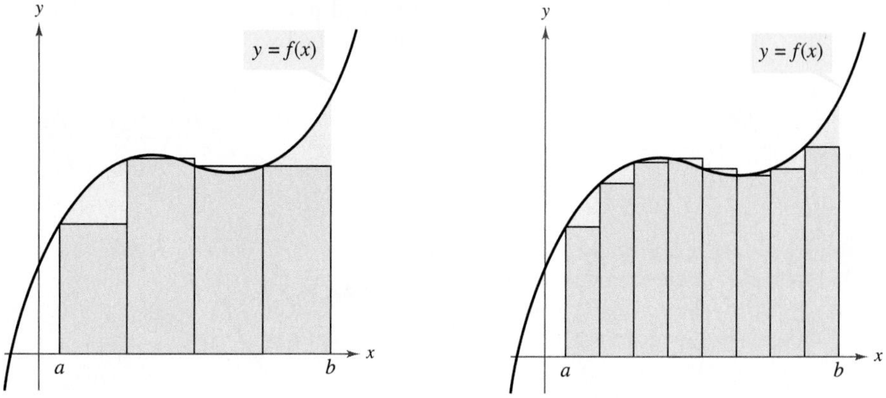

Approximation using four rectangles
Approximation using eight rectangles
Figure 1.4

EXPLORATION

Consider the region bounded by the graphs of $f(x) = x^2$, $y = 0$, and $x = 1$, as shown in part (a) of the figure. The area of the region can be approximated by two sets of rectangles—one set inscribed within the region and the other set circumscribed over the region, as shown in parts (b) and (c). Find the sum of the areas of each set of rectangles. Then use your results to approximate the area of the region.

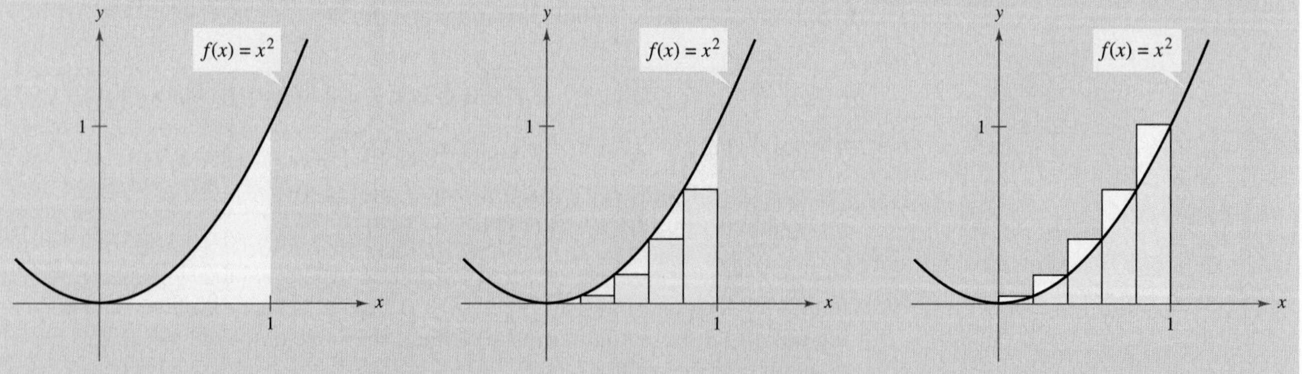

(a) Bounded region
(b) Inscribed rectangles
(c) Circumscribed rectangles

Exercises for Section 1.1

See www.CalcChat.com for worked-out solutions to odd-numbered exercises.

In Exercises 1–6, decide whether the problem can be solved using precalculus, or whether calculus is required. If the problem can be solved using precalculus, solve it. If the problem seems to require calculus, explain your reasoning and use a graphical or numerical approach to estimate the solution.

1. Find the distance traveled in 15 seconds by an object traveling at a constant velocity of 20 feet per second.
2. Find the distance traveled in 15 seconds by an object moving with a velocity of $v(t) = 20 + 7 \cos t$ feet per second.
3. A bicyclist is riding on a path modeled by the function $f(x) = 0.04(8x - x^2)$, where x and $f(x)$ are measured in miles. Find the rate of change of elevation when $x = 2$.

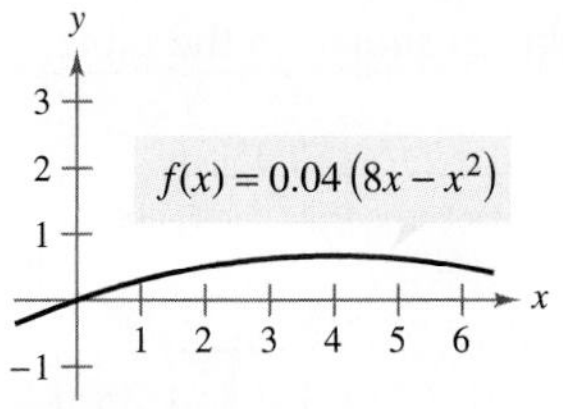

Figure for 3

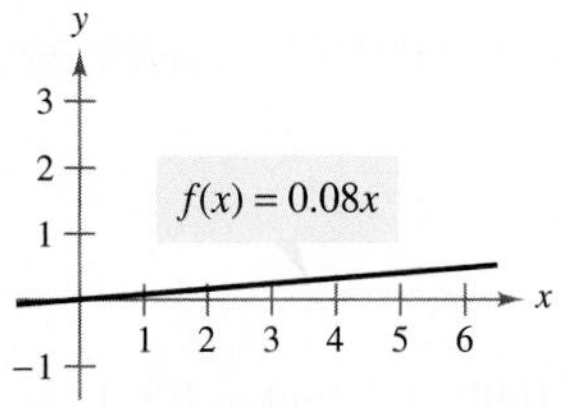

Figure for 4

4. A bicyclist is riding on a path modeled by the function $f(x) = 0.08x$, where x and $f(x)$ are measured in miles. Find the rate of change of elevation when $x = 2$.
5. Find the area of the shaded region.

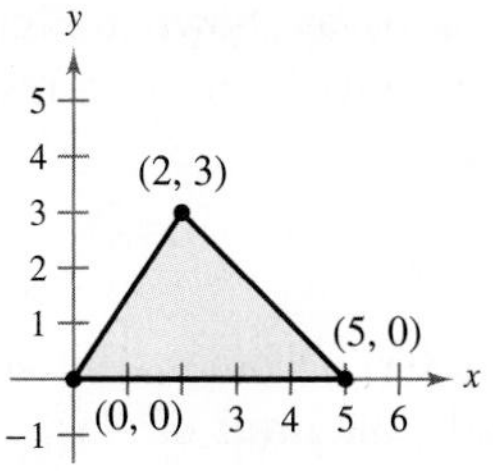

Figure for 5

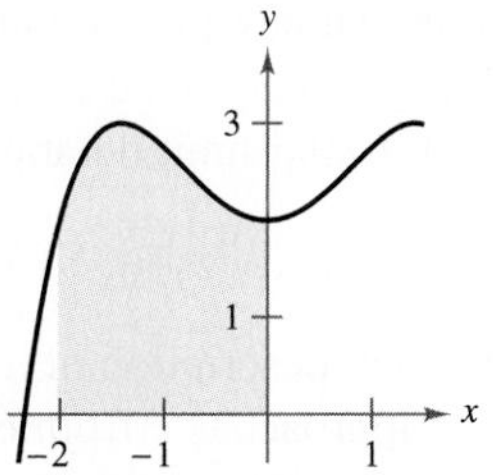

Figure for 6

6. Find the area of the shaded region.
7. ***Secant Lines*** Consider the function $f(x) = 4x - x^2$ and the point $P(1, 3)$ on the graph of f.
   (a) Graph f and the secant lines passing through $P(1, 3)$ and $Q(x, f(x))$ for x-values of 2, 1.5, and 0.5.
   (b) Find the slope of each secant line.
   (c) Use the results of part (b) to estimate the slope of the tangent line of f at $P(1, 3)$. Describe how to improve your approximation of the slope.
8. ***Secant Lines*** Consider the function $f(x) = \sqrt{x}$ and the point $P(4, 2)$ on the graph of f.
   (a) Graph f and the secant lines passing through $P(4, 2)$ and $Q(x, f(x))$ for x-values of 1, 3, and 5.
   (b) Find the slope of each secant line.
   (c) Use the results of part (b) to estimate the slope of the tangent line of f at $P(4, 2)$. Describe how to improve your approximation of the slope.
9. (a) Use the rectangles in each graph to approximate the area of the region bounded by $y = 5/x$, $y = 0$, $x = 1$, and $x = 5$.

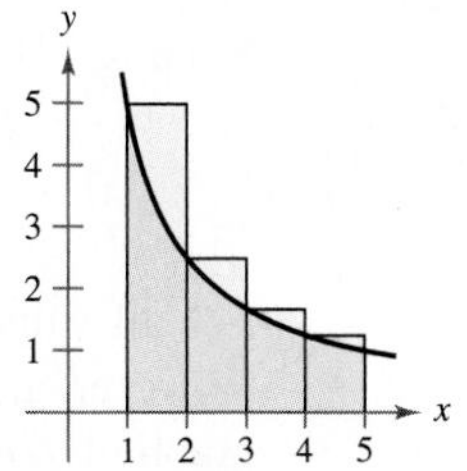

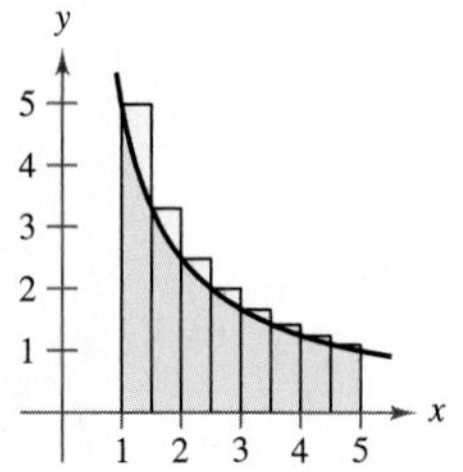

   (b) Describe how you could continue this process to obtain a more accurate approximation of the area.
10. (a) Use the rectangles in each graph to approximate the area of the region bounded by $y = \sin x$, $y = 0$, $x = 0$, and $x = \pi$.

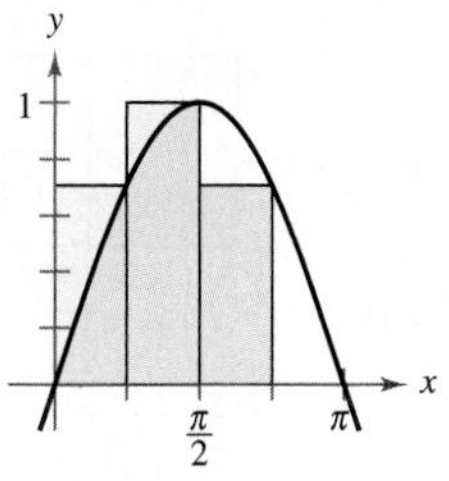

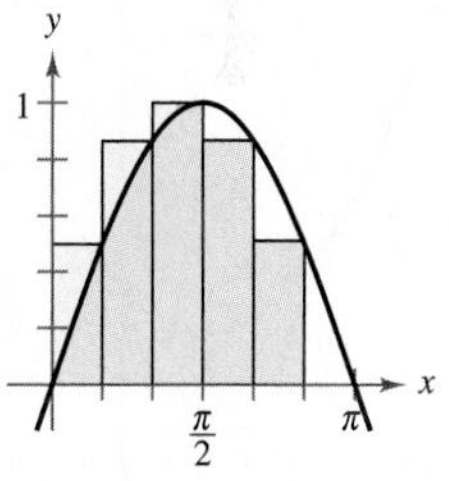

   (b) Describe how you could continue this process to obtain a more accurate approximation of the area.

Writing About Concepts

11. Consider the length of the graph of $f(x) = 5/x$ from $(1, 5)$ to $(5, 1)$.

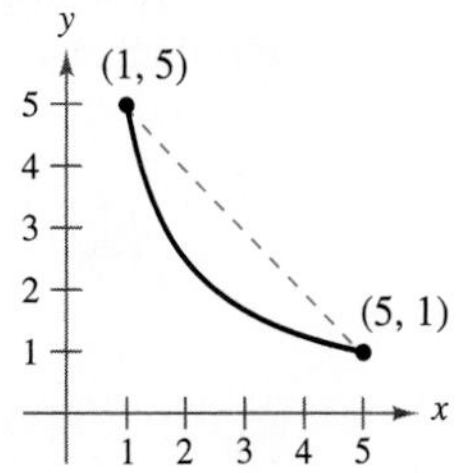

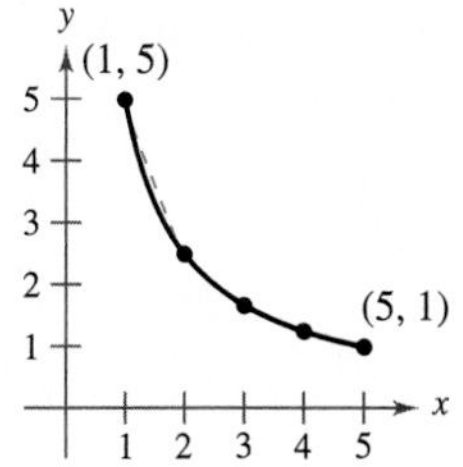

   (a) Approximate the length of the curve by finding the distance between its two endpoints, as shown in the first figure.
   (b) Approximate the length of the curve by finding the sum of the lengths of four line segments, as shown in the second figure.
   (c) Describe how you could continue this process to obtain a more accurate approximation of the length of the curve.

Section 1.2 Finding Limits Graphically and Numerically

- **Estimate a limit using a numerical or graphical approach.**
- **Learn different ways that a limit can fail to exist.**
- **Study and use a formal definition of limit.**

An Introduction to Limits

Suppose you are asked to sketch the graph of the function f given by

$$f(x) = \frac{x^3 - 1}{x - 1}, \qquad x \neq 1.$$

For all values other than $x = 1$, you can use standard curve-sketching techniques. However, at $x = 1$, it is not clear what to expect. To get an idea of the behavior of the graph of f near $x = 1$, you can use two sets of x-values—one set that approaches 1 from the left and one set that approaches 1 from the right, as shown in the table.

x approaches 1 from the left. → ← x approaches 1 from the right.

x	0.75	0.9	0.99	0.999	1	1.001	1.01	1.1	1.25
$f(x)$	2.313	2.710	2.970	2.997	?	3.003	3.030	3.310	3.813

$f(x)$ approaches 3. → ← $f(x)$ approaches 3.

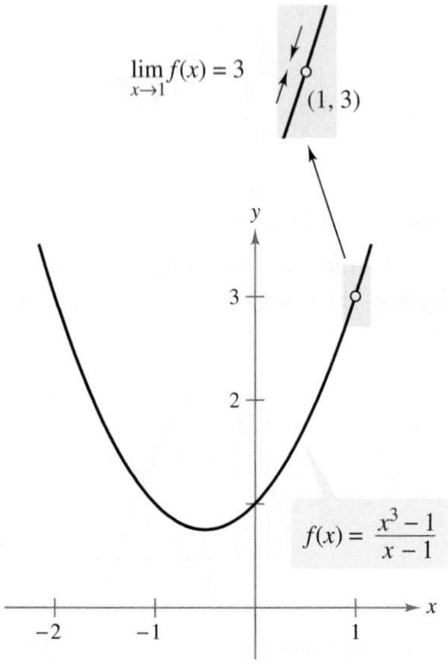

The limit of $f(x)$ as x approaches 1 is 3.
Figure 1.5

The graph of f is a parabola that has a gap at the point (1, 3), as shown in Figure 1.5. Although x cannot equal 1, you can move arbitrarily close to 1, and as a result $f(x)$ moves arbitrarily close to 3. Using limit notation, you can write

$$\lim_{x \to 1} f(x) = 3. \qquad \text{This is read as "the limit of } f(x) \text{ as } x \text{ approaches 1 is 3."}$$

This discussion leads to an informal description of a limit. If $f(x)$ becomes arbitrarily close to a single number L as x approaches c from either side, the **limit** of $f(x)$, as x approaches c, is L. This limit is written as

$$\lim_{x \to c} f(x) = L.$$

EXPLORATION

The discussion above gives an example of how you can estimate a limit *numerically* by constructing a table and *graphically* by drawing a graph. Estimate the following limit numerically by completing the table.

$$\lim_{x \to 2} \frac{x^2 - 3x + 2}{x - 2}$$

x	1.75	1.9	1.99	1.999	2	2.001	2.01	2.1	2.25
$f(x)$	?	?	?	?	?	?	?	?	?

Then use a graphing utility to estimate the limit graphically.

EXAMPLE 1 Estimating a Limit Numerically

Evaluate the function $f(x) = x/\left(\sqrt{x+1} - 1\right)$ at several points near $x = 0$ and use the results to estimate the limit

$$\lim_{x \to 0} \frac{x}{\sqrt{x+1} - 1}.$$

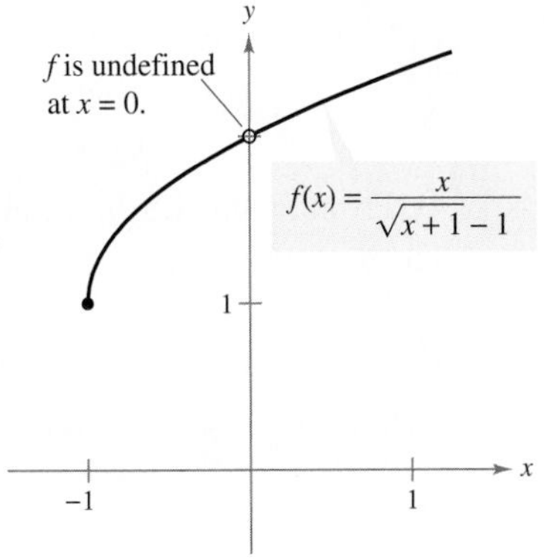

The limit of $f(x)$ as x approaches 0 is 2.
Figure 1.6

Solution The table lists the values of $f(x)$ for several x-values near 0.

x approaches 0 from the left. → ← x approaches 0 from the right.

x	−0.01	−0.001	−0.0001	0	0.0001	0.001	0.01
$f(x)$	1.99499	1.99950	1.99995	?	2.00005	2.00050	2.00499

$f(x)$ approaches 2. → ← $f(x)$ approaches 2.

From the results shown in the table, you can estimate the limit to be 2. This limit is reinforced by the graph of f (see Figure 1.6).

In Example 1, note that the function is undefined at $x = 0$ and yet $f(x)$ appears to be approaching a limit as x approaches 0. This often happens, and it is important to realize that *the existence or nonexistence of $f(x)$ at $x = c$ has no bearing on the existence of the limit of $f(x)$ as x approaches c.*

EXAMPLE 2 Finding a Limit

Find the limit of $f(x)$ as x approaches 2 where f is defined as

$$f(x) = \begin{cases} 1, & x \neq 2 \\ 0, & x = 2. \end{cases}$$

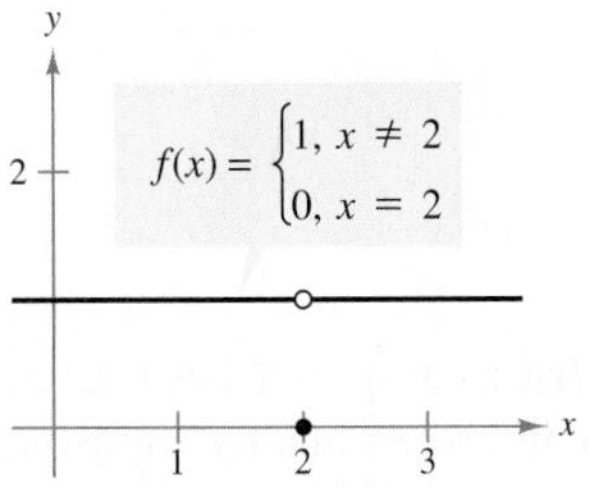

The limit of $f(x)$ as x approaches 2 is 1.
Figure 1.7

Solution Because $f(x) = 1$ for all x other than $x = 2$, you can conclude that the limit is 1, as shown in Figure 1.7. So, you can write

$$\lim_{x \to 2} f(x) = 1.$$

The fact that $f(2) = 0$ has no bearing on the existence or value of the limit as x approaches 2. For instance, if the function were defined as

$$f(x) = \begin{cases} 1, & x \neq 2 \\ 2, & x = 2 \end{cases}$$

the limit would be the same.

So far in this section, you have been estimating limits numerically and graphically. Each of these approaches produces an estimate of the limit. In Section 1.3, you will study analytic techniques for evaluating limits. Throughout the course, try to develop a habit of using this three-pronged approach to problem solving.

1. Numerical approach — Construct a table of values.
2. Graphical approach — Draw a graph by hand or using technology.
3. Analytic approach — Use algebra or calculus.

Limits That Fail to Exist

In the next three examples you will examine some limits that fail to exist.

EXAMPLE 3 Behavior That Differs from the Right and Left

Show that the limit does not exist.

$$\lim_{x\to 0} \frac{|x|}{x}$$

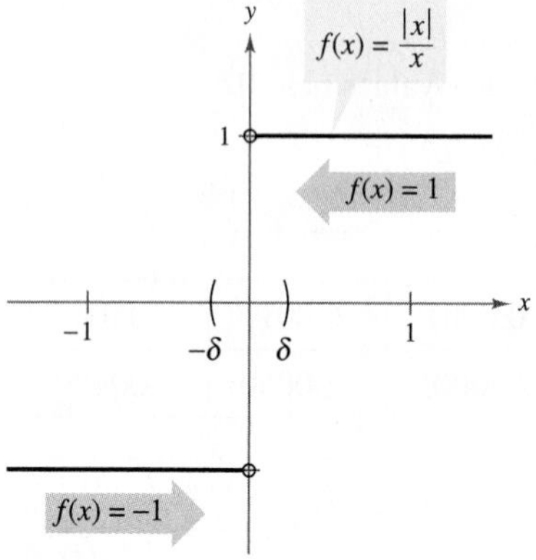

$\lim_{x\to 0} f(x)$ does not exist.

Figure 1.8

Solution Consider the graph of the function $f(x) = |x|/x$. From Figure 1.8, you can see that for positive x-values

$$\frac{|x|}{x} = 1, \qquad x > 0$$

and for negative x-values

$$\frac{|x|}{x} = -1, \qquad x < 0.$$

This means that no matter how close x gets to 0, there will be both positive and negative x-values that yield $f(x) = 1$ and $f(x) = -1$. Specifically, if δ (the lowercase Greek letter *delta*) is a positive number, then for x-values satisfying the inequality $0 < |x| < \delta$, you can classify the values of $|x|/x$ as shown.

$(-\delta, 0)$ — Negative x-values yield $|x|/x = -1$.

$(0, \delta)$ — Positive x-values yield $|x|/x = 1$.

This implies that the limit does not exist.

EXAMPLE 4 Unbounded Behavior

Discuss the existence of the limit

$$\lim_{x\to 0} \frac{1}{x^2}.$$

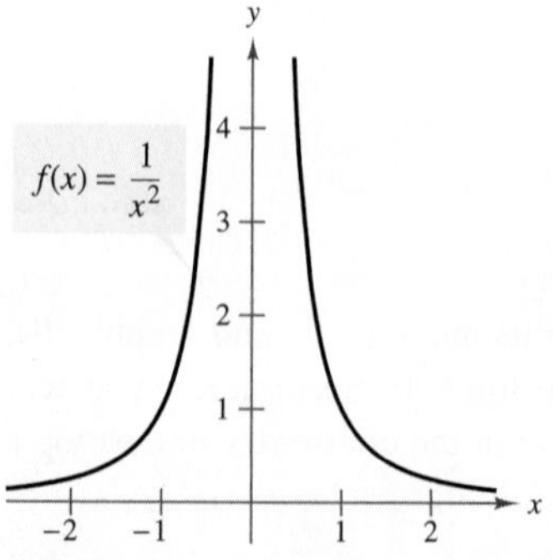

$\lim_{x\to 0} f(x)$ does not exist.

Figure 1.9

Solution Let $f(x) = 1/x^2$. In Figure 1.9, you can see that as x approaches 0 from either the right or the left, $f(x)$ increases without bound. This means that by choosing x close enough to 0, you can force $f(x)$ to be as large as you want. For instance, $f(x)$ will be larger than 100 if you choose x that is within $\frac{1}{10}$ of 0. That is,

$$0 < |x| < \frac{1}{10} \implies f(x) = \frac{1}{x^2} > 100.$$

Similarly, you can force $f(x)$ to be larger than 1,000,000, as follows.

$$0 < |x| < \frac{1}{1000} \implies f(x) = \frac{1}{x^2} > 1{,}000{,}000$$

Because $f(x)$ is not approaching a real number L as x approaches 0, you can conclude that the limit does not exist.

EXAMPLE 5 Oscillating Behavior

Discuss the existence of the limit $\lim_{x \to 0} \sin \frac{1}{x}$.

Solution Let $f(x) = \sin(1/x)$. In Figure 1.10, you can see that as x approaches 0, $f(x)$ oscillates between -1 and 1. So, the limit does not exist because no matter how small you choose δ, it is possible to choose x_1 and x_2 within δ units of 0 such that $\sin(1/x_1) = 1$ and $\sin(1/x_2) = -1$, as shown in the table.

x	$2/\pi$	$2/3\pi$	$2/5\pi$	$2/7\pi$	$2/9\pi$	$2/11\pi$	$x \to 0$
sin (1/x)	1	-1	1	-1	1	-1	Limit does not exist.

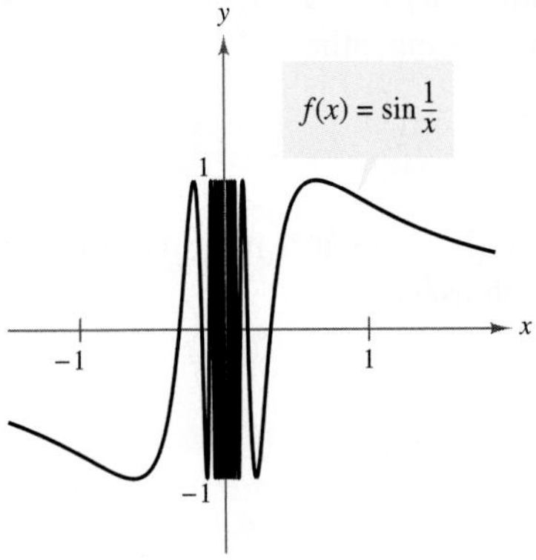

$\lim_{x \to 0} f(x)$ does not exist.
Figure 1.10

Common Types of Behavior Associated with Nonexistence of a Limit

1. $f(x)$ approaches a different number from the right side of c than it approaches from the left side.
2. $f(x)$ increases or decreases without bound as x approaches c.
3. $f(x)$ oscillates between two fixed values as x approaches c.

There are many other interesting functions that have unusual limit behavior. An often cited one is the *Dirichlet function*

$$f(x) = \begin{cases} 0, & \text{if } x \text{ is rational.} \\ 1, & \text{if } x \text{ is irrational.} \end{cases}$$

Because this function has *no limit* at any real number c, it is *not continuous* at any real number c. You will study continuity more closely in Section 1.4.

Peter Gustav Dirichlet (1805–1859)

In the early development of calculus, the definition of a function was much more restricted than it is today, and "functions" such as the Dirichlet function would not have been considered. The modern definition of function was given by the German mathematician Peter Gustav Dirichlet.

TECHNOLOGY PITFALL When you use a graphing utility to investigate the behavior of a function near the x-value at which you are trying to evaluate a limit, remember that you can't always trust the pictures that graphing utilities draw. If you use a graphing utility to graph the function in Example 5 over an interval containing 0, you will most likely obtain an incorrect graph such as that shown in Figure 1.11. The reason that a graphing utility can't show the correct graph is that the graph has infinitely many oscillations over any interval that contains 0.

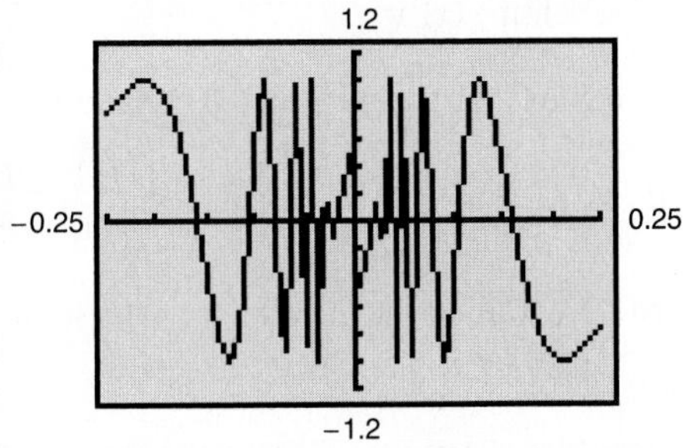

Incorrect graph of $f(x) = \sin(1/x)$.
Figure 1.11

indicates that in the HM mathSpace® *CD-ROM and the online* Eduspace® *system for this text, you will find an Open Exploration, which further explores this example using the computer algebra systems* Maple, Mathcad, Mathematica, *and* Derive.

A Formal Definition of Limit

Let's take another look at the informal description of a limit. If $f(x)$ becomes arbitrarily close to a single number L as x approaches c from either side, then the limit of $f(x)$ as x approaches c is L, written as

$$\lim_{x \to c} f(x) = L.$$

At first glance, this description looks fairly technical. Even so, it is informal because exact meanings have not yet been given to the two phrases

"$f(x)$ becomes arbitrarily close to L"

and

"x approaches c."

The first person to assign mathematically rigorous meanings to these two phrases was Augustin-Louis Cauchy. His **ε-δ definition of limit** is the standard used today.

In Figure 1.12, let ε (the lowercase Greek letter *epsilon*) represent a (small) positive number. Then the phrase "$f(x)$ becomes arbitrarily close to L" means that $f(x)$ lies in the interval $(L - \varepsilon, L + \varepsilon)$. Using absolute value, you can write this as

$$|f(x) - L| < \varepsilon.$$

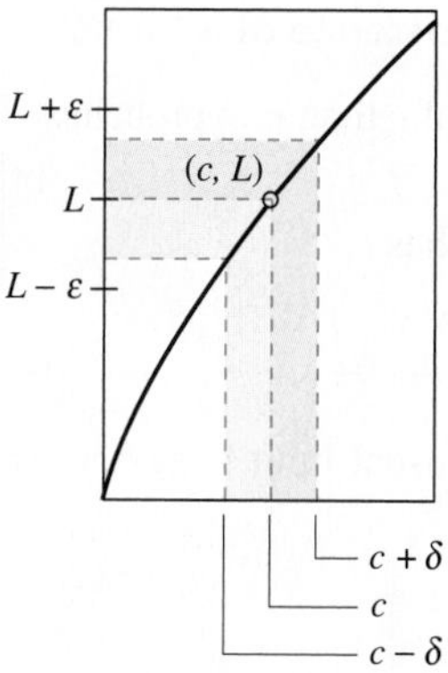

The ε-δ definition of the limit of $f(x)$ as x approaches c
Figure 1.12

Similarly, the phrase "x approaches c" means that there exists a positive number δ such that x lies in either the interval $(c - \delta, c)$ or the interval $(c, c + \delta)$. This fact can be concisely expressed by the double inequality

$$0 < |x - c| < \delta.$$

The first inequality

$$0 < |x - c| \qquad \text{The distance between } x \text{ and } c \text{ is more than 0.}$$

expresses the fact that $x \neq c$. The second inequality

$$|x - c| < \delta \qquad x \text{ is within } \delta \text{ units of } c.$$

says that x is within a distance δ of c.

Definition of Limit

Let f be a function defined on an open interval containing c (except possibly at c) and let L be a real number. The statement

$$\lim_{x \to c} f(x) = L$$

means that for each $\varepsilon > 0$ there exists a $\delta > 0$ such that if

$$0 < |x - c| < \delta, \quad \text{then} \quad |f(x) - L| < \varepsilon.$$

FOR FURTHER INFORMATION For more on the introduction of rigor to calculus, see "Who Gave You the Epsilon? Cauchy and the Origins of Rigorous Calculus" by Judith V. Grabiner in *The American Mathematical Monthly*. To view this article, go to the website *www.matharticles.com*.

NOTE Throughout this text, the expression

$$\lim_{x \to c} f(x) = L$$

implies two statements—the limit exists *and* the limit is L.

Some functions do not have limits as $x \to c$, but those that do cannot have two different limits as $x \to c$. That is, *if the limit of a function exists, it is unique* (see Exercise 69).

The next three examples should help you develop a better understanding of the ε-δ definition of limit.

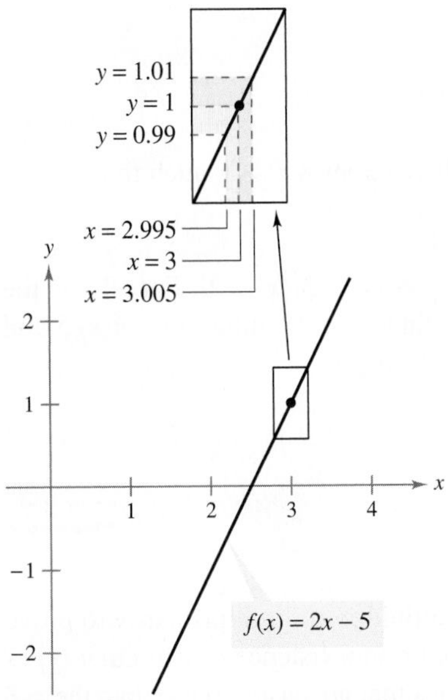

The limit of $f(x)$ as x approaches 3 is 1.
Figure 1.13

EXAMPLE 6 Finding a δ for a Given ε

Given the limit

$$\lim_{x\to 3} (2x - 5) = 1$$

find δ such that $|(2x - 5) - 1| < 0.01$ whenever $0 < |x - 3| < \delta$.

Solution In this problem, you are working with a given value of ε—namely, $\varepsilon = 0.01$. To find an appropriate δ, notice that

$$|(2x - 5) - 1| = |2x - 6| = 2|x - 3|.$$

Because the inequality $|(2x - 5) - 1| < 0.01$ is equivalent to $2|x - 3| < 0.01$, you can choose $\delta = \frac{1}{2}(0.01) = 0.005$. This choice works because

$$0 < |x - 3| < 0.005$$

implies that

$$|(2x - 5) - 1| = 2|x - 3| < 2(0.005) = 0.01$$

as shown in Figure 1.13.

NOTE In Example 6, note that 0.005 is the *largest* value of δ that will guarantee $|(2x - 5) - 1| < 0.01$ whenever $0 < |x - 3| < \delta$. Any *smaller* positive value of δ would also work.

In Example 6, you found a δ-value for a *given* ε. This does not prove the existence of the limit. To do that, you must prove that you can find a δ for any ε, as shown in the next example.

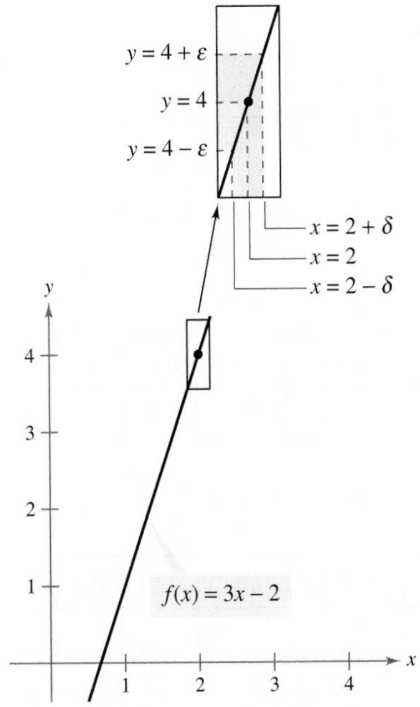

The limit of $f(x)$ as x approaches 2 is 4.
Figure 1.14

EXAMPLE 7 Using the ε-δ Definition of Limit

Use the ε-δ definition of limit to prove that

$$\lim_{x\to 2} (3x - 2) = 4.$$

Solution You must show that for each $\varepsilon > 0$, there exists a $\delta > 0$ such that $|(3x - 2) - 4| < \varepsilon$ whenever $0 < |x - 2| < \delta$. Because your choice of δ depends on ε, you need to establish a connection between the absolute values $|(3x - 2) - 4|$ and $|x - 2|$.

$$|(3x - 2) - 4| = |3x - 6| = 3|x - 2|$$

So, for a given $\varepsilon > 0$ you can choose $\delta = \varepsilon/3$. This choice works because

$$0 < |x - 2| < \delta = \frac{\varepsilon}{3}$$

implies that

$$|(3x - 2) - 4| = 3|x - 2| < 3\left(\frac{\varepsilon}{3}\right) = \varepsilon$$

as shown in Figure 1.14.

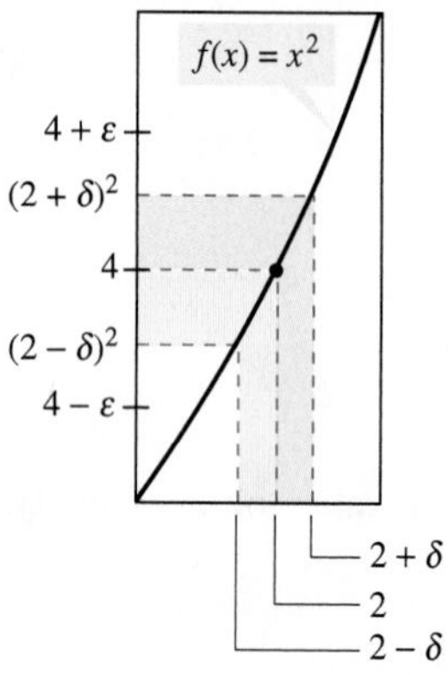

The limit of $f(x)$ as x approaches 2 is 4.
Figure 1.15

EXAMPLE 8 Using the ε-δ Definition of Limit

Use the ε-δ definition of limit to prove that

$$\lim_{x \to 2} x^2 = 4.$$

Solution You must show that for each $\varepsilon > 0$, there exists a $\delta > 0$ such that

$$|x^2 - 4| < \varepsilon \text{ whenever } 0 < |x - 2| < \delta.$$

To find an appropriate δ, begin by writing $|x^2 - 4| = |x - 2||x + 2|$. For all x in the interval $(1, 3)$, you know that $|x + 2| < 5$. So, letting δ be the minimum of $\varepsilon/5$ and 1, it follows that, whenever $0 < |x - 2| < \delta$, you have

$$|x^2 - 4| = |x - 2||x + 2| < \left(\frac{\varepsilon}{5}\right)(5) = \varepsilon$$

as shown in Figure 1.15.

Throughout this chapter you will use the ε-δ definition of limit primarily to prove theorems about limits and to establish the existence or nonexistence of particular types of limits. For *finding* limits, you will learn techniques that are easier to use than the ε-δ definition of limit.

Exercises for Section 1.2

See www.CalcChat.com for worked-out solutions to odd-numbered exercises.

In Exercises 1–8, complete the table and use the result to estimate the limit. Use a graphing utility to graph the function to confirm your result.

1. $\displaystyle\lim_{x \to 2} \frac{x - 2}{x^2 - x - 2}$

x	1.9	1.99	1.999	2.001	2.01	2.1
$f(x)$						

2. $\displaystyle\lim_{x \to 2} \frac{x - 2}{x^2 - 4}$

x	1.9	1.99	1.999	2.001	2.01	2.1
$f(x)$						

3. $\displaystyle\lim_{x \to 0} \frac{\sqrt{x + 3} - \sqrt{3}}{x}$

x	-0.1	-0.01	-0.001	0.001	0.01	0.1
$f(x)$						

4. $\displaystyle\lim_{x \to -3} \frac{\sqrt{1 - x} - 2}{x + 3}$

x	-3.1	-3.01	-3.001	-2.999	-2.99	-2.9
$f(x)$						

5. $\displaystyle\lim_{x \to 3} \frac{[1/(x + 1)] - (1/4)}{x - 3}$

x	2.9	2.99	2.999	3.001	3.01	3.1
$f(x)$						

6. $\displaystyle\lim_{x \to 4} \frac{[x/(x + 1)] - (4/5)}{x - 4}$

x	3.9	3.99	3.999	4.001	4.01	4.1
$f(x)$						

7. $\displaystyle\lim_{x \to 0} \frac{\sin x}{x}$

x	-0.1	-0.01	-0.001	0.001	0.01	0.1
$f(x)$						

8. $\displaystyle\lim_{x \to 0} \frac{\cos x - 1}{x}$

x	-0.1	-0.01	-0.001	0.001	0.01	0.1
$f(x)$						

In Exercises 9–18, use the graph to find the limit (if it exists). If the limit does not exist, explain why.

9. $\lim_{x\to 3}(4 - x)$

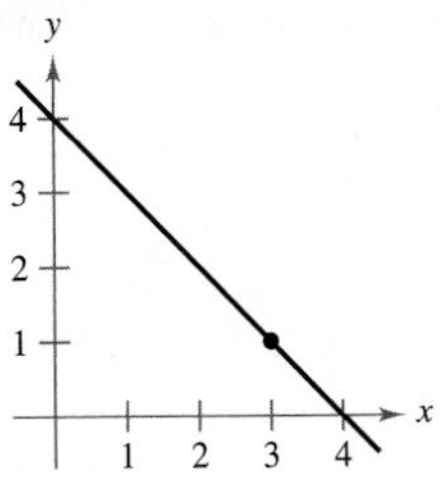

10. $\lim_{x\to 1}(x^2 + 2)$

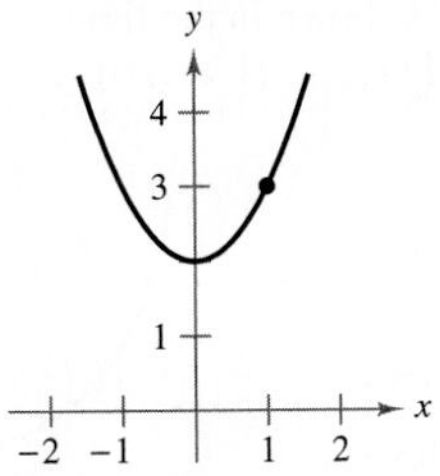

11. $\lim_{x\to 2} f(x)$

$$f(x) = \begin{cases} 4 - x, & x \neq 2 \\ 0, & x = 2 \end{cases}$$

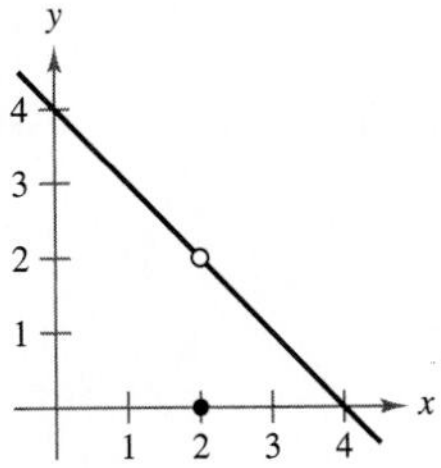

12. $\lim_{x\to 1} f(x)$

$$f(x) = \begin{cases} x^2 + 2, & x \neq 1 \\ 1, & x = 1 \end{cases}$$

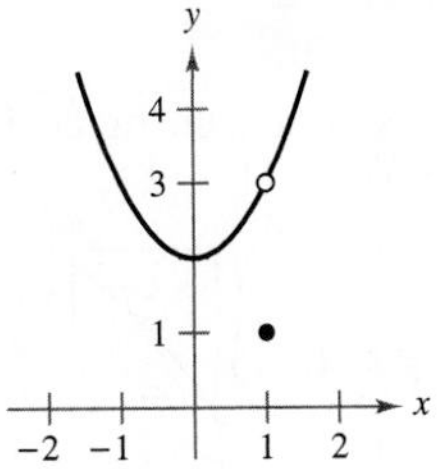

13. $\lim_{x\to 5}\dfrac{|x - 5|}{x - 5}$

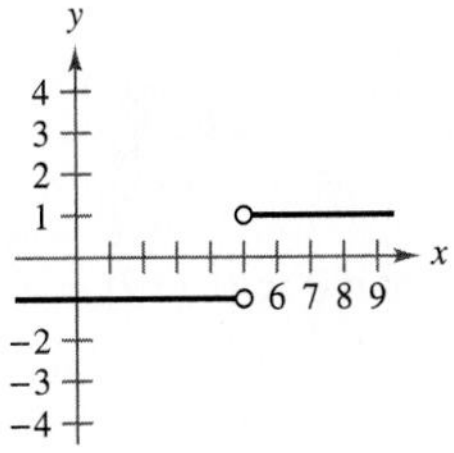

14. $\lim_{x\to 3}\dfrac{1}{x - 3}$

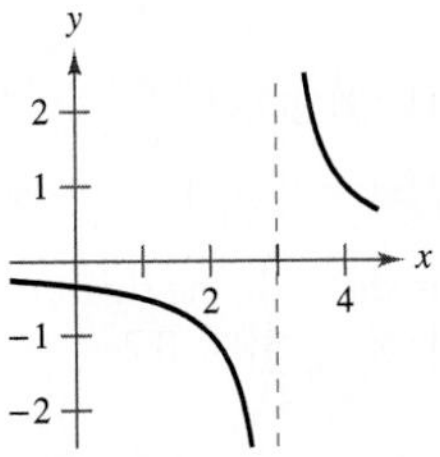

15. $\lim_{x\to 1}\sin \pi x$

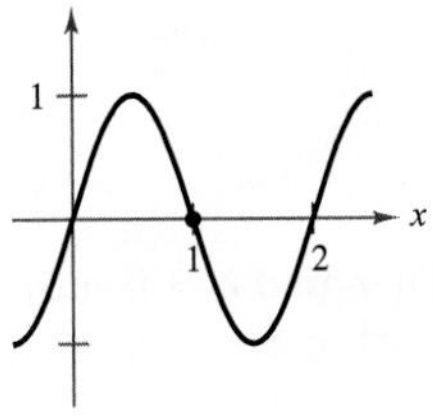

16. $\lim_{x\to 0}\sec x$

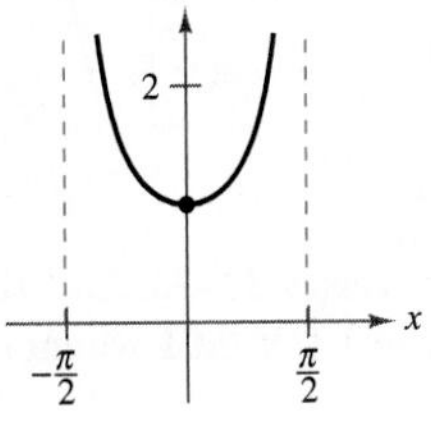

17. $\lim_{x\to 0}\cos\dfrac{1}{x}$

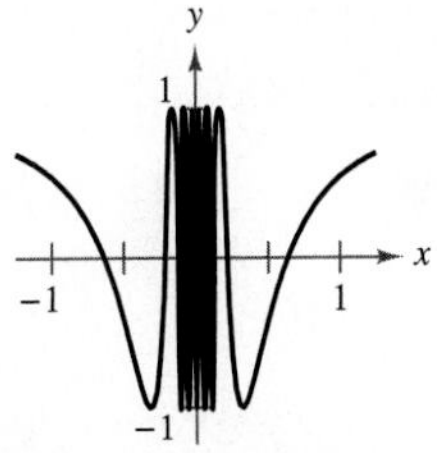

18. $\lim_{x\to \pi/2}\tan x$

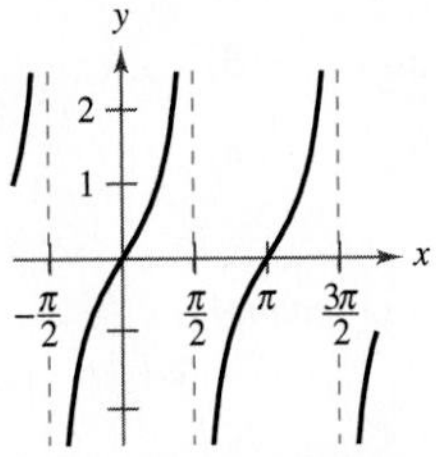

In Exercises 19 and 20, use the graph of the function f to decide whether the value of the given quantity exists. If it does, find it. If not, explain why.

19. (a) $f(1)$

(b) $\lim_{x\to 1} f(x)$

(c) $f(4)$

(d) $\lim_{x\to 4} f(x)$

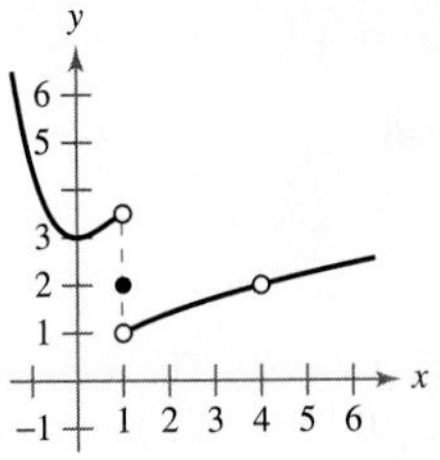

20. (a) $f(-2)$

(b) $\lim_{x\to -2} f(x)$

(c) $f(0)$

(d) $\lim_{x\to 0} f(x)$

(e) $f(2)$

(f) $\lim_{x\to 2} f(x)$

(g) $f(4)$

(h) $\lim_{x\to 4} f(x)$

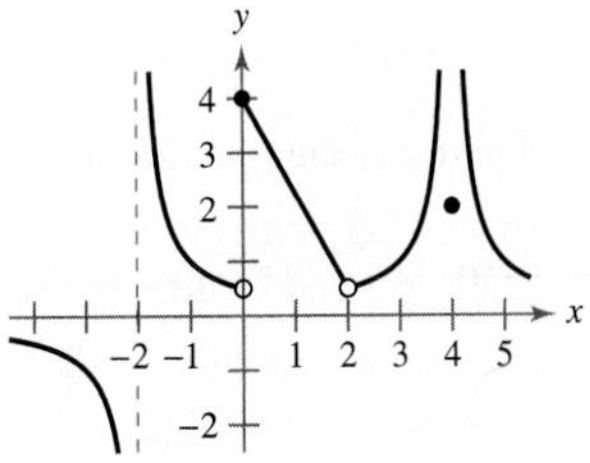

In Exercises 21 and 22, use the graph of f to identify the values of c for which $\lim_{x\to c} f(x)$ exists.

21.

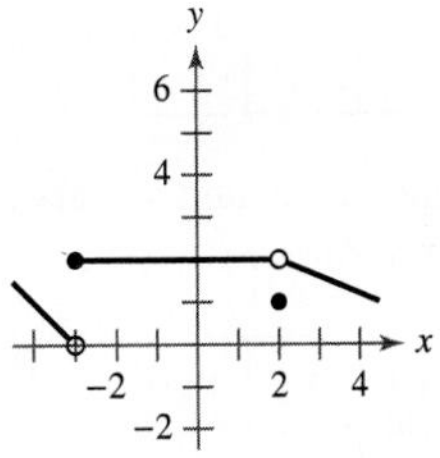

22.

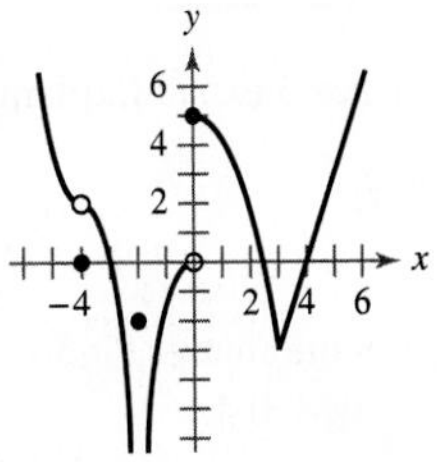

In Exercises 23 and 24, sketch the graph of f. Then identify the values of c for which $\lim_{x\to c} f(x)$ exists.

23. $f(x) = \begin{cases} x^2, & x \le 2 \\ 8 - 2x, & 2 < x < 4 \\ 4, & x \ge 4 \end{cases}$

24. $f(x) = \begin{cases} \sin x, & x < 0 \\ 1 - \cos x, & 0 \le x \le \pi \\ \cos x, & x > \pi \end{cases}$

In Exercises 25 and 26, sketch a graph of a function f that satisfies the given values. (There are many correct answers.)

25. $f(0)$ is undefined.

$\lim_{x\to 0} f(x) = 4$

$f(2) = 6$

$\lim_{x\to 2} f(x) = 3$

26. $f(-2) = 0$

$f(2) = 0$

$\lim_{x\to -2} f(x) = 0$

$\lim_{x\to 2} f(x)$ does not exist.

27. ***Modeling Data*** The cost of a telephone call between two cities is \$0.75 for the first minute and \$0.50 for each additional minute or fraction thereof. A formula for the cost is given by

$C(t) = 0.75 - 0.50[\![-(t-1)]\!]$

where t is the time in minutes.

(*Note:* $[\![x]\!]$ = greatest integer n such that $n \le x$. For example, $[\![3.2]\!] = 3$ and $[\![-1.6]\!] = -2$.)

(a) Use a graphing utility to graph the cost function for $0 < t \le 5$.

(b) Use the graph to complete the table and observe the behavior of the function as t approaches 3.5. Use the graph and the table to find

$\lim_{t\to 3.5} C(t)$.

t	3	3.3	3.4	3.5	3.6	3.7	4
C				?			

(c) Use the graph to complete the table and observe the behavior of the function as t approaches 3.

t	2	2.5	2.9	3	3.1	3.5	4
C				?			

Does the limit of $C(t)$ as t approaches 3 exist? Explain.

28. Repeat Exercise 27 for

$C(t) = 0.35 - 0.12[\![-(t-1)]\!]$.

29. The graph of $f(x) = x + 1$ is shown in the figure. Find δ such that if $0 < |x - 2| < \delta$ then $|f(x) - 3| < 0.4$.

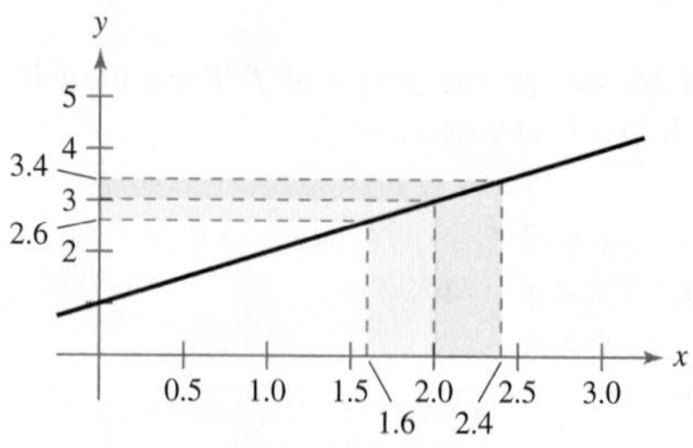

30. The graph of

$$f(x) = \frac{1}{x-1}$$

is shown in the figure. Find δ such that if $0 < |x - 2| < \delta$ then $|f(x) - 1| < 0.01$.

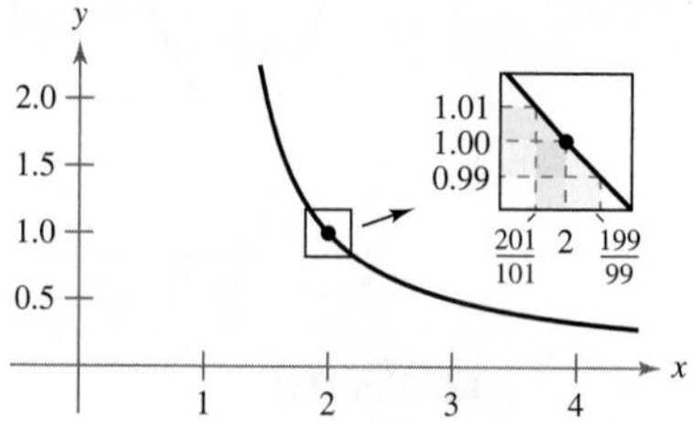

31. The graph of

$$f(x) = 2 - \frac{1}{x}$$

is shown in the figure. Find δ such that if $0 < |x - 1| < \delta$ then $|f(x) - 1| < 0.1$.

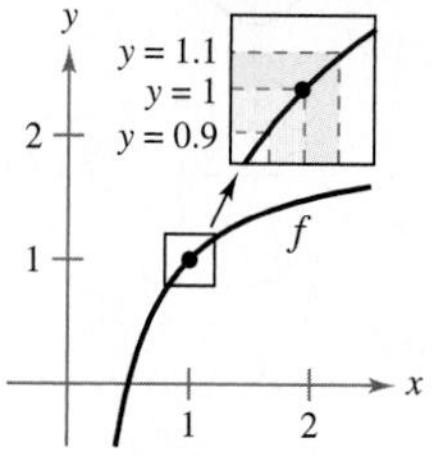

32. The graph of

$$f(x) = x^2 - 1$$

is shown in the figure. Find δ such that if $0 < |x - 2| < \delta$ then $|f(x) - 3| < 0.2$.

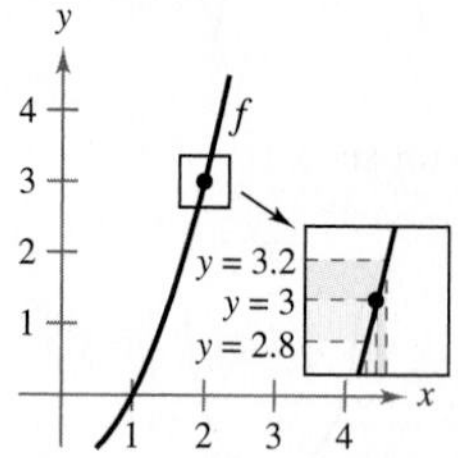

In Exercises 33–36, find the limit L. Then find $\delta > 0$ such that $|f(x) - L| < 0.01$ whenever $0 < |x - c| < \delta$.

33. $\lim_{x\to 2} (3x + 2)$

34. $\lim_{x\to 4} \left(4 - \frac{x}{2}\right)$

35. $\lim_{x\to 2} (x^2 - 3)$

36. $\lim_{x\to 5} (x^2 + 4)$

The symbol indicates an exercise in which you are instructed to use graphing technology or a symbolic computer algebra system. The solutions of other exercises may also be facilitated by use of appropriate technology.

In Exercises 37–48, find the limit L. Then use the ε-δ definition to prove that the limit is L.

37. $\lim_{x\to 2} (x + 3)$

38. $\lim_{x\to -3} (2x + 5)$

39. $\lim_{x\to -4} \left(\frac{1}{2}x - 1\right)$

40. $\lim_{x\to 1} \left(\frac{2}{3}x + 9\right)$

41. $\lim_{x\to 6} 3$

42. $\lim_{x\to 2} (-1)$

43. $\lim_{x\to 0} \sqrt[3]{x}$

44. $\lim_{x\to 4} \sqrt{x}$

45. $\lim_{x\to -2} |x - 2|$

46. $\lim_{x\to 3} |x - 3|$

47. $\lim_{x\to 1} (x^2 + 1)$

48. $\lim_{x\to -3} (x^2 + 3x)$

Writing **In Exercises 49–52, use a graphing utility to graph the function and estimate the limit (if it exists). What is the domain of the function? Can you detect a possible error in determining the domain of a function solely by analyzing the graph generated by a graphing utility? Write a short paragraph about the importance of examining a function analytically as well as graphically.**

49. $f(x) = \dfrac{\sqrt{x + 5} - 3}{x - 4}$

$\lim_{x\to 4} f(x)$

50. $f(x) = \dfrac{x - 3}{x^2 - 4x + 3}$

$\lim_{x\to 3} f(x)$

51. $f(x) = \dfrac{x - 9}{\sqrt{x} - 3}$

$\lim_{x\to 9} f(x)$

52. $f(x) = \dfrac{x - 3}{x^2 - 9}$

$\lim_{x\to 3} f(x)$

Writing About Concepts

53. Write a brief description of the meaning of the notation $\lim_{x\to 8} f(x) = 25$.

54. If $f(2) = 4$, can you conclude anything about the limit of $f(x)$ as x approaches 2? Explain your reasoning.

55. If the limit of $f(x)$ as x approaches 2 is 4, can you conclude anything about $f(2)$? Explain your reasoning.

Writing About Concepts *(continued)*

56. Identify three types of behavior associated with the nonexistence of a limit. Illustrate each type with a graph of a function.

57. ***Jewelry*** A jeweler resizes a ring so that its inner circumference is 6 centimeters.

(a) What is the radius of the ring?

(b) If the ring's inner circumference can vary between 5.5 centimeters and 6.5 centimeters, how can the radius vary?

(c) Use the ε-δ definition of limit to describe this situation. Identify ε and δ.

58. ***Sports*** A sporting goods manufacturer designs a golf ball having a volume of 2.48 cubic inches.

(a) What is the radius of the golf ball?

(b) If the ball's volume can vary between 2.45 cubic inches and 2.51 cubic inches, how can the radius vary?

(c) Use the ε-δ definition of limit to describe this situation. Identify ε and δ.

59. Consider the function $f(x) = (1 + x)^{1/x}$. Estimate the limit

$$\lim_{x\to 0} (1 + x)^{1/x}$$

by evaluating f at x-values near 0. Sketch the graph of f.

60. Consider the function

$$f(x) = \frac{|x + 1| - |x - 1|}{x}.$$

Estimate

$$\lim_{x\to 0} \frac{|x + 1| - |x - 1|}{x}$$

by evaluating f at x-values near 0. Sketch the graph of f.

61. ***Graphical Analysis*** The statement

$$\lim_{x\to 2} \frac{x^2 - 4}{x - 2} = 4$$

means that for each $\varepsilon > 0$ there corresponds a $\delta > 0$ such that if $0 < |x - 2| < \delta$, then

$$\left|\frac{x^2 - 4}{x - 2} - 4\right| < \varepsilon.$$

If $\varepsilon = 0.001$, then

$$\left|\frac{x^2 - 4}{x - 2} - 4\right| < 0.001.$$

Use a graphing utility to graph each side of this inequality. Use the *zoom* feature to find an interval $(2 - \delta, 2 + \delta)$ such that the graph of the left side is below the graph of the right side of the inequality.

62. ***Graphical Analysis*** The statement

$$\lim_{x \to 3} \frac{x^2 - 3x}{x - 3}$$

means that for each $\varepsilon > 0$ there corresponds a $\delta > 0$ such that if $0 < |x - 3| < \delta$, then

$$\left|\frac{x^2 - 3x}{x - 3} - 3\right| < \varepsilon.$$

If $\varepsilon = 0.001$, then

$$\left|\frac{x^2 - 3x}{x - 3} - 3\right| < 0.001.$$

Use a graphing utility to graph each side of this inequality. Use the *zoom* feature to find an interval $(3 - \delta, 3 + \delta)$ such that the graph of the left side is below the graph of the right side of the inequality.

True or False? **In Exercises 63–66, determine whether the statement is true or false. If it is false, explain why or give an example that shows it is false.**

63. If f is undefined at $x = c$, then the limit of $f(x)$ as x approaches c does not exist.

64. If the limit of $f(x)$ as x approaches c is 0, then there must exist a number k such that $f(k) < 0.001$.

65. If $f(c) = L$, then $\lim_{x \to c} f(x) = L$.

66. If $\lim_{x \to c} f(x) = L$, then $f(c) = L$.

67. Consider the function $f(x) = \sqrt{x}$.

(a) Is $\lim_{x \to 0.25} \sqrt{x} = 0.5$ a true statement? Explain.

(b) Is $\lim_{x \to 0} \sqrt{x} = 0$ a true statement? Explain.

68. ***Writing*** The definition of limit on page 52 requires that f is a function defined on an open interval containing c, except possibly at c. Why is this requirement necessary?

69. Prove that if the limit of $f(x)$ as $x \to c$ exists, then the limit must be unique. [*Hint:* Let

$$\lim_{x \to c} f(x) = L_1 \quad \text{and} \quad \lim_{x \to c} f(x) = L_2$$

and prove that $L_1 = L_2$.]

70. Consider the line $f(x) = mx + b$, where $m \neq 0$. Use the ε-δ definition of limit to prove that $\lim_{x \to c} f(x) = mc + b$.

71. Prove that $\lim_{x \to c} f(x) = L$ is equivalent to $\lim_{x \to c} [f(x) - L] = 0$.

72. (a) Given that

$$\lim_{x \to 0} (3x + 1)(3x - 1)x^2 + 0.01 = 0.01$$

prove that there exists an open interval (a, b) containing 0 such that $(3x + 1)(3x - 1)x^2 + 0.01 > 0$ for all $x \neq 0$ in (a, b).

(b) Given that $\lim_{x \to c} g(x) = L$, where $L > 0$, prove that there exists an open interval (a, b) containing c such that $g(x) > 0$ for all $x \neq c$ in (a, b).

73. ***Programming*** Use the programming capabilities of a graphing utility to write a program for approximating $\lim_{x \to c} f(x)$.

Assume the program will be applied only to functions whose limits exist as x approaches c. Let $y_1 = f(x)$ and generate two lists whose entries form the ordered pairs

$$(c \pm [0.1]^n, f(c \pm [0.1]^n))$$

for $n = 0, 1, 2, 3$, and 4.

74. ***Programming*** Use the program you created in Exercise 73 to approximate the limit

$$\lim_{x \to 4} \frac{x^2 - x - 12}{x - 4}.$$

Putnam Exam Challenge

75. Inscribe a rectangle of base b and height h and an isosceles triangle of base b in a circle of radius one as shown. For what value of h do the rectangle and triangle have the same area?

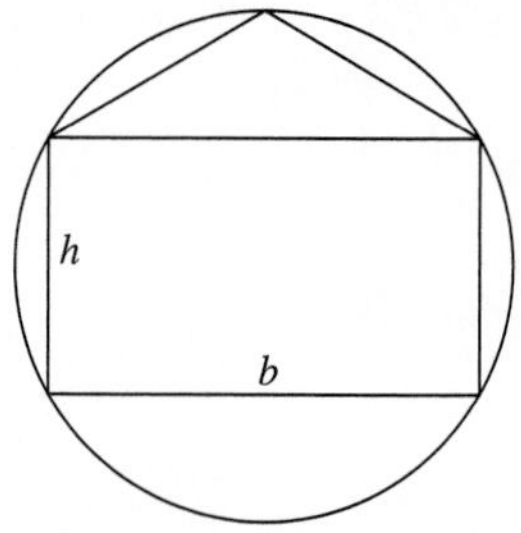

76. A right circular cone has base of radius 1 and height 3. A cube is inscribed in the cone so that one face of the cube is contained in the base of the cone. What is the side-length of the cube?

These problems were composed by the Committee on the Putnam Prize Competition.

Section 1.3 Evaluating Limits Analytically

- **Evaluate a limit using properties of limits.**
- **Develop and use a strategy for finding limits.**
- **Evaluate a limit using dividing out and rationalizing techniques.**
- **Evaluate a limit using the Squeeze Theorem.**

Properties of Limits

In Section 1.2, you learned that the limit of $f(x)$ as x approaches c does not depend on the value of f at $x = c$. It may happen, however, that the limit is precisely $f(c)$. In such cases, the limit can be evaluated by **direct substitution.** That is,

$$\lim_{x\to c} f(x) = f(c). \qquad \text{Substitute } c \text{ for } x.$$

Such *well-behaved* functions are **continuous at c.** You will examine this concept more closely in Section 1.4.

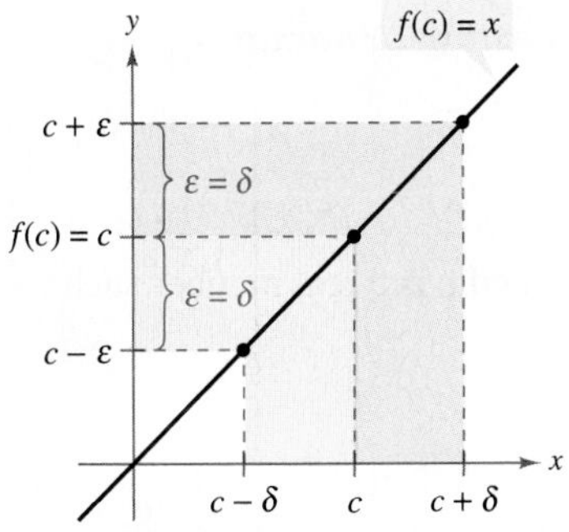

Figure 1.16

THEOREM 1.1 Some Basic Limits

Let b and c be real numbers and let n be a positive integer.

1. $\lim_{x\to c} b = b$ **2.** $\lim_{x\to c} x = c$ **3.** $\lim_{x\to c} x^n = c^n$

Proof To prove Property 2 of Theorem 1.1, you need to show that for each $\varepsilon > 0$ there exists a $\delta > 0$ such that $|x - c| < \varepsilon$ whenever $0 < |x - c| < \delta$. To do this, choose $\delta = \varepsilon$. The second inequality then implies the first, as shown in Figure 1.16. This completes the proof. (Proofs of the other properties of limits in this section are listed in Appendix A or are discussed in the exercises.)

NOTE When you encounter new notations or symbols in mathematics, be sure you know how the notations are read. For instance, the limit in Example 1(c) is read as "the limit of x^2 as x approaches 2 is 4."

EXAMPLE 1 Evaluating Basic Limits

a. $\lim_{x\to 2} 3 = 3$ **b.** $\lim_{x\to -4} x = -4$ **c.** $\lim_{x\to 2} x^2 = 2^2 = 4$

THEOREM 1.2 Properties of Limits

Let b and c be real numbers, let n be a positive integer, and let f and g be functions with the following limits.

$$\lim_{x\to c} f(x) = L \quad \text{and} \quad \lim_{x\to c} g(x) = K$$

1. Scalar multiple: $\lim_{x\to c} [b f(x)] = bL$

2. Sum or difference: $\lim_{x\to c} [f(x) \pm g(x)] = L \pm K$

3. Product: $\lim_{x\to c} [f(x)g(x)] = LK$

4. Quotient: $\lim_{x\to c} \dfrac{f(x)}{g(x)} = \dfrac{L}{K}$, provided $K \neq 0$

5. Power: $\lim_{x\to c} [f(x)]^n = L^n$

EXAMPLE 2 The Limit of a Polynomial

$$\begin{aligned}
\lim_{x\to 2} (4x^2 + 3) &= \lim_{x\to 2} 4x^2 + \lim_{x\to 2} 3 && \text{Property 2}\\
&= 4\left(\lim_{x\to 2} x^2\right) + \lim_{x\to 2} 3 && \text{Property 1}\\
&= 4(2^2) + 3 && \text{Example 1}\\
&= 19 && \text{Simplify.}
\end{aligned}$$

In Example 2, note that the limit (as $x \to 2$) of the *polynomial function* $p(x) = 4x^2 + 3$ is simply the value of p at $x = 2$.

$$\lim_{x\to 2} p(x) = p(2) = 4(2^2) + 3 = 19$$

This *direct substitution* property is valid for all polynomial and rational functions with nonzero denominators.

THEOREM 1.3 Limits of Polynomial and Rational Functions

If p is a polynomial function and c is a real number, then

$$\lim_{x\to c} p(x) = p(c).$$

If r is a rational function given by $r(x) = p(x)/q(x)$ and c is a real number such that $q(c) \neq 0$, then

$$\lim_{x\to c} r(x) = r(c) = \frac{p(c)}{q(c)}.$$

EXAMPLE 3 The Limit of a Rational Function

Find the limit: $\displaystyle\lim_{x\to 1} \frac{x^2 + x + 2}{x + 1}$.

Solution Because the denominator is not 0 when $x = 1$, you can apply Theorem 1.3 to obtain

$$\lim_{x\to 1} \frac{x^2 + x + 2}{x + 1} = \frac{1^2 + 1 + 2}{1 + 1} = \frac{4}{2} = 2.$$

Polynomial functions and rational functions are two of the three basic types of algebraic functions. The following theorem deals with the limit of the third type of algebraic function—one that involves a radical. See Appendix A for a proof of this theorem.

THE SQUARE ROOT SYMBOL

The first use of a symbol to denote the square root can be traced to the sixteenth century. Mathematicians first used the symbol $\surd$, which had only two strokes. This symbol was chosen because it resembled a lowercase *r*, to stand for the Latin word *radix*, meaning root.

THEOREM 1.4 The Limit of a Function Involving a Radical

Let n be a positive integer. The following limit is valid for all c if n is odd, and is valid for $c > 0$ if n is even.

$$\lim_{x\to c} \sqrt[n]{x} = \sqrt[n]{c}$$

The following theorem greatly expands your ability to evaluate limits because it shows how to analyze the limit of a composite function. See Appendix A for a proof of this theorem.

THEOREM 1.5 The Limit of a Composite Function

If f and g are functions such that $\lim_{x\to c} g(x) = L$ and $\lim_{x\to L} f(x) = f(L)$, then

$$\lim_{x\to c} f(g(x)) = f\left(\lim_{x\to c} g(x)\right) = f(L).$$

EXAMPLE 4 The Limit of a Composite Function

a. Because

$$\lim_{x\to 0} (x^2 + 4) = 0^2 + 4 = 4 \qquad \text{and} \qquad \lim_{x\to 4} \sqrt{x} = 2$$

it follows that

$$\lim_{x\to 0} \sqrt{x^2 + 4} = \sqrt{4} = 2.$$

b. Because

$$\lim_{x\to 3} (2x^2 - 10) = 2(3^2) - 10 = 8 \quad \text{and} \quad \lim_{x\to 8} \sqrt[3]{x} = 2$$

it follows that

$$\lim_{x\to 3} \sqrt[3]{2x^2 - 10} = \sqrt[3]{8} = 2.$$

You have seen that the limits of many algebraic functions can be evaluated by direct substitution. The six basic trigonometric functions also exhibit this desirable quality, as shown in the next theorem (presented without proof).

THEOREM 1.6 Limits of Trigonometric Functions

Let c be a real number in the domain of the given trigonometric function.

1. $\lim_{x\to c} \sin x = \sin c$ **2.** $\lim_{x\to c} \cos x = \cos c$

3. $\lim_{x\to c} \tan x = \tan c$ **4.** $\lim_{x\to c} \cot x = \cot c$

5. $\lim_{x\to c} \sec x = \sec c$ **6.** $\lim_{x\to c} \csc x = \csc c$

EXAMPLE 5 Limits of Trigonometric Functions

a. $\lim_{x\to 0} \tan x = \tan(0) = 0$

b. $\lim_{x\to \pi} (x \cos x) = \left(\lim_{x\to \pi} x\right)\left(\lim_{x\to \pi} \cos x\right) = \pi \cos(\pi) = -\pi$

c. $\lim_{x\to 0} \sin^2 x = \lim_{x\to 0} (\sin x)^2 = 0^2 = 0$

A Strategy for Finding Limits

On the previous three pages, you studied several types of functions whose limits can be evaluated by direct substitution. This knowledge, together with the following theorem, can be used to develop a strategy for finding limits. A proof of this theorem is given in Appendix A.

> **THEOREM 1.7 Functions That Agree at All But One Point**
>
> Let c be a real number and let $f(x) = g(x)$ for all $x \neq c$ in an open interval containing c. If the limit of $g(x)$ as x approaches c exists, then the limit of $f(x)$ also exists and
>
> $$\lim_{x \to c} f(x) = \lim_{x \to c} g(x).$$

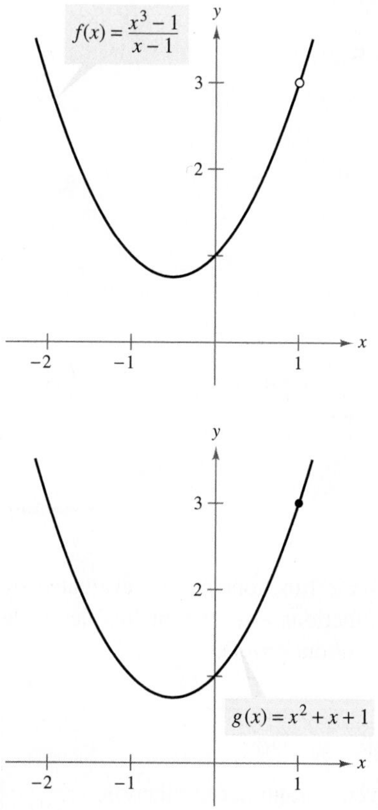

f and *g* agree at all but one point.
Figure 1.17

EXAMPLE 6 Finding the Limit of a Function

Find the limit: $\lim_{x \to 1} \frac{x^3 - 1}{x - 1}$.

Solution Let $f(x) = (x^3 - 1)/(x - 1)$. By factoring and dividing out like factors, you can rewrite f as

$$f(x) = \frac{\cancel{(x - 1)}(x^2 + x + 1)}{\cancel{(x - 1)}} = x^2 + x + 1 = g(x), \qquad x \neq 1.$$

So, for all x-values other than $x = 1$, the functions f and g agree, as shown in Figure 1.17. Because $\lim_{x \to 1} g(x)$ exists, you can apply Theorem 1.7 to conclude that f and g have the same limit at $x = 1$.

$$\begin{aligned}
\lim_{x \to 1} \frac{x^3 - 1}{x - 1} &= \lim_{x \to 1} \frac{(x - 1)(x^2 + x + 1)}{x - 1} && \text{Factor.} \\
&= \lim_{x \to 1} \frac{\cancel{(x - 1)}(x^2 + x + 1)}{\cancel{x - 1}} && \text{Divide out like factors.} \\
&= \lim_{x \to 1} (x^2 + x + 1) && \text{Apply Theorem 1.7.} \\
&= 1^2 + 1 + 1 && \text{Use direct substitution.} \\
&= 3 && \text{Simplify.}
\end{aligned}$$

STUDY TIP When applying this strategy for finding a limit, remember that some functions do not have a limit (as x approaches c). For instance, the following limit does not exist.

$$\lim_{x \to 1} \frac{x^3 + 1}{x - 1}$$

> **A Strategy for Finding Limits**
>
> 1. Learn to recognize which limits can be evaluated by direct substitution. (These limits are listed in Theorems 1.1 through 1.6.)
> 2. If the limit of $f(x)$ as x approaches c *cannot* be evaluated by direct substitution, try to find a function g that agrees with f for all x other than $x = c$. [Choose g such that the limit of $g(x)$ *can* be evaluated by direct substitution.]
> 3. Apply Theorem 1.7 to conclude *analytically* that
>
> $$\lim_{x \to c} f(x) = \lim_{x \to c} g(x) = g(c).$$
> 4. Use a *graph* or *table* to reinforce your conclusion.

Dividing Out and Rationalizing Techniques

Two techniques for finding limits analytically are shown in Examples 7 and 8. The first technique involves dividing out common factors, and the second technique involves rationalizing the numerator of a fractional expression.

EXAMPLE 7 Dividing Out Technique

Find the limit: $\lim_{x\to -3} \dfrac{x^2 + x - 6}{x + 3}$.

Solution Although you are taking the limit of a rational function, you *cannot* apply Theorem 1.3 because the limit of the denominator is 0.

$$\lim_{x\to -3} \frac{x^2 + x - 6}{x + 3} \quad \begin{cases} \lim_{x\to -3} (x^2 + x - 6) = 0 \\ \lim_{x\to -3} (x + 3) = 0 \end{cases}$$

Direct substitution fails.

Because the limit of the numerator is also 0, the numerator and denominator have a *common factor* of $(x + 3)$. So, for all $x \neq -3$, you can divide out this factor to obtain

$$f(x) = \frac{x^2 + x - 6}{x + 3} = \frac{\cancel{(x + 3)}(x - 2)}{\cancel{x + 3}} = x - 2 = g(x), \qquad x \neq -3.$$

Using Theorem 1.7, it follows that

$$\lim_{x\to -3} \frac{x^2 + x - 6}{x + 3} = \lim_{x\to -3} (x - 2) \qquad \text{Apply Theorem 1.7.}$$
$$= -5. \qquad \text{Use direct substitution.}$$

This result is shown graphically in Figure 1.18. Note that the graph of the function f coincides with the graph of the function $g(x) = x - 2$, except that the graph of f has a gap at the point $(-3, -5)$.

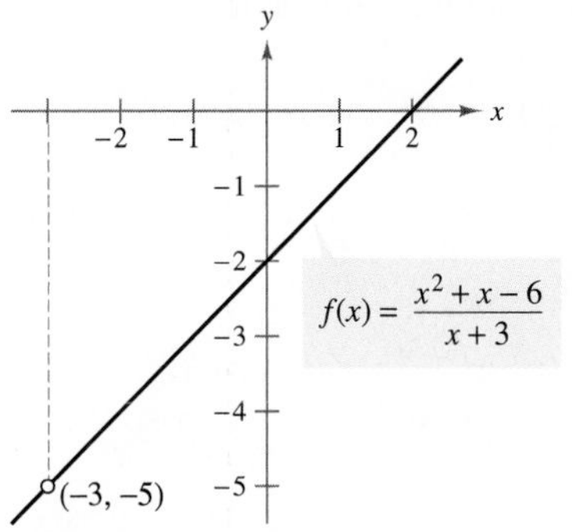

f is undefined when $x = -3$.
Figure 1.18

NOTE In the solution of Example 7, be sure you see the usefulness of the Factor Theorem of Algebra. This theorem states that if c is a zero of a polynomial function, $(x - c)$ is a factor of the polynomial. So, if you apply direct substitution to a rational function and obtain

$$r(c) = \frac{p(c)}{q(c)} = \frac{0}{0}$$

you can conclude that $(x - c)$ must be a common factor to both $p(x)$ and $q(x)$.

In Example 7, direct substitution produced the meaningless fractional form $0/0$. An expression such as $0/0$ is called an **indeterminate form** because you cannot (from the form alone) determine the limit. When you try to evaluate a limit and encounter this form, remember that you must rewrite the fraction so that the new denominator does not have 0 as its limit. One way to do this is to *divide out like factors*, as shown in Example 7. A second way is to *rationalize the numerator*, as shown in Example 8.

TECHNOLOGY PITFALL Because the graphs of

$$f(x) = \frac{x^2 + x - 6}{x + 3} \quad \text{and} \quad g(x) = x - 2$$

differ only at the point $(-3, -5)$, a standard graphing utility setting may not distinguish clearly between these graphs. However, because of the pixel configuration and rounding error of a graphing utility, it may be possible to find screen settings that distinguish between the graphs. Specifically, by repeatedly zooming in near the point $(-3, -5)$ on the graph of f, your graphing utility may show glitches or irregularities that do not exist on the actual graph. (See Figure 1.19.) By changing the screen settings on your graphing utility you may obtain the correct graph of f.

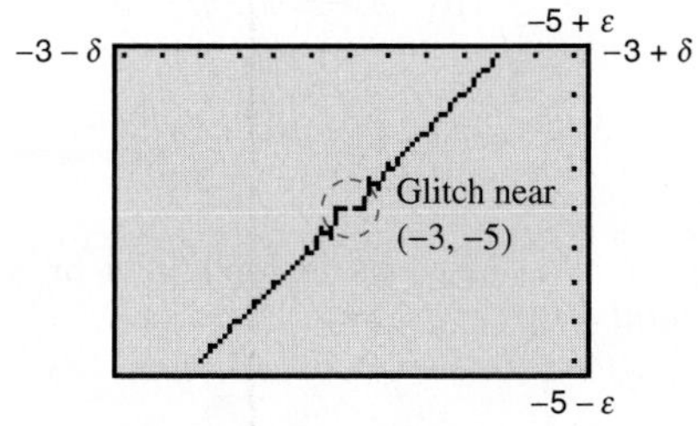

Incorrect graph of f
Figure 1.19

EXAMPLE 8 Rationalizing Technique

Find the limit: $\lim_{x \to 0} \frac{\sqrt{x+1}-1}{x}$.

Solution By direct substitution, you obtain the indeterminate form 0/0.

$$\lim_{x \to 0} \left(\sqrt{x+1}-1\right) = 0$$

$$\lim_{x \to 0} \frac{\sqrt{x+1}-1}{x}$$

Direct substitution fails.

$$\lim_{x \to 0} x = 0$$

In this case, you can rewrite the fraction by rationalizing the numerator.

$$\begin{aligned}
\frac{\sqrt{x+1}-1}{x} &= \left(\frac{\sqrt{x+1}-1}{x}\right)\left(\frac{\sqrt{x+1}+1}{\sqrt{x+1}+1}\right) \\
&= \frac{(x+1)-1}{x\left(\sqrt{x+1}+1\right)} \\
&= \frac{\cancel{x}}{\cancel{x}\left(\sqrt{x+1}+1\right)} \\
&= \frac{1}{\sqrt{x+1}+1}, \quad x \neq 0
\end{aligned}$$

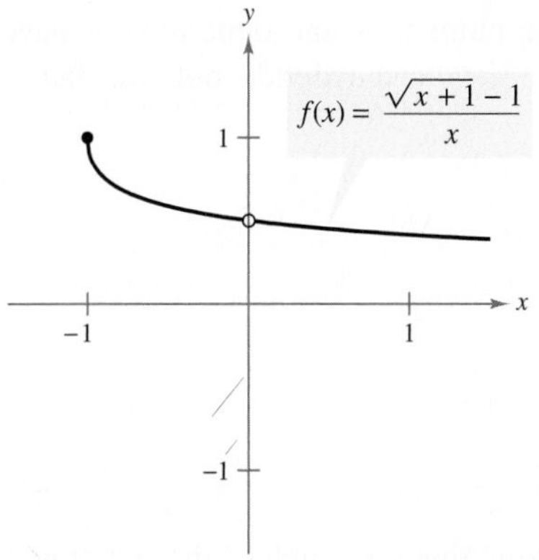

The limit of $f(x)$ as x approaches 0 is $\frac{1}{2}$.
Figure 1.20

Now, using Theorem 1.7, you can evaluate the limit as shown.

$$\begin{aligned}
\lim_{x \to 0} \frac{\sqrt{x+1}-1}{x} &= \lim_{x \to 0} \frac{1}{\sqrt{x+1}+1} \\
&= \frac{1}{1+1} \\
&= \frac{1}{2}
\end{aligned}$$

A table or a graph can reinforce your conclusion that the limit is $\frac{1}{2}$. (See Figure 1.20.)

x approaches 0 from the left. → ← *x* approaches 0 from the right.

x	-0.25	-0.1	-0.01	-0.001	0	0.001	0.01	0.1	0.25
$f(x)$	0.5359	0.5132	0.5013	0.5001	?	0.4999	0.4988	0.4881	0.4721

$f(x)$ approaches 0.5. → ← $f(x)$ approaches 0.5.

NOTE The rationalizing technique for evaluating limits is based on multiplication by a convenient form of 1. In Example 8, the convenient form is

$$1 = \frac{\sqrt{x+1}+1}{\sqrt{x+1}+1}.$$

The Squeeze Theorem

The next theorem concerns the limit of a function that is squeezed between two other functions, each of which has the same limit at a given x-value, as shown in Figure 1.21. (The proof of this theorem is given in Appendix A.)

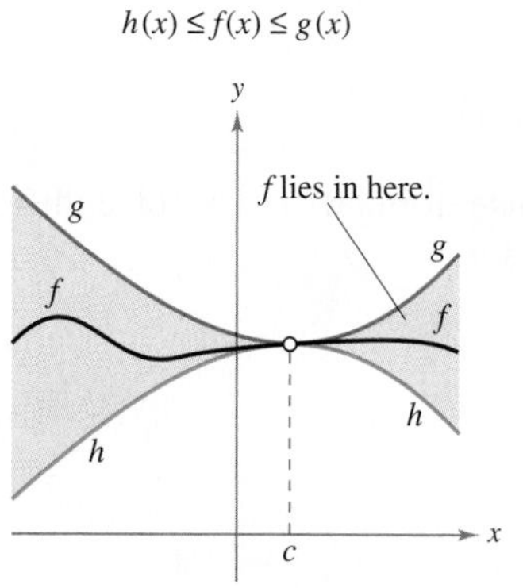

The Squeeze Theorem
Figure 1.21

> **THEOREM 1.8 The Squeeze Theorem**
>
> If $h(x) \le f(x) \le g(x)$ for all x in an open interval containing c, except possibly at c itself, and if
>
> $$\lim_{x \to c} h(x) = L = \lim_{x \to c} g(x)$$
>
> then $\lim_{x \to c} f(x)$ exists and is equal to L.

You can see the usefulness of the Squeeze Theorem in the proof of Theorem 1.9.

> **THEOREM 1.9 Two Special Trigonometric Limits**
>
> **1.** $\lim_{x \to 0} \dfrac{\sin x}{x} = 1$ **2.** $\lim_{x \to 0} \dfrac{1 - \cos x}{x} = 0$

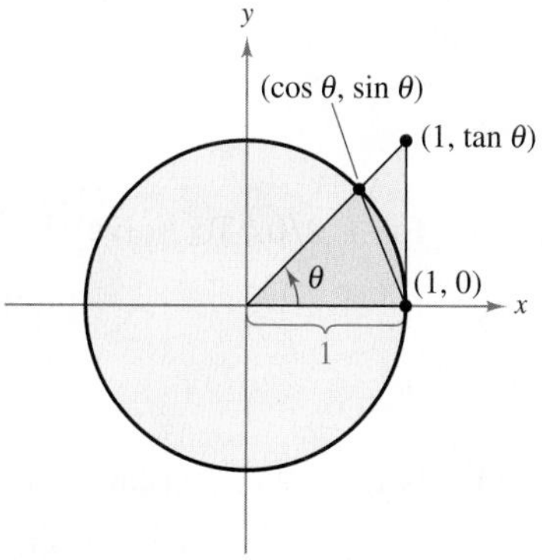

A circular sector is used to prove Theorem 1.9.
Figure 1.22

Proof To avoid the confusion of two different uses of x, the proof is presented using the variable θ, where θ is an acute positive angle *measured in radians*. Figure 1.22 shows a circular sector that is squeezed between two triangles.

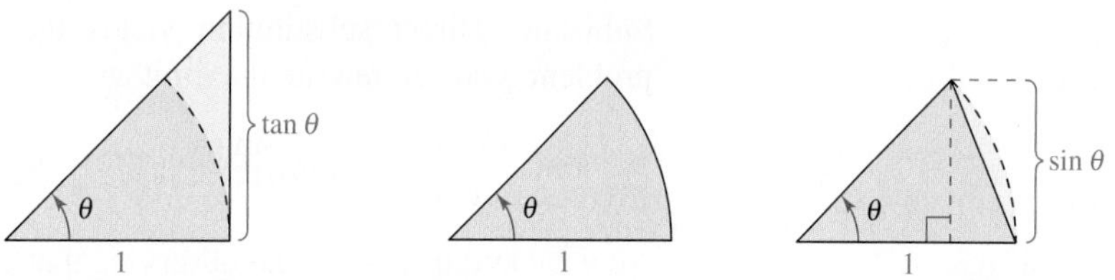

Area of triangle	$\ge$	Area of sector	$\ge$	Area of triangle
$\dfrac{\tan \theta}{2}$	$\ge$	$\dfrac{\theta}{2}$	$\ge$	$\dfrac{\sin \theta}{2}$

Multiplying each expression by $2/\sin \theta$ produces

$$\frac{1}{\cos \theta} \ge \frac{\theta}{\sin \theta} \ge 1$$

and taking reciprocals and reversing the inequalities yields

$$\cos \theta \le \frac{\sin \theta}{\theta} \le 1.$$

Because $\cos \theta = \cos(-\theta)$ and $(\sin \theta)/\theta = [\sin(-\theta)]/(-\theta)$, you can conclude that this inequality is valid for *all* nonzero θ in the open interval $(-\pi/2, \pi/2)$. Finally, because $\lim_{\theta \to 0} \cos \theta = 1$ and $\lim_{\theta \to 0} 1 = 1$, you can apply the Squeeze Theorem to conclude that $\lim_{\theta \to 0} (\sin \theta)/\theta = 1$. The proof of the second limit is left as an exercise (see Exercise 120).

FOR FURTHER INFORMATION
For more information on the function $f(x) = (\sin x)/x$, see the article "The Function $(\sin x)/x$" by William B. Gearhart and Harris S. Shultz in *The College Mathematics Journal*. To view this article, go to the website *www.matharticles.com.*

EXAMPLE 9 A Limit Involving a Trigonometric Function

Find the limit: $\lim_{x\to 0} \frac{\tan x}{x}$.

Solution Direct substitution yields the indeterminate form 0/0. To solve this problem, you can write $\tan x$ as $(\sin x)/(\cos x)$ and obtain

$$\lim_{x\to 0} \frac{\tan x}{x} = \lim_{x\to 0} \left(\frac{\sin x}{x}\right)\left(\frac{1}{\cos x}\right).$$

Now, because

$$\lim_{x\to 0} \frac{\sin x}{x} = 1 \quad \text{and} \quad \lim_{x\to 0} \frac{1}{\cos x} = 1$$

you can obtain

$$\begin{aligned}\lim_{x\to 0} \frac{\tan x}{x} &= \left(\lim_{x\to 0} \frac{\sin x}{x}\right)\left(\lim_{x\to 0} \frac{1}{\cos x}\right)\\ &= (1)(1)\\ &= 1.\end{aligned}$$

(See Figure 1.23.)

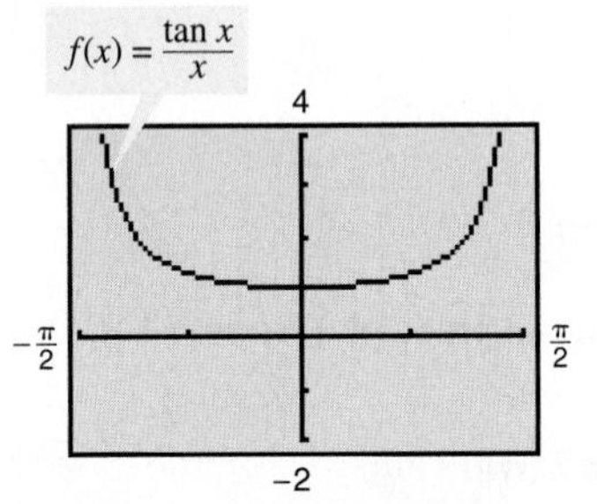

The limit of $f(x)$ as x approaches 0 is 1.
Figure 1.23

EXAMPLE 10 A Limit Involving a Trigonometric Function

Find the limit: $\lim_{x\to 0} \frac{\sin 4x}{x}$.

Solution Direct substitution yields the indeterminate form 0/0. To solve this problem, you can rewrite the limit as

$$\lim_{x\to 0} \frac{\sin 4x}{x} = 4\left(\lim_{x\to 0} \frac{\sin 4x}{4x}\right). \qquad \text{Multiply and divide by 4.}$$

Now, by letting $y = 4x$ and observing that $x \to 0$ if and only if $y \to 0$, you can write

$$\begin{aligned}\lim_{x\to 0} \frac{\sin 4x}{x} &= 4\left(\lim_{x\to 0} \frac{\sin 4x}{4x}\right)\\ &= 4\left(\lim_{y\to 0} \frac{\sin y}{y}\right)\\ &= 4(1)\\ &= 4.\end{aligned}$$

(See Figure 1.24.)

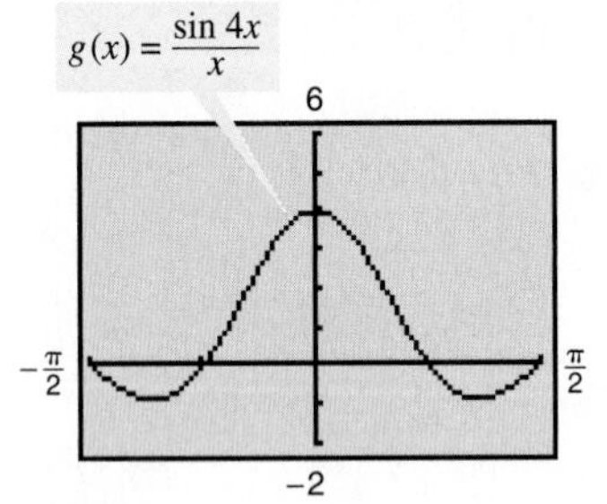

The limit of $g(x)$ as x approaches 0 is 4.
Figure 1.24

TECHNOLOGY Use a graphing utility to confirm the limits in the examples and exercise set. For instance, Figures 1.23 and 1.24 show the graphs of

$$f(x) = \frac{\tan x}{x} \quad \text{and} \quad g(x) = \frac{\sin 4x}{x}.$$

Note that the first graph appears to contain the point (0, 1) and the second graph appears to contain the point (0, 4), which lends support to the conclusions obtained in Examples 9 and 10.

Exercises for Section 1.3

See www.CalcChat.com for worked-out solutions to odd-numbered exercises.

In Exercises 1–4, use a graphing utility to graph the function and visually estimate the limits.

1. $h(x) = x^2 - 5x$

(a) $\lim_{x\to5} h(x)$

(b) $\lim_{x\to-1} h(x)$

2. $g(x) = \dfrac{12(\sqrt{x} - 3)}{x - 9}$

(a) $\lim_{x\to4} g(x)$

(b) $\lim_{x\to0} g(x)$

3. $f(x) = x \cos x$

(a) $\lim_{x\to0} f(x)$

(b) $\lim_{x\to\pi/3} f(x)$

4. $f(t) = t|t - 4|$

(a) $\lim_{t\to4} f(t)$

(b) $\lim_{t\to-1} f(t)$

In Exercises 5–22, find the limit.

5. $\lim_{x\to2} x^4$

6. $\lim_{x\to-2} x^3$

7. $\lim_{x\to0} (2x - 1)$

8. $\lim_{x\to-3} (3x + 2)$

9. $\lim_{x\to-3} (x^2 + 3x)$

10. $\lim_{x\to1} (-x^2 + 1)$

11. $\lim_{x\to-3} (2x^2 + 4x + 1)$

12. $\lim_{x\to1} (3x^3 - 2x^2 + 4)$

13. $\lim_{x\to2} \dfrac{1}{x}$

14. $\lim_{x\to-3} \dfrac{2}{x + 2}$

15. $\lim_{x\to1} \dfrac{x - 3}{x^2 + 4}$

16. $\lim_{x\to3} \dfrac{2x - 3}{x + 5}$

17. $\lim_{x\to7} \dfrac{5x}{\sqrt{x + 2}}$

18. $\lim_{x\to3} \dfrac{\sqrt{x + 1}}{x - 4}$

19. $\lim_{x\to3} \sqrt{x + 1}$

20. $\lim_{x\to4} \sqrt[3]{x + 4}$

21. $\lim_{x\to-4} (x + 3)^2$

22. $\lim_{x\to0} (2x - 1)^3$

In Exercises 23–26, find the limits.

23. $f(x) = 5 - x,\ g(x) = x^3$

(a) $\lim_{x\to1} f(x)$ (b) $\lim_{x\to4} g(x)$ (c) $\lim_{x\to1} g(f(x))$

24. $f(x) = x + 7,\ g(x) = x^2$

(a) $\lim_{x\to-3} f(x)$ (b) $\lim_{x\to4} g(x)$ (c) $\lim_{x\to-3} g(f(x))$

25. $f(x) = 4 - x^2,\ g(x) = \sqrt{x + 1}$

(a) $\lim_{x\to1} f(x)$ (b) $\lim_{x\to3} g(x)$ (c) $\lim_{x\to1} g(f(x))$

26. $f(x) = 2x^2 - 3x + 1,\ g(x) = \sqrt[3]{x + 6}$

(a) $\lim_{x\to4} f(x)$ (b) $\lim_{x\to21} g(x)$ (c) $\lim_{x\to4} g(f(x))$

In Exercises 27–36, find the limit of the trigonometric function.

27. $\lim_{x\to\pi/2} \sin x$

28. $\lim_{x\to\pi} \tan x$

29. $\lim_{x\to2} \cos \dfrac{\pi x}{3}$

30. $\lim_{x\to1} \sin \dfrac{\pi x}{2}$

31. $\lim_{x\to0} \sec 2x$

32. $\lim_{x\to\pi} \cos 3x$

33. $\lim_{x\to5\pi/6} \sin x$

34. $\lim_{x\to5\pi/3} \cos x$

35. $\lim_{x\to3} \tan\left(\dfrac{\pi x}{4}\right)$

36. $\lim_{x\to7} \sec\left(\dfrac{\pi x}{6}\right)$

In Exercises 37–40, use the information to evaluate the limits.

37. $\lim_{x\to c} f(x) = 2$

$\lim_{x\to c} g(x) = 3$

(a) $\lim_{x\to c} [5g(x)]$

(b) $\lim_{x\to c} [f(x) + g(x)]$

(c) $\lim_{x\to c} [f(x)g(x)]$

(d) $\lim_{x\to c} \dfrac{f(x)}{g(x)}$

38. $\lim_{x\to c} f(x) = \frac{3}{2}$

$\lim_{x\to c} g(x) = \frac{1}{2}$

(a) $\lim_{x\to c} [4f(x)]$

(b) $\lim_{x\to c} [f(x) + g(x)]$

(c) $\lim_{x\to c} [f(x)g(x)]$

(d) $\lim_{x\to c} \dfrac{f(x)}{g(x)}$

39. $\lim_{x\to c} f(x) = 4$

(a) $\lim_{x\to c} [f(x)]^3$

(b) $\lim_{x\to c} \sqrt{f(x)}$

(c) $\lim_{x\to c} [3f(x)]$

(d) $\lim_{x\to c} [f(x)]^{3/2}$

40. $\lim_{x\to c} f(x) = 27$

(a) $\lim_{x\to c} \sqrt[3]{f(x)}$

(b) $\lim_{x\to c} \dfrac{f(x)}{18}$

(c) $\lim_{x\to c} [f(x)]^2$

(d) $\lim_{x\to c} [f(x)]^{2/3}$

In Exercises 41–44, use the graph to determine the limit visually (if it exists). Write a simpler function that agrees with the given function at all but one point.

41. $g(x) = \dfrac{-2x^2 + x}{x}$

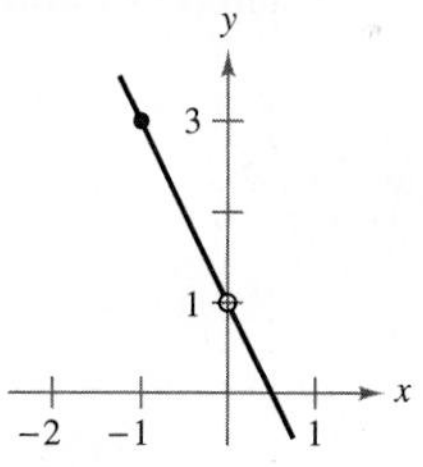

(a) $\lim_{x\to0} g(x)$

(b) $\lim_{x\to-1} g(x)$

42. $h(x) = \dfrac{x^2 - 3x}{x}$

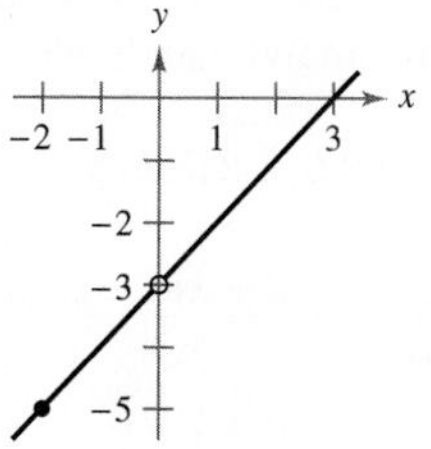

(a) $\lim_{x\to-2} h(x)$

(b) $\lim_{x\to0} h(x)$

43. $g(x) = \dfrac{x^3 - x}{x - 1}$

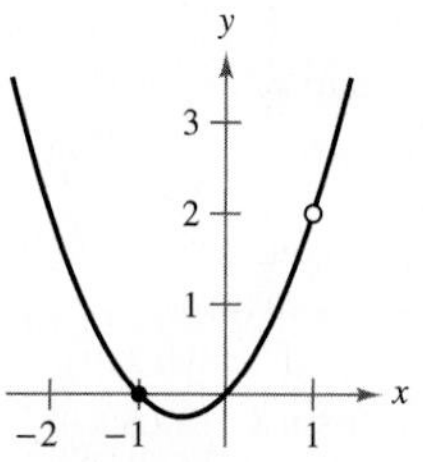

(a) $\lim_{x\to1} g(x)$

(b) $\lim_{x\to-1} g(x)$

44. $f(x) = \dfrac{x}{x^2 - x}$

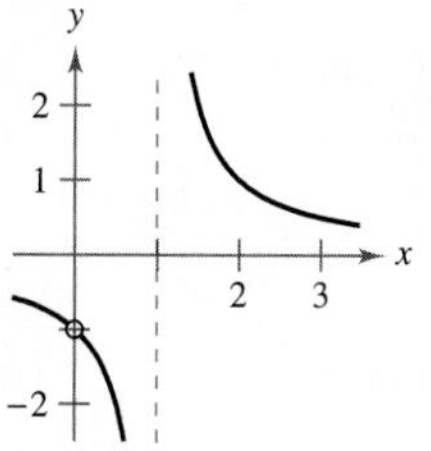

(a) $\lim_{x\to1} f(x)$

(b) $\lim_{x\to0} f(x)$

In Exercises 45–48, find the limit of the function (if it exists). Write a simpler function that agrees with the given function at all but one point. Use a graphing utility to confirm your result.

45. $\lim\limits_{x\to -1} \dfrac{x^2-1}{x+1}$

46. $\lim\limits_{x\to -1} \dfrac{2x^2-x-3}{x+1}$

47. $\lim\limits_{x\to 2} \dfrac{x^3-8}{x-2}$

48. $\lim\limits_{x\to -1} \dfrac{x^3+1}{x+1}$

In Exercises 49–62, find the limit (if it exists).

49. $\lim\limits_{x\to 5} \dfrac{x-5}{x^2-25}$

50. $\lim\limits_{x\to 2} \dfrac{2-x}{x^2-4}$

51. $\lim\limits_{x\to -3} \dfrac{x^2+x-6}{x^2-9}$

52. $\lim\limits_{x\to 4} \dfrac{x^2-5x+4}{x^2-2x-8}$

53. $\lim\limits_{x\to 0} \dfrac{\sqrt{x+5}-\sqrt{5}}{x}$

54. $\lim\limits_{x\to 0} \dfrac{\sqrt{2+x}-\sqrt{2}}{x}$

55. $\lim\limits_{x\to 4} \dfrac{\sqrt{x+5}-3}{x-4}$

56. $\lim\limits_{x\to 3} \dfrac{\sqrt{x+1}-2}{x-3}$

57. $\lim\limits_{x\to 0} \dfrac{[1/(3+x)]-(1/3)}{x}$

58. $\lim\limits_{x\to 0} \dfrac{[1/(x+4)]-(1/4)}{x}$

59. $\lim\limits_{\Delta x\to 0} \dfrac{2(x+\Delta x)-2x}{\Delta x}$

60. $\lim\limits_{\Delta x\to 0} \dfrac{(x+\Delta x)^2-x^2}{\Delta x}$

61. $\lim\limits_{\Delta x\to 0} \dfrac{(x+\Delta x)^2-2(x+\Delta x)+1-(x^2-2x+1)}{\Delta x}$

62. $\lim\limits_{\Delta x\to 0} \dfrac{(x+\Delta x)^3-x^3}{\Delta x}$

Graphical, Numerical, and Analytic Analysis **In Exercises 63–66, use a graphing utility to graph the function and estimate the limit. Use a table to reinforce your conclusion. Then find the limit by analytic methods.**

63. $\lim\limits_{x\to 0} \dfrac{\sqrt{x+2}-\sqrt{2}}{x}$

64. $\lim\limits_{x\to 16} \dfrac{4-\sqrt{x}}{x-16}$

65. $\lim\limits_{x\to 0} \dfrac{[1/(2+x)]-(1/2)}{x}$

66. $\lim\limits_{x\to 2} \dfrac{x^5-32}{x-2}$

In Exercises 67–78, determine the limit of the trigonometric function (if it exists).

67. $\lim\limits_{x\to 0} \dfrac{\sin x}{5x}$

68. $\lim\limits_{x\to 0} \dfrac{3(1-\cos x)}{x}$

69. $\lim\limits_{x\to 0} \dfrac{\sin x(1-\cos x)}{2x^2}$

70. $\lim\limits_{\theta\to 0} \dfrac{\cos\theta\tan\theta}{\theta}$

71. $\lim\limits_{x\to 0} \dfrac{\sin^2 x}{x}$

72. $\lim\limits_{x\to 0} \dfrac{\tan^2 x}{x}$

73. $\lim\limits_{h\to 0} \dfrac{(1-\cos h)^2}{h}$

74. $\lim\limits_{\phi\to \pi} \phi\sec\phi$

75. $\lim\limits_{x\to \pi/2} \dfrac{\cos x}{\cot x}$

76. $\lim\limits_{x\to \pi/4} \dfrac{1-\tan x}{\sin x-\cos x}$

77. $\lim\limits_{t\to 0} \dfrac{\sin 3t}{2t}$

78. $\lim\limits_{x\to 0} \dfrac{\sin 2x}{\sin 3x}$ $\left[\textit{Hint: } \text{Find } \lim\limits_{x\to 0}\left(\dfrac{2\sin 2x}{2x}\right)\left(\dfrac{3x}{3\sin 3x}\right).\right]$

Graphical, Numerical, and Analytic Analysis **In Exercises 79–82, use a graphing utility to graph the function and estimate the limit. Use a table to reinforce your conclusion. Then find the limit by analytic methods.**

79. $\lim\limits_{t\to 0} \dfrac{\sin 3t}{t}$

80. $\lim\limits_{x\to 0} \dfrac{\cos x-1}{2x^2}$

81. $\lim\limits_{x\to 0} \dfrac{\sin x^2}{x}$

82. $\lim\limits_{x\to 0} \dfrac{\sin x}{\sqrt[3]{x}}$

In Exercises 83–86, find $\lim\limits_{\Delta x\to 0} \dfrac{f(x+\Delta x)-f(x)}{\Delta x}$.

83. $f(x)=2x+3$

84. $f(x)=\sqrt{x}$

85. $f(x)=\dfrac{4}{x}$

86. $f(x)=x^2-4x$

In Exercises 87 and 88, use the Squeeze Theorem to find $\lim\limits_{x\to c} f(x)$.

87. $c=0$

$4-x^2\le f(x)\le 4+x^2$

88. $c=a$

$b-|x-a|\le f(x)\le b+|x-a|$

In Exercises 89–94, use a graphing utility to graph the given function and the equations $y=|x|$ **and** $y=-|x|$ **in the same viewing window. Using the graphs to observe the Squeeze Theorem visually, find** $\lim\limits_{x\to 0} f(x)$.

89. $f(x)=x\cos x$

90. $f(x)=|x\sin x|$

91. $f(x)=|x|\sin x$

92. $f(x)=|x|\cos x$

93. $f(x)=x\sin\dfrac{1}{x}$

94. $h(x)=x\cos\dfrac{1}{x}$

Writing About Concepts

95. In the context of finding limits, discuss what is meant by two functions that agree at all but one point.

96. Give an example of two functions that agree at all but one point.

97. What is meant by an indeterminate form?

98. In your own words, explain the Squeeze Theorem.

99. ***Writing*** Use a graphing utility to graph

$$f(x)=x,\quad g(x)=\sin x,\quad \text{and}\quad h(x)=\frac{\sin x}{x}$$

in the same viewing window. Compare the magnitudes of $f(x)$ and $g(x)$ when x is close to 0. Use the comparison to write a short paragraph explaining why

$$\lim_{x\to 0} h(x)=1.$$

100. ***Writing*** Use a graphing utility to graph

$$f(x) = x, \quad g(x) = \sin^2 x, \quad \text{and } h(x) = \frac{\sin^2 x}{x}$$

in the same viewing window. Compare the magnitudes of $f(x)$ and $g(x)$ when x is close to 0. Use the comparison to write a short paragraph explaining why

$$\lim_{x\to 0} h(x) = 0.$$

Free-Falling Object **In Exercises 101 and 102, use the position function $s(t) = -16t^2 + 1000$, which gives the height (in feet) of an object that has fallen for t seconds from a height of 1000 feet. The velocity at time $t = a$ seconds is given by**

$$\lim_{t\to a} \frac{s(a) - s(t)}{a - t}.$$

101. If a construction worker drops a wrench from a height of 1000 feet, how fast will the wrench be falling after 5 seconds?

102. If a construction worker drops a wrench from a height of 1000 feet, when will the wrench hit the ground? At what velocity will the wrench impact the ground?

Free-Falling Object **In Exercises 103 and 104, use the position function $s(t) = -4.9t^2 + 150$, which gives the height (in meters) of an object that has fallen from a height of 150 meters. The velocity at time $t = a$ seconds is given by**

$$\lim_{t\to a} \frac{s(a) - s(t)}{a - t}.$$

103. Find the velocity of the object when $t = 3$.

104. At what velocity will the object impact the ground?

105. Find two functions f and g such that $\lim_{x\to 0} f(x)$ and $\lim_{x\to 0} g(x)$ do not exist, but $\lim_{x\to 0} [f(x) + g(x)]$ does exist.

106. Prove that if $\lim_{x\to c} f(x)$ exists and $\lim_{x\to c} [f(x) + g(x)]$ does not exist, then $\lim_{x\to c} g(x)$ does not exist.

107. Prove Property 1 of Theorem 1.1.

108. Prove Property 3 of Theorem 1.1. (You may use Property 3 of Theorem 1.2.)

109. Prove Property 1 of Theorem 1.2.

110. Prove that if $\lim_{x\to c} f(x) = 0$, then $\lim_{x\to c} |f(x)| = 0$.

111. Prove that if $\lim_{x\to c} f(x) = 0$ and $|g(x)| \le M$ for a fixed number M and all $x \ne c$, then $\lim_{x\to c} f(x)g(x) = 0$.

112. (a) Prove that if $\lim_{x\to c} |f(x)| = 0$, then $\lim_{x\to c} f(x) = 0$.

(*Note:* This is the converse of Exercise 110.)

(b) Prove that if $\lim_{x\to c} f(x) = L$, then $\lim_{x\to c} |f(x)| = |L|$.

[*Hint:* Use the inequality $\big||f(x)| - |L|\big| \le |f(x) - L|$.]

True or False? **In Exercises 113–118, determine whether the statement is true or false. If it is false, explain why or give an example that shows it is false.**

113. $\lim_{x\to 0} \frac{|x|}{x} = 1$

114. $\lim_{x\to \pi} \frac{\sin x}{x} = 1$

115. If $f(x) = g(x)$ for all real numbers other than $x = 0$, and

$$\lim_{x\to 0} f(x) = L, \quad \text{then} \quad \lim_{x\to 0} g(x) = L.$$

116. If $\lim_{x\to c} f(x) = L$, then $f(c) = L$.

117. $\lim_{x\to 2} f(x) = 3$, where $f(x) = \begin{cases} 3, & x \le 2 \\ 0, & x > 2 \end{cases}$

118. If $f(x) < g(x)$ for all $x \ne a$, then

$$\lim_{x\to a} f(x) < \lim_{x\to a} g(x).$$

119. ***Think About It*** Find a function f to show that the converse of Exercise 112(b) is not true. [*Hint:* Find a function f such that $\lim_{x\to c} |f(x)| = |L|$ but $\lim_{x\to c} f(x)$ does not exist.]

120. Prove the second part of Theorem 1.9 by proving that

$$\lim_{x\to 0} \frac{1 - \cos x}{x} = 0.$$

121. Let $f(x) = \begin{cases} 0, & \text{if } x \text{ is rational} \\ 1, & \text{if } x \text{ is irrational} \end{cases}$

and

$$g(x) = \begin{cases} 0, & \text{if } x \text{ is rational} \\ x, & \text{if } x \text{ is irrational.} \end{cases}$$

Find (if possible) $\lim_{x\to 0} f(x)$ and $\lim_{x\to 0} g(x)$.

122. ***Graphical Reasoning*** Consider $f(x) = \dfrac{\sec x - 1}{x^2}$.

(a) Find the domain of f.

(b) Use a graphing utility to graph f. Is the domain of f obvious from the graph? If not, explain.

(c) Use the graph of f to approximate $\lim_{x\to 0} f(x)$.

(d) Confirm the answer in part (c) analytically.

123. ***Approximation***

(a) Find $\lim_{x\to 0} \dfrac{1 - \cos x}{x^2}$.

(b) Use the result in part (a) to derive the approximation $\cos x \approx 1 - \frac{1}{2}x^2$ for x near 0.

(c) Use the result in part (b) to approximate $\cos(0.1)$.

(d) Use a calculator to approximate $\cos(0.1)$ to four decimal places. Compare the result with part (c).

124. ***Think About It*** When using a graphing utility to generate a table to approximate $\lim_{x\to 0} [(\sin x)/x]$, a student concluded that the limit was 0.01745 rather than 1. Determine the probable cause of the error.

Section 1.4 Continuity and One-Sided Limits

- **Determine continuity at a point and continuity on an open interval.**
- **Determine one-sided limits and continuity on a closed interval.**
- **Use properties of continuity.**
- **Understand and use the Intermediate Value Theorem.**

EXPLORATION

Informally, you might say that a function is *continuous* on an open interval if its graph can be drawn with a pencil without lifting the pencil from the paper. Use a graphing utility to graph each function on the given interval. From the graphs, which functions would you say are continuous on the interval? Do you think you can trust the results you obtained graphically? Explain your reasoning.

Function	*Interval*
a. $y = x^2 + 1$	$(-3, 3)$
b. $y = \dfrac{1}{x-2}$	$(-3, 3)$
c. $y = \dfrac{\sin x}{x}$	$(-\pi, \pi)$
d. $y = \dfrac{x^2-4}{x+2}$	$(-3, 3)$
e. $y = \begin{cases} 2x-4, & x \le 0 \\ x+1, & x > 0 \end{cases}$	$(-3, 3)$

Continuity at a Point and on an Open Interval

In mathematics, the term *continuous* has much the same meaning as it has in everyday usage. Informally, to say that a function f is continuous at $x = c$ means that there is no interruption in the graph of f at c. That is, its graph is unbroken at c and there are no holes, jumps, or gaps. Figure 1.25 identifies three values of x at which the graph of f is *not* continuous. At all other points in the interval (a, b), the graph of f is uninterrupted and **continuous.**

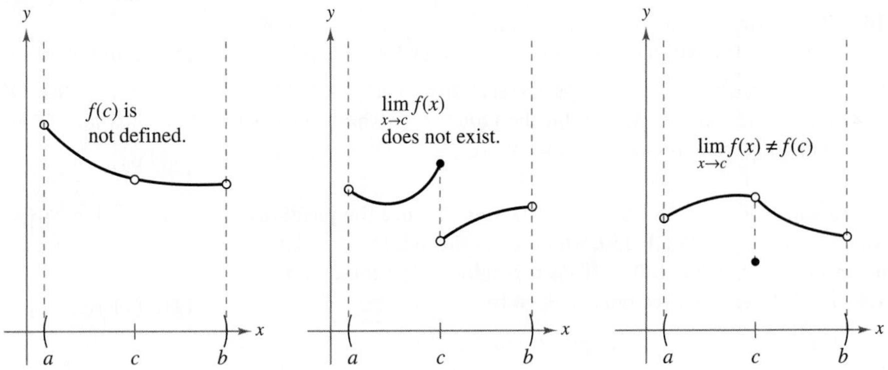

Three conditions exist for which the graph of f is not continuous at $x = c$.
Figure 1.25

In Figure 1.25, it appears that continuity at $x = c$ can be destroyed by any one of the following conditions.

1. The function is not defined at $x = c$.
2. The limit of $f(x)$ does not exist at $x = c$.
3. The limit of $f(x)$ exists at $x = c$, but it is not equal to $f(c)$.

If *none* of the above three conditions is true, the function f is called **continuous at c,** as indicated in the following important definition.

Definition of Continuity

Continuity at a Point: A function f is **continuous at c** if the following three conditions are met.

1. $f(c)$ is defined.
2. $\lim_{x\to c} f(x)$ exists.
3. $\lim_{x\to c} f(x) = f(c)$.

Continuity on an Open Interval: A function is **continuous on an open interval (a, b)** if it is continuous at each point in the interval. A function that is continuous on the entire real line $(-\infty, \infty)$ is **everywhere continuous.**

FOR FURTHER INFORMATION For more information on the concept of continuity, see the article "Leibniz and the Spell of the Continuous" by Hardy Grant in *The College Mathematics Journal*. To view this article, go to the website *www.matharticles.com*.

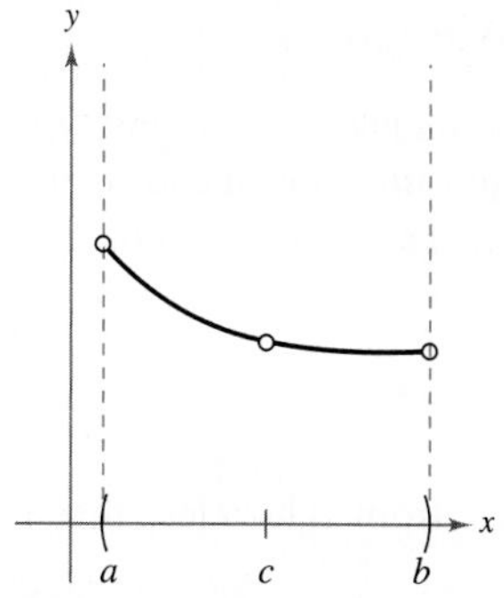

(a) Removable discontinuity

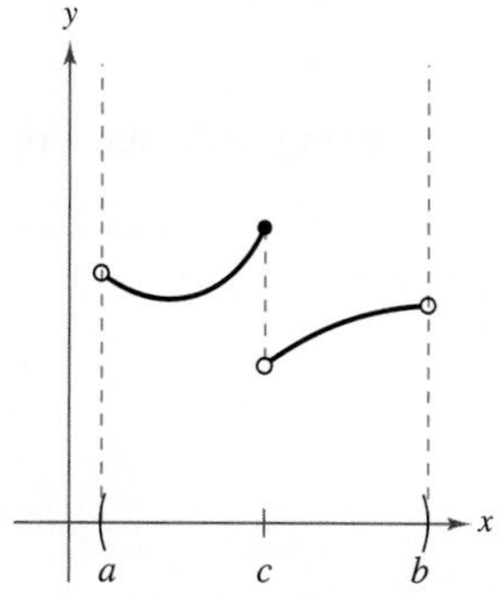

(b) Nonremovable discontinuity

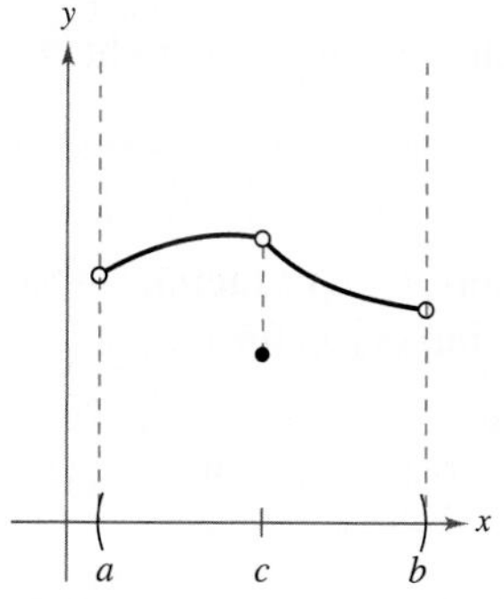

(c) Removable discontinuity
Figure 1.26

Consider an open interval I that contains a real number c. If a function f is defined on I (except possibly at c), and f is not continuous at c, then f is said to have a **discontinuity** at c. Discontinuities fall into two categories: **removable** and **nonremovable.** A discontinuity at c is called removable if f can be made continuous by appropriately defining (or redefining) $f(c)$. For instance, the functions shown in Figure 1.26(a) and (c) have removable discontinuities at c and the function shown in Figure 1.26(b) has a nonremovable discontinuity at c.

EXAMPLE 1 Continuity of a Function

Discuss the continuity of each function.

a. $f(x) = \dfrac{1}{x}$ **b.** $g(x) = \dfrac{x^2 - 1}{x - 1}$ **c.** $h(x) = \begin{cases} x + 1, & x \le 0 \\ x^2 + 1, & x > 0 \end{cases}$ **d.** $y = \sin x$

Solution

a. The domain of f is all nonzero real numbers. From Theorem 1.3, you can conclude that f is continuous at every x-value in its domain. At $x = 0$, f has a nonremovable discontinuity, as shown in Figure 1.27(a). In other words, there is no way to define $f(0)$ so as to make the function continuous at $x = 0$.

b. The domain of g is all real numbers except $x = 1$. From Theorem 1.3, you can conclude that g is continuous at every x-value in its domain. At $x = 1$, the function has a removable discontinuity, as shown in Figure 1.27(b). If $g(1)$ is defined as 2, the "newly defined" function is continuous for all real numbers.

c. The domain of h is all real numbers. The function h is continuous on $(-\infty, 0)$ and $(0, \infty)$, and, because $\lim_{x \to 0} h(x) = 1$, h is continuous on the entire real line, as shown in Figure 1.27(c).

d. The domain of y is all real numbers. From Theorem 1.6, you can conclude that the function is continuous on its entire domain, $(-\infty, \infty)$, as shown in Figure 1.27(d).

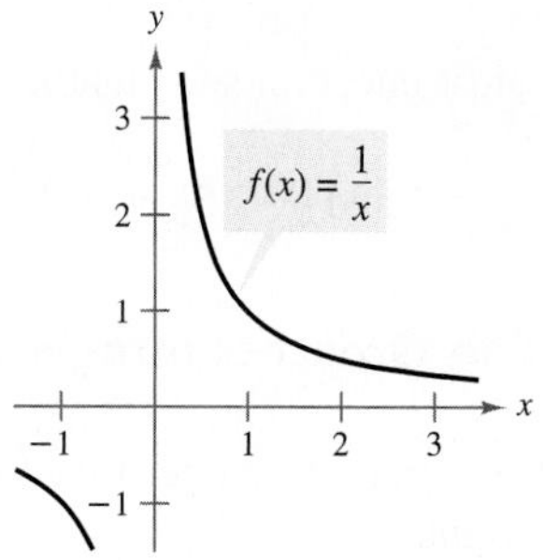

(a) Nonremovable discontinuity at $x = 0$

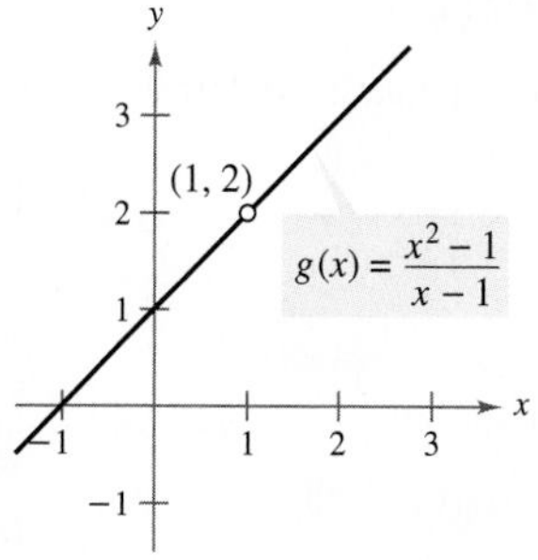

(b) Removable discontinuity at $x = 1$

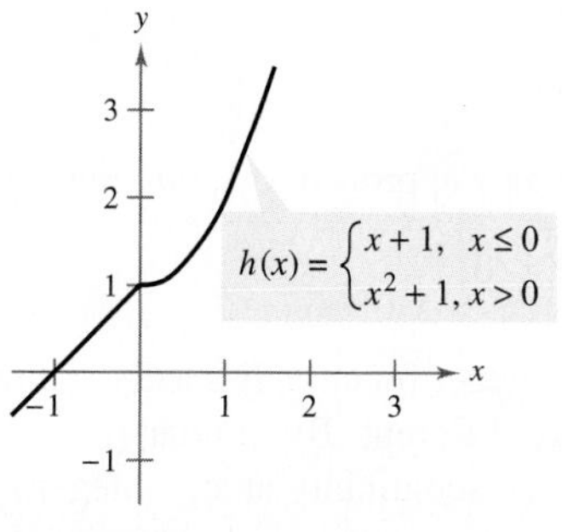

(c) Continuous on entire real line
Figure 1.27

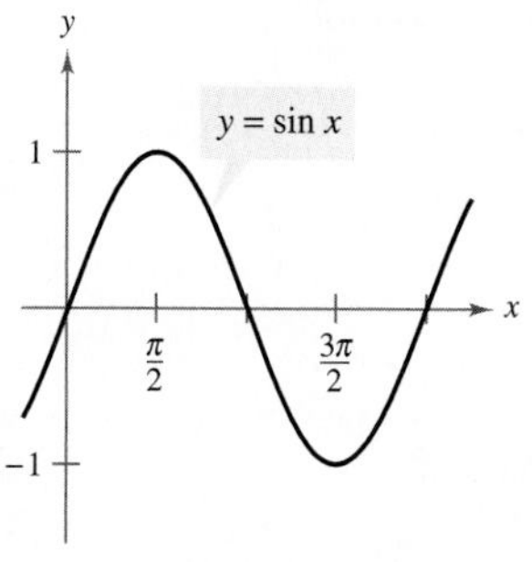

(d) Continuous on entire real line

STUDY TIP Some people may refer to the function in Example 1(a) as "discontinuous." We have found that this terminology can be confusing. Rather than saying the function is discontinuous, we prefer to say that it has a discontinuity at $x = 0$.

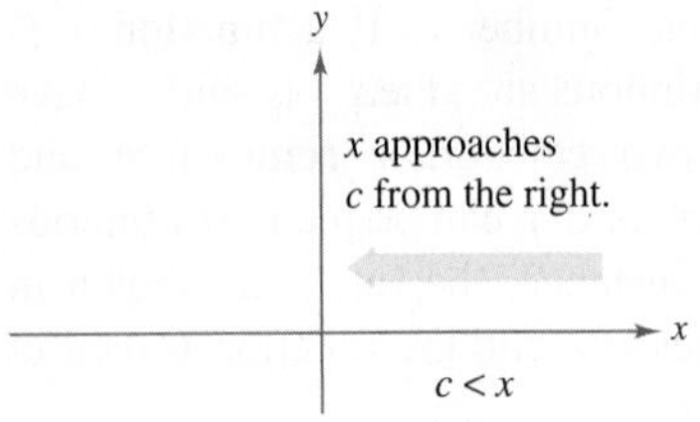

(a) Limit from right

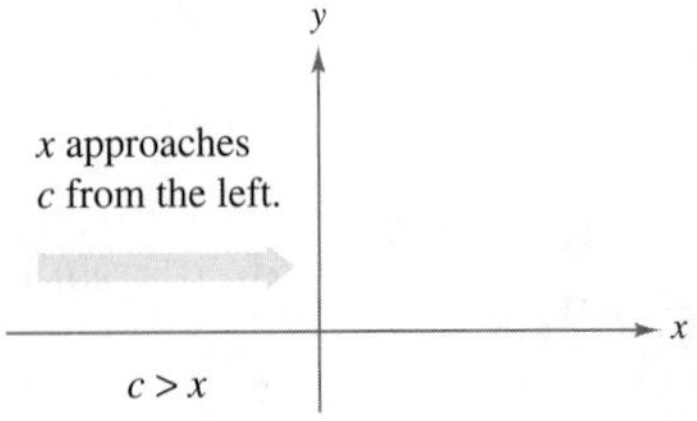

(b) Limit from left
Figure 1.28

One-Sided Limits and Continuity on a Closed Interval

To understand continuity on a closed interval, you first need to look at a different type of limit called a **one-sided limit.** For example, the **limit from the right** means that x approaches c from values greater than c [see Figure 1.28(a)]. This limit is denoted as

$$\lim_{x \to c^+} f(x) = L. \qquad \text{Limit from the right}$$

Similarly, the **limit from the left** means that x approaches c from values less than c [see Figure 1.28(b)]. This limit is denoted as

$$\lim_{x \to c^-} f(x) = L. \qquad \text{Limit from the left}$$

One-sided limits are useful in taking limits of functions involving radicals. For instance, if n is an even integer,

$$\lim_{x \to 0^+} \sqrt[n]{x} = 0.$$

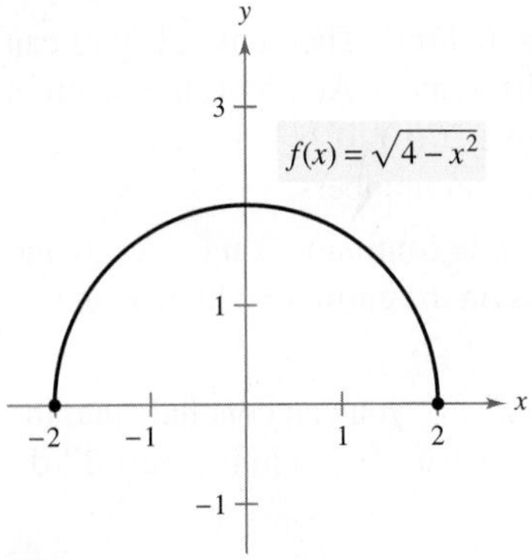

The limit of $f(x)$ as x approaches -2 from the right is 0.
Figure 1.29

EXAMPLE 2 A One-Sided Limit

Find the limit of $f(x) = \sqrt{4 - x^2}$ as x approaches -2 from the right.

Solution As shown in Figure 1.29, the limit as x approaches -2 from the right is

$$\lim_{x \to -2^+} \sqrt{4 - x^2} = 0.$$

One-sided limits can be used to investigate the behavior of **step functions.** One common type of step function is the **greatest integer function** $[\![x]\!]$, defined by

$$[\![x]\!] = \text{greatest integer } n \text{ such that } n \leq x. \qquad \text{Greatest integer function}$$

For instance, $[\![2.5]\!] = 2$ and $[\![-2.5]\!] = -3$.

EXAMPLE 3 The Greatest Integer Function

Find the limit of the greatest integer function $f(x) = [\![x]\!]$ as x approaches 0 from the left and from the right.

Solution As shown in Figure 1.30, the limit as x approaches 0 *from the left* is given by

$$\lim_{x \to 0^-} [\![x]\!] = -1$$

and the limit as x approaches 0 *from the right* is given by

$$\lim_{x \to 0^+} [\![x]\!] = 0.$$

The greatest integer function has a discontinuity at zero because the left and right limits at zero are different. By similar reasoning, you can see that the greatest integer function has a discontinuity at any integer n.

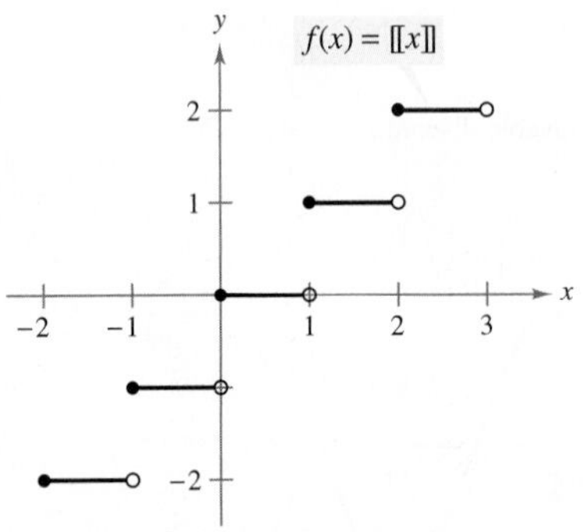

Greatest integer function
Figure 1.30

When the limit from the left is not equal to the limit from the right, the (two-sided) limit *does not exist*. The next theorem makes this more explicit. The proof of this theorem follows directly from the definition of a one-sided limit.

THEOREM 1.10 The Existence of a Limit

Let f be a function and let c and L be real numbers. The limit of $f(x)$ as x approaches c is L if and only if

$$\lim_{x\to c^-} f(x) = L \quad \text{and} \quad \lim_{x\to c^+} f(x) = L.$$

The concept of a one-sided limit allows you to extend the definition of continuity to closed intervals. Basically, a function is continuous on a closed interval if it is continuous in the interior of the interval and exhibits one-sided continuity at the endpoints. This is stated formally as follows.

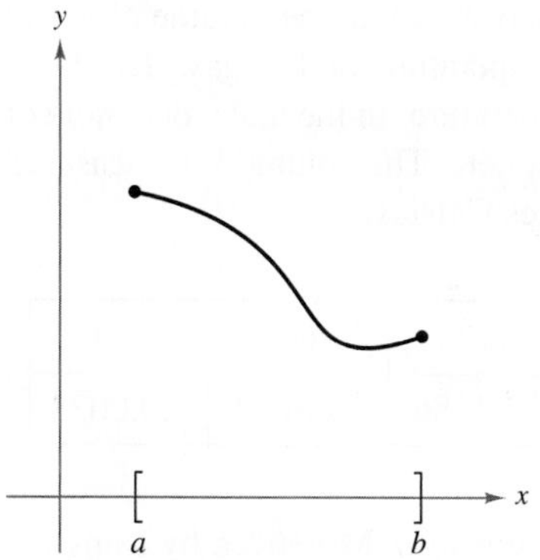

Continuous function on a closed interval
Figure 1.31

Definition of Continuity on a Closed Interval

A function f is **continuous on the closed interval $[a, b]$** if it is continuous on the open interval (a, b) and

$$\lim_{x\to a^+} f(x) = f(a) \quad \text{and} \quad \lim_{x\to b^-} f(x) = f(b).$$

The function f is **continuous from the right** at a and **continuous from the left** at b (see Figure 1.31).

Similar definitions can be made to cover continuity on intervals of the form $(a, b]$ and $[a, b)$ that are neither open nor closed, or on infinite intervals. For example, the function

$$f(x) = \sqrt{x}$$

is continuous on the infinite interval $[0, \infty)$, and the function

$$g(x) = \sqrt{2 - x}$$

is continuous on the infinite interval $(-\infty, 2]$.

EXAMPLE 4 Continuity on a Closed Interval

Discuss the continuity of $f(x) = \sqrt{1 - x^2}$.

Solution The domain of f is the closed interval $[-1, 1]$. At all points in the open interval $(-1, 1)$, the continuity of f follows from Theorems 1.4 and 1.5. Moreover, because

$$\lim_{x\to -1^+} \sqrt{1 - x^2} = 0 = f(-1) \qquad \text{Continuous from the right}$$

and

$$\lim_{x\to 1^-} \sqrt{1 - x^2} = 0 = f(1) \qquad \text{Continuous from the left}$$

you can conclude that f is continuous on the closed interval $[-1, 1]$, as shown in Figure 1.32.

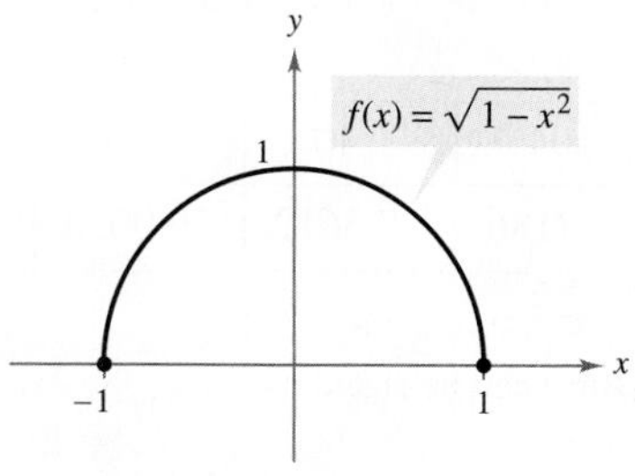

f is continuous on $[-1, 1]$.
Figure 1.32

The next example shows how a one-sided limit can be used to determine the value of absolute zero on the Kelvin scale.

EXAMPLE 5 Charles's Law and Absolute Zero

On the Kelvin scale, *absolute zero* is the temperature 0 K. Although temperatures of approximately 0.0001 K have been produced in laboratories, absolute zero has never been attained. In fact, evidence suggests that absolute zero *cannot* be attained. How did scientists determine that 0 K is the "lower limit" of the temperature of matter? What is absolute zero on the Celsius scale?

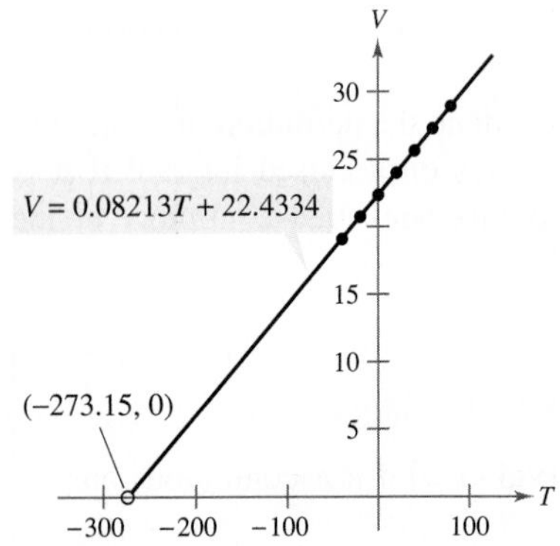

The volume of hydrogen gas depends on its temperature.
Figure 1.33

Solution The determination of absolute zero stems from the work of the French physicist Jacques Charles (1746–1823). Charles discovered that the volume of gas at a constant pressure increases linearly with the temperature of the gas. The table illustrates this relationship between volume and temperature. In the table, one mole of hydrogen is held at a constant pressure of one atmosphere. The volume V is measured in liters and the temperature T is measured in degrees Celsius.

T	-40	-20	0	20	40	60	80
V	19.1482	20.7908	22.4334	24.0760	25.7186	27.3612	29.0038

The points represented by the table are shown in Figure 1.33. Moreover, by using the points in the table, you can determine that T and V are related by the linear equation

$$V = 0.08213T + 22.4334 \qquad \text{or} \qquad T = \frac{V - 22.4334}{0.08213}.$$

By reasoning that the volume of the gas can approach 0 (but never equal or go below 0) you can determine that the "least possible temperature" is given by

$$\begin{aligned}\lim_{V\to 0^+} T &= \lim_{V\to 0^+} \frac{V - 22.4334}{0.08213}\\ &= \frac{0 - 22.4334}{0.08213} \qquad \text{Use direct substitution.}\\ &\approx -273.15.\end{aligned}$$

So, absolute zero on the Kelvin scale (0 K) is approximately $-273.15°$ on the Celsius scale.

University of Colorado at Boulder, Office of News Services

In 1995, physicists Carl Wieman and Eric Cornell of the University of Colorado at Boulder used lasers and evaporation to produce a supercold gas in which atoms overlap. This gas is called a Bose-Einstein condensate. "We get to within a billionth of a degree of absolute zero," reported Wieman. (*Source:* Time *magazine, April 10, 2000*)

The following table shows the temperatures in Example 5, converted to the Fahrenheit scale. Try repeating the solution shown in Example 5 using these temperatures and volumes. Use the result to find the value of absolute zero on the Fahrenheit scale.

T	-40	-4	32	68	104	140	176
V	19.1482	20.7908	22.4334	24.0760	25.7186	27.3612	29.0038

NOTE Charles's Law for gases (assuming constant pressure) can be stated as

$$V = RT \qquad \text{Charles's Law}$$

where V is volume, R is constant, and T is temperature. In the statement of this law, what property must the temperature scale have?

AUGUSTIN-LOUIS CAUCHY (1789–1857)

The concept of a continuous function was first introduced by Augustin-Louis Cauchy in 1821. The definition given in his text *Cours d'Analyse* stated that indefinite small changes in y were the result of indefinite small changes in x. "$\ldots f(x)$ will be called a *continuous* function if $\ldots$ the numerical values of the difference $f(x+\alpha)-f(x)$ decrease indefinitely with those of $\alpha \ldots$"

Properties of Continuity

In Section 1.3, you studied several properties of limits. Each of those properties yields a corresponding property pertaining to the continuity of a function. For instance, Theorem 1.11 follows directly from Theorem 1.2.

THEOREM 1.11 Properties of Continuity

If b is a real number and f and g are continuous at $x = c$, then the following functions are also continuous at c.

1. Scalar multiple: bf
2. Sum and difference: $f \pm g$
3. Product: fg
4. Quotient: $\dfrac{f}{g}$, if $g(c) \neq 0$

The following types of functions are continuous at every point in their domains.

1. Polynomial functions: $p(x) = a_n x^n + a_{n-1}x^{n-1} + \cdot\cdot\cdot + a_1 x + a_0$
2. Rational functions: $r(x) = \dfrac{p(x)}{q(x)}, \quad q(x) \neq 0$
3. Radical functions: $f(x) = \sqrt[n]{x}$
4. Trigonometric functions: $\sin x$, $\cos x$, $\tan x$, $\cot x$, $\sec x$, $\csc x$

By combining Theorem 1.11 with this summary, you can conclude that a wide variety of elementary functions are continuous at every point in their domains.

EXAMPLE 6 Applying Properties of Continuity

By Theorem 1.11, it follows that each of the following functions is continuous at every point in its domain.

$$f(x) = x + \sin x, \qquad f(x) = 3\tan x, \qquad f(x) = \frac{x^2+1}{\cos x}$$

The next theorem, which is a consequence of Theorem 1.5, allows you to determine the continuity of *composite* functions such as

$$f(x) = \sin 3x, \qquad f(x) = \sqrt{x^2+1}, \qquad f(x) = \tan\frac{1}{x}.$$

THEOREM 1.12 Continuity of a Composite Function

If g is continuous at c and f is continuous at $g(c)$, then the composite function given by $(f \circ g)(x) = f(g(x))$ is continuous at c.

One consequence of Theorem 1.12 is that if f and g satisfy the given conditions, you can determine the limit of $f(g(x))$ as x approaches c to be

$$\lim_{x \to c} f(g(x)) = f(g(c)).$$

EXAMPLE 7 Testing for Continuity

Describe the interval(s) on which each function is continuous.

a. $f(x) = \tan x$ **b.** $g(x) = \begin{cases} \sin \dfrac{1}{x}, & x \neq 0 \\ 0, & x = 0 \end{cases}$ **c.** $h(x) = \begin{cases} x \sin \dfrac{1}{x}, & x \neq 0 \\ 0, & x = 0 \end{cases}$

Solution

a. The tangent function $f(x) = \tan x$ is undefined at

$$x = \frac{\pi}{2} + n\pi, \qquad n \text{ is an integer.}$$

At all other points it is continuous. So, $f(x) = \tan x$ is continuous on the open intervals

$$\ldots, \left(-\frac{3\pi}{2}, -\frac{\pi}{2}\right), \left(-\frac{\pi}{2}, \frac{\pi}{2}\right), \left(\frac{\pi}{2}, \frac{3\pi}{2}\right), \ldots$$

as shown in Figure 1.34(a).

b. Because $y = 1/x$ is continuous except at $x = 0$ and the sine function is continuous for all real values of x, it follows that $y = \sin(1/x)$ is continuous at all real values except $x = 0$. At $x = 0$, the limit of $g(x)$ does not exist (see Example 5, Section 1.2). So, g is continuous on the intervals $(-\infty, 0)$ and $(0, \infty)$, as shown in Figure 1.34(b).

c. This function is similar to that in part (b) except that the oscillations are damped by the factor x. Using the Squeeze Theorem, you obtain

$$-|x| \leq x \sin \frac{1}{x} \leq |x|, \qquad x \neq 0$$

and you can conclude that

$$\lim_{x \to 0} h(x) = 0.$$

So, h is continuous on the entire real line, as shown in Figure 1.34(c).

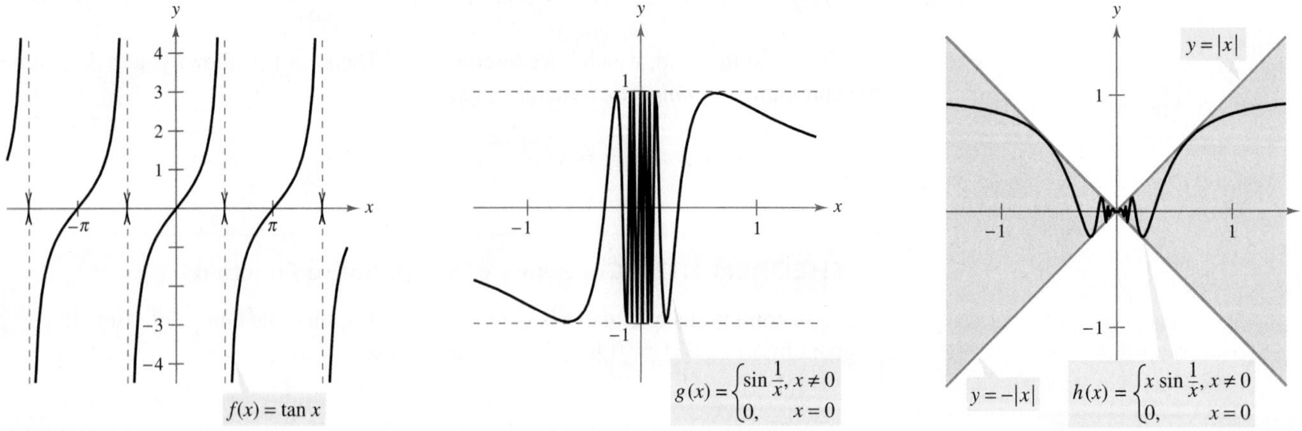

(a) f is continuous on each open interval in its domain.

(b) g is continuous on $(-\infty, 0)$ and $(0, \infty)$.

(c) h is continuous on the entire real line.

Figure 1.34

The Intermediate Value Theorem

Theorem 1.13 is an important theorem concerning the behavior of functions that are continuous on a closed interval.

THEOREM 1.13 Intermediate Value Theorem

If f is continuous on the closed interval $[a, b]$ and k is any number between $f(a)$ and $f(b)$, then there is at least one number c in $[a, b]$ such that

$$f(c) = k.$$

NOTE The Intermediate Value Theorem tells you that at least one c exists, but it does not give a method for finding c. Such theorems are called **existence theorems.**

By referring to a text on advanced calculus, you will find that a proof of this theorem is based on a property of real numbers called *completeness*. The Intermediate Value Theorem states that for a continuous function f, if x takes on all values between a and b, $f(x)$ must take on all values between $f(a)$ and $f(b)$.

As a simple example of this theorem, consider a person's height. Suppose that a girl is 5 feet tall on her thirteenth birthday and 5 feet 7 inches tall on her fourteenth birthday. Then, for any height h between 5 feet and 5 feet 7 inches, there must have been a time t when her height was exactly h. This seems reasonable because human growth is continuous and a person's height does not abruptly change from one value to another.

The Intermediate Value Theorem guarantees the existence of *at least one* number c in the closed interval $[a, b]$. There may, of course, be more than one number c such that $f(c) = k$, as shown in Figure 1.35. A function that is not continuous does not necessarily exhibit the intermediate value property. For example, the graph of the function shown in Figure 1.36 jumps over the horizontal line given by $y = k$, and for this function there is no value of c in $[a, b]$ such that $f(c) = k$.

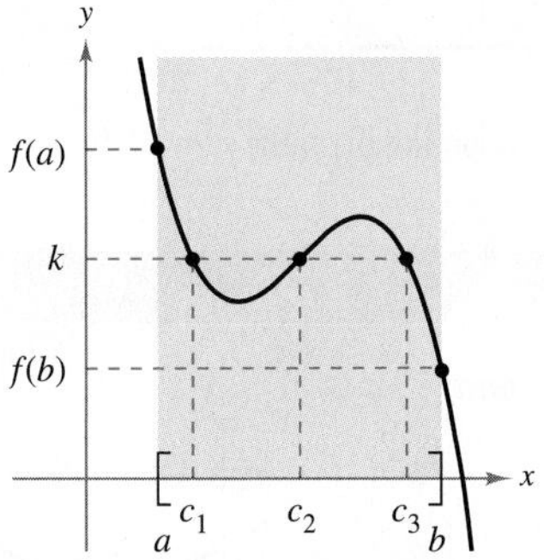

f is continuous on $[a, b]$.
[There exist three c's such that $f(c) = k$.]
Figure 1.35

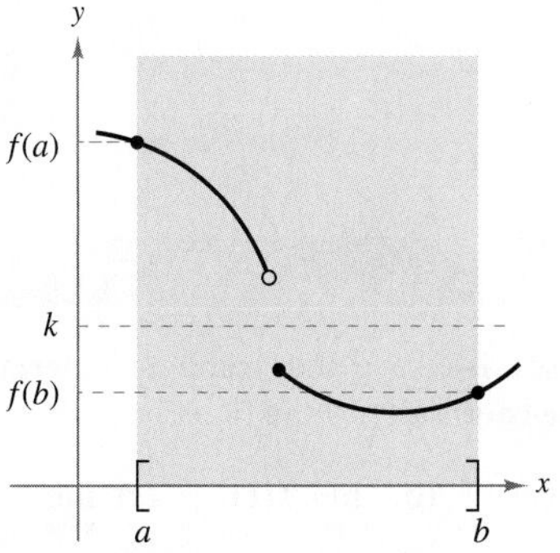

f is not continuous on $[a, b]$.
[There are no c's such that $f(c) = k$.]
Figure 1.36

The Intermediate Value Theorem often can be used to locate the zeros of a function that is continuous on a closed interval. Specifically, if f is continuous on $[a, b]$ and $f(a)$ and $f(b)$ differ in sign, the Intermediate Value Theorem guarantees the existence of at least one zero of f in the closed interval $[a, b]$.

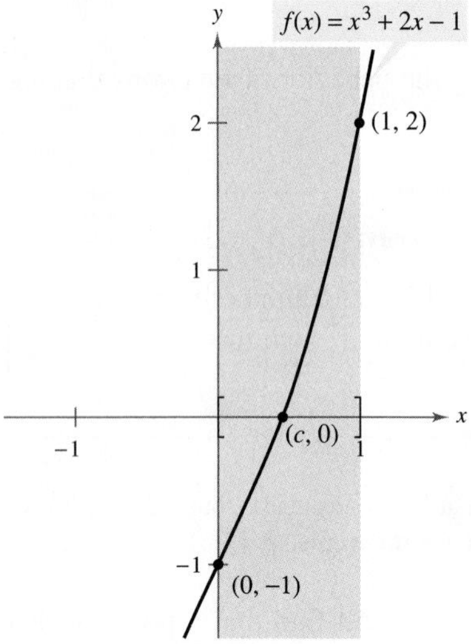

f is continuous on $[0, 1]$ with $f(0) < 0$ and $f(1) > 0$.
Figure 1.37

EXAMPLE 8 An Application of the Intermediate Value Theorem

Use the Intermediate Value Theorem to show that the polynomial function $f(x) = x^3 + 2x - 1$ has a zero in the interval $[0, 1]$.

Solution Note that f is continuous on the closed interval $[0, 1]$. Because

$$f(0) = 0^3 + 2(0) - 1 = -1 \quad \text{and} \quad f(1) = 1^3 + 2(1) - 1 = 2$$

it follows that $f(0) < 0$ and $f(1) > 0$. You can therefore apply the Intermediate Value Theorem to conclude that there must be some c in $[0, 1]$ such that

$$f(c) = 0 \qquad f \text{ has a zero in the closed interval } [0, 1].$$

as shown in Figure 1.37.

The **bisection method** for approximating the real zeros of a continuous function is similar to the method used in Example 8. If you know that a zero exists in the closed interval $[a, b]$, the zero must lie in the interval $[a, (a + b)/2]$ or $[(a + b)/2, b]$. From the sign of $f([a + b]/2)$, you can determine which interval contains the zero. By repeatedly bisecting the interval, you can "close in" on the zero of the function.

TECHNOLOGY You can also use the *zoom* feature of a graphing utility to approximate the real zeros of a continuous function. By repeatedly zooming in on the point where the graph crosses the x-axis, and adjusting the x-axis scale, you can approximate the zero of the function to any desired accuracy. The zero of $x^3 + 2x - 1$ is approximately 0.453, as shown in Figure 1.38.

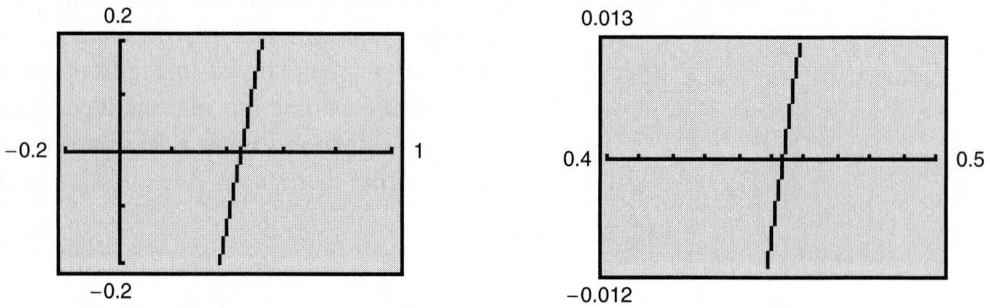

Figure 1.38 Zooming in on the zero of $f(x) = x^3 + 2x - 1$

Exercises for Section 1.4

See www.CalcChat.com for worked-out solutions to odd-numbered exercises.

In Exercises 1–6, use the graph to determine the limit, and discuss the continuity of the function.

(a) $\lim_{x \to c^+} f(x)$ **(b)** $\lim_{x \to c^-} f(x)$ **(c)** $\lim_{x \to c} f(x)$

1.

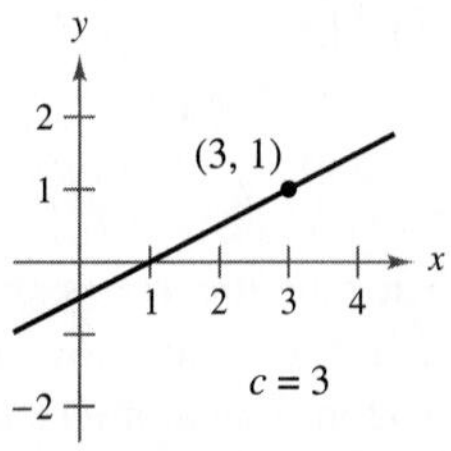

2.

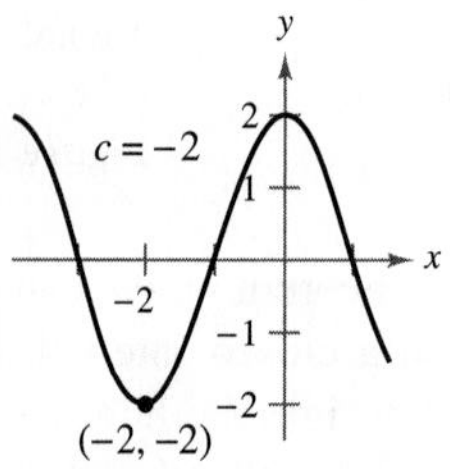

3.

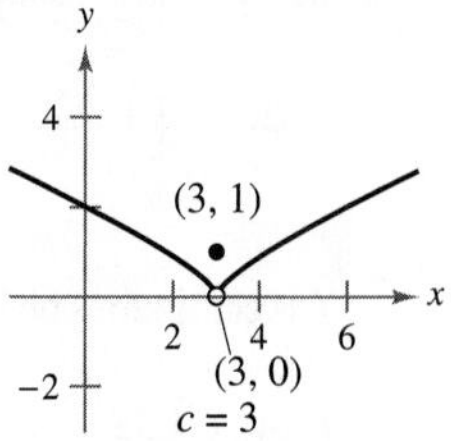

4.

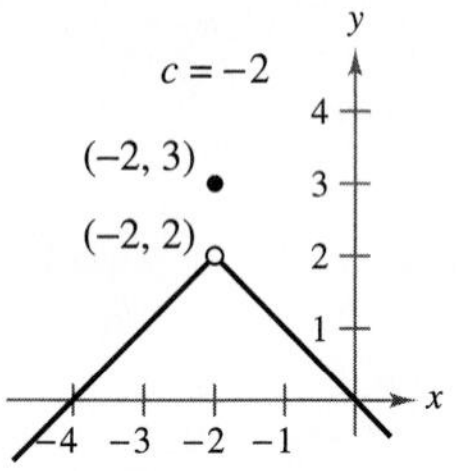

5.

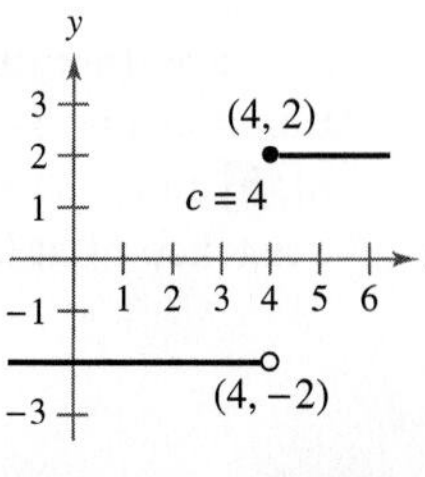

6.

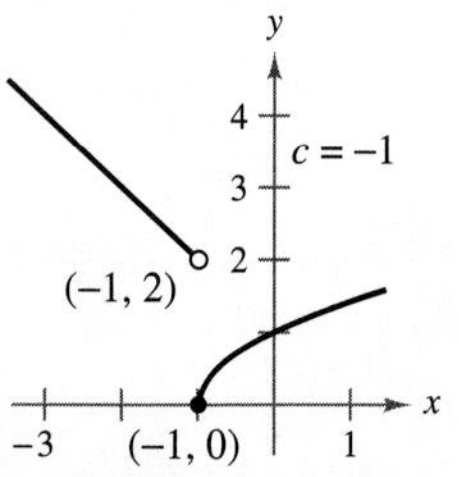

In Exercises 7–24, find the limit (if it exists). If it does not exist, explain why.

7. $\lim_{x \to 5^+} \frac{x - 5}{x^2 - 25}$

8. $\lim_{x \to 2^+} \frac{2 - x}{x^2 - 4}$

9. $\lim_{x \to -3^-} \frac{x}{\sqrt{x^2 - 9}}$

10. $\lim_{x \to 4^-} \frac{\sqrt{x} - 2}{x - 4}$

11. $\lim_{x \to 0^-} \frac{|x|}{x}$

12. $\lim_{x \to 2^+} \frac{|x - 2|}{x - 2}$

13. $\lim_{\Delta x \to 0^-} \frac{\frac{1}{x + \Delta x} - \frac{1}{x}}{\Delta x}$

14. $\lim_{\Delta x \to 0^+} \frac{(x + \Delta x)^2 + x + \Delta x - (x^2 + x)}{\Delta x}$

15. $\lim_{x \to 3^-} f(x)$, where $f(x) = \begin{cases} \frac{x + 2}{2}, & x \le 3 \\ \frac{12 - 2x}{3}, & x > 3 \end{cases}$

16. $\lim_{x \to 2} f(x)$, where $f(x) = \begin{cases} x^2 - 4x + 6, & x < 2 \\ -x^2 + 4x - 2, & x \ge 2 \end{cases}$

17. $\lim_{x \to 1} f(x)$, where $f(x) = \begin{cases} x^3 + 1, & x < 1 \\ x + 1, & x \ge 1 \end{cases}$

18. $\lim_{x \to 1^+} f(x)$, where $f(x) = \begin{cases} x, & x \le 1 \\ 1 - x, & x > 1 \end{cases}$

19. $\lim_{x \to \pi} \cot x$

20. $\lim_{x \to \pi/2} \sec x$

21. $\lim_{x \to 4^-} (3[\![x]\!] - 5)$

22. $\lim_{x \to 2^+} (2x - [\![x]\!])$

23. $\lim_{x \to 3} (2 - [\![-x]\!])$

24. $\lim_{x \to 1} \left(1 - \left[\!\left[-\frac{x}{2}\right]\!\right]\right)$

In Exercises 25–28, discuss the continuity of each function.

25. $f(x) = \frac{1}{x^2 - 4}$

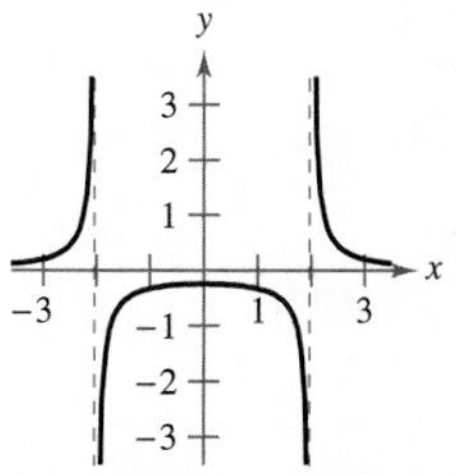

26. $f(x) = \frac{x^2 - 1}{x + 1}$

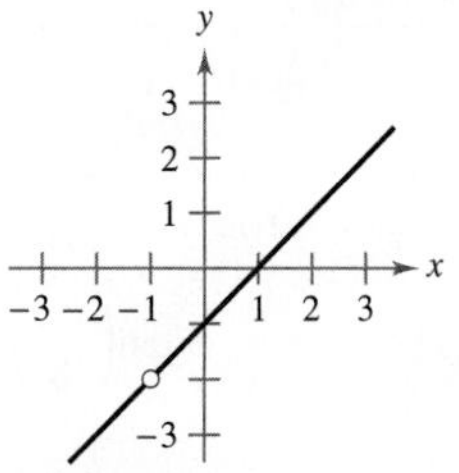

27. $f(x) = \frac{1}{2}[\![x]\!] + x$

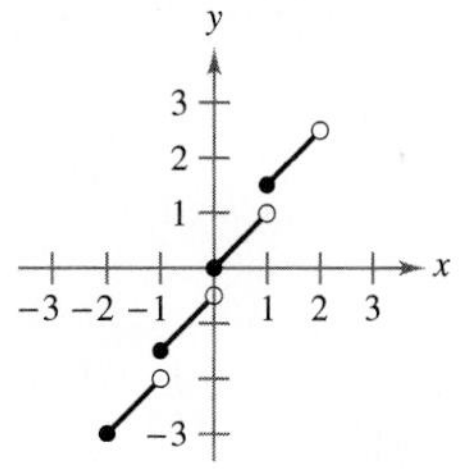

28. $f(x) = \begin{cases} x, & x < 1 \\ 2, & x = 1 \\ 2x - 1, & x > 1 \end{cases}$

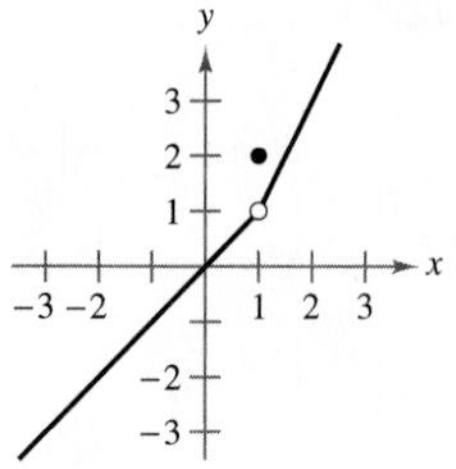

In Exercises 29–32, discuss the continuity of the function on the closed interval.

	Function	*Interval*
29.	$g(x) = \sqrt{25 - x^2}$	$[-5, 5]$
30.	$f(t) = 3 - \sqrt{9 - t^2}$	$[-3, 3]$
31.	$f(x) = \begin{cases} 3 - x, & x \le 0 \\ 3 + \frac{1}{2}x, & x > 0 \end{cases}$	$[-1, 4]$
32.	$g(x) = \frac{1}{x^2 - 4}$	$[-1, 2]$

In Exercises 33–54, find the x-values (if any) at which f is not continuous. Which of the discontinuities are removable?

33. $f(x) = x^2 - 2x + 1$

34. $f(x) = \frac{1}{x^2 + 1}$

35. $f(x) = 3x - \cos x$

36. $f(x) = \cos \frac{\pi x}{2}$

37. $f(x) = \frac{x}{x^2 - x}$

38. $f(x) = \frac{x}{x^2 - 1}$

39. $f(x) = \frac{x}{x^2 + 1}$

40. $f(x) = \frac{x - 3}{x^2 - 9}$

41. $f(x) = \frac{x + 2}{x^2 - 3x - 10}$

42. $f(x) = \frac{x - 1}{x^2 + x - 2}$

43. $f(x) = \frac{|x + 2|}{x + 2}$

44. $f(x) = \frac{|x - 3|}{x - 3}$

45. $f(x) = \begin{cases} x, & x \le 1 \\ x^2, & x > 1 \end{cases}$

46. $f(x) = \begin{cases} -2x + 3, & x < 1 \\ x^2, & x \ge 1 \end{cases}$

47. $f(x) = \begin{cases} \frac{1}{2}x + 1, & x \le 2 \\ 3 - x, & x > 2 \end{cases}$

48. $f(x) = \begin{cases} -2x, & x \le 2 \\ x^2 - 4x + 1, & x > 2 \end{cases}$

49. $f(x) = \begin{cases} \tan \dfrac{\pi x}{4}, & |x| < 1 \\ x, & |x| \ge 1 \end{cases}$

50. $f(x) = \begin{cases} \csc \dfrac{\pi x}{6}, & |x - 3| \le 2 \\ 2, & |x - 3| > 2 \end{cases}$

51. $f(x) = \csc 2x$

52. $f(x) = \tan \dfrac{\pi x}{2}$

53. $f(x) = [\![x - 1]\!]$

54. $f(x) = 3 - [\![x]\!]$

In Exercises 55 and 56, use a graphing utility to graph the function. From the graph, estimate

$$\lim_{x\to 0^+} f(x) \quad \text{and} \quad \lim_{x\to 0^-} f(x).$$

Is the function continuous on the entire real line? Explain.

55. $f(x) = \dfrac{|x^2 - 4|x}{x + 2}$

56. $f(x) = \dfrac{|x^2 + 4x|(x + 2)}{x + 4}$

In Exercises 57–60, find the constant *a*, or the constants *a* and *b*, such that the function is continuous on the entire real line.

57. $f(x) = \begin{cases} x^3, & x \le 2 \\ ax^2, & x > 2 \end{cases}$

58. $g(x) = \begin{cases} \dfrac{4 \sin x}{x}, & x < 0 \\ a - 2x, & x \ge 0 \end{cases}$

59. $f(x) = \begin{cases} 2, & x \le -1 \\ ax + b, & -1 < x < 3 \\ -2, & x \ge 3 \end{cases}$

60. $g(x) = \begin{cases} \dfrac{x^2 - a^2}{x - a}, & x \ne a \\ 8, & x = a \end{cases}$

In Exercises 61–64, discuss the continuity of the composite function $h(x) = f(g(x))$.

61. $f(x) = x^2$
$g(x) = x - 1$

62. $f(x) = \dfrac{1}{\sqrt{x}}$
$g(x) = x - 1$

63. $f(x) = \dfrac{1}{x - 6}$
$g(x) = x^2 + 5$

64. $f(x) = \sin x$
$g(x) = x^2$

In Exercises 65–68, use a graphing utility to graph the function. Use the graph to determine any *x*-values at which the function is not continuous.

65. $f(x) = [\![x]\!] - x$

66. $h(x) = \dfrac{1}{x^2 - x - 2}$

67. $g(x) = \begin{cases} 2x - 4, & x \le 3 \\ x^2 - 2x, & x > 3 \end{cases}$

68. $f(x) = \begin{cases} \dfrac{\cos x - 1}{x}, & x < 0 \\ 5x, & x \ge 0 \end{cases}$

In Exercises 69–72, describe the interval(s) on which the function is continuous.

69. $f(x) = \dfrac{x}{x^2 + 1}$

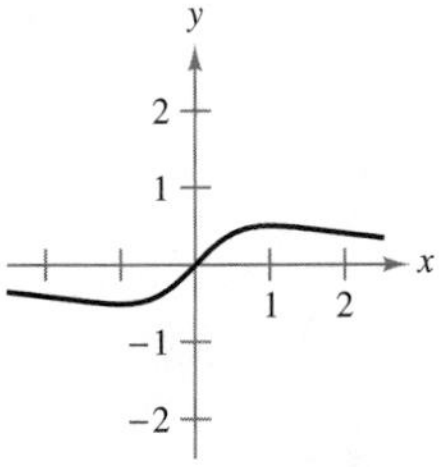

70. $f(x) = x\sqrt{x + 3}$

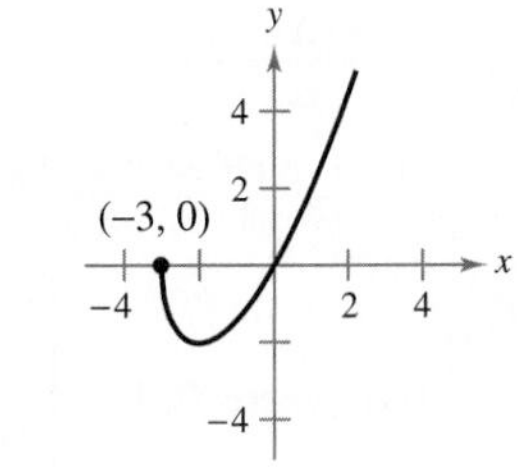

71. $f(x) = \sec \dfrac{\pi x}{4}$

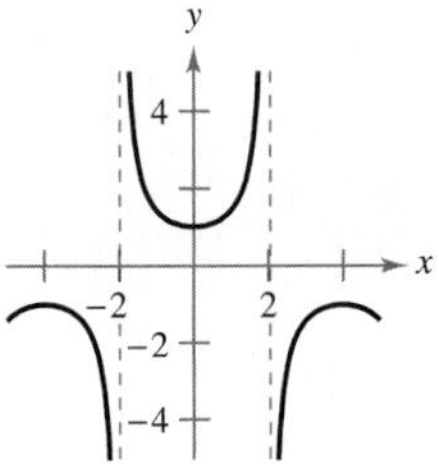

72. $f(x) = \dfrac{x + 1}{\sqrt{x}}$

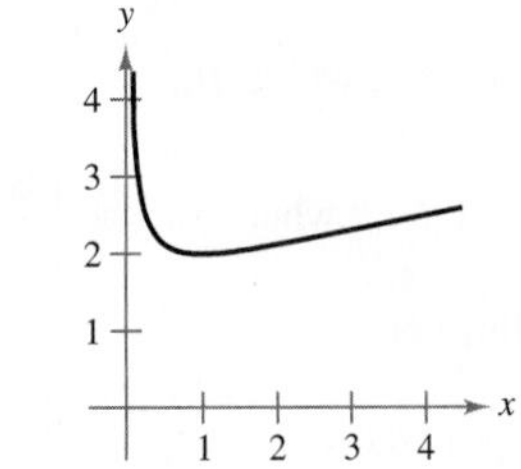

Writing **In Exercises 73 and 74, use a graphing utility to graph the function on the interval $[-4, 4]$. Does the graph of the function appear continuous on this interval? Is the function continuous on $[-4, 4]$? Write a short paragraph about the importance of examining a function analytically as well as graphically.**

73. $f(x) = \dfrac{\sin x}{x}$

74. $f(x) = \dfrac{x^3 - 8}{x - 2}$

Writing **In Exercises 75–78, explain why the function has a zero in the given interval.**

Function	*Interval*
75. $f(x) = \frac{1}{16}x^4 - x^3 + 3$	$[1, 2]$
76. $f(x) = x^3 + 3x - 2$	$[0, 1]$
77. $f(x) = x^2 - 2 - \cos x$	$[0, \pi]$
78. $f(x) = -\dfrac{4}{x} + \tan\left(\dfrac{\pi x}{8}\right)$	$[1, 3]$

In Exercises 79–82, use the Intermediate Value Theorem and a graphing utility to approximate the zero of the function in the interval [0, 1]. Repeatedly "zoom in" on the graph of the function to approximate the zero accurate to two decimal places. Use the *zero* or *root* feature of the graphing utility to approximate the zero accurate to four decimal places.

79. $f(x) = x^3 + x - 1$

80. $f(x) = x^3 + 3x - 2$

81. $g(t) = 2 \cos t - 3t$

82. $h(\theta) = 1 + \theta - 3 \tan \theta$

In Exercises 83–86, verify that the Intermediate Value Theorem applies to the indicated interval and find the value of c guaranteed by the theorem.

83. $f(x) = x^2 + x - 1, \quad [0, 5], \quad f(c) = 11$

84. $f(x) = x^2 - 6x + 8, \quad [0, 3], \quad f(c) = 0$

85. $f(x) = x^3 - x^2 + x - 2, \quad [0, 3], \quad f(c) = 4$

86. $f(x) = \dfrac{x^2 + x}{x - 1}, \quad \left[\dfrac{5}{2}, 4\right], \quad f(c) = 6$

Writing About Concepts

87. State how continuity is destroyed at $x = c$ for each of the following graphs.

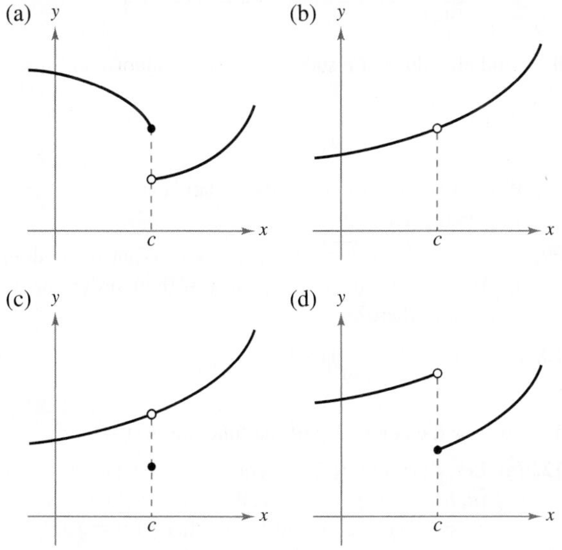

88. Describe the difference between a discontinuity that is removable and one that is nonremovable. In your explanation, give examples of the following descriptions.

(a) A function with a nonremovable discontinuity at $x = 2$

(b) A function with a removable discontinuity at $x = -2$

(c) A function that has both of the characteristics described in parts (a) and (b)

Writing About Concepts *(continued)*

89. Sketch the graph of any function f such that

$$\lim_{x \to 3^+} f(x) = 1 \quad \text{and} \quad \lim_{x \to 3^-} f(x) = 0.$$

Is the function continuous at $x = 3$? Explain.

90. If the functions f and g are continuous for all real x, is $f + g$ always continuous for all real x? Is f/g always continuous for all real x? If either is not continuous, give an example to verify your conclusion.

True or False? **In Exercises 91–94, determine whether the statement is true or false. If it is false, explain why or give an example that shows it is false.**

91. If $\lim_{x \to c} f(x) = L$ and $f(c) = L$, then f is continuous at c.

92. If $f(x) = g(x)$ for $x \neq c$ and $f(c) \neq g(c)$, then either f or g is not continuous at c.

93. A rational function can have infinitely many x-values at which it is not continuous.

94. The function $f(x) = |x - 1|/(x - 1)$ is continuous on $(-\infty, \infty)$.

95. ***Swimming Pool*** Every day you dissolve 28 ounces of chlorine in a swimming pool. The graph shows the amount of chlorine $f(t)$ in the pool after t days.

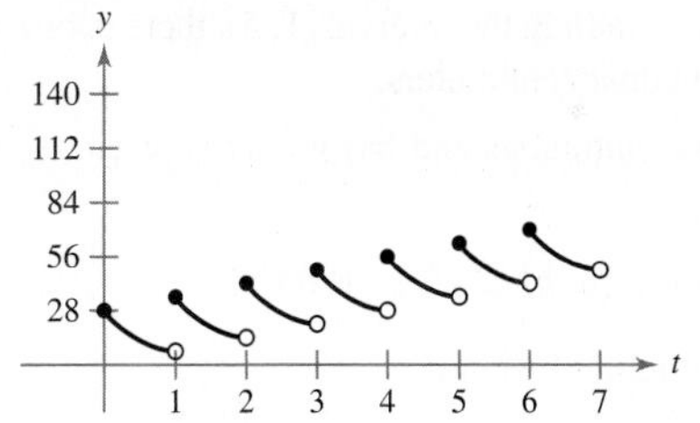

Estimate and interpret $\lim_{t \to 4^-} f(t)$ and $\lim_{t \to 4^+} f(t)$.

96. ***Think About It*** Describe how the functions

$f(x) = 3 + [\![x]\!]$

and

$g(x) = 3 - [\![-x]\!]$

differ.

97. ***Telephone Charges*** A dial-direct long distance call between two cities costs \$1.04 for the first 2 minutes and \$0.36 for each additional minute or fraction thereof. Use the greatest integer function to write the cost C of a call in terms of time t (in minutes). Sketch the graph of this function and discuss its continuity.

98. ***Inventory Management*** The number of units in inventory in a small company is given by

$$N(t) = 25\left(2\left[\!\left[\frac{t+2}{2}\right]\!\right] - t\right)$$

where t is the time in months. Sketch the graph of this function and discuss its continuity. How often must this company replenish its inventory?

99. ***Déjà Vu*** At 8:00 A.M. on Saturday a man begins running up the side of a mountain to his weekend campsite (see figure). On Sunday morning at 8:00 A.M. he runs back down the mountain. It takes him 20 minutes to run up, but only 10 minutes to run down. At some point on the way down, he realizes that he passed the same place at exactly the same time on Saturday. Prove that he is correct. [*Hint:* Let $s(t)$ and $r(t)$ be the position functions for the runs up and down, and apply the Intermediate Value Theorem to the function $f(t) = s(t) - r(t)$.]

100. ***Volume*** Use the Intermediate Value Theorem to show that for all spheres with radii in the interval $[1, 5]$, there is one with a volume of 275 cubic centimeters.

101. Prove that if f is continuous and has no zeros on $[a, b]$, then either

$f(x) > 0$ for all x in $[a, b]$ or $f(x) < 0$ for all x in $[a, b]$.

102. Show that the Dirichlet function

$$f(x) = \begin{cases} 0, & \text{if } x \text{ is rational} \\ 1, & \text{if } x \text{ is irrational} \end{cases}$$

is not continuous at any real number.

103. Show that the function

$$f(x) = \begin{cases} 0, & \text{if } x \text{ is rational} \\ kx, & \text{if } x \text{ is irrational} \end{cases}$$

is continuous only at $x = 0$. (Assume that k is any nonzero real number.)

104. The **signum function** is defined by

$$\operatorname{sgn}(x) = \begin{cases} -1, & x < 0 \\ 0, & x = 0 \\ 1, & x > 0. \end{cases}$$

Sketch a graph of $\operatorname{sgn}(x)$ and find the following (if possible).

(a) $\lim_{x \to 0^-} \operatorname{sgn}(x)$ (b) $\lim_{x \to 0^+} \operatorname{sgn}(x)$ (c) $\lim_{x \to 0} \operatorname{sgn}(x)$

105. ***Modeling Data*** After an object falls for t seconds, the speed S (in feet per second) of the object is recorded in the table.

t	0	5	10	15	20	25	30
S	0	48.2	53.5	55.2	55.9	56.2	56.3

(a) Create a line graph of the data.

(b) Does there appear to be a limiting speed of the object? If there is a limiting speed, identify a possible cause.

106. ***Creating Models*** A swimmer crosses a pool of width b by swimming in a straight line from $(0, 0)$ to $(2b, b)$. (See figure.)

(a) Let f be a function defined as the y-coordinate of the point on the long side of the pool that is nearest the swimmer at any given time during the swimmer's path across the pool. Determine the function f and sketch its graph. Is it continuous? Explain.

(b) Let g be the minimum distance between the swimmer and the long sides of the pool. Determine the function g and sketch its graph. Is it continuous? Explain.

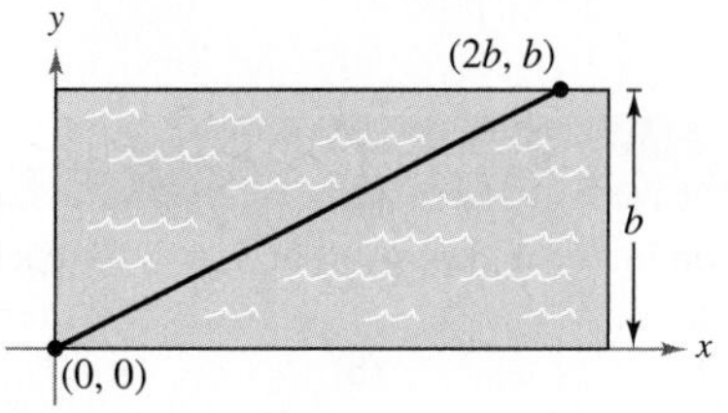

107. Find all values of c such that f is continuous on $(-\infty, \infty)$.

$$f(x) = \begin{cases} 1 - x^2, & x \le c \\ x, & x > c \end{cases}$$

108. Prove that for any real number y there exists x in $(-\pi/2, \pi/2)$ such that $\tan x = y$.

109. Let $f(x) = \left(\sqrt{x + c^2} - c\right)/x$, $c > 0$. What is the domain of f? How can you define f at $x = 0$ in order for f to be continuous there?

110. Prove that if $\lim_{\Delta x \to 0} f(c + \Delta x) = f(c)$, then f is continuous at c.

111. Discuss the continuity of the function $h(x) = x[\![x]\!]$.

112. (a) Let $f_1(x)$ and $f_2(x)$ be continuous on the closed interval $[a, b]$. If $f_1(a) < f_2(a)$ and $f_1(b) > f_2(b)$, prove that there exists c between a and b such that $f_1(c) = f_2(c)$.

(b) Show that there exists c in $\left[0, \frac{\pi}{2}\right]$ such that $\cos x = x$. Use a graphing utility to approximate c to three decimal places.

Putnam Exam Challenge

113. Prove or disprove: if x and y are real numbers with $y \ge 0$ and $y(y + 1) \le (x + 1)^2$, then $y(y - 1) \le x^2$.

114. Determine all polynomials $P(x)$ such that $P(x^2 + 1) = (P(x))^2 + 1$ and $P(0) = 0$.

These problems were composed by the Committee on the Putnam Prize Competition.

Section 1.5 Infinite Limits

- **Determine infinite limits from the left and from the right.**
- **Find and sketch the vertical asymptotes of the graph of a function.**

Infinite Limits

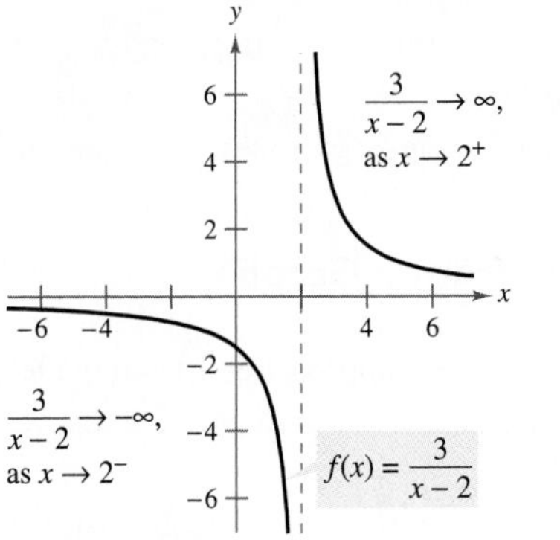

$f(x)$ increases and decreases without bound as x approaches 2.
Figure 1.39

Let f be the function given by

$$f(x) = \frac{3}{x-2}.$$

From Figure 1.39 and the table, you can see that $f(x)$ *decreases without bound* as x approaches 2 from the left, and $f(x)$ *increases without bound* as x approaches 2 from the right. This behavior is denoted as

$$\lim_{x \to 2^-} \frac{3}{x-2} = -\infty \qquad f(x) \text{ decreases without bound as } x \text{ approaches 2 from the left.}$$

and

$$\lim_{x \to 2^+} \frac{3}{x-2} = \infty \qquad f(x) \text{ increases without bound as } x \text{ approaches 2 from the right.}$$

x approaches 2 from the left. → ← x approaches 2 from the right.

x	1.5	1.9	1.99	1.999	2	2.001	2.01	2.1	2.5
$f(x)$	-6	-30	-300	-3000	?	3000	300	30	6

$f(x)$ decreases without bound. → ← $f(x)$ increases without bound.

A limit in which $f(x)$ increases or decreases without bound as x approaches c is called an **infinite limit.**

Definition of Infinite Limits

Let f be a function that is defined at every real number in some open interval containing c (except possibly at c itself). The statement

$$\lim_{x \to c} f(x) = \infty$$

means that for each $M > 0$ there exists a $\delta > 0$ such that $f(x) > M$ whenever $0 < |x - c| < \delta$ (see Figure 1.40). Similarly, the statement

$$\lim_{x \to c} f(x) = -\infty$$

means that for each $N < 0$ there exists a $\delta > 0$ such that $f(x) < N$ whenever $0 < |x - c| < \delta$. To define the **infinite limit from the left,** replace $0 < |x - c| < \delta$ by $c - \delta < x < c$. To define the **infinite limit from the right,** replace $0 < |x - c| < \delta$ by $c < x < c + \delta$.

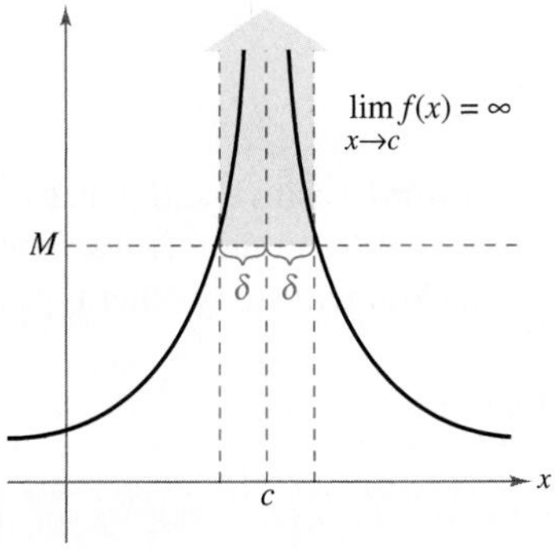

Infinite limits
Figure 1.40

Be sure you see that the equal sign in the statement $\lim f(x) = \infty$ does not mean that the limit exists! On the contrary, it tells you how the limit *fails to exist* by denoting the unbounded behavior of $f(x)$ as x approaches c.

EXPLORATION

Use a graphing utility to graph each function. For each function, analytically find the single real number c that is not in the domain. Then graphically find the limit of $f(x)$ as x approaches c from the left and from the right.

a. $f(x) = \dfrac{3}{x - 4}$ **b.** $f(x) = \dfrac{1}{2 - x}$

c. $f(x) = \dfrac{2}{(x - 3)^2}$ **d.** $f(x) = \dfrac{-3}{(x + 2)^2}$

EXAMPLE 1 Determining Infinite Limits from a Graph

Use Figure 1.41 to determine the limit of each function as x approaches 1 from the left and from the right.

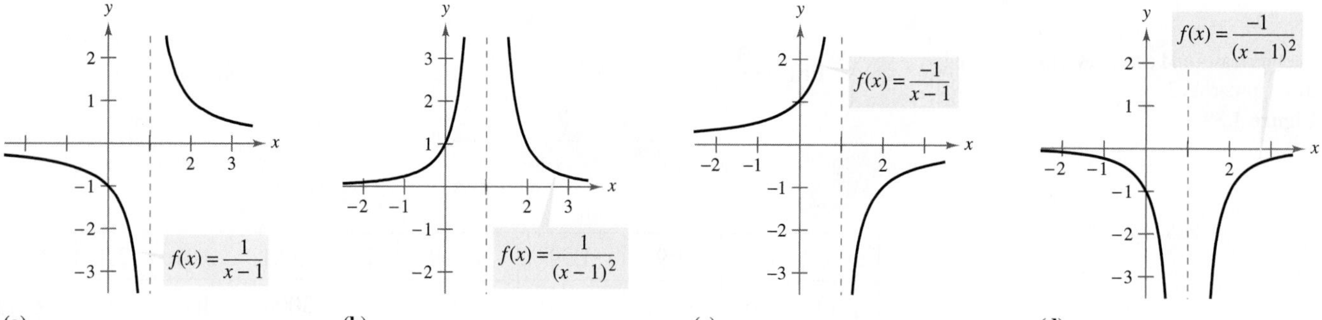

Figure 1.41 Each graph has an asymptote at $x = 1$.

Solution

a. $\displaystyle\lim_{x\to1^-} \frac{1}{x-1} = -\infty$ and $\displaystyle\lim_{x\to1^+} \frac{1}{x-1} = \infty$

b. $\displaystyle\lim_{x\to1} \frac{1}{(x-1)^2} = \infty$ Limit from each side is ∞.

c. $\displaystyle\lim_{x\to1^-} \frac{-1}{x-1} = \infty$ and $\displaystyle\lim_{x\to1^+} \frac{-1}{x-1} = -\infty$

d. $\displaystyle\lim_{x\to1} \frac{-1}{(x-1)^2} = -\infty$ Limit from each side is $-\infty$.

Vertical Asymptotes

If it were possible to extend the graphs in Figure 1.41 toward positive and negative infinity, you would see that each graph becomes arbitrarily close to the vertical line $x = 1$. This line is a **vertical asymptote** of the graph of f. (You will study other types of asymptotes in Sections 3.5 and 3.6.)

NOTE If the graph of a function f has a vertical asymptote at $x = c$, then f is *not continuous* at c.

Definition of Vertical Asymptote

If $f(x)$ approaches infinity (or negative infinity) as x approaches c from the right or the left, then the line $x = c$ is a **vertical asymptote** of the graph of f.

In Example 1, note that each of the functions is a *quotient* and that the vertical asymptote occurs at a number where the denominator is 0 (and the numerator is not 0). The next theorem generalizes this observation. (A proof of this theorem is given in Appendix A.)

THEOREM 1.14 Vertical Asymptotes

Let f and g be continuous on an open interval containing c. If $f(c) \neq 0$, $g(c) = 0$, and there exists an open interval containing c such that $g(x) \neq 0$ for all $x \neq c$ in the interval, then the graph of the function given by

$$h(x) = \frac{f(x)}{g(x)}$$

has a vertical asymptote at $x = c$.

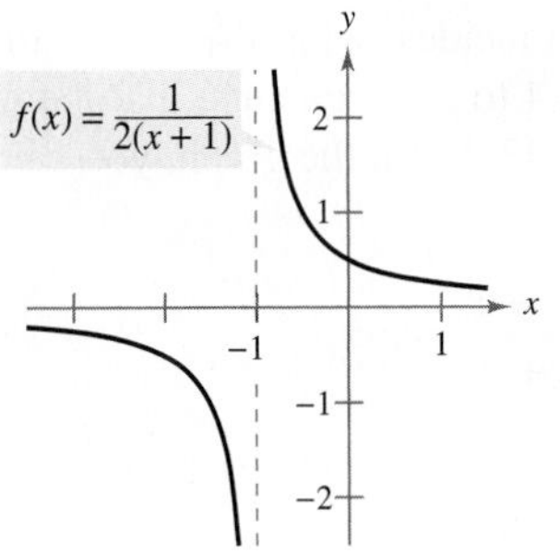

(a)

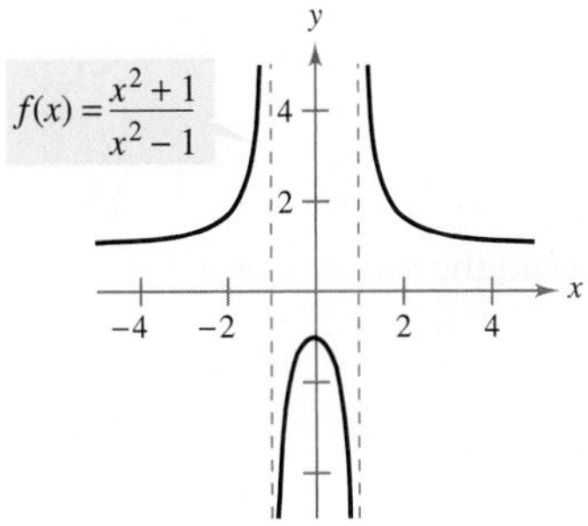

(b)

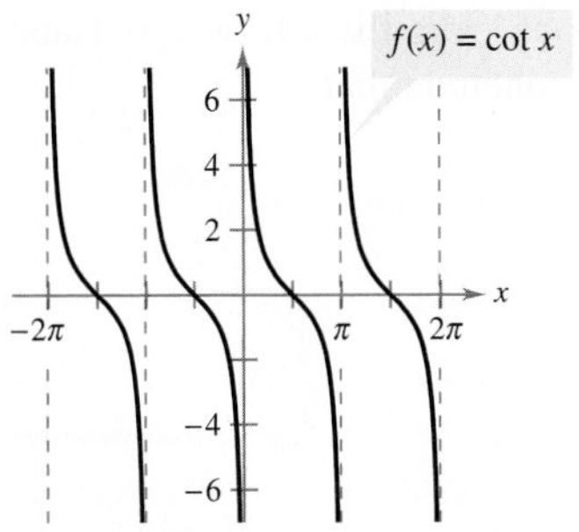

(c)

Functions with vertical asymptotes

Figure 1.42

EXAMPLE 2 Finding Vertical Asymptotes

Determine all vertical asymptotes of the graph of each function.

a. $f(x) = \dfrac{1}{2(x+1)}$ **b.** $f(x) = \dfrac{x^2+1}{x^2-1}$ **c.** $f(x) = \cot x$

Solution

a. When $x = -1$, the denominator of

$$f(x) = \frac{1}{2(x+1)}$$

is 0 and the numerator is not 0. So, by Theorem 1.14, you can conclude that $x = -1$ is a vertical asymptote, as shown in Figure 1.42(a).

b. By factoring the denominator as

$$f(x) = \frac{x^2+1}{x^2-1} = \frac{x^2+1}{(x-1)(x+1)}$$

you can see that the denominator is 0 at $x = -1$ and $x = 1$. Moreover, because the numerator is not 0 at these two points, you can apply Theorem 1.14 to conclude that the graph of f has two vertical asymptotes, as shown in Figure 1.42(b).

c. By writing the cotangent function in the form

$$f(x) = \cot x = \frac{\cos x}{\sin x}$$

you can apply Theorem 1.14 to conclude that vertical asymptotes occur at all values of x such that $\sin x = 0$ and $\cos x \neq 0$, as shown in Figure 1.42(c). So, the graph of this function has infinitely many vertical asymptotes. These asymptotes occur when $x = n\pi$, where n is an integer.

Theorem 1.14 requires that the value of the numerator at $x = c$ be nonzero. If both the numerator and the denominator are 0 at $x = c$, you obtain the *indeterminate form* $0/0$, and you cannot determine the limit behavior at $x = c$ without further investigation, as illustrated in Example 3.

EXAMPLE 3 A Rational Function with Common Factors

Determine all vertical asymptotes of the graph of

$$f(x) = \frac{x^2 + 2x - 8}{x^2 - 4}.$$

Solution Begin by simplifying the expression, as shown.

$$f(x) = \frac{x^2 + 2x - 8}{x^2 - 4}$$
$$= \frac{(x + 4)(x - 2)}{(x + 2)(x - 2)}$$
$$= \frac{x + 4}{x + 2}, \quad x \neq 2$$

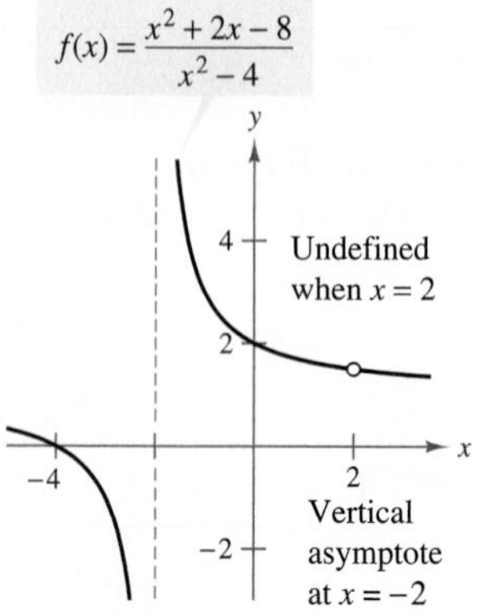

$f(x)$ increases and decreases without bound as x approaches -2.
Figure 1.43

At all x-values other than $x = 2$, the graph of f coincides with the graph of $g(x) = (x + 4)/(x + 2)$. So, you can apply Theorem 1.14 to g to conclude that there is a vertical asymptote at $x = -2$, as shown in Figure 1.43. From the graph, you can see that

$$\lim_{x \to -2^-} \frac{x^2 + 2x - 8}{x^2 - 4} = -\infty \quad \text{and} \quad \lim_{x \to -2^+} \frac{x^2 + 2x - 8}{x^2 - 4} = \infty.$$

Note that $x = 2$ is *not* a vertical asymptote.

EXAMPLE 4 Determining Infinite Limits

Find each limit.

$$\lim_{x \to 1^-} \frac{x^2 - 3x}{x - 1} \quad \text{and} \quad \lim_{x \to 1^+} \frac{x^2 - 3x}{x - 1}$$

Solution Because the denominator is 0 when $x = 1$ (and the numerator is not zero), you know that the graph of

$$f(x) = \frac{x^2 - 3x}{x - 1}$$

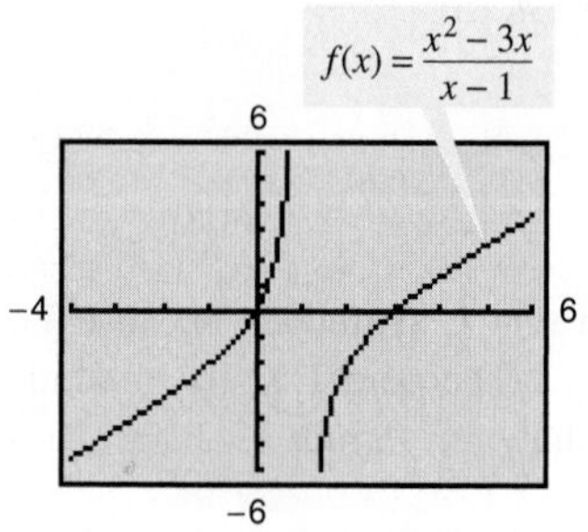

f has a vertical asymptote at $x = 1$.
Figure 1.44

has a vertical asymptote at $x = 1$. This means that each of the given limits is either ∞ or $-\infty$. A graphing utility can help determine the result. From the graph of f shown in Figure 1.44, you can see that the graph approaches ∞ from the left of $x = 1$ and approaches $-\infty$ from the right of $x = 1$. So, you can conclude that

$$\lim_{x \to 1^-} \frac{x^2 - 3x}{x - 1} = \infty \qquad \text{The limit from the left is infinity.}$$

and

$$\lim_{x \to 1^+} \frac{x^2 - 3x}{x - 1} = -\infty. \qquad \text{The limit from the right is negative infinity.}$$

TECHNOLOGY PITFALL When using a graphing calculator or graphing software, be careful to interpret correctly the graph of a function with a vertical asymptote—graphing utilities often have difficulty drawing this type of graph.

THEOREM 1.15 Properties of Infinite Limits

Let c and L be real numbers and let f and g be functions such that

$$\lim_{x \to c} f(x) = \infty \quad \text{and} \quad \lim_{x \to c} g(x) = L.$$

1. Sum or difference: $\lim_{x \to c} [f(x) \pm g(x)] = \infty$

2. Product: $\lim_{x \to c} [f(x)g(x)] = \infty, \quad L > 0$

$\lim_{x \to c} [f(x)g(x)] = -\infty, \quad L < 0$

3. Quotient: $\lim_{x \to c} \dfrac{g(x)}{f(x)} = 0$

Similar properties hold for one-sided limits and for functions for which the limit of $f(x)$ as x approaches c is $-\infty$.

Proof To show that the limit of $f(x) + g(x)$ is infinite, choose $M > 0$. You then need to find $\delta > 0$ such that

$$[f(x) + g(x)] > M$$

whenever $0 < |x - c| < \delta$. For simplicity's sake, you can assume L is positive. Let $M_1 = M + 1$. Because the limit of $f(x)$ is infinite, there exists δ_1 such that $f(x) > M_1$ whenever $0 < |x - c| < \delta_1$. Also, because the limit of $g(x)$ is L, there exists δ_2 such that $|g(x) - L| < 1$ whenever $0 < |x - c| < \delta_2$. By letting δ be the smaller of δ_1 and δ_2, you can conclude that $0 < |x - c| < \delta$ implies $f(x) > M + 1$ and $|g(x) - L| < 1$. The second of these two inequalities implies that $g(x) > L - 1$, and, adding this to the first inequality, you can write

$$f(x) + g(x) > (M + 1) + (L - 1) = M + L > M.$$

So, you can conclude that

$$\lim_{x \to c} [f(x) + g(x)] = \infty.$$

The proofs of the remaining properties are left as exercises (see Exercise 72).

EXAMPLE 5 Determining Limits

a. Because $\lim_{x \to 0} 1 = 1$ and $\lim_{x \to 0} \dfrac{1}{x^2} = \infty$, you can write

$$\lim_{x \to 0} \left(1 + \frac{1}{x^2}\right) = \infty. \qquad \text{Property 1, Theorem 1.15}$$

b. Because $\lim_{x \to 1^-} (x^2 + 1) = 2$ and $\lim_{x \to 1^-} (\cot \pi x) = -\infty$, you can write

$$\lim_{x \to 1^-} \frac{x^2 + 1}{\cot \pi x} = 0. \qquad \text{Property 3, Theorem 1.15}$$

c. Because $\lim_{x \to 0^+} 3 = 3$ and $\lim_{x \to 0^+} \cot x = \infty$, you can write

$$\lim_{x \to 0^+} 3 \cot x = \infty. \qquad \text{Property 2, Theorem 1.15}$$

Exercises for Section 1.5

See www.CalcChat.com for worked-out solutions to odd-numbered exercises.

In Exercises 1–4, determine whether $f(x)$ approaches ∞ or $-\infty$ as x approaches -2 from the left and from the right.

1. $f(x) = 2\left|\dfrac{x}{x^2 - 4}\right|$

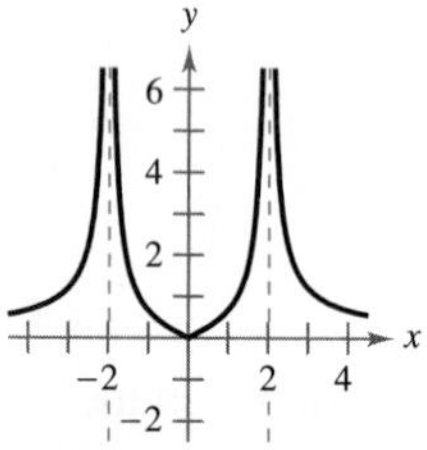

2. $f(x) = \dfrac{1}{x + 2}$

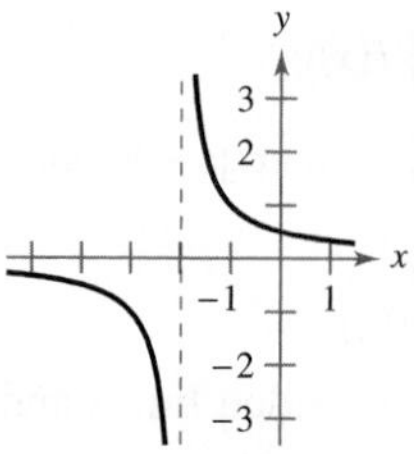

3. $f(x) = \tan \dfrac{\pi x}{4}$

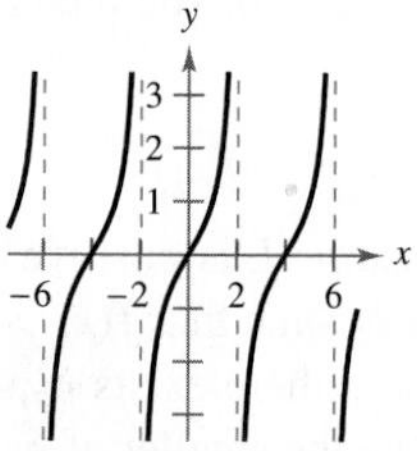

4. $f(x) = \sec \dfrac{\pi x}{4}$

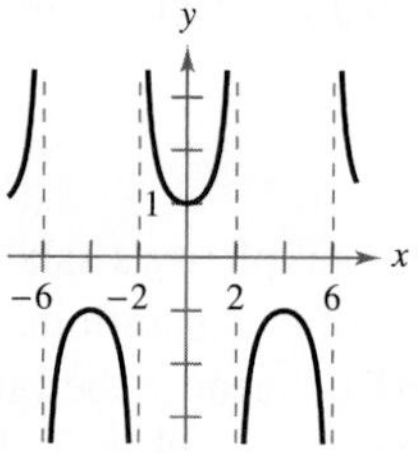

Numerical and Graphical Analysis **In Exercises 5–8, determine whether $f(x)$ approaches ∞ or $-\infty$ as x approaches -3 from the left and from the right by completing the table. Use a graphing utility to graph the function and confirm your answer.**

x	-3.5	-3.1	-3.01	-3.001
$f(x)$				

x	-2.999	-2.99	-2.9	-2.5
$f(x)$				

5. $f(x) = \dfrac{1}{x^2 - 9}$

6. $f(x) = \dfrac{x}{x^2 - 9}$

7. $f(x) = \dfrac{x^2}{x^2 - 9}$

8. $f(x) = \sec \dfrac{\pi x}{6}$

In Exercises 9–28, find the vertical asymptotes (if any) of the graph of the function.

9. $f(x) = \dfrac{1}{x^2}$

10. $f(x) = \dfrac{4}{(x - 2)^3}$

11. $h(x) = \dfrac{x^2 - 2}{x^2 - x - 2}$

12. $g(x) = \dfrac{2 + x}{x^2(1 - x)}$

13. $f(x) = \dfrac{x^2}{x^2 - 4}$

14. $f(x) = \dfrac{-4x}{x^2 + 4}$

15. $g(t) = \dfrac{t - 1}{t^2 + 1}$

16. $h(s) = \dfrac{2s - 3}{s^2 - 25}$

17. $f(x) = \tan 2x$

18. $f(x) = \sec \pi x$

19. $T(t) = 1 - \dfrac{4}{t^2}$

20. $g(x) = \dfrac{\frac{1}{2}x^3 - x^2 - 4x}{3x^2 - 6x - 24}$

21. $f(x) = \dfrac{x}{x^2 + x - 2}$

22. $f(x) = \dfrac{4x^2 + 4x - 24}{x^4 - 2x^3 - 9x^2 + 18x}$

23. $g(x) = \dfrac{x^3 + 1}{x + 1}$

24. $h(x) = \dfrac{x^2 - 4}{x^3 + 2x^2 + x + 2}$

25. $f(x) = \dfrac{x^2 - 2x - 15}{x^3 - 5x^2 + x - 5}$

26. $h(t) = \dfrac{t^2 - 2t}{t^4 - 16}$

27. $s(t) = \dfrac{t}{\sin t}$

28. $g(\theta) = \dfrac{\tan \theta}{\theta}$

In Exercises 29–32, determine whether the graph of the function has a vertical asymptote or a removable discontinuity at $x = -1$. Graph the function using a graphing utility to confirm your answer.

29. $f(x) = \dfrac{x^2 - 1}{x + 1}$

30. $f(x) = \dfrac{x^2 - 6x - 7}{x + 1}$

31. $f(x) = \dfrac{x^2 + 1}{x + 1}$

32. $f(x) = \dfrac{\sin(x + 1)}{x + 1}$

In Exercises 33–48, find the limit.

33. $\displaystyle\lim_{x \to 2^+} \frac{x - 3}{x - 2}$

34. $\displaystyle\lim_{x \to 1^+} \frac{2 + x}{1 - x}$

35. $\displaystyle\lim_{x \to 3^+} \frac{x^2}{x^2 - 9}$

36. $\displaystyle\lim_{x \to 4^-} \frac{x^2}{x^2 + 16}$

37. $\displaystyle\lim_{x \to -3^-} \frac{x^2 + 2x - 3}{x^2 + x - 6}$

38. $\displaystyle\lim_{x \to (-1/2)^+} \frac{6x^2 + x - 1}{4x^2 - 4x - 3}$

39. $\displaystyle\lim_{x \to 1} \frac{x^2 - x}{(x^2 + 1)(x - 1)}$

40. $\displaystyle\lim_{x \to 3} \frac{x - 2}{x^2}$

41. $\displaystyle\lim_{x \to 0^-} \left(1 + \frac{1}{x}\right)$

42. $\displaystyle\lim_{x \to 0^-} \left(x^2 - \frac{1}{x}\right)$

43. $\displaystyle\lim_{x \to 0^+} \frac{2}{\sin x}$

44. $\displaystyle\lim_{x \to (\pi/2)^+} \frac{-2}{\cos x}$

45. $\displaystyle\lim_{x \to \pi} \frac{\sqrt{x}}{\csc x}$

46. $\displaystyle\lim_{x \to 0} \frac{x + 2}{\cot x}$

47. $\displaystyle\lim_{x \to 1/2} x \sec \pi x$

48. $\displaystyle\lim_{x \to 1/2} x^2 \tan \pi x$

In Exercises 49–52, use a graphing utility to graph the function and determine the one-sided limit.

49. $f(x) = \dfrac{x^2 + x + 1}{x^3 - 1}$

$\displaystyle\lim_{x \to 1^+} f(x)$

50. $f(x) = \dfrac{x^3 - 1}{x^2 + x + 1}$

$\displaystyle\lim_{x \to 1^-} f(x)$

51. $f(x) = \dfrac{1}{x^2 - 25}$

$\displaystyle\lim_{x \to 5^-} f(x)$

52. $f(x) = \sec \dfrac{\pi x}{6}$

$\displaystyle\lim_{x \to 3^+} f(x)$

Writing About Concepts

53. In your own words, describe the meaning of an infinite limit. Is ∞ a real number?

54. In your own words, describe what is meant by an asymptote of a graph.

55. Write a rational function with vertical asymptotes at $x = 6$ and $x = -2$, and with a zero at $x = 3$.

56. Does the graph of every rational function have a vertical asymptote? Explain.

57. Use the graph of the function f (see figure) to sketch the graph of $g(x) = 1/f(x)$ on the interval $[-2, 3]$. To print an enlarged copy of the graph, go to the website *www.mathgraphs.com.*

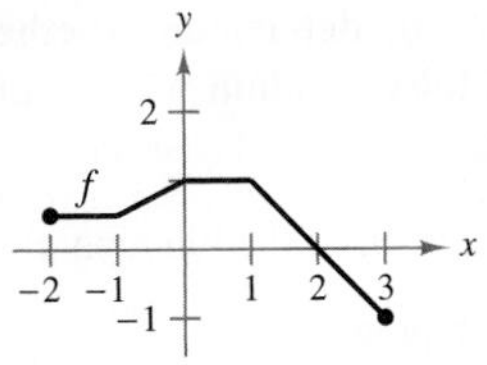

58. ***Boyle's Law*** For a quantity of gas at a constant temperature, the pressure P is inversely proportional to the volume V. Find the limit of P as $V \to 0^+$.

59. ***Rate of Change*** A patrol car is parked 50 feet from a long warehouse (see figure). The revolving light on top of the car turns at a rate of $\frac{1}{2}$ revolution per second. The rate at which the light beam moves along the wall is

$$r = 50\pi \sec^2 \theta \text{ ft/sec}.$$

(a) Find the rate r when θ is $\pi/6$.

(b) Find the rate r when θ is $\pi/3$.

(c) Find the limit of r as $\theta \to (\pi/2)^-$.

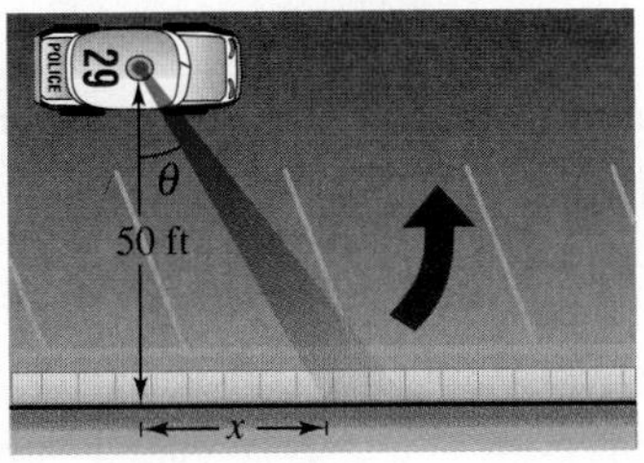

60. ***Illegal Drugs*** The cost in millions of dollars for a governmental agency to seize $x\%$ of an illegal drug is

$$C = \frac{528x}{100 - x}, \quad 0 \le x < 100.$$

(a) Find the cost of seizing 25% of the drug.

(b) Find the cost of seizing 50% of the drug.

(c) Find the cost of seizing 75% of the drug.

(d) Find the limit of C as $x \to 100^-$ and interpret its meaning.

61. ***Relativity*** According to the theory of relativity, the mass m of a particle depends on its velocity v. That is,

$$m = \frac{m_0}{\sqrt{1 - (v^2/c^2)}}$$

where m_0 is the mass when the particle is at rest and c is the speed of light. Find the limit of the mass as v approaches c^-.

62. ***Rate of Change*** A 25-foot ladder is leaning against a house (see figure). If the base of the ladder is pulled away from the house at a rate of 2 feet per second, the top will move down the wall at a rate of

$$r = \frac{2x}{\sqrt{625 - x^2}} \text{ ft/sec}$$

where x is the distance between the base of the ladder and the house.

(a) Find the rate r when x is 7 feet.

(b) Find the rate r when x is 15 feet.

(c) Find the limit of r as $x \to 25^-$.

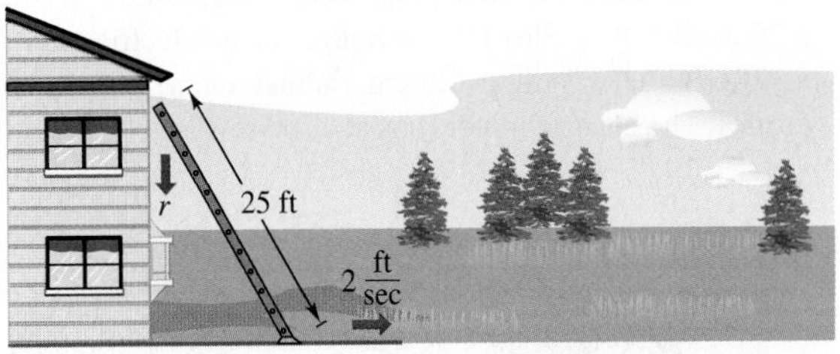

63. ***Average Speed*** On a trip of d miles to another city, a truck driver's average speed was x miles per hour. On the return trip the average speed was y miles per hour. The average speed for the round trip was 50 miles per hour.

(a) Verify that $y = \dfrac{25x}{x - 25}$. What is the domain?

(b) Complete the table.

x	30	40	50	60
y				

Are the values of y different than you expected? Explain.

(c) Find the limit of y as $x \to 25^+$ and interpret its meaning.

64. ***Numerical and Graphical Analysis*** Use a graphing utility to complete the table for each function and graph each function to estimate the limit. What is the value of the limit when the power on x in the denominator is greater than 3?

x	1	0.5	0.2	0.1	0.01	0.001	0.0001
$f(x)$							

(a) $\displaystyle\lim_{x \to 0^+} \frac{x - \sin x}{x}$ (b) $\displaystyle\lim_{x \to 0^+} \frac{x - \sin x}{x^2}$

(c) $\displaystyle\lim_{x \to 0^+} \frac{x - \sin x}{x^3}$ (d) $\displaystyle\lim_{x \to 0^+} \frac{x - \sin x}{x^4}$

65. ***Numerical and Graphical Analysis*** Consider the shaded region outside the sector of a circle of radius 10 meters and inside a right triangle (see figure).

(a) Write the area $A = f(\theta)$ of the region as a function of θ. Determine the domain of the function.

(b) Use a graphing utility to complete the table and graph the function over the appropriate domain.

θ	0.3	0.6	0.9	1.2	1.5
$f(\theta)$					

(c) Find the limit of A as $\theta \to (\pi/2)^-$.

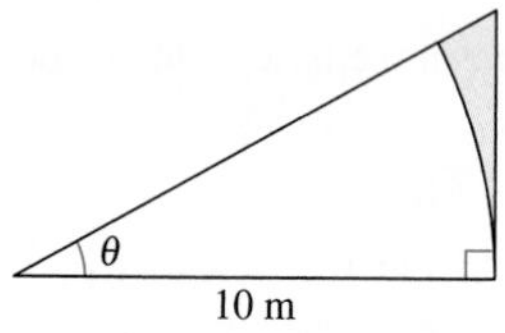

66. ***Numerical and Graphical Reasoning*** A crossed belt connects a 20-centimeter pulley (10-cm radius) on an electric motor with a 40-centimeter pulley (20-cm radius) on a saw arbor (see figure). The electric motor runs at 1700 revolutions per minute.

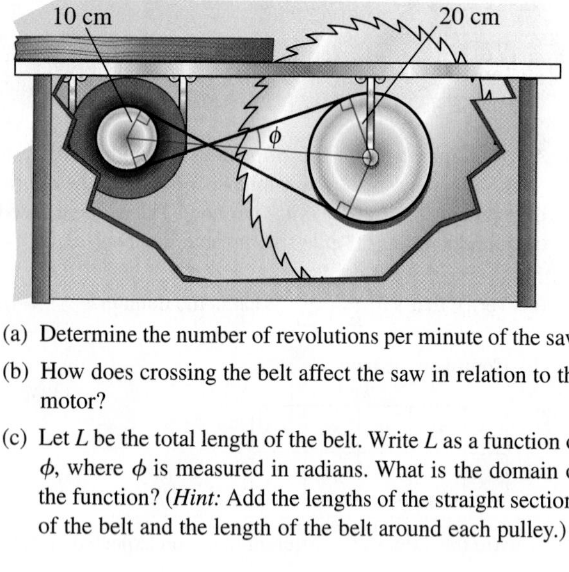

(a) Determine the number of revolutions per minute of the saw.

(b) How does crossing the belt affect the saw in relation to the motor?

(c) Let L be the total length of the belt. Write L as a function of ϕ, where ϕ is measured in radians. What is the domain of the function? (*Hint:* Add the lengths of the straight sections of the belt and the length of the belt around each pulley.)

(d) Use a graphing utility to complete the table.

ϕ	0.3	0.6	0.9	1.2	1.5
L					

(e) Use a graphing utility to graph the function over the appropriate domain.

(f) Find $\displaystyle\lim_{\phi \to (\pi/2)^-} L$. Use a geometric argument as the basis of a second method of finding this limit.

(g) Find $\displaystyle\lim_{\phi \to 0^+} L$.

True or False? **In Exercises 67–70, determine whether the statement is true or false. If it is false, explain why or give an example that shows it is false.**

67. If $p(x)$ is a polynomial, then the graph of the function given by $f(x) = \dfrac{p(x)}{x-1}$ has a vertical asymptote at $x = 1$.

68. The graph of a rational function has at least one vertical asymptote.

69. The graphs of polynomial functions have no vertical asymptotes.

70. If f has a vertical asymptote at $x = 0$, then f is undefined at $x = 0$.

71. Find functions f and g such that $\displaystyle\lim_{x \to c} f(x) = \infty$ and $\displaystyle\lim_{x \to c} g(x) = \infty$ but $\displaystyle\lim_{x \to c} [f(x) - g(x)] \neq 0$.

72. Prove the remaining properties of Theorem 1.15.

73. Prove that if $\displaystyle\lim_{x \to c} f(x) = \infty$, then $\displaystyle\lim_{x \to c} \frac{1}{f(x)} = 0$.

74. Prove that if $\displaystyle\lim_{x \to c} \frac{1}{f(x)} = 0$, then $\displaystyle\lim_{x \to c} f(x)$ does not exist.

Infinite Limits **In Exercises 75 and 76, use the ε-δ definition of infinite limits to prove the statement.**

75. $\displaystyle\lim_{x \to 3^+} \frac{1}{x-3} = \infty$

76. $\displaystyle\lim_{x \to 4^-} \frac{1}{x-4} = -\infty$

Section Project: Graphs and Limits of Trigonometric Functions

Recall from Theorem 1.9 that the limit of $f(x) = (\sin x)/x$ as x approaches 0 is 1.

(a) Use a graphing utility to graph the function f on the interval $-\pi \leq 0 \leq \pi$. Explain how the graph helps confirm this theorem.

(b) Explain how you could use a table of values to confirm the value of this limit numerically.

(c) Graph $g(x) = \sin x$ by hand. Sketch a tangent line at the point $(0, 0)$ and visually estimate the slope of this tangent line.

(d) Let $(x, \sin x)$ be a point on the graph of g near $(0, 0)$, and write a formula for the slope of the secant line joining $(x, \sin x)$ and $(0, 0)$. Evaluate this formula for $x = 0.1$ and $x = 0.01$. Then find the exact slope of the tangent line to g at the point $(0, 0)$.

(e) Sketch the graph of the cosine function $h(x) = \cos x$. What is the slope of the tangent line at the point $(0, 1)$? Use limits to find this slope analytically.

(f) Find the slope of the tangent line to $k(x) = \tan x$ at $(0, 0)$.

Review Exercises for Chapter 1

See www.CalcChat.com for worked-out solutions to odd-numbered exercises.

In Exercises 1 and 2, determine whether the problem can be solved using precalculus or if calculus is required. If the problem can be solved using precalculus, solve it. If the problem seems to require calculus, explain your reasoning. Use a graphical or numerical approach to estimate the solution.

1. Find the distance between the points $(1, 1)$ and $(3, 9)$ along the curve $y = x^2$.

2. Find the distance between the points $(1, 1)$ and $(3, 9)$ along the line $y = 4x - 3$.

In Exercises 3 and 4, complete the table and use the result to estimate the limit. Use a graphing utility to graph the function to confirm your result.

x	-0.1	-0.01	-0.001	0.001	0.01	0.1
$f(x)$						

3. $\lim_{x\to 0} \dfrac{[4/(x+2)] - 2}{x}$

4. $\lim_{x\to 0} \dfrac{4(\sqrt{x+2} - \sqrt{2})}{x}$

In Exercises 5 and 6, use the graph to determine each limit.

5. $h(x) = \dfrac{x^2 - 2x}{x}$

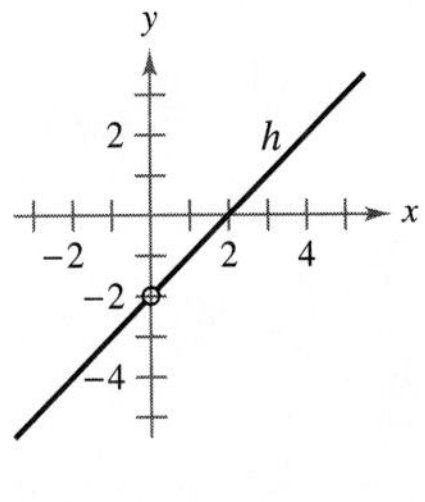

(a) $\lim_{x\to 0} h(x)$ (b) $\lim_{x\to -1} h(x)$

6. $g(x) = \dfrac{3x}{x - 2}$

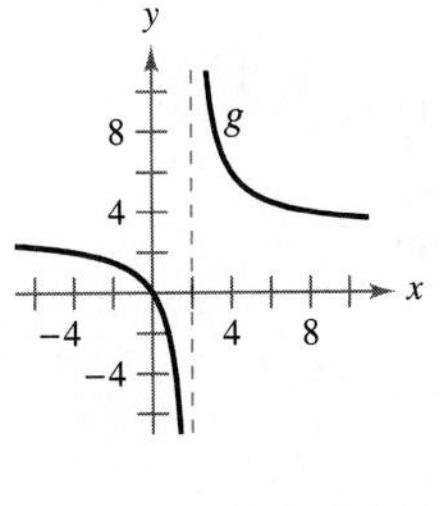

(a) $\lim_{x\to 2} g(x)$ (b) $\lim_{x\to 0} g(x)$

In Exercises 7–10, find the limit L. Then use the ε-δ definition to prove that the limit is L.

7. $\lim_{x\to 1} (3 - x)$

8. $\lim_{x\to 9} \sqrt{x}$

9. $\lim_{x\to 2} (x^2 - 3)$

10. $\lim_{x\to 5} 9$

In Exercises 11–24, find the limit (if it exists).

11. $\lim_{t\to 4} \sqrt{t+2}$

12. $\lim_{y\to 4} 3|y - 1|$

13. $\lim_{t\to -2} \dfrac{t+2}{t^2 - 4}$

14. $\lim_{t\to 3} \dfrac{t^2 - 9}{t - 3}$

15. $\lim_{x\to 4} \dfrac{\sqrt{x} - 2}{x - 4}$

16. $\lim_{x\to 0} \dfrac{\sqrt{4+x} - 2}{x}$

17. $\lim_{x\to 0} \dfrac{[1/(x+1)] - 1}{x}$

18. $\lim_{s\to 0} \dfrac{(1/\sqrt{1+s}) - 1}{s}$

19. $\lim_{x\to -5} \dfrac{x^3 + 125}{x + 5}$

20. $\lim_{x\to -2} \dfrac{x^2 - 4}{x^3 + 8}$

21. $\lim_{x\to 0} \dfrac{1 - \cos x}{\sin x}$

22. $\lim_{x\to \pi/4} \dfrac{4x}{\tan x}$

23. $\lim_{\Delta x\to 0} \dfrac{\sin[(\pi/6) + \Delta x] - (1/2)}{\Delta x}$

[*Hint:* $\sin(\theta + \phi) = \sin\theta\cos\phi + \cos\theta\sin\phi$]

24. $\lim_{\Delta x\to 0} \dfrac{\cos(\pi + \Delta x) + 1}{\Delta x}$

[*Hint:* $\cos(\theta + \phi) = \cos\theta\cos\phi - \sin\theta\sin\phi$]

In Exercises 25 and 26, evaluate the limit given $\lim_{x\to c} f(x) = -\frac{3}{4}$ and $\lim_{x\to c} g(x) = \frac{2}{3}$.

25. $\lim_{x\to c} [f(x)g(x)]$

26. $\lim_{x\to c} [f(x) + 2g(x)]$

Numerical, Graphical, and Analytic Analysis **In Exercises 27 and 28, consider**

$$\lim_{x\to 1^+} f(x).$$

(a) Complete the table to estimate the limit.

(b) Use a graphing utility to graph the function and use the graph to estimate the limit.

(c) Rationalize the numerator to find the exact value of the limit analytically.

x	1.1	1.01	1.001	1.0001
$f(x)$				

27. $f(x) = \dfrac{\sqrt{2x+1} - \sqrt{3}}{x - 1}$

28. $f(x) = \dfrac{1 - \sqrt[3]{x}}{x - 1}$

[*Hint:* $a^3 - b^3 = (a - b)(a^2 + ab + b^2)$]

Free-Falling Object **In Exercises 29 and 30, use the position function $s(t) = -4.9t^2 + 200$, which gives the height (in meters) of an object that has fallen from a height of 200 meters. The velocity at time $t = a$ seconds is given by**

$$\lim_{t\to a} \frac{s(a) - s(t)}{a - t}.$$

29. Find the velocity of the object when $t = 4$.

30. At what velocity will the object impact the ground?

In Exercises 31–36, find the limit (if it exists). If the limit does not exist, explain why.

31. $\lim_{x \to 3^-} \frac{|x - 3|}{x - 3}$

32. $\lim_{x \to 4} [\![x - 1]\!]$

33. $\lim_{x \to 2} f(x)$, where $f(x) = \begin{cases} (x - 2)^2, & x \le 2 \\ 2 - x, & x > 2 \end{cases}$

34. $\lim_{x \to 1^+} g(x)$, where $g(x) = \begin{cases} \sqrt{1 - x}, & x \le 1 \\ x + 1, & x > 1 \end{cases}$

35. $\lim_{t \to 1} h(t)$, where $h(t) = \begin{cases} t^3 + 1, & t < 1 \\ \frac{1}{2}(t + 1), & t \ge 1 \end{cases}$

36. $\lim_{s \to -2} f(s)$, where $f(s) = \begin{cases} -s^2 - 4s - 2, & s \le -2 \\ s^2 + 4s + 6, & s > -2 \end{cases}$

In Exercises 37–46, determine the intervals on which the function is continuous.

37. $f(x) = [\![x + 3]\!]$

38. $f(x) = \frac{3x^2 - x - 2}{x - 1}$

39. $f(x) = \begin{cases} \frac{3x^2 - x - 2}{x - 1}, & x \ne 1 \\ 0, & x = 1 \end{cases}$

40. $f(x) = \begin{cases} 5 - x, & x \le 2 \\ 2x - 3, & x > 2 \end{cases}$

41. $f(x) = \frac{1}{(x - 2)^2}$

42. $f(x) = \sqrt{\frac{x + 1}{x}}$

43. $f(x) = \frac{3}{x + 1}$

44. $f(x) = \frac{x + 1}{2x + 2}$

45. $f(x) = \csc \frac{\pi x}{2}$

46. $f(x) = \tan 2x$

47. Determine the value of c such that the function is continuous on the entire real line.

$$f(x) = \begin{cases} x + 3, & x \le 2 \\ cx + 6, & x > 2 \end{cases}$$

48. Determine the values of b and c such that the function is continuous on the entire real line.

$$f(x) = \begin{cases} x + 1, & 1 < x < 3 \\ x^2 + bx + c, & |x - 2| \ge 1 \end{cases}$$

49. Use the Intermediate Value Theorem to show that $f(x) = 2x^3 - 3$ has a zero in the interval $[1, 2]$.

50. ***Delivery Charges*** The cost of sending an overnight package from New York to Atlanta is \$9.80 for the first pound and \$2.50 for each additional pound or fraction thereof. Use the greatest integer function to create a model for the cost C of overnight delivery of a package weighing x pounds. Use a graphing utility to graph the function and discuss its continuity.

51. Let $f(x) = \frac{x^2 - 4}{|x - 2|}$. Find each limit (if possible).

(a) $\lim_{x \to 2^-} f(x)$

(b) $\lim_{x \to 2^+} f(x)$

(c) $\lim_{x \to 2} f(x)$

52. Let $f(x) = \sqrt{x(x - 1)}$.

(a) Find the domain of f.

(b) Find $\lim_{x \to 0^-} f(x)$.

(c) Find $\lim_{x \to 1^+} f(x)$.

In Exercises 53–56, find the vertical asymptotes (if any) of the graphs of the function.

53. $g(x) = 1 + \frac{2}{x}$

54. $h(x) = \frac{4x}{4 - x^2}$

55. $f(x) = \frac{8}{(x - 10)^2}$

56. $f(x) = \csc \pi x$

In Exercises 57–68, find the one-sided limit.

57. $\lim_{x \to -2^-} \frac{2x^2 + x + 1}{x + 2}$

58. $\lim_{x \to (1/2)^+} \frac{x}{2x - 1}$

59. $\lim_{x \to -1^+} \frac{x + 1}{x^3 + 1}$

60. $\lim_{x \to -1^-} \frac{x + 1}{x^4 - 1}$

61. $\lim_{x \to 1^-} \frac{x^2 + 2x + 1}{x - 1}$

62. $\lim_{x \to -1^+} \frac{x^2 - 2x + 1}{x + 1}$

63. $\lim_{x \to 0^+} \left(x - \frac{1}{x^3}\right)$

64. $\lim_{x \to 2^-} \frac{1}{\sqrt[3]{x^2 - 4}}$

65. $\lim_{x \to 0^+} \frac{\sin 4x}{5x}$

66. $\lim_{x \to 0^+} \frac{\sec x}{x}$

67. $\lim_{x \to 0^+} \frac{\csc 2x}{x}$

68. $\lim_{x \to 0^-} \frac{\cos^2 x}{x}$

69. ***Environment*** A utility company burns coal to generate electricity. The cost C in dollars of removing $p\%$ of the air pollutants in the stack emissions is

$$C = \frac{80{,}000p}{100 - p}, \quad 0 \le p < 100.$$

Find the cost of removing (a) 15%, (b) 50%, and (c) 90% of the pollutants. (d) Find the limit of C as $p \to 100^-$.

70. The function f is defined as shown.

$$f(x) = \frac{\tan 2x}{x}, \quad x \ne 0$$

(a) Find $\lim_{x \to 0} \frac{\tan 2x}{x}$ (if it exists).

(b) Can the function f be defined at $x = 0$ such that it is continuous at $x = 0$?

P.S. Problem Solving

See www.CalcChat.com for worked-out solutions to odd-numbered exercises.

1. Let $P(x, y)$ be a point on the parabola $y = x^2$ in the first quadrant. Consider the triangle $\triangle PAO$ formed by P, $A(0, 1)$, and the origin $O(0, 0)$, and the triangle $\triangle PBO$ formed by P, $B(1, 0)$, and the origin.

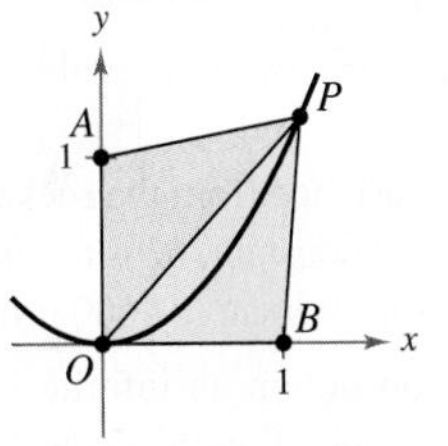

(a) Write the perimeter of each triangle in terms of x.

(b) Let $r(x)$ be the ratio of the perimeters of the two triangles,

$$r(x) = \frac{\text{Perimeter } \triangle PAO}{\text{Perimeter } \triangle PBO}.$$

Complete the table.

x	4	2	1	0.1	0.01
Perimeter $\triangle PAO$					
Perimeter $\triangle PBO$					
$r(x)$					

(c) Calculate $\lim_{x \to 0^+} r(x)$.

2. Let $P(x, y)$ be a point on the parabola $y = x^2$ in the first quadrant. Consider the triangle $\triangle PAO$ formed by P, $A(0, 1)$, and the origin $O(0, 0)$, and the triangle $\triangle PBO$ formed by P, $B(1, 0)$, and the origin.

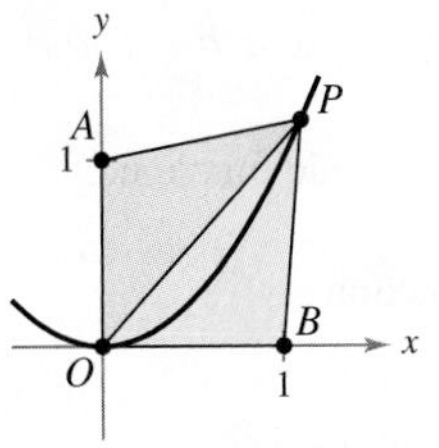

(a) Write the area of each triangle in terms of x.

(b) Let $a(x)$ be the ratio of the areas of the two triangles,

$$a(x) = \frac{\text{Area } \triangle PBO}{\text{Area } \triangle PAO}.$$

Complete the table.

x	4	2	1	0.1	0.01
Area $\triangle PAO$					
Area $\triangle PBO$					
$a(x)$					

(c) Calculate $\lim_{x \to 0^+} a(x)$.

3. (a) Find the area of a regular hexagon inscribed in a circle of radius 1. How close is this area to that of the circle?

(b) Find the area A_n of an n-sided regular polygon inscribed in a circle of radius 1. Write your answer as a function of n.

(c) Complete the table.

n	6	12	24	48	96
A_n					

(d) What number does A_n approach as n gets larger and larger?

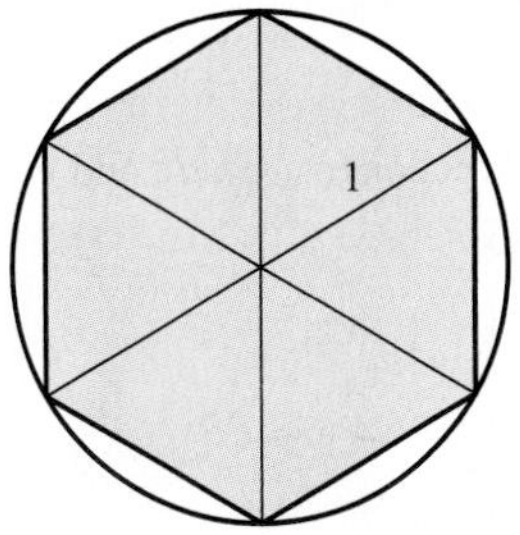

Figure for 3

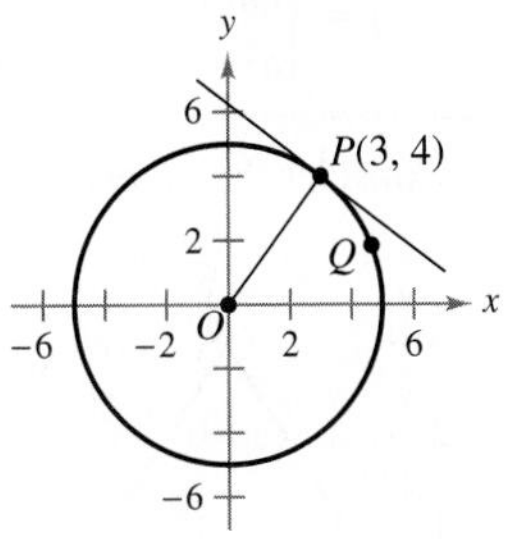

Figure for 4

4. Let $P(3, 4)$ be a point on the circle $x^2 + y^2 = 25$.

(a) What is the slope of the line joining P and $O(0, 0)$?

(b) Find an equation of the tangent line to the circle at P.

(c) Let $Q(x, y)$ be another point on the circle in the first quadrant. Find the slope m_x of the line joining P and Q in terms of x.

(d) Calculate $\lim_{x \to 3} m_x$. How does this number relate to your answer in part (b)?

5. Let $P(5, -12)$ be a point on the circle $x^2 + y^2 = 169$.

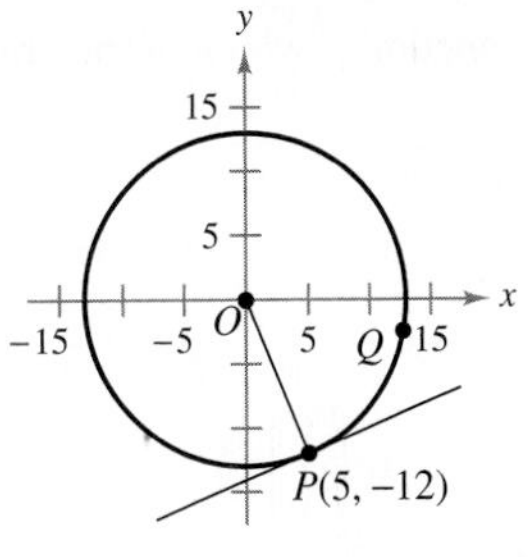

(a) What is the slope of the line joining P and $O(0, 0)$?

(b) Find an equation of the tangent line to the circle at P.

(c) Let $Q(x, y)$ be another point on the circle in the fourth quadrant. Find the slope m_x of the line joining P and Q in terms of x.

(d) Calculate $\lim_{x \to 5} m_x$. How does this number relate to your answer in part (b)?

6. Find the values of the constants a and b such that

$$\lim_{x \to 0} \frac{\sqrt{a + bx} - \sqrt{3}}{x} = \sqrt{3}.$$

7. Consider the function $f(x) = \dfrac{\sqrt{3 + x^{1/3}} - 2}{x - 1}$.

(a) Find the domain of f.

(b) Use a graphing utility to graph the function.

(c) Calculate $\lim\limits_{x \to -27^+} f(x)$.

(d) Calculate $\lim\limits_{x \to 1} f(x)$.

8. Determine all values of the constant a such that the following function is continuous for all real numbers.

$$f(x) = \begin{cases} \dfrac{ax}{\tan x}, & x \ge 0 \\ a^2 - 2, & x < 0 \end{cases}$$

9. Consider the graphs of the four functions g_1, g_2, g_3, and g_4.

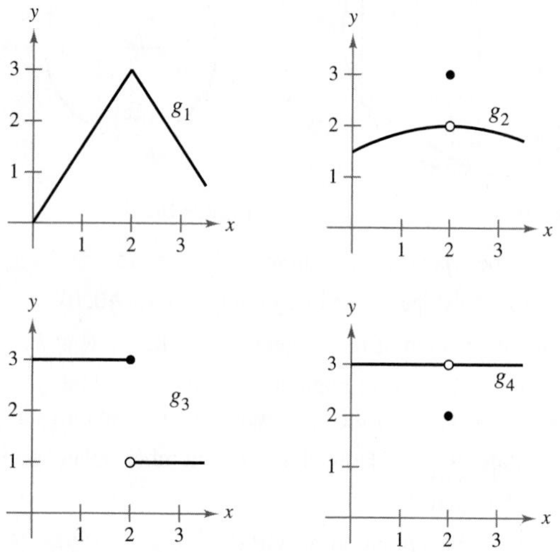

For each given condition of the function f, which of the graphs could be the graph of f?

(a) $\lim\limits_{x \to 2} f(x) = 3$

(b) f is continuous at 2.

(c) $\lim\limits_{x \to 2^-} f(x) = 3$

10. Sketch the graph of the function $f(x) = \left[\!\left[\dfrac{1}{x} \right]\!\right]$.

(a) Evaluate $f\left(\frac{1}{4}\right)$, $f(3)$, and $f(1)$.

(b) Evaluate the limits $\lim\limits_{x \to 1^-} f(x)$, $\lim\limits_{x \to 1^+} f(x)$, $\lim\limits_{x \to 0^-} f(x)$, and $\lim\limits_{x \to 0^+} f(x)$.

(c) Discuss the continuity of the function.

11. Sketch the graph of the function $f(x) = [\![x]\!] + [\![-x]\!]$.

(a) Evaluate $f(1)$, $f(0)$, $f\left(\frac{1}{2}\right)$, and $f(-2.7)$.

(b) Evaluate the limits $\lim\limits_{x \to 1^-} f(x)$, $\lim\limits_{x \to 1^+} f(x)$, and $\lim\limits_{x \to \frac{1}{2}} f(x)$.

(c) Discuss the continuity of the function.

12. To escape Earth's gravitational field, a rocket must be launched with an initial velocity called the **escape velocity.** A rocket launched from the surface of Earth has velocity v (in miles per second) given by

$$v = \sqrt{\frac{2GM}{r} + v_0^2 - \frac{2GM}{R}} \approx \sqrt{\frac{192{,}000}{r} + v_0^2 - 48}$$

where v_0 is the initial velocity, r is the distance from the rocket to the center of Earth, G is the gravitational constant, M is the mass of Earth, and R is the radius of Earth (approximately 4000 miles).

(a) Find the value of v_0 for which you obtain an infinite limit for r as v tends to zero. This value of v_0 is the escape velocity for Earth.

(b) A rocket launched from the surface of the moon has velocity v (in miles per second) given by

$$v = \sqrt{\frac{1920}{r} + v_0^2 - 2.17}.$$

Find the escape velocity for the moon.

(c) A rocket launched from the surface of a planet has velocity v (in miles per second) given by

$$v = \sqrt{\frac{10{,}600}{r} + v_0^2 - 6.99}.$$

Find the escape velocity for this planet. Is the mass of this planet larger or smaller than that of Earth? (Assume that the mean density of this planet is the same as that of Earth.)

13. For positive numbers $a < b$, the **pulse function** is defined as

$$P_{a,b}(x) = H(x - a) - H(x - b) = \begin{cases} 0, & x < a \\ 1, & a \le x < b \\ 0, & x \ge b \end{cases}$$

where $H(x) = \begin{cases} 1, & x \ge 0 \\ 0, & x < 0 \end{cases}$ is the Heaviside function.

(a) Sketch the graph of the pulse function.

(b) Find the following limits:

(i) $\lim\limits_{x \to a^+} P_{a,b}(x)$ (ii) $\lim\limits_{x \to a^-} P_{a,b}(x)$

(iii) $\lim\limits_{x \to b^+} P_{a,b}(x)$ (iv) $\lim\limits_{x \to b^-} P_{a,b}(x)$

(c) Discuss the continuity of the pulse function.

(d) Why is

$$U(x) = \frac{1}{b - a} P_{a,b}(x)$$

called the **unit** pulse function?

14. Let a be a nonzero constant. Prove that if $\lim\limits_{x \to 0} f(x) = L$, then $\lim\limits_{x \to 0} f(ax) = L$. Show by means of an example that a must be nonzero.

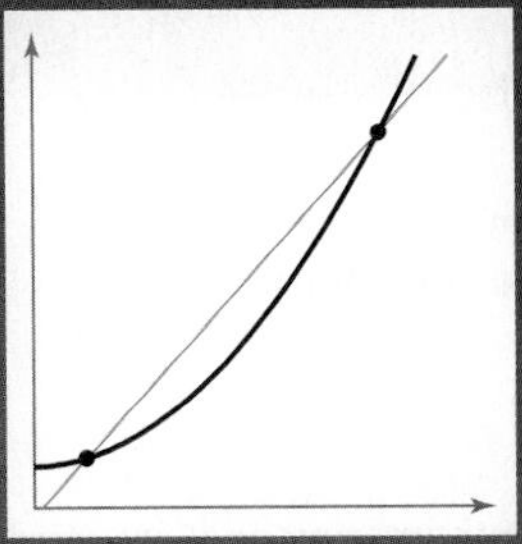
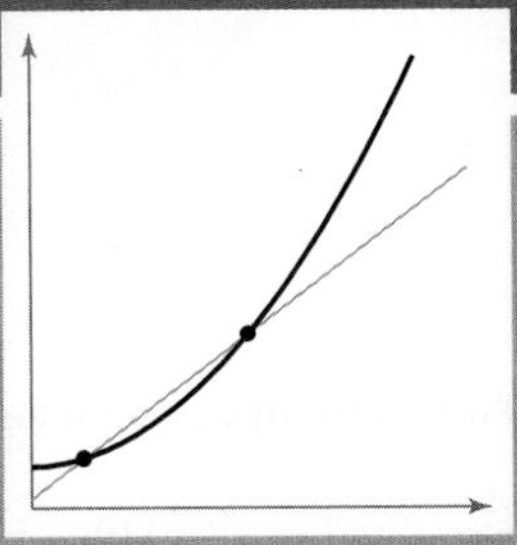
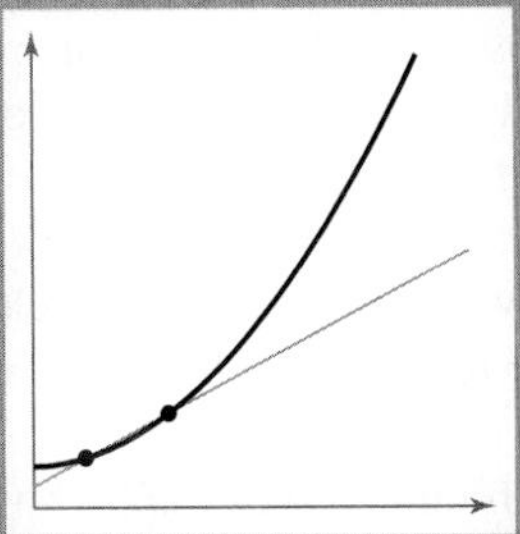
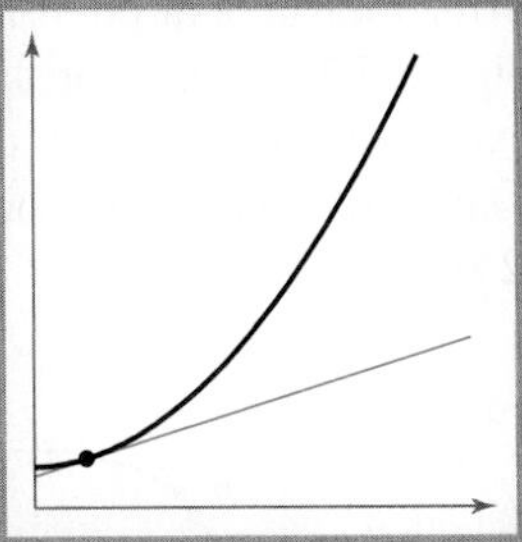

2 Differentiation

You pump air at a steady rate into a deflated balloon until the balloon bursts. Does the diameter of the balloon change faster when you first start pumping the air, or just before the balloon bursts? Why?

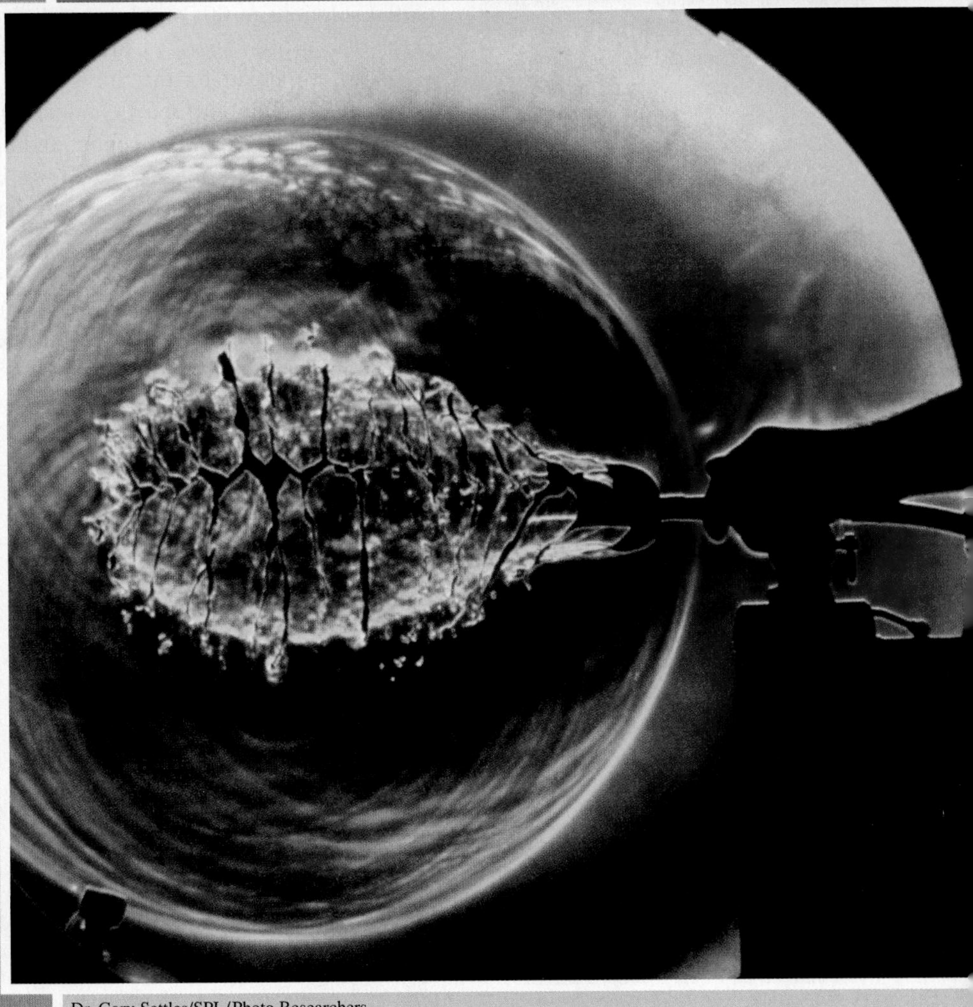

Dr. Gary Settles/SPL/Photo Researchers

To approximate the slope of a tangent line to a graph at a given point, find the slope of the secant line through the given point and a second point on the graph. As the second point approaches the given point, the approximation tends to become more accurate. In Section 2.1, you will use limits to find slopes of tangent lines to graphs. This process is called *differentiation*.

Section 2.1 The Derivative and the Tangent Line Problem

- **Find the slope of the tangent line to a curve at a point.**
- **Use the limit definition to find the derivative of a function.**
- **Understand the relationship between differentiability and continuity.**

ISAAC NEWTON (1642–1727)

In addition to his work in calculus, Newton made revolutionary contributions to physics, including the Law of Universal Gravitation and his three laws of motion.

The Tangent Line Problem

Calculus grew out of four major problems that European mathematicians were working on during the seventeenth century.

1. The tangent line problem (Section 1.1 and this section)
2. The velocity and acceleration problem (Sections 2.2 and 2.3)
3. The minimum and maximum problem (Section 3.1)
4. The area problem (Sections 1.1 and 4.2)

Each problem involves the notion of a limit, and calculus can be introduced with any of the four problems.

A brief introduction to the tangent line problem is given in Section 1.1. Although partial solutions to this problem were given by Pierre de Fermat (1601–1665), René Descartes (1596–1650), Christian Huygens (1629–1695), and Isaac Barrow (1630–1677), credit for the first general solution is usually given to Isaac Newton (1642–1727) and Gottfried Leibniz (1646–1716). Newton's work on this problem stemmed from his interest in optics and light refraction.

What does it mean to say that a line is tangent to a curve at a point? For a circle, the tangent line at a point P is the line that is perpendicular to the radial line at point P, as shown in Figure 2.1.

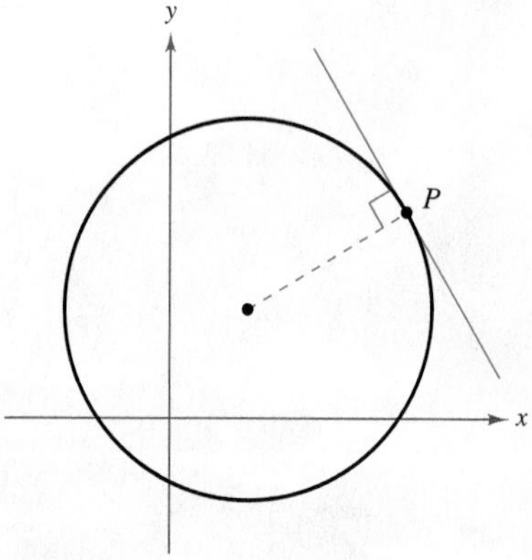

Tangent line to a circle
Figure 2.1

For a general curve, however, the problem is more difficult. For example, how would you define the tangent lines shown in Figure 2.2? You might say that a line is tangent to a curve at a point P if it touches, but does not cross, the curve at point P. This definition would work for the first curve shown in Figure 2.2, but not for the second. *Or* you might say that a line is tangent to a curve if the line touches or intersects the curve at exactly one point. This definition would work for a circle but not for more general curves, as the third curve in Figure 2.2 shows.

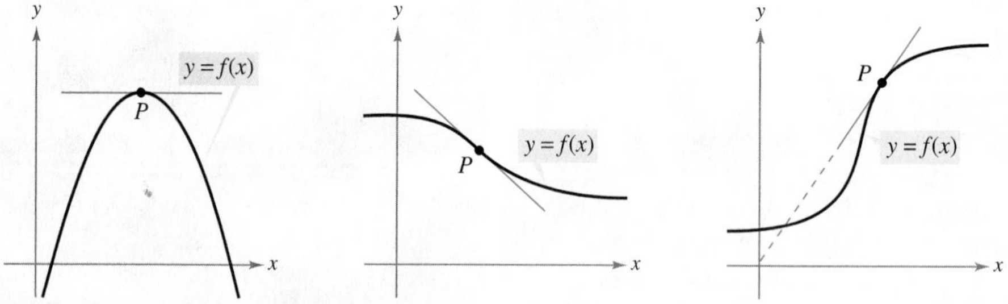

Tangent line to a curve at a point
Figure 2.2

For Further Information For more information on the crediting of mathematical discoveries to the first "discoverer," see the article "Mathematical Firsts—Who Done It?" by Richard H. Williams and Roy D. Mazzagatti in *Mathematics Teacher*. To view this article, go to the website *www.matharticles.com*.

EXPLORATION

Identifying a Tangent Line Use a graphing utility to graph the function $f(x) = 2x^3 - 4x^2 + 3x - 5$. On the same screen, graph $y = x - 5$, $y = 2x - 5$, and $y = 3x - 5$. Which of these lines, if any, appears to be tangent to the graph of f at the point $(0, -5)$? Explain your reasoning.

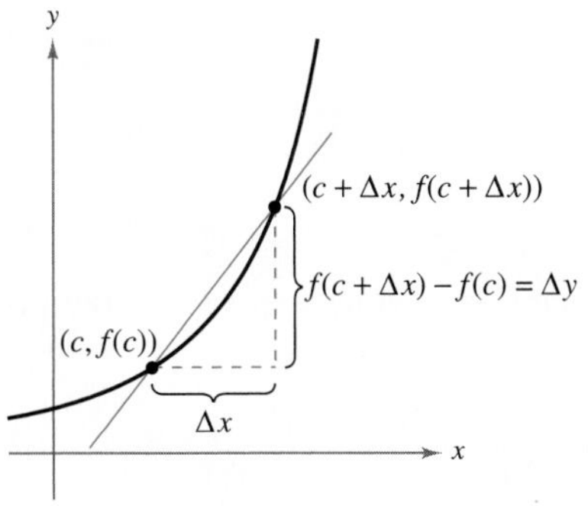

The secant line through $(c, f(c))$ and $(c + \Delta x, f(c + \Delta x))$
Figure 2.3

Essentially, the problem of finding the tangent line at a point P boils down to the problem of finding the *slope* of the tangent line at point P. You can approximate this slope using a **secant line*** through the point of tangency and a second point on the curve, as shown in Figure 2.3. If $(c, f(c))$ is the point of tangency and $(c + \Delta x, f(c + \Delta x))$ is a second point on the graph of f, the slope of the secant line through the two points is given by substitution into the slope formula

$$m = \frac{y_2 - y_1}{x_2 - x_1}$$

$$m_{\text{sec}} = \frac{f(c + \Delta x) - f(c)}{(c + \Delta x) - c} \qquad \frac{\text{Change in } y}{\text{Change in } x}$$

$$m_{\text{sec}} = \frac{f(c + \Delta x) - f(c)}{\Delta x}. \qquad \text{Slope of secant line}$$

The right-hand side of this equation is a **difference quotient.** The denominator Δx is the **change in** $\boldsymbol{x}$**,** and the numerator $\Delta y = f(c + \Delta x) - f(c)$ is the **change in** $\boldsymbol{y}$**.**

The beauty of this procedure is that you can obtain more and more accurate approximations of the slope of the tangent line by choosing points closer and closer to the point of tangency, as shown in Figure 2.4.

THE TANGENT LINE PROBLEM

In 1637, mathematician René Descartes stated this about the tangent line problem:

"And I dare say that this is not only the most useful and general problem in geometry that I know, but even that I ever desire to know."

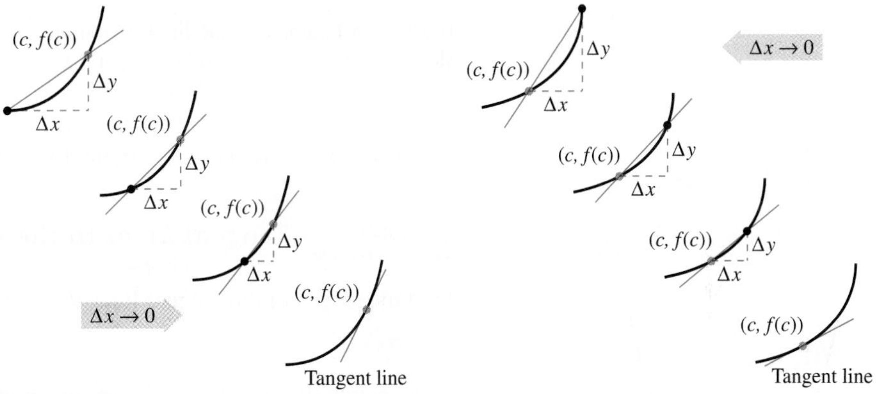

Tangent line approximations
Figure 2.4

Definition of Tangent Line with Slope m

If f is defined on an open interval containing c, and if the limit

$$\lim_{\Delta x \to 0} \frac{\Delta y}{\Delta x} = \lim_{\Delta x \to 0} \frac{f(c + \Delta x) - f(c)}{\Delta x} = m$$

exists, then the line passing through $(c, f(c))$ with slope m is the **tangent line** to the graph of f at the point $(c, f(c))$.

The slope of the tangent line to the graph of f at the point $(c, f(c))$ is also called the **slope of the graph of** $\boldsymbol{f}$ **at** $\boldsymbol{x = c}$**.**

* *This use of the word* secant *comes from the Latin* secare, *meaning to cut, and is not a reference to the trigonometric function of the same name.*

EXAMPLE 1 The Slope of the Graph of a Linear Function

Find the slope of the graph of

$$f(x) = 2x - 3$$

at the point (2, 1).

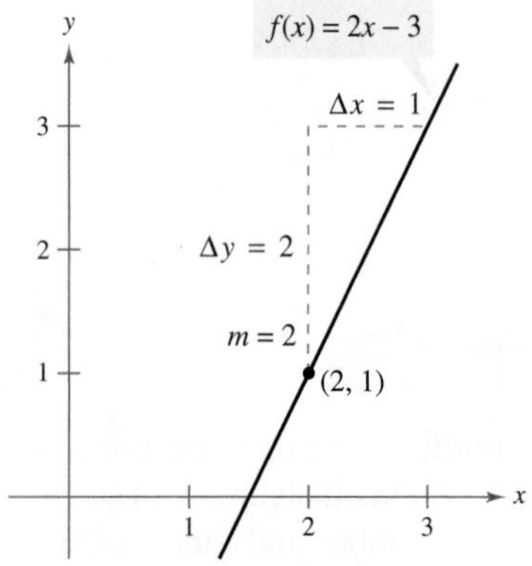

The slope of f at (2, 1) is $m = 2$.
Figure 2.5

Solution To find the slope of the graph of f when $c = 2$, you can apply the definition of the slope of a tangent line, as shown.

$$\begin{aligned}\lim_{\Delta x\to 0}\frac{f(2+\Delta x)-f(2)}{\Delta x} &= \lim_{\Delta x\to 0}\frac{[2(2+\Delta x)-3]-[2(2)-3]}{\Delta x}\\ &= \lim_{\Delta x\to 0}\frac{4+2\Delta x-3-4+3}{\Delta x}\\ &= \lim_{\Delta x\to 0}\frac{2\cancel{\Delta x}}{\cancel{\Delta x}}\\ &= \lim_{\Delta x\to 0} 2\\ &= 2\end{aligned}$$

The slope of f at $(c, f(c)) = (2, 1)$ is $m = 2$, as shown in Figure 2.5.

NOTE In Example 1, the limit definition of the slope of f agrees with the definition of the slope of a line as discussed in Section P.2.

The graph of a linear function has the same slope at any point. This is not true of nonlinear functions, as shown in the following example.

EXAMPLE 2 Tangent Lines to the Graph of a Nonlinear Function

Find the slopes of the tangent lines to the graph of

$$f(x) = x^2 + 1$$

at the points (0, 1) and $(-1, 2)$, as shown in Figure 2.6.

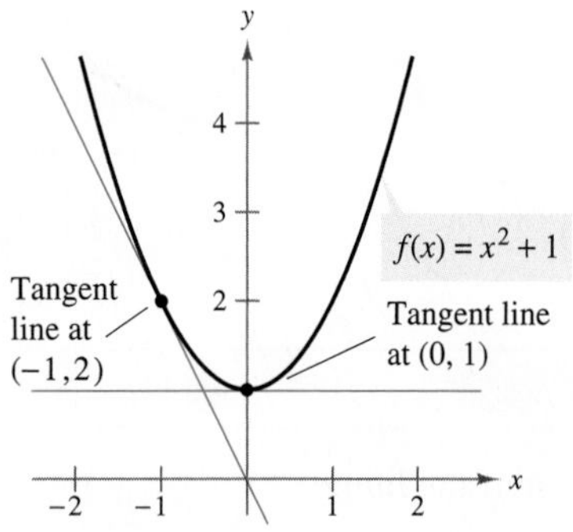

The slope of f at any point $(c, f(c))$ is $m = 2c$.
Figure 2.6

Solution Let $(c, f(c))$ represent an arbitrary point on the graph of f. Then the slope of the tangent line at $(c, f(c))$ is given by

$$\begin{aligned}\lim_{\Delta x\to 0}\frac{f(c+\Delta x)-f(c)}{\Delta x} &= \lim_{\Delta x\to 0}\frac{[(c+\Delta x)^2+1]-(c^2+1)}{\Delta x}\\ &= \lim_{\Delta x\to 0}\frac{c^2+2c(\Delta x)+(\Delta x)^2+1-c^2-1}{\Delta x}\\ &= \lim_{\Delta x\to 0}\frac{2c(\Delta x)+(\Delta x)^2}{\Delta x}\\ &= \lim_{\Delta x\to 0}(2c+\Delta x)\\ &= 2c.\end{aligned}$$

So, the slope at *any* point $(c, f(c))$ on the graph of f is $m = 2c$. At the point (0, 1), the slope is $m = 2(0) = 0$, and at $(-1, 2)$, the slope is $m = 2(-1) = -2$.

NOTE In Example 2, note that c is held constant in the limit process (as $\Delta x \to 0$).

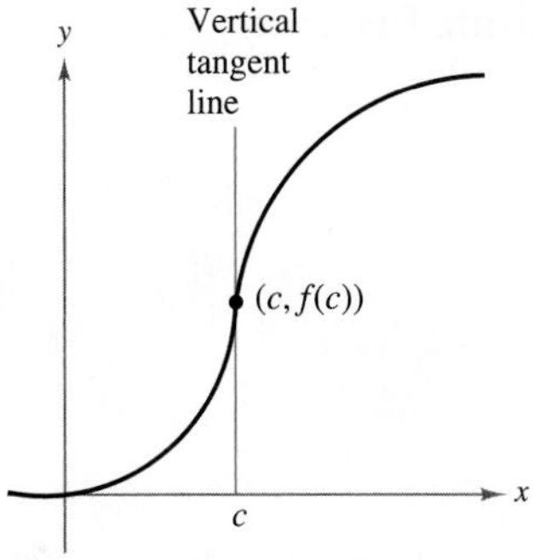

The graph of f has a vertical tangent line at $(c, f(c))$.
Figure 2.7

The definition of a tangent line to a curve does not cover the possibility of a vertical tangent line. For vertical tangent lines, you can use the following definition. If f is continuous at c and

$$\lim_{\Delta x \to 0} \frac{f(c + \Delta x) - f(c)}{\Delta x} = \infty \quad \text{or} \quad \lim_{\Delta x \to 0} \frac{f(c + \Delta x) - f(c)}{\Delta x} = -\infty$$

the vertical line $x = c$ passing through $(c, f(c))$ is a **vertical tangent line** to the graph of f. For example, the function shown in Figure 2.7 has a vertical tangent line at $(c, f(c))$. If the domain of f is the closed interval $[a, b]$, you can extend the definition of a vertical tangent line to include the endpoints by considering continuity and limits from the right (for $x = a$) and from the left (for $x = b$).

The Derivative of a Function

You have now arrived at a crucial point in the study of calculus. The limit used to define the slope of a tangent line is also used to define one of the two fundamental operations of calculus—**differentiation.**

Definition of the Derivative of a Function

The **derivative** of f at x is given by

$$f'(x) = \lim_{\Delta x \to 0} \frac{f(x + \Delta x) - f(x)}{\Delta x}$$

provided the limit exists. For all x for which this limit exists, f' is a function of x.

Be sure you see that the derivative of a function of x is also a function of x. This "new" function gives the slope of the tangent line to the graph of f at the point $(x, f(x))$, provided that the graph has a tangent line at this point.

The process of finding the derivative of a function is called **differentiation.** A function is **differentiable** at x if its derivative exists at x and is **differentiable on an open interval (a, b)** if it is differentiable at every point in the interval.

In addition to $f'(x)$, which is read as "f prime of x," other notations are used to denote the derivative of $y = f(x)$. The most common are

$$f'(x), \quad \frac{dy}{dx}, \quad y', \quad \frac{d}{dx}[f(x)], \quad D_x[y]. \qquad \text{Notation for derivatives}$$

The notation dy/dx is read as "the derivative of y *with respect to* x" or simply "$dy - dx$". Using limit notation, you can write

$$\begin{aligned} \frac{dy}{dx} &= \lim_{\Delta x \to 0} \frac{\Delta y}{\Delta x} \\ &= \lim_{\Delta x \to 0} \frac{f(x + \Delta x) - f(x)}{\Delta x} \\ &= f'(x). \end{aligned}$$

EXAMPLE 3 Finding the Derivative by the Limit Process

Find the derivative of $f(x) = x^3 + 2x$.

Solution

$$\begin{aligned} f'(x) &= \lim_{\Delta x \to 0} \frac{f(x+\Delta x) - f(x)}{\Delta x} && \text{Definition of derivative} \\ &= \lim_{\Delta x \to 0} \frac{(x+\Delta x)^3 + 2(x+\Delta x) - (x^3+2x)}{\Delta x} \\ &= \lim_{\Delta x \to 0} \frac{x^3 + 3x^2\Delta x + 3x(\Delta x)^2 + (\Delta x)^3 + 2x + 2\Delta x - x^3 - 2x}{\Delta x} \\ &= \lim_{\Delta x \to 0} \frac{3x^2\Delta x + 3x(\Delta x)^2 + (\Delta x)^3 + 2\Delta x}{\Delta x} \\ &= \lim_{\Delta x \to 0} \frac{\Delta x[3x^2 + 3x\Delta x + (\Delta x)^2 + 2]}{\Delta x} \\ &= \lim_{\Delta x \to 0} [3x^2 + 3x\Delta x + (\Delta x)^2 + 2] \\ &= 3x^2 + 2 \end{aligned}$$

STUDY TIP When using the definition to find a derivative of a function, the key is to rewrite the difference quotient so that Δx does not occur as a factor of the denominator.

Remember that the derivative of a function f is itself a function, which can be used to find the slope of the tangent line at the point $(x, f(x))$ on the graph of f.

EXAMPLE 4 Using the Derivative to Find the Slope at a Point

Find $f'(x)$ for $f(x) = \sqrt{x}$. Then find the slope of the graph of f at the points $(1, 1)$ and $(4, 2)$. Discuss the behavior of f at $(0, 0)$.

Solution Use the procedure for rationalizing numerators, as discussed in Section 1.3.

$$\begin{aligned} f'(x) &= \lim_{\Delta x \to 0} \frac{f(x+\Delta x) - f(x)}{\Delta x} && \text{Definition of derivative} \\ &= \lim_{\Delta x \to 0} \frac{\sqrt{x+\Delta x} - \sqrt{x}}{\Delta x} \\ &= \lim_{\Delta x \to 0} \left(\frac{\sqrt{x+\Delta x} - \sqrt{x}}{\Delta x}\right)\left(\frac{\sqrt{x+\Delta x} + \sqrt{x}}{\sqrt{x+\Delta x} + \sqrt{x}}\right) \\ &= \lim_{\Delta x \to 0} \frac{(x+\Delta x) - x}{\Delta x\left(\sqrt{x+\Delta x} + \sqrt{x}\right)} \\ &= \lim_{\Delta x \to 0} \frac{\Delta x}{\Delta x\left(\sqrt{x+\Delta x} + \sqrt{x}\right)} \\ &= \lim_{\Delta x \to 0} \frac{1}{\sqrt{x+\Delta x} + \sqrt{x}} \\ &= \frac{1}{2\sqrt{x}}, \quad x > 0 \end{aligned}$$

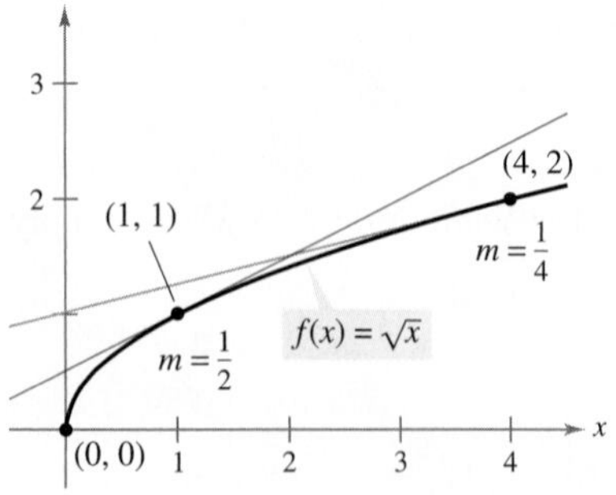

The slope of f at $(x, f(x)), x > 0$, is $m = 1/\left(2\sqrt{x}\right)$.
Figure 2.8

At the point $(1, 1)$, the slope is $f'(1) = \frac{1}{2}$. At the point $(4, 2)$, the slope is $f'(4) = \frac{1}{4}$. See Figure 2.8. At the point $(0, 0)$, the slope is undefined. Moreover, the graph of f has a vertical tangent line at $(0, 0)$.

WWW *indicates that in the* HM mathSpace® *CD-ROM and the online* Eduspace® *system for this text, you will find an Open Exploration, which further explores this example using the computer algebra systems* Maple, Mathcad, Mathematica, *and* Derive.

In many applications, it is convenient to use a variable other than x as the independent variable, as shown in Example 5.

EXAMPLE 5 Finding the Derivative of a Function

Find the derivative with respect to t for the function $y = 2/t$.

Solution Considering $y = f(t)$, you obtain

$$\frac{dy}{dt} = \lim_{\Delta t \to 0} \frac{f(t + \Delta t) - f(t)}{\Delta t} \qquad \text{Definition of derivative}$$

$$= \lim_{\Delta t \to 0} \frac{\dfrac{2}{t + \Delta t} - \dfrac{2}{t}}{\Delta t} \qquad f(t + \Delta t) = 2/(t + \Delta t) \text{ and } f(t) = 2/t$$

$$= \lim_{\Delta t \to 0} \frac{\dfrac{2t - 2(t + \Delta t)}{t(t + \Delta t)}}{\Delta t} \qquad \text{Combine fractions in numerator.}$$

$$= \lim_{\Delta t \to 0} \frac{-2\Delta t}{\Delta t(t)(t + \Delta t)} \qquad \text{Divide out common factor of } \Delta t.$$

$$= \lim_{\Delta t \to 0} \frac{-2}{t(t + \Delta t)} \qquad \text{Simplify.}$$

$$= -\frac{2}{t^2}. \qquad \text{Evaluate limit as } \Delta t \to 0.$$

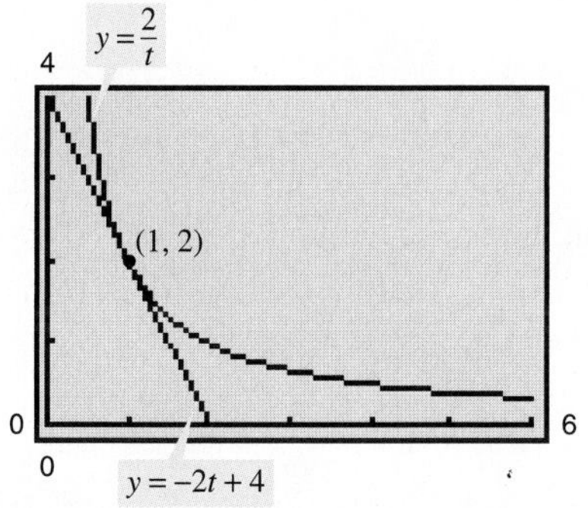

At the point $(1, 2)$ the line $y = -2t + 4$ is tangent to the graph of $y = 2/t$.
Figure 2.9

TECHNOLOGY A graphing utility can be used to reinforce the result given in Example 5. For instance, using the formula $dy/dt = -2/t^2$, you know that the slope of the graph of $y = 2/t$ at the point $(1, 2)$ is $m = -2$. This implies that an equation of the tangent line to the graph at $(1, 2)$ is

$$y - 2 = -2(t - 1) \quad \text{or} \quad y = -2t + 4$$

as shown in Figure 2.9.

Differentiability and Continuity

The following alternative limit form of the derivative is useful in investigating the relationship between differentiability and continuity. The derivative of f at c is

$$f'(c) = \lim_{x \to c} \frac{f(x) - f(c)}{x - c} \qquad \text{Alternative form of derivative}$$

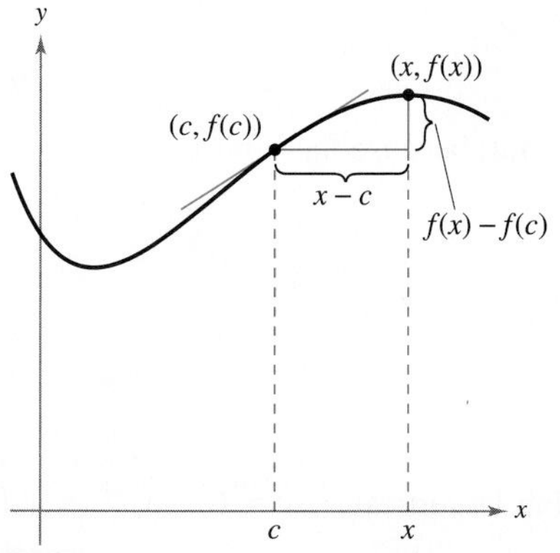

As x approaches c, the secant line approaches the tangent line.
Figure 2.10

provided this limit exists (see Figure 2.10). (A proof of the equivalence of this form is given in Appendix A.) Note that the existence of the limit in this alternative form requires that the one-sided limits

$$\lim_{x \to c^-} \frac{f(x) - f(c)}{x - c} \quad \text{and} \quad \lim_{x \to c^+} \frac{f(x) - f(c)}{x - c}$$

exist and are equal. These one-sided limits are called the **derivatives from the left and from the right,** respectively. It follows that f is **differentiable on the closed interval** $[a, b]$ if it is differentiable on (a, b) and if the derivative from the right at a and the derivative from the left at b both exist.

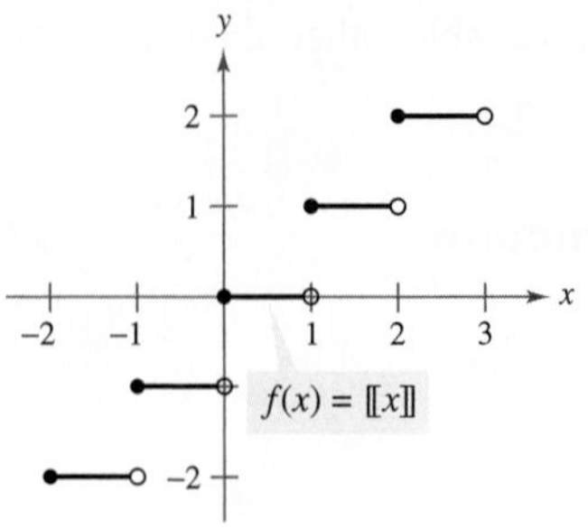

The greatest integer function is not differentiable at $x = 0$, because it is not continuous at $x = 0$.
Figure 2.11

If a function is not continuous at $x = c$, it is also not differentiable at $x = c$. For instance, the greatest integer function

$$f(x) = [\![x]\!]$$

is not continuous at $x = 0$, and so it is not differentiable at $x = 0$ (see Figure 2.11). You can verify this by observing that

$$\lim_{x \to 0^-} \frac{f(x) - f(0)}{x - 0} = \lim_{x \to 0^-} \frac{[\![x]\!] - 0}{x} = \infty \qquad \text{Derivative from the left}$$

and

$$\lim_{x \to 0^+} \frac{f(x) - f(0)}{x - 0} = \lim_{x \to 0^+} \frac{[\![x]\!] - 0}{x} = 0. \qquad \text{Derivative from the right}$$

Although it is true that differentiability implies continuity (as shown in Theorem 2.1 on the next page), the converse is not true. That is, it is possible for a function to be continuous at $x = c$ and *not* differentiable at $x = c$. Examples 6 and 7 illustrate this possibility.

EXAMPLE 6 A Graph with a Sharp Turn

The function

$$f(x) = |x - 2|$$

shown in Figure 2.12 is continuous at $x = 2$. But, the one-sided limits

$$\lim_{x \to 2^-} \frac{f(x) - f(2)}{x - 2} = \lim_{x \to 2^-} \frac{|x - 2| - 0}{x - 2} = -1 \qquad \text{Derivative from the left}$$

and

$$\lim_{x \to 2^+} \frac{f(x) - f(2)}{x - 2} = \lim_{x \to 2^+} \frac{|x - 2| - 0}{x - 2} = 1 \qquad \text{Derivative from the right}$$

are not equal. So, f is not differentiable at $x = 2$ and the graph of f does not have a tangent line at the point $(2, 0)$.

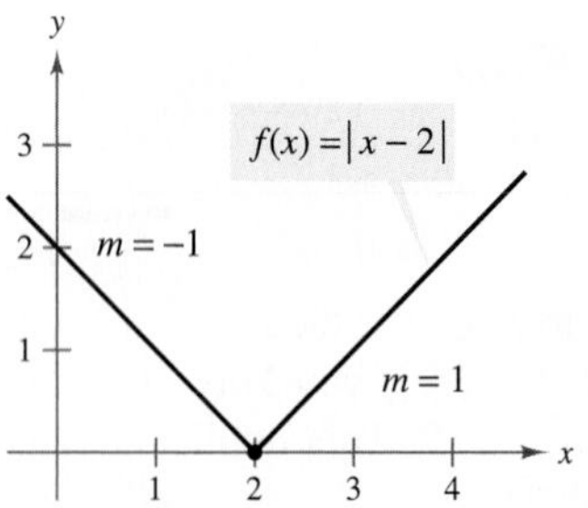

f is not differentiable at $x = 2$, because the derivatives from the left and from the right are not equal.
Figure 2.12

EXAMPLE 7 A Graph with a Vertical Tangent Line

The function

$$f(x) = x^{1/3}$$

is continuous at $x = 0$, as shown in Figure 2.13. But, because the limit

$$\begin{aligned}\lim_{x \to 0} \frac{f(x) - f(0)}{x - 0} &= \lim_{x \to 0} \frac{x^{1/3} - 0}{x} \\ &= \lim_{x \to 0} \frac{1}{x^{2/3}} \\ &= \infty\end{aligned}$$

is infinite, you can conclude that the tangent line is vertical at $x = 0$. So, f is not differentiable at $x = 0$.

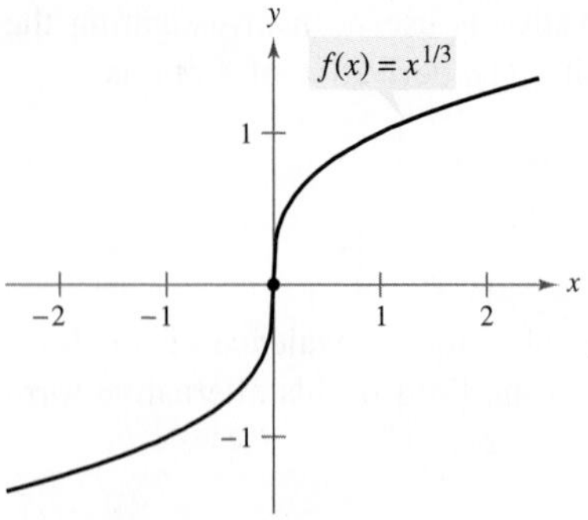

f is not differentiable at $x = 0$, because f has a vertical tangent at $x = 0$.
Figure 2.13

From Examples 6 and 7, you can see that a function is not differentiable at a point at which its graph has a sharp turn *or* a vertical tangent.

TECHNOLOGY Some graphing utilities, such as *Derive*, *Maple*, *Mathcad*, *Mathematica*, and the *TI-89*, perform symbolic differentiation. Others perform *numerical differentiation* by finding values of derivatives using the formula

$$f'(x) \approx \frac{f(x + \Delta x) - f(x - \Delta x)}{2\Delta x}$$

where Δx is a small number such as 0.001. Can you see any problems with this definition? For instance, using this definition, what is the value of the derivative of $f(x) = |x|$ when $x = 0$?

THEOREM 2.1 Differentiability Implies Continuity

If f is differentiable at $x = c$, then f is continuous at $x = c$.

Proof You can prove that f is continuous at $x = c$ by showing that $f(x)$ approaches $f(c)$ as $x \to c$. To do this, use the differentiability of f at $x = c$ and consider the following limit.

$$\begin{aligned}
\lim_{x\to c} [f(x) - f(c)] &= \lim_{x\to c} \left[(x - c)\left(\frac{f(x) - f(c)}{x - c}\right)\right] \\
&= \left[\lim_{x\to c} (x - c)\right]\left[\lim_{x\to c} \frac{f(x) - f(c)}{x - c}\right] \\
&= (0)[f'(c)] \\
&= 0
\end{aligned}$$

Because the difference $f(x) - f(c)$ approaches zero as $x \to c$, you can conclude that $\lim_{x\to c} f(x) = f(c)$. So, f is continuous at $x = c$.

The following statements summarize the relationship between continuity and differentiability.

1. If a function is differentiable at $x = c$, then it is continuous at $x = c$. So, differentiability implies continuity.
2. It is possible for a function to be continuous at $x = c$ and not be differentiable at $x = c$. So, continuity does not imply differentiability.

Exercises for Section 2.1

See www.CalcChat.com for worked-out solutions to odd-numbered exercises.

In Exercises 1 and 2, estimate the slope of the graph at the points (x_1, y_1) and (x_2, y_2).

1. (a)

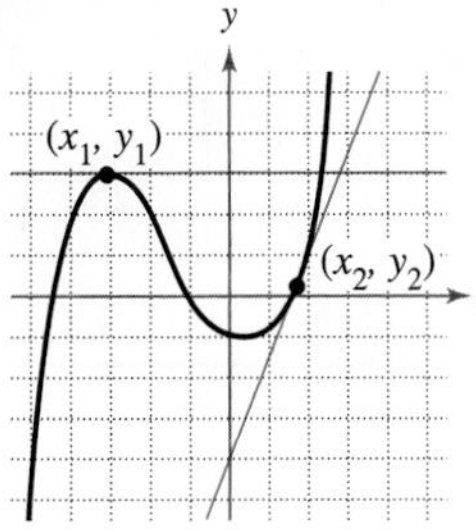

(b)

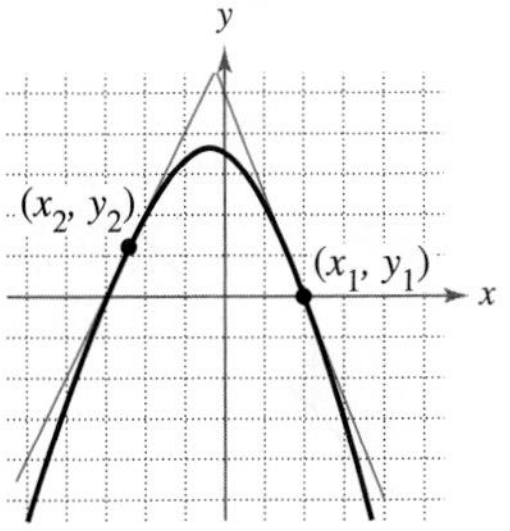

2. (a)

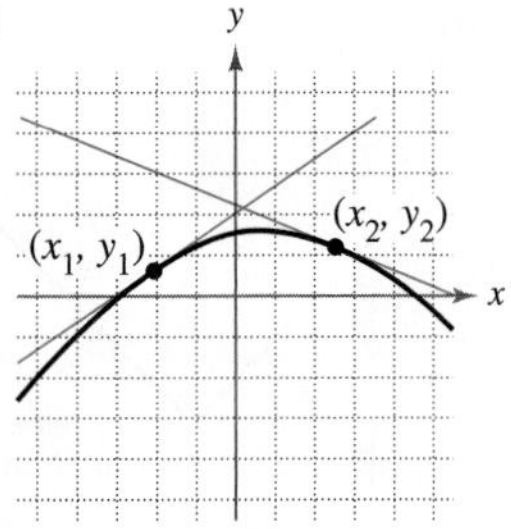

(b)

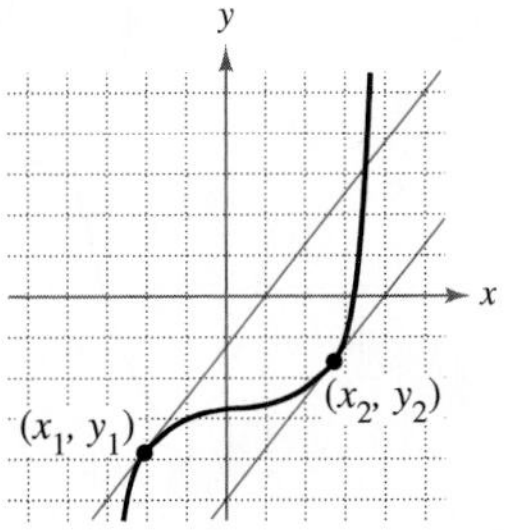

In Exercises 3 and 4, use the graph shown in the figure. To print an enlarged copy of the graph, go to the website *www.mathgraphs.com*.

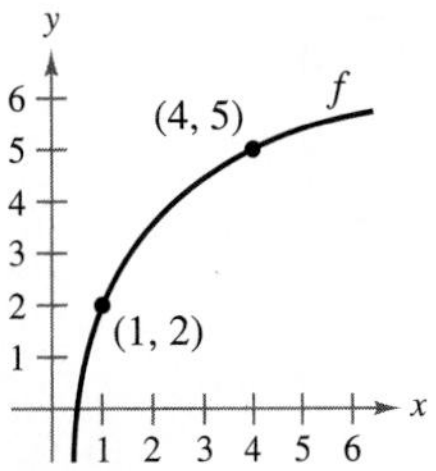

3. Identify or sketch each of the quantities on the figure.
   (a) $f(1)$ and $f(4)$ (b) $f(4) - f(1)$
   (c) $y = \dfrac{f(4) - f(1)}{4 - 1}(x - 1) + f(1)$
4. Insert the proper inequality symbol (< or >) between the given quantities.
   (a) $\dfrac{f(4) - f(1)}{4 - 1}$ ▒ $\dfrac{f(4) - f(3)}{4 - 3}$
   (b) $\dfrac{f(4) - f(1)}{4 - 1}$ ▒ $f'(1)$

In Exercises 5–10, find the slope of the tangent line to the graph of the function at the given point.

5. $f(x) = 3 - 2x, \quad (-1, 5)$

6. $g(x) = \frac{3}{2}x + 1, \quad (-2, -2)$

7. $g(x) = x^2 - 4, \quad (1, -3)$

8. $g(x) = 5 - x^2, \quad (2, 1)$

9. $f(t) = 3t - t^2, \quad (0, 0)$

10. $h(t) = t^2 + 3, \quad (-2, 7)$

In Exercises 11–24, find the derivative by the limit process.

11. $f(x) = 3$

12. $g(x) = -5$

13. $f(x) = -5x$

14. $f(x) = 3x + 2$

15. $h(s) = 3 + \frac{2}{3}s$

16. $f(x) = 9 - \frac{1}{2}x$

17. $f(x) = 2x^2 + x - 1$

18. $f(x) = 1 - x^2$

19. $f(x) = x^3 - 12x$

20. $f(x) = x^3 + x^2$

21. $f(x) = \dfrac{1}{x - 1}$

22. $f(x) = \dfrac{1}{x^2}$

23. $f(x) = \sqrt{x + 1}$

24. $f(x) = \dfrac{4}{\sqrt{x}}$

In Exercises 25–32, (a) find an equation of the tangent line to the graph of f at the given point, (b) use a graphing utility to graph the function and its tangent line at the point, and (c) use the *derivative* feature of a graphing utility to confirm your results.

25. $f(x) = x^2 + 1, \quad (2, 5)$

26. $f(x) = x^2 + 2x + 1, \quad (-3, 4)$

27. $f(x) = x^3, \quad (2, 8)$

28. $f(x) = x^3 + 1, \quad (1, 2)$

29. $f(x) = \sqrt{x}, \quad (1, 1)$

30. $f(x) = \sqrt{x - 1}, \quad (5, 2)$

31. $f(x) = x + \dfrac{4}{x}, \quad (4, 5)$

32. $f(x) = \dfrac{1}{x + 1}, \quad (0, 1)$

In Exercises 33–36, find an equation of the line that is tangent to the graph of f *and* parallel to the given line.

Function	*Line*
33. $f(x) = x^3$	$3x - y + 1 = 0$
34. $f(x) = x^3 + 2$	$3x - y - 4 = 0$
35. $f(x) = \dfrac{1}{\sqrt{x}}$	$x + 2y - 6 = 0$
36. $f(x) = \dfrac{1}{\sqrt{x - 1}}$	$x + 2y + 7 = 0$

In Exercises 37–40, the graph of f is given. Select the graph of f'.

37.

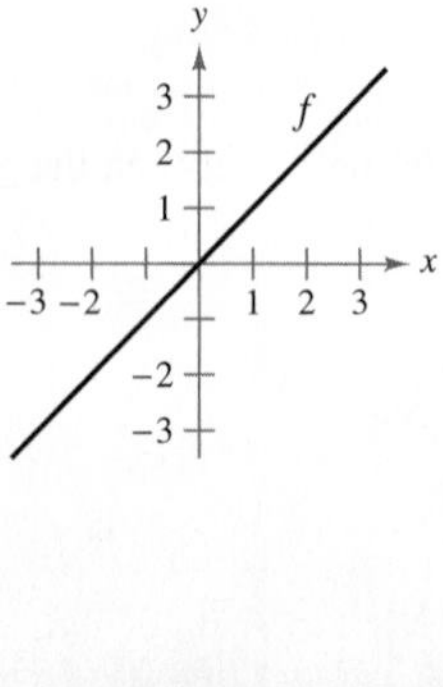

38.

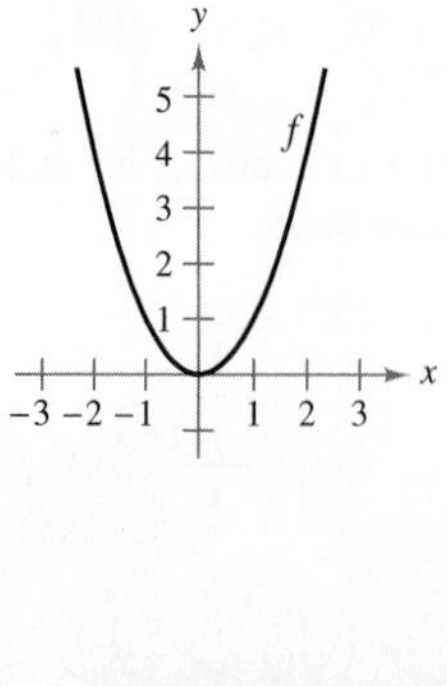

39.

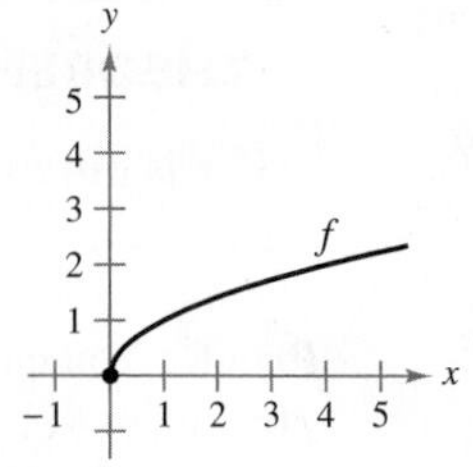

40.

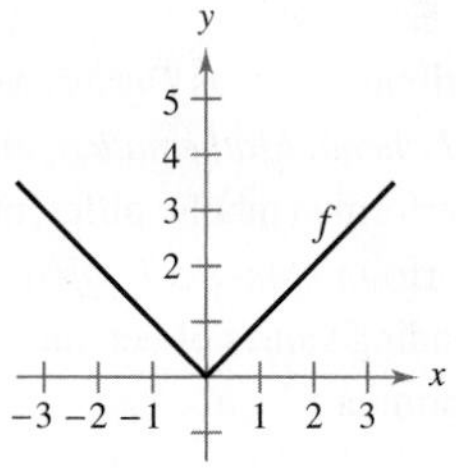

(a)

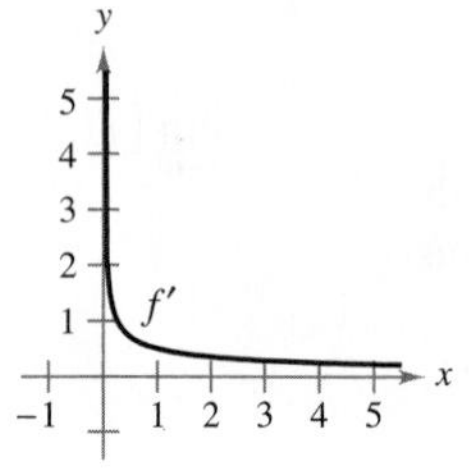

(b)

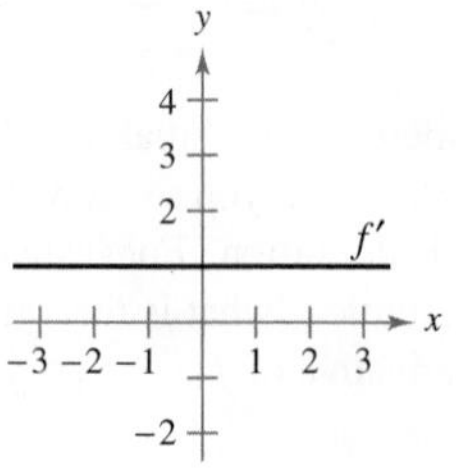

(c)

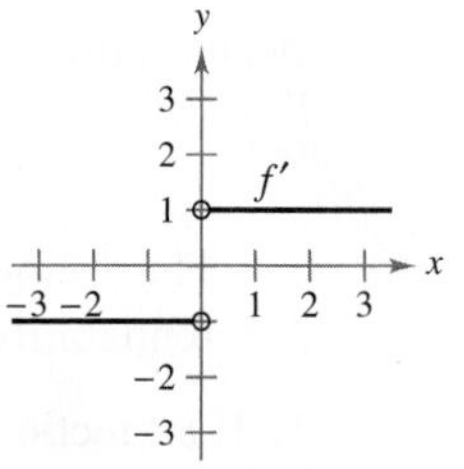

(d)

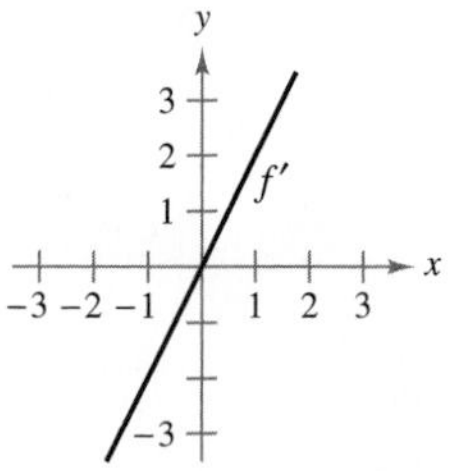

41. The tangent line to the graph of $y = g(x)$ at the point $(5, 2)$ passes through the point $(9, 0)$. Find $g(5)$ and $g'(5)$.

42. The tangent line to the graph of $y = h(x)$ at the point $(-1, 4)$ passes through the point $(3, 6)$. Find $h(-1)$ and $h'(-1)$.

Writing About Concepts

In Exercises 43–46, sketch the graph of f'. Explain how you found your answer.

43.

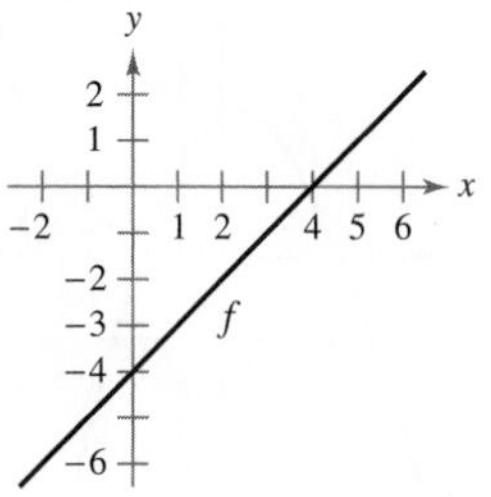

44.

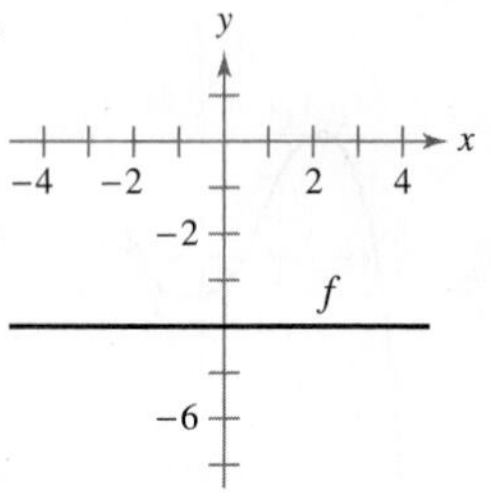

45.

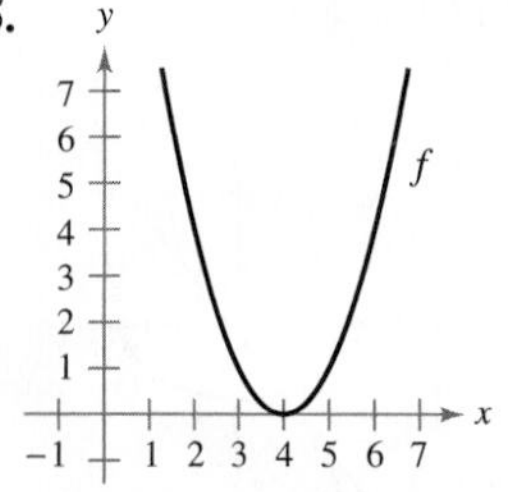

46.

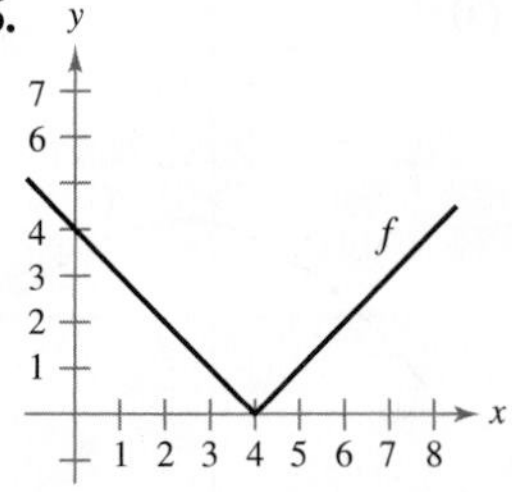

47. Sketch a graph of a function whose derivative is always negative.

Writing About Concepts *(continued)*

48. Sketch a graph of a function whose derivative is always positive.

In Exercises 49–52, the limit represents $f'(c)$ for a function f and a number c. Find f and c.

49. $\displaystyle\lim_{\Delta x \to 0} \frac{[5 - 3(1 + \Delta x)] - 2}{\Delta x}$ **50.** $\displaystyle\lim_{\Delta x \to 0} \frac{(-2 + \Delta x)^3 + 8}{\Delta x}$

51. $\displaystyle\lim_{x \to 6} \frac{-x^2 + 36}{x - 6}$ **52.** $\displaystyle\lim_{x \to 9} \frac{2\sqrt{x} - 6}{x - 9}$

In Exercises 53–55, identify a function f that has the following characteristics. Then sketch the function.

53. $f(0) = 2$; $f'(x) = -3, -\infty < x < \infty$

54. $f(0) = 4; f'(0) = 0$; $f'(x) < 0$ for $x < 0$; $f'(x) > 0$ for $x > 0$

55. $f(0) = 0; f'(0) = 0; f'(x) > 0$ if $x \neq 0$

56. Assume that $f'(c) = 3$. Find $f'(-c)$ if (a) f is an odd function and if (b) f is an even function.

In Exercises 57 and 58, find equations of the two tangent lines to the graph of f that pass through the indicated point.

57. $f(x) = 4x - x^2$ **58.** $f(x) = x^2$

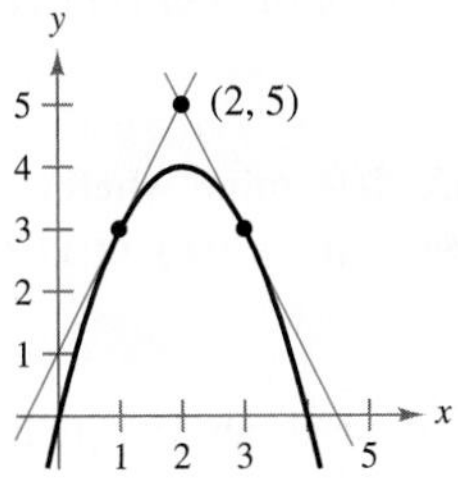

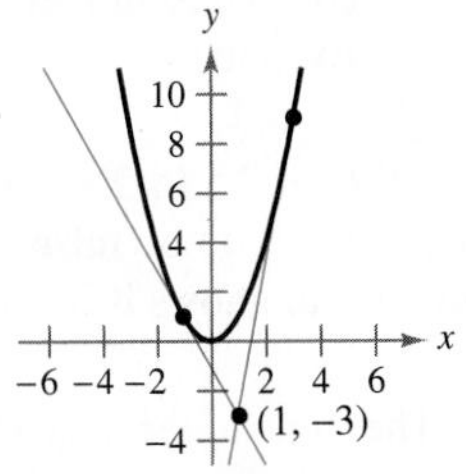

59. ***Graphical Reasoning*** The figure shows the graph of g'.

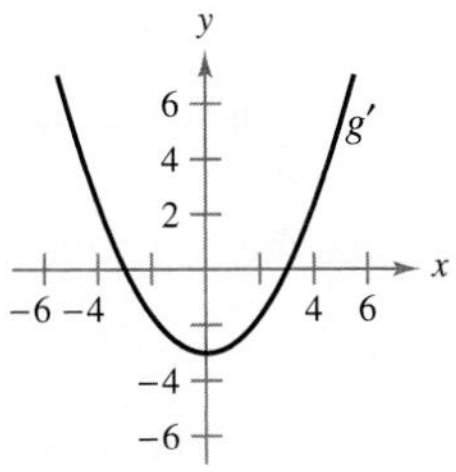

(a) $g'(0) = $ ▭ (b) $g'(3) = $ ▭

(c) What can you conclude about the graph of g knowing that $g'(1) = -\frac{8}{3}$?

(d) What can you conclude about the graph of g knowing that $g'(-4) = \frac{7}{3}$?

(e) Is $g(6) - g(4)$ positive or negative? Explain.

(f) Is it possible to find $g(2)$ from the graph? Explain.

60. ***Graphical Reasoning*** Use a graphing utility to graph each function and its tangent lines at $x = -1$, $x = 0$, and $x = 1$. Based on the results, determine whether the slopes of tangent lines to the graph of a function at different values of x are always distinct.

(a) $f(x) = x^2$ (b) $g(x) = x^3$

Graphical, Numerical, and Analytic Analysis **In Exercises 61 and 62, use a graphing utility to graph f on the interval $[-2, 2]$. Complete the table by graphically estimating the slopes of the graph at the indicated points. Then evaluate the slopes analytically and compare your results with those obtained graphically.**

x	-2	-1.5	-1	-0.5	0	0.5	1	1.5	2
$f(x)$									
$f'(x)$									

61. $f(x) = \frac{1}{4}x^3$ **62.** $f(x) = \frac{1}{2}x^2$

Graphical Reasoning **In Exercises 63 and 64, use a graphing utility to graph the functions f and g in the same viewing window where**

$$g(x) = \frac{f(x + 0.01) - f(x)}{0.01}.$$

Label the graphs and describe the relationship between them.

63. $f(x) = 2x - x^2$ **64.** $f(x) = 3\sqrt{x}$

In Exercises 65 and 66, evaluate $f(2)$ and $f(2.1)$ and use the results to approximate $f'(2)$.

65. $f(x) = x(4 - x)$ **66.** $f(x) = \frac{1}{4}x^3$

Graphical Reasoning **In Exercises 67 and 68, use a graphing utility to graph the function and its derivative in the same viewing window. Label the graphs and describe the relationship between them.**

67. $f(x) = \dfrac{1}{\sqrt{x}}$ **68.** $f(x) = \dfrac{x^3}{4} - 3x$

Writing **In Exercises 69 and 70, consider the functions f and $S_{\Delta x}$ where**

$$S_{\Delta x}(x) = \frac{f(2 + \Delta x) - f(2)}{\Delta x}(x - 2) + f(2).$$

(a) Use a graphing utility to graph f and $S_{\Delta x}$ in the same viewing window for $\Delta x = 1, 0.5$, and 0.1.

(b) Give a written description of the graphs of S for the different values of Δx in part (a).

69. $f(x) = 4 - (x - 3)^2$ **70.** $f(x) = x + \dfrac{1}{x}$

In Exercises 71–80, use the alternative form of the derivative to find the derivative at $x = c$ (if it exists).

71. $f(x) = x^2 - 1,\ c = 2$
72. $g(x) = x(x - 1),\ c = 1$
73. $f(x) = x^3 + 2x^2 + 1,\ c = -2$
74. $f(x) = x^3 + 2x,\ c = 1$
75. $g(x) = \sqrt{|x|},\ c = 0$
76. $f(x) = 1/x,\ c = 3$
77. $f(x) = (x - 6)^{2/3},\ c = 6$
78. $g(x) = (x + 3)^{1/3},\ c = -3$
79. $h(x) = |x + 5|,\ c = -5$
80. $f(x) = |x - 4|,\ c = 4$

In Exercises 81–86, describe the x-values at which f is differentiable.

81. $f(x) = \dfrac{1}{x + 1}$

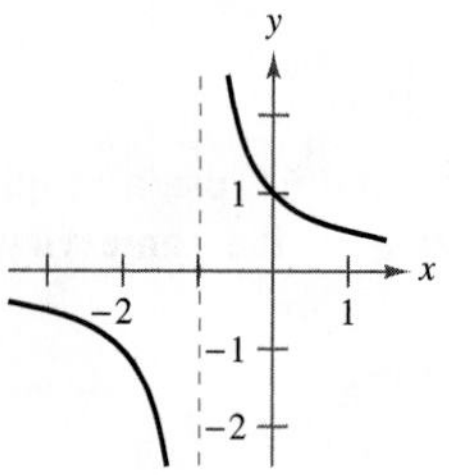

82. $f(x) = |x^2 - 9|$

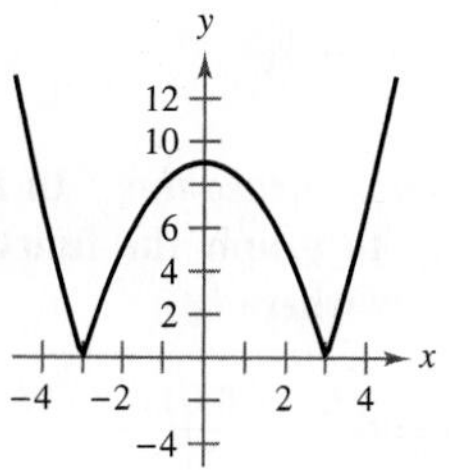

83. $f(x) = (x - 3)^{2/3}$

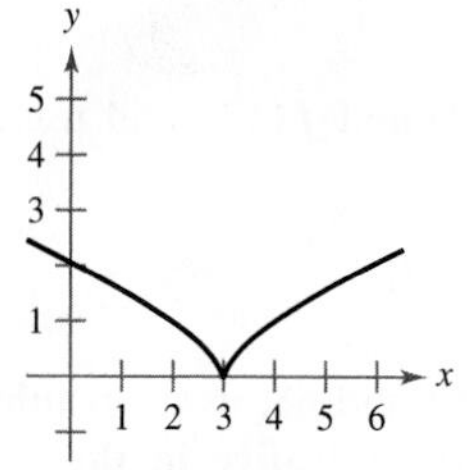

84. $f(x) = \dfrac{x^2}{x^2 - 4}$

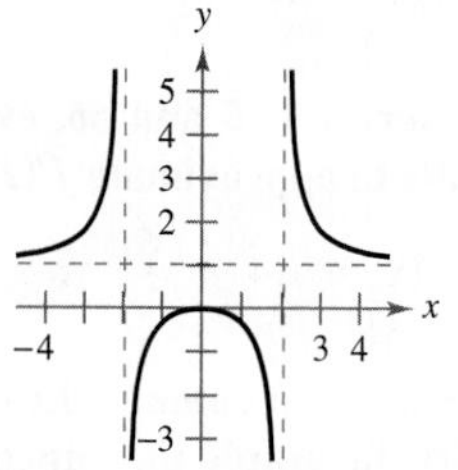

85. $f(x) = \sqrt{x - 1}$

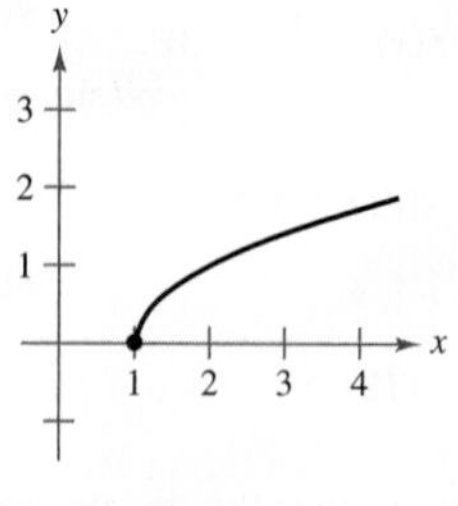

86. $f(x) = \begin{cases} x^2 - 4, & x \le 0 \\ 4 - x^2, & x > 0 \end{cases}$

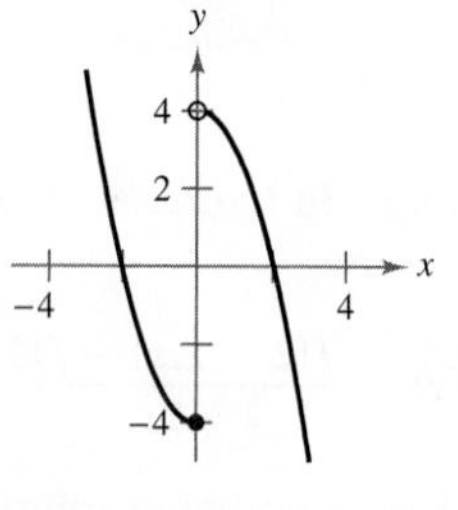

Graphical Analysis **In Exercises 87–90, use a graphing utility to find the x-values at which f is differentiable.**

87. $f(x) = |x + 3|$
88. $f(x) = \dfrac{2x}{x - 1}$
89. $f(x) = x^{2/5}$
90. $f(x) = \begin{cases} x^3 - 3x^2 + 3x, & x \le 1 \\ x^2 - 2x, & x > 1 \end{cases}$

In Exercises 91–94, find the derivatives from the left and from the right at $x = 1$ (if they exist). Is the function differentiable at $x = 1$?

91. $f(x) = |x - 1|$
92. $f(x) = \sqrt{1 - x^2}$
93. $f(x) = \begin{cases} (x - 1)^3, & x \le 1 \\ (x - 1)^2, & x > 1 \end{cases}$
94. $f(x) = \begin{cases} x, & x \le 1 \\ x^2, & x > 1 \end{cases}$

In Exercises 95 and 96, determine whether the function is differentiable at $x = 2$.

95. $f(x) = \begin{cases} x^2 + 1, & x \le 2 \\ 4x - 3, & x > 2 \end{cases}$
96. $f(x) = \begin{cases} \frac{1}{2}x + 1, & x < 2 \\ \sqrt{2x}, & x \ge 2 \end{cases}$

97. ***Graphical Reasoning*** A line with slope m passes through the point $(0, 4)$ and has the equation $y = mx + 4$.
(a) Write the distance d between the line and the point $(3, 1)$ as a function of m.
(b) Use a graphing utility to graph the function d in part (a). Based on the graph, is the function differentiable at every value of m? If not, where is it not differentiable?

98. ***Conjecture*** Consider the functions $f(x) = x^2$ and $g(x) = x^3$.
(a) Graph f and f' on the same set of axes.
(b) Graph g and g' on the same set of axes.
(c) Identify a pattern between f and g and their respective derivatives. Use the pattern to make a conjecture about $h'(x)$ if $h(x) = x^n$, where n is an integer and $n \ge 2$.
(d) Find $f'(x)$ if $f(x) = x^4$. Compare the result with the conjecture in part (c). Is this a proof of your conjecture? Explain.

True or False? **In Exercises 99–102, determine whether the statement is true or false. If it is false, explain why or give an example that shows it is false.**

99. The slope of the tangent line to the differentiable function f at the point $(2, f(2))$ is $\dfrac{f(2 + \Delta x) - f(2)}{\Delta x}$.

100. If a function is continuous at a point, then it is differentiable at that point.

101. If a function has derivatives from both the right and the left at a point, then it is differentiable at that point.

102. If a function is differentiable at a point, then it is continuous at that point.

103. Let $f(x) = \begin{cases} x \sin \frac{1}{x}, & x \ne 0 \\ 0, & x = 0 \end{cases}$ and $g(x) = \begin{cases} x^2 \sin \frac{1}{x}, & x \ne 0 \\ 0, & x = 0 \end{cases}$.

Show that f is continuous, but not differentiable, at $x = 0$. Show that g is differentiable at 0, and find $g'(0)$.

104. ***Writing*** Use a graphing utility to graph the two functions $f(x) = x^2 + 1$ and $g(x) = |x| + 1$ in the same viewing window. Use the *zoom* and *trace* features to analyze the graphs near the point $(0, 1)$. What do you observe? Which function is differentiable at this point? Write a short paragraph describing the geometric significance of differentiability at a point.

Section 2.2 Basic Differentiation Rules and Rates of Change

- **Find the derivative of a function using the Constant Rule.**
- **Find the derivative of a function using the Power Rule.**
- **Find the derivative of a function using the Constant Multiple Rule.**
- **Find the derivative of a function using the Sum and Difference Rules.**
- **Find the derivatives of the sine function and of the cosine function.**
- **Use derivatives to find rates of change.**

The Constant Rule

In Section 2.1 you used the limit definition to find derivatives. In this and the next two sections you will be introduced to several "differentiation rules" that allow you to find derivatives without the *direct* use of the limit definition.

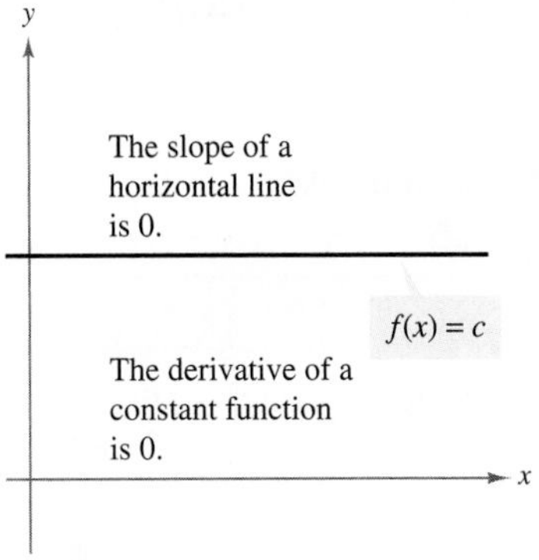

The Constant Rule
Figure 2.14

NOTE In Figure 2.14, note that the Constant Rule is equivalent to saying that the slope of a horizontal line is 0. This demonstrates the relationship between slope and derivative.

THEOREM 2.2 The Constant Rule

The derivative of a constant function is 0. That is, if c is a real number, then

$$\frac{d}{dx}[c] = 0.$$

Proof Let $f(x) = c$. Then, by the limit definition of the derivative,

$$\begin{aligned}\frac{d}{dx}[c] &= f'(x)\\ &= \lim_{\Delta x \to 0} \frac{f(x + \Delta x) - f(x)}{\Delta x}\\ &= \lim_{\Delta x \to 0} \frac{c - c}{\Delta x}\\ &= \lim_{\Delta x \to 0} 0\\ &= 0.\end{aligned}$$

EXAMPLE 1 Using the Constant Rule

	Function	*Derivative*
a.	$y = 7$	$\dfrac{dy}{dx} = 0$
b.	$f(x) = 0$	$f'(x) = 0$
c.	$s(t) = -3$	$s'(t) = 0$
d.	$y = k\pi^2$, k is constant	$y' = 0$

EXPLORATION

Writing a Conjecture Use the definition of the derivative given in Section 2.1 to find the derivative of each function. What patterns do you see? Use your results to write a conjecture about the derivative of $f(x) = x^n$.

a. $f(x) = x^1$ **b.** $f(x) = x^2$ **c.** $f(x) = x^3$

d. $f(x) = x^4$ **e.** $f(x) = x^{1/2}$ **f.** $f(x) = x^{-1}$

The Power Rule

Before proving the next rule, it is important to review the procedure for expanding a binomial.

$$(x + \Delta x)^2 = x^2 + 2x\Delta x + (\Delta x)^2$$
$$(x + \Delta x)^3 = x^3 + 3x^2\Delta x + 3x(\Delta x)^2 + (\Delta x)^3$$

The general binomial expansion for a positive integer n is

$$(x + \Delta x)^n = x^n + nx^{n-1}(\Delta x) + \underbrace{\frac{n(n-1)x^{n-2}}{2}(\Delta x)^2 + \cdot \cdot \cdot + (\Delta x)^n}_{(\Delta x)^2 \text{ is a factor of these terms.}}.$$

This binomial expansion is used in proving a special case of the Power Rule.

THEOREM 2.3 The Power Rule

If n is a rational number, then the function $f(x) = x^n$ is differentiable and

$$\frac{d}{dx}[x^n] = nx^{n-1}.$$

For f to be differentiable at $x = 0$, n must be a number such that x^{n-1} is defined on an interval containing 0.

Proof If n is a positive integer greater than 1, then the binomial expansion produces

$$\begin{aligned}\frac{d}{dx}[x^n] &= \lim_{\Delta x \to 0} \frac{(x + \Delta x)^n - x^n}{\Delta x}\\ &= \lim_{\Delta x \to 0} \frac{x^n + nx^{n-1}(\Delta x) + \frac{n(n-1)x^{n-2}}{2}(\Delta x)^2 + \cdot \cdot \cdot + (\Delta x)^n - x^n}{\Delta x}\\ &= \lim_{\Delta x \to 0} \left[nx^{n-1} + \frac{n(n-1)x^{n-2}}{2}(\Delta x) + \cdot \cdot \cdot + (\Delta x)^{n-1}\right]\\ &= nx^{n-1} + 0 + \cdot \cdot \cdot + 0\\ &= nx^{n-1}.\end{aligned}$$

This proves the case for which n is a positive integer greater than 1. You will prove the case for $n = 1$. Example 7 in Section 2.3 proves the case for which n is a negative integer. In Exercise 75 in Section 2.5 you are asked to prove the case for which n is rational. (In Section 5.5, the Power Rule will be extended to cover irrational values of n.)

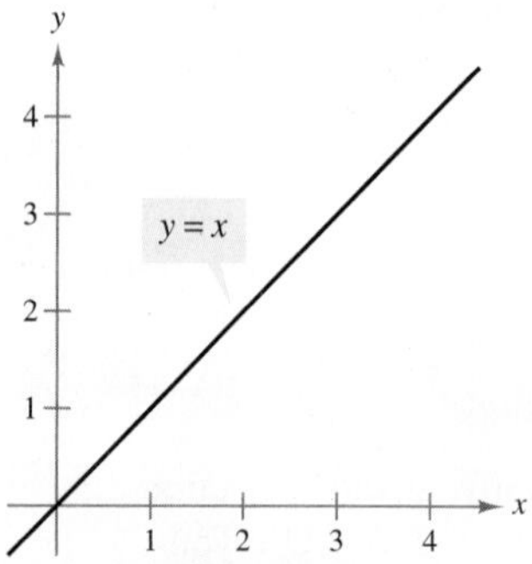

The slope of the line $y = x$ is 1.
Figure 2.15

When using the Power Rule, the case for which $n = 1$ is best thought of as a separate differentiation rule. That is,

$$\frac{d}{dx}[x] = 1. \qquad \text{Power Rule when } n = 1$$

This rule is consistent with the fact that the slope of the line $y = x$ is 1, as shown in Figure 2.15.

EXAMPLE 2 Using the Power Rule

Function	*Derivative*
a. $f(x) = x^3$	$f'(x) = 3x^2$
b. $g(x) = \sqrt[3]{x}$	$g'(x) = \dfrac{d}{dx}[x^{1/3}] = \dfrac{1}{3}x^{-2/3} = \dfrac{1}{3x^{2/3}}$
c. $y = \dfrac{1}{x^2}$	$\dfrac{dy}{dx} = \dfrac{d}{dx}[x^{-2}] = (-2)x^{-3} = -\dfrac{2}{x^3}$

In Example 2(c), note that *before* differentiating, $1/x^2$ was rewritten as x^{-2}. Rewriting is the first step in *many* differentiation problems.

Given:		Rewrite:		Differentiate:		Simplify:
$y = \dfrac{1}{x^2}$	⇨	$y = x^{-2}$	⇨	$\dfrac{dy}{dx} = (-2)x^{-3}$	⇨	$\dfrac{dy}{dx} = -\dfrac{2}{x^3}$

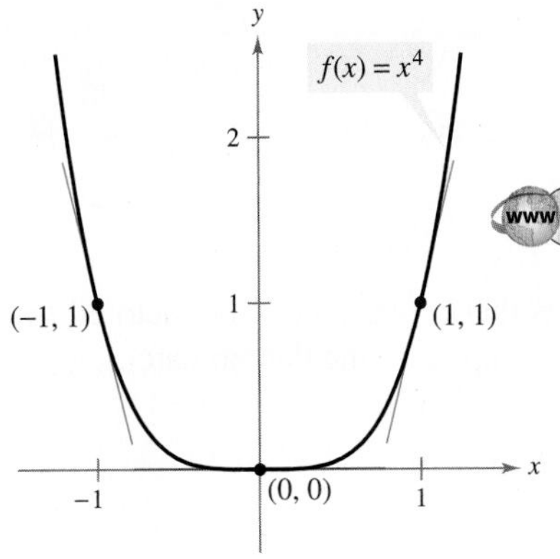

Note that the slope of the graph is negative at the point $(-1, 1)$, the slope is zero at the point $(0, 0)$, and the slope is positive at the point $(1, 1)$.
Figure 2.16

EXAMPLE 3 Finding the Slope of a Graph

Find the slope of the graph of $f(x) = x^4$ when

a. $x = -1$ **b.** $x = 0$ **c.** $x = 1$.

Solution The slope of a graph at a point is the value of the derivative at that point. The derivative of f is $f'(x) = 4x^3$.

a. When $x = -1$, the slope is $f'(-1) = 4(-1)^3 = -4$. Slope is negative.

b. When $x = 0$, the slope is $f'(0) = 4(0)^3 = 0$. Slope is zero.

c. When $x = 1$, the slope is $f'(1) = 4(1)^3 = 4$. Slope is positive.

See Figure 2.16.

EXAMPLE 4 Finding an Equation of a Tangent Line

Find an equation of the tangent line to the graph of $f(x) = x^2$ when $x = -2$.

Solution To find the *point* on the graph of f, evaluate the original function at $x = -2$.

$$(-2, f(-2)) = (-2, 4) \qquad \text{Point on graph}$$

To find the *slope* of the graph when $x = -2$, evaluate the derivative, $f'(x) = 2x$, at $x = -2$.

$$m = f'(-2) = -4 \qquad \text{Slope of graph at } (-2, 4)$$

Now, using the point-slope form of the equation of a line, you can write

$$y - y_1 = m(x - x_1) \qquad \text{Point-slope form}$$
$$y - 4 = -4[x - (-2)] \qquad \text{Substitute for } y_1, m, \text{ and } x_1.$$
$$y = -4x - 4. \qquad \text{Simplify.}$$

See Figure 2.17.

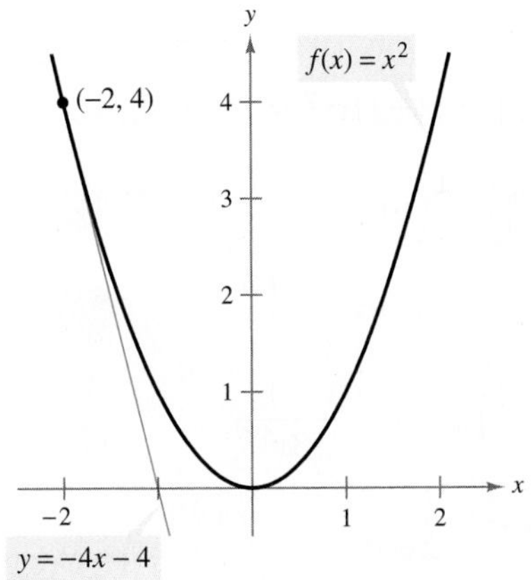

The line $y = -4x - 4$ is tangent to the graph of $f(x) = x^2$ at the point $(-2, 4)$.
Figure 2.17

The Constant Multiple Rule

THEOREM 2.4 The Constant Multiple Rule

If f is a differentiable function and c is a real number, then cf is also differentiable and $\frac{d}{dx}[cf(x)] = cf'(x)$.

Proof

$$\begin{aligned}\frac{d}{dx}[cf(x)] &= \lim_{\Delta x\to 0}\frac{cf(x+\Delta x) - cf(x)}{\Delta x} && \text{Definition of derivative}\\ &= \lim_{\Delta x\to 0} c\left[\frac{f(x+\Delta x)-f(x)}{\Delta x}\right]\\ &= c\left[\lim_{\Delta x\to 0}\frac{f(x+\Delta x)-f(x)}{\Delta x}\right] && \text{Apply Theorem 1.2.}\\ &= cf'(x)\end{aligned}$$

Informally, the Constant Multiple Rule states that constants can be factored out of the differentiation process, even if the constants appear in the denominator.

$$\frac{d}{dx}[cf(x)] = c\frac{d}{dx}[\;f(x)] = cf'(x)$$

$$\begin{aligned}\frac{d}{dx}\left[\frac{f(x)}{c}\right] &= \frac{d}{dx}\left[\left(\frac{1}{c}\right)f(x)\right]\\ &= \left(\frac{1}{c}\right)\frac{d}{dx}[\;f(x)] = \left(\frac{1}{c}\right)f'(x)\end{aligned}$$

EXAMPLE 5 Using the Constant Multiple Rule

Function	Derivative
a. $y = \frac{2}{x}$	$\frac{dy}{dx} = \frac{d}{dx}[2x^{-1}] = 2\frac{d}{dx}[x^{-1}] = 2(-1)x^{-2} = -\frac{2}{x^2}$
b. $f(t) = \frac{4t^2}{5}$	$f'(t) = \frac{d}{dt}\left[\frac{4}{5}t^2\right] = \frac{4}{5}\frac{d}{dt}[t^2] = \frac{4}{5}(2t) = \frac{8}{5}t$
c. $y = 2\sqrt{x}$	$\frac{dy}{dx} = \frac{d}{dx}[2x^{1/2}] = 2\left(\frac{1}{2}x^{-1/2}\right) = x^{-1/2} = \frac{1}{\sqrt{x}}$
d. $y = \frac{1}{2\sqrt[3]{x^2}}$	$\frac{dy}{dx} = \frac{d}{dx}\left[\frac{1}{2}x^{-2/3}\right] = \frac{1}{2}\left(-\frac{2}{3}\right)x^{-5/3} = -\frac{1}{3x^{5/3}}$
e. $y = -\frac{3x}{2}$	$y' = \frac{d}{dx}\left[-\frac{3}{2}x\right] = -\frac{3}{2}(1) = -\frac{3}{2}$

The Constant Multiple Rule and the Power Rule can be combined into one rule. The combination rule is

$$D_x[cx^n] = cnx^{n-1}.$$

EXAMPLE 6 Using Parentheses When Differentiating

	Original Function	*Rewrite*	*Differentiate*	*Simplify*
a.	$y = \dfrac{5}{2x^3}$	$y = \dfrac{5}{2}(x^{-3})$	$y' = \dfrac{5}{2}(-3x^{-4})$	$y' = -\dfrac{15}{2x^4}$
b.	$y = \dfrac{5}{(2x)^3}$	$y = \dfrac{5}{8}(x^{-3})$	$y' = \dfrac{5}{8}(-3x^{-4})$	$y' = -\dfrac{15}{8x^4}$
c.	$y = \dfrac{7}{3x^{-2}}$	$y = \dfrac{7}{3}(x^2)$	$y' = \dfrac{7}{3}(2x)$	$y' = \dfrac{14x}{3}$
d.	$y = \dfrac{7}{(3x)^{-2}}$	$y = 63(x^2)$	$y' = 63(2x)$	$y' = 126x$

The Sum and Difference Rules

THEOREM 2.5 The Sum and Difference Rules

The sum (or difference) of two differentiable functions f and g is itself differentiable. Moreover, the derivative of $f + g$ (or $f - g$) is the sum (or difference) of the derivatives of f and g.

$$\frac{d}{dx}[f(x) + g(x)] = f'(x) + g'(x) \qquad \text{Sum Rule}$$

$$\frac{d}{dx}[f(x) - g(x)] = f'(x) - g'(x) \qquad \text{Difference Rule}$$

Proof A proof of the Sum Rule follows from Theorem 1.2. (The Difference Rule can be proved in a similar way.)

$$\begin{aligned}
\frac{d}{dx}[f(x) + g(x)] &= \lim_{\Delta x \to 0} \frac{[f(x + \Delta x) + g(x + \Delta x)] - [f(x) + g(x)]}{\Delta x} \\
&= \lim_{\Delta x \to 0} \frac{f(x + \Delta x) + g(x + \Delta x) - f(x) - g(x)}{\Delta x} \\
&= \lim_{\Delta x \to 0} \left[\frac{f(x + \Delta x) - f(x)}{\Delta x} + \frac{g(x + \Delta x) - g(x)}{\Delta x}\right] \\
&= \lim_{\Delta x \to 0} \frac{f(x + \Delta x) - f(x)}{\Delta x} + \lim_{\Delta x \to 0} \frac{g(x + \Delta x) - g(x)}{\Delta x} \\
&= f'(x) + g'(x)
\end{aligned}$$

The Sum and Difference Rules can be extended to any finite number of functions. For instance, if $F(x) = f(x) + g(x) - h(x)$, then $F'(x) = f'(x) + g'(x) - h'(x)$.

EXAMPLE 7 Using the Sum and Difference Rules

	Function	*Derivative*
a.	$f(x) = x^3 - 4x + 5$	$f'(x) = 3x^2 - 4$
b.	$g(x) = -\dfrac{x^4}{2} + 3x^3 - 2x$	$g'(x) = -2x^3 + 9x^2 - 2$

FOR FURTHER INFORMATION For the outline of a geometric proof of the derivatives of the sine and cosine functions, see the article "The Spider's Spacewalk Derivation of sin′ and cos′" by Tim Hesterberg in *The College Mathematics Journal*. To view this article, go to the website *www.matharticles.com*.

Derivatives of Sine and Cosine Functions

In Section 1.3, you studied the following limits.

$$\lim_{\Delta x \to 0} \frac{\sin \Delta x}{\Delta x} = 1 \quad \text{and} \quad \lim_{\Delta x \to 0} \frac{1 - \cos \Delta x}{\Delta x} = 0$$

These two limits can be used to prove differentiation rules for the sine and cosine functions. (The derivatives of the other four trigonometric functions are discussed in Section 2.3.)

THEOREM 2.6 Derivatives of Sine and Cosine Functions

$$\frac{d}{dx}[\sin x] = \cos x \qquad \frac{d}{dx}[\cos x] = -\sin x$$

Proof

$$\begin{aligned}
\frac{d}{dx}[\sin x] &= \lim_{\Delta x \to 0} \frac{\sin(x + \Delta x) - \sin x}{\Delta x} && \text{Definition of derivative} \\
&= \lim_{\Delta x \to 0} \frac{\sin x \cos \Delta x + \cos x \sin \Delta x - \sin x}{\Delta x} \\
&= \lim_{\Delta x \to 0} \frac{\cos x \sin \Delta x - (\sin x)(1 - \cos \Delta x)}{\Delta x} \\
&= \lim_{\Delta x \to 0} \left[(\cos x)\left(\frac{\sin \Delta x}{\Delta x}\right) - (\sin x)\left(\frac{1 - \cos \Delta x}{\Delta x}\right) \right] \\
&= \cos x \left(\lim_{\Delta x \to 0} \frac{\sin \Delta x}{\Delta x} \right) - \sin x \left(\lim_{\Delta x \to 0} \frac{1 - \cos \Delta x}{\Delta x} \right) \\
&= (\cos x)(1) - (\sin x)(0) \\
&= \cos x
\end{aligned}$$

This differentiation rule is shown graphically in Figure 2.18. Note that for each x, the *slope* of the sine curve is equal to the value of the cosine. The proof of the second rule is left as an exercise (see Exercise 116).

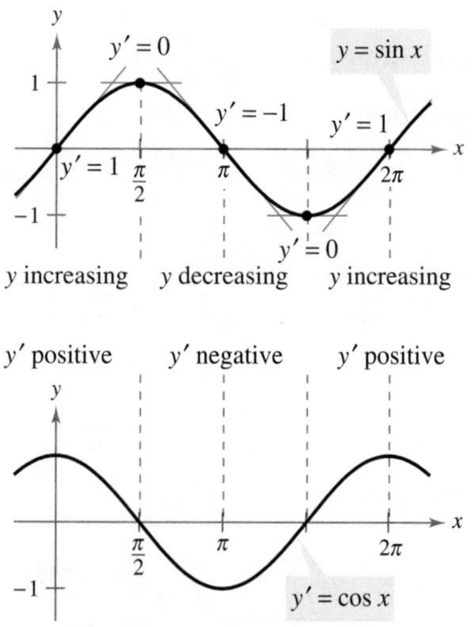

The derivative of the sine function is the cosine function.
Figure 2.18

EXAMPLE 8 Derivatives Involving Sines and Cosines

Function	*Derivative*
a. $y = 2 \sin x$	$y' = 2 \cos x$
b. $y = \dfrac{\sin x}{2} = \dfrac{1}{2} \sin x$	$y' = \dfrac{1}{2} \cos x = \dfrac{\cos x}{2}$
c. $y = x + \cos x$	$y' = 1 - \sin x$

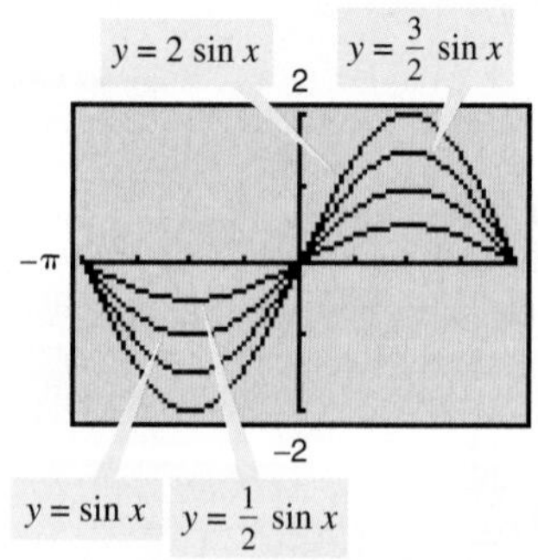

$\dfrac{d}{dx}[a \sin x] = a \cos x$
Figure 2.19

TECHNOLOGY A graphing utility can provide insight into the interpretation of a derivative. For instance, Figure 2.19 shows the graphs of

$$y = a \sin x$$

for $a = \frac{1}{2}, 1, \frac{3}{2}$, and 2. Estimate the slope of each graph at the point (0, 0). Then verify your estimates analytically by evaluating the derivative of each function when $x = 0$.

Rates of Change

You have seen how the derivative is used to determine slope. The derivative can also be used to determine the rate of change of one variable with respect to another. Applications involving rates of change occur in a wide variety of fields. A few examples are population growth rates, production rates, water flow rates, velocity, and acceleration.

A common use for rate of change is to describe the motion of an object moving in a straight line. In such problems, it is customary to use either a horizontal or a vertical line with a designated origin to represent the line of motion. On such lines, movement to the right (or upward) is considered to be in the positive direction, and movement to the left (or downward) is considered to be in the negative direction.

The function s that gives the position (relative to the origin) of an object as a function of time t is called a **position function.** If, over a period of time Δt, the object changes its position by the amount $\Delta s = s(t + \Delta t) - s(t)$, then, by the familiar formula

$$\text{Rate} = \frac{\text{distance}}{\text{time}}$$

the **average velocity** is

$$\frac{\text{Change in distance}}{\text{Change in time}} = \frac{\Delta s}{\Delta t}.$$ Average velocity

EXAMPLE 9 Finding Average Velocity of a Falling Object

If a billiard ball is dropped from a height of 100 feet, its height s at time t is given by the position function

$$s = -16t^2 + 100$$ Position function

where s is measured in feet and t is measured in seconds. Find the average velocity over each of the following time intervals.

a. $[1, 2]$ **b.** $[1, 1.5]$ **c.** $[1, 1.1]$

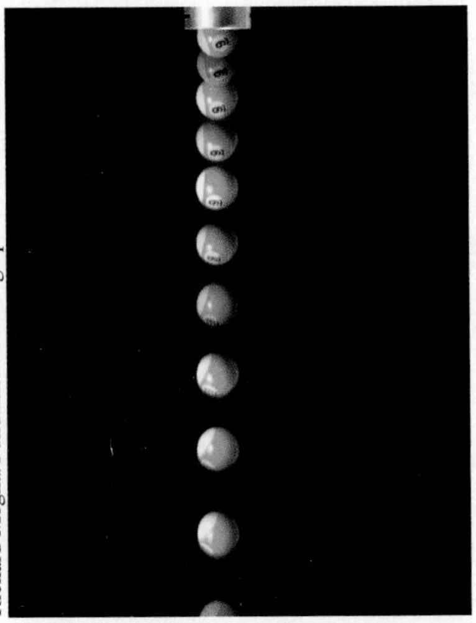
Richard Megna/Fundamental Photographs

Time-lapse photograph of a free-falling billiard ball

Solution

a. For the interval $[1, 2]$, the object falls from a height of $s(1) = -16(1)^2 + 100 = 84$ feet to a height of $s(2) = -16(2)^2 + 100 = 36$ feet. The average velocity is

$$\frac{\Delta s}{\Delta t} = \frac{36 - 84}{2 - 1} = \frac{-48}{1} = -48 \text{ feet per second.}$$

b. For the interval $[1, 1.5]$, the object falls from a height of 84 feet to a height of 64 feet. The average velocity is

$$\frac{\Delta s}{\Delta t} = \frac{64 - 84}{1.5 - 1} = \frac{-20}{0.5} = -40 \text{ feet per second.}$$

c. For the interval $[1, 1.1]$, the object falls from a height of 84 feet to a height of 80.64 feet. The average velocity is

$$\frac{\Delta s}{\Delta t} = \frac{80.64 - 84}{1.1 - 1} = \frac{-3.36}{0.1} = -33.6 \text{ feet per second.}$$

Note that the average velocities are *negative*, indicating that the object is moving downward.

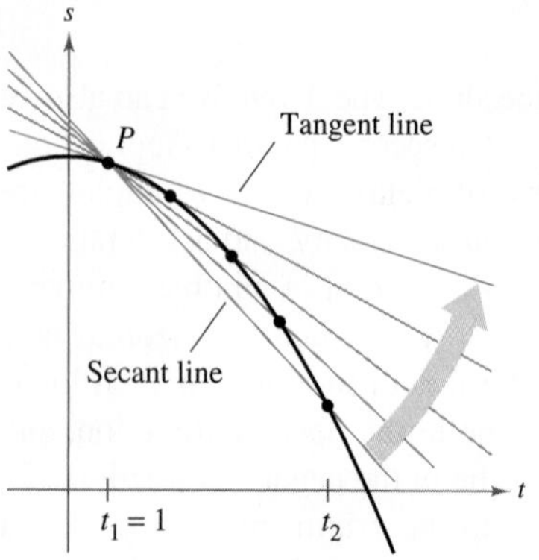

The average velocity between t_1 and t_2 is the slope of the secant line, and the instantaneous velocity at t_1 is the slope of the tangent line.
Figure 2.20

Suppose that in Example 9 you wanted to find the *instantaneous* velocity (or simply the velocity) of the object when $t = 1$. Just as you can approximate the slope of the tangent line by calculating the slope of the secant line, you can approximate the velocity at $t = 1$ by calculating the average velocity over a small interval $[1, 1 + \Delta t]$ (see Figure 2.20). By taking the limit as Δt approaches zero, you obtain the velocity when $t = 1$. Try doing this—you will find that the velocity when $t = 1$ is -32 feet per second.

In general, if $s = s(t)$ is the position function for an object moving along a straight line, the **velocity** of the object at time t is

$$v(t) = \lim_{\Delta t \to 0} \frac{s(t + \Delta t) - s(t)}{\Delta t} = s'(t).$$

Velocity function

In other words, the velocity function is the derivative of the position function. Velocity can be negative, zero, or positive. The **speed** of an object is the absolute value of its velocity. Speed cannot be negative.

The position of a free-falling object (neglecting air resistance) under the influence of gravity can be represented by the equation

$$s(t) = \frac{1}{2}gt^2 + v_0 t + s_0$$

Position function

where s_0 is the initial height of the object, v_0 is the initial velocity of the object, and g is the acceleration due to gravity. On Earth, the value of g is approximately -32 feet per second per second or -9.8 meters per second per second.

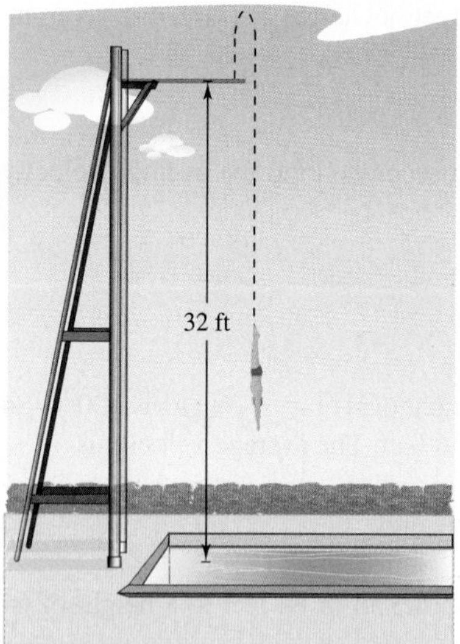

Velocity is positive when an object is rising, and is negative when an object is falling.
Figure 2.21

NOTE In Figure 2.21, note that the diver moves upward for the first half-second because the velocity is positive for $0 < t < \frac{1}{2}$. When the velocity is 0, the diver has reached the maximum height of the dive.

EXAMPLE 10 Using the Derivative to Find Velocity

At time $t = 0$, a diver jumps from a platform diving board that is 32 feet above the water (see Figure 2.21). The position of the diver is given by

$$s(t) = -16t^2 + 16t + 32$$

Position function

where s is measured in feet and t is measured in seconds.

a. When does the diver hit the water?

b. What is the diver's velocity at impact?

Solution

a. To find the time t when the diver hits the water, let $s = 0$ and solve for t.

$$-16t^2 + 16t + 32 = 0$$ Set position function equal to 0.

$$-16(t + 1)(t - 2) = 0$$ Factor.

$$t = -1 \text{ or } 2$$ Solve for t.

Because $t \geq 0$, choose the positive value to conclude that the diver hits the water at $t = 2$ seconds.

b. The velocity at time t is given by the derivative $s'(t) = -32t + 16$. So, the velocity at time $t = 2$ is

$$s'(2) = -32(2) + 16 = -48 \text{ feet per second.}$$

Exercises for Section 2.2

See www.CalcChat.com for worked-out solutions to odd-numbered exercises.

In Exercises 1 and 2, use the graph to estimate the slope of the tangent line to $y = x^n$ at the point (1, 1). Verify your answer analytically. To print an enlarged copy of the graph, go to the website *www.mathgraphs.com*.

1. (a) $y = x^{1/2}$ (b) $y = x^3$

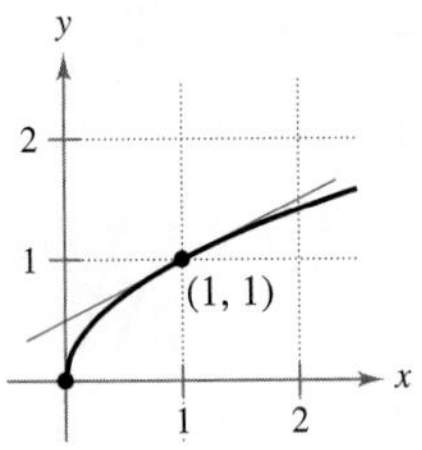

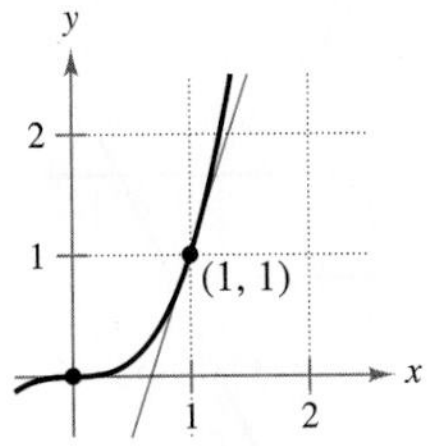

2. (a) $y = x^{-1/2}$ (b) $y = x^{-1}$

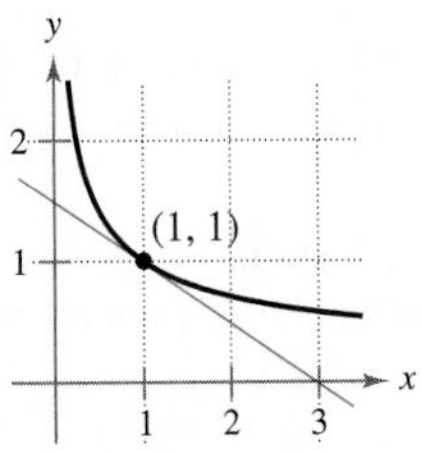

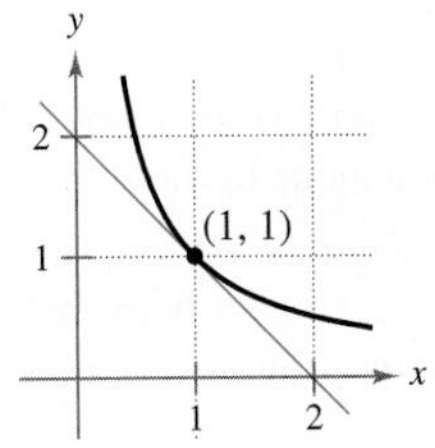

In Exercises 3–24, find the derivative of the function.

3. $y = 8$

4. $f(x) = -2$

5. $y = x^6$

6. $y = x^8$

7. $y = \dfrac{1}{x^7}$

8. $y = \dfrac{1}{x^8}$

9. $f(x) = \sqrt[5]{x}$

10. $g(x) = \sqrt[4]{x}$

11. $f(x) = x + 1$

12. $g(x) = 3x - 1$

13. $f(t) = -2t^2 + 3t - 6$

14. $y = t^2 + 2t - 3$

15. $g(x) = x^2 + 4x^3$

16. $y = 8 - x^3$

17. $s(t) = t^3 - 2t + 4$

18. $f(x) = 2x^3 - x^2 + 3x$

19. $y = \dfrac{\pi}{2}\sin\theta - \cos\theta$

20. $g(t) = \pi\cos t$

21. $y = x^2 - \frac{1}{2}\cos x$

22. $y = 5 + \sin x$

23. $y = \dfrac{1}{x} - 3\sin x$

24. $y = \dfrac{5}{(2x)^3} + 2\cos x$

In Exercises 25–30, complete the table.

	Original Function	*Rewrite*	*Differentiate*	*Simplify*
25.	$y = \dfrac{5}{2x^2}$			
26.	$y = \dfrac{2}{3x^2}$			
27.	$y = \dfrac{3}{(2x)^3}$			
28.	$y = \dfrac{\pi}{(3x)^2}$			
29.	$y = \dfrac{\sqrt{x}}{x}$			
30.	$y = \dfrac{4}{x^{-3}}$			

In Exercises 31–38, find the slope of the graph of the function at the given point. Use the *derivative* feature of a graphing utility to confirm your results.

	Function	*Point*
31.	$f(x) = \dfrac{3}{x^2}$	$(1, 3)$
32.	$f(t) = 3 - \dfrac{3}{5t}$	$(\frac{3}{5}, 2)$
33.	$f(x) = -\frac{1}{2} + \frac{7}{5}x^3$	$(0, -\frac{1}{2})$
34.	$y = 3x^3 - 6$	$(2, 18)$
35.	$y = (2x + 1)^2$	$(0, 1)$
36.	$f(x) = 3(5 - x)^2$	$(5, 0)$
37.	$f(\theta) = 4\sin\theta - \theta$	$(0, 0)$
38.	$g(t) = 2 + 3\cos t$	$(\pi, -1)$

In Exercises 39–52, find the derivative of the function.

39. $f(x) = x^2 + 5 - 3x^{-2}$

40. $f(x) = x^2 - 3x - 3x^{-2}$

41. $g(t) = t^2 - \dfrac{4}{t^3}$

42. $f(x) = x + \dfrac{1}{x^2}$

43. $f(x) = \dfrac{x^3 - 3x^2 + 4}{x^2}$

44. $h(x) = \dfrac{2x^2 - 3x + 1}{x}$

45. $y = x(x^2 + 1)$

46. $y = 3x(6x - 5x^2)$

47. $f(x) = \sqrt{x} - 6\sqrt[3]{x}$

48. $f(x) = \sqrt[3]{x} + \sqrt[5]{x}$

49. $h(s) = s^{4/5} - s^{2/3}$

50. $f(t) = t^{2/3} - t^{1/3} + 4$

51. $f(x) = 6\sqrt{x} + 5\cos x$

52. $f(x) = \dfrac{2}{\sqrt[3]{x}} + 3\cos x$

In Exercises 53–56, (a) find an equation of the tangent line to the graph of f at the given point, (b) use a graphing utility to graph the function and its tangent line at the point, and (c) use the *derivative* feature of a graphing utility to confirm your results.

	Function	*Point*
53.	$y = x^4 - 3x^2 + 2$	$(1, 0)$
54.	$y = x^3 + x$	$(-1, -2)$
55.	$f(x) = \dfrac{2}{\sqrt[4]{x^3}}$	$(1, 2)$
56.	$y = (x^2 + 2x)(x + 1)$	$(1, 6)$

In Exercises 57–62, determine the point(s) (if any) at which the graph of the function has a horizontal tangent line.

57. $y = x^4 - 8x^2 + 2$

58. $y = x^3 + x$

59. $y = \dfrac{1}{x^2}$

60. $y = x^2 + 1$

61. $y = x + \sin x, \quad 0 \le x < 2\pi$

62. $y = \sqrt{3}x + 2\cos x, \quad 0 \le x < 2\pi$

In Exercises 63–66, find k such that the line is tangent to the graph of the function.

	Function	*Line*
63.	$f(x) = x^2 - kx$	$y = 4x - 9$
64.	$f(x) = k - x^2$	$y = -4x + 7$
65.	$f(x) = \dfrac{k}{x}$	$y = -\dfrac{3}{4}x + 3$
66.	$f(x) = k\sqrt{x}$	$y = x + 4$

Writing About Concepts

67. Use the graph of f to answer each question. To print an enlarged copy of the graph, go to the website *www.mathgraphs.com.*

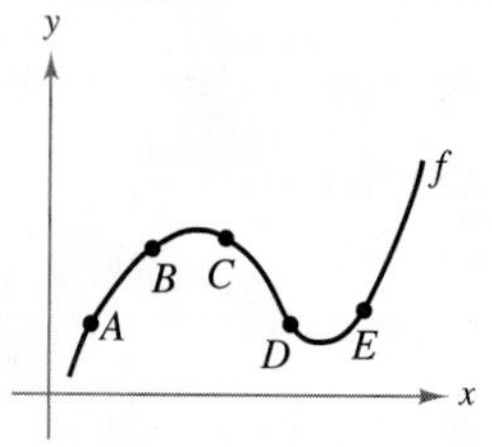

(a) Between which two consecutive points is the average rate of change of the function greatest?

(b) Is the average rate of change of the function between A and B greater than or less than the instantaneous rate of change at B?

(c) Sketch a tangent line to the graph between C and D such that the slope of the tangent line is the same as the average rate of change of the function between C and D.

68. Sketch the graph of a function f such that $f' > 0$ for all x and the rate of change of the function is decreasing.

In Exercises 69 and 70, the relationship between f and g is given. Explain the relationship between f' and g'.

69. $g(x) = f(x) + 6$

70. $g(x) = -5f(x)$

Writing About Concepts *(continued)*

In Exercises 71 and 72, the graphs of a function f and its derivative f' are shown on the same set of coordinate axes. Label the graphs as f or f' and write a short paragraph stating the criteria used in making the selection. To print an enlarged copy of the graph, go to the website *www.mathgraphs.com.*

71.

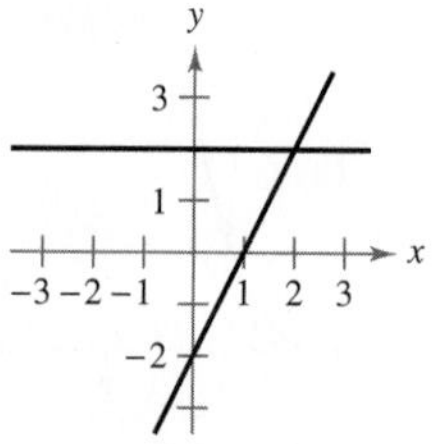

72.

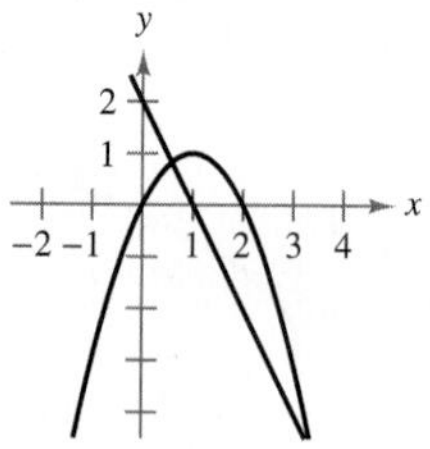

73. Sketch the graphs of $y = x^2$ and $y = -x^2 + 6x - 5$, and sketch the two lines that are tangent to both graphs. Find equations of these lines.

74. Show that the graphs of the two equations $y = x$ and $y = 1/x$ have tangent lines that are perpendicular to each other at their point of intersection.

75. Show that the graph of the function

$$f(x) = 3x + \sin x + 2$$

does not have a horizontal tangent line.

76. Show that the graph of the function

$$f(x) = x^5 + 3x^3 + 5x$$

does not have a tangent line with a slope of 3.

In Exercises 77 and 78, find an equation of the tangent line to the graph of the function f through the point (x_0, y_0) not on the graph. To find the point of tangency (x, y) on the graph of f, solve the equation

$$f'(x) = \frac{y_0 - y}{x_0 - x}.$$

77. $f(x) = \sqrt{x}$

$(x_0, y_0) = (-4, 0)$

78. $f(x) = \dfrac{2}{x}$

$(x_0, y_0) = (5, 0)$

79. ***Linear Approximation*** Use a graphing utility, with a square window setting, to zoom in on the graph of

$$f(x) = 4 - \tfrac{1}{2}x^2$$

to approximate $f'(1)$. Use the derivative to find $f'(1)$.

80. ***Linear Approximation*** Use a graphing utility, with a square window setting, to zoom in on the graph of

$$f(x) = 4\sqrt{x} + 1$$

to approximate $f'(4)$. Use the derivative to find $f'(4)$.

81. ***Linear Approximation*** Consider the function $f(x) = x^{3/2}$ with the solution point $(4, 8)$.

(a) Use a graphing utility to graph f. Use the *zoom* feature to obtain successive magnifications of the graph in the neighborhood of the point $(4, 8)$. After zooming in a few times, the graph should appear nearly linear. Use the *trace* feature to determine the coordinates of a point near $(4, 8)$. Find an equation of the secant line $S(x)$ through the two points.

(b) Find the equation of the line

$$T(x) = f'(4)(x - 4) + f(4)$$

tangent to the graph of f passing through the given point. Why are the linear functions S and T nearly the same?

(c) Use a graphing utility to graph f and T on the same set of coordinate axes. Note that T is a good approximation of f when x is close to 4. What happens to the accuracy of the approximation as you move farther away from the point of tangency?

(d) Demonstrate the conclusion in part (c) by completing the table.

Δx	-3	-2	-1	-0.5	-0.1	0
$f(4 + \Delta x)$						
$T(4 + \Delta x)$						

Δx	0.1	0.5	1	2	3
$f(4 + \Delta x)$					
$T(4 + \Delta x)$					

82. ***Linear Approximation*** Repeat Exercise 81 for the function $f(x) = x^3$ where $T(x)$ is the line tangent to the graph at the point $(1, 1)$. Explain why the accuracy of the linear approximation decreases more rapidly than in Exercise 81.

True or False? **In Exercises 83–88, determine whether the statement is true or false. If it is false, explain why or give an example that shows it is false.**

83. If $f'(x) = g'(x)$, then $f(x) = g(x)$.

84. If $f(x) = g(x) + c$, then $f'(x) = g'(x)$.

85. If $y = \pi^2$, then $dy/dx = 2\pi$.

86. If $y = x/\pi$, then $dy/dx = 1/\pi$.

87. If $g(x) = 3f(x)$, then $g'(x) = 3f'(x)$.

88. If $f(x) = 1/x^n$, then $f'(x) = 1/(nx^{n-1})$.

In Exercises 89–92, find the average rate of change of the function over the given interval. Compare this average rate of change with the instantaneous rates of change at the endpoints of the interval.

89. $f(t) = 2t + 7, \quad [1, 2]$

90. $f(t) = t^2 - 3, \quad [2, 2.1]$

91. $f(x) = \dfrac{-1}{x}, \quad [1, 2]$

92. $f(x) = \sin x, \quad \left[0, \dfrac{\pi}{6}\right]$

Vertical Motion **In Exercises 93 and 94, use the position function $s(t) = -16t^2 + v_0 t + s_0$ for free-falling objects.**

93. A silver dollar is dropped from the top of a building that is 1362 feet tall.

(a) Determine the position and velocity functions for the coin.

(b) Determine the average velocity on the interval $[1, 2]$.

(c) Find the instantaneous velocities when $t = 1$ and $t = 2$.

(d) Find the time required for the coin to reach ground level.

(e) Find the velocity of the coin at impact.

94. A ball is thrown straight down from the top of a 220-foot building with an initial velocity of -22 feet per second. What is its velocity after 3 seconds? What is its velocity after falling 108 feet?

Vertical Motion **In Exercises 95 and 96, use the position function $s(t) = -4.9t^2 + v_0 t + s_0$ for free-falling objects.**

95. A projectile is shot upward from the surface of Earth with an initial velocity of 120 meters per second. What is its velocity after 5 seconds? After 10 seconds?

96. To estimate the height of a building, a stone is dropped from the top of the building into a pool of water at ground level. How high is the building if the splash is seen 6.8 seconds after the stone is dropped?

Think About It **In Exercises 97 and 98, the graph of a position function is shown. It represents the distance in miles that a person drives during a 10-minute trip to work. Make a sketch of the corresponding velocity function.**

97.

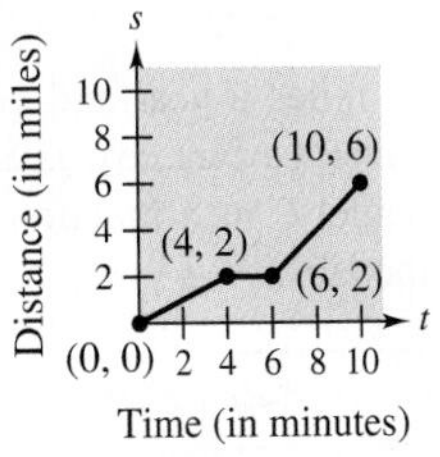

98.

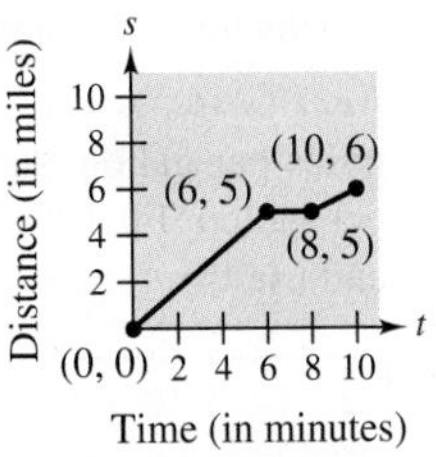

Think About It **In Exercises 99 and 100, the graph of a velocity function is shown. It represents the velocity in miles per hour during a 10-minute drive to work. Make a sketch of the corresponding position function.**

99.

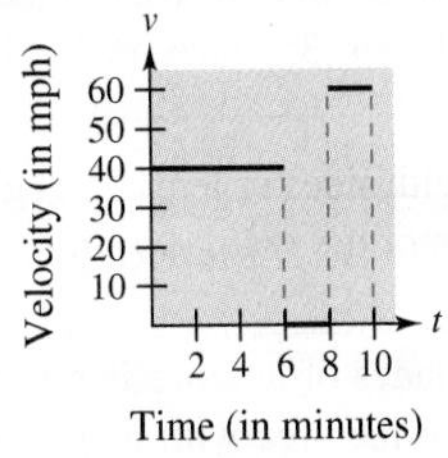

100.

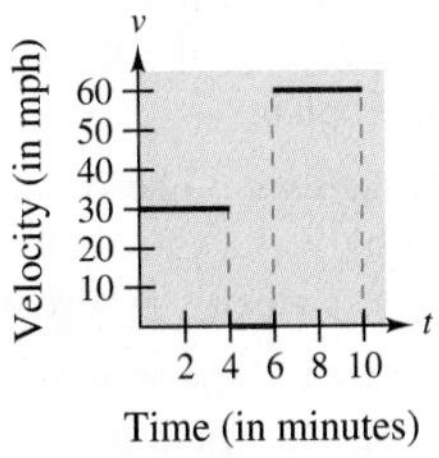

101. ***Modeling Data*** The stopping distance of an automobile, on dry, level pavement, traveling at a speed v (kilometers per hour) is the distance R (meters) the car travels during the reaction time of the driver plus the distance B (meters) the car travels after the brakes are applied (see figure). The table shows the results of an experiment.

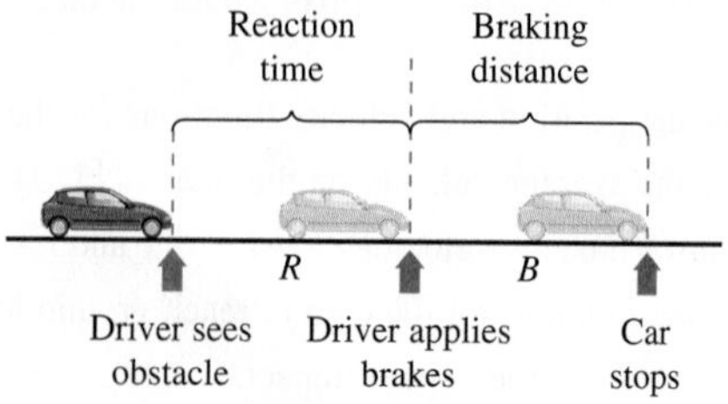

Speed, v	20	40	60	80	100
Reaction Time Distance, R	8.3	16.7	25.0	33.3	41.7
Braking Time Distance, B	2.3	9.0	20.2	35.8	55.9

(a) Use the regression capabilities of a graphing utility to find a linear model for reaction time distance.

(b) Use the regression capabilities of a graphing utility to find a quadratic model for braking distance.

(c) Determine the polynomial giving the total stopping distance T.

(d) Use a graphing utility to graph the functions R, B, and T in the same viewing window.

(e) Find the derivative of T and the rates of change of the total stopping distance for $v = 40$, $v = 80$, and $v = 100$.

(f) Use the results of this exercise to draw conclusions about the total stopping distance as speed increases.

102. ***Fuel Cost*** A car is driven 15,000 miles a year and gets x miles per gallon. Assume that the average fuel cost is \$1.55 per gallon. Find the annual cost of fuel C as a function of x and use this function to complete the table.

x	10	15	20	25	30	35	40
C							
dC/dx							

Who would benefit more from a one-mile-per-gallon increase in fuel efficiency—the driver of a car that gets 15 miles per gallon or the driver of a car that gets 35 miles per gallon? Explain.

103. ***Volume*** The volume of a cube with sides of length s is given by $V = s^3$. Find the rate of change of the volume with respect to s when $s = 4$ centimeters.

104. ***Area*** The area of a square with sides of length s is given by $A = s^2$. Find the rate of change of the area with respect to s when $s = 4$ meters.

105. ***Velocity*** Verify that the average velocity over the time interval $[t_0 - \Delta t, t_0 + \Delta t]$ is the same as the instantaneous velocity at $t = t_0$ for the position function

$$s(t) = -\tfrac{1}{2}at^2 + c.$$

106. ***Inventory Management*** The annual inventory cost C for a manufacturer is

$$C = \frac{1{,}008{,}000}{Q} + 6.3Q$$

where Q is the order size when the inventory is replenished. Find the change in annual cost when Q is increased from 350 to 351, and compare this with the instantaneous rate of change when $Q = 350$.

107. ***Writing*** The number of gallons N of regular unleaded gasoline sold by a gasoline station at a price of p dollars per gallon is given by $N = f(p)$.

(a) Describe the meaning of $f'(1.479)$.

(b) Is $f'(1.479)$ usually positive or negative? Explain.

108. ***Newton's Law of Cooling*** This law states that the rate of change of the temperature of an object is proportional to the difference between the object's temperature T and the temperature T_a of the surrounding medium. Write an equation for this law.

109. Find an equation of the parabola $y = ax^2 + bx + c$ that passes through $(0, 1)$ and is tangent to the line $y = x - 1$ at $(1, 0)$.

110. Let (a, b) be an arbitrary point on the graph of $y = 1/x$, $x > 0$. Prove that the area of the triangle formed by the tangent line through (a, b) and the coordinate axes is 2.

111. Find the tangent line(s) to the curve $y = x^3 - 9x$ through the point $(1, -9)$.

112. Find the equation(s) of the tangent line(s) to the parabola $y = x^2$ through the given point.

(a) $(0, a)$ (b) $(a, 0)$

Are there any restrictions on the constant a?

In Exercises 113 and 114, find *a* and *b* such that *f* is differentiable everywhere.

113. $f(x) = \begin{cases} ax^3, & x \le 2 \\ x^2 + b, & x > 2 \end{cases}$

114. $f(x) = \begin{cases} \cos x, & x < 0 \\ ax + b, & x \ge 0 \end{cases}$

115. Where are the functions $f_1(x) = |\sin x|$ and $f_2(x) = \sin |x|$ differentiable?

116. Prove that $\dfrac{d}{dx}[\cos x] = -\sin x$.

FOR FURTHER INFORMATION For a geometric interpretation of the derivatives of trigonometric functions, see the article "Sines and Cosines of the Times" by Victor J. Katz in *Math Horizons*. To view this article, go to the website *www.matharticles.com*.

Section 2.3 Product and Quotient Rules and Higher-Order Derivatives

- **Find the derivative of a function using the Product Rule.**
- **Find the derivative of a function using the Quotient Rule.**
- **Find the derivative of a trigonometric function.**
- **Find a higher-order derivative of a function.**

The Product Rule

In Section 2.2 you learned that the derivative of the sum of two functions is simply the sum of their derivatives. The rules for the derivatives of the product and quotient of two functions are not as simple.

NOTE A version of the Product Rule that some people prefer is

$$\frac{d}{dx}[f(x)g(x)] = f'(x)g(x) + f(x)g'(x).$$

The advantage of this form is that it generalizes easily to products involving three or more factors.

THEOREM 2.7 The Product Rule

The product of two differentiable functions f and g is itself differentiable. Moreover, the derivative of fg is the first function times the derivative of the second, plus the second function times the derivative of the first.

$$\frac{d}{dx}[f(x)g(x)] = f(x)g'(x) + g(x)f'(x)$$

Proof Some mathematical proofs, such as the proof of the Sum Rule, are straightforward. Others involve clever steps that may appear unmotivated to a reader. This proof involves such a step—subtracting and adding the same quantity—which is shown in color.

$$\begin{aligned}
\frac{d}{dx}[f(x)g(x)] &= \lim_{\Delta x\to 0} \frac{f(x+\Delta x)g(x+\Delta x) - f(x)g(x)}{\Delta x}\\
&= \lim_{\Delta x\to 0} \frac{f(x+\Delta x)g(x+\Delta x) - f(x+\Delta x)g(x) + f(x+\Delta x)g(x) - f(x)g(x)}{\Delta x}\\
&= \lim_{\Delta x\to 0} \left[f(x+\Delta x)\frac{g(x+\Delta x)-g(x)}{\Delta x} + g(x)\frac{f(x+\Delta x)-f(x)}{\Delta x}\right]\\
&= \lim_{\Delta x\to 0} \left[f(x+\Delta x)\frac{g(x+\Delta x)-g(x)}{\Delta x}\right] + \lim_{\Delta x\to 0}\left[g(x)\frac{f(x+\Delta x)-f(x)}{\Delta x}\right]\\
&= \lim_{\Delta x\to 0} f(x+\Delta x)\cdot \lim_{\Delta x\to 0}\frac{g(x+\Delta x)-g(x)}{\Delta x} + \lim_{\Delta x\to 0} g(x)\cdot \lim_{\Delta x\to 0}\frac{f(x+\Delta x)-f(x)}{\Delta x}\\
&= f(x)g'(x) + g(x)f'(x)
\end{aligned}$$

Note that $\lim_{\Delta x\to 0} f(x+\Delta x) = f(x)$ because f is given to be differentiable and therefore is continuous.

The Product Rule can be extended to cover products involving more than two factors. For example, if f, g, and h are differentiable functions of x, then

$$\frac{d}{dx}[f(x)g(x)h(x)] = f'(x)g(x)h(x) + f(x)g'(x)h(x) + f(x)g(x)h'(x).$$

For instance, the derivative of $y = x^2 \sin x \cos x$ is

$$\begin{aligned}
\frac{dy}{dx} &= 2x \sin x \cos x + x^2 \cos x \cos x + x^2 \sin x(-\sin x)\\
&= 2x \sin x \cos x + x^2(\cos^2 x - \sin^2 x).
\end{aligned}$$

THE PRODUCT RULE

When Leibniz originally wrote a formula for the Product Rule, he was motivated by the expression

$$(x + dx)(y + dy) - xy$$

from which he subtracted $dx\,dy$ (as being negligible) and obtained the differential form $x\,dy + y\,dx$. This derivation resulted in the traditional form of the Product Rule. *(Source: The History of Mathematics by David M. Burton)*

The derivative of a product of two functions is not (in general) given by the product of the derivatives of the two functions. To see this, try comparing the product of the derivatives of $f(x) = 3x - 2x^2$ and $g(x) = 5 + 4x$ with the derivative in Example 1.

EXAMPLE 1 Using the Product Rule

Find the derivative of $h(x) = (3x - 2x^2)(5 + 4x)$.

Solution

$$h'(x) = \underbrace{(3x - 2x^2)}_{\text{First}}\underbrace{\frac{d}{dx}[5 + 4x]}_{\text{Derivative of second}} + \underbrace{(5 + 4x)}_{\text{Second}}\underbrace{\frac{d}{dx}[3x - 2x^2]}_{\text{Derivative of first}} \qquad \text{Apply Product Rule.}$$

$$= (3x - 2x^2)(4) + (5 + 4x)(3 - 4x)$$
$$= (12x - 8x^2) + (15 - 8x - 16x^2)$$
$$= -24x^2 + 4x + 15$$

In Example 1, you have the option of finding the derivative with or without the Product Rule. To find the derivative without the Product Rule, you can write

$$D_x[(3x - 2x^2)(5 + 4x)] = D_x[-8x^3 + 2x^2 + 15x]$$
$$= -24x^2 + 4x + 15.$$

In the next example, you must use the Product Rule.

EXAMPLE 2 Using the Product Rule

Find the derivative of $y = 3x^2 \sin x$.

Solution

$$\frac{d}{dx}[3x^2 \sin x] = 3x^2 \frac{d}{dx}[\sin x] + \sin x \frac{d}{dx}[3x^2] \qquad \text{Apply Product Rule.}$$
$$= 3x^2 \cos x + (\sin x)(6x)$$
$$= 3x^2 \cos x + 6x \sin x$$
$$= 3x(x \cos x + 2 \sin x)$$

EXAMPLE 3 Using the Product Rule

Find the derivative of $y = 2x \cos x - 2 \sin x$.

Solution

$$\frac{dy}{dx} = \underbrace{(2x)\left(\frac{d}{dx}[\cos x]\right) + (\cos x)\left(\frac{d}{dx}[2x]\right)}_{\text{Product Rule}} - \underbrace{2\frac{d}{dx}[\sin x]}_{\text{Constant Multiple Rule}}$$
$$= (2x)(-\sin x) + (\cos x)(2) - 2(\cos x)$$
$$= -2x \sin x$$

NOTE In Example 3, notice that you use the Product Rule when both factors of the product are variable, and you use the Constant Multiple Rule when one of the factors is a constant.

The Quotient Rule

THEOREM 2.8 The Quotient Rule

The quotient f/g of two differentiable functions f and g is itself differentiable at all values of x for which $g(x) \neq 0$. Moreover, the derivative of f/g is given by the denominator times the derivative of the numerator minus the numerator times the derivative of the denominator, all divided by the square of the denominator.

$$\frac{d}{dx}\left[\frac{f(x)}{g(x)}\right] = \frac{g(x)f'(x) - f(x)g'(x)}{[g(x)]^2}, \qquad g(x) \neq 0$$

Proof As with the proof of Theorem 2.7, the key to this proof is subtracting and adding the same quantity.

$$\frac{d}{dx}\left[\frac{f(x)}{g(x)}\right] = \lim_{\Delta x \to 0} \frac{\dfrac{f(x+\Delta x)}{g(x+\Delta x)} - \dfrac{f(x)}{g(x)}}{\Delta x} \qquad \text{Definition of derivative}$$

$$= \lim_{\Delta x \to 0} \frac{g(x)f(x+\Delta x) - f(x)g(x+\Delta x)}{\Delta x g(x) g(x+\Delta x)}$$

$$= \lim_{\Delta x \to 0} \frac{g(x)f(x+\Delta x) - f(x)g(x) + f(x)g(x) - f(x)g(x+\Delta x)}{\Delta x g(x) g(x+\Delta x)}$$

$$= \frac{\displaystyle\lim_{\Delta x \to 0} \frac{g(x)[f(x+\Delta x) - f(x)]}{\Delta x} - \lim_{\Delta x \to 0} \frac{f(x)[g(x+\Delta x) - g(x)]}{\Delta x}}{\displaystyle\lim_{\Delta x \to 0} [g(x)g(x+\Delta x)]}$$

$$= \frac{g(x)\left[\displaystyle\lim_{\Delta x \to 0} \frac{f(x+\Delta x) - f(x)}{\Delta x}\right] - f(x)\left[\displaystyle\lim_{\Delta x \to 0} \frac{g(x+\Delta x) - g(x)}{\Delta x}\right]}{\displaystyle\lim_{\Delta x \to 0} [g(x)g(x+\Delta x)]}$$

$$= \frac{g(x)f'(x) - f(x)g'(x)}{[g(x)]^2}$$

Note that $\lim_{\Delta x \to 0} g(x + \Delta x) = g(x)$ because g is given to be differentiable and therefore is continuous.

TECHNOLOGY A graphing utility can be used to compare the graph of a function with the graph of its derivative. For instance, in Figure 2.22, the graph of the function in Example 4 appears to have two points that have horizontal tangent lines. What are the values of y' at these two points?

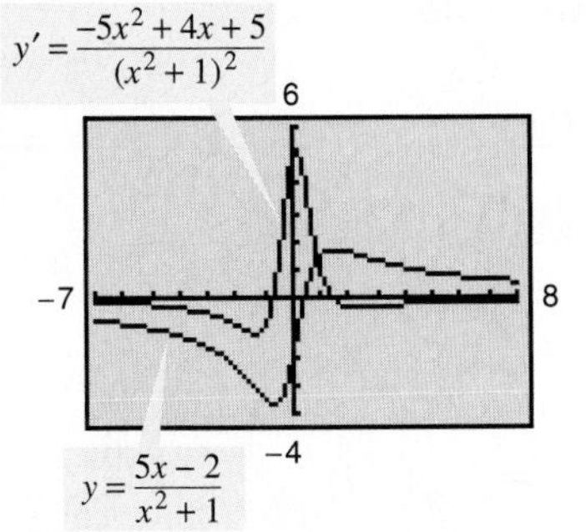

Graphical comparison of a function and its derivative
Figure 2.22

EXAMPLE 4 Using the Quotient Rule

Find the derivative of $y = \dfrac{5x - 2}{x^2 + 1}$.

Solution

$$\frac{d}{dx}\left[\frac{5x - 2}{x^2 + 1}\right] = \frac{(x^2+1)\frac{d}{dx}[5x - 2] - (5x - 2)\frac{d}{dx}[x^2 + 1]}{(x^2+1)^2} \qquad \text{Apply Quotient Rule.}$$

$$= \frac{(x^2+1)(5) - (5x - 2)(2x)}{(x^2+1)^2}$$

$$= \frac{(5x^2 + 5) - (10x^2 - 4x)}{(x^2+1)^2}$$

$$= \frac{-5x^2 + 4x + 5}{(x^2+1)^2}$$

Note the use of parentheses in Example 4. A liberal use of parentheses is recommended for *all* types of differentiation problems. For instance, with the Quotient Rule, it is a good idea to enclose all factors and derivatives in parentheses, and to pay special attention to the subtraction required in the numerator.

When differentiation rules were introduced in the preceding section, the need for rewriting *before* differentiating was emphasized. The next example illustrates this point with the Quotient Rule.

EXAMPLE 5 Rewriting Before Differentiating

Find an equation of the tangent line to the graph of $f(x) = \dfrac{3 - (1/x)}{x + 5}$ at $(-1, 1)$.

Solution Begin by rewriting the function.

$$f(x) = \frac{3 - (1/x)}{x + 5} \qquad \text{Write original function.}$$

$$= \frac{x\left(3 - \dfrac{1}{x}\right)}{x(x + 5)} \qquad \text{Multiply numerator and denominator by } x.$$

$$= \frac{3x - 1}{x^2 + 5x} \qquad \text{Rewrite.}$$

$$f'(x) = \frac{(x^2 + 5x)(3) - (3x - 1)(2x + 5)}{(x^2 + 5x)^2} \qquad \text{Quotient Rule}$$

$$= \frac{(3x^2 + 15x) - (6x^2 + 13x - 5)}{(x^2 + 5x)^2}$$

$$= \frac{-3x^2 + 2x + 5}{(x^2 + 5x)^2} \qquad \text{Simplify.}$$

To find the slope at $(-1, 1)$, evaluate $f'(-1)$.

$$f'(-1) = 0 \qquad \text{Slope of graph at } (-1, 1)$$

Then, using the point-slope form of the equation of a line, you can determine that the equation of the tangent line at $(-1, 1)$ is $y = 1$. See Figure 2.23.

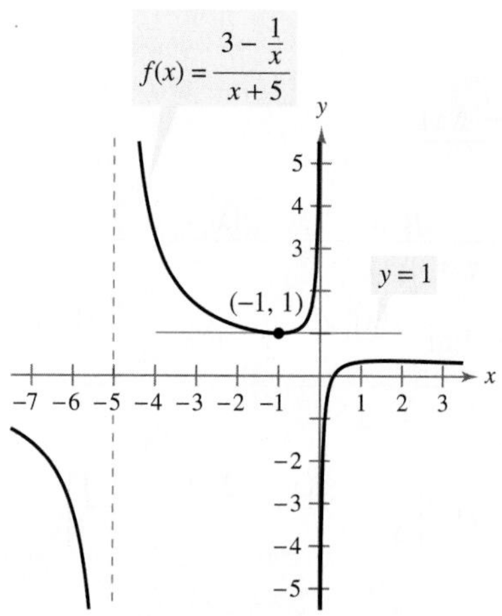

The line $y = 1$ is tangent to the graph of $f(x)$ at the point $(-1, 1)$.
Figure 2.23

Not every quotient needs to be differentiated by the Quotient Rule. For example, each quotient in the next example can be considered as the product of a constant times a function of x. In such cases it is more convenient to use the Constant Multiple Rule.

EXAMPLE 6 Using the Constant Multiple Rule

	Original Function	*Rewrite*	*Differentiate*	*Simplify*
a.	$y = \dfrac{x^2 + 3x}{6}$	$y = \dfrac{1}{6}(x^2 + 3x)$	$y' = \dfrac{1}{6}(2x + 3)$	$y' = \dfrac{2x + 3}{6}$
b.	$y = \dfrac{5x^4}{8}$	$y = \dfrac{5}{8}x^4$	$y' = \dfrac{5}{8}(4x^3)$	$y' = \dfrac{5}{2}x^3$
c.	$y = \dfrac{-3(3x - 2x^2)}{7x}$	$y = -\dfrac{3}{7}(3 - 2x)$	$y' = -\dfrac{3}{7}(-2)$	$y' = \dfrac{6}{7}$
d.	$y = \dfrac{9}{5x^2}$	$y = \dfrac{9}{5}(x^{-2})$	$y' = \dfrac{9}{5}(-2x^{-3})$	$y' = -\dfrac{18}{5x^3}$

NOTE To see the benefit of using the Constant Multiple Rule for some quotients, try using the Quotient Rule to differentiate the functions in Example 6—you should obtain the same results, but with more work.

In Section 2.2, the Power Rule was proved only for the case where the exponent n is a positive integer greater than 1. The next example extends the proof to include negative integer exponents.

EXAMPLE 7 Proof of the Power Rule (Negative Integer Exponents)

If n is a negative integer, there exists a positive integer k such that $n = -k$. So, by the Quotient Rule, you can write

$$\begin{aligned}\frac{d}{dx}[x^n] &= \frac{d}{dx}\left[\frac{1}{x^k}\right]\\ &= \frac{x^k(0) - (1)(kx^{k-1})}{(x^k)^2} && \text{Quotient Rule and Power Rule}\\ &= \frac{0 - kx^{k-1}}{x^{2k}}\\ &= -kx^{-k-1}\\ &= nx^{n-1}. && n = -k\end{aligned}$$

So, the Power Rule

$$D_x[x^n] = nx^{n-1} \qquad \text{Power Rule}$$

is valid for any integer. In Exercise 75 in Section 2.5, you are asked to prove the case for which n is any rational number.

Derivatives of Trigonometric Functions

Knowing the derivatives of the sine and cosine functions, you can use the Quotient Rule to find the derivatives of the four remaining trigonometric functions.

THEOREM 2.9 Derivatives of Trigonometric Functions

$$\frac{d}{dx}[\tan x] = \sec^2 x \qquad \frac{d}{dx}[\cot x] = -\csc^2 x$$

$$\frac{d}{dx}[\sec x] = \sec x \tan x \qquad \frac{d}{dx}[\csc x] = -\csc x \cot x$$

Proof Considering $\tan x = (\sin x)/(\cos x)$ and applying the Quotient Rule, you obtain

$$\begin{aligned}\frac{d}{dx}[\tan x] &= \frac{(\cos x)(\cos x) - (\sin x)(-\sin x)}{\cos^2 x} && \text{Apply Quotient Rule.}\\ &= \frac{\cos^2 x + \sin^2 x}{\cos^2 x}\\ &= \frac{1}{\cos^2 x}\\ &= \sec^2 x.\end{aligned}$$

The proofs of the other three parts of the theorem are left as an exercise (see Exercise 89).

EXAMPLE 8 Differentiating Trigonometric Functions

NOTE Because of trigonometric identities, the derivative of a trigonometric function can take many forms. This presents a challenge when you are trying to match your answers to those given in the back of the text.

Function	*Derivative*
a. $y = x - \tan x$	$\dfrac{dy}{dx} = 1 - \sec^2 x$
b. $y = x \sec x$	$y' = x(\sec x \tan x) + (\sec x)(1)$ $= (\sec x)(1 + x \tan x)$

EXAMPLE 9 Different Forms of a Derivative

Differentiate both forms of $y = \dfrac{1 - \cos x}{\sin x} = \csc x - \cot x$.

Solution

First form: $y = \dfrac{1 - \cos x}{\sin x}$

$$y' = \frac{(\sin x)(\sin x) - (1 - \cos x)(\cos x)}{\sin^2 x}$$

$$= \frac{\sin^2 x + \cos^2 x - \cos x}{\sin^2 x}$$

$$= \frac{1 - \cos x}{\sin^2 x}$$

Second form: $y = \csc x - \cot x$

$$y' = -\csc x \cot x + \csc^2 x$$

To show that the two derivatives are equal, you can write

$$\frac{1 - \cos x}{\sin^2 x} = \frac{1}{\sin^2 x} - \left(\frac{1}{\sin x}\right)\left(\frac{\cos x}{\sin x}\right)$$

$$= \csc^2 x - \csc x \cot x.$$

The summary below shows that much of the work in obtaining a simplified form of a derivative occurs *after* differentiating. Note that two characteristics of a simplified form are the absence of negative exponents and the combining of like terms.

	$f'(x)$ After Differentiating	*$f'(x)$ After Simplifying*
Example 1	$(3x - 2x^2)(4) + (5 + 4x)(3 - 4x)$	$-24x^2 + 4x + 15$
Example 3	$(2x)(-\sin x) + (\cos x)(2) - 2(\cos x)$	$-2x \sin x$
Example 4	$\dfrac{(x^2 + 1)(5) - (5x - 2)(2x)}{(x^2 + 1)^2}$	$\dfrac{-5x^2 + 4x + 5}{(x^2 + 1)^2}$
Example 5	$\dfrac{(x^2 + 5x)(3) - (3x - 1)(2x + 5)}{(x^2 + 5x)^2}$	$\dfrac{-3x^2 + 2x + 5}{(x^2 + 5x)^2}$
Example 9	$\dfrac{(\sin x)(\sin x) - (1 - \cos x)(\cos x)}{\sin^2 x}$	$\dfrac{1 - \cos x}{\sin^2 x}$

Higher-Order Derivatives

Just as you can obtain a velocity function by differentiating a position function, you can obtain an **acceleration** function by differentiating a velocity function. Another way of looking at this is that you can obtain an acceleration function by differentiating a position function *twice*.

$$s(t) \qquad \text{Position function}$$
$$v(t) = s'(t) \qquad \text{Velocity function}$$
$$a(t) = v'(t) = s''(t) \qquad \text{Acceleration function}$$

NOTE: The second derivative of f is the derivative of the first derivative of f.

The function given by $a(t)$ is the **second derivative** of $s(t)$ and is denoted by $s''(t)$.

The second derivative is an example of a **higher-order derivative.** You can define derivatives of any positive integer order. For instance, the **third derivative** is the derivative of the second derivative. Higher-order derivatives are denoted as follows.

First derivative:	y',	$f'(x)$,	$\frac{dy}{dx}$,	$\frac{d}{dx}[f(x)]$,	$D_x[y]$
Second derivative:	y'',	$f''(x)$,	$\frac{d^2y}{dx^2}$,	$\frac{d^2}{dx^2}[f(x)]$,	$D_x^2[y]$
Third derivative:	y''',	$f'''(x)$,	$\frac{d^3y}{dx^3}$,	$\frac{d^3}{dx^3}[f(x)]$,	$D_x^3[y]$
Fourth derivative:	$y^{(4)}$,	$f^{(4)}(x)$,	$\frac{d^4y}{dx^4}$,	$\frac{d^4}{dx^4}[f(x)]$,	$D_x^4[y]$
$\vdots$					
nth derivative:	$y^{(n)}$,	$f^{(n)}(x)$,	$\frac{d^ny}{dx^n}$,	$\frac{d^n}{dx^n}[f(x)]$,	$D_x^n[y]$

NASA

THE MOON

The moon's mass is 7.349×10^{22} kilograms, and Earth's mass is 5.976×10^{24} kilograms. The moon's radius is 1737 kilometers, and Earth's radius is 6378 kilometers. Because the gravitational force on the surface of a planet is directly proportional to its mass and inversely proportional to the square of its radius, the ratio of the gravitational force on Earth to the gravitational force on the moon is

$$\frac{(5.976 \times 10^{24})/6378^2}{(7.349 \times 10^{22})/1737^2} \approx 6.03.$$

EXAMPLE 10 Finding the Acceleration Due to Gravity

Because the moon has no atmosphere, a falling object on the moon encounters no air resistance. In 1971, astronaut David Scott demonstrated that a feather and a hammer fall at the same rate on the moon. The position function for each of these falling objects is given by

$$s(t) = -0.81t^2 + 2$$

where $s(t)$ is the height in meters and t is the time in seconds. What is the ratio of Earth's gravitational force to the moon's?

Solution To find the acceleration, differentiate the position function twice.

$$s(t) = -0.81t^2 + 2 \qquad \text{Position function}$$
$$s'(t) = -1.62t \qquad \text{Velocity function}$$
$$s''(t) = -1.62 \qquad \text{Acceleration function}$$

So, the acceleration due to gravity on the moon is -1.62 meters per second per second. Because the acceleration due to gravity on Earth is -9.8 meters per second per second, the ratio of Earth's gravitational force to the moon's is

$$\frac{\text{Earth's gravitational force}}{\text{Moon's gravitational force}} = \frac{-9.8}{-1.62} \approx 6.05.$$

Exercises for Section 2.3

See www.CalcChat.com for worked-out solutions to odd-numbered exercises.

In Exercises 1–6, use the Product Rule to differentiate the function.

1. $g(x) = (x^2 + 1)(x^2 - 2x)$
2. $f(x) = (6x + 5)(x^3 - 2)$
3. $h(t) = \sqrt[3]{t}(t^2 + 4)$
4. $g(s) = \sqrt{s}(4 - s^2)$
5. $f(x) = x^3 \cos x$
6. $g(x) = \sqrt{x} \sin x$

In Exercises 7–12, use the Quotient Rule to differentiate the function.

7. $f(x) = \dfrac{x}{x^2 + 1}$
8. $g(t) = \dfrac{t^2 + 2}{2t - 7}$
9. $h(x) = \dfrac{\sqrt[3]{x}}{x^3 + 1}$
10. $h(s) = \dfrac{s}{\sqrt{s} - 1}$
11. $g(x) = \dfrac{\sin x}{x^2}$
12. $f(t) = \dfrac{\cos t}{t^3}$

In Exercises 13–18, find $f'(x)$ and $f'(c)$.

Function	*Value of c*
13. $f(x) = (x^3 - 3x)(2x^2 + 3x + 5)$	$c = 0$
14. $f(x) = (x^2 - 2x + 1)(x^3 - 1)$	$c = 1$
15. $f(x) = \dfrac{x^2 - 4}{x - 3}$	$c = 1$
16. $f(x) = \dfrac{x + 1}{x - 1}$	$c = 2$
17. $f(x) = x \cos x$	$c = \dfrac{\pi}{4}$
18. $f(x) = \dfrac{\sin x}{x}$	$c = \dfrac{\pi}{6}$

In Exercises 19–24, complete the table without using the Quotient Rule.

Function	*Rewrite*	*Differentiate*	*Simplify*
19. $y = \dfrac{x^2 + 2x}{3}$			
20. $y = \dfrac{5x^2 - 3}{4}$			
21. $y = \dfrac{7}{3x^3}$			
22. $y = \dfrac{4}{5x^2}$			
23. $y = \dfrac{4x^{3/2}}{x}$			
24. $y = \dfrac{3x^2 - 5}{7}$			

In Exercises 25–38, find the derivative of the algebraic function.

25. $f(x) = \dfrac{3 - 2x - x^2}{x^2 - 1}$
26. $f(x) = \dfrac{x^3 + 3x + 2}{x^2 - 1}$
27. $f(x) = x\left(1 - \dfrac{4}{x + 3}\right)$
28. $f(x) = x^4\left(1 - \dfrac{2}{x + 1}\right)$
29. $f(x) = \dfrac{2x + 5}{\sqrt{x}}$
30. $f(x) = \sqrt[3]{x}(\sqrt{x} + 3)$
31. $h(s) = (s^3 - 2)^2$
32. $h(x) = (x^2 - 1)^2$
33. $f(x) = \dfrac{2 - \dfrac{1}{x}}{x - 3}$
34. $g(x) = x^2\left(\dfrac{2}{x} - \dfrac{1}{x + 1}\right)$
35. $f(x) = (3x^3 + 4x)(x - 5)(x + 1)$
36. $f(x) = (x^2 - x)(x^2 + 1)(x^2 + x + 1)$
37. $f(x) = \dfrac{x^2 + c^2}{x^2 - c^2}$, c is a constant
38. $f(x) = \dfrac{c^2 - x^2}{c^2 + x^2}$, c is a constant

In Exercises 39–54, find the derivative of the trigonometric function.

39. $f(t) = t^2 \sin t$
40. $f(\theta) = (\theta + 1) \cos \theta$
41. $f(t) = \dfrac{\cos t}{t}$
42. $f(x) = \dfrac{\sin x}{x}$
43. $f(x) = -x + \tan x$
44. $y = x + \cot x$
45. $g(t) = \sqrt[4]{t} + 8 \sec t$
46. $h(s) = \dfrac{1}{s} - 10 \csc s$
47. $y = \dfrac{3(1 - \sin x)}{2 \cos x}$
48. $y = \dfrac{\sec x}{x}$
49. $y = -\csc x - \sin x$
50. $y = x \sin x + \cos x$
51. $f(x) = x^2 \tan x$
52. $f(x) = \sin x \cos x$
53. $y = 2x \sin x + x^2 \cos x$
54. $h(\theta) = 5\theta \sec \theta + \theta \tan \theta$

In Exercises 55–58, use a computer algebra system to differentiate the function.

55. $g(x) = \left(\dfrac{x + 1}{x + 2}\right)(2x - 5)$
56. $f(x) = \left(\dfrac{x^2 - x - 3}{x^2 + 1}\right)(x^2 + x + 1)$
57. $g(\theta) = \dfrac{\theta}{1 - \sin \theta}$
58. $f(\theta) = \dfrac{\sin \theta}{1 - \cos \theta}$

In Exercises 59–62, evaluate the derivative of the function at the given point. Use a graphing utility to verify your result.

Function	*Point*
59. $y = \dfrac{1 + \csc x}{1 - \csc x}$	$\left(\dfrac{\pi}{6}, -3\right)$
60. $f(x) = \tan x \cot x$	$(1, 1)$
61. $h(t) = \dfrac{\sec t}{t}$	$\left(\pi, -\dfrac{1}{\pi}\right)$
62. $f(x) = \sin x(\sin x + \cos x)$	$\left(\dfrac{\pi}{4}, 1\right)$

In Exercises 63–68, (a) find an equation of the tangent line to the graph of f at the given point, (b) use a graphing utility to graph the function and its tangent line at the point, and (c) use the *derivative* feature of a graphing utility to confirm your results.

63. $f(x) = (x^3 - 3x + 1)(x + 2)$, $(1, -3)$

64. $f(x) = (x - 1)(x^2 - 2)$, $(0, 2)$

65. $f(x) = \dfrac{x}{x - 1}$, $(2, 2)$

66. $f(x) = \dfrac{(x - 1)}{(x + 1)}$, $\left(2, \dfrac{1}{3}\right)$

67. $f(x) = \tan x$, $\left(\dfrac{\pi}{4}, 1\right)$

68. $f(x) = \sec x$, $\left(\dfrac{\pi}{3}, 2\right)$

Famous Curves **In Exercises 69–72, find an equation of the tangent line to the graph at the given point. (The graphs in Exercises 69 and 70 are called *witches of Agnesi*. The graphs in Exercises 71 and 72 are called *serpentines*.**

69.

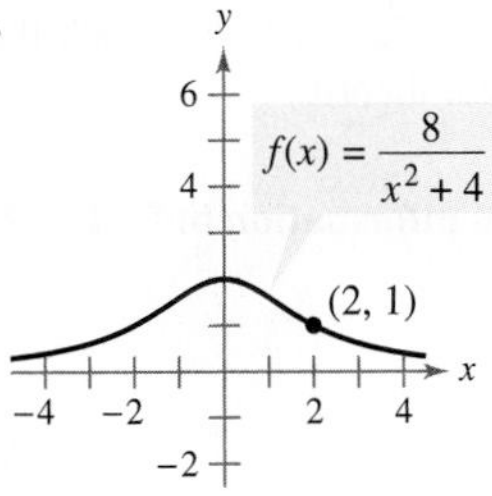

70.

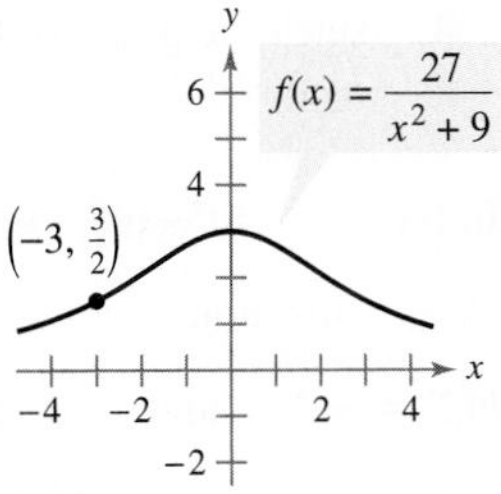

71.

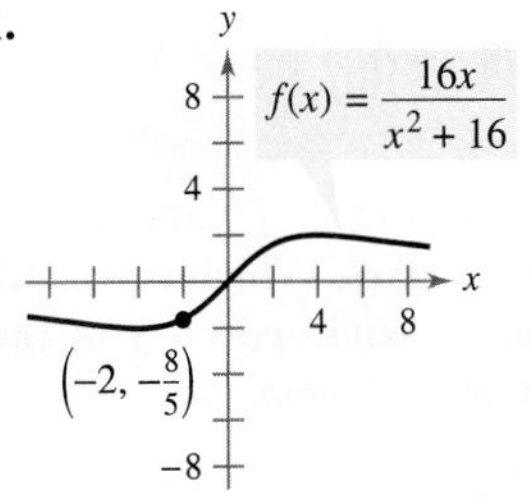

72.

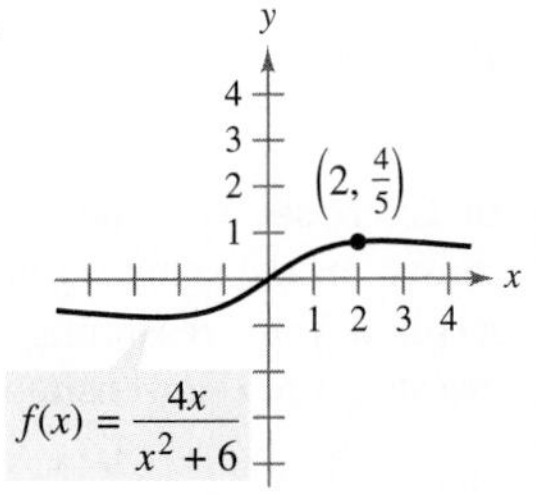

In Exercises 73–76, determine the point(s) at which the graph of the function has a horizontal tangent line.

73. $f(x) = \dfrac{x^2}{x - 1}$

74. $f(x) = \dfrac{x^2}{x^2 + 1}$

75. $f(x) = \dfrac{4x - 2}{x^2}$

76. $f(x) = \dfrac{x - 4}{x^2 - 7}$

77. ***Tangent Lines*** Find equations of the tangent lines to the graph of $f(x) = \dfrac{x + 1}{x - 1}$ that are parallel to the line $2y + x = 6$. Then graph the function and the tangent lines.

78. ***Tangent Lines*** Find equations of the tangent lines to the graph of $f(x) = \dfrac{x}{x - 1}$ that pass through the point $(-1, 5)$. Then graph the function and the tangent lines.

In Exercises 79 and 80, verify that $f'(x) = g'(x)$, and explain the relationship between f and g.

79. $f(x) = \dfrac{3x}{x + 2}$, $g(x) = \dfrac{5x + 4}{x + 2}$

80. $f(x) = \dfrac{\sin x - 3x}{x}$, $g(x) = \dfrac{\sin x + 2x}{x}$

In Exercises 81 and 82, use the graphs of f and g. Let $p(x) = f(x)g(x)$ and $q(x) = \dfrac{f(x)}{g(x)}$.

81. (a) Find $p'(1)$.
(b) Find $q'(4)$.

82. (a) Find $p'(4)$.
(b) Find $q'(7)$.

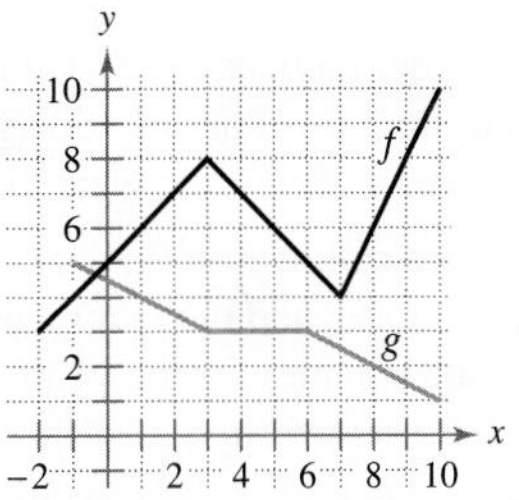

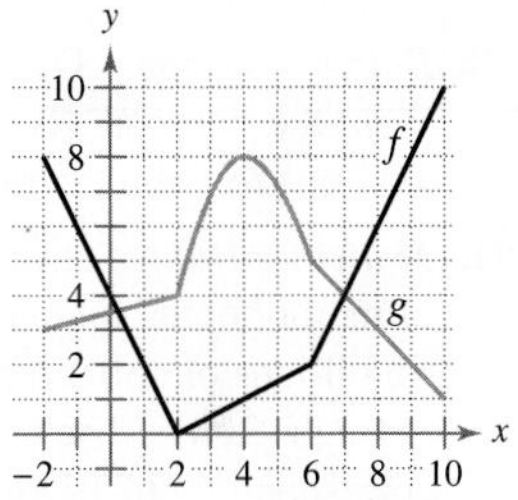

83. ***Area*** The length of a rectangle is given by $2t + 1$ and its height is $\sqrt{t}$, where t is time in seconds and the dimensions are in centimeters. Find the rate of change of the area with respect to time.

84. ***Volume*** The radius of a right circular cylinder is given by $\sqrt{t + 2}$ and its height is $\frac{1}{2}\sqrt{t}$, where t is time in seconds and the dimensions are in inches. Find the rate of change of the volume with respect to time.

85. ***Inventory Replenishment*** The ordering and transportation cost C for the components used in manufacturing a product is

$$C = 100\left(\frac{200}{x^2} + \frac{x}{x + 30}\right), \quad x \geq 1$$

where C is measured in thousands of dollars and x is the order size in hundreds. Find the rate of change of C with respect to x when (a) $x = 10$, (b) $x = 15$, and (c) $x = 20$. What do these rates of change imply about increasing order size?

86. ***Boyle's Law*** This law states that if the temperature of a gas remains constant, its pressure is inversely proportional to its volume. Use the derivative to show that the rate of change of the pressure is inversely proportional to the square of the volume.

87. ***Population Growth*** A population of 500 bacteria is introduced into a culture and grows in number according to the equation

$$P(t) = 500\left(1 + \frac{4t}{50 + t^2}\right)$$

where t is measured in hours. Find the rate at which the population is growing when $t = 2$.

88. ***Gravitational Force*** Newton's Law of Universal Gravitation states that the force F between two masses, m_1 and m_2, is

$$F = \frac{Gm_1m_2}{d^2}$$

where G is a constant and d is the distance between the masses. Find an equation that gives an instantaneous rate of change of F with respect to d. (Assume m_1 and m_2 represent moving points.)

89. Prove the following differentiation rules.

(a) $\frac{d}{dx}[\sec x] = \sec x \tan x$ (b) $\frac{d}{dx}[\csc x] = -\csc x \cot x$

(c) $\frac{d}{dx}[\cot x] = -\csc^2 x$

90. ***Rate of Change*** Determine whether there exist any values of x in the interval $[0, 2\pi)$ such that the rate of change of $f(x) = \sec x$ and the rate of change of $g(x) = \csc x$ are equal.

91. ***Modeling Data*** The table shows the numbers n (in thousands) of motor homes sold in the United States and the retail values v (in billions of dollars) of these motor homes for the years 1996 through 2001. The year is represented by t, with $t = 6$ corresponding to 1996. *(Source: Recreation Vehicle Industry Association)*

Year, t	6	7	8	9	10	11
n	247.5	254.5	292.7	321.2	300.1	256.8
v	6.3	6.9	8.4	10.4	9.5	8.6

(a) Use a graphing utility to find cubic models for the number of motor homes sold $n(t)$ and the total retail value $v(t)$ of the motor homes.

(b) Graph each model found in part (a).

(c) Find $A = v(t)/n(t)$, then graph A. What does this function represent?

(d) Interpret $A'(t)$ in the context of these data.

92. ***Satellites*** When satellites observe Earth, they can scan only part of Earth's surface. Some satellites have sensors that can measure the angle θ shown in the figure. Let h represent the satellite's distance from Earth's surface and let r represent Earth's radius.

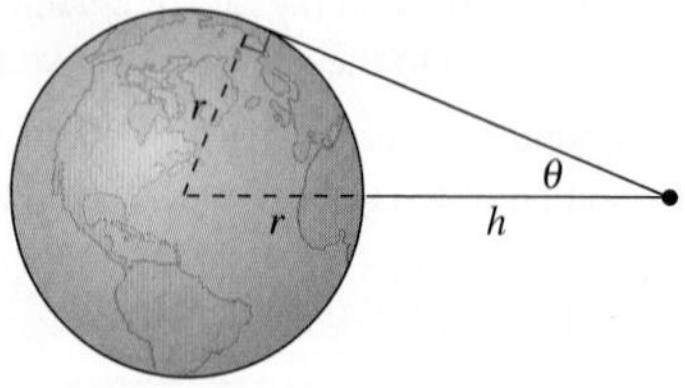

(a) Show that $h = r(\csc\theta - 1)$.

(b) Find the rate at which h is changing with respect to θ when $\theta = 30°$. (Assume $r = 3960$ miles.)

In Exercises 93–98, find the second derivative of the function.

93. $f(x) = 4x^{3/2}$ **94.** $f(x) = x + 32x^{-2}$

95. $f(x) = \frac{x}{x-1}$ **96.** $f(x) = \frac{x^2 + 2x - 1}{x}$

97. $f(x) = 3 \sin x$ **98.** $f(x) = \sec x$

In Exercises 99–102, find the given higher-order derivative.

99. $f'(x) = x^2$, $f''(x)$ **100.** $f''(x) = 2 - \frac{2}{x}$, $f'''(x)$

101. $f'''(x) = 2\sqrt{x}$, $f^{(4)}(x)$ **102.** $f^{(4)}(x) = 2x + 1$, $f^{(6)}(x)$

Writing About Concepts

103. Sketch the graph of a differentiable function f such that $f(2) = 0$, $f' < 0$ for $-\infty < x < 2$, and $f' > 0$ for $2 < x < \infty$.

104. Sketch the graph of a differentiable function f such that $f > 0$ and $f' < 0$ for all real numbers x.

In Exercises 105–108, use the given information to find $f'(2)$.

$g(2) = 3$ and $g'(2) = -2$

$h(2) = -1$ and $h'(2) = 4$

105. $f(x) = 2g(x) + h(x)$ **106.** $f(x) = 4 - h(x)$

107. $f(x) = \frac{g(x)}{h(x)}$ **108.** $f(x) = g(x)h(x)$

In Exercises 109 and 110, the graphs of f, f', and f'' are shown on the same set of coordinate axes. Which is which? Explain your reasoning. To print an enlarged copy of the graph, go to the website *www.mathgraphs.com*.

109.

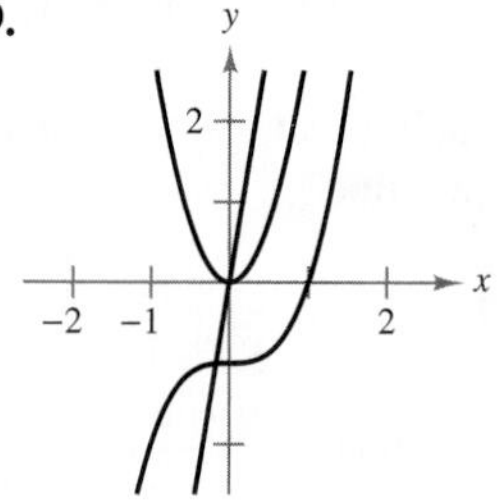

110.

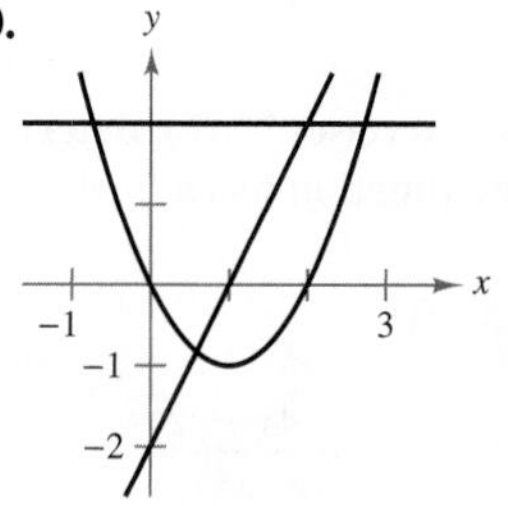

In Exercises 111–114, the graph of f is shown. Sketch the graphs of f' and f''. To print an enlarged copy of the graph, go to the website *www.mathgraphs.com*.

111.

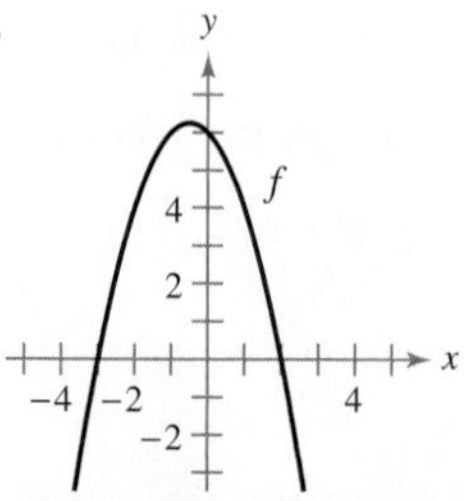

112.

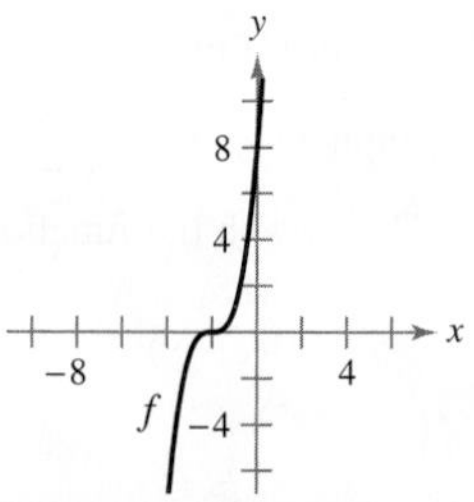

113.

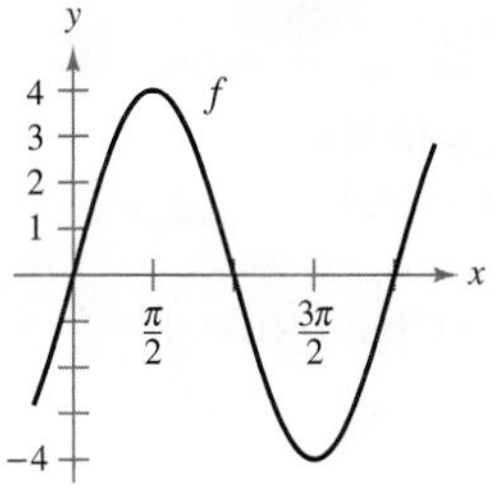

114.

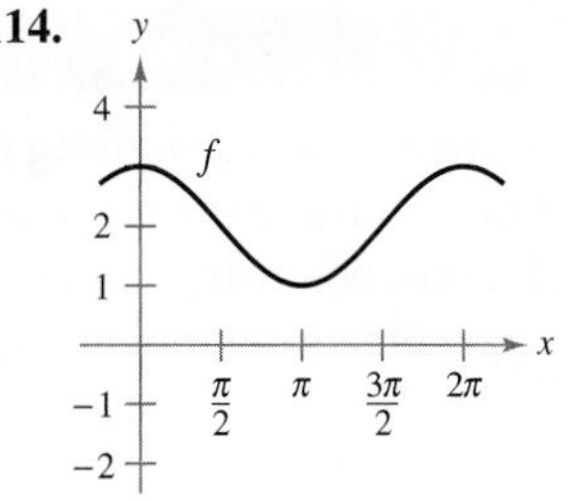

115. ***Acceleration*** The velocity of an object in meters per second is $v(t) = 36 - t^2, 0 \le t \le 6$. Find the velocity and acceleration of the object when $t = 3$. What can be said about the speed of the object when the velocity and acceleration have opposite signs?

116. ***Acceleration*** An automobile's velocity starting from rest is

$$v(t) = \frac{100t}{2t + 15}$$

where v is measured in feet per second. Find the acceleration at (a) 5 seconds, (b) 10 seconds, and (c) 20 seconds.

117. ***Stopping Distance*** A car is traveling at a rate of 66 feet per second (45 miles per hour) when the brakes are applied. The position function for the car is $s(t) = -8.25t^2 + 66t$, where s is measured in feet and t is measured in seconds. Use this function to complete the table, and find the average velocity during each time interval.

t	0	1	2	3	4
$s(t)$					
$v(t)$					
$a(t)$					

118. ***Particle Motion*** The figure shows the graphs of the position, velocity, and acceleration functions of a particle.

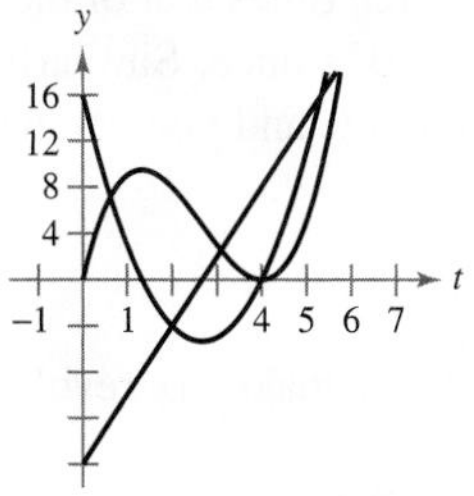

(a) Copy the graphs of the functions shown. Identify each graph. Explain your reasoning. To print an enlarged copy of the graph, go to the website *www.mathgraphs.com*.

(b) On your sketch, identify when the particle speeds up and when it slows down. Explain your reasoning.

Finding a Pattern **In Exercises 119 and 120, develop a general rule for $f^{(n)}(x)$ given $f(x)$.**

119. $f(x) = x^n$

120. $f(x) = \dfrac{1}{x}$

121. ***Finding a Pattern*** Consider the function $f(x) = g(x)h(x)$.

(a) Use the Product Rule to generate rules for finding $f''(x)$, $f'''(x)$, and $f^{(4)}(x)$.

(b) Use the results in part (a) to write a general rule for $f^{(n)}(x)$.

122. ***Finding a Pattern*** Develop a general rule for $[xf(x)]^{(n)}$ where f is a differentiable function of x.

In Exercises 123 and 124, find the derivatives of the function f for $n = 1, 2, 3$, and 4. Use the results to write a general rule for $f'(x)$ in terms of n.

123. $f(x) = x^n \sin x$

124. $f(x) = \dfrac{\cos x}{x^n}$

Differential Equations **In Exercises 125–128, verify that the function satisfies the differential equation.**

	Function	*Differential Equation*
125.	$y = \dfrac{1}{x}, x > 0$	$x^3 y'' + 2x^2 y' = 0$
126.	$y = 2x^3 - 6x + 10$	$-y''' - xy'' - 2y' = -24x^2$
127.	$y = 2 \sin x + 3$	$y'' + y = 3$
128.	$y = 3 \cos x + \sin x$	$y'' + y = 0$

True or False? **In Exercises 129–134, determine whether the statement is true or false. If it is false, explain why or give an example that shows it is false.**

129. If $y = f(x)g(x)$, then $dy/dx = f'(x)g'(x)$.

130. If $y = (x + 1)(x + 2)(x + 3)(x + 4)$, then $d^5y/dx^5 = 0$.

131. If $f'(c)$ and $g'(c)$ are zero and $h(x) = f(x)g(x)$, then $h'(c) = 0$.

132. If $f(x)$ is an nth-degree polynomial, then $f^{(n+1)}(x) = 0$.

133. The second derivative represents the rate of change of the first derivative.

134. If the velocity of an object is constant, then its acceleration is zero.

135. Find a second-degree polynomial $f(x) = ax^2 + bx + c$ such that its graph has a tangent line with slope 10 at the point $(2, 7)$ and an x-intercept at $(1, 0)$.

136. Consider the third-degree polynomial

$$f(x) = ax^3 + bx^2 + cx + d, \quad a \ne 0.$$

Determine conditions for a, b, c, and d if the graph of f has (a) no horizontal tangents, (b) exactly one horizontal tangent, and (c) exactly two horizontal tangents. Give an example for each case.

137. Find the derivative of $f(x) = x|x|$. Does $f''(0)$ exist?

138. ***Think About It*** Let f and g be functions whose first and second derivatives exist on an interval I. Which of the following formulas is (are) true?

(a) $fg'' - f''g = (fg' - f'g)'$

(b) $fg'' + f''g = (fg)''$

Section 2.4 The Chain Rule

- **Find the derivative of a composite function using the Chain Rule.**
- **Find the derivative of a function using the General Power Rule.**
- **Simplify the derivative of a function using algebra.**
- **Find the derivative of a trigonometric function using the Chain Rule.**

The Chain Rule

This text has yet to discuss one of the most powerful differentiation rules—the **Chain Rule.** This rule deals with composite functions and adds a surprising versatility to the rules discussed in the two previous sections. For example, compare the functions shown below. Those on the left can be differentiated without the Chain Rule, and those on the right are best done with the Chain Rule.

Without the Chain Rule	*With the Chain Rule*
$y = x^2 + 1$	$y = \sqrt{x^2 + 1}$
$y = \sin x$	$y = \sin 6x$
$y = 3x + 2$	$y = (3x + 2)^5$
$y = x + \tan x$	$y = x + \tan x^2$

Basically, the Chain Rule states that if y changes dy/du times as fast as u, and u changes du/dx times as fast as x, then y changes $(dy/du)(du/dx)$ times as fast as x.

EXAMPLE 1 The Derivative of a Composite Function

A set of gears is constructed, as shown in Figure 2.24, such that the second and third gears are on the same axle. As the first axle revolves, it drives the second axle, which in turn drives the third axle. Let y, u, and x represent the numbers of revolutions per minute of the first, second, and third axles. Find dy/du, du/dx, and dy/dx, and show that

$$\frac{dy}{dx} = \frac{dy}{du} \cdot \frac{du}{dx}.$$

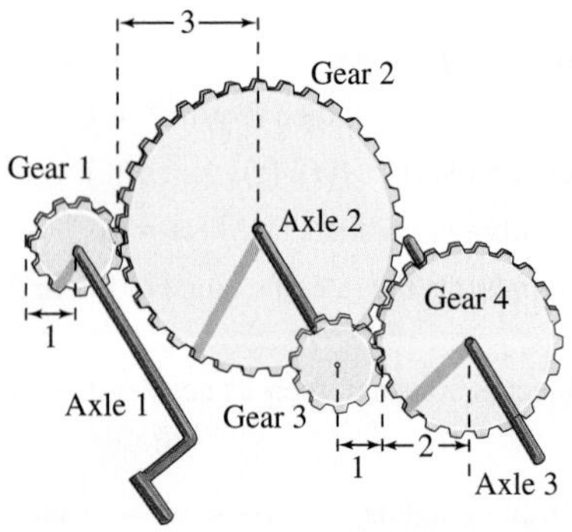

Axle 1: y revolutions per minute
Axle 2: u revolutions per minute
Axle 3: x revolutions per minute
Figure 2.24

Solution Because the circumference of the second gear is three times that of the first, the first axle must make three revolutions to turn the second axle once. Similarly, the second axle must make two revolutions to turn the third axle once, and you can write

$$\frac{dy}{du} = 3 \quad \text{and} \quad \frac{du}{dx} = 2.$$

Combining these two results, you know that the first axle must make six revolutions to turn the third axle once. So, you can write

$$\begin{aligned}\frac{dy}{dx} &= \boxed{\begin{matrix}\text{Rate of change of first axle}\\ \text{with respect to second axle}\end{matrix}} \cdot \boxed{\begin{matrix}\text{Rate of change of second axle}\\ \text{with respect to third axle}\end{matrix}}\\ &= \frac{dy}{du} \cdot \frac{du}{dx} = 3 \cdot 2 = 6\\ &= \boxed{\begin{matrix}\text{Rate of change of first axle}\\ \text{with respect to third axle}\end{matrix}}.\end{aligned}$$

In other words, the rate of change of y with respect to x is the product of the rate of change of y with respect to u and the rate of change of u with respect to x.

> **EXPLORATION**
>
> ***Using the Chain Rule*** Each of the following functions can be differentiated using rules that you studied in Sections 2.2 and 2.3. For each function, find the derivative using those rules. Then find the derivative using the Chain Rule. Compare your results. Which method is simpler?
>
> **a.** $\dfrac{2}{3x + 1}$
>
> **b.** $(x + 2)^3$
>
> **c.** $\sin 2x$

Example 1 illustrates a simple case of the Chain Rule. The general rule is stated below.

> **THEOREM 2.10 The Chain Rule**
>
> If $y = f(u)$ is a differentiable function of u and $u = g(x)$ is a differentiable function of x, then $y = f(g(x))$ is a differentiable function of x and
>
> $$\frac{dy}{dx} = \frac{dy}{du} \cdot \frac{du}{dx}$$
>
> or, equivalently,
>
> $$\frac{d}{dx}[f(g(x))] = f'(g(x))g'(x).$$

Proof Let $h(x) = f(g(x))$. Then, using the alternative form of the derivative, you need to show that, for $x = c$,

$$h'(c) = f'(g(c))g'(c).$$

An important consideration in this proof is the behavior of g as x approaches c. A problem occurs if there are values of x, other than c, such that $g(x) = g(c)$. Appendix A shows how to use the differentiability of f and g to overcome this problem. For now, assume that $g(x) \neq g(c)$ for values of x other than c. In the proofs of the Product Rule and the Quotient Rule, the same quantity was added and subtracted to obtain the desired form. This proof uses a similar technique—multiplying and dividing by the same (nonzero) quantity. Note that because g is differentiable, it is also continuous, and it follows that $g(x) \to g(c)$ as $x \to c$.

$$\begin{aligned} h'(c) &= \lim_{x \to c} \frac{f(g(x)) - f(g(c))}{x - c} \\ &= \lim_{x \to c} \left[\frac{f(g(x)) - f(g(c))}{g(x) - g(c)} \cdot \frac{g(x) - g(c)}{x - c} \right], \quad g(x) \neq g(c) \\ &= \left[\lim_{x \to c} \frac{f(g(x)) - f(g(c))}{g(x) - g(c)} \right] \left[\lim_{x \to c} \frac{g(x) - g(c)}{x - c} \right] \\ &= f'(g(c))g'(c) \end{aligned}$$

When applying the Chain Rule, it is helpful to think of the composite function $f \circ g$ as having two parts—an inner part and an outer part.

Outer function

$$y = f(g(x)) = f(u)$$

Inner function

The derivative of $y = f(u)$ is the derivative of the outer function (at the inner function u) *times* the derivative of the inner function.

$$y' = f'(u) \cdot u'$$

EXAMPLE 2 Decomposition of a Composite Function

$y = f(g(x))$	$u = g(x)$	$y = f(u)$
a. $y = \frac{1}{x+1}$	$u = x + 1$	$y = \frac{1}{u}$
b. $y = \sin 2x$	$u = 2x$	$y = \sin u$
c. $y = \sqrt{3x^2 - x + 1}$	$u = 3x^2 - x + 1$	$y = \sqrt{u}$
d. $y = \tan^2 x$	$u = \tan x$	$y = u^2$

EXAMPLE 3 Using the Chain Rule

Find dy/dx for $y = (x^2 + 1)^3$.

STUDY TIP You could also solve the problem in Example 3 without using the Chain Rule by observing that

$$y = x^6 + 3x^4 + 3x^2 + 1$$

and

$$y' = 6x^5 + 12x^3 + 6x.$$

Verify that this is the same as the derivative in Example 3. Which method would you use to find

$$\frac{d}{dx}(x^2 + 1)^{50}?$$

Solution For this function, you can consider the inside function to be $u = x^2 + 1$. By the Chain Rule, you obtain

$$\frac{dy}{dx} = \underbrace{3(x^2 + 1)^2}_{\frac{dy}{du}}\underbrace{(2x)}_{\frac{du}{dx}} = 6x(x^2 + 1)^2.$$

The General Power Rule

The function in Example 3 is an example of one of the most common types of composite functions, $y = [u(x)]^n$. The rule for differentiating such functions is called the **General Power Rule,** and it is a special case of the Chain Rule.

THEOREM 2.11 The General Power Rule

If $y = [u(x)]^n$, where u is a differentiable function of x and n is a rational number, then

$$\frac{dy}{dx} = n[u(x)]^{n-1}\frac{du}{dx}$$

or, equivalently,

$$\frac{d}{dx}[u^n] = n u^{n-1} u'.$$

Proof Because $y = u^n$, you apply the Chain Rule to obtain

$$\frac{dy}{dx} = \left(\frac{dy}{du}\right)\left(\frac{du}{dx}\right)$$
$$= \frac{d}{du}[u^n]\frac{du}{dx}.$$

By the (Simple) Power Rule in Section 2.2, you have $D_u[u^n] = nu^{n-1}$, and it follows that

$$\frac{dy}{dx} = n[u(x)]^{n-1}\frac{du}{dx}.$$

EXAMPLE 4 Applying the General Power Rule

Find the derivative of $f(x) = (3x - 2x^2)^3$.

Solution Let $u = 3x - 2x^2$. Then

$$f(x) = (3x - 2x^2)^3 = u^3$$

and, by the General Power Rule, the derivative is

$$f'(x) = \overset{n}{3}\overbrace{(3x - 2x^2)^2}^{u^{n-1}} \overbrace{\frac{d}{dx}[3x - 2x^2]}^{u'} \qquad \text{Apply General Power Rule.}$$

$$= 3(3x - 2x^2)^2(3 - 4x). \qquad \text{Differentiate } 3x - 2x^2.$$

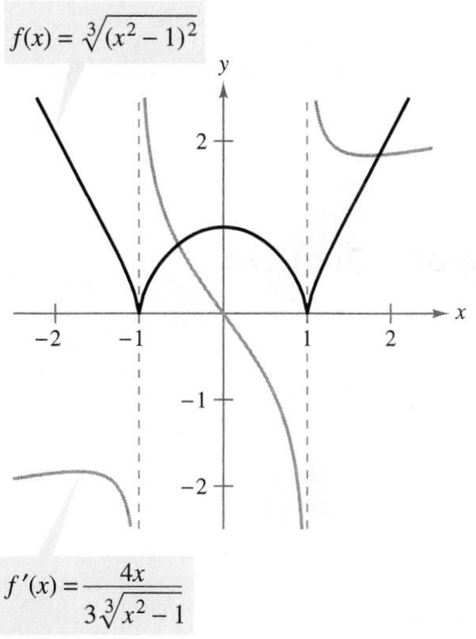

The derivative of f is 0 at $x = 0$ and is undefined at $x = \pm 1$.
Figure 2.25

EXAMPLE 5 Differentiating Functions Involving Radicals

Find all points on the graph of $f(x) = \sqrt[3]{(x^2 - 1)^2}$ for which $f'(x) = 0$ and those for which $f'(x)$ does not exist.

Solution Begin by rewriting the function as

$$f(x) = (x^2 - 1)^{2/3}.$$

Then, applying the General Power Rule (with $u = x^2 - 1$) produces

$$f'(x) = \overset{n}{\frac{2}{3}}\overbrace{(x^2 - 1)^{-1/3}}^{u^{n-1}}\overbrace{(2x)}^{u'} \qquad \text{Apply General Power Rule.}$$

$$= \frac{4x}{3\sqrt[3]{x^2 - 1}}. \qquad \text{Write in radical form.}$$

So, $f'(x) = 0$ when $x = 0$ and $f'(x)$ does not exist when $x = \pm 1$, as shown in Figure 2.25.

EXAMPLE 6 Differentiating Quotients with Constant Numerators

Differentiate $g(t) = \dfrac{-7}{(2t - 3)^2}$.

Solution Begin by rewriting the function as

$$g(t) = -7(2t - 3)^{-2}.$$

Then, applying the General Power Rule produces

$$g'(t) = \underbrace{(-7)}_{\substack{\text{Constant} \\ \text{Multiple Rule}}}\overbrace{(-2)}^{n}\overbrace{(2t - 3)^{-3}}^{u^{n-1}}\overbrace{(2)}^{u'} \qquad \text{Apply General Power Rule.}$$

$$= 28(2t - 3)^{-3} \qquad \text{Simplify.}$$

$$= \frac{28}{(2t - 3)^3}. \qquad \text{Write with positive exponent.}$$

NOTE Try differentiating the function in Example 6 using the Quotient Rule. You should obtain the same result, but using the Quotient Rule is less efficient than using the General Power Rule.

Simplifying Derivatives

The next three examples illustrate some techniques for simplifying the "raw derivatives" of functions involving products, quotients, and composites.

EXAMPLE 7 Simplifying by Factoring Out the Least Powers

$$
\begin{aligned}
f(x) &= x^2\sqrt{1-x^2} && \text{Original function}\\
&= x^2(1-x^2)^{1/2} && \text{Rewrite.}\\
f'(x) &= x^2\frac{d}{dx}\left[(1-x^2)^{1/2}\right] + (1-x^2)^{1/2}\frac{d}{dx}\left[x^2\right] && \text{Product Rule}\\
&= x^2\left[\frac{1}{2}(1-x^2)^{-1/2}(-2x)\right] + (1-x^2)^{1/2}(2x) && \text{General Power Rule}\\
&= -x^3(1-x^2)^{-1/2} + 2x(1-x^2)^{1/2} && \text{Simplify.}\\
&= x(1-x^2)^{-1/2}\left[-x^2(1) + 2(1-x^2)\right] && \text{Factor.}\\
&= \frac{x(2-3x^2)}{\sqrt{1-x^2}} && \text{Simplify.}
\end{aligned}
$$

EXAMPLE 8 Simplifying the Derivative of a Quotient

TECHNOLOGY Symbolic differentiation utilities are capable of differentiating very complicated functions. Often, however, the result is given in unsimplified form. If you have access to such a utility, use it to find the derivatives of the functions given in Examples 7, 8, and 9. Then compare the results with those given on this page.

$$
\begin{aligned}
f(x) &= \frac{x}{\sqrt[3]{x^2+4}} && \text{Original function}\\
&= \frac{x}{(x^2+4)^{1/3}} && \text{Rewrite.}\\
f'(x) &= \frac{(x^2+4)^{1/3}(1) - x(1/3)(x^2+4)^{-2/3}(2x)}{(x^2+4)^{2/3}} && \text{Quotient Rule}\\
&= \frac{1}{3}(x^2+4)^{-2/3}\left[\frac{3(x^2+4) - (2x^2)(1)}{(x^2+4)^{2/3}}\right] && \text{Factor.}\\
&= \frac{x^2+12}{3(x^2+4)^{4/3}} && \text{Simplify.}
\end{aligned}
$$

EXAMPLE 9 Simplifying the Derivative of a Power

$$
y = \left(\frac{3x-1}{x^2+3}\right)^2 \qquad \text{Original function}
$$

$$
\begin{aligned}
y' &= \underset{n}{2}\underbrace{\left(\frac{3x-1}{x^2+3}\right)}_{u^{n-1}}\underbrace{\frac{d}{dx}\left[\frac{3x-1}{x^2+3}\right]}_{u'} && \text{General Power Rule}\\
&= \left[\frac{2(3x-1)}{x^2+3}\right]\left[\frac{(x^2+3)(3) - (3x-1)(2x)}{(x^2+3)^2}\right] && \text{Quotient Rule}\\
&= \frac{2(3x-1)(3x^2+9-6x^2+2x)}{(x^2+3)^3} && \text{Multiply.}\\
&= \frac{2(3x-1)(-3x^2+2x+9)}{(x^2+3)^3} && \text{Simplify.}
\end{aligned}
$$

Trigonometric Functions and the Chain Rule

The "Chain Rule versions" of the derivatives of the six trigonometric functions are as shown.

$$\frac{d}{dx}[\sin u] = (\cos u)\, u' \qquad \frac{d}{dx}[\cos u] = -(\sin u)\, u'$$

$$\frac{d}{dx}[\tan u] = (\sec^2 u)\, u' \qquad \frac{d}{dx}[\cot u] = -(\csc^2 u)\, u'$$

$$\frac{d}{dx}[\sec u] = (\sec u \tan u)\, u' \qquad \frac{d}{dx}[\csc u] = -(\csc u \cot u)\, u'$$

EXAMPLE 10 Applying the Chain Rule to Trigonometric Functions

a. $y = \sin \overbrace{2x}^{u}$ $\qquad y' = \overbrace{\cos 2x}^{\cos u}\, \overbrace{\frac{d}{dx}[2x]}^{u'} = (\cos 2x)(2) = 2 \cos 2x$

b. $y = \cos(x - 1)$ $\qquad y' = -\sin(x - 1)$

c. $y = \tan 3x$ $\qquad y' = 3 \sec^2 3x$

Be sure that you understand the mathematical conventions regarding parentheses and trigonometric functions. For instance, in Example 10(a), $\sin 2x$ is written to mean $\sin(2x)$.

EXAMPLE 11 Parentheses and Trigonometric Functions

a. $y = \cos 3x^2 = \cos(3x^2)$ $\qquad y' = (-\sin 3x^2)(6x) = -6x \sin 3x^2$

b. $y = (\cos 3)x^2$ $\qquad y' = (\cos 3)(2x) = 2x \cos 3$

c. $y = \cos(3x)^2 = \cos(9x^2)$ $\qquad y' = (-\sin 9x^2)(18x) = -18x \sin 9x^2$

d. $y = \cos^2 x = (\cos x)^2$ $\qquad y' = 2(\cos x)(-\sin x) = -2 \cos x \sin x$

e. $y = \sqrt{\cos x} = (\cos x)^{1/2}$ $\qquad y' = \frac{1}{2}(\cos x)^{-1/2}(-\sin x) = -\frac{\sin x}{2\sqrt{\cos x}}$

To find the derivative of a function of the form $k(x) = f(g(h(x)))$, you need to apply the Chain Rule twice, as shown in Example 12.

EXAMPLE 12 Repeated Application of the Chain Rule

$$\begin{aligned}
f(t) &= \sin^3 4t && \text{Original function} \\
&= (\sin 4t)^3 && \text{Rewrite.} \\
f'(t) &= 3(\sin 4t)^2 \frac{d}{dt}[\sin 4t] && \text{Apply Chain Rule once.} \\
&= 3(\sin 4t)^2(\cos 4t) \frac{d}{dt}[4t] && \text{Apply Chain Rule a second time.} \\
&= 3(\sin 4t)^2(\cos 4t)(4) && \\
&= 12 \sin^2 4t \cos 4t && \text{Simplify.}
\end{aligned}$$

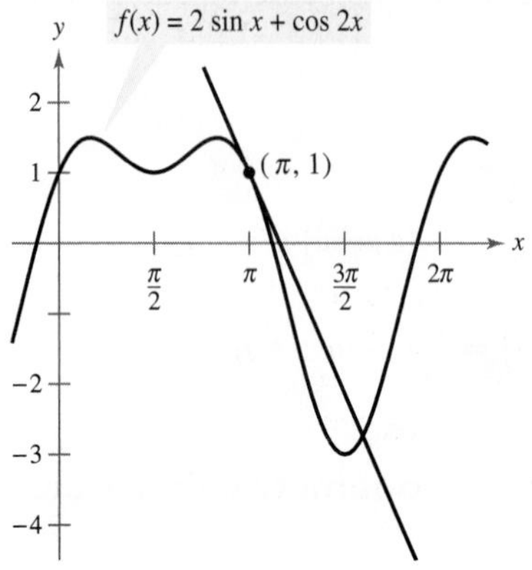

Figure 2.26

EXAMPLE 13 Tangent Line of a Trigonometric Function

Find an equation of the tangent line to the graph of

$$f(x) = 2\sin x + \cos 2x$$

at the point $(\pi, 1)$, as shown in Figure 2.26. Then determine all values of x in the interval $(0, 2\pi)$ at which the graph of f has a horizontal tangent.

Solution Begin by finding $f'(x)$.

$$f(x) = 2\sin x + \cos 2x \qquad \text{Write original function.}$$
$$f'(x) = 2\cos x + (-\sin 2x)(2) \qquad \text{Apply Chain Rule to } \cos 2x.$$
$$= 2\cos x - 2\sin 2x \qquad \text{Simplify.}$$

To find the equation of the tangent line at $(\pi, 1)$, evaluate $f'(\pi)$.

$$f'(\pi) = 2\cos\pi - 2\sin 2\pi \qquad \text{Substitute.}$$
$$= -2 \qquad \text{Slope of graph at } (\pi, 1)$$

Now, using the point-slope form of the equation of a line, you can write

$$y - y_1 = m(x - x_1) \qquad \text{Point-slope form}$$
$$y - 1 = -2(x - \pi) \qquad \text{Substitute for } y_1, m, \text{ and } x_1.$$
$$y = 1 - 2x + 2\pi. \qquad \text{Equation of tangent line at } (\pi, 1)$$

You can then determine that $f'(x) = 0$ when $x = \frac{\pi}{6}, \frac{\pi}{2}, \frac{5\pi}{6}$, and $\frac{3\pi}{2}$. So, f has a horizontal tangent at $x = \frac{\pi}{6}, \frac{\pi}{2}, \frac{5\pi}{6}$, and $\frac{3\pi}{2}$.

STUDY TIP To become skilled at differentiation, you should memorize each rule. As an aid to memorization, note that the cofunctions (cosine, cotangent, and cosecant) require a negative sign as part of their derivatives.

This section concludes with a summary of the differentiation rules studied so far.

Summary of Differentiation Rules

General Differentiation Rules Let f, g, and u be differentiable functions of x.

Constant Multiple Rule: $\frac{d}{dx}[cf] = cf'$

Sum or Difference Rule: $\frac{d}{dx}[f \pm g] = f' \pm g'$

Product Rule: $\frac{d}{dx}[fg] = fg' + gf'$

Quotient Rule: $\frac{d}{dx}\left[\frac{f}{g}\right] = \frac{gf' - fg'}{g^2}$

Derivatives of Algebraic Functions

Constant Rule: $\frac{d}{dx}[c] = 0$

(Simple) Power Rule: $\frac{d}{dx}[x^n] = nx^{n-1}, \quad \frac{d}{dx}[x] = 1$

Derivatives of Trigonometric Functions

$\frac{d}{dx}[\sin x] = \cos x$ $\quad \frac{d}{dx}[\tan x] = \sec^2 x$ $\quad \frac{d}{dx}[\sec x] = \sec x \tan x$

$\frac{d}{dx}[\cos x] = -\sin x$ $\quad \frac{d}{dx}[\cot x] = -\csc^2 x$ $\quad \frac{d}{dx}[\csc x] = -\csc x \cot x$

Chain Rule

Chain Rule: $\frac{d}{dx}[f(u)] = f'(u)\,u'$

General Power Rule: $\frac{d}{dx}[u^n] = nu^{n-1}u'$

Exercises for Section 2.4

See www.CalcChat.com for worked-out solutions to odd-numbered exercises.

In Exercises 1–6, complete the table.

$y = f(g(x))$	$u = g(x)$	$y = f(u)$
1. $y = (6x - 5)^4$		
2. $y = \dfrac{1}{\sqrt{x+1}}$		
3. $y = \sqrt{x^2 - 1}$		
4. $y = 3\tan(\pi x^2)$		
5. $y = \csc^3 x$		
6. $y = \cos\dfrac{3x}{2}$		

In Exercises 7–32, find the derivative of the function.

7. $y = (2x - 7)^3$

8. $y = 3(4 - x^2)^5$

9. $g(x) = 3(4 - 9x)^4$

10. $f(t) = (9t + 2)^{2/3}$

11. $f(t) = \sqrt{1 - t}$

12. $g(x) = \sqrt{5 - 3x}$

13. $y = \sqrt[3]{9x^2 + 4}$

14. $g(x) = \sqrt{x^2 - 2x + 1}$

15. $y = 2\sqrt[4]{4 - x^2}$

16. $f(x) = -3\sqrt[4]{2 - 9x}$

17. $y = \dfrac{1}{x - 2}$

18. $s(t) = \dfrac{1}{t^2 + 3t - 1}$

19. $f(t) = \left(\dfrac{1}{t - 3}\right)^2$

20. $y = -\dfrac{5}{(t + 3)^3}$

21. $y = \dfrac{1}{\sqrt{x + 2}}$

22. $g(t) = \sqrt{\dfrac{1}{t^2 - 2}}$

23. $f(x) = x^2(x - 2)^4$

24. $f(x) = x(3x - 9)^3$

25. $y = x\sqrt{1 - x^2}$

26. $y = \frac{1}{2}x^2\sqrt{16 - x^2}$

27. $y = \dfrac{x}{\sqrt{x^2 + 1}}$

28. $y = \dfrac{x}{\sqrt{x^4 + 4}}$

29. $g(x) = \left(\dfrac{x + 5}{x^2 + 2}\right)^2$

30. $h(t) = \left(\dfrac{t^2}{t^3 + 2}\right)^2$

31. $f(v) = \left(\dfrac{1 - 2v}{1 + v}\right)^3$

32. $g(x) = \left(\dfrac{3x^2 - 2}{2x + 3}\right)^3$

In Exercises 33–38, use a computer algebra system to find the derivative of the function. Then use the utility to graph the function and its derivative on the same set of coordinate axes. Describe the behavior of the function that corresponds to any zeros of the graph of the derivative.

33. $y = \dfrac{\sqrt{x} + 1}{x^2 + 1}$

34. $y = \sqrt{\dfrac{2x}{x + 1}}$

35. $y = \sqrt{\dfrac{x + 1}{x}}$

36. $g(x) = \sqrt{x - 1} + \sqrt{x + 1}$

37. $y = \dfrac{\cos \pi x + 1}{x}$

38. $y = x^2 \tan\dfrac{1}{x}$

In Exercises 39 and 40, find the slope of the tangent line to the sine function at the origin. Compare this value with the number of complete cycles in the interval $[0, 2\pi]$. What can you conclude about the slope of the sine function $\sin ax$ at the origin?

39. (a) (b)

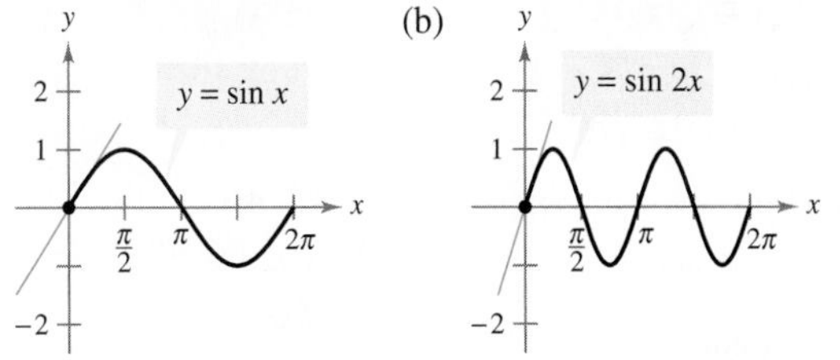

40. (a) (b)

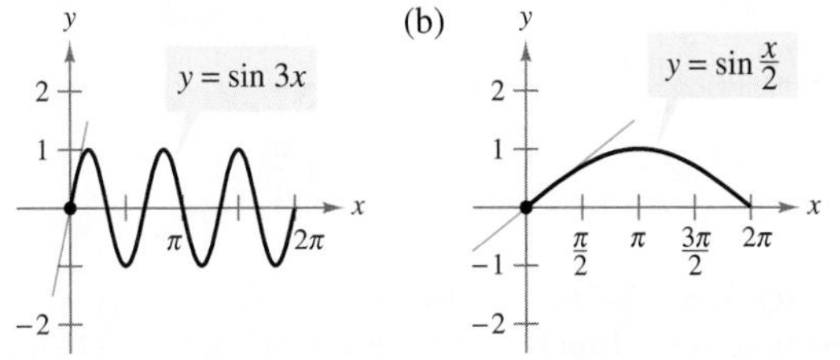

In Exercises 41–58, find the derivative of the function.

41. $y = \cos 3x$

42. $y = \sin \pi x$

43. $g(x) = 3\tan 4x$

44. $h(x) = \sec x^2$

45. $y = \sin(\pi x)^2$

46. $y = \cos(1 - 2x)^2$

47. $h(x) = \sin 2x \cos 2x$

48. $g(\theta) = \sec\left(\frac{1}{2}\theta\right)\tan\left(\frac{1}{2}\theta\right)$

49. $f(x) = \dfrac{\cot x}{\sin x}$

50. $g(v) = \dfrac{\cos v}{\csc v}$

51. $y = 4\sec^2 x$

52. $g(t) = 5\cos^2 \pi t$

53. $f(\theta) = \frac{1}{4}\sin^2 2\theta$

54. $h(t) = 2\cot^2(\pi t + 2)$

55. $f(t) = 3\sec^2(\pi t - 1)$

56. $y = 3x - 5\cos(\pi x)^2$

57. $y = \sqrt{x} + \frac{1}{4}\sin(2x)^2$

58. $y = \sin\sqrt[3]{x} + \sqrt[3]{\sin x}$

In Exercises 59–66, evaluate the derivative of the function at the given point. Use a graphing utility to verify your result.

Function	*Point*
59. $s(t) = \sqrt{t^2 + 2t + 8}$	$(2, 4)$
60. $y = \sqrt[5]{3x^3 + 4x}$	$(2, 2)$
61. $f(x) = \dfrac{3}{x^3 - 4}$	$\left(-1, -\dfrac{3}{5}\right)$
62. $f(x) = \dfrac{1}{(x^2 - 3x)^2}$	$\left(4, \dfrac{1}{16}\right)$
63. $f(t) = \dfrac{3t + 2}{t - 1}$	$(0, -2)$
64. $f(x) = \dfrac{x + 1}{2x - 3}$	$(2, 3)$
65. $y = 37 - \sec^3(2x)$	$(0, 36)$
66. $y = \dfrac{1}{x} + \sqrt{\cos x}$	$\left(\dfrac{\pi}{2}, \dfrac{2}{\pi}\right)$

In Exercises 67–74, (a) find an equation of the tangent line to the graph of f at the given point, (b) use a graphing utility to graph the function and its tangent line at the point, and (c) use the *derivative* feature of the graphing utility to confirm your results.

Function	*Point*
67. $f(x) = \sqrt{3x^2 - 2}$	$(3, 5)$
68. $f(x) = \frac{1}{3}x\sqrt{x^2 + 5}$	$(2, 2)$
69. $y = (2x^3 + 1)^2$	$(-1, 1)$
70. $f(x) = (9 - x^2)^{2/3}$	$(1, 4)$
71. $f(x) = \sin 2x$	$(\pi, 0)$
72. $y = \cos 3x$	$\left(\frac{\pi}{4}, -\frac{\sqrt{2}}{2}\right)$
73. $f(x) = \tan^2 x$	$\left(\frac{\pi}{4}, 1\right)$
74. $y = 2\tan^3 x$	$\left(\frac{\pi}{4}, 2\right)$

In Exercises 75–78, (a) use a graphing utility to find the derivative of the function at the given point, (b) find an equation of the tangent line to the graph of the function at the given point, and (c) use the utility to graph the function and its tangent line in the same viewing window.

75. $g(t) = \dfrac{3t^2}{\sqrt{t^2 + 2t - 1}}, \quad \left(\dfrac{1}{2}, \dfrac{3}{2}\right)$

76. $f(x) = \sqrt{x}(2 - x)^2, \quad (4, 8)$

77. $s(t) = \dfrac{(4 - 2t)\sqrt{1 + t}}{3}, \quad \left(0, \dfrac{4}{3}\right)$

78. $y = (t^2 - 9)\sqrt{t + 2}, \quad (2, -10)$

Famous Curves **In Exercises 79 and 80, find an equation of the tangent line to the graph at the given point. Then use a graphing utility to graph the function and its tangent line in the same viewing window.**

79. Top half of circle

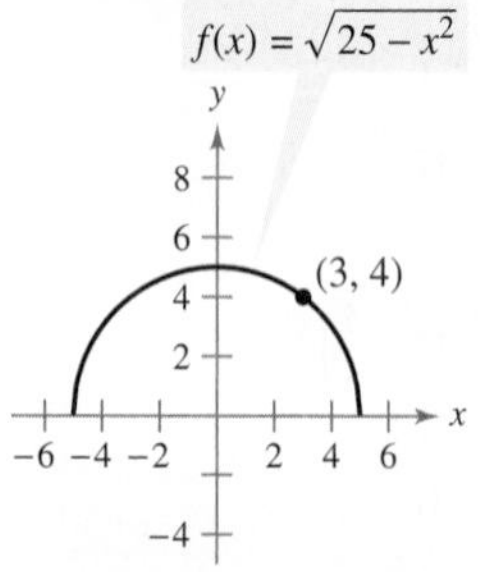

80. Bullet-nose curve

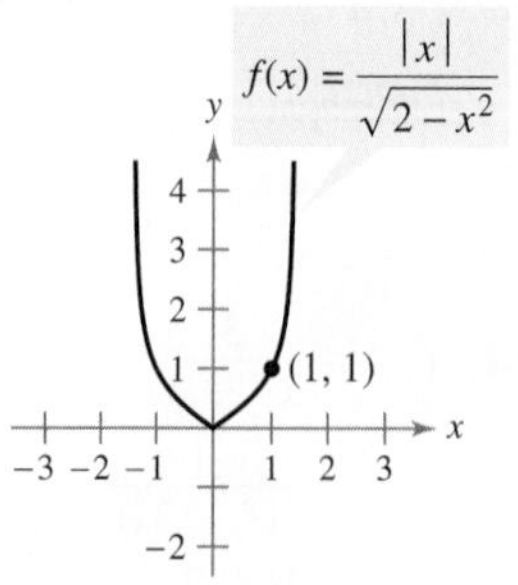

81. ***Horizontal Tangent Line*** Determine the point(s) in the interval $(0, 2\pi)$ at which the graph of $f(x) = 2\cos x + \sin 2x$ has a horizontal tangent.

82. ***Horizontal Tangent Line*** Determine the point(s) at which the graph of $f(x) = \dfrac{x}{\sqrt{2x - 1}}$ has a horizontal tangent.

In Exercises 83–86, find the second derivative of the function.

83. $f(x) = 2(x^2 - 1)^3$

84. $f(x) = \dfrac{1}{x - 2}$

85. $f(x) = \sin x^2$

86. $f(x) = \sec^2 \pi x$

In Exercises 87–90, evaluate the second derivative of the function at the given point. Use a computer algebra system to verify your result.

87. $h(x) = \frac{1}{9}(3x + 1)^3, \quad \left(1, \frac{64}{9}\right)$

88. $f(x) = \dfrac{1}{\sqrt{x + 4}}, \quad \left(0, \dfrac{1}{2}\right)$

89. $f(x) = \cos(x^2), \quad (0, 1)$

90. $g(t) = \tan 2t, \quad \left(\dfrac{\pi}{6}, \sqrt{3}\right)$

Writing About Concepts

In Exercises 91–94, the graphs of a function f and its derivative f' are shown. Label the graphs as f or f' and write a short paragraph stating the criteria used in making the selection. To print an enlarged copy of the graph, go to the website *www.mathgraphs.com*.

91.

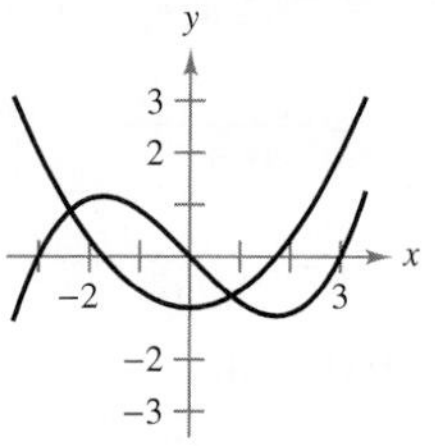

92.

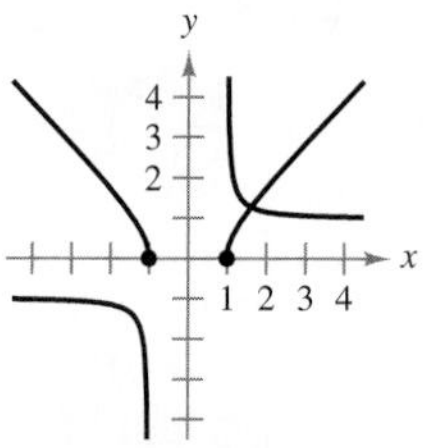

93.

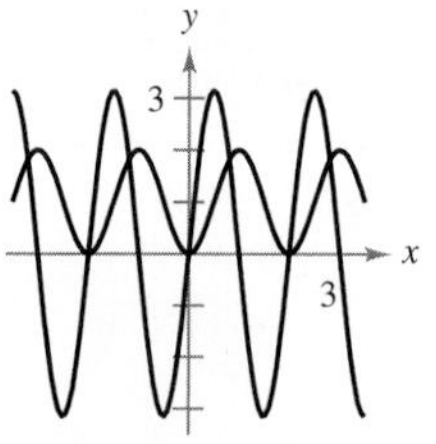

94.

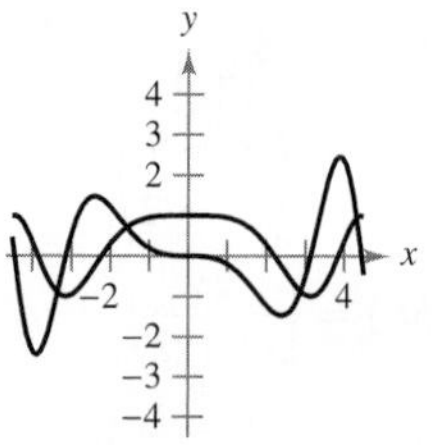

In Exercises 95 and 96, the relationship between f and g is given. Explain the relationship between f' and g'.

95. $g(x) = f(3x)$

96. $g(x) = f(x^2)$

97. Given that $g(5) = -3$, $g'(5) = 6$, $h(5) = 3$, and $h'(5) = -2$, find $f'(5)$ (if possible) for each of the following. If it is not possible, state what additional information is required.

(a) $f(x) = g(x)h(x)$

(b) $f(x) = g(h(x))$

(c) $f(x) = \dfrac{g(x)}{h(x)}$

(d) $f(x) = [g(x)]^3$

98. ***Think About It*** The table shows some values of the derivative of an unknown function f. Complete the table by finding (if possible) the derivative of each transformation of f.

(a) $g(x) = f(x) - 2$ (b) $h(x) = 2f(x)$

(c) $r(x) = f(-3x)$ (d) $s(x) = f(x + 2)$

x	-2	-1	0	1	2	3
$f'(x)$	4	$\frac{2}{3}$	$-\frac{1}{3}$	-1	-2	-4
$g'(x)$						
$h'(x)$						
$r'(x)$						
$s'(x)$						

In Exercises 99 and 100, the graphs of f and g are shown. Let $h(x) = f(g(x))$ and $s(x) = g(f(x))$. Find each derivative, if it exists. If the derivative does not exist, explain why.

99. (a) Find $h'(1)$.

(b) Find $s'(5)$.

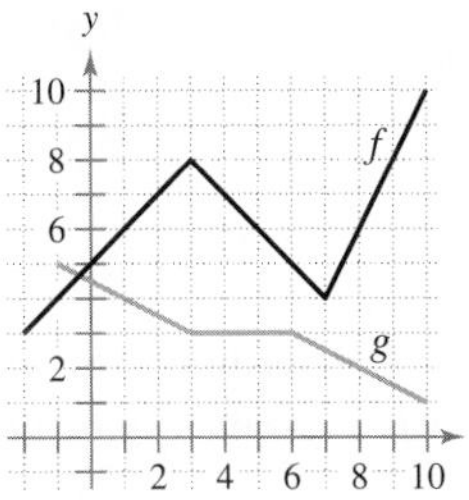

100. (a) Find $h'(3)$.

(b) Find $s'(9)$.

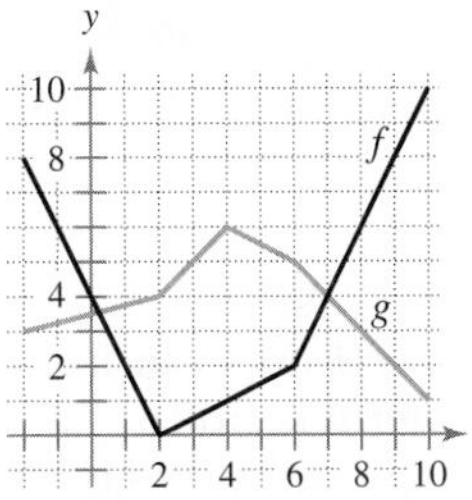

101. ***Doppler Effect*** The frequency F of a fire truck siren heard by a stationary observer is

$$F = \frac{132,400}{331 \pm v}$$

where $\pm v$ represents the velocity of the accelerating fire truck in meters per second (see figure). Find the rate of change of F with respect to v when

(a) the fire truck is approaching at a velocity of 30 meters per second (use $-v$).

(b) the fire truck is moving away at a velocity of 30 meters per second (use $+v$).

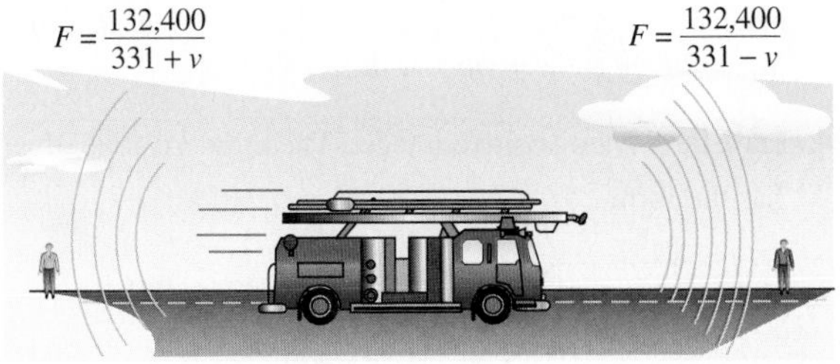

102. ***Harmonic Motion*** The displacement from equilibrium of an object in harmonic motion on the end of a spring is

$$y = \frac{1}{3}\cos 12t - \frac{1}{4}\sin 12t$$

where y is measured in feet and t is the time in seconds. Determine the position and velocity of the object when $t = \pi/8$.

103. ***Pendulum*** A 15-centimeter pendulum moves according to the equation $\theta = 0.2 \cos 8t$, where θ is the angular displacement from the vertical in radians and t is the time in seconds. Determine the maximum angular displacement and the rate of change of θ when $t = 3$ seconds.

104. ***Wave Motion*** A buoy oscillates in simple harmonic motion $y = A \cos \omega t$ as waves move past it. The buoy moves a total of 3.5 feet (vertically) from its low point to its high point. It returns to its high point every 10 seconds.

(a) Write an equation describing the motion of the buoy if it is at its high point at $t = 0$.

(b) Determine the velocity of the buoy as a function of t.

105. ***Circulatory System*** The speed S of blood that is r centimeters from the center of an artery is

$$S = C(R^2 - r^2)$$

where C is a constant, R is the radius of the artery, and S is measured in centimeters per second. Suppose a drug is administered and the artery begins to dilate at a rate of dR/dt. At a constant distance r, find the rate at which S changes with respect to t for $C = 1.76 \times 10^5$, $R = 1.2 \times 10^{-2}$, and $dR/dt = 10^{-5}$.

106. ***Modeling Data*** The normal daily maximum temperatures T (in degrees Fahrenheit) for Denver, Colorado, are shown in the table. *(Source: National Oceanic and Atmospheric Administration)*

Month	Jan	Feb	Mar	Apr	May	Jun
Temperature	43.2	47.2	53.7	60.9	70.5	82.1

Month	Jul	Aug	Sep	Oct	Nov	Dec
Temperature	88.0	86.0	77.4	66.0	51.5	44.1

(a) Use a graphing utility to plot the data and find a model for the data of the form

$$T(t) = a + b \sin(\pi t/6 - c)$$

where T is the temperature and t is the time in months, with $t = 1$ corresponding to January.

(b) Use a graphing utility to graph the model. How well does the model fit the data?

(c) Find T' and use a graphing utility to graph the derivative.

(d) Based on the graph of the derivative, during what times does the temperature change most rapidly? Most slowly? Do your answers agree with your observations of the temperature changes? Explain.

107. ***Modeling Data*** The cost of producing x units of a product is $C = 60x + 1350$. For one week management determined the number of units produced at the end of t hours during an eight-hour shift. The average values of x for the week are shown in the table.

t	0	1	2	3	4	5	6	7	8
x	0	16	60	130	205	271	336	384	392

(a) Use a graphing utility to fit a cubic model to the data.

(b) Use the Chain Rule to find dC/dt.

(c) Explain why the cost function is not increasing at a constant rate during the eight-hour shift.

108. ***Finding a Pattern*** Consider the function $f(x) = \sin \beta x$, where β is a constant.

(a) Find the first-, second-, third-, and fourth-order derivatives of the function.

(b) Verify that the function and its second derivative satisfy the equation $f''(x) + \beta^2 f(x) = 0$.

(c) Use the results in part (a) to write general rules for the even- and odd-order derivatives

$$f^{(2k)}(x) \text{ and } f^{(2k-1)}(x).$$

[*Hint:* $(-1)^k$ is positive if k is even and negative if k is odd.]

109. ***Conjecture*** Let f be a differentiable function of period p.

(a) Is the function f' periodic? Verify your answer.

(b) Consider the function $g(x) = f(2x)$. Is the function $g'(x)$ periodic? Verify your answer.

110. ***Think About It*** Let $r(x) = f(g(x))$ and $s(x) = g(f(x))$ where f and g are shown in the figure. Find (a) $r'(1)$ and (b) $s'(4)$.

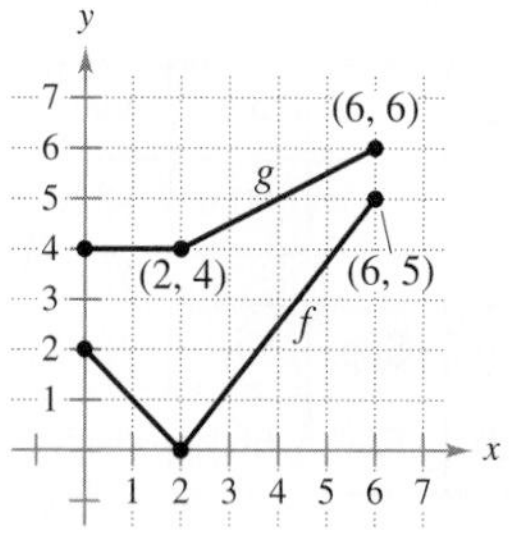

111. (a) Find the derivative of the function $g(x) = \sin^2 x + \cos^2 x$ in two ways.

(b) For $f(x) = \sec^2 x$ and $g(x) = \tan^2 x$, show that

$$f'(x) = g'(x).$$

112. (a) Show that the derivative of an odd function is even. That is, if $f(-x) = -f(x)$, then $f'(-x) = f'(x)$.

(b) Show that the derivative of an even function is odd. That is, if $f(-x) = f(x)$, then $f'(-x) = -f'(x)$.

113. Let u be a differentiable function of x. Use the fact that $|u| = \sqrt{u^2}$ to prove that

$$\frac{d}{dx}[|u|] = u'\frac{u}{|u|}, \quad u \neq 0.$$

In Exercises 114–117, use the result of Exercise 113 to find the derivative of the function.

114. $g(x) = |2x - 3|$

115. $f(x) = |x^2 - 4|$

116. $h(x) = |x| \cos x$

117. $f(x) = |\sin x|$

Linear and Quadratic Approximations **The linear and quadratic approximations of a function f at $x = a$ are**

$$P_1(x) = f'(a)(x - a) + f(a) \text{ and}$$

$$P_2(x) = \tfrac{1}{2}f''(a)(x - a)^2 + f'(a)(x - a) + f(a).$$

In Exercises 118 and 119, (a) find the specified linear and quadratic approximations of f, (b) use a graphing utility to graph f and the approximations, (c) determine whether P_1 or P_2 is the better approximation, and (d) state how the accuracy changes as you move farther from $x = a$.

118. $f(x) = \tan \dfrac{\pi x}{4}$

$a = 1$

119. $f(x) = \sec 2x$

$a = \dfrac{\pi}{6}$

True or False? **In Exercises 120–122, determine whether the statement is true or false. If it is false, explain why or give an example that shows it is false.**

120. If $y = (1 - x)^{1/2}$, then $y' = \frac{1}{2}(1 - x)^{-1/2}$.

121. If $f(x) = \sin^2(2x)$, then $f'(x) = 2(\sin 2x)(\cos 2x)$.

122. If y is a differentiable function of u, u is a differentiable function of v, and v is a differentiable function of x, then

$$\frac{dy}{dx} = \frac{dy}{du}\frac{du}{dv}\frac{dv}{dx}.$$

Putnam Exam Challenge

123. Let $f(x) = a_1 \sin x + a_2 \sin 2x + \cdots + a_n \sin nx$, where $a_1, a_2, \ldots, a_n$ are real numbers and where n is a positive integer. Given that $|f(x)| \leq |\sin x|$ for all real x, prove that $|a_1 + 2a_2 + \cdots + na_n| \leq 1$.

124. Let k be a fixed positive integer. The nth derivative of $\dfrac{1}{x^k - 1}$ has the form

$$\frac{P_n(x)}{(x^k - 1)^{n+1}}$$

where $P_n(x)$ is a polynomial. Find $P_n(1)$.

These problems were composed by the Committee on the Putnam Prize Competition.

Section 2.5 Implicit Differentiation

- **Distinguish between functions written in implicit form and explicit form.**
- **Use implicit differentiation to find the derivative of a function.**

Implicit and Explicit Functions

Up to this point in the text, most functions have been expressed in **explicit form.** For example, in the equation

$$y = 3x^2 - 5 \qquad \text{Explicit form}$$

the variable y is explicitly written as a function of x. Some functions, however, are only implied by an equation. For instance, the function $y = 1/x$ is defined **implicitly** by the equation $xy = 1$. Suppose you were asked to find dy/dx for this equation. You could begin by writing y explicitly as a function of x and then differentiating.

Implicit Form	*Explicit Form*	*Derivative*
$xy = 1$	$y = \frac{1}{x} = x^{-1}$	$\frac{dy}{dx} = -x^{-2} = -\frac{1}{x^2}$

This strategy works whenever you can solve for the function explicitly. You cannot, however, use this procedure when you are unable to solve for y as a function of x. For instance, how would you find dy/dx for the equation

$$x^2 - 2y^3 + 4y = 2$$

where it is very difficult to express y as a function of x explicitly? To do this, you can use **implicit differentiation.**

To understand how to find dy/dx implicitly, you must realize that the differentiation is taking place *with respect to x.* This means that when you differentiate terms involving x alone, you can differentiate as usual. However, when you differentiate terms involving y, you must apply the Chain Rule, because you are assuming that y is defined implicitly as a differentiable function of x.

EXPLORATION

Graphing an Implicit Equation
How could you use a graphing utility to sketch the graph of the equation

$$x^2 - 2y^3 + 4y = 2?$$

Here are two possible approaches.

a. Solve the equation for x. Switch the roles of x and y and graph the two resulting equations. The combined graphs will show a 90° rotation of the graph of the original equation.

b. Set the graphing utility to *parametric* mode and graph the equations

$$x = -\sqrt{2t^3 - 4t + 2}$$
$$y = t$$

and

$$x = \sqrt{2t^3 - 4t + 2}$$
$$y = t.$$

From either of these two approaches, can you decide whether the graph has a tangent line at the point $(0, 1)$? Explain your reasoning.

EXAMPLE 1 Differentiating with Respect to x

a. $\frac{d}{dx}[x^3] = 3x^2$ — Variables agree: use Simple Power Rule.

Variables agree

b. $\frac{d}{dx}[\underbrace{y^3}_{u^n}] = \underbrace{3y^2}_{nu^{n-1}}\underbrace{\frac{dy}{dx}}_{u'}$ — Variables disagree: use Chain Rule.

Variables disagree

c. $\frac{d}{dx}[x + 3y] = 1 + 3\frac{dy}{dx}$ — Chain Rule: $\frac{d}{dx}[3y] = 3y'$

d. $\frac{d}{dx}[xy^2] = x\frac{d}{dx}[y^2] + y^2\frac{d}{dx}[x]$ — Product Rule

$= x\left(2y\frac{dy}{dx}\right) + y^2(1)$ — Chain Rule

$= 2xy\frac{dy}{dx} + y^2$ — Simplify.

Implicit Differentiation

Guidelines for Implicit Differentiation

1. Differentiate both sides of the equation *with respect to x*.
2. Collect all terms involving dy/dx on the left side of the equation and move all other terms to the right side of the equation.
3. Factor dy/dx out of the left side of the equation.
4. Solve for dy/dx.

EXAMPLE 2 Implicit Differentiation

Find dy/dx given that $y^3 + y^2 - 5y - x^2 = -4$.

Solution

NOTE In Example 2, note that implicit differentiation can produce an expression for dy/dx that contains both x and y.

1. Differentiate both sides of the equation with respect to x.

$$\frac{d}{dx}[y^3 + y^2 - 5y - x^2] = \frac{d}{dx}[-4]$$

$$\frac{d}{dx}[y^3] + \frac{d}{dx}[y^2] - \frac{d}{dx}[5y] - \frac{d}{dx}[x^2] = \frac{d}{dx}[-4]$$

$$3y^2\frac{dy}{dx} + 2y\frac{dy}{dx} - 5\frac{dy}{dx} - 2x = 0$$

2. Collect the dy/dx terms on the left side of the equation and move all other terms to the right side of the equation.

$$3y^2\frac{dy}{dx} + 2y\frac{dy}{dx} - 5\frac{dy}{dx} = 2x$$

3. Factor dy/dx out of the left side of the equation.

$$\frac{dy}{dx}(3y^2 + 2y - 5) = 2x$$

4. Solve for dy/dx by dividing by $(3y^2 + 2y - 5)$.

$$\frac{dy}{dx} = \frac{2x}{3y^2 + 2y - 5}$$

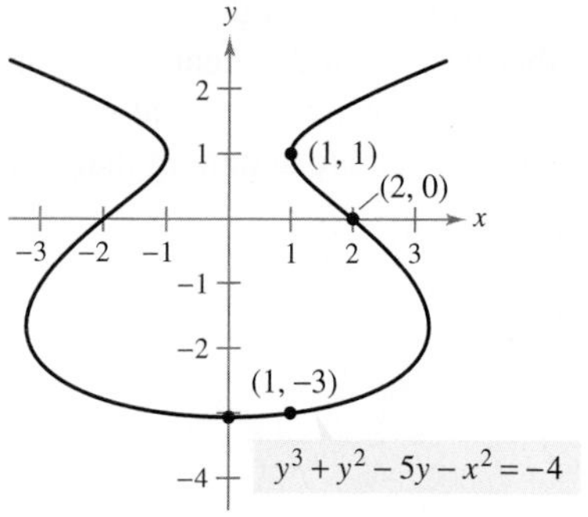

Point on Graph	*Slope of Graph*
(2, 0)	$-\frac{4}{5}$
(1, − 3)	$\frac{1}{8}$
$x = 0$	0
(1, 1)	Undefined

The implicit equation

$$y^3 + y^2 - 5y - x^2 = -4$$

has the derivative

$$\frac{dy}{dx} = \frac{2x}{3y^2 + 2y - 5}.$$

Figure 2.27

To see how you can use an *implicit derivative*, consider the graph shown in Figure 2.27. From the graph, you can see that y is not a function of x. Even so, the derivative found in Example 2 gives a formula for the slope of the tangent line at a point on this graph. The slopes at several points on the graph are shown below the graph.

TECHNOLOGY With most graphing utilities, it is easy to graph an equation that explicitly represents y as a function of x. Graphing other equations, however, can require some ingenuity. For instance, to graph the equation given in Example 2, use a graphing utility, set in *parametric* mode, to graph the parametric representations $x = \sqrt{t^3 + t^2 - 5t + 4}$, $y = t$, and $x = -\sqrt{t^3 + t^2 - 5t + 4}$, $y = t$, for $-5 \le t \le 5$. How does the result compare with the graph shown in Figure 2.27?

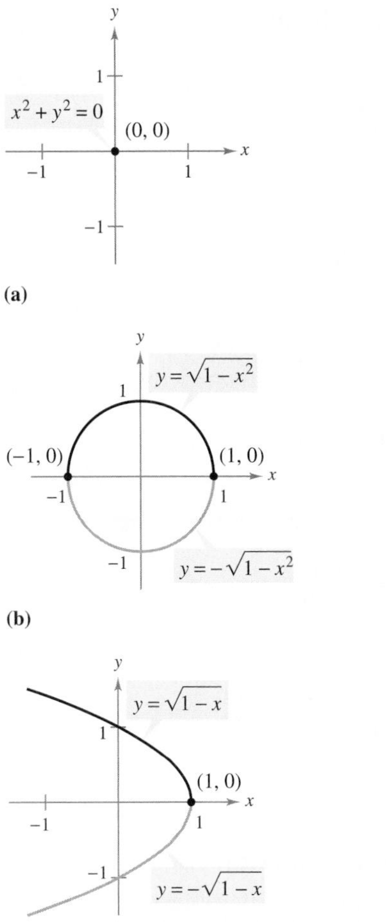

(c)

Some graph segments can be represented by differentiable functions.

Figure 2.28

It is meaningless to solve for dy/dx in an equation that has no solution points. (For example, $x^2 + y^2 = -4$ has no solution points.) If, however, a segment of a graph can be represented by a differentiable function, dy/dx will have meaning as the slope at each point on the segment. Recall that a function is not differentiable at (a) points with vertical tangents and (b) points at which the function is not continuous.

EXAMPLE 3 Representing a Graph by Differentiable Functions

If possible, represent y as a differentiable function of x.

a. $x^2 + y^2 = 0$ **b.** $x^2 + y^2 = 1$ **c.** $x + y^2 = 1$

Solution

a. The graph of this equation is a single point. So, it does not define y as a differentiable function of x. See Figure 2.28(a).

b. The graph of this equation is the unit circle, centered at $(0, 0)$. The upper semicircle is given by the differentiable function

$$y = \sqrt{1 - x^2}, \quad -1 < x < 1$$

and the lower semicircle is given by the differentiable function

$$y = -\sqrt{1 - x^2}, \quad -1 < x < 1.$$

At the points $(-1, 0)$ and $(1, 0)$, the slope of the graph is undefined. See Figure 2.28(b).

c. The upper half of this parabola is given by the differentiable function

$$y = \sqrt{1 - x}, \quad x < 1$$

and the lower half of this parabola is given by the differentiable function

$$y = -\sqrt{1 - x}, \quad x < 1.$$

At the point $(1, 0)$, the slope of the graph is undefined. See Figure 2.28(c).

EXAMPLE 4 Finding the Slope of a Graph Implicitly

Determine the slope of the tangent line to the graph of

$$x^2 + 4y^2 = 4$$

at the point $\left(\sqrt{2}, -1/\sqrt{2}\right)$. See Figure 2.29.

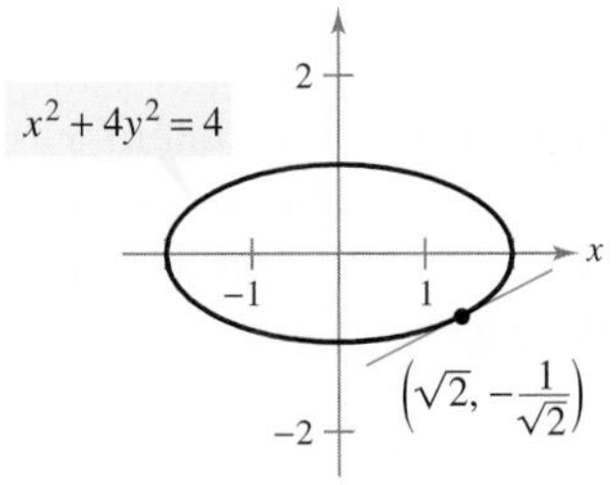

Figure 2.29

Solution

$$x^2 + 4y^2 = 4 \qquad \text{Write original equation.}$$

$$2x + 8y\frac{dy}{dx} = 0 \qquad \text{Differentiate with respect to } x.$$

$$\frac{dy}{dx} = \frac{-2x}{8y} = \frac{-x}{4y} \qquad \text{Solve for } \frac{dy}{dx}.$$

So, at $\left(\sqrt{2}, -1/\sqrt{2}\right)$, the slope is

$$\frac{dy}{dx} = \frac{-\sqrt{2}}{-4/\sqrt{2}} = \frac{1}{2}. \qquad \text{Evaluate } \frac{dy}{dx} \text{ when } x = \sqrt{2} \text{ and } y = -\frac{1}{\sqrt{2}}.$$

NOTE To see the benefit of implicit differentiation, try doing Example 4 using the explicit function $y = -\frac{1}{2}\sqrt{4 - x^2}$.

EXAMPLE 5 Finding the Slope of a Graph Implicitly

Determine the slope of the graph of $3(x^2 + y^2)^2 = 100xy$ at the point $(3, 1)$.

Solution

$$\frac{d}{dx}[3(x^2 + y^2)^2] = \frac{d}{dx}[100xy]$$

$$3(2)(x^2 + y^2)\left(2x + 2y\frac{dy}{dx}\right) = 100\left[x\frac{dy}{dx} + y(1)\right]$$

$$12y(x^2 + y^2)\frac{dy}{dx} - 100x\frac{dy}{dx} = 100y - 12x(x^2 + y^2)$$

$$[12y(x^2 + y^2) - 100x]\frac{dy}{dx} = 100y - 12x(x^2 + y^2)$$

$$\frac{dy}{dx} = \frac{100y - 12x(x^2 + y^2)}{-100x + 12y(x^2 + y^2)}$$

$$= \frac{25y - 3x(x^2 + y^2)}{-25x + 3y(x^2 + y^2)}$$

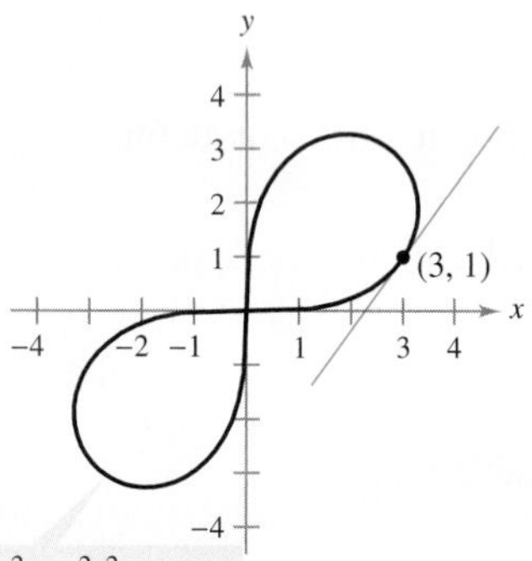

$3(x^2 + y^2)^2 = 100xy$

Lemniscate

Figure 2.30

At the point $(3, 1)$, the slope of the graph is

$$\frac{dy}{dx} = \frac{25(1) - 3(3)(3^2 + 1^2)}{-25(3) + 3(1)(3^2 + 1^2)} = \frac{25 - 90}{-75 + 30} = \frac{-65}{-45} = \frac{13}{9}$$

as shown in Figure 2.30. This graph is called a **lemniscate.**

EXAMPLE 6 Determining a Differentiable Function

Find dy/dx implicitly for the equation $\sin y = x$. Then find the largest interval of the form $-a < y < a$ on which y is a differentiable function of x (see Figure 2.31).

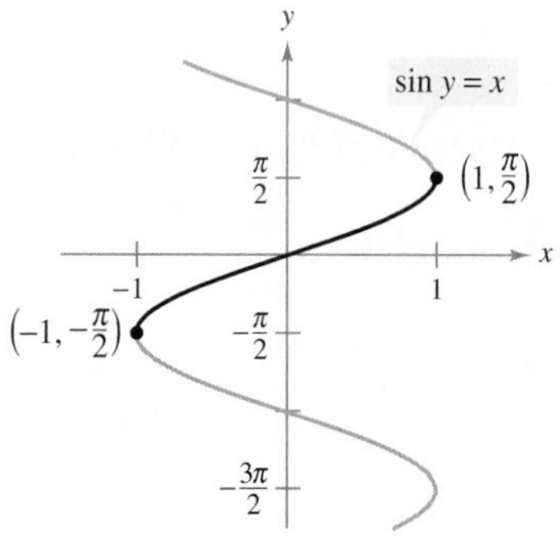

The derivative is $\frac{dy}{dx} = \frac{1}{\sqrt{1 - x^2}}$.

Figure 2.31

Solution

$$\frac{d}{dx}[\sin y] = \frac{d}{dx}[x]$$

$$\cos y\frac{dy}{dx} = 1$$

$$\frac{dy}{dx} = \frac{1}{\cos y}$$

The largest interval about the origin for which y is a differentiable function of x is $-\pi/2 < y < \pi/2$. To see this, note that $\cos y$ is positive for all y in this interval and is 0 at the endpoints. If you restrict y to the interval $-\pi/2 < y < \pi/2$, you should be able to write dy/dx explicitly as a function of x. To do this, you can use

$$\cos y = \sqrt{1 - \sin^2 y}$$

$$= \sqrt{1 - x^2}, \quad -\frac{\pi}{2} < y < \frac{\pi}{2}$$

and conclude that

$$\frac{dy}{dx} = \frac{1}{\sqrt{1 - x^2}}.$$

ISAAC BARROW (1630–1677)

The graph in Figure 2.32 is called the **kappa curve** because it resembles the Greek letter kappa, κ. The general solution for the tangent line to this curve was discovered by the English mathematician Isaac Barrow. Newton was Barrow's student, and they corresponded frequently regarding their work in the early development of calculus.

With implicit differentiation, the form of the derivative often can be simplified (as in Example 6) by an appropriate use of the *original* equation. A similar technique can be used to find and simplify higher-order derivatives obtained implicitly.

EXAMPLE 7 Finding the Second Derivative Implicitly

Given $x^2 + y^2 = 25$, find $\dfrac{d^2y}{dx^2}$.

Solution Differentiating each term with respect to x produces

$$2x + 2y\frac{dy}{dx} = 0$$
$$2y\frac{dy}{dx} = -2x$$
$$\frac{dy}{dx} = \frac{-2x}{2y} = -\frac{x}{y}.$$

Differentiating a second time with respect to x yields

$$\frac{d^2y}{dx^2} = -\frac{(y)(1) - (x)(dy/dx)}{y^2} \qquad \text{Quotient Rule}$$
$$= -\frac{y - (x)(-x/y)}{y^2} \qquad \text{Substitute } -x/y \text{ for } \frac{dy}{dx}.$$
$$= -\frac{y^2 + x^2}{y^3} \qquad \text{Simplify.}$$
$$= -\frac{25}{y^3}. \qquad \text{Substitute 25 for } x^2 + y^2.$$

EXAMPLE 8 Finding a Tangent Line to a Graph

Find the tangent line to the graph given by $x^2(x^2 + y^2) = y^2$ at the point $(\sqrt{2}/2, \sqrt{2}/2)$, as shown in Figure 2.32.

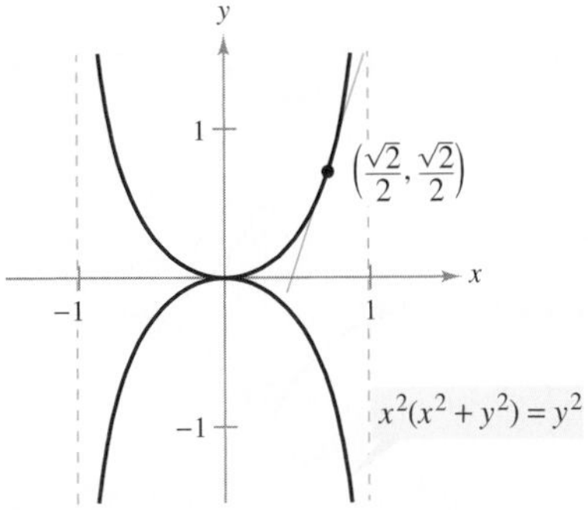

The kappa curve
Figure 2.32

Solution By rewriting and differentiating implicitly, you obtain

$$x^4 + x^2y^2 - y^2 = 0$$
$$4x^3 + x^2\left(2y\frac{dy}{dx}\right) + 2xy^2 - 2y\frac{dy}{dx} = 0$$
$$2y(x^2 - 1)\frac{dy}{dx} = -2x(2x^2 + y^2)$$
$$\frac{dy}{dx} = \frac{x(2x^2 + y^2)}{y(1 - x^2)}.$$

At the point $(\sqrt{2}/2, \sqrt{2}/2)$, the slope is

$$\frac{dy}{dx} = \frac{(\sqrt{2}/2)[2(1/2) + (1/2)]}{(\sqrt{2}/2)[1 - (1/2)]} = \frac{3/2}{1/2} = 3$$

and the equation of the tangent line at this point is

$$y - \frac{\sqrt{2}}{2} = 3\left(x - \frac{\sqrt{2}}{2}\right)$$
$$y = 3x - \sqrt{2}.$$

Exercises for Section 2.5

See www.CalcChat.com for worked-out solutions to odd-numbered exercises.

In Exercises 1–16, find dy/dx by implicit differentiation.

1. $x^2 + y^2 = 36$
2. $x^2 - y^2 = 16$
3. $x^{1/2} + y^{1/2} = 9$
4. $x^3 + y^3 = 8$
5. $x^3 - xy + y^2 = 4$
6. $x^2y + y^2x = -2$
7. $x^3y^3 - y = x$
8. $\sqrt{xy} = x - 2y$
9. $x^3 - 3x^2y + 2xy^2 = 12$
10. $2 \sin x \cos y = 1$
11. $\sin x + 2 \cos 2y = 1$
12. $(\sin \pi x + \cos \pi y)^2 = 2$
13. $\sin x = x(1 + \tan y)$
14. $\cot y = x - y$
15. $y = \sin(xy)$
16. $x = \sec \dfrac{1}{y}$

In Exercises 17–20, (a) find two explicit functions by solving the equation for y in terms of x, (b) sketch the graph of the equation and label the parts given by the corresponding explicit functions, (c) differentiate the explicit functions, and (d) find dy/dx and show that the result is equivalent to that of part (c).

17. $x^2 + y^2 = 16$
18. $x^2 + y^2 - 4x + 6y + 9 = 0$
19. $9x^2 + 16y^2 = 144$
20. $9y^2 - x^2 = 9$

In Exercises 21–28, find dy/dx by implicit differentiation and evaluate the derivative at the given point.

21. $xy = 4, \quad (-4, -1)$
22. $x^2 - y^3 = 0, \quad (1, 1)$
23. $y^2 = \dfrac{x^2 - 4}{x^2 + 4}, \quad (2, 0)$
24. $(x + y)^3 = x^3 + y^3, \quad (-1, 1)$
25. $x^{2/3} + y^{2/3} = 5, \quad (8, 1)$
26. $x^3 + y^3 = 4xy + 1, \quad (2, 1)$
27. $\tan(x + y) = x, \quad (0, 0)$
28. $x \cos y = 1, \quad \left(2, \dfrac{\pi}{3}\right)$

Famous Curves **In Exercises 29–32, find the slope of the tangent line to the graph at the given point.**

29. Witch of Agnesi:
$(x^2 + 4)y = 8$
Point: $(2, 1)$

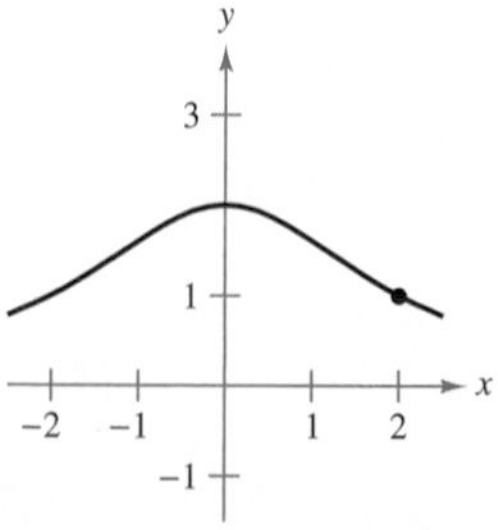

30. Cissoid:
$(4 - x)y^2 = x^3$
Point: $(2, 2)$

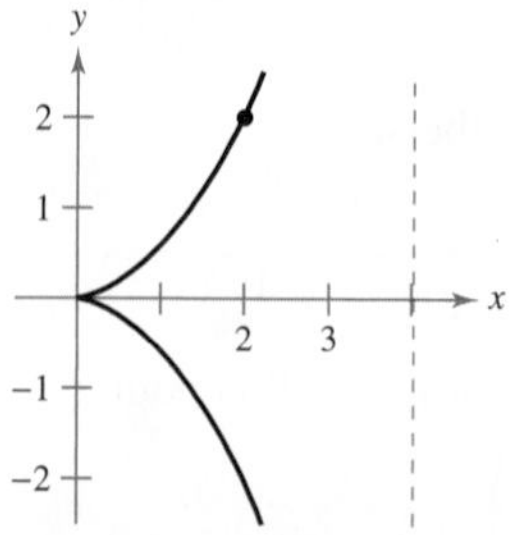

31. Bifolium:
$(x^2 + y^2)^2 = 4x^2y$
Point: $(1, 1)$

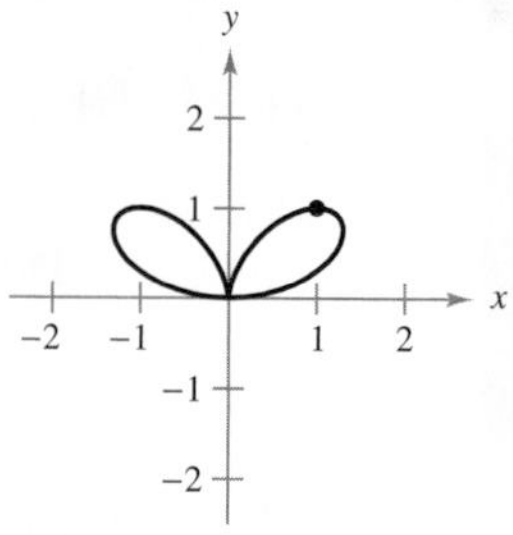

32. Folium of Descartes:
$x^3 + y^3 - 6xy = 0$
Point: $\left(\frac{4}{3}, \frac{8}{3}\right)$

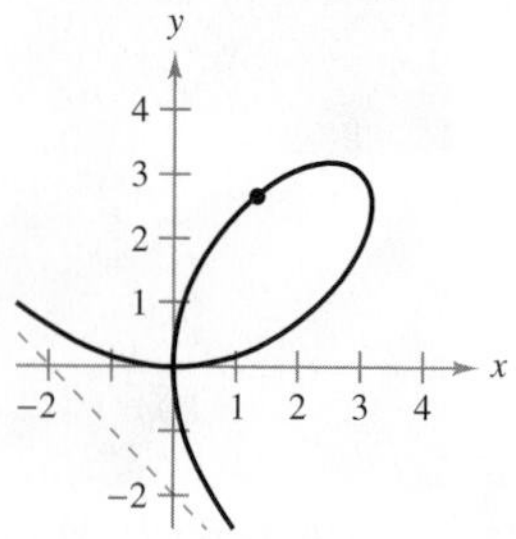

Famous Curves **In Exercises 33–40, find an equation of the tangent line to the graph at the given point. To print an enlarged copy of the graph, go to the website *www.mathgraphs.com*.**

33. Parabola

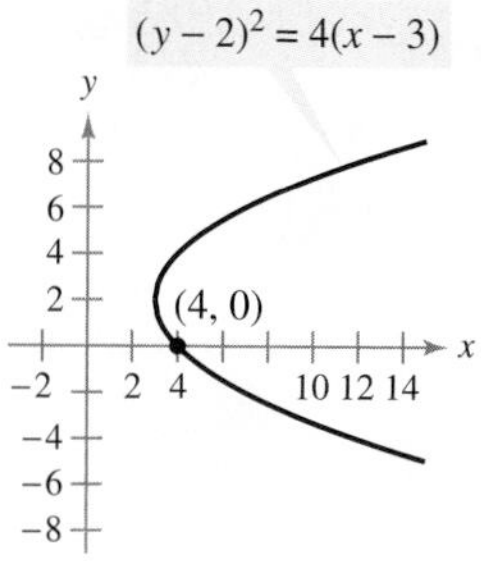

34. Circle

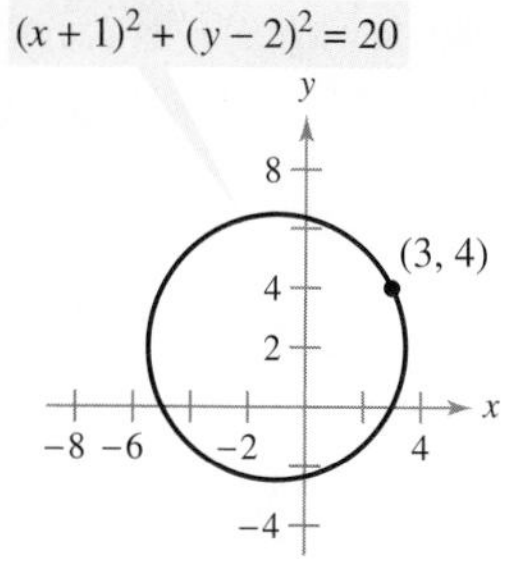

35. Rotated hyperbola

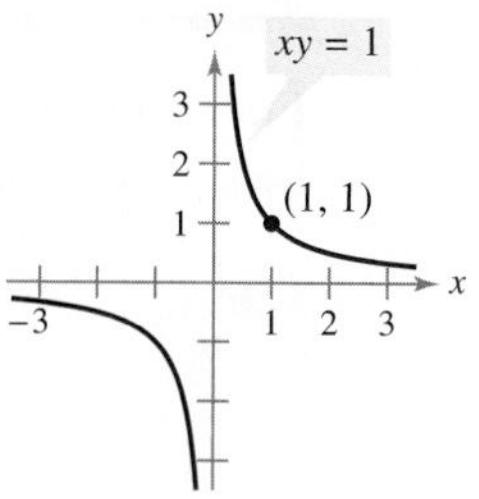

36. Rotated ellipse

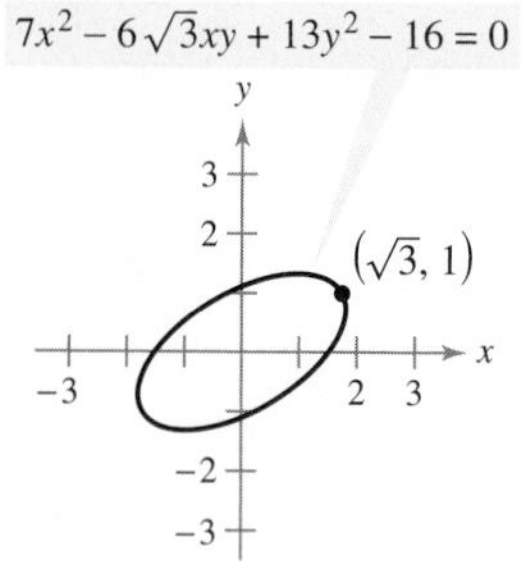

37. Cruciform

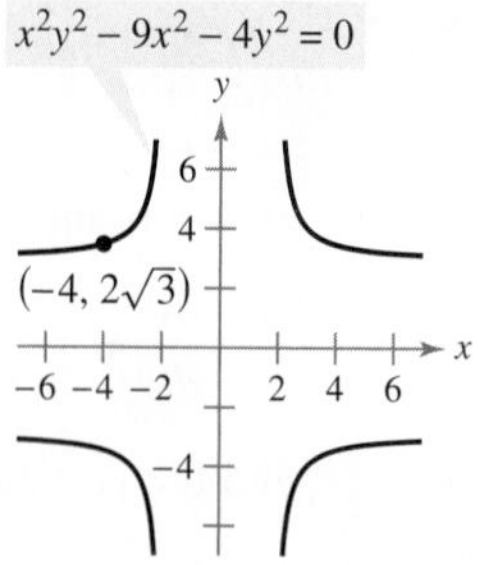

38. Astroid

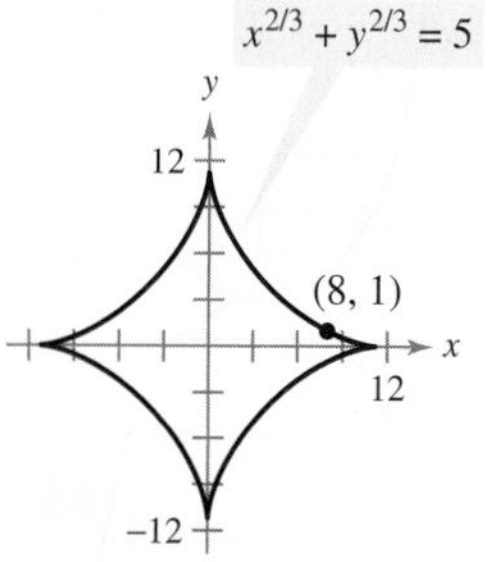

39. Lemniscate

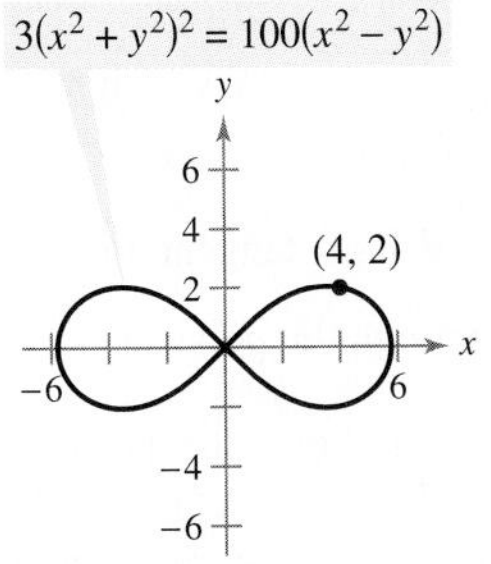

40. Kappa curve

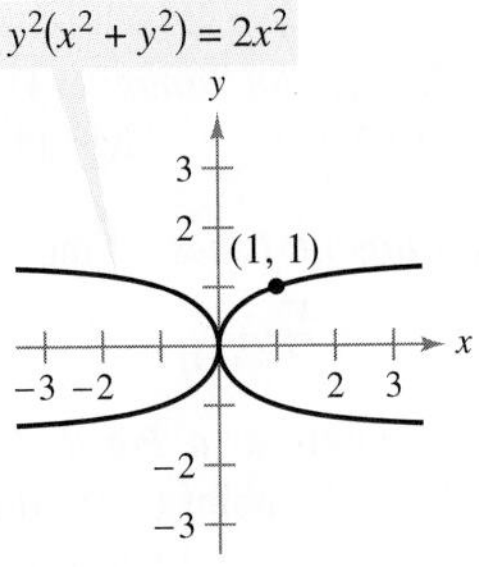

41. (a) Use implicit differentiation to find an equation of the tangent line to the ellipse $\frac{x^2}{2} + \frac{y^2}{8} = 1$ at $(1, 2)$.

(b) Show that the equation of the tangent line to the ellipse $\frac{x^2}{a^2} + \frac{y^2}{b^2} = 1$ at (x_0, y_0) is $\frac{x_0 x}{a^2} + \frac{y_0 y}{b^2} = 1$.

42. (a) Use implicit differentiation to find an equation of the tangent line to the hyperbola $\frac{x^2}{6} - \frac{y^2}{8} = 1$ at $(3, -2)$.

(b) Show that the equation of the tangent line to the hyperbola $\frac{x^2}{a^2} - \frac{y^2}{b^2} = 1$ at (x_0, y_0) is $\frac{x_0 x}{a^2} - \frac{y_0 y}{b^2} = 1$.

In Exercises 43 and 44, find dy/dx implicitly and find the largest interval of the form $-a < y < a$ or $0 < y < a$ such that y is a differentiable function of x. Write dy/dx as a function of x.

43. $\tan y = x$

44. $\cos y = x$

In Exercises 45–50, find d^2y/dx^2 in terms of x and y.

45. $x^2 + y^2 = 36$

46. $x^2y^2 - 2x = 3$

47. $x^2 - y^2 = 16$

48. $1 - xy = x - y$

49. $y^2 = x^3$

50. $y^2 = 4x$

In Exercises 51 and 52, use a graphing utility to graph the equation. Find an equation of the tangent line to the graph at the given point and graph the tangent line in the same viewing window.

51. $\sqrt{x} + \sqrt{y} = 4, \quad (9, 1)$

52. $y^2 = \frac{x - 1}{x^2 + 1}, \quad \left(2, \frac{\sqrt{5}}{5}\right)$

In Exercises 53 and 54, find equations for the tangent line and normal line to the circle at the given points. (The *normal line* at a point is perpendicular to the tangent line at the point.) Use a graphing utility to graph the equation, tangent line, and normal line.

53. $x^2 + y^2 = 25$
$(4, 3), (-3, 4)$

54. $x^2 + y^2 = 9$
$(0, 3), (2, \sqrt{5})$

55. Show that the normal line at any point on the circle $x^2 + y^2 = r^2$ passes through the origin.

56. Two circles of radius 4 are tangent to the graph of $y^2 = 4x$ at the point $(1, 2)$. Find equations of these two circles.

In Exercises 57 and 58, find the points at which the graph of the equation has a vertical or horizontal tangent line.

57. $25x^2 + 16y^2 + 200x - 160y + 400 = 0$

58. $4x^2 + y^2 - 8x + 4y + 4 = 0$

Orthogonal Trajectories **In Exercises 59–62, use a graphing utility to sketch the intersecting graphs of the equations and show that they are orthogonal. [Two graphs are *orthogonal* if at their point(s) of intersection their tangent lines are perpendicular to each other.]**

59. $2x^2 + y^2 = 6$
$y^2 = 4x$

60. $y^2 = x^3$
$2x^2 + 3y^2 = 5$

61. $x + y = 0$
$x = \sin y$

62. $x^3 = 3(y - 1)$
$x(3y - 29) = 3$

Orthogonal Trajectories **In Exercises 63 and 64, verify that the two families of curves are orthogonal where C and K are real numbers. Use a graphing utility to graph the two families for two values of C and two values of K.**

63. $xy = C, \quad x^2 - y^2 = K$

64. $x^2 + y^2 = C^2, \quad y = Kx$

In Exercises 65–68, differentiate (a) with respect to x (y is a function of x) and (b) with respect to t (x and y are functions of t).

65. $2y^2 - 3x^4 = 0$

66. $x^2 - 3xy^2 + y^3 = 10$

67. $\cos \pi y - 3 \sin \pi x = 1$

68. $4 \sin x \cos y = 1$

Writing About Concepts

69. Describe the difference between the explicit form of a function and an implicit equation. Give an example of each.

70. In your own words, state the guidelines for implicit differentiation.

71. ***Orthogonal Trajectories*** The figure below shows the topographic map carried by a group of hikers. The hikers are in a wooded area on top of the hill shown on the map and they decide to follow a path of steepest descent (orthogonal trajectories to the contours on the map). Draw their routes if they start from point A and if they start from point B. If their goal is to reach the road along the top of the map, which starting point should they use? To print an enlarged copy of the graph, go to the website *www.mathgraphs.com*.

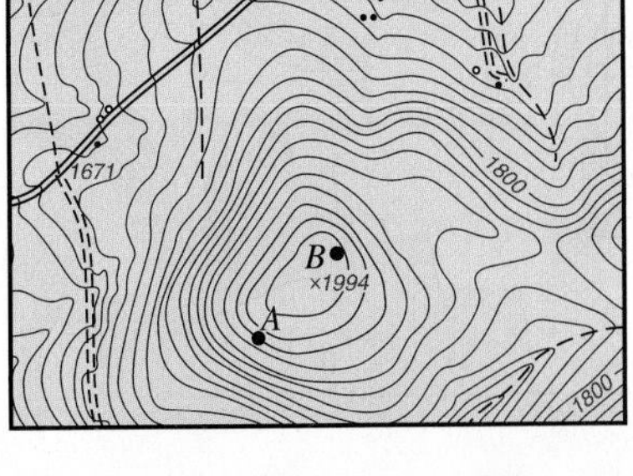

72. ***Weather Map*** The weather map shows several *isobars*—curves that represent areas of constant air pressure. Three high pressures H and one low pressure L are shown on the map. Given that wind speed is greatest along the orthogonal trajectories of the isobars, use the map to determine the areas having high wind speed.

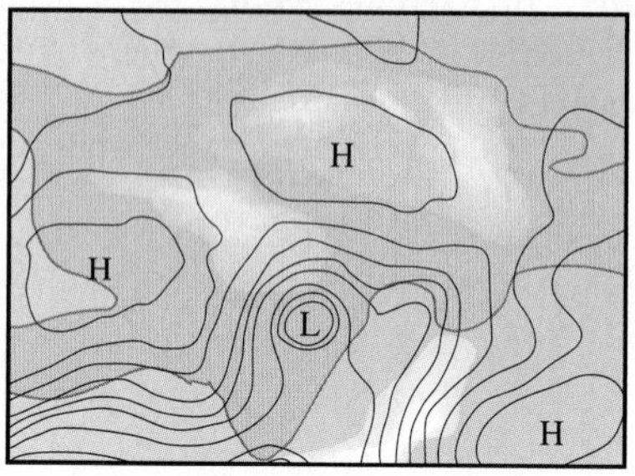

73. Consider the equation $x^4 = 4(4x^2 - y^2)$.

(a) Use a graphing utility to graph the equation.

(b) Find and graph the four tangent lines to the curve for $y = 3$.

(c) Find the exact coordinates of the point of intersection of the two tangent lines in the first quadrant.

74. Let L be any tangent line to the curve $\sqrt{x} + \sqrt{y} = \sqrt{c}$. Show that the sum of the x- and y-intercepts of L is c.

75. Prove (Theorem 2.3) that

$$\frac{d}{dx}[x^n] = nx^{n-1}$$

for the case in which n is a rational number. (*Hint:* Write $y = x^{p/q}$ in the form $y^q = x^p$ and differentiate implicitly. Assume that p and q are integers, where $q > 0$.)

76. ***Slope*** Find all points on the circle $x^2 + y^2 = 25$ where the slope is $\frac{3}{4}$.

77. ***Horizontal Tangent*** Determine the point(s) at which the graph of $y^4 = y^2 - x^2$ has a horizontal tangent.

78. ***Tangent Lines*** Find equations of both tangent lines to the ellipse $\frac{x^2}{4} + \frac{y^2}{9} = 1$ that passes through the point $(4, 0)$.

79. ***Normals to a Parabola*** The graph shows the normal lines from the point $(2, 0)$ to the graph of the parabola $x = y^2$. How many normal lines are there from the point $(x_0, 0)$ to the graph of the parabola if (a) $x_0 = \frac{1}{4}$, (b) $x_0 = \frac{1}{2}$, and (c) $x_0 = 1$? For what value of x_0 are two of the normal lines perpendicular to each other?

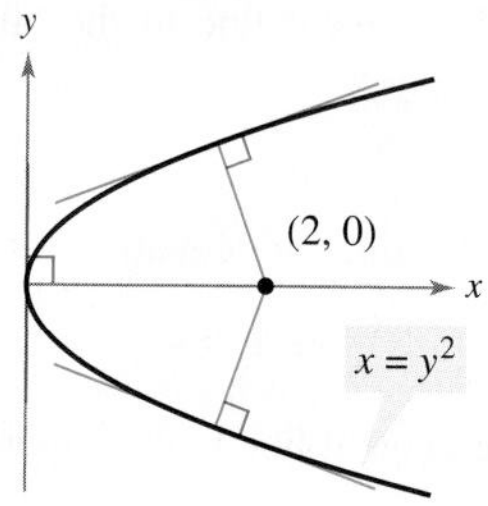

80. ***Normal Lines*** (a) Find an equation of the normal line to the ellipse

$$\frac{x^2}{32} + \frac{y^2}{8} = 1$$

at the point $(4, 2)$. (b) Use a graphing utility to graph the ellipse and the normal line. (c) At what other point does the normal line intersect the ellipse?

Section Project: Optical Illusions

In each graph below, an optical illusion is created by having lines intersect a family of curves. In each case, the lines appear to be curved. Find the value of dy/dx for the given values of x and y.

(a) Circles: $x^2 + y^2 = C^2$

$x = 3, y = 4, C = 5$

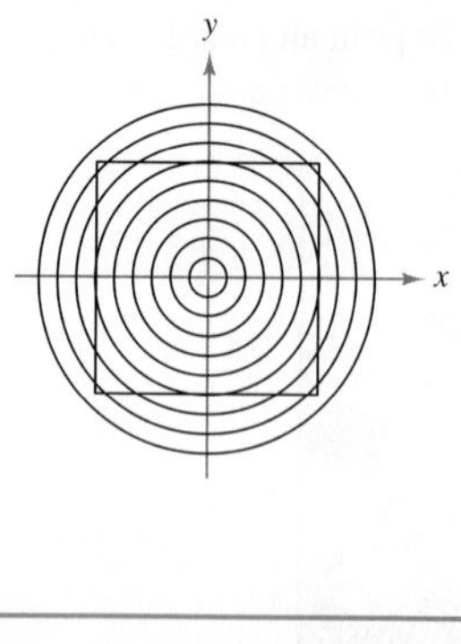

(b) Hyperbolas: $xy = C$

$x = 1, y = 4, C = 4$

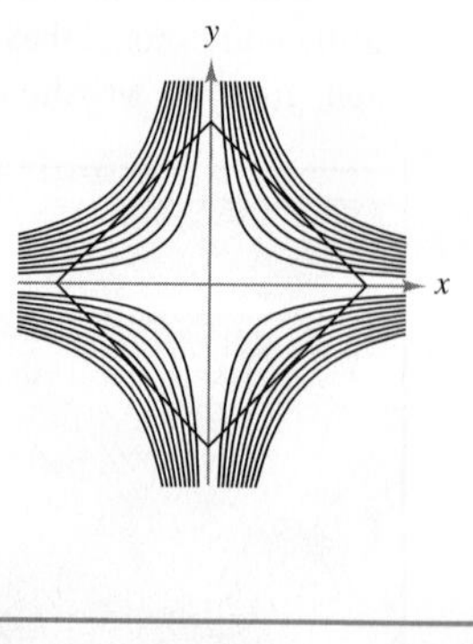

(c) Lines: $ax = by$

$x = \sqrt{3}, y = 3,$

$a = \sqrt{3}, b = 1$

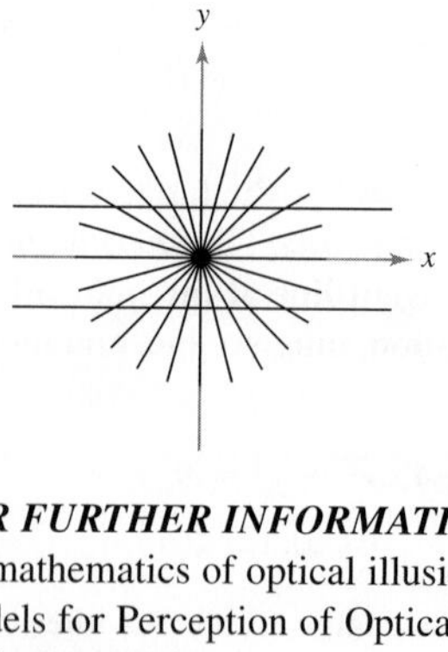

(d) Cosine curves: $y = C \cos x$

$x = \frac{\pi}{3}, y = \frac{1}{3}, C = \frac{2}{3}$

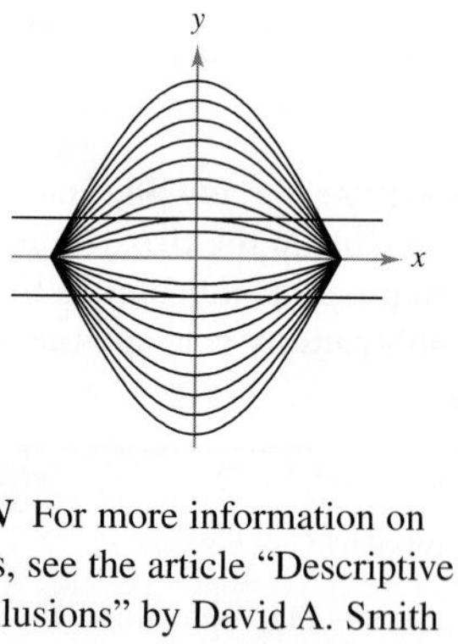

FOR FURTHER INFORMATION For more information on the mathematics of optical illusions, see the article "Descriptive Models for Perception of Optical Illusions" by David A. Smith in *The UMAP Journal*.

Section 2.6 Related Rates

- **Find a related rate.**
- **Use related rates to solve real-life problems.**

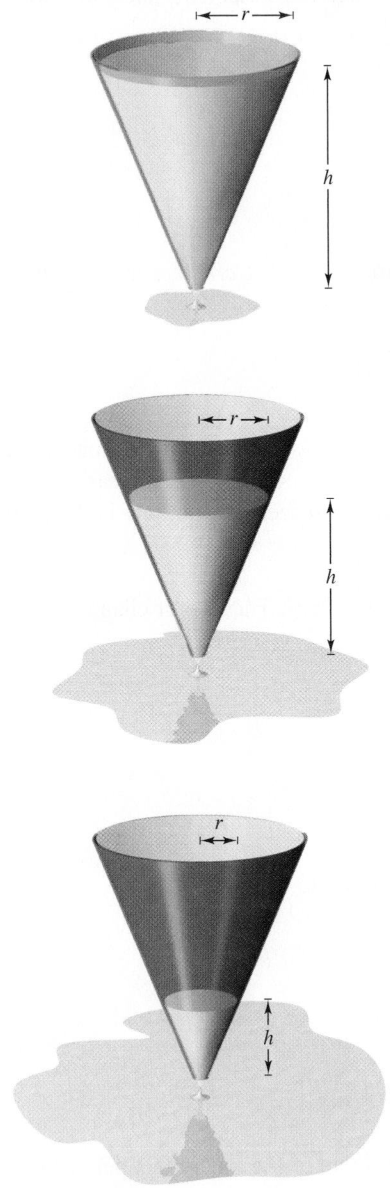

Volume is related to radius and height.
Figure 2.33

Finding Related Rates

You have seen how the Chain Rule can be used to find dy/dx implicitly. Another important use of the Chain Rule is to find the rates of change of two or more related variables that are changing with respect to *time*.

For example, when water is drained out of a conical tank (see Figure 2.33), the volume V, the radius r, and the height h of the water level are all functions of time t. Knowing that these variables are related by the equation

$$V = \frac{\pi}{3}r^2h \qquad \text{Original equation}$$

you can differentiate implicitly with respect to t to obtain the **related-rate** equation

$$\frac{d}{dt}(V) = \frac{d}{dt}\left(\frac{\pi}{3}r^2h\right)$$

$$\frac{dV}{dt} = \frac{\pi}{3}\left[r^2\frac{dh}{dt} + h\left(2r\frac{dr}{dt}\right)\right] \qquad \text{Differentiate with respect to } t.$$

$$= \frac{\pi}{3}\left(r^2\frac{dh}{dt} + 2rh\frac{dr}{dt}\right).$$

From this equation you can see that the rate of change of V is related to the rates of change of both h and r.

EXPLORATION

Finding a Related Rate In the conical tank shown in Figure 2.33, suppose that the height is changing at a rate of -0.2 foot per minute and the radius is changing at a rate of -0.1 foot per minute. What is the rate of change in the volume when the radius is $r = 1$ foot and the height is $h = 2$ feet? Does the rate of change in the volume depend on the values of r and h? Explain.

FOR FURTHER INFORMATION To learn more about the history of related-rate problems, see the article "The Lengthening Shadow: The Story of Related Rates" by Bill Austin, Don Barry, and David Berman in *Mathematics Magazine*. To view this article, go to the website *www.matharticles.com*.

EXAMPLE 1 Two Rates That Are Related

Suppose x and y are both differentiable functions of t and are related by the equation $y = x^2 + 3$. Find dy/dt when $x = 1$, given that $dx/dt = 2$ when $x = 1$.

Solution Using the Chain Rule, you can differentiate both sides of the equation *with respect to t.*

$$y = x^2 + 3 \qquad \text{Write original equation.}$$

$$\frac{d}{dt}[y] = \frac{d}{dt}[x^2 + 3] \qquad \text{Differentiate with respect to } t.$$

$$\frac{dy}{dt} = 2x\frac{dx}{dt} \qquad \text{Chain Rule}$$

When $x = 1$ and $dx/dt = 2$, you have

$$\frac{dy}{dt} = 2(1)(2) = 4.$$

Problem Solving with Related Rates

In Example 1, you were *given* an equation that related the variables x and y and were asked to find the rate of change of y when $x = 1$.

Equation: $y = x^2 + 3$

Given rate: $\frac{dx}{dt} = 2$ when $x = 1$

Find: $\frac{dy}{dt}$ when $x = 1$

In each of the remaining examples in this section, you must *create* a mathematical model from a verbal description.

EXAMPLE 2 Ripples in a Pond

W. Cody/Corbis

Total area increases as the outer radius increases.
Figure 2.34

A pebble is dropped into a calm pond, causing ripples in the form of concentric circles, as shown in Figure 2.34. The radius r of the outer ripple is increasing at a constant rate of 1 foot per second. When the radius is 4 feet, at what rate is the total area A of the disturbed water changing?

Solution The variables r and A are related by $A = \pi r^2$. The rate of change of the radius r is $dr/dt = 1$.

Equation: $A = \pi r^2$

Given rate: $\frac{dr}{dt} = 1$

Find: $\frac{dA}{dt}$ when $r = 4$

With this information, you can proceed as in Example 1.

$$\frac{d}{dt}[A] = \frac{d}{dt}[\pi r^2] \qquad \text{Differentiate with respect to } t.$$

$$\frac{dA}{dt} = 2\pi r \frac{dr}{dt} \qquad \text{Chain Rule}$$

$$\frac{dA}{dt} = 2\pi(4)(1) = 8\pi \qquad \text{Substitute 4 for } r \text{ and 1 for } dr/dt.$$

When the radius is 4 feet, the area is changing at a rate of 8π square feet per second.

Guidelines For Solving Related-Rate Problems

1. Identify all *given* quantities and quantities *to be determined.* Make a sketch and label the quantities.
2. Write an equation involving the variables whose rates of change either are given or are to be determined.
3. Using the Chain Rule, implicitly differentiate both sides of the equation *with respect to time t.*
4. *After* completing Step 3, substitute into the resulting equation all known values for the variables and their rates of change. Then solve for the required rate of change.

NOTE When using these guidelines, be sure you perform Step 3 before Step 4. Substituting the known values of the variables before differentiating will produce an inappropriate derivative.

The table below lists examples of mathematical models involving rates of change. For instance, the rate of change in the first example is the velocity of a car.

Verbal Statement	Mathematical Model
The velocity of a car after traveling for 1 hour is 50 miles per hour.	$x =$ distance traveled $\dfrac{dx}{dt} = 50$ when $t = 1$
Water is being pumped into a swimming pool at a rate of 10 cubic meters per hour.	$V =$ volume of water in pool $\dfrac{dV}{dt} = 10\ \text{m}^3/\text{hr}$
A gear is revolving at a rate of 25 revolutions per minute (1 revolution $= 2\pi$ rad).	$\theta =$ angle of revolution $\dfrac{d\theta}{dt} = 25(2\pi)$ rad/min

EXAMPLE 3 An Inflating Balloon

Air is being pumped into a spherical balloon (see Figure 2.35) at a rate of 4.5 cubic feet per minute. Find the rate of change of the radius when the radius is 2 feet.

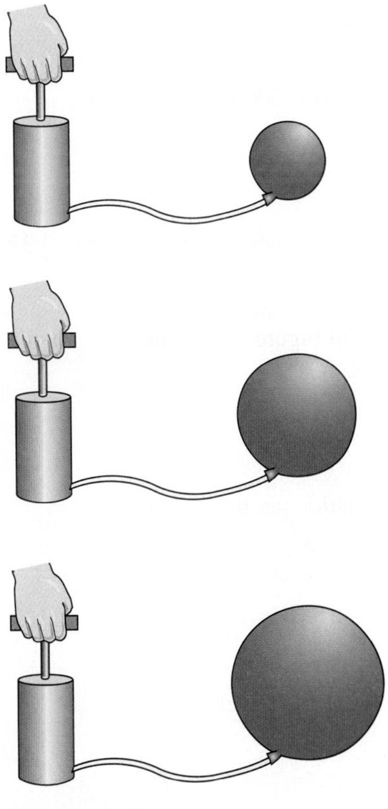

Inflating a balloon
Figure 2.35

Solution Let V be the volume of the balloon and let r be its radius. Because the volume is increasing at a rate of 4.5 cubic feet per minute, you know that at time t the rate of change of the volume is $dV/dt = \frac{9}{2}$. So, the problem can be stated as shown.

Given rate: $\dfrac{dV}{dt} = \dfrac{9}{2}$ (constant rate)

Find: $\dfrac{dr}{dt}$ when $r = 2$

To find the rate of change of the radius, you must find an equation that relates the radius r to the volume V.

Equation: $V = \dfrac{4}{3}\pi r^3$ Volume of a sphere

Differentiating both sides of the equation with respect to t produces

$$\frac{dV}{dt} = 4\pi r^2 \frac{dr}{dt}$$ Differentiate with respect to t.

$$\frac{dr}{dt} = \frac{1}{4\pi r^2}\left(\frac{dV}{dt}\right).$$ Solve for dr/dt.

Finally, when $r = 2$, the rate of change of the radius is

$$\frac{dr}{dt} = \frac{1}{16\pi}\left(\frac{9}{2}\right) \approx 0.09 \text{ foot per minute.}$$

In Example 3, note that the volume is increasing at a *constant* rate but the radius is increasing at a *variable* rate. Just because two rates are related does not mean that they are proportional. In this particular case, the radius is growing more and more slowly as t increases. Do you see why?

EXAMPLE 4 The Speed of an Airplane Tracked by Radar

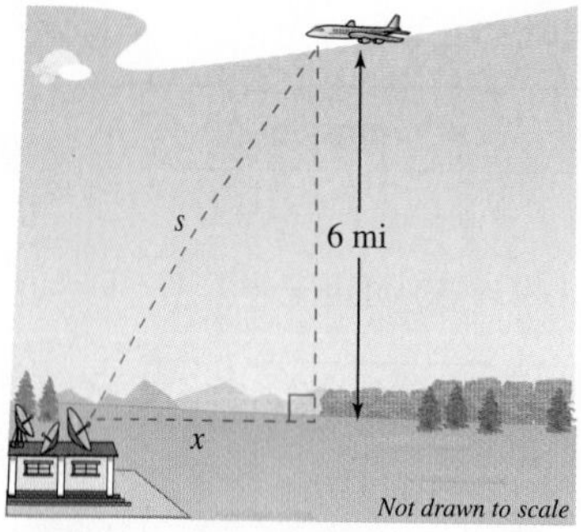

An airplane is flying at an altitude of 6 miles, s miles from the station.
Figure 2.36

An airplane is flying on a flight path that will take it directly over a radar tracking station, as shown in Figure 2.36. If s is decreasing at a rate of 400 miles per hour when $s = 10$ miles, what is the speed of the plane?

Solution Let x be the horizontal distance from the station, as shown in Figure 2.36. Notice that when $s = 10$, $x = \sqrt{10^2 - 36} = 8$.

Given rate: $ds/dt = -400$ when $s = 10$

Find: dx/dt when $s = 10$ and $x = 8$

You can find the velocity of the plane as shown.

Equation: $x^2 + 6^2 = s^2$ — Pythagorean Theorem

$$2x\frac{dx}{dt} = 2s\frac{ds}{dt}$$ Differentiate with respect to t.

$$\frac{dx}{dt} = \frac{s}{x}\left(\frac{ds}{dt}\right)$$ Solve for dx/dt.

$$\frac{dx}{dt} = \frac{10}{8}(-400)$$ Substitute for s, x, and ds/dt.

$$= -500 \text{ miles per hour}$$ Simplify.

Because the velocity is -500 miles per hour, the *speed* is 500 miles per hour.

EXAMPLE 5 A Changing Angle of Elevation

Find the rate of change in the angle of elevation of the camera shown in Figure 2.37 at 10 seconds after lift-off.

Solution Let θ be the angle of elevation, as shown in Figure 2.37. When $t = 10$, the height s of the rocket is $s = 50t^2 = 50(10)^2 = 5000$ feet.

Given rate: $ds/dt = 100t =$ velocity of rocket

Find: $d\theta/dt$ when $t = 10$ and $s = 5000$

Using Figure 2.37, you can relate s and θ by the equation $\tan\theta = s/2000$.

Equation: $$\tan\theta = \frac{s}{2000}$$ See Figure 2.37.

$$(\sec^2\theta)\frac{d\theta}{dt} = \frac{1}{2000}\left(\frac{ds}{dt}\right)$$ Differentiate with respect to t.

$$\frac{d\theta}{dt} = \cos^2\theta\,\frac{100t}{2000}$$ Substitute $100t$ for ds/dt.

$$= \left(\frac{2000}{\sqrt{s^2 + 2000^2}}\right)^2 \frac{100t}{2000}$$ $\cos\theta = 2000/\sqrt{s^2 + 2000^2}$

When $t = 10$ and $s = 5000$, you have

$$\frac{d\theta}{dt} = \frac{2000(100)(10)}{5000^2 + 2000^2} = \frac{2}{29} \text{ radian per second.}$$

So, when $t = 10$, θ is changing at a rate of $\frac{2}{29}$ radian per second.

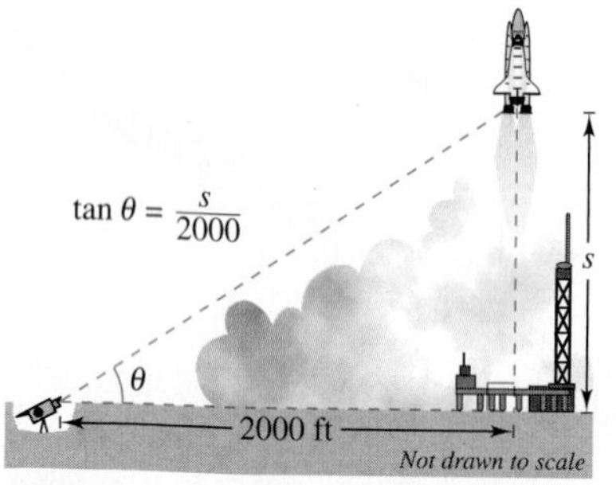

A television camera at ground level is filming the lift-off of a space shuttle that is rising vertically according to the position equation $s = 50t^2$, where s is measured in feet and t is measured in seconds. The camera is 2000 feet from the launch pad.
Figure 2.37

EXAMPLE 6 The Velocity of a Piston

In the engine shown in Figure 2.38, a 7-inch connecting rod is fastened to a crank of radius 3 inches. The crankshaft rotates counterclockwise at a constant rate of 200 revolutions per minute. Find the velocity of the piston when $\theta = \pi/3$.

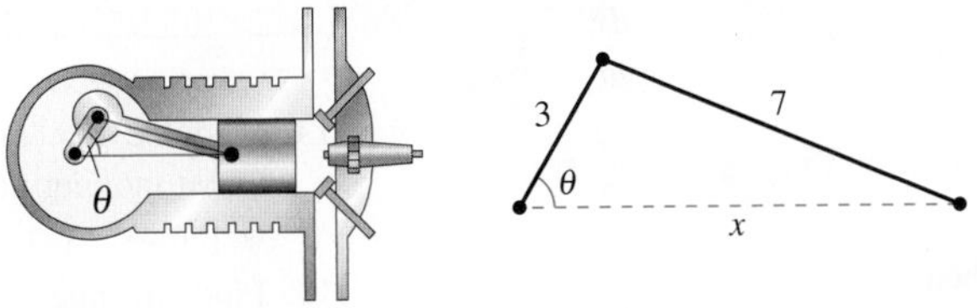

The velocity of a piston is related to the angle of the crankshaft.
Figure 2.38

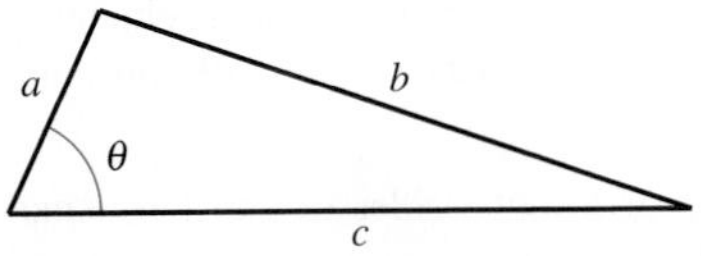

Law of Cosines:
$b^2 = a^2 + c^2 - 2ac\cos\theta$
Figure 2.39

Solution Label the distances as shown in Figure 2.38. Because a complete revolution corresponds to 2π radians, it follows that $d\theta/dt = 200(2\pi) = 400\pi$ radians per minute.

Given rate: $\dfrac{d\theta}{dt} = 400\pi$ (constant rate)

Find: $\dfrac{dx}{dt}$ when $\theta = \dfrac{\pi}{3}$

You can use the Law of Cosines (Figure 2.39) to find an equation that relates x and θ.

Equation:

$$7^2 = 3^2 + x^2 - 2(3)(x)\cos\theta$$

$$0 = 2x\frac{dx}{dt} - 6\left(-x\sin\theta\frac{d\theta}{dt} + \cos\theta\frac{dx}{dt}\right)$$

$$(6\cos\theta - 2x)\frac{dx}{dt} = 6x\sin\theta\frac{d\theta}{dt}$$

$$\frac{dx}{dt} = \frac{6x\sin\theta}{6\cos\theta - 2x}\left(\frac{d\theta}{dt}\right)$$

When $\theta = \pi/3$, you can solve for x as shown.

$$7^2 = 3^2 + x^2 - 2(3)(x)\cos\frac{\pi}{3}$$

$$49 = 9 + x^2 - 6x\left(\frac{1}{2}\right)$$

$$0 = x^2 - 3x - 40$$

$$0 = (x - 8)(x + 5)$$

$$x = 8$$ Choose positive solution.

So, when $x = 8$ and $\theta = \pi/3$, the velocity of the piston is

$$\frac{dx}{dt} = \frac{6(8)(\sqrt{3}/2)}{6(1/2) - 16}(400\pi)$$

$$= \frac{9600\pi\sqrt{3}}{-13}$$

$$\approx -4018 \text{ inches per minute.}$$

NOTE Note that the velocity in Example 6 is negative because x represents a distance that is decreasing.

Exercises for Section 2.6

See www.CalcChat.com for worked-out solutions to odd-numbered exercises.

In Exercises 1–4, assume that x and y are both differentiable functions of t and find the required values of dy/dt and dx/dt.

Equation	Find	Given
1. $y = \sqrt{x}$	(a) $\dfrac{dy}{dt}$ when $x = 4$	$\dfrac{dx}{dt} = 3$
	(b) $\dfrac{dx}{dt}$ when $x = 25$	$\dfrac{dy}{dt} = 2$
2. $y = 2(x^2 - 3x)$	(a) $\dfrac{dy}{dt}$ when $x = 3$	$\dfrac{dx}{dt} = 2$
	(b) $\dfrac{dx}{dt}$ when $x = 1$	$\dfrac{dy}{dt} = 5$
3. $xy = 4$	(a) $\dfrac{dy}{dt}$ when $x = 8$	$\dfrac{dx}{dt} = 10$
	(b) $\dfrac{dx}{dt}$ when $x = 1$	$\dfrac{dy}{dt} = -6$
4. $x^2 + y^2 = 25$	(a) $\dfrac{dy}{dt}$ when $x = 3, y = 4$	$\dfrac{dx}{dt} = 8$
	(b) $\dfrac{dx}{dt}$ when $x = 4, y = 3$	$\dfrac{dy}{dt} = -2$

In Exercises 5–8, a point is moving along the graph of the given function such that dx/dt is 2 centimeters per second. Find dy/dt for the given values of x.

5. $y = x^2 + 1$	(a) $x = -1$	(b) $x = 0$	(c) $x = 1$
6. $y = \dfrac{1}{1 + x^2}$	(a) $x = -2$	(b) $x = 0$	(c) $x = 2$
7. $y = \tan x$	(a) $x = -\dfrac{\pi}{3}$	(b) $x = -\dfrac{\pi}{4}$	(c) $x = 0$
8. $y = \sin x$	(a) $x = \dfrac{\pi}{6}$	(b) $x = \dfrac{\pi}{4}$	(c) $x = \dfrac{\pi}{3}$

Writing About Concepts

In Exercises 9 and 10, using the graph of f, (a) determine whether dy/dt is positive or negative given that dx/dt is negative, and (b) determine whether dx/dt is positive or negative given that dy/dt is positive.

9.

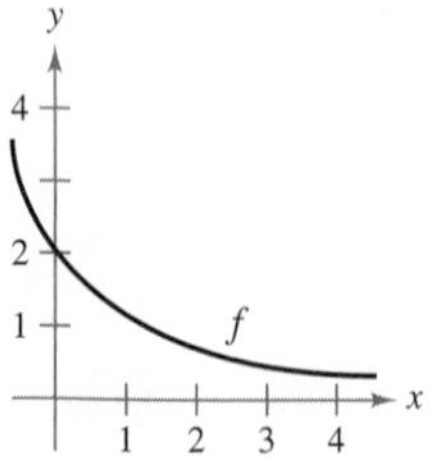

10.

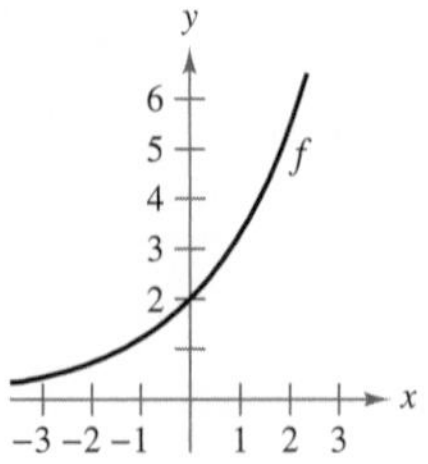

11. Consider the linear function $y = ax + b$. If x changes at a constant rate, does y change at a constant rate? If so, does it change at the same rate as x? Explain.

Writing About Concepts *(continued)*

12. In your own words, state the guidelines for solving related-rate problems.

13. Find the rate of change of the distance between the origin and a moving point on the graph of $y = x^2 + 1$ if $dx/dt = 2$ centimeters per second.

14. Find the rate of change of the distance between the origin and a moving point on the graph of $y = \sin x$ if $dx/dt = 2$ centimeters per second.

15. ***Area*** The radius r of a circle is increasing at a rate of 3 centimeters per minute. Find the rates of change of the area when (a) $r = 6$ centimeters and (b) $r = 24$ centimeters.

16. ***Area*** Let A be the area of a circle of radius r that is changing with respect to time. If dr/dt is constant, is dA/dt constant? Explain.

17. ***Area*** The included angle of the two sides of constant equal length s of an isosceles triangle is θ.

(a) Show that the area of the triangle is given by $A = \frac{1}{2}s^2 \sin \theta$.

(b) If θ is increasing at the rate of $\frac{1}{2}$ radian per minute, find the rates of change of the area when $\theta = \pi/6$ and $\theta = \pi/3$.

(c) Explain why the rate of change of the area of the triangle is not constant even though $d\theta/dt$ is constant.

18. ***Volume*** The radius r of a sphere is increasing at a rate of 2 inches per minute.

(a) Find the rate of change of the volume when $r = 6$ inches and $r = 24$ inches.

(b) Explain why the rate of change of the volume of the sphere is not constant even though dr/dt is constant.

19. ***Volume*** A spherical balloon is inflated with gas at the rate of 800 cubic centimeters per minute. How fast is the radius of the balloon increasing at the instant the radius is (a) 30 centimeters and (b) 60 centimeters?

20. ***Volume*** All edges of a cube are expanding at a rate of 3 centimeters per second. How fast is the volume changing when each edge is (a) 1 centimeter and (b) 10 centimeters?

21. ***Surface Area*** The conditions are the same as in Exercise 20. Determine how fast the *surface area* is changing when each edge is (a) 1 centimeter and (b) 10 centimeters.

22. ***Volume*** The formula for the volume of a cone is $V = \frac{1}{3}\pi r^2 h$. Find the rate of change of the volume if dr/dt is 2 inches per minute and $h = 3r$ when (a) $r = 6$ inches and (b) $r = 24$ inches.

23. ***Volume*** At a sand and gravel plant, sand is falling off a conveyor and onto a conical pile at a rate of 10 cubic feet per minute. The diameter of the base of the cone is approximately three times the altitude. At what rate is the height of the pile changing when the pile is 15 feet high?

24. ***Depth*** A conical tank (with vertex down) is 10 feet across the top and 12 feet deep. If water is flowing into the tank at a rate of 10 cubic feet per minute, find the rate of change of the depth of the water when the water is 8 feet deep.

25. ***Depth*** A swimming pool is 12 meters long, 6 meters wide, 1 meter deep at the shallow end, and 3 meters deep at the deep end (see figure). Water is being pumped into the pool at $\frac{1}{4}$ cubic meter per minute, and there is 1 meter of water at the deep end.

(a) What percent of the pool is filled?

(b) At what rate is the water level rising?

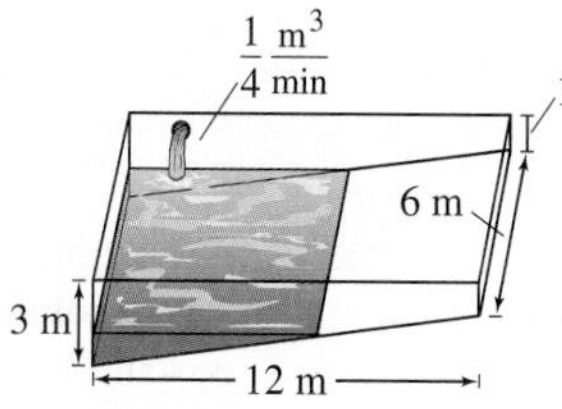

Figure for 25

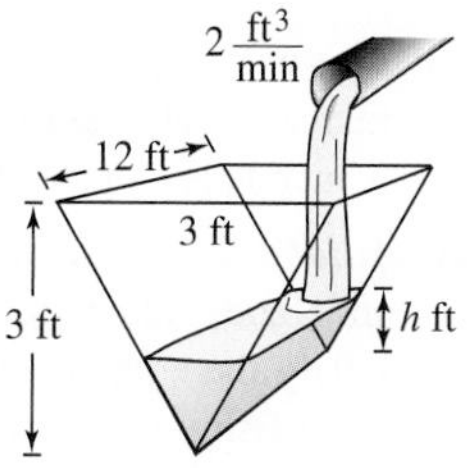

Figure for 26

26. ***Depth*** A trough is 12 feet long and 3 feet across the top (see figure). Its ends are isosceles triangles with altitudes of 3 feet.

(a) If water is being pumped into the trough at 2 cubic feet per minute, how fast is the water level rising when h is 1 foot deep?

(b) If the water is rising at a rate of $\frac{3}{8}$ inch per minute when $h = 2$, determine the rate at which water is being pumped into the trough.

27. ***Moving Ladder*** A ladder 25 feet long is leaning against the wall of a house (see figure). The base of the ladder is pulled away from the wall at a rate of 2 feet per second.

(a) How fast is the top of the ladder moving down the wall when its base is 7 feet, 15 feet, and 24 feet from the wall?

(b) Consider the triangle formed by the side of the house, the ladder, and the ground. Find the rate at which the area of the triangle is changing when the base of the ladder is 7 feet from the wall.

(c) Find the rate at which the angle between the ladder and the wall of the house is changing when the base of the ladder is 7 feet from the wall.

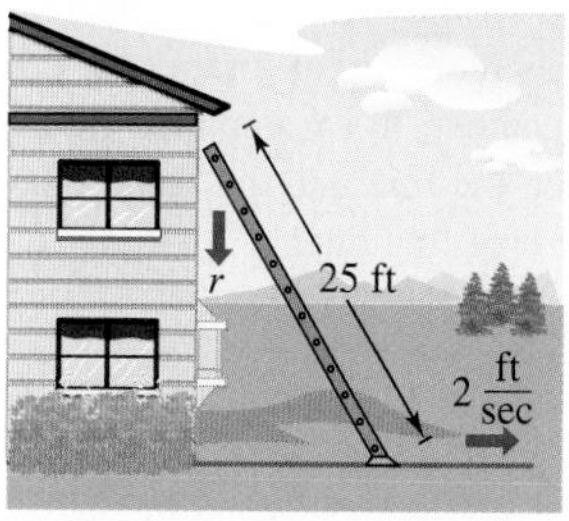

Figure for 27

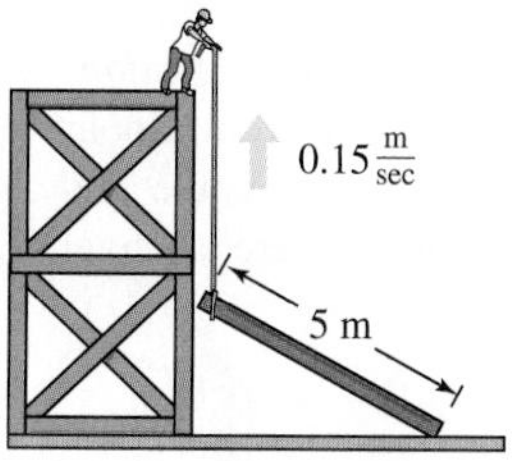

Figure for 28

FOR FURTHER INFORMATION For more information on the mathematics of moving ladders, see the article "The Falling Ladder Paradox" by Paul Scholten and Andrew Simoson in *The College Mathematics Journal*. To view this article, go to the website *www.matharticles.com*.

28. ***Construction*** A construction worker pulls a five-meter plank up the side of a building under construction by means of a rope tied to one end of the plank (see figure). Assume the opposite end of the plank follows a path perpendicular to the wall of the building and the worker pulls the rope at a rate of 0.15 meter per second. How fast is the end of the plank sliding along the ground when it is 2.5 meters from the wall of the building?

29. ***Construction*** A winch at the top of a 12-meter building pulls a pipe of the same length to a vertical position, as shown in the figure. The winch pulls in rope at a rate of -0.2 meter per second. Find the rate of vertical change and the rate of horizontal change at the end of the pipe when $y = 6$.

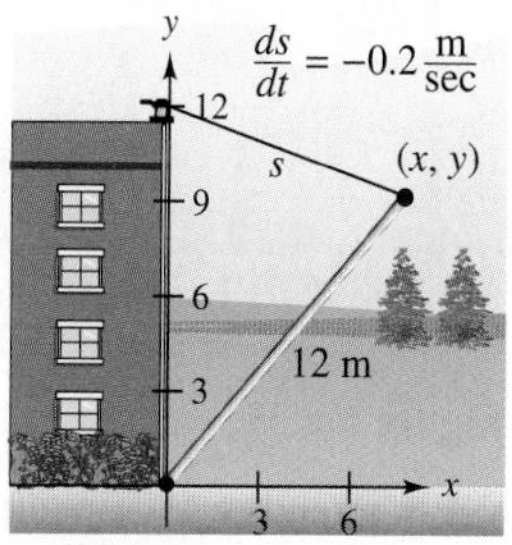

Figure for 29

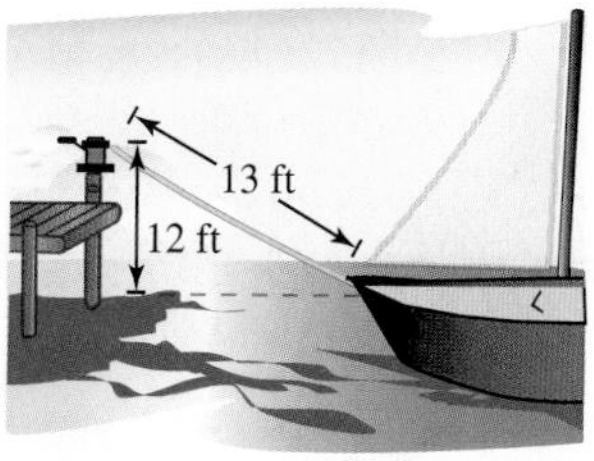

Not drawn to scale

Figure for 30

30. ***Boating*** A boat is pulled into a dock by means of a winch 12 feet above the deck of the boat (see figure).

(a) The winch pulls in rope at a rate of 4 feet per second. Determine the speed of the boat when there is 13 feet of rope out. What happens to the speed of the boat as it gets closer to the dock?

(b) Suppose the boat is moving at a constant rate of 4 feet per second. Determine the speed at which the winch pulls in rope when there is a total of 13 feet of rope out. What happens to the speed at which the winch pulls in rope as the boat gets closer to the dock?

31. ***Air Traffic Control*** An air traffic controller spots two planes at the same altitude converging on a point as they fly at right angles to each other (see figure). One plane is 150 miles from the point moving at 450 miles per hour. The other plane is 200 miles from the point moving at 600 miles per hour.

(a) At what rate is the distance between the planes decreasing?

(b) How much time does the air traffic controller have to get one of the planes on a different flight path?

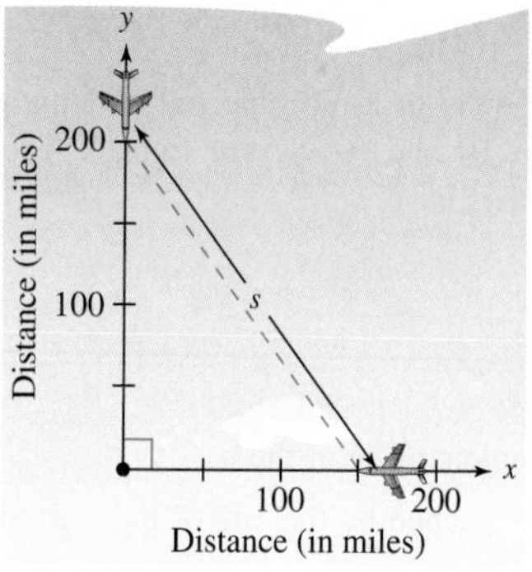

Figure for 31

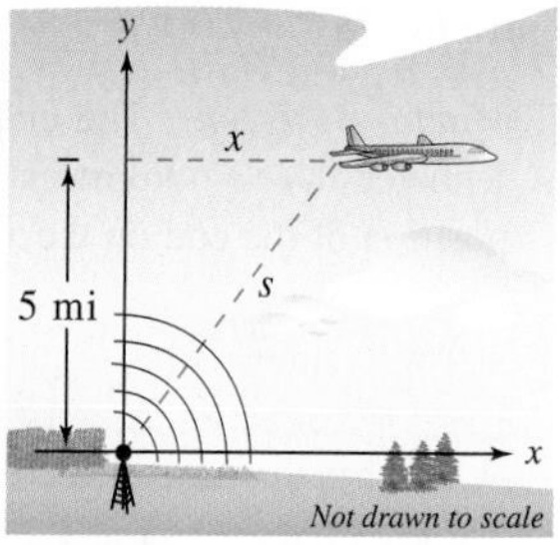

Figure for 32

32. ***Air Traffic Control*** An airplane is flying at an altitude of 5 miles and passes directly over a radar antenna (see figure on previous page). When the plane is 10 miles away ($s = 10$), the radar detects that the distance s is changing at a rate of 240 miles per hour. What is the speed of the plane?

33. ***Sports*** A baseball diamond has the shape of a square with sides 90 feet long (see figure). A player running from second base to third base at a speed of 28 feet per second is 30 feet from third base. At what rate is the player's distance s from home plate changing?

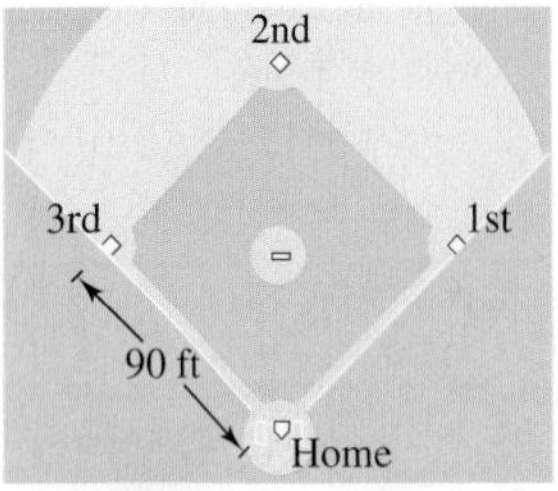

Figure for 33 and 34

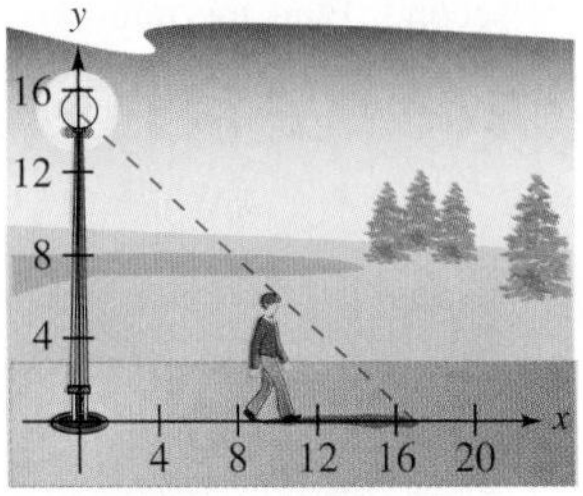

Figure for 35

34. ***Sports*** For the baseball diamond in Exercise 33, suppose the player is running from first to second at a speed of 28 feet per second. Find the rate at which the distance from home plate is changing when the player is 30 feet from second base.

35. ***Shadow Length*** A man 6 feet tall walks at a rate of 5 feet per second away from a light that is 15 feet above the ground (see figure). When he is 10 feet from the base of the light,

(a) at what rate is the tip of his shadow moving?

(b) at what rate is the length of his shadow changing?

36. ***Shadow Length*** Repeat Exercise 35 for a man 6 feet tall walking at a rate of 5 feet per second *toward* a light that is 20 feet above the ground (see figure).

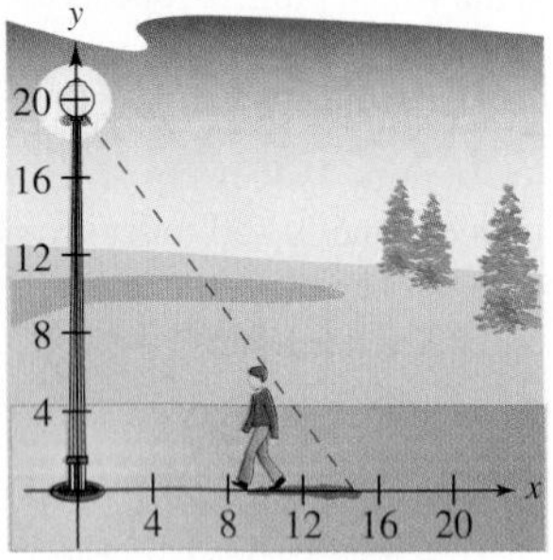

Figure for 36

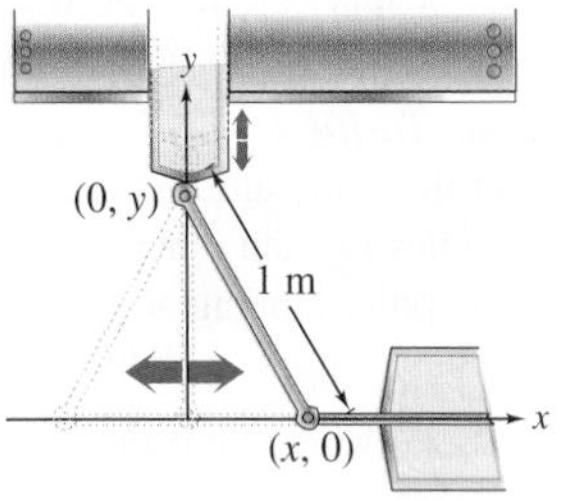

Figure for 37

37. ***Machine Design*** The endpoints of a movable rod of length 1 meter have coordinates $(x, 0)$ and $(0, y)$ (see figure). The position of the end on the x-axis is

$$x(t) = \frac{1}{2}\sin\frac{\pi t}{6}$$

where t is the time in seconds.

(a) Find the time of one complete cycle of the rod.

(b) What is the lowest point reached by the end of the rod on the y-axis?

(c) Find the speed of the y-axis endpoint when the x-axis endpoint is $\left(\frac{1}{4}, 0\right)$.

38. ***Machine Design*** Repeat Exercise 37 for a position function of $x(t) = \frac{3}{5}\sin \pi t$. Use the point $\left(\frac{3}{10}, 0\right)$ for part (c).

39. ***Evaporation*** As a spherical raindrop falls, it reaches a layer of dry air and begins to evaporate at a rate that is proportional to its surface area ($S = 4\pi r^2$). Show that the radius of the raindrop decreases at a constant rate.

40. ***Electricity*** The combined electrical resistance R of R_1 and R_2, connected in parallel, is given by

$$\frac{1}{R} = \frac{1}{R_1} + \frac{1}{R_2}$$

where R, R_1, and R_2 are measured in ohms. R_1 and R_2 are increasing at rates of 1 and 1.5 ohms per second, respectively. At what rate is R changing when $R_1 = 50$ ohms and $R_2 = 75$ ohms?

41. ***Adiabatic Expansion*** When a certain polyatomic gas undergoes adiabatic expansion, its pressure p and volume V satisfy the equation $pV^{1.3} = k$, where k is a constant. Find the relationship between the related rates dp/dt and dV/dt.

42. ***Roadway Design*** Cars on a certain roadway travel on a circular arc of radius r. In order not to rely on friction alone to overcome the centrifugal force, the road is banked at an angle of magnitude θ from the horizontal (see figure). The banking angle must satisfy the equation $rg \tan\theta = v^2$, where v is the velocity of the cars and $g = 32$ feet per second per second is the acceleration due to gravity. Find the relationship between the related rates dv/dt and $d\theta/dt$.

43. ***Angle of Elevation*** A balloon rises at a rate of 3 meters per second from a point on the ground 30 meters from an observer. Find the rate of change of the angle of elevation of the balloon from the observer when the balloon is 30 meters above the ground.

44. ***Angle of Elevation*** A fish is reeled in at a rate of 1 foot per second from a point 10 feet above the water (see figure). At what rate is the angle between the line and the water changing when there is a total of 25 feet of line out?

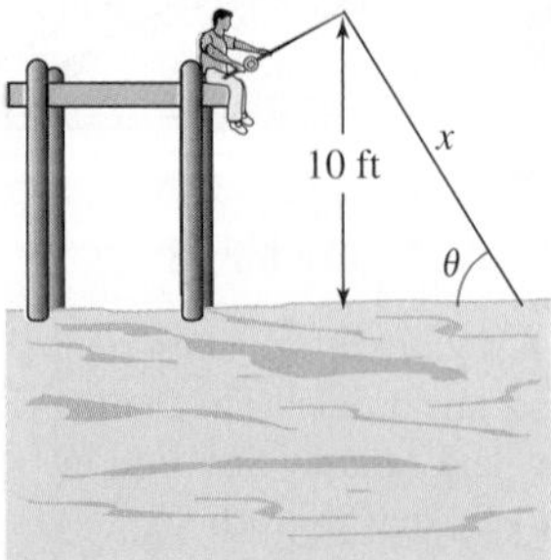

45. ***Angle of Elevation*** An airplane flies at an altitude of 5 miles toward a point directly over an observer (see figure). The speed of the plane is 600 miles per hour. Find the rates at which the angle of elevation θ is changing when the angle is (a) $\theta = 30°$, (b) $\theta = 60°$, and (c) $\theta = 75°$.

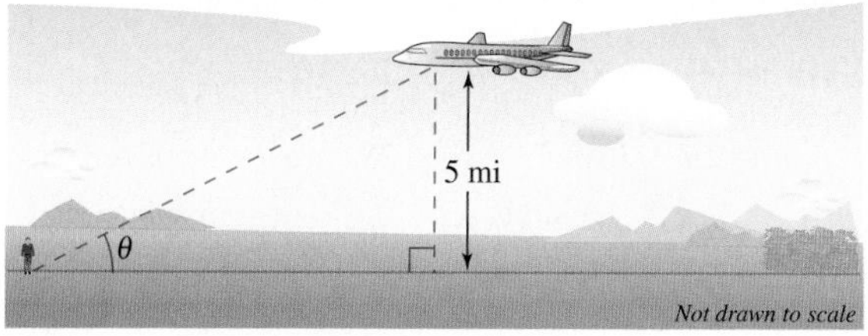

46. ***Linear vs. Angular Speed*** A patrol car is parked 50 feet from a long warehouse (see figure). The revolving light on top of the car turns at a rate of 30 revolutions per minute. How fast is the light beam moving along the wall when the beam makes angles of (a) $\theta = 30°$, (b) $\theta = 60°$, and (c) $\theta = 70°$ with the line perpendicular from the light to the wall?

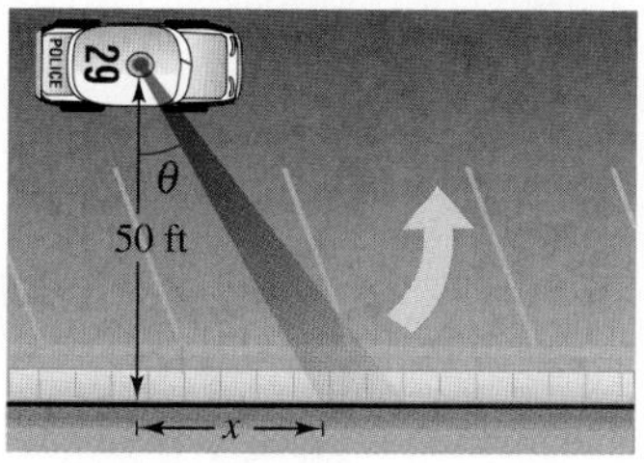

Figure for 46

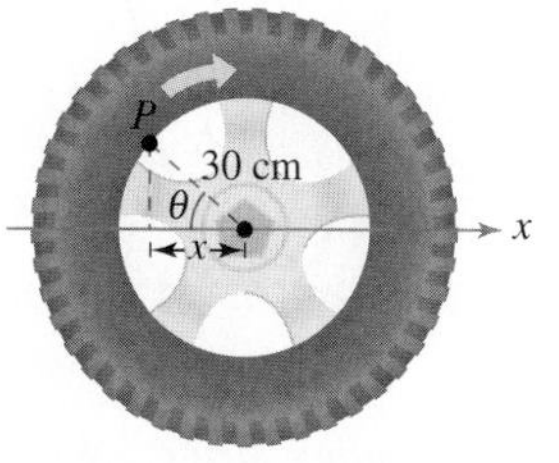

Figure for 47

47. ***Linear vs. Angular Speed*** A wheel of radius 30 centimeters revolves at a rate of 10 revolutions per second. A dot is painted at a point P on the rim of the wheel (see figure).

(a) Find dx/dt as a function of θ.

(b) Use a graphing utility to graph the function in part (a).

(c) When is the absolute value of the rate of change of x greatest? When is it least?

(d) Find dx/dt when $\theta = 30°$ and $\theta = 60°$.

48. ***Flight Control*** An airplane is flying in still air with an airspeed of 240 miles per hour. If it is climbing at an angle of 22°, find the rate at which it is gaining altitude.

49. ***Security Camera*** A security camera is centered 50 feet above a 100-foot hallway (see figure). It is easiest to design the camera with a constant angular rate of rotation, but this results in a variable rate at which the images of the surveillance area are recorded. So, it is desirable to design a system with a variable rate of rotation and a constant rate of movement of the scanning beam along the hallway. Find a model for the variable rate of rotation if $|dx/dt| = 2$ feet per second.

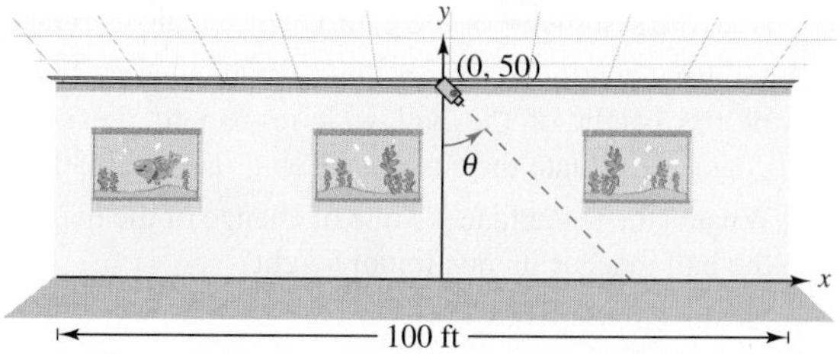

50. ***Think About It*** Describe the relationship between the rate of change of y and the rate of change of x in each expression. Assume all variables and derivatives are positive.

(a) $\dfrac{dy}{dt} = 3\dfrac{dx}{dt}$ (b) $\dfrac{dy}{dt} = x(L - x)\dfrac{dx}{dt}, \quad 0 \le x \le L$

Acceleration **In Exercises 51 and 52, find the acceleration of the specified object. (*Hint:* Recall that if a variable is changing at a constant rate, its acceleration is zero.)**

51. Find the acceleration of the top of the ladder described in Exercise 27 when the base of the ladder is 7 feet from the wall.

52. Find the acceleration of the boat in Exercise 30(a) when there is a total of 13 feet of rope out.

53. ***Modeling Data*** The table shows the numbers (in millions) of single women (never married) s and married women m in the civilian work force in the United States for the years 1993 through 2001. *(Source: U.S. Bureau of Labor Statistics)*

Year	1993	1994	1995	1996	1997	1998	1999	2000	2001
s	15.0	15.3	15.5	15.8	16.5	17.1	17.6	17.8	18.0
m	32.0	32.9	33.4	33.6	33.8	33.9	34.4	34.6	34.7

(a) Use the regression capabilities of a graphing utility to find a model of the form $m(s) = as^3 + bs^2 + cs + d$ for the data, where t is the time in years, with $t = 3$ corresponding to 1993.

(b) Find dm/dt. Then use the model to estimate dm/dt for $t = 10$ if it is predicted that the number of single women in the work force will increase at the rate of 0.75 million per year.

54. ***Moving Shadow*** A ball is dropped from a height of 20 meters, 12 meters away from the top of a 20-meter lamppost (see figure). The ball's shadow, caused by the light at the top of the lamppost, is moving along the level ground. How fast is the shadow moving 1 second after the ball is released? *(Submitted by Dennis Gittinger, St. Philips College, San Antonio, TX)*

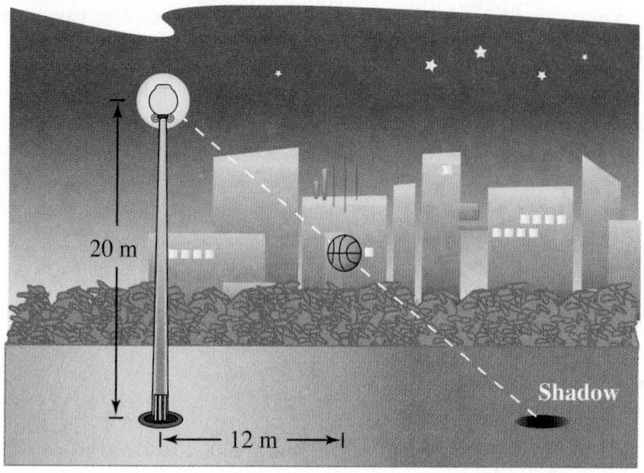

Review Exercises for Chapter 2

See www.CalcChat.com for worked-out solutions to odd-numbered exercises.

In Exercises 1–4, find the derivative of the function by using the definition of the derivative.

1. $f(x) = x^2 - 2x + 3$
2. $f(x) = \sqrt{x} + 1$
3. $f(x) = \dfrac{x+1}{x-1}$
4. $f(x) = \dfrac{2}{x}$

In Exercises 5 and 6, describe the x-values at which f is differentiable.

5. $f(x) = (x+1)^{2/3}$

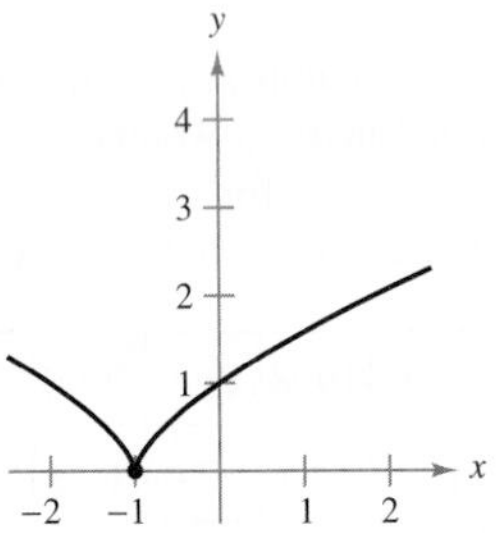

6. $f(x) = \dfrac{4x}{x+3}$

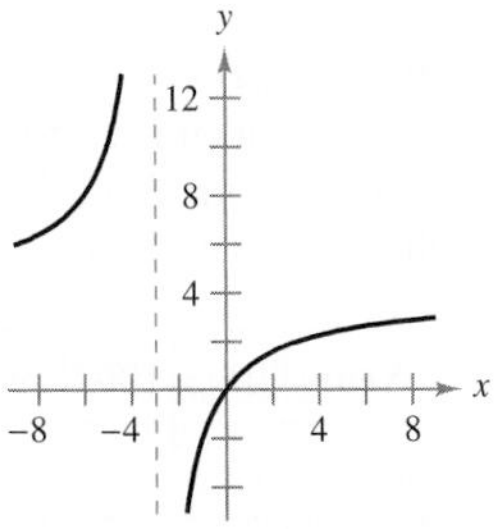

7. Sketch the graph of $f(x) = 4 - |x - 2|$.
   (a) Is f continuous at $x = 2$?
   (b) Is f differentiable at $x = 2$? Explain.
8. Sketch the graph of $f(x) = \begin{cases} x^2 + 4x + 2, & x < -2 \\ 1 - 4x - x^2, & x \geq -2. \end{cases}$
   (a) Is f continuous at $x = -2$?
   (b) Is f differentiable at $x = -2$? Explain.

In Exercises 9 and 10, find the slope of the tangent line to the graph of the function at the given point.

9. $g(x) = \dfrac{2}{3}x^2 - \dfrac{x}{6}, \quad \left(-1, \dfrac{5}{6}\right)$
10. $h(x) = \dfrac{3x}{8} - 2x^2, \quad \left(-2, -\dfrac{35}{4}\right)$

In Exercises 11 and 12, (a) find an equation of the tangent line to the graph of f at the given point, (b) use a graphing utility to graph the function and its tangent line at the point, and (c) use the *derivative* feature of the graphing utility to confirm your results.

11. $f(x) = x^3 - 1, \quad (-1, -2)$
12. $f(x) = \dfrac{2}{x+1}, \quad (0, 2)$

In Exercises 13 and 14, use the alternative form of the derivative to find the derivative at $x = c$ (if it exists).

13. $g(x) = x^2(x-1), \quad c = 2$
14. $f(x) = \dfrac{1}{x+1}, \quad c = 2$

In Exercises 15–30, find the derivative of the function.

15. $y = 25$
16. $y = -12$
17. $f(x) = x^8$
18. $g(x) = x^{12}$
19. $h(t) = 3t^4$
20. $f(t) = -8t^5$
21. $f(x) = x^3 - 3x^2$
22. $g(s) = 4s^4 - 5s^2$
23. $h(x) = 6\sqrt{x} + 3\sqrt[3]{x}$
24. $f(x) = x^{1/2} - x^{-1/2}$
25. $g(t) = \dfrac{2}{3t^2}$
26. $h(x) = \dfrac{2}{(3x)^2}$
27. $f(\theta) = 2\theta - 3\sin\theta$
28. $g(\alpha) = 4\cos\alpha + 6$
29. $f(\theta) = 3\cos\theta - \dfrac{\sin\theta}{4}$
30. $g(\alpha) = \dfrac{5\sin\alpha}{3} - 2\alpha$

Writing **In Exercises 31 and 32, the figure shows the graphs of a function and its derivative. Label the graphs as f or f' and write a short paragraph stating the criteria used in making the selection. To print an enlarged copy of the graph, go to the website *www.mathgraphs.com*.**

31.

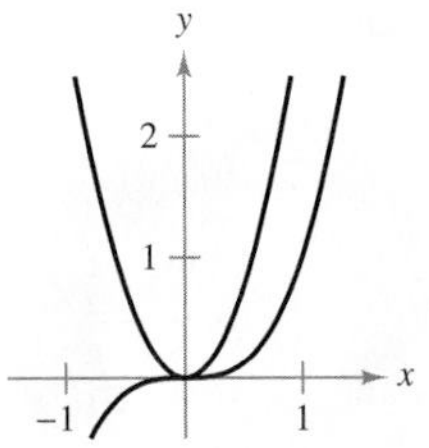

32.

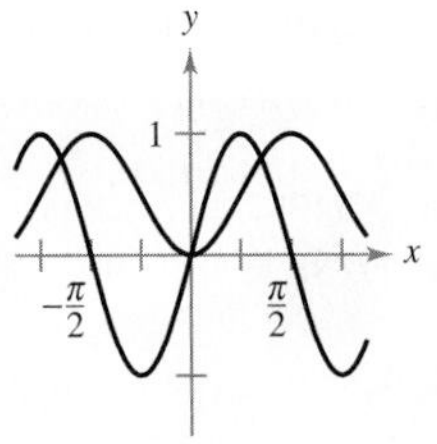

33. ***Vibrating String*** When a guitar string is plucked, it vibrates with a frequency of $F = 200\sqrt{T}$, where F is measured in vibrations per second and the tension T is measured in pounds. Find the rates of change of F when (a) $T = 4$ and (b) $T = 9$.
34. ***Vertical Motion*** A ball is dropped from a height of 100 feet. One second later, another ball is dropped from a height of 75 feet. Which ball hits the ground first?
35. ***Vertical Motion*** To estimate the height of a building, a weight is dropped from the top of the building into a pool at ground level. How high is the building if the splash is seen 9.2 seconds after the weight is dropped?
36. ***Vertical Motion*** A bomb is dropped from an airplane at an altitude of 14,400 feet. How long will it take for the bomb to reach the ground? (Because of the motion of the plane, the fall will not be vertical, but the time will be the same as that for a vertical fall.) The plane is moving at 600 miles per hour. How far will the bomb move horizontally after it is released from the plane?
37. ***Projectile Motion*** A ball thrown follows a path described by $y = x - 0.02x^2$.
    (a) Sketch a graph of the path.
    (b) Find the total horizontal distance the ball is thrown.
    (c) At what x-value does the ball reach its maximum height? (Use the symmetry of the path.)
    (d) Find an equation that gives the instantaneous rate of change of the height of the ball with respect to the horizontal change. Evaluate the equation at $x = 0, 10, 25, 30,$ and 50.
    (e) What is the instantaneous rate of change of the height when the ball reaches its maximum height?

38. ***Projectile Motion*** The path of a projectile thrown at an angle of 45° with level ground is

$$y = x - \frac{32}{v_0^2}(x^2)$$

where the initial velocity is v_0 feet per second.

(a) Find the x-coordinate of the point where the projectile strikes the ground. Use the symmetry of the path of the projectile to locate the x-coordinate of the point where the projectile reaches its maximum height.

(b) What is the instantaneous rate of change of the height when the projectile is at its maximum height?

(c) Show that doubling the initial velocity of the projectile multiplies both the maximum height and the range by a factor of 4.

(d) Find the maximum height and range of a projectile thrown with an initial velocity of 70 feet per second. Use a graphing utility to graph the path of the projectile.

39. ***Horizontal Motion*** The position function of a particle moving along the x-axis is

$$x(t) = t^2 - 3t + 2 \qquad \text{for} \qquad -\infty < t < \infty.$$

(a) Find the velocity of the particle.

(b) Find the open t-interval(s) in which the particle is moving to the left.

(c) Find the position of the particle when the velocity is 0.

(d) Find the speed of the particle when the position is 0.

40. ***Modeling Data*** The speed of a car in miles per hour and the stopping distance in feet are recorded in the table.

Speed, x	20	30	40	50	60
Stopping Distance, y	25	55	105	188	300

(a) Use the regression capabilities of a graphing utility to find a quadratic model for the data.

(b) Use a graphing utility to plot the data and graph the model.

(c) Use a graphing utility to graph dy/dx.

(d) Use the model to approximate the stopping distance at a speed of 65 miles per hour.

(e) Use the graphs in parts (b) and (c) to explain the change in stopping distance as the speed increases.

In Exercises 41–54, find the derivative of the function.

41. $f(x) = (3x^2 + 7)(x^2 - 2x + 3)$

42. $g(x) = (x^3 - 3x)(x + 2)$

43. $h(x) = \sqrt{x}\sin x$

44. $f(t) = t^3 \cos t$

45. $f(x) = \dfrac{x^2 + x - 1}{x^2 - 1}$

46. $f(x) = \dfrac{6x - 5}{x^2 + 1}$

47. $f(x) = \dfrac{1}{4 - 3x^2}$

48. $f(x) = \dfrac{9}{3x^2 - 2x}$

49. $y = \dfrac{x^2}{\cos x}$

50. $y = \dfrac{\sin x}{x^2}$

51. $y = 3x^2 \sec x$

52. $y = 2x - x^2 \tan x$

53. $y = x \cos x - \sin x$

54. $g(x) = 3x \sin x + x^2 \cos x$

In Exercises 55–58, find an equation of the tangent line to the graph of f at the given point.

55. $f(x) = \dfrac{2x^3 - 1}{x^2}, \quad (1, 1)$

56. $f(x) = \dfrac{x + 1}{x - 1}, \quad \left(\dfrac{1}{2}, -3\right)$

57. $f(x) = -x \tan x, \quad (0, 0)$

58. $f(x) = \dfrac{1 + \sin x}{1 - \sin x}, \quad (\pi, 1)$

59. ***Acceleration*** The velocity of an object in meters per second is $v(t) = 36 - t^2$, $0 \le t \le 6$. Find the velocity and acceleration of the object when $t = 4$.

60. ***Acceleration*** An automobile's velocity starting from rest is

$$v(t) = \frac{90t}{4t + 10}$$

where v is measured in feet per second. Find the vehicle's velocity and acceleration at each of the following times.

(a) 1 second (b) 5 seconds (c) 10 seconds

In Exercises 61–64, find the second derivative of the function.

61. $g(t) = t^3 - 3t + 2$

62. $f(x) = 12\sqrt[4]{x}$

63. $f(\theta) = 3 \tan \theta$

64. $h(t) = 4 \sin t - 5 \cos t$

In Exercises 65 and 66, show that the function satisfies the equation.

	Function	*Equation*
65.	$y = 2 \sin x + 3 \cos x$	$y'' + y = 0$
66.	$y = \dfrac{10 - \cos x}{x}$	$xy' + y = \sin x$

In Exercises 67–78, find the derivative of the function.

67. $h(x) = \left(\dfrac{x - 3}{x^2 + 1}\right)^2$

68. $f(x) = \left(x^2 + \dfrac{1}{x}\right)^5$

69. $f(s) = (s^2 - 1)^{5/2}(s^3 + 5)$

70. $h(\theta) = \dfrac{\theta}{(1 - \theta)^3}$

71. $y = 3\cos(3x + 1)$

72. $y = 1 - \cos 2x + 2\cos^2 x$

73. $y = \dfrac{x}{2} - \dfrac{\sin 2x}{4}$

74. $y = \dfrac{\sec^7 x}{7} - \dfrac{\sec^5 x}{5}$

75. $y = \dfrac{2}{3}\sin^{3/2} x - \dfrac{2}{7}\sin^{7/2} x$

76. $f(x) = \dfrac{3x}{\sqrt{x^2 + 1}}$

77. $y = \dfrac{\sin \pi x}{x + 2}$

78. $y = \dfrac{\cos(x - 1)}{x - 1}$

In Exercises 79–82, find the derivative of the function at the given point.

79. $f(x) = \sqrt{1 - x^3}, \quad (-2, 3)$

80. $f(x) = \sqrt[3]{x^2 - 1}, \quad (3, 2)$

81. $y = \frac{1}{2}\csc 2x, \quad \left(\frac{\pi}{4}, \frac{1}{2}\right)$

82. $y = \csc 3x + \cot 3x, \quad \left(\frac{\pi}{6}, 1\right)$

In Exercises 83–86, use a computer algebra system to find the derivative of the function. Use the utility to graph the function and its derivative on the same set of coordinate axes. Describe the behavior of the function that corresponds to any zeros of the graph of the derivative.

83. $g(x) = \frac{2x}{\sqrt{x+1}}$ **84.** $f(x) = [(x-2)(x+4)]^2$

85. $f(t) = \sqrt{t+1}\,\sqrt[3]{t+1}$ **86.** $y = \sqrt{3x}\,(x+2)^3$

In Exercises 87–90, (a) use a computer algebra system to find the derivative of the function at the given point, (b) find an equation of the tangent line to the graph of the function at the point, and (c) graph the function and its tangent line on the same set of coordinate axes.

87. $f(t) = t^2(t-1)^5, \quad (2, 4)$

88. $g(x) = x\sqrt{x^2+1}, \quad \left(3, 3\sqrt{10}\right)$

89. $y = \tan\sqrt{1-x}, \quad \left(-2, \tan\sqrt{3}\right)$

90. $y = 2\csc^3\left(\sqrt{x}\right), \quad (1, 2\csc^3 1)$

In Exercises 91–94, find the second derivative of the function.

91. $y = 2x^2 + \sin 2x$ **92.** $y = \frac{1}{x} + \tan x$

93. $f(x) = \cot x$ **94.** $y = \sin^2 x$

In Exercises 95–98, use a computer algebra system to find the second derivative of the function.

95. $f(t) = \frac{t}{(1-t)^2}$ **96.** $g(x) = \frac{6x-5}{x^2+1}$

97. $g(\theta) = \tan 3\theta - \sin(\theta - 1)$ **98.** $h(x) = x\sqrt{x^2-1}$

99. ***Refrigeration*** The temperature T of food put in a freezer is

$$T = \frac{700}{t^2 + 4t + 10}$$

where t is the time in hours. Find the rate of change of T with respect to t at each of the following times.

(a) $t = 1$ (b) $t = 3$ (c) $t = 5$ (d) $t = 10$

100. ***Fluid Flow*** The emergent velocity v of a liquid flowing from a hole in the bottom of a tank is given by $v = \sqrt{2gh}$, where g is the acceleration due to gravity (32 feet per second per second) and h is the depth of the liquid in the tank. Find the rate of change of v with respect to h when (a) $h = 9$ and (b) $h = 4$. (Note that $g = +32$ feet per second per second. The sign of g depends on how a problem is modeled. In this case, letting g be negative would produce an imaginary value for v.)

In Exercises 101–106, use implicit differentiation to find dy/dx.

101. $x^2 + 3xy + y^3 = 10$ **102.** $x^2 + 9y^2 - 4x + 3y = 0$

103. $y\sqrt{x} - x\sqrt{y} = 16$ **104.** $y^2 = (x-y)(x^2+y)$

105. $x \sin y = y \cos x$ **106.** $\cos(x+y) = x$

In Exercises 107 and 108, find the equations of the tangent line and the normal line to the graph of the equation at the given point. Use a graphing utility to graph the equation, the tangent line, and the normal line.

107. $x^2 + y^2 = 20, \quad (2, 4)$ **108.** $x^2 - y^2 = 16, \quad (5, 3)$

109. A point moves along the curve $y = \sqrt{x}$ in such a way that the y-value is increasing at a rate of 2 units per second. At what rate is x changing for each of the following values?

(a) $x = \frac{1}{2}$ (b) $x = 1$ (c) $x = 4$

110. ***Surface Area*** The edges of a cube are expanding at a rate of 5 centimeters per second. How fast is the surface area changing when each edge is 4.5 centimeters?

111. ***Depth*** The cross section of a five-meter trough is an isosceles trapezoid with a two-meter lower base, a three-meter upper base, and an altitude of 2 meters. Water is running into the trough at a rate of 1 cubic meter per minute. How fast is the water level rising when the water is 1 meter deep?

112. ***Linear and Angular Velocity*** A rotating beacon is located 1 kilometer off a straight shoreline (see figure). If the beacon rotates at a rate of 3 revolutions per minute, how fast (in kilometers per hour) does the beam of light appear to be moving to a viewer who is $\frac{1}{2}$ kilometer down the shoreline?

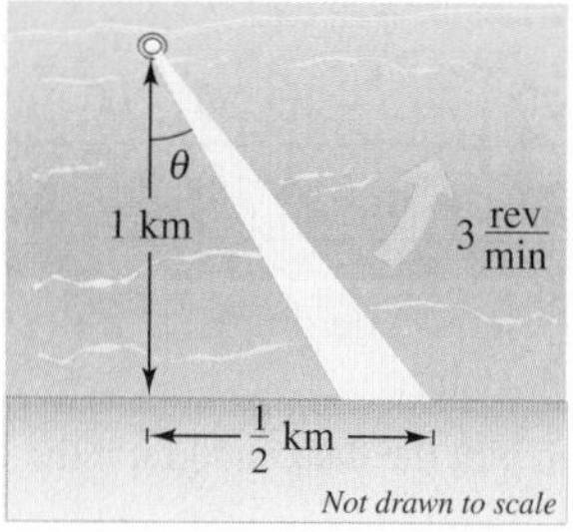

113. ***Moving Shadow*** A sandbag is dropped from a balloon at a height of 60 meters when the angle of elevation to the sun is 30° (see figure). Find the rate at which the shadow of the sandbag is traveling along the ground when the sandbag is at a height of 35 meters. [*Hint:* The position of the sandbag is given by $s(t) = 60 - 4.9t^2$.]

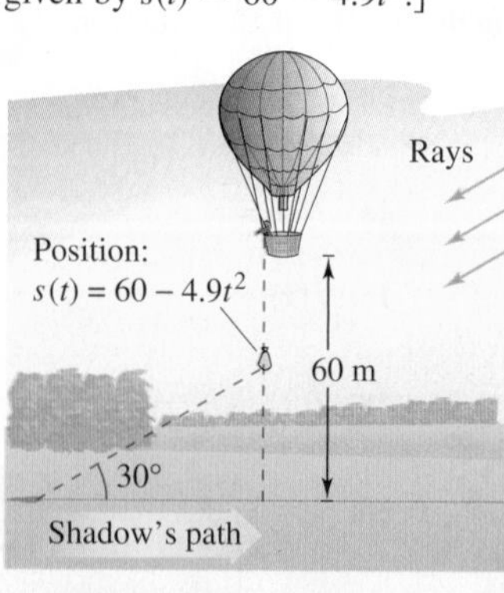

P.S. Problem Solving

See www.CalcChat.com for worked-out solutions to odd-numbered exercises.

1. Consider the graph of the parabola $y = x^2$.

(a) Find the radius r of the largest possible circle centered on the y-axis that is tangent to the parabola at the origin, as shown in the figure. This circle is called the **circle of curvature** (see Section 12.5). Find the equation of this circle. Use a graphing utility to graph the circle and parabola in the same viewing window to verify your answer.

(b) Find the center $(0, b)$ of the circle of radius 1 centered on the y-axis that is tangent to the parabola at two points, as shown in the figure. Find the equation of this circle. Use a graphing utility to graph the circle and parabola in the same viewing window to verify your answer.

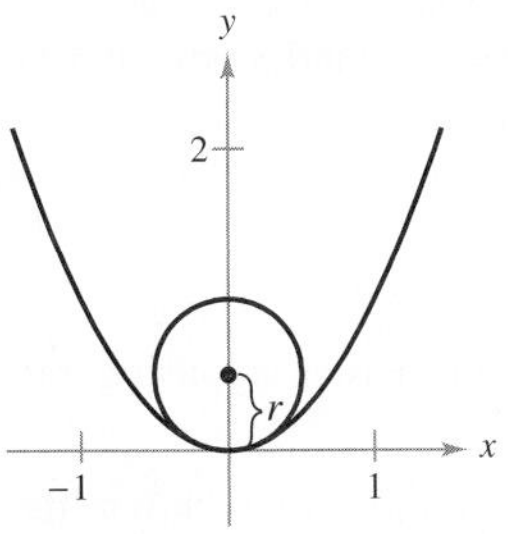

Figure for 1(a)

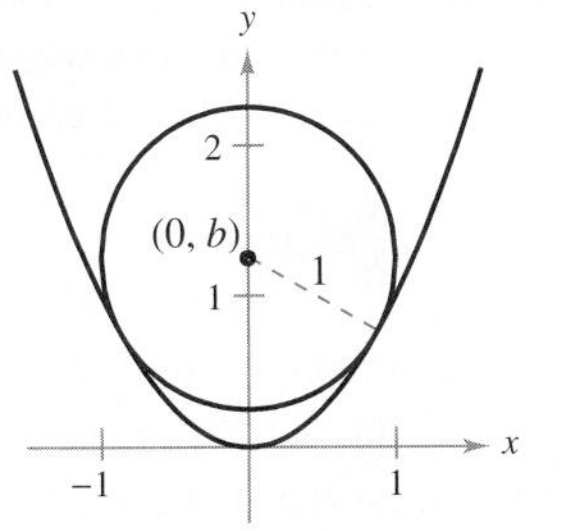

Figure for 1(b)

2. Graph the two parabolas $y = x^2$ and $y = -x^2 + 2x - 5$ in the same coordinate plane. Find equations of the two lines simultaneously tangent to both parabolas.

3. (a) Find the polynomial $P_1(x) = a_0 + a_1x$ whose value and slope agree with the value and slope of $f(x) = \cos x$ at the point $x = 0$.

(b) Find the polynomial $P_2(x) = a_0 + a_1x + a_2x^2$ whose value and first two derivatives agree with the value and first two derivatives of $f(x) = \cos x$ at the point $x = 0$. This polynomial is called the second-degree **Taylor polynomial** of $f(x) = \cos x$ at $x = 0$.

(c) Complete the table comparing the values of f and P_2. What do you observe?

x	-1.0	-0.1	-0.001	0	0.001	0.1	1.0
$\cos x$							
$P_2(x)$							

(d) Find the third-degree Taylor polynomial of $f(x) = \sin x$ at $x = 0$.

4. (a) Find an equation of the tangent line to the parabola $y = x^2$ at the point $(2, 4)$.

(b) Find an equation of the normal line to $y = x^2$ at the point $(2, 4)$. (The normal line is perpendicular to the tangent line.) Where does this line intersect the parabola a second time?

(c) Find equations of the tangent line and normal line to $y = x^2$ at the point $(0, 0)$.

(d) Prove that for any point $(a, b) \neq (0, 0)$ on the parabola $y = x^2$, the normal line intersects the graph a second time.

5. Find a third-degree polynomial $p(x)$ that is tangent to the line $y = 14x - 13$ at the point $(1, 1)$, and tangent to the line $y = -2x - 5$ at the point $(-1, -3)$.

6. Find a function of the form $f(x) = a + b \cos cx$ that is tangent to the line $y = 1$ at the point $(0, 1)$, and tangent to the line

$$y = x + \frac{3}{2} - \frac{\pi}{4}$$

at the point $\left(\frac{\pi}{4}, \frac{3}{2}\right)$.

7. The graph of the **eight curve,**

$$x^4 = a^2(x^2 - y^2),\ a \neq 0,$$

is shown below.

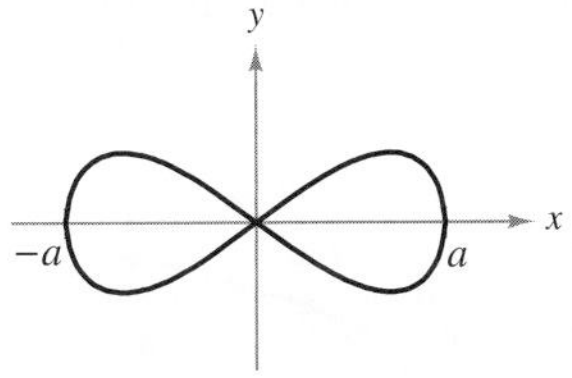

(a) Explain how you could use a graphing utility to graph this curve.

(b) Use a graphing utility to graph the curve for various values of the constant a. Describe how a affects the shape of the curve.

(c) Determine the points on the curve where the tangent line is horizontal.

8. The graph of the **pear-shaped quartic,**

$$b^2y^2 = x^3(a - x),\ a, b > 0,$$

is shown below.

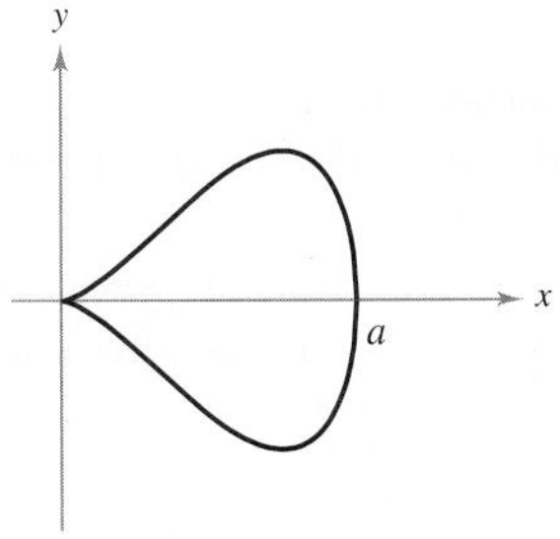

(a) Explain how you could use a graphing utility to graph this curve.

(b) Use a graphing utility to graph the curve for various values of the constants a and b. Describe how a and b affect the shape of the curve.

(c) Determine the points on the curve where the tangent line is horizontal.

9. A man 6 feet tall walks at a rate of 5 feet per second toward a streetlight that is 30 feet high (see figure). The man's 3-foot-tall child follows at the same speed, but 10 feet behind the man. At times, the shadow behind the child is caused by the man, and at other times, by the child.

(a) Suppose the man is 90 feet from the streetlight. Show that the man's shadow extends beyond the child's shadow.

(b) Suppose the man is 60 feet from the streetlight. Show that the child's shadow extends beyond the man's shadow.

(c) Determine the distance d from the man to the streetlight at which the tips of the two shadows are exactly the same distance from the streetlight.

(d) Determine how fast the tip of the shadow is moving as a function of x, the distance between the man and the street light. Discuss the continuity of this shadow speed function.

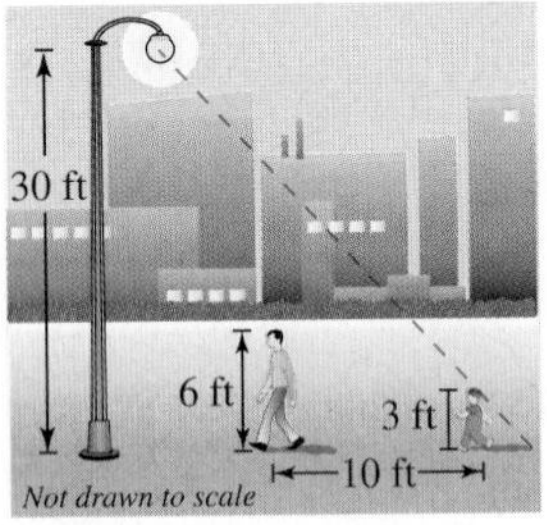

Figure for 9

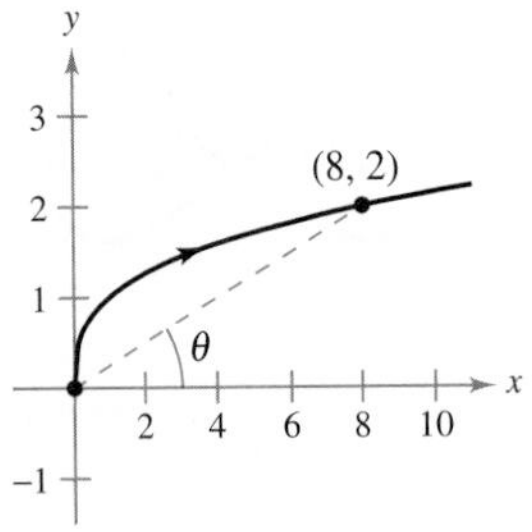

Figure for 10

10. A particle is moving along the graph of $y = \sqrt[3]{x}$ (see figure). When $x = 8$, the y-component of its position is increasing at the rate of 1 centimeter per second.

(a) How fast is the x-component changing at this moment?

(b) How fast is the distance from the origin changing at this moment?

(c) How fast is the angle of inclination θ changing at this moment?

11. Let L be a differentiable function for all x. Prove that if $L(a + b) = L(a) + L(b)$ for all a and b, then $L'(x) = L'(0)$ for all x. What does the graph of L look like?

12. Let E be a function satisfying $E(0) = E'(0) = 1$. Prove that if $E(a + b) = E(a)E(b)$ for all a and b, then E is differentiable and $E'(x) = E(x)$ for all x. Find an example of a function satisfying $E(a + b) = E(a)E(b)$.

13. The fundamental limit $\lim_{x \to 0} \frac{\sin x}{x} = 1$ assumes that x is measured in radians. What happens if you assume that x is measured in degrees instead of radians?

(a) Set your calculator to *degree* mode and complete the table.

z (in degrees)	0.1	0.01	0.0001
$\frac{\sin z}{z}$			

(b) Use the table to estimate

$$\lim_{z \to 0} \frac{\sin z}{z}$$

for z in degrees. What is the exact value of this limit? (*Hint:* $180° = \pi$ radians)

(c) Use the limit definition of the derivative to find

$$\frac{d}{dz} \sin z$$

for z in degrees.

(d) Define the new functions $S(z) = \sin(cz)$ and $C(z) = \cos(cz)$, where $c = \pi/180$. Find $S(90)$ and $C(180)$. Use the Chain Rule to calculate

$$\frac{d}{dz} S(z).$$

(e) Explain why calculus is made easier by using radians instead of degrees.

14. An astronaut standing on the moon throws a rock into the air. The height of the rock is

$$s = -\frac{27}{10}t^2 + 27t + 6$$

where s is measured in feet and t is measured in seconds.

(a) Find expressions for the velocity and acceleration of the rock.

(b) Find the time when the rock is at its highest point by finding the time when the velocity is zero. What is the height of the rock at this time?

(c) How does the acceleration of the rock compare with the acceleration due to gravity on Earth?

15. If a is the acceleration of an object, the *jerk* j is defined by $j = a'(t)$.

(a) Use this definition to give a physical interpretation of j.

(b) Find j for the slowing vehicle in Exercise 117 in Section 2.3 and interpret the result.

(c) The figure shows the graph of the position, velocity, acceleration, and jerk functions of a vehicle. Identify each graph and explain your reasoning.

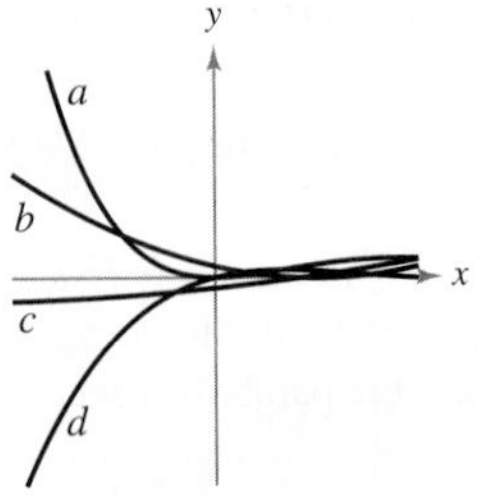

3 Applications of Differentiation

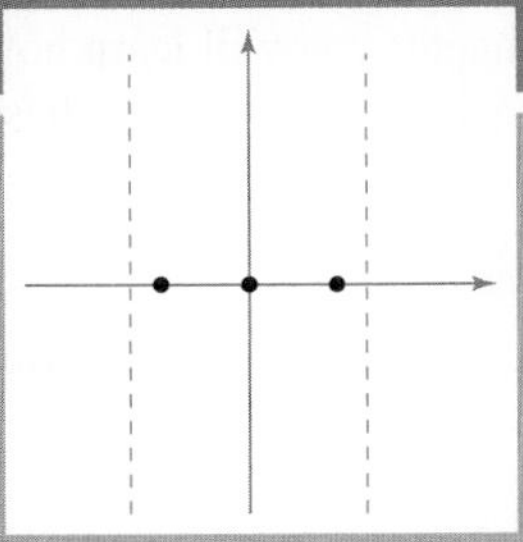

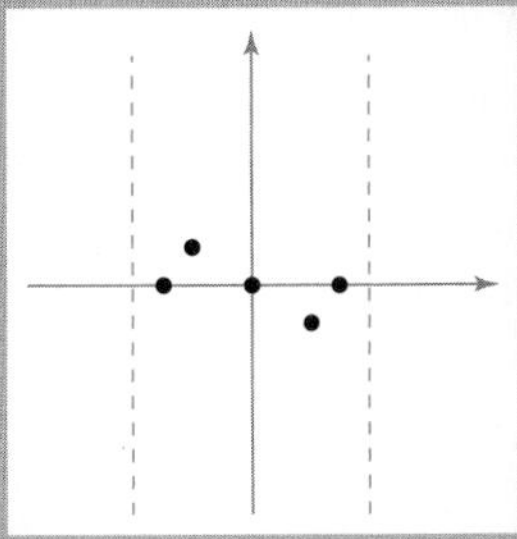

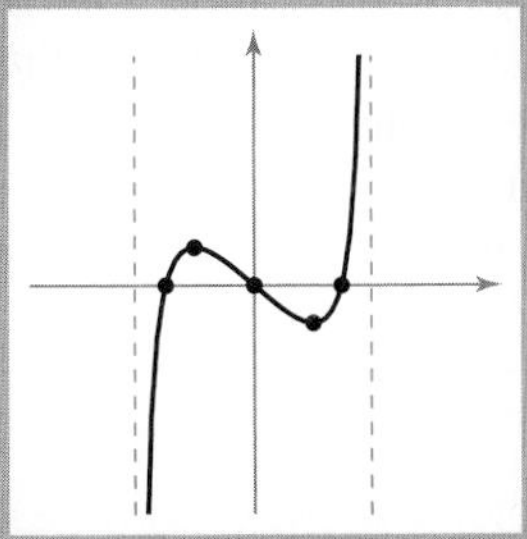

In Chapter 3, you will use calculus to analyze graphs of functions. For example, you can use the derivative of a function to determine the function's maximum and minimum values. You can use limits to identify any asymptotes of the function's graph. In Section 3.6, you will combine these techniques to sketch the graph of a function.

When a glassblower removes a glowing "blob" of molten glass from a kiln, its temperature is about 1700°F. At first, the molten glass cools rapidly. Then, as the temperature of the glass approaches room temperature, it cools more and more slowly. Will the temperature of the glass ever actually reach room temperature? Why?

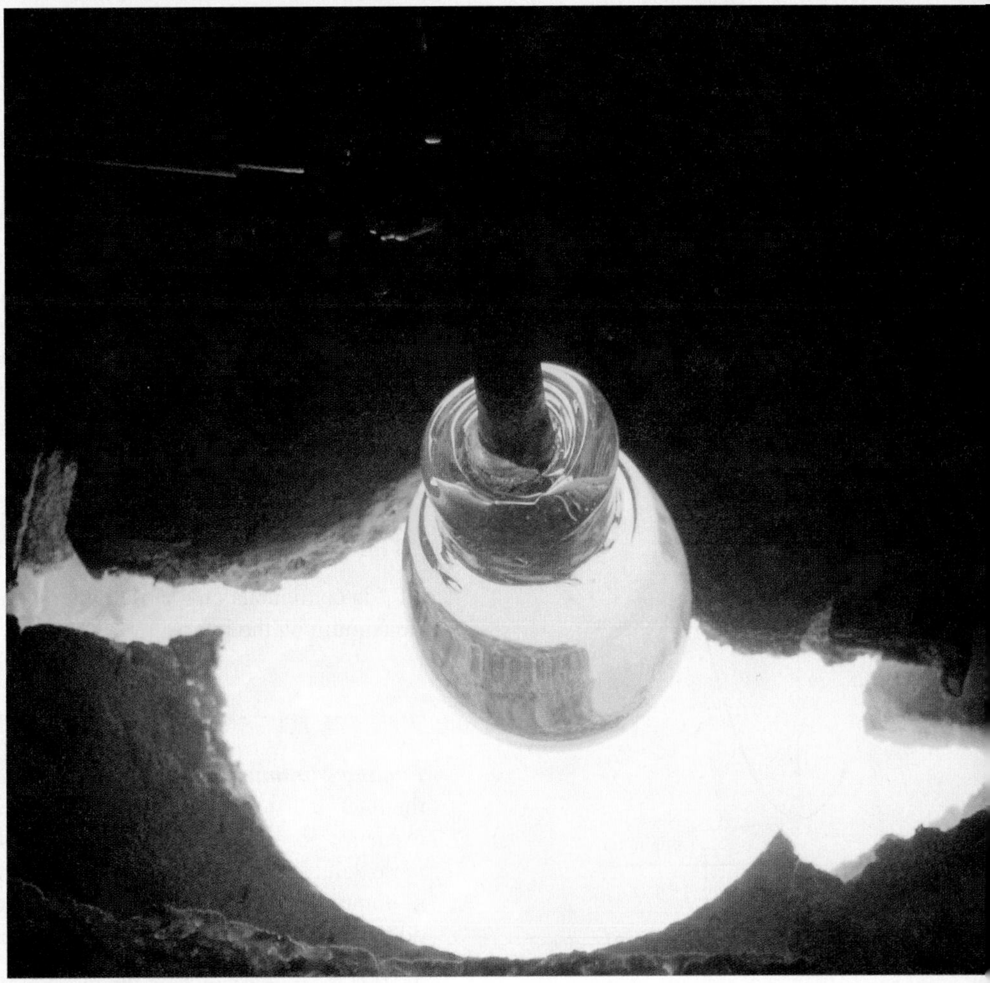

www.shawnolson.net/a/507

Section 3.1 Extrema on an Interval

- **Understand the definition of extrema of a function on an interval.**
- **Understand the definition of relative extrema of a function on an open interval.**
- **Find extrema on a closed interval.**

Extrema of a Function

In calculus, much effort is devoted to determining the behavior of a function f on an interval I. Does f have a maximum value on I? Does it have a minimum value? Where is the function increasing? Where is it decreasing? In this chapter you will learn how derivatives can be used to answer these questions. You will also see why these questions are important in real-life applications.

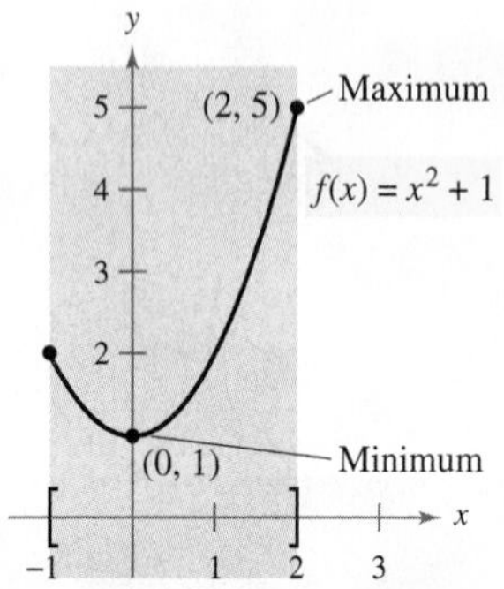

(a) f is continuous, $[-1, 2]$ is closed.

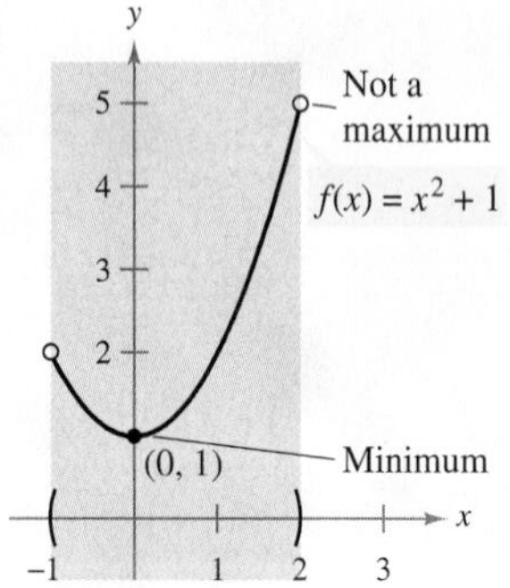

(b) f is continuous, $(-1, 2)$ is open.

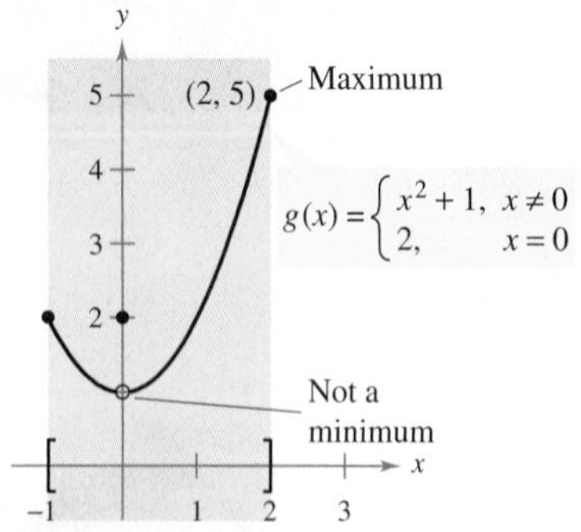

(c) g is not continuous, $[-1, 2]$ is closed.
Extrema can occur at interior points or endpoints of an interval. Extrema that occur at the endpoints are called **endpoint extrema.**
Figure 3.1

Definition of Extrema

Let f be defined on an interval I containing c.

1. $f(c)$ is the **minimum of f on I** if $f(c) \leq f(x)$ for all x in I.
2. $f(c)$ is the **maximum of f on I** if $f(c) \geq f(x)$ for all x in I.

The minimum and maximum of a function on an interval are the **extreme values,** or **extrema** (the singular form of extrema is extremum), of the function on the interval. The minimum and maximum of a function on an interval are also called the **absolute minimum** and **absolute maximum** on the interval.

A function need not have a minimum or a maximum on an interval. For instance, in Figure 3.1(a) and (b), you can see that the function $f(x) = x^2 + 1$ has both a minimum and a maximum on the closed interval $[-1, 2]$, but does not have a maximum on the open interval $(-1, 2)$. Moreover, in Figure 3.1(c), you can see that continuity (or the lack of it) can affect the existence of an extremum on the interval. This suggests the theorem below. (Although the Extreme Value Theorem is intuitively plausible, a proof of this theorem is not within the scope of this text.)

THEOREM 3.1 The Extreme Value Theorem

If f is continuous on a closed interval $[a, b]$, then f has both a minimum and a maximum on the interval.

EXPLORATION

Finding Minimum and Maximum Values The Extreme Value Theorem (like the Intermediate Value Theorem) is an *existence theorem* because it tells of the existence of minimum and maximum values but does not show how to find these values. Use the extreme-value capability of a graphing utility to find the minimum and maximum values of each of the following functions. In each case, do you think the x-values are exact or approximate? Explain your reasoning.

a. $f(x) = x^2 - 4x + 5$ on the closed interval $[-1, 3]$

b. $f(x) = x^3 - 2x^2 - 3x - 2$ on the closed interval $[-1, 3]$

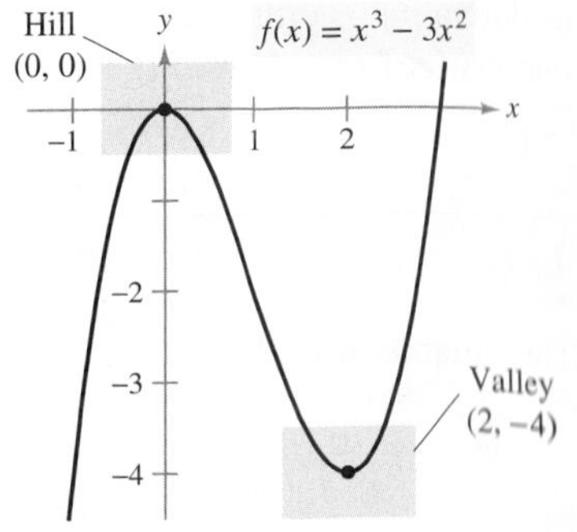

f has a relative maximum at $(0, 0)$ and a relative minimum at $(2, -4)$.
Figure 3.2

Critical Numbers

occu... the hill (... the high poi... represents a func... $x^3 - 3x^2$ has a **relative maximum** at the point ...e point $(2, -4)$. Informally, you can think of a ...ill" on the graph, and a relative minimum as ... a hill and valley can occur in two ways. If ...e graph has a horizontal tangent line at ...alley) is sharp and peaked, the graph ...e high point (or low point).

Definition of Relative Ex...

1. If there is an open interval contai... is called a **relative maximum** of f, ...a maximum, then $f(c)$ ...at f has a **relative** **maximum at $(c, f(c))$.**
2. If there is an open interval containing c on which ...um, then $f(c)$ is called a **relative minimum** of f, or you can sa... **relative** **minimum at $(c, f(c))$.**

The plural of relative maximum is relative maxima, and the plural of relative minimum is relative minima.

Example 1 examines the derivatives of functions at *given* relative extrema. (Much more is said about *finding* the relative extrema of a function in Section 3.3.)

EXAMPLE 1 The Value of the Derivative at Relative Extrema

Find the value of the derivative at each of the relative extrema shown in Figure 3.3.

Solution

a. The derivative of $f(x) = \dfrac{9(x^2 - 3)}{x^3}$ is

$$f'(x) = \frac{x^3(18x) - (9)(x^2 - 3)(3x^2)}{(x^3)^2} \qquad \text{Differentiate using Quotient Rule.}$$

$$= \frac{9(9 - x^2)}{x^4}. \qquad \text{Simplify.}$$

At the point $(3, 2)$, the value of the derivative is $f'(3) = 0$ [see Figure 3.3(a)].

b. At $x = 0$, the derivative of $f(x) = |x|$ *does not exist* because the following one-sided limits differ [see Figure 3.3(b)].

$$\lim_{x \to 0^-} \frac{f(x) - f(0)}{x - 0} = \lim_{x \to 0^-} \frac{|x|}{x} = -1 \qquad \text{Limit from the left}$$

$$\lim_{x \to 0^+} \frac{f(x) - f(0)}{x - 0} = \lim_{x \to 0^+} \frac{|x|}{x} = 1 \qquad \text{Limit from the right}$$

c. The derivative of $f(x) = \sin x$ is

$$f'(x) = \cos x.$$

At the point $(\pi/2, 1)$, the value of the derivative is $f'(\pi/2) = \cos(\pi/2) = 0$. At the point $(3\pi/2, -1)$, the value of the derivative is $f'(3\pi/2) = \cos(3\pi/2) = 0$ [see Figure 3.3(c)].

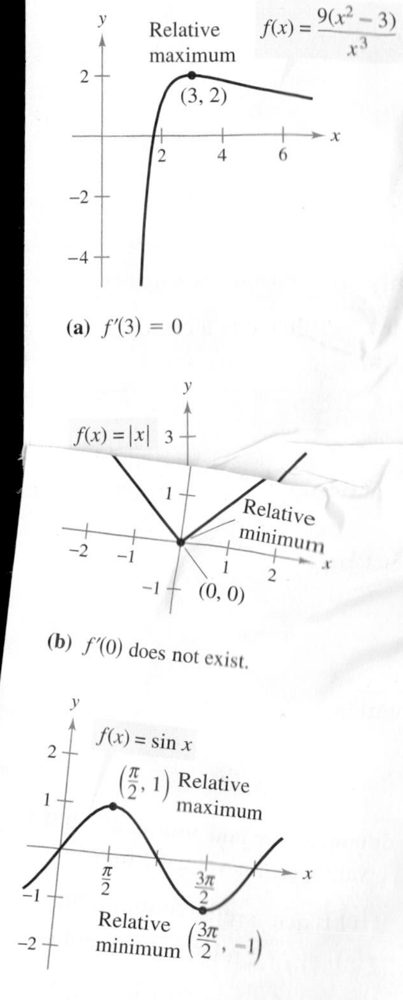

(a) $f'(3) = 0$

(b) $f'(0)$ does not exist.

(c) $f'\left(\frac{\pi}{2}\right) = 0$; $f'\left(\frac{3\pi}{2}\right) = 0$

Figure 3.3

relative extrema, the derivative is either zero or does … special points are called **critical numbers.** Figure 3.4 … critical numbers.

… of a Critical Number

… defined at c. If $f'(c) = 0$ or if f is not differentiable at c, then c is a …cal number** of f.

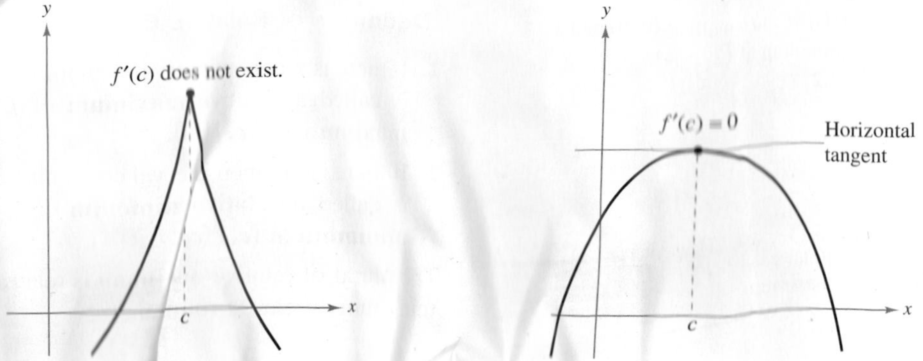

c is a critical number of f.
Figure 3.4

THEOREM 3.2 Relative Extrema Occur Only at Critical Numbers

If f has a relative minimum or relative maximum at $x = c$, then c is a critical number of f.

Mary Evans Picture Library

PIERRE DE FERMAT (1601–1665)

For Fermat, who was trained as a lawyer, mathematics was more of a hobby than a profession. Nevertheless, Fermat made many contributions to analytic geometry, number theory, calculus, and probability. In letters to friends, he wrote of many of the fundamental ideas of calculus, long before Newton or Leibniz. For instance, the theorem at the right is sometimes attributed to Fermat.

Proof

Case 1: If f is *not* differentiable at $x = c$, then, by definitio… of f and the theorem is valid.

Case 2: If f is differentiable at $x = c$, then $f'(c)$ must be positive, negative, or 0. Suppose $f'(c)$ is positive. Then

$$f'(c) = \lim_{x \to c} \frac{f(x) - f(c)}{x - c} > 0$$

which implies that there exists an interval (a, b) containing c such that

$$\frac{f(x) - f(c)}{x - c} > 0, \text{ for all } x \neq c \text{ in } (a, b).$$

[See Exercise 72(b), Section 1.2.]

Because this quotient is positive, the signs of the denominator and numerator must agree. This produces the following inequalities for x-values in the interval (a, b).

Left of c: $x < c$ and $f(x) < f(c)$ ⇒ $f(c)$ is not a relative minimum

Right of c: $x > c$ and $f(x) > f(c)$ ⇒ $f(c)$ is not a relative maximum

So, the assumption that $f'(c) > 0$ contradicts the hypothesis that $f(c)$ is a relative extremum. Assuming that $f'(c) < 0$ produces a similar contradiction, you are left with only one possibility—namely, $f'(c) = 0$. So, by definition, c is a critical number of f and the theorem is valid.

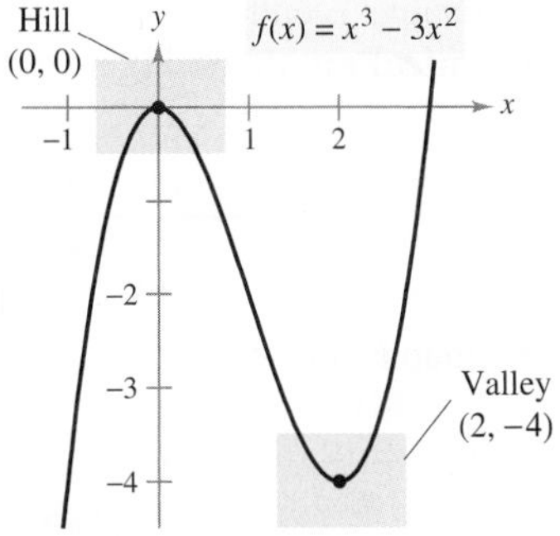

f has a relative maximum at $(0, 0)$ and a relative minimum at $(2, -4)$.
Figure 3.2

Relative Extrema and Critical Numbers

In Figure 3.2, the graph of $f(x) = x^3 - 3x^2$ has a **relative maximum** at the point $(0, 0)$ and a **relative minimum** at the point $(2, -4)$. Informally, you can think of a relative maximum as occurring on a "hill" on the graph, and a relative minimum as occurring in a "valley" on the graph. Such a hill and valley can occur in two ways. If the hill (or valley) is smooth and rounded, the graph has a horizontal tangent line at the high point (or low point). If the hill (or valley) is sharp and peaked, the graph represents a function that is not differentiable at the high point (or low point).

Definition of Relative Extrema

1. If there is an open interval containing c on which $f(c)$ is a maximum, then $f(c)$ is called a **relative maximum** of f, or you can say that f has a **relative maximum at $(c, f(c))$.**
2. If there is an open interval containing c on which $f(c)$ is a minimum, then $f(c)$ is called a **relative minimum** of f, or you can say that f has a **relative minimum at $(c, f(c))$.**

The plural of relative maximum is relative maxima, and the plural of relative minimum is relative minima.

Example 1 examines the derivatives of functions at *given* relative extrema. (Much more is said about *finding* the relative extrema of a function in Section 3.3.)

EXAMPLE 1 The Value of the Derivative at Relative Extrema

Find the value of the derivative at each of the relative extrema shown in Figure 3.3.

Solution

a. The derivative of $f(x) = \dfrac{9(x^2 - 3)}{x^3}$ is

$$f'(x) = \frac{x^3(18x) - (9)(x^2 - 3)(3x^2)}{(x^3)^2} \qquad \text{Differentiate using Quotient Rule.}$$

$$= \frac{9(9 - x^2)}{x^4}. \qquad \text{Simplify.}$$

At the point $(3, 2)$, the value of the derivative is $f'(3) = 0$ [see Figure 3.3(a)].

b. At $x = 0$, the derivative of $f(x) = |x|$ *does not exist* because the following one-sided limits differ [see Figure 3.3(b)].

$$\lim_{x \to 0^-} \frac{f(x) - f(0)}{x - 0} = \lim_{x \to 0^-} \frac{|x|}{x} = -1 \qquad \text{Limit from the left}$$

$$\lim_{x \to 0^+} \frac{f(x) - f(0)}{x - 0} = \lim_{x \to 0^+} \frac{|x|}{x} = 1 \qquad \text{Limit from the right}$$

c. The derivative of $f(x) = \sin x$ is

$$f'(x) = \cos x.$$

At the point $(\pi/2, 1)$, the value of the derivative is $f'(\pi/2) = \cos(\pi/2) = 0$. At the point $(3\pi/2, -1)$, the value of the derivative is $f'(3\pi/2) = \cos(3\pi/2) = 0$ [see Figure 3.3(c)].

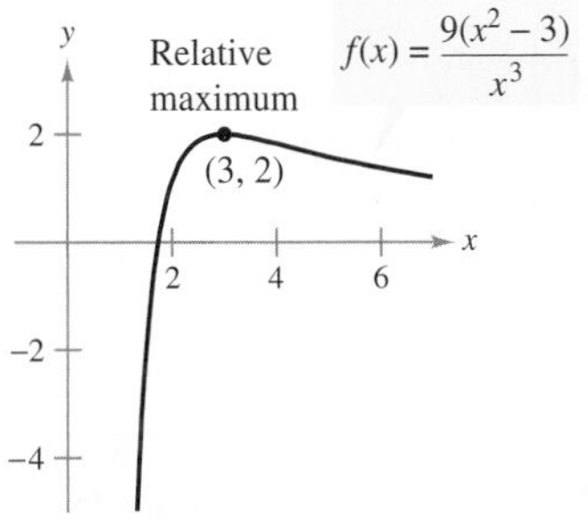

(a) $f'(3) = 0$

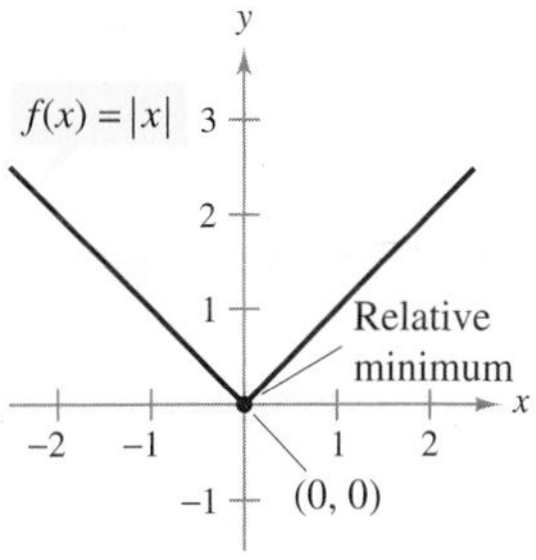

(b) $f'(0)$ does not exist.

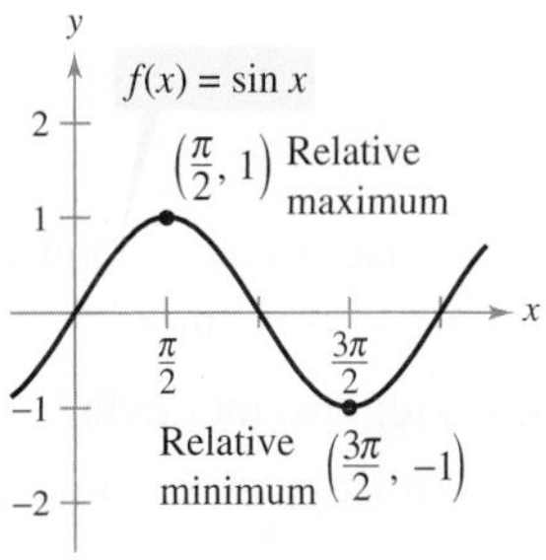

(c) $f'\left(\frac{\pi}{2}\right) = 0$; $f'\left(\frac{3\pi}{2}\right) = 0$
Figure 3.3

Note in Example 1 that at the relative extrema, the derivative is either zero or does not exist. The x-values at these special points are called **critical numbers.** Figure 3.4 illustrates the two types of critical numbers.

Definition of a Critical Number

Let f be defined at c. If $f'(c) = 0$ or if f is not differentiable at c, then c is a **critical number** of f.

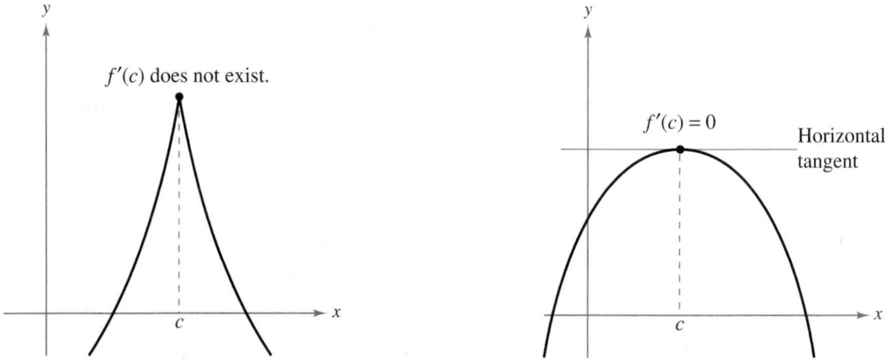

c is a critical number of f.
Figure 3.4

THEOREM 3.2 Relative Extrema Occur Only at Critical Numbers

If f has a relative minimum or relative maximum at $x = c$, then c is a critical number of f.

Pierre de Fermat (1601–1665)

For Fermat, who was trained as a lawyer, mathematics was more of a hobby than a profession. Nevertheless, Fermat made many contributions to analytic geometry, number theory, calculus, and probability. In letters to friends, he wrote of many of the fundamental ideas of calculus, long before Newton or Leibniz. For instance, the theorem at the right is sometimes attributed to Fermat.

Proof

Case 1: If f is *not* differentiable at $x = c$, then, by definition, c is a critical number of f and the theorem is valid.

Case 2: If f is differentiable at $x = c$, then $f'(c)$ must be positive, negative, or 0. Suppose $f'(c)$ is positive. Then

$$f'(c) = \lim_{x \to c} \frac{f(x) - f(c)}{x - c} > 0$$

which implies that there exists an interval (a, b) containing c such that

$$\frac{f(x) - f(c)}{x - c} > 0, \text{ for all } x \neq c \text{ in } (a, b).$$ [See Exercise 72(b), Section 1.2.]

Because this quotient is positive, the signs of the denominator and numerator must agree. This produces the following inequalities for x-values in the interval (a, b).

Left of c: $x < c$ and $f(x) < f(c)$ ⇨ $f(c)$ is not a relative minimum

Right of c: $x > c$ and $f(x) > f(c)$ ⇨ $f(c)$ is not a relative maximum

So, the assumption that $f'(c) > 0$ contradicts the hypothesis that $f(c)$ is a relative extremum. Assuming that $f'(c) < 0$ produces a similar contradiction, you are left with only one possibility—namely, $f'(c) = 0$. So, by definition, c is a critical number of f and the theorem is valid.

Finding Extrema on a Closed Interval

Theorem 3.2 states that the relative extrema of a function can occur *only* at the critical numbers of the function. Knowing this, you can use the following guidelines to find extrema on a closed interval.

Guidelines for Finding Extrema on a Closed Interval

To find the extrema of a continuous function f on a closed interval $[a, b]$, use the following steps.

1. Find the critical numbers of f in (a, b).
2. Evaluate f at each critical number in (a, b).
3. Evaluate f at each endpoint of $[a, b]$.
4. The least of these values is the minimum. The greatest is the maximum.

The next three examples show how to apply these guidelines. Be sure you see that finding the critical numbers of the function is only part of the procedure. Evaluating the function at the critical numbers *and* the endpoints is the other part.

EXAMPLE 2 Finding Extrema on a Closed Interval

Find the extrema of $f(x) = 3x^4 - 4x^3$ on the interval $[-1, 2]$.

Solution Begin by differentiating the function.

$$f(x) = 3x^4 - 4x^3 \quad \text{Write original function.}$$

$$f'(x) = 12x^3 - 12x^2 \quad \text{Differentiate.}$$

To find the critical numbers of f, you must find all x-values for which $f'(x) = 0$ and all x-values for which $f'(x)$ does not exist.

$$f'(x) = 12x^3 - 12x^2 = 0 \quad \text{Set } f'(x) \text{ equal to 0.}$$

$$12x^2(x - 1) = 0 \quad \text{Factor.}$$

$$x = 0, 1 \quad \text{Critical numbers}$$

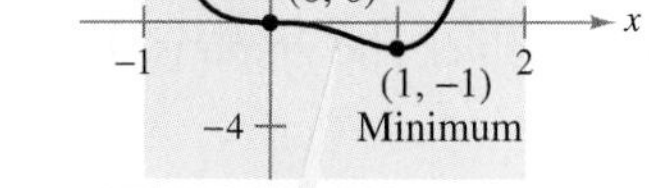

On the closed interval $[-1, 2]$, f has a minimum at $(1, -1)$ and a maximum at $(2, 16)$.
Figure 3.5

Because f' is defined for all x, you can conclude that these are the only critical numbers of f. By evaluating f at these two critical numbers and at the endpoints of $[-1, 2]$, you can determine that the maximum is $f(2) = 16$ and the minimum is $f(1) = -1$, as shown in the table. The graph of f is shown in Figure 3.5.

Left Endpoint	Critical Number	Critical Number	Right Endpoint
$f(-1) = 7$	$f(0) = 0$	$f(1) = -1$ Minimum	$f(2) = 16$ Maximum

In Figure 3.5, note that the critical number $x = 0$ does not yield a relative minimum or a relative maximum. This tells you that the converse of Theorem 3.2 is not true. In other words, *the critical numbers of a function need not produce relative extrema.*

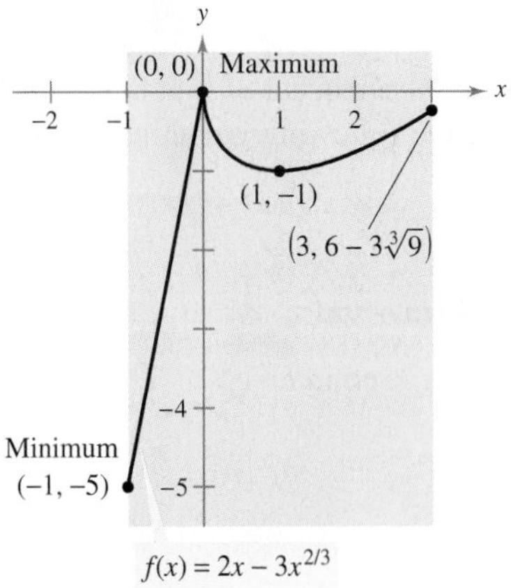

On the closed interval $[-1, 3]$, f has a minimum at $(-1, -5)$ and a maximum at $(0, 0)$.
Figure 3.6

EXAMPLE 3 Finding Extrema on a Closed Interval

Find the extrema of $f(x) = 2x - 3x^{2/3}$ on the interval $[-1, 3]$.

Solution Begin by differentiating the function.

$$f(x) = 2x - 3x^{2/3} \qquad \text{Write original function.}$$

$$f'(x) = 2 - \frac{2}{x^{1/3}} = 2\left(\frac{x^{1/3} - 1}{x^{1/3}}\right) \qquad \text{Differentiate.}$$

From this derivative, you can see that the function has two critical numbers in the interval $[-1, 3]$. The number 1 is a critical number because $f'(1) = 0$, and the number 0 is a critical number because $f'(0)$ does not exist. By evaluating f at these two numbers and at the endpoints of the interval, you can conclude that the minimum is $f(-1) = -5$ and the maximum is $f(0) = 0$, as shown in the table. The graph of f is shown in Figure 3.6.

Left Endpoint	Critical Number	Critical Number	Right Endpoint
$f(-1) = -5$ Minimum	$f(0) = 0$ Maximum	$f(1) = -1$	$f(3) = 6 - 3\sqrt[3]{9} \approx -0.24$

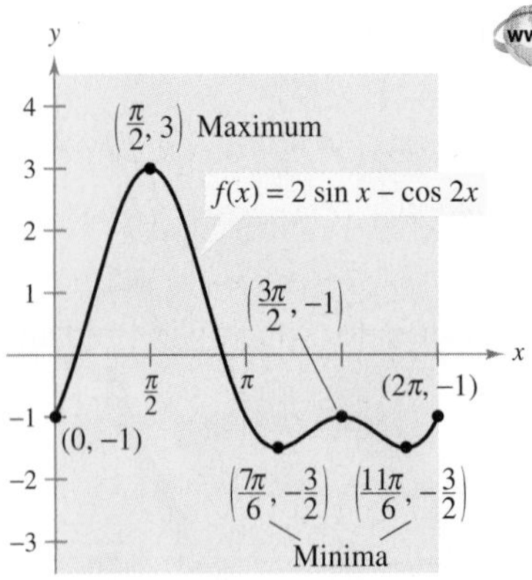

On the closed interval $[0, 2\pi]$, f has two minima at $(7\pi/6, -3/2)$ and $(11\pi/6, -3/2)$ and a maximum at $(\pi/2, 3)$.
Figure 3.7

EXAMPLE 4 Finding Extrema on a Closed Interval

Find the extrema of $f(x) = 2 \sin x - \cos 2x$ on the interval $[0, 2\pi]$.

Solution This function is differentiable for all real x, so you can find all critical numbers by differentiating the function and setting $f'(x)$ equal to zero, as shown.

$$f(x) = 2 \sin x - \cos 2x \qquad \text{Write original function.}$$

$$f'(x) = 2 \cos x + 2 \sin 2x = 0 \qquad \text{Set } f'(x) \text{ equal to 0.}$$

$$2 \cos x + 4 \cos x \sin x = 0 \qquad \sin 2x = 2 \cos x \sin x$$

$$2(\cos x)(1 + 2 \sin x) = 0 \qquad \text{Factor.}$$

In the interval $[0, 2\pi]$, the factor $\cos x$ is zero when $x = \pi/2$ and when $x = 3\pi/2$. The factor $(1 + 2 \sin x)$ is zero when $x = 7\pi/6$ and when $x = 11\pi/6$. By evaluating f at these four critical numbers and at the endpoints of the interval, you can conclude that the maximum is $f(\pi/2) = 3$ and the minimum occurs at *two* points, $f(7\pi/6) = -3/2$ and $f(11\pi/6) = -3/2$, as shown in the table. The graph is shown in Figure 3.7.

Left Endpoint	Critical Number	Critical Number	Critical Number	Critical Number	Right Endpoint
$f(0) = -1$	$f\left(\frac{\pi}{2}\right) = 3$ Maximum	$f\left(\frac{7\pi}{6}\right) = -\frac{3}{2}$ Minimum	$f\left(\frac{3\pi}{2}\right) = -1$	$f\left(\frac{11\pi}{6}\right) = -\frac{3}{2}$ Minimum	$f(2\pi) = -1$

indicates that in the HM mathSpace® *CD-ROM and the online* Eduspace® *system for this text, you will find an Open Exploration, which further explores this example using the computer algebra systems* Maple, Mathcad, Mathematica, *and* Derive.

Exercises for Section 3.1

See www.CalcChat.com for worked-out solutions to odd-numbered exercises.

In Exercises 1 and 2, decide whether each labeled point is an absolute maximum or minimum, a relative maximum or minimum, or neither.

1.

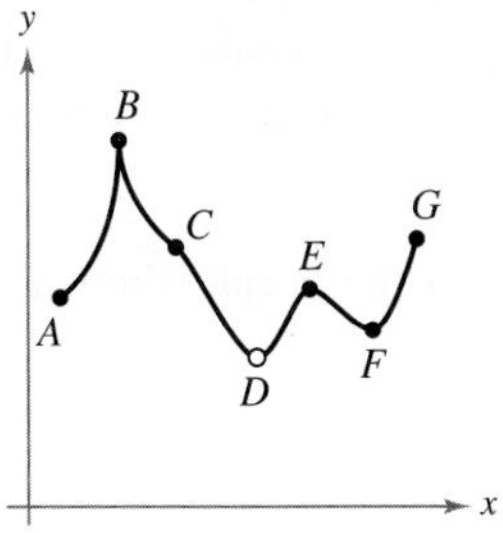

2.

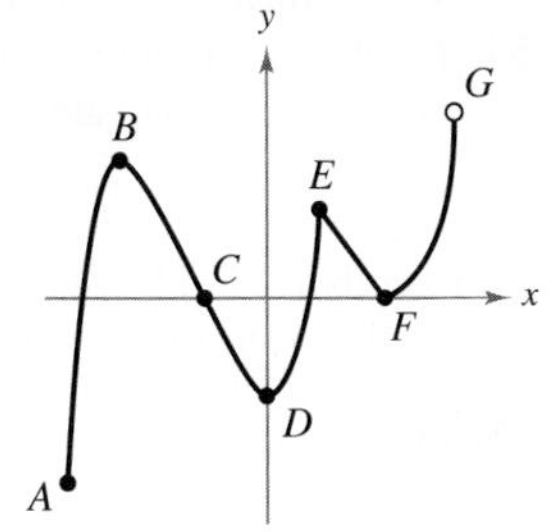

In Exercises 3–8, find the value of the derivative (if it exists) at each indicated extremum.

3. $f(x) = \dfrac{x^2}{x^2 + 4}$

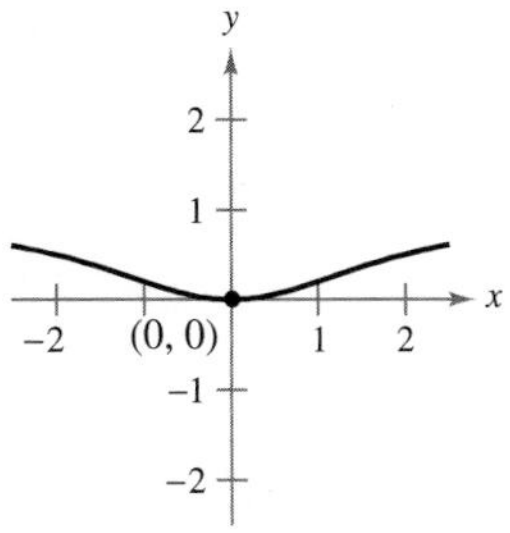

4. $f(x) = \cos \dfrac{\pi x}{2}$

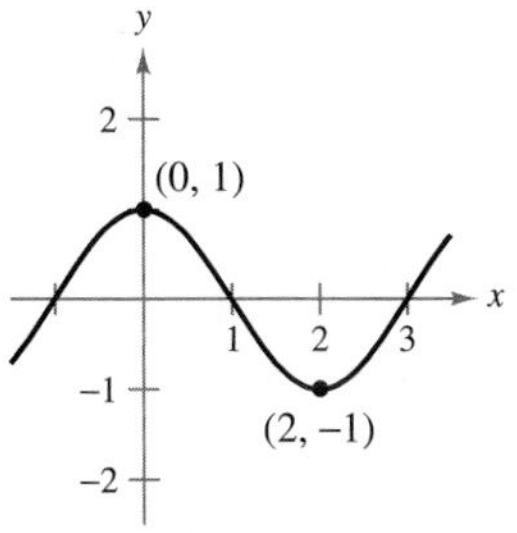

5. $f(x) = x + \dfrac{27}{2x^2}$

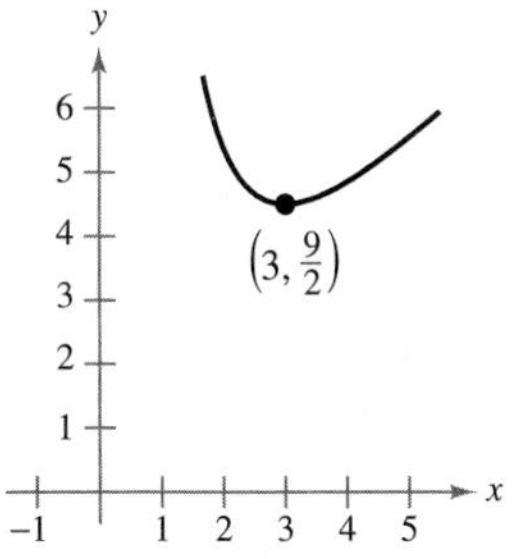

6. $f(x) = -3x\sqrt{x + 1}$

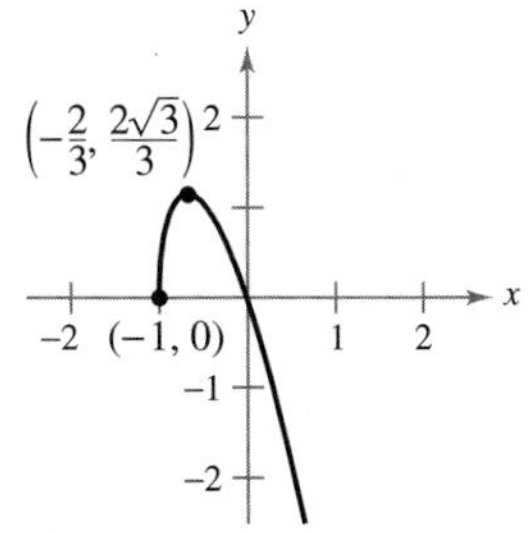

7. $f(x) = (x + 2)^{2/3}$

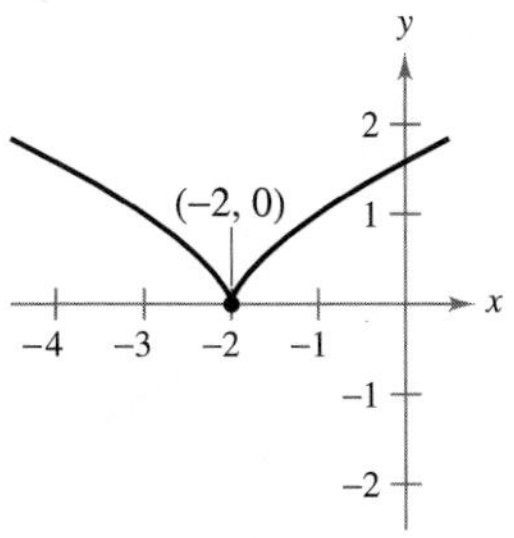

8. $f(x) = 4 - |x|$

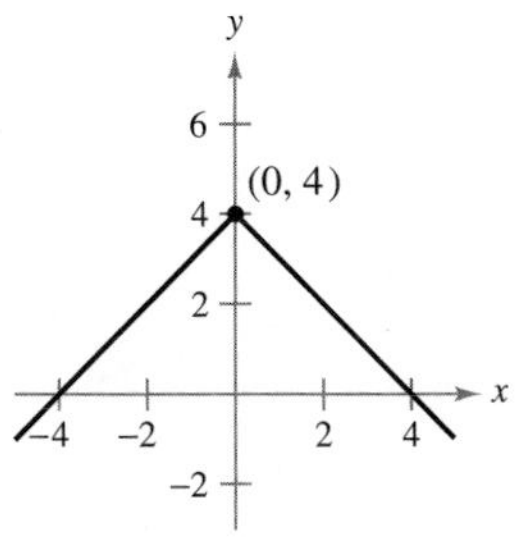

In Exercises 9–12, approximate the critical numbers of the function shown in the graph. Determine whether the function has a relative maximum, relative minimum, absolute maximum, absolute minimum, or none of these at each critical number on the interval shown.

9.

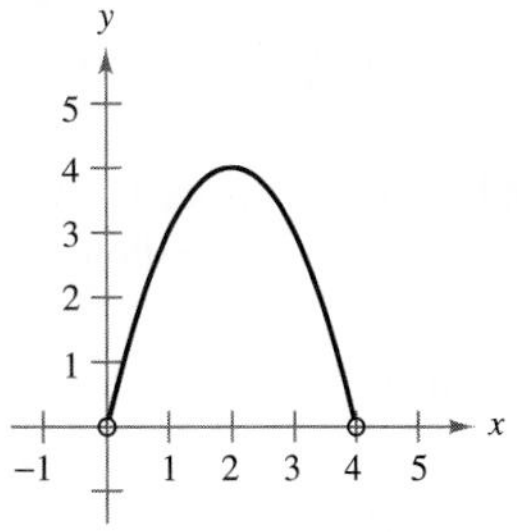

10.

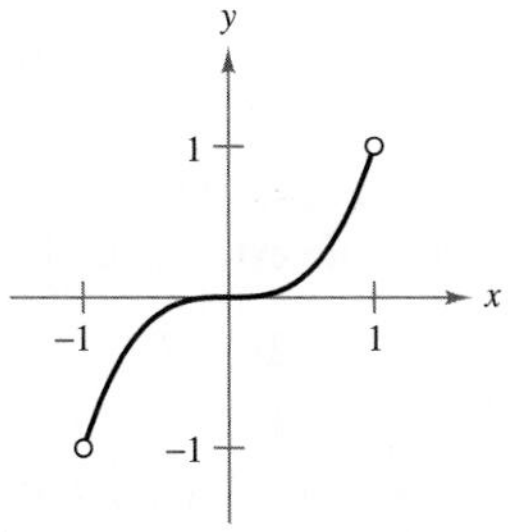

11.

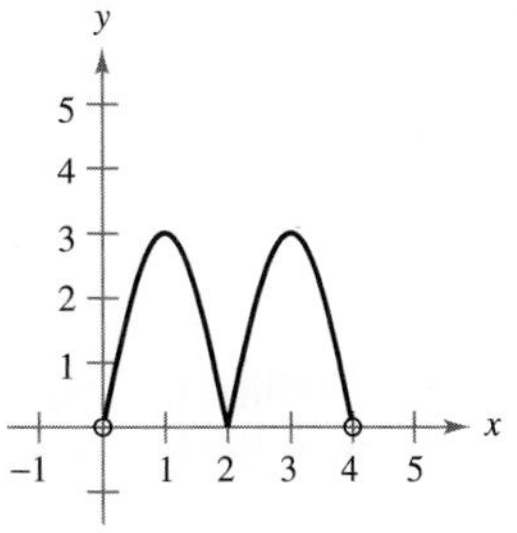

12.

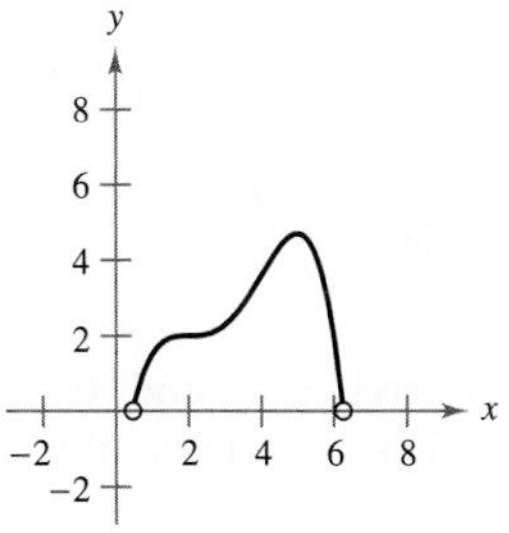

In Exercises 13–18, find any critical numbers of the function.

13. $f(x) = x^2(x - 3)$

14. $g(x) = x^2(x^2 - 4)$

15. $g(t) = t\sqrt{4 - t},\ t < 3$

16. $f(x) = \dfrac{4x}{x^2 + 1}$

17. $h(x) = \sin^2 x + \cos x$
$0 < x < 2\pi$

18. $f(\theta) = 2 \sec \theta + \tan \theta$
$0 < \theta < 2\pi$

In Exercises 19–36, locate the absolute extrema of the function on the closed interval.

19. $f(x) = 2(3 - x),\ [-1, 2]$

20. $f(x) = \dfrac{2x + 5}{3},\ [0, 5]$

21. $f(x) = -x^2 + 3x,\ [0, 3]$

22. $f(x) = x^2 + 2x - 4,\ [-1, 1]$

23. $f(x) = x^3 - \dfrac{3}{2}x^2,\ [-1, 2]$

24. $f(x) = x^3 - 12x,\ [0, 4]$

25. $y = 3x^{2/3} - 2x,\ [-1, 1]$

26. $g(x) = \sqrt[3]{x},\ [-1, 1]$

27. $g(t) = \dfrac{t^2}{t^2 + 3},\ [-1, 1]$

28. $f(x) = \dfrac{2x}{x^2 + 1},\ [-2, 2]$

29. $h(s) = \dfrac{1}{s - 2},\ [0, 1]$

30. $h(t) = \dfrac{t}{t - 2},\ [3, 5]$

31. $y = 3 - |t - 3|,\ [-1, 5]$

32. $f(x) = [\![x]\!],\ [-2, 2]$

33. $f(x) = \cos \pi x,\ \left[0, \dfrac{1}{6}\right]$

34. $g(x) = \sec x,\ \left[-\dfrac{\pi}{6}, \dfrac{\pi}{3}\right]$

35. $y = \dfrac{4}{x} + \tan\left(\dfrac{\pi x}{8}\right),\ [1, 2]$

36. $y = x^2 - 2 - \cos x,\ [-1, 3]$

In Exercises 37–40, locate the absolute extrema of the function (if any exist) over each interval.

37. $f(x) = 2x - 3$
(a) $[0, 2]$ (b) $[0, 2)$
(c) $(0, 2]$ (d) $(0, 2)$

38. $f(x) = 5 - x$
(a) $[1, 4]$ (b) $[1, 4)$
(c) $(1, 4]$ (d) $(1, 4)$

39. $f(x) = x^2 - 2x$
(a) $[-1, 2]$ (b) $(1, 3]$
(c) $(0, 2)$ (d) $[1, 4)$

40. $f(x) = \sqrt{4 - x^2}$
(a) $[-2, 2]$ (b) $[-2, 0)$
(c) $(-2, 2)$ (d) $[1, 2)$

In Exercises 41–44, sketch the graph of the function. Then locate the absolute extrema of the function over the given interval.

41. $f(x) = \begin{cases} 2x + 2, & 0 \le x \le 1 \\ 4x^2, & 1 < x \le 3 \end{cases}, \quad [0, 3]$

42. $f(x) = \begin{cases} 2 - x^2, & 1 \le x < 3 \\ 2 - 3x, & 3 \le x \le 5 \end{cases}, \quad [1, 5]$

43. $f(x) = \dfrac{3}{x - 1}, \quad (1, 4]$

44. $f(x) = \dfrac{2}{2 - x}, \quad [0, 2)$

In Exercises 45 and 46, use a graphing utility to graph the function. Then locate the absolute extrema of the function over the given interval.

45. $f(x) = x^4 - 2x^3 + x + 1, \quad [-1, 3]$

46. $f(x) = \sqrt{x} + \cos\dfrac{x}{2}, \quad [0, 2\pi]$

In Exercises 47 and 48, (a) use a computer algebra system to graph the function and approximate any absolute extrema on the given interval. (b) Use the utility to find any critical numbers, and use them to find any absolute extrema not located at the endpoints. Compare the results with those in part (a).

47. $f(x) = 3.2x^5 + 5x^3 - 3.5x, \quad [0, 1]$

48. $f(x) = \dfrac{4}{3}x\sqrt{3 - x}, \quad [0, 3]$

In Exercises 49 and 50, use a computer algebra system to find the maximum value of $|f''(x)|$ on the closed interval. (This value is used in the error estimate for the Trapezoidal Rule, as discussed in Section 4.6.)

49. $f(x) = \sqrt{1 + x^3}, \quad [0, 2]$

50. $f(x) = \dfrac{1}{x^2 + 1}, \quad \left[\dfrac{1}{2}, 3\right]$

In Exercises 51 and 52, use a computer algebra system to find the maximum value of $|f^{4}(x)|$ on the closed interval. (This value is used in the error estimate for Simpson's Rule, as discussed in Section 4.6.)

51. $f(x) = (x + 1)^{2/3}, \quad [0, 2]$

52. $f(x) = \dfrac{1}{x^2 + 1}, \quad [-1, 1]$

Writing About Concepts

In Exercises 53 and 54, graph a function on the interval $[-2, 5]$ having the given characteristics.

53. Absolute maximum at $x = -2$, absolute minimum at $x = 1$, relative maximum at $x = 3$

54. Relative minimum at $x = -1$, critical number at $x = 0$, but no extrema, absolute maximum at $x = 2$, absolute minimum at $x = 5$

In Exercises 55–58, determine from the graph whether f has a minimum in the open interval (a, b).

55. (a) (b)

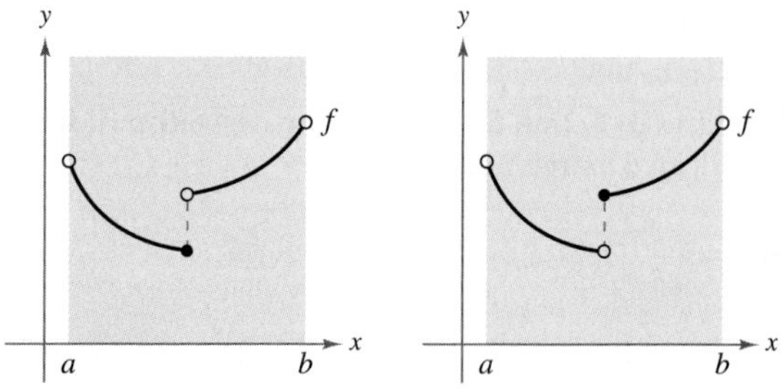

56. (a) (b)

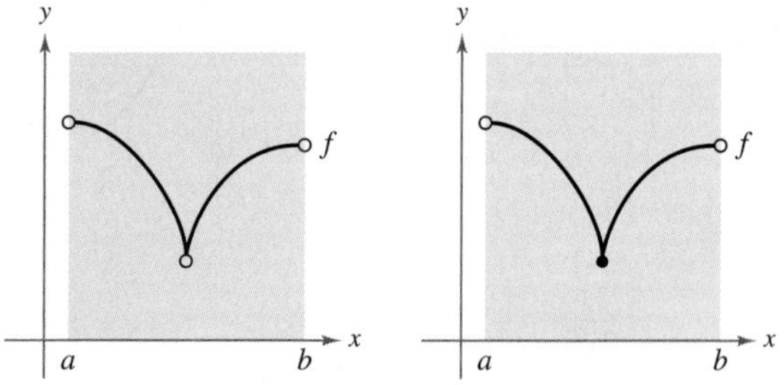

57. (a) (b)

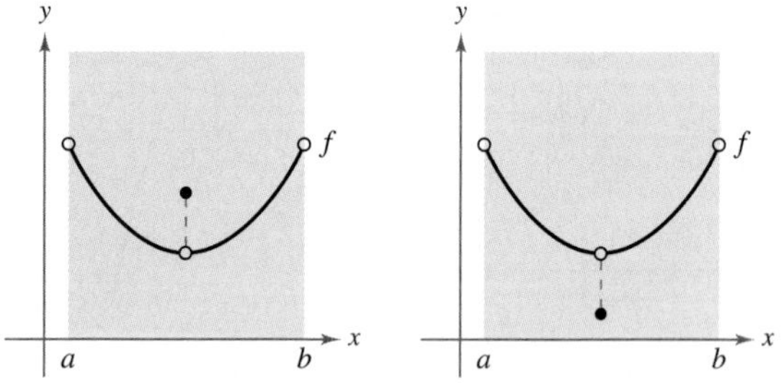

58. (a) (b)

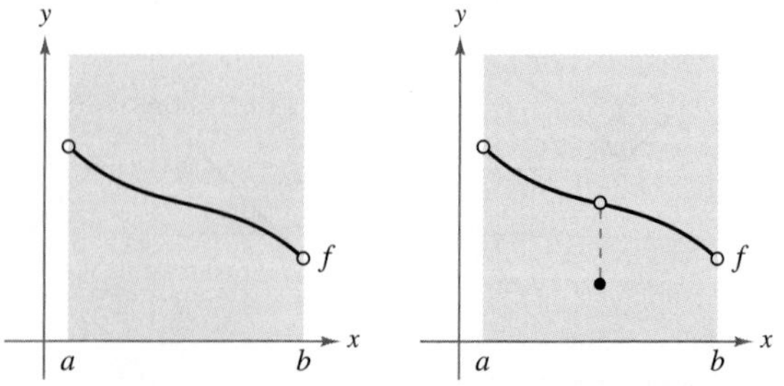

59. ***Power*** The formula for the power output P of a battery is $P = VI - RI^2$, where V is the electromotive force in volts, R is the resistance, and I is the current. Find the current (measured in amperes) that corresponds to a maximum value of P in a battery for which $V = 12$ volts and $R = 0.5$ ohm. Assume that a 15-ampere fuse bounds the output in the interval $0 \le I \le 15$. Could the power output be increased by replacing the 15-ampere fuse with a 20-ampere fuse? Explain.

60. ***Inventory Cost*** A retailer has determined that the cost C of ordering and storing x units of a product is

$$C = 2x + \frac{300{,}000}{x}, \quad 1 \le x \le 300.$$

The delivery truck can bring at most 300 units per order. Find the order size that will minimize cost. Could the cost be decreased if the truck were replaced with one that could bring at most 400 units? Explain.

61. ***Lawn Sprinkler*** A lawn sprinkler is constructed in such a way that $d\theta/dt$ is constant, where θ ranges between 45° and 135° (see figure). The distance the water travels horizontally is

$$x = \frac{v^2 \sin 2\theta}{32}, \quad 45° \le \theta \le 135°$$

where v is the speed of the water. Find dx/dt and explain why this lawn sprinkler does not water evenly. What part of the lawn receives the most water?

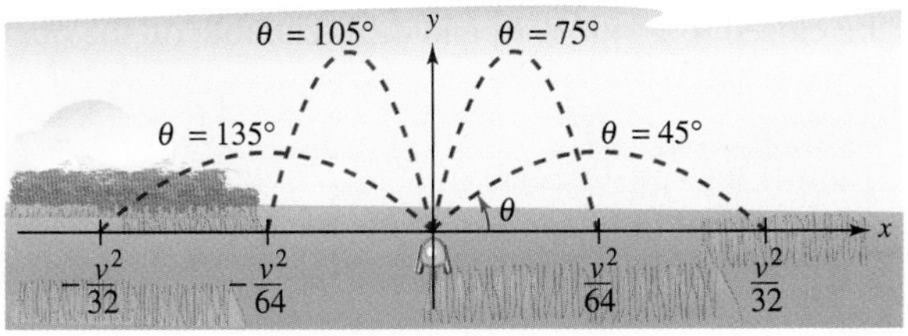

Water sprinkler: $45° \le \theta \le 135°$

FOR FURTHER INFORMATION For more information on the "calculus of lawn sprinklers," see the article "Design of an Oscillating Sprinkler" by Bart Braden in *Mathematics Magazine*. To view this article, go to the website *www.matharticles.com.*

62. ***Honeycomb*** The surface area of a cell in a honeycomb is

$$S = 6hs + \frac{3s^2}{2}\left(\frac{\sqrt{3} - \cos\theta}{\sin\theta}\right)$$

where h and s are positive constants and θ is the angle at which the upper faces meet the altitude of the cell (see figure). Find the angle θ $(\pi/6 \le \theta \le \pi/2)$ that minimizes the surface area S.

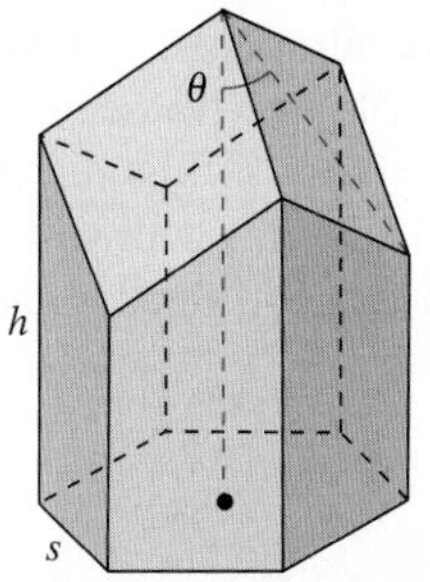

FOR FURTHER INFORMATION For more information on the geometric structure of a honeycomb cell, see the article "The Design of Honeycombs" by Anthony L. Peressini in UMAP Module 502, published by COMAP, Inc., Suite 210, 57 Bedford Street, Lexington, MA.

True or False? **In Exercises 63–66, determine whether the statement is true or false. If it is false, explain why or give an example that shows it is false.**

63. The maximum of a function that is continuous on a closed interval can occur at two different values in the interval.

64. If a function is continuous on a closed interval, then it must have a minimum on the interval.

65. If $x = c$ is a critical number of the function f, then it is also a critical number of the function $g(x) = f(x) + k$, where k is a constant.

66. If $x = c$ is a critical number of the function f, then it is also a critical number of the function $g(x) = f(x - k)$, where k is a constant.

67. Let the function f be differentiable on an interval I containing c. If f has a maximum value at $x = c$, show that $-f$ has a minimum value at $x = c$.

68. Consider the cubic function $f(x) = ax^3 + bx^2 + cx + d$ where $a \ne 0$. Show that f can have zero, one, or two critical numbers and give an example of each case.

69. ***Highway Design*** In order to build a highway, it is necessary to fill a section of a valley where the grades (slopes) of the sides are 9% and 6% (see figure). The top of the filled region will have the shape of a parabolic arc that is tangent to the two slopes at the points A and B. The horizontal distance between the points A and B is 1000 feet.

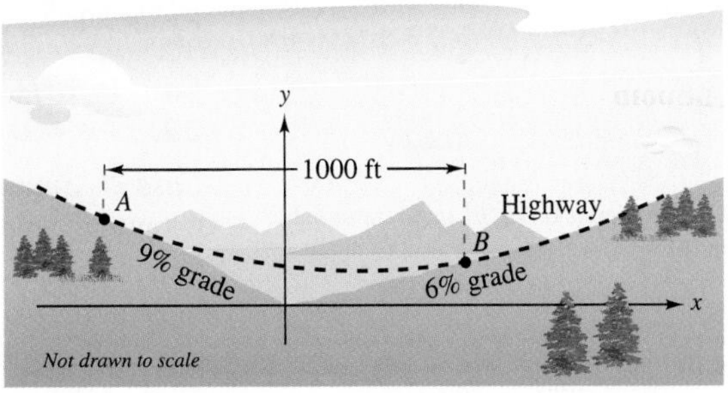

(a) Find a quadratic function $y = ax^2 + bx + c$, $-500 \le x \le 500$, that describes the top of the filled region.

(b) Construct a table giving the depths d of the fill for $x = -500, -400, -300, -200, -100, 0, 100, 200, 300, 400$, and 500.

(c) What will be the lowest point on the completed highway? Will it be directly over the point where the two hillsides come together?

Section 3.2 Rolle's Theorem and the Mean Value Theorem

- **Understand and use Rolle's Theorem.**
- **Understand and use the Mean Value Theorem.**

Rolle's Theorem

ROLLE'S THEOREM

French mathematician Michel Rolle first published the theorem that bears his name in 1691. Before this time, however, Rolle was one of the most vocal critics of calculus, stating that it gave erroneous results and was based on unsound reasoning. Later in life, Rolle came to see the usefulness of calculus.

The Extreme Value Theorem (Section 3.1) states that a continuous function on a closed interval $[a, b]$ must have both a minimum and a maximum on the interval. Both of these values, however, can occur at the endpoints. **Rolle's Theorem,** named after the French mathematician Michel Rolle (1652–1719), gives conditions that guarantee the existence of an extreme value in the *interior* of a closed interval.

EXPLORATION

Extreme Values in a Closed Interval Sketch a rectangular coordinate plane on a piece of paper. Label the points $(1, 3)$ and $(5, 3)$. Using a pencil or pen, draw the graph of a differentiable function f that starts at $(1, 3)$ and ends at $(5, 3)$. Is there at least one point on the graph for which the derivative is zero? Would it be possible to draw the graph so that there *isn't* a point for which the derivative is zero? Explain your reasoning.

THEOREM 3.3 Rolle's Theorem

Let f be continuous on the closed interval $[a, b]$ and differentiable on the open interval (a, b). If

$$f(a) = f(b)$$

then there is at least one number c in (a, b) such that $f'(c) = 0$.

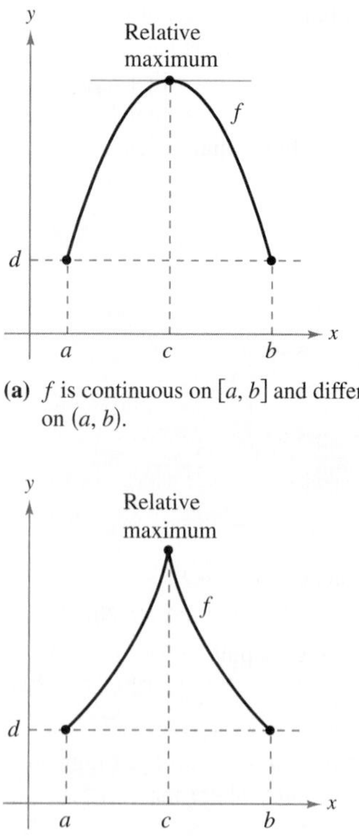

(a) f is continuous on $[a, b]$ and differentiable on (a, b).

(b) f is continuous on $[a, b]$.

Figure 3.8

Proof Let $f(a) = d = f(b)$.

Case 1: If $f(x) = d$ for all x in $[a, b]$, f is constant on the interval and, by Theorem 2.2, $f'(x) = 0$ for all x in (a, b).

Case 2: Suppose $f(x) > d$ for some x in (a, b). By the Extreme Value Theorem, you know that f has a maximum at some c in the interval. Moreover, because $f(c) > d$, this maximum does not occur at either endpoint. So, f has a maximum in the *open* interval (a, b). This implies that $f(c)$ is a *relative* maximum and, by Theorem 3.2, c is a critical number of f. Finally, because f is differentiable at c, you can conclude that $f'(c) = 0$.

Case 3: If $f(x) < d$ for some x in (a, b), you can use an argument similar to that in Case 2, but involving the minimum instead of the maximum.

From Rolle's Theorem, you can see that if a function f is continuous on $[a, b]$ and differentiable on (a, b), and if $f(a) = f(b)$, there must be at least one x-value between a and b at which the graph of f has a horizontal tangent, as shown in Figure 3.8(a). If the differentiability requirement is dropped from Rolle's Theorem, f will still have a critical number in (a, b), but it may not yield a horizontal tangent. Such a case is shown in Figure 3.8(b).

EXAMPLE 1 Illustrating Rolle's Theorem

Find the two x-intercepts of

$$f(x) = x^2 - 3x + 2$$

and show that $f'(x) = 0$ at some point between the two x-intercepts.

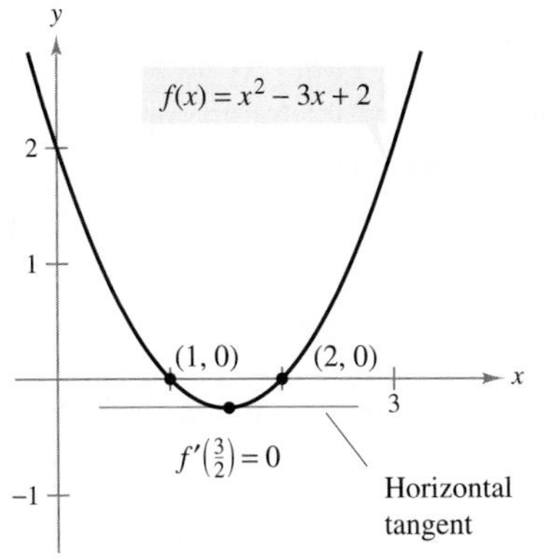

The x-value for which $f'(x) = 0$ is between the two x-intercepts.
Figure 3.9

Solution Note that f is differentiable on the entire real line. Setting $f(x)$ equal to 0 produces

$$x^2 - 3x + 2 = 0 \qquad \text{Set } f(x) \text{ equal to 0.}$$
$$(x - 1)(x - 2) = 0. \qquad \text{Factor.}$$

So, $f(1) = f(2) = 0$, and from Rolle's Theorem you know that there *exists* at least one c in the interval $(1, 2)$ such that $f'(c) = 0$. To *find* such a c, you can solve the equation

$$f'(x) = 2x - 3 = 0 \qquad \text{Set } f'(x) \text{ equal to 0.}$$

and determine that $f'(x) = 0$ when $x = \frac{3}{2}$. Note that the x-value lies in the open interval $(1, 2)$, as shown in Figure 3.9.

Rolle's Theorem states that if f satisfies the conditions of the theorem, there must be *at least* one point between a and b at which the derivative is 0. There may of course be more than one such point, as shown in the next example.

EXAMPLE 2 Illustrating Rolle's Theorem

Let $f(x) = x^4 - 2x^2$. Find all values of c in the interval $(-2, 2)$ such that $f'(c) = 0$.

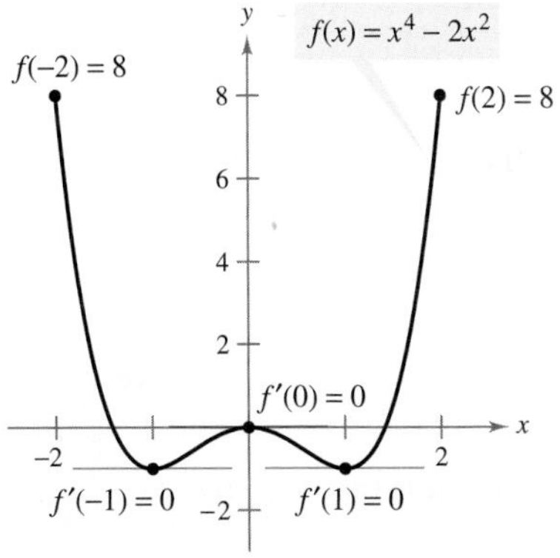

$f'(x) = 0$ for more than one x-value in the interval $(-2, 2)$.
Figure 3.10

Solution To begin, note that the function satisfies the conditions of Rolle's Theorem. That is, f is continuous on the interval $[-2, 2]$ and differentiable on the interval $(-2, 2)$. Moreover, because $f(-2) = f(2) = 8$, you can conclude that there exists at least one c in $(-2, 2)$ such that $f'(c) = 0$. Setting the derivative equal to 0 produces

$$f'(x) = 4x^3 - 4x = 0 \qquad \text{Set } f'(x) \text{ equal to 0.}$$
$$4x(x - 1)(x + 1) = 0 \qquad \text{Factor.}$$
$$x = 0, 1, -1. \qquad x\text{-values for which } f'(x) = 0$$

So, in the interval $(-2, 2)$, the derivative is zero at three different values of x, as shown in Figure 3.10.

TECHNOLOGY PITFALL A graphing utility can be used to indicate whether the points on the graphs in Examples 1 and 2 are relative minima or relative maxima of the functions. When using a graphing utility, however, you should keep in mind that it can give misleading pictures of graphs. For example, use a graphing utility to graph

$$f(x) = 1 - (x - 1)^2 - \frac{1}{1000(x - 1)^{1/7} + 1}.$$

With most viewing windows, it appears that the function has a maximum of 1 when $x = 1$ (see Figure 3.11). By evaluating the function at $x = 1$, however, you can see that $f(1) = 0$. To determine the behavior of this function near $x = 1$, you need to examine the graph analytically to get the complete picture.

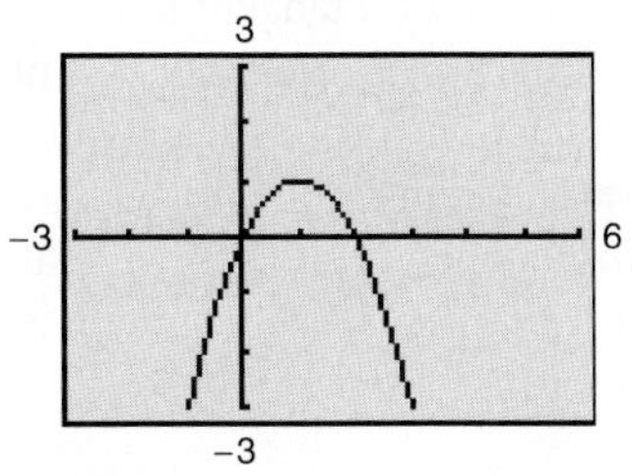

Figure 3.11

The Mean Value Theorem

Rolle's Theorem can be used to prove another theorem—the **Mean Value Theorem.**

THEOREM 3.4 The Mean Value Theorem

If f is continuous on the closed interval $[a, b]$ and differentiable on the open interval (a, b), then there exists a number c in (a, b) such that

$$f'(c) = \frac{f(b) - f(a)}{b - a}.$$

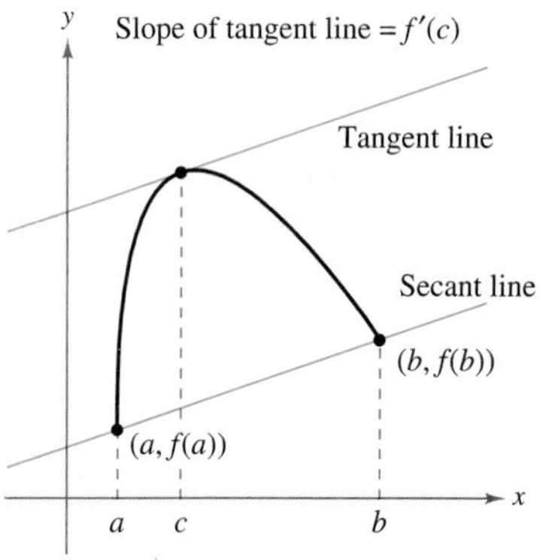

Figure 3.12

Proof Refer to Figure 3.12. The equation of the secant line containing the points $(a, f(a))$ and $(b, f(b))$ is

$$y = \left[\frac{f(b) - f(a)}{b - a}\right](x - a) + f(a).$$

Let $g(x)$ be the difference between $f(x)$ and y. Then

$$\begin{aligned} g(x) &= f(x) - y \\ &= f(x) - \left[\frac{f(b) - f(a)}{b - a}\right](x - a) - f(a). \end{aligned}$$

By evaluating g at a and b, you can see that $g(a) = 0 = g(b)$. Because f is continuous on $[a, b]$ it follows that g is also continuous on $[a, b]$. Furthermore, because f is differentiable, g is also differentiable, and you can apply Rolle's Theorem to the function g. So, there exists a number c in (a, b) such that $g'(c) = 0$, which implies that

$$\begin{aligned} 0 &= g'(c) \\ &= f'(c) - \frac{f(b) - f(a)}{b - a}. \end{aligned}$$

So, there exists a number c in (a, b) such that

$$f'(c) = \frac{f(b) - f(a)}{b - a}.$$

NOTE The "mean" in the Mean Value Theorem refers to the mean (or average) rate of change of f in the interval $[a, b]$.

JOSEPH-LOUIS LAGRANGE (1736–1813)

The Mean Value Theorem was first proved by the famous mathematician Joseph-Louis Lagrange. Born in Italy, Lagrange held a position in the court of Frederick the Great in Berlin for 20 years. Afterward, he moved to France, where he met emperor Napoleon Bonaparte, who is quoted as saying, "Lagrange is the lofty pyramid of the mathematical sciences."

Although the Mean Value Theorem can be used directly in problem solving, it is used more often to prove other theorems. In fact, some people consider this to be the most important theorem in calculus—it is closely related to the Fundamental Theorem of Calculus discussed in Chapter 4. For now, you can get an idea of the versatility of this theorem by looking at the results stated in Exercises 77–85 in this section.

The Mean Value Theorem has implications for both basic interpretations of the derivative. Geometrically, the theorem guarantees the existence of a tangent line that is parallel to the secant line through the points $(a, f(a))$ and $(b, f(b))$, as shown in Figure 3.12. Example 3 illustrates this geometric interpretation of the Mean Value Theorem. In terms of rates of change, the Mean Value Theorem implies that there must be a point in the open interval (a, b) at which the instantaneous rate of change is equal to the average rate of change over the interval $[a, b]$. This is illustrated in Example 4.

EXAMPLE 3 Finding a Tangent Line

Given $f(x) = 5 - (4/x)$, find all values of c in the open interval $(1, 4)$ such that

$$f'(c) = \frac{f(4) - f(1)}{4 - 1}.$$

Solution The slope of the secant line through $(1, f(1))$ and $(4, f(4))$ is

$$\frac{f(4) - f(1)}{4 - 1} = \frac{4 - 1}{4 - 1} = 1.$$

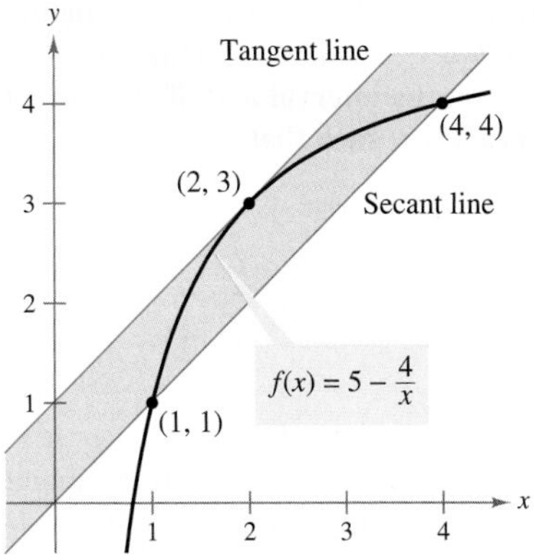

The tangent line at $(2, 3)$ is parallel to the secant line through $(1, 1)$ and $(4, 4)$.
Figure 3.13

Because f satisfies the conditions of the Mean Value Theorem, there exists at least one number c in $(1, 4)$ such that $f'(c) = 1$. Solving the equation $f'(x) = 1$ yields

$$f'(x) = \frac{4}{x^2} = 1$$

which implies that $x = \pm 2$. So, in the interval $(1, 4)$, you can conclude that $c = 2$, as shown in Figure 3.13.

EXAMPLE 4 Finding an Instantaneous Rate of Change

Two stationary patrol cars equipped with radar are 5 miles apart on a highway, as shown in Figure 3.14. As a truck passes the first patrol car, its speed is clocked at 55 miles per hour. Four minutes later, when the truck passes the second patrol car, its speed is clocked at 50 miles per hour. Prove that the truck must have exceeded the speed limit (of 55 miles per hour) at some time during the 4 minutes.

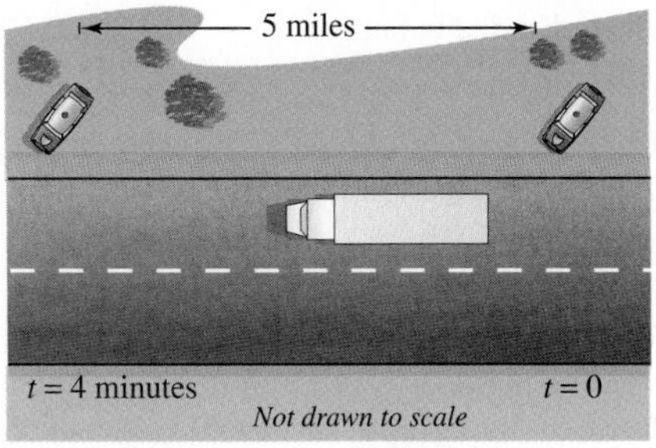

At some time t, the instantaneous velocity is equal to the average velocity over 4 minutes.
Figure 3.14

Solution Let $t = 0$ be the time (in hours) when the truck passes the first patrol car. The time when the truck passes the second patrol car is

$$t = \frac{4}{60} = \frac{1}{15} \text{ hour.}$$

By letting $s(t)$ represent the distance (in miles) traveled by the truck, you have $s(0) = 0$ and $s(\frac{1}{15}) = 5$. So, the average velocity of the truck over the five-mile stretch of highway is

$$\begin{aligned}\text{Average velocity} &= \frac{s(1/15) - s(0)}{(1/15) - 0} \\ &= \frac{5}{1/15} = 75 \text{ miles per hour.}\end{aligned}$$

Assuming that the position function is differentiable, you can apply the Mean Value Theorem to conclude that the truck must have been traveling at a rate of 75 miles per hour sometime during the 4 minutes.

A useful alternative form of the Mean Value Theorem is as follows: If f is continuous on $[a, b]$ and differentiable on (a, b), then there exists a number c in (a, b) such that

$$f(b) = f(a) + (b - a)f'(c).$$ Alternative form of Mean Value Theorem

NOTE When doing the exercises for this section, keep in mind that polynomial functions, rational functions, and trigonometric functions are differentiable at all points in their domains.

Exercises for Section 3.2

See www.CalcChat.com for worked-out solutions to odd-numbered exercises.

In Exercises 1–4, explain why Rolle's Theorem does not apply to the function even though there exist a and b such that $f(a) = f(b)$.

1. $f(x) = 1 - |x - 1|$

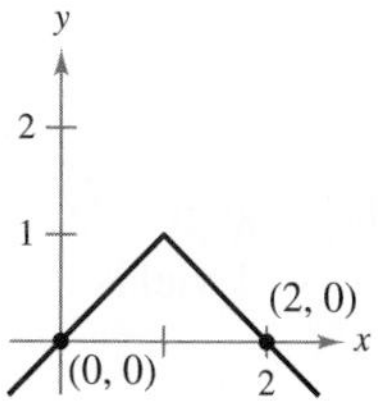

2. $f(x) = \cot \frac{x}{2}$

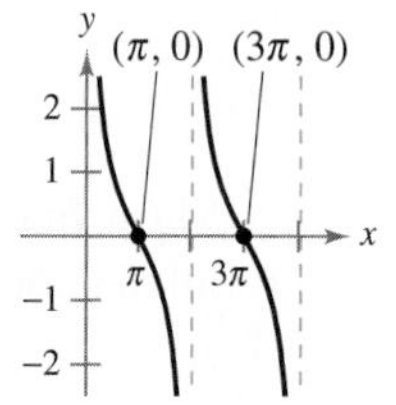

3. $f(x) = \left|\frac{1}{x}\right|$, $[-1, 1]$

4. $f(x) = \sqrt{(2 - x^{2/3})^3}$, $[-1, 1]$

In Exercises 5–8, find the two x-intercepts of the function f and show that $f'(x) = 0$ at some point between the two x-intercepts.

5. $f(x) = x^2 - x - 2$
6. $f(x) = x(x - 3)$
7. $f(x) = x\sqrt{x + 4}$
8. $f(x) = -3x\sqrt{x + 1}$

Rolle's Theorem **In Exercises 9 and 10, the graph of f is shown. Apply Rolle's Theorem and find all values of c such that $f'(c) = 0$ at some point between the labeled intercepts.**

9.

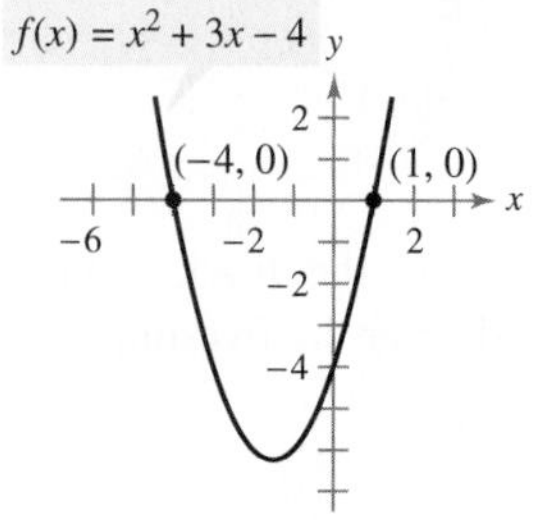

10.

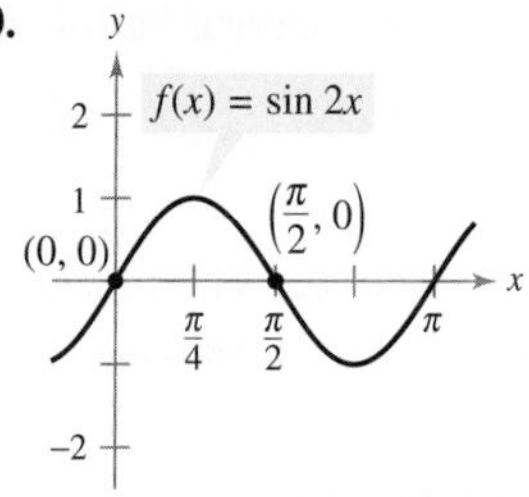

In Exercises 11–24, determine whether Rolle's Theorem can be applied to f on the closed interval $[a, b]$. If Rolle's Theorem can be applied, find all values of c in the open interval (a, b) such that $f'(c) = 0$.

11. $f(x) = x^2 - 2x, [0, 2]$
12. $f(x) = x^2 - 5x + 4, [1, 4]$
13. $f(x) = (x - 1)(x - 2)(x - 3), [1, 3]$
14. $f(x) = (x - 3)(x + 1)^2, [-1, 3]$
15. $f(x) = x^{2/3} - 1, [-8, 8]$
16. $f(x) = 3 - |x - 3|, [0, 6]$
17. $f(x) = \dfrac{x^2 - 2x - 3}{x + 2}, [-1, 3]$
18. $f(x) = \dfrac{x^2 - 1}{x}, [-1, 1]$
19. $f(x) = \sin x, [0, 2\pi]$
20. $f(x) = \cos x, [0, 2\pi]$
21. $f(x) = \dfrac{6x}{\pi} - 4\sin^2 x, \left[0, \dfrac{\pi}{6}\right]$
22. $f(x) = \cos 2x, \left[-\dfrac{\pi}{12}, \dfrac{\pi}{6}\right]$
23. $f(x) = \tan x, [0, \pi]$
24. $f(x) = \sec x, \left[-\dfrac{\pi}{4}, \dfrac{\pi}{4}\right]$

In Exercises 25–28, use a graphing utility to graph the function on the closed interval $[a, b]$. Determine whether Rolle's Theorem can be applied to f on the interval and, if so, find all values of c in the open interval (a, b) such that $f'(c) = 0$.

25. $f(x) = |x| - 1, [-1, 1]$
26. $f(x) = x - x^{1/3}, [0, 1]$
27. $f(x) = 4x - \tan \pi x, \left[-\frac{1}{4}, \frac{1}{4}\right]$
28. $f(x) = \dfrac{x}{2} - \sin \dfrac{\pi x}{6}, [-1, 0]$

29. ***Vertical Motion*** The height of a ball t seconds after it is thrown upward from a height of 32 feet and with an initial velocity of 48 feet per second is $f(t) = -16t^2 + 48t + 32$.
(a) Verify that $f(1) = f(2)$.
(b) According to Rolle's Theorem, what must be the velocity at some time in the interval $(1, 2)$? Find that time.

30. ***Reorder Costs*** The ordering and transportation cost C for components used in a manufacturing process is approximated by $C(x) = 10\left(\dfrac{1}{x} + \dfrac{x}{x + 3}\right)$, where C is measured in thousands of dollars and x is the order size in hundreds.
(a) Verify that $C(3) = C(6)$.
(b) According to Rolle's Theorem, the rate of change of the cost must be 0 for some order size in the interval $(3, 6)$. Find that order size.

In Exercises 31 and 32, copy the graph and sketch the secant line to the graph through the points $(a, f(a))$ and $(b, f(b))$. Then sketch any tangent lines to the graph for each value of c guaranteed by the Mean Value Theorem. To print an enlarged copy of the graph, go to the website *www.mathgraphs.com*.

31.

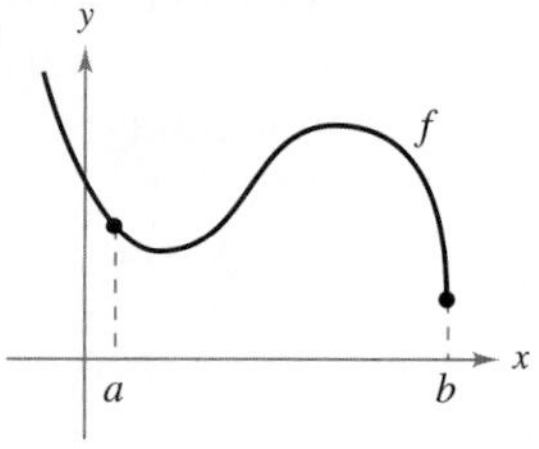

32.

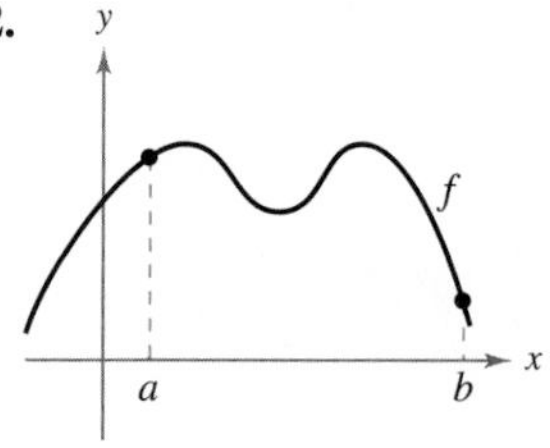

Writing **In Exercises 33–36, explain why the Mean Value Theorem does not apply to the function f on the interval $[0, 6]$.**

33.

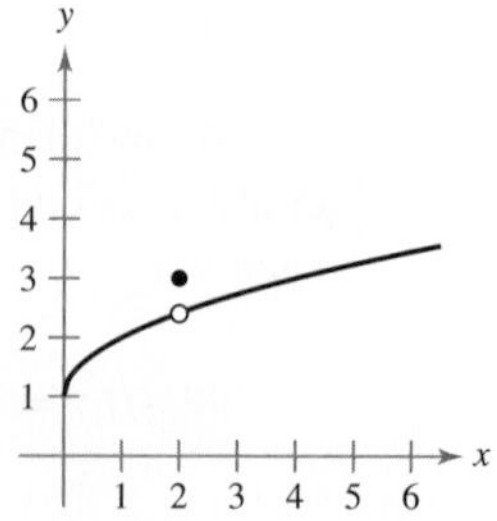

34.

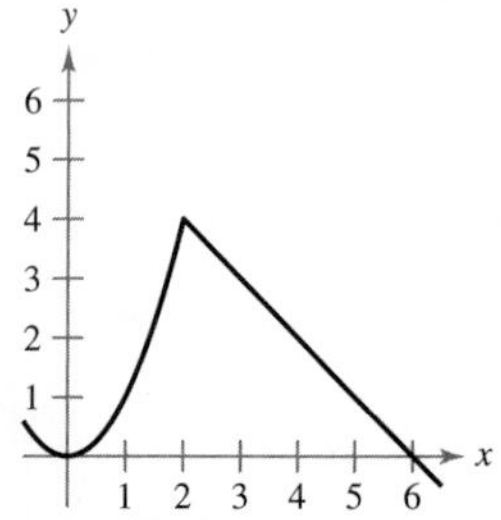

35. $f(x) = \dfrac{1}{x - 3}$

36. $f(x) = |x - 3|$

37. ***Mean Value Theorem*** Consider the graph of the function $f(x) = x^2 + 1$. (a) Find the equation of the secant line joining the points $(-1, 2)$ and $(2, 5)$. (b) Use the Mean Value Theorem to determine a point c in the interval $(-1, 2)$ such that the tangent line at c is parallel to the secant line. (c) Find the equation of the tangent line through c. (d) Then use a graphing utility to graph f, the secant line, and the tangent line.

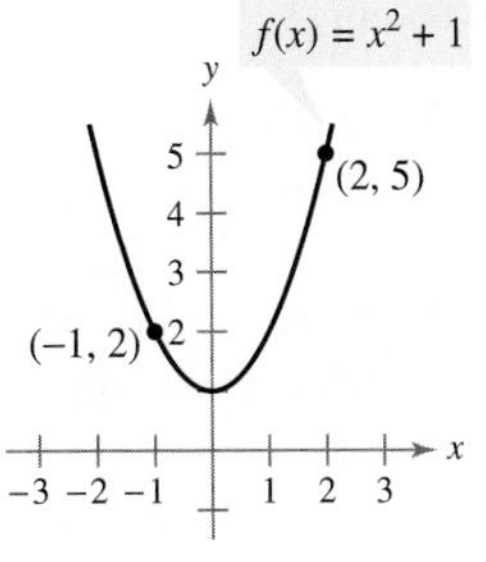

Figure for 37

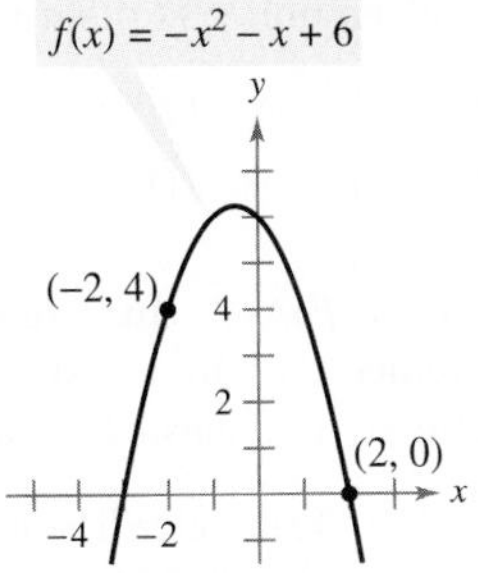

Figure for 38

38. ***Mean Value Theorem*** Consider the graph of the function $f(x) = -x^2 - x + 6$. (a) Find the equation of the secant line joining the points $(-2, 4)$ and $(2, 0)$. (b) Use the Mean Value Theorem to determine a point c in the interval $(-2, 2)$ such that the tangent line at c is parallel to the secant line. (c) Find the equation of the tangent line through c. (d) Then use a graphing utility to graph f, the secant line, and the tangent line.

In Exercises 39–46, determine whether the Mean Value Theorem can be applied to f on the closed interval $[a, b]$. If the Mean Value Theorem can be applied, find all values of c in the open interval (a, b) such that $f'(c) = \dfrac{f(b) - f(a)}{b - a}$.

39. $f(x) = x^2,\ [-2, 1]$

40. $f(x) = x(x^2 - x - 2),\ [-1, 1]$

41. $f(x) = x^{2/3},\ [0, 1]$

42. $f(x) = \dfrac{x + 1}{x},\ \left[\frac{1}{2}, 2\right]$

43. $f(x) = \sqrt{2 - x},\ [-7, 2]$

44. $f(x) = x^3,\ [0, 1]$

45. $f(x) = \sin x,\ [0, \pi]$

46. $f(x) = 2 \sin x + \sin 2x,\ [0, \pi]$

In Exercises 47–50, use a graphing utility to (a) graph the function f on the given interval, (b) find and graph the secant line through points on the graph of f at the endpoints of the given interval, and (c) find and graph any tangent lines to the graph of f that are parallel to the secant line.

47. $f(x) = \dfrac{x}{x + 1},\ \left[-\frac{1}{2}, 2\right]$

48. $f(x) = x - 2 \sin x,\ [-\pi, \pi]$

49. $f(x) = \sqrt{x},\ [1, 9]$

50. $f(x) = -x^4 + 4x^3 + 8x^2 + 5,\ [0, 5]$

51. ***Vertical Motion*** The height of an object t seconds after it is dropped from a height of 500 meters is $s(t) = -4.9t^2 + 500$.

(a) Find the average velocity of the object during the first 3 seconds.

(b) Use the Mean Value Theorem to verify that at some time during the first 3 seconds of fall the instantaneous velocity equals the average velocity. Find that time.

52. ***Sales*** A company introduces a new product for which the number of units sold S is

$$S(t) = 200\left(5 - \frac{9}{2 + t}\right)$$

where t is the time in months.

(a) Find the average value of $S(t)$ during the first year.

(b) During what month does $S'(t)$ equal the average value during the first year?

Writing About Concepts

53. Let f be continuous on $[a, b]$ and differentiable on (a, b). If there exists c in (a, b) such that $f'(c) = 0$, does it follow that $f(a) = f(b)$? Explain.

54. Let f be continuous on the closed interval $[a, b]$ and differentiable on the open interval (a, b). Also, suppose that $f(a) = f(b)$ and that c is a real number in the interval such that $f'(c) = 0$. Find an interval for the function g over which Rolle's Theorem can be applied, and find the corresponding critical number of g (k is a constant).

(a) $g(x) = f(x) + k$ (b) $g(x) = f(x - k)$

(c) $g(x) = f(kx)$

55. The function

$$f(x) = \begin{cases} 0, & x = 0 \\ 1 - x, & 0 < x \le 1 \end{cases}$$

is differentiable on $(0, 1)$ and satisfies $f(0) = f(1)$. However, its derivative is never zero on $(0, 1)$. Does this contradict Rolle's Theorem? Explain.

56. Can you find a function f such that $f(-2) = -2$, $f(2) = 6$, and $f'(x) < 1$ for all x. Why or why not?

57. ***Speed*** A plane begins its takeoff at 2:00 P.M. on a 2500-mile flight. The plane arrives at its destination at 7:30 P.M. Explain why there are at least two times during the flight when the speed of the plane is 400 miles per hour.

58. ***Temperature*** When an object is removed from a furnace and placed in an environment with a constant temperature of 90°F, its core temperature is 1500°F. Five hours later the core temperature is 390°F. Explain why there must exist a time in the interval when the temperature is decreasing at a rate of 222°F per hour.

59. ***Velocity*** Two bicyclists begin a race at 8:00 A.M. They both finish the race 2 hours and 15 minutes later. Prove that at some time during the race, the bicyclists are traveling at the same velocity.

60. ***Acceleration*** At 9:13 A.M., a sports car is traveling 35 miles per hour. Two minutes later, the car is traveling 85 miles per hour. Prove that at some time during this two-minute interval, the car's acceleration is exactly 1500 miles per hour squared.

61. ***Graphical Reasoning*** The figure shows two parts of the graph of a continuous differentiable function f on $[-10, 4]$. The derivative f' is also continuous. To print an enlarged copy of the graph, go to the website *www.mathgraphs.com.*

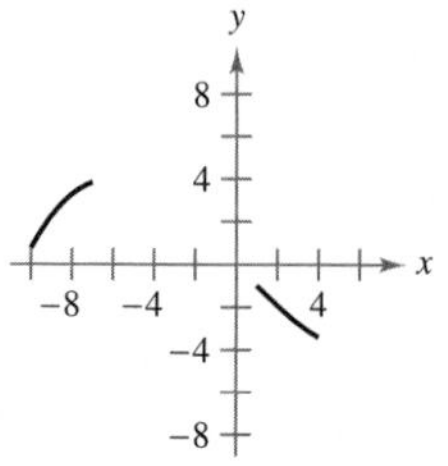

(a) Explain why f must have at least one zero in $[-10, 4]$.

(b) Explain why f' must also have at least one zero in the interval $[-10, 4]$. What are these zeros called?

(c) Make a possible sketch of the function with one zero of f' on the interval $[-10, 4]$.

(d) Make a possible sketch of the function with two zeros of f' on the interval $[-10, 4]$.

(e) Were the conditions of continuity of f and f' necessary to do parts (a) through (d)? Explain.

62. Consider the function $f(x) = 3\cos^2\left(\frac{\pi x}{2}\right)$.

(a) Use a graphing utility to graph f and f'.

(b) Is f a continuous function? Is f' a continuous function?

(c) Does Rolle's Theorem apply on the interval $[-1, 1]$? Does it apply on the interval $[1, 2]$? Explain.

(d) Evaluate, if possible, $\lim_{x\to 3^-} f'(x)$ and $\lim_{x\to 3^+} f'(x)$.

Think About It **In Exercises 63 and 64, sketch the graph of an arbitrary function f that satisfies the given condition but does not satisfy the conditions of the Mean Value Theorem on the interval $[-5, 5]$.**

63. f is continuous on $[-5, 5]$.

64. f is not continuous on $[-5, 5]$.

In Exercises 65 and 66, use the Intermediate Value Theorem and Rolle's Theorem to prove that the equation has exactly one real solution.

65. $x^5 + x^3 + x + 1 = 0$

66. $2x - 2 - \cos x = 0$

67. Determine the values a, b, and c such that the function f satisfies the hypotheses of the Mean Value Theorem on the interval $[0, 3]$.

$$f(x) = \begin{cases} 1, & x = 0 \\ ax + b, & 0 < x \le 1 \\ x^2 + 4x + c, & 1 < x \le 3 \end{cases}$$

68. Determine the values a, b, c, and d so that the function f satisfies the hypotheses of the Mean Value Theorem on the interval $[-1, 2]$.

$$f(x) = \begin{cases} a, & x = -1 \\ 2, & -1 < x \le 0 \\ bx^2 + c, & 0 < x \le 1 \\ dx + 4, & 1 < x \le 2 \end{cases}$$

Differential Equations **In Exercises 69–72, find a function f that has the derivative $f'(x)$ and whose graph passes through the given point. Explain your reasoning.**

69. $f'(x) = 0$, $(2, 5)$

70. $f'(x) = 4$, $(0, 1)$

71. $f'(x) = 2x$, $(1, 0)$

72. $f'(x) = 2x + 3$, $(1, 0)$

True or False? **In Exercises 73–76, determine whether the statement is true or false. If it is false, explain why or give an example that shows it is false.**

73. The Mean Value Theorem can be applied to $f(x) = 1/x$ on the interval $[-1, 1]$.

74. If the graph of a function has three x-intercepts, then it must have at least two points at which its tangent line is horizontal.

75. If the graph of a polynomial function has three x-intercepts, then it must have at least two points at which its tangent line is horizontal.

76. If $f'(x) = 0$ for all x in the domain of f, then f is a constant function.

77. Prove that if $a > 0$ and n is any positive integer, then the polynomial function $p(x) = x^{2n+1} + ax + b$ cannot have two real roots.

78. Prove that if $f'(x) = 0$ for all x in an interval (a, b), then f is constant on (a, b).

79. Let $p(x) = Ax^2 + Bx + C$. Prove that for any interval $[a, b]$, the value c guaranteed by the Mean Value Theorem is the midpoint of the interval.

80. (a) Let $f(x) = x^2$ and $g(x) = -x^3 + x^2 + 3x + 2$. Then $f(-1) = g(-1)$ and $f(2) = g(2)$. Show that there is at least one value c in the interval $(-1, 2)$ where the tangent line to f at $(c, f(c))$ is parallel to the tangent line to g at $(c, g(c))$. Identify c.

(b) Let f and g be differentiable functions on $[a, b]$ where $f(a) = g(a)$ and $f(b) = g(b)$. Show that there is at least one value c in the interval (a, b) where the tangent line to f at $(c, f(c))$ is parallel to the tangent line to g at $(c, g(c))$.

81. Prove that if f is differentiable on $(-\infty, \infty)$ and $f'(x) < 1$ for all real numbers, then f has at most one fixed point. A fixed point of a function f is a real number c such that $f(c) = c$.

82. Use the result of Exercise 81 to show that $f(x) = \frac{1}{2}\cos x$ has at most one fixed point.

83. Prove that $|\cos a - \cos b| \le |a - b|$ for all a and b.

84. Prove that $|\sin a - \sin b| \le |a - b|$ for all a and b.

85. Let $0 < a < b$. Use the Mean Value Theorem to show that

$$\sqrt{b} - \sqrt{a} < \frac{b - a}{2\sqrt{a}}.$$

Section 3.3 Increasing and Decreasing Functions and the First Derivative Test

- **Determine intervals on which a function is increasing or decreasing.**
- **Apply the First Derivative Test to find relative extrema of a function.**

Increasing and Decreasing Functions

In this section you will learn how derivatives can be used to *classify* relative extrema as either relative minima or relative maxima. First, it is important to define increasing and decreasing functions.

Definitions of Increasing and Decreasing Functions

A function f is **increasing** on an interval if for any two numbers x_1 and x_2 in the interval, $x_1 < x_2$ implies $f(x_1) < f(x_2)$.

A function f is **decreasing** on an interval if for any two numbers x_1 and x_2 in the interval, $x_1 < x_2$ implies $f(x_1) > f(x_2)$.

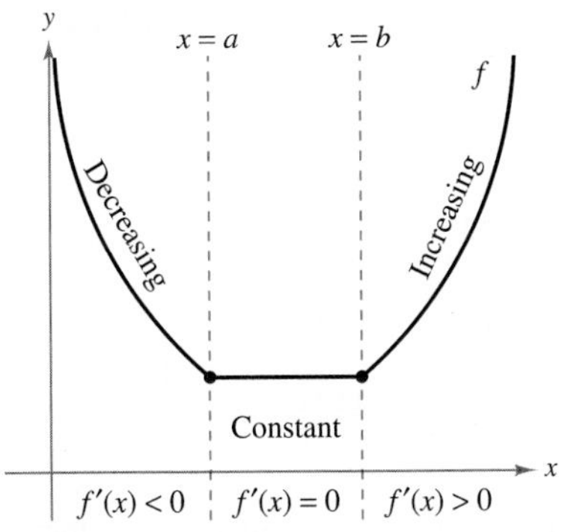

The derivative is related to the slope of a function.
Figure 3.15

A function is increasing if, *as x moves to the right*, its graph moves up, and is decreasing if its graph moves down. For example, the function in Figure 3.15 is decreasing on the interval $(-\infty, a)$, is constant on the interval (a, b), and is increasing on the interval (b, ∞). As shown in Theorem 3.5 below, a positive derivative implies that the function is increasing; a negative derivative implies that the function is decreasing; and a zero derivative on an entire interval implies that the function is constant on that interval.

THEOREM 3.5 Test for Increasing and Decreasing Functions

Let f be a function that is continuous on the closed interval $[a, b]$ and differentiable on the open interval (a, b).

1. If $f'(x) > 0$ for all x in (a, b), then f is increasing on $[a, b]$.
2. If $f'(x) < 0$ for all x in (a, b), then f is decreasing on $[a, b]$.
3. If $f'(x) = 0$ for all x in (a, b), then f is constant on $[a, b]$.

Proof To prove the first case, assume that $f'(x) > 0$ for all x in the interval (a, b) and let $x_1 < x_2$ be any two points in the interval. By the Mean Value Theorem, you know that there exists a number c such that $x_1 < c < x_2$, and

$$f'(c) = \frac{f(x_2) - f(x_1)}{x_2 - x_1}.$$

Because $f'(c) > 0$ and $x_2 - x_1 > 0$, you know that

$$f(x_2) - f(x_1) > 0$$

which implies that $f(x_1) < f(x_2)$. So, f is increasing on the interval. The second case has a similar proof (see Exercise 101), and the third case was given as Exercise 78 in Section 3.2.

NOTE The conclusions in the first two cases of Theorem 3.5 are valid even if $f'(x) = 0$ at a finite number of x-values in (a, b).

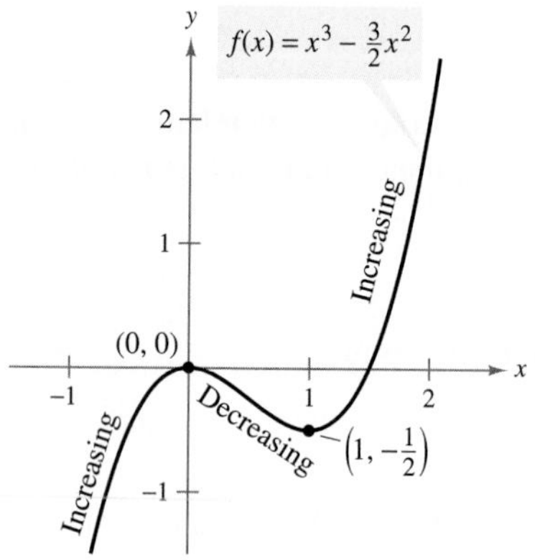

Figure 3.16

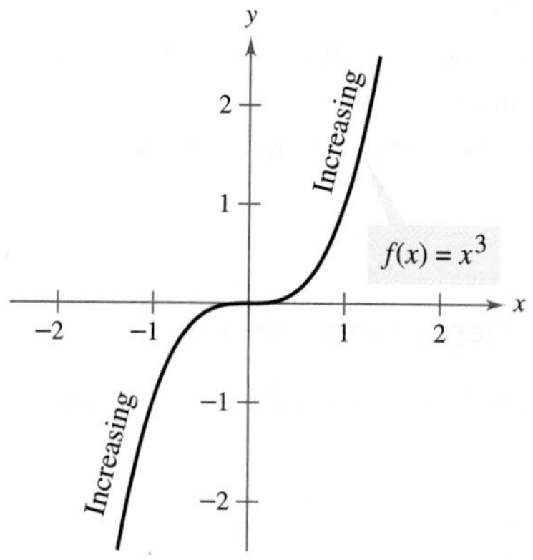

(a) Strictly monotonic function

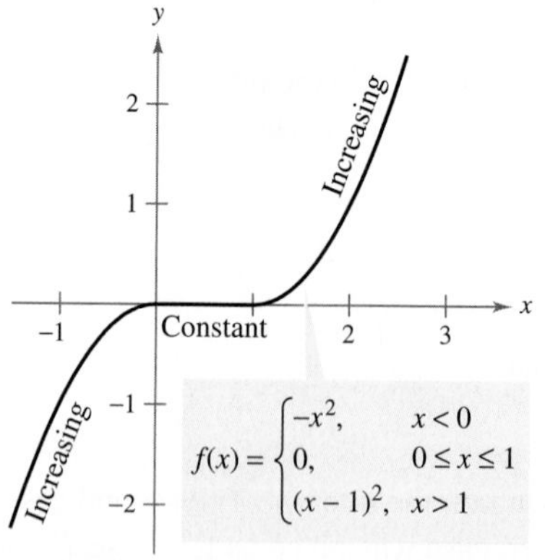

(b) Not strictly monotonic
Figure 3.17

EXAMPLE 1 Intervals on Which f Is Increasing or Decreasing

Find the open intervals on which $f(x) = x^3 - \frac{3}{2}x^2$ is increasing or decreasing.

Solution Note that f is differentiable on the entire real number line. To determine the critical numbers of f, set $f'(x)$ equal to zero.

$$f(x) = x^3 - \frac{3}{2}x^2 \qquad \text{Write original function.}$$

$$f'(x) = 3x^2 - 3x = 0 \qquad \text{Differentiate and set } f'(x) \text{ equal to 0.}$$

$$3(x)(x-1) = 0 \qquad \text{Factor.}$$

$$x = 0, 1 \qquad \text{Critical numbers}$$

Because there are no points for which f' does not exist, you can conclude that $x = 0$ and $x = 1$ are the only critical numbers. The table summarizes the testing of the three intervals determined by these two critical numbers.

Interval	$-\infty < x < 0$	$0 < x < 1$	$1 < x < \infty$
Test Value	$x = -1$	$x = \frac{1}{2}$	$x = 2$
Sign of $f'(x)$	$f'(-1) = 6 > 0$	$f'\left(\frac{1}{2}\right) = -\frac{3}{4} < 0$	$f'(2) = 6 > 0$
Conclusion	Increasing	Decreasing	Increasing

So, f is increasing on the intervals $(-\infty, 0)$ and $(1, \infty)$ and decreasing on the interval $(0, 1)$, as shown in Figure 3.16.

Example 1 gives you one example of how to find intervals on which a function is increasing or decreasing. The guidelines below summarize the steps followed in the example.

Guidelines for Finding Intervals on Which a Function Is Increasing or Decreasing

Let f be continuous on the interval (a, b). To find the open intervals on which f is increasing or decreasing, use the following steps.

1. Locate the critical numbers of f in (a, b), and use these numbers to determine test intervals.
2. Determine the sign of $f'(x)$ at one test value in each of the intervals.
3. Use Theorem 3.5 to determine whether f is increasing or decreasing on each interval.

These guidelines are also valid if the interval (a, b) is replaced by an interval of the form $(-\infty, b)$, (a, ∞), or $(-\infty, \infty)$.

A function is **strictly monotonic** on an interval if it is either increasing on the entire interval or decreasing on the entire interval. For instance, the function $f(x) = x^3$ is strictly monotonic on the entire real line because it is increasing on the entire real line, as shown in Figure 3.17(a). The function shown in Figure 3.17(b) is not strictly monotonic on the entire real line because it is constant on the interval $[0, 1]$.

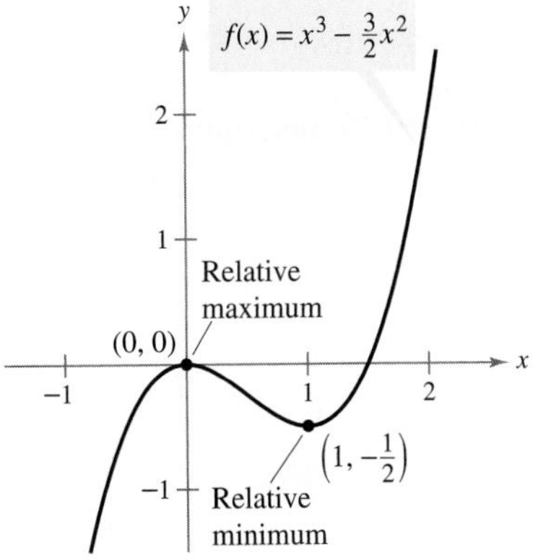

Relative extrema of f
Figure 3.18

The First Derivative Test

After you have determined the intervals on which a function is increasing or decreasing, it is not difficult to locate the relative extrema of the function. For instance, in Figure 3.18 (from Example 1), the function

$$f(x) = x^3 - \frac{3}{2}x^2$$

has a relative maximum at the point $(0, 0)$ because f is increasing immediately to the left of $x = 0$ and decreasing immediately to the right of $x = 0$. Similarly, f has a relative minimum at the point $\left(1, -\frac{1}{2}\right)$ because f is decreasing immediately to the left of $x = 1$ and increasing immediately to the right of $x = 1$. The following theorem, called the First Derivative Test, makes this more explicit.

THEOREM 3.6 The First Derivative Test

Let c be a critical number of a function f that is continuous on an open interval I containing c. If f is differentiable on the interval, except possibly at c, then $f(c)$ can be classified as follows.

1. If $f'(x)$ changes from negative to positive at c, then f has a *relative minimum* at $(c, f(c))$.
2. If $f'(x)$ changes from positive to negative at c, then f has a *relative maximum* at $(c, f(c))$.
3. If $f'(x)$ is positive on both sides of c or negative on both sides of c, then $f(c)$ is neither a relative minimum nor a relative maximum.

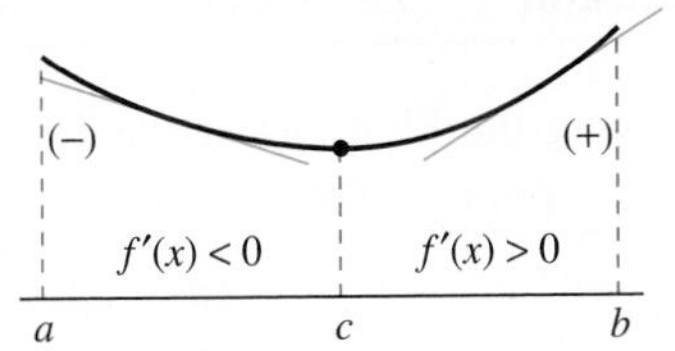

Relative minimum

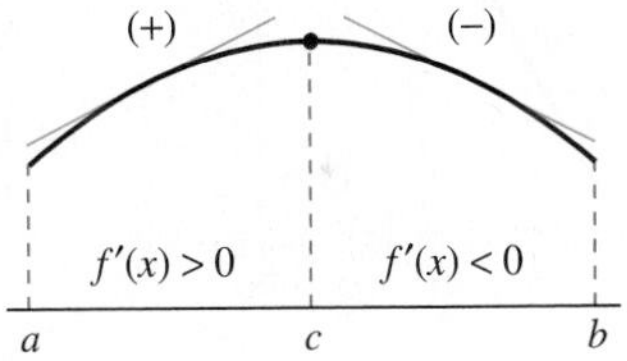

Relative maximum

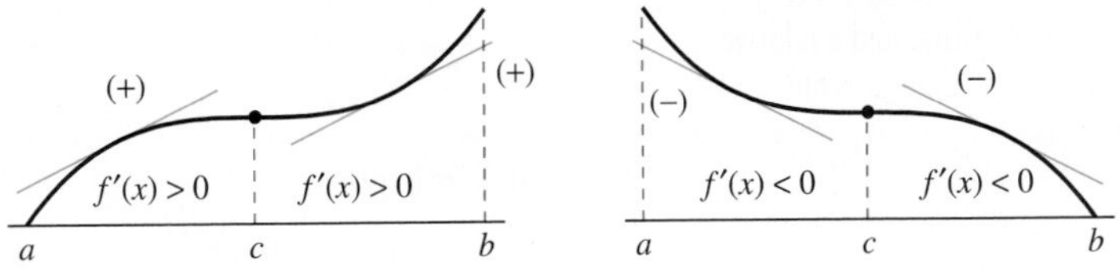

Neither relative minimum nor relative maximum

Proof Assume that $f'(x)$ changes from negative to positive at c. Then there exist a and b in I such that

$$f'(x) < 0 \text{ for all } x \text{ in } (a, c)$$

and

$$f'(x) > 0 \text{ for all } x \text{ in } (c, b).$$

By Theorem 3.5, f is decreasing on (a, c) and increasing on (c, b). So, $f(c)$ is a minimum of f on the open interval (a, b) and, consequently, a relative minimum of f. This proves the first case of the theorem. The second case can be proved in a similar way (see Exercise 102).

EXAMPLE 2 Applying the First Derivative Test

Find the relative extrema of the function $f(x) = \frac{1}{2}x - \sin x$ in the interval $(0, 2\pi)$.

Solution Note that f is continuous on the interval $(0, 2\pi)$. To determine the critical numbers of f in this interval, set $f'(x)$ equal to 0.

$$f'(x) = \frac{1}{2} - \cos x = 0 \qquad \text{Set } f'(x) \text{ equal to 0.}$$

$$\cos x = \frac{1}{2}$$

$$x = \frac{\pi}{3}, \frac{5\pi}{3} \qquad \text{Critical numbers}$$

Because there are no points for which f' does not exist, you can conclude that $x = \pi/3$ and $x = 5\pi/3$ are the only critical numbers. The table summarizes the testing of the three intervals determined by these two critical numbers.

Interval	$0 < x < \frac{\pi}{3}$	$\frac{\pi}{3} < x < \frac{5\pi}{3}$	$\frac{5\pi}{3} < x < 2\pi$
Test Value	$x = \frac{\pi}{4}$	$x = \pi$	$x = \frac{7\pi}{4}$
Sign of $f'(x)$	$f'\left(\frac{\pi}{4}\right) < 0$	$f'(\pi) > 0$	$f'\left(\frac{7\pi}{4}\right) < 0$
Conclusion	Decreasing	Increasing	Decreasing

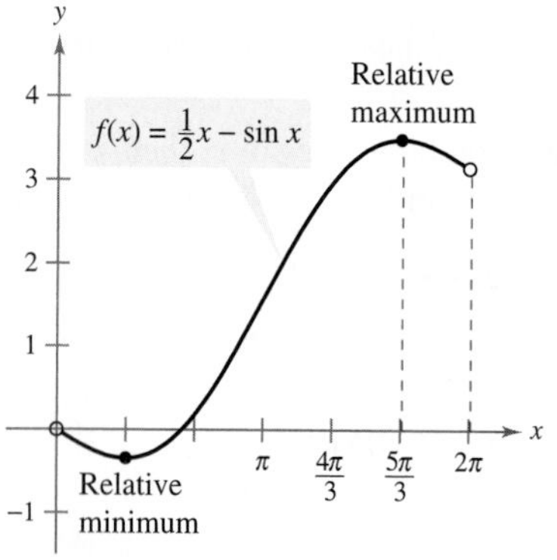

A relative minimum occurs where f changes from decreasing to increasing, and a relative maximum occurs where f changes from increasing to decreasing.
Figure 3.19

By applying the First Derivative Test, you can conclude that f has a relative minimum at the point where

$$x = \frac{\pi}{3} \qquad x\text{-value where relative minimum occurs}$$

and a relative maximum at the point where

$$x = \frac{5\pi}{3} \qquad x\text{-value where relative maximum occurs}$$

as shown in Figure 3.19.

EXPLORATION

Comparing Graphical and Analytic Approaches From Section 3.2, you know that, *by itself*, a graphing utility can give misleading information about the relative extrema of a graph. *Used in conjunction with an analytic approach,* however, a graphing utility can provide a good way to reinforce your conclusions. Use a graphing utility to graph the function in Example 2. Then use the *zoom* and *trace* features to estimate the relative extrema. How close are your graphical approximations?

Note that in Examples 1 and 2 the given functions are differentiable on the entire real line. For such functions, the only critical numbers are those for which $f'(x) = 0$. Example 3 concerns a function that has two types of critical numbers—those for which $f'(x) = 0$ and those for which f is not differentiable.

EXAMPLE 3 Applying the First Derivative Test

Find the relative extrema of

$$f(x) = (x^2 - 4)^{2/3}.$$

Solution Begin by noting that f is continuous on the entire real line. The derivative of f

$$f'(x) = \frac{2}{3}(x^2 - 4)^{-1/3}(2x) \qquad \text{General Power Rule}$$

$$= \frac{4x}{3(x^2 - 4)^{1/3}} \qquad \text{Simplify.}$$

is 0 when $x = 0$ and does not exist when $x = \pm 2$. So, the critical numbers are $x = -2$, $x = 0$, and $x = 2$. The table summarizes the testing of the four intervals determined by these three critical numbers.

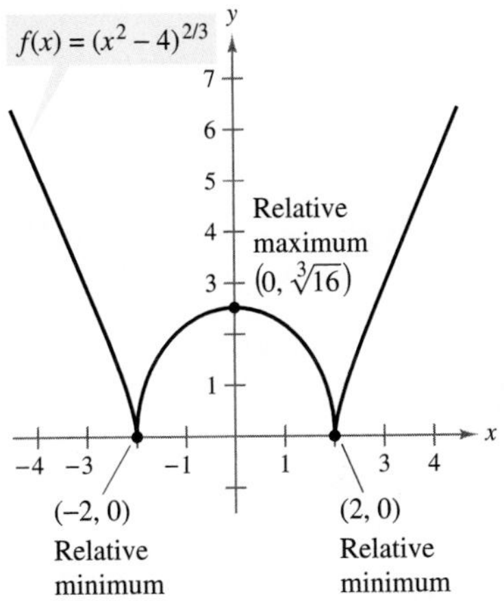

You can apply the First Derivative Test to find relative extrema.
Figure 3.20

Interval	$-\infty < x < -2$	$-2 < x < 0$	$0 < x < 2$	$2 < x < \infty$
Test Value	$x = -3$	$x = -1$	$x = 1$	$x = 3$
Sign of $f'(x)$	$f'(-3) < 0$	$f'(-1) > 0$	$f'(1) < 0$	$f'(3) > 0$
Conclusion	Decreasing	Increasing	Decreasing	Increasing

By applying the First Derivative Test, you can conclude that f has a relative minimum at the point $(-2, 0)$, a relative maximum at the point $(0, \sqrt[3]{16})$, and another relative minimum at the point $(2, 0)$, as shown in Figure 3.20.

TECHNOLOGY PITFALL When using a graphing utility to graph a function involving radicals or rational exponents, be sure you understand the way the utility evaluates radical expressions. For instance, even though

$$f(x) = (x^2 - 4)^{2/3}$$

and

$$g(x) = [(x^2 - 4)^2]^{1/3}$$

are the same algebraically, some graphing utilities distinguish between these two functions. Which of the graphs shown in Figure 3.21 is incorrect? Why did the graphing utility produce an incorrect graph?

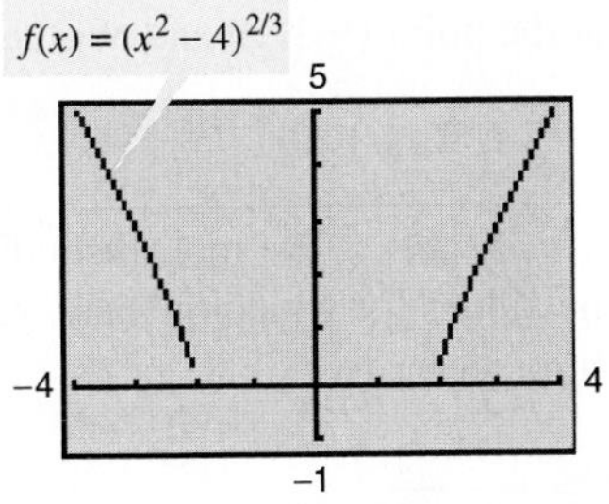

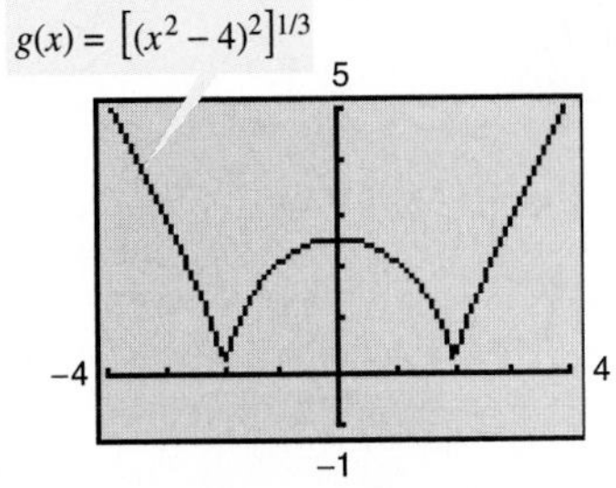

Which graph is incorrect?
Figure 3.21

When using the First Derivative Test, be sure to consider the domain of the function. For instance, in the next example, the function

$$f(x) = \frac{x^4 + 1}{x^2}$$

is not defined when $x = 0$. This x-value must be used with the critical numbers to determine the test intervals.

EXAMPLE 4 Applying the First Derivative Test

Find the relative extrema of $f(x) = \dfrac{x^4 + 1}{x^2}$.

Solution

$$f(x) = x^2 + x^{-2} \qquad \text{Rewrite original function.}$$

$$f'(x) = 2x - 2x^{-3} \qquad \text{Differentiate.}$$

$$= 2x - \frac{2}{x^3} \qquad \text{Rewrite with positive exponent.}$$

$$= \frac{2(x^4 - 1)}{x^3} \qquad \text{Simplify.}$$

$$= \frac{2(x^2 + 1)(x - 1)(x + 1)}{x^3} \qquad \text{Factor.}$$

So, $f'(x)$ is zero at $x = \pm 1$. Moreover, because $x = 0$ is not in the domain of f, you should use this x-value along with the critical numbers to determine the test intervals.

$$x = \pm 1 \qquad \text{Critical numbers, } f'(\pm 1) = 0$$

$$x = 0 \qquad 0 \text{ is not in the domain of } f.$$

The table summarizes the testing of the four intervals determined by these three x-values.

Interval	$-\infty < x < -1$	$-1 < x < 0$	$0 < x < 1$	$1 < x < \infty$
Test Value	$x = -2$	$x = -\frac{1}{2}$	$x = \frac{1}{2}$	$x = 2$
Sign of $f'(x)$	$f'(-2) < 0$	$f'(-\frac{1}{2}) > 0$	$f'(\frac{1}{2}) < 0$	$f'(2) > 0$
Conclusion	Decreasing	Increasing	Decreasing	Increasing

By applying the First Derivative Test, you can conclude that f has one relative minimum at the point $(-1, 2)$ and another at the point $(1, 2)$, as shown in Figure 3.22.

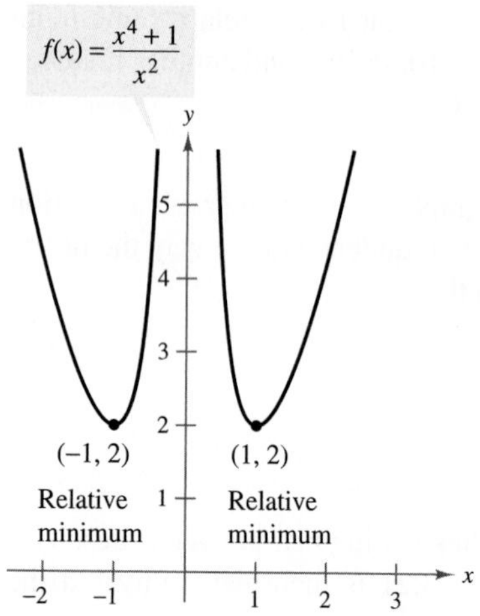

x-values that are not in the domain of f, as well as critical numbers, determine test intervals for f'.
Figure 3.22

TECHNOLOGY The most difficult step in applying the First Derivative Test is finding the values for which the derivative is equal to 0. For instance, the values of x for which the derivative of

$$f(x) = \frac{x^4 + 1}{x^2 + 1}$$

is equal to zero are $x = 0$ and $x = \pm\sqrt{\sqrt{2} - 1}$. If you have access to technology that can perform symbolic differentiation and solve equations, use it to apply the First Derivative Test to this function.

EXAMPLE 5 The Path of a Projectile

Michael Kevin Daly/Corbis

If a projectile is propelled from ground level and air resistance is neglected, the object will travel farthest with an initial angle of 45°. If, however, the projectile is propelled from a point above ground level, the angle that yields a maximum horizontal distance is not 45° (see Example 5).

Neglecting air resistance, the path of a projectile that is propelled at an angle θ is

$$y = \frac{g \sec^2 \theta}{2v_0^2}x^2 + (\tan \theta)x + h, \quad 0 \le \theta \le \frac{\pi}{2}$$

where y is the height, x is the horizontal distance, g is the acceleration due to gravity, v_0 is the initial velocity, and h is the initial height. (This equation is derived in Section 12.3.) Let $g = -32$ feet per second per second, $v_0 = 24$ feet per second, and $h = 9$ feet. What value of θ will produce a maximum horizontal distance?

Solution To find the distance the projectile travels, let $y = 0$, and use the Quadratic Formula to solve for x.

$$\frac{g \sec^2 \theta}{2v_0^2}x^2 + (\tan \theta)x + h = 0$$

$$\frac{-32 \sec^2 \theta}{2(24^2)}x^2 + (\tan \theta)x + 9 = 0$$

$$-\frac{\sec^2 \theta}{36}x^2 + (\tan \theta)x + 9 = 0$$

$$x = \frac{-\tan \theta \pm \sqrt{\tan^2 \theta + \sec^2 \theta}}{-\sec^2 \theta/18}$$

$$x = 18 \cos \theta\left(\sin \theta + \sqrt{\sin^2 \theta + 1}\right), \quad x \ge 0$$

At this point, you need to find the value of θ that produces a maximum value of x. Applying the First Derivative Test by hand would be very tedious. Using technology to solve the equation $dx/d\theta = 0$, however, eliminates most of the messy computations. The result is that the maximum value of x occurs when

$$\theta \approx 0.61548 \text{ radian, or } 35.3°.$$

This conclusion is reinforced by sketching the path of the projectile for different values of θ, as shown in Figure 3.23. Of the three paths shown, note that the distance traveled is greatest for $\theta = 35°$.

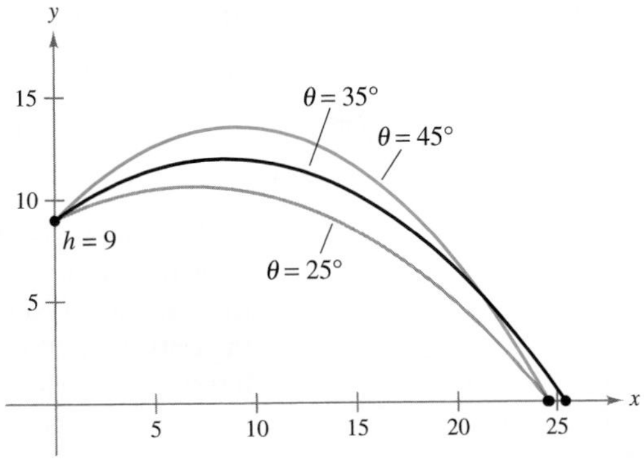

The path of a projectile with initial angle θ
Figure 3.23

NOTE A computer simulation of this example is given in the *HM mathSpace®* CD-ROM and the online *Eduspace®* system for this text. Using that simulation, you can experimentally discover that the maximum value of x occurs when $\theta \approx 35.3°$.

Exercises for Section 3.3

See www.CalcChat.com for worked-out solutions to odd-numbered exercises.

In Exercises 1 and 2, use the graph of f to find (a) the largest open interval on which f is increasing, and (b) the largest open interval on which f is decreasing.

1.

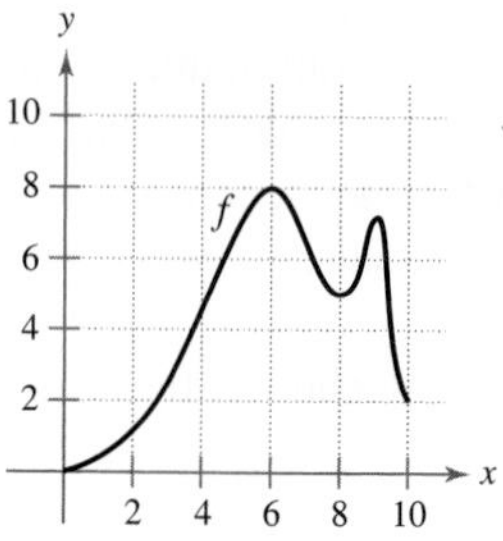

2.

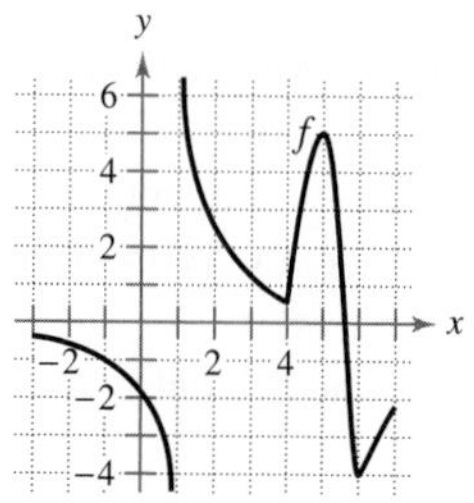

In Exercises 3–16, identify the open intervals on which the function is increasing or decreasing.

3. $f(x) = x^2 - 6x + 8$

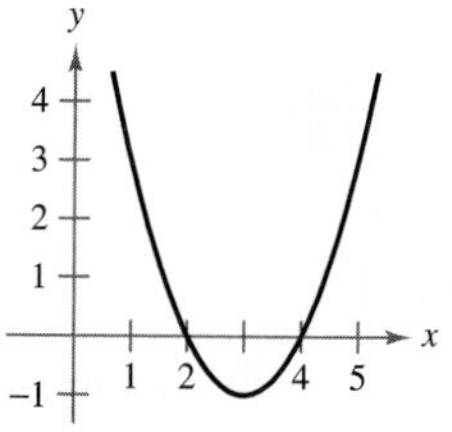

4. $y = -(x + 1)^2$

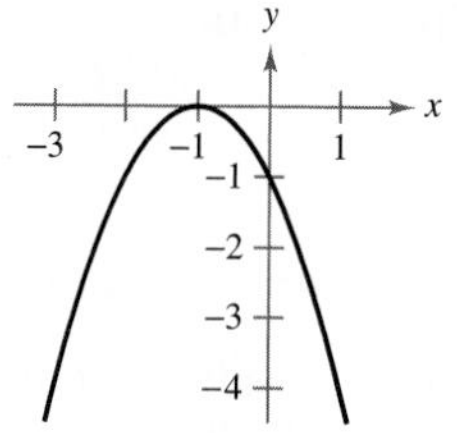

5. $y = \dfrac{x^3}{4} - 3x$

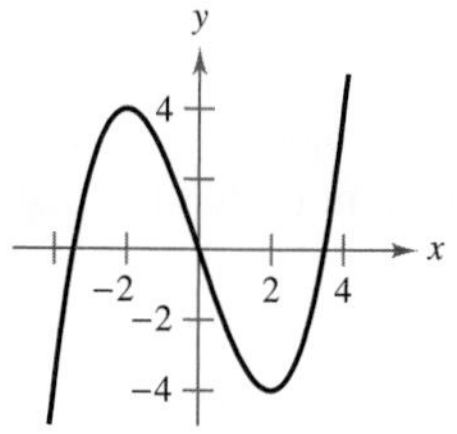

6. $f(x) = x^4 - 2x^2$

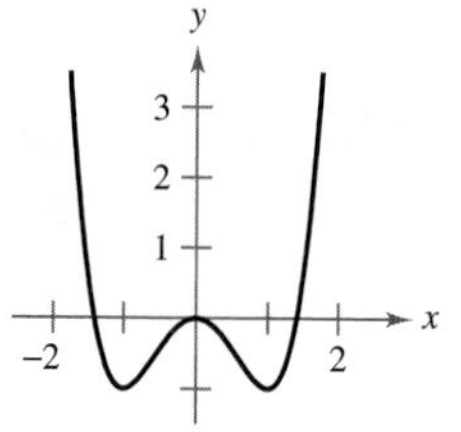

7. $f(x) = \sin x + 2, 0 < x < 2\pi$

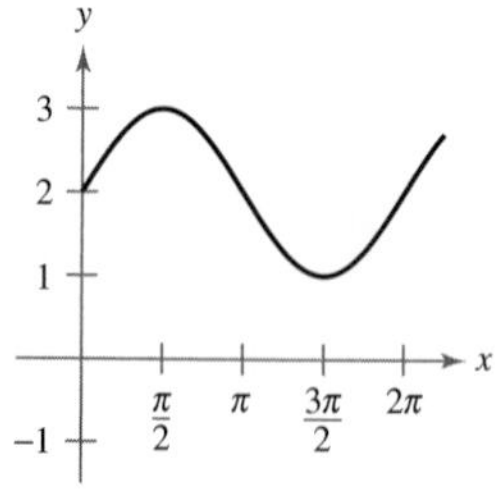

8. $h(x) = \cos\dfrac{x}{2}, 0 < x < 2\pi$

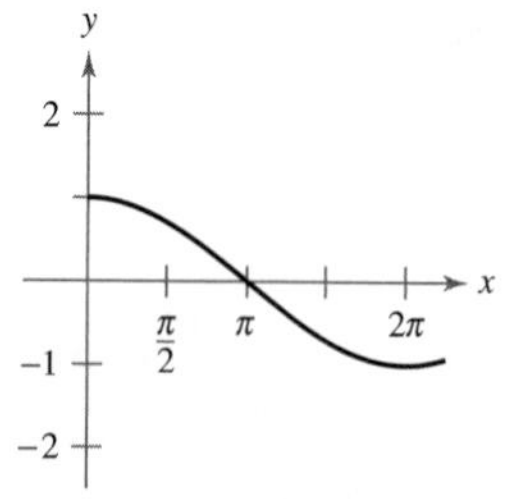

9. $f(x) = \dfrac{1}{x^2}$

10. $y = \dfrac{x^2}{x + 1}$

11. $g(x) = x^2 - 2x - 8$

12. $h(x) = 27x - x^3$

13. $y = x\sqrt{16 - x^2}$

14. $y = x + \dfrac{4}{x}$

15. $y = x - 2\cos x, \quad 0 < x < 2\pi$

16. $f(x) = \cos^2 x - \cos x, \quad 0 < x < 2\pi$

In Exercises 17–38, (a) find the critical numbers of f (if any), (b) find the open interval(s) on which the function is increasing or decreasing, (c) apply the First Derivative Test to identify all relative extrema, and (d) use a graphing utility to confirm your results.

17. $f(x) = x^2 - 6x$

18. $f(x) = x^2 + 8x + 10$

19. $f(x) = -2x^2 + 4x + 3$

20. $f(x) = -(x^2 + 8x + 12)$

21. $f(x) = 2x^3 + 3x^2 - 12x$

22. $f(x) = x^3 - 6x^2 + 15$

23. $f(x) = x^2(3 - x)$

24. $f(x) = (x + 2)^2(x - 1)$

25. $f(x) = \dfrac{x^5 - 5x}{5}$

26. $f(x) = x^4 - 32x + 4$

27. $f(x) = x^{1/3} + 1$

28. $f(x) = x^{2/3} - 4$

29. $f(x) = (x - 1)^{2/3}$

30. $f(x) = (x - 1)^{1/3}$

31. $f(x) = 5 - |x - 5|$

32. $f(x) = |x + 3| - 1$

33. $f(x) = x + \dfrac{1}{x}$

34. $f(x) = \dfrac{x}{x + 1}$

35. $f(x) = \dfrac{x^2}{x^2 - 9}$

36. $f(x) = \dfrac{x + 3}{x^2}$

37. $f(x) = \dfrac{x^2 - 2x + 1}{x + 1}$

38. $f(x) = \dfrac{x^2 - 3x - 4}{x - 2}$

In Exercises 39–46, consider the function on the interval $(0, 2\pi)$. For each function, (a) find the open interval(s) on which the function is increasing or decreasing, (b) apply the First Derivative Test to identify all relative extrema, and (c) use a graphing utility to confirm your results.

39. $f(x) = \dfrac{x}{2} + \cos x$

40. $f(x) = \sin x \cos x$

41. $f(x) = \sin x + \cos x$

42. $f(x) = x + 2\sin x$

43. $f(x) = \cos^2(2x)$

44. $f(x) = \sqrt{3}\sin x + \cos x$

45. $f(x) = \sin^2 x + \sin x$

46. $f(x) = \dfrac{\sin x}{1 + \cos^2 x}$

In Exercises 47–52, (a) use a computer algebra system to differentiate the function, (b) sketch the graphs of f and f' on the same set of coordinate axes over the given interval, (c) find the critical numbers of f in the open interval, and (d) find the interval(s) on which f' is positive and the interval(s) on which it is negative. Compare the behavior of f and the sign of f'.

47. $f(x) = 2x\sqrt{9 - x^2}, \quad [-3, 3]$

48. $f(x) = 10\left(5 - \sqrt{x^2 - 3x + 16}\right), \quad [0, 5]$

49. $f(t) = t^2 \sin t, \quad [0, 2\pi]$

50. $f(x) = \dfrac{x}{2} + \cos\dfrac{x}{2}, \quad [0, 4\pi]$

51. $f(x) = -3\sin\dfrac{x}{3}, \quad [0, 6\pi]$

52. $f(x) = 2\sin 3x + 4\cos 3x, \quad [0, \pi]$

In Exercises 53 and 54, use symmetry, extrema, and zeros to sketch the graph of f. How do the functions f and g differ?

53. $f(x) = \dfrac{x^5 - 4x^3 + 3x}{x^2 - 1}, \quad g(x) = x(x^2 - 3)$

54. $f(t) = \cos^2 t - \sin^2 t, \quad g(t) = 1 - 2\sin^2 t$

Think About It **In Exercises 55–60, the graph of f is shown in the figure. Sketch a graph of the derivative of f. To print an enlarged copy of the graph, go to the website *www.mathgraphs.com*.**

55.

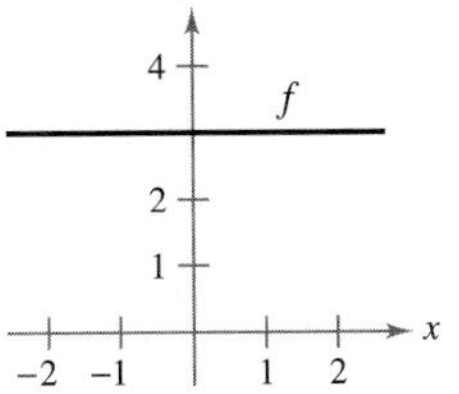

56.

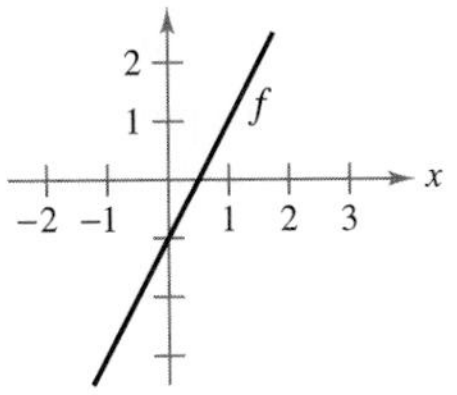

57.

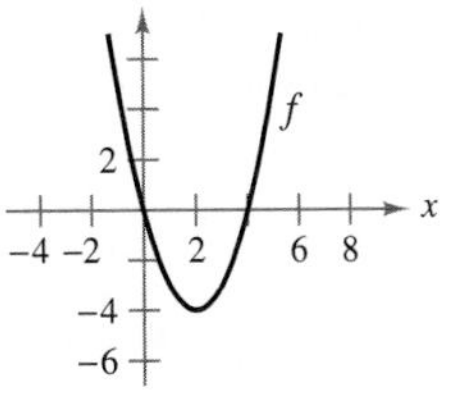

58.

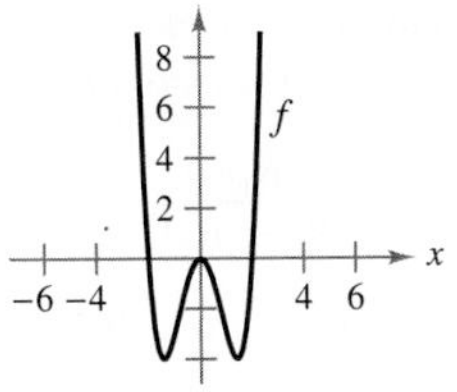

59.

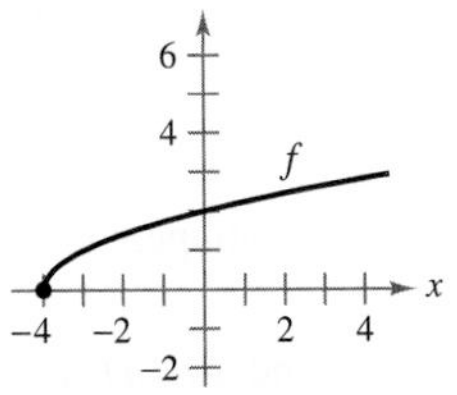

60.

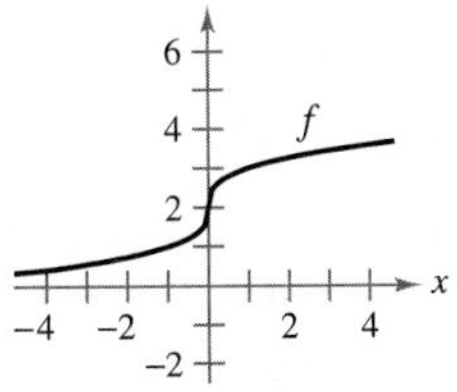

In Exercises 61–64, use the graph of f' to (a) identify the interval(s) on which f is increasing or decreasing, and (b) estimate the values of x at which f has a relative maximum or minimum.

61.

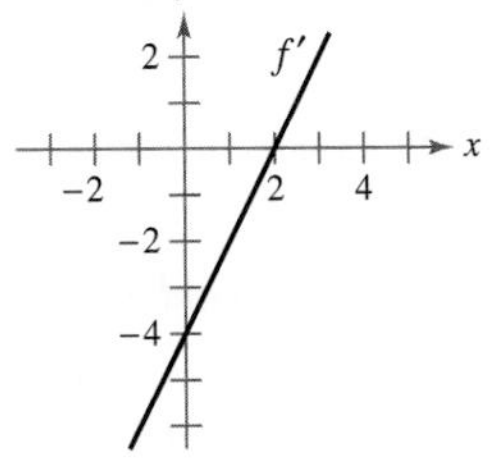

62.

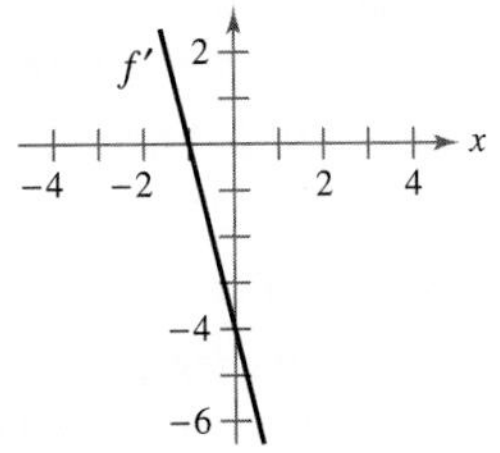

63.

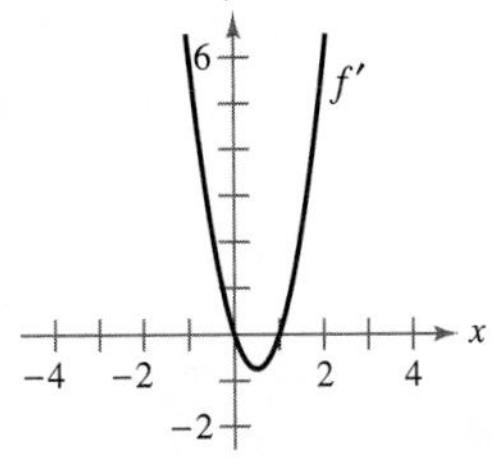

64.

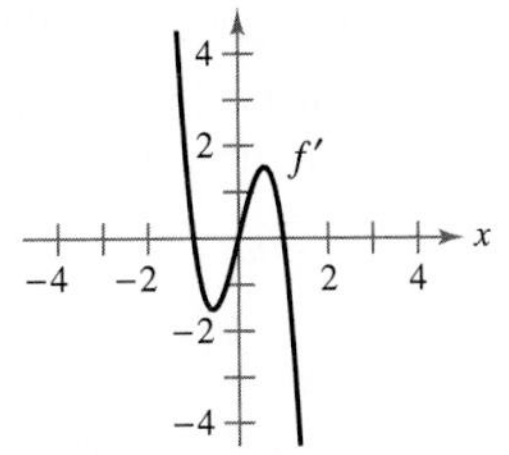

Writing About Concepts

In Exercises 65–70, assume that f is differentiable for all x. The signs of f' are as follows.

$f'(x) > 0$ **on** $(-\infty, -4)$

$f'(x) < 0$ **on** $(-4, 6)$

$f'(x) > 0$ **on** $(6, \infty)$

Supply the appropriate inequality for the indicated value of c.

	Function	*Sign of $g'(c)$*
65.	$g(x) = f(x) + 5$	$g'(0)$ ▒ 0
66.	$g(x) = 3f(x) - 3$	$g'(-5)$ ▒ 0
67.	$g(x) = -f(x)$	$g'(-6)$ ▒ 0
68.	$g(x) = -f(x)$	$g'(0)$ ▒ 0
69.	$g(x) = f(x - 10)$	$g'(0)$ ▒ 0
70.	$g(x) = f(x - 10)$	$g'(8)$ ▒ 0

71. Sketch the graph of the arbitrary function f such that

$$f'(x) \begin{cases} > 0, & x < 4 \\ \text{undefined}, & x = 4. \\ < 0, & x > 4 \end{cases}$$

72. A differentiable function f has one critical number at $x = 5$. Identify the relative extrema of f at the critical number if $f'(4) = -2.5$ and $f'(6) = 3$.

73. *Think About It* The function f is differentiable on the interval $[-1, 1]$. The table shows the values of f' for selected values of x. Sketch the graph of f, approximate the critical numbers, and identify the relative extrema.

x	-1	-0.75	-0.50	-0.25
$f'(x)$	-10	-3.2	-0.5	0.8

x	0	0.25	0.50	0.75	1
$f'(x)$	5.6	3.6	-0.2	-6.7	-20.1

74. *Think About It* The function f is differentiable on the interval $[0, \pi]$. The table shows the values of f' for selected values of x. Sketch the graph of f, approximate the critical numbers, and identify the relative extrema.

x	0	$\pi/6$	$\pi/4$	$\pi/3$	$\pi/2$
$f'(x)$	3.14	-0.23	-2.45	-3.11	0.69

x	$2\pi/3$	$3\pi/4$	$5\pi/6$	π
$f'(x)$	3.00	1.37	-1.14	-2.84

75. *Rolling a Ball Bearing* A ball bearing is placed on an inclined plane and begins to roll. The angle of elevation of the plane is θ. The distance (in meters) the ball bearing rolls in t seconds is $s(t) = 4.9(\sin \theta)t^2$.

(a) Determine the speed of the ball bearing after t seconds.

(b) Complete the table and use it to determine the value of θ that produces the maximum speed at a particular time.

θ	0	$\pi/4$	$\pi/3$	$\pi/2$	$2\pi/3$	$3\pi/4$	π
$s'(t)$							

76. *Numerical, Graphical, and Analytic Analysis* The concentration C of a chemical in the bloodstream t hours after injection into muscle tissue is

$$C(t) = \frac{3t}{27 + t^3}, \quad t \geq 0.$$

(a) Complete the table and use it to approximate the time when the concentration is greatest.

t	0	0.5	1	1.5	2	2.5	3
$C(t)$							

(b) Use a graphing utility to graph the concentration function and use the graph to approximate the time when the concentration is greatest.

(c) Use calculus to determine analytically the time when the concentration is greatest.

77. *Numerical, Graphical, and Analytic Analysis* Consider the functions $f(x) = x$ and $g(x) = \sin x$ on the interval $(0, \pi)$.

(a) Complete the table and make a conjecture about which is the greater function on the interval $(0, \pi)$.

x	0.5	1	1.5	2	2.5	3
$f(x)$						
$g(x)$						

(b) Use a graphing utility to graph the functions and use the graphs to make a conjecture about which is the greater function on the interval $(0, \pi)$.

(c) Prove that $f(x) > g(x)$ on the interval $(0, \pi)$. [*Hint:* Show that $h'(x) > 0$ where $h = f - g$.]

78. *Numerical, Graphical, and Analytic Analysis* Consider the functions $f(x) = x$ and $g(x) = \tan x$ on the interval $(0, \pi/2)$.

(a) Complete the table and make a conjecture about which is the greater function on the interval $(0, \pi/2)$.

x	0.25	0.5	0.75	1	1.25	1.5
$f(x)$						
$g(x)$						

(b) Use a graphing utility to graph the functions and use the graphs to make a conjecture about which is the greater function on the interval $(0, \pi/2)$.

(c) Prove that $f(x) < g(x)$ on the interval $(0, \pi/2)$. [*Hint:* Show that $h'(x) > 0$, where $h = g - f$.]

79. *Trachea Contraction* Coughing forces the trachea (windpipe) to contract, which affects the velocity v of the air passing through the trachea. The velocity of the air during coughing is

$$v = k(R - r)r^2, \quad 0 \leq r < R$$

where k is constant, R is the normal radius of the trachea, and r is the radius during coughing. What radius will produce the maximum air velocity?

80. *Profit* The profit P (in dollars) made by a fast-food restaurant selling x hamburgers is

$$P = 2.44x - \frac{x^2}{20{,}000} - 5000, \quad 0 \leq x \leq 35{,}000.$$

Find the open intervals on which P is increasing or decreasing.

81. *Power* The electric power P in watts in a direct-current circuit with two resistors R_1 and R_2 connected in parallel is

$$P = \frac{vR_1R_2}{(R_1 + R_2)^2}$$

where v is the voltage. If v and R_1 are held constant, what resistance R_2 produces maximum power?

82. *Electrical Resistance* The resistance R of a certain type of resistor is

$$R = \sqrt{0.001T^4 - 4T + 100}$$

where R is measured in ohms and the temperature T is measured in degrees Celsius.

(a) Use a computer algebra system to find dR/dT and the critical number of the function. Determine the minimum resistance for this type of resistor.

(b) Use a graphing utility to graph the function R and use the graph to approximate the minimum resistance for this type of resistor.

83. *Modeling Data* The end-of-year assets for the Medicare Hospital Insurance Trust Fund (in billions of dollars) for the years 1995 through 2001 are shown.

1995: 130.3; 1996: 124.9; 1997: 115.6; 1998: 120.4;

1999: 141.4; 2000: 177.5; 2001: 208.7

(Source: U.S. Centers for Medicare and Medicaid Services)

(a) Use the regression capabilities of a graphing utility to find a model of the form $M = at^2 + bt + c$ for the data. (Let $t = 5$ represent 1995.)

(b) Use a graphing utility to plot the data and graph the model.

(c) Analytically find the minimum of the model and compare the result with the actual data.

84. ***Modeling Data*** The number of bankruptcies (in thousands) for the years 1988 through 2001 are shown.

1988: 594.6; 1989: 643.0; 1990: 725.5; 1991: 880.4;

1992: 972.5; 1993: 918.7; 1994: 845.3; 1995: 858.1;

1996: 1042.1; 1997: 1317.0; 1998: 1429.5;

1999: 1392.0; 2000: 1277.0; 2001: 1386.6

(Source: Administrative Office of the U.S. Courts)

(a) Use the regression capabilities of a graphing utility to find a model of the form $B = at^4 + bt^3 + ct^2 + dt + e$ for the data. (Let $t = 8$ represent 1988.)

(b) Use a graphing utility to plot the data and graph the model.

(c) Find the maximum of the model and compare the result with the actual data.

Motion Along a Line **In Exercises 85–88, the function $s(t)$ describes the motion of a particle moving along a line. For each function, (a) find the velocity function of the particle at any time $t \geq 0$, (b) identify the time interval(s) when the particle is moving in a positive direction, (c) identify the time interval(s) when the particle is moving in a negative direction, and (d) identify the time(s) when the particle changes its direction.**

85. $s(t) = 6t - t^2$

86. $s(t) = t^2 - 7t + 10$

87. $s(t) = t^3 - 5t^2 + 4t$

88. $s(t) = t^3 - 20t^2 + 128t - 280$

Motion Along a Line **In Exercise 89 and 90, the graph shows the position of a particle moving along a line. Describe how the particle's position changes with respect to time.**

89.

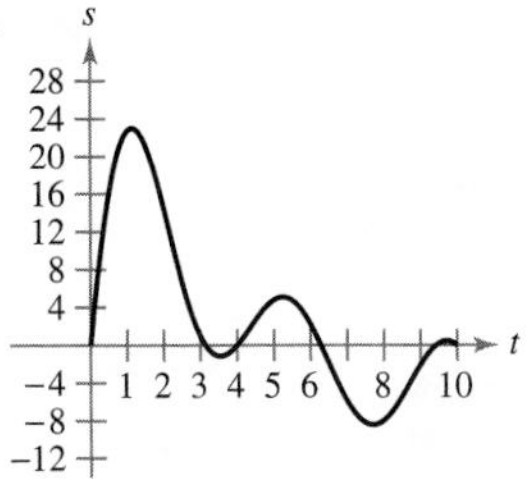

90.

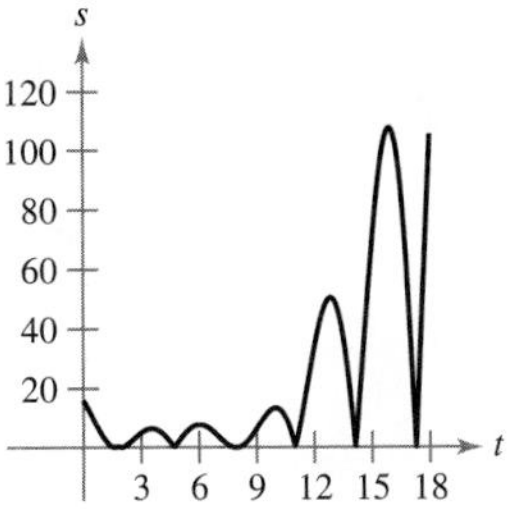

Creating Polynomial Functions **In Exercises 91–94, find a polynomial function**

$$f(x) = a_nx^n + a_{n-1}x^{n-1} + \cdots + a_2x^2 + a_1x + a_0$$

that has only the specified extrema. (a) Determine the minimum degree of the function and give the criteria you used in determining the degree. (b) Using the fact that the coordinates of the extrema are solution points of the function, and that the x-coordinates are critical numbers, determine a system of linear equations whose solution yields the coefficients of the required function. (c) Use a graphing utility to solve the system of equations and determine the function. (d) Use a graphing utility to confirm your result graphically.

91. Relative minimum: $(0, 0)$; Relative maximum: $(2, 2)$

92. Relative minimum: $(0, 0)$; Relative maximum: $(4, 1000)$

93. Relative minima: $(0, 0)$, $(4, 0)$

Relative maximum: $(2, 4)$

94. Relative minimum: $(1, 2)$

Relative maxima: $(-1, 4)$, $(3, 4)$

True or False? **In Exercises 95–100, determine whether the statement is true or false. If it is false, explain why or give an example that shows it is false.**

95. The sum of two increasing functions is increasing.

96. The product of two increasing functions is increasing.

97. Every nth-degree polynomial has $(n - 1)$ critical numbers.

98. An nth-degree polynomial has at most $(n - 1)$ critical numbers.

99. There is a relative maximum or minimum at each critical number.

100. The relative maxima of the function f are $f(1) = 4$ and $f(3) = 10$. So, f has at least one minimum for some x in the interval $(1, 3)$.

101. Prove the second case of Theorem 3.5.

102. Prove the second case of Theorem 3.6.

103. Let $x > 0$ and $n > 1$ be real numbers. Prove that $(1 + x)^n > 1 + nx$.

104. Use the definitions of increasing and decreasing functions to prove that $f(x) = x^3$ is increasing on $(-\infty, \infty)$.

105. Use the definitions of increasing and decreasing functions to prove that $f(x) = 1/x$ is decreasing on $(0, \infty)$.

Section Project: Rainbows

Rainbows are formed when light strikes raindrops and is reflected and refracted, as shown in the figure. (This figure shows a cross section of a spherical raindrop.) The Law of Refraction states that $(\sin \alpha)/(\sin \beta) = k$, where $k \approx 1.33$ (for water). The angle of deflection is given by $D = \pi + 2\alpha - 4\beta$.

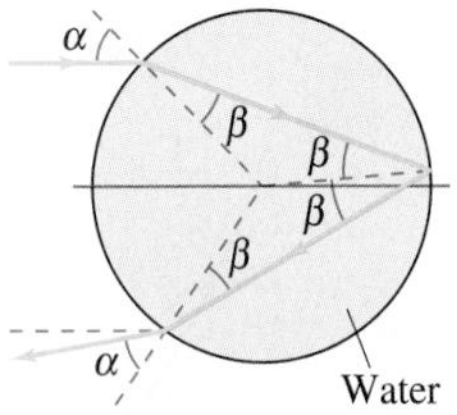

(a) Use a graphing utility to graph $D = \pi + 2\alpha - 4\sin^{-1}(1/k \sin \alpha)$, $0 \leq \alpha \leq \pi/2$.

(b) Prove that the minimum angle of deflection occurs when

$$\cos \alpha = \sqrt{\frac{k^2 - 1}{3}}.$$

For water, what is the minimum angle of deflection, $D_{\min}$? (The angle $\pi - D_{\min}$ is called the *rainbow angle.*) What value of α produces this minimum angle? (A ray of sunlight that strikes a raindrop at this angle, α, is called a *rainbow ray.*)

FOR FURTHER INFORMATION For more information about the mathematics of rainbows, see the article "Somewhere Within the Rainbow" by Steven Janke in *The UMAP Journal.*

Section 3.4 Concavity and the Second Derivative Test

- **Determine intervals on which a function is concave upward or concave downward.**
- **Find any points of inflection of the graph of a function.**
- **Apply the Second Derivative Test to find relative extrema of a function.**

Concavity

You have already seen that locating the intervals in which a function f increases or decreases helps to describe its graph. In this section, you will see how locating the intervals in which f' increases or decreases can be used to determine where the graph of f is *curving upward* or *curving downward.*

Definition of Concavity

Let f be differentiable on an open interval I. The graph of f is **concave upward** on I if f' is increasing on the interval and **concave downward** on I if f' is decreasing on the interval.

The following graphical interpretation of concavity is useful. (See Appendix A for a proof of these results.)

1. Let f be differentiable on an open interval I. If the graph of f is concave *upward* on I, then the graph of f lies *above* all of its tangent lines on I. [See Figure 3.24(a).]
2. Let f be differentiable on an open interval I. If the graph of f is concave *downward* on I, then the graph of f lies *below* all of its tangent lines on I. [See Figure 3.24(b).]

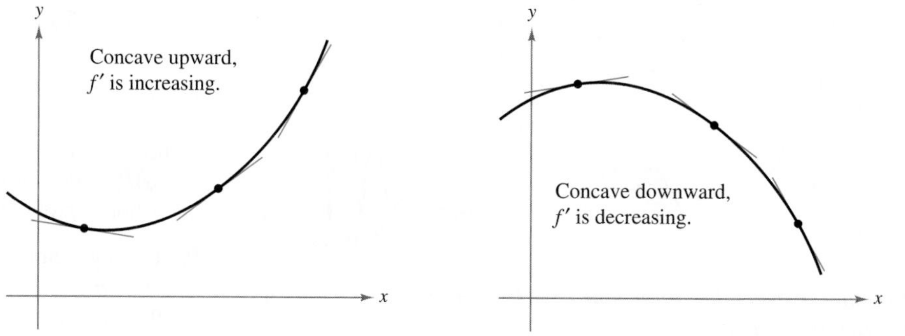

(a) The graph of f lies above its tangent lines. **(b)** The graph of f lies below its tangent lines.
Figure 3.24

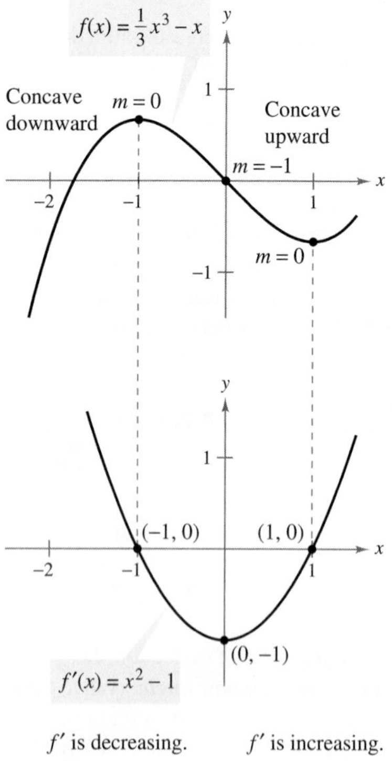

The concavity of f is related to the slope of the derivative.
Figure 3.25

To find the open intervals on which the graph of a function f is concave upward or downward, you need to find the intervals on which f' is increasing or decreasing. For instance, the graph of

$$f(x) = \frac{1}{3}x^3 - x$$

is concave downward on the open interval $(-\infty, 0)$ because $f'(x) = x^2 - 1$ is decreasing there. (See Figure 3.25.) Similarly, the graph of f is concave upward on the interval $(0, \infty)$ because f' is increasing on $(0, \infty)$.

The following theorem shows how to use the *second* derivative of a function f to determine intervals on which the graph of f is concave upward or downward. A proof of this theorem follows directly from Theorem 3.5 and the definition of concavity.

> **THEOREM 3.7 Test for Concavity**
>
> Let f be a function whose second derivative exists on an open interval I.
>
> **1.** If $f''(x) > 0$ for all x in I, then the graph of f is concave upward in I.
>
> **2.** If $f''(x) < 0$ for all x in I, then the graph of f is concave downward in I.

NOTE A third case of Theorem 3.7 could be that if $f''(x) = 0$ for all x in I, then f is linear. Note, however, that concavity is not defined for a line. In other words, a straight line is neither concave upward nor concave downward.

To apply Theorem 3.7, locate the x-values at which $f''(x) = 0$ or f'' does not exist. Second, use these x-values to determine test intervals. Finally, test the sign of $f''(x)$ in each of the test intervals.

EXAMPLE 1 Determining Concavity

Determine the open intervals on which the graph of

$$f(x) = \frac{6}{x^2 + 3}$$

is concave upward or downward.

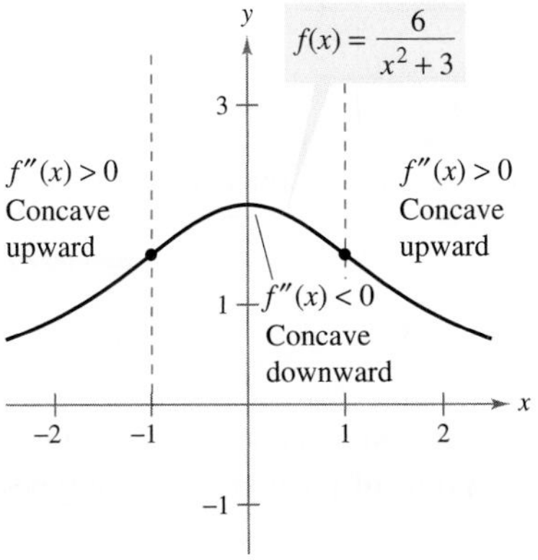

From the sign of f'' you can determine the concavity of the graph of f.
Figure 3.26

Solution Begin by observing that f is continuous on the entire real line. Next, find the second derivative of f.

$$f(x) = 6(x^2 + 3)^{-1} \quad \text{Rewrite original function.}$$

$$f'(x) = (-6)(x^2 + 3)^{-2}(2x) \quad \text{Differentiate.}$$

$$= \frac{-12x}{(x^2 + 3)^2} \quad \text{First derivative}$$

$$f''(x) = \frac{(x^2 + 3)^2(-12) - (-12x)(2)(x^2 + 3)(2x)}{(x^2 + 3)^4} \quad \text{Differentiate.}$$

$$= \frac{36(x^2 - 1)}{(x^2 + 3)^3} \quad \text{Second derivative}$$

Because $f''(x) = 0$ when $x = \pm 1$ and f'' is defined on the entire real line, you should test f'' in the intervals $(-\infty, -1)$, $(-1, 1)$, and $(1, \infty)$. The results are shown in the table and in Figure 3.26.

Interval	$-\infty < x < -1$	$-1 < x < 1$	$1 < x < \infty$
Test Value	$x = -2$	$x = 0$	$x = 2$
Sign of $f''(x)$	$f''(-2) > 0$	$f''(0) < 0$	$f''(2) > 0$
Conclusion	Concave upward	Concave downward	Concave upward

The function given in Example 1 is continuous on the entire real line. If there are x-values at which the function is not continuous, these values should be used along with the points at which $f''(x) = 0$ or $f''(x)$ does not exist to form the test intervals.

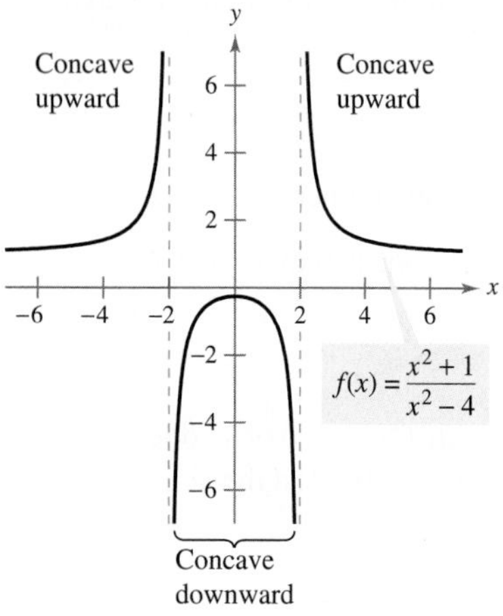

Figure 3.27

EXAMPLE 2 Determining Concavity

Determine the open intervals on which the graph of $f(x) = \dfrac{x^2+1}{x^2-4}$ is concave upward or downward.

Solution Differentiating twice produces the following.

$$f(x) = \frac{x^2+1}{x^2-4} \qquad \text{Write original function.}$$

$$f'(x) = \frac{(x^2-4)(2x) - (x^2+1)(2x)}{(x^2-4)^2} \qquad \text{Differentiate.}$$

$$= \frac{-10x}{(x^2-4)^2} \qquad \text{First derivative}$$

$$f''(x) = \frac{(x^2-4)^2(-10) - (-10x)(2)(x^2-4)(2x)}{(x^2-4)^4} \qquad \text{Differentiate.}$$

$$= \frac{10(3x^2+4)}{(x^2-4)^3} \qquad \text{Second derivative}$$

There are no points at which $f''(x) = 0$, but at $x = \pm 2$ the function f is not continuous, so test for concavity in the intervals $(-\infty, -2)$, $(-2, 2)$, and $(2, \infty)$, as shown in the table. The graph of f is shown in Figure 3.27.

Interval	$-\infty < x < -2$	$-2 < x < 2$	$2 < x < \infty$
Test Value	$x = -3$	$x = 0$	$x = 3$
Sign of $f''(x)$	$f''(-3) > 0$	$f''(0) < 0$	$f''(3) > 0$
Conclusion	Concave upward	Concave downward	Concave upward

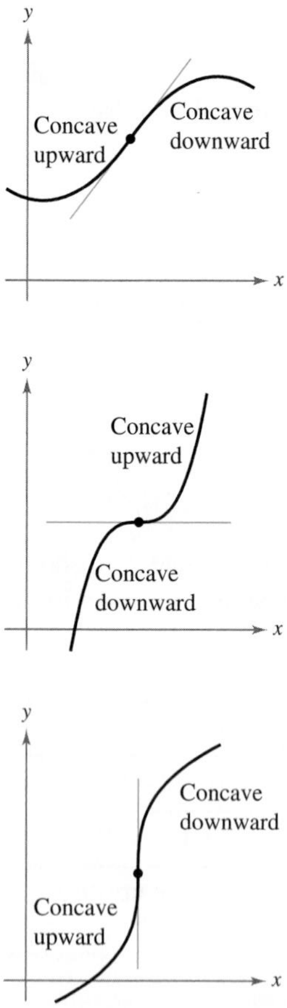

The concavity of f changes at a point of inflection. Note that a graph crosses its tangent line at a point of inflection.
Figure 3.28

Points of Inflection

The graph in Figure 3.26 has two points at which the concavity changes. If the tangent line to the graph exists at such a point, that point is a **point of inflection.** Three types of points of inflection are shown in Figure 3.28.

Definition of Point of Inflection

Let f be a function that is continuous on an open interval and let c be a point in the interval. If the graph of f has a tangent line at this point $(c, f(c))$, then this point is a **point of inflection** of the graph of f if the concavity of f changes from upward to downward (or downward to upward) at the point.

NOTE The definition of *point of inflection* given in this book requires that the tangent line exists at the point of inflection. Some books do not require this. For instance, we do not consider the function

$$f(x) = \begin{cases} x^3, & x < 0 \\ x^2 + 2x, & x \ge 0 \end{cases}$$

to have a point of inflection at the origin, even though the concavity of the graph changes from concave downward to concave upward.

To locate *possible* points of inflection, you can determine the values of x for which $f''(x) = 0$ or $f''(x)$ does not exist. This is similar to the procedure for locating relative extrema of f.

THEOREM 3.8 Points of Inflection

If $(c, f(c))$ is a point of inflection of the graph of f, then either $f''(c) = 0$ or f'' does not exist at $x = c$.

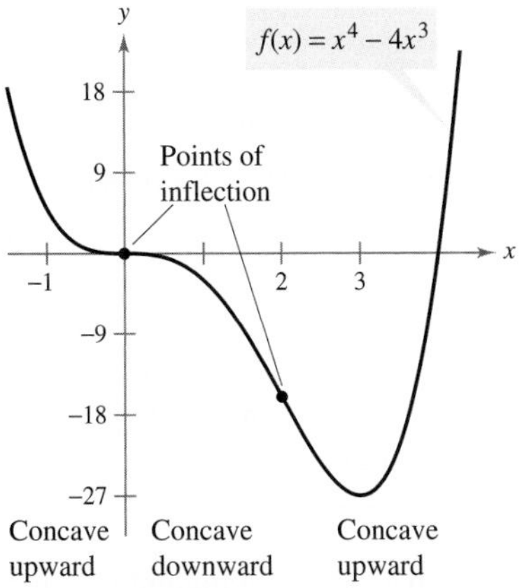

Points of inflection can occur where $f''(x) = 0$ or f'' does not exist.
Figure 3.29

EXAMPLE 3 Finding Points of Inflection

Determine the points of inflection and discuss the concavity of the graph of $f(x) = x^4 - 4x^3$.

Solution Differentiating twice produces the following.

$$f(x) = x^4 - 4x^3 \quad \text{Write original function.}$$

$$f'(x) = 4x^3 - 12x^2 \quad \text{Find first derivative.}$$

$$f''(x) = 12x^2 - 24x = 12x(x - 2) \quad \text{Find second derivative.}$$

Setting $f''(x) = 0$, you can determine that the possible points of inflection occur at $x = 0$ and $x = 2$. By testing the intervals determined by these x-values, you can conclude that they both yield points of inflection. A summary of this testing is shown in the table, and the graph of f is shown in Figure 3.29.

Interval	$-\infty < x < 0$	$0 < x < 2$	$2 < x < \infty$
Test Value	$x = -1$	$x = 1$	$x = 3$
Sign of $f''(x)$	$f''(-1) > 0$	$f''(1) < 0$	$f''(3) > 0$
Conclusion	Concave upward	Concave downward	Concave upward

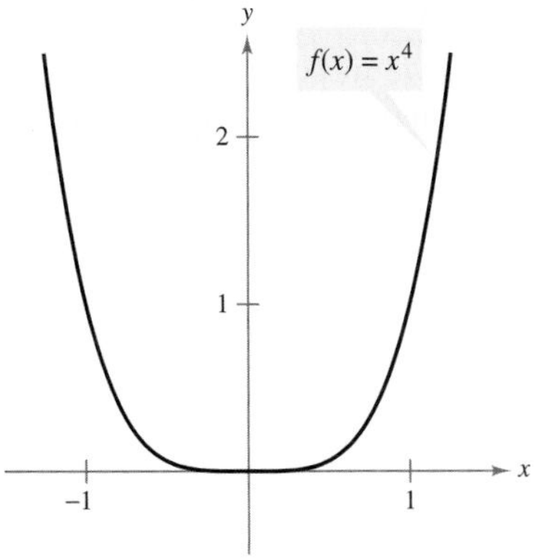

$f''(0) = 0$, but $(0, 0)$ is not a point of inflection.
Figure 3.30

The converse of Theorem 3.8 is not generally true. That is, it is possible for the second derivative to be 0 at a point that is *not* a point of inflection. For instance, the graph of $f(x) = x^4$ is shown in Figure 3.30. The second derivative is 0 when $x = 0$, but the point $(0, 0)$ is not a point of inflection because the graph of f is concave upward in both intervals $-\infty < x < 0$ and $0 < x < \infty$.

EXPLORATION

Consider a general cubic function of the form

$$f(x) = ax^3 + bx^2 + cx + d.$$

You know that the value of d has a bearing on the location of the graph but has no bearing on the value of the first derivative at given values of x. Graphically, this is true because changes in the value of d shift the graph up or down but do not change its basic shape. Use a graphing utility to graph several cubics with different values of c. Then give a graphical explanation of why changes in c do not affect the values of the second derivative.

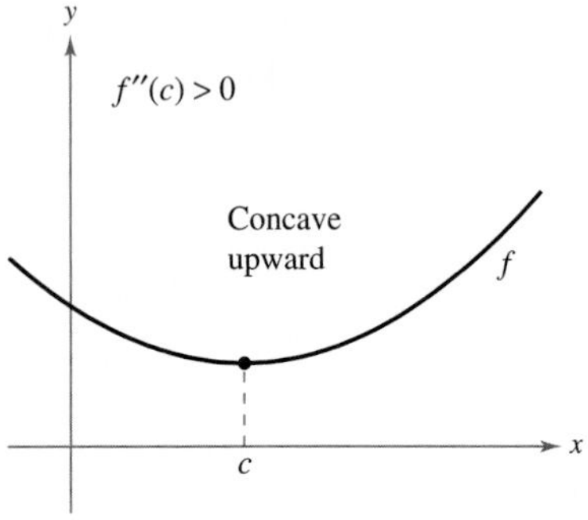

If $f'(c) = 0$ and $f''(c) > 0$, $f(c)$ is a relative minimum.

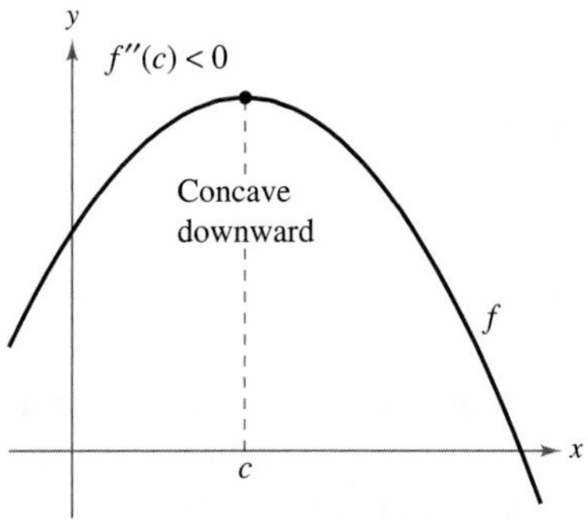

If $f'(c) = 0$ and $f''(c) < 0$, $f(c)$ is a relative maximum.
Figure 3.31

The Second Derivative Test

In addition to testing for concavity, the second derivative can be used to perform a simple test for relative maxima and minima. The test is based on the fact that if the graph of a function f is concave upward on an open interval containing c, and $f'(c) = 0$, $f(c)$ must be a relative minimum of f. Similarly, if the graph of a function f is concave downward on an open interval containing c, and $f'(c) = 0$, $f(c)$ must be a relative maximum of f (see Figure 3.31).

THEOREM 3.9 Second Derivative Test

Let f be a function such that $f'(c) = 0$ and the second derivative of f exists on an open interval containing c.

1. If $f''(c) > 0$, then f has a relative minimum at $(c, f(c))$.
2. If $f''(c) < 0$, then f has a relative maximum at $(c, f(c))$.

If $f''(c) = 0$, the test fails. That is, f may have a relative maximum, a relative minimum, or neither. In such cases, you can use the First Derivative Test.

Proof If $f'(c) = 0$ and $f''(c) > 0$, there exists an open interval I containing c for which

$$\frac{f'(x) - f'(c)}{x - c} = \frac{f'(x)}{x - c} > 0$$

for all $x \neq c$ in I. If $x < c$, then $x - c < 0$ and $f'(x) < 0$. Also, if $x > c$, then $x - c > 0$ and $f'(x) > 0$. So, $f'(x)$ changes from negative to positive at c, and the First Derivative Test implies that $f(c)$ is a relative minimum. A proof of the second case is left to you.

EXAMPLE 4 Using the Second Derivative Test

Find the relative extrema for $f(x) = -3x^5 + 5x^3$.

Solution Begin by finding the critical numbers of f.

$$f'(x) = -15x^4 + 15x^2 = 15x^2(1 - x^2) = 0 \qquad \text{Set } f'(x) \text{ equal to 0.}$$

$$x = -1, 0, 1 \qquad \text{Critical numbers}$$

Using

$$f''(x) = -60x^3 + 30x = 30(-2x^3 + x)$$

you can apply the Second Derivative Test as shown below.

Point	$(-1, -2)$	$(1, 2)$	$(0, 0)$
Sign of $f''(x)$	$f''(-1) > 0$	$f''(1) < 0$	$f''(0) = 0$
Conclusion	Relative minimum	Relative maximum	Test fails

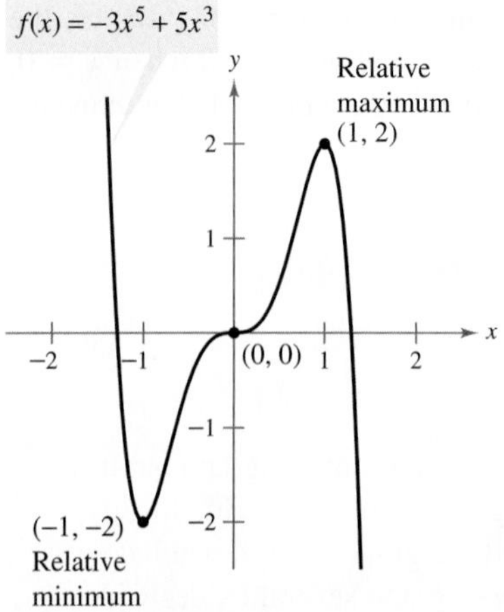

$(0, 0)$ is neither a relative minimum nor a relative maximum.
Figure 3.32

Because the Second Derivative Test fails at $(0, 0)$, you can use the First Derivative Test and observe that f increases to the left and right of $x = 0$. So, $(0, 0)$ is neither a relative minimum nor a relative maximum (even though the graph has a horizontal tangent line at this point). The graph of f is shown in Figure 3.32.

Exercises for Section 3.4

See www.CalcChat.com for worked-out solutions to odd-numbered exercises.

In Exercises 1–10, determine the open intervals on which the graph is concave upward or concave downward.

1. $y = x^2 - x - 2$

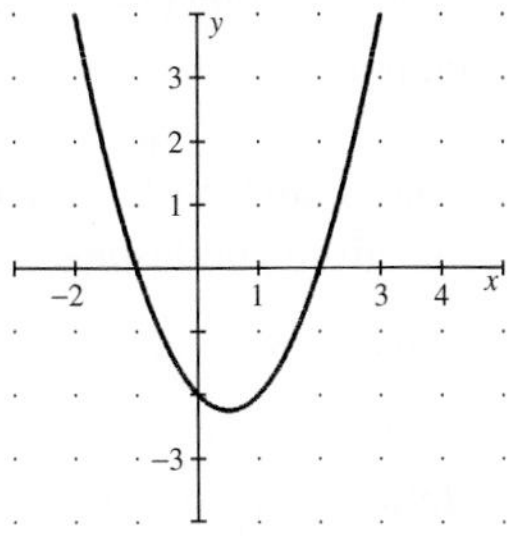

Generated by Derive

2. $y = -x^3 + 3x^2 - 2$

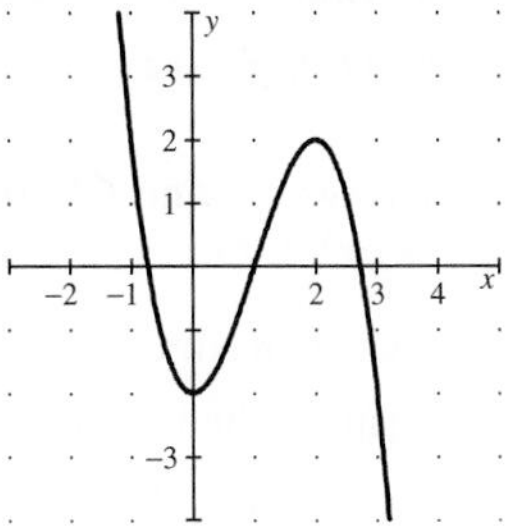

Generated by Derive

3. $f(x) = \dfrac{24}{x^2 + 12}$

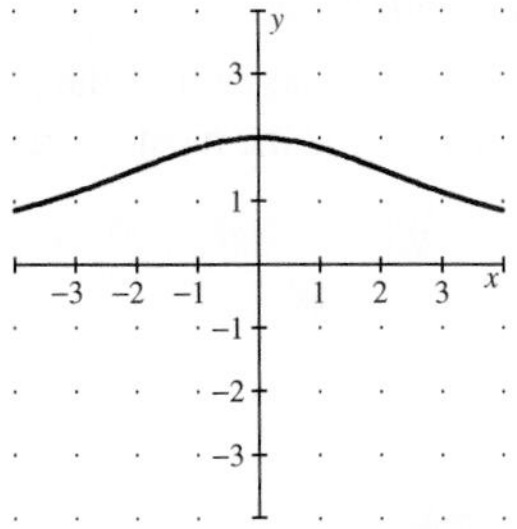

Generated by Derive

4. $f(x) = \dfrac{x^2 - 1}{2x + 1}$

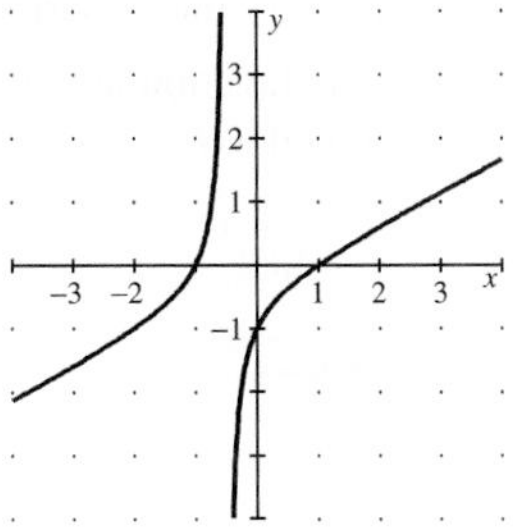

Generated by Derive

5. $f(x) = \dfrac{x^2 + 1}{x^2 - 1}$

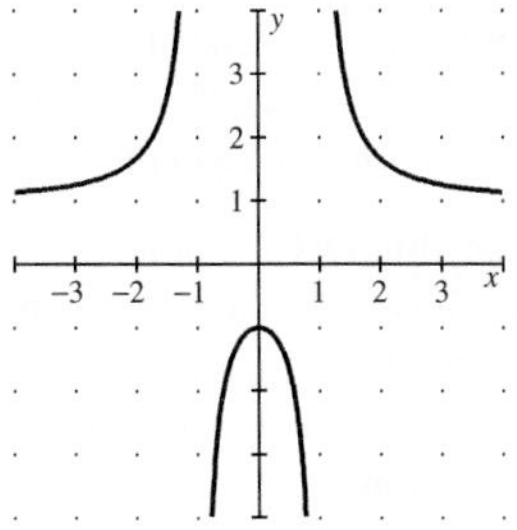

Generated by Derive

6. $y = \dfrac{-3x^5 + 40x^3 + 135x}{270}$

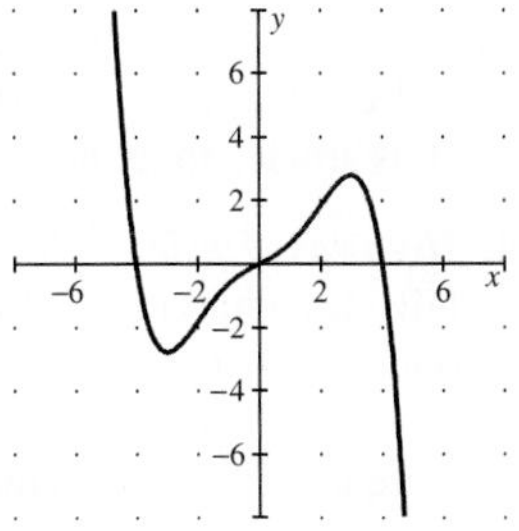

Generated by Derive

7. $g(x) = 3x^2 - x^3$
8. $h(x) = x^5 - 5x + 2$
9. $y = 2x - \tan x, \quad \left(-\dfrac{\pi}{2}, \dfrac{\pi}{2}\right)$
10. $y = x + \dfrac{2}{\sin x}, \quad (-\pi, \pi)$

In Exercises 11–26, find the points of inflection and discuss the concavity of the graph of the function.

11. $f(x) = x^3 - 6x^2 + 12x$
12. $f(x) = 2x^3 - 3x^2 - 12x + 5$
13. $f(x) = \frac{1}{4}x^4 - 2x^2$
14. $f(x) = 2x^4 - 8x + 3$
15. $f(x) = x(x - 4)^3$
16. $f(x) = x^3(x - 4)$
17. $f(x) = x\sqrt{x + 3}$
18. $f(x) = x\sqrt{x + 1}$
19. $f(x) = \dfrac{x}{x^2 + 1}$
20. $f(x) = \dfrac{x + 1}{\sqrt{x}}$
21. $f(x) = \sin\dfrac{x}{2}, \quad [0, 4\pi]$
22. $f(x) = 2\csc\dfrac{3x}{2}, \quad (0, 2\pi)$
23. $f(x) = \sec\left(x - \dfrac{\pi}{2}\right), \quad (0, 4\pi)$
24. $f(x) = \sin x + \cos x, \quad [0, 2\pi]$
25. $f(x) = 2\sin x + \sin 2x, \quad [0, 2\pi]$
26. $f(x) = x + 2\cos x, \quad [0, 2\pi]$

In Exercises 27–40, find all relative extrema. Use the Second Derivative Test where applicable.

27. $f(x) = x^4 - 4x^3 + 2$
28. $f(x) = x^2 + 3x - 8$
29. $f(x) = (x - 5)^2$
30. $f(x) = -(x - 5)^2$
31. $f(x) = x^3 - 3x^2 + 3$
32. $f(x) = x^3 - 9x^2 + 27x$
33. $g(x) = x^2(6 - x)^3$
34. $g(x) = -\frac{1}{8}(x + 2)^2(x - 4)^2$
35. $f(x) = x^{2/3} - 3$
36. $f(x) = \sqrt{x^2 + 1}$
37. $f(x) = x + \dfrac{4}{x}$
38. $f(x) = \dfrac{x}{x - 1}$
39. $f(x) = \cos x - x, \quad [0, 4\pi]$
40. $f(x) = 2\sin x + \cos 2x, \quad [0, 2\pi]$

In Exercises 41–44, use a computer algebra system to analyze the function over the given interval. (a) Find the first and second derivatives of the function. (b) Find any relative extrema and points of inflection. (c) Graph f, f', and f'' on the same set of coordinate axes and state the relationship between the behavior of f and the signs of f' and f''.

41. $f(x) = 0.2x^2(x - 3)^3, \quad [-1, 4]$
42. $f(x) = x^2\sqrt{6 - x^2}, \quad \left[-\sqrt{6}, \sqrt{6}\right]$
43. $f(x) = \sin x - \frac{1}{3}\sin 3x + \frac{1}{5}\sin 5x, \quad [0, \pi]$
44. $f(x) = \sqrt{2x}\sin x, \quad [0, 2\pi]$

Writing About Concepts

45. Consider a function f such that f' is increasing. Sketch graphs of f for (a) $f' < 0$ and (b) $f' > 0$.
46. Consider a function f such that f' is decreasing. Sketch graphs of f for (a) $f' < 0$ and (b) $f' > 0$.
47. Sketch the graph of a function f that does *not* have a point of inflection at $(c, f(c))$ even though $f''(c) = 0$.
48. S represents weekly sales of a product. What can be said of S' and S'' for each of the following statements?
    (a) The rate of change of sales is increasing.
    (b) Sales are increasing at a slower rate.
    (c) The rate of change of sales is constant.
    (d) Sales are steady.
    (e) Sales are declining, but at a slower rate.
    (f) Sales have bottomed out and have started to rise.

In Exercises 49–52, the graph of f is shown. Graph f, f', and f'' on the same set of coordinate axes. To print an enlarged copy of the graph, go to the website *www.mathgraphs.com*.

49.

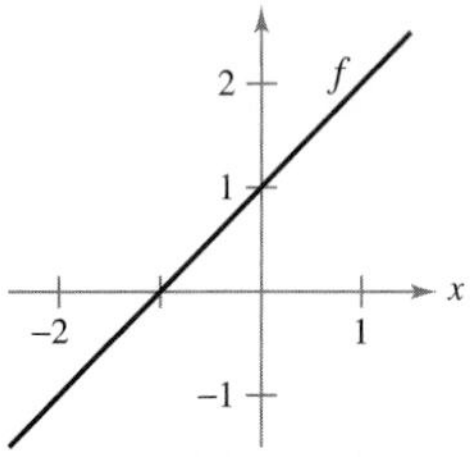

50.

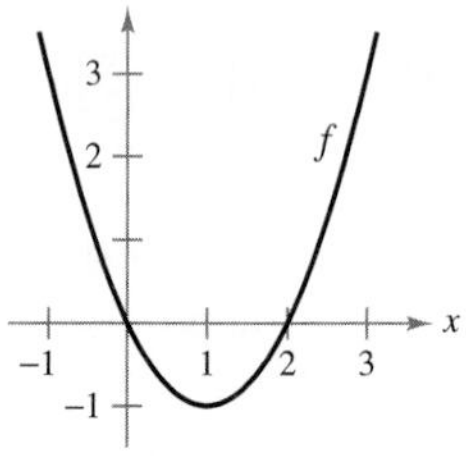

51.

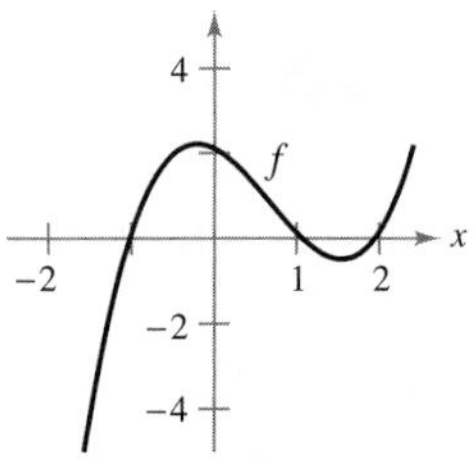

52.

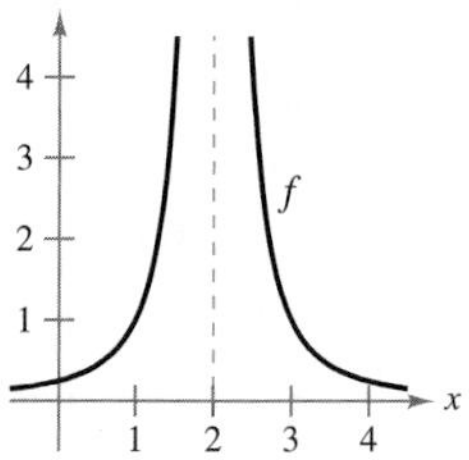

Think About It **In Exercises 53–56, sketch the graph of a function f having the given characteristics.**

53. $f(2) = f(4) = 0$
$f(3)$ is defined.
$f'(x) < 0$ if $x < 3$
$f'(3)$ does not exist.
$f'(x) > 0$ if $x > 3$
$f''(x) < 0, x \neq 3$

54. $f(0) = f(2) = 0$
$f'(x) > 0$ if $x < 1$
$f'(1) = 0$
$f'(x) < 0$ if $x > 1$
$f''(x) < 0$

55. $f(2) = f(4) = 0$
$f'(x) > 0$ if $x < 3$
$f'(3)$ does not exist.
$f'(x) < 0$ if $x > 3$
$f''(x) > 0, x \neq 3$

56. $f(0) = f(2) = 0$
$f'(x) < 0$ if $x < 1$
$f'(1) = 0$
$f'(x) > 0$ if $x > 1$
$f''(x) > 0$

57. ***Think About It*** The figure shows the graph of f''. Sketch a graph of f. (The answer is not unique.) To print an enlarged copy of the graph, go to the website *www.mathgraphs.com*.

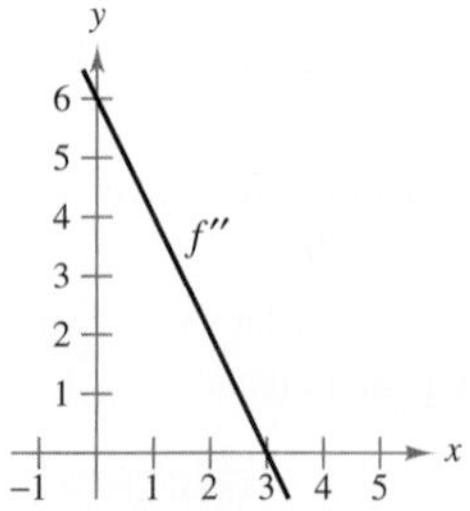

Figure for 57

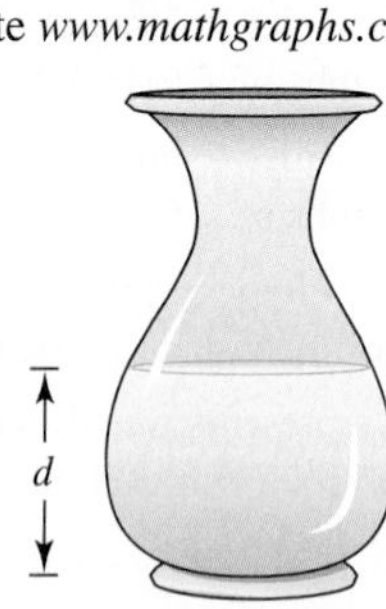

Figure for 58

58. ***Think About It*** Water is running into the vase shown in the figure at a constant rate.

(a) Graph the depth d of water in the vase as a function of time.

(b) Does the function have any extrema? Explain.

(c) Interpret the inflection points of the graph of d.

59. ***Conjecture*** Consider the function $f(x) = (x - 2)^n$.

(a) Use a graphing utility to graph f for $n = 1, 2, 3$, and 4. Use the graphs to make a conjecture about the relationship between n and any inflection points of the graph of f.

(b) Verify your conjecture in part (a).

60. (a) Graph $f(x) = \sqrt[3]{x}$ and identify the inflection point.

(b) Does $f''(x)$ exist at the inflection point? Explain.

In Exercises 61 and 62, find a, b, c, and d such that the cubic $f(x) = ax^3 + bx^2 + cx + d$ satisfies the given conditions.

61. Relative maximum: $(3, 3)$
Relative minimum: $(5, 1)$
Inflection point: $(4, 2)$

62. Relative maximum: $(2, 4)$
Relative minimum: $(4, 2)$
Inflection point: $(3, 3)$

63. ***Aircraft Glide Path*** A small aircraft starts its descent from an altitude of 1 mile, 4 miles west of the runway (see figure).

(a) Find the cubic $f(x) = ax^3 + bx^2 + cx + d$ on the interval $[-4, 0]$ that describes a smooth glide path for the landing.

(b) The function in part (a) models the glide path of the plane. When would the plane be descending at the most rapid rate?

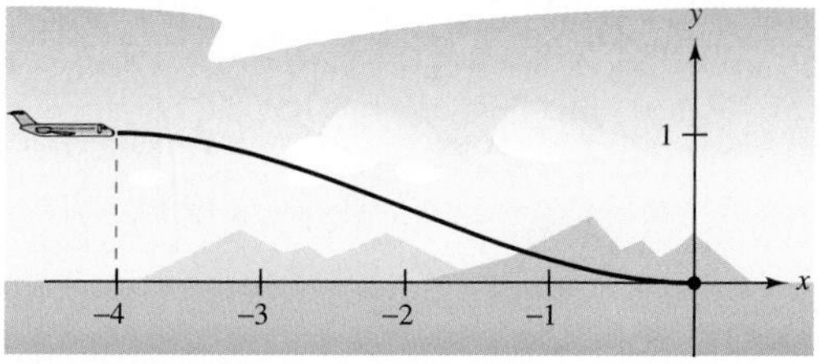

FOR FURTHER INFORMATION For more information on this type of modeling, see the article "How Not to Land at Lake Tahoe!" by Richard Barshinger in *The American Mathematical Monthly*. To view this article, go to the website *www.matharticles.com*.

64. ***Highway Design*** A section of highway connecting two hillsides with grades of 6% and 4% is to be built between two points that are separated by a horizontal distance of 2000 feet (see figure). At the point where the two hillsides come together, there is a 50-foot difference in elevation.

(a) Design a section of highway connecting the hillsides modeled by the function $f(x) = ax^3 + bx^2 + cx + d$ $(-1000 \le x \le 1000)$. At the points A and B, the slope of the model must match the grade of the hillside.

(b) Use a graphing utility to graph the model.

(c) Use a graphing utility to graph the derivative of the model.

(d) Determine the grade at the steepest part of the transitional section of the highway.

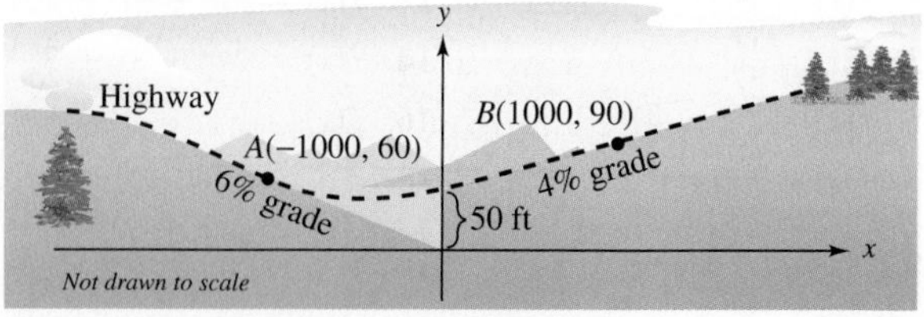

65. ***Beam Deflection*** The deflection D of a beam of length L is $D = 2x^4 - 5Lx^3 + 3L^2x^2$, where x is the distance from one end of the beam. Find the value of x that yields the maximum deflection.

66. ***Specific Gravity*** A model for the specific gravity of water S is

$$S = \frac{5.755}{10^8}T^3 - \frac{8.521}{10^6}T^2 + \frac{6.540}{10^5}T + 0.99987, \; 0 < T < 25$$

where T is the water temperature in degrees Celsius.

(a) Use a computer algebra system to find the coordinates of the maximum value of the function.

(b) Sketch a graph of the function over the specified domain. (Use a setting in which $0.996 \le S \le 1.001$.)

(c) Estimate the specific gravity of water when $T = 20°$.

67. ***Average Cost*** A manufacturer has determined that the total cost C of operating a factory is $C = 0.5x^2 + 15x + 5000$, where x is the number of units produced. At what level of production will the average cost per unit be minimized? (The average cost per unit is C/x.)

68. ***Inventory Cost*** The total cost C for ordering and storing x units is $C = 2x + 300{,}000/x$. What order size will produce a minimum cost?

69. ***Sales Growth*** The annual sales S of a new product are given by $S = \dfrac{5000t^2}{8 + t^2}$, $0 \le t \le 3$, where t is time in years.

(a) Complete the table. Then use it to estimate when the annual sales are increasing at the greatest rate.

t	0.5	1	1.5	2	2.5	3
S						

(b) Use a graphing utility to graph the function S. Then use the graph to estimate when the annual sales are increasing at the greatest rate.

(c) Find the exact time when the annual sales are increasing at the greatest rate.

70. ***Modeling Data*** The average typing speed S (words per minute) of a typing student after t weeks of lessons is shown in the table.

t	5	10	15	20	25	30
S	38	56	79	90	93	94

A model for the data is $S = \dfrac{100t^2}{65 + t^2}$, $t > 0$.

(a) Use a graphing utility to plot the data and graph the model.

(b) Use the second derivative to determine the concavity of S. Compare the result with the graph in part (a).

(c) What is the sign of the first derivative for $t > 0$? By combining this information with the concavity of the model, what inferences can be made about the typing speed as t increases?

Linear and Quadratic Approximations **In Exercises 71–74, use a graphing utility to graph the function. Then graph the linear and quadratic approximations**

$$P_1(x) = f(a) + f'(a)(x - a)$$

and

$$P_2(x) = f(a) + f'(a)(x - a) + \tfrac{1}{2}f''(a)(x - a)^2$$

in the same viewing window. Compare the values of f, P_1, and P_2 and their first derivatives at $x = a$. How do the approximations change as you move farther away from $x = a$?

Function	*Value of a*
71. $f(x) = 2(\sin x + \cos x)$	$a = \dfrac{\pi}{4}$
72. $f(x) = 2(\sin x + \cos x)$	$a = 0$
73. $f(x) = \sqrt{1 - x}$	$a = 0$
74. $f(x) = \dfrac{\sqrt{x}}{x - 1}$	$a = 2$

75. Use a graphing utility to graph $y = x\sin(1/x)$. Show that the graph is concave downward to the right of $x = 1/\pi$.

76. Show that the point of inflection of $f(x) = x(x - 6)^2$ lies midway between the relative extrema of f.

77. Prove that every cubic function with three distinct real zeros has a point of inflection whose x-coordinate is the average of the three zeros.

78. Show that the cubic polynomial $p(x) = ax^3 + bx^2 + cx + d$ has exactly one point of inflection (x_0, y_0), where

$$x_0 = \frac{-b}{3a} \quad \text{and} \quad y_0 = \frac{2b^3}{27a^2} - \frac{bc}{3a} + d.$$

Use this formula to find the point of inflection of $p(x) = x^3 - 3x^2 + 2$.

True or False? **In Exercises 79–82, determine whether the statement is true or false. If it is false, explain why or give an example that shows it is false.**

79. The graph of every cubic polynomial has precisely one point of inflection.

80. The graph of $f(x) = 1/x$ is concave downward for $x < 0$ and concave upward for $x > 0$, and thus it has a point of inflection at $x = 0$.

81. If $f'(c) > 0$, then f is concave upward at $x = c$.

82. If $f''(2) = 0$, then the graph of f must have a point of inflection at $x = 2$.

In Exercises 83 and 84, let f and g represent differentiable functions such that $f'' \neq 0$ and $g'' \neq 0$.

83. Show that if f and g are concave upward on the interval (a, b), then $f + g$ is also concave upward on (a, b).

84. Prove that if f and g are positive, increasing, and concave upward on the interval (a, b), then fg is also concave upward on (a, b).

Section 3.5 Limits at Infinity

- **Determine (finite) limits at infinity.**
- **Determine the horizontal asymptotes, if any, of the graph of a function.**
- **Determine infinite limits at infinity.**

Limits at Infinity

This section discusses the "end behavior" of a function on an *infinite* interval. Consider the graph of

$$f(x) = \frac{3x^2}{x^2 + 1}$$

as shown in Figure 3.33. Graphically, you can see that the values of $f(x)$ appear to approach 3 as x increases without bound or decreases without bound. You can come to the same conclusions numerically, as shown in the table.

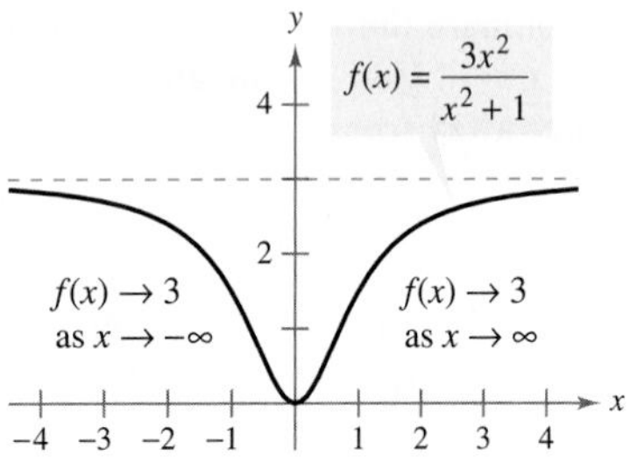

The limit of $f(x)$ as x approaches $-\infty$ or ∞ is 3.
Figure 3.33

← x decreases without bound. | x increases without bound. →

x	$-\infty \leftarrow$	-100	-10	-1	0	1	10	100	$\rightarrow \infty$
$f(x)$	$3 \leftarrow$	2.9997	2.97	1.5	0	1.5	2.97	2.9997	$\rightarrow 3$

← $f(x)$ approaches 3. | $f(x)$ approaches 3. →

The table suggests that the value of $f(x)$ approaches 3 as x increases without bound ($x \to \infty$). Similarly, $f(x)$ approaches 3 as x decreases without bound ($x \to -\infty$). These **limits at infinity** are denoted by

$$\lim_{x \to -\infty} f(x) = 3 \qquad \text{Limit at negative infinity}$$

and

$$\lim_{x \to \infty} f(x) = 3. \qquad \text{Limit at positive infinity}$$

NOTE The statement $\lim_{x \to -\infty} f(x) = L$ or $\lim_{x \to \infty} f(x) = L$, means that the limit exists *and* the limit is equal to L.

To say that a statement is true as x increases *without bound* means that for some (large) real number M, the statement is true for *all* x in the interval $\{x: x > M\}$. The following definition uses this concept.

Definition of Limits at Infinity

Let L be a real number.

1. The statement $\lim_{x \to \infty} f(x) = L$ means that for each $\varepsilon > 0$ there exists an $M > 0$ such that $|f(x) - L| < \varepsilon$ whenever $x > M$.
2. The statement $\lim_{x \to -\infty} f(x) = L$ means that for each $\varepsilon > 0$ there exists an $N < 0$ such that $|f(x) - L| < \varepsilon$ whenever $x < N$.

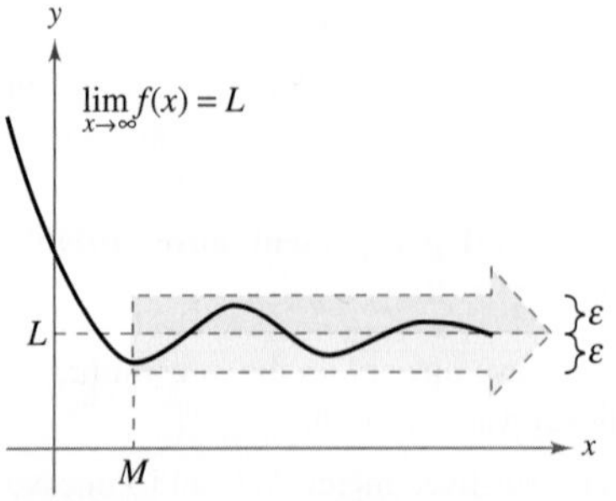

$f(x)$ is within ε units of L as $x \to \infty$.
Figure 3.34

The definition of a limit at infinity is shown in Figure 3.34. In this figure, note that for a given positive number ε there exists a positive number M such that, for $x > M$, the graph of f will lie between the horizontal lines given by $y = L + \varepsilon$ and $y = L - \varepsilon$.

EXPLORATION

Use a graphing utility to graph

$$f(x) = \frac{2x^2 + 4x - 6}{3x^2 + 2x - 16}.$$

Describe all the important features of the graph. Can you find a single viewing window that shows all of these features clearly? Explain your reasoning.

What are the horizontal asymptotes of the graph? How far to the right do you have to move on the graph so that the graph is within 0.001 unit of its horizontal asymptote? Explain your reasoning.

Horizontal Asymptotes

In Figure 3.34, the graph of f approaches the line $y = L$ as x increases without bound. The line $y = L$ is called a **horizontal asymptote** of the graph of f.

Definition of a Horizontal Asymptote

The line $y = L$ is a **horizontal asymptote** of the graph of f if

$$\lim_{x\to-\infty} f(x) = L \quad \text{or} \quad \lim_{x\to\infty} f(x) = L.$$

Note that from this definition, it follows that the graph of a *function* of x can have at most two horizontal asymptotes—one to the right and one to the left.

Limits at infinity have many of the same properties of limits discussed in Section 1.3. For example, if $\lim_{x\to\infty} f(x)$ and $\lim_{x\to\infty} g(x)$ both exist, then

$$\lim_{x\to\infty} [f(x) + g(x)] = \lim_{x\to\infty} f(x) + \lim_{x\to\infty} g(x)$$

and

$$\lim_{x\to\infty} [f(x)g(x)] = \left[\lim_{x\to\infty} f(x)\right]\left[\lim_{x\to\infty} g(x)\right].$$

Similar properties hold for limits at $-\infty$.

When evaluating limits at infinity, the following theorem is helpful. (A proof of this theorem is given in Appendix A.)

THEOREM 3.10 Limits at Infinity

If r is a positive rational number and c is any real number, then

$$\lim_{x\to\infty} \frac{c}{x^r} = 0.$$

Furthermore, if x^r is defined when $x < 0$, then

$$\lim_{x\to-\infty} \frac{c}{x^r} = 0.$$

EXAMPLE 1 Finding a Limit at Infinity

Find the limit: $\lim_{x\to\infty} \left(5 - \frac{2}{x^2}\right)$.

Solution Using Theorem 3.10, you can write

$$\begin{aligned} \lim_{x\to\infty} \left(5 - \frac{2}{x^2}\right) &= \lim_{x\to\infty} 5 - \lim_{x\to\infty} \frac{2}{x^2} && \text{Property of limits} \\ &= 5 - 0 \\ &= 5. \end{aligned}$$

EXAMPLE 2 Finding a Limit at Infinity

Find the limit: $\lim_{x\to\infty} \frac{2x-1}{x+1}$.

Solution Note that both the numerator and the denominator approach infinity as x approaches infinity.

$$\lim_{x\to\infty} \frac{2x-1}{x+1} \quad \begin{cases} \lim_{x\to\infty} (2x-1) \to \infty \\ \lim_{x\to\infty} (x+1) \to \infty \end{cases}$$

NOTE When you encounter an indeterminate form such as the one in Example 2, you should divide the numerator and denominator by the highest power of x in the *denominator*.

This results in $\frac{\infty}{\infty}$, an **indeterminate form.** To resolve this problem, you can divide both the numerator and the denominator by x. After dividing, the limit may be evaluated as shown.

$$\lim_{x\to\infty} \frac{2x-1}{x+1} = \lim_{x\to\infty} \frac{\frac{2x-1}{x}}{\frac{x+1}{x}} \qquad \text{Divide numerator and denominator by } x.$$

$$= \lim_{x\to\infty} \frac{2-\frac{1}{x}}{1+\frac{1}{x}} \qquad \text{Simplify.}$$

$$= \frac{\lim_{x\to\infty} 2 - \lim_{x\to\infty} \frac{1}{x}}{\lim_{x\to\infty} 1 + \lim_{x\to\infty} \frac{1}{x}} \qquad \text{Take limits of numerator and denominator.}$$

$$= \frac{2-0}{1+0} \qquad \text{Apply Theorem 3.10.}$$

$$= 2$$

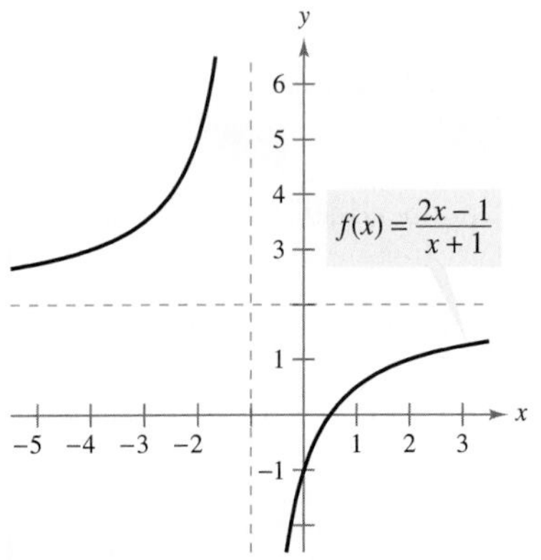

$y = 2$ is a horizontal asymptote.
Figure 3.35

So, the line $y = 2$ is a horizontal asymptote to the right. By taking the limit as $x \to -\infty$, you can see that $y = 2$ is also a horizontal asymptote to the left. The graph of the function is shown in Figure 3.35.

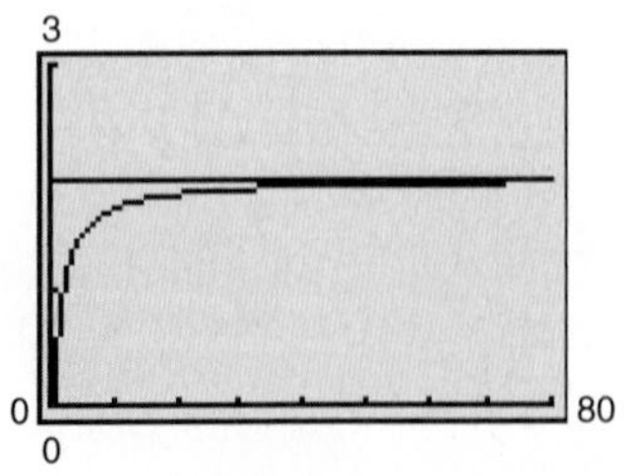

As x increases, the graph of f moves closer and closer to the line $y = 2$.
Figure 3.36

TECHNOLOGY You can test the reasonableness of the limit found in Example 2 by evaluating $f(x)$ for a few large positive values of x. For instance,

$$f(100) \approx 1.9703, \quad f(1000) \approx 1.9970, \quad \text{and} \quad f(10{,}000) \approx 1.9997.$$

Another way to test the reasonableness of the limit is to use a graphing utility. For instance, in Figure 3.36, the graph of

$$f(x) = \frac{2x-1}{x+1}$$

is shown with the horizontal line $y = 2$. Note that as x increases, the graph of f moves closer and closer to its horizontal asymptote.

EXAMPLE 3 A Comparison of Three Rational Functions

Maria Agnesi (1718–1799)

Agnesi was one of a handful of women to receive credit for significant contributions to mathematics before the twentieth century. In her early twenties, she wrote the first text that included both differential and integral calculus. By age 30, she was an honorary member of the faculty at the University of Bologna.

Find each limit.

a. $\displaystyle\lim_{x\to\infty}\frac{2x+5}{3x^2+1}$ **b.** $\displaystyle\lim_{x\to\infty}\frac{2x^2+5}{3x^2+1}$ **c.** $\displaystyle\lim_{x\to\infty}\frac{2x^3+5}{3x^2+1}$

Solution In each case, attempting to evaluate the limit produces the indeterminate form ∞/∞.

a. Divide both the numerator and the denominator by x^2.

$$\lim_{x\to\infty}\frac{2x+5}{3x^2+1}=\lim_{x\to\infty}\frac{(2/x)+(5/x^2)}{3+(1/x^2)}=\frac{0+0}{3+0}=\frac{0}{3}=0$$

b. Divide both the numerator and the denominator by x^2.

$$\lim_{x\to\infty}\frac{2x^2+5}{3x^2+1}=\lim_{x\to\infty}\frac{2+(5/x^2)}{3+(1/x^2)}=\frac{2+0}{3+0}=\frac{2}{3}$$

c. Divide both the numerator and the denominator by x^2.

$$\lim_{x\to\infty}\frac{2x^3+5}{3x^2+1}=\lim_{x\to\infty}\frac{2x+(5/x^2)}{3+(1/x^2)}=\frac{\infty}{3}$$

You can conclude that the limit *does not exist* because the numerator increases without bound while the denominator approaches 3.

Guidelines for Finding Limits at $\pm\infty$ of Rational Functions

1. If the degree of the numerator is *less than* the degree of the denominator, then the limit of the rational function is 0.
2. If the degree of the numerator is *equal to* the degree of the denominator, then the limit of the rational function is the ratio of the leading coefficients.
3. If the degree of the numerator is *greater than* the degree of the denominator, then the limit of the rational function does not exist.

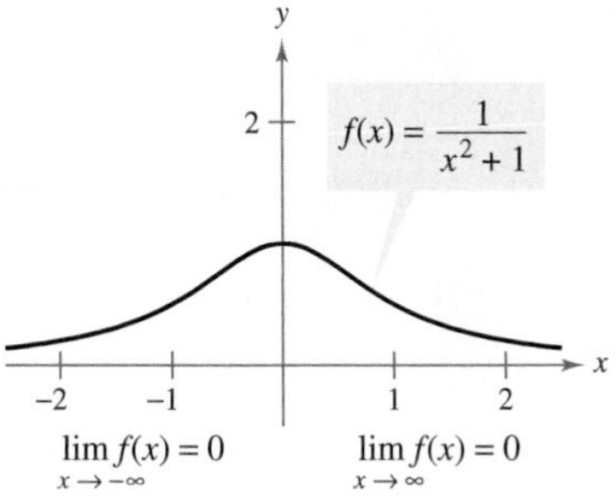

f has a horizontal asymptote at $y=0$.
Figure 3.37

FOR FURTHER INFORMATION For more information on the contributions of women to mathematics, see the article "Why Women Succeed in Mathematics" by Mona Fabricant, Sylvia Svitak, and Patricia Clark Kenschaft in *Mathematics Teacher*. To view this article, go to the website *www.matharticles.com*.

Use these guidelines to check the results in Example 3. These limits seem reasonable when you consider that for large values of x, the highest-power term of the rational function is the most "influential" in determining the limit. For instance, the limit as x approaches infinity of the function

$$f(x)=\frac{1}{x^2+1}$$

is 0 because the denominator overpowers the numerator as x increases or decreases without bound, as shown in Figure 3.37.

The function shown in Figure 3.37 is a special case of a type of curve studied by the Italian mathematician Maria Gaetana Agnesi. The general form of this function is

$$f(x)=\frac{8a^3}{x^2+4a^2} \qquad \text{Witch of Agnesi}$$

and, through a mistranslation of the Italian word *vertéré*, the curve has come to be known as the Witch of Agnesi. Agnesi's work with this curve first appeared in a comprehensive text on calculus that was published in 1748.

In Figure 3.37, you can see that the function $f(x) = 1/(x^2 + 1)$ approaches the same horizontal asymptote to the right and to the left. This is always true of rational functions. Functions that are not rational, however, may approach different horizontal asymptotes to the right and to the left. This is demonstrated in Example 4.

EXAMPLE 4 A Function with Two Horizontal Asymptotes

Find each limit.

a. $\displaystyle\lim_{x\to\infty} \frac{3x - 2}{\sqrt{2x^2 + 1}}$ **b.** $\displaystyle\lim_{x\to-\infty} \frac{3x - 2}{\sqrt{2x^2 + 1}}$

Solution

a. For $x > 0$, you can write $x = \sqrt{x^2}$. So, dividing both the numerator and the denominator by x produces

$$\frac{3x - 2}{\sqrt{2x^2 + 1}} = \frac{\frac{3x - 2}{x}}{\frac{\sqrt{2x^2 + 1}}{\sqrt{x^2}}} = \frac{3 - \frac{2}{x}}{\sqrt{\frac{2x^2 + 1}{x^2}}} = \frac{3 - \frac{2}{x}}{\sqrt{2 + \frac{1}{x^2}}}$$

and you can take the limit as follows.

$$\lim_{x\to\infty} \frac{3x - 2}{\sqrt{2x^2 + 1}} = \lim_{x\to\infty} \frac{3 - \frac{2}{x}}{\sqrt{2 + \frac{1}{x^2}}} = \frac{3 - 0}{\sqrt{2 + 0}} = \frac{3}{\sqrt{2}}$$

b. For $x < 0$, you can write $x = -\sqrt{x^2}$. So, dividing both the numerator and the denominator by x produces

$$\frac{3x - 2}{\sqrt{2x^2 + 1}} = \frac{\frac{3x - 2}{x}}{\frac{\sqrt{2x^2 + 1}}{-\sqrt{x^2}}} = \frac{3 - \frac{2}{x}}{-\sqrt{\frac{2x^2 + 1}{x^2}}} = \frac{3 - \frac{2}{x}}{-\sqrt{2 + \frac{1}{x^2}}}$$

and you can take the limit as follows.

$$\lim_{x\to-\infty} \frac{3x - 2}{\sqrt{2x^2 + 1}} = \lim_{x\to-\infty} \frac{3 - \frac{2}{x}}{-\sqrt{2 + \frac{1}{x^2}}} = \frac{3 - 0}{-\sqrt{2 + 0}} = -\frac{3}{\sqrt{2}}$$

The graph of $f(x) = (3x - 2)/\sqrt{2x^2 + 1}$ is shown in Figure 3.38.

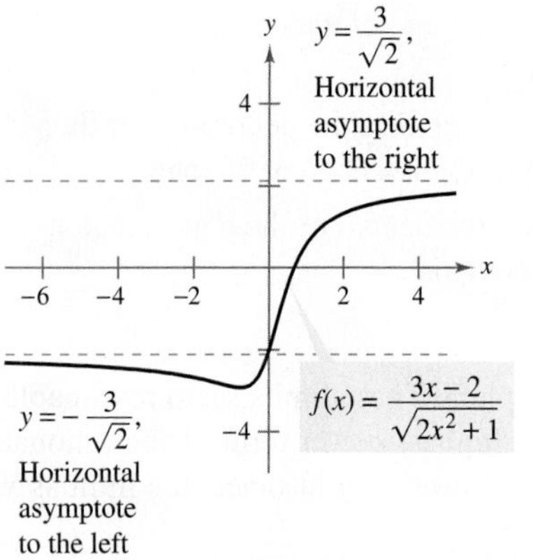

Functions that are not rational may have different right and left horizontal asymptotes.
Figure 3.38

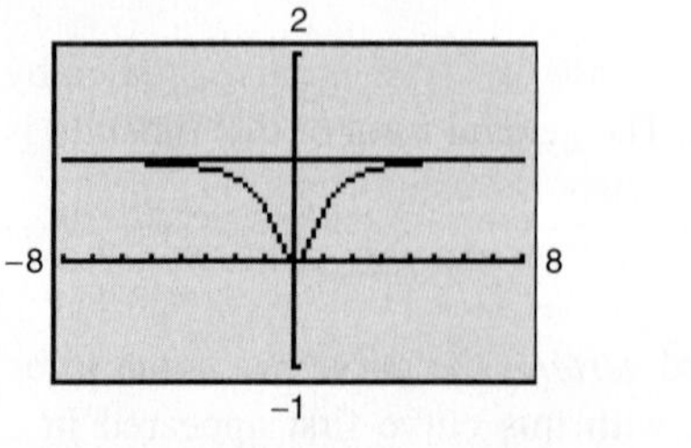

The horizontal asymptote appears to be the line $y = 1$ but it is actually the line $y = 2$.
Figure 3.39

TECHNOLOGY PITFALL If you use a graphing utility to help estimate a limit, be sure that you also confirm the estimate analytically—the pictures shown by a graphing utility can be misleading. For instance, Figure 3.39 shows one view of the graph of

$$y = \frac{2x^3 + 1000x^2 + x}{x^3 + 1000x^2 + x + 1000}.$$

From this view, one could be convinced that the graph has $y = 1$ as a horizontal asymptote. An analytical approach shows that the horizontal asymptote is actually $y = 2$. Confirm this by enlarging the viewing window on the graphing utility.

In Section 1.3 (Example 9), you saw how the Squeeze Theorem can be used to evaluate limits involving trigonometric functions. This theorem is also valid for limits at infinity.

EXAMPLE 5 Limits Involving Trigonometric Functions

Find each limit.

a. $\lim_{x\to\infty} \sin x$ **b.** $\lim_{x\to\infty} \dfrac{\sin x}{x}$

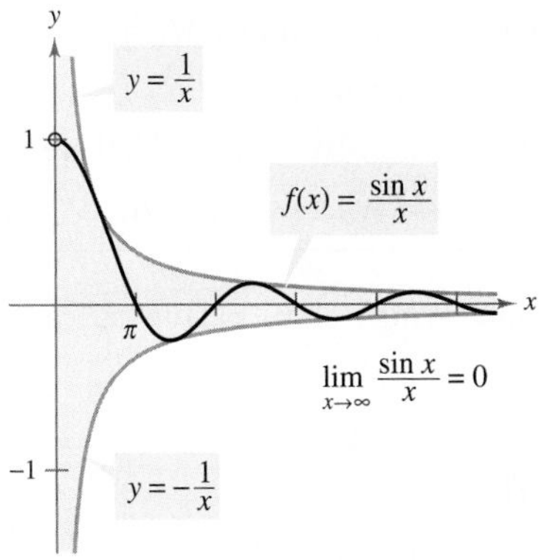

As x increases without bound, $f(x)$ approaches 0.
Figure 3.40

Solution

a. As x approaches infinity, the sine function oscillates between 1 and -1. So, this limit does not exist.

b. Because $-1 \le \sin x \le 1$, it follows that for $x > 0$,

$$-\frac{1}{x} \le \frac{\sin x}{x} \le \frac{1}{x}$$

where $\lim_{x\to\infty} (-1/x) = 0$ and $\lim_{x\to\infty} (1/x) = 0$. So, by the Squeeze Theorem, you can obtain

$$\lim_{x\to\infty} \frac{\sin x}{x} = 0$$

as shown in Figure 3.40.

EXAMPLE 6 Oxygen Level in a Pond

Suppose that $f(t)$ measures the level of oxygen in a pond, where $f(t) = 1$ is the normal (unpolluted) level and the time t is measured in weeks. When $t = 0$, organic waste is dumped into the pond, and as the waste material oxidizes, the level of oxygen in the pond is

$$f(t) = \frac{t^2 - t + 1}{t^2 + 1}.$$

What percent of the normal level of oxygen exists in the pond after 1 week? After 2 weeks? After 10 weeks? What is the limit as t approaches infinity?

Solution When $t = 1$, 2, and 10, the levels of oxygen are as shown.

$$f(1) = \frac{1^2 - 1 + 1}{1^2 + 1} = \frac{1}{2} = 50\% \qquad \text{1 week}$$

$$f(2) = \frac{2^2 - 2 + 1}{2^2 + 1} = \frac{3}{5} = 60\% \qquad \text{2 weeks}$$

$$f(10) = \frac{10^2 - 10 + 1}{10^2 + 1} = \frac{91}{101} \approx 90.1\% \qquad \text{10 weeks}$$

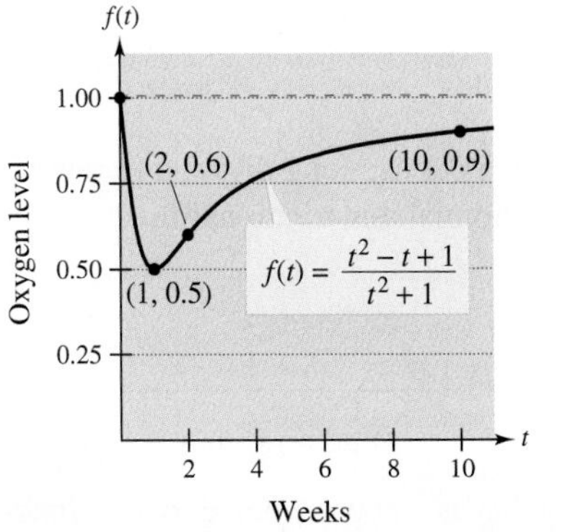

The level of oxygen in a pond approaches the normal level of 1 as t approaches ∞.
Figure 3.41

To find the limit as t approaches infinity, divide the numerator and the denominator by t^2 to obtain

$$\lim_{t\to\infty} \frac{t^2 - t + 1}{t^2 + 1} = \lim_{t\to\infty} \frac{1 - (1/t) + (1/t^2)}{1 + (1/t^2)} = \frac{1 - 0 + 0}{1 + 0} = 1 = 100\%.$$

See Figure 3.41.

Infinite Limits at Infinity

Many functions do not approach a finite limit as x increases (or decreases) without bound. For instance, no polynomial function has a finite limit at infinity. The following definition is used to describe the behavior of polynomial and other functions at infinity.

NOTE Determining whether a function has an infinite limit at infinity is useful in analyzing the "end behavior" of its graph. You will see examples of this in Section 3.6 on curve sketching.

Definition of Infinite Limits at Infinity

Let f be a function defined on the interval (a, ∞).

1. The statement $\lim_{x\to\infty} f(x) = \infty$ means that for each positive number M, there is a corresponding number $N > 0$ such that $f(x) > M$ whenever $x > N$.
2. The statement $\lim_{x\to\infty} f(x) = -\infty$ means that for each negative number M, there is a corresponding number $N > 0$ such that $f(x) < M$ whenever $x > N$.

Similar definitions can be given for the statements $\lim_{x\to-\infty} f(x) = \infty$ and $\lim_{x\to-\infty} f(x) = -\infty$.

EXAMPLE 7 Finding Infinite Limits at Infinity

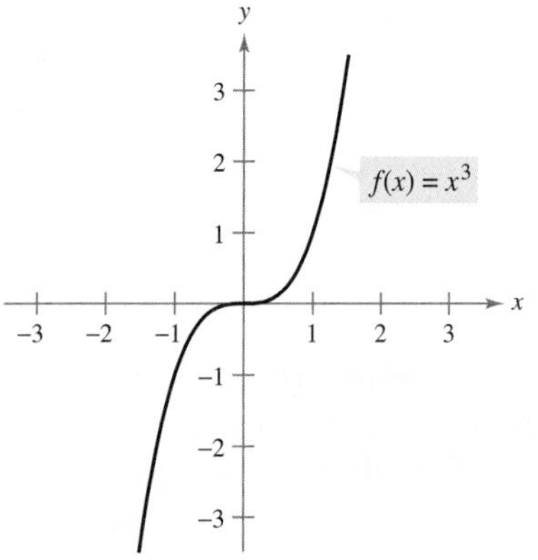

Figure 3.42

Find each limit.

a. $\lim_{x\to\infty} x^3$ **b.** $\lim_{x\to-\infty} x^3$

Solution

a. As x increases without bound, x^3 also increases without bound. So, you can write $\lim_{x\to\infty} x^3 = \infty$.

b. As x decreases without bound, x^3 also decreases without bound. So, you can write $\lim_{x\to-\infty} x^3 = -\infty$.

The graph of $f(x) = x^3$ in Figure 3.42 illustrates these two results. These results agree with the Leading Coefficient Test for polynomial functions as described in Section P.3.

EXAMPLE 8 Finding Infinite Limits at Infinity

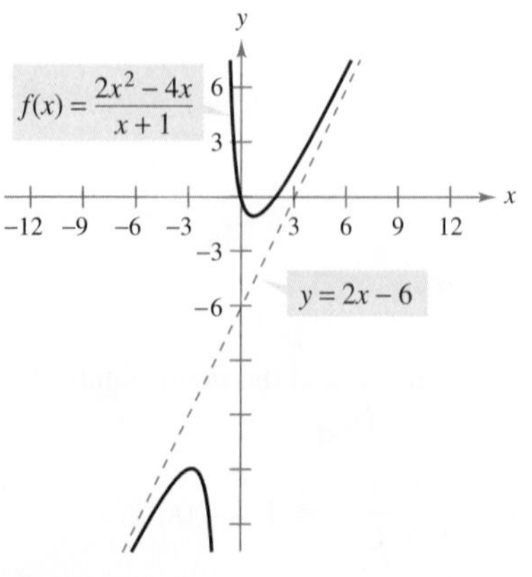

Figure 3.43

Find each limit.

a. $\lim_{x\to\infty} \dfrac{2x^2 - 4x}{x + 1}$ **b.** $\lim_{x\to-\infty} \dfrac{2x^2 - 4x}{x + 1}$

Solution One way to evaluate each of these limits is to use long division to rewrite the improper rational function as the sum of a polynomial and a rational function.

a. $$\lim_{x\to\infty} \frac{2x^2 - 4x}{x + 1} = \lim_{x\to\infty}\left(2x - 6 + \frac{6}{x + 1}\right) = \infty$$

b. $$\lim_{x\to-\infty} \frac{2x^2 - 4x}{x + 1} = \lim_{x\to-\infty}\left(2x - 6 + \frac{6}{x + 1}\right) = -\infty$$

The statements above can be interpreted as saying that as x approaches $\pm\infty$, the function $f(x) = (2x^2 - 4x)/(x + 1)$ behaves like the function $g(x) = 2x - 6$. In Section 3.6, you will see that this is graphically described by saying that the line $y = 2x - 6$ is a slant asymptote of the graph of f, as shown in Figure 3.43.

Exercises for Section 3.5

See www.CalcChat.com for worked-out solutions to odd-numbered exercises.

In Exercises 1 and 2, describe in your own words what the statement means.

1. $\lim_{x\to\infty} f(x) = 4$

2. $\lim_{x\to-\infty} f(x) = 2$

In Exercises 3–8, match the function with one of the graphs [(a), (b), (c), (d), (e), or (f)] using horizontal asymptotes as an aid.

(a)

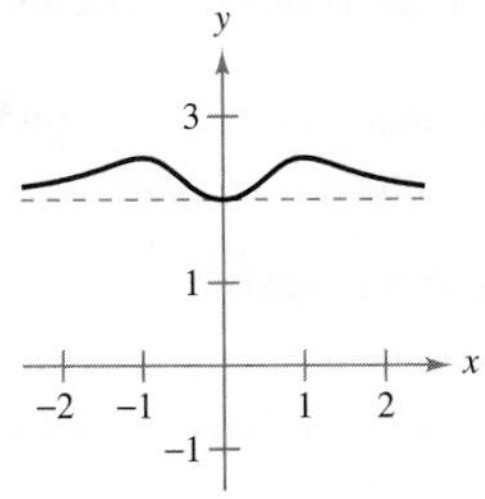

(b)

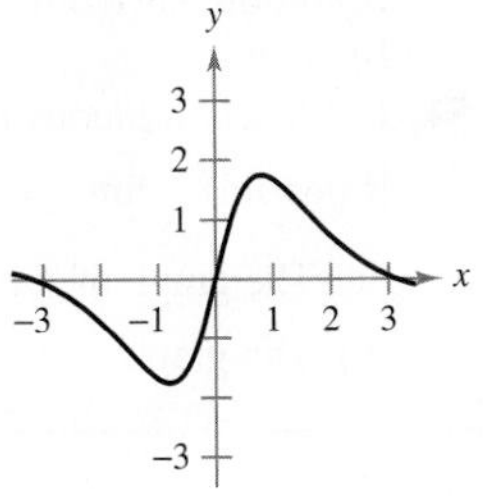

(c)

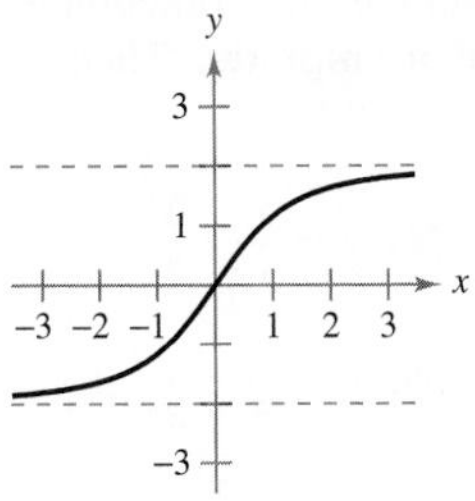

(d)

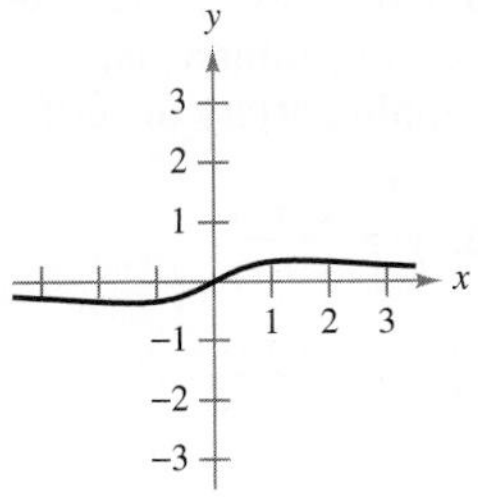

(e)

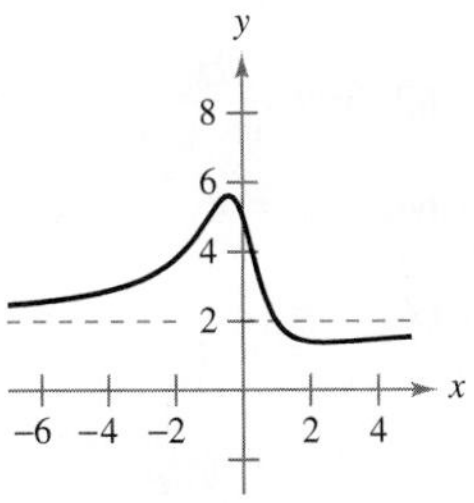

(f)

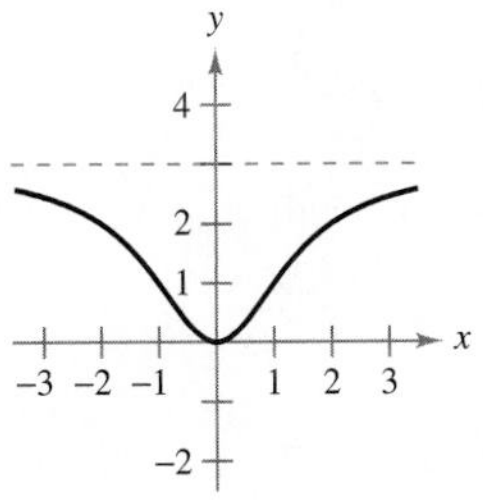

3. $f(x) = \dfrac{3x^2}{x^2+2}$

4. $f(x) = \dfrac{2x}{\sqrt{x^2+2}}$

5. $f(x) = \dfrac{x}{x^2+2}$

6. $f(x) = 2 + \dfrac{x^2}{x^4+1}$

7. $f(x) = \dfrac{4\sin x}{x^2+1}$

8. $f(x) = \dfrac{2x^2-3x+5}{x^2+1}$

Numerical and Graphical Analysis **In Exercises 9–14, use a graphing utility to complete the table and estimate the limit as x approaches infinity. Then use a graphing utility to graph the function and estimate the limit graphically.**

x	10^0	10^1	10^2	10^3	10^4	10^5	10^6
$f(x)$							

9. $f(x) = \dfrac{4x+3}{2x-1}$

10. $f(x) = \dfrac{2x^2}{x+1}$

11. $f(x) = \dfrac{-6x}{\sqrt{4x^2+5}}$

12. $f(x) = \dfrac{8x}{\sqrt{x^2-3}}$

13. $f(x) = 5 - \dfrac{1}{x^2+1}$

14. $f(x) = 4 + \dfrac{3}{x^2+2}$

In Exercises 15 and 16, find $\lim_{x\to\infty} h(x)$, if possible.

15. $f(x) = 5x^3 - 3x^2 + 10$

(a) $h(x) = \dfrac{f(x)}{x^2}$ (b) $h(x) = \dfrac{f(x)}{x^3}$

(c) $h(x) = \dfrac{f(x)}{x^4}$

16. $f(x) = 5x^2 - 3x + 7$

(a) $h(x) = \dfrac{f(x)}{x}$ (b) $h(x) = \dfrac{f(x)}{x^2}$

(c) $h(x) = \dfrac{f(x)}{x^3}$

In Exercises 17–20, find each limit, if possible.

17. (a) $\lim_{x\to\infty} \dfrac{x^2+2}{x^3-1}$

(b) $\lim_{x\to\infty} \dfrac{x^2+2}{x^2-1}$

(c) $\lim_{x\to\infty} \dfrac{x^2+2}{x-1}$

18. (a) $\lim_{x\to\infty} \dfrac{3-2x}{3x^3-1}$

(b) $\lim_{x\to\infty} \dfrac{3-2x}{3x-1}$

(c) $\lim_{x\to\infty} \dfrac{3-2x^2}{3x-1}$

19. (a) $\lim_{x\to\infty} \dfrac{5-2x^{3/2}}{3x^2-4}$

(b) $\lim_{x\to\infty} \dfrac{5-2x^{3/2}}{3x^{3/2}-4}$

(c) $\lim_{x\to\infty} \dfrac{5-2x^{3/2}}{3x-4}$

20. (a) $\lim_{x\to\infty} \dfrac{5x^{3/2}}{4x^2+1}$

(b) $\lim_{x\to\infty} \dfrac{5x^{3/2}}{4x^{3/2}+1}$

(c) $\lim_{x\to\infty} \dfrac{5x^{3/2}}{4\sqrt{x}+1}$

In Exercises 21–34, find the limit.

21. $\lim_{x\to\infty} \dfrac{2x-1}{3x+2}$

22. $\lim_{x\to\infty} \dfrac{3x^3+2}{9x^3-2x^2+7}$

23. $\lim_{x\to\infty} \dfrac{x}{x^2-1}$

24. $\lim_{x\to\infty} \left(4 + \dfrac{3}{x}\right)$

25. $\lim_{x\to-\infty} \dfrac{5x^2}{x+3}$

26. $\lim_{x\to-\infty} \left(\dfrac{1}{2}x - \dfrac{4}{x^2}\right)$

27. $\lim_{x\to-\infty} \dfrac{x}{\sqrt{x^2-x}}$

28. $\lim_{x\to-\infty} \dfrac{x}{\sqrt{x^2+1}}$

29. $\lim_{x\to-\infty} \dfrac{2x+1}{\sqrt{x^2-x}}$

30. $\lim_{x\to-\infty} \dfrac{-3x+1}{\sqrt{x^2+x}}$

31. $\lim_{x\to\infty} \dfrac{\sin 2x}{x}$

32. $\lim_{x\to\infty} \dfrac{x-\cos x}{x}$

33. $\lim_{x\to\infty} \dfrac{1}{2x+\sin x}$

34. $\lim_{x\to\infty} \cos\dfrac{1}{x}$

In Exercises 35–38, use a graphing utility to graph the function and identify any horizontal asymptotes.

35. $f(x) = \dfrac{|x|}{x+1}$

36. $f(x) = \dfrac{|3x+2|}{x-2}$

37. $f(x) = \dfrac{3x}{\sqrt{x^2+2}}$

38. $f(x) = \dfrac{\sqrt{9x^2-2}}{2x+1}$

In Exercises 39 and 40, find the limit. (*Hint:* Let $x = 1/t$ and find the limit as $t \to 0^+$.)

39. $\displaystyle\lim_{x\to\infty} x \sin\frac{1}{x}$

40. $\displaystyle\lim_{x\to\infty} x \tan\frac{1}{x}$

In Exercises 41–46, find the limit. (*Hint:* Treat the expression as a fraction whose denominator is 1, and rationalize the numerator.) Use a graphing utility to verify your result.

41. $\displaystyle\lim_{x\to-\infty} \left(x + \sqrt{x^2+3}\right)$

42. $\displaystyle\lim_{x\to\infty} \left(2x - \sqrt{4x^2+1}\right)$

43. $\displaystyle\lim_{x\to\infty} \left(x - \sqrt{x^2+x}\right)$

44. $\displaystyle\lim_{x\to-\infty} \left(3x + \sqrt{9x^2-x}\right)$

45. $\displaystyle\lim_{x\to\infty} \left(4x - \sqrt{16x^2-x}\right)$

46. $\displaystyle\lim_{x\to-\infty} \left(\frac{x}{2} + \sqrt{\frac{1}{4}x^2+x}\right)$

Numerical, Graphical, and Analytic Analysis **In Exercises 47–50, use a graphing utility to complete the table and estimate the limit as x approaches infinity. Then use a graphing utility to graph the function and estimate the limit. Finally, find the limit analytically and compare your results with the estimates.**

x	10^0	10^1	10^2	10^3	10^4	10^5	10^6
$f(x)$							

47. $f(x) = x - \sqrt{x(x-1)}$

48. $f(x) = x^2 - x\sqrt{x(x-1)}$

49. $f(x) = x \sin\dfrac{1}{2x}$

50. $f(x) = \dfrac{x+1}{x\sqrt{x}}$

Writing About Concepts

51. The graph of a function f is shown below. To print an enlarged copy of the graph, go to the website *www.mathgraphs.com.*

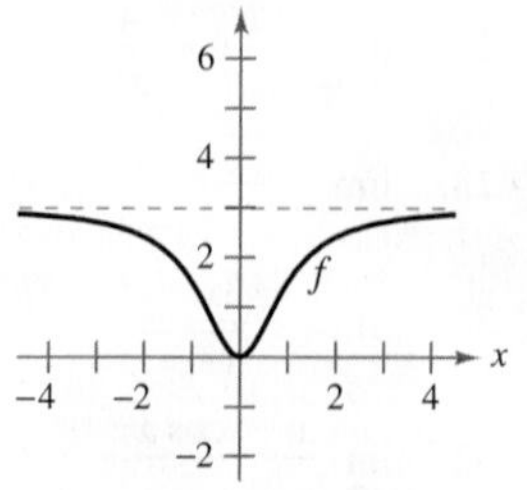

(a) Sketch f'.

(b) Use the graphs to estimate $\displaystyle\lim_{x\to\infty} f(x)$ and $\displaystyle\lim_{x\to\infty} f'(x)$.

(c) Explain the answers you gave in part (b).

Writing About Concepts (*continued*)

52. Sketch a graph of a differentiable function f that satisfies the following conditions and has $x = 2$ as its only critical number.

$f'(x) < 0$ for $x < 2$ $\quad$ $f'(x) > 0$ for $x > 2$

$\displaystyle\lim_{x\to-\infty} f(x) = \lim_{x\to\infty} f(x) = 6$

53. Is it possible to sketch a graph of a function that satisfies the conditions of Exercise 52 and has *no* points of inflection? Explain.

54. If f is a continuous function such that $\displaystyle\lim_{x\to\infty} f(x) = 5$, find, if possible, $\displaystyle\lim_{x\to-\infty} f(x)$ for each specified condition.

(a) The graph of f is symmetric to the y-axis.

(b) The graph of f is symmetric to the origin.

In Exercises 55–72, sketch the graph of the function using extrema, intercepts, symmetry, and asymptotes. Then use a graphing utility to verify your result.

55. $y = \dfrac{2+x}{1-x}$

56. $y = \dfrac{x-3}{x-2}$

57. $y = \dfrac{x}{x^2-4}$

58. $y = \dfrac{2x}{9-x^2}$

59. $y = \dfrac{x^2}{x^2+9}$

60. $y = \dfrac{x^2}{x^2-9}$

61. $y = \dfrac{2x^2}{x^2-4}$

62. $y = \dfrac{2x^2}{x^2+4}$

63. $xy^2 = 4$

64. $x^2y = 4$

65. $y = \dfrac{2x}{1-x}$

66. $y = \dfrac{2x}{1-x^2}$

67. $y = 2 - \dfrac{3}{x^2}$

68. $y = 1 + \dfrac{1}{x}$

69. $y = 3 + \dfrac{2}{x}$

70. $y = 4\left(1 - \dfrac{1}{x^2}\right)$

71. $y = \dfrac{x^3}{\sqrt{x^2-4}}$

72. $y = \dfrac{x}{\sqrt{x^2-4}}$

In Exercises 73–82, use a computer algebra system to analyze the graph of the function. Label any extrema and/or asymptotes that exist.

73. $f(x) = 5 - \dfrac{1}{x^2}$

74. $f(x) = \dfrac{x^2}{x^2-1}$

75. $f(x) = \dfrac{x}{x^2-4}$

76. $f(x) = \dfrac{1}{x^2-x-2}$

77. $f(x) = \dfrac{x-2}{x^2-4x+3}$

78. $f(x) = \dfrac{x+1}{x^2+x+1}$

79. $f(x) = \dfrac{3x}{\sqrt{4x^2+1}}$

80. $g(x) = \dfrac{2x}{\sqrt{3x^2+1}}$

81. $g(x) = \sin\left(\dfrac{x}{x-2}\right),\quad x > 3$

82. $f(x) = \dfrac{2\sin 2x}{x}$

In Exercises 83 and 84, (a) use a graphing utility to graph f and g in the same viewing window, (b) verify algebraically that f and g represent the same function, and (c) zoom out sufficiently far so that the graph appears as a line. What equation does this line appear to have? (Note that the points at which the function is not continuous are not readily seen when you zoom out.)

83. $f(x) = \dfrac{x^3 - 3x^2 + 2}{x(x - 3)}$

$g(x) = x + \dfrac{2}{x(x - 3)}$

84. $f(x) = -\dfrac{x^3 - 2x^2 + 2}{2x^2}$

$g(x) = -\dfrac{1}{2}x + 1 - \dfrac{1}{x^2}$

85. ***Average Cost*** A business has a cost of $C = 0.5x + 500$ for producing x units. The average cost per unit is

$$\overline{C} = \frac{C}{x}.$$

Find the limit of $\overline{C}$ as x approaches infinity.

86. ***Engine Efficiency*** The efficiency of an internal combustion engine is

$$\text{Efficiency } (\%) = 100\left[1 - \frac{1}{(v_1/v_2)^c}\right]$$

where v_1/v_2 is the ratio of the uncompressed gas to the compressed gas and c is a positive constant dependent on the engine design. Find the limit of the efficiency as the compression ratio approaches infinity.

87. ***Physics*** Newton's First Law of Motion and Einstein's Special Theory of Relativity differ concerning a particle's behavior as its velocity approaches the speed of light, c. Functions N and E represent the predicted velocity, v, with respect to time, t, for a particle accelerated by a constant force. Write a limit statement that describes each theory.

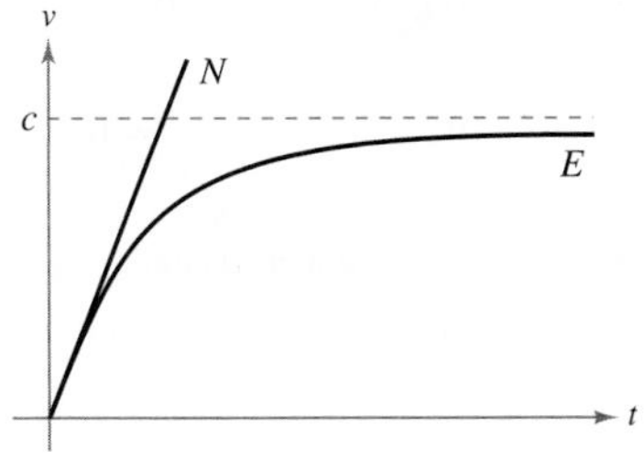

88. ***Temperature*** The graph shows the temperature T, in degrees Fahrenheit, of an apple pie t seconds after it is removed from an oven and placed on a cooling rack.

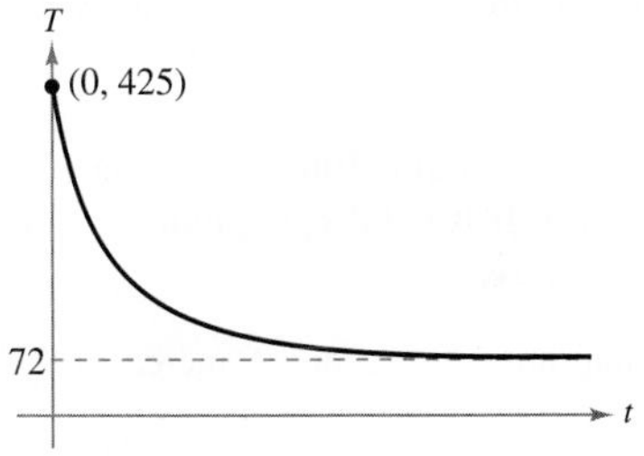

(a) Find $\lim_{t\to 0^+} T$. What does this limit represent?

(b) Find $\lim_{t\to\infty} T$. What does this limit represent?

89. ***Modeling Data*** The table shows the world record times for running 1 mile, where t represents the year, with $t = 0$ corresponding to 1900, and y is the time in minutes and seconds.

t	23	33	45	54
y	4:10.4	4:07.6	4:01.3	3:59.4

t	58	66	79	85	99
y	3:54.5	3:51.3	3:48.9	3:46.3	3:43.1

A model for the data is

$$y = \frac{3.351t^2 + 42.461t - 543.730}{t^2}$$

where the seconds have been changed to decimal parts of a minute.

(a) Use a graphing utility to plot the data and graph the model.

(b) Does there appear to be a limiting time for running 1 mile? Explain.

90. ***Modeling Data*** The average typing speeds S (words per minute) of a typing student after t weeks of lessons are shown in the table.

t	5	10	15	20	25	30
S	28	56	79	90	93	94

A model for the data is $S = \dfrac{100t^2}{65 + t^2}$, $t > 0$.

(a) Use a graphing utility to plot the data and graph the model.

(b) Does there appear to be a limiting typing speed? Explain.

91. ***Modeling Data*** A heat probe is attached to the heat exchanger of a heating system. The temperature T (degrees Celsius) is recorded t seconds after the furnace is started. The results for the first 2 minutes are recorded in the table.

t	0	15	30	45	60
T	25.2°	36.9°	45.5°	51.4°	56.0°

t	75	90	105	120
T	59.6°	62.0°	64.0°	65.2°

(a) Use the regression capabilities of a graphing utility to find a model of the form $T_1 = at^2 + bt + c$ for the data.

(b) Use a graphing utility to graph T_1.

(c) A rational model for the data is $T_2 = \dfrac{1451 + 86t}{58 + t}$. Use a graphing utility to graph the model.

(d) Find $T_1(0)$ and $T_2(0)$.

(e) Find $\lim_{t\to\infty} T_2$.

(f) Interpret the result in part (e) in the context of the problem. Is it possible to do this type of analysis using T_1? Explain.

92. ***Modeling Data*** A container contains 5 liters of a 25% brine solution. The table shows the concentrations C of the mixture after adding x liters of a 75% brine solution to the container.

x	0	0.5	1	1.5	2
C	0.25	0.295	0.333	0.365	0.393

x	2.5	3	3.5	4
C	0.417	0.438	0.456	0.472

(a) Use the regression features of a graphing utility to find a model of the form $C_1 = ax^2 + bx + c$ for the data.

(b) Use a graphing utility to graph C_1.

(c) A rational model for these data is $C_2 = \dfrac{5 + 3x}{20 + 4x}$. Use a graphing utility to graph C_2.

(d) Find $\lim_{x\to\infty} C_1$ and $\lim_{x\to\infty} C_2$. Which model do you think best represents the concentration of the mixture? Explain.

(e) What is the limiting concentration?

93. A line with slope m passes through the point $(0, 4)$.

(a) Write the distance d between the line and the point $(3, 1)$ as a function of m.

(b) Use a graphing utility to graph the equation in part (a).

(c) Find $\lim_{m\to\infty} d(m)$ and $\lim_{m\to-\infty} d(m)$. Interpret the results geometrically.

94. A line with slope m passes through the point $(0, -2)$.

(a) Write the distance d between the line and the point $(4, 2)$ as a function of m.

(b) Use a graphing utility to graph the equation in part (a).

(c) Find $\lim_{m\to\infty} d(m)$ and $\lim_{m\to-\infty} d(m)$. Interpret the results geometrically.

95. The graph of $f(x) = \dfrac{2x^2}{x^2 + 2}$ is shown.

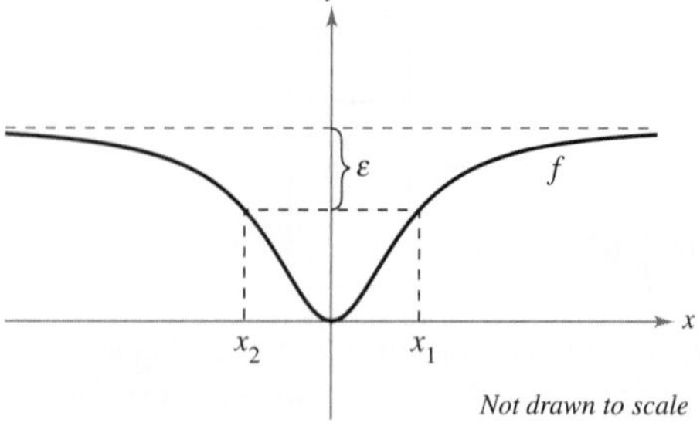

(a) Find $L = \lim_{x\to\infty} f(x)$.

(b) Determine x_1 and x_2 in terms of ε.

(c) Determine M, where $M > 0$, such that $|f(x) - L| < \varepsilon$ for $x > M$.

(d) Determine N, where $N < 0$, such that $|f(x) - L| < \varepsilon$ for $x < N$.

96. The graph of $f(x) = \dfrac{6x}{\sqrt{x^2 + 2}}$ is shown.

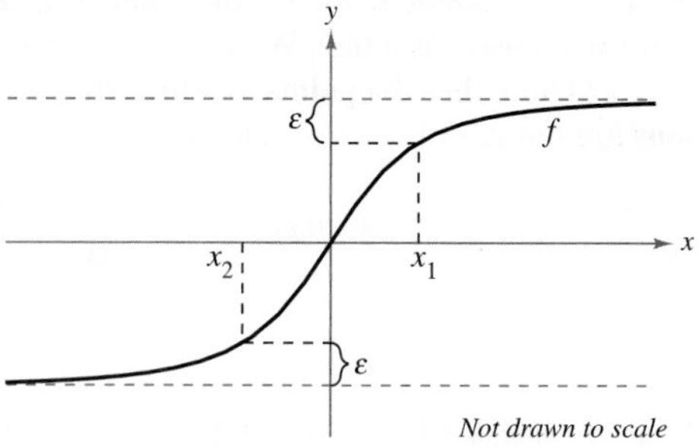

(a) Find $L = \lim_{x\to\infty} f(x)$ and $K = \lim_{x\to-\infty} f(x)$.

(b) Determine x_1 and x_2 in terms of ε.

(c) Determine M, where $M > 0$, such that $|f(x) - L| < \varepsilon$ for $x > M$.

(d) Determine N, where $N < 0$, such that $|f(x) - K| < \varepsilon$ for $x < N$.

97. Consider $\lim_{x\to\infty} \dfrac{3x}{\sqrt{x^2 + 3}}$. Use the definition of limits at infinity to find values of M that correspond to (a) $\varepsilon = 0.5$ and (b) $\varepsilon = 0.1$.

98. Consider $\lim_{x\to-\infty} \dfrac{3x}{\sqrt{x^2 + 3}}$. Use the definition of limits at infinity to find values of N that correspond to (a) $\varepsilon = 0.5$ and (b) $\varepsilon = 0.1$.

In Exercises 99–102, use the definition of limits at infinity to prove the limit.

99. $\lim_{x\to\infty} \dfrac{1}{x^2} = 0$

100. $\lim_{x\to\infty} \dfrac{2}{\sqrt{x}} = 0$

101. $\lim_{x\to-\infty} \dfrac{1}{x^3} = 0$

102. $\lim_{x\to-\infty} \dfrac{1}{x - 2} = 0$

103. Prove that if $p(x) = a_n x^n + \cdots + a_1 x + a_0$ and $q(x) = b_m x^m + \cdots + b_1 x + b_0$ $(a_n \neq 0, b_m \neq 0)$, then

$$\lim_{x\to\infty} \frac{p(x)}{q(x)} = \begin{cases} 0, & n < m \\ \dfrac{a_n}{b_m}, & n = m. \\ \pm\infty, & n > m \end{cases}$$

104. Use the definition of infinite limits at infinity to prove that $\lim_{x\to\infty} x^3 = \infty$.

True or False? **In Exercises 105 and 106, determine whether the statement is true or false. If it is false, explain why or give an example that shows it is false.**

105. If $f'(x) > 0$ for all real numbers x, then f increases without bound.

106. If $f''(x) < 0$ for all real numbers x, then f decreases without bound.

Section 3.6 A Summary of Curve Sketching

- **Analyze and sketch the graph of a function.**

Analyzing the Graph of a Function

It would be difficult to overstate the importance of using graphs in mathematics. Descartes's introduction of analytic geometry contributed significantly to the rapid advances in calculus that began during the mid-seventeenth century. In the words of Lagrange, "As long as algebra and geometry traveled separate paths their advance was slow and their applications limited. But when these two sciences joined company, they drew from each other fresh vitality and thenceforth marched on at a rapid pace toward perfection."

So far, you have studied several concepts that are useful in analyzing the graph of a function.

- x-intercepts and y-intercepts (Section P.1)
- Symmetry (Section P.1)
- Domain and range (Section P.3)
- Continuity (Section 1.4)
- Vertical asymptotes (Section 1.5)
- Differentiability (Section 2.1)
- Relative extrema (Section 3.1)
- Concavity (Section 3.4)
- Points of inflection (Section 3.4)
- Horizontal asymptotes (Section 3.5)
- Infinite limits at infinity (Section 3.5)

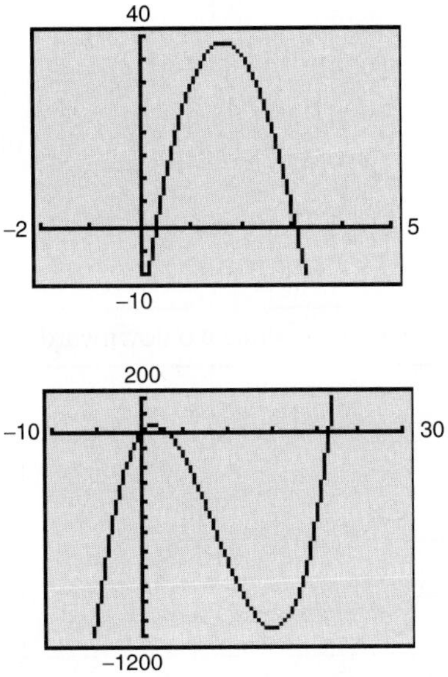

Different viewing windows for the graph of $f(x) = x^3 - 25x^2 + 74x - 20$
Figure 3.44

When you are sketching the graph of a function, either by hand or with a graphing utility, remember that normally you cannot show the *entire* graph. The decision as to which part of the graph you choose to show is often crucial. For instance, which of the viewing windows in Figure 3.44 better represents the graph of

$$f(x) = x^3 - 25x^2 + 74x - 20?$$

By seeing both views, it is clear that the second viewing window gives a more complete representation of the graph. But would a third viewing window reveal other interesting portions of the graph? To answer this, you need to use calculus to interpret the first and second derivatives. Here are some guidelines for determining a good viewing window for the graph of a function.

Guidelines for Analyzing the Graph of a Function

1. Determine the domain and range of the function.
2. Determine the intercepts, asymptotes, and symmetry of the graph.
3. Locate the x-values for which $f'(x)$ and $f''(x)$ either are zero or do not exist. Use the results to determine relative extrema and points of inflection.

NOTE In these guidelines, note the importance of *algebra* (as well as calculus) for solving the equations $f(x) = 0$, $f'(x) = 0$, and $f''(x) = 0$.

EXAMPLE 1 Sketching the Graph of a Rational Function

Analyze and sketch the graph of $f(x) = \dfrac{2(x^2 - 9)}{x^2 - 4}$.

Solution

First derivative:	$f'(x) = \dfrac{20x}{(x^2 - 4)^2}$
Second derivative:	$f''(x) = \dfrac{-20(3x^2 + 4)}{(x^2 - 4)^3}$
x-intercepts:	$(-3, 0), (3, 0)$
y-intercept:	$\left(0, \frac{9}{2}\right)$
Vertical asymptotes:	$x = -2, x = 2$
Horizontal asymptote:	$y = 2$
Critical number:	$x = 0$
Possible points of inflection:	None
Domain:	All real numbers except $x = \pm 2$
Symmetry:	With respect to y-axis
Test intervals:	$(-\infty, -2), (-2, 0), (0, 2), (2, \infty)$

The table shows how the test intervals are used to determine several characteristics of the graph. The graph of f is shown in Figure 3.45.

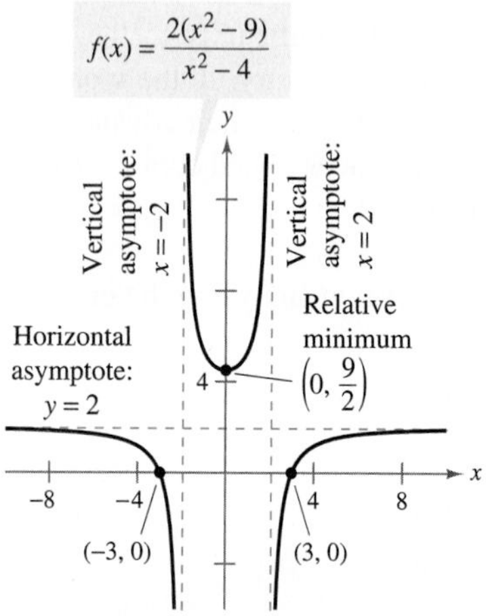

Using calculus, you can be certain that you have determined all characteristics of the graph of f.
Figure 3.45

	$f(x)$	$f'(x)$	$f''(x)$	Characteristic of Graph
$-\infty < x < -2$		$-$	$-$	Decreasing, concave downward
$x = -2$	Undef.	Undef.	Undef.	Vertical asymptote
$-2 < x < 0$		$-$	$+$	Decreasing, concave upward
$x = 0$	$\frac{9}{2}$	0	$+$	Relative minimum
$0 < x < 2$		$+$	$+$	Increasing, concave upward
$x = 2$	Undef.	Undef.	Undef.	Vertical asymptote
$2 < x < \infty$		$+$	$-$	Increasing, concave downward

FOR FURTHER INFORMATION For more information on the use of technology to graph rational functions, see the article "Graphs of Rational Functions for Computer Assisted Calculus" by Stan Byrd and Terry Walters in *The College Mathematics Journal*. To view this article, go to the website *www.matharticles.com*.

Be sure you understand all of the implications of creating a table such as that shown in Example 1. Because of the use of calculus, you can *be sure* that the graph has no relative extrema or points of inflection other than those shown in Figure 3.45.

TECHNOLOGY PITFALL Without using the type of analysis outlined in Example 1, it is easy to obtain an incomplete view of a graph's basic characteristics. For instance, Figure 3.46 shows a view of the graph of

$$g(x) = \frac{2(x^2 - 9)(x - 20)}{(x^2 - 4)(x - 21)}.$$

From this view, it appears that the graph of g is about the same as the graph of f shown in Figure 3.45. The graphs of these two functions, however, differ significantly. Try enlarging the viewing window to see the differences.

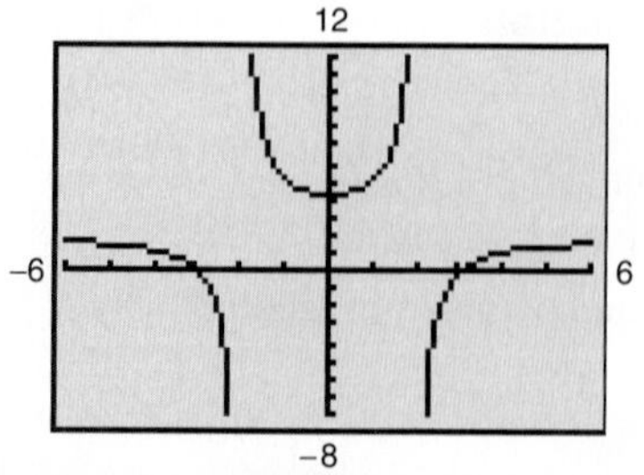

By not using calculus you may overlook important characteristics of the graph of g.
Figure 3.46

EXAMPLE 2 Sketching the Graph of a Rational Function

Analyze and sketch the graph of $f(x) = \dfrac{x^2 - 2x + 4}{x - 2}$.

Solution

First derivative:	$f'(x) = \dfrac{x(x-4)}{(x-2)^2}$
Second derivative:	$f''(x) = \dfrac{8}{(x-2)^3}$
x-intercepts:	None
y-intercept:	$(0, -2)$
Vertical asymptote:	$x = 2$
Horizontal asymptotes:	None
End behavior:	$\lim_{x\to-\infty} f(x) = -\infty, \ \lim_{x\to\infty} f(x) = \infty$
Critical numbers:	$x = 0, x = 4$
Possible points of inflection:	None
Domain:	All real numbers except $x = 2$
Test intervals:	$(-\infty, 0), (0, 2), (2, 4), (4, \infty)$

The analysis of the graph of f is shown in the table, and the graph is shown in Figure 3.47.

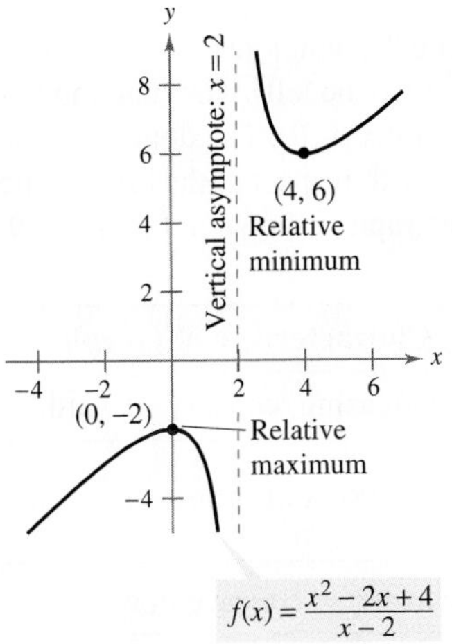

Figure 3.47

	$f(x)$	$f'(x)$	$f''(x)$	Characteristic of Graph
$-\infty < x < 0$		+	−	Increasing, concave downward
$x = 0$	−2	0	−	Relative maximum
$0 < x < 2$		−	−	Decreasing, concave downward
$x = 2$	Undef.	Undef.	Undef.	Vertical asymptote
$2 < x < 4$		−	+	Decreasing, concave upward
$x = 4$	6	0	+	Relative minimum
$4 < x < \infty$		+	+	Increasing, concave upward

Although the graph of the function in Example 2 has no horizontal asymptote, it does have a slant asymptote. The graph of a rational function (having no common factors and whose denominator is of degree 1 or greater) has a **slant asymptote** if the degree of the numerator exceeds the degree of the denominator by exactly 1. To find the slant asymptote, use long division to rewrite the rational function as the sum of a first-degree polynomial and another rational function.

$$f(x) = \frac{x^2 - 2x + 4}{x - 2} \qquad \text{Write original equation.}$$

$$= x + \frac{4}{x - 2} \qquad \text{Rewrite using long division.}$$

In Figure 3.48, note that the graph of f approaches the slant asymptote $y = x$ as x approaches $-\infty$ or ∞.

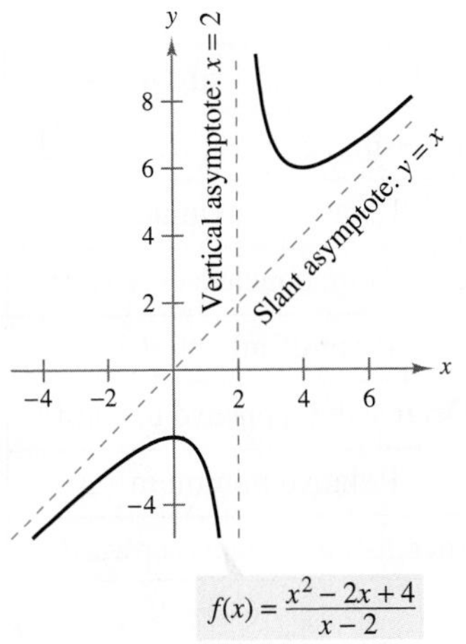

A slant asymptote
Figure 3.48

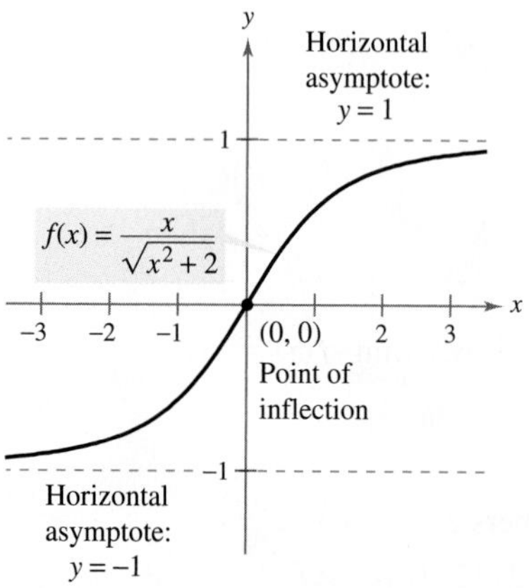

Figure 3.49

EXAMPLE 3 Sketching the Graph of a Radical Function

Analyze and sketch the graph of $f(x) = \dfrac{x}{\sqrt{x^2 + 2}}$.

Solution

$$f'(x) = \frac{2}{(x^2 + 2)^{3/2}} \qquad f''(x) = -\frac{6x}{(x^2 + 2)^{5/2}}$$

The graph has only one intercept, $(0, 0)$. It has no vertical asymptotes, but it has two horizontal asymptotes: $y = 1$ (to the right) and $y = -1$ (to the left). The function has no critical numbers and one possible point of inflection (at $x = 0$). The domain of the function is all real numbers, and the graph is symmetric with respect to the origin. The analysis of the graph of f is shown in the table, and the graph is shown in Figure 3.49.

	$f(x)$	$f'(x)$	$f''(x)$	Characteristic of Graph
$-\infty < x < 0$		$+$	$+$	Increasing, concave upward
$x = 0$	0	$\dfrac{1}{\sqrt{2}}$	0	Point of inflection
$0 < x < \infty$		$+$	$-$	Increasing, concave downward

EXAMPLE 4 Sketching the Graph of a Radical Function

Analyze and sketch the graph of $f(x) = 2x^{5/3} - 5x^{4/3}$.

Solution

$$f'(x) = \frac{10}{3}x^{1/3}(x^{1/3} - 2) \qquad f''(x) = \frac{20(x^{1/3} - 1)}{9x^{2/3}}$$

The function has two intercepts: $(0, 0)$ and $\left(\frac{125}{8}, 0\right)$. There are no horizontal or vertical asymptotes. The function has two critical numbers ($x = 0$ and $x = 8$) and two possible points of inflection ($x = 0$ and $x = 1$). The domain is all real numbers. The analysis of the graph of f is shown in the table, and the graph is shown in Figure 3.50.

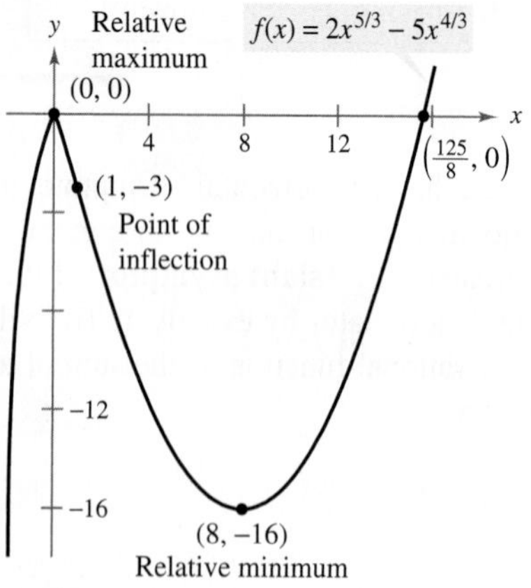

Figure 3.50

	$f(x)$	$f'(x)$	$f''(x)$	Characteristic of Graph
$-\infty < x < 0$		$+$	$-$	Increasing, concave downward
$x = 0$	0	0	Undef.	Relative maximum
$0 < x < 1$		$-$	$-$	Decreasing, concave downward
$x = 1$	-3	$-$	0	Point of inflection
$1 < x < 8$		$-$	$+$	Decreasing, concave upward
$x = 8$	-16	0	$+$	Relative minimum
$8 < x < \infty$		$+$	$+$	Increasing, concave upward

EXAMPLE 5 Sketching the Graph of a Polynomial Function

Analyze and sketch the graph of $f(x) = x^4 - 12x^3 + 48x^2 - 64x$.

Solution Begin by factoring to obtain

$$f(x) = x^4 - 12x^3 + 48x^2 - 64x$$
$$= x(x - 4)^3.$$

Then, using the factored form of $f(x)$, you can perform the following analysis.

First derivative:	$f'(x) = 4(x - 1)(x - 4)^2$
Second derivative:	$f''(x) = 12(x - 4)(x - 2)$
x-intercepts:	$(0, 0), (4, 0)$
y-intercept:	$(0, 0)$
Vertical asymptotes:	None
Horizontal asymptotes:	None
End behavior:	$\lim_{x \to -\infty} f(x) = \infty, \ \lim_{x \to \infty} f(x) = \infty$
Critical numbers:	$x = 1, x = 4$
Possible points of inflection:	$x = 2, x = 4$
Domain:	All real numbers
Test intervals:	$(-\infty, 1), (1, 2), (2, 4), (4, \infty)$

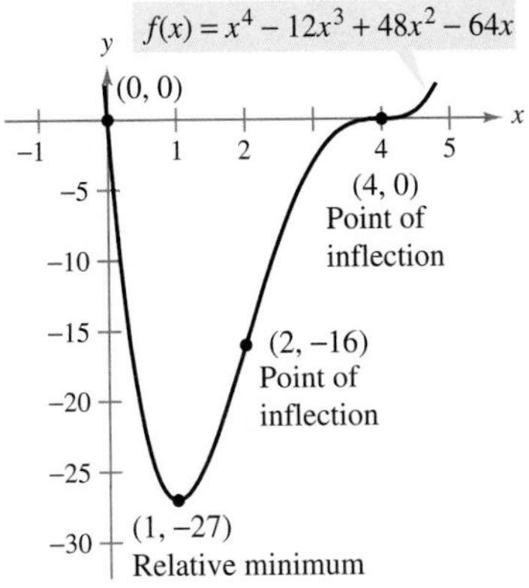

(a)

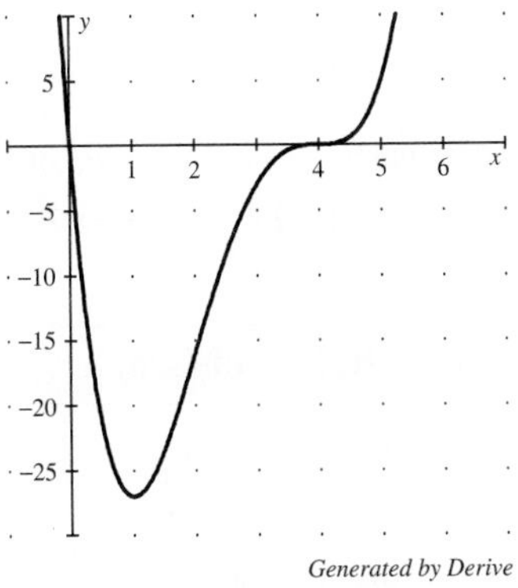

(b)

A polynomial function of even degree must have at least one relative extremum.

Figure 3.51

The analysis of the graph of f is shown in the table, and the graph is shown in Figure 3.51(a). Using a computer algebra system such as *Derive* [see Figure 3.51(b)] can help you verify your analysis.

	$f(x)$	$f'(x)$	$f''(x)$	Characteristic of Graph
$-\infty < x < 1$		$-$	$+$	Decreasing, concave upward
$x = 1$	-27	0	$+$	Relative minimum
$1 < x < 2$		$+$	$+$	Increasing, concave upward
$x = 2$	-16	$+$	0	Point of inflection
$2 < x < 4$		$+$	$-$	Increasing, concave downward
$x = 4$	0	0	0	Point of inflection
$4 < x < \infty$		$+$	$+$	Increasing, concave upward

The fourth-degree polynomial function in Example 5 has one relative minimum and no relative maxima. In general, a polynomial function of degree n can have *at most* $n - 1$ relative extrema, and *at most* $n - 2$ points of inflection. Moreover, polynomial functions of even degree must have *at least* one relative extremum.

Remember from the Leading Coefficient Test described in Section P.3 that the "end behavior" of the graph of a polynomial function is determined by its leading coefficient and its degree. For instance, because the polynomial in Example 5 has a positive leading coefficient, the graph rises to the right. Moreover, because the degree is even, the graph also rises to the left.

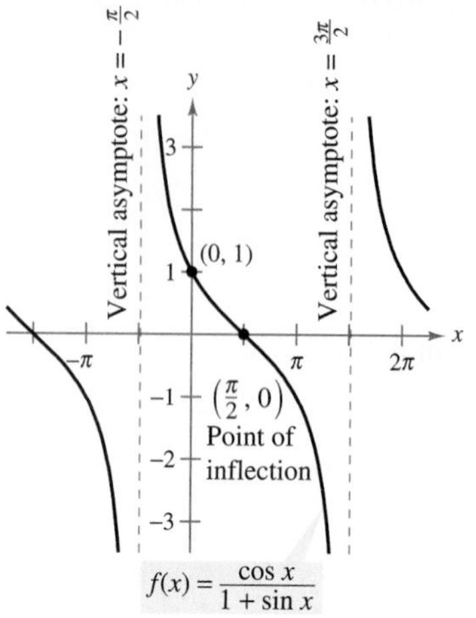

(a)

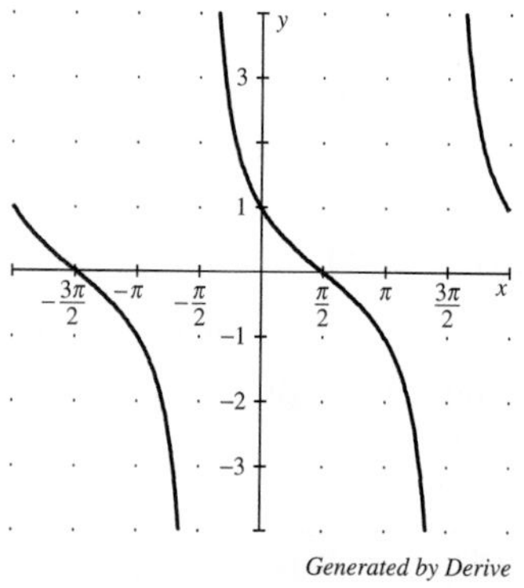

(b)

Figure 3.52

EXAMPLE 6 Sketching the Graph of a Trigonometric Function

Analyze and sketch the graph of $f(x) = \dfrac{\cos x}{1 + \sin x}$.

Solution Because the function has a period of 2π, you can restrict the analysis of the graph to any interval of length 2π. For convenience, choose $(-\pi/2, 3\pi/2)$.

First derivative: $f'(x) = -\dfrac{1}{1 + \sin x}$

Second derivative: $f''(x) = \dfrac{\cos x}{(1 + \sin x)^2}$

Period: 2π

x-intercept: $\left(\dfrac{\pi}{2}, 0\right)$

y-intercept: $(0, 1)$

Vertical asymptotes: $x = -\dfrac{\pi}{2},\ x = \dfrac{3\pi}{2}$ See Note below.

Horizontal asymptotes: None

Critical numbers: None

Possible points of inflection: $x = \dfrac{\pi}{2}$

Domain: All real numbers except $x = \dfrac{3 + 4n}{2}\pi$

Test intervals: $\left(-\dfrac{\pi}{2}, \dfrac{\pi}{2}\right), \left(\dfrac{\pi}{2}, \dfrac{3\pi}{2}\right)$

The analysis of the graph of f on the interval $(-\pi/2, 3\pi/2)$ is shown in the table, and the graph is shown in Figure 3.52(a). Compare this with the graph generated by the computer algebra system *Derive* in Figure 3.52(b).

	$f(x)$	$f'(x)$	$f''(x)$	Characteristic of Graph
$x = -\frac{\pi}{2}$	Undef.	Undef.	Undef.	Vertical asymptote
$-\frac{\pi}{2} < x < \frac{\pi}{2}$		$-$	$+$	Decreasing, concave upward
$x = \frac{\pi}{2}$	0	$-\frac{1}{2}$	0	Point of inflection
$\frac{\pi}{2} < x < \frac{3\pi}{2}$		$-$	$-$	Decreasing, concave downward
$x = \frac{3\pi}{2}$	Undef.	Undef.	Undef.	Vertical asymptote

NOTE By substituting $-\pi/2$ or $3\pi/2$ into the function, you obtain the form $0/0$. This is called an indeterminate form and you will study this in Section 8.7. To determine that the function has vertical asymptotes at these two values, you can rewrite the function as follows.

$$f(x) = \frac{\cos x}{1 + \sin x} = \frac{(\cos x)(1 - \sin x)}{(1 + \sin x)(1 - \sin x)} = \frac{(\cos x)(1 - \sin x)}{\cos^2 x} = \frac{1 - \sin x}{\cos x}$$

In this form, it is clear that the graph of f has vertical asymptotes when $x = -\pi/2$ and $3\pi/2$.

Exercises for Section 3.6

See www.CalcChat.com for worked-out solutions to odd-numbered exercises.

In Exercises 1–4, match the graph of f in the left column with that of its derivative in the right column.

Graph of f *Graph of f′*

1.

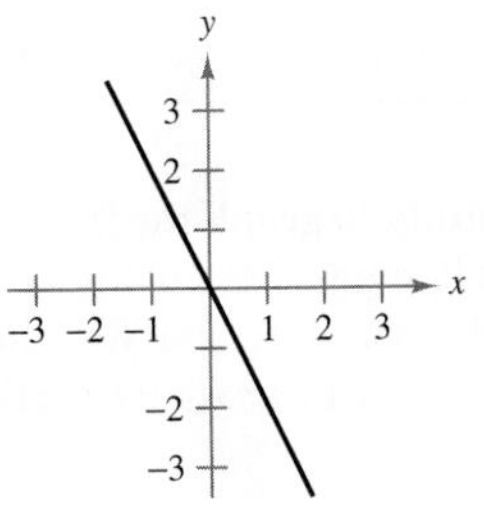

(a)

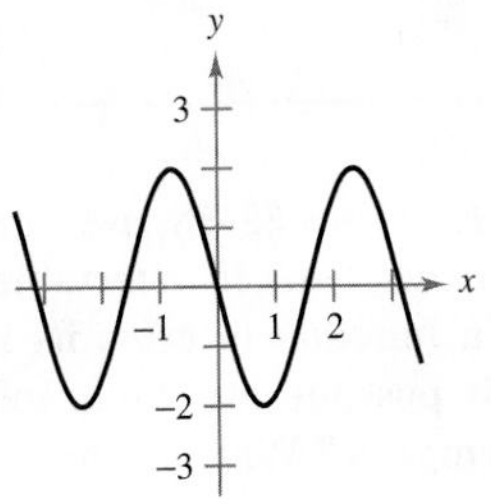

2.

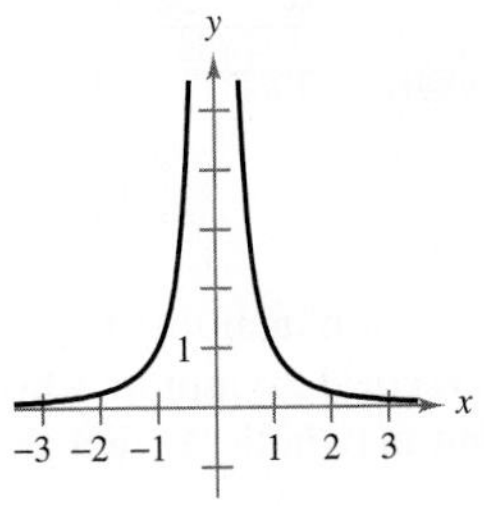

(b)

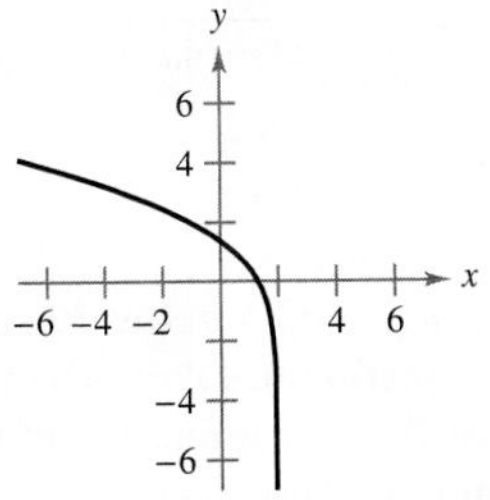

3.

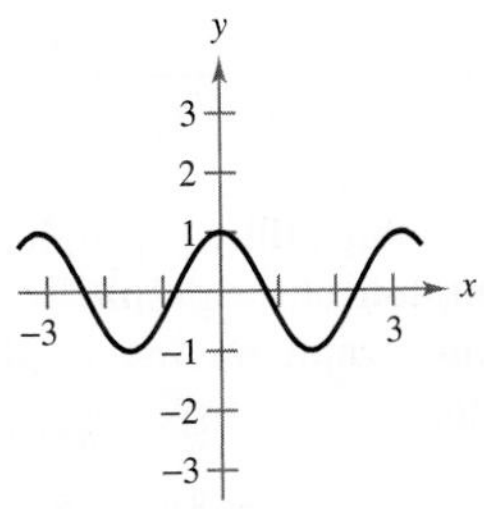

(c)

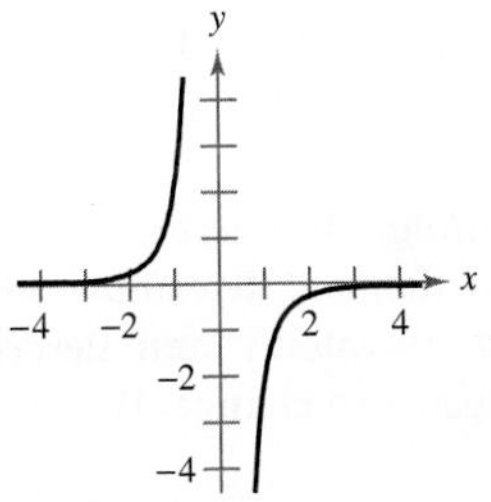

4.

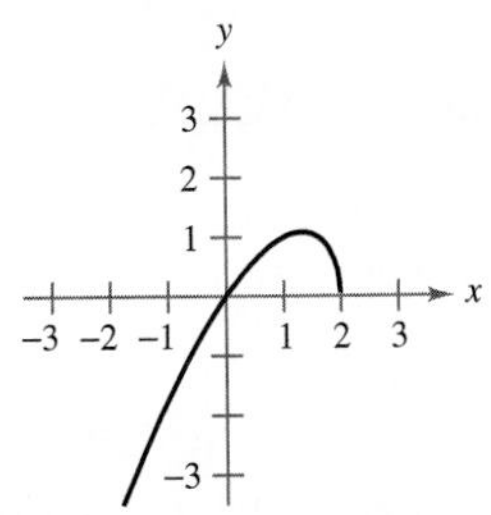

(d)

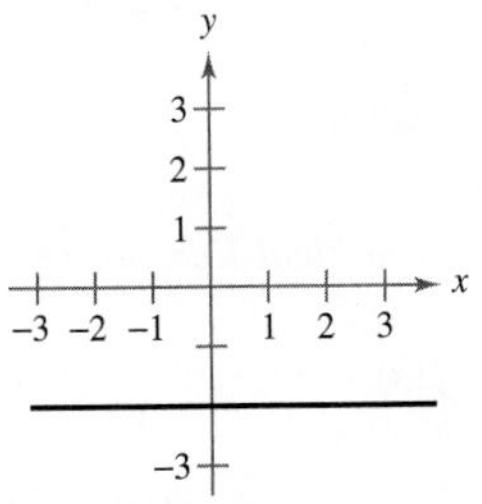

5. ***Graphical Reasoning*** The graph of f is shown in the figure.

(a) For which values of x is $f'(x)$ zero? Positive? Negative?

(b) For which values of x is $f''(x)$ zero? Positive? Negative?

(c) On what interval is f' an increasing function?

(d) For which value of x is $f'(x)$ minimum? For this value of x, how does the rate of change of f compare with the rate of change of f for other values of x? Explain.

6. ***Graphical Reasoning*** Identify the real numbers x_0, x_1, x_2, x_3, and x_4 in the figure such that each of the following is true.

(a) $f'(x) = 0$ (b) $f''(x) = 0$

(c) $f'(x)$ does not exist. (d) f has a relative maximum.

(e) f has a point of inflection.

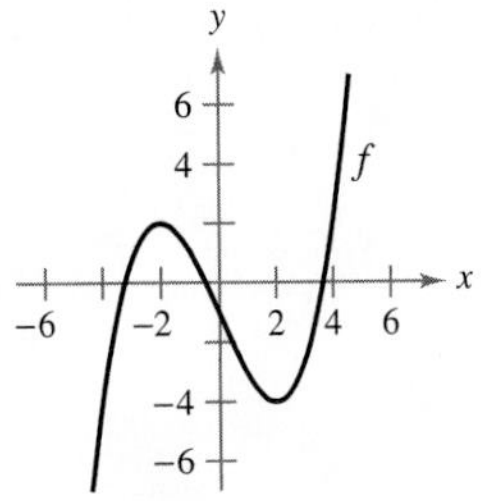

Figure for 5

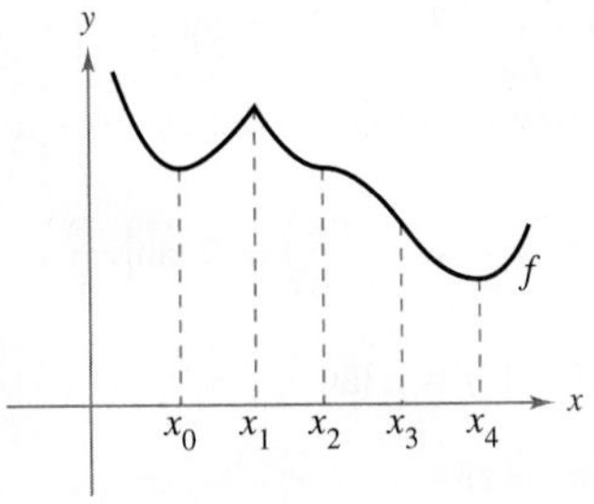

Figure for 6

In Exercises 7–34, analyze and sketch a graph of the function. Label any intercepts, relative extrema, points of inflection, and asymptotes. Use a graphing utility to verify your results.

7. $y = \dfrac{x^2}{x^2 + 3}$

8. $y = \dfrac{x}{x^2 + 1}$

9. $y = \dfrac{1}{x - 2} - 3$

10. $y = \dfrac{x^2 + 1}{x^2 - 9}$

11. $y = \dfrac{2x}{x^2 - 1}$

12. $f(x) = \dfrac{x + 2}{x}$

13. $g(x) = x + \dfrac{4}{x^2 + 1}$

14. $f(x) = x + \dfrac{32}{x^2}$

15. $f(x) = \dfrac{x^2 + 1}{x}$

16. $f(x) = \dfrac{x^3}{x^2 - 4}$

17. $y = \dfrac{x^2 - 6x + 12}{x - 4}$

18. $y = \dfrac{2x^2 - 5x + 5}{x - 2}$

19. $y = x\sqrt{4 - x}$

20. $g(x) = x\sqrt{9 - x}$

21. $h(x) = x\sqrt{9 - x^2}$

22. $y = x\sqrt{16 - x^2}$

23. $y = 3x^{2/3} - 2x$

24. $y = 3(x - 1)^{2/3} - (x - 1)^2$

25. $y = x^3 - 3x^2 + 3$

26. $y = -\frac{1}{3}(x^3 - 3x + 2)$

27. $y = 2 - x - x^3$

28. $f(x) = \frac{1}{3}(x - 1)^3 + 2$

29. $y = 3x^4 + 4x^3$

30. $y = 3x^4 - 6x^2 + \frac{5}{3}$

31. $y = x^5 - 5x$

32. $y = (x - 1)^5$

33. $y = |2x - 3|$

34. $y = |x^2 - 6x + 5|$

In Exercises 35–38, use a computer algebra system to analyze and graph the function. Identify any relative extrema, points of inflection, and asymptotes.

35. $f(x) = \dfrac{20x}{x^2 + 1} - \dfrac{1}{x}$

36. $f(x) = 5\left(\dfrac{1}{x - 4} - \dfrac{1}{x + 2}\right)$

37. $f(x) = \dfrac{x}{\sqrt{x^2 + 7}}$

38. $f(x) = \dfrac{4x}{\sqrt{x^2 + 15}}$

In Exercises 39–46, sketch a graph of the function over the given interval. Use a graphing utility to verify your graph.

39. $y = \sin x - \frac{1}{18}\sin 3x,\quad 0 \le x \le 2\pi$

40. $y = \cos x - \frac{1}{2}\cos 2x,\quad 0 \le x \le 2\pi$

41. $y = 2x - \tan x, \quad -\frac{\pi}{2} < x < \frac{\pi}{2}$

42. $y = 2(x - 2) + \cot x, \quad 0 < x < \pi$

43. $y = 2(\csc x + \sec x), \quad 0 < x < \frac{\pi}{2}$

44. $y = \sec^2\left(\frac{\pi x}{8}\right) - 2\tan\left(\frac{\pi x}{8}\right) - 1, \quad -3 < x < 3$

45. $g(x) = x \tan x, \quad -\frac{3\pi}{2} < x < \frac{3\pi}{2}$

46. $g(x) = x \cot x, \quad -2\pi < x < 2\pi$

Writing About Concepts

In Exercises 47 and 48, the graphs of f, f', and f'' are shown on the same set of coordinate axes. Which is which? Explain your reasoning. To print an enlarged copy of the graph, go to the website *www.mathgraphs.com*.

47.

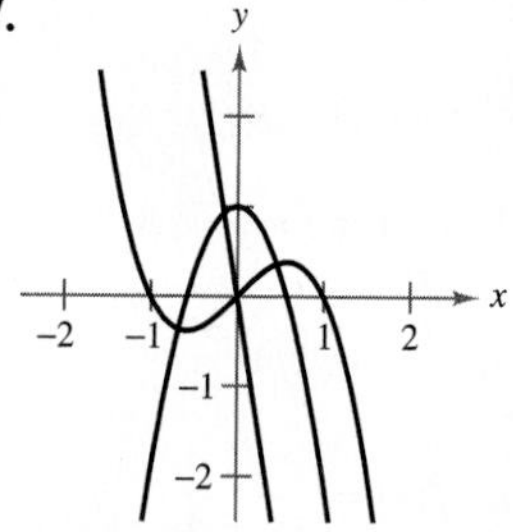

48.

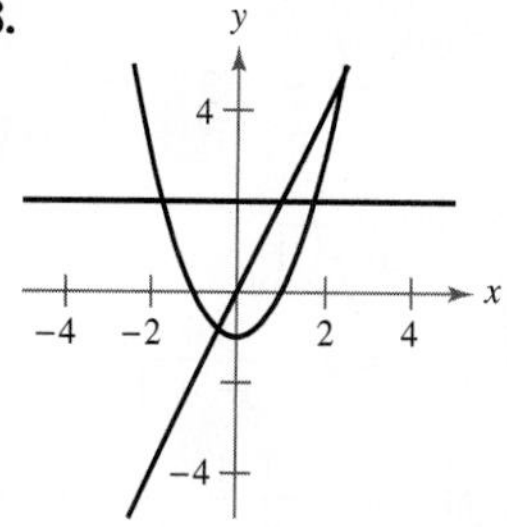

In Exercises 49–52, use the graph of f' to sketch a graph of f and the graph of f''. To print an enlarged copy of the graph, go to the website *www.mathgraphs.com*.

49.

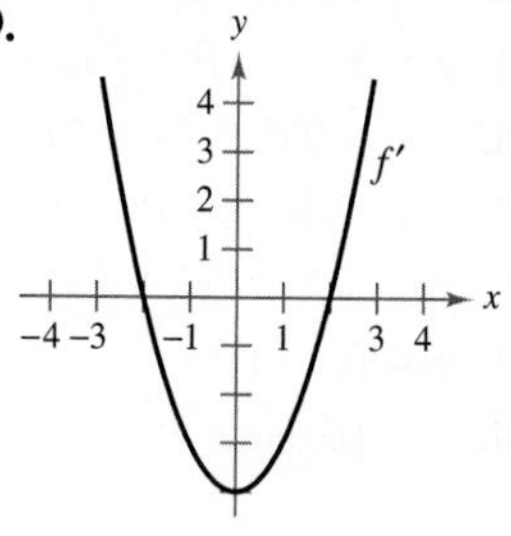

50.

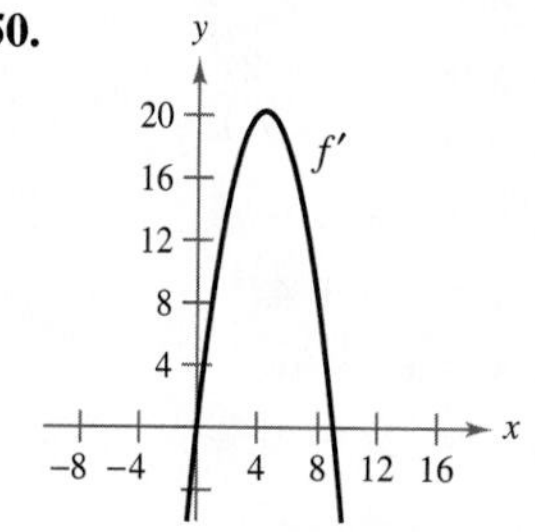

51.

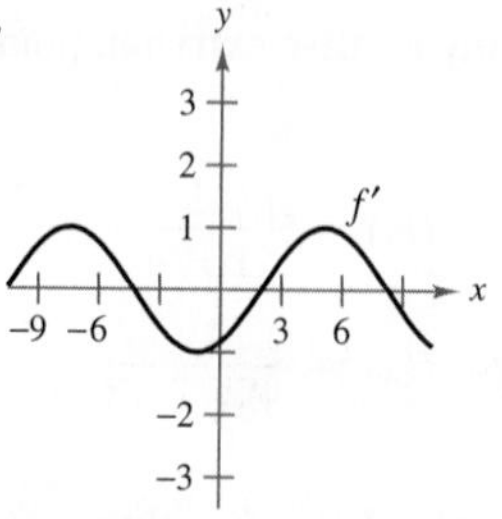

52.

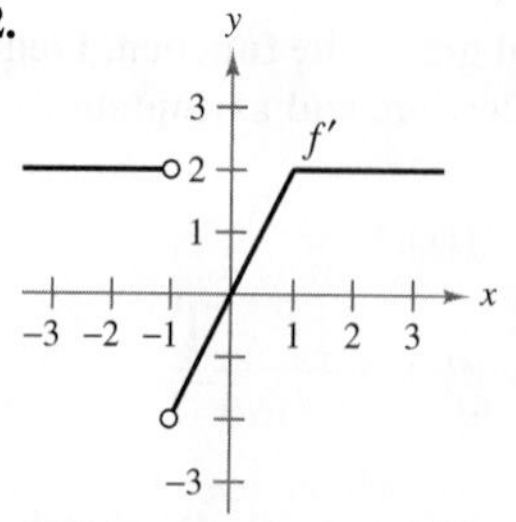

(Submitted by Bill Fox, Moberly Area Community College, Moberly, MO)

Writing About Concepts *(continued)*

53. Suppose $f'(t) < 0$ for all t in the interval $(2, 8)$. Explain why $f(3) > f(5)$.

54. Suppose $f(0) = 3$ and $2 \le f'(x) \le 4$ for all x in the interval $[-5, 5]$. Determine the greatest and least possible values of $f(2)$.

In Exercises 55–58, use a graphing utility to graph the function. Use the graph to determine whether it is possible for the graph of a function to cross its horizontal asymptote. Do you think it is possible for the graph of a function to cross its vertical asymptote? Why or why not?

55. $f(x) = \frac{4(x - 1)^2}{x^2 - 4x + 5}$

56. $g(x) = \frac{3x^4 - 5x + 3}{x^4 + 1}$

57. $h(x) = \frac{\sin 2x}{x}$

58. $f(x) = \frac{\cos 3x}{4x}$

Writing **In Exercises 59 and 60, use a graphing utility to graph the function. Explain why there is no vertical asymptote when a superficial examination of the function may indicate that there should be one.**

59. $h(x) = \frac{6 - 2x}{3 - x}$

60. $g(x) = \frac{x^2 + x - 2}{x - 1}$

Writing **In Exercises 61–64, use a graphing utility to graph the function and determine the slant asymptote of the graph. Zoom out repeatedly and describe how the graph on the display appears to change. Why does this occur?**

61. $f(x) = -\frac{x^2 - 3x - 1}{x - 2}$

62. $g(x) = \frac{2x^2 - 8x - 15}{x - 5}$

63. $f(x) = \frac{x^3}{x^2 + 1}$

64. $h(x) = \frac{-x^3 + x^2 + 4}{x^2}$

65. ***Graphical Reasoning*** Consider the function

$$f(x) = \frac{\cos^2 \pi x}{\sqrt{x^2 + 1}}, \quad 0 < x < 4.$$

(a) Use a computer algebra system to graph the function and use the graph to approximate the critical numbers visually.

(b) Use a computer algebra system to find f' and approximate the critical numbers. Are the results the same as the visual approximation in part (a)? Explain.

66. ***Graphical Reasoning*** Consider the function

$$f(x) = \tan(\sin \pi x).$$

(a) Use a graphing utility to graph the function.

(b) Identify any symmetry of the graph.

(c) Is the function periodic? If so, what is the period?

(d) Identify any extrema on $(-1, 1)$.

(e) Use a graphing utility to determine the concavity of the graph on $(0, 1)$.

Think About It **In Exercises 67–70, create a function whose graph has the given characteristics. (There is more than one correct answer.)**

67. Vertical asymptote: $x = 5$
Horizontal asymptote: $y = 0$

68. Vertical asymptote: $x = -3$
Horizontal asymptote: None

69. Vertical asymptote: $x = 5$
Slant asymptote: $y = 3x + 2$

70. Vertical asymptote: $x = 0$
Slant asymptote: $y = -x$

71. ***Graphical Reasoning*** Consider the function

$$f(x) = \frac{ax}{(x-b)^2}.$$

(a) Determine the effect on the graph of f if $b \neq 0$ and a is varied. Consider cases where a is positive and a is negative.

(b) Determine the effect on the graph of f if $a \neq 0$ and b is varied.

72. Consider the function $f(x) = \frac{1}{2}(ax)^2 - (ax),\quad a \neq 0.$

(a) Determine the changes (if any) in the intercepts, extrema, and concavity of the graph of f when a is varied.

(b) In the same viewing window, use a graphing utility to graph the function for four different values of a.

73. ***Investigation*** Consider the function

$$f(x) = \frac{3x^n}{x^4 + 1}$$

for nonnegative integer values of n.

(a) Discuss the relationship between the value of n and the symmetry of the graph.

(b) For which values of n will the x-axis be the horizontal asymptote?

(c) For which value of n will $y = 3$ be the horizontal asymptote?

(d) What is the asymptote of the graph when $n = 5$?

(e) Use a graphing utility to graph f for the indicated values of n in the table. Use the graph to determine the number of extrema M and the number of inflection points N of the graph.

n	0	1	2	3	4	5
M						
N						

74. ***Investigation*** Let $P(x_0, y_0)$ be an arbitrary point on the graph of f such that $f'(x_0) \neq 0$, as shown in the figure. Verify each statement.

(a) The x-intercept of the tangent line is $\left(x_0 - \dfrac{f(x_0)}{f'(x_0)}, 0\right)$.

(b) The y-intercept of the tangent line is $(0, f(x_0) - x_0 f'(x_0))$.

(c) The x-intercept of the normal line is $(x_0 + f(x_0)f'(x_0), 0)$.

(d) The y-intercept of the normal line is $\left(0, y_0 + \dfrac{x_0}{f'(x_0)}\right)$.

(e) $|BC| = \left|\dfrac{f(x_0)}{f'(x_0)}\right|$ (f) $|PC| = \left|\dfrac{f(x_0)\sqrt{1 + [f'(x_0)]^2}}{f'(x_0)}\right|$

(g) $|AB| = |f(x_0)f'(x_0)|$

(h) $|AP| = |f(x_0)|\sqrt{1 + [f'(x_0)]^2}$

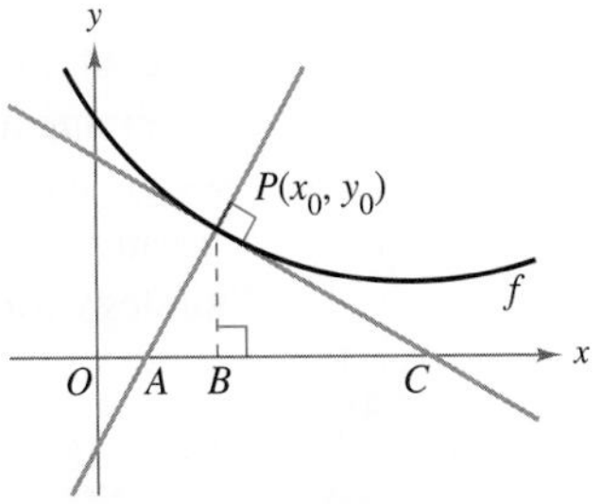

75. ***Modeling Data*** The data in the table show the number N of bacteria in a culture at time t, where t is measured in days.

t	1	2	3	4	5	6	7	8
N	25	200	804	1756	2296	2434	2467	2473

A model for these data is given by

$$N = \frac{24{,}670 - 35{,}153t + 13{,}250t^2}{100 - 39t + 7t^2}, \quad 1 \le t \le 8.$$

(a) Use a graphing utility to plot the data and graph the model.

(b) Use the model to estimate the number of bacteria when $t = 10$.

(c) Approximate the day when the number of bacteria is greatest.

(d) Use a computer algebra system to determine the time when the rate of increase in the number of bacteria is greatest.

(e) Find $\lim_{t\to\infty} N(t)$.

Slant Asymptotes **In Exercises 76 and 77, the graph of the function has two slant asymptotes. Identify each slant asymptote. Then graph the function and its asymptotes.**

76. $y = \sqrt{4 + 16x^2}$

77. $y = \sqrt{x^2 + 6x}$

Putnam Exam Challenge

78. Let $f(x)$ be defined for $a \le x \le b$. Assuming appropriate properties of continuity and derivability, prove for $a < x < b$ that

$$\frac{\dfrac{f(x) - f(a)}{x - a} - \dfrac{f(b) - f(a)}{b - a}}{x - b} = \frac{1}{2}f''(\beta)$$

where β is some number between a and b.

This problem was composed by the Committee on the Putnam Prize Competition.

Section 3.7 Optimization Problems

- **Solve applied minimum and maximum problems.**

Applied Minimum and Maximum Problems

One of the most common applications of calculus involves the determination of minimum and maximum values. Consider how frequently you hear or read terms such as greatest profit, least cost, least time, greatest voltage, optimum size, least size, greatest strength, and greatest distance. Before outlining a general problem-solving strategy for such problems, let's look at an example.

EXAMPLE 1 Finding Maximum Volume

A manufacturer wants to design an open box having a square base and a surface area of 108 square inches, as shown in Figure 3.53. What dimensions will produce a box with maximum volume?

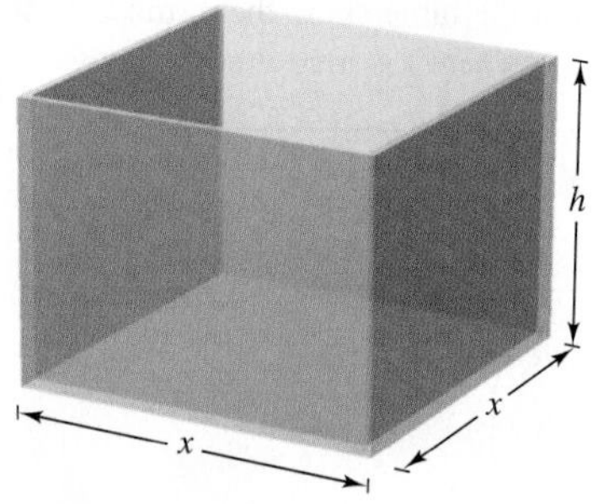

Open box with square base:
$S = x^2 + 4xh = 108$
Figure 3.53

Solution Because the box has a square base, its volume is

$$V = x^2h. \qquad \text{Primary equation}$$

This equation is called the **primary equation** because it gives a formula for the quantity to be optimized. The surface area of the box is

$$S = (\text{area of base}) + (\text{area of four sides})$$
$$S = x^2 + 4xh = 108. \qquad \text{Secondary equation}$$

Because V is to be maximized, you want to write V as a function of just one variable. To do this, you can solve the equation $x^2 + 4xh = 108$ for h in terms of x to obtain $h = (108 - x^2)/(4x)$. Substituting into the primary equation produces

$$V = x^2h \qquad \text{Function of two variables}$$
$$= x^2\left(\frac{108 - x^2}{4x}\right) \qquad \text{Substitute for } h.$$
$$= 27x - \frac{x^3}{4}. \qquad \text{Function of one variable}$$

Before finding which x-value will yield a maximum value of V, you should determine the *feasible domain.* That is, what values of x make sense in this problem? You know that $V \geq 0$. You also know that x must be nonnegative and that the area of the base ($A = x^2$) is at most 108. So, the feasible domain is

$$0 \leq x \leq \sqrt{108}. \qquad \text{Feasible domain}$$

To maximize V, find the critical numbers of the volume function.

$$\frac{dV}{dx} = 27 - \frac{3x^2}{4} = 0 \qquad \text{Set derivative equal to 0.}$$
$$3x^2 = 108 \qquad \text{Simplify.}$$
$$x = \pm 6 \qquad \text{Critical numbers}$$

So, the critical numbers are $x = \pm 6$. You do not need to consider $x = -6$ because it is outside the domain. Evaluating V at the critical number $x = 6$ and at the endpoints of the domain produces $V(0) = 0$, $V(6) = 108$, and $V(\sqrt{108}) = 0$. So, V is maximum when $x = 6$ and the dimensions of the box are $6 \times 6 \times 3$ inches.

TECHNOLOGY You can verify your answer by using a graphing utility to graph the volume function

$$V = 27x - \frac{x^3}{4}.$$

Use a viewing window in which $0 \leq x \leq \sqrt{108} \approx 10.4$ and $0 \leq y \leq 120$, and the *trace* feature to determine the maximum value of V.

In Example 1, you should realize that there are infinitely many open boxes having 108 square inches of surface area. To begin solving the problem, you might ask yourself which basic shape would seem to yield a maximum volume. Should the box be tall, squat, or nearly cubical?

You might even try calculating a few volumes, as shown in Figure 3.54, to see if you can get a better feeling for what the optimum dimensions should be. Remember that you are not ready to begin solving a problem until you have clearly identified what the problem is.

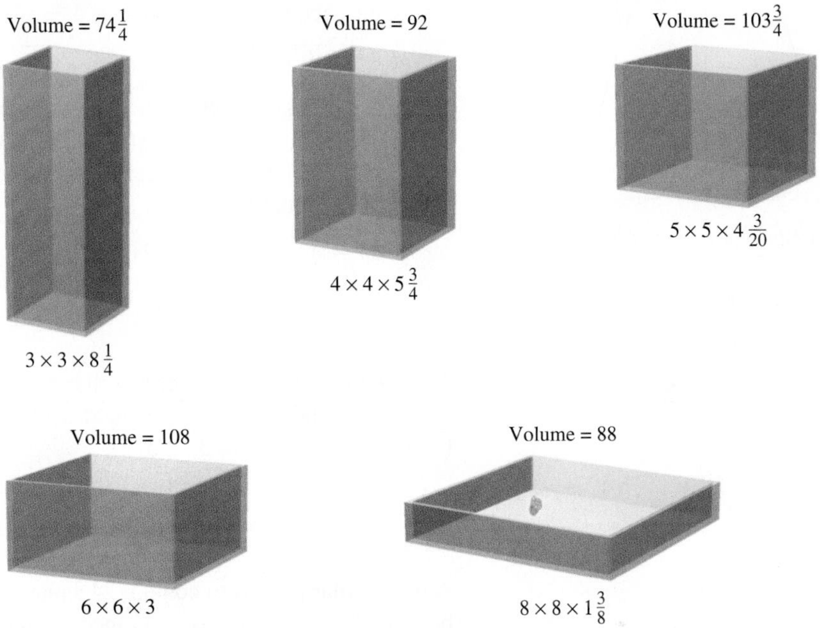

Which box has the greatest volume?
Figure 3.54

Example 1 illustrates the following guidelines for solving applied minimum and maximum problems.

Guidelines for Solving Applied Minimum and Maximum Problems

1. Identify all *given* quantities and quantities *to be determined.* If possible, make a sketch.
2. Write a **primary equation** for the quantity that is to be maximized or minimized. (A review of several useful formulas from geometry is presented inside the front cover.)
3. Reduce the primary equation to one having a *single independent variable.* This may involve the use of **secondary equations** relating the independent variables of the primary equation.
4. Determine the feasible domain of the primary equation. That is, determine the values for which the stated problem makes sense.
5. Determine the desired maximum or minimum value by the calculus techniques discussed in Sections 3.1 through 3.4.

NOTE When performing Step 5, recall that to determine the maximum or minimum value of a continuous function f on a closed interval, you should compare the values of f at its critical numbers with the values of f at the endpoints of the interval.

EXAMPLE 2 Finding Minimum Distance

Which points on the graph of $y = 4 - x^2$ are closest to the point $(0, 2)$?

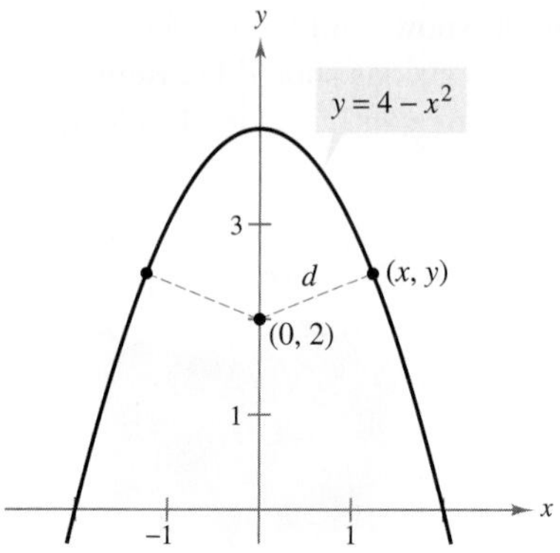

The quantity to be minimized is distance: $d = \sqrt{(x-0)^2 + (y-2)^2}$.
Figure 3.55

Solution Figure 3.55 shows that there are two points at a minimum distance from the point $(0, 2)$. The distance between the point $(0, 2)$ and a point (x, y) on the graph of $y = 4 - x^2$ is given by

$$d = \sqrt{(x-0)^2 + (y-2)^2}. \qquad \text{Primary equation}$$

Using the secondary equation $y = 4 - x^2$, you can rewrite the primary equation as

$$d = \sqrt{x^2 + (4 - x^2 - 2)^2} = \sqrt{x^4 - 3x^2 + 4}.$$

Because d is smallest when the expression inside the radical is smallest, you need only find the critical numbers of $f(x) = x^4 - 3x^2 + 4$. Note that the domain of f is the entire real line. So, there are no endpoints of the domain to consider. Moreover, setting $f'(x)$ equal to 0 yields

$$f'(x) = 4x^3 - 6x = 2x(2x^2 - 3) = 0$$

$$x = 0, \sqrt{\frac{3}{2}}, -\sqrt{\frac{3}{2}}.$$

The First Derivative Test verifies that $x = 0$ yields a relative maximum, whereas both $x = \sqrt{3/2}$ and $x = -\sqrt{3/2}$ yield a minimum distance. So, the closest points are $\left(\sqrt{3/2}, 5/2\right)$ and $\left(-\sqrt{3/2}, 5/2\right)$.

EXAMPLE 3 Finding Minimum Area

A rectangular page is to contain 24 square inches of print. The margins at the top and bottom of the page are to be $1\frac{1}{2}$ inches, and the margins on the left and right are to be 1 inch (see Figure 3.56). What should the dimensions of the page be so that the least amount of paper is used?

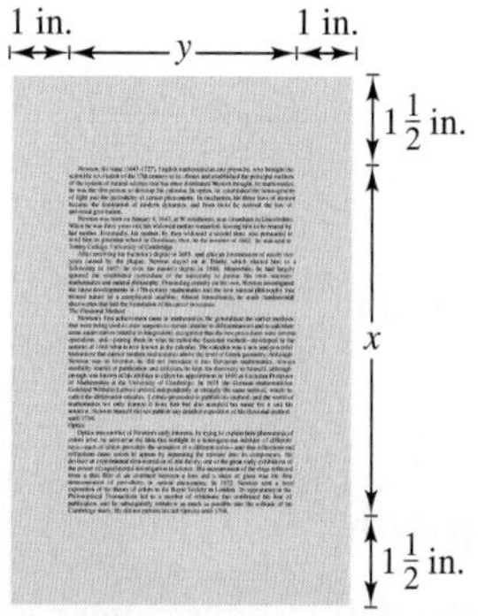

The quantity to be minimized is area: $A = (x + 3)(y + 2)$.
Figure 3.56

Solution Let A be the area to be minimized.

$$A = (x + 3)(y + 2) \qquad \text{Primary equation}$$

The printed area inside the margins is given by

$$24 = xy. \qquad \text{Secondary equation}$$

Solving this equation for y produces $y = 24/x$. Substitution into the primary equation produces

$$A = (x + 3)\left(\frac{24}{x} + 2\right) = 30 + 2x + \frac{72}{x}. \qquad \text{Function of one variable}$$

Because x must be positive, you are interested only in values of A for $x > 0$. To find the critical numbers, differentiate with respect to x.

$$\frac{dA}{dx} = 2 - \frac{72}{x^2} = 0 \quad \Longrightarrow \quad x^2 = 36$$

So, the critical numbers are $x = \pm 6$. You do not have to consider $x = -6$ because it is outside the domain. The First Derivative Test confirms that A is a minimum when $x = 6$. So, $y = \frac{24}{6} = 4$ and the dimensions of the page should be $x + 3 = 9$ inches by $y + 2 = 6$ inches.

EXAMPLE 4 Finding Minimum Length

Two posts, one 12 feet high and the other 28 feet high, stand 30 feet apart. They are to be stayed by two wires, attached to a single stake, running from ground level to the top of each post. Where should the stake be placed to use the least amount of wire?

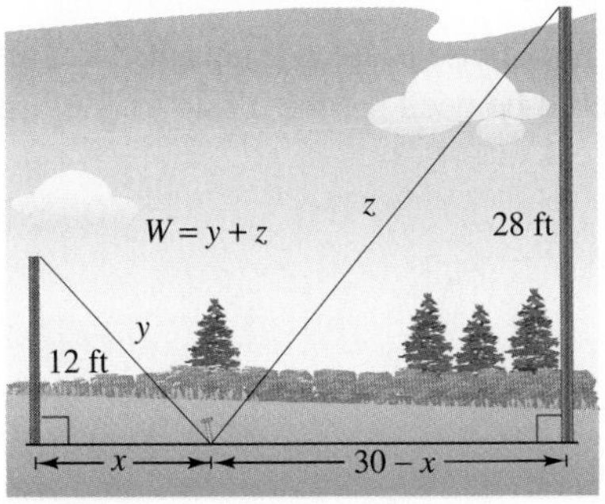

The quantity to be minimized is length. From the diagram, you can see that x varies between 0 and 30.
Figure 3.57

Solution Let W be the wire length to be minimized. Using Figure 3.57, you can write

$$W = y + z. \qquad \text{Primary equation}$$

In this problem, rather than solving for y in terms of z (or vice versa), you can solve for both y and z in terms of a third variable x, as shown in Figure 3.57. From the Pythagorean Theorem, you obtain

$$x^2 + 12^2 = y^2$$
$$(30 - x)^2 + 28^2 = z^2$$

which implies that

$$y = \sqrt{x^2 + 144}$$
$$z = \sqrt{x^2 - 60x + 1684}.$$

So, W is given by

$$\begin{aligned} W &= y + z \\ &= \sqrt{x^2 + 144} + \sqrt{x^2 - 60x + 1684}, \qquad 0 \le x \le 30. \end{aligned}$$

Differentiating W with respect to x yields

$$\frac{dW}{dx} = \frac{x}{\sqrt{x^2 + 144}} + \frac{x - 30}{\sqrt{x^2 - 60x + 1684}}.$$

By letting $dW/dx = 0$, you obtain

$$\begin{aligned} \frac{x}{\sqrt{x^2 + 144}} + \frac{x - 30}{\sqrt{x^2 - 60x + 1684}} &= 0 \\ x\sqrt{x^2 - 60x + 1684} &= (30 - x)\sqrt{x^2 + 144} \\ x^2(x^2 - 60x + 1684) &= (30 - x)^2(x^2 + 144) \\ x^4 - 60x^3 + 1684x^2 &= x^4 - 60x^3 + 1044x^2 - 8640x + 129{,}600 \\ 640x^2 + 8640x - 129{,}600 &= 0 \\ 320(x - 9)(2x + 45) &= 0 \\ x &= 9, -22.5. \end{aligned}$$

Because $x = -22.5$ is not in the domain and

$$W(0) \approx 53.04, \qquad W(9) = 50, \qquad \text{and} \qquad W(30) \approx 60.31$$

you can conclude that the wire should be staked at 9 feet from the 12-foot pole.

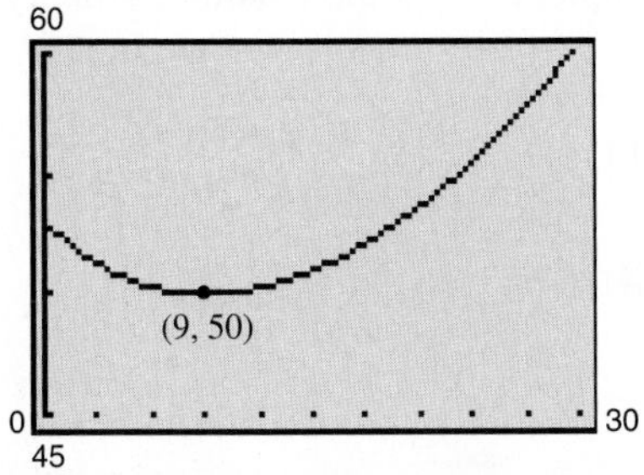

You can confirm the minimum value of W with a graphing utility.
Figure 3.58

TECHNOLOGY From Example 4, you can see that applied optimization problems can involve a lot of algebra. If you have access to a graphing utility, you can confirm that $x = 9$ yields a minimum value of W by graphing

$$W = \sqrt{x^2 + 144} + \sqrt{x^2 - 60x + 1684}$$

as shown in Figure 3.58.

In each of the first four examples, the extreme value occurred at a critical number. Although this happens often, remember that an extreme value can also occur at an endpoint of an interval, as shown in Example 5.

EXAMPLE 5 An Endpoint Maximum

Four feet of wire is to be used to form a square and a circle. How much of the wire should be used for the square and how much should be used for the circle to enclose the maximum total area?

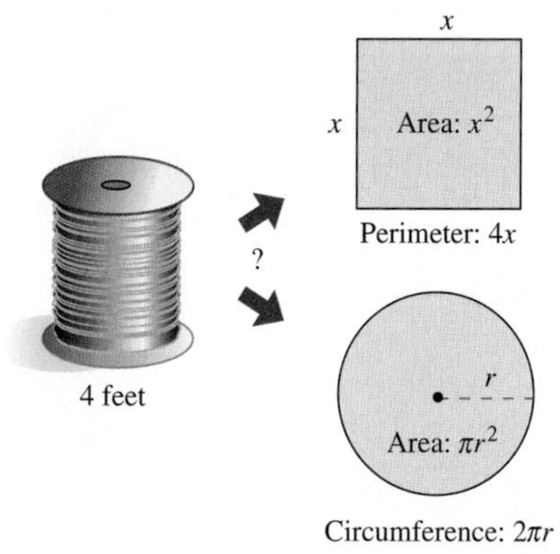

The quantity to be maximized is area: $A = x^2 + \pi r^2$.
Figure 3.59

Solution The total area (see Figure 3.59) is given by

$$A = (\text{area of square}) + (\text{area of circle})$$

$$A = x^2 + \pi r^2. \qquad \text{Primary equation}$$

Because the total length of wire is 4 feet, you obtain

$$4 = (\text{perimeter of square}) + (\text{circumference of circle})$$

$$4 = 4x + 2\pi r.$$

So, $r = 2(1 - x)/\pi$, and by substituting into the primary equation you have

$$\begin{aligned} A &= x^2 + \pi\left[\frac{2(1 - x)}{\pi}\right]^2 \\ &= x^2 + \frac{4(1 - x)^2}{\pi} \\ &= \frac{1}{\pi}[(\pi + 4)x^2 - 8x + 4]. \end{aligned}$$

The feasible domain is $0 \le x \le 1$ restricted by the square's perimeter. Because

$$\frac{dA}{dx} = \frac{2(\pi + 4)x - 8}{\pi}$$

the only critical number in $(0, 1)$ is $x = 4/(\pi + 4) \approx 0.56$. So, using

$$A(0) \approx 1.273, \quad A(0.56) \approx 0.56, \quad \text{and} \quad A(1) = 1$$

you can conclude that the maximum area occurs when $x = 0$. That is, *all* the wire is used for the circle.

EXPLORATION

What would the answer be if Example 5 asked for the dimensions needed to enclose the *minimum* total area?

Let's review the primary equations developed in the first five examples. As applications go, these five examples are fairly simple, and yet the resulting primary equations are quite complicated.

$$V = 27x - \frac{x^3}{4} \qquad W = \sqrt{x^2 + 144} + \sqrt{x^2 - 60x + 1684}$$

$$d = \sqrt{x^4 - 3x^2 + 4} \qquad A = \frac{1}{\pi}[(\pi + 4)x^2 - 8x + 4]$$

$$A = 30 + 2x + \frac{72}{x}$$

You must expect that real-life applications often involve equations that are *at least as complicated* as these five. Remember that one of the main goals of this course is to learn to use calculus to analyze equations that initially seem formidable.

Exercises for Section 3.7

1. ***Numerical, Graphical, and Analytic Analysis*** Find two positive numbers whose sum is 110 and whose product is a maximum.

   (a) Analytically complete six rows of a table such as the one below. (The first two rows are shown.)

First Number x	Second Number	Product P
10	$110 - 10$	$10(110 - 10) = 1000$
20	$110 - 20$	$20(110 - 20) = 1800$

   (b) Use a graphing utility to generate additional rows of the table. Use the table to estimate the solution. (*Hint:* Use the *table* feature of the graphing utility.)

   (c) Write the product P as a function of x.

   (d) Use a graphing utility to graph the function in part (c) and estimate the solution from the graph.

   (e) Use calculus to find the critical number of the function in part (c). Then find the two numbers.

2. ***Numerical, Graphical, and Analytic Analysis*** An open box of maximum volume is to be made from a square piece of material, 24 inches on a side, by cutting equal squares from the corners and turning up the sides (see figure).

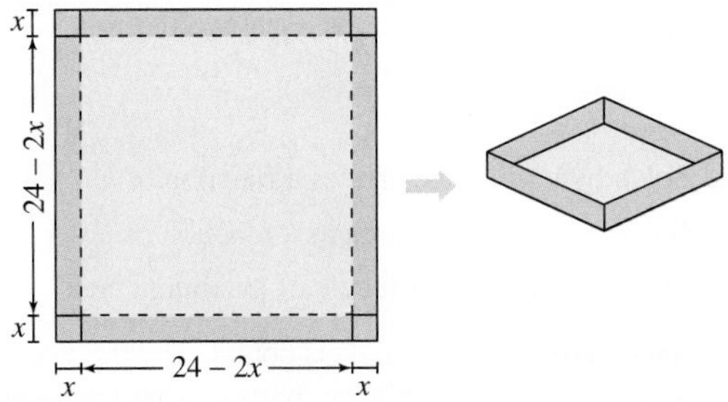

   (a) Analytically complete six rows of a table such as the one below. (The first two rows are shown.) Use the table to guess the maximum volume.

Height	Length and Width	Volume
1	$24 - 2(1)$	$1[24 - 2(1)]^2 = 484$
2	$24 - 2(2)$	$2[24 - 2(2)]^2 = 800$

   (b) Write the volume V as a function of x.

   (c) Use calculus to find the critical number of the function in part (b) and find the maximum value.

   (d) Use a graphing utility to graph the function in part (b) and verify the maximum volume from the graph.

In Exercises 3–8, find two positive numbers that satisfy the given requirements.

3. The sum is S and the product is a maximum.
4. The product is 192 and the sum is a minimum.
5. The product is 192 and the sum of the first plus three times the second is a minimum.
6. The second number is the reciprocal of the first and the sum is a minimum.
7. The sum of the first and twice the second is 100 and the product is a maximum.
8. The sum of the first number squared and the second is 27 and the product is a maximum.

In Exercises 9 and 10, find the length and width of a rectangle that has the given perimeter and a maximum area.

9. Perimeter: 100 meters
10. Perimeter: P units

In Exercises 11 and 12, find the length and width of a rectangle that has the given area and a minimum perimeter.

11. Area: 64 square feet
12. Area: A square centimeters

In Exercises 13–16, find the point on the graph of the function that is closest to the given point.

	Function	*Point*		*Function*	*Point*
13.	$f(x) = \sqrt{x}$	$(4, 0)$	14.	$f(x) = \sqrt{x - 8}$	$(2, 0)$
15.	$f(x) = x^2$	$\left(2, \frac{1}{2}\right)$	16.	$f(x) = (x + 1)^2$	$(5, 3)$

17. ***Chemical Reaction*** In an autocatalytic chemical reaction, the product formed is a catalyst for the reaction. If Q_0 is the amount of the original substance and x is the amount of catalyst formed, the rate of chemical reaction is

$$\frac{dQ}{dx} = kx(Q_0 - x).$$

For what value of x will the rate of chemical reaction be greatest?

18. ***Traffic Control*** On a given day, the flow rate F (cars per hour) on a congested roadway is

$$F = \frac{v}{22 + 0.02v^2}$$

where v is the speed of the traffic in miles per hour. What speed will maximize the flow rate on the road?

19. ***Area*** A farmer plans to fence a rectangular pasture adjacent to a river. The pasture must contain 180,000 square meters in order to provide enough grass for the herd. What dimensions would require the least amount of fencing if no fencing is needed along the river?

20. ***Maximum Area*** A rancher has 200 feet of fencing with which to enclose two adjacent rectangular corrals (see figure). What dimensions should be used so that the enclosed area will be a maximum?

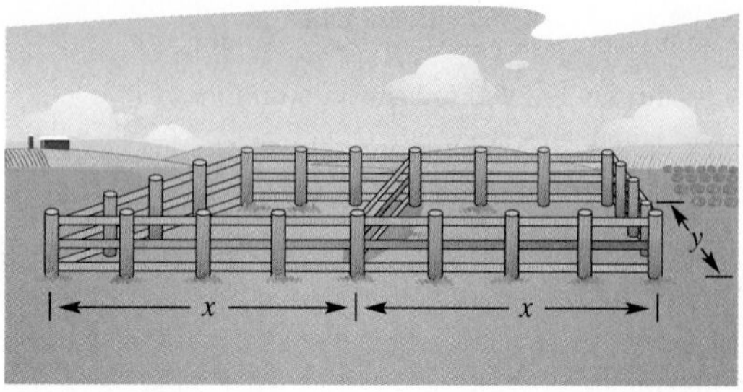

21. ***Maximum Volume***

(a) Verify that each of the rectangular solids shown in the figure has a surface area of 150 square inches.

(b) Find the volume of each solid.

(c) Determine the dimensions of a rectangular solid (with a square base) of maximum volume if its surface area is 150 square inches.

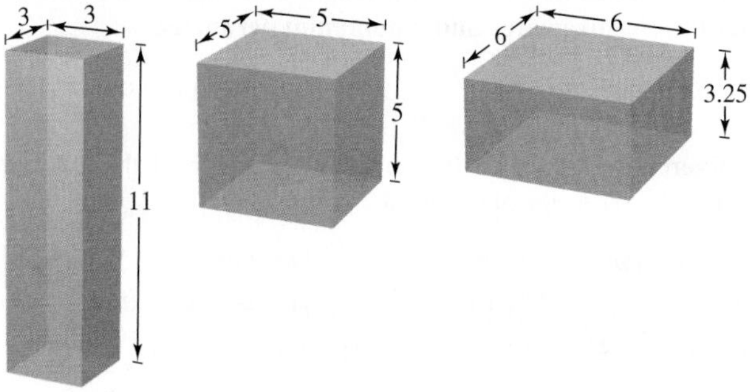

22. ***Maximum Volume*** Determine the dimensions of a rectangular solid (with a square base) with maximum volume if its surface area is 337.5 square centimeters.

23. ***Maximum Area*** A Norman window is constructed by adjoining a semicircle to the top of an ordinary rectangular window (see figure). Find the dimensions of a Norman window of maximum area if the total perimeter is 16 feet.

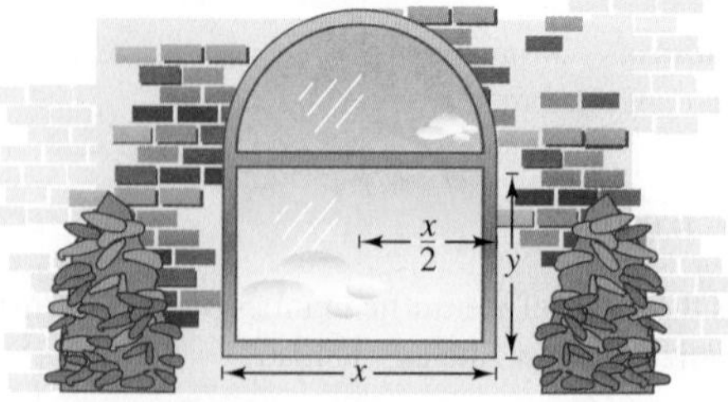

24. ***Maximum Area*** A rectangle is bounded by the x- and y-axes and the graph of $y = (6 - x)/2$ (see figure). What length and width should the rectangle have so that its area is a maximum?

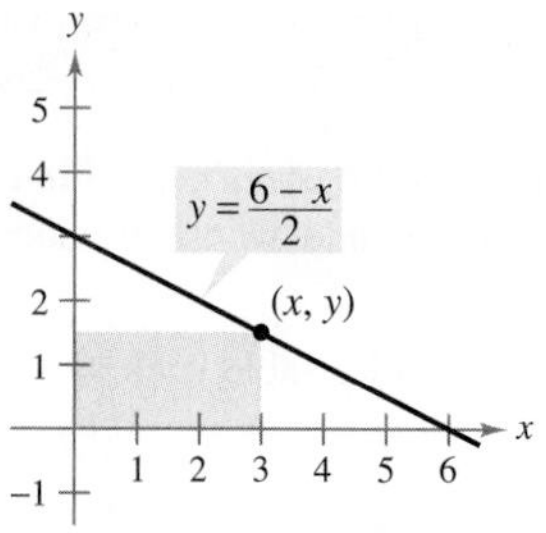

Figure for 24

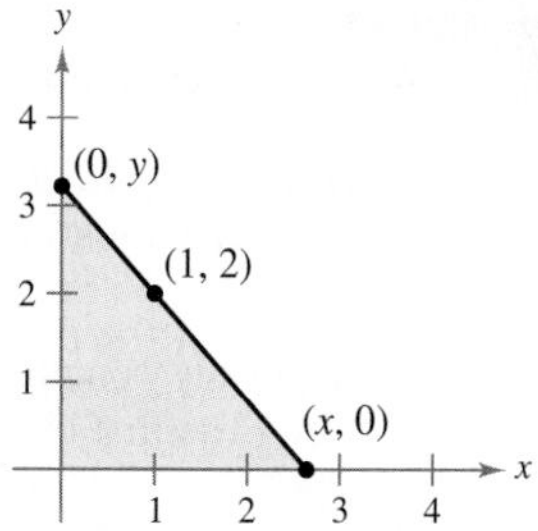

Figure for 25

25. ***Minimum Length*** A right triangle is formed in the first quadrant by the x- and y-axes and a line through the point $(1, 2)$ (see figure).

(a) Write the length L of the hypotenuse as a function of x.

(b) Use a graphing utility to approximate x graphically such that the length of the hypotenuse is a minimum.

(c) Find the vertices of the triangle such that its area is a minimum.

26. ***Maximum Area*** Find the area of the largest isosceles triangle that can be inscribed in a circle of radius 4 (see figure).

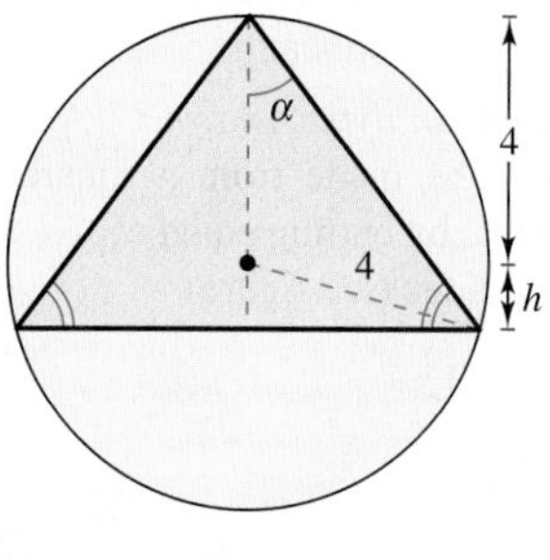

(a) Solve by writing the area as a function of h.

(b) Solve by writing the area as a function of α.

(c) Identify the type of triangle of maximum area.

27. ***Maximum Area*** A rectangle is bounded by the x-axis and the semicircle $y = \sqrt{25 - x^2}$ (see figure). What length and width should the rectangle have so that its area is a maximum?

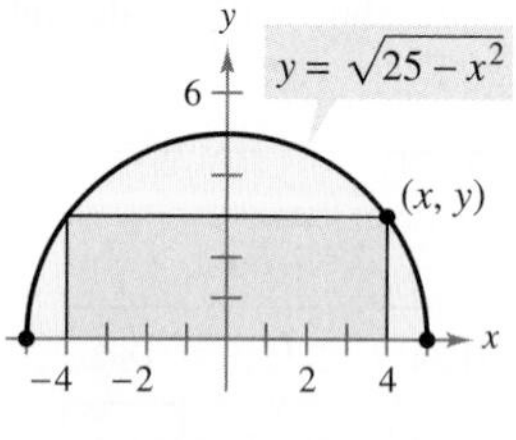

28. ***Area*** Find the dimensions of the largest rectangle that can be inscribed in a semicircle of radius r (see Exercise 27).

29. ***Area*** A rectangular page is to contain 30 square inches of print. The margins on each side are 1 inch. Find the dimensions of the page such that the least amount of paper is used.

30. ***Area*** A rectangular page is to contain 36 square inches of print. The margins on each side are to be $1\frac{1}{2}$ inches. Find the dimensions of the page such that the least amount of paper is used.

31. ***Numerical, Graphical, and Analytic Analysis*** An exercise room consists of a rectangle with a semicircle on each end. A 200-meter running track runs around the outside of the room.

(a) Draw a figure to represent the problem. Let x and y represent the length and width of the rectangle.

(b) Analytically complete six rows of a table such as the one below. (The first two rows are shown.) Use the table to guess the maximum area of the rectangular region.

Length x	Width y	Area
10	$\frac{2}{\pi}(100 - 10)$	$(10)\frac{2}{\pi}(100 - 10) \approx 573$
20	$\frac{2}{\pi}(100 - 20)$	$(20)\frac{2}{\pi}(100 - 20) \approx 1019$

(c) Write the area A as a function of x.

(d) Use calculus to find the critical number of the function in part (c) and find the maximum value.

(e) Use a graphing utility to graph the function in part (c) and verify the maximum area from the graph.

32. ***Numerical, Graphical, and Analytic Analysis*** A right circular cylinder is to be designed to hold 22 cubic inches of a soft drink (approximately 12 fluid ounces).

(a) Analytically complete six rows of a table such as the one below. (The first two rows are shown.)

Radius r	Height	Surface Area
0.2	$\frac{22}{\pi(0.2)^2}$	$2\pi(0.2)\left[0.2 + \frac{22}{\pi(0.2)^2}\right] \approx 220.3$
0.4	$\frac{22}{\pi(0.4)^2}$	$2\pi(0.4)\left[0.4 + \frac{22}{\pi(0.4)^2}\right] \approx 111.0$

(b) Use a graphing utility to generate additional rows of the table. Use the table to estimate the minimum surface area. (*Hint:* Use the *table* feature of the graphing utility.)

(c) Write the surface area S as a function of r.

(d) Use a graphing utility to graph the function in part (c) and estimate the minimum surface area from the graph.

(e) Use calculus to find the critical number of the function in part (c) and find dimensions that will yield the minimum surface area.

33. ***Maximum Volume*** A rectangular package to be sent by a postal service can have a maximum combined length and girth (perimeter of a cross section) of 108 inches (see figure). Find the dimensions of the package of maximum volume that can be sent. (Assume the cross section is square.)

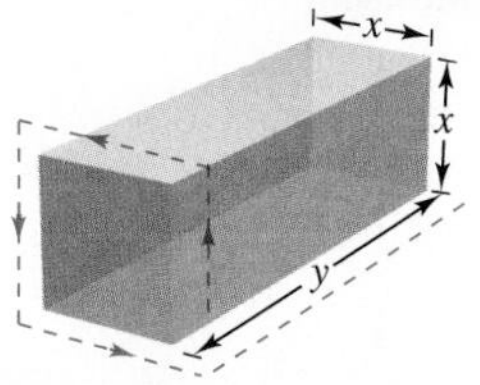

34. ***Maximum Volume*** Rework Exercise 33 for a cylindrical package. (The cross section is circular.)

35. ***Maximum Volume*** Find the volume of the largest right circular cone that can be inscribed in a sphere of radius r.

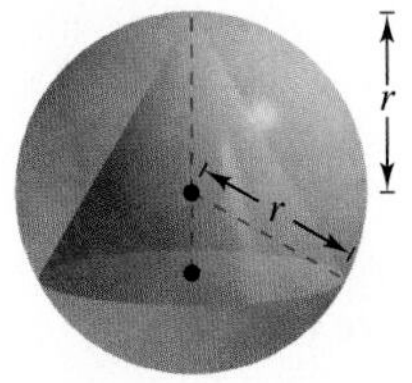

36. ***Maximum Volume*** Find the volume of the largest right circular cylinder that can be inscribed in a sphere of radius r.

Writing About Concepts

37. The perimeter of a rectangle is 20 feet. Of all possible dimensions, the maximum area is 25 square feet when its length and width are both 5 feet. Are there dimensions that yield a minimum area? Explain.

38. A shampoo bottle is a right circular cylinder. Because the surface area of the bottle does not change when it is squeezed, is it true that the volume remains the same? Explain.

39. ***Minimum Surface Area*** A solid is formed by adjoining two hemispheres to the ends of a right circular cylinder. The total volume of the solid is 12 cubic centimeters. Find the radius of the cylinder that produces the minimum surface area.

40. ***Minimum Cost*** An industrial tank of the shape described in Exercise 39 must have a volume of 3000 cubic feet. The hemispherical ends cost twice as much per square foot of surface area as the sides. Find the dimensions that will minimize cost.

41. ***Minimum Area*** The sum of the perimeters of an equilateral triangle and a square is 10. Find the dimensions of the triangle and the square that produce a minimum total area.

42. ***Maximum Area*** Twenty feet of wire is to be used to form two figures. In each of the following cases, how much wire should be used for each figure so that the total enclosed area is maximum?

(a) Equilateral triangle and square

(b) Square and regular pentagon

(c) Regular pentagon and regular hexagon

(d) Regular hexagon and circle

What can you conclude from this pattern? {*Hint:* The area of a regular polygon with n sides of length x is $A = (n/4)[\cot(\pi/n)]x^2$.}

43. ***Beam Strength*** A wooden beam has a rectangular cross section of height h and width w (see figure on the next page). The strength S of the beam is directly proportional to the width and the square of the height. What are the dimensions of the strongest beam that can be cut from a round log of diameter 24 inches? (*Hint:* $S = kh^2w$, where k is the proportionality constant.)

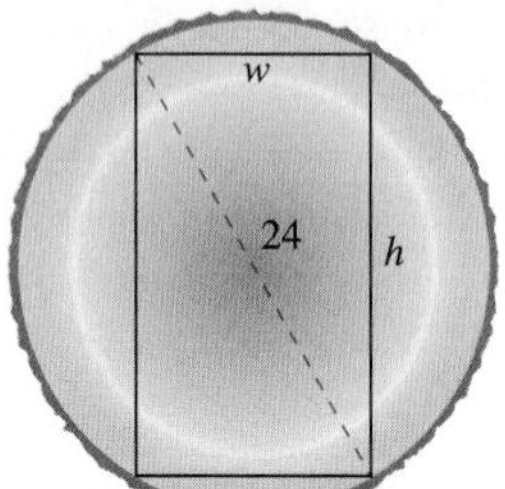

Figure for 43

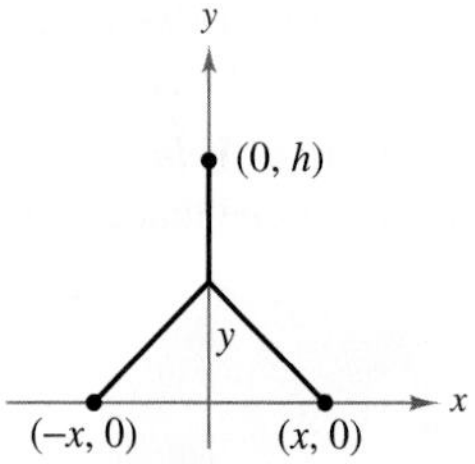

Figure for 44

44. ***Minimum Length*** Two factories are located at the coordinates $(-x, 0)$ and $(x, 0)$ with their power supply located at $(0, h)$ (see figure). Find y such that the total length of power line from the power supply to the factories is a minimum.

45. ***Projectile Range*** The range R of a projectile fired with an initial velocity v_0 at an angle θ with the horizontal is

$R = \dfrac{v_0^2 \sin 2\theta}{g}$, where g is the acceleration due to gravity. Find the angle θ such that the range is a maximum.

46. ***Conjecture*** Consider the functions $f(x) = \frac{1}{2}x^2$ and $g(x) = \frac{1}{16}x^4 - \frac{1}{2}x^2$ on the domain $[0, 4]$.

(a) Use a graphing utility to graph the functions on the specified domain.

(b) Write the vertical distance d between the functions as a function of x and use calculus to find the value of x for which d is maximum.

(c) Find the equations of the tangent lines to the graphs of f and g at the critical number found in part (b). Graph the tangent lines. What is the relationship between the lines?

(d) Make a conjecture about the relationship between tangent lines to the graphs of two functions at the value of x at which the vertical distance between the functions is greatest, and prove your conjecture.

47. ***Illumination*** A light source is located over the center of a circular table of diameter 4 feet (see figure). Find the height h of the light source such that the illumination I at the perimeter of the table is maximum if $I = k(\sin \alpha)/s^2$, where s is the slant height, α is the angle at which the light strikes the table, and k is a constant.

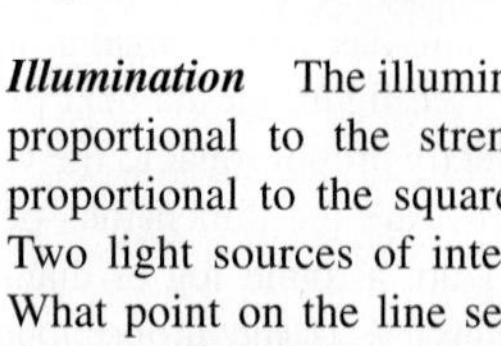

48. ***Illumination*** The illumination from a light source is directly proportional to the strength of the source and inversely proportional to the square of the distance from the source. Two light sources of intensities I_1 and I_2 are d units apart. What point on the line segment joining the two sources has the least illumination?

49. ***Minimum Time*** A man is in a boat 2 miles from the nearest point on the coast. He is to go to a point Q, located 3 miles down the coast and 1 mile inland (see figure). He can row at 2 miles per hour and walk at 4 miles per hour. Toward what point on the coast should he row in order to reach point Q in the least time?

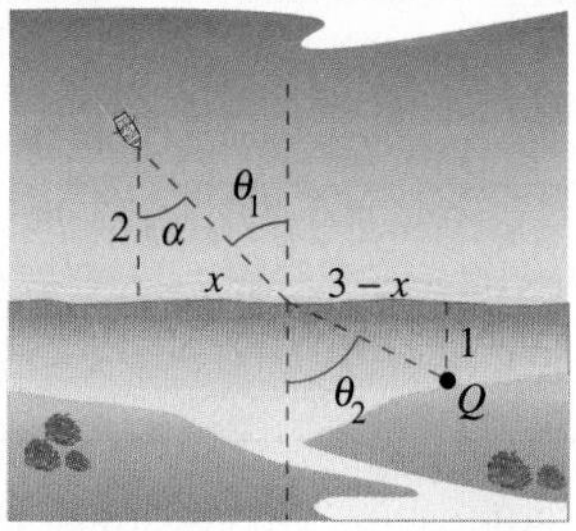

50. ***Minimum Time*** Consider Exercise 49 if the point Q is on the shoreline rather than 1 mile inland.

(a) Write the travel time T as a function of α.

(b) Use the result of part (a) to find the minimum time to reach Q.

(c) The man can row at v_1 miles per hour and walk at v_2 miles per hour. Write the time T as a function of α. Show that the critical number of T depends only on v_1 and v_2 and not the distances. Explain how this result would be more beneficial to the man than the result of Exercise 49.

(d) Describe how to apply the result of part (c) to minimizing the cost of constructing a power transmission cable that costs c_1 dollars per mile under water and c_2 dollars per mile over land.

51. ***Minimum Time*** The conditions are the same as in Exercise 49 except that the man can row at v_1 miles per hour and walk at v_2 miles per hour. If θ_1 and θ_2 are the magnitudes of the angles, show that the man will reach point Q in the least time when

$$\frac{\sin \theta_1}{v_1} = \frac{\sin \theta_2}{v_2}.$$

52. ***Minimum Time*** When light waves, traveling in a transparent medium, strike the surface of a second transparent medium, they change direction. This change of direction is called *refraction* and is defined by **Snell's Law of Refraction,**

$$\frac{\sin \theta_1}{v_1} = \frac{\sin \theta_2}{v_2}$$

where θ_1 and θ_2 are the magnitudes of the angles shown in the figure and v_1 and v_2 are the velocities of light in the two media. Show that this problem is equivalent to that of Exercise 51, and that light waves traveling from P to Q follow the path of minimum time.

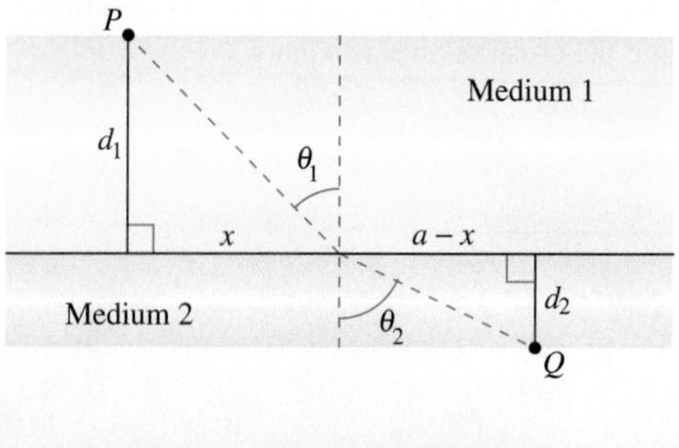

53. Sketch the graph of $f(x) = 2 - 2 \sin x$ on the interval $[0, \pi/2]$.

(a) Find the distance from the origin to the y-intercept and the distance from the origin to the x-intercept.

(b) Write the distance d from the origin to a point on the graph of f as a function of x. Use your graphing utility to graph d and find the minimum distance.

(c) Use calculus and the *zero* or *root* feature of a graphing utility to find the value of x that minimizes the function d on the interval $[0, \pi/2]$. What is the minimum distance?

(Submitted by Tim Chapell, Penn Valley Community College, Kansas City, MO.)

54. ***Minimum Cost*** An offshore oil well is 2 kilometers off the coast. The refinery is 4 kilometers down the coast. Laying pipe in the ocean is twice as expensive as on land. What path should the pipe follow in order to minimize the cost?

55. ***Minimum Force*** A component is designed to slide a block of steel with weight W across a table and into a chute (see figure). The motion of the block is resisted by a frictional force proportional to its apparent weight. (Let k be the constant of proportionality.) Find the minimum force F needed to slide the block, and find the corresponding value of θ. (*Hint:* $F \cos \theta$ is the force in the direction of motion, and $F \sin \theta$ is the amount of force tending to lift the block. So, the apparent weight of the block is $W - F \sin \theta$.)

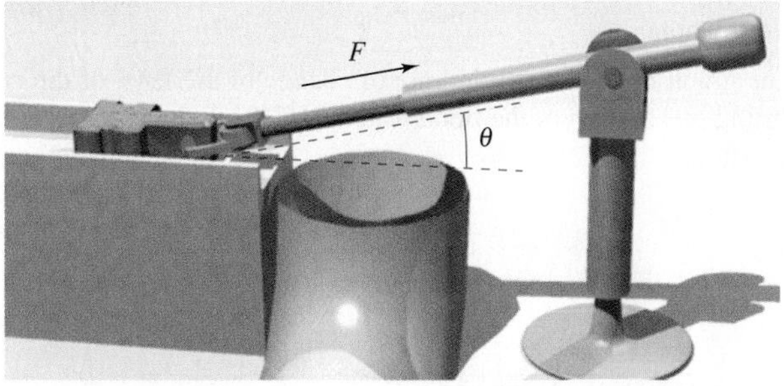

56. ***Maximum Volume*** A sector with central angle θ is cut from a circle of radius 12 inches (see figure), and the edges of the sector are brought together to form a cone. Find the magnitude of θ such that the volume of the cone is a maximum.

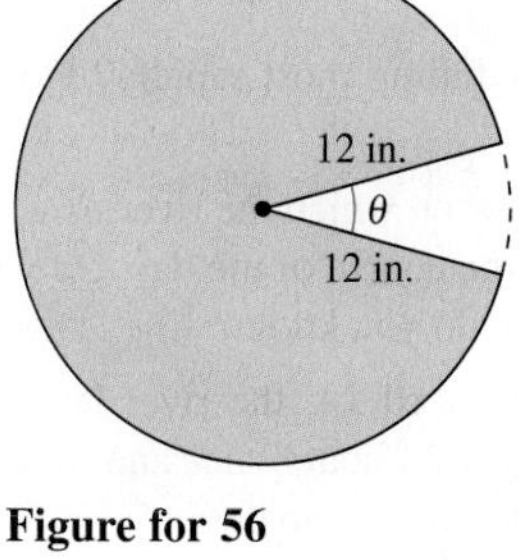

Figure for 56

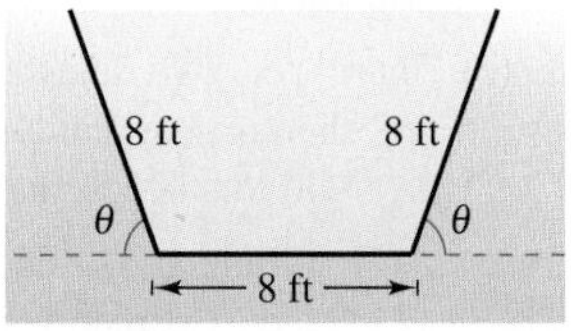

Figure for 57

57. ***Numerical, Graphical, and Analytic Analysis*** The cross sections of an irrigation canal are isosceles trapezoids of which three sides are 8 feet long (see figure). Determine the angle of elevation θ of the sides such that the area of the cross section is a maximum by completing the following.

(a) Analytically complete six rows of a table such as the one below. (The first two rows are shown.)

Base 1	Base 2	Altitude	Area
8	$8 + 16 \cos 10°$	$8 \sin 10°$	≈ 22.1
8	$8 + 16 \cos 20°$	$8 \sin 20°$	≈ 42.5

(b) Use a graphing utility to generate additional rows of the table and estimate the maximum cross-sectional area. (*Hint:* Use the *table* feature of the graphing utility.)

(c) Write the cross-sectional area A as a function of θ.

(d) Use calculus to find the critical number of the function in part (c) and find the angle that will yield the maximum cross-sectional area.

(e) Use a graphing utility to graph the function in part (c) and verify the maximum cross-sectional area.

58. ***Maximum Profit*** Assume that the amount of money deposited in a bank is proportional to the square of the interest rate the bank pays on this money. Furthermore, the bank can reinvest this money at 12%. Find the interest rate the bank should pay to maximize profit. (Use the simple interest formula.)

59. ***Minimum Cost*** The ordering and transportation cost C of the components used in manufacturing a product is

$$C = 100\left(\frac{200}{x^2} + \frac{x}{x + 30}\right), \quad x \geq 1$$

where C is measured in thousands of dollars and x is the order size in hundreds. Find the order size that minimizes the cost. (*Hint:* Use the *root* feature of a graphing utility.)

60. ***Diminishing Returns*** The profit P (in thousands of dollars) for a company spending an amount s (in thousands of dollars) on advertising is

$$P = -\tfrac{1}{10}s^3 + 6s^2 + 400.$$

(a) Find the amount of money the company should spend on advertising in order to yield a maximum profit.

(b) The *point of diminishing returns* is the point at which the rate of growth of the profit function begins to decline. Find the point of diminishing returns.

Minimum Distance **In Exercises 61–63, consider a fuel distribution center located at the origin of the rectangular coordinate system (units in miles; see figures on next page). The center supplies three factories with coordinates (4, 1), (5, 6), and (10, 3). A trunk line will run from the distribution center along the line $y = mx$, and feeder lines will run to the three factories. The objective is to find m such that the lengths of the feeder lines are minimized.**

61. Minimize the sum of the squares of the lengths of vertical feeder lines given by

$$S_1 = (4m - 1)^2 + (5m - 6)^2 + (10m - 3)^2.$$

Find the equation for the trunk line by this method and then determine the sum of the lengths of the feeder lines.

62. Minimize the sum of the absolute values of the lengths of vertical feeder lines given by

$$S_2 = |4m - 1| + |5m - 6| + |10m - 3|.$$

Find the equation for the trunk line by this method and then determine the sum of the lengths of the feeder lines. (*Hint:* Use a graphing utility to graph the function S_2 and approximate the required critical number.)

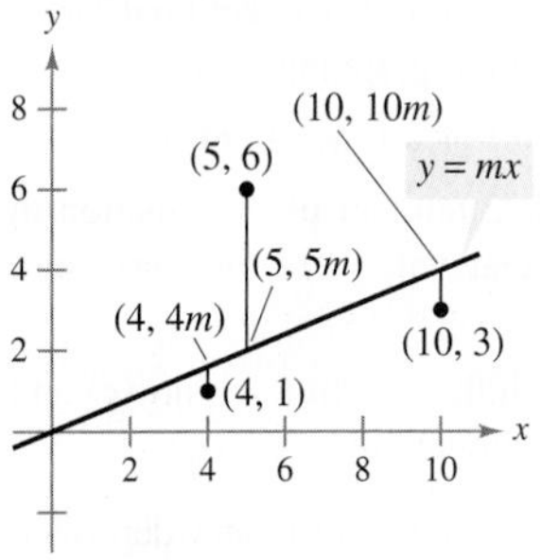

Figure for 61 and 62

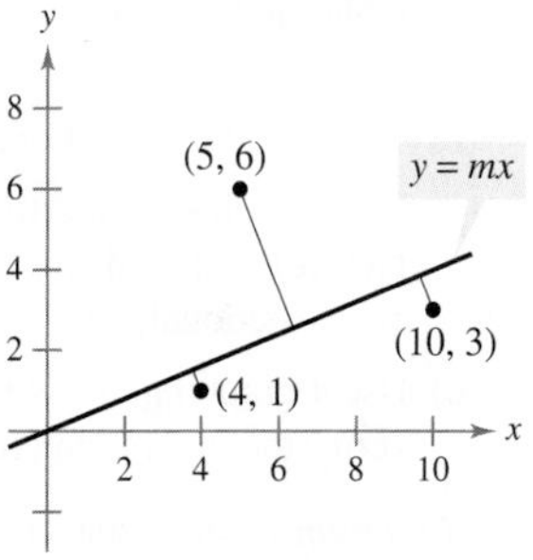

Figure for 63

63. Minimize the sum of the perpendicular distances (see Exercises 85–90 in Section P.2) from the trunk line to the factories given by

$$S_3 = \frac{|4m - 1|}{\sqrt{m^2 + 1}} + \frac{|5m - 6|}{\sqrt{m^2 + 1}} + \frac{|10m - 3|}{\sqrt{m^2 + 1}}.$$

Find the equation for the trunk line by this method and then determine the sum of the lengths of the feeder lines. (*Hint:* Use a graphing utility to graph the function S_3 and approximate the required critical number.)

64. ***Maximum Area*** Consider a symmetric cross inscribed in a circle of radius r (see figure).

(a) Write the area A of the cross as a function of x and find the value of x that maximizes the area.

(b) Write the area A of the cross as a function of θ and find the value of θ that maximizes the area.

(c) Show that the critical numbers of parts (a) and (b) yield the same maximum area. What is that area?

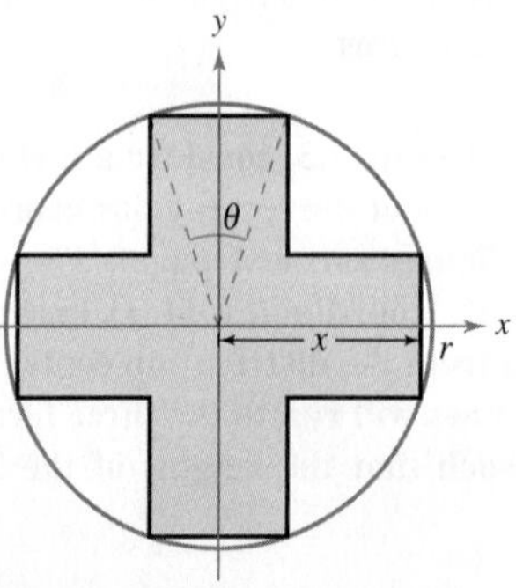

Putnam Exam Challenge

65. Find the maximum value of $f(x) = x^3 - 3x$ on the set of all real numbers x satisfying $x^4 + 36 \le 13x^2$. Explain your reasoning.

66. Find the minimum value of

$$\frac{(x + 1/x)^6 - (x^6 + 1/x^6) - 2}{(x + 1/x)^3 + (x^3 + 1/x^3)}$$

for $x > 0$.

These problems were composed by the Committee on the Putnam Prize Competition.

Section Project: Connecticut River

Whenever the Connecticut River reaches a level of 105 feet above sea level, two Northampton, Massachusetts flood control station operators begin a round-the-clock river watch. Every 2 hours, they check the height of the river, using a scale marked off in tenths of a foot, and record the data in a log book. In the spring of 1996, the flood watch lasted from April 4, when the river reached 105 feet and was rising at 0.2 foot per hour, until April 25, when the level subsided again to 105 feet. Between those dates, their log shows that the river rose and fell several times, at one point coming close to the 115-foot mark. If the river had reached 115 feet, the city would have closed down Mount Tom Road (Route 5, south of Northampton).

The graph below shows the rate of change of the level of the river during one portion of the flood watch. Use the graph to answer each question.

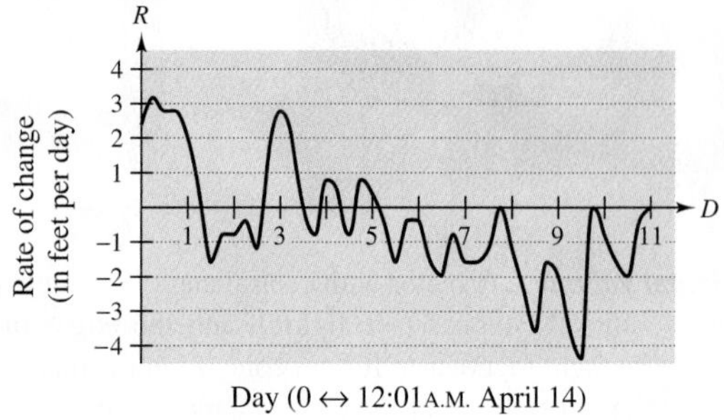

(a) On what date was the river rising most rapidly? How do you know?

(b) On what date was the river falling most rapidly? How do you know?

(c) There were two dates in a row on which the river rose, then fell, then rose again during the course of the day. On which days did this occur, and how do you know?

(d) At 1 minute past midnight, April 14, the river level was 111.0 feet. Estimate its height 24 hours later and 48 hours later. Explain how you made your estimates.

(e) The river crested at 114.4 feet. On what date do you think this occurred?

(Submitted by Mary Murphy, Smith College, Northampton, MA)

Section 3.8 Newton's Method

- **Approximate a zero of a function using Newton's Method.**

Newton's Method

In this section you will study a technique for approximating the real zeros of a function. The technique is called **Newton's Method,** and it uses tangent lines to approximate the graph of the function near its x-intercepts.

To see how Newton's Method works, consider a function f that is continuous on the interval $[a, b]$ and differentiable on the interval (a, b). If $f(a)$ and $f(b)$ differ in sign, then, by the Intermediate Value Theorem, f must have at least one zero in the interval (a, b). Suppose you estimate this zero to occur at

$$x = x_1 \qquad \text{First estimate}$$

as shown in Figure 3.60(a). Newton's Method is based on the assumption that the graph of f and the tangent line at $(x_1, f(x_1))$ both cross the x-axis at *about* the same point. Because you can easily calculate the x-intercept for this tangent line, you can use it as a second (and, usually, better) estimate for the zero of f. The tangent line passes through the point $(x_1, f(x_1))$ with a slope of $f'(x_1)$. In point-slope form, the equation of the tangent line is therefore

$$y - f(x_1) = f'(x_1)(x - x_1)$$
$$y = f'(x_1)(x - x_1) + f(x_1).$$

Letting $y = 0$ and solving for x produces

$$x = x_1 - \frac{f(x_1)}{f'(x_1)}.$$

So, from the initial estimate x_1 you obtain a new estimate

$$x_2 = x_1 - \frac{f(x_1)}{f'(x_1)}. \qquad \text{Second estimate [see Figure 3.60(b)]}$$

You can improve on x_2 and calculate yet a third estimate

$$x_3 = x_2 - \frac{f(x_2)}{f'(x_2)}. \qquad \text{Third estimate}$$

Repeated application of this process is called Newton's Method.

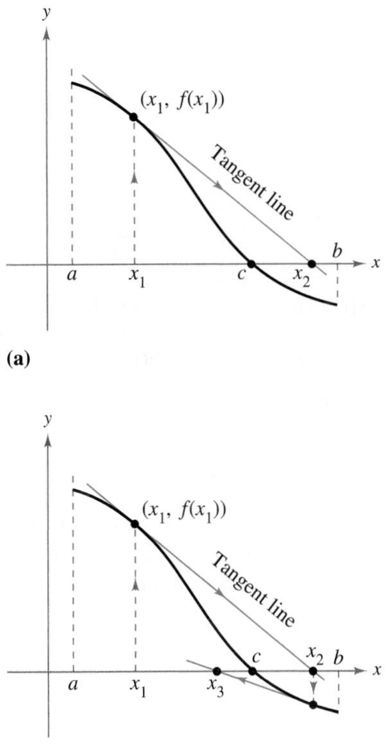

(b)
The x-intercept of the tangent line approximates the zero of f.
Figure 3.60

> **Newton's Method for Approximating the Zeros of a Function**
>
> Let $f(c) = 0$, where f is differentiable on an open interval containing c. Then, to approximate c, use the following steps.
>
> 1. Make an initial estimate x_1 that is close to c. (A graph is helpful.)
> 2. Determine a new approximation
>
> $$x_{n+1} = x_n - \frac{f(x_n)}{f'(x_n)}.$$
>
> 3. If $|x_n - x_{n+1}|$ is within the desired accuracy, let x_{n+1} serve as the final approximation. Otherwise, return to Step 2 and calculate a new approximation.
>
> Each successive application of this procedure is called an **iteration.**

NEWTON'S METHOD

Isaac Newton first described the method for approximating the real zeros of a function in his text *Method of Fluxions*. Although the book was written in 1671, it was not published until 1736. Meanwhile, in 1690, Joseph Raphson (1648–1715) published a paper describing a method for approximating the real zeros of a function that was very similar to Newton's. For this reason, the method is often referred to as the Newton-Raphson method.

NOTE For many functions, just a few iterations of Newton's Method will produce approximations having very small errors, as shown in Example 1.

EXAMPLE 1 Using Newton's Method

Calculate three iterations of Newton's Method to approximate a zero of $f(x) = x^2 - 2$. Use $x_1 = 1$ as the initial guess.

Solution Because $f(x) = x^2 - 2$, you have $f'(x) = 2x$, and the iterative process is given by the formula

$$x_{n+1} = x_n - \frac{f(x_n)}{f'(x_n)} = x_n - \frac{x_n^2 - 2}{2x_n}.$$

The calculations for three iterations are shown in the table.

n	x_n	$f(x_n)$	$f'(x_n)$	$\frac{f(x_n)}{f'(x_n)}$	$x_n - \frac{f(x_n)}{f'(x_n)}$
1	1.000000	−1.000000	2.000000	−0.500000	1.500000
2	1.500000	0.250000	3.000000	0.083333	1.416667
3	1.416667	0.006945	2.833334	0.002451	1.414216
4	1.414216				

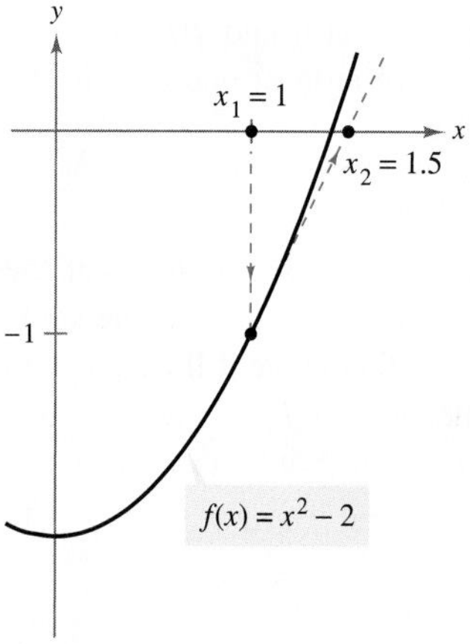

The first iteration of Newton's Method
Figure 3.61

Of course, in this case you know that the two zeros of the function are $\pm\sqrt{2}$. To six decimal places, $\sqrt{2} = 1.414214$. So, after only three iterations of Newton's Method, you have obtained an approximation that is within 0.000002 of an actual root. The first iteration of this process is shown in Figure 3.61.

EXAMPLE 2 Using Newton's Method

Use Newton's Method to approximate the zeros of

$$f(x) = 2x^3 + x^2 - x + 1.$$

Continue the iterations until two successive approximations differ by less than 0.0001.

Solution Begin by sketching a graph of f, as shown in Figure 3.62. From the graph, you can observe that the function has only one zero, which occurs near $x = -1.2$. Next, differentiate f and form the iterative formula

$$x_{n+1} = x_n - \frac{f(x_n)}{f'(x_n)} = x_n - \frac{2x_n^3 + x_n^2 - x_n + 1}{6x_n^2 + 2x_n - 1}.$$

The calculations are shown in the table.

n	x_n	$f(x_n)$	$f'(x_n)$	$\frac{f(x_n)}{f'(x_n)}$	$x_n - \frac{f(x_n)}{f'(x_n)}$
1	−1.20000	0.18400	5.24000	0.03511	−1.23511
2	−1.23511	−0.00771	5.68276	−0.00136	−1.23375
3	−1.23375	0.00001	5.66533	0.00000	−1.23375
4	−1.23375				

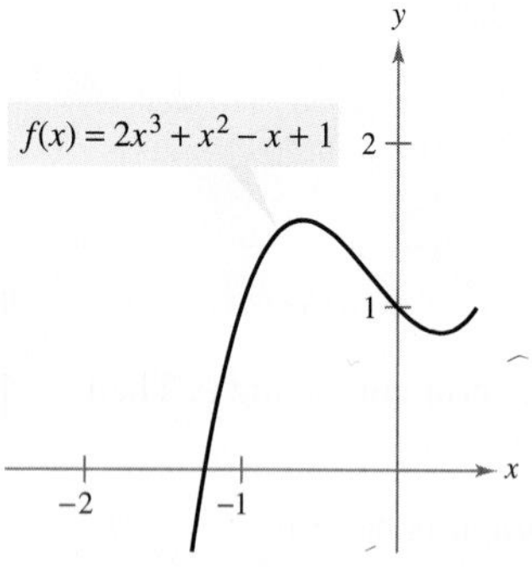

After three iterations of Newton's Method, the zero of f is approximated to the desired accuracy.
Figure 3.62

Because two successive approximations differ by less than the required 0.0001, you can estimate the zero of f to be -1.23375.

When, as in Examples 1 and 2, the approximations approach a limit, the sequence $x_1, x_2, x_3, \ldots, x_n, \ldots$ is said to **converge.** Moreover, if the limit is c, it can be shown that c must be a zero of f.

Newton's Method does not always yield a convergent sequence. One way it can fail to do so is shown in Figure 3.63. Because Newton's Method involves division by $f'(x_n)$, it is clear that the method will fail if the derivative is zero for any x_n in the sequence. When you encounter this problem, you can usually overcome it by choosing a different value for x_1. Another way Newton's Method can fail is shown in the next example.

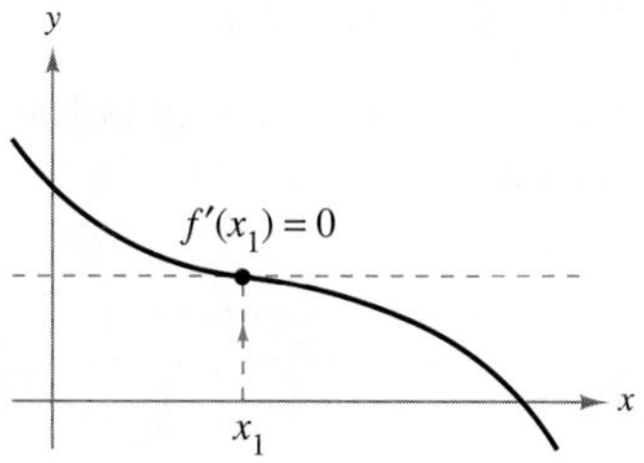

Newton's Method fails to converge if $f'(x_n) = 0$.
Figure 3.63

EXAMPLE 3 An Example in Which Newton's Method Fails

The function $f(x) = x^{1/3}$ is not differentiable at $x = 0$. Show that Newton's Method fails to converge using $x_1 = 0.1$.

Solution Because $f'(x) = \frac{1}{3}x^{-2/3}$, the iterative formula is

$$\begin{aligned} x_{n+1} &= x_n - \frac{f(x_n)}{f'(x_n)} \\ &= x_n - \frac{x_n^{1/3}}{\frac{1}{3}x_n^{-2/3}} \\ &= x_n - 3x_n \\ &= -2x_n. \end{aligned}$$

The calculations are shown in the table. This table and Figure 3.64 indicate that x_n continues to increase in magnitude as $n \to \infty$, and so the limit of the sequence does not exist.

n	x_n	$f(x_n)$	$f'(x_n)$	$\frac{f(x_n)}{f'(x_n)}$	$x_n - \frac{f(x_n)}{f'(x_n)}$
1	0.10000	0.46416	1.54720	0.30000	−0.20000
2	−0.20000	−0.58480	0.97467	−0.60000	0.40000
3	0.40000	0.73681	0.61401	1.20000	−0.80000
4	−0.80000	−0.92832	0.38680	−2.40000	1.60000

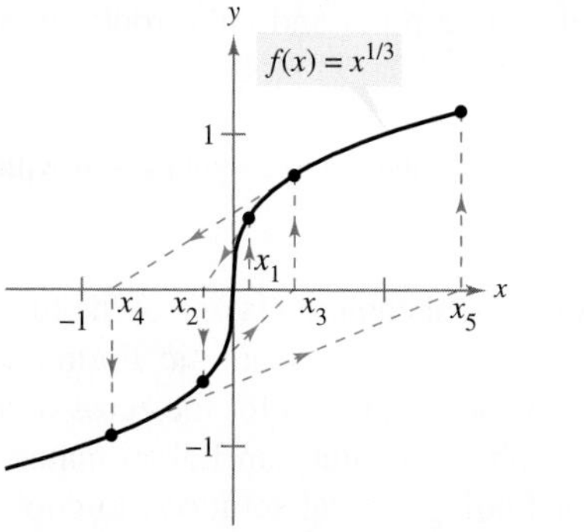

Newton's Method fails to converge for every x-value other than the actual zero of f.
Figure 3.64

NOTE In Example 3, the initial estimate $x_1 = 0.1$ fails to produce a convergent sequence. Try showing that Newton's Method also fails for every other choice of x_1 (other than the actual zero).

It can be shown that a condition sufficient to produce convergence of Newton's Method to a zero of f is that

$$\left|\frac{f(x)f''(x)}{[f'(x)]^2}\right| < 1$$

Condition for convergence

on an open interval containing the zero. For instance, in Example 1 this test would yield $f(x) = x^2 - 2, f'(x) = 2x, f''(x) = 2$, and

$$\left|\frac{f(x)f''(x)}{[f'(x)]^2}\right| = \left|\frac{(x^2-2)(2)}{4x^2}\right| = \left|\frac{1}{2} - \frac{1}{x^2}\right|.$$

Example 1

On the interval $(1, 3)$, this quantity is less than 1 and therefore the convergence of Newton's Method is guaranteed. On the other hand, in Example 3, you have $f(x) = x^{1/3}, f'(x) = \frac{1}{3}x^{-2/3}, f''(x) = -\frac{2}{9}x^{-5/3}$, and

$$\left|\frac{f(x)f''(x)}{[f'(x)]^2}\right| = \left|\frac{x^{1/3}(-2/9)(x^{-5/3})}{(1/9)(x^{-4/3})}\right| = 2$$

Example 3

which is not less than 1 for any value of x, so you cannot conclude that Newton's Method will converge.

The Granger Collection

NIELS HENRIK ABEL (1802–1829)

The Granger Collection

EVARISTE GALOIS (1811–1832)

Although the lives of both Abel and Galois were brief, their work in the fields of analysis and abstract algebra was far-reaching.

Algebraic Solutions of Polynomial Equations

The zeros of some functions, such as

$$f(x) = x^3 - 2x^2 - x + 2$$

can be found by simple algebraic techniques, such as factoring. The zeros of other functions, such as

$$f(x) = x^3 - x + 1$$

cannot be found by *elementary* algebraic methods. This particular function has only one real zero, and by using more advanced algebraic techniques you can determine the zero to be

$$x = -\sqrt[3]{\frac{3 - \sqrt{23/3}}{6}} - \sqrt[3]{\frac{3 + \sqrt{23/3}}{6}}.$$

Because the *exact* solution is written in terms of square roots and cube roots, it is called a **solution by radicals.**

NOTE Try approximating the real zero of $f(x) = x^3 - x + 1$ and compare your result with the exact solution shown above.

The determination of radical solutions of a polynomial equation is one of the fundamental problems of algebra. The earliest such result is the Quadratic Formula, which dates back at least to Babylonian times. The general formula for the zeros of a cubic function was developed much later. In the sixteenth century an Italian mathematician, Jerome Cardan, published a method for finding radical solutions to cubic and quartic equations. Then, for 300 years, the problem of finding a general quintic formula remained open. Finally, in the nineteenth century, the problem was answered independently by two young mathematicians. Niels Henrik Abel, a Norwegian mathematician, and Evariste Galois, a French mathematician, proved that it is not possible to solve a *general* fifth- (or higher-) degree polynomial equation by radicals. Of course, you can solve particular fifth-degree equations such as $x^5 - 1 = 0$, but Abel and Galois were able to show that no general *radical* solution exists.

Exercises for Section 3.8

See www.CalcChat.com for worked-out solutions to odd-numbered exercises.

In Exercises 1–4, complete two iterations of Newton's Method for the function using the given initial guess.

1. $f(x) = x^2 - 3,\ \ x_1 = 1.7$

2. $f(x) = 2x^2 - 3,\ \ x_1 = 1$

3. $f(x) = \sin x,\ \ x_1 = 3$

4. $f(x) = \tan x,\ \ x_1 = 0.1$

In Exercises 5–14, approximate the zero(s) of the function. Use Newton's Method and continue the process until two successive approximations differ by less than 0.001. Then find the zero(s) using a graphing utility and compare the results.

5. $f(x) = x^3 + x - 1$

6. $f(x) = x^5 + x - 1$

7. $f(x) = 3\sqrt{x-1} - x$

8. $f(x) = x - 2\sqrt{x+1}$

9. $f(x) = x^3 + 3$

10. $f(x) = 1 - 2x^3$

11. $f(x) = x^3 - 3.9x^2 + 4.79x - 1.881$

12. $f(x) = \frac{1}{2}x^4 - 3x - 3$

13. $f(x) = x + \sin(x + 1)$

14. $f(x) = x^3 - \cos x$

In Exercises 15–18, apply Newton's Method to approximate the x-value(s) of the indicated point(s) of intersection of the two graphs. Continue the process until two successive approximations differ by less than 0.001. [*Hint:* Let $h(x) = f(x) - g(x)$.]

15. $f(x) = 2x + 1$
$g(x) = \sqrt{x + 4}$

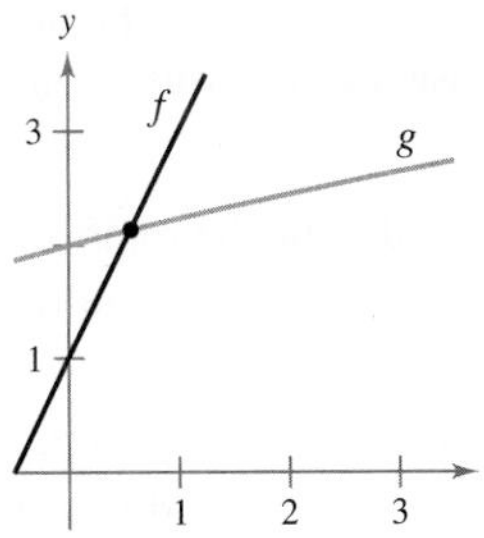

16. $f(x) = 3 - x$
$g(x) = 1/(x^2 + 1)$

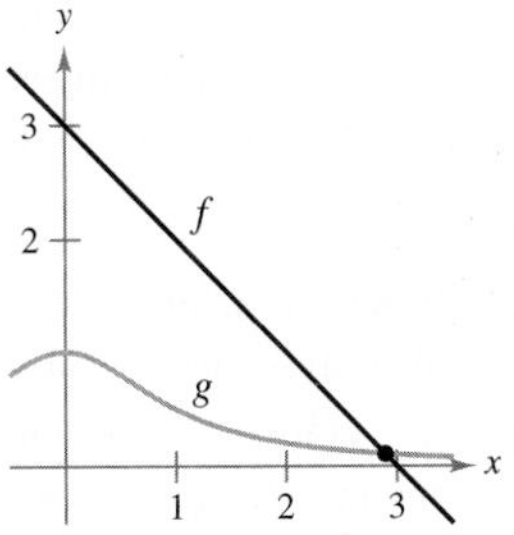

17. $f(x) = x$
$g(x) = \tan x$

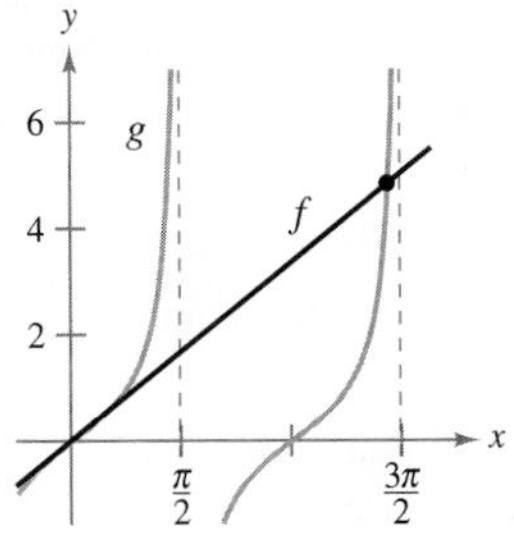

18. $f(x) = x^2$
$g(x) = \cos x$

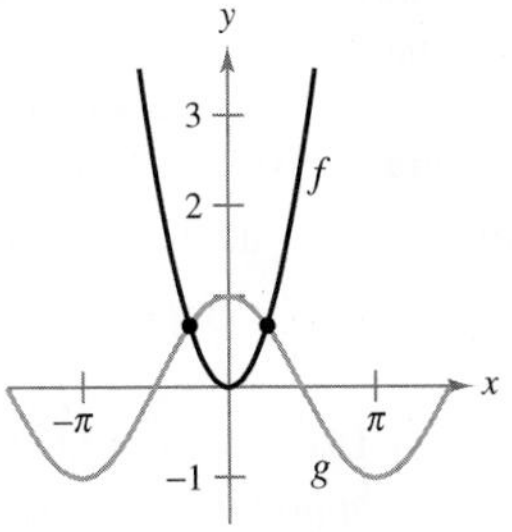

19. ***Mechanic's Rule*** The Mechanic's Rule for approximating $\sqrt{a}$, $a > 0$, is

$$x_{n+1} = \frac{1}{2}\left(x_n + \frac{a}{x_n}\right), \quad n = 1, 2, 3 \ldots$$

where x_1 is an approximation of $\sqrt{a}$.

(a) Use Newton's Method and the function $f(x) = x^2 - a$ to derive the Mechanic's Rule.

(b) Use the Mechanic's Rule to approximate $\sqrt{5}$ and $\sqrt{7}$ to three decimal places.

20. (a) Use Newton's Method and the function $f(x) = x^n - a$ to obtain a general rule for approximating $x = \sqrt[n]{a}$.

(b) Use the general rule found in part (a) to approximate $\sqrt[4]{6}$ and $\sqrt[3]{15}$ to three decimal places.

In Exercises 21–24, apply Newton's Method using the given initial guess, and explain why the method fails.

21. $y = 2x^3 - 6x^2 + 6x - 1,\ \ x_1 = 1$

22. $y = 4x^3 - 12x^2 + 12x - 3,\ \ x_1 = \frac{3}{2}$

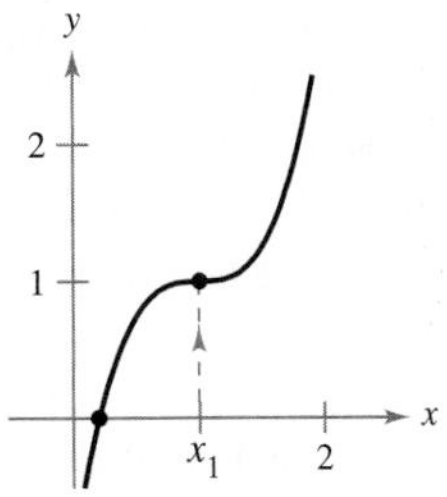

Figure for 21

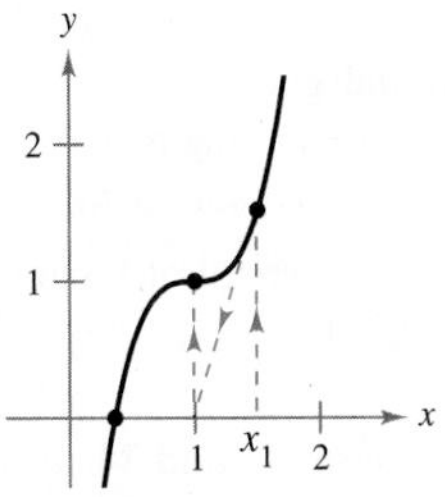

Figure for 22

23. $f(x) = -x^3 + 6x^2 - 10x + 6,\ \ x_1 = 2$

24. $f(x) = 2\sin x + \cos 2x,\ \ x_1 = \dfrac{3\pi}{2}$

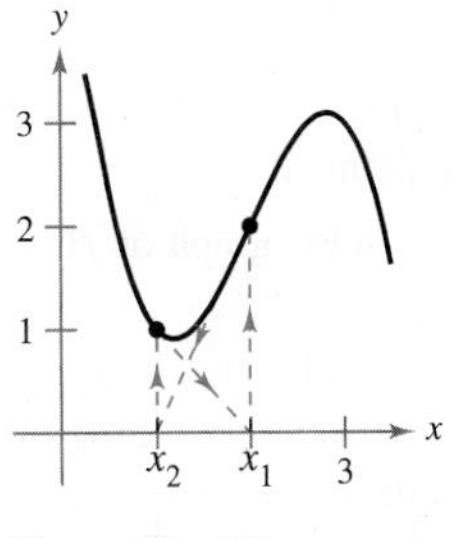

Figure for 23

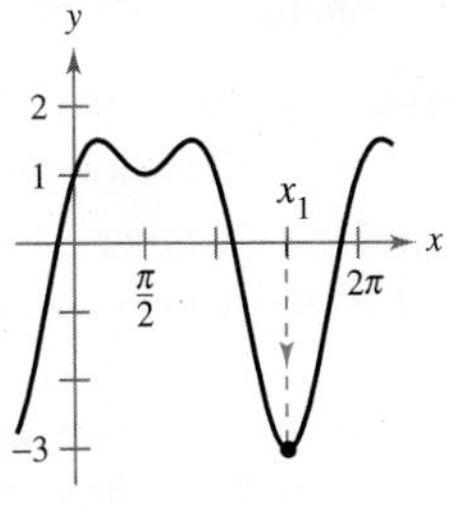

Figure for 24

Writing About Concepts

25. In your own words and using a sketch, describe Newton's Method for approximating the zeros of a function.

26. Under what conditions will Newton's Method fail?

Fixed Point **In Exercises 27 and 28, approximate the fixed point of the function to two decimal places. [A *fixed point* x_0 of a function f is a value of x such that $f(x_0) = x_0$.]**

27. $f(x) = \cos x$

28. $f(x) = \cot x,\ \ 0 < x < \pi$

29. ***Writing*** Consider the function $f(x) = x^3 - 3x^2 + 3$.

(a) Use a graphing utility to graph f.

(b) Use Newton's Method with $x_1 = 1$ as an initial guess.

(c) Repeat part (b) using $x_1 = \frac{1}{4}$ as an initial guess and observe that the result is different.

(d) To understand why the results in parts (b) and (c) are different, sketch the tangent lines to the graph of f at the points $(1, f(1))$ and $\left(\frac{1}{4}, f\left(\frac{1}{4}\right)\right)$. Find the x-intercept of each tangent line and compare the intercepts with the first iteration of Newton's Method using the respective initial guesses.

(e) Write a short paragraph summarizing how Newton's Method works. Use the results of this exercise to describe why it is important to select the initial guess carefully.

30. ***Writing*** Repeat the steps in Exercise 29 for the function $f(x) = \sin x$ with initial guesses of $x_1 = 1.8$ and $x_1 = 3$.

31. Use Newton's Method to show that the equation $x_{n+1} = x_n(2 - ax_n)$ can be used to approximate $1/a$ if x_1 is an initial guess of the reciprocal of a. Note that this method of approximating reciprocals uses only the operations of multiplication and subtraction. [*Hint:* Consider $f(x) = (1/x) - a$.]

32. Use the result of Exercise 31 to approximate (a) $\frac{1}{3}$ and (b) $\frac{1}{11}$ to three decimal places.

In Exercises 33 and 34, approximate the critical number of f on the interval $(0, \pi)$. Sketch the graph of f, labeling any extrema.

33. $f(x) = x \cos x$

34. $f(x) = x \sin x$

In Exercises 35–38, some typical problems from the previous sections of this chapter are given. In each case, use Newton's Method to approximate the solution.

35. ***Minimum Distance*** Find the point on the graph of $f(x) = 4 - x^2$ that is closest to the point $(1, 0)$.

36. ***Minimum Distance*** Find the point on the graph of $f(x) = x^2$ that is closest to the point $(4, -3)$.

37. ***Minimum Time*** You are in a boat 2 miles from the nearest point on the coast (see figure). You are to go to a point Q, which is 3 miles down the coast and 1 mile inland. You can row at 3 miles per hour and walk at 4 miles per hour. Toward what point on the coast should you row in order to reach Q in the least time?

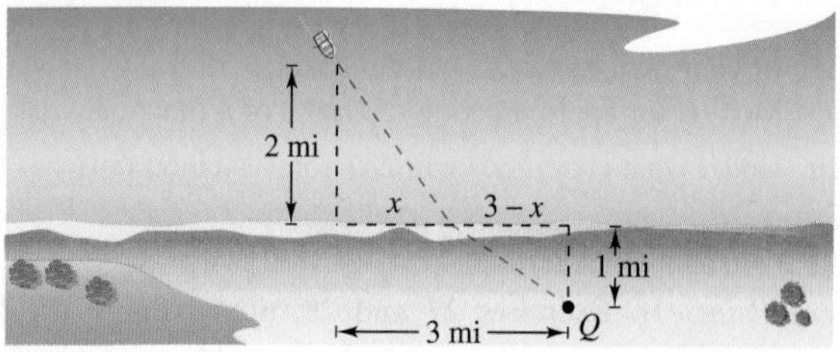

38. ***Medicine*** The concentration C of a chemical in the bloodstream t hours after injection into muscle tissue is given by $C = (3t^2 + t)/(50 + t^3)$. When is the concentration greatest?

39. ***Advertising Costs*** A company that produces portable CD players estimates that the profit for selling a particular model is

$$P = -76x^3 + 4830x^2 - 320{,}000, \quad 0 \le x \le 60$$

where P is the profit in dollars and x is the advertising expense in 10,000s of dollars (see figure). According to this model, find the smaller of two advertising amounts that yield a profit P of \$2,500,000.

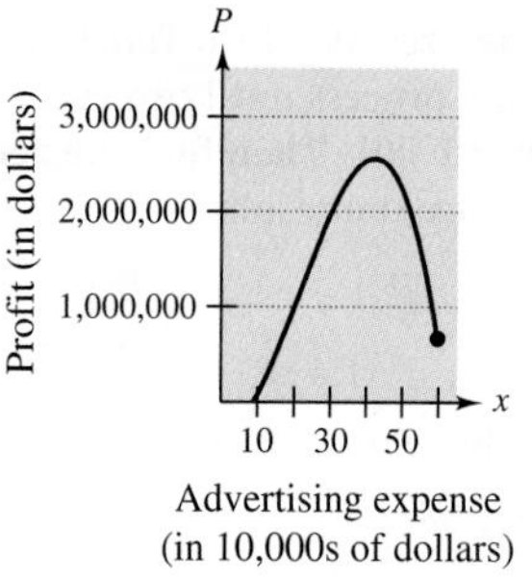

Figure for 39

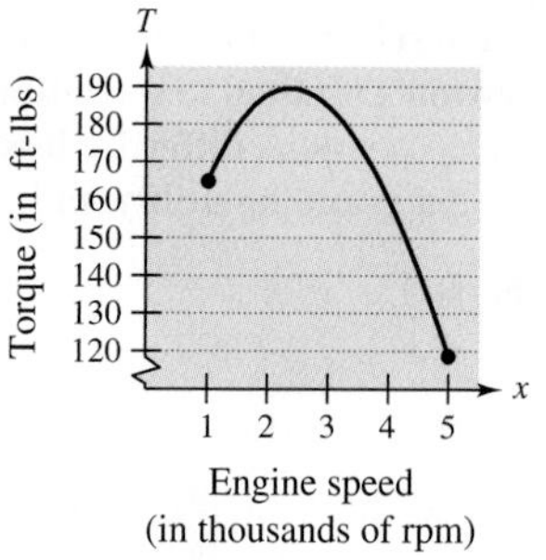

Figure for 40

40. ***Engine Power*** The torque produced by a compact automobile engine is approximated by the model

$$T = 0.808x^3 - 17.974x^2 + 71.248x + 110.843, \quad 1 \le x \le 5$$

where T is the torque in foot-pounds and x is the engine speed in thousands of revolutions per minute (see figure). Approximate the two engine speeds that yield a torque T of 170 foot-pounds.

True or False? **In Exercises 41–44, determine whether the statement is true or false. If it is false, explain why or give an example that shows it is false.**

41. The zeros of $f(x) = p(x)/q(x)$ coincide with the zeros of $p(x)$.

42. If the coefficients of a polynomial function are all positive, then the polynomial has no positive zeros.

43. If $f(x)$ is a cubic polynomial such that $f'(x)$ is never zero, then any initial guess will force Newton's Method to converge to the zero of f.

44. The roots of $\sqrt{f(x)} = 0$ coincide with the roots of $f(x) = 0$.

45. ***Tangent Lines*** The graph of $f(x) = -\sin x$ has infinitely many tangent lines that pass through the origin. Use Newton's Method to approximate the slope of the tangent line having the greatest slope to three decimal places.

46. Consider the function $f(x) = 2x^3 - 20x^2 - 12x - 24$.

(a) Use a graphing utility to determine the number of zeros of f.

(b) Use Newton's Method with an initial estimate of $x_1 = 2$ to approximate the zero of f to four decimal places.

(c) Repeat part (b) using initial estimates of $x_1 = 10$ and $x_1 = 100$.

(d) Discuss the results of parts (b) and (c). What can you conclude?

Section 3.9 Differentials

- **Understand the concept of a tangent line approximation.**
- **Compare the value of the differential, dy, with the actual change in y, Δy.**
- **Estimate a propagated error using a differential.**
- **Find the differential of a function using differentiation formulas.**

EXPLORATION

Tangent Line Approximation Use a graphing utility to graph

$$f(x) = x^2.$$

In the same viewing window, graph the tangent line to the graph of f at the point $(1, 1)$. Zoom in twice on the point of tangency. Does your graphing utility distinguish between the two graphs? Use the *trace* feature to compare the two graphs. As the x-values get closer to 1, what can you say about the y-values?

Tangent Line Approximations

Newton's Method (Section 3.8) is an example of the use of a tangent line to a graph to approximate the graph. In this section, you will study other situations in which the graph of a function can be approximated by a straight line.

To begin, consider a function f that is differentiable at c. The equation for the tangent line at the point $(c, f(c))$ is given by

$$y - f(c) = f'(c)(x - c)$$
$$y = f(c) + f'(c)(x - c)$$

and is called the **tangent line approximation** (or **linear approximation**) **of f at c.** Because c is a constant, y is a linear function of x. Moreover, by restricting the values of x to be sufficiently close to c, the values of y can be used as approximations (to any desired accuracy) of the values of the function f. In other words, as $x \to c$, the limit of y is $f(c)$.

EXAMPLE 1 Using a Tangent Line Approximation

Find the tangent line approximation of

$$f(x) = 1 + \sin x$$

at the point $(0, 1)$. Then use a table to compare the y-values of the linear function with those of $f(x)$ on an open interval containing $x = 0$.

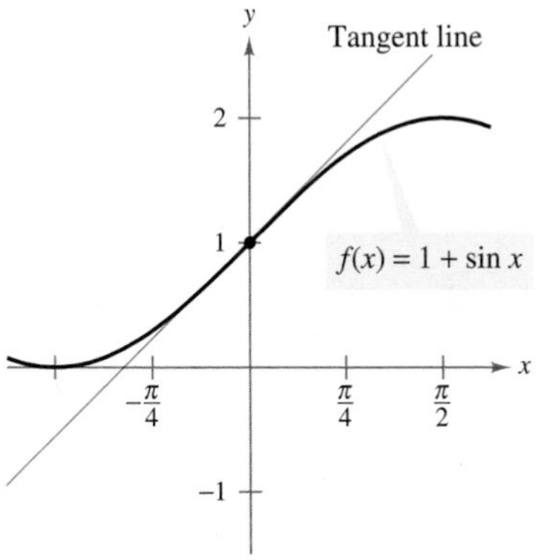

The tangent line approximation of f at the point $(0, 1)$
Figure 3.65

Solution The derivative of f is

$$f'(x) = \cos x. \quad \text{First derivative}$$

So, the equation of the tangent line to the graph of f at the point $(0, 1)$ is

$$y - f(0) = f'(0)(x - 0)$$
$$y - 1 = (1)(x - 0)$$
$$y = 1 + x. \quad \text{Tangent line approximation}$$

The table compares the values of y given by this linear approximation with the values of $f(x)$ near $x = 0$. Notice that the closer x is to 0, the better the approximation is. This conclusion is reinforced by the graph shown in Figure 3.65.

x	-0.5	-0.1	-0.01	0	0.01	0.1	0.5
$f(x) = 1 + \sin x$	0.521	0.9002	0.9900002	1	1.0099998	1.0998	1.479
$y = 1 + x$	0.5	0.9	0.99	1	1.01	1.1	1.5

NOTE Be sure you see that this linear approximation of $f(x) = 1 + \sin x$ depends on the point of tangency. At a different point on the graph of f, you would obtain a different tangent line approximation.

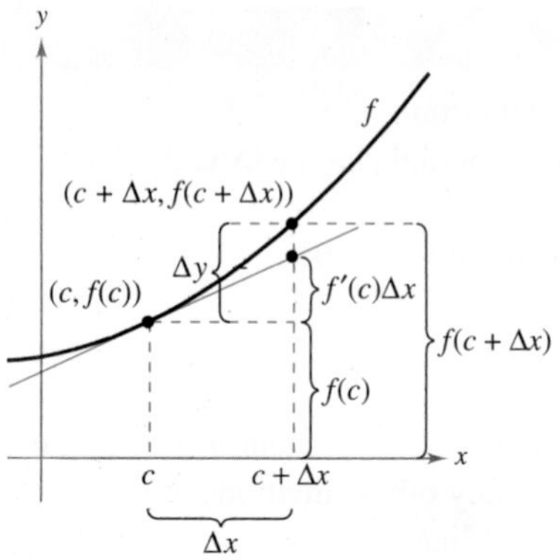

When Δx is small, $\Delta y = f(c + \Delta x) - f(c)$ is approximated by $f'(c)\Delta x$.
Figure 3.66

Differentials

When the tangent line to the graph of f at the point $(c, f(c))$

$$y = f(c) + f'(c)(x - c) \quad \text{Tangent line at } (c, f(c))$$

is used as an approximation of the graph of f, the quantity $x - c$ is called the change in x, and is denoted by Δx, as shown in Figure 3.66. When Δx is small, the change in y (denoted by Δy) can be approximated as shown.

$$\Delta y = f(c + \Delta x) - f(c) \quad \text{Actual change in } y$$
$$\approx f'(c)\Delta x \quad \text{Approximate change in } y$$

For such an approximation, the quantity Δx is traditionally denoted by dx, and is called the **differential of *x*.** The expression $f'(x)\,dx$ is denoted by dy, and is called the **differential of *y*.**

Definition of Differentials

Let $y = f(x)$ represent a function that is differentiable on an open interval containing x. The **differential of *x*** (denoted by dx) is any nonzero real number. The **differential of *y*** (denoted by dy) is

$$dy = f'(x)\,dx.$$

In many types of applications, the differential of y can be used as an approximation of the change in y. That is,

$$\Delta y \approx dy \quad \text{or} \quad \Delta y \approx f'(x)dx.$$

EXAMPLE 2 Comparing Δy and dy

Let $y = x^2$. Find dy when $x = 1$ and $dx = 0.01$. Compare this value with Δy for $x = 1$ and $\Delta x = 0.01$.

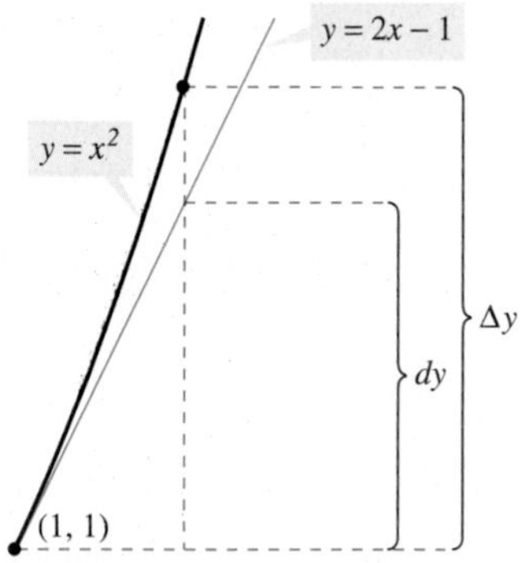

The change in y, Δy, is approximated by the differential of y, dy.
Figure 3.67

Solution Because $y = f(x) = x^2$, you have $f'(x) = 2x$, and the differential dy is given by

$$dy = f'(x)\,dx = f'(1)(0.01) = 2(0.01) = 0.02. \quad \text{Differential of } y$$

Now, using $\Delta x = 0.01$, the change in y is

$$\Delta y = f(x + \Delta x) - f(x) = f(1.01) - f(1) = (1.01)^2 - 1^2 = 0.0201.$$

Figure 3.67 shows the geometric comparison of dy and Δy. Try comparing other values of dy and Δy. You will see that the values become closer to each other as dx (or Δx) approaches 0.

In Example 2, the tangent line to the graph of $f(x) = x^2$ at $x = 1$ is

$$y = 2x - 1 \quad \text{or} \quad g(x) = 2x - 1. \quad \text{Tangent line to the graph of } f \text{ at } x = 1.$$

For x-values near 1, this line is close to the graph of f, as shown in Figure 3.67. For instance,

$$f(1.01) = 1.01^2 = 1.0201 \quad \text{and} \quad g(1.01) = 2(1.01) - 1 = 1.02.$$

Error Propagation

Physicists and engineers tend to make liberal use of the approximation of Δy by dy. One way this occurs in practice is in the estimation of errors propagated by physical measuring devices. For example, if you let x represent the measured value of a variable and let $x + \Delta x$ represent the exact value, then Δx is the *error in measurement*. Finally, if the measured value x is used to compute another value $f(x)$, the difference between $f(x + \Delta x)$ and $f(x)$ is the **propagated error.**

$$\underbrace{f(x + \overbrace{\Delta x}^{\text{Measurement error}})}_{\text{Exact value}} - \underbrace{f(x)}_{\text{Measured value}} = \overbrace{\Delta y}^{\text{Propagated error}}$$

EXAMPLE 3 Estimation of Error

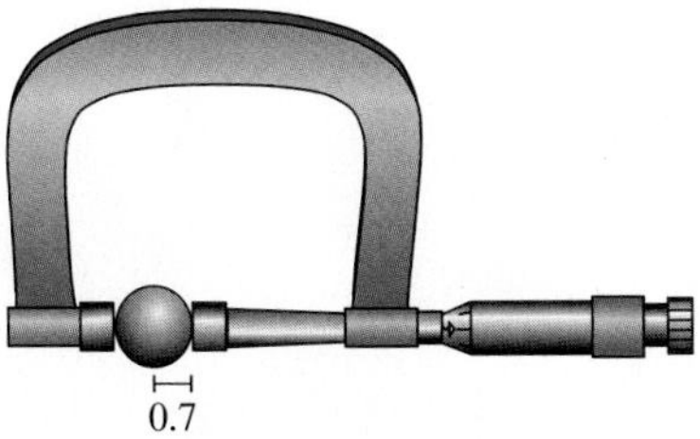

Ball bearing with measured radius that is correct to within 0.01 inch
Figure 3.68

The radius of a ball bearing is measured to be 0.7 inch, as shown in Figure 3.68. If the measurement is correct to within 0.01 inch, estimate the propagated error in the volume V of the ball bearing.

Solution The formula for the volume of a sphere is $V = \frac{4}{3}\pi r^3$, where r is the radius of the sphere. So, you can write

$$r = 0.7 \qquad \text{Measured radius}$$

and

$$-0.01 \le \Delta r \le 0.01. \qquad \text{Possible error}$$

To approximate the propagated error in the volume, differentiate V to obtain $dV/dr = 4\pi r^2$ and write

$$\begin{aligned} \Delta V &\approx dV && \text{Approximate } \Delta V \text{ by } dV. \\ &= 4\pi r^2\, dr \\ &= 4\pi(0.7)^2(\pm 0.01) && \text{Substitute for } r \text{ and } dr. \\ &\approx \pm 0.06158 \text{ cubic inch.} \end{aligned}$$

So, the volume has a propagated error of about 0.06 cubic inch.

Would you say that the propagated error in Example 3 is large or small? The answer is best given in *relative* terms by comparing dV with V. The ratio

$$\begin{aligned} \frac{dV}{V} &= \frac{4\pi r^2\, dr}{\frac{4}{3}\pi r^3} && \text{Ratio of } dV \text{ to } V \\ &= \frac{3\, dr}{r} && \text{Simplify.} \\ &\approx \frac{3}{0.7}(\pm 0.01) && \text{Substitute for } dr \text{ and } r. \\ &\approx \pm 0.0429 \end{aligned}$$

is called the **relative error.** The corresponding **percent error** is approximately 4.29%.

Calculating Differentials

Each of the differentiation rules that you studied in Chapter 2 can be written in **differential form.** For example, suppose u and v are differentiable functions of x. By the definition of differentials, you have

$$du = u'\,dx \quad \text{and} \quad dv = v'\,dx.$$

So, you can write the differential form of the Product Rule as shown below.

$$\begin{aligned} d[uv] &= \frac{d}{dx}[uv]\,dx && \text{Differential of } uv \\ &= [uv' + vu']\,dx && \text{Product Rule} \\ &= uv'\,dx + vu'\,dx \\ &= u\,dv + v\,du \end{aligned}$$

Differential Formulas

Let u and v be differentiable functions of x.

Constant multiple: $d[cu] = c\,du$

Sum or difference: $d[u \pm v] = du \pm dv$

Product: $d[uv] = u\,dv + v\,du$

Quotient: $d\left[\dfrac{u}{v}\right] = \dfrac{v\,du - u\,dv}{v^2}$

EXAMPLE 4 Finding Differentials

Function	*Derivative*	*Differential*
a. $y = x^2$	$\dfrac{dy}{dx} = 2x$	$dy = 2x\,dx$
b. $y = 2\sin x$	$\dfrac{dy}{dx} = 2\cos x$	$dy = 2\cos x\,dx$
c. $y = x\cos x$	$\dfrac{dy}{dx} = -x\sin x + \cos x$	$dy = (-x\sin x + \cos x)\,dx$
d. $y = \dfrac{1}{x}$	$\dfrac{dy}{dx} = -\dfrac{1}{x^2}$	$dy = -\dfrac{dx}{x^2}$

GOTTFRIED WILHELM LEIBNIZ (1646–1716)

Both Leibniz and Newton are credited with creating calculus. It was Leibniz, however, who tried to broaden calculus by developing rules and formal notation. He often spent days choosing an appropriate notation for a new concept.

The notation in Example 4 is called the **Leibniz notation** for derivatives and differentials, named after the German mathematician Gottfried Wilhelm Leibniz. The beauty of this notation is that it provides an easy way to remember several important calculus formulas by making it seem as though the formulas were derived from algebraic manipulations of differentials. For instance, in Leibniz notation, the *Chain Rule*

$$\frac{dy}{dx} = \frac{dy}{du}\frac{du}{dx}$$

would appear to be true because the du's divide out. Even though this reasoning is *incorrect*, the notation does help one remember the Chain Rule.

EXAMPLE 5 Finding the Differential of a Composite Function

$$y = f(x) = \sin 3x \qquad \text{Original function}$$

$$f'(x) = 3 \cos 3x \qquad \text{Apply Chain Rule.}$$

$$dy = f'(x)\,dx = 3 \cos 3x\,dx \qquad \text{Differential form}$$

EXAMPLE 6 Finding the Differential of a Composite Function

$$y = f(x) = (x^2 + 1)^{1/2} \qquad \text{Original function}$$

$$f'(x) = \frac{1}{2}(x^2 + 1)^{-1/2}(2x) = \frac{x}{\sqrt{x^2 + 1}} \qquad \text{Apply Chain Rule.}$$

$$dy = f'(x)\,dx = \frac{x}{\sqrt{x^2 + 1}}\,dx \qquad \text{Differential form}$$

Differentials can be used to approximate function values. To do this for the function given by $y = f(x)$, you use the formula

$$f(x + \Delta x) \approx f(x) + dy = f(x) + f'(x)\,dx$$

which is derived from the approximation $\Delta y = f(x + \Delta x) - f(x) \approx dy$. The key to using this formula is to choose a value for x that makes the calculations easier, as shown in Example 7.

EXAMPLE 7 Approximating Function Values

Use differentials to approximate $\sqrt{16.5}$.

Solution Using $f(x) = \sqrt{x}$, you can write

$$f(x + \Delta x) \approx f(x) + f'(x)\,dx = \sqrt{x} + \frac{1}{2\sqrt{x}}\,dx.$$

Now, choosing $x = 16$ and $dx = 0.5$, you obtain the following approximation.

$$f(x + \Delta x) = \sqrt{16.5} \approx \sqrt{16} + \frac{1}{2\sqrt{16}}(0.5) = 4 + \left(\frac{1}{8}\right)\left(\frac{1}{2}\right) = 4.0625$$

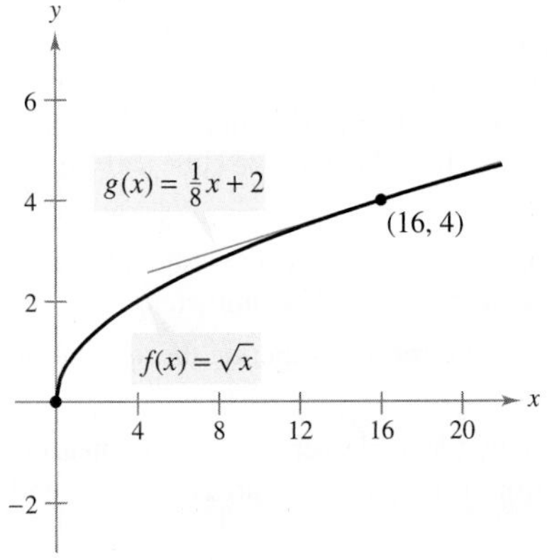

Figure 3.69

The tangent line approximation to $f(x) = \sqrt{x}$ at $x = 16$ is the line $g(x) = \frac{1}{8}x + 2$. For x-values near 16, the graphs of f and g are close together, as shown in Figure 3.69. For instance,

$$f(16.5) = \sqrt{16.5} \approx 4.0620 \quad \text{and} \quad g(16.5) = \frac{1}{8}(16.5) + 2 = 4.0625.$$

In fact, if you use a graphing utility to zoom in near the point of tangency (16, 4), you will see that the two graphs appear to coincide. Notice also that as you move farther away from the point of tangency, the linear approximation is less accurate.

Exercises for Section 3.9

See www.CalcChat.com for worked-out solutions to odd-numbered exercises.

In Exercises 1–6, find the equation of the tangent line T to the graph of f at the given point. Use this linear approximation to complete the table.

x	1.9	1.99	2	2.01	2.1
$f(x)$					
$T(x)$					

1. $f(x) = x^2, \quad (2, 4)$

2. $f(x) = \dfrac{6}{x^2}, \quad \left(2, \dfrac{3}{2}\right)$

3. $f(x) = x^5, \quad (2, 32)$

4. $f(x) = \sqrt{x}, \quad (2, \sqrt{2})$

5. $f(x) = \sin x, \quad (2, \sin 2)$

6. $f(x) = \csc x, \quad (2, \csc 2)$

In Exercises 7–10, use the information to evaluate and compare Δy and dy.

7. $y = \frac{1}{2}x^3 \qquad x = 2 \qquad \Delta x = dx = 0.1$

8. $y = 1 - 2x^2 \qquad x = 0 \qquad \Delta x = dx = -0.1$

9. $y = x^4 + 1 \qquad x = -1 \qquad \Delta x = dx = 0.01$

10. $y = 2x + 1 \qquad x = 2 \qquad \Delta x = dx = 0.01$

In Exercises 11–20, find the differential dy of the given function.

11. $y = 3x^2 - 4$

12. $y = 3x^{2/3}$

13. $y = \dfrac{x + 1}{2x - 1}$

14. $y = \sqrt{9 - x^2}$

15. $y = x\sqrt{1 - x^2}$

16. $y = \sqrt{x} + \dfrac{1}{\sqrt{x}}$

17. $y = 2x - \cot^2 x$

18. $y = x \sin x$

19. $y = \dfrac{1}{3}\cos\left(\dfrac{6\pi x - 1}{2}\right)$

20. $y = \dfrac{\sec^2 x}{x^2 + 1}$

In Exercises 21–24, use differentials and the graph of f to approximate (a) $f(1.9)$ and (b) $f(2.04)$. To print an enlarged copy of the graph, go to the website *www.mathgraphs.com*.

21.

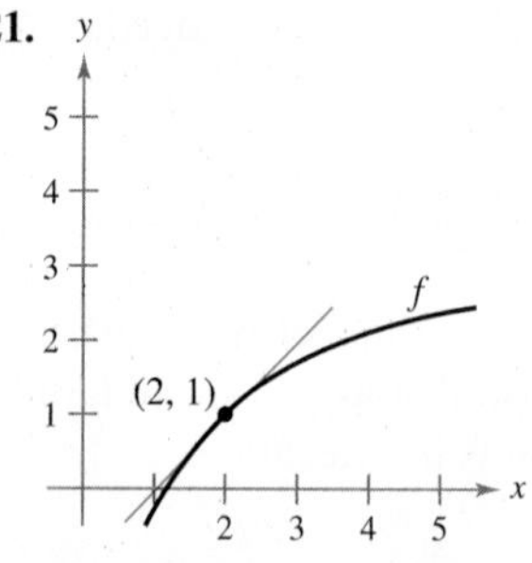

22.

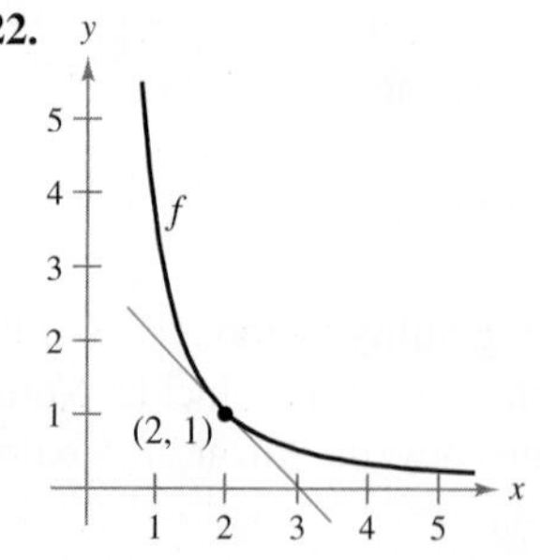

23.

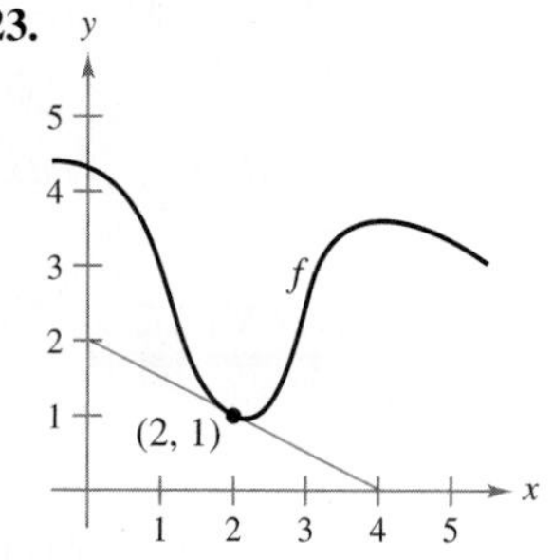

24.

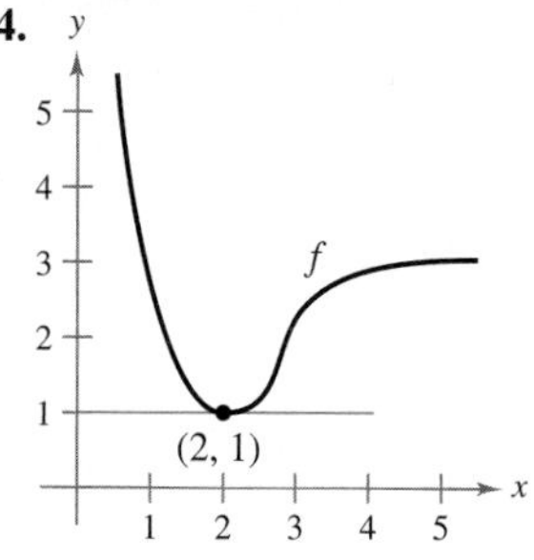

In Exercises 25 and 26, use differentials and the graph of g' to approximate (a) $g(2.93)$ and (b) $g(3.1)$ given that $g(3) = 8$.

25.

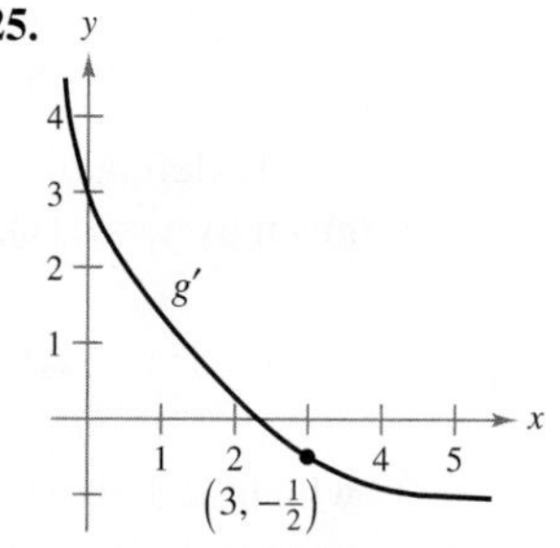

26.

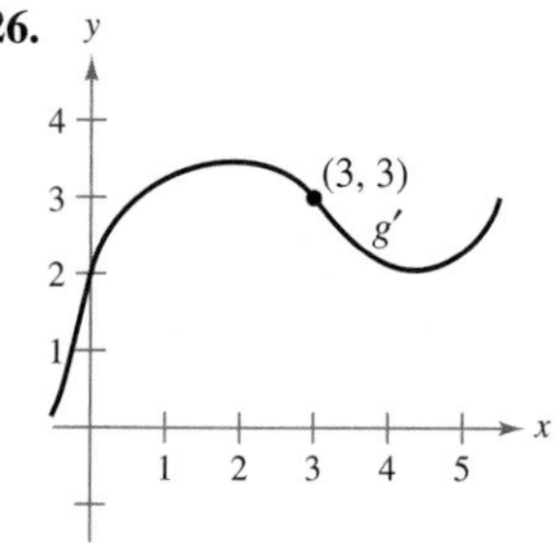

27. ***Area*** The measurement of the side of a square is found to be 12 inches, with a possible error of $\frac{1}{64}$ inch. Use differentials to approximate the possible propagated error in computing the area of the square.

28. ***Area*** The measurements of the base and altitude of a triangle are found to be 36 and 50 centimeters, respectively. The possible error in each measurement is 0.25 centimeter. Use differentials to approximate the possible propagated error in computing the area of the triangle.

29. ***Area*** The measurement of the radius of the end of a log is found to be 14 inches, with a possible error of $\frac{1}{4}$ inch. Use differentials to approximate the possible propagated error in computing the area of the end of the log.

30. ***Volume and Surface Area*** The measurement of the edge of a cube is found to be 12 inches, with a possible error of 0.03 inch. Use differentials to approximate the maximum possible propagated error in computing (a) the volume of the cube and (b) the surface area of the cube.

31. ***Area*** The measurement of a side of a square is found to be 15 centimeters, with a possible error of 0.05 centimeter.

(a) Approximate the percent error in computing the area of the square.

(b) Estimate the maximum allowable percent error in measuring the side if the error in computing the area cannot exceed 2.5%.

32. ***Circumference*** The measurement of the circumference of a circle is found to be 56 inches, with a possible error of 1.2 inches.

(a) Approximate the percent error in computing the area of the circle.

(b) Estimate the maximum allowable percent error in measuring the circumference if the error in computing the area cannot exceed 3%.

33. ***Volume and Surface Area*** The radius of a sphere is measured to be 6 inches, with a possible error of 0.02 inch. Use differentials to approximate the maximum possible error in calculating (a) the volume of the sphere, (b) the surface area of the sphere, and (c) the relative errors in parts (a) and (b).

34. ***Profit*** The profit P for a company is given by

$$P = (500x - x^2) - \left(\tfrac{1}{2}x^2 - 77x + 3000\right).$$

Approximate the change and percent change in profit as production changes from $x = 115$ to $x = 120$ units.

Volume **In Exercises 35 and 36, the thickness of each shell is 0.2 centimeter. Use differentials to approximate the volume of each shell.**

35.

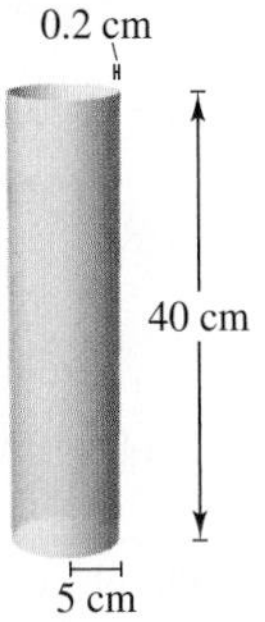

36.

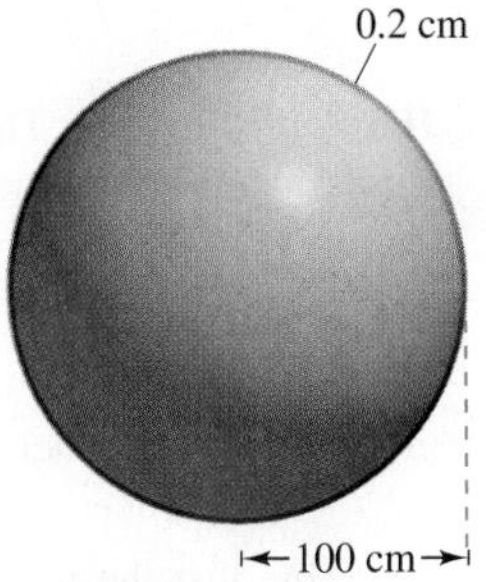

37. ***Pendulum*** The period of a pendulum is given by

$$T = 2\pi\sqrt{\frac{L}{g}}$$

where L is the length of the pendulum in feet, g is the acceleration due to gravity, and T is the time in seconds. The pendulum has been subjected to an increase in temperature such that the length has increased by $\frac{1}{2}\%$.

(a) Find the approximate percent change in the period.

(b) Using the result in part (a), find the approximate error in this pendulum clock in 1 day.

38. ***Ohm's Law*** A current of I amperes passes through a resistor of R ohms. **Ohm's Law** states that the voltage E applied to the resistor is $E = IR$. If the voltage is constant, show that the magnitude of the relative error in R caused by a change in I is equal in magnitude to the relative error in I.

39. ***Triangle Measurements*** The measurement of one side of a right triangle is found to be 9.5 inches, and the angle opposite that side is $26°45'$ with a possible error of $15'$.

(a) Approximate the percent error in computing the length of the hypotenuse.

(b) Estimate the maximum allowable percent error in measuring the angle if the error in computing the length of the hypotenuse cannot exceed 2%.

40. ***Area*** Approximate the percent error in computing the area of the triangle in Exercise 39.

41. ***Projectile Motion*** The range R of a projectile is

$$R = \frac{v_0^2}{32}(\sin 2\theta)$$

where v_0 is the initial velocity in feet per second and θ is the angle of elevation. If $v_0 = 2200$ feet per second and θ is changed from 10° to 11°, use differentials to approximate the change in the range.

42. ***Surveying*** A surveyor standing 50 feet from the base of a large tree measures the angle of elevation to the top of the tree as 71.5°. How accurately must the angle be measured if the percent error in estimating the height of the tree is to be less than 6%?

In Exercises 43–46, use differentials to approximate the value of the expression. Compare your answer with that of a calculator.

43. $\sqrt{99.4}$ **44.** $\sqrt[3]{26}$

45. $\sqrt[4]{624}$ **46.** $(2.99)^3$

Writing **In Exercises 47 and 48, give a short explanation of why the approximation is valid.**

47. $\sqrt{4.02} \approx 2 + \frac{1}{4}(0.02)$ **48.** $\tan 0.05 \approx 0 + 1(0.05)$

In Exercises 49–52, verify the tangent line approximation of the function at the given point. Then use a graphing utility to graph the function and its approximation in the same viewing window.

Function	*Approximation*	*Point*
49. $f(x) = \sqrt{x+4}$	$y = 2 + \dfrac{x}{4}$	$(0, 2)$
50. $f(x) = \sqrt{x}$	$y = \dfrac{1}{2} + \dfrac{x}{2}$	$(1, 1)$
51. $f(x) = \tan x$	$y = x$	$(0, 0)$
52. $f(x) = \dfrac{1}{1-x}$	$y = 1 + x$	$(0, 1)$

Writing About Concepts

53. Describe the change in accuracy of dy as an approximation for Δy when Δx is decreased.

54. When using differentials, what is meant by the terms *propagated error*, *relative error*, and *percent error*?

True or False? **In Exercises 55–58, determine whether the statement is true or false. If it is false, explain why or give an example that shows it is false.**

55. If $y = x + c$, then $dy = dx$.

56. If $y = ax + b$, then $\Delta y/\Delta x = dy/dx$.

57. If y is differentiable, then $\lim_{\Delta x \to 0} (\Delta y - dy) = 0$.

58. If $y = f(x)$, f is increasing and differentiable, and $\Delta x > 0$, then $\Delta y \geq dy$.

Review Exercises for Chapter 3

See www.CalcChat.com for worked-out solutions to odd-numbered exercises.

1. Give the definition of a critical number, and graph a function f showing the different types of critical numbers.

2. Consider the odd function f that is continuous and differentiable and has the functional values shown in the table.

x	-5	-4	-1	0	2	3	6
$f(x)$	1	3	2	0	-1	-4	0

(a) Determine $f(4)$.

(b) Determine $f(-3)$.

(c) Plot the points and make a possible sketch of the graph of f on the interval $[-6, 6]$. What is the smallest number of critical points in the interval? Explain.

(d) Does there exist at least one real number c in the interval $(-6, 6)$ where $f'(c) = -1$? Explain.

(e) Is it possible that $\lim_{x\to 0} f(x)$ does not exist? Explain.

(f) Is it necessary that $f'(x)$ exists at $x = 2$? Explain.

In Exercises 3 and 4, find the absolute extrema of the function on the closed interval. Use a graphing utility to graph the function over the given interval to confirm your results.

3. $g(x) = 2x + 5\cos x, \quad [0, 2\pi]$

4. $f(x) = \dfrac{x}{\sqrt{x^2 + 1}}, \quad [0, 2]$

In Exercises 5 and 6, determine whether Rolle's Theorem can be applied to f on the closed interval $[a, b]$. If Rolle's Theorem can be applied, find all values of c in the open interval (a, b) such that $f'(c) = 0$.

5. $f(x) = (x - 2)(x + 3)^2, \quad [-3, 2]$

6. $f(x) = |x - 2| - 2, \quad [0, 4]$

7. Consider the function $f(x) = 3 - |x - 4|$.

(a) Graph the function and verify that $f(1) = f(7)$.

(b) Note that $f'(x)$ is not equal to zero for any x in $[1, 7]$. Explain why this does not contradict Rolle's Theorem.

8. Can the Mean Value Theorem be applied to the function $f(x) = 1/x^2$ on the interval $[-2, 1]$? Explain.

In Exercises 9–12, find the point(s) guaranteed by the Mean Value Theorem for the closed interval $[a, b]$.

9. $f(x) = x^{2/3}, \quad [1, 8]$

10. $f(x) = \dfrac{1}{x}, \quad [1, 4]$

11. $f(x) = x - \cos x, \quad \left[-\dfrac{\pi}{2}, \dfrac{\pi}{2}\right]$

12. $f(x) = \sqrt{x} - 2x, \quad [0, 4]$

13. For the function $f(x) = Ax^2 + Bx + C$, determine the value of c guaranteed by the Mean Value Theorem on the interval $[x_1, x_2]$.

14. Demonstrate the result of Exercise 13 for $f(x) = 2x^2 - 3x + 1$ on the interval $[0, 4]$.

In Exercises 15–18, find the critical numbers (if any) and the open intervals on which the function is increasing or decreasing.

15. $f(x) = (x - 1)^2(x - 3)$

16. $g(x) = (x + 1)^3$

17. $h(x) = \sqrt{x}(x - 3), \quad x > 0$

18. $f(x) = \sin x + \cos x, \quad [0, 2\pi]$

In Exercises 19 and 20, use the First Derivative Test to find any relative extrema of the function. Use a graphing utility to verify your results.

19. $h(t) = \dfrac{1}{4}t^4 - 8t$

20. $g(x) = \dfrac{3}{2}\sin\left(\dfrac{\pi x}{2} - 1\right), \quad [0, 4]$

21. ***Harmonic Motion*** The height of an object attached to a spring is given by the harmonic equation

$$y = \tfrac{1}{3}\cos 12t - \tfrac{1}{4}\sin 12t$$

where y is measured in inches and t is measured in seconds.

(a) Calculate the height and velocity of the object when $t = \pi/8$ second.

(b) Show that the maximum displacement of the object is $\frac{5}{12}$ inch.

(c) Find the period P of y. Also, find the frequency f (number of oscillations per second) if $f = 1/P$.

22. ***Writing*** The general equation giving the height of an oscillating object attached to a spring is

$$y = A\sin\sqrt{\frac{k}{m}}\,t + B\cos\sqrt{\frac{k}{m}}\,t$$

where k is the spring constant and m is the mass of the object.

(a) Show that the maximum displacement of the object is $\sqrt{A^2 + B^2}$.

(b) Show that the object oscillates with a frequency of

$$f = \frac{1}{2\pi}\sqrt{\frac{k}{m}}.$$

In Exercises 23 and 24, determine the points of inflection and discuss the concavity of the graph of the function.

23. $f(x) = x + \cos x, \quad [0, 2\pi]$

24. $f(x) = (x + 2)^2(x - 4)$

In Exercises 25 and 26, use the Second Derivative Test to find all relative extrema.

25. $g(x) = 2x^2(1 - x^2)$

26. $h(t) = t - 4\sqrt{t + 1}$

Think About It **In Exercises 27 and 28, sketch the graph of a function f having the given characteristics.**

27. $f(0) = f(6) = 0$

$f'(3) = f'(5) = 0$

$f'(x) > 0$ if $x < 3$

$f'(x) > 0$ if $3 < x < 5$

$f'(x) < 0$ if $x > 5$

$f''(x) < 0$ if $x < 3$ or $x > 4$

$f''(x) > 0$ if $3 < x < 4$

28. $f(0) = 4,\ f(6) = 0$

$f'(x) < 0$ if $x < 2$ or $x > 4$

$f'(2)$ does not exist.

$f'(4) = 0$

$f'(x) > 0$ if $2 < x < 4$

$f''(x) < 0$ if $x \neq 2$

29. ***Writing*** A newspaper headline states that "The rate of growth of the national deficit is decreasing." What does this mean? What does it imply about the graph of the deficit as a function of time?

30. ***Inventory Cost*** The cost of inventory depends on the ordering and storage costs according to the inventory model

$$C = \left(\frac{Q}{x}\right)s + \left(\frac{x}{2}\right)r.$$

Determine the order size that will minimize the cost, assuming that sales occur at a constant rate, Q is the number of units sold per year, r is the cost of storing one unit for 1 year, s is the cost of placing an order, and x is the number of units per order.

31. ***Modeling Data*** Outlays for national defense D (in billions of dollars) for selected years from 1970 through 1999 are shown in the table, where t is time in years, with $t = 0$ corresponding to 1970. *(Source: U.S. Office of Management and Budget)*

t	0	5	10	15	20
D	90.4	103.1	155.1	279.0	328.3

t	25	26	27	28	29
D	309.9	302.7	309.8	310.3	320.2

(a) Use the regression capabilities of a graphing utility to fit a model of the form

$$D = at^4 + bt^3 + ct^2 + dt + e$$

to the data.

(b) Use a graphing utility to plot the data and graph the model.

(c) For the years shown in the table, when does the model indicate that the outlay for national defense is at a maximum? When is it at a minimum?

(d) For the years shown in the table, when does the model indicate that the outlay for national defense is increasing at the greatest rate?

32. ***Modeling Data*** The manager of a store recorded the annual sales S (in thousands of dollars) of a product over a period of 7 years, as shown in the table, where t is the time in years, with $t = 7$ corresponding to 1997.

t	7	8	9	10	11	12	13
S	5.4	6.9	11.5	15.5	19.0	22.0	23.6

(a) Use the regression capabilities of a graphing utility to find a model of the form $S = at^3 + bt^2 + ct + d$ for the data.

(b) Use a graphing utility to plot the data and graph the model.

(c) Use calculus to find the time t when sales were increasing at the greatest rate.

(d) Do you think the model would be accurate for predicting future sales? Explain.

In Exercises 33–40, find the limit.

33. $\displaystyle\lim_{x\to\infty} \frac{2x^2}{3x^2 + 5}$ **34.** $\displaystyle\lim_{x\to\infty} \frac{2x}{3x^2 + 5}$

35. $\displaystyle\lim_{x\to-\infty} \frac{3x^2}{x + 5}$ **36.** $\displaystyle\lim_{x\to-\infty} \frac{\sqrt{x^2 + x}}{-2x}$

37. $\displaystyle\lim_{x\to\infty} \frac{5\cos x}{x}$ **38.** $\displaystyle\lim_{x\to\infty} \frac{3x}{\sqrt{x^2 + 4}}$

39. $\displaystyle\lim_{x\to-\infty} \frac{6x}{x + \cos x}$ **40.** $\displaystyle\lim_{x\to-\infty} \frac{x}{2\sin x}$

In Exercises 41–44, find any vertical and horizontal asymptotes of the graph of the function. Use a graphing utility to verify your results.

41. $h(x) = \dfrac{2x + 3}{x - 4}$ **42.** $g(x) = \dfrac{5x^2}{x^2 + 2}$

43. $f(x) = \dfrac{3}{x} - 2$ **44.** $f(x) = \dfrac{3x}{\sqrt{x^2 + 2}}$

In Exercises 45–48, use a graphing utility to graph the function. Use the graph to approximate any relative extrema or asymptotes.

45. $f(x) = x^3 + \dfrac{243}{x}$ **46.** $f(x) = |x^3 - 3x^2 + 2x|$

47. $f(x) = \dfrac{x - 1}{1 + 3x^2}$ **48.** $g(x) = \dfrac{\pi^2}{3} - 4\cos x + \cos 2x$

In Exercises 49–66, analyze and sketch the graph of the function.

49. $f(x) = 4x - x^2$ **50.** $f(x) = 4x^3 - x^4$

51. $f(x) = x\sqrt{16 - x^2}$ **52.** $f(x) = (x^2 - 4)^2$

53. $f(x) = (x - 1)^3(x - 3)^2$ **54.** $f(x) = (x - 3)(x + 2)^3$

55. $f(x) = x^{1/3}(x + 3)^{2/3}$ **56.** $f(x) = (x - 2)^{1/3}(x + 1)^{2/3}$

57. $f(x) = \dfrac{x + 1}{x - 1}$ **58.** $f(x) = \dfrac{2x}{1 + x^2}$

59. $f(x) = \dfrac{4}{1 + x^2}$

60. $f(x) = \dfrac{x^2}{1 + x^4}$

61. $f(x) = x^3 + x + \dfrac{4}{x}$

62. $f(x) = x^2 + \dfrac{1}{x}$

63. $f(x) = |x^2 - 9|$

64. $f(x) = |x - 1| + |x - 3|$

65. $f(x) = x + \cos x, \quad 0 \le x \le 2\pi$

66. $f(x) = \dfrac{1}{\pi}(2 \sin \pi x - \sin 2\pi x), \quad -1 \le x \le 1$

67. Find the maximum and minimum points on the graph of

$$x^2 + 4y^2 - 2x - 16y + 13 = 0$$

(a) without using calculus.

(b) using calculus.

68. Consider the function $f(x) = x^n$ for positive integer values of n.

(a) For what values of n does the function have a relative minimum at the origin?

(b) For what values of n does the function have a point of inflection at the origin?

69. ***Distance*** At noon, ship A is 100 kilometers due east of ship B. Ship A is sailing west at 12 kilometers per hour, and ship B is sailing south at 10 kilometers per hour. At what time will the ships be nearest to each other, and what will this distance be?

70. ***Maximum Area*** Find the dimensions of the rectangle of maximum area, with sides parallel to the coordinate axes, that can be inscribed in the ellipse given by

$$\frac{x^2}{144} + \frac{y^2}{16} = 1.$$

71. ***Minimum Length*** A right triangle in the first quadrant has the coordinate axes as sides, and the hypotenuse passes through the point $(1, 8)$. Find the vertices of the triangle such that the length of the hypotenuse is minimum.

72. ***Minimum Length*** The wall of a building is to be braced by a beam that must pass over a parallel fence 5 feet high and 4 feet from the building. Find the length of the shortest beam that can be used.

73. ***Maximum Area*** Three sides of a trapezoid have the same length s. Of all such possible trapezoids, show that the one of maximum area has a fourth side of length $2s$.

74. ***Maximum Area*** Show that the greatest area of any rectangle inscribed in a triangle is one-half that of the triangle.

75. ***Distance*** Find the length of the longest pipe that can be carried level around a right-angle corner at the intersection of two corridors of widths 4 feet and 6 feet. (Do not use trigonometry.)

76. ***Distance*** Rework Exercise 75, given corridors of widths a meters and b meters.

77. ***Distance*** A hallway of width 6 feet meets a hallway of width 9 feet at right angles. Find the length of the longest pipe that can be carried level around this corner. [*Hint:* If L is the length of the pipe, show that

$$L = 6 \csc \theta + 9 \csc\left(\frac{\pi}{2} - \theta\right)$$

where θ is the angle between the pipe and the wall of the narrower hallway.]

78. ***Length*** Rework Exercise 77, given that one hallway is of width a meters and the other is of width b meters. Show that the result is the same as in Exercise 76.

Minimum Cost **In Exercises 79 and 80, find the speed v, in miles per hour, that will minimize costs on a 110-mile delivery trip. The cost per hour for fuel is C dollars, and the driver is paid W dollars per hour. (Assume there are no costs other than wages and fuel.)**

79. Fuel cost: $C = \dfrac{v^2}{600}$

Driver: $W = \$5$

80. Fuel cost: $C = \dfrac{v^2}{500}$

Driver: $W = \$7.50$

In Exercises 81 and 82, use Newton's Method to approximate any real zeros of the function accurate to three decimal places. Use the *zero* or *root* feature of a graphing utility to verify your results.

81. $f(x) = x^3 - 3x - 1$

82. $f(x) = x^3 + 2x + 1$

In Exercises 83 and 84, use Newton's Method to approximate, to three decimal places, the x-value(s) of the point(s) of intersection of the equations. Use a graphing utility to verify your results.

83. $y = x^4$

$y = x + 3$

84. $y = \sin \pi x$

$y = 1 - x$

In Exercises 85 and 86, find the differential dy.

85. $y = x(1 - \cos x)$

86. $y = \sqrt{36 - x^2}$

87. ***Surface Area and Volume*** The diameter of a sphere is measured to be 18 centimeters, with a maximum possible error of 0.05 centimeter. Use differentials to approximate the possible propagated error and percent error in calculating the surface area and the volume of the sphere.

88. ***Demand Function*** A company finds that the demand for its commodity is

$$p = 75 - \frac{1}{4}x.$$

If x changes from 7 to 8, find and compare the values of Δp and dp.

P.S. Problem Solving

See www.CalcChat.com for worked-out solutions to odd-numbered exercises.

1. Graph the fourth-degree polynomial $p(x) = x^4 + ax^2 + 1$ for various values of the constant a.

(a) Determine the values of a for which p has exactly one relative minimum.

(b) Determine the values of a for which p has exactly one relative maximum.

(c) Determine the values of a for which p has exactly two relative minima.

(d) Show that the graph of p cannot have exactly two relative extrema.

2. (a) Graph the fourth-degree polynomial $p(x) = ax^4 - 6x^2$ for $a = -3, -2, -1, 0, 1, 2,$ and 3. For what values of the constant a does p have a relative minimum or relative maximum?

(b) Show that p has a relative maximum for all values of the constant a.

(c) Determine analytically the values of a for which p has a relative minimum.

(d) Let $(x, y) = (x, p(x))$ be a relative extremum of p. Show that (x, y) lies on the graph of $y = -3x^2$. Verify this result graphically by graphing $y = -3x^2$ together with the seven curves from part (a).

3. Let $f(x) = \dfrac{c}{x} + x^2$. Determine all values of the constant c such that f has a relative minimum, but no relative maximum.

4. (a) Let $f(x) = ax^2 + bx + c$, $a \neq 0$, be a quadratic polynomial. How many points of inflection does the graph of f have?

(b) Let $f(x) = ax^3 + bx^2 + cx + d$, $a \neq 0$, be a cubic polynomial. How many points of inflection does the graph of f have?

(c) Suppose the function $y = f(x)$ satisfies the equation

$$\frac{dy}{dx} = ky\left(1 - \frac{y}{L}\right)$$ where k and L are positive constants.

Show that the graph of f has a point of inflection at the point where $y = \dfrac{L}{2}$. (This equation is called the **logistic differential equation.**)

5. Prove Darboux's Theorem: Let f be differentiable on the closed interval $[a, b]$ such that $f'(a) = y_1$ and $f'(b) = y_2$. If d lies between y_1 and y_2, then there exists c in (a, b) such that $f'(c) = d$.

6. Let f and g be functions that are continuous on $[a, b]$ and differentiable on (a, b). Prove that if $f(a) = g(a)$ and $g'(x) > f'(x)$ for all x in (a, b), then $g(b) > f(b)$.

7. Prove the following **Extended Mean Value Theorem.** If f and f' are continuous on the closed interval $[a, b]$, and if f'' exists in the open interval (a, b), then there exists a number c in (a, b) such that

$$f(b) = f(a) + f'(a)(b - a) + \frac{1}{2}f''(c)(b - a)^2.$$

8. (a) Let $V = x^3$. Find dV and ΔV. Show that for small values of x, the difference $\Delta V - dV$ is very small in the sense that there exists ε such that $\Delta V - dV = \varepsilon\Delta x$, where $\varepsilon \to 0$ as $\Delta x \to 0$.

(b) Generalize this result by showing that if $y = f(x)$ is a differentiable function, then $\Delta y - dy = \varepsilon\Delta x$, where $\varepsilon \to 0$ as $\Delta x \to 0$.

9. The amount of illumination of a surface is proportional to the intensity of the light source, inversely proportional to the square of the distance from the light source, and proportional to $\sin\theta$, where θ is the angle at which the light strikes the surface. A rectangular room measures 10 feet by 24 feet, with a 10-foot ceiling. Determine the height at which the light should be placed to allow the corners of the floor to receive as much light as possible.

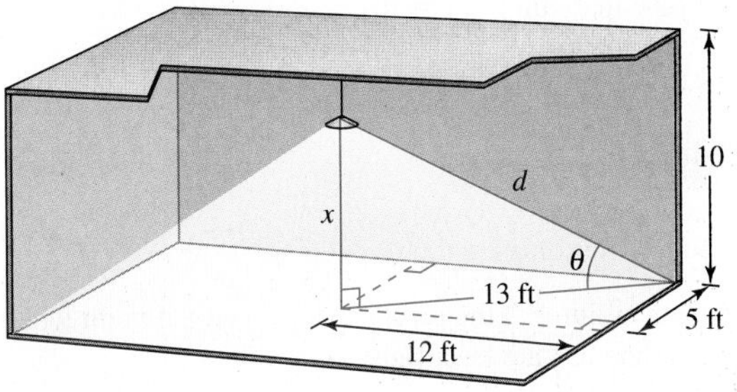

10. Consider a room in the shape of a cube, 4 meters on each side. A bug at point P wants to walk to point Q at the opposite corner, as shown in the figure. Use calculus to determine the shortest path. Can you solve the problem without calculus?

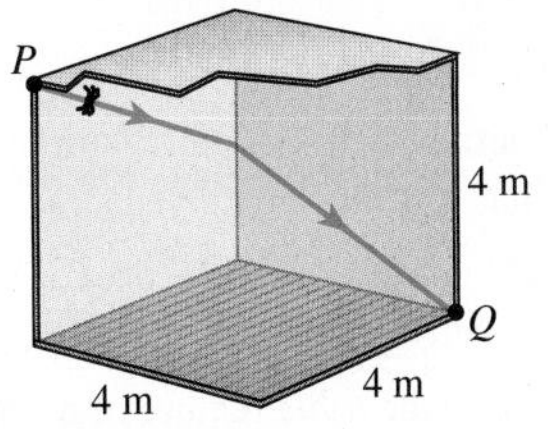

11. The line joining P and Q crosses the two parallel lines, as shown in the figure. The point R is d units from P. How far from Q should the point S be chosen so that the sum of the areas of the two shaded triangles is a minimum? So that the sum is a maximum?

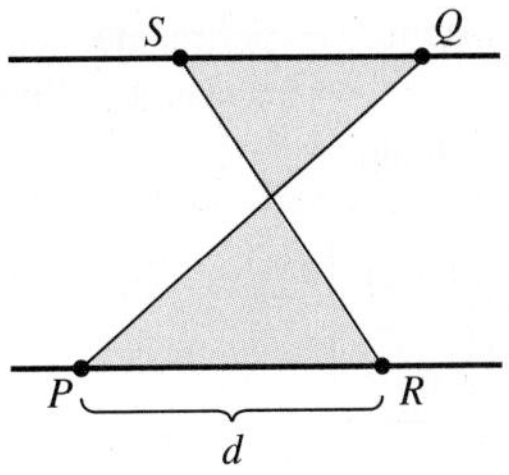

12. The figures show a rectangle, a circle, and a semicircle inscribed in a triangle bounded by the coordinate axes and the first-quadrant portion of the line with intercepts (3, 0) and (0, 4). Find the dimensions of each inscribed figure such that its area is maximum. State whether calculus was helpful in finding the required dimensions. Explain your reasoning.

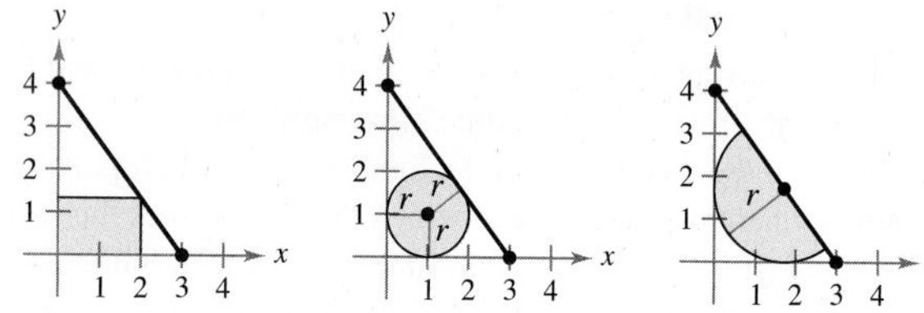

13. (a) Prove that $\lim_{x\to\infty} x^2 = \infty$.

(b) Prove that $\lim_{x\to\infty}\left(\frac{1}{x^2}\right) = 0$.

(c) Let L be a real number. Prove that if $\lim_{x\to\infty} f(x) = L$, then $\lim_{y\to 0^+} f\left(\frac{1}{y}\right) = L$.

14. Find the point on the graph of $y = \dfrac{1}{1 + x^2}$ (see figure) where the tangent line has the greatest slope, and the point where the tangent line has the least slope.

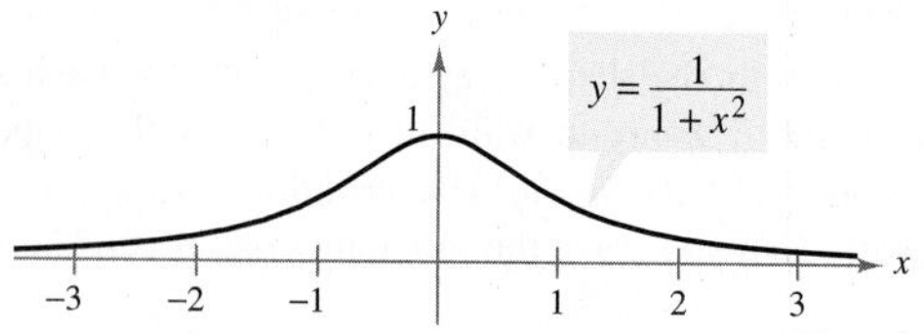

15. (a) Let x be a positive number. Use the *table* feature of a graphing utility to verify that $\sqrt{1 + x} < \frac{1}{2}x + 1$.

(b) Use the Mean Value Theorem to prove that $\sqrt{1 + x} < \frac{1}{2}x + 1$ for all positive real numbers x.

16. (a) Let x be a positive number. Use the *table* feature of a graphing utility to verify that $\sin x < x$.

(b) Use the Mean Value Theorem to prove that $\sin x < x$ for all positive real numbers x.

17. The police department must determine the speed limit on a bridge such that the flow rate of cars is maximum per unit time. The greater the speed limit, the farther apart the cars must be in order to keep a safe stopping distance. Experimental data on the stopping distance d (in meters) for various speeds v (in kilometers per hour) are shown in the table.

v	20	40	60	80	100
d	5.1	13.7	27.2	44.2	66.4

(a) Convert the speeds v in the table to the speeds s in meters per second. Use the regression capabilities of a graphing utility to find a model of the form $d(s) = as^2 + bs + c$ for the data.

(b) Consider two consecutive vehicles of average length 5.5 meters, traveling at a safe speed on the bridge. Let T be the difference between the times (in seconds) when the front bumpers of the vehicles pass a given point on the bridge. Verify that this difference in times is given by

$$T = \frac{d(s)}{s} + \frac{5.5}{s}.$$

(c) Use a graphing utility to graph the function T and estimate the speed s that minimizes the time between vehicles.

(d) Use calculus to determine the speed that minimizes T. What is the minimum value of T? Convert the required speed to kilometers per hour.

(e) Find the optimal distance between vehicles for the posted speed limit determined in part (d).

18. A legal-sized sheet of paper (8.5 inches by 14 inches) is folded so that corner P touches the opposite 14 inch edge at R. (Note: $PQ = \sqrt{C^2 - x^2}$.)

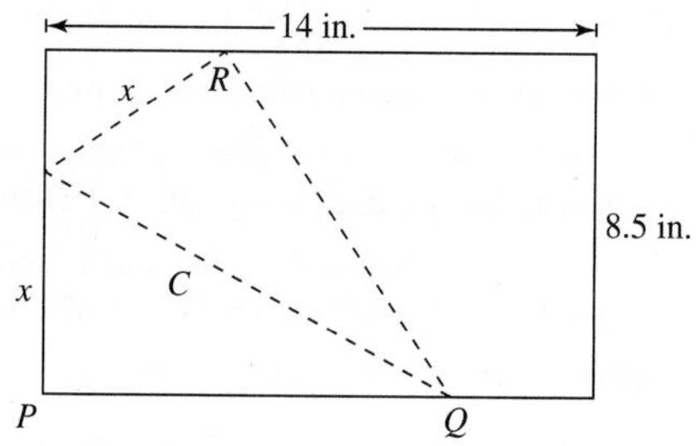

(a) Show that $C^2 = \dfrac{2x^3}{2x - 8.5}$.

(b) What is the domain of C?

(c) Determine the x-value that minimizes C.

(d) Determine the minimum length C.

19. The polynomial $P(x) = c_0 + c_1(x - a) + c_2(x - a)^2$ is the quadratic approximation of the function f at $(a, f(a))$ if $P(a) = f(a)$, $P'(a) = f'(a)$, and $P''(a) = f''(a)$.

(a) Find the quadratic approximation of

$$f(x) = \frac{x}{x + 1}$$

at (0, 0).

(b) Use a graphing utility to graph P and f in the same viewing window.

4 Integration

This photo of a jet breaking the sound barrier was taken by Ensign John Gay. At the time the photo was taken, was the jet's velocity constant or changing? Why?

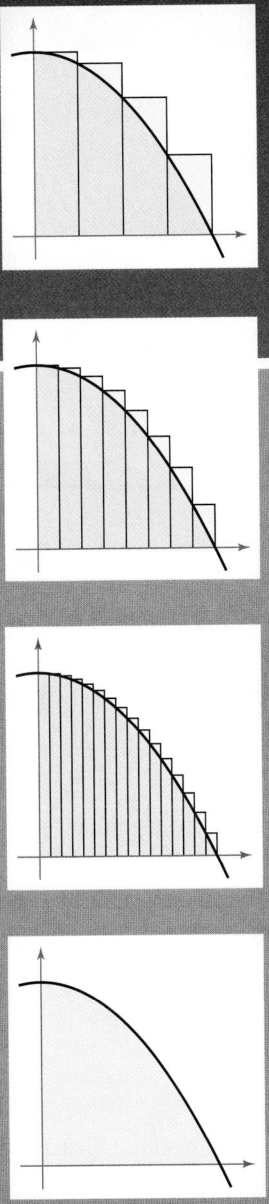

The area of a parabolic region can be approximated as the sum of the areas of rectangles. As you increase the number of rectangles, the approximation tends to become more and more accurate. In Section 4.2, you will learn how the limit process can be used to find areas of a wide variety of regions. This process is called *integration* and is closely related to differentiation.

Section 4.1 Antiderivatives and Indefinite Integration

- **Write the general solution of a differential equation.**
- **Use indefinite integral notation for antiderivatives.**
- **Use basic integration rules to find antiderivatives.**
- **Find a particular solution of a differential equation.**

EXPLORATION

Finding Antiderivatives For each derivative, describe the original function F.

a. $F'(x) = 2x$

b. $F'(x) = x$

c. $F'(x) = x^2$

d. $F'(x) = \dfrac{1}{x^2}$

e. $F'(x) = \dfrac{1}{x^3}$

f. $F'(x) = \cos x$

What strategy did you use to find F?

Antiderivatives

Suppose you were asked to find a function F whose derivative is $f(x) = 3x^2$. From your knowledge of derivatives, you would probably say that

$$F(x) = x^3 \text{ because } \frac{d}{dx}[x^3] = 3x^2.$$

The function F is an *antiderivative* of f.

Definition of an Antiderivative

A function F is an **antiderivative** of f on an interval I if $F'(x) = f(x)$ for all x in I.

Note that F is called *an* antiderivative of f, rather than *the* antiderivative of f. To see why, observe that

$$F_1(x) = x^3, \quad F_2(x) = x^3 - 5, \quad \text{and} \quad F_3(x) = x^3 + 97$$

are all antiderivatives of $f(x) = 3x^2$. In fact, for any constant C, the function given by $F(x) = x^3 + C$ is an antiderivative of f.

THEOREM 4.1 Representation of Antiderivatives

If F is an antiderivative of f on an interval I, then G is an antiderivative of f on the interval I if and only if G is of the form $G(x) = F(x) + C$, for all x in I where C is a constant.

Proof The proof of Theorem 4.1 in one direction is straightforward. That is, if $G(x) = F(x) + C$, $F'(x) = f(x)$, and C is a constant, then

$$G'(x) = \frac{d}{dx}[F(x) + C] = F'(x) + 0 = f(x).$$

To prove this theorem in the other direction, assume that G is an antiderivative of f. Define a function H such that

$$H(x) = G(x) - F(x).$$

If H is not constant on the interval I, there must exist a and b $(a < b)$ in the interval such that $H(a) \neq H(b)$. Moreover, because H is differentiable on (a, b), you can apply the Mean Value Theorem to conclude that there exists some c in (a, b) such that

$$H'(c) = \frac{H(b) - H(a)}{b - a}.$$

Because $H(b) \neq H(a)$, it follows that $H'(c) \neq 0$. However, because $G'(c) = F'(c)$, you know that $H'(c) = G'(c) - F'(c) = 0$, which contradicts the fact that $H'(c) \neq 0$. Consequently, you can conclude that $H(x)$ is a constant, C. So, $G(x) - F(x) = C$ and it follows that $G(x) = F(x) + C$.

Using Theorem 4.1, you can represent the entire family of antiderivatives of a function by adding a constant to a *known* antiderivative. For example, knowing that $D_x[x^2] = 2x$, you can represent the family of *all* antiderivatives of $f(x) = 2x$ by

$$G(x) = x^2 + C \qquad \text{Family of all antiderivatives of } f(x) = 2x$$

where C is a constant. The constant C is called the **constant of integration.** The family of functions represented by G is the **general antiderivative** of f, and $G(x) = x^2 + C$ is the **general solution** of the *differential equation*

$$G'(x) = 2x. \qquad \text{Differential equation}$$

A **differential equation** in x and y is an equation that involves x, y, and derivatives of y. For instance, $y' = 3x$ and $y' = x^2 + 1$ are examples of differential equations.

EXAMPLE 1 Solving a Differential Equation

Find the general solution of the differential equation $y' = 2$.

Solution To begin, you need to find a function whose derivative is 2. One such function is

$$y = 2x. \qquad 2x \text{ is } an \text{ antiderivative of 2.}$$

Now, you can use Theorem 4.1 to conclude that the general solution of the differential equation is

$$y = 2x + C. \qquad \text{General solution}$$

The graphs of several functions of the form $y = 2x + C$ are shown in Figure 4.1.

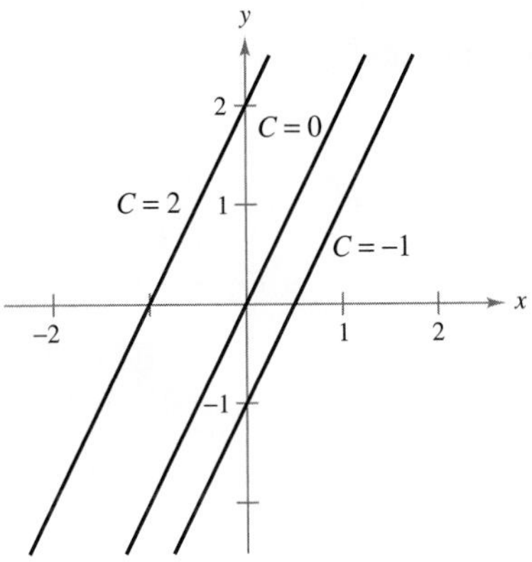

Functions of the form $y = 2x + C$
Figure 4.1

Notation for Antiderivatives

When solving a differential equation of the form

$$\frac{dy}{dx} = f(x)$$

it is convenient to write it in the equivalent differential form

$$dy = f(x)\,dx.$$

The operation of finding all solutions of this equation is called **antidifferentiation** (or **indefinite integration**) and is denoted by an integral sign $\int$. The general solution is denoted by

Variable of integration — Constant of integration

$$y = \int f(x)\,dx = F(x) + C.$$

Integrand

The expression $\int f(x)dx$ is read as the *antiderivative of f with respect to x.* So, the differential dx serves to identify x as the variable of integration. The term **indefinite integral** is a synonym for antiderivative.

NOTE In this text, the notation $\int f(x)\,dx = F(x) + C$ means that F is an antiderivative of f *on an interval.*

Basic Integration Rules

The inverse nature of integration and differentiation can be verified by substituting $F'(x)$ for $f(x)$ in the indefinite integration definition to obtain

$$\int F'(x)\,dx = F(x) + C.$$

Integration is the "inverse" of differentiation.

Moreover, if $\int f(x)\,dx = F(x) + C$, then

$$\frac{d}{dx}\left[\int f(x)\,dx\right] = f(x).$$

Differentiation is the "inverse" of integration.

These two equations allow you to obtain integration formulas directly from differentiation formulas, as shown in the following summary.

Basic Integration Rules

Differentiation Formula	*Integration Formula*	
$\frac{d}{dx}[C] = 0$	$\int 0\,dx = C$	
$\frac{d}{dx}[kx] = k$	$\int k\,dx = kx + C$	
$\frac{d}{dx}[kf(x)] = kf'(x)$	$\int kf(x)\,dx = k\int f(x)\,dx$	
$\frac{d}{dx}[f(x) \pm g(x)] = f'(x) \pm g'(x)$	$\int [f(x) \pm g(x)]\,dx = \int f(x)\,dx \pm \int g(x)\,dx$	
$\frac{d}{dx}[x^n] = nx^{n-1}$	$\int x^n\,dx = \frac{x^{n+1}}{n+1} + C, \quad n \neq -1$	Power Rule
$\frac{d}{dx}[\sin x] = \cos x$	$\int \cos x\,dx = \sin x + C$	
$\frac{d}{dx}[\cos x] = -\sin x$	$\int \sin x\,dx = -\cos x + C$	
$\frac{d}{dx}[\tan x] = \sec^2 x$	$\int \sec^2 x\,dx = \tan x + C$	
$\frac{d}{dx}[\sec x] = \sec x \tan x$	$\int \sec x \tan x\,dx = \sec x + C$	
$\frac{d}{dx}[\cot x] = -\csc^2 x$	$\int \csc^2 x\,dx = -\cot x + C$	
$\frac{d}{dx}[\csc x] = -\csc x \cot x$	$\int \csc x \cot x\,dx = -\csc x + C$	

NOTE Note that the Power Rule for Integration has the restriction that $n \neq -1$. The evaluation of $\int 1/x\,dx$ must wait until the introduction of the natural logarithm function in Chapter 5.

EXAMPLE 2 Applying the Basic Integration Rules

Describe the antiderivatives of $3x$.

Solution

$$\int 3x\,dx = 3\int x\,dx \qquad \text{Constant Multiple Rule}$$

$$= 3\int x^1\,dx \qquad \text{Rewrite } x \text{ as } x^1.$$

$$= 3\left(\frac{x^2}{2}\right) + C \qquad \text{Power Rule } (n = 1)$$

$$= \frac{3}{2}x^2 + C \qquad \text{Simplify.}$$

So, the antiderivatives of $3x$ are of the form $\frac{3}{2}x^2 + C$, where C is any constant.

When indefinite integrals are evaluated, a strict application of the basic integration rules tends to produce complicated constants of integration. For instance, in Example 2, you could have written

$$\int 3x\,dx = 3\int x\,dx = 3\left(\frac{x^2}{2} + C\right) = \frac{3}{2}x^2 + 3C.$$

However, because C represents *any* constant, it is both cumbersome and unnecessary to write $3C$ as the constant of integration. So, $\frac{3}{2}x^2 + 3C$ is written in the simpler form, $\frac{3}{2}x^2 + C$.

In Example 2, note that the general pattern of integration is similar to that of differentiation.

Original integral ⇨ Rewrite ⇨ Integrate ⇨ Simplify

EXAMPLE 3 Rewriting Before Integrating

	Original Integral	*Rewrite*	*Integrate*	*Simplify*
a.	$\int \frac{1}{x^3}\,dx$	$\int x^{-3}\,dx$	$\frac{x^{-2}}{-2} + C$	$-\frac{1}{2x^2} + C$
b.	$\int \sqrt{x}\,dx$	$\int x^{1/2}\,dx$	$\frac{x^{3/2}}{3/2} + C$	$\frac{2}{3}x^{3/2} + C$
c.	$\int 2\sin x\,dx$	$2\int \sin x\,dx$	$2(-\cos x) + C$	$-2\cos x + C$

TECHNOLOGY Some software programs, such as *Derive*, *Maple*, *Mathcad*, *Mathematica*, and the *TI-89*, are capable of performing integration symbolically. If you have access to such a symbolic integration utility, try using it to evaluate the indefinite integrals in Example 3.

Remember that you can check your answer to an antidifferentiation problem by differentiating. For instance, in Example 3(b), you can check that $\frac{2}{3}x^{3/2} + C$ is the correct antiderivative by differentiating the answer to obtain

$$D_x\left[\frac{2}{3}x^{3/2} + C\right] = \left(\frac{2}{3}\right)\left(\frac{3}{2}\right)x^{1/2} = \sqrt{x}. \qquad \text{Use differentiation to check antiderivative.}$$

indicates that in the HM mathSpace® *CD-ROM and the online* Eduspace® *system for this text, you will find an Open Exploration, which further explores this example using the computer algebra systems* Maple, Mathcad, Mathematica, *and* Derive.

The basic integration rules listed earlier in this section allow you to integrate any polynomial function, as shown in Example 4.

EXAMPLE 4 Integrating Polynomial Functions

a. $\int dx = \int 1\, dx$ — Integrand is understood to be 1.

$= x + C$ — Integrate.

b. $\int (x + 2)\, dx = \int x\, dx + \int 2\, dx$

$= \dfrac{x^2}{2} + C_1 + 2x + C_2$ — Integrate.

$= \dfrac{x^2}{2} + 2x + C$ — $C = C_1 + C_2$

The second line in the solution is usually omitted.

c. $\int (3x^4 - 5x^2 + x)\, dx = 3\left(\dfrac{x^5}{5}\right) - 5\left(\dfrac{x^3}{3}\right) + \dfrac{x^2}{2} + C$ — Integrate.

$= \dfrac{3}{5}x^5 - \dfrac{5}{3}x^3 + \dfrac{1}{2}x^2 + C$ — Simplify.

EXAMPLE 5 Rewriting Before Integrating

$\int \dfrac{x + 1}{\sqrt{x}}\, dx = \int \left(\dfrac{x}{\sqrt{x}} + \dfrac{1}{\sqrt{x}}\right) dx$ — Rewrite as two fractions.

$= \int (x^{1/2} + x^{-1/2})\, dx$ — Rewrite with fractional exponents.

$= \dfrac{x^{3/2}}{3/2} + \dfrac{x^{1/2}}{1/2} + C$ — Integrate.

$= \dfrac{2}{3}x^{3/2} + 2x^{1/2} + C$ — Simplify.

$= \dfrac{2}{3}\sqrt{x}(x + 3) + C$

NOTE When integrating quotients, do not integrate the numerator and denominator separately. This is no more valid in integration than it is in differentiation. For instance, in Example 5, be sure you understand that

$$\int \frac{x + 1}{\sqrt{x}}\, dx = \frac{2}{3}\sqrt{x}(x + 3) + C \text{ is not the same as } \frac{\int (x + 1)\, dx}{\int \sqrt{x}\, dx} = \frac{\frac{1}{2}x^2 + x + C_1}{\frac{2}{3}x\sqrt{x} + C_2}.$$

EXAMPLE 6 Rewriting Before Integrating

$\int \dfrac{\sin x}{\cos^2 x}\, dx = \int \left(\dfrac{1}{\cos x}\right)\left(\dfrac{\sin x}{\cos x}\right) dx$ — Rewrite as a product.

$= \int \sec x \tan x\, dx$ — Rewrite using trigonometric identities.

$= \sec x + C$ — Integrate.

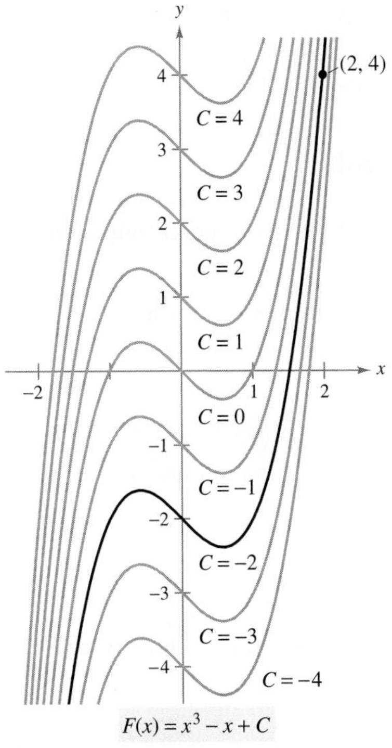

The particular solution that satisfies the initial condition $F(2) = 4$ is $F(x) = x^3 - x - 2$.
Figure 4.2

Initial Conditions and Particular Solutions

You have already seen that the equation $y = \int f(x)\,dx$ has many solutions (each differing from the others by a constant). This means that the graphs of any two antiderivatives of f are vertical translations of each other. For example, Figure 4.2 shows the graphs of several antiderivatives of the form

$$y = \int (3x^2 - 1)\,dx = x^3 - x + C \qquad \text{General solution}$$

for various integer values of C. Each of these antiderivatives is a solution of the differential equation

$$\frac{dy}{dx} = 3x^2 - 1.$$

In many applications of integration, you are given enough information to determine a **particular solution.** To do this, you need only know the value of $y = F(x)$ for one value of x. This information is called an **initial condition.** For example, in Figure 4.2, only one curve passes through the point (2, 4). To find this curve, you can use the following information.

$$F(x) = x^3 - x + C \qquad \text{General solution}$$
$$F(2) = 4 \qquad \text{Initial condition}$$

By using the initial condition in the general solution, you can determine that $F(2) = 8 - 2 + C = 4$, which implies that $C = -2$. So, you obtain

$$F(x) = x^3 - x - 2. \qquad \text{Particular solution}$$

EXAMPLE 7 Finding a Particular Solution

Find the general solution of

$$F'(x) = \frac{1}{x^2}, \quad x > 0$$

and find the particular solution that satisfies the initial condition $F(1) = 0$.

Solution To find the general solution, integrate to obtain

$$F(x) = \int \frac{1}{x^2}\,dx \qquad F(x) = \int F'(x)dx$$
$$= \int x^{-2}\,dx \qquad \text{Rewrite as a power.}$$
$$= \frac{x^{-1}}{-1} + C \qquad \text{Integrate.}$$
$$= -\frac{1}{x} + C, \quad x > 0. \qquad \text{General solution}$$

Using the initial condition $F(1) = 0$, you can solve for C as follows.

$$F(1) = -\frac{1}{1} + C = 0 \quad \Longrightarrow \quad C = 1$$

So, the particular solution, as shown in Figure 4.3, is

$$F(x) = -\frac{1}{x} + 1, \quad x > 0. \qquad \text{Particular solution}$$

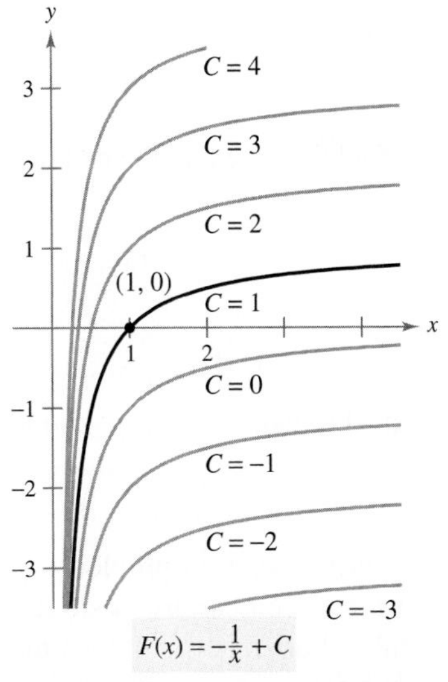

The particular solution that satisfies the initial condition $F(1) = 0$ is $F(x) = -(1/x) + 1, x > 0$.
Figure 4.3

So far in this section you have been using x as the variable of integration. In applications, it is often convenient to use a different variable. For instance, in the following example involving *time*, the variable of integration is t.

EXAMPLE 8 Solving a Vertical Motion Problem

A ball is thrown upward with an initial velocity of 64 feet per second from an initial height of 80 feet.

a. Find the position function giving the height s as a function of the time t.

b. When does the ball hit the ground?

Solution

a. Let $t = 0$ represent the initial time. The two given initial conditions can be written as follows.

$$s(0) = 80 \qquad \text{Initial height is 80 feet.}$$
$$s'(0) = 64 \qquad \text{Initial velocity is 64 feet per second.}$$

Using -32 feet per second per second as the acceleration due to gravity, you can write

$$s''(t) = -32$$
$$s'(t) = \int s''(t)\,dt = \int -32\,dt = -32t + C_1.$$

Using the initial velocity, you obtain $s'(0) = 64 = -32(0) + C_1$, which implies that $C_1 = 64$. Next, by integrating $s'(t)$, you obtain

$$s(t) = \int s'(t)\,dt = \int (-32t + 64)\,dt = -16t^2 + 64t + C_2.$$

Using the initial height, you obtain

$$s(0) = 80 = -16(0^2) + 64(0) + C_2$$

which implies that $C_2 = 80$. So, the position function is

$$s(t) = -16t^2 + 64t + 80. \qquad \text{See Figure 4.4.}$$

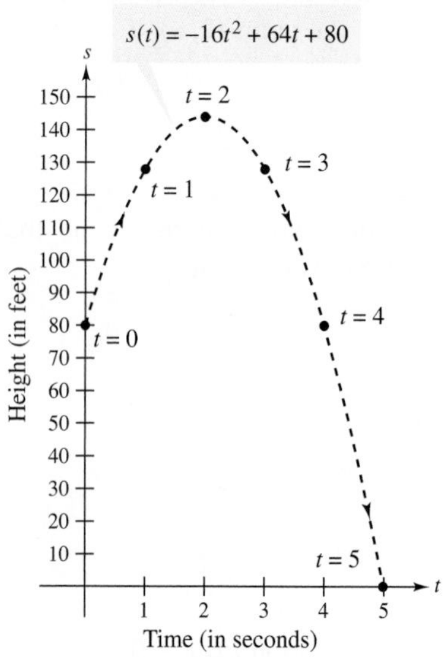

Height of a ball at time t
Figure 4.4

b. Using the position function found in part (a), you can find the time that the ball hits the ground by solving the equation $s(t) = 0$.

$$s(t) = -16t^2 + 64t + 80 = 0$$
$$-16(t + 1)(t - 5) = 0$$
$$t = -1, 5$$

Because t must be positive, you can conclude that the ball hits the ground 5 seconds after it was thrown.

NOTE In Example 8, note that the position function has the form

$$s(t) = \tfrac{1}{2}gt^2 + v_0t + s_0$$

where $g = -32$, v_0 is the initial velocity, and s_0 is the initial height, as presented in Section 2.2.

Example 8 shows how to use calculus to analyze vertical motion problems in which the acceleration is determined by a gravitational force. You can use a similar strategy to analyze other linear motion problems (vertical or horizontal) in which the acceleration (or deceleration) is the result of some other force, as you will see in Exercises 77–86.

Before you begin the exercise set, be sure you realize that one of the most important steps in integration is *rewriting the integrand* in a form that fits the basic integration rules. To illustrate this point further, here are some additional examples.

Original Integral	*Rewrite*	*Integrate*	*Simplify*
$\int \frac{2}{\sqrt{x}}\,dx$	$2\int x^{-1/2}\,dx$	$2\left(\frac{x^{1/2}}{1/2}\right) + C$	$4x^{1/2} + C$
$\int (t^2+1)^2\,dt$	$\int (t^4 + 2t^2 + 1)\,dt$	$\frac{t^5}{5} + 2\left(\frac{t^3}{3}\right) + t + C$	$\frac{1}{5}t^5 + \frac{2}{3}t^3 + t + C$
$\int \frac{x^3+3}{x^2}\,dx$	$\int (x + 3x^{-2})\,dx$	$\frac{x^2}{2} + 3\left(\frac{x^{-1}}{-1}\right) + C$	$\frac{1}{2}x^2 - \frac{3}{x} + C$
$\int \sqrt[3]{x}(x-4)\,dx$	$\int (x^{4/3} - 4x^{1/3})\,dx$	$\frac{x^{7/3}}{7/3} - 4\left(\frac{x^{4/3}}{4/3}\right) + C$	$\frac{3}{7}x^{7/3} - 3x^{4/3}$

Exercises for Section 4.1

See www.CalcChat.com for worked-out solutions to odd-numbered exercises.

In Exercises 1–4, verify the statement by showing that the derivative of the right side equals the integrand of the left side.

1. $\displaystyle\int \left(-\frac{9}{x^4}\right) dx = \frac{3}{x^3} + C$

2. $\displaystyle\int \left(4x^3 - \frac{1}{x^2}\right) dx = x^4 + \frac{1}{x} + C$

3. $\displaystyle\int (x-2)(x+2)\,dx = \tfrac{1}{3}x^3 - 4x + C$

4. $\displaystyle\int \frac{x^2-1}{x^{3/2}}\,dx = \frac{2(x^2+3)}{3\sqrt{x}} + C$

In Exercises 5–8, find the general solution of the differential equation and check the result by differentiation.

5. $\dfrac{dy}{dt} = 3t^2$

6. $\dfrac{dr}{d\theta} = \pi$

7. $\dfrac{dy}{dx} = x^{3/2}$

8. $\dfrac{dy}{dx} = 2x^{-3}$

In Exercises 9–14, complete the table.

Original Integral	*Rewrite*	*Integrate*	*Simplify*
9. $\int \sqrt[3]{x}\,dx$			
10. $\int \frac{1}{x^2}\,dx$			
11. $\int \frac{1}{x\sqrt{x}}\,dx$			
12. $\int x(x^2+3)\,dx$			
13. $\int \frac{1}{2x^3}\,dx$			
14. $\int \frac{1}{(3x)^2}\,dx$			

In Exercises 15–34, find the indefinite integral and check the result by differentiation.

15. $\displaystyle\int (x+3)\,dx$

16. $\displaystyle\int (5-x)\,dx$

17. $\displaystyle\int (2x - 3x^2)\,dx$

18. $\displaystyle\int (4x^3 + 6x^2 - 1)\,dx$

19. $\displaystyle\int (x^3 + 2)\,dx$

20. $\displaystyle\int (x^3 - 4x + 2)\,dx$

21. $\displaystyle\int (x^{3/2} + 2x + 1)\,dx$

22. $\displaystyle\int \left(\sqrt{x} + \frac{1}{2\sqrt{x}}\right) dx$

23. $\displaystyle\int \sqrt[3]{x^2}\,dx$

24. $\displaystyle\int \left(\sqrt[4]{x^3} + 1\right) dx$

25. $\displaystyle\int \frac{1}{x^3}\,dx$

26. $\displaystyle\int \frac{1}{x^4}\,dx$

27. $\displaystyle\int \frac{x^2+x+1}{\sqrt{x}}\,dx$

28. $\displaystyle\int \frac{x^2+2x-3}{x^4}\,dx$

29. $\displaystyle\int (x+1)(3x-2)\,dx$

30. $\displaystyle\int (2t^2-1)^2\,dt$

31. $\displaystyle\int y^2\sqrt{y}\,dy$

32. $\displaystyle\int (1+3t)t^2\,dt$

33. $\displaystyle\int dx$

34. $\displaystyle\int 3\,dt$

In Exercises 35–42, find the indefinite integral and check the result by differentiation.

35. $\displaystyle\int (2\sin x + 3\cos x)\,dx$

36. $\displaystyle\int (t^2 - \sin t)\,dt$

37. $\displaystyle\int (1 - \csc t \cot t)\,dt$

38. $\displaystyle\int (\theta^2 + \sec^2\theta)\,d\theta$

39. $\displaystyle\int (\sec^2\theta - \sin\theta)\,d\theta$

40. $\displaystyle\int \sec y(\tan y - \sec y)\,dy$

41. $\displaystyle\int (\tan^2 y + 1)\,dy$

42. $\displaystyle\int \frac{\cos x}{1 - \cos^2 x}\,dx$

In Exercises 43–46, the graph of the derivative of a function is given. Sketch the graphs of *two* functions that have the given derivative. (There is more than one correct answer.) To print an enlarged copy of the graph, go to the website *www.mathgraphs.com*.

43.

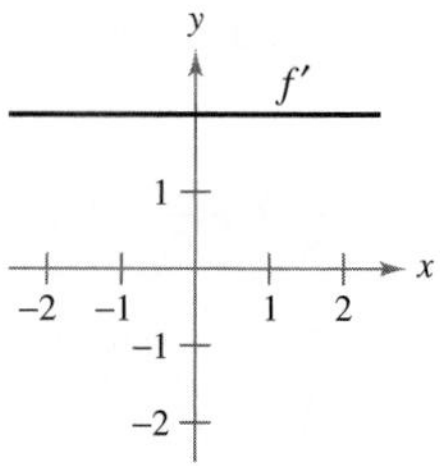

44.

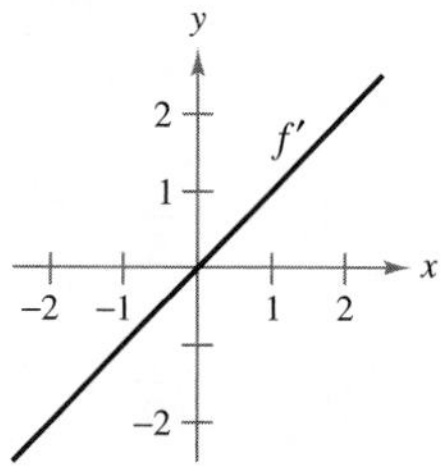

45.

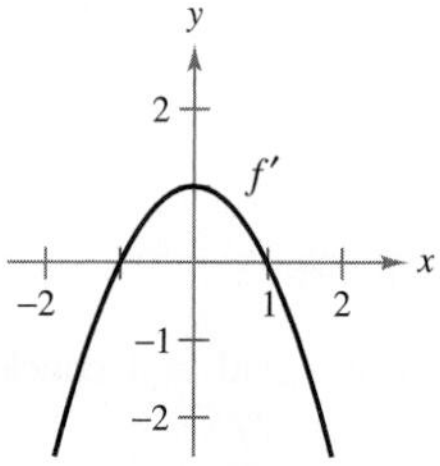

46.

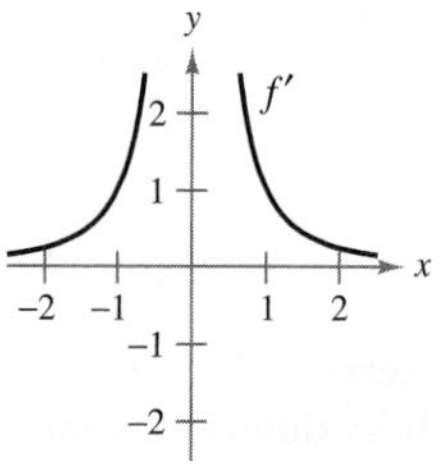

In Exercises 47 and 48, find the equation for *y*, given the derivative and the indicated point on the curve.

47. $\dfrac{dy}{dx} = 2x - 1$

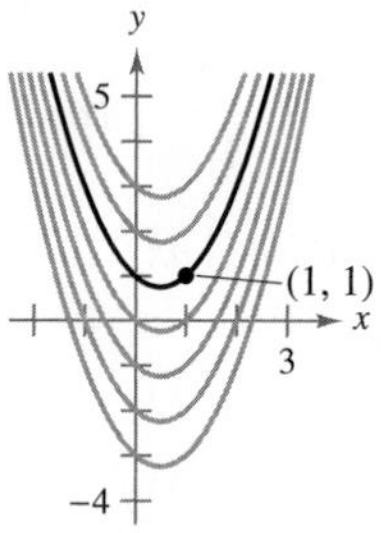

48. $\dfrac{dy}{dx} = 2(x - 1)$

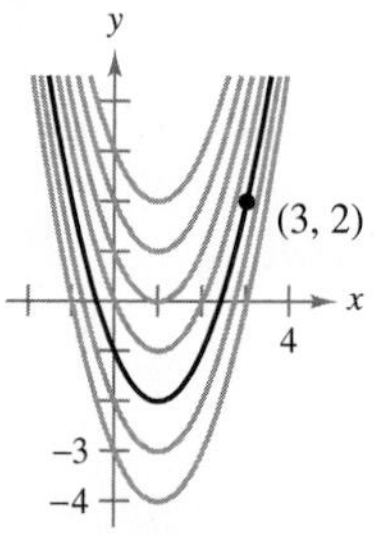

Slope Fields **In Exercises 49–52, a differential equation, a point, and a slope field are given. A *slope field* (or *direction field*) consists of line segments with slopes given by the differential equation. These line segments give a visual perspective of the slopes of the solutions of the differential equation. (a) Sketch two approximate solutions of the differential equation on the slope field, one of which passes through the indicated point. (To print an enlarged copy of the graph, go to the website *www.mathgraphs.com*.) (b) Use integration to find the particular solution of the differential equation and use a graphing utility to graph the solution. Compare the result with the sketches in part (a).**

49. $\dfrac{dy}{dx} = \dfrac{1}{2}x - 1, \quad (4, 2)$

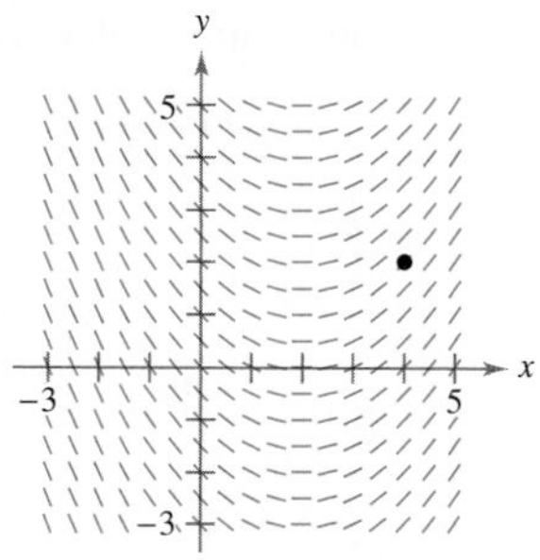

50. $\dfrac{dy}{dx} = x^2 - 1, \quad (-1, 3)$

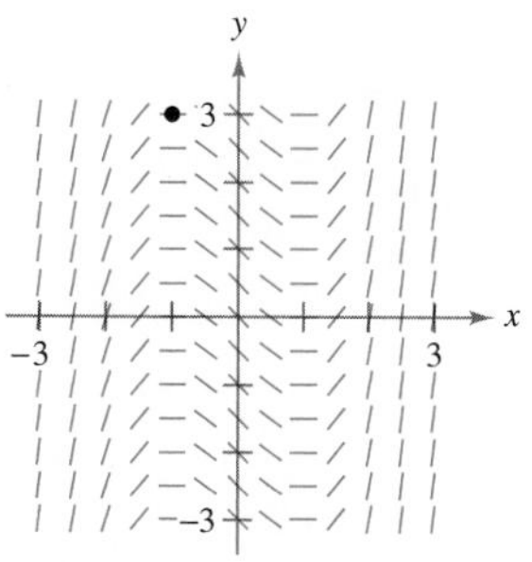

51. $\dfrac{dy}{dx} = \cos x, \ (0, 4)$

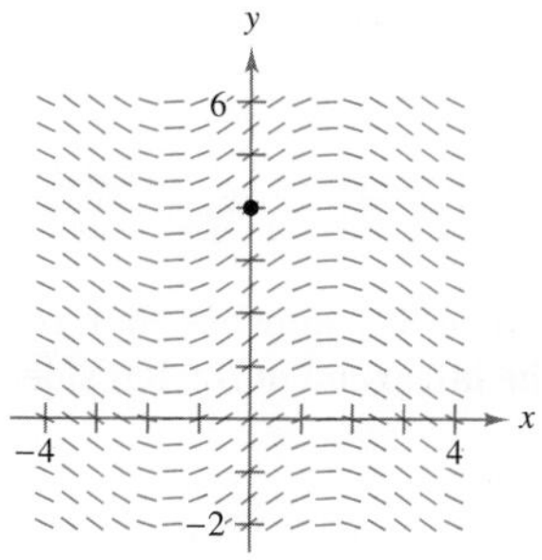

52. $\dfrac{dy}{dx} = -\dfrac{1}{x^2}, \ x > 0, \ (1, 3)$

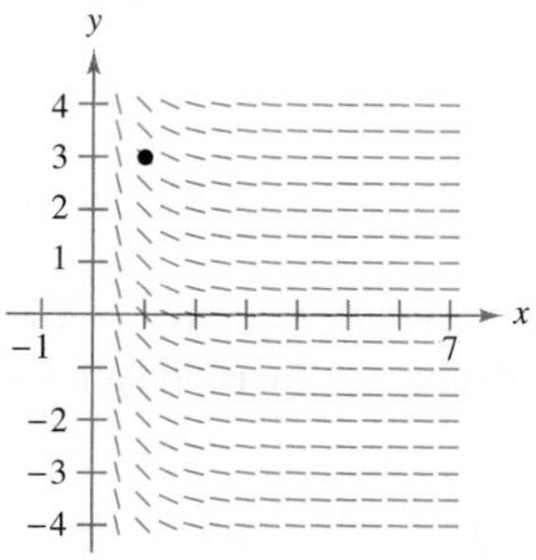

Slope Fields **In Exercises 53 and 54, (a) use a graphing utility to graph a slope field for the differential equation, (b) use integration and the given point to find the particular solution of the differential equation, and (c) graph the solution and the slope field in the same viewing window.**

53. $\dfrac{dy}{dx} = 2x, \ (-2, -2)$

54. $\dfrac{dy}{dx} = 2\sqrt{x}, \ (4, 12)$

In Exercises 55–62, solve the differential equation.

55. $f'(x) = 4x, f(0) = 6$

56. $g'(x) = 6x^2, g(0) = -1$

57. $h'(t) = 8t^3 + 5, h(1) = -4$

58. $f'(s) = 6s - 8s^3, f(2) = 3$

59. $f''(x) = 2, \ f'(2) = 5, \ f(2) = 10$

60. $f''(x) = x^2, \ f'(0) = 6, \ f(0) = 3$

61. $f''(x) = x^{-3/2}, \ f'(4) = 2, \ f(0) = 0$

62. $f''(x) = \sin x, \ f'(0) = 1, \ f(0) = 6$

63. *Tree Growth* An evergreen nursery usually sells a certain shrub after 6 years of growth and shaping. The growth rate during those 6 years is approximated by $dh/dt = 1.5t + 5$, where t is the time in years and h is the height in centimeters. The seedlings are 12 centimeters tall when planted ($t = 0$).

(a) Find the height after t years.

(b) How tall are the shrubs when they are sold?

64. *Population Growth* The rate of growth dP/dt of a population of bacteria is proportional to the square root of t, where P is the population size and t is the time in days ($0 \le t \le 10$). That is, $dP/dt = k\sqrt{t}$. The initial size of the population is 500. After 1 day the population has grown to 600. Estimate the population after 7 days.

Writing About Concepts

65. Use the graph of f' shown in the figure to answer the following, given that $f(0) = -4$.

(a) Approximate the slope of f at $x = 4$. Explain.

(b) Is it possible that $f(2) = -1$? Explain.

(c) Is $f(5) - f(4) > 0$? Explain.

(d) Approximate the value of x where f is maximum. Explain.

(e) Approximate any intervals in which the graph of f is concave upward and any intervals in which it is concave downward. Approximate the x-coordinates of any points of inflection.

(f) Approximate the x-coordinate of the minimum of $f''(x)$.

(g) Sketch an approximate graph of f. To print an enlarged copy of the graph, go to the website *www.mathgraphs.com.*

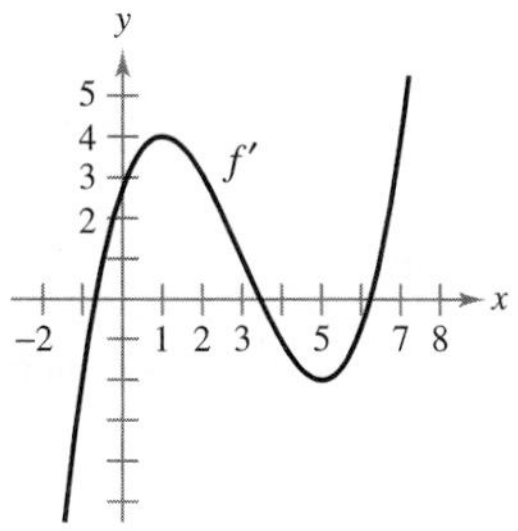

Figure for 65

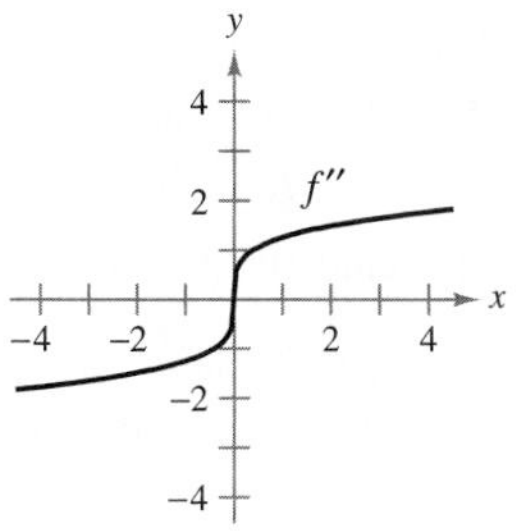

Figure for 66

66. The graphs of f and f' each pass through the origin. Use the graph of f'' shown in the figure to sketch the graphs of f and f'. To print an enlarged copy of the graph, go to the website *www.mathgraphs.com.*

Vertical Motion **In Exercises 67–70, use $a(t) = -32$ feet per second per second as the acceleration due to gravity. (Neglect air resistance.)**

67. A ball is thrown vertically upward from a height of 6 feet with an initial velocity of 60 feet per second. How high will the ball go?

68. Show that the height above the ground of an object thrown upward from a point s_0 feet above the ground with an initial velocity of v_0 feet per second is given by the function

$$f(t) = -16t^2 + v_0t + s_0.$$

69. With what initial velocity must an object be thrown upward (from ground level) to reach the top of the Washington Monument (approximately 550 feet)?

70. A balloon, rising vertically with a velocity of 16 feet per second, releases a sandbag at the instant it is 64 feet above the ground.

(a) How many seconds after its release will the bag strike the ground?

(b) At what velocity will it hit the ground?

Vertical Motion **In Exercises 71–74, use $a(t) = -9.8$ meters per second per second as the acceleration due to gravity. (Neglect air resistance.)**

71. Show that the height above the ground of an object thrown upward from a point s_0 meters above the ground with an initial velocity of v_0 meters per second is given by the function

$$f(t) = -4.9t^2 + v_0t + s_0.$$

72. The Grand Canyon is 1800 meters deep at its deepest point. A rock is dropped from the rim above this point. Write the height of the rock as a function of the time t in seconds. How long will it take the rock to hit the canyon floor?

73. A baseball is thrown upward from a height of 2 meters with an initial velocity of 10 meters per second. Determine its maximum height.

74. With what initial velocity must an object be thrown upward (from a height of 2 meters) to reach a maximum height of 200 meters?

75. ***Lunar Gravity*** On the moon, the acceleration due to gravity is -1.6 meters per second per second. A stone is dropped from a cliff on the moon and hits the surface of the moon 20 seconds later. How far did it fall? What was its velocity at impact?

76. ***Escape Velocity*** The minimum velocity required for an object to escape Earth's gravitational pull is obtained from the solution of the equation

$$\int v\,dv = -GM\int \frac{1}{y^2}\,dy$$

where v is the velocity of the object projected from Earth, y is the distance from the center of Earth, G is the gravitational constant, and M is the mass of Earth. Show that v and y are related by the equation

$$v^2 = v_0^2 + 2GM\left(\frac{1}{y} - \frac{1}{R}\right)$$

where v_0 is the initial velocity of the object and R is the radius of Earth.

Rectilinear Motion **In Exercises 77–80, consider a particle moving along the x-axis where $x(t)$ is the position of the particle at time t, $x'(t)$ is its velocity, and $x''(t)$ is its acceleration.**

77. $x(t) = t^3 - 6t^2 + 9t - 2, \quad 0 \le t \le 5$

(a) Find the velocity and acceleration of the particle.

(b) Find the open t-intervals on which the particle is moving to the right.

(c) Find the velocity of the particle when the acceleration is 0.

78. Repeat Exercise 77 for the position function

$$x(t) = (t-1)(t-3)^2, \quad 0 \le t \le 5.$$

79. A particle moves along the x-axis at a velocity of $v(t) = 1/\sqrt{t}$, $t > 0$. At time $t = 1$, its position is $x = 4$. Find the acceleration and position functions for the particle.

80. A particle, initially at rest, moves along the x-axis such that its acceleration at time $t > 0$ is given by $a(t) = \cos t$. At the time $t = 0$, its position is $x = 3$.

(a) Find the velocity and position functions for the particle.

(b) Find the values of t for which the particle is at rest.

81. ***Acceleration*** The maker of an automobile advertises that it takes 13 seconds to accelerate from 25 kilometers per hour to 80 kilometers per hour. Assuming constant acceleration, compute the following.

(a) The acceleration in meters per second per second

(b) The distance the car travels during the 13 seconds

82. ***Deceleration*** A car traveling at 45 miles per hour is brought to a stop, at constant deceleration, 132 feet from where the brakes are applied.

(a) How far has the car moved when its speed has been reduced to 30 miles per hour?

(b) How far has the car moved when its speed has been reduced to 15 miles per hour?

(c) Draw the real number line from 0 to 132, and plot the points found in parts (a) and (b). What can you conclude?

83. ***Acceleration*** At the instant the traffic light turns green, a car that has been waiting at an intersection starts with a constant acceleration of 6 feet per second per second. At the same instant, a truck traveling with a constant velocity of 30 feet per second passes the car.

(a) How far beyond its starting point will the car pass the truck?

(b) How fast will the car be traveling when it passes the truck?

84. ***Modeling Data*** The table shows the velocities (in miles per hour) of two cars on an entrance ramp to an interstate highway. The time t is in seconds.

t	0	5	10	15	20	25	30
v_1	0	2.5	7	16	29	45	65
v_2	0	21	38	51	60	64	65

(a) Rewrite the table converting miles per hour to feet per second.

(b) Use the regression capabilities of a graphing utility to find quadratic models for the data in part (a).

(c) Approximate the distance traveled by each car during the 30 seconds. Explain the difference in the distances.

85. ***Acceleration*** Assume that a fully loaded plane starting from rest has a constant acceleration while moving down a runway. The plane requires 0.7 mile of runway and a speed of 160 miles per hour in order to lift off. What is the plane's acceleration?

86. ***Airplane Separation*** Two airplanes are in a straight-line landing pattern and, according to FAA regulations, must keep at least a three-mile separation. Airplane A is 10 miles from touchdown and is gradually decreasing its speed from 150 miles per hour to a landing speed of 100 miles per hour. Airplane B is 17 miles from touchdown and is gradually decreasing its speed from 250 miles per hour to a landing speed of 115 miles per hour.

(a) Assuming the deceleration of each airplane is constant, find the position functions s_1 and s_2 for airplane A and airplane B. Let $t = 0$ represent the times when the airplanes are 10 and 17 miles from the airport.

(b) Use a graphing utility to graph the position functions.

(c) Find a formula for the magnitude of the distance d between the two airplanes as a function of t. Use a graphing utility to graph d. Is $d < 3$ for some time prior to the landing of airplane A? If so, find that time.

True or False? **In Exercises 87–92, determine whether the statement is true or false. If it is false, explain why or give an example that shows it is false.**

87. Each antiderivative of an nth-degree polynomial function is an $(n + 1)$th-degree polynomial function.

88. If $p(x)$ is a polynomial function, then p has exactly one antiderivative whose graph contains the origin.

89. If $F(x)$ and $G(x)$ are antiderivatives of $f(x)$, then $F(x) = G(x) + C$.

90. If $f'(x) = g(x)$, then $\int g(x)\,dx = f(x) + C$.

91. $\int f(x)g(x)\,dx = \int f(x)\,dx \int g(x)\,dx$

92. The antiderivative of $f(x)$ is unique.

93. Find a function f such that the graph of f has a horizontal tangent at $(2, 0)$ and $f''(x) = 2x$.

94. The graph of f' is shown. Sketch the graph of f given that f is continuous and $f(0) = 1$.

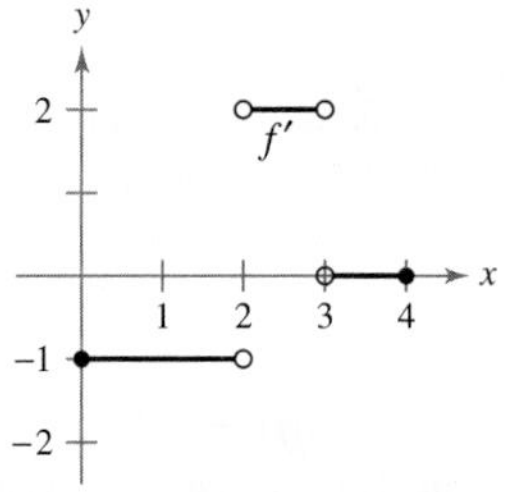

95. If $f'(x) = \begin{cases} 1, & 0 \le x < 2 \\ 3x, & 2 \le x \le 5 \end{cases}$, f is continuous, and $f(1) = 3$, find f. Is f differentiable at $x = 2$?

96. Let $s(x)$ and $c(x)$ be two functions satisfying $s'(x) = c(x)$ and $c'(x) = -s(x)$ for all x. If $s(0) = 0$ and $c(0) = 1$, prove that $[s(x)]^2 + [c(x)]^2 = 1$.

Putnam Exam Challenge

97. Suppose f and g are nonconstant, differentiable, real-valued functions on R. Furthermore, suppose that for each pair of real numbers x and y, $f(x + y) = f(x)f(y) - g(x)g(y)$ and $g(x + y) = f(x)g(y) + g(x)f(y)$. If $f'(0) = 0$, prove that $(f(x))^2 + (g(x))^2 = 1$ for all x.

This problem was composed by the Committee on the Putnam Prize Competition.

Section 4.2 Area

- **Use sigma notation to write and evaluate a sum.**
- **Understand the concept of area.**
- **Approximate the area of a plane region.**
- **Find the area of a plane region using limits.**

Sigma Notation

In the preceding section, you studied antidifferentiation. In this section, you will look further into a problem introduced in Section 1.1—that of finding the area of a region in the plane. At first glance, these two ideas may seem unrelated, but you will discover in Section 4.4 that they are closely related by an extremely important theorem called the Fundamental Theorem of Calculus.

This section begins by introducing a concise notation for sums. This notation is called **sigma notation** because it uses the uppercase Greek letter sigma, written as Σ.

> **Sigma Notation**
>
> The sum of n terms $a_1, a_2, a_3, \ldots, a_n$ is written as
>
> $$\sum_{i=1}^{n} a_i = a_1 + a_2 + a_3 + \cdots + a_n$$
>
> where i is the **index of summation,** a_i is the ***i*th term** of the sum, and the **upper and lower bounds of summation** are n and 1.

NOTE The upper and lower bounds must be constant with respect to the index of summation. However, the lower bound doesn't have to be 1. Any integer less than or equal to the upper bound is legitimate.

EXAMPLE 1 Examples of Sigma Notation

a. $\displaystyle\sum_{i=1}^{6} i = 1 + 2 + 3 + 4 + 5 + 6$

b. $\displaystyle\sum_{i=0}^{5} (i + 1) = 1 + 2 + 3 + 4 + 5 + 6$

c. $\displaystyle\sum_{j=3}^{7} j^2 = 3^2 + 4^2 + 5^2 + 6^2 + 7^2$

d. $\displaystyle\sum_{k=1}^{n} \frac{1}{n}(k^2 + 1) = \frac{1}{n}(1^2 + 1) + \frac{1}{n}(2^2 + 1) + \cdots + \frac{1}{n}(n^2 + 1)$

e. $\displaystyle\sum_{i=1}^{n} f(x_i)\,\Delta x = f(x_1)\,\Delta x + f(x_2)\,\Delta x + \cdots + f(x_n)\,\Delta x$

From parts (a) and (b), notice that the same sum can be represented in different ways using sigma notation.

FOR FURTHER INFORMATION For a geometric interpretation of summation formulas, see the article, "Looking at $\sum_{k=1}^{n} k$ and $\sum_{k=1}^{n} k^2$ Geometrically" by Eric Hegblom in *Mathematics Teacher*. To view this article, go to the website *www.matharticles.com.*

Although any variable can be used as the index of summation i, j, and k are often used. Notice in Example 1 that the index of summation does not appear in the terms of the expanded sum.

The Sum of the First 100 Integers

Carl Friedrich Gauss's (1777–1855) teacher asked him to add all the integers from 1 to 100. When Gauss returned with the correct answer after only a few moments, the teacher could only look at him in astounded silence. This is what Gauss did:

$$\begin{array}{rcrcrcccr} 1 & + & 2 & + & 3 & + & \cdots & + & 100 \\ 100 & + & 99 & + & 98 & + & \cdots & + & 1 \\ \hline 101 & + & 101 & + & 101 & + & \cdots & + & 101 \end{array}$$

$$\frac{100 \times 101}{2} = 5050$$

This is generalized by Theorem 4.2, where

$$\sum_{i=1}^{100} i = \frac{100(101)}{2} = 5050.$$

The following properties of summation can be derived using the associative and commutative properties of addition and the distributive property of addition over multiplication. (In the first property, k is a constant.)

1. $\displaystyle\sum_{i=1}^{n} ka_i = k\sum_{i=1}^{n} a_i$

2. $\displaystyle\sum_{i=1}^{n} (a_i \pm b_i) = \sum_{i=1}^{n} a_i \pm \sum_{i=1}^{n} b_i$

The next theorem lists some useful formulas for sums of powers. A proof of this theorem is given in Appendix A.

THEOREM 4.2 Summation Formulas

1. $\displaystyle\sum_{i=1}^{n} c = cn$

2. $\displaystyle\sum_{i=1}^{n} i = \frac{n(n+1)}{2}$

3. $\displaystyle\sum_{i=1}^{n} i^2 = \frac{n(n+1)(2n+1)}{6}$

4. $\displaystyle\sum_{i=1}^{n} i^3 = \frac{n^2(n+1)^2}{4}$

EXAMPLE 2 Evaluating a Sum

Evaluate $\displaystyle\sum_{i=1}^{n} \frac{i+1}{n^2}$ for $n = 10$, 100, 1000, and 10,000.

Solution Applying Theorem 4.2, you can write

$$\sum_{i=1}^{n} \frac{i+1}{n^2} = \frac{1}{n^2}\sum_{i=1}^{n}(i+1) \qquad \text{Factor constant } 1/n^2 \text{ out of sum.}$$

$$= \frac{1}{n^2}\left(\sum_{i=1}^{n} i + \sum_{i=1}^{n} 1\right) \qquad \text{Write as two sums.}$$

$$= \frac{1}{n^2}\left[\frac{n(n+1)}{2} + n\right] \qquad \text{Apply Theorem 4.2.}$$

$$= \frac{1}{n^2}\left[\frac{n^2+3n}{2}\right] \qquad \text{Simplify.}$$

$$= \frac{n+3}{2n}. \qquad \text{Simplify.}$$

Now you can evaluate the sum by substituting the appropriate values of n, as shown in the table at the left.

n	$\displaystyle\sum_{i=1}^{n} \frac{i+1}{n^2} = \frac{n+3}{2n}$
10	0.65000
100	0.51500
1,000	0.50150
10,000	0.50015

In the table, note that the sum appears to approach a limit as n increases. Although the discussion of limits at infinity in Section 3.5 applies to a variable x, where x can be any real number, many of the same results hold true for limits involving the variable n, where n is restricted to positive integer values. So, to find the limit of $(n+3)/2n$ as n approaches infinity, you can write

$$\lim_{n\to\infty} \frac{n+3}{2n} = \frac{1}{2}.$$

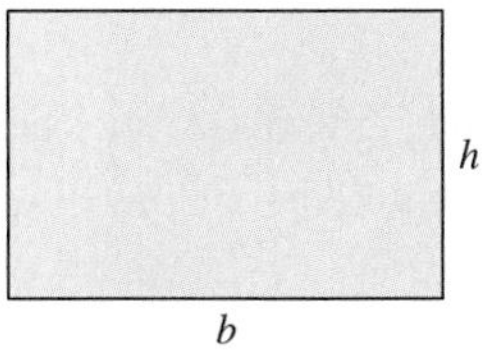

Rectangle: $A = bh$
Figure 4.5

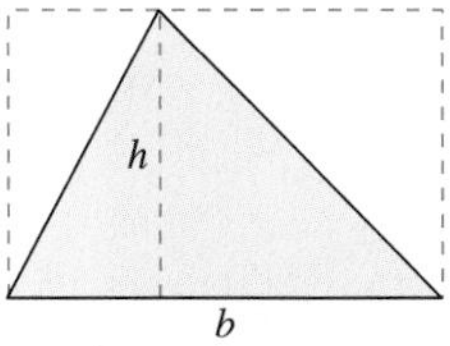

Triangle: $A = \frac{1}{2}bh$
Figure 4.6

Area

In Euclidean geometry, the simplest type of plane region is a rectangle. Although people often say that the *formula* for the area of a rectangle is $A = bh$, as shown in Figure 4.5, it is actually more proper to say that this is the *definition* of the **area of a rectangle.**

From this definition, you can develop formulas for the areas of many other plane regions. For example, to determine the area of a triangle, you can form a rectangle whose area is twice that of the triangle, as shown in Figure 4.6. Once you know how to find the area of a triangle, you can determine the area of any polygon by subdividing the polygon into triangular regions, as shown in Figure 4.7.

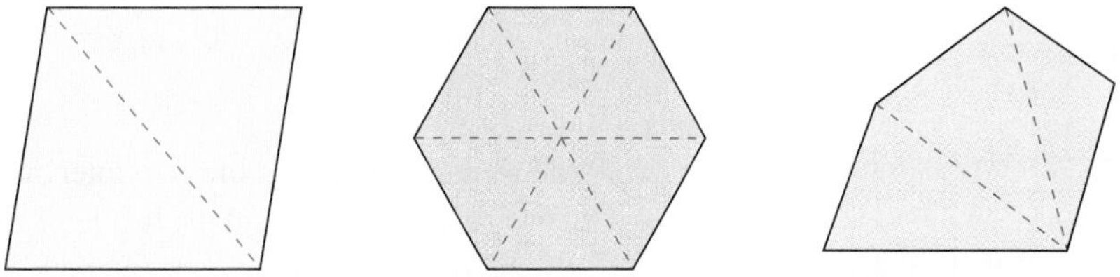

Parallelogram Hexagon Polygon
Figure 4.7

ARCHIMEDES (287–212 B.C.)

Archimedes used the method of exhaustion to derive formulas for the areas of ellipses, parabolic segments, and sectors of a spiral. He is considered to have been the greatest applied mathematician of antiquity.

Finding the areas of regions other than polygons is more difficult. The ancient Greeks were able to determine formulas for the areas of some general regions (principally those bounded by conics) by the *exhaustion* method. The clearest description of this method was given by Archimedes. Essentially, the method is a limiting process in which the area is squeezed between two polygons—one inscribed in the region and one circumscribed about the region.

For instance, in Figure 4.8 the area of a circular region is approximated by an n-sided inscribed polygon and an n-sided circumscribed polygon. For each value of n the area of the inscribed polygon is less than the area of the circle, and the area of the circumscribed polygon is greater than the area of the circle. Moreover, as n increases, the areas of both polygons become better and better approximations of the area of the circle.

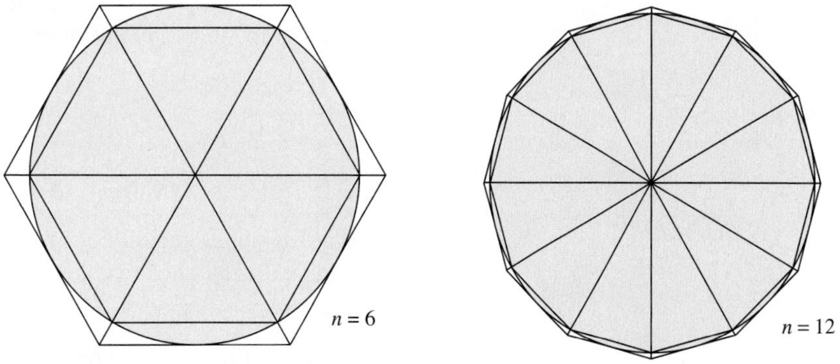

The exhaustion method for finding the area of a circular region
Figure 4.8

FOR FURTHER INFORMATION For an alternative development of the formula for the area of a circle, see the article "Proof Without Words: Area of a Disk is πR^2" by Russell Jay Hendel in *Mathematics Magazine*. To view this article, go to the website *www.matharticles.com.*

A process that is similar to that used by Archimedes to determine the area of a plane region is used in the remaining examples in this section.

The Area of a Plane Region

Recall from Section 1.1 that the origins of calculus are connected to two classic problems: the tangent line problem and the area problem. Example 3 begins the investigation of the area problem.

EXAMPLE 3 Approximating the Area of a Plane Region

Use the five rectangles in Figure 4.9(a) and (b) to find *two* approximations of the area of the region lying between the graph of

$$f(x) = -x^2 + 5$$

and the x-axis between $x = 0$ and $x = 2$.

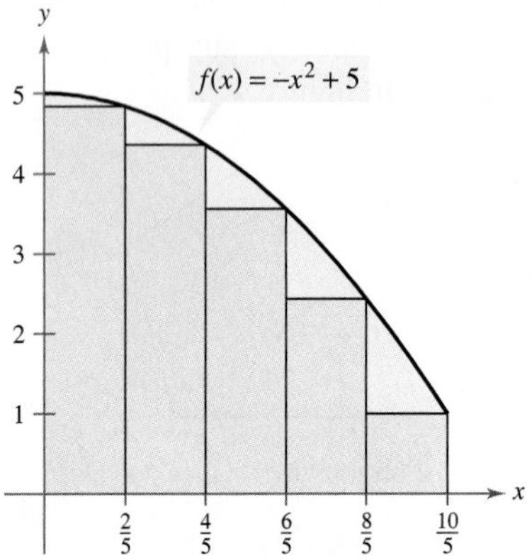

(a) The area of the parabolic region is greater than the area of the rectangles.

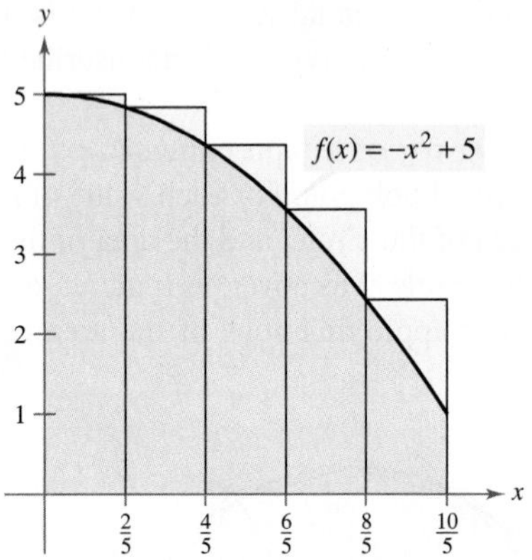

(b) The area of the parabolic region is less than the area of the rectangles.

Figure 4.9

Solution

a. The right endpoints of the five intervals are $\frac{2}{5}i$, where $i = 1, 2, 3, 4, 5$. The width of each rectangle is $\frac{2}{5}$, and the height of each rectangle can be obtained by evaluating f at the right endpoint of each interval.

$$\left[0, \frac{2}{5}\right], \left[\frac{2}{5}, \frac{4}{5}\right], \left[\frac{4}{5}, \frac{6}{5}\right], \left[\frac{6}{5}, \frac{8}{5}\right], \left[\frac{8}{5}, \frac{10}{5}\right]$$

Evaluate f at the right endpoints of these intervals.

The sum of the areas of the five rectangles is

$$\sum_{i=1}^{5} \underbrace{f\left(\frac{2i}{5}\right)}_{\text{Height}} \underbrace{\left(\frac{2}{5}\right)}_{\text{Width}} = \sum_{i=1}^{5} \left[-\left(\frac{2i}{5}\right)^2 + 5\right]\left(\frac{2}{5}\right) = \frac{162}{25} = 6.48.$$

Because each of the five rectangles lies inside the parabolic region, you can conclude that the area of the parabolic region is greater than 6.48.

b. The left endpoints of the five intervals are $\frac{2}{5}(i - 1)$, where $i = 1, 2, 3, 4, 5$. The width of each rectangle is $\frac{2}{5}$, and the height of each rectangle can be obtained by evaluating f at the left endpoint of each interval.

$$\sum_{i=1}^{5} \underbrace{f\left(\frac{2i-2}{5}\right)}_{\text{Height}} \underbrace{\left(\frac{2}{5}\right)}_{\text{Width}} = \sum_{i=1}^{5} \left[-\left(\frac{2i-2}{5}\right)^2 + 5\right]\left(\frac{2}{5}\right) = \frac{202}{25} = 8.08.$$

Because the parabolic region lies within the union of the five rectangular regions, you can conclude that the area of the parabolic region is less than 8.08.

By combining the results in parts (a) and (b), you can conclude that

$$6.48 < (\text{Area of region}) < 8.08.$$

NOTE By increasing the number of rectangles used in Example 3, you can obtain closer and closer approximations of the area of the region. For instance, using 25 rectangles of width $\frac{2}{25}$ each, you can conclude that

$$7.17 < (\text{Area of region}) < 7.49.$$

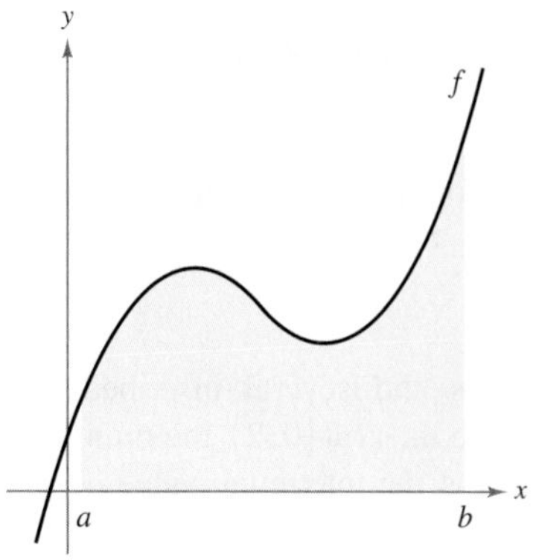

The region under a curve
Figure 4.10

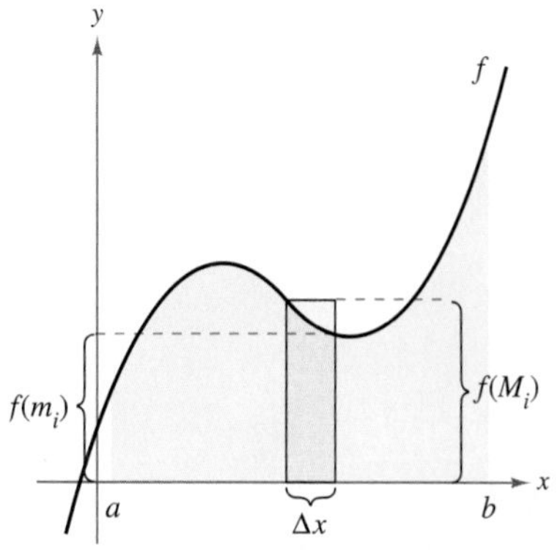

The interval $[a, b]$ is divided into n subintervals of width $\Delta x = \dfrac{b - a}{n}$.
Figure 4.11

Upper and Lower Sums

The procedure used in Example 3 can be generalized as follows. Consider a plane region bounded above by the graph of a nonnegative, continuous function $y = f(x)$, as shown in Figure 4.10. The region is bounded below by the x-axis, and the left and right boundaries of the region are the vertical lines $x = a$ and $x = b$.

To approximate the area of the region, begin by subdividing the interval $[a, b]$ into n subintervals, each of width $\Delta x = (b - a)/n$, as shown in Figure 4.11. The endpoints of the intervals are as follows.

$$\underbrace{a + 0(\Delta x)}_{a = x_0} < \underbrace{a + 1(\Delta x)}_{x_1} < \underbrace{a + 2(\Delta x)}_{x_2} < \cdot \cdot \cdot < \underbrace{a + n(\Delta x)}_{x_n = b}$$

Because f is continuous, the Extreme Value Theorem guarantees the existence of a minimum and a maximum value of $f(x)$ in *each* subinterval.

$f(m_i)$ = Minimum value of $f(x)$ in ith subinterval

$f(M_i)$ = Maximum value of $f(x)$ in ith subinterval

Next, define an **inscribed rectangle** lying *inside* the ith subregion and a **circumscribed rectangle** extending *outside* the ith subregion. The height of the ith inscribed rectangle is $f(m_i)$ and the height of the ith circumscribed rectangle is $f(M_i)$. For *each i*, the area of the inscribed rectangle is less than or equal to the area of the circumscribed rectangle.

$$\begin{pmatrix}\text{Area of inscribed}\\ \text{rectangle}\end{pmatrix} = f(m_i)\,\Delta x \le f(M_i)\,\Delta x = \begin{pmatrix}\text{Area of circumscribed}\\ \text{rectangle}\end{pmatrix}$$

The sum of the areas of the inscribed rectangles is called a **lower sum,** and the sum of the areas of the circumscribed rectangles is called an **upper sum.**

$$\text{Lower sum} = s(n) = \sum_{i=1}^{n} f(m_i)\,\Delta x \qquad \text{Area of inscribed rectangles}$$

$$\text{Upper sum} = S(n) = \sum_{i=1}^{n} f(M_i)\,\Delta x \qquad \text{Area of circumscribed rectangles}$$

From Figure 4.12, you can see that the lower sum $s(n)$ is less than or equal to the upper sum $S(n)$. Moreover, the actual area of the region lies between these two sums.

$$s(n) \le (\text{Area of region}) \le S(n)$$

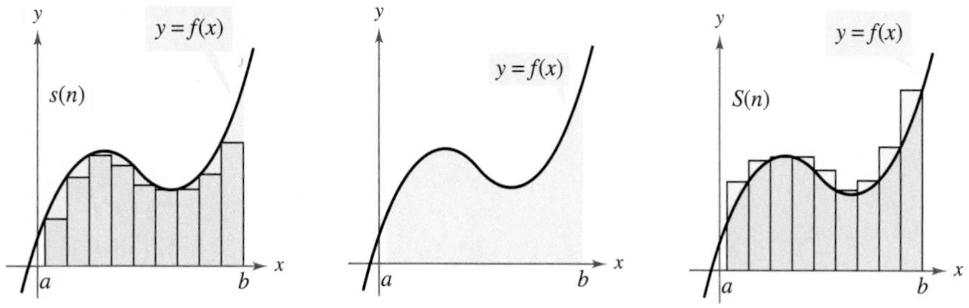

Area of inscribed rectangles is less than area of region.

Area of region

Area of circumscribed rectangles is greater than area of region.

Figure 4.12

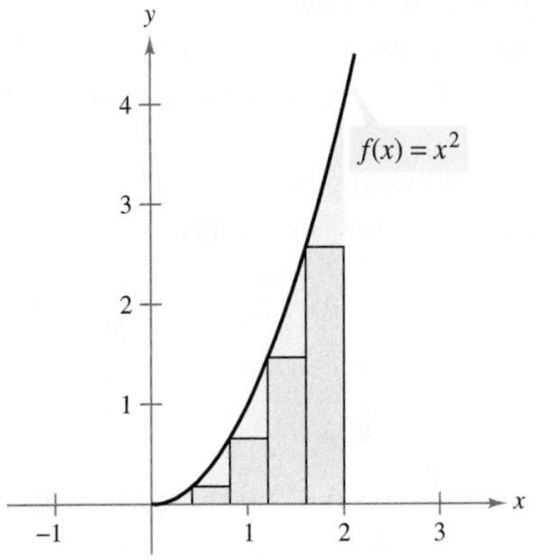

Inscribed rectangles

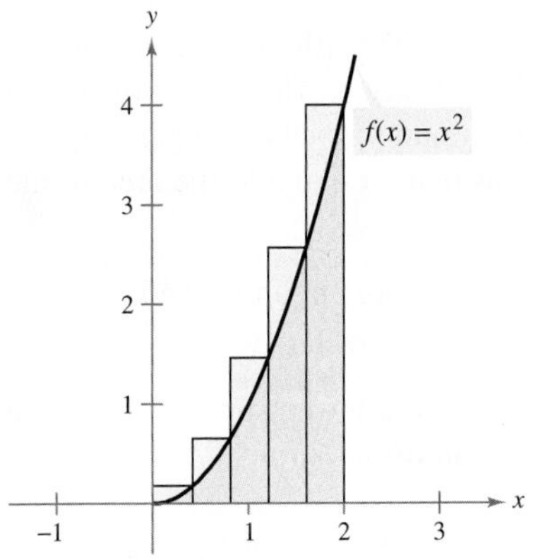

Circumscribed rectangles
Figure 4.13

EXAMPLE 4 Finding Upper and Lower Sums for a Region

Find the upper and lower sums for the region bounded by the graph of $f(x) = x^2$ and the x-axis between $x = 0$ and $x = 2$.

Solution To begin, partition the interval $[0, 2]$ into n subintervals, each of width

$$\Delta x = \frac{b-a}{n} = \frac{2-0}{n} = \frac{2}{n}.$$

Figure 4.13 shows the endpoints of the subintervals and several inscribed and circumscribed rectangles. Because f is increasing on the interval $[0, 2]$, the minimum value on each subinterval occurs at the left endpoint, and the maximum value occurs at the right endpoint.

Left Endpoints	*Right Endpoints*
$m_i = 0 + (i-1)\left(\frac{2}{n}\right) = \frac{2(i-1)}{n}$	$M_i = 0 + i\left(\frac{2}{n}\right) = \frac{2i}{n}$

Using the left endpoints, the lower sum is

$$\begin{aligned}
s(n) = \sum_{i=1}^{n} f(m_i)\,\Delta x &= \sum_{i=1}^{n} f\left[\frac{2(i-1)}{n}\right]\left(\frac{2}{n}\right)\\
&= \sum_{i=1}^{n} \left[\frac{2(i-1)}{n}\right]^2\left(\frac{2}{n}\right)\\
&= \sum_{i=1}^{n} \left(\frac{8}{n^3}\right)(i^2 - 2i + 1)\\
&= \frac{8}{n^3}\left(\sum_{i=1}^{n} i^2 - 2\sum_{i=1}^{n} i + \sum_{i=1}^{n} 1\right)\\
&= \frac{8}{n^3}\left\{\frac{n(n+1)(2n+1)}{6} - 2\left[\frac{n(n+1)}{2}\right] + n\right\}\\
&= \frac{4}{3n^3}(2n^3 - 3n^2 + n)\\
&= \frac{8}{3} - \frac{4}{n} + \frac{4}{3n^2}. \qquad \text{Lower sum}
\end{aligned}$$

Using the right endpoints, the upper sum is

$$\begin{aligned}
S(n) = \sum_{i=1}^{n} f(M_i)\,\Delta x &= \sum_{i=1}^{n} f\left(\frac{2i}{n}\right)\left(\frac{2}{n}\right)\\
&= \sum_{i=1}^{n} \left(\frac{2i}{n}\right)^2\left(\frac{2}{n}\right)\\
&= \sum_{i=1}^{n} \left(\frac{8}{n^3}\right) i^2\\
&= \frac{8}{n^3}\left[\frac{n(n+1)(2n+1)}{6}\right]\\
&= \frac{4}{3n^3}(2n^3 + 3n^2 + n)\\
&= \frac{8}{3} + \frac{4}{n} + \frac{4}{3n^2}. \qquad \text{Upper sum}
\end{aligned}$$

EXPLORATION

For the region given in Example 4, evaluate the lower sum

$$s(n) = \frac{8}{3} - \frac{4}{n} + \frac{4}{3n^2}$$

and the upper sum

$$S(n) = \frac{8}{3} + \frac{4}{n} + \frac{4}{3n^2}$$

for $n = 10, 100,$ and 1000. Use your results to determine the area of the region.

Example 4 illustrates some important things about lower and upper sums. First, notice that for any value of n, the lower sum is less than (or equal to) the upper sum.

$$s(n) = \frac{8}{3} - \frac{4}{n} + \frac{4}{3n^2} < \frac{8}{3} + \frac{4}{n} + \frac{4}{3n^2} = S(n)$$

Second, the difference between these two sums lessens as n increases. In fact, if you take the limits as $n \to \infty$, both the upper sum and the lower sum approach $\frac{8}{3}$.

$$\lim_{n\to\infty} s(n) = \lim_{n\to\infty}\left(\frac{8}{3} - \frac{4}{n} + \frac{4}{3n^2}\right) = \frac{8}{3} \qquad \text{Lower sum limit}$$

$$\lim_{n\to\infty} S(n) = \lim_{n\to\infty}\left(\frac{8}{3} + \frac{4}{n} + \frac{4}{3n^2}\right) = \frac{8}{3} \qquad \text{Upper sum limit}$$

The next theorem shows that the equivalence of the limits (as $n \to \infty$) of the upper and lower sums is not mere coincidence. It is true for all functions that are continuous and nonnegative on the closed interval $[a, b]$. The proof of this theorem is best left to a course in advanced calculus.

THEOREM 4.3 Limits of the Lower and Upper Sums

Let f be continuous and nonnegative on the interval $[a, b]$. The limits as $n \to \infty$ of both the lower and upper sums exist and are equal to each other. That is,

$$\begin{aligned}\lim_{n\to\infty} s(n) &= \lim_{n\to\infty} \sum_{i=1}^{n} f(m_i)\,\Delta x \\ &= \lim_{n\to\infty} \sum_{i=1}^{n} f(M_i)\,\Delta x \\ &= \lim_{n\to\infty} S(n)\end{aligned}$$

where $\Delta x = (b - a)/n$ and $f(m_i)$ and $f(M_i)$ are the minimum and maximum values of f on the subinterval.

Because the same limit is attained for both the minimum value $f(m_i)$ and the maximum value $f(M_i)$, it follows from the Squeeze Theorem (Theorem 1.8) that the choice of x in the ith subinterval does not affect the limit. This means that you are free to choose an *arbitrary* x-value in the ith subinterval, as in the following *definition of the area of a region in the plane.*

Definition of the Area of a Region in the Plane

Let f be continuous and nonnegative on the interval $[a, b]$. The area of the region bounded by the graph of f, the x-axis, and the vertical lines $x = a$ and $x = b$ is

$$\text{Area} = \lim_{n\to\infty} \sum_{i=1}^{n} f(c_i)\,\Delta x, \qquad x_{i-1} \le c_i \le x_i$$

where $\Delta x = (b - a)/n$ (see Figure 4.14).

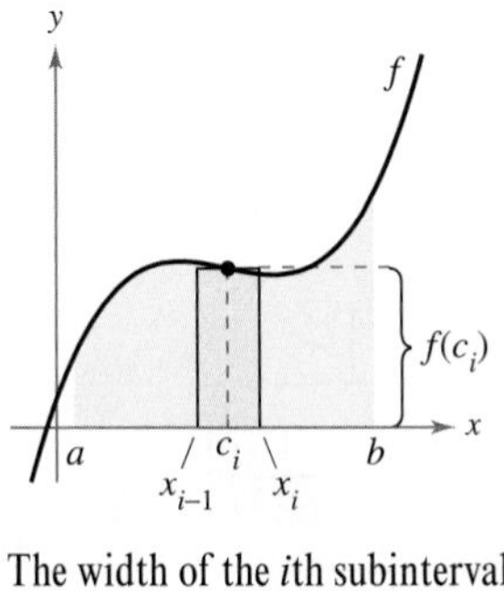

The width of the ith subinterval is $\Delta x = x_i - x_{i-1}$.
Figure 4.14

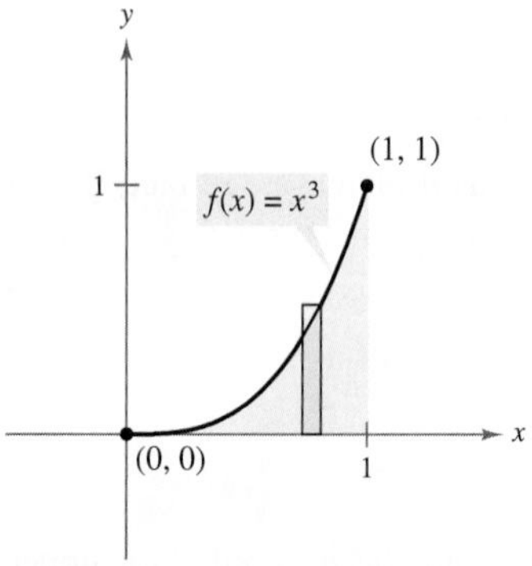

The area of the region bounded by the graph of f, the x-axis, $x = 0$, and $x = 1$ is $\frac{1}{4}$.
Figure 4.15

EXAMPLE 5 Finding Area by the Limit Definition

Find the area of the region bounded by the graph $f(x) = x^3$, the x-axis, and the vertical lines $x = 0$ and $x = 1$, as shown in Figure 4.15.

Solution Begin by noting that f is continuous and nonnegative on the interval $[0, 1]$. Next, partition the interval $[0, 1]$ into n subintervals, each of width $\Delta x = 1/n$. According to the definition of area, you can choose any x-value in the ith subinterval. For this example, the right endpoints $c_i = i/n$ are convenient.

$$\begin{aligned}
\text{Area} = \lim_{n\to\infty}\sum_{i=1}^{n} f(c_i)\,\Delta x &= \lim_{n\to\infty}\sum_{i=1}^{n}\left(\frac{i}{n}\right)^3\left(\frac{1}{n}\right) \qquad \text{Right endpoints: } c_i = \frac{i}{n}\\
&= \lim_{n\to\infty}\frac{1}{n^4}\sum_{i=1}^{n} i^3\\
&= \lim_{n\to\infty}\frac{1}{n^4}\left[\frac{n^2(n+1)^2}{4}\right]\\
&= \lim_{n\to\infty}\left(\frac{1}{4}+\frac{1}{2n}+\frac{1}{4n^2}\right)\\
&= \frac{1}{4}
\end{aligned}$$

The area of the region is $\frac{1}{4}$.

EXAMPLE 6 Finding Area by the Limit Definition

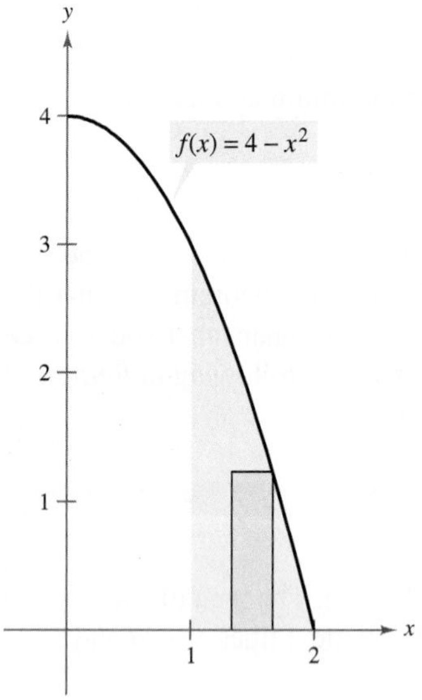

The area of the region bounded by the graph of f, the x-axis, $x = 1$, and $x = 2$ is $\frac{5}{3}$.
Figure 4.16

Find the area of the region bounded by the graph of $f(x) = 4 - x^2$, the x-axis, and the vertical lines $x = 1$ and $x = 2$, as shown in Figure 4.16.

Solution The function f is continuous and nonnegative on the interval $[1, 2]$, and so begin by partitioning the interval into n subintervals, each of width $\Delta x = 1/n$. Choosing the right endpoint

$$c_i = a + i\Delta x = 1 + \frac{i}{n} \qquad \text{Right endpoints}$$

of each subinterval, you obtain

$$\begin{aligned}
\text{Area} = \lim_{n\to\infty}\sum_{i=1}^{n} f(c_i)\,\Delta x &= \lim_{n\to\infty}\sum_{i=1}^{n}\left[4-\left(1+\frac{i}{n}\right)^2\right]\left(\frac{1}{n}\right)\\
&= \lim_{n\to\infty}\sum_{i=1}^{n}\left(3-\frac{2i}{n}-\frac{i^2}{n^2}\right)\left(\frac{1}{n}\right)\\
&= \lim_{n\to\infty}\left(\frac{1}{n}\sum_{i=1}^{n}3-\frac{2}{n^2}\sum_{i=1}^{n}i-\frac{1}{n^3}\sum_{i=1}^{n}i^2\right)\\
&= \lim_{n\to\infty}\left[3-\left(1+\frac{1}{n}\right)-\left(\frac{1}{3}+\frac{1}{2n}+\frac{1}{6n^2}\right)\right]\\
&= 3-1-\frac{1}{3}\\
&= \frac{5}{3}.
\end{aligned}$$

The area of the region is $\frac{5}{3}$.

The last example in this section looks at a region that is bounded by the y-axis (rather than by the x-axis).

EXAMPLE 7 A Region Bounded by the y-axis

Find the area of the region bounded by the graph of $f(y) = y^2$ and the y-axis for $0 \le y \le 1$, as shown in Figure 4.17.

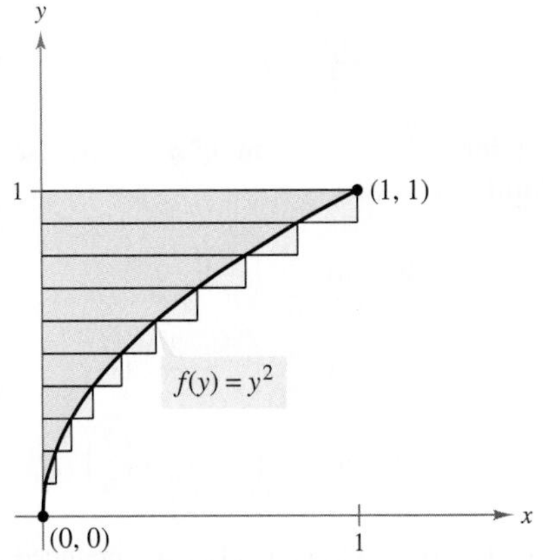

The area of the region bounded by the graph of f and the y-axis for $0 \le y \le 1$ is $\frac{1}{3}$.
Figure 4.17

Solution When f is a continuous, nonnegative function of y, you still can use the same basic procedure shown in Examples 5 and 6. Begin by partitioning the interval $[0, 1]$ into n subintervals, each of width $\Delta y = 1/n$. Then, using the upper endpoints $c_i = i/n$, you obtain

$$\begin{aligned}
\text{Area} &= \lim_{n\to\infty} \sum_{i=1}^{n} f(c_i)\,\Delta y = \lim_{n\to\infty} \sum_{i=1}^{n} \left(\frac{i}{n}\right)^2\left(\frac{1}{n}\right) \qquad \text{Upper endpoints: } c_i = \frac{i}{n}\\
&= \lim_{n\to\infty} \frac{1}{n^3}\sum_{i=1}^{n} i^2\\
&= \lim_{n\to\infty} \frac{1}{n^3}\left[\frac{n(n+1)(2n+1)}{6}\right]\\
&= \lim_{n\to\infty} \left(\frac{1}{3} + \frac{1}{2n} + \frac{1}{6n^2}\right)\\
&= \frac{1}{3}.
\end{aligned}$$

The area of the region is $\frac{1}{3}$.

Exercises for Section 4.2

See www.CalcChat.com for worked-out solutions to odd-numbered exercises.

In Exercises 1–6, find the sum. Use the summation capabilities of a graphing utility to verify your result.

1. $\sum_{i=1}^{5} (2i + 1)$
2. $\sum_{k=3}^{6} k(k - 2)$
3. $\sum_{k=0}^{4} \frac{1}{k^2 + 1}$
4. $\sum_{j=3}^{5} \frac{1}{j}$
5. $\sum_{k=1}^{4} c$
6. $\sum_{i=1}^{4} [(i - 1)^2 + (i + 1)^3]$

In Exercises 7–14, use sigma notation to write the sum.

7. $\frac{1}{3(1)} + \frac{1}{3(2)} + \frac{1}{3(3)} + \cdots + \frac{1}{3(9)}$
8. $\frac{5}{1+1} + \frac{5}{1+2} + \frac{5}{1+3} + \cdots + \frac{5}{1+15}$
9. $\left[5\left(\frac{1}{8}\right) + 3\right] + \left[5\left(\frac{2}{8}\right) + 3\right] + \cdots + \left[5\left(\frac{8}{8}\right) + 3\right]$
10. $\left[1 - \left(\frac{1}{4}\right)^2\right] + \left[1 - \left(\frac{2}{4}\right)^2\right] + \cdots + \left[1 - \left(\frac{4}{4}\right)^2\right]$
11. $\left[\left(\frac{2}{n}\right)^3 - \frac{2}{n}\right]\left(\frac{2}{n}\right) + \cdots + \left[\left(\frac{2n}{n}\right)^3 - \frac{2n}{n}\right]\left(\frac{2}{n}\right)$
12. $\left[1 - \left(\frac{2}{n} - 1\right)^2\right]\left(\frac{2}{n}\right) + \cdots + \left[1 - \left(\frac{2n}{n} - 1\right)^2\right]\left(\frac{2}{n}\right)$
13. $\left[2\left(1 + \frac{3}{n}\right)^2\right]\left(\frac{3}{n}\right) + \cdots + \left[2\left(1 + \frac{3n}{n}\right)^2\right]\left(\frac{3}{n}\right)$
14. $\left(\frac{1}{n}\right)\sqrt{1 - \left(\frac{0}{n}\right)^2} + \cdots + \left(\frac{1}{n}\right)\sqrt{1 - \left(\frac{n-1}{n}\right)^2}$

In Exercises 15–20, use the properties of summation and Theorem 4.2 to evaluate the sum. Use the summation capabilities of a graphing utility to verify your result.

15. $\sum_{i=1}^{20} 2i$
16. $\sum_{i=1}^{15} (2i - 3)$
17. $\sum_{i=1}^{20} (i - 1)^2$
18. $\sum_{i=1}^{10} (i^2 - 1)$
19. $\sum_{i=1}^{15} i(i - 1)^2$
20. $\sum_{i=1}^{10} i(i^2 + 1)$

In Exercises 21 and 22, use the summation capabilities of a graphing utility to evaluate the sum. Then use the properties of summation and Theorem 4.2 to verify the sum.

21. $\sum_{i=1}^{20} (i^2 + 3)$
22. $\sum_{i=1}^{15} (i^3 - 2i)$

In Exercises 23–26, bound the area of the shaded region by approximating the upper and lower sums. Use rectangles of width 1.

23.

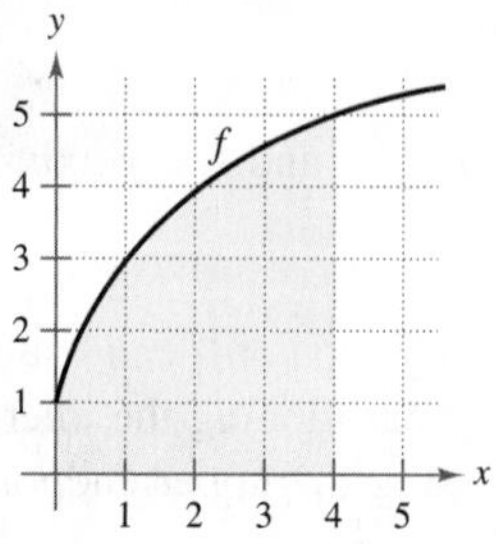

24.

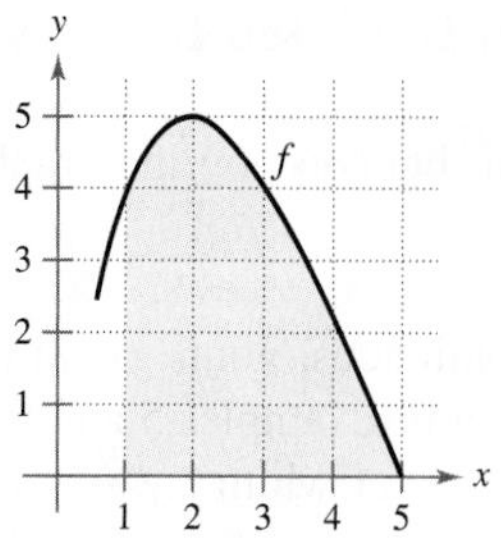

25.

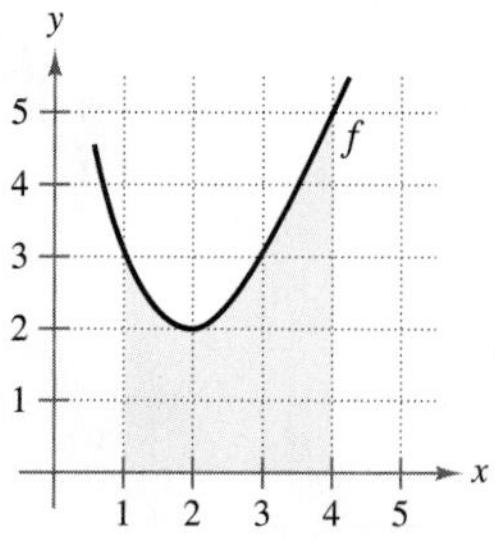

26.

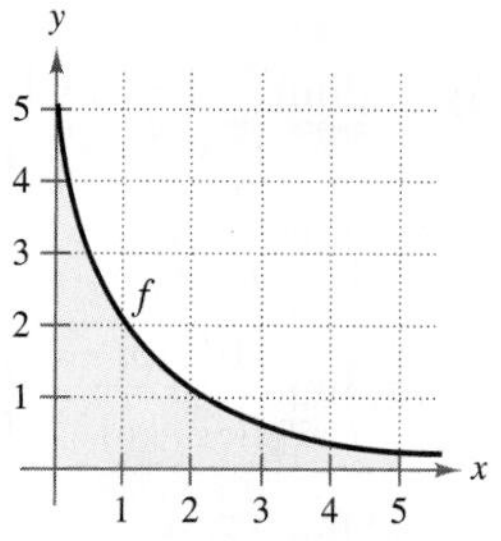

In Exercises 27–30, use upper and lower sums to approximate the area of the region using the given number of subintervals (of equal width).

27. $y = \sqrt{x}$

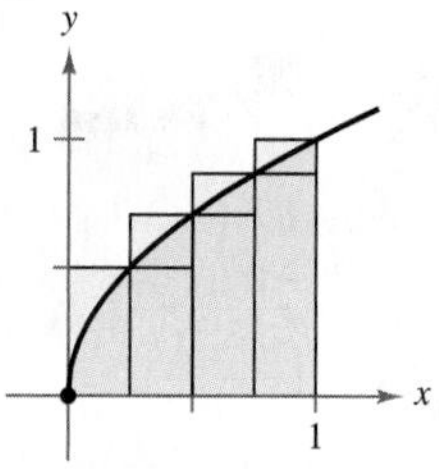

28. $y = \sqrt{x} + 2$

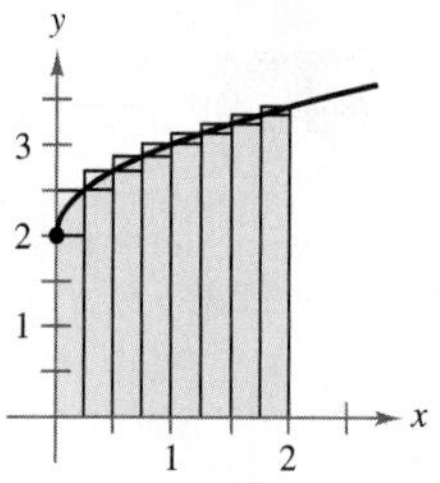

29. $y = \dfrac{1}{x}$

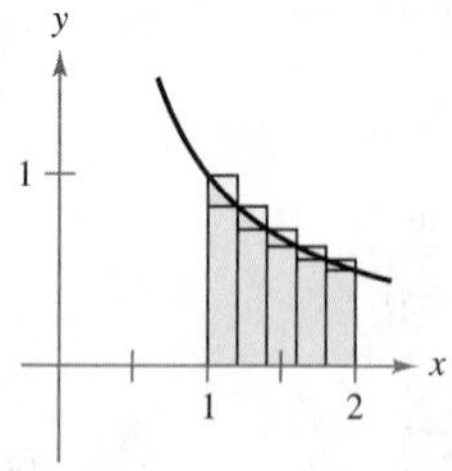

30. $y = \sqrt{1 - x^2}$

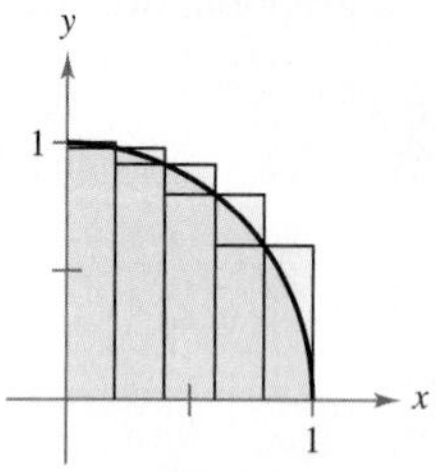

In Exercises 31–34, find the limit of $s(n)$ as $n \to \infty$.

31. $s(n) = \dfrac{81}{n^4}\left[\dfrac{n^2(n+1)^2}{4}\right]$

32. $s(n) = \dfrac{64}{n^3}\left[\dfrac{n(n+1)(2n+1)}{6}\right]$

33. $s(n) = \dfrac{18}{n^2}\left[\dfrac{n(n+1)}{2}\right]$

34. $s(n) = \dfrac{1}{n^2}\left[\dfrac{n(n+1)}{2}\right]$

In Exercises 35–38, use the summation formulas to rewrite the expression without the summation notation. Use the result to find the sum for $n = 10, 100, 1000$, and $10{,}000$.

35. $\displaystyle\sum_{i=1}^{n} \frac{2i+1}{n^2}$

36. $\displaystyle\sum_{j=1}^{n} \frac{4j+3}{n^2}$

37. $\displaystyle\sum_{k=1}^{n} \frac{6k(k-1)}{n^3}$

38. $\displaystyle\sum_{i=1}^{n} \frac{4i^2(i-1)}{n^4}$

In Exercises 39–44, find a formula for the sum of n terms. Use the formula to find the limit as $n \to \infty$.

39. $\displaystyle\lim_{n\to\infty} \sum_{i=1}^{n} \frac{16i}{n^2}$

40. $\displaystyle\lim_{n\to\infty} \sum_{i=1}^{n} \left(\frac{2i}{n}\right)\left(\frac{2}{n}\right)$

41. $\displaystyle\lim_{n\to\infty} \sum_{i=1}^{n} \frac{1}{n^3}(i-1)^2$

42. $\displaystyle\lim_{n\to\infty} \sum_{i=1}^{n} \left(1 + \frac{2i}{n}\right)^2\left(\frac{2}{n}\right)$

43. $\displaystyle\lim_{n\to\infty} \sum_{i=1}^{n} \left(1 + \frac{i}{n}\right)\left(\frac{2}{n}\right)$

44. $\displaystyle\lim_{n\to\infty} \sum_{i=1}^{n} \left(1 + \frac{2i}{n}\right)^3\left(\frac{2}{n}\right)$

45. ***Numerical Reasoning*** Consider a triangle of area 2 bounded by the graphs of $y = x$, $y = 0$, and $x = 2$.

(a) Sketch the region.

(b) Divide the interval $[0, 2]$ into n subintervals of equal width and show that the endpoints are

$$0 < 1\left(\frac{2}{n}\right) < \cdots < (n-1)\left(\frac{2}{n}\right) < n\left(\frac{2}{n}\right).$$

(c) Show that $s(n) = \displaystyle\sum_{i=1}^{n}\left[(i-1)\left(\frac{2}{n}\right)\right]\left(\frac{2}{n}\right)$.

(d) Show that $S(n) = \displaystyle\sum_{i=1}^{n}\left[i\left(\frac{2}{n}\right)\right]\left(\frac{2}{n}\right)$.

(e) Complete the table.

n	5	10	50	100
$s(n)$				
$S(n)$				

(f) Show that $\displaystyle\lim_{n\to\infty} s(n) = \lim_{n\to\infty} S(n) = 2$.

46. ***Numerical Reasoning*** Consider a trapezoid of area 4 bounded by the graphs of $y = x$, $y = 0$, $x = 1$, and $x = 3$.

(a) Sketch the region.

(b) Divide the interval $[1, 3]$ into n subintervals of equal width and show that the endpoints are

$$1 < 1 + 1\left(\frac{2}{n}\right) < \cdots < 1 + (n-1)\left(\frac{2}{n}\right) < 1 + n\left(\frac{2}{n}\right).$$

(c) Show that $s(n) = \displaystyle\sum_{i=1}^{n}\left[1 + (i-1)\left(\frac{2}{n}\right)\right]\left(\frac{2}{n}\right)$.

(d) Show that $S(n) = \displaystyle\sum_{i=1}^{n}\left[1 + i\left(\frac{2}{n}\right)\right]\left(\frac{2}{n}\right)$.

(e) Complete the table.

n	5	10	50	100
$s(n)$				
$S(n)$				

(f) Show that $\displaystyle\lim_{n\to\infty} s(n) = \lim_{n\to\infty} S(n) = 4$.

In Exercises 47–56, use the limit process to find the area of the region between the graph of the function and the x-axis over the given interval. Sketch the region.

47. $y = -2x + 3,\ [0, 1]$
48. $y = 3x - 4,\ [2, 5]$
49. $y = x^2 + 2,\ [0, 1]$
50. $y = x^2 + 1,\ [0, 3]$
51. $y = 16 - x^2,\ [1, 3]$
52. $y = 1 - x^2,\ [-1, 1]$
53. $y = 64 - x^3,\ [1, 4]$
54. $y = 2x - x^3,\ [0, 1]$
55. $y = x^2 - x^3,\ [-1, 1]$
56. $y = x^2 - x^3,\ [-1, 0]$

In Exercises 57–62, use the limit process to find the area of the region between the graph of the function and the y-axis over the given y-interval. Sketch the region.

57. $f(y) = 3y, 0 \le y \le 2$
58. $g(y) = \frac{1}{2}y, 2 \le y \le 4$
59. $f(y) = y^2, 0 \le y \le 3$
60. $f(y) = 4y - y^2, 1 \le y \le 2$
61. $g(y) = 4y^2 - y^3, 1 \le y \le 3$
62. $h(y) = y^3 + 1, 1 \le y \le 2$

In Exercises 63–66, use the *Midpoint Rule*

$$\text{Area} \approx \sum_{i=1}^{n} f\left(\frac{x_i + x_{i-1}}{2}\right)\Delta x$$

with $n = 4$ to approximate the area of the region bounded by the graph of the function and the x-axis over the given interval.

63. $f(x) = x^2 + 3,\ [0, 2]$
64. $f(x) = x^2 + 4x,\ [0, 4]$
65. $f(x) = \tan x,\ \left[0, \frac{\pi}{4}\right]$
66. $f(x) = \sin x,\ \left[0, \frac{\pi}{2}\right]$

Programming **Write a program for a graphing utility to approximate areas by using the Midpoint Rule. Assume that the function is positive over the given interval and the subintervals are of equal width. In Exercises 67–70, use the program to approximate the area of the region between the graph of the function and the x-axis over the given interval, and complete the table.**

n	4	8	12	16	20
Approximate Area					

67. $f(x) = \sqrt{x},\ [0, 4]$
68. $f(x) = \dfrac{8}{x^2 + 1},\ [2, 6]$
69. $f(x) = \tan\left(\dfrac{\pi x}{8}\right),\ [1, 3]$
70. $f(x) = \cos \sqrt{x},\ [0, 2]$

Writing About Concepts

Approximation **In Exercises 71 and 72, determine which value best approximates the area of the region between the x-axis and the graph of the function over the given interval. (Make your selection on the basis of a sketch of the region and not by performing calculations.)**

71. $f(x) = 4 - x^2,\ [0, 2]$

(a) -2 (b) 6 (c) 10 (d) 3 (e) 8

Writing About Concepts (continued)

72. $f(x) = \sin \dfrac{\pi x}{4},\ [0, 4]$

(a) 3 (b) 1 (c) -2 (d) 8 (e) 6

73. In your own words and using appropriate figures, describe the methods of upper sums and lower sums in approximating the area of a region.

74. Give the definition of the area of a region in the plane.

75. ***Graphical Reasoning*** Consider the region bounded by the graphs of

$$f(x) = \frac{8x}{x + 1},$$

$x = 0$, $x = 4$, and $y = 0$, as shown in the figure. To print an enlarged copy of the graph, go to the website *www.mathgraphs.com.*

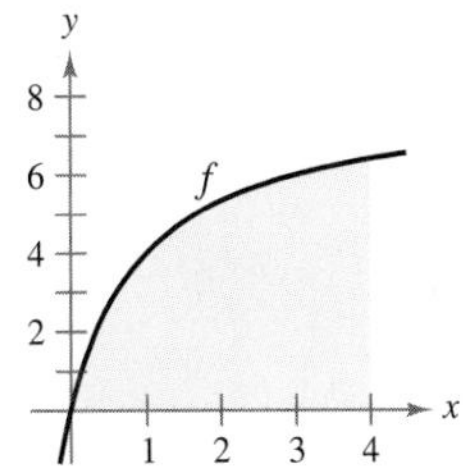

(a) Redraw the figure, and complete and shade the rectangles representing the lower sum when $n = 4$. Find this lower sum.

(b) Redraw the figure, and complete and shade the rectangles representing the upper sum when $n = 4$. Find this upper sum.

(c) Redraw the figure, and complete and shade the rectangles whose heights are determined by the functional values at the midpoint of each subinterval when $n = 4$. Find this sum using the Midpoint Rule.

(d) Verify the following formulas for approximating the area of the region using n subintervals of equal width.

Lower sum: $s(n) = \displaystyle\sum_{i=1}^{n} f\left[(i - 1)\frac{4}{n}\right]\left(\frac{4}{n}\right)$

Upper sum: $S(n) = \displaystyle\sum_{i=1}^{n} f\left[(i)\frac{4}{n}\right]\left(\frac{4}{n}\right)$

Midpoint Rule: $M(n) = \displaystyle\sum_{i=1}^{n} f\left[\left(i - \frac{1}{2}\right)\frac{4}{n}\right]\left(\frac{4}{n}\right)$

(e) Use a graphing utility and the formulas in part (d) to complete the table.

n	4	8	20	100	200
$s(n)$					
$S(n)$					
$M(n)$					

(f) Explain why $s(n)$ increases and $S(n)$ decreases for increasing values of n, as shown in the table in part (e).

76. ***Monte Carlo Method*** The following computer program approximates the area of the region under the graph of a monotonic function and above the x-axis between $x = a$ and $x = b$. Run the program for $a = 0$ and $b = \pi/2$ for several values of N2. Explain why the Monte Carlo Method works. [*Adaptation of Monte Carlo Method program from James M. Sconyers, "Approximation of Area Under a Curve," MATHEMATICS TEACHER 77, no. 2 (February 1984). Copyright © 1984 by the National Council of Teachers of Mathematics. Reprinted with permission.*]

```
10  DEF FNF(X)=SIN(X)
20  A=0
30  B=π/2
40  PRINT "Input Number of Random Points"
50  INPUT N2
60  N1=0
70  IF FNF(A)>FNF(B) THEN YMAX=FNF(A) ELSE
    YMAX=FNF(B)
80  FOR I=1 TO N2
90  X=A+(B-A)*RND(1)
100 Y=YMAX*RND(1)
110 IF Y>=FNF(X) THEN GOTO 130
120 N1=N1+1
130 NEXT I
140 AREA=(N1/N2)*(B-A)*YMAX
150 PRINT "Approximate Area:"; AREA
160 END
```

True or False? **In Exercises 77 and 78, determine whether the statement is true or false. If it is false, explain why or give an example that shows it is false.**

77. The sum of the first n positive integers is $n(n + 1)/2$.

78. If f is continuous and nonnegative on $[a, b]$, then the limits as $n \to \infty$ of its lower sum $s(n)$ and upper sum $S(n)$ both exist and are equal.

79. ***Writing*** Use the figure to write a short paragraph explaining why the formula $1 + 2 + \cdots + n = \frac{1}{2}n(n + 1)$ is valid for all positive integers n.

Figure for 79

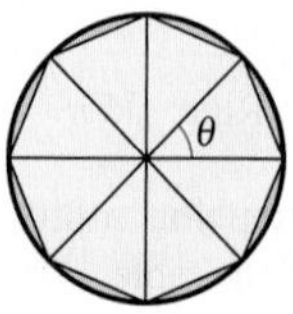

Figure for 80

80. ***Graphical Reasoning*** Consider an n-sided regular polygon inscribed in a circle of radius r. Join the vertices of the polygon to the center of the circle, forming n congruent triangles (see figure).

(a) Determine the central angle θ in terms of n.

(b) Show that the area of each triangle is $\frac{1}{2}r^2 \sin \theta$.

(c) Let A_n be the sum of the areas of the n triangles. Find $\lim_{n\to\infty} A_n$.

81. ***Modeling Data*** The table lists the measurements of a lot bounded by a stream and two straight roads that meet at right angles, where x and y are measured in feet (see figure).

x	0	50	100	150	200	250	300
y	450	362	305	268	245	156	0

(a) Use the regression capabilities of a graphing utility to find a model of the form $y = ax^3 + bx^2 + cx + d$.

(b) Use a graphing utility to plot the data and graph the model.

(c) Use the model in part (a) to estimate the area of the lot.

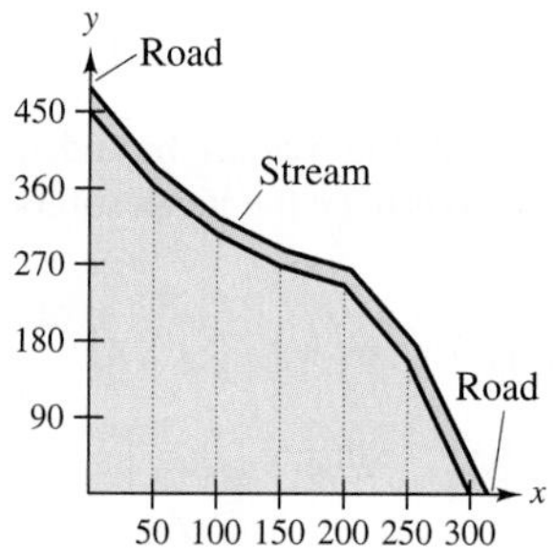

Figure for 81

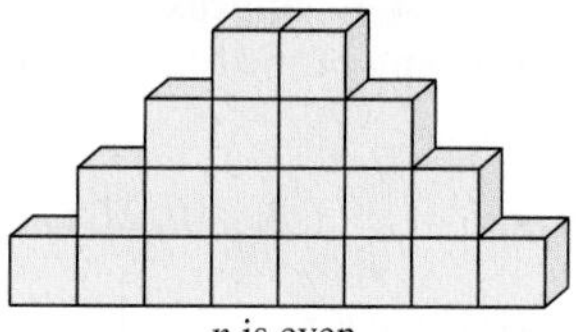

Figure for 82

82. ***Building Blocks*** A child places n cubic building blocks in a row to form the base of a triangular design (see figure). Each successive row contains two fewer blocks than the preceding row. Find a formula for the number of blocks used in the design. (*Hint:* The number of building blocks in the design depends on whether n is odd or even.)

83. Prove each formula by mathematical induction. (You may need to review the method of proof by induction from a precalculus text.)

(a) $\sum_{i=1}^{n} 2i = n(n + 1)$

(b) $\sum_{i=1}^{n} i^3 = \frac{n^2(n + 1)^2}{4}$

Putnam Exam Challenge

84. A dart, thrown at random, hits a square target. Assuming that any two parts of the target of equal area are equally likely to be hit, find the probability that the point hit is nearer to the center than to any edge. Write your answer in the form $(a\sqrt{b} + c)/d$, where a, b, c, and d are positive integers.

This problem was composed by the Committee on the Putnam Prize Competition.

Section 4.3 Riemann Sums and Definite Integrals

- **Understand the definition of a Riemann sum.**
- **Evaluate a definite integral using limits.**
- **Evaluate a definite integral using properties of definite integrals.**

Riemann Sums

In the definition of area given in Section 4.2, the partitions have subintervals of *equal width.* This was done only for computational convenience. The following example shows that it is not necessary to have subintervals of equal width.

EXAMPLE 1 A Partition with Subintervals of Unequal Widths

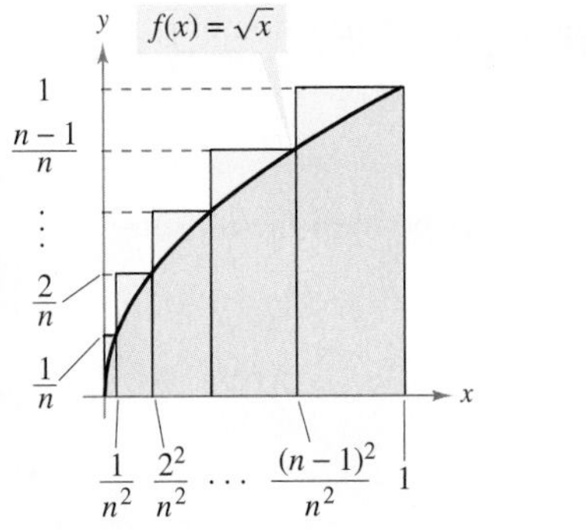

The subintervals do not have equal widths.
Figure 4.18

Consider the region bounded by the graph of $f(x) = \sqrt{x}$ and the x-axis for $0 \le x \le 1$, as shown in Figure 4.18. Evaluate the limit

$$\lim_{n\to\infty} \sum_{i=1}^{n} f(c_i)\,\Delta x_i$$

where c_i is the right endpoint of the partition given by $c_i = i^2/n^2$ and Δx_i is the width of the ith interval.

Solution The width of the ith interval is given by

$$\begin{aligned}\Delta x_i &= \frac{i^2}{n^2} - \frac{(i-1)^2}{n^2}\\ &= \frac{i^2 - i^2 + 2i - 1}{n^2}\\ &= \frac{2i-1}{n^2}.\end{aligned}$$

So, the limit is

$$\begin{aligned}\lim_{n\to\infty} \sum_{i=1}^{n} f(c_i)\,\Delta x_i &= \lim_{n\to\infty} \sum_{i=1}^{n} \sqrt{\frac{i^2}{n^2}}\left(\frac{2i-1}{n^2}\right)\\ &= \lim_{n\to\infty} \frac{1}{n^3} \sum_{i=1}^{n} (2i^2 - i)\\ &= \lim_{n\to\infty} \frac{1}{n^3}\left[2\left(\frac{n(n+1)(2n+1)}{6}\right) - \frac{n(n+1)}{2}\right]\\ &= \lim_{n\to\infty} \frac{4n^3 + 3n^2 - n}{6n^3}\\ &= \frac{2}{3}.\end{aligned}$$

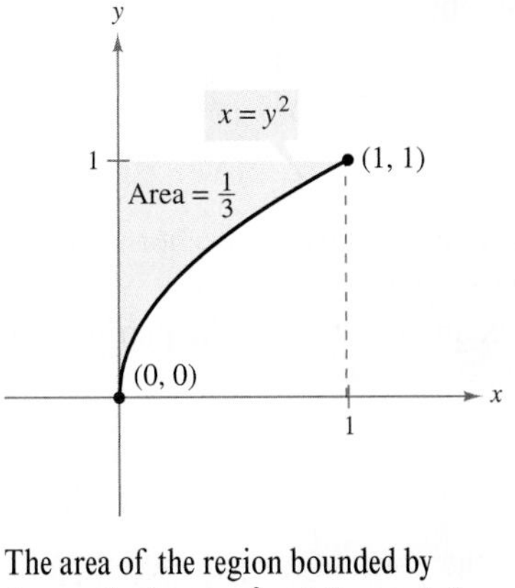

The area of the region bounded by the graph of $x = y^2$ and the y-axis for $0 \le y \le 1$ is $\frac{1}{3}$.
Figure 4.19

From Example 7 in Section 4.2, you know that the region shown in Figure 4.19 has an area of $\frac{1}{3}$. Because the square bounded by $0 \le x \le 1$ and $0 \le y \le 1$ has an area of 1, you can conclude that the area of the region shown in Figure 4.18 has an area of $\frac{2}{3}$. This agrees with the limit found in Example 1, even though that example used a partition having subintervals of unequal widths. The reason this particular partition gave the proper area is that as n increases, the *width of the largest subinterval approaches zero.* This is a key feature of the development of definite integrals.

The Granger Collection

GEORG FRIEDRICH BERNHARD RIEMANN (1826–1866)

German mathematician Riemann did his most famous work in the areas of non-Euclidean geometry, differential equations, and number theory. It was Riemann's results in physics and mathematics that formed the structure on which Einstein's theory of general relativity is based.

In the preceding section, the limit of a sum was used to define the area of a region in the plane. Finding area by this means is only one of *many* applications involving the limit of a sum. A similar approach can be used to determine quantities as diverse as arc lengths, average values, centroids, volumes, work, and surface areas. The following definition is named after Georg Friedrich Bernhard Riemann. Although the definite integral had been defined and used long before the time of Riemann, he generalized the concept to cover a broader category of functions.

In the following definition of a Riemann sum, note that the function f has no restrictions other than being defined on the interval $[a, b]$. (In the preceding section, the function f was assumed to be continuous and nonnegative because we were dealing with the area under a curve.)

Definition of a Riemann Sum

Let f be defined on the closed interval $[a, b]$, and let Δ be a partition of $[a, b]$ given by

$$a = x_0 < x_1 < x_2 < \cdot \cdot \cdot < x_{n-1} < x_n = b$$

where Δx_i is the width of the ith subinterval. If c_i is *any* point in the ith subinterval, then the sum

$$\sum_{i=1}^{n} f(c_i)\,\Delta x_i, \qquad x_{i-1} \le c_i \le x_i$$

is called a **Riemann sum** of f for the partition Δ.

NOTE The sums in Section 4.2 are examples of Riemann sums, but there are more general Riemann sums than those covered there.

The width of the largest subinterval of a partition Δ is the **norm** of the partition and is denoted by $\|\Delta\|$. If every subinterval is of equal width, the partition is **regular** and the norm is denoted by

$$\|\Delta\| = \Delta x = \frac{b-a}{n}. \qquad \text{Regular partition}$$

For a general partition, the norm is related to the number of subintervals of $[a, b]$ in the following way.

$$\frac{b-a}{\|\Delta\|} \le n \qquad \text{General partition}$$

So, the number of subintervals in a partition approaches infinity as the norm of the partition approaches 0. That is, $\|\Delta\| \to 0$ implies that $n \to \infty$.

The converse of this statement is not true. For example, let Δ_n be the partition of the interval $[0, 1]$ given by

$$0 < \frac{1}{2^n} < \frac{1}{2^{n-1}} < \cdot \cdot \cdot < \frac{1}{8} < \frac{1}{4} < \frac{1}{2} < 1.$$

As shown in Figure 4.20, for any positive value of n, the norm of the partition Δ_n is $\frac{1}{2}$. So, letting n approach infinity does not force $\|\Delta\|$ to approach 0. In a regular partition, however, the statements $\|\Delta\| \to 0$ and $n \to \infty$ are equivalent.

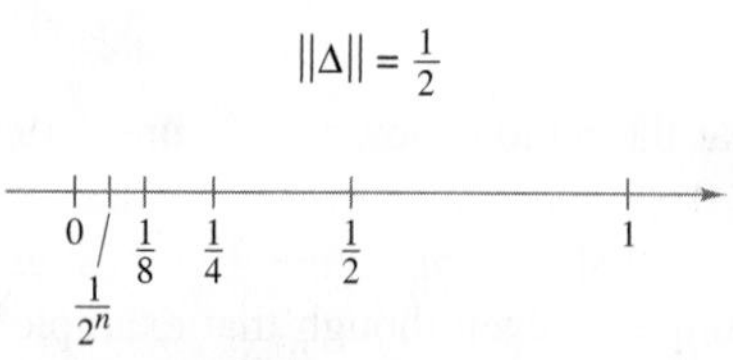

$n \to \infty$ does not imply that $\|\Delta\| \to 0$.
Figure 4.20

Definite Integrals

To define the definite integral, consider the following limit.

$$\lim_{\|\Delta\| \to 0} \sum_{i=1}^{n} f(c_i)\,\Delta x_i = L$$

To say that this limit exists means that for $\varepsilon > 0$ there exists a $\delta > 0$ such that for every partition with $\|\Delta\| < \delta$ it follows that

$$\left| L - \sum_{i=1}^{n} f(c_i)\,\Delta x_i \right| < \varepsilon.$$

(This must be true for any choice of c_i in the ith subinterval of Δ.)

FOR FURTHER INFORMATION For insight into the history of the definite integral, see the article "The Evolution of Integration" by A. Shenitzer and J. Steprāns in The *American Mathematical Monthly*. To view this article, go to the website *www.matharticles.com.*

Definition of a Definite Integral

If f is defined on the closed interval $[a, b]$ and the limit

$$\lim_{\|\Delta\| \to 0} \sum_{i=1}^{n} f(c_i)\,\Delta x_i$$

exists (as described above), then f is **integrable** on $[a, b]$ and the limit is denoted by

$$\lim_{\|\Delta\| \to 0} \sum_{i=1}^{n} f(c_i)\,\Delta x_i = \int_a^b f(x)\,dx.$$

The limit is called the **definite integral** of f from a to b. The number a is the **lower limit** of integration, and the number b is the **upper limit** of integration.

It is not a coincidence that the notation for definite integrals is similar to that used for indefinite integrals. You will see why in the next section when the Fundamental Theorem of Calculus is introduced. For now it is important to see that definite integrals and indefinite integrals are different identities. A definite integral is a *number*, whereas an indefinite integral is a *family of functions*.

A sufficient condition for a function f to be integrable on $[a, b]$ is that it is continuous on $[a, b]$. A proof of this theorem is beyond the scope of this text.

THEOREM 4.4 Continuity Implies Integrability

If a function f is continuous on the closed interval $[a, b]$, then f is integrable on $[a, b]$.

EXPLORATION

The Converse of Theorem 4.4 Is the converse of Theorem 4.4 true? That is, if a function is integrable, does it have to be continuous? Explain your reasoning and give examples.

Describe the relationships among continuity, differentiability, and integrability. Which is the strongest condition? Which is the weakest? Which conditions imply other conditions?

EXAMPLE 2 Evaluating a Definite Integral as a Limit

Evaluate the definite integral $\int_{-2}^{1} 2x\,dx$.

Solution The function $f(x) = 2x$ is integrable on the interval $[-2, 1]$ because it is continuous on $[-2, 1]$. Moreover, the definition of integrability implies that any partition whose norm approaches 0 can be used to determine the limit. For computational convenience, define Δ by subdividing $[-2, 1]$ into n subintervals of equal width

$$\Delta x_i = \Delta x = \frac{b-a}{n} = \frac{3}{n}.$$

Choosing c_i as the right endpoint of each subinterval produces

$$c_i = a + i(\Delta x) = -2 + \frac{3i}{n}.$$

So, the definite integral is given by

$$\begin{aligned}
\int_{-2}^{1} 2x\,dx &= \lim_{\|\Delta\|\to 0} \sum_{i=1}^{n} f(c_i)\,\Delta x_i \\
&= \lim_{n\to\infty} \sum_{i=1}^{n} f(c_i)\,\Delta x \\
&= \lim_{n\to\infty} \sum_{i=1}^{n} 2\left(-2 + \frac{3i}{n}\right)\left(\frac{3}{n}\right) \\
&= \lim_{n\to\infty} \frac{6}{n} \sum_{i=1}^{n} \left(-2 + \frac{3i}{n}\right) \\
&= \lim_{n\to\infty} \frac{6}{n}\left\{-2n + \frac{3}{n}\left[\frac{n(n+1)}{2}\right]\right\} \\
&= \lim_{n\to\infty} \left(-12 + 9 + \frac{9}{n}\right) \\
&= -3.
\end{aligned}$$

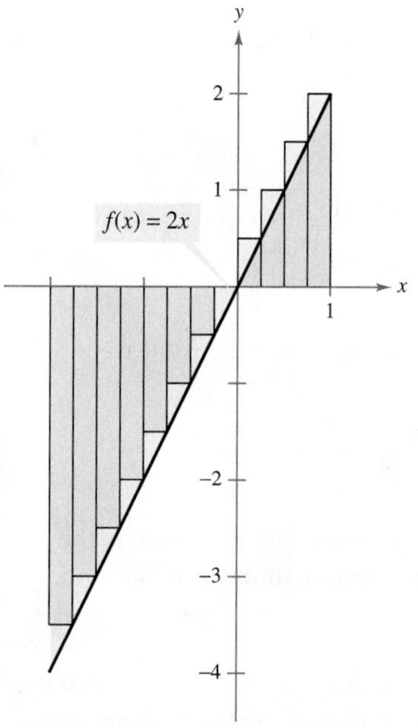

Because the definite integral is negative, it does not represent the area of the region.
Figure 4.21

Because the definite integral in Example 2 is negative, it *does not* represent the area of the region shown in Figure 4.21. Definite integrals can be positive, negative, or zero. For a definite integral to be interpreted as an area (as defined in Section 4.2), the function f must be continuous and nonnegative on $[a, b]$, as stated in the following theorem. (The proof of this theorem is straightforward—you simply use the definition of area given in Section 4.2.)

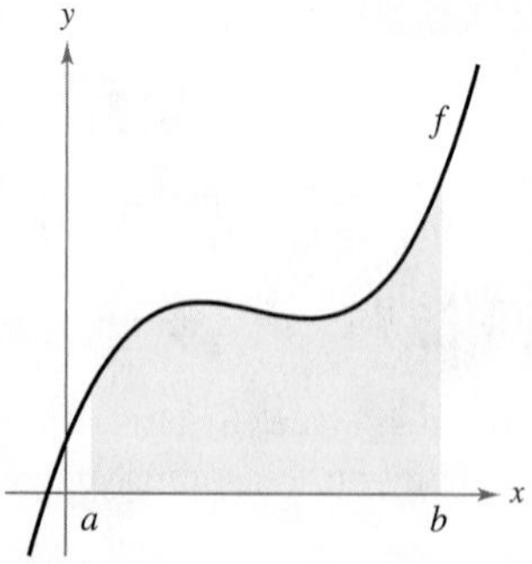

You can use a definite integral to find the area of the region bounded by the graph of f, the x-axis, $x = a$, and $x = b$.
Figure 4.22

THEOREM 4.5 The Definite Integral as the Area of a Region

If f is continuous and nonnegative on the closed interval $[a, b]$, then the area of the region bounded by the graph of f, the x-axis, and the vertical lines $x = a$ and $x = b$ is given by

$$\text{Area} = \int_a^b f(x)\,dx.$$

(See Figure 4.22.)

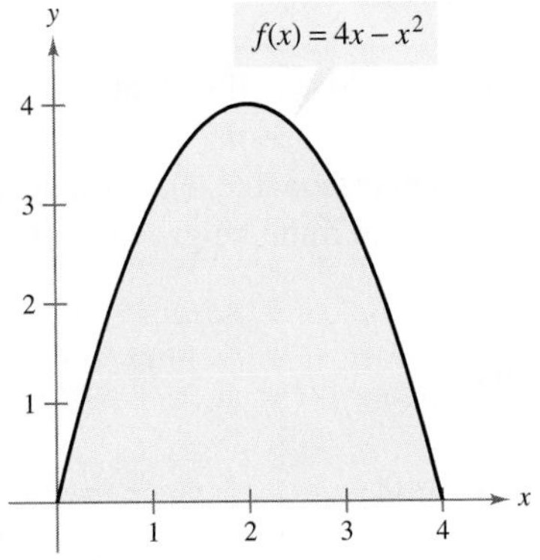

$\text{Area} = \int_0^4 (4x - x^2)\,dx$

Figure 4.23

As an example of Theorem 4.5, consider the region bounded by the graph of

$$f(x) = 4x - x^2$$

and the x-axis, as shown in Figure 4.23. Because f is continuous and nonnegative on the closed interval $[0, 4]$, the area of the region is

$$\text{Area} = \int_0^4 (4x - x^2)\,dx.$$

A straightforward technique for evaluating a definite integral such as this will be discussed in Section 4.4. For now, however, you can evaluate a definite integral in two ways—you can use the limit definition *or* you can check to see whether the definite integral represents the area of a common geometric region such as a rectangle, triangle, or semicircle.

EXAMPLE 3 Areas of Common Geometric Figures

Sketch the region corresponding to each definite integral. Then evaluate each integral using a geometric formula.

a. $\int_1^3 4\,dx$ **b.** $\int_0^3 (x + 2)\,dx$ **c.** $\int_{-2}^2 \sqrt{4 - x^2}\,dx$

Solution A sketch of each region is shown in Figure 4.24.

a. This region is a rectangle of height 4 and width 2.

$$\int_1^3 4\,dx = (\text{Area of rectangle}) = 4(2) = 8$$

b. This region is a trapezoid with an altitude of 3 and parallel bases of lengths 2 and 5. The formula for the area of a trapezoid is $\frac{1}{2}h(b_1 + b_2)$.

$$\int_0^3 (x + 2)\,dx = (\text{Area of trapezoid}) = \frac{1}{2}(3)(2 + 5) = \frac{21}{2}$$

c. This region is a semicircle of radius 2. The formula for the area of a semicircle is $\frac{1}{2}\pi r^2$.

$$\int_{-2}^2 \sqrt{4 - x^2}\,dx = (\text{Area of semicircle}) = \frac{1}{2}\pi(2^2) = 2\pi$$

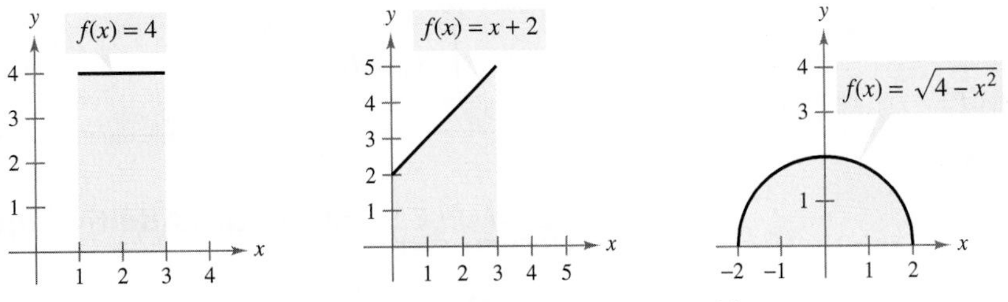

Figure 4.24

NOTE The variable of integration in a definite integral is sometimes called a *dummy variable* because it can be replaced by any other variable without changing the value of the integral. For instance, the definite integrals

$$\int_0^3 (x + 2)\,dx$$

and

$$\int_0^3 (t + 2)\,dt$$

have the same value.

Properties of Definite Integrals

The definition of the definite integral of f on the interval $[a, b]$ specifies that $a < b$. Now, however, it is convenient to extend the definition to cover cases in which $a = b$ or $a > b$. Geometrically, the following two definitions seem reasonable. For instance, it makes sense to define the area of a region of zero width and finite height to be 0.

Definitions of Two Special Definite Integrals

1. If f is defined at $x = a$, then we define $\int_a^a f(x)\,dx = 0$.

2. If f is integrable on $[a, b]$, then we define $\int_b^a f(x)\,dx = -\int_a^b f(x)\,dx$.

EXAMPLE 4 Evaluating Definite Integrals

a. Because the sine function is defined at $x = \pi$, and the upper and lower limits of integration are equal, you can write

$$\int_\pi^\pi \sin x\,dx = 0.$$

b. The integral $\int_3^0 (x + 2)\,dx$ is the same as that given in Example 3(b) except that the upper and lower limits are interchanged. Because the integral in Example 3(b) has a value of $\frac{21}{2}$, you can write

$$\int_3^0 (x + 2)\,dx = -\int_0^3 (x + 2)\,dx = -\frac{21}{2}.$$

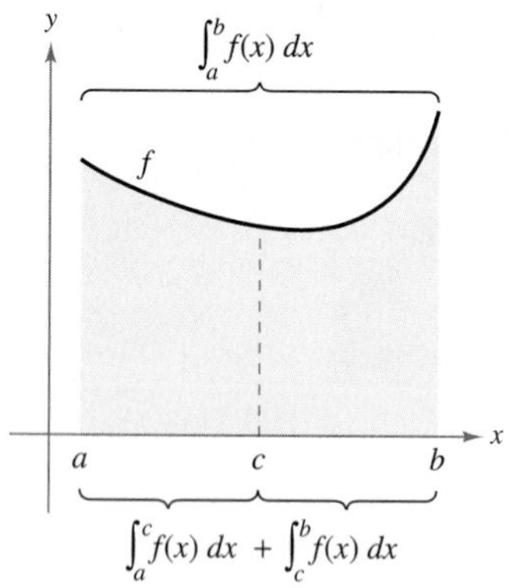

Figure 4.25

In Figure 4.25, the larger region can be divided at $x = c$ into two subregions whose intersection is a line segment. Because the line segment has zero area, it follows that the area of the larger region is equal to the sum of the areas of the two smaller regions.

THEOREM 4.6 Additive Interval Property

If f is integrable on the three closed intervals determined by a, b, and c, then

$$\int_a^b f(x)\,dx = \int_a^c f(x)\,dx + \int_c^b f(x)\,dx.$$

EXAMPLE 5 Using the Additive Interval Property

$$\int_{-1}^1 |x|\,dx = \int_{-1}^0 -x\,dx + \int_0^1 x\,dx \qquad \text{Theorem 4.6}$$

$$= \frac{1}{2} + \frac{1}{2} \qquad \text{Area of a triangle}$$

$$= 1$$

Because the definite integral is defined as the limit of a sum, it inherits the properties of summation given at the top of page 260.

THEOREM 4.7 Properties of Definite Integrals

If f and g are integrable on $[a, b]$ and k is a constant, then the functions of kf and $f \pm g$ are integrable on $[a, b]$, and

1. $\displaystyle\int_a^b kf(x)\,dx = k\int_a^b f(x)\,dx$

2. $\displaystyle\int_a^b [f(x) \pm g(x)]\,dx = \int_a^b f(x)\,dx \pm \int_a^b g(x)\,dx.$

Note that Property 2 of Theorem 4.7 can be extended to cover any finite number of functions. For example,

$$\int_a^b [f(x) + g(x) + h(x)]\,dx = \int_a^b f(x)\,dx + \int_a^b g(x)\,dx + \int_a^b h(x)\,dx.$$

EXAMPLE 6 Evaluation of a Definite Integral

Evaluate $\displaystyle\int_1^3 (-x^2 + 4x - 3)\,dx$ using each of the following values.

$$\int_1^3 x^2\,dx = \frac{26}{3}, \qquad \int_1^3 x\,dx = 4, \qquad \int_1^3 dx = 2$$

Solution

$$\begin{aligned}\int_1^3 (-x^2 + 4x - 3)\,dx &= \int_1^3 (-x^2)\,dx + \int_1^3 4x\,dx + \int_1^3 (-3)\,dx\\ &= -\int_1^3 x^2\,dx + 4\int_1^3 x\,dx - 3\int_1^3 dx\\ &= -\left(\frac{26}{3}\right) + 4(4) - 3(2)\\ &= \frac{4}{3}\end{aligned}$$

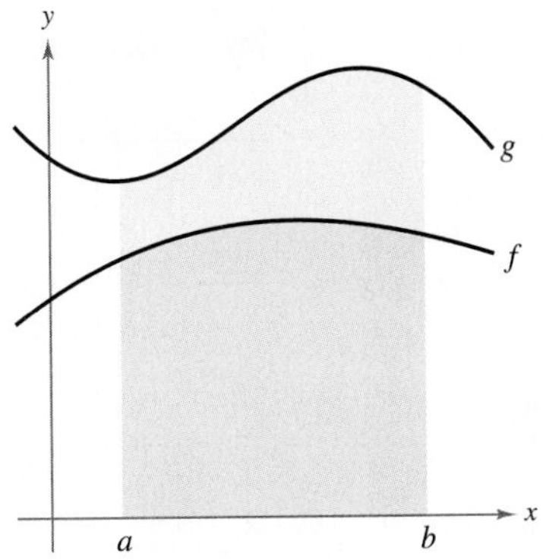

$\displaystyle\int_a^b f(x)\,dx \le \int_a^b g(x)\,dx$

Figure 4.26

If f and g are continuous on the closed interval $[a, b]$ and

$0 \le f(x) \le g(x)$

for $a \le x \le b$, the following properties are true. First, the area of the region bounded by the graph of f and the x-axis (between a and b) must be nonnegative. Second, this area must be less than or equal to the area of the region bounded by the graph of g and the x-axis (between a and b), as shown in Figure 4.26. These two results are generalized in Theorem 4.8. (A proof of this theorem is given in Appendix A.)

THEOREM 4.8 Preservation of Inequality

1. If f is integrable and nonnegative on the closed interval $[a, b]$, then

$$0 \le \int_a^b f(x)\, dx.$$

2. If f and g are integrable on the closed interval $[a, b]$ and $f(x) \le g(x)$ for every x in $[a, b]$, then

$$\int_a^b f(x)\, dx \le \int_a^b g(x)\, dx.$$

Exercises for Section 4.3

See www.CalcChat.com for worked-out solutions to odd-numbered exercises.

In Exercises 1 and 2, use Example 1 as a model to evaluate the limit

$$\lim_{n\to\infty} \sum_{i=1}^{n} f(c_i)\,\Delta x_i$$

over the region bounded by the graphs of the equations.

1. $f(x) = \sqrt{x}, \quad y = 0, \quad x = 0, \quad x = 3$
   (*Hint:* Let $c_i = 3i^2/n^2$.)
2. $f(x) = \sqrt[3]{x}, \quad y = 0, \quad x = 0, \quad x = 1$
   (*Hint:* Let $c_i = i^3/n^3$.)

In Exercises 3–8, evaluate the definite integral by the limit definition.

3. $\displaystyle\int_4^{10} 6\, dx$
4. $\displaystyle\int_{-2}^{3} x\, dx$
5. $\displaystyle\int_{-1}^{1} x^3\, dx$
6. $\displaystyle\int_1^3 3x^2\, dx$
7. $\displaystyle\int_1^2 (x^2 + 1)\, dx$
8. $\displaystyle\int_{-1}^{2} (3x^2 + 2)\, dx$

In Exercises 9–12, write the limit as a definite integral on the interval $[a, b]$, where c_i is any point in the ith subinterval.

Limit	*Interval*
9. $\displaystyle\lim_{\|\Delta\|\to 0} \sum_{i=1}^{n} (3c_i + 10)\,\Delta x_i$	$[-1, 5]$
10. $\displaystyle\lim_{\|\Delta\|\to 0} \sum_{i=1}^{n} 6c_i(4 - c_i)^2\,\Delta x_i$	$[0, 4]$
11. $\displaystyle\lim_{\|\Delta\|\to 0} \sum_{i=1}^{n} \sqrt{c_i^2 + 4}\,\Delta x_i$	$[0, 3]$
12. $\displaystyle\lim_{\|\Delta\|\to 0} \sum_{i=1}^{n} \left(\frac{3}{c_i^2}\right)\Delta x_i$	$[1, 3]$

In Exercises 13–22, set up a definite integral that yields the area of the region. (Do not evaluate the integral.)

13. $f(x) = 3$

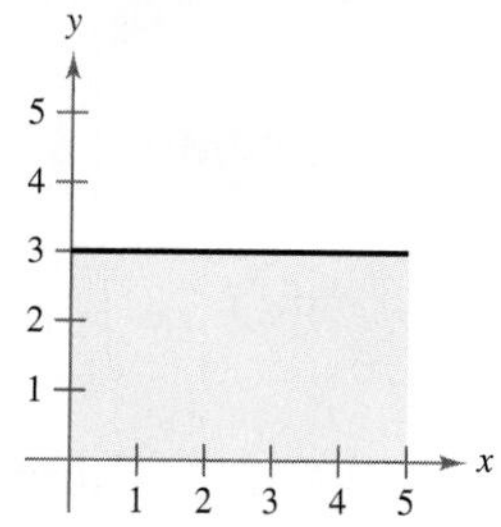

14. $f(x) = 4 - 2x$

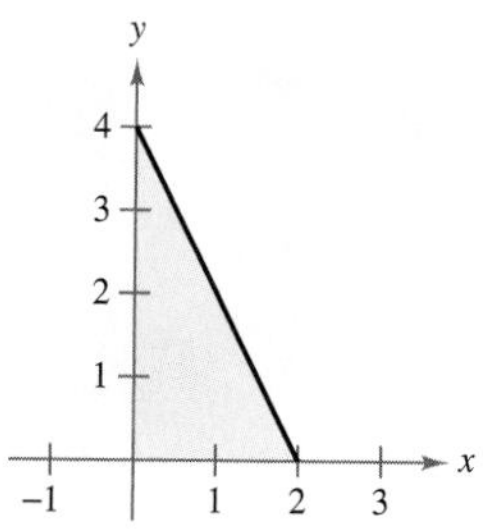

15. $f(x) = 4 - |x|$

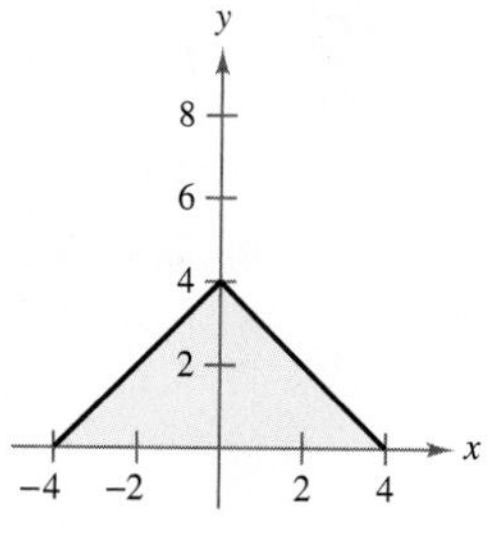

16. $f(x) = x^2$

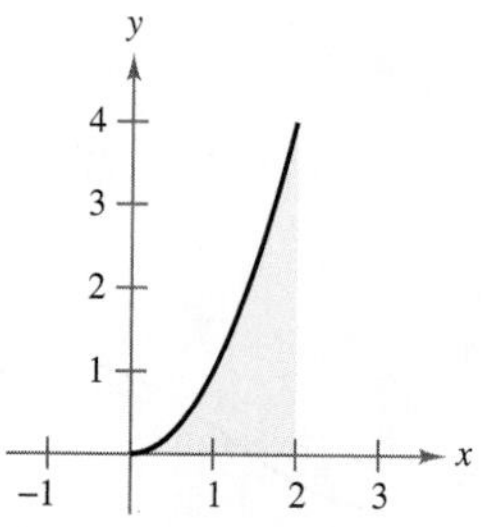

17. $f(x) = 4 - x^2$

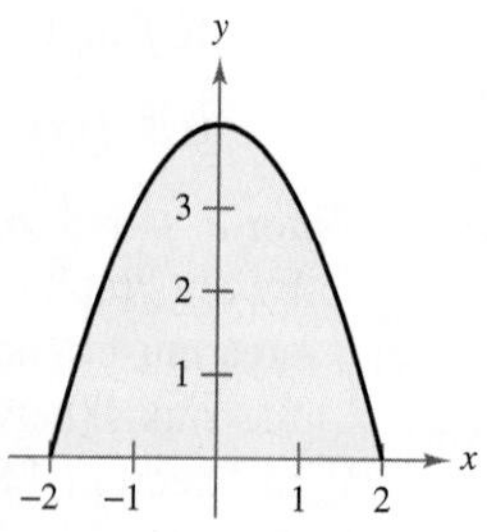

18. $f(x) = \dfrac{1}{x^2 + 1}$

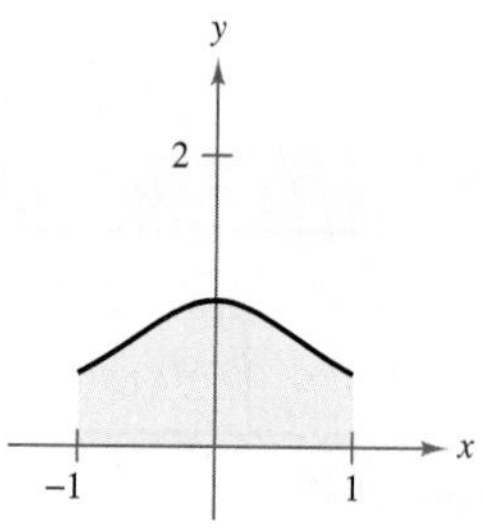

19. $f(x) = \sin x$

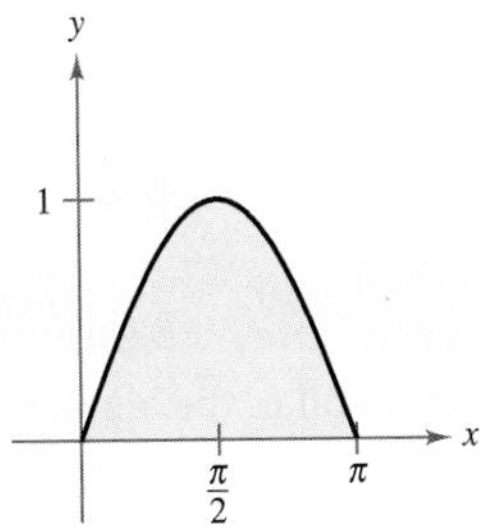

20. $f(x) = \tan x$

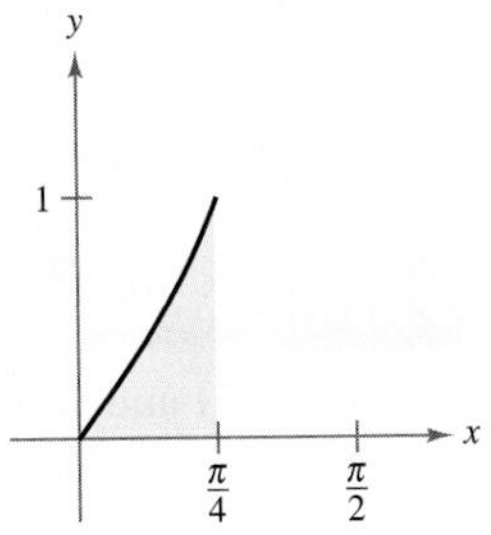

21. $g(y) = y^3$

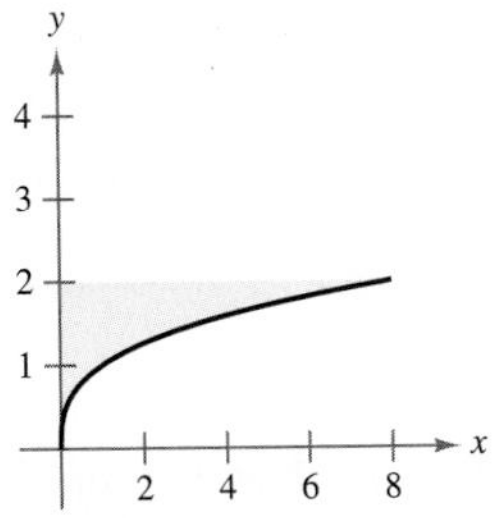

22. $f(y) = (y - 2)^2$

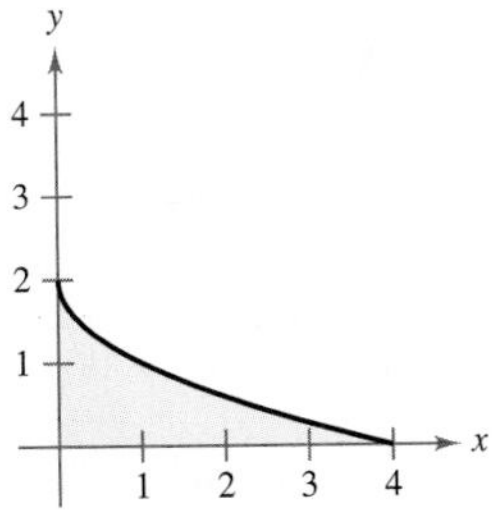

In Exercises 23–32, sketch the region whose area is given by the definite integral. Then use a geometric formula to evaluate the integral ($a > 0, r > 0$).

23. $\displaystyle\int_0^3 4\, dx$

24. $\displaystyle\int_{-a}^{a} 4\, dx$

25. $\displaystyle\int_0^4 x\, dx$

26. $\displaystyle\int_0^4 \frac{x}{2}\, dx$

27. $\displaystyle\int_0^2 (2x + 5)\, dx$

28. $\displaystyle\int_0^8 (8 - x)\, dx$

29. $\displaystyle\int_{-1}^{1} (1 - |x|)\, dx$

30. $\displaystyle\int_{-a}^{a} (a - |x|)\, dx$

31. $\displaystyle\int_{-3}^{3} \sqrt{9 - x^2}\, dx$

32. $\displaystyle\int_{-r}^{r} \sqrt{r^2 - x^2}\, dx$

In Exercises 33–40, evaluate the integral using the following values.

$$\int_2^4 x^3\, dx = 60, \qquad \int_2^4 x\, dx = 6, \qquad \int_2^4 dx = 2$$

33. $\displaystyle\int_4^2 x\, dx$

34. $\displaystyle\int_2^2 x^3\, dx$

35. $\displaystyle\int_2^4 4x\, dx$

36. $\displaystyle\int_2^4 15\, dx$

37. $\displaystyle\int_2^4 (x - 8)\, dx$

38. $\displaystyle\int_2^4 (x^3 + 4)\, dx$

39. $\displaystyle\int_2^4 \left(\tfrac{1}{2}x^3 - 3x + 2\right) dx$

40. $\displaystyle\int_2^4 (6 + 2x - x^3)\, dx$

41. Given $\displaystyle\int_0^5 f(x)\, dx = 10$ and $\displaystyle\int_5^7 f(x)\, dx = 3$, evaluate

(a) $\displaystyle\int_0^7 f(x)\, dx.$ (b) $\displaystyle\int_5^0 f(x)\, dx.$

(c) $\displaystyle\int_5^5 f(x)\, dx.$ (d) $\displaystyle\int_0^5 3f(x)\, dx.$

42. Given $\displaystyle\int_0^3 f(x)\, dx = 4$ and $\displaystyle\int_3^6 f(x)\, dx = -1$, evaluate

(a) $\displaystyle\int_0^6 f(x)\, dx.$ (b) $\displaystyle\int_6^3 f(x)\, dx.$

(c) $\displaystyle\int_3^3 f(x)\, dx.$ (d) $\displaystyle\int_3^6 -5f(x)\, dx.$

43. Given $\displaystyle\int_2^6 f(x)\, dx = 10$ and $\displaystyle\int_2^6 g(x)\, dx = -2$, evaluate

(a) $\displaystyle\int_2^6 [f(x) + g(x)]\, dx.$ (b) $\displaystyle\int_2^6 [g(x) - f(x)]\, dx.$

(c) $\displaystyle\int_2^6 2g(x)\, dx.$ (d) $\displaystyle\int_2^6 3f(x)\, dx.$

44. Given $\displaystyle\int_{-1}^{1} f(x)\, dx = 0$ and $\displaystyle\int_0^1 f(x)\, dx = 5$, evaluate

(a) $\displaystyle\int_{-1}^{0} f(x)\, dx.$ (b) $\displaystyle\int_0^1 f(x)\, dx - \int_{-1}^{0} f(x)\, dx.$

(c) $\displaystyle\int_{-1}^{1} 3f(x)\, dx.$ (d) $\displaystyle\int_0^1 3f(x)\, dx.$

45. Use the table of values to find lower and upper estimates of

$$\int_0^{10} f(x)\, dx.$$

Assume that f is a decreasing function.

x	0	2	4	6	8	10
$f(x)$	32	24	12	-4	-20	-36

46. Use the table of values to estimate

$$\int_0^6 f(x)\, dx.$$

Use three equal subintervals and the (a) left endpoints, (b) right endpoints, and (c) midpoints. If f is an increasing function, how does each estimate compare with the actual value? Explain your reasoning.

x	0	1	2	3	4	5	6
$f(x)$	-6	0	8	18	30	50	80

47. ***Think About It*** The graph of f consists of line segments and a semicircle, as shown in the figure. Evaluate each definite integral by using geometric formulas.

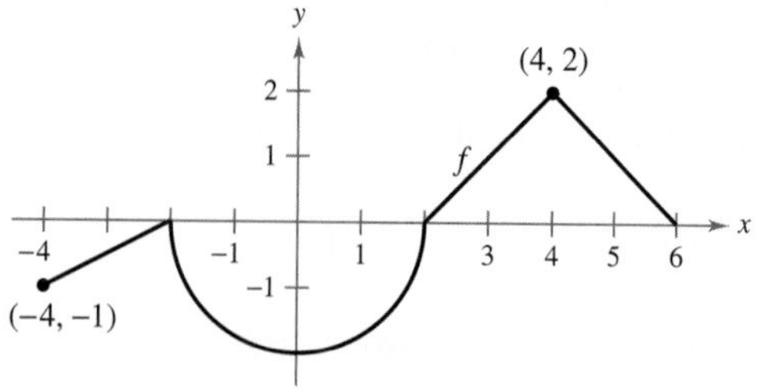

(a) $\int_0^2 f(x)\,dx$ (b) $\int_2^6 f(x)\,dx$

(c) $\int_{-4}^2 f(x)\,dx$ (d) $\int_{-4}^6 f(x)\,dx$

(e) $\int_{-4}^6 |f(x)|\,dx$ (f) $\int_{-4}^6 [f(x) + 2]\,dx$

48. ***Think About It*** The graph of f consists of line segments, as shown in the figure. Evaluate each definite integral by using geometric formulas.

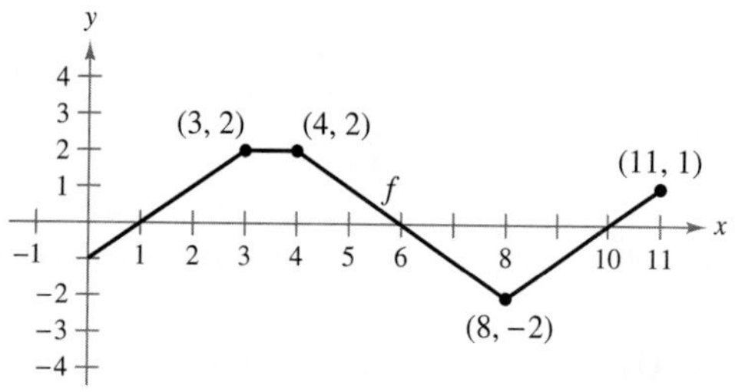

(a) $\int_0^1 -f(x)\,dx$ (b) $\int_3^4 3f(x)\,dx$

(c) $\int_0^7 f(x)\,dx$ (d) $\int_5^{11} f(x)\,dx$

(e) $\int_0^{11} f(x)\,dx$ (f) $\int_4^{10} f(x)\,dx$

49. ***Think About It*** Consider the function f that is continuous on the interval $[-5, 5]$ and for which

$$\int_0^5 f(x)\,dx = 4.$$

Evaluate each integral.

(a) $\int_0^5 [f(x) + 2]\,dx$

(b) $\int_{-2}^3 f(x + 2)\,dx$

(c) $\int_{-5}^5 f(x)\,dx$ (f is even.)

(d) $\int_{-5}^5 f(x)\,dx$ (f is odd.)

50. ***Think About It*** A function f is defined below. Use geometric formulas to find $\int_0^8 f(x)\,dx$.

$$f(x) = \begin{cases} 4, & x < 4 \\ x, & x \geq 4 \end{cases}$$

Writing About Concepts

In Exercises 51 and 52, use the figure to fill in the blank with the symbol <, >, or =.

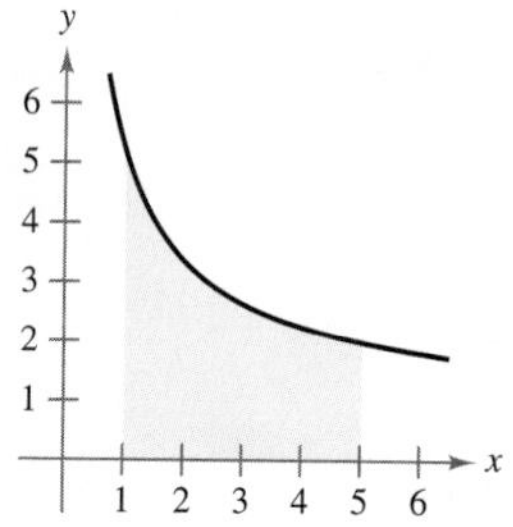

51. The interval $[1, 5]$ is partitioned into n subintervals of equal width Δx, and x_i is the left endpoint of the ith subinterval.

$$\sum_{i=1}^{n} f(x_i)\,\Delta x \quad \boxed{} \quad \int_1^5 f(x)\,dx$$

52. The interval $[1, 5]$ is partitioned into n subintervals of equal width Δx, and x_i is the right endpoint of the ith subinterval.

$$\sum_{i=1}^{n} f(x_i)\,\Delta x \quad \boxed{} \quad \int_1^5 f(x)\,dx$$

53. Determine whether the function $f(x) = \dfrac{1}{x - 4}$ is integrable on the interval $[3, 5]$. Explain.

54. Give an example of a function that is integrable on the interval $[-1, 1]$, but not continuous on $[-1, 1]$.

In Exercises 55–58, determine which value best approximates the definite integral. Make your selection on the basis of a sketch.

55. $\int_0^4 \sqrt{x}\,dx$

(a) 5 (b) -3 (c) 10 (d) 2 (e) 8

56. $\int_0^{1/2} 4 \cos \pi x\,dx$

(a) 4 (b) $\frac{4}{3}$ (c) 16 (d) 2π (e) -6

57. $\int_0^1 2 \sin \pi x\,dx$

(a) 6 (b) $\frac{1}{2}$ (c) 4 (d) $\frac{5}{4}$

58. $\int_0^9 \left(1 + \sqrt{x}\right) dx$

(a) -3 (b) 9 (c) 27 (d) 3

Programming **Write a program for your graphing utility to approximate a definite integral using the Riemann sum**

$$\sum_{i=1}^{n} f(c_i)\,\Delta x_i$$

where the subintervals are of equal width. The output should give three approximations of the integral where c_i is the left-hand endpoint $L(n)$, midpoint $M(n)$, and right-hand endpoint $R(n)$ of each subinterval. In Exercises 59–62, use the program to approximate the definite integral and complete the table.

n	4	8	12	16	20
$L(n)$					
$M(n)$					
$R(n)$					

59. $\displaystyle\int_0^3 x\sqrt{3-x}\,dx$

60. $\displaystyle\int_0^3 \frac{5}{x^2+1}\,dx$

61. $\displaystyle\int_0^{\pi/2} \sin^2 x\,dx$

62. $\displaystyle\int_0^3 x\sin x\,dx$

True or False? **In Exercises 63–68, determine whether the statement is true or false. If it is false, explain why or give an example that shows it is false.**

63. $\displaystyle\int_a^b [f(x) + g(x)]\,dx = \int_a^b f(x)\,dx + \int_a^b g(x)\,dx$

64. $\displaystyle\int_a^b f(x)g(x)\,dx = \left[\int_a^b f(x)\,dx\right]\left[\int_a^b g(x)\,dx\right]$

65. If the norm of a partition approaches zero, then the number of subintervals approaches infinity.

66. If f is increasing on $[a, b]$, then the minimum value of $f(x)$ on $[a, b]$ is $f(a)$.

67. The value of

$$\int_a^b f(x)\,dx$$

must be positive.

68. The value of

$$\int_2^2 \sin(x^2)\,dx$$

is 0.

69. Find the Riemann sum for

$$f(x) = x^2 + 3x$$

over the interval $[0, 8]$, where $x_0 = 0$, $x_1 = 1$, $x_2 = 3$, $x_3 = 7$, and $x_4 = 8$, and where $c_1 = 1$, $c_2 = 2$, $c_3 = 5$, and $c_4 = 8$.

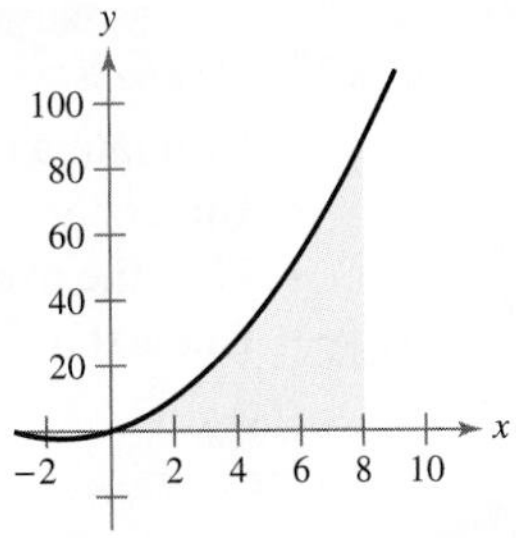

Figure for 69

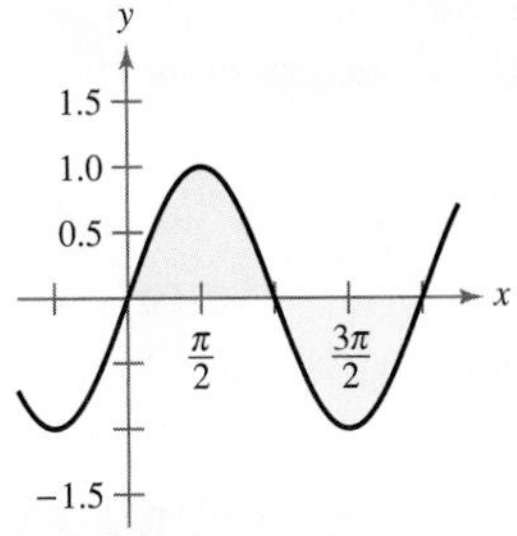

Figure for 70

70. Find the Riemann sum for $f(x) = \sin x$ over the interval $[0, 2\pi]$, where $x_0 = 0$, $x_1 = \pi/4$, $x_2 = \pi/3$, $x_3 = \pi$, and $x_4 = 2\pi$, and where $c_1 = \pi/6$, $c_2 = \pi/3$, $c_3 = 2\pi/3$, and $c_4 = 3\pi/2$.

71. Prove that $\displaystyle\int_a^b x\,dx = \frac{b^2 - a^2}{2}$.

72. Prove that $\displaystyle\int_a^b x^2\,dx = \frac{b^3 - a^3}{3}$.

73. ***Think About It*** Determine whether the Dirichlet function

$$f(x) = \begin{cases} 1, & x \text{ is rational} \\ 0, & x \text{ is irrational} \end{cases}$$

is integrable on the interval $[0, 1]$. Explain.

74. Suppose the function f is defined on $[0, 1]$, as shown in the figure.

$$f(x) = \begin{cases} 0, & x = 0 \\ \dfrac{1}{x}, & 0 < x \le 1 \end{cases}$$

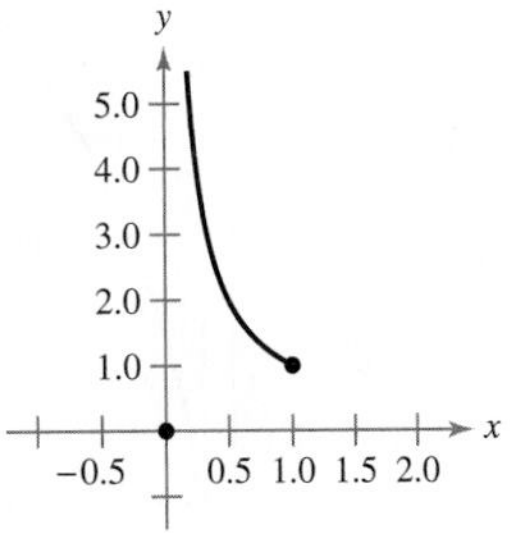

Show that $\int_0^1 f(x)\,dx$ does not exist. Why doesn't this contradict Theorem 4.4?

75. Find the constants a and b that maximize the value of

$$\int_a^b (1 - x^2)\,dx.$$

Explain your reasoning.

76. Evaluate, if possible, the integral $\displaystyle\int_0^2 [\![x]\!]\,dx$.

77. Determine

$$\lim_{n\to\infty} \frac{1}{n^3}[1^2 + 2^2 + 3^2 + \cdots + n^2]$$

by using an appropriate Riemann sum.

Section 4.4 The Fundamental Theorem of Calculus

- **Evaluate a definite integral using the Fundamental Theorem of Calculus.**
- **Understand and use the Mean Value Theorem for Integrals.**
- **Find the average value of a function over a closed interval.**
- **Understand and use the Second Fundamental Theorem of Calculus.**

EXPLORATION

Integration and Antidifferentiation Throughout this chapter, you have been using the integral sign to denote an antiderivative (a family of functions) and a definite integral (a number).

Antidifferentiation: $\int f(x)\,dx$

Definite integration: $\int_a^b f(x)\,dx$

The use of this same symbol for both operations makes it appear that they are related. In the early work with calculus, however, it was not known that the two operations were related. Do you think the symbol $\int$ was first applied to antidifferentiation or to definite integration? Explain your reasoning. *(Hint:* The symbol was first used by Leibniz and was derived from the letter *S*.)

The Fundamental Theorem of Calculus

You have now been introduced to the two major branches of calculus: differential calculus (introduced with the tangent line problem) and integral calculus (introduced with the area problem). At this point, these two problems might seem unrelated—but there is a very close connection. The connection was discovered independently by Isaac Newton and Gottfried Leibniz and is stated in a theorem that is appropriately called the **Fundamental Theorem of Calculus.**

Informally, the theorem states that differentiation and (definite) integration are inverse operations, in the same sense that division and multiplication are inverse operations. To see how Newton and Leibniz might have anticipated this relationship, consider the approximations shown in Figure 4.27. The slope of the tangent line was defined using the *quotient* $\Delta y/\Delta x$ (the slope of the secant line). Similarly, the area of a region under a curve was defined using the *product* $\Delta y \Delta x$ (the area of a rectangle). So, at least in the primitive approximation stage, the operations of differentiation and definite integration appear to have an inverse relationship in the same sense that division and multiplication are inverse operations. The Fundamental Theorem of Calculus states that the limit processes (used to define the derivative and definite integral) preserve this inverse relationship.

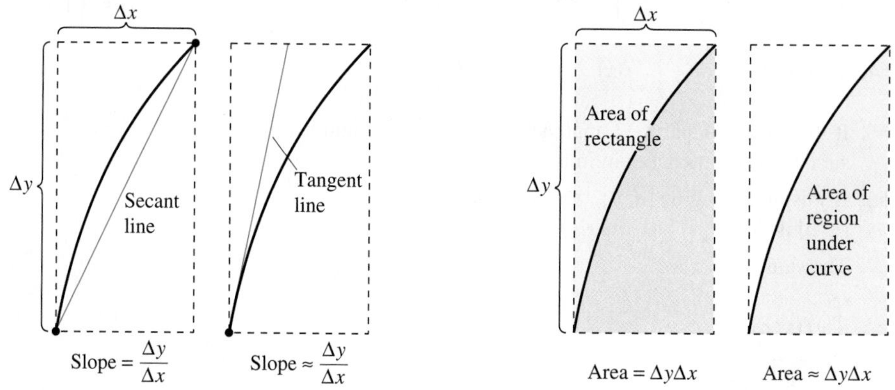

(a) Differentiation **(b)** Definite integration

Differentiation and definite integration have an "inverse" relationship.
Figure 4.27

> **THEOREM 4.9 The Fundamental Theorem of Calculus**
>
> If a function f is continuous on the closed interval $[a, b]$ and F is an antiderivative of f on the interval $[a, b]$, then
>
> $$\int_a^b f(x)\,dx = F(b) - F(a).$$

Proof The key to the proof is in writing the difference $F(b) - F(a)$ in a convenient form. Let Δ be the following partition of $[a, b]$.

$$a = x_0 < x_1 < x_2 < \cdot \cdot \cdot < x_{n-1} < x_n = b$$

By pairwise subtraction and addition of like terms, you can write

$$\begin{aligned} F(b) - F(a) &= F(x_n) - F(x_{n-1}) + F(x_{n-1}) - \cdot \cdot \cdot - F(x_1) + F(x_1) - F(x_0) \\ &= \sum_{i=1}^{n} [F(x_i) - F(x_{i-1})]. \end{aligned}$$

By the Mean Value Theorem, you know that there exists a number c_i in the ith subinterval such that

$$F'(c_i) = \frac{F(x_i) - F(x_{i-1})}{x_i - x_{i-1}}.$$

Because $F'(c_i) = f(c_i)$, you can let $\Delta x_i = x_i - x_{i-1}$ and obtain

$$F(b) - F(a) = \sum_{i=1}^{n} f(c_i)\,\Delta x_i.$$

This important equation tells you that by applying the Mean Value Theorem you can always find a collection of c_i's such that the *constant* $F(b) - F(a)$ is a Riemann sum of f on $[a, b]$. Taking the limit (as $\|\Delta\| \to 0$) produces

$$F(b) - F(a) = \int_a^b f(x)\,dx.$$

The following guidelines can help you understand the use of the Fundamental Theorem of Calculus.

Guidelines for Using the Fundamental Theorem of Calculus

1. *Provided you can find* an antiderivative of f, you now have a way to evaluate a definite integral without having to use the limit of a sum.
2. When applying the Fundamental Theorem of Calculus, the following notation is convenient.

$$\begin{aligned} \int_a^b f(x)\,dx &= F(x)\Big]_a^b \\ &= F(b) - F(a) \end{aligned}$$

For instance, to evaluate $\int_1^3 x^3\,dx$, you can write

$$\int_1^3 x^3\,dx = \frac{x^4}{4}\Big]_1^3 = \frac{3^4}{4} - \frac{1^4}{4} = \frac{81}{4} - \frac{1}{4} = 20.$$

3. It is not necessary to include a constant of integration C in the antiderivative because

$$\begin{aligned} \int_a^b f(x)\,dx &= \Big[F(x) + C\Big]_a^b \\ &= [F(b) + C] - [F(a) + C] \\ &= F(b) - F(a). \end{aligned}$$

EXAMPLE 1 Evaluating a Definite Integral

Evaluate each definite integral.

a. $\int_1^2 (x^2 - 3)\, dx$ **b.** $\int_1^4 3\sqrt{x}\, dx$ **c.** $\int_0^{\pi/4} \sec^2 x\, dx$

Solution

a. $$\int_1^2 (x^2 - 3)\, dx = \left[\frac{x^3}{3} - 3x\right]_1^2 = \left(\frac{8}{3} - 6\right) - \left(\frac{1}{3} - 3\right) = -\frac{2}{3}$$

b. $$\int_1^4 3\sqrt{x}\, dx = 3\int_1^4 x^{1/2}\, dx = 3\left[\frac{x^{3/2}}{3/2}\right]_1^4 = 2(4)^{3/2} - 2(1)^{3/2} = 14$$

c. $$\int_0^{\pi/4} \sec^2 x\, dx = \tan x\Big]_0^{\pi/4} = 1 - 0 = 1$$

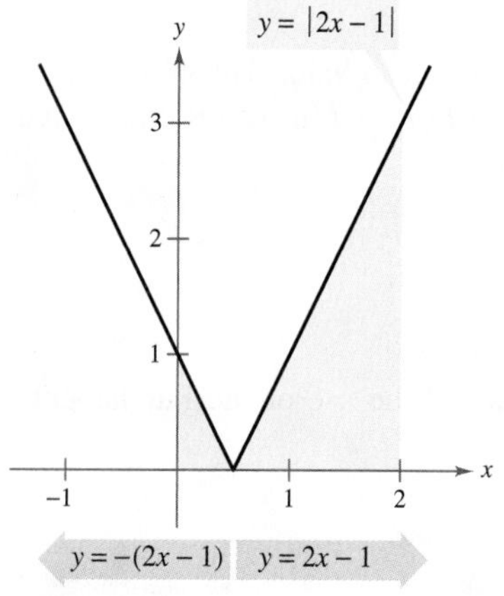

The definite integral of y on $[0, 2]$ is $\frac{5}{2}$.
Figure 4.28

EXAMPLE 2 A Definite Integral Involving Absolute Value

Evaluate $\int_0^2 |2x - 1|\, dx$.

Solution Using Figure 4.28 and the definition of absolute value, you can rewrite the integrand as shown.

$$|2x - 1| = \begin{cases} -(2x - 1), & x < \frac{1}{2} \\ 2x - 1, & x \geq \frac{1}{2} \end{cases}$$

From this, you can rewrite the integral in two parts.

$$\begin{aligned} \int_0^2 |2x - 1|\, dx &= \int_0^{1/2} -(2x - 1)\, dx + \int_{1/2}^2 (2x - 1)\, dx \\ &= \Big[-x^2 + x\Big]_0^{1/2} + \Big[x^2 - x\Big]_{1/2}^2 \\ &= \left(-\frac{1}{4} + \frac{1}{2}\right) - (0 + 0) + (4 - 2) - \left(\frac{1}{4} - \frac{1}{2}\right) = \frac{5}{2} \end{aligned}$$

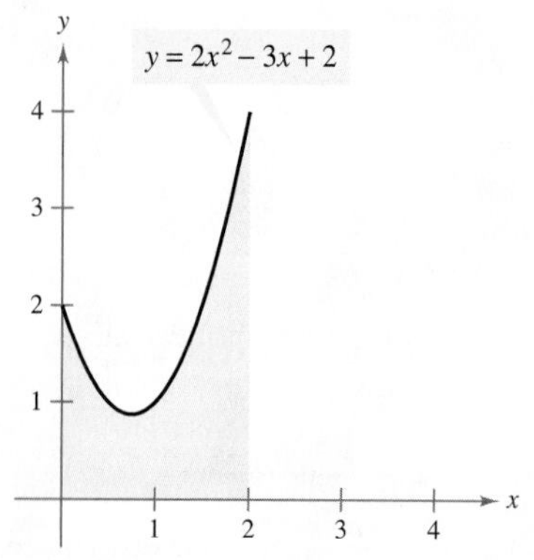

The area of the region bounded by the graph of y, the x-axis, $x = 0$, and $x = 2$ is $\frac{10}{3}$.
Figure 4.29

EXAMPLE 3 Using the Fundamental Theorem to Find Area

Find the area of the region bounded by the graph of $y = 2x^2 - 3x + 2$, the x-axis, and the vertical lines $x = 0$ and $x = 2$, as shown in Figure 4.29.

Solution Note that $y > 0$ on the interval $[0, 2]$.

$$\text{Area} = \int_0^2 (2x^2 - 3x + 2)\, dx \qquad \text{Integrate between } x = 0 \text{ and } x = 2.$$

$$= \left[\frac{2x^3}{3} - \frac{3x^2}{2} + 2x\right]_0^2 \qquad \text{Find antiderivative.}$$

$$= \left(\frac{16}{3} - 6 + 4\right) - (0 - 0 + 0) \qquad \text{Apply Fundamental Theorem.}$$

$$= \frac{10}{3} \qquad \text{Simplify.}$$

The Mean Value Theorem for Integrals

In Section 4.2, you saw that the area of a region under a curve is greater than the area of an inscribed rectangle and less than the area of a circumscribed rectangle. The Mean Value Theorem for Integrals states that somewhere "between" the inscribed and circumscribed rectangles there is a rectangle whose area is precisely equal to the area of the region under the curve, as shown in Figure 4.30.

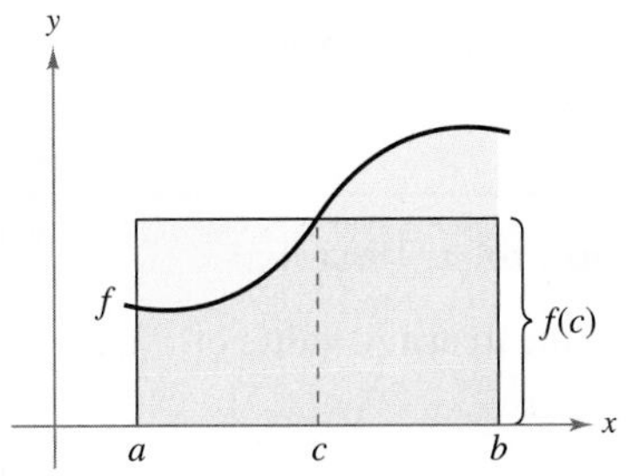

Mean value rectangle:

$$f(c)(b-a) = \int_a^b f(x)\,dx$$

Figure 4.30

THEOREM 4.10 Mean Value Theorem for Integrals

If f is continuous on the closed interval $[a, b]$, then there exists a number c in the closed interval $[a, b]$ such that

$$\int_a^b f(x)\,dx = f(c)(b-a).$$

Proof

Case 1: If f is constant on the interval $[a, b]$, the theorem is clearly valid because c can be any point in $[a, b]$.

Case 2: If f is not constant on $[a, b]$, then, by the Extreme Value Theorem, you can choose $f(m)$ and $f(M)$ to be the minimum and maximum values of f on $[a, b]$. Because $f(m) \le f(x) \le f(M)$ for all x in $[a, b]$, you can apply Theorem 4.8 to write the following.

$$\int_a^b f(m)\,dx \le \int_a^b f(x)\,dx \le \int_a^b f(M)\,dx \qquad \text{See Figure 4.31.}$$

$$f(m)(b-a) \le \int_a^b f(x)\,dx \le f(M)(b-a)$$

$$f(m) \le \frac{1}{b-a}\int_a^b f(x)\,dx \le f(M)$$

From the third inequality, you can apply the Intermediate Value Theorem to conclude that there exists some c in $[a, b]$ such that

$$f(c) = \frac{1}{b-a}\int_a^b f(x)\,dx \qquad \text{or} \qquad f(c)(b-a) = \int_a^b f(x)\,dx.$$

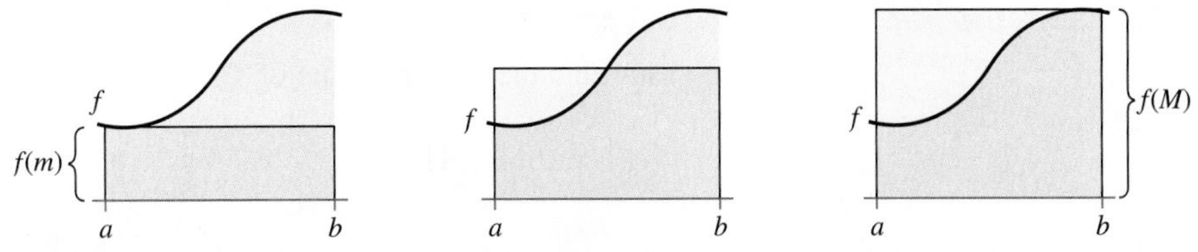

Inscribed rectangle (less than actual area)

$$\int_a^b f(m)\,dx = f(m)(b-a)$$

Mean value rectangle (equal to actual area)

$$\int_a^b f(x)\,dx$$

Circumscribed rectangle (greater than actual area)

$$\int_a^b f(M)\,dx = f(M)(b-a)$$

Figure 4.31

NOTE Notice that Theorem 4.10 does not specify how to determine c. It merely guarantees the existence of at least one number c in the interval.

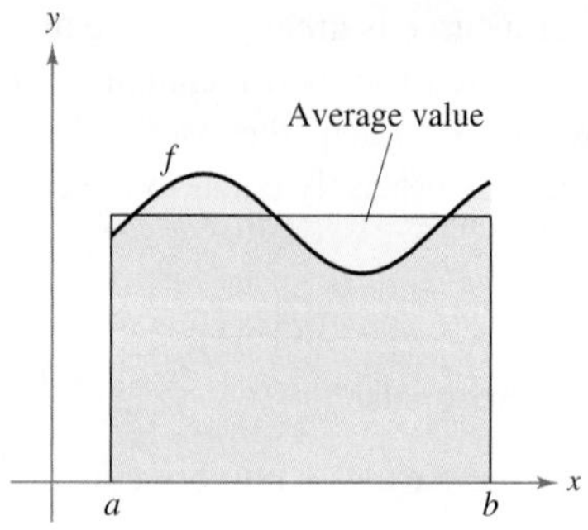

Average value $= \dfrac{1}{b-a}\displaystyle\int_a^b f(x)\,dx$

Figure 4.32

Average Value of a Function

The value of $f(c)$ given in the Mean Value Theorem for Integrals is called the **average value** of f on the interval $[a, b]$.

> **Definition of the Average Value of a Function on an Interval**
>
> If f is integrable on the closed interval $[a, b]$, then the **average value** of f on the interval is
>
> $$\frac{1}{b-a}\int_a^b f(x)\,dx.$$

NOTE Notice in Figure 4.32 that the area of the region under the graph of f is equal to the area of the rectangle whose height is the average value.

To see why the average value of f is defined in this way, suppose that you partition $[a, b]$ into n subintervals of equal width $\Delta x = (b-a)/n$. If c_i is any point in the ith subinterval, the arithmetic average (or mean) of the function values at the c_i's is given by

$$a_n = \frac{1}{n}[f(c_1) + f(c_2) + \cdot \cdot \cdot + f(c_n)]. \qquad \text{Average of } f(c_1), \ldots, f(c_n)$$

By multiplying and dividing by $(b-a)$, you can write the average as

$$a_n = \frac{1}{n}\sum_{i=1}^{n} f(c_i)\left(\frac{b-a}{b-a}\right) = \frac{1}{b-a}\sum_{i=1}^{n} f(c_i)\left(\frac{b-a}{n}\right)$$
$$= \frac{1}{b-a}\sum_{i=1}^{n} f(c_i)\,\Delta x.$$

Finally, taking the limit as $n \to \infty$ produces the average value of f on the interval $[a, b]$, as given in the definition above.

This development of the average value of a function on an interval is only one of many practical uses of definite integrals to represent summation processes. In Chapter 7, you will study other applications, such as volume, arc length, centers of mass, and work.

EXAMPLE 4 Finding the Average Value of a Function

Find the average value of $f(x) = 3x^2 - 2x$ on the interval $[1, 4]$.

Solution The average value is given by

$$\frac{1}{b-a}\int_a^b f(x)\,dx = \frac{1}{3}\int_1^4 (3x^2 - 2x)\,dx$$
$$= \frac{1}{3}\Big[x^3 - x^2\Big]_1^4$$
$$= \frac{1}{3}[64 - 16 - (1 - 1)] = \frac{48}{3} = 16.$$

(See Figure 4.33.)

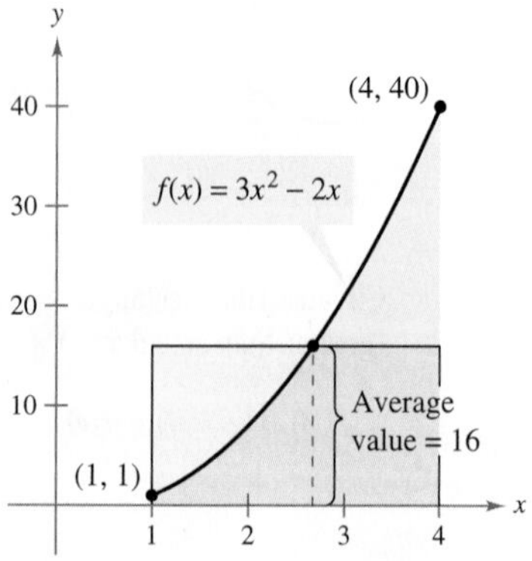

Figure 4.33

The first person to fly at a speed greater than the speed of sound was Charles Yeager. On October 14, 1947, Yeager was clocked at 295.9 meters per second at an altitude of 12.2 kilometers. If Yeager had been flying at an altitude below 11.275 kilometers, this speed would not have "broken the sound barrier." The photo above shows an F-14 *Tomcat*, a supersonic, twin-engine strike fighter. Currently, the *Tomcat* can reach heights of 15.24 kilometers and speeds up to 2 mach (707.78 meters per second).

EXAMPLE 5 The Speed of Sound

At different altitudes in Earth's atmosphere, sound travels at different speeds. The speed of sound $s(x)$ (in meters per second) can be modeled by

$$s(x) = \begin{cases} -4x + 341, & 0 \le x < 11.5 \\ 295, & 11.5 \le x < 22 \\ \frac{3}{4}x + 278.5, & 22 \le x < 32 \\ \frac{3}{2}x + 254.5, & 32 \le x < 50 \\ -\frac{3}{2}x + 404.5, & 50 \le x \le 80 \end{cases}$$

where x is the altitude in kilometers (see Figure 4.34). What is the average speed of sound over the interval $[0, 80]$?

Solution Begin by integrating $s(x)$ over the interval $[0, 80]$. To do this, you can break the integral into five parts.

$$\int_0^{11.5} s(x)\,dx = \int_0^{11.5} (-4x + 341)\,dx = \Big[-2x^2 + 341x\Big]_0^{11.5} = 3657$$

$$\int_{11.5}^{22} s(x)\,dx = \int_{11.5}^{22} (295)\,dx = \Big[295x\Big]_{11.5}^{22} = 3097.5$$

$$\int_{22}^{32} s(x)\,dx = \int_{22}^{32} \left(\tfrac{3}{4}x + 278.5\right)dx = \Big[\tfrac{3}{8}x^2 + 278.5x\Big]_{22}^{32} = 2987.5$$

$$\int_{32}^{50} s(x)\,dx = \int_{32}^{50} \left(\tfrac{3}{2}x + 254.5\right)dx = \Big[\tfrac{3}{4}x^2 + 254.5x\Big]_{32}^{50} = 5688$$

$$\int_{50}^{80} s(x)\,dx = \int_{50}^{80} \left(-\tfrac{3}{2}x + 404.5\right)dx = \Big[-\tfrac{3}{4}x^2 + 404.5x\Big]_{50}^{80} = 9210$$

By adding the values of the five integrals, you have

$$\int_0^{80} s(x)\,dx = 24{,}640.$$

So, the average speed of sound from an altitude of 0 kilometers to an altitude of 80 kilometers is

$$\text{Average speed} = \frac{1}{80}\int_0^{80} s(x)\,dx = \frac{24{,}640}{80} = 308 \text{ meters per second.}$$

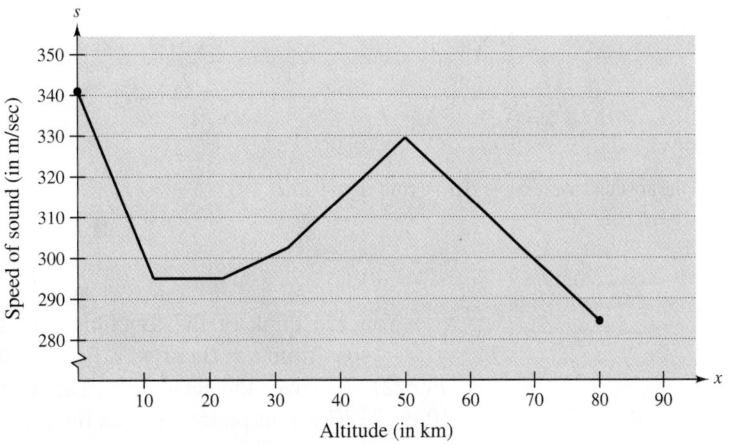

Speed of sound depends on altitude.
Figure 4.34

The Second Fundamental Theorem of Calculus

Earlier you saw that the definite integral of f on the interval $[a, b]$ was defined using the constant b as the upper limit of integration and x as the variable of integration. However, a slightly different situation may arise in which the variable x is used as the upper limit of integration. To avoid the confusion of using x in two different ways, t is temporarily used as the variable of integration. (Remember that the definite integral is *not* a function of its variable of integration.)

The Definite Integral as a Number

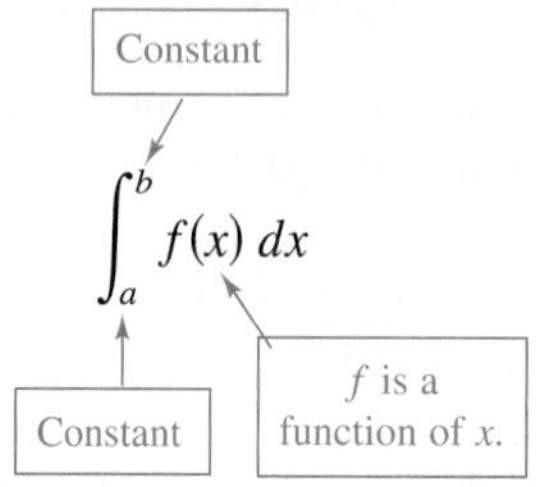

The Definite Integral as a Function of x

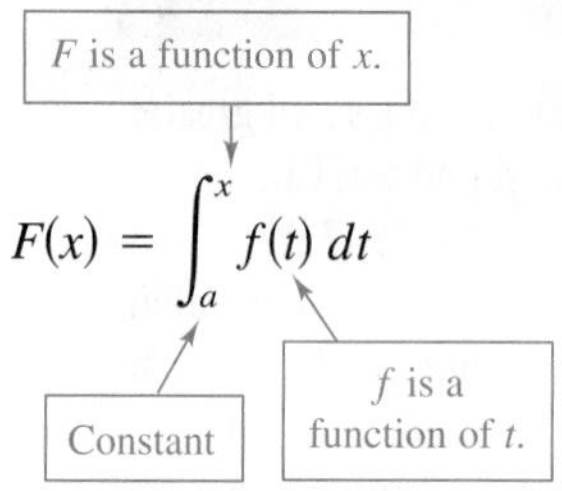

EXPLORATION

Use a graphing utility to graph the function

$$F(x) = \int_0^x \cos t\,dt$$

for $0 \le x \le \pi$. Do you recognize this graph? Explain.

EXAMPLE 6 The Definite Integral as a Function

Evaluate the function

$$F(x) = \int_0^x \cos t\,dt$$

at $x = 0, \pi/6, \pi/4, \pi/3$, and $\pi/2$.

Solution You could evaluate five different definite integrals, one for each of the given upper limits. However, it is much simpler to fix x (as a constant) temporarily to obtain

$$\int_0^x \cos t\,dt = \sin t\Big]_0^x = \sin x - \sin 0 = \sin x.$$

Now, using $F(x) = \sin x$, you can obtain the results shown in Figure 4.35.

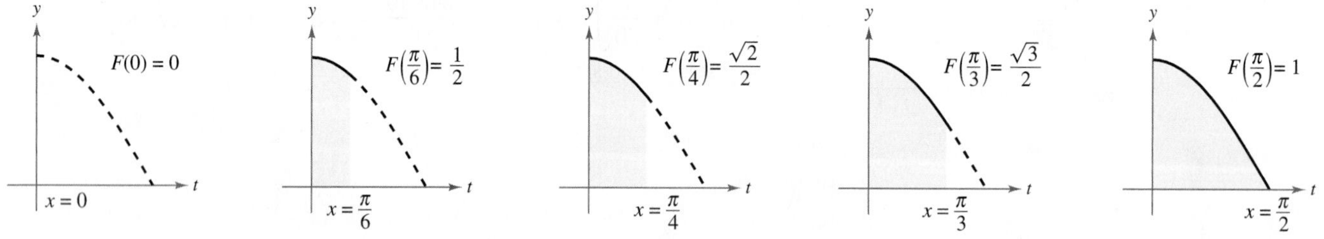

$F(x) = \int_0^x \cos t\,dt$ is the area under the curve $f(t) = \cos t$ from 0 to x.

Figure 4.35

You can think of the function $F(x)$ as *accumulating* the area under the curve $f(t) = \cos t$ from $t = 0$ to $t = x$. For $x = 0$, the area is 0 and $F(0) = 0$. For $x = \pi/2$, $F(\pi/2) = 1$ gives the accumulated area under the cosine curve on the entire interval $[0, \pi/2]$. This interpretation of an integral as an **accumulation function** is used often in applications of integration.

In Example 6, note that the derivative of F is the original integrand (with only the variable changed). That is,

$$\frac{d}{dx}[F(x)] = \frac{d}{dx}[\sin x] = \frac{d}{dx}\left[\int_0^x \cos t\, dt\right] = \cos x.$$

This result is generalized in the following theorem, called the **Second Fundamental Theorem of Calculus.**

THEOREM 4.11 The Second Fundamental Theorem of Calculus

If f is continuous on an open interval I containing a, then, for every x in the interval,

$$\frac{d}{dx}\left[\int_a^x f(t)\, dt\right] = f(x).$$

Proof Begin by defining F as

$$F(x) = \int_a^x f(t)\, dt.$$

Then, by the definition of the derivative, you can write

$$\begin{aligned} F'(x) &= \lim_{\Delta x \to 0} \frac{F(x+\Delta x) - F(x)}{\Delta x} \\ &= \lim_{\Delta x \to 0} \frac{1}{\Delta x}\left[\int_a^{x+\Delta x} f(t)\, dt - \int_a^x f(t)\, dt\right] \\ &= \lim_{\Delta x \to 0} \frac{1}{\Delta x}\left[\int_a^{x+\Delta x} f(t)\, dt + \int_x^a f(t)\, dt\right] \\ &= \lim_{\Delta x \to 0} \frac{1}{\Delta x}\left[\int_x^{x+\Delta x} f(t)\, dt\right]. \end{aligned}$$

From the Mean Value Theorem for Integrals (assuming $\Delta x > 0$), you know there exists a number c in the interval $[x, x + \Delta x]$ such that the integral in the expression above is equal to $f(c)\,\Delta x$. Moreover, because $x \le c \le x + \Delta x$, it follows that $c \to x$ as $\Delta x \to 0$. So, you obtain

$$\begin{aligned} F'(x) &= \lim_{\Delta x \to 0}\left[\frac{1}{\Delta x} f(c)\,\Delta x\right] \\ &= \lim_{\Delta x \to 0} f(c) \\ &= f(x). \end{aligned}$$

A similar argument can be made for $\Delta x < 0$.

NOTE Using the area model for definite integrals, you can view the approximation

$$f(x)\,\Delta x \approx \int_x^{x+\Delta x} f(t)\, dt$$

as saying that the area of the rectangle of height $f(x)$ and width Δx is approximately equal to the area of the region lying between the graph of f and the x-axis on the interval $[x, x + \Delta x]$, as shown in Figure 4.36.

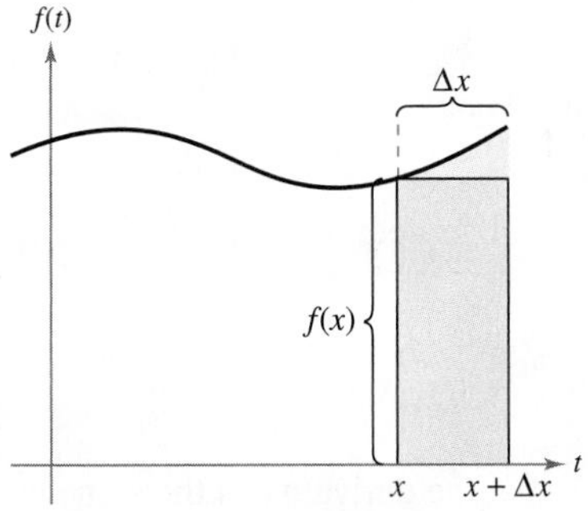

$$f(x)\,\Delta x \approx \int_x^{x+\Delta x} f(t)\, dt$$

Figure 4.36

Note that the Second Fundamental Theorem of Calculus tells you that if a function is continuous, you can be sure that it has an antiderivative. This antiderivative need not, however, be an elementary function. (Recall the discussion of elementary functions in Section P.3.)

EXAMPLE 7 Using the Second Fundamental Theorem of Calculus

Evaluate $\frac{d}{dx}\left[\int_0^x \sqrt{t^2+1}\, dt\right]$.

Solution Note that $f(t) = \sqrt{t^2+1}$ is continuous on the entire real line. So, using the Second Fundamental Theorem of Calculus, you can write

$$\frac{d}{dx}\left[\int_0^x \sqrt{t^2+1}\, dt\right] = \sqrt{x^2+1}.$$

The differentiation shown in Example 7 is a straightforward application of the Second Fundamental Theorem of Calculus. The next example shows how this theorem can be combined with the Chain Rule to find the derivative of a function.

EXAMPLE 8 Using the Second Fundamental Theorem of Calculus

Find the derivative of $F(x) = \int_{\pi/2}^{x^3} \cos t\, dt$.

Solution Using $u = x^3$, you can apply the Second Fundamental Theorem of Calculus with the Chain Rule as shown.

$$\begin{aligned}
F'(x) &= \frac{dF}{du}\frac{du}{dx} && \text{Chain Rule} \\
&= \frac{d}{du}[F(x)]\frac{du}{dx} && \text{Definition of } \frac{dF}{du} \\
&= \frac{d}{du}\left[\int_{\pi/2}^{x^3} \cos t\, dt\right]\frac{du}{dx} && \text{Substitute } \int_{\pi/2}^{x^3} \cos t\, dt \text{ for } F(x). \\
&= \frac{d}{du}\left[\int_{\pi/2}^{u} \cos t\, dt\right]\frac{du}{dx} && \text{Substitute } u \text{ for } x^3. \\
&= (\cos u)(3x^2) && \text{Apply Second Fundamental Theorem of Calculus.} \\
&= (\cos x^3)(3x^2) && \text{Rewrite as function of } x.
\end{aligned}$$

Because the integrand in Example 8 is easily integrated, you can verify the derivative as follows.

$$F(x) = \int_{\pi/2}^{x^3} \cos t\, dt = \sin t\Big]_{\pi/2}^{x^3} = \sin x^3 - \sin\frac{\pi}{2} = (\sin x^3) - 1$$

In this form, you can apply the Power Rule to verify that the derivative is the same as that obtained in Example 8.

$$F'(x) = (\cos x^3)(3x^2)$$

Exercises for Section 4.4

See www.CalcChat.com for worked-out solutions to odd-numbered exercises.

Graphical Reasoning **In Exercises 1–4, use a graphing utility to graph the integrand. Use the graph to determine whether the definite integral is positive, negative, or zero.**

1. $\displaystyle\int_0^{\pi} \frac{4}{x^2+1}\,dx$ **2.** $\displaystyle\int_0^{\pi} \cos x\,dx$

3. $\displaystyle\int_{-2}^{2} x\sqrt{x^2+1}\,dx$ **4.** $\displaystyle\int_{-2}^{2} x\sqrt{2-x}\,dx$

In Exercises 5–26, evaluate the definite integral of the algebraic function. Use a graphing utility to verify your result.

5. $\displaystyle\int_0^1 2x\,dx$ **6.** $\displaystyle\int_2^7 3\,dv$

7. $\displaystyle\int_{-1}^0 (x-2)\,dx$ **8.** $\displaystyle\int_2^5 (-3v+4)\,dv$

9. $\displaystyle\int_{-1}^1 (t^2-2)\,dt$ **10.** $\displaystyle\int_1^3 (3x^2+5x-4)\,dx$

11. $\displaystyle\int_0^1 (2t-1)^2\,dt$ **12.** $\displaystyle\int_{-1}^1 (t^3-9t)\,dt$

13. $\displaystyle\int_1^2 \left(\frac{3}{x^2}-1\right)dx$ **14.** $\displaystyle\int_{-2}^{-1} \left(u-\frac{1}{u^2}\right)du$

15. $\displaystyle\int_1^4 \frac{u-2}{\sqrt{u}}\,du$ **16.** $\displaystyle\int_{-3}^3 v^{1/3}\,dv$

17. $\displaystyle\int_{-1}^1 (\sqrt[3]{t}-2)\,dt$ **18.** $\displaystyle\int_1^8 \sqrt{\frac{2}{x}}\,dx$

19. $\displaystyle\int_0^1 \frac{x-\sqrt{x}}{3}\,dx$ **20.** $\displaystyle\int_0^2 (2-t)\sqrt{t}\,dt$

21. $\displaystyle\int_{-1}^0 \left(t^{1/3}-t^{2/3}\right)dt$ **22.** $\displaystyle\int_{-8}^{-1} \frac{x-x^2}{2\sqrt[3]{x}}\,dx$

23. $\displaystyle\int_0^3 |2x-3|\,dx$ **24.** $\displaystyle\int_1^4 (3-|x-3|)\,dx$

25. $\displaystyle\int_0^3 |x^2-4|\,dx$ **26.** $\displaystyle\int_0^4 |x^2-4x+3|\,dx$

In Exercises 27–32, evaluate the definite integral of the trigonometric function. Use a graphing utility to verify your result.

27. $\displaystyle\int_0^{\pi} (1+\sin x)\,dx$

28. $\displaystyle\int_0^{\pi/4} \frac{1-\sin^2\theta}{\cos^2\theta}\,d\theta$

29. $\displaystyle\int_{-\pi/6}^{\pi/6} \sec^2 x\,dx$

30. $\displaystyle\int_{\pi/4}^{\pi/2} (2-\csc^2 x)\,dx$

31. $\displaystyle\int_{-\pi/3}^{\pi/3} 4\sec\theta\tan\theta\,d\theta$

32. $\displaystyle\int_{-\pi/2}^{\pi/2} (2t+\cos t)\,dt$

In Exercises 33–38, determine the area of the given region.

33. $y = x - x^2$

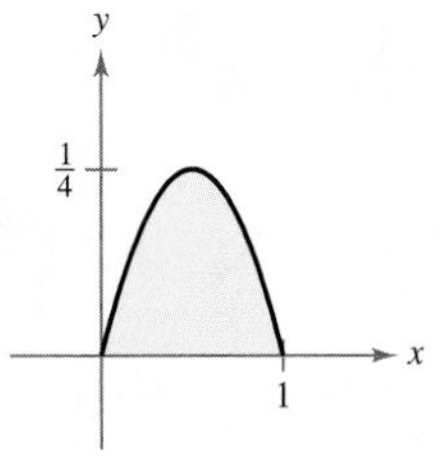

34. $y = 1 - x^4$

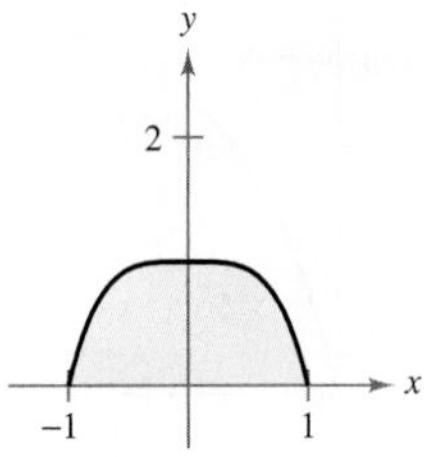

35. $y = (3-x)\sqrt{x}$

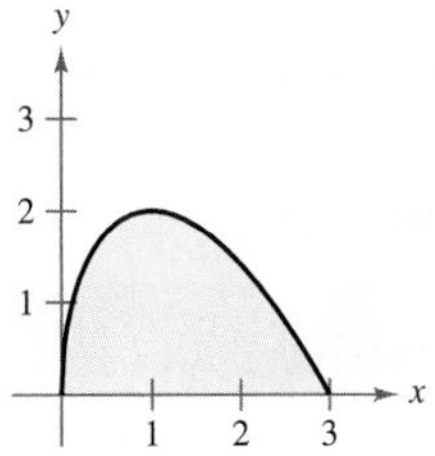

36. $y = \dfrac{1}{x^2}$

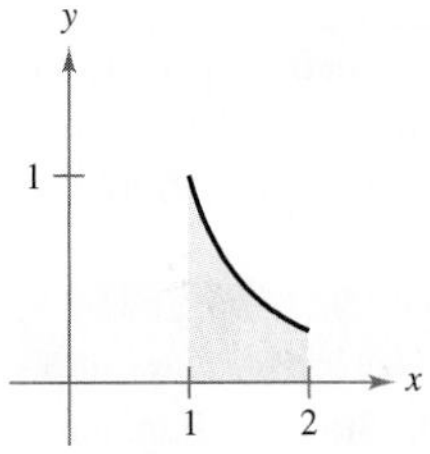

37. $y = \cos x$

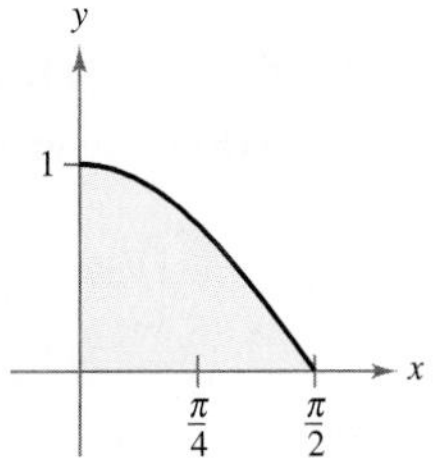

38. $y = x + \sin x$

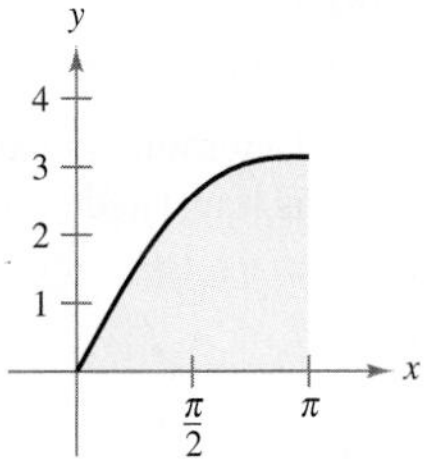

In Exercises 39–42, find the area of the region bounded by the graphs of the equations.

39. $y = 3x^2+1,\quad x=0,\quad x=2,\quad y=0$

40. $y = 1+\sqrt[3]{x},\quad x=0,\quad x=8,\quad y=0$

41. $y = x^3+x,\quad x=2,\quad y=0$ **42.** $y = -x^2+3x,\quad y=0$

In Exercises 43–46, find the value(s) of *c* guaranteed by the Mean Value Theorem for Integrals for the function over the given interval.

43. $f(x) = x - 2\sqrt{x},\quad [0, 2]$ **44.** $f(x) = \dfrac{9}{x^3},\quad [1, 3]$

45. $f(x) = 2\sec^2 x,\quad [-\pi/4, \pi/4]$

46. $f(x) = \cos x,\quad [-\pi/3, \pi/3]$

In Exercises 47–50, find the average value of the function over the given interval and all values of *x* in the interval for which the function equals its average value.

47. $f(x) = 4 - x^2,\quad [-2, 2]$ **48.** $f(x) = \dfrac{4(x^2+1)}{x^2},\quad [1, 3]$

49. $f(x) = \sin x,\quad [0, \pi]$ **50.** $f(x) = \cos x,\quad [0, \pi/2]$

51. ***Velocity*** The graph shows the velocity, in feet per second, of a car accelerating from rest. Use the graph to estimate the distance the car travels in 8 seconds.

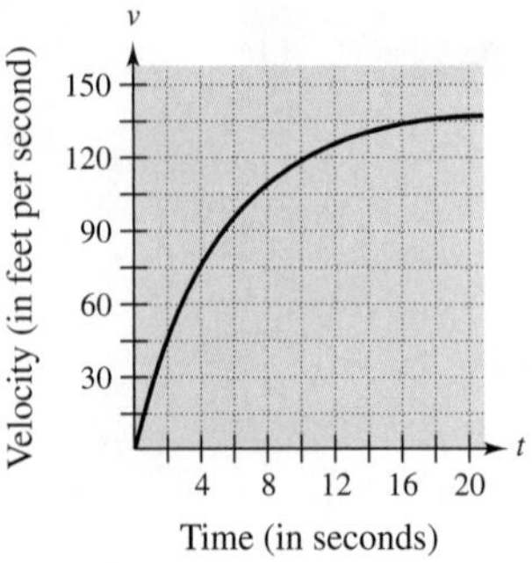

Figure for 51

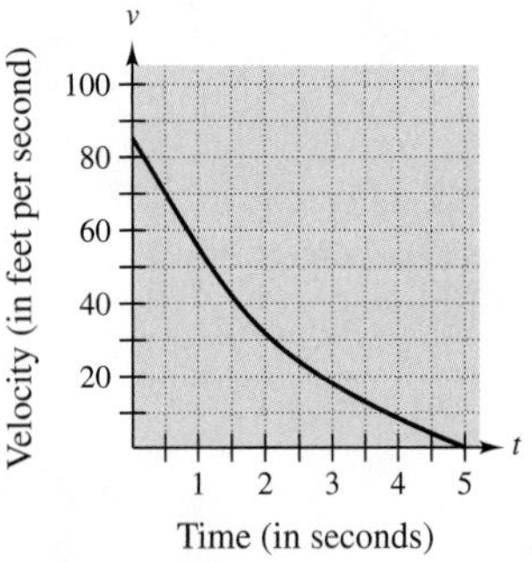

Figure for 52

52. ***Velocity*** The graph shows the velocity of a car as soon as the driver applies the brakes. Use the graph to estimate how far the car travels before it comes to a stop.

Writing About Concepts

53. State the Fundamental Theorem of Calculus.

54. The graph of f is shown in the figure.

(a) Evaluate $\int_1^7 f(x)\,dx$.

(b) Determine the average value of f on the interval $[1, 7]$.

(c) Determine the answers to parts (a) and (b) if the graph is translated two units upward.

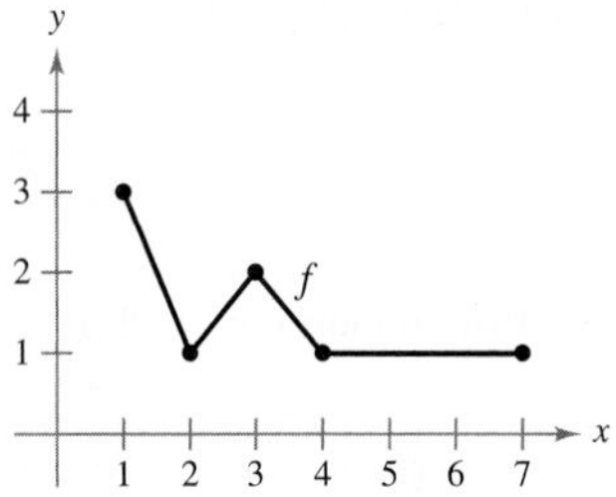

Figure for 54

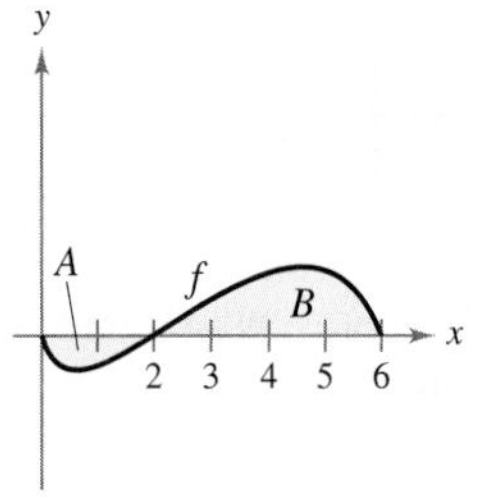

Figure for 55–60

In Exercises 55–60, use the graph of f shown in the figure. The shaded region A has an area of 1.5, and $\int_0^6 f(x)\,dx = 3.5$. Use this information to fill in the blanks.

55. $\int_0^2 f(x)\,dx =$ ____

56. $\int_2^6 f(x)\,dx =$ ____

57. $\int_0^6 |f(x)|\,dx =$ ____

58. $\int_0^2 -2f(x)\,dx =$ ____

59. $\int_0^6 [2 + f(x)]\,dx =$ ____

60. The average value of f over the interval $[0, 6]$ is ____.

61. ***Force*** The force F (in newtons) of a hydraulic cylinder in a press is proportional to the square of $\sec x$, where x is the distance (in meters) that the cylinder is extended in its cycle. The domain of F is $[0, \pi/3]$, and $F(0) = 500$.

(a) Find F as a function of x.

(b) Find the average force exerted by the press over the interval $[0, \pi/3]$.

62. ***Blood Flow*** The velocity v of the flow of blood at a distance r from the central axis of an artery of radius R is

$$v = k(R^2 - r^2)$$

where k is the constant of proportionality. Find the average rate of flow of blood along a radius of the artery. (Use 0 and R as the limits of integration.)

63. ***Respiratory Cycle*** The volume V in liters of air in the lungs during a five-second respiratory cycle is approximated by the model

$$V = 0.1729t + 0.1522t^2 - 0.0374t^3$$

where t is the time in seconds. Approximate the average volume of air in the lungs during one cycle.

64. ***Average Sales*** A company fits a model to the monthly sales data of a seasonal product. The model is

$$S(t) = \frac{t}{4} + 1.8 + 0.5\sin\left(\frac{\pi t}{6}\right), \quad 0 \le t \le 24$$

where S is sales (in thousands) and t is time in months.

(a) Use a graphing utility to graph $f(t) = 0.5\sin(\pi t/6)$ for $0 \le t \le 24$. Use the graph to explain why the average value of $f(t)$ is 0 over the interval.

(b) Use a graphing utility to graph $S(t)$ and the line $g(t) = t/4 + 1.8$ in the same viewing window. Use the graph and the result of part (a) to explain why g is called the *trend line*.

65. ***Modeling Data*** An experimental vehicle is tested on a straight track. It starts from rest, and its velocity v (meters per second) is recorded in the table every 10 seconds for 1 minute.

t	0	10	20	30	40	50	60
v	0	5	21	40	62	78	83

(a) Use a graphing utility to find a model of the form $v = at^3 + bt^2 + ct + d$ for the data.

(b) Use a graphing utility to plot the data and graph the model.

(c) Use the Fundamental Theorem of Calculus to approximate the distance traveled by the vehicle during the test.

66. *Modeling Data* A department store manager wants to estimate the number of customers that enter the store from noon until closing at 9 P.M. The table shows the number of customers N entering the store during a randomly selected minute each hour from $t - 1$ to t, with $t = 0$ corresponding to noon.

t	1	2	3	4	5	6	7	8	9
N	6	7	9	12	15	14	11	7	2

(a) Draw a histogram of the data.

(b) Estimate the total number of customers entering the store between noon and 9 P.M.

(c) Use the regression capabilities of a graphing utility to find a model of the form $N(t) = at^3 + bt^2 + ct + d$ for the data.

(d) Use a graphing utility to plot the data and graph the model.

(e) Use a graphing utility to evaluate $\int_0^9 N(t)\,dt$, and use the result to estimate the number of customers entering the store between noon and 9 P.M. Compare this with your answer in part (b).

(f) Estimate the average number of customers entering the store per minute between 3 P.M. and 7 P.M.

In Exercises 67–72, find F as a function of x and evaluate it at $x = 2$, $x = 5$, and $x = 8$.

67. $F(x) = \int_0^x (t - 5)\,dt$

68. $F(x) = \int_2^x (t^3 + 2t - 2)\,dt$

69. $F(x) = \int_1^x \frac{10}{v^2}\,dv$

70. $F(x) = \int_2^x -\frac{2}{t^3}\,dt$

71. $F(x) = \int_1^x \cos\theta\,d\theta$

72. $F(x) = \int_0^x \sin\theta\,d\theta$

73. Let $g(x) = \int_0^x f(t)\,dt$, where f is a function whose graph is shown.

(a) Estimate $g(0)$, $g(2)$, $g(4)$, $g(6)$, and $g(8)$.

(b) Find the largest open interval on which g is increasing. Find the largest open interval on which g is decreasing.

(c) Identify any extrema of g.

(d) Sketch a rough graph of g.

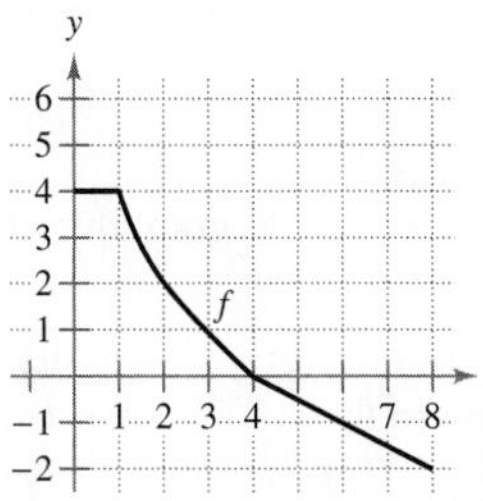

Figure for 73

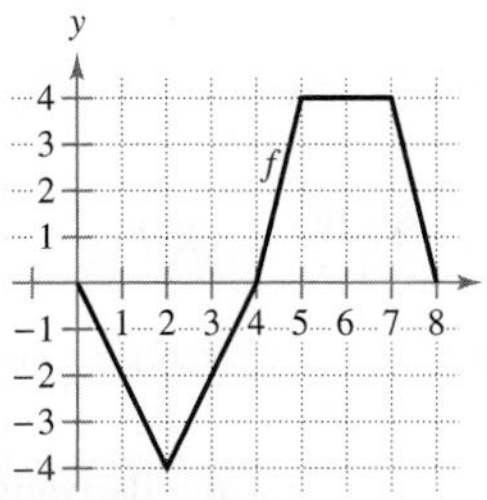

Figure for 74

74. Let $g(x) = \int_0^x f(t)\,dt$, where f is a function whose graph is shown.

(a) Estimate $g(0)$, $g(2)$, $g(4)$, $g(6)$, and $g(8)$.

(b) Find the largest open interval on which g is increasing. Find the largest open interval on which g is decreasing.

(c) Identify any extrema of g.

(d) Sketch a rough graph of g.

In Exercises 75–80, (a) integrate to find F as a function of x and (b) demonstrate the Second Fundamental Theorem of Calculus by differentiating the result in part (a).

75. $F(x) = \int_0^x (t + 2)\,dt$

76. $F(x) = \int_0^x t(t^2 + 1)\,dt$

77. $F(x) = \int_8^x \sqrt[3]{t}\,dt$

78. $F(x) = \int_4^x \sqrt{t}\,dt$

79. $F(x) = \int_{\pi/4}^x \sec^2 t\,dt$

80. $F(x) = \int_{\pi/3}^x \sec t \tan t\,dt$

In Exercises 81–86, use the Second Fundamental Theorem of Calculus to find $F'(x)$.

81. $F(x) = \int_{-2}^x (t^2 - 2t)\,dt$

82. $F(x) = \int_1^x \frac{t^2}{t^2 + 1}\,dt$

83. $F(x) = \int_{-1}^x \sqrt{t^4 + 1}\,dt$

84. $F(x) = \int_1^x \sqrt[4]{t}\,dt$

85. $F(x) = \int_0^x t \cos t\,dt$

86. $F(x) = \int_0^x \sec^3 t\,dt$

In Exercises 87–92, find $F'(x)$.

87. $F(x) = \int_x^{x+2} (4t + 1)\,dt$

88. $F(x) = \int_{-x}^x t^3\,dt$

89. $F(x) = \int_0^{\sin x} \sqrt{t}\,dt$

90. $F(x) = \int_2^{x^2} \frac{1}{t^3}\,dt$

91. $F(x) = \int_0^{x^3} \sin t^2\,dt$

92. $F(x) = \int_0^{x^2} \sin\theta^2\,d\theta$

93. *Graphical Analysis* Approximate the graph of g on the interval $0 \le x \le 4$, where $g(x) = \int_0^x f(t)\,dt$. Identify the x-coordinate of an extremum of g. To print an enlarged copy of the graph, go to the website *www.mathgraphs.com*.

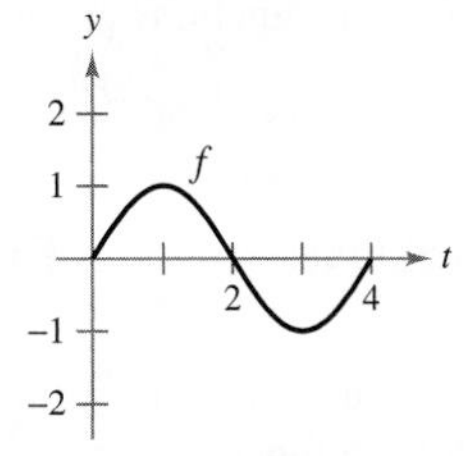

94. Use the graph of the function f shown in the figure on the next page and the function g defined by $g(x) = \int_0^x f(t)\,dt$.

(a) Complete the table.

x	1	2	3	4	5	6	7	8	9	10
$g(x)$										

(b) Plot the points from the table in part (a) and graph g.

(c) Where does g have its minimum? Explain.

(d) Where does g have a maximum? Explain.

(e) On what interval does g increase at the greatest rate? Explain.

(f) Identify the zeros of g.

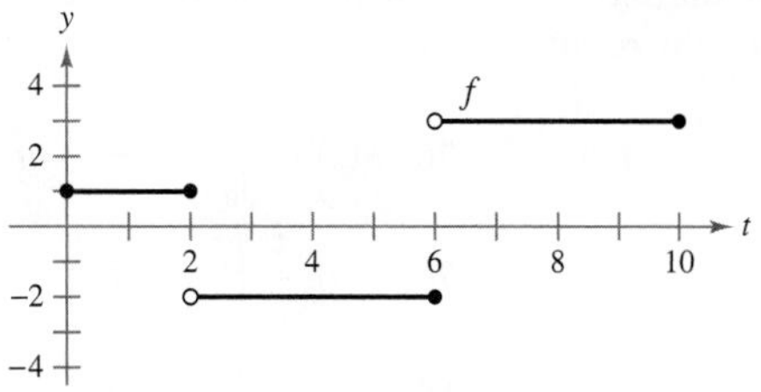

95. ***Cost*** The total cost C (in dollars) of purchasing and maintaining a piece of equipment for x years is

$$C(x) = 5000\left(25 + 3\int_0^x t^{1/4}\,dt\right).$$

(a) Perform the integration to write C as a function of x.

(b) Find $C(1)$, $C(5)$, and $C(10)$.

96. ***Area*** The area A between the graph of the function $g(t) = 4 - 4/t^2$ and the t-axis over the interval $[1, x]$ is

$$A(x) = \int_1^x \left(4 - \frac{4}{t^2}\right) dt.$$

(a) Find the horizontal asymptote of the graph of g.

(b) Integrate to find A as a function of x. Does the graph of A have a horizontal asymptote? Explain.

Rectilinear Motion **In Exercises 97–99, consider a particle moving along the x-axis where $x(t)$ is the position of the particle at time t, $x'(t)$ is its velocity, and $\int_a^b |x'(t)|\,dt$ is the distance the particle travels in the interval of time.**

97. The position function is given by $x(t) = t^3 - 6t^2 + 9t - 2$, $0 \le t \le 5$. Find the total distance the particle travels in 5 units of time.

98. Repeat Exercise 97 for the position function given by $x(t) = (t - 1)(t - 3)^2$, $0 \le t \le 5$.

99. A particle moves along the x-axis with velocity $v(t) = 1/\sqrt{t}$, $t > 0$. At time $t = 1$, its position is $x = 4$. Find the total distance traveled by the particle on the interval $1 \le t \le 4$.

100. ***Buffon's Needle Experiment*** A horizontal plane is ruled with parallel lines 2 inches apart. A two-inch needle is tossed randomly onto the plane. The probability that the needle will touch a line is

$$P = \frac{2}{\pi}\int_0^{\pi/2} \sin\theta\,d\theta$$

where θ is the acute angle between the needle and any one of the parallel lines. Find this probability.

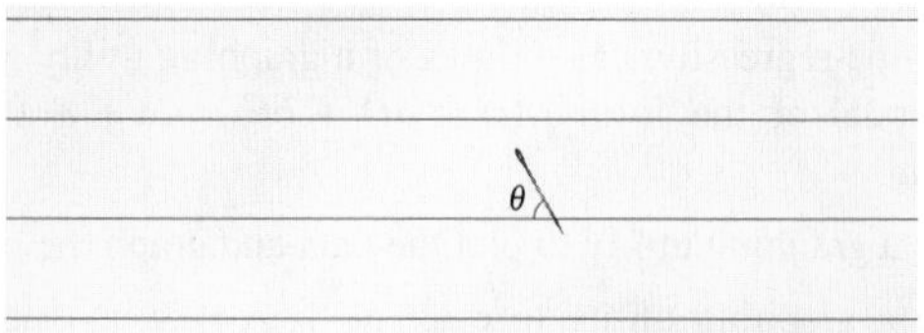

True or False? **In Exercises 101 and 102, determine whether the statement is true or false. If it is false, explain why or give an example that shows it is false.**

101. If $F'(x) = G'(x)$ on the interval $[a, b]$, then $F(b) - F(a) = G(b) - G(a)$.

102. If f is continuous on $[a, b]$, then f is integrable on $[a, b]$.

103. ***Find the Error*** Describe why the statement is incorrect.

$$\int_{-1}^{1} x^{-2}\,dx = \left[-x^{-1}\right]_{-1}^{1} = (-1) - 1 = -2$$

104. Prove that $\dfrac{d}{dx}\left[\displaystyle\int_{u(x)}^{v(x)} f(t)\,dt\right] = f(v(x))v'(x) - f(u(x))u'(x)$.

105. Show that the function

$$f(x) = \int_0^{1/x} \frac{1}{t^2 + 1}\,dt + \int_0^x \frac{1}{t^2 + 1}\,dt$$

is constant for $x > 0$.

106. Let $G(x) = \displaystyle\int_0^x \left[s\int_0^s f(t)dt\right] ds$, where f is continuous for all real t. Find (a) $G(0)$, (b) $G'(0)$, (c) $G''(x)$, and (d) $G''(0)$.

Section Project: Demonstrating the Fundamental Theorem

Use a graphing utility to graph the function $y_1 = \sin^2 t$ on the interval $0 \le t \le \pi$. Let $F(x)$ be the following function of x.

$$F(x) = \int_0^x \sin^2 t\,dt$$

(a) Complete the table. Explain why the values of F are increasing.

x	0	$\pi/6$	$\pi/3$	$\pi/2$	$2\pi/3$	$5\pi/6$	π
$F(x)$							

(b) Use the integration capabilities of a graphing utility to graph F.

(c) Use the differentiation capabilities of a graphing utility to graph $F'(x)$. How is this graph related to the graph in part (b)?

(d) Verify that the derivative of $y = (1/2)t - (\sin 2t)/4$ is $\sin^2 t$. Graph y and write a short paragraph about how this graph is related to those in parts (b) and (c).

Section 4.5 Integration by Substitution

- **Use pattern recognition to find an indefinite integral.**
- **Use a change of variables to find an indefinite integral.**
- **Use the General Power Rule for Integration to find an indefinite integral.**
- **Use a change of variables to evaluate a definite integral.**
- **Evaluate a definite integral involving an even or odd function.**

Pattern Recognition

In this section you will study techniques for integrating composite functions. The discussion is split into two parts—*pattern recognition* and *change of variables*. Both techniques involve a ***u*-substitution.** With pattern recognition you perform the substitution mentally, and with change of variables you write the substitution steps.

The role of substitution in integration is comparable to the role of the Chain Rule in differentiation. Recall that for differentiable functions given by $y = F(u)$ and $u = g(x)$, the Chain Rule states that

$$\frac{d}{dx}[F(g(x))] = F'(g(x))g'(x).$$

From the definition of an antiderivative, it follows that

$$\begin{aligned}\int F'(g(x))g'(x)\,dx &= F(g(x)) + C\\ &= F(u) + C.\end{aligned}$$

These results are summarized in the following theorem.

NOTE The statement of Theorem 4.12 doesn't tell how to distinguish between $f(g(x))$ and $g'(x)$ in the integrand. As you become more experienced at integration, your skill in doing this will increase. Of course, part of the key is familiarity with derivatives.

THEOREM 4.12 Antidifferentiation of a Composite Function

Let g be a function whose range is an interval I, and let f be a function that is continuous on I. If g is differentiable on its domain and F is an antiderivative of f on I, then

$$\int f(g(x))g'(x)\,dx = F(g(x)) + C.$$

If $u = g(x)$, then $du = g'(x)\,dx$ and

$$\int f(u)\,du = F(u) + C.$$

STUDY TIP There are several techniques for applying substitution, each differing slightly from the others. However, you should remember that the goal is the same with every technique—*you are trying to find an antiderivative of the integrand.*

EXPLORATION

Recognizing Patterns The integrand in each of the following integrals fits the pattern $f(g(x))g'(x)$. Identify the pattern and use the result to evaluate the integral.

a. $\int 2x(x^2 + 1)^4\,dx$ **b.** $\int 3x^2\sqrt{x^3 + 1}\,dx$ **c.** $\int \sec^2 x(\tan x + 3)\,dx$

The next three integrals are similar to the first three. Show how you can multiply and divide by a constant to evaluate these integrals.

d. $\int x(x^2 + 1)^4\,dx$ **e.** $\int x^2\sqrt{x^3 + 1}\,dx$ **f.** $\int 2\sec^2 x(\tan x + 3)\,dx$

Examples 1 and 2 show how to apply Theorem 4.12 *directly*, by recognizing the presence of $f(g(x))$ and $g'(x)$. Note that the composite function in the integrand has an *outside function* f and an *inside function* g. Moreover, the derivative $g'(x)$ is present as a factor of the integrand.

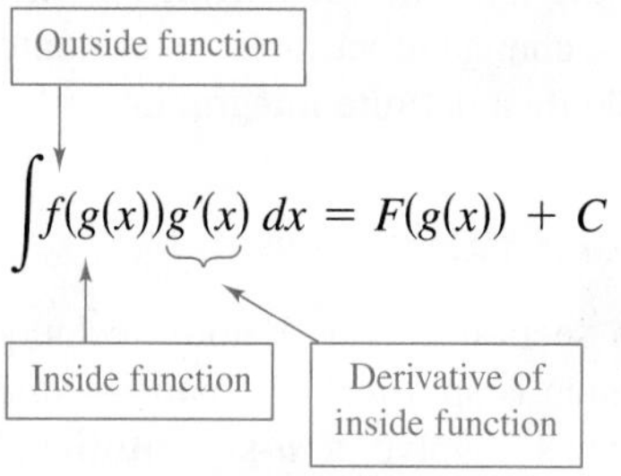

EXAMPLE 1 Recognizing the $f(g(x))g'(x)$ Pattern

Find $\int (x^2 + 1)^2(2x)\, dx$.

Solution Letting $g(x) = x^2 + 1$, you obtain

$$g'(x) = 2x$$

and

$$f(g(x)) = f(x^2 + 1) = (x^2 + 1)^2.$$

From this, you can recognize that the integrand follows the $f(g(x))g'(x)$ pattern. Using the Power Rule for Integration and Theorem 4.12, you can write

$$\int \overbrace{(x^2 + 1)^2}^{f(g(x))}\overbrace{(2x)}^{g'(x)}\, dx = \frac{1}{3}(x^2 + 1)^3 + C.$$

Try using the Chain Rule to check that the derivative of $\frac{1}{3}(x^2 + 1)^3 + C$ is the integrand of the original integral.

TECHNOLOGY Try using a computer algebra system, such as *Maple*, *Derive*, *Mathematica*, *Mathcad*, or the *TI-89*, to solve the integrals given in Examples 1 and 2. Do you obtain the same antiderivatives that are listed in the examples?

EXAMPLE 2 Recognizing the $f(g(x))g'(x)$ Pattern

Find $\int 5 \cos 5x\, dx$.

Solution Letting $g(x) = 5x$, you obtain

$$g'(x) = 5$$

and

$$f(g(x)) = f(5x) = \cos 5x.$$

From this, you can recognize that the integrand follows the $f(g(x))g'(x)$ pattern. Using the Cosine Rule for Integration and Theorem 4.12, you can write

$$\int \overbrace{(\cos 5x)}^{f(g(x))}\overbrace{(5)}^{g'(x)}\, dx = \sin 5x + C.$$

You can check this by differentiating $\sin 5x + C$ to obtain the original integrand.

The integrands in Examples 1 and 2 fit the $f(g(x))g'(x)$ pattern exactly—you only had to recognize the pattern. You can extend this technique considerably with the Constant Multiple Rule

$$\int kf(x)\,dx = k\int f(x)\,dx.$$

Many integrands contain the essential part (the variable part) of $g'(x)$ but are missing a constant multiple. In such cases, you can multiply and divide by the necessary constant multiple, as shown in Example 3.

EXAMPLE 3 Multiplying and Dividing by a Constant

Find $\int x(x^2+1)^2\,dx$.

Solution This is similar to the integral given in Example 1, except that the integrand is missing a factor of 2. Recognizing that $2x$ is the derivative of x^2+1, you can let $g(x) = x^2+1$ and supply the $2x$ as follows.

$$\int x(x^2+1)^2\,dx = \int (x^2+1)^2\left(\frac{1}{2}\right)(2x)\,dx \qquad \text{Multiply and divide by 2.}$$

$$= \frac{1}{2}\int \overbrace{(x^2+1)^2}^{f(g(x))}\,\overbrace{(2x)}^{g'(x)}\,dx \qquad \text{Constant Multiple Rule}$$

$$= \frac{1}{2}\left[\frac{(x^2+1)^3}{3}\right] + C \qquad \text{Integrate.}$$

$$= \frac{1}{6}(x^2+1)^3 + C \qquad \text{Simplify.}$$

In practice, most people would not write as many steps as are shown in Example 3. For instance, you could evaluate the integral by simply writing

$$\int x(x^2+1)^2\,dx = \frac{1}{2}\int (x^2+1)^2\,2x\,dx$$

$$= \frac{1}{2}\left[\frac{(x^2+1)^3}{3}\right] + C$$

$$= \frac{1}{6}(x^2+1)^3 + C.$$

NOTE Be sure you see that the *Constant* Multiple Rule applies only to *constants*. You cannot multiply and divide by a variable and then move the variable outside the integral sign. For instance,

$$\int (x^2+1)^2\,dx \neq \frac{1}{2x}\int (x^2+1)^2(2x)\,dx.$$

After all, if it were legitimate to move variable quantities outside the integral sign, you could move the entire integrand out and simplify the whole process. But the result would be incorrect.

Change of Variables

With a formal **change of variables,** you completely rewrite the integral in terms of u and du (or any other convenient variable). Although this procedure can involve more written steps than the pattern recognition illustrated in Examples 1 to 3, it is useful for complicated integrands. The change of variable technique uses the Leibniz notation for the differential. That is, if $u = g(x)$, then $du = g'(x)\,dx$, and the integral in Theorem 4.12 takes the form

$$\int f(g(x))g'(x)\,dx = \int f(u)\,du = F(u) + C.$$

EXAMPLE 4 Change of Variables

Find $\int \sqrt{2x-1}\,dx$.

Solution First, let u be the inner function, $u = 2x - 1$. Then calculate the differential du to be $du = 2\,dx$. Now, using $\sqrt{2x-1} = \sqrt{u}$ and $dx = du/2$, substitute to obtain

$$\begin{aligned}
\int \sqrt{2x-1}\,dx &= \int \sqrt{u}\left(\frac{du}{2}\right) && \text{Integral in terms of } u\\
&= \frac{1}{2}\int u^{1/2}\,du && \text{Constant Multiple Rule}\\
&= \frac{1}{2}\left(\frac{u^{3/2}}{3/2}\right) + C && \text{Antiderivative in terms of } u\\
&= \frac{1}{3}u^{3/2} + C && \text{Simplify.}\\
&= \frac{1}{3}(2x-1)^{3/2} + C. && \text{Antiderivative in terms of } x
\end{aligned}$$

STUDY TIP Because integration is usually more difficult than differentiation, you should always check your answer to an integration problem by differentiating. For instance, in Example 4 you should differentiate $\frac{1}{3}(2x-1)^{3/2} + C$ to verify that you obtain the original integrand.

EXAMPLE 5 Change of Variables

Find $\int x\sqrt{2x-1}\,dx$.

Solution As in the previous example, let $u = 2x - 1$ and obtain $dx = du/2$. Because the integrand contains a factor of x, you must also solve for x in terms of u, as shown.

$$u = 2x - 1 \quad \Longrightarrow \quad x = (u+1)/2 \qquad \text{Solve for } x \text{ in terms of } u.$$

Now, using substitution, you obtain

$$\begin{aligned}
\int x\sqrt{2x-1}\,dx &= \int \left(\frac{u+1}{2}\right)u^{1/2}\left(\frac{du}{2}\right)\\
&= \frac{1}{4}\int (u^{3/2} + u^{1/2})\,du\\
&= \frac{1}{4}\left(\frac{u^{5/2}}{5/2} + \frac{u^{3/2}}{3/2}\right) + C\\
&= \frac{1}{10}(2x-1)^{5/2} + \frac{1}{6}(2x-1)^{3/2} + C.
\end{aligned}$$

To complete the change of variables in Example 5, you solved for x in terms of u. Sometimes this is very difficult. Fortunately it is not always necessary, as shown in the next example.

EXAMPLE 6 Change of Variables

Find $\int \sin^2 3x \cos 3x \, dx$.

Solution Because $\sin^2 3x = (\sin 3x)^2$, you can let $u = \sin 3x$. Then

$$du = (\cos 3x)(3)\, dx.$$

Now, because $\cos 3x\, dx$ is part of the original integral, you can write

$$\frac{du}{3} = \cos 3x \, dx.$$

Substituting u and $du/3$ in the original integral yields

$$\begin{aligned}\int \sin^2 3x \cos 3x \, dx &= \int u^2 \frac{du}{3}\\ &= \frac{1}{3}\int u^2\, du\\ &= \frac{1}{3}\left(\frac{u^3}{3}\right) + C\\ &= \frac{1}{9}\sin^3 3x + C.\end{aligned}$$

You can check this by differentiating.

$$\begin{aligned}\frac{d}{dx}\left[\frac{1}{9}\sin^3 3x\right] &= \left(\frac{1}{9}\right)(3)(\sin 3x)^2(\cos 3x)(3)\\ &= \sin^2 3x \cos 3x\end{aligned}$$

Because differentiation produces the original integrand, you know that you have obtained the correct antiderivative.

STUDY TIP When making a change of variables, be sure that your answer is written using the same variables as in the original integrand. For instance, in Example 6, you should not leave your answer as

$$\tfrac{1}{9}u^3 + C$$

but rather, replace u by $\sin 3x$.

The steps used for integration by substitution are summarized in the following guidelines.

Guidelines for Making a Change of Variables

1. Choose a substitution $u = g(x)$. Usually, it is best to choose the *inner* part of a composite function, such as a quantity raised to a power.
2. Compute $du = g'(x)\, dx$.
3. Rewrite the integral in terms of the variable u.
4. Find the resulting integral in terms of u.
5. Replace u by $g(x)$ to obtain an antiderivative in terms of x.
6. Check your answer by differentiating.

The General Power Rule for Integration

One of the most common u-substitutions involves quantities in the integrand that are raised to a power. Because of the importance of this type of substitution, it is given a special name—the **General Power Rule for Integration.** A proof of this rule follows directly from the (simple) Power Rule for Integration, together with Theorem 4.12.

> **THEOREM 4.13 The General Power Rule for Integration**
>
> If g is a differentiable function of x, then
>
> $$\int [g(x)]^n g'(x)\,dx = \frac{[g(x)]^{n+1}}{n+1} + C, \qquad n \neq -1.$$
>
> Equivalently, if $u = g(x)$, then
>
> $$\int u^n\,du = \frac{u^{n+1}}{n+1} + C, \qquad n \neq -1.$$

EXAMPLE 7 Substitution and the General Power Rule

a. $$\int 3(3x-1)^4\,dx = \int \overbrace{(3x-1)^4}^{u^4}\overbrace{(3)\,dx}^{du} = \overbrace{\frac{(3x-1)^5}{5}}^{u^5/5} + C$$

b. $$\int (2x+1)(x^2+x)\,dx = \int \overbrace{(x^2+x)^1}^{u^1}\overbrace{(2x+1)\,dx}^{du} = \overbrace{\frac{(x^2+x)^2}{2}}^{u^2/2} + C$$

c. $$\int 3x^2\sqrt{x^3-2}\,dx = \int \overbrace{(x^3-2)^{1/2}}^{u^{1/2}}\overbrace{(3x^2)\,dx}^{du} = \overbrace{\frac{(x^3-2)^{3/2}}{3/2}}^{u^{3/2}/(3/2)} + C = \frac{2}{3}(x^3-2)^{3/2} + C$$

d. $$\int \frac{-4x}{(1-2x^2)^2}\,dx = \int \overbrace{(1-2x^2)^{-2}}^{u^{-2}}\overbrace{(-4x)\,dx}^{du} = \overbrace{\frac{(1-2x^2)^{-1}}{-1}}^{u^{-1}/(-1)} + C = -\frac{1}{1-2x^2} + C$$

e. $$\int \cos^2 x \sin x\,dx = -\int \overbrace{(\cos x)^2}^{u^2}\overbrace{(-\sin x)\,dx}^{du} = -\overbrace{\frac{(\cos x)^3}{3}}^{u^3/3} + C$$

Some integrals whose integrands involve quantities raised to powers cannot be found by the General Power Rule. Consider the two integrals

$$\int x(x^2+1)^2\,dx \quad \text{and} \quad \int (x^2+1)^2\,dx.$$

The substitution $u = x^2 + 1$ works in the first integral but not in the second. In the second, the substitution fails because the integrand lacks the factor x needed for du. Fortunately, *for this particular integral*, you can expand the integrand as $(x^2+1)^2 = x^4 + 2x^2 + 1$ and use the (simple) Power Rule to integrate each term.

EXPLORATION

Suppose you were asked to find one of the following integrals. Which one would you choose? Explain your reasoning.

a. $\int \sqrt{x^3+1}\,dx$ or

$\int x^2\sqrt{x^3+1}\,dx$

b. $\int \tan(3x)\sec^2(3x)\,dx$ or

$\int \tan(3x)\,dx$

Change of Variables for Definite Integrals

When using u-substitution with a definite integral, it is often convenient to determine the limits of integration for the variable u rather than to convert the antiderivative back to the variable x and evaluate at the original limits. This change of variables is stated explicitly in the next theorem. The proof follows from Theorem 4.12 combined with the Fundamental Theorem of Calculus.

THEOREM 4.14 Change of Variables for Definite Integrals

If the function $u = g(x)$ has a continuous derivative on the closed interval $[a, b]$ and f is continuous on the range of g, then

$$\int_a^b f(g(x))g'(x)\,dx = \int_{g(a)}^{g(b)} f(u)\,du.$$

EXAMPLE 8 Change of Variables

Evaluate $\int_0^1 x(x^2 + 1)^3\,dx$.

Solution To evaluate this integral, let $u = x^2 + 1$. Then, you obtain

$$u = x^2 + 1 \Rightarrow du = 2x\,dx.$$

Before substituting, determine the new upper and lower limits of integration.

Lower Limit	*Upper Limit*
When $x = 0$, $u = 0^2 + 1 = 1$.	When $x = 1$, $u = 1^2 + 1 = 2$.

Now, you can substitute to obtain

$$\int_0^1 x(x^2 + 1)^3\,dx = \frac{1}{2}\int_0^1 (x^2 + 1)^3(2x)\,dx \qquad \text{Integration limits for } x$$

$$= \frac{1}{2}\int_1^2 u^3\,du \qquad \text{Integration limits for } u$$

$$= \frac{1}{2}\left[\frac{u^4}{4}\right]_1^2$$

$$= \frac{1}{2}\left(4 - \frac{1}{4}\right)$$

$$= \frac{15}{8}.$$

Try rewriting the antiderivative $\frac{1}{2}(u^4/4)$ in terms of the variable x and evaluate the definite integral at the original limits of integration, as shown.

$$\frac{1}{2}\left[\frac{u^4}{4}\right]_1^2 = \frac{1}{2}\left[\frac{(x^2 + 1)^4}{4}\right]_0^1$$

$$= \frac{1}{2}\left(4 - \frac{1}{4}\right) = \frac{15}{8}$$

Notice that you obtain the same result.

EXAMPLE 9 Change of Variables

Evaluate $A = \displaystyle\int_1^5 \frac{x}{\sqrt{2x - 1}}\, dx$.

Solution To evaluate this integral, let $u = \sqrt{2x - 1}$. Then, you obtain

$$u^2 = 2x - 1$$

$$u^2 + 1 = 2x$$

$$\frac{u^2 + 1}{2} = x$$

$$u\, du = dx. \qquad \text{Differentiate each side.}$$

Before substituting, determine the new upper and lower limits of integration.

Lower Limit	*Upper Limit*
When $x = 1$, $u = \sqrt{2 - 1} = 1$.	When $x = 5$, $u = \sqrt{10 - 1} = 3$.

Now, substitute to obtain

$$\begin{aligned}\int_1^5 \frac{x}{\sqrt{2x - 1}}\, dx &= \int_1^3 \frac{1}{u}\left(\frac{u^2 + 1}{2}\right) u\, du \\ &= \frac{1}{2}\int_1^3 (u^2 + 1)\, du \\ &= \frac{1}{2}\left[\frac{u^3}{3} + u\right]_1^3 \\ &= \frac{1}{2}\left(9 + 3 - \frac{1}{3} - 1\right) \\ &= \frac{16}{3}.\end{aligned}$$

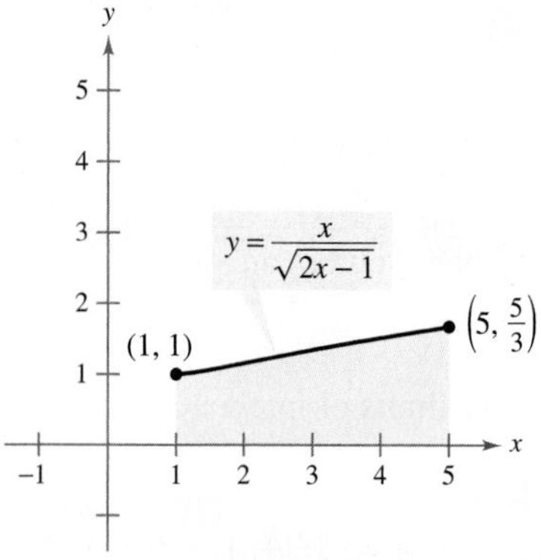

The region before substitution has an area of $\frac{16}{3}$.
Figure 4.37

Geometrically, you can interpret the equation

$$\int_1^5 \frac{x}{\sqrt{2x - 1}}\, dx = \int_1^3 \frac{u^2 + 1}{2}\, du$$

to mean that the two *different* regions shown in Figures 4.37 and 4.38 have the *same* area.

When evaluating definite integrals by substitution, it is possible for the upper limit of integration of the u-variable form to be smaller than the lower limit. If this happens, don't rearrange the limits. Simply evaluate as usual. For example, after substituting $u = \sqrt{1 - x}$ in the integral

$$\int_0^1 x^2(1 - x)^{1/2}\, dx$$

you obtain $u = \sqrt{1 - 1} = 0$ when $x = 1$, and $u = \sqrt{1 - 0} = 1$ when $x = 0$. So, the correct u-variable form of this integral is

$$-2\int_1^0 (1 - u^2)^2 u^2\, du.$$

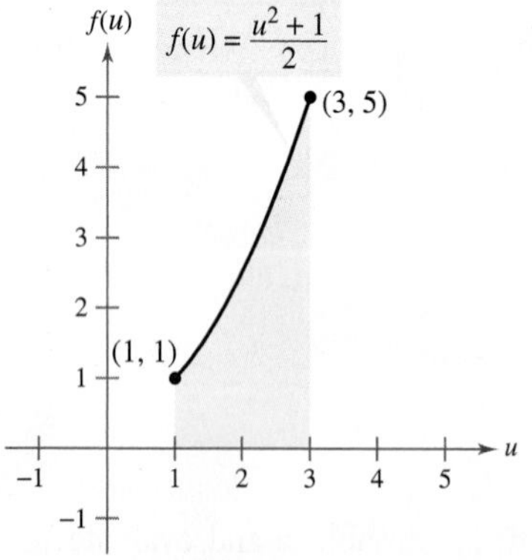

The region after substitution has an area of $\frac{16}{3}$.
Figure 4.38

Integration of Even and Odd Functions

Even with a change of variables, integration can be difficult. Occasionally, you can simplify the evaluation of a definite integral (over an interval that is symmetric about the y-axis or about the origin) by recognizing the integrand to be an even or odd function (see Figure 4.39).

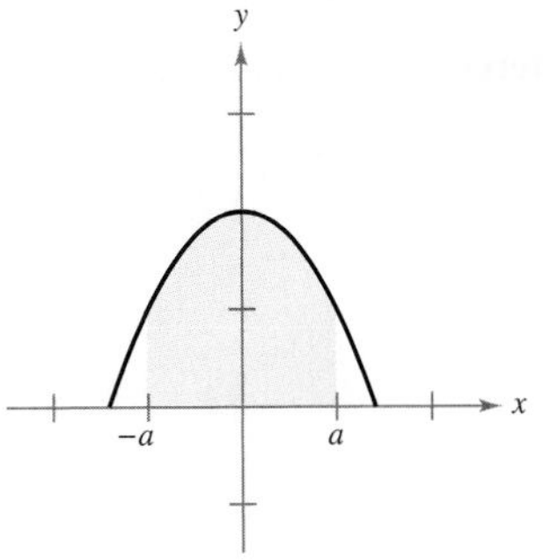

Even function

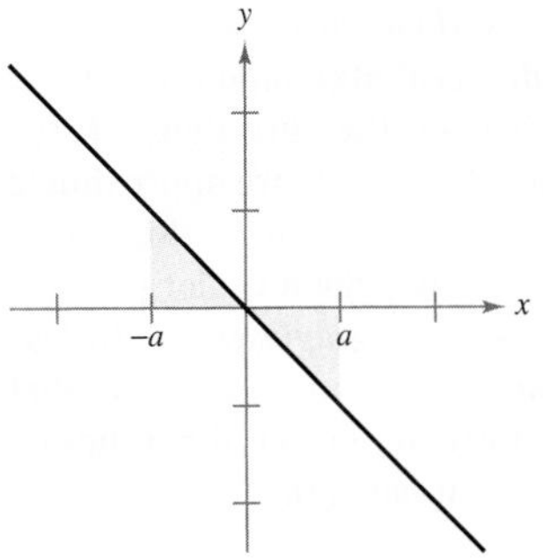

Odd function
Figure 4.39

> **THEOREM 4.15 Integration of Even and Odd Functions**
>
> Let f be integrable on the closed interval $[-a, a]$.
>
> **1.** If f is an *even* function, then $\displaystyle\int_{-a}^{a} f(x)\,dx = 2\int_{0}^{a} f(x)\,dx.$
>
> **2.** If f is an *odd* function, then $\displaystyle\int_{-a}^{a} f(x)\,dx = 0.$

Proof Because f is even, you know that $f(x) = f(-x)$. Using Theorem 4.12 with the substitution $u = -x$ produces

$$\int_{-a}^{0} f(x)\,dx = \int_{a}^{0} f(-u)(-du) = -\int_{a}^{0} f(u)\,du = \int_{0}^{a} f(u)\,du = \int_{0}^{a} f(x)\,dx.$$

Finally, using Theorem 4.6, you obtain

$$\begin{aligned}\int_{-a}^{a} f(x)\,dx &= \int_{-a}^{0} f(x)\,dx + \int_{0}^{a} f(x)\,dx\\ &= \int_{0}^{a} f(x)\,dx + \int_{0}^{a} f(x)\,dx = 2\int_{0}^{a} f(x)\,dx.\end{aligned}$$

This proves the first property. The proof of the second property is left to you (see Exercise 133).

EXAMPLE 10 Integration of an Odd Function

Evaluate $\displaystyle\int_{-\pi/2}^{\pi/2} (\sin^3 x \cos x + \sin x \cos x)\,dx.$

Solution Letting $f(x) = \sin^3 x \cos x + \sin x \cos x$ produces

$$\begin{aligned}f(-x) &= \sin^3(-x)\cos(-x) + \sin(-x)\cos(-x)\\ &= -\sin^3 x \cos x - \sin x \cos x = -f(x).\end{aligned}$$

So, f is an odd function, and because f is symmetric about the origin over $[-\pi/2, \pi/2]$, you can apply Theorem 4.15 to conclude that

$$\int_{-\pi/2}^{\pi/2} (\sin^3 x \cos x + \sin x \cos x)\,dx = 0.$$

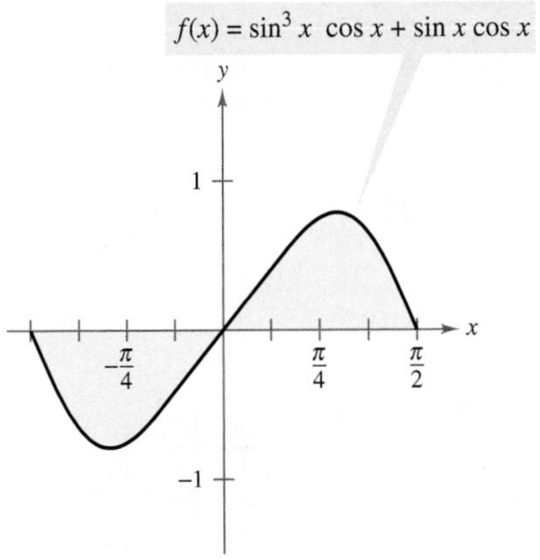

Because f is an odd function,

$\displaystyle\int_{-\pi/2}^{\pi/2} f(x)\,dx = 0.$

Figure 4.40

NOTE From Figure 4.40 you can see that the two regions on either side of the y-axis have the same area. However, because one lies below the x-axis and one lies above it, integration produces a cancellation effect. (More will be said about this in Section 7.1.)

Exercises for Section 4.5

See www.CalcChat.com for worked-out solutions to odd-numbered exercises.

In Exercises 1–6, complete the table by identifying u and du for the integral.

$\int f(g(x))g'(x)\,dx$	$u = g(x)$	$du = g'(x)\,dx$
1. $\int (5x^2 + 1)^2(10x)\,dx$		
2. $\int x^2\sqrt{x^3 + 1}\,dx$		
3. $\int \frac{x}{\sqrt{x^2 + 1}}\,dx$		
4. $\int \sec 2x \tan 2x\,dx$		
5. $\int \tan^2 x \sec^2 x\,dx$		
6. $\int \frac{\cos x}{\sin^2 x}\,dx$		

In Exercises 7–34, find the indefinite integral and check the result by differentiation.

7. $\int (1 + 2x)^4(2)\,dx$

8. $\int (x^2 - 9)^3(2x)\,dx$

9. $\int \sqrt{9 - x^2}\,(-2x)\,dx$

10. $\int \sqrt[3]{(1 - 2x^2)}(-4x)\,dx$

11. $\int x^3(x^4 + 3)^2\,dx$

12. $\int x^2(x^3 + 5)^4\,dx$

13. $\int x^2(x^3 - 1)^4\,dx$

14. $\int x(4x^2 + 3)^3\,dx$

15. $\int t\sqrt{t^2 + 2}\,dt$

16. $\int t^3\sqrt{t^4 + 5}\,dt$

17. $\int 5x\,\sqrt[3]{1 - x^2}\,dx$

18. $\int u^2\sqrt{u^3 + 2}\,du$

19. $\int \frac{x}{(1 - x^2)^3}\,dx$

20. $\int \frac{x^3}{(1 + x^4)^2}\,dx$

21. $\int \frac{x^2}{(1 + x^3)^2}\,dx$

22. $\int \frac{x^2}{(16 - x^3)^2}\,dx$

23. $\int \frac{x}{\sqrt{1 - x^2}}\,dx$

24. $\int \frac{x^3}{\sqrt{1 + x^4}}\,dx$

25. $\int \left(1 + \frac{1}{t}\right)^3\left(\frac{1}{t^2}\right)dt$

26. $\int \left[x^2 + \frac{1}{(3x)^2}\right]dx$

27. $\int \frac{1}{\sqrt{2x}}\,dx$

28. $\int \frac{1}{2\sqrt{x}}\,dx$

29. $\int \frac{x^2 + 3x + 7}{\sqrt{x}}\,dx$

30. $\int \frac{t + 2t^2}{\sqrt{t}}\,dt$

31. $\int t^2\left(t - \frac{2}{t}\right)dt$

32. $\int \left(\frac{t^3}{3} + \frac{1}{4t^2}\right)dt$

33. $\int (9 - y)\sqrt{y}\,dy$

34. $\int 2\pi y(8 - y^{3/2})\,dy$

In Exercises 35–38, solve the differential equation.

35. $\frac{dy}{dx} = 4x + \frac{4x}{\sqrt{16 - x^2}}$

36. $\frac{dy}{dx} = \frac{10x^2}{\sqrt{1 + x^3}}$

37. $\frac{dy}{dx} = \frac{x + 1}{(x^2 + 2x - 3)^2}$

38. $\frac{dy}{dx} = \frac{x - 4}{\sqrt{x^2 - 8x + 1}}$

Slope Fields **In Exercises 39–42, a differential equation, a point, and a slope field are given. A *slope field* consists of line segments with slopes given by the differential equation. These line segments give a visual perspective of the directions of the solutions of the differential equation. (a) Sketch two approximate solutions of the differential equation on the slope field, one of which passes through the given point. (To print an enlarged copy of the graph, go to the website *www.mathgraphs.com*.) (b) Use integration to find the particular solution of the differential equation and use a graphing utility to graph the solution. Compare the result with the sketches in part (a).**

39. $\frac{dy}{dx} = x\sqrt{4 - x^2}$

$(2, 2)$

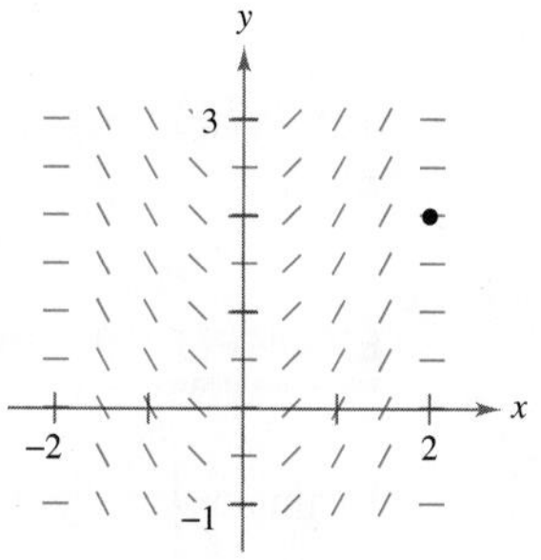

40. $\frac{dy}{dx} = x^2(x^3 - 1)^2$

$(1, 0)$

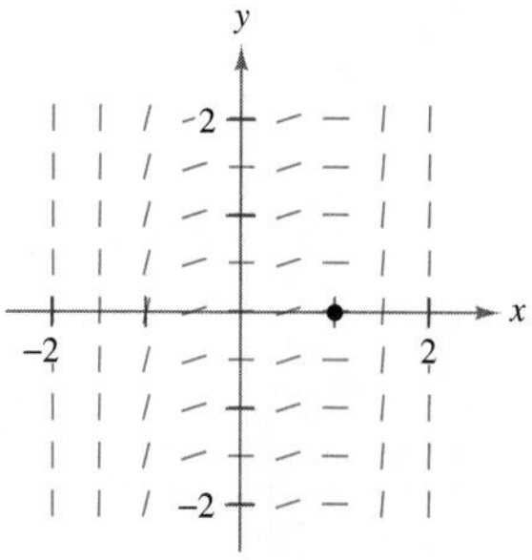

41. $\frac{dy}{dx} = x\cos x^2$

$(0, 1)$

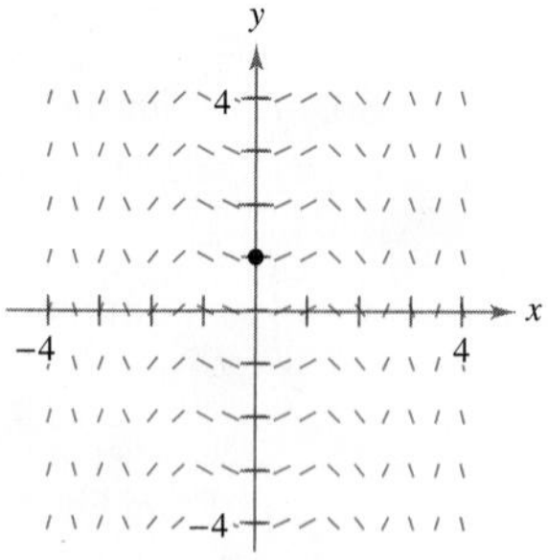

42. $\frac{dy}{dx} = -2\sec(2x)\tan(2x)$

$(0, -1)$

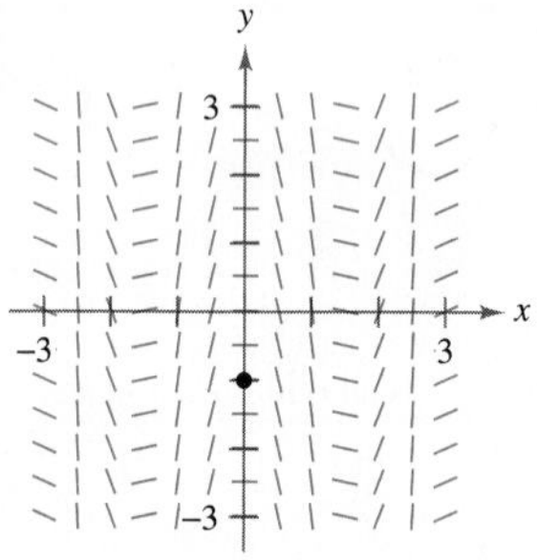

In Exercises 43–56, find the indefinite integral.

43. $\int \pi \sin \pi x \, dx$

44. $\int 4x^3 \sin x^4 \, dx$

45. $\int \sin 2x \, dx$

46. $\int \cos 6x \, dx$

47. $\int \frac{1}{\theta^2} \cos \frac{1}{\theta} \, d\theta$

48. $\int x \sin x^2 \, dx$

49. $\int \sin 2x \cos 2x \, dx$

50. $\int \sec(1 - x) \tan(1 - x) \, dx$

51. $\int \tan^4 x \sec^2 x \, dx$

52. $\int \sqrt{\tan x} \sec^2 x \, dx$

53. $\int \frac{\csc^2 x}{\cot^3 x} \, dx$

54. $\int \frac{\sin x}{\cos^3 x} \, dx$

55. $\int \cot^2 x \, dx$

56. $\int \csc^2\left(\frac{x}{2}\right) dx$

In Exercises 57–62, find an equation for the function f that has the given derivative and whose graph passes through the given point.

Derivative	*Point*
57. $f'(x) = \cos \frac{x}{2}$	$(0, 3)$
58. $f'(x) = \pi \sec \pi x \tan \pi x$	$(\frac{1}{3}, 1)$
59. $f'(x) = \sin 4x$	$\left(\frac{\pi}{4}, -\frac{3}{4}\right)$
60. $f'(x) = \sec^2(2x)$	$\left(\frac{\pi}{2}, 2\right)$
61. $f'(x) = 2x(4x^2 - 10)^2$	$(2, 10)$
62. $f'(x) = -2x\sqrt{8 - x^2}$	$(2, 7)$

In Exercises 63–70, find the indefinite integral by the method shown in Example 5.

63. $\int x\sqrt{x + 2} \, dx$

64. $\int x\sqrt{2x + 1} \, dx$

65. $\int x^2\sqrt{1 - x} \, dx$

66. $\int (x + 1)\sqrt{2 - x} \, dx$

67. $\int \frac{x^2 - 1}{\sqrt{2x - 1}} \, dx$

68. $\int \frac{2x + 1}{\sqrt{x + 4}} \, dx$

69. $\int \frac{-x}{(x + 1) - \sqrt{x + 1}} \, dx$

70. $\int t \sqrt[3]{t - 4} \, dt$

In Exercises 71–82, evaluate the definite integral. Use a graphing utility to verify your result.

71. $\int_{-1}^{1} x(x^2 + 1)^3 \, dx$

72. $\int_{-2}^{4} x^2(x^3 + 8)^2 \, dx$

73. $\int_{1}^{2} 2x^2\sqrt{x^3 + 1} \, dx$

74. $\int_{0}^{1} x\sqrt{1 - x^2} \, dx$

75. $\int_{0}^{4} \frac{1}{\sqrt{2x + 1}} \, dx$

76. $\int_{0}^{2} \frac{x}{\sqrt{1 + 2x^2}} \, dx$

77. $\int_{1}^{9} \frac{1}{\sqrt{x}\left(1 + \sqrt{x}\right)^2} \, dx$

78. $\int_{0}^{2} x \sqrt[3]{4 + x^2} \, dx$

79. $\int_{1}^{2} (x - 1)\sqrt{2 - x} \, dx$

80. $\int_{1}^{5} \frac{x}{\sqrt{2x - 1}} \, dx$

81. $\int_{0}^{\pi/2} \cos\left(\frac{2x}{3}\right) dx$

82. $\int_{\pi/3}^{\pi/2} (x + \cos x) \, dx$

Differential Equations **In Exercises 83–86, the graph of a function f is shown. Use the differential equation and the given point to find an equation of the function.**

83. $\frac{dy}{dx} = 18x^2(2x^3 + 1)^2$

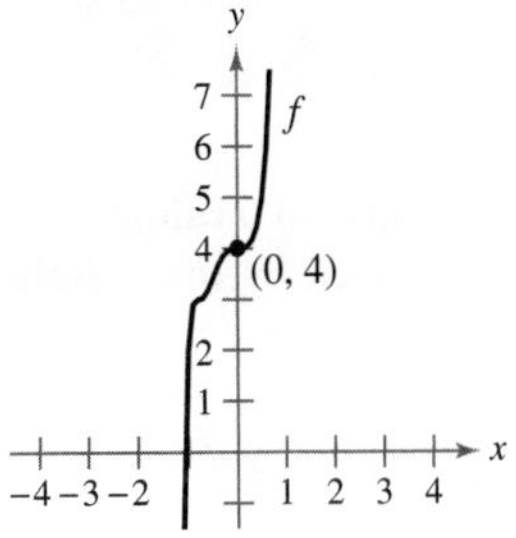

84. $\frac{dy}{dx} = \frac{-48}{(3x + 5)^3}$

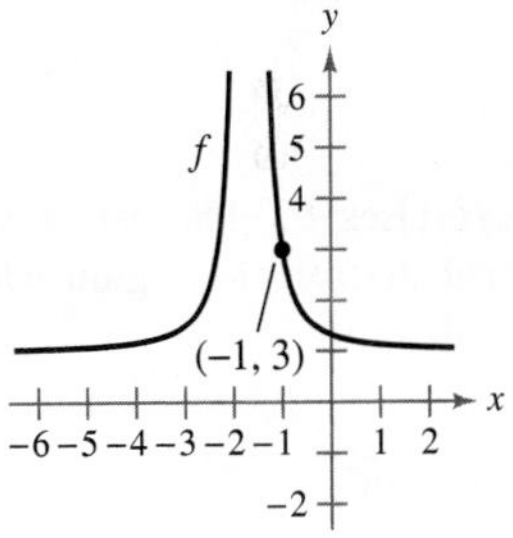

85. $\frac{dy}{dx} = \frac{2x}{\sqrt{2x^2 - 1}}$

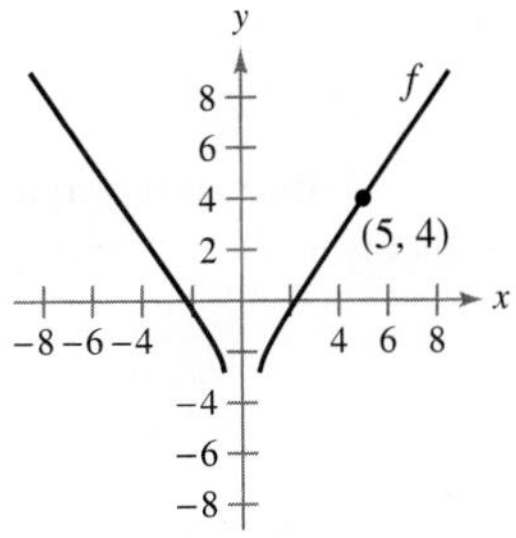

86. $\frac{dy}{dx} = 4x + \frac{9x^2}{(3x^3 + 1)^{(3/2)}}$

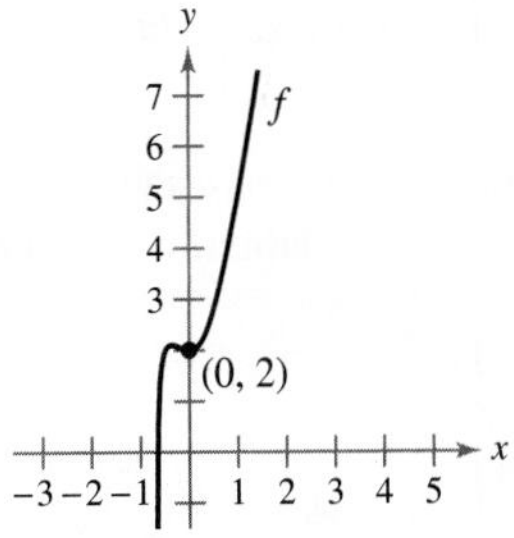

In Exercises 87–92, find the area of the region. Use a graphing utility to verify your result.

87. $\int_{0}^{7} x \sqrt[3]{x + 1} \, dx$

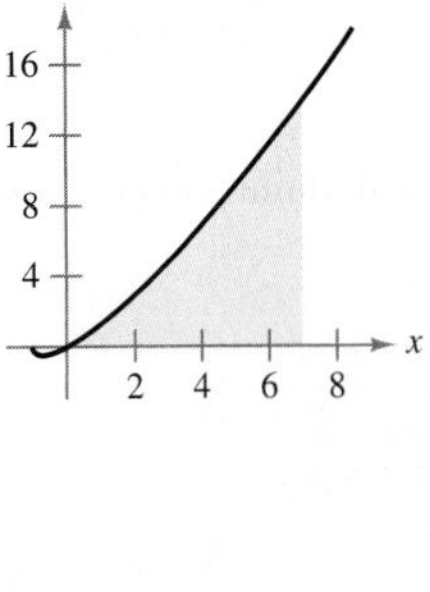

88. $\int_{-2}^{6} x^2 \sqrt[3]{x + 2} \, dx$

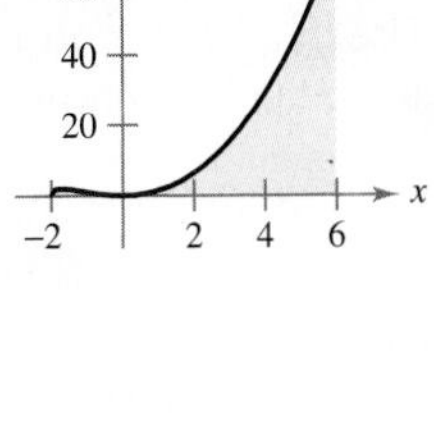

89. $y = 2\sin x + \sin 2x$

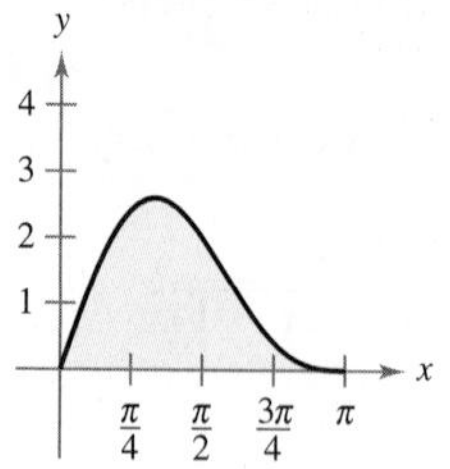

90. $y = \sin x + \cos 2x$

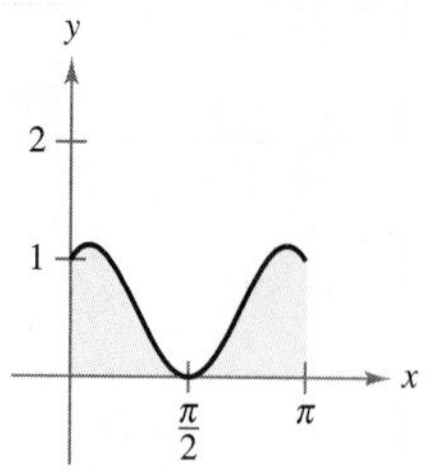

91. $\displaystyle\int_{\pi/2}^{2\pi/3} \sec^2\left(\frac{x}{2}\right) dx$

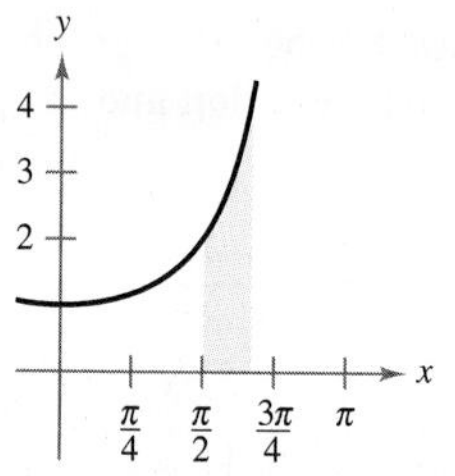

92. $\displaystyle\int_{\pi/12}^{\pi/4} \csc 2x \cot 2x \, dx$

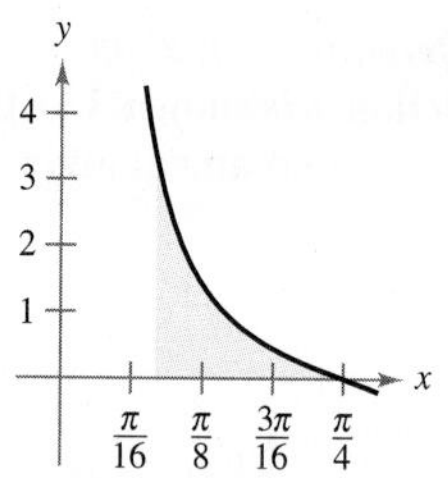

In Exercises 93–98, use a graphing utility to evaluate the integral. Graph the region whose area is given by the definite integral.

93. $\displaystyle\int_0^4 \frac{x}{\sqrt{2x+1}}\, dx$ **94.** $\displaystyle\int_0^2 x^3\sqrt{x+2}\, dx$

95. $\displaystyle\int_3^7 x\sqrt{x-3}\, dx$ **96.** $\displaystyle\int_1^5 x^2\sqrt{x-1}\, dx$

97. $\displaystyle\int_0^3 \left(\theta + \cos\frac{\theta}{6}\right) d\theta$ **98.** $\displaystyle\int_0^{\pi/2} \sin 2x\, dx$

Writing **In Exercises 99 and 100, find the indefinite integral in two ways. Explain any difference in the forms of the answers.**

99. $\displaystyle\int (2x-1)^2\, dx$

100. $\displaystyle\int \sin x \cos x\, dx$

In Exercises 101–104, evaluate the integral using the properties of even and odd functions as an aid.

101. $\displaystyle\int_{-2}^{2} x^2(x^2+1)\, dx$ **102.** $\displaystyle\int_{-2}^{2} x(x^2+1)^3\, dx$

103. $\displaystyle\int_{-\pi/2}^{\pi/2} \sin^2 x \cos x\, dx$ **104.** $\displaystyle\int_{-\pi/2}^{\pi/2} \sin x \cos x\, dx$

105. Use $\int_0^2 x^2\, dx = \frac{8}{3}$ to evaluate each definite integral without using the Fundamental Theorem of Calculus.

(a) $\displaystyle\int_{-2}^{0} x^2\, dx$ (b) $\displaystyle\int_{-2}^{2} x^2\, dx$

(c) $\displaystyle\int_{0}^{2} -x^2\, dx$ (d) $\displaystyle\int_{-2}^{0} 3x^2\, dx$

106. Use the symmetry of the graphs of the sine and cosine functions as an aid in evaluating each definite integral.

(a) $\displaystyle\int_{-\pi/4}^{\pi/4} \sin x\, dx$ (b) $\displaystyle\int_{-\pi/4}^{\pi/4} \cos x\, dx$

(c) $\displaystyle\int_{-\pi/2}^{\pi/2} \cos x\, dx$ (d) $\displaystyle\int_{-\pi/2}^{\pi/2} \sin x \cos x\, dx$

In Exercises 107 and 108, write the integral as the sum of the integral of an odd function and the integral of an even function. Use this simplification to evaluate the integral.

107. $\displaystyle\int_{-4}^{4} (x^3 + 6x^2 - 2x - 3)\, dx$ **108.** $\displaystyle\int_{-\pi}^{\pi} (\sin 3x + \cos 3x)\, dx$

Writing About Concepts

109. Describe why

$$\int x(5-x^2)^3\, dx \neq \int u^3\, du$$

where $u = 5 - x^2$.

110. Without integrating, explain why

$$\int_{-2}^{2} x(x^2+1)^2\, dx = 0.$$

111. ***Cash Flow*** The rate of disbursement dQ/dt of a 2 million dollar federal grant is proportional to the square of $100 - t$. Time t is measured in days $(0 \le t \le 100)$, and Q is the amount that remains to be disbursed. Find the amount that remains to be disbursed after 50 days. Assume that all the money will be disbursed in 100 days.

112. ***Depreciation*** The rate of depreciation dV/dt of a machine is inversely proportional to the square of $t + 1$, where V is the value of the machine t years after it was purchased. The initial value of the machine was \$500,000, and its value decreased \$100,000 in the first year. Estimate its value after 4 years.

113. ***Rainfall*** The normal monthly rainfall at the Seattle-Tacoma airport can be approximated by the model

$$R = 3.121 + 2.399\sin(0.524t + 1.377)$$

where R is measured in inches and t is the time in months, with $t = 1$ corresponding to January. *(Source: U.S. National Oceanic and Atmospheric Administration)*

(a) Determine the extrema of the function over a one-year period.

(b) Use integration to approximate the normal annual rainfall. (*Hint:* Integrate over the interval [0, 12].)

(c) Approximate the average monthly rainfall during the months of October, November, and December.

114. *Sales* The sales S (in thousands of units) of a seasonal product are given by the model

$$S = 74.50 + 43.75 \sin \frac{\pi t}{6}$$

where t is the time in months, with $t = 1$ corresponding to January. Find the average sales for each time period.

(a) The first quarter $(0 \le t \le 3)$

(b) The second quarter $(3 \le t \le 6)$

(c) The entire year $(0 \le t \le 12)$

115. *Water Supply* A model for the flow rate of water at a pumping station on a given day is

$$R(t) = 53 + 7 \sin\left(\frac{\pi t}{6} + 3.6\right) + 9 \cos\left(\frac{\pi t}{12} + 8.9\right)$$

where $0 \le t \le 24$. R is the flow rate in thousands of gallons per hour, and t is the time in hours.

(a) Use a graphing utility to graph the rate function and approximate the maximum flow rate at the pumping station.

(b) Approximate the total volume of water pumped in 1 day.

116. *Electricity* The oscillating current in an electrical circuit is

$$I = 2 \sin(60\pi t) + \cos(120\pi t)$$

where I is measured in amperes and t is measured in seconds. Find the average current for each time interval.

(a) $0 \le t \le \frac{1}{60}$

(b) $0 \le t \le \frac{1}{240}$

(c) $0 \le t \le \frac{1}{30}$

Probability **In Exercises 117 and 118, the function**

$$f(x) = kx^n(1 - x)^m, \quad 0 \le x \le 1$$

where $n > 0$, $m > 0$, and k is a constant, can be used to represent various probability distributions. If k is chosen such that

$$\int_0^1 f(x)\,dx = 1$$

the probability that x will fall between a and b $(0 \le a \le b \le 1)$ is

$$P_{a,b} = \int_a^b f(x)\,dx.$$

117. The probability that a person will remember between $a\%$ and $b\%$ of material learned in an experiment is

$$P_{a,b} = \int_a^b \frac{15}{4} x\sqrt{1 - x}\,dx$$

where x represents the percent remembered. (See figure.)

(a) For a randomly chosen individual, what is the probability that he or she will recall between 50% and 75% of the material?

(b) What is the median percent recall? That is, for what value of b is it true that the probability of recalling 0 to b is 0.5?

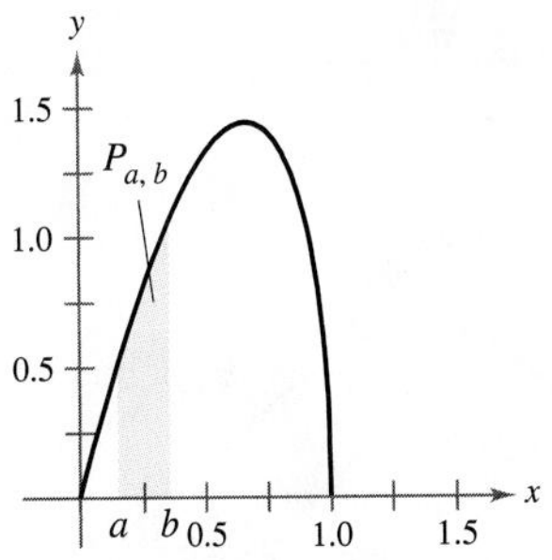

Figure for 117

118. The probability that ore samples taken from a region contain between $a\%$ and $b\%$ iron is

$$P_{a,b} = \int_a^b \frac{1155}{32} x^3(1 - x)^{3/2}\,dx$$

where x represents the percent of iron. (See figure.) What is the probability that a sample will contain between

(a) 0% and 25% iron?

(b) 50% and 100% iron?

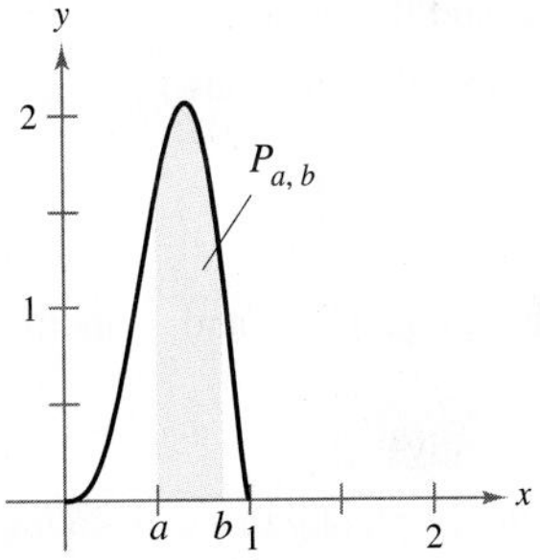

119. *Temperature* The temperature in degrees Fahrenheit in a house is

$$T = 72 + 12 \sin\left[\frac{\pi(t - 8)}{12}\right]$$

where t is time in hours, with $t = 0$ representing midnight. The hourly cost of cooling a house is \$0.10 per degree.

(a) Find the cost C of cooling the house if its thermostat is set at 72°F by evaluating the integral

$$C = 0.1\int_8^{20} \left[72 + 12 \sin \frac{\pi(t - 8)}{12} - 72\right] dt. \text{ (See figure.)}$$

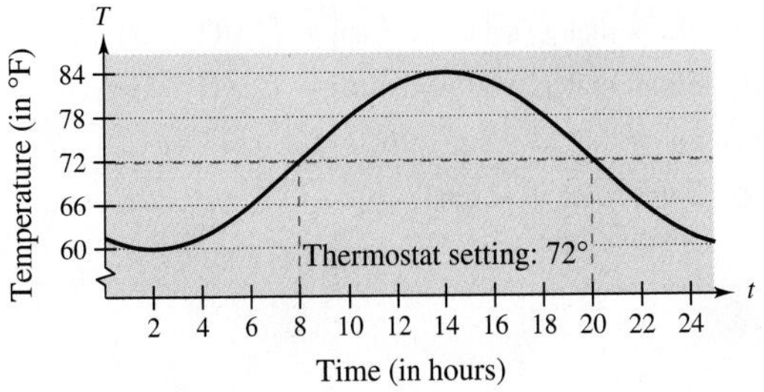

(b) Find the savings from resetting the thermostat to 78°F by evaluating the integral

$$C = 0.1\int_{10}^{18}\left[72 + 12\sin\frac{\pi(t-8)}{12} - 78\right]dt.$$

(See figure.)

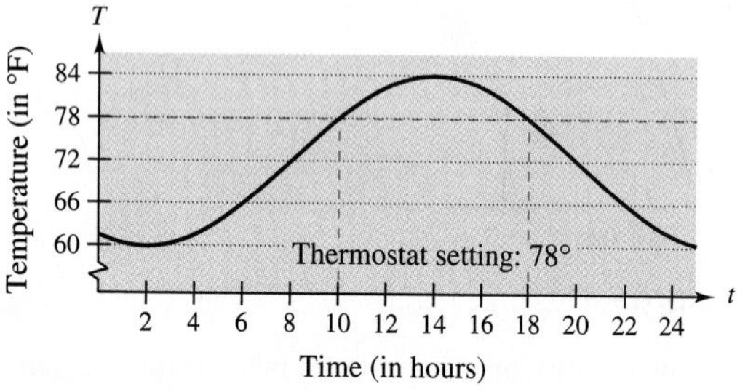

120. ***Manufacturing*** A manufacturer of fertilizer finds that national sales of fertilizer follow the seasonal pattern

$$F = 100{,}000\left[1 + \sin\frac{2\pi(t-60)}{365}\right]$$

where F is measured in pounds and t represents the time in days, with $t = 1$ corresponding to January 1. The manufacturer wants to set up a schedule to produce a uniform amount of fertilizer each day. What should this amount be?

121. ***Graphical Analysis*** Consider the functions f and g, where

$$f(x) = 6\sin x\cos^2 x \quad \text{and} \quad g(t) = \int_0^t f(x)\,dx.$$

(a) Use a graphing utility to graph f and g in the same viewing window.

(b) Explain why g is nonnegative.

(c) Identify the points on the graph of g that correspond to the extrema of f.

(d) Does each of the zeros of f correspond to an extremum of g? Explain.

(e) Consider the function

$$h(t) = \int_{\pi/2}^t f(x)\,dx.$$

Use a graphing utility to graph h. What is the relationship between g and h? Verify your conjecture.

122. Find $\displaystyle\lim_{n\to+\infty}\sum_{i=1}^{n}\frac{\sin(i\pi/n)}{n}$ by evaluating an appropriate definite integral over the interval $[0, 1]$.

123. (a) Show that $\int_0^1 x^2(1-x)^5\,dx = \int_0^1 x^5(1-x)^2\,dx$.

(b) Show that $\int_0^1 x^a(1-x)^b\,dx = \int_0^1 x^b(1-x)^a\,dx$.

124. (a) Show that $\int_0^{\pi/2}\sin^2 x\,dx = \int_0^{\pi/2}\cos^2 x\,dx$.

(b) Show that $\int_0^{\pi/2}\sin^n x\,dx = \int_0^{\pi/2}\cos^n x\,dx$, where n is a positive integer.

True or False? **In Exercises 125–130, determine whether the statement is true or false. If it is false, explain why or give an example that shows it is false.**

125. $\displaystyle\int(2x+1)^2\,dx = \tfrac{1}{3}(2x+1)^3 + C$

126. $\displaystyle\int x(x^2+1)\,dx = \tfrac{1}{2}x^2\left(\tfrac{1}{3}x^3 + x\right) + C$

127. $\displaystyle\int_{-10}^{10}(ax^3 + bx^2 + cx + d)\,dx = 2\int_0^{10}(bx^2 + d)\,dx$

128. $\displaystyle\int_a^b \sin x\,dx = \int_a^{b+2\pi}\sin x\,dx$

129. $\displaystyle 4\int\sin x\cos x\,dx = -\cos 2x + C$

130. $\displaystyle\int\sin^2 2x\cos 2x\,dx = \tfrac{1}{3}\sin^3 2x + C$

131. Assume that f is continuous everywhere and that c is a constant. Show that

$$\int_{ca}^{cb} f(x)\,dx = c\int_a^b f(cx)\,dx.$$

132. (a) Verify that $\sin u - u\cos u + C = \int u\sin u\,du$.

(b) Use part (a) to show that $\int_0^{\pi^2}\sin\sqrt{x}\,dx = 2\pi$.

133. Complete the proof of Theorem 4.15.

134. Show that if f is continuous on the entire real number line, then

$$\int_a^b f(x+h)\,dx = \int_{a+h}^{b+h} f(x)\,dx.$$

Putnam Exam Challenge

135. If $a_0, a_1, \ldots, a_n$ are real numbers satisfying

$$\frac{a_0}{1} + \frac{a_1}{2} + \cdots + \frac{a_n}{n+1} = 0$$

show that the equation $a_0 + a_1x + a_2x^2 + \cdots + a_nx^n = 0$ has at least one real zero.

136. Find all the continuous positive functions $f(x)$, for $0 \le x \le 1$, such that

$$\int_0^1 f(x)\,dx = 1$$

$$\int_0^1 f(x)x\,dx = \alpha$$

$$\int_0^1 f(x)x^2\,dx = \alpha^2$$

where α is a real number.

These problems were composed by the Committee on the Putnam Prize Competition.

Section 4.6 Numerical Integration

- **Approximate a definite integral using the Trapezoidal Rule.**
- **Approximate a definite integral using Simpson's Rule.**
- **Analyze the approximate errors in the Trapezoidal Rule and Simpson's Rule.**

The Trapezoidal Rule

Some elementary functions simply do not have antiderivatives that are elementary functions. For example, there is no elementary function that has any of the following functions as its derivative.

$$\sqrt[3]{x}\sqrt{1-x}, \qquad \sqrt{x}\cos x, \qquad \frac{\cos x}{x}, \qquad \sqrt{1-x^3}, \qquad \sin x^2$$

If you need to evaluate a definite integral involving a function whose antiderivative cannot be found, the Fundamental Theorem of Calculus cannot be applied, and you must resort to an approximation technique. Two such techniques are described in this section.

One way to approximate a definite integral is to use n trapezoids, as shown in Figure 4.41. In the development of this method, assume that f is continuous and positive on the interval $[a, b]$. So, the definite integral

$$\int_a^b f(x)\,dx$$

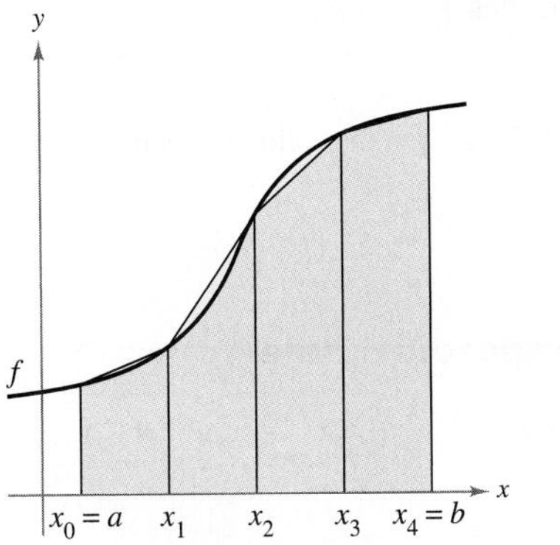

The area of the region can be approximated using four trapezoids.
Figure 4.41

represents the area of the region bounded by the graph of f and the x-axis, from $x = a$ to $x = b$. First, partition the interval $[a, b]$ into n subintervals, each of width $\Delta x = (b-a)/n$, such that

$$a = x_0 < x_1 < x_2 < \cdot \cdot \cdot < x_n = b.$$

Then form a trapezoid for each subinterval (see Figure 4.42). The area of the ith trapezoid is

$$\text{Area of } i\text{th trapezoid} = \left[\frac{f(x_{i-1}) + f(x_i)}{2}\right]\left(\frac{b-a}{n}\right).$$

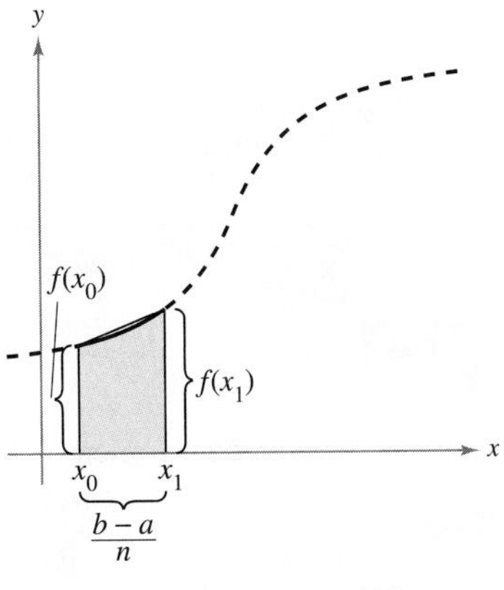

The area of the first trapezoid is $\left[\frac{f(x_0) + f(x_1)}{2}\right]\left(\frac{b-a}{n}\right)$.
Figure 4.42

This implies that the sum of the areas of the n trapezoids is

$$\begin{aligned}
\text{Area} &= \left(\frac{b-a}{n}\right)\left[\frac{f(x_0) + f(x_1)}{2} + \cdot \cdot \cdot + \frac{f(x_{n-1}) + f(x_n)}{2}\right] \\
&= \left(\frac{b-a}{2n}\right)[f(x_0) + f(x_1) + f(x_1) + f(x_2) + \cdot \cdot \cdot + f(x_{n-1}) + f(x_n)] \\
&= \left(\frac{b-a}{2n}\right)[f(x_0) + 2f(x_1) + 2f(x_2) + \cdot \cdot \cdot + 2f(x_{n-1}) + f(x_n)].
\end{aligned}$$

Letting $\Delta x = (b-a)/n$, you can take the limit as $n \to \infty$ to obtain

$$\begin{aligned}
&\lim_{n\to\infty}\left(\frac{b-a}{2n}\right)[f(x_0) + 2f(x_1) + \cdot \cdot \cdot + 2f(x_{n-1}) + f(x_n)] \\
&= \lim_{n\to\infty}\left[\frac{[f(a) - f(b)]\,\Delta x}{2} + \sum_{i=1}^{n} f(x_i)\,\Delta x\right] \\
&= \lim_{n\to\infty}\frac{[f(a) - f(b)](b-a)}{2n} + \lim_{n\to\infty}\sum_{i=1}^{n} f(x_i)\,\Delta x \\
&= 0 + \int_a^b f(x)\,dx.
\end{aligned}$$

The result is summarized in the following theorem.

THEOREM 4.16 The Trapezoidal Rule

Let f be continuous on $[a, b]$. The Trapezoidal Rule for approximating $\int_a^b f(x)\,dx$ is given by

$$\int_a^b f(x)\,dx \approx \frac{b-a}{2n}\left[f(x_0) + 2f(x_1) + 2f(x_2) + \cdots + 2f(x_{n-1}) + f(x_n)\right].$$

Moreover, as $n \to \infty$, the right-hand side approaches $\int_a^b f(x)\,dx$.

NOTE Observe that the coefficients in the Trapezoidal Rule have the following pattern.

1 2 2 2 . . . 2 2 1

EXAMPLE 1 Approximation with the Trapezoidal Rule

Use the Trapezoidal Rule to approximate

$$\int_0^{\pi} \sin x\,dx.$$

Compare the results for $n = 4$ and $n = 8$, as shown in Figure 4.43.

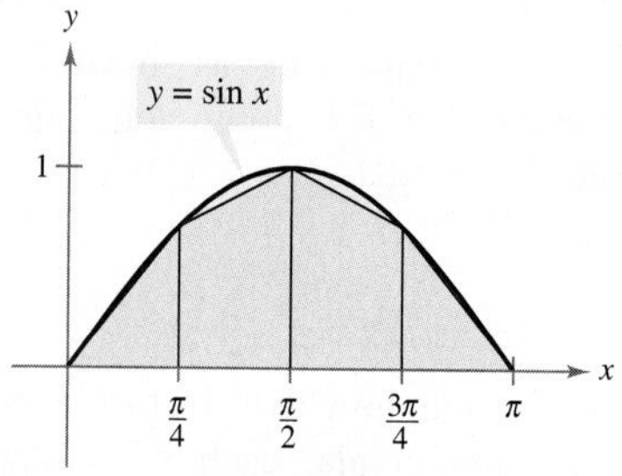

Four subintervals

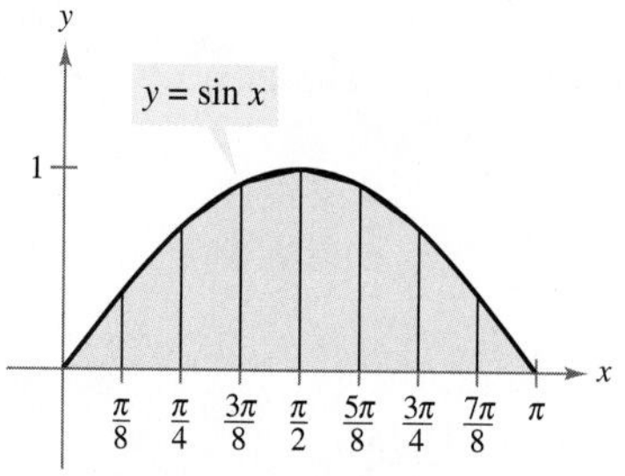

Eight subintervals

Trapezoidal approximations
Figure 4.43

Solution When $n = 4$, $\Delta x = \pi/4$, and you obtain

$$\int_0^{\pi} \sin x\,dx \approx \frac{\pi}{8}\left(\sin 0 + 2\sin\frac{\pi}{4} + 2\sin\frac{\pi}{2} + 2\sin\frac{3\pi}{4} + \sin\pi\right)$$
$$= \frac{\pi}{8}\left(0 + \sqrt{2} + 2 + \sqrt{2} + 0\right) = \frac{\pi\left(1+\sqrt{2}\right)}{4} \approx 1.896.$$

When $n = 8$, $\Delta x = \pi/8$, and you obtain

$$\int_0^{\pi} \sin x\,dx \approx \frac{\pi}{16}\left(\sin 0 + 2\sin\frac{\pi}{8} + 2\sin\frac{\pi}{4} + 2\sin\frac{3\pi}{8} + 2\sin\frac{\pi}{2}\right.$$
$$\left. + 2\sin\frac{5\pi}{8} + 2\sin\frac{3\pi}{4} + 2\sin\frac{7\pi}{8} + \sin\pi\right)$$
$$= \frac{\pi}{16}\left(2 + 2\sqrt{2} + 4\sin\frac{\pi}{8} + 4\sin\frac{3\pi}{8}\right) \approx 1.974.$$

For this particular integral, you could have found an antiderivative and determined that the exact area of the region is 2.

TECHNOLOGY Most graphing utilities and computer algebra systems have built-in programs that can be used to approximate the value of a definite integral. Try using such a program to approximate the integral in Example 1. How close is your approximation?

When you use such a program, you need to be aware of its limitations. Often, you are given no indication of the degree of accuracy of the approximation. Other times, you may be given an approximation that is completely wrong. For instance, try using a built-in numerical integration program to evaluate

$$\int_{-1}^{2} \frac{1}{x}\,dx.$$

Your calculator should give an error message. Does yours?

It is interesting to compare the Trapezoidal Rule with the Midpoint Rule given in Section 4.2 (Exercises 63–66). For the Trapezoidal Rule, you average the function values at the endpoints of the subintervals, but for the Midpoint Rule you take the function values of the subinterval midpoints.

$$\int_a^b f(x)\,dx \approx \sum_{i=1}^{n} f\left(\frac{x_i + x_{i-1}}{2}\right)\Delta x \qquad \text{Midpoint Rule}$$

$$\int_a^b f(x)\,dx \approx \sum_{i=1}^{n} \left(\frac{f(x_i) + f(x_{i-1})}{2}\right)\Delta x \qquad \text{Trapezoidal Rule}$$

NOTE There are two important points that should be made concerning the Trapezoidal Rule (or the Midpoint Rule). First, the approximation tends to become more accurate as n increases. For instance, in Example 1, if $n = 16$, the Trapezoidal Rule yields an approximation of 1.994. Second, although you could have used the Fundamental Theorem to evaluate the integral in Example 1, this theorem cannot be used to evaluate an integral as simple as $\int_0^\pi \sin x^2\,dx$ because $\sin x^2$ has no elementary antiderivative. Yet, the Trapezoidal Rule can be applied easily to this integral.

Simpson's Rule

One way to view the trapezoidal approximation of a definite integral is to say that on each subinterval you approximate f by a *first*-degree polynomial. In Simpson's Rule, named after the English mathematician Thomas Simpson (1710–1761), you take this procedure one step further and approximate f by *second*-degree polynomials.

Before presenting Simpson's Rule, we list a theorem for evaluating integrals of polynomials of degree 2 (or less).

THEOREM 4.17 Integral of $p(x) = Ax^2 + Bx + C$

If $p(x) = Ax^2 + Bx + C$, then

$$\int_a^b p(x)\,dx = \left(\frac{b-a}{6}\right)\left[p(a) + 4p\left(\frac{a+b}{2}\right) + p(b)\right].$$

Proof

$$\begin{aligned}\int_a^b p(x)\,dx &= \int_a^b (Ax^2 + Bx + C)\,dx\\ &= \left[\frac{Ax^3}{3} + \frac{Bx^2}{2} + Cx\right]_a^b\\ &= \frac{A(b^3 - a^3)}{3} + \frac{B(b^2 - a^2)}{2} + C(b - a)\\ &= \left(\frac{b-a}{6}\right)[2A(a^2 + ab + b^2) + 3B(b + a) + 6C]\end{aligned}$$

By expansion and collection of terms, the expression inside the brackets becomes

$$\underbrace{(Aa^2 + Ba + C)}_{p(a)} + \underbrace{4\left[A\left(\frac{b+a}{2}\right)^2 + B\left(\frac{b+a}{2}\right) + C\right]}_{4p\left(\frac{a+b}{2}\right)} + \underbrace{(Ab^2 + Bb + C)}_{p(b)}$$

and you can write

$$\int_a^b p(x)\,dx = \left(\frac{b-a}{6}\right)\left[p(a) + 4p\left(\frac{a+b}{2}\right) + p(b)\right].$$

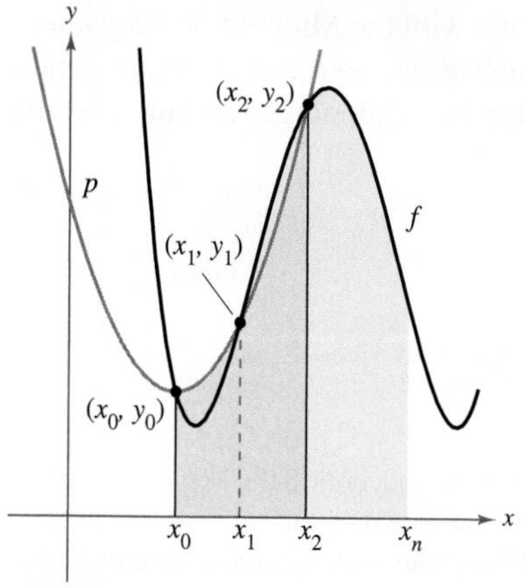

$$\int_{x_0}^{x_2} p(x)\,dx \approx \int_{x_0}^{x_2} f(x)\,dx$$

Figure 4.44

To develop Simpson's Rule for approximating a definite integral, you again partition the interval $[a, b]$ into n subintervals, each of width $\Delta x = (b - a)/n$. This time, however, n is required to be even, and the subintervals are grouped in pairs such that

$$a = \underbrace{x_0 < x_1 < x_2}_{[x_0,\, x_2]} < \underbrace{x_3 < x_4}_{[x_2,\, x_4]} < \cdot \cdot \cdot < \underbrace{x_{n-2} < x_{n-1} < x_n}_{[x_{n-2},\, x_n]} = b.$$

On each (double) subinterval $[x_{i-2}, x_i]$, you can approximate f by a polynomial p of degree less than or equal to 2. (See Exercise 55.) For example, on the subinterval $[x_0, x_2]$, choose the polynomial of least degree passing through the points (x_0, y_0), (x_1, y_1), and (x_2, y_2), as shown in Figure 4.44. Now, using p as an approximation of f on this subinterval, you have, by Theorem 4.17,

$$\begin{aligned}\int_{x_0}^{x_2} f(x)\,dx \approx \int_{x_0}^{x_2} p(x)\,dx &= \frac{x_2 - x_0}{6}\left[p(x_0) + 4p\left(\frac{x_0 + x_2}{2}\right) + p(x_2)\right] \\ &= \frac{2[(b-a)/n]}{6}\left[p(x_0) + 4p(x_1) + p(x_2)\right] \\ &= \frac{b-a}{3n}\left[f(x_0) + 4f(x_1) + f(x_2)\right].\end{aligned}$$

Repeating this procedure on the entire interval $[a, b]$ produces the following theorem.

THEOREM 4.18 Simpson's Rule (n is even)

Let f be continuous on $[a, b]$. Simpson's Rule for approximating $\int_a^b f(x)\,dx$ is

$$\int_a^b f(x)\,dx \approx \frac{b-a}{3n}\left[f(x_0) + 4f(x_1) + 2f(x_2) + 4f(x_3) + \cdot \cdot \cdot + 4f(x_{n-1}) + f(x_n)\right].$$

Moreover, as $n \to \infty$, the right-hand side approaches $\int_a^b f(x)\,dx$.

NOTE Observe that the coefficients in Simpson's Rule have the following pattern.

1 4 2 4 2 4 . . . 4 2 4 1

In Example 1, the Trapezoidal Rule was used to estimate $\int_0^{\pi} \sin x\,dx$. In the next example, Simpson's Rule is applied to the same integral.

EXAMPLE 2 Approximation with Simpson's Rule

NOTE In Example 1, the Trapezoidal Rule with $n = 8$ approximated $\int_0^{\pi} \sin x\,dx$ as 1.974. In Example 2, Simpson's Rule with $n = 8$ gave an approximation of 2.0003. The antiderivative would produce the true value of 2.

Use Simpson's Rule to approximate

$$\int_0^{\pi} \sin x\,dx.$$

Compare the results for $n = 4$ and $n = 8$.

Solution When $n = 4$, you have

$$\int_0^{\pi} \sin x\,dx \approx \frac{\pi}{12}\left(\sin 0 + 4\sin\frac{\pi}{4} + 2\sin\frac{\pi}{2} + 4\sin\frac{3\pi}{4} + \sin\pi\right) \approx 2.005.$$

When $n = 8$, you have $\int_0^{\pi} \sin x\,dx \approx 2.0003$.

Error Analysis

If you must use an approximation technique, it is important to know how accurate you can expect the approximation to be. The following theorem, which is listed without proof, gives the formulas for estimating the errors involved in the use of Simpson's Rule and the Trapezoidal Rule.

THEOREM 4.19 Errors in the Trapezoidal Rule and Simpson's Rule

If f has a continuous second derivative on $[a, b]$, then the error E in approximating $\int_a^b f(x)\,dx$ by the Trapezoidal Rule is

$$E \le \frac{(b-a)^3}{12n^2}\left[\max|f''(x)|\right], \quad a \le x \le b. \qquad \text{Trapezoidal Rule}$$

Moreover, if f has a continuous fourth derivative on $[a, b]$, then the error E in approximating $\int_a^b f(x)\,dx$ by Simpson's Rule is

$$E \le \frac{(b-a)^5}{180n^4}\left[\max|f^{(4)}(x)|\right], \quad a \le x \le b. \qquad \text{Simpson's Rule}$$

TECHNOLOGY If you have access to a computer algebra system, use it to evaluate the definite integral in Example 3. You should obtain a value of

$$\int_0^1 \sqrt{1+x^2}\,dx = \tfrac{1}{2}\left[\sqrt{2} + \ln\left(1+\sqrt{2}\right)\right] \approx 1.14779.$$

("ln" represents the natural logarithmic function, which you will study in Section 5.1.)

Theorem 4.19 states that the errors generated by the Trapezoidal Rule and Simpson's Rule have upper bounds dependent on the extreme values of $f''(x)$ and $f^{(4)}(x)$ in the interval $[a, b]$. Furthermore, these errors can be made arbitrarily small by *increasing* n, provided that f'' and $f^{(4)}$ are continuous and therefore bounded in $[a, b]$.

EXAMPLE 3 The Approximate Error in the Trapezoidal Rule

Determine a value of n such that the Trapezoidal Rule will approximate the value of $\int_0^1 \sqrt{1+x^2}\,dx$ with an error that is less than 0.01.

Solution Begin by letting $f(x) = \sqrt{1+x^2}$ and finding the second derivative of f.

$$f'(x) = x(1+x^2)^{-1/2} \quad \text{and} \quad f''(x) = (1+x^2)^{-3/2}$$

The maximum value of $|f''(x)|$ on the interval $[0, 1]$ is $|f''(0)| = 1$. So, by Theorem 4.19, you can write

$$E \le \frac{(b-a)^3}{12n^2}|f''(0)| = \frac{1}{12n^2}(1) = \frac{1}{12n^2}.$$

To obtain an error E that is less than 0.01, you must choose n such that $1/(12n^2) \le 1/100$.

$$100 \le 12n^2 \quad \Rightarrow \quad n \ge \sqrt{\tfrac{100}{12}} \approx 2.89$$

So, you can choose $n = 3$ (because n must be greater than or equal to 2.89) and apply the Trapezoidal Rule, as shown in Figure 4.45, to obtain

$$\int_0^1 \sqrt{1+x^2}\,dx \approx \frac{1}{6}\left[\sqrt{1+0^2} + 2\sqrt{1+\left(\tfrac{1}{3}\right)^2} + 2\sqrt{1+\left(\tfrac{2}{3}\right)^2} + \sqrt{1+1^2}\right] \approx 1.154.$$

So, with an error no larger than 0.01, you know that

$$1.144 \le \int_0^1 \sqrt{1+x^2}\,dx \le 1.164.$$

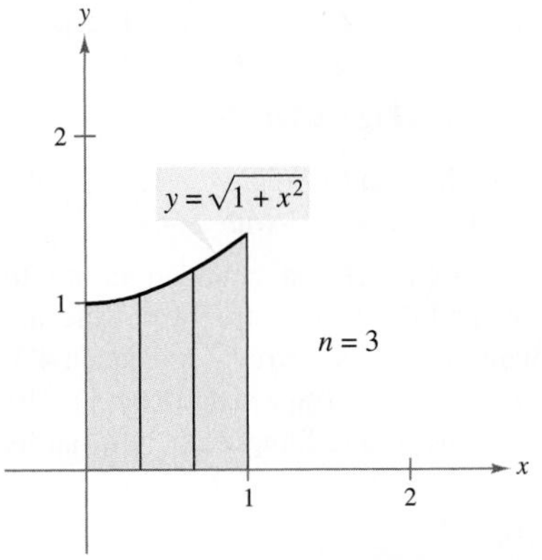

$1.144 \le \int_0^1 \sqrt{1+x^2}\,dx \le 1.164$

Figure 4.45

Exercises for Section 4.6

See www.CalcChat.com for worked-out solutions to odd-numbered exercises.

In Exercises 1–10, use the Trapezoidal Rule and Simpson's Rule to approximate the value of the definite integral for the given value of n. Round your answer to four decimal places and compare the results with the exact value of the definite integral.

1. $\int_0^2 x^2\,dx,\quad n = 4$

2. $\int_0^1 \left(\frac{x^2}{2} + 1\right) dx,\quad n = 4$

3. $\int_0^2 x^3\,dx,\quad n = 4$

4. $\int_1^2 \frac{1}{x^2}\,dx,\quad n = 4$

5. $\int_0^2 x^3\,dx,\quad n = 8$

6. $\int_0^8 \sqrt[3]{x}\,dx,\quad n = 8$

7. $\int_4^9 \sqrt{x}\,dx,\quad n = 8$

8. $\int_1^3 (4 - x^2)\,dx,\quad n = 4$

9. $\int_1^2 \frac{1}{(x+1)^2}\,dx,\quad n = 4$

10. $\int_0^2 x\sqrt{x^2+1}\,dx,\quad n = 4$

In Exercises 11–20, approximate the definite integral using the Trapezoidal Rule and Simpson's Rule with $n = 4$. Compare these results with the approximation of the integral using a graphing utility.

11. $\int_0^2 \sqrt{1 + x^3}\,dx$

12. $\int_0^2 \frac{1}{\sqrt{1 + x^3}}\,dx$

13. $\int_0^1 \sqrt{x}\,\sqrt{1 - x}\,dx$

14. $\int_{\pi/2}^{\pi} \sqrt{x}\sin x\,dx$

15. $\int_0^{\sqrt{\pi/2}} \cos x^2\,dx$

16. $\int_0^{\sqrt{\pi/4}} \tan x^2\,dx$

17. $\int_1^{1.1} \sin x^2\,dx$

18. $\int_0^{\pi/2} \sqrt{1 + \cos^2 x}\,dx$

19. $\int_0^{\pi/4} x\tan x\,dx$

20. $\int_0^{\pi} f(x)\,dx,\quad f(x) = \begin{cases} \frac{\sin x}{x}, & x > 0 \\ 1, & x = 0 \end{cases}$

Writing About Concepts

21. If the function f is concave upward on the interval $[a, b]$, will the Trapezoidal Rule yield a result greater than or less than $\int_a^b f(x)\,dx$? Explain.

22. The Trapezoidal Rule and Simpson's Rule yield approximations of a definite integral $\int_a^b f(x)\,dx$ based on polynomial approximations of f. What degree polynomial is used for each?

In Exercises 23–28, use the error formulas in Theorem 4.19 to estimate the error in approximating the integral, with $n = 4$, using (a) the Trapezoidal Rule and (b) Simpson's Rule.

23. $\int_0^2 x^3\,dx$

24. $\int_1^3 (2x + 3)\,dx$

25. $\int_0^1 \frac{1}{x+1}\,dx$

26. $\int_2^4 \frac{1}{(x-1)^2}\,dx$

27. $\int_0^{\pi} \cos x\,dx$

28. $\int_0^1 \sin(\pi x)\,dx$

In Exercises 29–34, use the error formulas in Theorem 4.19 to find n such that the error in the approximation of the definite integral is less than 0.00001 using (a) the Trapezoidal Rule and (b) Simpson's Rule.

29. $\int_1^3 \frac{1}{x}\,dx$

30. $\int_0^1 \frac{1}{1+x}\,dx$

31. $\int_0^2 \sqrt{x+2}\,dx$

32. $\int_1^3 \frac{1}{\sqrt{x}}\,dx$

33. $\int_0^1 \cos(\pi x)\,dx$

34. $\int_0^{\pi/2} \sin x\,dx$

In Exercises 35–38, use a computer algebra system and the error formulas to find n such that the error in the approximation of the definite integral is less than 0.00001 using (a) the Trapezoidal Rule and (b) Simpson's Rule.

35. $\int_0^2 \sqrt{1+x}\,dx$

36. $\int_0^2 (x+1)^{2/3}\,dx$

37. $\int_0^1 \tan x^2\,dx$

38. $\int_0^1 \sin x^2\,dx$

39. Approximate the area of the shaded region using (a) the Trapezoidal Rule and (b) Simpson's Rule with $n = 4$.

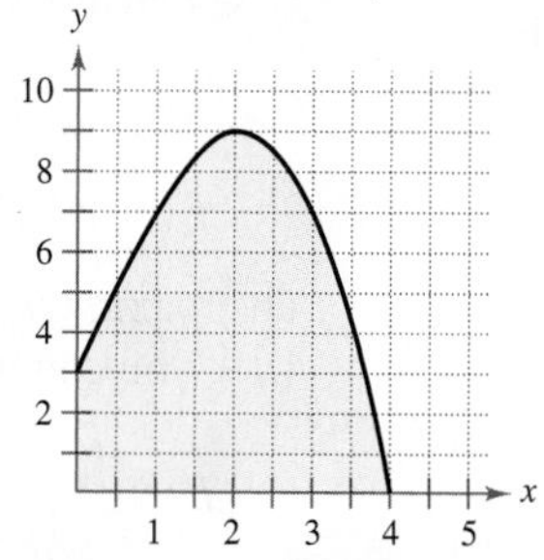

Figure for 39

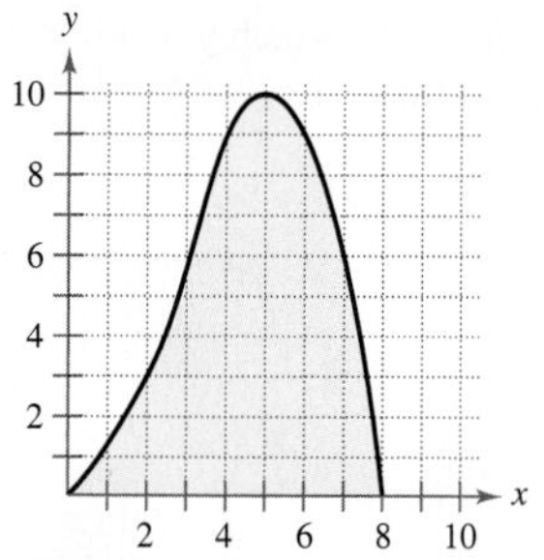

Figure for 40

40. Approximate the area of the shaded region using (a) the Trapezoidal Rule and (b) Simpson's Rule with $n = 8$.

41. ***Programming*** Write a program for a graphing utility to approximate a definite integral using the Trapezoidal Rule and Simpson's Rule. Start with the program written in Section 4.3, Exercises 59–62, and note that the Trapezoidal Rule can be written as $T(n) = \frac{1}{2}[L(n) + R(n)]$ and Simpson's Rule can be written as

$S(n) = \frac{1}{3}[T(n/2) + 2M(n/2)]$.

[Recall that $L(n)$, $M(n)$, and $R(n)$ represent the Riemann sums using the left-hand endpoints, midpoints, and right-hand endpoints of subintervals of equal width.]

Programming **In Exercises 42–44, use the program in Exercise 41 to approximate the definite integral and complete the table.**

n	$L(n)$	$M(n)$	$R(n)$	$T(n)$	$S(n)$
4					
8					
10					
12					
16					
20					

42. $\displaystyle\int_0^4 \sqrt{2+3x^2}\,dx$ **43.** $\displaystyle\int_0^1 \sqrt{1-x^2}\,dx$ **44.** $\displaystyle\int_0^4 \sin\sqrt{x}\,dx$

45. ***Area*** Use Simpson's Rule with $n = 14$ to approximate the area of the region bounded by the graphs of $y = \sqrt{x}\cos x$, $y = 0$, $x = 0$, and $x = \pi/2$.

46. ***Circumference*** The **elliptic integral**

$$8\sqrt{3}\int_0^{\pi/2} \sqrt{1-\tfrac{2}{3}\sin^2\theta}\,d\theta$$

gives the circumference of an ellipse. Use Simpson's Rule with $n = 8$ to approximate the circumference.

47. ***Work*** To determine the size of the motor required to operate a press, a company must know the amount of work done when the press moves an object linearly 5 feet. The variable force to move the object is

$$F(x) = 100x\sqrt{125 - x^3}$$

where F is given in pounds and x gives the position of the unit in feet. Use Simpson's Rule with $n = 12$ to approximate the work W (in foot-pounds) done through one cycle if

$$W = \int_0^5 F(x)\,dx.$$

48. The table lists several measurements gathered in an experiment to approximate an unknown continuous function $y = f(x)$.

(a) Approximate the integral $\int_0^2 f(x)\,dx$ using the Trapezoidal Rule and Simpson's Rule.

x	0.00	0.25	0.50	0.75	1.00
y	4.32	4.36	4.58	5.79	6.14

x	1.25	1.50	1.75	2.00
y	7.25	7.64	8.08	8.14

(b) Use a graphing utility to find a model of the form $y = ax^3 + bx^2 + cx + d$ for the data. Integrate the resulting polynomial over $[0, 2]$ and compare the result with part (a).

Approximation of Pi **In Exercises 49 and 50, use Simpson's Rule with $n = 6$ to approximate π using the given equation. (In Section 5.7, you will be able to evaluate the integral using inverse trigonometric functions.)**

49. $\displaystyle\pi = \int_0^{1/2} \frac{6}{\sqrt{1-x^2}}\,dx$ **50.** $\displaystyle\pi = \int_0^1 \frac{4}{1+x^2}\,dx$

Area **In Exercises 51 and 52, use the Trapezoidal Rule to estimate the number of square meters of land in a lot where x and y are measured in meters, as shown in the figures. The land is bounded by a stream and two straight roads that meet at right angles.**

51.

x	y
0	125
100	125
200	120
300	112
400	90
500	90
600	95
700	88
800	75
900	35
1000	0

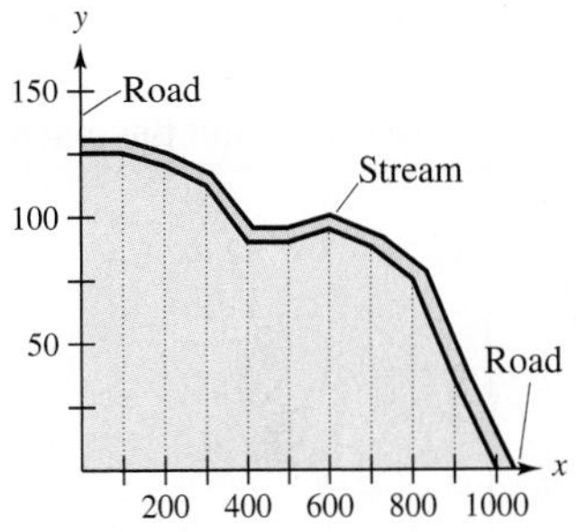

52.

x	y
0	75
10	81
20	84
30	76
40	67
50	68
60	69
70	72
80	68
90	56
100	42
110	23
120	0

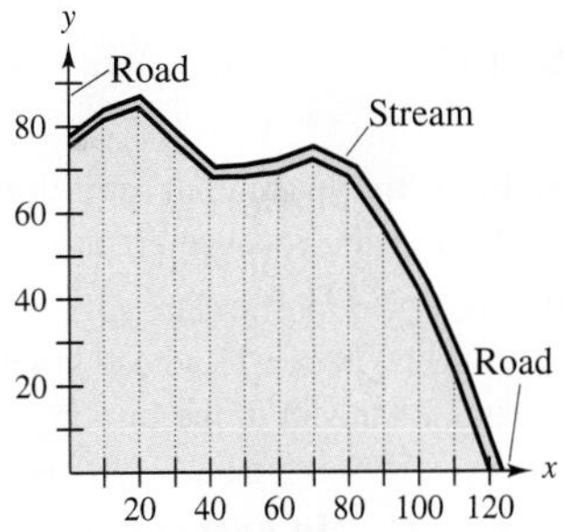

53. Prove that Simpson's Rule is exact when approximating the integral of a cubic polynomial function, and demonstrate the result for $\int_0^1 x^3\,dx$, $n = 2$.

54. Use Simpson's Rule with $n = 10$ and a computer algebra system to approximate t in the integral equation

$$\int_0^t \sin\sqrt{x}\,dx = 2.$$

55. Prove that you can find a polynomial $p(x) = Ax^2 + Bx + C$ that passes through any three points (x_1, y_1), (x_2, y_2), and (x_3, y_3), where the x_i's are distinct.

Review Exercises for Chapter 4

See www.CalcChat.com for worked-out solutions to odd-numbered exercises.

In Exercises 1 and 2, use the graph of f' to sketch a graph of f. To print an enlarged copy of the graph, go to the website *www.mathgraphs.com*.

1. 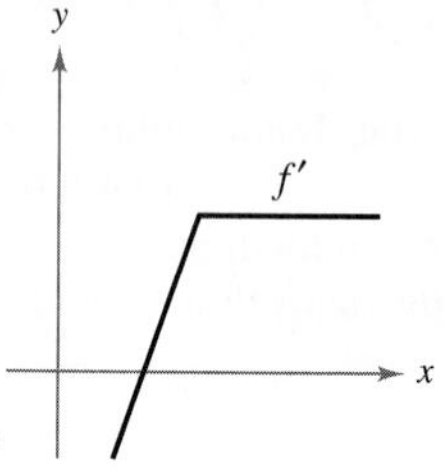

2. 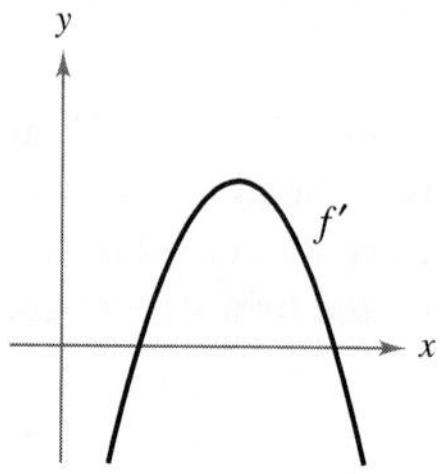

In Exercises 3–8, find the indefinite integral.

3. $\int (2x^2 + x - 1)\, dx$

4. $\int \frac{2}{\sqrt[3]{3x}}\, dx$

5. $\int \frac{x^3 + 1}{x^2}\, dx$

6. $\int \frac{x^3 - 2x^2 + 1}{x^2}\, dx$

7. $\int (4x - 3 \sin x)\, dx$

8. $\int (5 \cos x - 2 \sec^2 x)\, dx$

9. Find the particular solution of the differential equation $f'(x) = -2x$ whose graph passes through the point $(-1, 1)$.

10. Find the particular solution of the differential equation $f''(x) = 6(x - 1)$ whose graph passes through the point $(2, 1)$ and is tangent to the line $3x - y - 5 = 0$ at that point.

Slope Fields **In Exercises 11 and 12, a differential equation, a point, and a slope field are given. (a) Sketch two approximate solutions of the differential equation on the slope field, one of which passes through the given point. (To print an enlarged copy of the graph, go to the website *www.mathgraphs.com*.) (b) Use integration to find the particular solution of the differential equation and use a graphing utility to graph the solution.**

11. $\frac{dy}{dx} = 2x - 4, \quad (4, -2)$

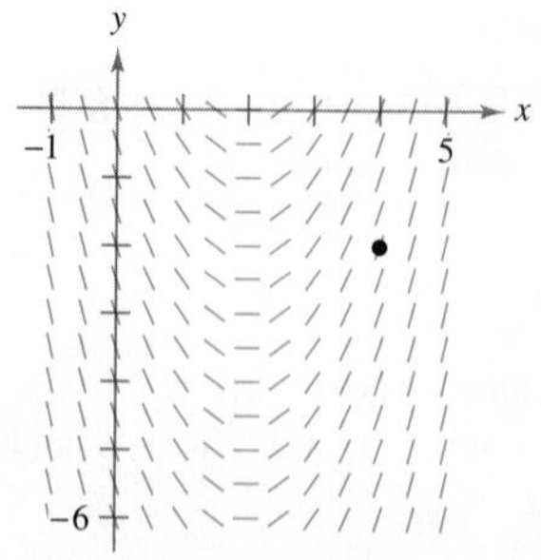

12. $\frac{dy}{dx} = \frac{1}{2}x^2 - 2x, \quad (6, 2)$

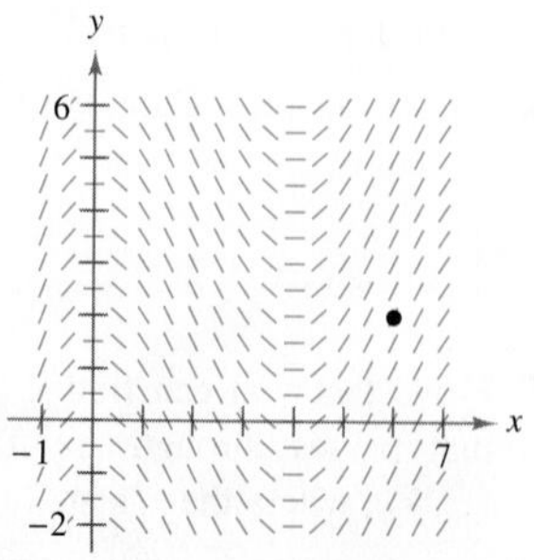

13. ***Velocity and Acceleration*** An airplane taking off from a runway travels 3600 feet before lifting off. The airplane starts from rest, moves with constant acceleration, and makes the run in 30 seconds. With what speed does it lift off?

14. ***Velocity and Acceleration*** The speed of a car traveling in a straight line is reduced from 45 to 30 miles per hour in a distance of 264 feet. Find the distance in which the car can be brought to rest from 30 miles per hour, assuming the same constant deceleration.

15. ***Velocity and Acceleration*** A ball is thrown vertically upward from ground level with an initial velocity of 96 feet per second.
    (a) How long will it take the ball to rise to its maximum height?
    (b) What is the maximum height?
    (c) When is the velocity of the ball one-half the initial velocity?
    (d) What is the height of the ball when its velocity is one-half the initial velocity?

16. ***Velocity and Acceleration*** Repeat Exercise 15 for an initial velocity of 40 meters per second.

In Exercises 17–20, use sigma notation to write the sum.

17. $\frac{1}{4(1)} + \frac{1}{4(2)} + \frac{1}{4(3)} + \cdots + \frac{1}{4(8)}$

18. $\frac{1+2}{2(1)} + \frac{2+2}{2(2)} + \frac{3+2}{2(3)} + \cdots + \frac{12+2}{2(12)}$

19. $\left(\frac{3}{n}\right)\left(\frac{1+1}{n}\right)^2 + \left(\frac{3}{n}\right)\left(\frac{2+1}{n}\right)^2 + \cdots + \left(\frac{3}{n}\right)\left(\frac{n+1}{n}\right)^2$

20. $3\left(2 + \frac{4}{n}\right) + \cdots + 3n\left(2 + \frac{(n+1)^2}{n}\right)$

In Exercises 21–24, use the properties of summation and Theorem 4.2 to evaluate the sum.

21. $\sum_{i=1}^{10} 3i$

22. $\sum_{i=1}^{20} (4i - 1)$

23. $\sum_{i=1}^{20} (i + 1)^2$

24. $\sum_{i=1}^{12} i(i^2 - 1)$

25. Write in sigma notation (a) the sum of the first ten positive odd integers, (b) the sum of the cubes of the first n positive integers, and (c) $6 + 10 + 14 + 18 + \cdots + 42$.

26. Evaluate each sum for $x_1 = 2, x_2 = -1, x_3 = 5, x_4 = 3,$ and $x_5 = 7$.
    (a) $\frac{1}{5}\sum_{i=1}^{5} x_i$
    (b) $\sum_{i=1}^{5} \frac{1}{x_i}$
    (c) $\sum_{i=1}^{5} (2x_i - x_i^2)$
    (d) $\sum_{i=2}^{5} (x_i - x_{i-1})$

In Exercises 27 and 28, use upper and lower sums to approximate the area of the region using the indicated number of subintervals of equal width.

27. $y = \dfrac{10}{x^2 + 1}$

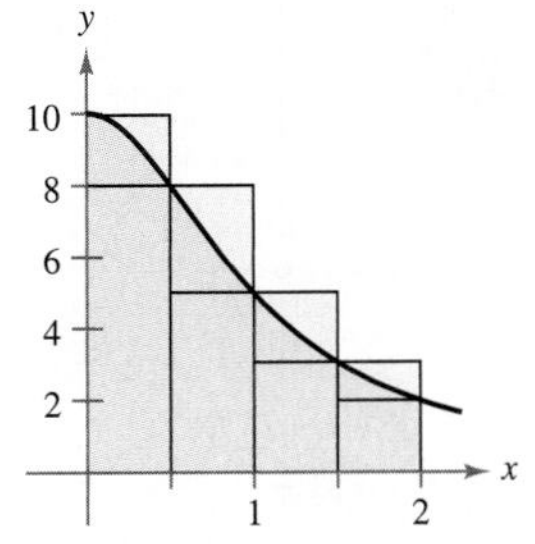

28. $y = 9 - \frac{1}{4}x^2$

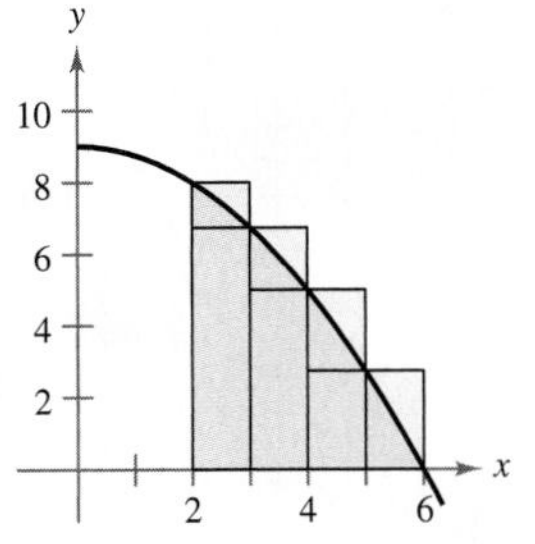

In Exercises 29–32, use the limit process to find the area of the region between the graph of the function and the x-axis over the given interval. Sketch the region.

29. $y = 6 - x, \quad [0, 4]$

30. $y = x^2 + 3, \quad [0, 2]$

31. $y = 5 - x^2, \quad [-2, 1]$

32. $y = \frac{1}{4}x^3, \quad [2, 4]$

33. Use the limit process to find the area of the region bounded by $x = 5y - y^2$, $x = 0$, $y = 2$, and $y = 5$.

34. Consider the region bounded by $y = mx$, $y = 0$, $x = 0$, and $x = b$.

(a) Find the upper and lower sums to approximate the area of the region when $\Delta x = b/4$.

(b) Find the upper and lower sums to approximate the area of the region when $\Delta x = b/n$.

(c) Find the area of the region by letting n approach infinity in both sums in part (b). Show that in each case you obtain the formula for the area of a triangle.

In Exercises 35 and 36, write the limit as a definite integral on the interval $[a, b]$, where c_i is any point in the ith subinterval.

Limit	*Interval*
35. $\displaystyle\lim_{\|\Delta\| \to 0} \sum_{i=1}^{n} (2c_i - 3)\,\Delta x_i$	$[4, 6]$
36. $\displaystyle\lim_{\|\Delta\| \to 0} \sum_{i=1}^{n} 3c_i(9 - c_i^2)\,\Delta x_i$	$[1, 3]$

In Exercises 37 and 38, set up a definite integral that yields the area of the region. (Do not evaluate the integral.)

37. $f(x) = 3x + 6$

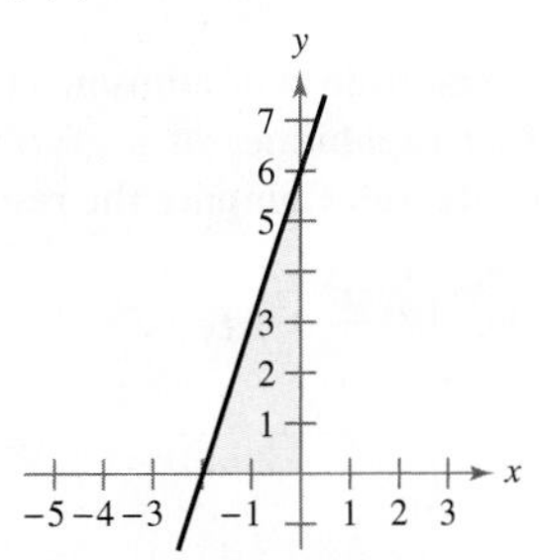

38. $f(x) = 9 - x^2$

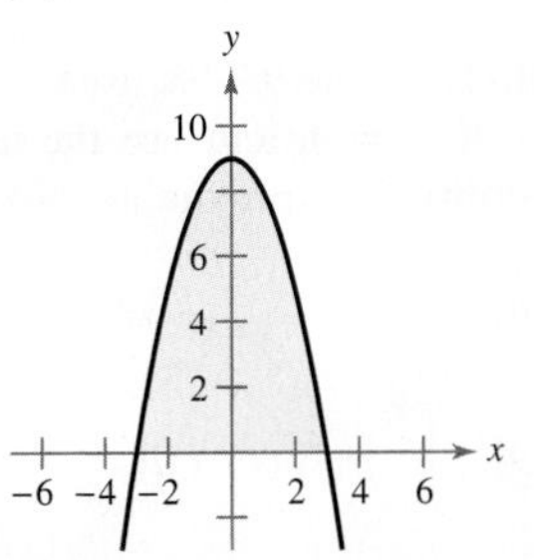

In Exercises 39 and 40, sketch the region whose area is given by the definite integral. Then use a geometric formula to evaluate the integral.

39. $\displaystyle\int_0^5 (5 - |x - 5|)\,dx$

40. $\displaystyle\int_{-4}^4 \sqrt{16 - x^2}\,dx$

41. Given $\displaystyle\int_2^6 f(x)\,dx = 10$ and $\displaystyle\int_2^6 g(x)\,dx = 3$, evaluate

(a) $\displaystyle\int_2^6 [f(x) + g(x)]\,dx.$

(b) $\displaystyle\int_2^6 [f(x) - g(x)]\,dx.$

(c) $\displaystyle\int_2^6 [2f(x) - 3g(x)]\,dx.$

(d) $\displaystyle\int_2^6 5f(x)\,dx.$

42. Given $\displaystyle\int_0^3 f(x)\,dx = 4$ and $\displaystyle\int_3^6 f(x)\,dx = -1$, evaluate

(a) $\displaystyle\int_0^6 f(x)\,dx.$

(b) $\displaystyle\int_6^3 f(x)\,dx.$

(c) $\displaystyle\int_4^4 f(x)\,dx.$

(d) $\displaystyle\int_3^6 -10f(x)\,dx.$

In Exercises 43–50, use the Fundamental Theorem of Calculus to evaluate the definite integral.

43. $\displaystyle\int_0^4 (2 + x)\,dx$

44. $\displaystyle\int_{-1}^1 (t^2 + 2)\,dt$

45. $\displaystyle\int_{-1}^1 (4t^3 - 2t)\,dt$

46. $\displaystyle\int_{-2}^2 (x^4 + 2x^2 - 5)\,dx$

47. $\displaystyle\int_4^9 x\sqrt{x}\,dx$

48. $\displaystyle\int_1^2 \left(\frac{1}{x^2} - \frac{1}{x^3}\right) dx$

49. $\displaystyle\int_0^{3\pi/4} \sin\theta\,d\theta$

50. $\displaystyle\int_{-\pi/4}^{\pi/4} \sec^2 t\,dt$

In Exercises 51–56, sketch the graph of the region whose area is given by the integral, and find the area.

51. $\displaystyle\int_1^3 (2x - 1)\,dx$

52. $\displaystyle\int_0^2 (x + 4)\,dx$

53. $\displaystyle\int_3^4 (x^2 - 9)\,dx$

54. $\displaystyle\int_{-1}^2 (-x^2 + x + 2)\,dx$

55. $\displaystyle\int_0^1 (x - x^3)\,dx$

56. $\displaystyle\int_0^1 \sqrt{x}\,(1 - x)\,dx$

In Exercises 57 and 58, determine the area of the given region.

57. $y = \sin x$

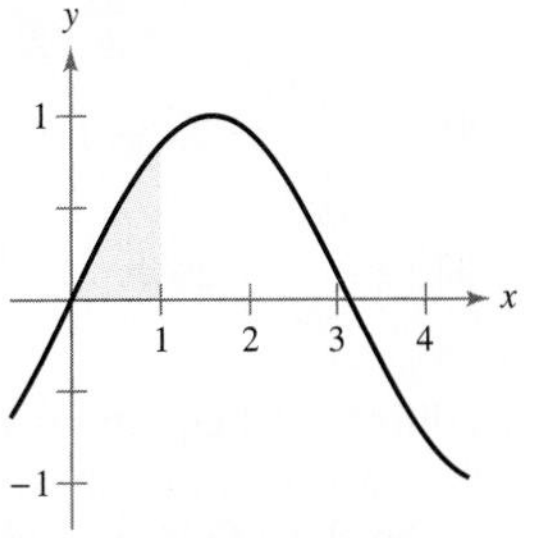

58. $y = x + \cos x$

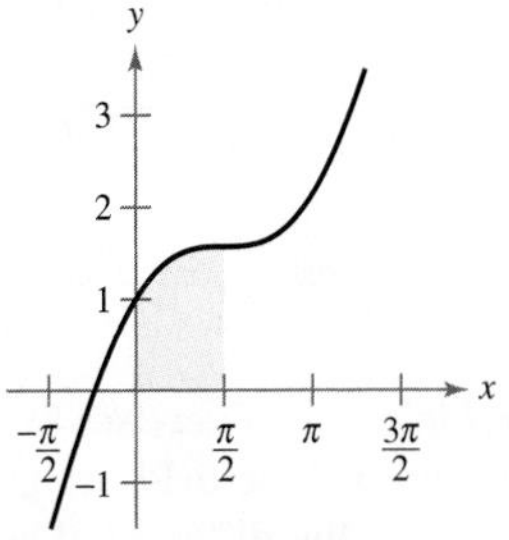

In Exercises 59 and 60, sketch the region bounded by the graphs of the equations, and determine its area.

59. $y = \dfrac{4}{\sqrt{x}}, \quad y = 0, \quad x = 1, \quad x = 9$

60. $y = \sec^2 x, \quad y = 0, \quad x = 0, \quad x = \dfrac{\pi}{3}$

In Exercises 61 and 62, find the average value of the function over the given interval. Find the values of x at which the function assumes its average value, and graph the function.

61. $f(x) = \dfrac{1}{\sqrt{x}}, \quad [4, 9]$

62. $f(x) = x^3, \quad [0, 2]$

In Exercises 63–66, use the Second Fundamental Theorem of Calculus to find $F'(x)$.

63. $F(x) = \displaystyle\int_0^x t^2\sqrt{1 + t^3}\, dt$

64. $F(x) = \displaystyle\int_1^x \frac{1}{t^2}\, dt$

65. $F(x) = \displaystyle\int_{-3}^x (t^2 + 3t + 2)\, dt$

66. $F(x) = \displaystyle\int_0^x \csc^2 t\, dt$

In Exercises 67–80, find the indefinite integral.

67. $\displaystyle\int (x^2 + 1)^3\, dx$

68. $\displaystyle\int \left(x + \frac{1}{x}\right)^2 dx$

69. $\displaystyle\int \frac{x^2}{\sqrt{x^3 + 3}}\, dx$

70. $\displaystyle\int x^2\sqrt{x^3 + 3}\, dx$

71. $\displaystyle\int x(1 - 3x^2)^4\, dx$

72. $\displaystyle\int \frac{x + 3}{(x^2 + 6x - 5)^2}\, dx$

73. $\displaystyle\int \sin^3 x \cos x\, dx$

74. $\displaystyle\int x \sin 3x^2\, dx$

75. $\displaystyle\int \frac{\sin\theta}{\sqrt{1 - \cos\theta}}\, d\theta$

76. $\displaystyle\int \frac{\cos x}{\sqrt{\sin x}}\, dx$

77. $\displaystyle\int \tan^n x \sec^2 x\, dx, \quad n \neq -1$

78. $\displaystyle\int \sec 2x \tan 2x\, dx$

79. $\displaystyle\int (1 + \sec \pi x)^2 \sec \pi x \tan \pi x\, dx$

80. $\displaystyle\int \cot^4 \alpha \csc^2 \alpha\, d\alpha$

In Exercises 81–88, evaluate the definite integral. Use a graphing utility to verify your result.

81. $\displaystyle\int_{-1}^{2} x(x^2 - 4)\, dx$

82. $\displaystyle\int_0^1 x^2(x^3 + 1)^3\, dx$

83. $\displaystyle\int_0^3 \frac{1}{\sqrt{1 + x}}\, dx$

84. $\displaystyle\int_3^6 \frac{x}{3\sqrt{x^2 - 8}}\, dx$

85. $2\pi \displaystyle\int_0^1 (y + 1)\sqrt{1 - y}\, dy$

86. $2\pi \displaystyle\int_{-1}^0 x^2\sqrt{x + 1}\, dx$

87. $\displaystyle\int_0^\pi \cos\frac{x}{2}\, dx$

88. $\displaystyle\int_{-\pi/4}^{\pi/4} \sin 2x\, dx$

Slope Fields **In Exercises 89 and 90, a differential equation, a point, and a slope field are given. (a) Sketch two approximate solutions of the differential equation on the slope field, one of which passes through the given point. (To print an enlarged copy of the graph, go to the website *www.mathgraphs.com*.) (b) Use integration to find the particular solution of the differential equation and use a graphing utility to graph the solution.**

89. $\dfrac{dy}{dx} = x\sqrt{9 - x^2}, \quad (0, -4)$

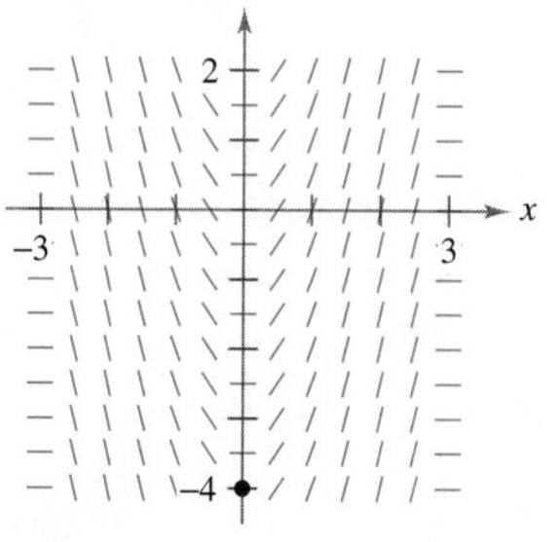

90. $\dfrac{dy}{dx} = -\dfrac{1}{2}x\sin(x^2), \quad (0, 0)$

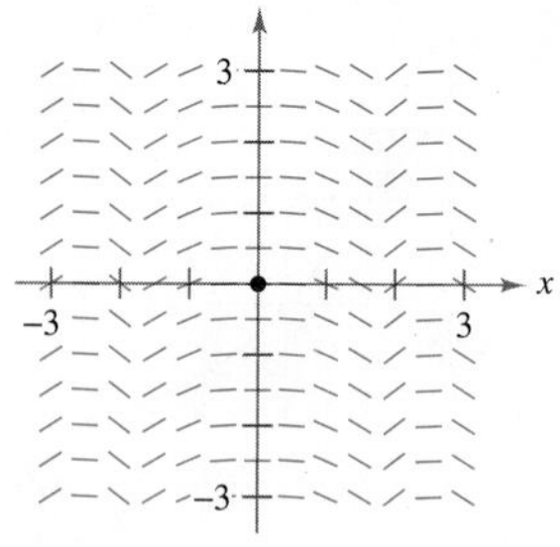

In Exercises 91 and 92, find the area of the region. Use a graphing utility to verify your result.

91. $\displaystyle\int_1^9 x\sqrt[3]{x - 1}\, dx$

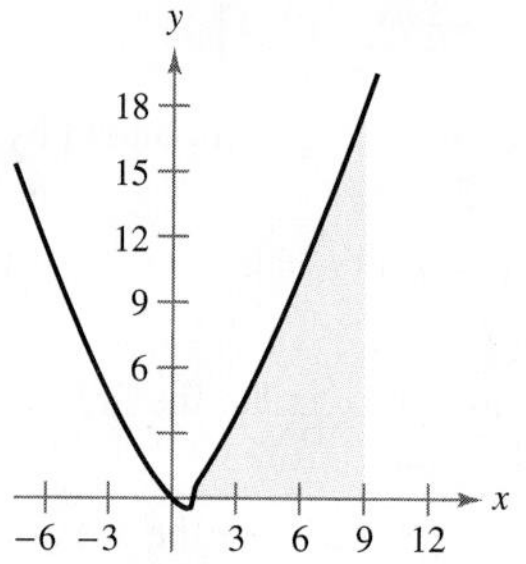

92. $\displaystyle\int_0^{\pi/2} [\cos x + \sin(2x)]\, dx$

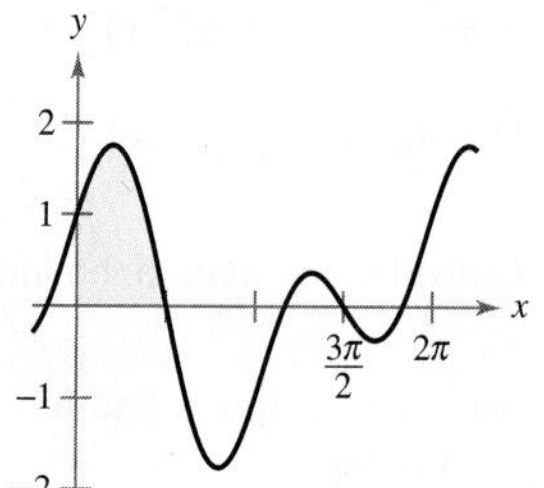

93. ***Fuel Cost*** Gasoline is increasing in price according to the equation $p = 1.20 + 0.04t$, where p is the dollar price per gallon and t is the time in years, with $t = 0$ representing 1990. An automobile is driven 15,000 miles a year and gets M miles per gallon. The annual fuel cost is

$$C = \frac{15{,}000}{M}\int_t^{t+1} p\, dt.$$

Estimate the annual fuel cost in (a) 2000 and (b) 2005.

94. ***Respiratory Cycle*** After exercising for a few minutes, a person has a respiratory cycle for which the rate of air intake is

$$v = 1.75 \sin\frac{\pi t}{2}.$$

Find the volume, in liters, of air inhaled during one cycle by integrating the function over the interval $[0, 2]$.

In Exercises 95–98, use the Trapezoidal Rule and Simpson's Rule with $n = 4$, and use the integration capabilities of a graphing utility, to approximate the definite integral. Compare the results.

95. $\displaystyle\int_1^2 \frac{1}{1 + x^3}\, dx$

96. $\displaystyle\int_0^1 \frac{x^{3/2}}{3 - x^2}\, dx$

97. $\displaystyle\int_0^{\pi/2} \sqrt{x}\cos x\, dx$

98. $\displaystyle\int_0^\pi \sqrt{1 + \sin^2 x}\, dx$

P.S. Problem Solving

See www.CalcChat.com for worked-out solutions to odd-numbered exercises.

1. Let $L(x) = \int_1^x \frac{1}{t}\,dt,\ x > 0.$

(a) Find $L(1)$.

(b) Find $L'(x)$ and $L'(1)$.

(c) Use a graphing utility to approximate the value of x (to three decimal places) for which $L(x) = 1$.

(d) Prove that $L(x_1x_2) = L(x_1) + L(x_2)$ for all positive values of x_1 and x_2.

2. Let $F(x) = \int_2^x \sin t^2\,dt.$

(a) Use a graphing utility to complete the table.

x	0	1.0	1.5	1.9	2.0
$F(x)$					
x	2.1	2.5	3.0	4.0	5.0
$F(x)$					

(b) Let $G(x) = \frac{1}{x-2}F(x) = \frac{1}{x-2}\int_2^x \sin t^2\,dt$. Use a graphing utility to complete the table and estimate $\lim_{x\to 2} G(x)$.

x	1.9	1.95	1.99	2.01	2.1
$G(x)$					

(c) Use the definition of the derivative to find the exact value of the limit $\lim_{x\to 2} G(x)$.

In Exercises 3 and 4, (a) write the area under the graph of the given function defined on the given interval as a limit. Then use a computer algebra system to (b) evaluate the sum in part (a), and (c) evaluate the limit using the result of part (b).

3. $y = x^4 - 4x^3 + 4x^2,\quad [0, 2]$

$\left(\textit{Hint: } \sum_{i=1}^{n} i^4 = \frac{n(n+1)(2n+1)(3n^2+3n-1)}{30}\right)$

4. $y = \frac{1}{2}x^5 + 2x^3,\quad [0, 2]$

$\left(\textit{Hint: } \sum_{i=1}^{n} i^5 = \frac{n^2(n+1)^2(2n^2+2n-1)}{12}\right)$

5. The **Fresnel function** S is defined by the integral

$$S(x) = \int_0^x \sin\left(\frac{\pi t^2}{2}\right) dt.$$

(a) Graph the function $y = \sin\left(\frac{\pi x^2}{2}\right)$ on the interval $[0, 3]$.

(b) Use the graph in part (a) to sketch the graph of S on the interval $[0, 3]$.

(c) Locate all relative extrema of S on the interval $(0, 3)$.

(d) Locate all points of inflection of S on the interval $(0, 3)$.

6. The **Two-Point Gaussian Quadrature Approximation** for f is

$$\int_{-1}^{1} f(x)\,dx \approx f\left(-\frac{1}{\sqrt{3}}\right) + f\left(\frac{1}{\sqrt{3}}\right).$$

(a) Use this formula to approximate $\int_{-1}^{1} \cos x\,dx$. Find the error of the approximation.

(b) Use this formula to approximate $\int_{-1}^{1} \frac{1}{1+x^2}\,dx$.

(c) Prove that the Two-Point Gaussian Quadrature Approximation is exact for all polynomials of degree 3 or less.

7. Archimedes showed that the area of a parabolic arch is equal to $\frac{2}{3}$ the product of the base and the height (see figure).

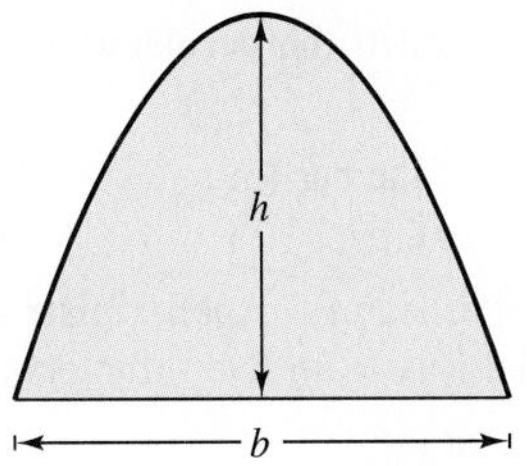

(a) Graph the parabolic arch bounded by $y = 9 - x^2$ and the x-axis. Use an appropriate integral to find the area A.

(b) Find the base and height of the arch and verify Archimedes' formula.

(c) Prove Archimedes' formula for a general parabola.

8. Galileo Galilei (1564–1642) stated the following proposition concerning falling objects:

The time in which any space is traversed by a uniformly accelerating body is equal to the time in which that same space would be traversed by the same body moving at a uniform speed whose value is the mean of the highest speed of the accelerating body and the speed just before acceleration began.

Use the techniques of this chapter to verify this proposition.

9. The graph of the function f consists of the three line segments joining the points $(0, 0)$, $(2, -2)$, $(6, 2)$, and $(8, 3)$. The function F is defined by the integral

$$F(x) = \int_0^x f(t)\,dt.$$

(a) Sketch the graph of f.

(b) Complete the table.

x	0	1	2	3	4	5	6	7	8
$F(x)$									

(c) Find the extrema of F on the interval $[0, 8]$.

(d) Determine all points of inflection of F on the interval $(0, 8)$.

10. A car is traveling in a straight line for 1 hour. Its velocity v in miles per hour at six-minute intervals is shown in the table.

t (hours)	0	0.1	0.2	0.3	0.4	0.5
v (mi/h)	0	10	20	40	60	50

t (hours)	0.6	0.7	0.8	0.9	1.0
v (mi/h)	40	35	40	50	65

(a) Produce a reasonable graph of the velocity function v by graphing these points and connecting them with a smooth curve.

(b) Find the open intervals over which the acceleration a is positive.

(c) Find the average acceleration of the car (in miles per hour squared) over the interval $[0, 0.4]$.

(d) What does the integral $\int_0^1 v(t)\,dt$ signify? Approximate this integral using the Trapezoidal Rule with five subintervals.

(e) Approximate the acceleration at $t = 0.8$.

11. Prove $\displaystyle\int_0^x f(t)(x - t)\,dt = \int_0^x \left(\int_0^t f(v)\,dv\right) dt.$

12. Prove $\displaystyle\int_a^b f(x)f'(x)\,dx = \tfrac{1}{2}([f(b)]^2 - [f(a)]^2).$

13. Use an appropriate Riemann sum to evaluate the limit

$$\lim_{n\to\infty} \frac{\sqrt{1} + \sqrt{2} + \sqrt{3} + \cdot\cdot\cdot + \sqrt{n}}{n^{3/2}}.$$

14. Use an appropriate Riemann sum to evaluate the limit

$$\lim_{n\to\infty} \frac{1^5 + 2^5 + 3^5 + \cdot\cdot\cdot + n^5}{n^6}.$$

15. Suppose that f is integrable on $[a, b]$ and $0 < m \le f(x) \le M$ for all x in the interval $[a, b]$. Prove that

$$m(a - b) \le \int_a^b f(x)\,dx \le M(b - a).$$

Use this result to estimate $\int_0^1 \sqrt{1 + x^4}\,dx.$

16. Let f be continuous on the interval $[0, b]$ where $f(x) + f(b - x) \ne 0$ on $[0, b]$.

(a) Show that $\displaystyle\int_0^b \frac{f(x)}{f(x) + f(b - x)}\,dx = \frac{b}{2}.$

(b) Use the result in part (a) to evaluate

$$\int_0^1 \frac{\sin x}{\sin(1 - x) + \sin x}\,dx.$$

(c) Use the result in part (a) to evaluate

$$\int_0^3 \frac{\sqrt{x}}{\sqrt{x} + \sqrt{3 - x}}\,dx.$$

17. Verify that

$$\sum_{i=1}^{n} i^2 = \frac{n(n + 1)(2n + 1)}{6}$$

by showing the following.

(a) $(1 + i)^3 - i^3 = 3i^2 + 3i + 1$

(b) $(n + 1)^3 = \displaystyle\sum_{i=1}^{n} (3i^2 + 3i + 1) + 1$

(c) $\displaystyle\sum_{i=1}^{n} i^2 = \frac{n(n + 1)(2n + 1)}{6}$

18. Prove that if f is a continuous function on a closed interval $[a, b]$, then

$$\left|\int_a^b f(x)\,dx\right| \le \int_a^b |f(x)|\,dx.$$

19. Let

$$I = \int_0^4 f(x)\,dx$$

where f is shown in the figure. Let $L(n)$ and $R(n)$ represent the Riemann sums using the left-hand endpoints and right-hand endpoints of n subintervals of equal width. (Assume n is even.) Let $T(n)$ and $S(n)$ be the corresponding values of the Trapezoidal Rule and Simpson's Rule.

(a) For any n, list $L(n)$, $R(n)$, $T(n)$, and I in increasing order.

(b) Approximate $S(4)$.

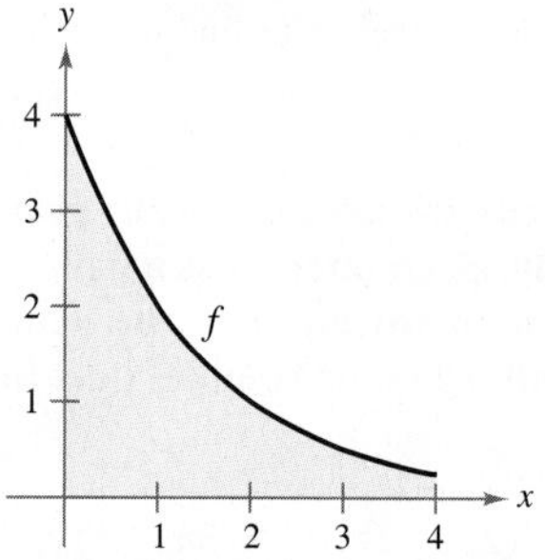

20. The **sine integral function**

$$\mathrm{Si}(x) = \int_0^x \frac{\sin t}{t}\,dt$$

is often used in engineering. The function $f(t) = \dfrac{\sin t}{t}$ is not defined at $t = 0$, but its limit is 1 as $t \to 0$. So, define $f(0) = 1$. Then f is continuous everywhere.

(a) Use a graphing utility to graph $\mathrm{Si}(x)$.

(b) At what values of x does $\mathrm{Si}(x)$ have relative maxima?

(c) Find the coordinates of the first inflection point where $x > 0$.

(d) Decide whether $\mathrm{Si}(x)$ has any horizontal asymptotes. If so, identify each.

5 Logarithmic, Exponential, and Other Transcendental Functions

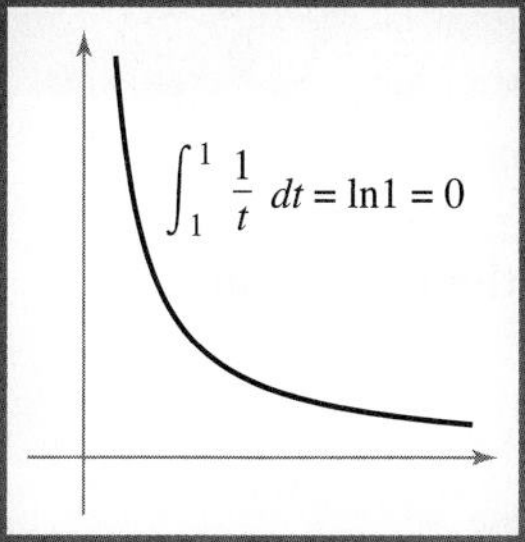

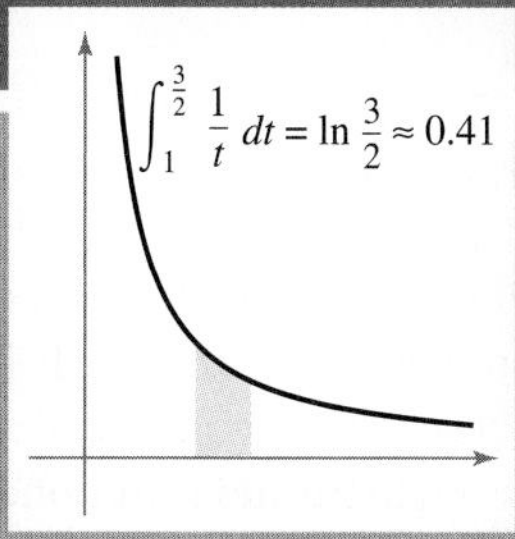

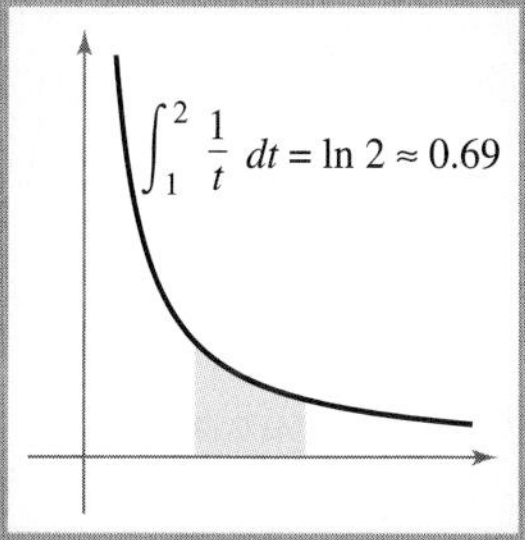

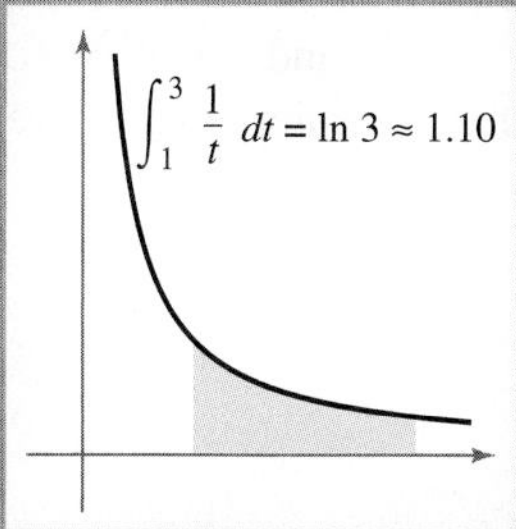

In Chapter 5, you will see how the function $f(x) = 1/x$ can be used to define the natural logarithmic function. To do this, consider the definite integral

$$\int_1^x \frac{1}{t}\,dt.$$

When $x < 1$, the value of this definite integral is negative. When $x = 1$, the value is 0. When $x > 1$, the value is positive.

A geyser is a hot spring that erupts periodically when groundwater in a confined space boils and produces steam. The steam forces overlying water up and out through an opening on Earth's surface. The temperature at which water boils is affected by pressure. Do you think an *increase* or a *decrease* in pressure causes water to boil at a lower temperature? Why?

Brian Maslyar/Index Stock

Section 5.1 The Natural Logarithmic Function: Differentiation

- **Develop and use properties of the natural logarithmic function.**
- **Understand the definition of the number *e*.**
- **Find derivatives of functions involving the natural logarithmic function.**

The Granger Collection

John Napier (1550–1617)

Logarithms were invented by the Scottish mathematician John Napier. Although he did not introduce the *natural* logarithmic function, it is sometimes called the *Napierian* logarithm.

The Natural Logarithmic Function

Recall that the General Power Rule

$$\int x^n \, dx = \frac{x^{n+1}}{n+1} + C, \quad n \neq -1 \qquad \text{General Power Rule}$$

has an important disclaimer—it doesn't apply when $n = -1$. Consequently, you have not yet found an antiderivative for the function $f(x) = 1/x$. In this section, you will use the Second Fundamental Theorem of Calculus to *define* such a function. This antiderivative is a function that you have not encountered previously in the text. It is neither algebraic nor trigonometric, but falls into a new class of functions called *logarithmic functions*. This particular function is the **natural logarithmic function.**

Definition of the Natural Logarithmic Function

The **natural logarithmic function** is defined by

$$\ln x = \int_1^x \frac{1}{t} \, dt, \quad x > 0.$$

The domain of the natural logarithmic function is the set of all positive real numbers.

From the definition, you can see that $\ln x$ is positive for $x > 1$ and negative for $0 < x < 1$, as shown in Figure 5.1. Moreover, $\ln(1) = 0$, because the upper and lower limits of integration are equal when $x = 1$.

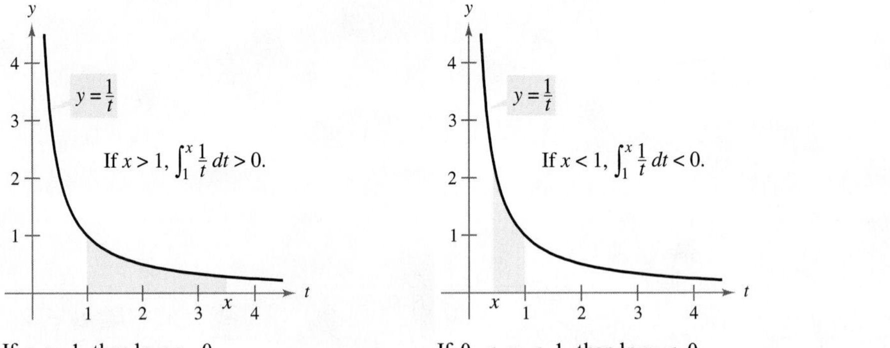

If $x > 1$, then $\ln x > 0$.

If $0 < x < 1$, then $\ln x < 0$.

Figure 5.1

EXPLORATION

Graphing the Natural Logarithmic Function Using *only* the definition of the natural logarithmic function, sketch a graph of the function. Explain your reasoning.

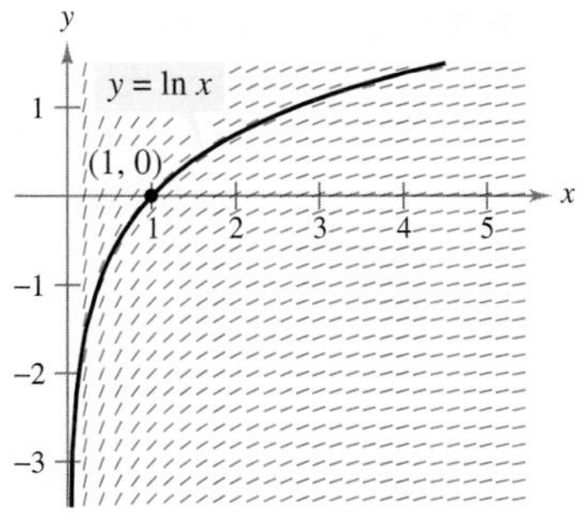

Each small line segment has a slope of $\frac{1}{x}$.
Figure 5.2

NOTE Slope fields can be helpful in getting a visual perspective of the directions of the solutions of a differential equation.

To sketch the graph of $y = \ln x$, you can think of the natural logarithmic function as an *antiderivative* given by the differential equation

$$\frac{dy}{dx} = \frac{1}{x}.$$

Figure 5.2 is a computer-generated graph, called a *slope (or direction) field*, showing small line segments of slope $1/x$. The graph of $y = \ln x$ is the solution that passes through the point $(1, 0)$. You will study slope fields in Section 6.1.

The following theorem lists some basic properties of the natural logarithmic function.

THEOREM 5.1 Properties of the Natural Logarithmic Function

The natural logarithmic function has the following properties.

1. The domain is $(0, \infty)$ and the range is $(-\infty, \infty)$.
2. The function is continuous, increasing, and one-to-one.
3. The graph is concave downward.

Proof The domain of $f(x) = \ln x$ is $(0, \infty)$ by definition. Moreover, the function is continuous because it is differentiable. It is increasing because its derivative

$$f'(x) = \frac{1}{x} \qquad \text{First derivative}$$

is positive for $x > 0$, as shown in Figure 5.3. It is concave downward because

$$f''(x) = -\frac{1}{x^2} \qquad \text{Second derivative}$$

is negative for $x > 0$. The proof that f is one-to-one is left as an exercise (see Exercise 111). The following limits imply that its range is the entire real line.

$$\lim_{x \to 0^+} \ln x = -\infty \quad \text{and} \quad \lim_{x \to \infty} \ln x = \infty$$

Verification of these two limits is given in Appendix A.

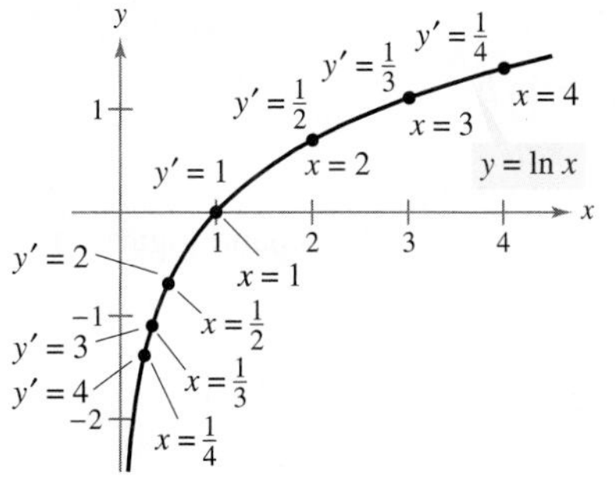

The natural logarithmic function is increasing, and its graph is concave downward.
Figure 5.3

Using the definition of the natural logarithmic function, you can prove several important properties involving operations with natural logarithms. If you are already familiar with logarithms, you will recognize that these properties are characteristic of all logarithms.

LOGARITHMS

Napier coined the term *logarithm*, from the two Greek words *logos* (or ratio) and *arithmos* (or number), to describe the theory that he spent 20 years developing and that first appeared in the book *Mirifici Logarithmorum canonis descriptio* (A Description of the Marvelous Rule of Logarithms).

THEOREM 5.2 Logarithmic Properties

If a and b are positive numbers and n is rational, then the following properties are true.

1. $\ln(1) = 0$
2. $\ln(ab) = \ln a + \ln b$
3. $\ln(a^n) = n \ln a$
4. $\ln\left(\frac{a}{b}\right) = \ln a - \ln b$

Proof The first property has already been discussed. The proof of the second property follows from the fact that two antiderivatives of the same function differ at most by a constant. From the Second Fundamental Theorem of Calculus and the definition of the natural logarithmic function, you know that

$$\frac{d}{dx}[\ln x] = \frac{d}{dx}\left[\int_1^x \frac{1}{t}\,dt\right] = \frac{1}{x}.$$

So, consider the two derivatives

$$\frac{d}{dx}[\ln(ax)] = \frac{a}{ax} = \frac{1}{x}$$

and

$$\frac{d}{dx}[\ln a + \ln x] = 0 + \frac{1}{x} = \frac{1}{x}.$$

Because $\ln(ax)$ and $(\ln a + \ln x)$ are both antiderivatives of $1/x$, they must differ at most by a constant.

$$\ln(ax) = \ln a + \ln x + C$$

By letting $x = 1$, you can see that $C = 0$. The third property can be proved similarly by comparing the derivatives of $\ln(x^n)$ and $n \ln x$. Finally, using the second and third properties, you can prove the fourth property.

$$\ln\left(\frac{a}{b}\right) = \ln[a(b^{-1})] = \ln a + \ln(b^{-1}) = \ln a - \ln b$$

Example 1 shows how logarithmic properties can be used to expand logarithmic expressions.

EXAMPLE 1 Expanding Logarithmic Expressions

a. $\ln\frac{10}{9} = \ln 10 - \ln 9$ — Property 4

b. $\ln\sqrt{3x+2} = \ln(3x+2)^{1/2}$ — Rewrite with rational exponent.

$= \frac{1}{2}\ln(3x+2)$ — Property 3

c. $\ln\frac{6x}{5} = \ln(6x) - \ln 5$ — Property 4

$= \ln 6 + \ln x - \ln 5$ — Property 2

d. $\ln\frac{(x^2+3)^2}{x\sqrt[3]{x^2+1}} = \ln(x^2+3)^2 - \ln\left(x\sqrt[3]{x^2+1}\right)$

$= 2\ln(x^2+3) - [\ln x + \ln(x^2+1)^{1/3}]$

$= 2\ln(x^2+3) - \ln x - \ln(x^2+1)^{1/3}$

$= 2\ln(x^2+3) - \ln x - \frac{1}{3}\ln(x^2+1)$

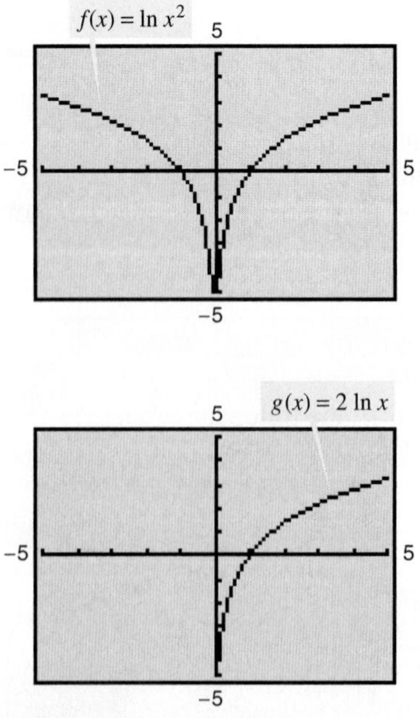

Figure 5.4

When using the properties of logarithms to rewrite logarithmic functions, you must check to see whether the domain of the rewritten function is the same as the domain of the original. For instance, the domain of $f(x) = \ln x^2$ is all real numbers except $x = 0$, and the domain of $g(x) = 2 \ln x$ is all positive real numbers. (See Figure 5.4.)

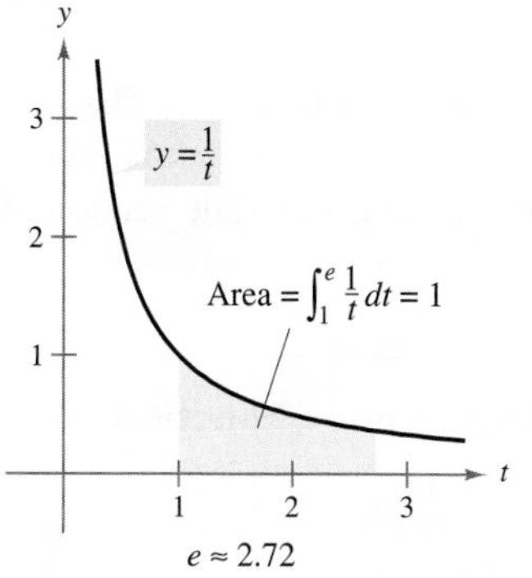

e is the base for the natural logarithm because $\ln e = 1$.
Figure 5.5

The Number e

It is likely that you have studied logarithms in an algebra course. There, without the benefit of calculus, logarithms would have been defined in terms of a **base** number. For example, common logarithms have a base of 10 and therefore $\log_{10}10 = 1$. (You will learn more about this in Section 5.5.)

The **base for the natural logarithm** is defined using the fact that the natural logarithmic function is continuous, is one-to-one, and has a range of $(-\infty, \infty)$. So, there must be a unique real number x such that $\ln x = 1$, as shown in Figure 5.5. This number is denoted by the letter e. It can be shown that e is irrational and has the following decimal approximation.

$$e \approx 2.71828182846$$

Definition of e

The letter e denotes the positive real number such that

$$\ln e = \int_1^e \frac{1}{t}\,dt = 1.$$

FOR FURTHER INFORMATION To learn more about the number e, see the article "Unexpected Occurrences of the Number e" by Harris S. Shultz and Bill Leonard in *Mathematics Magazine*. To view this article, go to the website *www.matharticles.com*.

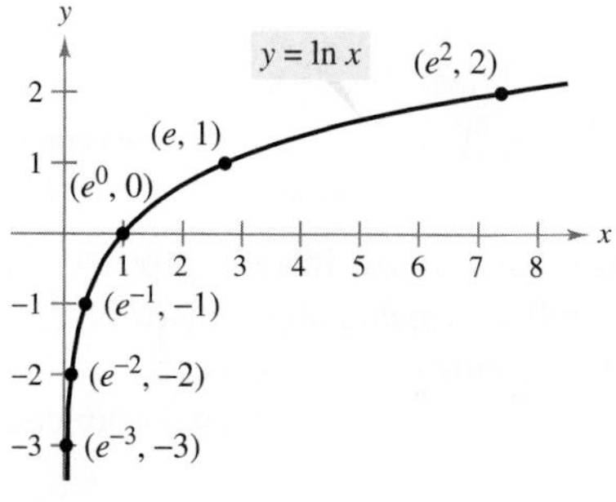

If $x = e^n$, then $\ln x = n$.
Figure 5.6

Once you know that $\ln e = 1$, you can use logarithmic properties to evaluate the natural logarithms of several other numbers. For example, by using the property

$$\begin{aligned}\ln(e^n) &= n \ln e \\ &= n(1) \\ &= n\end{aligned}$$

you can evaluate $\ln(e^n)$ for various values of n, as shown in the table and in Figure 5.6.

x	$\frac{1}{e^3} \approx 0.050$	$\frac{1}{e^2} \approx 0.135$	$\frac{1}{e} \approx 0.368$	$e^0 = 1$	$e \approx 2.718$	$e^2 \approx 7.389$
ln x	-3	-2	-1	0	1	2

The logarithms shown in the table above are convenient because the x-values are integer powers of e. Most logarithmic expressions are, however, best evaluated with a calculator.

EXAMPLE 2 Evaluating Natural Logarithmic Expressions

a. $\ln 2 \approx 0.693$

b. $\ln 32 \approx 3.466$

c. $\ln 0.1 \approx -2.303$

The Derivative of the Natural Logarithmic Function

The derivative of the natural logarithmic function is given in Theorem 5.3. The first part of the theorem follows from the definition of the natural logarithmic function as an antiderivative. The second part of the theorem is simply the Chain Rule version of the first part.

THEOREM 5.3 Derivative of the Natural Logarithmic Function

Let u be a differentiable function of x.

1. $\frac{d}{dx}[\ln x] = \frac{1}{x}, \quad x > 0$ **2.** $\frac{d}{dx}[\ln u] = \frac{1}{u}\frac{du}{dx} = \frac{u'}{u}, \quad u > 0$

EXAMPLE 3 Differentiation of Logarithmic Functions

a. $\frac{d}{dx}[\ln(2x)] = \frac{u'}{u} = \frac{2}{2x} = \frac{1}{x}$ $u = 2x$

b. $\frac{d}{dx}[\ln(x^2+1)] = \frac{u'}{u} = \frac{2x}{x^2+1}$ $u = x^2 + 1$

c. $\frac{d}{dx}[x \ln x] = x\left(\frac{d}{dx}[\ln x]\right) + (\ln x)\left(\frac{d}{dx}[x]\right)$ Product Rule

$= x\left(\frac{1}{x}\right) + (\ln x)(1) = 1 + \ln x$

d. $\frac{d}{dx}[(\ln x)^3] = 3(\ln x)^2 \frac{d}{dx}[\ln x]$ Chain Rule

$= 3(\ln x)^2 \frac{1}{x}$

EXPLORATION

Use a graphing utility to graph

$$y_1 = \frac{1}{x}$$

and

$$y_2 = \frac{d}{dx}[\ln x]$$

in the same viewing window, in which $0.1 \le x \le 5$ and $-2 \le y \le 8$. Explain why the graphs appear to be identical.

Napier used logarithmic properties to simplify *calculations* involving products, quotients, and powers. Of course, given the availability of calculators, there is now little need for this particular application of logarithms. However, there is great value in using logarithmic properties to simplify *differentiation* involving products, quotients, and powers.

EXAMPLE 4 Logarithmic Properties as Aids to Differentiation

Differentiate $f(x) = \ln\sqrt{x+1}$.

Solution Because

$$f(x) = \ln\sqrt{x+1} = \ln(x+1)^{1/2} = \frac{1}{2}\ln(x+1)$$ Rewrite before differentiating.

you can write

$$f'(x) = \frac{1}{2}\left(\frac{1}{x+1}\right) = \frac{1}{2(x+1)}.$$ Differentiate.

indicates that in the HM mathSpace® *CD-ROM and the online* Eduspace® *system for this text, you will find an Open Exploration, which further explores this example using the computer algebra systems* Maple, Mathcad, Mathematica, *and* Derive.

EXAMPLE 5 Logarithmic Properties as Aids to Differentiation

Differentiate $f(x) = \ln \frac{x(x^2 + 1)^2}{\sqrt{2x^3 - 1}}$.

Solution

$$f(x) = \ln \frac{x(x^2 + 1)^2}{\sqrt{2x^3 - 1}}$$ Write original function.

$$= \ln x + 2\ln(x^2 + 1) - \frac{1}{2}\ln(2x^3 - 1)$$ Rewrite before differentiating.

$$f'(x) = \frac{1}{x} + 2\left(\frac{2x}{x^2 + 1}\right) - \frac{1}{2}\left(\frac{6x^2}{2x^3 - 1}\right)$$ Differentiate.

$$= \frac{1}{x} + \frac{4x}{x^2 + 1} - \frac{3x^2}{2x^3 - 1}$$ Simplify.

NOTE In Examples 4 and 5, be sure you see the benefit of applying logarithmic properties *before* differentiating. Consider, for instance, the difficulty of direct differentiation of the function given in Example 5.

On occasion, it is convenient to use logarithms as aids in differentiating *nonlogarithmic* functions. This procedure is called **logarithmic differentiation.**

EXAMPLE 6 Logarithmic Differentiation

Find the derivative of

$$y = \frac{(x - 2)^2}{\sqrt{x^2 + 1}}, \quad x \neq 2.$$

Solution Note that $y > 0$ for all $x \neq 2$. So, $\ln y$ is defined. Begin by taking the natural logarithm of each side of the equation. Then apply logarithmic properties and differentiate implicitly. Finally, solve for y'.

$$y = \frac{(x - 2)^2}{\sqrt{x^2 + 1}}, \quad x \neq 2$$ Write original equation.

$$\ln y = \ln \frac{(x - 2)^2}{\sqrt{x^2 + 1}}$$ Take natural log of each side.

$$\ln y = 2\ln(x - 2) - \frac{1}{2}\ln(x^2 + 1)$$ Logarithmic properties

$$\frac{y'}{y} = 2\left(\frac{1}{x - 2}\right) - \frac{1}{2}\left(\frac{2x}{x^2 + 1}\right)$$ Differentiate.

$$= \frac{2}{x - 2} - \frac{x}{x^2 + 1}$$ Simplify.

$$y' = y\left(\frac{2}{x - 2} - \frac{x}{x^2 + 1}\right)$$ Solve for y'.

$$= \frac{(x - 2)^2}{\sqrt{x^2 + 1}}\left[\frac{x^2 + 2x + 2}{(x - 2)(x^2 + 1)}\right]$$ Substitute for y.

$$= \frac{(x - 2)(x^2 + 2x + 2)}{(x^2 + 1)^{3/2}}$$ Simplify.

Because the natural logarithm is undefined for negative numbers, you will often encounter expressions of the form $\ln|u|$. The following theorem states that you can differentiate functions of the form $y = \ln|u|$ as if the absolute value sign were not present.

THEOREM 5.4 Derivative Involving Absolute Value

If u is a differentiable function of x such that $u \neq 0$, then

$$\frac{d}{dx}[\ln|u|] = \frac{u'}{u}.$$

Proof If $u > 0$, then $|u| = u$, and the result follows from Theorem 5.3. If $u < 0$, then $|u| = -u$, and you have

$$\begin{aligned}\frac{d}{dx}[\ln|u|] &= \frac{d}{dx}[\ln(-u)]\\ &= \frac{-u'}{-u}\\ &= \frac{u'}{u}.\end{aligned}$$

EXAMPLE 7 Derivative Involving Absolute Value

Find the derivative of

$$f(x) = \ln|\cos x|.$$

Solution Using Theorem 5.4, let $u = \cos x$ and write

$$\begin{aligned}\frac{d}{dx}[\ln|\cos x|] &= \frac{u'}{u} && \frac{d}{dx}[\ln|u|] = \frac{u'}{u}\\ &= \frac{-\sin x}{\cos x} && u = \cos x\\ &= -\tan x. && \text{Simplify.}\end{aligned}$$

EXAMPLE 8 Finding Relative Extrema

Locate the relative extrema of

$$y = \ln(x^2 + 2x + 3).$$

Solution Differentiating y, you obtain

$$\frac{dy}{dx} = \frac{2x + 2}{x^2 + 2x + 3}.$$

Because $dy/dx = 0$ when $x = -1$, you can apply the First Derivative Test and conclude that the point $(-1, \ln 2)$ is a relative minimum. Because there are no other critical points, it follows that this is the only relative extremum (see Figure 5.7).

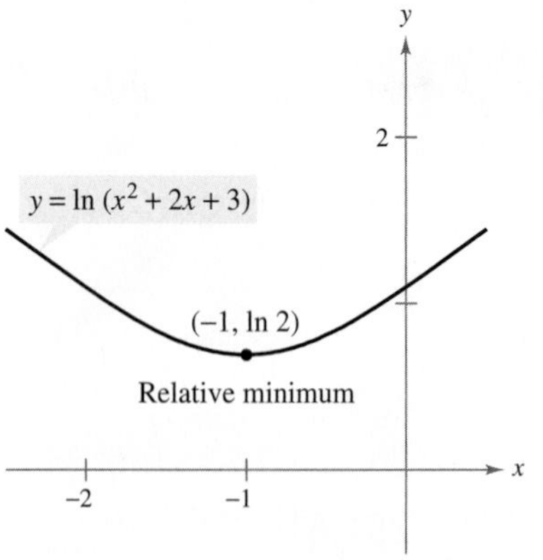

The derivative of y changes from negative to positive at $x = -1$.
Figure 5.7

Exercises for Section 5.1

See www.CalcChat.com for worked-out solutions to odd-numbered exercises.

1. Complete the table below. Use a graphing utility and Simpson's Rule with $n = 10$ to approximate the integral $\int_1^x (1/t)\,dt$.

x	0.5	1.5	2	2.5	3	3.5	4
$\int_1^x (1/t)\,dt$							

2. (a) Plot the points generated in Exercise 1 and connect them with a smooth curve. Compare the result with the graph of $y = \ln x$.

(b) Use a graphing utility to graph $y = \int_1^x (1/t)\,dt$ for $0.2 \le x \le 4$. Compare the result with the graph of $y = \ln x$.

In Exercises 3–6, use a graphing utility to evaluate the logarithm by (a) using the natural logarithm key and (b) using the integration capabilities to evaluate the integral $\int_1^x (1/t)\,dt$.

3. $\ln 45$ **4.** $\ln 8.3$

5. $\ln 0.8$ **6.** $\ln 0.6$

In Exercises 7–10, match the function with its graph. [The graphs are labeled (a), (b), (c), and (d).]

(a)

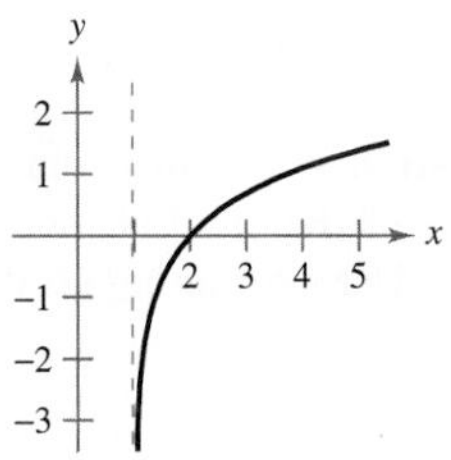

(b)

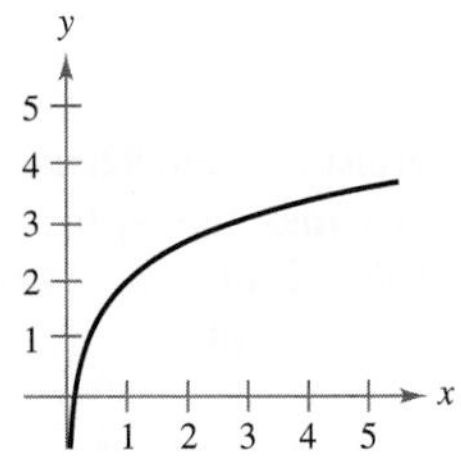

(c)

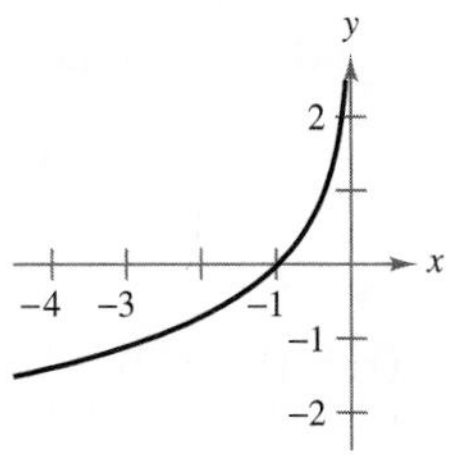

(d)

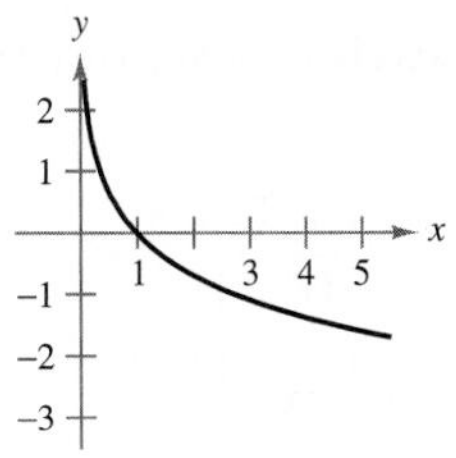

7. $f(x) = \ln x + 2$ **8.** $f(x) = -\ln x$

9. $f(x) = \ln(x - 1)$ **10.** $f(x) = -\ln(-x)$

In Exercises 11–16, sketch the graph of the function and state its domain.

11. $f(x) = 3 \ln x$ **12.** $f(x) = -2 \ln x$

13. $f(x) = \ln 2x$ **14.** $f(x) = \ln|x|$

15. $f(x) = \ln(x - 1)$ **16.** $g(x) = 2 + \ln x$

In Exercises 17 and 18, use the properties of logarithms to approximate the indicated logarithms, given that $\ln 2 \approx 0.6931$ and $\ln 3 \approx 1.0986$.

17. (a) $\ln 6$ (b) $\ln \frac{2}{3}$ (c) $\ln 81$ (d) $\ln \sqrt{3}$

18. (a) $\ln 0.25$ (b) $\ln 24$ (c) $\ln \sqrt[3]{12}$ (d) $\ln \frac{1}{72}$

In Exercises 19–28, use the properties of logarithms to expand the logarithmic expression.

19. $\ln \frac{2}{3}$ **20.** $\ln \sqrt{2^3}$

21. $\ln \dfrac{xy}{z}$ **22.** $\ln(xyz)$

23. $\ln \sqrt[3]{a^2 + 1}$ **24.** $\ln \sqrt{a - 1}$

25. $\ln\left(\dfrac{x^2 - 1}{x^3}\right)^3$ **26.** $\ln(3e^2)$

27. $\ln z(z - 1)^2$ **28.** $\ln \dfrac{1}{e}$

In Exercises 29–34, write the expression as a logarithm of a single quantity.

29. $\ln(x - 2) - \ln(x + 2)$ **30.** $3 \ln x + 2 \ln y - 4 \ln z$

31. $\frac{1}{3}[2 \ln(x + 3) + \ln x - \ln(x^2 - 1)]$

32. $2[\ln x - \ln(x + 1) - \ln(x - 1)]$

33. $2 \ln 3 - \frac{1}{2} \ln(x^2 + 1)$

34. $\frac{3}{2}[\ln(x^2 + 1) - \ln(x + 1) - \ln(x - 1)]$

In Exercises 35 and 36, (a) verify that $f = g$ by using a graphing utility to graph f and g in the same viewing window. (b) Then verify that $f = g$ algebraically.

35. $f(x) = \ln \dfrac{x^2}{4}$, $x > 0$, $g(x) = 2 \ln x - \ln 4$

36. $f(x) = \ln \sqrt{x(x^2 + 1)}$, $g(x) = \frac{1}{2}[\ln x + \ln(x^2 + 1)]$

In Exercises 37–40, find the limit.

37. $\lim\limits_{x \to 3^+} \ln(x - 3)$ **38.** $\lim\limits_{x \to 6^-} \ln(6 - x)$

39. $\lim\limits_{x \to 2^-} \ln[x^2(3 - x)]$ **40.** $\lim\limits_{x \to 5^+} \ln \dfrac{x}{\sqrt{x - 4}}$

In Exercises 41–44, find an equation of the tangent line to the graph of the logarithmic function at the point (1, 0).

41. $y = \ln x^3$

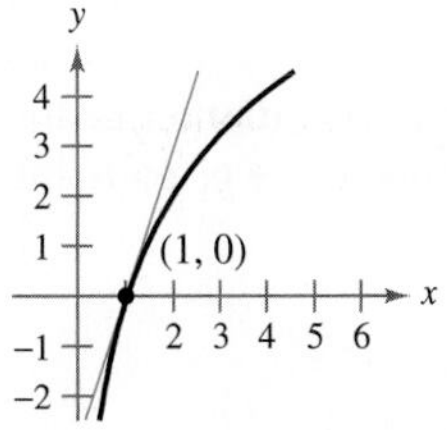

42. $y = \ln x^{3/2}$

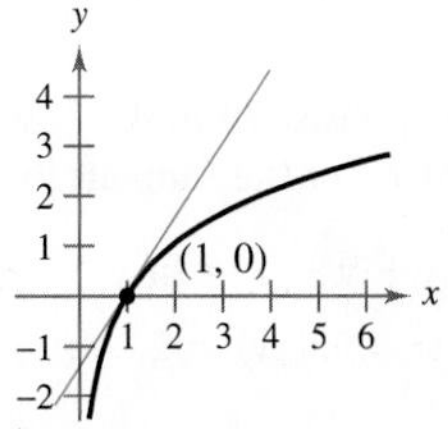

43. $y = \ln x^2$

44. $y = \ln x^{1/2}$

In Exercises 45–70, find the derivative of the function.

45. $g(x) = \ln x^2$

46. $h(x) = \ln(2x^2 + 1)$

47. $y = (\ln x)^4$

48. $y = x \ln x$

49. $y = \ln\left(x\sqrt{x^2 - 1}\right)$

50. $y = \ln\sqrt{x^2 - 4}$

51. $f(x) = \ln\left(\dfrac{x}{x^2 + 1}\right)$

52. $f(x) = \ln\left(\dfrac{2x}{x + 3}\right)$

53. $g(t) = \dfrac{\ln t}{t^2}$

54. $h(t) = \dfrac{\ln t}{t}$

55. $y = \ln(\ln x^2)$

56. $y = \ln(\ln x)$

57. $y = \ln\sqrt{\dfrac{x + 1}{x - 1}}$

58. $y = \ln\sqrt[3]{\dfrac{x - 1}{x + 1}}$

59. $f(x) = \ln\left(\dfrac{\sqrt{4 + x^2}}{x}\right)$

60. $f(x) = \ln\left(x + \sqrt{4 + x^2}\right)$

61. $y = \dfrac{-\sqrt{x^2 + 1}}{x} + \ln\left(x + \sqrt{x^2 + 1}\right)$

62. $y = \dfrac{-\sqrt{x^2 + 4}}{2x^2} - \dfrac{1}{4}\ln\left(\dfrac{2 + \sqrt{x^2 + 4}}{x}\right)$

63. $y = \ln|\sin x|$

64. $y = \ln|\csc x|$

65. $y = \ln\left|\dfrac{\cos x}{\cos x - 1}\right|$

66. $y = \ln|\sec x + \tan x|$

67. $y = \ln\left|\dfrac{-1 + \sin x}{2 + \sin x}\right|$

68. $y = \ln\sqrt{2 + \cos^2 x}$

69. $f(x) = \displaystyle\int_2^{\ln(2x)} (t + 1)\, dt$

70. $g(x) = \displaystyle\int_1^{\ln x} (t^2 + 3)\, dt$

In Exercises 71–76, (a) find an equation of the tangent line to the graph of f at the given point, (b) use a graphing utility to graph the function and its tangent line at the point, and (c) use the *derivative* feature of a graphing utility to confirm your results.

71. $f(x) = 3x^2 - \ln x, \quad (1, 3)$

72. $f(x) = 4 - x^2 - \ln\left(\frac{1}{2}x + 1\right), \quad (0, 4)$

73. $f(x) = \ln\sqrt{1 + \sin^2 x}, \quad \left(\dfrac{\pi}{4}, \ln\sqrt{\dfrac{3}{2}}\right)$

74. $f(x) = \sin(2x)\ln(x^2), \quad (1, 0)$

75. $f(x) = x^3 \ln x, \quad (1, 0)$

76. $f(x) = \dfrac{1}{2}x\ln(x^2), \quad (-1, 0)$

In Exercises 77 and 78, use implicit differentiation to find dy/dx.

77. $x^2 - 3\ln y + y^2 = 10$

78. $\ln xy + 5x = 30$

In Exercises 79 and 80, use implicit differentiation to find an equation of the tangent line to the graph at the given point.

79. $x + y - 1 = \ln(x^2 + y^2), \quad (1, 0)$

80. $y^2 + \ln(xy) = 2, \quad (e, 1)$

In Exercises 81 and 82, show that the function is a solution of the differential equation.

Function	*Differential Equation*
81. $y = 2\ln x + 3$	$xy'' + y' = 0$
82. $y = x\ln x - 4x$	$x + y - xy' = 0$

In Exercises 83–88, locate any relative extrema and inflection points. Use a graphing utility to confirm your results.

83. $y = \dfrac{x^2}{2} - \ln x$

84. $y = x - \ln x$

85. $y = x\ln x$

86. $y = \dfrac{\ln x}{x}$

87. $y = \dfrac{x}{\ln x}$

88. $y = x^2 \ln\dfrac{x}{4}$

Linear and Quadratic Approximations **In Exercises 89 and 90, use a graphing utility to graph the function. Then graph**

$$P_1(x) = f(1) + f'(1)(x - 1)$$

and

$$P_2(x) = f(1) + f'(1)(x - 1) + \tfrac{1}{2}f''(1)(x - 1)^2$$

in the same viewing window. Compare the values of f, P_1, and P_2 and their first derivatives at $x = 1$.

89. $f(x) = \ln x$

90. $f(x) = x\ln x$

In Exercises 91 and 92, use Newton's Method to approximate, to three decimal places, the x-coordinate of the point of intersection of the graphs of the two equations. Use a graphing utility to verify your result.

91. $y = \ln x, \quad y = -x$

92. $y = \ln x, \quad y = 3 - x$

In Exercises 93–98, use logarithmic differentiation to find dy/dx.

93. $y = x\sqrt{x^2 - 1}$

94. $y = \sqrt{(x - 1)(x - 2)(x - 3)}$

95. $y = \dfrac{x^2\sqrt{3x - 2}}{(x - 1)^2}$

96. $y = \sqrt{\dfrac{x^2 - 1}{x^2 + 1}}$

97. $y = \dfrac{x(x - 1)^{3/2}}{\sqrt{x + 1}}$

98. $y = \dfrac{(x + 1)(x + 2)}{(x - 1)(x - 2)}$

Writing About Concepts

99. In your own words, state the properties of the natural logarithmic function.

100. Define the base for the natural logarithmic function.

101. Let f be a function that is positive and differentiable on the entire real line. Let $g(x) = \ln f(x)$.

(a) If g is increasing, must f be increasing? Explain.

(b) If the graph of f is concave upward, must the graph of g be concave upward? Explain.

Writing About Concepts *(continued)*

102. Consider the function $f(x) = x - 2 \ln x$ on $[1, 3]$.

(a) Explain why Rolle's Theorem (Section 3.2) does not apply.

(b) Do you think the conclusion of Rolle's Theorem is true for f? Explain.

True or False? **In Exercises 103 and 104, determine whether the statement is true or false. If it is false, explain why or give an example that shows it is false.**

103. $\ln(x + 25) = \ln x + \ln 25$

104. If $y = \ln \pi$, then $y' = 1/\pi$.

105. ***Home Mortgage*** The term t (in years) of a \$120,000 home mortgage at 10% interest can be approximated by

$$t = \frac{5.315}{-6.7968 + \ln x}, \quad x > 1000$$

where x is the monthly payment in dollars.

(a) Use a graphing utility to graph the model.

(b) Use the model to approximate the term of a home mortgage for which the monthly payment is \$1167.41. What is the total amount paid?

(c) Use the model to approximate the term of a home mortgage for which the monthly payment is \$1068.45. What is the total amount paid?

(d) Find the instantaneous rate of change of t with respect to x when $x = 1167.41$ and $x = 1068.45$.

(e) Write a short paragraph describing the benefit of the higher monthly payment.

106. ***Sound Intensity*** The relationship between the number of decibels β and the intensity of a sound I in watts per centimeter squared is

$$\beta = 10 \log_{10}\left(\frac{I}{10^{-16}}\right).$$

Use the properties of logarithms to write the formula in simpler form, and determine the number of decibels of a sound with an intensity of 10^{-10} watts per square centimeter.

107. ***Modeling Data*** The table shows the temperature T (°F) at which water boils at selected pressures p (pounds per square inch). *(Source: Standard Handbook of Mechanical Engineers)*

p	5	10	14.696 (1 atm)	20
T	162.24°	193.21°	212.00°	227.96°

p	30	40	60	80	100
T	250.33°	267.25°	292.71°	312.03°	327.81°

A model that approximates the data is

$$T = 87.97 + 34.96 \ln p + 7.91\sqrt{p}.$$

(a) Use a graphing utility to plot the data and graph the model.

(b) Find the rate of change of T with respect to p when $p = 10$ and $p = 70$.

(c) Use a graphing utility to graph T'. Find $\lim_{p \to \infty} T'(p)$ and interpret the result in the context of the problem.

108. ***Modeling Data*** The atmospheric pressure decreases with increasing altitude. At sea level, the average air pressure is one atmosphere (1.033227 kilograms per square centimeter). The table shows the pressures p (in atmospheres) at selected altitudes h (in kilometers).

h	0	5	10	15	20	25
p	1	0.55	0.25	0.12	0.06	0.02

(a) Use a graphing utility to find a model of the form $p = a + b \ln h$ for the data. Explain why the result is an error message.

(b) Use a graphing utility to find the logarithmic model $h = a + b \ln p$ for the data.

(c) Use a graphing utility to plot the data and graph the model.

(d) Use the model to estimate the altitude when $p = 0.75$.

(e) Use the model to estimate the pressure when $h = 13$.

(f) Use the model to find the rate of change of pressure when $h = 5$ and $h = 20$. Interpret the results.

109. ***Tractrix*** A person walking along a dock drags a boat by a 10-meter rope. The boat travels along a path known as a *tractrix* (see figure). The equation of this path is

$$y = 10 \ln\left(\frac{10 + \sqrt{100 - x^2}}{x}\right) - \sqrt{100 - x^2}.$$

(a) Use a graphing utility to graph the function.

(b) What is the slope of this path when $x = 5$ and $x = 9$?

(c) What does the slope of the path approach as $x \to 10$?

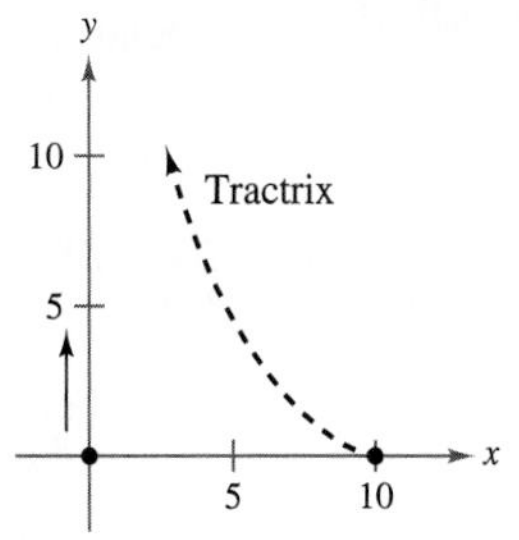

110. ***Conjecture*** Use a graphing utility to graph f and g in the same viewing window and determine which is increasing at the greater rate for large values of x. What can you conclude about the rate of growth of the natural logarithmic function?

(a) $f(x) = \ln x, \quad g(x) = \sqrt{x}$ (b) $f(x) = \ln x, \quad g(x) = \sqrt[4]{x}$

111. Prove that the natural logarithmic function is one-to-one.

112. (a) Use a graphing utility to graph $y = \sqrt{x} - 4 \ln x$.

(b) Use the graph to identify any relative minima and inflection points.

(c) Use calculus to verify your answer to part (b).

Section 5.2 The Natural Logarithmic Function: Integration

- **Use the Log Rule for Integration to integrate a rational function.**
- **Integrate trigonometric functions.**

EXPLORATION

Integrating Rational Functions
Early in Chapter 4, you learned rules that allowed you to integrate *any* polynomial function. The Log Rule presented in this section goes a long way toward enabling you to integrate rational functions. For instance, each of the following functions can be integrated with the Log Rule.

$\frac{2}{x}$	Example 1
$\frac{1}{4x-1}$	Example 2
$\frac{x}{x^2+1}$	Example 3
$\frac{3x^2+1}{x^3+x}$	Example 4(a)
$\frac{x+1}{x^2+2x}$	Example 4(c)
$\frac{1}{3x+2}$	Example 4(d)
$\frac{x^2+x+1}{x^2+1}$	Example 5
$\frac{2x}{(x+1)^2}$	Example 6

There are still some rational functions that cannot be integrated using the Log Rule. Give examples of these functions, and explain your reasoning.

Log Rule for Integration

The differentiation rules

$$\frac{d}{dx}[\ln|x|] = \frac{1}{x} \quad \text{and} \quad \frac{d}{dx}[\ln|u|] = \frac{u'}{u}$$

that you studied in the preceding section produce the following integration rule.

THEOREM 5.5 Log Rule for Integration

Let u be a differentiable function of x.

1. $\int \frac{1}{x}\,dx = \ln|x| + C$ **2.** $\int \frac{1}{u}\,du = \ln|u| + C$

Because $du = u'\,dx$, the second formula can also be written as

$$\int \frac{u'}{u}\,dx = \ln|u| + C.$$ Alternative form of Log Rule

EXAMPLE 1 Using the Log Rule for Integration

$$\int \frac{2}{x}\,dx = 2\int \frac{1}{x}\,dx$$ Constant Multiple Rule

$$= 2\ln|x| + C$$ Log Rule for Integration

$$= \ln(x^2) + C$$ Property of logarithms

Because x^2 cannot be negative, the absolute value is unnecessary in the final form of the antiderivative.

EXAMPLE 2 Using the Log Rule with a Change of Variables

Find $\int \frac{1}{4x-1}\,dx$.

Solution If you let $u = 4x - 1$, then $du = 4\,dx$.

$$\int \frac{1}{4x-1}\,dx = \frac{1}{4}\int \left(\frac{1}{4x-1}\right)4\,dx$$ Multiply and divide by 4.

$$= \frac{1}{4}\int \frac{1}{u}\,du$$ Substitute: $u = 4x - 1$.

$$= \frac{1}{4}\ln|u| + C$$ Apply Log Rule.

$$= \frac{1}{4}\ln|4x-1| + C$$ Back-substitute.

Example 3 uses the alternative form of the Log Rule. To apply this rule, look for quotients in which the numerator is the derivative of the denominator.

EXAMPLE 3 Finding Area with the Log Rule

Find the area of the region bounded by the graph of

$$y = \frac{x}{x^2 + 1}$$

the x-axis, and the line $x = 3$.

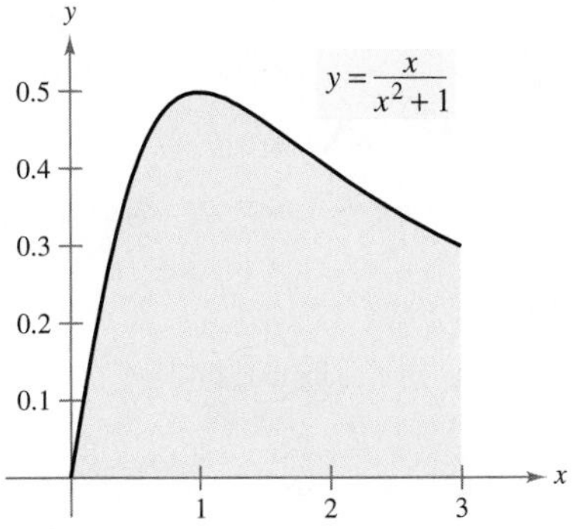

$$\text{Area} = \int_0^3 \frac{x}{x^2 + 1}\, dx$$

The area of the region bounded by the graph of y, the x-axis, and $x = 3$ is $\frac{1}{2}\ln 10$.
Figure 5.8

Solution From Figure 5.8, you can see that the area of the region is given by the definite integral

$$\int_0^3 \frac{x}{x^2 + 1}\, dx.$$

If you let $u = x^2 + 1$, then $u' = 2x$. To apply the Log Rule, multiply and divide by 2 as shown.

$$\begin{aligned}
\int_0^3 \frac{x}{x^2 + 1}\, dx &= \frac{1}{2}\int_0^3 \frac{2x}{x^2 + 1}\, dx && \text{Multiply and divide by 2.}\\
&= \frac{1}{2}\Big[\ln(x^2 + 1)\Big]_0^3 && \int \frac{u'}{u}\, dx = \ln|u| + C\\
&= \frac{1}{2}(\ln 10 - \ln 1)\\
&= \frac{1}{2}\ln 10 && \ln 1 = 0\\
&\approx 1.151
\end{aligned}$$

EXAMPLE 4 Recognizing Quotient Forms of the Log Rule

a. $\displaystyle\int \frac{3x^2 + 1}{x^3 + x}\, dx = \ln|x^3 + x| + C$ $\qquad u = x^3 + x$

b. $\displaystyle\int \frac{\sec^2 x}{\tan x}\, dx = \ln|\tan x| + C$ $\qquad u = \tan x$

c. $\displaystyle\int \frac{x + 1}{x^2 + 2x}\, dx = \frac{1}{2}\int \frac{2x + 2}{x^2 + 2x}\, dx$ $\qquad u = x^2 + 2x$

$$= \frac{1}{2}\ln|x^2 + 2x| + C$$

d. $\displaystyle\int \frac{1}{3x + 2}\, dx = \frac{1}{3}\int \frac{3}{3x + 2}\, dx$ $\qquad u = 3x + 2$

$$= \frac{1}{3}\ln|3x + 2| + C$$

With antiderivatives involving logarithms, it is easy to obtain forms that look quite different but are still equivalent. For instance, which of the following are equivalent to the antiderivative listed in Example 4(d)?

$$\ln|(3x + 2)^{1/3}| + C, \quad \frac{1}{3}\ln\left|x + \tfrac{2}{3}\right| + C, \quad \ln|3x + 2|^{1/3} + C$$

Integrals to which the Log Rule can be applied often appear in disguised form. For instance, if a rational function has a *numerator of degree greater than or equal to that of the denominator*, division may reveal a form to which you can apply the Log Rule. This is shown in Example 5.

EXAMPLE 5 Using Long Division Before Integrating

Find $\int \frac{x^2 + x + 1}{x^2 + 1}\, dx$.

Solution Begin by using long division to rewrite the integrand.

$$\frac{x^2 + x + 1}{x^2 + 1} \quad \Rightarrow \quad \begin{array}{r} 1 \\ x^2 + 1 \overline{\smash{)}\, x^2 + x + 1} \\ \underline{x^2 \quad\;\; + 1} \\ x \end{array} \quad \Rightarrow \quad 1 + \frac{x}{x^2 + 1}$$

Now, you can integrate to obtain

$$\int \frac{x^2 + x + 1}{x^2 + 1}\, dx = \int \left(1 + \frac{x}{x^2 + 1}\right) dx \qquad \text{Rewrite using long division.}$$

$$= \int dx + \frac{1}{2}\int \frac{2x}{x^2 + 1}\, dx \qquad \text{Rewrite as two integrals.}$$

$$= x + \frac{1}{2}\ln(x^2 + 1) + C. \qquad \text{Integrate.}$$

Check this result by differentiating to obtain the original integrand.

The next example gives another instance in which the use of the Log Rule is disguised. In this case, a change of variables helps you recognize the Log Rule.

EXAMPLE 6 Change of Variables with the Log Rule

Find $\int \frac{2x}{(x + 1)^2}\, dx$.

Solution If you let $u = x + 1$, then $du = dx$ and $x = u - 1$.

$$\int \frac{2x}{(x + 1)^2}\, dx = \int \frac{2(u - 1)}{u^2}\, du \qquad \text{Substitute.}$$

$$= 2\int \left(\frac{u}{u^2} - \frac{1}{u^2}\right) du \qquad \text{Rewrite as two fractions.}$$

$$= 2\int \frac{du}{u} - 2\int u^{-2}\, du \qquad \text{Rewrite as two integrals.}$$

$$= 2\ln|u| - 2\left(\frac{u^{-1}}{-1}\right) + C \qquad \text{Integrate.}$$

$$= 2\ln|u| + \frac{2}{u} + C \qquad \text{Simplify.}$$

$$= 2\ln|x + 1| + \frac{2}{x + 1} + C \qquad \text{Back-substitute.}$$

Check this result by differentiating to obtain the original integrand.

TECHNOLOGY If you have access to a computer algebra system, use it to find the indefinite integrals in Examples 5 and 6. How does the form of the antiderivative that it gives you compare with that given in Examples 5 and 6?

As you study the methods shown in Examples 5 and 6, be aware that both methods involve rewriting a disguised integrand so that it fits one or more of the basic integration formulas. Throughout the remaining sections of Chapter 5 and in Chapter 8, much time will be devoted to integration techniques. To master these techniques, you must recognize the "form-fitting" nature of integration. In this sense, integration is not nearly as straightforward as differentiation. Differentiation takes the form

"Here is the question; what is the answer?"

Integration is more like

"Here is the answer; what is the question?"

The following are guidelines you can use for integration.

STUDY TIP Keep in mind that you can check your answer to an integration problem by differentiating the answer. For instance, in Example 7, the derivative of $y = \ln|\ln x| + C$ is $y' = 1/(x \ln x)$.

Guidelines for Integration

1. Learn a basic list of integration formulas. (Including those given in this section, you now have 12 formulas: the Power Rule, the Log Rule, and ten trigonometric rules. By the end of Section 5.7, this list will have expanded to 20 basic rules.)
2. Find an integration formula that resembles all or part of the integrand, and, by trial and error, find a choice of u that will make the integrand conform to the formula.
3. If you cannot find a u-substitution that works, try altering the integrand. You might try a trigonometric identity, multiplication and division by the same quantity, or addition and subtraction of the same quantity. Be creative.
4. If you have access to computer software that will find antiderivatives symbolically, use it.

EXAMPLE 7 *u*-Substitution and the Log Rule

Solve the differential equation $\dfrac{dy}{dx} = \dfrac{1}{x \ln x}$.

Solution The solution can be written as an indefinite integral.

$$y = \int \frac{1}{x \ln x}\, dx$$

Because the integrand is a quotient whose denominator is raised to the first power, you should try the Log Rule. There are three basic choices for u. The choices $u = x$ and $u = x \ln x$ fail to fit the u'/u form of the Log Rule. However, the third choice does fit. Letting $u = \ln x$ produces $u' = 1/x$, and you obtain the following.

$$\int \frac{1}{x \ln x}\, dx = \int \frac{1/x}{\ln x}\, dx \qquad \text{Divide numerator and denominator by } x.$$

$$= \int \frac{u'}{u}\, dx \qquad \text{Substitute: } u = \ln x.$$

$$= \ln|u| + C \qquad \text{Apply Log Rule.}$$

$$= \ln|\ln x| + C \qquad \text{Back-substitute.}$$

So, the solution is $y = \ln|\ln x| + C$.

Integrals of Trigonometric Functions

In Section 4.1, you looked at six trigonometric integration rules—the six that correspond directly to differentiation rules. With the Log Rule, you can now complete the set of basic trigonometric integration formulas.

EXAMPLE 8 Using a Trigonometric Identity

Find $\int \tan x \, dx$.

Solution This integral does not seem to fit any formulas on our basic list. However, by using a trigonometric identity, you obtain

$$\int \tan x \, dx = \int \frac{\sin x}{\cos x} \, dx.$$

Knowing that $D_x[\cos x] = -\sin x$, you can let $u = \cos x$ and write

$$\int \tan x \, dx = -\int \frac{-\sin x}{\cos x} \, dx \qquad \text{Trigonometric identity}$$

$$= -\int \frac{u'}{u} \, dx \qquad \text{Substitute: } u = \cos x.$$

$$= -\ln|u| + C \qquad \text{Apply Log Rule.}$$

$$= -\ln|\cos x| + C. \qquad \text{Back-substitute.}$$

Example 8 uses a trigonometric identity to derive an integration rule for the tangent function. The next example takes a rather unusual step (multiplying and dividing by the same quantity) to derive an integration rule for the secant function.

EXAMPLE 9 Derivation of the Secant Formula

Find $\int \sec x \, dx$.

Solution Consider the following procedure.

$$\int \sec x \, dx = \int \sec x \left(\frac{\sec x + \tan x}{\sec x + \tan x}\right) dx$$

$$= \int \frac{\sec^2 x + \sec x \tan x}{\sec x + \tan x} \, dx$$

Letting u be the denominator of this quotient produces

$$u = \sec x + \tan x \quad \Longrightarrow \quad u' = \sec x \tan x + \sec^2 x.$$

So, you can conclude that

$$\int \sec x \, dx = \int \frac{\sec^2 x + \sec x \tan x}{\sec x + \tan x} \, dx \qquad \text{Rewrite integrand.}$$

$$= \int \frac{u'}{u} \, dx \qquad \text{Substitute: } u = \sec x + \tan x.$$

$$= \ln|u| + C \qquad \text{Apply Log Rule.}$$

$$= \ln|\sec x + \tan x| + C. \qquad \text{Back-substitute.}$$

With the results of Examples 8 and 9, you now have integration formulas for $\sin x$, $\cos x$, $\tan x$, and $\sec x$. All six trigonometric rules are summarized below.

NOTE Using trigonometric identities and properties of logarithms, you could rewrite these six integration rules in other forms. For instance, you could write

$$\int \csc u\,du = \ln|\csc u - \cot u| + C.$$

(See Exercises 83–86.)

Integrals of the Six Basic Trigonometric Functions

$$\int \sin u\,du = -\cos u + C \qquad \int \cos u\,du = \sin u + C$$

$$\int \tan u\,du = -\ln|\cos u| + C \qquad \int \cot u\,du = \ln|\sin u| + C$$

$$\int \sec u\,du = \ln|\sec u + \tan u| + C \qquad \int \csc u\,du = -\ln|\csc u + \cot u| + C$$

EXAMPLE 10 Integrating Trigonometric Functions

Evaluate $\displaystyle\int_0^{\pi/4} \sqrt{1 + \tan^2 x}\,dx$.

Solution Using $1 + \tan^2 x = \sec^2 x$, you can write

$$\begin{aligned}
\int_0^{\pi/4} \sqrt{1 + \tan^2 x}\,dx &= \int_0^{\pi/4} \sqrt{\sec^2 x}\,dx \\
&= \int_0^{\pi/4} \sec x\,dx && \sec x \ge 0 \text{ for } 0 \le x \le \frac{\pi}{4}. \\
&= \ln|\sec x + \tan x|\Big]_0^{\pi/4} \\
&= \ln(\sqrt{2} + 1) - \ln 1 \\
&\approx 0.881.
\end{aligned}$$

EXAMPLE 11 Finding an Average Value

Find the average value of $f(x) = \tan x$ on the interval $\left[0, \frac{\pi}{4}\right]$.

Solution

$$\begin{aligned}
\text{Average value} &= \frac{1}{(\pi/4) - 0}\int_0^{\pi/4} \tan x\,dx && \text{Average value} = \frac{1}{b-a}\int_a^b f(x)\,dx \\
&= \frac{4}{\pi}\int_0^{\pi/4} \tan x\,dx && \text{Simplify.} \\
&= \frac{4}{\pi}\Big[-\ln|\cos x|\Big]_0^{\pi/4} && \text{Integrate.} \\
&= -\frac{4}{\pi}\left[\ln\left(\frac{\sqrt{2}}{2}\right) - \ln(1)\right] \\
&= -\frac{4}{\pi}\ln\left(\frac{\sqrt{2}}{2}\right) \\
&\approx 0.441
\end{aligned}$$

The average value is about 0.441, as shown in Figure 5.9.

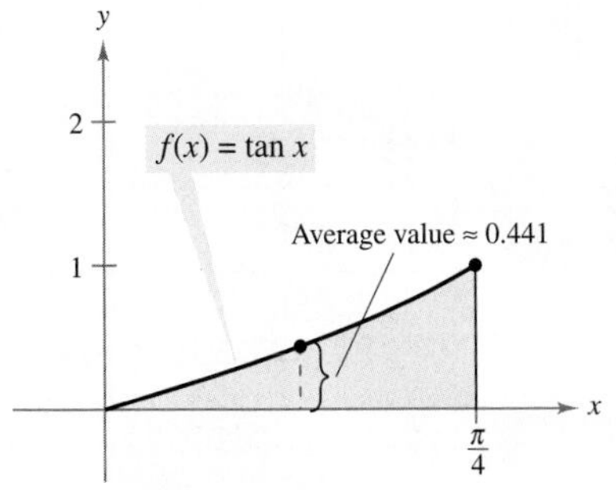

Figure 5.9

Exercises for Section 5.2

See www.CalcChat.com for worked-out solutions to odd-numbered exercises.

In Exercises 1–24, find the indefinite integral.

1. $\displaystyle\int \frac{5}{x}\,dx$
2. $\displaystyle\int \frac{10}{x}\,dx$
3. $\displaystyle\int \frac{1}{x+1}\,dx$
4. $\displaystyle\int \frac{1}{x-5}\,dx$
5. $\displaystyle\int \frac{1}{3-2x}\,dx$
6. $\displaystyle\int \frac{1}{3x+2}\,dx$
7. $\displaystyle\int \frac{x}{x^2+1}\,dx$
8. $\displaystyle\int \frac{x^2}{3-x^3}\,dx$
9. $\displaystyle\int \frac{x^2-4}{x}\,dx$
10. $\displaystyle\int \frac{x}{\sqrt{9-x^2}}\,dx$
11. $\displaystyle\int \frac{x^2+2x+3}{x^3+3x^2+9x}\,dx$
12. $\displaystyle\int \frac{x(x+2)}{x^3+3x^2-4}\,dx$
13. $\displaystyle\int \frac{x^2-3x+2}{x+1}\,dx$
14. $\displaystyle\int \frac{2x^2+7x-3}{x-2}\,dx$
15. $\displaystyle\int \frac{x^3-3x^2+5}{x-3}\,dx$
16. $\displaystyle\int \frac{x^3-6x-20}{x+5}\,dx$
17. $\displaystyle\int \frac{x^4+x-4}{x^2+2}\,dx$
18. $\displaystyle\int \frac{x^3-3x^2+4x-9}{x^2+3}\,dx$
19. $\displaystyle\int \frac{(\ln x)^2}{x}\,dx$
20. $\displaystyle\int \frac{1}{x\ln(x^3)}\,dx$
21. $\displaystyle\int \frac{1}{\sqrt{x}+1}\,dx$
22. $\displaystyle\int \frac{1}{x^{2/3}(1+x^{1/3})}\,dx$
23. $\displaystyle\int \frac{2x}{(x-1)^2}\,dx$
24. $\displaystyle\int \frac{x(x-2)}{(x-1)^3}\,dx$

In Exercises 25–28, find the indefinite integral by *u*-substitution. (*Hint:* Let *u* be the denominator of the integrand.)

25. $\displaystyle\int \frac{1}{1+\sqrt{2x}}\,dx$
26. $\displaystyle\int \frac{1}{1+\sqrt{3x}}\,dx$
27. $\displaystyle\int \frac{\sqrt{x}}{\sqrt{x}-3}\,dx$
28. $\displaystyle\int \frac{\sqrt[3]{x}}{\sqrt[3]{x}-1}\,dx$

In Exercises 29–36, find the indefinite integral.

29. $\displaystyle\int \frac{\cos\theta}{\sin\theta}\,d\theta$
30. $\displaystyle\int \tan 5\theta\,d\theta$
31. $\displaystyle\int \csc 2x\,dx$
32. $\displaystyle\int \sec\frac{x}{2}\,dx$
33. $\displaystyle\int \frac{\cos t}{1+\sin t}\,dt$
34. $\displaystyle\int \frac{\csc^2 t}{\cot t}\,dt$
35. $\displaystyle\int \frac{\sec x\tan x}{\sec x-1}\,dx$
36. $\displaystyle\int (\sec t+\tan t)\,dt$

In Exercises 37–40, solve the differential equation. Use a graphing utility to graph three solutions, one of which passes through the given point.

37. $\dfrac{dy}{dx}=\dfrac{3}{2-x},\quad (1, 0)$
38. $\dfrac{dy}{dx}=\dfrac{2x}{x^2-9},\quad (0, 4)$
39. $\dfrac{ds}{d\theta}=\tan 2\theta,\quad (0, 2)$
40. $\dfrac{dr}{dt}=\dfrac{\sec^2 t}{\tan t+1},\quad (\pi, 4)$

41. Determine the function f if $f''(x)=\dfrac{2}{x^2}$, $f(1)=1$, $f'(1)=1$, $x>0$.

42. Determine the function f if $f''(x)=-\dfrac{4}{(x-1)^2}-2$, $f(2)=3$, $f'(2)=0$, $x>1$.

Slope Fields **In Exercises 43–46, a differential equation, a point, and a slope field are given. (a) Sketch two approximate solutions of the differential equation on the slope field, one of which passes through the given point. (b) Use integration to find the particular solution of the differential equation and use a graphing utility to graph the solution. Compare the result with the sketches in part (a). To print an enlarged copy of the graph, go to the website *www.mathgraphs.com*.**

43. $\dfrac{dy}{dx}=\dfrac{1}{x+2},\quad (0, 1)$

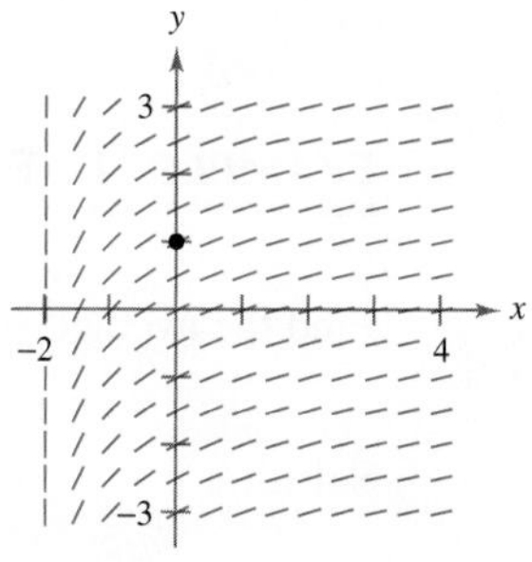

44. $\dfrac{dy}{dx}=\dfrac{\ln x}{x},\quad (1, -2)$

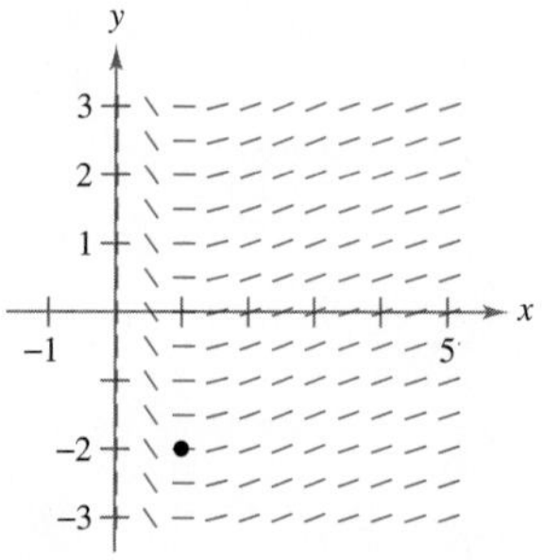

45. $\dfrac{dy}{dx}=1+\dfrac{1}{x},\quad (1, 4)$

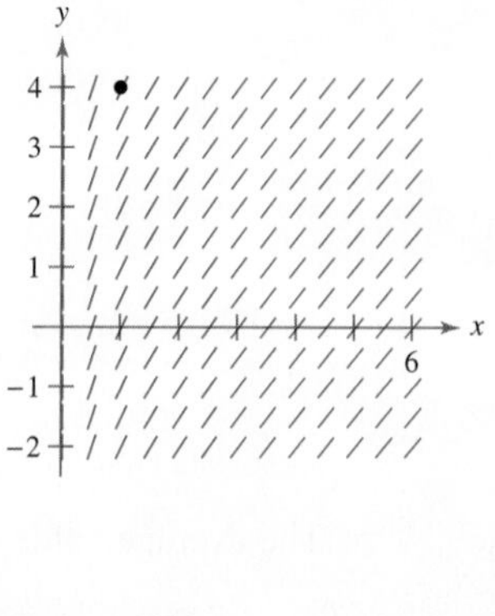

46. $\dfrac{dy}{dx}=\sec x,\quad (0, 1)$

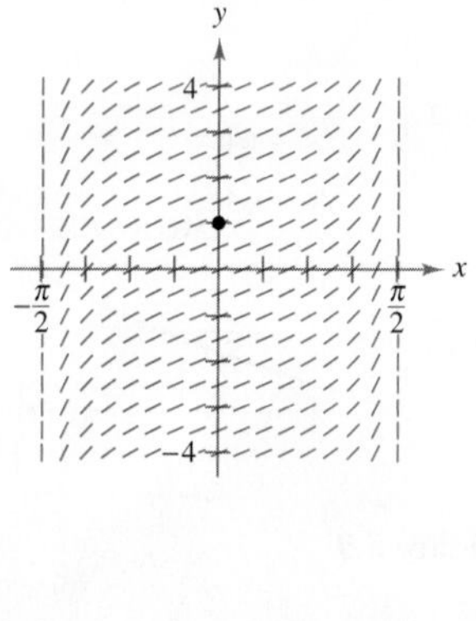

In Exercises 47–54, evaluate the definite integral. Use a graphing utility to verify your result.

47. $\int_0^4 \frac{5}{3x + 1}\,dx$

48. $\int_{-1}^1 \frac{1}{x + 2}\,dx$

49. $\int_1^e \frac{(1 + \ln x)^2}{x}\,dx$

50. $\int_e^{e^2} \frac{1}{x \ln x}\,dx$

51. $\int_0^2 \frac{x^2 - 2}{x + 1}\,dx$

52. $\int_0^1 \frac{x - 1}{x + 1}\,dx$

53. $\int_1^2 \frac{1 - \cos\theta}{\theta - \sin\theta}\,d\theta$

54. $\int_{0.1}^{0.2} (\csc 2\theta - \cot 2\theta)^2\,d\theta$

In Exercises 55–60, use a computer algebra system to find or evaluate the integral.

55. $\int \frac{1}{1 + \sqrt{x}}\,dx$

56. $\int \frac{1 - \sqrt{x}}{1 + \sqrt{x}}\,dx$

57. $\int \frac{\sqrt{x}}{x - 1}\,dx$

58. $\int \frac{x^2}{x - 1}\,dx$

59. $\int_{\pi/4}^{\pi/2} (\csc x - \sin x)\,dx$

60. $\int_{-\pi/4}^{\pi/4} \frac{\sin^2 x - \cos^2 x}{\cos x}\,dx$

In Exercises 61–64, find $F'(x)$.

61. $F(x) = \int_1^x \frac{1}{t}\,dt$

62. $F(x) = \int_0^x \tan t\,dt$

63. $F(x) = \int_1^{3x} \frac{1}{t}\,dt$

64. $F(x) = \int_1^{x^2} \frac{1}{t}\,dt$

Approximation **In Exercises 65 and 66, determine which value best approximates the area of the region between the x-axis and the graph of the function over the given interval. (Make your selection on the basis of a sketch of the region and not by performing any calculations.)**

65. $f(x) = \sec x,\ [0, 1]$

(a) 6 (b) -6 (c) $\frac{1}{2}$ (d) 1.25 (e) 3

66. $f(x) = \frac{2x}{x^2 + 1},\ [0, 4]$

(a) 3 (b) 7 (c) -2 (d) 5 (e) 1

Area **In Exercises 67–70, find the area of the given region. Use a graphing utility to verify your result.**

67. $y = \frac{4}{x}$

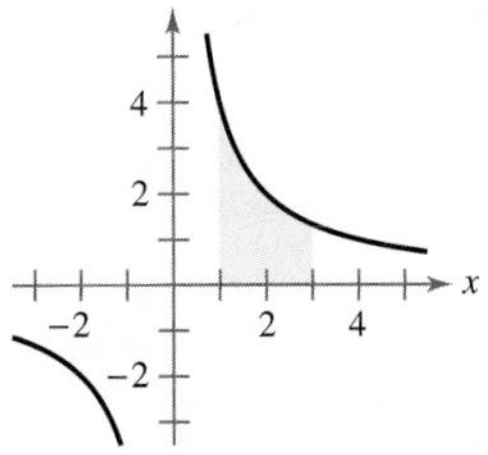

68. $y = \frac{2}{x \ln x}$

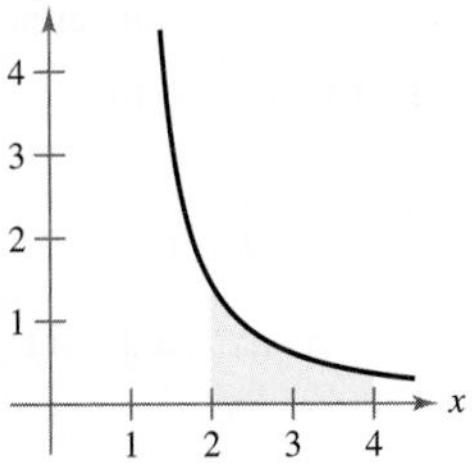

69. $y = \tan x$

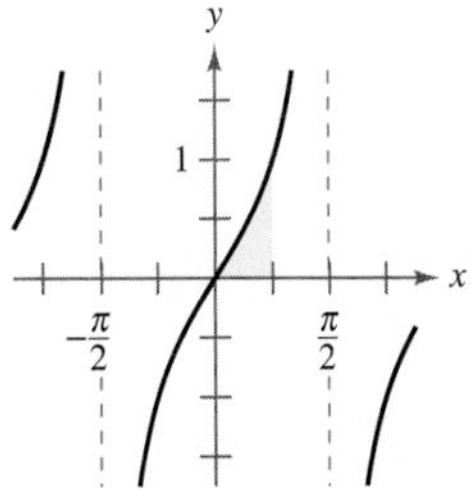

70. $y = \frac{\sin x}{1 + \cos x}$

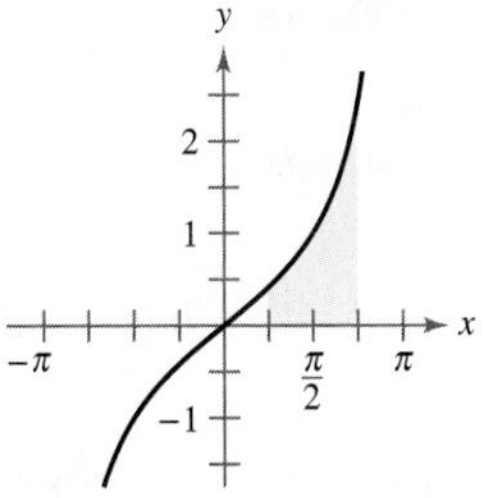

Area **In Exercises 71–74, find the area of the region bounded by the graphs of the equations. Use a graphing utility to verify your result.**

71. $y = \frac{x^2 + 4}{x},\ x = 1,\ x = 4,\ y = 0$

72. $y = \frac{x + 4}{x},\ x = 1,\ x = 4,\ y = 0$

73. $y = 2 \sec \frac{\pi x}{6},\ x = 0,\ x = 2,\ y = 0$

74. $y = 2x - \tan(0.3x),\ x = 1,\ x = 4,\ y = 0$

Numerical Integration **In Exercises 75–78, use the Trapezoidal Rule and Simpson's Rule to approximate the value of the definite integral. Let $n = 4$ and round your answer to four decimal places. Use a graphing utility to verify your result.**

75. $\int_1^5 \frac{12}{x}\,dx$

76. $\int_0^4 \frac{8x}{x^2 + 4}\,dx$

77. $\int_2^6 \ln x\,dx$

78. $\int_{-\pi/3}^{\pi/3} \sec x\,dx$

Writing About Concepts

In Exercises 79–82, state the integration formula you would use to perform the integration. Do not integrate.

79. $\int \sqrt[3]{x}\,dx$

80. $\int \frac{x}{(x^2 + 4)^3}\,dx$

81. $\int \frac{x}{x^2 + 4}\,dx$

82. $\int \frac{\sec^2 x}{\tan x}\,dx$

In Exercises 83–86, show that the two formulas are equivalent.

83. $\int \tan x\,dx = -\ln|\cos x| + C$

$\int \tan x\,dx = \ln|\sec x| + C$

84. $\int \cot x\,dx = \ln|\sin x| + C$

$\int \cot x\,dx = -\ln|\csc x| + C$

85. $\int \sec x\,dx = \ln|\sec x + \tan x| + C$

$\int \sec x\,dx = -\ln|\sec x - \tan x| + C$

86. $\int \csc x\,dx = -\ln|\csc x + \cot x| + C$

$\int \csc x\,dx = \ln|\csc x - \cot x| + C$

In Exercises 87–90, find the average value of the function over the given interval.

87. $f(x) = \dfrac{8}{x^2}$, $[2, 4]$

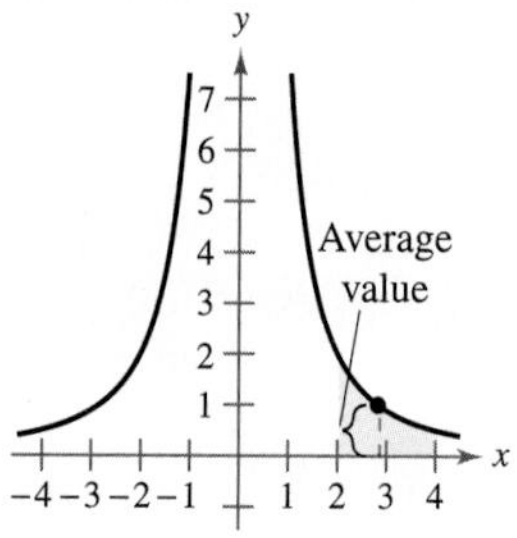

88. $f(x) = \dfrac{4(x+1)}{x^2}$, $[2, 4]$

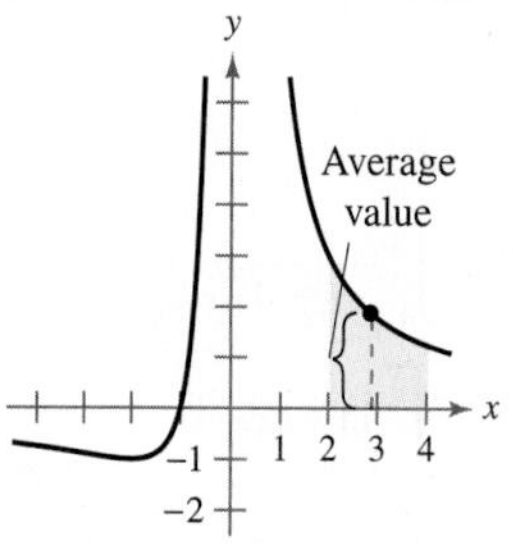

89. $f(x) = \dfrac{\ln x}{x}$, $[1, e]$

90. $f(x) = \sec \dfrac{\pi x}{6}$, $[0, 2]$

91. ***Population Growth*** A population of bacteria is changing at a rate of

$$\frac{dP}{dt} = \frac{3000}{1 + 0.25t}$$

where t is the time in days. The initial population (when $t = 0$) is 1000. Write an equation that gives the population at any time t, and find the population when $t = 3$ days.

92. ***Heat Transfer*** Find the time required for an object to cool from 300°F to 250°F by evaluating

$$t = \frac{10}{\ln 2}\int_{250}^{300} \frac{1}{T - 100}\,dT$$

where t is time in minutes.

93. ***Average Price*** The demand equation for a product is

$$p = \frac{90{,}000}{400 + 3x}.$$

Find the *average* price p on the interval $40 \le x \le 50$.

94. ***Sales*** The rate of change in sales S is inversely proportional to time t $(t > 1)$ measured in weeks. Find S as a function of t if sales after 2 and 4 weeks are 200 units and 300 units, respectively.

95. ***Orthogonal Trajectory***

(a) Use a graphing utility to graph the equation $2x^2 - y^2 = 8$.

(b) Evaluate the integral to find y^2 in terms of x.

$$y^2 = e^{-\int (1/x)\,dx}$$

For a particular value of the constant of integration, graph the result in the same viewing window used in part (a).

(c) Verify that the tangents to the graphs of parts (a) and (b) are perpendicular at the points of intersection.

96. Graph the function

$$f_k(x) = \frac{x^k - 1}{k}$$

for $k = 1, 0.5$, and 0.1 on $[0, 10]$. Find $\lim_{k\to 0^+} f_k(x)$.

True or False? **In Exercises 97–100, determine whether the statement is true or false. If it is false, explain why or give an example that shows it is false.**

97. $(\ln x)^{1/2} = \frac{1}{2}(\ln x)$

98. $\int \ln x\,dx = (1/x) + C$

99. $\int \frac{1}{x}\,dx = \ln|cx|, \quad c \ne 0$

100. $\int_{-1}^{2} \frac{1}{x}\,dx = \Big[\ln|x|\Big]_{-1}^{2} = \ln 2 - \ln 1 = \ln 2$

101. Graph the function

$$f(x) = \frac{x}{1 + x^2}$$

on the interval $[0, \infty)$.

(a) Find the area bounded by the graph of f and the line $y = \frac{1}{2}x$.

(b) Determine the values of the slope m such that the line $y = mx$ and the graph of f enclose a finite region.

(c) Calculate the area of this region as a function of m.

102. Prove that the function

$$F(x) = \int_x^{2x} \frac{1}{t}\,dt$$

is constant on the interval $(0, \infty)$.

Section 5.3 Inverse Functions

- **Verify that one function is the inverse function of another function.**
- **Determine whether a function has an inverse function.**
- **Find the derivative of an inverse function.**

Inverse Functions

Recall from Section P.3 that a function can be represented by a set of ordered pairs. For instance, the function $f(x) = x + 3$ from $A = \{1, 2, 3, 4\}$ to $B = \{4, 5, 6, 7\}$ can be written as

$$f: \{(1, 4), (2, 5), (3, 6), (4, 7)\}.$$

By interchanging the first and second coordinates of each ordered pair, you can form the **inverse function** of f. This function is denoted by f^{-1}. It is a function from B to A, and can be written as

$$f^{-1}: \{(4, 1), (5, 2), (6, 3), (7, 4)\}.$$

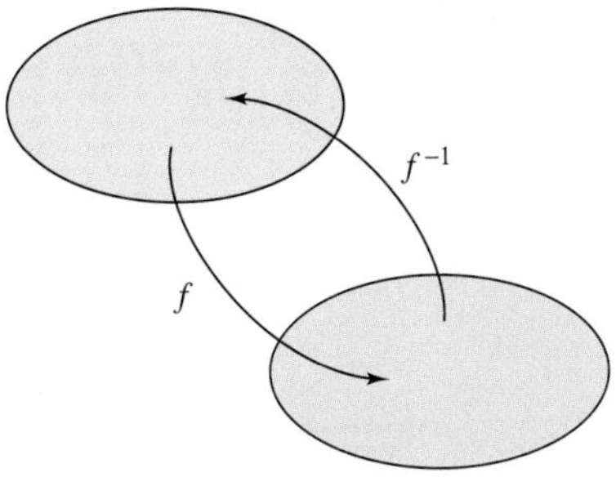

Domain of f = range of f^{-1}
Domain of f^{-1} = range of f
Figure 5.10

Note that the domain of f is equal to the range of f^{-1}, and vice versa, as shown in Figure 5.10. The functions f and f^{-1} have the effect of "undoing" each other. That is, when you form the composition of f with f^{-1} or the composition of f^{-1} with f, you obtain the identity function.

$$f(f^{-1}(x)) = x \qquad \text{and} \qquad f^{-1}(f(x)) = x$$

EXPLORATION

Finding Inverse Functions Explain how to "undo" each of the following functions. Then use your explanation to write the inverse function of f.

a. $f(x) = x - 5$

b. $f(x) = 6x$

c. $f(x) = \dfrac{x}{2}$

d. $f(x) = 3x + 2$

e. $f(x) = x^3$

f. $f(x) = 4(x - 2)$

Use a graphing utility to graph each function and its inverse function in the same "square" viewing window. What observation can you make about each pair of graphs?

Definition of Inverse Function

A function g is the **inverse function** of the function f if

$$f(g(x)) = x \quad \text{for each } x \text{ in the domain of } g$$

and

$$g(f(x)) = x \quad \text{for each } x \text{ in the domain of } f.$$

The function g is denoted by f^{-1} (read "f inverse").

NOTE Although the notation used to denote an inverse function resembles *exponential notation*, it is a different use of -1 as a superscript. That is, in general, $f^{-1}(x) \neq 1/f(x)$.

Here are some important observations about inverse functions.

1. If g is the inverse function of f, then f is the inverse function of g.
2. The domain of f^{-1} is equal to the range of f, and the range of f^{-1} is equal to the domain of f.
3. A function need not have an inverse function, but if it does, the inverse function is unique (see Exercise 99).

You can think of f^{-1} as undoing what has been done by f. For example, subtraction can be used to undo addition, and division can be used to undo multiplication. Use the definition of an inverse function to check the following.

$f(x) = x + c$ and $f^{-1}(x) = x - c$ are inverse functions of each other.

$f(x) = cx$ and $f^{-1}(x) = \dfrac{x}{c}$, $c \neq 0$, are inverse functions of each other.

EXAMPLE 1 Verifying Inverse Functions

Show that the functions are inverse functions of each other.

$$f(x) = 2x^3 - 1 \qquad \text{and} \qquad g(x) = \sqrt[3]{\frac{x+1}{2}}$$

Solution Because the domains and ranges of both f and g consist of all real numbers, you can conclude that both composite functions exist for all x. The composition of f with g is given by

$$\begin{aligned} f(g(x)) &= 2\left(\sqrt[3]{\frac{x+1}{2}}\right)^3 - 1 \\ &= 2\left(\frac{x+1}{2}\right) - 1 \\ &= x + 1 - 1 \\ &= x. \end{aligned}$$

The composition of g with f is given by

$$\begin{aligned} g(f(x)) &= \sqrt[3]{\frac{(2x^3 - 1) + 1}{2}} \\ &= \sqrt[3]{\frac{2x^3}{2}} \\ &= \sqrt[3]{x^3} \\ &= x. \end{aligned}$$

Because $f(g(x)) = x$ and $g(f(x)) = x$, you can conclude that f and g are inverse functions of each other (see Figure 5.11).

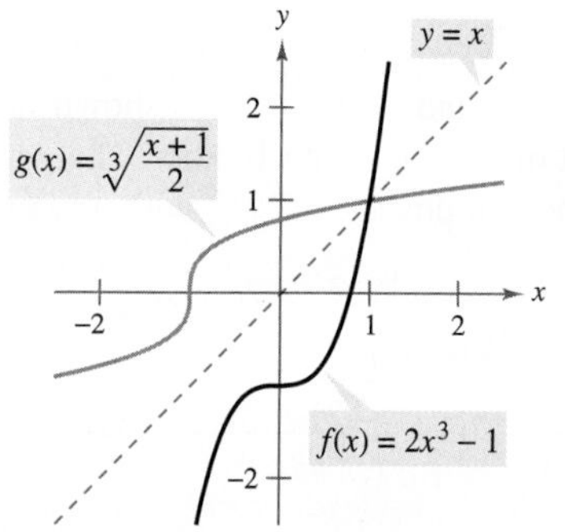

f and g are inverse functions of each other.
Figure 5.11

STUDY TIP In Example 1, try comparing the functions f and g verbally.

For f: First cube x, then multiply by 2, then subtract 1.

For g: First add 1, then divide by 2, then take the cube root.

Do you see the "undoing pattern"?

In Figure 5.11, the graphs of f and $g = f^{-1}$ appear to be mirror images of each other with respect to the line $y = x$. The graph of f^{-1} is a **reflection** of the graph of f in the line $y = x$. This idea is generalized in the following theorem.

THEOREM 5.6 Reflective Property of Inverse Functions

The graph of f contains the point (a, b) if and only if the graph of f^{-1} contains the point (b, a).

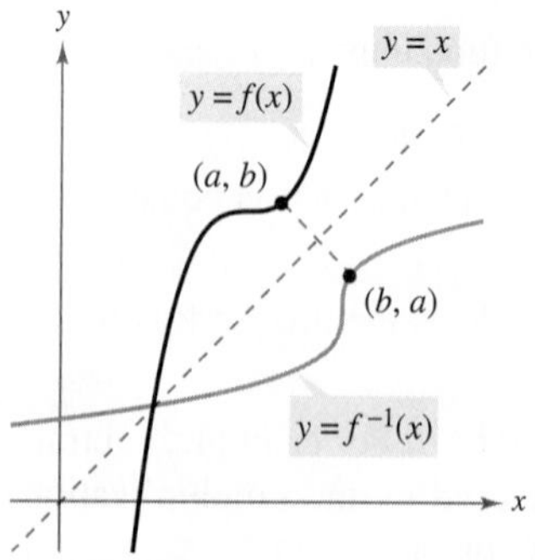

The graph of f^{-1} is a reflection of the graph of f in the line $y = x$.
Figure 5.12

Proof If (a, b) is on the graph of f, then $f(a) = b$ and you can write

$$f^{-1}(b) = f^{-1}(f(a)) = a.$$

So, (b, a) is on the graph of f^{-1}, as shown in Figure 5.12. A similar argument will prove the theorem in the other direction.

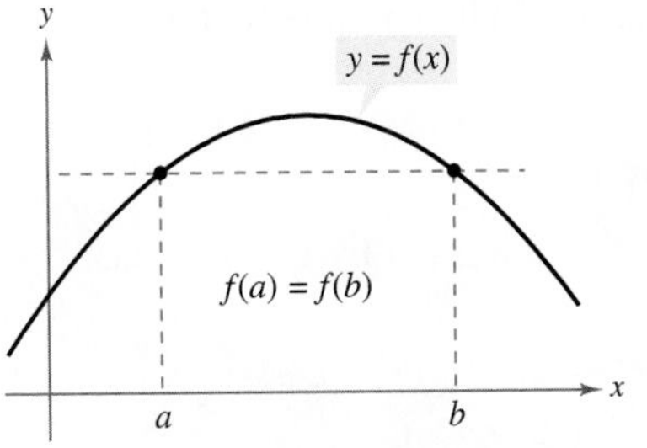

If a horizontal line intersects the graph of f twice, then f is not one-to-one.
Figure 5.13

Existence of an Inverse Function

Not every function has an inverse function, and Theorem 5.6 suggests a graphical test for those that do—the **Horizontal Line Test** for an inverse function. This test states that a function f has an inverse function if and only if every horizontal line intersects the graph of f at most once (see Figure 5.13). The following theorem formally states why the horizontal line test is valid. (Recall from Section 3.3 that a function is *strictly monotonic* if it is either increasing on its entire domain or decreasing on its entire domain.)

> **THEOREM 5.7 The Existence of an Inverse Function**
>
> **1.** A function has an inverse function if and only if it is one-to-one.
>
> **2.** If f is strictly monotonic on its entire domain, then it is one-to-one and therefore has an inverse function.

Proof To prove the second part of the theorem, recall from Section P.3 that f is one-to-one if for x_1 and x_2 in its domain

$$f(x_1) = f(x_2) \quad \Rightarrow \quad x_1 = x_2.$$

The *contrapositive* of this implication is logically equivalent and states that

$$x_1 \neq x_2 \quad \Rightarrow \quad f(x_1) \neq f(x_2).$$

Now, choose x_1 and x_2 in the domain of f. If $x_1 \neq x_2$, then, because f is strictly monotonic, it follows that either

$$f(x_1) < f(x_2) \qquad \text{or} \qquad f(x_1) > f(x_2).$$

In either case, $f(x_1) \neq f(x_2)$. So, f is one-to-one on the interval. The proof of the first part of the theorem is left as an exercise (see Exercise 100).

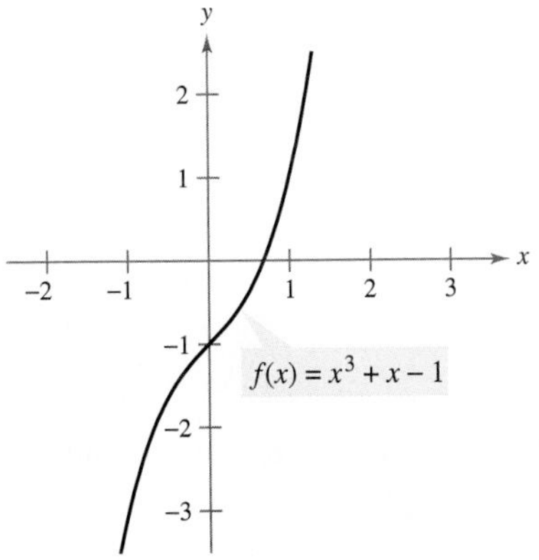

(a) Because f is increasing over its entire domain, it has an inverse function.

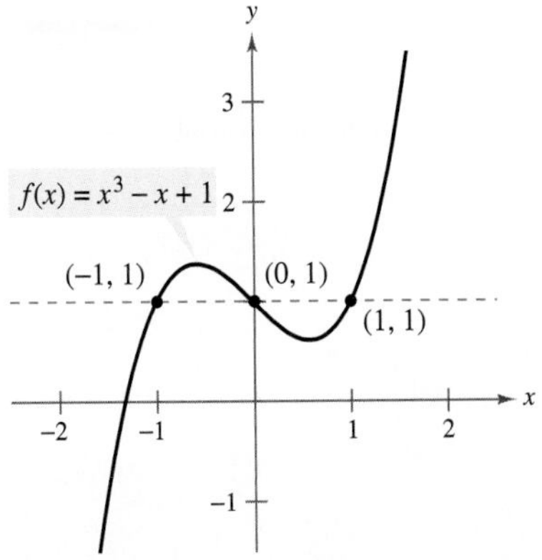

(b) Because f is not one-to-one, it does not have an inverse function.
Figure 5.14

EXAMPLE 2 The Existence of an Inverse Function

Which of the functions has an inverse function?

a. $f(x) = x^3 + x - 1$ **b.** $f(x) = x^3 - x + 1$

Solution

a. From the graph of f shown in Figure 5.14(a), it appears that f is increasing over its entire domain. To verify this, note that the derivative, $f'(x) = 3x^2 + 1$, is positive for all real values of x. So, f is strictly monotonic and it must have an inverse function.

b. From the graph of f shown in Figure 5.14(b), you can see that the function does not pass the horizontal line test. In other words, it is not one-to-one. For instance, f has the same value when $x = -1, 0,$ and 1.

$$f(-1) = f(1) = f(0) = 1 \qquad \text{Not one-to-one}$$

So, by Theorem 5.7, f does not have an inverse function.

NOTE Often it is easier to prove that a function *has* an inverse function than to find the inverse function. For instance, it would be difficult algebraically to find the inverse function of the function in Example 2(a).

The following guidelines suggest a procedure for finding an inverse function.

Guidelines for Finding an Inverse Function

1. Use Theorem 5.7 to determine whether the function given by $y = f(x)$ has an inverse function.
2. Solve for x as a function of y: $x = g(y) = f^{-1}(y)$.
3. Interchange x and y. The resulting equation is $y = f^{-1}(x)$.
4. Define the domain of f^{-1} to be the range of f.
5. Verify that $f(f^{-1}(x)) = x$ and $f^{-1}(f(x)) = x$.

EXAMPLE 3 Finding an Inverse Function

Find the inverse function of

$$f(x) = \sqrt{2x - 3}.$$

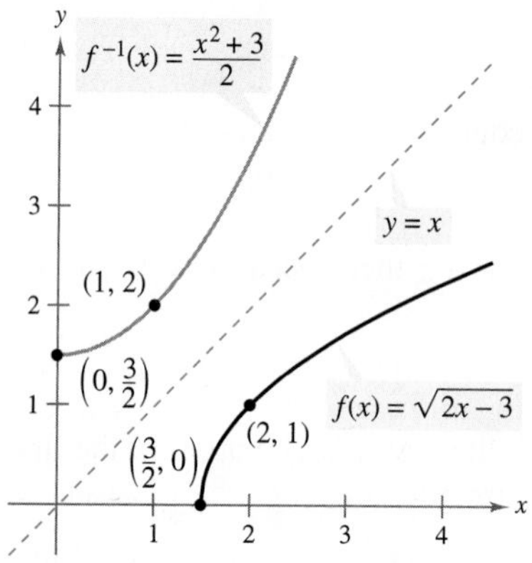

The domain of f^{-1}, $[0, \infty)$, is the range of f.
Figure 5.15

Solution The function has an inverse function because it is increasing on its entire domain (see Figure 5.15). To find an equation for the inverse function, let $y = f(x)$ and solve for x in terms of y.

$$\sqrt{2x - 3} = y \qquad \text{Let } y = f(x).$$

$$2x - 3 = y^2 \qquad \text{Square each side.}$$

$$x = \frac{y^2 + 3}{2} \qquad \text{Solve for } x.$$

$$y = \frac{x^2 + 3}{2} \qquad \text{Interchange } x \text{ and } y.$$

$$f^{-1}(x) = \frac{x^2 + 3}{2} \qquad \text{Replace } y \text{ by } f^{-1}(x).$$

The domain of f^{-1} is the range of f, which is $[0, \infty)$. You can verify this result as shown.

$$f(f^{-1}(x)) = \sqrt{2\left(\frac{x^2 + 3}{2}\right) - 3} = \sqrt{x^2} = x, \quad x \ge 0$$

$$f^{-1}(f(x)) = \frac{\left(\sqrt{2x - 3}\right)^2 + 3}{2} = \frac{2x - 3 + 3}{2} = x, \quad x \ge \frac{3}{2}$$

NOTE Remember that any letter can be used to represent the independent variable. So,

$$f^{-1}(y) = \frac{y^2 + 3}{2}$$

$$f^{-1}(x) = \frac{x^2 + 3}{2}$$

$$f^{-1}(s) = \frac{s^2 + 3}{2}$$

all represent the same function.

Theorem 5.7 is useful in the following type of problem. Suppose you are given a function that is *not* one-to-one on its domain. By restricting the domain to an interval on which the function is strictly monotonic, you can conclude that the new function *is* one-to-one on the restricted domain.

EXAMPLE 4 Testing Whether a Function Is One-to-One

Show that the sine function

$$f(x) = \sin x$$

is not one-to-one on the entire real line. Then show that $[-\pi/2, \pi/2]$ is the largest interval, centered at the origin, for which f is strictly monotonic.

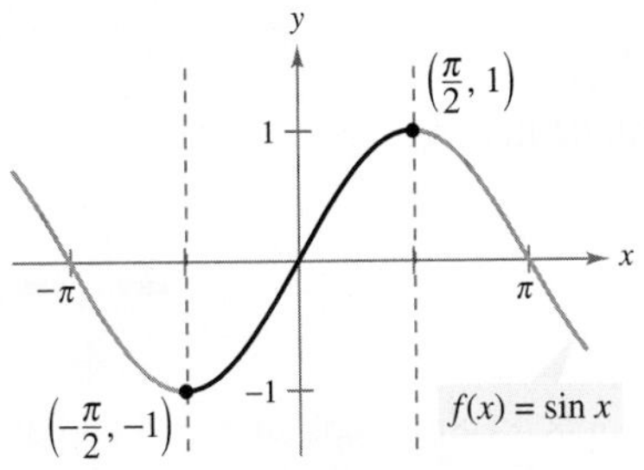

f is one-to-one on the interval $[-\pi/2, \pi/2]$.
Figure 5.16

Solution It is clear that f is not one-to-one, because many different x-values yield the same y-value. For instance,

$$\sin(0) = 0 = \sin(\pi).$$

Moreover, f is increasing on the open interval $(-\pi/2, \pi/2)$, because its derivative

$$f'(x) = \cos x$$

is positive there. Finally, because the left and right endpoints correspond to relative extrema of the sine function, you can conclude that f is increasing on the closed interval $[-\pi/2, \pi/2]$ *and* that in any larger interval the function is not strictly monotonic (see Figure 5.16).

Derivative of an Inverse Function

The next two theorems discuss the derivative of an inverse function. The reasonableness of Theorem 5.8 follows from the reflective property of inverse functions as shown in Figure 5.12. Proofs of the two theorems are given in Appendix A.

THEOREM 5.8 Continuity and Differentiability of Inverse Functions

Let f be a function whose domain is an interval I. If f has an inverse function, then the following statements are true.

1. If f is continuous on its domain, then f^{-1} is continuous on its domain.
2. If f is increasing on its domain, then f^{-1} is increasing on its domain.
3. If f is decreasing on its domain, then f^{-1} is decreasing on its domain.
4. If f is differentiable at c and $f'(c) \neq 0$, then f^{-1} is differentiable at $f(c)$.

EXPLORATION

Graph the inverse functions

$$f(x) = x^3$$

and

$$g(x) = x^{1/3}.$$

Calculate the slope of f at $(1, 1)$, $(2, 8)$, and $(3, 27)$, and the slope of g at $(1, 1)$, $(8, 2)$, and $(27, 3)$. What do you observe? What happens at $(0, 0)$?

THEOREM 5.9 The Derivative of an Inverse Function

Let f be a function that is differentiable on an interval I. If f has an inverse function g, then g is differentiable at any x for which $f'(g(x)) \neq 0$. Moreover,

$$g'(x) = \frac{1}{f'(g(x))}, \quad f'(g(x)) \neq 0.$$

EXAMPLE 5 Evaluating the Derivative of an Inverse Function

Let $f(x) = \frac{1}{4}x^3 + x - 1$.

a. What is the value of $f^{-1}(x)$ when $x = 3$?

b. What is the value of $(f^{-1})'(x)$ when $x = 3$?

Solution Notice that f is one-to-one and therefore has an inverse function.

a. Because $f(x) = 3$ when $x = 2$, you know that $f^{-1}(3) = 2$.

b. Because the function f is differentiable and has an inverse function, you can apply Theorem 5.9 (with $g = f^{-1}$) to write

$$(f^{-1})'(3) = \frac{1}{f'(f^{-1}(3))} = \frac{1}{f'(2)}.$$

Moreover, using $f'(x) = \frac{3}{4}x^2 + 1$, you can conclude that

$$(f^{-1})'(3) = \frac{1}{f'(2)} = \frac{1}{\frac{3}{4}(2^2) + 1} = \frac{1}{4}.$$

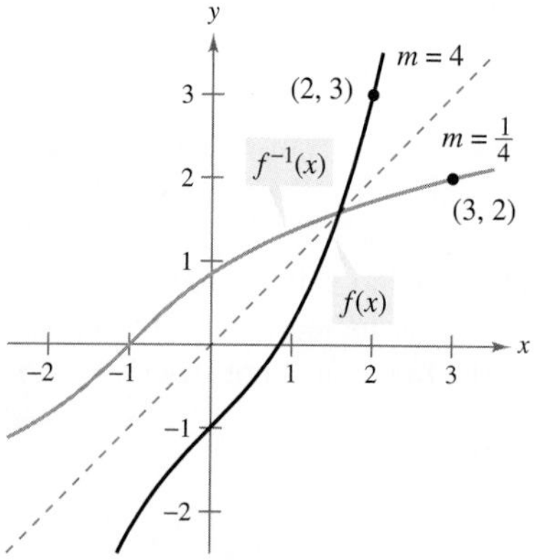

The graphs of the inverse functions f and f^{-1} have reciprocal slopes at points (a, b) and (b, a).
Figure 5.17

In Example 5, note that at the point (2, 3) the slope of the graph of f is 4 and at the point (3, 2) the slope of the graph of f^{-1} is $\frac{1}{4}$ (see Figure 5.17). This reciprocal relationship (which follows from Theorem 5.9) can be written as shown below. If $y = g(x) = f^{-1}(x)$, then $f(y) = x$ and $f'(y) = \frac{dx}{dy}$. Theorem 5.9 says that

$$g'(x) = \frac{dy}{dx} = \frac{1}{f'(g(x))} = \frac{1}{f'(y)} = \frac{1}{(dx/dy)}.$$

So, $$\frac{dy}{dx} = \frac{1}{dx/dy}.$$

EXAMPLE 6 Graphs of Inverse Functions Have Reciprocal Slopes

Let $f(x) = x^2$ (for $x \geq 0$) and let $f^{-1}(x) = \sqrt{x}$. Show that the slopes of the graphs of f and f^{-1} are reciprocals at each of the following points.

a. (2, 4) and (4, 2) **b.** (3, 9) and (9, 3)

Solution The derivatives of f and f^{-1} are given by

$$f'(x) = 2x \quad \text{and} \quad (f^{-1})'(x) = \frac{1}{2\sqrt{x}}.$$

a. At (2, 4), the slope of the graph of f is $f'(2) = 2(2) = 4$. At (4, 2), the slope of the graph of f^{-1} is

$$(f^{-1})'(4) = \frac{1}{2\sqrt{4}} = \frac{1}{2(2)} = \frac{1}{4}.$$

b. At (3, 9), the slope of the graph of f is $f'(3) = 2(3) = 6$. At (9, 3), the slope of the graph of f^{-1} is

$$(f^{-1})'(9) = \frac{1}{2\sqrt{9}} = \frac{1}{2(3)} = \frac{1}{6}.$$

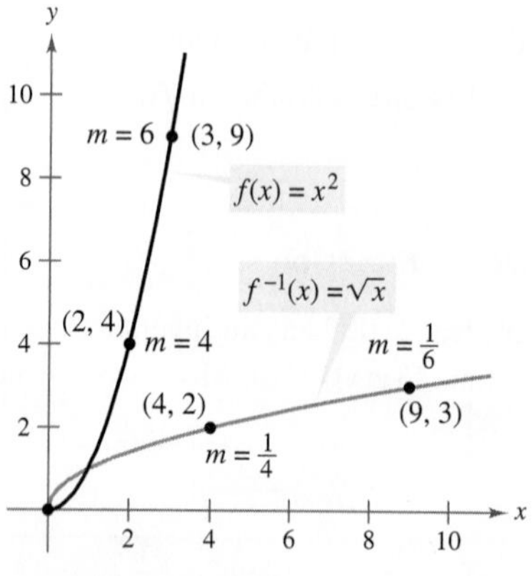

At (0, 0), the derivative of f is 0, and the derivative of f^{-1} does not exist.
Figure 5.18

So, in both cases, the slopes are reciprocals, as shown in Figure 5.18.

Exercises for Section 5.3

See www.CalcChat.com for worked-out solutions to odd-numbered exercises.

In Exercises 1–8, show that f and g are inverse functions (a) analytically and (b) graphically.

1. $f(x) = 5x + 1$, $g(x) = (x - 1)/5$
2. $f(x) = 3 - 4x$, $g(x) = (3 - x)/4$
3. $f(x) = x^3$, $g(x) = \sqrt[3]{x}$
4. $f(x) = 1 - x^3$, $g(x) = \sqrt[3]{1 - x}$
5. $f(x) = \sqrt{x - 4}$, $g(x) = x^2 + 4$, $x \ge 0$
6. $f(x) = 16 - x^2$, $x \ge 0$, $g(x) = \sqrt{16 - x}$
7. $f(x) = 1/x$, $g(x) = 1/x$
8. $f(x) = \dfrac{1}{1 + x}$, $x \ge 0$, $g(x) = \dfrac{1 - x}{x}$, $0 < x \le 1$

In Exercises 9–12, match the graph of the function with the graph of its inverse function. [The graphs of the inverse functions are labeled (a), (b), (c), and (d).]

(a)

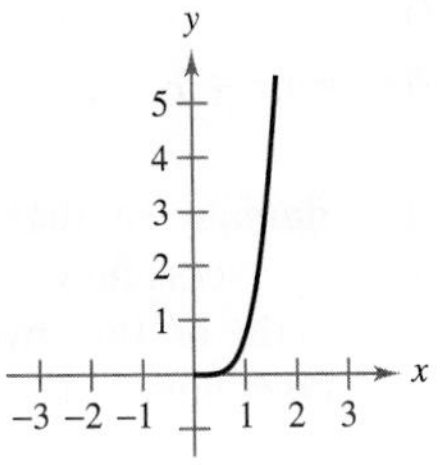

(b)

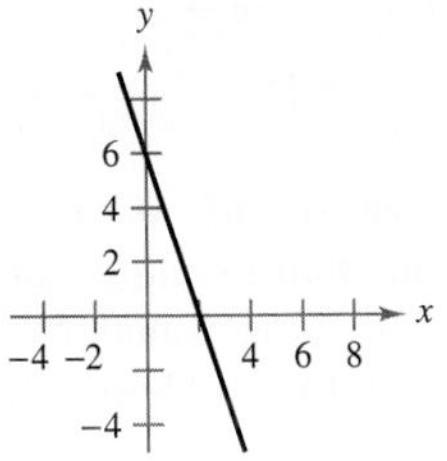

(c)

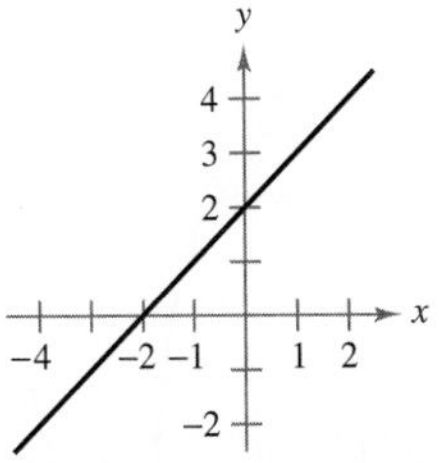

(d)

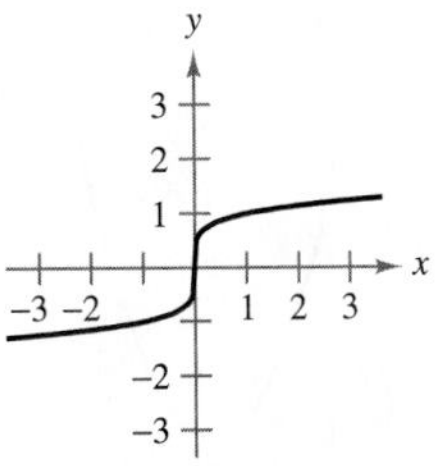

9.

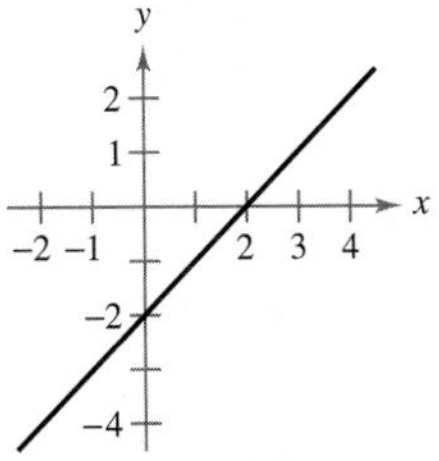

10.

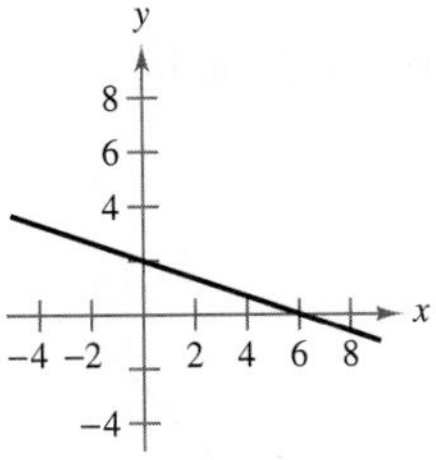

11.

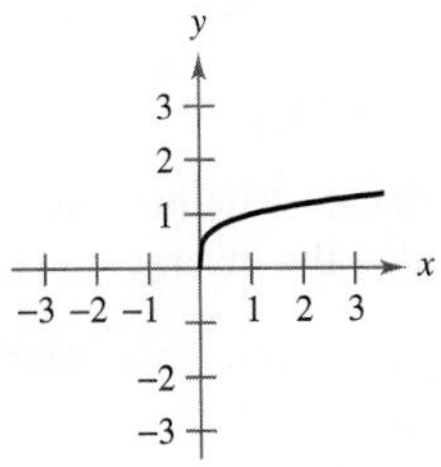

12.

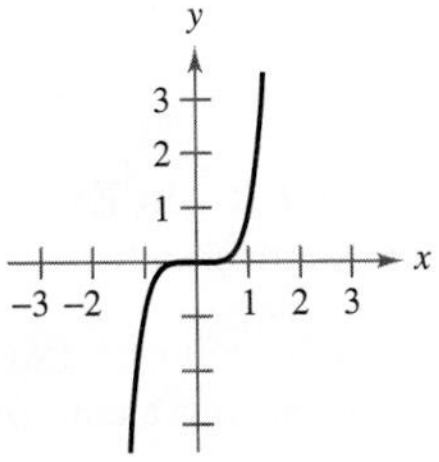

In Exercises 13–16, use the Horizontal Line Test to determine whether the function is one-to-one on its entire domain and therefore has an inverse function. To print an enlarged copy of the graph, go to the website *www.mathgraphs.com*.

13. $f(x) = \frac{3}{4}x + 6$

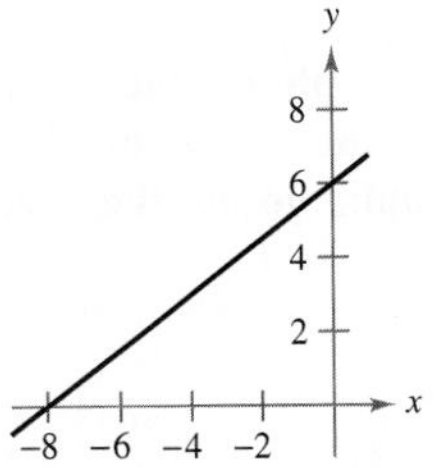

14. $f(x) = 5x - 3$

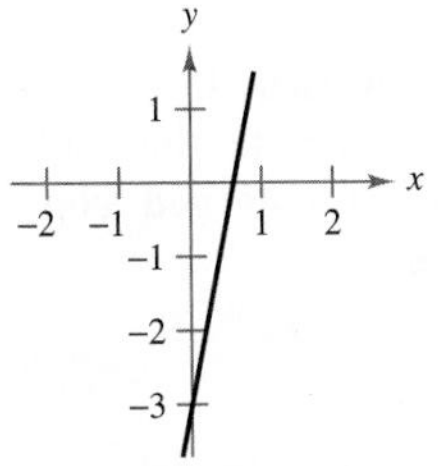

15. $f(\theta) = \sin\theta$

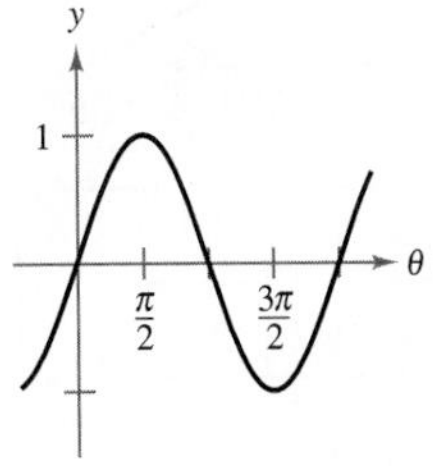

16. $f(x) = \dfrac{x^2}{x^2 + 4}$

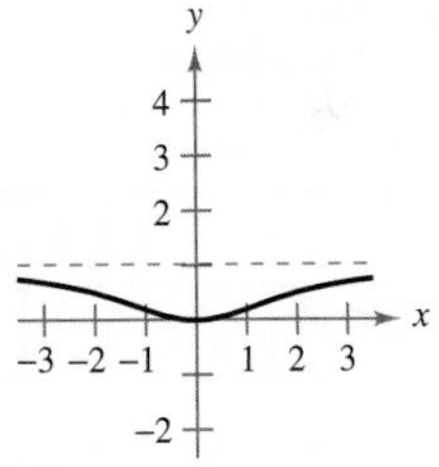

In Exercises 17–22, use a graphing utility to graph the function. Determine whether the function is one-to-one on its entire domain.

17. $h(s) = \dfrac{1}{s - 2} - 3$
18. $g(t) = \dfrac{1}{\sqrt{t^2 + 1}}$
19. $f(x) = \ln x$
20. $f(x) = 5x\sqrt{x - 1}$
21. $g(x) = (x + 5)^3$
22. $h(x) = |x + 4| - |x - 4|$

In Exercises 23–28, use the derivative to determine whether the function is strictly monotonic on its entire domain and therefore has an inverse function.

23. $f(x) = \ln(x - 3)$
24. $f(x) = \cos\dfrac{3x}{2}$
25. $f(x) = \dfrac{x^4}{4} - 2x^2$
26. $f(x) = x^3 - 6x^2 + 12x$
27. $f(x) = 2 - x - x^3$
28. $f(x) = (x + a)^3 + b$

In Exercises 29–36, find the inverse function of f. Graph (by hand) f and f^{-1}. Describe the relationship between the graphs.

29. $f(x) = 2x - 3$
30. $f(x) = 3x$
31. $f(x) = x^5$
32. $f(x) = x^3 - 1$
33. $f(x) = \sqrt{x}$
34. $f(x) = x^2$, $x \ge 0$
35. $f(x) = \sqrt{4 - x^2}$, $x \ge 0$
36. $f(x) = \sqrt{x^2 - 4}$, $x \ge 2$

In Exercises 37–42, find the inverse function of f. Use a graphing utility to graph f and f^{-1} in the same viewing window. Describe the relationship between the graphs.

37. $f(x) = \sqrt[3]{x - 1}$ **38.** $f(x) = 3\sqrt[5]{2x - 1}$

39. $f(x) = x^{2/3}, \quad x \geq 0$ **40.** $f(x) = x^{3/5}$

41. $f(x) = \dfrac{x}{\sqrt{x^2 + 7}}$ **42.** $f(x) = \dfrac{x + 2}{x}$

In Exercises 43 and 44, use the graph of the function f to complete the table and sketch the graph of f^{-1}. To print an enlarged copy of the graph, go to the website *www.mathgraphs.com*.

43.

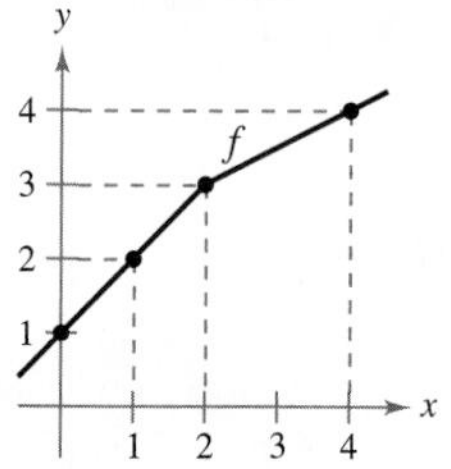

x	1	2	3	4
$f^{-1}(x)$				

44.

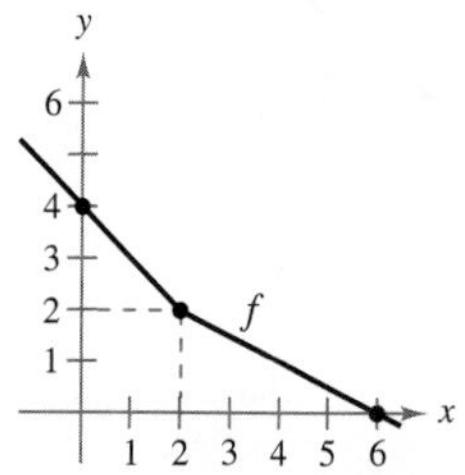

x	0	2	4
$f^{-1}(x)$			

45. ***Cost*** You need 50 pounds of two commodities costing \$1.25 and \$1.60 per pound.

(a) Verify that the total cost is $y = 1.25x + 1.60(50 - x)$, where x is the number of pounds of the less expensive commodity.

(b) Find the inverse function of the cost function. What does each variable represent in the inverse function?

(c) Use the context of the problem to determine the domain of the inverse function.

(d) Determine the number of pounds of the less expensive commodity purchased if the total cost is \$73.

46. ***Temperature*** The formula $C = \frac{5}{9}(F - 32)$, where $F \geq -459.6$, represents Celsius temperature C as a function of Fahrenheit temperature F.

(a) Find the inverse function of C.

(b) What does the inverse function represent?

(c) Determine the domain of the inverse function.

(d) The temperature is 22°C. What is the corresponding temperature in degrees Fahrenheit?

In Exercises 47–52, show that f is strictly monotonic on the given interval and therefore has an inverse function on that interval.

47. $f(x) = (x - 4)^2, \quad [4, \infty)$ **48.** $f(x) = |x + 2|, \quad [-2, \infty)$

49. $f(x) = \dfrac{4}{x^2}, \quad (0, \infty)$ **50.** $f(x) = \cot x, \quad (0, \pi)$

51. $f(x) = \cos x, \quad [0, \pi]$ **52.** $f(x) = \sec x, \quad \left[0, \dfrac{\pi}{2}\right)$

In Exercises 53 and 54, find the inverse function of f over the given interval. Use a graphing utility to graph f and f^{-1} in the same viewing window. Describe the relationship between the graphs.

53. $f(x) = \dfrac{x}{x^2 - 4}, \quad (-2, 2)$ **54.** $f(x) = 2 - \dfrac{3}{x^2}, \quad (0, 10)$

Graphical Reasoning **In Exercises 55–58, (a) use a graphing utility to graph the function, (b) use the *drawing* feature of a graphing utility to draw the inverse function of the function, and (c) determine whether the graph of the inverse relation is an inverse function. Explain your reasoning.**

55. $f(x) = x^3 + x + 4$ **56.** $h(x) = x\sqrt{4 - x^2}$

57. $g(x) = \dfrac{3x^2}{x^2 + 1}$ **58.** $f(x) = \dfrac{4x}{\sqrt{x^2 + 15}}$

In Exercises 59–62, determine whether the function is one-to-one. If it is, find its inverse function.

59. $f(x) = \sqrt{x - 2}$ **60.** $f(x) = -3$

61. $f(x) = |x - 2|, \quad x \leq 2$ **62.** $f(x) = ax + b, \quad a \neq 0$

In Exercises 63–66, delete part of the domain so that the function that remains is one-to-one. Find the inverse function of the remaining function and give the domain of the inverse function. (*Note:* There is more than one correct answer.)

63. $f(x) = (x - 3)^2$

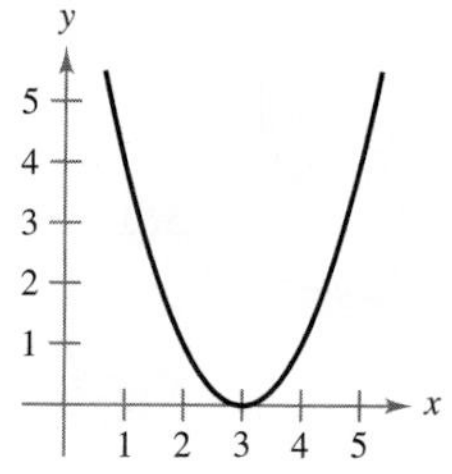

64. $f(x) = 16 - x^4$

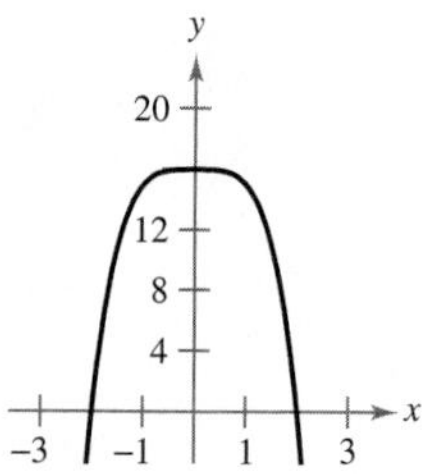

65. $f(x) = |x + 3|$

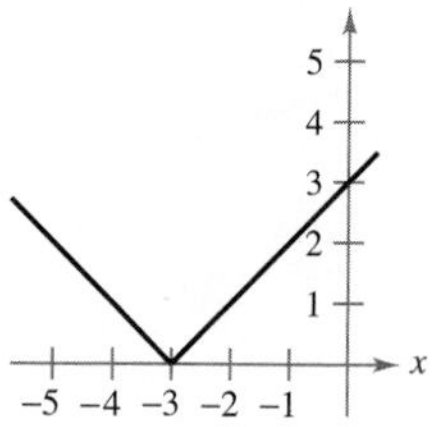

66. $f(x) = |x - 3|$

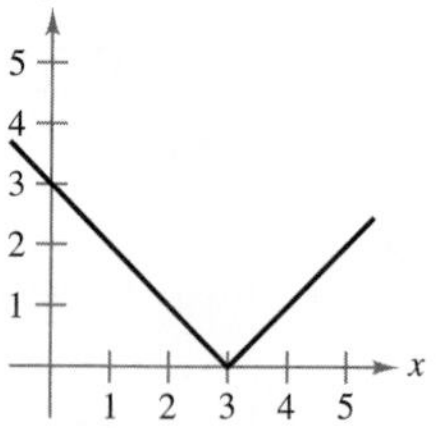

Think About It **In Exercises 67–70, decide whether the function has an inverse function. If so, what is the inverse function?**

67. $g(t)$ is the volume of water that has passed through a water line t minutes after a control valve is opened.

68. $h(t)$ is the height of the tide t hours after midnight, where $0 \leq t < 24$.

69. $C(t)$ is the cost of a long distance call lasting t minutes.

70. $A(r)$ is the area of a circle of radius r.

In Exercises 71–76, find $(f^{-1})'(a)$ for the function f and the given real number a.

71. $f(x) = x^3 + 2x - 1, \quad a = 2$

72. $f(x) = \frac{1}{27}(x^5 + 2x^3), \quad a = -11$

73. $f(x) = \sin x, \quad -\frac{\pi}{2} \le x \le \frac{\pi}{2}, \quad a = \frac{1}{2}$

74. $f(x) = \cos 2x, \quad 0 \le x \le \frac{\pi}{2}, \quad a = 1$

75. $f(x) = x^3 - \frac{4}{x}, \quad a = 6$

76. $f(x) = \sqrt{x - 4}, \quad a = 2$

In Exercises 77–80, (a) find the domains of f and f^{-1}, (b) find the ranges of f and f^{-1}, (c) graph f and f^{-1}, and (d) show that the slopes of the graphs of f and f^{-1} are reciprocals at the given points.

	Functions	*Point*
77.	$f(x) = x^3$	$\left(\frac{1}{2}, \frac{1}{8}\right)$
	$f^{-1}(x) = \sqrt[3]{x}$	$\left(\frac{1}{8}, \frac{1}{2}\right)$
78.	$f(x) = 3 - 4x$	$(1, -1)$
	$f^{-1}(x) = \frac{3 - x}{4}$	$(-1, 1)$
79.	$f(x) = \sqrt{x - 4}$	$(5, 1)$
	$f^{-1}(x) = x^2 + 4, \quad x \ge 0$	$(1, 5)$
80.	$f(x) = \frac{4}{1 + x^2}, \quad x \ge 0$	$(1, 2)$
	$f^{-1}(x) = \sqrt{\frac{4 - x}{x}}$	$(2, 1)$

In Exercises 81 and 82, find dy/dx at the given point for the equation.

81. $x = y^3 - 7y^2 + 2, \quad (-4, 1)$

82. $x = 2\ln(y^2 - 3), \quad (0, 4)$

In Exercises 83–86, use the functions $f(x) = \frac{1}{8}x - 3$ and $g(x) = x^3$ to find the given value.

83. $(f^{-1} \circ g^{-1})(1)$

84. $(g^{-1} \circ f^{-1})(-3)$

85. $(f^{-1} \circ f^{-1})(6)$

86. $(g^{-1} \circ g^{-1})(-4)$

In Exercises 87–90, use the functions $f(x) = x + 4$ and $g(x) = 2x - 5$ to find the given function.

87. $g^{-1} \circ f^{-1}$

88. $f^{-1} \circ g^{-1}$

89. $(f \circ g)^{-1}$

90. $(g \circ f)^{-1}$

Writing About Concepts

91. Describe how to find the inverse function of a one-to-one function given by an equation in x and y. Give an example.

92. Describe the relationship between the graph of a function and the graph of its inverse function.

Writing About Concepts *(continued)*

In Exercises 93 and 94, the derivative of the function has the same sign for all x in its domain, but the function is not one-to-one. Explain.

93. $f(x) = \tan x$

94. $f(x) = \frac{x}{x^2 - 4}$

95. ***Think About It*** The function $f(x) = k(2 - x - x^3)$ is one-to-one and $f^{-1}(3) = -2$. Find k.

96. (a) Show that $f(x) = 2x^3 + 3x^2 - 36x$ is not one-to-one on $(-\infty, \infty)$.

(b) Determine the greatest value c such that f is one-to-one on $(-c, c)$.

97. Let f and g be one-to-one functions. Prove that (a) $f \circ g$ is one-to-one and (b) $(f \circ g)^{-1}(x) = (g^{-1} \circ f^{-1})(x)$.

98. Prove that if f has an inverse function, then $(f^{-1})^{-1} = f$.

99. Prove that if a function has an inverse function, then the inverse function is unique.

100. Prove that a function has an inverse function if and only if it is one-to-one.

True or False? **In Exercises 101–104, determine whether the statement is true or false. If it is false, explain why or give an example that shows it is false.**

101. If f is an even function, then f^{-1} exists.

102. If the inverse function of f exists, then the y-intercept of f is an x-intercept of f^{-1}.

103. If $f(x) = x^n$ where n is odd, then f^{-1} exists.

104. There exists no function f such that $f = f^{-1}$.

105. Is the converse of the second part of Theorem 5.7 true? That is, if a function is one-to-one (and therefore has an inverse function), then must the function be strictly monotonic? If so, prove it. If not, give a counterexample.

106. Let f be twice-differentiable and one-to-one on an open interval I. Show that its inverse function g satisfies

$$g''(x) = -\frac{f''(g(x))}{[f'(g(x))]^3}.$$

If f is increasing and concave downward, what is the concavity of $f^{-1} = g$?

107. If $f(x) = \int_2^x \frac{dt}{\sqrt{1 + t^4}}$, find $(f^{-1})'(0)$.

108. Show that $f(x) = \int_2^x \sqrt{1 + t^2}\, dt$ is one-to-one and find $(f^{-1})'(0)$.

109. Let $y = \frac{x - 2}{x - 1}$. Show that y is its own inverse function. What can you conclude about the graph of f? Explain.

Section 5.4 Exponential Functions: Differentiation and Integration

- **Develop properties of the natural exponential function.**
- **Differentiate natural exponential functions.**
- **Integrate natural exponential functions.**

The Natural Exponential Function

The function $f(x) = \ln x$ is increasing on its entire domain, and therefore it has an inverse function f^{-1}. The domain of f^{-1} is the set of all reals, and the range is the set of positive reals, as shown in Figure 5.19. So, for any real number x,

$$f(f^{-1}(x)) = \ln[f^{-1}(x)] = x. \qquad x \text{ is any real number.}$$

If x happens to be rational, then

$$\ln(e^x) = x \ln e = x(1) = x. \qquad x \text{ is a rational number.}$$

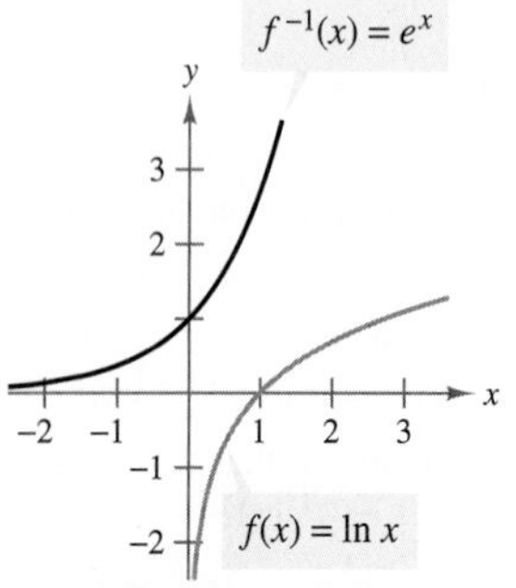

The inverse function of the natural logarithmic function is the natural exponential function.
Figure 5.19

Because the natural logarithmic function is one-to-one, you can conclude that $f^{-1}(x)$ and e^x agree for *rational* values of x. The following definition extends the meaning of e^x to include *all* real values of x.

> **Definition of the Natural Exponential Function**
>
> The inverse function of the natural logarithmic function $f(x) = \ln x$ is called the **natural exponential function** and is denoted by
>
> $$f^{-1}(x) = e^x.$$
>
> That is,
>
> $$y = e^x \quad \text{if and only if} \quad x = \ln y.$$

> **THE NUMBER *e***
>
> The symbol *e* was first used by mathematician Leonhard Euler to represent the base of natural logarithms in a letter to another mathematician, Christian Goldbach, in 1731.

The inverse relationship between the natural logarithmic function and the natural exponential function can be summarized as follows.

$$\ln(e^x) = x \quad \text{and} \quad e^{\ln x} = x \qquad \text{Inverse relationship}$$

EXAMPLE 1 Solving Exponential Equations

Solve $7 = e^{x+1}$.

Solution You can convert from exponential form to logarithmic form by *taking the natural logarithim of each side* of the equation.

$$7 = e^{x+1} \qquad \text{Write original equation.}$$
$$\ln 7 = \ln(e^{x+1}) \qquad \text{Take natural logarithm of each side.}$$
$$\ln 7 = x + 1 \qquad \text{Apply inverse property.}$$
$$-1 + \ln 7 = x \qquad \text{Solve for } x.$$
$$0.946 \approx x \qquad \text{Use a calculator.}$$

Check this solution in the original equation.

EXAMPLE 2 Solving a Logarithmic Equation

Solve $\ln(2x - 3) = 5$.

Solution To convert from logarithmic form to exponential form, you can *exponentiate each side* of the logarithmic equation.

$$\ln(2x - 3) = 5 \quad \text{Write original equation.}$$
$$e^{\ln(2x-3)} = e^5 \quad \text{Exponentiate each side.}$$
$$2x - 3 = e^5 \quad \text{Apply inverse property.}$$
$$x = \tfrac{1}{2}(e^5 + 3) \quad \text{Solve for } x.$$
$$x \approx 75.707 \quad \text{Use a calculator.}$$

The familiar rules for operating with rational exponents can be extended to the natural exponential function, as shown in the following theorem.

THEOREM 5.10 Operations with Exponential Functions

Let a and b be any real numbers.

1. $e^a e^b = e^{a+b}$

2. $\dfrac{e^a}{e^b} = e^{a-b}$

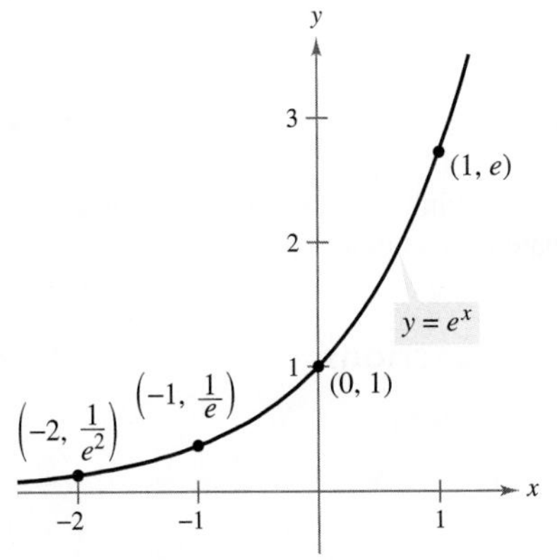

The natural exponential function is increasing, and its graph is concave upward.
Figure 5.20

Proof To prove Property 1, you can write

$$\ln(e^a e^b) = \ln(e^a) + \ln(e^b) = a + b = \ln(e^{a+b}).$$

Because the natural logarithmic function is one-to-one, you can conclude that

$$e^a e^b = e^{a+b}.$$

The proof of the second property is left to you (see Exercise 129).

In Section 5.3, you learned that an inverse function f^{-1} shares many properties with f. So, the natural exponential function inherits the following properties from the natural logarithmic function (see Figure 5.20).

Properties of the Natural Exponential Function

1. The domain of $f(x) = e^x$ is $(-\infty, \infty)$, and the range is $(0, \infty)$.
2. The function $f(x) = e^x$ is continuous, increasing, and one-to-one on its entire domain.
3. The graph of $f(x) = e^x$ is concave upward on its entire domain.
4. $\displaystyle\lim_{x\to-\infty} e^x = 0$ and $\displaystyle\lim_{x\to\infty} e^x = \infty$

Derivatives of Exponential Functions

One of the most intriguing (and useful) characteristics of the natural exponential function is that *it is its own derivative.* In other words, it is a solution to the differential equation $y' = y$. This result is stated in the next theorem.

FOR FURTHER INFORMATION To find out about derivatives of exponential functions of order 1/2, see the article "A Child's Garden of Fractional Derivatives" by Marcia Kleinz and Thomas J. Osler in *The College Mathematics Journal.* To view this article, go to the website *www.matharticles.com.*

THEOREM 5.11 Derivative of the Natural Exponential Function

Let u be a differentiable function of x.

1. $\dfrac{d}{dx}[e^x] = e^x$

2. $\dfrac{d}{dx}[e^u] = e^u \dfrac{du}{dx}$

Proof To prove Property 1, use the fact that $\ln e^x = x$, and differentiate each side of the equation.

$$\ln e^x = x \qquad \text{Definition of exponential function}$$

$$\frac{d}{dx}[\ln e^x] = \frac{d}{dx}[x] \qquad \text{Differentiate each side with respect to } x.$$

$$\frac{1}{e^x}\frac{d}{dx}[e^x] = 1$$

$$\frac{d}{dx}[e^x] = e^x$$

The derivative of e^u follows from the Chain Rule.

NOTE You can interpret this theorem geometrically by saying that the slope of the graph of $f(x) = e^x$ at any point (x, e^x) is equal to the y-coordinate of the point.

EXAMPLE 3 Differentiating Exponential Functions

a. $\dfrac{d}{dx}[e^{2x-1}] = e^u \dfrac{du}{dx} = 2e^{2x-1}$ $\qquad u = 2x - 1$

b. $\dfrac{d}{dx}[e^{-3/x}] = e^u \dfrac{du}{dx} = \left(\dfrac{3}{x^2}\right)e^{-3/x} = \dfrac{3e^{-3/x}}{x^2}$ $\qquad u = -\dfrac{3}{x}$

EXAMPLE 4 Locating Relative Extrema

Find the relative extrema of $f(x) = xe^x$.

Solution The derivative of f is given by

$$f'(x) = x(e^x) + e^x(1) \qquad \text{Product Rule}$$
$$= e^x(x + 1).$$

Because e^x is never 0, the derivative is 0 only when $x = -1$. Moreover, by the First Derivative Test, you can determine that this corresponds to a relative minimum, as shown in Figure 5.21. Because the derivative $f'(x) = e^x(x + 1)$ is defined for all x, there are no other critical points.

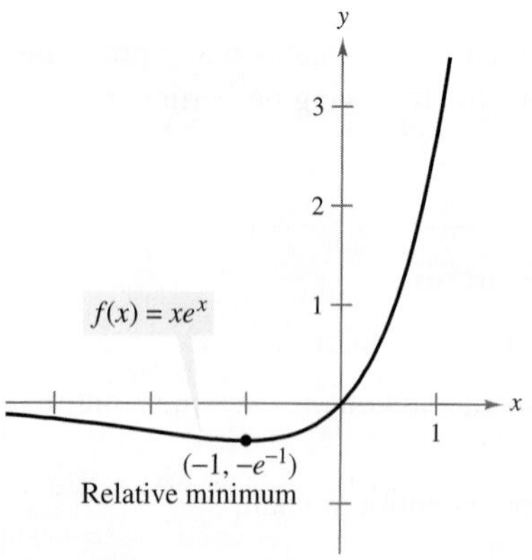

The derivative of f changes from negative to positive at $x = -1$.
Figure 5.21

EXAMPLE 5 The Standard Normal Probability Density Function

Show that the *standard normal probability density function*

$$f(x) = \frac{1}{\sqrt{2\pi}}e^{-x^2/2}$$

has points of inflection when $x = \pm 1$.

Solution To locate possible points of inflection, find the x-values for which the second derivative is 0.

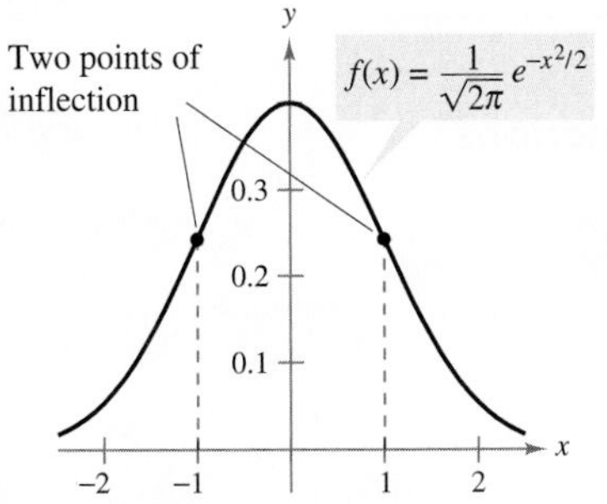

The bell-shaped curve given by a standard normal probability density function
Figure 5.22

$$f(x) = \frac{1}{\sqrt{2\pi}}e^{-x^2/2}$$ Write original function.

$$f'(x) = \frac{1}{\sqrt{2\pi}}(-x)e^{-x^2/2}$$ First derivative

$$f''(x) = \frac{1}{\sqrt{2\pi}}[(-x)(-x)e^{-x^2/2} + (-1)e^{-x^2/2}]$$ Product Rule

$$= \frac{1}{\sqrt{2\pi}}(e^{-x^2/2})(x^2 - 1)$$ Second derivative

So, $f''(x) = 0$ when $x = \pm 1$, and you can apply the techniques of Chapter 3 to conclude that these values yield the two points of inflection shown in Figure 5.22.

NOTE The general form of a normal probability density function (whose mean is 0) is given by

$$f(x) = \frac{1}{\sigma\sqrt{2\pi}}e^{-x^2/2\sigma^2}$$

where σ is the standard deviation (σ is the lowercase Greek letter sigma). This "bell-shaped curve" has points of inflection when $x = \pm\sigma$.

EXAMPLE 6 Shares Traded

The number y of shares traded (in millions) on the New York Stock Exchange from 1990 through 2002 can be modeled by

$$y = 36{,}663e^{0.1902t}$$

where t represents the year, with $t = 0$ corresponding to 1990. At what rate was the number of shares traded changing in 1998? *(Source: New York Stock Exchange, Inc.)*

Solution The derivative of the given model is

$$y' = (0.1902)(36{,}663)e^{0.1902t}$$
$$\approx 6973e^{0.1902t}.$$

By evaluating the derivative when $t = 8$, you can conclude that the rate of change in 1998 was about

31,933 million shares per year.

The graph of this model is shown in Figure 5.23.

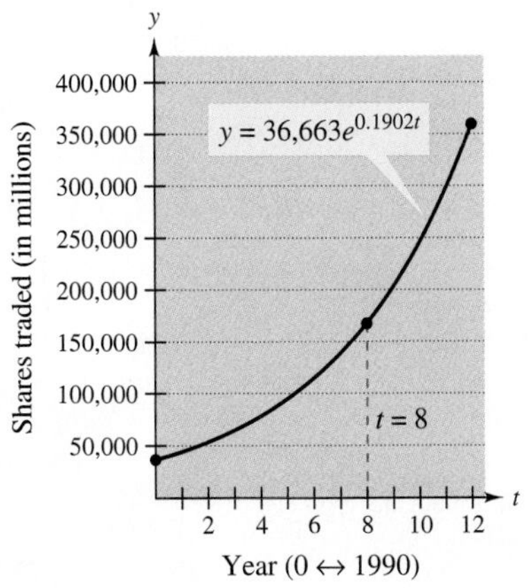

Figure 5.23

Integrals of Exponential Functions

Each differentiation formula in Theorem 5.11 has a corresponding integration formula.

THEOREM 5.12 Integration Rules for Exponential Functions

Let u be a differentiable function of x.

1. $\int e^x\,dx = e^x + C$ **2.** $\int e^u\,du = e^u + C$

EXAMPLE 7 Integrating Exponential Functions

Find $\int e^{3x+1}\,dx$.

Solution If you let $u = 3x + 1$, then $du = 3\,dx$.

$$\begin{aligned}
\int e^{3x+1}dx &= \frac{1}{3}\int e^{3x+1}(3)\,dx && \text{Multiply and divide by 3.}\\
&= \frac{1}{3}\int e^u\,du && \text{Substitute: } u = 3x + 1.\\
&= \frac{1}{3}e^u + C && \text{Apply Exponential Rule.}\\
&= \frac{e^{3x+1}}{3} + C && \text{Back-substitute.}
\end{aligned}$$

NOTE In Example 7, the missing *constant* factor 3 was introduced to create $du = 3\,dx$. However, remember that you cannot introduce a missing *variable* factor in the integrand. For instance,

$$\int e^{-x^2}\,dx \neq \frac{1}{x}\int e^{-x^2}\,(x\,dx).$$

EXAMPLE 8 Integrating Exponential Functions

Find $\int 5xe^{-x^2}\,dx$.

Solution If you let $u = -x^2$, then $du = -2x\,dx$ or $x\,dx = -du/2$.

$$\begin{aligned}
\int 5xe^{-x^2}dx &= \int 5e^{-x^2}(x\,dx) && \text{Regroup integrand.}\\
&= \int 5e^u\left(-\frac{du}{2}\right) && \text{Substitute: } u = -x^2.\\
&= -\frac{5}{2}\int e^u\,du && \text{Constant Multiple Rule}\\
&= -\frac{5}{2}e^u + C && \text{Apply Exponential Rule.}\\
&= -\frac{5}{2}e^{-x^2} + C && \text{Back-substitute.}
\end{aligned}$$

EXAMPLE 9 Integrating Exponential Functions

a. $$\int \frac{e^{1/x}}{x^2}\,dx = -\int \overbrace{e^{1/x}}^{e^u}\overbrace{\left(-\frac{1}{x^2}\right)dx}^{du} \qquad u = \frac{1}{x}$$
$$= -e^{1/x} + C$$

b. $$\int \sin x\, e^{\cos x}\,dx = -\int \overbrace{e^{\cos x}}^{e^u}\overbrace{(-\sin x\,dx)}^{du} \qquad u = \cos x$$
$$= -e^{\cos x} + C$$

EXAMPLE 10 Finding Areas Bounded by Exponential Functions

Evaluate each definite integral.

a. $\displaystyle\int_0^1 e^{-x}\,dx$ **b.** $\displaystyle\int_0^1 \frac{e^x}{1+e^x}\,dx$ **c.** $\displaystyle\int_{-1}^0 [e^x\cos(e^x)]\,dx$

Solution

a. $$\int_0^1 e^{-x}\,dx = -e^{-x}\Big]_0^1 \qquad \text{See Figure 5.24(a).}$$
$$= -e^{-1} - (-1)$$
$$= 1 - \frac{1}{e}$$
$$\approx 0.632$$

b. $$\int_0^1 \frac{e^x}{1+e^x}\,dx = \ln(1+e^x)\Big]_0^1 \qquad \text{See Figure 5.24(b).}$$
$$= \ln(1+e) - \ln 2$$
$$\approx 0.620$$

c. $$\int_{-1}^0 [e^x\cos(e^x)]\,dx = \sin(e^x)\Big]_{-1}^0 \qquad \text{See Figure 5.24(c).}$$
$$= \sin 1 - \sin(e^{-1})$$
$$\approx 0.482$$

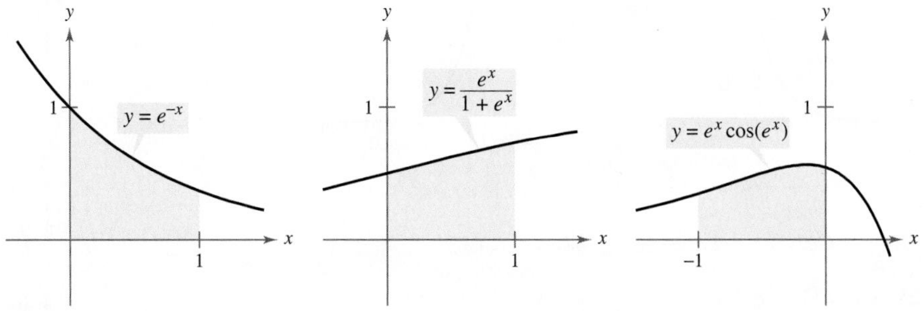

(a) (b) (c)

Figure 5.24

Exercises for Section 5.4

See www.CalcChat.com for worked-out solutions to odd-numbered exercises.

In Exercises 1–14, solve for x accurate to three decimal places.

1. $e^{\ln x} = 4$

2. $e^{\ln 2x} = 12$

3. $e^x = 12$

4. $4e^x = 83$

5. $9 - 2e^x = 7$

6. $-6 + 3e^x = 8$

7. $50e^{-x} = 30$

8. $200e^{-4x} = 15$

9. $\ln x = 2$

10. $\ln x^2 = 10$

11. $\ln(x - 3) = 2$

12. $\ln 4x = 1$

13. $\ln\sqrt{x + 2} = 1$

14. $\ln(x - 2)^2 = 12$

In Exercises 15–18, sketch the graph of the function.

15. $y = e^{-x}$

16. $y = \frac{1}{2}e^x$

17. $y = e^{-x^2}$

18. $y = e^{-x/2}$

19. Use a graphing utility to graph $f(x) = e^x$ and the given function in the same viewing window. How are the two graphs related?

(a) $g(x) = e^{x-2}$ (b) $h(x) = -\frac{1}{2}e^x$ (c) $q(x) = e^{-x} + 3$

20. Use a graphing utility to graph the function. Use the graph to determine any asymptotes of the function.

(a) $f(x) = \dfrac{8}{1 + e^{-0.5x}}$

(b) $g(x) = \dfrac{8}{1 + e^{-0.5/x}}$

In Exercises 21–24, match the equation with the correct graph. Assume that a and C are positive real numbers. [The graphs are labeled (a), (b), (c), and (d).]

(a)

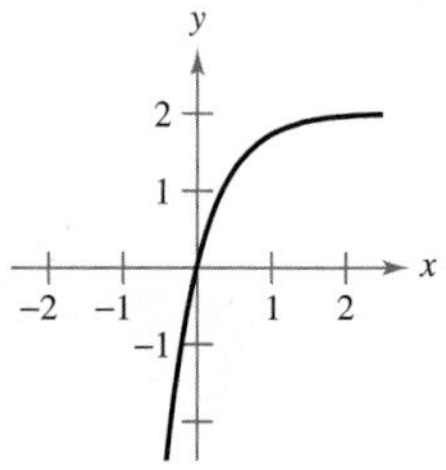

(b)

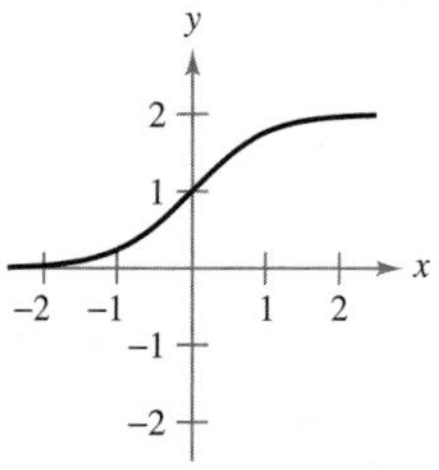

(c)

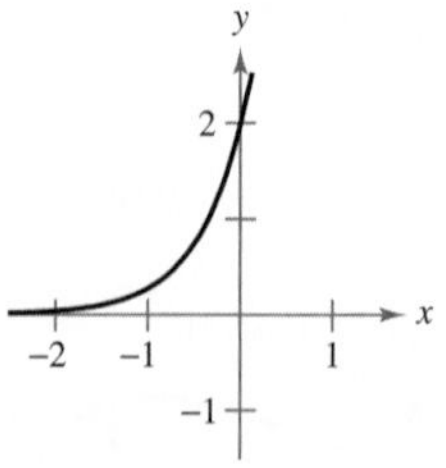

(d)

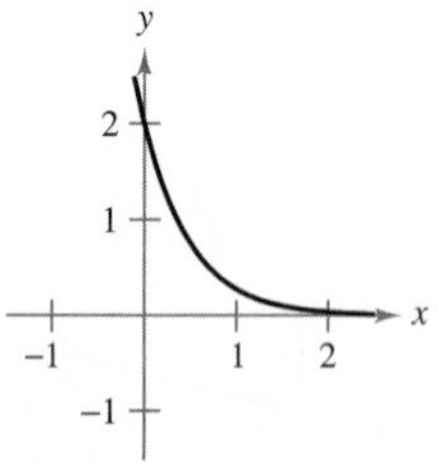

21. $y = Ce^{ax}$

22. $y = Ce^{-ax}$

23. $y = C(1 - e^{-ax})$

24. $y = \dfrac{C}{1 + e^{-ax}}$

In Exercises 25–28, illustrate that the functions are inverses of each other by graphing both functions on the same set of coordinate axes.

25. $f(x) = e^{2x}$
$g(x) = \ln\sqrt{x}$

26. $f(x) = e^{x/3}$
$g(x) = \ln x^3$

27. $f(x) = e^x - 1$
$g(x) = \ln(x + 1)$

28. $f(x) = e^{x-1}$
$g(x) = 1 + \ln x$

29. ***Graphical Analysis*** Use a graphing utility to graph

$$f(x) = \left(1 + \frac{0.5}{x}\right)^x \quad \text{and} \quad g(x) = e^{0.5}$$

in the same viewing window. What is the relationship between f and g as $x \to \infty$?

30. ***Conjecture*** Use the result of Exercise 29 to make a conjecture about the value of

$$\left(1 + \frac{r}{x}\right)^x$$

as $x \to \infty$.

In Exercises 31 and 32, compare the given number with the number e. Is the number less than or greater than e?

31. $\left(1 + \dfrac{1}{1{,}000{,}000}\right)^{1{,}000{,}000}$ (See Exercise 30.)

32. $1 + 1 + \dfrac{1}{2} + \dfrac{1}{6} + \dfrac{1}{24} + \dfrac{1}{120} + \dfrac{1}{720} + \dfrac{1}{5040}$

In Exercises 33 and 34, find an equation of the tangent line to the graph of the function at the point (0, 1).

33. (a) $y = e^{3x}$ (b) $y = e^{-3x}$

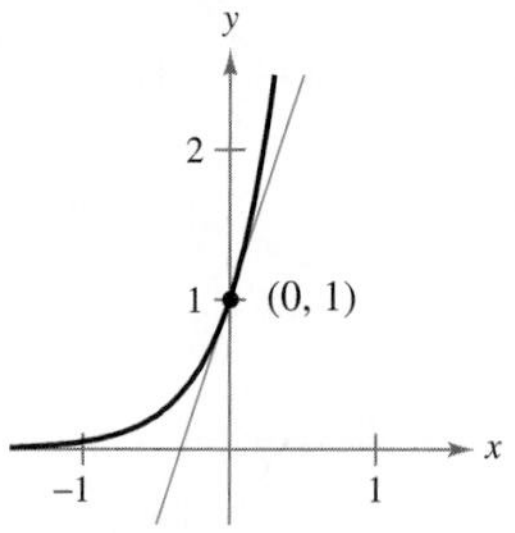

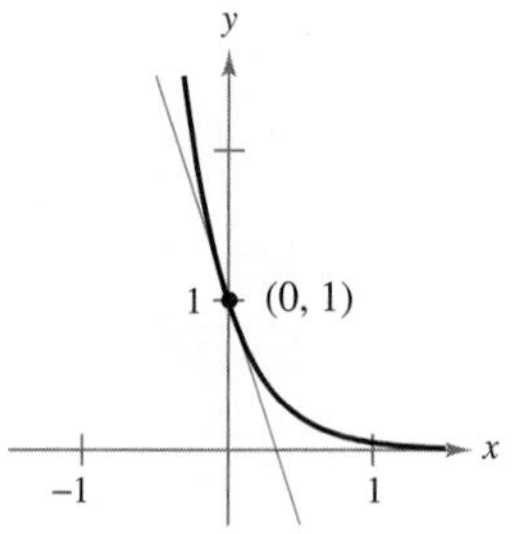

34. (a) $y = e^{2x}$ (b) $y = e^{-2x}$

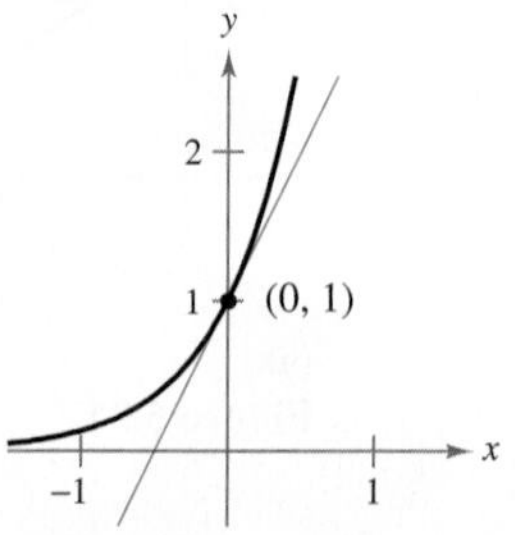

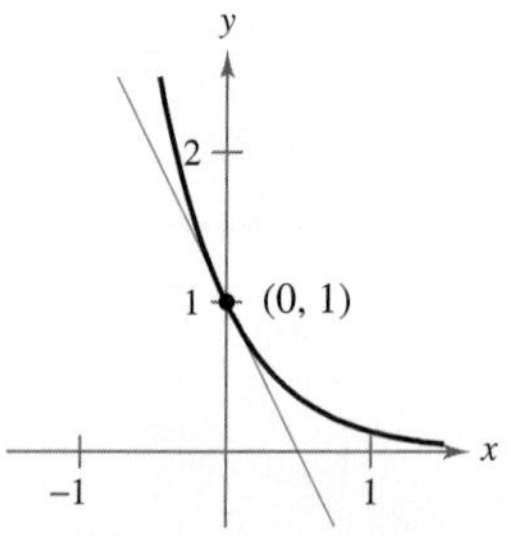

In Exercises 35–48, find the derivative.

35. $f(x) = e^{2x}$

36. $y = e^{-x^2}$

37. $y = e^{\sqrt{x}}$

38. $y = x^2e^{-x}$

39. $g(t) = (e^{-t} + e^t)^3$

40. $g(t) = e^{-3/t^2}$

41. $y = \ln(1 + e^{2x})$

42. $y = \ln\left(\dfrac{1 + e^x}{1 - e^x}\right)$

43. $y = \dfrac{2}{e^x + e^{-x}}$

44. $y = \dfrac{e^x - e^{-x}}{2}$

45. $y = e^x(\sin x + \cos x)$

46. $y = \ln e^x$

47. $F(x) = \displaystyle\int_{\pi}^{\ln x} \cos e^t\,dt$

48. $F(x) = \displaystyle\int_0^{e^{2x}} \ln(t + 1)\,dt$

In Exercises 49–56, find an equation of the tangent line to the graph of the function at the given point.

49. $f(x) = e^{1-x}$, $(1, 1)$

50. $y = e^{-2x+x^2}$, $(2, 1)$

51. $y = \ln(e^{x^2})$, $(-2, 4)$

52. $y = \ln\dfrac{e^x + e^{-x}}{2}$, $(0, 0)$

53. $y = x^2e^x - 2xe^x + 2e^x$, $(1, e)$

54. $y = xe^x - e^x$, $(1, 0)$

55. $f(x) = e^{-x}\ln x$, $(1, 0)$

56. $f(x) = e^3 \ln x$, $(1, 0)$

In Exercises 57 and 58, use implicit differentiation to find dy/dx.

57. $xe^y - 10x + 3y = 0$

58. $e^{xy} + x^2 - y^2 = 10$

In Exercises 59 and 60, find an equation of the tangent line to the graph of the function at the given point.

59. $xe^y + ye^x = 1$, $(0, 1)$

60. $1 + \ln xy = e^{x-y}$, $(1, 1)$

In Exercises 61 and 62, find the second derivative of the function.

61. $f(x) = (3 + 2x)e^{-3x}$

62. $g(x) = \sqrt{x} + e^x \ln x$

In Exercises 63 and 64, show that the function $y = f(x)$ is a solution of the differential equation.

63. $y = e^x\left(\cos\sqrt{2}x + \sin\sqrt{2}x\right)$

$y'' - 2y' + 3y = 0$

64. $y = e^x(3\cos 2x - 4\sin 2x)$

$y'' - 2y' + 5y = 0$

In Exercises 65–72, find the extrema and the points of inflection (if any exist) of the function. Use a graphing utility to graph the function and confirm your results.

65. $f(x) = \dfrac{e^x + e^{-x}}{2}$

66. $f(x) = \dfrac{e^x - e^{-x}}{2}$

67. $g(x) = \dfrac{1}{\sqrt{2\pi}}e^{-(x-2)^2/2}$

68. $g(x) = \dfrac{1}{\sqrt{2\pi}}e^{-(x-3)^2/2}$

69. $f(x) = x^2e^{-x}$

70. $f(x) = xe^{-x}$

71. $g(t) = 1 + (2 + t)e^{-t}$

72. $f(x) = -2 + e^{3x}(4 - 2x)$

73. ***Area*** Find the area of the largest rectangle that can be inscribed under the curve $y = e^{-x^2}$ in the first and second quadrants.

74. ***Area*** Perform the following steps to find the maximum area of the rectangle shown in the figure.

(a) Solve for c in the equation $f(c) = f(c + x)$.

(b) Use the result in part (a) to write the area A as a function of x. [*Hint*: $A = xf(c)$]

(c) Use a graphing utility to graph the area function. Use the graph to approximate the dimensions of the rectangle of maximum area. Determine the maximum area.

(d) Use a graphing utility to graph the expression for c found in part (a). Use the graph to approximate

$$\lim_{x\to 0^+} c \quad \text{and} \quad \lim_{x\to\infty} c.$$

Use this result to describe the changes in dimensions and position of the rectangle for $0 < x < \infty$.

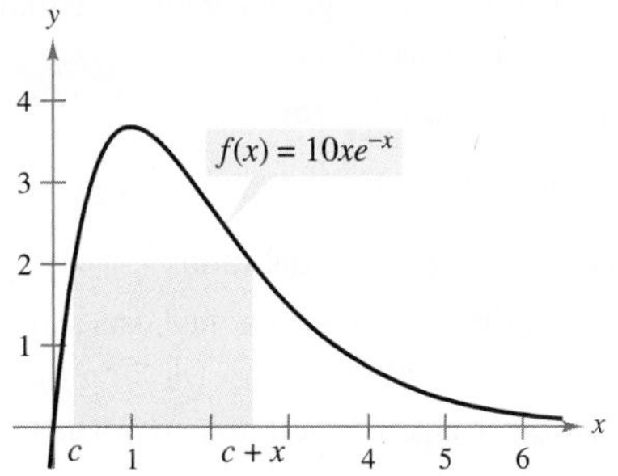

75. Verify that the function

$$y = \frac{L}{1 + ae^{-x/b}}, \quad a > 0, \quad b > 0, \quad L > 0$$

increases at a maximum rate when $y = L/2$.

76. ***Writing*** Consider the function $f(x) = \dfrac{2}{1 + e^{1/x}}$.

(a) Use a graphing utility to graph f.

(b) Write a short paragraph explaining why the graph has a horizontal asymptote at $y = 1$ and why the function has a nonremovable discontinuity at $x = 0$.

77. Find a point on the graph of the function $f(x) = e^{2x}$ such that the tangent line to the graph at that point passes through the origin. Use a graphing utility to graph f and the tangent line in the same viewing window.

78. Find the point on the graph of $y = e^{-x}$ where the normal line to the curve passes through the origin. (Use Newton's Method or the *zero* or *root* feature of a graphing utility.)

79. ***Depreciation*** The value V of an item t years after it is purchased is $V = 15{,}000e^{-0.6286t}$, $0 \le t \le 10$.

(a) Use a graphing utility to graph the function.

(b) Find the rate of change of V with respect to t when $t = 1$ and $t = 5$.

(c) Use a graphing utility to graph the tangent line to the function when $t = 1$ and $t = 5$.

80. ***Harmonic Motion*** The displacement from equilibrium of a mass oscillating on the end of a spring suspended from a ceiling is

$$y = 1.56e^{-0.22t} \cos 4.9t$$

where y is the displacement in feet and t is the time in seconds. Use a graphing utility to graph the displacement function on the interval $[0, 10]$. Find a value of t past which the displacement is less than 3 inches from equilibrium.

81. ***Modeling Data*** A meteorologist measures the atmospheric pressure P (in kilograms per square meter) at altitude h (in kilometers). The data are shown below.

h	0	5	10	15	20
P	10,332	5583	2376	1240	517

(a) Use a graphing utility to plot the points $(h, \ln P)$. Use the regression capabilities of the graphing utility to find a linear model for the revised data points.

(b) The line in part (a) has the form

$\ln P = ah + b.$

Write the equation in exponential form.

(c) Use a graphing utility to plot the original data and graph the exponential model in part (b).

(d) Find the rate of change of the pressure when $h = 5$ and $h = 18$.

82. ***Modeling Data*** The table lists the approximate value V of a mid-sized sedan for the years 1997 through 2003. The variable t represents the time in years, with $t = 7$ corresponding to 1997.

t	7	8	9	10
V	\$17,040	\$14,590	\$12,845	\$10,995

t	11	12	13
V	\$9220	\$8095	\$6835

(a) Use a computer algebra system to find linear and quadratic models for the data. Plot the data and graph the models.

(b) What does the slope represent in the linear model in part (a)?

(c) Use a computer algebra system to fit an exponential model to the data.

(d) Determine the horizontal asymptote of the exponential model found in part (c). Interpret its meaning in the context of the problem.

(e) Find the rate of decrease in the value of the sedan when $t = 8$ and $t = 12$ using the exponential model.

Linear and Quadratic Approximations **In Exercises 83 and 84, use a graphing utility to graph the function. Then graph**

$$P_1(x) = f(0) + f'(0)(x - 0)$$

and

$$P_2(x) = f(0) + f'(0)(x - 0) + \tfrac{1}{2}f''(0)(x - 0)^2$$

in the same viewing window. Compare the values of f, P_1, and P_2 and their first derivatives at $x = 0$.

83. $f(x) = e^{x/2}$

84. $f(x) = e^{-x^2/2}$

In Exercises 85–98, find the indefinite integral.

85. $\displaystyle\int e^{5x}(5)\,dx$

86. $\displaystyle\int e^{-x^4}(-4x^3)\,dx$

87. $\displaystyle\int \frac{e^{\sqrt{x}}}{\sqrt{x}}\,dx$

88. $\displaystyle\int \frac{e^{1/x^2}}{x^3}\,dx$

89. $\displaystyle\int \frac{e^{-x}}{1 + e^{-x}}\,dx$

90. $\displaystyle\int \frac{e^{2x}}{1 + e^{2x}}\,dx$

91. $\displaystyle\int e^x\sqrt{1 - e^x}\,dx$

92. $\displaystyle\int \frac{e^x - e^{-x}}{e^x + e^{-x}}\,dx$

93. $\displaystyle\int \frac{e^x + e^{-x}}{e^x - e^{-x}}\,dx$

94. $\displaystyle\int \frac{2e^x - 2e^{-x}}{(e^x + e^{-x})^2}\,dx$

95. $\displaystyle\int \frac{5 - e^x}{e^{2x}}\,dx$

96. $\displaystyle\int \frac{e^{2x} + 2e^x + 1}{e^x}\,dx$

97. $\displaystyle\int e^{-x}\tan(e^{-x})\,dx$

98. $\displaystyle\int \ln(e^{2x-1})\,dx$

In Exercises 99–106, evaluate the definite integral. Use a graphing utility to verify your result.

99. $\displaystyle\int_0^1 e^{-2x}\,dx$

100. $\displaystyle\int_3^4 e^{3-x}\,dx$

101. $\displaystyle\int_0^1 xe^{-x^2}\,dx$

102. $\displaystyle\int_{-2}^0 x^2 e^{x^3/2}\,dx$

103. $\displaystyle\int_1^3 \frac{e^{3/x}}{x^2}\,dx$

104. $\displaystyle\int_0^{\sqrt{2}} xe^{-(x^2/2)}\,dx$

105. $\displaystyle\int_0^{\pi/2} e^{\sin \pi x}\cos \pi x\,dx$

106. $\displaystyle\int_{\pi/3}^{\pi/2} e^{\sec 2x}\sec 2x \tan 2x\,dx$

Differential Equations **In Exercises 107 and 108, solve the differential equation.**

107. $\dfrac{dy}{dx} = xe^{ax^2}$

108. $\dfrac{dy}{dx} = (e^x - e^{-x})^2$

Differential Equations **In Exercises 109 and 110, find the particular solution that satisfies the initial conditions.**

109. $f''(x) = \frac{1}{2}(e^x + e^{-x})$, $f(0) = 1, f'(0) = 0$

110. $f''(x) = \sin x + e^{2x}$, $f(0) = \frac{1}{4}, f'(0) = \frac{1}{2}$

Slope Fields **In Exercises 111 and 112, a differential equation, a point, and a slope field are given. (a) Sketch two approximate solutions of the differential equation on the slope field, one of which passes through the given point. (b) Use integration to find the particular solution of the differential equation and use a graphing utility to graph the solution. Compare the result with the sketches in part (a). To print an enlarged copy of the graph, go to the website *www.mathgraphs.com*.**

111. $\frac{dy}{dx} = 2e^{-x/2}$, $(0, 1)$

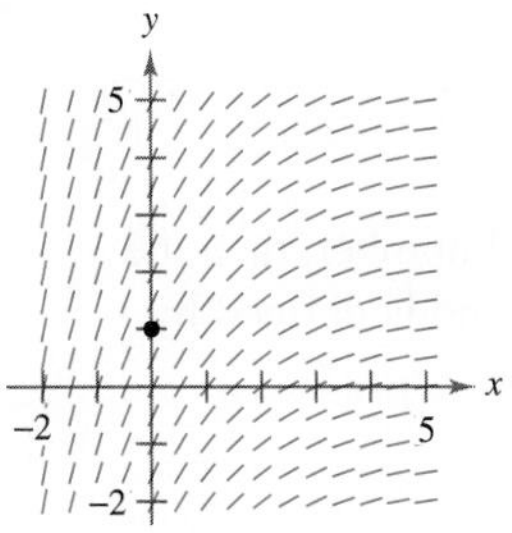

112. $\frac{dy}{dx} = xe^{-0.2x^2}$, $\left(0, -\frac{3}{2}\right)$

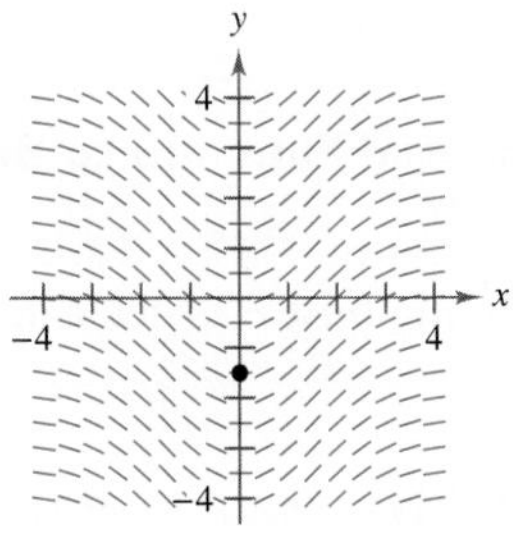

Area **In Exercises 113–116, find the area of the region bounded by the graphs of the equations. Use a graphing utility to graph the region and verify your result.**

113. $y = e^x,\ y = 0,\ x = 0,\ x = 5$

114. $y = e^{-x},\ y = 0,\ x = a,\ x = b$

115. $y = xe^{-x^2/4},\ y = 0,\ x = 0,\ x = \sqrt{6}$

116. $y = e^{-2x} + 2,\ y = 0,\ x = 0,\ x = 2$

Numerical Integration **In Exercises 117 and 118, approximate the integral using the Midpoint Rule, the Trapezoidal Rule, and Simpson's Rule with $n = 12$. Use a graphing utility to verify your results.**

117. $\int_0^4 \sqrt{x}\, e^x\, dx$

118. $\int_0^2 2xe^{-x}\, dx$

119. ***Probability*** A car battery has an average lifetime of 48 months with a standard deviation of 6 months. The battery lives are normally distributed. The probability that a given battery will last between 48 months and 60 months is

$$0.0665 \int_{48}^{60} e^{-0.0139(t-48)^2}\, dt.$$

Use the integration capabilities of a graphing utility to approximate the integral. Interpret the resulting probability.

120. ***Probability*** The median waiting time (in minutes) for people waiting for service in a convenience store is given by the solution of the equation

$$\int_0^x 0.3e^{-0.3t}\, dt = \frac{1}{2}.$$

Solve the equation.

121. Given $e^x \geq 1$ for $x \geq 0$, it follows that

$$\int_0^x e^t\, dt \geq \int_0^x 1\, dt.$$

Perform this integration to derive the inequality $e^x \geq 1 + x$ for $x \geq 0$.

122. ***Modeling Data*** A valve on a storage tank is opened for 4 hours to release a chemical in a manufacturing process. The flow rate R (in liters per hour) at time t (in hours) is given in the table.

t	0	1	2	3	4
R	425	240	118	71	36

(a) Use the regression capabilities of a graphing utility to find a linear model for the points $(t, \ln R)$. Write the resulting equation of the form $\ln R = at + b$ in exponential form.

(b) Use a graphing utility to plot the data and graph the exponential model.

(c) Use the definite integral to approximate the number of liters of chemical released during the 4 hours.

Writing About Concepts

123. In your own words, state the properties of the natural exponential function.

124. Describe the relationship between the graphs of $f(x) = \ln x$ and $g(x) = e^x$.

125. Is there a function f such that $f(x) = f'(x)$? If so, identify it.

126. Without integrating, state the integration formula you can use to integrate each of the following.

(a) $\int \frac{e^x}{e^x + 1}\, dx$ (b) $\int xe^{x^2}\, dx$

127. Find, to three decimal places, the value of x such that $e^{-x} = x$. (Use Newton's Method or the *zero* or *root* feature of a graphing utility.)

128. Find the value of a such that the area bounded by $y = e^{-x}$, the x-axis, $x = -a$, and $x = a$ is $\frac{8}{3}$.

129. Prove that $\frac{e^a}{e^b} = e^{a-b}$.

130. Let $f(x) = \frac{\ln x}{x}$.

(a) Graph f on $(0, \infty)$ and show that f is strictly decreasing on (e, ∞).

(b) Show that if $e \leq A < B$, then $A^B > B^A$.

(c) Use part (b) to show that $e^\pi > \pi^e$.

Section 5.5 Bases Other Than e and Applications

- **Define exponential functions that have bases other than *e*.**
- **Differentiate and integrate exponential functions that have bases other than *e*.**
- **Use exponential functions to model compound interest and exponential growth.**

Bases Other than e

The **base** of the natural exponential function is e. This "natural" base can be used to assign a meaning to a general base a.

Definition of Exponential Function to Base *a*

If a is a positive real number $(a \neq 1)$ and x is any real number, then the **exponential function to the base *a*** is denoted by a^x and is defined by

$$a^x = e^{(\ln a)x}.$$

If $a = 1$, then $y = 1^x = 1$ is a constant function.

These functions obey the usual laws of exponents. For instance, here are some familiar properties.

1. $a^0 = 1$ **2.** $a^x a^y = a^{x+y}$

3. $\dfrac{a^x}{a^y} = a^{x-y}$ **4.** $(a^x)^y = a^{xy}$

When modeling the half-life of a radioactive sample, it is convenient to use $\frac{1}{2}$ as the base of the exponential model.

EXAMPLE 1 Radioactive Half-Life Model

The half-life of carbon-14 is about 5715 years. A sample contains 1 gram of carbon-14. How much will be present in 10,000 years?

Solution Let $t = 0$ represent the present time and let y represent the amount (in grams) of carbon-14 in the sample. Using a base of $\frac{1}{2}$, you can model y by the equation

$$y = \left(\frac{1}{2}\right)^{t/5715}.$$

Notice that when $t = 5715$, the amount is reduced to half of the original amount.

$$y = \left(\frac{1}{2}\right)^{5715/5715} = \frac{1}{2} \text{ gram}$$

When $t = 11{,}430$, the amount is reduced to a quarter of the original amount, and so on. To find the amount of carbon-14 after 10,000 years, substitute 10,000 for t.

$$y = \left(\frac{1}{2}\right)^{10{,}000/5715} \approx 0.30 \text{ gram}$$

The graph of y is shown in Figure 5.25.

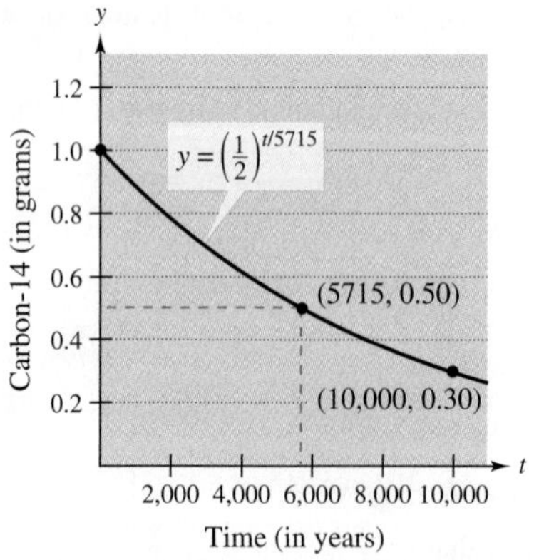

The half-life of carbon-14 is about 5715 years.
Figure 5.25

Logarithmic functions to bases other than e can be defined in much the same way as exponential functions to other bases are defined.

NOTE In precalculus, you learned that $\log_a x$ is the value to which a must be raised to produce x. This agrees with the definition given here because

$$\begin{aligned} a^{\log_a x} &= a^{(1/\ln a)\ln x} \\ &= (e^{\ln a})^{(1/\ln a)\ln x} \\ &= e^{(\ln a/\ln a)\ln x} \\ &= e^{\ln x} \\ &= x. \end{aligned}$$

Definition of Logarithmic Function to Base *a*

If a is a positive real number $(a \neq 1)$ and x is any positive real number, then the **logarithmic function to the base *a*** is denoted by $\log_a x$ and is defined as

$$\log_a x = \frac{1}{\ln a} \ln x.$$

Logarithmic functions to the base a have properties similar to those of the natural logarithmic function given in Theorem 5.2.

1. $\log_a 1 = 0$ — Log of 1
2. $\log_a xy = \log_a x + \log_a y$ — Log of a product
3. $\log_a x^n = n \log_a x$ — Log of a power
4. $\log_a \dfrac{x}{y} = \log_a x - \log_a y$ — Log of a quotient

From the definitions of the exponential and logarithmic functions to the base a, it follows that $f(x) = a^x$ and $g(x) = \log_a x$ are inverse functions of each other.

Properties of Inverse Functions

1. $y = a^x$ if and only if $x = \log_a y$
2. $a^{\log_a x} = x$, for $x > 0$
3. $\log_a a^x = x$, for all x

The logarithmic function to the base 10 is called the **common logarithmic function.** So, for common logarithms, $y = 10^x$ if and only if $x = \log_{10} y$.

EXAMPLE 2 Bases Other Than e

Solve for x in each equation.

a. $3^x = \dfrac{1}{81}$

b. $\log_2 x = -4$

Solution

a. To solve this equation, you can apply the logarithmic function to the base 3 to each side of the equation.

$$\begin{aligned} 3^x &= \frac{1}{81} \\ \log_3 3^x &= \log_3 \frac{1}{81} \\ x &= \log_3 3^{-4} \\ x &= -4 \end{aligned}$$

b. To solve this equation, you can apply the exponential function to the base 2 to each side of the equation.

$$\begin{aligned} \log_2 x &= -4 \\ 2^{\log_2 x} &= 2^{-4} \\ x &= \frac{1}{2^4} \\ x &= \frac{1}{16} \end{aligned}$$

Differentiation and Integration

To differentiate exponential and logarithmic functions to other bases, you have three options: (1) use the definitions of a^x and $\log_a x$ and differentiate using the rules for the natural exponential and logarithmic functions, (2) use logarithmic differentiation, or (3) use the following differentiation rules for bases other than e.

THEOREM 5.13 Derivatives for Bases Other Than e

Let a be a positive real number ($a \neq 1$) and let u be a differentiable function of x.

1. $\dfrac{d}{dx}[a^x] = (\ln a)a^x$

2. $\dfrac{d}{dx}[a^u] = (\ln a)a^u\dfrac{du}{dx}$

3. $\dfrac{d}{dx}[\log_a x] = \dfrac{1}{(\ln a)x}$

4. $\dfrac{d}{dx}[\log_a u] = \dfrac{1}{(\ln a)u}\dfrac{du}{dx}$

Proof By definition, $a^x = e^{(\ln a)x}$. So, you can prove the first rule by letting $u = (\ln a)x$ and differentiating with base e to obtain

$$\frac{d}{dx}[a^x] = \frac{d}{dx}[e^{(\ln a)x}] = e^u\frac{du}{dx} = e^{(\ln a)x}(\ln a) = (\ln a)a^x.$$

To prove the third rule, you can write

$$\frac{d}{dx}[\log_a x] = \frac{d}{dx}\left[\frac{1}{\ln a}\ln x\right] = \frac{1}{\ln a}\left(\frac{1}{x}\right) = \frac{1}{(\ln a)x}.$$

The second and fourth rules are simply the Chain Rule versions of the first and third rules.

NOTE These differentiation rules are similar to those for the natural exponential function and natural logarithmic function. In fact, they differ only by the constant factors $\ln a$ and $1/\ln a$. This points out one reason why, for calculus, e is the most convenient base.

EXAMPLE 3 Differentiating Functions to Other Bases

Find the derivative of each function.

a. $y = 2^x$

b. $y = 2^{3x}$

c. $y = \log_{10} \cos x$

Solution

a. $y' = \dfrac{d}{dx}[2^x] = (\ln 2)2^x$

b. $y' = \dfrac{d}{dx}[2^{3x}] = (\ln 2)2^{3x}(3) = (3 \ln 2)2^{3x}$

Try writing 2^{3x} as 8^x and differentiating to see that you obtain the same result.

c. $y' = \dfrac{d}{dx}[\log_{10} \cos x] = \dfrac{-\sin x}{(\ln 10)\cos x} = -\dfrac{1}{\ln 10}\tan x$

Occasionally, an integrand involves an exponential function to a base other than e. When this occurs, there are two options: (1) convert to base e using the formula $a^x = e^{(\ln a)x}$ and then integrate, or (2) integrate directly, using the integration formula

$$\int a^x\,dx = \left(\frac{1}{\ln a}\right)a^x + C$$

(which follows from Theorem 5.13).

EXAMPLE 4 Integrating an Exponential Function to Another Base

Find $\int 2^x\,dx$.

Solution

$$\int 2^x\,dx = \frac{1}{\ln 2}2^x + C$$

When the Power Rule, $D_x[x^n] = nx^{n-1}$, was introduced in Chapter 2, the exponent n was required to be a rational number. Now the rule is extended to cover any real value of n. Try to prove this theorem using logarithmic differentiation.

THEOREM 5.14 The Power Rule for Real Exponents

Let n be any real number and let u be a differentiable function of x.

1. $\dfrac{d}{dx}[x^n] = nx^{n-1}$

2. $\dfrac{d}{dx}[u^n] = nu^{n-1}\dfrac{du}{dx}$

The next example compares the derivatives of four types of functions. Each function uses a different differentiation formula, depending on whether the base and exponent are constants or variables.

EXAMPLE 5 Comparing Variables and Constants

a. $\dfrac{d}{dx}[e^e] = 0$ Constant Rule

b. $\dfrac{d}{dx}[e^x] = e^x$ Exponential Rule

c. $\dfrac{d}{dx}[x^e] = ex^{e-1}$ Power Rule

d. $y = x^x$ Logarithmic differentiation

$$\ln y = \ln x^x$$

$$\ln y = x \ln x$$

$$\frac{y'}{y} = x\left(\frac{1}{x}\right) + (\ln x)(1) = 1 + \ln x$$

$$y' = y(1 + \ln x) = x^x(1 + \ln x)$$

NOTE Be sure you see that there is no simple differentiation rule for calculating the derivative of $y = x^x$. In general, if $y = u(x)^{v(x)}$, you need to use logarithmic differentiation.

Applications of Exponential Functions

n	A
1	\$1080.00
2	\$1081.60
4	\$1082.43
12	\$1083.00
365	\$1083.28

Suppose P dollars is deposited in an account at an annual interest rate r (in decimal form). If interest accumulates in the account, what is the balance in the account at the end of 1 year? The answer depends on the number of times n the interest is compounded according to the formula

$$A = P\left(1 + \frac{r}{n}\right)^n.$$

For instance, the result for a deposit of \$1000 at 8% interest compounded n times a year is shown in the upper table at the left.

As n increases, the balance A approaches a limit. To develop this limit, use the following theorem. To test the reasonableness of this theorem, try evaluating $[(x + 1)/x]^x$ for several values of x, as shown in the lower table at the left. (A proof of this theorem is given in Appendix A.)

x	$\left(\frac{x+1}{x}\right)^x$
10	2.59374
100	2.70481
1000	2.71692
10,000	2.71815
100,000	2.71827
1,000,000	2.71828

THEOREM 5.15 A Limit Involving e

$$\lim_{x\to\infty}\left(1 + \frac{1}{x}\right)^x = \lim_{x\to\infty}\left(\frac{x+1}{x}\right)^x = e$$

Now, let's take another look at the formula for the balance A in an account in which the interest is compounded n times per year. By taking the limit as n approaches infinity, you obtain

$$\begin{aligned} A &= \lim_{n\to\infty} P\left(1 + \frac{r}{n}\right)^n && \text{Take limit as } n\to\infty. \\ &= P \lim_{n\to\infty}\left[\left(1 + \frac{1}{n/r}\right)^{n/r}\right]^r && \text{Rewrite.} \\ &= P\left[\lim_{x\to\infty}\left(1 + \frac{1}{x}\right)^x\right]^r && \text{Let } x = n/r. \text{ Then } x\to\infty \text{ as } n\to\infty. \\ &= Pe^r. && \text{Apply Theorem 5.15.} \end{aligned}$$

This limit produces the balance after 1 year of **continuous compounding.** So, for a deposit of \$1000 at 8% interest compounded continuously, the balance at the end of 1 year would be

$$\begin{aligned} A &= 1000e^{0.08} \\ &\approx \$1083.29. \end{aligned}$$

These results are summarized below.

Summary of Compound Interest Formulas

Let P = amount of deposit, t = number of years, A = balance after t years, r = annual interest rate (decimal form), and n = number of compoundings per year.

1. Compounded n times per year: $A = P\left(1 + \frac{r}{n}\right)^{nt}$

2. Compounded continuously: $A = Pe^{rt}$

EXAMPLE 6 Comparing Continuous and Quarterly Compounding

A deposit of \$2500 is made in an account that pays an annual interest rate of 5%. Find the balance in the account at the end of 5 years if the interest is compounded (a) quarterly, (b) monthly, and (c) continuously.

Solution

a. $A = P\left(1 + \frac{r}{n}\right)^{nt} = 2500\left(1 + \frac{0.05}{4}\right)^{4(5)}$ Compounded quarterly

$= 2500(1.0125)^{20}$

$\approx \$3205.09$

b. $A = P\left(1 + \frac{r}{n}\right)^{nt} = 2500\left(1 + \frac{0.05}{12}\right)^{12(5)}$ Compounded monthly

$\approx 2500(1.0041667)^{60}$

$\approx \$3208.40$

c. $A = Pe^{rt} = 2500[e^{0.05(5)}]$ Compounded continuously

$= 2500e^{0.25} \approx \$3210.06$

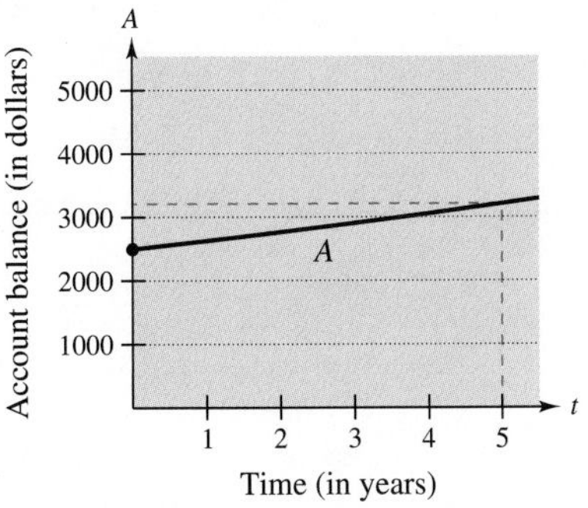

The balance in a savings account grows exponentially.
Figure 5.26

Figure 5.26 shows how the balance increases over the five-year period. Notice that the scale used in the figure does not graphically distinguish among the three types of exponential growth in (a), (b), and (c).

EXAMPLE 7 Bacterial Culture Growth

A bacterial culture is growing according to the *logistic growth function*

$$y = \frac{1.25}{1 + 0.25e^{-0.4t}}, \quad t \geq 0$$

where y is the weight of the culture in grams and t is the time in hours. Find the weight of the culture after (a) 0 hours, (b) 1 hour, and (c) 10 hours. (d) What is the limit as t approaches infinity?

Solution

a. When $t = 0$, $y = \frac{1.25}{1 + 0.25e^{-0.4(0)}}$

$= 1$ gram.

b. When $t = 1$, $y = \frac{1.25}{1 + 0.25e^{-0.4(1)}}$

≈ 1.071 grams.

c. When $t = 10$, $y = \frac{1.25}{1 + 0.25e^{-0.4(10)}}$

≈ 1.244 grams.

d. Finally, taking the limit as t approaches infinity, you obtain

$$\lim_{t\to\infty} \frac{1.25}{1 + 0.25e^{-0.4t}} = \frac{1.25}{1 + 0} = 1.25 \text{ grams.}$$

The graph of the function is shown in Figure 5.27.

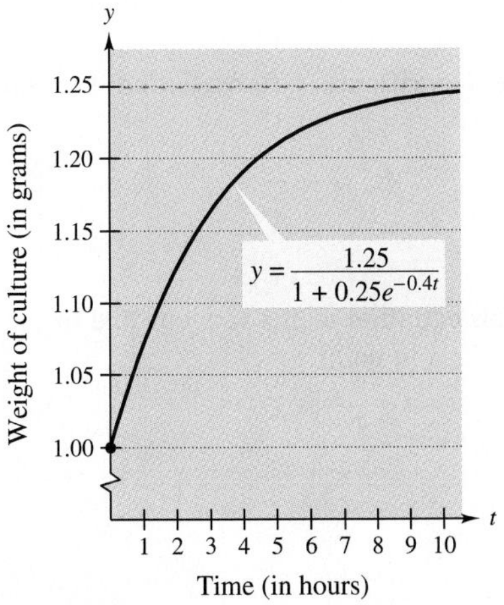

The limit of the weight of the culture as $t \to \infty$ is 1.25 grams.
Figure 5.27

Exercises for Section 5.5

See www.CalcChat.com for worked-out solutions to odd-numbered exercises.

In Exercises 1–4, evaluate the expression without using a calculator.

1. $\log_2 \frac{1}{8}$
2. $\log_{27} 9$
3. $\log_7 1$
4. $\log_a \dfrac{1}{a}$

In Exercises 5–8, write the exponential equation as a logarithmic equation or vice versa.

5. (a) $2^3 = 8$
   (b) $3^{-1} = \frac{1}{3}$
6. (a) $27^{2/3} = 9$
   (b) $16^{3/4} = 8$
7. (a) $\log_{10} 0.01 = -2$
   (b) $\log_{0.5} 8 = -3$
8. (a) $\log_3 \frac{1}{9} = -2$
   (b) $49^{1/2} = 7$

In Exercises 9–14, sketch the graph of the function by hand.

9. $y = 3^x$
10. $y = 3^{x-1}$
11. $y = \left(\frac{1}{3}\right)^x$
12. $y = 2^{x^2}$
13. $h(x) = 5^{x-2}$
14. $y = 3^{-|x|}$

In Exercises 15–20, solve for x or b.

15. (a) $\log_{10} 1000 = x$
   (b) $\log_{10} 0.1 = x$
16. (a) $\log_3 \frac{1}{81} = x$
   (b) $\log_6 36 = x$
17. (a) $\log_3 x = -1$
   (b) $\log_2 x = -4$
18. (a) $\log_b 27 = 3$
   (b) $\log_b 125 = 3$
19. (a) $x^2 - x = \log_5 25$
   (b) $3x + 5 = \log_2 64$
20. (a) $\log_3 x + \log_3(x - 2) = 1$
   (b) $\log_{10}(x + 3) - \log_{10} x = 1$

In Exercises 21–30, solve the equation accurate to three decimal places.

21. $3^{2x} = 75$
22. $5^{6x} = 8320$
23. $2^{3-z} = 625$
24. $3(5^{x-1}) = 86$
25. $\left(1 + \dfrac{0.09}{12}\right)^{12t} = 3$
26. $\left(1 + \dfrac{0.10}{365}\right)^{365t} = 2$
27. $\log_2(x - 1) = 5$
28. $\log_{10}(t - 3) = 2.6$
29. $\log_3 x^2 = 4.5$
30. $\log_5 \sqrt{x - 4} = 3.2$

In Exercises 31–34, use a graphing utility to graph the function and approximate its zero(s) accurate to three decimal places.

31. $g(x) = 6(2^{1-x}) - 25$
32. $f(t) = 300(1.0075^{12t}) - 735.41$
33. $h(s) = 32 \log_{10}(s - 2) + 15$
34. $g(x) = 1 - 2 \log_{10}[x(x - 3)]$

In Exercises 35 and 36, illustrate that the functions are inverse functions of each other by sketching their graphs on the same set of coordinate axes.

35. $f(x) = 4^x$
   $g(x) = \log_4 x$
36. $f(x) = 3^x$
   $g(x) = \log_3 x$

In Exercises 37–48, find the derivative of the function.

37. $f(x) = 4^x$
38. $y = x(6^{-2x})$
39. $g(t) = t^2 2^t$
40. $f(t) = \dfrac{3^{2t}}{t}$
41. $h(\theta) = 2^{-\theta} \cos \pi\theta$
42. $g(\alpha) = 5^{-\alpha/2} \sin 2\alpha$
43. $f(x) = \log_2 \dfrac{x^2}{x - 1}$
44. $h(x) = \log_3 \dfrac{x\sqrt{x - 1}}{2}$
45. $y = \log_5 \sqrt{x^2 - 1}$
46. $y = \log_{10} \dfrac{x^2 - 1}{x}$
47. $g(t) = \dfrac{10 \log_4 t}{t}$
48. $f(t) = t^{3/2} \log_2 \sqrt{t + 1}$

In Exercises 49–52, find an equation of the tangent line to the graph of the function at the given point.

49. $y = 2^{-x}$, $(-1, 2)$
50. $y = 5^{x-2}$, $(2, 1)$
51. $y = \log_3 x$, $(27, 3)$
52. $y = \log_{10} 2x$, $(5, 1)$

In Exercises 53–56, use logarithmic differentiation to find dy/dx.

53. $y = x^{2/x}$
54. $y = x^{x-1}$
55. $y = (x - 2)^{x+1}$
56. $y = (1 + x)^{1/x}$

In Exercises 57–60, find an equation of the tangent line to the graph of the function at the given point.

57. $y = x^{\sin x}$, $\left(\dfrac{\pi}{2}, \dfrac{\pi}{2}\right)$
58. $y = (\sin x)^{2x}$, $\left(\dfrac{\pi}{2}, 1\right)$
59. $y = (\ln x)^{\cos x}$, $(e, 1)$
60. $y = x^{1/x}$, $(1, 1)$

In Exercises 61–66, find the integral.

61. $\int 3^x \, dx$

62. $\int 5^{-x} \, dx$

63. $\int x(5^{-x^2}) \, dx$

64. $\int (3 - x)7^{(3-x)^2} \, dx$

65. $\int \frac{3^{2x}}{1 + 3^{2x}} \, dx$

66. $\int 2^{\sin x} \cos x \, dx$

In Exercises 67–70, evaluate the integral.

67. $\int_{-1}^{2} 2^x \, dx$

68. $\int_{-2}^{2} 4^{x/2} \, dx$

69. $\int_{0}^{1} (5^x - 3^x) \, dx$

70. $\int_{1}^{e} (6^x - 2^x) \, dx$

Area **In Exercises 71 and 72, find the area of the region bounded by the graphs of the equations.**

71. $y = 3^x, y = 0, x = 0, x = 3$

72. $y = 3^{\cos x} \sin x, y = 0, x = 0, x = \pi$

Slope Fields **In Exercises 73 and 74, a differential equation, a point, and a slope field are given. (a) Sketch two approximate solutions of the differential equation on the slope field, one of which passes through the given point. (b) Use integration to find the particular solution of the differential equation and use a graphing utility to graph the solution. Compare the result with the sketches in part (a). To print an enlarged copy of the graph, go to the website *www.mathgraphs.com*.**

73. $\frac{dy}{dx} = 0.4^{x/3}, \quad \left(0, \frac{1}{2}\right)$

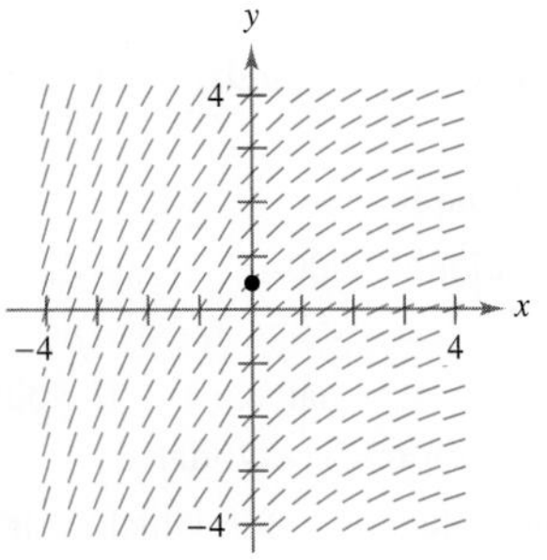

74. $\frac{dy}{dx} = e^{\sin x} \cos x, \quad (\pi, 2)$

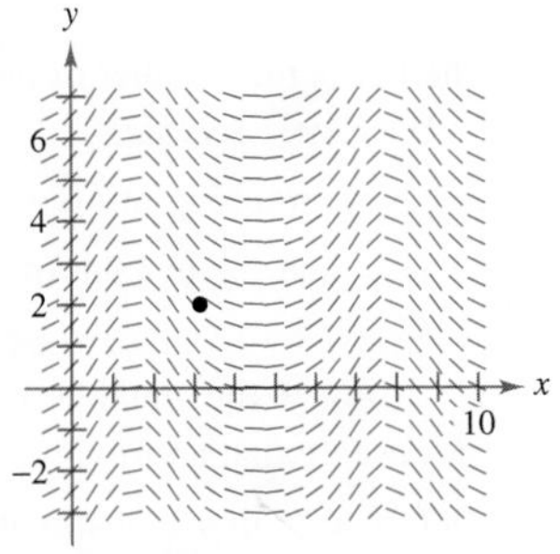

Writing About Concepts

75. The table of values below was obtained by evaluating a function. Determine which of the statements may be true and which must be false, and explain why.

(a) y is an exponential function of x.

(b) y is a logarithmic function of x.

(c) x is an exponential function of y.

(d) y is a linear function of x.

x	1	2	8
y	0	1	3

76. Consider the function $f(x) = \log_{10} x$.

(a) What is the domain of f?

(b) Find f^{-1}.

(c) If x is a real number between 1000 and 10,000, determine the interval in which $f(x)$ will be found.

(d) Determine the interval in which x will be found if $f(x)$ is negative.

(e) If $f(x)$ is increased by one unit, x must have been increased by what factor?

(f) Find the ratio of x_1 to x_2 given that $f(x_1) = 3n$ and $f(x_2) = n$.

77. Order the functions

$$f(x) = \log_2 x, \; g(x) = x^x, \; h(x) = x^2, \text{ and } k(x) = 2^x$$

from the one with the greatest rate of growth to the one with the smallest rate of growth for large values of x.

78. Find the derivative of each function, given that a is constant.

(a) $y = x^a$ (b) $y = a^x$

(c) $y = x^x$ (d) $y = a^a$

79. ***Inflation*** If the annual rate of inflation averages 5% over the next 10 years, the approximate cost C of goods or services during any year in that decade is

$$C(t) = P(1.05)^t$$

where t is the time in years and P is the present cost.

(a) The price of an oil change for your car is presently \$24.95. Estimate the price 10 years from now.

(b) Find the rates of change of C with respect to t when $t = 1$ and $t = 8$.

(c) Verify that the rate of change of C is proportional to C. What is the constant of proportionality?

80. *Depreciation* After t years, the value of a car purchased for \$20,000 is

$$V(t) = 20{,}000\left(\frac{3}{4}\right)^t.$$

(a) Use a graphing utility to graph the function and determine the value of the car 2 years after it was purchased.

(b) Find the rates of change of V with respect to t when $t = 1$ and $t = 4$.

(c) Use a graphing utility to graph $V'(t)$ and determine the horizontal asymptote of $V'(t)$. Interpret its meaning in the context of the problem.

Compound Interest **In Exercises 81–84, complete the table to determine the balance *A* for *P* dollars invested at rate *r* for *t* years and compounded *n* times per year.**

n	1	2	4	12	365	Continuous compounding
A						

81. $P = \$1000$
$r = 3\frac{1}{2}\%$
$t = 10$ years

82. $P = \$2500$
$r = 6\%$
$t = 20$ years

83. $P = \$1000$
$r = 5\%$
$t = 30$ years

84. $P = \$5000$
$r = 7\%$
$t = 25$ years

Compound Interest **In Exercises 85–88, complete the table to determine the amount of money *P* (present value) that should be invested at rate *r* to produce a balance of \$100,000 in *t* years.**

t	1	10	20	30	40	50
P						

85. $r = 5\%$
Compounded continuously

86. $r = 6\%$
Compounded continuously

87. $r = 5\%$
Compounded monthly

88. $r = 7\%$
Compounded daily

89. *Compound Interest* Assume that you can earn 6% on an investment, compounded daily. Which of the following options would yield the greatest balance after 8 years?

(a) \$20,000 now

(b) \$30,000 after 8 years

(c) \$8000 now and \$20,000 after 4 years

(d) \$9000 now, \$9000 after 4 years, and \$9000 after 8 years

90. *Compound Interest* Consider a deposit of \$100 placed in an account for 20 years at $r\%$ compounded continuously. Use a graphing utility to graph the exponential functions giving the growth of the investment over the 20 years for each of the following interest rates. Compare the ending balances for each of the rates.

(a) $r = 3\%$

(b) $r = 5\%$

(c) $r = 6\%$

91. *Timber Yield* The yield V (in millions of cubic feet per acre) for a stand of timber at age t is

$$V = 6.7e^{(-48.1)/t}$$

where t is measured in years.

(a) Find the limiting volume of wood per acre as t approaches infinity.

(b) Find the rates at which the yield is changing when $t = 20$ years and $t = 60$ years.

92. *Learning Theory* In a group project in learning theory, a mathematical model for the proportion P of correct responses after n trials was found to be

$$P = \frac{0.86}{1 + e^{-0.25n}}.$$

(a) Find the limiting proportion of correct responses as n approaches infinity.

(b) Find the rates at which P is changing after $n = 3$ trials and $n = 10$ trials.

93. *Forest Defoliation* To estimate the amount of defoliation caused by the gypsy moth during a year, a forester counts the number of egg masses on $\frac{1}{40}$ of an acre the preceding fall. The percent of defoliation y is approximated by

$$y = \frac{300}{3 + 17e^{-0.0625x}}$$

where x is the number of egg masses in thousands. *(Source: USDA Forest Service)*

(a) Use a graphing utility to graph the function.

(b) Estimate the percent of defoliation if 2000 egg masses are counted.

(c) Estimate the number of egg masses that existed if you observe that approximately $\frac{2}{3}$ of a forest is defoliated.

(d) Use calculus to estimate the value of x for which y is increasing most rapidly.

94. *Population Growth* A lake is stocked with 500 fish, and their population increases according to the logistic curve

$$p(t) = \frac{10,000}{1 + 19e^{-t/5}}$$

where t is measured in months.

(a) Use a graphing utility to graph the function.

(b) What is the limiting size of the fish population?

(c) At what rates is the fish population changing at the end of 1 month and at the end of 10 months?

(d) After how many months is the population increasing most rapidly?

95. *Modeling Data* The breaking strengths B (in tons) of a steel cable of various diameters d (in inches) are shown in the table.

d	0.50	0.75	1.00	1.25	1.50	1.75
B	9.85	21.8	38.3	59.2	84.4	114.0

(a) Use the regression capabilities of a graphing utility to fit an exponential model to the data.

(b) Use a graphing utility to plot the data and graph the model.

(c) Find the rates of growth of the model when $d = 0.8$ and $d = 1.5$.

96. *Comparing Models* The amounts y (in billions of dollars) given to philanthropy (from individuals, foundations, corporations, and charitable bequests) in the United States for the years 1995 through 2002 are shown in the table, with $x = 5$ corresponding to 1995. *(Source: AAFRC Trust for Philanthropy)*

x	5	6	7	8	9	10	11	12
y	124.0	138.6	157.1	174.8	199.0	210.9	212.0	240.9

(a) Use the regression capabilities of a graphing utility to find the following models for the data.

$y_1 = ax + b$

$y_2 = a + b \ln x$

$y_3 = ab^x$

$y_4 = ax^b$

(b) Use a graphing utility to plot the data and graph each of the models. Which model do you think best fits the data?

(c) Interpret the slope of the linear model in the context of the problem.

(d) Find the rate of change of each of the models for the year 1996. Which model is increasing at the greatest rate in 1996?

97. *Conjecture*

(a) Use a graphing utility to approximate the integrals of the functions

$$f(t) = 4\left(\frac{3}{8}\right)^{2t/3}, \quad g(t) = 4\left(\frac{\sqrt[3]{9}}{4}\right)^t, \text{ and } h(t) = 4e^{-0.653886t}$$

on the interval $[0, 4]$.

(b) Use a graphing utility to graph the three functions.

(c) Use the results in parts (a) and (b) to make a conjecture about the three functions. Could you make the conjecture using only part (a)? Explain. Prove your conjecture analytically.

98. Complete the table to demonstrate that e can also be defined as $\lim\limits_{x \to 0^+} (1 + x)^{1/x}$.

x	1	10^{-1}	10^{-2}	10^{-4}	10^{-6}
$(1 + x)^{1/x}$					

In Exercises 99 and 100, find an exponential function that fits the experimental data collected over time *t*.

99.

t	0	1	2	3	4
y	1200.00	720.00	432.00	259.20	155.52

100.

t	0	1	2	3	4
y	600.00	630.00	661.50	694.58	729.30

True or False? **In Exercises 101–106, determine whether the statement is true or false. If it is false, explain why or give an example that shows it is false.**

101. $e = \dfrac{271,801}{99,900}$

102. If $f(x) = \ln x$, then $f(e^{n+1}) - f(e^n) = 1$ for any value of n.

103. The functions $f(x) = 2 + e^x$ and $g(x) = \ln(x - 2)$ are inverse functions of each other.

104. The exponential function $y = Ce^x$ is a solution of the differential equation $d^n y/dx^n = y, n = 1, 2, 3, \ldots$.

105. The graphs of $f(x) = e^x$ and $g(x) = e^{-x}$ meet at right angles.

106. If $f(x) = g(x)e^x$, then the only zeros of f are the zeros of g.

107. Solve the logistic differential equation

$$\frac{dy}{dt} = \frac{8}{25}y\left(\frac{5}{4} - y\right), \quad y(0) = 1$$

and obtain the logistic growth function in Example 7.

$$\left[\textit{Hint: } \frac{1}{y\left(\frac{5}{4} - y\right)} = \frac{4}{5}\left(\frac{1}{y} + \frac{1}{\frac{5}{4} - y}\right)\right]$$

108. Given the exponential function $f(x) = a^x$, show that

(a) $f(u + v) = f(u) \cdot f(v)$.

(b) $f(2x) = [f(x)]^2$.

109. (a) Determine y' given $y^x = x^y$.

(b) Find the slope of the tangent line to the graph of $y^x = x^y$ at each of the following points.

(i) (c, c)

(ii) $(2, 4)$

(iii) $(4, 2)$

(c) At what point on the graph of $y^x = x^y$ does the tangent line not exist?

110. Consider the functions $f(x) = 1 + x$ and $g(x) = b^x$, $b > 1$.

(a) Given $b = 2$, use a graphing utility to graph f and g in the same viewing window. Identify the point(s) of intersection.

(b) Repeat part (a) using $b = 3$.

(c) Find all values of b such that $g(x) \geq f(x)$ for all x.

Putnam Exam Challenge

111. Which is greater

$$\left(\sqrt{n}\right)^{\sqrt{n+1}} \quad \text{or} \quad \left(\sqrt{n+1}\right)^{\sqrt{n}}$$

where $n > 8$?

112. Show that if x is positive, then

$$\log_e\left(1 + \frac{1}{x}\right) > \frac{1}{1 + x}.$$

These problems were composed by the Committee on the Putnam Prize Competition.

Section Project: Using Graphing Utilities to Estimate Slope

Let $f(x) = \begin{cases} |x|^x, & x \neq 0 \\ 1, & x = 0. \end{cases}$

(a) Use a graphing utility to graph f in the viewing window $-3 \leq x \leq 3$, $-2 \leq y \leq 2$. What is the domain of f?

(b) Use the *zoom* and *trace* features of a graphing utility to estimate

$$\lim_{x \to 0} f(x).$$

(c) Write a short paragraph explaining why the function f is continuous for all real numbers.

(d) Visually estimate the slope of f at the point $(0, 1)$.

(e) Explain why the derivative of a function can be approximated by the formula

$$\frac{f(x + \Delta x) - f(x - \Delta x)}{2\Delta x}$$

for small values of Δx. Use this formula to approximate the slope of f at the point $(0, 1)$.

$$f'(0) \approx \frac{f(0 + \Delta x) - f(0 - \Delta x)}{2\Delta x} = \frac{f(\Delta x) - f(-\Delta x)}{2\Delta x}$$

What do you think the slope of the graph of f is at $(0, 1)$?

(f) Find a formula for the derivative of f and determine $f'(0)$. Write a short paragraph explaining how a graphing utility might lead you to approximate the slope of a graph incorrectly.

(g) Use your formula for the derivative of f to find the relative extrema of f. Verify your answer with a graphing utility.

FOR FURTHER INFORMATION For more information on using graphing utilities to estimate slope, see the article "Computer-Aided Delusions" by Richard L. Hall in *The College Mathematics Journal*. To view this article, go to the website *www.matharticles.com*.

Section 5.6 Inverse Trigonometric Functions: Differentiation

- **Develop properties of the six inverse trigonometric functions.**
- **Differentiate an inverse trigonometric function.**
- **Review the basic differentiation rules for elementary functions.**

Inverse Trigonometric Functions

This section begins with a rather surprising statement: *None of the six basic trigonometric functions has an inverse function.* This statement is true because all six trigonometric functions are periodic and therefore are not one-to-one. In this section you will examine these six functions to see whether their domains can be redefined in such a way that they will have inverse functions on the *restricted domains.*

In Example 4 of Section 5.3, you saw that the sine function is increasing (and therefore is one-to-one) on the interval $[-\pi/2, \pi/2]$ (see Figure 5.28). On this interval you can define the inverse of the *restricted* sine function to be

$$y = \arcsin x \quad \text{if and only if} \quad \sin y = x$$

where $-1 \le x \le 1$ and $-\pi/2 \le \arcsin x \le \pi/2$.

Under suitable restrictions, each of the six trigonometric functions is one-to-one and so has an inverse function, as shown in the following definition.

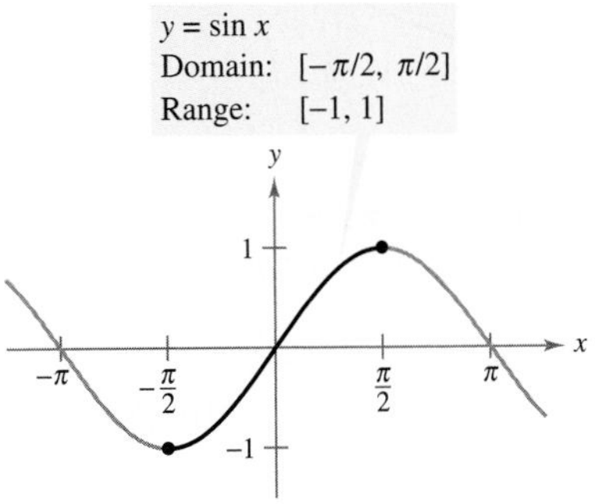

The sine function is one-to-one on $[-\pi/2, \pi/2]$.
Figure 5.28

NOTE The term "iff" is used to represent the phrase "if and only if."

Definitions of Inverse Trigonometric Functions

Function	*Domain*	*Range*
$y = \arcsin x$ iff $\sin y = x$	$-1 \le x \le 1$	$-\frac{\pi}{2} \le y \le \frac{\pi}{2}$
$y = \arccos x$ iff $\cos y = x$	$-1 \le x \le 1$	$0 \le y \le \pi$
$y = \arctan x$ iff $\tan y = x$	$-\infty < x < \infty$	$-\frac{\pi}{2} < y < \frac{\pi}{2}$
$y = \operatorname{arccot} x$ iff $\cot y = x$	$-\infty < x < \infty$	$0 < y < \pi$
$y = \operatorname{arcsec} x$ iff $\sec y = x$	$\lvert x \rvert \ge 1$	$0 \le y \le \pi, \quad y \ne \frac{\pi}{2}$
$y = \operatorname{arccsc} x$ iff $\csc y = x$	$\lvert x \rvert \ge 1$	$-\frac{\pi}{2} \le y \le \frac{\pi}{2}, \quad y \ne 0$

NOTE The term "arcsin x" is read as "the arcsine of x" or sometimes "the angle whose sine is x." An alternative notation for the inverse sine function is "$\sin^{-1} x$."

EXPLORATION

The Inverse Secant Function In the definition above, the inverse secant function is defined by restricting the domain of the secant function to the intervals $\left[0, \frac{\pi}{2}\right) \cup \left(\frac{\pi}{2}, \pi\right]$. Most other texts and reference books agree with this, but some disagree. What other domains might make sense? Explain your reasoning graphically. Most calculators do not have a key for the inverse secant function. How can you use a calculator to evaluate the inverse secant function?

The graphs of the six inverse trigonometric functions are shown in Figure 5.29.

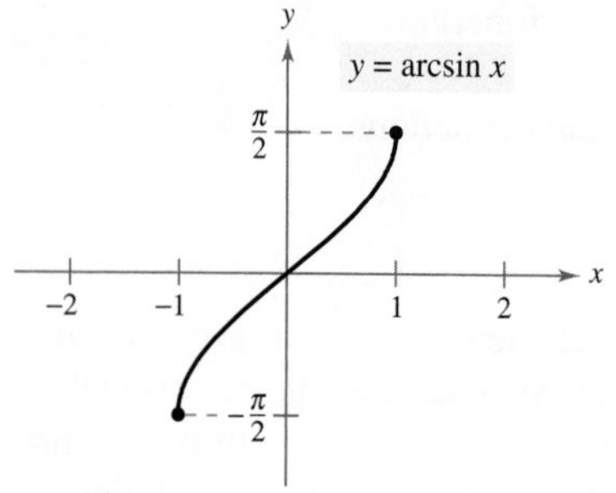

Domain: $[-1, 1]$
Range: $[-\pi/2, \pi/2]$

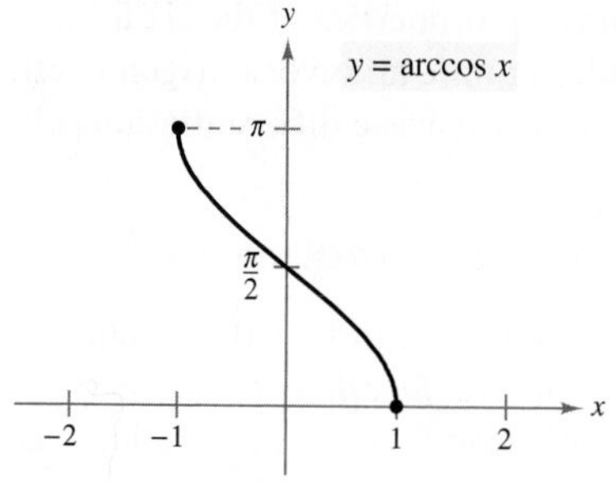

Domain: $[-1, 1]$
Range: $[0, \pi]$

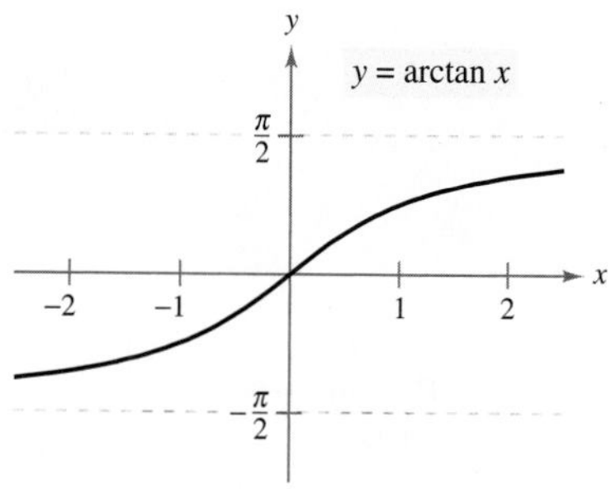

Domain: $(-\infty, \infty)$
Range: $(-\pi/2, \pi/2)$

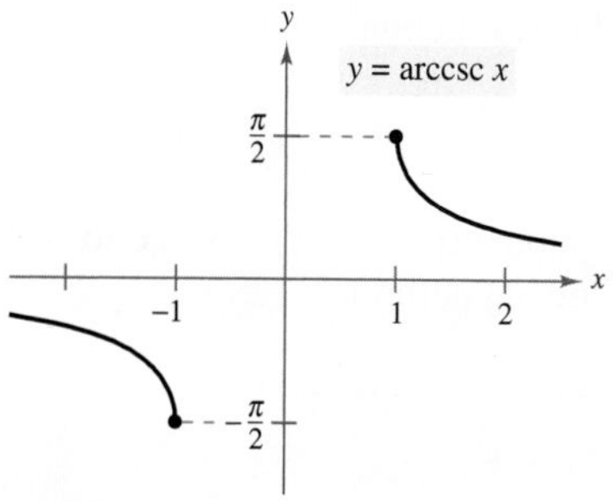

Domain: $(-\infty, -1] \cup [1, \infty)$
Range: $[-\pi/2, 0) \cup (0, \pi/2]$
Figure 5.29

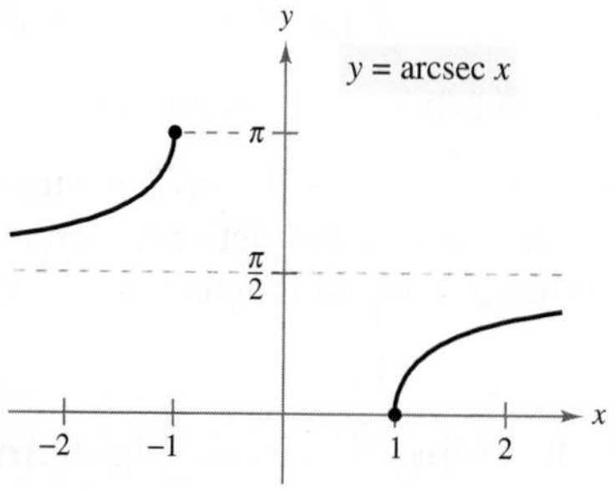

Domain: $(-\infty, -1] \cup [1, \infty)$
Range: $[0, \pi/2) \cup (\pi/2, \pi]$

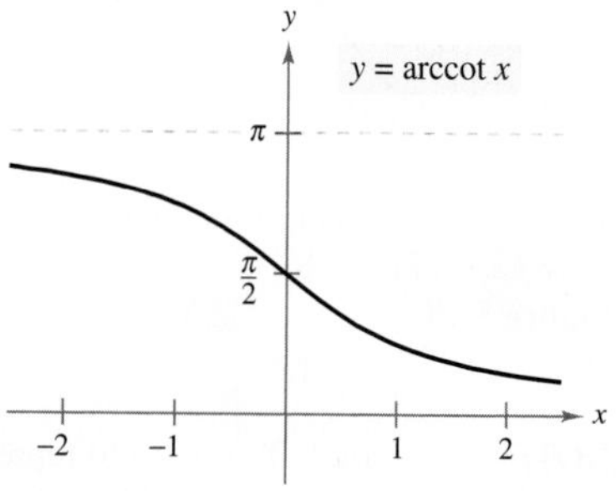

Domain: $(-\infty, \infty)$
Range: $(0, \pi)$

EXAMPLE 1 Evaluating Inverse Trigonometric Functions

Evaluate each function.

a. $\arcsin\left(-\dfrac{1}{2}\right)$ **b.** $\arccos 0$ **c.** $\arctan \sqrt{3}$ **d.** $\arcsin(0.3)$

NOTE When evaluating inverse trigonometric functions, remember that they denote *angles in radian measure.*

Solution

a. By definition, $y = \arcsin\left(-\frac{1}{2}\right)$ implies that $\sin y = -\frac{1}{2}$. In the interval $[-\pi/2, \pi/2]$, the correct value of y is $-\pi/6$.

$$\arcsin\left(-\frac{1}{2}\right) = -\frac{\pi}{6}$$

b. By definition, $y = \arccos 0$ implies that $\cos y = 0$. In the interval $[0, \pi]$, you have $y = \pi/2$.

$$\arccos 0 = \frac{\pi}{2}$$

c. By definition, $y = \arctan \sqrt{3}$ implies that $\tan y = \sqrt{3}$. In the interval $(-\pi/2, \pi/2)$, you have $y = \pi/3$.

$$\arctan \sqrt{3} = \frac{\pi}{3}$$

d. Using a calculator set in *radian* mode produces

$$\arcsin(0.3) \approx 0.305.$$

EXPLORATION

Graph $y = \arccos(\cos x)$ for $-4\pi \le x \le 4\pi$. Why isn't the graph the same as the graph of $y = x$?

Inverse functions have the properties

$$f(f^{-1}(x)) = x \quad \text{and} \quad f^{-1}(f(x)) = x.$$

When applying these properties to inverse trigonometric functions, remember that the trigonometric functions have inverse functions only in restricted domains. For x-values outside these domains, these two properties do not hold. For example, $\arcsin(\sin \pi)$ is equal to 0, not π.

Properties of Inverse Trigonometric Functions

If $-1 \le x \le 1$ and $-\pi/2 \le y \le \pi/2$, then

$$\sin(\arcsin x) = x \quad \text{and} \quad \arcsin(\sin y) = y.$$

If $-\pi/2 < y < \pi/2$, then

$$\tan(\arctan x) = x \quad \text{and} \quad \arctan(\tan y) = y.$$

If $|x| \ge 1$ and $0 \le y < \pi/2$ or $\pi/2 < y \le \pi$, then

$$\sec(\operatorname{arcsec} x) = x \quad \text{and} \quad \operatorname{arcsec}(\sec y) = y.$$

Similar properties hold for the other inverse trigonometric functions.

EXAMPLE 2 Solving an Equation

$$\arctan(2x - 3) = \frac{\pi}{4} \qquad \text{Original equation}$$

$$\tan[\arctan(2x - 3)] = \tan\frac{\pi}{4} \qquad \text{Take tangent of each side.}$$

$$2x - 3 = 1 \qquad \tan(\arctan x) = x$$

$$x = 2 \qquad \text{Solve for } x.$$

Some problems in calculus require that you evaluate expressions such as $\cos(\arcsin x)$, as shown in Example 3.

EXAMPLE 3 Using Right Triangles

a. Given $y = \arcsin x$, where $0 < y < \pi/2$, find $\cos y$.

b. Given $y = \operatorname{arcsec}(\sqrt{5}/2)$, find $\tan y$.

Solution

a. Because $y = \arcsin x$, you know that $\sin y = x$. This relationship between x and y can be represented by a right triangle, as shown in Figure 5.30.

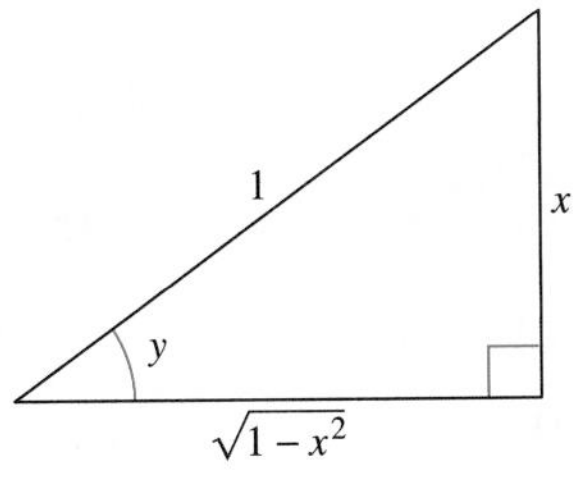

$y = \arcsin x$
Figure 5.30

$$\cos y = \cos(\arcsin x) = \frac{\text{adj.}}{\text{hyp.}} = \sqrt{1 - x^2}$$

(This result is also valid for $-\pi/2 < y < 0$.)

b. Use the right triangle shown in Figure 5.31.

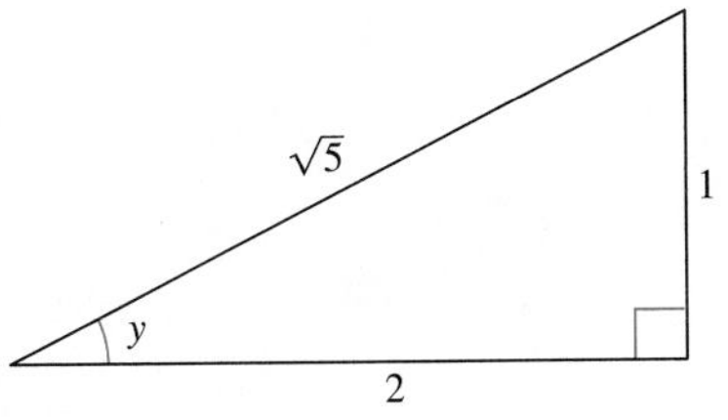

$y = \operatorname{arcsec}\dfrac{\sqrt{5}}{2}$
Figure 5.31

$$\tan y = \tan\left[\operatorname{arcsec}\left(\frac{\sqrt{5}}{2}\right)\right] = \frac{\text{opp.}}{\text{adj.}} = \frac{1}{2}$$

NOTE There is no common agreement on the definition of arcsec x (or arccsc x) for negative values of x. When we defined the range of the arcsecant, we chose to preserve the reciprocal identity

$$\operatorname{arcsec} x = \arccos \frac{1}{x}.$$

For example, to evaluate arcsec(-2), you can write

$$\operatorname{arcsec}(-2) = \arccos(-0.5) \approx 2.09.$$

One of the consequences of the definition of the inverse secant function given in this text is that its graph has a positive slope at every x-value in its domain. (See Figure 5.29.) This accounts for the absolute value sign in the formula for the derivative of arcsec x.

Derivatives of Inverse Trigonometric Functions

In Section 5.1 you saw that the derivative of the *transcendental* function $f(x) = \ln x$ is the *algebraic* function $f'(x) = 1/x$. You will now see that the derivatives of the inverse trigonometric functions also are algebraic (even though the inverse trigonometric functions are themselves transcendental).

The following theorem lists the derivatives of the six inverse trigonometric functions. Note that the derivatives of arccos u, arccot u, and arccsc u are the *negatives* of the derivatives of arcsin u, arctan u, and arcsec u, respectively.

THEOREM 5.16 Derivatives of Inverse Trigonometric Functions

Let u be a differentiable function of x.

$$\frac{d}{dx}[\arcsin u] = \frac{u'}{\sqrt{1-u^2}} \qquad \frac{d}{dx}[\arccos u] = \frac{-u'}{\sqrt{1-u^2}}$$

$$\frac{d}{dx}[\arctan u] = \frac{u'}{1+u^2} \qquad \frac{d}{dx}[\operatorname{arccot} u] = \frac{-u'}{1+u^2}$$

$$\frac{d}{dx}[\operatorname{arcsec} u] = \frac{u'}{|u|\sqrt{u^2-1}} \qquad \frac{d}{dx}[\operatorname{arccsc} u] = \frac{-u'}{|u|\sqrt{u^2-1}}$$

To derive these formulas, you can use implicit differentiation. For instance, if $y = \arcsin x$, then $\sin y = x$ and $(\cos y)y' = 1$. (See Exercise 94.)

TECHNOLOGY If your graphing utility does not have the arcsecant function, you can obtain its graph using

$$f(x) = \operatorname{arcsec} x = \arccos \frac{1}{x}.$$

EXAMPLE 4 Differentiating Inverse Trigonometric Functions

a. $\dfrac{d}{dx}[\arcsin(2x)] = \dfrac{2}{\sqrt{1-(2x)^2}} = \dfrac{2}{\sqrt{1-4x^2}}$

b. $\dfrac{d}{dx}[\arctan(3x)] = \dfrac{3}{1+(3x)^2} = \dfrac{3}{1+9x^2}$

c. $\dfrac{d}{dx}\left[\arcsin\sqrt{x}\right] = \dfrac{(1/2)\,x^{-1/2}}{\sqrt{1-x}} = \dfrac{1}{2\sqrt{x}\sqrt{1-x}} = \dfrac{1}{2\sqrt{x-x^2}}$

d. $\dfrac{d}{dx}[\operatorname{arcsec} e^{2x}] = \dfrac{2e^{2x}}{e^{2x}\sqrt{(e^{2x})^2-1}} = \dfrac{2e^{2x}}{e^{2x}\sqrt{e^{4x}-1}} = \dfrac{2}{\sqrt{e^{4x}-1}}$

The absolute value sign is not necessary because $e^{2x} > 0$.

EXAMPLE 5 A Derivative That Can Be Simplified

Differentiate $y = \arcsin x + x\sqrt{1-x^2}$.

Solution

$$\begin{aligned} y' &= \frac{1}{\sqrt{1-x^2}} + x\left(\frac{1}{2}\right)(-2x)(1-x^2)^{-1/2} + \sqrt{1-x^2} \\ &= \frac{1}{\sqrt{1-x^2}} - \frac{x^2}{\sqrt{1-x^2}} + \sqrt{1-x^2} \\ &= \sqrt{1-x^2} + \sqrt{1-x^2} \\ &= 2\sqrt{1-x^2} \end{aligned}$$

NOTE From Example 5, you can see one of the benefits of inverse trigonometric functions—they can be used to integrate common algebraic functions. For instance, from the result shown in the example, it follows that

$$\int \sqrt{1-x^2}\,dx = \frac{1}{2}\left(\arcsin x + x\sqrt{1-x^2}\right).$$

EXAMPLE 6 Analyzing an Inverse Trigonometric Graph

Analyze the graph of $y = (\arctan x)^2$.

Solution From the derivative

$$y' = 2(\arctan x)\left(\frac{1}{1+x^2}\right) = \frac{2 \arctan x}{1+x^2}$$

you can see that the only critical number is $x = 0$. By the First Derivative Test, this value corresponds to a relative minimum. From the second derivative

$$y'' = \frac{(1+x^2)\left(\frac{2}{1+x^2}\right) - (2 \arctan x)(2x)}{(1+x^2)^2} = \frac{2(1 - 2x \arctan x)}{(1+x^2)^2}$$

it follows that points of inflection occur when $2x \arctan x = 1$. Using Newton's Method, these points occur when $x \approx \pm 0.765$. Finally, because

$$\lim_{x \to \pm\infty} (\arctan x)^2 = \frac{\pi^2}{4}$$

it follows that the graph has a horizontal asymptote at $y = \pi^2/4$. The graph is shown in Figure 5.32.

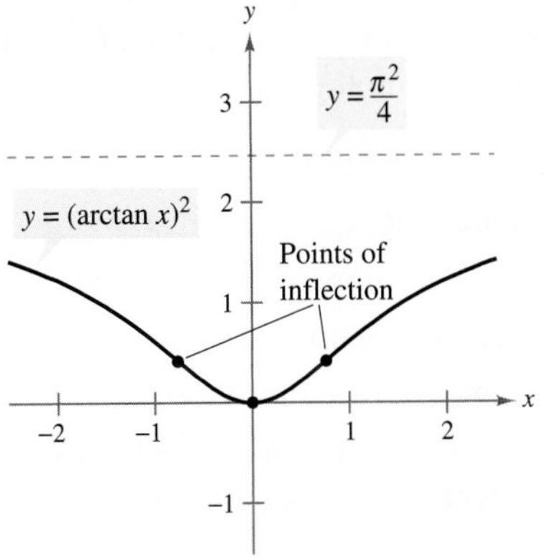

The graph of $y = (\arctan x)^2$ has a horizontal asymptote at $y = \pi^2/4$.
Figure 5.32

EXAMPLE 7 Maximizing an Angle

A photographer is taking a picture of a four-foot painting hung in an art gallery. The camera lens is 1 foot below the lower edge of the painting, as shown in Figure 5.33. How far should the camera be from the painting to maximize the angle subtended by the camera lens?

Solution In Figure 5.33, let β be the angle to be maximized.

$$\beta = \theta - \alpha = \operatorname{arccot} \frac{x}{5} - \operatorname{arccot} x$$

Differentiating produces

$$\frac{d\beta}{dx} = \frac{-1/5}{1 + (x^2/25)} - \frac{-1}{1+x^2} = \frac{-5}{25+x^2} + \frac{1}{1+x^2} = \frac{4(5 - x^2)}{(25+x^2)(1+x^2)}.$$

Because $d\beta/dx = 0$ when $x = \sqrt{5}$, you can conclude from the First Derivative Test that this distance yields a maximum value of β. So, the distance is $x \approx 2.236$ feet and the angle is $\beta \approx 0.7297$ radian $\approx 41.81°$.

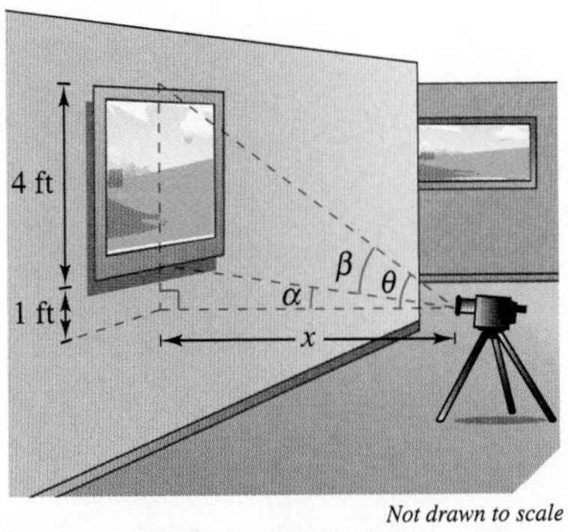

The camera should be 2.236 feet from the painting to maximize the angle β.
Figure 5.33

The Granger Collection

GALILEO GALILEI (1564–1642)

Galileo's approach to science departed from the accepted Aristotelian view that nature had describable *qualities*, such as "fluidity" and "potentiality." He chose to describe the physical world in terms of measurable *quantities*, such as time, distance, force, and mass.

Review of Basic Differentiation Rules

In the 1600s, Europe was ushered into the scientific age by such great thinkers as Descartes, Galileo, Huygens, Newton, and Kepler. These men believed that nature is governed by basic laws—laws that can, for the most part, be written in terms of mathematical equations. One of the most influential publications of this period—*Dialogue on the Great World Systems*, by Galileo Galilei—has become a classic description of modern scientific thought.

As mathematics has developed during the past few hundred years, a small number of elementary functions has proven sufficient for modeling most* phenomena in physics, chemistry, biology, engineering, economics, and a variety of other fields. An **elementary function** is a function from the following list or one that can be formed as the sum, product, quotient, or composition of functions in the list.

Algebraic Functions	*Transcendental Functions*
Polynomial functions	Logarithmic functions
Rational functions	Exponential functions
Functions involving radicals	Trigonometric functions
	Inverse trigonometric functions

With the differentiation rules introduced so far in the text, you can differentiate *any* elementary function. For convenience, these differentiation rules are summarized below.

Basic Differentiation Rules for Elementary Functions

1. $\dfrac{d}{dx}[cu] = cu'$
2. $\dfrac{d}{dx}[u \pm v] = u' \pm v'$
3. $\dfrac{d}{dx}[uv] = uv' + vu'$
4. $\dfrac{d}{dx}\left[\dfrac{u}{v}\right] = \dfrac{vu' - uv'}{v^2}$
5. $\dfrac{d}{dx}[c] = 0$
6. $\dfrac{d}{dx}[u^n] = nu^{n-1}u'$
7. $\dfrac{d}{dx}[x] = 1$
8. $\dfrac{d}{dx}[|u|] = \dfrac{u}{|u|}(u'), \quad u \neq 0$
9. $\dfrac{d}{dx}[\ln u] = \dfrac{u'}{u}$
10. $\dfrac{d}{dx}[e^u] = e^u u'$
11. $\dfrac{d}{dx}[\log_a u] = \dfrac{u'}{(\ln a)u}$
12. $\dfrac{d}{dx}[a^u] = (\ln a)a^u u'$
13. $\dfrac{d}{dx}[\sin u] = (\cos u)u'$
14. $\dfrac{d}{dx}[\cos u] = -(\sin u)u'$
15. $\dfrac{d}{dx}[\tan u] = (\sec^2 u)u'$
16. $\dfrac{d}{dx}[\cot u] = -(\csc^2 u)u'$
17. $\dfrac{d}{dx}[\sec u] = (\sec u \tan u)u'$
18. $\dfrac{d}{dx}[\csc u] = -(\csc u \cot u)u'$
19. $\dfrac{d}{dx}[\arcsin u] = \dfrac{u'}{\sqrt{1 - u^2}}$
20. $\dfrac{d}{dx}[\arccos u] = \dfrac{-u'}{\sqrt{1 - u^2}}$
21. $\dfrac{d}{dx}[\arctan u] = \dfrac{u'}{1 + u^2}$
22. $\dfrac{d}{dx}[\operatorname{arccot} u] = \dfrac{-u'}{1 + u^2}$
23. $\dfrac{d}{dx}[\operatorname{arcsec} u] = \dfrac{u'}{|u|\sqrt{u^2 - 1}}$
24. $\dfrac{d}{dx}[\operatorname{arccsc} u] = \dfrac{-u'}{|u|\sqrt{u^2 - 1}}$

* *Some important functions used in engineering and science (such as Bessel functions and gamma functions) are not elementary functions.*

Exercises for Section 5.6

See www.CalcChat.com for worked-out solutions to odd-numbered exercises.

Numerical and Graphical Analysis **In Exercises 1 and 2, (a) use a graphing utility to complete the table, (b) plot the points in the table and graph the function by hand, (c) use a graphing utility to graph the function and compare the result with your hand-drawn graph in part (b), and (d) determine any intercepts and symmetry of the graph.**

x	-1	-0.8	-0.6	-0.4	-0.2	0	0.2	0.4	0.6	0.8	1
y											

1. $y = \arcsin x$ **2.** $y = \arccos x$

3. Determine the missing coordinates of the points on the graph of the function.

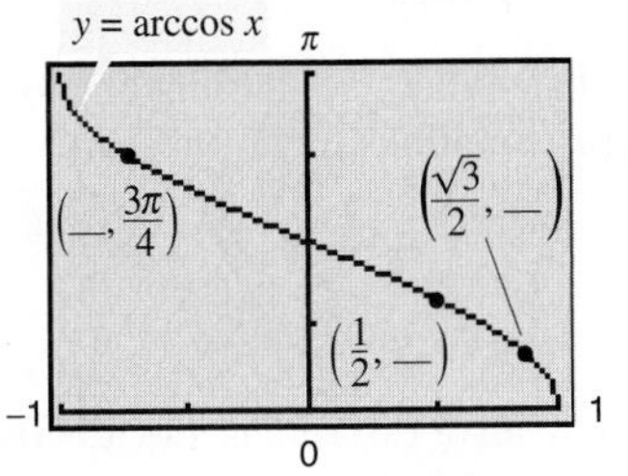

4. Determine the missing coordinates of the points on the graph of the function.

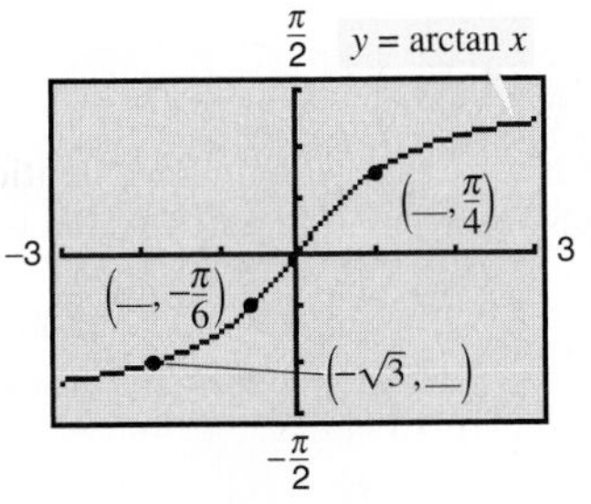

In Exercises 5–12, evaluate the expression without using a calculator.

5. $\arcsin \frac{1}{2}$ **6.** $\arcsin 0$

7. $\arccos \frac{1}{2}$ **8.** $\arccos 0$

9. $\arctan \frac{\sqrt{3}}{3}$ **10.** $\text{arccot}(-\sqrt{3})$

11. $\text{arccsc}(-\sqrt{2})$ **12.** $\arccos\left(-\frac{\sqrt{3}}{2}\right)$

In Exercises 13–16, use a calculator to approximate the value. Round your answer to two decimal places.

13. $\arccos(-0.8)$ **14.** $\arcsin(-0.39)$

15. $\text{arcsec } 1.269$ **16.** $\arctan(-3)$

In Exercises 17–20, evaluate each expression without using a calculator. (*Hint:* See Example 3.)

17. (a) $\sin\left(\arctan \frac{3}{4}\right)$ (b) $\sec\left(\arcsin \frac{4}{5}\right)$

18. (a) $\tan\left(\arccos \frac{\sqrt{2}}{2}\right)$ (b) $\cos\left(\arcsin \frac{5}{13}\right)$

19. (a) $\cot\left[\arcsin\left(-\frac{1}{2}\right)\right]$ (b) $\csc\left[\arctan\left(-\frac{5}{12}\right)\right]$

20. (a) $\sec\left[\arctan\left(-\frac{3}{5}\right)\right]$ (b) $\tan\left[\arcsin\left(-\frac{5}{6}\right)\right]$

In Exercises 21–28, write the expression in algebraic form.

21. $\cos(\arcsin 2x)$ **22.** $\sec(\arctan 4x)$

23. $\sin(\text{arcsec } x)$ **24.** $\cos(\text{arccot } x)$

25. $\tan\left(\text{arcsec } \frac{x}{3}\right)$ **26.** $\sec[\arcsin(x - 1)]$

27. $\csc\left(\arctan \frac{x}{\sqrt{2}}\right)$ **28.** $\cos\left(\arcsin \frac{x - h}{r}\right)$

In Exercises 29 and 30, (a) use a graphing utility to graph f and g in the same viewing window to verify that they are equal, (b) use algebra to verify that f and g are equal, and (c) identify any horizontal asymptotes of the graphs.

29. $f(x) = \sin(\arctan 2x), \quad g(x) = \frac{2x}{\sqrt{1 + 4x^2}}$

30. $f(x) = \tan\left(\arccos \frac{x}{2}\right), \quad g(x) = \frac{\sqrt{4 - x^2}}{x}$

In Exercises 31–34, solve the equation for x.

31. $\arcsin(3x - \pi) = \frac{1}{2}$ **32.** $\arctan(2x - 5) = -1$

33. $\arcsin \sqrt{2x} = \arccos \sqrt{x}$ **34.** $\arccos x = \text{arcsec } x$

In Exercises 35 and 36, verify each identity.

35. (a) $\text{arccsc } x = \arcsin \frac{1}{x}, \quad x \geq 1$

(b) $\arctan x + \arctan \frac{1}{x} = \frac{\pi}{2}, \quad x > 0$

36. (a) $\arcsin(-x) = -\arcsin x, \quad |x| \leq 1$

(b) $\arccos(-x) = \pi - \arccos x, \quad |x| \leq 1$

In Exercises 37–40, sketch the graph of the function. Use a graphing utility to verify your graph.

37. $f(x) = \arcsin(x - 1)$ **38.** $f(x) = \arctan x + \frac{\pi}{2}$

39. $f(x) = \text{arcsec } 2x$ **40.** $f(x) = \arccos \frac{x}{4}$

In Exercises 41–60, find the derivative of the function.

41. $f(x) = 2\arcsin(x - 1)$

42. $f(t) = \arcsin t^2$

43. $g(x) = 3\arccos\dfrac{x}{2}$

44. $f(x) = \operatorname{arcsec} 2x$

45. $f(x) = \arctan\dfrac{x}{a}$

46. $f(x) = \arctan\sqrt{x}$

47. $g(x) = \dfrac{\arcsin 3x}{x}$

48. $h(x) = x^2 \arctan x$

49. $h(t) = \sin(\arccos t)$

50. $f(x) = \arcsin x + \arccos x$

51. $y = x\arccos x - \sqrt{1 - x^2}$

52. $y = \ln(t^2 + 4) - \dfrac{1}{2}\arctan\dfrac{t}{2}$

53. $y = \dfrac{1}{2}\left(\dfrac{1}{2}\ln\dfrac{x + 1}{x - 1} + \arctan x\right)$

54. $y = \dfrac{1}{2}\left[x\sqrt{4 - x^2} + 4\arcsin\left(\dfrac{x}{2}\right)\right]$

55. $y = x\arcsin x + \sqrt{1 - x^2}$

56. $y = x\arctan 2x - \dfrac{1}{4}\ln(1 + 4x^2)$

57. $y = 8\arcsin\dfrac{x}{4} - \dfrac{x\sqrt{16 - x^2}}{2}$

58. $y = 25\arcsin\dfrac{x}{5} - x\sqrt{25 - x^2}$

59. $y = \arctan x + \dfrac{x}{1 + x^2}$

60. $y = \arctan\dfrac{x}{2} - \dfrac{1}{2(x^2 + 4)}$

In Exercises 61–66, find an equation of the tangent line to the graph of the function at the given point.

61. $y = 2\arcsin x$

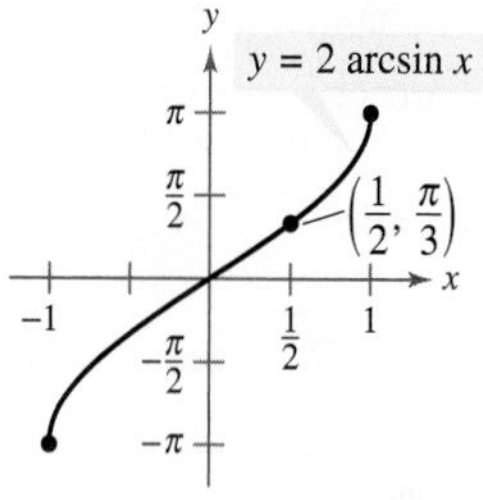

62. $y = \dfrac{1}{2}\arccos x$

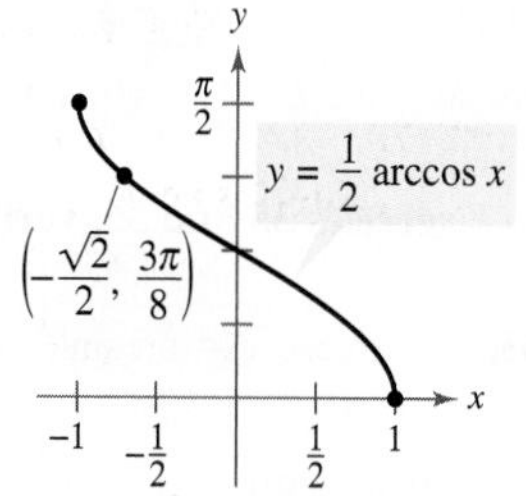

63. $y = \arctan\dfrac{x}{2}$

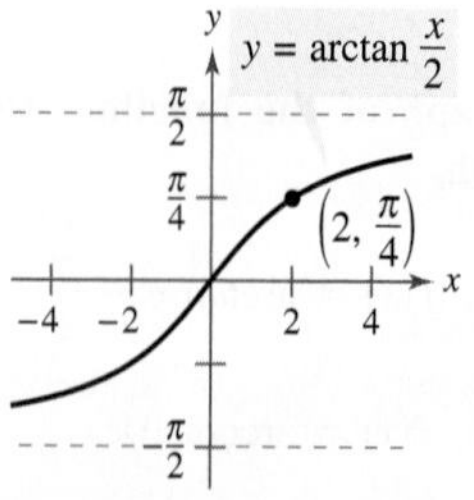

64. $y = \operatorname{arcsec} 4x$

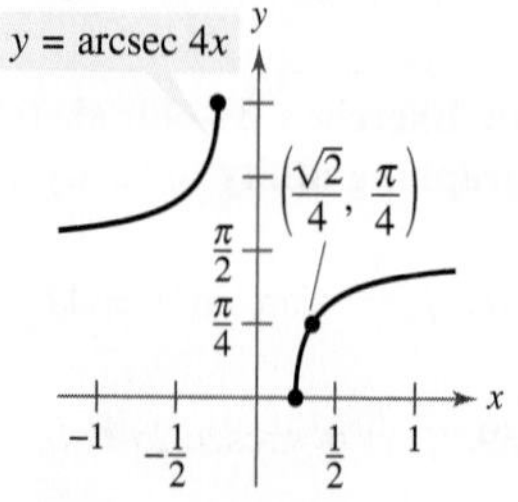

65. $y = 4x\arccos(x - 1)$

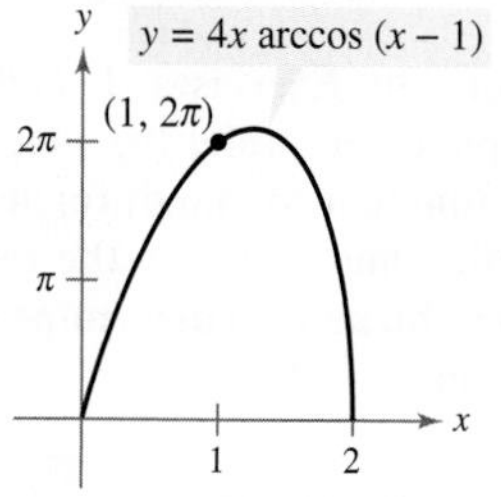

66. $y = 3x\arcsin x$

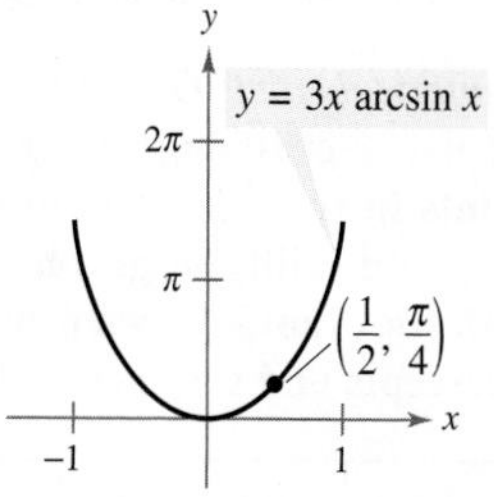

Linear and Quadratic Approximations **In Exercises 67–70, use a computer algebra system to find the linear approximation**

$$P_1(x) = f(a) + f'(a)(x - a)$$

and the quadratic approximation

$$P_2(x) = f(a) + f'(a)(x - a) + \tfrac{1}{2}f''(a)(x - a)^2$$

of the function f at $x = a$. Sketch the graph of the function and its linear and quadratic approximations.

67. $f(x) = \arctan x, \quad a = 0$

68. $f(x) = \arccos x, \quad a = 0$

69. $f(x) = \arcsin x, \quad a = \frac{1}{2}$

70. $f(x) = \arctan x, \quad a = 1$

In Exercises 71–74, find any relative extrema of the function.

71. $f(x) = \operatorname{arcsec} x - x$

72. $f(x) = \arcsin x - 2x$

73. $f(x) = \arctan x - \arctan(x - 4)$

74. $h(x) = \arcsin x - 2\arctan x$

Implicit Differentiation **In Exercises 75–78, find an equation of the tangent line to the graph of the equation at the given point.**

75. $x^2 + x\arctan y = y - 1, \quad \left(-\dfrac{\pi}{4}, 1\right)$

76. $\arctan(xy) = \arcsin(x + y), \quad (0, 0)$

77. $\arcsin x + \arcsin y = \dfrac{\pi}{2}, \quad \left(\dfrac{\sqrt{2}}{2}, \dfrac{\sqrt{2}}{2}\right)$

78. $\arctan(x + y) = y^2 + \dfrac{\pi}{4}, \quad (1, 0)$

Writing About Concepts

79. Explain why the domains of the trigonometric functions are restricted when finding the inverse trigonometric functions.

80. Explain why $\tan \pi = 0$ does not imply that $\arctan 0 = \pi$.

81. Explain how to graph $y = \operatorname{arccot} x$ on a graphing utility that does not have the arccotangent function.

82. Are the derivatives of the inverse trigonometric functions algebraic or transcendental functions? List the derivatives of the inverse trigonometric functions.

True or False? **In Exercises 83–88, determine whether the statement is true or false. If it is false, explain why or give an example that shows it is false.**

83. Because $\cos\left(-\frac{\pi}{3}\right) = \frac{1}{2}$, it follows that $\arccos \frac{1}{2} = -\frac{\pi}{3}$.

84. $\arcsin \frac{\pi}{4} = \frac{\sqrt{2}}{2}$

85. The slope of the graph of the inverse tangent function is positive for all x.

86. The range of $y = \arcsin x$ is $[0, \pi]$.

87. $\frac{d}{dx}[\arctan(\tan x)] = 1$ for all x in the domain.

88. $\arcsin^2 x + \arccos^2 x = 1$

89. ***Angular Rate of Change*** An airplane flies at an altitude of 5 miles toward a point directly over an observer. Consider θ and x as shown in the figure.

(a) Write θ as a function of x.

(b) The speed of the plane is 400 miles per hour. Find $d\theta/dt$ when $x = 10$ miles and $x = 3$ miles.

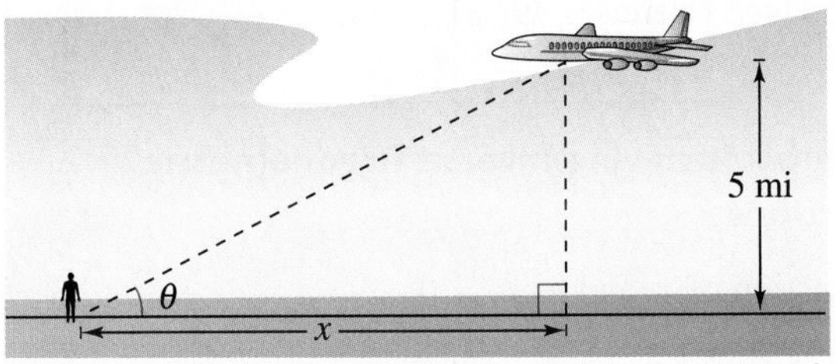

Not drawn to scale

90. ***Writing*** Repeat Exercise 89 if the altitude of the plane is 3 miles and describe how the altitude affects the rate of change of θ.

91. ***Angular Rate of Change*** In a free-fall experiment, an object is dropped from a height of 256 feet. A camera on the ground 500 feet from the point of impact records the fall of the object (see figure).

(a) Find the position function giving the height of the object at time t assuming the object is released at time $t = 0$. At what time will the object reach ground level?

(b) Find the rates of change of the angle of elevation of the camera when $t = 1$ and $t = 2$.

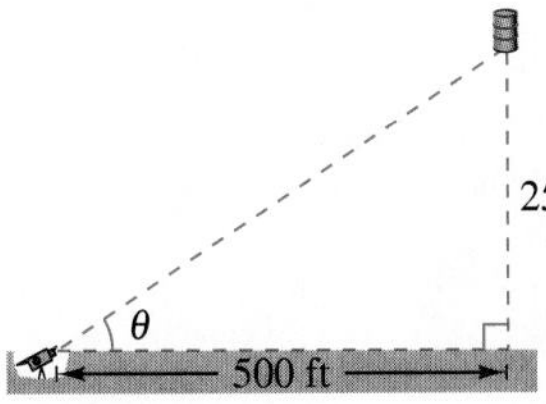

Figure for 91

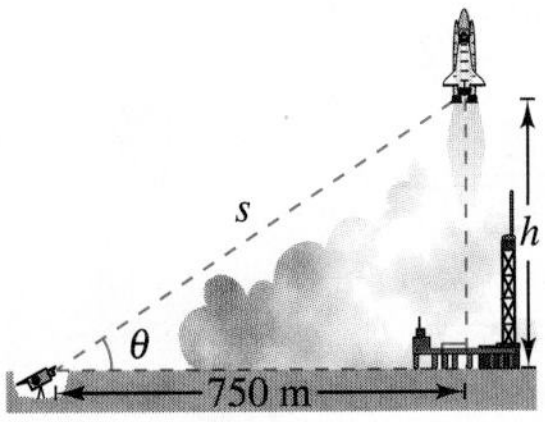

Figure for 92

92. ***Angular Rate of Change*** A television camera at ground level is filming the lift-off of a space shuttle at a point 750 meters from the launch pad. Let θ be the angle of elevation of the shuttle and let s be the distance between the camera and the shuttle (see figure). Write θ as a function of s for the period of time when the shuttle is moving vertically. Differentiate the result to find $d\theta/dt$ in terms of s and ds/dt.

93. (a) Prove that

$$\arctan x + \arctan y = \arctan \frac{x + y}{1 - xy}, \quad xy \neq 1.$$

(b) Use the formula in part (a) to show that

$$\arctan \frac{1}{2} + \arctan \frac{1}{3} = \frac{\pi}{4}.$$

94. Verify each differentiation formula.

(a) $\frac{d}{dx}[\arcsin u] = \frac{u'}{\sqrt{1 - u^2}}$ (b) $\frac{d}{dx}[\arctan u] = \frac{u'}{1 + u^2}$

(c) $\frac{d}{dx}[\operatorname{arcsec} u] = \frac{u'}{|u|\sqrt{u^2 - 1}}$

(d) $\frac{d}{dx}[\arccos u] = \frac{-u'}{\sqrt{1 - u^2}}$ (e) $\frac{d}{dx}[\operatorname{arccot} u] = \frac{-u'}{1 + u^2}$

(f) $\frac{d}{dx}[\operatorname{arccsc} u] = \frac{-u'}{|u|\sqrt{u^2 - 1}}$

95. ***Existence of an Inverse*** Determine the values of k such that the function $f(x) = kx + \sin x$ has an inverse function.

96. ***Think About It*** Use a graphing utility to graph $f(x) = \sin x$ and $g(x) = \arcsin(\sin x)$.

(a) Why isn't the graph of g the line $y = x$?

(b) Determine the extrema of g.

97. (a) Graph the function $f(x) = \arccos x + \arcsin x$ on the interval $[-1, 1]$. (b) Describe the graph of f. (c) Prove the result from part (b) analytically.

98. Prove that $\arcsin x = \arctan\left(\frac{x}{\sqrt{1 - x^2}}\right)$, $|x| < 1$.

99. Find the value of c in the interval $[0, 4]$ on the x-axis that maximizes angle θ.

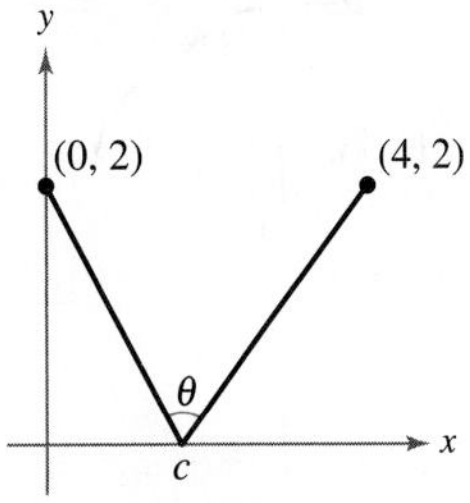

Figure for 99

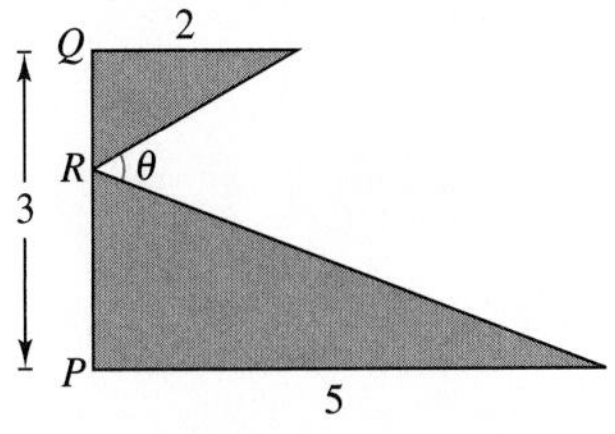

Figure for 100

100. Find PR such that $0 \leq PR \leq 3$ and $m \angle \theta$ is a maximum.

101. Some calculus textbooks define the inverse secant function using the range $[0, \pi/2) \cup [\pi, 3\pi/2)$.

(a) Sketch the graph $y = \operatorname{arcsec} x$ using this range.

(b) Show that $y' = \frac{1}{x\sqrt{x^2 - 1}}$.

Section 5.7 Inverse Trigonometric Functions: Integration

- **Integrate functions whose antiderivatives involve inverse trigonometric functions.**
- **Use the method of completing the square to integrate a function.**
- **Review the basic integration rules involving elementary functions.**

Integrals Involving Inverse Trigonometric Functions

The derivatives of the six inverse trigonometric functions fall into three pairs. In each pair, the derivative of one function is the negative of the other. For example,

$$\frac{d}{dx}[\arcsin x] = \frac{1}{\sqrt{1-x^2}}$$

and

$$\frac{d}{dx}[\arccos x] = -\frac{1}{\sqrt{1-x^2}}.$$

When listing the *antiderivative* that corresponds to each of the inverse trigonometric functions, you need to use only one member from each pair. It is conventional to use $\arcsin x$ as the antiderivative of $1/\sqrt{1-x^2}$, rather than $-\arccos x$. The next theorem gives one antiderivative formula for each of the three pairs. The proofs of these integration rules are left to you (see Exercises 79–81).

FOR FURTHER INFORMATION For a detailed proof of part 2 of Theorem 5.17, see the article "A Direct Proof of the Integral Formula for Arctangent" by Arnold J. Insel in *The College Mathematics Journal*. To view this article, go to the website *www.matharticles.com.*

THEOREM 5.17 Integrals Involving Inverse Trigonometric Functions

Let u be a differentiable function of x, and let $a > 0$.

1. $\displaystyle\int \frac{du}{\sqrt{a^2-u^2}} = \arcsin\frac{u}{a} + C$ **2.** $\displaystyle\int \frac{du}{a^2+u^2} = \frac{1}{a}\arctan\frac{u}{a} + C$

3. $\displaystyle\int \frac{du}{u\sqrt{u^2-a^2}} = \frac{1}{a}\operatorname{arcsec}\frac{|u|}{a} + C$

EXAMPLE 1 Integration with Inverse Trigonometric Functions

a. $\displaystyle\int \frac{dx}{\sqrt{4-x^2}} = \arcsin\frac{x}{2} + C$

b. $\displaystyle\int \frac{dx}{2+9x^2} = \frac{1}{3}\int \frac{3\,dx}{(\sqrt{2})^2+(3x)^2}$ $u = 3x,\ a = \sqrt{2}$

$$= \frac{1}{3\sqrt{2}}\arctan\frac{3x}{\sqrt{2}} + C$$

c. $\displaystyle\int \frac{dx}{x\sqrt{4x^2-9}} = \int \frac{2\,dx}{2x\sqrt{(2x)^2-3^2}}$ $u = 2x,\ a = 3$

$$= \frac{1}{3}\operatorname{arcsec}\frac{|2x|}{3} + C$$

The integrals in Example 1 are fairly straightforward applications of integration formulas. Unfortunately, this is not typical. The integration formulas for inverse trigonometric functions can be disguised in many ways.

TECHNOLOGY PITFALL

Computer software that can perform symbolic integration is useful for integrating functions such as the one in Example 2. When using such software, however, you must remember that it can fail to find an antiderivative for two reasons. First, some elementary functions simply do not have antiderivatives that are elementary functions. Second, every symbolic integration utility has limitations—you might have entered a function that the software was not programmed to handle. You should also remember that antiderivatives involving trigonometric functions or logarithmic functions can be written in many different forms. For instance, one symbolic integration utility found the integral in Example 2 to be

$$\int \frac{dx}{\sqrt{e^{2x}-1}} = \arctan \sqrt{e^{2x}-1} + C.$$

Try showing that this antiderivative is equivalent to that obtained in Example 2.

EXAMPLE 2 Integration by Substitution

Find $\displaystyle\int \frac{dx}{\sqrt{e^{2x}-1}}$.

Solution As it stands, this integral doesn't fit any of the three inverse trigonometric formulas. Using the substitution $u = e^x$, however, produces

$$u = e^x \quad \Rightarrow \quad du = e^x\,dx \quad \Rightarrow \quad dx = \frac{du}{e^x} = \frac{du}{u}.$$

With this substitution, you can integrate as follows.

$$\begin{aligned}
\int \frac{dx}{\sqrt{e^{2x}-1}} &= \int \frac{dx}{\sqrt{(e^x)^2-1}} && \text{Write } e^{2x} \text{ as } (e^x)^2.\\
&= \int \frac{du/u}{\sqrt{u^2-1}} && \text{Substitute.}\\
&= \int \frac{du}{u\sqrt{u^2-1}} && \text{Rewrite to fit Arcsecant Rule.}\\
&= \operatorname{arcsec} \frac{|u|}{1} + C && \text{Apply Arcsecant Rule.}\\
&= \operatorname{arcsec} e^x + C && \text{Back-substitute.}
\end{aligned}$$

EXAMPLE 3 Rewriting as the Sum of Two Quotients

Find $\displaystyle\int \frac{x+2}{\sqrt{4-x^2}}\,dx$.

Solution This integral does not appear to fit any of the basic integration formulas. By splitting the integrand into two parts, however, you can see that the first part can be found with the Power Rule and the second part yields an inverse sine function.

$$\begin{aligned}
\int \frac{x+2}{\sqrt{4-x^2}}\,dx &= \int \frac{x}{\sqrt{4-x^2}}\,dx + \int \frac{2}{\sqrt{4-x^2}}\,dx\\
&= -\frac{1}{2}\int (4-x^2)^{-1/2}(-2x)\,dx + 2\int \frac{1}{\sqrt{4-x^2}}\,dx\\
&= -\frac{1}{2}\left[\frac{(4-x^2)^{1/2}}{1/2}\right] + 2\arcsin\frac{x}{2} + C\\
&= -\sqrt{4-x^2} + 2\arcsin\frac{x}{2} + C
\end{aligned}$$

Completing the Square

Completing the square helps when quadratic functions are involved in the integrand. For example, the quadratic $x^2 + bx + c$ can be written as the difference of two squares by adding and subtracting $(b/2)^2$.

$$\begin{aligned}
x^2 + bx + c &= x^2 + bx + \left(\frac{b}{2}\right)^2 - \left(\frac{b}{2}\right)^2 + c\\
&= \left(x + \frac{b}{2}\right)^2 - \left(\frac{b}{2}\right)^2 + c
\end{aligned}$$

EXAMPLE 4 Completing the Square

Find $\int \frac{dx}{x^2 - 4x + 7}$.

Solution You can write the denominator as the sum of two squares as shown.

$$x^2 - 4x + 7 = (x^2 - 4x + 4) - 4 + 7$$
$$= (x - 2)^2 + 3 = u^2 + a^2$$

Now, in this completed square form, let $u = x - 2$ and $a = \sqrt{3}$.

$$\int \frac{dx}{x^2 - 4x + 7} = \int \frac{dx}{(x - 2)^2 + 3} = \frac{1}{\sqrt{3}} \arctan \frac{x - 2}{\sqrt{3}} + C$$

If the leading coefficient is not 1, it helps to factor before completing the square. For instance, you can complete the square of $2x^2 - 8x + 10$ by factoring first.

$$2x^2 - 8x + 10 = 2(x^2 - 4x + 5)$$
$$= 2(x^2 - 4x + 4 - 4 + 5)$$
$$= 2[(x - 2)^2 + 1]$$

To complete the square when the coefficient of x^2 is negative, use the same factoring process shown above. For instance, you can complete the square for $3x - x^2$ as shown.

$$3x - x^2 = -(x^2 - 3x)$$
$$= -\left[x^2 - 3x + \left(\tfrac{3}{2}\right)^2 - \left(\tfrac{3}{2}\right)^2\right]$$
$$= \left(\tfrac{3}{2}\right)^2 - \left(x - \tfrac{3}{2}\right)^2$$

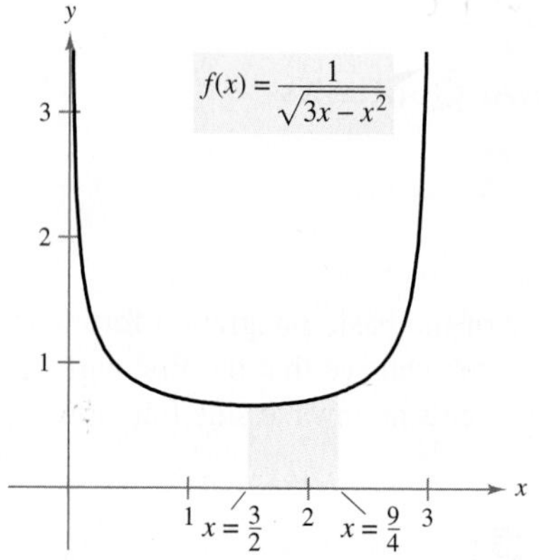

The area of the region bounded by the graph of f, the x-axis, $x = \frac{3}{2}$, and $x = \frac{9}{4}$ is $\pi/6$.
Figure 5.34

EXAMPLE 5 Completing the Square (Negative Leading Coefficient)

Find the area of the region bounded by the graph of

$$f(x) = \frac{1}{\sqrt{3x - x^2}}$$

the x-axis, and the lines $x = \frac{3}{2}$ and $x = \frac{9}{4}$.

Solution From Figure 5.34, you can see that the area is given by

$$\text{Area} = \int_{3/2}^{9/4} \frac{1}{\sqrt{3x - x^2}}\, dx.$$

Using the completed square form derived above, you can integrate as shown.

$$\int_{3/2}^{9/4} \frac{dx}{\sqrt{3x - x^2}} = \int_{3/2}^{9/4} \frac{dx}{\sqrt{(3/2)^2 - [x - (3/2)]^2}}$$
$$= \arcsin \frac{x - (3/2)}{3/2} \Bigg]_{3/2}^{9/4}$$
$$= \arcsin \frac{1}{2} - \arcsin 0$$
$$= \frac{\pi}{6}$$
$$\approx 0.524$$

TECHNOLOGY With definite integrals such as the one given in Example 5, remember that you can resort to a numerical solution. For instance, applying Simpson's Rule (with $n = 12$) to the integral in the example, you obtain

$$\int_{3/2}^{9/4} \frac{1}{\sqrt{3x - x^2}}\, dx \approx 0.523599.$$

This differs from the exact value of the integral ($\pi/6 \approx 0.5235988$) by less than one millionth.

Review of Basic Integration Rules

You have now completed the introduction of the **basic integration rules.** To be efficient at applying these rules, you should have practiced enough so that each rule is committed to memory.

Basic Integration Rules ($a > 0$)

1. $\displaystyle\int kf(u)\,du = k\int f(u)\,du$

2. $\displaystyle\int [f(u) \pm g(u)]\,du = \int f(u)\,du \pm \int g(u)\,du$

3. $\displaystyle\int du = u + C$

4. $\displaystyle\int u^n\,du = \frac{u^{n+1}}{n+1} + C, \quad n \neq -1$

5. $\displaystyle\int \frac{du}{u} = \ln|u| + C$

6. $\displaystyle\int e^u\,du = e^u + C$

7. $\displaystyle\int a^u\,du = \left(\frac{1}{\ln a}\right)a^u + C$

8. $\displaystyle\int \sin u\,du = -\cos u + C$

9. $\displaystyle\int \cos u\,du = \sin u + C$

10. $\displaystyle\int \tan u\,du = -\ln|\cos u| + C$

11. $\displaystyle\int \cot u\,du = \ln|\sin u| + C$

12. $\displaystyle\int \sec u\,du = \ln|\sec u + \tan u| + C$

13. $\displaystyle\int \csc u\,du = -\ln|\csc u + \cot u| + C$

14. $\displaystyle\int \sec^2 u\,du = \tan u + C$

15. $\displaystyle\int \csc^2 u\,du = -\cot u + C$

16. $\displaystyle\int \sec u \tan u\,du = \sec u + C$

17. $\displaystyle\int \csc u \cot u\,du = -\csc u + C$

18. $\displaystyle\int \frac{du}{\sqrt{a^2 - u^2}} = \arcsin\frac{u}{a} + C$

19. $\displaystyle\int \frac{du}{a^2 + u^2} = \frac{1}{a}\arctan\frac{u}{a} + C$

20. $\displaystyle\int \frac{du}{u\sqrt{u^2 - a^2}} = \frac{1}{a}\operatorname{arcsec}\frac{|u|}{a} + C$

You can learn a lot about the nature of integration by comparing this list with the summary of differentiation rules given in the preceding section. For differentiation, you now have rules that allow you to differentiate *any* elementary function. For integration, this is far from true.

The integration rules listed above are primarily those that were happened on when developing differentiation rules. So far, you have not learned any rules or techniques for finding the antiderivative of a general product or quotient, the natural logarithmic function, or the inverse trigonometric functions. More importantly, you cannot apply any of the rules in this list unless you can create the proper du corresponding to the u in the formula. The point is that you need to work more on integration techniques, which you will do in Chapter 8. The next two examples should give you a better feeling for the integration problems that you *can* and *cannot* do with the techniques and rules you now know.

EXAMPLE 6 Comparing Integration Problems

Find as many of the following integrals as you can using the formulas and techniques you have studied so far in the text.

a. $\displaystyle\int \frac{dx}{x\sqrt{x^2-1}}$ **b.** $\displaystyle\int \frac{x\,dx}{\sqrt{x^2-1}}$ **c.** $\displaystyle\int \frac{dx}{\sqrt{x^2-1}}$

Solution

a. You *can* find this integral (it fits the Arcsecant Rule).

$$\int \frac{dx}{x\sqrt{x^2-1}} = \operatorname{arcsec}|x| + C$$

b. You *can* find this integral (it fits the Power Rule).

$$\begin{aligned}\int \frac{x\,dx}{\sqrt{x^2-1}} &= \frac{1}{2}\int (x^2-1)^{-1/2}(2x)\,dx \\ &= \frac{1}{2}\left[\frac{(x^2-1)^{1/2}}{1/2}\right] + C \\ &= \sqrt{x^2-1} + C\end{aligned}$$

c. You *cannot* find this integral using present techniques. (You should scan the list of basic integration rules to verify this conclusion.)

EXAMPLE 7 Comparing Integration Problems

Find as many of the following integrals as you can using the formulas and techniques you have studied so far in the text.

a. $\displaystyle\int \frac{dx}{x\ln x}$ **b.** $\displaystyle\int \frac{\ln x\,dx}{x}$ **c.** $\displaystyle\int \ln x\,dx$

Solution

a. You *can* find this integral (it fits the Log Rule).

$$\begin{aligned}\int \frac{dx}{x\ln x} &= \int \frac{1/x}{\ln x}\,dx \\ &= \ln|\ln x| + C\end{aligned}$$

b. You *can* find this integral (it fits the Power Rule).

$$\begin{aligned}\int \frac{\ln x\,dx}{x} &= \int \left(\frac{1}{x}\right)(\ln x)^1\,dx \\ &= \frac{(\ln x)^2}{2} + C\end{aligned}$$

c. You *cannot* find this integral using present techniques.

NOTE Note in Examples 6 and 7 that the *simplest* functions are the ones that you cannot yet integrate.

Exercises for Section 5.7

See www.CalcChat.com for worked-out solutions to odd-numbered exercises.

In Exercises 1–20, find the integral.

1. $\int \frac{5}{\sqrt{9-x^2}}\,dx$
2. $\int \frac{3}{\sqrt{1-4x^2}}\,dx$
3. $\int \frac{7}{16+x^2}\,dx$
4. $\int \frac{4}{1+9x^2}\,dx$
5. $\int \frac{1}{x\sqrt{4x^2-1}}\,dx$
6. $\int \frac{1}{4+(x-1)^2}\,dx$
7. $\int \frac{x^3}{x^2+1}\,dx$
8. $\int \frac{x^4-1}{x^2+1}\,dx$
9. $\int \frac{1}{\sqrt{1-(x+1)^2}}\,dx$
10. $\int \frac{t}{t^4+16}\,dt$
11. $\int \frac{t}{\sqrt{1-t^4}}\,dt$
12. $\int \frac{1}{x\sqrt{x^4-4}}\,dx$
13. $\int \frac{e^{2x}}{4+e^{4x}}\,dx$
14. $\int \frac{1}{3+(x-2)^2}\,dx$
15. $\int \frac{1}{\sqrt{x}\sqrt{1-x}}\,dx$
16. $\int \frac{3}{2\sqrt{x}(1+x)}\,dx$
17. $\int \frac{x-3}{x^2+1}\,dx$
18. $\int \frac{4x+3}{\sqrt{1-x^2}}\,dx$
19. $\int \frac{x+5}{\sqrt{9-(x-3)^2}}\,dx$
20. $\int \frac{x-2}{(x+1)^2+4}\,dx$

In Exercises 21–30, evaluate the integral.

21. $\int_0^{1/6} \frac{1}{\sqrt{1-9x^2}}\,dx$
22. $\int_0^{1} \frac{dx}{\sqrt{4-x^2}}$
23. $\int_0^{\sqrt{3}/2} \frac{1}{1+4x^2}\,dx$
24. $\int_{\sqrt{3}}^{3} \frac{1}{9+x^2}\,dx$
25. $\int_0^{1/\sqrt{2}} \frac{\arcsin x}{\sqrt{1-x^2}}\,dx$
26. $\int_0^{1/\sqrt{2}} \frac{\arccos x}{\sqrt{1-x^2}}\,dx$
27. $\int_{-1/2}^{0} \frac{x}{\sqrt{1-x^2}}\,dx$
28. $\int_{-\sqrt{3}}^{0} \frac{x}{1+x^2}\,dx$
29. $\int_{\pi/2}^{\pi} \frac{\sin x}{1+\cos^2 x}\,dx$
30. $\int_0^{\pi/2} \frac{\cos x}{1+\sin^2 x}\,dx$

In Exercises 31–42, find or evaluate the integral. (Complete the square, if necessary.)

31. $\int_0^2 \frac{dx}{x^2-2x+2}$
32. $\int_{-2}^2 \frac{dx}{x^2+4x+13}$
33. $\int \frac{2x}{x^2+6x+13}\,dx$
34. $\int \frac{2x-5}{x^2+2x+2}\,dx$
35. $\int \frac{1}{\sqrt{-x^2-4x}}\,dx$
36. $\int \frac{2}{\sqrt{-x^2+4x}}\,dx$
37. $\int \frac{x+2}{\sqrt{-x^2-4x}}\,dx$
38. $\int \frac{x-1}{\sqrt{x^2-2x}}\,dx$
39. $\int_2^3 \frac{2x-3}{\sqrt{4x-x^2}}\,dx$
40. $\int \frac{1}{(x-1)\sqrt{x^2-2x}}\,dx$
41. $\int \frac{x}{x^4+2x^2+2}\,dx$
42. $\int \frac{x}{\sqrt{9+8x^2-x^4}}\,dx$

In Exercises 43–46, use the specified substitution to find or evaluate the integral.

43. $\int \sqrt{e^t-3}\,dt$

$u = \sqrt{e^t-3}$

44. $\int \frac{\sqrt{x-2}}{x+1}\,dx$

$u = \sqrt{x-2}$

45. $\int_1^3 \frac{dx}{\sqrt{x}(1+x)}$

$u = \sqrt{x}$

46. $\int_0^1 \frac{dx}{2\sqrt{3-x}\sqrt{x+1}}$

$u = \sqrt{x+1}$

Writing About Concepts

In Exercises 47–50, determine which of the integrals can be found using the basic integration formulas you have studied so far in the text.

47. (a) $\int \frac{1}{\sqrt{1-x^2}}\,dx$ (b) $\int \frac{x}{\sqrt{1-x^2}}\,dx$ (c) $\int \frac{1}{x\sqrt{1-x^2}}\,dx$

48. (a) $\int e^{x^2}\,dx$ (b) $\int xe^{x^2}\,dx$ (c) $\int \frac{1}{x^2}e^{1/x}\,dx$

49. (a) $\int \sqrt{x-1}\,dx$ (b) $\int x\sqrt{x-1}\,dx$ (c) $\int \frac{x}{\sqrt{x-1}}\,dx$

50. (a) $\int \frac{1}{1+x^4}\,dx$ (b) $\int \frac{x}{1+x^4}\,dx$ (c) $\int \frac{x^3}{1+x^4}\,dx$

51. Determine which value best approximates the area of the region between the x-axis and the function

$$f(x) = \frac{1}{\sqrt{1-x^2}}$$

over the interval $[-0.5, 0.5]$. (Make your selection on the basis of a sketch of the region and *not* by performing any calculations.)

(a) 4 (b) -3 (c) 1 (d) 2 (e) 3

52. Decide whether you can find the integral

$$\int \frac{2\,dx}{\sqrt{x^2+4}}$$

using the formulas and techniques you have studied so far. Explain your reasoning.

Differential Equations **In Exercises 53 and 54, use the differential equation and the specified initial condition to find y.**

53. $\frac{dy}{dx} = \frac{1}{\sqrt{4-x^2}}$

$y(0) = \pi$

54. $\frac{dy}{dx} = \frac{1}{4+x^2}$

$y(2) = \pi$

Slope Fields **In Exercises 55–58, a differential equation, a point, and a slope field are given. (a) Sketch two approximate solutions of the differential equation on the slope field, one of which passes through the given point. (b) Use integration to find the particular solution of the differential equation and use a graphing utility to graph the solution. Compare the result with the sketches in part (a). To print an enlarged copy of the graph, go to the website *www.mathgraphs.com*.**

55. $\dfrac{dy}{dx} = \dfrac{3}{1+x^2}$, $(0, 0)$

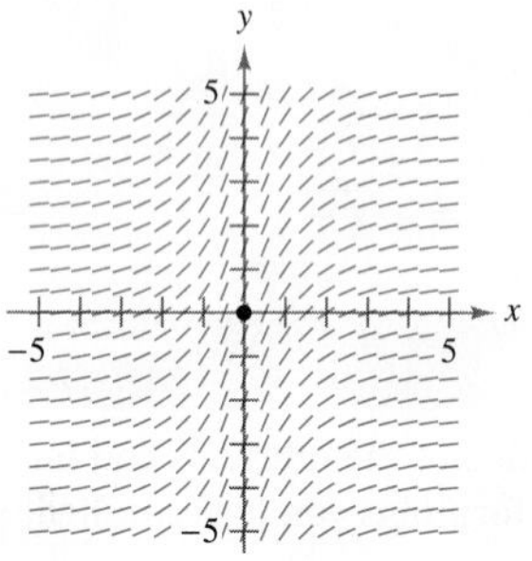

56. $\dfrac{dy}{dx} = \dfrac{2}{9+x^2}$, $(0, 2)$

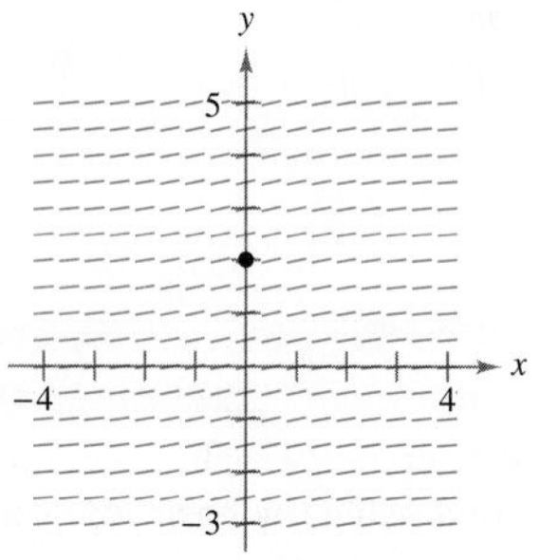

57. $\dfrac{dy}{dx} = \dfrac{1}{x\sqrt{x^2-4}}$, $(2, 1)$

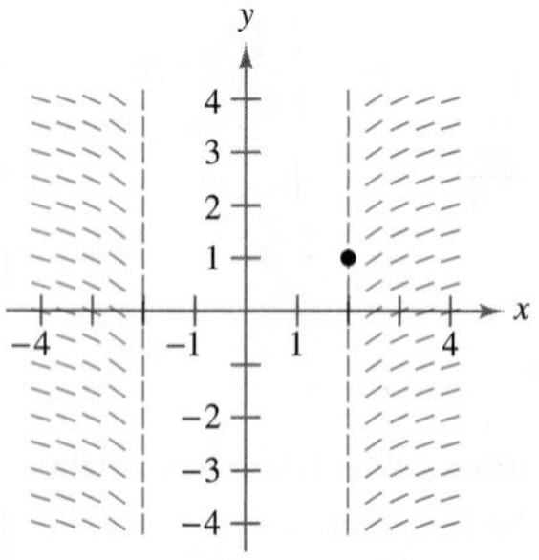

58. $\dfrac{dy}{dx} = \dfrac{2}{\sqrt{25-x^2}}$, $(5, \pi)$

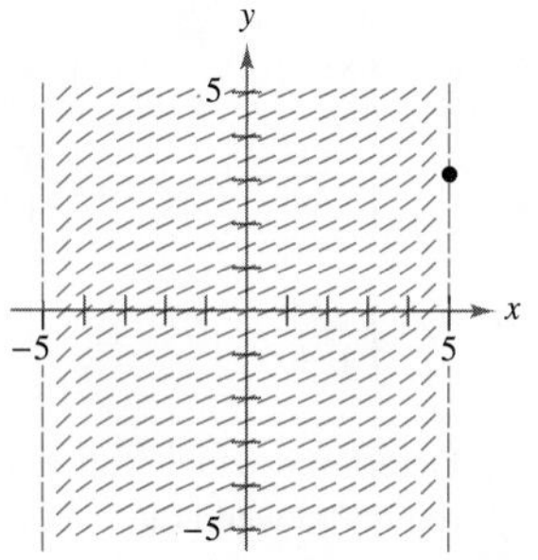

Slope Fields **In Exercises 59–62, use a computer algebra system to graph the slope field for the differential equation and graph the solution satisfying the specified initial condition.**

59. $\dfrac{dy}{dx} = \dfrac{10}{x\sqrt{x^2-1}}$

$y(3) = 0$

60. $\dfrac{dy}{dx} = \dfrac{1}{12+x^2}$

$y(4) = 2$

61. $\dfrac{dy}{dx} = \dfrac{2y}{\sqrt{16-x^2}}$

$y(0) = 2$

62. $\dfrac{dy}{dx} = \dfrac{\sqrt{y}}{1+x^2}$

$y(0) = 4$

Area **In Exercises 63–68, find the area of the region.**

63. $y = \dfrac{1}{x^2-2x+5}$

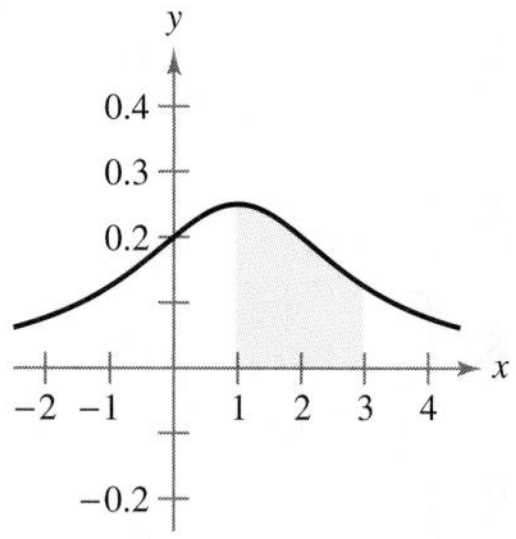

64. $y = \dfrac{2}{x^2+4x+8}$

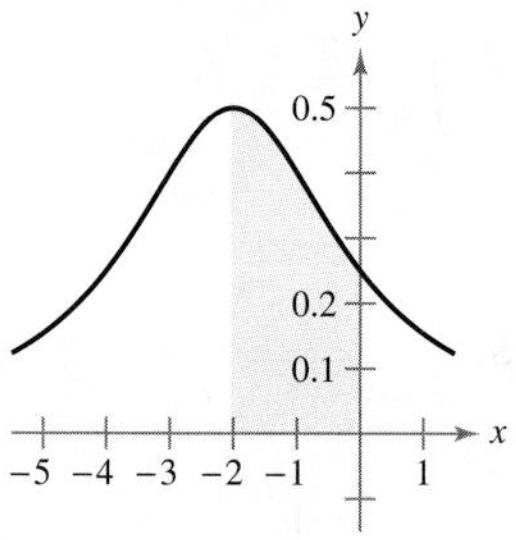

65. $y = \dfrac{1}{\sqrt{4-x^2}}$

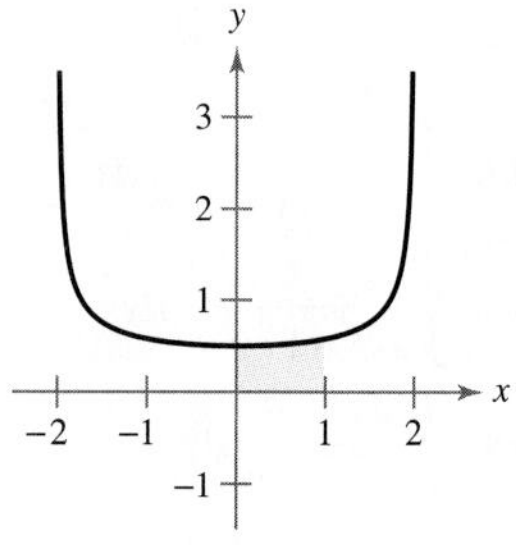

66. $y = \dfrac{1}{x\sqrt{x^2-1}}$

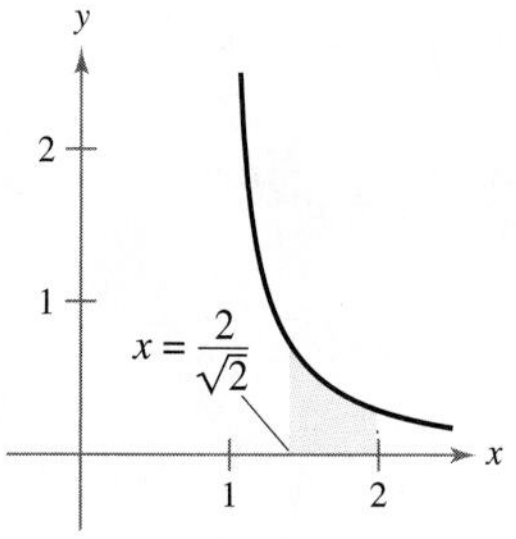

67. $y = \dfrac{3\cos x}{1+\sin^2 x}$

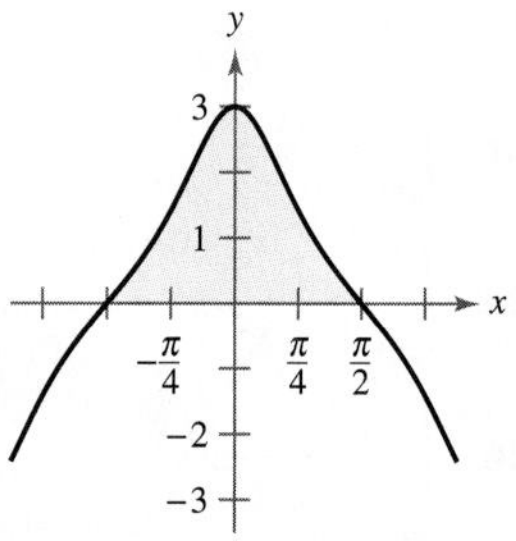

68. $y = \dfrac{e^x}{1+e^{2x}}$

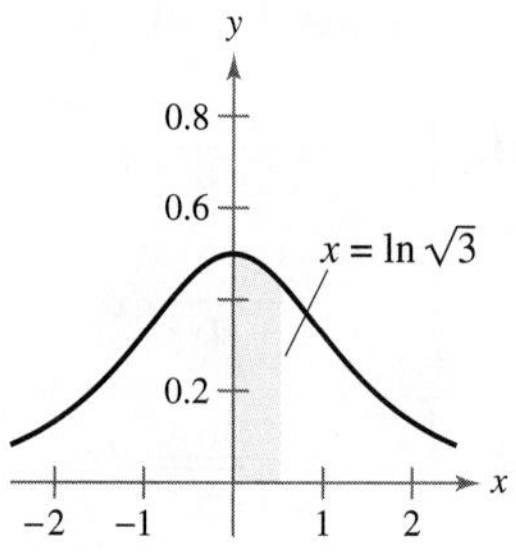

In Exercises 69 and 70, (a) verify the integration formula, then (b) use it to find the area of the region.

69. $\displaystyle\int \frac{\arctan x}{x^2}\,dx = \ln x - \frac{1}{2}\ln(1+x^2) - \frac{\arctan x}{x} + C$

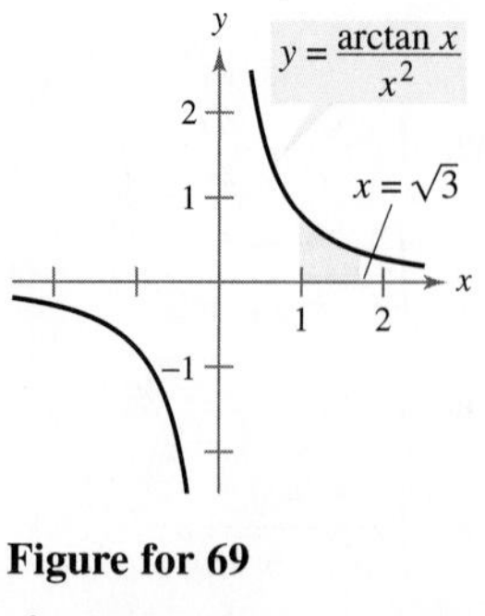

Figure for 69

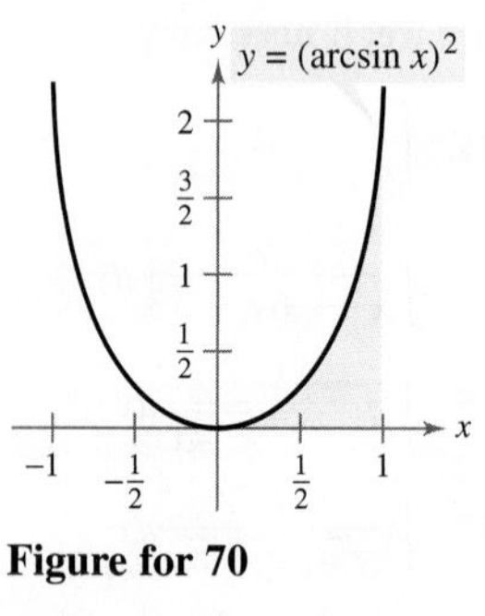

Figure for 70

70. $\displaystyle\int (\arcsin x)^2\,dx$

$= x(\arcsin x)^2 - 2x + 2\sqrt{1-x^2}\arcsin x + C$

71. (a) Sketch the region whose area is represented by

$$\int_0^1 \arcsin x\, dx.$$

(b) Use the integration capabilities of a graphing utility to approximate the area.

(c) Find the exact area analytically.

72. (a) Show that $\displaystyle\int_0^1 \frac{4}{1+x^2}\,dx = \pi.$

(b) Approximate the number π using Simpson's Rule (with $n = 6$) and the integral in part (a).

(c) Approximate the number π by using the integration capabilities of a graphing utility.

73. ***Investigation*** Consider the function $F(x) = \dfrac{1}{2}\displaystyle\int_x^{x+2} \frac{2}{t^2+1}\,dt.$

(a) Write a short paragraph giving a geometric interpretation of the function $F(x)$ relative to the function $f(x) = \dfrac{2}{x^2+1}$. Use what you have written to guess the value of x that will make F maximum.

(b) Perform the specified integration to find an alternative form of $F(x)$. Use calculus to locate the value of x that will make F maximum and compare the result with your guess in part (a).

74. Consider the integral $\displaystyle\int \frac{1}{\sqrt{6x - x^2}}\,dx.$

(a) Find the integral by completing the square of the radicand.

(b) Find the integral by making the substitution $u = \sqrt{x}$.

(c) The antiderivatives in parts (a) and (b) appear to be significantly different. Use a graphing utility to graph each antiderivative in the same viewing window and determine the relationship between them. Find the domain of each.

True or False? **In Exercises 75–78, determine whether the statement is true or false. If it is false, explain why or give an example that shows it is false.**

75. $\displaystyle\int \frac{dx}{3x\sqrt{9x^2 - 16}} = \frac{1}{4}\,\mathrm{arcsec}\,\frac{3x}{4} + C$

76. $\displaystyle\int \frac{dx}{25 + x^2} = \frac{1}{25}\arctan\frac{x}{25} + C$

77. $\displaystyle\int \frac{dx}{\sqrt{4 - x^2}} = -\arccos\frac{x}{2} + C$

78. One way to find $\displaystyle\int \frac{2e^{2x}}{\sqrt{9 - e^{2x}}}\,dx$ is to use the Arcsine Rule.

Verifying Integration Rules **In Exercises 79–81, verify each rule by differentiating. Let $a > 0$.**

79. $\displaystyle\int \frac{du}{\sqrt{a^2 - u^2}} = \arcsin\frac{u}{a} + C$

80. $\displaystyle\int \frac{du}{a^2 + u^2} = \frac{1}{a}\arctan\frac{u}{a} + C$

81. $\displaystyle\int \frac{du}{u\sqrt{u^2 - a^2}} = \frac{1}{a}\,\mathrm{arcsec}\,\frac{|u|}{a} + C$

82. ***Numerical Integration*** (a) Write an integral that represents the area of the region. (b) Then use the Trapezoidal Rule with $n = 8$ to estimate the area of the region. (c) Explain how you can use the results of parts (a) and (b) to estimate π.

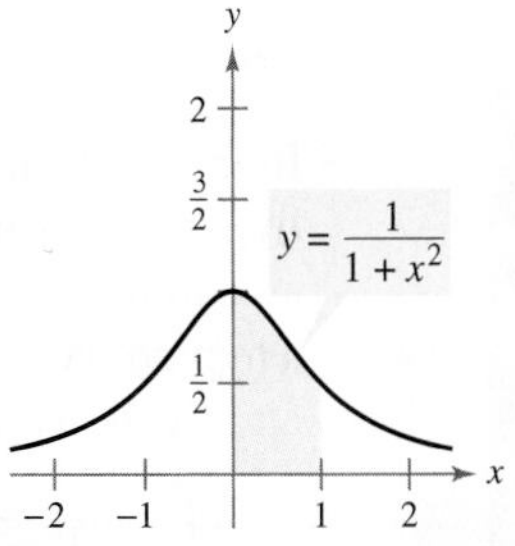

83. ***Vertical Motion*** An object is projected upward from ground level with an initial velocity of 500 feet per second. In this exercise, the goal is to analyze the motion of the object during its upward flight.

(a) If air resistance is neglected, find the velocity of the object as a function of time. Use a graphing utility to graph this function.

(b) Use the result in part (a) to find the position function and determine the maximum height attained by the object.

(c) If the air resistance is proportional to the square of the velocity, you obtain the equation

$$\frac{dv}{dt} = -(32 + kv^2)$$

where -32 feet per second per second is the acceleration due to gravity and k is a constant. Find the velocity as a function of time by solving the equation

$$\int \frac{dv}{32 + kv^2} = -\int dt.$$

(d) Use a graphing utility to graph the velocity function $v(t)$ in part (c) if $k = 0.001$. Use the graph to approximate the time t_0 at which the object reaches its maximum height.

(e) Use the integration capabilities of a graphing utility to approximate the integral

$$\int_0^{t_0} v(t)\,dt$$

where $v(t)$ and t_0 are those found in part (d). This is the approximation of the maximum height of the object.

(f) Explain the difference between the results in parts (b) and (e).

FOR FURTHER INFORMATION For more information on this topic, see "What Goes Up Must Come Down; Will Air Resistance Make It Return Sooner, or Later?" by John Lekner in *Mathematics Magazine*. To view this article, go to the website *www.matharticles.com*.

84. Graph $y_1 = \dfrac{x}{1+x^2}$, $y_2 = \arctan x$, and $y_3 = x$ on $[0, 10]$.

Prove that $\dfrac{x}{1+x^2} < \arctan x < x$ for $x > 0$.

Section 5.8 Hyperbolic Functions

- **Develop properties of hyperbolic functions.**
- **Differentiate and integrate hyperbolic functions.**
- **Develop properties of inverse hyperbolic functions.**
- **Differentiate and integrate functions involving inverse hyperbolic functions.**

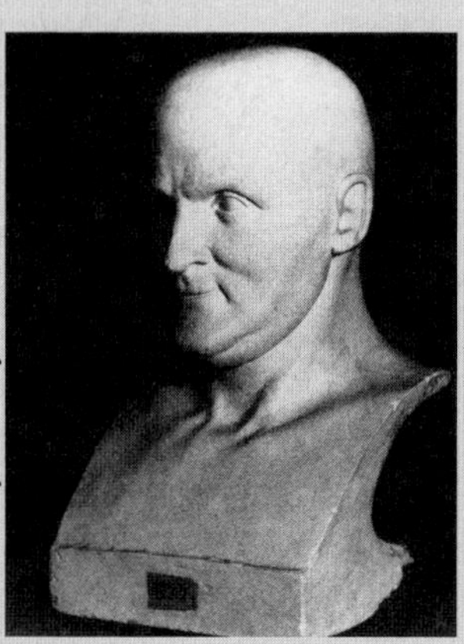

American Institute of Physics/Emilio Segre Visual Archives, Physics Today Collection

JOHANN HEINRICH LAMBERT (1728–1777)

The first person to publish a comprehensive study on hyperbolic functions was Johann Heinrich Lambert, a Swiss-German mathematician and colleague of Euler.

Hyperbolic Functions

In this section you will look briefly at a special class of exponential functions called **hyperbolic functions.** The name *hyperbolic function* arose from comparison of the area of a semicircular region, as shown in Figure 5.35, with the area of a region under a hyperbola, as shown in Figure 5.36. The integral for the semicircular region involves an inverse trigonometric (circular) function:

$$\int_{-1}^{1} \sqrt{1 - x^2}\, dx = \frac{1}{2}\left[x\sqrt{1 - x^2} + \arcsin x\right]_{-1}^{1} = \frac{\pi}{2} \approx 1.571.$$

The integral for the hyperbolic region involves an inverse hyperbolic function:

$$\int_{-1}^{1} \sqrt{1 + x^2}\, dx = \frac{1}{2}\left[x\sqrt{1 + x^2} + \sinh^{-1}x\right]_{-1}^{1} \approx 2.296.$$

This is only one of many ways in which the hyperbolic functions are similar to the trigonometric functions.

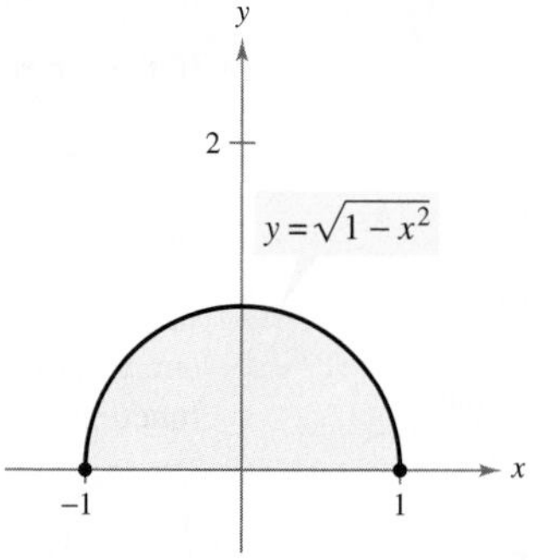

Circle: $x^2 + y^2 = 1$
Figure 5.35

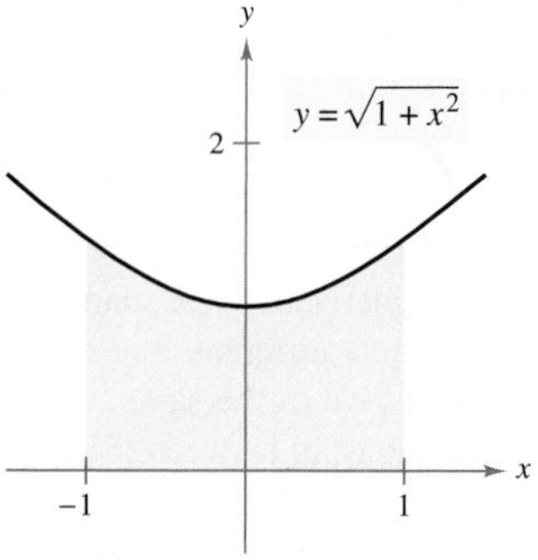

Hyperbola: $-x^2 + y^2 = 1$
Figure 5.36

FOR FURTHER INFORMATION For more information on the development of hyperbolic functions, see the article "An Introduction to Hyperbolic Functions in Elementary Calculus" by Jerome Rosenthal in *Mathematics Teacher*. To view this article, go to the website *www.matharticles.com.*

Definitions of the Hyperbolic Functions

$$\sinh x = \frac{e^x - e^{-x}}{2} \qquad \operatorname{csch} x = \frac{1}{\sinh x}, \quad x \neq 0$$

$$\cosh x = \frac{e^x + e^{-x}}{2} \qquad \operatorname{sech} x = \frac{1}{\cosh x}$$

$$\tanh x = \frac{\sinh x}{\cosh x} \qquad \coth x = \frac{1}{\tanh x}, \quad x \neq 0$$

NOTE $\sinh x$ is read as "the hyperbolic sine of x," $\cosh x$ as "the hyperbolic cosine of x," and so on.

The graphs of the six hyperbolic functions and their domains and ranges are shown in Figure 5.37. Note that the graph of $\sinh x$ can be obtained by *addition of ordinates* using the exponential functions $f(x) = \frac{1}{2}e^x$ and $g(x) = -\frac{1}{2}e^{-x}$. Likewise, the graph of $\cosh x$ can be obtained by *addition of ordinates* using the exponential functions $f(x) = \frac{1}{2}e^x$ and $h(x) = \frac{1}{2}e^{-x}$.

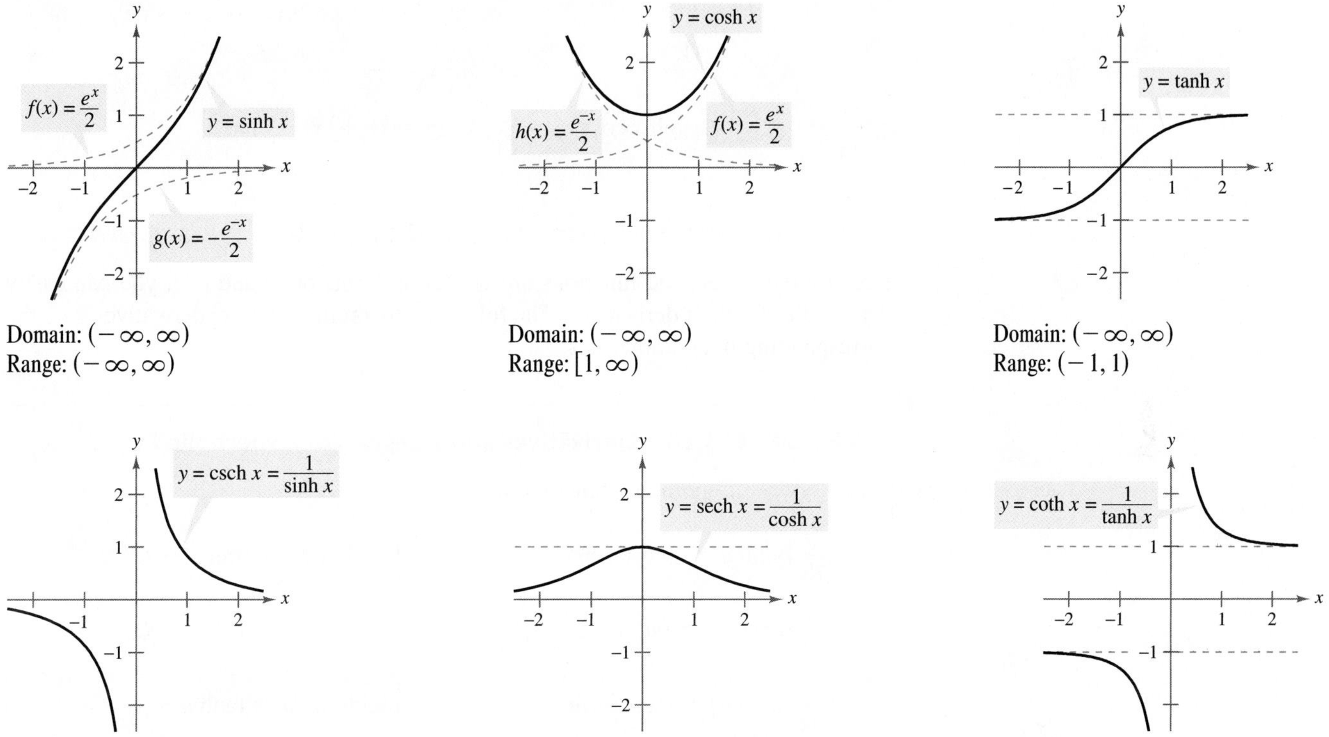

Domain: $(-\infty, \infty)$
Range: $(-\infty, \infty)$

Domain: $(-\infty, \infty)$
Range: $[1, \infty)$

Domain: $(-\infty, \infty)$
Range: $(-1, 1)$

Domain: $(-\infty, 0) \cup (0, \infty)$
Range: $(-\infty, 0) \cup (0, \infty)$

Domain: $(-\infty, \infty)$
Range: $(0, 1]$

Domain: $(-\infty, 0) \cup (0, \infty)$
Range: $(-\infty, -1) \cup (1, \infty)$

Figure 5.37

Many of the trigonometric identities have corresponding *hyperbolic identities.* For instance,

$$\begin{aligned}
\cosh^2 x - \sinh^2 x &= \left(\frac{e^x + e^{-x}}{2}\right)^2 - \left(\frac{e^x - e^{-x}}{2}\right)^2 \\
&= \frac{e^{2x} + 2 + e^{-2x}}{4} - \frac{e^{2x} - 2 + e^{-2x}}{4} \\
&= \frac{4}{4} \\
&= 1
\end{aligned}$$

and

$$\begin{aligned}
2 \sinh x \cosh x &= 2\left(\frac{e^x - e^{-x}}{2}\right)\left(\frac{e^x + e^{-x}}{2}\right) \\
&= \frac{e^{2x} - e^{-2x}}{2} \\
&= \sinh 2x.
\end{aligned}$$

FOR FURTHER INFORMATION To understand geometrically the relationship between the hyperbolic and exponential functions, see the article "A Short Proof Linking the Hyperbolic and Exponential Functions" by Michael J. Seery in *The AMATYC Review*.

Hyperbolic Identities

$\cosh^2 x - \sinh^2 x = 1$

$\tanh^2 x + \operatorname{sech}^2 x = 1$

$\coth^2 x - \operatorname{csch}^2 x = 1$

$\sinh(x + y) = \sinh x \cosh y + \cosh x \sinh y$

$\sinh(x - y) = \sinh x \cosh y - \cosh x \sinh y$

$\cosh(x + y) = \cosh x \cosh y + \sinh x \sinh y$

$\cosh(x - y) = \cosh x \cosh y - \sinh x \sinh y$

$\sinh^2 x = \dfrac{-1 + \cosh 2x}{2}$ $\qquad$ $\cosh^2 x = \dfrac{1 + \cosh 2x}{2}$

$\sinh 2x = 2 \sinh x \cosh x$ $\qquad$ $\cosh 2x = \cosh^2 x + \sinh^2 x$

Differentiation and Integration of Hyperbolic Functions

Because the hyperbolic functions are written in terms of e^x and e^{-x}, you can easily derive rules for their derivatives. The following theorem lists these derivatives with the corresponding integration rules.

THEOREM 5.18 Derivatives and Integrals of Hyperbolic Functions

Let u be a differentiable function of x.

$$\frac{d}{dx}[\sinh u] = (\cosh u)u' \qquad \int \cosh u\, du = \sinh u + C$$

$$\frac{d}{dx}[\cosh u] = (\sinh u)u' \qquad \int \sinh u\, du = \cosh u + C$$

$$\frac{d}{dx}[\tanh u] = (\operatorname{sech}^2 u)u' \qquad \int \operatorname{sech}^2 u\, du = \tanh u + C$$

$$\frac{d}{dx}[\coth u] = -(\operatorname{csch}^2 u)u' \qquad \int \operatorname{csch}^2 u\, du = -\coth u + C$$

$$\frac{d}{dx}[\operatorname{sech} u] = -(\operatorname{sech} u \tanh u)u' \qquad \int \operatorname{sech} u \tanh u\, du = -\operatorname{sech} u + C$$

$$\frac{d}{dx}[\operatorname{csch} u] = -(\operatorname{csch} u \coth u)u' \qquad \int \operatorname{csch} u \coth u\, du = -\operatorname{csch} u + C$$

Proof

$$\begin{aligned}\frac{d}{dx}[\sinh x] &= \frac{d}{dx}\left[\frac{e^x - e^{-x}}{2}\right]\\ &= \frac{e^x + e^{-x}}{2} = \cosh x\end{aligned}$$

$$\begin{aligned}\frac{d}{dx}[\tanh x] &= \frac{d}{dx}\left[\frac{\sinh x}{\cosh x}\right]\\ &= \frac{\cosh x(\cosh x) - \sinh x(\sinh x)}{\cosh^2 x}\\ &= \frac{1}{\cosh^2 x}\\ &= \operatorname{sech}^2 x\end{aligned}$$

In Exercises 98 and 102, you are asked to prove some of the other differentiation rules.

EXAMPLE 1 Differentiation of Hyperbolic Functions

a. $\dfrac{d}{dx}[\sinh(x^2 - 3)] = 2x\cosh(x^2 - 3)$ **b.** $\dfrac{d}{dx}[\ln(\cosh x)] = \dfrac{\sinh x}{\cosh x} = \tanh x$

c. $\dfrac{d}{dx}[x \sinh x - \cosh x] = x\cosh x + \sinh x - \sinh x = x\cosh x$

EXAMPLE 2 Finding Relative Extrema

Find the relative extrema of $f(x) = (x - 1)\cosh x - \sinh x$.

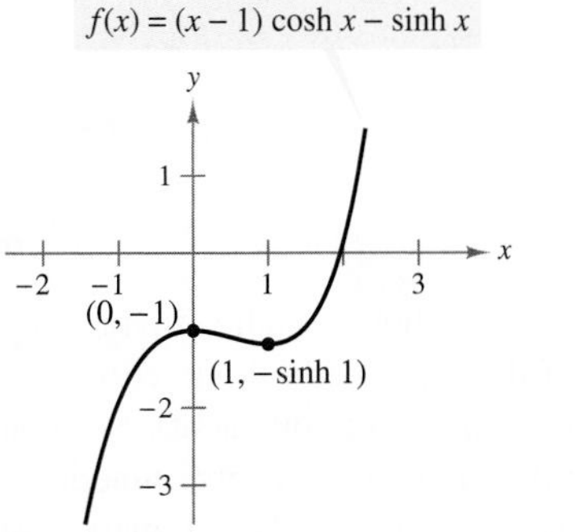

$f''(0) < 0$, so $(0, -1)$ is a relative maximum. $f''(1) > 0$, so $(1, -\sinh 1)$ is a relative minimum.

Figure 5.38

Solution Begin by setting the first derivative of f equal to 0.

$$f'(x) = (x - 1)\sinh x + \cosh x - \cosh x = 0$$
$$(x - 1)\sinh x = 0$$

So, the critical numbers are $x = 1$ and $x = 0$. Using the Second Derivative Test, you can verify that the point $(0, -1)$ yields a relative maximum and the point $(1, -\sinh 1)$ yields a relative minimum, as shown in Figure 5.38. Try using a graphing utility to confirm this result. If your graphing utility does not have hyperbolic functions, you can use exponential functions as follows.

$$\begin{aligned} f(x) &= (x - 1)\left(\tfrac{1}{2}\right)(e^x + e^{-x}) - \tfrac{1}{2}(e^x - e^{-x}) \\ &= \tfrac{1}{2}(xe^x + xe^{-x} - e^x - e^{-x} - e^x + e^{-x}) \\ &= \tfrac{1}{2}(xe^x + xe^{-x} - 2e^x) \end{aligned}$$

When a uniform flexible cable, such as a telephone wire, is suspended from two points, it takes the shape of a *catenary*, as discussed in Example 3.

EXAMPLE 3 Hanging Power Cables

Power cables are suspended between two towers, forming the catenary shown in Figure 5.39. The equation for this catenary is

$$y = a\cosh\frac{x}{a}.$$

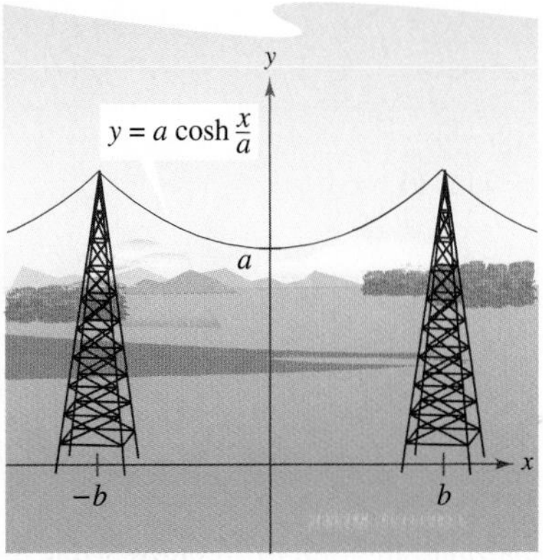

Catenary

Figure 5.39

The distance between the two towers is $2b$. Find the slope of the catenary at the point where the cable meets the right-hand tower.

Solution Differentiating produces

$$y' = a\left(\frac{1}{a}\right)\sinh\frac{x}{a} = \sinh\frac{x}{a}.$$

At the point $(b, a\cosh(b/a))$, the slope (from the left) is given by $m = \sinh\dfrac{b}{a}$.

FOR FURTHER INFORMATION In Example 3, the cable is a catenary between two supports at the same height. To learn about the shape of a cable hanging between supports of different heights, see the article "Reexamining the Catenary" by Paul Cella in *The College Mathematics Journal*. To view this article, go to the website *www.matharticles.com*.

EXAMPLE 4 Integrating a Hyperbolic Function

Find $\int \cosh 2x \sinh^2 2x\, dx$.

Solution

$$\int \cosh 2x \sinh^2 2x\, dx = \frac{1}{2}\int (\sinh 2x)^2(2\cosh 2x)\, dx \qquad u = \sinh 2x$$
$$= \frac{1}{2}\left[\frac{(\sinh 2x)^3}{3}\right] + C$$
$$= \frac{\sinh^3 2x}{6} + C$$

Inverse Hyperbolic Functions

Unlike trigonometric functions, hyperbolic functions are not periodic. In fact, by looking back at Figure 5.37, you can see that four of the six hyperbolic functions are actually one-to-one (the hyperbolic sine, tangent, cosecant, and cotangent). So, you can apply Theorem 5.7 to conclude that these four functions have inverse functions. The other two (the hyperbolic cosine and secant) are one-to-one if their domains are restricted to the positive real numbers, and for this restricted domain they also have inverse functions. Because the hyperbolic functions are defined in terms of exponential functions, it is not surprising to find that the inverse hyperbolic functions can be written in terms of logarithmic functions, as shown in Theorem 5.19.

THEOREM 5.19 Inverse Hyperbolic Functions

Function	*Domain*
$\sinh^{-1} x = \ln\left(x + \sqrt{x^2 + 1}\right)$	$(-\infty, \infty)$
$\cosh^{-1} x = \ln\left(x + \sqrt{x^2 - 1}\right)$	$[1, \infty)$
$\tanh^{-1} x = \frac{1}{2} \ln \frac{1 + x}{1 - x}$	$(-1, 1)$
$\coth^{-1} x = \frac{1}{2} \ln \frac{x + 1}{x - 1}$	$(-\infty, -1) \cup (1, \infty)$
$\operatorname{sech}^{-1} x = \ln \frac{1 + \sqrt{1 - x^2}}{x}$	$(0, 1]$
$\operatorname{csch}^{-1} x = \ln\left(\frac{1}{x} + \frac{\sqrt{1 + x^2}}{\lvert x\rvert}\right)$	$(-\infty, 0) \cup (0, \infty)$

Proof The proof of this theorem is a straightforward application of the properties of the exponential and logarithmic functions. For example, if

$$f(x) = \sinh x = \frac{e^x - e^{-x}}{2}$$

and

$$g(x) = \ln\left(x + \sqrt{x^2 + 1}\right)$$

you can show that $f(g(x)) = x$ and $g(f(x)) = x$, which implies that g is the inverse function of f.

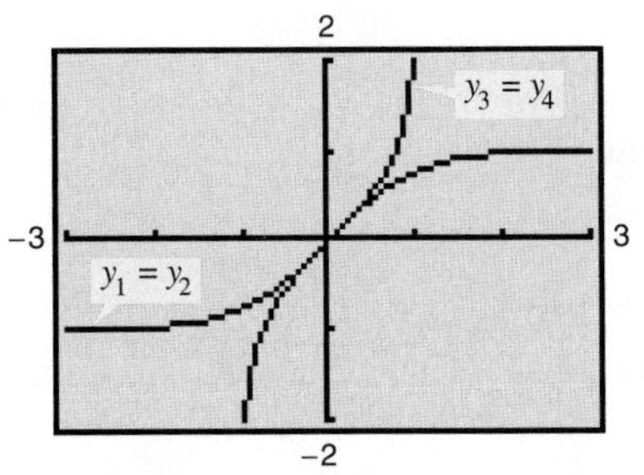

Graphs of the hyperbolic tangent function and the inverse hyperbolic tangent function
Figure 5.40

TECHNOLOGY You can use a graphing utility to confirm graphically the results of Theorem 5.19. For instance, graph the following functions.

$$y_1 = \tanh x \qquad \text{Hyperbolic tangent}$$

$$y_2 = \frac{e^x - e^{-x}}{e^x + e^{-x}} \qquad \text{Definition of hyperbolic tangent}$$

$$y_3 = \tanh^{-1} x \qquad \text{Inverse hyperbolic tangent}$$

$$y_4 = \frac{1}{2}\ln\frac{1+x}{1-x} \qquad \text{Definition of inverse hyperbolic tangent}$$

The resulting display is shown in Figure 5.40. As you watch the graphs being traced out, notice that $y_1 = y_2$ and $y_3 = y_4$. Also notice that the graph of y_1 is the reflection of the graph of y_3 in the line $y = x$.

The graphs of the inverse hyperbolic functions are shown in Figure 5.41.

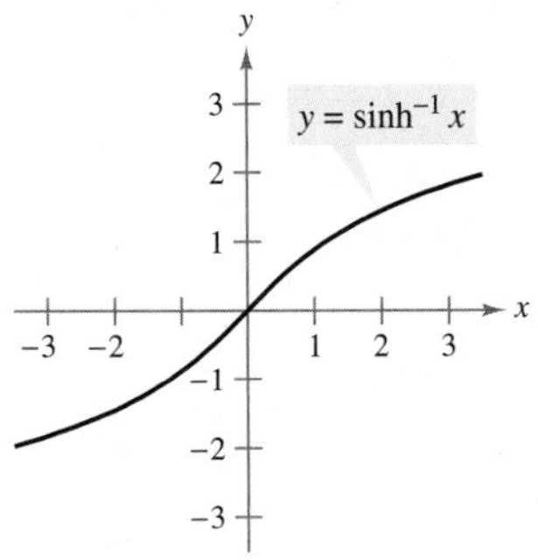

Domain: $(-\infty, \infty)$
Range: $(-\infty, \infty)$

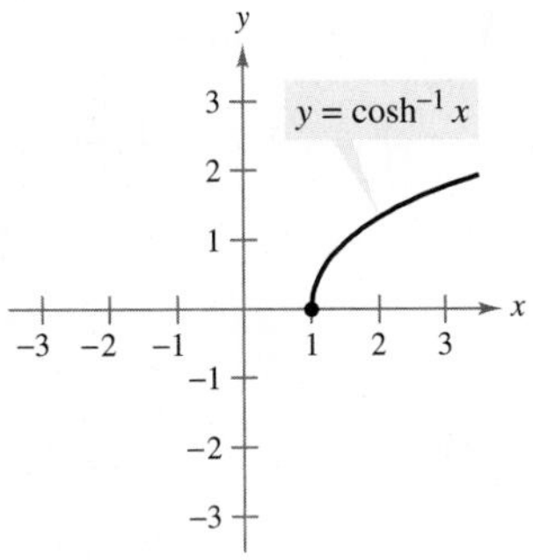

Domain: $[1, \infty)$
Range: $[0, \infty)$

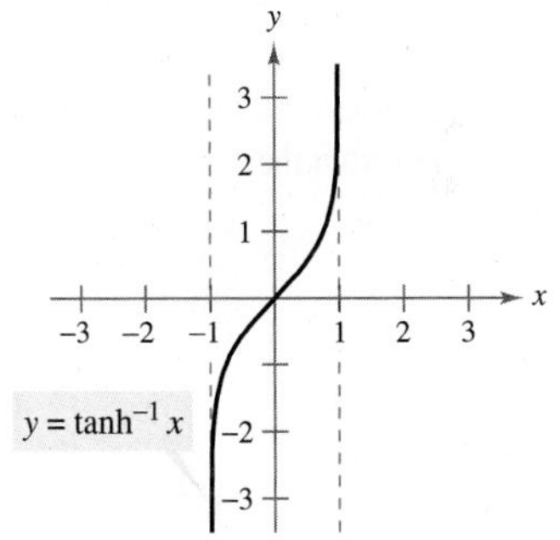

Domain: $(-1, 1)$
Range: $(-\infty, \infty)$

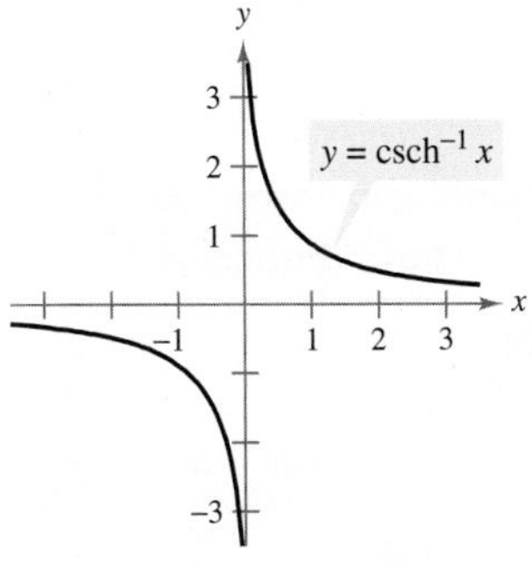

Domain: $(-\infty, 0) \cup (0, \infty)$
Range: $(-\infty, 0) \cup (0, \infty)$

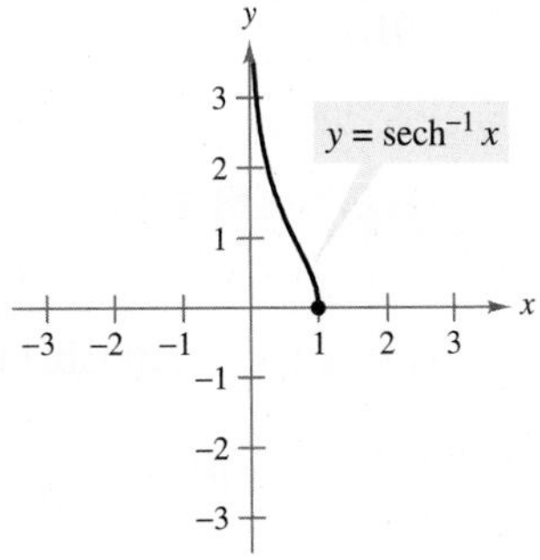

Domain: $(0, 1]$
Range: $[0, \infty)$

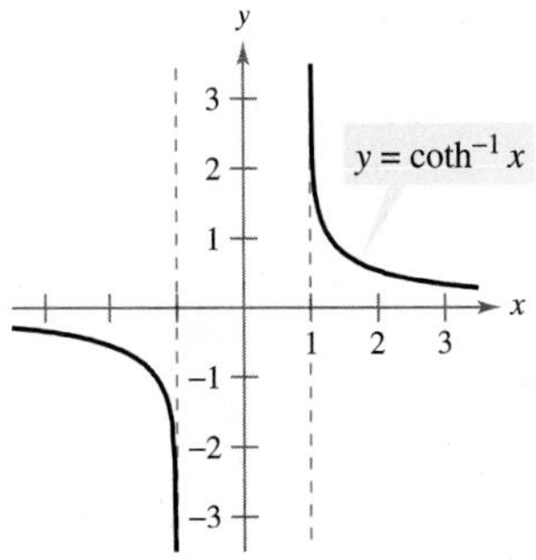

Domain: $(-\infty, -1) \cup (1, \infty)$
Range: $(-\infty, 0) \cup (0, \infty)$

Figure 5.41

The inverse hyperbolic secant can be used to define a curve called a *tractrix* or *pursuit curve*, as discussed in Example 5.

EXAMPLE 5 A Tractrix

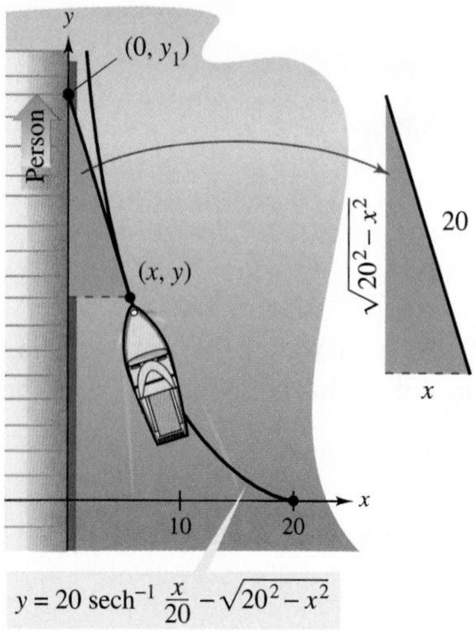

A person must walk 41.27 feet to bring the boat 5 feet from the dock.
Figure 5.42

A person is holding a rope that is tied to a boat, as shown in Figure 5.42. As the person walks along the dock, the boat travels along a **tractrix,** given by the equation

$$y = a \operatorname{sech}^{-1}\frac{x}{a} - \sqrt{a^2 - x^2}$$

where a is the length of the rope. If $a = 20$ feet, find the distance the person must walk to bring the boat 5 feet from the dock.

Solution In Figure 5.42, notice that the distance the person has walked is given by

$$\begin{aligned} y_1 = y + \sqrt{20^2 - x^2} &= \left(20 \operatorname{sech}^{-1}\frac{x}{20} - \sqrt{20^2 - x^2}\right) + \sqrt{20^2 - x^2} \\ &= 20 \operatorname{sech}^{-1}\frac{x}{20}. \end{aligned}$$

When $x = 5$, this distance is

$$\begin{aligned} y_1 = 20 \operatorname{sech}^{-1}\frac{5}{20} &= 20 \ln \frac{1 + \sqrt{1 - (1/4)^2}}{1/4} \\ &= 20 \ln\left(4 + \sqrt{15}\right) \\ &\approx 41.27 \text{ feet.} \end{aligned}$$

Differentiation and Integration of Inverse Hyperbolic Functions

The derivatives of the inverse hyperbolic functions, which resemble the derivatives of the inverse trigonometric functions, are listed in Theorem 5.20 with the corresponding integration formulas (in logarithmic form). You can verify each of these formulas by applying the logarithmic definitions of the inverse hyperbolic functions. (See Exercises 99–101.)

THEOREM 5.20 Differentiation and Integration Involving Inverse Hyperbolic Functions

Let u be a differentiable function of x.

$$\frac{d}{dx}[\sinh^{-1} u] = \frac{u'}{\sqrt{u^2 + 1}} \qquad \frac{d}{dx}[\cosh^{-1} u] = \frac{u'}{\sqrt{u^2 - 1}}$$

$$\frac{d}{dx}[\tanh^{-1} u] = \frac{u'}{1 - u^2} \qquad \frac{d}{dx}[\coth^{-1} u] = \frac{u'}{1 - u^2}$$

$$\frac{d}{dx}[\operatorname{sech}^{-1} u] = \frac{-u'}{u\sqrt{1 - u^2}} \qquad \frac{d}{dx}[\operatorname{csch}^{-1} u] = \frac{-u'}{|u|\sqrt{1 + u^2}}$$

$$\int \frac{du}{\sqrt{u^2 \pm a^2}} = \ln\left(u + \sqrt{u^2 \pm a^2}\right) + C$$

$$\int \frac{du}{a^2 - u^2} = \frac{1}{2a} \ln\left|\frac{a + u}{a - u}\right| + C$$

$$\int \frac{du}{u\sqrt{a^2 \pm u^2}} = -\frac{1}{a} \ln \frac{a + \sqrt{a^2 \pm u^2}}{|u|} + C$$

EXAMPLE 6 More About a Tractrix

For the tractrix given in Example 5, show that the boat is always pointing toward the person.

Solution For a point (x, y) on a tractrix, the slope of the graph gives the direction of the boat, as shown in Figure 5.42.

$$\begin{aligned}
y' &= \frac{d}{dx}\left[20 \operatorname{sech}^{-1} \frac{x}{20} - \sqrt{20^2 - x^2}\right] \\
&= -20\left(\frac{1}{20}\right)\left[\frac{1}{(x/20)\sqrt{1 - (x/20)^2}}\right] - \left(\frac{1}{2}\right)\left(\frac{-2x}{\sqrt{20^2 - x^2}}\right) \\
&= \frac{-20^2}{x\sqrt{20^2 - x^2}} + \frac{x}{\sqrt{20^2 - x^2}} \\
&= -\frac{\sqrt{20^2 - x^2}}{x}
\end{aligned}$$

However, from Figure 5.42, you can see that the slope of the line segment connecting the point $(0, y_1)$ with the point (x, y) is also

$$m = -\frac{\sqrt{20^2 - x^2}}{x}.$$

So, the boat is always pointing toward the person. (It is because of this property that a tractrix is called a *pursuit curve*.)

EXAMPLE 7 Integration Using Inverse Hyperbolic Functions

Find $\displaystyle\int \frac{dx}{x\sqrt{4 - 9x^2}}$.

Solution Let $a = 2$ and $u = 3x$.

$$\begin{aligned}
\int \frac{dx}{x\sqrt{4 - 9x^2}} &= \int \frac{3\,dx}{(3x)\sqrt{4 - 9x^2}} & \int \frac{du}{u\sqrt{a^2 - u^2}} \\
&= -\frac{1}{2} \ln \frac{2 + \sqrt{4 - 9x^2}}{|3x|} + C & -\frac{1}{a} \ln \frac{a + \sqrt{a^2 - u^2}}{|u|} + C
\end{aligned}$$

EXAMPLE 8 Integration Using Inverse Hyperbolic Functions

Find $\displaystyle\int \frac{dx}{5 - 4x^2}$.

Solution Let $a = \sqrt{5}$ and $u = 2x$.

$$\begin{aligned}
\int \frac{dx}{5 - 4x^2} &= \frac{1}{2} \int \frac{2\,dx}{(\sqrt{5})^2 - (2x)^2} & \int \frac{du}{a^2 - u^2} \\
&= \frac{1}{2}\left(\frac{1}{2\sqrt{5}} \ln\left|\frac{\sqrt{5} + 2x}{\sqrt{5} - 2x}\right|\right) + C & \frac{1}{2a} \ln\left|\frac{a + u}{a - u}\right| + C \\
&= \frac{1}{4\sqrt{5}} \ln\left|\frac{\sqrt{5} + 2x}{\sqrt{5} - 2x}\right| + C
\end{aligned}$$

Exercises for Section 5.8

See www.CalcChat.com for worked-out solutions to odd-numbered exercises.

In Exercises 1–6, evaluate the function. If the value is not a rational number, give the answer to three-decimal-place accuracy.

1. (a) $\sinh 3$
(b) $\tanh(-2)$

2. (a) $\cosh 0$
(b) $\operatorname{sech} 1$

3. (a) $\operatorname{csch}(\ln 2)$
(b) $\coth(\ln 5)$

4. (a) $\sinh^{-1} 0$
(b) $\tanh^{-1} 0$

5. (a) $\cosh^{-1} 2$
(b) $\operatorname{sech}^{-1} \frac{2}{3}$

6. (a) $\operatorname{csch}^{-1} 2$
(b) $\coth^{-1} 3$

In Exercises 7–12, verify the identity.

7. $\tanh^2 x + \operatorname{sech}^2 x = 1$

8. $\cosh^2 x = \dfrac{1 + \cosh 2x}{2}$

9. $\sinh(x + y) = \sinh x \cosh y + \cosh x \sinh y$

10. $\sinh 2x = 2 \sinh x \cosh x$

11. $\sinh 3x = 3 \sinh x + 4 \sinh^3 x$

12. $\cosh x + \cosh y = 2 \cosh \dfrac{x + y}{2} \cosh \dfrac{x - y}{2}$

In Exercises 13 and 14, use the value of the given hyperbolic function to find the values of the other hyperbolic functions at x.

13. $\sinh x = \dfrac{3}{2}$

14. $\tanh x = \dfrac{1}{2}$

In Exercises 15–24, find the derivative of the function.

15. $y = \operatorname{sech}(x + 1)$

16. $y = \coth 3x$

17. $f(x) = \ln(\sinh x)$

18. $g(x) = \ln(\cosh x)$

19. $y = \ln\left(\tanh \dfrac{x}{2}\right)$

20. $y = x \cosh x - \sinh x$

21. $h(x) = \dfrac{1}{4} \sinh 2x - \dfrac{x}{2}$

22. $h(t) = t - \coth t$

23. $f(t) = \arctan(\sinh t)$

24. $g(x) = \operatorname{sech}^2 3x$

In Exercises 25–28, find an equation of the tangent line to the graph of the function at the given point.

25. $y = \sinh(1 - x^2)$

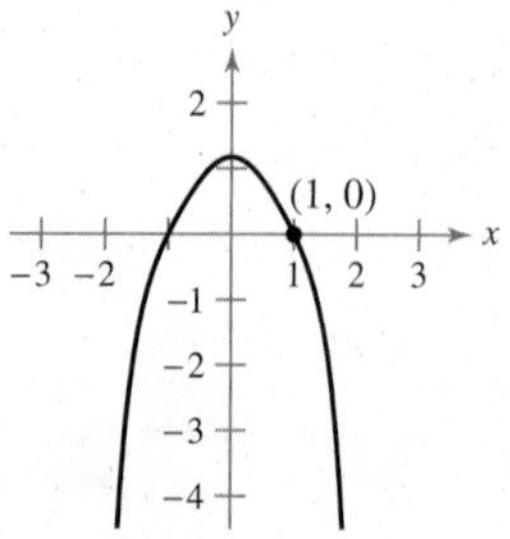

26. $y = x^{\cosh x}$

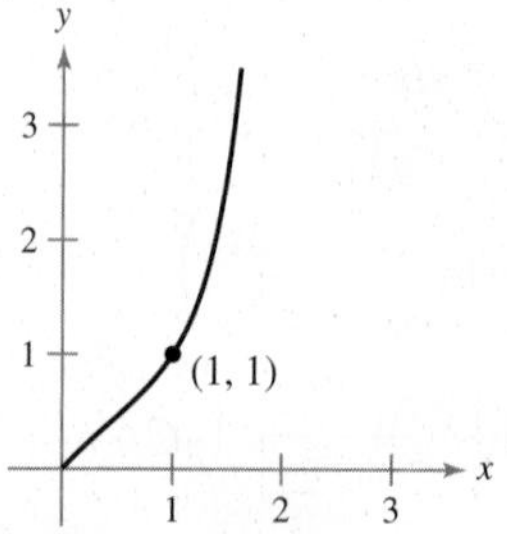

27. $y = (\cosh x - \sinh x)^2$

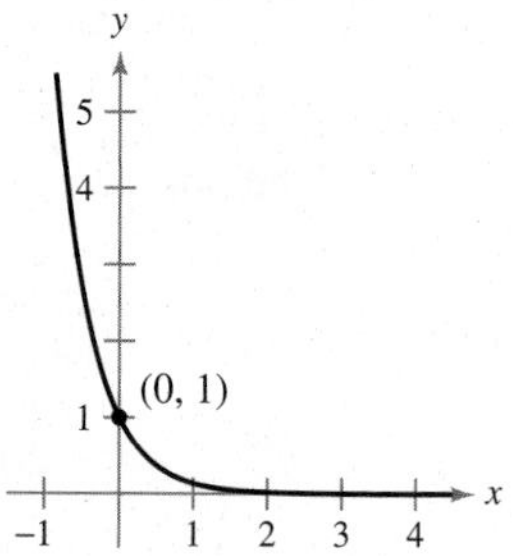

28. $y = e^{\sinh x}$

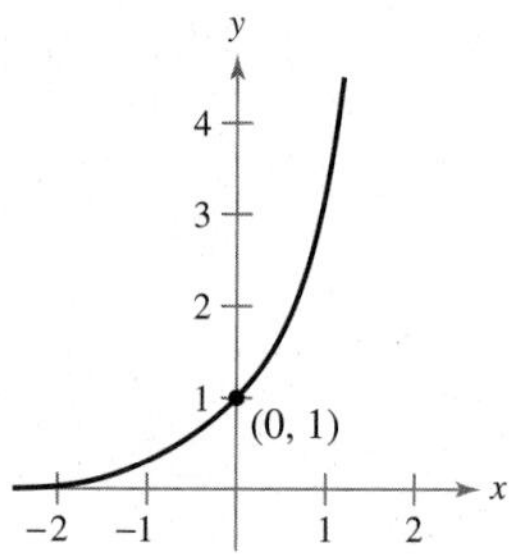

In Exercises 29–32, find any relative extrema of the function. Use a graphing utility to confirm your result.

29. $f(x) = \sin x \sinh x - \cos x \cosh x, \quad -4 \le x \le 4$

30. $f(x) = x \sinh(x - 1) - \cosh(x - 1)$

31. $g(x) = x \operatorname{sech} x$

32. $h(x) = 2 \tanh x - x$

In Exercises 33 and 34, show that the function satisfies the differential equation.

Function	*Differential Equation*
33. $y = a \sinh x$	$y''' - y' = 0$
34. $y = a \cosh x$	$y'' - y = 0$

Linear and Quadratic Approximations **In Exercises 35 and 36, use a computer algebra system to find the linear approximation**

$$P_1(x) = f(a) + f'(a)(x - a)$$

and the quadratic approximation

$$P_2(x) = f(a) + f'(a)(x - a) + \tfrac{1}{2}f''(a)(x - a)^2$$

of the function f at $x = a$. Use a graphing utility to graph the function and its linear and quadratic approximations.

35. $f(x) = \tanh x, \quad a = 0$

36. $f(x) = \cosh x, \quad a = 0$

Catenary **In Exercises 37 and 38, a model for a power cable suspended between two towers is given. (a) Graph the model, (b) find the heights of the cable at the towers and at the midpoint between the towers, and (c) find the slope of the model at the point where the cable meets the right-hand tower.**

37. $y = 10 + 15 \cosh \dfrac{x}{15}, \quad -15 \le x \le 15$

38. $y = 18 + 25 \cosh \dfrac{x}{25}, \quad -25 \le x \le 25$

In Exercises 39–50, find the integral.

39. $\displaystyle\int \sinh(1 - 2x)\, dx$

40. $\displaystyle\int \frac{\cosh \sqrt{x}}{\sqrt{x}}\, dx$

41. $\displaystyle\int \cosh^2(x - 1) \sinh(x - 1)\, dx$

42. $\displaystyle\int \frac{\sinh x}{1 + \sinh^2 x}\, dx$

43. $\displaystyle\int \frac{\cosh x}{\sinh x}\,dx$

44. $\displaystyle\int \operatorname{sech}^2(2x - 1)\,dx$

45. $\displaystyle\int x \operatorname{csch}^2 \frac{x^2}{2}\,dx$

46. $\displaystyle\int \operatorname{sech}^3 x \tanh x\,dx$

47. $\displaystyle\int \frac{\operatorname{csch}(1/x)\coth(1/x)}{x^2}\,dx$

48. $\displaystyle\int \frac{\cosh x}{\sqrt{9 - \sinh^2 x}}\,dx$

49. $\displaystyle\int \frac{x}{x^4 + 1}\,dx$

50. $\displaystyle\int \frac{2}{x\sqrt{1 + 4x^2}}\,dx$

In Exercises 51–56, evaluate the integral.

51. $\displaystyle\int_0^{\ln 2} \tanh x\,dx$

52. $\displaystyle\int_0^1 \cosh^2 x\,dx$

53. $\displaystyle\int_0^4 \frac{1}{25 - x^2}\,dx$

54. $\displaystyle\int_0^4 \frac{1}{\sqrt{25 - x^2}}\,dx$

55. $\displaystyle\int_0^{\sqrt{2}/4} \frac{2}{\sqrt{1 - 4x^2}}\,dx$

56. $\displaystyle\int_0^{\ln 2} 2e^{-x}\cosh x\,dx$

In Exercises 57–64, find the derivative of the function.

57. $y = \cosh^{-1}(3x)$

58. $y = \tanh^{-1}\dfrac{x}{2}$

59. $y = \sinh^{-1}(\tan x)$

60. $y = \operatorname{sech}^{-1}(\cos 2x), \quad 0 < x < \pi/4$

61. $y = \tanh^{-1}(\sin 2x)$

62. $y = (\operatorname{csch}^{-1} x)^2$

63. $y = 2x \sinh^{-1}(2x) - \sqrt{1 + 4x^2}$

64. $y = x \tanh^{-1} x + \ln\sqrt{1 - x^2}$

Writing About Concepts

65. Discuss several ways in which the hyperbolic functions are similar to the trigonometric functions.

66. Sketch the graph of each hyperbolic function. Then identify the domain and range of each function.

Limits **In Exercises 67–72, find the limit.**

67. $\displaystyle\lim_{x\to\infty} \sinh x$

68. $\displaystyle\lim_{x\to\infty} \tanh x$

69. $\displaystyle\lim_{x\to\infty} \operatorname{sech} x$

70. $\displaystyle\lim_{x\to-\infty} \operatorname{csch} x$

71. $\displaystyle\lim_{x\to 0} \frac{\sinh x}{x}$

72. $\displaystyle\lim_{x\to 0^-} \coth x$

In Exercises 73–80, find the indefinite integral using the formulas of Theorem 5.20.

73. $\displaystyle\int \frac{1}{\sqrt{1 + e^{2x}}}\,dx$

74. $\displaystyle\int \frac{x}{9 - x^4}\,dx$

75. $\displaystyle\int \frac{1}{\sqrt{x}\sqrt{1 + x}}\,dx$

76. $\displaystyle\int \frac{\sqrt{x}}{\sqrt{1 + x^3}}\,dx$

77. $\displaystyle\int \frac{-1}{4x - x^2}\,dx$

78. $\displaystyle\int \frac{dx}{(x + 2)\sqrt{x^2 + 4x + 8}}$

79. $\displaystyle\int \frac{1}{1 - 4x - 2x^2}\,dx$

80. $\displaystyle\int \frac{dx}{(x + 1)\sqrt{2x^2 + 4x + 8}}$

In Exercises 81–84, solve the differential equation.

81. $\dfrac{dy}{dx} = \dfrac{1}{\sqrt{80 + 8x - 16x^2}}$

82. $\dfrac{dy}{dx} = \dfrac{1}{(x - 1)\sqrt{-4x^2 + 8x - 1}}$

83. $\dfrac{dy}{dx} = \dfrac{x^3 - 21x}{5 + 4x - x^2}$

84. $\dfrac{dy}{dx} = \dfrac{1 - 2x}{4x - x^2}$

Area **In Exercises 85–88, find the area of the region.**

85. $y = \operatorname{sech}\dfrac{x}{2}$

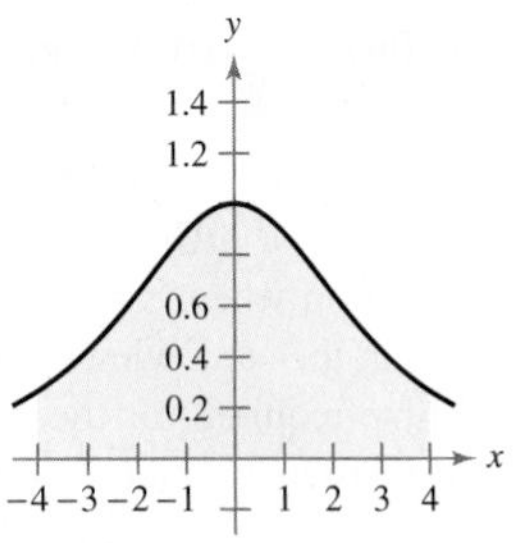

86. $y = \tanh 2x$

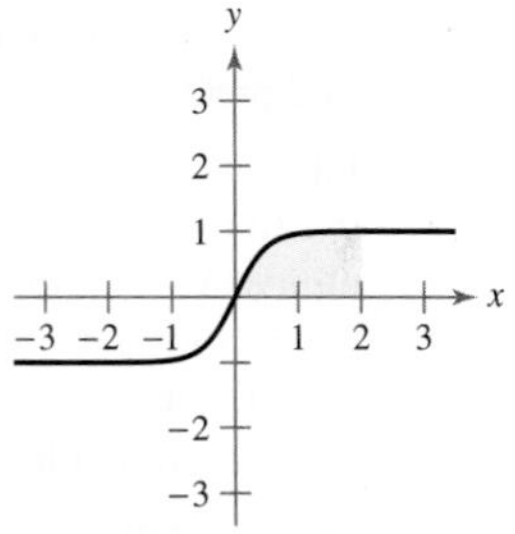

87. $y = \dfrac{5x}{\sqrt{x^4 + 1}}$

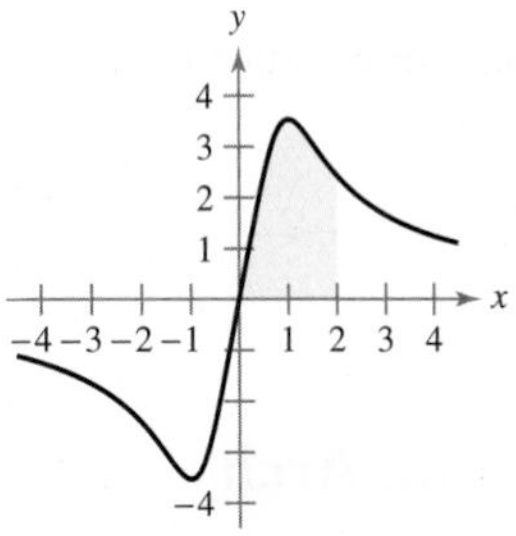

88. $y = \dfrac{6}{\sqrt{x^2 - 4}}$

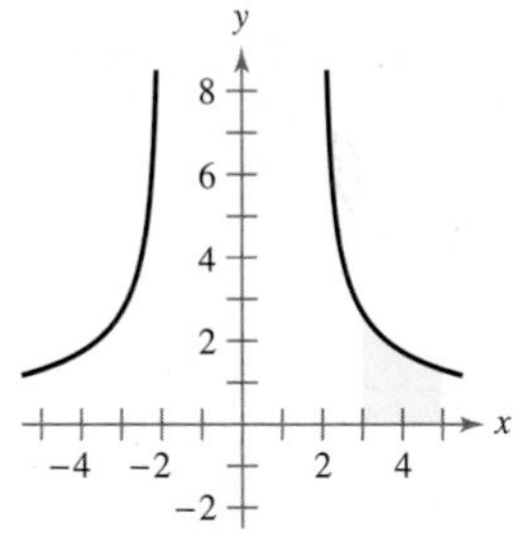

In Exercises 89 and 90, evaluate the integral in terms of (a) natural logarithms and (b) inverse hyperbolic functions.

89. $\displaystyle\int_0^{\sqrt{3}} \frac{dx}{\sqrt{x^2 + 1}}$

90. $\displaystyle\int_{-1/2}^{1/2} \frac{dx}{1 - x^2}$

91. ***Chemical Reactions*** Chemicals A and B combine in a 3-to-1 ratio to form a compound. The amount of compound x being produced at any time t is proportional to the unchanged amounts of A and B remaining in the solution. So, if 3 kilograms of A is mixed with 2 kilograms of B, you have

$$\frac{dx}{dt} = k\left(3 - \frac{3x}{4}\right)\left(2 - \frac{x}{4}\right) = \frac{3k}{16}(x^2 - 12x + 32).$$

One kilogram of the compound is formed after 10 minutes. Find the amount formed after 20 minutes by solving the equation

$$\int \frac{3k}{16}\,dt = \int \frac{dx}{x^2 - 12x + 32}.$$

92. ***Vertical Motion*** An object is dropped from a height of 400 feet.

(a) Find the velocity of the object as a function of time (neglect air resistance on the object).

(b) Use the result in part (a) to find the position function.

(c) If the air resistance is proportional to the square of the velocity, then $dv/dt = -32 + kv^2$, where -32 feet per second per second is the acceleration due to gravity and k is a constant. Show that the velocity v as a function of time is

$$v(t) = -\sqrt{\frac{32}{k}}\tanh\left(\sqrt{32k}\,t\right)$$

by performing the following integration and simplifying the result.

$$\int \frac{dv}{32 - kv^2} = -\int dt$$

(d) Use the result in part (c) to find $\lim_{t\to\infty} v(t)$ and give its interpretation.

(e) Integrate the velocity function in part (c) and find the position s of the object as a function of t. Use a graphing utility to graph the position function when $k = 0.01$ and the position function in part (b) in the same viewing window. Estimate the additional time required for the object to reach ground level when air resistance is not neglected.

(f) Give a written description of what you believe would happen if k were increased. Then test your assertion with a particular value of k.

Tractrix **In Exercises 93 and 94, use the equation of the tractrix** $y = a \operatorname{sech}^{-1}\frac{x}{a} - \sqrt{a^2 - x^2}$, $a > 0$.

93. Find dy/dx.

94. Let L be the tangent line to the tractrix at the point P. If L intersects the y-axis at the point Q, show that the distance between P and Q is a.

95. Prove that $\tanh^{-1} x = \frac{1}{2}\ln\left(\frac{1+x}{1-x}\right)$, $\quad -1 < x < 1$.

96. Show that $\arctan(\sinh x) = \arcsin(\tanh x)$.

97. Let $x > 0$ and $b > 0$. Show that $\displaystyle\int_{-b}^{b} e^{xt}\,dt = \frac{2\sinh bx}{x}$.

In Exercises 98–102, verify the differentiation formula.

98. $\frac{d}{dx}[\cosh x] = \sinh x$

99. $\frac{d}{dx}[\operatorname{sech}^{-1} x] = \frac{-1}{x\sqrt{1-x^2}}$

100. $\frac{d}{dx}[\cosh^{-1} x] = \frac{1}{\sqrt{x^2-1}}$

101. $\frac{d}{dx}[\sinh^{-1} x] = \frac{1}{\sqrt{x^2+1}}$

102. $\frac{d}{dx}[\operatorname{sech} x] = -\operatorname{sech} x \tanh x$

Putnam Exam Challenge

103. From the vertex $(0, c)$ of the catenary $y = c\cosh(x/c)$ a line L is drawn perpendicular to the tangent to the catenary at a point P. Prove that the length of L intercepted by the axes is equal to the ordinate y of the point P.

104. Prove or disprove that there is at least one straight line normal to the graph of $y = \cosh x$ at a point $(a, \cosh a)$ and also normal to the graph of $y = \sinh x$ at a point $(c, \sinh c)$.

[At a point on a graph, the normal line is the perpendicular to the tangent at that point. Also, $\cosh x = (e^x + e^{-x})/2$ and $\sinh x = (e^x - e^{-x})/2$.]

These problems were composed by the Committee on the Putnam Prize Competition.

Section Project: St. Louis Arch

Roy Ooms/Masterfile

The Gateway Arch in St. Louis, Missouri was constructed using the hyperbolic cosine function. The equation used for construction was

$y = 693.8597 - 68.7672 \cosh 0.0100333x,$

$-299.2239 \le x \le 299.2239$

where x and y are measured in feet. Cross sections of the arch are equilateral triangles, and (x, y) traces the path of the centers of mass of the cross-sectional triangles. For each value of x, the area of the cross-sectional triangle is $A = 125.1406 \cosh 0.0100333x$. (*Source:* Owner's Manual for the Gateway Arch, Saint Louis, MO, by William Thayer)

(a) How high above the ground is the center of the highest triangle? (At ground level, $y = 0$.)

(b) What is the height of the arch? (*Hint:* For an equilateral triangle, $A = \sqrt{3}c^2$, where c is one-half the base of the triangle, and the center of mass of the triangle is located at two-thirds the height of the triangle.)

(c) How wide is the arch at ground level?

Review Exercises for Chapter 5

See www.CalcChat.com for worked-out solutions to odd-numbered exercises.

In Exercises 1 and 2, sketch the graph of the function by hand. Identify any asymptotes of the graph.

1. $f(x) = \ln x + 3$

2. $f(x) = \ln(x - 3)$

In Exercises 3 and 4, use the properties of logarithms to expand the logarithmic function.

3. $\ln \sqrt[5]{\dfrac{4x^2 - 1}{4x^2 + 1}}$

4. $\ln[(x^2 + 1)(x - 1)]$

In Exercises 5 and 6, write the expression as the logarithm of a single quantity.

5. $\ln 3 + \frac{1}{3}\ln(4 - x^2) - \ln x$

6. $3[\ln x - 2\ln(x^2 + 1)] + 2\ln 5$

In Exercises 7 and 8, solve the equation for x.

7. $\ln \sqrt{x + 1} = 2$

8. $\ln x + \ln(x - 3) = 0$

In Exercises 9–14, find the derivative of the function.

9. $g(x) = \ln \sqrt{x}$

10. $h(x) = \ln \dfrac{x(x - 1)}{x - 2}$

11. $f(x) = x\sqrt{\ln x}$

12. $f(x) = \ln[x(x^2 - 2)^{2/3}]$

13. $y = \dfrac{1}{b^2}[a + bx - a\ln(a + bx)]$

14. $y = -\dfrac{1}{ax} + \dfrac{b}{a^2}\ln\dfrac{a + bx}{x}$

In Exercises 15 and 16, find an equation of the tangent line to the graph of the function at the given point.

15. $y = \ln(2 + x) + \dfrac{2}{2 + x}$

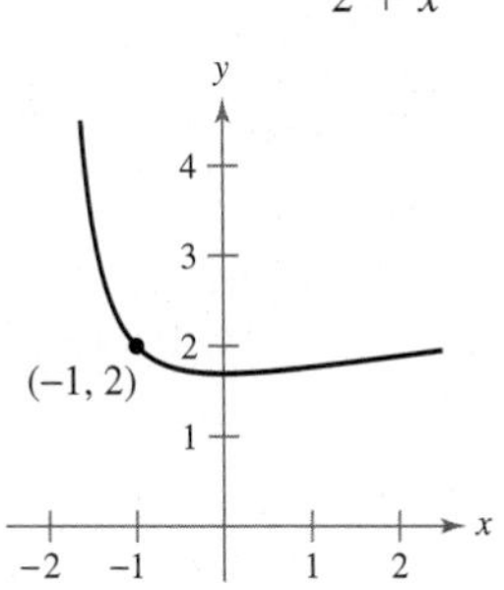

16. $y = \ln \dfrac{1 + x}{x}$

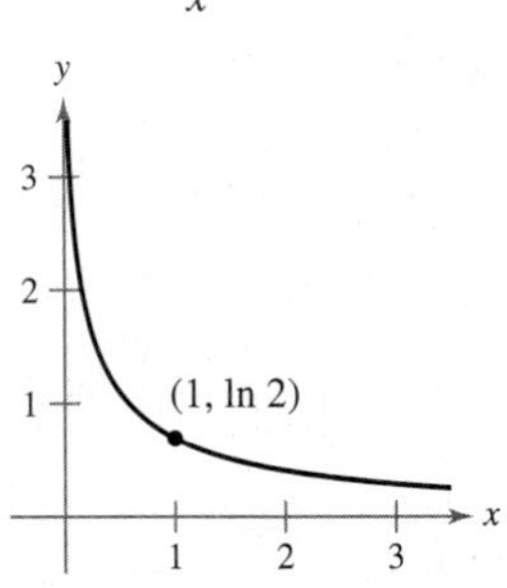

In Exercises 17–24, find or evaluate the integral.

17. $\displaystyle\int \frac{1}{7x - 2}\,dx$

18. $\displaystyle\int \frac{x}{x^2 - 1}\,dx$

19. $\displaystyle\int \frac{\sin x}{1 + \cos x}\,dx$

20. $\displaystyle\int \frac{\ln \sqrt{x}}{x}\,dx$

21. $\displaystyle\int_1^4 \frac{x + 1}{x}\,dx$

22. $\displaystyle\int_1^e \frac{\ln x}{x}\,dx$

23. $\displaystyle\int_0^{\pi/3} \sec\theta\,d\theta$

24. $\displaystyle\int_0^{\pi/4} \tan\left(\frac{\pi}{4} - x\right)dx$

In Exercises 25–30, (a) find the inverse function of f, (b) use a graphing utility to graph f and f^{-1} in the same viewing window, and (c) verify that $f^{-1}(f(x)) = x$ and $f(f^{-1}(x)) = x$.

25. $f(x) = \frac{1}{2}x - 3$

26. $f(x) = 5x - 7$

27. $f(x) = \sqrt{x + 1}$

28. $f(x) = x^3 + 2$

29. $f(x) = \sqrt[3]{x + 1}$

30. $f(x) = x^2 - 5, \quad x \geq 0$

In Exercises 31–34, find $(f^{-1})'(a)$ for the function f and the given real number a.

31. $f(x) = x^3 + 2, \quad a = -1$

32. $f(x) = x\sqrt{x - 3}, \quad a = 4$

33. $f(x) = \tan x, \quad -\dfrac{\pi}{4} \leq x \leq \dfrac{\pi}{4}, \quad a = \dfrac{\sqrt{3}}{3}$

34. $f(x) = \ln x, \quad a = 0$

In Exercises 35 and 36, (a) find the inverse function of f, (b) use a graphing utility to graph f and f^{-1} in the same viewing window, and (c) verify that $f^{-1}(f(x)) = x$ and $f(f^{-1}(x)) = x$.

35. $f(x) = \ln \sqrt{x}$

36. $f(x) = e^{1-x}$

In Exercises 37 and 38, graph the function without the aid of a graphing utility.

37. $y = e^{-x/2}$

38. $y = 4e^{-x^2}$

In Exercises 39–44, find the derivative of the function.

39. $g(t) = t^2 e^t$

40. $g(x) = \ln \dfrac{e^x}{1 + e^x}$

41. $y = \sqrt{e^{2x} + e^{-2x}}$

42. $h(z) = e^{-z^2/2}$

43. $g(x) = \dfrac{x^2}{e^x}$

44. $y = 3e^{-3/t}$

In Exercises 45 and 46, find an equation of the tangent line to the graph of the function at the given point.

45. $f(x) = \ln(e^{-x^2}), \quad (2, -4)$

46. $f(\theta) = \dfrac{1}{2}e^{\sin 2\theta}, \quad \left(0, \dfrac{1}{2}\right)$

In Exercises 47 and 48, use implicit differentiation to find dy/dx.

47. $y\ln x + y^2 = 0$

48. $\cos x^2 = xe^y$

In Exercises 49–56, find or evaluate the integral.

49. $\displaystyle\int_0^1 xe^{-3x^2}\,dx$

50. $\displaystyle\int_{1/2}^2 \frac{e^{1/x}}{x^2}\,dx$

51. $\displaystyle\int \frac{e^{4x} - e^{2x} + 1}{e^x}\,dx$

52. $\displaystyle\int \frac{e^{2x} - e^{-2x}}{e^{2x} + e^{-2x}}\,dx$

53. $\int xe^{1-x^2}\,dx$

54. $\int x^2e^{x^3+1}\,dx$

55. $\int_1^3 \frac{e^x}{e^x-1}\,dx$

56. $\int_0^2 \frac{e^{2x}}{e^{2x}+1}\,dx$

57. Show that $y = e^x(a\cos 3x + b\sin 3x)$ satisfies the differential equation $y'' - 2y' + 10y = 0$.

58. ***Depreciation*** The value V of an item t years after it is purchased is $V = 8000e^{-0.6t}$, $0 \le t \le 5$.

(a) Use a graphing utility to graph the function.

(b) Find the rates of change of V with respect to t when $t = 1$ and $t = 4$.

(c) Use a graphing utility to graph the tangent lines to the function when $t = 1$ and $t = 4$.

In Exercises 59 and 60, find the area of the region bounded by the graphs of the equations.

59. $y = xe^{-x^2}$, $y = 0$, $x = 0$, $x = 4$

60. $y = 2e^{-x}$, $y = 0$, $x = 0$, $x = 2$

In Exercises 61–64, sketch the graph of the function by hand.

61. $y = 3^{x/2}$

62. $y = 6(2^{-x^2})$

63. $y = \log_2(x-1)$

64. $y = \log_4 x^2$

In Exercises 65–70, find the derivative of the function.

65. $f(x) = 3^{x-1}$

66. $f(x) = (4e)^x$

67. $y = x^{2x+1}$

68. $y = x(4^{-x})$

69. $g(x) = \log_3 \sqrt{1-x}$

70. $h(x) = \log_5 \frac{x}{x-1}$

In Exercises 71 and 72, find the indefinite integral.

71. $\int (x+1)5^{(x+1)^2}\,dx$

72. $\int \frac{2^{-1/t}}{t^2}\,dt$

73. ***Climb Rate*** The time t (in minutes) for a small plane to climb to an altitude of h feet is

$$t = 50\log_{10}\frac{18{,}000}{18{,}000-h}$$

where 18,000 feet is the plane's absolute ceiling.

(a) Determine the domain of the function appropriate for the context of the problem.

(b) Use a graphing utility to graph the time function and identify any asymptotes.

(c) Find the time when the altitude is increasing at the greatest rate.

74. ***Compound Interest***

(a) How large a deposit, at 7% interest compounded continuously, must be made to obtain a balance of \$10,000 in 15 years?

(b) A deposit earns interest at a rate of r percent compounded continuously and doubles in value in 10 years. Find r.

In Exercises 75 and 76, sketch the graph of the function.

75. $f(x) = 2\arctan(x+3)$

76. $h(x) = -3\arcsin 2x$

In Exercises 77 and 78, evaluate the expression without using a calculator. (*Hint:* Make a sketch of a right triangle.)

77. (a) $\sin\left(\arcsin \frac{1}{2}\right)$
(b) $\cos\left(\arcsin \frac{1}{2}\right)$

78. (a) $\tan(\operatorname{arccot} 2)$
(b) $\cos\left(\operatorname{arcsec}\sqrt{5}\right)$

In Exercises 79–84, find the derivative of the function.

79. $y = \tan(\arcsin x)$

80. $y = \arctan(x^2 - 1)$

81. $y = x\operatorname{arcsec} x$

82. $y = \frac{1}{2}\arctan e^{2x}$

83. $y = x(\arcsin x)^2 - 2x + 2\sqrt{1-x^2}\arcsin x$

84. $y = \sqrt{x^2-4} - 2\operatorname{arcsec}\frac{x}{2}$, $2 < x < 4$

In Exercises 85–90, find the indefinite integral.

85. $\int \frac{1}{e^{2x}+e^{-2x}}\,dx$

86. $\int \frac{1}{3+25x^2}\,dx$

87. $\int \frac{x}{\sqrt{1-x^4}}\,dx$

88. $\int \frac{1}{16+x^2}\,dx$

89. $\int \frac{\arctan(x/2)}{4+x^2}\,dx$

90. $\int \frac{\arcsin x}{\sqrt{1-x^2}}\,dx$

In Exercises 91 and 92, find the area of the region.

91. $y = \frac{4-x}{\sqrt{4-x^2}}$

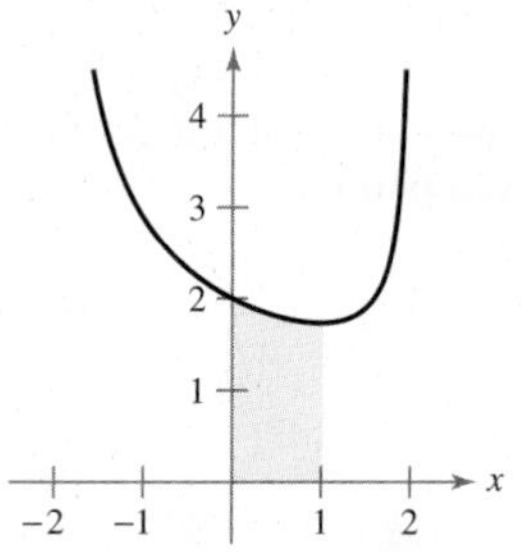

92. $y = \frac{x}{16+x^2}$

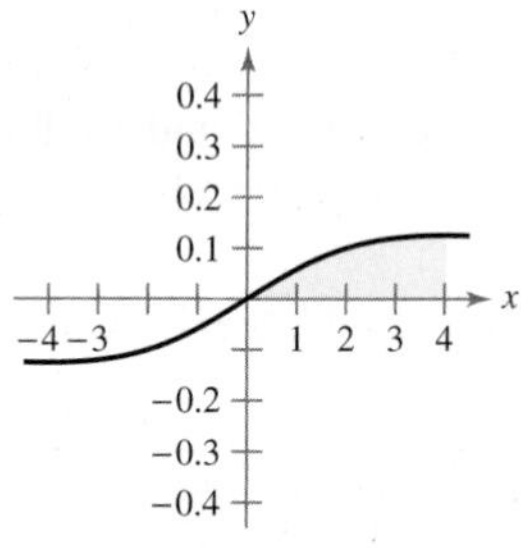

93. ***Harmonic Motion*** A weight of mass m is attached to a spring and oscillates with simple harmonic motion. By Hooke's Law, you can determine that

$$\int \frac{dy}{\sqrt{A^2-y^2}} = \int \sqrt{\frac{k}{m}}\,dt$$

where A is the maximum displacement, t is the time, and k is a constant. Find y as a function of t, given that $y = 0$ when $t = 0$.

In Exercises 94 and 95, find the derivative of the function.

94. $y = 2x - \cosh\sqrt{x}$

95. $y = x\tanh^{-1} 2x$

In Exercises 96 and 97, find the indefinite integral.

96. $\int \frac{x}{\sqrt{x^4-1}}\,dx$

97. $\int x^2\operatorname{sech}^2 x^3\,dx$

P.S. Problem Solving

See www.CalcChat.com for worked-out solutions to odd-numbered exercises.

1. Find the value of a that maximizes the angle θ shown in the figure. What is the approximate measure of this angle?

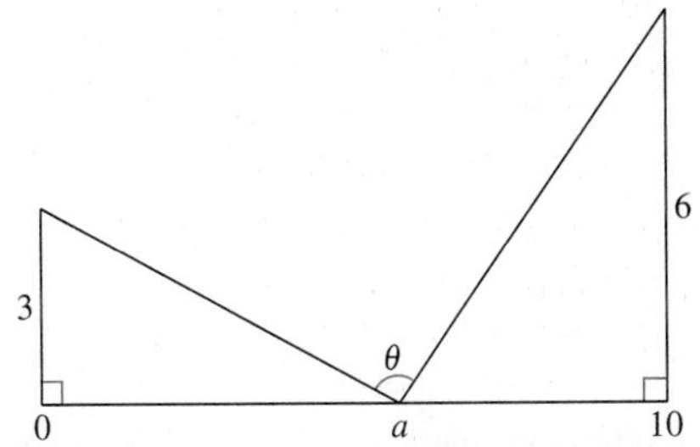

2. Recall that the graph of a function $y = f(x)$ is symmetric with respect to the origin if whenever (x, y) is a point on the graph, $(-x, -y)$ is also a point on the graph. The graph of the function $y = f(x)$ is **symmetric with respect to the point (a, b)** if, whenever $(a - x, b - y)$ is a point on the graph, $(a + x, b + y)$ is also a point on the graph, as shown in the figure.

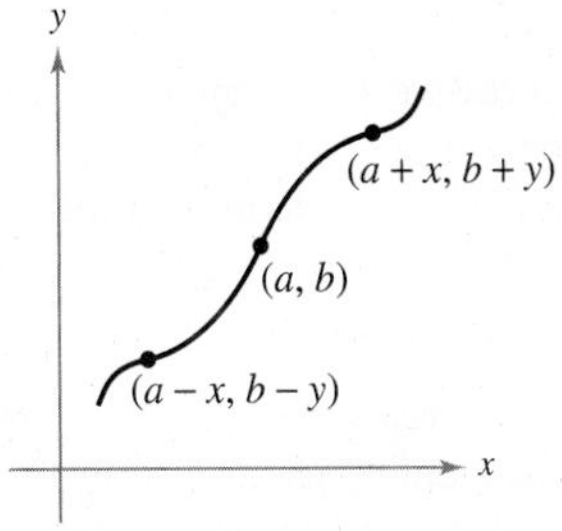

(a) Sketch the graph of $y = \sin x$ on the interval $[0, 2\pi]$. Write a short paragraph explaining how the symmetry of the graph with respect to the point $(0, \pi)$ allows you to conclude that

$$\int_0^{2\pi} \sin x \, dx = 0.$$

(b) Sketch the graph of $y = \sin x + 2$ on the interval $[0, 2\pi]$. Use the symmetry of the graph with respect to the point $(\pi, 2)$ to evaluate the integral

$$\int_0^{2\pi} (\sin x + 2) \, dx.$$

(c) Sketch the graph of $y = \arccos x$ on the interval $[-1, 1]$. Use the symmetry of the graph to evaluate the integral

$$\int_{-1}^{1} \arccos x \, dx.$$

(d) Evaluate the integral $\displaystyle\int_0^{\pi/2} \frac{1}{1 + (\tan x)^{\sqrt{2}}} \, dx.$

3. (a) Use a graphing utility to graph $f(x) = \dfrac{\ln(x + 1)}{x}$ on the interval $[-1, 1]$.

(b) Use the graph to estimate $\lim_{x \to 0} f(x)$.

(c) Use the definition of derivative to prove your answer to part (b).

4. Let $f(x) = \sin(\ln x)$.

(a) Determine the domain of the function f.

(b) Find two values of x satisfying $f(x) = 1$.

(c) Find two values of x satisfying $f(x) = -1$.

(d) What is the range of the function f?

(e) Calculate $f'(x)$ and use calculus to find the maximum value of f on the interval $[1, 10]$.

(f) Use a graphing utility to graph f in the viewing window $[0, 5] \times [-2, 2]$ and estimate $\lim_{x \to 0^+} f(x)$, if it exists.

(g) Determine $\lim_{x \to 0^+} f(x)$ analytically, if it exists.

5. Graph the exponential function $y = a^x$ for $a = 0.5$, 1.2, and 2.0. Which of these curves intersects the line $y = x$? Determine all positive numbers a for which the curve $y = a^x$ intersects the line $y = x$.

6. (a) Let $P(\cos t, \sin t)$ be a point on the unit circle $x^2 + y^2 = 1$ in the first quadrant (see figure). Show that t is equal to twice the area of the shaded circular sector AOP.

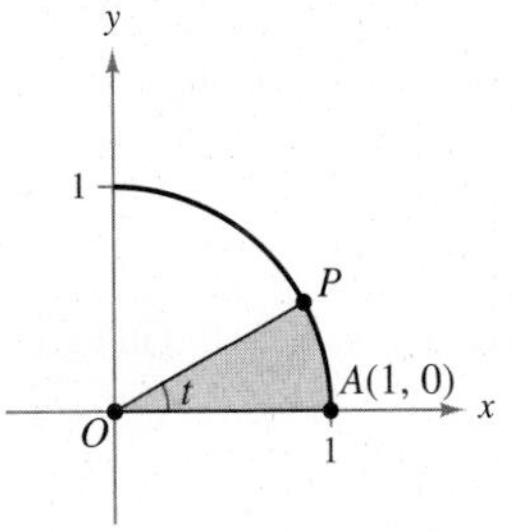

(b) Let $P(\cosh t, \sinh t)$ be a point on the unit hyperbola $x^2 - y^2 = 1$ in the first quadrant (see figure). Show that t is equal to twice the area of the shaded region AOP. Begin by showing that the area of the shaded region AOP is given by the formula

$$A(t) = \frac{1}{2} \cosh t \sinh t - \int_1^{\cosh t} \sqrt{x^2 - 1} \, dx.$$

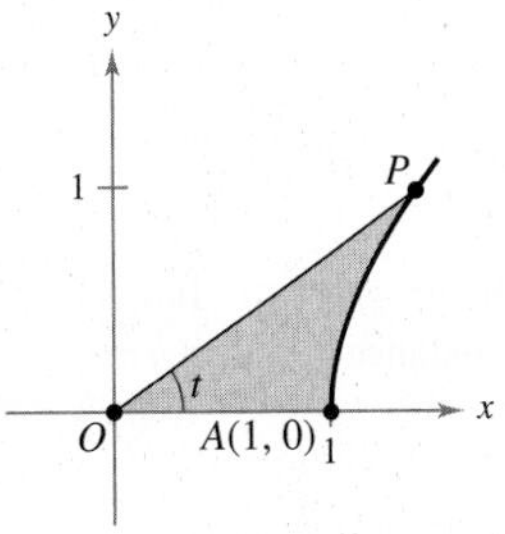

7. Consider the three regions A, B, and C determined by the graph of $f(x) = \arcsin x$, as shown in the figure.

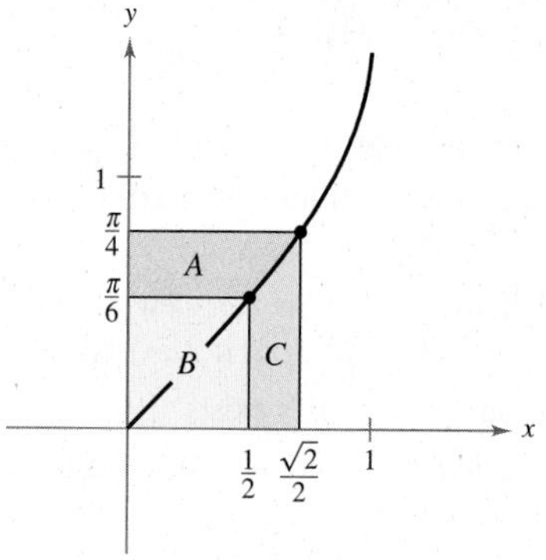

(a) Calculate the areas of regions A and B.

(b) Use your answers in part (a) to evaluate the integral

$$\int_{1/2}^{\sqrt{2}/2} \arcsin x \, dx.$$

(c) Use your answers in part (a) to evaluate the integral

$$\int_{1}^{3} \ln x \, dx.$$

(d) Use your answers in part (a) to evaluate the integral

$$\int_{1}^{\sqrt{3}} \arctan x \, dx.$$

8. Let L be the tangent line to the graph of the function $y = \ln x$ at the point (a, b). Show that the distance between b and c is always equal to 1.

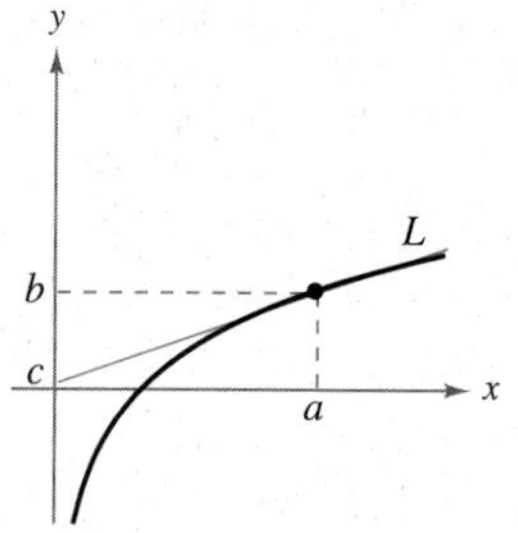

9. Let L be the tangent line to the graph of the function $y = e^x$ at the point (a, b). Show that the distance between a and c is always equal to 1.

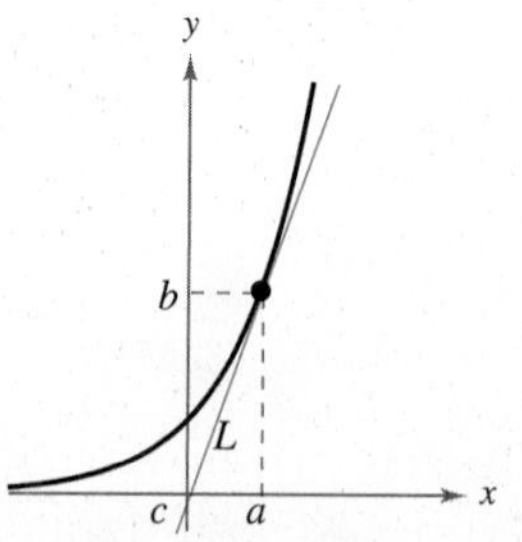

10. Use integration by substitution to find the area under the curve

$$y = \frac{1}{\sqrt{x} + x}$$

between $x = 1$ and $x = 4$.

11. Use integration by substitution to find the area under the curve

$$y = \frac{1}{\sin^2 x + 4\cos^2 x}$$

between $x = 0$ and $x = \pi/4$.

12. (a) Use a graphing utility to compare the graph of the function $y = e^x$ with the graphs of each of the given functions.

(i) $y_1 = 1 + \dfrac{x}{1!}$

(ii) $y_2 = 1 + \dfrac{x}{1!} + \dfrac{x^2}{2!}$

(iii) $y_3 = 1 + \dfrac{x}{1!} + \dfrac{x^2}{2!} + \dfrac{x^3}{3!}$

(b) Identify the pattern of successive polynomials in part (a) and extend the pattern one more term and compare the graph of the resulting polynomial function with the graph of $y = e^x$.

(c) What do you think this pattern implies?

13. A \$120,000 home mortgage for 35 years at $9\frac{1}{2}\%$ has a monthly payment of \$985.93. Part of the monthly payment goes for the interest charge on the unpaid balance and the remainder of the payment is used to reduce the principal. The amount that goes for interest is

$$u = M - \left(M - \frac{Pr}{12}\right)\left(1 + \frac{r}{12}\right)^{12t}$$

and the amount that goes toward reduction of the principal is

$$v = \left(M - \frac{Pr}{12}\right)\left(1 + \frac{r}{12}\right)^{12t}.$$

In these formulas, P is the amount of the mortgage, r is the interest rate, M is the monthly payment, and t is the time in years.

(a) Use a graphing utility to graph each function in the same viewing window. (The viewing window should show all 35 years of mortgage payments.)

(b) In the early years of the mortgage, the larger part of the monthly payment goes for what purpose? Approximate the time when the monthly payment is evenly divided between interest and principal reduction.

(c) Use the graphs in part (a) to make a conjecture about the relationship between the slopes of the tangent lines to the two curves for a specified value of t. Give an analytical argument to verify your conjecture. Find $u'(15)$ and $v'(15)$.

(d) Repeat parts (a) and (b) for a repayment period of 20 years ($M = \$1118.56$). What can you conclude?

6 Differential Equations

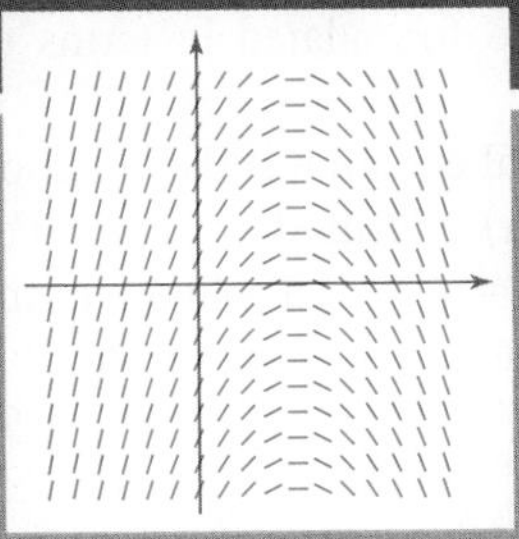

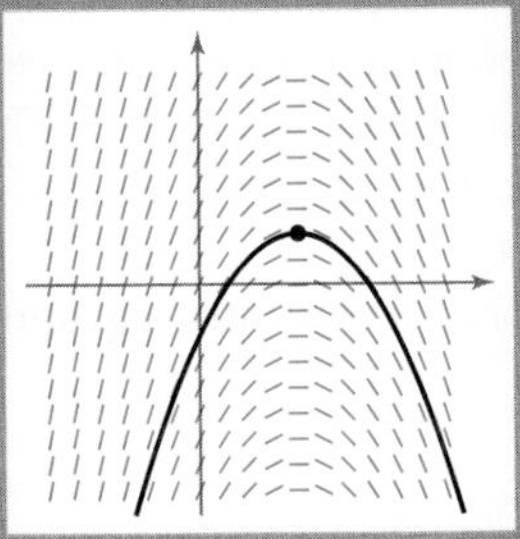

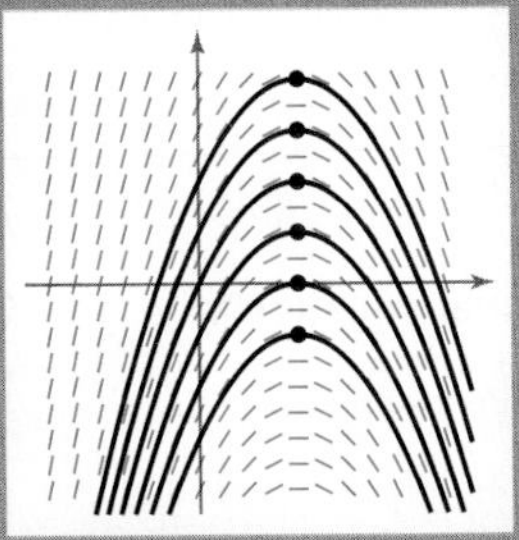

A function $y = f(x)$ is a solution of a differential equation if the equation is satisfied when y and its derivatives are replaced by $f(x)$ and its derivatives. One way to solve a differential equation is to use slope fields, which show the general shape of all solutions of a differential equation. In Chapter 6, you will learn how to sketch slope fields and solve differential equations.

The SkyDome in Toronto is an entertainment center that has a retractable roof. How do you think the noise levels in the stadium compare when the roof is open and when the roof is closed? Explain your reasoning.

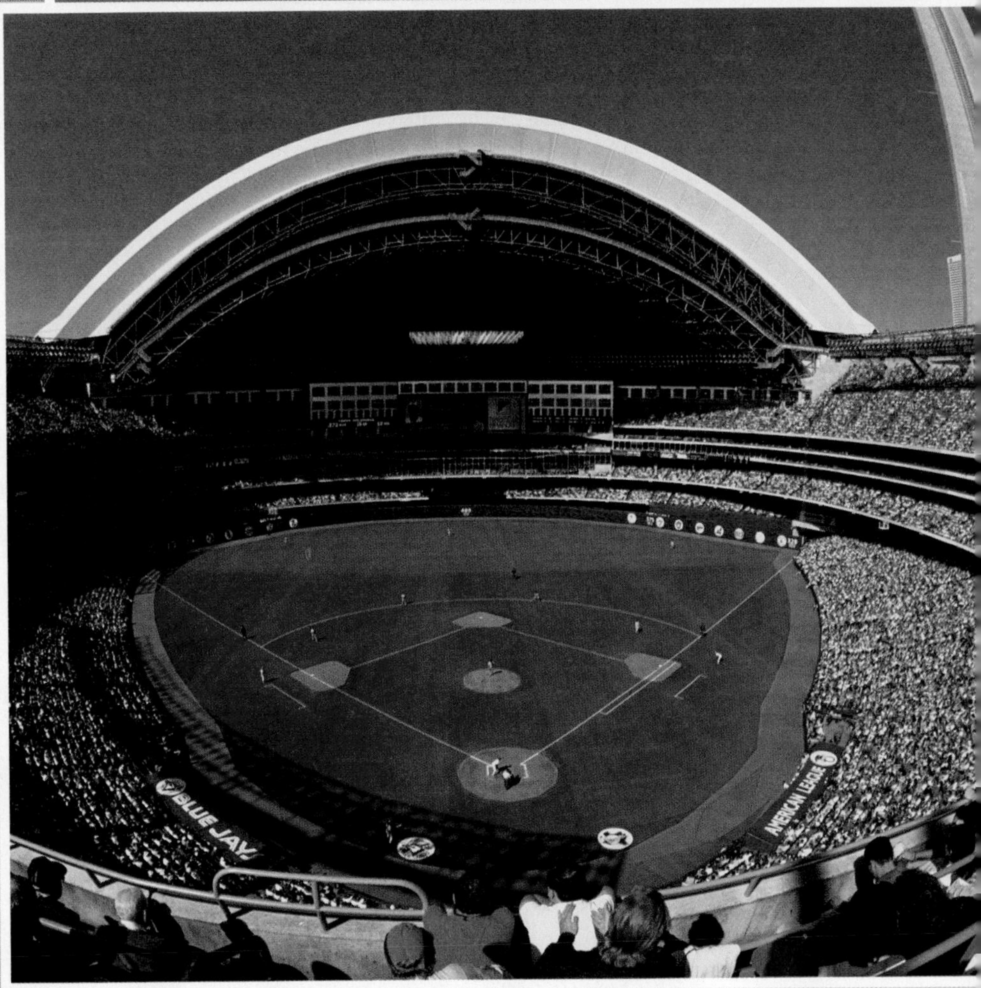

Image Gap/Alamy Images

Section 6.1 Slope Fields and Euler's Method

- **Use initial conditions to find particular solutions of differential equations.**
- **Use slope fields to approximate solutions of differential equations.**
- **Use Euler's Method to approximate solutions of differential equations.**

General and Particular Solutions

In this text, you will learn that physical phenomena can be described by differential equations. In Section 6.2, you will see that problems involving radioactive decay, population growth, and Newton's Law of Cooling can be formulated in terms of differential equations.

A function $y = f(x)$ is called a **solution** of a differential equation if the equation is satisfied when y and its derivatives are replaced by $f(x)$ and its derivatives. For example, differentiation and substitution would show that $y = e^{-2x}$ is a solution of the differential equation $y' + 2y = 0$. It can be shown that every solution of this differential equation is of the form

$$y = Ce^{-2x} \qquad \text{General solution of } y' + 2y = 0$$

where C is any real number. This solution is called the **general solution.** Some differential equations have **singular solutions** that cannot be written as special cases of the general solution. However, such solutions are not considered in this text. The **order** of a differential equation is determined by the highest-order derivative in the equation. For instance, $y' = 4y$ is a first-order differential equation.

NOTE First-order linear differential equations are discussed in Section 6.4.

In Section 4.1, Example 8, you saw that the second-order differential equation $s''(t) = -32$ has the general solution

$$s(t) = -16t^2 + C_1t + C_2 \qquad \text{General solution of } s''(t) = -32$$

which contains two arbitrary constants. It can be shown that a differential equation of order n has a general solution with n arbitrary constants.

EXAMPLE 1 Verifying Solutions

Determine whether the function is a solution of the differential equation $y'' - y = 0$.

a. $y = \sin x$ **b.** $y = 4e^{-x}$ **c.** $y = Ce^x$

Solution

a. Because $y = \sin x$, $y' = \cos x$, and $y'' = -\sin x$, it follows that

$$y'' - y = -\sin x - \sin x = -2 \sin x \neq 0.$$

So, $y = \sin x$ is *not* a solution.

b. Because $y = 4e^{-x}$, $y' = -4e^{-x}$, and $y'' = 4e^{-x}$, it follows that

$$y'' - y = 4e^{-x} - 4e^{-x} = 0.$$

So, $y = 4e^{-x}$ is a solution.

c. Because $y = Ce^x$, $y' = Ce^x$, and $y'' = Ce^x$, it follows that

$$y'' - y = Ce^x - Ce^x = 0.$$

So, $y = Ce^x$ is a solution for any value of C.

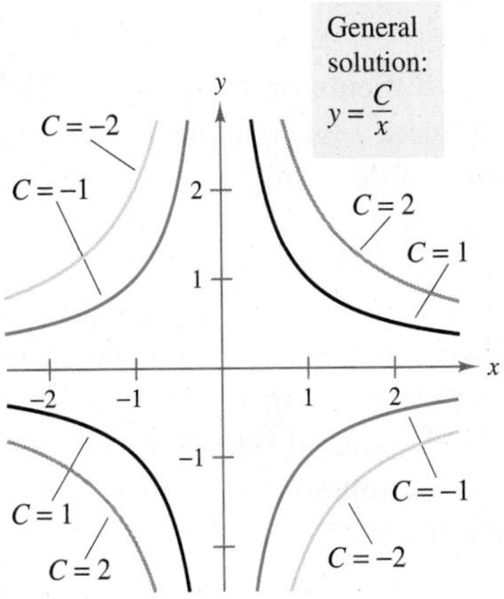

Solution curves for $xy' + y = 0$
Figure 6.1

Geometrically, the general solution of a first-order differential equation represents a family of curves known as **solution curves,** one for each value assigned to the arbitrary constant. For instance, you can verify that every function of the form

$$y = \frac{C}{x} \qquad \text{General solution of } xy' + y = 0$$

is a solution of the differential equation $xy' + y = 0$. Figure 6.1 shows four of the solution curves corresponding to different values of C.

As discussed in Section 4.1, **particular solutions** of a differential equation are obtained from **initial conditions** that give the value of the dependent variable or one of its derivatives for a particular value of the independent variable. The term "initial condition" stems from the fact that, often in problems involving time, the value of the dependent variable or one of its derivatives is known at the *initial* time $t = 0$. For instance, the second-order differential equation $s''(t) = -32$ having the general solution

$$s(t) = -16t^2 + C_1t + C_2 \qquad \text{General solution of } s''(t) = -32$$

might have the following initial conditions.

$$s(0) = 80, \qquad s'(0) = 64 \qquad \text{Initial conditions}$$

In this case, the initial conditions yield the particular solution

$$s(t) = -16t^2 + 64t + 80. \qquad \text{Particular solution}$$

EXAMPLE 2 Finding a Particular Solution

For the differential equation $xy' - 3y = 0$, verify that $y = Cx^3$ is a solution, and find the particular solution determined by the initial condition $y = 2$ when $x = -3$.

Solution You know that $y = Cx^3$ is a solution because $y' = 3Cx^2$ and

$$\begin{aligned} xy' - 3y &= x(3Cx^2) - 3(Cx^3) \\ &= 0. \end{aligned}$$

Furthermore, the initial condition $y = 2$ when $x = -3$ yields

$$y = Cx^3 \qquad \text{General solution}$$
$$2 = C(-3)^3 \qquad \text{Substitute initial condition.}$$
$$-\frac{2}{27} = C \qquad \text{Solve for } C.$$

and you can conclude that the particular solution is

$$y = -\frac{2x^3}{27}. \qquad \text{Particular solution}$$

Try checking this solution by substituting for y and y' in the original differential equation.

NOTE To determine a particular solution, the number of initial conditions must match the number of constants in the general solution.

indicates that in the HM mathSpace® *CD-ROM and the online* Eduspace® *system for this text, you will find an Open Exploration, which further explores this example using the computer algebra systems* Maple, Mathcad, Mathematica, *and* Derive.

Slope Fields

Solving a differential equation analytically can be difficult or even impossible. However, there is a graphical approach you can use to learn a lot about the solution of a differential equation. Consider a differential equation of the form

$$y' = F(x, y). \qquad \text{Differential equation}$$

At each point (x, y) in the xy-plane where F is defined, the differential equation determines the slope $y' = F(x, y)$ of the solution at that point. If you draw a short line segment with slope $F(x, y)$ at selected points (x, y) in the domain of F, then these line segments form a **slope field,** or a *direction field* for the differential equation $y' = F(x, y)$. Each line segment has the same slope as the solution curve through that point. A slope field shows the general shape of all the solutions.

EXAMPLE 3 Sketching a Slope Field

Sketch a slope field for the differential equation $y' = x - y$ for the points $(-1, 1)$, $(0, 1)$, and $(1, 1)$.

Solution

The slope of the solution curve at any point (x, y) is $F(x, y) = x - y$. So, the slope at $(-1, 1)$ is $y' = -1 - 1 = -2$, the slope at $(0, 1)$ is $y' = 0 - 1 = -1$, and the slope at $(1, 1)$ is $y' = 1 - 1 = 0$. Draw short line segments at the three points with their respective slopes, as shown in Figure 6.2.

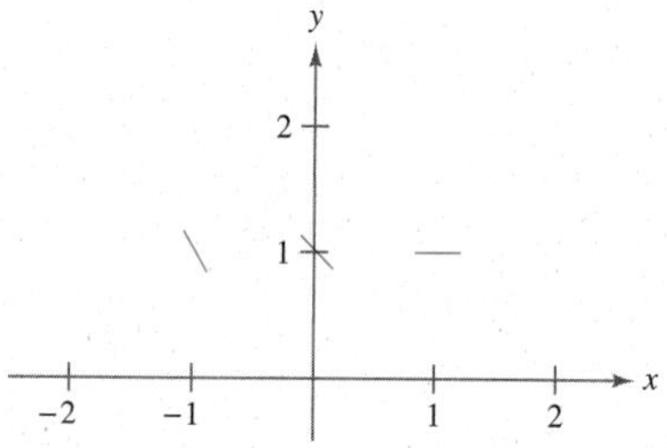

Figure 6.2

EXAMPLE 4 Identifying Slope Fields for Differential Equations

Match each slope field with its differential equation.

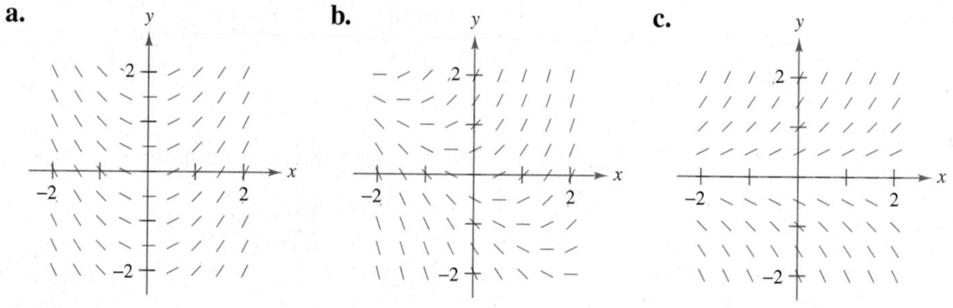

Figure 6.3

i. $y' = x + y$ **ii.** $y' = x$ **iii.** $y' = y$

Solution

a. From Figure 6.3(a), you can see that the slope at any point along the y-axis is 0. The only equation that satisfies this condition is $y' = x$. So, the graph matches (ii).

b. From Figure 6.3(b), you can see that the slope at the point $(1, -1)$ is 0. The only equation that satisfies this condition is $y' = x + y$. So, the graph matches (i).

c. From Figure 6.3(c), you can see that the slope at any point along the x-axis is 0. The only equation that satisfies this condition is $y' = y$. So, the graph matches (iii).

A solution curve of a differential equation $y' = F(x, y)$ is simply a curve in the xy-plane whose tangent line at each point (x, y) has slope equal to $F(x, y)$. This is illustrated in Example 5.

EXAMPLE 5 Sketching a Solution Using a Slope Field

Sketch a slope field for the differential equation

$$y' = 2x + y.$$

Use the slope field to sketch the solution that passes through the point $(1, 1)$.

Solution

Make a table showing the slopes at several points. The table shown is a small sample. The slopes at many other points should be calculated to get a representative slope field.

x	-2	-2	-1	-1	0	0	1	1	2	2
y	-1	1	-1	1	-1	1	-1	1	-1	1
$y' = 2x + y$	-5	-3	-3	-1	-1	1	1	3	3	5

Next draw line segments at the points with their respective slopes, as shown in Figure 6.4.

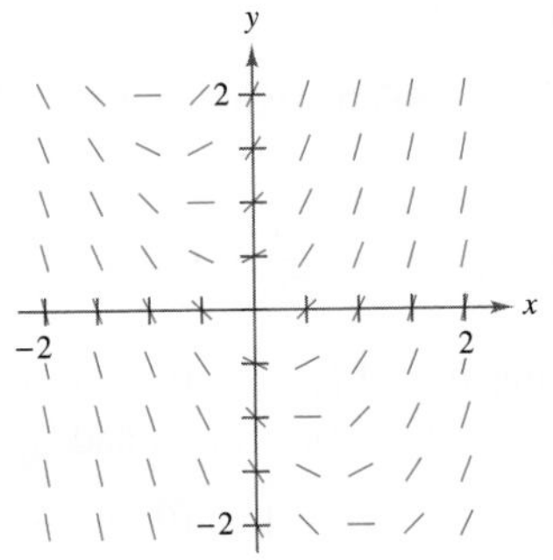

Slope field for $y' = 2x + y$
Figure 6.4

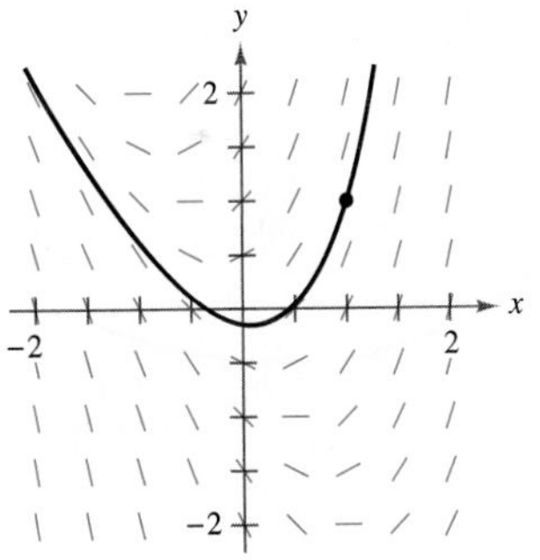

Particular solution for $y' = 2x + y$ passing through $(1, 1)$
Figure 6.5

After the slope field is drawn, start at the initial point $(1, 1)$ and move to the right in the direction of the line segment. Continue to draw the solution curve so that it moves parallel to the nearby line segments. Do the same to the left of $(1, 1)$. The resulting solution is shown in Figure 6.5.

From Example 5, note that the slope field shows that y' increases to infinity as x increases.

NOTE Drawing a slope field by hand is tedious. In practice, slope fields are usually drawn using a graphing utility.

Euler's Method

Euler's Method is a numerical approach to approximating the particular solution of the differential equation

$$y' = F(x, y)$$

that passes through the point (x_0, y_0). From the given information, you know that the graph of the solution passes through the point (x_0, y_0) and has a slope of $F(x_0, y_0)$ at this point. This gives you a "starting point" for approximating the solution.

From this starting point, you can proceed in the direction indicated by the slope. Using a small step h, move along the tangent line until you arrive at the point (x_1, y_1), where

$$x_1 = x_0 + h \quad \text{and} \quad y_1 = y_0 + hF(x_0, y_0)$$

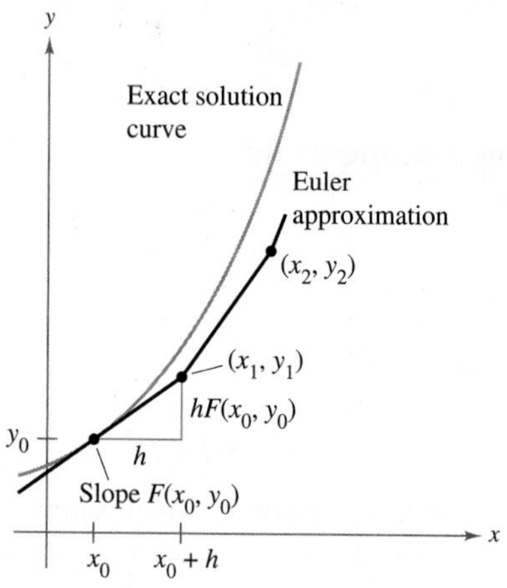

Figure 6.6

as shown in Figure 6.6. If you think of (x_1, y_1) as a new starting point, you can repeat the process to obtain a second point (x_2, y_2). The values of x_i and y_i are as follows.

$$\begin{aligned} x_1 &= x_0 + h & y_1 &= y_0 + hF(x_0, y_0) \\ x_2 &= x_1 + h & y_2 &= y_1 + hF(x_1, y_1) \\ &\vdots & &\vdots \\ x_n &= x_{n-1} + h & y_n &= y_{n-1} + hF(x_{n-1}, y_{n-1}) \end{aligned}$$

NOTE You can obtain better approximations of the exact solution by choosing smaller and smaller step sizes.

EXAMPLE 6 Approximating a Solution Using Euler's Method

Use Euler's Method to approximate the particular solution of the differential equation

$$y' = x - y$$

passing through the point (0, 1). Use a step of $h = 0.1$.

Solution Using $h = 0.1$, $x_0 = 0$, $y_0 = 1$, and $F(x, y) = x - y$, you have $x_0 = 0$, $x_1 = 0.1$, $x_2 = 0.2$, $x_3 = 0.3, \ldots,$ and

$$\begin{aligned} y_1 &= y_0 + hF(x_0, y_0) = 1 + (0.1)(0 - 1) = 0.9 \\ y_2 &= y_1 + hF(x_1, y_1) = 0.9 + (0.1)(0.1 - 0.9) = 0.82 \\ y_3 &= y_2 + hF(x_2, y_2) = 0.82 + (0.1)(0.2 - 0.82) = 0.758. \end{aligned}$$

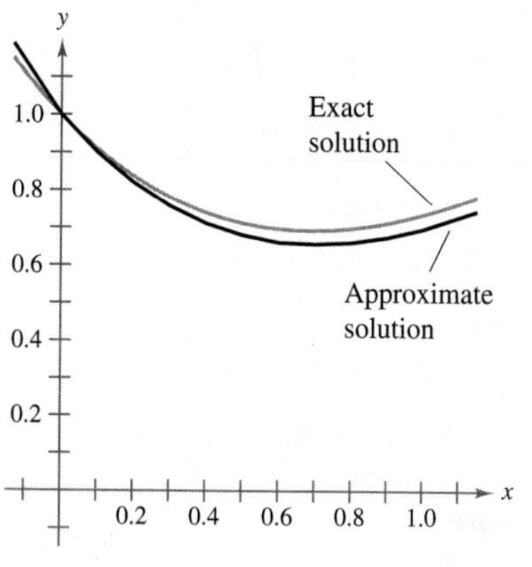

Figure 6.7

The first ten approximations are shown in the table. You can plot these values to see a graph of the approximate solution, as shown in Figure 6.7.

n	0	1	2	3	4	5	6	7	8	9	10
x_n	0	0.1	0.2	0.3	0.4	0.5	0.6	0.7	0.8	0.9	1.0
y_n	1	0.900	0.820	0.758	0.712	0.681	0.663	0.657	0.661	0.675	0.697

NOTE For the differential equation in Example 6, you can verify the exact solution to be $y = x - 1 + 2e^{-x}$. Figure 6.7 compares this exact solution with the approximate solution obtained in Example 6.

Exercises for Section 6.1

See www.CalcChat.com for worked-out solutions to odd-numbered exercises.

In Exercises 1–8, verify the solution of the differential equation.

Solution	*Differential Equation*
1. $y = Ce^{4x}$	$y' = 4y$
2. $y = e^{-x}$	$3y' + 4y = e^{-x}$
3. $x^2 + y^2 = Cy$	$y' = 2xy/(x^2 - y^2)$
4. $y^2 - 2 \ln y = x^2$	$\frac{dy}{dx} = \frac{xy}{y^2 - 1}$
5. $y = C_1 \cos x + C_2 \sin x$	$y'' + y = 0$
6. $y = C_1 e^{-x} \cos x + C_2 e^{-x} \sin x$	$y'' + 2y' + 2y = 0$
7. $y = -\cos x \ln\lvert\sec x + \tan x\rvert$	$y'' + y = \tan x$
8. $y = \frac{2}{5}(e^{-2x} + e^x)$	$y'' + 2y' = 2e^x$

In Exercises 9–12, verify the particular solution of the differential equation.

Solution	*Differential Equation and Initial Condition*
9. $y = \sin x \cos x - \cos^2 x$	$2y + y' = 2\sin(2x) - 1$ $y\left(\frac{\pi}{4}\right) = 0$
10. $y = \frac{1}{2}x^2 - 4\cos x + 2$	$y' = x + 4\sin x$ $y(0) = -2$
11. $y = 6e^{-2x^2}$	$y' = -4xy$ $y(0) = 6$
12. $y = e^{-\cos x}$	$y' = y \sin x$ $y\left(\frac{\pi}{2}\right) = 1$

In Exercises 13–18, determine whether the function is a solution of the differential equation $y^{(4)} - 16y = 0$.

13. $y = 3\cos x$

14. $y = 3\cos 2x$

15. $y = e^{-2x}$

16. $y = 5 \ln x$

17. $y = C_1e^{2x} + C_2e^{-2x} + C_3 \sin 2x + C_4 \cos 2x$

18. $y = 3e^{2x} - 4 \sin 2x$

In Exercises 19–24, determine whether the function is a solution of the differential equation $xy' - 2y = x^3e^x$.

19. $y = x^2$

20. $y = x^2e^x$

21. $y = x^2(2 + e^x)$

22. $y = \sin x$

23. $y = \ln x$

24. $y = x^2e^x - 5x^2$

In Exercises 25–28, some of the curves corresponding to different values of C in the general solution of the differential equation are given. Find the particular solution that passes through the point shown on the graph.

Solution	*Differential Equation*
25. $y = Ce^{-x/2}$	$2y' + y = 0$
26. $y(x^2 + y) = C$	$2xy + (x^2 + 2y)y' = 0$
27. $y^2 = Cx^3$	$2xy' - 3y = 0$
28. $2x^2 - y^2 = C$	$yy' - 2x = 0$

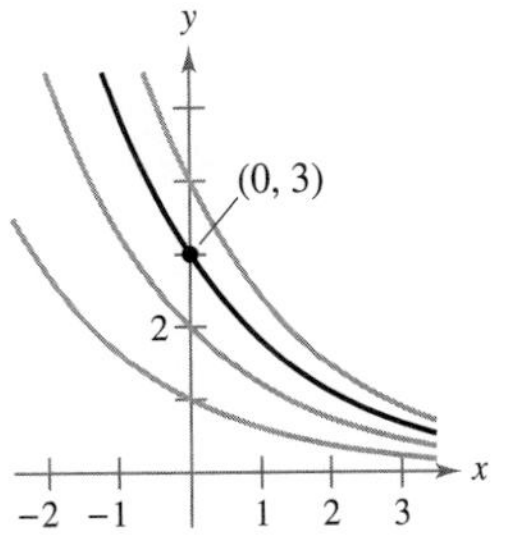

Figure for 25

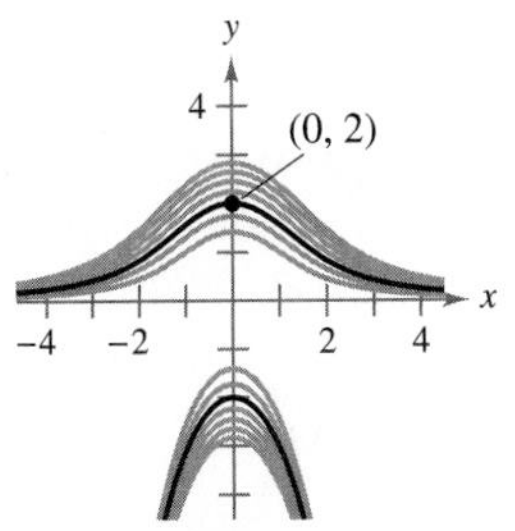

Figure for 26

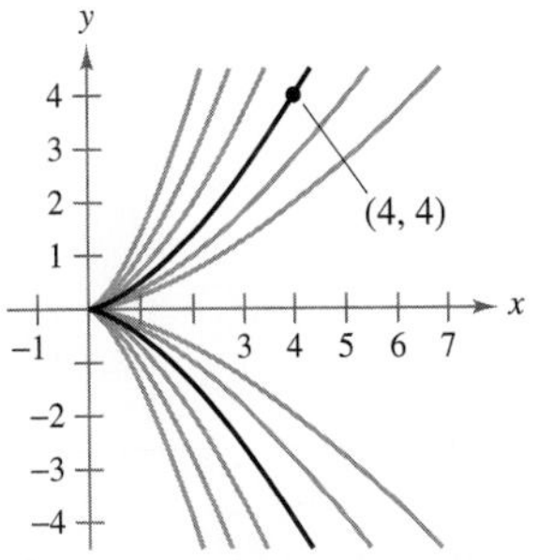

Figure for 27

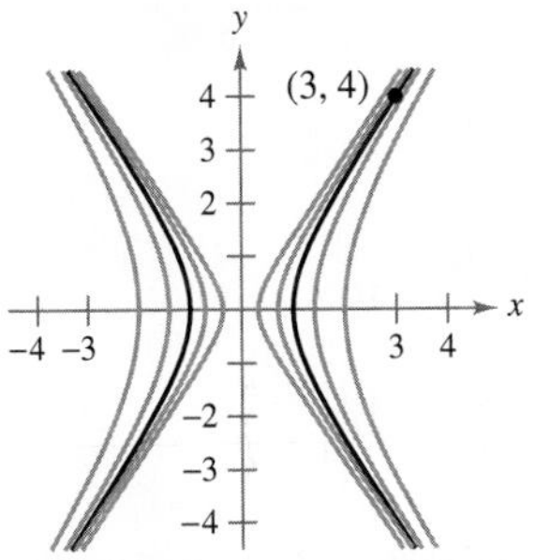

Figure for 28

In Exercises 29 and 30, the general solution of the differential equation is given. Use a graphing utility to graph the particular solutions for the given values of C.

29. $4yy' - x = 0$
$4y^2 - x^2 = C$
$C = 0, C = \pm 1, C = \pm 4$

30. $yy' + x = 0$
$x^2 + y^2 = C$
$C = 0, C = 1, C = 4$

In Exercises 31–36, verify that the general solution satisfies the differential equation. Then find the particular solution that satisfies the initial condition.

31. $y = Ce^{-2x}$
$y' + 2y = 0$
$y = 3$ when $x = 0$

32. $3x^2 + 2y^2 = C$
$3x + 2yy' = 0$
$y = 3$ when $x = 1$

33. $y = C_1 \sin 3x + C_2 \cos 3x$
$y'' + 9y = 0$
$y = 2$ when $x = \pi/6$
$y' = 1$ when $x = \pi/6$

34. $y = C_1 + C_2 \ln x$
$xy'' + y' = 0$
$y = 0$ when $x = 2$
$y' = \frac{1}{2}$ when $x = 2$

35. $y = C_1x + C_2x^3$
$x^2y'' - 3xy' + 3y = 0$
$y = 0$ when $x = 2$
$y' = 4$ when $x = 2$

36. $y = e^{2x/3}(C_1 + C_2x)$
$9y'' - 12y' + 4y = 0$
$y = 4$ when $x = 0$
$y = 0$ when $x = 3$

In Exercises 37–48, use integration to find a general solution of the differential equation.

37. $\dfrac{dy}{dx} = 3x^2$

38. $\dfrac{dy}{dx} = x^3 - 4x$

39. $\dfrac{dy}{dx} = \dfrac{x}{1 + x^2}$

40. $\dfrac{dy}{dx} = \dfrac{e^x}{1 + e^x}$

41. $\dfrac{dy}{dx} = \dfrac{x - 2}{x}$

42. $\dfrac{dy}{dx} = x \cos x^2$

43. $\dfrac{dy}{dx} = \sin 2x$

44. $\dfrac{dy}{dx} = \tan^2 x$

45. $\dfrac{dy}{dx} = x\sqrt{x - 3}$

46. $\dfrac{dy}{dx} = x\sqrt{5 - x}$

47. $\dfrac{dy}{dx} = xe^{x^2}$

48. $\dfrac{dy}{dx} = 5e^{-x/2}$

Slope Fields **In Exercises 49–52, a differential equation and its slope field are given. Determine the slopes (if possible) in the slope field at the points given in the table.**

x	-4	-2	0	2	4	8
y	2	0	4	4	6	8
dy/dx						

49. $\dfrac{dy}{dx} = \dfrac{x}{y}$

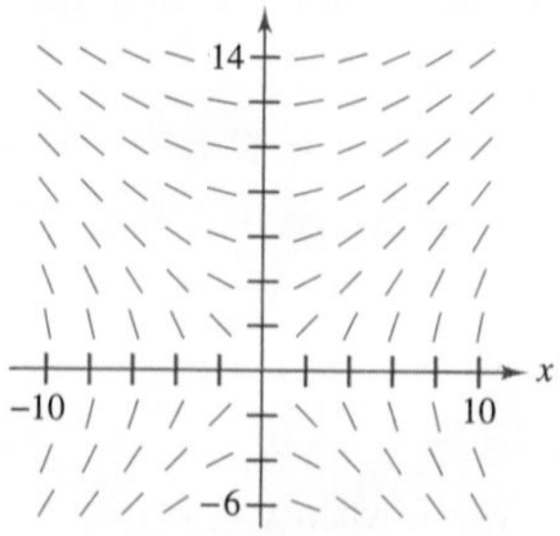

50. $\dfrac{dy}{dx} = x - y$

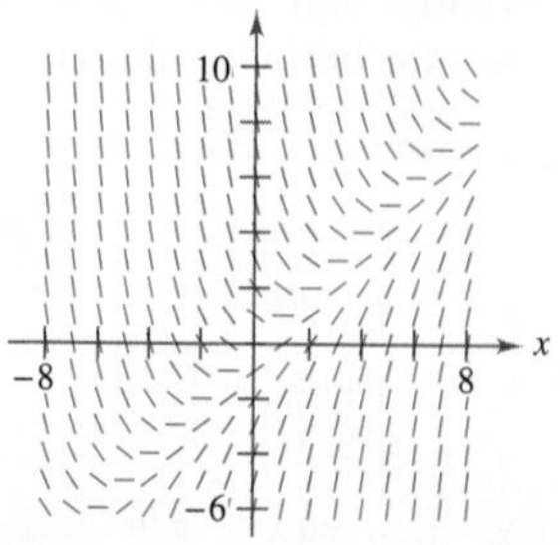

51. $\dfrac{dy}{dx} = x \cos \dfrac{\pi y}{8}$

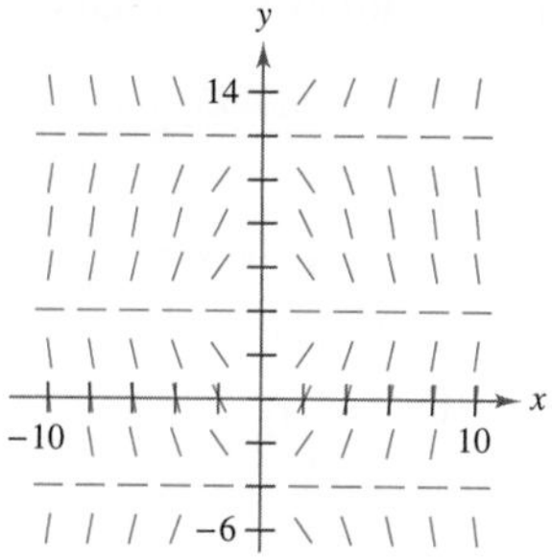

52. $\dfrac{dy}{dx} = \tan\left(\dfrac{\pi y}{6}\right)$

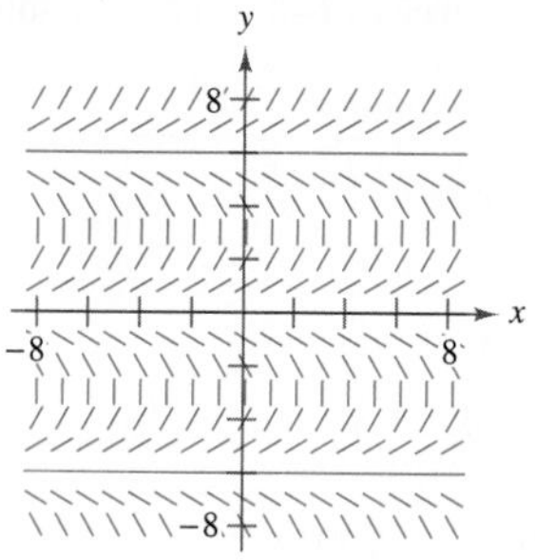

In Exercises 53–56, match the differential equation with its slope field. [The slope fields are labeled (a), (b), (c), and (d).]

(a)

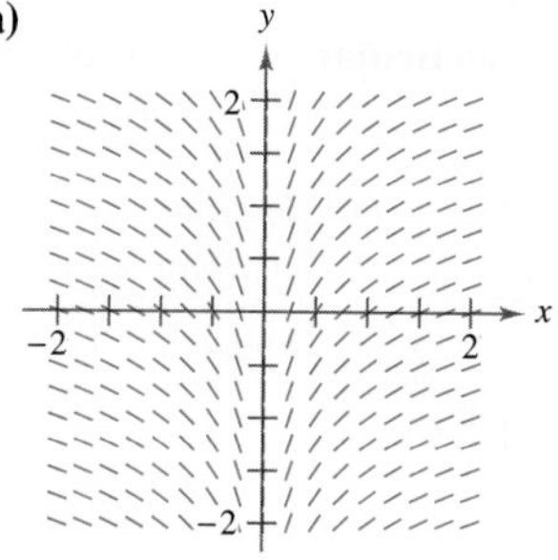

(b)

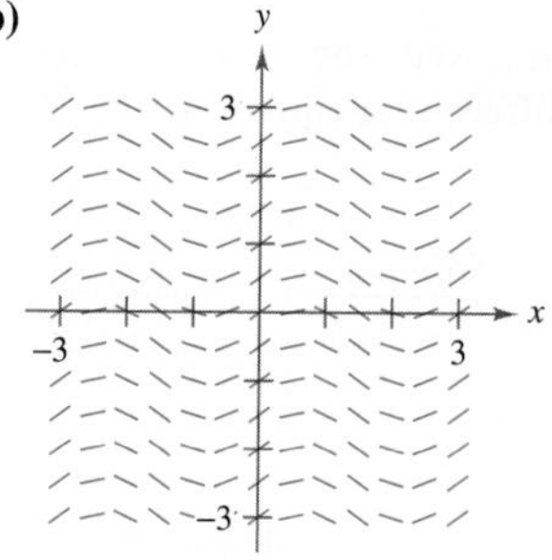

(c)

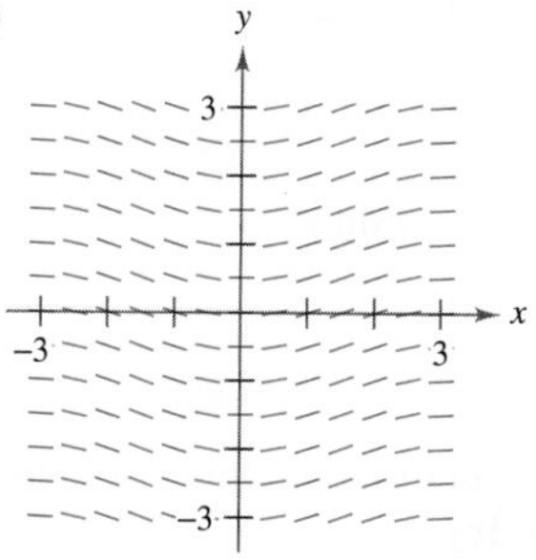

(d)

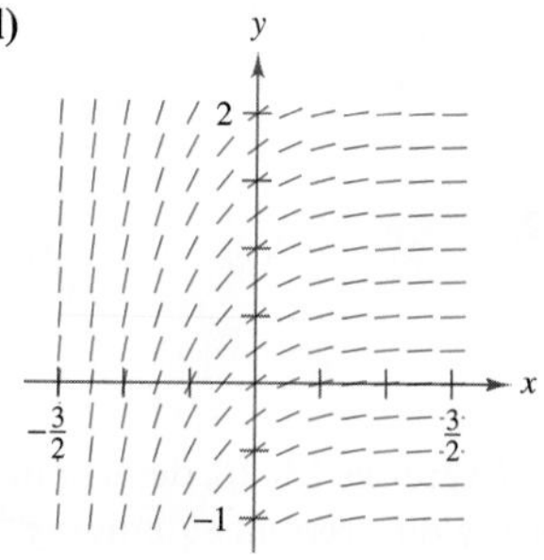

53. $\dfrac{dy}{dx} = \cos(2x)$

54. $\dfrac{dy}{dx} = \dfrac{1}{2} \sin x$

55. $\dfrac{dy}{dx} = e^{-2x}$

56. $\dfrac{dy}{dx} = \dfrac{1}{x}$

Slope Fields **In Exercises 57–60, (a) sketch the slope field for the differential equation, (b) use the slope field to sketch the solution that passes through the given point, and (c) discuss the graph of the solution as $x \to \infty$ and $x \to -\infty$.**

	Differential Equation	*Point*
57.	$y' = -x + 1$	$(2, 4)$
58.	$y' = \frac{1}{3}x^2 - \frac{1}{2}x$	$(1, 1)$
59.	$y' = y - 2x$	$(1, 1)$
60.	$y' = y + xy$	$(0, 4)$

61. ***Slope Field*** Use the slope field for the differential equation $y' = 1/x$, where $x > 0$, to sketch the graph of the solution that satisfies each given initial condition. Then make a conjecture about the behavior of a particular solution of $y' = 1/x$ as $x \to \infty$. To print an enlarged copy of the graph, go to the website *www.mathgraphs.com.*

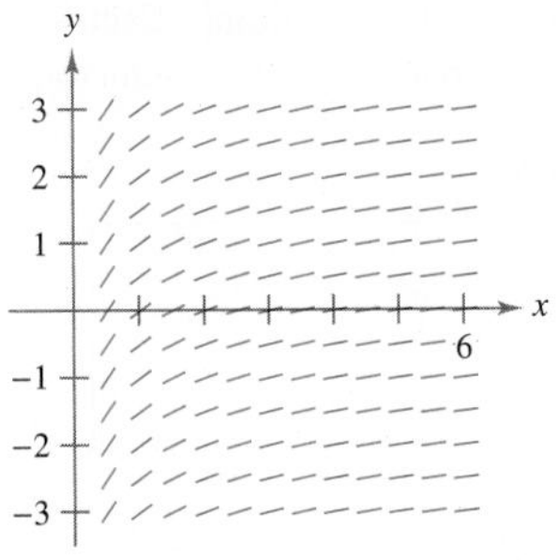

(a) $(1, 0)$ (b) $(2, -1)$

62. ***Slope Field*** Use the slope field for the differential equation $y' = 1/y$, where $y > 0$, to sketch the graph of the solution that satisfies each initial condition. Then make a conjecture about the behavior of a particular solution of $y' = 1/y$ as $x \to \infty$. To print an enlarged copy of the graph, go to the website *www.mathgraphs.com.*

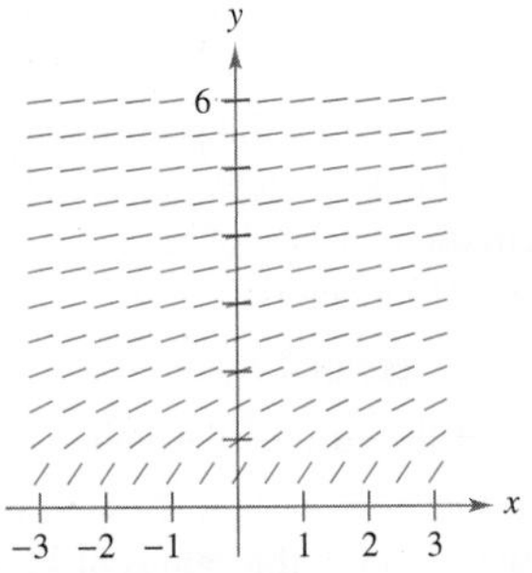

(a) $(0, 1)$ (b) $(1, 1)$

Slope Fields **In Exercises 63–68, use a computer algebra system to (a) graph the slope field for the differential equation and (b) graph the solution satisfying the specified initial condition.**

63. $\dfrac{dy}{dx} = 0.5y, \quad y(0) = 6$

64. $\dfrac{dy}{dx} = 2 - y, \quad y(0) = 4$

65. $\dfrac{dy}{dx} = 0.02y(10 - y), \quad y(0) = 2$

66. $\dfrac{dy}{dx} = 0.2x(2 - y), \quad y(0) = 9$

67. $\dfrac{dy}{dx} = 0.4y(3 - x), \quad y(0) = 1$

68. $\dfrac{dy}{dx} = \dfrac{1}{2}e^{-x/8} \sin \dfrac{\pi y}{4}, \quad y(0) = 2$

Euler's Method **In Exercises 69–74, use Euler's Method to make a table of values for the approximate solution of the differential equation with the specified initial value. Use n steps of size h.**

69. $y' = x + y, \quad y(0) = 2, \quad n = 10, \quad h = 0.1$

70. $y' = x + y, \quad y(0) = 2, \quad n = 20, \quad h = 0.05$

71. $y' = 3x - 2y, \quad y(0) = 3, \quad n = 10, \quad h = 0.05$

72. $y' = 0.5x(3 - y), \quad y(0) = 1, \quad n = 5, \quad h = 0.4$

73. $y' = e^{xy}, \quad y(0) = 1, \quad n = 10, \quad h = 0.1$

74. $y' = \cos x + \sin y, \quad y(0) = 5, \quad n = 10, \quad h = 0.1$

In Exercises 75–77, complete the table using the exact solution of the differential equation and two approximations obtained using Euler's Method to approximate the particular solution of the differential equation. Use $h = 0.2$ and 0.1 and compute each approximation to four decimal places.

x	0	0.2	0.4	0.6	0.8	1.0
$y(x)$ (exact)						
$y(x)$ ($h = 0.2$)						
$y(x)$ ($h = 0.1$)						

	Differential Equation	*Initial Condition*	*Exact Solution*
75.	$\dfrac{dy}{dx} = y$	$(0, 3)$	$y = 3e^x$
76.	$\dfrac{dy}{dx} = \dfrac{2x}{y}$	$(0, 2)$	$y = \sqrt{2x^2 + 4}$
77.	$\dfrac{dy}{dx} = y + \cos(x)$	$(0, 0)$	$y = \frac{1}{2}(\sin x - \cos x + e^x)$

78. Compare the values of the approximations in Exercises 75–77 with the values given by the exact solution. How does the error change as h increases?

79. ***Temperature*** At time $t = 0$ minutes, the temperature of an object is 140°F. The temperature of the object is changing at the rate given by the differential equation

$$\frac{dy}{dt} = -\frac{1}{2}(y - 72).$$

(a) Use a graphing utility and Euler's Method to approximate the particular solutions of this differential equation at $t = 1$, 2, and 3. Use a step size of $h = 0.1$. (A graphing utility program for Euler's Method is available on the website *college.hmco.com.*)

(b) Compare your results with the exact solution

$$y = 72 + 68e^{-t/2}.$$

80. ***Temperature*** Repeat Exercise 79 using a step size of $h = 0.05$. Compare the results.

Writing About Concepts

81. In your own words, describe the difference between a general solution of a differential equation and a particular solution.

82. Explain how to interpret a slope field.

83. Describe how to use Euler's Method to approximate the particular solution of a differential equation.

84. It is known that $y = Ce^{kx}$ is a solution of the differential equation $y' = 0.07y$. Is it possible to determine C or k from the information given? If so, find its value.

True or False? **In Exercises 85–88, determine whether the statement is true or false. If it is false, explain why or give an example that shows it is false.**

85. If $y = f(x)$ is a solution of a first-order differential equation, then $y = f(x) + C$ is also a solution.

86. The general solution of a differential equation is $y = -4.9x^2 + C_1x + C_2$. To find a particular solution, you must be given two initial conditions.

87. Slope fields represent the general solutions of differential equations.

88. A slope field shows that the slope at the point $(1, 1)$ is 6. This slope field represents the family of solutions for the differential equation $y' = 4x + 2y$.

89. ***Error and Euler's Method*** The exact solution of the differential equation

$$\frac{dy}{dx} = -2y$$

where $y(0) = 4$, is $y = 4e^{-2x}$.

(a) Use a graphing utility to complete the table, where y is the exact value of the solution, y_1 is the approximate solution using Euler's Method with $h = 0.1$, y_2 is the approximate solution using Euler's Method with $h = 0.2$, e_1 is the absolute error $|y - y_1|$, e_2 is the absolute error $|y - y_2|$, and r is the ratio e_1/e_2.

x	0	0.2	0.4	0.6	0.8	1
y						
y_1						
y_2						
e_1						
e_2						
r						

(b) What can you conclude about the ratio r as h changes?

(c) Predict the absolute error when $h = 0.05$.

90. ***Error and Euler's Method*** Repeat Exercise 89 where the exact solution of the differential equation

$$\frac{dy}{dx} = x - y$$

where $y(0) = 1$, is $y = x - 1 + 2e^{-x}$.

91. ***Electric Circuits*** The diagram shows a simple electric circuit consisting of a power source, a resistor, and an inductor.

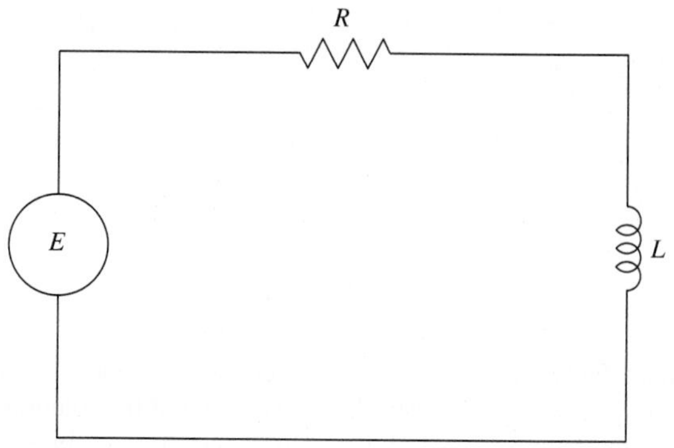

A model of the current I, in amperes (A), at time t is given by the first-order differential equation

$$L\frac{dI}{dt} + RI = E(t)$$

where $E(t)$ is the voltage (V) produced by the power source, R is the resistance, in ohms (Ω), and L is the inductance, in henrys (H). Suppose the electric circuit consists of a 24-V power source, a 12-Ω resistor, and a 4-H inductor.

(a) Sketch a slope field for the differential equation.

(b) What is the limiting value of the current? Explain.

92. ***Think About It*** It is known that $y = e^{kt}$ is a solution of the differential equation $y'' - 16y = 0$. Find the values of k.

93. ***Think About It*** It is known that $y = A \sin \omega t$ is a solution of the differential equation $y'' + 16y = 0$. Find the values of ω.

Putnam Exam Challenge

94. Let f be a twice-differentiable real-valued function satisfying

$$f(x) + f''(x) = -xg(x)f'(x)$$

where $g(x) \geq 0$ for all real x. Prove that $|f(x)|$ is bounded.

95. Prove that if the family of integral curves of the differential equation

$$\frac{dy}{dx} + p(x)y = q(x), \qquad p(x) \cdot q(x) \neq 0$$

is cut by the line $x = k$, the tangents at the points of intersection are concurrent.

Section 6.2 Differential Equations: Growth and Decay

- **Use separation of variables to solve a simple differential equation.**
- **Use exponential functions to model growth and decay in applied problems.**

Differential Equations

In the preceding section, you learned to analyze visually the solutions of differential equations using slope fields and to approximate solutions numerically using Euler's Method. Analytically, you have learned to solve only two types of differential equations—those of the forms

$$y' = f(x) \quad \text{and} \quad y'' = f(x).$$

In this section, you will learn how to solve a more general type of differential equation. The strategy is to rewrite the equation so that each variable occurs on only one side of the equation. This strategy is called *separation of variables.* (You will study this strategy in detail in Section 6.3.)

EXAMPLE 1 Solving a Differential Equation

Solve the differential equation $y' = 2x/y$.

Solution

$$y' = \frac{2x}{y} \qquad \text{Write original equation.}$$

$$yy' = 2x \qquad \text{Multiply both sides by } y.$$

$$\int yy'\,dx = \int 2x\,dx \qquad \text{Integrate with respect to } x.$$

$$\int y\,dy = \int 2x\,dx \qquad dy = y'\,dx$$

$$\frac{1}{2}y^2 = x^2 + C_1 \qquad \text{Apply Power Rule.}$$

$$y^2 - 2x^2 = C \qquad \text{Rewrite, letting } C = 2C_1.$$

So, the general solution is given by

$$y^2 - 2x^2 = C.$$

You can use implicit differentiation to check this result.

NOTE When you integrate both sides of the equation in Example 1, you don't need to add a constant of integration to both sides of the equation. If you did, you would obtain the same result as in Example 1.

$$\int y\,dy = \int 2x\,dx$$

$$\tfrac{1}{2}y^2 + C_2 = x^2 + C_3$$

$$\tfrac{1}{2}y^2 = x^2 + (C_3 - C_2)$$

$$\tfrac{1}{2}y^2 = x^2 + C_1$$

EXPLORATION

In Example 1, the general solution of the differential equation is

$$y^2 - 2x^2 = C.$$

Use a graphing utility to sketch several particular solutions—those given by $C = \pm 2$, $C = \pm 1$, and $C = 0$. Describe the solutions graphically. Is the following statement true of each solution?

The slope of the graph at the point (x, y) is equal to twice the ratio of x and y.

Explain your reasoning. Are all curves for which this statement is true represented by the general solution?

In practice, most people prefer to use Leibniz notation and differentials when applying separation of variables. The solution of Example 1 is shown below using this notation.

$$\frac{dy}{dx} = \frac{2x}{y}$$

$$y\,dy = 2x\,dx$$

$$\int y\,dy = \int 2x\,dx$$

$$\frac{1}{2}y^2 = x^2 + C_1$$

$$y^2 - 2x^2 = C$$

Growth and Decay Models

In many applications, the rate of change of a variable y is proportional to the value of y. If y is a function of time t, the proportion can be written as shown.

Rate of change of y is proportional to y.

$$\frac{dy}{dt} = ky$$

The general solution of this differential equation is given in the following theorem.

THEOREM 6.1 Exponential Growth and Decay Model

If y is a differentiable function of t such that $y > 0$ and $y' = ky$, for some constant k, then

$$y = Ce^{kt}.$$

C is the **initial value** of y, and k is the **proportionality constant. Exponential growth** occurs when $k > 0$, and **exponential decay** occurs when $k < 0$.

NOTE Differentiate the function $y = Ce^{kt}$ with respect to t, and verify that $y' = ky$.

Proof

$y' = ky$	Write original equation.
$\dfrac{y'}{y} = k$	Separate variables.
$\displaystyle\int \frac{y'}{y}\,dt = \int k\,dt$	Integrate with respect to t.
$\displaystyle\int \frac{1}{y}\,dy = \int k\,dt$	$dy = y'\,dt$
$\ln y = kt + C_1$	Find antiderivative of each side.
$y = e^{kt}e^{C_1}$	Solve for y.
$y = Ce^{kt}$	Let $C = e^{C_1}$.

So, all solutions of $y' = ky$ are of the form $y = Ce^{kt}$.

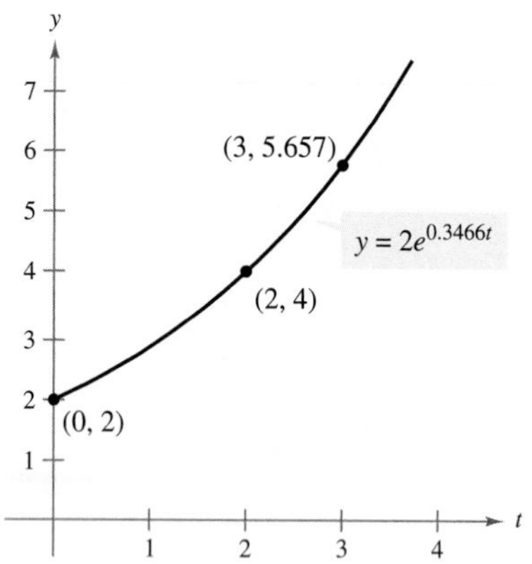

If the rate of change of y is proportional to y, then y follows an exponential model.
Figure 6.8

EXAMPLE 2 Using an Exponential Growth Model

The rate of change of y is proportional to y. When $t = 0$, $y = 2$. When $t = 2$, $y = 4$. What is the value of y when $t = 3$?

Solution Because $y' = ky$, you know that y and t are related by the equation $y = Ce^{kt}$. You can find the values of the constants C and k by applying the initial conditions.

$2 = Ce^0$ ⇨	$C = 2$	When $t = 0$, $y = 2$.
$4 = 2e^{2k}$ ⇨	$k = \frac{1}{2}\ln 2 \approx 0.3466$	When $t = 2$, $y = 4$.

So, the model is $y \approx 2e^{0.3466t}$. When $t = 3$, the value of y is $2e^{0.3466(3)} \approx 5.657$ (see Figure 6.8).

STUDY TIP Using logarithmic properties, note that the value of k in Example 2 can also be written as $\ln(\sqrt{2})$. So, the model becomes $y = 2e^{(\ln\sqrt{2})t}$, which can then be rewritten as $y = 2(\sqrt{2})^t$.

TECHNOLOGY Most graphing utilities have curve-fitting capabilities that can be used to find models that represent data. Use the *exponential regression* feature of a graphing utility and the information in Example 2 to find a model for the data. How does your model compare with the given model?

Radioactive decay is measured in terms of *half-life*—the number of years required for half of the atoms in a sample of radioactive material to decay. The half-lives of some common radioactive isotopes are shown below.

Uranium (^{238}U)	4,470,000,000 years
Plutonium (^{239}Pu)	24,100 years
Carbon (^{14}C)	5715 years
Radium (^{226}Ra)	1599 years
Einsteinium (^{254}Es)	276 days
Nobelium (^{257}No)	25 seconds

EXAMPLE 3 Radioactive Decay

Sergei Supinsky/AFP/Getty Images

Suppose that 10 grams of the plutonium isotope Pu-239 was released in the Chernobyl nuclear accident. How long will it take for the 10 grams to decay to 1 gram?

Solution Let y represent the mass (in grams) of the plutonium. Because the rate of decay is proportional to y, you know that

$$y = Ce^{kt}$$

where t is the time in years. To find the values of the constants C and k, apply the initial conditions. Using the fact that $y = 10$ when $t = 0$, you can write

$$10 = Ce^{k(0)} = Ce^0$$

which implies that $C = 10$. Next, using the fact that $y = 5$ when $t = 24{,}100$, you can write

$$5 = 10e^{k(24,100)}$$

$$\frac{1}{2} = e^{24,100k}$$

$$\frac{1}{24,100} \ln \frac{1}{2} = k$$

$$-0.000028761 \approx k.$$

So, the model is

$$y = 10e^{-0.000028761t}. \qquad \text{Half-life model}$$

To find the time it would take for 10 grams to decay to 1 gram, you can solve for t in the equation

$$1 = 10e^{-0.000028761t}.$$

The solution is approximately 80,059 years.

NOTE The exponential decay model in Example 3 could also be written as $y = 10\left(\frac{1}{2}\right)^{t/24,100}$. This model is much easier to derive, but for some applications it is not as convenient to use.

From Example 3, notice that in an exponential growth or decay problem, it is easy to solve for C when you are given the value of y at $t = 0$. The next example demonstrates a procedure for solving for C and k when you do not know the value of y at $t = 0$.

EXAMPLE 4 Population Growth

Suppose an experimental population of fruit flies increases according to the law of exponential growth. There were 100 flies after the second day of the experiment and 300 flies after the fourth day. Approximately how many flies were in the original population?

Solution Let $y = Ce^{kt}$ be the number of flies at time t, where t is measured in days. Because $y = 100$ when $t = 2$ and $y = 300$ when $t = 4$, you can write

$$100 = Ce^{2k} \quad \text{and} \quad 300 = Ce^{4k}.$$

From the first equation, you know that $C = 100e^{-2k}$. Substituting this value into the second equation produces the following.

$$\begin{aligned} 300 &= 100e^{-2k}e^{4k} \\ 300 &= 100e^{2k} \\ \ln 3 &= 2k \\ \frac{1}{2}\ln 3 &= k \\ 0.5493 &\approx k \end{aligned}$$

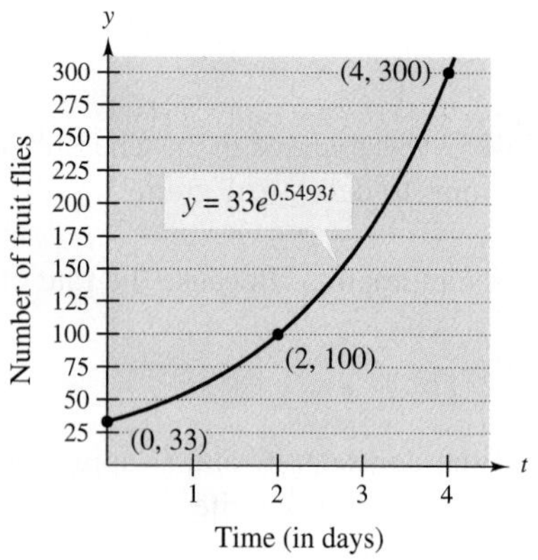

Figure 6.9

So, the exponential growth model is

$$y = Ce^{0.5493t}.$$

To solve for C, reapply the condition $y = 100$ when $t = 2$ and obtain

$$\begin{aligned} 100 &= Ce^{0.5493(2)} \\ C &= 100e^{-1.0986} \approx 33. \end{aligned}$$

So, the original population (when $t = 0$) consisted of approximately $y = C = 33$ flies, as shown in Figure 6.9.

EXAMPLE 5 Declining Sales

Four months after it stops advertising, a manufacturing company notices that its sales have dropped from 100,000 units per month to 80,000 units per month. If the sales follow an exponential pattern of decline, what will they be after another 2 months?

Solution Use the exponential decay model $y = Ce^{kt}$, where t is measured in months. From the initial condition ($t = 0$), you know that $C = 100{,}000$. Moreover, because $y = 80{,}000$ when $t = 4$, you have

$$\begin{aligned} 80{,}000 &= 100{,}000e^{4k} \\ 0.8 &= e^{4k} \\ \ln(0.8) &= 4k \\ -0.0558 &\approx k. \end{aligned}$$

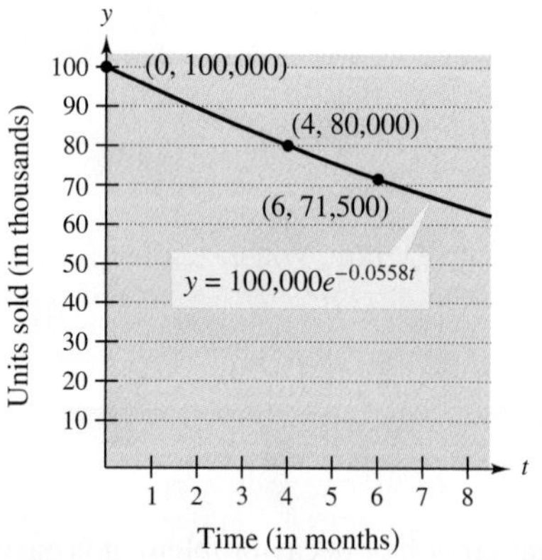

Figure 6.10

So, after 2 more months ($t = 6$), you can expect the monthly sales rate to be

$$\begin{aligned} y &\approx 100{,}000e^{-0.0558(6)} \\ &\approx 71{,}500 \text{ units.} \end{aligned}$$

See Figure 6.10.

In Examples 2 through 5, you did not actually have to solve the differential equation

$$y' = ky.$$

(This was done once in the proof of Theorem 6.1.) The next example demonstrates a problem whose solution involves the separation of variables technique. The example concerns **Newton's Law of Cooling,** which states that the rate of change in the temperature of an object is proportional to the difference between the object's temperature and the temperature of the surrounding medium.

EXAMPLE 6 Newton's Law of Cooling

Let y represent the temperature (in °F) of an object in a room whose temperature is kept at a constant 60°. If the object cools from 100° to 90° in 10 minutes, how much longer will it take for its temperature to decrease to 80°?

Solution From Newton's Law of Cooling, you know that the rate of change in y is proportional to the difference between y and 60. This can be written as

$$y' = k(y - 60), \qquad 80 \le y \le 100.$$

To solve this differential equation, use separation of variables, as shown.

$$\frac{dy}{dt} = k(y - 60) \qquad \text{Differential equation}$$

$$\left(\frac{1}{y - 60}\right) dy = k\,dt \qquad \text{Separate variables.}$$

$$\int \frac{1}{y - 60}\,dy = \int k\,dt \qquad \text{Integrate each side.}$$

$$\ln|y - 60| = kt + C_1 \qquad \text{Find antiderivative of each side.}$$

Because $y > 60$, $|y - 60| = y - 60$, and you can omit the absolute value signs. Using exponential notation, you have

$$y - 60 = e^{kt + C_1} \implies y = 60 + Ce^{kt}. \qquad C = e^{C_1}$$

Using $y = 100$ when $t = 0$, you obtain $100 = 60 + Ce^{k(0)} = 60 + C$, which implies that $C = 40$. Because $y = 90$ when $t = 10$,

$$90 = 60 + 40e^{k(10)}$$
$$30 = 40e^{10k}$$
$$k = \tfrac{1}{10} \ln \tfrac{3}{4} \approx -0.02877.$$

So, the model is

$$y = 60 + 40e^{-0.02877t} \qquad \text{Cooling model}$$

and finally, when $y = 80$, you obtain

$$80 = 60 + 40e^{-0.02877t}$$
$$20 = 40e^{-0.02877t}$$
$$\tfrac{1}{2} = e^{-0.02877t}$$
$$\ln \tfrac{1}{2} = -0.02877t$$
$$t \approx 24.09 \text{ minutes.}$$

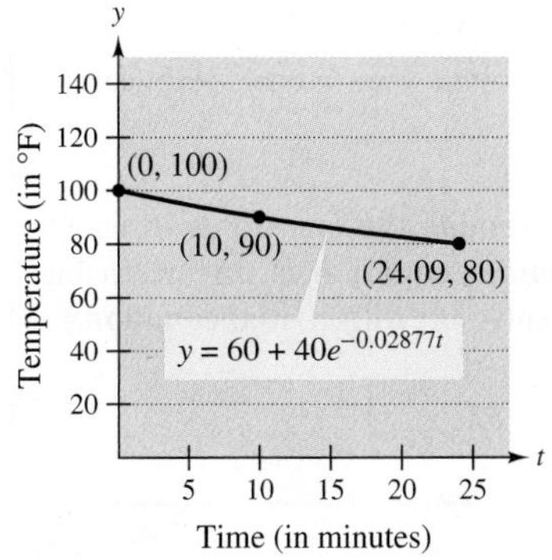

Figure 6.11

So, it will require about 14.09 *more* minutes for the object to cool to a temperature of 80° (see Figure 6.11).

Exercises for Section 6.2

See www.CalcChat.com for worked-out solutions to odd-numbered exercises.

In Exercises 1–10, solve the differential equation.

1. $\dfrac{dy}{dx} = x + 2$
2. $\dfrac{dy}{dx} = 4 - x$
3. $\dfrac{dy}{dx} = y + 2$
4. $\dfrac{dy}{dx} = 4 - y$
5. $y' = \dfrac{5x}{y}$
6. $y' = \dfrac{\sqrt{x}}{3y}$
7. $y' = \sqrt{x}\, y$
8. $y' = x(1 + y)$
9. $(1 + x^2)y' - 2xy = 0$
10. $xy + y' = 100x$

In Exercises 11–14, write and solve the differential equation that models the verbal statement.

11. The rate of change of Q with respect to t is inversely proportional to the square of t.
12. The rate of change of P with respect to t is proportional to $10 - t$.
13. The rate of change of N with respect to s is proportional to $250 - s$.
14. The rate of change of y with respect to x varies jointly as x and $L - y$.

Slope Fields **In Exercises 15 and 16, a differential equation, a point, and a slope field are given. (a) Sketch two approximate solutions of the differential equation on the slope field, one of which passes through the given point. (b) Use integration to find the particular solution of the differential equation and use a graphing utility to graph the solution. Compare the result with the sketch in part (a). To print an enlarged copy of the graph, go to the website *www.mathgraphs.com*.**

15. $\dfrac{dy}{dx} = x(6 - y)$, $(0, 0)$

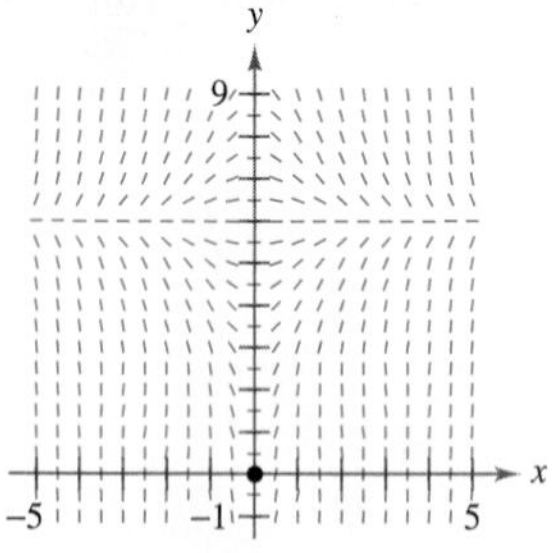

16. $\dfrac{dy}{dx} = xy$, $\left(0, \frac{1}{2}\right)$

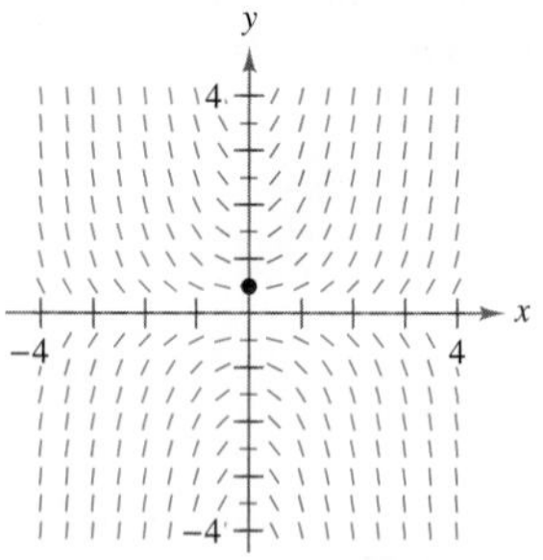

In Exercises 17–20, find the function $y = f(t)$ passing through the point $(0, 10)$ with the given first derivative. Use a graphing utility to graph the solution.

17. $\dfrac{dy}{dt} = \dfrac{1}{2}t$
18. $\dfrac{dy}{dt} = -\dfrac{3}{4}\sqrt{t}$
19. $\dfrac{dy}{dt} = -\dfrac{1}{2}y$
20. $\dfrac{dy}{dt} = \dfrac{3}{4}y$

In Exercises 21–24, write and solve the differential equation that models the verbal statement. Evaluate the solution at the specified value of the independent variable.

21. The rate of change of y is proportional to y. When $x = 0$, $y = 4$ and when $x = 3$, $y = 10$. What is the value of y when $x = 6$?
22. The rate of change of N is proportional to N. When $t = 0$, $N = 250$ and when $t = 1$, $N = 400$. What is the value of N when $t = 4$?
23. The rate of change of V is proportional to V. When $t = 0$, $V = 20{,}000$ and when $t = 4$, $V = 12{,}500$. What is the value of V when $t = 6$?
24. The rate of change of P is proportional to P. When $t = 0$, $P = 5000$ and when $t = 1$, $P = 4750$. What is the value of P when $t = 5$?

In Exercises 25–28, find the exponential function $y = Ce^{kt}$ that passes through the two given points.

25.

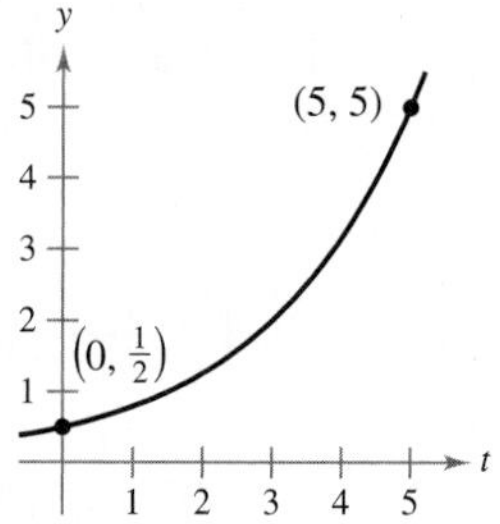

26.

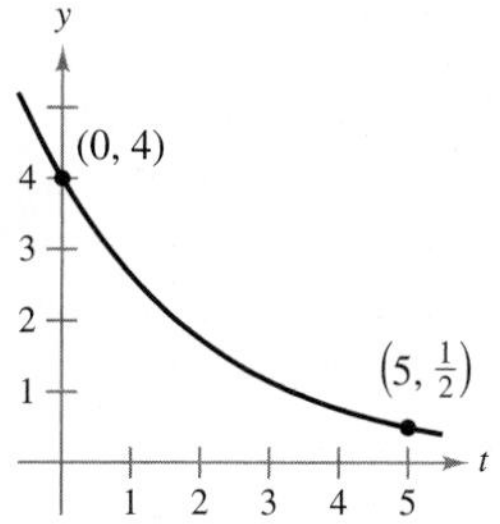

27.

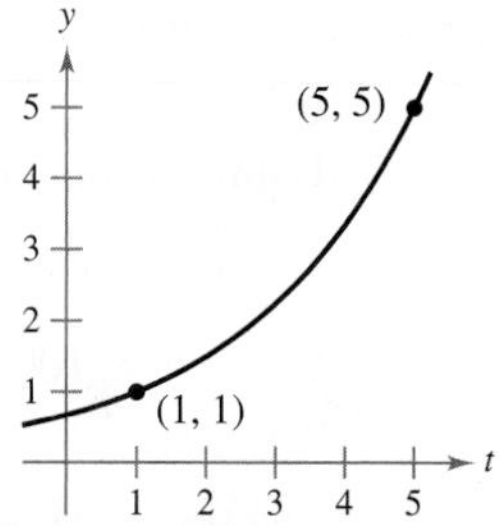

28.

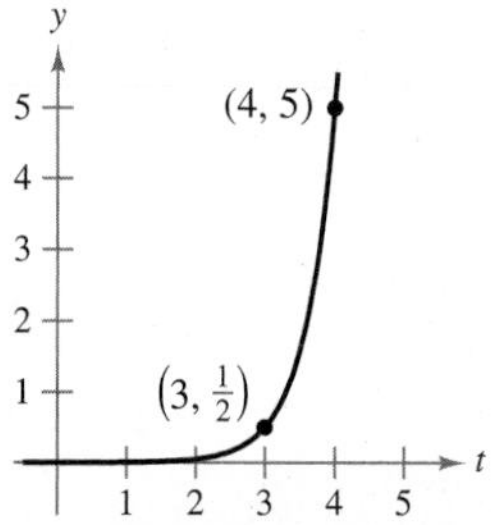

Writing About Concepts

29. Describe what the values of C and k represent in the exponential growth and decay model, $y = Ce^{kt}$.
30. Give the differential equation that models exponential growth and decay.

In Exercises 31 and 32, determine the quadrants in which the solution of the differential equation is an increasing function. Explain. (Do not solve the differential equation.)

31. $\dfrac{dy}{dx} = \dfrac{1}{2}xy$
32. $\dfrac{dy}{dx} = \dfrac{1}{2}x^2y$

Radioactive Decay **In Exercises 33–40, complete the table for the radioactive isotope.**

	Isotope	*Half-Life (in years)*	*Initial Quantity*	*Amount After 1000 Years*	*Amount After 10,000 Years*
33.	^{226}Ra	1599	10 g		
34.	^{226}Ra	1599		1.5 g	
35.	^{226}Ra	1599			0.5 g
36.	^{14}C	5715			2 g
37.	^{14}C	5715	5 g		
38.	^{14}C	5715		3.2 g	
39.	^{239}Pu	24,100		2.1 g	
40.	^{239}Pu	24,100			0.4 g

41. ***Radioactive Decay*** Radioactive radium has a half-life of approximately 1599 years. What percent of a given amount remains after 100 years?

42. ***Carbon Dating*** Carbon-14 dating assumes that the carbon dioxide on Earth today has the same radioactive content as it did centuries ago. If this is true, the amount of ^{14}C absorbed by a tree that grew several centuries ago should be the same as the amount of ^{14}C absorbed by a tree growing today. A piece of ancient charcoal contains only 15% as much of the radioactive carbon as a piece of modern charcoal. How long ago was the tree burned to make the ancient charcoal? (The half-life of ^{14}C is 5715 years.)

Compound Interest **In Exercises 43–48, complete the table for a savings account in which interest is compounded continuously.**

	Initial Investment	*Annual Rate*	*Time to Double*	*Amount After 10 Years*
43.	\$1000	6%		
44.	\$20,000	$5\frac{1}{2}\%$		
45.	\$750		$7\frac{3}{4}$ yr	
46.	\$10,000		5 yr	
47.	\$500			\$1292.85
48.	\$2000			\$5436.56

Compound Interest **In Exercises 49–52, find the principal P that must be invested at rate r, compounded monthly, so that \$500,000 will be available for retirement in t years.**

49. $r = 7\frac{1}{2}\%, \quad t = 20$

50. $r = 6\%, \quad t = 40$

51. $r = 8\%, \quad t = 35$

52. $r = 9\%, \quad t = 25$

Compound Interest **In Exercises 53–56, find the time necessary for \$1000 to double if it is invested at a rate of r compounded (a) annually, (b) monthly, (c) daily, and (d) continuously.**

53. $r = 7\%$

54. $r = 6\%$

55. $r = 8.5\%$

56. $r = 5.5\%$

Population **In Exercises 57–60, the population (in millions) of a country in 2001 and the expected continuous annual rate of change k of the population for the years 2000 through 2010 are given.** ***(Source: U.S. Census Bureau, International Data Base)***

(a) Find the exponential growth model $P = Ce^{kt}$ for the population by letting $t = 0$ correspond to 2000.

(b) Use the model to predict the population of the country in 2015.

(c) Discuss the relationship between the sign of k and the change in population for the country.

	Country	*2001 Population*	k
57.	Bulgaria	7.7	−0.009
58.	Cambodia	12.7	0.018
59.	Jordan	5.2	0.026
60.	Lithuania	3.6	−0.002

61. ***Modeling Data*** One hundred bacteria are started in a culture and the number N of bacteria is counted each hour for 5 hours. The results are shown in the table, where t is the time in hours.

t	0	1	2	3	4	5
N	100	126	151	198	243	297

(a) Use the regression capabilities of a graphing utility to find an exponential model for the data.

(b) Use the model to estimate the time required for the population to quadruple in size.

62. ***Bacteria Growth*** The number of bacteria in a culture is increasing according to the law of exponential growth. There are 125 bacteria in the culture after 2 hours and 350 bacteria after 4 hours.

(a) Find the initial population.

(b) Write an exponential growth model for the bacteria population. Let t represent time in hours.

(c) Use the model to determine the number of bacteria after 8 hours.

(d) After how many hours will the bacteria count be 25,000?

63. ***Learning Curve*** The management at a certain factory has found that a worker can produce at most 30 units in a day. The learning curve for the number of units N produced per day after a new employee has worked t days is $N = 30(1 - e^{kt})$. After 20 days on the job, a particular worker produces 19 units.

(a) Find the learning curve for this worker.

(b) How many days should pass before this worker is producing 25 units per day?

64. ***Learning Curve*** If in Exercise 63 management requires a new employee to produce at least 20 units per day after 30 days on the job, find (a) the learning curve that describes this minimum requirement and (b) the number of days before a minimal achiever is producing 25 units per day.

65. *Modeling Data* The table shows the population P (in millions) of the United States from 1960 to 2000. *(Source: U.S. Census Bureau)*

Year	1960	1970	1980	1990	2000
Population, P	181	205	228	250	282

(a) Use the 1960 and 1970 data to find an exponential model P_1 for the data. Let $t = 0$ represent 1960.

(b) Use a graphing utility to find an exponential model P_2 for the data. Let $t = 0$ represent 1960.

(c) Use a graphing utility to plot the data and graph both models in the same viewing window. Compare the actual data with the predictions. Which model better fits the data?

(d) Estimate when the population will be 320 million.

66. *Modeling Data* The table shows the net receipts and the amounts required to service the national debt (interest on Treasury debt securities) of the United States from 1992 through 2001. The monetary amounts are given in billions of dollars. *(Source: U.S. Office of Management and Budget)*

Year	1992	1993	1994	1995	1996
Receipts	1091.3	1154.4	1258.6	1351.8	1453.1
Interest	292.3	292.5	296.3	332.4	343.9

Year	1997	1998	1999	2000	2001
Receipts	1579.3	1721.8	1827.5	2025.2	1991.2
Interest	355.8	363.8	353.5	361.9	359.5

(a) Use the regression capabilities of a graphing utility to find an exponential model R for the receipts and a quartic model I for the amount required to service the debt. Let t represent the time in years, with $t = 2$ corresponding to 1992.

(b) Use a graphing utility to plot the points corresponding to the receipts, and graph the corresponding model. Based on the model, what is the continuous rate of growth of the receipts?

(c) Use a graphing utility to plot the points corresponding to the amount required to service the debt, and graph the quartic model.

(d) Find a function $P(t)$ that approximates the percent of the receipts that is required to service the national debt. Use a graphing utility to graph this function.

67. *Sound Intensity* The level of sound β (in decibels), with an intensity of I is

$$\beta(I) = 10 \log_{10} \frac{I}{I_0}$$

where I_0 is an intensity of 10^{-16} watts per square centimeter, corresponding roughly to the faintest sound that can be heard. Determine $\beta(I)$ for the following.

(a) $I = 10^{-14}$ watts per square centimeter (whisper)

(b) $I = 10^{-9}$ watts per square centimeter (busy street corner)

(c) $I = 10^{-6.5}$ watts per square centimeter (air hammer)

(d) $I = 10^{-4}$ watts per square centimeter (threshold of pain)

68. *Noise Level* With the installation of noise suppression materials, the noise level in an auditorium was reduced from 93 to 80 decibels. Use the function in Exercise 67 to find the percent decrease in the intensity level of the noise as a result of the installation of these materials.

69. *Forestry* The value of a tract of timber is

$$V(t) = 100{,}000e^{0.8\sqrt{t}}$$

where t is the time in years, with $t = 0$ corresponding to 1998. If money earns interest continuously at 10%, the present value of the timber at any time t is $A(t) = V(t)e^{-0.10t}$. Find the year in which the timber should be harvested to maximize the present value function.

70. *Earthquake Intensity* On the Richter scale, the magnitude R of an earthquake of intensity I is

$$R = \frac{\ln I - \ln I_0}{\ln 10}$$

where I_0 is the minimum intensity used for comparison. Assume that $I_0 = 1$.

(a) Find the intensity of the 1906 San Francisco earthquake ($R = 8.3$).

(b) Find the factor by which the intensity is increased if the Richter scale measurement is doubled.

(c) Find dR/dI.

71. *Newton's Law of Cooling* When an object is removed from a furnace and placed in an environment with a constant temperature of 80°F, its core temperature is 1500°F. One hour after it is removed, the core temperature is 1120°F. Find the core temperature 5 hours after the object is removed from the furnace.

72. *Newton's Law of Cooling* A container of hot liquid is placed in a freezer that is kept at a constant temperature of 20°F. The initial temperature of the liquid is 160°F. After 5 minutes, the liquid's temperature is 60°F. How much longer will it take for its temperature to decrease to 30°F?

True or False? **In Exercises 73–76, determine whether the statement is true or false. If it is false, explain why or give an example that shows it is false.**

73. In exponential growth, the rate of growth is constant.

74. In linear growth, the rate of growth is constant.

75. If prices are rising at a rate of 0.5% per month, then they are rising at a rate of 6% per year.

76. The differential equation modeling exponential growth is $dy/dx = ky$, where k is a constant.

Section 6.3 Separation of Variables and the Logistic Equation

- **Recognize and solve differential equations that can be solved by separation of variables.**
- **Recognize and solve homogeneous differential equations.**
- **Use differential equations to model and solve applied problems.**
- **Solve and analyze logistic differential equations.**

Separation of Variables

Consider a differential equation that can be written in the form

$$M(x) + N(y)\frac{dy}{dx} = 0$$

where M is a continuous function of x alone and N is a continuous function of y alone. As you saw in the preceding section, for this type of equation, all x terms can be collected with dx and all y terms with dy, and a solution can be obtained by integration. Such equations are said to be **separable,** and the solution procedure is called *separation of variables*. Below are some examples of differential equations that are separable.

Original Differential Equation	*Rewritten with Variables Separated*
$x^2 + 3y\dfrac{dy}{dx} = 0$	$3y\,dy = -x^2\,dx$
$(\sin x)y' = \cos x$	$dy = \cot x\,dx$
$\dfrac{xy'}{e^y + 1} = 2$	$\dfrac{1}{e^y + 1}\,dy = \dfrac{2}{x}\,dx$

EXAMPLE 1 Separation of Variables

Find the general solution of $(x^2 + 4)\dfrac{dy}{dx} = xy$.

Solution To begin, note that $y = 0$ is a solution. To find other solutions, assume that $y \neq 0$ and separate variables as shown.

$$(x^2 + 4)\,dy = xy\,dx \qquad \text{Differential form}$$

$$\frac{dy}{y} = \frac{x}{x^2 + 4}\,dx \qquad \text{Separate variables.}$$

Now, integrate to obtain

$$\int \frac{dy}{y} = \int \frac{x}{x^2 + 4}\,dx \qquad \text{Integrate.}$$

$$\ln|y| = \frac{1}{2}\ln(x^2 + 4) + C_1$$

$$\ln|y| = \ln\sqrt{x^2 + 4} + C_1$$

$$|y| = e^{C_1}\sqrt{x^2 + 4}$$

$$y = \pm e^{C_1}\sqrt{x^2 + 4}.$$

Because $y = 0$ is also a solution, you can write the general solution as

$$y = C\sqrt{x^2 + 4}. \qquad \text{General solution}$$

NOTE Be sure to check your solutions throughout this chapter. In Example 1, you can check the solution $y = C\sqrt{x^2 + 4}$ by differentiating and substituting into the original equation.

$$(x^2 + 4)\frac{dy}{dx} = xy$$

$$(x^2 + 4)\frac{Cx}{\sqrt{x^2 + 4}} \stackrel{?}{=} x\left(C\sqrt{x^2 + 4}\right)$$

$$Cx\sqrt{x^2 + 4} = Cx\sqrt{x^2 + 4}$$

So, the solution checks.

In some cases it is not feasible to write the general solution in the explicit form $y = f(x)$. The next example illustrates such a solution. Implicit differentiation can be used to verify this solution.

FOR FURTHER INFORMATION For an example (from engineering) of a differential equation that is separable, see the article "Designing a Rose Cutter" by J. S. Hartzler in *The College Mathematics Journal*. To view this article, go to the website *www.matharticles.com*.

EXAMPLE 2 Finding a Particular Solution

Given the initial condition $y(0) = 1$, find the particular solution of the equation

$$xy\,dx + e^{-x^2}(y^2 - 1)\,dy = 0.$$

Solution Note that $y = 0$ is a solution of the differential equation—but this solution does not satisfy the initial condition. So, you can assume that $y \neq 0$. To separate variables, you must rid the first term of y and the second term of e^{-x^2}. So, you should multiply by e^{x^2}/y and obtain the following.

$$xy\,dx + e^{-x^2}(y^2 - 1)\,dy = 0$$
$$e^{-x^2}(y^2 - 1)\,dy = -xy\,dx$$
$$\int\left(y - \frac{1}{y}\right)dy = \int -xe^{x^2}\,dx$$
$$\frac{y^2}{2} - \ln|y| = -\frac{1}{2}e^{x^2} + C$$

From the initial condition $y(0) = 1$, you have $\frac{1}{2} - 0 = -\frac{1}{2} + C$, which implies that $C = 1$. So, the particular solution has the implicit form

$$\frac{y^2}{2} - \ln|y| = -\frac{1}{2}e^{x^2} + 1$$
$$y^2 - \ln y^2 + e^{x^2} = 2.$$

You can check this by differentiating and rewriting to get the original equation.

EXAMPLE 3 Finding a Particular Solution Curve

Find the equation of the curve that passes through the point $(1, 3)$ and has a slope of y/x^2 at any point (x, y).

Solution Because the slope of the curve is given by y/x^2, you have

$$\frac{dy}{dx} = \frac{y}{x^2}$$

with the initial condition $y(1) = 3$. Separating variables and integrating produces

$$\int\frac{dy}{y} = \int\frac{dx}{x^2}, \quad y \neq 0$$
$$\ln|y| = -\frac{1}{x} + C_1$$
$$y = e^{-(1/x)+C_1} = Ce^{-1/x}.$$

Because $y = 3$ when $x = 1$, it follows that $3 = Ce^{-1}$ and $C = 3e$. So, the equation of the specified curve is

$$y = (3e)e^{-1/x} = 3e^{(x-1)/x}, \quad x > 0.$$

See Figure 6.12.

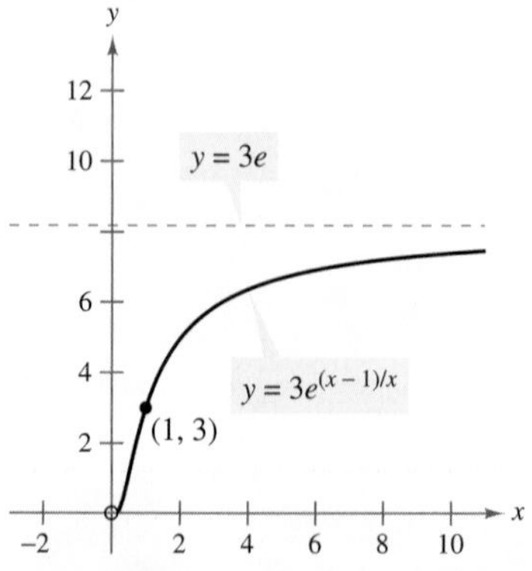

Figure 6.12

Homogeneous Differential Equations

Some differential equations that are not separable in x and y can be made separable by a change of variables. This is true for differential equations of the form $y' = f(x, y)$, where f is a **homogeneous function.** The function given by $f(x, y)$ is **homogeneous of degree *n*** if

NOTE The notation $f(x, y)$ is used to denote a function of two variables in much the same way as $f(x)$ denotes a function of one variable. You will study functions of two variables in detail in Chapter 13.

$$f(tx, ty) = t^n f(x, y)$$ Homogeneous function of degree n

where n is a real number.

EXAMPLE 4 Verifying Homogeneous Functions

a. $f(x, y) = x^2y - 4x^3 + 3xy^2$ is a homogeneous function of degree 3 because

$$\begin{aligned} f(tx, ty) &= (tx)^2(ty) - 4(tx)^3 + 3(tx)(ty)^2 \\ &= t^3(x^2y) - t^3(4x^3) + t^3(3xy^2) \\ &= t^3(x^2y - 4x^3 + 3xy^2) \\ &= t^3 f(x, y). \end{aligned}$$

b. $f(x, y) = xe^{x/y} + y \sin(y/x)$ is a homogeneous function of degree 1 because

$$\begin{aligned} f(tx, ty) &= txe^{tx/ty} + ty \sin \frac{ty}{tx} \\ &= t\left(xe^{x/y} + y \sin \frac{y}{x}\right) \\ &= tf(x, y). \end{aligned}$$

c. $f(x, y) = x + y^2$ is *not* a homogeneous function because

$$f(tx, ty) = tx + t^2y^2 = t(x + ty^2) \neq t^n(x + y^2).$$

d. $f(x, y) = x/y$ is a homogeneous function of degree 0 because

$$f(tx, ty) = \frac{tx}{ty} = t^0 \frac{x}{y}.$$

Definition of Homogeneous Differential Equation

A **homogeneous differential equation** is an equation of the form

$$M(x, y)\, dx + N(x, y) dy = 0$$

where M and N are homogeneous functions of the same degree.

EXAMPLE 5 Testing for Homogeneous Differential Equations

a. $(x^2 + xy)\, dx + y^2\, dy = 0$ is homogeneous of degree 2.

b. $x^3\, dx = y^3\, dy$ is homogeneous of degree 3.

c. $(x^2 + 1)\, dx + y^2\, dy = 0$ is *not* a homogeneous differential equation.

To solve a homogeneous differential equation by the method of separation of variables, use the following change of variables theorem.

THEOREM 6.2 Change of Variables for Homogeneous Equations

If $M(x, y)\,dx + N(x, y)\,dy = 0$ is homogeneous, then it can be transformed into a differential equation whose variables are separable by the substitution

$$y = vx$$

where v is a differentiable function of x.

EXAMPLE 6 Solving a Homogeneous Differential Equation

Find the general solution of

$$(x^2 - y^2)\,dx + 3xy\,dy = 0.$$

STUDY TIP The substitution $y = vx$ will yield a differential equation that is separable with respect to the variables x and v. You must write your final solution, however, in terms of x and y.

Solution Because $(x^2 - y^2)$ and $3xy$ are both homogeneous of degree 2, let $y = vx$ to obtain $dy = x\,dv + v\,dx$. Then, by substitution, you have

$$\begin{aligned} (x^2 - v^2x^2)\,dx + 3x(vx)\overbrace{(x\,dv + v\,dx)}^{dy} &= 0 \\ (x^2 + 2v^2x^2)\,dx + 3x^3v\,dv &= 0 \\ x^2(1 + 2v^2)\,dx + x^2(3vx)\,dv &= 0. \end{aligned}$$

Dividing by x^2 and separating variables produces

$$\begin{aligned} (1 + 2v^2)\,dx &= -3vx\,dv \\ \int \frac{dx}{x} &= \int \frac{-3v}{1 + 2v^2}\,dv \\ \ln|x| &= -\frac{3}{4}\ln(1 + 2v^2) + C_1 \\ 4\ln|x| &= -3\ln(1 + 2v^2) + \ln|C| \\ \ln x^4 &= \ln|C(1 + 2v^2)^{-3}| \\ x^4 &= C(1 + 2v^2)^{-3}. \end{aligned}$$

Substituting for v produces the following general solution.

$$x^4 = C\left[1 + 2\left(\frac{y}{x}\right)^2\right]^{-3}$$

$$\left(1 + \frac{2y^2}{x^2}\right)^3 x^4 = C$$

$$(x^2 + 2y^2)^3 = Cx^2 \qquad \text{General solution}$$

You can check this by differentiating and rewriting to get the original equation.

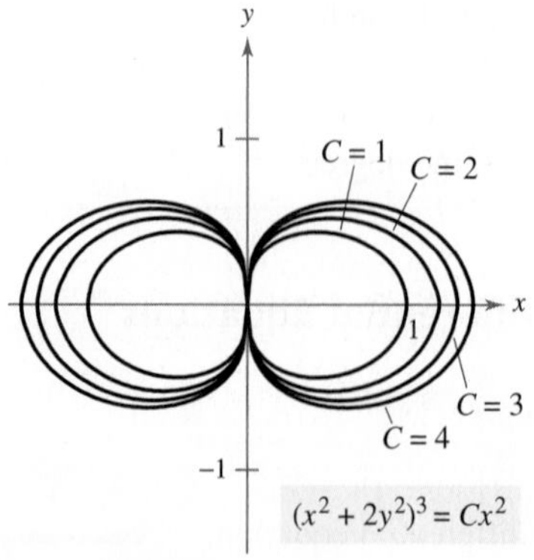

General solutions of $(x^2 - y^2)\,dx + 3xy\,dy = 0$
Figure 6.13

TECHNOLOGY If you have access to a graphing utility, try using it to graph several of the solutions in Example 6. For instance, Figure 6.13 shows the graphs of

$$(x^2 + 2y^2)^3 = Cx^2$$

for $C = 1, 2, 3,$ and 4.

Applications

EXAMPLE 7 Wildlife Population

Photodisc/Getty Images

The rate of change of the number of coyotes $N(t)$ in a population is directly proportional to $650 - N(t)$, where t is the time in years. When $t = 0$, the population is 300, and when $t = 2$, the population has increased to 500. Find the population when $t = 3$.

Solution Because the rate of change of the population is proportional to $650 - N(t)$, you can write the following differential equation.

$$\frac{dN}{dt} = k(650 - N)$$

You can solve this equation using separation of variables.

$$dN = k(650 - N)\,dt \qquad \text{Differential form}$$

$$\frac{dN}{650 - N} = k\,dt \qquad \text{Separate variables.}$$

$$-\ln|650 - N| = kt + C_1 \qquad \text{Integrate.}$$

$$\ln|650 - N| = -kt - C_1$$

$$650 - N = e^{-kt - C_1} \qquad \text{Assume } N < 650.$$

$$N = 650 - Ce^{-kt} \qquad \text{General solution}$$

Using $N = 300$ when $t = 0$, you can conclude that $C = 350$, which produces

$$N = 650 - 350e^{-kt}.$$

Then, using $N = 500$ when $t = 2$, it follows that

$$500 = 650 - 350e^{-2k} \quad \Longrightarrow \quad e^{-2k} = \frac{3}{7} \quad \Longrightarrow \quad k \approx 0.4236.$$

So, the model for the coyote population is

$$N = 650 - 350e^{-0.4236t}. \qquad \text{Model for population}$$

When $t = 3$, you can approximate the population to be

$$N = 650 - 350e^{-0.4236(3)} \approx 552 \text{ coyotes.}$$

The model for the population is shown in Figure 6.14.

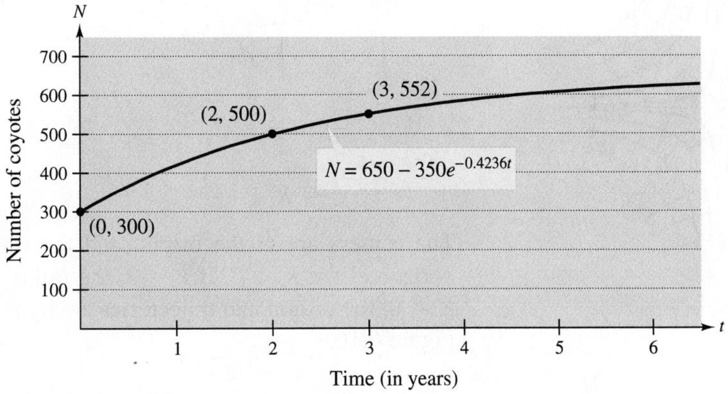

Figure 6.14

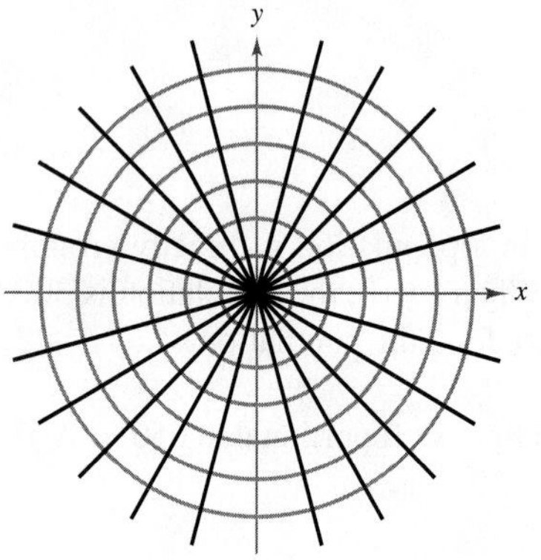

Each line $y = Kx$ is an orthogonal trajectory to the family of circles.
Figure 6.15

A common problem in electrostatics, thermodynamics, and hydrodynamics involves finding a family of curves, each of which is orthogonal to all members of a given family of curves. For example, Figure 6.15 shows a family of circles

$$x^2 + y^2 = C \qquad \text{Family of circles}$$

each of which intersects the lines in the family

$$y = Kx \qquad \text{Family of lines}$$

at right angles. Two such families of curves are said to be **mutually orthogonal,** and each curve in one of the families is called an **orthogonal trajectory** of the other family. In electrostatics, lines of force are orthogonal to the *equipotential curves.* In thermodynamics, the flow of heat across a plane surface is orthogonal to the *isothermal curves.* In hydrodynamics, the flow (stream) lines are orthogonal trajectories of the *velocity potential curves.*

EXAMPLE 8 Finding Orthogonal Trajectories

Describe the orthogonal trajectories for the family of curves given by

$$y = \frac{C}{x}$$

for $C \neq 0$. Sketch several members of each family.

Solution First, solve the given equation for C and write $xy = C$. Then, by differentiating implicitly with respect to x, you obtain the differential equation

$$xy' + y = 0 \qquad \text{Differential equation}$$

$$x\frac{dy}{dx} = -y$$

$$\frac{dy}{dx} = -\frac{y}{x}. \qquad \text{Slope of given family}$$

Because y' represents the slope of the given family of curves at (x, y), it follows that the orthogonal family has the negative reciprocal slope x/y. So,

$$\frac{dy}{dx} = \frac{x}{y}. \qquad \text{Slope of orthogonal family}$$

Now you can find the orthogonal family by separating variables and integrating.

$$\int y\,dy = \int x\,dx$$

$$\frac{y^2}{2} = \frac{x^2}{2} + C_1$$

$$y^2 - x^2 = K$$

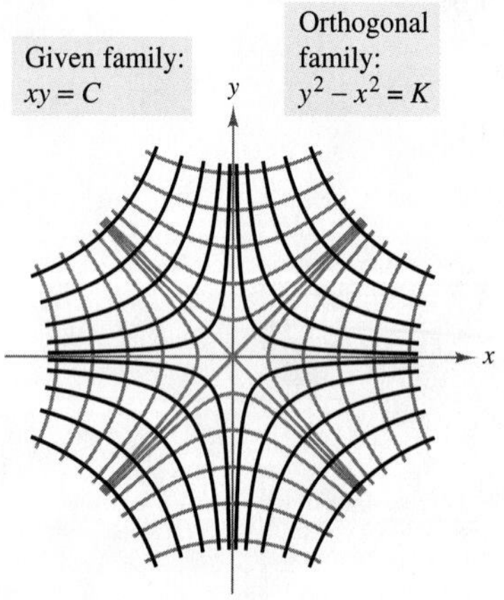

Orthogonal trajectories
Figure 6.16

The centers are at the origin, and the transverse axes are vertical for $K > 0$ and horizontal for $K < 0$. If $k = 0$, the orthogonal trajectories are the lines $y = \pm x$. If $K \neq 0$, the orthogonal trajectories are hyperbolas. Several trajectories are shown in Figure 6.16.

Logistic Differential Equation

In Section 6.2, the exponential growth model is derived from the fact that the rate of change of a variable y is proportional to the value of y. You observed that the differential equation $dy/dt = ky$ has the general solution $y = Ce^{kt}$. Exponential growth is unlimited, but when describing a population, there often exists some upper limit L past which growth cannot occur. This upper limit L is called the **carrying capacity,** which is the maximum population $y(t)$ that can be sustained or supported as time t increases. A model that is often used for this type of growth is the **logistic differential equation**

$$\frac{dy}{dt} = ky\left(1 - \frac{y}{L}\right) \qquad \text{Logistic differential equation}$$

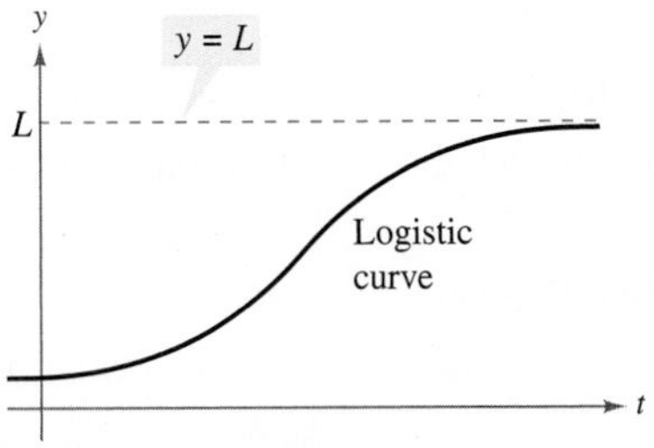

Note that as $t \to \infty$, $y \to L$.
Figure 6.17

where k and L are positive constants. A population that satisfies this equation does not grow without bound, but approaches the carrying capacity L as t increases.

From the equation, you can see that if y is between 0 and the carrying capacity L, then $dy/dt > 0$, and the population increases. If y is greater than L, then $dy/dt < 0$, and the population decreases. The graph of the function y is called the *logistic curve*, as shown in Figure 6.17.

EXAMPLE 9 Deriving the General Solution

Solve the logistic differential equation $\dfrac{dy}{dt} = ky\left(1 - \dfrac{y}{L}\right)$.

Solution Begin by separating variables.

$$\frac{dy}{dt} = ky\left(1 - \frac{y}{L}\right) \qquad \text{Write differential equation.}$$

$$\frac{1}{y(1 - y/L)}dy = kdt \qquad \text{Separate variables.}$$

$$\int \frac{1}{y(1 - y/L)}dy = \int kdt \qquad \text{Integrate each side.}$$

$$\int \left(\frac{1}{y} + \frac{1}{L - y}\right)dy = \int kdt \qquad \text{Rewrite left side using partial fractions.}$$

$$\ln|y| - \ln|L - y| = kt + C \qquad \text{Find antiderivative of each side.}$$

$$\ln\left|\frac{L - y}{y}\right| = -kt - C \qquad \text{Multiply each side by } -1 \text{ and simplify.}$$

$$\left|\frac{L - y}{y}\right| = e^{-kt-C} = e^{-C}e^{-kt} \qquad \text{Exponentiate each side.}$$

$$\frac{L - y}{y} = be^{-kt} \qquad \text{Let } \pm e^{-C} = b.$$

Solving this equation for y produces $y = \dfrac{L}{1 + be^{-kt}}$.

> **EXPLORATION**
>
> Use a graphing utility to investigate the effects of the values of L, b, and k on the graph of
>
> $$y = \frac{L}{1 + be^{-kt}}.$$
>
> Include some examples to support your results.

From Example 9, you can conclude that all solutions of the logistic differential equation are of the general form

$$y = \frac{L}{1 + be^{-kt}}.$$

EXPLORATION

Explain what happens if $p(0) = L$.

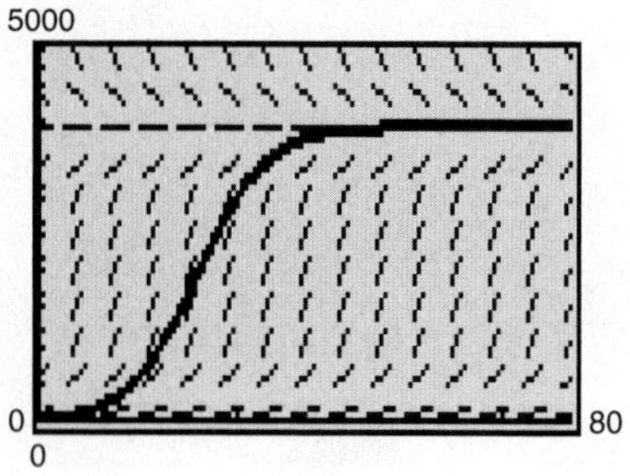

Figure 6.18

EXAMPLE 10 Solving a Logistic Differential Equation

A state game commission releases 40 elk into a game refuge. After 5 years, the elk population is 104. The commission believes that the environment can support no more than 4000 elk. The growth rate of the elk population p is

$$\frac{dp}{dt} = kp\left(1 - \frac{p}{4000}\right), \quad 40 \le p \le 4000$$

where t is the number of years.

a. Write a model for the elk population in terms of t.

b. Graph the slope field of the differential equation and the solution that passes through the point $(0, 40)$.

c. Use the model to estimate the elk population after 15 years.

d. Find the limit of the model as $t \to \infty$.

Solution

a. You know that $L = 4000$. So, the solution of the equation is of the form

$$p = \frac{4000}{1 + be^{-kt}}.$$

Because $p(0) = 40$, you can solve for b as shown.

$$40 = \frac{4000}{1 + be^{-k(0)}}$$

$$40 = \frac{4000}{1 + b} \quad \Longrightarrow \quad b = 99$$

Then, because $p = 104$ when $t = 5$, you can solve for k.

$$104 = \frac{4000}{1 + 99e^{-k(5)}} \quad \Longrightarrow \quad k \approx 0.194$$

So, a model for the elk population is given by $p = \dfrac{4000}{1 + 99e^{-0.194t}}$.

b. Using a graphing utility, you can graph the slope field of

$$\frac{dp}{dt} = 0.194p\left(1 - \frac{p}{4000}\right)$$

and the solution that passes through $(0, 40)$, as shown in Figure 6.18.

c. To estimate the elk population after 15 years, substitute 15 for t in the model

$$p = \frac{4000}{1 + 99e^{-0.194(15)}} \qquad \text{Substitute 15 for } t.$$

$$= \frac{4000}{1 + 99e^{-2.91}} \approx 626 \qquad \text{Simplify.}$$

d. As t increases without bound, the denominator of $\dfrac{4000}{1 + 99e^{-0.194t}}$ gets closer to 1.

So, $\displaystyle\lim_{t\to\infty} \frac{4000}{1 + 99e^{-0.194t}} = 4000.$

Exercises for Section 6.3

See www.CalcChat.com for worked-out solutions to odd-numbered exercises.

In Exercises 1–12, find the general solution of the differential equation.

1. $\dfrac{dy}{dx} = \dfrac{x}{y}$
2. $\dfrac{dy}{dx} = \dfrac{x^2 + 2}{3y^2}$
3. $\dfrac{dr}{ds} = 0.05r$
4. $\dfrac{dr}{ds} = 0.05s$
5. $(2 + x)y' = 3y$
6. $xy' = y$
7. $yy' = \sin x$
8. $yy' = 6\cos(\pi x)$
9. $\sqrt{1 - 4x^2}\, y' = x$
10. $\sqrt{x^2 - 9}\, y' = 5x$
11. $y \ln x - xy' = 0$
12. $4yy' - 3e^x = 0$

In Exercises 13–22, find the particular solution that satisfies the initial condition.

	Differential Equation	*Initial Condition*
13.	$yy' - e^x = 0$	$y(0) = 4$
14.	$\sqrt{x} + \sqrt{y}\, y' = 0$	$y(1) = 4$
15.	$y(x + 1) + y' = 0$	$y(-2) = 1$
16.	$2xy' - \ln x^2 = 0$	$y(1) = 2$
17.	$y(1 + x^2)y' - x(1 + y^2) = 0$	$y(0) = \sqrt{3}$
18.	$y\sqrt{1 - x^2}\, y' - x\sqrt{1 - y^2} = 0$	$y(0) = 1$
19.	$\dfrac{du}{dv} = uv \sin v^2$	$u(0) = 1$
20.	$\dfrac{dr}{ds} = e^{r-2s}$	$r(0) = 0$
21.	$dP - kP\, dt = 0$	$P(0) = P_0$
22.	$dT + k(T - 70)\, dt = 0$	$T(0) = 140$

In Exercises 23 and 24, find an equation of the graph that passes through the point and has the given slope.

23. $(1, 1),\quad y' = -\dfrac{9x}{16y}$
24. $(8, 2),\quad y' = \dfrac{2y}{3x}$

In Exercises 25 and 26, find all functions *f* having the indicated property.

25. The tangent to the graph of f at the point (x, y) intersects the x-axis at $(x + 2, 0)$.
26. All tangents to the graph of f pass through the origin.

In Exercises 27–34, determine whether the function is homogeneous, and if it is, determine its degree.

27. $f(x, y) = x^3 - 4xy^2 + y^3$
28. $f(x, y) = x^3 + 3x^2y^2 - 2y^2$
29. $f(x, y) = \dfrac{x^2y^2}{\sqrt{x^2 + y^2}}$
30. $f(x, y) = \dfrac{xy}{\sqrt{x^2 + y^2}}$
31. $f(x, y) = 2 \ln xy$
32. $f(x, y) = \tan(x + y)$
33. $f(x, y) = 2 \ln \dfrac{x}{y}$
34. $f(x, y) = \tan \dfrac{y}{x}$

In Exercises 35–40, solve the homogeneous differential equation.

35. $y' = \dfrac{x + y}{2x}$
36. $y' = \dfrac{x^3 + y^3}{xy^2}$
37. $y' = \dfrac{x - y}{x + y}$
38. $y' = \dfrac{x^2 + y^2}{2xy}$
39. $y' = \dfrac{xy}{x^2 - y^2}$
40. $y' = \dfrac{2x + 3y}{x}$

In Exercises 41–44, find the particular solution that satisfies the initial condition.

	Differential Equation	*Initial Condition*
41.	$x\, dy - (2xe^{-y/x} + y)\, dx = 0$	$y(1) = 0$
42.	$-y^2\, dx + x(x + y)\, dy = 0$	$y(1) = 1$
43.	$\left(x \sec \dfrac{y}{x} + y\right) dx - x\, dy = 0$	$y(1) = 0$
44.	$(2x^2 + y^2)\, dx + xy\, dy = 0$	$y(1) = 0$

Slope Fields **In Exercises 45–48, sketch a few solutions of the differential equation on the slope field and then find the general solution analytically. To print an enlarged copy of the graph, go to the website *www.mathgraphs.com*.**

45. $\dfrac{dy}{dx} = x$

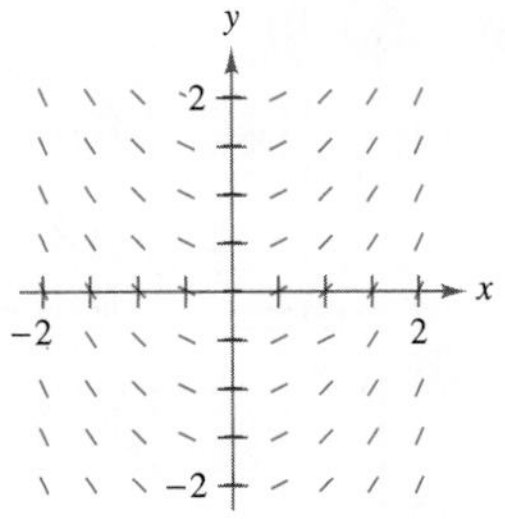

46. $\dfrac{dy}{dx} = -\dfrac{x}{y}$

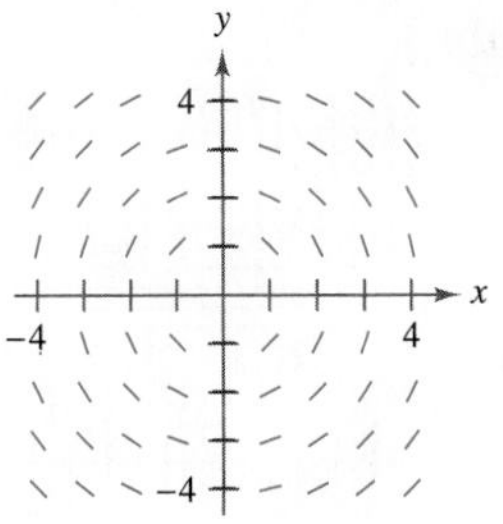

47. $\dfrac{dy}{dx} = 4 - y$

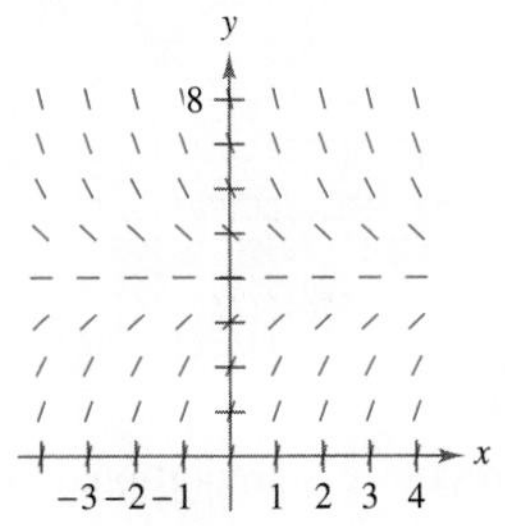

48. $\dfrac{dy}{dx} = 0.25x(4 - y)$

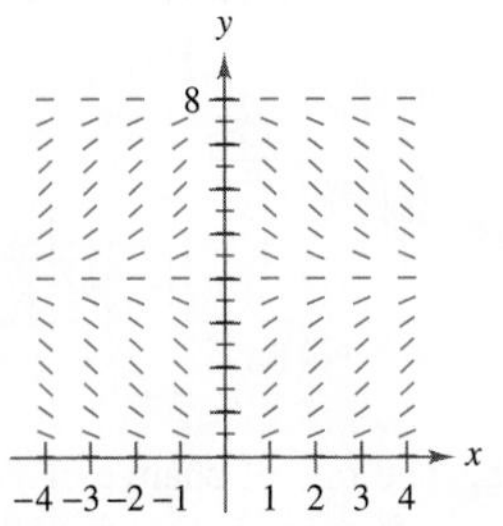

Euler's Method **In Exercises 49–52, (a) use Euler's Method with a step size of $h = 0.1$ to approximate the particular solution of the initial value problem at the given x-value, (b) find the exact solution of the differential equation analytically, and (c) compare the solutions at the given x-value.**

Differential Equation	*Initial Condition*	*x-value*
49. $\dfrac{dy}{dx} = -6xy$	$(0, 5)$	$x = 1$
50. $\dfrac{dy}{dx} + 6xy^2 = 0$	$(0, 3)$	$x = 1$
51. $\dfrac{dy}{dx} = \dfrac{2x + 12}{3y^2 - 4}$	$(1, 2)$	$x = 2$
52. $\dfrac{dy}{dx} = 2x(1 + y^2)$	$(1, 0)$	$x = 1.5$

53. ***Radioactive Decay*** The rate of decomposition of radioactive radium is proportional to the amount present at any time. The half-life of radioactive radium is 1599 years. What percent of a present amount will remain after 25 years?

54. ***Chemical Reaction*** In a chemical reaction, a certain compound changes into another compound at a rate proportional to the unchanged amount. If initially there are 20 grams of the original compound, and there is 16 grams after 1 hour, when will 75 percent of the compound be changed?

Slope Fields **In Exercises 55–58, (a) write a differential equation for the statement, (b) match the differential equation with a possible slope field, and (c) verify your result by using a graphing utility to graph a slope field for the differential equation. [The slope fields are labeled (a), (b), (c), and (d).] To print an enlarged copy of the graph, go to the website *www.mathgraphs.com*.**

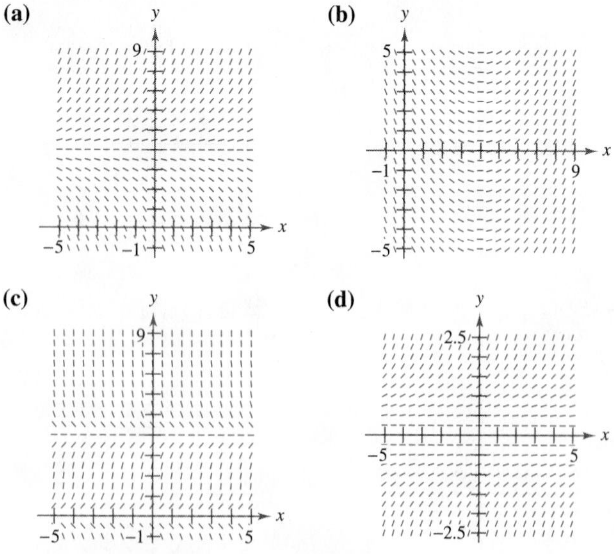

55. The rate of change of y with respect to x is proportional to the difference between y and 4.

56. The rate of change of y with respect to x is proportional to the difference between x and 4.

57. The rate of change of y with respect to x is proportional to the product of y and the difference between y and 4.

58. The rate of change of y with respect to x is proportional to y^2.

59. ***Weight Gain*** A calf that weighs 60 pounds at birth gains weight at the rate

$$\frac{dw}{dt} = k(1200 - w)$$

where w is weight in pounds and t is time in years. Solve the differential equation.

(a) Use a computer algebra system to solve the differential equation for $k = 0.8, 0.9$, and 1. Graph the three solutions.

(b) If the animal is sold when its weight reaches 800 pounds, find the time of sale for each of the models in part (a).

(c) What is the maximum weight of the animal for each of the models?

60. ***Weight Gain*** A calf that weighs w_0 pounds at birth gains weight at the rate

$$\frac{dw}{dt} = 1200 - w$$

where w is weight in pounds and t is time in years. Solve the differential equation.

In Exercises 61–66, find the orthogonal trajectories of the family. Use a graphing utility to graph several members of each family.

61. $x^2 + y^2 = C$

62. $x^2 - 2y^2 = C$

63. $x^2 = Cy$

64. $y^2 = 2Cx$

65. $y^2 = Cx^3$

66. $y = Ce^x$

In Exercises 67–70, match the logistic equation with its graph. [The graphs are labeled (a), (b), (c), and (d).]

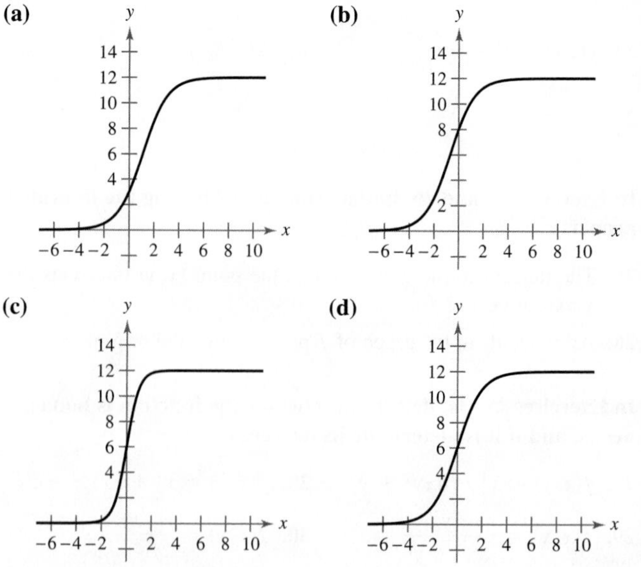

67. $y = \dfrac{12}{1 + e^{-x}}$

68. $y = \dfrac{12}{1 + 3e^{-x}}$

69. $y = \dfrac{12}{1 + \frac{1}{2}e^{-x}}$

70. $y = \dfrac{12}{1 + e^{-2x}}$

In Exercises 71 and 72, the logistic equation models the growth of a population. Use the equation to (a) find the value of k, (b) find the carrying capacity, (c) find the initial population, (d) determine when the population will reach 50% of its carrying capacity, and (e) write a logistic differential equation that has the solution $P(t)$.

71. $P(t) = \dfrac{1500}{1 + 24e^{-0.75t}}$

72. $P(t) = \dfrac{5000}{1 + 39e^{-0.2t}}$

In Exercises 73 and 74, the logistic differential equation models the growth rate of a population. Use the equation to (a) find the value of k, (b) find the carrying capacity, (c) use a computer algebra system to graph a slope field, and (d) determine the value of P at which the population growth rate is the greatest.

73. $\dfrac{dP}{dt} = 3P\left(1 - \dfrac{P}{100}\right)$

74. $\dfrac{dP}{dt} = 0.1P - 0.0004P^2$

In Exercises 75–78, find the logistic equation that satisfies the initial condition.

Logistic Differential Equation	*Initial Condition*
75. $\dfrac{dy}{dt} = y\left(1 - \dfrac{y}{40}\right)$	$(0, 8)$
76. $\dfrac{dy}{dt} = 1.2y\left(1 - \dfrac{y}{8}\right)$	$(0, 5)$
77. $\dfrac{dy}{dt} = \dfrac{4y}{5} - \dfrac{y^2}{150}$	$(0, 8)$
78. $\dfrac{dy}{dt} = \dfrac{3y}{20} - \dfrac{y^2}{1600}$	$(0, 15)$

79. ***Endangered Species*** A conservation organization releases 25 Florida panthers into a game preserve. After 2 years, there are 39 panthers in the preserve. The Florida preserve has a carrying capacity of 200 panthers.

(a) Write a logistic equation that models the population of panthers in the preserve.

(b) Find the population after 5 years.

(c) When will the population reach 100?

(d) Write a logistic differential equation that models the growth rate of the panther population. Then repeat part (b) using Euler's Method with a step size of $h = 1$. Compare the approximation with the exact answers.

(e) At what time is the panther population growing most rapidly? Explain.

80. ***Bacteria Growth*** At time $t = 0$, a bacterial culture weighs 1 gram. Two hours later, the culture weighs 2 grams. The maximum weight of the culture is 10 grams.

(a) Write a logistic equation that models the weight of the bacterial culture.

(b) Find the culture's weight after 5 hours.

(c) When will the culture's weight reach 8 grams?

(d) Write a logistic differential equation that models the growth rate of the culture's weight. Then repeat part (b) using Euler's Method with a step size of $h = 1$. Compare the approximation with the exact answers.

(e) At what time is the culture's weight increasing most rapidly? Explain.

Writing About Concepts

81. In your own words, describe how to recognize and solve differential equations that can be solved by separation of variables.

82. State the test for determining if a differential equation is homogeneous. Give an example.

83. In your own words, describe the relationship between two families of curves that are mutually orthogonal.

84. ***Sailing*** Ignoring resistance, a sailboat starting from rest accelerates (dv/dt) at a rate proportional to the difference between the velocities of the wind and the boat.

(a) The wind is blowing at 20 knots, and after 1 minute the boat is moving at 5 knots. Write the velocity v as a function of time t.

(b) Use the result of part (a) to write the distance traveled by the boat as a function of time.

True or False? **In Exercises 85–88, determine whether the statement is true or false. If it is false, explain why or give an example that shows it is false.**

85. The function $y = 0$ is always a solution of a differential equation that can be solved by separation of variables.

86. The differential equation $y' = xy - 2y + x - 2$ can be written in separated variables form.

87. The function $f(x, y) = x^2 + xy + 2$ is homogeneous.

88. The families $x^2 + y^2 = 2Cy$ and $x^2 + y^2 = 2Kx$ are mutually orthogonal.

89. Show that if $y = \dfrac{1}{1 + be^{-kt}}$, then $\dfrac{dy}{dt} = ky(1 - y)$.

Putnam Exam Challenge

90. A not uncommon calculus mistake is to believe that the product rule for derivatives says that $(fg)' = f'g'$. If $f(x) = e^{x^2}$, determine, with proof, whether there exists an open interval (a, b) and a nonzero function g defined on (a, b) such that this wrong product rule is true for x in (a, b).

This problem was composed by the Committee on the Putnam Prize Competition.

Section 6.4 First-Order Linear Differential Equations

- **Solve a first-order linear differential equation.**
- **Solve a Bernoulli differential equation.**
- **Use linear differential equations to solve applied problems.**

First-Order Linear Differential Equations

In this section, you will see how to solve a very important class of first-order differential equations—first-order linear differential equations.

Definition of First-Order Linear Differential Equation

A first-order linear differential equation is an equation of the form

$$\frac{dy}{dx} + P(x)y = Q(x)$$

where P and Q are continuous functions of x. This first-order linear differential equation is said to be in **standard form.**

NOTE It is instructive to see why the integrating factor helps solve a linear differential equation of the form $y' + P(x)y = Q(x)$. When both sides of the equation are multiplied by the integrating factor $u(x) = e^{\int P(x)\,dx}$, the left-hand side becomes the derivative of a product.

$$y'e^{\int P(x)\,dx} + P(x)ye^{\int P(x)\,dx} = Q(x)e^{\int P(x)\,dx}$$

$$\left[ye^{\int P(x)\,dx}\right]' = Q(x)e^{\int P(x)\,dx}$$

Integrating both sides of this second equation and dividing by $u(x)$ produces the general solution.

To solve a linear differential equation, write it in standard form to identify the functions $P(x)$ and $Q(x)$. Then integrate $P(x)$ and form the expression

$$u(x) = e^{\int P(x)\,dx} \qquad \text{Integrating factor}$$

which is called an **integrating factor.** The general solution of the equation is

$$y = \frac{1}{u(x)}\int Q(x)u(x)\,dx. \qquad \text{General solution}$$

EXAMPLE 1 Solving a Linear Differential Equation

Find the general solution of

$$y' + y = e^x.$$

Solution

For this equation, $P(x) = 1$ and $Q(x) = e^x$. So, the integrating factor is

$$\begin{aligned} u(x) &= e^{\int P(x)\,dx} && \text{Integrating factor} \\ &= e^{\int dx} \\ &= e^x. \end{aligned}$$

This implies that the general solution is

$$\begin{aligned} y &= \frac{1}{u(x)}\int Q(x)u(x)\,dx \\ &= \frac{1}{e^x}\int e^x(e^x)\,dx \\ &= e^{-x}\left(\frac{1}{2}e^{2x} + C\right) \\ &= \frac{1}{2}e^x + Ce^{-x}. && \text{General solution} \end{aligned}$$

THEOREM 6.3 Solution of a First-Order Linear Differential Equation

An integrating factor for the first-order linear differential equation

$$y' + P(x)y = Q(x)$$

is $u(x) = e^{\int P(x)\,dx}$. The solution of the differential equation is

$$ye^{\int P(x)\,dx} = \int Q(x)e^{\int P(x)\,dx}\,dx + C.$$

ANNA JOHNSON PELL WHEELER (1883–1966)

Anna Johnson Pell Wheeler was awarded a master's degree from the University of Iowa for her thesis *The Extension of Galois Theory to Linear Differential Equations* in 1904. Influenced by David Hilbert, she worked on integral equations while studying infinite linear spaces.

STUDY TIP Rather than memorizing the formula in Theorem 6.3, just remember that multiplication by the integrating factor $e^{\int P(x)\,dx}$ converts the left side of the differential equation into the derivative of the product $ye^{\int P(x)\,dx}$.

EXAMPLE 2 Solving a First-Order Linear Differential Equation

Find the general solution of

$$xy' - 2y = x^2.$$

Solution The standard form of the given equation is

$$y' + P(x)y = Q(x)$$

$$y' - \left(\frac{2}{x}\right)y = x. \qquad \text{Standard form}$$

So, $P(x) = -2/x$, and you have

$$\int P(x)\,dx = -\int \frac{2}{x}\,dx$$

$$= -\ln x^2$$

$$e^{\int P(x)\,dx} = e^{-\ln x^2}$$

$$= \frac{1}{e^{\ln x^2}}$$

$$= \frac{1}{x^2}. \qquad \text{Integrating factor}$$

So, multiplying each side of the standard form by $1/x^2$ yields

$$\frac{y'}{x^2} - \frac{2y}{x^3} = \frac{1}{x}$$

$$\frac{d}{dx}\left[\frac{y}{x^2}\right] = \frac{1}{x}$$

$$\frac{y}{x^2} = \int \frac{1}{x}\,dx$$

$$\frac{y}{x^2} = \ln|x| + C$$

$$y = x^2(\ln|x| + C). \qquad \text{General solution}$$

Several solution curves (for $C = -2, -1, 0, 1, 2, 3,$ and 4) are shown in Figure 6.19.

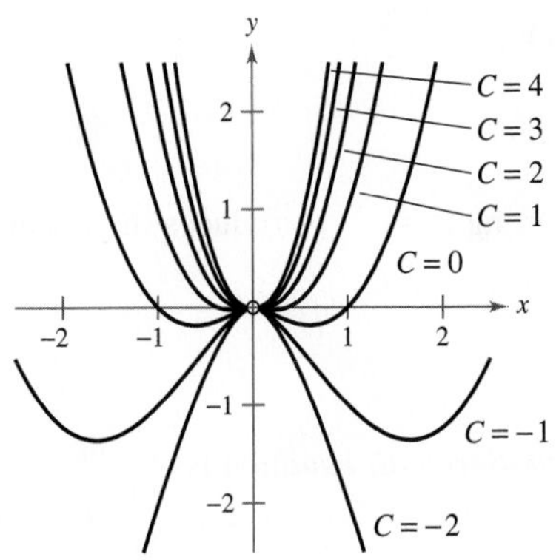

Figure 6.19

EXAMPLE 3 Solving a First-Order Linear Differential Equation

Find the general solution of

$$y' - y \tan t = 1, \quad -\frac{\pi}{2} < t < \frac{\pi}{2}.$$

Solution The equation is already in the standard form $y' + P(t)y = Q(t)$. So, $P(t) = -\tan t$, and

$$\int P(t)\,dt = -\int \tan t\,dt = \ln|\cos t|$$

which implies that the integrating factor is

$$\begin{aligned} e^{\int P(t)\,dt} &= e^{\ln|\cos t|} \\ &= |\cos t|. \end{aligned}$$ Integrating factor

A quick check shows that $\cos t$ is also an integrating factor. So, multiplying $y' - y \tan t = 1$ by $\cos t$ produces

$$\begin{aligned} \frac{d}{dt}[y \cos t] &= \cos t \\ y \cos t &= \int \cos t\,dt \\ y \cos t &= \sin t + C \\ y &= \tan t + C \sec t. \end{aligned}$$ General solution

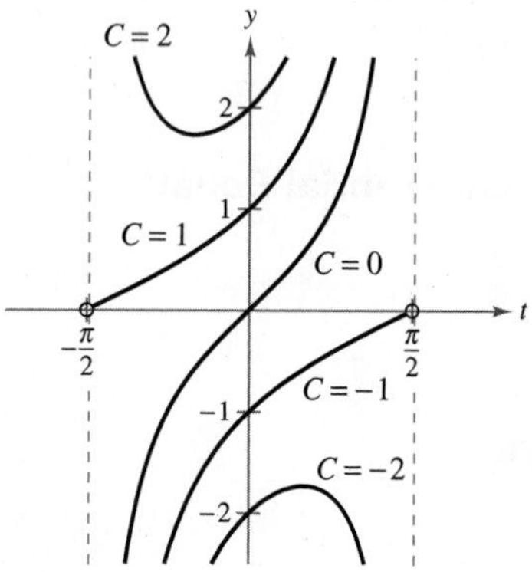

Figure 6.20

Several solution curves are shown in Figure 6.20.

Bernoulli Equation

A well-known nonlinear equation that reduces to a linear one with an appropriate substitution is the **Bernoulli equation,** named after James Bernoulli (1654–1705).

$$y' + P(x)y = Q(x)y^n$$ Bernoulli equation

This equation is linear if $n = 0$, and has separable variables if $n = 1$. So, in the following development, assume that $n \neq 0$ and $n \neq 1$. Begin by multiplying by y^{-n} and $(1 - n)$ to obtain

$$\begin{aligned} y^{-n}y' + P(x)y^{1-n} &= Q(x) \\ (1-n)y^{-n}y' + (1-n)P(x)y^{1-n} &= (1-n)Q(x) \\ \frac{d}{dx}[y^{1-n}] + (1-n)P(x)y^{1-n} &= (1-n)Q(x) \end{aligned}$$

which is a linear equation in the variable y^{1-n}. Letting $z = y^{1-n}$ produces the linear equation

$$\frac{dz}{dx} + (1-n)P(x)z = (1-n)Q(x).$$

Finally, by Theorem 6.3, the *general solution of the Bernoulli equation* is

$$y^{1-n}e^{\int(1-n)P(x)\,dx} = \int (1-n)Q(x)e^{\int(1-n)P(x)\,dx}\,dx + C.$$

EXAMPLE 4 Solving a Bernoulli Equation

Find the general solution of

$$y' + xy = xe^{-x^2}y^{-3}.$$

Solution For this Bernoulli equation, let $n = -3$, and use the substitution

$$z = y^4 \qquad \text{Let } z = y^{1-n} = y^{1-(-3)}.$$
$$z' = 4y^3y'. \qquad \text{Differentiate.}$$

Multiplying the original equation by $4y^3$ produces

$$y' + xy = xe^{-x^2}y^{-3} \qquad \text{Write original equation.}$$
$$4y^3y' + 4xy^4 = 4xe^{-x^2} \qquad \text{Multiply each side by } 4y^3.$$
$$z' + 4xz = 4xe^{-x^2}. \qquad \text{Linear equation: } z' + P(x)z = Q(x)$$

This equation is linear in z. Using $P(x) = 4x$ produces

$$\int P(x)\,dx = \int 4x\,dx$$
$$= 2x^2$$

which implies that e^{2x^2} is an integrating factor. Multiplying the linear equation by this factor produces

$$z' + 4xz = 4xe^{-x^2} \qquad \text{Linear equation}$$
$$z'e^{2x^2} + 4xze^{2x^2} = 4xe^{x^2} \qquad \text{Multiply by integrating factor.}$$
$$\frac{d}{dx}[ze^{2x^2}] = 4xe^{x^2} \qquad \text{Write left side as derivative.}$$
$$ze^{2x^2} = \int 4xe^{x^2}\,dx \qquad \text{Integrate each side.}$$
$$ze^{2x^2} = 2e^{x^2} + C$$
$$z = 2e^{-x^2} + Ce^{-2x^2}. \qquad \text{Divide each side by } e^{2x^2}.$$

Finally, substituting $z = y^4$, the general solution is

$$y^4 = 2e^{-x^2} + Ce^{-2x^2}. \qquad \text{General solution}$$

So far you have studied several types of first-order differential equations. Of these, the separable variables case is usually the simplest, and solution by an integrating factor is ordinarily used only as a last resort.

Summary of First-Order Differential Equations

Method	*Form of Equation*
1. Separable variables:	$M(x)\,dx + N(y)\,dy = 0$
2. Homogeneous:	$M(x, y)\,dx + N(x, y)\,dy = 0$, where M and N are nth-degree homogeneous
3. Linear:	$y' + P(x)y = Q(x)$
4. Bernoulli equation:	$y' + P(x)y = Q(x)y^n$

Applications

One type of problem that can be described in terms of a differential equation involves chemical mixtures, as illustrated in the next example.

EXAMPLE 5 A Mixture Problem

Figure 6.21

A tank contains 50 gallons of a solution composed of 90% water and 10% alcohol. A second solution containing 50% water and 50% alcohol is added to the tank at the rate of 4 gallons per minute. As the second solution is being added, the tank is being drained at a rate of 5 gallons per minute, as shown in Figure 6.21. Assuming the solution in the tank is stirred constantly, how much alcohol is in the tank after 10 minutes?

Solution Let y be the number of gallons of alcohol in the tank at any time t. You know that $y = 5$ when $t = 0$. Because the number of gallons of solution in the tank at any time is $50 - t$, and the tank loses 5 gallons of solution per minute, it must lose

$$\left(\frac{5}{50 - t}\right)y$$

gallons of alcohol per minute. Furthermore, because the tank is gaining 2 gallons of alcohol per minute, the rate of change of alcohol in the tank is given by

$$\frac{dy}{dt} = 2 - \left(\frac{5}{50 - t}\right)y \quad \Longrightarrow \quad \frac{dy}{dt} + \left(\frac{5}{50 - t}\right)y = 2.$$

To solve this linear equation, let $P(t) = 5/(50 - t)$ and obtain

$$\int P(t)\,dt = \int \frac{5}{50 - t}\,dt = -5 \ln|50 - t|.$$

Because $t < 50$, you can drop the absolute value signs and conclude that

$$e^{\int P(t)\,dt} = e^{-5\ln(50 - t)} = \frac{1}{(50 - t)^5}.$$

So, the general solution is

$$\frac{y}{(50 - t)^5} = \int \frac{2}{(50 - t)^5}\,dt = \frac{1}{2(50 - t)^4} + C$$

$$y = \frac{50 - t}{2} + C(50 - t)^5.$$

Because $y = 5$ when $t = 0$, you have

$$5 = \frac{50}{2} + C(50)^5 \quad \Longrightarrow \quad -\frac{20}{50^5} = C$$

which means that the particular solution is

$$y = \frac{50 - t}{2} - 20\left(\frac{50 - t}{50}\right)^5.$$

Finally, when $t = 10$, the amount of alcohol in the tank is

$$y = \frac{50 - 10}{2} - 20\left(\frac{50 - 10}{50}\right)^5 \approx 13.45 \text{ gal}$$

which represents a solution containing 33.6% alcohol.

In most falling-body problems discussed so far in the text, air resistance has been neglected. The next example includes this factor. In the example, the air resistance on the falling object is assumed to be proportional to its velocity v. If g is the gravitational constant, the downward force F on a falling object of mass m is given by the difference $mg - kv$. But by Newton's Second Law of Motion, you know that

$$\begin{aligned} F &= ma \\ &= m(dv/dt) \end{aligned}$$

which yields the following differential equation.

$$m\frac{dv}{dt} = mg - kv \quad \Longrightarrow \quad \frac{dv}{dt} + \frac{k}{m}v = g$$

EXAMPLE 6 A Falling Object with Air Resistance

An object of mass m is dropped from a hovering helicopter. Find its velocity as a function of time t, assuming that the air resistance is proportional to the velocity of the object.

Solution The velocity v satisfies the equation

$$\frac{dv}{dt} + \frac{kv}{m} = g$$

where g is the gravitational constant and k is the constant of proportionality. Letting $b = k/m$, you can *separate variables* to obtain

$$\begin{aligned} dv &= (g - bv)\,dt \\ \int \frac{dv}{g - bv} &= \int dt \\ -\frac{1}{b}\ln|g - bv| &= t + C_1 \\ \ln|g - bv| &= -bt - bC_1 \\ g - bv &= Ce^{-bt}. \end{aligned}$$

Because the object was dropped, $v = 0$ when $t = 0$; so $g = C$, and it follows that

$$-bv = -g + ge^{-bt} \quad \Longrightarrow \quad v = \frac{g - ge^{-bt}}{b} = \frac{mg}{k}(1 - e^{-kt/m}).$$

NOTE Notice in Example 6 that the velocity approaches a limit of mg/k as a result of the air resistance. For falling-body problems in which air resistance is neglected, the velocity increases without bound.

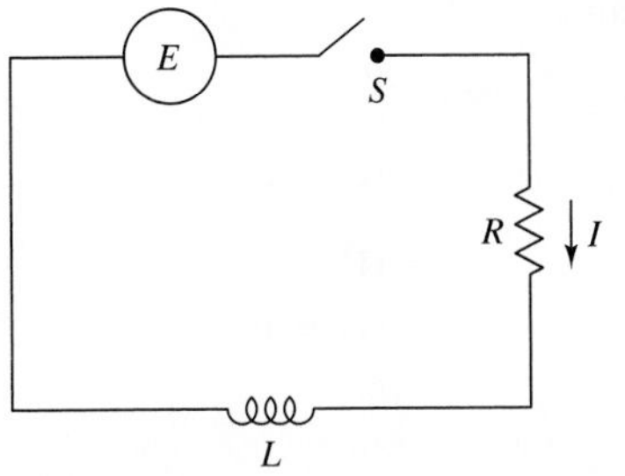

Figure 6.22

A simple electric circuit consists of electric current I (in amperes), a resistance R (in ohms), an inductance L (in henrys), and a constant electromotive force E (in volts), as shown in Figure 6.22. According to Kirchhoff's Second Law, if the switch S is closed when $t = 0$, the applied electromotive force (voltage) is equal to the sum of the voltage drops in the rest of the circuit. This in turn means that the current I satisfies the differential equation

$$L\frac{dI}{dt} + RI = E.$$

EXAMPLE 7 An Electric Circuit Problem

Find the current I as a function of time t (in seconds), given that I satisfies the differential equation $L(dI/dt) + RI = \sin 2t$, where R and L are nonzero constants.

Solution In standard form, the given linear equation is

$$\frac{dI}{dt} + \frac{R}{L}I = \frac{1}{L}\sin 2t.$$

Let $P(t) = R/L$, so that $e^{\int P(t)\,dt} = e^{(R/L)t}$, and, by Theorem 6.3,

$$Ie^{(R/L)t} = \frac{1}{L}\int e^{(R/L)t}\sin 2t\,dt$$

$$= \frac{1}{4L^2 + R^2}e^{(R/L)t}(R\sin 2t - 2L\cos 2t) + C.$$

So the general solution is

$$I = e^{-(R/L)t}\left[\frac{1}{4L^2 + R^2}e^{(R/L)t}(R\sin 2t - 2L\cos 2t) + C\right]$$

$$I = \frac{1}{4L^2 + R^2}(R\sin 2t - 2L\cos 2t) + Ce^{-(R/L)t}.$$

TECHNOLOGY The integral in Example 7 was found using symbolic algebra software. If you have access to *Derive*, *Maple*, *Mathcad*, *Mathematica*, or the *TI-89*, try using it to integrate

$$\frac{1}{L}\int e^{(R/L)t}\sin 2t\,dt.$$

In Chapter 8 you will learn how to integrate functions of this type using integration by parts.

Exercises for Section 6.4

See www.CalcChat.com for worked-out solutions to odd-numbered exercises.

In Exercises 1–4, determine whether the differential equation is linear. Explain your reasoning.

1. $x^3y' + xy = e^x + 1$
2. $2xy - y'\ln x = y$
3. $y' + y\cos x = xy^2$
4. $\dfrac{1 - y'}{y} = 3x$

In Exercises 5–14, solve the first-order linear differential equation.

5. $\dfrac{dy}{dx} + \left(\dfrac{1}{x}\right)y = 3x + 4$
6. $\dfrac{dy}{dx} + \left(\dfrac{2}{x}\right)y = 3x + 2$
7. $y' - y = 10$
8. $y' + 2xy = 4x$
9. $(y + 1)\cos x\,dx - dy = 0$
10. $(y - 1)\sin x\,dx - dy = 0$
11. $(x - 1)y' + y = x^2 - 1$
12. $y' + 3y = e^{3x}$
13. $y' - 3x^2y = e^{x^3}$
14. $y' - y = \cos x$

Slope Fields **In Exercises 15 and 16, (a) sketch an approximate solution of the differential equation satisfying the initial condition by hand on the slope field, (b) find the particular solution that satisfies the initial condition, and (c) use a graphing utility to graph the particular solution. Compare the graph with the hand-drawn graph of part (a). To print an enlarged copy of the graph, go to the website *www.mathgraphs.com*.**

Differential Equation	*Initial Condition*
15. $\dfrac{dy}{dx} = e^x - y$	$(0, 1)$
16. $y' + \left(\dfrac{1}{x}\right)y = \sin x^2$	$(\sqrt{\pi}, 0)$

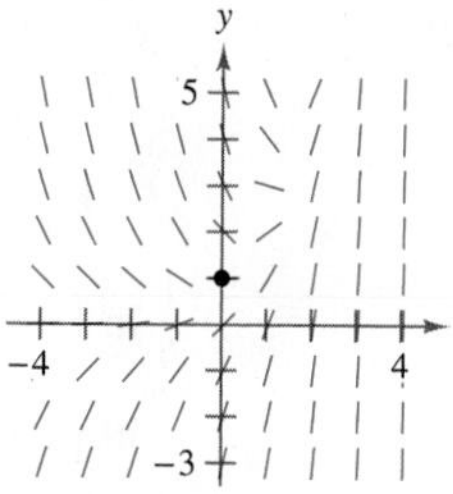

Figure for 15

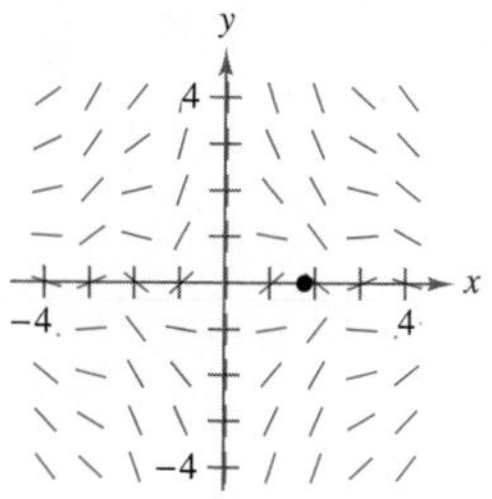

Figure for 16

In Exercises 17–24, find the particular solution of the differential equation that satisfies the boundary condition.

Differential Equation	*Boundary Condition*
17. $y'\cos^2 x + y - 1 = 0$	$y(0) = 5$
18. $x^3y' + 2y = e^{1/x^2}$	$y(1) = e$
19. $y' + y\tan x = \sec x + \cos x$	$y(0) = 1$
20. $y' + y\sec x = \sec x$	$y(0) = 4$
21. $y' + \left(\dfrac{1}{x}\right)y = 0$	$y(2) = 2$
22. $y' + (2x - 1)y = 0$	$y(1) = 2$
23. $x\,dy = (x + y + 2)\,dx$	$y(1) = 10$
24. $2xy' - y = x^3 - x$	$y(4) = 2$

In Exercises 25–30, solve the Bernoulli differential equation.

25. $y' + 3x^2y = x^2y^3$

26. $y' + xy = xy^{-1}$

27. $y' + \left(\frac{1}{x}\right)y = xy^2$

28. $y' + \left(\frac{1}{x}\right)y = x\sqrt{y}$

29. $y' - y = e^x\sqrt[3]{y}$

30. $yy' - 2y^2 = e^x$

Slope Fields **In Exercises 31–34, (a) use a graphing utility to graph the slope field for the differential equation, (b) find the particular solutions of the differential equation passing through the given points, and (c) use a graphing utility to graph the particular solutions on the slope field.**

Differential Equation	*Points*
31. $\frac{dy}{dx} - \frac{1}{x}y = x^2$	$(-2, 4)$, $(2, 8)$
32. $\frac{dy}{dx} + 4x^3y = x^3$	$\left(0, \frac{7}{2}\right)$, $\left(0, -\frac{1}{2}\right)$
33. $\frac{dy}{dx} + (\cot x)y = 2$	$(1, 1)$, $(3, -1)$
34. $\frac{dy}{dx} + 2xy = xy^2$	$(0, 3)$, $(0, 1)$

35. ***Population Growth*** When predicting population growth, demographers must consider birth and death rates as well as the net change caused by the difference between the rates of immigration and emigration. Let P be the population at time t and let N be the net increase per unit time resulting from the difference between immigration and emigration. So, the rate of growth of the population is given by

$$\frac{dP}{dt} = kP + N, \quad N \text{ is constant.}$$

Solve this differential equation to find P as a function of time if at time $t = 0$ the size of the population is P_0.

36. ***Investment Growth*** A large corporation starts at time $t = 0$ to invest part of its receipts continuously at a rate of P dollars per year in a fund for future corporate expansion. Assume that the fund earns r percent interest per year compounded continuously. So, the rate of growth of the amount A in the fund is given by

$$\frac{dA}{dt} = rA + P$$

where $A = 0$ when $t = 0$. Solve this differential equation for A as a function of t.

Investment Growth **In Exercises 37 and 38, use the result of Exercise 36.**

37. Find A for the following.

(a) $P = \$100{,}000$, $r = 6\%$, and $t = 5$ years

(b) $P = \$250{,}000$, $r = 5\%$, and $t = 10$ years

38. Find t if the corporation needs \$800,000 and it can invest \$75,000 per year in a fund earning 8% interest compounded continuously.

39. ***Intravenous Feeding*** Glucose is added intravenously to the bloodstream at the rate of q units per minute, and the body removes glucose from the bloodstream at a rate proportional to the amount present. Assume that $Q(t)$ is the amount of glucose in the bloodstream at time t.

(a) Determine the differential equation describing the rate of change of glucose in the bloodstream with respect to time.

(b) Solve the differential equation from part (a), letting $Q = Q_0$ when $t = 0$.

(c) Find the limit of $Q(t)$ as $t \to \infty$.

40. ***Learning Curve*** The management at a certain factory has found that the maximum number of units a worker can produce in a day is 40. The rate of increase in the number of units N produced with respect to time t in days by a new employee is proportional to $40 - N$.

(a) Determine the differential equation describing the rate of change of performance with respect to time.

(b) Solve the differential equation from part (a).

(c) Find the particular solution for a new employee who produced 10 units on the first day at the factory and 19 units on the twentieth day.

Mixture **In Exercises 41–46, consider a tank that at time $t = 0$ contains v_0 gallons of a solution of which, by weight, q_0 pounds is soluble concentrate. Another solution containing q_1 pounds of the concentrate per gallon is running into the tank at the rate of r_1 gallons per minute. The solution in the tank is kept well stirred and is withdrawn at the rate of r_2 gallons per minute.**

41. If Q is the amount of concentrate in the solution at any time t, show that

$$\frac{dQ}{dt} + \frac{r_2Q}{v_0 + (r_1 - r_2)t} = q_1r_1.$$

42. If Q is the amount of concentrate in the solution at any time t, write the differential equation for the rate of change of Q with respect to t if $r_1 = r_2 = r$.

43. A 200-gallon tank is full of a solution containing 25 pounds of concentrate. Starting at time $t = 0$, distilled water is admitted to the tank at a rate of 10 gallons per minute, and the well-stirred solution is withdrawn at the same rate.

(a) Find the amount of concentrate Q in the solution as a function of t.

(b) Find the time at which the amount of concentrate in the tank reaches 15 pounds.

(c) Find the quantity of the concentrate in the solution as $t \to \infty$.

44. Repeat Exercise 43, assuming that the solution entering the tank contains 0.04 pound of concentrate per gallon.

45. A 200-gallon tank is half full of distilled water. At time $t = 0$, a solution containing 0.5 pound of concentrate per gallon enters the tank at the rate of 5 gallons per minute, and the well-stirred mixture is withdrawn at the rate of 3 gallons per minute.

(a) At what time will the tank be full?

(b) At the time the tank is full, how many pounds of concentrate will it contain?

46. Repeat Exercise 45, assuming that the solution entering the tank contains 1 pound of concentrate per gallon.

Falling Object **In Exercises 47 and 48, consider an eight-pound object dropped from a height of 5000 feet, where the air resistance is proportional to the velocity.**

47. Write the velocity as a function of time if its velocity after 5 seconds is approximately -101 feet per second. What is the limiting value of the velocity function?

48. Use the result of Exercise 47 to write the position of the object as a function of time. Approximate the velocity of the object when it reaches ground level.

Electric Circuits **In Exercises 49 and 50, use the differential equation for electric circuits given by**

$$L\frac{dI}{dt} + RI = E.$$

In this equation, I is the current, R is the resistance, L is the inductance, and E is the electromotive force (voltage).

49. Solve the differential equation given a constant voltage E_0.

50. Use the result of Exercise 49 to find the equation for the current if $I(0) = 0$, $E_0 = 120$ volts, $R = 600$ ohms, and $L = 4$ henrys. When does the current reach 90% of its limiting value?

Writing About Concepts

51. Give the standard form of a first-order linear differential equation. What is its integrating factor?

52. Give the standard form of the Bernoulli equation. Describe how one reduces it to a linear equation.

In Exercises 53–56, match the differential equation with its solution.

Differential Equation	*Solution*
53. $y' - 2x = 0$	(a) $y = Ce^{x^2}$
54. $y' - 2y = 0$	(b) $y = -\frac{1}{2} + Ce^{x^2}$
55. $y' - 2xy = 0$	(c) $y = x^2 + C$
56. $y' - 2xy = x$	(d) $y = Ce^{2x}$

In Exercises 57–68, solve the first-order differential equation by any appropriate method.

57. $\dfrac{dy}{dx} = \dfrac{e^{2x+y}}{e^{x-y}}$

58. $\dfrac{dy}{dx} = \dfrac{x+1}{y(y+2)}$

59. $y\cos x - \cos x + \dfrac{dy}{dx} = 0$

60. $y' = 2x\sqrt{1-y^2}$

61. $(3y^2 + 4xy)\,dx + (2xy + x^2)\,dy = 0$

62. $(x + y)\,dx - x\,dy = 0$

63. $(2y - e^x)\,dx + x\,dy = 0$

64. $(y^2 + xy)\,dx - x^2\,dy = 0$

65. $(x^2y^4 - 1)\,dx + x^3y^3\,dy = 0$

66. $y\,dx + (3x + 4y)\,dy = 0$

67. $3(y - 4x^2)\,dx + x\,dy = 0$

68. $x\,dx + (y + e^y)(x^2 + 1)\,dy = 0$

True or False? **In Exercises 69 and 70, determine whether the statement is true or false. If it is false, explain why or give an example that shows it is false.**

69. $y' + x\sqrt{y} = x^2$ is a first-order linear differential equation.

70. $y' + xy = e^x y$ is a first-order linear differential equation.

Section Project: Weight Loss

A person's weight depends on both the number of calories consumed and the energy used. Moreover, the amount of energy used depends on a person's weight—the average amount of energy used by a person is 17.5 calories per pound per day. So, the more weight a person loses, the less energy a person uses (assuming that the person maintains a constant level of activity). An equation that can be used to model weight loss is

$$\left(\frac{dw}{dt}\right) = \frac{C}{3500} - \frac{17.5}{3500}w$$

where w is the person's weight (in pounds), t is the time in days, and C is the constant daily calorie consumption.

(a) Find the general solution of the differential equation.

(b) Consider a person who weighs 180 pounds and begins a diet of 2500 calories per day. How long will it take the person to lose 10 pounds? How long will it take the person to lose 35 pounds?

(c) Use a graphing utility to graph the solution. What is the "limiting" weight of the person?

(d) Repeat parts (b) and (c) for a person who weighs 200 pounds when the diet is started.

FOR FURTHER INFORMATION For more information on modeling weight loss, see the article "A Linear Diet Model" by Arthur C. Segal in *The College Mathematics Journal*.

Review Exercises for Chapter 6

See www.CalcChat.com for worked-out solutions to odd-numbered exercises.

1. Determine whether the function $y = x^3$ is a solution of the differential equation $x^2y' + 3y = 6x^3$.

2. Determine whether the function $y = 2 \sin 2x$ is a solution of the differential equation $y''' - 8y = 0$.

In Exercises 3–8, use integration to find a general solution of the differential equation.

3. $\dfrac{dy}{dx} = 2x^2 + 5$

4. $\dfrac{dy}{dx} = x^3 - 2x$

5. $\dfrac{dy}{dx} = \cos 2x$

6. $\dfrac{dy}{dx} = 2 \sin x$

7. $\dfrac{dy}{dx} = 2x\sqrt{x - 7}$

8. $\dfrac{dy}{dx} = 3e^{-x/3}$

Slope Fields **In Exercises 9 and 10, a differential equation and its slope field are given. Determine the slopes (if possible) in the slope field at the points given in the table.**

x	-4	-2	0	2	4	8
y	2	0	4	4	6	8
dy/dx						

9. $\dfrac{dy}{dx} = \dfrac{2x}{y}$

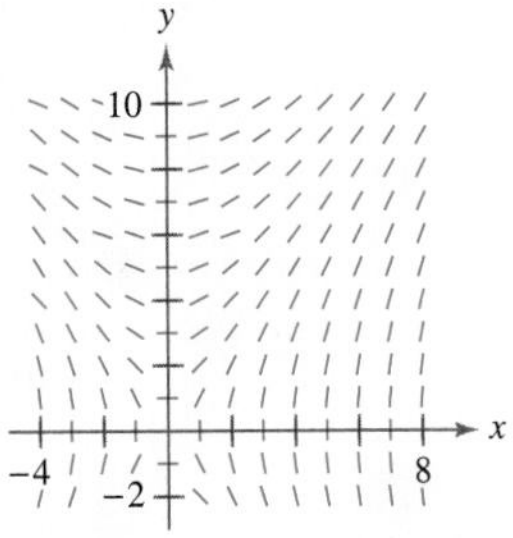

10. $\dfrac{dy}{dx} = x \sin\left(\dfrac{\pi y}{4}\right)$

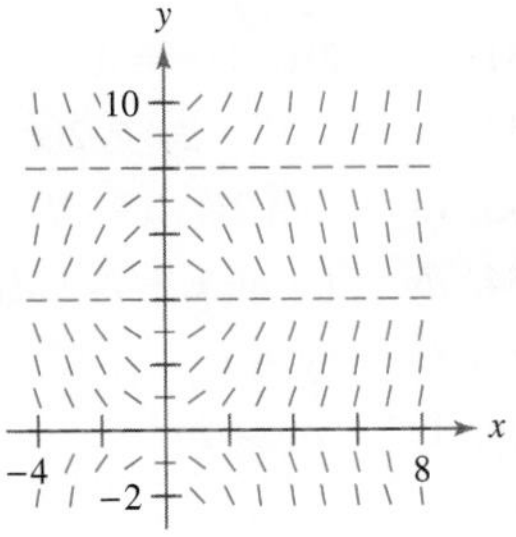

Slope Fields **In Exercises 11–16, (a) sketch the slope field for the differential equation, and (b) use the slope field to sketch the solution that passes through the given point.**

	Differential Equation	*Point*
11.	$y' = -x - 2$	$(-1, 1)$
12.	$y' = 2x^2 - x$	$(0, 2)$
13.	$y' = \frac{1}{4}x^2 - \frac{1}{3}x$	$(0, 3)$
14.	$y' = y + 3x$	$(2, 1)$
15.	$y' = \dfrac{xy}{x^2 + 4}$	$(0, 1)$
16.	$y' = \dfrac{y}{x^2 + 1}$	$(0, -2)$

In Exercises 17–22, solve the differential equation.

17. $\dfrac{dy}{dx} = 6 - x$

18. $\dfrac{dy}{dx} = y + 6$

19. $\dfrac{dy}{dx} = (3 + y)^2$

20. $\dfrac{dy}{dx} = 4\sqrt{y}$

21. $(2 + x)y' - xy = 0$

22. $xy' - (x + 1)y = 0$

In Exercises 23–26, find the exponential function $y = Ce^{kt}$ that passes through the two points.

23.

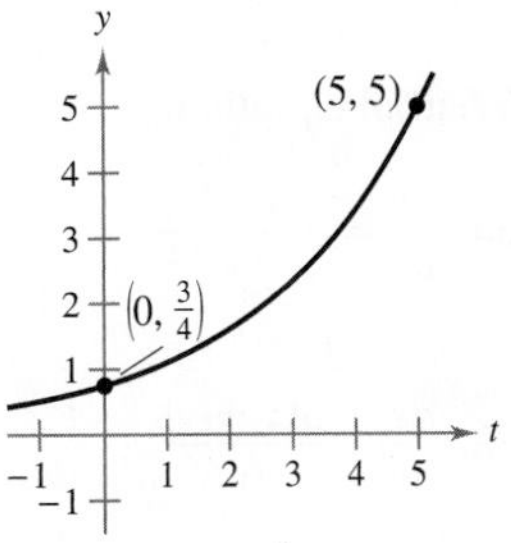

24.

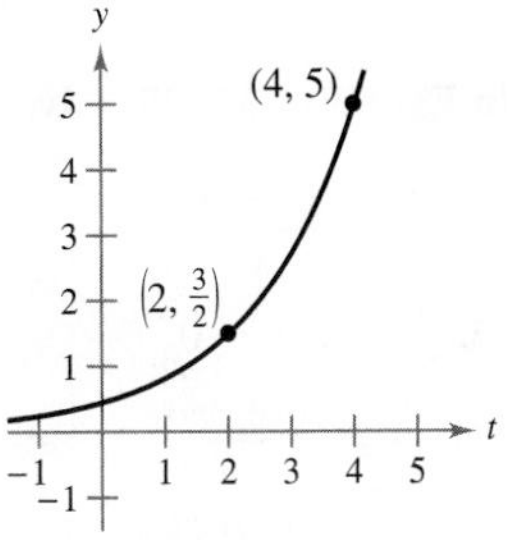

25. $(0, 5), \left(5, \dfrac{1}{6}\right)$

26. $(1, 9), (6, 2)$

27. ***Air Pressure*** Under ideal conditions, air pressure decreases continuously with the height above sea level at a rate proportional to the pressure at that height. The barometer reads 30 inches at sea level and 15 inches at 18,000 feet. Find the barometric pressure at 35,000 feet.

28. ***Radioactive Decay*** Radioactive radium has a half-life of approximately 1599 years. The initial quantity is 5 grams. How much remains after 600 years?

29. ***Sales*** The sales S (in thousands of units) of a new product after it has been on the market for t years is given by

$$S = Ce^{k/t}.$$

(a) Find S as a function of t if 5000 units have been sold after 1 year and the saturation point for the market is 30,000 units (that is, $\lim_{t\to\infty} S = 30$).

(b) How many units will have been sold after 5 years?

(c) Use a graphing utility to graph this sales function.

30. ***Sales*** The sales S (in thousands of units) of a new product after it has been on the market for t years is given by

$$S = 25(1 - e^{kt}).$$

(a) Find S as a function of t if 45,000 units have been sold after 1 year.

(b) How many units will saturate this market?

(c) How many units will have been sold after 5 years?

(d) Use a graphing utility to graph this sales function.

31. ***Population Growth*** A population grows continuously at the rate of 1.5%. How long will it take the population to double?

32. ***Fuel Economy*** An automobile gets 28 miles per gallon of gasoline for speeds up to 50 miles per hour. Over 50 miles per hour, the number of miles per gallon drops at the rate of 12 percent for each 10 miles per hour.

(a) s is the speed and y is the number of miles per gallon. Find y as a function of s by solving the differential equation

$$\frac{dy}{ds} = -0.012y, \quad s > 50.$$

(b) Use the function in part (a) to complete the table.

Speed	50	55	60	65	70
Miles per Gallon					

In Exercises 33–38, solve the differential equation.

33. $\dfrac{dy}{dx} = \dfrac{x^2 + 3}{x}$

34. $\dfrac{dy}{dx} = \dfrac{e^{-2x}}{1 + e^{-2x}}$

35. $y' - 2xy = 0$

36. $y' - e^y \sin x = 0$

37. $\dfrac{dy}{dx} = \dfrac{x^2 + y^2}{2xy}$

38. $\dfrac{dy}{dx} = \dfrac{3(x + y)}{x}$

39. Verify that the general solution $y = C_1x + C_2x^3$ satisfies the differential equation $x^2y'' - 3xy' + 3y = 0$. Then find the particular solution that satisfies the initial condition $y = 0$ and $y' = 4$ when $x = 2$.

40. ***Vertical Motion*** A falling object encounters air resistance that is proportional to its velocity. The acceleration due to gravity is -9.8 meters per second per second. The net change in velocity is $dv/dt = kv - 9.8$.

(a) Find the velocity of the object as a function of time if the initial velocity is v_0.

(b) Use the result of part (a) to find the limit of the velocity as t approaches infinity.

(c) Integrate the velocity function found in part (a) to find the position function s.

Slope Fields **In Exercises 41 and 42, sketch a few solutions of the differential equation on the slope field and then find the general solution analytically. To print an enlarged copy of the graph, go to the website *www.mathgraphs.com*.**

41. $\dfrac{dy}{dx} = -\dfrac{4x}{y}$

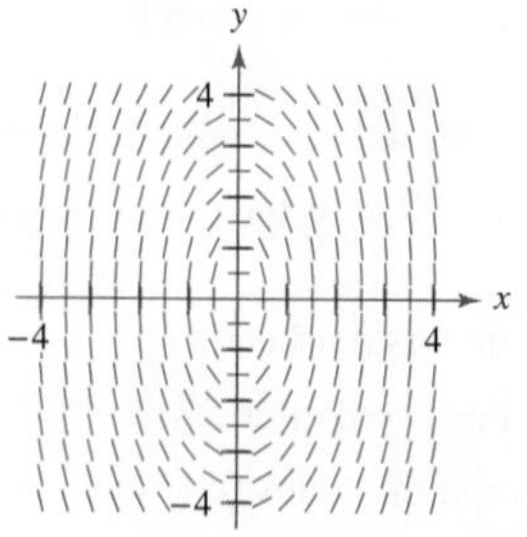

42. $\dfrac{dy}{dx} = 3 - 2y$

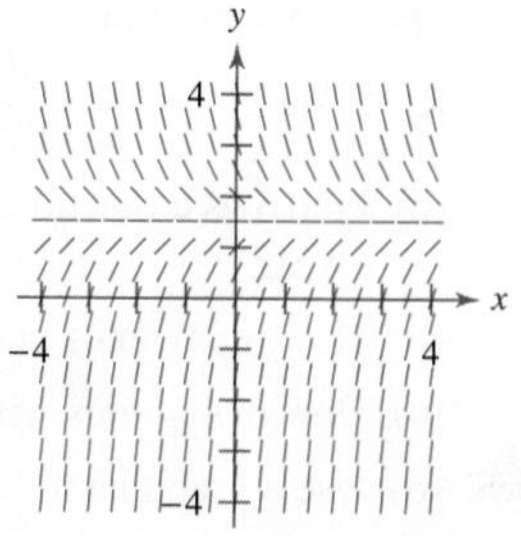

In Exercises 43 and 44, the logistic equation models the growth of a population. Use the equation to (a) find the value of k, (b) find the carrying capacity, (c) find the initial population, (d) determine when the population will reach 50% of its carrying capacity, and (e) write a logistic differential equation that has the solution $P(t)$.

43. $P(t) = \dfrac{7200}{1 + 44e^{-0.55t}}$

44. $P(t) = \dfrac{4800}{1 + 14e^{-0.15t}}$

45. ***Environment*** A conservation department releases 1200 brook trout into a lake. It is estimated that the carrying capacity of the lake for the species is 20,400. After the first year, there are 2000 brook trout in the lake.

(a) Write a logistic equation that models the number of brook trout in the lake.

(b) Find the number of brook trout in the lake after 8 years.

(c) When will the number of brook trout reach 10,000?

46. ***Environment*** Write a logistic differential equation that models the growth rate of the brook trout population in Exercise 45. Then repeat part (b) using Euler's Method with a step size of $h = 1$. Compare the approximation with the exact answers.

In Exercises 47–56, solve the first-order linear differential equation.

47. $y' - y = 8$

48. $e^x y' + 4e^x y = 1$

49. $4y' = e^{x/4} + y$

50. $\dfrac{dy}{dx} - \dfrac{5y}{x^2} = \dfrac{1}{x^2}$

51. $(x - 2)y' + y = 1$

52. $(x + 3)y' + 2y = 2(x + 3)^2$

53. $(3y + \sin 2x)\,dx - dy = 0$

54. $dy = (y \tan x + 2e^x)\,dx$

55. $y' + 5y = e^{5x}$

56. $xy' - ay = bx^4$

In Exercises 57–60, solve the Bernoulli differential equation.

57. $y' + y = xy^2$ $\left[\textit{Hint: } \int xe^{-x}\,dx = (-x - 1)e^{-x}\right]$

58. $y' + 2xy = xy^2$

59. $y' + \left(\dfrac{1}{x}\right)y = \dfrac{y^3}{x^2}$

60. $xy' + y = xy^2$

In Exercises 61–64, write an example of the given differential equation. Then solve your equation.

61. Homogeneous differential equation

62. Logistic differential equation

63. First-order linear differential equation

64. Bernoulli differential equation

P.S. Problem Solving

See www.CalcChat.com for worked-out solutions to odd-numbered exercises.

1. The differential equation

$$\frac{dy}{dt} = ky^{1+\varepsilon}$$

where k and ε are positive constants, is called the **doomsday equation.**

(a) Solve the doomsday equation

$$\frac{dy}{dt} = y^{1.01}$$

given that $y(0) = 1$. Find the time T at which

$$\lim_{t \to T^-} y(t) = \infty.$$

(b) Solve the doomsday equation

$$\frac{dy}{dt} = ky^{1+\varepsilon}$$

given that $y(0) = y_0$. Explain why this equation is called the doomsday equation.

2. A thermometer is taken from a room at 72°F to the outdoors, where the temperature is 20°F. The reading drops to 48°F after 1 minute. Determine the reading on the thermometer after 5 minutes.

3. Let S represent sales of a new product (in thousands of units), let L represent the maximum level of sales (in thousands of units), and let t represent time (in months). The rate of change of S with respect to t varies jointly as the product of S and $L - S$.

(a) Write the differential equation for the sales model if $L = 100$, $S = 10$ when $t = 0$, and $S = 20$ when $t = 1$. Verify that

$$S = \frac{L}{1 + Ce^{-kt}}.$$

(b) At what time is the growth in sales increasing most rapidly?

(c) Use a graphing utility to graph the sales function.

(d) Sketch the solution from part (a) on the slope field shown in the figure below. To print an enlarged copy of the graph, go to the website *www.mathgraphs.com*.

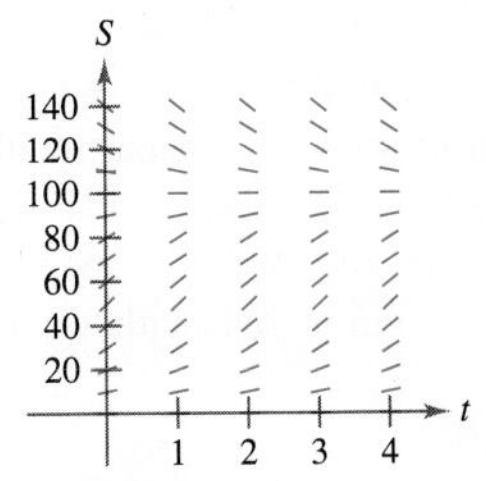

(e) If the estimated maximum level of sales is correct, use the slope field to describe the shape of the solution curves for sales if, at some period of time, sales exceed L.

4. Another model that can be used to represent population growth is the **Gompertz equation,** which is the solution of the differential equation

$$\frac{dy}{dt} = k \ln\left(\frac{L}{y}\right)y$$

where k is a constant and L is the carrying capacity.

(a) Solve the differential equation.

(b) Use a graphing utility to graph the slope field for the differential equation when $k = 0.05$ and $L = 1000$.

(c) Describe the behavior of the graph as $t \to \infty$.

(d) Graph the equation you found in part (a) for $L = 5000$, $y_0 = 500$, and $k = 0.02$. Determine the concavity of the graph and how it compares with the general solution of the logistical differential equation.

5. Show that the logistic equation

$$y = \frac{L}{1 + be^{-kt}}$$

can be written as

$$y = \frac{1}{2}L\left[1 + \tanh\left(\frac{1}{2}k\left(t - \frac{\ln b}{k}\right)\right)\right].$$

What can you conclude about the graph of the logistic equation?

6. **Torricelli's Law** states that water will flow from an opening at the bottom of a tank with the same speed that it would attain falling from the surface of the water to the opening. One of the forms of Torricelli's Law is

$$A(h)\frac{dh}{dt} = -k\sqrt{2gh}$$

where h is the height of the water in the tank, k is the area of the opening at the bottom of the tank, $A(h)$ is the horizontal cross-sectional area at height h, and g is the acceleration due to gravity ($g \approx 32$ feet per second per second). A hemispherical water tank has a radius of 6 feet. When the tank is full, a circular valve with a radius of 1 inch is opened at the bottom, as shown in the figure. How long will it take for the tank to drain completely?

7. The cylindrical water tank shown in the figure has a height of 18 feet. When the tank is full, a circular valve is opened at the bottom of the tank. After 30 minutes, the depth of the water is 12 feet.

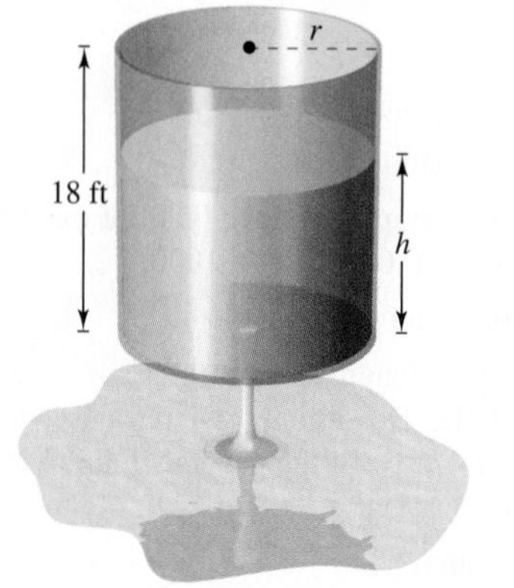

(a) How long will it take for the tank to drain completely?

(b) What is the depth of the water in the tank after 1 hour?

8. Suppose the tank in Exercise 7 has a height of 20 feet, a radius of 8 feet, and the valve is circular with a radius of 2 inches. The tank is full when the valve is opened. How long will it take for the tank to drain completely?

9. In hilly areas, radio reception may be poor. Consider a situation where an FM transmitter is located at the point $(-1, 1)$ behind a hill modeled by the graph of

$$y = x - x^2$$

and a radio receiver is on the opposite side of the hill. (Assume that the x-axis represents ground level at the base of the hill.)

(a) What is the closest position $(x, 0)$ the radio can be to the hill so that reception is unobstructed?

(b) Write the closest position $(x, 0)$ of the radio with x represented as a function of h if the transmitter is located at $(-1, h)$.

(c) Use a graphing utility to graph the function for x in part (b). Determine the vertical asymptote of the function and interpret the result.

10. Biomass is a measure of an amount of living matter in an ecosystem. Suppose the biomass $s(t)$ in a given ecosystem increases at a rate of about 3.5 tons per year, and decreases by about 1.9% per year. This situation can be modeled by the differential equation

$$\frac{ds}{dt} = 3.5 - 0.019s.$$

(a) Solve the differential equation.

(b) Use a graphing utility to graph the slope field for the differential equation. What do you notice?

(c) Explain what happens as $t \to \infty$.

In Exercises 11–13, a medical researcher wants to determine the concentration C (in moles per liter) of a tracer drug injected into a moving fluid. Solve this problem by considering a single-compartment dilution model (see figure). Assume that the fluid is continuously mixed and that the volume of the fluid in the compartment is constant.

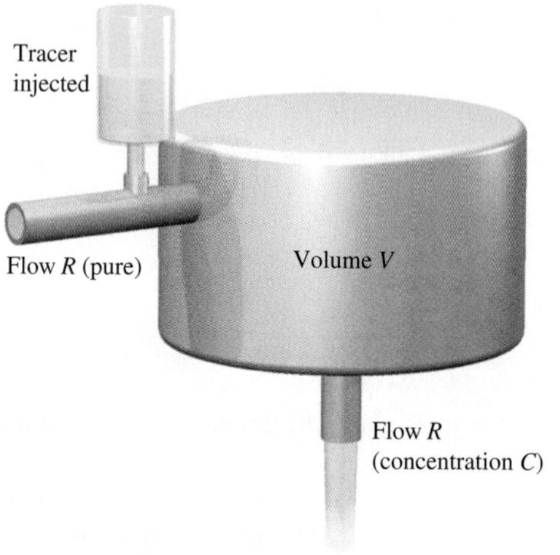

Figure for 11–13

11. If the tracer is injected instantaneously at time $t = 0$, then the concentration of the fluid in the compartment begins diluting according to the differential equation

$$\frac{dC}{dt} = \left(-\frac{R}{V}\right)C, \quad C = C_0 \text{ when } t = 0.$$

(a) Solve this differential equation to find the concentration C as a function of time t.

(b) Find the limit of C as $t \to \infty$.

12. Use the solution of the differential equation in Exercise 11 to find the concentration C as a function of time t, and use a graphing utility to graph the function.

(a) $V = 2$ liters, $R = 0.5$ liter per minute, and $C_0 = 0.6$ mole per liter

(b) $V = 2$ liters, $R = 1.5$ liters per minute, and $C_0 = 0.6$ mole per liter

13. In Exercises 11 and 12, it was assumed that there was a single initial injection of the tracer drug into the compartment. Now consider the case in which the tracer is continuously injected (beginning at $t = 0$) at the rate of Q moles per minute. Considering Q to be negligible compared with R, use the differential equation

$$\frac{dC}{dt} = \frac{Q}{V} - \left(\frac{R}{V}\right)C, \quad C = 0 \text{ when } t = 0.$$

(a) Solve this differential equation to find the concentration C as a function of time t.

(b) Find the limit of C as $t \to \infty$.

7 Applications of Integration

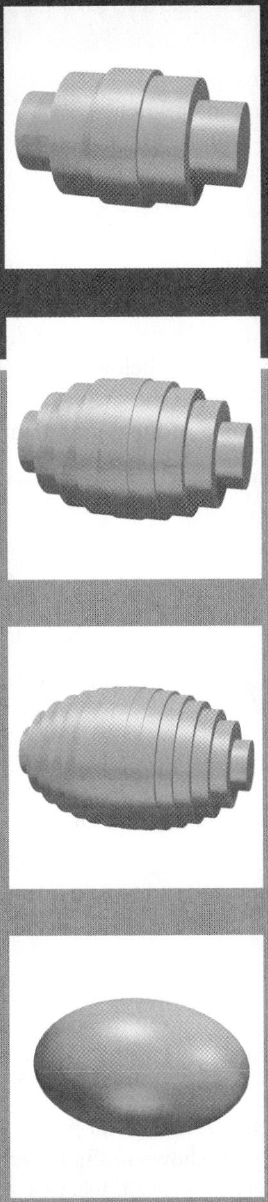

The *disk method* is one method that is used to find the volume of a solid. This method requires finding the sum of the volumes of representative disks to approximate the volume of the solid. As you increase the number of disks, the approximation tends to become more accurate. In Section 7.2, you will use limits to write the exact volume of the solid as a definite integral.

The Atomium, located in Belgium, represents an iron crystal molecule magnified 165 billion times. The structure contains nine spheres connected with cylindrical tubes. The central sphere has one tube passing directly through its center. Explain how to find the volume of the portion of the central sphere that does not include the tube.

Andre Jenny/Alamy Images

Section 7.1 Area of a Region Between Two Curves

- **Find the area of a region between two curves using integration.**
- **Find the area of a region between intersecting curves using integration.**
- **Describe integration as an accumulation process.**

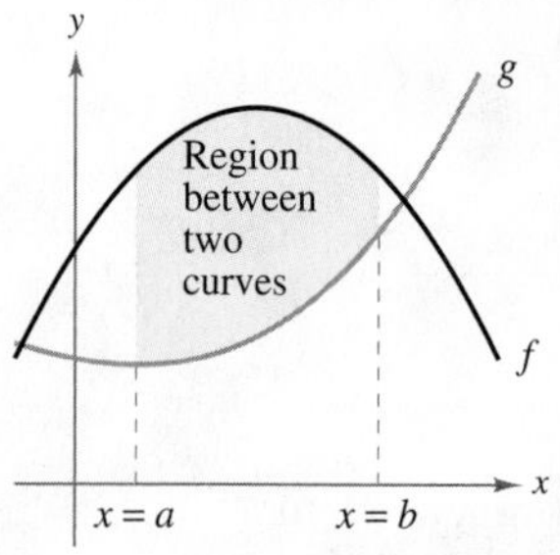

Figure 7.1

Area of a Region Between Two Curves

With a few modifications you can extend the application of definite integrals from the area of a region *under* a curve to the area of a region *between* two curves. Consider two functions f and g that are continuous on the interval $[a, b]$. If, as in Figure 7.1, the graphs of both f and g lie above the x-axis, and the graph of g lies below the graph of f, you can geometrically interpret the area of the region between the graphs as the area of the region under the graph of g subtracted from the area of the region under the graph of f, as shown in Figure 7.2.

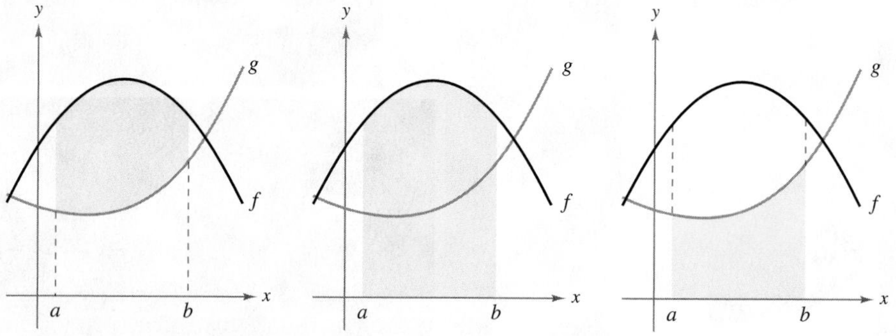

Area of region between f and g	=	Area of region under f	−	Area of region under g

$$\int_a^b [f(x) - g(x)]\,dx = \int_a^b f(x)\,dx - \int_a^b g(x)\,dx$$

Figure 7.2

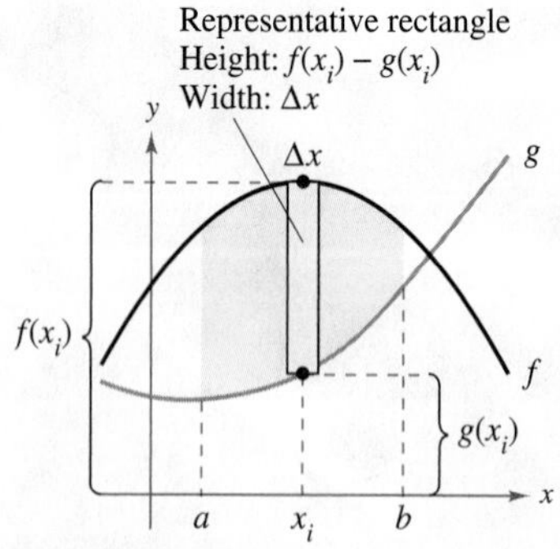

Figure 7.3

To verify the reasonableness of the result shown in Figure 7.2, you can partition the interval $[a, b]$ into n subintervals, each of width Δx. Then, as shown in Figure 7.3, sketch a **representative rectangle** of width Δx and height $f(x_i) - g(x_i)$, where x_i is in the ith interval. The area of this representative rectangle is

$$\Delta A_i = (\text{height})(\text{width}) = [f(x_i) - g(x_i)]\,\Delta x.$$

By adding the areas of the n rectangles and taking the limit as $\|\Delta\| \to 0$ $(n \to \infty)$, you obtain

$$\lim_{n\to\infty} \sum_{i=1}^{n} [f(x_i) - g(x_i)]\,\Delta x.$$

Because f and g are continuous on $[a, b]$, $f - g$ is also continuous on $[a, b]$ and the limit exists. So, the area of the given region is

$$\begin{aligned}\text{Area} &= \lim_{n\to\infty} \sum_{i=1}^{n} [f(x_i) - g(x_i)]\,\Delta x \\ &= \int_a^b [f(x) - g(x)]\,dx.\end{aligned}$$

Area of a Region Between Two Curves

If f and g are continuous on $[a, b]$ and $g(x) \leq f(x)$ for all x in $[a, b]$, then the area of the region bounded by the graphs of f and g and the vertical lines $x = a$ and $x = b$ is

$$A = \int_a^b [f(x) - g(x)]\, dx.$$

In Figure 7.1, the graphs of f and g are shown above the x-axis. This, however, is not necessary. The same integrand $[f(x) - g(x)]$ can be used as long as f and g are continuous and $g(x) \leq f(x)$ for all x in the interval $[a, b]$. This result is summarized graphically in Figure 7.4.

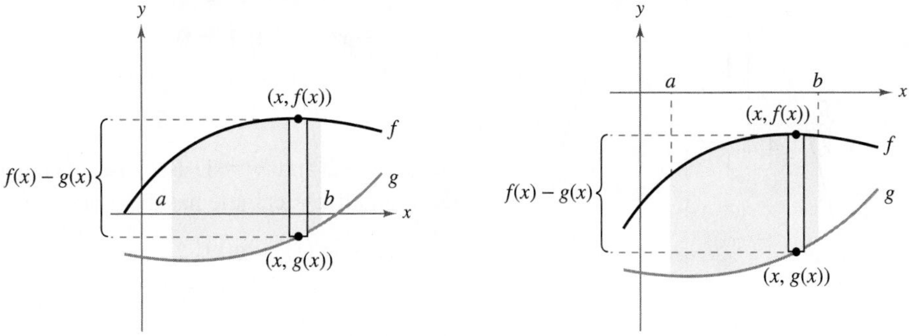

Figure 7.4

NOTE The height of a representative rectangle is $f(x) - g(x)$ regardless of the relative position of the x-axis, as shown in Figure 7.4.

Representative rectangles are used throughout this chapter in various applications of integration. A vertical rectangle (of width Δx) implies integration with respect to x, whereas a horizontal rectangle (of width Δy) implies integration with respect to y.

EXAMPLE 1 Finding the Area of a Region Between Two Curves

Find the area of the region bounded by the graphs of $y = x^2 + 2$, $y = -x$, $x = 0$, and $x = 1$.

Solution Let $g(x) = -x$ and $f(x) = x^2 + 2$. Then $g(x) \leq f(x)$ for all x in $[0, 1]$, as shown in Figure 7.5. So, the area of the representative rectangle is

$$\Delta A = [f(x) - g(x)]\,\Delta x$$
$$= [(x^2 + 2) - (-x)]\,\Delta x$$

and the area of the region is

$$A = \int_a^b [f(x) - g(x)]\, dx = \int_0^1 [(x^2 + 2) - (-x)]\, dx$$
$$= \left[\frac{x^3}{3} + \frac{x^2}{2} + 2x\right]_0^1$$
$$= \frac{1}{3} + \frac{1}{2} + 2$$
$$= \frac{17}{6}.$$

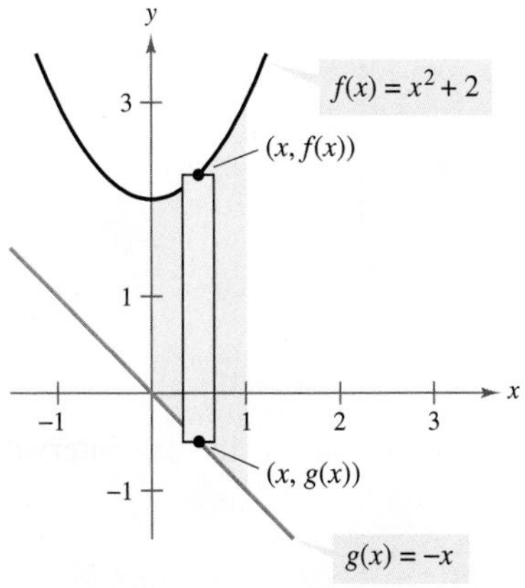

Region bounded by the graph of f, the graph of g, $x = 0$, and $x = 1$
Figure 7.5

Area of a Region Between Intersecting Curves

In Example 1, the graphs of $f(x) = x^2 + 2$ and $g(x) = -x$ do not intersect, and the values of a and b are given explicitly. A more common problem involves the area of a region bounded by two *intersecting* graphs, where the values of a and b must be calculated.

EXAMPLE 2 A Region Lying Between Two Intersecting Graphs

Find the area of the region bounded by the graphs of $f(x) = 2 - x^2$ and $g(x) = x$.

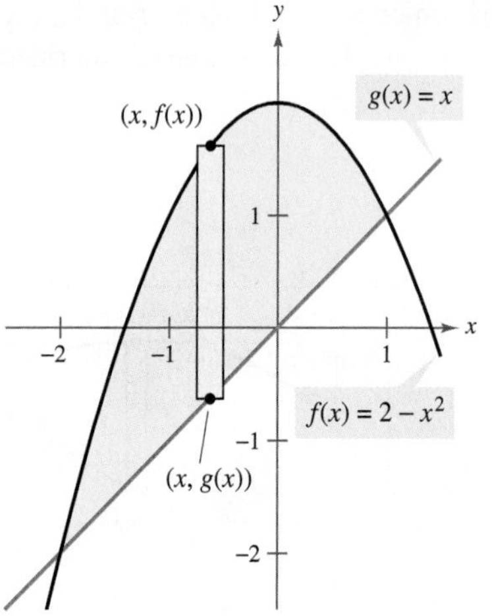

Region bounded by the graph of f and the graph of g
Figure 7.6

Solution In Figure 7.6, notice that the graphs of f and g have two points of intersection. To find the x-coordinates of these points, set $f(x)$ and $g(x)$ equal to each other and solve for x.

$$2 - x^2 = x \qquad \text{Set } f(x) \text{ equal to } g(x).$$
$$-x^2 - x + 2 = 0 \qquad \text{Write in general form.}$$
$$-(x + 2)(x - 1) = 0 \qquad \text{Factor.}$$
$$x = -2 \text{ or } 1 \qquad \text{Solve for } x.$$

So, $a = -2$ and $b = 1$. Because $g(x) \le f(x)$ for all x in the interval $[-2, 1]$, the representative rectangle has an area of

$$\Delta A = [f(x) - g(x)]\,\Delta x$$
$$= [(2 - x^2) - x]\,\Delta x$$

and the area of the region is

$$A = \int_{-2}^{1} [(2 - x^2) - x]\,dx = \left[-\frac{x^3}{3} - \frac{x^2}{2} + 2x\right]_{-2}^{1}$$
$$= \frac{9}{2}.$$

EXAMPLE 3 A Region Lying Between Two Intersecting Graphs

The sine and cosine curves intersect infinitely many times, bounding regions of equal areas, as shown in Figure 7.7. Find the area of one of these regions.

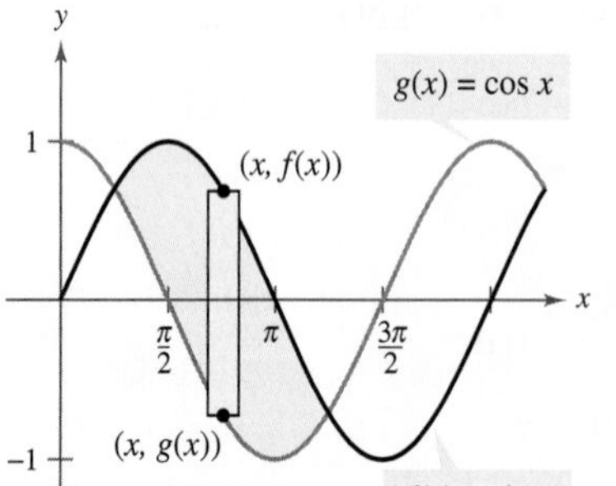

One of the regions bounded by the graphs of the sine and cosine functions
Figure 7.7

Solution

$$\sin x = \cos x \qquad \text{Set } f(x) \text{ equal to } g(x).$$
$$\frac{\sin x}{\cos x} = 1 \qquad \text{Divide each side by } \cos x.$$
$$\tan x = 1 \qquad \text{Trigonometric identity}$$
$$x = \frac{\pi}{4} \text{ or } \frac{5\pi}{4}, \quad 0 \le x \le 2\pi \qquad \text{Solve for } x.$$

So, $a = \pi/4$ and $b = 5\pi/4$. Because $\sin x \ge \cos x$ for all x in the interval $[\pi/4, 5\pi/4]$, the area of the region is

$$A = \int_{\pi/4}^{5\pi/4} [\sin x - \cos x]\,dx = \Big[-\cos x - \sin x\Big]_{\pi/4}^{5\pi/4}$$
$$= 2\sqrt{2}.$$

If two curves intersect at more than two points, then to find the area of the region between the curves, you must find all points of intersection and check to see which curve is above the other in each interval determined by these points.

EXAMPLE 4 Curves That Intersect at More Than Two Points

Find the area of the region between the graphs of $f(x) = 3x^3 - x^2 - 10x$ and $g(x) = -x^2 + 2x$.

Solution Begin by setting $f(x)$ and $g(x)$ equal to each other and solving for x. This yields the x-values at each point of intersection of the two graphs.

$$3x^3 - x^2 - 10x = -x^2 + 2x \quad \text{Set } f(x) \text{ equal to } g(x).$$
$$3x^3 - 12x = 0 \quad \text{Write in general form.}$$
$$3x(x-2)(x+2) = 0 \quad \text{Factor.}$$
$$x = -2, 0, 2 \quad \text{Solve for } x.$$

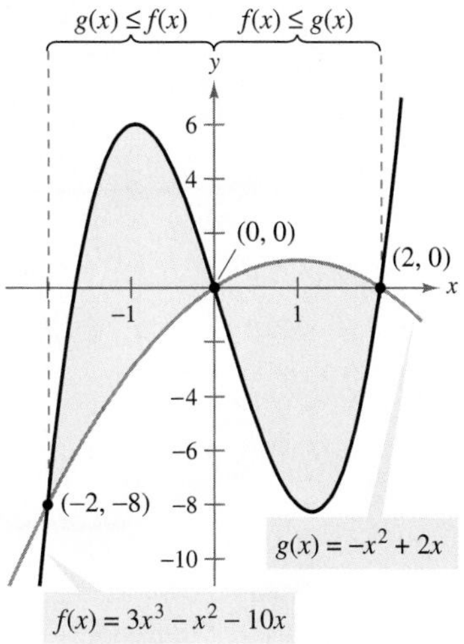

On $[-2, 0]$, $g(x) \le f(x)$, and on $[0, 2]$, $f(x) \le g(x)$
Figure 7.8

So, the two graphs intersect when $x = -2, 0,$ and 2. In Figure 7.8, notice that $g(x) \le f(x)$ on the interval $[-2, 0]$. However, the two graphs switch at the origin, and $f(x) \le g(x)$ on the interval $[0, 2]$. So, you need two integrals—one for the interval $[-2, 0]$ and one for the interval $[0, 2]$.

$$\begin{aligned} A &= \int_{-2}^{0} [f(x) - g(x)]\,dx + \int_{0}^{2} [g(x) - f(x)]\,dx \\ &= \int_{-2}^{0} (3x^3 - 12x)\,dx + \int_{0}^{2} (-3x^3 + 12x)\,dx \\ &= \left[\frac{3x^4}{4} - 6x^2\right]_{-2}^{0} + \left[\frac{-3x^4}{4} + 6x^2\right]_{0}^{2} \\ &= -(12 - 24) + (-12 + 24) = 24 \end{aligned}$$

NOTE In Example 4, notice that you obtain an incorrect result if you integrate from -2 to 2. Such integration produces

$$\int_{-2}^{2} [f(x) - g(x)]\,dx = \int_{-2}^{2} (3x^3 - 12x)\,dx = 0.$$

If the graph of a function of y is a boundary of a region, it is often convenient to use representative rectangles that are *horizontal* and find the area by integrating with respect to y. In general, to determine the area between two curves, you can use

$$A = \int_{x_1}^{x_2} \underbrace{[(\text{top curve}) - (\text{bottom curve})]}_{\text{in variable } x}\,dx \quad \text{Vertical rectangles}$$

$$A = \int_{y_1}^{y_2} \underbrace{[(\text{right curve}) - (\text{left curve})]}_{\text{in variable } y}\,dy \quad \text{Horizontal rectangles}$$

where (x_1, y_1) and (x_2, y_2) are either adjacent points of intersection of the two curves involved or points on the specified boundary lines.

---

indicates that in the HM mathSpace® *CD-ROM and the online* Eduspace® *system for this text, you will find an Open Exploration, which further explores this example using the computer algebra systems* Maple, Mathcad, Mathematica, *and* Derive.

EXAMPLE 5 Horizontal Representative Rectangles

Find the area of the region bounded by the graphs of $x = 3 - y^2$ and $x = y + 1$.

Solution Consider

$$g(y) = 3 - y^2 \quad \text{and} \quad f(y) = y + 1.$$

These two curves intersect when $y = -2$ and $y = 1$, as shown in Figure 7.9. Because $f(y) \le g(y)$ on this interval, you have

$$\Delta A = [g(y) - f(y)]\,\Delta y = [(3 - y^2) - (y + 1)]\,\Delta y.$$

So, the area is

$$\begin{aligned} A &= \int_{-2}^{1} [(3 - y^2) - (y + 1)]\,dy \\ &= \int_{-2}^{1} (-y^2 - y + 2)\,dy \\ &= \left[\frac{-y^3}{3} - \frac{y^2}{2} + 2y\right]_{-2}^{1} \\ &= \left(-\frac{1}{3} - \frac{1}{2} + 2\right) - \left(\frac{8}{3} - 2 - 4\right) \\ &= \frac{9}{2}. \end{aligned}$$

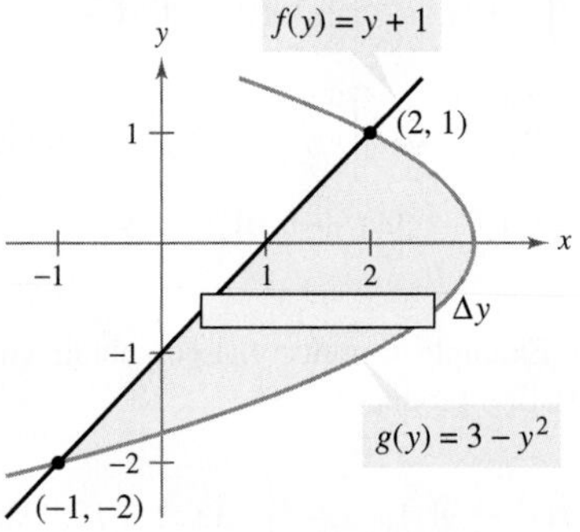

Horizontal rectangles (integration with respect to y)
Figure 7.9

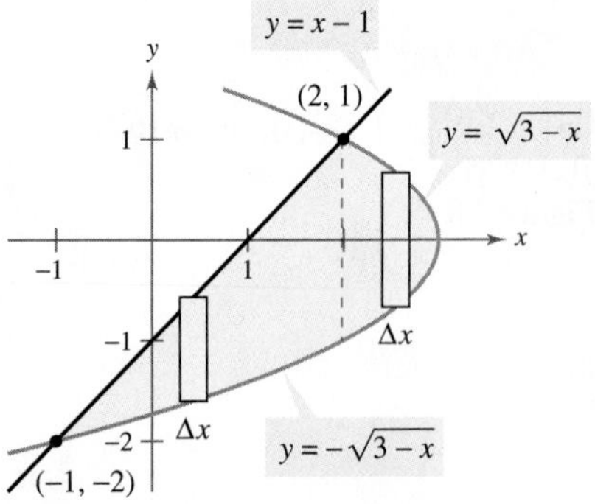

Vertical rectangles (integration with respect to x)
Figure 7.10

In Example 5, notice that by integrating with respect to y you need only one integral. If you had integrated with respect to x, you would have needed two integrals because the upper boundary would have changed at $x = 2$, as shown in Figure 7.10.

$$\begin{aligned} A &= \int_{-1}^{2} \left[(x - 1) + \sqrt{3 - x}\right] dx + \int_{2}^{3} \left(\sqrt{3 - x} + \sqrt{3 - x}\right) dx \\ &= \int_{-1}^{2} \left[x - 1 + (3 - x)^{1/2}\right] dx + 2\int_{2}^{3} (3 - x)^{1/2}\,dx \\ &= \left[\frac{x^2}{2} - x - \frac{(3 - x)^{3/2}}{3/2}\right]_{-1}^{2} - 2\left[\frac{(3 - x)^{3/2}}{3/2}\right]_{2}^{3} \\ &= \left(2 - 2 - \frac{2}{3}\right) - \left(\frac{1}{2} + 1 - \frac{16}{3}\right) - 2(0) + 2\left(\frac{2}{3}\right) \\ &= \frac{9}{2} \end{aligned}$$

Integration as an Accumulation Process

In this section, the integration formula for the area between two curves was developed by using a rectangle as the *representative element*. For each new application in the remaining sections of this chapter, an appropriate representative element will be constructed using precalculus formulas you already know. Each integration formula will then be obtained by summing or accumulating these representative elements.

Known precalculus formula ⇨ Representative element ⇨ New integration formula

For example, in this section the area formula was developed as follows.

$A = (\text{height})(\text{width})$ ⇨ $\Delta A = [f(x) - g(x)]\,\Delta x$ ⇨ $A = \int_a^b [f(x) - g(x)]\,dx$

EXAMPLE 6 Describing Integration as an Accumulation Process

Find the area of the region bounded by the graph of $y = 4 - x^2$ and the x-axis. Describe the integration as an accumulation process.

Solution The area of the region is given by

$$A = \int_{-2}^{2} (4 - x^2)\,dx.$$

You can think of the integration as an accumulation of the areas of the rectangles formed as the representative rectangle slides from $x = -2$ to $x = 2$, as shown in Figure 7.11.

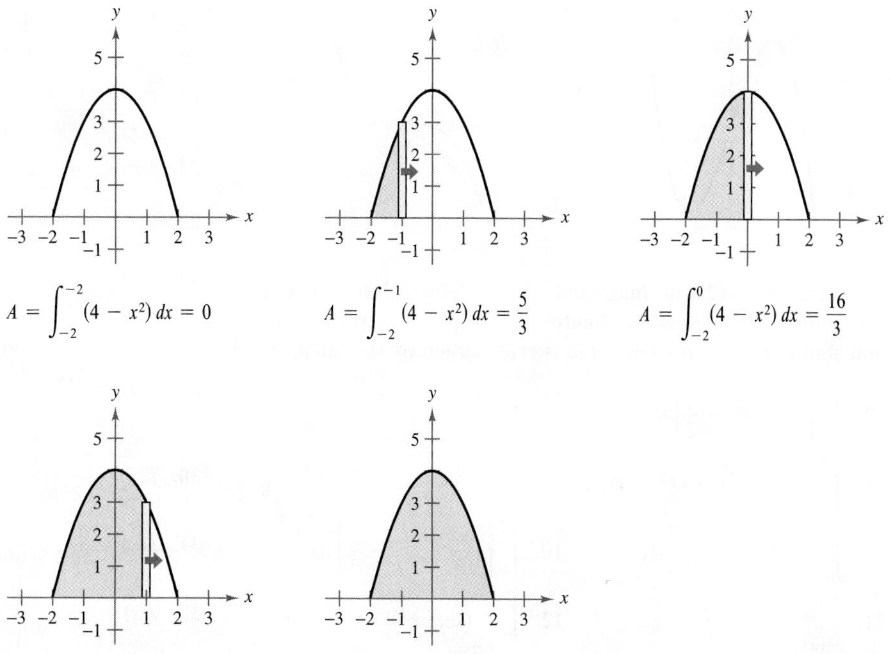

$A = \int_{-2}^{-2} (4 - x^2)\,dx = 0$ $A = \int_{-2}^{-1} (4 - x^2)\,dx = \frac{5}{3}$ $A = \int_{-2}^{0} (4 - x^2)\,dx = \frac{16}{3}$

$A = \int_{-2}^{1} (4 - x^2)\,dx = 9$ $A = \int_{-2}^{2} (4 - x^2)\,dx = \frac{32}{3}$

Figure 7.11

Exercises for Section 7.1

See www.CalcChat.com for worked-out solutions to odd-numbered exercises.

In Exercises 1–6, set up the definite integral that gives the area of the region.

1. $f(x) = x^2 - 6x$
   $g(x) = 0$

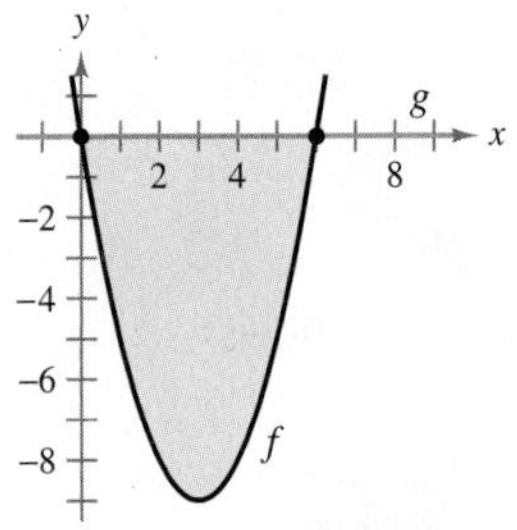

2. $f(x) = x^2 + 2x + 1$
   $g(x) = 2x + 5$

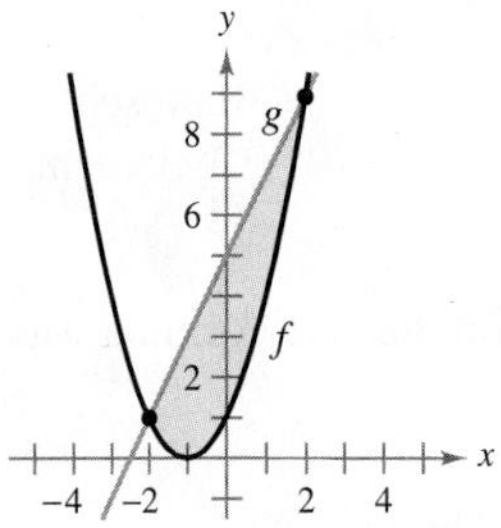

3. $f(x) = x^2 - 4x + 3$
   $g(x) = -x^2 + 2x + 3$

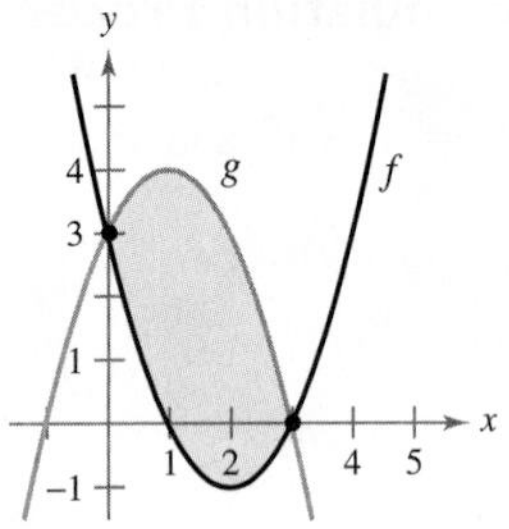

4. $f(x) = x^2$
   $g(x) = x^3$

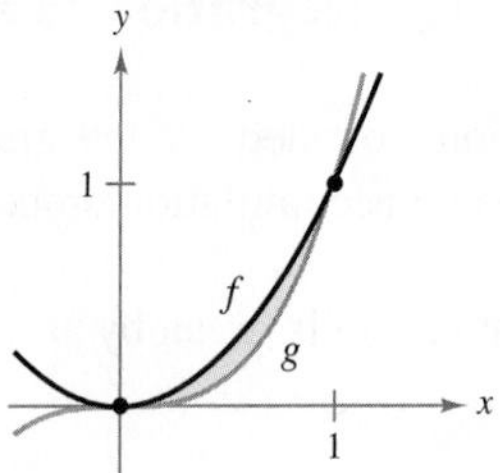

5. $f(x) = 3(x^3 - x)$
   $g(x) = 0$

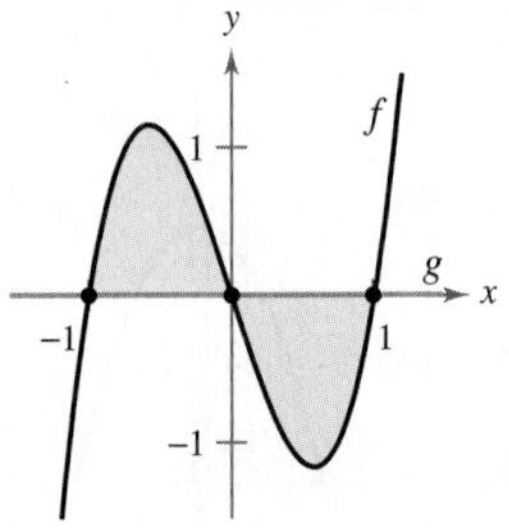

6. $f(x) = (x - 1)^3$
   $g(x) = x - 1$

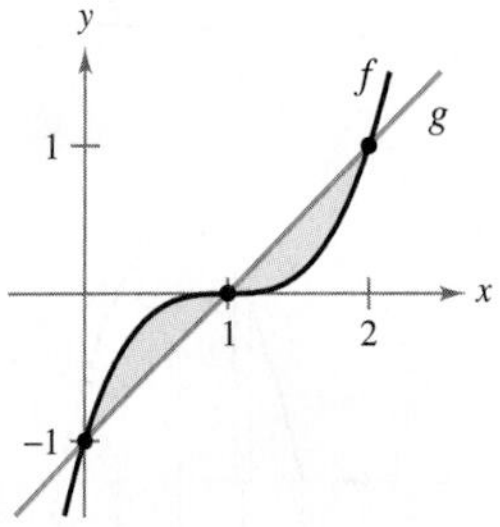

In Exercises 7–12, the integrand of the definite integral is a difference of two functions. Sketch the graph of each function and shade the region whose area is represented by the integral.

7. $\displaystyle\int_0^4 \left[(x + 1) - \frac{x}{2}\right] dx$

8. $\displaystyle\int_{-1}^1 [(1 - x^2) - (x^2 - 1)]\, dx$

9. $\displaystyle\int_0^6 \left[4(2^{-x/3}) - \frac{x}{6}\right] dx$

10. $\displaystyle\int_2^3 \left[\left(\frac{x^3}{3} - x\right) - \frac{x}{3}\right] dx$

11. $\displaystyle\int_{-\pi/3}^{\pi/3} (2 - \sec x)\, dx$

12. $\displaystyle\int_{-\pi/4}^{\pi/4} (\sec^2 x - \cos x)\, dx$

In Exercises 13 and 14, find the area of the region by integrating (a) with respect to x and (b) with respect to y.

13. $x = 4 - y^2$
    $x = y - 2$

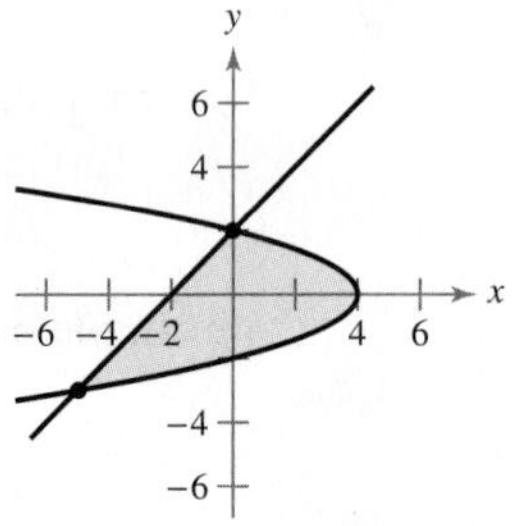

14. $y = x^2$
    $y = 6 - x$

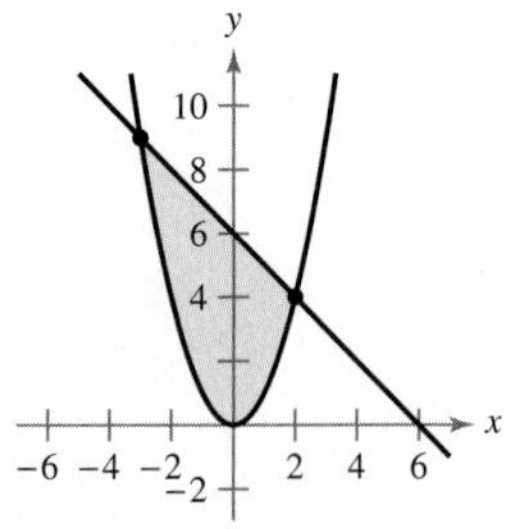

Think About It **In Exercises 15 and 16, determine which value best approximates the area of the region bounded by the graphs of f and g. (Make your selection on the basis of a sketch of the region and not by performing any calculations.)**

15. $f(x) = x + 1,\quad g(x) = (x - 1)^2$
    (a) -2 (b) 2 (c) 10 (d) 4 (e) 8

16. $f(x) = 2 - \frac{1}{2}x,\quad g(x) = 2 - \sqrt{x}$
    (a) 1 (b) 6 (c) -3 (d) 3 (e) 4

In Exercises 17–32, sketch the region bounded by the graphs of the algebraic functions and find the area of the region.

17. $y = \frac{1}{2}x^3 + 2,\ y = x + 1,\ x = 0,\ x = 2$

18. $y = -\frac{3}{8}x(x - 8),\ y = 10 - \frac{1}{2}x,\ x = 2,\ x = 8$

19. $f(x) = x^2 - 4x,\ g(x) = 0$

20. $f(x) = -x^2 + 4x + 1,\ g(x) = x + 1$

21. $f(x) = x^2 + 2x + 1,\ g(x) = 3x + 3$

22. $f(x) = -x^2 + 4x + 2,\ g(x) = x + 2$

23. $y = x,\ y = 2 - x,\ y = 0$

24. $y = \dfrac{1}{x^2},\ y = 0,\ x = 1,\ x = 5$

25. $f(x) = \sqrt{3x} + 1,\ g(x) = x + 1$

26. $f(x) = \sqrt[3]{x - 1},\ g(x) = x - 1$

27. $f(y) = y^2,\ g(y) = y + 2$

28. $f(y) = y(2 - y),\ g(y) = -y$

29. $f(y) = y^2 + 1,\ g(y) = 0,\ y = -1,\ y = 2$

30. $f(y) = \dfrac{y}{\sqrt{16 - y^2}},\ g(y) = 0,\ y = 3$

31. $f(x) = \dfrac{10}{x},\ x = 0,\ y = 2,\ y = 10$

32. $g(x) = \dfrac{4}{2 - x},\ y = 4,\ x = 0$

In Exercises 33–42, (a) use a graphing utility to graph the region bounded by the graphs of the equations, (b) find the area of the region, and (c) use the integration capabilities of the graphing utility to verify your results.

33. $f(x) = x(x^2 - 3x + 3),\; g(x) = x^2$

34. $f(x) = x^3 - 2x + 1,\; g(x) = -2x,\; x = 1$

35. $y = x^2 - 4x + 3,\; y = 3 + 4x - x^2$

36. $y = x^4 - 2x^2,\; y = 2x^2$

37. $f(x) = x^4 - 4x^2,\; g(x) = x^2 - 4$

38. $f(x) = x^4 - 4x^2,\; g(x) = x^3 - 4x$

39. $f(x) = 1/(1 + x^2),\; g(x) = \frac{1}{2}x^2$

40. $f(x) = 6x/(x^2 + 1),\; y = 0,\; 0 \le x \le 3$

41. $y = \sqrt{1 + x^3},\; y = \frac{1}{2}x + 2,\; x = 0$

42. $y = x\sqrt{\dfrac{4 - x}{4 + x}},\; y = 0,\; x = 4$

In Exercises 43–48, sketch the region bounded by the graphs of the functions, and find the area of the region.

43. $f(x) = 2\sin x,\; g(x) = \tan x,\; -\dfrac{\pi}{3} \le x \le \dfrac{\pi}{3}$

44. $f(x) = \sin x,\; g(x) = \cos 2x,\; -\dfrac{\pi}{2} \le x \le \dfrac{\pi}{6}$

45. $f(x) = \cos x,\; g(x) = 2 - \cos x,\; 0 \le x \le 2\pi$

46. $f(x) = \sec\dfrac{\pi x}{4}\tan\dfrac{\pi x}{4},\; g(x) = \left(\sqrt{2} - 4\right)x + 4,\; x = 0$

47. $f(x) = xe^{-x^2},\; y = 0,\; 0 \le x \le 1$

48. $f(x) = 3^x,\; g(x) = 2x + 1$

In Exercises 49–52, (a) use a graphing utility to graph the region bounded by the graphs of the equations, (b) find the area of the region, and (c) use the integration capabilities of the graphing utility to verify your results.

49. $f(x) = 2\sin x + \sin 2x,\; y = 0,\; 0 \le x \le \pi$

50. $f(x) = 2\sin x + \cos 2x,\; y = 0,\; 0 < x \le \pi$

51. $f(x) = \dfrac{1}{x^2}e^{1/x},\; y = 0,\; 1 \le x \le 3$

52. $g(x) = \dfrac{4\ln x}{x},\; y = 0,\; x = 5$

In Exercises 53–56, (a) use a graphing utility to graph the region bounded by the graphs of the equations, (b) explain why the area of the region is difficult to find by hand, and (c) use the integration capabilities of the graphing utility to approximate the area to four decimal places.

53. $y = \sqrt{\dfrac{x^3}{4 - x}},\; y = 0,\; x = 3$

54. $y = \sqrt{x}\,e^x,\; y = 0,\; x = 0,\; x = 1$

55. $y = x^2,\; y = 4\cos x$

56. $y = x^2,\; y = \sqrt{3 + x}$

In Exercises 57–60, find the accumulation function F. Then evaluate F at each value of the independent variable and graphically show the area given by each value of F.

57. $F(x) = \displaystyle\int_0^x \left(\tfrac{1}{2}t + 1\right) dt$ (a) $F(0)$ (b) $F(2)$ (c) $F(6)$

58. $F(x) = \displaystyle\int_0^x \left(\tfrac{1}{2}t^2 + 2\right) dt$ (a) $F(0)$ (b) $F(4)$ (c) $F(6)$

59. $F(\alpha) = \displaystyle\int_{-1}^{\alpha} \cos\frac{\pi\theta}{2}\, d\theta$ (a) $F(-1)$ (b) $F(0)$ (c) $F\left(\tfrac{1}{2}\right)$

60. $F(y) = \displaystyle\int_{-1}^{y} 4e^{x/2}\, dx$ (a) $F(-1)$ (b) $F(0)$ (c) $F(4)$

In Exercises 61–64, use integration to find the area of the figure having the given vertices.

61. $(2, -3), (4, 6), (6, 1)$

62. $(0, 0), (a, 0), (b, c)$

63. $(0, 2), (4, 2), (0, -2), (-4, -2)$

64. $(0, 0), (1, 2), (3, -2), (1, -3)$

65. ***Numerical Integration*** Estimate the surface area of the golf green using (a) the Trapezoidal Rule and (b) Simpson's Rule.

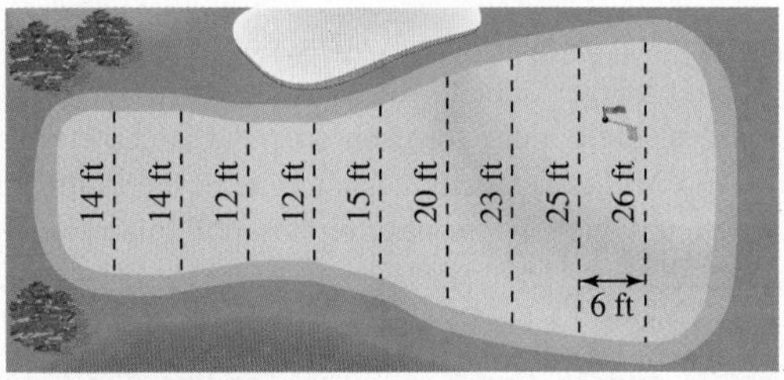

66. ***Numerical Integration*** Estimate the surface area of the oil spill using (a) the Trapezoidal Rule and (b) Simpson's Rule.

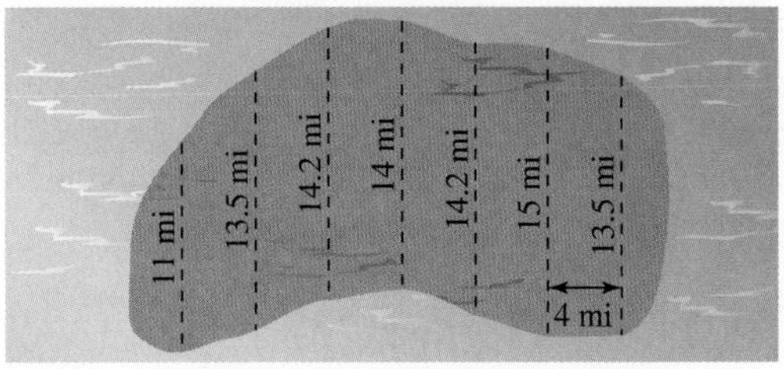

In Exercises 67–70, set up and evaluate the definite integral that gives the area of the region bounded by the graph of the function and the tangent line to the graph at the given point.

67. $f(x) = x^3,\; (1, 1)$

68. $y = x^3 - 2x,\; (-1, 1)$

69. $f(x) = \dfrac{1}{x^2 + 1},\; \left(1, \dfrac{1}{2}\right)$

70. $y = \dfrac{2}{1 + 4x^2},\; \left(\dfrac{1}{2}, 1\right)$

Writing About Concepts

71. The graphs of $y = x^4 - 2x^2 + 1$ and $y = 1 - x^2$ intersect at three points. However, the area between the curves *can* be found by a single integral. Explain why this is so, and write an integral for this area.

Writing About Concepts *(continued)*

72. The area of the region bounded by the graphs of $y = x^3$ and $y = x$ *cannot* be found by the single integral $\int_{-1}^{1} (x^3 - x)\,dx$. Explain why this is so. Use symmetry to write a single integral that does represent the area.

73. A college graduate has two job offers. The starting salary for each is \$32,000, and after 8 years of service each will pay \$54,000. The salary increase for each offer is shown in the figure. From a strictly monetary viewpoint, which is the better offer? Explain.

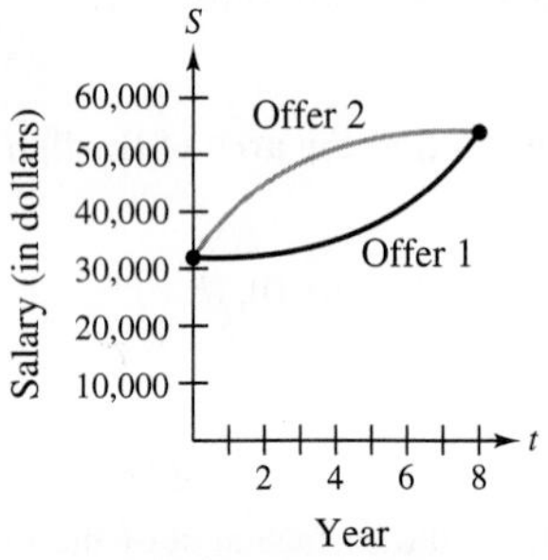

Figure for 73

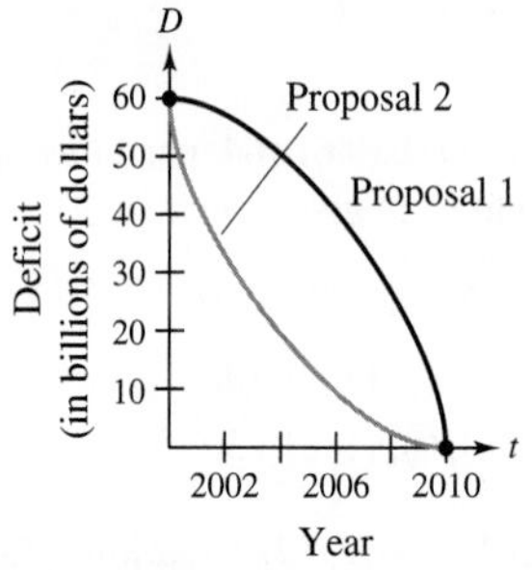

Figure for 74

74. A state legislature is debating two proposals for eliminating the annual budget deficits by the year 2010. The rate of decrease of the deficits for each proposal is shown in the figure. From the viewpoint of minimizing the cumulative state deficit, which is the better proposal? Explain.

In Exercises 75 and 76, find b such that the line $y = b$ divides the region bounded by the graphs of the two equations into two regions of equal area.

75. $y = 9 - x^2, \quad y = 0$

76. $y = 9 - |x|, \quad y = 0$

In Exercises 77 and 78, find a such that the line $x = a$ divides the region bounded by the graphs of the equations into two regions of equal area.

77. $y = x, \quad y = 4, \quad x = 0$

78. $y^2 = 4 - x, \quad x = 0$

In Exercises 79 and 80, evaluate the limit and sketch the graph of the region whose area is represented by the limit.

79. $\displaystyle\lim_{\|\Delta\| \to 0} \sum_{i=1}^{n} (x_i - x_i^2)\,\Delta x$, where $x_i = i/n$ and $\Delta x = 1/n$

80. $\displaystyle\lim_{\|\Delta\| \to 0} \sum_{i=1}^{n} (4 - x_i^2)\,\Delta x$, where $x_i = -2 + (4i/n)$ and $\Delta x = 4/n$

Revenue **In Exercises 81 and 82, two models R_1 and R_2 are given for revenue (in billions of dollars per year) for a large corporation. The model R_1 gives projected annual revenues from 2000 to 2005, with $t = 0$ corresponding to 2000, and R_2 gives projected revenues if there is a decrease in the rate of growth of corporate sales over the period. Approximate the total reduction in revenue if corporate sales are actually closer to the model R_2.**

81. $R_1 = 7.21 + 0.58t$
$R_2 = 7.21 + 0.45t$

82. $R_1 = 7.21 + 0.26t + 0.02t^2$
$R_2 = 7.21 + 0.1t + 0.01t^2$

83. ***Modeling Data*** The table shows the total receipts R and total expenditures E for the Old-Age and Survivors Insurance Trust Fund (Social Security Trust Fund) in billions of dollars. The time t is given in years, with $t = 1$ corresponding to 1991. *(Source: Social Security Administration)*

t	1	2	3	4	5	6
R	299.3	311.2	323.3	328.3	342.8	363.7
E	245.6	259.9	273.1	284.1	297.8	308.2

t	7	8	9	10	11
R	397.2	424.8	457.0	490.5	518.1
E	322.1	332.3	339.9	358.3	377.5

(a) Use a graphing utility to fit an exponential model to the data for receipts. Plot the data and graph the model.

(b) Use a graphing utility to fit an exponential model to the data for expenditures. Plot the data and graph the model.

(c) If the models are assumed to be true for the years 2002 through 2007, use integration to approximate the surplus revenue generated during those years.

(d) Will the models found in parts (a) and (b) intersect? Explain. Based on your answer and news reports about the fund, will these models be accurate for long-term analysis?

84. ***Lorenz Curve*** Economists use *Lorenz curves* to illustrate the distribution of income in a country. A Lorenz curve, $y = f(x)$, represents the actual income distribution in the country. In this model, x represents percents of families in the country and y represents percents of total income. The model $y = x$ represents a country in which each family has the same income. The area between these two models, where $0 \le x \le 100$, indicates a country's "income inequality." The table lists percents of income y for selected percents of families x in a country.

x	10	20	30	40	50
y	3.35	6.07	9.17	13.39	19.45

x	60	70	80	90
y	28.03	39.77	55.28	75.12

(a) Use a graphing utility to find a quadratic model for the Lorenz curve.

(b) Plot the data and graph the model.

(c) Graph the model $y = x$. How does this model compare with the model in part (a)?

(d) Use the integration capabilities of a graphing utility to approximate the "income inequality."

85. ***Profit*** The chief financial officer of a company reports that profits for the past fiscal year were \$893,000. The officer predicts that profits for the next 5 years will grow at a continuous annual rate somewhere between $3\frac{1}{2}\%$ and 5%. Estimate the cumulative difference in total profit over the 5 years based on the predicted range of growth rates.

86. ***Area*** The shaded region in the figure consists of all points whose distances from the center of the square are less than their distances from the edges of the square. Find the area of the region.

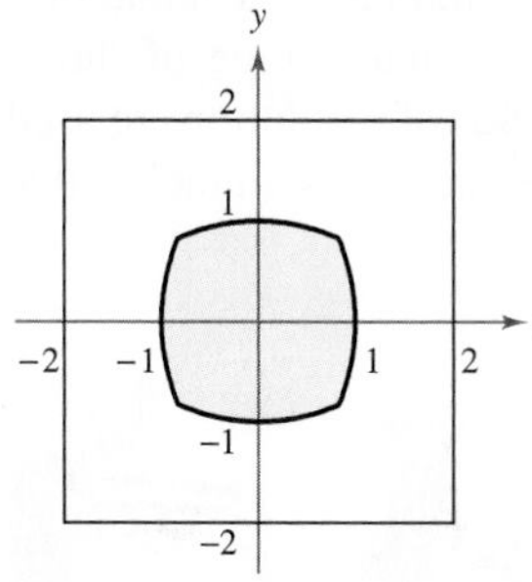

Figure for 86

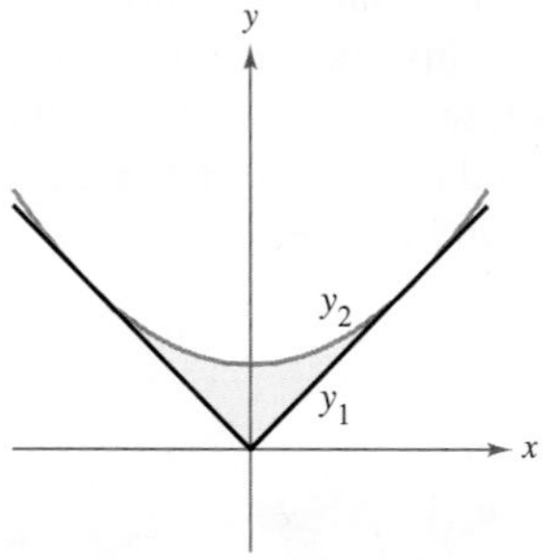

Figure for 87

87. ***Mechanical Design*** The surface of a machine part is the region between the graphs of $y_1 = |x|$ and $y_2 = 0.08x^2 + k$ (see figure).

(a) Find k if the parabola is tangent to the graph of y_1.

(b) Find the area of the surface of the machine part.

88. ***Building Design*** Concrete sections for a new building have the dimensions (in meters) and shape shown in the figure.

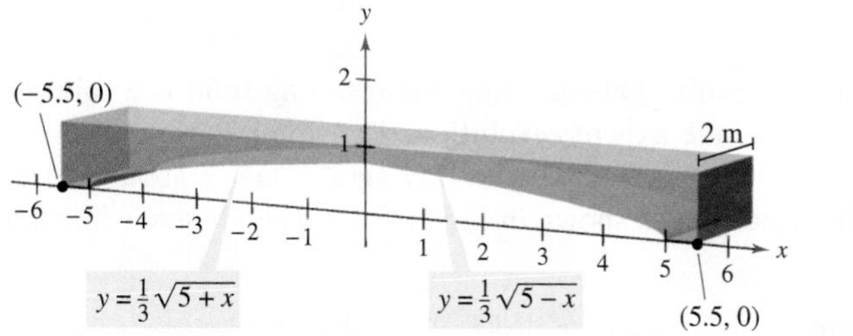

(a) Find the area of the face of the section superimposed on the rectangular coordinate system.

(b) Find the volume of concrete in one of the sections by multiplying the area in part (a) by 2 meters.

(c) One cubic meter of concrete weighs 5000 pounds. Find the weight of the section.

89. ***Building Design*** To decrease the weight and to aid in the hardening process, the concrete sections in Exercise 88 often are not solid. Rework Exercise 88 to allow for cylindrical openings such as those shown in the figure.

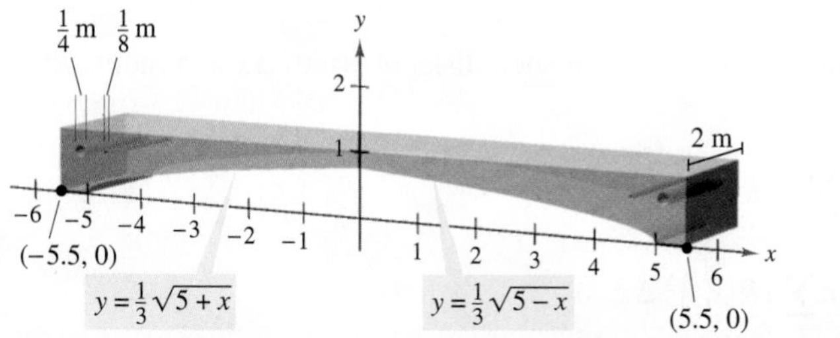

True or False? **In Exercises 90–92, determine whether the statement is true or false. If it is false, explain why or give an example that shows it is false.**

90. If the area of the region bounded by the graphs of f and g is 1, then the area of the region bounded by the graphs of $h(x) = f(x) + C$ and $k(x) = g(x) + C$ is also 1.

91. If $\int_a^b [f(x) - g(x)]\,dx = A$, then $\int_a^b [g(x) - f(x)]\,dx = -A$.

92. If the graphs of f and g intersect midway between $x = a$ and $x = b$, then

$$\int_a^b [f(x) - g(x)]\,dx = 0.$$

93. ***Area*** Find the area between the graph of $y = \sin x$ and the line segments joining the points $(0, 0)$ and $\left(\frac{7\pi}{6}, -\frac{1}{2}\right)$, as shown in the figure.

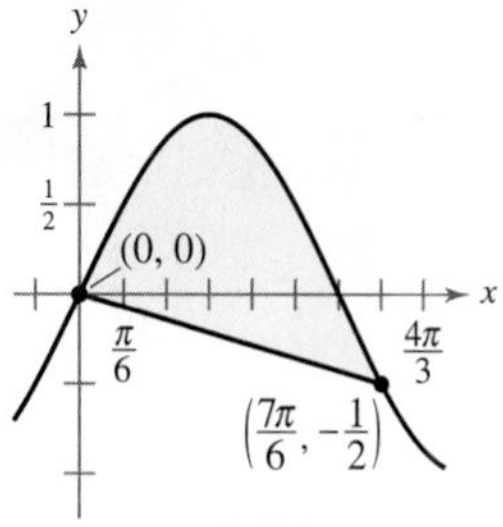

Figure for 93

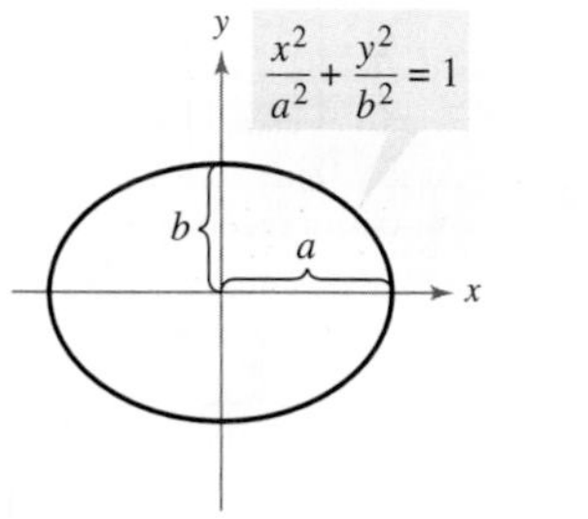

Figure for 94

94. ***Area*** Let $a > 0$ and $b > 0$. Show that the area of the ellipse

$$\frac{x^2}{a^2} + \frac{y^2}{b^2} = 1$$

is πab (see figure).

Putnam Exam Challenge

95. The horizontal line $y = c$ intersects the curve $y = 2x - 3x^3$ in the first quadrant as shown in the figure. Find c so that the areas of the two shaded regions are equal.

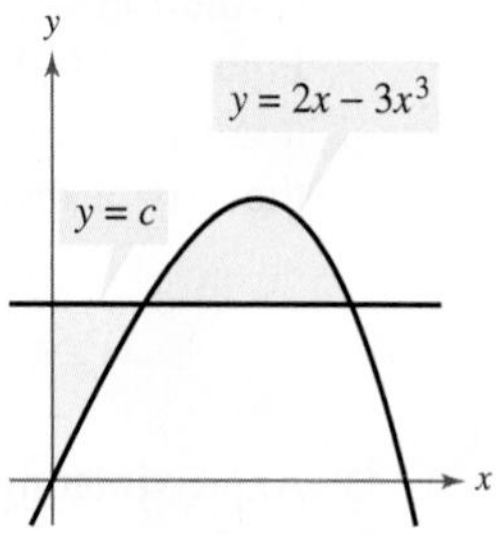

Section 7.2 Volume: The Disk Method

- **Find the volume of a solid of revolution using the disk method.**
- **Find the volume of a solid of revolution using the washer method.**
- **Find the volume of a solid with known cross sections.**

The Disk Method

In Chapter 4 we mentioned that area is only one of the *many* applications of the definite integral. Another important application is its use in finding the volume of a three-dimensional solid. In this section you will study a particular type of three-dimensional solid—one whose cross sections are similar. Solids of revolution are used commonly in engineering and manufacturing. Some examples are axles, funnels, pills, bottles, and pistons, as shown in Figure 7.12.

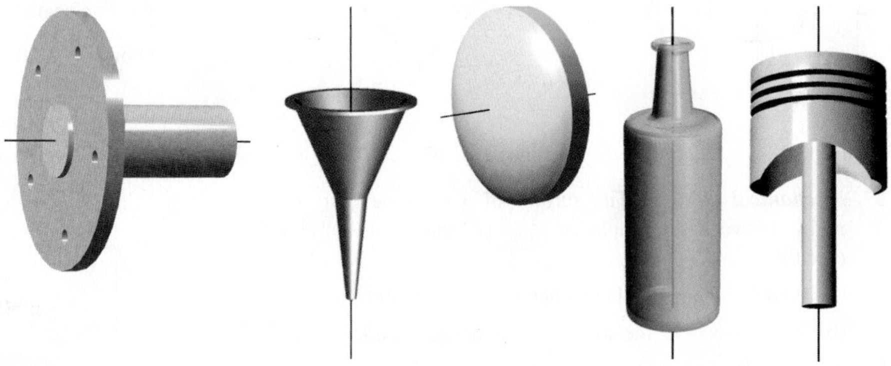

Solids of revolution
Figure 7.12

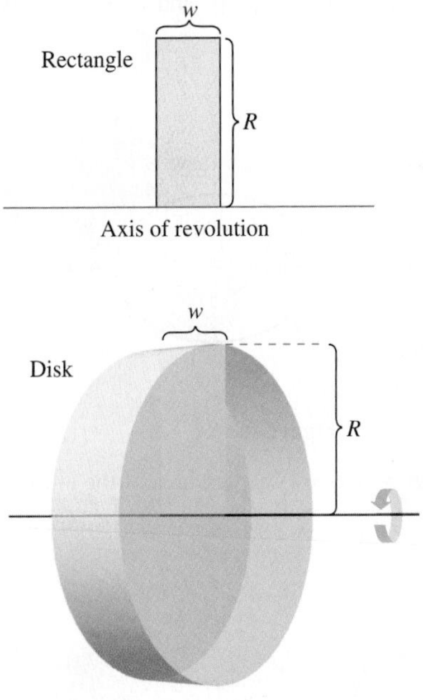

Volume of a disk: $\pi R^2 w$
Figure 7.13

If a region in the plane is revolved about a line, the resulting solid is a **solid of revolution,** and the line is called the **axis of revolution.** The simplest such solid is a right circular cylinder or **disk,** which is formed by revolving a rectangle about an axis adjacent to one side of the rectangle, as shown in Figure 7.13. The volume of such a disk is

$$\begin{aligned}\text{Volume of disk} &= (\text{area of disk})(\text{width of disk})\\ &= \pi R^2 w\end{aligned}$$

where R is the radius of the disk and w is the width.

To see how to use the volume of a disk to find the volume of a general solid of revolution, consider a solid of revolution formed by revolving the plane region in Figure 7.14 about the indicated axis. To determine the volume of this solid, consider a representative rectangle in the plane region. When this rectangle is revolved about the axis of revolution, it generates a representative disk whose volume is

$$\Delta V = \pi R^2 \Delta x.$$

Approximating the volume of the solid by n such disks of width Δx and radius $R(x_i)$ produces

$$\begin{aligned}\text{Volume of solid} &\approx \sum_{i=1}^{n} \pi[R(x_i)]^2 \,\Delta x\\ &= \pi \sum_{i=1}^{n} [R(x_i)]^2 \,\Delta x.\end{aligned}$$

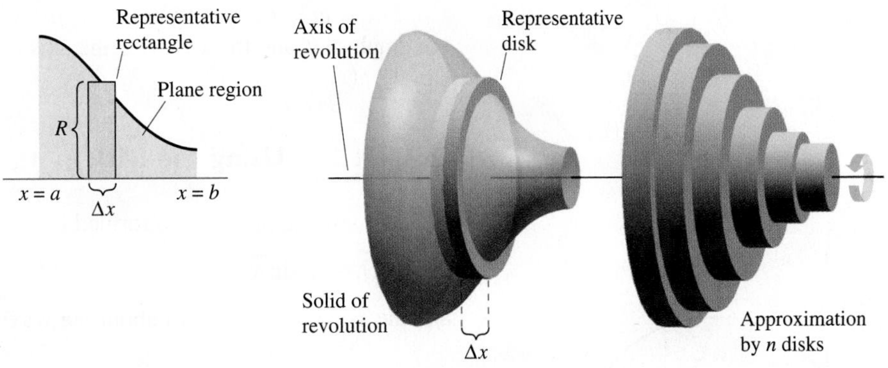

Disk method
Figure 7.14

This approximation appears to become better and better as $\|\Delta\| \to 0$ $(n \to \infty)$. So, you can define the volume of the solid as

$$\text{Volume of solid} = \lim_{\|\Delta\| \to 0} \pi \sum_{i=1}^{n} [R(x_i)]^2 \, \Delta x = \pi \int_a^b [R(x)]^2 \, dx.$$

Schematically, the disk method looks like this.

Known Precalculus Formula		*Representative Element*		*New Integration Formula*
Volume of disk $V = \pi R^2 w$	⇨	$\Delta V = \pi [R(x_i)]^2 \, \Delta x$	⇨	Solid of revolution $V = \pi \int_a^b [R(x)]^2 \, dx$

A similar formula can be derived if the axis of revolution is vertical.

The Disk Method

To find the volume of a solid of revolution with the **disk method,** use one of the following, as shown in Figure 7.15.

Horizontal Axis of Revolution	*Vertical Axis of Revolution*
$\text{Volume} = V = \pi \int_a^b [R(x)]^2 \, dx$	$\text{Volume} = V = \pi \int_c^d [R(y)]^2 \, dy$

NOTE In Figure 7.15, note that you can determine the variable of integration by placing a representative rectangle in the plane region "perpendicular" to the axis of revolution. If the width of the rectangle is Δx, integrate with respect to x, and if the width of the rectangle is Δy, integrate with respect to y.

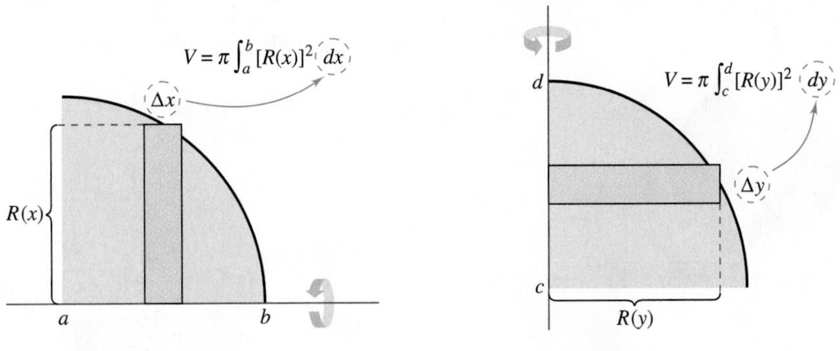

Figure 7.15

The simplest application of the disk method involves a plane region bounded by the graph of f and the x-axis. If the axis of revolution is the x-axis, the radius $R(x)$ is simply $f(x)$.

EXAMPLE 1 Using the Disk Method

Find the volume of the solid formed by revolving the region bounded by the graph of

$$f(x) = \sqrt{\sin x}$$

and the x-axis $(0 \le x \le \pi)$ about the x-axis.

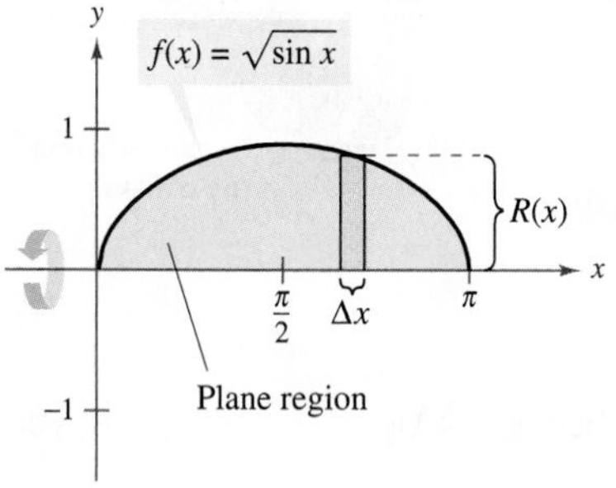

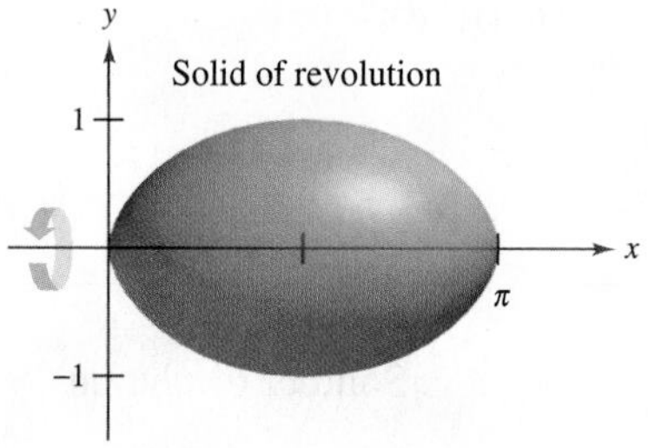

Figure 7.16

Solution From the representative rectangle in the upper graph in Figure 7.16, you can see that the radius of this solid is

$$\begin{aligned} R(x) &= f(x) \\ &= \sqrt{\sin x}. \end{aligned}$$

So, the volume of the solid of revolution is

$$\begin{aligned} V = \pi \int_a^b [R(x)]^2\,dx &= \pi \int_0^{\pi} \left(\sqrt{\sin x}\right)^2 dx && \text{Apply disk method.} \\ &= \pi \int_0^{\pi} \sin x\,dx && \text{Simplify.} \\ &= \pi \Big[-\cos x\Big]_0^{\pi} && \text{Integrate.} \\ &= \pi(1 + 1) \\ &= 2\pi. \end{aligned}$$

EXAMPLE 2 Revolving About a Line That Is Not a Coordinate Axis

Find the volume of the solid formed by revolving the region bounded by

$$f(x) = 2 - x^2$$

and $g(x) = 1$ about the line $y = 1$, as shown in Figure 7.17.

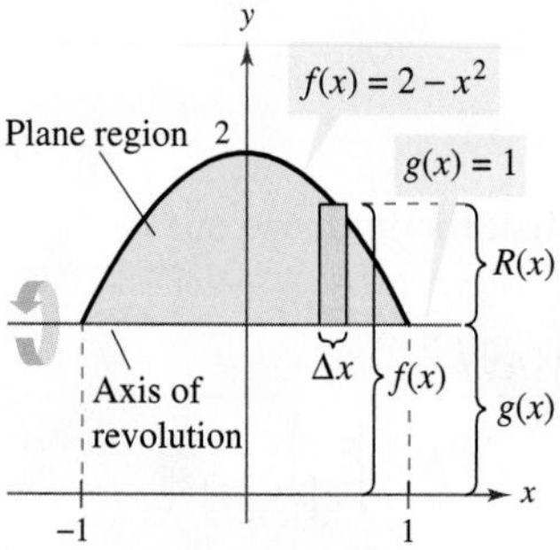

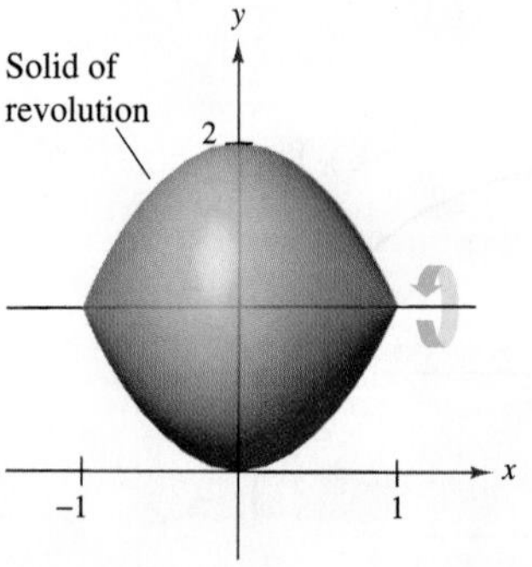

Figure 7.17

Solution By equating $f(x)$ and $g(x)$, you can determine that the two graphs intersect when $x = \pm 1$. To find the radius, subtract $g(x)$ from $f(x)$.

$$\begin{aligned} R(x) &= f(x) - g(x) \\ &= (2 - x^2) - 1 \\ &= 1 - x^2 \end{aligned}$$

Finally, integrate between -1 and 1 to find the volume.

$$\begin{aligned} V = \pi \int_a^b [R(x)]^2\,dx &= \pi \int_{-1}^{1} (1 - x^2)^2\,dx && \text{Apply disk method.} \\ &= \pi \int_{-1}^{1} (1 - 2x^2 + x^4)\,dx && \text{Simplify.} \\ &= \pi \left[x - \frac{2x^3}{3} + \frac{x^5}{5}\right]_{-1}^{1} && \text{Integrate.} \\ &= \frac{16\pi}{15} \end{aligned}$$

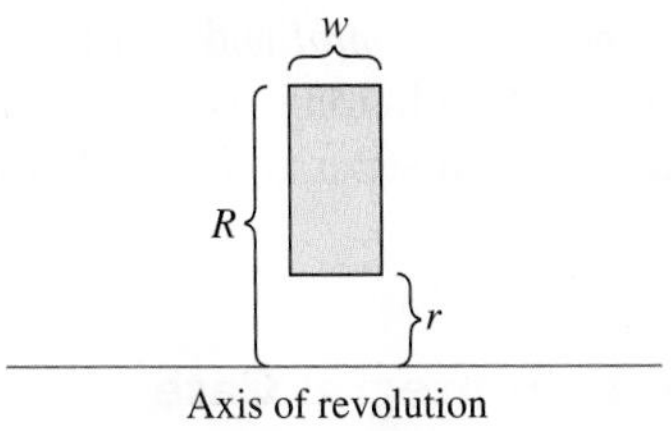

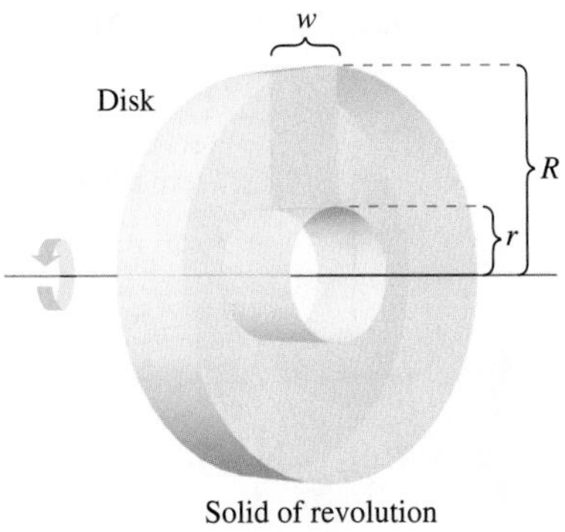

Figure 7.18

The Washer Method

The disk method can be extended to cover solids of revolution with holes by replacing the representative disk with a representative **washer.** The washer is formed by revolving a rectangle about an axis, as shown in Figure 7.18. If r and R are the inner and outer radii of the washer and w is the width of the washer, the volume is given by

$$\text{Volume of washer} = \pi(R^2 - r^2)w.$$

To see how this concept can be used to find the volume of a solid of revolution, consider a region bounded by an **outer radius** $R(x)$ and an **inner radius** $r(x)$, as shown in Figure 7.19. If the region is revolved about its axis of revolution, the volume of the resulting solid is given by

$$V = \pi\int_a^b ([R(x)]^2 - [r(x)]^2)\,dx. \qquad \text{Washer method}$$

Note that the integral involving the inner radius represents the volume of the hole and is *subtracted* from the integral involving the outer radius.

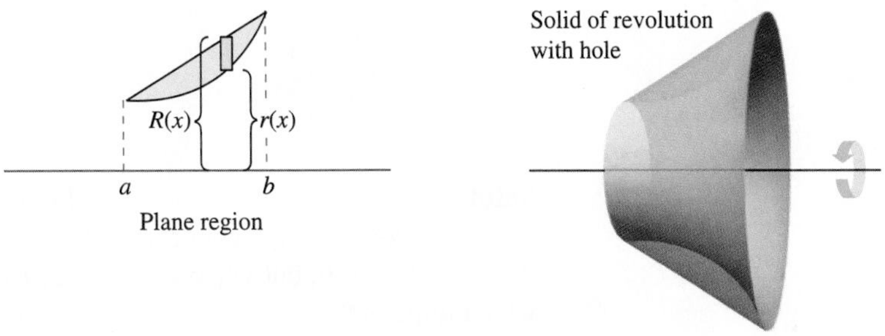

Figure 7.19

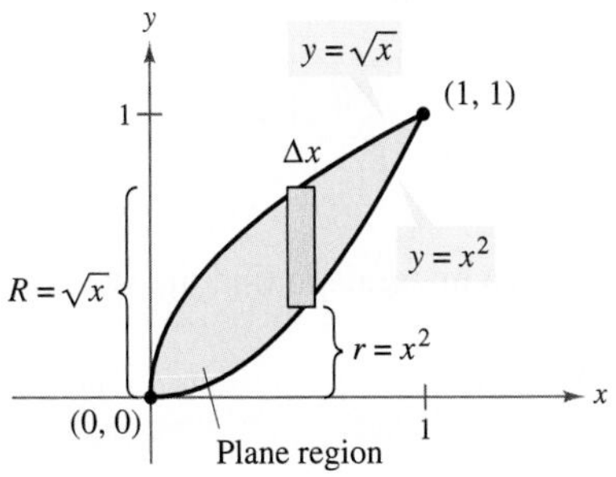

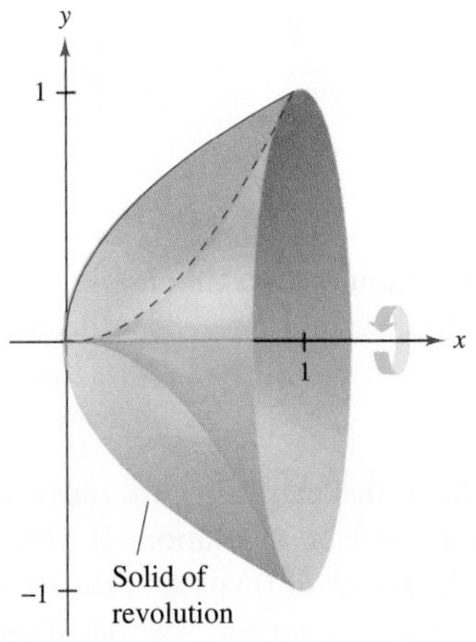

Solid of revolution
Figure 7.20

EXAMPLE 3 Using the Washer Method

Find the volume of the solid formed by revolving the region bounded by the graphs of $y = \sqrt{x}$ and $y = x^2$ about the x-axis, as shown in Figure 7.20.

Solution In Figure 7.20, you can see that the outer and inner radii are as follows.

$$R(x) = \sqrt{x} \qquad \text{Outer radius}$$
$$r(x) = x^2 \qquad \text{Inner radius}$$

Integrating between 0 and 1 produces

$$\begin{aligned} V &= \pi\int_a^b ([R(x)]^2 - [r(x)]^2)\,dx && \text{Apply washer method.} \\ &= \pi\int_0^1 \left[\left(\sqrt{x}\right)^2 - (x^2)^2\right] dx \\ &= \pi\int_0^1 (x - x^4)\,dx && \text{Simplify.} \\ &= \pi\left[\frac{x^2}{2} - \frac{x^5}{5}\right]_0^1 && \text{Integrate.} \\ &= \frac{3\pi}{10}. \end{aligned}$$

In each example so far, the axis of revolution has been *horizontal* and you have integrated with respect to x. In the next example, the axis of revolution is *vertical* and you integrate with respect to y. In this example, you need two separate integrals to compute the volume.

EXAMPLE 4 Integrating with Respect to y, Two-Integral Case

Find the volume of the solid formed by revolving the region bounded by the graphs of $y = x^2 + 1$, $y = 0$, $x = 0$, and $x = 1$ about the y-axis, as shown in Figure 7.21.

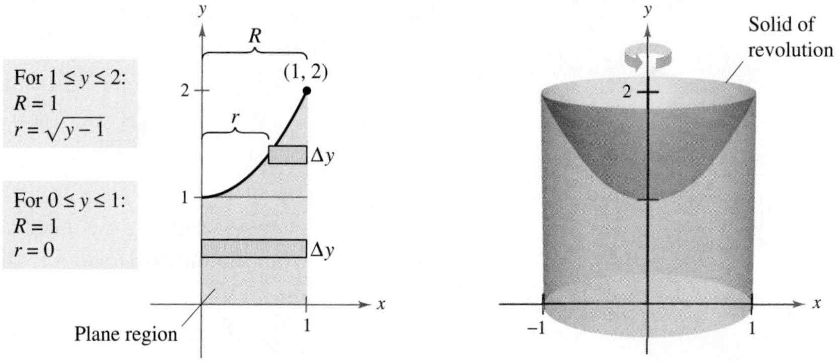

Figure 7.21

Solution For the region shown in Figure 7.21, the outer radius is simply $R = 1$. There is, however, no convenient formula that represents the inner radius. When $0 \le y \le 1$, $r = 0$, but when $1 \le y \le 2$, r is determined by the equation $y = x^2 + 1$, which implies that $r = \sqrt{y-1}$.

$$r(y) = \begin{cases} 0, & 0 \le y \le 1 \\ \sqrt{y-1}, & 1 \le y \le 2 \end{cases}$$

Using this definition of the inner radius, you can use two integrals to find the volume.

$$V = \pi \int_0^1 (1^2 - 0^2)\, dy + \pi \int_1^2 \left[1^2 - \left(\sqrt{y-1}\right)^2\right] dy \qquad \text{Apply washer method.}$$

$$= \pi \int_0^1 1\, dy + \pi \int_1^2 (2 - y)\, dy \qquad \text{Simplify.}$$

$$= \pi \Big[y \Big]_0^1 + \pi \left[2y - \frac{y^2}{2} \right]_1^2 \qquad \text{Integrate.}$$

$$= \pi + \pi\left(4 - 2 - 2 + \frac{1}{2}\right) = \frac{3\pi}{2}$$

Note that the first integral $\pi \int_0^1 1\, dy$ represents the volume of a right circular cylinder of radius 1 and height 1. This portion of the volume could have been determined without using calculus.

TECHNOLOGY Some graphing utilities have the capability to generate (or have built-in software capable of generating) a solid of revolution. If you have access to such a utility, use it to graph some of the solids of revolution described in this section. For instance, the solid in Example 4 might appear like that shown in Figure 7.22.

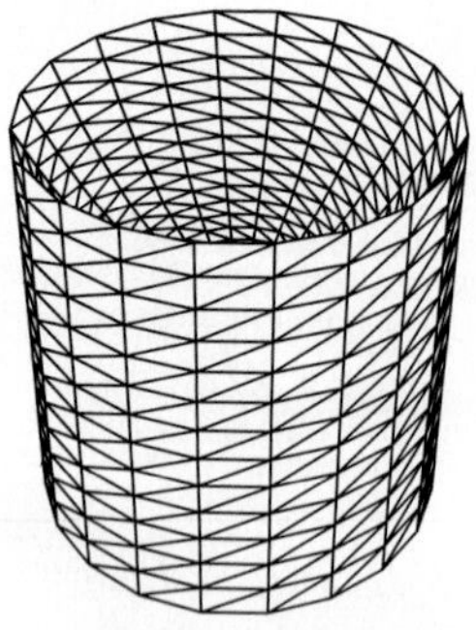

Generated by Mathematica

Figure 7.22

EXAMPLE 5 Manufacturing

A manufacturer drills a hole through the center of a metal sphere of radius 5 inches, as shown in Figure 7.23(a). The hole has a radius of 3 inches. What is the volume of the resulting metal ring?

Solution You can imagine the ring to be generated by a segment of the circle whose equation is $x^2 + y^2 = 25$, as shown in Figure 7.23(b). Because the radius of the hole is 3 inches, you can let $y = 3$ and solve the equation $x^2 + y^2 = 25$ to determine that the limits of integration are $x = \pm 4$. So, the inner and outer radii are $r(x) = 3$ and $R(x) = \sqrt{25 - x^2}$ and the volume is given by

$$\begin{aligned} V = \pi \int_a^b \left([R(x)]^2 - [r(x)]^2\right) dx &= \pi \int_{-4}^{4} \left[\left(\sqrt{25 - x^2}\right)^2 - (3)^2\right] dx \\ &= \pi \int_{-4}^{4} (16 - x^2)\, dx \\ &= \pi \left[16x - \frac{x^3}{3}\right]_{-4}^{4} \\ &= \frac{256\pi}{3} \text{ cubic inches.} \end{aligned}$$

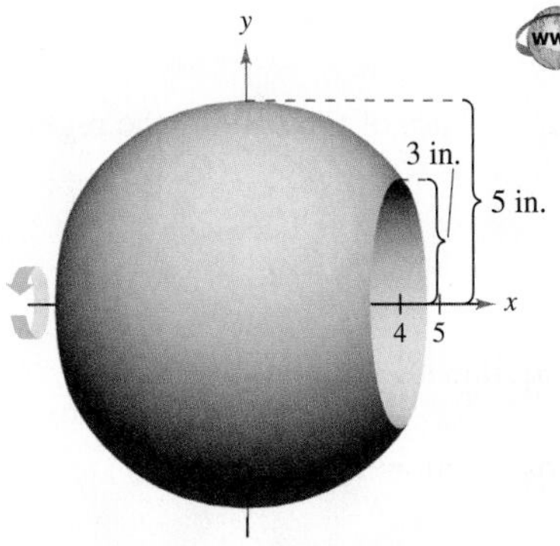

Solid of revolution

(a)

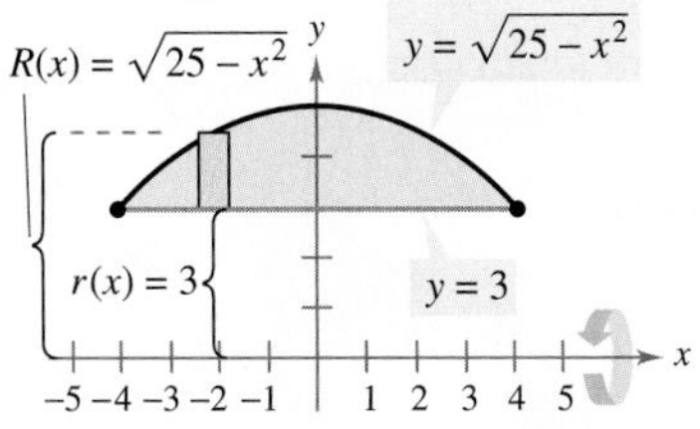

Plane region

(b)

Figure 7.23

Solids with Known Cross Sections

With the disk method, you can find the volume of a solid having a circular cross section whose area is $A = \pi R^2$. This method can be generalized to solids of any shape, as long as you know a formula for the area of an arbitrary cross section. Some common cross sections are squares, rectangles, triangles, semicircles, and trapezoids.

Volumes of Solids with Known Cross Sections

1. For cross sections of area $A(x)$ taken perpendicular to the x-axis,

$$\text{Volume} = \int_a^b A(x)\, dx.$$ See Figure 7.24(a).

2. For cross sections of area $A(y)$ taken perpendicular to the y-axis,

$$\text{Volume} = \int_c^d A(y)\, dy.$$ See Figure 7.24(b).

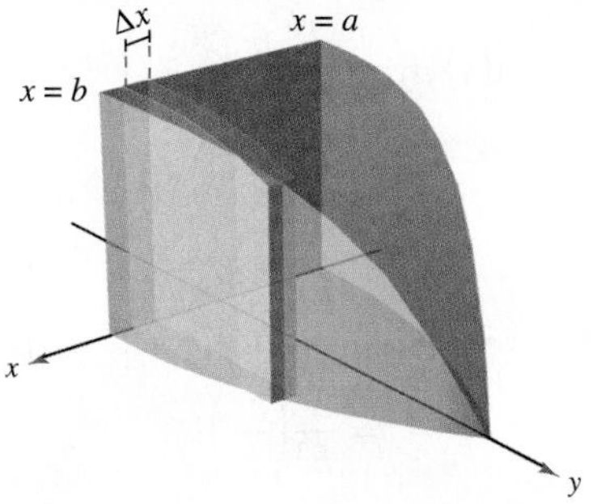

(a) Cross sections perpendicular to x-axis

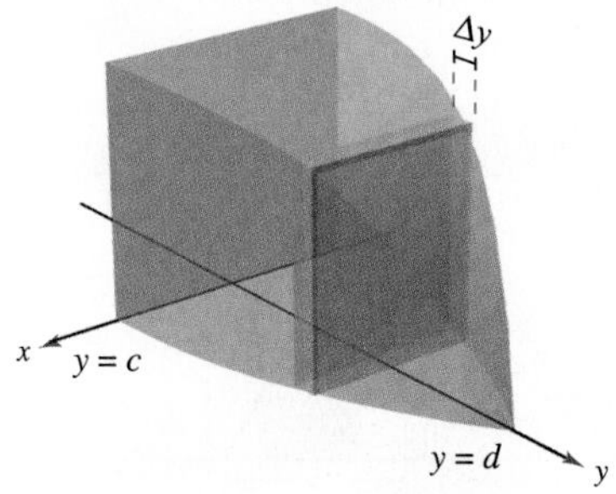

(b) Cross sections perpendicular to y-axis

Figure 7.24

EXAMPLE 6 Triangular Cross Sections

Find the volume of the solid shown in Figure 7.25. The base of the solid is the region bounded by the lines

$$f(x) = 1 - \frac{x}{2}, \qquad g(x) = -1 + \frac{x}{2}, \qquad \text{and} \qquad x = 0.$$

The cross sections perpendicular to the x-axis are equilateral triangles.

Solution The base and area of each triangular cross section are as follows.

$$\text{Base} = \left(1 - \frac{x}{2}\right) - \left(-1 + \frac{x}{2}\right) = 2 - x \qquad \text{Length of base}$$

$$\text{Area} = \frac{\sqrt{3}}{4}(\text{base})^2 \qquad \text{Area of equilateral triangle}$$

$$A(x) = \frac{\sqrt{3}}{4}(2 - x)^2 \qquad \text{Area of cross section}$$

Because x ranges from 0 to 2, the volume of the solid is

$$V = \int_a^b A(x)\,dx = \int_0^2 \frac{\sqrt{3}}{4}(2 - x)^2\,dx$$

$$= -\frac{\sqrt{3}}{4}\left[\frac{(2 - x)^3}{3}\right]_0^2 = \frac{2\sqrt{3}}{3}.$$

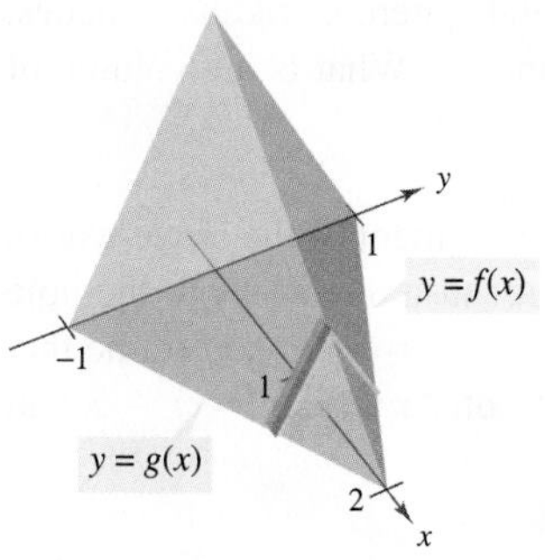

Cross sections are equilateral triangles.

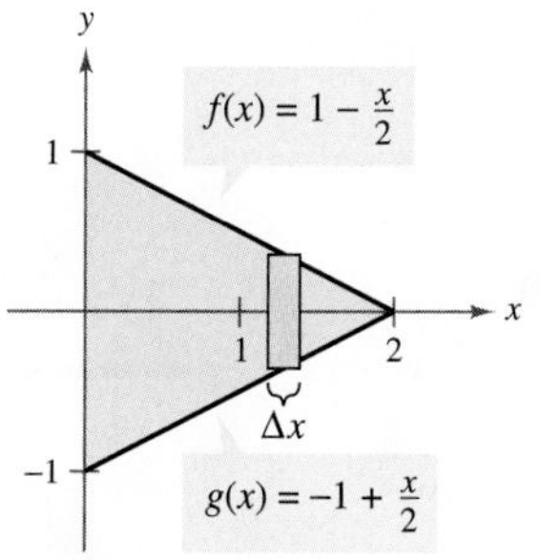

Triangular base in xy-plane
Figure 7.25

EXAMPLE 7 An Application to Geometry

Prove that the volume of a pyramid with a square base is $V = \frac{1}{3}hB$, where h is the height of the pyramid and B is the area of the base.

Solution As shown in Figure 7.26, you can intersect the pyramid with a plane parallel to the base at height y to form a square cross section whose sides are of length b'. Using similar triangles, you can show that

$$\frac{b'}{b} = \frac{h - y}{h} \qquad \text{or} \qquad b' = \frac{b}{h}(h - y)$$

where b is the length of the sides of the base of the pyramid. So,

$$A(y) = (b')^2 = \frac{b^2}{h^2}(h - y)^2.$$

Integrating between 0 and h produces

$$V = \int_0^h A(y)\,dy = \int_0^h \frac{b^2}{h^2}(h - y)^2\,dy$$

$$= \frac{b^2}{h^2}\int_0^h (h - y)^2\,dy$$

$$= -\left(\frac{b^2}{h^2}\right)\left[\frac{(h - y)^3}{3}\right]_0^h$$

$$= \frac{b^2}{h^2}\left(\frac{h^3}{3}\right)$$

$$= \frac{1}{3}hB. \qquad B = b^2$$

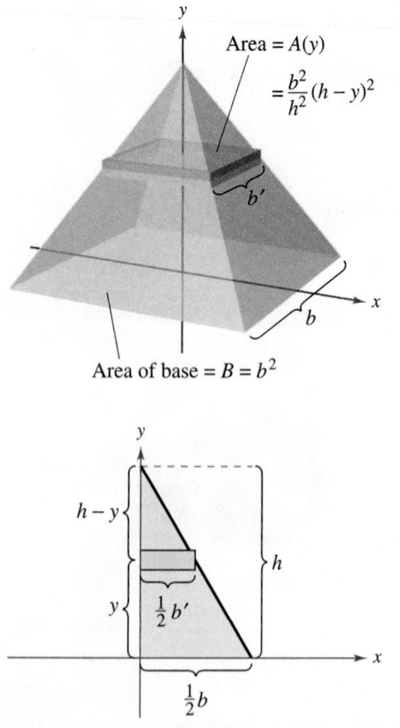

Figure 7.26

Exercises for Section 7.2

See www.CalcChat.com for worked-out solutions to odd-numbered exercises.

In Exercises 1–6, set up and evaluate the integral that gives the volume of the solid formed by revolving the region about the *x*-axis.

1. $y = -x + 1$

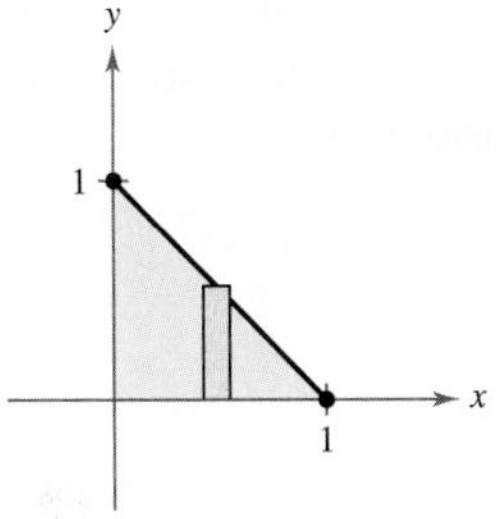

2. $y = 4 - x^2$

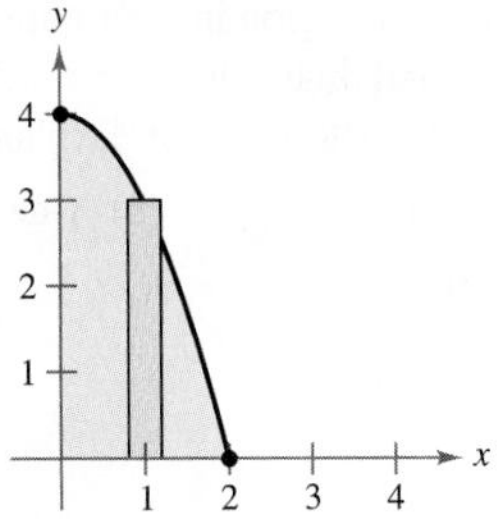

3. $y = \sqrt{x}$

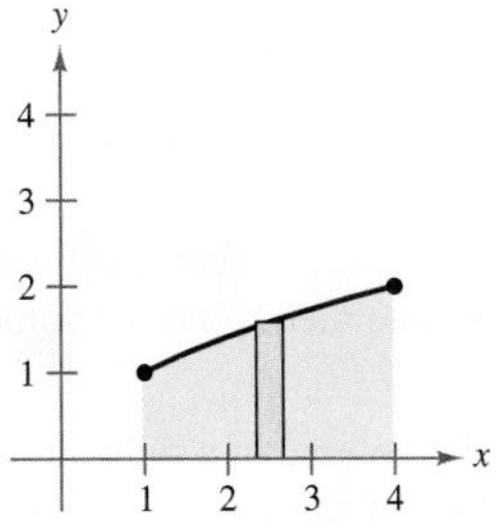

4. $y = \sqrt{9 - x^2}$

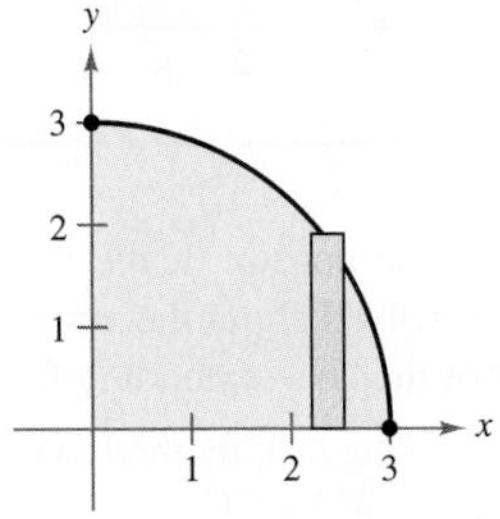

5. $y = x^2,\ y = x^3$

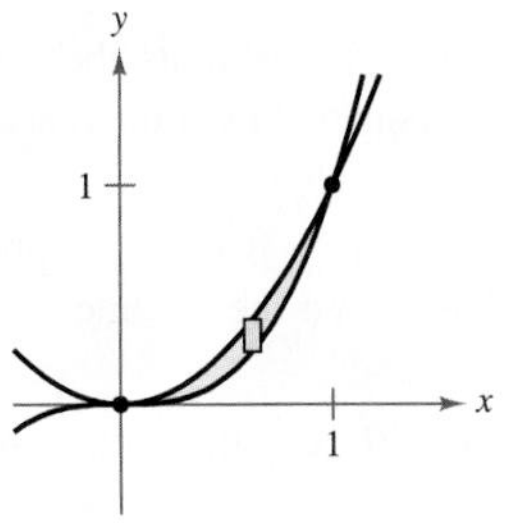

6. $y = 2,\ \ y = 4 - \dfrac{x^2}{4}$

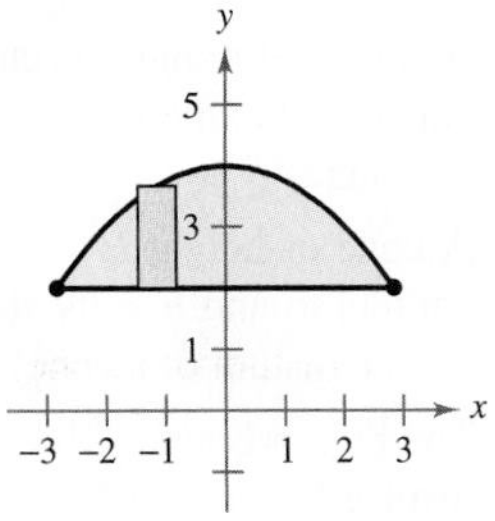

In Exercises 7–10, set up and evaluate the integral that gives the volume of the solid formed by revolving the region about the *y*-axis.

7. $y = x^2$

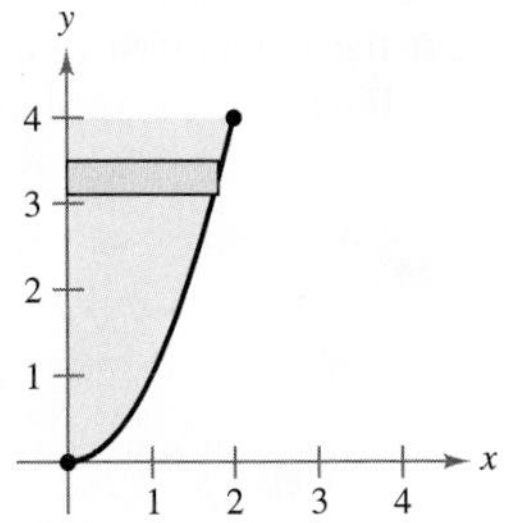

8. $y = \sqrt{16 - x^2}$

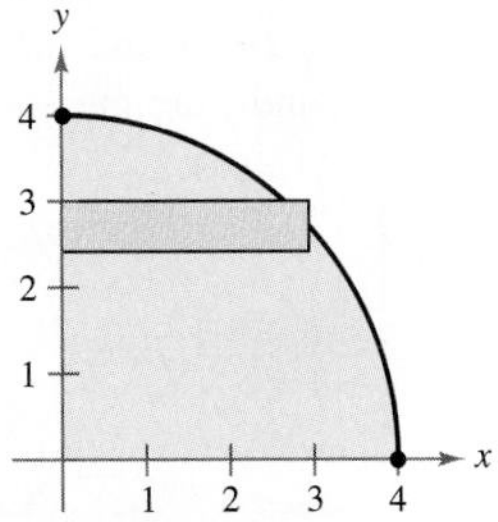

9. $y = x^{2/3}$

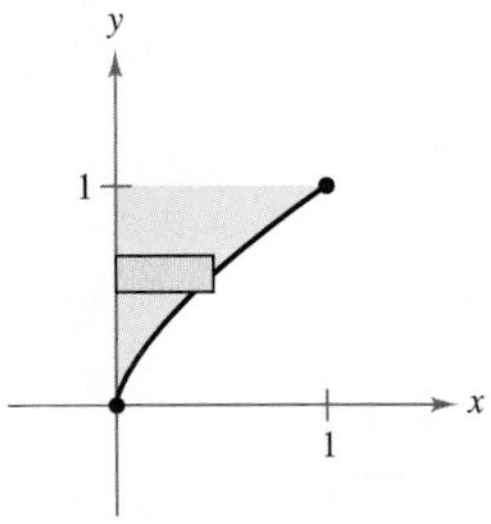

10. $x = -y^2 + 4y$

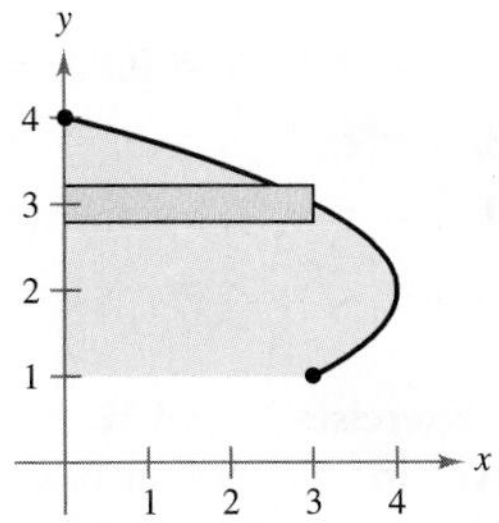

In Exercises 11–14, find the volume of the solid generated by revolving the region bounded by the graphs of the equations about the given lines.

11. $y = \sqrt{x},\ \ y = 0,\ \ x = 4$
    (a) the x-axis (b) the y-axis
    (c) the line $x = 4$ (d) the line $x = 6$

12. $y = 2x^2,\ \ y = 0,\ \ x = 2$
    (a) the y-axis (b) the x-axis
    (c) the line $y = 8$ (d) the line $x = 2$

13. $y = x^2,\ \ y = 4x - x^2$
    (a) the x-axis (b) the line $y = 6$

14. $y = 6 - 2x - x^2,\ \ y = x + 6$
    (a) the x-axis (b) the line $y = 3$

In Exercises 15–18, find the volume of the solid generated by revolving the region bounded by the graphs of the equations about the line $y = 4$.

15. $y = x,\ \ y = 3,\ \ x = 0$

16. $y = \frac{1}{2}x^3,\ \ y = 4,\ \ x = 0$

17. $y = \dfrac{1}{1 + x},\ \ y = 0,\ \ x = 0,\ \ x = 3$

18. $y = \sec x,\ \ y = 0,\ \ 0 \le x \le \dfrac{\pi}{3}$

In Exercises 19–22, find the volume of the solid generated by revolving the region bounded by the graphs of the equations about the line $x = 6$.

19. $y = x,\ \ y = 0,\ \ y = 4,\ \ x = 6$

20. $y = 6 - x,\ \ y = 0,\ \ y = 4,\ \ x = 0$

21. $x = y^2,\ \ x = 4$

22. $xy = 6,\ \ y = 2,\ \ y = 6,\ \ x = 6$

In Exercises 23–30, find the volume of the solid generated by revolving the region bounded by the graphs of the equations about the *x*-axis.

23. $y = \dfrac{1}{\sqrt{x + 1}},\ \ y = 0,\ \ x = 0,\ \ x = 3$

24. $y = x\sqrt{4 - x^2},\ \ y = 0$

25. $y = \frac{1}{x}, \quad y = 0, \quad x = 1, \quad x = 4$

26. $y = \frac{3}{x + 1}, \quad y = 0, \quad x = 0, \quad x = 8$

27. $y = e^{-x}, \quad y = 0, \quad x = 0, \quad x = 1$

28. $y = e^{x/2}, \quad y = 0, \quad x = 0, \quad x = 4$

29. $y = x^2 + 1, \quad y = -x^2 + 2x + 5, \quad x = 0, \quad x = 3$

30. $y = \sqrt{x}, \quad y = -\frac{1}{2}x + 4, \quad x = 0, \quad x = 8$

In Exercises 31 and 32, find the volume of the solid generated by revolving the region bounded by the graphs of the equations about the *y*-axis.

31. $y = 3(2 - x), \quad y = 0, \quad x = 0$

32. $y = 9 - x^2, \quad y = 0, \quad x = 2, \quad x = 3$

In Exercises 33–36, find the volume of the solid generated by revolving the region bounded by the graphs of the equations about the *x*-axis. Verify your results using the integration capabilities of a graphing utility.

33. $y = \sin x, \quad y = 0, \quad x = 0, \quad x = \pi$

34. $y = \cos x, \quad y = 0, \quad x = 0, \quad x = \frac{\pi}{2}$

35. $y = e^{x-1}, \quad y = 0, \quad x = 1, \quad x = 2$

36. $y = e^{x/2} + e^{-x/2}, \quad y = 0, \quad x = -1, \quad x = 2$

In Exercises 37–40, use the integration capabilities of a graphing utility to approximate the volume of the solid generated by revolving the region bounded by the graphs of the equations about the *x*-axis.

37. $y = e^{-x^2}, \quad y = 0, \quad x = 0, \quad x = 2$

38. $y = \ln x, \quad y = 0, \quad x = 1, \quad x = 3$

39. $y = 2\arctan(0.2x), \quad y = 0, \quad x = 0, \quad x = 5$

40. $y = \sqrt{2x}, \quad y = x^2$

Writing About Concepts

In Exercises 41 and 42, the integral represents the volume of a solid. Describe the solid.

41. $\pi \int_0^{\pi/2} \sin^2 x \, dx$

42. $\pi \int_2^4 y^4 \, dy$

Think About It **In Exercises 43 and 44, determine which value best approximates the volume of the solid generated by revolving the region bounded by the graphs of the equations about the *x*-axis. (Make your selection on the basis of a sketch of the solid and *not* by performing any calculations.)**

43. $y = e^{-x^2/2}, \quad y = 0, \quad x = 0, \quad x = 2$

(a) 3 (b) -5 (c) 10 (d) 7 (e) 20

44. $y = \arctan x, \quad y = 0, \quad x = 0, \quad x = 1$

(a) 10 (b) $\frac{3}{4}$ (c) 5 (d) -6 (e) 15

Writing About Concepts *(continued)*

45. A region bounded by the parabola $y = 4x - x^2$ and the x-axis is revolved about the x-axis. A second region bounded by the parabola $y = 4 - x^2$ and the x-axis is revolved about the x-axis. Without integrating, how do the volumes of the two solids compare? Explain.

46. The region in the figure is revolved about the indicated axes and line. Order the volumes of the resulting solids from least to greatest. Explain your reasoning.

(a) x-axis (b) y-axis (c) $x = 8$

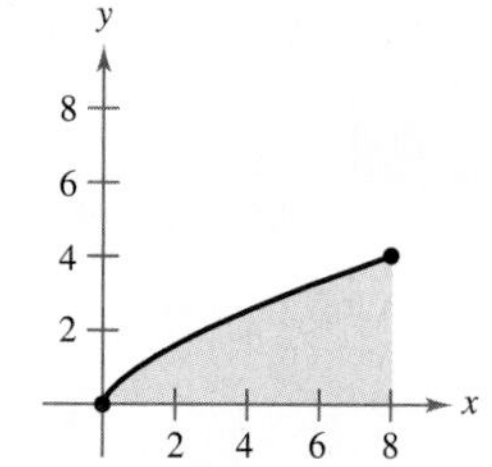

47. If the portion of the line $y = \frac{1}{2}x$ lying in the first quadrant is revolved about the x-axis, a cone is generated. Find the volume of the cone extending from $x = 0$ to $x = 6$.

48. Use the disk method to verify that the volume of a right circular cone is $\frac{1}{3}\pi r^2 h$, where r is the radius of the base and h is the height.

49. Use the disk method to verify that the volume of a sphere is $\frac{4}{3}\pi r^3$.

50. A sphere of radius r is cut by a plane h $(h < r)$ units above the equator. Find the volume of the solid (spherical segment) above the plane.

51. A cone of height H with a base of radius r is cut by a plane parallel to and h units above the base. Find the volume of the solid (frustum of a cone) below the plane.

52. The region bounded by $y = \sqrt{x}$, $y = 0$, $x = 0$, and $x = 4$ is revolved about the x-axis.

(a) Find the value of x in the interval $[0, 4]$ that divides the solid into two parts of equal volume.

(b) Find the values of x in the interval $[0, 4]$ that divide the solid into three parts of equal volume.

53. ***Volume of a Fuel Tank*** A tank on the wing of a jet aircraft is formed by revolving the region bounded by the graph of $y = \frac{1}{8}x^2\sqrt{2 - x}$ and the x-axis about the x-axis (see figure), where x and y are measured in meters. Find the tank's volume.

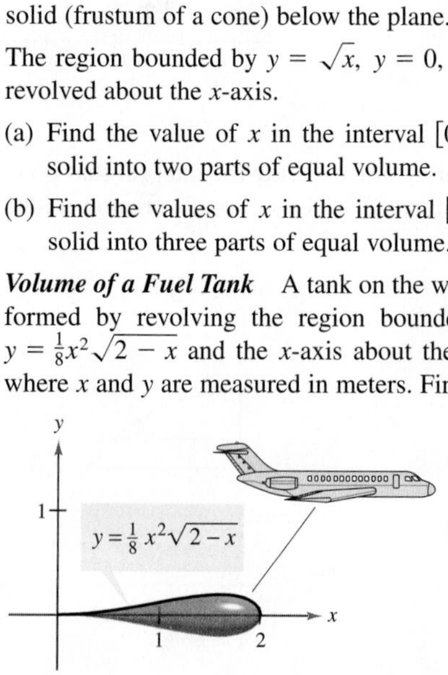

54. *Volume of a Lab Glass* A glass container can be modeled by revolving the graph of

$$y = \begin{cases} \sqrt{0.1x^3 - 2.2x^2 + 10.9x + 22.2}, & 0 \le x \le 11.5 \\ 2.95, & 11.5 < x \le 15 \end{cases}$$

about the x-axis, where x and y are measured in centimeters. Use a graphing utility to graph the function and find the volume of the container.

55. Find the volume of the solid generated if the upper half of the ellipse $9x^2 + 25y^2 = 225$ is revolved about (a) the x-axis to form a prolate spheroid (shaped like a football), and (b) the y-axis to form an oblate spheroid (shaped like half of a candy).

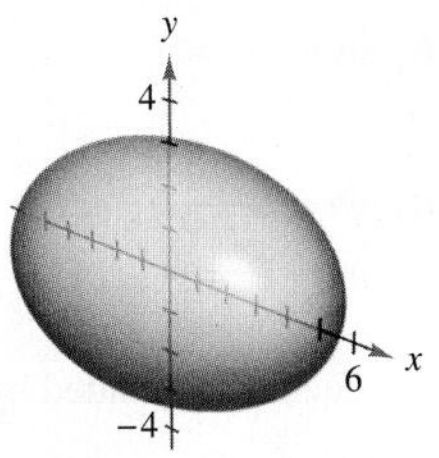

Figure for 55(a)

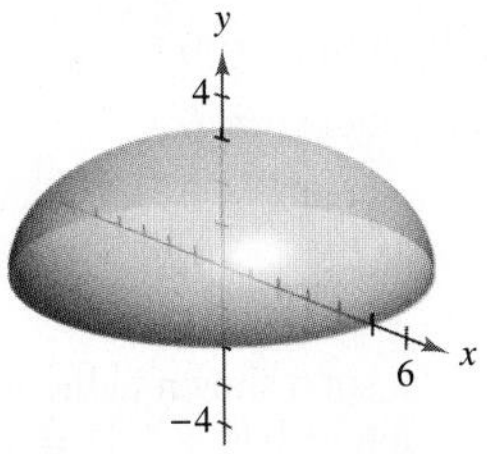

Figure for 55(b)

56. *Minimum Volume* The arc of

$$y = 4 - \frac{x^2}{4}$$

on the interval $[0, 4]$ is revolved about the line $y = b$ (see figure).

(a) Find the volume of the resulting solid as a function of b.

(b) Use a graphing utility to graph the function in part (a), and use the graph to approximate the value of b that minimizes the volume of the solid.

(c) Use calculus to find the value of b that minimizes the volume of the solid, and compare the result with the answer to part (b).

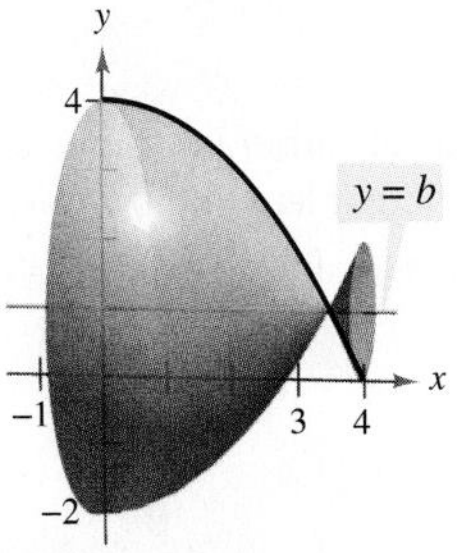

Figure for 56

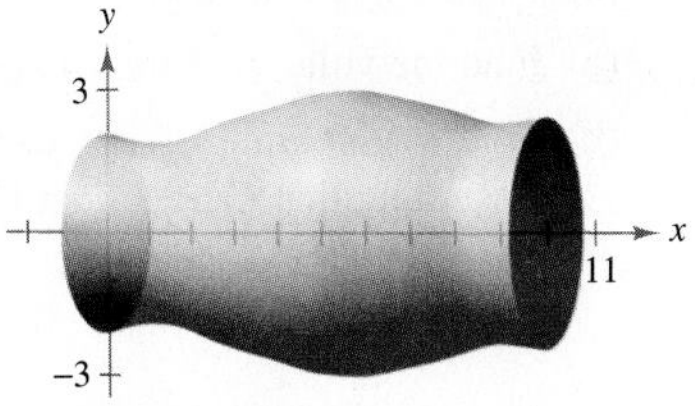

Figure for 58

57. *Water Depth in a Tank* A tank on a water tower is a sphere of radius 50 feet. Determine the depths of the water when the tank is filled to one-fourth and three-fourths of its total capacity. (*Note:* Use the *zero* or *root* feature of a graphing utility after evaluating the definite integral.)

58. *Modeling Data* A draftsman is asked to determine the amount of material required to produce a machine part (see figure in first column). The diameters d of the part at equally spaced points x are listed in the table. The measurements are listed in centimeters.

x	0	1	2	3	4	5
d	4.2	3.8	4.2	4.7	5.2	5.7

x	6	7	8	9	10
d	5.8	5.4	4.9	4.4	4.6

(a) Use these data with Simpson's Rule to approximate the volume of the part.

(b) Use the regression capabilities of a graphing utility to find a fourth-degree polynomial through the points representing the radius of the solid. Plot the data and graph the model.

(c) Use a graphing utility to approximate the definite integral yielding the volume of the part. Compare the result with the answer to part (a).

59. *Think About It* Match each integral with the solid whose volume it represents, and give the dimensions of each solid.

(a) Right circular cylinder (b) Ellipsoid

(c) Sphere (d) Right circular cone (e) Torus

(i) $\pi \displaystyle\int_0^h \left(\frac{rx}{h}\right)^2 dx$ (ii) $\pi \displaystyle\int_0^h r^2\, dx$

(iii) $\pi \displaystyle\int_{-r}^{r} \left(\sqrt{r^2 - x^2}\right)^2 dx$ (iv) $\pi \displaystyle\int_{-b}^{b} \left(a\sqrt{1 - \frac{x^2}{b^2}}\right)^2 dx$

(v) $\pi \displaystyle\int_{-r}^{r} \left[\left(R + \sqrt{r^2 - x^2}\right)^2 - \left(R - \sqrt{r^2 - x^2}\right)^2\right] dx$

60. *Cavalieri's Theorem* Prove that if two solids have equal altitudes and all plane sections parallel to their bases and at equal distances from their bases have equal areas, then the solids have the same volume (see figure).

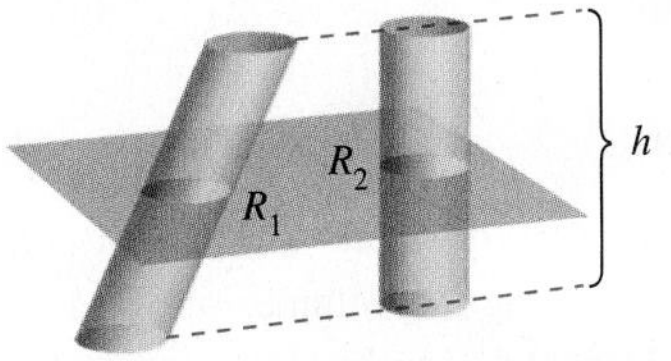

Area of R_1 = area of R_2

61. Find the volume of the solid whose base is bounded by the graphs of $y = x + 1$ and $y = x^2 - 1$, with the indicated cross sections taken perpendicular to the x-axis.

(a) Squares (b) Rectangles of height 1

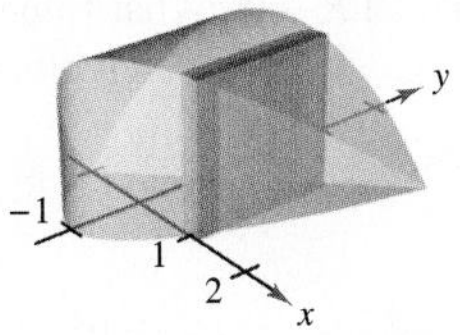

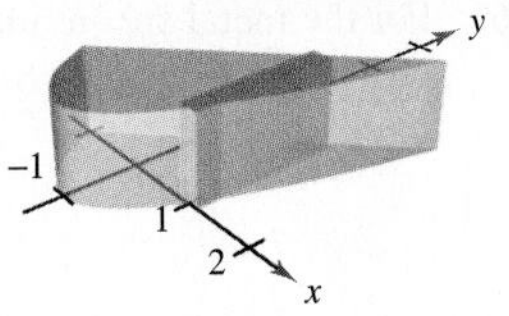

62. Find the volume of the solid whose base is bounded by the circle

$$x^2 + y^2 = 4$$

with the indicated cross sections taken perpendicular to the x-axis.

(a) Squares

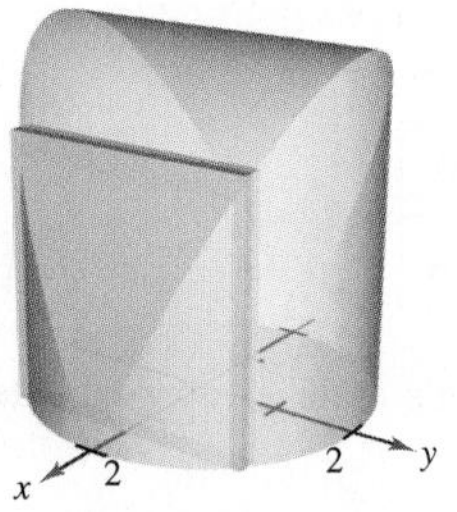

(b) Equilateral triangles

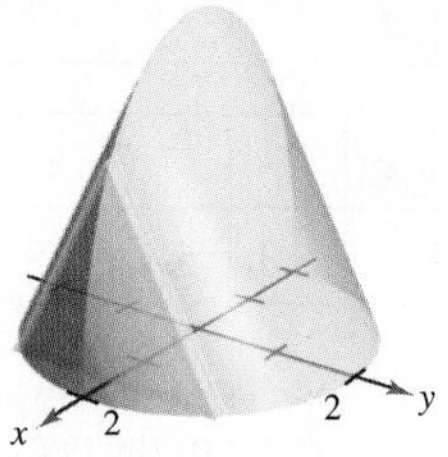

(c) Semicircles

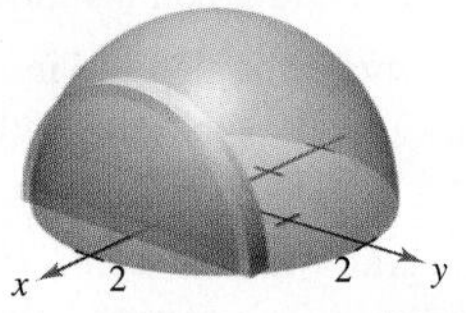

(d) Isosceles right triangles

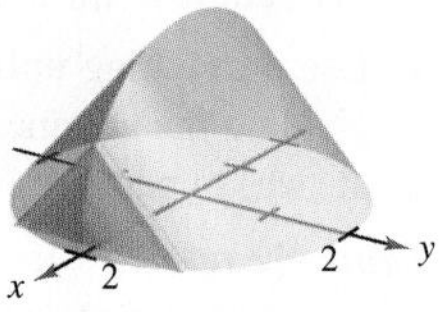

63. The base of a solid is bounded by $y = x^3$, $y = 0$, and $x = 1$. Find the volume of the solid for each of the following cross sections (taken perpendicular to the y-axis): (a) squares, (b) semicircles, (c) equilateral triangles, and (d) semiellipses whose heights are twice the lengths of their bases.

64. Find the volume of the solid of intersection (the solid common to both) of the two right circular cylinders of radius r whose axes meet at right angles (see figure).

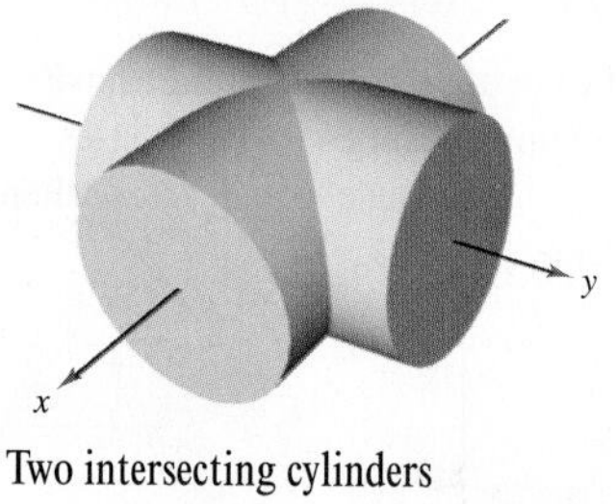

Two intersecting cylinders

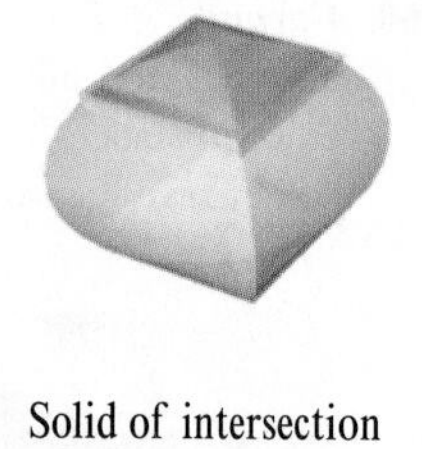

Solid of intersection

FOR FURTHER INFORMATION For more information on this problem, see the article "Estimating the Volumes of Solid Figures with Curved Surfaces" by Donald Cohen in *Mathematics Teacher*. To view this article, go to the website *www.matharticles.com*.

65. A manufacturer drills a hole through the center of a metal sphere of radius R. The hole has a radius r. Find the volume of the resulting ring.

66. For the metal sphere in Exercise 65, let $R = 5$. What value of r will produce a ring whose volume is exactly half the volume of the sphere?

In Exercises 67–74, find the volume generated by rotating the given region about the specified line.

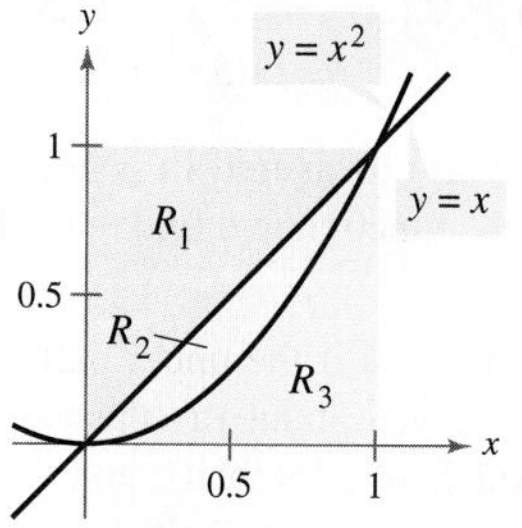

67. R_1 about $x = 0$

68. R_1 about $x = 1$

69. R_2 about $y = 0$

70. R_2 about $y = 1$

71. R_3 about $x = 0$

72. R_3 about $x = 1$

73. R_2 about $x = 0$

74. R_2 about $x = 1$

75. The solid shown in the figure has cross sections bounded by the graph of $|x|^a + |y|^a = 1$, where $1 \le a \le 2$.

(a) Describe the cross section when $a = 1$ and $a = 2$.

(b) Describe a procedure for approximating the volume of the solid.

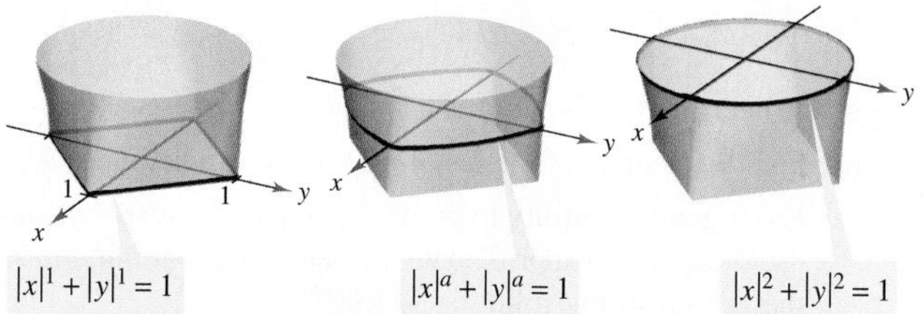

76. Two planes cut a right circular cylinder to form a wedge. One plane is perpendicular to the axis of the cylinder and the second makes an angle of θ degrees with the first (see figure).

(a) Find the volume of the wedge if $\theta = 45°$.

(b) Find the volume of the wedge for an arbitrary angle θ. Assuming that the cylinder has sufficient length, how does the volume of the wedge change as θ increases from 0° to 90°?

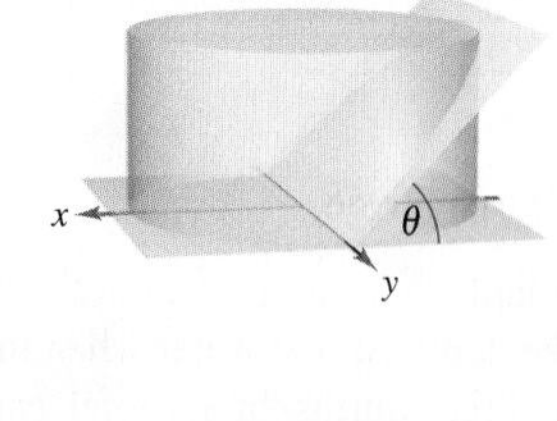

Figure for 76

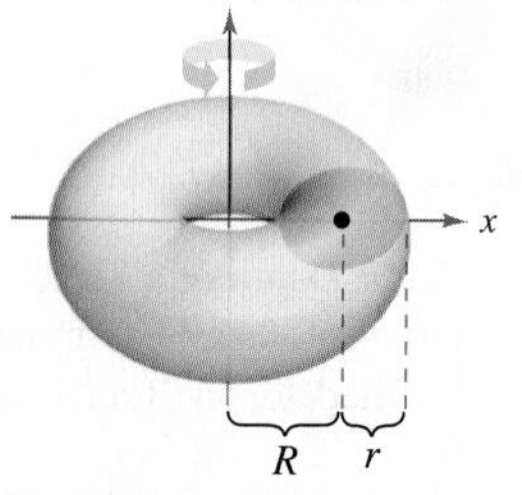

Figure for 77

77. (a) Show that the volume of the torus shown is given by the integral $8\pi R \int_0^r \sqrt{r^2 - y^2}\, dy$, where $R > r > 0$.

(b) Find the volume of the torus.

Section 7.3 Volume: The Shell Method

- **Find the volume of a solid of revolution using the shell method.**
- **Compare the uses of the disk method and the shell method.**

The Shell Method

In this section, you will study an alternative method for finding the volume of a solid of revolution. This method is called the **shell method** because it uses cylindrical shells. A comparison of the advantages of the disk and shell methods is given later in this section.

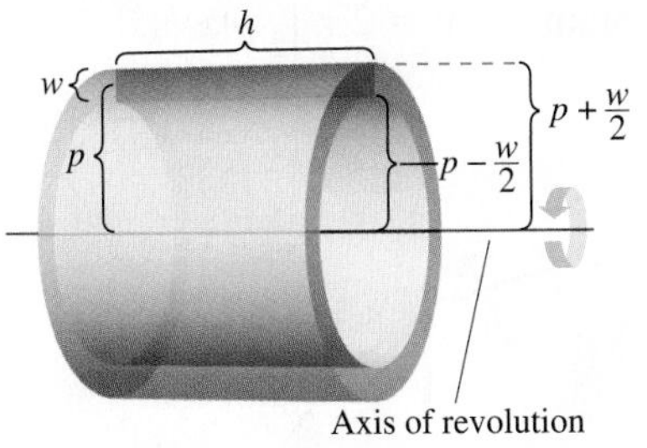

Figure 7.27

To begin, consider a representative rectangle as shown in Figure 7.27, where w is the width of the rectangle, h is the height of the rectangle, and p is the distance between the axis of revolution and the *center* of the rectangle. When this rectangle is revolved about its axis of revolution, it forms a cylindrical shell (or tube) of thickness w. To find the volume of this shell, consider two cylinders. The radius of the larger cylinder corresponds to the outer radius of the shell, and the radius of the smaller cylinder corresponds to the inner radius of the shell. Because p is the average radius of the shell, you know the outer radius is $p + (w/2)$ and the inner radius is $p - (w/2)$.

$$p + \frac{w}{2} \qquad \text{Outer radius}$$

$$p - \frac{w}{2} \qquad \text{Inner radius}$$

So, the volume of the shell is

$$\begin{aligned}\text{Volume of shell} &= (\text{volume of cylinder}) - (\text{volume of hole}) \\ &= \pi\left(p + \frac{w}{2}\right)^2 h - \pi\left(p - \frac{w}{2}\right)^2 h \\ &= 2\pi phw \\ &= 2\pi(\text{average radius})(\text{height})(\text{thickness}).\end{aligned}$$

You can use this formula to find the volume of a solid of revolution. Assume that the plane region in Figure 7.28 is revolved about a line to form the indicated solid. If you consider a horizontal rectangle of width Δy, then, as the plane region is revolved about a line parallel to the x-axis, the rectangle generates a representative shell whose volume is

$$\Delta V = 2\pi[p(y)h(y)]\,\Delta y.$$

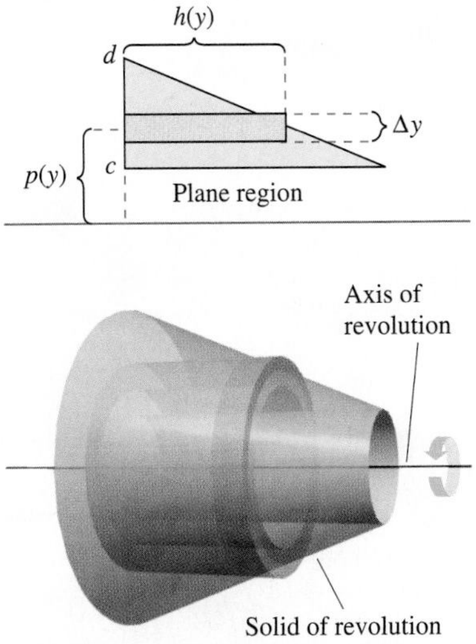

Figure 7.28

You can approximate the volume of the solid by n such shells of thickness Δy, height $h(y_i)$, and average radius $p(y_i)$.

$$\text{Volume of solid} \approx \sum_{i=1}^{n} 2\pi[p(y_i)h(y_i)]\,\Delta y = 2\pi\sum_{i=1}^{n}[p(y_i)h(y_i)]\,\Delta y$$

This approximation appears to become better and better as $\|\Delta\| \to 0$ $(n \to \infty)$. So, the volume of the solid is

$$\begin{aligned}\text{Volume of solid} &= \lim_{\|\Delta\|\to 0} 2\pi\sum_{i=1}^{n}[p(y_i)h(y_i)]\,\Delta y \\ &= 2\pi\int_c^d [p(y)h(y)]\,dy.\end{aligned}$$

The Shell Method

To find the volume of a solid of revolution with the **shell method,** use one of the following, as shown in Figure 7.29.

Horizontal Axis of Revolution	*Vertical Axis of Revolution*
$\text{Volume} = V = 2\pi \int_c^d p(y)h(y)\,dy$	$\text{Volume} = V = 2\pi \int_a^b p(x)h(x)\,dx$

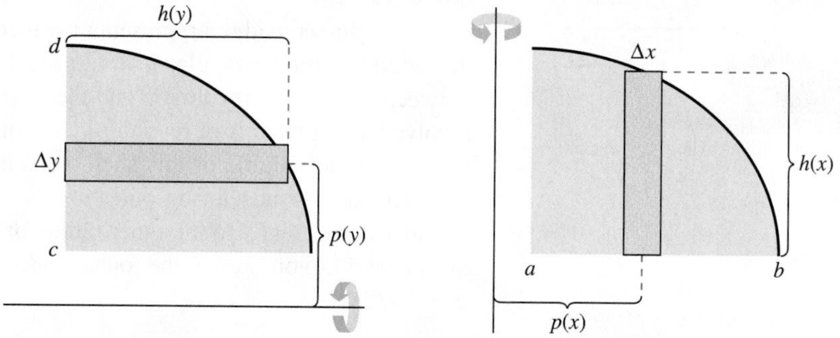

Horizontal axis of revolution
Vertical axis of revolution
Figure 7.29

EXAMPLE 1 Using the Shell Method to Find Volume

Find the volume of the solid of revolution formed by revolving the region bounded by

$$y = x - x^3$$

and the x-axis $(0 \le x \le 1)$ about the y-axis.

Solution Because the axis of revolution is vertical, use a vertical representative rectangle, as shown in Figure 7.30. The width Δx indicates that x is the variable of integration. The distance from the center of the rectangle to the axis of revolution is $p(x) = x$, and the height of the rectangle is

$$h(x) = x - x^3.$$

Because x ranges from 0 to 1, the volume of the solid is

$$\begin{aligned} V = 2\pi \int_a^b p(x)h(x)\,dx &= 2\pi \int_0^1 x(x - x^3)\,dx && \text{Apply shell method.} \\ &= 2\pi \int_0^1 (-x^4 + x^2)\,dx && \text{Simplify.} \\ &= 2\pi \left[-\frac{x^5}{5} + \frac{x^3}{3} \right]_0^1 && \text{Integrate.} \\ &= 2\pi \left(-\frac{1}{5} + \frac{1}{3} \right) \\ &= \frac{4\pi}{15}. \end{aligned}$$

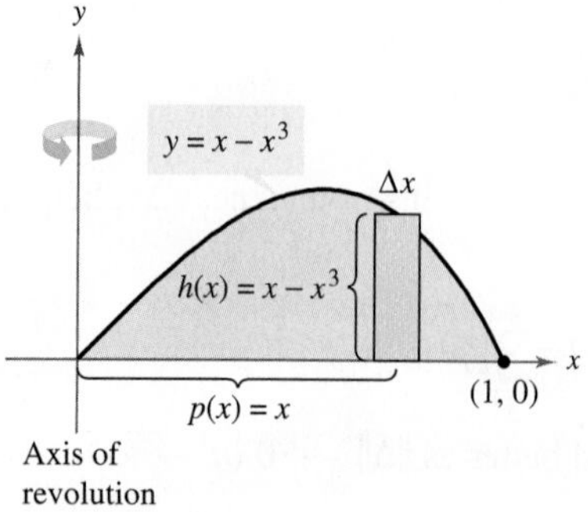

Figure 7.30

EXAMPLE 2 Using the Shell Method to Find Volume

Find the volume of the solid of revolution formed by revolving the region bounded by the graph of

$$x = e^{-y^2}$$

and the y-axis $(0 \le y \le 1)$ about the x-axis.

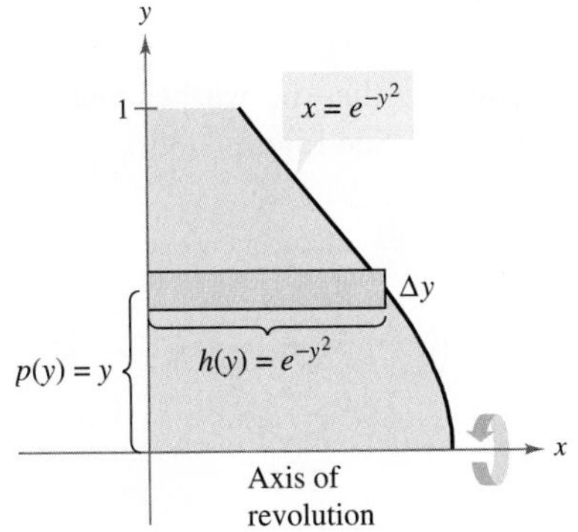

Figure 7.31

Solution Because the axis of revolution is horizontal, use a horizontal representative rectangle, as shown in Figure 7.31. The width Δy indicates that y is the variable of integration. The distance from the center of the rectangle to the axis of revolution is $p(y) = y$, and the height of the rectangle is $h(y) = e^{-y^2}$. Because y ranges from 0 to 1, the volume of the solid is

$$\begin{aligned} V &= 2\pi \int_c^d p(y)h(y)\,dy = 2\pi \int_0^1 ye^{-y^2}\,dy && \text{Apply shell method.} \\ &= -\pi\Big[e^{-y^2}\Big]_0^1 && \text{Integrate.} \\ &= \pi\left(1 - \frac{1}{e}\right) \\ &\approx 1.986. \end{aligned}$$

NOTE To see the advantage of using the shell method in Example 2, solve the equation $x = e^{-y^2}$ for y.

$$y = \begin{cases} 1, & 0 \le x \le 1/e \\ \sqrt{-\ln x}, & 1/e < x \le 1 \end{cases}$$

Then use this equation to find the volume using the disk method.

Comparison of Disk and Shell Methods

The disk and shell methods can be distinguished as follows. For the disk method, the representative rectangle is always *perpendicular* to the axis of revolution, whereas for the shell method, the representative rectangle is always *parallel* to the axis of revolution, as shown in Figure 7.32.

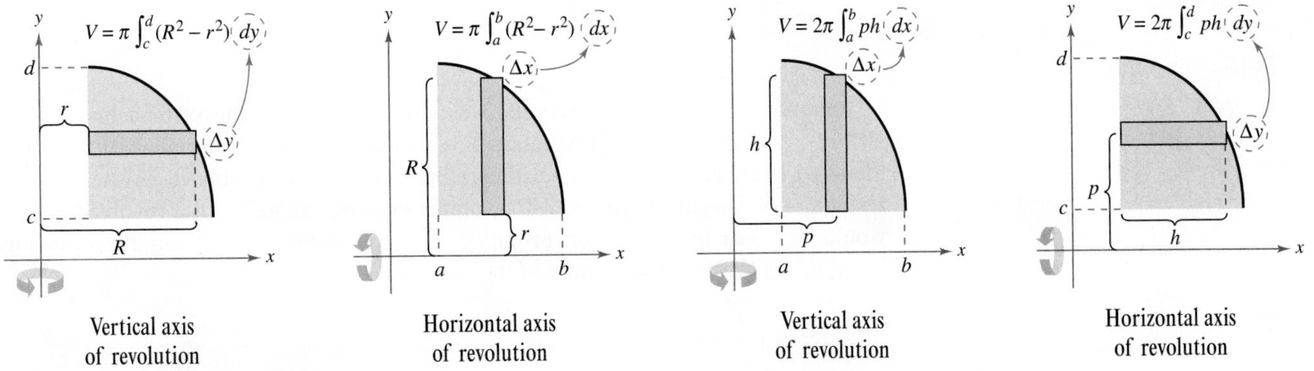

Disk method: Representative rectangle is perpendicular to the axis of revolution.

Shell method: Representative rectangle is parallel to the axis of revolution.

Figure 7.32

Often, one method is more convenient to use than the other. The following example illustrates a case in which the shell method is preferable.

EXAMPLE 3 Shell Method Preferable

Find the volume of the solid formed by revolving the region bounded by the graphs of

$$y = x^2 + 1, \quad y = 0, \quad x = 0, \quad \text{and} \quad x = 1$$

about the y-axis.

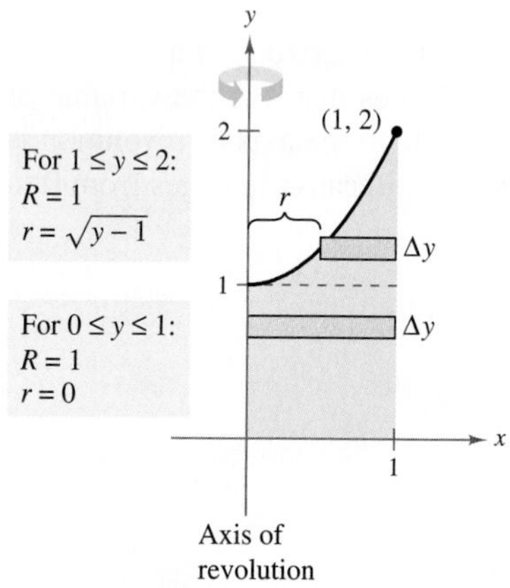

(a) Disk method

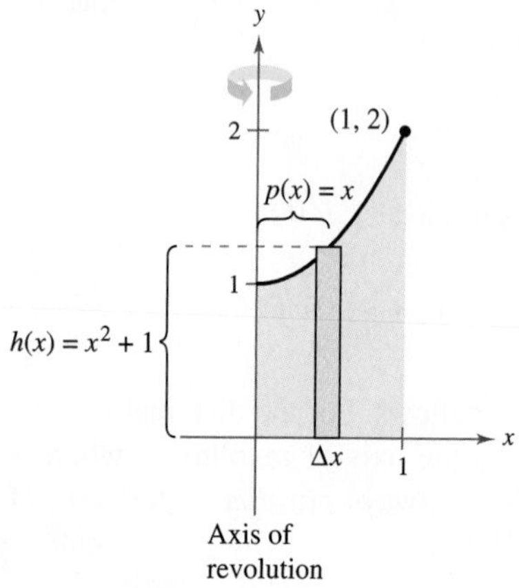

(b) Shell method

Figure 7.33

Solution In Example 4 in the preceding section, you saw that the washer method requires two integrals to determine the volume of this solid. See Figure 7.33(a).

$$\begin{aligned} V &= \pi \int_0^1 (1^2 - 0^2)\,dy + \pi \int_1^2 \left[1^2 - \left(\sqrt{y-1}\right)^2\right] dy && \text{Apply washer method.} \\ &= \pi \int_0^1 1\,dy + \pi \int_1^2 (2 - y)\,dy && \text{Simplify.} \\ &= \pi \Big[y\Big]_0^1 + \pi \left[2y - \frac{y^2}{2}\right]_1^2 && \text{Integrate.} \\ &= \pi + \pi\left(4 - 2 - 2 + \frac{1}{2}\right) \\ &= \frac{3\pi}{2} \end{aligned}$$

In Figure 7.33(b), you can see that the shell method requires only one integral to find the volume.

$$\begin{aligned} V &= 2\pi \int_a^b p(x)h(x)\,dx && \text{Apply shell method.} \\ &= 2\pi \int_0^1 x(x^2 + 1)\,dx \\ &= 2\pi \left[\frac{x^4}{4} + \frac{x^2}{2}\right]_0^1 && \text{Integrate.} \\ &= 2\pi\left(\frac{3}{4}\right) \\ &= \frac{3\pi}{2} \end{aligned}$$

Suppose the region in Example 3 were revolved about the vertical line $x = 1$. Would the resulting solid of revolution have a greater volume or a smaller volume than the solid in Example 3? Without integrating, you should be able to reason that the resulting solid would have a smaller volume because "more" of the revolved region would be closer to the axis of revolution. To confirm this, try solving the following integral, which gives the volume of the solid.

$$V = 2\pi \int_0^1 (1 - x)(x^2 + 1)\,dx \qquad p(x) = 1 - x$$

FOR FURTHER INFORMATION To learn more about the disk and shell methods, see the article "The Disk and Shell Method" by Charles A. Cable in *The American Mathematical Monthly*. To view this article, go to the website *www.matharticles.com*.

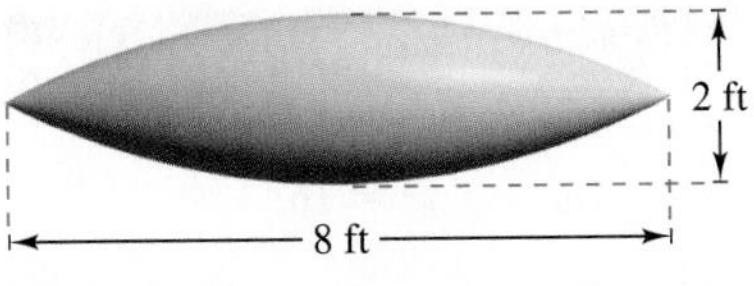

Figure 7.34

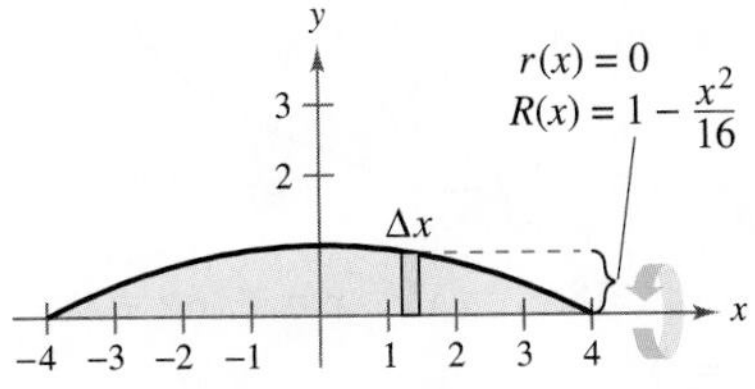

(a) Disk method

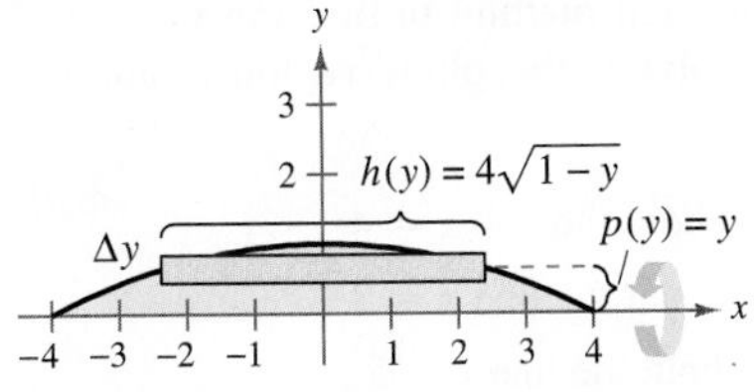

(b) Shell method
Figure 7.35

EXAMPLE 4 Volume of a Pontoon

A pontoon is to be made in the shape shown in Figure 7.34. The pontoon is designed by rotating the graph of

$$y = 1 - \frac{x^2}{16}, \qquad -4 \le x \le 4$$

about the x-axis, where x and y are measured in feet. Find the volume of the pontoon.

Solution Refer to Figure 7.35(a) and use the disk method as follows.

$$V = \pi \int_{-4}^{4} \left(1 - \frac{x^2}{16}\right)^2 dx \qquad \text{Apply disk method.}$$

$$= \pi \int_{-4}^{4} \left(1 - \frac{x^2}{8} + \frac{x^4}{256}\right) dx \qquad \text{Simplify.}$$

$$= \pi \left[x - \frac{x^3}{24} + \frac{x^5}{1280}\right]_{-4}^{4} \qquad \text{Integrate.}$$

$$= \frac{64\pi}{15} \approx 13.4 \text{ cubic feet}$$

Try using Figure 7.35(b) to set up the integral for the volume using the shell method. Does the integral seem more complicated?

For the shell method in Example 4, you would have to solve for x in terms of y in the equation

$$y = 1 - (x^2/16).$$

Sometimes, solving for x is very difficult (or even impossible). In such cases you must use a vertical rectangle (of width Δx), thus making x the variable of integration. The position (horizontal or vertical) of the axis of revolution then determines the method to be used. This is shown in Example 5.

EXAMPLE 5 Shell Method Necessary

Find the volume of the solid formed by revolving the region bounded by the graphs of $y = x^3 + x + 1$, $y = 1$, and $x = 1$ about the line $x = 2$, as shown in Figure 7.36.

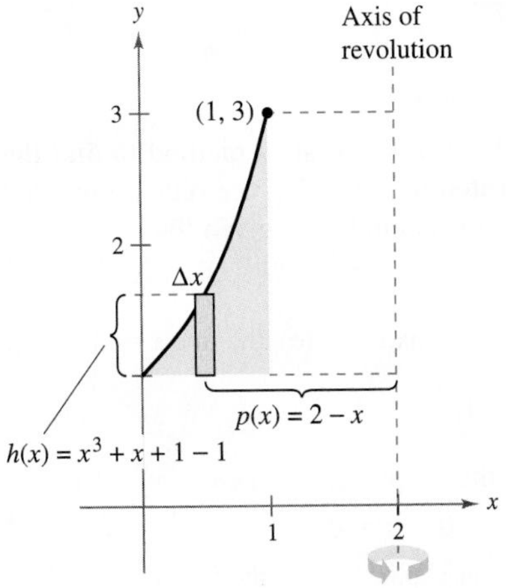

Figure 7.36

Solution In the equation $y = x^3 + x + 1$, you cannot easily solve for x in terms of y. (See Section 3.8 on Newton's Method.) Therefore, the variable of integration must be x, and you should choose a vertical representative rectangle. Because the rectangle is parallel to the axis of revolution, use the shell method and obtain

$$V = 2\pi \int_a^b p(x)h(x)\,dx = 2\pi \int_0^1 (2 - x)(x^3 + x + 1 - 1)\,dx \qquad \text{Apply shell method.}$$

$$= 2\pi \int_0^1 (-x^4 + 2x^3 - x^2 + 2x)\,dx \qquad \text{Simplify.}$$

$$= 2\pi \left[-\frac{x^5}{5} + \frac{x^4}{2} - \frac{x^3}{3} + x^2\right]_0^1 \qquad \text{Integrate.}$$

$$= 2\pi\left(-\frac{1}{5} + \frac{1}{2} - \frac{1}{3} + 1\right)$$

$$= \frac{29\pi}{15}.$$

Exercises for Section 7.3

See www.CalcChat.com for worked-out solutions to odd-numbered exercises.

In Exercises 1–12, use the shell method to set up and evaluate the integral that gives the volume of the solid generated by revolving the plane region about the y-axis.

1. $y = x$

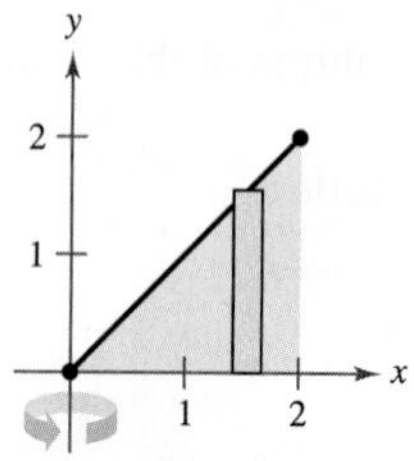

2. $y = 1 - x$

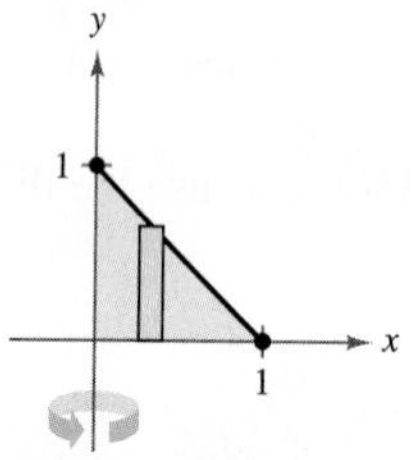

3. $y = \sqrt{x}$

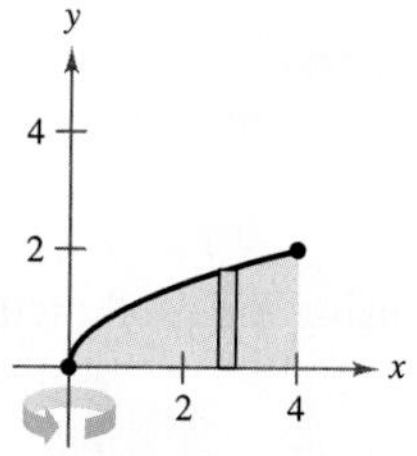

4. $y = x^2 + 4$

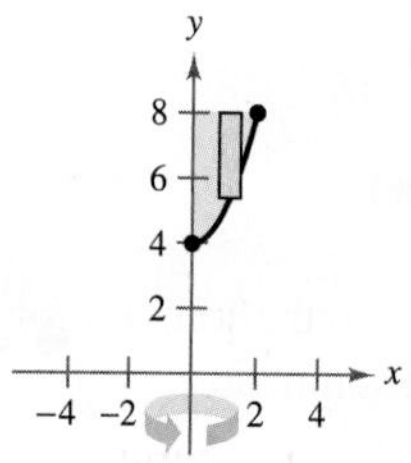

5. $y = x^2, \quad y = 0, \quad x = 2$

6. $y = \frac{1}{2}x^2, \quad y = 0, \quad x = 6$

7. $y = x^2, \quad y = 4x - x^2$

8. $y = 4 - x^2, \quad y = 0$

9. $y = 4x - x^2, \quad x = 0, \quad y = 4$

10. $y = 2x, \quad y = 4, \quad x = 0$

11. $y = \dfrac{1}{\sqrt{2\pi}} e^{-x^2/2}, \quad y = 0, \quad x = 0, \quad x = 1$

12. $y = \begin{cases} \dfrac{\sin x}{x}, & x > 0 \\ 1, & x = 0 \end{cases}, \quad y = 0, \quad x = 0, \quad x = \pi$

In Exercises 13–20, use the shell method to set up and evaluate the integral that gives the volume of the solid generated by revolving the plane region about the x-axis.

13. $y = x$

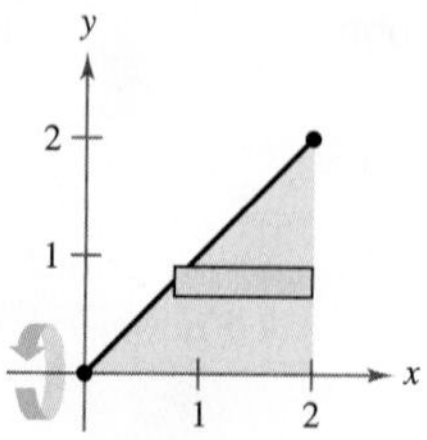

14. $y = 2 - x$

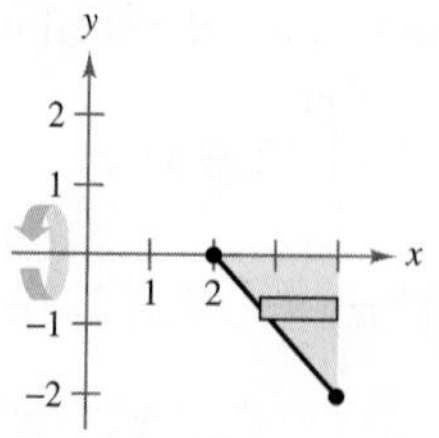

15. $y = \dfrac{1}{x}$

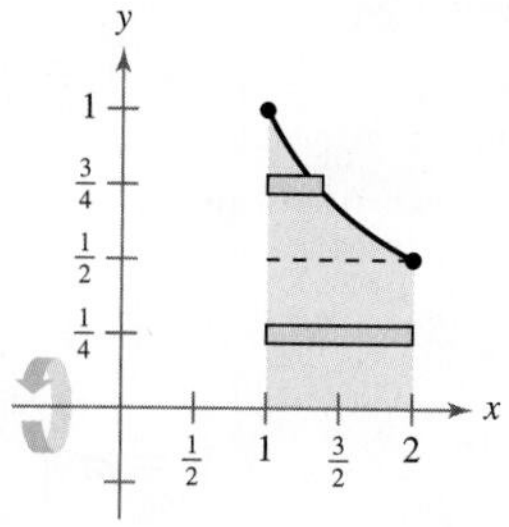

16. $x + y^2 = 16$

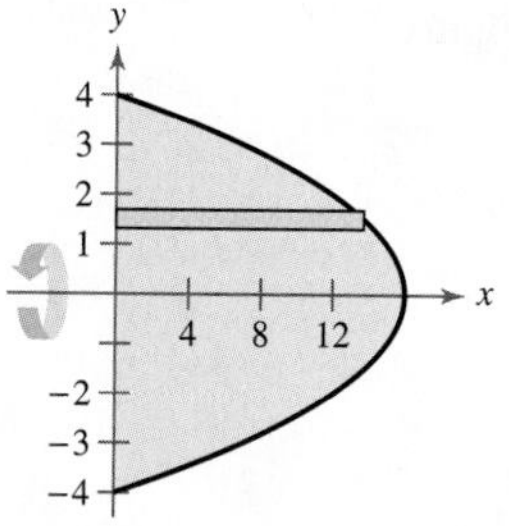

17. $y = x^3, \quad x = 0, \quad y = 8$

18. $y = x^2, \quad x = 0, \quad y = 9$

19. $x + y = 4, \quad y = x, \quad y = 0$

20. $y = \sqrt{x + 2}, \quad y = x, \quad y = 0$

In Exercises 21–24, use the shell method to find the volume of the solid generated by revolving the plane region about the given line.

21. $y = x^2, \quad y = 4x - x^2$, about the line $x = 4$

22. $y = x^2, \quad y = 4x - x^2$, about the line $x = 2$

23. $y = 4x - x^2, \quad y = 0$, about the line $x = 5$

24. $y = \sqrt{x}, \quad y = 0, \quad x = 4$, about the line $x = 6$

In Exercises 25 and 26, decide whether it is more convenient to use the disk method or the shell method to find the volume of the solid of revolution. Explain your reasoning. (Do not find the volume.)

25. $(y - 2)^2 = 4 - x$

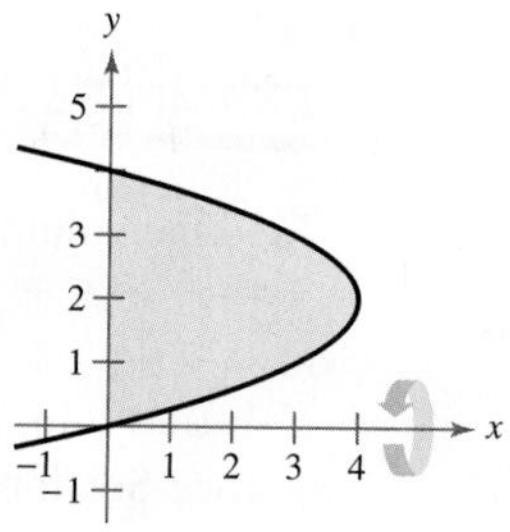

26. $y = 4 - e^x$

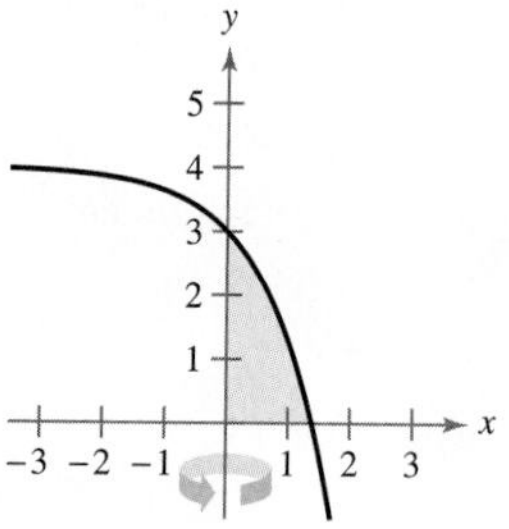

In Exercises 27–30, use the disk *or* the shell method to find the volume of the solid generated by revolving the region bounded by the graphs of the equations about each given line.

27. $y = x^3, \quad y = 0, \quad x = 2$

(a) the x-axis (b) the y-axis (c) the line $x = 4$

28. $y = \dfrac{10}{x^2}, \quad y = 0, \quad x = 1, \quad x = 5$

(a) the x-axis (b) the y-axis (c) the line $y = 10$

29. $x^{1/2} + y^{1/2} = a^{1/2}, \quad x = 0, \quad y = 0$

(a) the x-axis (b) the y-axis (c) the line $x = a$

30. $x^{2/3} + y^{2/3} = a^{2/3}$, $a > 0$ (hypocycloid)
(a) the x-axis (b) the y-axis

Writing About Concepts

31. Consider a solid that is generated by revolving a plane region about the y-axis. Describe the position of a representative rectangle when using (a) the shell method and (b) the disk method to find the volume of the solid.

32. The region in the figure is revolved about the indicated axes and line. Order the volumes of the resulting solids from least to greatest. Explain your reasoning.
(a) x-axis (b) y-axis (c) $x = 5$

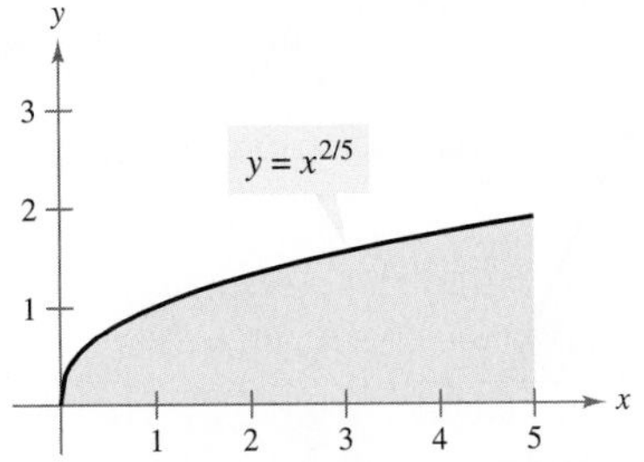

In Exercises 33 and 34, give a geometric argument that explains why the integrals have equal values.

33. $\pi\int_1^5 (x - 1)\,dx = 2\pi\int_0^2 y[5 - (y^2 + 1)]\,dy$

34. $\pi\int_0^2 [16 - (2y)^2]\,dy = 2\pi\int_0^4 x\left(\frac{x}{2}\right)dx$

In Exercises 35–38, (a) use a graphing utility to graph the plane region bounded by the graphs of the equations, and (b) use the integration capabilities of the graphing utility to approximate the volume of the solid generated by revolving the region about the y-axis.

35. $x^{4/3} + y^{4/3} = 1$, $x = 0$, $y = 0$, first quadrant

36. $y = \sqrt{1 - x^3}$, $y = 0$, $x = 0$

37. $y = \sqrt[3]{(x - 2)^2(x - 6)^2}$, $y = 0$, $x = 2$, $x = 6$

38. $y = \dfrac{2}{1 + e^{1/x}}$, $y = 0$, $x = 1$, $x = 3$

Think About It **In Exercises 39 and 40, determine which value best approximates the volume of the solid generated by revolving the region bounded by the graphs of the equations about the y-axis. (Make your selection on the basis of a sketch of the solid and *not* by performing any calculations.)**

39. $y = 2e^{-x}$, $y = 0$, $x = 0$, $x = 2$
(a) $\frac{3}{2}$ (b) -2 (c) 4 (d) 7.5 (e) 15

40. $y = \tan x$, $y = 0$, $x = 0$, $x = \dfrac{\pi}{4}$
(a) 3.5 (b) $-\frac{9}{4}$ (c) 8 (d) 10 (e) 1

41. ***Machine Part*** A solid is generated by revolving the region bounded by $y = \frac{1}{2}x^2$ and $y = 2$ about the y-axis. A hole, centered along the axis of revolution, is drilled through this solid so that one-fourth of the volume is removed. Find the diameter of the hole.

42. ***Machine Part*** A solid is generated by revolving the region bounded by $y = \sqrt{9 - x^2}$ and $y = 0$ about the y-axis. A hole, centered along the axis of revolution, is drilled through this solid so that one-third of the volume is removed. Find the diameter of the hole.

43. ***Volume of a Torus*** A torus is formed by revolving the region bounded by the circle $x^2 + y^2 = 1$ about the line $x = 2$ (see figure). Find the volume of this "doughnut-shaped" solid. (*Hint:* The integral $\int_{-1}^{1} \sqrt{1 - x^2}\,dx$ represents the area of a semicircle.)

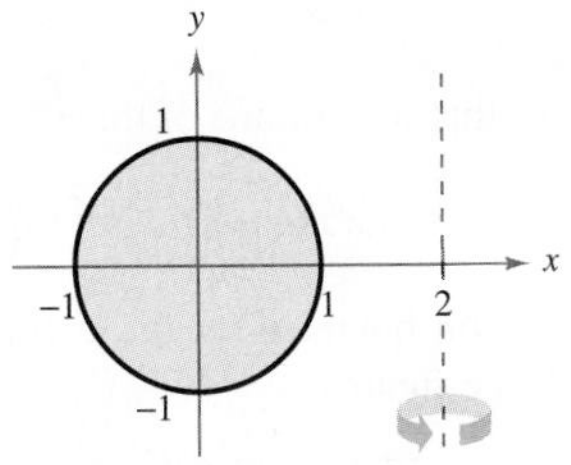

44. ***Volume of a Torus*** Repeat Exercise 43 for a torus formed by revolving the region bounded by the circle $x^2 + y^2 = r^2$ about the line $x = R$, where $r < R$.

45. (a) Use differentiation to verify that

$$\int x \sin x\,dx = \sin x - x\cos x + C.$$

(b) Use the result of part (a) to find the volume of the solid generated by revolving each plane region about the y-axis.

(i)
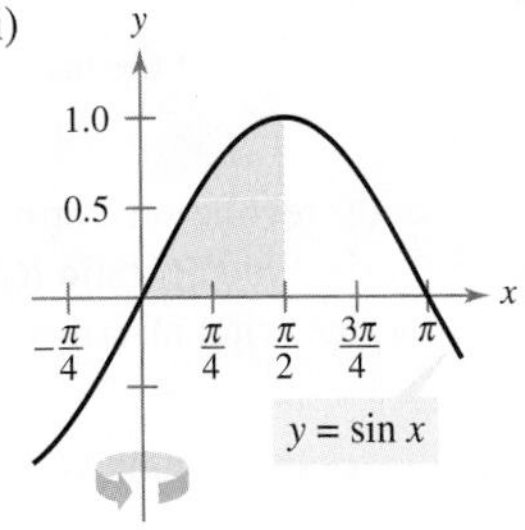

(ii)
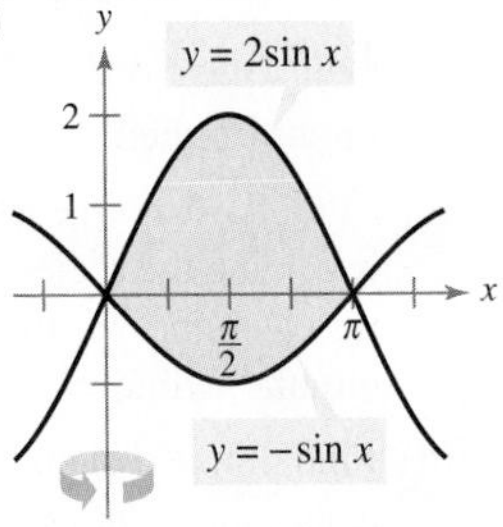

46. (a) Use differentiation to verify that

$$\int x\cos x\,dx = \cos x + x\sin x + C.$$

(b) Use the result of part (a) to find the volume of the solid generated by revolving each plane region about the y-axis. (*Hint:* Begin by approximating the points of intersection.)

(i)
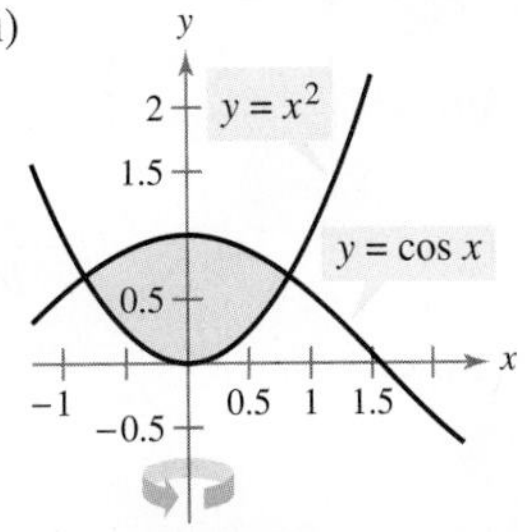

(ii)
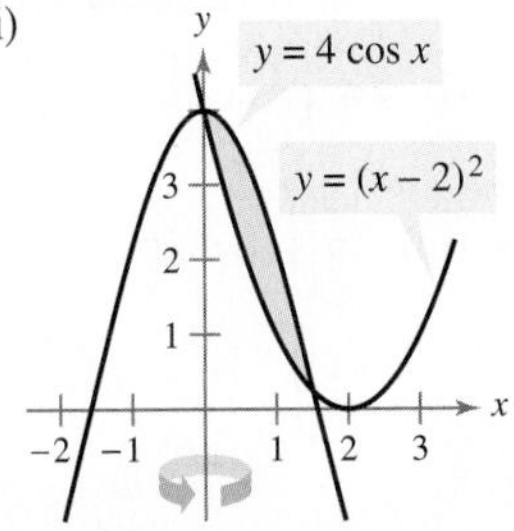

In Exercises 47–50, the integral represents the volume of a solid of revolution. Identify (a) the plane region that is revolved and (b) the axis of revolution.

47. $2\pi \int_0^2 x^3\,dx$

48. $2\pi \int_0^1 y - y^{3/2}\,dy$

49. $2\pi \int_0^6 (y+2)\sqrt{6-y}\,dy$

50. $2\pi \int_0^1 (4-x)e^x\,dx$

51. ***Volume of a Segment of a Sphere*** Let a sphere of radius r be cut by a plane, thereby forming a segment of height h. Show that the volume of this segment is $\frac{1}{3}\pi h^2(3r - h)$.

52. ***Volume of an Ellipsoid*** Consider the plane region bounded by the graph of

$$\left(\frac{x}{a}\right)^2 + \left(\frac{y}{b}\right)^2 = 1$$

where $a > 0$ and $b > 0$. Show that the volume of the ellipsoid formed when this region revolves about the y-axis is $\dfrac{4\pi a^2 b}{3}$.

53. ***Exploration*** Consider the region bounded by the graphs of $y = ax^n$, $y = ab^n$, and $x = 0$ (see figure).

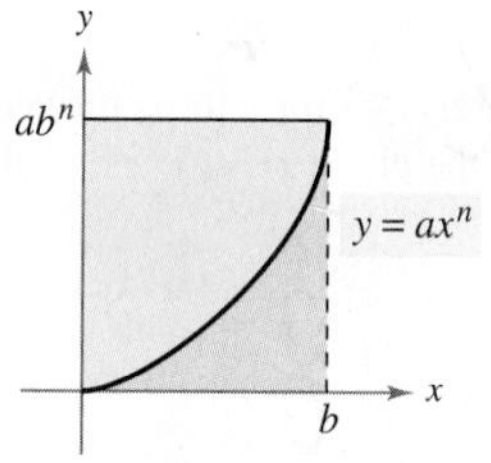

(a) Find the ratio $R_1(n)$ of the area of the region to the area of the circumscribed rectangle.

(b) Find $\lim_{n\to\infty} R_1(n)$ and compare the result with the area of the circumscribed rectangle.

(c) Find the volume of the solid of revolution formed by revolving the region about the y-axis. Find the ratio $R_2(n)$ of this volume to the volume of the circumscribed right circular cylinder.

(d) Find $\lim_{n\to\infty} R_2(n)$ and compare the result with the volume of the circumscribed cylinder.

(e) Use the results of parts (b) and (d) to make a conjecture about the shape of the graph of $y = ax^n$ $(0 \le x \le b)$ as $n \to \infty$.

54. ***Think About It*** Match each integral with the solid whose volume it represents, and give the dimensions of each solid.

(a) Right circular cone (b) Torus (c) Sphere

(d) Right circular cylinder (e) Ellipsoid

(i) $2\pi \int_0^r hx\,dx$

(ii) $2\pi \int_0^r hx\left(1 - \frac{x}{r}\right)dx$

(iii) $2\pi \int_0^r 2x\sqrt{r^2 - x^2}\,dx$

(iv) $2\pi \int_0^b 2ax\sqrt{1 - \frac{x^2}{b^2}}\,dx$

(v) $2\pi \int_{-r}^r (R - x)\left(2\sqrt{r^2 - x^2}\right)dx$

55. ***Volume of a Storage Shed*** A storage shed has a circular base of diameter 80 feet (see figure). Starting at the center, the interior height is measured every 10 feet and recorded in the table.

x	0	10	20	30	40
Height	50	45	40	20	0

(a) Use Simpson's Rule to approximate the volume of the shed.

(b) Note that the roof line consists of two line segments. Find the equations of the line segments and use integration to find the volume of the shed.

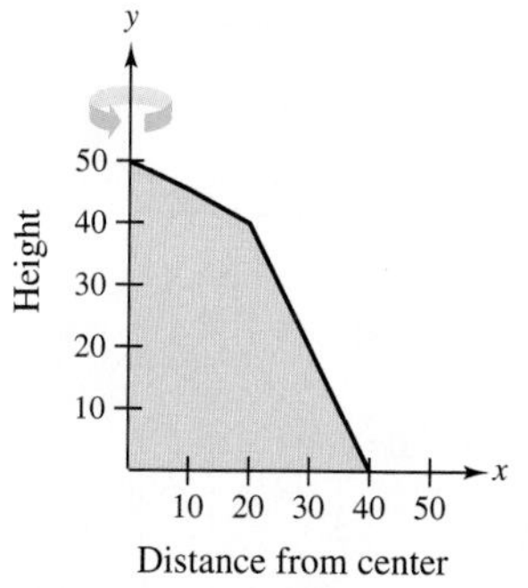

56. ***Modeling Data*** A pond is approximately circular, with a diameter of 400 feet (see figure). Starting at the center, the depth of the water is measured every 25 feet and recorded in the table.

x	0	25	50	75	100	125	150	175	200
Depth	20	19	19	17	15	14	10	6	0

(a) Use Simpson's Rule to approximate the volume of water in the pond.

(b) Use the regression capabilities of a graphing utility to find a quadratic model for the depths recorded in the table. Use the graphing utility to plot the depths and graph the model.

(c) Use the integration capabilities of a graphing utility and the model in part (b) to approximate the volume of water in the pond.

(d) Use the result of part (c) to approximate the number of gallons of water in the pond if 1 cubic foot of water is approximately 7.48 gallons.

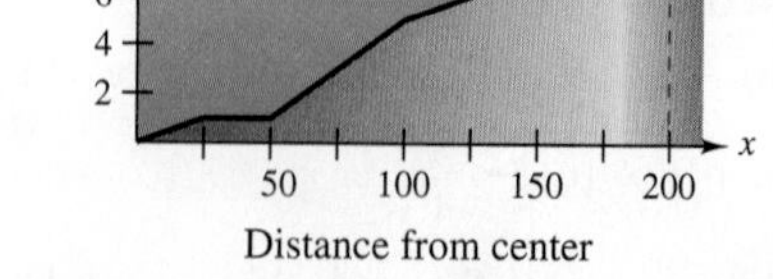

57. Consider the graph of $y^2 = x(4 - x)^2$ (see figure). Find the volumes of the solids that are generated when the loop of this graph is revolved around (a) the x-axis, (b) the y-axis, and (c) the line $x = 4$.

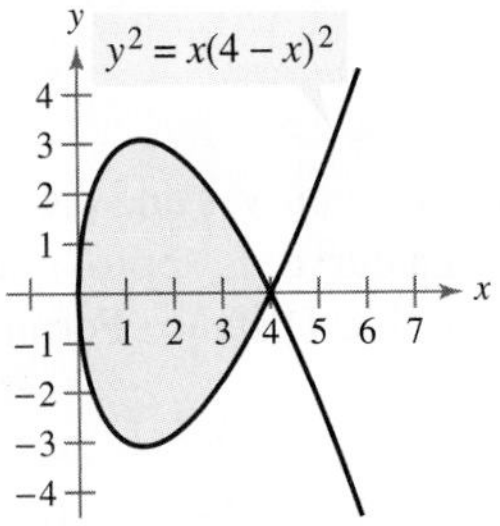

Figure for 57

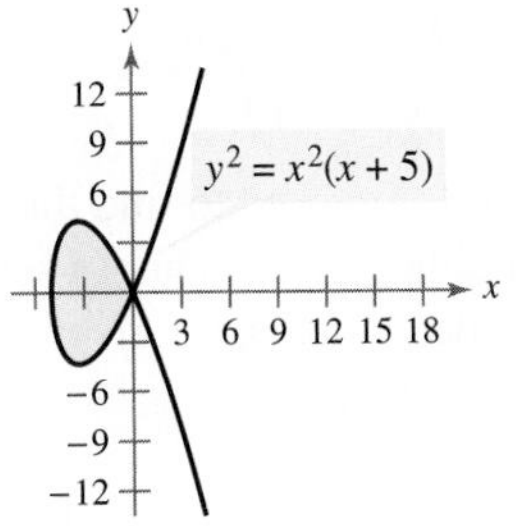

Figure for 58

58. Consider the graph of $y^2 = x^2(x + 5)$ (see figure). Find the volume of the solid that is generated when the loop of this graph is revolved around (a) the x-axis, (b) the y-axis, and (c) the line $x = -5$.

59. Let V_1 and V_2 be the volumes of the solids that result when the plane region bounded by $y = 1/x$, $y = 0$, $x = \frac{1}{4}$, and $x = c\left(c > \frac{1}{4}\right)$ is revolved about the x-axis and y-axis, respectively. Find the value of c for which $V_1 = V_2$.

Section Project: Saturn

The Oblateness of Saturn Saturn is the most oblate of the nine planets in our solar system. Its equatorial radius is 60,268 kilometers and its polar radius is 54,364 kilometers. The color enhanced photograph of Saturn was taken by Voyager 1. In the photograph, the oblateness of Saturn is clearly visible.

(a) Find the ratio of the volumes of the sphere and the oblate ellipsoid shown below.

(b) If a planet were spherical and had the same volume as Saturn, what would its radius be?

NSSDC

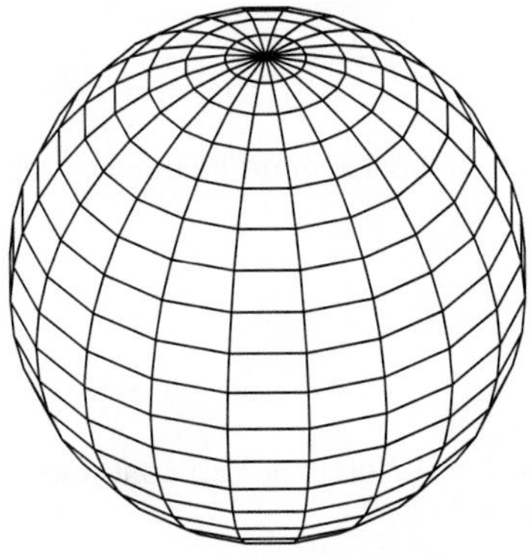

Computer model of "spherical Saturn," whose equatorial radius is equal to its polar radius. The equation of the cross section passing through the pole is

$x^2 + y^2 = 60{,}268^2.$

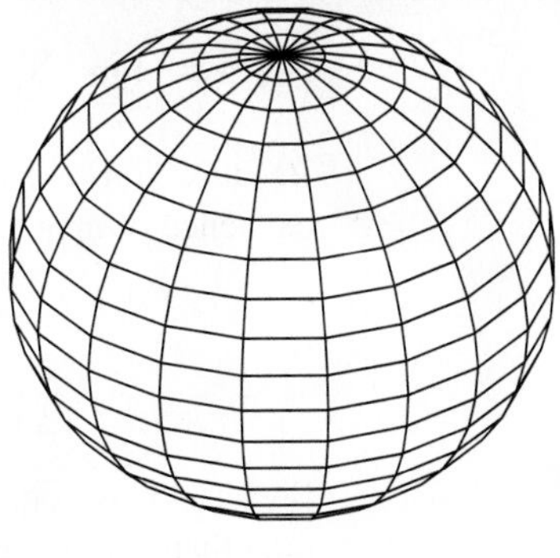

Computer model of "oblate Saturn," whose equatorial radius is greater than its polar radius. The equation of the cross section passing through the pole is

$$\frac{x^2}{60{,}268^2} + \frac{y^2}{54{,}364^2} = 1.$$

Section 7.4 Arc Length and Surfaces of Revolution

Bettmann/Corbis

Christian Huygens (1629–1695)

The Dutch mathematician Christian Huygens, who invented the pendulum clock, and James Gregory (1638–1675), a Scottish mathematician, both made early contributions to the problem of finding the length of a rectifiable curve.

- **Find the arc length of a smooth curve.**
- **Find the area of a surface of revolution.**

Arc Length

In this section, definite integrals are used to find the arc lengths of curves and the areas of surfaces of revolution. In either case, an arc (a segment of a curve) is approximated by straight line segments whose lengths are given by the familiar Distance Formula

$$d = \sqrt{(x_2 - x_1)^2 + (y_2 - y_1)^2}.$$

A **rectifiable** curve is one that has a finite arc length. You will see that a sufficient condition for the graph of a function f to be rectifiable between $(a, f(a))$ and $(b, f(b))$ is that f' be continuous on $[a, b]$. Such a function is **continuously differentiable** on $[a, b]$, and its graph on the interval $[a, b]$ is a **smooth curve.**

Consider a function $y = f(x)$ that is continuously differentiable on the interval $[a, b]$. You can approximate the graph of f by n line segments whose endpoints are determined by the partition

$$a = x_0 < x_1 < x_2 < \cdot \cdot \cdot < x_n = b$$

as shown in Figure 7.37. By letting $\Delta x_i = x_i - x_{i-1}$ and $\Delta y_i = y_i - y_{i-1}$, you can approximate the length of the graph by

$$\begin{aligned} s &\approx \sum_{i=1}^{n} \sqrt{(x_i - x_{i-1})^2 + (y_i - y_{i-1})^2} \\ &= \sum_{i=1}^{n} \sqrt{(\Delta x_i)^2 + (\Delta y_i)^2} \\ &= \sum_{i=1}^{n} \sqrt{(\Delta x_i)^2 + \left(\frac{\Delta y_i}{\Delta x_i}\right)^2 (\Delta x_i)^2} \\ &= \sum_{i=1}^{n} \sqrt{1 + \left(\frac{\Delta y_i}{\Delta x_i}\right)^2} (\Delta x_i). \end{aligned}$$

This approximation appears to become better and better as $\|\Delta\| \to 0$ $(n \to \infty)$. So, the length of the graph is

$$s = \lim_{\|\Delta\| \to 0} \sum_{i=1}^{n} \sqrt{1 + \left(\frac{\Delta y_i}{\Delta x_i}\right)^2} (\Delta x_i).$$

Because $f'(x)$ exists for each x in (x_{i-1}, x_i), the Mean Value Theorem guarantees the existence of c_i in (x_{i-1}, x_i) such that

$$\begin{aligned} f(x_i) - f(x_{i-1}) &= f'(c_i)(x_i - x_{i-1}) \\ \frac{\Delta y_i}{\Delta x_i} &= f'(c_i). \end{aligned}$$

Because f' is continuous on $[a, b]$, it follows that $\sqrt{1 + [f'(x)]^2}$ is also continuous (and therefore integrable) on $[a, b]$, which implies that

$$\begin{aligned} s &= \lim_{\|\Delta\| \to 0} \sum_{i=1}^{n} \sqrt{1 + [f'(c_i)]^2} (\Delta x_i) \\ &= \int_a^b \sqrt{1 + [f'(x)]^2}\, dx \end{aligned}$$

where s is called the **arc length** of f between a and b.

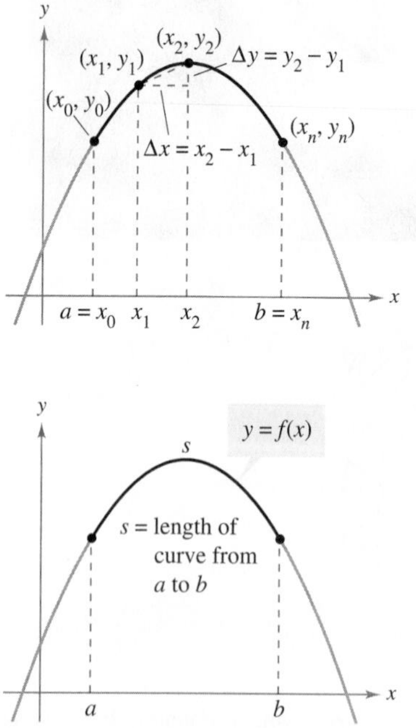

Figure 7.37

Definition of Arc Length

Let the function given by $y = f(x)$ represent a smooth curve on the interval $[a, b]$. The **arc length** of f between a and b is

$$s = \int_a^b \sqrt{1 + [f'(x)]^2}\,dx.$$

Similarly, for a smooth curve given by $x = g(y)$, the **arc length** of g between c and d is

$$s = \int_c^d \sqrt{1 + [g'(y)]^2}\,dy.$$

Because the definition of arc length can be applied to a linear function, you can check to see that this new definition agrees with the standard Distance Formula for the length of a line segment. This is shown in Example 1.

EXAMPLE 1 The Length of a Line Segment

Find the arc length from (x_1, y_1) to (x_2, y_2) on the graph of $f(x) = mx + b$, as shown in Figure 7.38.

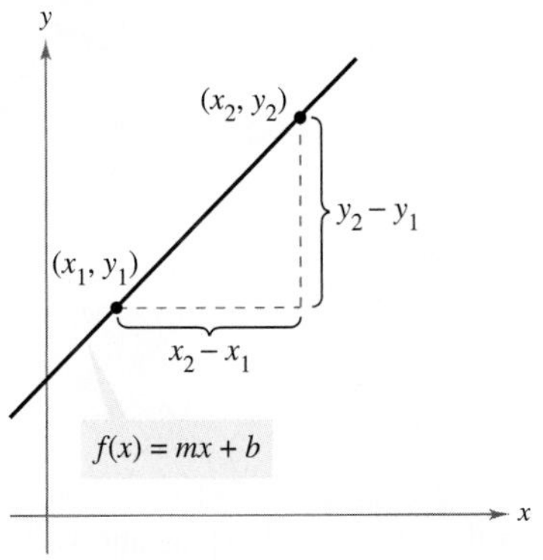

The arc length of the graph of f from (x_1, y_1) to (x_2, y_2) is the same as the standard Distance Formula.
Figure 7.38

Solution Because

$$m = f'(x) = \frac{y_2 - y_1}{x_2 - x_1}$$

it follows that

$$\begin{aligned} s &= \int_{x_1}^{x_2} \sqrt{1 + [f'(x)]^2}\,dx && \text{Formula for arc length} \\ &= \int_{x_1}^{x_2} \sqrt{1 + \left(\frac{y_2 - y_1}{x_2 - x_1}\right)^2}\,dx \\ &= \sqrt{\frac{(x_2 - x_1)^2 + (y_2 - y_1)^2}{(x_2 - x_1)^2}}\,(x)\Big]_{x_1}^{x_2} && \text{Integrate and simplify.} \\ &= \sqrt{\frac{(x_2 - x_1)^2 + (y_2 - y_1)^2}{(x_2 - x_1)^2}}\,(x_2 - x_1) \\ &= \sqrt{(x_2 - x_1)^2 + (y_2 - y_1)^2} \end{aligned}$$

which is the formula for the distance between two points in the plane.

TECHNOLOGY Definite integrals representing arc length often are very difficult to evaluate. In this section, a few examples are presented. In the next chapter, with more advanced integration techniques, you will be able to tackle more difficult arc length problems. In the meantime, remember that you can always use a numerical integration program to approximate an arc length. For instance, use the *numerical integration* feature of a graphing utility to approximate the arc lengths in Examples 2 and 3.

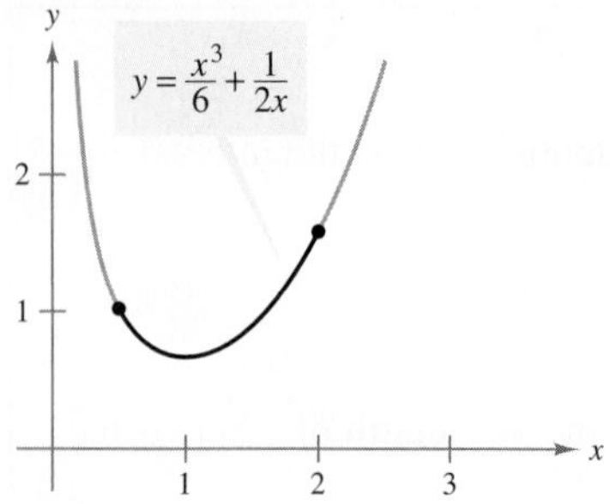

The arc length of the graph of y on $\left[\frac{1}{2}, 2\right]$
Figure 7.39

FOR FURTHER INFORMATION To see how arc length can be used to define trigonometric functions, see the article "Trigonometry Requires Calculus, Not Vice Versa" by Yves Nievergelt in *UMAP Modules*.

EXAMPLE 2 Finding Arc Length

Find the arc length of the graph of

$$y = \frac{x^3}{6} + \frac{1}{2x}$$

on the interval $\left[\frac{1}{2}, 2\right]$, as shown in Figure 7.39.

Solution Using

$$\frac{dy}{dx} = \frac{3x^2}{6} - \frac{1}{2x^2} = \frac{1}{2}\left(x^2 - \frac{1}{x^2}\right)$$

yields an arc length of

$$s = \int_a^b \sqrt{1 + \left(\frac{dy}{dx}\right)^2}\, dx = \int_{1/2}^{2} \sqrt{1 + \left[\frac{1}{2}\left(x^2 - \frac{1}{x^2}\right)\right]^2}\, dx \qquad \text{Formula for arc length}$$

$$= \int_{1/2}^{2} \sqrt{\frac{1}{4}\left(x^4 + 2 + \frac{1}{x^4}\right)}\, dx$$

$$= \int_{1/2}^{2} \frac{1}{2}\left(x^2 + \frac{1}{x^2}\right) dx \qquad \text{Simplify.}$$

$$= \frac{1}{2}\left[\frac{x^3}{3} - \frac{1}{x}\right]_{1/2}^{2} \qquad \text{Integrate.}$$

$$= \frac{1}{2}\left(\frac{13}{6} + \frac{47}{24}\right)$$

$$= \frac{33}{16}.$$

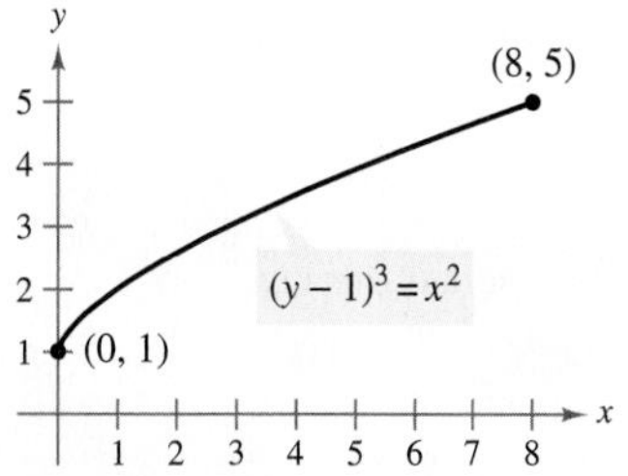

The arc length of the graph of y on $[0, 8]$
Figure 7.40

EXAMPLE 3 Finding Arc Length

Find the arc length of the graph of $(y - 1)^3 = x^2$ on the interval $[0, 8]$, as shown in Figure 7.40.

Solution Begin by solving for x in terms of y: $x = \pm(y - 1)^{3/2}$. Choosing the positive value of x produces

$$\frac{dx}{dy} = \frac{3}{2}(y - 1)^{1/2}.$$

The x-interval $[0, 8]$ corresponds to the y-interval $[1, 5]$, and the arc length is

$$s = \int_c^d \sqrt{1 + \left(\frac{dx}{dy}\right)^2}\, dy = \int_1^5 \sqrt{1 + \left[\frac{3}{2}(y - 1)^{1/2}\right]^2}\, dy \qquad \text{Formula for arc length}$$

$$= \int_1^5 \sqrt{\frac{9}{4}y - \frac{5}{4}}\, dy$$

$$= \frac{1}{2}\int_1^5 \sqrt{9y - 5}\, dy \qquad \text{Simplify.}$$

$$= \frac{1}{18}\left[\frac{(9y - 5)^{3/2}}{3/2}\right]_1^5 \qquad \text{Integrate.}$$

$$= \frac{1}{27}(40^{3/2} - 4^{3/2})$$

$$\approx 9.073.$$

EXAMPLE 4 Finding Arc Length

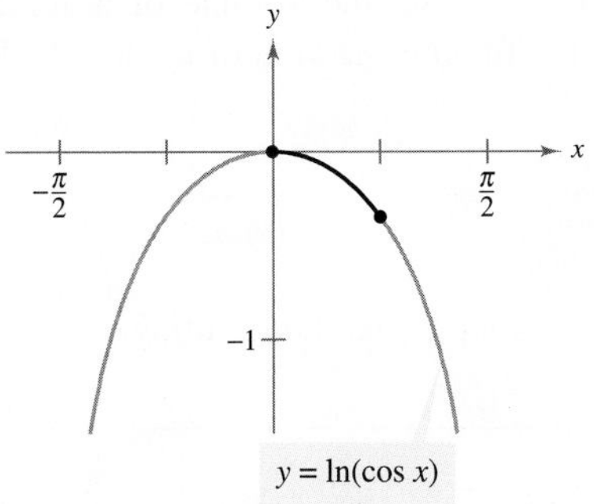

The arc length of the graph of y on $\left[0, \frac{\pi}{4}\right]$
Figure 7.41

Find the arc length of the graph of $y = \ln(\cos x)$ from $x = 0$ to $x = \pi/4$, as shown in Figure 7.41.

Solution Using

$$\frac{dy}{dx} = -\frac{\sin x}{\cos x} = -\tan x$$

yields an arc length of

$$s = \int_a^b \sqrt{1 + \left(\frac{dy}{dx}\right)^2}\, dx = \int_0^{\pi/4} \sqrt{1 + \tan^2 x}\, dx \qquad \text{Formula for arc length}$$

$$= \int_0^{\pi/4} \sqrt{\sec^2 x}\, dx \qquad \text{Trigonometric identity}$$

$$= \int_0^{\pi/4} \sec x\, dx \qquad \text{Simplify.}$$

$$= \Big[\ln|\sec x + \tan x|\Big]_0^{\pi/4} \qquad \text{Integrate.}$$

$$= \ln\left(\sqrt{2} + 1\right) - \ln 1$$

$$\approx 0.881.$$

EXAMPLE 5 Length of a Cable

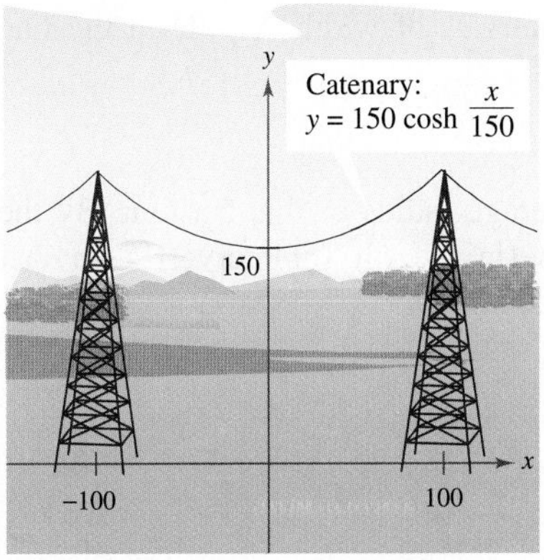

Figure 7.42

An electric cable is hung between two towers that are 200 feet apart, as shown in Figure 7.42. The cable takes the shape of a catenary whose equation is

$$y = 75(e^{x/150} + e^{-x/150}) = 150 \cosh \frac{x}{150}.$$

Find the arc length of the cable between the two towers.

Solution Because $y' = \frac{1}{2}(e^{x/150} - e^{-x/150})$, you can write

$$(y')^2 = \frac{1}{4}(e^{x/75} - 2 + e^{-x/75})$$

and

$$1 + (y')^2 = \frac{1}{4}(e^{x/75} + 2 + e^{-x/75}) = \left[\frac{1}{2}(e^{x/150} + e^{-x/150})\right]^2.$$

Therefore, the arc length of the cable is

$$s = \int_a^b \sqrt{1 + (y')^2}\, dx = \frac{1}{2}\int_{-100}^{100} (e^{x/150} + e^{-x/150})\, dx \qquad \text{Formula for arc length}$$

$$= 75\Big[e^{x/150} - e^{-x/150}\Big]_{-100}^{100} \qquad \text{Integrate.}$$

$$= 150(e^{2/3} - e^{-2/3})$$

$$\approx 215 \text{ feet.}$$

Area of a Surface of Revolution

In Sections 7.2 and 7.3, integration was used to calculate the volume of a solid of revolution. You will now look at a procedure for finding the area of a surface of revolution.

> **Definition of Surface of Revolution**
>
> If the graph of a continuous function is revolved about a line, the resulting surface is a **surface of revolution.**

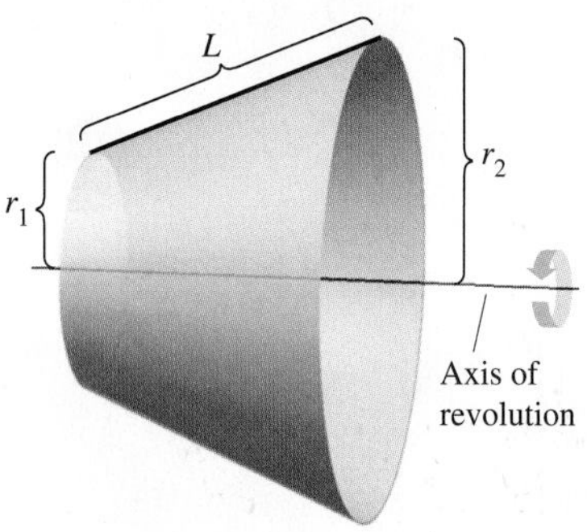

Figure 7.43

The area of a surface of revolution is derived from the formula for the lateral surface area of the frustum of a right circular cone. Consider the line segment in Figure 7.43, where L is the length of the line segment, r_1 is the radius at the left end of the line segment, and r_2 is the radius at the right end of the line segment. When the line segment is revolved about its axis of revolution, it forms a frustum of a right circular cone, with

$$S = 2\pi r L \qquad \text{Lateral surface area of frustum}$$

where

$$r = \frac{1}{2}(r_1 + r_2). \qquad \text{Average radius of frustum}$$

(In Exercise 60, you are asked to verify the formula for S.)

Suppose the graph of a function f, having a continuous derivative on the interval $[a, b]$, is revolved about the x-axis to form a surface of revolution, as shown in Figure 7.44. Let Δ be a partition of $[a, b]$, with subintervals of width Δx_i. Then the line segment of length

$$\Delta L_i = \sqrt{\Delta x_i^2 + \Delta y_i^2}$$

generates a frustum of a cone. Let r_i be the average radius of this frustum. By the Intermediate Value Theorem, a point d_i exists (in the ith subinterval) such that $r_i = f(d_i)$. The lateral surface area ΔS_i of the frustum is

$$\begin{aligned}\Delta S_i &= 2\pi r_i \Delta L_i\\ &= 2\pi f(d_i)\sqrt{\Delta x_i^2 + \Delta y_i^2}\\ &= 2\pi f(d_i)\sqrt{1 + \left(\frac{\Delta y_i}{\Delta x_i}\right)^2}\,\Delta x_i.\end{aligned}$$

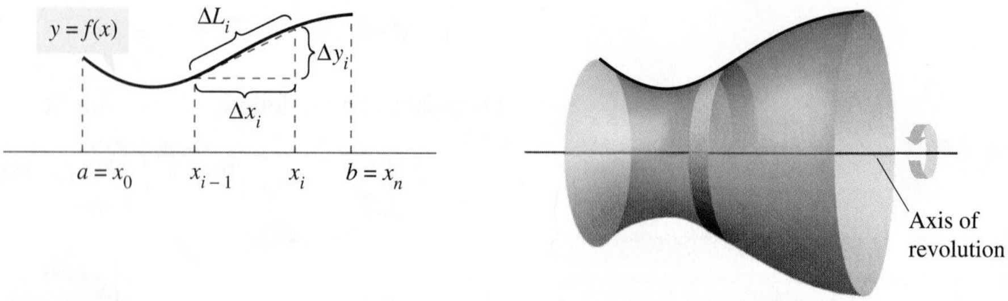

Figure 7.44

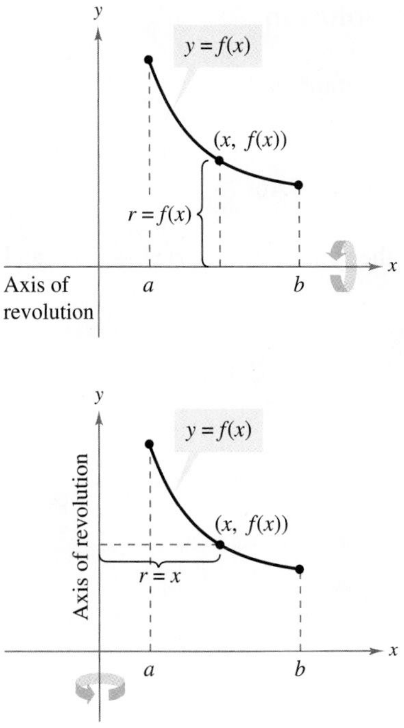

Figure 7.45

By the Mean Value Theorem, a point c_i exists in (x_{i-1}, x_i) such that

$$f'(c_i) = \frac{f(x_i) - f(x_{i-1})}{x_i - x_{i-1}} = \frac{\Delta y_i}{\Delta x_i}.$$

So, $\Delta S_i = 2\pi f(d_i)\sqrt{1 + [f'(c_i)]^2}\,\Delta x_i$, and the total surface area can be approximated by

$$S \approx 2\pi \sum_{i=1}^{n} f(d_i)\sqrt{1 + [f'(c_i)]^2}\,\Delta x_i.$$

It can be shown that the limit of the right side as $\|\Delta\| \to 0\ (n \to \infty)$ is

$$S = 2\pi \int_a^b f(x)\sqrt{1 + [f'(x)]^2}\,dx.$$

In a similar manner, if the graph of f is revolved about the y-axis, then S is

$$S = 2\pi \int_a^b x\sqrt{1 + [f'(x)]^2}\,dx.$$

In both formulas for S, you can regard the products $2\pi f(x)$ and $2\pi x$ as the circumference of the circle traced by a point (x, y) on the graph of f as it is revolved about the x- or y-axis (Figure 7.45). In one case the radius is $r = f(x)$, and in the other case the radius is $r = x$. Moreover, by appropriately adjusting r, you can generalize the formula for surface area to cover *any* horizontal or vertical axis of revolution, as indicated in the following definition.

Definition of the Area of a Surface of Revolution

Let $y = f(x)$ have a continuous derivative on the interval $[a, b]$. The area S of the surface of revolution formed by revolving the graph of f about a horizontal or vertical axis is

$$S = 2\pi \int_a^b r(x)\sqrt{1 + [f'(x)]^2}\,dx \qquad y \text{ is a function of } x.$$

where $r(x)$ is the distance between the graph of f and the axis of revolution. If $x = g(y)$ on the interval $[c, d]$, then the surface area is

$$S = 2\pi \int_c^d r(y)\sqrt{1 + [g'(y)]^2}\,dy \qquad x \text{ is a function of } y.$$

where $r(y)$ is the distance between the graph of g and the axis of revolution.

The formulas in this definition are sometimes written as

$$S = 2\pi \int_a^b r(x)\,ds \qquad y \text{ is a function of } x.$$

and

$$S = 2\pi \int_c^d r(y)\,ds \qquad x \text{ is a function of } y.$$

where $ds = \sqrt{1 + [f'(x)]^2}\,dx$ and $ds = \sqrt{1 + [g'(y)]^2}\,dy$, respectively.

EXAMPLE 6 The Area of a Surface of Revolution

Find the area of the surface formed by revolving the graph of

$$f(x) = x^3$$

on the interval $[0, 1]$ about the x-axis, as shown in Figure 7.46.

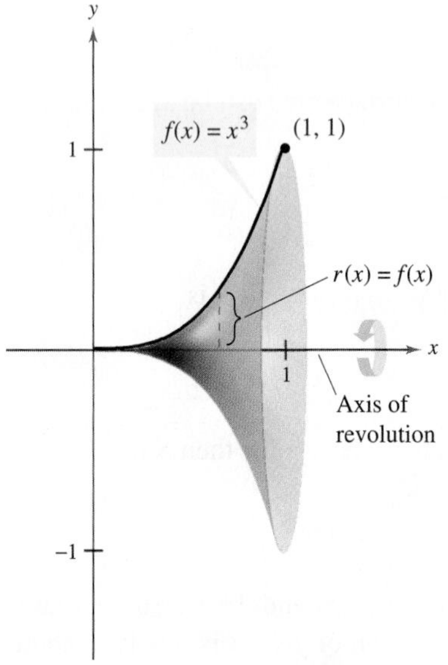

Figure 7.46

Solution The distance between the x-axis and the graph of f is $r(x) = f(x)$, and because $f'(x) = 3x^2$, the surface area is

$$\begin{aligned} S &= 2\pi \int_a^b r(x)\sqrt{1 + [f'(x)]^2}\,dx && \text{Formula for surface area} \\ &= 2\pi \int_0^1 x^3\sqrt{1 + (3x^2)^2}\,dx \\ &= \frac{2\pi}{36}\int_0^1 (36x^3)(1 + 9x^4)^{1/2}\,dx && \text{Simplify.} \\ &= \frac{\pi}{18}\left[\frac{(1 + 9x^4)^{3/2}}{3/2}\right]_0^1 && \text{Integrate.} \\ &= \frac{\pi}{27}(10^{3/2} - 1) \\ &\approx 3.563. \end{aligned}$$

EXAMPLE 7 The Area of a Surface of Revolution

Find the area of the surface formed by revolving the graph of

$$f(x) = x^2$$

on the interval $[0, \sqrt{2}]$ about the y-axis, as shown in Figure 7.47.

Solution In this case, the distance between the graph of f and the y-axis is $r(x) = x$. Using $f'(x) = 2x$, you can determine that the surface area is

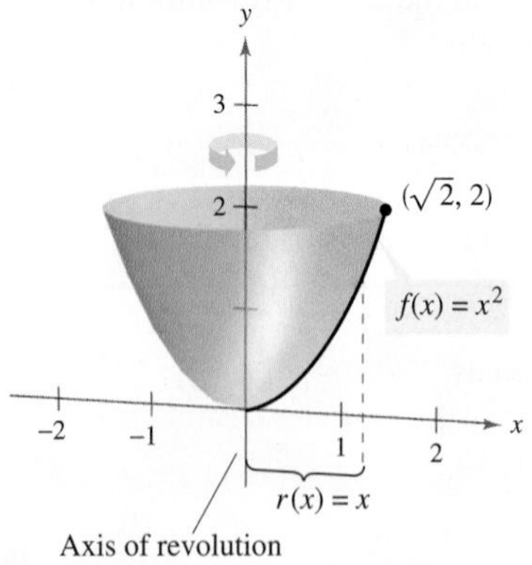

Figure 7.47

$$\begin{aligned} S &= 2\pi \int_a^b r(x)\sqrt{1 + [f'(x)]^2}\,dx && \text{Formula for surface area} \\ &= 2\pi \int_0^{\sqrt{2}} x\sqrt{1 + (2x)^2}\,dx \\ &= \frac{2\pi}{8}\int_0^{\sqrt{2}} (1 + 4x^2)^{1/2}(8x)\,dx && \text{Simplify.} \\ &= \frac{\pi}{4}\left[\frac{(1 + 4x^2)^{3/2}}{3/2}\right]_0^{\sqrt{2}} && \text{Integrate.} \\ &= \frac{\pi}{6}[(1 + 8)^{3/2} - 1] \\ &= \frac{13\pi}{3} \\ &\approx 13.614. \end{aligned}$$

Exercises for Section 7.4

See www.CalcChat.com for worked-out solutions to odd-numbered exercises.

In Exercises 1 and 2, find the distance between the points using (a) the Distance Formula and (b) integration.

1. $(0, 0), \quad (5, 12)$

2. $(1, 2), \quad (7, 10)$

In Exercises 3–14, find the arc length of the graph of the function over the indicated interval.

3. $y = \frac{2}{3}x^{3/2} + 1$

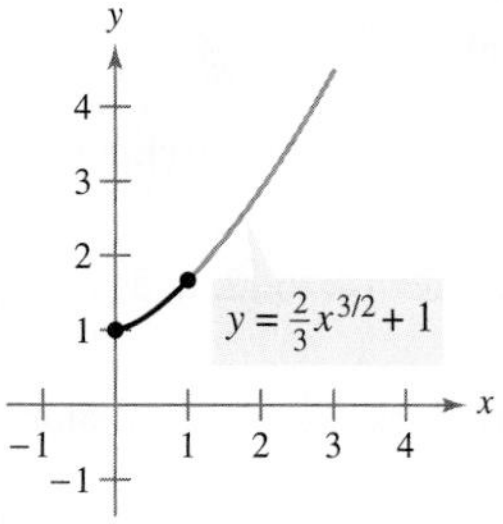

4. $y = 2x^{3/2} + 3$

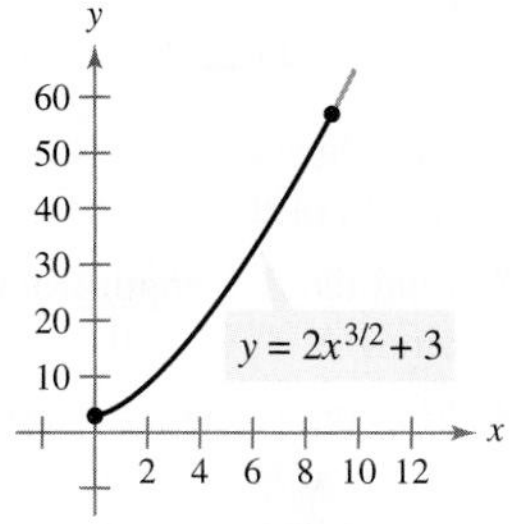

5. $y = \frac{3}{2}x^{2/3}$

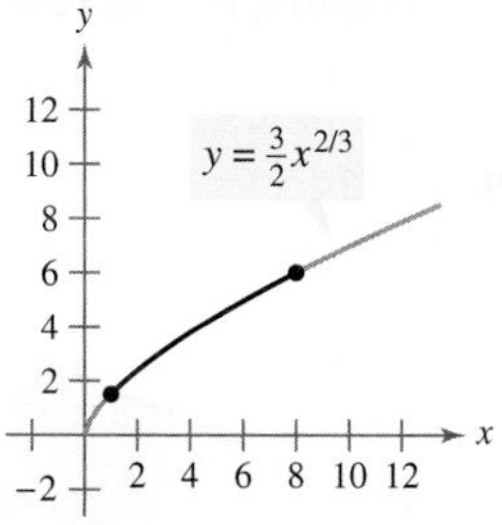

6. $y = \frac{x^4}{8} + \frac{1}{4x^2}$

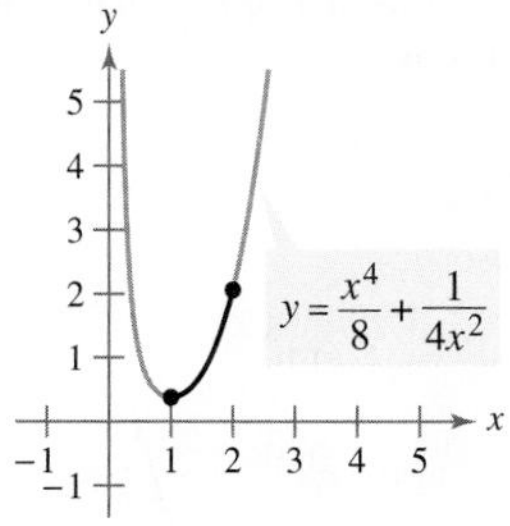

7. $y = \frac{x^5}{10} + \frac{1}{6x^3}, \quad [1, 2]$

8. $y = \frac{3}{2}x^{2/3} + 4, \quad [1, 27]$

9. $y = \ln(\sin x), \quad \left[\frac{\pi}{4}, \frac{3\pi}{4}\right]$

10. $y = \ln(\cos x), \quad \left[0, \frac{\pi}{3}\right]$

11. $y = \frac{1}{2}(e^x + e^{-x}), \quad [0, 2]$

12. $y = \ln\left(\frac{e^x + 1}{e^x - 1}\right), \quad [\ln 2, \ln 3]$

13. $x = \frac{1}{3}(y^2 + 2)^{3/2}, \quad 0 \le y \le 4$

14. $x = \frac{1}{3}\sqrt{y}(y - 3), \quad 1 \le y \le 4$

In Exercises 15–24, (a) graph the function, highlighting the part indicated by the given interval, (b) find a definite integral that represents the arc length of the curve over the indicated interval and observe that the integral cannot be evaluated with the techniques studied so far, and (c) use the integration capabilities of a graphing utility to approximate the arc length.

15. $y = 4 - x^2, \quad 0 \le x \le 2$

16. $y = x^2 + x - 2, \quad -2 \le x \le 1$

17. $y = \frac{1}{x}, \quad 1 \le x \le 3$

18. $y = \frac{1}{x + 1}, \quad 0 \le x \le 1$

19. $y = \sin x, \quad 0 \le x \le \pi$

20. $y = \cos x, \quad -\frac{\pi}{2} \le x \le \frac{\pi}{2}$

21. $x = e^{-y}, \quad 0 \le y \le 2$

22. $y = \ln x, \quad 1 \le x \le 5$

23. $y = 2 \arctan x, \quad 0 \le x \le 1$

24. $x = \sqrt{36 - y^2}, \quad 0 \le y \le 3$

Approximation **In Exercises 25 and 26, determine which value best approximates the length of the arc represented by the integral. (Make your selection on the basis of a sketch of the arc and *not* by performing any calculations.)**

25. $\displaystyle\int_0^2 \sqrt{1 + \left[\frac{d}{dx}\left(\frac{5}{x^2 + 1}\right)\right]^2}\, dx$

(a) 25 (b) 5 (c) 2 (d) -4 (e) 3

26. $\displaystyle\int_0^{\pi/4} \sqrt{1 + \left[\frac{d}{dx}(\tan x)\right]^2}\, dx$

(a) 3 (b) -2 (c) 4 (d) $\frac{4\pi}{3}$ (e) 1

Approximation **In Exercises 27 and 28, approximate the arc length of the graph of the function over the interval [0, 4] in four ways. (a) Use the Distance Formula to find the distance between the endpoints of the arc. (b) Use the Distance Formula to find the lengths of the four line segments connecting the points on the arc when $x = 0$, $x = 1$, $x = 2$, $x = 3$, and $x = 4$. Find the sum of the four lengths. (c) Use Simpson's Rule with $n = 10$ to approximate the integral yielding the indicated arc length. (d) Use the integration capabilities of a graphing utility to approximate the integral yielding the indicated arc length.**

27. $f(x) = x^3$

28. $f(x) = (x^2 - 4)^2$

29. (a) Use a graphing utility to graph the function $f(x) = x^{2/3}$.

(b) Can you integrate with respect to x to find the arc length of the graph of f on the interval $[-1, 8]$? Explain.

(c) Find the arc length of the graph of f on the interval $[-1, 8]$.

30. ***Astroid*** Find the total length of the graph of the astroid $x^{2/3} + y^{2/3} = 4$.

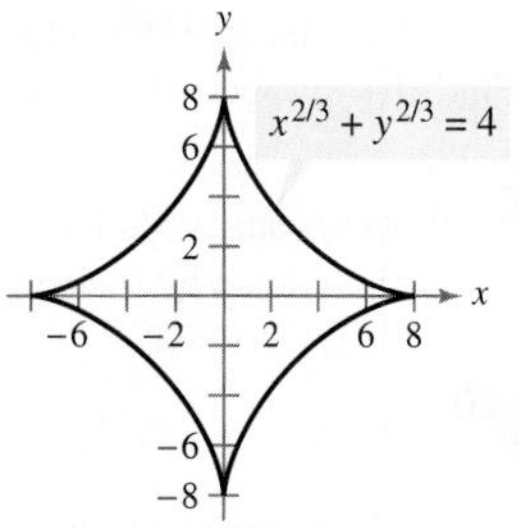

31. ***Think About It*** The figure shows the graphs of the functions $y_1 = x$, $y_2 = \frac{1}{2}x^{3/2}$, $y_3 = \frac{1}{4}x^2$, and $y_4 = \frac{1}{8}x^{5/2}$ on the interval $[0, 4]$. To print an enlarged copy of the graph, go to the website *www.mathgraphs.com.*

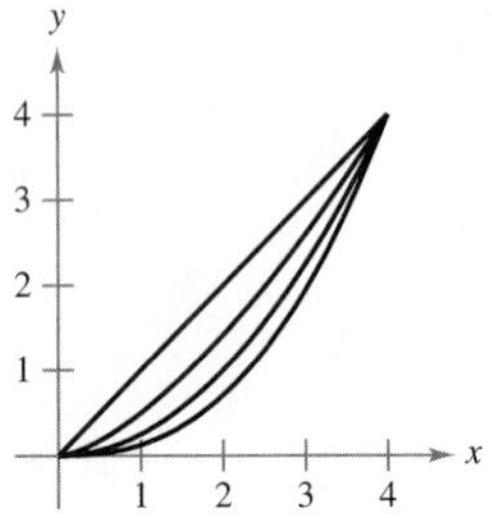

(a) Label the functions.

(b) List the functions in order of increasing arc length.

(c) Verify your answer in part (b) by approximating each arc length accurate to three decimal places.

32. ***Think About It*** Explain why the two integrals are equal.

$$\int_1^e \sqrt{1 + \frac{1}{x^2}}\, dx = \int_0^1 \sqrt{1 + e^{2x}}\, dx$$

Use the integration capabilities of a graphing utility to verify that the integrals are equal.

33. ***Length of Pursuit*** A fleeing object leaves the origin and moves up the y-axis (see figure). At the same time, a pursuer leaves the point (1, 0) and always moves toward the fleeing object. The pursuer's speed is twice that of the fleeing object. The equation of the path is modeled by

$$y = \frac{1}{3}(x^{3/2} - 3x^{1/2} + 2).$$

How far has the fleeing object traveled when it is caught? Show that the pursuer has traveled twice as far.

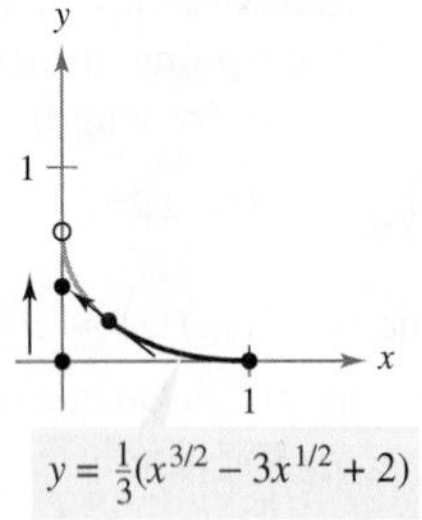

Figure for 33

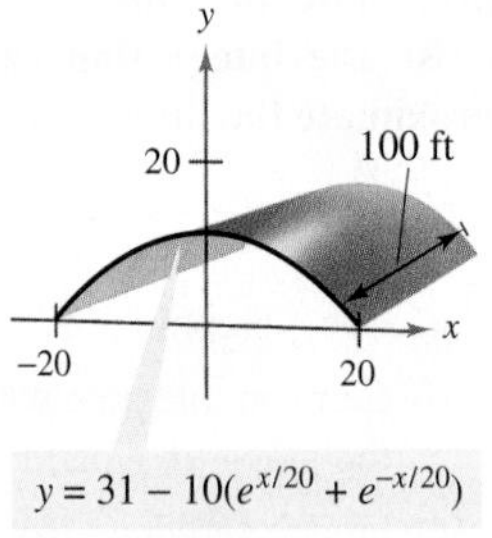

Figure for 34

34. ***Roof Area*** A barn is 100 feet long and 40 feet wide (see figure). A cross section of the roof is the inverted catenary $y = 31 - 10(e^{x/20} + e^{-x/20})$. Find the number of square feet of roofing on the barn.

35. ***Length of a Catenary*** Electrical wires suspended between two towers form a catenary (see figure) modeled by the equation

$$y = 20 \cosh \frac{x}{20}, \quad -20 \le x \le 20$$

where x and y are measured in meters. The towers are 40 meters apart. Find the length of the suspended cable.

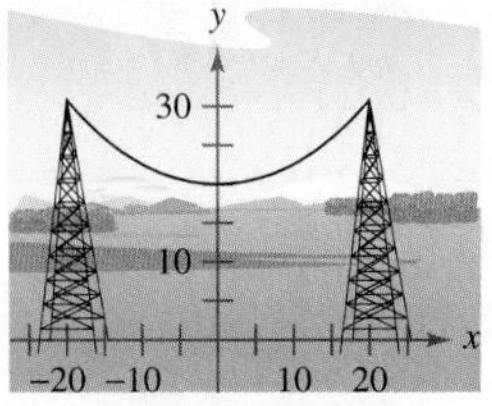

Figure for 35

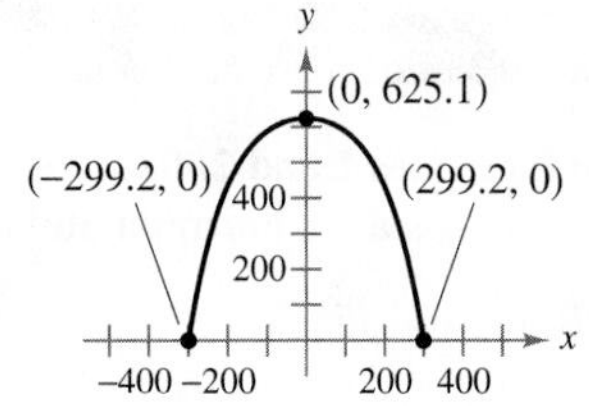

Figure for 36

36. ***Length of Gateway Arch*** The Gateway Arch in St. Louis, Missouri, is modeled by

$$y = 693.8597 - 68.7672 \cosh 0.0100333x,$$
$$-299.2239 \le x \le 299.2239.$$

(See Section 5.8, Section Project: St. Louis Arch.) Find the length of this curve (see figure).

37. Find the arc length from (0, 3) clockwise to $(2, \sqrt{5})$ along the circle $x^2 + y^2 = 9$.

38. Find the arc length from $(-3, 4)$ clockwise to (4, 3) along the circle $x^2 + y^2 = 25$. Show that the result is one-fourth the circumference of the circle.

In Exercises 39–42, set up and evaluate the definite integral for the area of the surface generated by revolving the curve about the x-axis.

39. $y = \dfrac{1}{3}x^3$

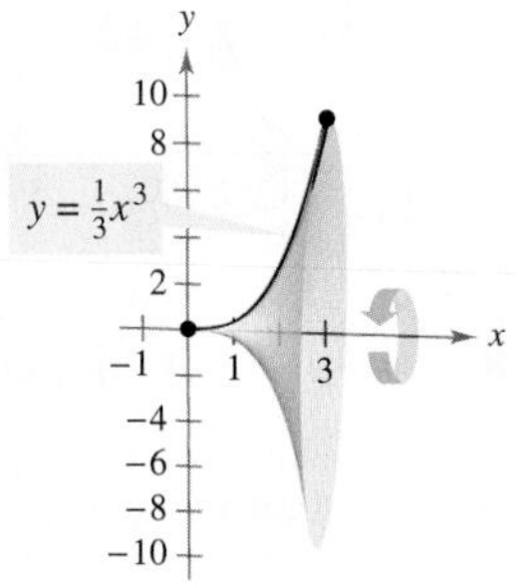

40. $y = 2\sqrt{x}$

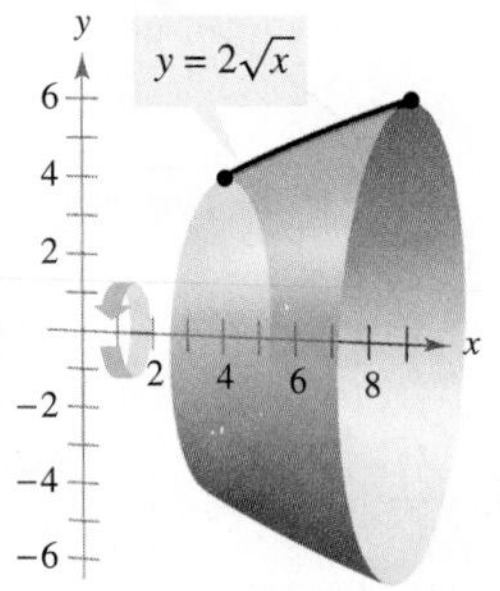

41. $y = \dfrac{x^3}{6} + \dfrac{1}{2x}, \quad 1 \le x \le 2$

42. $y = \dfrac{x}{2}, \quad 0 \le x \le 6$

In Exercises 43 and 44, set up and evaluate the definite integral for the area of the surface generated by revolving the curve about the y-axis.

43. $y = \sqrt[3]{x} + 2$

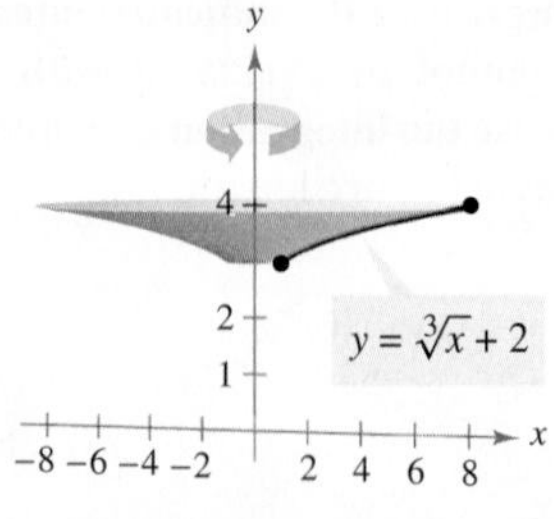

44. $y = 9 - x^2$

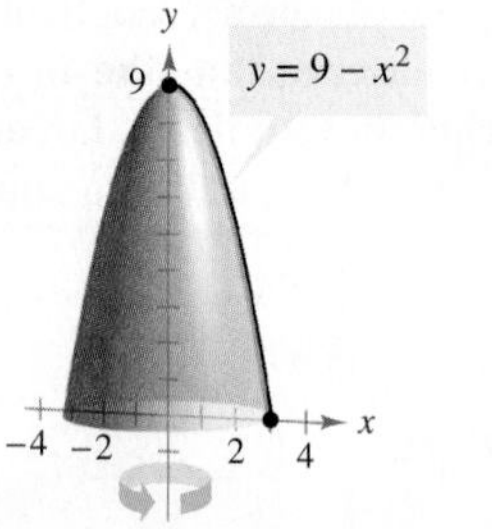

In Exercises 45 and 46, use the integration capabilities of a graphing utility to approximate the surface area of the solid of revolution.

Function	*Interval*
45. $y = \sin x$ revolved about the x-axis	$[0, \pi]$
46. $y = \ln x$ revolved about the y-axis	$[1, e]$

Writing About Concepts

47. Define a rectifiable curve.

48. What precalculus formula and representative element are used to develop the integration formula for arc length?

49. What precalculus formula and representative element are used to develop the integration formula for the area of a surface of revolution?

50. The graphs of the functions f_1 and f_2 on the interval $[a, b]$ are shown in the figure. The graph of each is revolved about the x-axis. Which surface of revolution has the greater surface area? Explain.

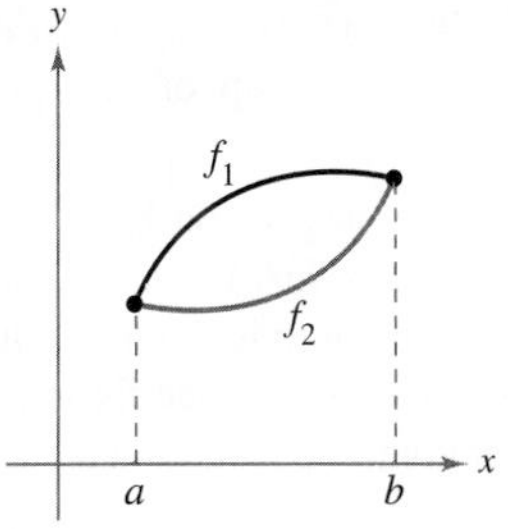

51. A right circular cone is generated by revolving the region bounded by $y = hx/r$, $y = h$, and $x = 0$ about the y-axis. Verify that the lateral surface area of the cone is

$$S = \pi r\sqrt{r^2 + h^2}.$$

52. A sphere of radius r is generated by revolving the graph of $y = \sqrt{r^2 - x^2}$ about the x-axis. Verify that the surface area of the sphere is $4\pi r^2$.

53. Find the area of the zone of a sphere formed by revolving the graph of $y = \sqrt{9 - x^2}$, $0 \le x \le 2$, about the y-axis.

54. Find the area of the zone of a sphere formed by revolving the graph of $y = \sqrt{r^2 - x^2}$, $0 \le x \le a$, about the y-axis. Assume that $a < r$.

55. *Bulb Design* An ornamental light bulb is designed by revolving the graph of

$$y = \tfrac{1}{3}x^{1/2} - x^{3/2}, \quad 0 \le x \le \tfrac{1}{3}$$

about the x-axis, where x and y are measured in feet (see figure). Find the surface area of the bulb and use the result to approximate the amount of glass needed to make the bulb. (Assume that the glass is 0.015 inch thick.)

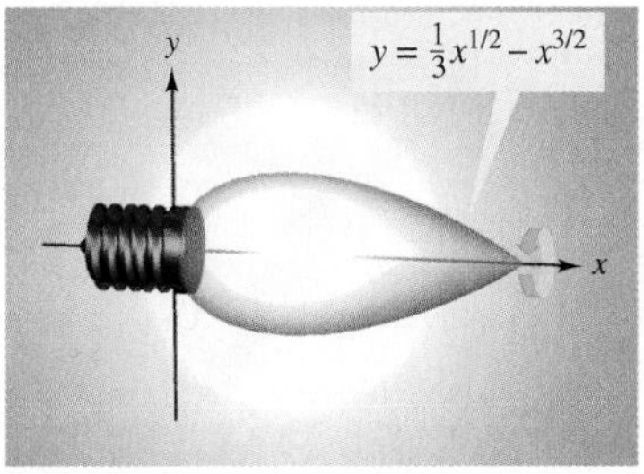

Figure for 55

56. *Think About It* Consider the equation $\dfrac{x^2}{9} + \dfrac{y^2}{4} = 1$.

(a) Use a graphing utility to graph the equation.

(b) Set up the definite integral for finding the first quadrant arc length of the graph in part (a).

(c) Compare the interval of integration in part (b) and the domain of the integrand. Is it possible to evaluate the definite integral? Is it possible to use Simpson's Rule to evaluate the definite integral? Explain. (You will learn how to evaluate this type of integral in Section 8.8.)

57. *Modeling Data* The circumference C (in inches) of a vase is measured at three-inch intervals starting at its base. The measurements are shown in the table, where y is the vertical distance in inches from the base.

y	0	3	6	9	12	15	18
C	50	65.5	70	66	58	51	48

(a) Use the data to approximate the volume of the vase by summing the volumes of approximating disks.

(b) Use the data to approximate the outside surface area (excluding the base) of the vase by summing the outside surface areas of approximating frustums of right circular cones.

(c) Use the regression capabilities of a graphing utility to find a cubic model for the points (y, r) where $r = C/(2\pi)$. Use the graphing utility to plot the points and graph the model.

(d) Use the model in part (c) and the integration capabilities of a graphing utility to approximate the volume and outside surface area of the vase. Compare the results with your answers in parts (a) and (b).

58. *Modeling Data* Property bounded by two perpendicular roads and a stream is shown in the figure on the next page. All distances are measured in feet.

(a) Use the regression capabilities of a graphing utility to fit a fourth-degree polynomial to the path of the stream.

(b) Use the model in part (a) to approximate the area of the property in acres.

(c) Use the integration capabilities of a graphing utility to find the length of the stream that bounds the property.

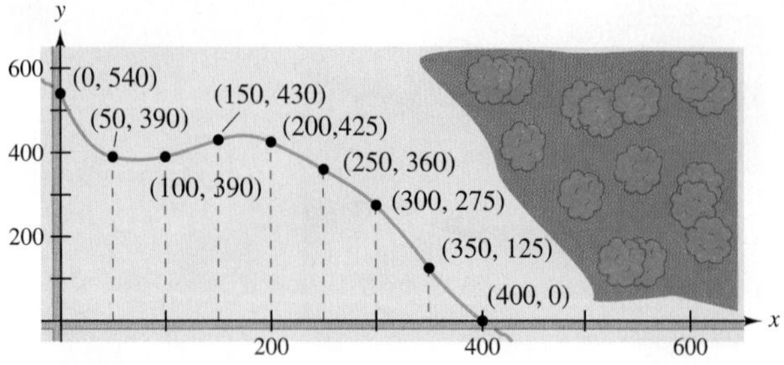

Figure for 58

59. Let R be the region bounded by $y = 1/x$, the x-axis, $x = 1$, and $x = b$, where $b > 1$. Let D be the solid formed when R is revolved about the x-axis.

(a) Find the volume V of D.

(b) Write the surface area S as an integral.

(c) Show that V approaches a finite limit as $b \to \infty$.

(d) Show that $S \to \infty$ as $b \to \infty$.

60. (a) Given a circular sector with radius L and central angle θ (see figure), show that the area of the sector is given by

$$S = \frac{1}{2}L^2\theta.$$

(b) By joining the straight line edges of the sector in part (a), a right circular cone is formed (see figure) and the lateral surface area of the cone is the same as the area of the sector. Show that the area is $S = \pi rL$, where r is the radius of the base of the cone. (*Hint:* The arc length of the sector equals the circumference of the base of the cone.)

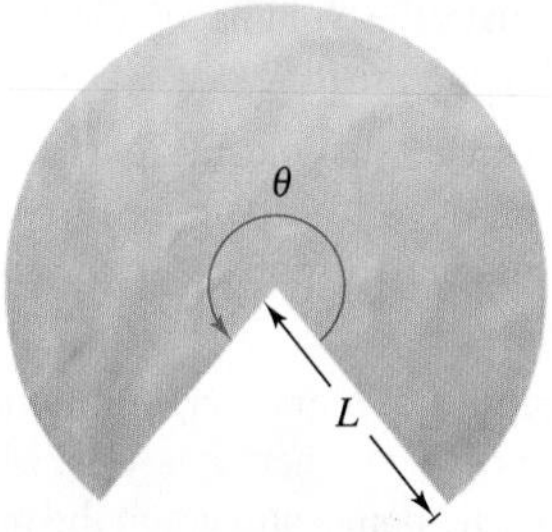

Figure for 60(a)

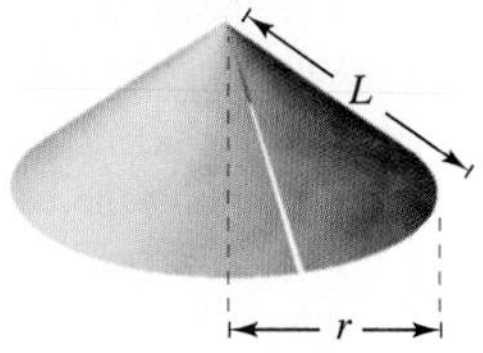

Figure for 60(b)

(c) Use the result of part (b) to verify that the formula for the lateral surface area of the frustum of a cone with slant height L and radii r_1 and r_2 (see figure) is $S = \pi(r_1 + r_2)L$. (*Note:* This formula was used to develop the integral for finding the surface area of a surface of revolution.)

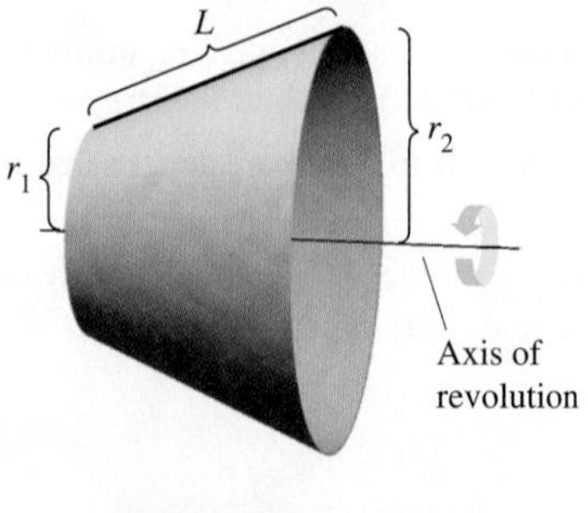

61. ***Individual Project*** Select a solid of revolution from everyday life. Measure the radius of the solid at a minimum of seven points along its axis. Use the data to approximate the volume of the solid and the surface area of the lateral sides of the solid.

62. ***Writing*** Read the article "Arc Length, Area and the Arcsine Function" by Andrew M. Rockett in *Mathematics Magazine*. Then write a paragraph explaining how the arcsine function can be defined in terms of an arc length. (To view this article, go to the website *www.matharticles.com*.)

63. ***Astroid*** Find the area of the surface formed by revolving the portion in the first quadrant of the graph of $x^{2/3} + y^{2/3} = 4$, $0 \le y \le 8$ about the y-axis.

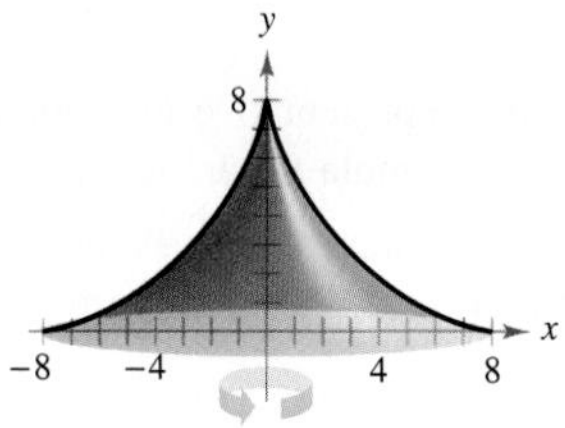

Figure for 63

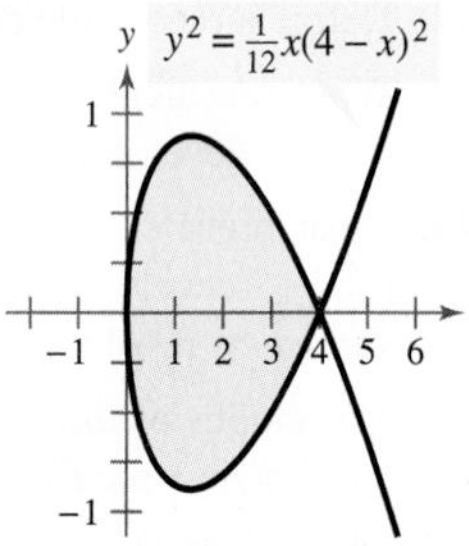

Figure for 64

64. Consider the graph of $y^2 = \frac{1}{12}x(4 - x)^2$ (see figure). Find the area of the surface formed when the loop of this graph is revolved around the x-axis.

65. ***Suspension Bridge*** A cable for a suspension bridge has the shape of a parabola with equation $y = kx^2$. Let h represent the height of the cable from its lowest point to its highest point and let $2w$ represent the total span of the bridge (see figure). Show that the length C of the cable is given by

$$C = 2\int_0^w \sqrt{1 + \frac{4h^2}{w^4}x^2}\,dx.$$

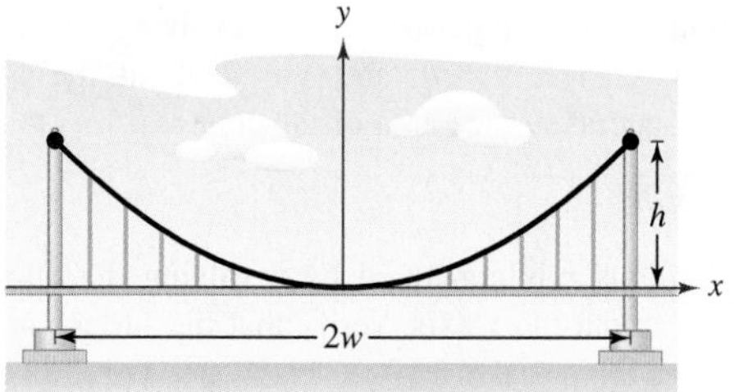

66. ***Suspension Bridge*** The Humber Bridge, located in the United Kingdom and opened in 1981, has a main span of about 1400 meters. Each of its towers has a height of about 155 meters. Use these dimensions, the integral in Exercise 65, and the integration capabilities of a graphing utility to approximate the length of a parabolic cable along the main span.

Putnam Exam Challenge

67. Find the length of the curve $y^2 = x^3$ from the origin to the point where the tangent makes an angle of 45° with the x-axis.

This problem was composed by the Committee on the Putnam Prize Competition.

Section 7.5 Work

- **Find the work done by a constant force.**
- **Find the work done by a variable force.**

Work Done by a Constant Force

The concept of work is important to scientists and engineers for determining the energy needed to perform various jobs. For instance, it is useful to know the amount of work done when a crane lifts a steel girder, when a spring is compressed, when a rocket is propelled into the air, or when a truck pulls a load along a highway.

In general, **work** is done by a force when it moves an object. If the force applied to the object is *constant*, then the definition of work is as follows.

Definition of Work Done by a Constant Force

If an object is moved a distance D in the direction of an applied constant force F, then the **work** W done by the force is defined as $W = FD$.

There are many types of forces—centrifugal, electromotive, and gravitational, to name a few. A **force** can be thought of as a *push* or a *pull*; a force changes the state of rest or state of motion of a body. For gravitational forces on Earth, it is common to use units of measure corresponding to the weight of an object.

EXAMPLE 1 Lifting an Object

Determine the work done in lifting a 50-pound object 4 feet.

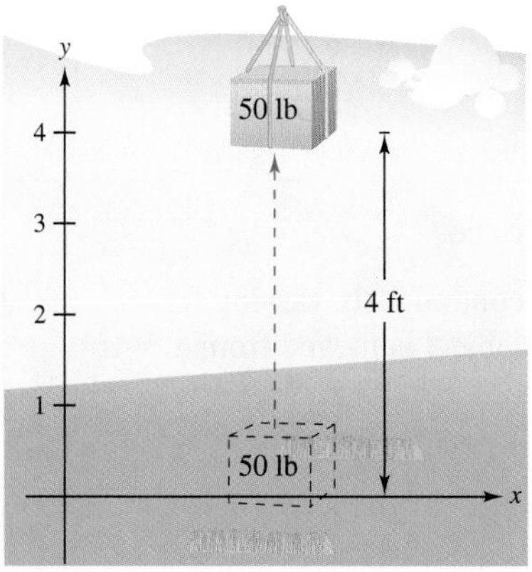

The work done in lifting a 50-pound object 4 feet is 200 foot-pounds.
Figure 7.48

Solution The magnitude of the required force F is the weight of the object, as shown in Figure 7.48. So, the work done in lifting the object 4 feet is

$$\begin{aligned} W &= FD && \text{Work} = \text{(force)(distance)} \\ &= 50(4) && \text{Force} = 50 \text{ pounds, distance} = 4 \text{ feet} \\ &= 200 \text{ foot-pounds.} \end{aligned}$$

In the U.S. measurement system, work is typically expressed in foot-pounds (ft-lb), inch-pounds, or foot-tons. In the centimeter-gram-second (C-G-S) system, the basic unit of force is the **dyne**—the force required to produce an acceleration of 1 centimeter per second per second on a mass of 1 gram. In this system, work is typically expressed in dyne-centimeters (ergs) or newton-meters (joules), where 1 joule $= 10^7$ ergs.

EXPLORATION

How Much Work? In Example 1, 200 foot-pounds of work was needed to lift the 50-pound object 4 feet vertically off the ground. Suppose that once you lifted the object, you held it and walked a horizontal distance of 4 feet. Would this require an additional 200 foot-pounds of work? Explain your reasoning.

Work Done by a Variable Force

In Example 1, the force involved was *constant*. If a *variable* force is applied to an object, calculus is needed to determine the work done, because the amount of force changes as the object changes position. For instance, the force required to compress a spring increases as the spring is compressed.

Suppose that an object is moved along a straight line from $x = a$ to $x = b$ by a continuously varying force $F(x)$. Let Δ be a partition that divides the interval $[a, b]$ into n subintervals determined by

$$a = x_0 < x_1 < x_2 < \cdot \cdot \cdot < x_n = b$$

and let $\Delta x_i = x_i - x_{i-1}$. For each i, choose c_i such that $x_{i-1} \le c_i \le x_i$. Then at c_i the force is given by $F(c_i)$. Because F is continuous, you can approximate the work done in moving the object through the ith subinterval by the increment

$$\Delta W_i = F(c_i)\,\Delta x_i$$

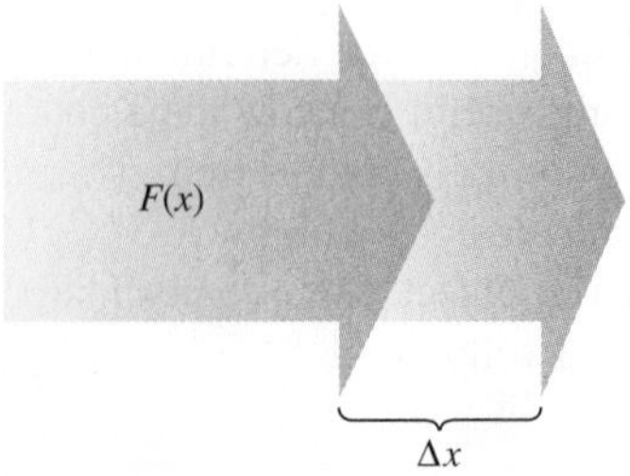

The amount of force changes as an object changes position (Δx).
Figure 7.49

as shown in Figure 7.49. So, the total work done as the object moves from a to b is approximated by

$$\begin{aligned} W &\approx \sum_{i=1}^{n} \Delta W_i \\ &= \sum_{i=1}^{n} F(c_i)\,\Delta x_i. \end{aligned}$$

This approximation appears to become better and better as $\|\Delta\| \to 0$ $(n \to \infty)$. So, the work done is

$$\begin{aligned} W &= \lim_{\|\Delta\|\to 0} \sum_{i=1}^{n} F(c_i)\,\Delta x_i \\ &= \int_a^b F(x)\,dx. \end{aligned}$$

Definition of Work Done by a Variable Force

If an object is moved along a straight line by a continuously varying force $F(x)$, then the **work** W done by the force as the object is moved from $x = a$ to $x = b$ is

$$\begin{aligned} W &= \lim_{\|\Delta\|\to 0} \sum_{i=1}^{n} \Delta W_i \\ &= \int_a^b F(x)\,dx. \end{aligned}$$

Bettmann/Corbis

EMILIE DE BRETEUIL (1706–1749)
Another major work by de Breteuil was the translation of Newton's "Philosophiae Naturalis Principia Mathematica" into French. Her translation and commentary greatly contributed to the acceptance of Newtonian science in Europe.

The remaining examples in this section use some well-known physical laws. The discoveries of many of these laws occurred during the same period in which calculus was being developed. In fact, during the seventeenth and eighteenth centuries, there was little difference between physicists and mathematicians. One such physicist-mathematician was Emilie de Breteuil. Breteuil was instrumental in synthesizing the work of many other scientists, including Newton, Leibniz, Huygens, Kepler, and Descartes. Her physics text *Institutions* was widely used for many years.

The following three laws of physics were developed by Robert Hooke (1635–1703), Isaac Newton (1642–1727), and Charles Coulomb (1736–1806).

1. **Hooke's Law:** The force F required to compress or stretch a spring (within its elastic limits) is proportional to the distance d that the spring is compressed or stretched from its original length. That is,

$$F = kd$$

where the constant of proportionality k (the spring constant) depends on the specific nature of the spring.

2. **Newton's Law of Universal Gravitation:** The force F of attraction between two particles of masses m_1 and m_2 is proportional to the product of the masses and inversely proportional to the square of the distance d between the two particles. That is,

$$F = k\frac{m_1 m_2}{d^2}.$$

If m_1 and m_2 are given in grams and d in centimeters, F will be in dynes for a value of $k = 6.670 \times 10^{-8}$ cubic centimeter per gram-second squared.

3. **Coulomb's Law:** The force between two charges q_1 and q_2 in a vacuum is proportional to the product of the charges and inversely proportional to the square of the distance d between the two charges. That is,

$$F = k\frac{q_1 q_2}{d^2}.$$

If q_1 and q_2 are given in electrostatic units and d in centimeters, F will be in dynes for a value of $k = 1$.

EXPLORATION

The work done in compressing the spring in Example 2 from $x = 3$ inches to $x = 6$ inches is 3375 inch-pounds. Should the work done in compressing the spring from $x = 0$ inches to $x = 3$ inches be more than, the same as, or less than this? Explain.

EXAMPLE 2 Compressing a Spring

A force of 750 pounds compresses a spring 3 inches from its natural length of 15 inches. Find the work done in compressing the spring an additional 3 inches.

Solution By Hooke's Law, the force $F(x)$ required to compress the spring x units (from its natural length) is $F(x) = kx$. Using the given data, it follows that $F(3) = 750 = (k)(3)$ and so $k = 250$ and $F(x) = 250x$, as shown in Figure 7.50. To find the increment of work, assume that the force required to compress the spring over a small increment Δx is nearly constant. So, the increment of work is

$$\Delta W = (\text{force})(\text{distance increment}) = (250x)\,\Delta x.$$

Because the spring is compressed from $x = 3$ to $x = 6$ inches less than its natural length, the work required is

$$W = \int_a^b F(x)\,dx = \int_3^6 250x\,dx \qquad \text{Formula for work}$$

$$= 125x^2\Big]_3^6 = 4500 - 1125 = 3375 \text{ inch-pounds.}$$

Note that you do *not* integrate from $x = 0$ to $x = 6$ because you were asked to determine the work done in compressing the spring an *additional* 3 inches (not including the first 3 inches).

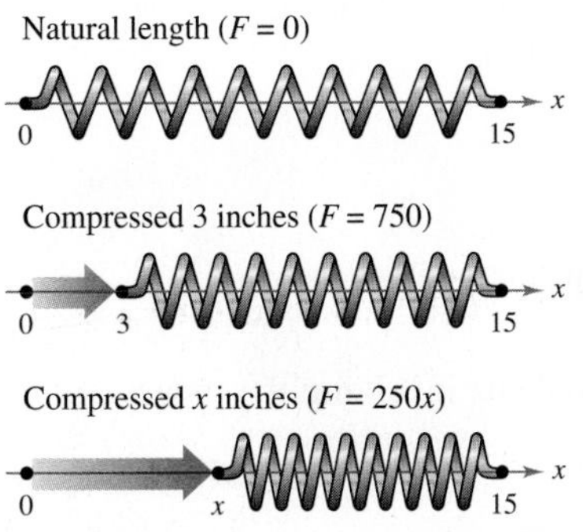

Figure 7.50

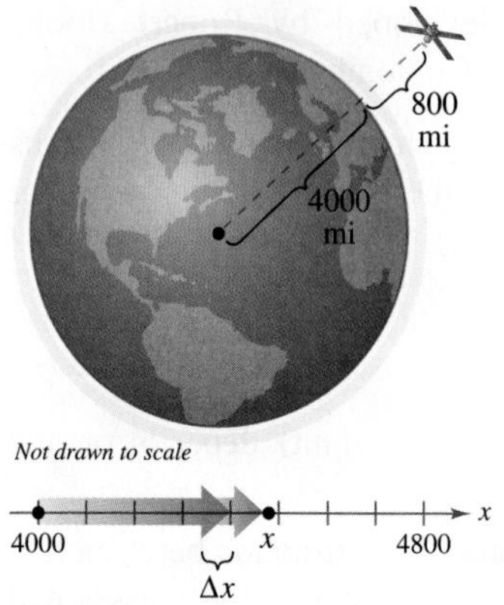

Figure 7.51

EXAMPLE 3 Moving a Space Module into Orbit

A space module weighs 15 metric tons on the surface of Earth. How much work is done in propelling the module to a height of 800 miles above Earth, as shown in Figure 7.51? (Use 4000 miles as the radius of Earth. Do not consider the effect of air resistance or the weight of the propellant.)

Solution Because the weight of a body varies inversely as the square of its distance from the center of Earth, the force $F(x)$ exerted by gravity is

$$F(x) = \frac{C}{x^2}. \qquad \text{C is the constant of proportionality.}$$

Because the module weighs 15 metric tons on the surface of Earth and the radius of Earth is approximately 4000 miles, you have

$$15 = \frac{C}{(4000)^2}$$
$$240{,}000{,}000 = C.$$

So, the increment of work is

$$\begin{aligned}\Delta W &= (\text{force})(\text{distance increment})\\ &= \frac{240{,}000{,}000}{x^2}\Delta x.\end{aligned}$$

Finally, because the module is propelled from $x = 4000$ to $x = 4800$ miles, the total work done is

$$\begin{aligned}W &= \int_a^b F(x)\,dx = \int_{4000}^{4800} \frac{240{,}000{,}000}{x^2}\,dx && \text{Formula for work}\\ &= \frac{-240{,}000{,}000}{x}\Bigg]_{4000}^{4800} && \text{Integrate.}\\ &= -50{,}000 + 60{,}000\\ &= 10{,}000 \text{ mile-tons}\\ &\approx 1.164 \times 10^{11} \text{ foot-pounds.}\end{aligned}$$

In the C-G-S system, using a conversion factor of 1 foot-pound $\approx$ 1.35582 joules, the work done is

$$W \approx 1.578 \times 10^{11} \text{ joules.}$$

The solutions to Examples 2 and 3 conform to our development of work as the summation of increments in the form

$$\Delta W = (\text{force})(\text{distance increment}) = (F)(\Delta x).$$

Another way to formulate the increment of work is

$$\Delta W = (\text{force increment})(\text{distance}) = (\Delta F)(x).$$

This second interpretation of ΔW is useful in problems involving the movement of nonrigid substances such as fluids and chains.

EXAMPLE 4 Emptying a Tank of Oil

A spherical tank of radius 8 feet is half full of oil that weighs 50 pounds per cubic foot. Find the work required to pump oil out through a hole in the top of the tank.

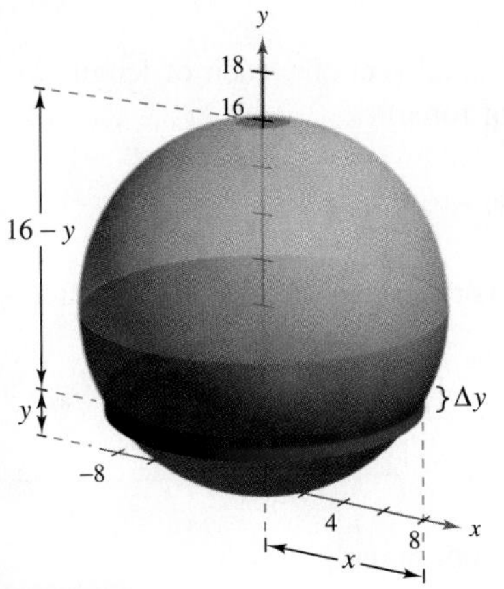

Figure 7.52

Solution Consider the oil to be subdivided into disks of thickness Δy and radius x, as shown in Figure 7.52. Because the increment of force for each disk is given by its weight, you have

$$\begin{aligned}\Delta F &= \text{weight}\\ &= \left(\frac{50 \text{ pounds}}{\text{cubic foot}}\right)(\text{volume})\\ &= 50(\pi x^2 \Delta y) \text{ pounds.}\end{aligned}$$

For a circle of radius 8 and center at (0, 8), you have

$$\begin{aligned}x^2 + (y-8)^2 &= 8^2\\ x^2 &= 16y - y^2\end{aligned}$$

and you can write the force increment as

$$\begin{aligned}\Delta F &= 50(\pi x^2 \Delta y)\\ &= 50\pi(16y - y^2)\,\Delta y.\end{aligned}$$

In Figure 7.52, note that a disk y feet from the bottom of the tank must be moved a distance of $(16 - y)$ feet. So, the increment of work is

$$\begin{aligned}\Delta W &= \Delta F(16-y)\\ &= 50\pi(16y - y^2)\,\Delta y(16-y)\\ &= 50\pi(256y - 32y^2 + y^3)\,\Delta y.\end{aligned}$$

Because the tank is half full, y ranges from 0 to 8, and the work required to empty the tank is

$$\begin{aligned}W &= \int_0^8 50\pi(256y - 32y^2 + y^3)\,dy\\ &= 50\pi\left[128y^2 - \frac{32}{3}y^3 + \frac{y^4}{4}\right]_0^8\\ &= 50\pi\left(\frac{11{,}264}{3}\right)\\ &\approx 589{,}782 \text{ foot-pounds.}\end{aligned}$$

To estimate the reasonableness of the result in Example 4, consider that the weight of the oil in the tank is

$$\left(\frac{1}{2}\right)(\text{volume})(\text{density}) = \frac{1}{2}\left(\frac{4}{3}\pi 8^3\right)(50)$$
$$\approx 53{,}616.5 \text{ pounds.}$$

Lifting the entire half-tank of oil 8 feet would involve work of $8(53{,}616.5) \approx 428{,}932$ foot-pounds. Because the oil is actually lifted between 8 and 16 feet, it seems reasonable that the work done is 589,782 foot-pounds.

Work required to raise one end of the chain
Figure 7.53

EXAMPLE 5 Lifting a Chain

A 20-foot chain weighing 5 pounds per foot is lying coiled on the ground. How much work is required to raise one end of the chain to a height of 20 feet so that it is fully extended, as shown in Figure 7.53?

Solution Imagine that the chain is divided into small sections, each of length Δy. Then the weight of each section is the increment of force

$$\Delta F = (\text{weight}) = \left(\frac{5 \text{ pounds}}{\text{foot}}\right)(\text{length}) = 5\Delta y.$$

Because a typical section (initially on the ground) is raised to a height of y, the increment of work is

$$\Delta W = (\text{force increment})(\text{distance}) = (5\ \Delta y)y = 5y\ \Delta y.$$

Because y ranges from 0 to 20, the total work is

$$W = \int_0^{20} 5y\, dy = \frac{5y^2}{2}\bigg]_0^{20} = \frac{5(400)}{2} = 1000 \text{ foot-pounds.}$$

In the next example you will consider a piston of radius r in a cylindrical casing, as shown in Figure 7.54. As the gas in the cylinder expands, the piston moves and work is done. If p represents the pressure of the gas (in pounds per square foot) against the piston head and V represents the volume of the gas (in cubic feet), the work increment involved in moving the piston Δx feet is

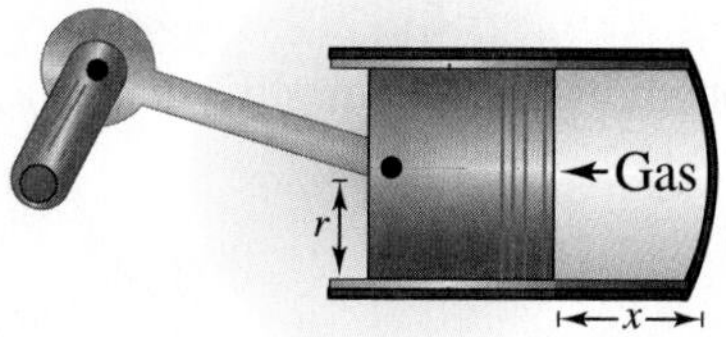

Work done by expanding gas
Figure 7.54

$$\Delta W = (\text{force})(\text{distance increment}) = F(\Delta x) = p(\pi r^2)\,\Delta x = p\ \Delta V.$$

So, as the volume of the gas expands from V_0 to V_1, the work done in moving the piston is

$$W = \int_{V_0}^{V_1} p\, dV.$$

Assuming the pressure of the gas to be inversely proportional to its volume, you have $p = k/V$ and the integral for work becomes

$$W = \int_{V_0}^{V_1} \frac{k}{V}\, dV.$$

EXAMPLE 6 Work Done by an Expanding Gas

A quantity of gas with an initial volume of 1 cubic foot and a pressure of 500 pounds per square foot expands to a volume of 2 cubic feet. Find the work done by the gas. (Assume that the pressure is inversely proportional to the volume.)

Solution Because $p = k/V$ and $p = 500$ when $V = 1$, you have $k = 500$. So, the work is

$$\begin{aligned} W &= \int_{V_0}^{V_1} \frac{k}{V}\, dV \\ &= \int_1^2 \frac{500}{V}\, dV \\ &= 500 \ln|V|\bigg]_1^2 \approx 346.6 \text{ foot-pounds.} \end{aligned}$$

Exercises for Section 7.5

See www.CalcChat.com for worked-out solutions to odd-numbered exercises.

Constant Force **In Exercises 1–4, determine the work done by the constant force.**

1. A 100-pound bag of sugar is lifted 10 feet.

2. An electric hoist lifts a 2800-pound car 4 feet.

3. A force of 112 newtons is required to slide a cement block 4 meters in a construction project.

4. The locomotive of a freight train pulls its cars with a constant force of 9 tons a distance of one-half mile.

Writing About Concepts

5. State the definition of work done by a constant force.

6. State the definition of work done by a variable force.

7. The graphs show the force F_i (in pounds) required to move an object 9 feet along the x-axis. Order the force functions from the one that yields the least work to the one that yields the most work without doing any calculations. Explain your reasoning.

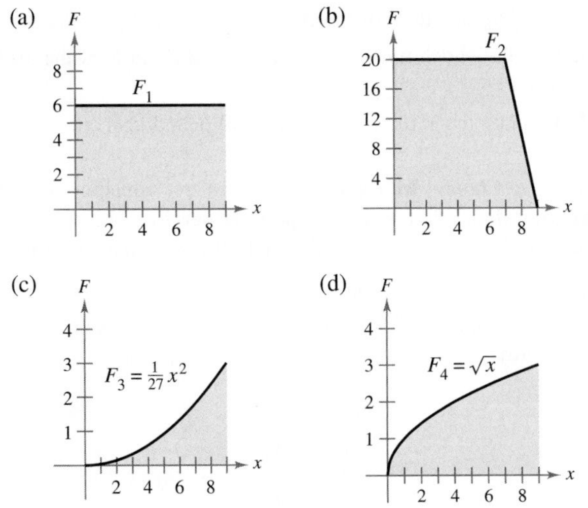

8. Verify your answer to Exercise 7 by calculating the work for each force function.

Hooke's Law **In Exercises 9–16, use Hooke's Law to determine the variable force in the spring problem.**

9. A force of 5 pounds compresses a 15-inch spring a total of 4 inches. How much work is done in compressing the spring 7 inches?

10. How much work is done in compressing the spring in Exercise 9 from a length of 10 inches to a length of 6 inches?

11. A force of 250 newtons stretches a spring 30 centimeters. How much work is done in stretching the spring from 20 centimeters to 50 centimeters?

12. A force of 800 newtons stretches a spring 70 centimeters on a mechanical device for driving fence posts. Find the work done in stretching the spring the required 70 centimeters.

13. A force of 20 pounds stretches a spring 9 inches in an exercise machine. Find the work done in stretching the spring 1 foot from its natural position.

14. An overhead garage door has two springs, one on each side of the door. A force of 15 pounds is required to stretch each spring 1 foot. Because of the pulley system, the springs stretch only one-half the distance the door travels. The door moves a total of 8 feet and the springs are at their natural length when the door is open. Find the work done by the pair of springs.

15. Eighteen foot-pounds of work is required to stretch a spring 4 inches from its natural length. Find the work required to stretch the spring an additional 3 inches.

16. Seven and one-half foot-pounds of work is required to compress a spring 2 inches from its natural length. Find the work required to compress the spring an additional one-half inch.

17. ***Propulsion*** Neglecting air resistance and the weight of the propellant, determine the work done in propelling a five-ton satellite to a height of

(a) 100 miles above Earth.

(b) 300 miles above Earth.

18. ***Propulsion*** Use the information in Exercise 17 to write the work W of the propulsion system as a function of the height h of the satellite above Earth. Find the limit (if it exists) of W as h approaches infinity.

19. ***Propulsion*** Neglecting air resistance and the weight of the propellant, determine the work done in propelling a 10-ton satellite to a height of

(a) 11,000 miles above Earth.

(b) 22,000 miles above Earth.

20. ***Propulsion*** A lunar module weighs 12 tons on the surface of Earth. How much work is done in propelling the module from the surface of the moon to a height of 50 miles? Consider the radius of the moon to be 1100 miles and its force of gravity to be one-sixth that of Earth.

21. ***Pumping Water*** A rectangular tank with a base 4 feet by 5 feet and a height of 4 feet is full of water (see figure). The water weighs 62.4 pounds per cubic foot. How much work is done in pumping water out over the top edge in order to empty (a) half of the tank? (b) all of the tank?

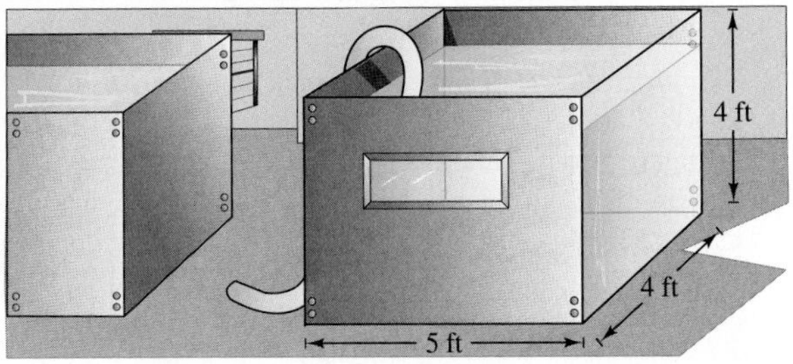

22. ***Think About It*** Explain why the answer in part (b) of Exercise 21 is not twice the answer in part (a).

23. ***Pumping Water*** A cylindrical water tank 4 meters high with a radius of 2 meters is buried so that the top of the tank is 1 meter below ground level (see figure). How much work is done in pumping a full tank of water up to ground level? (The water weighs 9800 newtons per cubic meter.)

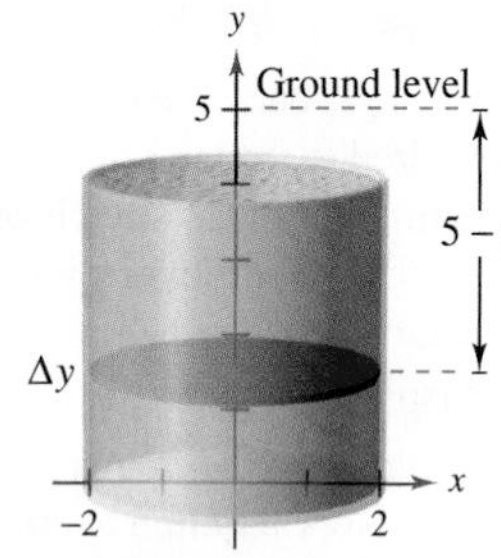

Figure for 23

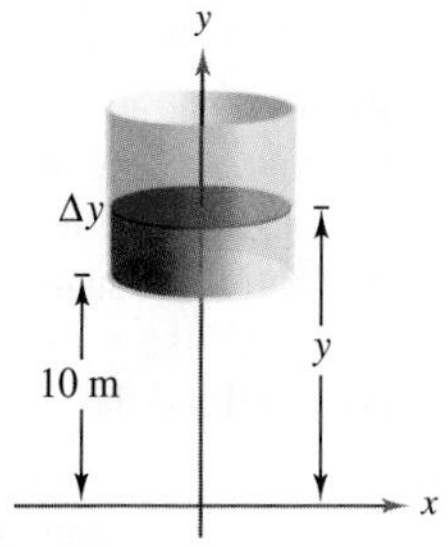

Figure for 24

24. ***Pumping Water*** Suppose the tank in Exercise 23 is located on a tower so that the bottom of the tank is 10 meters above the level of a stream (see figure). How much work is done in filling the tank half full of water through a hole in the bottom, using water from the stream?

25. ***Pumping Water*** An open tank has the shape of a right circular cone (see figure). The tank is 8 feet across the top and 6 feet high. How much work is done in emptying the tank by pumping the water over the top edge?

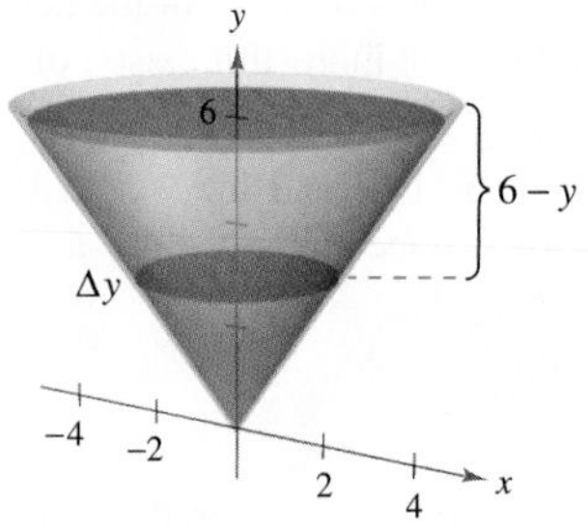

Figure for 25

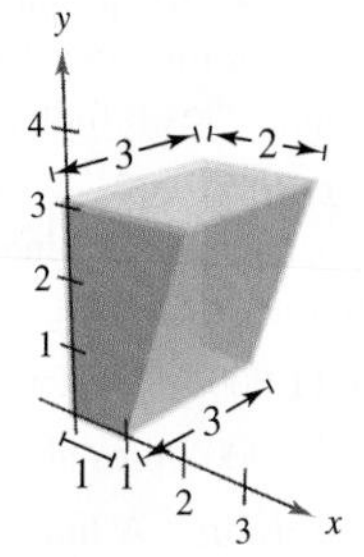

Figure for 28

26. ***Pumping Water*** Water is pumped in through the bottom of the tank in Exercise 25. How much work is done to fill the tank

(a) to a depth of 2 feet?

(b) from a depth of 4 feet to a depth of 6 feet?

27. ***Pumping Water*** A hemispherical tank of radius 6 feet is positioned so that its base is circular. How much work is required to fill the tank with water through a hole in the base if the water source is at the base?

28. ***Pumping Diesel Fuel*** The fuel tank on a truck has trapezoidal cross sections with dimensions (in feet) shown in the figure. Assume that an engine is approximately 3 feet above the top of the fuel tank and that diesel fuel weighs approximately 53.1 pounds per cubic foot. Find the work done by the fuel pump in raising a full tank of fuel to the level of the engine.

Pumping Gasoline **In Exercises 29 and 30, find the work done in pumping gasoline that weighs 42 pounds per cubic foot. (*Hint:* Evaluate one integral by a geometric formula and the other by observing that the integrand is an odd function.)**

29. A cylindrical gasoline tank 3 feet in diameter and 4 feet long is carried on the back of a truck and is used to fuel tractors. The axis of the tank is horizontal. The opening on the tractor tank is 5 feet above the top of the tank in the truck. Find the work done in pumping the entire contents of the fuel tank into a tractor.

30. The top of a cylindrical storage tank for gasoline at a service station is 4 feet below ground level. The axis of the tank is horizontal and its diameter and length are 5 feet and 12 feet, respectively. Find the work done in pumping the entire contents of the full tank to a height of 3 feet above ground level.

Lifting a Chain **In Exercises 31–34, consider a 15-foot chain that weighs 3 pounds per foot hanging from a winch 15 feet above ground level. Find the work done by the winch in winding up the specified amount of chain.**

31. Wind up the entire chain.

32. Wind up one-third of the chain.

33. Run the winch until the bottom of the chain is at the 10-foot level.

34. Wind up the entire chain with a 500-pound load attached to it.

Lifting a Chain **In Exercises 35 and 36, consider a 15-foot hanging chain that weighs 3 pounds per foot. Find the work done in lifting the chain vertically to the indicated position.**

35. Take the bottom of the chain and raise it to the 15-foot level, leaving the chain doubled and still hanging vertically (see figure).

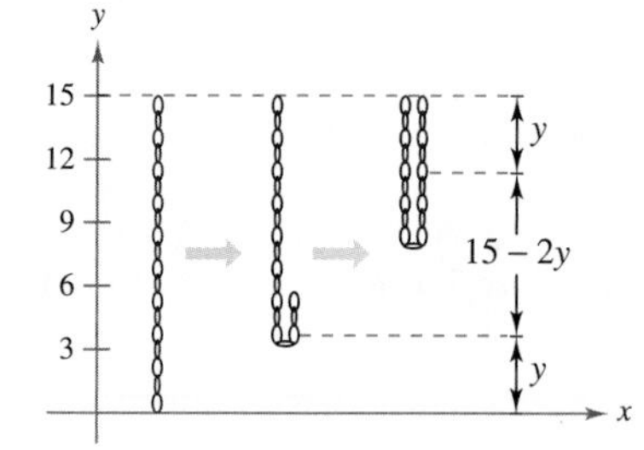

36. Repeat Exercise 35 raising the bottom of the chain to the 12-foot level.

Demolition Crane **In Exercises 37 and 38, consider a demolition crane with a 500-pound ball suspended from a 40-foot cable that weighs 1 pound per foot.**

37. Find the work required to wind up 15 feet of the apparatus.

38. Find the work required to wind up all 40 feet of the apparatus.

Boyle's Law **In Exercises 39 and 40, find the work done by the gas for the given volume and pressure. Assume that the pressure is inversely proportional to the volume. (See Example 6.)**

39. A quantity of gas with an initial volume of 2 cubic feet and a pressure of 1000 pounds per square foot expands to a volume of 3 cubic feet.

40. A quantity of gas with an initial volume of 1 cubic foot and a pressure of 2500 pounds per square foot expands to a volume of 3 cubic feet.

41. ***Electric Force*** Two electrons repel each other with a force that varies inversely as the square of the distance between them. One electron is fixed at the point (2, 4). Find the work done in moving the second electron from (−2, 4) to (1, 4).

42. ***Modeling Data*** The hydraulic cylinder on a woodsplitter has a four-inch bore (diameter) and a stroke of 2 feet. The hydraulic pump creates a maximum pressure of 2000 pounds per square inch. Therefore, the maximum force created by the cylinder is $2000(\pi 2^2) = 8000\pi$ pounds.

(a) Find the work done through one extension of the cylinder given that the maximum force is required.

(b) The force exerted in splitting a piece of wood is variable. Measurements of the force obtained when a piece of wood was split are shown in the table. The variable x measures the extension of the cylinder in feet, and F is the force in pounds. Use Simpson's Rule to approximate the work done in splitting the piece of wood.

x	0	$\frac{1}{3}$	$\frac{2}{3}$	1	$\frac{4}{3}$	$\frac{5}{3}$	2
$F(x)$	0	20,000	22,000	15,000	10,000	5000	0

Table for 42(b)

(c) Use the regression capabilities of a graphing utility to find a fourth-degree polynomial model for the data. Plot the data and graph the model.

(d) Use the model in part (c) to approximate the extension of the cylinder when the force is maximum.

(e) Use the model in part (c) to approximate the work done in splitting the piece of wood.

Hydraulic Press **In Exercises 43–46, use the integration capabilities of a graphing utility to approximate the work done by a press in a manufacturing process. A model for the variable force F (in pounds) and the distance x (in feet) the press moves is given.**

	Force	*Interval*
43.	$F(x) = 1000[1.8 - \ln(x + 1)]$	$0 \le x \le 5$
44.	$F(x) = \dfrac{e^{x^2} - 1}{100}$	$0 \le x \le 4$
45.	$F(x) = 100x\sqrt{125 - x^3}$	$0 \le x \le 5$
46.	$F(x) = 1000 \sinh x$	$0 \le x \le 2$

Section Project: Tidal Energy

Tidal power plants use "tidal energy" to produce electrical energy. To construct a tidal power plant, a dam is built to separate a basin from the sea. Electrical energy is produced as the water flows back and forth between the basin and the sea. The amount of "natural energy" produced depends on the volume of the basin and the tidal range—the vertical distance between high and low tides. (Several natural basin have tidal ranges in excess of 15 feet; the Bay of Fundy in Nova Scotia has a tidal range of 53 feet.)

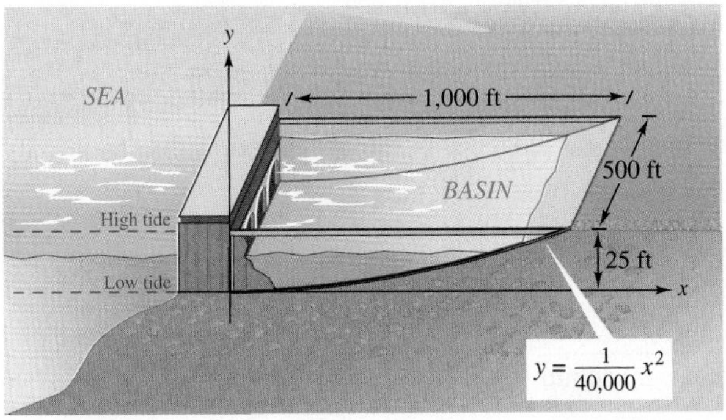

(a) Consider a basin with a rectangular base, as shown in the figure. The basin has a tidal range of 25 feet, with low tide corresponding to $y = 0$. How much water does the basin hold at high tide?

(b) The amount of energy produced during the filling (or the emptying) of the basin is proportional to the amount of work required to fill (or empty) the basin. How much work is required to fill the basin with seawater? (Use a seawater density of 64 pounds per cubic foot.)

Francois Gohier/Photo Researchers, Inc.

Francois Gohier/Photo Researchers, Inc.

The Bay of Fundy in Nova Scotia has an extreme tidal range, as displayed in the greatly contrasting photos above.

FOR FURTHER INFORMATION For more information on tidal power, see the article "LaRance: Six Years of Operating a Tidal Power Plant in France" by J. Cotillon in *Water Power Magazine*.

Section 7.6 Moments, Centers of Mass, and Centroids

- **Understand the definition of mass.**
- **Find the center of mass in a one-dimensional system.**
- **Find the center of mass in a two-dimensional system.**
- **Find the center of mass of a planar lamina.**
- **Use the Theorem of Pappus to find the volume of a solid of revolution.**

Mass

In this section you will study several important applications of integration that are related to **mass.** Mass is a measure of a body's resistance to changes in motion, and is independent of the particular gravitational system in which the body is located. However, because so many applications involving mass occur on Earth's surface, an object's mass is sometimes equated with its weight. This is not technically correct. Weight is a type of force and as such is dependent on gravity. Force and mass are related by the equation

$$\text{Force} = (\text{mass})(\text{acceleration}).$$

The table below lists some commonly used measures of mass and force, together with their conversion factors.

System of Measurement	Measure of Mass	Measure of Force
U.S.	Slug	Pound = (slug)(ft/sec^2)
International	Kilogram	Newton = (kilogram)(m/sec^2)
C-G-S	Gram	Dyne = (gram)(cm/sec^2)
Conversions: 1 pound = 4.448 newtons 1 newton = 0.2248 pound 1 dyne = 0.000002248 pound 1 dyne = 0.00001 newton		 1 slug = 14.59 kilograms 1 kilogram = 0.06852 slug 1 gram = 0.00006852 slug 1 foot = 0.3048 meter

EXAMPLE 1 **Mass on the Surface of Earth**

Find the mass (in slugs) of an object whose weight at sea level is 1 pound.

Solution Using 32 feet per second per second as the acceleration due to gravity produces

$$\begin{aligned}\text{Mass} &= \frac{\text{force}}{\text{acceleration}} && \text{Force} = (\text{mass})(\text{acceleration})\\ &= \frac{1 \text{ pound}}{32 \text{ feet per second per second}}\\ &= 0.03125 \frac{\text{pound}}{\text{foot per second per second}}\\ &= 0.03125 \text{ slug.}\end{aligned}$$

Because many applications involving mass occur on Earth's surface, this amount of mass is called a **pound mass.**

Center of Mass in a One-Dimensional System

You will now consider two types of moments of a mass—the **moment about a point** and the **moment about a line.** To define these two moments, consider an idealized situation in which a mass m is concentrated at a point. If x is the distance between this point mass and another point P, the **moment of m about the point P is**

$$\text{Moment} = mx$$

and x is the **length of the moment arm.**

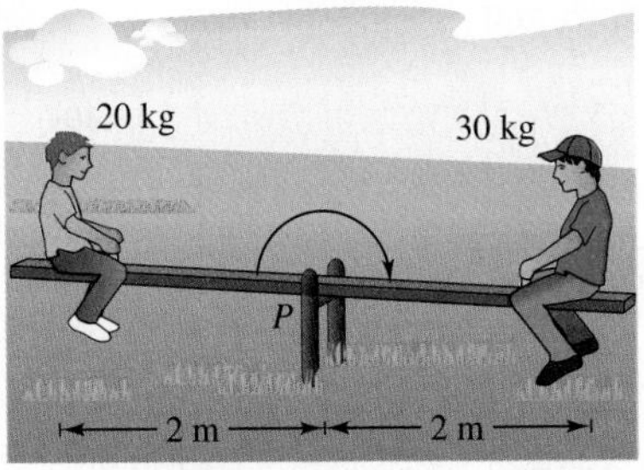

The seesaw will balance when the left and the right moments are equal.
Figure 7.55

The concept of moment can be demonstrated simply by a seesaw, as shown in Figure 7.55. A child of mass 20 kilograms sits 2 meters to the left of fulcrum P, and an older child of mass 30 kilograms sits 2 meters to the right of P. From experience, you know that the seesaw will begin to rotate clockwise, moving the larger child down. This rotation occurs because the moment produced by the child on the left is less than the moment produced by the child on the right.

$$\text{Left moment} = (20)(2) = 40 \text{ kilogram-meters}$$
$$\text{Right moment} = (30)(2) = 60 \text{ kilogram-meters}$$

To balance the seesaw, the two moments must be equal. For example, if the larger child moved to a position $\frac{4}{3}$ meters from the fulcrum, the seesaw would balance, because each child would produce a moment of 40 kilogram-meters.

To generalize this, you can introduce a coordinate line on which the origin corresponds to the fulcrum, as shown in Figure 7.56. Suppose several point masses are located on the x-axis. The measure of the tendency of this system to rotate about the origin is the **moment about the origin,** and it is defined as the sum of the n products $m_i x_i$.

$$M_0 = m_1x_1 + m_2x_2 + \cdot\cdot\cdot + m_nx_n$$

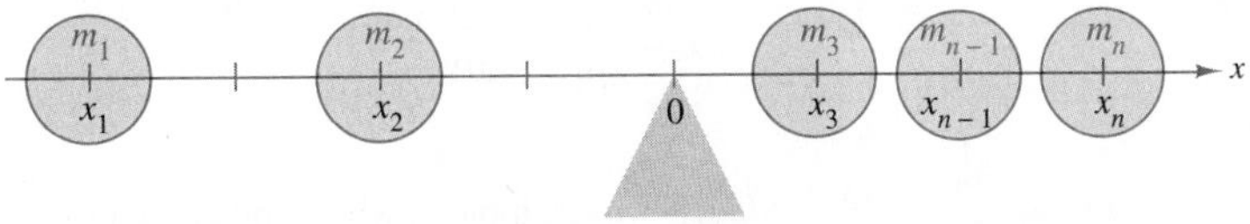

If $m_1x_1 + m_2x_2 + \cdot\cdot\cdot + m_nx_n = 0$, the system is in equilibrium.
Figure 7.56

If M_0 is 0, the system is said to be in **equilibrium.**

For a system that is not in equilibrium, the **center of mass** is defined as the point $\bar{x}$ at which the fulcrum could be relocated to attain equilibrium. If the system were translated $\bar{x}$ units, each coordinate x_i would become $(x_i - \bar{x})$, and because the moment of the translated system is 0, you have

$$\sum_{i=1}^{n} m_i(x_i - \bar{x}) = \sum_{i=1}^{n} m_ix_i - \sum_{i=1}^{n} m_i\bar{x} = 0.$$

Solving for $\bar{x}$ produces

$$\bar{x} = \frac{\sum_{i=1}^{n} m_ix_i}{\sum_{i=1}^{n} m_i} = \frac{\text{moment of system about origin}}{\text{total mass of system}}.$$

If $m_1x_1 + m_2x_2 + \cdot\cdot\cdot + m_nx_n = 0$, the system is in equilibrium.

Moments and Center of Mass: One-Dimensional System

Let the point masses $m_1, m_2, \ldots, m_n$ be located at $x_1, x_2, \ldots, x_n$.

1. The **moment about the origin** is $M_0 = m_1x_1 + m_2x_2 + \cdots + m_nx_n$.
2. The **center of mass** is $\bar{x} = \dfrac{M_0}{m}$, where $m = m_1 + m_2 + \cdots + m_n$ is the **total mass** of the system.

EXAMPLE 2 The Center of Mass of a Linear System

Find the center of mass of the linear system shown in Figure 7.57.

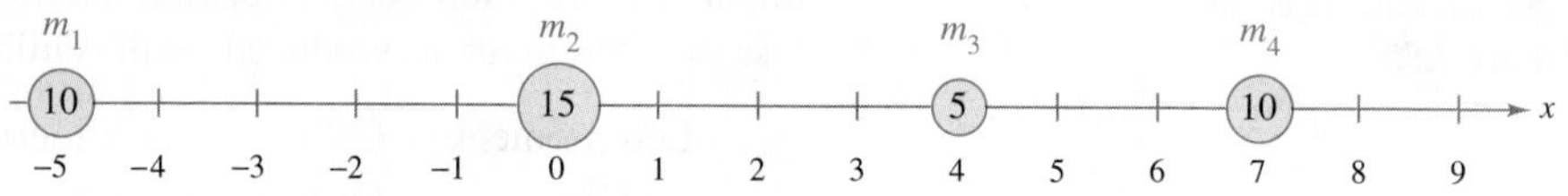

Figure 7.57

Solution The moment about the origin is

$$\begin{aligned} M_0 &= m_1x_1 + m_2x_2 + m_3x_3 + m_4x_4 \\ &= 10(-5) + 15(0) + 5(4) + 10(7) \\ &= -50 + 0 + 20 + 70 \\ &= 40. \end{aligned}$$

Because the total mass of the system is $m = 10 + 15 + 5 + 10 = 40$, the center of mass is

$$\bar{x} = \frac{M_0}{m} = \frac{40}{40} = 1.$$

NOTE In Example 2, where should you locate the fulcrum so that the point masses will be in equilibrium?

Rather than define the moment of a mass, you could define the moment of a *force.* In this context, the center of mass is called the **center of gravity.** Suppose that a system of point masses $m_1, m_2, \ldots, m_n$ is located at $x_1, x_2, \ldots, x_n$. Then, because force = (mass)(acceleration), the total force of the system is

$$\begin{aligned} F &= m_1a + m_2a + \cdots + m_na \\ &= ma. \end{aligned}$$

The **torque** (moment) about the origin is

$$\begin{aligned} T_0 &= (m_1a)x_1 + (m_2a)x_2 + \cdots + (m_na)x_n \\ &= M_0a \end{aligned}$$

and the **center of gravity** is

$$\frac{T_0}{F} = \frac{M_0a}{ma} = \frac{M_0}{m} = \bar{x}.$$

So, the center of gravity and the center of mass have the same location.

Center of Mass in a Two-Dimensional System

You can extend the concept of moment to two dimensions by considering a system of masses located in the xy-plane at the points $(x_1, y_1), (x_2, y_2), \ldots, (x_n, y_n)$, as shown in Figure 7.58. Rather than defining a single moment (with respect to the origin), two moments are defined—one with respect to the x-axis and one with respect to the y-axis.

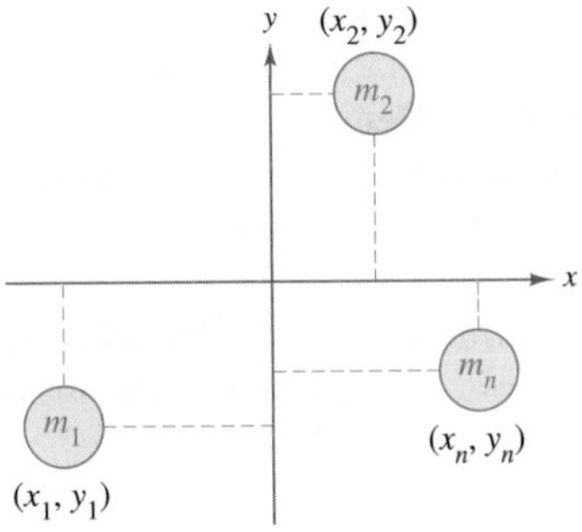

In a two-dimensional system, there is a moment about the y-axis, M_y, and a moment about the x-axis, M_x.
Figure 7.58

Moments and Center of Mass: Two-Dimensional System

Let the point masses $m_1, m_2, \ldots, m_n$ be located at $(x_1, y_1), (x_2, y_2), \ldots, (x_n, y_n)$.

1. The **moment about the y-axis** is $M_y = m_1x_1 + m_2x_2 + \cdots + m_nx_n$.
2. The **moment about the x-axis** is $M_x = m_1y_1 + m_2y_2 + \cdots + m_ny_n$.
3. The **center of mass** $(\bar{x}, \bar{y})$ (or **center of gravity**) is

$$\bar{x} = \frac{M_y}{m} \quad \text{and} \quad \bar{y} = \frac{M_x}{m}$$

where $m = m_1 + m_2 + \cdots + m_n$ is the **total mass** of the system.

The moment of a system of masses in the plane can be taken about any horizontal or vertical line. In general, the moment about a line is the sum of the product of the masses and the *directed distances* from the points to the line.

$$\text{Moment} = m_1(y_1 - b) + m_2(y_2 - b) + \cdots + m_n(y_n - b) \qquad \text{Horizontal line } y = b$$
$$\text{Moment} = m_1(x_1 - a) + m_2(x_2 - a) + \cdots + m_n(x_n - a) \qquad \text{Vertical line } x = a$$

EXAMPLE 3 The Center of Mass of a Two-Dimensional System

Find the center of mass of a system of point masses $m_1 = 6$, $m_2 = 3$, $m_3 = 2$, and $m_4 = 9$, located at

$$(3, -2),\ (0, 0),\ (-5, 3),\ \text{and}\ (4, 2)$$

as shown in Figure 7.59.

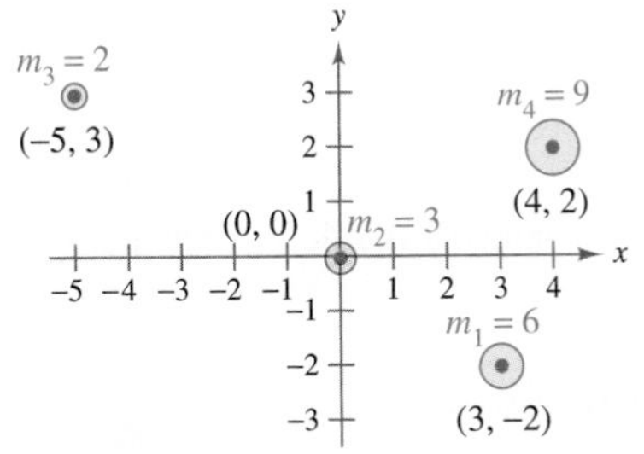

Figure 7.59

Solution

$$\begin{aligned} m &= 6 + 3 + 2 + 9 = 20 && \text{Mass} \\ M_y &= 6(3) + 3(0) + 2(-5) + 9(4) = 44 && \text{Moment about } y\text{-axis} \\ M_x &= 6(-2) + 3(0) + 2(3) + 9(2) = 12 && \text{Moment about } x\text{-axis} \end{aligned}$$

So,

$$\bar{x} = \frac{M_y}{m} = \frac{44}{20} = \frac{11}{5}$$

and

$$\bar{y} = \frac{M_x}{m} = \frac{12}{20} = \frac{3}{5}$$

and so the center of mass is $\left(\frac{11}{5}, \frac{3}{5}\right)$.

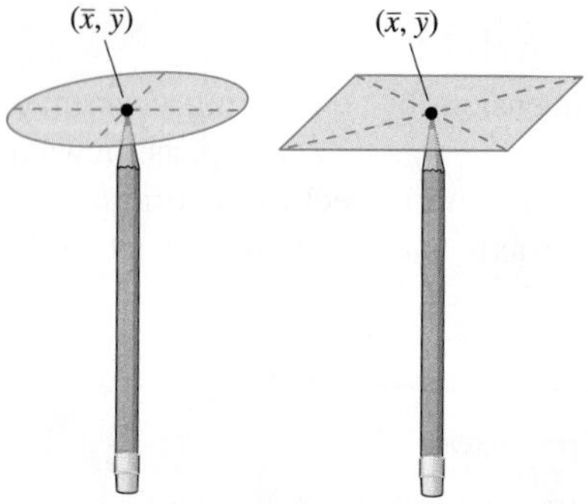

You can think of the center of mass $(\bar{x}, \bar{y})$ of a lamina as its balancing point. For a circular lamina, the center of mass is the center of the circle. For a rectangular lamina, the center of mass is the center of the rectangle.
Figure 7.60

Center of Mass of a Planar Lamina

So far in this section you have assumed the total mass of a system to be distributed at discrete points in a plane or on a line. Now consider a thin, flat plate of material of constant density called a **planar lamina** (see Figure 7.60). **Density** is a measure of mass per unit of volume, such as grams per cubic centimeter. For planar laminas, however, density is considered to be a measure of mass per unit of area. Density is denoted by ρ, the lowercase Greek letter rho.

Consider an irregularly shaped planar lamina of uniform density ρ, bounded by the graphs of $y = f(x)$, $y = g(x)$, and $a \le x \le b$, as shown in Figure 7.61. The mass of this region is given by

$$\begin{aligned} m &= (\text{density})(\text{area}) \\ &= \rho \int_a^b [f(x) - g(x)]\,dx \\ &= \rho A \end{aligned}$$

where A is the area of the region. To find the center of mass of this lamina, partition the interval $[a, b]$ into n subintervals of equal width Δx. Let x_i be the center of the ith subinterval. You can approximate the portion of the lamina lying in the ith subinterval by a rectangle whose height is $h = f(x_i) - g(x_i)$. Because the density of the rectangle is ρ, its mass is

$$\begin{aligned} m_i &= (\text{density})(\text{area}) \\ &= \underbrace{\rho}_{\text{Density}}\underbrace{[f(x_i) - g(x_i)]}_{\text{Height}}\underbrace{\Delta x}_{\text{Width}}. \end{aligned}$$

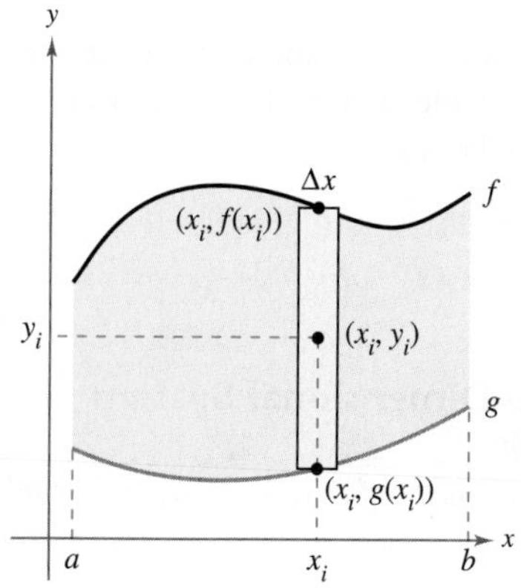

Planar lamina of uniform density ρ
Figure 7.61

Now, considering this mass to be located at the center (x_i, y_i) of the rectangle, the directed distance from the x-axis to (x_i, y_i) is $y_i = [f(x_i) + g(x_i)]/2$. So, the moment of m_i about the x-axis is

$$\begin{aligned} \text{Moment} &= (\text{mass})(\text{distance}) \\ &= m_i y_i \\ &= \rho[f(x_i) - g(x_i)]\,\Delta x\left[\frac{f(x_i) + g(x_i)}{2}\right]. \end{aligned}$$

Summing the moments and taking the limit as $n \to \infty$ suggest the definitions below.

Moments and Center of Mass of a Planar Lamina

Let f and g be continuous functions such that $f(x) \ge g(x)$ on $[a, b]$, and consider the planar lamina of uniform density ρ bounded by the graphs of $y = f(x)$, $y = g(x)$, and $a \le x \le b$.

1. The **moments about the x- and y-axes** are

$$M_x = \rho \int_a^b \left[\frac{f(x) + g(x)}{2}\right][f(x) - g(x)]\,dx$$

$$M_y = \rho \int_a^b x[f(x) - g(x)]\,dx.$$

2. The **center of mass** $(\bar{x}, \bar{y})$ is given by $\bar{x} = \dfrac{M_y}{m}$ and $\bar{y} = \dfrac{M_x}{m}$, where $m = \rho \int_a^b [f(x) - g(x)]\,dx$ is the mass of the lamina.

EXAMPLE 4 The Center of Mass of a Planar Lamina

Find the center of mass of the lamina of uniform density ρ bounded by the graph of $f(x) = 4 - x^2$ and the x-axis.

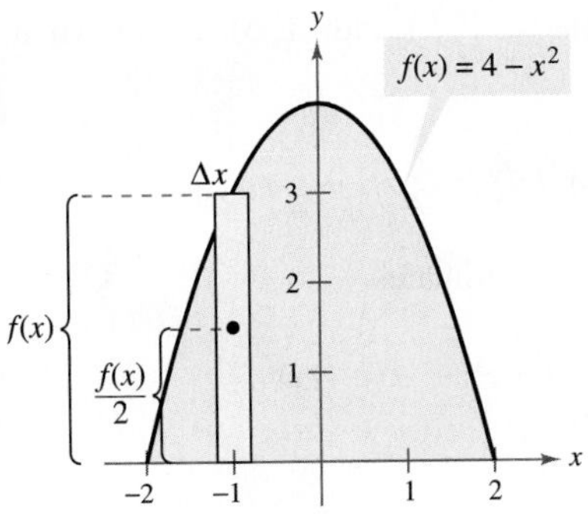

Figure 7.62

Solution Because the center of mass lies on the axis of symmetry, you know that $\bar{x} = 0$. Moreover, the mass of the lamina is

$$m = \rho \int_{-2}^{2} (4 - x^2)\, dx$$
$$= \rho \left[4x - \frac{x^3}{3} \right]_{-2}^{2}$$
$$= \frac{32\rho}{3}.$$

To find the moment about the x-axis, place a representative rectangle in the region, as shown in Figure 7.62. The distance from the x-axis to the center of this rectangle is

$$y_i = \frac{f(x)}{2} = \frac{4 - x^2}{2}.$$

Because the mass of the representative rectangle is

$$\rho f(x)\, \Delta x = \rho(4 - x^2)\, \Delta x$$

you have

$$M_x = \rho \int_{-2}^{2} \frac{4 - x^2}{2} (4 - x^2)\, dx$$
$$= \frac{\rho}{2} \int_{-2}^{2} (16 - 8x^2 + x^4)\, dx$$
$$= \frac{\rho}{2} \left[16x - \frac{8x^3}{3} + \frac{x^5}{5} \right]_{-2}^{2}$$
$$= \frac{256\rho}{15}$$

and $\bar{y}$ is given by

$$\bar{y} = \frac{M_x}{m} = \frac{256\rho/15}{32\rho/3} = \frac{8}{5}.$$

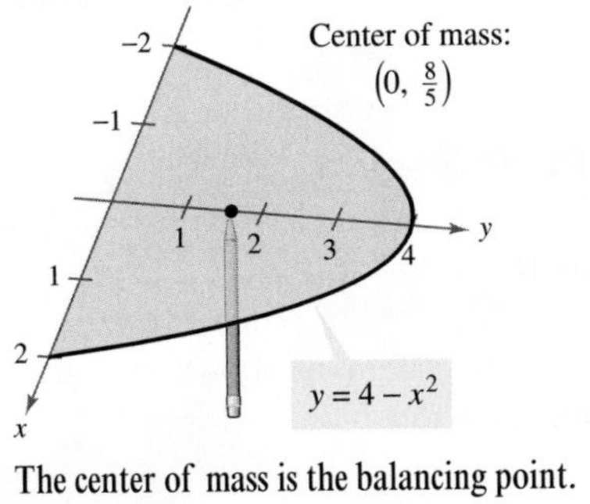

The center of mass is the balancing point.
Figure 7.63

So, the center of mass (the balancing point) of the lamina is $\left(0, \frac{8}{5}\right)$, as shown in Figure 7.63.

The density ρ in Example 4 is a common factor of both the moments and the mass, and as such divides out of the quotients representing the coordinates of the center of mass. So, the center of mass of a lamina of *uniform* density depends only on the shape of the lamina and not on its density. For this reason, the point

$$(\bar{x}, \bar{y}) \qquad \text{Center of mass or centroid}$$

is sometimes called the center of mass of a *region* in the plane, or the **centroid** of the region. In other words, to find the centroid of a region in the plane, you simply assume that the region has a constant density of $\rho = 1$ and compute the corresponding center of mass.

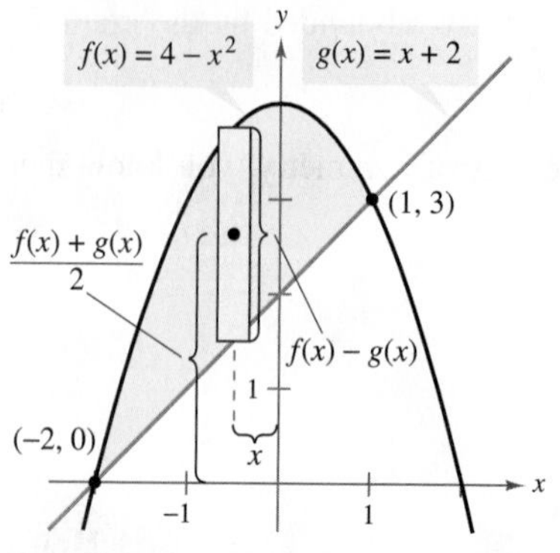

Figure 7.64

EXPLORATION

Cut an irregular shape from a piece of cardboard.

a. Hold a pencil vertically and move the object on the pencil point until the centroid is located.

b. Divide the object into representative elements. Make the necessary measurements and numerically approximate the centroid. Compare your result with the result in part (a).

EXAMPLE 5 The Centroid of a Plane Region

Find the centroid of the region bounded by the graphs of $f(x) = 4 - x^2$ and $g(x) = x + 2$.

Solution The two graphs intersect at the points $(-2, 0)$ and $(1, 3)$, as shown in Figure 7.64. So, the area of the region is

$$A = \int_{-2}^{1} [f(x) - g(x)]\,dx = \int_{-2}^{1} (2 - x - x^2)\,dx = \frac{9}{2}.$$

The centroid $(\bar{x}, \bar{y})$ of the region has the following coordinates.

$$\bar{x} = \frac{1}{A}\int_{-2}^{1} x[(4 - x^2) - (x + 2)]\,dx = \frac{2}{9}\int_{-2}^{1} (-x^3 - x^2 + 2x)\,dx$$

$$= \frac{2}{9}\left[-\frac{x^4}{4} - \frac{x^3}{3} + x^2\right]_{-2}^{1} = -\frac{1}{2}$$

$$\bar{y} = \frac{1}{A}\int_{-2}^{1} \left[\frac{(4 - x^2) + (x + 2)}{2}\right][(4 - x^2) - (x + 2)]\,dx$$

$$= \frac{2}{9}\left(\frac{1}{2}\right)\int_{-2}^{1} (-x^2 + x + 6)(-x^2 - x + 2)\,dx$$

$$= \frac{1}{9}\int_{-2}^{1} (x^4 - 9x^2 - 4x + 12)\,dx$$

$$= \frac{1}{9}\left[\frac{x^5}{5} - 3x^3 - 2x^2 + 12x\right]_{-2}^{1} = \frac{12}{5}.$$

So, the centroid of the region is $(\bar{x}, \bar{y}) = \left(-\frac{1}{2}, \frac{12}{5}\right)$.

For simple plane regions, you may be able to find the centroids without resorting to integration.

EXAMPLE 6 The Centroid of a Simple Plane Region

Find the centroid of the region shown in Figure 7.65(a).

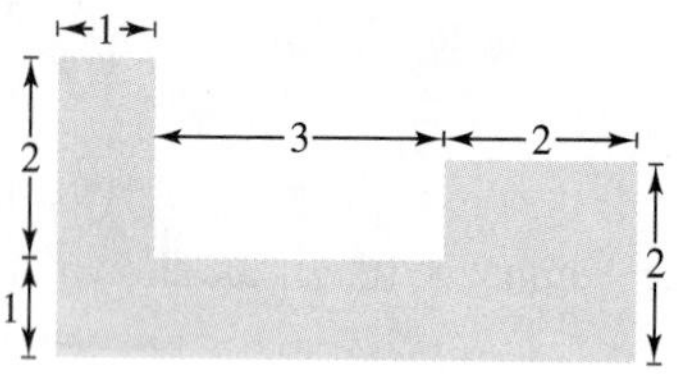

(a) Original region

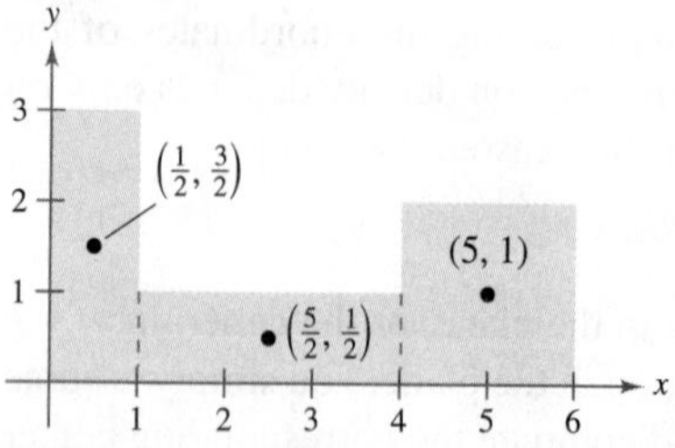

(b) The centroids of the three rectangles

Figure 7.65

Solution By superimposing a coordinate system on the region, as shown in Figure 7.65(b), you can locate the centroids of the three rectangles at

$$\left(\frac{1}{2}, \frac{3}{2}\right), \quad \left(\frac{5}{2}, \frac{1}{2}\right), \quad \text{and} \quad (5, 1).$$

Using these three points, you can find the centroid of the region.

$$A = \text{area of region} = 3 + 3 + 4 = 10$$

$$\bar{x} = \frac{(1/2)(3) + (5/2)(3) + (5)(4)}{10} = \frac{29}{10} = 2.9$$

$$\bar{y} = \frac{(3/2)(3) + (1/2)(3) + (1)(4)}{10} = \frac{10}{10} = 1$$

So, the centroid of the region is (2.9, 1).

NOTE In Example 6, notice that (2.9, 1) is not the "average" of $\left(\frac{1}{2}, \frac{3}{2}\right)$, $\left(\frac{5}{2}, \frac{1}{2}\right)$, and (5, 1).

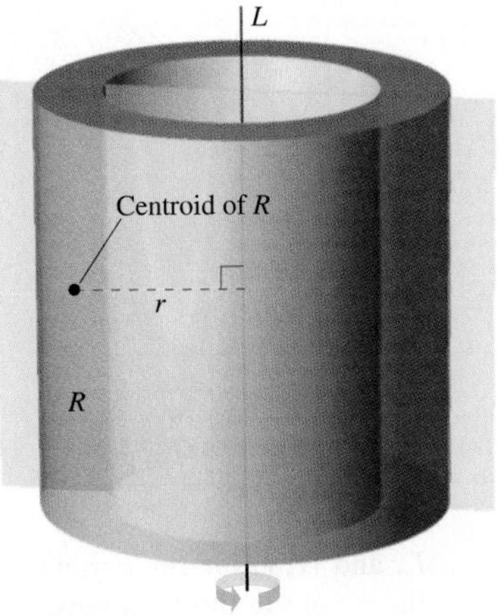

The volume V is $2\pi rA$, where A is the area of region R.
Figure 7.66

Theorem of Pappus

The final topic in this section is a useful theorem credited to Pappus of Alexandria (ca. 300 A.D.), a Greek mathematician whose eight-volume *Mathematical Collection* is a record of much of classical Greek mathematics. The proof of this theorem is given in Section 14.4.

THEOREM 7.1 The Theorem of Pappus

Let R be a region in a plane and let L be a line in the same plane such that L does not intersect the interior of R, as shown in Figure 7.66. If r is the distance between the centroid of R and the line, then the volume V of the solid of revolution formed by revolving R about the line is

$$V = 2\pi rA$$

where A is the area of R. (Note that $2\pi r$ is the distance traveled by the centroid as the region is revolved about the line.)

The Theorem of Pappus can be used to find the volume of a torus, as shown in the following example. Recall that a torus is a doughnut-shaped solid formed by revolving a circular region about a line that lies in the same plane as the circle (but does not intersect the circle).

EXAMPLE 7 Finding Volume by the Theorem of Pappus

Find the volume of the torus shown in Figure 7.67(a), which was formed by revolving the circular region bounded by

$$(x - 2)^2 + y^2 = 1$$

about the y-axis, as shown in Figure 7.67(b).

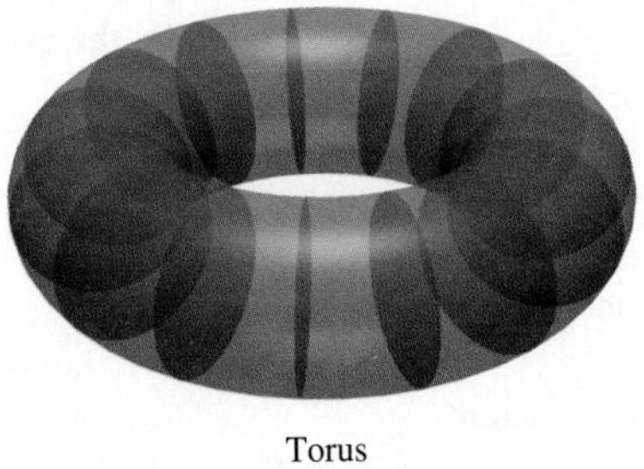

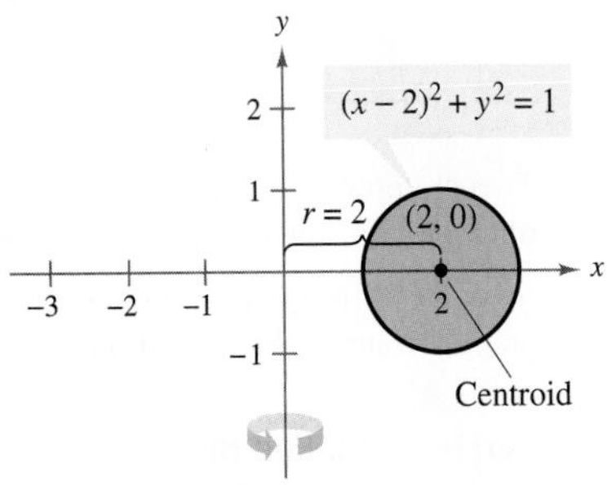

(a) **(b)**
Figure 7.67

EXPLORATION

Use the shell method to show that the volume of the torus is given by

$$V = \int_1^3 4\pi x\sqrt{1 - (x - 2)^2}\, dx.$$

Evaluate this integral using a graphing utility. Does your answer agree with the one in Example 7?

Solution In Figure 7.67(b), you can see that the centroid of the circular region is (2, 0). So, the distance between the centroid and the axis of revolution is $r = 2$. Because the area of the circular region is $A = \pi$, the volume of the torus is

$$\begin{aligned} V &= 2\pi rA \\ &= 2\pi(2)(\pi) \\ &= 4\pi^2 \\ &\approx 39.5. \end{aligned}$$

Exercises for Section 7.6

See www.CalcChat.com for worked-out solutions to odd-numbered exercises.

In Exercises 1–4, find the center of mass of the point masses lying on the x-axis.

1. $m_1 = 6, m_2 = 3, m_3 = 5$

$x_1 = -5, x_2 = 1, x_3 = 3$

2. $m_1 = 7, m_2 = 4, m_3 = 3, m_4 = 8$

$x_1 = -3, x_2 = -2, x_3 = 5, x_4 = 6$

3. $m_1 = 1, m_2 = 1, m_3 = 1, m_4 = 1, m_5 = 1$

$x_1 = 7, x_2 = 8, x_3 = 12, x_4 = 15, x_5 = 18$

4. $m_1 = 12, m_2 = 1, m_3 = 6, m_4 = 3, m_5 = 11$

$x_1 = -6, x_2 = -4, x_3 = -2, x_4 = 0, x_5 = 8$

5. ***Graphical Reasoning***

(a) Translate each point mass in Exercise 3 to the right five units and determine the resulting center of mass.

(b) Translate each point mass in Exercise 4 to the left three units and determine the resulting center of mass.

6. ***Conjecture*** Use the result of Exercise 5 to make a conjecture about the change in the center of mass that results when each point mass is translated k units horizontally.

Statics Problems **In Exercises 7 and 8, consider a beam of length L with a fulcrum x feet from one end (see figure). There are objects with weights W_1 and W_2 placed on opposite ends of the beam. Find x such that the system is in equilibrium.**

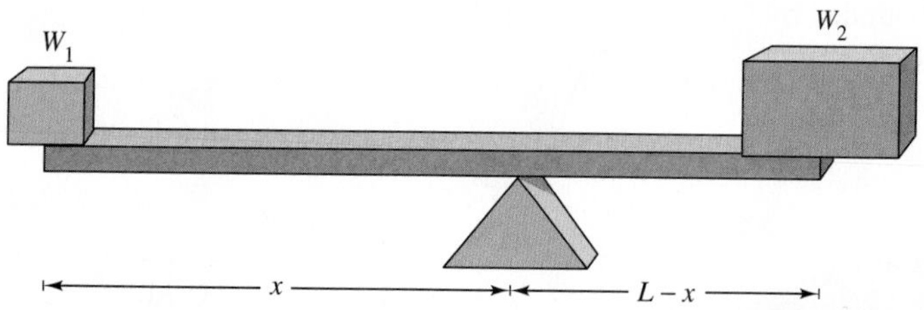

7. Two children weighing 50 pounds and 75 pounds are going to play on a seesaw that is 10 feet long.

8. In order to move a 550-pound rock, a person weighing 200 pounds wants to balance it on a beam that is 5 feet long.

In Exercise 9–12, find the center of mass of the given system of point masses.

9.

m_i	5	1	3
(x_1, y_1)	$(2, 2)$	$(-3, 1)$	$(1, -4)$

10.

m_i	10	2	5
(x_1, y_1)	$(1, -1)$	$(5, 5)$	$(-4, 0)$

11.

m_i	3	4
(x_1, y_1)	$(-2, -3)$	$(5, 5)$

m_i	2	1	6
(x_1, y_1)	$(7, 1)$	$(0, 0)$	$(-3, 0)$

12.

m_i	12	6	$\frac{15}{2}$	15
(x_1, y_1)	$(2, 3)$	$(-1, 5)$	$(6, 8)$	$(2, -2)$

In Exercises 13–24, find M_x, M_y, and $(\bar{x}, \bar{y})$ for the laminas of uniform density ρ bounded by the graphs of the equations.

13. $y = \sqrt{x}, y = 0, x = 4$

14. $y = \frac{1}{2}x^2, y = 0, x = 2$

15. $y = x^2, y = x^3$

16. $y = \sqrt{x}, y = x$

17. $y = -x^2 + 4x + 2, y = x + 2$

18. $y = \sqrt{x} + 1, y = \frac{1}{3}x + 1$

19. $y = x^{2/3}, y = 0, x = 8$

20. $y = x^{2/3}, y = 4$

21. $x = 4 - y^2, x = 0$

22. $x = 2y - y^2, x = 0$

23. $x = -y, x = 2y - y^2$

24. $x = y + 2, x = y^2$

In Exercises 25–28, set up and evaluate the integrals for finding the area and moments about the x- and y-axes for the region bounded by the graphs of the equations. (Assume $\rho = 1$.)

25. $y = x^2, y = x$

26. $y = \dfrac{1}{x}, y = 0, 1 \le x \le 4$

27. $y = 2x + 4, y = 0, 0 \le x \le 3$

28. $y = x^2 - 4, y = 0$

In Exercises 29–32, use a graphing utility to graph the region bounded by the graphs of the equations. Use the integration capabilities of the graphing utility to approximate the centroid of the region.

29. $y = 10x\sqrt{125 - x^3}, y = 0$

30. $y = xe^{-x/2}, y = 0, x = 0, x = 4$

31. ***Prefabricated End Section of a Building***

$y = 5\sqrt[3]{400 - x^2}, y = 0$

32. ***Witch of Agnesi***

$y = 8/(x^2 + 4), y = 0, x = -2, x = 2$

In Exercises 33–38, find and/or verify the centroid of the common region used in engineering.

33. ***Triangle*** Show that the centroid of the triangle with vertices $(-a, 0)$, $(a, 0)$, and (b, c) is the point of intersection of the medians (see figure).

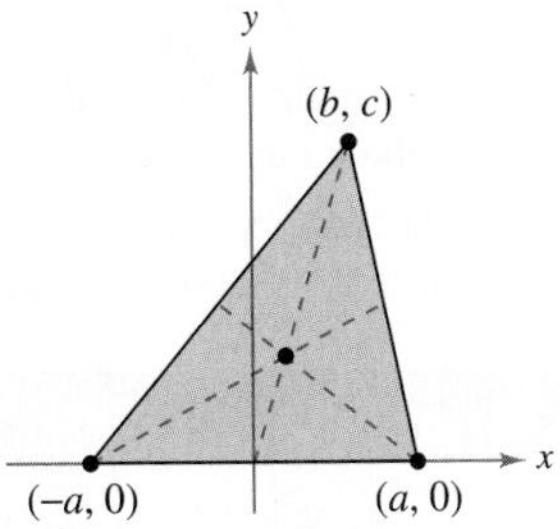

Figure for 33

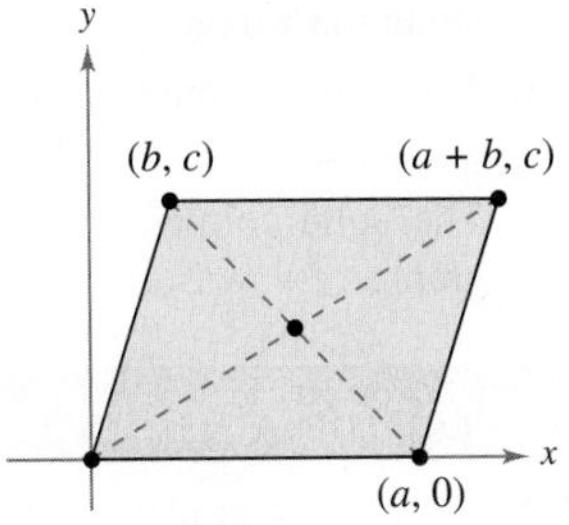

Figure for 34

34. ***Parallelogram*** Show that the centroid of the parallelogram with vertices $(0, 0)$, $(a, 0)$, (b, c), and $(a + b, c)$ is the point of intersection of the diagonals (see figure).

35. ***Trapezoid*** Find the centroid of the trapezoid with vertices $(0, 0)$, $(0, a)$, (c, b), and $(c, 0)$. Show that it is the intersection of the line connecting the midpoints of the parallel sides and the line connecting the extended parallel sides, as shown in the figure.

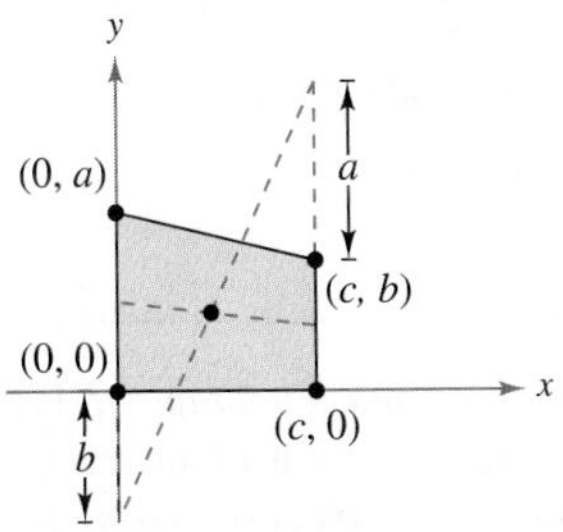

Figure for 35

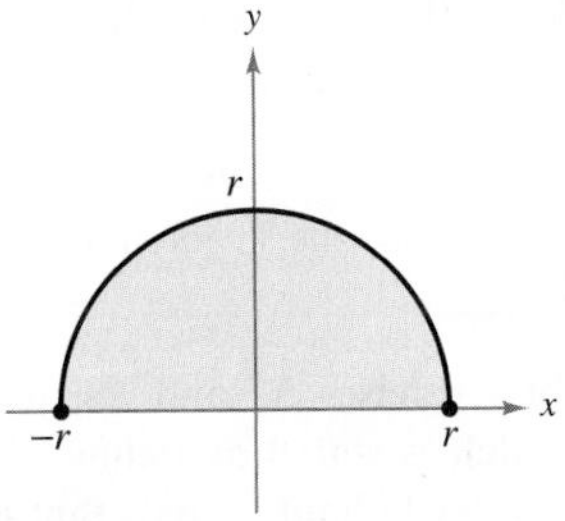

Figure for 36

36. ***Semicircle*** Find the centroid of the region bounded by the graphs of $y = \sqrt{r^2 - x^2}$ and $y = 0$ (see figure).

37. ***Semiellipse*** Find the centroid of the region bounded by the graphs of $y = \dfrac{b}{a}\sqrt{a^2 - x^2}$ and $y = 0$ (see figure).

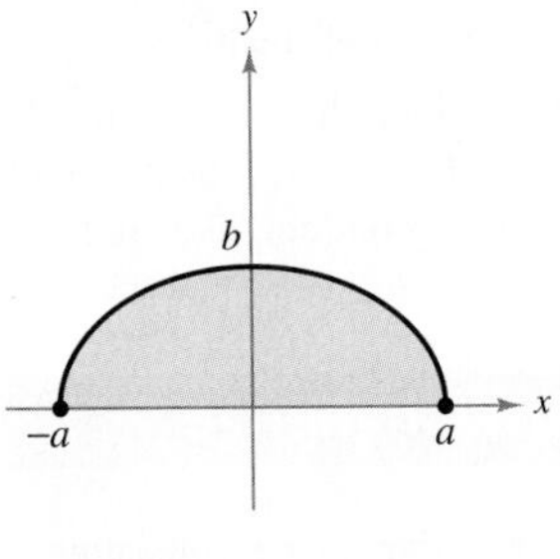

Figure for 37

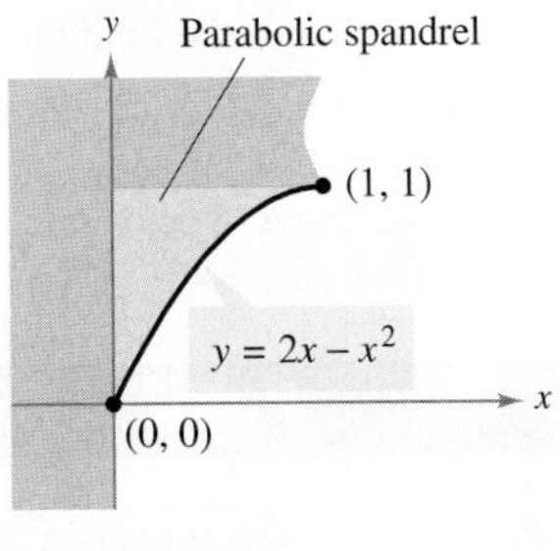

Figure for 38

38. ***Parabolic Spandrel*** Find the centroid of the **parabolic spandrel** shown in the figure.

39. ***Graphical Reasoning*** Consider the region bounded by the graphs of $y = x^2$ and $y = b$, where $b > 0$.

(a) Sketch a graph of the region.

(b) Use the graph in part (a) to determine $\bar{x}$. Explain.

(c) Set up the integral for finding M_y. Because of the form of the integrand, the value of the integral can be obtained without integrating. What is the form of the integrand and what is the value of the integral? Compare with the result in part (b).

(d) Use the graph in part (a) to determine whether $\bar{y} > \dfrac{b}{2}$ or $\bar{y} < \dfrac{b}{2}$. Explain.

(e) Use integration to verify your answer in part (d).

40. ***Graphical and Numerical Reasoning*** Consider the region bounded by the graphs of $y = x^{2n}$ and $y = b$, where $b > 0$ and n is a positive integer.

(a) Set up the integral for finding M_y. Because of the form of the integrand, the value of the integral can be obtained without integrating. What is the form of the integrand and what is the value of the integral? Compare with the result in part (b).

(b) Is $\bar{y} > \dfrac{b}{2}$ or $\bar{y} < \dfrac{b}{2}$? Explain.

(c) Use integration to find $\bar{y}$ as a function of n.

(d) Use the result of part (c) to complete the table.

n	1	2	3	4
$\bar{y}$				

(e) Find $\lim_{n\to\infty} \bar{y}$.

(f) Give a geometric explanation of the result in part (e).

41. ***Modeling Data*** The manufacturer of glass for a window in a conversion van needs to approximate its center of mass. A coordinate system is superimposed on a prototype of the glass (see figure). The measurements (in centimeters) for the right half of the symmetric piece of glass are shown in the table.

x	0	10	20	30	40
y	30	29	26	20	0

(a) Use Simpson's Rule to approximate the center of mass of the glass.

(b) Use the regression capabilities of a graphing utility to find a fourth-degree polynomial model for the data.

(c) Use the integration capabilities of a graphing utility and the model to approximate the center of mass of the glass. Compare with the result in part (a).

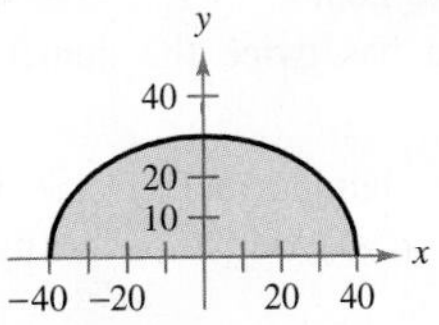

42. ***Modeling Data*** The manufacturer of a boat needs to approximate the center of mass of a section of the hull. A coordinate system is superimposed on a prototype (see figure). The measurements (in feet) for the right half of the symmetric prototype are listed in the table.

x	0	0.5	1.0	1.5	2
l	1.50	1.45	1.30	0.99	0
d	0.50	0.48	0.43	0.33	0

(a) Use Simpson's Rule to approximate the center of mass of the hull section.

(b) Use the regression capabilities of a graphing utility to find fourth-degree polynomial models for both curves shown in the figure. Plot the data and graph the models.

(c) Use the integration capabilities of a graphing utility and the model to approximate the center of mass of the hull section. Compare with the result in part (a).

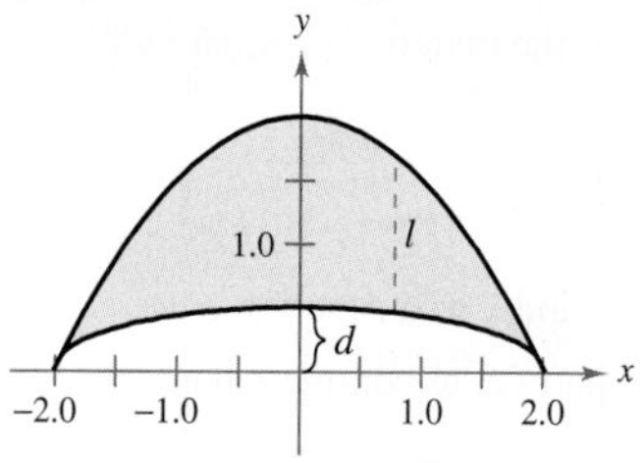

In Exercises 43–46, introduce an appropriate coordinate system and find the coordinates of the center of mass of the planar lamina. (The answer depends on the position of the coordinate system.)

43.

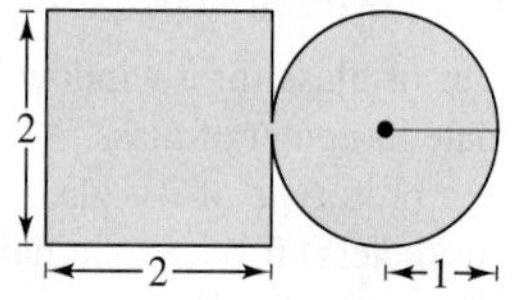

44.

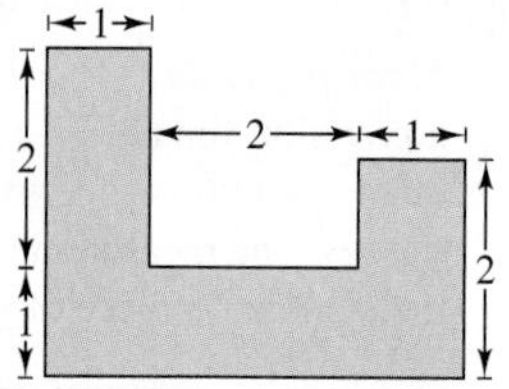

45.

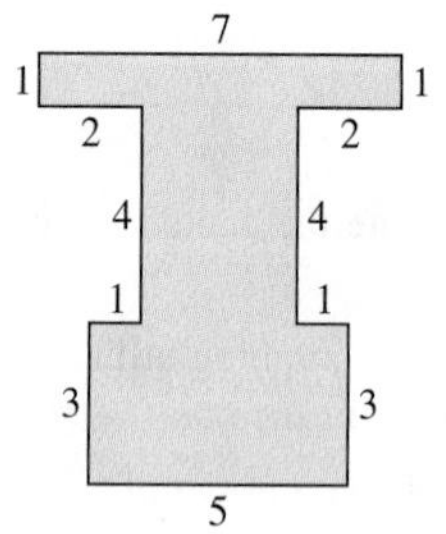

46.

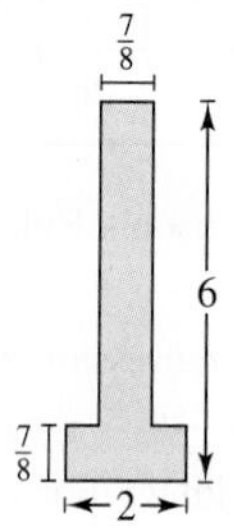

47. Find the center of mass of the lamina in Exercise 43 if the circular portion of the lamina has twice the density of the square portion of the lamina.

48. Find the center of mass of the lamina in Exercise 43 if the square portion of the lamina has twice the density of the circular portion of the lamina.

In Exercises 49–52, use the Theorem of Pappus to find the volume of the solid of revolution.

49. The torus formed by revolving the circle $(x-5)^2+y^2=16$ about the y-axis

50. The torus formed by revolving the circle $x^2+(y-3)^2=4$ about the x-axis

51. The solid formed by revolving the region bounded by the graphs of $y=x$, $y=4$, and $x=0$ about the x-axis

52. The solid formed by revolving the region bounded by the graphs of $y=2\sqrt{x-2}$, $y=0$, and $x=6$ about the y-axis

Writing About Concepts

53. Let the point masses $m_1, m_2, \ldots, m_n$ be located at (x_1, y_1), $(x_2, y_2), \ldots, (x_n, y_n)$. Define the center of mass $(\bar{x}, \bar{y})$.

54. What is a planar lamina? Describe what is meant by the center of mass $(\bar{x}, \bar{y})$ of a planar lamina.

55. The centroid of the plane region bounded by the graphs of $y=f(x)$, $y=0$, $x=0$, and $x=1$ is $\left(\frac{5}{6}, \frac{5}{18}\right)$. Is it possible to find the centroid of each of the regions bounded by the graphs of the following sets of equations? If so, identify the centroid and explain your answer.

(a) $y=f(x)+2$, $y=2$, $x=0$, and $x=1$

(b) $y=f(x-2)$, $y=0$, $x=2$, and $x=3$

(c) $y=-f(x)$, $y=0$, $x=0$, and $x=1$

(d) $y=f(x)$, $y=0$, $x=-1$, and $x=1$

56. State the Theorem of Pappus.

In Exercises 57 and 58, use the *Second Theorem of Pappus*, which is stated as follows. If a segment of a plane curve C is revolved about an axis that does not intersect the curve (except possibly at its endpoints), the area S of the resulting surface of revolution is given by the product of the length of C times the distance d traveled by the centroid of C.

57. A sphere is formed by revolving the graph of $y=\sqrt{r^2-x^2}$ about the x-axis. Use the formula for surface area, $S=4\pi r^2$, to find the centroid of the semicircle $y=\sqrt{r^2-x^2}$.

58. A torus is formed by revolving the graph of $(x-1)^2+y^2=1$ about the y-axis. Find the surface area of the torus.

59. Let $n \geq 1$ be constant, and consider the region bounded by $f(x)=x^n$, the x-axis, and $x=1$. Find the centroid of this region. As $n \to \infty$, what does the region look like, and where is its centroid?

Putnam Exam Challenge

60. Let V be the region in the cartesian plane consisting of all points (x, y) satisfying the simultaneous conditions

$$|x| \leq y \leq |x|+3 \quad \text{and} \quad y \leq 4.$$

Find the centroid $(\bar{x}, \bar{y})$ of V.

Section 7.7 Fluid Pressure and Fluid Force

- **Find fluid pressure and fluid force.**

Fluid Pressure and Fluid Force

Swimmers know that the deeper an object is submerged in a fluid, the greater the pressure on the object. **Pressure** is defined as the force per unit of area over the surface of a body. For example, because a column of water that is 10 feet in height and 1 inch square weighs 4.3 pounds, the *fluid pressure* at a depth of 10 feet of water is 4.3 pounds per square inch.* At 20 feet, this would increase to 8.6 pounds per square inch, and in general the pressure is proportional to the depth of the object in the fluid.

BLAISE PASCAL (1623–1662)

Pascal is well known for his work in many areas of mathematics and physics, and also for his influence on Leibniz. Although much of Pascal's work in calculus was intuitive and lacked the rigor of modern mathematics, he nevertheless anticipated many important results.

Definition of Fluid Pressure

The **pressure** on an object at depth h in a liquid is

Pressure $= P = wh$

where w is the weight-density of the liquid per unit of volume.

Below are some common weight-densities of fluids in pounds per cubic foot.

Ethyl alcohol	49.4
Gasoline	41.0–43.0
Glycerin	78.6
Kerosene	51.2
Mercury	849.0
Seawater	64.0
Water	62.4

When calculating fluid pressure, you can use an important (and rather surprising) physical law called **Pascal's Principle,** named after the French mathematician Blaise Pascal. Pascal's Principle states that the pressure exerted by a fluid at a depth h is transmitted equally *in all directions*. For example, in Figure 7.68, the pressure at the indicated depth is the same for all three objects. Because fluid pressure is given in terms of force per unit area ($P = F/A$), the fluid force on a *submerged horizontal* surface of area A is

Fluid force $= F = PA =$ (pressure)(area).

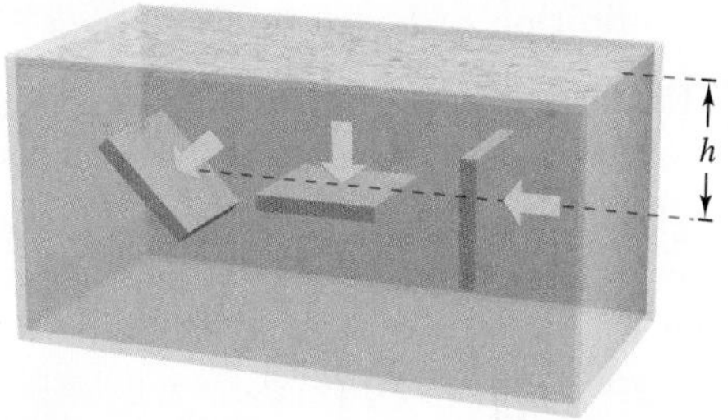

The pressure at h is the same for all three objects.
Figure 7.68

* *The total pressure on an object in 10 feet of water would also include the pressure due to Earth's atmosphere. At sea level, atmospheric pressure is approximately 14.7 pounds per square inch.*

The fluid force on a horizontal metal sheet is equal to the fluid pressure times the area.
Figure 7.69

EXAMPLE 1 Fluid Force on a Submerged Sheet

Find the fluid force on a rectangular metal sheet measuring 3 feet by 4 feet that is submerged in 6 feet of water, as shown in Figure 7.69.

Solution Because the weight-density of water is 62.4 pounds per cubic foot and the sheet is submerged in 6 feet of water, the fluid pressure is

$$\begin{aligned} P &= (62.4)(6) && P = wh \\ &= 374.4 \text{ pounds per square foot.} \end{aligned}$$

Because the total area of the sheet is $A = (3)(4) = 12$ square feet, the fluid force is

$$\begin{aligned} F = PA &= \left(374.4 \frac{\text{pounds}}{\text{square foot}}\right)(12 \text{ square feet}) \\ &= 4492.8 \text{ pounds.} \end{aligned}$$

This result is independent of the size of the body of water. The fluid force would be the same in a swimming pool or lake.

In Example 1, the fact that the sheet is rectangular and horizontal means that you do not need the methods of calculus to solve the problem. Consider a surface that is submerged vertically in a fluid. This problem is more difficult because the pressure is not constant over the surface.

Suppose a vertical plate is submerged in a fluid of weight-density w (per unit of volume), as shown in Figure 7.70. To determine the total force against *one side* of the region from depth c to depth d, you can subdivide the interval $[c, d]$ into n subintervals, each of width Δy. Next, consider the representative rectangle of width Δy and length $L(y_i)$, where y_i is in the ith subinterval. The force against this representative rectangle is

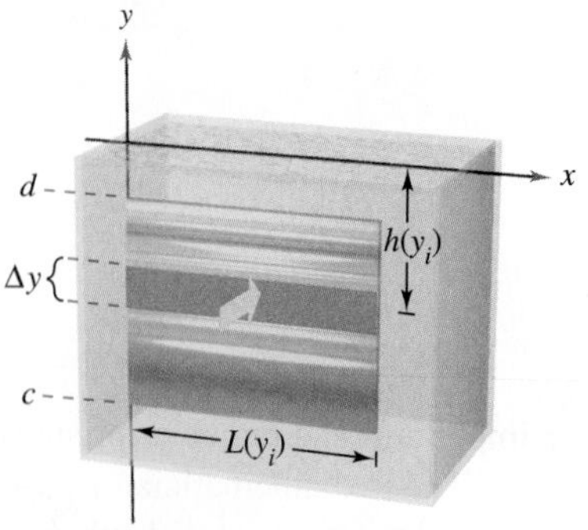

Calculus methods must be used to find the fluid force on a vertical metal plate.
Figure 7.70

$$\begin{aligned} \Delta F_i &= w(\text{depth})(\text{area}) \\ &= wh(y_i)L(y_i)\,\Delta y. \end{aligned}$$

The force against n such rectangles is

$$\sum_{i=1}^{n} \Delta F_i = w\sum_{i=1}^{n} h(y_i)L(y_i)\,\Delta y.$$

Note that w is considered to be constant and is factored out of the summation. Therefore, taking the limit as $\|\Delta\| \to 0$ $(n \to \infty)$ suggests the following definition.

Definition of Force Exerted by a Fluid

The **force F exerted by a fluid** of constant weight-density w (per unit of volume) against a submerged vertical plane region from $y = c$ to $y = d$ is

$$\begin{aligned} F &= w \lim_{\|\Delta\| \to 0} \sum_{i=1}^{n} h(y_i)L(y_i)\,\Delta y \\ &= w \int_c^d h(y)L(y)\,dy \end{aligned}$$

where $h(y)$ is the depth of the fluid at y and $L(y)$ is the horizontal length of the region at y.

EXAMPLE 2 Fluid Force on a Vertical Surface

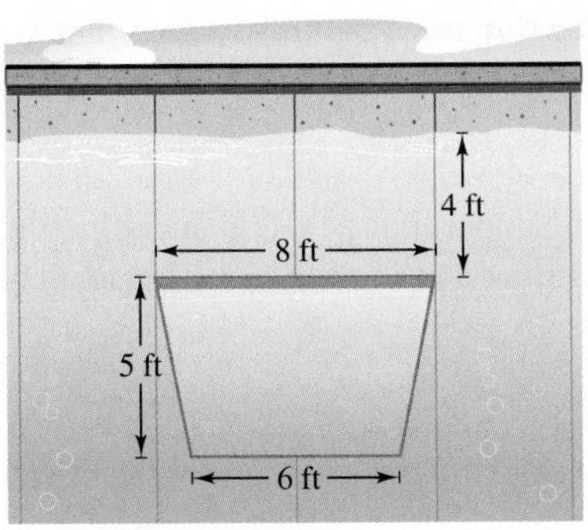

(a) Water gate in a dam

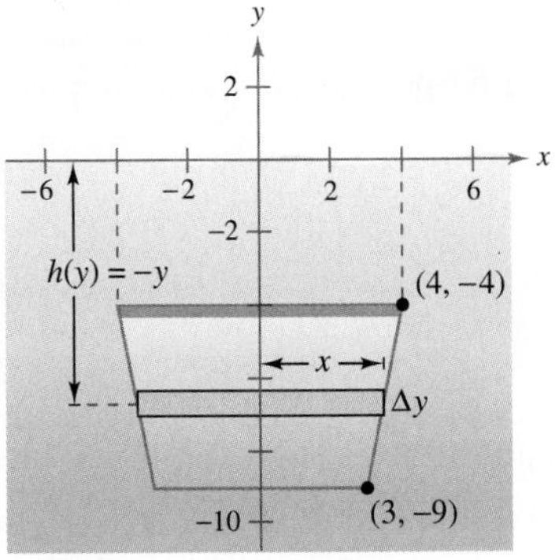

(b) The fluid force against the gate
Figure 7.71

A vertical gate in a dam has the shape of an isosceles trapezoid 8 feet across the top and 6 feet across the bottom, with a height of 5 feet, as shown in Figure 7.71(a). What is the fluid force on the gate when the top of the gate is 4 feet below the surface of the water?

Solution In setting up a mathematical model for this problem, you are at liberty to locate the x- and y-axes in several different ways. A convenient approach is to let the y-axis bisect the gate and place the x-axis at the surface of the water, as shown in Figure 7.71(b). So, the depth of the water at y in feet is

$$\text{Depth} = h(y) = -y.$$

To find the length $L(y)$ of the region at y, find the equation of the line forming the right side of the gate. Because this line passes through the points $(3, -9)$ and $(4, -4)$, its equation is

$$\begin{aligned} y - (-9) &= \frac{-4 - (-9)}{4 - 3}(x - 3) \\ y + 9 &= 5(x - 3) \\ y &= 5x - 24 \\ x &= \frac{y + 24}{5}. \end{aligned}$$

In Figure 7.71(b) you can see that the length of the region at y is

$$\begin{aligned} \text{Length} &= 2x \\ &= \frac{2}{5}(y + 24) \\ &= L(y). \end{aligned}$$

Finally, by integrating from $y = -9$ to $y = -4$, you can calculate the fluid force to be

$$\begin{aligned} F &= w\int_c^d h(y)L(y)\,dy \\ &= 62.4\int_{-9}^{-4} (-y)\left(\frac{2}{5}\right)(y + 24)\,dy \\ &= -62.4\left(\frac{2}{5}\right)\int_{-9}^{-4} (y^2 + 24y)\,dy \\ &= -62.4\left(\frac{2}{5}\right)\left[\frac{y^3}{3} + 12y^2\right]_{-9}^{-4} \\ &= -62.4\left(\frac{2}{5}\right)\left(\frac{-1675}{3}\right) \\ &= 13{,}936 \text{ pounds.} \end{aligned}$$

NOTE In Example 2, the x-axis coincided with the surface of the water. This was convenient, but arbitrary. In choosing a coordinate system to represent a physical situation, you should consider various possibilities. Often you can simplify the calculations in a problem by locating the coordinate system to take advantage of special characteristics of the problem, such as symmetry.

EXAMPLE 3 Fluid Force on a Vertical Surface

A circular observation window on a marine science ship has a radius of 1 foot, and the center of the window is 8 feet below water level, as shown in Figure 7.72. What is the fluid force on the window?

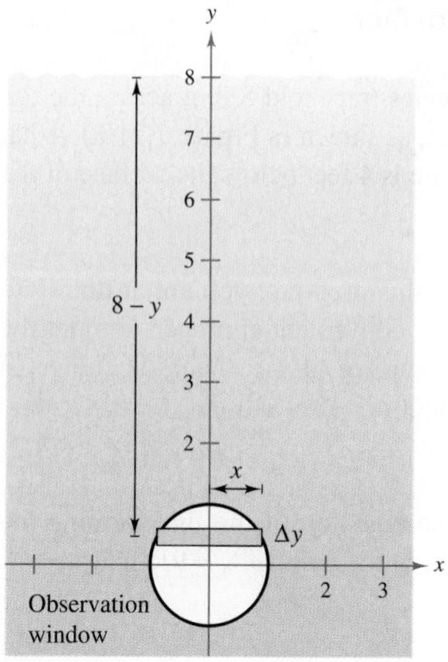

The fluid force on the window
Figure 7.72

Solution To take advantage of symmetry, locate a coordinate system such that the origin coincides with the center of the window, as shown in Figure 7.72. The depth at y is then

$$\text{Depth} = h(y) = 8 - y.$$

The horizontal length of the window is $2x$, and you can use the equation for the circle, $x^2 + y^2 = 1$, to solve for x as follows.

$$\begin{aligned}\text{Length} &= 2x \\ &= 2\sqrt{1 - y^2} = L(y)\end{aligned}$$

Finally, because y ranges from -1 to 1, and using 64 pounds per cubic foot as the weight-density of seawater, you have

$$\begin{aligned}F &= w\int_c^d h(y)L(y)\,dy \\ &= 64\int_{-1}^{1} (8 - y)(2)\sqrt{1 - y^2}\,dy.\end{aligned}$$

Initially it looks as if this integral would be difficult to solve. However, if you break the integral into two parts and apply symmetry, the solution is simple.

$$F = 64(16)\int_{-1}^{1} \sqrt{1 - y^2}\,dy - 64(2)\int_{-1}^{1} y\sqrt{1 - y^2}\,dy$$

The second integral is 0 (because the integrand is odd and the limits of integration are symmetric to the origin). Moreover, by recognizing that the first integral represents the area of a semicircle of radius 1, you obtain

$$\begin{aligned}F &= 64(16)\left(\frac{\pi}{2}\right) - 64(2)(0) \\ &= 512\pi \\ &\approx 1608.5 \text{ pounds.}\end{aligned}$$

So, the fluid force on the window is 1608.5 pounds.

TECHNOLOGY To confirm the result obtained in Example 3, you might have considered using Simpson's Rule to approximate the value of

$$128\int_{-1}^{1} (8 - x)\sqrt{1 - x^2}\,dx.$$

From the graph of

$$f(x) = (8 - x)\sqrt{1 - x^2}$$

however, you can see that f is not differentiable when $x = \pm 1$ (see Figure 7.73). This means that you cannot apply Theorem 4.19 from Section 4.6 to determine the potential error in Simpson's Rule. Without knowing the potential error, the approximation is of little value. Use a graphing utility to approximate the integral.

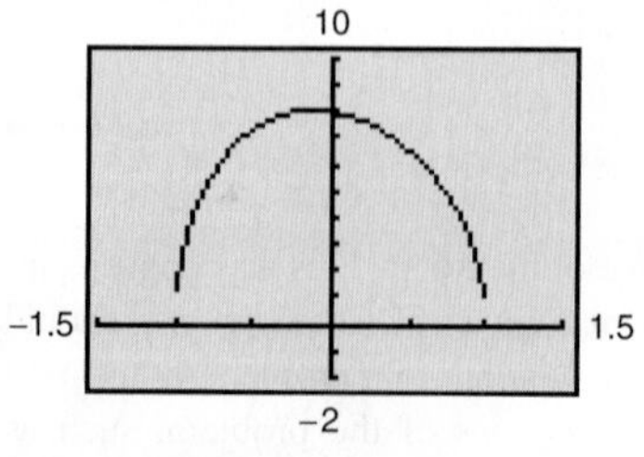

f is not differentiable at $x = \pm 1$.
Figure 7.73

Exercises for Section 7.7

See www.CalcChat.com for worked-out solutions to odd-numbered exercises.

Force on a Submerged Sheet **In Exercises 1 and 2, the area of the top side of a piece of sheet metal is given. The sheet metal is submerged horizontally in 5 feet of water. Find the fluid force on the top side.**

1. 3 square feet

2. 16 square feet

Buoyant Force **In Exercises 3 and 4, find the buoyant force of a rectangular solid of the given dimensions submerged in water so that the top side is parallel to the surface of the water. The buoyant force is the difference between the fluid forces on the top and bottom sides of the solid.**

3.

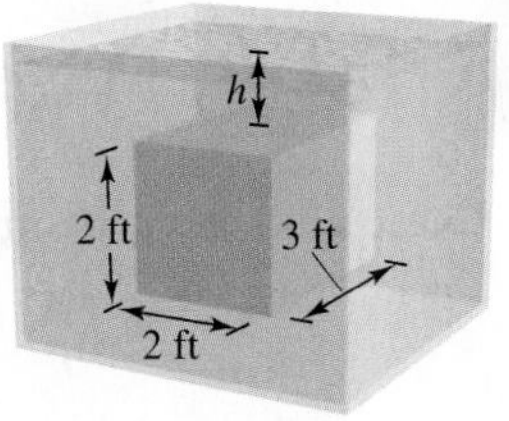

4.

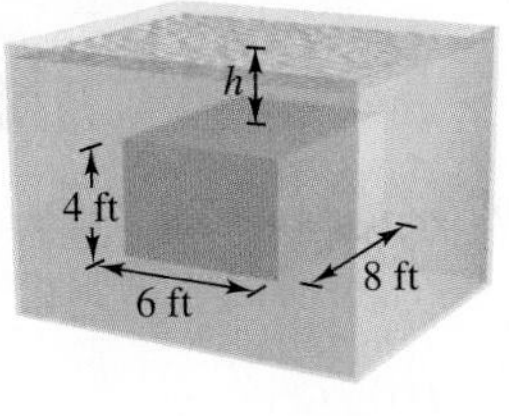

Fluid Force on a Tank Wall **In Exercises 5–10, find the fluid force on the vertical side of the tank, where the dimensions are given in feet. Assume that the tank is full of water.**

5. Rectangle

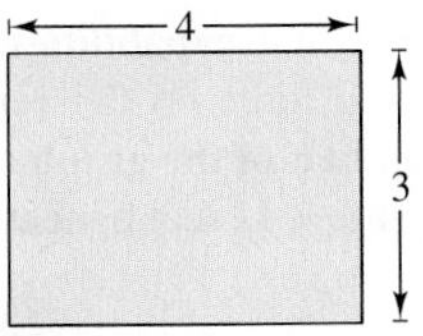

6. Triangle

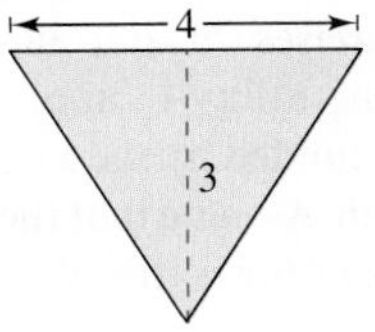

7. Trapezoid

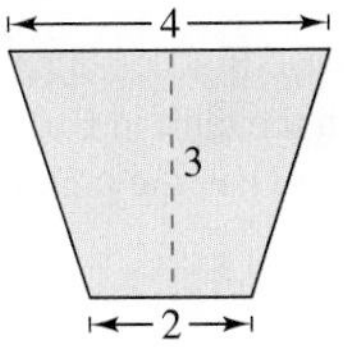

8. Semicircle

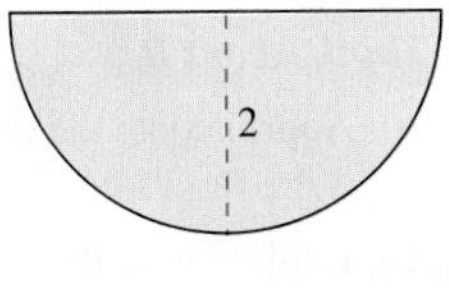

9. Parabola, $y = x^2$

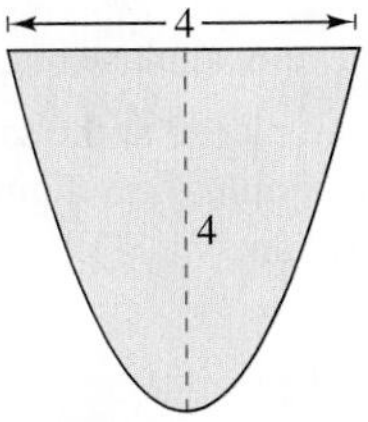

10. Semiellipse, $y = -\frac{1}{2}\sqrt{36 - 9x^2}$

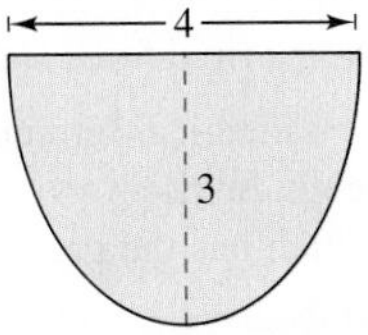

Fluid Force of Water **In Exercises 11–14, find the fluid force on the vertical plate submerged in water, where the dimensions are given in meters and the weight-density of water is 9800 newtons per cubic meter.**

11. Square

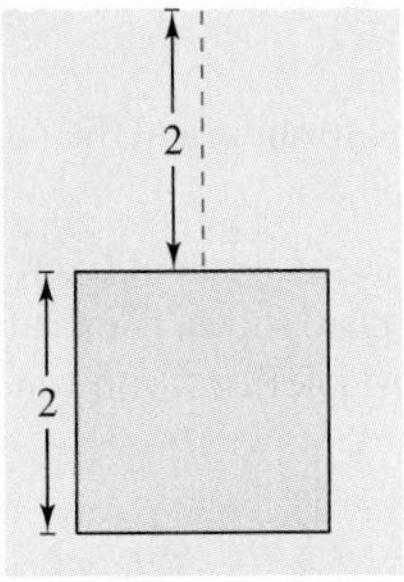

12. Square

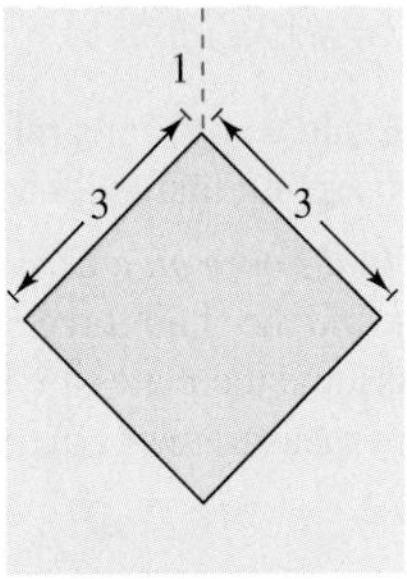

13. Triangle

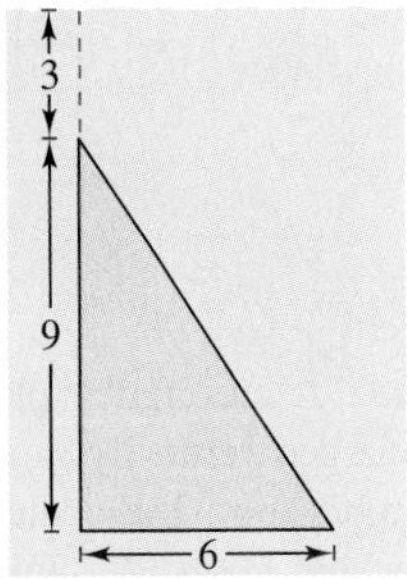

14. Rectangle

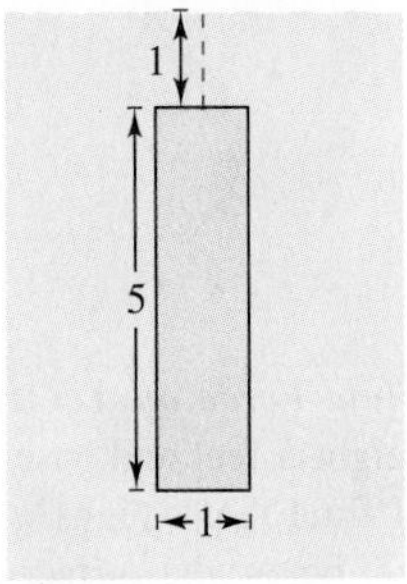

Force on a Concrete Form **In Exercises 15–18, the figure is the vertical side of a form for poured concrete that weighs 140.7 pounds per cubic foot. Determine the force on this part of the concrete form.**

15. Rectangle

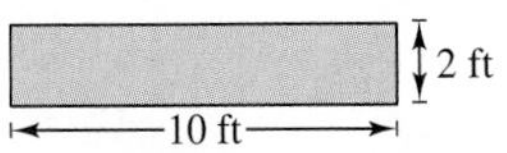

16. Semiellipse, $y = -\frac{3}{4}\sqrt{16 - x^2}$

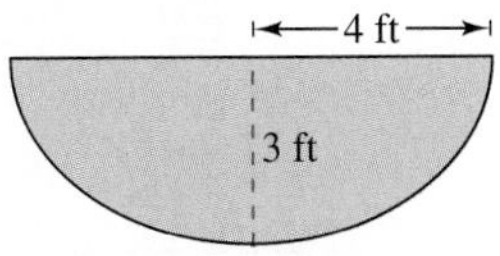

17. Rectangle

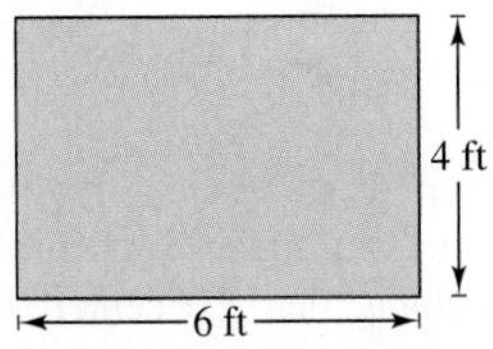

18. Triangle

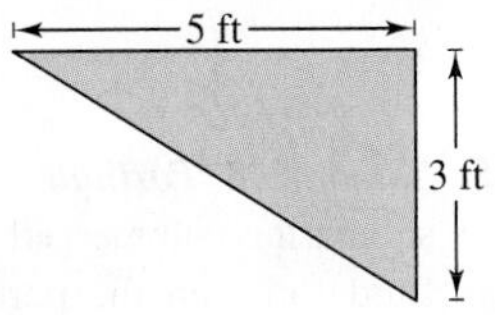

19. ***Fluid Force of Gasoline*** A cylindrical gasoline tank is placed so that the axis of the cylinder is horizontal. Find the fluid force on a circular end of the tank if the tank is half full, assuming that the diameter is 3 feet and the gasoline weighs 42 pounds per cubic foot.

20. ***Fluid Force of Gasoline*** Repeat Exercise 19 for a tank that is full. (Evaluate one integral by a geometric formula and the other by observing that the integrand is an odd function.)

21. ***Fluid Force on a Circular Plate*** A circular plate of radius r feet is submerged vertically in a tank of fluid that weighs w pounds per cubic foot. The center of the circle is k $(k > r)$ feet below the surface of the fluid. Show that the fluid force on the surface of the plate is

$$F = wk(\pi r^2).$$

(Evaluate one integral by a geometric formula and the other by observing that the integrand is an odd function.)

22. ***Fluid Force on a Circular Plate*** Use the result of Exercise 21 to find the fluid force on the circular plate shown in each figure. Assume the plates are in the wall of a tank filled with water and the measurements are given in feet.

(a)

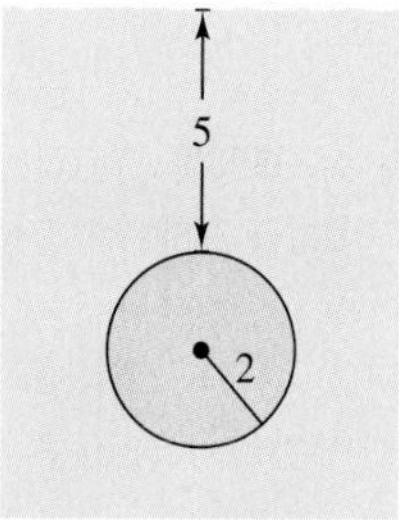

(b)

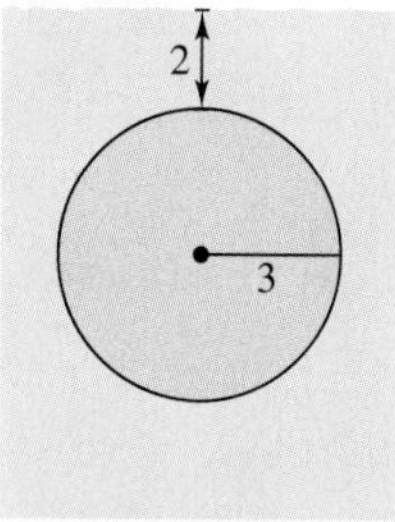

23. ***Fluid Force on a Rectangular Plate*** A rectangular plate of height h feet and base b feet is submerged vertically in a tank of fluid that weighs w pounds per cubic foot. The center is k feet below the surface of the fluid, where $h \le k/2$. Show that the fluid force on the surface of the plate is

$$F = wkhb.$$

24. ***Fluid Force on a Rectangular Plate*** Use the result of Exercise 23 to find the fluid force on the rectangular plate shown in each figure. Assume the plates are in the wall of a tank filled with water and the measurements are given in feet.

(a)

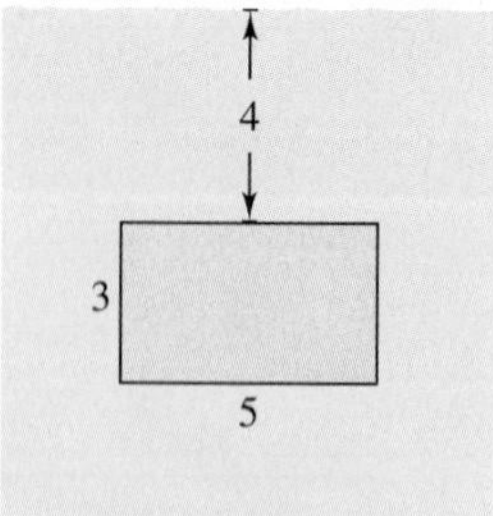

(b)

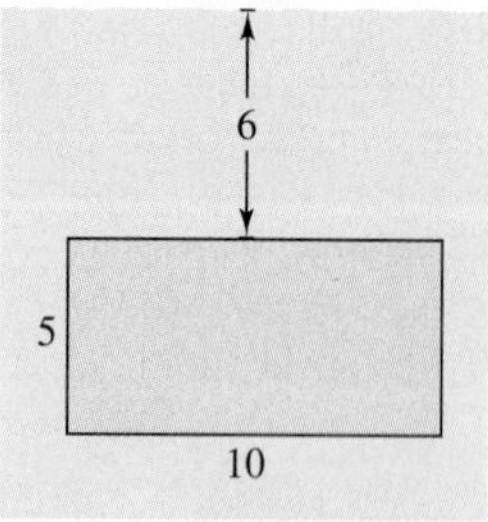

25. ***Submarine Porthole*** A porthole on a vertical side of a submarine (submerged in seawater) is 1 square foot. Find the fluid force on the porthole, assuming that the center of the square is 15 feet below the surface.

26. ***Submarine Porthole*** Repeat Exercise 25 for a circular porthole that has a diameter of 1 foot. The center is 15 feet below the surface.

27. ***Modeling Data*** The vertical stern of a boat with a superimposed coordinate system is shown in the figure. The table shows the width w of the stern at indicated values of y. Find the fluid force against the stern if the measurements are given in feet.

y	0	$\frac{1}{2}$	1	$\frac{3}{2}$	2	$\frac{5}{2}$	3	$\frac{7}{2}$	4
w	0	3	5	8	9	10	10.25	10.5	10.5

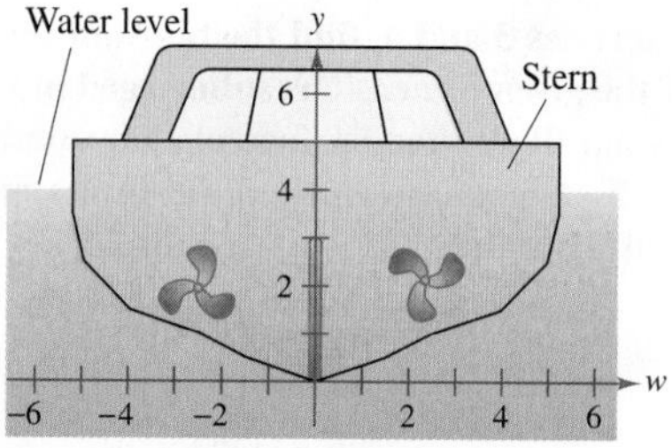

28. ***Irrigation Canal Gate*** The vertical cross section of an irrigation canal is modeled by

$$f(x) = \frac{5x^2}{x^2 + 4}$$

where x is measured in feet and $x = 0$ corresponds to the center of the canal. Use the integration capabilities of a graphing utility to approximate the fluid force against a vertical gate used to stop the flow of water if the water is 3 feet deep.

In Exercises 29 and 30, use the integration capabilities of a graphing utility to approximate the fluid force on the vertical plate bounded by the x-axis and the top half of the graph of the equation. Assume that the base of the plate is 12 feet beneath the surface of the water.

29. $x^{2/3} + y^{2/3} = 4^{2/3}$

30. $\dfrac{x^2}{28} + \dfrac{y^2}{16} = 1$

31. ***Think About It***

(a) Approximate the depth of the water in the tank in Exercise 5 if the fluid force is one-half as great as when the tank is full.

(b) Explain why the answer in part (a) is not $\frac{3}{2}$.

Writing About Concepts

32. Define fluid pressure.

33. Define fluid force against a submerged vertical plane region.

34. Two identical semicircular windows are placed at the same depth in the vertical wall of an aquarium (see figure). Which has the greater fluid force? Explain.

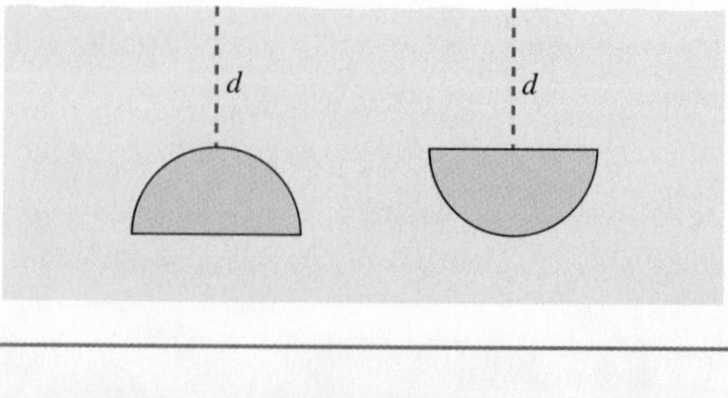

Review Exercises for Chapter 7

See www.CalcChat.com for worked-out solutions to odd-numbered exercises.

In Exercises 1–10, sketch the region bounded by the graphs of the equations, and determine the area of the region.

1. $y = \dfrac{1}{x^2}, \; y = 0, \; x = 1, \; x = 5$

2. $y = \dfrac{1}{x^2}, \; y = 4, \; x = 5$

3. $y = \dfrac{1}{x^2 + 1}, \; y = 0, \; x = -1, \; x = 1$

4. $x = y^2 - 2y, \; x = -1, \; y = 0$

5. $y = x, \; y = x^3$

6. $x = y^2 + 1, \; x = y + 3$

7. $y = e^x, \; y = e^2, \; x = 0$

8. $y = \csc x, \; y = 2$ (one region)

9. $y = \sin x, \; y = \cos x, \; \dfrac{\pi}{4} \le x \le \dfrac{5\pi}{4}$

10. $x = \cos y, \; x = \dfrac{1}{2}, \; \dfrac{\pi}{3} \le y \le \dfrac{7\pi}{3}$

In Exercises 11–14, use a graphing utility to graph the region bounded by the graphs of the functions, and use the integration capabilities of the graphing utility to find the area of the region.

11. $y = x^2 - 8x + 3, \; y = 3 + 8x - x^2$

12. $y = x^2 - 4x + 3, \; y = x^3, \; x = 0$

13. $\sqrt{x} + \sqrt{y} = 1, \; y = 0, \; x = 0$

14. $y = x^4 - 2x^2, \; y = 2x^2$

In Exercises 15–18, use vertical and horizontal representative rectangles to set up integrals for finding the area of the region bounded by the graphs of the equations. Find the area of the region by evaluating the easier of the two integrals.

15. $x = y^2 - 2y, \; x = 0$

16. $y = \sqrt{x - 1}, \; y = \dfrac{x - 1}{2}$

17. $y = 1 - \dfrac{x}{2}, \; y = x - 2, \; y = 1$

18. $y = \sqrt{x - 1}, \; y = 2, \; y = 0, \; x = 0$

19. ***Think About It*** A person has two job offers. The starting salary for each is \$30,000, and after 10 years of service each will pay \$56,000. The salary increases for each offer are shown in the figure. From a strictly monetary viewpoint, which is the better offer? Explain.

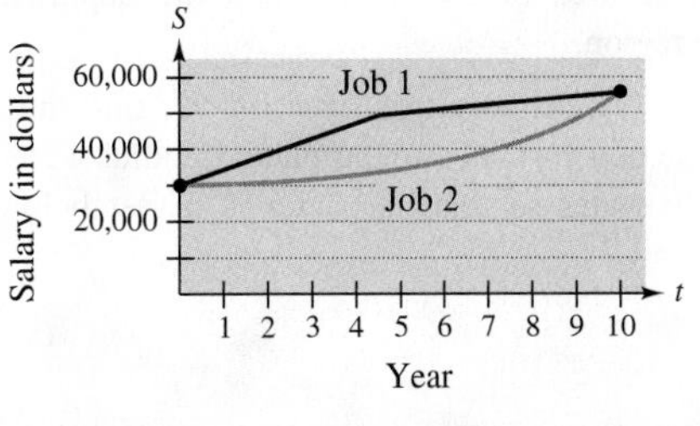

20. ***Modeling Data*** The table shows the annual service revenue R_1 in billions of dollars for the cellular telephone industry for the years 1995 through 2001. *(Source: Cellular Telecommunications & Internet Association)*

Year	1995	1996	1997	1998	1999	2000	2001
R_1	19.1	23.6	27.5	33.1	40.0	52.5	65.0

(a) Use the regression capabilities of a graphing utility to find an exponential model for the data. Let t represent the year, with $t = 5$ corresponding to 1995. Use the graphing utility to plot the data and graph the model in the same viewing window.

(b) A financial consultant believes that a model for service revenue for the years 2005 through 2010 is

$R_2 = 5 + 6.83e^{0.2t}$.

What is the difference in total service revenue between the two models for the years 2005 through 2010?

In Exercises 21–28, find the volume of the solid generated by revolving the plane region bounded by the equations about the indicated line(s).

21. $y = x, \; y = 0, \; x = 4$

(a) the x-axis (b) the y-axis
(c) the line $x = 4$ (d) the line $x = 6$

22. $y = \sqrt{x}, \; y = 2, \; x = 0$

(a) the x-axis (b) the line $y = 2$
(c) the y-axis (d) the line $x = -1$

23. $\dfrac{x^2}{16} + \dfrac{y^2}{9} = 1$

(a) the y-axis (oblate spheroid)
(b) the x-axis (prolate spheroid)

24. $\dfrac{x^2}{a^2} + \dfrac{y^2}{b^2} = 1$

(a) the y-axis (oblate spheroid)
(b) the x-axis (prolate spheroid)

25. $y = \dfrac{1}{x^4 + 1}, \; y = 0, \; x = 0, \; x = 1$

revolved about the y-axis

26. $y = \dfrac{1}{\sqrt{1 + x^2}}, \; y = 0, \; x = -1, \; x = 1$

revolved about the x-axis

27. $y = 1/(1 + \sqrt{x - 2}), \; y = 0, \; x = 2, \; x = 6$

revolved about the y-axis

28. $y = e^{-x}, \; y = 0, \; x = 0, \; x = 1$

revolved about the x-axis

In Exercises 29 and 30, consider the region bounded by the graphs of the equations $y = x\sqrt{x + 1}$ and $y = 0$.

29. ***Area*** Find the area of the region.

30. ***Volume*** Find the volume of the solid generated by revolving the region about (a) the x-axis and (b) the y-axis.

31. ***Depth of Gasoline in a Tank*** A gasoline tank is an oblate spheroid generated by revolving the region bounded by the graph of $(x^2/16) + (y^2/9) = 1$ about the y-axis, where x and y are measured in feet. Find the depth of the gasoline in the tank when it is filled to one-fourth its capacity.

32. ***Magnitude of a Base*** The base of a solid is a circle of radius a, and its vertical cross sections are equilateral triangles. The volume of the solid is 10 cubic meters. Find the radius of the circle.

In Exercises 33 and 34, find the arc length of the graph of the function over the given interval.

33. $f(x) = \frac{4}{5}x^{5/4}$, $[0, 4]$

34. $y = \frac{1}{6}x^3 + \frac{1}{2x}$, $[1, 3]$

35. ***Length of a Catenary*** A cable of a suspension bridge forms a catenary modeled by the equation

$$y = 300\cosh\left(\frac{x}{2000}\right) - 280, \quad -2000 \le x \le 2000$$

where x and y are measured in feet. Use a graphing utility to approximate the length of the cable.

36. ***Approximation*** Determine which value best approximates the length of the arc represented by the integral

$$\int_0^{\pi/4} \sqrt{1 + (\sec^2 x)^2}\, dx.$$

(Make your selection on the basis of a sketch of the arc and *not* by performing any calculations.)

(a) -2 (b) 1 (c) π (d) 4 (e) 3

37. ***Surface Area*** Use integration to find the lateral surface area of a right circular cone of height 4 and radius 3.

38. ***Surface Area*** The region bounded by the graphs of $y = 2\sqrt{x}$, $y = 0$, and $x = 3$ is revolved about the x-axis. Find the surface area of the solid generated.

39. ***Work*** A force of 4 pounds is needed to stretch a spring 1 inch from its natural position. Find the work done in stretching the spring from its natural length of 10 inches to a length of 15 inches.

40. ***Work*** The force required to stretch a spring is 50 pounds. Find the work done in stretching the spring from its natural length of 9 inches to double that length.

41. ***Work*** A water well has an eight-inch casing (diameter) and is 175 feet deep. The water is 25 feet from the top of the well. Determine the amount of work done in pumping the well dry, assuming that no water enters it while it is being pumped.

42. ***Work*** Repeat Exercise 41, assuming that water enters the well at a rate of 4 gallons per minute and the pump works at a rate of 12 gallons per minute. How many gallons are pumped in this case?

43. ***Work*** A chain 10 feet long weighs 5 pounds per foot and is hung from a platform 20 feet above the ground. How much work is required to raise the entire chain to the 20-foot level?

44. ***Work*** A windlass, 200 feet above ground level on the top of a building, uses a cable weighing 4 pounds per foot. Find the work done in winding up the cable if

(a) one end is at ground level.

(b) there is a 300-pound load attached to the end of the cable.

45. ***Work*** The work done by a variable force in a press is 80 foot-pounds. The press moves a distance of 4 feet and the force is a quadratic of the form $F = ax^2$. Find a.

46. ***Work*** Find the work done by the force F shown in the figure.

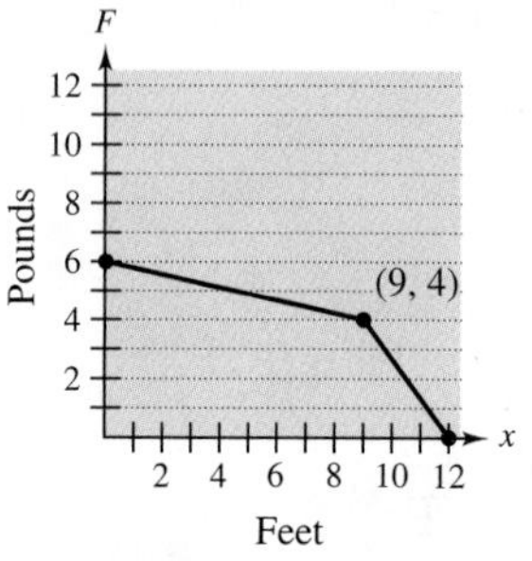

In Exercises 47–50, find the centroid of the region bounded by the graphs of the equations.

47. $\sqrt{x} + \sqrt{y} = \sqrt{a}$, $x = 0$, $y = 0$

48. $y = x^2$, $y = 2x + 3$

49. $y = a^2 - x^2$, $y = 0$

50. $y = x^{2/3}$, $y = \frac{1}{2}x$

51. ***Centroid*** A blade on an industrial fan has the configuration of a semicircle attached to a trapezoid (see figure). Find the centroid of the blade.

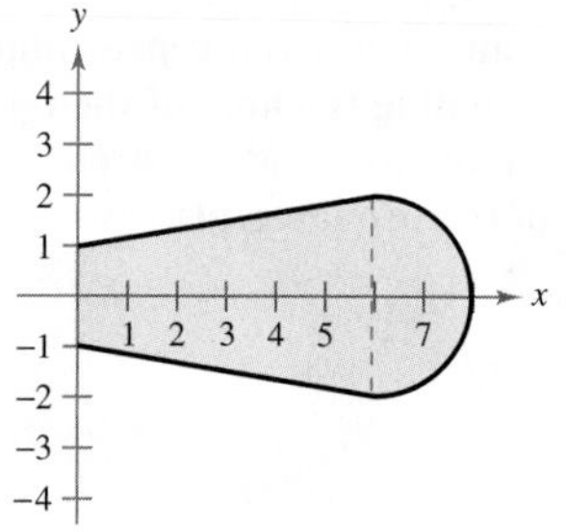

52. ***Fluid Force*** A swimming pool is 5 feet deep at one end and 10 feet deep at the other, and the bottom is an inclined plane. The length and width of the pool are 40 feet and 20 feet. If the pool is full of water, what is the fluid force on each of the vertical walls?

53. ***Fluid Force*** Show that the fluid force against any vertical region in a liquid is the product of the weight per cubic volume of the liquid, the area of the region, and the depth of the centroid of the region.

54. ***Fluid Force*** Using the result of Exercise 53, find the fluid force on one side of a vertical circular plate of radius 4 feet that is submerged in water so that its center is 5 feet below the surface.

P.S. Problem Solving

See www.CalcChat.com for worked-out solutions to odd-numbered exercises.

1. Let R be the area of the region in the first quadrant bounded by the parabola $y = x^2$ and the line $y = cx$, $c > 0$. Let T be the area of the triangle AOB. Calculate the limit

$$\lim_{c \to 0^+} \frac{T}{R}.$$

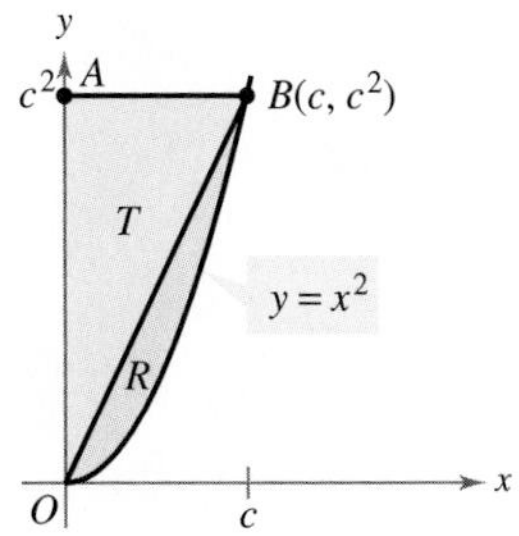

2. Let R be the region bounded by the parabola $y = x - x^2$ and the x-axis. Find the equation of the line $y = mx$ that divides this region into two regions of equal area.

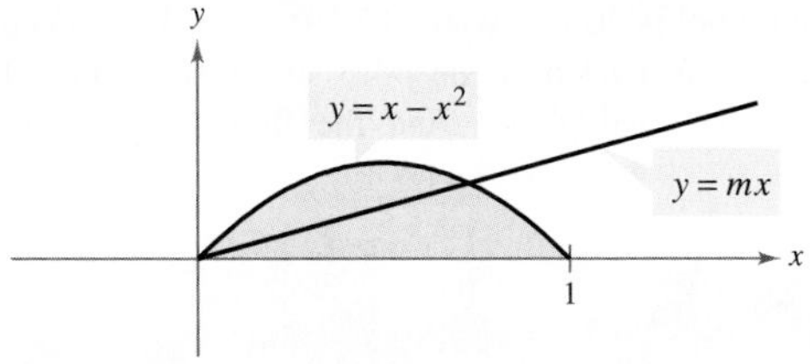

3. (a) A torus is formed by revolving the region bounded by the circle

$$(x - 2)^2 + y^2 = 1$$

about the y-axis (see figure). Use the disk method to calculate the volume of the torus.

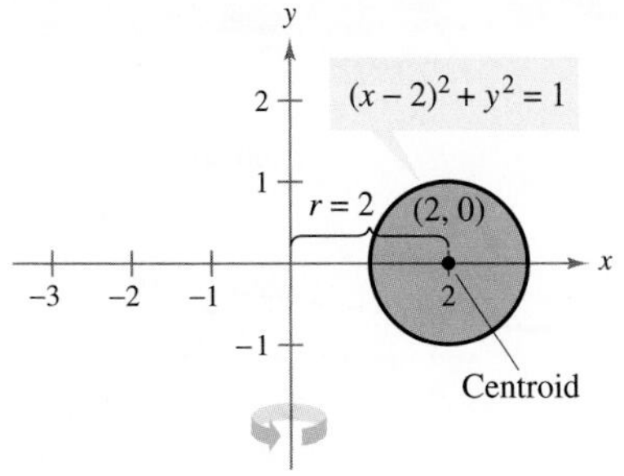

(b) Use the disk method to find the volume of the general torus if the circle has radius r and its center is R units from the axis of rotation.

4. Graph the curve

$$8y^2 = x^2(1 - x^2).$$

Use a computer algebra system to find the surface area of the solid of revolution obtained by revolving the curve about the x-axis.

5. A hole is cut through the center of a sphere of radius r (see figure). The height of the remaining spherical ring is h. Find the volume of the ring and show that it is independent of the radius of the sphere.

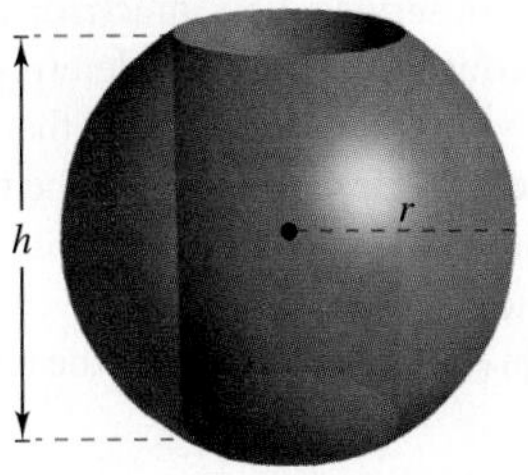

6. A rectangle R of length l and width w is revolved about the line L (see figure). Find the volume of the resulting solid of revolution.

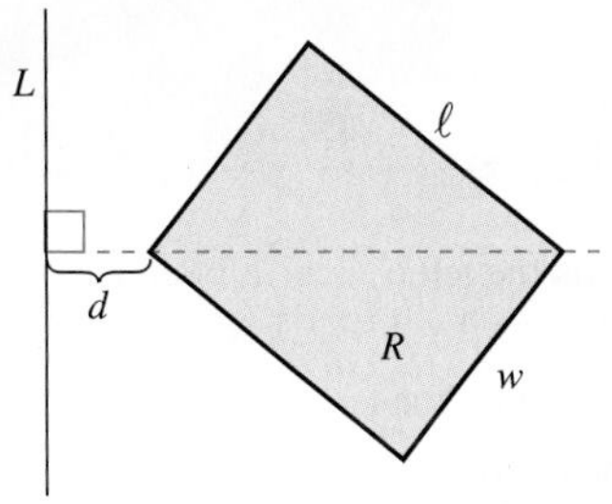

Figure for 6

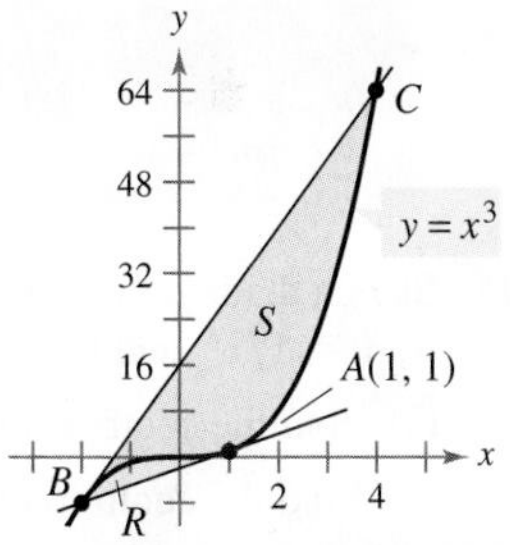

Figure for 7

7. (a) The tangent line to the curve $y = x^3$ at the point $A(1, 1)$ intersects the curve at another point B. Let R be the area of the region bounded by the curve and the tangent line. The tangent line at B intersects the curve at another point C (see figure). Let S be the area of the region bounded by the curve and this second tangent line. How are the areas R and S related?

(b) Repeat the construction in part (a) by selecting an arbitrary point A on the curve $y = x^3$. Show that the two areas R and S are always related in the same way.

8. The graph of $y = f(x)$ passes through the origin. The arc length of the curve from $(0, 0)$ to $(x, f(x))$ is given by

$$s(x) = \int_0^x \sqrt{1 + e^t}\, dt.$$

Identify the function f.

9. Let f be rectifiable on the interval $[a, b]$, and let

$$s(x) = \int_a^x \sqrt{1 + [f'(t)]^2}\, dt.$$

(a) Find $\dfrac{ds}{dx}$.

(b) Find ds and $(ds)^2$.

(c) If $f(t) = t^{3/2}$, find $s(x)$ on $[1, 3]$.

(d) Calculate $s(2)$ and describe what it signifies.

10. The **Archimedes Principle** states that the upward or buoyant force on an object within a fluid is equal to the weight of the fluid that the object displaces. For a partially submerged object, you can obtain information about the relative densities of the floating object and the fluid by observing how much of the object is above and below the surface. You can also determine the size of a floating object if you know the amount that is above the surface and the relative densities. You can see the top of a floating iceberg (see figure). The density of ocean water is 1.03×10^3 kilograms per cubic meter, and that of ice is 0.92×10^3 kilograms per cubic meter. What percent of the total iceberg is below the surface?

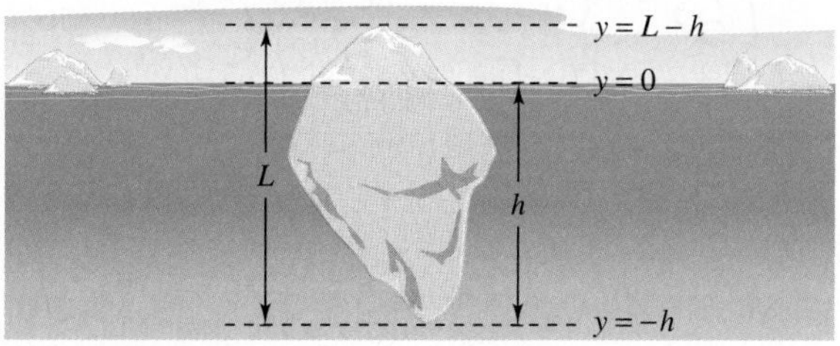

11. Sketch the region bounded on the left by $x = 1$, bounded above by $y = 1/x^3$, and bounded below by $y = -1/x^3$.

(a) Find the centroid of the region for $1 \le x \le 6$.

(b) Find the centroid of the region for $1 \le x \le b$.

(c) Where is the centroid as $b \to \infty$?

12. Sketch the region to the right of the y-axis, bounded above by $y = 1/x^4$ and bounded below by $y = -1/x^4$.

(a) Find the centroid of the region for $1 \le x \le 6$.

(b) Find the centroid of the region for $1 \le x \le b$.

(c) Where is the centroid as $b \to \infty$?

13. Find the work done by each force F.

(a) 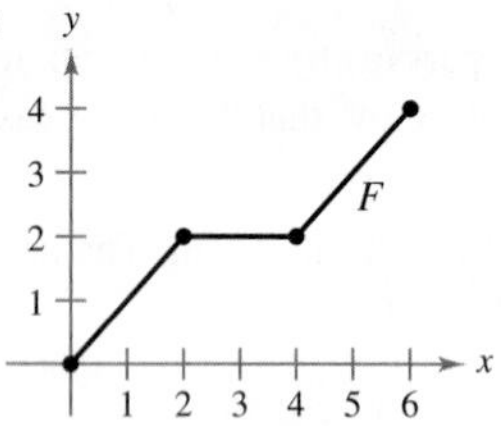

(b) 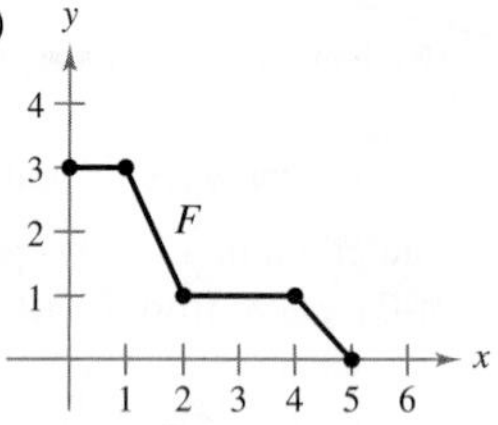

14. Estimate the surface area of the pond using (a) the Trapezoidal Rule and (b) Simpson's Rule.

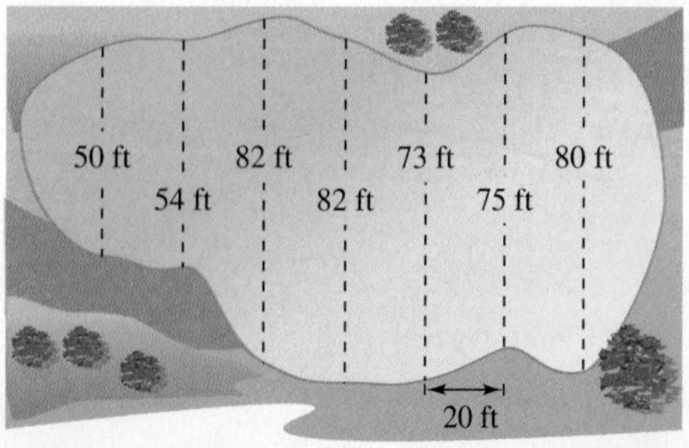

In Exercises 15 and 16, find the consumer surplus and producer surplus for the given demand $[p_1(x)]$ and supply $[p_2(x)]$ curves. The consumer surplus and producer surplus are represented by the areas shown in the figure.

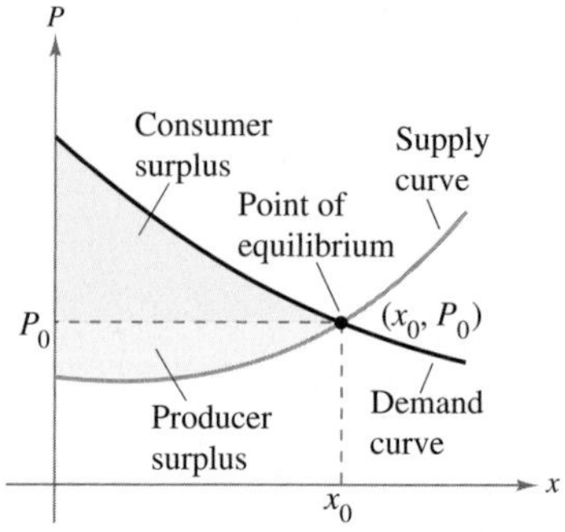

15. $p_1(x) = 50 - 0.5x, \quad p_2(x) = 0.125x$

16. $p_1(x) = 1000 - 0.4x^2, \quad p_2(x) = 42x$

17. A swimming pool is 20 feet wide, 40 feet long, 4 feet deep at one end, and 8 feet deep at the other end (see figure). The bottom is an inclined plane. Find the fluid force on each vertical wall.

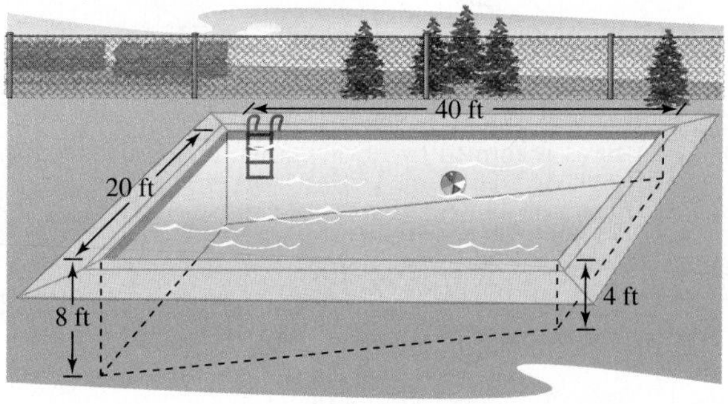

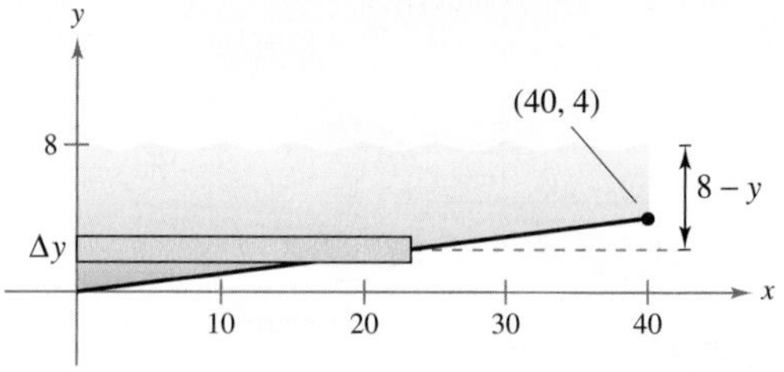

18. (a) Find at least two continuous functions f that satisfy each condition.

(i) $f(x) \ge 0$ on $[0, 1]$ (ii) $f(0) = 0$ and $f(1) = 0$

(iii) The area bounded by the graph of f and the x-axis for $0 \le x \le 1$ equals 1.

(b) For each function found in part (a), approximate the arc length of the graph of the function on the interval $[0, 1]$. (Use a graphing utility if necessary.)

(c) Can you find a function f that satisfies the conditions in part (a) and whose graph has an arc length of less than 3 on the interval $[0, 1]$?

8 Integration Techniques, L'Hôpital's Rule, and Improper Integrals

The NASA Hubble Space Telescope image of a planetary nebula nicknamed the "Cat's Eye Nebula" gives just a glimpse of the kinds of things you might see if you could travel through space. Would it be possible to propel a spacecraft an unlimited distance away from Earth's surface? Why?

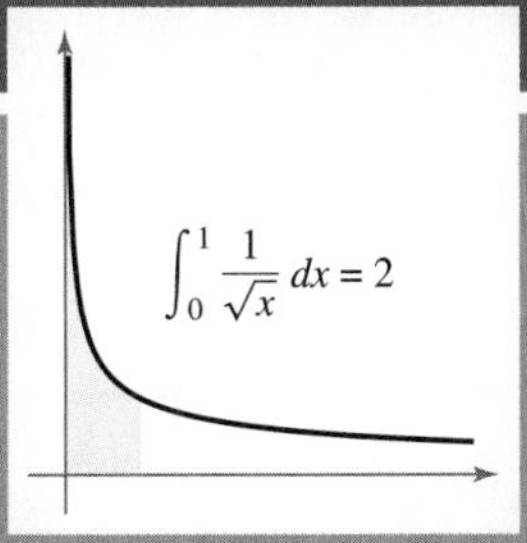

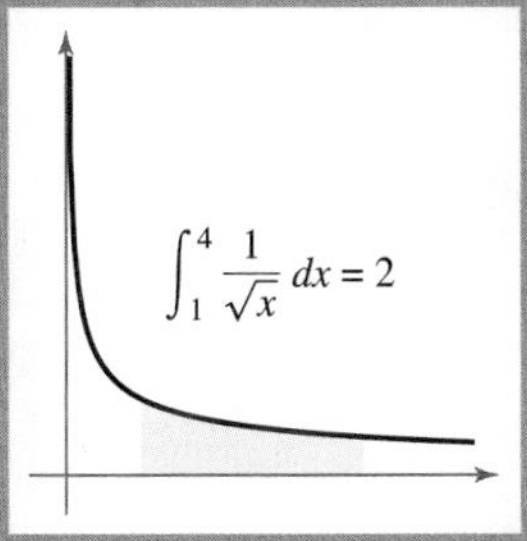

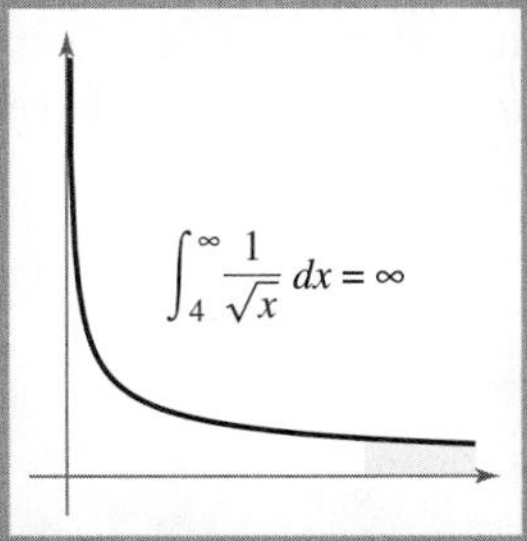

From your studies of calculus thus far, you know that a definite integral has finite limits of integration and a continuous integrand. In Chapter 8, you will study *improper integrals*. Improper integrals have at least one infinite limit of integration or have an integrand with an infinite discontinuity. You will see that improper integrals either converge or diverge.

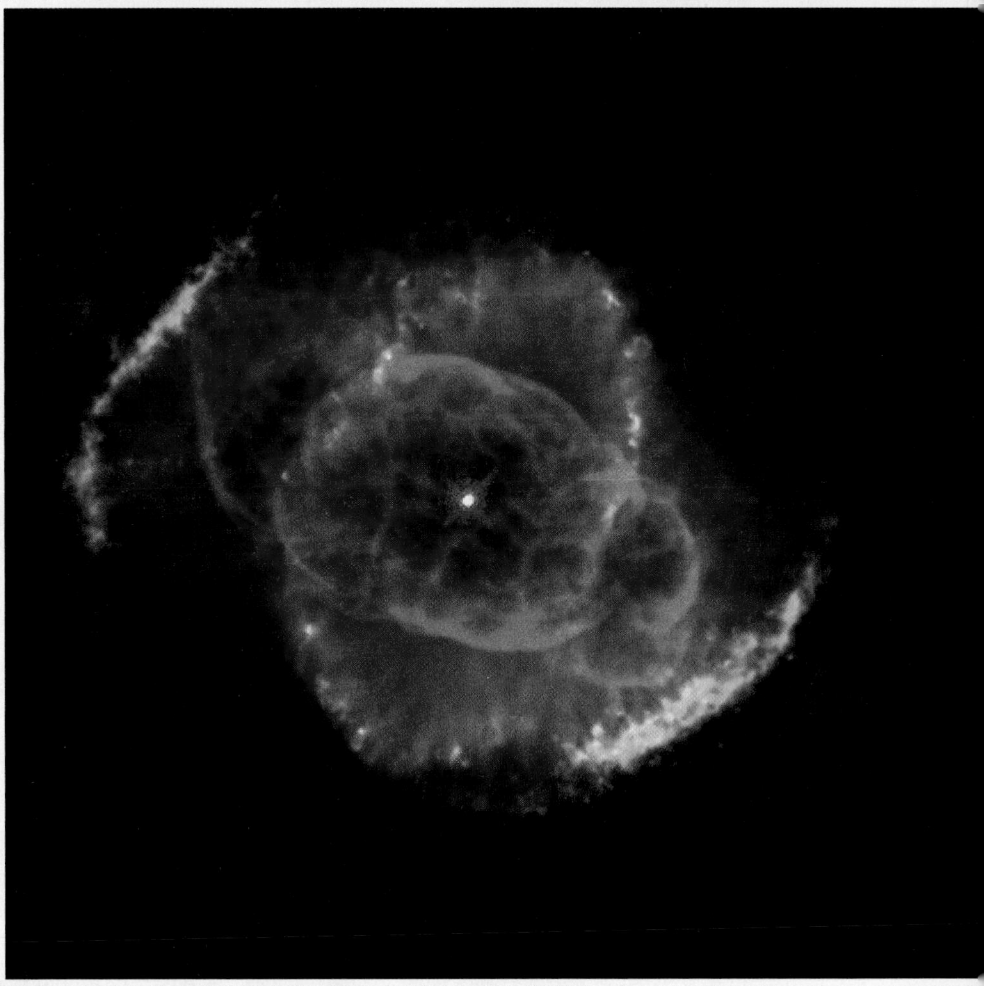

P. Harrington and K.J. Borkowski (University of Maryland), and NASA

Section 8.1 Basic Integration Rules

- **Review procedures for fitting an integrand to one of the basic integration rules.**

Fitting Integrands to Basic Rules

In this chapter, you will study several integration techniques that greatly expand the set of integrals to which the basic integration rules can be applied. These rules are reviewed on page 520. A major step in solving any integration problem is recognizing which basic integration rule to use. As shown in Example 1, slight differences in the integrand can lead to very different solution techniques.

EXPLORATION

A Comparison of Three Similar Integrals Which, if any, of the following integrals can be evaluated using the 20 basic integration rules? For any that can be evaluated, do so. For any that can't, explain why.

a. $\displaystyle\int \frac{3}{\sqrt{1-x^2}}\,dx$

b. $\displaystyle\int \frac{3x}{\sqrt{1-x^2}}\,dx$

c. $\displaystyle\int \frac{3x^2}{\sqrt{1-x^2}}\,dx$

EXAMPLE 1 A Comparison of Three Similar Integrals

Find each integral.

a. $\displaystyle\int \frac{4}{x^2+9}\,dx$ **b.** $\displaystyle\int \frac{4x}{x^2+9}\,dx$ **c.** $\displaystyle\int \frac{4x^2}{x^2+9}\,dx$

Solution

a. Use the Arctangent Rule and let $u = x$ and $a = 3$.

$$\int \frac{4}{x^2+9}\,dx = 4\int \frac{1}{x^2+3^2}\,dx \qquad \text{Constant Multiple Rule}$$

$$= 4\left(\frac{1}{3}\arctan\frac{x}{3}\right) + C \qquad \text{Arctangent Rule}$$

$$= \frac{4}{3}\arctan\frac{x}{3} + C \qquad \text{Simplify.}$$

b. Here the Arctangent Rule does not apply because the numerator contains a factor of x. Consider the Log Rule and let $u = x^2 + 9$. Then $du = 2x\,dx$, and you have

$$\int \frac{4x}{x^2+9}\,dx = 2\int \frac{2x\,dx}{x^2+9} \qquad \text{Constant Multiple Rule}$$

$$= 2\int \frac{du}{u} \qquad \text{Substitution: } u = x^2+9$$

$$= 2\ln|u| + C = 2\ln(x^2+9) + C. \qquad \text{Log Rule}$$

c. Because the degree of the numerator is equal to the degree of the denominator, you should first use division to rewrite the improper rational function as the sum of a polynomial and a proper rational function.

$$\int \frac{4x^2}{x^2+9}\,dx = \int \left(4 - \frac{36}{x^2+9}\right)dx \qquad \text{Rewrite using long division.}$$

$$= \int 4\,dx - 36\int \frac{1}{x^2+9}\,dx \qquad \text{Write as two integrals.}$$

$$= 4x - 36\left(\frac{1}{3}\arctan\frac{x}{3}\right) + C \qquad \text{Integrate.}$$

$$= 4x - 12\arctan\frac{x}{3} + C \qquad \text{Simplify.}$$

NOTE Notice in Example 1(c) that some preliminary algebra is required before applying the rules for integration, and that subsequently more than one rule is needed to evaluate the resulting integral.

(www) indicates that in the HM mathSpace® *CD-ROM and the online* Eduspace® *system for this text, you will find an Open Exploration, which further explores this example using the computer algebra systems* Maple, Mathcad, Mathematica, *and* Derive.

EXAMPLE 2 Using Two Basic Rules to Solve a Single Integral

Evaluate $\displaystyle\int_0^1 \frac{x+3}{\sqrt{4-x^2}}\,dx.$

Solution Begin by writing the integral as the sum of two integrals. Then apply the Power Rule and the Arcsine Rule as follows.

$$\begin{aligned}\int_0^1 \frac{x+3}{\sqrt{4-x^2}}\,dx &= \int_0^1 \frac{x}{\sqrt{4-x^2}}\,dx + \int_0^1 \frac{3}{\sqrt{4-x^2}}\,dx\\ &= -\frac{1}{2}\int_0^1 (4-x^2)^{-1/2}(-2x)\,dx + 3\int_0^1 \frac{1}{\sqrt{2^2-x^2}}\,dx\\ &= \left[-(4-x^2)^{1/2} + 3\arcsin\frac{x}{2}\right]_0^1\\ &= \left(-\sqrt{3}+\frac{\pi}{2}\right) - (-2+0)\\ &\approx 1.839\end{aligned}$$

See Figure 8.1.

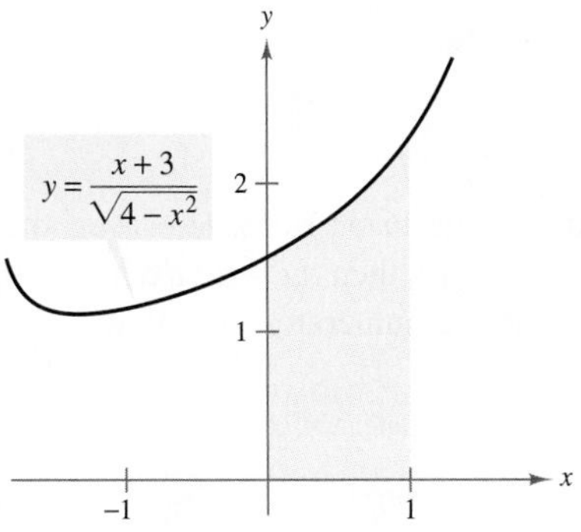

The area of the region is approximately 1.839.
Figure 8.1

TECHNOLOGY Simpson's Rule can be used to give a good approximation of the value of the integral in Example 2 (for $n = 10$, the approximation is 1.839). When using numerical integration, however, you should be aware that Simpson's Rule does not always give good approximations when one or both of the limits of integration are near a vertical asymptote. For instance, using the Fundamental Theorem of Calculus, you can obtain

$$\int_0^{1.99} \frac{x+3}{\sqrt{4-x^2}}\,dx \approx 6.213.$$

Applying Simpson's Rule (with $n = 10$) to this integral produces an approximation of 6.889.

EXAMPLE 3 A Substitution Involving $a^2 - u^2$

Find $\displaystyle\int \frac{x^2}{\sqrt{16-x^6}}\,dx.$

Solution Because the radical in the denominator can be written in the form

$$\sqrt{a^2-u^2} = \sqrt{4^2-(x^3)^2}$$

you can try the substitution $u = x^3$. Then $du = 3x^2\,dx$, and you have

$$\begin{aligned}\int \frac{x^2}{\sqrt{16-x^6}}\,dx &= \frac{1}{3}\int \frac{3x^2\,dx}{\sqrt{16-(x^3)^2}} && \text{Rewrite integral.}\\ &= \frac{1}{3}\int \frac{du}{\sqrt{4^2-u^2}} && \text{Substitution: } u = x^3\\ &= \frac{1}{3}\arcsin\frac{u}{4} + C && \text{Arcsine Rule}\\ &= \frac{1}{3}\arcsin\frac{x^3}{4} + C. && \text{Rewrite as a function of } x.\end{aligned}$$

STUDY TIP Rules 18, 19, and 20 of the basic integration rules on the next page all have expressions involving the sum or difference of two squares:

$a^2 - u^2$

$a^2 + u^2$

$u^2 - a^2$

With such an expression, consider the substitution $u = f(x)$, as in Example 3.

Surprisingly, two of the most commonly overlooked integration rules are the Log Rule and the Power Rule. Notice in the next two examples how these two integration rules can be disguised.

EXAMPLE 4 A Disguised Form of the Log Rule

Find $\int \frac{1}{1+e^x}\,dx$.

Solution The integral does not appear to fit any of the basic rules. However, the quotient form suggests the Log Rule. If you let $u = 1 + e^x$, then $du = e^x\,dx$. You can obtain the required du by adding and subtracting e^x in the numerator, as follows.

$$\int \frac{1}{1+e^x}\,dx = \int \frac{1+e^x-e^x}{1+e^x}\,dx \qquad \text{Add and subtract } e^x \text{ in numerator.}$$

$$= \int \left(\frac{1+e^x}{1+e^x} - \frac{e^x}{1+e^x}\right)dx \qquad \text{Rewrite as two fractions.}$$

$$= \int dx - \int \frac{e^x\,dx}{1+e^x} \qquad \text{Rewrite as two integrals.}$$

$$= x - \ln(1+e^x) + C \qquad \text{Integrate.}$$

NOTE There is usually more than one way to solve an integration problem. For instance, in Example 4, try integrating by multiplying the numerator and denominator by e^{-x} to obtain an integral of the form $-\int du/u$. See if you can get the same answer by this procedure. (Be careful: the answer will appear in a different form.)

EXAMPLE 5 A Disguised Form of the Power Rule

Find $\int (\cot x)[\ln(\sin x)]\,dx$.

Solution Again, the integral does not appear to fit any of the basic rules. However, considering the two primary choices for u $[u = \cot x$ and $u = \ln(\sin x)]$, you can see that the second choice is the appropriate one because

$$u = \ln(\sin x) \quad \text{and} \quad du = \frac{\cos x}{\sin x}\,dx = \cot x\,dx.$$

So,

$$\int (\cot x)[\ln(\sin x)]\,dx = \int u\,du \qquad \text{Substitution: } u = \ln(\sin x)$$

$$= \frac{u^2}{2} + C \qquad \text{Integrate.}$$

$$= \frac{1}{2}[\ln(\sin x)]^2 + C. \qquad \text{Rewrite as a function of } x.$$

NOTE In Example 5, try *checking* that the derivative of

$$\frac{1}{2}[\ln(\sin x)]^2 + C$$

is the integrand of the original integral.

Review of Basic Integration Rules ($a > 0$)

1. $\int kf(u)\,du = k\int f(u)\,du$
2. $\int [f(u) \pm g(u)]\,du = \int f(u)\,du \pm \int g(u)\,du$
3. $\int du = u + C$
4. $\int u^n\,du = \frac{u^{n+1}}{n+1} + C,\ n \neq -1$
5. $\int \frac{du}{u} = \ln|u| + C$
6. $\int e^u\,du = e^u + C$
7. $\int a^u\,du = \left(\frac{1}{\ln a}\right)a^u + C$
8. $\int \sin u\,du = -\cos u + C$
9. $\int \cos u\,du = \sin u + C$
10. $\int \tan u\,du = -\ln|\cos u| + C$
11. $\int \cot u\,du = \ln|\sin u| + C$
12. $\int \sec u\,du = \ln|\sec u + \tan u| + C$
13. $\int \csc u\,du = -\ln|\csc u + \cot u| + C$
14. $\int \sec^2 u\,du = \tan u + C$
15. $\int \csc^2 u\,du = -\cot u + C$
16. $\int \sec u \tan u\,du = \sec u + C$
17. $\int \csc u \cot u\,du = -\csc u + C$
18. $\int \frac{du}{\sqrt{a^2-u^2}} = \arcsin\frac{u}{a} + C$
19. $\int \frac{du}{a^2+u^2} = \frac{1}{a}\arctan\frac{u}{a} + C$
20. $\int \frac{du}{u\sqrt{u^2-a^2}} = \frac{1}{a}\operatorname{arcsec}\frac{|u|}{a} + C$

Trigonometric identities can often be used to fit integrals to one of the basic integration rules.

EXAMPLE 6 Using Trigonometric Identities

Find $\int \tan^2 2x\,dx$.

TECHNOLOGY If you have access to a computer algebra system, try using it to evaluate the integrals in this section. Compare the *form* of the antiderivative given by the software with the form obtained by hand. Sometimes the forms will be the same, but often they will differ. For instance, why is the antiderivative $\ln 2x + C$ equivalent to the antiderivative $\ln x + C$?

Solution Note that $\tan^2 u$ is not in the list of basic integration rules. However, $\sec^2 u$ is in the list. This suggests the trigonometric identity $\tan^2 u = \sec^2 u - 1$. If you let $u = 2x$, then $du = 2\,dx$ and

$$\int \tan^2 2x\,dx = \frac{1}{2}\int \tan^2 u\,du \qquad \text{Substitution: } u = 2x$$

$$= \frac{1}{2}\int (\sec^2 u - 1)\,du \qquad \text{Trigonometric identity}$$

$$= \frac{1}{2}\int \sec^2 u\,du - \frac{1}{2}\int du \qquad \text{Rewrite as two integrals.}$$

$$= \frac{1}{2}\tan u - \frac{u}{2} + C \qquad \text{Integrate.}$$

$$= \frac{1}{2}\tan 2x - x + C. \qquad \text{Rewrite as a function of } x.$$

This section concludes with a summary of the common procedures for fitting integrands to the basic integration rules.

Procedures for Fitting Integrands to Basic Rules

Technique	*Example*
Expand (numerator).	$(1 + e^x)^2 = 1 + 2e^x + e^{2x}$
Separate numerator.	$\dfrac{1 + x}{x^2 + 1} = \dfrac{1}{x^2 + 1} + \dfrac{x}{x^2 + 1}$
Complete the square.	$\dfrac{1}{\sqrt{2x - x^2}} = \dfrac{1}{\sqrt{1 - (x - 1)^2}}$
Divide improper rational function.	$\dfrac{x^2}{x^2 + 1} = 1 - \dfrac{1}{x^2 + 1}$
Add and subtract terms in numerator.	$\dfrac{2x}{x^2 + 2x + 1} = \dfrac{2x + 2 - 2}{x^2 + 2x + 1} = \dfrac{2x + 2}{x^2 + 2x + 1} - \dfrac{2}{(x + 1)^2}$
Use trigonometric identities.	$\cot^2 x = \csc^2 x - 1$
Multiply and divide by Pythagorean conjugate.	$\dfrac{1}{1 + \sin x} = \left(\dfrac{1}{1 + \sin x}\right)\left(\dfrac{1 - \sin x}{1 - \sin x}\right) = \dfrac{1 - \sin x}{1 - \sin^2 x}$ $= \dfrac{1 - \sin x}{\cos^2 x} = \sec^2 x - \dfrac{\sin x}{\cos^2 x}$

NOTE Remember that you can separate numerators but not denominators. Watch out for this common error when fitting integrands to basic rules.

$$\frac{1}{x^2 + 1} \neq \frac{1}{x^2} + \frac{1}{1} \qquad \text{Do not separate denominators.}$$

Exercises for Section 8.1

See www.CalcChat.com for worked-out solutions to odd-numbered exercises.

In Exercises 1–4, select the correct antiderivative.

1. $\dfrac{dy}{dx} = \dfrac{x}{\sqrt{x^2+1}}$

(a) $2\sqrt{x^2+1} + C$ (b) $\sqrt{x^2+1} + C$
(c) $\frac{1}{2}\sqrt{x^2+1} + C$ (d) $\ln(x^2+1) + C$

2. $\dfrac{dy}{dx} = \dfrac{x}{x^2+1}$

(a) $\ln\sqrt{x^2+1} + C$ (b) $\dfrac{2x}{(x^2+1)^2} + C$
(c) $\arctan x + C$ (d) $\ln(x^2+1) + C$

3. $\dfrac{dy}{dx} = \dfrac{1}{x^2+1}$

(a) $\ln\sqrt{x^2+1} + C$ (b) $\dfrac{2x}{(x^2+1)^2} + C$
(c) $\arctan x + C$ (d) $\ln(x^2+1) + C$

4. $\dfrac{dy}{dx} = x\cos(x^2+1)$

(a) $2x\sin(x^2+1) + C$ (b) $-\frac{1}{2}\sin(x^2+1) + C$
(c) $\frac{1}{2}\sin(x^2+1) + C$ (d) $-2x\sin(x^2+1) + C$

In Exercises 5–14, select the basic integration formula you can use to find the integral, and identify *u* and *a* when appropriate.

5. $\displaystyle\int (3x-2)^4\,dx$
6. $\displaystyle\int \frac{2t-1}{t^2-t+2}\,dt$
7. $\displaystyle\int \frac{1}{\sqrt{x}\left(1-2\sqrt{x}\right)}\,dx$
8. $\displaystyle\int \frac{2}{(2t-1)^2+4}\,dt$
9. $\displaystyle\int \frac{3}{\sqrt{1-t^2}}\,dt$
10. $\displaystyle\int \frac{-2x}{\sqrt{x^2-4}}\,dx$
11. $\displaystyle\int t\sin t^2\,dt$
12. $\displaystyle\int \sec 3x\tan 3x\,dx$
13. $\displaystyle\int (\cos x)e^{\sin x}\,dx$
14. $\displaystyle\int \frac{1}{x\sqrt{x^2-4}}\,dx$

In Exercises 15–50, find the indefinite integral.

15. $\displaystyle\int 6(x-4)^5\,dx$
16. $\displaystyle\int \frac{2}{(t-9)^2}\,dt$
17. $\displaystyle\int \frac{5}{(z-4)^5}\,dz$
18. $\displaystyle\int t^2\sqrt[3]{t^3-1}\,dt$
19. $\displaystyle\int \left[v + \frac{1}{(3v-1)^3}\right]dv$
20. $\displaystyle\int \left[x - \frac{3}{(2x+3)^2}\right]dx$
21. $\displaystyle\int \frac{t^2-3}{-t^3+9t+1}\,dt$
22. $\displaystyle\int \frac{x+1}{\sqrt{x^2+2x-4}}\,dx$
23. $\displaystyle\int \frac{x^2}{x-1}\,dx$
24. $\displaystyle\int \frac{2x}{x-4}\,dx$
25. $\displaystyle\int \frac{e^x}{1+e^x}\,dx$
26. $\displaystyle\int \left(\frac{1}{3x-1} - \frac{1}{3x+1}\right)dx$
27. $\displaystyle\int (1+2x^2)^2\,dx$
28. $\displaystyle\int x\left(1+\frac{1}{x}\right)^3 dx$
29. $\displaystyle\int x\cos 2\pi x^2\,dx$
30. $\displaystyle\int \sec 4x\,dx$
31. $\displaystyle\int \csc \pi x\cot \pi x\,dx$
32. $\displaystyle\int \frac{\sin x}{\sqrt{\cos x}}\,dx$
33. $\displaystyle\int e^{5x}\,dx$
34. $\displaystyle\int \csc^2 x e^{\cot x}\,dx$
35. $\displaystyle\int \frac{2}{e^{-x}+1}\,dx$
36. $\displaystyle\int \frac{5}{3e^x-2}\,dx$
37. $\displaystyle\int \frac{\ln x^2}{x}\,dx$
38. $\displaystyle\int (\tan x)[\ln(\cos x)]\,dx$
39. $\displaystyle\int \frac{1+\sin x}{\cos x}\,dx$
40. $\displaystyle\int \frac{1+\cos\alpha}{\sin\alpha}\,d\alpha$
41. $\displaystyle\int \frac{1}{\cos\theta - 1}\,d\theta$
42. $\displaystyle\int \frac{2}{3(\sec x - 1)}\,dx$
43. $\displaystyle\int \frac{-1}{\sqrt{1-(2t-1)^2}}\,dt$
44. $\displaystyle\int \frac{1}{4+3x^2}\,dx$
45. $\displaystyle\int \frac{\tan(2/t)}{t^2}\,dt$
46. $\displaystyle\int \frac{e^{1/t}}{t^2}\,dt$
47. $\displaystyle\int \frac{3}{\sqrt{6x-x^2}}\,dx$
48. $\displaystyle\int \frac{1}{(x-1)\sqrt{4x^2-8x+3}}\,dx$
49. $\displaystyle\int \frac{4}{4x^2+4x+65}\,dx$
50. $\displaystyle\int \frac{1}{\sqrt{1-4x-x^2}}\,dx$

Slope Fields **In Exercises 51–54, a differential equation, a point, and a slope field are given. (a) Sketch two approximate solutions of the differential equation on the slope field, one of which passes through the given point. (b) Use integration to find the particular solution of the differential equation and use a graphing utility to graph the solution. Compare the result with the sketches in part (a). To print an enlarged copy of the graph, go to the website *www.mathgraphs.com*.**

51. $\dfrac{ds}{dt} = \dfrac{t}{\sqrt{1-t^4}}$

$\left(0, -\dfrac{1}{2}\right)$

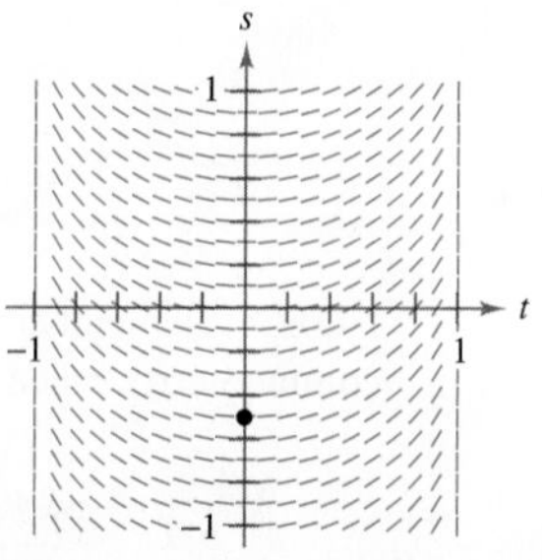

52. $\dfrac{dy}{dx} = \tan^2(2x)$

$(0, 0)$

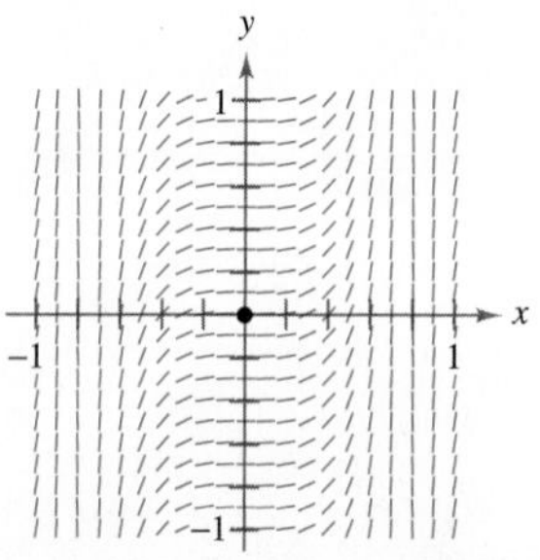

53. $\dfrac{dy}{dx} = (\sec x + \tan x)^2$

$(0, 1)$

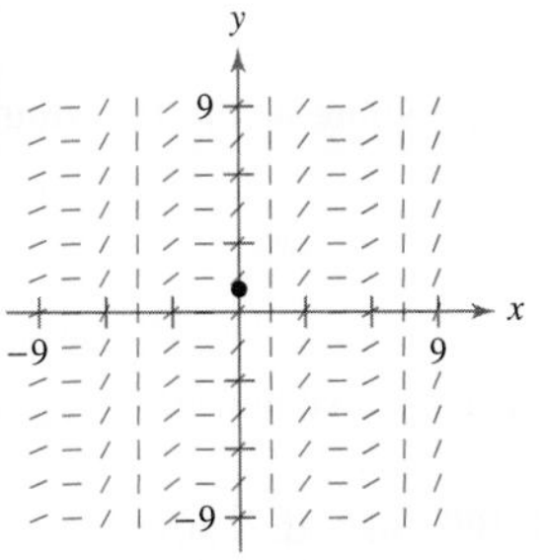

54. $\dfrac{dy}{dx} = \dfrac{1}{\sqrt{4x - x^2}}$

$\left(2, \dfrac{1}{2}\right)$

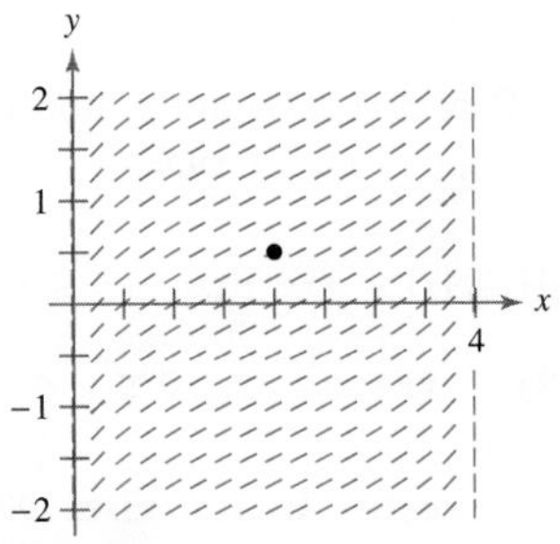

Slope Fields **In Exercises 55 and 56, use a computer algebra system to graph the slope field for the differential equation and graph the solution through the specified initial condition.**

55. $\dfrac{dy}{dx} = 0.2y,\ y(0) = 3$

56. $\dfrac{dy}{dx} = 5 - y,\ y(0) = 1$

In Exercises 57–60, solve the differential equation.

57. $\dfrac{dy}{dx} = (1 + e^x)^2$

58. $\dfrac{dr}{dt} = \dfrac{(1 + e^t)^2}{e^t}$

59. $(4 + \tan^2 x)y' = \sec^2 x$

60. $y' = \dfrac{1}{x\sqrt{4x^2 - 1}}$

In Exercises 61–68, evaluate the definite integral. Use the integration capabilities of a graphing utility to verify your result.

61. $\displaystyle\int_0^{\pi/4} \cos 2x\, dx$

62. $\displaystyle\int_0^{\pi} \sin^2 t \cos t\, dt$

63. $\displaystyle\int_0^1 xe^{-x^2}\, dx$

64. $\displaystyle\int_1^e \frac{1 - \ln x}{x}\, dx$

65. $\displaystyle\int_0^4 \frac{2x}{\sqrt{x^2 + 9}}\, dx$

66. $\displaystyle\int_1^2 \frac{x - 2}{x}\, dx$

67. $\displaystyle\int_0^{2/\sqrt{3}} \frac{1}{4 + 9x^2}\, dx$

68. $\displaystyle\int_0^4 \frac{1}{\sqrt{25 - x^2}}\, dx$

Area **In Exercises 69–74, find the area of the region.**

69. $y = (-2x + 5)^{3/2}$

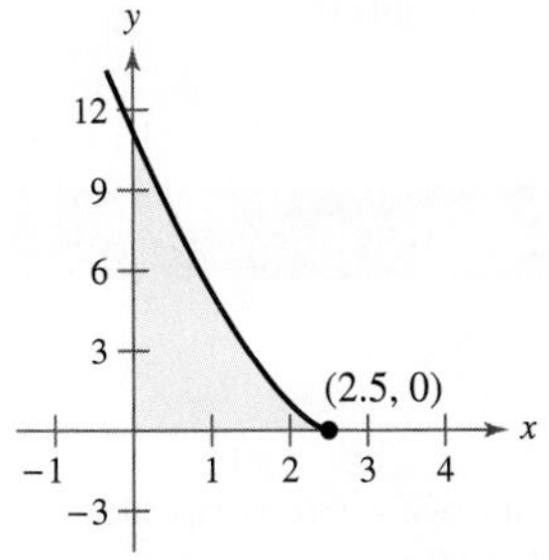

70. $y = x\sqrt{8 - 2x^2}$

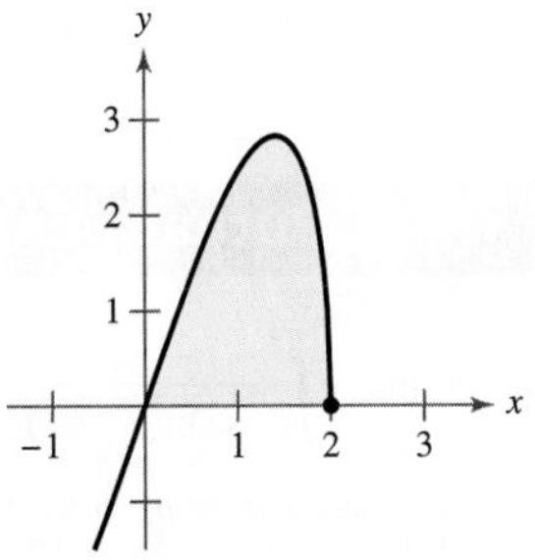

71. $y = \dfrac{3x + 2}{x^2 + 9}$

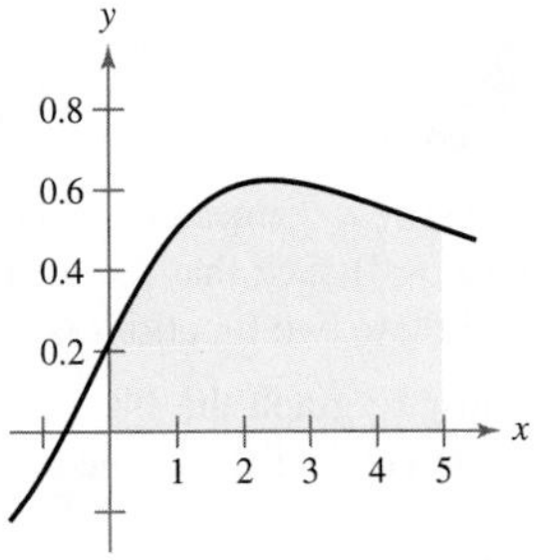

72. $y = \dfrac{3}{x^2 + 1}$

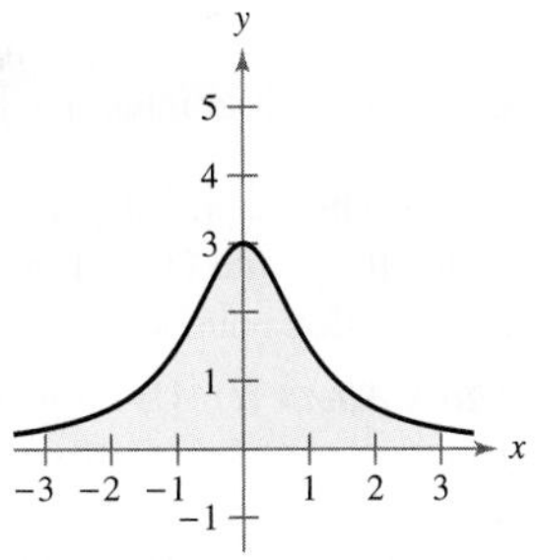

73. $y^2 = x^2(1 - x^2)$

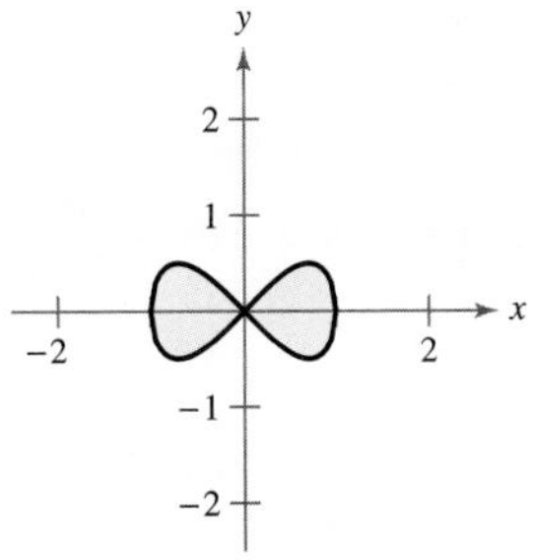

74. $y = \sin 2x$

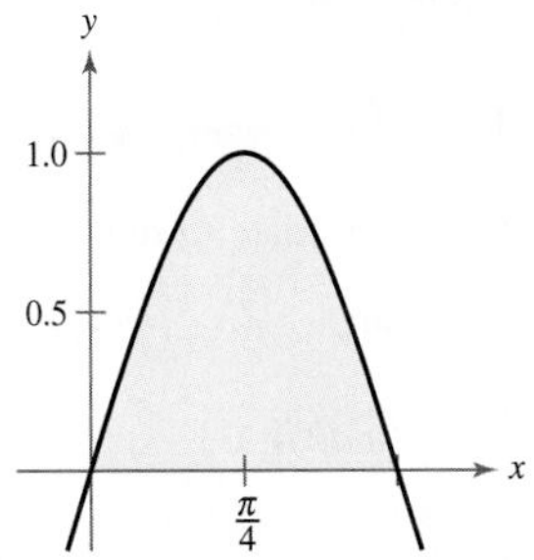

In Exercises 75–78, use a computer algebra system to find the integral. Use the computer algebra system to graph two antiderivatives. Describe the relationship between the two graphs of the antiderivatives.

75. $\displaystyle\int \frac{1}{x^2 + 4x + 13}\, dx$

76. $\displaystyle\int \frac{x - 2}{x^2 + 4x + 13}\, dx$

77. $\displaystyle\int \frac{1}{1 + \sin\theta}\, d\theta$

78. $\displaystyle\int \left(\frac{e^x + e^{-x}}{2}\right)^3 dx$

Writing About Concepts

In Exercises 79–82, state the integration formula you would use to perform the integration. Explain why you chose that formula. Do not integrate.

79. $\displaystyle\int x(x^2 + 1)^3\, dx$

80. $\displaystyle\int x \sec(x^2 + 1)\tan(x^2 + 1)\, dx$

81. $\displaystyle\int \frac{x}{x^2 + 1}\, dx$

82. $\displaystyle\int \frac{1}{x^2 + 1}\, dx$

83. Explain why the antiderivative $y_1 = e^{x + C_1}$ is equivalent to the antiderivative $y_2 = Ce^x$.

84. Explain why the antiderivative $y_1 = \sec^2 x + C_1$ is equivalent to the antiderivative $y_2 = \tan^2 x + C$.

85. Determine the constants a and b such that

$$\sin x + \cos x = a \sin(x + b).$$

Use this result to integrate $\displaystyle\int \frac{dx}{\sin x + \cos x}$.

86. ***Area*** The graphs of $f(x) = x$ and $g(x) = ax^2$ intersect at the points $(0, 0)$ and $(1/a, 1/a)$. Find a $(a > 0)$ such that the area of the region bounded by the graphs of these two functions is $\frac{2}{3}$.

87. ***Think About It*** Use a graphing utility to graph the function $f(x) = \frac{1}{5}(x^3 - 7x^2 + 10x)$. Use the graph to determine whether $\displaystyle\int_0^5 f(x)\,dx$ is positive or negative. Explain.

88. ***Think About It*** When evaluating

$$\int_{-1}^{1} x^2\,dx$$

is it appropriate to substitute $u = x^2$, $x = \sqrt{u}$, and $dx = \dfrac{du}{2\sqrt{u}}$ to obtain

$$\frac{1}{2}\int_1^1 \sqrt{u}\,du = 0?$$

Explain.

Approximation **In Exercises 89 and 90, determine which value best approximates the area of the region between the x-axis and the function over the given interval. (Make your selection on the basis of a sketch of the region and *not* by integrating.)**

89. $f(x) = \dfrac{4x}{x^2 + 1}$, $[0, 2]$

(a) 3 (b) 1 (c) -8 (d) 8 (e) 10

90. $f(x) = \dfrac{4}{x^2 + 1}$, $[0, 2]$

(a) 3 (b) 1 (c) -4 (d) 4 (e) 10

Interpreting Integrals **In Exercises 91 and 92, (a) sketch the region whose area is given by the integral, (b) sketch the solid whose volume is given by the integral if the disk method is used, and (c) sketch the solid whose volume is given by the integral if the shell method is used. (There is more than one correct answer for each part.)**

91. $\displaystyle\int_0^2 2\pi x^2\,dx$

92. $\displaystyle\int_0^4 \pi y\,dy$

93. ***Volume*** The region bounded by $y = e^{-x^2}$, $y = 0$, $x = 0$, and $x = b$ $(b > 0)$ is revolved about the y-axis.

(a) Find the volume of the solid generated if $b = 1$.

(b) Find b such that the volume of the generated solid is $\frac{4}{3}$ cubic units.

94. ***Arc Length*** Find the arc length of the graph of $y = \ln(\sin x)$ from $x = \pi/4$ to $x = \pi/2$.

95. ***Surface Area*** Find the area of the surface formed by revolving the graph of $y = 2\sqrt{x}$ on the interval $[0, 9]$ about the x-axis.

96. ***Centroid*** Find the x-coordinate of the centroid of the region bounded by the graphs of

$$y = \frac{5}{\sqrt{25 - x^2}}, \quad y = 0, \quad x = 0, \quad \text{and} \quad x = 4.$$

In Exercises 97 and 98, find the average value of the function over the given interval.

97. $f(x) = \dfrac{1}{1 + x^2}$, $-3 \le x \le 3$

98. $f(x) = \sin nx$, $0 \le x \le \pi/n$, n is a positive integer.

Arc Length **In Exercises 99 and 100, use the integration capabilities of a graphing utility to approximate the arc length of the curve over the given interval.**

99. $y = \tan \pi x$, $\left[0, \frac{1}{4}\right]$

100. $y = x^{2/3}$, $[1, 8]$

101. ***Finding a Pattern***

(a) Find $\displaystyle\int \cos^3 x\,dx$.

(b) Find $\displaystyle\int \cos^5 x\,dx$.

(c) Find $\displaystyle\int \cos^7 x\,dx$.

(d) Explain how to find $\int \cos^{15} x\,dx$ without actually integrating.

102. ***Finding a Pattern***

(a) Write $\int \tan^3 x\,dx$ in terms of $\int \tan x\,dx$. Then find $\int \tan^3 x\,dx$.

(b) Write $\int \tan^5 x\,dx$ in terms of $\int \tan^3 x\,dx$.

(c) Write $\int \tan^{2k+1} x\,dx$, where k is a positive integer, in terms of $\int \tan^{2k-1} x\,dx$.

(d) Explain how to find $\int \tan^{15} x\,dx$ without actually integrating.

103. ***Methods of Integration*** Show that the following results are equivalent.

Integration by tables:

$$\int \sqrt{x^2 + 1}\,dx = \frac{1}{2}\left(x\sqrt{x^2 + 1} + \ln\left|x + \sqrt{x^2 + 1}\right|\right) + C$$

Integration by computer algebra system:

$$\int \sqrt{x^2 + 1}\,dx = \frac{1}{2}\left(x\sqrt{x^2 + 1} + \operatorname{arcsinh}(x)\right) + C$$

Putnam Exam Challenge

104. Evaluate $\displaystyle\int_2^4 \frac{\sqrt{\ln(9 - x)}\,dx}{\sqrt{\ln(9 - x)} + \sqrt{\ln(x + 3)}}$.

This problem was composed by the Committee on the Putnam Prize Competition.

Section 8.2 Integration by Parts

- **Find an antiderivative using integration by parts.**
- **Use a tabular method to perform integration by parts.**

EXPLORATION

Proof Without Words Here is a different approach to proving the formula for integration by parts. Exercise taken from "Proof Without Words: Integration by Parts" by Roger B. Nelsen, *Mathematics Magazine*, April 1991, by permission of the author.

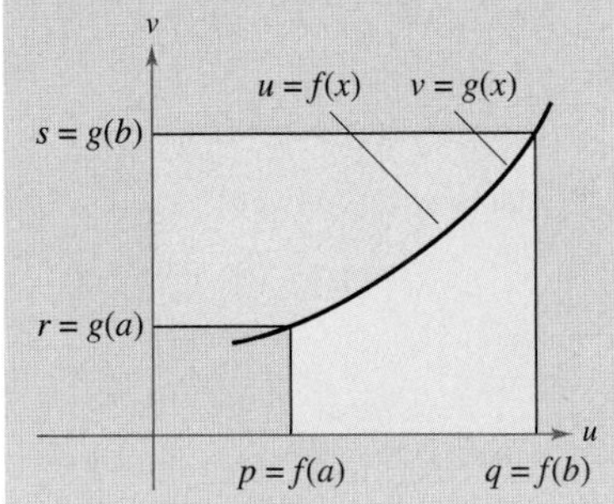

Area + Area $= qs - pr$

$$\int_r^s u\,dv + \int_q^p v\,du = \Big[uv\Big]_{(p,r)}^{(q,s)}$$

$$\int_r^s u\,dv = \Big[uv\Big]_{(p,r)}^{(q,s)} - \int_q^p v\,du$$

Explain how this graph proves the theorem. Which notation in this proof is unfamiliar? What do you think it means?

Integration by Parts

In this section you will study an important integration technique called **integration by parts.** This technique can be applied to a wide variety of functions and is particularly useful for integrands involving *products* of algebraic and transcendental functions. For instance, integration by parts works well with integrals such as

$$\int x \ln x\,dx, \quad \int x^2 e^x\,dx, \quad \text{and} \quad \int e^x \sin x\,dx.$$

Integration by parts is based on the formula for the derivative of a product

$$\begin{aligned}\frac{d}{dx}[uv] &= u\frac{dv}{dx} + v\frac{du}{dx}\\ &= uv' + vu'\end{aligned}$$

where both u and v are differentiable functions of x. If u' and v' are continuous, you can integrate both sides of this equation to obtain

$$\begin{aligned}uv &= \int uv'\,dx + \int vu'\,dx\\ &= \int u\,dv + \int v\,du.\end{aligned}$$

By rewriting this equation, you obtain the following theorem.

THEOREM 8.1 Integration by Parts

If u and v are functions of x and have continuous derivatives, then

$$\int u\,dv = uv - \int v\,du.$$

This formula expresses the original integral in terms of another integral. Depending on the choices of u and dv, it may be easier to evaluate the second integral than the original one. Because the choices of u and dv are critical in the integration by parts process, the following guidelines are provided.

Guidelines for Integration by Parts

1. Try letting dv be the most complicated portion of the integrand that fits a basic integration rule. Then u will be the remaining factor(s) of the integrand.
2. Try letting u be the portion of the integrand whose derivative is a function simpler than u. Then dv will be the remaining factor(s) of the integrand.

EXAMPLE 1 Integration by Parts

Find $\int xe^x\,dx$.

Solution To apply integration by parts, you need to write the integral in the form $\int u\,dv$. There are several ways to do this.

$$\int \underbrace{(x)}_{u}\underbrace{(e^x dx)}_{dv}, \quad \int \underbrace{(e^x)}_{u}\underbrace{(x\,dx)}_{dv}, \quad \int \underbrace{(1)}_{u}\underbrace{(xe^x\,dx)}_{dv}, \quad \int \underbrace{(xe^x)}_{u}\underbrace{(dx)}_{dv}$$

The guidelines on page 525 suggest choosing the first option because the derivative of $u = x$ is simpler than x, and $dv = e^x\,dx$ is the most complicated portion of the integrand that fits a basic integration formula.

$$dv = e^x\,dx \quad \Longrightarrow \quad v = \int dv = \int e^x\,dx = e^x$$

$$u = x \quad \Longrightarrow \quad du = dx$$

Now, integration by parts produces

$$\int u\,dv = uv - \int v\,du \qquad \text{Integration by parts formula}$$

$$\int xe^x\,dx = xe^x - \int e^x\,dx \qquad \text{Substitute.}$$

$$= xe^x - e^x + C. \qquad \text{Integrate.}$$

To check this, differentiate $xe^x - e^x + C$ to see that you obtain the original integrand.

NOTE In Example 1, note that it is not necessary to include a constant of integration when solving

$$v = \int e^x\,dx = e^x + C_1.$$

To illustrate this, replace $v = e^x$ by $v = e^x + C_1$ and apply integration by parts to see that you obtain the same result.

EXAMPLE 2 Integration by Parts

Find $\int x^2 \ln x\,dx$.

Solution In this case, x^2 is more easily integrated than $\ln x$. Furthermore, the derivative of $\ln x$ is simpler than $\ln x$. So, you should let $dv = x^2\,dx$.

$$dv = x^2\,dx \quad \Longrightarrow \quad v = \int x^2\,dx = \frac{x^3}{3}$$

$$u = \ln x \quad \Longrightarrow \quad du = \frac{1}{x}\,dx$$

Integration by parts produces

$$\int u\,dv = uv - \int v\,du \qquad \text{Integration by parts formula}$$

$$\int x^2 \ln x\,dx = \frac{x^3}{3}\ln x - \int \left(\frac{x^3}{3}\right)\left(\frac{1}{x}\right)dx \qquad \text{Substitute.}$$

$$= \frac{x^3}{3}\ln x - \frac{1}{3}\int x^2\,dx \qquad \text{Simplify.}$$

$$= \frac{x^3}{3}\ln x - \frac{x^3}{9} + C. \qquad \text{Integrate.}$$

You can check this result by differentiating.

$$\frac{d}{dx}\left[\frac{x^3}{3}\ln x - \frac{x^3}{9}\right] = \frac{x^3}{3}\left(\frac{1}{x}\right) + (\ln x)(x^2) - \frac{x^2}{3} = x^2 \ln x$$

FOR FURTHER INFORMATION To see how integration by parts is used to prove Stirling's approximation

$$\ln(n!) = n \ln n - n$$

see the article "The Validity of Stirling's Approximation: A Physical Chemistry Project" by A. S. Wallner and K. A. Brandt in *Journal of Chemical Education*.

TECHNOLOGY Try graphing

$$\int x^2 \ln x\,dx \quad \text{and} \quad \frac{x^3}{3}\ln x - \frac{x^3}{9}$$

on your graphing utility. Do you get the same graph? (This will take a while, so be patient.)

One surprising application of integration by parts involves integrands consisting of a single term, such as $\int \ln x\, dx$ or $\int \arcsin x\, dx$. In these cases, try letting $dv = dx$, as shown in the next example.

EXAMPLE 3 An Integrand with a Single Term

Evaluate $\displaystyle\int_0^1 \arcsin x\, dx$.

Solution Let $dv = dx$.

$$dv = dx \quad \Rightarrow \quad v = \int dx = x$$

$$u = \arcsin x \quad \Rightarrow \quad du = \frac{1}{\sqrt{1-x^2}}\, dx$$

Integration by parts now produces

$$\int u\, dv = uv - \int v\, du \qquad \text{Integration by parts formula}$$

$$\int \arcsin x\, dx = x \arcsin x - \int \frac{x}{\sqrt{1-x^2}}\, dx \qquad \text{Substitute.}$$

$$= x \arcsin x + \frac{1}{2}\int (1-x^2)^{-1/2}(-2x)\, dx \qquad \text{Rewrite.}$$

$$= x \arcsin x + \sqrt{1-x^2} + C. \qquad \text{Integrate.}$$

Using this antiderivative, you can evaluate the definite integral as follows.

$$\int_0^1 \arcsin x\, dx = \Big[x \arcsin x + \sqrt{1-x^2}\Big]_0^1$$

$$= \frac{\pi}{2} - 1$$

$$\approx 0.571$$

The area represented by this definite integral is shown in Figure 8.2.

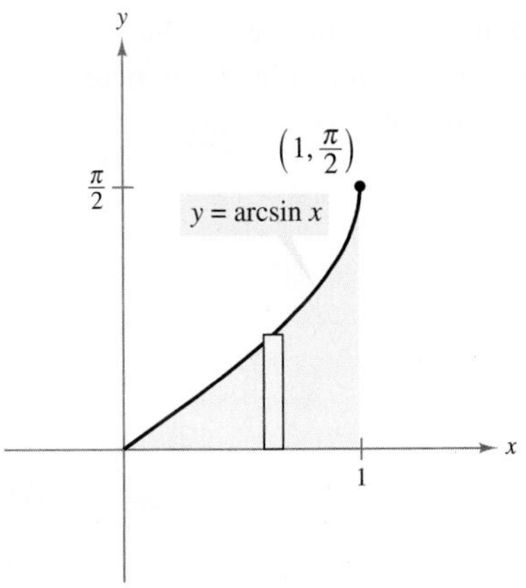

The area of the region is approximately 0.571.
Figure 8.2

TECHNOLOGY Remember that there are two ways to use technology to evaluate a definite integral: (1) you can use a numerical approximation such as the Trapezoidal Rule or Simpson's Rule, or (2) you can use a computer algebra system to find the antiderivative and then apply the Fundamental Theorem of Calculus. Both methods have shortcomings. To find the possible error when using a numerical method, the integrand must have a second derivative (Trapezoidal Rule) or a fourth derivative (Simpson's Rule) in the interval of integration: the integrand in Example 3 fails to meet either of these requirements. To apply the Fundamental Theorem of Calculus, the symbolic integration utility must be able to find the antiderivative.

Which method would you use to evaluate

$$\int_0^1 \arctan x\, dx?$$

Which method would you use to evaluate

$$\int_0^1 \arctan x^2\, dx?$$

Some integrals require repeated use of the integration by parts formula.

EXAMPLE 4 Repeated Use of Integration by Parts

Find $\int x^2 \sin x\, dx$.

Solution The factors x^2 and $\sin x$ are equally easy to integrate. However, the derivative of x^2 becomes simpler, whereas the derivative of $\sin x$ does not. So, you should let $u = x^2$.

$$dv = \sin x\, dx \quad \Rightarrow \quad v = \int \sin x\, dx = -\cos x$$

$$u = x^2 \quad \Rightarrow \quad du = 2x\, dx$$

Now, integration by parts produces

$$\int x^2 \sin x\, dx = -x^2 \cos x + \int 2x \cos x\, dx. \qquad \text{First use of integration by parts}$$

This first use of integration by parts has succeeded in simplifying the original integral, but the integral on the right still doesn't fit a basic integration rule. To evaluate that integral, you can apply integration by parts again. This time, let $u = 2x$.

$$dv = \cos x\, dx \quad \Rightarrow \quad v = \int \cos x\, dx = \sin x$$

$$u = 2x \quad \Rightarrow \quad du = 2\, dx$$

Now, integration by parts produces

$$\int 2x \cos x\, dx = 2x \sin x - \int 2 \sin x\, dx \qquad \text{Second use of integration by parts}$$

$$= 2x \sin x + 2 \cos x + C.$$

Combining these two results, you can write

$$\int x^2 \sin x\, dx = -x^2 \cos x + 2x \sin x + 2 \cos x + C.$$

When making repeated applications of integration by parts, you need to be careful not to interchange the substitutions in successive applications. For instance, in Example 4, the first substitution was $u = x^2$ and $dv = \sin x\, dx$. If, in the second application, you had switched the substitution to $u = \cos x$ and $dv = 2x$, you would have obtained

$$\int x^2 \sin x\, dx = -x^2 \cos x + \int 2x \cos x\, dx$$

$$= -x^2 \cos x + x^2 \cos x + \int x^2 \sin x\, dx$$

$$= \int x^2 \sin x\, dx$$

thereby undoing the previous integration and returning to the *original* integral. When making repeated applications of integration by parts, you should also watch for the appearance of a *constant multiple* of the original integral. For instance, this occurs when you use integration by parts to evaluate $\int e^x \cos 2x\, dx$, and also occurs in the next example.

EXPLORATION

Try to find

$$\int e^x \cos 2x\, dx$$

by letting $u = \cos 2x$ and $dv = e^x\, dx$ in the first substitution. For the second substitution, let $u = \sin 2x$ and $dv = e^x\, dx$.

EXAMPLE 5 Integration by Parts

Find $\int \sec^3 x\, dx$.

NOTE The integral in Example 5 is an important one. In Section 8.4 (Example 5), you will see that it is used to find the arc length of a parabolic segment.

Solution The most complicated portion of the integrand that can be easily integrated is $\sec^2 x$, so you should let $dv = \sec^2 x\, dx$ and $u = \sec x$.

$$dv = \sec^2 x\, dx \quad \Rightarrow \quad v = \int \sec^2 x\, dx = \tan x$$

$$u = \sec x \quad \Rightarrow \quad du = \sec x \tan x\, dx$$

Integration by parts produces

$$\int u\, dv = uv - \int v\, du \qquad \text{Integration by parts formula}$$

$$\int \sec^3 x\, dx = \sec x \tan x - \int \sec x \tan^2 x\, dx \qquad \text{Substitute.}$$

$$\int \sec^3 x\, dx = \sec x \tan x - \int \sec x(\sec^2 x - 1)\, dx \qquad \text{Trigonometric identity}$$

$$\int \sec^3 x\, dx = \sec x \tan x - \int \sec^3 x\, dx + \int \sec x\, dx \qquad \text{Rewrite.}$$

$$2\int \sec^3 x\, dx = \sec x \tan x + \int \sec x\, dx \qquad \text{Collect like integrals.}$$

$$\int \sec^3 x\, dx = \frac{1}{2}\sec x \tan x + \frac{1}{2}\ln|\sec x + \tan x| + C. \qquad \text{Integrate and divide by 2.}$$

STUDY TIP The trigonometric identities

$$\sin^2 x = \frac{1 - \cos 2x}{2}$$

$$\cos^2 x = \frac{1 + \cos 2x}{2}$$

play an important role in this chapter.

EXAMPLE 6 Finding a Centroid

A machine part is modeled by the region bounded by the graph of $y = \sin x$ and the x-axis, $0 \le x \le \pi/2$, as shown in Figure 8.3. Find the centroid of this region.

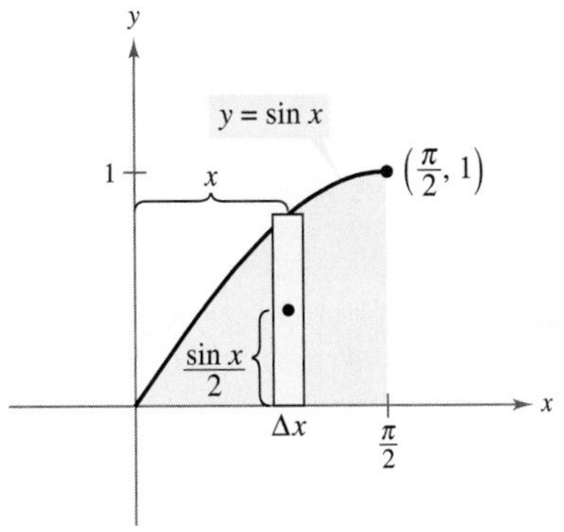

Figure 8.3

Solution Begin by finding the area of the region.

$$A = \int_0^{\pi/2} \sin x\, dx = \Big[-\cos x\Big]_0^{\pi/2} = 1$$

Now, you can find the coordinates of the centroid as follows.

$$\bar{y} = \frac{1}{A}\int_0^{\pi/2} \frac{\sin x}{2}(\sin x)\, dx = \frac{1}{4}\int_0^{\pi/2} (1 - \cos 2x)\, dx = \frac{1}{4}\left[x - \frac{\sin 2x}{2}\right]_0^{\pi/2} = \frac{\pi}{8}$$

You can evaluate the integral for $\bar{x}$, $(1/A)\int_0^{\pi/2} x \sin x\, dx$, with integration by parts. To do this, let $dv = \sin x\, dx$ and $u = x$. This produces $v = -\cos x$ and $du = dx$, and you can write

$$\int x \sin x\, dx = -x\cos x + \int \cos x\, dx$$

$$= -x\cos x + \sin x + C.$$

Finally, you can determine $\bar{x}$ to be

$$\bar{x} = \frac{1}{A}\int_0^{\pi/2} x \sin x\, dx = \Big[-x\cos x + \sin x\Big]_0^{\pi/2} = 1.$$

So, the centroid of the region is $(1, \pi/8)$.

As you gain experience in using integration by parts, your skill in determining u and dv will increase. The following summary lists several common integrals with suggestions for the choices of u and dv.

STUDY TIP You can use the acronym LIATE as a guideline for choosing u in integration by parts. In order, check the integrand for the following.

Is there a Logarithmic part?

Is there an Inverse trigonometric part?

Is there an Algebraic part?

Is there a Trigonometric part?

Is there an Exponential part?

Summary of Common Integrals Using Integration by Parts

1. For integrals of the form

$$\int x^n e^{ax}\,dx, \qquad \int x^n \sin ax\,dx, \qquad \text{or} \qquad \int x^n \cos ax\,dx$$

let $u = x^n$ and let $dv = e^{ax}\,dx$, $\sin ax\,dx$, or $\cos ax\,dx$.

2. For integrals of the form

$$\int x^n \ln x\,dx, \qquad \int x^n \arcsin ax\,dx, \qquad \text{or} \qquad \int x^n \arctan ax\,dx$$

let $u = \ln x$, $\arcsin ax$, or $\arctan ax$ and let $dv = x^n\,dx$.

3. For integrals of the form

$$\int e^{ax} \sin bx\,dx \qquad \text{or} \qquad \int e^{ax} \cos bx\,dx$$

let $u = \sin bx$ or $\cos bx$ and let $dv = e^{ax}\,dx$.

Tabular Method

In problems involving repeated applications of integration by parts, a tabular method, illustrated in Example 7, can help to organize the work. This method works well for integrals of the form $\int x^n \sin ax\,dx$, $\int x^n \cos ax\,dx$, and $\int x^n e^{ax}\,dx$.

EXAMPLE 7 Using the Tabular Method

Find $\displaystyle\int x^2 \sin 4x\,dx$.

Solution Begin as usual by letting $u = x^2$ and $dv = v'\,dx = \sin 4x\,dx$. Next, create a table consisting of three columns, as shown.

Alternate Signs	*u and Its Derivatives*	*v′ and Its Antiderivatives*
$+$	x^2	$\sin 4x$
$-$	$2x$	$-\frac{1}{4}\cos 4x$
$+$	2	$-\frac{1}{16}\sin 4x$
$-$	0	$\frac{1}{64}\cos 4x$

Differentiate until you obtain 0 as a derivative.

The solution is obtained by adding the signed products of the diagonal entries:

$$\int x^2 \sin 4x\,dx = -\frac{1}{4}x^2 \cos 4x + \frac{1}{8}x \sin 4x + \frac{1}{32}\cos 4x + C.$$

FOR FURTHER INFORMATION For more information on the tabular method, see the article "Tabular Integration by Parts" by David Horowitz in *The College Mathematics Journal*, and the article "More on Tabular Integration by Parts" by Leonard Gillman in *The College Mathematics Journal*. To view these articles, go to the website *www.matharticles.com*.

Exercises for Section 8.2

See www.CalcChat.com for worked-out solutions to odd-numbered exercises.

In Exercises 1–4, match the antiderivative with the correct integral. [Integrals are labeled (a), (b), (c), and (d).]

(a) $\int \ln x\,dx$ (b) $\int x \sin x\,dx$

(c) $\int x^2 e^x\,dx$ (d) $\int x^2 \cos x\,dx$

1. $y = \sin x - x \cos x$
2. $y = x^2 \sin x + 2x \cos x - 2 \sin x$
3. $y = x^2 e^x - 2xe^x + 2e^x$
4. $y = -x + x \ln x$

In Exercises 5–10, identify u and dv for finding the integral using integration by parts. (Do not evaluate the integral.)

5. $\int xe^{2x}\,dx$
6. $\int x^2 e^{2x}\,dx$
7. $\int (\ln x)^2\,dx$
8. $\int \ln 3x\,dx$
9. $\int x \sec^2 x\,dx$
10. $\int x^2 \cos x\,dx$

In Exercises 11–36, find the integral. (*Note:* Solve by the simplest method—not all require integration by parts.)

11. $\int xe^{-2x}\,dx$
12. $\int \frac{2x}{e^x}\,dx$
13. $\int x^3 e^x\,dx$
14. $\int \frac{e^{1/t}}{t^2}\,dt$
15. $\int x^2 e^{x^3}\,dx$
16. $\int x^4 \ln x\,dx$
17. $\int t \ln(t+1)\,dt$
18. $\int \frac{1}{x(\ln x)^3}\,dx$
19. $\int \frac{(\ln x)^2}{x}\,dx$
20. $\int \frac{\ln x}{x^2}\,dx$
21. $\int \frac{xe^{2x}}{(2x+1)^2}\,dx$
22. $\int \frac{x^3 e^{x^2}}{(x^2+1)^2}\,dx$
23. $\int (x^2 - 1)e^x\,dx$
24. $\int \frac{\ln 2x}{x^2}\,dx$
25. $\int x\sqrt{x-1}\,dx$
26. $\int \frac{x}{\sqrt{2+3x}}\,dx$
27. $\int x \cos x\,dx$
28. $\int x \sin x\,dx$
29. $\int x^3 \sin x\,dx$
30. $\int x^2 \cos x\,dx$
31. $\int t \csc t \cot t\,dt$
32. $\int \theta \sec \theta \tan \theta\,d\theta$
33. $\int \arctan x\,dx$
34. $\int 4 \arccos x\,dx$
35. $\int e^{2x} \sin x\,dx$
36. $\int e^x \cos 2x\,dx$

In Exercises 37–42, solve the differential equation.

37. $y' = xe^{x^2}$
38. $y' = \ln x$
39. $\frac{dy}{dt} = \frac{t^2}{\sqrt{2+3t}}$
40. $\frac{dy}{dx} = x^2\sqrt{x-1}$
41. $(\cos y)y' = 2x$
42. $y' = \arctan \frac{x}{2}$

Slope Fields **In Exercises 43 and 44, a differential equation, a point, and a slope field are given. (a) Sketch two approximate solutions of the differential equation on the slope field, one of which passes through the given point. (b) Use integration to find the particular solution of the differential equation and use a graphing utility to graph the solution. Compare the result with the sketches in part (a). To print an enlarged copy of the graph, go to the website *www.mathgraphs.com*.**

43. $\frac{dy}{dx} = x\sqrt{y} \cos x,\ (0, 4)$

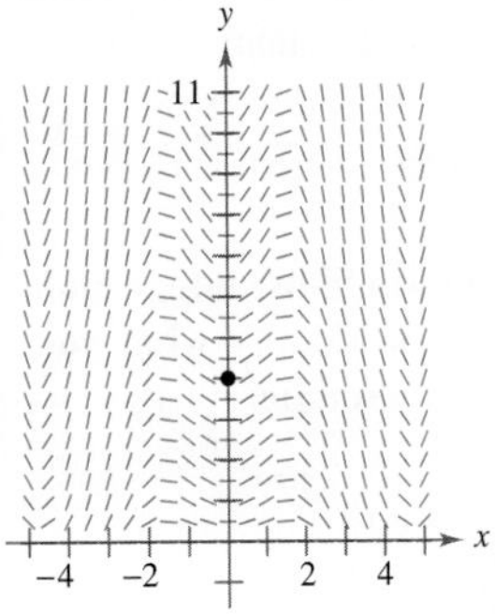

44. $\frac{dy}{dx} = e^{-x/3} \sin 2x,\ \left(0, -\frac{18}{37}\right)$

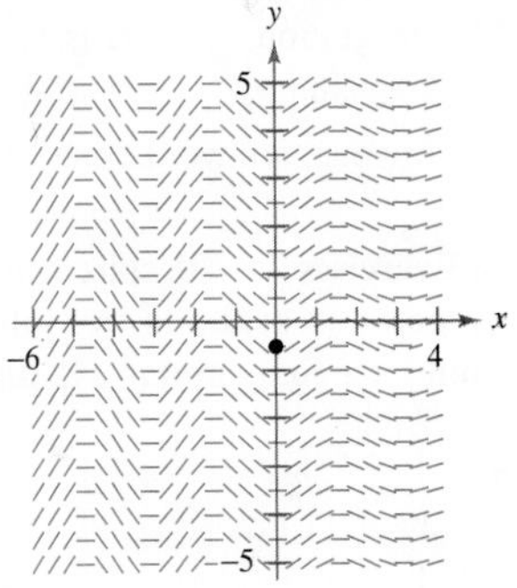

Slope Fields **In Exercises 45 and 46, use a computer algebra system to graph the slope field for the differential equation and graph the solution through the specified initial condition.**

45. $\frac{dy}{dx} = \frac{x}{y}e^{x/8}$, $y(0) = 2$
46. $\frac{dy}{dx} = \frac{x}{y} \sin x$, $y(0) = 4$

In Exercises 47–58, evaluate the definite integral. Use a graphing utility to confirm your result.

47. $\int_0^4 xe^{-x/2}\,dx$
48. $\int_0^1 x^2 e^x\,dx$
49. $\int_0^{\pi/2} x \cos x\,dx$
50. $\int_0^{\pi} x \sin 2x\,dx$
51. $\int_0^{1/2} \arccos x\,dx$
52. $\int_0^1 x \arcsin x^2\,dx$
53. $\int_0^1 e^x \sin x\,dx$
54. $\int_0^2 e^{-x} \cos x\,dx$
55. $\int_1^2 x^2 \ln x\,dx$
56. $\int_0^1 \ln(1 + x^2)\,dx$
57. $\int_2^4 x \operatorname{arcsec} x\,dx$
58. $\int_0^{\pi/4} x \sec^2 x\,dx$

In Exercises 59–64, use the tabular method to find the integral.

59. $\int x^2 e^{2x}\, dx$

60. $\int x^3 e^{-2x}\, dx$

61. $\int x^3 \sin x\, dx$

62. $\int x^3 \cos 2x\, dx$

63. $\int x \sec^2 x\, dx$

64. $\int x^2 (x-2)^{3/2}\, dx$

In Exercises 65–70, find or evaluate the integral using substitution first, then using integration by parts.

65. $\int \sin \sqrt{x}\, dx$

66. $\int 2x^3 \cos x^2\, dx$

67. $\int_0^4 x\sqrt{4-x}\, dx$

68. $\int_0^2 e^{\sqrt{2x}}\, dx$

69. $\int \cos(\ln x)\, dx$

70. $\int \ln(x^2+1)\, dx$

Writing About Concepts

71. Integration by parts is based on what differentiation rule? Explain.

72. In your own words, state guidelines for integration by parts.

In Exercises 73–78, state whether you would use integration by parts to evaluate the integral. If so, identify what you would use for u and dv. Explain your reasoning.

73. $\int \frac{\ln x}{x}\, dx$

74. $\int x \ln x\, dx$

75. $\int x^2 e^{2x}\, dx$

76. $\int 2x e^{x^2}\, dx$

77. $\int \frac{x}{\sqrt{x+1}}\, dx$

78. $\int \frac{x}{\sqrt{x^2+1}}\, dx$

In Exercises 79–82, use a computer algebra system to (a) find or evaluate the integral and (b) graph two antiderivatives. (c) Describe the relationship between the graphs of the antiderivatives.

79. $\int t^3 e^{-4t}\, dt$

80. $\int \alpha^4 \sin \pi\alpha\, d\alpha$

81. $\int_0^{\pi/2} e^{-2x} \sin 3x\, dx$

82. $\int_0^5 x^4 (25-x^2)^{3/2}\, dx$

83. Integrate $\int 2x\sqrt{2x-3}\, dx$

(a) by parts, letting $dv = \sqrt{2x-3}\, dx$.

(b) by substitution, letting $u = 2x - 3$.

84. Integrate $\int x\sqrt{4+x}\, dx$

(a) by parts, letting $dv = \sqrt{4+x}\, dx$.

(b) by substitution, letting $u = 4 + x$.

85. Integrate $\int \frac{x^3}{\sqrt{4+x^2}}\, dx$

(a) by parts, letting $dv = \left(x/\sqrt{4+x^2}\right) dx$.

(b) by substitution, letting $u = 4 + x^2$.

86. Integrate $\int x\sqrt{4-x}\, dx$

(a) by parts, letting $dv = \sqrt{4-x}\, dx$.

(b) by substitution, letting $u = 4 - x$.

In Exercises 87 and 88, use a computer algebra system to find the integral for $n = 0, 1, 2,$ and 3. Use the result to obtain a general rule for the integral for any positive integer n and test your results for $n = 4$.

87. $\int x^n \ln x\, dx$

88. $\int x^n e^x\, dx$

In Exercises 89–94, use integration by parts to verify the formula. (For Exercises 89–92, assume that n is a positive integer.)

89. $\int x^n \sin x\, dx = -x^n \cos x + n \int x^{n-1} \cos x\, dx$

90. $\int x^n \cos x\, dx = x^n \sin x - n \int x^{n-1} \sin x\, dx$

91. $\int x^n \ln x\, dx = \frac{x^{n+1}}{(n+1)^2}[-1 + (n+1)\ln x] + C$

92. $\int x^n e^{ax}\, dx = \frac{x^n e^{ax}}{a} - \frac{n}{a} \int x^{n-1} e^{ax}\, dx$

93. $\int e^{ax} \sin bx\, dx = \frac{e^{ax}(a \sin bx - b \cos bx)}{a^2 + b^2} + C$

94. $\int e^{ax} \cos bx\, dx = \frac{e^{ax}(a \cos bx + b \sin bx)}{a^2 + b^2} + C$

In Exercises 95–98, find the integral by using the appropriate formula from Exercises 89–94.

95. $\int x^3 \ln x\, dx$

96. $\int x^2 \cos x\, dx$

97. $\int e^{2x} \cos 3x\, dx$

98. $\int x^3 e^{2x}\, dx$

Area **In Exercises 99–102, use a graphing utility to graph the region bounded by the graphs of the equations, and find the area of the region.**

99. $y = xe^{-x},\ y = 0,\ x = 4$

100. $y = \frac{1}{9} x e^{-x/3},\ y = 0,\ x = 0,\ x = 3$

101. $y = e^{-x} \sin \pi x,\ y = 0,\ x = 0,\ x = 1$

102. $y = x \sin x,\ y = 0,\ x = 0,\ x = \pi$

103. ***Area, Volume, and Centroid*** Given the region bounded by the graphs of $y = \ln x$, $y = 0$, and $x = e$, find

(a) the area of the region.

(b) the volume of the solid generated by revolving the region about the x-axis.

(c) the volume of the solid generated by revolving the region about the y-axis.

(d) the centroid of the region.

104. ***Volume and Centroid*** Given the region bounded by the graphs of $y = x \sin x$, $y = 0$, $x = 0$, and $x = \pi$, find

(a) the volume of the solid generated by revolving the region about the x-axis.

(b) the volume of the solid generated by revolving the region about the y-axis.

(c) the centroid of the region.

105. ***Centroid*** Find the centroid of the region bounded by the graphs of $y = \arcsin x$, $x = 0$, and $y = \pi/2$. How is this problem related to Example 6 in this section?

106. ***Centroid*** Find the centroid of the region bounded by the graphs of $f(x) = x^2$, $g(x) = 2^x$, $x = 2$, and $x = 4$.

107. ***Average Displacement*** A damping force affects the vibration of a spring so that the displacement of the spring is given by $y = e^{-4t}(\cos 2t + 5 \sin 2t)$. Find the average value of y on the interval from $t = 0$ to $t = \pi$.

108. ***Memory Model*** A model for the ability M of a child to memorize, measured on a scale from 0 to 10, is given by $M = 1 + 1.6t \ln t$, $0 < t \le 4$, where t is the child's age in years. Find the average value of this model

(a) between the child's first and second birthdays.

(b) between the child's third and fourth birthdays.

Present Value **In Exercises 109 and 110, find the present value *P* of a continuous income flow of *c*(*t*) dollars per year if**

$$P = \int_0^{t_1} c(t)e^{-rt}\,dt$$

where t_1 is the time in years and *r* is the annual interest rate compounded continuously.

109. $c(t) = 100{,}000 + 4000t$, $r = 5\%$, $t_1 = 10$

110. $c(t) = 30{,}000 + 500t$, $r = 7\%$, $t_1 = 5$

Integrals Used to Find Fourier Coefficients **In Exercises 111 and 112, verify the value of the definite integral, where *n* is a positive integer.**

111. $$\int_{-\pi}^{\pi} x \sin nx\,dx = \begin{cases} \dfrac{2\pi}{n}, & n \text{ is odd} \\ -\dfrac{2\pi}{n}, & n \text{ is even} \end{cases}$$

112. $$\int_{-\pi}^{\pi} x^2 \cos nx\,dx = \frac{(-1)^n 4\pi}{n^2}$$

113. ***Vibrating String*** A string stretched between the two points (0, 0) and (2, 0) is plucked by displacing the string h units at its midpoint. The motion of the string is modeled by a **Fourier Sine Series** whose coefficients are given by

$$b_n = h\int_0^1 x \sin\frac{n\pi x}{2}\,dx + h\int_1^2 (-x + 2)\sin\frac{n\pi x}{2}\,dx.$$

Find b_n.

114. Find the fallacy in the following argument that $0 = 1$.

$$dv = dx \;\Rightarrow\; v = \int dx = x$$

$$u = \frac{1}{x} \;\Rightarrow\; du = -\frac{1}{x^2}\,dx$$

$$0 + \int \frac{dx}{x} = \left(\frac{1}{x}\right)(x) - \int \left(-\frac{1}{x^2}\right)(x)\,dx = 1 + \int \frac{dx}{x}$$

So, $0 = 1$.

115. Let $y = f(x)$ be positive and strictly increasing on the interval $0 < a \le x \le b$. Consider the region R bounded by the graphs of $y = f(x)$, $y = 0$, $x = a$, and $x = b$. If R is revolved about the y-axis, show that the disk method and shell method yield the same volume.

116. ***Euler's Method*** Consider the differential equation $f'(x) = xe^{-x}$ with the initial condition $f(0) = 0$.

(a) Use integration to solve the differential equation.

(b) Use a graphing utility to graph the solution of the differential equation.

(c) Use Euler's Method with $h = 0.05$, and the recursive capabilities of a graphing utility, to generate the first 80 points of the graph of the approximate solution. Use the graphing utility to plot the points. Compare the result with the graph in part (b).

(d) Repeat part (c) using $h = 0.1$ and generate the first 40 points.

(e) Why is the result in part (c) a better approximation of the solution than the result in part (d)?

Euler's Method **In Exercises 117 and 118, consider the differential equation and repeat parts (a)–(d) of Exercise 116.**

117. $f'(x) = 3x \sin(2x)$, $f(0) = 0$

118. $f'(x) = \cos\sqrt{x}$, $f(1) = 1$

119. ***Think About It*** Give a geometric explanation to explain why

$$\int_0^{\pi/2} x \sin x\,dx \le \int_0^{\pi/2} x\,dx.$$

Verify the inequality by evaluating the integrals.

120. ***Finding a Pattern*** Find the area bounded by the graphs of $y = x \sin x$ and $y = 0$ over each interval.

(a) $[0, \pi]$ (b) $[\pi, 2\pi]$ (c) $[2\pi, 3\pi]$

Describe any patterns that you notice. What is the area between the graphs of $y = x \sin x$ and $y = 0$ over the interval $[n\pi, (n + 1)\pi]$, where n is any nonnegative integer? Explain.

Section 8.3 Trigonometric Integrals

- **Solve trigonometric integrals involving powers of sine and cosine.**
- **Solve trigonometric integrals involving powers of secant and tangent.**
- **Solve trigonometric integrals involving sine-cosine products with different angles.**

Integrals Involving Powers of Sine and Cosine

In this section you will study techniques for evaluating integrals of the form

$$\int \sin^m x \cos^n x\, dx \qquad \text{and} \qquad \int \sec^m x \tan^n x\, dx$$

where either m or n is a positive integer. To find antiderivatives for these forms, try to break them into combinations of trigonometric integrals to which you can apply the Power Rule.

For instance, you can evaluate $\int \sin^5 x \cos x\, dx$ with the Power Rule by letting $u = \sin x$. Then, $du = \cos x\, dx$ and you have

$$\int \sin^5 x \cos x\, dx = \int u^5\, du = \frac{u^6}{6} + C = \frac{\sin^6 x}{6} + C.$$

To break up $\int \sin^m x \cos^n x\, dx$ into forms to which you can apply the Power Rule, use the following identities.

$$\sin^2 x + \cos^2 x = 1 \qquad \text{Pythagorean identity}$$

$$\sin^2 x = \frac{1 - \cos 2x}{2} \qquad \text{Half-angle identity for } \sin^2 x$$

$$\cos^2 x = \frac{1 + \cos 2x}{2} \qquad \text{Half-angle identity for } \cos^2 x$$

SHEILA SCOTT MACINTYRE (1910–1960)

Sheila Scott Macintyre published her first paper on the asymptotic periods of integral functions in 1935. She completed her doctorate work at Aberdeen University, where she taught. In 1958 she accepted a visiting research fellowship at the University of Cincinnati.

Guidelines for Evaluating Integrals Involving Sine and Cosine

1. If the power of the sine is odd and positive, save one sine factor and convert the remaining factors to cosines. Then, expand and integrate.

$$\int \overbrace{\sin^{2k+1}}^{\text{Odd}} x \cos^n x\, dx = \int \overbrace{(\sin^2 x)^k}^{\text{Convert to cosines}} \cos^n x \overbrace{\sin x\, dx}^{\text{Save for } du} = \int (1 - \cos^2 x)^k \cos^n x \sin x\, dx$$

2. If the power of the cosine is odd and positive, save one cosine factor and convert the remaining factors to sines. Then, expand and integrate.

$$\int \sin^m x \overbrace{\cos^{2k+1}}^{\text{Odd}} x\, dx = \int \sin^m x \overbrace{(\cos^2 x)^k}^{\text{Convert to sines}} \overbrace{\cos x\, dx}^{\text{Save for } du} = \int \sin^m x (1 - \sin^2 x)^k \cos x\, dx$$

3. If the powers of both the sine and cosine are even and nonnegative, make repeated use of the identities

$$\sin^2 x = \frac{1 - \cos 2x}{2} \qquad \text{and} \qquad \cos^2 x = \frac{1 + \cos 2x}{2}$$

to convert the integrand to odd powers of the cosine. Then proceed as in guideline 2.

TECHNOLOGY Use a computer algebra system to find the integral in Example 1. You should obtain

$$\int \sin^3 x \cos^4 x \, dx =$$

$$-\cos^5 x\left(\frac{1}{7}\sin^2 x + \frac{2}{35}\right) + C.$$

Is this equivalent to the result obtained in Example 1?

EXAMPLE 1 Power of Sine Is Odd and Positive

Find $\int \sin^3 x \cos^4 x \, dx$.

Solution Because you expect to use the Power Rule with $u = \cos x$, *save one sine factor* to form du and convert the remaining sine factors to cosines.

$$\int \sin^3 x \cos^4 x \, dx = \int \sin^2 x \cos^4 x (\sin x) \, dx \qquad \text{Rewrite.}$$

$$= \int (1 - \cos^2 x) \cos^4 x \sin x \, dx \qquad \text{Trigonometric identity}$$

$$= \int (\cos^4 x - \cos^6 x) \sin x \, dx \qquad \text{Multiply.}$$

$$= \int \cos^4 x \sin x \, dx - \int \cos^6 x \sin x \, dx \qquad \text{Rewrite.}$$

$$= -\int \cos^4 x(-\sin x) \, dx + \int \cos^6 x(-\sin x) \, dx$$

$$= -\frac{\cos^5 x}{5} + \frac{\cos^7 x}{7} + C \qquad \text{Integrate.}$$

In Example 1, *both* of the powers m and n happened to be positive integers. However, the same strategy will work as long as either m or n is odd and positive. For instance, in the next example the power of the cosine is 3, but the power of the sine is $-\frac{1}{2}$.

EXAMPLE 2 Power of Cosine Is Odd and Positive

Evaluate $\displaystyle\int_{\pi/6}^{\pi/3} \frac{\cos^3 x}{\sqrt{\sin x}} \, dx$.

Solution Because you expect to use the Power Rule with $u = \sin x$, *save one cosine factor* to form du and convert the remaining cosine factors to sines.

$$\int_{\pi/6}^{\pi/3} \frac{\cos^3 x}{\sqrt{\sin x}} \, dx = \int_{\pi/6}^{\pi/3} \frac{\cos^2 x \cos x}{\sqrt{\sin x}} \, dx$$

$$= \int_{\pi/6}^{\pi/3} \frac{(1 - \sin^2 x)(\cos x)}{\sqrt{\sin x}} \, dx$$

$$= \int_{\pi/6}^{\pi/3} \left[(\sin x)^{-1/2} \cos x - (\sin x)^{3/2} \cos x\right] dx$$

$$= \left[\frac{(\sin x)^{1/2}}{1/2} - \frac{(\sin x)^{5/2}}{5/2}\right]_{\pi/6}^{\pi/3}$$

$$= 2\left(\frac{\sqrt{3}}{2}\right)^{1/2} - \frac{2}{5}\left(\frac{\sqrt{3}}{2}\right)^{5/2} - \sqrt{2} + \frac{\sqrt{32}}{80}$$

$$\approx 0.239$$

Figure 8.4 shows the region whose area is represented by this integral.

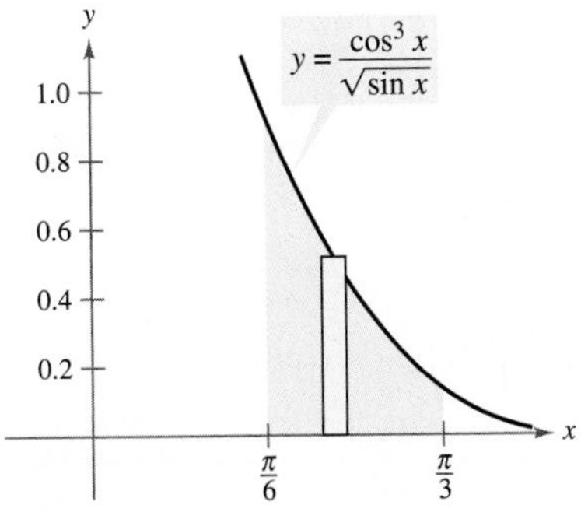

The area of the region is approximately 0.239.
Figure 8.4

EXAMPLE 3 Power of Cosine Is Even and Nonnegative

Find $\int \cos^4 x\, dx.$

Solution Because m and n are both even and nonnegative ($m = 0$), you can replace $\cos^4 x$ by $[(1 + \cos 2x)/2]^2$.

$$\begin{aligned}\int \cos^4 x\, dx &= \int \left(\frac{1+\cos 2x}{2}\right)^2 dx\\ &= \int \left(\frac{1}{4} + \frac{\cos 2x}{2} + \frac{\cos^2 2x}{4}\right) dx\\ &= \int \left[\frac{1}{4} + \frac{\cos 2x}{2} + \frac{1}{4}\left(\frac{1+\cos 4x}{2}\right)\right] dx\\ &= \frac{3}{8}\int dx + \frac{1}{4}\int 2\cos 2x\, dx + \frac{1}{32}\int 4\cos 4x\, dx\\ &= \frac{3x}{8} + \frac{\sin 2x}{4} + \frac{\sin 4x}{32} + C\end{aligned}$$

Use a symbolic differentiation utility to verify this. Can you simplify the derivative to obtain the original integrand?

In Example 3, if you were to evaluate the definite integral from 0 to $\pi/2$, you would obtain

$$\begin{aligned}\int_0^{\pi/2} \cos^4 x\, dx &= \left[\frac{3x}{8} + \frac{\sin 2x}{4} + \frac{\sin 4x}{32}\right]_0^{\pi/2}\\ &= \left(\frac{3\pi}{16} + 0 + 0\right) - (0 + 0 + 0)\\ &= \frac{3\pi}{16}.\end{aligned}$$

Note that the only term that contributes to the solution is $3x/8$. This observation is generalized in the following formulas developed by John Wallis.

Bettmann/Corbis

JOHN WALLIS (1616–1703)
Wallis did much of his work in calculus prior to Newton and Leibniz, and he influenced the thinking of both of these men. Wallis is also credited with introducing the present symbol (∞) for infinity.

Wallis's Formulas

1. If n is odd ($n \geq 3$), then

$$\int_0^{\pi/2} \cos^n x\, dx = \left(\frac{2}{3}\right)\left(\frac{4}{5}\right)\left(\frac{6}{7}\right) \cdots \left(\frac{n-1}{n}\right).$$

2. If n is even ($n \geq 2$), then

$$\int_0^{\pi/2} \cos^n x\, dx = \left(\frac{1}{2}\right)\left(\frac{3}{4}\right)\left(\frac{5}{6}\right) \cdots \left(\frac{n-1}{n}\right)\left(\frac{\pi}{2}\right).$$

These formulas are also valid if $\cos^n x$ is replaced by $\sin^n x$. (You are asked to prove both formulas in Exercise 104.)

Integrals Involving Powers of Secant and Tangent

The following guidelines can help you evaluate integrals of the form

$$\int \sec^m x \tan^n x \, dx.$$

Guidelines for Evaluating Integrals Involving Secant and Tangent

1. If the power of the secant is even and positive, save a secant-squared factor and convert the remaining factors to tangents. Then expand and integrate.

$$\int \overbrace{\sec^{2k}}^{\text{Even}} x \tan^n x \, dx = \int \overbrace{(\sec^2 x)^{k-1}}^{\text{Convert to tangents}} \tan^n x \overbrace{\sec^2 x \, dx}^{\text{Save for } du} = \int (1 + \tan^2 x)^{k-1} \tan^n x \sec^2 x \, dx$$

2. If the power of the tangent is odd and positive, save a secant-tangent factor and convert the remaining factors to secants. Then expand and integrate.

$$\int \sec^m x \overbrace{\tan^{2k+1}}^{\text{Odd}} x \, dx = \int \sec^{m-1} x \overbrace{(\tan^2 x)^k}^{\text{Convert to secants}} \overbrace{\sec x \tan x \, dx}^{\text{Save for } du} = \int \sec^{m-1} x (\sec^2 x - 1)^k \sec x \tan x \, dx$$

3. If there are no secant factors and the power of the tangent is even and positive, convert a tangent-squared factor to a secant-squared factor, then expand and repeat if necessary.

$$\int \tan^n x \, dx = \int \tan^{n-2} x \overbrace{(\tan^2 x)}^{\text{Convert to secants}} \, dx = \int \tan^{n-2} x (\sec^2 x - 1) \, dx$$

4. If the integral is of the form $\int \sec^m x \, dx$, where m is odd and positive, use integration by parts, as illustrated in Example 5 in the preceding section.
5. If none of the first four guidelines applies, try converting to sines and cosines.

EXAMPLE 4 Power of Tangent Is Odd and Positive

Find $\displaystyle\int \frac{\tan^3 x}{\sqrt{\sec x}} \, dx$.

Solution Because you expect to use the Power Rule with $u = \sec x$, *save a factor of* $(\sec x \tan x)$ to form du and convert the remaining tangent factors to secants.

$$\begin{aligned}
\int \frac{\tan^3 x}{\sqrt{\sec x}} \, dx &= \int (\sec x)^{-1/2} \tan^3 x \, dx \\
&= \int (\sec x)^{-3/2} (\tan^2 x)(\sec x \tan x) \, dx \\
&= \int (\sec x)^{-3/2} (\sec^2 x - 1)(\sec x \tan x) \, dx \\
&= \int [(\sec x)^{1/2} - (\sec x)^{-3/2}](\sec x \tan x) \, dx \\
&= \frac{2}{3}(\sec x)^{3/2} + 2(\sec x)^{-1/2} + C
\end{aligned}$$

NOTE In Example 5, the power of the tangent is odd and positive. So, you could also find the integral using the procedure described in guideline 2 on page 537. In Exercise 85, you are asked to show that the results obtained by these two procedures differ only by a constant.

EXAMPLE 5 Power of Secant Is Even and Positive

Find $\int \sec^4 3x \tan^3 3x\, dx$.

Solution Let $u = \tan 3x$, then $du = 3 \sec^2 3x\, dx$ and you can write

$$\begin{aligned}\int \sec^4 3x \tan^3 3x\, dx &= \int \sec^2 3x \tan^3 3x(\sec^2 3x)\, dx \\ &= \int (1 + \tan^2 3x) \tan^3 3x(\sec^2 3x)\, dx \\ &= \frac{1}{3}\int (\tan^3 3x + \tan^5 3x)(3 \sec^2 3x)\, dx \\ &= \frac{1}{3}\left(\frac{\tan^4 3x}{4} + \frac{\tan^6 3x}{6}\right) + C \\ &= \frac{\tan^4 3x}{12} + \frac{\tan^6 3x}{18} + C.\end{aligned}$$

EXAMPLE 6 Power of Tangent Is Even

Evaluate $\int_0^{\pi/4} \tan^4 x\, dx$.

Solution Because there are no secant factors, you can begin by converting a tangent-squared factor to a secant-squared factor.

$$\begin{aligned}\int \tan^4 x\, dx &= \int \tan^2 x(\tan^2 x)\, dx \\ &= \int \tan^2 x(\sec^2 x - 1)\, dx \\ &= \int \tan^2 x \sec^2 x\, dx - \int \tan^2 x\, dx \\ &= \int \tan^2 x \sec^2 x\, dx - \int (\sec^2 x - 1)\, dx \\ &= \frac{\tan^3 x}{3} - \tan x + x + C\end{aligned}$$

You can evaluate the definite integral as follows.

$$\begin{aligned}\int_0^{\pi/4} \tan^4 x\, dx &= \left[\frac{\tan^3 x}{3} - \tan x + x\right]_0^{\pi/4} \\ &= \frac{\pi}{4} - \frac{2}{3} \\ &\approx 0.119\end{aligned}$$

The area represented by the definite integral is shown in Figure 8.5. Try using Simpson's Rule to approximate this integral. With $n = 18$, you should obtain an approximation that is within 0.00001 of the actual value.

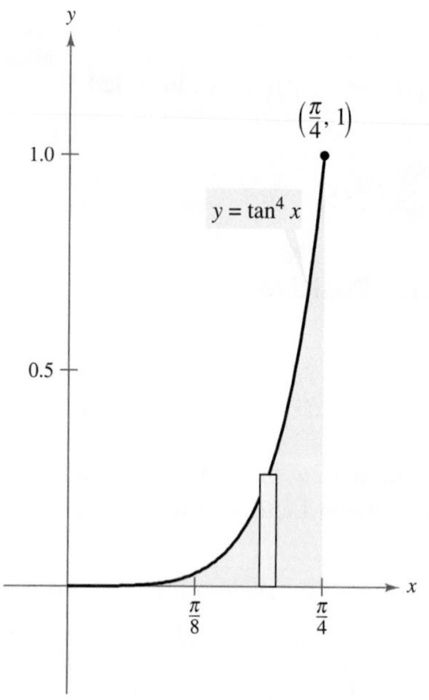

The area of the region is approximately 0.119.
Figure 8.5

For integrals involving powers of cotangents and cosecants, you can follow a strategy similar to that used for powers of tangents and secants. Also, when integrating trigonometric functions, remember that it sometimes helps to convert the entire integrand to powers of sines and cosines.

EXAMPLE 7 Converting to Sines and Cosines

Find $\displaystyle\int \frac{\sec x}{\tan^2 x}\, dx$.

Solution Because the first four guidelines on page 537 do not apply, try converting the integrand to sines and cosines. In this case, you are able to integrate the resulting powers of sine and cosine as follows.

$$\begin{aligned}\int \frac{\sec x}{\tan^2 x}\, dx &= \int \left(\frac{1}{\cos x}\right)\left(\frac{\cos x}{\sin x}\right)^2 dx \\ &= \int (\sin x)^{-2}(\cos x)\, dx \\ &= -(\sin x)^{-1} + C \\ &= -\csc x + C\end{aligned}$$

Integrals Involving Sine-Cosine Products with Different Angles

FOR FURTHER INFORMATION To learn more about integrals involving sine-cosine products with different angles, see the article "Integrals of Products of Sine and Cosine with Different Arguments" by Sherrie J. Nicol in *The College Mathematics Journal*. To view this article, go to the website *www.matharticles.com*.

Integrals involving the products of sines and cosines of two *different* angles occur in many applications. In such instances you can use the following product-to-sum identities.

$$\sin mx \sin nx = \frac{1}{2}(\cos[(m-n)x] - \cos[(m+n)x])$$

$$\sin mx \cos nx = \frac{1}{2}(\sin[(m-n)x] + \sin[(m+n)x])$$

$$\cos mx \cos nx = \frac{1}{2}(\cos[(m-n)x] + \cos[(m+n)x])$$

EXAMPLE 8 Using Product-to-Sum Identities

Find $\displaystyle\int \sin 5x \cos 4x\, dx$.

Solution Considering the second product-to-sum identity above, you can write

$$\begin{aligned}\int \sin 5x \cos 4x\, dx &= \frac{1}{2}\int (\sin x + \sin 9x)\, dx \\ &= \frac{1}{2}\left(-\cos x - \frac{\cos 9x}{9}\right) + C \\ &= -\frac{\cos x}{2} - \frac{\cos 9x}{18} + C.\end{aligned}$$

Exercises for Section 8.3

See www.CalcChat.com for worked-out solutions to odd-numbered exercises.

In Exercises 1–4, use differentiation to match the antiderivative with the correct integral. [Integrals are labeled (a), (b), (c), and (d).]

(a) $\int \sin x \tan^2 x\, dx$ (b) $8\int \cos^4 x\, dx$

(c) $\int \sin x \sec^2 x\, dx$ (d) $\int \tan^4 x\, dx$

1. $y = \sec x$
2. $y = \cos x + \sec x$
3. $y = x - \tan x + \frac{1}{3}\tan^3 x$
4. $y = 3x + 2\sin x \cos^3 x + 3\sin x \cos x$

In Exercises 5–18, find the integral.

5. $\int \cos^3 x \sin x\, dx$
6. $\int \cos^3 x \sin^4 x\, dx$
7. $\int \sin^5 2x \cos 2x\, dx$
8. $\int \sin^3 x\, dx$
9. $\int \sin^5 x \cos^2 x\, dx$
10. $\int \cos^3 \frac{x}{3}\, dx$
11. $\int \cos^3 \theta \sqrt{\sin \theta}\, d\theta$
12. $\int \frac{\sin^5 t}{\sqrt{\cos t}}\, dt$
13. $\int \cos^2 3x\, dx$
14. $\int \sin^2 2x\, dx$
15. $\int \sin^2 \alpha \cos^2 \alpha\, d\alpha$
16. $\int \sin^4 2\theta\, d\theta$
17. $\int x \sin^2 x\, dx$
18. $\int x^2 \sin^2 x\, dx$

In Exercises 19–24, use Wallis's Formulas to evaluate the integral.

19. $\int_0^{\pi/2} \cos^3 x\, dx$
20. $\int_0^{\pi/2} \cos^5 x\, dx$
21. $\int_0^{\pi/2} \cos^7 x\, dx$
22. $\int_0^{\pi/2} \sin^2 x\, dx$
23. $\int_0^{\pi/2} \sin^6 x\, dx$
24. $\int_0^{\pi/2} \sin^7 x\, dx$

In Exercises 25–42, find the integral involving secant and tangent.

25. $\int \sec 3x\, dx$
26. $\int \sec^2(2x - 1)\, dx$
27. $\int \sec^4 5x\, dx$
28. $\int \sec^6 3x\, dx$
29. $\int \sec^3 \pi x\, dx$
30. $\int \tan^2 x\, dx$
31. $\int \tan^5 \frac{x}{4}\, dx$
32. $\int \tan^3 \frac{\pi x}{2} \sec^2 \frac{\pi x}{2}\, dx$
33. $\int \sec^2 x \tan x\, dx$
34. $\int \tan^3 2t \sec^3 2t\, dt$
35. $\int \tan^2 x \sec^2 x\, dx$
36. $\int \tan^5 2x \sec^2 2x\, dx$
37. $\int \sec^6 4x \tan 4x\, dx$
38. $\int \sec^2 \frac{x}{2} \tan \frac{x}{2}\, dx$
39. $\int \sec^3 x \tan x\, dx$
40. $\int \tan^3 3x\, dx$
41. $\int \frac{\tan^2 x}{\sec x}\, dx$
42. $\int \frac{\tan^2 x}{\sec^5 x}\, dx$

In Exercises 43–46, solve the differential equation.

43. $\frac{dr}{d\theta} = \sin^4 \pi\theta$
44. $\frac{ds}{d\alpha} = \sin^2 \frac{\alpha}{2} \cos^2 \frac{\alpha}{2}$
45. $y' = \tan^3 3x \sec 3x$
46. $y' = \sqrt{\tan x}\, \sec^4 x$

Slope Fields **In Exercises 47 and 48, a differential equation, a point, and a slope field are given. (a) Sketch two approximate solutions of the differential equation on the slope field, one of which passes through the given point. (b) Use integration to find the particular solution of the differential equation and use a graphing utility to graph the solution. Compare the result with the sketches in part (a). To print an enlarged copy of the graph, go to the website *www.mathgraphs.com*.**

47. $\frac{dy}{dx} = \sin^2 x,\ (0, 0)$
48. $\frac{dy}{dx} = \sec^2 x \tan^2 x,\ \left(0, -\frac{1}{4}\right)$

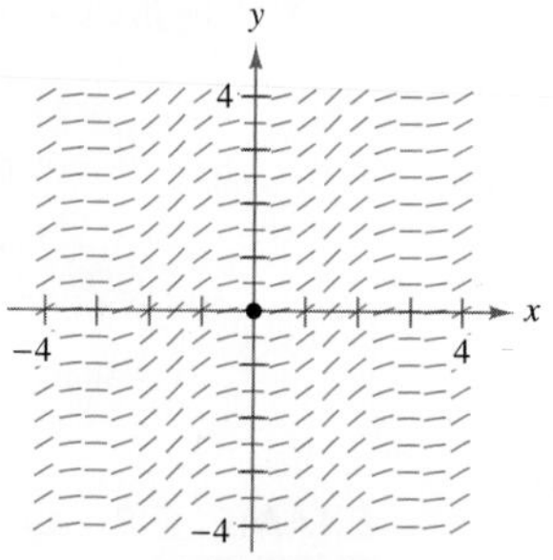

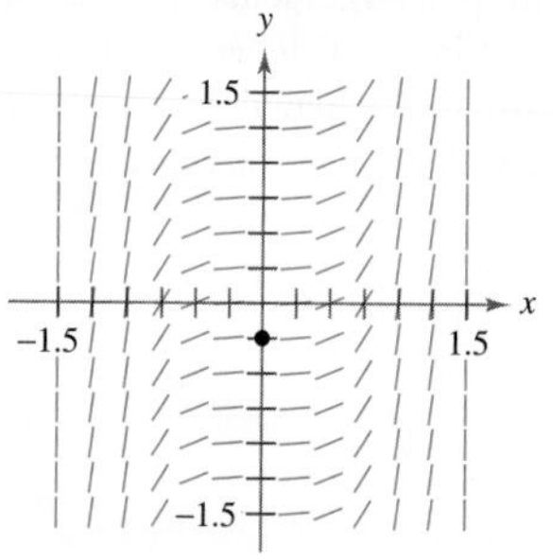

Slope Fields **In Exercises 49 and 50, use a computer algebra system to graph the slope field for the differential equation, and graph the solution through the specified initial condition.**

49. $\frac{dy}{dx} = \frac{3 \sin x}{y},\ y(0) = 2$
50. $\frac{dy}{dx} = 3\sqrt{y} \tan^2 x,\ y(0) = 3$

In Exercises 51–54, find the integral.

51. $\int \sin 3x \cos 2x\, dx$
52. $\int \cos 4\theta \cos(-3\theta)\, d\theta$
53. $\int \sin \theta \sin 3\theta\, d\theta$
54. $\int \sin(-4x) \cos 3x\, dx$

In Exercises 55–64, find the integral. Use a computer algebra system to confirm your result.

55. $\int \cot^3 2x\, dx$

56. $\int \tan^4 \frac{x}{2} \sec^4 \frac{x}{2}\, dx$

57. $\int \csc^4 \theta\, d\theta$

58. $\int \csc^2 3x \cot 3x\, dx$

59. $\int \frac{\cot^2 t}{\csc t}\, dt$

60. $\int \frac{\cot^3 t}{\csc t}\, dt$

61. $\int \frac{1}{\sec x \tan x}\, dx$

62. $\int \frac{\sin^2 x - \cos^2 x}{\cos x}\, dx$

63. $\int (\tan^4 t - \sec^4 t)\, dt$

64. $\int \frac{1 - \sec t}{\cos t - 1}\, dt$

In Exercises 65–72, evaluate the definite integral.

65. $\int_{-\pi}^{\pi} \sin^2 x\, dx$

66. $\int_0^{\pi/3} \tan^2 x\, dx$

67. $\int_0^{\pi/4} \tan^3 x\, dx$

68. $\int_0^{\pi/4} \sec^2 t \sqrt{\tan t}\, dt$

69. $\int_0^{\pi/2} \frac{\cos t}{1 + \sin t}\, dt$

70. $\int_{-\pi}^{\pi} \sin 3\theta \cos \theta\, d\theta$

71. $\int_{-\pi/2}^{\pi/2} \cos^3 x\, dx$

72. $\int_{-\pi/2}^{\pi/2} (\sin^2 x + 1)\, dx$

In Exercises 73–78, use a computer algebra system to find the integral. Graph the antiderivatives for two different values of the constant of integration.

73. $\int \cos^4 \frac{x}{2}\, dx$

74. $\int \sin^2 x \cos^2 x\, dx$

75. $\int \sec^5 \pi x\, dx$

76. $\int \tan^3(1 - x)\, dx$

77. $\int \sec^5 \pi x \tan \pi x\, dx$

78. $\int \sec^4(1 - x) \tan(1 - x)\, dx$

In Exercises 79–82, use a computer algebra system to evaluate the definite integral.

79. $\int_0^{\pi/4} \sin 2\theta \sin 3\theta\, d\theta$

80. $\int_0^{\pi/2} (1 - \cos \theta)^2\, d\theta$

81. $\int_0^{\pi/2} \sin^4 x\, dx$

82. $\int_0^{\pi/2} \sin^6 x\, dx$

Writing About Concepts

83. In your own words, describe how you would integrate $\int \sin^m x \cos^n x\, dx$ for each condition.
    (a) m is positive and odd.
    (b) n is positive and odd.
    (c) m and n are both positive and even.

Writing About Concepts *(continued)*

84. In your own words, describe how you would integrate $\int \sec^m x \tan^n x\, dx$ for each condition.
    (a) m is positive and even.
    (b) n is positive and odd.
    (c) n is positive and even, and there are no secant factors.
    (d) m is positive and odd, and there are no tangent factors.

In Exercises 85 and 86, (a) find the indefinite integral in two different ways. (b) Use a graphing utility to graph the antiderivative (without the constant of integration) obtained by each method to show that the results differ only by a constant. (c) Verify analytically that the results differ only by a constant.

85. $\int \sec^4 3x \tan^3 3x\, dx$

86. $\int \sec^2 x \tan x\, dx$

Area **In Exercises 87–90, find the area of the region bounded by the graphs of the equations.**

87. $y = \sin x,\quad y = \sin^3 x,\quad x = 0,\quad x = \pi/2$

88. $y = \sin^2 \pi x,\quad y = 0,\quad x = 0,\quad x = 1$

89. $y = \cos^2 x,\quad y = \sin^2 x,\quad x = -\pi/4,\quad x = \pi/4$

90. $y = \cos^2 x,\quad y = \sin x \cos x,\quad x = -\pi/2,\quad x = \pi/4$

Volume **In Exercises 91 and 92, find the volume of the solid generated by revolving the region bounded by the graphs of the equations about the *x*-axis.**

91. $y = \tan x,\quad y = 0,\quad x = -\pi/4\quad x = \pi/4$

92. $y = \cos \frac{x}{2},\quad y = \sin \frac{x}{2},\quad x = 0,\quad x = \pi/2$

Volume and Centroid **In Exercises 93 and 94, for the region bounded by the graphs of the equations, find (a) the volume of the solid formed by revolving the region about the *x*-axis and (b) the centroid of the region.**

93. $y = \sin x, y = 0, x = 0, x = \pi$

94. $y = \cos x, y = 0, x = 0, x = \pi/2$

In Exercises 95–98, use integration by parts to verify the reduction formula.

95. $\int \sin^n x\, dx = -\frac{\sin^{n-1} x \cos x}{n} + \frac{n-1}{n} \int \sin^{n-2} x\, dx$

96. $\int \cos^n x\, dx = \frac{\cos^{n-1} x \sin x}{n} + \frac{n-1}{n} \int \cos^{n-2} x\, dx$

97. $\int \cos^m x \sin^n x\, dx = -\frac{\cos^{m+1} x \sin^{n-1} x}{m+n} + \frac{n-1}{m+n} \int \cos^m x \sin^{n-2} x\, dx$

98. $\int \sec^n x\, dx = \frac{1}{n-1} \sec^{n-2} x \tan x + \frac{n-2}{n-1} \int \sec^{n-2} x\, dx$

In Exercises 99–102, use the results of Exercises 95–98 to find the integral.

99. $\displaystyle\int \sin^5 x\,dx$

100. $\displaystyle\int \cos^4 x\,dx$

101. $\displaystyle\int \sec^4 \frac{2\pi x}{5}\,dx$

102. $\displaystyle\int \sin^4 x \cos^2 x\,dx$

103. ***Modeling Data*** The table shows the normal maximum (high) and minimum (low) temperatures (in degrees Fahrenheit) for Erie, Pennsylvania for each month of the year. *(Source: NOAA)*

Month	Jan	Feb	Mar	Apr	May	Jun
Max	33.5	35.4	44.7	55.6	67.4	76.2
Min	20.3	20.9	28.2	37.9	48.7	58.5

Month	Jul	Aug	Sep	Oct	Nov	Dec
Max	80.4	79.0	72.0	61.0	49.3	38.6
Min	63.7	62.7	55.9	45.5	36.4	26.8

The maximum and minimum temperatures can be modeled by

$$f(t) = a_0 + a_1 \cos\frac{\pi t}{6} + b_1 \sin\frac{\pi t}{6}$$

where $t = 0$ corresponds to January and a_0, a_1, and b_1 are as follows.

$$a_0 = \frac{1}{12}\int_0^{12} f(t)\,dt$$

$$a_1 = \frac{1}{6}\int_0^{12} f(t)\cos\frac{\pi t}{6}\,dt$$

$$b_1 = \frac{1}{6}\int_0^{12} f(t)\sin\frac{\pi t}{6}\,dt$$

(a) Approximate the model $H(t)$ for the maximum temperatures. (*Hint:* Use Simpson's Rule to approximate the integrals and use the January data twice.)

(b) Repeat part (a) for a model $L(t)$ for the minimum temperature data.

(c) Use a graphing utility to compare each model with the actual data. During what part of the year is the difference between the maximum and minimum temperatures greatest?

104. ***Wallis's Formulas*** Use the result of Exercise 96 to prove the following versions of Wallis's Formulas.

(a) If n is odd $(n \ge 3)$, then

$$\int_0^{\pi/2} \cos^n x\,dx = \left(\frac{2}{3}\right)\left(\frac{4}{5}\right)\left(\frac{6}{7}\right)\cdots\left(\frac{n-1}{n}\right).$$

(b) If n is even $(n \ge 2)$, then

$$\int_0^{\pi/2} \cos^n x\,dx = \left(\frac{1}{2}\right)\left(\frac{3}{4}\right)\left(\frac{5}{6}\right)\cdots\left(\frac{n-1}{n}\right)\left(\frac{\pi}{2}\right).$$

105. The **inner product** of two functions f and g on $[a, b]$ is given by $\langle f, g\rangle = \int_a^b f(x)g(x)\,dx$. Two distinct functions f and g are said to be **orthogonal** if $\langle f, g\rangle = 0$. Show that the following set of functions is orthogonal on $[-\pi, \pi]$.

$$\{\sin x, \sin 2x, \sin 3x, \ldots, \cos x, \cos 2x, \cos 3x, \ldots\}$$

106. ***Fourier Series*** The following sum is a *finite Fourier series.*

$$f(x) = \sum_{i=1}^{N} a_i \sin ix$$

$$= a_1 \sin x + a_2 \sin 2x + a_3 \sin 3x + \cdots + a_N \sin Nx$$

(a) Use Exercise 105 to show that the nth coefficient a_n is given by $a_n = \dfrac{1}{\pi}\displaystyle\int_{-\pi}^{\pi} f(x)\sin nx\,dx$.

(b) Let $f(x) = x$. Find a_1, a_2, and a_3.

Section Project: Power Lines

Power lines are constructed by stringing wire between supports and adjusting the tension on each span. The wire hangs between supports in the shape of a catenary, as shown in the figure.

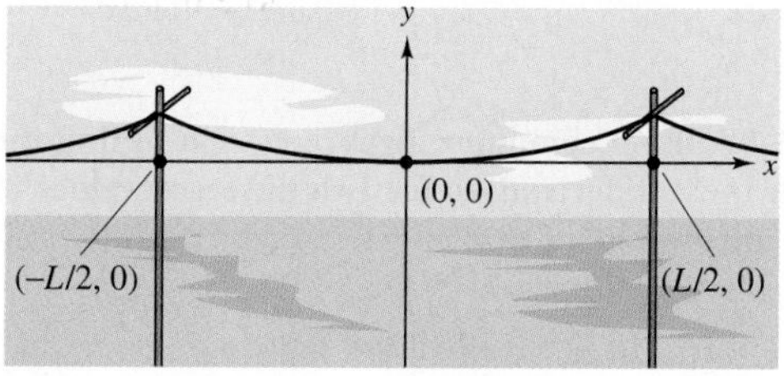

Let T be the tension (in pounds) on a span of wire, let u be the density (in pounds per foot), let $g \approx 32.2$ be the acceleration due to gravity (in feet per second per second), and let L be the distance (in feet) between the supports. Then the equation of the catenary is $y = \dfrac{T}{ug}\left(\cosh\dfrac{ugx}{T} - 1\right)$, where x and y are measured in feet.

(a) Find the length of the wire between two spans.

(b) To measure the tension in a span, power line workers use the *return wave method.* The wire is struck at one support, creating a wave in the line, and the time t (in seconds) it takes for the wave to make a round trip is measured. The velocity v (in feet per second) is given by $v = \sqrt{T/u}$. How long does it take the wave to make a round trip between supports?

(c) The sag s (in inches) can be obtained by evaluating y when $x = L/2$ in the equation for the catenary (and multiplying by 12). In practice, however, power line workers use the "lineman's equation" given by $s \approx 12.075t^2$. Use the fact that $[\cosh(ugL/2T) + 1] \approx 2$ to derive this equation.

FOR FURTHER INFORMATION To learn more about the mathematics of power lines, see the article "Constructing Power Lines" by Thomas O'Neil in *The UMAP Journal.*

Section 8.4 Trigonometric Substitution

- **Use trigonometric substitution to solve an integral.**
- **Use integrals to model and solve real-life applications.**

EXPLORATION

Integrating a Radical Function
Up to this point in the text, you have not evaluated the following integral.

$$\int_{-1}^{1} \sqrt{1 - x^2}\, dx$$

From geometry, you should be able to find the exact value of this integral—what is it? Using numerical integration, with Simpson's Rule or the Trapezoidal Rule, you can't be sure of the accuracy of the approximation. Why?

Try finding the exact value using the substitution

$$x = \sin\theta \text{ and } dx = \cos\theta\, d\theta.$$

Does your answer agree with the value you obtained using geometry?

Trigonometric Substitution

Now that you can evaluate integrals involving powers of trigonometric functions, you can use **trigonometric substitution** to evaluate integrals involving the radicals

$$\sqrt{a^2 - u^2}, \qquad \sqrt{a^2 + u^2}, \qquad \text{and} \qquad \sqrt{u^2 - a^2}.$$

The objective with trigonometric substitution is to eliminate the radical in the integrand. You do this with the Pythagorean identities

$$\cos^2\theta = 1 - \sin^2\theta, \quad \sec^2\theta = 1 + \tan^2\theta, \quad \text{and} \quad \tan^2\theta = \sec^2\theta - 1.$$

For example, if $a > 0$, let $u = a \sin\theta$, where $-\pi/2 \le \theta \le \pi/2$. Then

$$\begin{aligned}\sqrt{a^2 - u^2} &= \sqrt{a^2 - a^2\sin^2\theta}\\ &= \sqrt{a^2(1 - \sin^2\theta)}\\ &= \sqrt{a^2\cos^2\theta}\\ &= a\cos\theta.\end{aligned}$$

Note that $\cos\theta \ge 0$, because $-\pi/2 \le \theta \le \pi/2$.

Trigonometric Substitution ($a > 0$)

1. For integrals involving $\sqrt{a^2 - u^2}$, let

$$u = a\sin\theta.$$

Then $\sqrt{a^2 - u^2} = a\cos\theta$, where $-\pi/2 \le \theta \le \pi/2$.

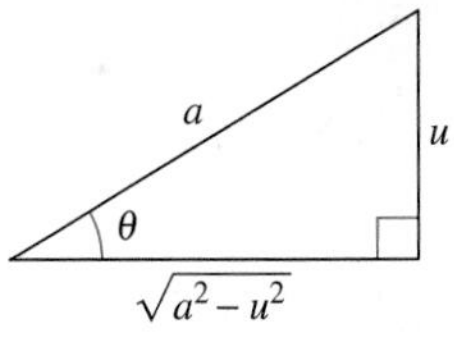

2. For integrals involving $\sqrt{a^2 + u^2}$, let

$$u = a\tan\theta.$$

Then $\sqrt{a^2 + u^2} = a\sec\theta$, where $-\pi/2 < \theta < \pi/2$.

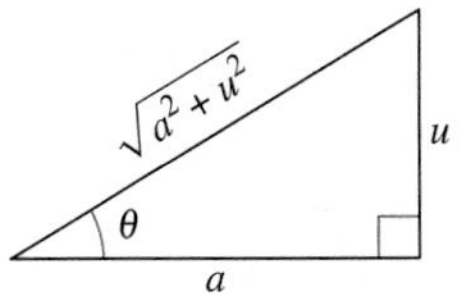

3. For integrals involving $\sqrt{u^2 - a^2}$, let

$$u = a\sec\theta.$$

Then $\sqrt{u^2 - a^2} = \pm a\tan\theta$, where $0 \le \theta < \pi/2$ or $\pi/2 < \theta \le \pi$. Use the positive value if $u > a$ and the negative value if $u < -a$.

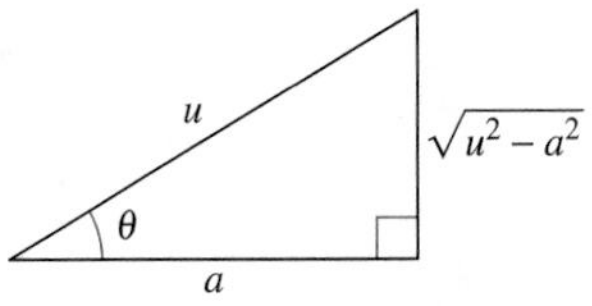

NOTE The restrictions on θ ensure that the function that defines the substitution is one-to-one. In fact, these are the same intervals over which the arcsine, arctangent, and arcsecant are defined.

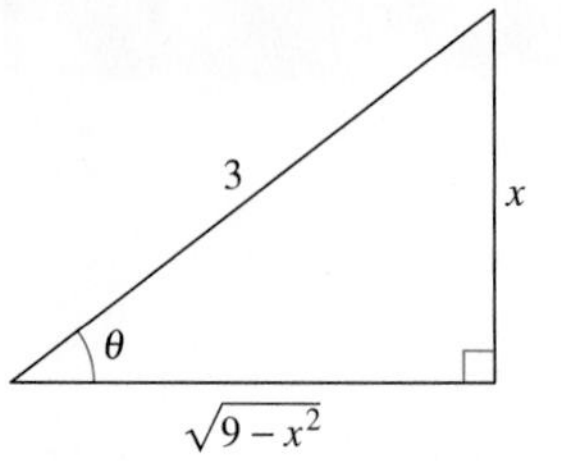

$\sin\theta = \frac{x}{3}, \cot\theta = \frac{\sqrt{9-x^2}}{x}$

Figure 8.6

EXAMPLE 1 Trigonometric Substitution: $u = a\sin\theta$

Find $\displaystyle\int \frac{dx}{x^2\sqrt{9-x^2}}$.

Solution First, note that none of the basic integration rules applies. To use trigonometric substitution, you should observe that $\sqrt{9-x^2}$ is of the form $\sqrt{a^2-u^2}$. So, you can use the substitution

$$x = a\sin\theta = 3\sin\theta.$$

Using differentiation and the triangle shown in Figure 8.6, you obtain

$$dx = 3\cos\theta\, d\theta, \qquad \sqrt{9-x^2} = 3\cos\theta, \qquad \text{and} \qquad x^2 = 9\sin^2\theta.$$

So, trigonometric substitution yields

$$\begin{aligned}
\int \frac{dx}{x^2\sqrt{9-x^2}} &= \int \frac{3\cos\theta\, d\theta}{(9\sin^2\theta)(3\cos\theta)} && \text{Substitute.}\\
&= \frac{1}{9}\int \frac{d\theta}{\sin^2\theta} && \text{Simplify.}\\
&= \frac{1}{9}\int \csc^2\theta\, d\theta && \text{Trigonometric identity}\\
&= -\frac{1}{9}\cot\theta + C && \text{Apply Cosecant Rule.}\\
&= -\frac{1}{9}\left(\frac{\sqrt{9-x^2}}{x}\right) + C && \text{Substitute for } \cot\theta.\\
&= -\frac{\sqrt{9-x^2}}{9x} + C.
\end{aligned}$$

Note that the triangle in Figure 8.6 can be used to convert the θ's back to x's as follows.

$$\begin{aligned}
\cot\theta &= \frac{\text{adj.}}{\text{opp.}}\\
&= \frac{\sqrt{9-x^2}}{x}
\end{aligned}$$

TECHNOLOGY Use a computer algebra system to find each definite integral.

$$\int \frac{dx}{\sqrt{9-x^2}} \qquad \int \frac{dx}{x\sqrt{9-x^2}} \qquad \int \frac{dx}{x^2\sqrt{9-x^2}} \qquad \int \frac{dx}{x^3\sqrt{9-x^2}}$$

Then use trigonometric substitution to duplicate the results obtained with the computer algebra system.

In an earlier chapter, you saw how the inverse hyperbolic functions can be used to evaluate the integrals

$$\int \frac{du}{\sqrt{u^2 \pm a^2}}, \qquad \int \frac{du}{a^2-u^2}, \qquad \text{and} \qquad \int \frac{du}{u\sqrt{a^2 \pm u^2}}.$$

You can also evaluate these integrals using trigonometric substitution. This is shown in the next example.

EXAMPLE 2 Trigonometric Substitution: $u = a \tan \theta$

Find $\displaystyle\int \frac{dx}{\sqrt{4x^2 + 1}}$.

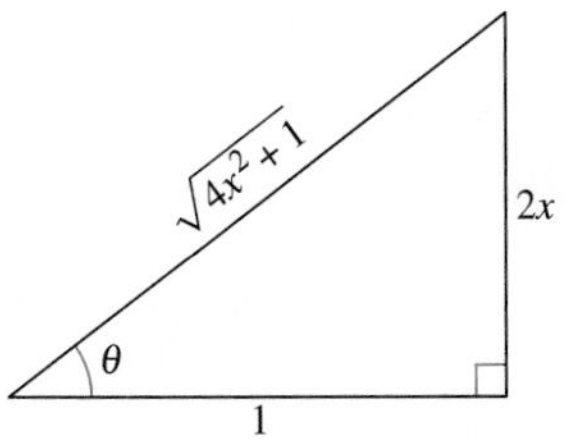

$\tan \theta = 2x$, $\sec \theta = \sqrt{4x^2 + 1}$
Figure 8.7

Solution Let $u = 2x$, $a = 1$, and $2x = \tan \theta$, as shown in Figure 8.7. Then,

$$dx = \frac{1}{2} \sec^2 \theta \, d\theta \qquad \text{and} \qquad \sqrt{4x^2 + 1} = \sec \theta.$$

Trigonometric substitution produces

$$\begin{aligned}
\int \frac{1}{\sqrt{4x^2 + 1}} \, dx &= \frac{1}{2} \int \frac{\sec^2 \theta \, d\theta}{\sec \theta} && \text{Substitute.} \\
&= \frac{1}{2} \int \sec \theta \, d\theta && \text{Simplify.} \\
&= \frac{1}{2} \ln\left|\sec \theta + \tan \theta\right| + C && \text{Apply Secant Rule.} \\
&= \frac{1}{2} \ln\left|\sqrt{4x^2 + 1} + 2x\right| + C. && \text{Back-substitute.}
\end{aligned}$$

Try checking this result with a computer algebra system. Is the result given in this form or in the form of an inverse hyperbolic function?

You can extend the use of trigonometric substitution to cover integrals involving expressions such as $(a^2 - u^2)^{n/2}$ by writing the expression as

$$(a^2 - u^2)^{n/2} = \left(\sqrt{a^2 - u^2}\right)^n.$$

EXAMPLE 3 Trigonometric Substitution: Rational Powers

Find $\displaystyle\int \frac{dx}{(x^2 + 1)^{3/2}}$.

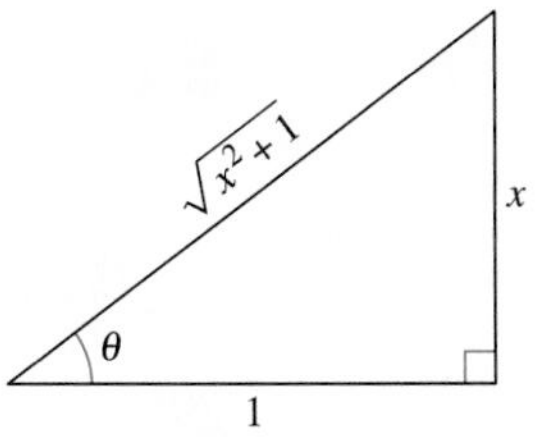

$\tan \theta = x$, $\sin \theta = \dfrac{x}{\sqrt{x^2 + 1}}$
Figure 8.8

Solution Begin by writing $(x^2 + 1)^{3/2}$ as $\left(\sqrt{x^2 + 1}\right)^3$. Then, let $a = 1$ and $u = x = \tan \theta$, as shown in Figure 8.8. Using

$$dx = \sec^2 \theta \, d\theta \qquad \text{and} \qquad \sqrt{x^2 + 1} = \sec \theta$$

you can apply trigonometric substitution as follows.

$$\begin{aligned}
\int \frac{dx}{(x^2 + 1)^{3/2}} &= \int \frac{dx}{\left(\sqrt{x^2 + 1}\right)^3} && \text{Rewrite denominator.} \\
&= \int \frac{\sec^2 \theta \, d\theta}{\sec^3 \theta} && \text{Substitute.} \\
&= \int \frac{d\theta}{\sec \theta} && \text{Simplify.} \\
&= \int \cos \theta \, d\theta && \text{Trigonometric identity} \\
&= \sin \theta + C && \text{Apply Cosine Rule.} \\
&= \frac{x}{\sqrt{x^2 + 1}} + C && \text{Back-substitute.}
\end{aligned}$$

For definite integrals, it is often convenient to determine the integration limits for θ that avoid converting back to x. You might want to review this procedure in Section 4.5, Examples 8 and 9.

EXAMPLE 4 Converting the Limits of Integration

Evaluate $\displaystyle\int_{\sqrt{3}}^{2} \frac{\sqrt{x^2-3}}{x}\,dx.$

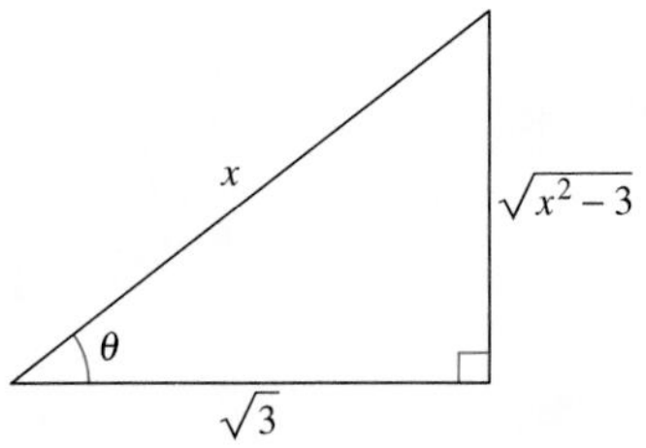

$\sec\theta = \dfrac{x}{\sqrt{3}}, \tan\theta = \dfrac{\sqrt{x^2-3}}{\sqrt{3}}$

Figure 8.9

Solution Because $\sqrt{x^2-3}$ has the form $\sqrt{u^2-a^2}$, you can consider

$$u = x, \quad a = \sqrt{3}, \quad \text{and} \quad x = \sqrt{3}\sec\theta$$

as shown in Figure 8.9. Then,

$$dx = \sqrt{3}\sec\theta\tan\theta\,d\theta \qquad \text{and} \qquad \sqrt{x^2-3} = \sqrt{3}\tan\theta.$$

To determine the upper and lower limits of integration, use the substitution $x = \sqrt{3}\sec\theta$, as follows.

Lower Limit	*Upper Limit*
When $x = \sqrt{3}$, $\sec\theta = 1$ and $\theta = 0$.	When $x = 2$, $\sec\theta = \dfrac{2}{\sqrt{3}}$ and $\theta = \dfrac{\pi}{6}$.

So, you have

Integration limits for x — Integration limits for θ

$$\begin{aligned}
\int_{\sqrt{3}}^{2} \frac{\sqrt{x^2-3}}{x}\,dx &= \int_{0}^{\pi/6} \frac{(\sqrt{3}\tan\theta)(\sqrt{3}\sec\theta\tan\theta)\,d\theta}{\sqrt{3}\sec\theta}\\
&= \int_{0}^{\pi/6} \sqrt{3}\tan^2\theta\,d\theta\\
&= \sqrt{3}\int_{0}^{\pi/6} (\sec^2\theta - 1)\,d\theta\\
&= \sqrt{3}\Big[\tan\theta - \theta\Big]_{0}^{\pi/6}\\
&= \sqrt{3}\left(\frac{1}{\sqrt{3}} - \frac{\pi}{6}\right)\\
&= 1 - \frac{\sqrt{3}\pi}{6}\\
&\approx 0.0931.
\end{aligned}$$

In Example 4, try converting back to the variable x and evaluating the antiderivative at the original limits of integration. You should obtain

$$\int_{\sqrt{3}}^{2} \frac{\sqrt{x^2-3}}{x}\,dx = \sqrt{3}\,\frac{\sqrt{x^2-3}}{\sqrt{3}} - \operatorname{arcsec}\frac{x}{\sqrt{3}}\Bigg]_{\sqrt{3}}^{2}.$$

When using trigonometric substitution to evaluate definite integrals, you must be careful to check that the values of θ lie in the intervals discussed at the beginning of this section. For instance, if in Example 4 you had been asked to evaluate the definite integral

$$\int_{-2}^{-\sqrt{3}} \frac{\sqrt{x^2 - 3}}{x}\, dx$$

then using $u = x$ and $a = \sqrt{3}$ in the interval $\left[-2, -\sqrt{3}\right]$ would imply that $u < -a$. So, when determining the upper and lower limits of integration, you would have to choose θ such that $\pi/2 < \theta \leq \pi$. In this case the integral would be evaluated as follows.

$$\begin{aligned}
\int_{-2}^{-\sqrt{3}} \frac{\sqrt{x^2 - 3}}{x}\, dx &= \int_{5\pi/6}^{\pi} \frac{\left(-\sqrt{3}\tan\theta\right)\left(\sqrt{3}\sec\theta\tan\theta\right) d\theta}{\sqrt{3}\sec\theta} \\
&= \int_{5\pi/6}^{\pi} -\sqrt{3}\tan^2\theta\, d\theta \\
&= -\sqrt{3}\int_{5\pi/6}^{\pi} (\sec^2\theta - 1)\, d\theta \\
&= -\sqrt{3}\Big[\tan\theta - \theta\Big]_{5\pi/6}^{\pi} \\
&= -\sqrt{3}\left[(0 - \pi) - \left(-\frac{1}{\sqrt{3}} - \frac{5\pi}{6}\right)\right] \\
&= -1 + \frac{\sqrt{3}\pi}{6} \\
&\approx -0.0931
\end{aligned}$$

Trigonometric substitution can be used with completing the square. For instance, try evaluating the following integral.

$$\int \sqrt{x^2 - 2x}\, dx$$

To begin, you could complete the square and write the integral as

$$\int \sqrt{(x - 1)^2 - 1^2}\, dx.$$

Trigonometric substitution can be used to evaluate the three integrals listed in the following theorem. These integrals will be encountered several times in the remainder of the text. When this happens, we will simply refer to this theorem. (In Exercise 85, you are asked to verify the formulas given in the theorem.)

THEOREM 8.2 Special Integration Formulas ($a > 0$)

1. $\displaystyle\int \sqrt{a^2 - u^2}\, du = \frac{1}{2}\left(a^2 \arcsin\frac{u}{a} + u\sqrt{a^2 - u^2}\right) + C$

2. $\displaystyle\int \sqrt{u^2 - a^2}\, du = \frac{1}{2}\left(u\sqrt{u^2 - a^2} - a^2 \ln\left|u + \sqrt{u^2 - a^2}\right|\right) + C, \quad u > a$

3. $\displaystyle\int \sqrt{u^2 + a^2}\, du = \frac{1}{2}\left(u\sqrt{u^2 + a^2} + a^2 \ln\left|u + \sqrt{u^2 + a^2}\right|\right) + C$

Applications

EXAMPLE 5 Finding Arc Length

Find the arc length of the graph of $f(x) = \frac{1}{2}x^2$ from $x = 0$ to $x = 1$ (see Figure 8.10).

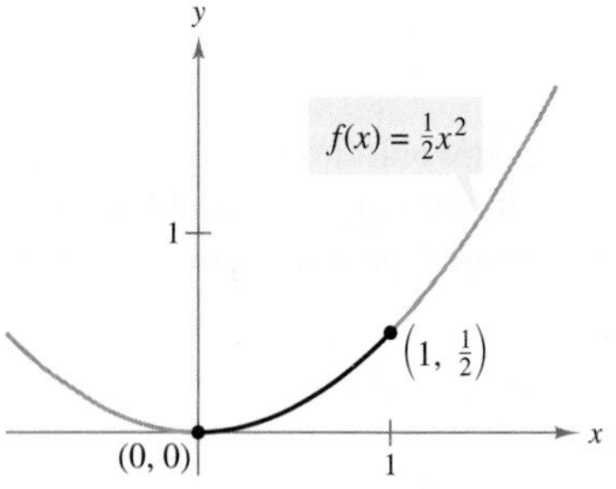

The arc length of the curve from (0, 0) to $(1, \frac{1}{2})$

Figure 8.10

Solution Refer to the arc length formula in Section 7.4.

$$
\begin{aligned}
s &= \int_0^1 \sqrt{1 + [f'(x)]^2}\,dx && \text{Formula for arc length} \\
&= \int_0^1 \sqrt{1 + x^2}\,dx && f'(x) = x \\
&= \int_0^{\pi/4} \sec^3\theta\,d\theta && \text{Let } a = 1 \text{ and } x = \tan\theta. \\
&= \frac{1}{2}\Big[\sec\theta\tan\theta + \ln|\sec\theta + \tan\theta|\Big]_0^{\pi/4} && \text{Example 5, Section 8.2} \\
&= \frac{1}{2}\left[\sqrt{2} + \ln\left(\sqrt{2} + 1\right)\right] \approx 1.148
\end{aligned}
$$

EXAMPLE 6 Comparing Two Fluid Forces

A sealed barrel of oil (weighing 48 pounds per cubic foot) is floating in seawater (weighing 64 pounds per cubic foot), as shown in Figures 8.11 and 8.12. (The barrel is not completely full of oil—on its side, the top 0.2 foot of the barrel is empty.) Compare the fluid forces against one end of the barrel from the inside and from the outside.

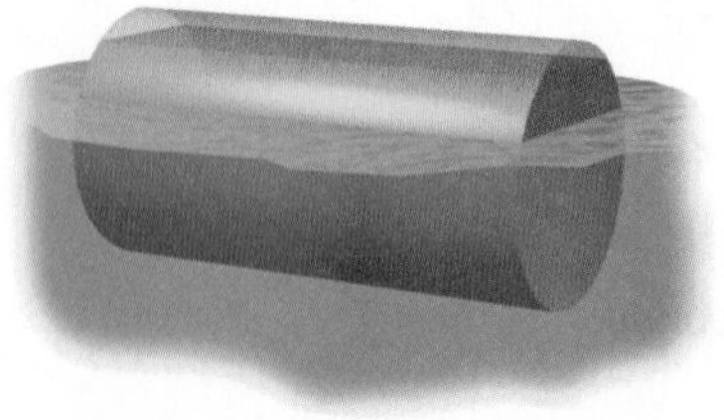

The barrel is not quite full of oil—the top 0.2 foot of the barrel is empty.

Figure 8.11

Solution In Figure 8.12, locate the coordinate system with the origin at the center of the circle given by $x^2 + y^2 = 1$. To find the fluid force against an end of the barrel *from the inside*, integrate between -1 and 0.8 (using a weight of $w = 48$).

$$F = w\int_c^d h(y)L(y)\,dy \qquad \text{General equation (see Section 7.7)}$$

$$
\begin{aligned}
F_{\text{inside}} &= 48\int_{-1}^{0.8} (0.8 - y)(2)\sqrt{1 - y^2}\,dy \\
&= 76.8\int_{-1}^{0.8} \sqrt{1 - y^2}\,dy - 96\int_{-1}^{0.8} y\sqrt{1 - y^2}\,dy
\end{aligned}
$$

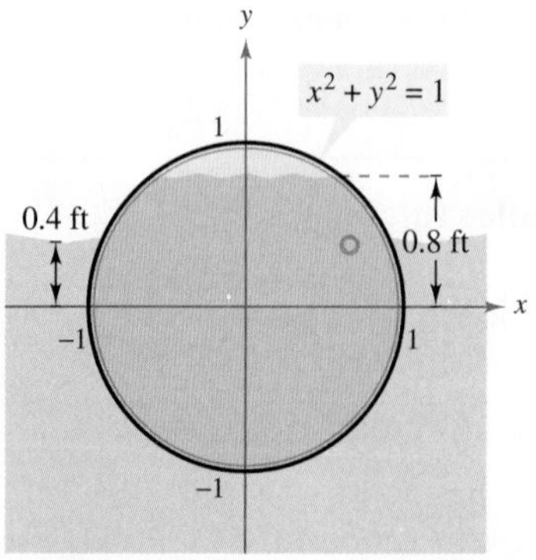

Figure 8.12

To find the fluid force *from the outside*, integrate between -1 and 0.4 (using a weight of $w = 64$).

$$
\begin{aligned}
F_{\text{outside}} &= 64\int_{-1}^{0.4} (0.4 - y)(2)\sqrt{1 - y^2}\,dy \\
&= 51.2\int_{-1}^{0.4} \sqrt{1 - y^2}\,dy - 128\int_{-1}^{0.4} y\sqrt{1 - y^2}\,dy
\end{aligned}
$$

The details of integration are left for you to complete in Exercise 84. Intuitively, would you say that the force from the oil (the inside) or the force from the seawater (the outside) is greater? By evaluating these two integrals, you can determine that

$$F_{\text{inside}} \approx 121.3 \text{ pounds} \qquad \text{and} \qquad F_{\text{outside}} \approx 93.0 \text{ pounds.}$$

Exercises for Section 8.4

See www.CalcChat.com for worked-out solutions to odd-numbered exercises.

In Exercises 1–4, use differentiation to match the antiderivative with the correct integral. [Integrals are labeled (a), (b), (c), and (d).]

(a) $\displaystyle\int \frac{x^2}{\sqrt{16-x^2}}\,dx$ (b) $\displaystyle\int \frac{\sqrt{x^2+16}}{x}\,dx$

(c) $\displaystyle\int \sqrt{7+6x-x^2}\,dx$ (d) $\displaystyle\int \frac{x^2}{\sqrt{x^2-16}}\,dx$

1. $4\ln\left|\frac{\sqrt{x^2+16}-4}{x}\right| + \sqrt{x^2+16} + C$

2. $8\ln\left|\sqrt{x^2-16}+x\right| + \frac{x\sqrt{x^2-16}}{2} + C$

3. $8\arcsin\frac{x}{4} - \frac{x\sqrt{16-x^2}}{2} + C$

4. $8\arcsin\frac{x-3}{4} + \frac{(x-3)\sqrt{7+6x-x^2}}{2} + C$

In Exercises 5–8, find the indefinite integral using the substitution $x = 5\sin\theta$.

5. $\displaystyle\int \frac{1}{(25-x^2)^{3/2}}\,dx$

6. $\displaystyle\int \frac{10}{x^2\sqrt{25-x^2}}\,dx$

7. $\displaystyle\int \frac{\sqrt{25-x^2}}{x}\,dx$

8. $\displaystyle\int \frac{x^2}{\sqrt{25-x^2}}\,dx$

In Exercises 9–12, find the indefinite integral using the substitution $x = 2\sec\theta$.

9. $\displaystyle\int \frac{1}{\sqrt{x^2-4}}\,dx$

10. $\displaystyle\int \frac{\sqrt{x^2-4}}{x}\,dx$

11. $\displaystyle\int x^3\sqrt{x^2-4}\,dx$

12. $\displaystyle\int \frac{x^3}{\sqrt{x^2-4}}\,dx$

In Exercises 13–16, find the indefinite integral using the substitution $x = \tan\theta$.

13. $\displaystyle\int x\sqrt{1+x^2}\,dx$

14. $\displaystyle\int \frac{9x^3}{\sqrt{1+x^2}}\,dx$

15. $\displaystyle\int \frac{1}{(1+x^2)^2}\,dx$

16. $\displaystyle\int \frac{x^2}{(1+x^2)^2}\,dx$

In Exercises 17–20, use the Special Integration Formulas (Theorem 8.2) to find the integral.

17. $\displaystyle\int \sqrt{4+9x^2}\,dx$

18. $\displaystyle\int \sqrt{1+x^2}\,dx$

19. $\displaystyle\int \sqrt{25-4x^2}\,dx$

20. $\displaystyle\int \sqrt{2x^2-1}\,dx$

In Exercises 21–42, find the integral.

21. $\displaystyle\int \frac{x}{\sqrt{x^2+9}}\,dx$

22. $\displaystyle\int \frac{x}{\sqrt{9-x^2}}\,dx$

23. $\displaystyle\int \frac{1}{\sqrt{16-x^2}}\,dx$

24. $\displaystyle\int \frac{1}{\sqrt{25-x^2}}\,dx$

25. $\displaystyle\int \sqrt{16-4x^2}\,dx$

26. $\displaystyle\int x\sqrt{16-4x^2}\,dx$

27. $\displaystyle\int \frac{1}{\sqrt{x^2-9}}\,dx$

28. $\displaystyle\int \frac{t}{(1-t^2)^{3/2}}\,dt$

29. $\displaystyle\int \frac{\sqrt{1-x^2}}{x^4}\,dx$

30. $\displaystyle\int \frac{\sqrt{4x^2+9}}{x^4}\,dx$

31. $\displaystyle\int \frac{1}{x\sqrt{4x^2+9}}\,dx$

32. $\displaystyle\int \frac{1}{x\sqrt{4x^2+16}}\,dx$

33. $\displaystyle\int \frac{-5x}{(x^2+5)^{3/2}}\,dx$

34. $\displaystyle\int \frac{1}{(x^2+3)^{3/2}}\,dx$

35. $\displaystyle\int e^{2x}\sqrt{1+e^{2x}}\,dx$

36. $\displaystyle\int (x+1)\sqrt{x^2+2x+2}\,dx$

37. $\displaystyle\int e^x\sqrt{1-e^{2x}}\,dx$

38. $\displaystyle\int \frac{\sqrt{1-x}}{\sqrt{x}}\,dx$

39. $\displaystyle\int \frac{1}{4+4x^2+x^4}\,dx$

40. $\displaystyle\int \frac{x^3+x+1}{x^4+2x^2+1}\,dx$

41. $\displaystyle\int \operatorname{arcsec} 2x\,dx,\quad x > \tfrac{1}{2}$

42. $\displaystyle\int x\arcsin x\,dx$

In Exercises 43–46, complete the square and find the integral.

43. $\displaystyle\int \frac{1}{\sqrt{4x-x^2}}\,dx$

44. $\displaystyle\int \frac{x^2}{\sqrt{2x-x^2}}\,dx$

45. $\displaystyle\int \frac{x}{\sqrt{x^2+4x+8}}\,dx$

46. $\displaystyle\int \frac{x}{\sqrt{x^2-6x+5}}\,dx$

In Exercises 47–52, evaluate the integral using (a) the given integration limits and (b) the limits obtained by trigonometric substitution.

47. $\displaystyle\int_0^{\sqrt{3}/2} \frac{t^2}{(1-t^2)^{3/2}}\,dt$

48. $\displaystyle\int_0^{\sqrt{3}/2} \frac{1}{(1-t^2)^{5/2}}\,dt$

49. $\displaystyle\int_0^3 \frac{x^3}{\sqrt{x^2+9}}\,dx$

50. $\displaystyle\int_0^{3/5} \sqrt{9-25x^2}\,dx$

51. $\displaystyle\int_4^6 \frac{x^2}{\sqrt{x^2-9}}\,dx$

52. $\displaystyle\int_3^6 \frac{\sqrt{x^2-9}}{x^2}\,dx$

In Exercises 53 and 54, find the particular solution of the differential equation.

53. $x\dfrac{dy}{dx} = \sqrt{x^2-9},\quad x \ge 3,\quad y(3) = 1$

54. $\sqrt{x^2+4}\,\dfrac{dy}{dx} = 1,\quad x \ge -2,\quad y(0) = 4$

In Exercises 55–58, use a computer algebra system to find the integral. Verify the result by differentiation.

55. $\displaystyle\int \frac{x^2}{\sqrt{x^2+10x+9}}\,dx$

56. $\displaystyle\int (x^2+2x+11)^{3/2}\,dx$

57. $\displaystyle\int \frac{x^2}{\sqrt{x^2-1}}\,dx$

58. $\displaystyle\int x^2\sqrt{x^2-4}\,dx$

Writing About Concepts

59. State the substitution you would make if you used trigonometric substitution and the integral involving the given radical, where $a > 0$. Explain your reasoning.

(a) $\sqrt{a^2-u^2}$ (b) $\sqrt{a^2+u^2}$ (c) $\sqrt{u^2-a^2}$

60. State the method of integration you would use to perform each integration. Explain why you chose that method. Do not integrate.

(a) $\displaystyle\int x\sqrt{x^2+1}\,dx$ (b) $\displaystyle\int x^2\sqrt{x^2-1}\,dx$

61. Evaluate the integral $\displaystyle\int \frac{x}{x^2+9}\,dx$ using (a) u-substitution and (b) trigonometric substitution. Discuss the results.

62. Evaluate the integral $\displaystyle\int \frac{x^2}{x^2+9}\,dx$ (a) algebraically using $x^2 = (x^2+9) - 9$ and (b) using trigonometric substitution Discuss the results.

True or False? **In Exercises 63–66, determine whether the statement is true or false. If it is false, explain why or give an example that shows it is false.**

63. If $x = \sin\theta$, then $\displaystyle\int \frac{dx}{\sqrt{1-x^2}} = \int d\theta.$

64. If $x = \sec\theta$, then $\displaystyle\int \frac{\sqrt{x^2-1}}{x}\,dx = \int \sec\theta\tan\theta\,d\theta.$

65. If $x = \tan\theta$, then $\displaystyle\int_0^{\sqrt{3}} \frac{dx}{(1+x^2)^{3/2}} = \int_0^{4\pi/3} \cos\theta\,d\theta.$

66. If $x = \sin\theta$, then $\displaystyle\int_{-1}^{1} x^2\sqrt{1-x^2}\,dx = 2\int_0^{\pi/2} \sin^2\theta\cos^2\theta\,d\theta.$

67. *Area* Find the area enclosed by the ellipse shown in the figure.

$$\frac{x^2}{a^2} + \frac{y^2}{b^2} = 1$$

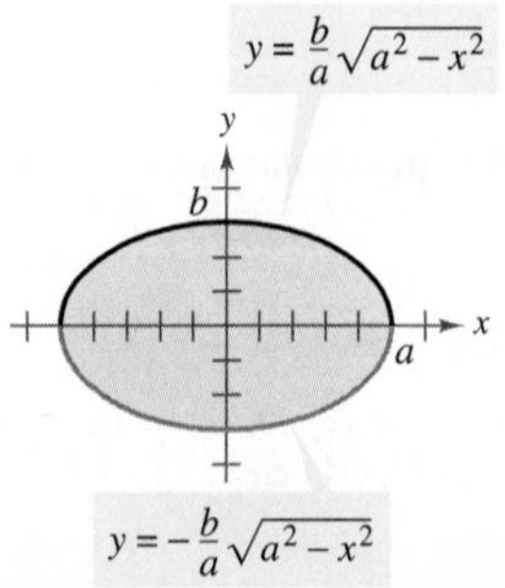

68. *Area* Find the area of the shaded region of the circle of radius a, if the chord is h units $(0 < h < a)$ from the center of the circle (see figure).

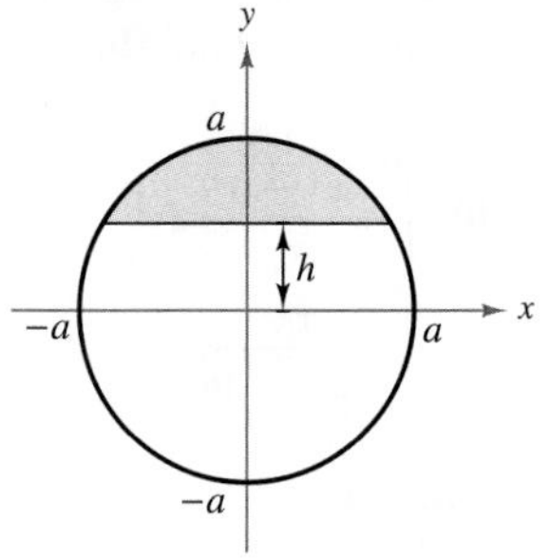

69. *Mechanical Design* The surface of a machine part is the region between the graphs of $y = |x|$ and $x^2 + (y-k)^2 = 25$ (see figure).

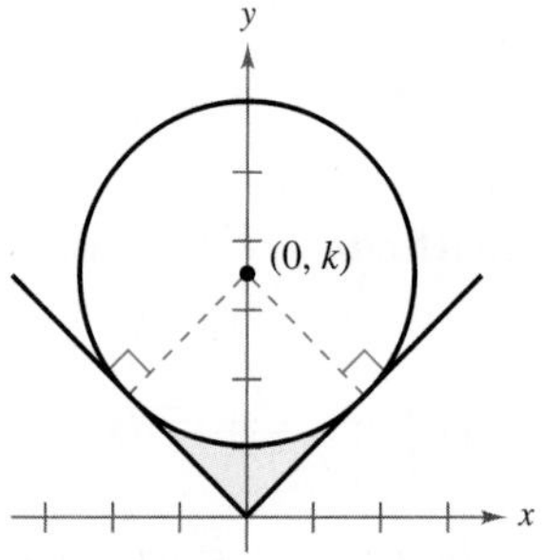

(a) Find k if the circle is tangent to the graph of $y = |x|$.

(b) Find the area of the surface of the machine part.

(c) Find the area of the surface of the machine part as a function of the radius r of the circle.

70. *Volume* The axis of a storage tank in the form of a right circular cylinder is horizontal (see figure). The radius and length of the tank are 1 meter and 3 meters, respectively.

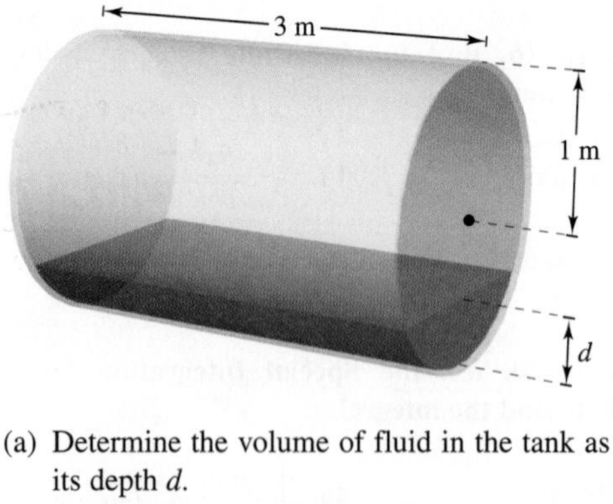

(a) Determine the volume of fluid in the tank as a function of its depth d.

(b) Use a graphing utility to graph the function in part (a).

(c) Design a dip stick for the tank with markings of $\frac{1}{4}$, $\frac{1}{2}$, and $\frac{3}{4}$.

(d) Fluid is entering the tank at a rate of $\frac{1}{4}$ cubic meter per minute. Determine the rate of change of the depth of the fluid as a function of its depth d.

(e) Use a graphing utility to graph the function in part (d). When will the rate of change of the depth be minimum? Does this agree with your intuition? Explain.

Volume of a Torus **In Exercises 71 and 72, find the volume of the torus generated by revolving the region bounded by the graph of the circle about the y-axis.**

71. $(x - 3)^2 + y^2 = 1$ (see figure)

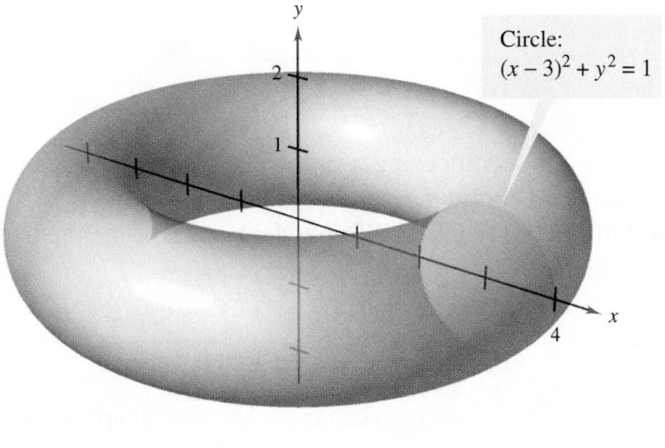

72. $(x - h)^2 + y^2 = r^2, \; h > r$

Arc Length **In Exercises 73 and 74, find the arc length of the curve over the given interval.**

73. $y = \ln x, \quad [1, 5]$

74. $y = \frac{1}{2}x^2, \quad [0, 4]$

75. ***Arc Length*** Show that the length of one arch of the sine curve is equal to the length of one arch of the cosine curve.

76. ***Conjecture***

(a) Find formulas for the distance between $(0, 0)$ and (a, a^2) along the line between these points and along the parabola $y = x^2$.

(b) Use the formulas from part (a) to find the distances for $a = 1$ and $a = 10$.

(c) Make a conjecture about the difference between the two distances as a increases.

Projectile Motion **In Exercises 77 and 78, (a) use a graphing utility to graph the path of a projectile that follows the path given by the graph of the equation, (b) determine the range of the projectile, and (c) use the integration capabilities of a graphing utility to determine the distance the projectile travels.**

77. $y = x - 0.005x^2$

78. $y = x - \dfrac{x^2}{72}$

Centroid **In Exercises 79 and 80, find the centroid of the region determined by the graphs of the inequalities.**

79. $y \leq 3/\sqrt{x^2 + 9}, y \geq 0, x \geq -4, x \leq 4$

80. $y \leq \frac{1}{4}x^2, (x - 4)^2 + y^2 \leq 16, y \geq 0$

81. ***Surface Area*** Find the surface area of the solid generated by revolving the region bounded by the graphs of $y = x^2$, $y = 0$, $x = 0$, and $x = \sqrt{2}$ about the x-axis.

82. ***Field Strength*** The field strength H of a magnet of length $2L$ on a particle r units from the center of the magnet is

$$H = \frac{2mL}{(r^2 + L^2)^{3/2}}$$

where $\pm m$ are the poles of the magnet (see figure). Find the average field strength as the particle moves from 0 to R units from the center by evaluating the integral

$$\frac{1}{R}\int_0^R \frac{2mL}{(r^2 + L^2)^{3/2}}\, dr.$$

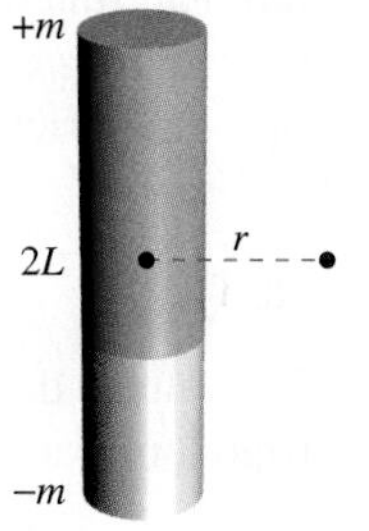

Figure for 82

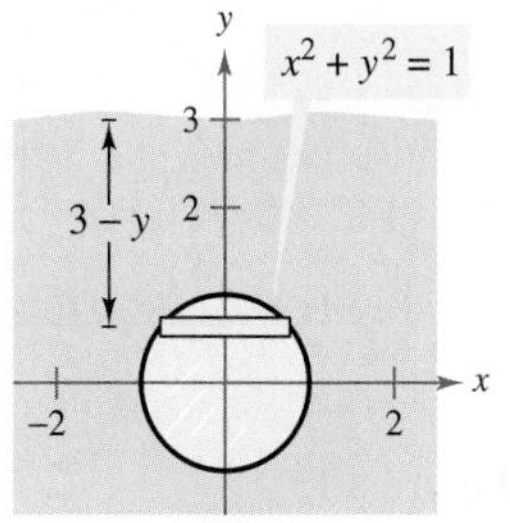

Figure for 83

83. ***Fluid Force*** Find the fluid force on a circular observation window of radius 1 foot in a vertical wall of a large water-filled tank at a fish hatchery when the center of the window is (a) 3 feet and (b) d feet $(d > 1)$ below the water's surface (see figure). Use trigonometric substitution to evaluate the one integral. (Recall that in Section 7.7 in a similar problem, you evaluated one integral by a geometric formula and the other by observing that the integrand was odd.)

84. ***Fluid Force*** Evaluate the following two integrals, which yield the fluid forces given in Example 6.

(a) $F_{\text{inside}} = 48\displaystyle\int_{-1}^{0.8} (0.8 - y)(2)\sqrt{1 - y^2}\, dy$

(b) $F_{\text{outside}} = 64\displaystyle\int_{-1}^{0.4} (0.4 - y)(2)\sqrt{1 - y^2}\, dy$

85. Use trigonometric substitution to verify the integration formulas given in Theorem 8.2.

86. ***Arc Length*** Show that the arc length of the graph of $y = \sin x$ on the interval $[0, 2\pi]$ is equal to the circumference of the ellipse $x^2 + 2y^2 = 2$ (see figure).

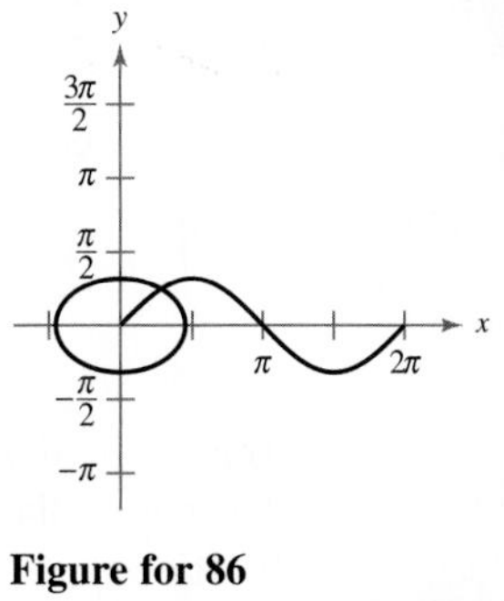

Figure for 86

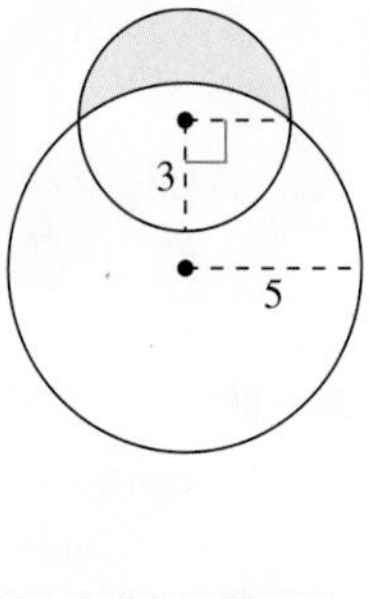

Figure for 87

87. ***Area of a Lune*** The crescent-shaped region bounded by two circles forms a *lune* (see figure). Find the area of the lune given that the radius of the smaller circle is 3 and the radius of the larger circle is 5.

Section 8.5 Partial Fractions

- **Understand the concept of a partial fraction decomposition.**
- **Use partial fraction decomposition with linear factors to integrate rational functions.**
- **Use partial fraction decomposition with quadratic factors to integrate rational functions.**

Partial Fractions

This section examines a procedure for decomposing a rational function into simpler rational functions to which you can apply the basic integration formulas. This procedure is called the **method of partial fractions.** To see the benefit of the method of partial fractions, consider the integral

$$\int \frac{1}{x^2 - 5x + 6}\,dx.$$

To evaluate this integral *without* partial fractions, you can complete the square and use trigonometric substitution (see Figure 8.13) to obtain

$$\begin{aligned}\int \frac{1}{x^2 - 5x + 6}\,dx &= \int \frac{dx}{(x - 5/2)^2 - (1/2)^2} && a = \tfrac{1}{2},\ x - \tfrac{5}{2} = \tfrac{1}{2}\sec\theta \\ &= \int \frac{(1/2)\sec\theta\tan\theta\,d\theta}{(1/4)\tan^2\theta} && dx = \tfrac{1}{2}\sec\theta\tan\theta\,d\theta \\ &= 2\int \csc\theta\,d\theta \\ &= 2\ln|\csc\theta - \cot\theta| + C \\ &= 2\ln\left|\frac{2x - 5}{2\sqrt{x^2 - 5x + 6}} - \frac{1}{2\sqrt{x^2 - 5x + 6}}\right| + C \\ &= 2\ln\left|\frac{x - 3}{\sqrt{x^2 - 5x + 6}}\right| + C \\ &= 2\ln\left|\frac{\sqrt{x - 3}}{\sqrt{x - 2}}\right| + C \\ &= \ln\left|\frac{x - 3}{x - 2}\right| + C \\ &= \ln|x - 3| - \ln|x - 2| + C.\end{aligned}$$

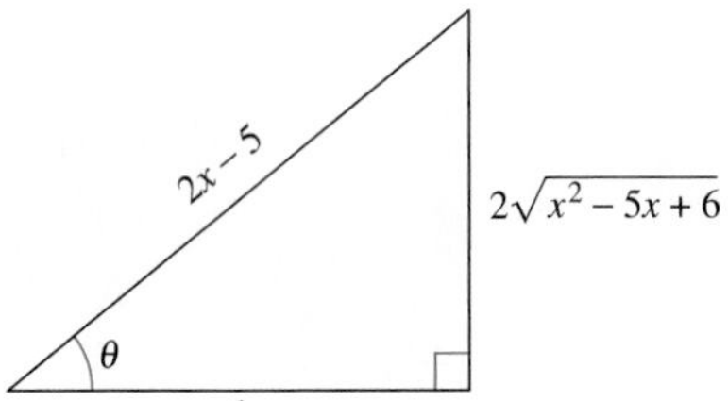

$\sec\theta = 2x - 5$
Figure 8.13

Now, suppose you had observed that

$$\frac{1}{x^2 - 5x + 6} = \frac{1}{x - 3} - \frac{1}{x - 2}. \qquad \text{Partial fraction decomposition}$$

Then you could evaluate the integral easily, as follows.

$$\begin{aligned}\int \frac{1}{x^2 - 5x + 6}\,dx &= \int \left(\frac{1}{x - 3} - \frac{1}{x - 2}\right) dx \\ &= \ln|x - 3| - \ln|x - 2| + C\end{aligned}$$

This method is clearly preferable to trigonometric substitution. However, its use depends on the ability to factor the denominator, $x^2 - 5x + 6$, and to find the **partial fractions**

$$\frac{1}{x - 3} \quad \text{and} \quad -\frac{1}{x - 2}.$$

In this section, you will study techniques for finding partial fraction decompositions.

Mary Evans Picture Library

JOHN BERNOULLI (1667–1748)

The method of partial fractions was introduced by John Bernoulli, a Swiss mathematician who was instrumental in the early development of calculus. John Bernoulli was a professor at the University of Basel and taught many outstanding students, the most famous of whom was Leonhard Euler.

STUDY TIP In precalculus you learned how to combine functions such as

$$\frac{1}{x-2} + \frac{-1}{x+3} = \frac{5}{(x-2)(x+3)}.$$

The method of partial fractions shows you how to reverse this process.

$$\frac{5}{(x-2)(x+3)} = \frac{?}{x-2} + \frac{?}{x+3}$$

Recall from algebra that every polynomial with real coefficients can be factored into linear and irreducible quadratic factors.* For instance, the polynomial

$$x^5 + x^4 - x - 1$$

can be written as

$$\begin{aligned} x^5 + x^4 - x - 1 &= x^4(x+1) - (x+1) \\ &= (x^4 - 1)(x+1) \\ &= (x^2+1)(x^2-1)(x+1) \\ &= (x^2+1)(x+1)(x-1)(x+1) \\ &= (x-1)(x+1)^2(x^2+1) \end{aligned}$$

where $(x - 1)$ is a linear factor, $(x + 1)^2$ is a repeated linear factor, and $(x^2 + 1)$ is an irreducible quadratic factor. Using this factorization, you can write the partial fraction decomposition of the rational expression

$$\frac{N(x)}{x^5 + x^4 - x - 1}$$

where $N(x)$ is a polynomial of degree less than 5, as follows.

$$\frac{N(x)}{(x-1)(x+1)^2(x^2+1)} = \frac{A}{x-1} + \frac{B}{x+1} + \frac{C}{(x+1)^2} + \frac{Dx+E}{x^2+1}$$

Decomposition of $N(x)/D(x)$ into Partial Fractions

1. **Divide if improper:** If $N(x)/D(x)$ is an improper fraction (that is, if the degree of the numerator is greater than or equal to the degree of the denominator), divide the denominator into the numerator to obtain

$$\frac{N(x)}{D(x)} = (\text{a polynomial}) + \frac{N_1(x)}{D(x)}$$

where the degree of $N_1(x)$ is less than the degree of $D(x)$. Then apply Steps 2, 3, and 4 to the proper rational expression $N_1(x)/D(x)$.

2. **Factor denominator:** Completely factor the denominator into factors of the form

$$(px + q)^m \quad \text{and} \quad (ax^2 + bx + c)^n$$

where $ax^2 + bx + c$ is irreducible.

3. **Linear factors:** For each factor of the form $(px + q)^m$, the partial fraction decomposition must include the following sum of m fractions.

$$\frac{A_1}{(px+q)} + \frac{A_2}{(px+q)^2} + \cdots + \frac{A_m}{(px+q)^m}$$

4. **Quadratic factors:** For each factor of the form $(ax^2 + bx + c)^n$, the partial fraction decomposition must include the following sum of n fractions.

$$\frac{B_1x + C_1}{ax^2+bx+c} + \frac{B_2x + C_2}{(ax^2+bx+c)^2} + \cdots + \frac{B_nx + C_n}{(ax^2+bx+c)^n}$$

* *For a review of factorization techniques, see* Precalculus, *6th edition, by Larson and Hostetler or* Precalculus: A Graphing Approach, *4th edition, by Larson, Hostetler, and Edwards (Boston, Massachusetts: Houghton Mifflin, 2004 and 2005, respectively).*

Linear Factors

Algebraic techniques for determining the constants in the numerators of a partial decomposition with linear or repeated linear factors are shown in Examples 1 and 2.

EXAMPLE 1 Distinct Linear Factors

Write the partial fraction decomposition for $\dfrac{1}{x^2 - 5x + 6}$.

Solution Because $x^2 - 5x + 6 = (x - 3)(x - 2)$, you should include one partial fraction for each factor and write

$$\frac{1}{x^2 - 5x + 6} = \frac{A}{x - 3} + \frac{B}{x - 2}$$

where A and B are to be determined. Multiplying this equation by the least common denominator $(x - 3)(x - 2)$ yields the **basic equation**

$$1 = A(x - 2) + B(x - 3). \qquad \text{Basic equation}$$

Because this equation is to be true for all x, you can substitute any *convenient* values for x to obtain equations in A and B. The most convenient values are the ones that make particular factors equal to 0.

NOTE Note that the substitutions for x in Example 1 are chosen for their convenience in determining values for A and B; $x = 2$ is chosen to eliminate the term $A(x - 2)$, and $x = 3$ is chosen to eliminate the term $B(x - 3)$. The goal is to make *convenient* substitutions whenever possible.

To solve for A, let $x = 3$ and obtain

$$1 = A(3 - 2) + B(3 - 3) \qquad \text{Let } x = 3 \text{ in basic equation.}$$
$$1 = A(1) + B(0)$$
$$A = 1.$$

To solve for B, let $x = 2$ and obtain

$$1 = A(2 - 2) + B(2 - 3) \qquad \text{Let } x = 2 \text{ in basic equation.}$$
$$1 = A(0) + B(-1)$$
$$B = -1.$$

So, the decomposition is

$$\frac{1}{x^2 - 5x + 6} = \frac{1}{x - 3} - \frac{1}{x - 2}$$

as shown at the beginning of this section.

FOR FURTHER INFORMATION To learn a different method for finding the partial fraction decomposition, called the Heavyside Method, see the article "Calculus to Algebra Connections in Partial Fraction Decomposition" by Joseph Wiener and Will Watkins in *The AMATYC Review*.

Be sure you see that the method of partial fractions is practical only for integrals of rational functions whose denominators factor "nicely." For instance, if the denominator in Example 1 were changed to $x^2 - 5x + 5$, its factorization as

$$x^2 - 5x + 5 = \left[x + \frac{5 + \sqrt{5}}{2}\right]\left[x - \frac{5 - \sqrt{5}}{2}\right]$$

would be too cumbersome to use with partial fractions. In such cases, you should use completing the square or a computer algebra system to perform the integration. If you do this, you should obtain

$$\int \frac{1}{x^2 - 5x + 5}\,dx = \frac{\sqrt{5}}{5}\ln\left|2x - \sqrt{5} - 5\right| - \frac{\sqrt{5}}{5}\ln\left|2x + \sqrt{5} - 5\right| + C.$$

EXAMPLE 2 Repeated Linear Factors

Find $\displaystyle\int \frac{5x^2 + 20x + 6}{x^3 + 2x^2 + x}\,dx.$

Solution Because

$$\begin{aligned} x^3 + 2x^2 + x &= x(x^2 + 2x + 1) \\ &= x(x + 1)^2 \end{aligned}$$

you should include one fraction for *each power* of x and $(x + 1)$ and write

$$\frac{5x^2 + 20x + 6}{x(x + 1)^2} = \frac{A}{x} + \frac{B}{x + 1} + \frac{C}{(x + 1)^2}.$$

Multiplying by the least common denominator $x(x + 1)^2$ yields the *basic equation*

$$5x^2 + 20x + 6 = A(x + 1)^2 + Bx(x + 1) + Cx. \qquad \text{Basic equation}$$

To solve for A, let $x = 0$. This eliminates the B and C terms and yields

$$\begin{aligned} 6 &= A(1) + 0 + 0 \\ A &= 6. \end{aligned}$$

To solve for C, let $x = -1$. This eliminates the A and B terms and yields

$$\begin{aligned} 5 - 20 + 6 &= 0 + 0 - C \\ C &= 9. \end{aligned}$$

The most convenient choices for x have been used, so to find the value of B, you can use *any other value* of x along with the calculated values of A and C. Using $x = 1$, $A = 6$, and $C = 9$ produces

$$\begin{aligned} 5 + 20 + 6 &= A(4) + B(2) + C \\ 31 &= 6(4) + 2B + 9 \\ -2 &= 2B \\ B &= -1. \end{aligned}$$

So, it follows that

$$\begin{aligned} \int \frac{5x^2 + 20x + 6}{x(x + 1)^2}\,dx &= \int \left(\frac{6}{x} - \frac{1}{x + 1} + \frac{9}{(x + 1)^2}\right) dx \\ &= 6 \ln|x| - \ln|x + 1| + 9\frac{(x + 1)^{-1}}{-1} + C \\ &= \ln\left|\frac{x^6}{x + 1}\right| - \frac{9}{x + 1} + C. \end{aligned}$$

Try checking this result by differentiating. Include algebra in your check, simplifying the derivative until you have obtained the original integrand.

NOTE It is necessary to make as many substitutions for x as there are unknowns $(A, B, C, \ldots)$ to be determined. For instance, in Example 2, three substitutions ($x = 0$, $x = -1$, and $x = 1$) were made to solve for A, B, and C.

FOR FURTHER INFORMATION For an alternative approach to using partial fractions, see the article "A Shortcut in Partial Fractions" by Xun-Cheng Huang in *The College Mathematics Journal.*

TECHNOLOGY Most computer algebra systems, such as *Derive*, *Maple*, *Mathcad*, *Mathematica*, and the *TI-89*, can be used to convert a rational function to its partial fraction decomposition. For instance, using *Maple*, you obtain the following.

$$> \text{convert}\left(\frac{5x^2 + 20x + 6}{x^3 + 2x^2 + x}, \text{parfrac}, x\right)$$

$$\frac{6}{x} + \frac{9}{(x + 1)^2} - \frac{1}{x + 1}$$

Quadratic Factors

When using the method of partial fractions with *linear* factors, a convenient choice of x immediately yields a value for one of the coefficients. With *quadratic* factors, a system of linear equations usually has to be solved, regardless of the choice of x.

EXAMPLE 3 Distinct Linear and Quadratic Factors

Find $\displaystyle\int \frac{2x^3 - 4x - 8}{(x^2 - x)(x^2 + 4)}\,dx.$

Solution Because

$$(x^2 - x)(x^2 + 4) = x(x - 1)(x^2 + 4)$$

you should include one partial fraction for each factor and write

$$\frac{2x^3 - 4x - 8}{x(x - 1)(x^2 + 4)} = \frac{A}{x} + \frac{B}{x - 1} + \frac{Cx + D}{x^2 + 4}.$$

Multiplying by the least common denominator $x(x - 1)(x^2 + 4)$ yields the *basic equation*

$$2x^3 - 4x - 8 = A(x - 1)(x^2 + 4) + Bx(x^2 + 4) + (Cx + D)(x)(x - 1).$$

To solve for A, let $x = 0$ and obtain

$$-8 = A(-1)(4) + 0 + 0 \quad \Rightarrow \quad 2 = A.$$

To solve for B, let $x = 1$ and obtain

$$-10 = 0 + B(5) + 0 \quad \Rightarrow \quad -2 = B.$$

At this point, C and D are yet to be determined. You can find these remaining constants by choosing two other values for x and solving the resulting system of linear equations. If $x = -1$, then, using $A = 2$ and $B = -2$, you can write

$$-6 = (2)(-2)(5) + (-2)(-1)(5) + (-C + D)(-1)(-2)$$
$$2 = -C + D.$$

If $x = 2$, you have

$$0 = (2)(1)(8) + (-2)(2)(8) + (2C + D)(2)(1)$$
$$8 = 2C + D.$$

Solving the linear system by subtracting the first equation from the second

$$-C + D = 2$$
$$2C + D = 8$$

yields $C = 2$. Consequently, $D = 4$, and it follows that

$$\int \frac{2x^3 - 4x - 8}{x(x - 1)(x^2 + 4)}\,dx = \int \left(\frac{2}{x} - \frac{2}{x - 1} + \frac{2x}{x^2 + 4} + \frac{4}{x^2 + 4}\right) dx$$

$$= 2\ln|x| - 2\ln|x - 1| + \ln(x^2 + 4) + 2\arctan\frac{x}{2} + C.$$

In Examples 1, 2, and 3, the solution of the basic equation began with substituting values of x that made the linear factors equal to 0. This method works well when the partial fraction decomposition involves linear factors. However, if the decomposition involves only quadratic factors, an alternative procedure is often more convenient.

EXAMPLE 4 Repeated Quadratic Factors

Find $\displaystyle\int \frac{8x^3 + 13x}{(x^2 + 2)^2}\, dx.$

Solution Include one partial fraction for each power of $(x^2 + 2)$ and write

$$\frac{8x^3 + 13x}{(x^2 + 2)^2} = \frac{Ax + B}{x^2 + 2} + \frac{Cx + D}{(x^2 + 2)^2}.$$

Multiplying by the least common denominator $(x^2 + 2)^2$ yields the *basic equation*

$$8x^3 + 13x = (Ax + B)(x^2 + 2) + Cx + D.$$

Expanding the basic equation and collecting like terms produces

$$8x^3 + 13x = Ax^3 + 2Ax + Bx^2 + 2B + Cx + D$$
$$8x^3 + 13x = Ax^3 + Bx^2 + (2A + C)x + (2B + D).$$

Now, you can equate the coefficients of like terms on opposite sides of the equation.

$8 = A$ $\qquad$ $0 = 2B + D$

$$8x^3 + 0x^2 + 13x + 0 = Ax^3 + Bx^2 + (2A + C)x + (2B + D)$$

$0 = B$ $\qquad$ $13 = 2A + C$

Using the known values $A = 8$ and $B = 0$, you can write

$$13 = 2A + C = 2(8) + C \quad \Rightarrow \quad C = -3$$
$$0 = 2B + D = 2(0) + D \quad \Rightarrow \quad D = 0.$$

Finally, you can conclude that

$$\int \frac{8x^3 + 13x}{(x^2 + 2)^2}\, dx = \int \left(\frac{8x}{x^2 + 2} + \frac{-3x}{(x^2 + 2)^2}\right) dx$$
$$= 4 \ln(x^2 + 2) + \frac{3}{2(x^2 + 2)} + C.$$

TECHNOLOGY Use a computer algebra system to evaluate the integral in Example 4—you might find that the form of the antiderivative is different. For instance, when you use a computer algebra system to work Example 4, you obtain

$$\int \frac{8x^3 + 13x}{(x^2 + 2)^2}\, dx = \ln(x^8 + 8x^6 + 24x^4 + 32x^2 + 16) + \frac{3}{2(x^2 + 2)} + C.$$

Is this result equivalent to that obtained in Example 4?

When integrating rational expressions, keep in mind that for *improper* rational expressions such as

$$\frac{N(x)}{D(x)} = \frac{2x^3 + x^2 - 7x + 7}{x^2 + x - 2}$$

you must first divide to obtain

$$\frac{N(x)}{D(x)} = 2x - 1 + \frac{-2x + 5}{x^2 + x - 2}.$$

The proper rational expression is then decomposed into its partial fractions by the usual methods. Here are some guidelines for solving the basic equation that is obtained in a partial fraction decomposition.

Guidelines for Solving the Basic Equation

Linear Factors

1. Substitute the roots of the distinct linear factors into the basic equation.
2. For repeated linear factors, use the coefficients determined in guideline 1 to rewrite the basic equation. Then substitute other convenient values of x and solve for the remaining coefficients.

Quadratic Factors

1. Expand the basic equation.
2. Collect terms according to powers of x.
3. Equate the coefficients of like powers to obtain a system of linear equations involving A, B, C, and so on.
4. Solve the system of linear equations.

Before concluding this section, here are a few things you should remember. First, it is not necessary to use the partial fractions technique on all rational functions. For instance, the following integral is evaluated more easily by the Log Rule.

$$\int \frac{x^2 + 1}{x^3 + 3x - 4}\,dx = \frac{1}{3}\int \frac{3x^2 + 3}{x^3 + 3x - 4}\,dx$$
$$= \frac{1}{3}\ln\left|x^3 + 3x - 4\right| + C$$

Second, if the integrand is not in reduced form, reducing it may eliminate the need for partial fractions, as shown in the following integral.

$$\int \frac{x^2 - x - 2}{x^3 - 2x - 4}\,dx = \int \frac{(x + 1)(x - 2)}{(x - 2)(x^2 + 2x + 2)}\,dx$$
$$= \int \frac{x + 1}{x^2 + 2x + 2}\,dx$$
$$= \frac{1}{2}\ln\left|x^2 + 2x + 2\right| + C$$

Finally, partial fractions can be used with some quotients involving transcendental functions. For instance, the substitution $u = \sin x$ allows you to write

$$\int \frac{\cos x}{\sin x(\sin x - 1)}\,dx = \int \frac{du}{u(u - 1)}. \qquad u = \sin x,\ du = \cos x\,dx$$

Exercises for Section 8.5

See www.CalcChat.com for worked-out solutions to odd-numbered exercises.

In Exercises 1–6, write the form of the partial fraction decomposition of the rational expression. Do not solve for the constants.

1. $\dfrac{5}{x^2 - 10x}$

2. $\dfrac{4x^2 + 3}{(x - 5)^3}$

3. $\dfrac{2x - 3}{x^3 + 10x}$

4. $\dfrac{x - 2}{x^2 + 4x + 3}$

5. $\dfrac{16}{x^2 - 10x}$

6. $\dfrac{2x - 1}{x(x^2 + 1)^2}$

In Exercises 7–28, use partial fractions to find the integral.

7. $\displaystyle\int \frac{1}{x^2 - 1}\,dx$

8. $\displaystyle\int \frac{1}{4x^2 - 9}\,dx$

9. $\displaystyle\int \frac{3}{x^2 + x - 2}\,dx$

10. $\displaystyle\int \frac{x + 1}{x^2 + 4x + 3}\,dx$

11. $\displaystyle\int \frac{5 - x}{2x^2 + x - 1}\,dx$

12. $\displaystyle\int \frac{5x^2 - 12x - 12}{x^3 - 4x}\,dx$

13. $\displaystyle\int \frac{x^2 + 12x + 12}{x^3 - 4x}\,dx$

14. $\displaystyle\int \frac{x^3 - x + 3}{x^2 + x - 2}\,dx$

15. $\displaystyle\int \frac{2x^3 - 4x^2 - 15x + 5}{x^2 - 2x - 8}\,dx$

16. $\displaystyle\int \frac{x + 2}{x^2 - 4x}\,dx$

17. $\displaystyle\int \frac{4x^2 + 2x - 1}{x^3 + x^2}\,dx$

18. $\displaystyle\int \frac{2x - 3}{(x - 1)^2}\,dx$

19. $\displaystyle\int \frac{x^2 + 3x - 4}{x^3 - 4x^2 + 4x}\,dx$

20. $\displaystyle\int \frac{4x^2}{x^3 + x^2 - x - 1}\,dx$

21. $\displaystyle\int \frac{x^2 - 1}{x^3 + x}\,dx$

22. $\displaystyle\int \frac{6x}{x^3 - 8}\,dx$

23. $\displaystyle\int \frac{x^2}{x^4 - 2x^2 - 8}\,dx$

24. $\displaystyle\int \frac{x^2 - x + 9}{(x^2 + 9)^2}\,dx$

25. $\displaystyle\int \frac{x}{16x^4 - 1}\,dx$

26. $\displaystyle\int \frac{x^2 - 4x + 7}{x^3 - x^2 + x + 3}\,dx$

27. $\displaystyle\int \frac{x^2 + 5}{x^3 - x^2 + x + 3}\,dx$

28. $\displaystyle\int \frac{x^2 + x + 3}{x^4 + 6x^2 + 9}\,dx$

In Exercises 29–32, evaluate the definite integral. Use a graphing utility to verify your result.

29. $\displaystyle\int_0^1 \frac{3}{2x^2 + 5x + 2}\,dx$

30. $\displaystyle\int_1^5 \frac{x - 1}{x^2(x + 1)}\,dx$

31. $\displaystyle\int_1^2 \frac{x + 1}{x(x^2 + 1)}\,dx$

32. $\displaystyle\int_0^1 \frac{x^2 - x}{x^2 + x + 1}\,dx$

In Exercises 33–40, use a computer algebra system to determine the antiderivative that passes through the given point. Use the system to graph the resulting antiderivative.

33. $\displaystyle\int \frac{3x}{x^2 - 6x + 9}\,dx,\quad (4, 0)$

34. $\displaystyle\int \frac{6x^2 + 1}{x^2(x - 1)^3}\,dx,\quad (2, 1)$

35. $\displaystyle\int \frac{x^2 + x + 2}{(x^2 + 2)^2}\,dx,\quad (0, 1)$

36. $\displaystyle\int \frac{x^3}{(x^2 - 4)^2}\,dx,\quad (3, 4)$

37. $\displaystyle\int \frac{2x^2 - 2x + 3}{x^3 - x^2 - x - 2}\,dx,\quad (3, 10)$

38. $\displaystyle\int \frac{x(2x - 9)}{x^3 - 6x^2 + 12x - 8}\,dx,\quad (3, 2)$

39. $\displaystyle\int \frac{1}{x^2 - 4}\,dx,\quad (6, 4)$

40. $\displaystyle\int \frac{x^2 - x + 2}{x^3 - x^2 + x - 1}\,dx,\quad (2, 6)$

In Exercises 41–46, use substitution to find the integral.

41. $\displaystyle\int \frac{\sin x}{\cos x(\cos x - 1)}\,dx$

42. $\displaystyle\int \frac{\sin x}{\cos x + \cos^2 x}\,dx$

43. $\displaystyle\int \frac{3\cos x}{\sin^2 x + \sin x - 2}\,dx$

44. $\displaystyle\int \frac{\sec^2 x}{\tan x(\tan x + 1)}\,dx$

45. $\displaystyle\int \frac{e^x}{(e^x - 1)(e^x + 4)}\,dx$

46. $\displaystyle\int \frac{e^x}{(e^{2x} + 1)(e^x - 1)}\,dx$

In Exercises 47–50, use the method of partial fractions to verify the integration formula.

47. $\displaystyle\int \frac{1}{x(a + bx)}\,dx = \frac{1}{a}\ln\left|\frac{x}{a + bx}\right| + C$

48. $\displaystyle\int \frac{1}{a^2 - x^2}\,dx = \frac{1}{2a}\ln\left|\frac{a + x}{a - x}\right| + C$

49. $\displaystyle\int \frac{x}{(a + bx)^2}\,dx = \frac{1}{b^2}\left(\frac{a}{a + bx} + \ln|a + bx|\right) + C$

50. $\displaystyle\int \frac{1}{x^2(a + bx)}\,dx = -\frac{1}{ax} - \frac{b}{a^2}\ln\left|\frac{x}{a + bx}\right| + C$

Slope Fields **In Exercises 51 and 52, use a computer algebra system to graph the slope field for the differential equation and graph the solution through the given initial condition.**

51. $\dfrac{dy}{dx} = \dfrac{6}{4 - x^2}$

$y(0) = 3$

52. $\dfrac{dy}{dx} = \dfrac{4}{x^2 - 2x - 3}$

$y(0) = 5$

Writing About Concepts

53. What is the first step when integrating $\displaystyle\int \frac{x^3}{x - 5}\,dx$? Explain.

54. Describe the decomposition of the proper rational function $N(x)/D(x)$ (a) if $D(x) = (px + q)^m$, and (b) if $D(x) = (ax^2 + bx + c)^n$, where $ax^2 + bx + c$ is irreducible. Explain why you chose that method.

55. State the method you would use to evaluate each integral. Explain why you chose that method. Do not integrate.

(a) $\displaystyle\int \frac{x + 1}{x^2 + 2x - 8}\,dx$

(b) $\displaystyle\int \frac{7x + 4}{x^2 + 2x - 8}\,dx$

(c) $\displaystyle\int \frac{4}{x^2 + 2x + 5}\,dx$

Writing About Concepts (continued)

56. Determine which value best approximates the area of the region between the x-axis and the graph of $f(x) = 10/[x(x^2 + 1)]$ over the interval $[1, 3]$. Make your selection on the basis of a sketch of the region and not by performing any calculations. Explain your reasoning.

(a) -6 (b) 6 (c) 3 (d) 5 (e) 8

57. ***Area*** Find the area of the region bounded by the graphs of $y = 12/(x^2 + 5x + 6)$, $y = 0$, $x = 0$, and $x = 1$.

58. ***Area*** Find the area of the region bounded by the graphs of $y = 7/(16 - x^2)$ and $y = 1$.

59. ***Modeling Data*** The predicted cost C (in hundreds of thousands of dollars) for a company to remove $p\%$ of a chemical from its waste water is shown in the table.

p	0	10	20	30	40	50	60	70	80	90
C	0	0.7	1.0	1.3	1.7	2.0	2.7	3.6	5.5	11.2

A model for the data is given by

$$C = \frac{124p}{(10 + p)(100 - p)}, \quad 0 \le p < 100.$$

Use the model to find the average cost for removing between 75% and 80% of the chemical.

60. ***Logistic Growth*** In Chapter 6, the exponential growth equation was derived from the assumption that the rate of growth was proportional to the existing quantity. In practice, there often exists some upper limit L past which growth cannot occur. In such cases, you assume the rate of growth to be proportional not only to the existing quantity, but also to the difference between the existing quantity y and the upper limit L. That is, $dy/dt = ky(L - y)$. In integral form, you can write this relationship as

$$\int \frac{dy}{y(L - y)} = \int k\,dt.$$

(a) A slope field for the differential equation $dy/dt = y(3 - y)$ is shown. Draw a possible solution to the differential equation if $y(0) = 5$, and another if $y(0) = \frac{1}{2}$. To print an enlarged copy of the graph, go to the website *www.mathgraphs.com.*

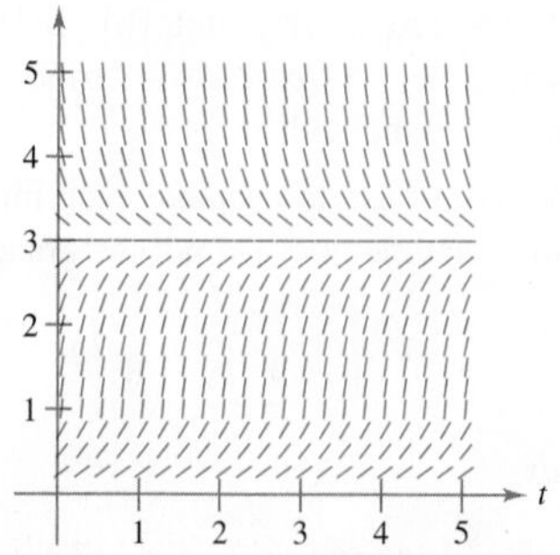

(b) Where $y(0)$ is greater than 3, what is the sign of the slope of the solution?

(c) For $y > 0$, find $\lim_{t \to \infty} y(t)$.

(d) Evaluate the two given integrals and solve for y as a function of t, where y_0 is the initial quantity.

(e) Use the result of part (d) to find and graph the solutions in part (a). Use a graphing utility to graph the solutions and compare the results with the solutions in part (a).

(f) The graph of the function y is a **logistic curve.** Show that the rate of growth is maximum at the point of inflection, and that this occurs when $y = L/2$.

61. ***Volume and Centroid*** Consider the region bounded by the graphs of $y = 2x/(x^2 + 1)$, $y = 0$, $x = 0$, and $x = 3$. Find the volume of the solid generated by revolving the region about the x-axis. Find the centroid of the region.

62. ***Volume*** Consider the region bounded by the graph of $y^2 = (2 - x)^2/(1 + x)^2$ on the interval $[0, 1]$. Find the volume of the solid generated by revolving this region about the x-axis.

63. ***Epidemic Model*** A single infected individual enters a community of n susceptible individuals. Let x be the number of newly infected individuals at time t. The common epidemic model assumes that the disease spreads at a rate proportional to the product of the total number infected and the number not yet infected. So, $dx/dt = k(x + 1)(n - x)$ and you obtain

$$\int \frac{1}{(x + 1)(n - x)}\,dx = \int k\,dt.$$

Solve for x as a function of t.

64. ***Chemical Reactions*** In a chemical reaction, one unit of compound Y and one unit of compound Z are converted into a single unit of compound X. x is the amount of compound X formed, and the rate of formation of X is proportional to the product of the amounts of unconverted compounds Y and Z. So, $dx/dt = k(y_0 - x)(z_0 - x)$, where y_0 and z_0 are the initial amounts of compounds Y and Z. From this equation you obtain

$$\int \frac{1}{(y_0 - x)(z_0 - x)}\,dx = \int k\,dt.$$

(a) Perform the two integrations and solve for x in terms of t.

(b) Use the result of part (a) to find x as $t \to \infty$ if (1) $y_0 < z_0$, (2) $y_0 > z_0$, and (3) $y_0 = z_0$.

65. Evaluate

$$\int_0^1 \frac{x}{1 + x^4}\,dx$$

in two different ways, one of which is partial fractions.

Putnam Exam Challenge

66. Prove $\dfrac{22}{7} - \pi = \displaystyle\int_0^1 \frac{x^4(1 - x)^4}{1 + x^2}\,dx.$

This problem was composed by the Committee on the Putnam Prize Competition.

Section 8.6 Integration by Tables and Other Integration Techniques

- **Evaluate an indefinite integral using a table of integrals.**
- **Evaluate an indefinite integral using reduction formulas.**
- **Evaluate an indefinite integral involving rational functions of sine and cosine.**

Integration by Tables

So far in this chapter you have studied several integration techniques that can be used with the basic integration rules. But merely knowing *how* to use the various techniques is not enough. You also need to know *when* to use them. Integration is first and foremost a problem of recognition. That is, you must recognize which rule or technique to apply to obtain an antiderivative. Frequently, a slight alteration of an integrand will require a different integration technique (or produce a function whose antiderivative is not an elementary function), as shown below.

$$\int x \ln x \, dx = \frac{x^2}{2} \ln x - \frac{x^2}{4} + C$$ Integration by parts

$$\int \frac{\ln x}{x} \, dx = \frac{(\ln x)^2}{2} + C$$ Power Rule

$$\int \frac{1}{x \ln x} \, dx = \ln|\ln x| + C$$ Log Rule

$$\int \frac{x}{\ln x} \, dx = ?$$ Not an elementary function

TECHNOLOGY A computer algebra system consists, in part, of a database of integration formulas. The primary difference between using a computer algebra system and using tables of integrals is that with a computer algebra system the computer searches through the database to find a fit. With integration tables, *you* must do the searching.

Many people find tables of integrals to be a valuable supplement to the integration techniques discussed in this chapter. Tables of common integrals can be found in Appendix B. **Integration by tables** is not a "cure-all" for all of the difficulties that can accompany integration—using tables of integrals requires considerable thought and insight and often involves substitution.

Each integration formula in Appendix B can be developed using one or more of the techniques in this chapter. You should try to verify several of the formulas. For instance, Formula 4

$$\int \frac{u}{(a + bu)^2} \, du = \frac{1}{b^2}\left(\frac{a}{a + bu} + \ln|a + bu|\right) + C$$ Formula 4

can be verified using the method of partial fractions, and Formula 19

$$\int \frac{\sqrt{a + bu}}{u} \, du = 2\sqrt{a + bu} + a \int \frac{du}{u\sqrt{a + bu}}$$ Formula 19

can be verified using integration by parts. Note that the integrals in Appendix B are classified according to forms involving the following.

u^n	$(a + bu)$
$(a + bu + cu^2)$	$\sqrt{a + bu}$
$(a^2 \pm u^2)$	$\sqrt{u^2 \pm a^2}$
$\sqrt{a^2 - u^2}$	Trigonometric functions
Inverse trigonometric functions	Exponential functions
Logarithmic functions	

EXPLORATION

Use the tables of integrals in Appendix B and the substitution

$$u = \sqrt{x-1}$$

to evaluate the integral in Example 1. If you do this, you should obtain

$$\int \frac{dx}{x\sqrt{x-1}} = \int \frac{2\,du}{u^2+1}.$$

Does this produce the same result as that obtained in Example 1?

EXAMPLE 1 Integration by Tables

Find $\displaystyle\int \frac{dx}{x\sqrt{x-1}}$.

Solution Because the expression inside the radical is linear, you should consider forms involving $\sqrt{a+bu}$.

$$\int \frac{du}{u\sqrt{a+bu}} = \frac{2}{\sqrt{-a}} \arctan \sqrt{\frac{a+bu}{-a}} + C \qquad \text{Formula 17 } (a < 0)$$

Let $a = -1$, $b = 1$, and $u = x$. Then $du = dx$, and you can write

$$\int \frac{dx}{x\sqrt{x-1}} = 2 \arctan \sqrt{x-1} + C.$$

EXAMPLE 2 Integration by Tables

Find $\displaystyle\int x\sqrt{x^4-9}\,dx$.

Solution Because the radical has the form $\sqrt{u^2-a^2}$, you should consider Formula 26.

$$\int \sqrt{u^2-a^2}\,du = \frac{1}{2}\left(u\sqrt{u^2-a^2} - a^2 \ln\left|u + \sqrt{u^2-a^2}\right|\right) + C$$

Let $u = x^2$ and $a = 3$. Then $du = 2x\,dx$, and you have

$$\int x\sqrt{x^4-9}\,dx = \frac{1}{2}\int \sqrt{(x^2)^2 - 3^2}\,(2x)\,dx$$

$$= \frac{1}{4}\left(x^2\sqrt{x^4-9} - 9 \ln\left|x^2 + \sqrt{x^4-9}\right|\right) + C.$$

EXAMPLE 3 Integration by Tables

Find $\displaystyle\int \frac{x}{1+e^{-x^2}}\,dx$.

Solution Of the forms involving e^u, consider the following formula.

$$\int \frac{du}{1+e^u} = u - \ln(1+e^u) + C \qquad \text{Formula 84}$$

Let $u = -x^2$. Then $du = -2x\,dx$, and you have

$$\int \frac{x}{1+e^{-x^2}}\,dx = -\frac{1}{2}\int \frac{-2x\,dx}{1+e^{-x^2}}$$

$$= -\frac{1}{2}\left[-x^2 - \ln(1+e^{-x^2})\right] + C$$

$$= \frac{1}{2}\left[x^2 + \ln(1+e^{-x^2})\right] + C.$$

TECHNOLOGY Example 3 shows the importance of having several solution techniques at your disposal. This integral is not difficult to solve with a table, but when it was entered into a well-known computer algebra system, the utility was unable to find the antiderivative.

Reduction Formulas

Several of the integrals in the integration tables have the form $\int f(x)\,dx = g(x) + \int h(x)\,dx$. Such integration formulas are called **reduction formulas** because they reduce a given integral to the sum of a function and a simpler integral.

EXAMPLE 4 Using a Reduction Formula

Find $\displaystyle\int x^3 \sin x\,dx$.

Solution Consider the following three formulas.

$$\int u \sin u\,du = \sin u - u\cos u + C \qquad \text{Formula 52}$$

$$\int u^n \sin u\,du = -u^n \cos u + n\int u^{n-1}\cos u\,du \qquad \text{Formula 54}$$

$$\int u^n \cos u\,du = u^n \sin u - n\int u^{n-1}\sin u\,du \qquad \text{Formula 55}$$

Using Formula 54, Formula 55, and then Formula 52 produces

$$\begin{aligned}\int x^3 \sin x\,dx &= -x^3\cos x + 3\int x^2\cos x\,dx\\ &= -x^3\cos x + 3\left(x^2\sin x - 2\int x\sin x\,dx\right)\\ &= -x^3\cos x + 3x^2\sin x + 6x\cos x - 6\sin x + C.\end{aligned}$$

TECHNOLOGY Sometimes when you use computer algebra systems you obtain results that look very different, but are actually equivalent. Here is how several different systems evaluated the integral in Example 5.

Maple

$\sqrt{3-5x} -$
$\sqrt{3}\,\text{arctanh}\left(\frac{1}{3}\sqrt{3-5x}\sqrt{3}\right)$

Derive

$\sqrt{3}\ln\left[\dfrac{\sqrt{(3-5x)}-\sqrt{3}}{\sqrt{x}}\right] +$
$\sqrt{(3-5x)}$

Mathematica

Sqrt[3 − 5x] −
$\text{Sqrt[3] ArcTanh}\left[\dfrac{\text{Sqrt[3 − 5x]}}{\text{Sqrt[3]}}\right]$

Mathcad

$\sqrt{3-5x} +$
$\frac{1}{2}\sqrt{3}\ln\left[-\frac{1}{5}\dfrac{(-6+5x+2\sqrt{3}\sqrt{3-5x})}{x}\right]$

Notice that computer algebra systems do not include a constant of integration.

EXAMPLE 5 Using a Reduction Formula

Find $\displaystyle\int \frac{\sqrt{3-5x}}{2x}\,dx$.

Solution Consider the following two formulas.

$$\int \frac{du}{u\sqrt{a+bu}} = \frac{1}{\sqrt{a}}\ln\left|\frac{\sqrt{a+bu}-\sqrt{a}}{\sqrt{a+bu}+\sqrt{a}}\right| + C \qquad \text{Formula 17 } (a > 0)$$

$$\int \frac{\sqrt{a+bu}}{u}\,du = 2\sqrt{a+bu} + a\int\frac{du}{u\sqrt{a+bu}} \qquad \text{Formula 19}$$

Using Formula 19, with $a = 3$, $b = -5$, and $u = x$, produces

$$\begin{aligned}\frac{1}{2}\int\frac{\sqrt{3-5x}}{x}\,dx &= \frac{1}{2}\left(2\sqrt{3-5x} + 3\int\frac{dx}{x\sqrt{3-5x}}\right)\\ &= \sqrt{3-5x} + \frac{3}{2}\int\frac{dx}{x\sqrt{3-5x}}.\end{aligned}$$

Using Formula 17, with $a = 3$, $b = -5$, and $u = x$, produces

$$\begin{aligned}\int\frac{\sqrt{3-5x}}{2x}\,dx &= \sqrt{3-5x} + \frac{3}{2}\left(\frac{1}{\sqrt{3}}\ln\left|\frac{\sqrt{3-5x}-\sqrt{3}}{\sqrt{3-5x}+\sqrt{3}}\right|\right) + C\\ &= \sqrt{3-5x} + \frac{\sqrt{3}}{2}\ln\left|\frac{\sqrt{3-5x}-\sqrt{3}}{\sqrt{3-5x}+\sqrt{3}}\right| + C.\end{aligned}$$

Rational Functions of Sine and Cosine

EXAMPLE 6 Integration by Tables

Find $\displaystyle\int \frac{\sin 2x}{2 + \cos x}\, dx.$

Solution Substituting $2 \sin x \cos x$ for $\sin 2x$ produces

$$\int \frac{\sin 2x}{2 + \cos x}\, dx = 2 \int \frac{\sin x \cos x}{2 + \cos x}\, dx.$$

A check of the forms involving $\sin u$ or $\cos u$ in Appendix B shows that none of those listed applies. So, you can consider forms involving $a + bu$. For example,

$$\int \frac{u\, du}{a + bu} = \frac{1}{b^2}(bu - a \ln|a + bu|) + C. \qquad \text{Formula 3}$$

Let $a = 2$, $b = 1$, and $u = \cos x$. Then $du = -\sin x\, dx$, and you have

$$\begin{aligned} 2 \int \frac{\sin x \cos x}{2 + \cos x}\, dx &= -2 \int \frac{\cos x(-\sin x\, dx)}{2 + \cos x} \\ &= -2(\cos x - 2 \ln|2 + \cos x|) + C \\ &= -2 \cos x + 4 \ln|2 + \cos x| + C. \end{aligned}$$

Example 6 involves a rational expression of $\sin x$ and $\cos x$. If you are unable to find an integral of this form in the integration tables, try using the following special substitution to convert the trigonometric expression to a standard rational expression.

Substitution for Rational Functions of Sine and Cosine

For integrals involving rational functions of sine and cosine, the substitution

$$u = \frac{\sin x}{1 + \cos x} = \tan \frac{x}{2}$$

yields

$$\cos x = \frac{1 - u^2}{1 + u^2}, \quad \sin x = \frac{2u}{1 + u^2}, \quad \text{and} \quad dx = \frac{2\, du}{1 + u^2}.$$

Proof From the substitution for u, it follows that

$$u^2 = \frac{\sin^2 x}{(1 + \cos x)^2} = \frac{1 - \cos^2 x}{(1 + \cos x)^2} = \frac{1 - \cos x}{1 + \cos x}.$$

Solving for $\cos x$ produces $\cos x = (1 - u^2)/(1 + u^2)$. To find $\sin x$, write $u = \sin x/(1 + \cos x)$ as

$$\sin x = u(1 + \cos x) = u\left(1 + \frac{1 - u^2}{1 + u^2}\right) = \frac{2u}{1 + u^2}.$$

Finally, to find dx, consider $u = \tan(x/2)$. Then you have $\arctan u = x/2$ and $dx = (2\, du)/(1 + u^2)$.

Exercises for Section 8.6

See www.CalcChat.com for worked-out solutions to odd-numbered exercises.

In Exercises 1 and 2, use a table of integrals with forms involving $a + bu$ to find the integral.

1. $\displaystyle\int \frac{x^2}{1+x}\,dx$

2. $\displaystyle\int \frac{2}{3x^2(2x-5)^2}\,dx$

In Exercises 3 and 4, use a table of integrals with forms involving $\sqrt{u^2 \pm a^2}$ to find the integral.

3. $\displaystyle\int e^x\sqrt{1+e^{2x}}\,dx$

4. $\displaystyle\int \frac{\sqrt{x^2-9}}{3x}\,dx$

In Exercises 5 and 6, use a table of integrals with forms involving $\sqrt{a^2 - u^2}$ to find the integral.

5. $\displaystyle\int \frac{1}{x^2\sqrt{1-x^2}}\,dx$

6. $\displaystyle\int \frac{x}{\sqrt{9-x^4}}\,dx$

In Exercises 7–10, use a table of integrals with forms involving the trigonometric functions to find the integral.

7. $\displaystyle\int \sin^4 2x\,dx$

8. $\displaystyle\int \frac{\cos^3\sqrt{x}}{\sqrt{x}}\,dx$

9. $\displaystyle\int \frac{1}{\sqrt{x}\left(1-\cos\sqrt{x}\right)}\,dx$

10. $\displaystyle\int \frac{1}{1-\tan 5x}\,dx$

In Exercises 11 and 12, use a table of integrals with forms involving e^u to find the integral.

11. $\displaystyle\int \frac{1}{1+e^{2x}}\,dx$

12. $\displaystyle\int e^{-x/2}\sin 2x\,dx$

In Exercises 13 and 14, use a table of integrals with forms involving $\ln u$ to find the integral.

13. $\displaystyle\int x^3\ln x\,dx$

14. $\displaystyle\int (\ln x)^3\,dx$

In Exercises 15–18, find the indefinite integral (a) using integration tables and (b) using the given method.

	Integral	*Method*
15.	$\displaystyle\int x^2 e^x\,dx$	Integration by parts
16.	$\displaystyle\int x^4\ln x\,dx$	Integration by parts
17.	$\displaystyle\int \frac{1}{x^2(x+1)}\,dx$	Partial fractions
18.	$\displaystyle\int \frac{1}{x^2-75}\,dx$	Partial fractions

In Exercises 19–42, use integration tables to find the integral.

19. $\displaystyle\int x\,\text{arcsec}(x^2+1)\,dx$

20. $\displaystyle\int \text{arcsec}\,2x\,dx$

21. $\displaystyle\int \frac{1}{x^2\sqrt{x^2-4}}\,dx$

22. $\displaystyle\int \frac{1}{x^2+2x+2}\,dx$

23. $\displaystyle\int \frac{2x}{(1-3x)^2}\,dx$

24. $\displaystyle\int \frac{\theta^2}{1-\sin\theta^3}\,d\theta$

25. $\displaystyle\int e^x\arccos e^x\,dx$

26. $\displaystyle\int \frac{e^x}{1-\tan e^x}\,dx$

27. $\displaystyle\int \frac{x}{1-\sec x^2}\,dx$

28. $\displaystyle\int \frac{1}{t[1+(\ln t)^2]}\,dt$

29. $\displaystyle\int \frac{\cos\theta}{3+2\sin\theta+\sin^2\theta}\,d\theta$

30. $\displaystyle\int x^2\sqrt{2+9x^2}\,dx$

31. $\displaystyle\int \frac{1}{x^2\sqrt{2+9x^2}}\,dx$

32. $\displaystyle\int \sqrt{x}\arctan x^{3/2}\,dx$

33. $\displaystyle\int \frac{\ln x}{x(3+2\ln x)}\,dx$

34. $\displaystyle\int \frac{e^x}{(1-e^{2x})^{3/2}}\,dx$

35. $\displaystyle\int \frac{x}{(x^2-6x+10)^2}\,dx$

36. $\displaystyle\int (2x-3)^2\sqrt{(2x-3)^2+4}\,dx$

37. $\displaystyle\int \frac{x}{\sqrt{x^4-6x^2+5}}\,dx$

38. $\displaystyle\int \frac{\cos x}{\sqrt{\sin^2 x+1}}\,dx$

39. $\displaystyle\int \frac{x^3}{\sqrt{4-x^2}}\,dx$

40. $\displaystyle\int \sqrt{\frac{3-x}{3+x}}\,dx$

41. $\displaystyle\int \frac{e^{3x}}{(1+e^x)^3}\,dx$

42. $\displaystyle\int \tan^3\theta\,d\theta$

In Exercises 43–50, use integration tables to evaluate the integral.

43. $\displaystyle\int_0^1 xe^{x^2}\,dx$

44. $\displaystyle\int_0^3 \frac{x}{\sqrt{1+x}}\,dx$

45. $\displaystyle\int_1^3 x^2\ln x\,dx$

46. $\displaystyle\int_0^{\pi} x\sin x\,dx$

47. $\displaystyle\int_{-\pi/2}^{\pi/2} \frac{\cos x}{1+\sin^2 x}\,dx$

48. $\displaystyle\int_2^4 \frac{x^2}{(3x-5)^2}\,dx$

49. $\displaystyle\int_0^{\pi/2} t^3\cos t\,dt$

50. $\displaystyle\int_0^1 \sqrt{3+x^2}\,dx$

In Exercises 51–56, verify the integration formula.

51. $\displaystyle\int \frac{u^2}{(a+bu)^2}\,du = \frac{1}{b^3}\left(bu - \frac{a^2}{a+bu} - 2a\ln|a+bu|\right) + C$

52. $\displaystyle\int \frac{u^n}{\sqrt{a+bu}}\,du = \frac{2}{(2n+1)b}\left(u^n\sqrt{a+bu} - na\int \frac{u^{n-1}}{\sqrt{a+bu}}\,du\right)$

53. $\displaystyle\int \frac{1}{(u^2 \pm a^2)^{3/2}}\,du = \frac{\pm u}{a^2\sqrt{u^2 \pm a^2}} + C$

54. $\displaystyle\int u^n\cos u\,du = u^n\sin u - n\int u^{n-1}\sin u\,du$

55. $\displaystyle\int \arctan u\,du = u\arctan u - \ln\sqrt{1+u^2} + C$

56. $\int (\ln u)^n \, du = u(\ln u)^n - n\int (\ln u)^{n-1}\, du$

In Exercises 57–62, use a computer algebra system to determine the antiderivative that passes through the given point. Use the system to graph the resulting antiderivative.

57. $\int \frac{1}{x^{3/2}\sqrt{1-x}}\, dx,\ \left(\frac{1}{2}, 5\right)$

58. $\int x\sqrt{x^2+2x}\, dx,\ (0, 0)$

59. $\int \frac{1}{(x^2-6x+10)^2}\, dx,\ (3, 0)$

60. $\int \frac{\sqrt{2-2x-x^2}}{x+1}\, dx,\ \left(0, \sqrt{2}\right)$

61. $\int \frac{1}{\sin\theta \tan\theta}\, d\theta,\ \left(\frac{\pi}{4}, 2\right)$

62. $\int \frac{\sin\theta}{(\cos\theta)(1+\sin\theta)}\, d\theta,\ (0, 1)$

In Exercises 63–70, find or evaluate the integral.

63. $\int \frac{1}{2-3\sin\theta}\, d\theta$

64. $\int \frac{\sin\theta}{1+\cos^2\theta}\, d\theta$

65. $\int_0^{\pi/2} \frac{1}{1+\sin\theta+\cos\theta}\, d\theta$

66. $\int_0^{\pi/2} \frac{1}{3-2\cos\theta}\, d\theta$

67. $\int \frac{\sin\theta}{3-2\cos\theta}\, d\theta$

68. $\int \frac{\cos\theta}{1+\cos\theta}\, d\theta$

69. $\int \frac{\cos\sqrt{\theta}}{\sqrt{\theta}}\, d\theta$

70. $\int \frac{1}{\sec\theta-\tan\theta}\, d\theta$

Area **In Exercises 71 and 72, find the area of the region bounded by the graphs of the equations.**

71. $y = \frac{x}{\sqrt{x+1}},\ y = 0,\ x = 8$

72. $y = \frac{x}{1+e^{x^2}},\ y = 0,\ x = 2$

Writing About Concepts

In Exercises 73–78, state (if possible) the method or integration formula you would use to find the antiderivative. Explain why you chose that method or formula. Do not integrate.

73. $\int \frac{e^x}{e^{2x}+1}\, dx$

74. $\int \frac{e^x}{e^x+1}\, dx$

75. $\int x\, e^{x^2}\, dx$

76. $\int x\, e^x\, dx$

77. $\int e^{x^2}\, dx$

78. $\int e^{2x}\sqrt{e^{2x}+1}\, dx$

79. (a) Evaluate $\int x^n \ln x\, dx$ for $n = 1, 2,$ and 3. Describe any patterns you notice.

(b) Write a general rule for evaluating the integral in part (a), for an integer $n \geq 1$.

80. Describe what is meant by a reduction formula. Give an example.

True or False? **In Exercises 81 and 82, determine whether the statement is true or false. If it is false, explain why or give an example that shows it is false.**

81. To use a table of integrals, the integral you are evaluating must appear in the table.

82. When using a table of integrals, you may have to make substitutions to rewrite your integral in the form in which it appears in the table.

83. ***Work*** A hydraulic cylinder on an industrial machine pushes a steel block a distance of x feet $(0 \leq x \leq 5)$, where the variable force required is $F(x) = 2000xe^{-x}$ pounds. Find the work done in pushing the block the full 5 feet through the machine.

84. ***Work*** Repeat Exercise 83, using $F(x) = \frac{500x}{\sqrt{26-x^2}}$ pounds.

85. ***Building Design*** The cross section of a precast concrete beam for a building is bounded by the graphs of the equations

$$x = \frac{2}{\sqrt{1+y^2}},\ x = \frac{-2}{\sqrt{1+y^2}},\ y = 0, \text{ and } y = 3$$

where x and y are measured in feet. The length of the beam is 20 feet (see figure). (a) Find the volume V and the weight W of the beam. Assume the concrete weighs 148 pounds per cubic foot. (b) Then find the centroid of a cross section of the beam.

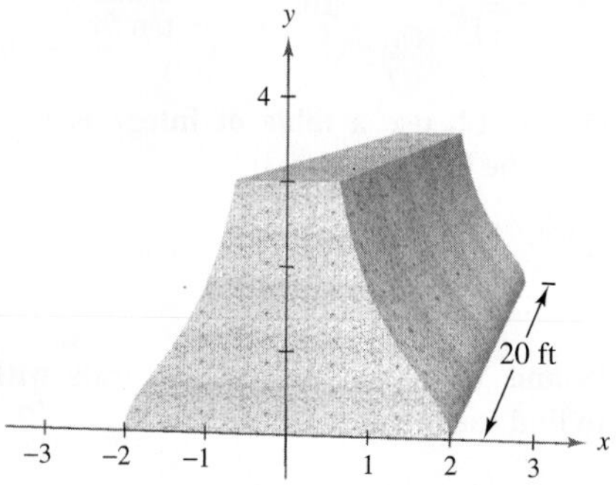

86. ***Population*** A population is growing according to the logistic model $N = \frac{5000}{1+e^{4.8-1.9t}}$ where t is the time in days. Find the average population over the interval $[0, 2]$.

In Exercises 87 and 88, use a graphing utility to (a) solve the integral equation for the constant k and (b) graph the region whose area is given by the integral.

87. $\int_0^4 \frac{k}{2+3x}\, dx = 10$

88. $\int_0^k 6x^2 e^{-x/2}\, dx = 50$

Putnam Exam Challenge

89. Evaluate $\int_0^{\pi/2} \frac{dx}{1+(\tan x)^{\sqrt{2}}}$.

This problem was composed by the Committee on the Putnam Prize Competition.

Section 8.7 Indeterminate Forms and L'Hôpital's Rule

- **Recognize limits that produce indeterminate forms.**
- **Apply L'Hôpital's Rule to evaluate a limit.**

Indeterminate Forms

Recall from Chapters 1 and 3 that the forms $0/0$ and ∞/∞ are called *indeterminate* because they do not guarantee that a limit exists, nor do they indicate what the limit is, if one does exist. When you encountered one of these indeterminate forms earlier in the text, you attempted to rewrite the expression by using various algebraic techniques.

Indeterminate Form	*Limit*	*Algebraic Technique*
$\frac{0}{0}$	$\lim_{x\to -1} \frac{2x^2-2}{x+1} = \lim_{x\to -1} 2(x-1) = -4$	Divide numerator and denominator by $(x+1)$.
$\frac{\infty}{\infty}$	$\lim_{x\to\infty} \frac{3x^2-1}{2x^2+1} = \lim_{x\to\infty} \frac{3-(1/x^2)}{2+(1/x^2)} = \frac{3}{2}$	Divide numerator and denominator by x^2.

Occasionally, you can extend these algebraic techniques to find limits of transcendental functions. For instance, the limit

$$\lim_{x\to 0} \frac{e^{2x}-1}{e^x-1}$$

produces the indeterminate form $0/0$. Factoring and then dividing produces

$$\lim_{x\to 0} \frac{e^{2x}-1}{e^x-1} = \lim_{x\to 0} \frac{(e^x+1)(e^x-1)}{e^x-1} = \lim_{x\to 0} (e^x+1) = 2.$$

However, not all indeterminate forms can be evaluated by algebraic manipulation. This is often true when *both* algebraic and transcendental functions are involved. For instance, the limit

$$\lim_{x\to 0} \frac{e^{2x}-1}{x}$$

produces the indeterminate form $0/0$. Rewriting the expression to obtain

$$\lim_{x\to 0} \left(\frac{e^{2x}}{x} - \frac{1}{x}\right)$$

merely produces another indeterminate form, $\infty - \infty$. Of course, you could use technology to estimate the limit, as shown in the table and in Figure 8.14. From the table and the graph, the limit appears to be 2. (This limit will be verified in Example 1.)

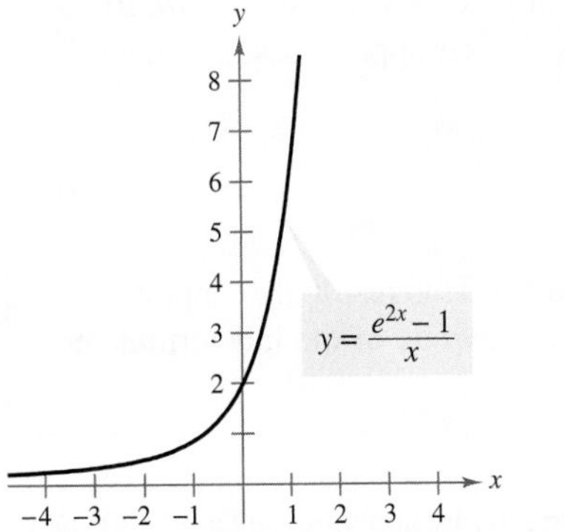

The limit as x approaches 0 appears to be 2.
Figure 8.14

x	-1	-0.1	-0.01	-0.001	0	0.001	0.01	0.1	1
$\frac{e^{2x}-1}{x}$	0.865	1.813	1.980	1.998	?	2.002	2.020	2.214	6.389

The Granger Collection

GUILLAUME L'HÔPITAL (1661–1704)

L'Hôpital's Rule is named after the French mathematician Guillaume François Antoine de L'Hôpital. L'Hôpital is credited with writing the first text on differential calculus (in 1696) in which the rule publicly appeared. It was recently discovered that the rule and its proof were written in a letter from John Bernoulli to L'Hôpital. "… I acknowledge that I owe very much to the bright minds of the Bernoulli brothers. … I have made free use of their discoveries …," said L'Hôpital.

L'Hôpital's Rule

To find the limit illustrated in Figure 8.14, you can use a theorem called **L'Hôpital's Rule.** This theorem states that under certain conditions the limit of the quotient $f(x)/g(x)$ is determined by the limit of the quotient of the derivatives

$$\frac{f'(x)}{g'(x)}.$$

To prove this theorem, you can use a more general result called the **Extended Mean Value Theorem.**

THEOREM 8.3 The Extended Mean Value Theorem

If f and g are differentiable on an open interval (a, b) and continuous on $[a, b]$ such that $g'(x) \neq 0$ for any x in (a, b), then there exists a point c in (a, b) such that

$$\frac{f'(c)}{g'(c)} = \frac{f(b) - f(a)}{g(b) - g(a)}.$$

NOTE To see why this is called the Extended Mean Value Theorem, consider the special case in which $g(x) = x$. For this case, you obtain the "standard" Mean Value Theorem as presented in Section 3.2.

The Extended Mean Value Theorem and L'Hôpital's Rule are both proved in Appendix A.

THEOREM 8.4 L'Hôpital's Rule

Let f and g be functions that are differentiable on an open interval (a, b) containing c, except possibly at c itself. Assume that $g'(x) \neq 0$ for all x in (a, b), except possibly at c itself. If the limit of $f(x)/g(x)$ as x approaches c produces the indeterminate form $0/0$, then

$$\lim_{x \to c} \frac{f(x)}{g(x)} = \lim_{x \to c} \frac{f'(x)}{g'(x)}$$

provided the limit on the right exists (or is infinite). This result also applies if the limit of $f(x)/g(x)$ as x approaches c produces any one of the indeterminate forms ∞/∞, $(-\infty)/\infty$, $\infty/(-\infty)$, or $(-\infty)/(-\infty)$.

NOTE People occasionally use L'Hôpital's Rule incorrectly by applying the Quotient Rule to $f(x)/g(x)$. Be sure you see that the rule involves $f'(x)/g'(x)$, not the derivative of $f(x)/g(x)$.

FOR FURTHER INFORMATION To enhance your understanding of the necessity of the restriction that $g'(x)$ be nonzero for all x in (a, b), except possibly at c, see the article "Counterexamples to L'Hôpital's Rule" by R. P. Boas in *The American Mathematical Monthly.* To view this article, go to the website *www.matharticles.com.*

L'Hôpital's Rule can also be applied to one-sided limits. For instance, if the limit of $f(x)/g(x)$ as x approaches c *from the right* produces the indeterminate form $0/0$, then

$$\lim_{x \to c^+} \frac{f(x)}{g(x)} = \lim_{x \to c^+} \frac{f'(x)}{g'(x)}$$

provided the limit exists (or is infinite).

TECHNOLOGY *Numerical and Graphical Approaches* Use a numerical or a graphical approach to approximate each limit.

a. $\displaystyle\lim_{x\to0}\frac{2^{2x}-1}{x}$

b. $\displaystyle\lim_{x\to0}\frac{3^{2x}-1}{x}$

c. $\displaystyle\lim_{x\to0}\frac{4^{2x}-1}{x}$

d. $\displaystyle\lim_{x\to0}\frac{5^{2x}-1}{x}$

What pattern do you observe? Does an analytic approach have an advantage for these limits? If so, explain your reasoning.

EXAMPLE 1 Indeterminate Form 0/0

Evaluate $\displaystyle\lim_{x\to0}\frac{e^{2x}-1}{x}$.

Solution Because direct substitution results in the indeterminate form $0/0$

$$\lim_{x\to0}\frac{e^{2x}-1}{x} \quad \begin{array}{l}\nearrow \lim_{x\to0}(e^{2x}-1)=0\\ \searrow \lim_{x\to0}x=0\end{array}$$

you can apply L'Hôpital's Rule as shown below.

$$\lim_{x\to0}\frac{e^{2x}-1}{x}=\lim_{x\to0}\frac{\frac{d}{dx}[e^{2x}-1]}{\frac{d}{dx}[x]}$$ Apply L'Hôpital's Rule.

$$=\lim_{x\to0}\frac{2e^{2x}}{1}$$ Differentiate numerator and denominator.

$$=2$$ Evaluate the limit.

NOTE In writing the string of equations in Example 1, you actually do not know that the first limit is equal to the second until you have shown that the second limit exists. In other words, if the second limit had not existed, it would not have been permissible to apply L'Hôpital's Rule.

Another form of L'Hôpital's Rule states that if the limit of $f(x)/g(x)$ as x approaches ∞ (or $-\infty$) produces the indeterminate form $0/0$ or ∞/∞, then

$$\lim_{x\to\infty}\frac{f(x)}{g(x)}=\lim_{x\to\infty}\frac{f'(x)}{g'(x)}$$

provided the limit on the right exists.

EXAMPLE 2 Indeterminate Form ∞/∞

Evaluate $\displaystyle\lim_{x\to\infty}\frac{\ln x}{x}$.

Solution Because direct substitution results in the indeterminate form ∞/∞, you can apply L'Hôpital's Rule to obtain

$$\lim_{x\to\infty}\frac{\ln x}{x}=\lim_{x\to\infty}\frac{\frac{d}{dx}[\ln x]}{\frac{d}{dx}[x]}$$ Apply L'Hôpital's Rule.

$$=\lim_{x\to\infty}\frac{1}{x}$$ Differentiate numerator and denominator.

$$=0.$$ Evaluate the limit.

NOTE Try graphing $y_1=\ln x$ and $y_2=x$ in the same viewing window. Which function grows faster as x approaches ∞? How is this observation related to Example 2?

Occasionally it is necessary to apply L'Hôpital's Rule more than once to remove an indeterminate form, as shown in Example 3.

EXAMPLE 3 Applying L'Hôpital's Rule More Than Once

Evaluate $\lim\limits_{x\to-\infty} \dfrac{x^2}{e^{-x}}$.

Solution Because direct substitution results in the indeterminate form ∞/∞, you can apply L'Hôpital's Rule.

$$\lim_{x\to-\infty} \frac{x^2}{e^{-x}} = \lim_{x\to-\infty} \frac{\frac{d}{dx}[x^2]}{\frac{d}{dx}[e^{-x}]} = \lim_{x\to-\infty} \frac{2x}{-e^{-x}}$$

This limit yields the indeterminate form $(-\infty)/(-\infty)$, so you can apply L'Hôpital's Rule again to obtain

$$\lim_{x\to-\infty} \frac{2x}{-e^{-x}} = \lim_{x\to-\infty} \frac{\frac{d}{dx}[2x]}{\frac{d}{dx}[-e^{-x}]} = \lim_{x\to-\infty} \frac{2}{e^{-x}} = 0.$$

In addition to the forms $0/0$ and ∞/∞, there are other indeterminate forms such as $0 \cdot \infty$, 1^∞, ∞^0, 0^0, and $\infty - \infty$. For example, consider the following four limits that lead to the indeterminate form $0 \cdot \infty$.

$$\underbrace{\lim_{x\to0}\,(x)\left(\frac{1}{x}\right)}_{\text{Limit is 1.}}, \quad \underbrace{\lim_{x\to0}\,(x)\left(\frac{2}{x}\right)}_{\text{Limit is 2.}}, \quad \underbrace{\lim_{x\to\infty}\,(x)\left(\frac{1}{e^x}\right)}_{\text{Limit is 0.}}, \quad \underbrace{\lim_{x\to\infty}\,(e^x)\left(\frac{1}{x}\right)}_{\text{Limit is }\infty.}$$

Because each limit is different, it is clear that the form $0 \cdot \infty$ is indeterminate in the sense that it does not determine the value (or even the existence) of the limit. The following examples indicate methods for evaluating these forms. Basically, you attempt to convert each of these forms to $0/0$ or ∞/∞ so that L'Hôpital's Rule can be applied.

EXAMPLE 4 Indeterminate Form $0 \cdot \infty$

Evaluate $\lim\limits_{x\to\infty} e^{-x}\sqrt{x}$.

Solution Because direct substitution produces the indeterminate form $0 \cdot \infty$, you should try to rewrite the limit to fit the form $0/0$ or ∞/∞. In this case, you can rewrite the limit to fit the second form.

$$\lim_{x\to\infty} e^{-x}\sqrt{x} = \lim_{x\to\infty} \frac{\sqrt{x}}{e^x}$$

Now, by L'Hôpital's Rule, you have

$$\lim_{x\to\infty} \frac{\sqrt{x}}{e^x} = \lim_{x\to\infty} \frac{1/(2\sqrt{x})}{e^x} = \lim_{x\to\infty} \frac{1}{2\sqrt{x}\,e^x} = 0.$$

If rewriting a limit in one of the forms $0/0$ or ∞/∞ does not seem to work, try the other form. For instance, in Example 4 you can write the limit as

$$\lim_{x\to\infty} e^{-x}\sqrt{x} = \lim_{x\to\infty} \frac{e^{-x}}{x^{-1/2}}$$

which yields the indeterminate form $0/0$. As it happens, applying L'Hôpital's Rule to this limit produces

$$\lim_{x\to\infty} \frac{e^{-x}}{x^{-1/2}} = \lim_{x\to\infty} \frac{-e^{-x}}{-1/(2x^{3/2})}$$

which also yields the indeterminate form $0/0$.

The indeterminate forms 1^∞, ∞^0, and 0^0 arise from limits of functions that have variable bases and variable exponents. When you previously encountered this type of function, you used logarithmic differentiation to find the derivative. You can use a similar procedure when taking limits, as shown in the next example.

EXAMPLE 5 Indeterminate Form 1^∞

Evaluate $\displaystyle\lim_{x\to\infty}\left(1+\frac{1}{x}\right)^x$.

Solution Because direct substitution yields the indeterminate form 1^∞, you can proceed as follows. To begin, assume that the limit exists and is equal to y.

$$y = \lim_{x\to\infty}\left(1+\frac{1}{x}\right)^x$$

Taking the natural logarithm of each side produces

$$\ln y = \ln\left[\lim_{x\to\infty}\left(1+\frac{1}{x}\right)^x\right].$$

Because the natural logarithmic function is continuous, you can write

$$\begin{aligned}
\ln y &= \lim_{x\to\infty}\left[x\ln\left(1+\frac{1}{x}\right)\right] && \text{Indeterminate form } \infty\cdot 0\\
&= \lim_{x\to\infty}\left(\frac{\ln[1+(1/x)]}{1/x}\right) && \text{Indeterminate form } 0/0\\
&= \lim_{x\to\infty}\left(\frac{(-1/x^2)\{1/[1+(1/x)]\}}{-1/x^2}\right) && \text{L'Hôpital's Rule}\\
&= \lim_{x\to\infty}\frac{1}{1+(1/x)}\\
&= 1.
\end{aligned}$$

Now, because you have shown that $\ln y = 1$, you can conclude that $y = e$ and obtain

$$\lim_{x\to\infty}\left(1+\frac{1}{x}\right)^x = e.$$

You can use a graphing utility to confirm this result, as shown in Figure 8.15.

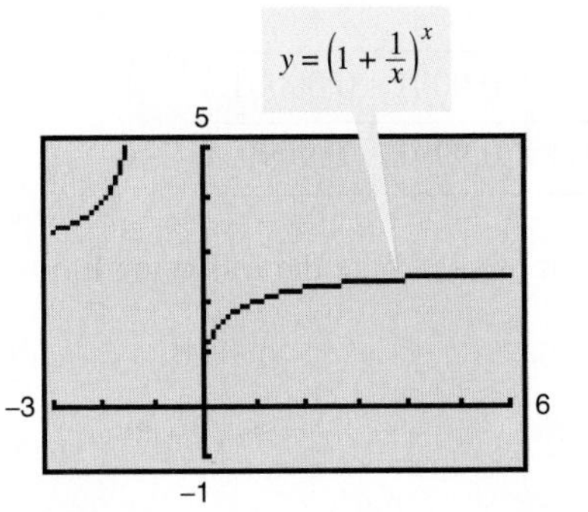

The limit of $[1+(1/x)]^x$ as x approaches infinity is e.
Figure 8.15

L'Hôpital's Rule can also be applied to one-sided limits, as demonstrated in Examples 6 and 7.

EXAMPLE 6 Indeterminate Form 0^0

Find $\lim_{x\to 0^+} (\sin x)^x$.

Solution Because direct substitution produces the indeterminate form 0^0, you can proceed as shown below. To begin, assume that the limit exists and is equal to y.

$$y = \lim_{x\to 0^+} (\sin x)^x \qquad \text{Indeterminate form } 0^0$$

$$\ln y = \ln\left[\lim_{x\to 0^+} (\sin x)^x\right] \qquad \text{Take natural log of each side.}$$

$$= \lim_{x\to 0^+} [\ln(\sin x)^x] \qquad \text{Continuity}$$

$$= \lim_{x\to 0^+} [x \ln(\sin x)] \qquad \text{Indeterminate form } 0 \cdot (-\infty)$$

$$= \lim_{x\to 0^+} \frac{\ln(\sin x)}{1/x} \qquad \text{Indeterminate form } -\infty/\infty$$

$$= \lim_{x\to 0^+} \frac{\cot x}{-1/x^2} \qquad \text{L'Hôpital's Rule}$$

$$= \lim_{x\to 0^+} \frac{-x^2}{\tan x} \qquad \text{Indeterminate form } 0/0$$

$$= \lim_{x\to 0^+} \frac{-2x}{\sec^2 x} = 0 \qquad \text{L'Hôpital's Rule}$$

Now, because $\ln y = 0$, you can conclude that $y = e^0 = 1$, and it follows that

$$\lim_{x\to 0^+} (\sin x)^x = 1.$$

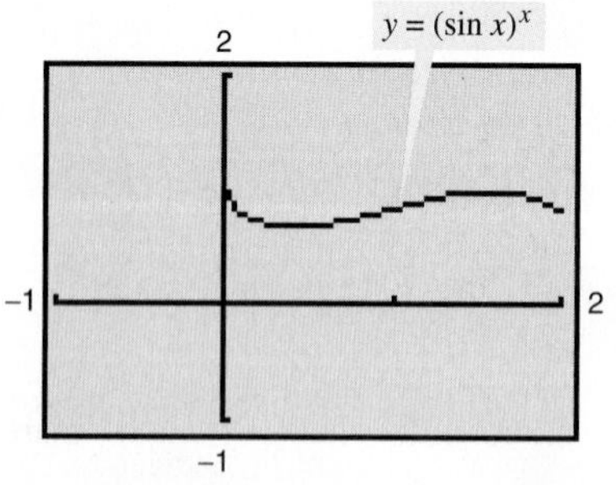

The limit of $(\sin x)^x$ is 1 as x approaches 0 from the right.
Figure 8.16

TECHNOLOGY When evaluating complicated limits such as the one in Example 6, it is helpful to check the reasonableness of the solution with a computer or with a graphing utility. For instance, the calculations in the following table and the graph in Figure 8.16 are consistent with the conclusion that $(\sin x)^x$ approaches 1 as x approaches 0 from the right.

x	1.0	0.1	0.01	0.001	0.0001	0.00001
$(\sin x)^x$	0.8415	0.7942	0.9550	0.9931	0.9991	0.9999

Use a computer algebra system or graphing utility to estimate the following limits:

$$\lim_{x\to 0} (1 - \cos x)^x$$

and

$$\lim_{x\to 0^+} (\tan x)^x.$$

Then see if you can verify your estimates analytically.

EXAMPLE 7 Indeterminate Form $\infty - \infty$

Evaluate $\lim_{x\to 1^+}\left(\frac{1}{\ln x} - \frac{1}{x-1}\right)$.

Solution Because direct substitution yields the indeterminate form $\infty - \infty$, you should try to rewrite the expression to produce a form to which you can apply L'Hôpital's Rule. In this case, you can combine the two fractions to obtain

$$\lim_{x\to 1^+}\left(\frac{1}{\ln x} - \frac{1}{x-1}\right) = \lim_{x\to 1^+}\left[\frac{x-1-\ln x}{(x-1)\ln x}\right].$$

Now, because direct substitution produces the indeterminate form $0/0$, you can apply L'Hôpital's Rule to obtain

$$\begin{aligned}
\lim_{x\to 1^+}\left(\frac{1}{\ln x} - \frac{1}{x-1}\right) &= \lim_{x\to 1^+}\frac{\frac{d}{dx}[x-1-\ln x]}{\frac{d}{dx}[(x-1)\ln x]}\\
&= \lim_{x\to 1^+}\left[\frac{1-(1/x)}{(x-1)(1/x)+\ln x}\right]\\
&= \lim_{x\to 1^+}\left(\frac{x-1}{x-1+x\ln x}\right).
\end{aligned}$$

This limit also yields the indeterminate form $0/0$, so you can apply L'Hôpital's Rule again to obtain

$$\begin{aligned}
\lim_{x\to 1^+}\left(\frac{1}{\ln x} - \frac{1}{x-1}\right) &= \lim_{x\to 1^+}\left[\frac{1}{1+x(1/x)+\ln x}\right]\\
&= \frac{1}{2}.
\end{aligned}$$

STUDY TIP In each of the examples presented in this section, L'Hôpital's Rule is used to find a limit that exists. It can also be used to conclude that a limit is infinite. For instance, try using L'Hôpital's Rule to show that

$$\lim_{x\to\infty}\frac{e^x}{x} = \infty.$$

The forms $0/0$, ∞/∞, $\infty - \infty$, $0 \cdot \infty$, 0^0, 1^∞, and ∞^0 have been identified as *indeterminate*. There are similar forms that you should recognize as "determinate."

$\infty + \infty \to \infty$ Limit is positive infinity.

$-\infty - \infty \to -\infty$ Limit is negative infinity.

$0^\infty \to 0$ Limit is zero.

$0^{-\infty} \to \infty$ Limit is positive infinity.

(You are asked to verify two of these in Exercises 106 and 107.)

As a final comment, remember that L'Hôpital's Rule can be applied only to quotients leading to the indeterminate forms $0/0$ and ∞/∞. For instance, the following application of L'Hôpital's Rule is *incorrect*.

$$\lim_{x\to 0}\frac{e^x}{x} \;\textcircled{=}\; \lim_{x\to 0}\frac{e^x}{1} = 1$$ Incorrect use of L'Hôpital's Rule

The reason this application is incorrect is that, even though the limit of the denominator is 0, the limit of the numerator is 1, which means that the hypotheses of L'Hôpital's Rule have not been satisfied.

Exercises for Section 8.7

See www.CalcChat.com for worked-out solutions to odd-numbered exercises.

Numerical and Graphical Analysis **In Exercises 1–4, complete the table and use the result to estimate the limit. Use a graphing utility to graph the function to support your result.**

1. $\lim_{x\to 0} \dfrac{\sin 5x}{\sin 2x}$

x	-0.1	-0.01	-0.001	0.001	0.01	0.1
$f(x)$						

2. $\lim_{x\to 0} \dfrac{1 - e^x}{x}$

x	-0.1	-0.01	-0.001	0.001	0.01	0.1
$f(x)$						

3. $\lim_{x\to\infty} x^5 e^{-x/100}$

x	1	10	10^2	10^3	10^4	10^5
$f(x)$						

4. $\lim_{x\to\infty} \dfrac{6x}{\sqrt{3x^2 - 2x}}$

x	1	10	10^2	10^3	10^4	10^5
$f(x)$						

In Exercises 5–10, evaluate the limit (a) using techniques from Chapters 1 and 3 and (b) using L'Hôpital's Rule.

5. $\lim_{x\to 3} \dfrac{2(x-3)}{x^2 - 9}$

6. $\lim_{x\to -1} \dfrac{2x^2 - x - 3}{x + 1}$

7. $\lim_{x\to 3} \dfrac{\sqrt{x+1} - 2}{x - 3}$

8. $\lim_{x\to 0} \dfrac{\sin 4x}{2x}$

9. $\lim_{x\to\infty} \dfrac{5x^2 - 3x + 1}{3x^2 - 5}$

10. $\lim_{x\to\infty} \dfrac{2x + 1}{4x^2 + x}$

In Exercises 11–36, evaluate the limit, using L'Hôpital's Rule if necessary. (In Exercise 18, *n* is a positive integer.)

11. $\lim_{x\to 2} \dfrac{x^2 - x - 2}{x - 2}$

12. $\lim_{x\to -1} \dfrac{x^2 - x - 2}{x + 1}$

13. $\lim_{x\to 0} \dfrac{\sqrt{4 - x^2} - 2}{x}$

14. $\lim_{x\to 2^-} \dfrac{\sqrt{4 - x^2}}{x - 2}$

15. $\lim_{x\to 0} \dfrac{e^x - (1 - x)}{x}$

16. $\lim_{x\to 1} \dfrac{\ln x^2}{x^2 - 1}$

17. $\lim_{x\to 0^+} \dfrac{e^x - (1 + x)}{x^3}$

18. $\lim_{x\to 0^+} \dfrac{e^x - (1 + x)}{x^n}$

19. $\lim_{x\to 0} \dfrac{\sin 2x}{\sin 3x}$

20. $\lim_{x\to 0} \dfrac{\sin ax}{\sin bx}$

21. $\lim_{x\to 0} \dfrac{\arcsin x}{x}$

22. $\lim_{x\to 1} \dfrac{\arctan x - (\pi/4)}{x - 1}$

23. $\lim_{x\to\infty} \dfrac{3x^2 - 2x + 1}{2x^2 + 3}$

24. $\lim_{x\to\infty} \dfrac{x - 1}{x^2 + 2x + 3}$

25. $\lim_{x\to\infty} \dfrac{x^2 + 2x + 3}{x - 1}$

26. $\lim_{x\to\infty} \dfrac{x^3}{x + 2}$

27. $\lim_{x\to\infty} \dfrac{x^3}{e^{x/2}}$

28. $\lim_{x\to\infty} \dfrac{x^2}{e^x}$

29. $\lim_{x\to\infty} \dfrac{x}{\sqrt{x^2 + 1}}$

30. $\lim_{x\to\infty} \dfrac{x^2}{\sqrt{x^2 + 1}}$

31. $\lim_{x\to\infty} \dfrac{\cos x}{x}$

32. $\lim_{x\to\infty} \dfrac{\sin x}{x - \pi}$

33. $\lim_{x\to\infty} \dfrac{\ln x}{x^2}$

34. $\lim_{x\to\infty} \dfrac{\ln x^4}{x^3}$

35. $\lim_{x\to\infty} \dfrac{e^x}{x^2}$

36. $\lim_{x\to\infty} \dfrac{e^{x/2}}{x}$

In Exercises 37–54, (a) describe the type of indeterminate form (if any) that is obtained by direct substitution. (b) Evaluate the limit, using L'Hôpital's Rule if necessary. (c) Use a graphing utility to graph the function and verify the result in part (b).

37. $\lim_{x\to\infty} x \ln x$

38. $\lim_{x\to 0^+} x^3 \cot x$

39. $\lim_{x\to\infty} \left(x \sin \dfrac{1}{x}\right)$

40. $\lim_{x\to\infty} x \tan \dfrac{1}{x}$

41. $\lim_{x\to 0^+} x^{1/x}$

42. $\lim_{x\to 0^+} (e^x + x)^{2/x}$

43. $\lim_{x\to\infty} x^{1/x}$

44. $\lim_{x\to\infty} \left(1 + \dfrac{1}{x}\right)^x$

45. $\lim_{x\to 0^+} (1 + x)^{1/x}$

46. $\lim_{x\to\infty} (1 + x)^{1/x}$

47. $\lim_{x\to 0^+} [3(x)^{x/2}]$

48. $\lim_{x\to 4^+} [3(x - 4)]^{x-4}$

49. $\lim_{x\to 1^+} (\ln x)^{x-1}$

50. $\lim_{x\to 0^+} \left[\cos\left(\dfrac{\pi}{2} - x\right)\right]^x$

51. $\lim_{x\to 2^+} \left(\dfrac{8}{x^2 - 4} - \dfrac{x}{x - 2}\right)$

52. $\lim_{x\to 2^+} \left(\dfrac{1}{x^2 - 4} - \dfrac{\sqrt{x - 1}}{x^2 - 4}\right)$

53. $\lim_{x\to 1^+} \left(\dfrac{3}{\ln x} - \dfrac{2}{x - 1}\right)$

54. $\lim_{x\to 0^+} \left(\dfrac{10}{x} - \dfrac{3}{x^2}\right)$

In Exercises 55–58, use a graphing utility to (a) graph the function and (b) find the required limit (if it exists).

55. $\lim_{x\to 3} \dfrac{x - 3}{\ln(2x - 5)}$

56. $\lim_{x\to 0^+} (\sin x)^x$

57. $\lim_{x\to\infty} \left(\sqrt{x^2 + 5x + 2} - x\right)$

58. $\lim_{x\to\infty} \dfrac{x^3}{e^{2x}}$

Writing About Concepts

59. List six different indeterminate forms.

60. State L'Hôpital's Rule.

61. Find the differentiable functions f and g that satisfy the specified condition such that

$\lim_{x\to 5} f(x) = 0$ and $\lim_{x\to 5} g(x) = 0$.

Explain how you obtained your answers. (*Note:* There are many correct answers.)

(a) $\displaystyle\lim_{x\to 5}\frac{f(x)}{g(x)} = 10$ (b) $\displaystyle\lim_{x\to 5}\frac{f(x)}{g(x)} = 0$

(c) $\displaystyle\lim_{x\to 5}\frac{f(x)}{g(x)} = \infty$

62. Find differentiable functions f and g such that

$\lim_{x\to\infty} f(x) = \lim_{x\to\infty} g(x) = \infty$ and

$\lim_{x\to\infty} [f(x) - g(x)] = 25.$

Explain how you obtained your answers. (*Note:* There are many correct answers.)

63. ***Numerical Approach*** Complete the table to show that x eventually "overpowers" $(\ln x)^4$.

x	10	10^2	10^4	10^6	10^8	10^{10}
$\dfrac{(\ln x)^4}{x}$						

64. ***Numerical Approach*** Complete the table to show that e^x eventually "overpowers" x^5.

x	1	5	10	20	30	40	50	100
$\dfrac{e^x}{x^5}$								

Comparing Functions **In Exercises 65–70, use L'Hôpital's Rule to determine the comparative rates of increase of the functions**

$$f(x) = x^m,\quad g(x) = e^{nx},\quad \text{and}\quad h(x) = (\ln x)^n$$

where $n > 0$, $m > 0$, and $x \to \infty$.

65. $\displaystyle\lim_{x\to\infty}\frac{x^2}{e^{5x}}$ **66.** $\displaystyle\lim_{x\to\infty}\frac{x^3}{e^{2x}}$

67. $\displaystyle\lim_{x\to\infty}\frac{(\ln x)^3}{x}$ **68.** $\displaystyle\lim_{x\to\infty}\frac{(\ln x)^2}{x^3}$

69. $\displaystyle\lim_{x\to\infty}\frac{(\ln x)^n}{x^m}$ **70.** $\displaystyle\lim_{x\to\infty}\frac{x^m}{e^{nx}}$

In Exercises 71–74, find any asymptotes and relative extrema that may exist and use a graphing utility to graph the function. (*Hint:* Some of the limits required in finding asymptotes have been found in preceding exercises.)

71. $y = x^{1/x},\quad x > 0$ **72.** $y = x^x,\quad x > 0$

73. $y = 2xe^{-x}$ **74.** $y = \dfrac{\ln x}{x}$

Think About It **In Exercises 75–78, L'Hôpital's Rule is used incorrectly. Describe the error.**

75. $\displaystyle\lim_{x\to 0}\frac{e^{2x}-1}{e^x} = \lim_{x\to 0}\frac{2e^{2x}}{e^x} = \lim_{x\to 0} 2e^x = 2$

76. $\displaystyle\lim_{x\to 0}\frac{\sin \pi x - 1}{x} = \lim_{x\to 0}\frac{\pi\cos \pi x}{1} = \pi$

77. $\displaystyle\lim_{x\to\infty} x\cos\frac{1}{x} = \lim_{x\to\infty}\frac{\cos(1/x)}{1/x}$

$\displaystyle = \lim_{x\to\infty}\frac{[-\sin(1/x)](1/x^2)}{-1/x^2}$

$= 0$

78. $\displaystyle\lim_{x\to\infty}\frac{e^{-x}}{1+e^{-x}} = \lim_{x\to\infty}\frac{-e^{-x}}{-e^{-x}}$

$\displaystyle = \lim_{x\to\infty} 1$

$= 1$

Analytical Approach **In Exercises 79 and 80, (a) explain why L'Hôpital's Rule cannot be used to find the limit, (b) find the limit analytically, and (c) use a graphing utility to graph the function and approximate the limit from the graph. Compare the result with that in part (b).**

79. $\displaystyle\lim_{x\to\infty}\frac{x}{\sqrt{x^2+1}}$ **80.** $\displaystyle\lim_{x\to\pi/2^-}\frac{\tan x}{\sec x}$

Graphical Analysis **In Exercises 81 and 82, graph $f(x)/g(x)$ and $f'(x)/g'(x)$ near $x = 0$. What do you notice about these ratios as $x \to 0$? How does this illustrate L'Hôpital's Rule?**

81. $f(x) = \sin 3x,\quad g(x) = \sin 4x$

82. $f(x) = e^{3x} - 1,\quad g(x) = x$

83. ***Velocity in a Resisting Medium*** The velocity v of an object falling through a resisting medium such as air or water is given by

$$v = \frac{32}{k}\left(1 - e^{-kt} + \frac{v_0 k e^{-kt}}{32}\right)$$

where v_0 is the initial velocity, t is the time in seconds, and k is the resistance constant of the medium. Use L'Hôpital's Rule to find the formula for the velocity of a falling body in a vacuum by fixing v_0 and t and letting k approach zero. (Assume that the downward direction is positive.)

84. ***Compound Interest*** The formula for the amount A in a savings account compounded n times per year for t years at an interest rate r and an initial deposit of P is given by

$$A = P\left(1 + \frac{r}{n}\right)^{nt}.$$

Use L'Hôpital's Rule to show that the limiting formula as the number of compoundings per year becomes infinite is given by $A = Pe^{rt}$.

85. ***The Gamma Function*** The Gamma Function $\Gamma(n)$ is defined in terms of the integral of the function given by $f(x) = x^{n-1}e^{-x}$, $n > 0$. Show that for any fixed value of n, the limit of $f(x)$ as x approaches infinity is zero.

86. ***Tractrix*** A person moves from the origin along the positive y-axis pulling a weight at the end of a 12-meter rope (see figure). Initially, the weight is located at the point $(12, 0)$.

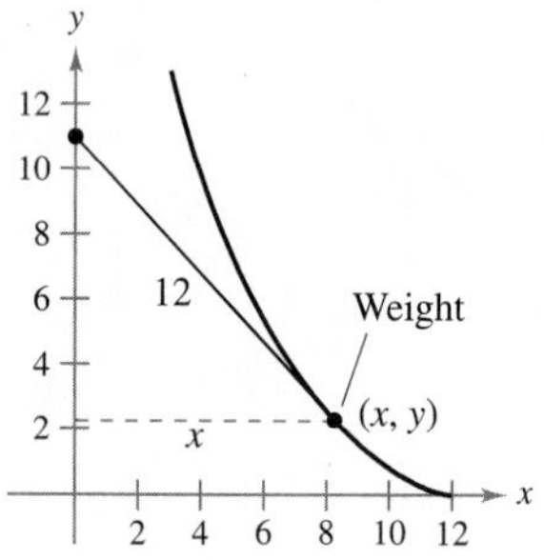

(a) Show that the slope of the tangent line of the path of the weight is

$$\frac{dy}{dx} = -\frac{\sqrt{144 - x^2}}{x}.$$

(b) Use the result of part (a) to find the equation of the path of the weight. Use a graphing utility to graph the path and compare it with the figure.

(c) Find any vertical asymptotes of the graph in part (b).

(d) When the person has reached the point $(0, 12)$, how far has the weight moved?

In Exercises 87–90, apply the Extended Mean Value Theorem to the functions f and g on the given interval. Find all values c in the interval (a, b) such that

$$\frac{f'(c)}{g'(c)} = \frac{f(b) - f(a)}{g(b) - g(a)}.$$

Functions	*Interval*
87. $f(x) = x^3$, $g(x) = x^2 + 1$	$[0, 1]$
88. $f(x) = \frac{1}{x}$, $g(x) = x^2 - 4$	$[1, 2]$
89. $f(x) = \sin x$, $g(x) = \cos x$	$\left[0, \frac{\pi}{2}\right]$
90. $f(x) = \ln x$, $g(x) = x^3$	$[1, 4]$

True or False? **In Exercises 91–94, determine whether the statement is true or false. If it is false, explain why or give an example that shows it is false.**

91. $\lim_{x\to 0}\left[\frac{x^2 + x + 1}{x}\right] = \lim_{x\to 0}\left[\frac{2x + 1}{1}\right] = 1$

92. If $y = e^x/x^2$, then $y' = e^x/2x$.

93. If $p(x)$ is a polynomial, then $\lim_{x\to\infty}[p(x)/e^x] = 0$.

94. If $\lim_{x\to\infty}\frac{f(x)}{g(x)} = 1$, then $\lim_{x\to\infty}[f(x) - g(x)] = 0$.

95. ***Area*** Find the limit, as x approaches 0, of the ratio of the area of the triangle to the total shaded area in the figure.

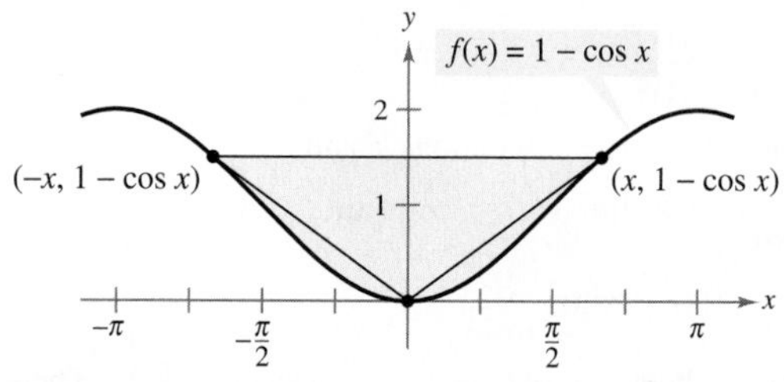

96. In Section 1.3, a geometric argument (see figure) was used to prove that

$$\lim_{\theta\to 0}\frac{\sin\theta}{\theta} = 1.$$

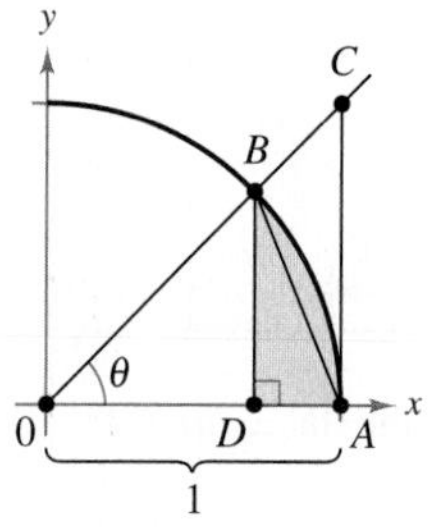

(a) Write the area of $\triangle ABD$ in terms of θ.

(b) Write the area of the shaded region in terms of θ.

(c) Write the ratio R of the area of $\triangle ABD$ to that of the shaded region.

(d) Find $\lim_{\theta\to 0} R$.

Continuous Functions **In Exercises 97 and 98, find the value of c that makes the function continuous at $x = 0$.**

97. $f(x) = \begin{cases} \dfrac{4x - 2\sin 2x}{2x^3}, & x \neq 0 \\ c, & x = 0 \end{cases}$

98. $f(x) = \begin{cases} (e^x + x)^{1/x}, & x \neq 0 \\ c, & x = 0 \end{cases}$

99. Find the values of a and b such that $\lim_{x\to 0}\frac{a - \cos bx}{x^2} = 2$.

100. Show that $\lim_{x\to\infty}\frac{x^n}{e^x} = 0$ for any integer $n > 0$.

101. (a) Let $f'(x)$ be continuous. Show that

$$\lim_{h\to 0} \frac{f(x+h) - f(x-h)}{2h} = f'(x).$$

(b) Explain the result of part (a) graphically.

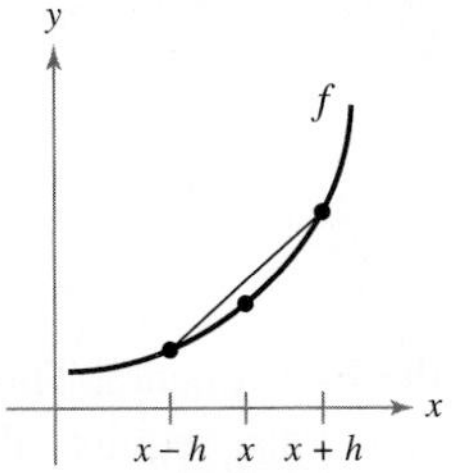

102. Let $f''(x)$ be continuous. Show that

$$\lim_{h\to 0} \frac{f(x+h) - 2f(x) + f(x-h)}{h^2} = f''(x).$$

103. Sketch the graph of

$$g(x) = \begin{cases} e^{-1/x^2}, & x \neq 0 \\ 0, & x = 0 \end{cases}$$

and determine $g'(0)$.

104. Use a graphing utility to graph

$$f(x) = \frac{x^k - 1}{k}$$

for $k = 1, 0.1$, and 0.01. Then evaluate the limit

$$\lim_{k\to 0^+} \frac{x^k - 1}{k}.$$

105. Consider the limit $\lim_{x\to 0^+} (-x \ln x)$.

(a) Describe the type of indeterminate form that is obtained by direct substitution.

(b) Evaluate the limit.

(c) Use a graphing utility to verify the result of part (b).

FOR FURTHER INFORMATION For a geometric approach to this exercise, see the article "A Geometric Proof of $\lim_{d\to 0^+} (-d \ln d) = 0$" by John H. Mathews in the *College Mathematics Journal*. To view this article, go to the website *www.matharticles.com*.

106. Prove that if $f(x) \geq 0$, $\lim_{x\to a} f(x) = 0$, and $\lim_{x\to a} g(x) = \infty$, then $\lim_{x\to a} f(x)^{g(x)} = 0$.

107. Prove that if $f(x) \geq 0$, $\lim_{x\to a} f(x) = 0$, and $\lim_{x\to a} g(x) = -\infty$, then $\lim_{x\to a} f(x)^{g(x)} = \infty$.

108. Prove the following generalization of the Mean Value Theorem. If f is twice differentiable on the closed interval $[a, b]$, then

$$f(b) - f(a) = f'(a)(b - a) - \int_a^b f''(t)(t - b)\, dt.$$

109. ***Indeterminate Forms*** Show that the indeterminate forms 0^0, ∞^0, and 1^∞ do not always have a value of 1 by evaluating each limit.

(a) $\lim_{x\to 0^+} x^{\ln 2/(1 + \ln x)}$

(b) $\lim_{x\to \infty} x^{\ln 2/(1 + \ln x)}$

(c) $\lim_{x\to 0} (x + 1)^{(\ln 2)/x}$

110. ***Calculus History*** In L'Hôpital's 1696 calculus textbook, he illustrated his rule using the limit of the function

$$f(x) = \frac{\sqrt{2a^3x - x^4} - a\sqrt[3]{a^2x}}{a - \sqrt[4]{ax^3}}$$

as x approaches a, $a > 0$. Find this limit.

111. Consider the function

$$h(x) = \frac{x + \sin x}{x}.$$

(a) Use a graphing utility to graph the function. Then use the *zoom* and *trace* features to investigate $\lim_{x\to\infty} h(x)$.

(b) Find $\lim_{x\to\infty} h(x)$ analytically by writing

$$h(x) = \frac{x}{x} + \frac{\sin x}{x}.$$

(c) Can you use L'Hôpital's Rule to find $\lim_{x\to\infty} h(x)$? Explain your reasoning.

Putnam Exam Challenge

112. Evaluate

$$\lim_{x\to\infty} \left[\frac{1}{x} \cdot \frac{a^x - 1}{a - 1}\right]^{1/x}$$

where $a > 0$, $a \neq 1$.

Section 8.8 Improper Integrals

- **Evaluate an improper integral that has an infinite limit of integration.**
- **Evaluate an improper integral that has an infinite discontinuity.**

Improper Integrals with Infinite Limits of Integration

The definition of a definite integral

$$\int_a^b f(x)\,dx$$

requires that the interval $[a, b]$ be finite. Furthermore, the Fundamental Theorem of Calculus, by which you have been evaluating definite integrals, requires that f be continuous on $[a, b]$. In this section you will study a procedure for evaluating integrals that do not satisfy these requirements—usually because either one or both of the limits of integration are infinite, or f has a finite number of infinite discontinuities in the interval $[a, b]$. Integrals that possess either property are **improper integrals.** Note that a function f is said to have an **infinite discontinuity** at c if, *from the right or left,*

$$\lim_{x\to c} f(x) = \infty \quad \text{or} \quad \lim_{x\to c} f(x) = -\infty.$$

To get an idea of how to evaluate an improper integral, consider the integral

$$\int_1^b \frac{dx}{x^2} = -\frac{1}{x}\Big]_1^b = -\frac{1}{b} + 1 = 1 - \frac{1}{b}$$

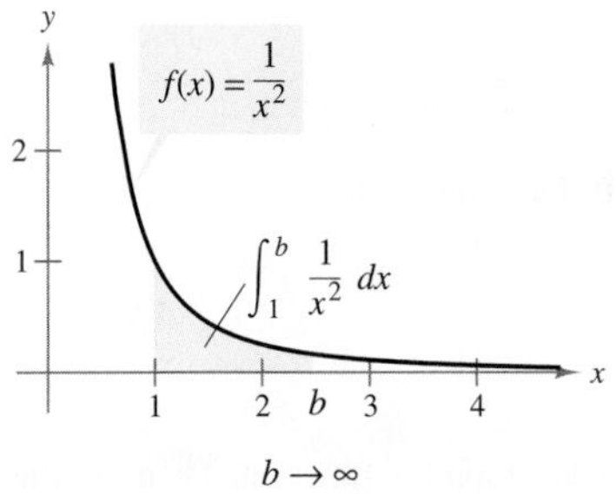

The unbounded region has an area of 1.
Figure 8.17

which can be interpreted as the area of the shaded region shown in Figure 8.17. Taking the limit as $b \to \infty$ produces

$$\int_1^\infty \frac{dx}{x^2} = \lim_{b\to\infty}\left(\int_1^b \frac{dx}{x^2}\right) = \lim_{b\to\infty}\left(1 - \frac{1}{b}\right) = 1.$$

This improper integral can be interpreted as the area of the *unbounded* region between the graph of $f(x) = 1/x^2$ and the x-axis (to the right of $x = 1$).

Definition of Improper Integrals with Infinite Integration Limits

1. If f is continuous on the interval $[a, \infty)$, then

$$\int_a^\infty f(x)\,dx = \lim_{b\to\infty}\int_a^b f(x)\,dx.$$

2. If f is continuous on the interval $(-\infty, b]$, then

$$\int_{-\infty}^b f(x)\,dx = \lim_{a\to-\infty}\int_a^b f(x)\,dx.$$

3. If f is continuous on the interval $(-\infty, \infty)$, then

$$\int_{-\infty}^\infty f(x)\,dx = \int_{-\infty}^c f(x)\,dx + \int_c^\infty f(x)\,dx$$

where c is any real number (see Exercise 110).

In the first two cases, the improper integral **converges** if the limit exists—otherwise, the improper integral **diverges.** In the third case, the improper integral on the left diverges if either of the improper integrals on the right diverges.

EXAMPLE 1 An Improper Integral That Diverges

Evaluate $\int_1^{\infty} \frac{dx}{x}$.

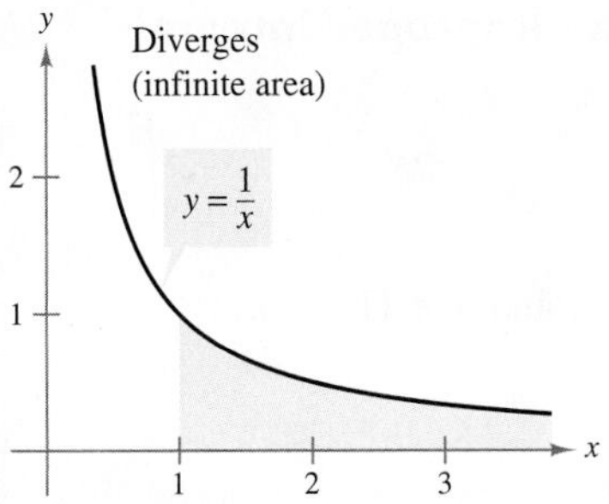

This unbounded region has an infinite area.
Figure 8.18

Solution

$$\begin{aligned}
\int_1^{\infty} \frac{dx}{x} &= \lim_{b\to\infty} \int_1^b \frac{dx}{x} && \text{Take limit as } b \to \infty. \\
&= \lim_{b\to\infty} \Big[\ln x\Big]_1^b && \text{Apply Log Rule.} \\
&= \lim_{b\to\infty} (\ln b - 0) && \text{Apply Fundamental Theorem of Calculus.} \\
&= \infty && \text{Evaluate limit.}
\end{aligned}$$

See Figure 8.18.

NOTE Try comparing the regions shown in Figures 8.17 and 8.18. They look similar, yet the region in Figure 8.17 has a finite area of 1 and the region in Figure 8.18 has an infinite area.

EXAMPLE 2 Improper Integrals That Converge

Evaluate each improper integral.

a. $\int_0^{\infty} e^{-x}\,dx$ **b.** $\int_0^{\infty} \frac{1}{x^2+1}\,dx$

Solution

a.
$$\begin{aligned}
\int_0^{\infty} e^{-x}\,dx &= \lim_{b\to\infty} \int_0^b e^{-x}\,dx \\
&= \lim_{b\to\infty} \Big[-e^{-x}\Big]_0^b \\
&= \lim_{b\to\infty} (-e^{-b} + 1) \\
&= 1
\end{aligned}$$

See Figure 8.19.

b.
$$\begin{aligned}
\int_0^{\infty} \frac{1}{x^2+1}\,dx &= \lim_{b\to\infty} \int_0^b \frac{1}{x^2+1}\,dx \\
&= \lim_{b\to\infty} \Big[\arctan x\Big]_0^b \\
&= \lim_{b\to\infty} \arctan b \\
&= \frac{\pi}{2}
\end{aligned}$$

See Figure 8.20.

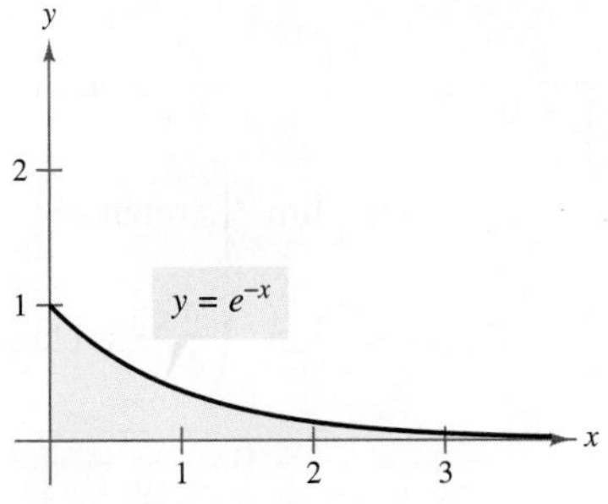

The area of the unbounded region is 1.
Figure 8.19

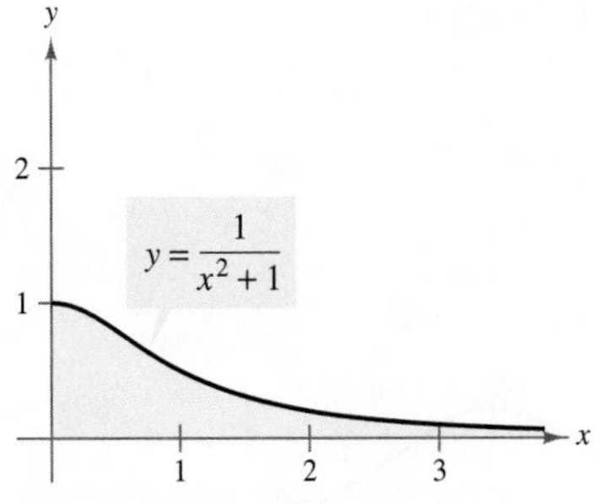

The area of the unbounded region is $\pi/2$.
Figure 8.20

In the following example, note how L'Hôpital's Rule can be used to evaluate an improper integral.

EXAMPLE 3 Using L'Hôpital's Rule with an Improper Integral

Evaluate $\int_1^{\infty} (1 - x)e^{-x}\, dx$.

Solution Use integration by parts, with $dv = e^{-x}\, dx$ and $u = (1 - x)$.

$$\begin{aligned}\int (1 - x)e^{-x}\, dx &= -e^{-x}(1 - x) - \int e^{-x}\, dx \\ &= -e^{-x} + xe^{-x} + e^{-x} + C \\ &= xe^{-x} + C\end{aligned}$$

Now, apply the definition of an improper integral.

$$\begin{aligned}\int_1^{\infty} (1 - x)e^{-x}\, dx &= \lim_{b\to\infty} \Big[xe^{-x} \Big]_1^b \\ &= \left(\lim_{b\to\infty} \frac{b}{e^b} \right) - \frac{1}{e}\end{aligned}$$

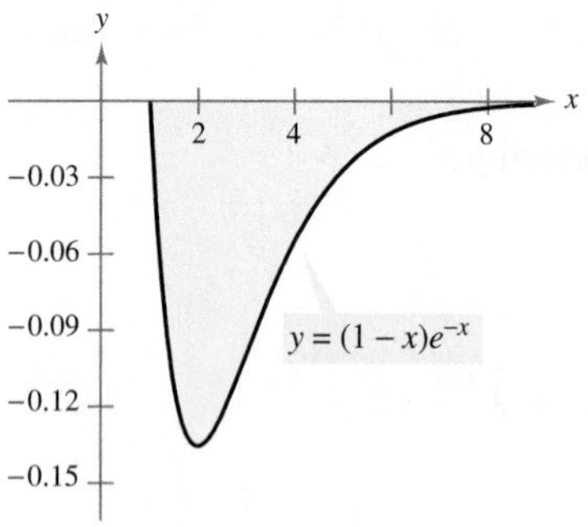

The area of the unbounded region is $|-1/e|$.
Figure 8.21

Finally, using L'Hôpital's Rule on the right-hand limit produces

$$\lim_{b\to\infty} \frac{b}{e^b} = \lim_{b\to\infty} \frac{1}{e^b} = 0$$

from which you can conclude that

$$\int_1^{\infty} (1 - x)e^{-x}\, dx = -\frac{1}{e}.$$

See Figure 8.21.

EXAMPLE 4 Infinite Upper and Lower Limits of Integration

Evaluate $\displaystyle\int_{-\infty}^{\infty} \frac{e^x}{1 + e^{2x}}\, dx$.

Solution Note that the integrand is continuous on $(-\infty, \infty)$. To evaluate the integral, you can break it into two parts, choosing $c = 0$ as a convenient value.

$$\begin{aligned}\int_{-\infty}^{\infty} \frac{e^x}{1 + e^{2x}}\, dx &= \int_{-\infty}^{0} \frac{e^x}{1 + e^{2x}}\, dx + \int_{0}^{\infty} \frac{e^x}{1 + e^{2x}}\, dx \\ &= \lim_{b\to-\infty} \Big[\arctan e^x \Big]_b^0 + \lim_{b\to\infty} \Big[\arctan e^x \Big]_0^b \\ &= \lim_{b\to-\infty} \left(\frac{\pi}{4} - \arctan e^b \right) + \lim_{b\to\infty} \left(\arctan e^b - \frac{\pi}{4} \right) \\ &= \frac{\pi}{4} - 0 + \frac{\pi}{2} - \frac{\pi}{4} \\ &= \frac{\pi}{2}\end{aligned}$$

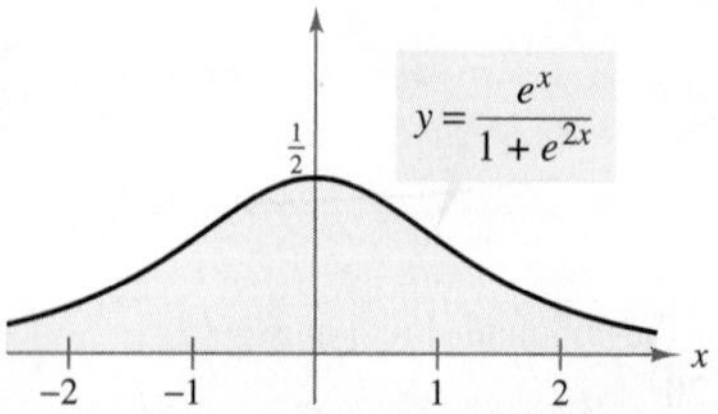

The area of the unbounded region is $\pi/2$.
Figure 8.22

See Figure 8.22.

EXAMPLE 5 Sending a Space Module into Orbit

In Example 3 of Section 7.5, you found that it would require 10,000 mile-tons of work to propel a 15-metric-ton space module to a height of 800 miles above Earth. How much work is required to propel the module an unlimited distance away from Earth's surface?

Solution At first you might think that an infinite amount of work would be required. But if this were the case, it would be impossible to send rockets into outer space. Because this has been done, the work required must be finite. You can determine the work in the following manner. Using the integral of Example 3, Section 7.5, replace the upper bound of 4800 miles by ∞ and write

$$\begin{aligned} W &= \int_{4000}^{\infty} \frac{240{,}000{,}000}{x^2}\,dx \\ &= \lim_{b\to\infty}\left[-\frac{240{,}000{,}000}{x}\right]_{4000}^{b} \\ &= \lim_{b\to\infty}\left(-\frac{240{,}000{,}000}{b} + \frac{240{,}000{,}000}{4000}\right) \\ &= 60{,}000 \text{ mile-tons} \\ &\approx 6.984 \times 10^{11} \text{ foot-pounds.} \end{aligned}$$

See Figure 8.23.

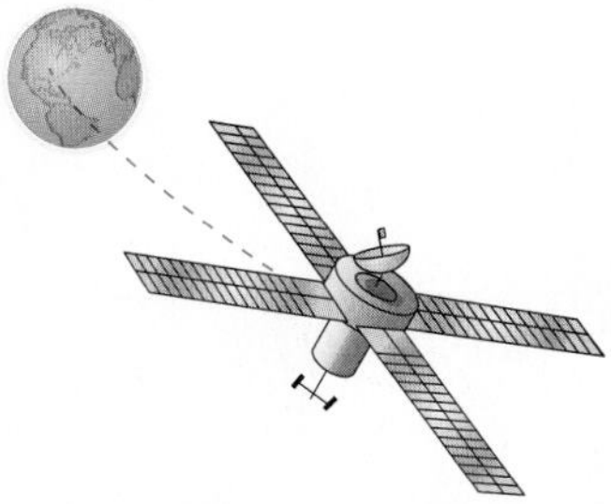

The work required to move a space module an unlimited distance away from Earth is approximately 6.984×10^{11} foot-pounds.
Figure 8.23

Improper Integrals with Infinite Discontinuities

The second basic type of improper integral is one that has an infinite discontinuity *at or between* the limits of integration.

Definition of Improper Integrals with Infinite Discontinuities

1. If f is continuous on the interval $[a, b)$ and has an infinite discontinuity at b, then

$$\int_a^b f(x)\,dx = \lim_{c\to b^-}\int_a^c f(x)\,dx.$$

2. If f is continuous on the interval $(a, b]$ and has an infinite discontinuity at a, then

$$\int_a^b f(x)\,dx = \lim_{c\to a^+}\int_c^b f(x)\,dx.$$

3. If f is continuous on the interval $[a, b]$, except for some c in (a, b) at which f has an infinite discontinuity, then

$$\int_a^b f(x)\,dx = \int_a^c f(x)\,dx + \int_c^b f(x)\,dx.$$

In the first two cases, the improper integral **converges** if the limit exists—otherwise, the improper integral **diverges.** In the third case, the improper integral on the left diverges if either of the improper integrals on the right diverges.

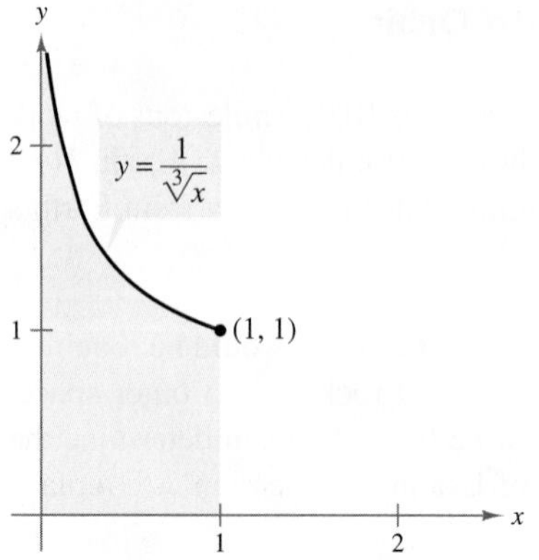

Infinite discontinuity at $x = 0$
Figure 8.24

EXAMPLE 6 An Improper Integral with an Infinite Discontinuity

Evaluate $\int_0^1 \frac{dx}{\sqrt[3]{x}}$.

Solution The integrand has an infinite discontinuity at $x = 0$, as shown in Figure 8.24. You can evaluate this integral as shown below.

$$\begin{aligned}\int_0^1 x^{-1/3}\,dx &= \lim_{b\to 0^+}\left[\frac{x^{2/3}}{2/3}\right]_b^1\\ &= \lim_{b\to 0^+}\frac{3}{2}(1 - b^{2/3})\\ &= \frac{3}{2}\end{aligned}$$

EXAMPLE 7 An Improper Integral That Diverges

Evaluate $\int_0^2 \frac{dx}{x^3}$.

Solution Because the integrand has an infinite discontinuity at $x = 0$, you can write

$$\begin{aligned}\int_0^2 \frac{dx}{x^3} &= \lim_{b\to 0^+}\left[-\frac{1}{2x^2}\right]_b^2\\ &= \lim_{b\to 0^+}\left(-\frac{1}{8} + \frac{1}{2b^2}\right)\\ &= \infty.\end{aligned}$$

So, you can conclude that the improper integral diverges.

EXAMPLE 8 An Improper Integral with an Interior Discontinuity

Evaluate $\int_{-1}^2 \frac{dx}{x^3}$.

Solution This integral is improper because the integrand has an infinite discontinuity at the interior point $x = 0$, as shown in Figure 8.25. So, you can write

$$\int_{-1}^2 \frac{dx}{x^3} = \int_{-1}^0 \frac{dx}{x^3} + \int_0^2 \frac{dx}{x^3}.$$

From Example 7 you know that the second integral diverges. So, the original improper integral also diverges.

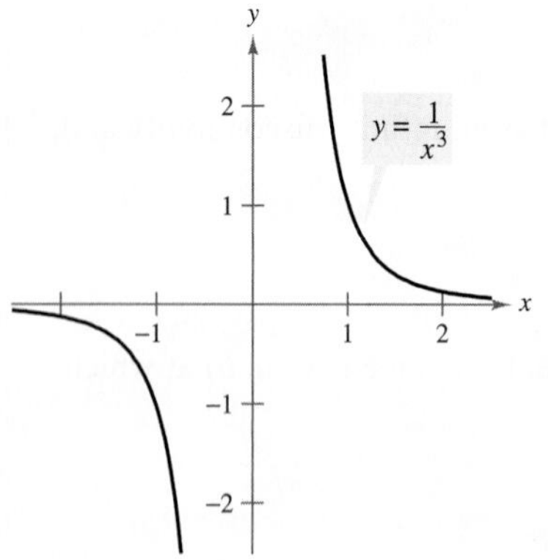

The improper integral $\int_{-1}^2 1/x^3\,dx$ diverges.
Figure 8.25

NOTE Remember to check for infinite discontinuities at interior points as well as endpoints when determining whether an integral is improper. For instance, if you had not recognized that the integral in Example 8 was improper, you would have obtained the *incorrect* result

$$\int_{-1}^2 \frac{dx}{x^3} = \left[\frac{-1}{2x^2}\right]_{-1}^2 = -\frac{1}{8} + \frac{1}{2} = \frac{3}{8}.$$ Incorrect evaluation

The integral in the next example is improper for *two* reasons. One limit of integration is infinite, and the integrand has an infinite discontinuity at the outer limit of integration.

EXAMPLE 9 A Doubly Improper Integral

Evaluate $\displaystyle\int_0^{\infty} \frac{dx}{\sqrt{x}(x+1)}$.

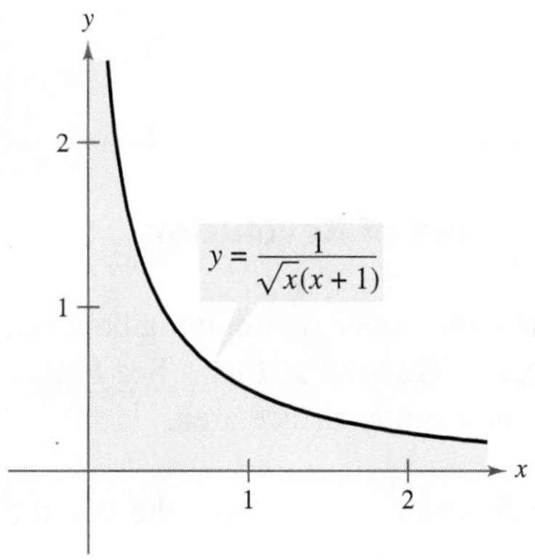

The area of the unbounded region is π.
Figure 8.26

Solution To evaluate this integral, split it at a convenient point (say, $x = 1$) and write

$$\begin{aligned}
\int_0^{\infty} \frac{dx}{\sqrt{x}(x+1)} &= \int_0^{1} \frac{dx}{\sqrt{x}(x+1)} + \int_1^{\infty} \frac{dx}{\sqrt{x}(x+1)} \\
&= \lim_{b\to 0^+}\Big[2\arctan\sqrt{x}\Big]_b^1 + \lim_{c\to\infty}\Big[2\arctan\sqrt{x}\Big]_1^c \\
&= 2\left(\frac{\pi}{4}\right) - 0 + 2\left(\frac{\pi}{2}\right) - 2\left(\frac{\pi}{4}\right) \\
&= \pi.
\end{aligned}$$

See Figure 8.26.

EXAMPLE 10 An Application Involving Arc Length

Use the formula for arc length to show that the circumference of the circle $x^2 + y^2 = 1$ is 2π.

Solution To simplify the work, consider the quarter circle given by $y = \sqrt{1 - x^2}$, where $0 \le x \le 1$. The function y is differentiable for any x in this interval except $x = 1$. Therefore, the arc length of the quarter circle is given by the improper integral

$$\begin{aligned}
s &= \int_0^1 \sqrt{1 + (y')^2}\, dx \\
&= \int_0^1 \sqrt{1 + \left(\frac{-x}{\sqrt{1-x^2}}\right)^2}\, dx \\
&= \int_0^1 \frac{dx}{\sqrt{1-x^2}}.
\end{aligned}$$

This integral is improper because it has an infinite discontinuity at $x = 1$. So, you can write

$$\begin{aligned}
s &= \int_0^1 \frac{dx}{\sqrt{1-x^2}} \\
&= \lim_{b\to 1^-}\Big[\arcsin x\Big]_0^b \\
&= \frac{\pi}{2} - 0 \\
&= \frac{\pi}{2}.
\end{aligned}$$

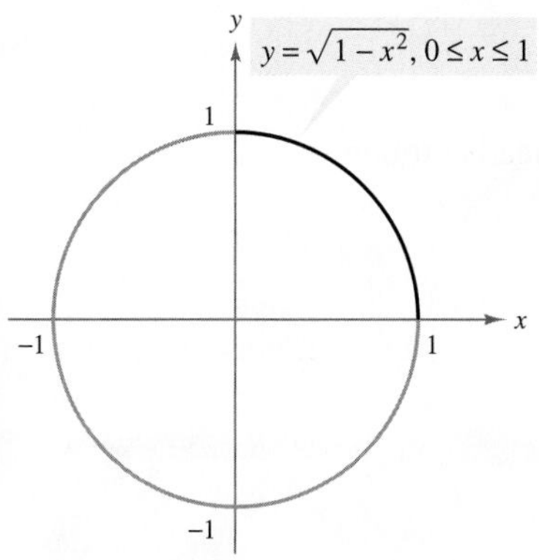

The circumference of the circle is 2π.
Figure 8.27

Finally, multiplying by 4, you can conclude that the circumference of the circle is $4s = 2\pi$, as shown in Figure 8.27.

This section concludes with a useful theorem describing the convergence or divergence of a common type of improper integral. The proof of this theorem is left as an exercise (see Exercise 49).

THEOREM 8.5 A Special Type of Improper Integral

$$\int_1^{\infty} \frac{dx}{x^p} = \begin{cases} \dfrac{1}{p-1}, & \text{if } p > 1 \\ \text{diverges}, & \text{if } p \le 1 \end{cases}$$

EXAMPLE 11 An Application Involving A Solid of Revolution

FOR FURTHER INFORMATION For further investigation of solids that have finite volumes and infinite surface areas, see the article "Supersolids: Solids Having Finite Volume and Infinite Surfaces" by William P. Love in *Mathematics Teacher*. To view this article, go to the website *www.matharticles.com.*

The solid formed by revolving (about the x-axis) the *unbounded* region lying between the graph of $f(x) = 1/x$ and the x-axis $(x \ge 1)$ is called **Gabriel's Horn.** (See Figure 8.28.) Show that this solid has a finite volume and an infinite surface area.

Solution Using the disk method and Theorem 8.5, you can determine the volume to be

$$\begin{aligned} V &= \pi \int_1^{\infty} \left(\frac{1}{x}\right)^2 dx && \text{Theorem 8.5, } p = 2 > 1 \\ &= \pi\left(\frac{1}{2-1}\right) = \pi. \end{aligned}$$

The surface area is given by

$$S = 2\pi \int_1^{\infty} f(x)\sqrt{1 + [f'(x)]^2}\, dx = 2\pi \int_1^{\infty} \frac{1}{x}\sqrt{1 + \frac{1}{x^4}}\, dx.$$

Because

$$\sqrt{1 + \frac{1}{x^4}} > 1$$

on the interval $[1, \infty)$, and the improper integral

$$\int_1^{\infty} \frac{1}{x}\, dx$$

diverges, you can conclude that the improper integral

$$\int_1^{\infty} \frac{1}{x}\sqrt{1 + \frac{1}{x^4}}\, dx$$

also diverges. (See Exercise 52.) So, the surface area is infinite.

FOR FURTHER INFORMATION To learn about another function that has a finite volume and an infinite surface area, see the article "Gabriel's Wedding Cake" by Julian F. Fleron in *The College Mathematics Journal*. To view this article, go to the website *www.matharticles.com.*

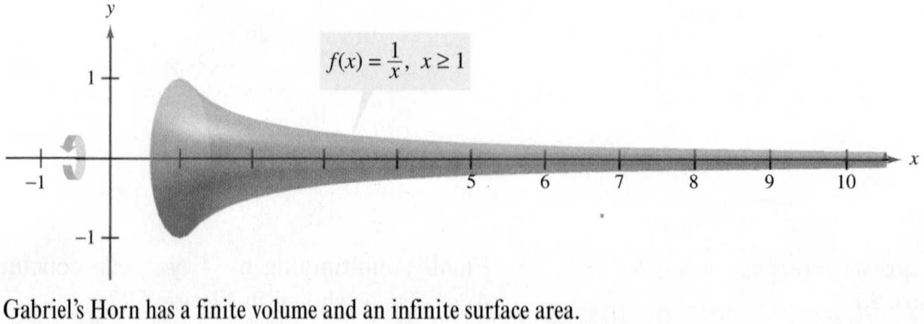

Gabriel's Horn has a finite volume and an infinite surface area.
Figure 8.28

Exercises for Section 8.8

See www.CalcChat.com for worked-out solutions to odd-numbered exercises.

In Exercises 1–4, decide whether the integral is improper. Explain your reasoning.

1. $\displaystyle\int_0^1 \frac{dx}{3x-2}$

2. $\displaystyle\int_1^3 \frac{dx}{x^2}$

3. $\displaystyle\int_0^1 \frac{2x-5}{x^2-5x+6}\,dx$

4. $\displaystyle\int_1^\infty \ln(x^2)\,dx$

In Exercises 5–10, explain why the integral is improper and determine whether it diverges or converges. Evaluate the integral if it converges.

5. $\displaystyle\int_0^4 \frac{1}{\sqrt{x}}\,dx$

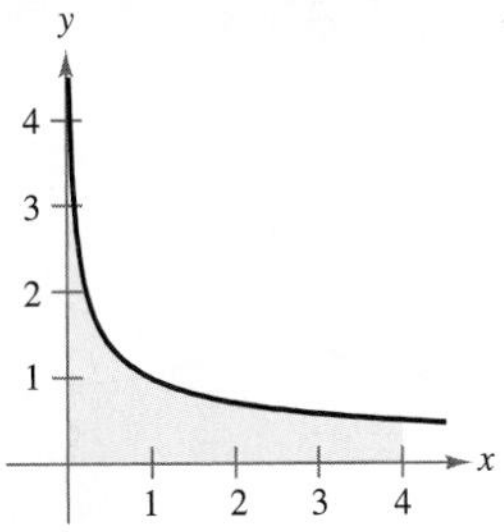

6. $\displaystyle\int_3^4 \frac{1}{(x-3)^{3/2}}\,dx$

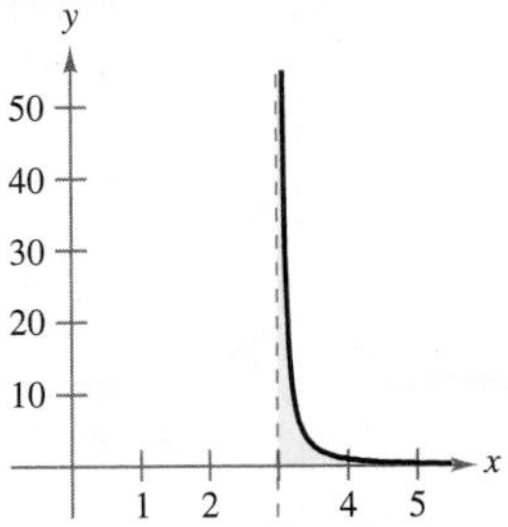

7. $\displaystyle\int_0^2 \frac{1}{(x-1)^2}\,dx$

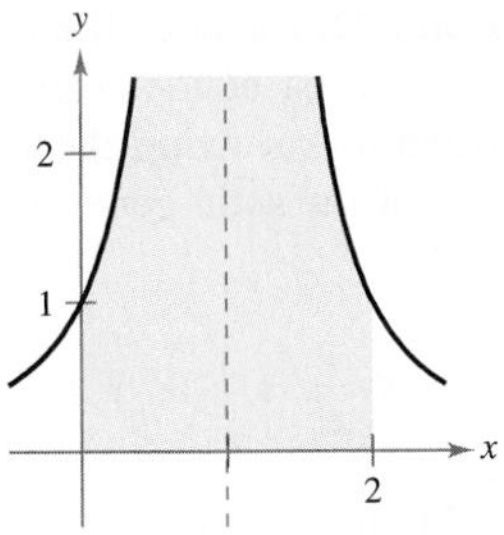

8. $\displaystyle\int_0^2 \frac{1}{(x-1)^{2/3}}\,dx$

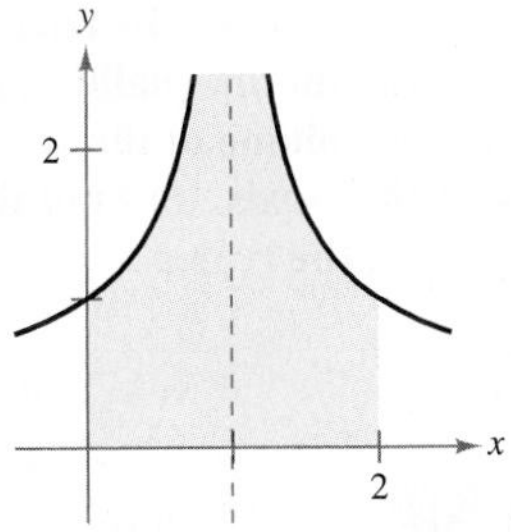

9. $\displaystyle\int_0^\infty e^{-x}\,dx$

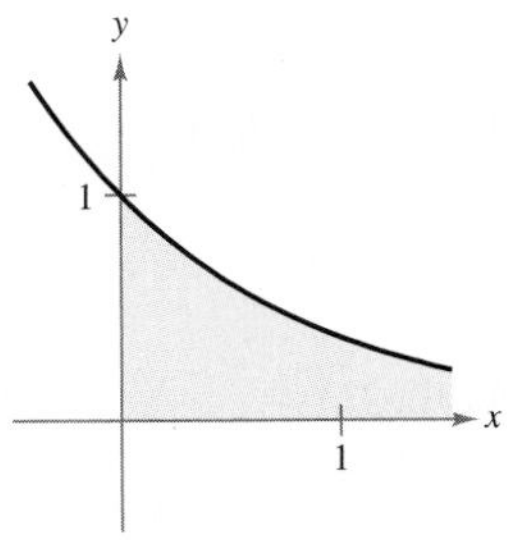

10. $\displaystyle\int_{-\infty}^0 e^{2x}\,dx$

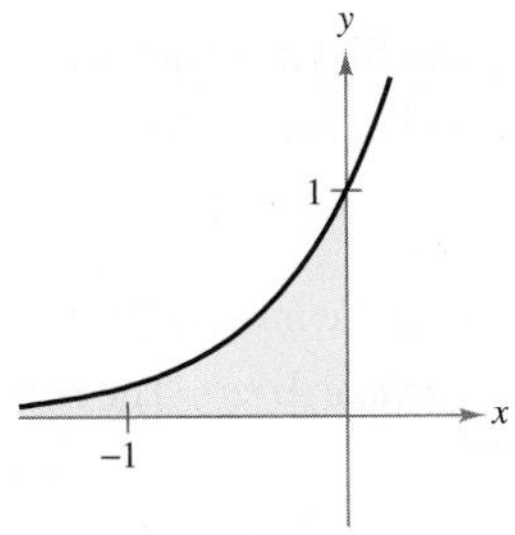

Writing **In Exercises 11–14, explain why the evaluation of the integral is *incorrect*. Use the integration capabilities of a graphing utility to attempt to evaluate the integral. Determine whether the utility gives the correct answer.**

11. $\displaystyle\int_{-1}^1 \frac{1}{x^2}\,dx = -2$

12. $\displaystyle\int_{-2}^2 \frac{-2}{(x-1)^3}\,dx = \frac{8}{9}$

13. $\displaystyle\int_0^\infty e^{-x}\,dx = 0$

14. $\displaystyle\int_0^\pi \sec x\,dx = 0$

In Exercises 15–32, determine whether the improper integral diverges or converges. Evaluate the integral if it converges.

15. $\displaystyle\int_1^\infty \frac{1}{x^2}\,dx$

16. $\displaystyle\int_1^\infty \frac{5}{x^3}\,dx$

17. $\displaystyle\int_1^\infty \frac{3}{\sqrt[3]{x}}\,dx$

18. $\displaystyle\int_1^\infty \frac{4}{\sqrt[4]{x}}\,dx$

19. $\displaystyle\int_{-\infty}^0 xe^{-2x}\,dx$

20. $\displaystyle\int_0^\infty xe^{-x/2}\,dx$

21. $\displaystyle\int_0^\infty x^2e^{-x}\,dx$

22. $\displaystyle\int_0^\infty (x-1)e^{-x}\,dx$

23. $\displaystyle\int_0^\infty e^{-x}\cos x\,dx$

24. $\displaystyle\int_0^\infty e^{-ax}\sin bx\,dx, \quad a > 0$

25. $\displaystyle\int_4^\infty \frac{1}{x(\ln x)^3}\,dx$

26. $\displaystyle\int_1^\infty \frac{\ln x}{x}\,dx$

27. $\displaystyle\int_{-\infty}^\infty \frac{2}{4+x^2}\,dx$

28. $\displaystyle\int_0^\infty \frac{x^3}{(x^2+1)^2}\,dx$

29. $\displaystyle\int_0^\infty \frac{1}{e^x+e^{-x}}\,dx$

30. $\displaystyle\int_0^\infty \frac{e^x}{1+e^x}\,dx$

31. $\displaystyle\int_0^\infty \cos \pi x\,dx$

32. $\displaystyle\int_0^\infty \sin\frac{x}{2}\,dx$

In Exercises 33–48, determine whether the improper integral diverges or converges. Evaluate the integral if it converges, and check your results with the results obtained by using the integration capabilities of a graphing utility.

33. $\displaystyle\int_0^1 \frac{1}{x^2}\,dx$

34. $\displaystyle\int_0^4 \frac{8}{x}\,dx$

35. $\displaystyle\int_0^8 \frac{1}{\sqrt[3]{8-x}}\,dx$

36. $\displaystyle\int_0^6 \frac{4}{\sqrt{6-x}}\,dx$

37. $\displaystyle\int_0^1 x\ln x\,dx$

38. $\displaystyle\int_0^e \ln x^2\,dx$

39. $\displaystyle\int_0^{\pi/2} \tan\theta\,d\theta$

40. $\displaystyle\int_0^{\pi/2} \sec\theta\,d\theta$

41. $\displaystyle\int_2^4 \frac{2}{x\sqrt{x^2-4}}\,dx$

42. $\displaystyle\int_0^2 \frac{1}{\sqrt{4-x^2}}\,dx$

43. $\displaystyle\int_2^4 \frac{1}{\sqrt{x^2-4}}\,dx$

44. $\displaystyle\int_0^2 \frac{1}{4-x^2}\,dx$

45. $\displaystyle\int_0^2 \frac{1}{\sqrt[3]{x-1}}\,dx$

46. $\displaystyle\int_1^3 \frac{2}{(x-2)^{8/3}}\,dx$

47. $\displaystyle\int_0^\infty \frac{4}{\sqrt{x}(x+6)}\,dx$

48. $\displaystyle\int_1^\infty \frac{1}{x\ln x}\,dx$

In Exercises 49 and 50, determine all values of p for which the improper integral converges.

49. $\int_1^{\infty} \frac{1}{x^p}\,dx$

50. $\int_0^{1} \frac{1}{x^p}\,dx$

51. Use mathematical induction to verify that the following integral converges for any positive integer n.

$$\int_0^{\infty} x^n e^{-x}\,dx$$

52. Given continuous functions f and g such that $0 \le f(x) \le g(x)$ on the interval $[a, \infty)$, prove the following.

(a) If $\int_a^{\infty} g(x)\,dx$ converges, then $\int_a^{\infty} f(x)\,dx$ converges.

(b) If $\int_a^{\infty} f(x)\,dx$ diverges, then $\int_a^{\infty} g(x)\,dx$ diverges.

In Exercises 53–62, use the results of Exercises 49–52 to determine whether the improper integral converges or diverges.

53. $\int_0^{1} \frac{1}{x^3}\,dx$

54. $\int_0^{1} \frac{1}{\sqrt[3]{x}}\,dx$

55. $\int_1^{\infty} \frac{1}{x^3}\,dx$

56. $\int_0^{\infty} x^4 e^{-x}\,dx$

57. $\int_1^{\infty} \frac{1}{x^2+5}\,dx$

58. $\int_2^{\infty} \frac{1}{\sqrt{x-1}}\,dx$

59. $\int_2^{\infty} \frac{1}{\sqrt[3]{x(x-1)}}\,dx$

60. $\int_1^{\infty} \frac{1}{\sqrt{x}(x+1)}\,dx$

61. $\int_0^{\infty} e^{-x^2}\,dx$

62. $\int_2^{\infty} \frac{1}{\sqrt{x}\ln x}\,dx$

Writing About Concepts

63. Describe the different types of improper integrals.

64. Define the terms *converges* and *diverges* when working with improper integrals.

65. Explain why $\int_{-1}^{1} \frac{1}{x^3}\,dx \neq 0$.

66. Consider the integral

$$\int_0^{3} \frac{10}{x^2-2x}\,dx.$$

To determine the convergence or divergence of the integral, how many improper integrals must be analyzed? What must be true of each of these integrals if the given integral converges?

Area **In Exercises 67–70, find the area of the unbounded shaded region.**

67. $y = e^x, \quad -\infty < x \le 1$

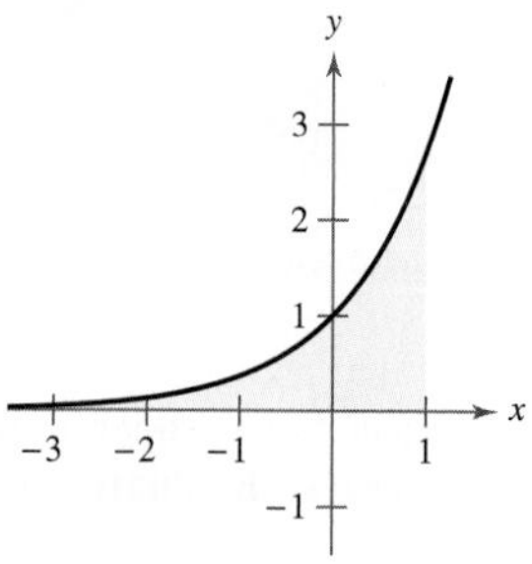

68. $y = -\ln x$

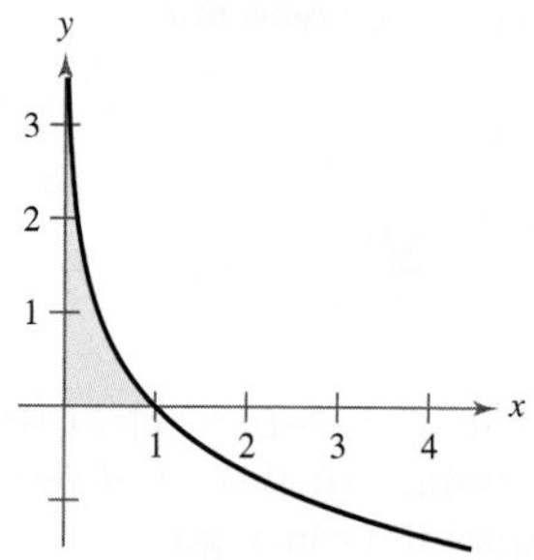

69. Witch of Agnesi:

$$y = \frac{1}{x^2+1}$$

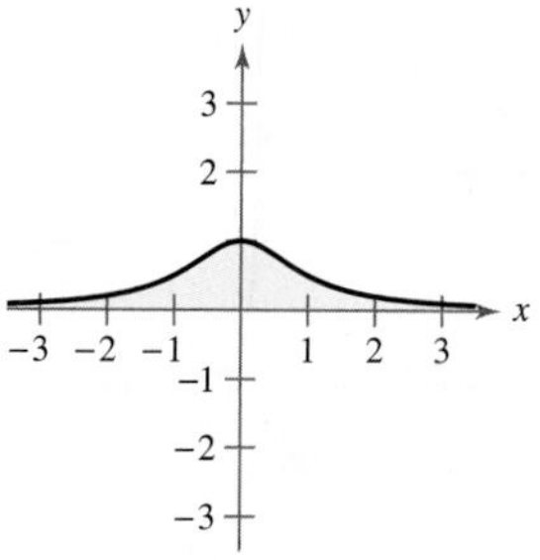

70. Witch of Agnesi:

$$y = \frac{8}{x^2+4}$$

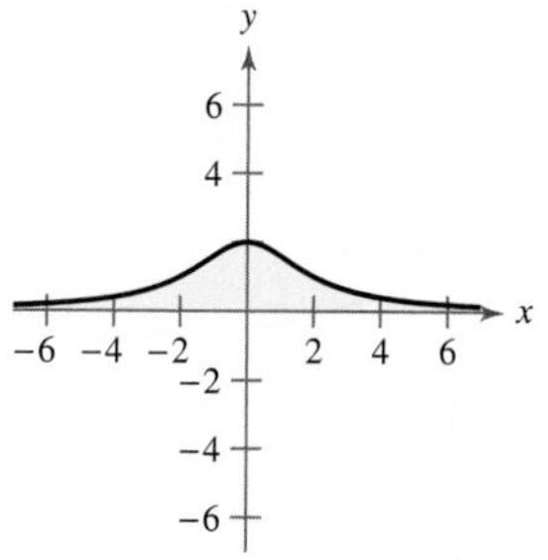

Area and Volume **In Exercises 71 and 72, consider the region satisfying the inequalities. (a) Find the area of the region. (b) Find the volume of the solid generated by revolving the region about the x-axis. (c) Find the volume of the solid generated by revolving the region about the y-axis.**

71. $y \le e^{-x}, \ y \ge 0, \ x \ge 0$

72. $y \le \frac{1}{x^2}, \ y \ge 0, \ x \ge 1$

73. ***Arc Length*** Sketch the graph of the hypocycloid of four cusps

$$x^{2/3} + y^{2/3} = 4$$

and find its perimeter.

74. ***Arc Length*** Find the arc length of the graph of

$$y = \sqrt{16 - x^2}$$

over the interval $[0, 4]$.

75. ***Surface Area*** The region bounded by

$$(x-2)^2 + y^2 = 1$$

is revolved about the y-axis to form a torus. Find the surface area of the torus.

76. ***Surface Area*** Find the area of the surface formed by revolving the graph of $y = 2e^{-x}$ on the interval $[0, \infty)$ about the x-axis.

Propulsion **In Exercises 77 and 78, use the weight of the rocket to answer each question. (Use 4000 miles as the radius of Earth and do not consider the effect of air resistance.)**

(a) How much work is required to propel the rocket an unlimited distance away from Earth's surface?

(b) How far has the rocket traveled when half the total work has occurred?

77. 5-ton rocket

78. 10-ton rocket

Probability **A nonnegative function f is called a *probability density function* if**

$$\int_{-\infty}^{\infty} f(t)\,dt = 1.$$

The probability that x lies between a and b is given by

$$P(a \le x \le b) = \int_a^b f(t)\,dt.$$

The expected value of x is given by

$$E(x) = \int_{-\infty}^{\infty} t f(t)\,dt.$$

In Exercises 79 and 80, (a) show that the nonnegative function is a probability density function, (b) find $P(0 \le x \le 4)$, and (c) find $E(x)$.

79. $f(t) = \begin{cases} \frac{1}{7}e^{-t/7}, & t \ge 0 \\ 0, & t < 0 \end{cases}$

80. $f(t) = \begin{cases} \frac{2}{5}e^{-2t/5}, & t \ge 0 \\ 0, & t < 0 \end{cases}$

Capitalized Cost **In Exercises 81 and 82, find the capitalized cost C of an asset (a) for $n = 5$ years, (b) for $n = 10$ years, and (c) forever. The capitalized cost is given by**

$$C = C_0 + \int_0^n c(t)e^{-rt}\,dt$$

where C_0 is the original investment, t is the time in years, r is the annual interest rate compounded continuously, and $c(t)$ is the annual cost of maintenance.

81. $C_0 = \$650{,}000$
$c(t) = \$25{,}000$
$r = 0.06$

82. $C_0 = \$650{,}000$
$c(t) = \$25{,}000(1 + 0.08t)$
$r = 0.06$

83. ***Electromagnetic Theory*** The magnetic potential P at a point on the axis of a circular coil is given by

$$P = \frac{2\pi NIr}{k}\int_c^{\infty} \frac{1}{(r^2 + x^2)^{3/2}}\,dx$$

where N, I, r, k, and c are constants. Find P.

84. ***Gravitational Force*** A "semi-infinite" uniform rod occupies the nonnegative x-axis. The rod has a linear density δ which means that a segment of length dx has a mass of $\delta\,dx$. A particle of mass m is located at the point $(-a, 0)$. The gravitational force F that the rod exerts on the mass is given by

$$F = \int_0^{\infty} \frac{GM\delta}{(a + x)^2}\,dx$$

where G is the gravitational constant. Find F.

True or False? **In Exercises 85–88, determine whether the statement is true or false. If it is false, explain why or give an example that shows it is false.**

85. If f is continuous on $[0, \infty)$ and $\lim_{x\to\infty} f(x) = 0$, then $\int_0^{\infty} f(x)\,dx$ converges.

86. If f is continuous on $[0, \infty)$ and $\int_0^{\infty} f(x)\,dx$ diverges, then $\lim_{x\to\infty} f(x) \ne 0$.

87. If f' is continuous on $[0, \infty)$ and $\lim_{x\to\infty} f(x) = 0$, then $\int_0^{\infty} f'(x)\,dx = -f(0)$.

88. If the graph of f is symmetric with respect to the origin or the y-axis, then $\int_0^{\infty} f(x)\,dx$ converges if and only if $\int_{-\infty}^{\infty} f(x)\,dx$ converges.

89. ***Writing***

(a) The improper integrals

$$\int_1^{\infty} \frac{1}{x}\,dx \quad \text{and} \quad \int_1^{\infty} \frac{1}{x^2}\,dx$$

diverge and converge, respectively. Describe the essential differences between the integrands that cause one integral to converge and the other to diverge.

(b) Sketch a graph of the function $y = \sin x/x$ over the interval $(1, \infty)$. Use your knowledge of the definite integral to make an inference as to whether or not the integral

$$\int_1^{\infty} \frac{\sin x}{x}\,dx$$

converges. Give reasons for your answer.

(c) Use one iteration of integration by parts on the integral in part (b) to determine its divergence or convergence.

90. ***Exploration*** Consider the integral

$$\int_0^{\pi/2} \frac{4}{1 + (\tan x)^n}\,dx$$

where n is a positive integer.

(a) Is the integral improper? Explain.

(b) Use a graphing utility to graph the integrand for $n = 2$, 4, 8, and 12.

(c) Use the graphs to approximate the integral as $n \to \infty$.

(d) Use a computer algebra system to evaluate the integral for the values of n in part (b). Make a conjecture about the value of the integral for any positive integer n. Compare your results with your answer in part (c).

91. ***The Gamma Function*** The Gamma Function $\Gamma(n)$ is defined by

$$\Gamma(n) = \int_0^{\infty} x^{n-1}e^{-x}\,dx, \quad n > 0.$$

(a) Find $\Gamma(1)$, $\Gamma(2)$, and $\Gamma(3)$.

(b) Use integration by parts to show that $\Gamma(n+1) = n\Gamma(n)$.

(c) Write $\Gamma(n)$ using factorial notation where n is a positive integer.

92. Prove that $I_n = \left(\dfrac{n-1}{n+2}\right)I_{n-1}$, where

$$I_n = \int_0^{\infty} \frac{x^{2n-1}}{(x^2+1)^{n+3}}\,dx, \quad n \geq 1.$$

Then evaluate each integral.

(a) $\displaystyle\int_0^{\infty} \frac{x}{(x^2+1)^4}\,dx$

(b) $\displaystyle\int_0^{\infty} \frac{x^3}{(x^2+1)^5}\,dx$

(c) $\displaystyle\int_0^{\infty} \frac{x^5}{(x^2+1)^6}\,dx$

Laplace Transforms **Let $f(t)$ be a function defined for all positive values of t. The Laplace Transform of $f(t)$ is defined by**

$$F(s) = \int_0^{\infty} e^{-st} f(t)\,dt$$

if the improper integral exists. Laplace Transforms are used to solve differential equations. In Exercises 93–100, find the Laplace Transform of the function.

93. $f(t) = 1$

94. $f(t) = t$

95. $f(t) = t^2$

96. $f(t) = e^{at}$

97. $f(t) = \cos at$

98. $f(t) = \sin at$

99. $f(t) = \cosh at$

100. $f(t) = \sinh at$

101. ***Normal Probability*** The mean height of American men between 18 and 24 years old is 70 inches, and the standard deviation is 3 inches. An 18- to 24-year-old man is chosen at random from the population. The probability that he is 6 feet tall or taller is

$$P(72 \leq x < \infty) = \int_{72}^{\infty} \frac{1}{3\sqrt{2\pi}} e^{-(x-70)^2/18}\,dx.$$

(Source: National Center for Health Statistics)

(a) Use a graphing utility to graph the integrand. Use the graphing utility to convince yourself that the area between the x-axis and the integrand is 1.

(b) Use a graphing utility to approximate $P(72 \leq x < \infty)$.

(c) Approximate $0.5 - P(70 \leq x \leq 72)$ using a graphing utility. Use the graph in part (a) to explain why this result is the same as the answer in part (b).

102. (a) Sketch the semicircle $y = \sqrt{4 - x^2}$.

(b) Explain why

$$\int_{-2}^{2} \frac{2\,dx}{\sqrt{4-x^2}} = \int_{-2}^{2} \sqrt{4-x^2}\,dx$$

without evaluating either integral.

103. For what value of c does the integral

$$\int_0^{\infty} \left(\frac{1}{\sqrt{x^2+1}} - \frac{c}{x+1}\right) dx$$

converge? Evaluate the integral for this value of c.

104. For what value of c does the integral

$$\int_1^{\infty} \left(\frac{cx}{x^2+2} - \frac{1}{3x}\right) dx$$

converge? Evaluate the integral for this value of c.

105. ***Volume*** Find the volume of the solid generated by revolving the region bounded by the graph of f about the x-axis.

$$f(x) = \begin{cases} x \ln x, & 0 < x \leq 2 \\ 0, & x = 0 \end{cases}$$

106. ***Volume*** Find the volume of the solid generated by revolving the unbounded region lying between $y = -\ln x$ and the y-axis $(y \geq 0)$ about the x-axis.

u-Substitution **In Exercises 107 and 108, rewrite the improper integral as a proper integral using the given u-substitution. Then use the Trapezoidal Rule with $n = 5$ to approximate the integral.**

107. $\displaystyle\int_0^1 \frac{\sin x}{\sqrt{x}}\,dx, \quad u = \sqrt{x}$

108. $\displaystyle\int_0^1 \frac{\cos x}{\sqrt{1-x}}\,dx, \quad u = \sqrt{1-x}$

109. (a) Use a graphing utility to graph the function $y = e^{-x^2}$.

(b) Show that $\displaystyle\int_0^{\infty} e^{-x^2}\,dx = \int_0^1 \sqrt{-\ln y}\,dy$.

110. Let $\displaystyle\int_{-\infty}^{\infty} f(x)\,dx$ be convergent and let a and b be real numbers where $a \neq b$. Show that

$$\int_{-\infty}^{a} f(x)\,dx + \int_a^{\infty} f(x)\,dx = \int_{-\infty}^{b} f(x)\,dx + \int_b^{\infty} f(x)\,dx.$$

Review Exercises for Chapter 8

See www.CalcChat.com for worked-out solutions to odd-numbered exercises.

In Exercises 1–8, use the basic integration rules to find or evaluate the integral.

1. $\int x\sqrt{x^2 - 1}\,dx$

2. $\int xe^{x^2-1}\,dx$

3. $\int \frac{x}{x^2 - 1}\,dx$

4. $\int \frac{x}{\sqrt{1 - x^2}}\,dx$

5. $\int_1^e \frac{\ln(2x)}{x}\,dx$

6. $\int_{3/2}^{2} 2x\sqrt{2x - 3}\,dx$

7. $\int \frac{16}{\sqrt{16 - x^2}}\,dx$

8. $\int \frac{x^4 + 2x^2 + x + 1}{(x^2 + 1)^2}\,dx$

In Exercises 9–16, use integration by parts to find the integral.

9. $\int e^{2x} \sin 3x\,dx$

10. $\int (x^2 - 1)e^x\,dx$

11. $\int x\sqrt{x - 5}\,dx$

12. $\int \arctan 2x\,dx$

13. $\int x^2 \sin 2x\,dx$

14. $\int \ln\sqrt{x^2 - 1}\,dx$

15. $\int x \arcsin 2x\,dx$

16. $\int e^x \arctan e^x\,dx$

In Exercises 17–22, find the trigonometric integral.

17. $\int \cos^3(\pi x - 1)\,dx$

18. $\int \sin^2 \frac{\pi x}{2}\,dx$

19. $\int \sec^4 \frac{x}{2}\,dx$

20. $\int \tan\theta \sec^4\theta\,d\theta$

21. $\int \frac{1}{1 - \sin\theta}\,d\theta$

22. $\int \cos 2\theta(\sin\theta + \cos\theta)^2\,d\theta$

Area **In Exercises 23 and 24, find the area of the region.**

23. $y = \sin^4 x$

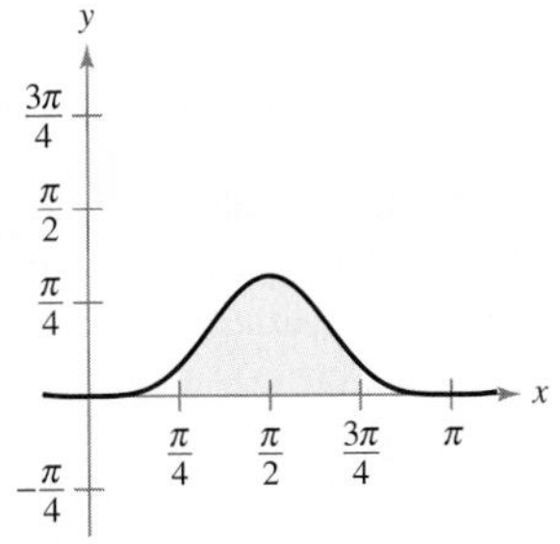

24. $y = \cos(3x)\cos x$

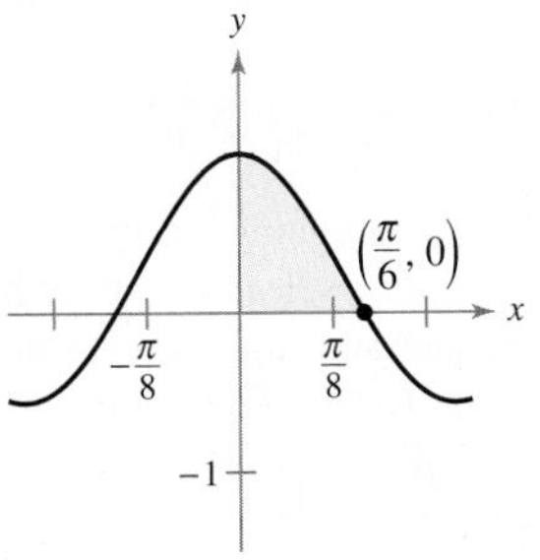

In Exercises 25–30, use trigonometric substitution to find or evaluate the integral.

25. $\int \frac{-12}{x^2\sqrt{4 - x^2}}\,dx$

26. $\int \frac{\sqrt{x^2 - 9}}{x}\,dx,\quad x > 3$

27. $\int \frac{x^3}{\sqrt{4 + x^2}}\,dx$

28. $\int \sqrt{9 - 4x^2}\,dx$

29. $\int_{-2}^{0} \sqrt{4 - x^2}\,dx$

30. $\int_0^{\pi/2} \frac{\sin\theta}{1 + 2\cos^2\theta}\,d\theta$

In Exercises 31 and 32, find the integral using each method.

31. $\int \frac{x^3}{\sqrt{4 + x^2}}\,dx$

(a) Trigonometric substitution

(b) Substitution: $u^2 = 4 + x^2$

(c) Integration by parts: $dv = \left(x/\sqrt{4 + x^2}\right)dx$

32. $\int x\sqrt{4 + x}\,dx$

(a) Trigonometric substitution

(b) Substitution: $u^2 = 4 + x$

(c) Substitution: $u = 4 + x$

(d) Integration by parts: $dv = \sqrt{4 + x}\,dx$

In Exercises 33–38, use partial fractions to find the integral.

33. $\int \frac{x - 28}{x^2 - x - 6}\,dx$

34. $\int \frac{2x^3 - 5x^2 + 4x - 4}{x^2 - x}\,dx$

35. $\int \frac{x^2 + 2x}{x^3 - x^2 + x - 1}\,dx$

36. $\int \frac{4x - 2}{3(x - 1)^2}\,dx$

37. $\int \frac{x^2}{x^2 + 2x - 15}\,dx$

38. $\int \frac{\sec^2\theta}{\tan\theta(\tan\theta - 1)}\,d\theta$

In Exercises 39–46, use integration tables to find or evaluate the integral.

39. $\int \frac{x}{(2 + 3x)^2}\,dx$

40. $\int \frac{x}{\sqrt{2 + 3x}}\,dx$

41. $\int_0^{\sqrt{\pi}/2} \frac{x}{1 + \sin x^2}\,dx$

42. $\int_0^1 \frac{x}{1 + e^{x^2}}\,dx$

43. $\int \frac{x}{x^2 + 4x + 8}\,dx$

44. $\int \frac{3}{2x\sqrt{9x^2 - 1}}\,dx,\quad x > \frac{1}{3}$

45. $\int \frac{1}{\sin \pi x \cos \pi x}\,dx$

46. $\int \frac{1}{1 + \tan \pi x}\,dx$

47. Verify the reduction formula

$$\int (\ln x)^n\,dx = x(\ln x)^n - n\int (\ln x)^{n-1}\,dx.$$

48. Verify the reduction formula

$$\int \tan^n x\,dx = \frac{1}{n - 1}\tan^{n-1} x - \int \tan^{n-2} x\,dx.$$

In Exercises 49–56, find the integral using any method.

49. $\int \theta \sin \theta \cos \theta \, d\theta$

50. $\int \frac{\csc \sqrt{2x}}{\sqrt{x}} \, dx$

51. $\int \frac{x^{1/4}}{1 + x^{1/2}} \, dx$

52. $\int \sqrt{1 + \sqrt{x}} \, dx$

53. $\int \sqrt{1 + \cos x} \, dx$

54. $\int \frac{3x^3 + 4x}{(x^2 + 1)^2} \, dx$

55. $\int \cos x \ln(\sin x) \, dx$

56. $\int (\sin \theta + \cos \theta)^2 \, d\theta$

In Exercises 57–60, solve the differential equation using any method.

57. $\frac{dy}{dx} = \frac{9}{x^2 - 9}$

58. $\frac{dy}{dx} = \frac{\sqrt{4 - x^2}}{2x}$

59. $y' = \ln(x^2 + x)$

60. $y' = \sqrt{1 - \cos \theta}$

In Exercises 61–66, evaluate the definite integral using any method. Use a graphing utility to verify your result.

61. $\int_2^{\sqrt{5}} x(x^2 - 4)^{3/2} \, dx$

62. $\int_0^1 \frac{x}{(x - 2)(x - 4)} \, dx$

63. $\int_1^4 \frac{\ln x}{x} \, dx$

64. $\int_0^2 xe^{3x} \, dx$

65. $\int_0^{\pi} x \sin x \, dx$

66. $\int_0^3 \frac{x}{\sqrt{1 + x}} \, dx$

Area **In Exercises 67 and 68, find the area of the region.**

67. $y = x\sqrt{4 - x}$

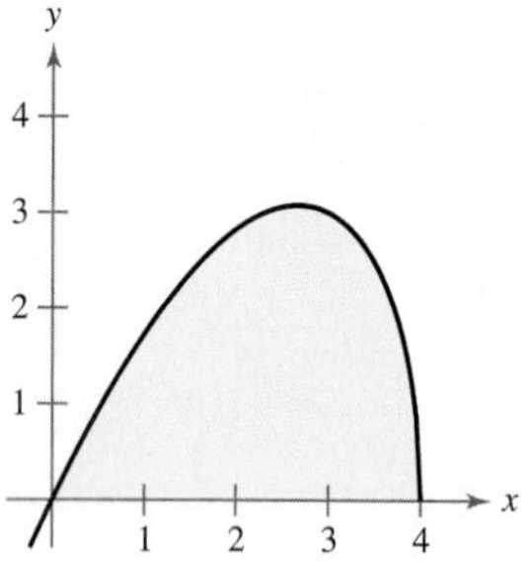

68. $y = \frac{1}{25 - x^2}$

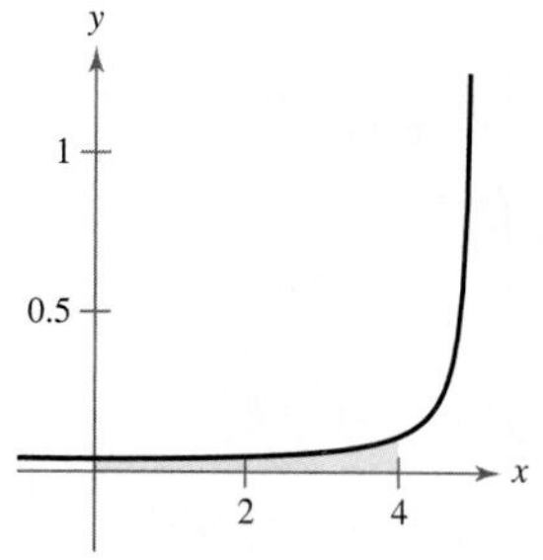

Centroid **In Exercises 69 and 70, find the centroid of the region bounded by the graphs of the equations.**

69. $y = \sqrt{1 - x^2}, \quad y = 0$

70. $(x - 1)^2 + y^2 = 1, \quad (x - 4)^2 + y^2 = 4$

Arc Length **In Exercises 71 and 72, approximate to two decimal places the arc length of the curve over the given interval.**

	Function	*Interval*
71.	$y = \sin x$	$[0, \pi]$
72.	$y = \sin^2 x$	$[0, \pi]$

In Exercises 73–80, use L'Hôpital's Rule to evaluate the limit.

73. $\lim_{x \to 1} \frac{(\ln x)^2}{x - 1}$

74. $\lim_{x \to 0} \frac{\sin \pi x}{\sin 2\pi x}$

75. $\lim_{x \to \infty} \frac{e^{2x}}{x^2}$

76. $\lim_{x \to \infty} xe^{-x^2}$

77. $\lim_{x \to \infty} (\ln x)^{2/x}$

78. $\lim_{x \to 1^+} (x - 1)^{\ln x}$

79. $\lim_{n \to \infty} 1000\left(1 + \frac{0.09}{n}\right)^n$

80. $\lim_{x \to 1^+} \left(\frac{2}{\ln x} - \frac{2}{x - 1}\right)$

In Exercises 81–86, determine whether the improper integral diverges or converges. Evaluate the integral if it converges.

81. $\int_0^{16} \frac{1}{\sqrt[4]{x}} \, dx$

82. $\int_0^1 \frac{6}{x - 1} \, dx$

83. $\int_1^{\infty} x^2 \ln x \, dx$

84. $\int_0^{\infty} \frac{e^{-1/x}}{x^2} \, dx$

85. $\int_1^{\infty} \frac{\ln x}{x^2} \, dx$

86. $\int_1^{\infty} \frac{1}{\sqrt[4]{x}} \, dx$

87. ***Present Value*** The board of directors of a corporation is calculating the price to pay for a business that is forecast to yield a continuous flow of profit of \$500,000 per year. If money will earn a nominal rate of 5% per year compounded continuously, what is the present value of the business

(a) for 20 years?

(b) forever (in perpetuity)?

(*Note:* The present value for t_0 years is $\int_0^{t_0} 500{,}000e^{-0.05t} \, dt$.)

88. ***Volume*** Find the volume of the solid generated by revolving the region bounded by the graphs of $y = xe^{-x}$, $y = 0$, and $x = 0$ about the x-axis.

89. ***Probability*** The average lengths (from beak to tail) of different species of warblers in the eastern United States are approximately normally distributed with a mean of 12.9 centimeters and a standard deviation of 0.95 centimeter (see figure). The probability that a randomly selected warbler has a length between a and b centimeters is

$$P(a \leq x \leq b) = \frac{1}{0.95\sqrt{2\pi}} \int_a^b e^{-(x - 12.9)^2/2(0.95)^2} \, dx.$$

Use a graphing utility to approximate the probability that a randomly selected warbler has a length of (a) 13 centimeters or greater and (b) 15 centimeters or greater. *(Source: Peterson's Field Guide: Eastern Birds)*

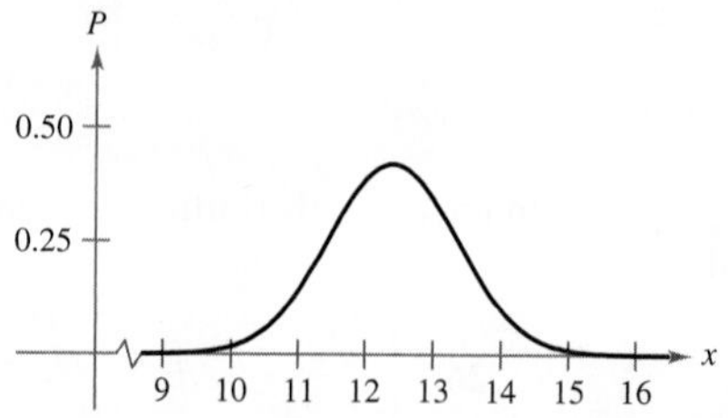

P.S. Problem Solving

See www.CalcChat.com for worked-out solutions to odd-numbered exercises.

1. (a) Evaluate the integrals

$$\int_{-1}^{1} (1 - x^2)\, dx \quad \text{and} \quad \int_{-1}^{1} (1 - x^2)^2\, dx.$$

(b) Use Wallis's Formulas to prove that

$$\int_{-1}^{1} (1 - x^2)^n\, dx = \frac{2^{2n+1}(n!)^2}{(2n + 1)!}$$

for all positive integers n.

2. (a) Evaluate the integrals $\int_0^1 \ln x\, dx$ and $\int_0^1 (\ln x)^2\, dx$.

(b) Prove that

$$\int_0^1 (\ln x)^n\, dx = (-1)^n n!$$

for all positive integers n.

3. Find the value of the positive constant c such that

$$\lim_{x\to\infty} \left(\frac{x + c}{x - c}\right)^x = 9.$$

4. Find the value of the positive constant c such that

$$\lim_{x\to\infty} \left(\frac{x - c}{x + c}\right)^x = \frac{1}{4}.$$

5. In the figure, the line $x = 1$ is tangent to the unit circle at A. The length of segment QA equals the length of the circular arc $\widehat{PA}$. Show that the length of segment OR approaches 2 as P approaches A.

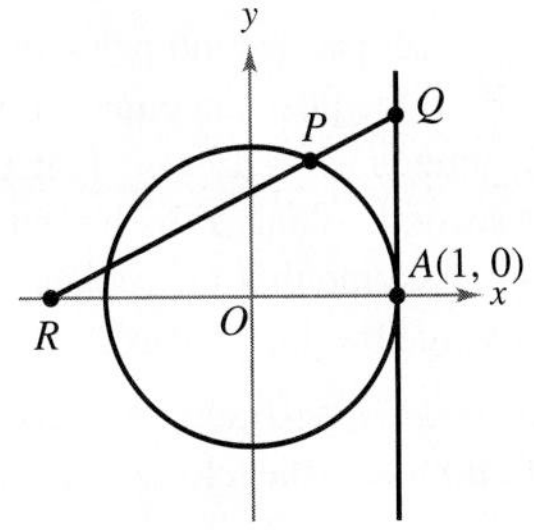

6. In the figure, the segment BD is the height of $\triangle OAB$. Let R be the ratio of the area of $\triangle DAB$ to that of the shaded region formed by deleting $\triangle OAB$ from the circular sector subtended by angle θ. Find $\lim_{\theta\to 0^+} R$.

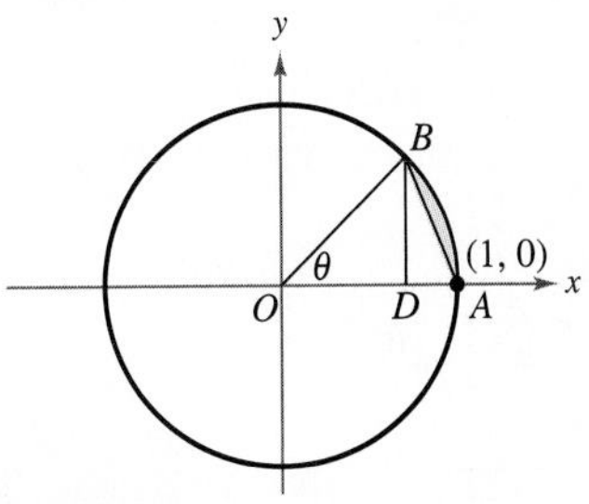

7. Consider the problem of finding the area of the region bounded by the x-axis, the line $x = 4$, and the curve

$$y = \frac{x^2}{(x^2 + 9)^{3/2}}.$$

(a) Use a graphing utility to graph the region and approximate its area.

(b) Use an appropriate trigonometric substitution to find the exact area.

(c) Use the substitution $x = 3 \sinh u$ to find the exact area and verify that you obtain the same answer as in part (b).

8. Use the substitution $u = \tan \frac{x}{2}$ to find the area of the shaded region under the graph of $y = \dfrac{1}{2 + \cos x}$, $0 \le x \le \pi/2$ (see figure).

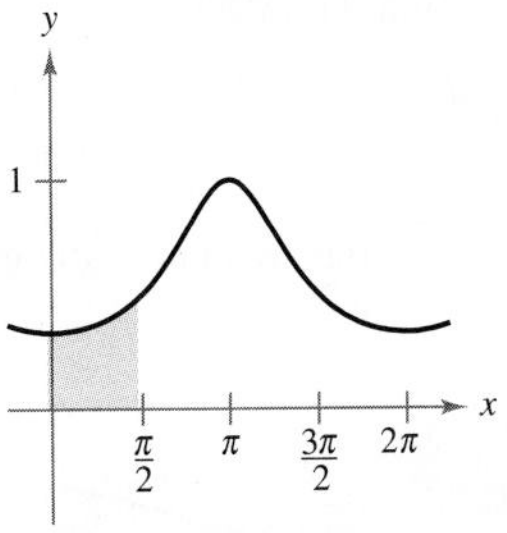

9. Find the arc length of the graph of the function $y = \ln(1 - x^2)$ on the interval $0 \le x \le \frac{1}{2}$ (see figure).

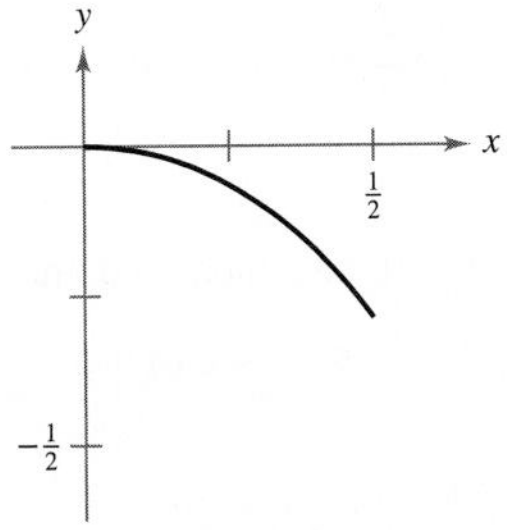

10. Find the centroid of the region above the x-axis and bounded above by the curve $y = e^{-c^2x^2}$, where c is a positive constant (see figure).

$\left(\textit{Hint:} \text{ Show that } \int_0^\infty e^{-c^2x^2}\, dx = \frac{1}{c}\int_0^\infty e^{-x^2}\, dx.\right)$

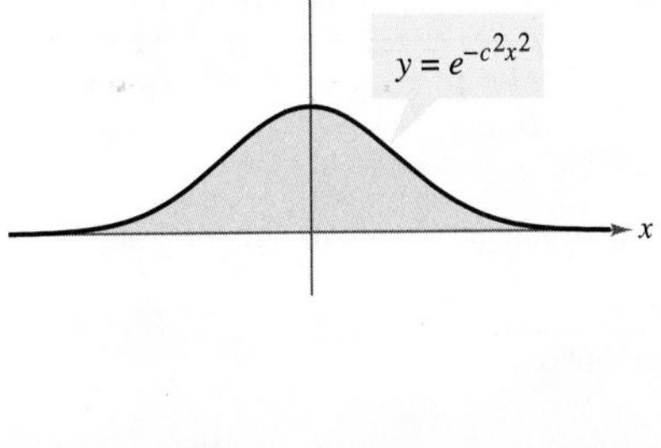

11. Some elementary functions, such as $f(x) = \sin(x^2)$, do not have antiderivatives that are elementary functions. Joseph Liouville proved that

$$\int \frac{e^x}{x}\,dx$$

does not have an elementary antiderivative. Use this fact to prove that

$$\int \frac{1}{\ln x}\,dx$$

is not elementary.

12. (a) Let $y = f^{-1}(x)$ be the inverse function of f. Use integration by parts to derive the formula

$$\int f^{-1}(x)\,dx = xf^{-1}(x) - \int f(y)\,dy.$$

(b) Use the formula in part (a) to find the integral

$$\int \arcsin x\,dx.$$

(c) Use the formula in part (a) to find the area under the graph of $y = \ln x$, $1 \le x \le e$ (see figure).

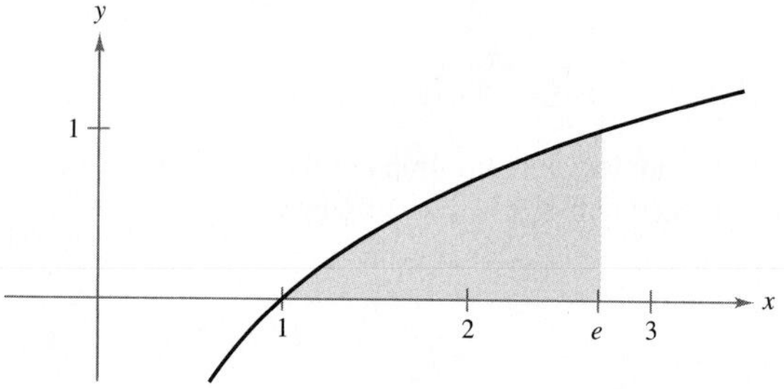

13. Factor the polynomial $p(x) = x^4 + 1$ and then find the area under the graph of $y = \dfrac{1}{x^4 + 1}$, $0 \le x \le 1$ (see figure).

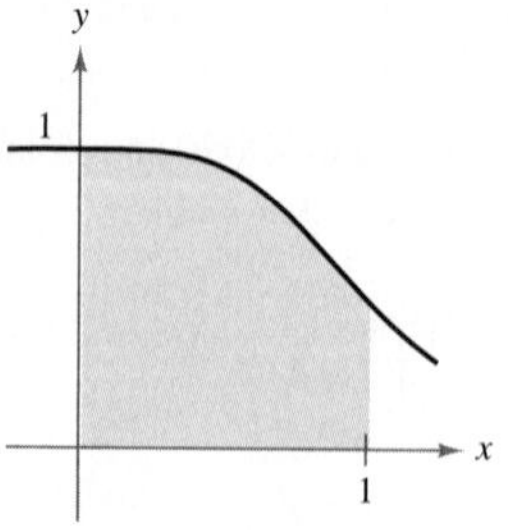

14. (a) Use the substitution $u = \dfrac{\pi}{2} - x$ to evaluate the integral

$$\int_0^{\pi/2} \frac{\sin x}{\cos x + \sin x}\,dx.$$

(b) Let n be a positive integer. Evaluate the integral

$$\int_0^{\pi/2} \frac{\sin^n x}{\cos^n x + \sin^n x}\,dx.$$

15. Use a graphing utility to estimate each limit. Then calculate each limit using L'Hôpital's Rule. What can you conclude about the indeterminate form $0 \cdot \infty$?

(a) $\displaystyle\lim_{x\to 0^+}\left(\cot x + \frac{1}{x}\right)$ (b) $\displaystyle\lim_{x\to 0^+}\left(\cot x - \frac{1}{x}\right)$

(c) $\displaystyle\lim_{x\to 0^+}\left[\left(\cot x + \frac{1}{x}\right)\left(\cot x - \frac{1}{x}\right)\right]$

16. Suppose the denominator of a rational function can be factored into distinct linear factors

$$D(x) = (x - c_1)(x - c_2) \cdot \cdot \cdot (x - c_n)$$

for a positive integer n and distinct real numbers $c_1, c_2, \ldots, c_n$. If N is a polynomial of degree less than n, show that

$$\frac{N(x)}{D(x)} = \frac{P_1}{x - c_1} + \frac{P_2}{x - c_2} + \cdot \cdot \cdot + \frac{P_n}{x - c_n}$$

where $P_k = N(c_k)/D'(c_k)$ for $k = 1, 2, \ldots, n$. Note that this is the partial fraction decomposition of $N(x)/D(x)$.

17. Use the results of Exercise 16 to find the partial fraction decomposition of

$$\frac{x^3 - 3x^2 + 1}{x^4 - 13x^2 + 12x}.$$

18. The velocity v (in feet per second) of a rocket whose initial mass (including fuel) is m is given by

$$v = gt + u \ln \frac{m}{m - rt}, \quad t < \frac{m}{r}$$

where u is the expulsion speed of the fuel, r is the rate at which the fuel is consumed, and $g = -32$ feet per second per second is the acceleration due to gravity. Find the position equation for a rocket for which $m = 50{,}000$ pounds, $u = 12{,}000$ feet per second, and $r = 400$ pounds per second. What is the height of the rocket when $t = 100$ seconds? (Assume that the rocket was fired from ground level and is moving straight upward.)

19. Suppose that $f(a) = f(b) = g(a) = g(b) = 0$ and the second derivatives of f and g are continuous on the closed interval $[a, b]$. Prove that

$$\int_a^b f(x)g''(x)\,dx = \int_a^b f''(x)g(x)\,dx.$$

20. Suppose that $f(a) = f(b) = 0$ and the second derivatives of f exist on the closed interval $[a, b]$. Prove that

$$\int_a^b (x - a)(x - b)f''(x)\,dx = 2\int_a^b f(x)\,dx.$$

21. Using the inequality

$$\frac{1}{x^5} + \frac{1}{x^{10}} + \frac{1}{x^{15}} < \frac{1}{x^5 - 1} < \frac{1}{x^5} + \frac{1}{x^{10}} + \frac{2}{x^{15}}$$

for $x \ge 2$, approximate $\displaystyle\int_2^{\infty} \frac{1}{x^5 - 1}\,dx$.

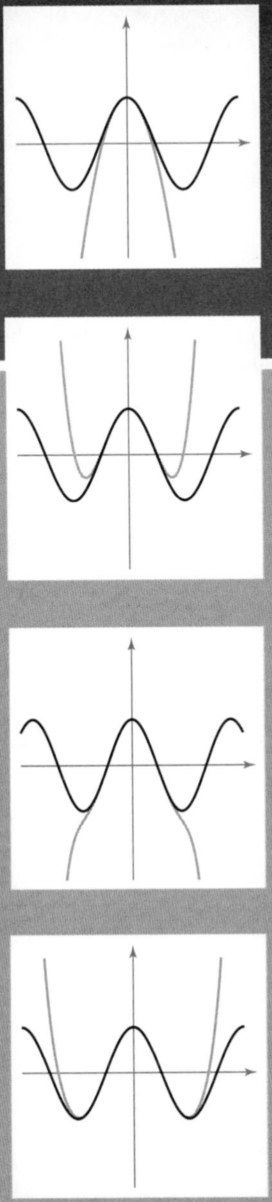

9 Infinite Series

A *ballista* was used as a portable rock-throwing machine. The propulsion mechanism was similar in appearance to a crossbow. Skilled artillerymen aimed and fired the ballista entirely by eye. What type of projectile path do you think these artillerymen preferred–a high, arching trajectory or a low, relatively level trajectory? Why?

Charles & Josette Lenars/Corbis

Maclaurin *polynomials* approximate a given function in an interval around $x = 0$. As you add terms to the Maclaurin polynomial, it becomes a better approximation of the given function near $x = 0$. In Section 9.10, you will see that a Maclaurin *series* is equivalent to the given function (under suitable conditions).

Section 9.1 Sequences

- **List the terms of a sequence.**
- **Determine whether a sequence converges or diverges.**
- **Write a formula for the *n*th term of a sequence.**
- **Use properties of monotonic sequences and bounded sequences.**

Sequences

In mathematics, the word "sequence" is used in much the same way as in ordinary English. To say that a collection of objects or events is *in sequence* usually means that the collection is ordered so that it has an identified first member, second member, third member, and so on.

Mathematically, a **sequence** is defined as a function whose domain is the set of positive integers. Although a sequence is a function, it is common to represent sequences by subscript notation rather than by the standard function notation. For instance, in the sequence

$$\begin{array}{ccccccc} 1, & 2, & 3, & 4, & \ldots, & n, & \ldots \\ \downarrow & \downarrow & \downarrow & \downarrow & \downarrow & \downarrow & \downarrow \\ a_1, & a_2, & a_3, & a_4, & \ldots, & a_n, & \ldots \end{array} \qquad \text{Sequence}$$

1 is mapped onto a_1, 2 is mapped onto a_2, and so on. The numbers $a_1, a_2, a_3, \ldots, a_n, \ldots$ are the **terms** of the sequence. The number a_n is the ***n*th term** of the sequence, and the entire sequence is denoted by $\{a_n\}$.

EXPLORATION

Finding Patterns Describe a pattern for each of the following sequences. Then use your description to write a formula for the *n*th term of each sequence. As *n* increases, do the terms appear to be approaching a limit? Explain your reasoning.

a. $1, \frac{1}{2}, \frac{1}{4}, \frac{1}{8}, \frac{1}{16}, \ldots$

b. $1, \frac{1}{2}, \frac{1}{6}, \frac{1}{24}, \frac{1}{120}, \ldots$

c. $10, \frac{10}{3}, \frac{10}{6}, \frac{10}{10}, \frac{10}{15}, \ldots$

d. $\frac{1}{4}, \frac{4}{9}, \frac{9}{16}, \frac{16}{25}, \frac{25}{36}, \ldots$

e. $\frac{3}{7}, \frac{5}{10}, \frac{7}{13}, \frac{9}{16}, \frac{11}{19}, \ldots$

NOTE Occasionally, it is convenient to begin a sequence with a_0, so that the terms of the sequence become $a_0, a_1, a_2, a_3, \ldots, a_n, \ldots$

EXAMPLE 1 **Listing the Terms of a Sequence**

a. The terms of the sequence $\{a_n\} = \{3 + (-1)^n\}$ are

$$\begin{array}{cccc} 3 + (-1)^1, & 3 + (-1)^2, & 3 + (-1)^3, & 3 + (-1)^4, \ldots \\ 2, & 4, & 2, & 4, \quad \ldots. \end{array}$$

b. The terms of the sequence $\{b_n\} = \left\{\dfrac{n}{1 - 2n}\right\}$ are

$$\begin{array}{cccc} \dfrac{1}{1 - 2 \cdot 1}, & \dfrac{2}{1 - 2 \cdot 2}, & \dfrac{3}{1 - 2 \cdot 3}, & \dfrac{4}{1 - 2 \cdot 4}, \ldots \\ -1, & -\dfrac{2}{3}, & -\dfrac{3}{5}, & -\dfrac{4}{7}, \quad \ldots. \end{array}$$

c. The terms of the sequence $\{c_n\} = \left\{\dfrac{n^2}{2^n - 1}\right\}$ are

$$\begin{array}{cccc} \dfrac{1^2}{2^1 - 1}, & \dfrac{2^2}{2^2 - 1}, & \dfrac{3^2}{2^3 - 1}, & \dfrac{4^2}{2^4 - 1}, \ldots \\ \dfrac{1}{1}, & \dfrac{4}{3}, & \dfrac{9}{7}, & \dfrac{16}{15}, \quad \ldots. \end{array}$$

d. The terms of the **recursively defined** sequence $\{d_n\}$, where $d_1 = 25$ and $d_{n+1} = d_n - 5$ are

$$25, \quad 25 - 5 = 20, \quad 20 - 5 = 15, \quad 15 - 5 = 10, \ldots.$$

STUDY TIP Some sequences are defined recursively. To define a sequence recursively, you need to be given one or more of the first few terms. All other terms of the sequence are then defined using previous terms, as shown in Example 1(d).

Limit of a Sequence

The primary focus of this chapter concerns sequences whose terms approach limiting values. Such sequences are said to **converge.** For instance, the sequence $\{1/2^n\}$

$$\frac{1}{2}, \frac{1}{4}, \frac{1}{8}, \frac{1}{16}, \frac{1}{32}, \ldots$$

converges to 0, as indicated in the following definition.

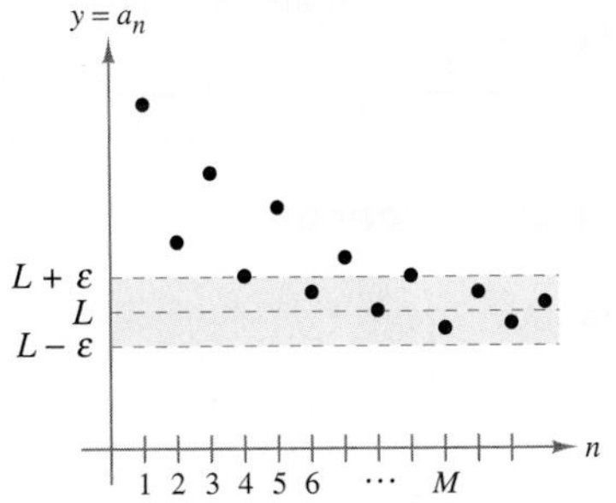

For $n > M$, the terms of the sequence all lie within ε units of L.
Figure 9.1

Definition of the Limit of a Sequence

Let L be a real number. The **limit** of a sequence $\{a_n\}$ is L, written as

$$\lim_{n\to\infty} a_n = L$$

if for each $\varepsilon > 0$, there exists $M > 0$ such that $|a_n - L| < \varepsilon$ whenever $n > M$. If the limit L of a sequence exists, then the sequence **converges** to L. If the limit of a sequence does not exist, then the sequence **diverges.**

Graphically, this definition says that eventually (for $n > M$ and $\varepsilon > 0$) the terms of a sequence that converges to L will lie within the band between the lines $y = L + \varepsilon$ and $y = L - \varepsilon$, as shown in Figure 9.1.

If a sequence $\{a_n\}$ agrees with a function f at every positive integer, and if $f(x)$ approaches a limit L as $x \to \infty$, the sequence must converge to the same limit L.

THEOREM 9.1 Limit of a Sequence

Let L be a real number. Let f be a function of a real variable such that

$$\lim_{x\to\infty} f(x) = L.$$

If $\{a_n\}$ is a sequence such that $f(n) = a_n$ for every positive integer n, then

$$\lim_{n\to\infty} a_n = L.$$

EXAMPLE 2 Finding the Limit of a Sequence

Find the limit of the sequence whose nth term is

$$a_n = \left(1 + \frac{1}{n}\right)^n.$$

Solution In Theorem 5.15, you learned that

$$\lim_{x\to\infty} \left(1 + \frac{1}{x}\right)^x = e.$$

So, you can apply Theorem 9.1 to conclude that

$$\lim_{n\to\infty} a_n = \lim_{n\to\infty} \left(1 + \frac{1}{n}\right)^n$$
$$= e.$$

NOTE There are different ways in which a sequence can fail to have a limit. One way is that the terms of the sequence increase without bound or decrease without bound. These cases are written symbolically as follows.

Terms increase without bound:

$$\lim_{n\to\infty} a_n = \infty$$

Terms decrease without bound:

$$\lim_{n\to\infty} a_n = -\infty$$

The following properties of limits of sequences parallel those given for limits of functions of a real variable in Section 1.3.

THEOREM 9.2 Properties of Limits of Sequences

Let $\lim_{n\to\infty} a_n = L$ and $\lim_{n\to\infty} b_n = K$.

1. $\lim_{n\to\infty} (a_n \pm b_n) = L \pm K$
2. $\lim_{n\to\infty} ca_n = cL$, c is any real number
3. $\lim_{n\to\infty} (a_n b_n) = LK$
4. $\lim_{n\to\infty} \frac{a_n}{b_n} = \frac{L}{K}$, $b_n \neq 0$ and $K \neq 0$

EXAMPLE 3 Determining Convergence or Divergence

a. Because the sequence $\{a_n\} = \{3 + (-1)^n\}$ has terms

$$2, 4, 2, 4, \ldots$$

See Example 1(a), page 594.

that alternate between 2 and 4, the limit

$$\lim_{n\to\infty} a_n$$

does not exist. So, the sequence diverges.

b. For $\{b_n\} = \left\{\frac{n}{1-2n}\right\}$, divide the numerator and denominator by n to obtain

$$\lim_{n\to\infty} \frac{n}{1-2n} = \lim_{n\to\infty} \frac{1}{(1/n) - 2} = -\frac{1}{2}$$

See Example 1(b), page 594.

which implies that the sequence converges to $-\frac{1}{2}$.

EXAMPLE 4 Using L'Hôpital's Rule to Determine Convergence

Show that the sequence whose nth term is $a_n = \frac{n^2}{2^n - 1}$ converges.

Solution Consider the function of a real variable

$$f(x) = \frac{x^2}{2^x - 1}.$$

Applying L'Hôpital's Rule twice produces

$$\lim_{x\to\infty} \frac{x^2}{2^x - 1} = \lim_{x\to\infty} \frac{2x}{(\ln 2)2^x} = \lim_{x\to\infty} \frac{2}{(\ln 2)^2 2^x} = 0.$$

Because $f(n) = a_n$ for every positive integer, you can apply Theorem 9.1 to conclude that

$$\lim_{n\to\infty} \frac{n^2}{2^n - 1} = 0.$$

See Example 1(c), page 594.

So, the sequence converges to 0.

TECHNOLOGY Use a graphing utility to graph the function in Example 4. Notice that as x approaches infinity, the value of the function gets closer and closer to 0. If you have access to a graphing utility that can generate terms of a sequence, try using it to calculate the first 20 terms of the sequence in Example 4. Then view the terms to observe numerically that the sequence converges to 0.

www indicates that in the HM mathSpace® *CD-ROM and the online* Eduspace® *system for this text, you will find an Open Exploration, which further explores this example using the computer algebra systems* Maple, Mathcad, Mathematica, *and* Derive.

The symbol $n!$ (read "n factorial") is used to simplify some of the formulas developed in this chapter. Let n be a positive integer; then ***n* factorial** is defined as

$$n! = 1 \cdot 2 \cdot 3 \cdot 4 \cdot \cdot \cdot (n-1) \cdot n.$$

As a special case, **zero factorial** is defined as $0! = 1$. From this definition, you can see that $1! = 1$, $2! = 1 \cdot 2 = 2$, $3! = 1 \cdot 2 \cdot 3 = 6$, and so on. Factorials follow the same conventions for order of operations as exponents. That is, just as $2x^3$ and $(2x)^3$ imply different orders of operations, $2n!$ and $(2n)!$ imply the following orders.

$$2n! = 2(n!) = 2(1 \cdot 2 \cdot 3 \cdot 4 \cdot \cdot \cdot n)$$

and

$$(2n)! = 1 \cdot 2 \cdot 3 \cdot 4 \cdot \cdot \cdot n \cdot (n+1) \cdot \cdot \cdot 2n$$

Another useful limit theorem that can be rewritten for sequences is the Squeeze Theorem from Section 1.3.

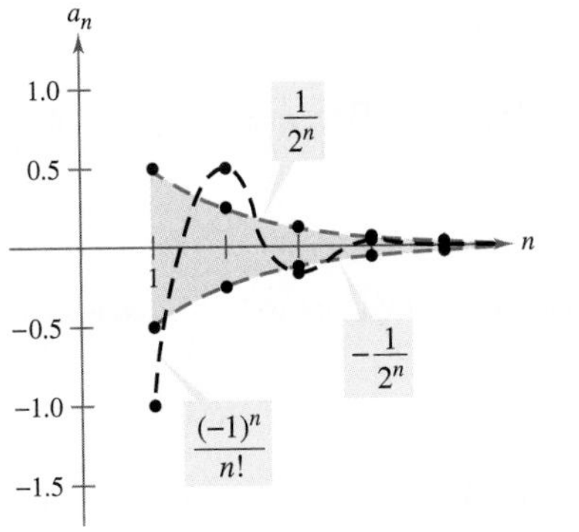

For $n \geq 4$, $(-1)^n/n!$ is squeezed between $-1/2^n$ and $1/2^n$.
Figure 9.2

THEOREM 9.3 Squeeze Theorem for Sequences

If

$$\lim_{n\to\infty} a_n = L = \lim_{n\to\infty} b_n$$

and there exists an integer N such that $a_n \leq c_n \leq b_n$ for all $n > N$, then

$$\lim_{n\to\infty} c_n = L.$$

EXAMPLE 5 Using the Squeeze Theorem

Show that the sequence $\{c_n\} = \left\{(-1)^n \frac{1}{n!}\right\}$ converges, and find its limit.

Solution To apply the Squeeze Theorem, you must find two convergent sequences that can be related to the given sequence. Two possibilities are $a_n = -1/2^n$ and $b_n = 1/2^n$, both of which converge to 0. By comparing the term $n!$ with 2^n, you can see that

$$n! = 1 \cdot 2 \cdot 3 \cdot 4 \cdot 5 \cdot 6 \cdot \cdot \cdot n = 24 \cdot \underbrace{5 \cdot 6 \cdot \cdot \cdot n}_{n-4 \text{ factors}} \qquad (n \geq 4)$$

and

$$2^n = 2 \cdot 2 \cdot 2 \cdot 2 \cdot 2 \cdot 2 \cdot \cdot \cdot 2 = 16 \cdot \underbrace{2 \cdot 2 \cdot \cdot \cdot 2}_{n-4 \text{ factors}}. \qquad (n \geq 4)$$

This implies that for $n \geq 4$, $2^n < n!$, and you have

$$\frac{-1}{2^n} \leq (-1)^n \frac{1}{n!} \leq \frac{1}{2^n}, \quad n \geq 4$$

as shown in Figure 9.2. So, by the Squeeze Theorem it follows that

$$\lim_{n\to\infty} (-1)^n \frac{1}{n!} = 0.$$

NOTE Example 5 suggests something about the rate at which $n!$ increases as $n \to \infty$. As Figure 9.2 suggests, both $1/2^n$ and $1/n!$ approach 0 as $n \to \infty$. Yet $1/n!$ approaches 0 so much faster than $1/2^n$ does that

$$\lim_{n\to\infty} \frac{1/n!}{1/2^n} = \lim_{n\to\infty} \frac{2^n}{n!} = 0.$$

In fact, it can be shown that for any fixed number k,

$$\lim_{n\to\infty} \frac{k^n}{n!} = 0.$$

This means that *the factorial function grows faster than any exponential function.*

In Example 5, the sequence $\{c_n\}$ has both positive and negative terms. For this sequence, it happens that the sequence of absolute values, $\{|c_n|\}$, also converges to 0. You can show this by the Squeeze Theorem using the inequality

$$0 \leq \frac{1}{n!} \leq \frac{1}{2^n}, \quad n \geq 4.$$

In such cases, it is often convenient to consider the sequence of absolute values—and then apply Theorem 9.4, which states that if the absolute value sequence converges to 0, the original signed sequence also converges to 0.

THEOREM 9.4 Absolute Value Theorem

For the sequence $\{a_n\}$, if

$$\lim_{n\to\infty} |a_n| = 0 \qquad \text{then} \qquad \lim_{n\to\infty} a_n = 0.$$

Proof Consider the two sequences $\{|a_n|\}$ and $\{-|a_n|\}$. Because both of these sequences converge to 0 and

$$-|a_n| \leq a_n \leq |a_n|$$

you can use the Squeeze Theorem to conclude that $\{a_n\}$ converges to 0.

Pattern Recognition for Sequences

Sometimes the terms of a sequence are generated by some rule that does not explicitly identify the *n*th term of the sequence. In such cases, you may be required to discover a *pattern* in the sequence and to describe the *n*th term. Once the *n*th term has been specified, you can investigate the convergence or divergence of the sequence.

EXAMPLE 6 Finding the *n*th Term of a Sequence

Find a sequence $\{a_n\}$ whose first five terms are

$$\frac{2}{1}, \frac{4}{3}, \frac{8}{5}, \frac{16}{7}, \frac{32}{9}, \ldots$$

and then determine whether the particular sequence you have chosen converges or diverges.

Solution First, note that the numerators are successive powers of 2, and the denominators form the sequence of positive odd integers. By comparing a_n with n, you have the following pattern.

$$\frac{2^1}{1}, \frac{2^2}{3}, \frac{2^3}{5}, \frac{2^4}{7}, \frac{2^5}{9}, \ldots, \frac{2^n}{2n-1}$$

Using L'Hôpital's Rule to evaluate the limit of $f(x) = 2^x/(2x - 1)$, you obtain

$$\lim_{x\to\infty} \frac{2^x}{2x-1} = \lim_{x\to\infty} \frac{2^x(\ln 2)}{2} = \infty \quad \Longrightarrow \quad \lim_{n\to\infty} \frac{2^n}{2n-1} = \infty.$$

So, the sequence diverges.

Without a specific rule for generating the terms of a sequence or some knowledge of the context in which the terms of the sequence are obtained, it is not possible to determine the convergence or divergence of the sequence merely from its first several terms. For instance, although the first three terms of the following four sequences are identical, the first two sequences converge to 0, the third sequence converges to $\frac{1}{9}$, and the fourth sequence diverges.

$$\{a_n\}: \frac{1}{2}, \frac{1}{4}, \frac{1}{8}, \frac{1}{16}, \ldots, \frac{1}{2^n}, \ldots$$

$$\{b_n\}: \frac{1}{2}, \frac{1}{4}, \frac{1}{8}, \frac{1}{15}, \ldots, \frac{6}{(n+1)(n^2-n+6)}, \ldots$$

$$\{c_n\}: \frac{1}{2}, \frac{1}{4}, \frac{1}{8}, \frac{7}{62}, \ldots, \frac{n^2-3n+3}{9n^2-25n+18}, \ldots$$

$$\{d_n\}: \frac{1}{2}, \frac{1}{4}, \frac{1}{8}, 0, \ldots, \frac{-n(n+1)(n-4)}{6(n^2+3n-2)}, \ldots$$

The process of determining an nth term from the pattern observed in the first several terms of a sequence is an example of *inductive reasoning*.

EXAMPLE 7 Finding the nth Term of a Sequence

Determine an nth term for a sequence whose first five terms are

$$-\frac{2}{1}, \frac{8}{2}, -\frac{26}{6}, \frac{80}{24}, -\frac{242}{120}, \ldots$$

and then decide whether the sequence converges or diverges.

Solution Note that the numerators are 1 less than 3^n. So, you can reason that the numerators are given by the rule $3^n - 1$. Factoring the denominators produces

$$\begin{aligned} 1 &= 1 \\ 2 &= 1 \cdot 2 \\ 6 &= 1 \cdot 2 \cdot 3 \\ 24 &= 1 \cdot 2 \cdot 3 \cdot 4 \\ 120 &= 1 \cdot 2 \cdot 3 \cdot 4 \cdot 5 \cdot \cdot \cdot . \end{aligned}$$

This suggests that the denominators are represented by $n!$. Finally, because the signs alternate, you can write the nth term as

$$a_n = (-1)^n \left(\frac{3^n - 1}{n!}\right).$$

From the discussion about the growth of $n!$, it follows that

$$\lim_{n \to \infty} |a_n| = \lim_{n \to \infty} \frac{3^n - 1}{n!} = 0.$$

Applying Theorem 9.4, you can conclude that

$$\lim_{n \to \infty} a_n = 0.$$

So, the sequence $\{a_n\}$ converges to 0.

Monotonic Sequences and Bounded Sequences

So far you have determined the convergence of a sequence by finding its limit. Even if you cannot determine the limit of a particular sequence, it still may be useful to know whether the sequence converges. Theorem 9.5 provides a test for convergence of sequences without determining the limit. First, some preliminary definitions are given.

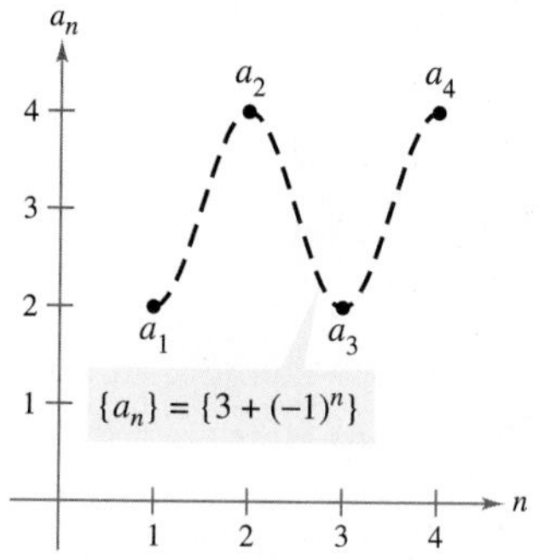

(a) Not monotonic

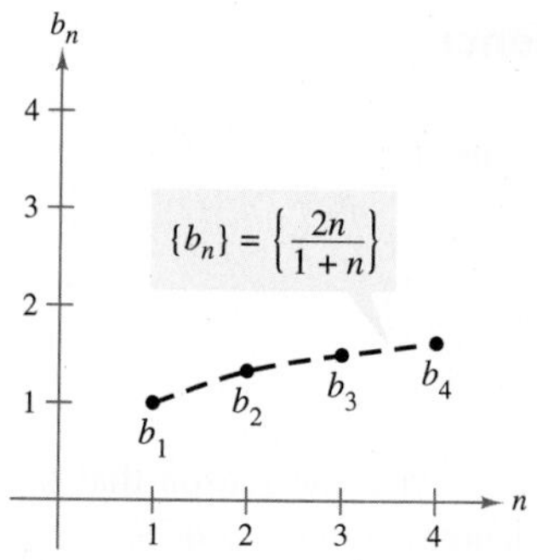

(b) Monotonic

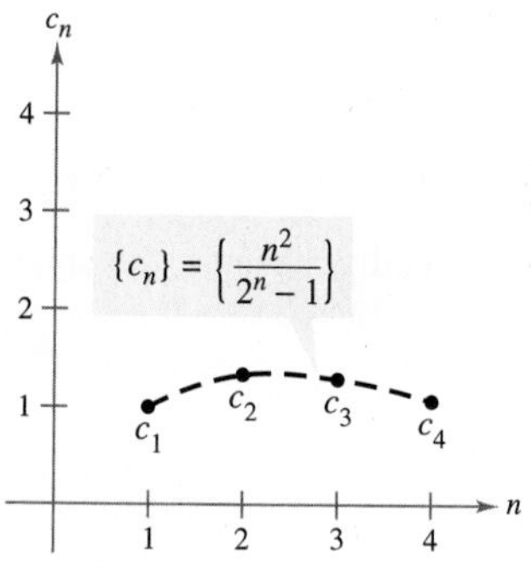

(c) Not monotonic

Figure 9.3

Definition of a Monotonic Sequence

A sequence $\{a_n\}$ is **monotonic** if its terms are nondecreasing

$$a_1 \le a_2 \le a_3 \le \cdot \cdot \cdot \le a_n \le \cdot \cdot \cdot$$

or if its terms are nonincreasing

$$a_1 \ge a_2 \ge a_3 \ge \cdot \cdot \cdot \ge a_n \ge \cdot \cdot \cdot.$$

EXAMPLE 8 Determining Whether a Sequence Is Monotonic

Determine whether each sequence having the given nth term is monotonic.

a. $a_n = 3 + (-1)^n$ **b.** $b_n = \dfrac{2n}{1+n}$ **c.** $c_n = \dfrac{n^2}{2^n - 1}$

Solution

a. This sequence alternates between 2 and 4. So, it is not monotonic.

b. This sequence is monotonic because each successive term is larger than its predecessor. To see this, compare the terms b_n and b_{n+1}. [Note that, because n is positive, you can multiply each side of the inequality by $(1 + n)$ and $(2 + n)$ without reversing the inequality sign.]

$$b_n = \frac{2n}{1+n} \overset{?}{<} \frac{2(n+1)}{1+(n+1)} = b_{n+1}$$
$$2n(2+n) \overset{?}{<} (1+n)(2n+2)$$
$$4n + 2n^2 \overset{?}{<} 2 + 4n + 2n^2$$
$$0 < 2$$

Starting with the final inequality, which is valid, you can reverse the steps to conclude that the original inequality is also valid.

c. This sequence is not monotonic, because the second term is larger than the first term, and larger than the third. (Note that if you drop the first term, the remaining sequence $c_2, c_3, c_4, \ldots$ is monotonic.)

Figure 9.3 graphically illustrates these three sequences.

NOTE In Example 8(b), another way to see that the sequence is monotonic is to argue that the derivative of the corresponding differentiable function $f(x) = 2x/(1 + x)$ is positive for all x. This implies that f is increasing, which in turn implies that $\{a_n\}$ is increasing.

NOTE All three sequences shown in Figure 9.3 are bounded. To see this, consider the following.

$$2 \le a_n \le 4$$
$$1 \le b_n \le 2$$
$$0 \le c_n \le \frac{4}{3}$$

Definition of a Bounded Sequence

1. A sequence $\{a_n\}$ is **bounded above** if there is a real number M such that $a_n \le M$ for all n. The number M is called an **upper bound** of the sequence.
2. A sequence $\{a_n\}$ is **bounded below** if there is a real number N such that $N \le a_n$ for all n. The number N is called a **lower bound** of the sequence.
3. A sequence $\{a_n\}$ is **bounded** if it is bounded above and bounded below.

One important property of the real numbers is that they are **complete.** Informally, this means that there are no holes or gaps on the real number line. (The set of rational numbers does not have the completeness property.) The completeness axiom for real numbers can be used to conclude that if a sequence has an upper bound, it must have a **least upper bound** (an upper bound that is smaller than all other upper bounds for the sequence). For example, the least upper bound of the sequence $\{a_n\} = \{n/(n + 1)\}$,

$$\frac{1}{2}, \frac{2}{3}, \frac{3}{4}, \frac{4}{5}, \ldots, \frac{n}{n+1}, \ldots$$

is 1. The completeness axiom is used in the proof of Theorem 9.5.

THEOREM 9.5 Bounded Monotonic Sequences

If a sequence $\{a_n\}$ is bounded and monotonic, then it converges.

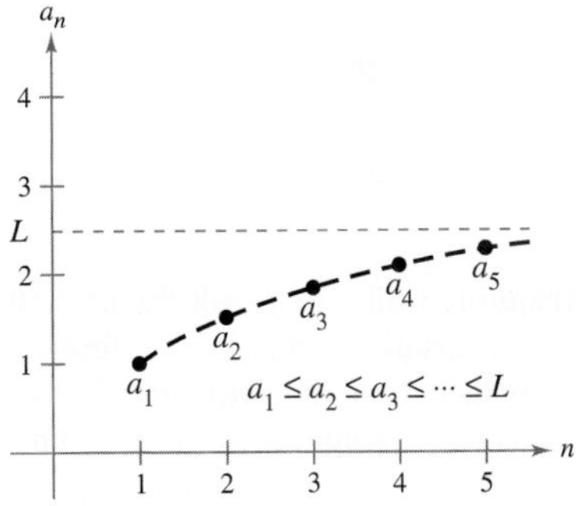

Every bounded nondecreasing sequence converges.
Figure 9.4

Proof Assume that the sequence is nondecreasing, as shown in Figure 9.4. For the sake of simplicity, also assume that each term in the sequence is positive. Because the sequence is bounded, there must exist an upper bound M such that

$$a_1 \le a_2 \le a_3 \le \cdots \le a_n \le \cdots \le M.$$

From the completeness axiom, it follows that there is a least upper bound L such that

$$a_1 \le a_2 \le a_3 \le \cdots \le a_n \le \cdots \le L.$$

For $\varepsilon > 0$, it follows that $L - \varepsilon < L$, and therefore $L - \varepsilon$ cannot be an upper bound for the sequence. Consequently, at least one term of $\{a_n\}$ is greater than $L - \varepsilon$. That is, $L - \varepsilon < a_N$ for some positive integer N. Because the terms of $\{a_n\}$ are nondecreasing, it follows that $a_N \le a_n$ for $n > N$. You now know that $L - \varepsilon < a_N \le a_n \le L < L + \varepsilon$, for every $n > N$. It follows that $|a_n - L| < \varepsilon$ for $n > N$, which by definition means that $\{a_n\}$ converges to L. The proof for a nonincreasing sequence is similar.

EXAMPLE 9 Bounded and Monotonic Sequences

a. The sequence $\{a_n\} = \{1/n\}$ is both bounded and monotonic and so, by Theorem 9.5, must converge.

b. The divergent sequence $\{b_n\} = \{n^2/(n + 1)\}$ is monotonic, but not bounded. (It is bounded below.)

c. The divergent sequence $\{c_n\} = \{(-1)^n\}$ is bounded, but not monotonic.

Exercises for Section 9.1

See www.CalcChat.com for worked-out solutions to odd-numbered exercises.

In Exercises 1–10, write the first five terms of the sequence.

1. $a_n = 2^n$

2. $a_n = \dfrac{3^n}{n!}$

3. $a_n = \left(-\frac{1}{2}\right)^n$

4. $a_n = \left(-\frac{2}{3}\right)^n$

5. $a_n = \sin \dfrac{n\pi}{2}$

6. $a_n = \dfrac{2n}{n+3}$

7. $a_n = \dfrac{(-1)^{n(n+1)/2}}{n^2}$

8. $a_n = (-1)^{n+1}\left(\dfrac{2}{n}\right)$

9. $a_n = 5 - \dfrac{1}{n} + \dfrac{1}{n^2}$

10. $a_n = 10 + \dfrac{2}{n} + \dfrac{6}{n^2}$

In Exercises 11–14, write the first five terms of the recursively defined sequence.

11. $a_1 = 3,\ a_{k+1} = 2(a_k - 1)$

12. $a_1 = 4,\ a_{k+1} = \left(\dfrac{k+1}{2}\right)a_k$

13. $a_1 = 32,\ a_{k+1} = \frac{1}{2}a_k$

14. $a_1 = 6,\ a_{k+1} = \frac{1}{3}a_k^2$

In Exercises 15–20, match the sequence with its graph. [The graphs are labeled (a), (b), (c), (d), (e), and (f).]

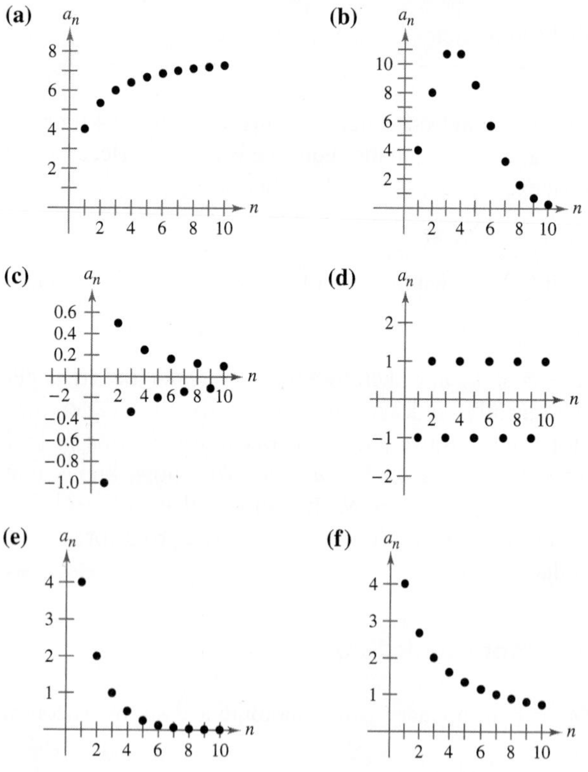

15. $a_n = \dfrac{8}{n+1}$

16. $a_n = \dfrac{8n}{n+1}$

17. $a_n = 4(0.5)^{n-1}$

18. $a_n = \dfrac{4^n}{n!}$

19. $a_n = (-1)^n$

20. $a_n = \dfrac{(-1)^n}{n}$

In Exercises 21–24, use a graphing utility to graph the first 10 terms of the sequence.

21. $a_n = \dfrac{2}{3}n$

22. $a_n = 2 - \dfrac{4}{n}$

23. $a_n = 16(-0.5)^{n-1}$

24. $a_n = \dfrac{2n}{n+1}$

In Exercises 25–30, write the next two apparent terms of the sequence. Describe the pattern you used to find these terms.

25. $2, 5, 8, 11, \ldots$

26. $\frac{7}{2}, 4, \frac{9}{2}, 5, \ldots$

27. $5, 10, 20, 40, \ldots$

28. $1, -\frac{1}{2}, \frac{1}{4}, -\frac{1}{8}, \ldots$

29. $3, -\frac{3}{2}, \frac{3}{4}, -\frac{3}{8}, \ldots$

30. $1, -\frac{3}{2}, \frac{9}{4}, -\frac{27}{8}, \ldots$

In Exercises 31–36, simplify the ratio of factorials.

31. $\dfrac{10!}{8!}$

32. $\dfrac{25!}{23!}$

33. $\dfrac{(n+1)!}{n!}$

34. $\dfrac{(n+2)!}{n!}$

35. $\dfrac{(2n-1)!}{(2n+1)!}$

36. $\dfrac{(2n+2)!}{(2n)!}$

In Exercises 37–42, find the limit (if possible) of the sequence.

37. $a_n = \dfrac{5n^2}{n^2+2}$

38. $a_n = 5 - \dfrac{1}{n^2}$

39. $a_n = \dfrac{2n}{\sqrt{n^2+1}}$

40. $a_n = \dfrac{5n}{\sqrt{n^2+4}}$

41. $a_n = \sin \dfrac{1}{n}$

42. $a_n = \cos \dfrac{2}{n}$

In Exercises 43–46, use a graphing utility to graph the first 10 terms of the sequence. Use the graph to make an inference about the convergence or divergence of the sequence. Verify your inference analytically and, if the sequence converges, find its limit.

43. $a_n = \dfrac{n+1}{n}$

44. $a_n = \dfrac{1}{n^{3/2}}$

45. $a_n = \cos \dfrac{n\pi}{2}$

46. $a_n = 3 - \dfrac{1}{2^n}$

In Exercises 47–68, determine the convergence or divergence of the sequence with the given nth term. If the sequence converges, find its limit.

47. $a_n = (-1)^n\left(\dfrac{n}{n+1}\right)$

48. $a_n = 1 + (-1)^n$

49. $a_n = \dfrac{3n^2 - n + 4}{2n^2 + 1}$

50. $a_n = \dfrac{\sqrt[3]{n}}{\sqrt[3]{n} + 1}$

51. $a_n = \dfrac{1 \cdot 3 \cdot 5 \cdot \cdots \cdot (2n-1)}{(2n)^n}$

52. $a_n = \dfrac{1 \cdot 3 \cdot 5 \cdot \cdots \cdot (2n-1)}{n!}$

53. $a_n = \dfrac{1 + (-1)^n}{n}$

54. $a_n = \dfrac{1 + (-1)^n}{n^2}$

55. $a_n = \dfrac{\ln(n^3)}{2n}$

56. $a_n = \dfrac{\ln \sqrt{n}}{n}$

57. $a_n = \dfrac{3^n}{4^n}$

58. $a_n = (0.5)^n$

59. $a_n = \dfrac{(n + 1)!}{n!}$

60. $a_n = \dfrac{(n - 2)!}{n!}$

61. $a_n = \dfrac{n - 1}{n} - \dfrac{n}{n - 1}, \quad n \geq 2$

62. $a_n = \dfrac{n^2}{2n + 1} - \dfrac{n^2}{2n - 1}$

63. $a_n = \dfrac{n^p}{e^n}, \quad p > 0$

64. $a_n = n \sin \dfrac{1}{n}$

65. $a_n = \left(1 + \dfrac{k}{n}\right)^n$

66. $a_n = 2^{1/n}$

67. $a_n = \dfrac{\sin n}{n}$

68. $a_n = \dfrac{\cos \pi n}{n^2}$

In Exercises 69–82, write an expression for the *n*th term of the sequence. (There is more than one correct answer.)

69. $1, 4, 7, 10, \ldots$

70. $3, 7, 11, 15, \ldots$

71. $-1, 2, 7, 14, 23, \ldots$

72. $1, -\frac{1}{4}, \frac{1}{9}, -\frac{1}{16}, \ldots$

73. $\frac{2}{3}, \frac{3}{4}, \frac{4}{5}, \frac{5}{6}, \ldots$

74. $2, -1, \frac{1}{2}, -\frac{1}{4}, \frac{1}{8}, \ldots$

75. $2, 1 + \frac{1}{2}, 1 + \frac{1}{3}, 1 + \frac{1}{4}, 1 + \frac{1}{5}, \ldots$

76. $1 + \frac{1}{2}, 1 + \frac{3}{4}, 1 + \frac{7}{8}, 1 + \frac{15}{16}, 1 + \frac{31}{32}, \ldots$

77. $\dfrac{1}{2 \cdot 3}, \dfrac{2}{3 \cdot 4}, \dfrac{3}{4 \cdot 5}, \dfrac{4}{5 \cdot 6}, \ldots$

78. $1, \frac{1}{2}, \frac{1}{6}, \frac{1}{24}, \frac{1}{120}, \ldots$

79. $1, -\dfrac{1}{1 \cdot 3}, \dfrac{1}{1 \cdot 3 \cdot 5}, -\dfrac{1}{1 \cdot 3 \cdot 5 \cdot 7}, \ldots$

80. $1, x, \dfrac{x^2}{2}, \dfrac{x^3}{6}, \dfrac{x^4}{24}, \dfrac{x^5}{120}, \ldots$

81. $2, 24, 720, 40{,}320, 3{,}628{,}800, \ldots$

82. $1, 6, 120, 5040, 362{,}880, \ldots$

In Exercises 83–94, determine whether the sequence with the given *n*th term is monotonic. Discuss the boundedness of the sequence. Use a graphing utility to confirm your results.

83. $a_n = 4 - \dfrac{1}{n}$

84. $a_n = \dfrac{3n}{n + 2}$

85. $a_n = \dfrac{n}{2^{n+2}}$

86. $a_n = ne^{-n/2}$

87. $a_n = (-1)^n\left(\dfrac{1}{n}\right)$

88. $a_n = \left(-\dfrac{2}{3}\right)^n$

89. $a_n = \left(\dfrac{2}{3}\right)^n$

90. $a_n = \left(\dfrac{3}{2}\right)^n$

91. $a_n = \sin \dfrac{n\pi}{6}$

92. $a_n = \cos\left(\dfrac{n\pi}{2}\right)$

93. $a_n = \dfrac{\cos n}{n}$

94. $a_n = \dfrac{\sin \sqrt{n}}{n}$

In Exercises 95–98, (a) use Theorem 9.5 to show that the sequence with the given *n*th term converges and (b) use a graphing utility to graph the first 10 terms of the sequence and find its limit.

95. $a_n = 5 + \dfrac{1}{n}$

96. $a_n = 4 - \dfrac{3}{n}$

97. $a_n = \dfrac{1}{3}\left(1 - \dfrac{1}{3^n}\right)$

98. $a_n = 4 + \dfrac{1}{2^n}$

99. Let $\{a_n\}$ be an increasing sequence such that $2 \leq a_n \leq 4$. Explain why $\{a_n\}$ has a limit. What can you conclude about the limit?

100. Let $\{a_n\}$ be a monotonic sequence such that $a_n \leq 1$. Discuss the convergence of $\{a_n\}$. If $\{a_n\}$ converges, what can you conclude about its limit?

101. ***Compound Interest*** Consider the sequence $\{A_n\}$ whose *n*th term is given by

$$A_n = P\left(1 + \frac{r}{12}\right)^n$$

where P is the principal, A_n is the account balance after n months, and r is the interest rate compounded annually.

(a) Is $\{A_n\}$ a convergent sequence? Explain.

(b) Find the first 10 terms of the sequence if $P = \$9000$ and $r = 0.055$.

102. ***Compound Interest*** A deposit of \$100 is made at the beginning of each month in an account at an annual interest rate of 3% compounded monthly. The balance in the account after n months is $A_n = 100(401)(1.0025^n - 1)$.

(a) Compute the first six terms of the sequence $\{A_n\}$.

(b) Find the balance in the account after 5 years by computing the 60th term of the sequence.

(c) Find the balance in the account after 20 years by computing the 240th term of the sequence.

Writing About Concepts

103. In your own words, define each of the following.

(a) Sequence (b) Convergence of a sequence

(c) Monotonic sequence (d) Bounded sequence

104. The graphs of two sequences are shown in the figures. Which graph represents the sequence with alternating signs? Explain.

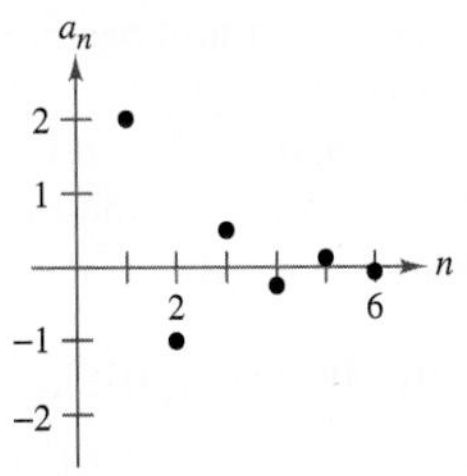

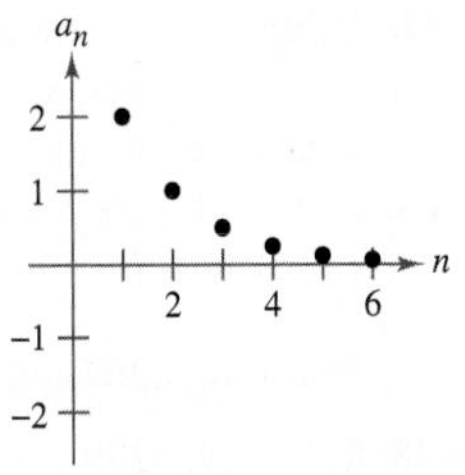

Writing About Concepts (continued)

In Exercises 105–108, give an example of a sequence satisfying the condition or explain why no such sequence exists. (Examples are not unique.)

105. A monotonically increasing sequence that converges to 10

106. A monotonically increasing bounded sequence that does not converge

107. A sequence that converges to $\frac{3}{4}$

108. An unbounded sequence that converges to 100

109. ***Government Expenditures*** A government program that currently costs taxpayers \$2.5 billion per year is cut back by 20 percent per year.

(a) Write an expression for the amount budgeted for this program after n years.

(b) Compute the budgets for the first 4 years.

(c) Determine the convergence or divergence of the sequence of reduced budgets. If the sequence converges, find its limit.

110. ***Inflation*** If the rate of inflation is $4\frac{1}{2}\%$ per year and the average price of a car is currently \$16,000, the average price after n years is

$$P_n = \$16{,}000(1.045)^n.$$

Compute the average prices for the next 5 years.

111. ***Modeling Data*** The number a_n of endangered and threatened species in the United States from 1996 through 2002 is shown in the table, where n represents the year, with $n = 6$ corresponding to 1996. *(Source: U.S. Fish and Wildlife Service)*

n	6	7	8	9	10	11	12
a_n	1053	1132	1194	1205	1244	1254	1263

(a) Use the regression capabilities of a graphing utility to find a model of the form

$$a_n = bn^2 + cn + d, \qquad n = 6, 7, . . . , 12$$

for the data. Use the graphing utility to plot the points and graph the model.

(b) Use the model to predict the number of endangered and threatened species in the year 2008.

112. ***Modeling Data*** The annual sales a_n (in millions of dollars) for Avon Products, Inc. from 1993 through 2002 are given below as ordered pairs of the form (n, a_n), where n represents the year, with $n = 3$ corresponding to 1993. *(Source: 2002 Avon Products, Inc. Annual Report)*

(3, 3844), (4, 4267), (5, 4492), (6, 4814), (7, 5079),
(8, 5213), (9, 5289), (10, 5682), (11, 5958), (12, 6171)

(a) Use the regression capabilities of a graphing utility to find a model of the form

$$a_n = bn + c, \quad n = 3, 4, . . . , 12$$

for the data. Graphically compare the points and the model.

(b) Use the model to predict sales in the year 2008.

113. ***Comparing Exponential and Factorial Growth*** Consider the sequence $a_n = 10^n/n!$.

(a) Find two consecutive terms that are equal in magnitude.

(b) Are the terms following those found in part (a) increasing or decreasing?

(c) In Section 8.7, Exercises 65–70, it was shown that for "large" values of the independent variable an exponential function increases more rapidly than a polynomial function. From the result in part (b), what inference can you make about the rate of growth of an exponential function versus a factorial function for "large" integer values of n?

114. Compute the first six terms of the sequence

$$\{a_n\} = \left\{\left(1 + \frac{1}{n}\right)^n\right\}.$$

If the sequence converges, find its limit.

115. Compute the first six terms of the sequence $\{a_n\} = \{\sqrt[n]{n}\}$. If the sequence converges, find its limit.

116. Prove that if $\{s_n\}$ converges to L and $L > 0$, then there exists a number N such that $s_n > 0$ for $n > N$.

True or False? **In Exercises 117–120, determine whether the statement is true or false. If it is false, explain why or give an example that shows it is false.**

117. If $\{a_n\}$ converges to 3 and $\{b_n\}$ converges to 2, then $\{a_n + b_n\}$ converges to 5.

118. If $\{a_n\}$ converges, then $\lim_{n\to\infty} (a_n - a_{n+1}) = 0$.

119. If $n > 1$, then $n! = n(n - 1)!$.

120. If $\{a_n\}$ converges, then $\{a_n/n\}$ converges to 0.

121. ***Fibonacci Sequence*** In a study of the progeny of rabbits, Fibonacci (ca. 1170–ca. 1240) encountered the sequence now bearing his name. It is defined recursively by

$$a_{n+2} = a_n + a_{n+1}, \qquad \text{where} \quad a_1 = 1 \text{ and } a_2 = 1.$$

(a) Write the first 12 terms of the sequence.

(b) Write the first 10 terms of the sequence defined by

$$b_n = \frac{a_{n+1}}{a_n}, \quad n \ge 1.$$

(c) Using the definition in part (b), show that

$$b_n = 1 + \frac{1}{b_{n-1}}.$$

(d) The **golden ratio** ρ can be defined by $\lim_{n\to\infty} b_n = \rho$. Show that $\rho = 1 + 1/\rho$ and solve this equation for ρ.

122. ***Conjecture*** Let $x_0 = 1$ and consider the sequence x_n given by the formula

$$x_n = \frac{1}{2}x_{n-1} + \frac{1}{x_{n-1}}, \qquad n = 1, 2, \ldots .$$

Use a graphing utility to compute the first 10 terms of the sequence and make a conjecture about the limit of the sequence.

123. Consider the sequence

$$\sqrt{2}, \sqrt{2 + \sqrt{2}}, \sqrt{2 + \sqrt{2 + \sqrt{2}}}, \ldots$$

(a) Compute the first five terms of this sequence.

(b) Write a recursion formula for a_n, for $n \geq 2$.

(c) Find $\lim_{n\to\infty} a_n$.

124. Consider the sequence

$$\sqrt{6}, \sqrt{6 + \sqrt{6}}, \sqrt{6 + \sqrt{6 + \sqrt{6}}}, \ldots$$

(a) Compute the first five terms of this sequence.

(b) Write a recursion formula for a_n, for $n \geq 2$.

(c) Find $\lim_{n\to\infty} a_n$.

125. Consider the sequence $\{a_n\}$ where $a_1 = \sqrt{k}$, $a_{n+1} = \sqrt{k + a_n}$, and $k > 0$.

(a) Show that $\{a_n\}$ is increasing and bounded.

(b) Prove that $\lim_{n\to\infty} a_n$ exists.

(c) Find $\lim_{n\to\infty} a_n$.

126. ***Arithmetic-Geometric Mean*** Let $a_0 > b_0 > 0$. Let a_1 be the arithmetic mean of a_0 and b_0 and let b_1 be the geometric mean of a_0 and b_0.

$$a_1 = \frac{a_0 + b_0}{2} \qquad \text{Arithmetic mean}$$

$$b_1 = \sqrt{a_0 b_0} \qquad \text{Geometric mean}$$

Now define the sequences $\{a_n\}$ and $\{b_n\}$ as follows.

$$a_n = \frac{a_{n-1} + b_{n-1}}{2} \qquad b_n = \sqrt{a_{n-1}b_{n-1}}$$

(a) Let $a_0 = 10$ and $b_0 = 3$. Write out the first five terms of $\{a_n\}$ and $\{b_n\}$. Compare the terms of $\{b_n\}$. Compare a_n and b_n. What do you notice?

(b) Use induction to show that $a_n > a_{n+1} > b_{n+1} > b_n$, for $a_0 > b_0 > 0$.

(c) Explain why $\{a_n\}$ and $\{b_n\}$ are both convergent.

(d) Show that $\lim_{n\to\infty} a_n = \lim_{n\to\infty} b_n$.

127. (a) Let $f(x) = \sin x$ and $a_n = n \sin 1/n$. Show that

$$\lim_{n\to\infty} a_n - f'(0) = 1.$$

(b) Let $f(x)$ be differentiable on the interval $[0, 1]$ and $f(0) = 0$. Consider the sequence $\{a_n\}$, where $a_n = n f(1/n)$. Show that $\lim_{n\to\infty} a_n = f'(0)$.

128. Consider the sequence $\{a_n\} = \{nr^n\}$. Decide whether $\{a_n\}$ converges for each value of r.

(a) $r = \frac{1}{2}$ (b) $r = 1$ (c) $r = \frac{3}{2}$

(d) For what values or r does the sequence $\{nr^n\}$ converge?

129. (a) Show that $\int_1^n \ln x\, dx < \ln(n!)$ for $n \geq 2$.

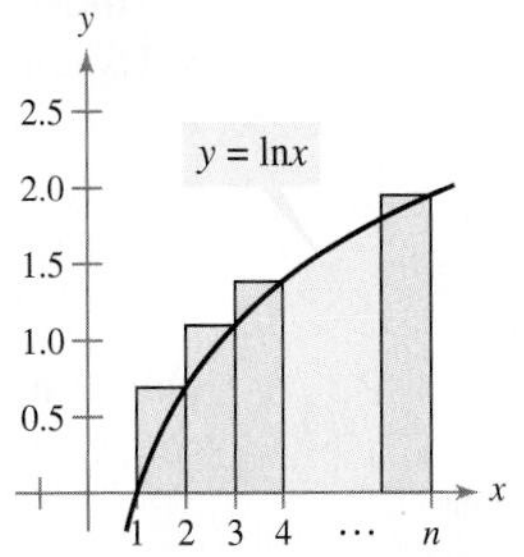

(b) Draw a graph similar to the one above that shows

$$\ln(n!) < \int_1^{n+1} \ln x\, dx.$$

(c) Use the results of parts (a) and (b) to show that

$$\frac{n^n}{e^{n-1}} < n! < \frac{(n+1)^{n+1}}{e^n}, \text{ for } n > 1.$$

(d) Use the Squeeze Theorem for Sequences and the result of part (c) to show that

$$\lim_{n\to\infty} \frac{\sqrt[n]{n!}}{n} = \frac{1}{e}.$$

(e) Test the result of part (d) for $n = 20$, 50, and 100.

130. Consider the sequence $\{a_n\} = \left\{\frac{1}{n}\sum_{k=1}^{n} \frac{1}{1 + (k/n)}\right\}$.

(a) Write the first five terms of $\{a_n\}$.

(b) Show that $\lim_{n\to\infty} a_n = \ln 2$ by interpreting a_n as a Riemann sum of a definite integral.

131. Prove, using the definition of the limit of a sequence, that

$$\lim_{n\to\infty} \frac{1}{n^3} = 0.$$

132. Prove, using the definition of the limit of a sequence, that

$$\lim_{n\to\infty} r^n = 0 \text{ for } -1 < r < 1.$$

133. Complete the proof of Theorem 9.5.

Putnam Exam Challenge

134. Let $\{x_n\}$, $n \geq 0$, be a sequence of nonzero real numbers such that $x_n^2 - x_{n-1}x_{n+1} = 1$ for $n = 1, 2, 3, \ldots$. Prove that there exists a real number a such that $x_{n+1} = ax_n - x_{n-1}$, for all $n \geq 1$.

135. Let $T_0 = 2$, $T_1 = 3$, $T_2 = 6$, and, for $n \geq 3$,

$$T_n = (n + 4)T_{n-1} - 4nT_{n-2} + (4n - 8)T_{n-3}.$$

The first 10 terms of the sequence are

2, 3, 6, 14, 40, 152, 784, 5168, 40,576, 363,392.

Find, with proof, a formula for T_n of the form $T_n = A_n + B_n$, where $\{A_n\}$ and $\{B_n\}$ are well-known sequences.

These problems were composed by the Committee on the Putnam Prize Competition.

Section 9.2 Series and Convergence

- **Understand the definition of a convergent infinite series.**
- **Use properties of infinite geometric series.**
- **Use the *n*th-Term Test for Divergence of an infinite series.**

Infinite Series

One important application of infinite sequences is in representing "infinite summations." Informally, if $\{a_n\}$ is an infinite sequence, then

$$\sum_{n=1}^{\infty} a_n = a_1 + a_2 + a_3 + \cdot \cdot \cdot + a_n + \cdot \cdot \cdot \qquad \text{Infinite series}$$

is an **infinite series** (or simply a **series**). The numbers a_1, a_2, a_3, are the **terms** of the series. For some series it is convenient to begin the index at $n = 0$ (or some other integer). As a typesetting convention, it is common to represent an infinite series as simply $\Sigma\, a_n$. In such cases, the starting value for the index must be taken from the context of the statement.

To find the sum of an infinite series, consider the following **sequence of partial sums.**

$$S_1 = a_1$$
$$S_2 = a_1 + a_2$$
$$S_3 = a_1 + a_2 + a_3$$
$$\vdots$$
$$S_n = a_1 + a_2 + a_3 + \cdot \cdot \cdot + a_n$$

If this sequence of partial sums converges, the series is said to converge and has the sum indicated in the following definition.

INFINITE SERIES

The study of infinite series was considered a novelty in the fourteenth century. Logician Richard Suiseth, whose nickname was Calculator, solved this problem.

If throughout the first half of a given time interval a variation continues at a certain intensity, throughout the next quarter of the interval at double the intensity, throughout the following eighth at triple the intensity and so ad infinitum; then the average intensity for the whole interval will be the intensity of the variation during the second subinterval (or double the intensity).

This is the same as saying that the sum of the infinite series

$$\frac{1}{2} + \frac{2}{4} + \frac{3}{8} + \cdot \cdot \cdot + \frac{n}{2^n} + \cdot \cdot \cdot$$

is 2.

Definitions of Convergent and Divergent Series

For the infinite series $\sum_{n=1}^{\infty} a_n$, the ***n*th partial sum** is given by

$$S_n = a_1 + a_2 + \cdot \cdot \cdot + a_n.$$

If the sequence of partial sums $\{S_n\}$ converges to S, then the series $\sum_{n=1}^{\infty} a_n$ **converges.** The limit S is called the **sum of the series.**

$$S = a_1 + a_2 + \cdot \cdot \cdot + a_n + \cdot \cdot \cdot$$

If $\{S_n\}$ diverges, then the series **diverges.**

STUDY TIP As you study this chapter, you will see that there are two basic questions involving infinite series. Does a series converge or does it diverge? If a series converges, what is its sum? These questions are not always easy to answer, especially the second one.

EXPLORATION

Finding the Sum of an Infinite Series Find the sum of each infinite series. Explain your reasoning.

a. $0.1 + 0.01 + 0.001 + 0.0001 + \cdot \cdot \cdot$

b. $\frac{3}{10} + \frac{3}{100} + \frac{3}{1000} + \frac{3}{10{,}000} + \cdot \cdot \cdot$

c. $1 + \frac{1}{2} + \frac{1}{4} + \frac{1}{8} + \frac{1}{16} + \cdot \cdot \cdot$

d. $\frac{15}{100} + \frac{15}{10{,}000} + \frac{15}{1{,}000{,}000} + \cdot \cdot \cdot$

TECHNOLOGY Figure 9.5 shows the first 15 partial sums of the infinite series in Example 1(a). Notice how the values appear to approach the line $y = 1$.

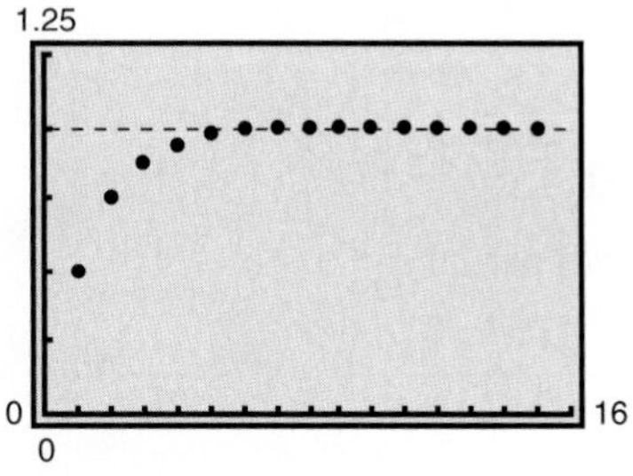

Figure 9.5

EXAMPLE 1 Convergent and Divergent Series

a. The series

$$\sum_{n=1}^{\infty} \frac{1}{2^n} = \frac{1}{2} + \frac{1}{4} + \frac{1}{8} + \frac{1}{16} + \cdot \cdot \cdot$$

has the following partial sums.

$$S_1 = \frac{1}{2}$$

$$S_2 = \frac{1}{2} + \frac{1}{4} = \frac{3}{4}$$

$$S_3 = \frac{1}{2} + \frac{1}{4} + \frac{1}{8} = \frac{7}{8}$$

$$\vdots$$

$$S_n = \frac{1}{2} + \frac{1}{4} + \frac{1}{8} + \cdot \cdot \cdot + \frac{1}{2^n} = \frac{2^n - 1}{2^n}$$

Because

$$\lim_{n\to\infty} \frac{2^n - 1}{2^n} = 1$$

it follows that the series converges and its sum is 1.

b. The nth partial sum of the series

$$\sum_{n=1}^{\infty} \left(\frac{1}{n} - \frac{1}{n+1}\right) = \left(1 - \frac{1}{2}\right) + \left(\frac{1}{2} - \frac{1}{3}\right) + \left(\frac{1}{3} - \frac{1}{4}\right) + \cdot \cdot \cdot$$

is given by

$$S_n = 1 - \frac{1}{n+1}.$$

Because the limit of S_n is 1, the series converges and its sum is 1.

c. The series

$$\sum_{n=1}^{\infty} 1 = 1 + 1 + 1 + 1 + \cdot \cdot \cdot$$

diverges because $S_n = n$ and the sequence of partial sums diverges.

NOTE You can geometrically determine the partial sums of the series in Example 1(a) using Figure 9.6.

Figure 9.6

The series in Example 1(b) is a **telescoping series** of the form

$$(b_1 - b_2) + (b_2 - b_3) + (b_3 - b_4) + (b_4 - b_5) + \cdot \cdot \cdot .$$ Telescoping series

Note that b_2 is canceled by the second term, b_3 is canceled by the third term, and so on. Because the nth partial sum of this series is

$$S_n = b_1 - b_{n+1}$$

it follows that a telescoping series will converge if and only if b_n approaches a finite number as $n \to \infty$. Moreover, if the series converges, its sum is

$$S = b_1 - \lim_{n\to\infty} b_{n+1}.$$

FOR FURTHER INFORMATION To learn more about the partial sums of infinite series, see the article "Six Ways to Sum a Series" by Dan Kalman in *The College Mathematics Journal*. To view this article, go to the website *www.matharticles.com*.

EXAMPLE 2 Writing a Series in Telescoping Form

Find the sum of the series $\displaystyle\sum_{n=1}^{\infty} \frac{2}{4n^2 - 1}$.

Solution
Using partial fractions, you can write

$$a_n = \frac{2}{4n^2 - 1} = \frac{2}{(2n-1)(2n+1)} = \frac{1}{2n-1} - \frac{1}{2n+1}.$$

From this telescoping form, you can see that the nth partial sum is

$$S_n = \left(\frac{1}{1} - \frac{1}{3}\right) + \left(\frac{1}{3} - \frac{1}{5}\right) + \cdot \cdot \cdot + \left(\frac{1}{2n-1} - \frac{1}{2n+1}\right) = 1 - \frac{1}{2n+1}.$$

So, the series converges and its sum is 1. That is,

$$\sum_{n=1}^{\infty} \frac{2}{4n^2 - 1} = \lim_{n\to\infty} S_n = \lim_{n\to\infty}\left(1 - \frac{1}{2n+1}\right) = 1.$$

EXPLORATION

In "Proof Without Words," by Benjamin G. Klein and Irl C. Bivens, the authors present the following diagram. Explain why the final statement below the diagram is valid. How is this result related to Theorem 9.6?

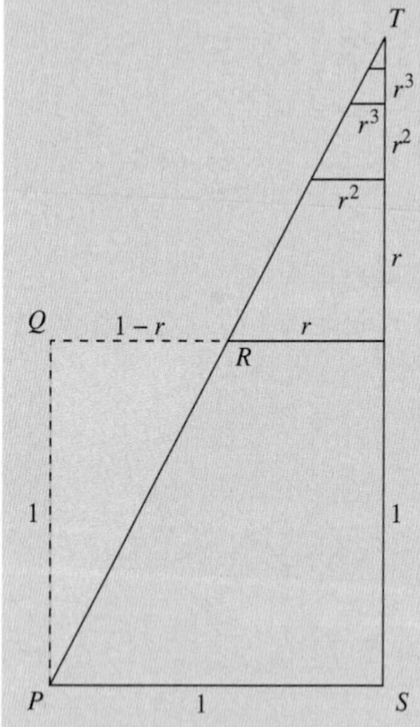

$$\Delta PQR \approx \Delta TSP$$

$$1 + r + r^2 + r^3 + \cdot \cdot \cdot = \frac{1}{1 - r}$$

Exercise taken from "Proof Without Words" by Benjamin G. Klein and Irl C. Bivens, *Mathematics Magazine*, October 1988, by permission of the authors.

Geometric Series

The series given in Example 1(a) is a **geometric series.** In general, the series given by

$$\sum_{n=0}^{\infty} ar^n = a + ar + ar^2 + \cdot \cdot \cdot + ar^n + \cdot \cdot \cdot, \quad a \neq 0 \qquad \text{Geometric series}$$

is a **geometric series** with ratio r.

THEOREM 9.6 Convergence of a Geometric Series

A geometric series with ratio r diverges if $|r| \geq 1$. If $0 < |r| < 1$, then the series converges to the sum

$$\sum_{n=0}^{\infty} ar^n = \frac{a}{1-r}, \quad 0 < |r| < 1.$$

Proof It is easy to see that the series diverges if $r = \pm 1$. If $r \neq \pm 1$, then $S_n = a + ar + ar^2 + \cdot \cdot \cdot + ar^{n-1}$. Multiplication by r yields

$$rS_n = ar + ar^2 + ar^3 + \cdot \cdot \cdot + ar^n.$$

Subtracting the second equation from the first produces $S_n - rS_n = a - ar^n$. Therefore, $S_n(1 - r) = a(1 - r^n)$, and the nth partial sum is

$$S_n = \frac{a}{1-r}(1 - r^n).$$

If $0 < |r| < 1$, it follows that $r^n \to 0$ as $n \to \infty$, and you obtain

$$\lim_{n\to\infty} S_n = \lim_{n\to\infty}\left[\frac{a}{1-r}(1 - r^n)\right] = \frac{a}{1-r}\left[\lim_{n\to\infty}(1 - r^n)\right] = \frac{a}{1-r}$$

which means that the series *converges* and its sum is $a/(1 - r)$. It is left to you to show that the series diverges if $|r| > 1$.

TECHNOLOGY Try using a graphing utility or writing a computer program to compute the sum of the first 20 terms of the sequence in Example 3(a). You should obtain a sum of about 5.999994.

EXAMPLE 3 Convergent and Divergent Geometric Series

a. The geometric series

$$\sum_{n=0}^{\infty} \frac{3}{2^n} = \sum_{n=0}^{\infty} 3\left(\frac{1}{2}\right)^n$$
$$= 3(1) + 3\left(\frac{1}{2}\right) + 3\left(\frac{1}{2}\right)^2 + \cdots$$

has a ratio of $r = \frac{1}{2}$ with $a = 3$. Because $0 < |r| < 1$, the series converges and its sum is

$$S = \frac{a}{1-r} = \frac{3}{1-(1/2)} = 6.$$

b. The geometric series

$$\sum_{n=0}^{\infty} \left(\frac{3}{2}\right)^n = 1 + \frac{3}{2} + \frac{9}{4} + \frac{27}{8} + \cdots$$

has a ratio of $r = \frac{3}{2}$. Because $|r| \geq 1$, the series diverges.

The formula for the sum of a geometric series can be used to write a repeating decimal as the ratio of two integers, as demonstrated in the next example.

EXAMPLE 4 A Geometric Series for a Repeating Decimal

Use a geometric series to write $0.\overline{08}$ as the ratio of two integers.

Solution For the repeating decimal $0.\overline{08}$, you can write

$$0.080808\ldots = \frac{8}{10^2} + \frac{8}{10^4} + \frac{8}{10^6} + \frac{8}{10^8} + \cdots$$
$$= \sum_{n=0}^{\infty} \left(\frac{8}{10^2}\right)\left(\frac{1}{10^2}\right)^n.$$

For this series, you have $a = 8/10^2$ and $r = 1/10^2$. So,

$$0.080808\ldots = \frac{a}{1-r} = \frac{8/10^2}{1-(1/10^2)} = \frac{8}{99}.$$

Try dividing 8 by 99 on a calculator to see that it produces $0.\overline{08}$.

The convergence of a series is not affected by removal of a finite number of terms from the beginning of the series. For instance, the geometric series

$$\sum_{n=4}^{\infty} \left(\frac{1}{2}\right)^n \quad \text{and} \quad \sum_{n=0}^{\infty} \left(\frac{1}{2}\right)^n$$

both converge. Furthermore, because the sum of the second series is $a/(1-r) = 2$, you can conclude that the sum of the first series is

$$S = 2 - \left[\left(\frac{1}{2}\right)^0 + \left(\frac{1}{2}\right)^1 + \left(\frac{1}{2}\right)^2 + \left(\frac{1}{2}\right)^3\right]$$
$$= 2 - \frac{15}{8} = \frac{1}{8}.$$

STUDY TIP As you study this chapter, it is important to distinguish between an infinite series and a sequence. A sequence is an ordered collection of numbers

$$a_1, a_2, a_3, \ldots, a_n, \ldots$$

whereas a series is an infinite sum of terms from a sequence

$$a_1 + a_2 + \cdots + a_n + \cdots.$$

The following properties are direct consequences of the corresponding properties of limits of sequences.

THEOREM 9.7 Properties of Infinite Series

If $\Sigma\, a_n = A$, $\Sigma\, b_n = B$, and c is a real number, then the following series converge to the indicated sums.

1. $\displaystyle\sum_{n=1}^{\infty} ca_n = cA$

2. $\displaystyle\sum_{n=1}^{\infty} (a_n + b_n) = A + B$

3. $\displaystyle\sum_{n=1}^{\infty} (a_n - b_n) = A - B$

*n*th-Term Test for Divergence

The following theorem states that if a series converges, the limit of its nth term must be 0.

NOTE Be sure you see that the converse of Theorem 9.8 is generally not true. That is, if the sequence $\{a_n\}$ converges to 0, then the series $\Sigma\, a_n$ may either converge or diverge.

THEOREM 9.8 Limit of *n*th Term of a Convergent Series

If $\displaystyle\sum_{n=1}^{\infty} a_n$ converges, then $\displaystyle\lim_{n\to\infty} a_n = 0$.

Proof Assume that

$$\sum_{n=1}^{\infty} a_n = \lim_{n\to\infty} S_n = L.$$

Then, because $S_n = S_{n-1} + a_n$ and

$$\lim_{n\to\infty} S_n = \lim_{n\to\infty} S_{n-1} = L$$

it follows that

$$\begin{aligned} L = \lim_{n\to\infty} S_n &= \lim_{n\to\infty} (S_{n-1} + a_n) \\ &= \lim_{n\to\infty} S_{n-1} + \lim_{n\to\infty} a_n \\ &= L + \lim_{n\to\infty} a_n \end{aligned}$$

which implies that $\{a_n\}$ converges to 0.

The contrapositive of Theorem 9.8 provides a useful test for *divergence.* This ***n*th-Term Test for Divergence** states that if the limit of the nth term of a series does *not* converge to 0, the series must diverge.

THEOREM 9.9 *n*th-Term Test for Divergence

If $\displaystyle\lim_{n\to\infty} a_n \neq 0$, then $\displaystyle\sum_{n=1}^{\infty} a_n$ diverges.

EXAMPLE 5 Using the *n*th-Term Test for Divergence

a. For the series $\sum_{n=0}^{\infty} 2^n$, you have

$$\lim_{n\to\infty} 2^n = \infty.$$

So, the limit of the nth term is not 0, and the series diverges.

b. For the series $\sum_{n=1}^{\infty} \frac{n!}{2n! + 1}$, you have

$$\lim_{n\to\infty} \frac{n!}{2n! + 1} = \frac{1}{2}.$$

So, the limit of the nth term is not 0, and the series diverges.

c. For the series $\sum_{n=1}^{\infty} \frac{1}{n}$, you have

$$\lim_{n\to\infty} \frac{1}{n} = 0.$$

Because the limit of the nth term is 0, the nth-Term Test for Divergence does *not* apply and you can draw no conclusions about convergence or divergence. (In the next section, you will see that this particular series diverges.)

STUDY TIP The series in Example 5(c) will play an important role in this chapter.

$$\sum_{n=1}^{\infty} \frac{1}{n} = 1 + \frac{1}{2} + \frac{1}{3} + \frac{1}{4} + \cdot \cdot \cdot$$

You will see that this series diverges even though the nth term approaches 0 as n approaches ∞.

EXAMPLE 6 Bouncing Ball Problem

A ball is dropped from a height of 6 feet and begins bouncing, as shown in Figure 9.7. The height of each bounce is three-fourths the height of the previous bounce. Find the total vertical distance traveled by the ball.

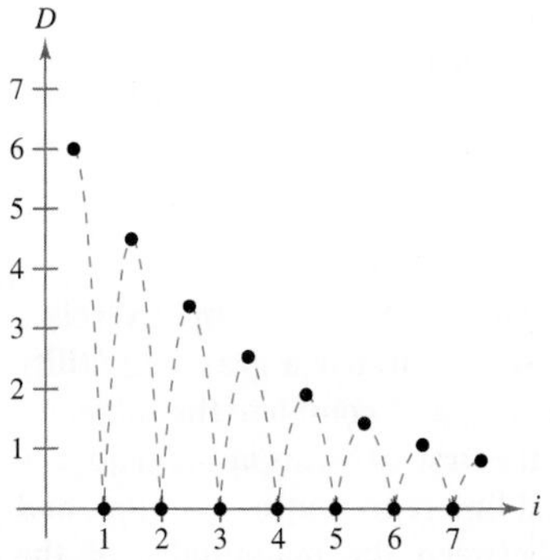

The height of each bounce is three-fourths the height of the preceding bounce.
Figure 9.7

Solution When the ball hits the ground for the first time, it has traveled a distance of $D_1 = 6$ feet. For subsequent bounces, let D_i be the distance traveled up and down. For example, D_2 and D_3 are as follows.

$$D_2 = \underbrace{6\left(\tfrac{3}{4}\right)}_{\text{Up}} + \underbrace{6\left(\tfrac{3}{4}\right)}_{\text{Down}} = 12\left(\tfrac{3}{4}\right)$$

$$D_3 = \underbrace{6\left(\tfrac{3}{4}\right)\left(\tfrac{3}{4}\right)}_{\text{Up}} + \underbrace{6\left(\tfrac{3}{4}\right)\left(\tfrac{3}{4}\right)}_{\text{Down}} = 12\left(\tfrac{3}{4}\right)^2$$

By continuing this process, it can be determined that the total vertical distance is

$$\begin{aligned} D &= 6 + 12\left(\tfrac{3}{4}\right) + 12\left(\tfrac{3}{4}\right)^2 + 12\left(\tfrac{3}{4}\right)^3 + \cdot \cdot \cdot \\ &= 6 + 12 \sum_{n=0}^{\infty} \left(\tfrac{3}{4}\right)^{n+1} \\ &= 6 + 12\left(\tfrac{3}{4}\right) \sum_{n=0}^{\infty} \left(\tfrac{3}{4}\right)^n \\ &= 6 + 9\left(\frac{1}{1 - \frac{3}{4}}\right) \\ &= 6 + 9(4) \\ &= 42 \text{ feet.} \end{aligned}$$

Exercises for Section 9.2

See www.CalcChat.com for worked-out solutions to odd-numbered exercises.

In Exercises 1–6, find the first five terms of the sequence of partial sums.

1. $1 + \frac{1}{4} + \frac{1}{9} + \frac{1}{16} + \frac{1}{25} + \cdots$

2. $\frac{1}{2 \cdot 3} + \frac{2}{3 \cdot 4} + \frac{3}{4 \cdot 5} + \frac{4}{5 \cdot 6} + \frac{5}{6 \cdot 7} + \cdots$

3. $3 - \frac{9}{2} + \frac{27}{4} - \frac{81}{8} + \frac{243}{16} - \cdots$

4. $\frac{1}{1} + \frac{1}{3} + \frac{1}{5} + \frac{1}{7} + \frac{1}{9} + \frac{1}{11} + \cdots$

5. $\sum_{n=1}^{\infty} \frac{3}{2^{n-1}}$

6. $\sum_{n=1}^{\infty} \frac{(-1)^{n+1}}{n!}$

In Exercises 7–16, verify that the infinite series diverges.

7. $\sum_{n=0}^{\infty} 3\left(\frac{3}{2}\right)^n$

8. $\sum_{n=0}^{\infty} \left(\frac{4}{3}\right)^n$

9. $\sum_{n=0}^{\infty} 1000(1.055)^n$

10. $\sum_{n=0}^{\infty} 2(-1.03)^n$

11. $\sum_{n=1}^{\infty} \frac{n}{n+1}$

12. $\sum_{n=1}^{\infty} \frac{n}{2n+3}$

13. $\sum_{n=1}^{\infty} \frac{n^2}{n^2+1}$

14. $\sum_{n=1}^{\infty} \frac{n}{\sqrt{n^2+1}}$

15. $\sum_{n=1}^{\infty} \frac{2^n+1}{2^{n+1}}$

16. $\sum_{n=1}^{\infty} \frac{n!}{2^n}$

In Exercises 17–22, match the series with the graph of its sequence of partial sums. [The graphs are labeled (a), (b), (c), (d), (e), and (f).] Use the graph to estimate the sum of the series. Confirm your answer analytically.

(a)

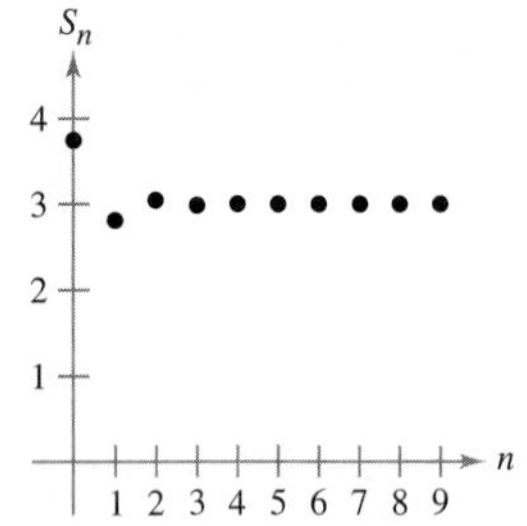

(b)

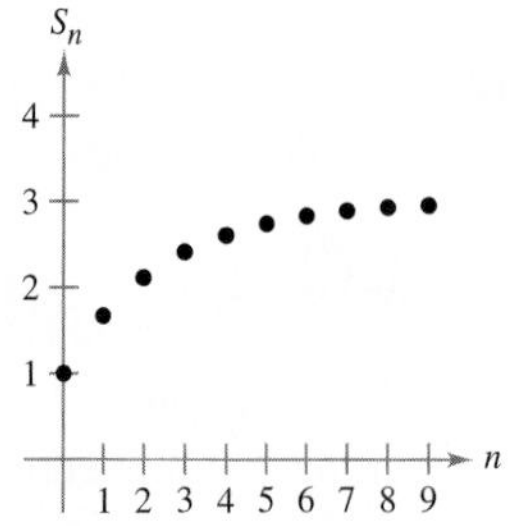

(c)

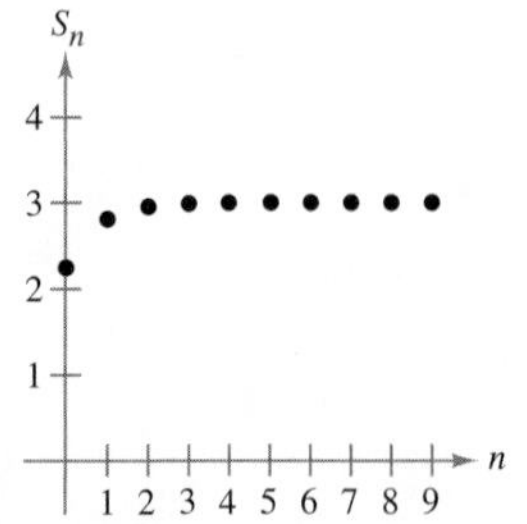

(d)

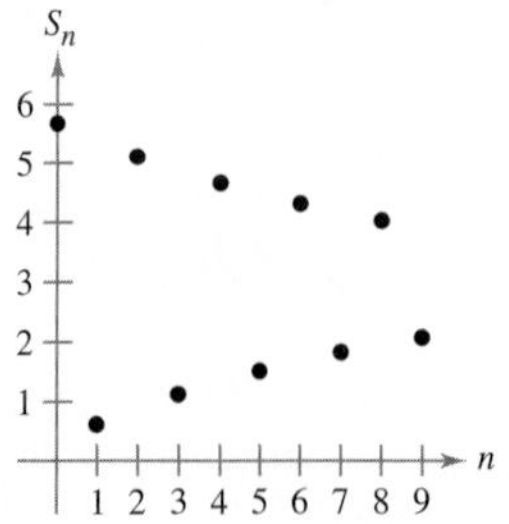

(e)

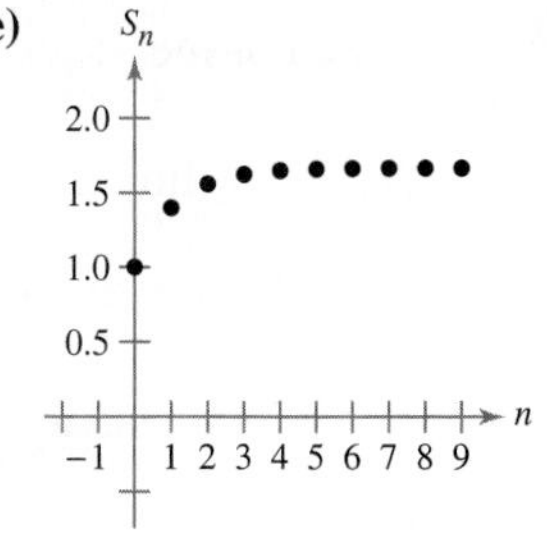

(f)

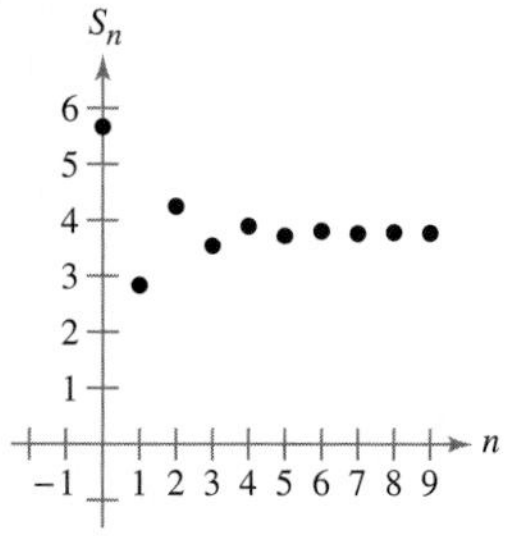

17. $\sum_{n=0}^{\infty} \frac{9}{4}\left(\frac{1}{4}\right)^n$

18. $\sum_{n=0}^{\infty} \left(\frac{2}{3}\right)^n$

19. $\sum_{n=0}^{\infty} \frac{15}{4}\left(-\frac{1}{4}\right)^n$

20. $\sum_{n=0}^{\infty} \frac{17}{3}\left(-\frac{8}{9}\right)^n$

21. $\sum_{n=0}^{\infty} \frac{17}{3}\left(-\frac{1}{2}\right)^n$

22. $\sum_{n=0}^{\infty} \left(\frac{2}{5}\right)^n$

In Exercises 23–28, verify that the infinite series converges.

23. $\sum_{n=1}^{\infty} \frac{1}{n(n+1)}$ (Use partial fractions.)

24. $\sum_{n=1}^{\infty} \frac{1}{n(n+2)}$ (Use partial fractions.)

25. $\sum_{n=0}^{\infty} 2\left(\frac{3}{4}\right)^n$

26. $\sum_{n=1}^{\infty} 2\left(-\frac{1}{2}\right)^n$

27. $\sum_{n=0}^{\infty} (0.9)^n = 1 + 0.9 + 0.81 + 0.729 + \cdots$

28. $\sum_{n=0}^{\infty} (-0.6)^n = 1 - 0.6 + 0.36 - 0.216 + \cdots$

Numerical, Graphical, and Analytic Analysis **In Exercises 29–34, (a) find the sum of the series, (b) use a graphing utility to find the indicated partial sum S_n and complete the table, (c) use a graphing utility to graph the first 10 terms of the sequence of partial sums and a horizontal line representing the sum, and (d) explain the relationship between the magnitudes of the terms of the series and the rate at which the sequence of partial sums approaches the sum of the series.**

n	5	10	20	50	100
S_n					

29. $\sum_{n=1}^{\infty} \frac{6}{n(n+3)}$

30. $\sum_{n=1}^{\infty} \frac{4}{n(n+4)}$

31. $\sum_{n=1}^{\infty} 2(0.9)^{n-1}$

32. $\sum_{n=1}^{\infty} 3(0.85)^{n-1}$

33. $\sum_{n=1}^{\infty} 10(0.25)^{n-1}$

34. $\sum_{n=1}^{\infty} 5\left(-\frac{1}{3}\right)^{n-1}$

In Exercises 35–50, find the sum of the convergent series.

35. $\sum_{n=2}^{\infty} \frac{1}{n^2-1}$

36. $\sum_{n=1}^{\infty} \frac{4}{n(n+2)}$

37. $\sum_{n=1}^{\infty} \frac{8}{(n+1)(n+2)}$

38. $\sum_{n=1}^{\infty} \frac{1}{(2n+1)(2n+3)}$

39. $\sum_{n=0}^{\infty} \left(\frac{1}{2}\right)^n$

40. $\sum_{n=0}^{\infty} 6\left(\frac{4}{5}\right)^n$

41. $\sum_{n=0}^{\infty} \left(-\frac{1}{2}\right)^n$

42. $\sum_{n=0}^{\infty} 2\left(-\frac{2}{3}\right)^n$

43. $1 + 0.1 + 0.01 + 0.001 + \cdots$

44. $8 + 6 + \frac{9}{2} + \frac{27}{8} + \cdots$

45. $3 - 1 + \frac{1}{3} - \frac{1}{9} + \cdots$

46. $4 - 2 + 1 - \frac{1}{2} + \cdots$

47. $\sum_{n=0}^{\infty} \left(\frac{1}{2^n} - \frac{1}{3^n}\right)$

48. $\sum_{n=1}^{\infty} [(0.7)^n + (0.9)^n]$

49. $\sum_{n=1}^{\infty} (\sin 1)^n$

50. $\sum_{n=1}^{\infty} \frac{1}{9n^2 + 3n - 2}$

In Exercises 51–56, (a) write the repeating decimal as a geometric series and (b) write its sum as the ratio of two integers.

51. $0.\overline{4}$

52. $0.\overline{9}$

53. $0.\overline{81}$

54. $0.\overline{01}$

55. $0.0\overline{75}$

56. $0.2\overline{15}$

In Exercises 57–72, determine the convergence or divergence of the series.

57. $\sum_{n=1}^{\infty} \frac{n+10}{10n+1}$

58. $\sum_{n=1}^{\infty} \frac{n+1}{2n-1}$

59. $\sum_{n=1}^{\infty} \left(\frac{1}{n} - \frac{1}{n+2}\right)$

60. $\sum_{n=1}^{\infty} \frac{1}{n(n+3)}$

61. $\sum_{n=1}^{\infty} \frac{3n-1}{2n+1}$

62. $\sum_{n=1}^{\infty} \frac{3^n}{n^3}$

63. $\sum_{n=0}^{\infty} \frac{4}{2^n}$

64. $\sum_{n=0}^{\infty} \frac{1}{4^n}$

65. $\sum_{n=0}^{\infty} (1.075)^n$

66. $\sum_{n=1}^{\infty} \frac{2^n}{100}$

67. $\sum_{n=2}^{\infty} \frac{n}{\ln n}$

68. $\sum_{n=1}^{\infty} \ln \frac{1}{n}$

69. $\sum_{n=1}^{\infty} \left(1 + \frac{k}{n}\right)^n$

70. $\sum_{n=1}^{\infty} e^{-n}$

71. $\sum_{n=1}^{\infty} \arctan n$

72. $\sum_{n=1}^{\infty} \ln\left(\frac{n+1}{n}\right)$

Writing About Concepts

73. State the definitions of convergent and divergent series.

74. Describe the difference between $\lim_{n \to \infty} a_n = 5$ and $\sum_{n=1}^{\infty} a_n = 5$.

75. Define a geometric series, state when it converges, and give the formula for the sum of a convergent geometric series.

Writing About Concepts *(continued)*

76. State the nth-Term Test for Divergence.

77. Let $a_n = \frac{n+1}{n}$. Discuss the convergence of $\{a_n\}$ and $\sum_{n=1}^{\infty} a_n$.

78. Explain any differences among the following series.

(a) $\sum_{n=1}^{\infty} a_n$ (b) $\sum_{k=1}^{\infty} a_k$ (c) $\sum_{n=1}^{\infty} a_k$

In Exercises 79–86, find all values of x for which the series converges. For these values of x, write the sum of the series as a function of x.

79. $\sum_{n=1}^{\infty} \frac{x^n}{2^n}$

80. $\sum_{n=1}^{\infty} (3x)^n$

81. $\sum_{n=1}^{\infty} (x-1)^n$

82. $\sum_{n=0}^{\infty} 4\left(\frac{x-3}{4}\right)^n$

83. $\sum_{n=0}^{\infty} (-1)^n x^n$

84. $\sum_{n=0}^{\infty} (-1)^n x^{2n}$

85. $\sum_{n=0}^{\infty} \left(\frac{1}{x}\right)^n$

86. $\sum_{n=1}^{\infty} \left(\frac{x^2}{x^2+4}\right)^n$

87. (a) You delete a finite number of terms from a divergent series. Will the new series still diverge? Explain your reasoning.

(b) You add a finite number of terms to a convergent series. Will the new series still converge? Explain your reasoning.

88. ***Think About It*** Consider the formula

$$\frac{1}{x-1} = 1 + x + x^2 + x^3 + \cdots.$$

Given $x = -1$ and $x = 2$, can you conclude that either of the following statements is true? Explain your reasoning.

(a) $\frac{1}{2} = 1 - 1 + 1 - 1 + \cdots$

(b) $-1 = 1 + 2 + 4 + 8 + \cdots$

In Exercises 89 and 90, (a) find the common ratio of the geometric series, (b) write the function that gives the sum of the series, and (c) use a graphing utility to graph the function and the partial sums S_3 and S_5. What do you notice?

89. $1 + x + x^2 + x^3 + \cdots$

90. $1 - \frac{x}{2} + \frac{x^2}{4} - \frac{x^3}{8} + \cdots$

In Exercises 91 and 92, use a graphing utility to graph the function. Identify the horizontal asymptote of the graph and determine its relationship to the sum of the series.

	Function	*Series*
91.	$f(x) = 3\left[\frac{1-(0.5)^x}{1-0.5}\right]$	$\sum_{n=0}^{\infty} 3\left(\frac{1}{2}\right)^n$
92.	$f(x) = 2\left[\frac{1-(0.8)^x}{1-0.8}\right]$	$\sum_{n=0}^{\infty} 2\left(\frac{4}{5}\right)^n$

Writing **In Exercises 93 and 94, use a graphing utility to determine the first term that is less than 0.0001 in each of the convergent series. Note that the answers are very different. Explain how this will affect the rate at which the series converges.**

93. $\displaystyle\sum_{n=1}^{\infty} \frac{1}{n(n+1)}, \quad \sum_{n=1}^{\infty} \left(\frac{1}{8}\right)^n$ **94.** $\displaystyle\sum_{n=1}^{\infty} \frac{1}{2^n}, \quad \sum_{n=1}^{\infty} (0.01)^n$

95. ***Marketing*** An electronic games manufacturer producing a new product estimates the annual sales to be 8000 units. Each year 10% of the units that have been sold will become inoperative. So, 8000 units will be in use after 1 year, $[8000 + 0.9(8000)]$ units will be in use after 2 years, and so on. How many units will be in use after n years?

96. ***Depreciation*** A company buys a machine for \$225,000 that depreciates at a rate of 30% per year. Find a formula for the value of the machine after n years. What is its value after 5 years?

97. ***Multiplier Effect*** The annual spending by tourists in a resort city is \$100 million. Approximately 75% of that revenue is again spent in the resort city, and of that amount approximately 75% is again spent in the same city, and so on. Write the geometric series that gives the total amount of spending generated by the \$100 million and find the sum of the series.

98. ***Multiplier Effect*** Repeat Exercise 97 if the percent of the revenue that is spent again in the city decreases to 60%.

99. ***Distance*** A ball is dropped from a height of 16 feet. Each time it drops h feet, it rebounds $0.81h$ feet. Find the total distance traveled by the ball.

100. ***Time*** The ball in Exercise 99 takes the following times for each fall.

$$
\begin{aligned}
s_1 &= -16t^2 + 16, & s_1 &= 0 \text{ if } t = 1\\
s_2 &= -16t^2 + 16(0.81), & s_2 &= 0 \text{ if } t = 0.9\\
s_3 &= -16t^2 + 16(0.81)^2, & s_3 &= 0 \text{ if } t = (0.9)^2\\
s_4 &= -16t^2 + 16(0.81)^3, & s_4 &= 0 \text{ if } t = (0.9)^3\\
&\vdots & &\vdots\\
s_n &= -16t^2 + 16(0.81)^{n-1}, & s_n &= 0 \text{ if } t = (0.9)^{n-1}
\end{aligned}
$$

Beginning with s_2, the ball takes the same amount of time to bounce up as it does to fall, and so the total time elapsed before it comes to rest is given by

$$t = 1 + 2\sum_{n=1}^{\infty} (0.9)^n.$$

Find this total time.

Probability **In Exercises 101 and 102, the random variable n represents the number of units of a product sold per day in a store. The probability distribution of n is given by $P(n)$. Find the probability that two units are sold in a given day $[P(2)]$ and show that $P(1) + P(2) + P(3) + \cdots = 1$.**

101. $P(n) = \dfrac{1}{2}\left(\dfrac{1}{2}\right)^n$ **102.** $P(n) = \dfrac{1}{3}\left(\dfrac{2}{3}\right)^n$

103. ***Probability*** A fair coin is tossed repeatedly. The probability that the first head occurs on the nth toss is given by $P(n) = \left(\frac{1}{2}\right)^n$, where $n \geq 1$.

(a) Show that $\displaystyle\sum_{n=1}^{\infty} \left(\frac{1}{2}\right)^n = 1.$

(b) The expected number of tosses required until the first head occurs in the experiment is given by

$$\sum_{n=1}^{\infty} n\left(\frac{1}{2}\right)^n.$$

Is this series geometric?

(c) Use a computer algebra system to find the sum in part (b).

104. ***Probability*** In an experiment, three people toss a fair coin one at a time until one of them tosses a head. Determine, for each person, the probability that he or she tosses the first head. Verify that the sum of the three probabilities is 1.

105. ***Area*** The sides of a square are 16 inches in length. A new square is formed by connecting the midpoints of the sides of the original square, and two of the triangles outside the second square are shaded (see figure). Determine the area of the shaded regions (a) if this process is continued five more times and (b) if this pattern of shading is continued infinitely.

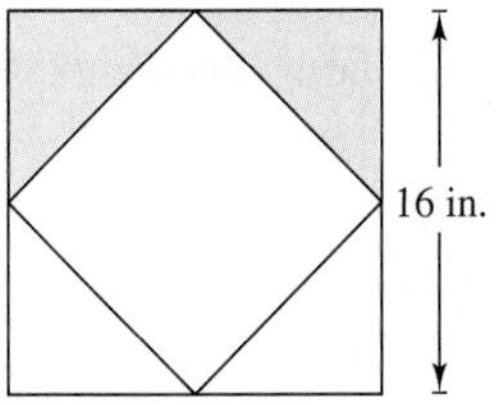

Figure for 105

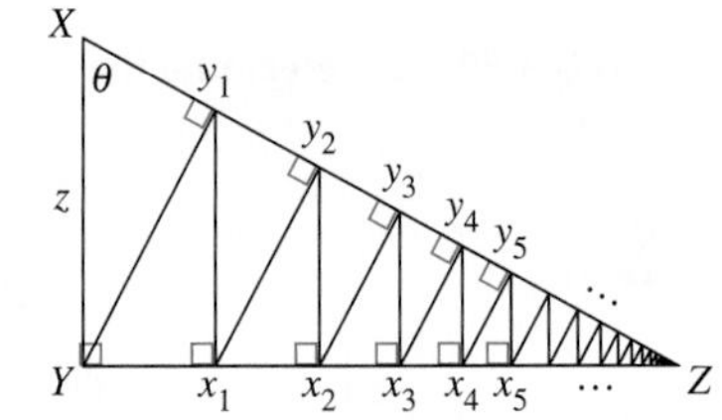

Figure for 106

106. ***Length*** A right triangle XYZ is shown above where $|XY| = z$ and $\angle X = \theta$. Line segments are continually drawn to be perpendicular to the triangle, as shown in the figure.

(a) Find the total length of the perpendicular line segments $|Yy_1| + |x_1y_1| + |x_1y_2| + \cdots$ in terms of z and θ.

(b) If $z = 1$ and $\theta = \pi/6$, find the total length of the perpendicular line segments.

In Exercises 107–110, use the formula for the nth partial sum of a geometric series

$$\sum_{i=0}^{n-1} ar^i = \frac{a(1-r^n)}{1-r}.$$

107. ***Present Value*** The winner of a \$1,000,000 sweepstakes will be paid \$50,000 per year for 20 years. The money earns 6% interest per year. The present value of the winnings is

$$\sum_{n=1}^{20} 50{,}000\left(\frac{1}{1.06}\right)^n.$$

Compute the present value and interpret its meaning.

108. ***Sphereflake*** A sphereflake shown below is a computer-generated fractal that was created by Eric Haines. The radius of the large sphere is 1. To the large sphere, nine spheres of radius $\frac{1}{3}$ are attached. To each of these, nine spheres of radius $\frac{1}{9}$ are attached. This process is continued infinitely. Prove that the sphereflake has an infinite surface area.

Eric Haines

109. ***Salary*** You go to work at a company that pays \$0.01 for the first day, \$0.02 for the second day, \$0.04 for the third day, and so on. If the daily wage keeps doubling, what would your total income be for working (a) 29 days, (b) 30 days, and (c) 31 days?

110. ***Annuities*** When an employee receives a paycheck at the end of each month, P dollars is invested in a retirement account. These deposits are made each month for t years and the account earns interest at the annual percentage rate r. If the interest is compounded monthly, the amount A in the account at the end of t years is

$$A = P + P\left(1 + \frac{r}{12}\right) + \cdot \cdot \cdot + P\left(1 + \frac{r}{12}\right)^{12t-1}$$
$$= P\left(\frac{12}{r}\right)\left[\left(1 + \frac{r}{12}\right)^{12t} - 1\right].$$

If the interest is compounded continuously, the amount A in the account after t years is

$$A = P + Pe^{r/12} + Pe^{2r/12} + Pe^{(12t-1)r/12}$$
$$= \frac{P(e^{rt} - 1)}{e^{r/12} - 1}.$$

Verify the formulas for the sums given above.

Annuities **In Exercises 111–114, consider making monthly deposits of P dollars in a savings account at an annual interest rate r. Use the results of Exercise 110 to find the balance A after t years if the interest is compounded (a) monthly and (b) continuously.**

111. $P = \$50, \quad r = 3\%, \quad t = 20$ years

112. $P = \$75, \quad r = 5\%, \quad t = 25$ years

113. $P = \$100, \quad r = 4\%, \quad t = 40$ years

114. $P = \$20, \quad r = 6\%, \quad t = 50$ years

115. ***Modeling Data*** The annual sales a_n (in millions of dollars) for Avon Products, Inc. from 1993 through 2002 are given below as ordered pairs of the form (n, a_n), where n represents the year, with $n = 3$ corresponding to 1993. *(Source: 2002 Avon Products, Inc. Annual Report)*

(3, 3844), (4, 4267), (5, 4492), (6, 4814), (7, 5079), (8, 5213), (9, 5289), (10, 5682), (11, 5958), (12, 6171)

(a) Use the regression capabilities of a graphing utility to find a model of the form

$$a_n = ce^{kn}, \quad n = 3, 4, 5, \ldots, 12$$

for the data. Graphically compare the points and the model.

(b) Use the data to find the total sales for the 10-year period.

(c) Approximate the total sales for the 10-year period using the formula for the sum of a geometric series. Compare the result with that in part (b).

116. ***Salary*** You accept a job that pays a salary of \$40,000 for the first year. During the next 39 years you receive a 4% raise each year. What would be your total compensation over the 40-year period?

True or False? **In Exercises 117–122, determine whether the statement is true or false. If it is false, explain why or give an example that shows it is false.**

117. If $\lim_{n\to\infty} a_n = 0$, then $\sum_{n=1}^{\infty} a_n$ converges.

118. If $\sum_{n=1}^{\infty} a_n = L$, then $\sum_{n=0}^{\infty} a_n = L + a_0$.

119. If $|r| < 1$, then $\sum_{n=1}^{\infty} ar^n = \frac{a}{(1 - r)}$.

120. The series $\sum_{n=1}^{\infty} \frac{n}{1000(n + 1)}$ diverges.

121. $0.75 = 0.749999 \ldots$.

122. Every decimal with a repeating pattern of digits is a rational number.

123. Show that the series $\sum_{n=1}^{\infty} a_n$ can be written in the telescoping form

$$\sum_{n=1}^{\infty} [(c - S_{n-1}) - (c - S_n)]$$

where $S_0 = 0$ and S_n is the nth partial sum.

124. Let $\Sigma\, a_n$ be a convergent series, and let

$$R_N = a_{N+1} + a_{N+2} + \cdot \cdot \cdot$$

be the remainder of the series after the first N terms. Prove that $\lim_{N\to\infty} R_N = 0$.

125. Find two divergent series $\Sigma\, a_n$ and $\Sigma\, b_n$ such that $\Sigma(a_n + b_n)$ converges.

126. Given two infinite series $\Sigma\, a_n$ and $\Sigma\, b_n$ such that $\Sigma\, a_n$ converges and $\Sigma\, b_n$ diverges, prove that $\Sigma(a_n + b_n)$ diverges.

127. Suppose that $\Sigma\, a_n$ diverges and c is a nonzero constant. Prove that $\Sigma\, ca_n$ diverges.

128. If $\sum_{n=1}^{\infty} a_n$ converges where a_n is nonzero, show that $\sum_{n=1}^{\infty} \frac{1}{a_n}$ diverges.

129. The Fibonacci sequence is defined recursively by $a_{n+2} = a_n + a_{n+1}$, where $a_1 = 1$ and $a_2 = 1$.

(a) Show that $\dfrac{1}{a_{n+1}\,a_{n+3}} = \dfrac{1}{a_{n+1}\,a_{n+2}} - \dfrac{1}{a_{n+2}\,a_{n+3}}$.

(b) Show that $\displaystyle\sum_{n=0}^{\infty} \frac{1}{a_{n+1}\,a_{n+3}} = 1$.

130. Find the values of x for which the infinite series

$$1 + 2x + x^2 + 2x^3 + x^4 + 2x^5 + x^6 + \cdot\cdot\cdot$$

converges. What is the sum when the series converges?

131. Prove that $\dfrac{1}{r} + \dfrac{1}{r^2} + \dfrac{1}{r^3} + \cdot\cdot\cdot = \dfrac{1}{r-1}$, for $|r| > 1$.

132. ***Writing*** The figure below represents an informal way of showing that $\displaystyle\sum_{n=1}^{\infty} \frac{1}{n^2} < 2$. Explain how the figure implies this conclusion.

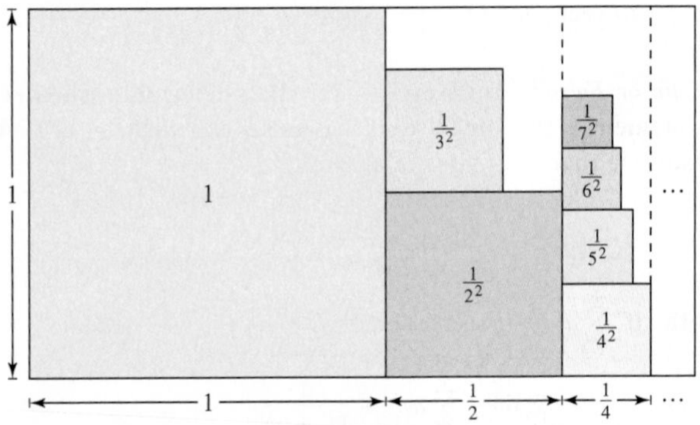

FOR FURTHER INFORMATION For more on this exercise, see the article "Convergence with Pictures" by P.J. Rippon in *American Mathematical Monthly.*

133. ***Writing*** Read the article "The Exponential-Decay Law Applied to Medical Dosages" by Gerald M. Armstrong and Calvin P. Midgley in *Mathematics Teacher*. (To view this article, go to the website *www.matharticles.com*.) Then write a paragraph on how a geometric sequence can be used to find the total amount of a drug that remains in a patient's system after n equal doses have been administered (at equal time intervals).

Putnam Exam Challenge

134. Write $\displaystyle\sum_{k=1}^{\infty} \frac{6^k}{(3^{k+1} - 2^{k+1})(3^k - 2^k)}$ as a rational number.

135. Let $f(n)$ be the sum of the first n terms of the sequence 0, 1, 1, 2, 2, 3, 3, 4, . . . , where the nth term is given by

$$a_n = \begin{cases} n/2, & \text{if } n \text{ is even} \\ (n-1)/2, & \text{if } n \text{ is odd} \end{cases}.$$

Show that if x and y are positive integers and $x > y$ then $xy = f(x + y) - f(x - y)$.

These problems were composed by the Committee on the Putnam Prize Competition.

Section Project: Cantor's Disappearing Table

The following procedure shows how to make a table disappear by removing only half of the table!

(a) Original table has a length of L.

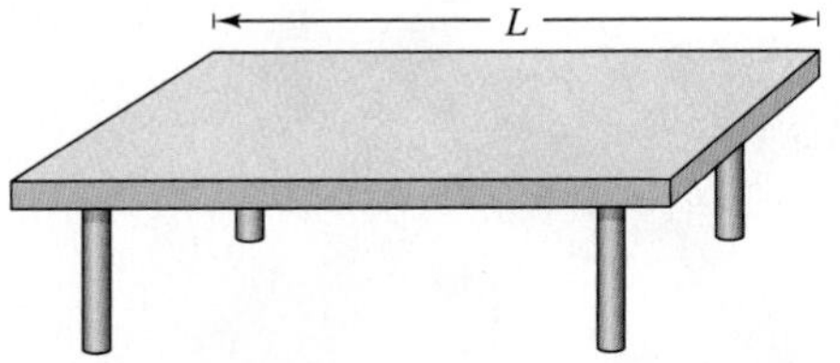

(b) Remove $\frac{1}{4}$ of the table centered at the midpoint. Each remaining piece has a length that is less than $\frac{1}{2}L$.

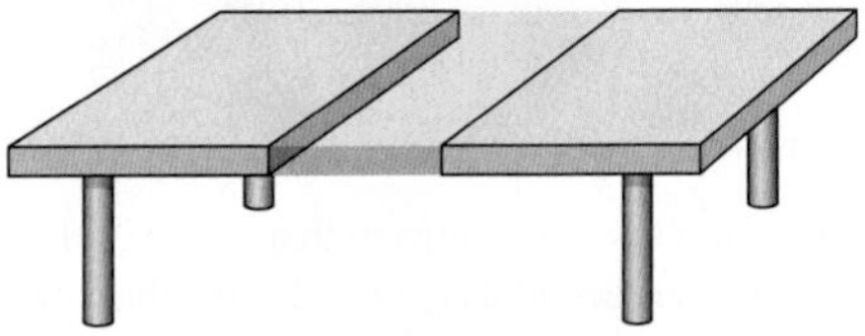

(c) Remove $\frac{1}{8}$ of the table by taking sections of length $\frac{1}{16}L$ from the centers of each of the two remaining pieces. Now, you have removed $\frac{1}{4} + \frac{1}{8}$ of the table. Each remaining piece has a length that is less than $\frac{1}{4}L$.

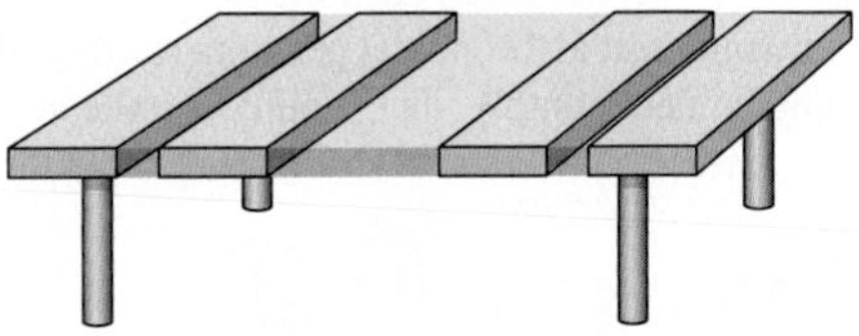

(d) Remove $\frac{1}{16}$ of the table by taking sections of length $\frac{1}{64}L$ from the centers of each of the four remaining pieces. Now, you have removed $\frac{1}{4} + \frac{1}{8} + \frac{1}{16}$ of the table. Each remaining piece has a length that is less than $\frac{1}{8}L$.

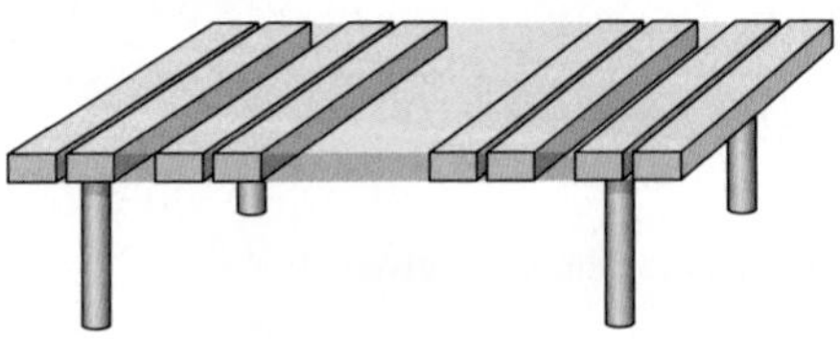

Will continuing this process cause the table to disappear, even though you have only removed half of the table? Why?

FOR FURTHER INFORMATION Read the article "Cantor's Disappearing Table" by Larry E. Knop in *The College Mathematics Journal.* To view this article, go to the website *www.matharticles.com.*

Section 9.3 The Integral Test and p-Series

- **Use the Integral Test to determine whether an infinite series converges or diverges.**
- **Use properties of p-series and harmonic series.**

The Integral Test

In this and the following section, you will study several convergence tests that apply to series with *positive* terms.

> **THEOREM 9.10 The Integral Test**
>
> If f is positive, continuous, and decreasing for $x \geq 1$ and $a_n = f(n)$, then
>
> $$\sum_{n=1}^{\infty} a_n \quad \text{and} \quad \int_1^{\infty} f(x)\,dx$$
>
> either both converge or both diverge.

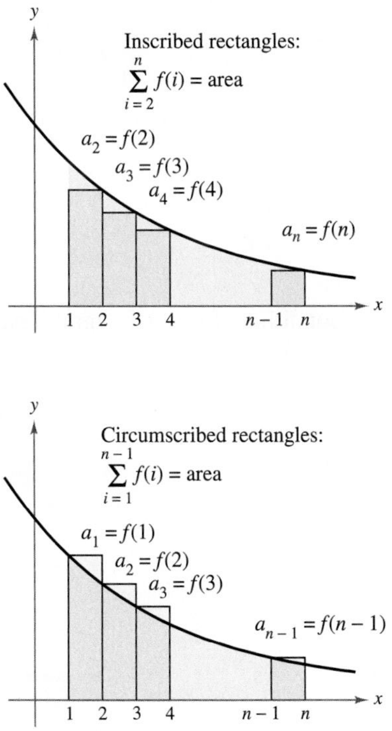

Figure 9.8

Proof Begin by partitioning the interval $[1, n]$ into $n - 1$ unit intervals, as shown in Figure 9.8. The total areas of the inscribed rectangles and the circumscribed rectangles are as follows.

$$\sum_{i=2}^{n} f(i) = f(2) + f(3) + \cdot \cdot \cdot + f(n) \qquad \text{Inscribed area}$$

$$\sum_{i=1}^{n-1} f(i) = f(1) + f(2) + \cdot \cdot \cdot + f(n-1) \qquad \text{Circumscribed area}$$

The exact area under the graph of f from $x = 1$ to $x = n$ lies between the inscribed and circumscribed areas.

$$\sum_{i=2}^{n} f(i) \leq \int_1^n f(x)\,dx \leq \sum_{i=1}^{n-1} f(i)$$

Using the nth partial sum, $S_n = f(1) + f(2) + \cdot \cdot \cdot + f(n)$, you can write this inequality as

$$S_n - f(1) \leq \int_1^n f(x)\,dx \leq S_{n-1}.$$

Now, assuming that $\int_1^{\infty} f(x)\,dx$ converges to L, it follows that for $n \geq 1$

$$S_n - f(1) \leq L \quad \Longrightarrow \quad S_n \leq L + f(1).$$

Consequently, $\{S_n\}$ is bounded and monotonic, and by Theorem 9.5 it converges. So, $\Sigma\, a_n$ converges. For the other direction of the proof, assume that the improper integral diverges. Then $\int_1^n f(x)\,dx$ approaches infinity as $n \to \infty$, and the inequality $S_{n-1} \geq \int_1^n f(x)\,dx$ implies that $\{S_n\}$ diverges. So, $\Sigma\, a_n$ diverges.

NOTE Remember that the convergence or divergence of $\Sigma\, a_n$ is not affected by deleting the first N terms. Similarly, if the conditions for the Integral Test are satisfied for all $x \geq N > 1$, you can simply use the integral $\int_N^{\infty} f(x)\,dx$ to test for convergence or divergence. (This is illustrated in Example 4.)

EXAMPLE 1 Using the Integral Test

Apply the Integral Test to the series $\sum_{n=1}^{\infty} \frac{n}{n^2+1}$.

Solution The function $f(x) = x/(x^2+1)$ is positive and continuous for $x \geq 1$. To determine whether f is decreasing, find the derivative.

$$f'(x) = \frac{(x^2+1)(1) - x(2x)}{(x^2+1)^2} = \frac{-x^2+1}{(x^2+1)^2}$$

So, $f'(x) < 0$ for $x > 1$ and it follows that f satisfies the conditions for the Integral Test. You can integrate to obtain

$$\begin{aligned}\int_1^{\infty} \frac{x}{x^2+1}\,dx &= \frac{1}{2}\int_1^{\infty} \frac{2x}{x^2+1}\,dx \\ &= \frac{1}{2}\lim_{b\to\infty}\int_1^{b} \frac{2x}{x^2+1}\,dx \\ &= \frac{1}{2}\lim_{b\to\infty}\Big[\ln(x^2+1)\Big]_1^b \\ &= \frac{1}{2}\lim_{b\to\infty}[\ln(b^2+1) - \ln 2] \\ &= \infty.\end{aligned}$$

So, the series *diverges*.

EXAMPLE 2 Using the Integral Test

Apply the Integral Test to the series $\sum_{n=1}^{\infty} \frac{1}{n^2+1}$.

Solution Because $f(x) = 1/(x^2+1)$ satisfies the conditions for the Integral Test (check this), you can integrate to obtain

$$\begin{aligned}\int_1^{\infty} \frac{1}{x^2+1}\,dx &= \lim_{b\to\infty}\int_1^{b} \frac{1}{x^2+1}\,dx \\ &= \lim_{b\to\infty}\Big[\arctan x\Big]_1^b \\ &= \lim_{b\to\infty}(\arctan b - \arctan 1) \\ &= \frac{\pi}{2} - \frac{\pi}{4} = \frac{\pi}{4}.\end{aligned}$$

So, the series *converges* (see Figure 9.9).

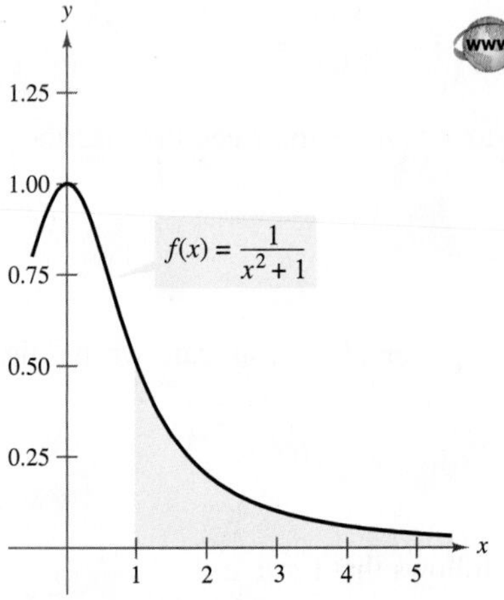

Because the improper integral converges, the infinite series also converges.
Figure 9.9

TECHNOLOGY
In Example 2, the fact that the improper integral converges to $\pi/4$ does not imply that the infinite series converges to $\pi/4$. To approximate the sum of the series, you can use the inequality

$$\sum_{n=1}^{N} \frac{1}{n^2+1} \leq \sum_{n=1}^{\infty} \frac{1}{n^2+1} \leq \sum_{n=1}^{N} \frac{1}{n^2+1} + \int_N^{\infty} \frac{1}{x^2+1}\,dx.$$

(See Exercise 60.) The larger the value of N, the better the approximation. For instance, using $N = 200$ produces $1.072 \leq \Sigma 1/(n^2+1) \leq 1.077$.

HARMONIC SERIES

Pythagoras and his students paid close attention to the development of music as an abstract science. This led to the discovery of the relationship between the tone and the length of the vibrating string. It was observed that the most beautiful musical harmonies corresponded to the simplest ratios of whole numbers. Later mathematicians developed this idea into the harmonic series, where the terms in the harmonic series correspond to the nodes on a vibrating string that produce multiples of the fundamental frequency. For example, $\frac{1}{2}$ is twice the fundamental frequency, $\frac{1}{3}$ is three times the fundamental frequency, and so on.

p-Series and Harmonic Series

In the remainder of this section, you will investigate a second type of series that has a simple arithmetic test for convergence or divergence. A series of the form

$$\sum_{n=1}^{\infty} \frac{1}{n^p} = \frac{1}{1^p} + \frac{1}{2^p} + \frac{1}{3^p} + \cdot \cdot \cdot \qquad p\text{-series}$$

is a ***p*-series,** where p is a positive constant. For $p = 1$, the series

$$\sum_{n=1}^{\infty} \frac{1}{n} = 1 + \frac{1}{2} + \frac{1}{3} + \cdot \cdot \cdot \qquad \text{Harmonic series}$$

is the **harmonic** series. A **general harmonic series** is of the form $\Sigma 1/(an + b)$. In music, strings of the same material, diameter, and tension, whose lengths form a harmonic series, produce harmonic tones.

The Integral Test is convenient for establishing the convergence or divergence of p-series. This is shown in the proof of Theorem 9.11.

THEOREM 9.11 Convergence of p-Series

The p-series

$$\sum_{n=1}^{\infty} \frac{1}{n^p} = \frac{1}{1^p} + \frac{1}{2^p} + \frac{1}{3^p} + \frac{1}{4^p} + \cdot \cdot \cdot$$

1. converges if $p > 1$, and

2. diverges if $0 < p \leq 1$.

Proof The proof follows from the Integral Test and from Theorem 8.5, which states that

$$\int_1^{\infty} \frac{1}{x^p}\,dx$$

converges if $p > 1$ and diverges if $0 < p \leq 1$.

NOTE The sum of the series in Example 3(b) can be shown to be $\pi^2/6$. (This was proved by Leonhard Euler, but the proof is too difficult to present here.) Be sure you see that the Integral Test does not tell you that the sum of the series is equal to the value of the integral. For instance, the sum of the series in Example 3(b) is

$$\sum_{n=1}^{\infty} \frac{1}{n^2} = \frac{\pi^2}{6} \approx 1.645$$

but the value of the corresponding improper integral is

$$\int_1^{\infty} \frac{1}{x^2}\,dx = 1.$$

EXAMPLE 3 Convergent and Divergent p-Series

Discuss the convergence or divergence of (a) the harmonic series and (b) the p-series with $p = 2$.

Solution

a. From Theorem 9.11, it follows that the harmonic series

$$\sum_{n=1}^{\infty} \frac{1}{n} = \frac{1}{1} + \frac{1}{2} + \frac{1}{3} + \cdot \cdot \cdot \qquad p = 1$$

diverges.

b. From Theorem 9.11, it follows that the p-series

$$\sum_{n=1}^{\infty} \frac{1}{n^2} = \frac{1}{1^2} + \frac{1}{2^2} + \frac{1}{3^2} + \cdot \cdot \cdot \qquad p = 2$$

converges.

EXAMPLE 4 Testing a Series for Convergence

Determine whether the following series converges or diverges.

$$\sum_{n=2}^{\infty} \frac{1}{n \ln n}$$

Solution This series is similar to the divergent harmonic series. If its terms were larger than those of the harmonic series, you would expect it to diverge. However, because its terms are smaller, you are not sure what to expect. The function $f(x) = 1/(x \ln x)$ is positive and continuous for $x \geq 2$. To determine whether f is decreasing, first rewrite f as $f(x) = (x \ln x)^{-1}$ and then find its derivative.

$$f'(x) = (-1)(x \ln x)^{-2}(1 + \ln x) = -\frac{1 + \ln x}{x^2(\ln x)^2}$$

So, $f'(x) < 0$ for $x > 2$ and it follows that f satisfies the conditions for the Integral Test.

$$\begin{aligned}\int_2^{\infty} \frac{1}{x \ln x}\,dx &= \int_2^{\infty} \frac{1/x}{\ln x}\,dx \\ &= \lim_{b\to\infty} \Big[\ln(\ln x)\Big]_2^b \\ &= \lim_{b\to\infty} [\ln(\ln b) - \ln(\ln 2)] = \infty\end{aligned}$$

The series diverges.

NOTE The infinite series in Example 4 diverges very slowly. For instance, the sum of the first 10 terms is approximately 1.6878196, whereas the sum of the first 100 terms is just slightly larger: 2.3250871. In fact, the sum of the first 10,000 terms is approximately 3.015021704. You can see that although the infinite series "adds up to infinity," it does so very slowly.

Exercises for Section 9.3

See www.CalcChat.com for worked-out solutions to odd-numbered exercises.

In Exercises 1–18, use the Integral Test to determine the convergence or divergence of the series.

1. $\displaystyle\sum_{n=1}^{\infty} \frac{1}{n+1}$

2. $\displaystyle\sum_{n=1}^{\infty} \frac{2}{3n+5}$

3. $\displaystyle\sum_{n=1}^{\infty} e^{-n}$

4. $\displaystyle\sum_{n=1}^{\infty} ne^{-n/2}$

5. $\frac{1}{2} + \frac{1}{5} + \frac{1}{10} + \frac{1}{17} + \frac{1}{26} + \cdots$

6. $\frac{1}{3} + \frac{1}{5} + \frac{1}{7} + \frac{1}{9} + \frac{1}{11} + \cdots$

7. $\frac{\ln 2}{2} + \frac{\ln 3}{3} + \frac{\ln 4}{4} + \frac{\ln 5}{5} + \frac{\ln 6}{6} + \cdots$

8. $\frac{\ln 2}{\sqrt{2}} + \frac{\ln 3}{\sqrt{3}} + \frac{\ln 4}{\sqrt{4}} + \frac{\ln 5}{\sqrt{5}} + \frac{\ln 6}{\sqrt{6}} + \cdots$

9. $\frac{1}{\sqrt{1}(\sqrt{1}+1)} + \frac{1}{\sqrt{2}(\sqrt{2}+1)} + \frac{1}{\sqrt{3}(\sqrt{3}+1)} + \cdots + \frac{1}{\sqrt{n}(\sqrt{n}+1)} + \cdots$

10. $\frac{1}{4} + \frac{2}{7} + \frac{3}{12} + \cdots + \frac{n}{n^2+3} + \cdots$

11. $\displaystyle\sum_{n=1}^{\infty} \frac{1}{\sqrt{n+1}}$

12. $\displaystyle\sum_{n=2}^{\infty} \frac{\ln n}{n^3}$

13. $\displaystyle\sum_{n=1}^{\infty} \frac{\ln n}{n^2}$

14. $\displaystyle\sum_{n=2}^{\infty} \frac{1}{n\sqrt{\ln n}}$

15. $\displaystyle\sum_{n=1}^{\infty} \frac{\arctan n}{n^2+1}$

16. $\displaystyle\sum_{n=3}^{\infty} \frac{1}{n \ln n \ln(\ln n)}$

17. $\displaystyle\sum_{n=1}^{\infty} \frac{2n}{n^2+1}$

18. $\displaystyle\sum_{n=1}^{\infty} \frac{n}{n^4+1}$

In Exercises 19 and 20, use the Integral Test to determine the convergence or divergence of the series, where k is a positive integer.

19. $\displaystyle\sum_{n=1}^{\infty} \frac{n^{k-1}}{n^k+c}$

20. $\displaystyle\sum_{n=1}^{\infty} n^k e^{-n}$

In Exercises 21–24, explain why the Integral Test does not apply to the series.

21. $\sum_{n=1}^{\infty} \frac{(-1)^n}{n}$

22. $\sum_{n=1}^{\infty} e^{-n} \cos n$

23. $\sum_{n=1}^{\infty} \frac{2 + \sin n}{n}$

24. $\sum_{n=1}^{\infty} \left(\frac{\sin n}{n}\right)^2$

In Exercises 25–28, use the Integral Test to determine the convergence or divergence of the p-series.

25. $\sum_{n=1}^{\infty} \frac{1}{n^3}$

26. $\sum_{n=1}^{\infty} \frac{1}{n^{1/3}}$

27. $\sum_{n=1}^{\infty} \frac{1}{\sqrt{n}}$

28. $\sum_{n=1}^{\infty} \frac{1}{n^2}$

In Exercises 29–36, use Theorem 9.11 to determine the convergence or divergence of the p-series.

29. $\sum_{n=1}^{\infty} \frac{1}{\sqrt[5]{n}}$

30. $\sum_{n=1}^{\infty} \frac{3}{n^{5/3}}$

31. $1 + \frac{1}{\sqrt{2}} + \frac{1}{\sqrt{3}} + \frac{1}{\sqrt{4}} + \cdots$

32. $1 + \frac{1}{4} + \frac{1}{9} + \frac{1}{16} + \frac{1}{25} + \cdots$

33. $1 + \frac{1}{2\sqrt{2}} + \frac{1}{3\sqrt{3}} + \frac{1}{4\sqrt{4}} + \frac{1}{5\sqrt{5}} + \cdots$

34. $1 + \frac{1}{\sqrt[3]{4}} + \frac{1}{\sqrt[3]{9}} + \frac{1}{\sqrt[3]{16}} + \frac{1}{\sqrt[3]{25}} + \cdots$

35. $\sum_{n=1}^{\infty} \frac{1}{n^{1.04}}$

36. $\sum_{n=1}^{\infty} \frac{1}{n^{\pi}}$

In Exercises 37–42, match the series with the graph of its sequence of partial sums. [The graphs are labeled (a), (b), (c), (d), (e), and (f).] Determine the convergence or divergence of the series.

(a)

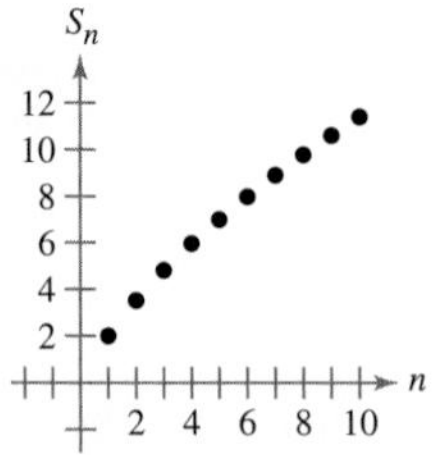

(b)

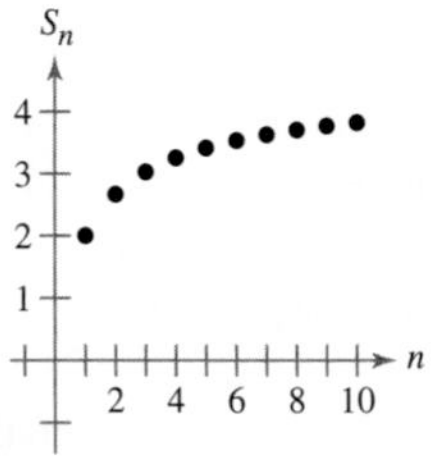

(c)

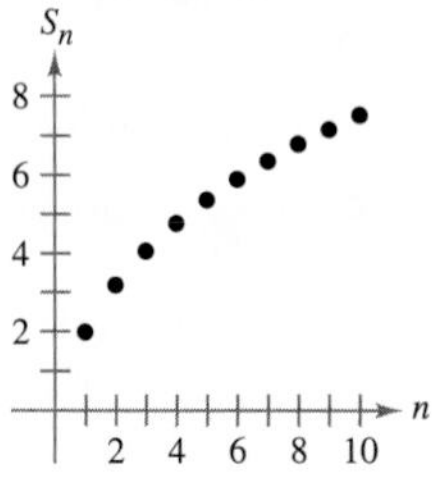

(d)

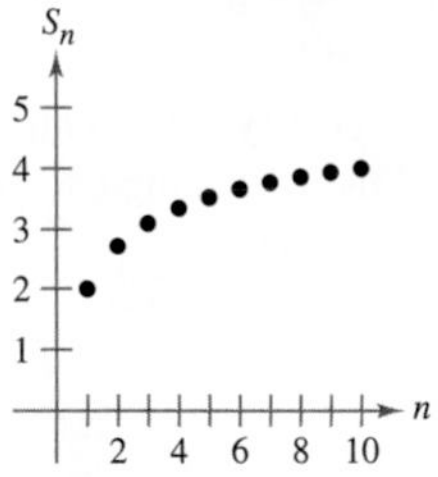

(e)

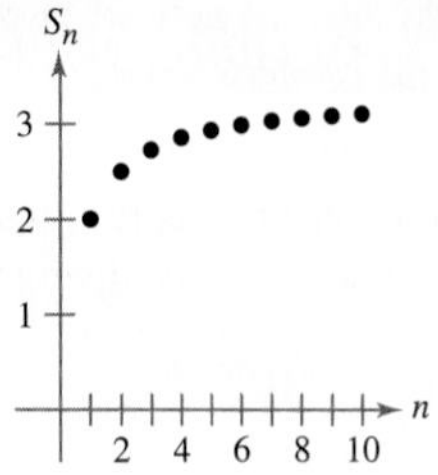

(f)

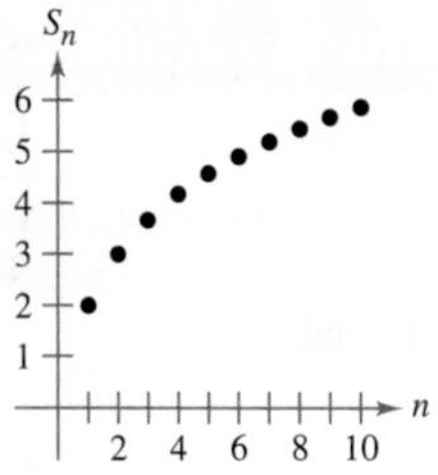

37. $\sum_{n=1}^{\infty} \frac{2}{\sqrt[4]{n^3}}$

38. $\sum_{n=1}^{\infty} \frac{2}{n}$

39. $\sum_{n=1}^{\infty} \frac{2}{\sqrt{n^{\pi}}}$

40. $\sum_{n=1}^{\infty} \frac{2}{\sqrt[5]{n^2}}$

41. $\sum_{n=1}^{\infty} \frac{2}{n\sqrt{n}}$

42. $\sum_{n=1}^{\infty} \frac{2}{n^2}$

43. ***Numerical and Graphical Analysis*** Use a graphing utility to find the indicated partial sum S_n and complete the table. Then use a graphing utility to graph the first 10 terms of the sequence of partial sums. Compare the rate at which the sequence of partial sums approaches the sum of the series for each series.

n	5	10	20	50	100
S_n					

(a) $\sum_{n=1}^{\infty} 3\left(\frac{1}{5}\right)^{n-1} = \frac{15}{4}$ (b) $\sum_{n=1}^{\infty} \frac{1}{n^2} = \frac{\pi^2}{6}$

44. ***Numerical Reasoning*** Because the harmonic series diverges, it follows that for any positive real number M there exists a positive integer N such that the partial sum

$$\sum_{n=1}^{N} \frac{1}{n} > M.$$

(a) Use a graphing utility to complete the table.

M	2	4	6	8
N				

(b) As the real number M increases in equal increments, does the number N increase in equal increments? Explain.

Writing About Concepts

45. State the Integral Test and give an example of its use.

46. Define a p-series and state the requirements for its convergence.

47. A friend in your calculus class tells you that the following series converges because the terms are very small and approach 0 rapidly. Is your friend correct? Explain.

$$\frac{1}{10{,}000} + \frac{1}{10{,}001} + \frac{1}{10{,}002} + \cdots$$

Writing About Concepts (continued)

48. In Exercises 37–42, $\lim_{n\to\infty} a_n = 0$ for each series but they do not all converge. Is this a contradiction of Theorem 9.9? Why do you think some converge and others diverge? Explain.

49. Use a graph to show that

$$\sum_{n=1}^{\infty} \frac{1}{\sqrt{n}} > \int_1^{\infty} \frac{1}{\sqrt{x}}\,dx.$$

What can you conclude about the convergence or divergence of the series? Explain.

50. Let f be a positive, continuous, and decreasing function for $x \geq 1$, such that $a_n = f(n)$. Use a graph to rank the following quantities in decreasing order. Explain your reasoning.

(a) $\displaystyle\sum_{n=2}^{7} a_n$ (b) $\displaystyle\int_1^7 f(x)\,dx$ (c) $\displaystyle\sum_{n=1}^{6} a_n$

In Exercises 51–54, find the positive values of p for which the series converges.

51. $\displaystyle\sum_{n=2}^{\infty} \frac{1}{n(\ln n)^p}$

52. $\displaystyle\sum_{n=2}^{\infty} \frac{\ln n}{n^p}$

53. $\displaystyle\sum_{n=1}^{\infty} \frac{n}{(1+n^2)^p}$

54. $\displaystyle\sum_{n=1}^{\infty} n(1+n^2)^p$

In Exercises 55–58, use the result of Exercise 51 to determine the convergence or divergence of the series.

55. $\displaystyle\sum_{n=2}^{\infty} \frac{1}{n \ln n}$

56. $\displaystyle\sum_{n=2}^{\infty} \frac{1}{n\sqrt[3]{(\ln n)^2}}$

57. $\displaystyle\sum_{n=2}^{\infty} \frac{1}{n(\ln n)^2}$

58. $\displaystyle\sum_{n=2}^{\infty} \frac{1}{n \ln(n^2)}$

59. Let f be a positive, continuous, and decreasing function for $x \geq 1$, such that $a_n = f(n)$. Prove that if the series

$$\sum_{n=1}^{\infty} a_n$$

converges to S, then the remainder $R_N = S - S_N$ is bounded by

$$0 \leq R_N \leq \int_N^{\infty} f(x)\,dx.$$

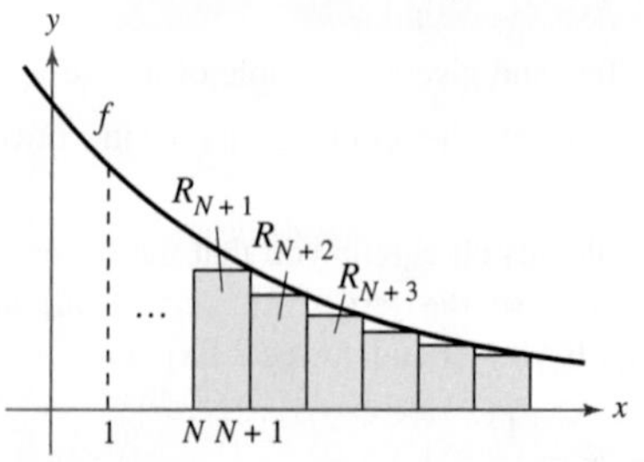

60. Show that the result of Exercise 59 can be written as

$$\sum_{n=1}^{N} a_n \leq \sum_{n=1}^{\infty} a_n \leq \sum_{n=1}^{N} a_n + \int_N^{\infty} f(x)\,dx.$$

In Exercises 61–66, use the result of Exercise 59 to approximate the sum of the convergent series using the indicated number of terms. Include an estimate of the maximum error for your approximation.

61. $\displaystyle\sum_{n=1}^{\infty} \frac{1}{n^4}$, six terms

62. $\displaystyle\sum_{n=1}^{\infty} \frac{1}{n^5}$, four terms

63. $\displaystyle\sum_{n=1}^{\infty} \frac{1}{n^2+1}$, ten terms

64. $\displaystyle\sum_{n=1}^{\infty} \frac{1}{(n+1)[\ln(n+1)]^3}$, ten terms

65. $\displaystyle\sum_{n=1}^{\infty} ne^{-n^2}$, four terms

66. $\displaystyle\sum_{n=1}^{\infty} e^{-n}$, four terms

In Exercises 67–72, use the result of Exercise 59 to find N such that $R_N \leq 0.001$ for the convergent series.

67. $\displaystyle\sum_{n=1}^{\infty} \frac{1}{n^4}$

68. $\displaystyle\sum_{n=1}^{\infty} \frac{1}{n^{3/2}}$

69. $\displaystyle\sum_{n=1}^{\infty} e^{-5n}$

70. $\displaystyle\sum_{n=1}^{\infty} e^{-n/2}$

71. $\displaystyle\sum_{n=1}^{\infty} \frac{1}{n^2+1}$

72. $\displaystyle\sum_{n=1}^{\infty} \frac{2}{n^2+5}$

73. (a) Show that $\displaystyle\sum_{n=2}^{\infty} \frac{1}{n^{1.1}}$ converges and $\displaystyle\sum_{n=2}^{\infty} \frac{1}{n \ln n}$ diverges.

(b) Compare the first five terms of each series in part (a).

(c) Find $n > 3$ such that

$$\frac{1}{n^{1.1}} < \frac{1}{n \ln n}.$$

74. Ten terms are used to approximate a convergent p-series. Therefore, the remainder is a function of p and is

$$0 \leq R_{10}(p) \leq \int_{10}^{\infty} \frac{1}{x^p}\,dx, \quad p > 1.$$

(a) Perform the integration in the inequality.

(b) Use a graphing utility to represent the inequality graphically.

(c) Identify any asymptotes of the error function and interpret their meaning.

75. *Euler's Constant* Let

$$S_n = \sum_{k=1}^{n} \frac{1}{k} = 1 + \frac{1}{2} + \cdot \cdot \cdot + \frac{1}{n}.$$

(a) Show that $\ln(n + 1) \le S_n \le 1 + \ln n$.

(b) Show that the sequence $\{a_n\} = \{S_n - \ln n\}$ is bounded.

(c) Show that the sequence $\{a_n\}$ is decreasing.

(d) Show that a_n converges to a limit γ (called Euler's constant).

(e) Approximate γ using a_{100}.

76. Find the sum of the series $\sum_{n=2}^{\infty} \ln\left(1 - \frac{1}{n^2}\right)$.

77. Consider the series

$$\sum_{n=2}^{\infty} x^{\ln n}.$$

(a) Determine the convergence or divergence of the series for $x = 1$.

(b) Determine the convergence or divergence of the series for $x = 1/e$.

(c) Find the positive values of x for which the series converges.

78. The **Riemann zeta function** for real numbers is defined for all x for which the series

$$\zeta(x) = \sum_{n=1}^{\infty} n^{-x}$$

converges. Find the domain of the function.

Review **In Exercises 79–90, determine the convergence or divergence of the series.**

79. $\sum_{n=1}^{\infty} \frac{1}{2n - 1}$

80. $\sum_{n=2}^{\infty} \frac{1}{n\sqrt{n^2 - 1}}$

81. $\sum_{n=1}^{\infty} \frac{1}{n\sqrt[4]{n}}$

82. $3\sum_{n=1}^{\infty} \frac{1}{n^{0.95}}$

83. $\sum_{n=0}^{\infty} \left(\frac{2}{3}\right)^n$

84. $\sum_{n=0}^{\infty} (1.075)^n$

85. $\sum_{n=1}^{\infty} \frac{n}{\sqrt{n^2 + 1}}$

86. $\sum_{n=1}^{\infty} \left(\frac{1}{n^2} - \frac{1}{n^3}\right)$

87. $\sum_{n=1}^{\infty} \left(1 + \frac{1}{n}\right)^n$

88. $\sum_{n=2}^{\infty} \ln n$

89. $\sum_{n=2}^{\infty} \frac{1}{n(\ln n)^3}$

90. $\sum_{n=2}^{\infty} \frac{\ln n}{n^3}$

Section Project: The Harmonic Series

The harmonic series

$$\sum_{n=1}^{\infty} \frac{1}{n} = 1 + \frac{1}{2} + \frac{1}{3} + \frac{1}{4} + \cdot \cdot \cdot + \frac{1}{n} + \cdot \cdot \cdot$$

is one of the most important series in this chapter. Even though its terms tend to zero as n increases,

$$\lim_{n\to\infty} \frac{1}{n} = 0$$

the harmonic series diverges. In other words, even though the terms are getting smaller and smaller, the sum "adds up to infinity."

(a) One way to show that the harmonic series diverges is attributed to Jakob Bernoulli. He grouped the terms of the harmonic series as follows:

$$1 + \frac{1}{2} + \underbrace{\frac{1}{3} + \frac{1}{4}}_{> \frac{1}{2}} + \underbrace{\frac{1}{5} + \cdot \cdot \cdot + \frac{1}{8}}_{> \frac{1}{2}} + \underbrace{\frac{1}{9} + \cdot \cdot \cdot + \frac{1}{16}}_{> \frac{1}{2}} + \underbrace{\frac{1}{17} + \cdot \cdot \cdot + \frac{1}{32}}_{> \frac{1}{2}} + \cdot \cdot \cdot$$

Write a short paragraph explaining how you can use this grouping to show that the harmonic series diverges.

(b) Use the proof of the Integral Test, Theorem 9.10, to show that

$$\ln(n + 1) \le 1 + \frac{1}{2} + \frac{1}{3} + \frac{1}{4} + \cdot \cdot \cdot + \frac{1}{n} \le 1 + \ln n.$$

(c) Use part (b) to determine how many terms M you would need so that

$$\sum_{n=1}^{M} \frac{1}{n} > 50.$$

(d) Show that the sum of the first million terms of the harmonic series is less than 15.

(e) Show that the following inequalities are valid.

$$\ln \tfrac{21}{10} \le \tfrac{1}{10} + \tfrac{1}{11} + \cdot \cdot \cdot + \tfrac{1}{20} \le \ln \tfrac{20}{9}$$

$$\ln \tfrac{201}{100} \le \tfrac{1}{100} + \tfrac{1}{101} + \cdot \cdot \cdot + \tfrac{1}{200} \le \ln \tfrac{200}{99}$$

(f) Use the ideas in part (e) to find the limit

$$\lim_{m\to\infty} \sum_{n=m}^{2m} \frac{1}{n}.$$

Section 9.4 Comparisons of Series

- **Use the Direct Comparison Test to determine whether a series converges or diverges.**
- **Use the Limit Comparison Test to determine whether a series converges or diverges.**

Direct Comparison Test

For the convergence tests developed so far, the terms of the series have to be fairly simple and the series must have special characteristics in order for the convergence tests to be applied. A slight deviation from these special characteristics can make a test nonapplicable. For example, in the following pairs, the second series cannot be tested by the same convergence test as the first series even though it is similar to the first.

1. $\sum_{n=0}^{\infty} \frac{1}{2^n}$ is geometric, but $\sum_{n=0}^{\infty} \frac{n}{2^n}$ is not.

2. $\sum_{n=1}^{\infty} \frac{1}{n^3}$ is a p-series, but $\sum_{n=1}^{\infty} \frac{1}{n^3 + 1}$ is not.

3. $a_n = \frac{n}{(n^2 + 3)^2}$ is easily integrated, but $b_n = \frac{n^2}{(n^2 + 3)^2}$ is not.

In this section you will study two additional tests for positive-term series. These two tests greatly expand the variety of series you are able to test for convergence or divergence. They allow you to *compare* a series having complicated terms with a simpler series whose convergence or divergence is known.

THEOREM 9.12 Direct Comparison Test

Let $0 < a_n \le b_n$ for all n.

1. If $\sum_{n=1}^{\infty} b_n$ converges, then $\sum_{n=1}^{\infty} a_n$ converges.

2. If $\sum_{n=1}^{\infty} a_n$ diverges, then $\sum_{n=1}^{\infty} b_n$ diverges.

Proof To prove the first property, let $L = \sum_{n=1}^{\infty} b_n$ and let

$$S_n = a_1 + a_2 + \cdot \cdot \cdot + a_n.$$

Because $0 < a_n \le b_n$, the sequence $S_1, S_2, S_3, . . .$ is nondecreasing and bounded above by L; so, it must converge. Because

$$\lim_{n\to\infty} S_n = \sum_{n=1}^{\infty} a_n$$

it follows that $\Sigma\, a_n$ converges. The second property is logically equivalent to the first.

NOTE As stated, the Direct Comparison Test requires that $0 < a_n \le b_n$ for all n. Because the convergence of a series is not dependent on its first several terms, you could modify the test to require only that $0 < a_n \le b_n$ for all n greater than some integer N.

EXAMPLE 1 Using the Direct Comparison Test

Determine the convergence or divergence of

$$\sum_{n=1}^{\infty} \frac{1}{2+3^n}.$$

Solution This series resembles

$$\sum_{n=1}^{\infty} \frac{1}{3^n}. \qquad \text{Convergent geometric series}$$

Term-by-term comparison yields

$$a_n = \frac{1}{2+3^n} < \frac{1}{3^n} = b_n, \quad n \ge 1.$$

So, by the Direct Comparison Test, the series converges.

EXAMPLE 2 Using the Direct Comparison Test

Determine the convergence or divergence of

$$\sum_{n=1}^{\infty} \frac{1}{2+\sqrt{n}}.$$

Solution This series resembles

$$\sum_{n=1}^{\infty} \frac{1}{n^{1/2}}. \qquad \text{Divergent } p\text{-series}$$

Term-by-term comparison yields

$$\frac{1}{2+\sqrt{n}} \le \frac{1}{\sqrt{n}}, \quad n \ge 1$$

which *does not* meet the requirements for divergence. (Remember that if term-by-term comparison reveals a series that is *smaller* than a divergent series, the Direct Comparison Test tells you nothing.) Still expecting the series to diverge, you can compare the given series with

$$\sum_{n=1}^{\infty} \frac{1}{n}. \qquad \text{Divergent harmonic series}$$

In this case, term-by-term comparison yields

$$a_n = \frac{1}{n} \le \frac{1}{2+\sqrt{n}} = b_n, \quad n \ge 4$$

and, by the Direct Comparison Test, the given series diverges.

NOTE To verify the last inequality in Example 2, try showing that $2 + \sqrt{n} \le n$ whenever $n \ge 4$.

Remember that both parts of the Direct Comparison Test require that $0 < a_n \le b_n$. Informally, the test says the following about the two series with nonnegative terms.

1. If the "larger" series converges, the "smaller" series must also converge.
2. If the "smaller" series diverges, the "larger" series must also diverge.

Limit Comparison Test

Often a given series closely resembles a p-series or a geometric series, yet you cannot establish the term-by-term comparison necessary to apply the Direct Comparison Test. Under these circumstances you may be able to apply a second comparison test, called the **Limit Comparison Test.**

THEOREM 9.13 Limit Comparison Test

Suppose that $a_n > 0$, $b_n > 0$, and

$$\lim_{n\to\infty}\left(\frac{a_n}{b_n}\right) = L$$

where L is *finite and positive*. Then the two series $\sum a_n$ and $\sum b_n$ either both converge or both diverge.

NOTE As with the Direct Comparison Test, the Limit Comparison Test could be modified to require only that a_n and b_n be positive for all n greater than some integer N.

Proof Because $a_n > 0$, $b_n > 0$, and

$$\lim_{n\to\infty}\left(\frac{a_n}{b_n}\right) = L$$

there exists $N > 0$ such that

$$0 < \frac{a_n}{b_n} < L + 1, \quad \text{for } n \geq N.$$

This implies that

$$0 < a_n < (L + 1)b_n.$$

So, by the Direct Comparison Test, the convergence of $\sum b_n$ implies the convergence of $\sum a_n$. Similarly, the fact that

$$\lim_{n\to\infty}\left(\frac{b_n}{a_n}\right) = \frac{1}{L}$$

can be used to show that the convergence of $\sum a_n$ implies the convergence of $\sum b_n$.

EXAMPLE 3 Using the Limit Comparison Test

Show that the following general harmonic series diverges.

$$\sum_{n=1}^{\infty} \frac{1}{an + b}, \quad a > 0, \quad b > 0$$

Solution By comparison with

$$\sum_{n=1}^{\infty} \frac{1}{n} \qquad \text{Divergent harmonic series}$$

you have

$$\lim_{n\to\infty} \frac{1/(an + b)}{1/n} = \lim_{n\to\infty} \frac{n}{an + b} = \frac{1}{a}.$$

Because this limit is greater than 0, you can conclude from the Limit Comparison Test that the given series diverges.

The Limit Comparison Test works well for comparing a "messy" algebraic series with a p-series. In choosing an appropriate p-series, you must choose one with an nth term of the same magnitude as the nth term of the given series.

Given Series	*Comparison Series*	*Conclusion*
$\displaystyle\sum_{n=1}^{\infty} \frac{1}{3n^2 - 4n + 5}$	$\displaystyle\sum_{n=1}^{\infty} \frac{1}{n^2}$	Both series converge.
$\displaystyle\sum_{n=1}^{\infty} \frac{1}{\sqrt{3n - 2}}$	$\displaystyle\sum_{n=1}^{\infty} \frac{1}{\sqrt{n}}$	Both series diverge.
$\displaystyle\sum_{n=1}^{\infty} \frac{n^2 - 10}{4n^5 + n^3}$	$\displaystyle\sum_{n=1}^{\infty} \frac{n^2}{n^5} = \sum_{n=1}^{\infty} \frac{1}{n^3}$	Both series converge.

In other words, when choosing a series for comparison, you can disregard all but the *highest powers of n* in both the numerator and the denominator.

EXAMPLE 4 Using the Limit Comparison Test

Determine the convergence or divergence of

$$\sum_{n=1}^{\infty} \frac{\sqrt{n}}{n^2 + 1}.$$

Solution Disregarding all but the highest powers of n in the numerator and the denominator, you can compare the series with

$$\sum_{n=1}^{\infty} \frac{\sqrt{n}}{n^2} = \sum_{n=1}^{\infty} \frac{1}{n^{3/2}}. \qquad \text{Convergent } p\text{-series}$$

Because

$$\lim_{n\to\infty} \frac{a_n}{b_n} = \lim_{n\to\infty} \left(\frac{\sqrt{n}}{n^2 + 1}\right)\left(\frac{n^{3/2}}{1}\right)$$
$$= \lim_{n\to\infty} \frac{n^2}{n^2 + 1} = 1$$

you can conclude by the Limit Comparison Test that the given series converges.

EXAMPLE 5 Using the Limit Comparison Test

Determine the convergence or divergence of

$$\sum_{n=1}^{\infty} \frac{n2^n}{4n^3 + 1}.$$

Solution A reasonable comparison would be with the series

$$\sum_{n=1}^{\infty} \frac{2^n}{n^2}. \qquad \text{Divergent series}$$

Note that this series diverges by the nth-Term Test. From the limit

$$\lim_{n\to\infty} \frac{a_n}{b_n} = \lim_{n\to\infty} \left(\frac{n2^n}{4n^3 + 1}\right)\left(\frac{n^2}{2^n}\right)$$
$$= \lim_{n\to\infty} \frac{1}{4 + (1/n^3)} = \frac{1}{4}$$

you can conclude that the given series diverges.

Exercises for Section 9.4

See www.CalcChat.com for worked-out solutions to odd-numbered exercises.

1. ***Graphical Analysis*** The figures show the graphs of the first 10 terms, and the graphs of the first 10 terms of the sequence of partial sums, of each series.

$$\sum_{n=1}^{\infty} \frac{6}{n^{3/2}}, \quad \sum_{n=1}^{\infty} \frac{6}{n^{3/2}+3}, \quad \text{and} \quad \sum_{n=1}^{\infty} \frac{6}{n\sqrt{n^2+0.5}}$$

(a) Identify the series in each figure.

(b) Which series is a p-series? Does it converge or diverge?

(c) For the series that are not p-series, how do the magnitudes of the terms compare with the magnitudes of the terms of the p-series? What conclusion can you draw about the convergence or divergence of the series?

(d) Explain the relationship between the magnitudes of the terms of the series and the magnitudes of the terms of the partial sums.

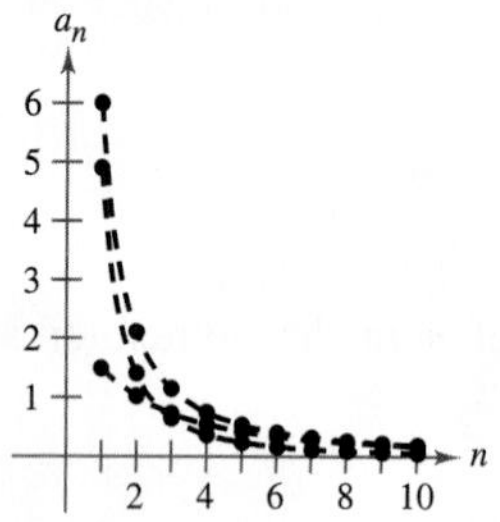

Graphs of terms

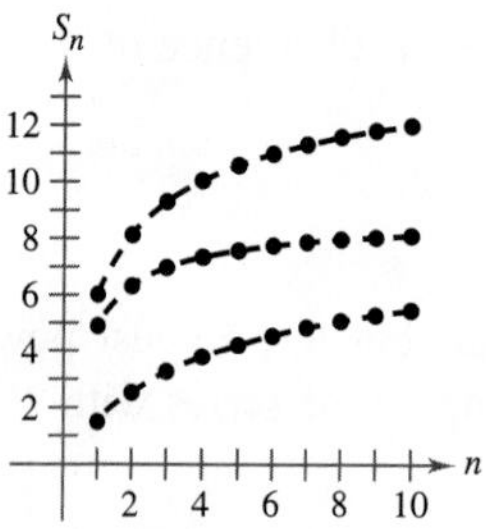

Graphs of partial sums

2. ***Graphical Analysis*** The figures show the graphs of the first 10 terms, and the graphs of the first 10 terms of the sequence of partial sums, of each series.

$$\sum_{n=1}^{\infty} \frac{2}{\sqrt{n}}, \quad \sum_{n=1}^{\infty} \frac{2}{\sqrt{n}-0.5}, \quad \text{and} \quad \sum_{n=1}^{\infty} \frac{4}{\sqrt{n+0.5}}$$

(a) Identify the series in each figure.

(b) Which series is a p-series? Does it converge or diverge?

(c) For the series that are not p-series, how do the magnitudes of the terms compare with the magnitudes of the terms of the p-series? What conclusion can you draw about the convergence or divergence of the series?

(d) Explain the relationship between the magnitudes of the terms of the series and the magnitudes of the terms of the partial sums.

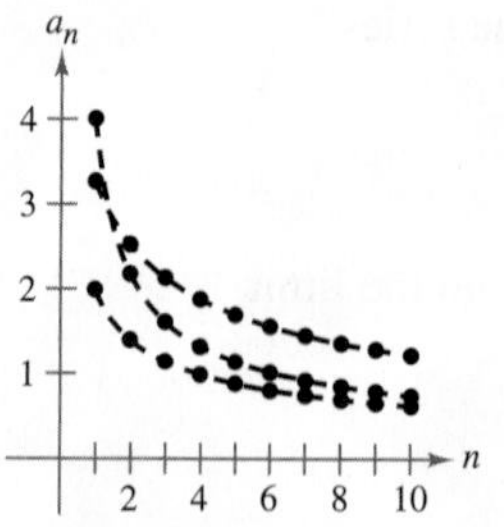

Graphs of terms

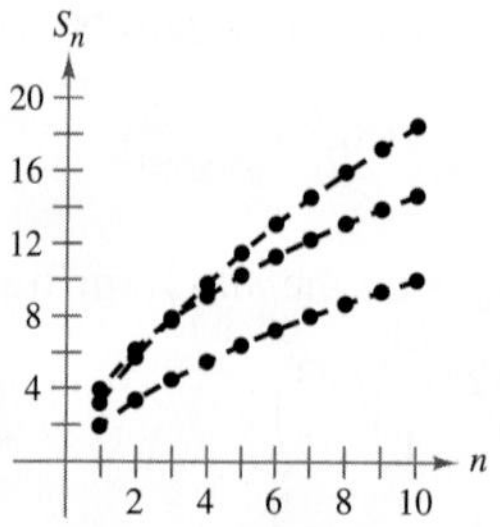

Graphs of partial sums

In Exercises 3–14, use the Direct Comparison Test to determine the convergence or divergence of the series.

3. $\displaystyle\sum_{n=1}^{\infty} \frac{1}{n^2+1}$

4. $\displaystyle\sum_{n=1}^{\infty} \frac{1}{3n^2+2}$

5. $\displaystyle\sum_{n=2}^{\infty} \frac{1}{n-1}$

6. $\displaystyle\sum_{n=2}^{\infty} \frac{1}{\sqrt{n}-1}$

7. $\displaystyle\sum_{n=0}^{\infty} \frac{1}{3^n+1}$

8. $\displaystyle\sum_{n=0}^{\infty} \frac{3^n}{4^n+5}$

9. $\displaystyle\sum_{n=2}^{\infty} \frac{\ln n}{n+1}$

10. $\displaystyle\sum_{n=1}^{\infty} \frac{1}{\sqrt{n^3+1}}$

11. $\displaystyle\sum_{n=0}^{\infty} \frac{1}{n!}$

12. $\displaystyle\sum_{n=1}^{\infty} \frac{1}{4\sqrt[3]{n}-1}$

13. $\displaystyle\sum_{n=0}^{\infty} e^{-n^2}$

14. $\displaystyle\sum_{n=1}^{\infty} \frac{4^n}{3^n-1}$

In Exercises 15–28, use the Limit Comparison Test to determine the convergence or divergence of the series.

15. $\displaystyle\sum_{n=1}^{\infty} \frac{n}{n^2+1}$

16. $\displaystyle\sum_{n=1}^{\infty} \frac{2}{3^n-5}$

17. $\displaystyle\sum_{n=0}^{\infty} \frac{1}{\sqrt{n^2+1}}$

18. $\displaystyle\sum_{n=3}^{\infty} \frac{3}{\sqrt{n^2-4}}$

19. $\displaystyle\sum_{n=1}^{\infty} \frac{2n^2-1}{3n^5+2n+1}$

20. $\displaystyle\sum_{n=1}^{\infty} \frac{5n-3}{n^2-2n+5}$

21. $\displaystyle\sum_{n=1}^{\infty} \frac{n+3}{n(n+2)}$

22. $\displaystyle\sum_{n=1}^{\infty} \frac{1}{n(n^2+1)}$

23. $\displaystyle\sum_{n=1}^{\infty} \frac{1}{n\sqrt{n^2+1}}$

24. $\displaystyle\sum_{n=1}^{\infty} \frac{n}{(n+1)2^{n-1}}$

25. $\displaystyle\sum_{n=1}^{\infty} \frac{n^{k-1}}{n^k+1}, \quad k>2$

26. $\displaystyle\sum_{n=1}^{\infty} \frac{5}{n+\sqrt{n^2+4}}$

27. $\displaystyle\sum_{n=1}^{\infty} \sin\frac{1}{n}$

28. $\displaystyle\sum_{n=1}^{\infty} \tan\frac{1}{n}$

In Exercises 29–36, test for convergence or divergence, using each test at least once. Identify which test was used.

(a) nth-Term Test
(b) Geometric Series Test
(c) p-Series Test
(d) Telescoping Series Test
(e) Integral Test
(f) Direct Comparison Test
(g) Limit Comparison Test

29. $\displaystyle\sum_{n=1}^{\infty} \frac{\sqrt[3]{n}}{n}$

30. $\displaystyle\sum_{n=0}^{\infty} 5\left(-\frac{1}{5}\right)^n$

31. $\displaystyle\sum_{n=1}^{\infty} \frac{1}{3^n+2}$

32. $\displaystyle\sum_{n=4}^{\infty} \frac{1}{3n^2-2n-15}$

33. $\displaystyle\sum_{n=1}^{\infty} \frac{n}{2n+3}$

34. $\displaystyle\sum_{n=1}^{\infty} \left(\frac{1}{n+1}-\frac{1}{n+2}\right)$

35. $\displaystyle\sum_{n=1}^{\infty} \frac{n}{(n^2+1)^2}$

36. $\displaystyle\sum_{n=1}^{\infty} \frac{3}{n(n+3)}$

37. Use the Limit Comparison Test with the harmonic series to show that the series $\sum a_n$ (where $0 < a_n < a_{n-1}$) diverges if $\lim_{n\to\infty} na_n$ is finite and nonzero.

38. Prove that, if $P(n)$ and $Q(n)$ are polynomials of degree j and k, respectively, then the series

$$\sum_{n=1}^{\infty} \frac{P(n)}{Q(n)}$$

converges if $j < k - 1$ and diverges if $j \geq k - 1$.

In Exercises 39–42, use the polynomial test given in Exercise 38 to determine whether the series converges or diverges.

39. $\frac{1}{2} + \frac{2}{5} + \frac{3}{10} + \frac{4}{17} + \frac{5}{26} + \cdots$

40. $\frac{1}{3} + \frac{1}{8} + \frac{1}{15} + \frac{1}{24} + \frac{1}{35} + \cdots$

41. $\displaystyle\sum_{n=1}^{\infty} \frac{1}{n^3 + 1}$

42. $\displaystyle\sum_{n=1}^{\infty} \frac{n^2}{n^3 + 1}$

In Exercises 43 and 44, use the divergence test given in Exercise 37 to show that the series diverges.

43. $\displaystyle\sum_{n=1}^{\infty} \frac{n^3}{5n^4 + 3}$

44. $\displaystyle\sum_{n=2}^{\infty} \frac{1}{\ln n}$

In Exercises 45–48, determine the convergence or divergence of the series.

45. $\frac{1}{200} + \frac{1}{400} + \frac{1}{600} + \frac{1}{800} + \cdots$

46. $\frac{1}{200} + \frac{1}{210} + \frac{1}{220} + \frac{1}{230} + \cdots$

47. $\frac{1}{201} + \frac{1}{204} + \frac{1}{209} + \frac{1}{216} + \cdots$

48. $\frac{1}{201} + \frac{1}{208} + \frac{1}{227} + \frac{1}{264} + \cdots$

Writing About Concepts

49. Review the results of Exercises 45–48. Explain why careful analysis is required to determine the convergence or divergence of a series and why only considering the magnitudes of the terms of a series could be misleading.

50. State the Direct Comparison Test and give an example of its use.

51. State the Limit Comparison Test and give an example of its use.

52. It appears that the terms of the series

$$\frac{1}{1000} + \frac{1}{1001} + \frac{1}{1002} + \frac{1}{1003} + \cdots$$

are less than the corresponding terms of the convergent series

$$1 + \frac{1}{4} + \frac{1}{9} + \frac{1}{16} + \cdots.$$

If the statement above is correct, the first series converges. Is this correct? Why or why not? Make a statement about how the divergence or convergence of a series is affected by inclusion or exclusion of the first finite number of terms.

Writing About Concepts *(continued)*

53. The figure shows the first 20 terms of the convergent series $\sum_{n=1}^{\infty} a_n$ and the first 20 terms of the series $\sum_{n=1}^{\infty} a_n^2$. Identify the two series and explain your reasoning in making the selection.

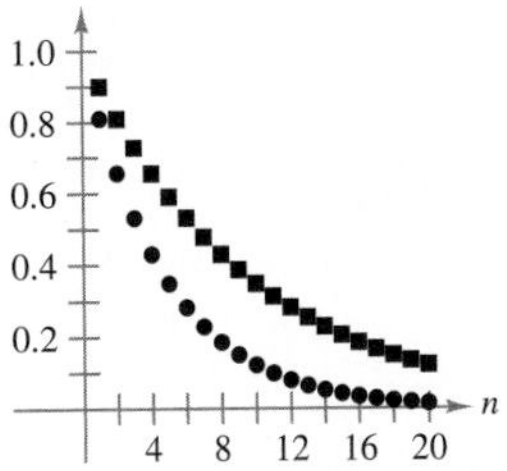

54. Consider the series $\displaystyle\sum_{n=1}^{\infty} \frac{1}{(2n-1)^2}$.

(a) Verify that the series converges.

(b) Use a graphing utility to complete the table.

n	5	10	20	50	100
S_n					

(c) The sum of the series is $\pi^2/8$. Find the sum of the series

$$\sum_{n=3}^{\infty} \frac{1}{(2n-1)^2}.$$

(d) Use a graphing utility to find the sum of the series

$$\sum_{n=10}^{\infty} \frac{1}{(2n-1)^2}.$$

True or False? **In Exercises 55–60, determine whether the statement is true or false. If it is false, explain why or give an example that shows it is false.**

55. If $0 < a_n \leq b_n$ and $\sum_{n=1}^{\infty} a_n$ converges, then $\sum_{n=1}^{\infty} b_n$ diverges.

56. If $0 < a_{n+10} \leq b_n$ and $\sum_{n=1}^{\infty} b_n$ converges, then $\sum_{n=1}^{\infty} a_n$ converges.

57. If $a_n + b_n \leq c_n$ and $\sum_{n=1}^{\infty} c_n$ converges, then the series $\sum_{n=1}^{\infty} a_n$ and $\sum_{n=1}^{\infty} b_n$ both converge. (Assume that the terms of all three series are positive.)

58. If $a_n \leq b_n + c_n$ and $\sum_{n=1}^{\infty} a_n$ diverges, then the series $\sum_{n=1}^{\infty} b_n$ and $\sum_{n=1}^{\infty} c_n$ both diverge. (Assume that the terms of all three series are positive.)

59. If $0 < a_n \leq b_n$ and $\sum_{n=1}^{\infty} a_n$ diverges, then $\sum_{n=1}^{\infty} b_n$ diverges.

60. If $0 < a_n \le b_n$ and $\sum_{n=1}^{\infty} b_n$ diverges, then $\sum_{n=1}^{\infty} a_n$ diverges.

61. Prove that if the nonnegative series

$$\sum_{n=1}^{\infty} a_n \quad \text{and} \quad \sum_{n=1}^{\infty} b_n$$

converge, then so does the series

$$\sum_{n=1}^{\infty} a_n b_n.$$

62. Use the result of Exercise 61 to prove that if the nonnegative series

$$\sum_{n=1}^{\infty} a_n$$

converges, then so does the series

$$\sum_{n=1}^{\infty} a_n^2.$$

63. Find two series that demonstrate the result of Exercise 61.

64. Find two series that demonstrate the result of Exercise 62.

65. Suppose that $\Sigma\, a_n$ and $\Sigma\, b_n$ are series with positive terms. Prove that if $\lim_{n\to\infty} \frac{a_n}{b_n} = 0$ and $\Sigma\, b_n$ converges, $\Sigma\, a_n$ also converges.

66. Suppose that $\Sigma\, a_n$ and $\Sigma\, b_n$ are series with positive terms. Prove that if $\lim_{n\to\infty} \frac{a_n}{b_n} = \infty$ and $\Sigma\, b_n$ diverges, $\Sigma\, a_n$ also diverges.

67. Use the result of Exercise 65 to show that each series converges.

(a) $\sum_{n=1}^{\infty} \frac{1}{(n+1)^3}$ (b) $\sum_{n=1}^{\infty} \frac{1}{\sqrt{n}\pi^n}$

68. Use the result of Exercise 66 to show that each series diverges.

(a) $\sum_{n=1}^{\infty} \frac{\ln n}{n}$ (b) $\sum_{n=2}^{\infty} \frac{1}{\ln n}$

69. Suppose that $\Sigma\, a_n$ is a series with positive terms. Prove that if $\Sigma\, a_n$ converges, then $\Sigma \sin a_n$ also converges.

70. Prove that the series $\sum_{n=1}^{\infty} \frac{1}{1+2+3+\cdots+n}$ converges.

Putnam Exam Challenge

71. Is the infinite series

$$\sum_{n=1}^{\infty} \frac{1}{n^{(n+1)/n}}$$

convergent? Prove your statement.

72. Prove that if $\sum_{n=1}^{\infty} a_n$ is a convergent series of positive real numbers, then so is

$$\sum_{n=1}^{\infty} (a_n)^{n/(n+1)}.$$

These problems were composed by the Committee on the Putnam Prize Competition.

Section Project: Solera Method

Most wines are produced entirely from grapes grown in a single year. Sherry, however, is a complex mixture of older wines with new wines. This is done with a sequence of barrels (called a solera) stacked on top of each other, as shown in the photo.

Everton/The Image Works

The oldest wine is in the bottom tier of barrels, and the newest is in the top tier. Each year, half of each barrel in the bottom tier is bottled as sherry. The bottom barrels are then refilled with the wine from the barrels above. This process is repeated throughout the solera, with new wine being added to the top barrels. A mathematical model for the amount of n-year-old wine that is removed from a solera (with k tiers) each year is

$$f(n, k) = \binom{n-1}{k-1}\left(\frac{1}{2}\right)^{n+1}, \quad k \le n.$$

(a) Consider a solera that has five tiers, numbered $k = 1, 2, 3, 4,$ and 5. In 1990 ($n = 0$), half of each barrel in the top tier (tier 1) was refilled with new wine. How much of this wine was removed from the solera in 1991? In 1992? In 1993? . . . In 2005? During which year(s) was the greatest amount of the 1990 wine removed from the solera?

(b) In part (a), let a_n be the amount of 1990 wine that is removed from the solera in year n. Evaluate

$$\sum_{n=0}^{\infty} a_n.$$

FOR FURTHER INFORMATION See the article "Finding Vintage Concentrations in a Sherry Solera" by Rhodes Peele and John T. MacQueen in the *UMAP Modules*.

Section 9.5 Alternating Series

- **Use the Alternating Series Test to determine whether an infinite series converges.**
- **Use the Alternating Series Remainder to approximate the sum of an alternating series.**
- **Classify a convergent series as absolutely or conditionally convergent.**
- **Rearrange an infinite series to obtain a different sum.**

Alternating Series

So far, most series you have dealt with have had positive terms. In this section and the following section, you will study series that contain both positive and negative terms. The simplest such series is an **alternating series,** whose terms alternate in sign. For example, the geometric series

$$\sum_{n=0}^{\infty}\left(-\frac{1}{2}\right)^n = \sum_{n=0}^{\infty}(-1)^n\frac{1}{2^n}$$
$$= 1 - \frac{1}{2} + \frac{1}{4} - \frac{1}{8} + \frac{1}{16} - \cdot\cdot\cdot$$

is an *alternating geometric series* with $r = -\frac{1}{2}$. Alternating series occur in two ways: either the odd terms are negative or the even terms are negative.

> **THEOREM 9.14 Alternating Series Test**
>
> Let $a_n > 0$. The alternating series
>
> $$\sum_{n=1}^{\infty}(-1)^n a_n \quad \text{and} \quad \sum_{n=1}^{\infty}(-1)^{n+1} a_n$$
>
> converge if the following two conditions are met.
>
> **1.** $\lim_{n\to\infty} a_n = 0$ **2.** $a_{n+1} \le a_n$, for all n

Proof Consider the alternating series $\Sigma(-1)^{n+1}a_n$. For this series, the partial sum (where $2n$ is even)

$$S_{2n} = (a_1 - a_2) + (a_3 - a_4) + (a_5 - a_6) + \cdot\cdot\cdot + (a_{2n-1} - a_{2n})$$

has all nonnegative terms, and therefore $\{S_{2n}\}$ is a nondecreasing sequence. But you can also write

$$S_{2n} = a_1 - (a_2 - a_3) - (a_4 - a_5) - \cdot\cdot\cdot - (a_{2n-2} - a_{2n-1}) - a_{2n}$$

which implies that $S_{2n} \le a_1$ for every integer n. So, $\{S_{2n}\}$ is a bounded, nondecreasing sequence that converges to some value L. Because $S_{2n-1} - a_{2n} = S_{2n}$ and $a_{2n} \to 0$, you have

$$\lim_{n\to\infty} S_{2n-1} = \lim_{n\to\infty} S_{2n} + \lim_{n\to\infty} a_{2n}$$
$$= L + \lim_{n\to\infty} a_{2n} = L.$$

Because both S_{2n} and S_{2n-1} converge to the same limit L, it follows that $\{S_n\}$ also converges to L. Consequently, the given alternating series converges.

NOTE The second condition in the Alternating Series Test can be modified to require only that $0 < a_{n+1} \le a_n$ for all n greater than some integer N.

EXAMPLE 1 Using the Alternating Series Test

NOTE The series in Example 1 is called the *alternating harmonic series*—more is said about this series in Example 7.

Determine the convergence or divergence of $\sum_{n=1}^{\infty} (-1)^{n+1}\frac{1}{n}$.

Solution Note that $\lim_{n\to\infty} a_n = \lim_{n\to\infty} \frac{1}{n} = 0$. So, the first condition of Theorem 9.14 is satisfied. Also note that the second condition of Theorem 9.14 is satisfied because

$$a_{n+1} = \frac{1}{n+1} \le \frac{1}{n} = a_n$$

for all n. So, applying the Alternating Series Test, you can conclude that the series converges.

EXAMPLE 2 Using the Alternating Series Test

Determine the convergence or divergence of $\sum_{n=1}^{\infty} \frac{n}{(-2)^{n-1}}$.

Solution To apply the Alternating Series Test, note that, for $n \ge 1$,

$$\frac{1}{2} \le \frac{n}{n+1}$$

$$\frac{2^{n-1}}{2^n} \le \frac{n}{n+1}$$

$$(n+1)2^{n-1} \le n2^n$$

$$\frac{n+1}{2^n} \le \frac{n}{2^{n-1}}.$$

So, $a_{n+1} = (n+1)/2^n \le n/2^{n-1} = a_n$ for all n. Furthermore, by L'Hôpital's Rule,

$$\lim_{x\to\infty} \frac{x}{2^{x-1}} = \lim_{x\to\infty} \frac{1}{2^{x-1}(\ln 2)} = 0 \quad \Longrightarrow \quad \lim_{n\to\infty} \frac{n}{2^{n-1}} = 0.$$

Therefore, by the Alternating Series Test, the series converges.

EXAMPLE 3 Cases for Which the Alternating Series Test Fails

NOTE In Example 3(a), remember that whenever a series does not pass the first condition of the Alternating Series Test, you can use the nth-Term Test for Divergence to conclude that the series diverges.

a. The alternating series

$$\sum_{n=1}^{\infty} \frac{(-1)^{n+1}(n+1)}{n} = \frac{2}{1} - \frac{3}{2} + \frac{4}{3} - \frac{5}{4} + \frac{6}{5} - \cdots$$

passes the second condition of the Alternating Series Test because $a_{n+1} \le a_n$ for all n. You cannot apply the Alternating Series Test, however, because the series does not pass the first condition. In fact, the series diverges.

b. The alternating series

$$\frac{2}{1} - \frac{1}{1} + \frac{2}{2} - \frac{1}{2} + \frac{2}{3} - \frac{1}{3} + \frac{2}{4} - \frac{1}{4} + \cdots$$

passes the first condition because a_n approaches 0 as $n \to \infty$. You cannot apply the Alternating Series Test, however, because the series does not pass the second condition. To conclude that the series diverges, you can argue that S_{2N} equals the Nth partial sum of the divergent harmonic series. This implies that the sequence of partial sums diverges. So, the series diverges.

Alternating Series Remainder

For a convergent alternating series, the partial sum S_N can be a useful approximation for the sum S of the series. The error involved in using $S \approx S_N$ is the remainder $R_N = S - S_N$.

THEOREM 9.15 Alternating Series Remainder

If a convergent alternating series satisfies the condition $a_{n+1} \le a_n$, then the absolute value of the remainder R_N involved in approximating the sum S by S_N is less than (or equal to) the first neglected term. That is,

$$|S - S_N| = |R_N| \le a_{N+1}.$$

Proof The series obtained by deleting the first N terms of the given series satisfies the conditions of the Alternating Series Test and has a sum of R_N.

$$\begin{aligned} R_N = S - S_N &= \sum_{n=1}^{\infty} (-1)^{n+1} a_n - \sum_{n=1}^{N} (-1)^{n+1} a_n \\ &= (-1)^N a_{N+1} + (-1)^{N+1} a_{N+2} + (-1)^{N+2} a_{N+3} + \cdots \\ &= (-1)^N (a_{N+1} - a_{N+2} + a_{N+3} - \cdots) \\ |R_N| &= a_{N+1} - a_{N+2} + a_{N+3} - a_{N+4} + a_{N+5} - \cdots \\ &= a_{N+1} - (a_{N+2} - a_{N+3}) - (a_{N+4} - a_{N+5}) - \cdots \le a_{N+1} \end{aligned}$$

Consequently, $|S - S_N| = |R_N| \le a_{N+1}$, which establishes the theorem.

EXAMPLE 4 Approximating the Sum of an Alternating Series

Approximate the sum of the following series by its first six terms.

$$\sum_{n=1}^{\infty} (-1)^{n+1} \left(\frac{1}{n!}\right) = \frac{1}{1!} - \frac{1}{2!} + \frac{1}{3!} - \frac{1}{4!} + \frac{1}{5!} - \frac{1}{6!} + \cdots$$

Solution The series converges by the Alternating Series Test because

$$\frac{1}{(n+1)!} \le \frac{1}{n!} \quad \text{and} \quad \lim_{n \to \infty} \frac{1}{n!} = 0.$$

The sum of the first six terms is

$$S_6 = 1 - \frac{1}{2} + \frac{1}{6} - \frac{1}{24} + \frac{1}{120} - \frac{1}{720} = \frac{91}{144} \approx 0.63194$$

and, by the Alternating Series Remainder, you have

$$|S - S_6| = |R_6| \le a_7 = \frac{1}{5040} \approx 0.0002.$$

So, the sum S lies between $0.63194 - 0.0002$ and $0.63194 + 0.0002$, and you have

$$0.63174 \le S \le 0.63214.$$

TECHNOLOGY Later, in Section 9.10, you will be able to show that the series in Example 4 converges to

$$\frac{e-1}{e} \approx 0.63212.$$

For now, try using a computer to obtain an approximation of the sum of the series. How many terms do you need to obtain an approximation that is within 0.00001 unit of the actual sum?

Absolute and Conditional Convergence

Occasionally, a series may have both positive and negative terms and not be an alternating series. For instance, the series

$$\sum_{n=1}^{\infty} \frac{\sin n}{n^2} = \frac{\sin 1}{1} + \frac{\sin 2}{4} + \frac{\sin 3}{9} + \cdots$$

has both positive and negative terms, yet it is not an alternating series. One way to obtain some information about the convergence of this series is to investigate the convergence of the series

$$\sum_{n=1}^{\infty} \left|\frac{\sin n}{n^2}\right|.$$

By direct comparison, you have $|\sin n| \leq 1$ for all n, so

$$\left|\frac{\sin n}{n^2}\right| \leq \frac{1}{n^2}, \quad n \geq 1.$$

Therefore, by the Direct Comparison Test, the series $\sum \left|\frac{\sin n}{n^2}\right|$ converges. The next theorem tells you that the original series also converges.

THEOREM 9.16 Absolute Convergence

If the series $\Sigma|a_n|$ converges, then the series $\Sigma\, a_n$ also converges.

Proof Because $0 \leq a_n + |a_n| \leq 2|a_n|$ for all n, the series

$$\sum_{n=1}^{\infty} (a_n + |a_n|)$$

converges by comparison with the convergent series

$$\sum_{n=1}^{\infty} 2|a_n|.$$

Furthermore, because $a_n = (a_n + |a_n|) - |a_n|$, you can write

$$\sum_{n=1}^{\infty} a_n = \sum_{n=1}^{\infty} (a_n + |a_n|) - \sum_{n=1}^{\infty} |a_n|$$

where both series on the right converge. So, it follows that $\Sigma\, a_n$ converges.

The converse of Theorem 9.16 is not true. For instance, the **alternating harmonic series**

$$\sum_{n=1}^{\infty} \frac{(-1)^{n+1}}{n} = \frac{1}{1} - \frac{1}{2} + \frac{1}{3} - \frac{1}{4} + \cdots$$

converges by the Alternating Series Test. Yet the harmonic series diverges. This type of convergence is called **conditional.**

Definitions of Absolute and Conditional Convergence

1. $\Sigma\, a_n$ is **absolutely convergent** if $\Sigma\, |a_n|$ converges.
2. $\Sigma\, a_n$ is **conditionally convergent** if $\Sigma\, a_n$ converges but $\Sigma\, |a_n|$ diverges.

EXAMPLE 5 **Absolute and Conditional Convergence**

Determine whether each of the series is convergent or divergent. Classify any convergent series as absolutely or conditionally convergent.

a. $\displaystyle\sum_{n=0}^{\infty} \frac{(-1)^n\, n!}{2^n} = \frac{0!}{2^0} - \frac{1!}{2^1} + \frac{2!}{2^2} - \frac{3!}{2^3} + \cdots$

b. $\displaystyle\sum_{n=1}^{\infty} \frac{(-1)^n}{\sqrt{n}} = -\frac{1}{\sqrt{1}} + \frac{1}{\sqrt{2}} - \frac{1}{\sqrt{3}} + \frac{1}{\sqrt{4}} - \cdots$

Solution

a. By the nth-Term Test for Divergence, you can conclude that this series diverges.

b. The given series can be shown to be convergent by the Alternating Series Test. Moreover, because the p-series

$$\sum_{n=1}^{\infty} \left|\frac{(-1)^n}{\sqrt{n}}\right| = \frac{1}{\sqrt{1}} + \frac{1}{\sqrt{2}} + \frac{1}{\sqrt{3}} + \frac{1}{\sqrt{4}} + \cdots$$

diverges, the given series is *conditionally* convergent.

EXAMPLE 6 **Absolute and Conditional Convergence**

Determine whether each of the series is convergent or divergent. Classify any convergent series as absolutely or conditionally convergent.

a. $\displaystyle\sum_{n=1}^{\infty} \frac{(-1)^{n(n+1)/2}}{3^n} = -\frac{1}{3} - \frac{1}{9} + \frac{1}{27} + \frac{1}{81} - \cdots$

b. $\displaystyle\sum_{n=1}^{\infty} \frac{(-1)^n}{\ln(n+1)} = -\frac{1}{\ln 2} + \frac{1}{\ln 3} - \frac{1}{\ln 4} + \frac{1}{\ln 5} - \cdots$

Solution

a. This is *not* an alternating series. However, because

$$\sum_{n=1}^{\infty} \left|\frac{(-1)^{n(n+1)/2}}{3^n}\right| = \sum_{n=1}^{\infty} \frac{1}{3^n}$$

is a convergent geometric series, you can apply Theorem 9.16 to conclude that the given series is *absolutely* convergent (and therefore convergent).

b. In this case, the Alternating Series Test indicates that the given series converges. However, the series

$$\sum_{n=1}^{\infty} \left|\frac{(-1)^n}{\ln(n+1)}\right| = \frac{1}{\ln 2} + \frac{1}{\ln 3} + \frac{1}{\ln 4} + \cdots$$

diverges by direct comparison with the terms of the harmonic series. Therefore, the given series is *conditionally* convergent.

Rearrangement of Series

A finite sum such as $(1 + 3 - 2 + 5 - 4)$ can be rearranged without changing the value of the sum. This is not necessarily true of an infinite series—it depends on whether the series is absolutely convergent (every rearrangement has the same sum) or conditionally convergent.

FOR FURTHER INFORMATION Georg Friedrich Riemann (1826–1866) proved that if $\Sigma\, a_n$ is conditionally convergent and S is any real number, the terms of the series can be rearranged to converge to S. For more on this topic, see the article "Riemann's Rearrangement Theorem" by Stewart Galanor in *Mathematics Teacher*. To view this article, go to the website *www.matharticles.com*.

EXAMPLE 7 Rearrangement of a Series

The alternating harmonic series converges to ln 2. That is,

$$\sum_{n=1}^{\infty} (-1)^{n+1}\frac{1}{n} = \frac{1}{1} - \frac{1}{2} + \frac{1}{3} - \frac{1}{4} + \cdot \cdot \cdot = \ln 2. \qquad \text{(See Exercise 49, Section 9.10.)}$$

Rearrange the series to produce a different sum.

Solution Consider the following rearrangement.

$$1 - \frac{1}{2} - \frac{1}{4} + \frac{1}{3} - \frac{1}{6} - \frac{1}{8} + \frac{1}{5} - \frac{1}{10} - \frac{1}{12} + \frac{1}{7} - \frac{1}{14} - \cdot \cdot \cdot$$

$$= \left(1 - \frac{1}{2}\right) - \frac{1}{4} + \left(\frac{1}{3} - \frac{1}{6}\right) - \frac{1}{8} + \left(\frac{1}{5} - \frac{1}{10}\right) - \frac{1}{12} + \left(\frac{1}{7} - \frac{1}{14}\right) - \cdot \cdot \cdot$$

$$= \frac{1}{2} - \frac{1}{4} + \frac{1}{6} - \frac{1}{8} + \frac{1}{10} - \frac{1}{12} + \frac{1}{14} - \cdot \cdot \cdot$$

$$= \frac{1}{2}\left(1 - \frac{1}{2} + \frac{1}{3} - \frac{1}{4} + \frac{1}{5} - \frac{1}{6} + \frac{1}{7} - \cdot \cdot \cdot\right) = \frac{1}{2}(\ln 2)$$

By rearranging the terms, you obtain a sum that is half the original sum.

Exercises for Section 9.5

See www.CalcChat.com for worked-out solutions to odd-numbered exercises.

In Exercises 1–6, match the series with the graph of its sequence of partial sums. [The graphs are labeled (a), (b), (c), (d), (e), and (f).]

(a)

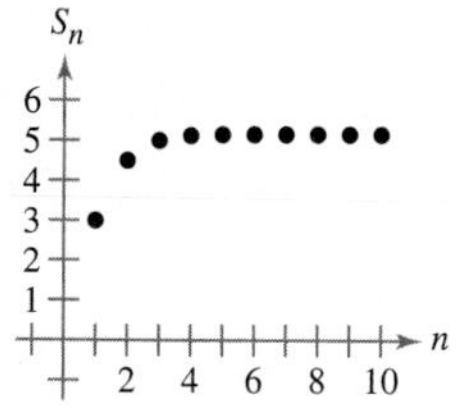

(b)

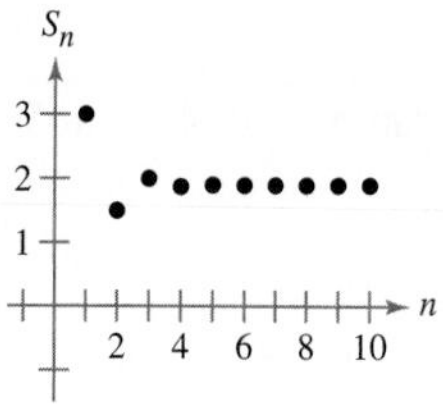

(c)

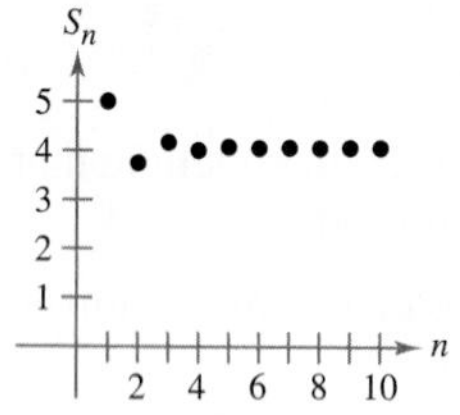

(d)

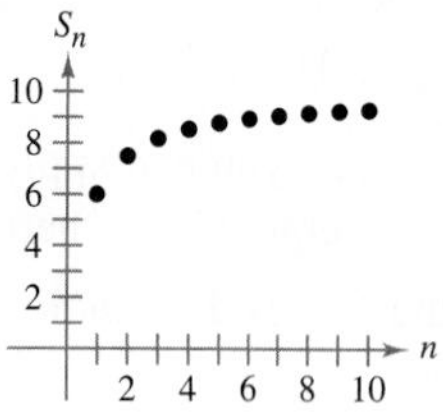

(e)

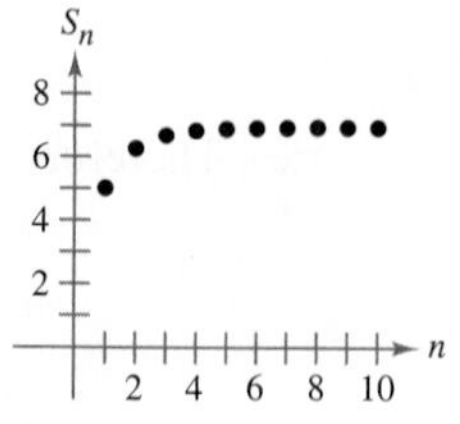

(f)

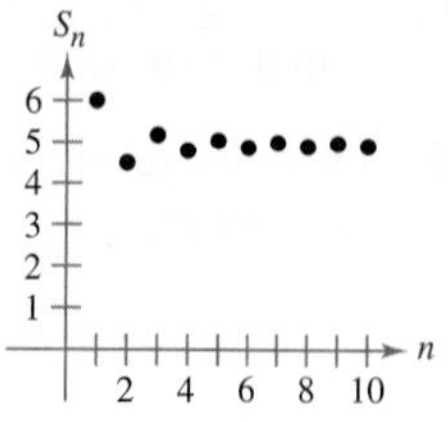

1. $\displaystyle\sum_{n=1}^{\infty} \frac{6}{n^2}$

2. $\displaystyle\sum_{n=1}^{\infty} \frac{(-1)^{n-1}6}{n^2}$

3. $\displaystyle\sum_{n=1}^{\infty} \frac{3}{n!}$

4. $\displaystyle\sum_{n=1}^{\infty} \frac{(-1)^{n-1}3}{n!}$

5. $\displaystyle\sum_{n=1}^{\infty} \frac{10}{n2^n}$

6. $\displaystyle\sum_{n=1}^{\infty} \frac{(-1)^{n-1}10}{n2^n}$

Numerical and Graphical Analysis **In Exercises 7–10, explore the Alternating Series Remainder.**

(a) Use a graphing utility to find the indicated partial sum S_n and complete the table.

n	1	2	3	4	5	6	7	8	9	10
S_n										

(b) Use a graphing utility to graph the first 10 terms of the sequence of partial sums and a horizontal line representing the sum.

(c) What pattern exists between the plot of the successive points in part (b) relative to the horizontal line representing the sum of the series? Do the distances between the successive points and the horizontal line increase or decrease?

(d) Discuss the relationship between the answers in part (c) and the Alternating Series Remainder as given in Theorem 9.15.

7. $\displaystyle\sum_{n=1}^{\infty} \frac{(-1)^{n-1}}{2n-1} = \frac{\pi}{4}$

8. $\displaystyle\sum_{n=1}^{\infty} \frac{(-1)^{n-1}}{(n-1)!} = \frac{1}{e}$

9. $\displaystyle\sum_{n=1}^{\infty} \frac{(-1)^{n-1}}{n^2} = \frac{\pi^2}{12}$

10. $\displaystyle\sum_{n=1}^{\infty} \frac{(-1)^{n-1}}{(2n-1)!} = \sin 1$

In Exercises 11–32, determine the convergence or divergence of the series.

11. $\sum_{n=1}^{\infty} \frac{(-1)^{n+1}}{n}$

12. $\sum_{n=1}^{\infty} \frac{(-1)^{n+1}n}{2n-1}$

13. $\sum_{n=1}^{\infty} \frac{(-1)^{n+1}}{2n-1}$

14. $\sum_{n=1}^{\infty} \frac{(-1)^{n}}{\ln(n+1)}$

15. $\sum_{n=1}^{\infty} \frac{(-1)^{n}n^2}{n^2+1}$

16. $\sum_{n=1}^{\infty} \frac{(-1)^{n+1}n}{n^2+1}$

17. $\sum_{n=1}^{\infty} \frac{(-1)^{n}}{\sqrt{n}}$

18. $\sum_{n=1}^{\infty} \frac{(-1)^{n+1}n^2}{n^2+5}$

19. $\sum_{n=1}^{\infty} \frac{(-1)^{n+1}(n+1)}{\ln(n+1)}$

20. $\sum_{n=1}^{\infty} \frac{(-1)^{n+1}\ln(n+1)}{n+1}$

21. $\sum_{n=1}^{\infty} \sin\frac{(2n-1)\pi}{2}$

22. $\sum_{n=1}^{\infty} \frac{1}{n}\sin\frac{(2n-1)\pi}{2}$

23. $\sum_{n=1}^{\infty} \cos n\pi$

24. $\sum_{n=1}^{\infty} \frac{1}{n}\cos n\pi$

25. $\sum_{n=0}^{\infty} \frac{(-1)^{n}}{n!}$

26. $\sum_{n=0}^{\infty} \frac{(-1)^{n}}{(2n+1)!}$

27. $\sum_{n=1}^{\infty} \frac{(-1)^{n+1}\sqrt{n}}{n+2}$

28. $\sum_{n=1}^{\infty} \frac{(-1)^{n+1}\sqrt{n}}{\sqrt[3]{n}}$

29. $\sum_{n=1}^{\infty} \frac{(-1)^{n+1}n!}{1\cdot 3\cdot 5\cdot\cdot\cdot(2n-1)}$

30. $\sum_{n=1}^{\infty} (-1)^{n+1}\frac{1\cdot 3\cdot 5\cdot\cdot\cdot(2n-1)}{1\cdot 4\cdot 7\cdot\cdot\cdot(3n-2)}$

31. $\sum_{n=1}^{\infty} \frac{2(-1)^{n+1}}{e^n-e^{-n}} = \sum_{n=1}^{\infty}(-1)^{n+1}\operatorname{csch} n$

32. $\sum_{n=1}^{\infty} \frac{2(-1)^{n+1}}{e^n+e^{-n}} = \sum_{n=1}^{\infty}(-1)^{n+1}\operatorname{sech} n$

In Exercises 33–36, approximate the sum of the series by using the first six terms. (See Example 4.)

33. $\sum_{n=1}^{\infty} \frac{(-1)^{n+1}3}{n^2}$

34. $\sum_{n=1}^{\infty} \frac{(-1)^{n+1}4}{\ln(n+1)}$

35. $\sum_{n=0}^{\infty} \frac{(-1)^{n}2}{n!}$

36. $\sum_{n=1}^{\infty} \frac{(-1)^{n+1}n}{2^n}$

In Exercises 37–42, (a) use Theorem 9.15 to determine the number of terms required to approximate the sum of the convergent series with an error of less than 0.001, and (b) use a graphing utility to approximate the sum of the series with an error of less than 0.001.

37. $\sum_{n=0}^{\infty} \frac{(-1)^{n}}{n!} = \frac{1}{e}$

38. $\sum_{n=0}^{\infty} \frac{(-1)^{n}}{2^n n!} = \frac{1}{\sqrt{e}}$

39. $\sum_{n=0}^{\infty} \frac{(-1)^{n}}{(2n+1)!} = \sin 1$

40. $\sum_{n=0}^{\infty} \frac{(-1)^{n}}{(2n)!} = \cos 1$

41. $\sum_{n=1}^{\infty} \frac{(-1)^{n+1}}{n} = \ln 2$

42. $\sum_{n=1}^{\infty} \frac{(-1)^{n+1}}{n4^n} = \ln\frac{5}{4}$

In Exercises 43–46, use Theorem 9.15 to determine the number of terms required to approximate the sum of the series with an error of less than 0.001.

43. $\sum_{n=1}^{\infty} \frac{(-1)^{n+1}}{n^3}$

44. $\sum_{n=1}^{\infty} \frac{(-1)^{n+1}}{n^2}$

45. $\sum_{n=1}^{\infty} \frac{(-1)^{n+1}}{2n^3-1}$

46. $\sum_{n=1}^{\infty} \frac{(-1)^{n+1}}{n^4}$

In Exercises 47–62, determine whether the series converges conditionally or absolutely, or diverges.

47. $\sum_{n=1}^{\infty} \frac{(-1)^{n+1}}{(n+1)^2}$

48. $\sum_{n=1}^{\infty} \frac{(-1)^{n+1}}{n+1}$

49. $\sum_{n=1}^{\infty} \frac{(-1)^{n+1}}{\sqrt{n}}$

50. $\sum_{n=1}^{\infty} \frac{(-1)^{n+1}}{n\sqrt{n}}$

51. $\sum_{n=1}^{\infty} \frac{(-1)^{n+1}n^2}{(n+1)^2}$

52. $\sum_{n=1}^{\infty} \frac{(-1)^{n+1}(2n+3)}{n+10}$

53. $\sum_{n=2}^{\infty} \frac{(-1)^{n}}{\ln n}$

54. $\sum_{n=0}^{\infty} (-1)^{n}e^{-n^2}$

55. $\sum_{n=2}^{\infty} \frac{(-1)^{n}n}{n^3-1}$

56. $\sum_{n=1}^{\infty} \frac{(-1)^{n+1}}{n^{1.5}}$

57. $\sum_{n=0}^{\infty} \frac{(-1)^{n}}{(2n+1)!}$

58. $\sum_{n=0}^{\infty} \frac{(-1)^{n}}{\sqrt{n+4}}$

59. $\sum_{n=0}^{\infty} \frac{\cos n\pi}{n+1}$

60. $\sum_{n=1}^{\infty} (-1)^{n+1}\arctan n$

61. $\sum_{n=1}^{\infty} \frac{\cos n\pi}{n^2}$

62. $\sum_{n=1}^{\infty} \frac{\sin[(2n-1)\pi/2]}{n}$

Writing About Concepts

63. Define an alternating series and state the Alternating Series Test.

64. Give the remainder after N terms of a convergent alternating series.

65. In your own words, state the difference between absolute and conditional convergence of an alternating series.

66. The graphs of the sequences of partial sums of two series are shown in the figures. Which graph represents the partial sums of an alternating series? Explain.

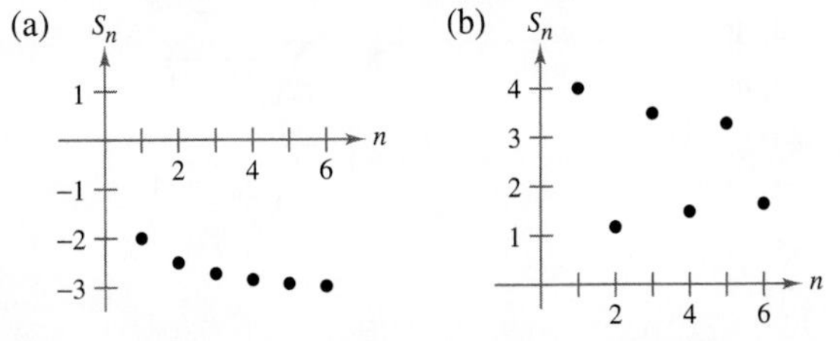

True or False? **In Exercises 67–70, determine whether the statement is true or false. If it is false, explain why or give an example that shows it is false.**

67. If both $\Sigma\, a_n$ and $\Sigma\,(-a_n)$ converge, then $\Sigma\, |a_n|$ converges.

68. If $\Sigma\, a_n$ diverges, then $\Sigma\, |a_n|$ diverges.

69. For the alternating series $\displaystyle\sum_{n=1}^{\infty} \frac{(-1)^n}{n}$, the partial sum S_{100} is an overestimate of the sum of the series.

70. If $\Sigma\, a_n$ and $\Sigma\, b_n$ both converge, then $\Sigma\, a_n b_n$ converges.

In Exercises 71 and 72, find the values of p for which the series converges.

71. $\displaystyle\sum_{n=1}^{\infty} (-1)^n\left(\frac{1}{n^p}\right)$

72. $\displaystyle\sum_{n=1}^{\infty} (-1)^n\left(\frac{1}{n+p}\right)$

73. Prove that if $\Sigma\, |a_n|$ converges, then $\Sigma\, a_n^2$ converges. Is the converse true? If not, give an example that shows it is false.

74. Use the result of Exercise 71 to give an example of an alternating p-series that converges, but whose corresponding p-series diverges.

75. Give an example of a series that demonstrates the statement you proved in Exercise 73.

76. Find all values of x for which the series $\Sigma\,(x^n/n)$ (a) converges absolutely and (b) converges conditionally.

77. Consider the following series.

$$\frac{1}{2} - \frac{1}{3} + \frac{1}{4} - \frac{1}{9} + \frac{1}{8} - \frac{1}{27} + \cdots + \frac{1}{2^n} - \frac{1}{3^n} + \cdots$$

(a) Does the series meet the conditions of Theorem 9.14? Explain why or why not.

(b) Does the series converge? If so, what is the sum?

78. Consider the following series.

$$\sum_{n=1}^{\infty} (-1)^{n+1} a_n,\quad a_n = \begin{cases} \dfrac{1}{\sqrt{n}}, & \text{if } n \text{ is odd} \\[2ex] \dfrac{1}{n^3}, & \text{if } n \text{ is even} \end{cases}$$

(a) Does the series meet the conditions of Theorem 9.14? Explain why or why not.

(b) Does the series converge? If so, what is the sum?

Review **In Exercises 79–88, test for convergence or divergence and identify the test used.**

79. $\displaystyle\sum_{n=1}^{\infty} \frac{10}{n^{3/2}}$

80. $\displaystyle\sum_{n=1}^{\infty} \frac{3}{n^2+5}$

81. $\displaystyle\sum_{n=1}^{\infty} \frac{3^n}{n^2}$

82. $\displaystyle\sum_{n=1}^{\infty} \frac{1}{2^n+1}$

83. $\displaystyle\sum_{n=0}^{\infty} 5\left(\frac{7}{8}\right)^n$

84. $\displaystyle\sum_{n=1}^{\infty} \frac{3n^2}{2n^2+1}$

85. $\displaystyle\sum_{n=1}^{\infty} 100e^{-n/2}$

86. $\displaystyle\sum_{n=0}^{\infty} \frac{(-1)^n}{n+4}$

87. $\displaystyle\sum_{n=1}^{\infty} \frac{(-1)^{n+1}4}{3n^2-1}$

88. $\displaystyle\sum_{n=2}^{\infty} \frac{\ln n}{n}$

89. The following argument, that $0 = 1$, is *incorrect*. Describe the error.

$$\begin{aligned} 0 &= 0 + 0 + 0 + \cdots \\ &= (1-1) + (1-1) + (1-1) + \cdots \\ &= 1 + (-1+1) + (-1+1) + \cdots \\ &= 1 + 0 + 0 + \cdots \\ &= 1 \end{aligned}$$

90. The following argument, $2 = 1$, is *incorrect*. Describe the error. Multiply each side of the alternating harmonic series

$$S = 1 - \frac{1}{2} + \frac{1}{3} - \frac{1}{4} + \frac{1}{5} - \frac{1}{6} + \frac{1}{7} - \frac{1}{8} + \frac{1}{9} - \frac{1}{10} + \cdots$$

by 2 to get

$$2S = 2 - 1 + \frac{2}{3} - \frac{1}{2} + \frac{2}{5} - \frac{1}{3} + \frac{2}{7} - \frac{1}{4} + \frac{2}{9} - \frac{1}{5} + \cdots$$

Now collect terms with like denominators (as indicated by the arrows) to get

$$2S = 1 - \frac{1}{2} + \frac{1}{3} - \frac{1}{4} + \frac{1}{5} + \cdots$$

The resulting series is the same one that you started with. So, $2S = S$ and divide each side by S to get $2 = 1$.

FOR FURTHER INFORMATION For more on this exercise, see the article "Riemann's Rearrangement Theorem" by Stewart Galanor in *Mathematics Teacher*. To view this article, go to the website *www.matharticles.com*.

Putnam Exam Challenge

91. Assume as known the (true) fact that the alternating harmonic series

(1) $1 - \frac{1}{2} + \frac{1}{3} - \frac{1}{4} + \frac{1}{5} - \frac{1}{6} + \frac{1}{7} - \frac{1}{8} + \cdots$

is convergent, and denote its sum by s. Rearrange the series (1) as follows:

(2) $1 + \frac{1}{3} - \frac{1}{2} + \frac{1}{5} + \frac{1}{7} - \frac{1}{4} + \frac{1}{9} + \frac{1}{11} - \frac{1}{6} + \cdots$

Assume as known the (true) fact that the series (2) is also convergent, and denote its sum by S. Denote by s_k, S_k the kth partial sum of the series (1) and (2), respectively. Prove each statement.

(i) $S_{3n} = s_{4n} + \frac{1}{2}s_{2n}$, (ii) $S \neq s$

Section 9.6 The Ratio and Root Tests

- **Use the Ratio Test to determine whether a series converges or diverges.**
- **Use the Root Test to determine whether a series converges or diverges.**
- **Review the tests for convergence and divergence of an infinite series.**

EXPLORATION

Writing a Series One of the following conditions guarantees that a series will diverge, two conditions guarantee that a series will converge, and one has no guarantee—the series can either converge or diverge. Which is which? Explain your reasoning.

a. $\lim_{n\to\infty} \left|\frac{a_{n+1}}{a_n}\right| = 0$

b. $\lim_{n\to\infty} \left|\frac{a_{n+1}}{a_n}\right| = \frac{1}{2}$

c. $\lim_{n\to\infty} \left|\frac{a_{n+1}}{a_n}\right| = 1$

d. $\lim_{n\to\infty} \left|\frac{a_{n+1}}{a_n}\right| = 2$

The Ratio Test

This section begins with a test for absolute convergence—the **Ratio Test.**

THEOREM 9.17 Ratio Test

Let $\Sigma\, a_n$ be a series with nonzero terms.

1. $\Sigma\, a_n$ converges absolutely if $\lim_{n\to\infty} \left|\frac{a_{n+1}}{a_n}\right| < 1$.

2. $\Sigma\, a_n$ diverges if $\lim_{n\to\infty} \left|\frac{a_{n+1}}{a_n}\right| > 1$ or $\lim_{n\to\infty} \left|\frac{a_{n+1}}{a_n}\right| = \infty$.

3. The Ratio Test is inconclusive if $\lim_{n\to\infty} \left|\frac{a_{n+1}}{a_n}\right| = 1$.

Proof To prove Property 1, assume that

$$\lim_{n\to\infty} \left|\frac{a_{n+1}}{a_n}\right| = r < 1$$

and choose R such that $0 \le r < R < 1$. By the definition of the limit of a sequence, there exists some $N > 0$ such that $|a_{n+1}/a_n| < R$ for all $n > N$. Therefore, you can write the following inequalities.

$$\begin{aligned}
|a_{N+1}| &< |a_N|R \\
|a_{N+2}| &< |a_{N+1}|R < |a_N|R^2 \\
|a_{N+3}| &< |a_{N+2}|R < |a_{N+1}|R^2 < |a_N|R^3 \\
&\vdots
\end{aligned}$$

The geometric series $\Sigma\, |a_N|R^n = |a_N|R + |a_N|R^2 + \cdot \cdot \cdot + |a_N|R^n + \cdot \cdot \cdot$ converges, and so, by the Direct Comparison Test, the series

$$\sum_{n=1}^{\infty} |a_{N+n}| = |a_{N+1}| + |a_{N+2}| + \cdot \cdot \cdot + |a_{N+n}| + \cdot \cdot \cdot$$

also converges. This in turn implies that the series $\Sigma\, |a_n|$ converges, because discarding a finite number of terms $(n = N - 1)$ does not affect convergence. Consequently, by Theorem 9.16, the series $\Sigma\, a_n$ converges absolutely. The proof of Property 2 is similar and is left as an exercise (see Exercise 98).

NOTE The fact that the Ratio Test is inconclusive when $|a_{n+1}/a_n| \to 1$ can be seen by comparing the two series $\Sigma\,(1/n)$ and $\Sigma\,(1/n^2)$. The first series diverges and the second one converges, but in both cases

$$\lim_{n\to\infty} \left|\frac{a_{n+1}}{a_n}\right| = 1.$$

Although the Ratio Test is not a cure for all ills related to tests for convergence, it is particularly useful for series that *converge rapidly*. Series involving factorials or exponentials are frequently of this type.

EXAMPLE 1 Using the Ratio Test

Determine the convergence or divergence of

$$\sum_{n=0}^{\infty} \frac{2^n}{n!}.$$

Solution Because $a_n = 2^n/n!$, you can write the following.

$$\begin{aligned}
\lim_{n\to\infty}\left|\frac{a_{n+1}}{a_n}\right| &= \lim_{n\to\infty}\left[\frac{2^{n+1}}{(n+1)!} \div \frac{2^n}{n!}\right] \\
&= \lim_{n\to\infty}\left[\frac{2^{n+1}}{(n+1)!}\cdot\frac{n!}{2^n}\right] \\
&= \lim_{n\to\infty}\frac{2}{n+1} \\
&= 0
\end{aligned}$$

Therefore, the series converges.

STUDY TIP A step frequently used in applications of the Ratio Test involves simplifying quotients of factorials. In Example 1, for instance, notice that

$$\frac{n!}{(n+1)!} = \frac{n!}{(n+1)n!} = \frac{1}{n+1}.$$

EXAMPLE 2 Using the Ratio Test

Determine whether each series converges or diverges.

a. $\displaystyle\sum_{n=0}^{\infty} \frac{n^2 2^{n+1}}{3^n}$ **b.** $\displaystyle\sum_{n=1}^{\infty} \frac{n^n}{n!}$

Solution

a. This series converges because the limit of $|a_{n+1}/a_n|$ is less than 1.

$$\begin{aligned}
\lim_{n\to\infty}\left|\frac{a_{n+1}}{a_n}\right| &= \lim_{n\to\infty}\left[(n+1)^2\left(\frac{2^{n+2}}{3^{n+1}}\right)\left(\frac{3^n}{n^2 2^{n+1}}\right)\right] \\
&= \lim_{n\to\infty}\frac{2(n+1)^2}{3n^2} \\
&= \frac{2}{3} < 1
\end{aligned}$$

b. This series diverges because the limit of $|a_{n+1}/a_n|$ is greater than 1.

$$\begin{aligned}
\lim_{n\to\infty}\left|\frac{a_{n+1}}{a_n}\right| &= \lim_{n\to\infty}\left[\frac{(n+1)^{n+1}}{(n+1)!}\left(\frac{n!}{n^n}\right)\right] \\
&= \lim_{n\to\infty}\left[\frac{(n+1)^{n+1}}{(n+1)}\left(\frac{1}{n^n}\right)\right] \\
&= \lim_{n\to\infty}\frac{(n+1)^n}{n^n} \\
&= \lim_{n\to\infty}\left(1+\frac{1}{n}\right)^n \\
&= e > 1
\end{aligned}$$

EXAMPLE 3 A Failure of the Ratio Test

Determine the convergence or divergence of $\sum_{n=1}^{\infty} (-1)^n \frac{\sqrt{n}}{n+1}$.

Solution The limit of $|a_{n+1}/a_n|$ is equal to 1.

$$\begin{aligned}\lim_{n\to\infty} \left|\frac{a_{n+1}}{a_n}\right| &= \lim_{n\to\infty} \left[\left(\frac{\sqrt{n+1}}{n+2}\right)\left(\frac{n+1}{\sqrt{n}}\right)\right] \\ &= \lim_{n\to\infty} \left[\sqrt{\frac{n+1}{n}}\left(\frac{n+1}{n+2}\right)\right] \\ &= \sqrt{1}(1) \\ &= 1\end{aligned}$$

NOTE The Ratio Test is also inconclusive for any p-series.

So, the Ratio Test is inconclusive. To determine whether the series converges, you need to try a different test. In this case, you can apply the Alternating Series Test. To show that $a_{n+1} \le a_n$, let

$$f(x) = \frac{\sqrt{x}}{x+1}.$$

Then the derivative is

$$f'(x) = \frac{-x+1}{2\sqrt{x}(x+1)^2}.$$

Because the derivative is negative for $x > 1$, you know that f is a decreasing function. Also, by L'Hôpital's Rule,

$$\begin{aligned}\lim_{x\to\infty} \frac{\sqrt{x}}{x+1} &= \lim_{x\to\infty} \frac{1/(2\sqrt{x})}{1} \\ &= \lim_{x\to\infty} \frac{1}{2\sqrt{x}} \\ &= 0.\end{aligned}$$

Therefore, by the Alternating Series Test, the series converges.

The series in Example 3 is *conditionally convergent*. This follows from the fact that the series

$$\sum_{n=1}^{\infty} |a_n|$$

diverges (by the Limit Comparison Test with $\Sigma\, 1/\sqrt{n}$), but the series

$$\sum_{n=1}^{\infty} a_n$$

converges.

TECHNOLOGY A computer or programmable calculator can reinforce the conclusion that the series in Example 3 converges *conditionally*. By adding the first 100 terms of the series, you obtain a sum of about -0.2. (The sum of the first 100 terms of the series $\Sigma\, |a_n|$ is about 17.)

The Root Test

The next test for convergence or divergence of series works especially well for series involving nth powers. The proof of this theorem is similar to that given for the Ratio Test, and is left as an exercise (see Exercise 99).

THEOREM 9.18 Root Test

Let $\Sigma\, a_n$ be a series.

1. $\Sigma\, a_n$ converges absolutely if $\lim_{n\to\infty} \sqrt[n]{|a_n|} < 1$.
2. $\Sigma\, a_n$ diverges if $\lim_{n\to\infty} \sqrt[n]{|a_n|} > 1$ or $\lim_{n\to\infty} \sqrt[n]{|a_n|} = \infty$.
3. The Root Test is inconclusive if $\lim_{n\to\infty} \sqrt[n]{|a_n|} = 1$.

EXAMPLE 4 Using the Root Test

Determine the convergence or divergence of

$$\sum_{n=1}^{\infty} \frac{e^{2n}}{n^n}.$$

Solution You can apply the Root Test as follows.

$$\begin{aligned}\lim_{n\to\infty} \sqrt[n]{|a_n|} &= \lim_{n\to\infty} \sqrt[n]{\frac{e^{2n}}{n^n}} \\ &= \lim_{n\to\infty} \frac{e^{2n/n}}{n^{n/n}} \\ &= \lim_{n\to\infty} \frac{e^2}{n} \\ &= 0 < 1\end{aligned}$$

NOTE The Root Test is always inconclusive for any p-series.

Because this limit is less than 1, you can conclude that the series converges absolutely (and therefore converges).

FOR FURTHER INFORMATION For more information on the usefulness of the Root Test, see the article "N! and the Root Test" by Charles C. Mumma II in *The American Mathematical Monthly*. To view this article, go to the website *www.matharticles.com.*

To see the usefulness of the Root Test for the series in Example 4, try applying the Ratio Test to that series. When you do this, you obtain the following.

$$\begin{aligned}\lim_{n\to\infty} \left|\frac{a_{n+1}}{a_n}\right| &= \lim_{n\to\infty} \left[\frac{e^{2(n+1)}}{(n+1)^{n+1}} \div \frac{e^{2n}}{n^n}\right] \\ &= \lim_{n\to\infty} e^2 \frac{n^n}{(n+1)^{n+1}} \\ &= \lim_{n\to\infty} e^2 \left(\frac{n}{n+1}\right)^n \left(\frac{1}{n+1}\right) \\ &= 0\end{aligned}$$

Note that this limit is not as easily evaluated as the limit obtained by the Root Test in Example 4.

Strategies for Testing Series

You have now studied 10 tests for determining the convergence or divergence of an infinite series. (See the summary in the table on page 644.) Skill in choosing and applying the various tests will come only with practice. Below is a set of guidelines for choosing an appropriate test.

Guidelines for Testing a Series for Convergence or Divergence

1. Does the nth term approach 0? If not, the series diverges.
2. Is the series one of the special types—geometric, p-series, telescoping, or alternating?
3. Can the Integral Test, the Root Test, or the Ratio Test be applied?
4. Can the series be compared favorably to one of the special types?

In some instances, more than one test is applicable. However, your objective should be to learn to choose the most efficient test.

EXAMPLE 5 Applying the Strategies for Testing Series

Determine the convergence or divergence of each series.

a. $\displaystyle\sum_{n=1}^{\infty} \frac{n+1}{3n+1}$ **b.** $\displaystyle\sum_{n=1}^{\infty} \left(\frac{\pi}{6}\right)^n$ **c.** $\displaystyle\sum_{n=1}^{\infty} ne^{-n^2}$

d. $\displaystyle\sum_{n=1}^{\infty} \frac{1}{3n+1}$ **e.** $\displaystyle\sum_{n=1}^{\infty} (-1)^n \frac{3}{4n+1}$ **f.** $\displaystyle\sum_{n=1}^{\infty} \frac{n!}{10^n}$

g. $\displaystyle\sum_{n=1}^{\infty} \left(\frac{n+1}{2n+1}\right)^n$

Solution

a. For this series, the limit of the nth term is not 0 $\left(a_n \rightarrow \frac{1}{3} \text{ as } n \rightarrow \infty\right)$. So, by the nth-Term Test, the series diverges.

b. This series is geometric. Moreover, because the ratio $r = \pi/6$ of the terms is less than 1 in absolute value, you can conclude that the series converges.

c. Because the function $f(x) = xe^{-x^2}$ is easily integrated, you can use the Integral Test to conclude that the series converges.

d. The nth term of this series can be compared to the nth term of the harmonic series. After using the Limit Comparison Test, you can conclude that the series diverges.

e. This is an alternating series whose nth term approaches 0. Because $a_{n+1} \leq a_n$, you can use the Alternating Series Test to conclude that the series converges.

f. The nth term of this series involves a factorial, which indicates that the Ratio Test may work well. After applying the Ratio Test, you can conclude that the series diverges.

g. The nth term of this series involves a variable that is raised to the nth power, which indicates that the Root Test may work well. After applying the Root Test, you can conclude that the series converges.

Summary of Tests for Series

Test	Series	Condition(s) of Convergence	Condition(s) of Divergence	Comment
nth-Term	$\sum_{n=1}^{\infty} a_n$		$\lim_{n\to\infty} a_n \neq 0$	This test cannot be used to show convergence.
Geometric Series	$\sum_{n=0}^{\infty} ar^n$	$\lvert r\rvert < 1$	$\lvert r\rvert \geq 1$	Sum: $S = \frac{a}{1-r}$
Telescoping Series	$\sum_{n=1}^{\infty} (b_n - b_{n+1})$	$\lim_{n\to\infty} b_n = L$		Sum: $S = b_1 - L$
p-Series	$\sum_{n=1}^{\infty} \frac{1}{n^p}$	$p > 1$	$p \leq 1$	
Alternating Series	$\sum_{n=1}^{\infty} (-1)^{n-1}a_n$	$0 < a_{n+1} \leq a_n$ and $\lim_{n\to\infty} a_n = 0$		Remainder: $\lvert R_N\rvert \leq a_{N+1}$
Integral (f is continuous, positive, and decreasing)	$\sum_{n=1}^{\infty} a_n$, $a_n = f(n) \geq 0$	$\int_1^{\infty} f(x)\,dx$ converges	$\int_1^{\infty} f(x)\,dx$ diverges	Remainder: $0 < R_N < \int_N^{\infty} f(x)\,dx$
Root	$\sum_{n=1}^{\infty} a_n$	$\lim_{n\to\infty} \sqrt[n]{\lvert a_n\rvert} < 1$	$\lim_{n\to\infty} \sqrt[n]{\lvert a_n\rvert} > 1$	Test is inconclusive if $\lim_{n\to\infty} \sqrt[n]{\lvert a_n\rvert} = 1$.
Ratio	$\sum_{n=1}^{\infty} a_n$	$\lim_{n\to\infty} \left\lvert\frac{a_{n+1}}{a_n}\right\rvert < 1$	$\lim_{n\to\infty} \left\lvert\frac{a_{n+1}}{a_n}\right\rvert > 1$	Test is inconclusive if $\lim_{n\to\infty} \left\lvert\frac{a_{n+1}}{a_n}\right\rvert = 1$.
Direct Comparison ($a_n, b_n > 0$)	$\sum_{n=1}^{\infty} a_n$	$0 < a_n \leq b_n$ and $\sum_{n=1}^{\infty} b_n$ converges	$0 < b_n \leq a_n$ and $\sum_{n=1}^{\infty} b_n$ diverges	
Limit Comparison ($a_n, b_n > 0$)	$\sum_{n=1}^{\infty} a_n$	$\lim_{n\to\infty} \frac{a_n}{b_n} = L > 0$ and $\sum_{n=1}^{\infty} b_n$ converges	$\lim_{n\to\infty} \frac{a_n}{b_n} = L > 0$ and $\sum_{n=1}^{\infty} b_n$ diverges	

Exercises for Section 9.6

See www.CalcChat.com for worked-out solutions to odd-numbered exercises.

In Exercises 1–4, verify the formula.

1. $\dfrac{(n+1)!}{(n-2)!} = (n+1)(n)(n-1)$

2. $\dfrac{(2k-2)!}{(2k)!} = \dfrac{1}{(2k)(2k-1)}$

3. $1 \cdot 3 \cdot 5 \cdot \cdot \cdot (2k-1) = \dfrac{(2k)!}{2^k k!}$

4. $\dfrac{1}{1 \cdot 3 \cdot 5 \cdot \cdot \cdot (2k-5)} = \dfrac{2^k k!(2k-3)(2k-1)}{(2k)!}, \quad k \geq 3$

In Exercises 5–10, match the series with the graph of its sequence of partial sums. [The graphs are labeled (a), (b), (c), (d), (e), and (f).]

(a)

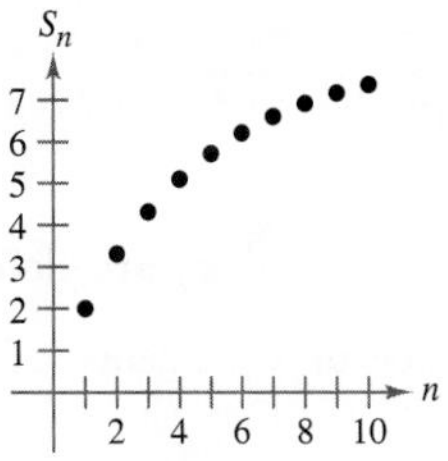

(b)

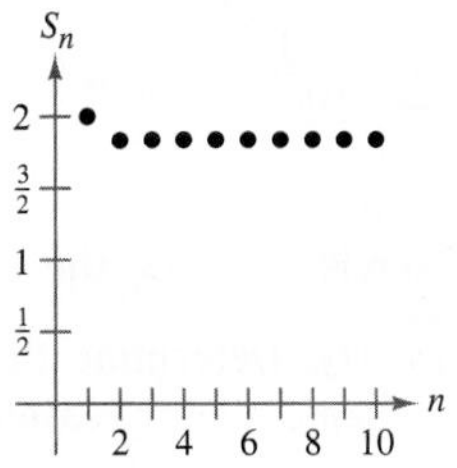

(c)

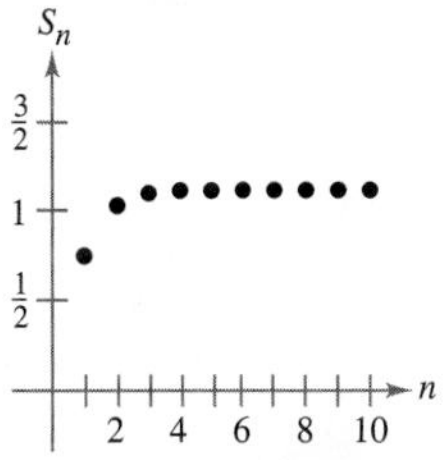

(d)

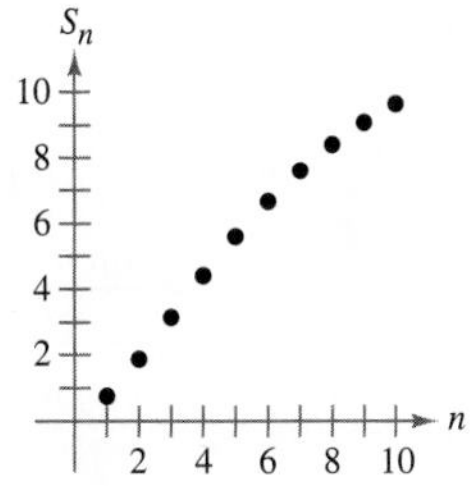

(e)

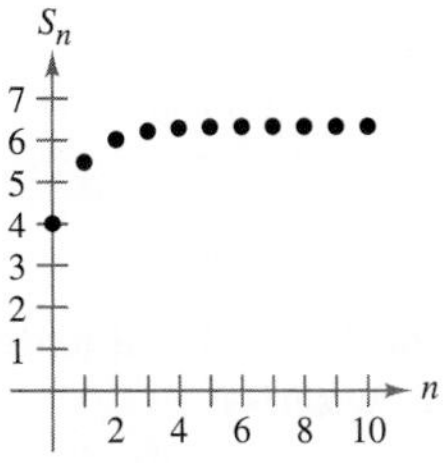

(f)

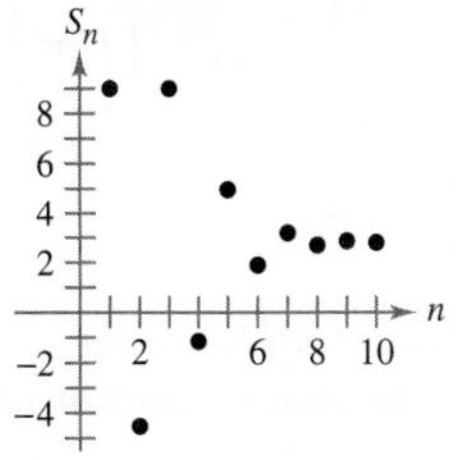

5. $\displaystyle\sum_{n=1}^{\infty} n\left(\frac{3}{4}\right)^n$

6. $\displaystyle\sum_{n=1}^{\infty} \left(\frac{3}{4}\right)^n\left(\frac{1}{n!}\right)$

7. $\displaystyle\sum_{n=1}^{\infty} \frac{(-3)^{n+1}}{n!}$

8. $\displaystyle\sum_{n=1}^{\infty} \frac{(-1)^{n-1}4}{(2n)!}$

9. $\displaystyle\sum_{n=1}^{\infty} \left(\frac{4n}{5n-3}\right)^n$

10. $\displaystyle\sum_{n=0}^{\infty} 4e^{-n}$

Numerical, Graphical, and Analytic Analysis **In Exercises 11 and 12, (a) verify that the series converges. (b) Use a graphing utility to find the indicated partial sum S_n and complete the table. (c) Use a graphing utility to graph the first 10 terms of the sequence of partial sums. (d) Use the table to estimate the sum of the series. (e) Explain the relationship between the magnitudes of the terms of the series and the rate at which the sequence of partial sums approaches the sum of the series.**

n	5	10	15	20	25
S_n					

11. $\displaystyle\sum_{n=1}^{\infty} n^2\left(\frac{5}{8}\right)^n$

12. $\displaystyle\sum_{n=1}^{\infty} \frac{n^2+1}{n!}$

In Exercises 13–32, use the Ratio Test to determine the convergence or divergence of the series.

13. $\displaystyle\sum_{n=0}^{\infty} \frac{n!}{3^n}$

14. $\displaystyle\sum_{n=0}^{\infty} \frac{3^n}{n!}$

15. $\displaystyle\sum_{n=1}^{\infty} n\left(\frac{3}{4}\right)^n$

16. $\displaystyle\sum_{n=1}^{\infty} n\left(\frac{3}{2}\right)^n$

17. $\displaystyle\sum_{n=1}^{\infty} \frac{n}{2^n}$

18. $\displaystyle\sum_{n=1}^{\infty} \frac{n^3}{2^n}$

19. $\displaystyle\sum_{n=1}^{\infty} \frac{2^n}{n^2}$

20. $\displaystyle\sum_{n=1}^{\infty} \frac{(-1)^{n+1}(n+2)}{n(n+1)}$

21. $\displaystyle\sum_{n=0}^{\infty} \frac{(-1)^n 2^n}{n!}$

22. $\displaystyle\sum_{n=1}^{\infty} \frac{(-1)^{n-1}(3/2)^n}{n^2}$

23. $\displaystyle\sum_{n=1}^{\infty} \frac{n!}{n3^n}$

24. $\displaystyle\sum_{n=1}^{\infty} \frac{(2n)!}{n^5}$

25. $\displaystyle\sum_{n=0}^{\infty} \frac{4^n}{n!}$

26. $\displaystyle\sum_{n=1}^{\infty} \frac{n^n}{n!}$

27. $\displaystyle\sum_{n=0}^{\infty} \frac{3^n}{(n+1)^n}$

28. $\displaystyle\sum_{n=0}^{\infty} \frac{(n!)^2}{(3n)!}$

29. $\displaystyle\sum_{n=0}^{\infty} \frac{4^n}{3^n+1}$

30. $\displaystyle\sum_{n=0}^{\infty} \frac{(-1)^n 2^{4n}}{(2n+1)!}$

31. $\displaystyle\sum_{n=0}^{\infty} \frac{(-1)^{n+1}n!}{1 \cdot 3 \cdot 5 \cdot \cdot \cdot (2n+1)}$

32. $\displaystyle\sum_{n=1}^{\infty} \frac{(-1)^n[2 \cdot 4 \cdot 6 \cdot \cdot \cdot (2n)]}{2 \cdot 5 \cdot 8 \cdot \cdot \cdot (3n-1)}$

In Exercises 33–36, verify that the Ratio Test is inconclusive for the p-series.

33. $\displaystyle\sum_{n=1}^{\infty} \frac{1}{n^{3/2}}$

34. $\displaystyle\sum_{n=1}^{\infty} \frac{1}{n^{1/2}}$

35. $\displaystyle\sum_{n=1}^{\infty} \frac{1}{n^4}$

36. $\displaystyle\sum_{n=1}^{\infty} \frac{1}{n^p}$

In Exercises 37–50, use the Root Test to determine the convergence or divergence of the series.

37. $\sum_{n=1}^{\infty} \left(\frac{n}{2n+1}\right)^n$

38. $\sum_{n=1}^{\infty} \left(\frac{2n}{n+1}\right)^n$

39. $\sum_{n=2}^{\infty} \left(\frac{2n+1}{n-1}\right)^n$

40. $\sum_{n=1}^{\infty} \left(\frac{4n+3}{2n-1}\right)^n$

41. $\sum_{n=2}^{\infty} \frac{(-1)^n}{(\ln n)^n}$

42. $\sum_{n=1}^{\infty} \left(\frac{-3n}{2n+1}\right)^{3n}$

43. $\sum_{n=1}^{\infty} \left(2\sqrt[n]{n}+1\right)^n$

44. $\sum_{n=0}^{\infty} e^{-n}$

45. $\sum_{n=1}^{\infty} \frac{n}{4^n}$

46. $\sum_{n=1}^{\infty} \left(\frac{n}{500}\right)^n$

47. $\sum_{n=1}^{\infty} \left(\frac{1}{n}-\frac{1}{n^2}\right)^n$

48. $\sum_{n=1}^{\infty} \left(\frac{\ln n}{n}\right)^n$

49. $\sum_{n=2}^{\infty} \frac{n}{(\ln n)^n}$

50. $\sum_{n=1}^{\infty} \frac{(n!)^n}{(n^n)^2}$

In Exercises 51–68, determine the convergence or divergence of the series using any appropriate test from this chapter. Identify the test used.

51. $\sum_{n=1}^{\infty} \frac{(-1)^{n+1}5}{n}$

52. $\sum_{n=1}^{\infty} \frac{5}{n}$

53. $\sum_{n=1}^{\infty} \frac{3}{n\sqrt{n}}$

54. $\sum_{n=1}^{\infty} \left(\frac{\pi}{4}\right)^n$

55. $\sum_{n=1}^{\infty} \frac{2n}{n+1}$

56. $\sum_{n=1}^{\infty} \frac{n}{2n^2+1}$

57. $\sum_{n=1}^{\infty} \frac{(-1)^n 3^{n-2}}{2^n}$

58. $\sum_{n=1}^{\infty} \frac{10}{3\sqrt{n^3}}$

59. $\sum_{n=1}^{\infty} \frac{10n+3}{n2^n}$

60. $\sum_{n=1}^{\infty} \frac{2^n}{4n^2-1}$

61. $\sum_{n=1}^{\infty} \frac{\cos n}{2^n}$

62. $\sum_{n=2}^{\infty} \frac{(-1)^n}{n \ln n}$

63. $\sum_{n=1}^{\infty} \frac{n7^n}{n!}$

64. $\sum_{n=1}^{\infty} \frac{\ln n}{n^2}$

65. $\sum_{n=1}^{\infty} \frac{(-1)^n 3^{n-1}}{n!}$

66. $\sum_{n=1}^{\infty} \frac{(-1)^n 3^n}{n2^n}$

67. $\sum_{n=1}^{\infty} \frac{(-3)^n}{3 \cdot 5 \cdot 7 \cdot \cdot \cdot (2n+1)}$

68. $\sum_{n=1}^{\infty} \frac{3 \cdot 5 \cdot 7 \cdot \cdot \cdot (2n+1)}{18^n(2n-1)n!}$

In Exercises 69–72, identify the two series that are the same.

69. (a) $\sum_{n=1}^{\infty} \frac{n5^n}{n!}$

(b) $\sum_{n=0}^{\infty} \frac{n5^n}{(n+1)!}$

(c) $\sum_{n=0}^{\infty} \frac{(n+1)5^{n+1}}{(n+1)!}$

70. (a) $\sum_{n=4}^{\infty} n\left(\frac{3}{4}\right)^n$

(b) $\sum_{n=0}^{\infty} (n+1)\left(\frac{3}{4}\right)^n$

(c) $\sum_{n=1}^{\infty} n\left(\frac{3}{4}\right)^{n-1}$

71. (a) $\sum_{n=0}^{\infty} \frac{(-1)^n}{(2n+1)!}$

(b) $\sum_{n=1}^{\infty} \frac{(-1)^{n-1}}{(2n-1)!}$

(c) $\sum_{n=1}^{\infty} \frac{(-1)^{n-1}}{(2n+1)!}$

72. (a) $\sum_{n=2}^{\infty} \frac{(-1)^n}{(n-1)2^{n-1}}$

(b) $\sum_{n=1}^{\infty} \frac{(-1)^{n+1}}{n2^n}$

(c) $\sum_{n=0}^{\infty} \frac{(-1)^{n+1}}{(n+1)2^n}$

In Exercises 73 and 74, write an equivalent series with the index of summation beginning at $n = 0$.

73. $\sum_{n=1}^{\infty} \frac{n}{4^n}$

74. $\sum_{n=2}^{\infty} \frac{2^n}{(n-2)!}$

In Exercises 75 and 76, (a) determine the number of terms required to approximate the sum of the series with an error less than 0.0001, and (b) use a graphing utility to approximate the sum of the series with an error less than 0.0001.

75. $\sum_{k=1}^{\infty} \frac{(-3)^k}{2^k k!}$

76. $\sum_{k=0}^{\infty} \frac{(-3)^k}{1 \cdot 3 \cdot 5 \cdot \cdot \cdot (2k+1)}$

In Exercises 77–82, the terms of a series $\sum_{n=1}^{\infty} a_n$ are defined recursively. Determine the convergence or divergence of the series. Explain your reasoning.

77. $a_1 = \frac{1}{2}, a_{n+1} = \frac{4n-1}{3n+2}a_n$

78. $a_1 = 2, a_{n+1} = \frac{2n+1}{5n-4}a_n$

79. $a_1 = 1, a_{n+1} = \frac{\sin n + 1}{\sqrt{n}}a_n$

80. $a_1 = \frac{1}{5}, a_{n+1} = \frac{\cos n + 1}{n}a_n$

81. $a_1 = \frac{1}{3}, a_{n+1} = \left(1 + \frac{1}{n}\right)a_n$

82. $a_1 = \frac{1}{4}, a_{n+1} = \sqrt[n]{a_n}$

In Exercises 83–86, use the Ratio Test or the Root Test to determine the convergence or divergence of the series.

83. $1 + \frac{1 \cdot 2}{1 \cdot 3} + \frac{1 \cdot 2 \cdot 3}{1 \cdot 3 \cdot 5} + \frac{1 \cdot 2 \cdot 3 \cdot 4}{1 \cdot 3 \cdot 5 \cdot 7} + \cdot \cdot \cdot$

84. $1 + \frac{2}{3} + \frac{3}{3^2} + \frac{4}{3^3} + \frac{5}{3^4} + \frac{6}{3^5} + \cdot \cdot \cdot$

85. $\frac{1}{(\ln 3)^3} + \frac{1}{(\ln 4)^4} + \frac{1}{(\ln 5)^5} + \frac{1}{(\ln 6)^6} + \cdot \cdot \cdot$

86. $1 + \frac{1 \cdot 3}{1 \cdot 2 \cdot 3} + \frac{1 \cdot 3 \cdot 5}{1 \cdot 2 \cdot 3 \cdot 4 \cdot 5} + \frac{1 \cdot 3 \cdot 5 \cdot 7}{1 \cdot 2 \cdot 3 \cdot 4 \cdot 5 \cdot 6 \cdot 7} + \cdot \cdot \cdot$

In Exercises 87–92, find the values of x for which the series converges.

87. $\sum_{n=0}^{\infty} 2\left(\frac{x}{3}\right)^n$

88. $\sum_{n=0}^{\infty} \left(\frac{x+1}{4}\right)^n$

89. $\sum_{n=1}^{\infty} \frac{(-1)^n(x+1)^n}{n}$

90. $\sum_{n=0}^{\infty} 2(x-1)^n$

91. $\sum_{n=0}^{\infty} n!\left(\frac{x}{2}\right)^n$

92. $\sum_{n=0}^{\infty} \frac{(x+1)^n}{n!}$

Writing About Concepts

93. State the Ratio Test.

94. State the Root Test.

95. You are told that the terms of a positive series appear to approach zero rapidly as n approaches infinity. In fact, $a_7 \le 0.0001$. Given no other information, does this imply that the series converges? Support your conclusion with examples.

96. The graph shows the first 10 terms of the sequence of partial sums of the convergent series

$$\sum_{n=1}^{\infty} \left(\frac{2n}{3n+2}\right)^n.$$

Find a series such that the terms of its sequence of partial sums are less than the corresponding terms of the sequence in the figure, but such that the series diverges. Explain your reasoning.

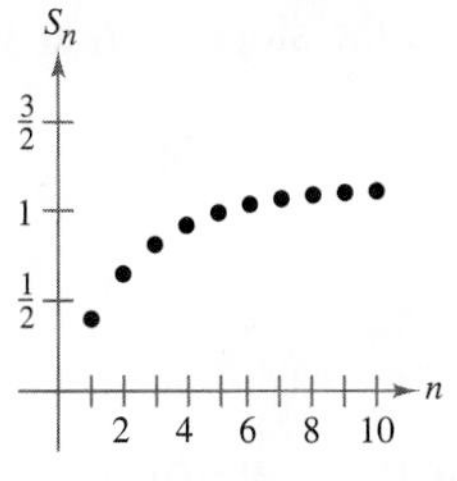

97. Using the Ratio Test, it is determined that an alternating series converges. Does the series converge conditionally or absolutely? Explain.

98. Prove Property 2 of Theorem 9.17.

99. Prove Theorem 9.18. (*Hint for Property 1:* If the limit equals $r < 1$, choose a real number R such that $r < R < 1$. By the definitions of the limit, there exists some $N > 0$ such that $\sqrt[n]{|a_n|} < R$ for $n > N$.)

100. Show that the Root Test is inconclusive for the p-series

$$\sum_{n=1}^{\infty} \frac{1}{n^p}.$$

101. Show that the Ratio Test and the Root Test are both inconclusive for the logarithmic p-series

$$\sum_{n=2}^{\infty} \frac{1}{n(\ln n)^p}.$$

102. Determine the convergence or divergence of the series

$$\sum_{n=1}^{\infty} \frac{(n!)^2}{(xn)!}$$

when (a) $x = 1$, (b) $x = 2$, (c) $x = 3$, and (d) x is a positive integer.

103. Show that if $\sum_{n=1}^{\infty} a_n$ is absolutely convergent, then

$$\left|\sum_{n=1}^{\infty} a_n\right| \le \sum_{n=1}^{\infty} |a_n|.$$

104. ***Writing*** Read the article "A Differentiation Test for Absolute Convergence" by Yaser S. Abu-Mostafa in *Mathematics Magazine.* Then write a paragraph that describes the test. Include examples of series that converge and examples of series that diverge.

Putnam Exam Challenge

105. Is the following series convergent or divergent?

$$1 + \frac{1}{2} \cdot \frac{19}{7} + \frac{2!}{3^2}\left(\frac{19}{7}\right)^2 + \frac{3!}{4^3}\left(\frac{19}{7}\right)^3 + \frac{4!}{5^4}\left(\frac{19}{7}\right)^4 + \cdots$$

106. Show that if the series

$$a_1 + a_2 + a_3 + \cdots + a_n + \cdots$$

converges, then the series

$$a_1 + \frac{a_2}{2} + \frac{a_3}{3} + \cdots + \frac{a_n}{n} + \cdots$$

converges also.

These problems were composed by the Committee on the Putnam Prize Competition.

Section 9.7 Taylor Polynomials and Approximations

- **Find polynomial approximations of elementary functions and compare them with the elementary functions.**
- **Find Taylor and Maclaurin polynomial approximations of elementary functions.**
- **Use the remainder of a Taylor polynomial.**

Polynomial Approximations of Elementary Functions

The goal of this section is to show how polynomial functions can be used as approximations for other elementary functions. To find a polynomial function P that approximates another function f, begin by choosing a number c in the domain of f at which f and P have the same value. That is,

$$P(c) = f(c). \qquad \text{Graphs of } f \text{ and } P \text{ pass through } (c, f(c)).$$

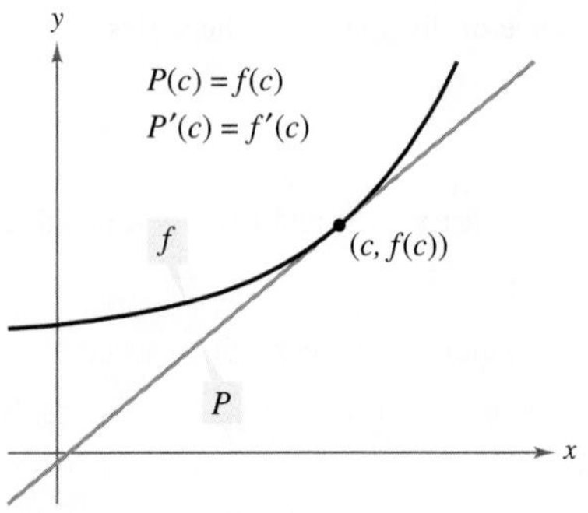

Near $(c, f(c))$, the graph of P can be used to approximate the graph of f.
Figure 9.10

The approximating polynomial is said to be **expanded about** c or **centered at** c. Geometrically, the requirement that $P(c) = f(c)$ means that the graph of P passes through the point $(c, f(c))$. Of course, there are many polynomials whose graphs pass through the point $(c, f(c))$. Your task is to find a polynomial whose graph resembles the graph of f near this point. One way to do this is to impose the additional requirement that the slope of the polynomial function be the same as the slope of the graph of f at the point $(c, f(c))$.

$$P'(c) = f'(c) \qquad \text{Graphs of } f \text{ and } P \text{ have the same slope at } (c, f(c)).$$

With these two requirements, you can obtain a simple linear approximation of f, as shown in Figure 9.10.

EXAMPLE 1 First-Degree Polynomial Approximation of $f(x) = e^x$

For the function $f(x) = e^x$, find a first-degree polynomial function

$$P_1(x) = a_0 + a_1 x$$

whose value and slope agree with the value and slope of f at $x = 0$.

Solution Because $f(x) = e^x$ and $f'(x) = e^x$, the value and the slope of f, at $x = 0$, are given by

$$f(0) = e^0 = 1$$

and

$$f'(0) = e^0 = 1.$$

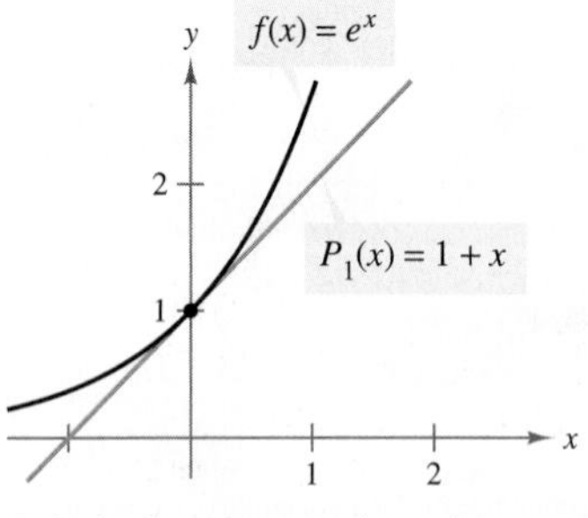

P_1 is the first-degree polynomial approximation of $f(x) = e^x$.
Figure 9.11

Because $P_1(x) = a_0 + a_1 x$, you can use the condition that $P_1(0) = f(0)$ to conclude that $a_0 = 1$. Moreover, because $P_1'(x) = a_1$, you can use the condition that $P_1'(0) = f'(0)$ to conclude that $a_1 = 1$. Therefore,

$$P_1(x) = 1 + x.$$

Figure 9.11 shows the graphs of $P_1(x) = 1 + x$ and $f(x) = e^x$.

NOTE Example 1 isn't the first time you have used a linear function to approximate another function. The same procedure was used as the basis for Newton's Method.

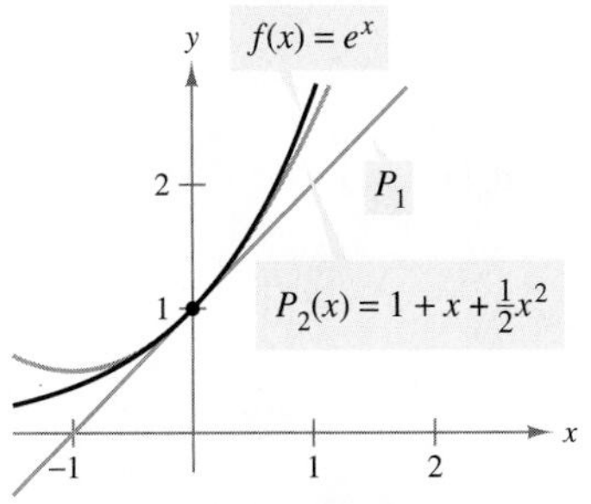

P_2 is the second-degree polynomial approximation of $f(x) = e^x$.
Figure 9.12

In Figure 9.12 you can see that, at points near (0, 1), the graph of

$$P_1(x) = 1 + x$$ 1st-degree approximation

is reasonably close to the graph of $f(x) = e^x$. However, as you move away from (0, 1), the graphs move farther from each other and the accuracy of the approximation decreases. To improve the approximation, you can impose yet another requirement—that the values of the second derivatives of P and f agree when $x = 0$. The polynomial, P_2, of least degree that satisfies all three requirements $P_2(0) = f(0)$, $P_2'(0) = f'(0)$, and $P_2''(0) = f''(0)$ can be shown to be

$$P_2(x) = 1 + x + \frac{1}{2}x^2.$$ 2nd-degree approximation

Moreover, in Figure 9.12, you can see that P_2 is a better approximation of f than P_1. If you continue this pattern, requiring that the values of $P_n(x)$ and its first n derivatives match those of $f(x) = e^x$ at $x = 0$, you obtain the following.

$$P_n(x) = 1 + x + \frac{1}{2}x^2 + \frac{1}{3!}x^3 + \cdot \cdot \cdot + \frac{1}{n!}x^n$$ nth-degree approximation

$$\approx e^x$$

EXAMPLE 2 Third-Degree Polynomial Approximation of $f(x) = e^x$

Construct a table comparing the values of the polynomial

$$P_3(x) = 1 + x + \frac{1}{2}x^2 + \frac{1}{3!}x^3$$ 3rd-degree approximation

with $f(x) = e^x$ for several values of x near 0.

Solution Using a calculator or a computer, you can obtain the results shown in the table. Note that for $x = 0$, the two functions have the same value, but that as x moves farther away from 0, the accuracy of the approximating polynomial $P_3(x)$ decreases.

x	−1.0	−0.2	−0.1	0	0.1	0.2	1.0
e^x	0.3679	0.81873	0.904837	1	1.105171	1.22140	2.7183
$P_3(x)$	0.3333	0.81867	0.904833	1	1.105167	1.22133	2.6667

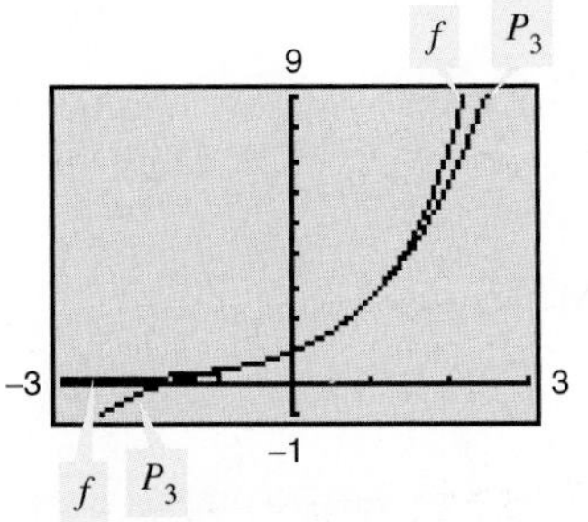

P_3 is the third-degree polynomial approximation of $f(x) = e^x$.
Figure 9.13

TECHNOLOGY A graphing utility can be used to compare the graph of the approximating polynomial with the graph of the function f. For instance, in Figure 9.13, the graph of

$$P_3(x) = 1 + x + \tfrac{1}{2}x^2 + \tfrac{1}{6}x^3$$ 3rd-degree approximation

is compared with the graph of $f(x) = e^x$. If you have access to a graphing utility, try comparing the graphs of

$$P_4(x) = 1 + x + \tfrac{1}{2}x^2 + \tfrac{1}{6}x^3 + \tfrac{1}{24}x^4$$ 4th-degree approximation

$$P_5(x) = 1 + x + \tfrac{1}{2}x^2 + \tfrac{1}{6}x^3 + \tfrac{1}{24}x^4 + \tfrac{1}{120}x^5$$ 5th-degree approximation

$$P_6(x) = 1 + x + \tfrac{1}{2}x^2 + \tfrac{1}{6}x^3 + \tfrac{1}{24}x^4 + \tfrac{1}{120}x^5 + \tfrac{1}{720}x^6$$ 6th-degree approximation

with the graph of f. What do you notice?

BROOK TAYLOR (1685–1731)

Although Taylor was not the first to seek polynomial approximations of transcendental functions, his account published in 1715 was one of the first comprehensive works on the subject.

Taylor and Maclaurin Polynomials

The polynomial approximation of $f(x) = e^x$ given in Example 2 is expanded about $c = 0$. For expansions about an arbitrary value of c, it is convenient to write the polynomial in the form

$$P_n(x) = a_0 + a_1(x - c) + a_2(x - c)^2 + a_3(x - c)^3 + \cdot \cdot \cdot + a_n(x - c)^n.$$

In this form, repeated differentiation produces

$$\begin{aligned} P_n'(x) &= a_1 + 2a_2(x - c) + 3a_3(x - c)^2 + \cdot \cdot \cdot + na_n(x - c)^{n-1} \\ P_n''(x) &= 2a_2 + 2(3a_3)(x - c) + \cdot \cdot \cdot + n(n-1)a_n(x - c)^{n-2} \\ P_n'''(x) &= 2(3a_3) + \cdot \cdot \cdot + n(n-1)(n-2)a_n(x - c)^{n-3} \\ &\vdots \\ P_n^{(n)}(x) &= n(n-1)(n-2) \cdot \cdot \cdot (2)(1)a_n. \end{aligned}$$

Letting $x = c$, you then obtain

$$P_n(c) = a_0, \quad P_n'(c) = a_1, \quad P_n''(c) = 2a_2, \quad . \ . \ . \ , \quad P_n^{(n)}(c) = n!a_n$$

and because the value of f and its first n derivatives must agree with the value of P_n and its first n derivatives at $x = c$, it follows that

$$f(c) = a_0, \quad f'(c) = a_1, \quad \frac{f''(c)}{2!} = a_2, \quad . \ . \ . \ , \quad \frac{f^{(n)}(c)}{n!} = a_n.$$

With these coefficients, you can obtain the following definition of **Taylor polynomials,** named after the English mathematician Brook Taylor, and **Maclaurin polynomials,** named after the English mathematician Colin Maclaurin (1698–1746).

NOTE Maclaurin polynomials are special types of Taylor polynomials for which $c = 0$.

Definitions of *n*th Taylor Polynomial and *n*th Maclaurin Polynomial

If f has n derivatives at c, then the polynomial

$$P_n(x) = f(c) + f'(c)(x - c) + \frac{f''(c)}{2!}(x - c)^2 + \cdot \cdot \cdot + \frac{f^{(n)}(c)}{n!}(x - c)^n$$

is called the ***n*th Taylor polynomial for *f* at *c*.** If $c = 0$, then

$$P_n(x) = f(0) + f'(0)x + \frac{f''(0)}{2!}x^2 + \frac{f'''(0)}{3!}x^3 + \cdot \cdot \cdot + \frac{f^{(n)}(0)}{n!}x^n$$

is also called the ***n*th Maclaurin polynomial for *f*.**

EXAMPLE 3 A Maclaurin Polynomial for $f(x) = e^x$

Find the nth Maclaurin polynomial for $f(x) = e^x$.

Solution From the discussion on page 649, the nth Maclaurin polynomial for

$$f(x) = e^x$$

is given by

$$P_n(x) = 1 + x + \frac{1}{2!}x^2 + \frac{1}{3!}x^3 + \cdot \cdot \cdot + \frac{1}{n!}x^n.$$

FOR FURTHER INFORMATION To see how to use series to obtain other approximations to e, see the article "Novel Series-based Approximations to ***e***" by John Knox and Harlan J. Brothers in *The College Mathematics Journal*. To view this article, go to the website *www.matharticles.com*.

EXAMPLE 4 Finding Taylor Polynomials for ln *x*

Find the Taylor polynomials P_0, P_1, P_2, P_3, and P_4 for $f(x) = \ln x$ centered at $c = 1$.

Solution Expanding about $c = 1$ yields the following.

$$f(x) = \ln x \qquad f(1) = \ln 1 = 0$$

$$f'(x) = \frac{1}{x} \qquad f'(1) = \frac{1}{1} = 1$$

$$f''(x) = -\frac{1}{x^2} \qquad f''(1) = -\frac{1}{1^2} = -1$$

$$f'''(x) = \frac{2!}{x^3} \qquad f'''(1) = \frac{2!}{1^3} = 2$$

$$f^{(4)}(x) = -\frac{3!}{x^4} \qquad f^{(4)}(1) = -\frac{3!}{1^4} = -6$$

Therefore, the Taylor polynomials are as follows.

$$P_0(x) = f(1) = 0$$

$$P_1(x) = f(1) + f'(1)(x-1) = (x-1)$$

$$P_2(x) = f(1) + f'(1)(x-1) + \frac{f''(1)}{2!}(x-1)^2$$

$$= (x-1) - \frac{1}{2}(x-1)^2$$

$$P_3(x) = f(1) + f'(1)(x-1) + \frac{f''(1)}{2!}(x-1)^2 + \frac{f'''(1)}{3!}(x-1)^3$$

$$= (x-1) - \frac{1}{2}(x-1)^2 + \frac{1}{3}(x-1)^3$$

$$P_4(x) = f(1) + f'(1)(x-1) + \frac{f''(1)}{2!}(x-1)^2 + \frac{f'''(1)}{3!}(x-1)^3 + \frac{f^{(4)}(1)}{4!}(x-1)^4$$

$$= (x-1) - \frac{1}{2}(x-1)^2 + \frac{1}{3}(x-1)^3 - \frac{1}{4}(x-1)^4$$

Figure 9.14 compares the graphs of P_1, P_2, P_3, and P_4 with the graph of $f(x) = \ln x$. Note that near $x = 1$ the graphs are nearly indistinguishable. For instance, $P_4(0.9) \approx -0.105358$ and $\ln(0.9) \approx -0.105361$.

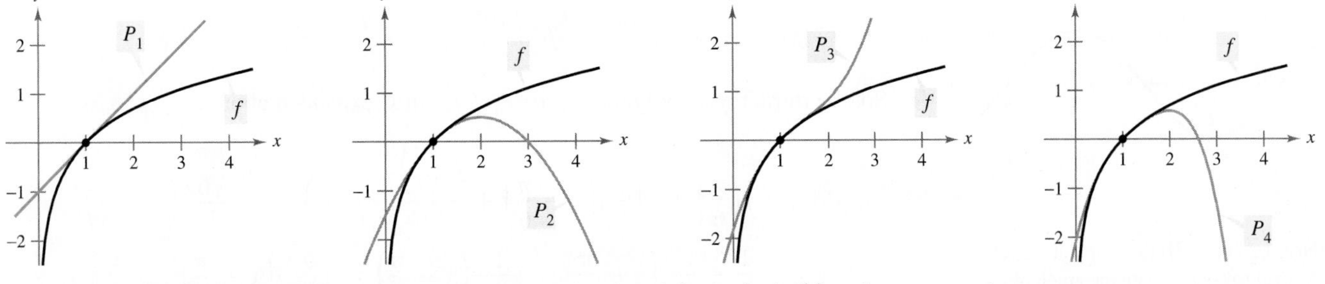

As n increases, the graph of P_n becomes a better and better approximation of the graph of $f(x) = \ln x$ near $x = 1$.
Figure 9.14

EXAMPLE 5 Finding Maclaurin Polynomials for cos x

Find the Maclaurin polynomials P_0, P_2, P_4, and P_6 for $f(x) = \cos x$. Use $P_6(x)$ to approximate the value of $\cos(0.1)$.

Solution Expanding about $c = 0$ yields the following.

$$\begin{aligned} f(x) &= \cos x & f(0) &= \cos 0 = 1 \\ f'(x) &= -\sin x & f'(0) &= -\sin 0 = 0 \\ f''(x) &= -\cos x & f''(0) &= -\cos 0 = -1 \\ f'''(x) &= \sin x & f'''(0) &= \sin 0 = 0 \end{aligned}$$

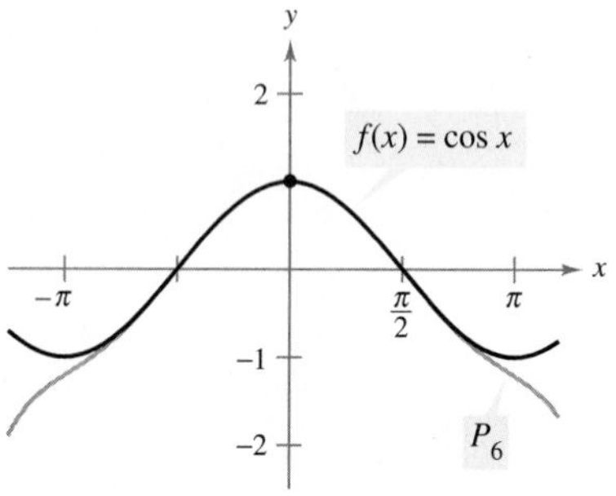

Near $(0, 1)$, the graph of P_6 can be used to approximate the graph of $f(x) = \cos x$.
Figure 9.15

Through repeated differentiation, you can see that the pattern $1, 0, -1, 0$ continues, and you obtain the following Maclaurin polynomials.

$$P_0(x) = 1, \quad P_2(x) = 1 - \frac{1}{2!}x^2,$$

$$P_4(x) = 1 - \frac{1}{2!}x^2 + \frac{1}{4!}x^4, \quad P_6(x) = 1 - \frac{1}{2!}x^2 + \frac{1}{4!}x^4 - \frac{1}{6!}x^6$$

Using $P_6(x)$, you obtain the approximation $\cos(0.1) \approx 0.995004165$, which coincides with the calculator value to nine decimal places. Figure 9.15 compares the graphs of $f(x) = \cos x$ and P_6.

Note in Example 5 that the Maclaurin polynomials for $\cos x$ have only even powers of x. Similarly, the Maclaurin polynomials for $\sin x$ have only odd powers of x (see Exercise 17). This is not generally true of the Taylor polynomials for $\sin x$ and $\cos x$ expanded about $c \neq 0$, as you can see in the next example.

EXAMPLE 6 Finding a Taylor Polynomial for sin x

Find the third Taylor polynomial for $f(x) = \sin x$, expanded about $c = \pi/6$.

Solution Expanding about $c = \pi/6$ yields the following.

$$\begin{aligned} f(x) &= \sin x & f\left(\frac{\pi}{6}\right) &= \sin\frac{\pi}{6} = \frac{1}{2} \\ f'(x) &= \cos x & f'\left(\frac{\pi}{6}\right) &= \cos\frac{\pi}{6} = \frac{\sqrt{3}}{2} \\ f''(x) &= -\sin x & f''\left(\frac{\pi}{6}\right) &= -\sin\frac{\pi}{6} = -\frac{1}{2} \\ f'''(x) &= -\cos x & f'''\left(\frac{\pi}{6}\right) &= -\cos\frac{\pi}{6} = -\frac{\sqrt{3}}{2} \end{aligned}$$

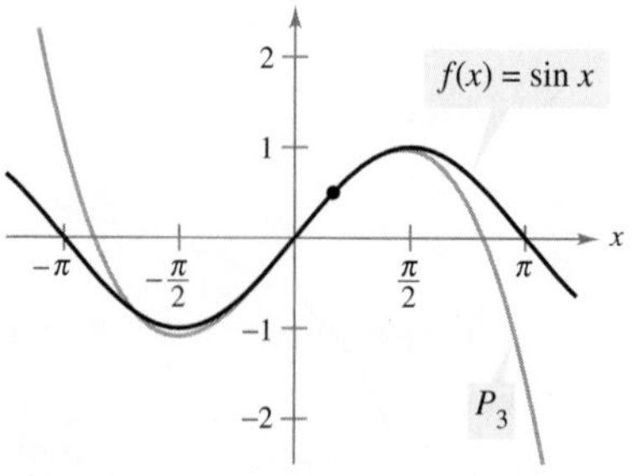

Near $(\pi/6, 1/2)$, the graph of P_3 can be used to approximate the graph of $f(x) = \sin x$.
Figure 9.16

So, the third Taylor polynomial for $f(x) = \sin x$, expanded about $c = \pi/6$, is

$$\begin{aligned} P_3(x) &= f\left(\frac{\pi}{6}\right) + f'\left(\frac{\pi}{6}\right)\left(x - \frac{\pi}{6}\right) + \frac{f''\left(\frac{\pi}{6}\right)}{2!}\left(x - \frac{\pi}{6}\right)^2 + \frac{f'''\left(\frac{\pi}{6}\right)}{3!}\left(x - \frac{\pi}{6}\right)^3 \\ &= \frac{1}{2} + \frac{\sqrt{3}}{2}\left(x - \frac{\pi}{6}\right) - \frac{1}{2(2!)}\left(x - \frac{\pi}{6}\right)^2 - \frac{\sqrt{3}}{2(3!)}\left(x - \frac{\pi}{6}\right)^3. \end{aligned}$$

Figure 9.16 compares the graphs of $f(x) = \sin x$ and P_3.

Taylor polynomials and Maclaurin polynomials can be used to approximate the value of a function at a specific point. For instance, to approximate the value of $\ln(1.1)$, you can use Taylor polynomials for $f(x) = \ln x$ expanded about $c = 1$, as shown in Example 4, or you can use Maclaurin polynomials, as shown in Example 7.

EXAMPLE 7 Approximation Using Maclaurin Polynomials

Use a fourth Maclaurin polynomial to approximate the value of $\ln(1.1)$.

Solution Because 1.1 is closer to 1 than to 0, you should consider Maclaurin polynomials for the function $g(x) = \ln(1 + x)$.

$$\begin{aligned} g(x) &= \ln(1+x) & g(0) &= \ln(1+0) = 0 \\ g'(x) &= (1+x)^{-1} & g'(0) &= (1+0)^{-1} = 1 \\ g''(x) &= -(1+x)^{-2} & g''(0) &= -(1+0)^{-2} = -1 \\ g'''(x) &= 2(1+x)^{-3} & g'''(0) &= 2(1+0)^{-3} = 2 \\ g^{(4)}(x) &= -6(1+x)^{-4} & g^{(4)}(0) &= -6(1+0)^{-4} = -6 \end{aligned}$$

Note that you obtain the same coefficients as in Example 4. Therefore, the fourth Maclaurin polynomial for $g(x) = \ln(1 + x)$ is

$$\begin{aligned} P_4(x) &= g(0) + g'(0)x + \frac{g''(0)}{2!}x^2 + \frac{g'''(0)}{3!}x^3 + \frac{g^{(4)}(0)}{4!}x^4 \\ &= x - \frac{1}{2}x^2 + \frac{1}{3}x^3 - \frac{1}{4}x^4. \end{aligned}$$

Consequently,

$$\ln(1.1) = \ln(1 + 0.1) \approx P_4(0.1) \approx 0.0953083.$$

Check to see that the fourth Taylor polynomial (from Example 4), evaluated at $x = 1.1$, yields the same result.

n	$P_n(0.1)$
1	0.1000000
2	0.0950000
3	0.0953333
4	0.0953083

The table at the left illustrates the accuracy of the Taylor polynomial approximation of the calculator value of $\ln(1.1)$. You can see that as n becomes larger, $P_n(0.1)$ approaches the calculator value of 0.0953102.

On the other hand, the table below illustrates that as you move away from the expansion point $c = 1$, the accuracy of the approximation decreases.

Fourth Taylor Polynomial Approximation of $\ln(1 + x)$

x	0	0.1	0.5	0.75	1.0
$\ln(1+x)$	0	0.0953102	0.4054651	0.5596158	0.6931472
$P_4(x)$	0	0.0953083	0.4010417	0.5302734	0.5833333

These two tables illustrate two very important points about the accuracy of Taylor (or Maclaurin) polynomials for use in approximations.

1. The approximation is usually better at x-values close to c than at x-values far from c.
2. The approximation is usually better for higher-degree Taylor (or Maclaurin) polynomials than for those of lower degree.

Remainder of a Taylor Polynomial

An approximation technique is of little value without some idea of its accuracy. To measure the accuracy of approximating a function value $f(x)$ by the Taylor polynomial $P_n(x)$, you can use the concept of a **remainder** $R_n(x)$, defined as follows.

$$f(x) = P_n(x) + R_n(x)$$

Exact value | Approximate Value | Remainder

So, $R_n(x) = f(x) - P_n(x)$. The absolute value of $R_n(x)$ is called the **error** associated with the approximation. That is,

$$\text{Error} = |R_n(x)| = |f(x) - P_n(x)|.$$

The next theorem gives a general procedure for estimating the remainder associated with a Taylor polynomial. This important theorem is called **Taylor's Theorem,** and the remainder given in the theorem is called the **Lagrange form of the remainder.** (The proof of the theorem is lengthy, and is given in Appendix A.)

THEOREM 9.19 Taylor's Theorem

If a function f is differentiable through order $n + 1$ in an interval I containing c, then, for each x in I, there exists z between x and c such that

$$f(x) = f(c) + f'(c)(x - c) + \frac{f''(c)}{2!}(x - c)^2 + \cdot \cdot \cdot + \frac{f^{(n)}(c)}{n!}(x - c)^n + R_n(x)$$

where

$$R_n(x) = \frac{f^{(n+1)}(z)}{(n + 1)!}(x - c)^{n+1}.$$

NOTE One useful consequence of Taylor's Theorem is that

$$|R_n(x)| \le \frac{|x - c|^{n+1}}{(n + 1)!} \max |f^{(n+1)}(z)|$$

where $\max|f^{(n+1)}(z)|$ is the maximum value of $f^{(n+1)}(z)$ between x and c.

For $n = 0$, Taylor's Theorem states that if f is differentiable in an interval I containing c, then, for each x in I, there exists z between x and c such that

$$f(x) = f(c) + f'(z)(x - c) \quad \text{or} \quad f'(z) = \frac{f(x) - f(c)}{x - c}.$$

Do you recognize this special case of Taylor's Theorem? (It is the Mean Value Theorem.)

When applying Taylor's Theorem, you should not expect to be able to find the exact value of z. (If you could do this, an approximation would not be necessary.) Rather, you try to find bounds for $f^{(n+1)}(z)$ from which you are able to tell how large the remainder $R_n(x)$ is.

EXAMPLE 8 Determining the Accuracy of an Approximation

The third Maclaurin polynomial for $\sin x$ is given by

$$P_3(x) = x - \frac{x^3}{3!}.$$

Use Taylor's Theorem to approximate $\sin(0.1)$ by $P_3(0.1)$ and determine the accuracy of the approximation.

Solution Using Taylor's Theorem, you have

$$\sin x = x - \frac{x^3}{3!} + R_3(x) = x - \frac{x^3}{3!} + \frac{f^{(4)}(z)}{4!}x^4$$

where $0 < z < 0.1$. Therefore,

$$\sin(0.1) \approx 0.1 - \frac{(0.1)^3}{3!} \approx 0.1 - 0.000167 = 0.099833.$$

Because $f^{(4)}(z) = \sin z$, it follows that the error $|R_3(0.1)|$ can be bounded as follows.

$$0 < R_3(0.1) = \frac{\sin z}{4!}(0.1)^4 < \frac{0.0001}{4!} \approx 0.000004$$

This implies that

$$0.099833 < \sin(0.1) = 0.099833 + R_3(x) < 0.099833 + 0.000004$$
$$0.099833 < \sin(0.1) < 0.099837.$$

NOTE Try using a calculator to verify the results obtained in Examples 8 and 9. For Example 8, you obtain

$$\sin(0.1) \approx 0.0998334.$$

For Example 9, you obtain

$$P_3(1.2) \approx 0.1827$$

and

$$\ln(1.2) \approx 0.1823.$$

EXAMPLE 9 Approximating a Value to a Desired Accuracy

Determine the degree of the Taylor polynomial $P_n(x)$ expanded about $c = 1$ that should be used to approximate $\ln(1.2)$ so that the error is less than 0.001.

Solution Following the pattern of Example 4, you can see that the $(n + 1)$st derivative of $f(x) = \ln x$ is given by

$$f^{(n+1)}(x) = (-1)^n \frac{n!}{x^{n+1}}.$$

Using Taylor's Theorem, you know that the error $|R_n(1.2)|$ is given by

$$|R_n(1.2)| = \left|\frac{f^{(n+1)}(z)}{(n+1)!}(1.2 - 1)^{n+1}\right| = \frac{n!}{z^{n+1}}\left[\frac{1}{(n+1)!}\right](0.2)^{n+1}$$
$$= \frac{(0.2)^{n+1}}{z^{n+1}(n+1)}$$

where $1 < z < 1.2$. In this interval, $(0.2)^{n+1}/[z^{n+1}(n + 1)]$ is less than $(0.2)^{n+1}/(n + 1)$. So, you are seeking a value of n such that

$$\frac{(0.2)^{n+1}}{(n+1)} < 0.001 \quad \Longrightarrow \quad 1000 < (n + 1)5^{n+1}.$$

By trial and error, you can determine that the smallest value of n that satisfies this inequality is $n = 3$. So, you would need the third Taylor polynomial to achieve the desired accuracy in approximating $\ln(1.2)$.

Exercises for Section 9.7

See www.CalcChat.com for worked-out solutions to odd-numbered exercises.

In Exercises 1–4, match the Taylor polynomial approximation of the function $f(x) = e^{-x^2/2}$ with the correct graph. [The graphs are labeled (a), (b), (c), and (d).]

(a)

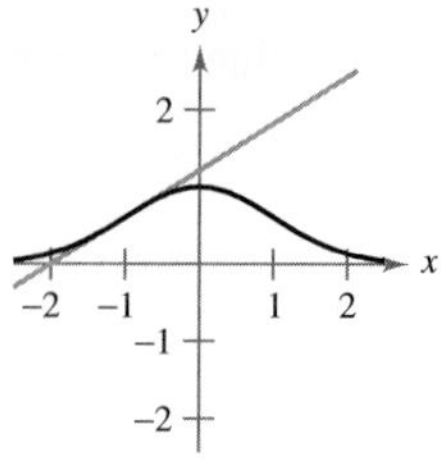

(b)

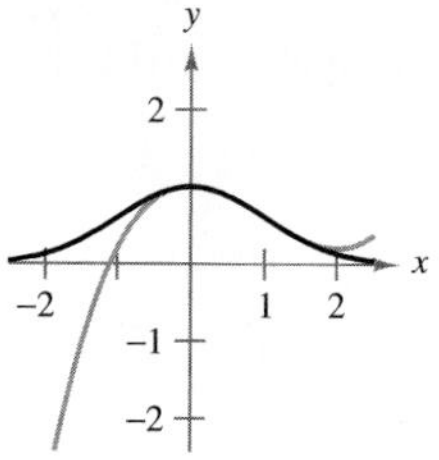

(c)

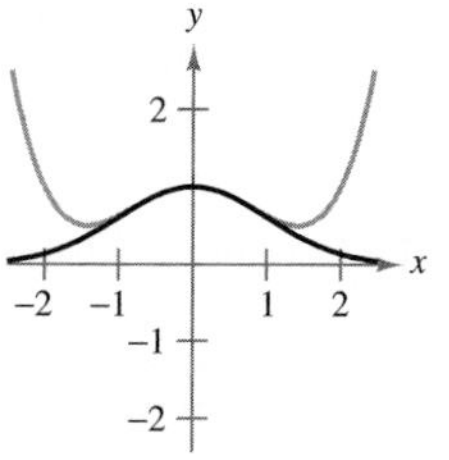

(d)

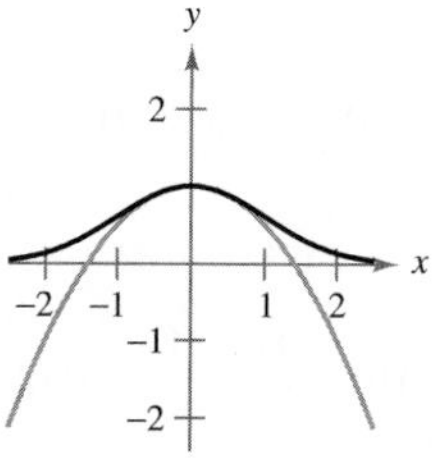

1. $g(x) = -\frac{1}{2}x^2 + 1$
2. $g(x) = \frac{1}{8}x^4 - \frac{1}{2}x^2 + 1$
3. $g(x) = e^{-1/2}[(x + 1) + 1]$
4. $g(x) = e^{-1/2}\left[\frac{1}{3}(x - 1)^3 - (x - 1) + 1\right]$

In Exercises 5–8, find a first-degree polynomial function P_1 whose value and slope agree with the value and slope of f at $x = c$. Use a graphing utility to graph f and P_1. What is P_1 called?

5. $f(x) = \dfrac{4}{\sqrt{x}}, \quad c = 1$
6. $f(x) = \dfrac{4}{\sqrt[3]{x}}, \quad c = 8$
7. $f(x) = \sec x, \quad c = \dfrac{\pi}{4}$
8. $f(x) = \tan x, \quad c = \dfrac{\pi}{4}$

Graphical and Numerical Analysis **In Exercises 9 and 10, use a graphing utility to graph f and its second-degree polynomial approximation P_2 at $x = c$. Complete the table comparing the values of f and P_2.**

9. $f(x) = \dfrac{4}{\sqrt{x}}, \quad c = 1$

$P_2(x) = 4 - 2(x - 1) + \frac{3}{2}(x - 1)^2$

x	0	0.8	0.9	1	1.1	1.2	2
$f(x)$							
$P_2(x)$							

10. $f(x) = \sec x, \quad c = \dfrac{\pi}{4}$

$P_2(x) = \sqrt{2} + \sqrt{2}\left(x - \dfrac{\pi}{4}\right) + \dfrac{3}{2}\sqrt{2}\left(x - \dfrac{\pi}{4}\right)^2$

x	−2.15	0.585	0.685	$\frac{\pi}{4}$	0.885	0.985	1.785
$f(x)$							
$P_2(x)$							

11. ***Conjecture*** Consider the function $f(x) = \cos x$ and its Maclaurin polynomials P_2, P_4, and P_6 (see Example 5).
   (a) Use a graphing utility to graph f and the indicated polynomial approximations.
   (b) Evaluate and compare the values of $f^{(n)}(0)$ and $P_n^{(n)}(0)$ for $n = 2, 4$, and 6.
   (c) Use the results in part (b) to make a conjecture about $f^{(n)}(0)$ and $P_n^{(n)}(0)$.

12. ***Conjecture*** Consider the function $f(x) = x^2e^x$.
   (a) Find the Maclaurin polynomials P_2, P_3, and P_4 for f.
   (b) Use a graphing utility to graph f, P_2, P_3, and P_4.
   (c) Evaluate and compare the values of $f^{(n)}(0)$ and $P_n^{(n)}(0)$ for $n = 2, 3$, and 4.
   (d) Use the results in part (c) to make a conjecture about $f^{(n)}(0)$ and $P_n^{(n)}(0)$.

In Exercises 13–24, find the Maclaurin polynomial of degree n for the function.

13. $f(x) = e^{-x}, \quad n = 3$
14. $f(x) = e^{-x}, \quad n = 5$
15. $f(x) = e^{2x}, \quad n = 4$
16. $f(x) = e^{3x}, \quad n = 4$
17. $f(x) = \sin x, \quad n = 5$
18. $f(x) = \sin \pi x, \quad n = 3$
19. $f(x) = xe^x, \quad n = 4$
20. $f(x) = x^2e^{-x}, \quad n = 4$
21. $f(x) = \dfrac{1}{x + 1}, \quad n = 4$
22. $f(x) = \dfrac{x}{x + 1}, \quad n = 4$
23. $f(x) = \sec x, \quad n = 2$
24. $f(x) = \tan x, \quad n = 3$

In Exercises 25–30, find the nth Taylor polynomial centered at c.

25. $f(x) = \dfrac{1}{x}, \quad n = 4, \quad c = 1$
26. $f(x) = \dfrac{2}{x^2}, \quad n = 4, \quad c = 2$
27. $f(x) = \sqrt{x}, \quad n = 4, \quad c = 1$
28. $f(x) = \sqrt[3]{x}, \quad n = 3, \quad c = 8$
29. $f(x) = \ln x, \quad n = 4, \quad c = 1$
30. $f(x) = x^2 \cos x, \quad n = 2, \quad c = \pi$

In Exercises 31 and 32, use a computer algebra system to find the indicated Taylor polynomials for the function f. Graph the function and the Taylor polynomials.

31. $f(x) = \tan x$

(a) $n = 3, \quad c = 0$

(b) $n = 3, \quad c = \pi/4$

32. $f(x) = \dfrac{1}{x^2 + 1}$

(a) $n = 4, \quad c = 0$

(b) $n = 4, \quad c = 1$

33. ***Numerical and Graphical Approximations***

(a) Use the Maclaurin polynomials $P_1(x)$, $P_3(x)$, and $P_5(x)$ for $f(x) = \sin x$ to complete the table.

x	0	0.25	0.50	0.75	1.00
$\sin x$	0	0.2474	0.4794	0.6816	0.8415
$P_1(x)$					
$P_3(x)$					
$P_5(x)$					

(b) Use a graphing utility to graph $f(x) = \sin x$ and the Maclaurin polynomials in part (a).

(c) Describe the change in accuracy of a polynomial approximation as the distance from the point where the polynomial is centered increases.

34. ***Numerical and Graphical Approximations***

(a) Use the Taylor polynomials $P_1(x)$, $P_2(x)$, and $P_4(x)$ for $f(x) = \ln x$ centered at $c = 1$ to complete the table.

x	1.00	1.25	1.50	1.75	2.00
$\ln x$	0	0.2231	0.4055	0.5596	0.6931
$P_1(x)$					
$P_2(x)$					
$P_4(x)$					

(b) Use a graphing utility to graph $f(x) = \ln x$ and the Taylor polynomials in part (a).

(c) Describe the change in accuracy of polynomial approximations as the degree increases.

Numerical and Graphical Approximations **In Exercises 35 and 36, (a) find the Maclaurin polynomial $P_3(x)$ for $f(x)$, (b) complete the table for $f(x)$ and $P_3(x)$, and (c) sketch the graphs of $f(x)$ and $P_3(x)$ on the same set of coordinate axes.**

x	-0.75	-0.50	-0.25	0	0.25	0.50	0.75
$f(x)$							
$P_3(x)$							

35. $f(x) = \arcsin x$

36. $f(x) = \arctan x$

In Exercises 37–40, the graph of $y = f(x)$ is shown with four of its Maclaurin polynomials. Identify the Maclaurin polynomials and use a graphing utility to confirm your results.

37.

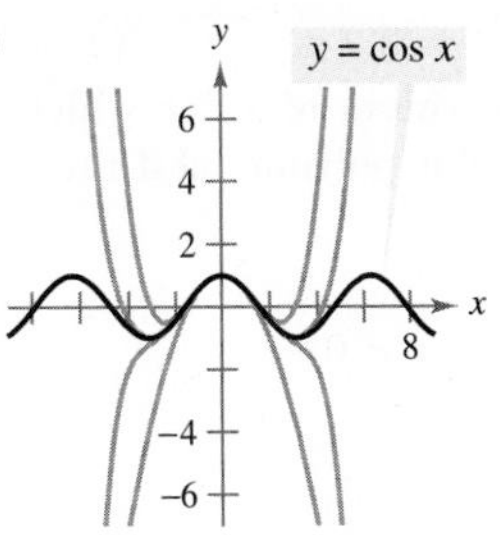

38.

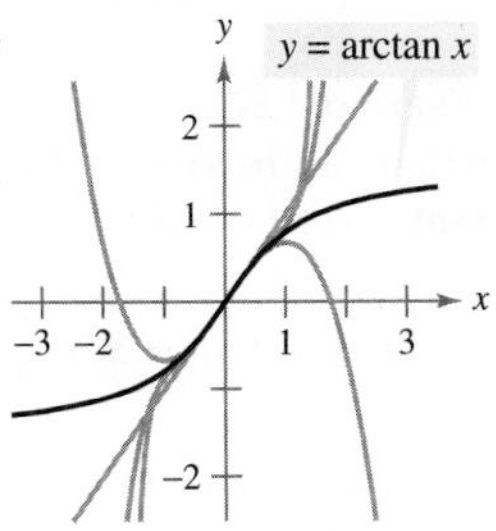

39.

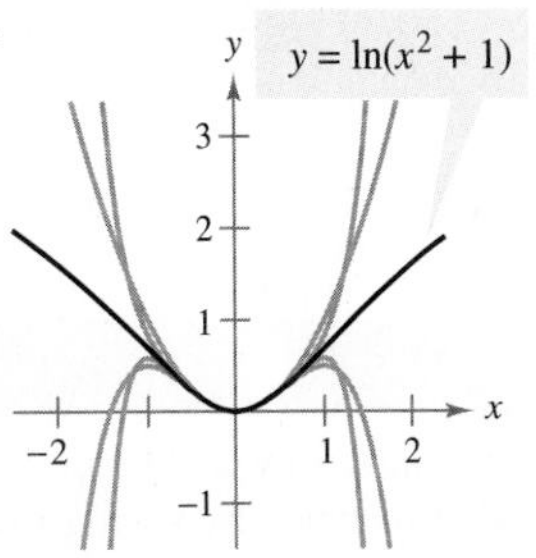

40.

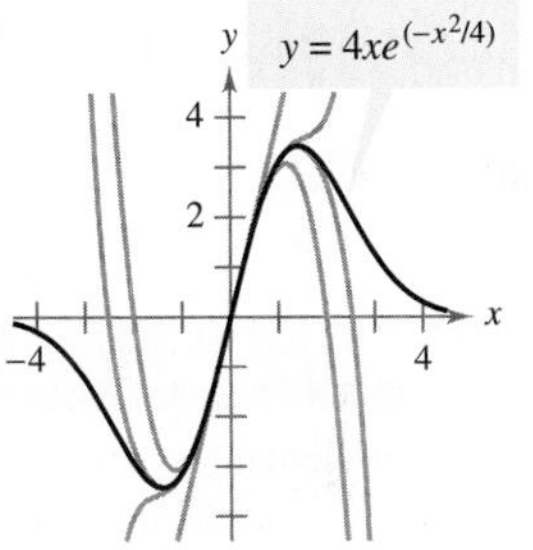

In Exercises 41–44, approximate the function at the given value of x, using the polynomial found in the indicated exercise.

41. $f(x) = e^{-x}, \quad f\left(\frac{1}{2}\right)$, Exercise 13

42. $f(x) = x^2e^{-x}, \quad f\left(\frac{1}{5}\right)$, Exercise 20

43. $f(x) = \ln x, \quad f(1.2)$, Exercise 29

44. $f(x) = x^2 \cos x, \quad f\left(\dfrac{7\pi}{8}\right)$, Exercise 30

In Exercises 45–48, use Taylor's Theorem to obtain an upper bound for the error of the approximation. Then calculate the exact value of the error.

45. $\cos(0.3) \approx 1 - \dfrac{(0.3)^2}{2!} + \dfrac{(0.3)^4}{4!}$

46. $e \approx 1 + 1 + \dfrac{1^2}{2!} + \dfrac{1^3}{3!} + \dfrac{1^4}{4!} + \dfrac{1^5}{5!}$

47. $\arcsin(0.4) \approx 0.4 + \dfrac{(0.4)^3}{2 \cdot 3}$

48. $\arctan(0.4) \approx 0.4 - \dfrac{(0.4)^3}{3}$

In Exercises 49–52, determine the degree of the Maclaurin polynomial required for the error in the approximation of the function at the indicated value of x to be less than 0.001.

49. $\sin(0.3)$

50. $\cos(0.1)$

51. $e^{0.6}$

52. $e^{0.3}$

In Exercises 53–56, determine the degree of the Maclaurin polynomial required for the error in the approximation of the function at the indicated value of x to be less than 0.0001. Use a computer algebra system to obtain and evaluate the required derivatives.

53. $f(x) = \ln(x + 1)$, approximate $f(0.5)$.

54. $f(x) = \cos(\pi x^2)$, approximate $f(0.6)$.

55. $f(x) = e^{-\pi x}$, approximate $f(1.3)$.

56. $f(x) = e^{-x}$, approximate $f(1)$.

In Exercises 57–60, determine the values of *x* for which the function can be replaced by the Taylor polynomial if the error cannot exceed 0.001.

57. $f(x) = e^x \approx 1 + x + \dfrac{x^2}{2!} + \dfrac{x^3}{3!}, \quad x < 0$

58. $f(x) = \sin x \approx x - \dfrac{x^3}{3!}$

59. $f(x) = \cos x \approx 1 - \dfrac{x^2}{2!} + \dfrac{x^4}{4!}$

60. $f(x) = e^{-2x} \approx 1 - 2x + 2x^2 - \dfrac{4}{3}x^3$

Writing About Concepts

61. An elementary function is approximated by a polynomial. In your own words, describe what is meant by saying that the polynomial is *expanded about c* or *centered at c*.

62. When an elementary function f is approximated by a second-degree polynomial P_2 centered at c, what is known about f and P_2 at c? Explain your reasoning.

63. State the definition of an nth-degree Taylor polynomial of f centered at c.

64. Describe the accuracy of the nth-degree Taylor polynomial of f centered at c as the distance between c and x increases.

65. In general, how does the accuracy of a Taylor polynomial change as the degree of the polynomial is increased? Explain your reasoning.

66. The graphs show first-, second-, and third-degree polynomial approximations P_1, P_2, and P_3 of a function f. Label the graphs of P_1, P_2, and P_3. To print an enlarged copy of the graph, go to the website *www.mathgraphs.com*.

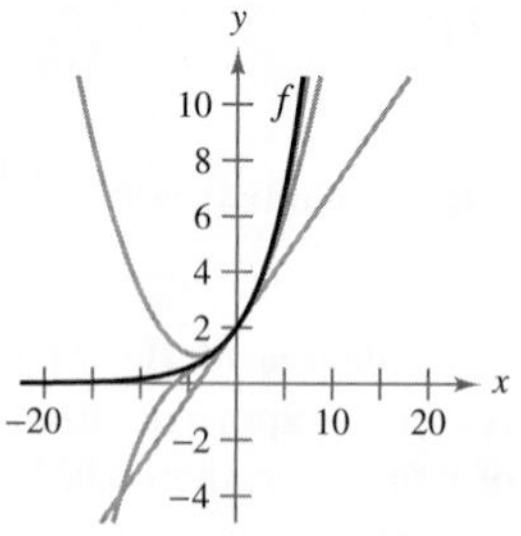

67. ***Comparing Maclaurin Polynomials***

(a) Compare the Maclaurin polynomials of degree 4 and degree 5, respectively, for the functions $f(x) = e^x$ and $g(x) = xe^x$. What is the relationship between them?

(b) Use the result in part (a) and the Maclaurin polynomial of degree 5 for $f(x) = \sin x$ to find a Maclaurin polynomial of degree 6 for the function $g(x) = x \sin x$.

(c) Use the result in part (a) and the Maclaurin polynomial of degree 5 for $f(x) = \sin x$ to find a Maclaurin polynomial of degree 4 for the function $g(x) = (\sin x)/x$.

68. ***Differentiating Maclaurin Polynomials***

(a) Differentiate the Maclaurin polynomial of degree 5 for $f(x) = \sin x$ and compare the result with the Maclaurin polynomial of degree 4 for $g(x) = \cos x$.

(b) Differentiate the Maclaurin polynomial of degree 6 for $f(x) = \cos x$ and compare the result with the Maclaurin polynomial of degree 5 for $g(x) = \sin x$.

(c) Differentiate the Maclaurin polynomial of degree 4 for $f(x) = e^x$. Describe the relationship between the two series.

69. ***Graphical Reasoning*** The figure shows the graph of the function

$$f(x) = \sin\left(\frac{\pi x}{4}\right)$$

and the second-degree Taylor polynomial

$$P_2(x) = 1 - \frac{\pi^2}{32}(x - 2)^2$$

centered at $x = 2$.

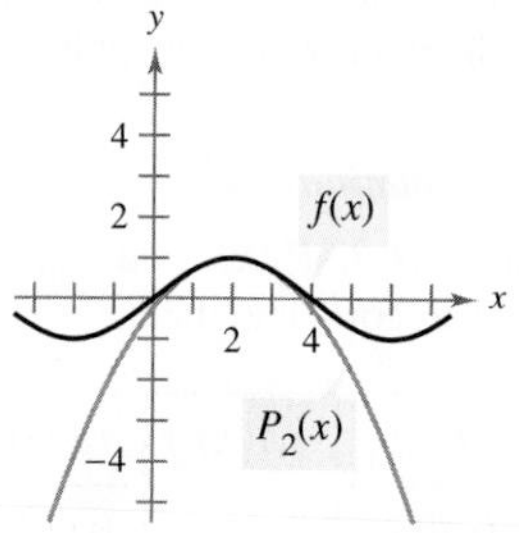

(a) Use the symmetry of the graph of f to write the second-degree Taylor polynomial for f centered at $x = -2$.

(b) Use a horizontal translation of the result in part (a) to find the second-degree Taylor polynomial for f centered at $x = 6$.

(c) Is it possible to use a horizontal translation of the result in part (a) to write a second-degree Taylor polynomial for f centered at $x = 4$? Explain.

70. Prove that if f is an odd function, then its nth Maclaurin polynomial contains only terms with odd powers of x.

71. Prove that if f is an even function, then its nth Maclaurin polynomial contains only terms with even powers of x.

72. Let $P_n(x)$ be the nth Taylor polynomial for f at c. Prove that $P_n(c) = f(c)$ and $P^{(k)}(c) = f^{(k)}(c)$ for $1 \le k \le n$. (See Exercises 9 and 10.)

73. ***Writing*** The proof in Exercise 72 guarantees that the Taylor polynomial and its derivatives agree with the function and its derivatives at $x = c$. Use the graphs and tables in Exercises 33–36 to discuss what happens to the accuracy of the Taylor polynomial as you move away from $x = c$.

Section 9.8 Power Series

- **Understand the definition of a power series.**
- **Find the radius and interval of convergence of a power series.**
- **Determine the endpoint convergence of a power series.**
- **Differentiate and integrate a power series.**

Power Series

In Section 9.7, you were introduced to the concept of approximating functions by Taylor polynomials. For instance, the function $f(x) = e^x$ can be *approximated* by its Maclaurin polynomials as follows.

$$e^x \approx 1 + x \qquad \text{1st-degree polynomial}$$

$$e^x \approx 1 + x + \frac{x^2}{2!} \qquad \text{2nd-degree polynomial}$$

$$e^x \approx 1 + x + \frac{x^2}{2!} + \frac{x^3}{3!} \qquad \text{3rd-degree polynomial}$$

$$e^x \approx 1 + x + \frac{x^2}{2!} + \frac{x^3}{3!} + \frac{x^4}{4!} \qquad \text{4th-degree polynomial}$$

$$e^x \approx 1 + x + \frac{x^2}{2!} + \frac{x^3}{3!} + \frac{x^4}{4!} + \frac{x^5}{5!} \qquad \text{5th-degree polynomial}$$

In that section, you saw that the higher the degree of the approximating polynomial, the better the approximation becomes.

In this and the next two sections, you will see that several important types of functions, including

$$f(x) = e^x$$

can be represented *exactly* by an infinite series called a **power series.** For example, the power series representation for e^x is

$$e^x = 1 + x + \frac{x^2}{2!} + \frac{x^3}{3!} + \cdot\cdot\cdot + \frac{x^n}{n!} + \cdot\cdot\cdot.$$

For each real number x, it can be shown that the infinite series on the right converges to the number e^x. Before doing this, however, some preliminary results dealing with power series will be discussed—beginning with the following definition.

EXPLORATION

Graphical Reasoning Use a graphing utility to approximate the graph of each power series near $x = 0$. (Use the first several terms of each series.) Each series represents a well-known function. What is the function?

a. $\displaystyle\sum_{n=0}^{\infty} \frac{(-1)^n x^n}{n!}$

b. $\displaystyle\sum_{n=0}^{\infty} \frac{(-1)^n x^{2n}}{(2n)!}$

c. $\displaystyle\sum_{n=0}^{\infty} \frac{(-1)^n x^{2n+1}}{(2n+1)!}$

d. $\displaystyle\sum_{n=0}^{\infty} \frac{(-1)^n x^{2n+1}}{2n+1}$

e. $\displaystyle\sum_{n=0}^{\infty} \frac{2^n x^n}{n!}$

Definition of Power Series

If x is a variable, then an infinite series of the form

$$\sum_{n=0}^{\infty} a_n x^n = a_0 + a_1 x + a_2 x^2 + a_3 x^3 + \cdot\cdot\cdot + a_n x^n + \cdot\cdot\cdot$$

is called a **power series.** More generally, an infinite series of the form

$$\sum_{n=0}^{\infty} a_n (x-c)^n = a_0 + a_1(x-c) + a_2(x-c)^2 + \cdot\cdot\cdot + a_n(x-c)^n + \cdot\cdot\cdot$$

is called a **power series centered at c,** where c is a constant.

NOTE To simplify the notation for power series, we agree that $(x - c)^0 = 1$, even if $x = c$.

EXAMPLE 1 Power Series

a. The following power series is centered at 0.

$$\sum_{n=0}^{\infty} \frac{x^n}{n!} = 1 + x + \frac{x^2}{2} + \frac{x^3}{3!} + \cdot \cdot \cdot$$

b. The following power series is centered at -1.

$$\sum_{n=0}^{\infty} (-1)^n (x+1)^n = 1 - (x+1) + (x+1)^2 - (x+1)^3 + \cdot \cdot \cdot$$

c. The following power series is centered at 1.

$$\sum_{n=1}^{\infty} \frac{1}{n}(x-1)^n = (x-1) + \frac{1}{2}(x-1)^2 + \frac{1}{3}(x-1)^3 + \cdot \cdot \cdot$$

Radius and Interval of Convergence

A power series in x can be viewed as a function of x

$$f(x) = \sum_{n=0}^{\infty} a_n(x-c)^n$$

where the *domain of f* is the set of all x for which the power series converges. Determination of the domain of a power series is the primary concern in this section. Of course, every power series converges at its center c because

$$\begin{aligned} f(c) &= \sum_{n=0}^{\infty} a_n(c-c)^n \\ &= a_0(1) + 0 + 0 + \cdot \cdot \cdot + 0 + \cdot \cdot \cdot \\ &= a_0. \end{aligned}$$

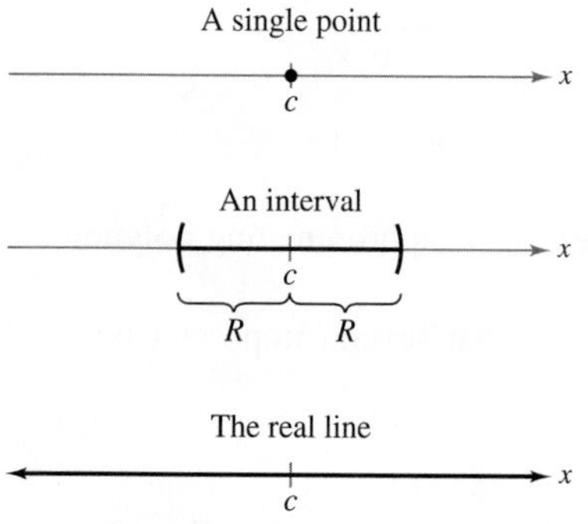

The domain of a power series has only three basic forms: a single point, an interval centered at c, or the entire real line.
Figure 9.17

So, c always lies in the domain of f. The following important theorem states that the domain of a power series can take three basic forms: a single point, an interval centered at c, or the entire real line, as shown in Figure 9.17. A proof is given in Appendix A.

THEOREM 9.20 Convergence of a Power Series

For a power series centered at c, precisely one of the following is true.

1. The series converges only at c.
2. There exists a real number $R > 0$ such that the series converges absolutely for $|x - c| < R$, and diverges for $|x - c| > R$.
3. The series converges absolutely for all x.

The number R is the **radius of convergence** of the power series. If the series converges only at c, the radius of convergence is $R = 0$, and if the series converges for all x, the radius of convergence is $R = \infty$. The set of all values of x for which the power series converges is the **interval of convergence** of the power series.

STUDY TIP To determine the radius of convergence of a power series, use the Ratio Test, as demonstrated in Examples 2, 3, and 4.

EXAMPLE 2 Finding the Radius of Convergence

Find the radius of convergence of $\sum_{n=0}^{\infty} n!x^n$.

Solution For $x = 0$, you obtain

$$f(0) = \sum_{n=0}^{\infty} n!0^n = 1 + 0 + 0 + \cdots = 1.$$

For any fixed value of x such that $|x| > 0$, let $u_n = n!x^n$. Then

$$\begin{aligned}\lim_{n\to\infty}\left|\frac{u_{n+1}}{u_n}\right| &= \lim_{n\to\infty}\left|\frac{(n+1)!x^{n+1}}{n!x^n}\right| \\ &= |x| \lim_{n\to\infty}(n+1) \\ &= \infty.\end{aligned}$$

Therefore, by the Ratio Test, the series diverges for $|x| > 0$ and converges only at its center, 0. So, the radius of convergence is $R = 0$.

EXAMPLE 3 Finding the Radius of Convergence

Find the radius of convergence of

$$\sum_{n=0}^{\infty} 3(x-2)^n.$$

Solution For $x \neq 2$, let $u_n = 3(x-2)^n$. Then

$$\begin{aligned}\lim_{n\to\infty}\left|\frac{u_{n+1}}{u_n}\right| &= \lim_{n\to\infty}\left|\frac{3(x-2)^{n+1}}{3(x-2)^n}\right| \\ &= \lim_{n\to\infty}|x-2| \\ &= |x-2|.\end{aligned}$$

By the Ratio Test, the series converges if $|x-2| < 1$ and diverges if $|x-2| > 1$. Therefore, the radius of convergence of the series is $R = 1$.

EXAMPLE 4 Finding the Radius of Convergence

Find the radius of convergence of

$$\sum_{n=0}^{\infty} \frac{(-1)^n x^{2n+1}}{(2n+1)!}.$$

Solution Let $u_n = (-1)^n x^{2n+1}/(2n+1)!$. Then

$$\begin{aligned}\lim_{n\to\infty}\left|\frac{u_{n+1}}{u_n}\right| &= \lim_{n\to\infty}\left|\frac{\dfrac{(-1)^{n+1}x^{2n+3}}{(2n+3)!}}{\dfrac{(-1)^n x^{2n+1}}{(2n+1)!}}\right| \\ &= \lim_{n\to\infty}\frac{x^2}{(2n+3)(2n+2)}.\end{aligned}$$

For any *fixed* value of x, this limit is 0. So, by the Ratio Test, the series converges for all x. Therefore, the radius of convergence is $R = \infty$.

Endpoint Convergence

Note that for a power series whose radius of convergence is a finite number R, Theorem 9.20 says nothing about the convergence at the *endpoints* of the interval of convergence. Each endpoint must be tested separately for convergence or divergence. As a result, the interval of convergence of a power series can take any one of the six forms shown in Figure 9.18.

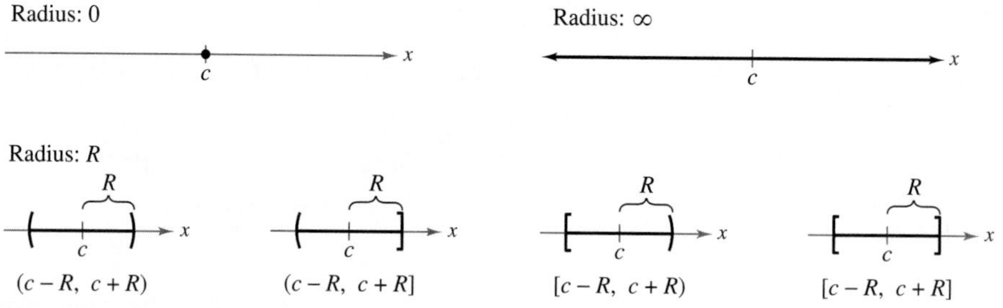

Intervals of convergence
Figure 9.18

EXAMPLE 5 Finding the Interval of Convergence

Find the interval of convergence of $\sum_{n=1}^{\infty} \frac{x^n}{n}$.

Solution Letting $u_n = x^n/n$ produces

$$\lim_{n\to\infty} \left|\frac{u_{n+1}}{u_n}\right| = \lim_{n\to\infty} \left|\frac{\frac{x^{n+1}}{(n+1)}}{\frac{x^n}{n}}\right|$$

$$= \lim_{n\to\infty} \left|\frac{nx}{n+1}\right|$$

$$= |x|.$$

So, by the Ratio Test, the radius of convergence is $R = 1$. Moreover, because the series is centered at 0, it converges in the interval $(-1, 1)$. This interval, however, is not necessarily the *interval of convergence*. To determine this, you must test for convergence at each endpoint. When $x = 1$, you obtain the *divergent* harmonic series

$$\sum_{n=1}^{\infty} \frac{1}{n} = \frac{1}{1} + \frac{1}{2} + \frac{1}{3} + \cdot \cdot \cdot . \qquad \text{Diverges when } x = 1$$

When $x = -1$, you obtain the *convergent* alternating harmonic series

$$\sum_{n=1}^{\infty} \frac{(-1)^n}{n} = -1 + \frac{1}{2} - \frac{1}{3} + \frac{1}{4} - \cdot \cdot \cdot . \qquad \text{Converges when } x = -1$$

So, the interval of convergence for the series is $[-1, 1)$, as shown in Figure 9.19.

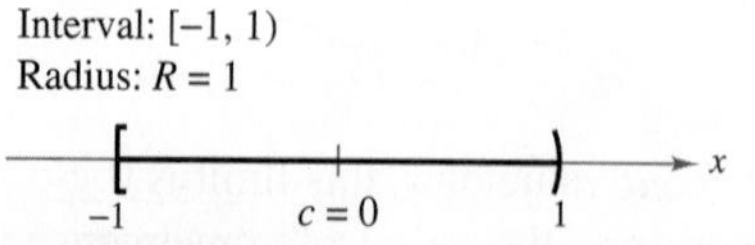

Figure 9.19

EXAMPLE 6 Finding the Interval of Convergence

Find the interval of convergence of $\displaystyle\sum_{n=0}^{\infty} \frac{(-1)^n(x+1)^n}{2^n}$.

Solution Letting $u_n = (-1)^n(x+1)^n/2^n$ produces

$$\lim_{n\to\infty}\left|\frac{u_{n+1}}{u_n}\right| = \lim_{n\to\infty}\left|\frac{\dfrac{(-1)^{n+1}(x+1)^{n+1}}{2^{n+1}}}{\dfrac{(-1)^n(x+1)^n}{2^n}}\right|$$

$$= \lim_{n\to\infty}\left|\frac{2^n(x+1)}{2^{n+1}}\right|$$

$$= \left|\frac{x+1}{2}\right|.$$

By the Ratio Test, the series converges if $|(x+1)/2| < 1$ or $|x+1| < 2$. So, the radius of convergence is $R = 2$. Because the series is centered at $x = -1$, it will converge in the interval $(-3, 1)$. Furthermore, at the endpoints you have

$$\sum_{n=0}^{\infty} \frac{(-1)^n(-2)^n}{2^n} = \sum_{n=0}^{\infty} \frac{2^n}{2^n} = \sum_{n=0}^{\infty} 1 \qquad \text{Diverges when } x = -3$$

and

$$\sum_{n=0}^{\infty} \frac{(-1)^n(2)^n}{2^n} = \sum_{n=0}^{\infty} (-1)^n \qquad \text{Diverges when } x = 1$$

both of which diverge. So, the interval of convergence is $(-3, 1)$, as shown in Figure 9.20.

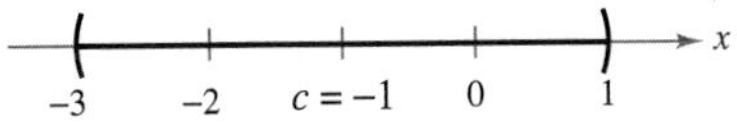

Figure 9.20

EXAMPLE 7 Finding the Interval of Convergence

Find the interval of convergence of

$$\sum_{n=1}^{\infty} \frac{x^n}{n^2}.$$

Solution Letting $u_n = x^n/n^2$ produces

$$\lim_{n\to\infty}\left|\frac{u_{n+1}}{u_n}\right| = \lim_{n\to\infty}\left|\frac{x^{n+1}/(n+1)^2}{x^n/n^2}\right|$$

$$= \lim_{n\to\infty}\left|\frac{n^2x}{(n+1)^2}\right| = |x|.$$

So, the radius of convergence is $R = 1$. Because the series is centered at $x = 0$, it converges in the interval $(-1, 1)$. When $x = 1$, you obtain the convergent p-series

$$\sum_{n=1}^{\infty} \frac{1}{n^2} = \frac{1}{1^2} + \frac{1}{2^2} + \frac{1}{3^2} + \frac{1}{4^2} + \cdots. \qquad \text{Converges when } x = 1$$

When $x = -1$, you obtain the convergent alternating series

$$\sum_{n=1}^{\infty} \frac{(-1)^n}{n^2} = -\frac{1}{1^2} + \frac{1}{2^2} - \frac{1}{3^2} + \frac{1}{4^2} - \cdots. \qquad \text{Converges when } x = -1$$

Therefore, the interval of convergence for the given series is $[-1, 1]$.

The Granger Collection

JAMES GREGORY (1638–1675)

One of the earliest mathematicians to work with power series was a Scotsman, James Gregory. He developed a power series method for interpolating table values—a method that was later used by Brook Taylor in the development of Taylor polynomials and Taylor series.

Differentiation and Integration of Power Series

Power series representation of functions has played an important role in the development of calculus. In fact, much of Newton's work with differentiation and integration was done in the context of power series—especially his work with complicated algebraic functions and transcendental functions. Euler, Lagrange, Leibniz, and the Bernoullis all used power series extensively in calculus.

Once you have defined a function with a power series, it is natural to wonder how you can determine the characteristics of the function. Is it continuous? Differentiable? Theorem 9.21, which is stated without proof, answers these questions.

THEOREM 9.21 Properties of Functions Defined by Power Series

If the function given by

$$\begin{aligned} f(x) &= \sum_{n=0}^{\infty} a_n(x-c)^n \\ &= a_0 + a_1(x-c) + a_2(x-c)^2 + a_3(x-c)^3 + \cdots \end{aligned}$$

has a radius of convergence of $R > 0$, then, on the interval $(c - R, c + R)$, f is differentiable (and therefore continuous). Moreover, the derivative and antiderivative of f are as follows.

1. $$\begin{aligned} f'(x) &= \sum_{n=1}^{\infty} na_n(x-c)^{n-1} \\ &= a_1 + 2a_2(x-c) + 3a_3(x-c)^2 + \cdots \end{aligned}$$

2. $$\begin{aligned} \int f(x)\,dx &= C + \sum_{n=0}^{\infty} a_n \frac{(x-c)^{n+1}}{n+1} \\ &= C + a_0(x-c) + a_1\frac{(x-c)^2}{2} + a_2\frac{(x-c)^3}{3} + \cdots \end{aligned}$$

The *radius of convergence* of the series obtained by differentiating or integrating a power series is the same as that of the original power series. The *interval of convergence*, however, may differ as a result of the behavior at the endpoints.

Theorem 9.21 states that, in many ways, a function defined by a power series behaves like a polynomial. It is continuous in its interval of convergence, and both its derivative and its antiderivative can be determined by differentiating and integrating each term of the given power series. For instance, the derivative of the power series

$$\begin{aligned} f(x) &= \sum_{n=0}^{\infty} \frac{x^n}{n!} \\ &= 1 + x + \frac{x^2}{2} + \frac{x^3}{3!} + \frac{x^4}{4!} + \cdots \end{aligned}$$

is

$$\begin{aligned} f'(x) &= 1 + (2)\frac{x}{2} + (3)\frac{x^2}{3!} + (4)\frac{x^3}{4!} + \cdots \\ &= 1 + x + \frac{x^2}{2} + \frac{x^3}{3!} + \frac{x^4}{4!} + \cdots \\ &= f(x). \end{aligned}$$

Notice that $f'(x) = f(x)$. Do you recognize this function?

EXAMPLE 8 Intervals of Convergence for $f(x)$, $f'(x)$, and $\int f(x)\,dx$

Consider the function given by

$$f(x) = \sum_{n=1}^{\infty} \frac{x^n}{n} = x + \frac{x^2}{2} + \frac{x^3}{3} + \cdots.$$

Find the intervals of convergence for each of the following.

a. $\int f(x)\,dx$ **b.** $f(x)$ **c.** $f'(x)$

Solution By Theorem 9.21, you have

$$\begin{aligned} f'(x) &= \sum_{n=1}^{\infty} x^{n-1} \\ &= 1 + x + x^2 + x^3 + \cdots \end{aligned}$$

and

$$\begin{aligned} \int f(x)\,dx &= C + \sum_{n=1}^{\infty} \frac{x^{n+1}}{n(n+1)} \\ &= C + \frac{x^2}{1 \cdot 2} + \frac{x^3}{2 \cdot 3} + \frac{x^4}{3 \cdot 4} + \cdots. \end{aligned}$$

By the Ratio Test, you can show that each series has a radius of convergence of $R = 1$. Considering the interval $(-1, 1)$, you have the following.

a. For $\int f(x)\,dx$, the series

$$\sum_{n=1}^{\infty} \frac{x^{n+1}}{n(n+1)} \qquad \text{Interval of convergence: } [-1, 1]$$

converges for $x = \pm 1$, and its interval of convergence is $[-1, 1]$. See Figure 9.21(a).

b. For $f(x)$, the series

$$\sum_{n=1}^{\infty} \frac{x^n}{n} \qquad \text{Interval of convergence: } [-1, 1)$$

converges for $x = -1$ and diverges for $x = 1$. So, its interval of convergence is $[-1, 1)$. See Figure 9.21(b).

c. For $f'(x)$, the series

$$\sum_{n=1}^{\infty} x^{n-1} \qquad \text{Interval of convergence: } (-1, 1)$$

diverges for $x = \pm 1$, and its interval of convergence is $(-1, 1)$. See Figure 9.21(c).

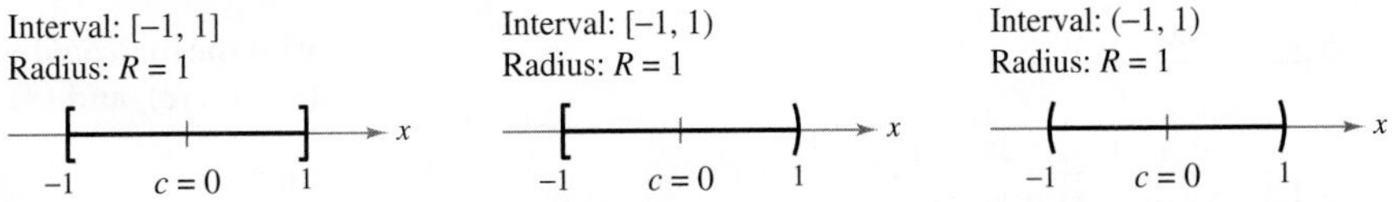

(a) (b) (c)

Figure 9.21

From Example 8, it appears that of the three series, the one for the derivative, $f'(x)$, is the least likely to converge at the endpoints. In fact, it can be shown that if the series for $f'(x)$ converges at the endpoints $x = c \pm R$, the series for $f(x)$ will also converge there.

Exercises for Section 9.8

See www.CalcChat.com for worked-out solutions to odd-numbered exercises.

In Exercises 1–4, state where the power series is centered.

1. $\sum_{n=0}^{\infty} nx^n$

2. $\sum_{n=1}^{\infty} \frac{(-1)^n 1 \cdot 3 \cdots (2n-1)}{2^n n!} x^n$

3. $\sum_{n=1}^{\infty} \frac{(x-2)^n}{n^3}$

4. $\sum_{n=0}^{\infty} \frac{(-1)^n (x-\pi)^{2n}}{(2n)!}$

In Exercises 5–10, find the radius of convergence of the power series.

5. $\sum_{n=0}^{\infty} (-1)^n \frac{x^n}{n+1}$

6. $\sum_{n=0}^{\infty} (2x)^n$

7. $\sum_{n=1}^{\infty} \frac{(2x)^n}{n^2}$

8. $\sum_{n=0}^{\infty} \frac{(-1)^n x^n}{2^n}$

9. $\sum_{n=0}^{\infty} \frac{(2x)^{2n}}{(2n)!}$

10. $\sum_{n=0}^{\infty} \frac{(2n)! x^{2n}}{n!}$

In Exercises 11–34, find the interval of convergence of the power series. (Be sure to include a check for convergence at the endpoints of the interval.)

11. $\sum_{n=0}^{\infty} \left(\frac{x}{2}\right)^n$

12. $\sum_{n=0}^{\infty} \left(\frac{x}{5}\right)^n$

13. $\sum_{n=1}^{\infty} \frac{(-1)^n x^n}{n}$

14. $\sum_{n=0}^{\infty} (-1)^{n+1}(n+1)x^n$

15. $\sum_{n=0}^{\infty} \frac{x^n}{n!}$

16. $\sum_{n=0}^{\infty} \frac{(3x)^n}{(2n)!}$

17. $\sum_{n=0}^{\infty} (2n)! \left(\frac{x}{2}\right)^n$

18. $\sum_{n=0}^{\infty} \frac{(-1)^n x^n}{(n+1)(n+2)}$

19. $\sum_{n=1}^{\infty} \frac{(-1)^{n+1} x^n}{4^n}$

20. $\sum_{n=0}^{\infty} \frac{(-1)^n n!(x-4)^n}{3^n}$

21. $\sum_{n=1}^{\infty} \frac{(-1)^{n+1}(x-5)^n}{n5^n}$

22. $\sum_{n=0}^{\infty} \frac{(x-2)^{n+1}}{(n+1)4^{n+1}}$

23. $\sum_{n=0}^{\infty} \frac{(-1)^{n+1}(x-1)^{n+1}}{n+1}$

24. $\sum_{n=1}^{\infty} \frac{(-1)^{n+1}(x-2)^n}{n2^n}$

25. $\sum_{n=1}^{\infty} \frac{(x-3)^{n-1}}{3^{n-1}}$

26. $\sum_{n=0}^{\infty} \frac{(-1)^n x^{2n+1}}{2n+1}$

27. $\sum_{n=1}^{\infty} \frac{n}{n+1}(-2x)^{n-1}$

28. $\sum_{n=0}^{\infty} \frac{(-1)^n x^{2n}}{n!}$

29. $\sum_{n=0}^{\infty} \frac{x^{2n+1}}{(2n+1)!}$

30. $\sum_{n=1}^{\infty} \frac{n! x^n}{(2n)!}$

31. $\sum_{n=1}^{\infty} \frac{2 \cdot 3 \cdot 4 \cdots (n+1)x^n}{n!}$

32. $\sum_{n=1}^{\infty} \left[\frac{2 \cdot 4 \cdot 6 \cdots 2n}{3 \cdot 5 \cdot 7 \cdots (2n+1)}\right] x^{2n+1}$

33. $\sum_{n=1}^{\infty} \frac{(-1)^{n+1} 3 \cdot 7 \cdot 11 \cdots (4n-1)(x-3)^n}{4^n}$

34. $\sum_{n=1}^{\infty} \frac{n!(x+1)^n}{1 \cdot 3 \cdot 5 \cdots (2n-1)}$

In Exercises 35 and 36, find the radius of convergence of the power series, where $c > 0$ and k is a positive integer.

35. $\sum_{n=1}^{\infty} \frac{(x-c)^{n-1}}{c^{n-1}}$

36. $\sum_{n=0}^{\infty} \frac{(n!)^k x^n}{(kn)!}$

In Exercises 37–40, find the interval of convergence of the power series. (Be sure to include a check for convergence at the endpoints of the interval.)

37. $\sum_{n=0}^{\infty} \left(\frac{x}{k}\right)^n, \quad k > 0$

38. $\sum_{n=1}^{\infty} \frac{(-1)^{n+1}(x-c)^n}{nc^n}$

39. $\sum_{n=1}^{\infty} \frac{k(k+1)(k+2)\cdots(k+n-1)x^n}{n!}, \quad k \geq 1$

40. $\sum_{n=1}^{\infty} \frac{n!(x-c)^n}{1 \cdot 3 \cdot 5 \cdots (2n-1)}$

In Exercises 41–44, write an equivalent series with the index of summation beginning at $n = 1$.

41. $\sum_{n=0}^{\infty} \frac{x^n}{n!}$

42. $\sum_{n=0}^{\infty} (-1)^{n+1}(n+1)x^n$

43. $\sum_{n=0}^{\infty} \frac{x^{2n+1}}{(2n+1)!}$

44. $\sum_{n=0}^{\infty} \frac{(-1)^n x^{2n+1}}{2n+1}$

In Exercises 45–48, find the intervals of convergence of (a) $f(x)$, (b) $f'(x)$, (c) $f''(x)$, and (d) $\int f(x)\,dx$. Include a check for convergence at the endpoints of the interval.

45. $f(x) = \sum_{n=0}^{\infty} \left(\frac{x}{2}\right)^n$

46. $f(x) = \sum_{n=1}^{\infty} \frac{(-1)^{n+1}(x-5)^n}{n5^n}$

47. $f(x) = \sum_{n=0}^{\infty} \frac{(-1)^{n+1}(x-1)^{n+1}}{n+1}$

48. $f(x) = \sum_{n=1}^{\infty} \frac{(-1)^{n+1}(x-2)^n}{n}$

Writing **In Exercises 49–52, match the graph of the first 10 terms of the sequence of partial sums of the series**

$$g(x) = \sum_{n=0}^{\infty} \left(\frac{x}{3}\right)^n$$

with the indicated value of the function. [The graphs are labeled (a), (b), (c), and (d).] Explain how you made your choice.

(a)

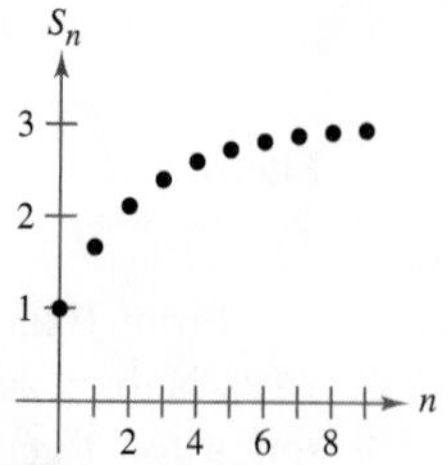

(b)

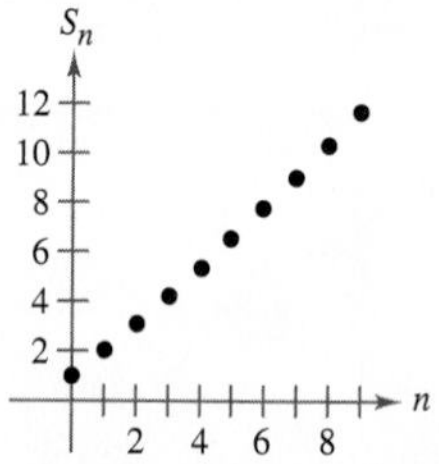

(c)
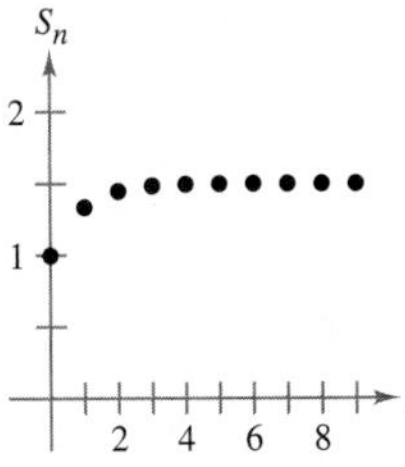

(d)
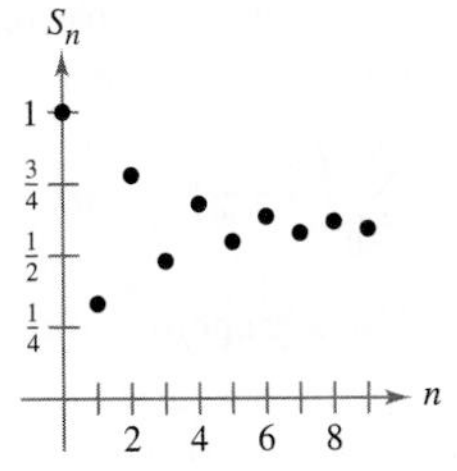

49. $g(1)$

50. $g(2)$

51. $g(3.1)$

52. $g(-2)$

Writing **In Exercises 53–56, match the graph of the first 10 terms of the sequence of partial sums of the series**

$$g(x) = \sum_{n=0}^{\infty} (2x)^n$$

with the indicated value of the function. [The graphs are labeled (a), (b), (c), and (d).] Explain how you made your choice.

(a)
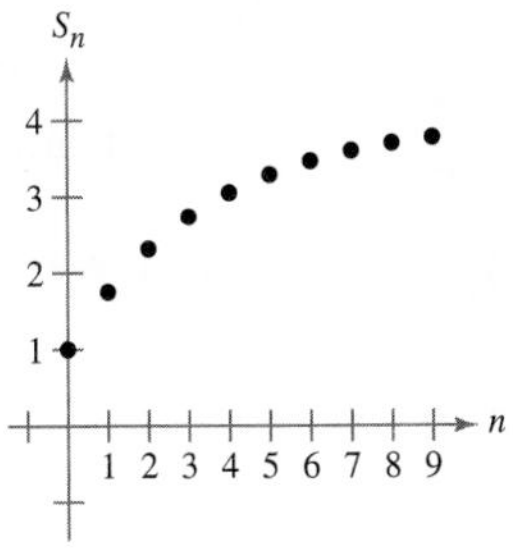

(b)
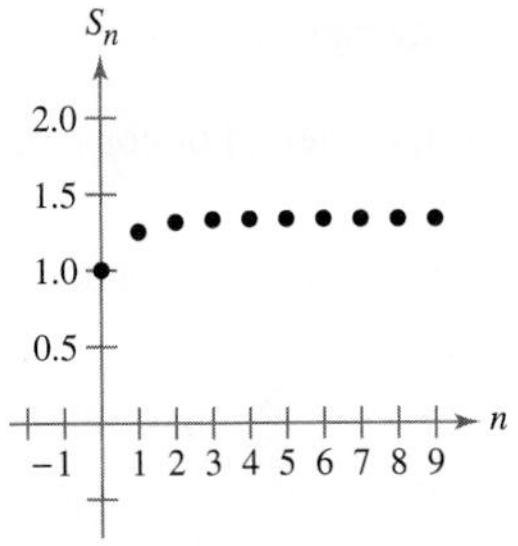

(c)
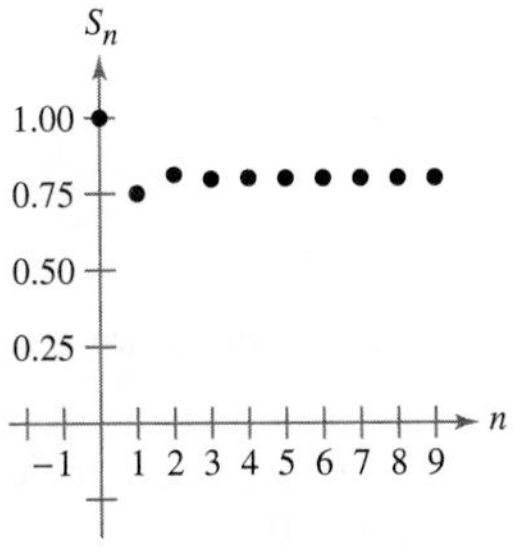

(d)
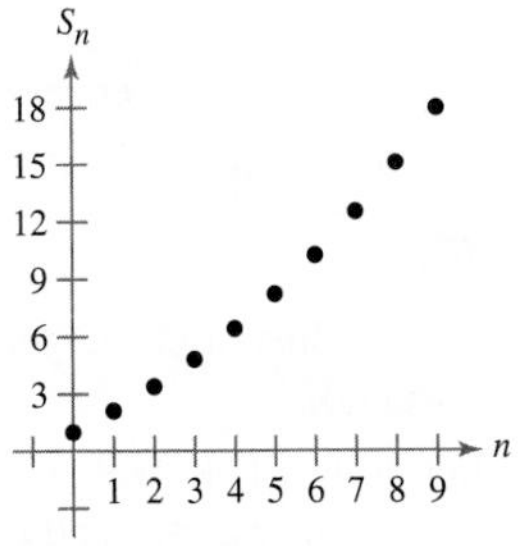

53. $g\left(\frac{1}{8}\right)$

54. $g\left(-\frac{1}{8}\right)$

55. $g\left(\frac{9}{16}\right)$

56. $g\left(\frac{3}{8}\right)$

Writing About Concepts

57. Define a power series centered at c.

58. Describe the radius of convergence of a power series. Describe the interval of convergence of a power series.

59. Describe the three basic forms of the domain of a power series.

Writing About Concepts *(continued)*

60. Describe how to differentiate and integrate a power series with a radius of convergence R. Will the series resulting from the operations of differentiation and integration have a different radius of convergence? Explain.

61. Give examples that show that the convergence of a power series at an endpoint of its interval of convergence may be either conditional or absolute. Explain your reasoning.

62. Write a power series that has the indicated interval of convergence. Explain your reasoning.

(a) $(-2, 2)$ (b) $(-1, 1]$ (c) $(-1, 0)$ (d) $[-2, 6)$

63. Let $f(x) = \displaystyle\sum_{n=0}^{\infty} \frac{(-1)^n x^{2n+1}}{(2n+1)!}$ and $g(x) = \displaystyle\sum_{n=0}^{\infty} \frac{(-1)^n x^{2n}}{(2n)!}$.

(a) Find the intervals of convergence of f and g.

(b) Show that $f'(x) = g(x)$.

(c) Show that $g'(x) = -f(x)$.

(d) Identify the functions f and g.

64. Let $f(x) = \displaystyle\sum_{n=0}^{\infty} \frac{x^n}{n!}$.

(a) Find the interval of convergence of f.

(b) Show that $f'(x) = f(x)$.

(c) Show that $f(0) = 1$.

(d) Identify the function f.

In Exercises 65–70, show that the function represented by the power series is a solution of the differential equation.

65. $y = \displaystyle\sum_{n=0}^{\infty} \frac{(-1)^n x^{2n+1}}{(2n+1)!}$, $y'' + y = 0$

66. $y = \displaystyle\sum_{n=0}^{\infty} \frac{(-1)^n x^{2n}}{(2n)!}$, $y'' + y = 0$

67. $y = \displaystyle\sum_{n=0}^{\infty} \frac{x^{2n+1}}{(2n+1)!}$, $y'' - y = 0$

68. $y = \displaystyle\sum_{n=0}^{\infty} \frac{x^{2n}}{(2n)!}$, $y'' - y = 0$

69. $y = \displaystyle\sum_{n=0}^{\infty} \frac{x^{2n}}{2^n n!}$, $y'' - xy' - y = 0$

70. $y = 1 + \displaystyle\sum_{n=1}^{\infty} \frac{(-1)^n x^{4n}}{2^{2n} n! \cdot 3 \cdot 7 \cdot 11 \cdots (4n-1)}$, $y'' + x^2 y = 0$

71. ***Bessel Function*** The Bessel function of order 0 is

$$J_0(x) = \sum_{k=0}^{\infty} \frac{(-1)^k x^{2k}}{2^{2k}(k!)^2}.$$

(a) Show that the series converges for all x.

(b) Show that the series is a solution of the differential equation $x^2 J_0'' + x J_0' + x^2 J_0 = 0$.

(c) Use a graphing utility to graph the polynomial composed of the first four terms of J_0.

(d) Approximate $\int_0^1 J_0\, dx$ accurate to two decimal places.

72. *Bessel Function* The Bessel function of order 1 is

$$J_1(x) = x\sum_{k=0}^{\infty} \frac{(-1)^k x^{2k}}{2^{2k+1} k!(k+1)!}.$$

(a) Show that the series converges for all x.

(b) Show that the series is a solution of the differential equation $x^2 J_1'' + x J_1' + (x^2 - 1) J_1 = 0$.

(c) Use a graphing utility to graph the polynomial composed of the first four terms of J_1.

(d) Show that $J_0'(x) = -J_1(x)$.

In Exercises 73–76, the series represents a well-known function. Use a computer algebra system to graph the partial sum S_{10} and identify the function from the graph.

73. $f(x) = \sum_{n=0}^{\infty} (-1)^n \frac{x^{2n}}{(2n)!}$ **74.** $f(x) = \sum_{n=0}^{\infty} (-1)^n \frac{x^{2n+1}}{(2n+1)!}$

75. $f(x) = \sum_{n=0}^{\infty} (-1)^n x^n, \quad -1 < x < 1$

76. $f(x) = \sum_{n=0}^{\infty} (-1)^n \frac{x^{2n+1}}{2n+1}, \quad -1 \le x \le 1$

77. *Investigation* In Exercise 11 you found that the interval of convergence of the geometric series $\sum_{n=0}^{\infty} \left(\frac{x}{2}\right)^n$ is $(-2, 2)$.

(a) Find the sum of the series when $x = \frac{3}{4}$. Use a graphing utility to graph the first six terms of the sequence of partial sums and the horizontal line representing the sum of the series.

(b) Repeat part (a) for $x = -\frac{3}{4}$.

(c) Write a short paragraph comparing the rate of convergence of the partial sums with the sum of the series in parts (a) and (b). How do the plots of the partial sums differ as they converge toward the sum of the series?

(d) Given any positive real number M, there exists a positive integer N such that the partial sum

$$\sum_{n=0}^{N} \left(\frac{3}{2}\right)^n > M.$$

Use a graphing utility to complete the table.

M	10	100	1000	10,000
N				

78. *Investigation* The interval of convergence of the series $\sum_{n=0}^{\infty} (3x)^n$ is $\left(-\frac{1}{3}, \frac{1}{3}\right)$.

(a) Find the sum of the series when $x = \frac{1}{6}$. Use a graphing utility to graph the first six terms of the sequence of partial sums and the horizontal line representing the sum of the series.

(b) Repeat part (a) for $x = -\frac{1}{6}$.

(c) Write a short paragraph comparing the rate of convergence of the partial sums with the sum of the series in parts (a) and (b). How do the plots of the partial sums differ as they converge toward the sum of the series?

(d) Given any positive real number M, there exists a positive integer N such that the partial sum

$$\sum_{n=0}^{N} \left(3 \cdot \frac{2}{3}\right)^n > M.$$

Use a graphing utility to complete the table.

M	10	100	1000	10,000
N				

True or False? **In Exercises 79–82, determine whether the statement is true or false. If it is false, explain why or give an example that shows it is false.**

79. If the power series $\sum_{n=0}^{\infty} a_n x^n$ converges for $x = 2$, then it also converges for $x = -2$.

80. If the power series $\sum_{n=0}^{\infty} a_n x^n$ converges for $x = 2$, then it also converges for $x = -1$.

81. If the interval of convergence for $\sum_{n=0}^{\infty} a_n x^n$ is $(-1, 1)$, then the interval of convergence for $\sum_{n=0}^{\infty} a_n (x-1)^n$ is $(0, 2)$.

82. If $f(x) = \sum_{n=0}^{\infty} a_n x^n$ converges for $|x| < 2$, then

$$\int_0^1 f(x)\, dx = \sum_{n=0}^{\infty} \frac{a_n}{n+1}.$$

83. Prove that the power series

$$\sum_{n=0}^{\infty} \frac{(n+p)!}{n!(n+q)!} x^n$$

has a radius of convergence of $R = \infty$ if p and q are positive integers.

84. Let $g(x) = 1 + 2x + x^2 + 2x^3 + x^4 + \cdots$, where the coefficients are $c_{2n} = 1$ and $c_{2n+1} = 2$ for $n \ge 0$.

(a) Find the interval of convergence of the series.

(b) Find an explicit formula for $g(x)$.

85. Let $f(x) = \sum_{n=0}^{\infty} c_n x^n$, where $c_{n+3} = c_n$ for $n \ge 0$.

(a) Find the interval of convergence of the series.

(b) Find an explicit formula for $f(x)$.

86. Prove that if the power series $\sum_{n=0}^{\infty} c_n x^n$ has a radius of convergence of R, then $\sum_{n=0}^{\infty} c_n x^{2n}$ has a radius of convergence of $\sqrt{R}$.

87. For $n > 0$, let $R > 0$ and $c_n > 0$. Prove that if the interval of convergence of the series $\sum_{n=0}^{\infty} c_n(x - x_0)^n$ is $(x_0 - R, x_0 + R]$, then the series converges conditionally at $x_0 + R$.

Section 9.9

Representation of Functions by Power Series

- **Find a geometric power series that represents a function.**
- **Construct a power series using series operations.**

The Granger Collection

JOSEPH FOURIER (1768–1830)

Some of the early work in representing functions by power series was done by the French mathematician Joseph Fourier. Fourier's work is important in the history of calculus, partly because it forced eighteenth century mathematicians to question the then-prevailing narrow concept of a function. Both Cauchy and Dirichlet were motivated by Fourier's work with series, and in 1837 Dirichlet published the general definition of a function that is used today.

Geometric Power Series

In this section and the next, you will study several techniques for finding a power series that represents a given function.

Consider the function given by $f(x) = 1/(1-x)$. The form of f closely resembles the sum of a geometric series

$$\sum_{n=0}^{\infty} ar^n = \frac{a}{1-r}, \quad |r| < 1.$$

In other words, if you let $a = 1$ and $r = x$, a power series representation for $1/(1-x)$, centered at 0, is

$$\frac{1}{1-x} = \sum_{n=0}^{\infty} x^n$$
$$= 1 + x + x^2 + x^3 + \cdots, \quad |x| < 1.$$

Of course, this series represents $f(x) = 1/(1-x)$ only on the interval $(-1, 1)$, whereas f is defined for all $x \neq 1$, as shown in Figure 9.22. To represent f in another interval, you must develop a different series. For instance, to obtain the power series centered at -1, you could write

$$\frac{1}{1-x} = \frac{1}{2-(x+1)} = \frac{1/2}{1-[(x+1)/2]} = \frac{a}{1-r}$$

which implies that $a = \frac{1}{2}$ and $r = (x+1)/2$. So, for $|x+1| < 2$, you have

$$\frac{1}{1-x} = \sum_{n=0}^{\infty} \left(\frac{1}{2}\right)\left(\frac{x+1}{2}\right)^n$$
$$= \frac{1}{2}\left[1 + \frac{(x+1)}{2} + \frac{(x+1)^2}{4} + \frac{(x+1)^3}{8} + \cdots\right], \quad |x+1| < 2$$

which converges on the interval $(-3, 1)$.

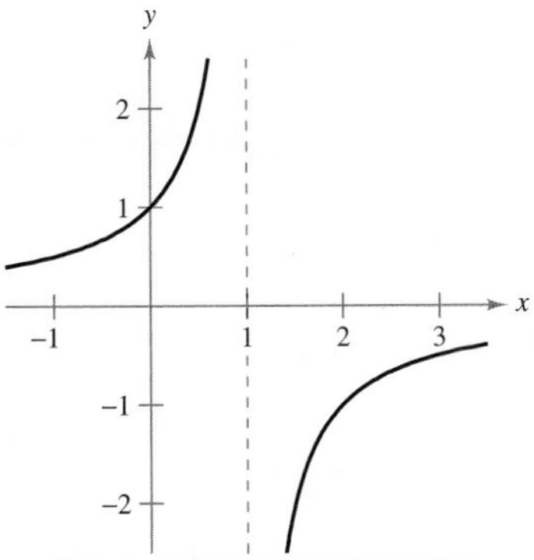

$f(x) = \dfrac{1}{1-x}$, Domain: all $x \neq 1$

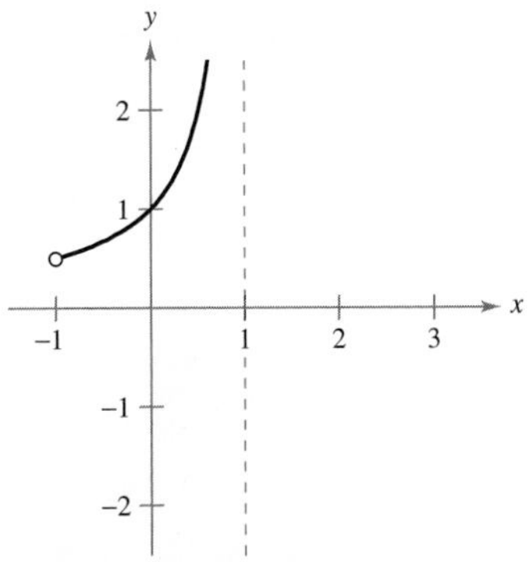

$f(x) = \sum_{n=0}^{\infty} x^n$, Domain: $-1 < x < 1$

Figure 9.22

EXAMPLE 1 Finding a Geometric Power Series Centered at 0

Find a power series for $f(x) = \dfrac{4}{x+2}$, centered at 0.

Solution Writing $f(x)$ in the form $a/(1-r)$ produces

$$\frac{4}{2+x} = \frac{2}{1-(-x/2)} = \frac{a}{1-r}$$

which implies that $a = 2$ and $r = -x/2$. So, the power series for $f(x)$ is

$$\begin{aligned}\frac{4}{x+2} &= \sum_{n=0}^{\infty} ar^n \\ &= \sum_{n=0}^{\infty} 2\left(-\frac{x}{2}\right)^n \\ &= 2\left(1 - \frac{x}{2} + \frac{x^2}{4} - \frac{x^3}{8} + \cdots\right).\end{aligned}$$

This power series converges when

$$\left|-\frac{x}{2}\right| < 1$$

which implies that the interval of convergence is $(-2, 2)$.

Long Division

$$\begin{array}{r} 2 - x + \frac{1}{2}x^2 - \frac{1}{4}x^3 + \cdots \\ 2 + x \,\big)\, 4 \qquad\qquad\qquad\qquad \\ \underline{4 + 2x} \qquad\qquad\qquad \\ -2x \qquad\qquad\qquad \\ \underline{-2x - x^2} \qquad\qquad \\ x^2 \qquad\qquad \\ \underline{x^2 + \frac{1}{2}x^3} \qquad \\ -\frac{1}{2}x^3 \qquad \\ \underline{-\frac{1}{2}x^3 - \frac{1}{4}x^4} \end{array}$$

Another way to determine a power series for a rational function such as the one in Example 1 is to use long division. For instance, by dividing $2 + x$ into 4, you obtain the result shown at the left.

EXAMPLE 2 Finding a Geometric Power Series Centered at 1

Find a power series for $f(x) = \dfrac{1}{x}$, centered at 1.

Solution Writing $f(x)$ in the form $a/(1-r)$ produces

$$\frac{1}{x} = \frac{1}{1-(-x+1)} = \frac{a}{1-r}$$

which implies that $a = 1$ and $r = 1 - x = -(x-1)$. So, the power series for $f(x)$ is

$$\begin{aligned}\frac{1}{x} &= \sum_{n=0}^{\infty} ar^n \\ &= \sum_{n=0}^{\infty} [-(x-1)]^n \\ &= \sum_{n=0}^{\infty} (-1)^n(x-1)^n \\ &= 1 - (x-1) + (x-1)^2 - (x-1)^3 + \cdots.\end{aligned}$$

This power series converges when

$$|x - 1| < 1$$

which implies that the interval of convergence is $(0, 2)$.

Operations with Power Series

The versatility of geometric power series will be shown later in this section, following a discussion of power series operations. These operations, used with differentiation and integration, provide a means of developing power series for a variety of elementary functions. (For simplicity, the following properties are stated for a series centered at 0.)

Operations with Power Series

Let $f(x) = \Sigma\, a_n x^n$ and $g(x) = \Sigma\, b_n x^n$.

1. $f(kx) = \displaystyle\sum_{n=0}^{\infty} a_n k^n x^n$
2. $f(x^N) = \displaystyle\sum_{n=0}^{\infty} a_n x^{nN}$
3. $f(x) \pm g(x) = \displaystyle\sum_{n=0}^{\infty} (a_n \pm b_n)x^n$

The operations described above can change the interval of convergence for the resulting series. For example, in the following addition, the interval of convergence for the sum is the *intersection* of the intervals of convergence of the two original series.

$$\underbrace{\sum_{n=0}^{\infty} x^n}_{(-1,\,1)} + \underbrace{\sum_{n=0}^{\infty}\left(\frac{x}{2}\right)^n}_{\cap\,(-2,\,2)} = \underbrace{\sum_{n=0}^{\infty}\left(1 + \frac{1}{2^n}\right)x^n}_{=\,(-1,\,1)}$$

EXAMPLE 3 Adding Two Power Series

Find a power series, centered at 0, for $f(x) = \dfrac{3x - 1}{x^2 - 1}$.

Solution Using partial fractions, you can write $f(x)$ as

$$\frac{3x - 1}{x^2 - 1} = \frac{2}{x + 1} + \frac{1}{x - 1}.$$

By adding the two geometric power series

$$\frac{2}{x + 1} = \frac{2}{1 - (-x)} = \sum_{n=0}^{\infty} 2(-1)^n x^n, \quad |x| < 1$$

and

$$\frac{1}{x - 1} = \frac{-1}{1 - x} = -\sum_{n=0}^{\infty} x^n, \quad |x| < 1$$

you obtain the following power series.

$$\frac{3x - 1}{x^2 - 1} = \sum_{n=0}^{\infty} [2(-1)^n - 1]x^n = 1 - 3x + x^2 - 3x^3 + x^4 - \cdots$$

The interval of convergence for this power series is $(-1, 1)$.

EXAMPLE 4 Finding a Power Series by Integration

Find a power series for $f(x) = \ln x$, centered at 1.

Solution From Example 2, you know that

$$\frac{1}{x} = \sum_{n=0}^{\infty} (-1)^n (x-1)^n. \qquad \text{Interval of convergence: } (0, 2)$$

Integrating this series produces

$$\ln x = \int \frac{1}{x}\,dx + C$$
$$= C + \sum_{n=0}^{\infty} (-1)^n \frac{(x-1)^{n+1}}{n+1}.$$

By letting $x = 1$, you can conclude that $C = 0$. Therefore,

$$\ln x = \sum_{n=0}^{\infty} (-1)^n \frac{(x-1)^{n+1}}{n+1}$$
$$= \frac{(x-1)}{1} - \frac{(x-1)^2}{2} + \frac{(x-1)^3}{3} - \frac{(x-1)^4}{4} + \cdots. \qquad \text{Interval of convergence: } (0, 2]$$

Note that the series converges at $x = 2$. This is consistent with the observation in the preceding section that integration of a power series may alter the convergence at the endpoints of the interval of convergence.

TECHNOLOGY In Section 9.7, the fourth-degree Taylor polynomial for the natural logarithmic function

$$\ln x \approx (x-1) - \frac{(x-1)^2}{2} + \frac{(x-1)^3}{3} - \frac{(x-1)^4}{4}$$

was used to approximate $\ln(1.1)$.

$$\ln(1.1) \approx (0.1) - \frac{1}{2}(0.1)^2 + \frac{1}{3}(0.1)^3 - \frac{1}{4}(0.1)^4$$
$$\approx 0.0953083$$

You now know from Example 4 that this polynomial represents the first four terms of the power series for $\ln x$. Moreover, using the Alternating Series Remainder, you can determine that the error in this approximation is less than

$$|R_4| \le |a_5|$$
$$= \frac{1}{5}(0.1)^5$$
$$= 0.000002.$$

During the seventeenth and eighteenth centuries, mathematical tables for logarithms and values of other transcendental functions were computed in this manner. Such numerical techniques are far from outdated, because it is precisely by such means that many modern calculating devices are programmed to evaluate transcendental functions.

EXAMPLE 5 Finding a Power Series by Integration

Find a power series for $g(x) = \arctan x$, centered at 0.

Solution Because $D_x[\arctan x] = 1/(1 + x^2)$, you can use the series

$$f(x) = \frac{1}{1 + x} = \sum_{n=0}^{\infty} (-1)^n x^n. \qquad \text{Interval of convergence: } (-1, 1)$$

Substituting x^2 for x produces

$$f(x^2) = \frac{1}{1 + x^2} = \sum_{n=0}^{\infty} (-1)^n x^{2n}.$$

Finally, by integrating, you obtain

$$\begin{aligned}
\arctan x &= \int \frac{1}{1 + x^2}\, dx + C \\
&= C + \sum_{n=0}^{\infty} (-1)^n \frac{x^{2n+1}}{2n + 1} \\
&= \sum_{n=0}^{\infty} (-1)^n \frac{x^{2n+1}}{2n + 1} && \text{Let } x = 0, \text{ then } C = 0. \\
&= x - \frac{x^3}{3} + \frac{x^5}{5} - \frac{x^7}{7} + \cdots. && \text{Interval of convergence: } (-1, 1)
\end{aligned}$$

The Granger Collection

Srinivasa Ramanujan (1887–1920)

Series that can be used to approximate π have interested mathematicians for the past 300 years. An amazing series for approximating $1/\pi$ was discovered by the Indian mathematician Srinivasa Ramanujan in 1914 (see Exercise 64). Each successive term of Ramanujan's series adds roughly eight more correct digits to the value of $1/\pi$. For more information about Ramanujan's work, see the article "Ramanujan and Pi" by Jonathan M. Borwein and Peter B. Borwein in *Scientific American*.

It can be shown that the power series developed for $\arctan x$ in Example 5 also converges (to $\arctan x$) for $x = \pm 1$. For instance, when $x = 1$, you can write

$$\begin{aligned}
\arctan 1 &= 1 - \frac{1}{3} + \frac{1}{5} - \frac{1}{7} + \cdots \\
&= \frac{\pi}{4}.
\end{aligned}$$

However, this series (developed by James Gregory in 1671) does not give us a practical way of approximating π because it converges so slowly that hundreds of terms would have to be used to obtain reasonable accuracy. Example 6 shows how to use *two* different arctangent series to obtain a very good approximation of π using only a few terms. This approximation was developed by John Machin in 1706.

EXAMPLE 6 Approximating π with a Series

Use the trigonometric identity

$$4 \arctan \frac{1}{5} - \arctan \frac{1}{239} = \frac{\pi}{4}$$

to approximate the number π [see Exercise 50(b)].

Solution By using only five terms from each of the series for $\arctan(1/5)$ and $\arctan(1/239)$, you obtain

$$4\left(4 \arctan \frac{1}{5} - \arctan \frac{1}{239}\right) \approx 3.1415926$$

which agrees with the exact value of π with an error of less than 0.0000001.

Exercises for Section 9.9

See www.CalcChat.com for worked-out solutions to odd-numbered exercises.

In Exercises 1–4, find a geometric power series for the function, centered at 0, (a) by the technique shown in Examples 1 and 2 and (b) by long division.

1. $f(x) = \dfrac{1}{2 - x}$
2. $f(x) = \dfrac{4}{5 - x}$
3. $f(x) = \dfrac{1}{2 + x}$
4. $f(x) = \dfrac{1}{1 + x}$

In Exercises 5–16, find a power series for the function, centered at c, and determine the interval of convergence.

5. $f(x) = \dfrac{1}{2 - x}, \quad c = 5$
6. $f(x) = \dfrac{4}{5 - x}, \quad c = -2$
7. $f(x) = \dfrac{3}{2x - 1}, \quad c = 0$
8. $f(x) = \dfrac{3}{2x - 1}, \quad c = 2$
9. $g(x) = \dfrac{1}{2x - 5}, \quad c = -3$
10. $h(x) = \dfrac{1}{2x - 5}, \quad c = 0$
11. $f(x) = \dfrac{3}{x + 2}, \quad c = 0$
12. $f(x) = \dfrac{4}{3x + 2}, \quad c = 2$
13. $g(x) = \dfrac{3x}{x^2 + x - 2}, \quad c = 0$
14. $g(x) = \dfrac{4x - 7}{2x^2 + 3x - 2}, \quad c = 0$
15. $f(x) = \dfrac{2}{1 - x^2}, \quad c = 0$
16. $f(x) = \dfrac{4}{4 + x^2}, \quad c = 0$

In Exercises 17–26, use the power series

$$\frac{1}{1 + x} = \sum_{n=0}^{\infty} (-1)^n x^n$$

to determine a power series, centered at 0, for the function. Identify the interval of convergence.

17. $h(x) = \dfrac{-2}{x^2 - 1} = \dfrac{1}{1 + x} + \dfrac{1}{1 - x}$
18. $h(x) = \dfrac{x}{x^2 - 1} = \dfrac{1}{2(1 + x)} - \dfrac{1}{2(1 - x)}$
19. $f(x) = -\dfrac{1}{(x + 1)^2} = \dfrac{d}{dx}\left[\dfrac{1}{x + 1}\right]$
20. $f(x) = \dfrac{2}{(x + 1)^3} = \dfrac{d^2}{dx^2}\left[\dfrac{1}{x + 1}\right]$
21. $f(x) = \ln(x + 1) = \displaystyle\int \frac{1}{x + 1}\,dx$
22. $f(x) = \ln(1 - x^2) = \displaystyle\int \frac{1}{1 + x}\,dx - \int \frac{1}{1 - x}\,dx$
23. $g(x) = \dfrac{1}{x^2 + 1}$
24. $f(x) = \ln(x^2 + 1)$
25. $h(x) = \dfrac{1}{4x^2 + 1}$
26. $f(x) = \arctan 2x$

Graphical and Numerical Analysis In Exercises 27 and 28, let

$$S_n = x - \frac{x^2}{2} + \frac{x^3}{3} - \frac{x^4}{4} + \cdots \pm \frac{x^n}{n}.$$

Use a graphing utility to confirm the inequality graphically. Then complete the table to confirm the inequality numerically.

x	0.0	0.2	0.4	0.6	0.8	1.0
S_n						
$\ln(x + 1)$						
S_{n+1}						

27. $S_2 \le \ln(x + 1) \le S_3$
28. $S_4 \le \ln(x + 1) \le S_5$

In Exercises 29 and 30, (a) graph several partial sums of the series, (b) find the sum of the series and its radius of convergence, (c) use 50 terms of the series to approximate the sum when $x = 0.5$, and (d) determine what the approximation represents and how good the approximation is.

29. $\displaystyle\sum_{n=1}^{\infty} \frac{(-1)^{n+1}(x - 1)^n}{n}$
30. $\displaystyle\sum_{n=0}^{\infty} \frac{(-1)^n x^{2n+1}}{(2n + 1)!}$

In Exercises 31–34, match the polynomial approximation of the function $f(x) = \arctan x$ with the correct graph. [The graphs are labeled (a), (b), (c), and (d).]

(a)
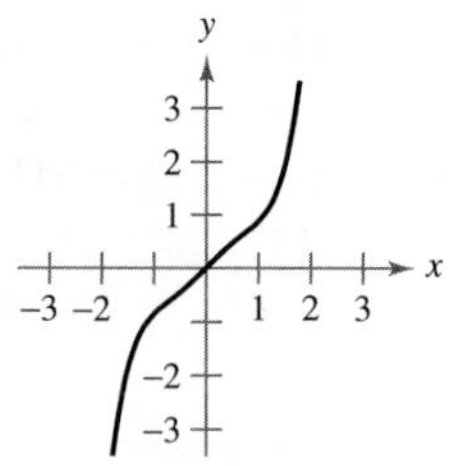

(b)
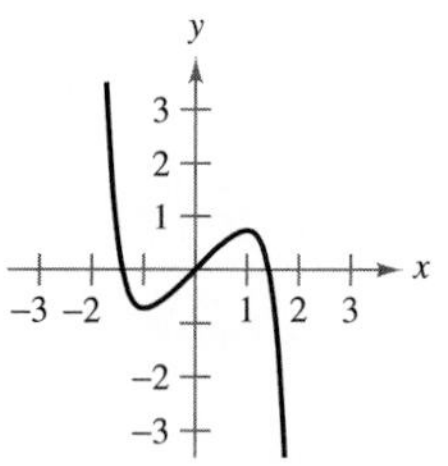

(c)
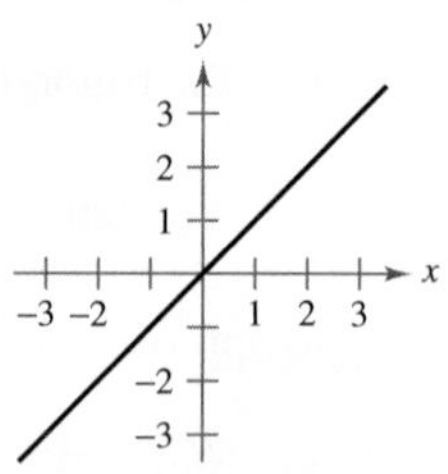

(d)
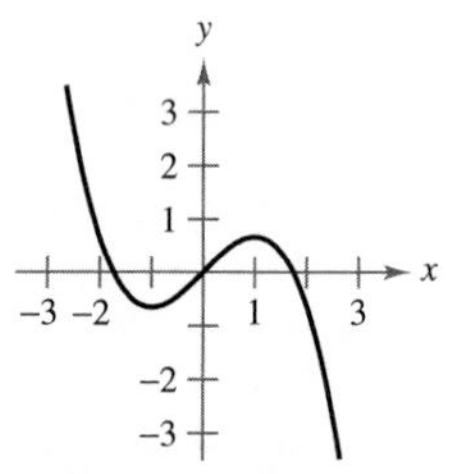

31. $g(x) = x$
32. $g(x) = x - \dfrac{x^3}{3}$
33. $g(x) = x - \dfrac{x^3}{3} + \dfrac{x^5}{5}$
34. $g(x) = x - \dfrac{x^3}{3} + \dfrac{x^5}{5} - \dfrac{x^7}{7}$

In Exercises 35–38, use the series for $f(x) = \arctan x$ to approximate the value, using $R_N \le 0.001$.

35. $\arctan \frac{1}{4}$

36. $\int_0^{3/4} \arctan x^2 \, dx$

37. $\int_0^{1/2} \frac{\arctan x^2}{x} \, dx$

38. $\int_0^{1/2} x^2 \arctan x \, dx$

In Exercises 39–42, use the power series

$$\frac{1}{1-x} = \sum_{n=0}^{\infty} x^n, \; |x| < 1.$$

Find the series representation of the function and determine its interval of convergence.

39. $f(x) = \frac{1}{(1-x)^2}$

40. $f(x) = \frac{x}{(1-x)^2}$

41. $f(x) = \frac{1+x}{(1-x)^2}$

42. $f(x) = \frac{x(1+x)}{(1-x)^2}$

43. ***Probability*** A fair coin is tossed repeatedly. The probability that the first head occurs on the nth toss is $P(n) = \left(\frac{1}{2}\right)^n$. When this game is repeated many times, the average number of tosses required until the first head occurs is

$$E(n) = \sum_{n=1}^{\infty} nP(n).$$

(This value is called the *expected value of n*.) Use the results of Exercises 39–42 to find $E(n)$. Is the answer what you expected? Why or why not?

44. Use the results of Exercises 39–42 to find the sum of each series.

(a) $\frac{1}{3}\sum_{n=1}^{\infty} n\left(\frac{2}{3}\right)^n$ (b) $\frac{1}{10}\sum_{n=1}^{\infty} n\left(\frac{9}{10}\right)^n$

Writing **In Exercises 45–48, explain how to use the geometric series**

$$g(x) = \frac{1}{1-x} = \sum_{n=0}^{\infty} x^n, \quad |x| < 1$$

to find the series for the function. Do not find the series.

45. $f(x) = \frac{1}{1+x}$

46. $f(x) = \frac{1}{1-x^2}$

47. $f(x) = \frac{5}{1+x}$

48. $f(x) = \ln(1-x)$

49. Prove that $\arctan x + \arctan y = \arctan \frac{x+y}{1-xy}$ for $xy \ne 1$ provided the value of the left side of the equation is between $-\pi/2$ and $\pi/2$.

50. Use the result of Exercise 49 to verify each identity.

(a) $\arctan \frac{120}{119} - \arctan \frac{1}{239} = \frac{\pi}{4}$

(b) $4 \arctan \frac{1}{5} - \arctan \frac{1}{239} = \frac{\pi}{4}$

[*Hint:* Use Exercise 49 twice to find $4 \arctan \frac{1}{5}$. Then use part (a).]

In Exercises 51 and 52, (a) verify the given equation and (b) use the equation and the series for the arctangent to approximate π to two-decimal-place accuracy.

51. $2 \arctan \frac{1}{2} - \arctan \frac{1}{7} = \frac{\pi}{4}$

52. $\arctan \frac{1}{2} + \arctan \frac{1}{3} = \frac{\pi}{4}$

In Exercises 53–58, find the sum of the convergent series by using a well-known function. Identify the function and explain how you obtained the sum.

53. $\sum_{n=1}^{\infty} (-1)^{n+1} \frac{1}{2^n n}$

54. $\sum_{n=1}^{\infty} (-1)^{n+1} \frac{1}{3^n n}$

55. $\sum_{n=1}^{\infty} (-1)^{n+1} \frac{2^n}{5^n n}$

56. $\sum_{n=0}^{\infty} (-1)^n \frac{1}{2n+1}$

57. $\sum_{n=0}^{\infty} (-1)^n \frac{1}{2^{2n+1}(2n+1)}$

58. $\sum_{n=1}^{\infty} (-1)^{n+1} \frac{1}{3^{2n-1}(2n-1)}$

Writing About Concepts

59. Use the results of Exercises 31–34 to make a geometric argument for why the series approximations of $f(x) = \arctan x$ have only odd powers of x.

60. Use the results of Exercises 31–34 to make a conjecture about the degrees of series approximations of $f(x) = \arctan x$ that have relative extrema.

61. One of the series in Exercises 53–58 converges to its sum at a much lower rate than the other five series. Which is it? Explain why this series converges so slowly. Use a graphing utility to illustrate the rate of convergence.

62. The radius of convergence of the power series $\sum_{n=0}^{\infty} a_n x^n$ is 3. What is the radius of convergence of the series $\sum_{n=1}^{\infty} n a_n x^{n-1}$? Explain.

63. The power series $\sum_{n=0}^{\infty} a_n x^n$ converges for $|x + 1| < 4$. What can you conclude about the series $\sum_{n=0}^{\infty} a_n \frac{x^{n+1}}{n+1}$? Explain.

64. Use a graphing utility to show that

$$\frac{\sqrt{8}}{9801} \sum_{n=0}^{\infty} \frac{(4n)!(1103 + 26{,}390n)}{(n!)396^{4n}} = \frac{1}{\pi}.$$

(*Note:* This series was discovered by the Indian mathematician Srinivasa Ramanujan in 1914.)

In Exercises 65 and 66, find the sum of the series.

65. $\sum_{n=0}^{\infty} \frac{(-1)^n}{3^n(2n+1)}$

66. $\sum_{n=0}^{\infty} \frac{(-1)^n \pi^{2n+1}}{3^{2n+1}(2n+1)!}$

Section 9.10 Taylor and Maclaurin Series

- **Find a Taylor or Maclaurin series for a function.**
- **Find a binomial series.**
- **Use a basic list of Taylor series to find other Taylor series.**

Bettmann/Corbis

COLIN MACLAURIN (1698–1746)

The development of power series to represent functions is credited to the combined work of many seventeenth and eighteenth century mathematicians. Gregory, Newton, John and James Bernoulli, Leibniz, Euler, Lagrange, Wallis, and Fourier all contributed to this work. However, the two names that are most commonly associated with power series are Brook Taylor (1685–1731) and Colin Maclaurin.

Taylor Series and Maclaurin Series

In Section 9.9, you derived power series for several functions using geometric series with term-by-term differentiation or integration. In this section you will study a *general* procedure for deriving the power series for a function that has derivatives of all orders. The following theorem gives the form that *every* convergent power series must take.

THEOREM 9.22 The Form of a Convergent Power Series

If f is represented by a power series $f(x) = \Sigma\, a_n(x - c)^n$ for all x in an open interval I containing c, then $a_n = f^{(n)}(c)/n!$ and

$$f(x) = f(c) + f'(c)(x - c) + \frac{f''(c)}{2!}(x - c)^2 + \cdot \cdot \cdot + \frac{f^{(n)}(c)}{n!}(x - c)^n + \cdot \cdot \cdot .$$

Proof Suppose the power series $\Sigma\, a_n(x - c)^n$ has a radius of convergence R. Then, by Theorem 9.21, you know that the nth derivative of f exists for $|x - c| < R$, and by successive differentiation you obtain the following.

$$\begin{aligned} f^{(0)}(x) &= a_0 + a_1(x - c) + a_2(x - c)^2 + a_3(x - c)^3 + a_4(x - c)^4 + \cdot \cdot \cdot \\ f^{(1)}(x) &= a_1 + 2a_2(x - c) + 3a_3(x - c)^2 + 4a_4(x - c)^3 + \cdot \cdot \cdot \\ f^{(2)}(x) &= 2a_2 + 3!a_3(x - c) + 4 \cdot 3a_4(x - c)^2 + \cdot \cdot \cdot \\ f^{(3)}(x) &= 3!a_3 + 4!a_4(x - c) + \cdot \cdot \cdot \\ &\vdots \\ f^{(n)}(x) &= n!a_n + (n + 1)!a_{n+1}(x - c) + \cdot \cdot \cdot \end{aligned}$$

Evaluating each of these derivatives at $x = c$ yields

$$\begin{aligned} f^{(0)}(c) &= 0!a_0 \\ f^{(1)}(c) &= 1!a_1 \\ f^{(2)}(c) &= 2!a_2 \\ f^{(3)}(c) &= 3!a_3 \end{aligned}$$

and, in general, $f^{(n)}(c) = n!a_n$. By solving for a_n, you find that the coefficients of the power series representation of $f(x)$ are

$$a_n = \frac{f^{(n)}(c)}{n!}.$$

NOTE Be sure you understand Theorem 9.22. The theorem says that *if a power series converges to* $f(x)$, the series must be a Taylor series. The theorem does *not* say that every series formed with the Taylor coefficients $a_n = f^{(n)}(c)/n!$ will converge to $f(x)$.

Notice that the coefficients of the power series in Theorem 9.22 are precisely the coefficients of the Taylor polynomials for $f(x)$ at c as defined in Section 9.7. For this reason, the series is called the **Taylor series** for $f(x)$ at c.

Definitions of Taylor and Maclaurin Series

If a function f has derivatives of all orders at $x = c$, then the series

$$\sum_{n=0}^{\infty} \frac{f^{(n)}(c)}{n!}(x-c)^n = f(c) + f'(c)(x-c) + \cdot \cdot \cdot + \frac{f^{(n)}(c)}{n!}(x-c)^n + \cdot \cdot \cdot$$

is called the **Taylor series for $f(x)$ at c.** Moreover, if $c = 0$, then the series is the **Maclaurin series for f.**

If you know the pattern for the coefficients of the Taylor polynomials for a function, you can extend the pattern easily to form the corresponding Taylor series. For instance, in Example 4 in Section 9.7, you found the fourth Taylor polynomial for $\ln x$, centered at 1, to be

$$P_4(x) = (x-1) - \frac{1}{2}(x-1)^2 + \frac{1}{3}(x-1)^3 - \frac{1}{4}(x-1)^4.$$

From this pattern, you can obtain the Taylor series for $\ln x$ centered at $c = 1$,

$$(x-1) - \frac{1}{2}(x-1)^2 + \cdot \cdot \cdot + \frac{(-1)^{n+1}}{n}(x-1)^n + \cdot \cdot \cdot .$$

EXAMPLE 1 Forming a Power Series

Use the function $f(x) = \sin x$ to form the Maclaurin series

$$\sum_{n=0}^{\infty} \frac{f^{(n)}(0)}{n!}x^n = f(0) + f'(0)x + \frac{f''(0)}{2!}x^2 + \frac{f^{(3)}(0)}{3!}x^3 + \frac{f^{(4)}(0)}{4!}x^4 + \cdot \cdot \cdot$$

and determine the interval of convergence.

Solution Successive differentiation of $f(x)$ yields

$$\begin{aligned} f(x) &= \sin x & f(0) &= \sin 0 = 0 \\ f'(x) &= \cos x & f'(0) &= \cos 0 = 1 \\ f''(x) &= -\sin x & f''(0) &= -\sin 0 = 0 \\ f^{(3)}(x) &= -\cos x & f^{(3)}(0) &= -\cos 0 = -1 \\ f^{(4)}(x) &= \sin x & f^{(4)}(0) &= \sin 0 = 0 \\ f^{(5)}(x) &= \cos x & f^{(5)}(0) &= \cos 0 = 1 \end{aligned}$$

and so on. The pattern repeats after the third derivative. So, the power series is as follows.

$$\begin{aligned} \sum_{n=0}^{\infty} \frac{f^{(n)}(0)}{n!}x^n &= f(0) + f'(0)x + \frac{f''(0)}{2!}x^2 + \frac{f^{(3)}(0)}{3!}x^3 + \frac{f^{(4)}(0)}{4!}x^4 + \cdot \cdot \cdot \\ \sum_{n=0}^{\infty} \frac{(-1)^n x^{2n+1}}{(2n+1)!} &= 0 + (1)x + \frac{0}{2!}x^2 + \frac{(-1)}{3!}x^3 + \frac{0}{4!}x^4 + \frac{1}{5!}x^5 + \frac{0}{6!}x^6 \\ &\quad + \frac{(-1)}{7!}x^7 + \cdot \cdot \cdot \\ &= x - \frac{x^3}{3!} + \frac{x^5}{5!} - \frac{x^7}{7!} + \cdot \cdot \cdot \end{aligned}$$

By the Ratio Test, you can conclude that this series converges for all x.

$$f(x) = \begin{cases} -1, & x < -\frac{\pi}{2} \\ \sin x, & |x| \le \frac{\pi}{2} \\ 1, & x > \frac{\pi}{2} \end{cases}$$

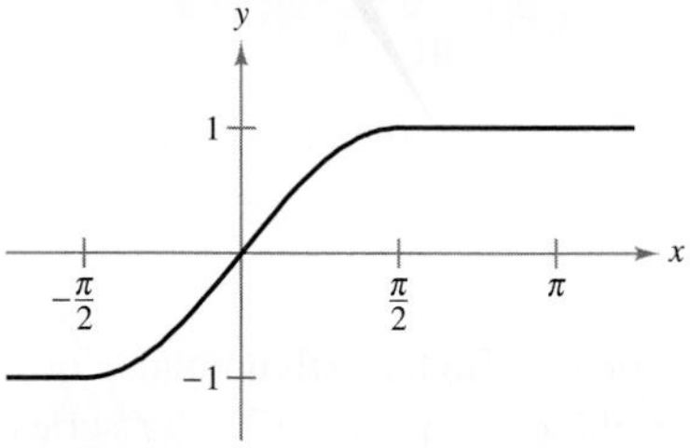

Figure 9.23

Notice that in Example 1 you cannot conclude that the power series converges to $\sin x$ for all x. You can simply conclude that the power series converges to some function, but you are not sure what function it is. This is a subtle, but important, point in dealing with Taylor or Maclaurin series. To persuade yourself that the series

$$f(c) + f'(c)(x - c) + \frac{f''(c)}{2!}(x - c)^2 + \cdot \cdot \cdot + \frac{f^{(n)}(c)}{n!}(x - c)^n + \cdot \cdot \cdot$$

might converge to a function other than f, remember that the derivatives are being evaluated at a single point. It can easily happen that another function will agree with the values of $f^{(n)}(x)$ when $x = c$ and disagree at other x-values. For instance, if you formed the power series (centered at 0) for the function shown in Figure 9.23, you would obtain the same series as in Example 1. You know that the series converges for all x, and yet it obviously cannot converge to both $f(x)$ and $\sin x$ for all x.

Let f have derivatives of all orders in an open interval I centered at c. The Taylor series for f may fail to converge for some x in I. Or, even if it is convergent, it may fail to have $f(x)$ as its sum. Nevertheless, Theorem 9.19 tells us that for each n,

$$f(x) = f(c) + f'(c)(x - c) + \frac{f''(c)}{2!}(x - c)^2 + \cdot \cdot \cdot + \frac{f^{(n)}(c)}{n!}(x - c)^n + R_n(x),$$

where

$$R_n(x) = \frac{f^{(n+1)}(z)}{(n + 1)!}(x - c)^{n+1}.$$

Note that in this remainder formula the particular value of z that makes the remainder formula true depends on the values of x and n. If $R_n \to 0$, then the following theorem tells us that the Taylor series for f actually converges to $f(x)$ for all x in I.

THEOREM 9.23 Convergence of Taylor Series

If $\lim_{n\to\infty} R_n = 0$ for all x in the interval I, then the Taylor series for f converges and equals $f(x)$,

$$f(x) = \sum_{n=0}^{\infty} \frac{f^{(n)}(c)}{n!}(x - c)^n.$$

Proof For a Taylor series, the nth partial sum coincides with the nth Taylor polynomial. That is, $S_n(x) = P_n(x)$. Moreover, because

$$P_n(x) = f(x) - R_n(x)$$

it follows that

$$\begin{aligned} \lim_{n\to\infty} S_n(x) &= \lim_{n\to\infty} P_n(x) \\ &= \lim_{n\to\infty} [f(x) - R_n(x)] \\ &= f(x) - \lim_{n\to\infty} R_n(x). \end{aligned}$$

So, for a given x, the Taylor series (the sequence of partial sums) converges to $f(x)$ if and only if $R_n(x) \to 0$ as $n \to \infty$.

NOTE Stated another way, Theorem 9.23 says that a power series formed with Taylor coefficients $a_n = f^{(n)}(c)/n!$ converges to the function from which it was derived at precisely those values for which the remainder approaches 0 as $n \to \infty$.

In Example 1, you derived the power series from the sine function and you also concluded that the series converges to some function on the entire real line. In Example 2, you will see that the series actually converges to sin x. The key observation is that although the value of z is not known, it is possible to obtain an upper bound for $|f^{(n+1)}(z)|$.

EXAMPLE 2 A Convergent Maclaurin Series

Show that the Maclaurin series for $f(x) = \sin x$ converges to $\sin x$ for all x.

Solution Using the result in Example 1, you need to show that

$$\sin x = x - \frac{x^3}{3!} + \frac{x^5}{5!} - \frac{x^7}{7!} + \cdot \cdot \cdot + \frac{(-1)^n x^{2n+1}}{(2n+1)!} + \cdot \cdot \cdot$$

is true for all x. Because

$$f^{(n+1)}(x) = \pm\sin x$$

or

$$f^{(n+1)}(x) = \pm\cos x$$

you know that $|f^{(n+1)}(z)| \le 1$ for every real number z. Therefore, for any fixed x, you can apply Taylor's Theorem (Theorem 9.19) to conclude that

$$0 \le |R_n(x)| = \left|\frac{f^{(n+1)}(z)}{(n+1)!}x^{n+1}\right| \le \frac{|x|^{n+1}}{(n+1)!}.$$

From the discussion in Section 9.1 regarding the relative rates of convergence of exponential and factorial sequences, it follows that for a fixed x

$$\lim_{n\to\infty} \frac{|x|^{n+1}}{(n+1)!} = 0.$$

Finally, by the Squeeze Theorem, it follows that for all x, $R_n(x) \to 0$ as $n \to \infty$. So, by Theorem 9.23, the Maclaurin series for sin x converges to sin x for all x.

Figure 9.24 visually illustrates the convergence of the Maclaurin series for sin x by comparing the graphs of the Maclaurin polynomials $P_1(x)$, $P_3(x)$, $P_5(x)$, and $P_7(x)$ with the graph of the sine function. Notice that as the degree of the polynomial increases, its graph more closely resembles that of the sine function.

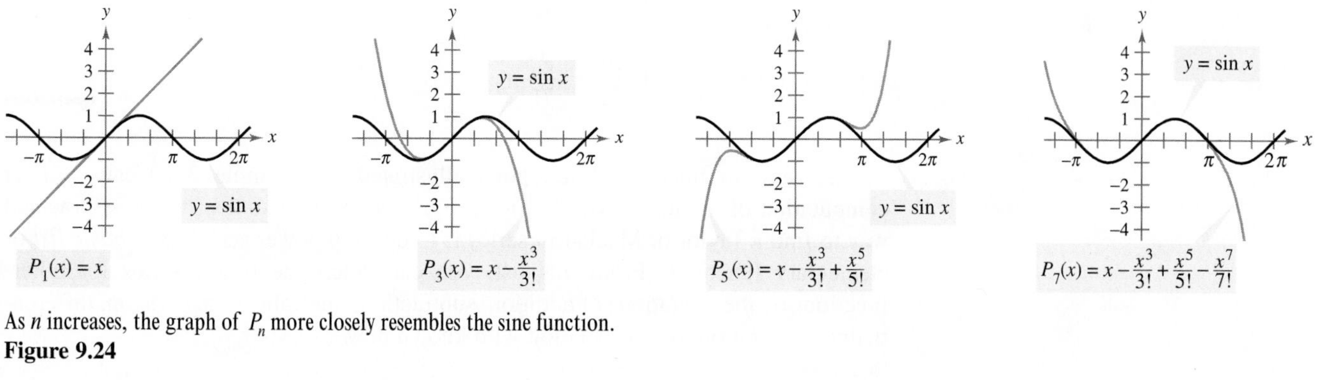

As n increases, the graph of P_n more closely resembles the sine function.
Figure 9.24

The guidelines for finding a Taylor series for $f(x)$ at c are summarized below.

Guidelines for Finding a Taylor Series

1. Differentiate $f(x)$ several times and evaluate each derivative at c.

$$f(c), f'(c), f''(c), f'''(c), \cdot \cdot \cdot, f^{(n)}(c), \cdot \cdot \cdot$$

Try to recognize a pattern in these numbers.

2. Use the sequence developed in the first step to form the Taylor coefficients $a_n = f^{(n)}(c)/n!$, and determine the interval of convergence for the resulting power series

$$f(c) + f'(c)(x - c) + \frac{f''(c)}{2!}(x - c)^2 + \cdot \cdot \cdot + \frac{f^{(n)}(c)}{n!}(x - c)^n + \cdot \cdot \cdot .$$

3. Within this interval of convergence, determine whether or not the series converges to $f(x)$.

The direct determination of Taylor or Maclaurin coefficients using successive differentiation can be difficult, and the next example illustrates a shortcut for finding the coefficients indirectly—using the coefficients of a known Taylor or Maclaurin series.

EXAMPLE 3 Maclaurin Series for a Composite Function

Find the Maclaurin series for $f(x) = \sin x^2$.

Solution To find the coefficients for this Maclaurin series directly, you must calculate successive derivatives of $f(x) = \sin x^2$. By calculating just the first two,

$$f'(x) = 2x \cos x^2 \quad \text{and} \quad f''(x) = -4x^2 \sin x^2 + 2 \cos x^2$$

you can see that this task would be quite cumbersome. Fortunately, there is an alternative. First consider the Maclaurin series for $\sin x$ found in Example 1.

$$\begin{aligned} g(x) &= \sin x \\ &= x - \frac{x^3}{3!} + \frac{x^5}{5!} - \frac{x^7}{7!} + \cdot \cdot \cdot \end{aligned}$$

Now, because $\sin x^2 = g(x^2)$, you can substitute x^2 for x in the series for $\sin x$ to obtain

$$\begin{aligned} \sin x^2 &= g(x^2) \\ &= x^2 - \frac{x^6}{3!} + \frac{x^{10}}{5!} - \frac{x^{14}}{7!} + \cdot \cdot \cdot . \end{aligned}$$

Be sure to understand the point illustrated in Example 3. Because direct computation of Taylor or Maclaurin coefficients can be tedious, the most practical way to find a Taylor or Maclaurin series is to develop power series for a *basic list* of elementary functions. From this list, you can determine power series for other functions by the operations of addition, subtraction, multiplication, division, differentiation, integration, or composition with known power series.

Binomial Series

Before presenting the basic list for elementary functions, you wll develop one more series—for a function of the form $f(x) = (1 + x)^k$. This produces the **binomial series.**

EXAMPLE 4 Binomial Series

Find the Maclaurin series for $f(x) = (1 + x)^k$ and determine its radius of convergence. Assume that k is not a positive integer.

Solution By successive differentiation, you have

$$\begin{aligned} f(x) &= (1+x)^k & f(0) &= 1 \\ f'(x) &= k(1+x)^{k-1} & f'(0) &= k \\ f''(x) &= k(k-1)(1+x)^{k-2} & f''(0) &= k(k-1) \\ f'''(x) &= k(k-1)(k-2)(1+x)^{k-3} & f'''(0) &= k(k-1)(k-2) \\ &\vdots & &\vdots \\ f^{(n)}(x) &= k\cdot\cdot\cdot(k-n+1)(1+x)^{k-n} & f^{(n)}(0) &= k(k-1)\cdot\cdot\cdot(k-n+1) \end{aligned}$$

which produces the series

$$1 + kx + \frac{k(k-1)x^2}{2} + \cdot\cdot\cdot + \frac{k(k-1)\cdot\cdot\cdot(k-n+1)x^n}{n!} + \cdot\cdot\cdot.$$

Because $a_{n+1}/a_n \rightarrow 1$, you can apply the Ratio Test to conclude that the radius of convergence is $R = 1$. So, the series converges to some function in the interval $(-1, 1)$.

Note that Example 4 shows that the Taylor series for $(1 + x)^k$ converges to some function in the interval $(-1, 1)$. However, the example does not show that the series actually converges to $(1 + x)^k$. To do this, you could show that the remainder $R_n(x)$ converges to 0, as illustrated in Example 2.

EXAMPLE 5 Finding a Binomial Series

Find the power series for $f(x) = \sqrt[3]{1 + x}$.

Solution Using the binomial series

$$(1+x)^k = 1 + kx + \frac{k(k-1)x^2}{2!} + \frac{k(k-1)(k-2)x^3}{3!} + \cdot\cdot\cdot$$

let $k = \frac{1}{3}$ and write

$$(1+x)^{1/3} = 1 + \frac{x}{3} - \frac{2x^2}{3^2 2!} + \frac{2\cdot 5x^3}{3^3 3!} - \frac{2\cdot 5\cdot 8x^4}{3^4 4!} + \cdot\cdot\cdot$$

which converges for $-1 \le x \le 1$.

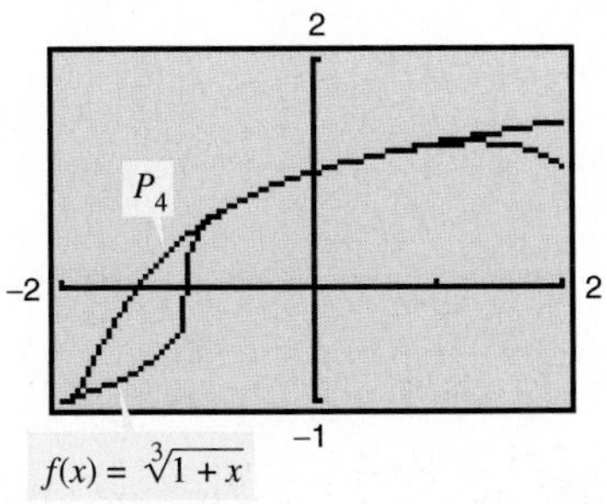

Figure 9.25

TECHNOLOGY Use a graphing utility to confirm the result in Example 5. When you graph the functions

$$f(x) = (1+x)^{1/3} \quad \text{and} \quad P_4(x) = 1 + \frac{x}{3} - \frac{x^2}{9} + \frac{5x^3}{81} - \frac{10x^4}{243}$$

in the same viewing window, you should obtain the result shown in Figure 9.25.

Deriving Taylor Series from a Basic List

The following list provides the power series for several elementary functions with the corresponding intervals of convergence.

Power Series for Elementary Functions

Function	*Interval of Convergence*
$\dfrac{1}{x} = 1 - (x-1) + (x-1)^2 - (x-1)^3 + (x-1)^4 - \cdots + (-1)^n(x-1)^n + \cdots$	$0 < x < 2$
$\dfrac{1}{1+x} = 1 - x + x^2 - x^3 + x^4 - x^5 + \cdots + (-1)^n x^n + \cdots$	$-1 < x < 1$
$\ln x = (x-1) - \dfrac{(x-1)^2}{2} + \dfrac{(x-1)^3}{3} - \dfrac{(x-1)^4}{4} + \cdots + \dfrac{(-1)^{n-1}(x-1)^n}{n} + \cdots$	$0 < x \le 2$
$e^x = 1 + x + \dfrac{x^2}{2!} + \dfrac{x^3}{3!} + \dfrac{x^4}{4!} + \dfrac{x^5}{5!} + \cdots + \dfrac{x^n}{n!} + \cdots$	$-\infty < x < \infty$
$\sin x = x - \dfrac{x^3}{3!} + \dfrac{x^5}{5!} - \dfrac{x^7}{7!} + \dfrac{x^9}{9!} - \cdots + \dfrac{(-1)^n x^{2n+1}}{(2n+1)!} + \cdots$	$-\infty < x < \infty$
$\cos x = 1 - \dfrac{x^2}{2!} + \dfrac{x^4}{4!} - \dfrac{x^6}{6!} + \dfrac{x^8}{8!} - \cdots + \dfrac{(-1)^n x^{2n}}{(2n)!} + \cdots$	$-\infty < x < \infty$
$\arctan x = x - \dfrac{x^3}{3} + \dfrac{x^5}{5} - \dfrac{x^7}{7} + \dfrac{x^9}{9} - \cdots + \dfrac{(-1)^n x^{2n+1}}{2n+1} + \cdots$	$-1 \le x \le 1$
$\arcsin x = x + \dfrac{x^3}{2 \cdot 3} + \dfrac{1 \cdot 3x^5}{2 \cdot 4 \cdot 5} + \dfrac{1 \cdot 3 \cdot 5x^7}{2 \cdot 4 \cdot 6 \cdot 7} + \cdots + \dfrac{(2n)!x^{2n+1}}{(2^n n!)^2(2n+1)} + \cdots$	$-1 \le x \le 1$
$(1+x)^k = 1 + kx + \dfrac{k(k-1)x^2}{2!} + \dfrac{k(k-1)(k-2)x^3}{3!} + \dfrac{k(k-1)(k-2)(k-3)x^4}{4!} + \cdots$	$-1 < x < 1$*

* *The convergence at $x = \pm 1$ depends on the value of k.*

NOTE The binomial series is valid for noninteger values of k. Moreover, if k happens to be a positive integer, the binomial series reduces to a simple binomial expansion.

EXAMPLE 6 Deriving a Power Series from a Basic List

Find the power series for $f(x) = \cos\sqrt{x}$.

Solution Using the power series

$$\cos x = 1 - \frac{x^2}{2!} + \frac{x^4}{4!} - \frac{x^6}{6!} + \frac{x^8}{8!} - \cdots$$

you can replace x by $\sqrt{x}$ to obtain the series

$$\cos\sqrt{x} = 1 - \frac{x}{2!} + \frac{x^2}{4!} - \frac{x^3}{6!} + \frac{x^4}{8!} - \cdots.$$

This series converges for all x in the domain of $\cos\sqrt{x}$—that is, for $x \ge 0$.

Power series can be multiplied and divided like polynomials. After finding the first few terms of the product (or quotient), you may be able to recognize a pattern.

EXAMPLE 7 Multiplication and Division of Power Series

Find the first three nonzero terms in each of the Maclaurin series.

a. $e^x \arctan x$ **b.** $\tan x$

Solution

a. Using the Maclaurin series for e^x and $\arctan x$ in the table, you have

$$e^x \arctan x = \left(1 + \frac{x}{1!} + \frac{x^2}{2!} + \frac{x^3}{3!} + \frac{x^4}{4!} + \cdots\right)\left(x - \frac{x^3}{3} + \frac{x^5}{5} - \cdots\right).$$

Multiply these expressions and collect like terms as you would for multiplying polynomials.

$$\begin{array}{lllllll}
1 + x + \frac{1}{2}x^2 + \frac{1}{6}x^3 + \frac{1}{24}x^4 + \cdots & & & & & & \\
x & - \frac{1}{3}x^3 & & + \frac{1}{5}x^5 - \cdots & & & \\
\hline
x + & x^2 + \frac{1}{2}x^3 + & \frac{1}{6}x^4 + & \frac{1}{24}x^5 + \cdots & & & \\
 & - \frac{1}{3}x^3 - & \frac{1}{3}x^4 - & \frac{1}{6}x^5 - \cdots & & & \\
 & & & + \frac{1}{5}x^5 + \cdots & & & \\
\hline
x + & x^2 + \frac{1}{6}x^3 - & \frac{1}{6}x^4 + & \frac{3}{40}x^5 + \cdots & & &
\end{array}$$

So, $e^x \arctan x = x + x^2 + \frac{1}{6}x^3 + \cdots$.

b. Using the Maclaurin series for $\sin x$ and $\cos x$ in the table, you have

$$\tan x = \frac{\sin x}{\cos x} = \frac{x - \dfrac{x^3}{3!} + \dfrac{x^5}{5!} - \cdots}{1 - \dfrac{x^2}{2!} + \dfrac{x^4}{4!} - \cdots}.$$

Divide using long division.

$$\begin{array}{r|l}
 & x + \frac{1}{3}x^3 + \frac{2}{15}x^5 + \cdots \\
\hline
1 - \frac{1}{2}x^2 + \frac{1}{24}x^4 - \cdots & x - \frac{1}{6}x^3 + \frac{1}{120}x^5 - \cdots \\
 & x - \frac{1}{2}x^3 + \frac{1}{24}x^5 - \cdots \\
\hline
 & \frac{1}{3}x^3 - \frac{1}{30}x^5 + \cdots \\
 & \frac{1}{3}x^3 - \frac{1}{6}x^5 + \cdots \\
\hline
 & \frac{2}{15}x^5 + \cdots
\end{array}$$

So, $\tan x = x + \frac{1}{3}x^3 + \frac{2}{15}x^5 + \cdots$.

EXAMPLE 8 A Power Series for $\sin^2 x$

Find the power series for $f(x) = \sin^2 x$.

Solution Consider rewriting $\sin^2 x$ as follows.

$$\sin^2 x = \frac{1 - \cos 2x}{2} = \frac{1}{2} - \frac{\cos 2x}{2}$$

Now, use the series for $\cos x$.

$$\cos x = 1 - \frac{x^2}{2!} + \frac{x^4}{4!} - \frac{x^6}{6!} + \frac{x^8}{8!} - \cdots$$

$$\cos 2x = 1 - \frac{2^2}{2!}x^2 + \frac{2^4}{4!}x^4 - \frac{2^6}{6!}x^6 + \frac{2^8}{8!}x^8 - \cdots$$

$$-\frac{1}{2}\cos 2x = -\frac{1}{2} + \frac{2}{2!}x^2 - \frac{2^3}{4!}x^4 + \frac{2^5}{6!}x^6 - \frac{2^7}{8!}x^8 + \cdots$$

$$\sin^2 x = \frac{1}{2} - \frac{1}{2}\cos 2x = \frac{1}{2} - \frac{1}{2} + \frac{2}{2!}x^2 - \frac{2^3}{4!}x^4 + \frac{2^5}{6!}x^6 - \frac{2^7}{8!}x^8 + \cdots$$

$$= \frac{2}{2!}x^2 - \frac{2^3}{4!}x^4 + \frac{2^5}{6!}x^6 - \frac{2^7}{8!}x^8 + \cdots$$

This series converges for $-\infty < x < \infty$.

As mentioned in the preceding section, power series can be used to obtain tables of values of transcendental functions. They are also useful for estimating the values of definite integrals for which antiderivatives cannot be found. The next example demonstrates this use.

EXAMPLE 9 Power Series Approximation of a Definite Integral

Use a power series to approximate

$$\int_0^1 e^{-x^2}\,dx$$

with an error of less than 0.01.

Solution Replacing x with $-x^2$ in the series for e^x produces the following.

$$e^{-x^2} = 1 - x^2 + \frac{x^4}{2!} - \frac{x^6}{3!} + \frac{x^8}{4!} - \cdots$$

$$\int_0^1 e^{-x^2}\,dx = \left[x - \frac{x^3}{3} + \frac{x^5}{5 \cdot 2!} - \frac{x^7}{7 \cdot 3!} + \frac{x^9}{9 \cdot 4!} - \cdots\right]_0^1$$

$$= 1 - \frac{1}{3} + \frac{1}{10} - \frac{1}{42} + \frac{1}{216} - \cdots$$

Summing the first four terms, you have

$$\int_0^1 e^{-x^2}\,dx \approx 0.74$$

which, by the Alternating Series Test, has an error of less than $\frac{1}{216} \approx 0.005$.

Exercises for Section 9.10

See www.CalcChat.com for worked-out solutions to odd-numbered exercises.

In Exercises 1–10, use the definition to find the Taylor series (centered at c) for the function.

1. $f(x) = e^{2x}, \quad c = 0$
2. $f(x) = e^{3x}, \quad c = 0$
3. $f(x) = \cos x, \quad c = \dfrac{\pi}{4}$
4. $f(x) = \sin x, \quad c = \dfrac{\pi}{4}$
5. $f(x) = \ln x, \quad c = 1$
6. $f(x) = e^x, \quad c = 1$
7. $f(x) = \sin 2x, \quad c = 0$
8. $f(x) = \ln(x^2 + 1), \quad c = 0$
9. $f(x) = \sec x, \quad c = 0$ (first three nonzero terms)
10. $f(x) = \tan x, \quad c = 0$ (first three nonzero terms)

In Exercises 11–14, prove that the Maclaurin series for the function converges to the function for all x.

11. $f(x) = \cos x$
12. $f(x) = e^{-2x}$
13. $f(x) = \sinh x$
14. $f(x) = \cosh x$

In Exercises 15–20, use the binomial series to find the Maclaurin series for the function.

15. $f(x) = \dfrac{1}{(1 + x)^2}$
16. $f(x) = \dfrac{1}{\sqrt{1 - x}}$
17. $f(x) = \dfrac{1}{\sqrt{4 + x^2}}$
18. $f(x) = \sqrt[4]{1 + x}$
19. $f(x) = \sqrt{1 + x^2}$
20. $f(x) = \sqrt{1 + x^3}$

In Exercises 21–30, find the Maclaurin series for the function. (Use the table of power series for elementary functions.)

21. $f(x) = e^{x^2/2}$
22. $g(x) = e^{-3x}$
23. $g(x) = \sin 3x$
24. $f(x) = \cos 4x$
25. $f(x) = \cos x^{3/2}$
26. $g(x) = 2 \sin x^3$
27. $f(x) = \frac{1}{2}(e^x - e^{-x}) = \sinh x$
28. $f(x) = e^x + e^{-x} = 2 \cosh x$
29. $f(x) = \cos^2 x$
30. $f(x) = \sinh^{-1} x = \ln\left(x + \sqrt{x^2 + 1}\right)$

    $\left(\textit{Hint:} \text{ Integrate the series for } \dfrac{1}{\sqrt{x^2 + 1}}.\right)$

In Exercises 31–34, find the Maclaurin series for the function. (See Example 7.)

31. $f(x) = x \sin x$
32. $h(x) = x \cos x$
33. $g(x) = \begin{cases} \dfrac{\sin x}{x}, & x \neq 0 \\ 1, & x = 0 \end{cases}$
34. $f(x) = \begin{cases} \dfrac{\arcsin x}{x}, & x \neq 0 \\ 1, & x = 0 \end{cases}$

In Exercises 35 and 36, use a power series and the fact that $i^2 = -1$ to verify the formula.

35. $g(x) = \dfrac{1}{2i}(e^{ix} - e^{-ix}) = \sin x$
36. $g(x) = \frac{1}{2}(e^{ix} + e^{-ix}) = \cos x$

In Exercises 37–42, find the first four nonzero terms of the Maclaurin series for the function by multiplying or dividing the appropriate power series. Use the table of power series for elementary functions on page 682. Use a graphing utility to graph the function and its corresponding polynomial approximation.

37. $f(x) = e^x \sin x$
38. $g(x) = e^x \cos x$
39. $h(x) = \cos x \ln(1 + x)$
40. $f(x) = e^x \ln(1 + x)$
41. $g(x) = \dfrac{\sin x}{1 + x}$
42. $f(x) = \dfrac{e^x}{1 + x}$

In Exercises 43–46, match the polynomial with its graph. [The graphs are labeled (a), (b), (c), and (d).] Factor a common factor from each polynomial and identify the function approximated by the remaining Taylor polynomial.

(a)

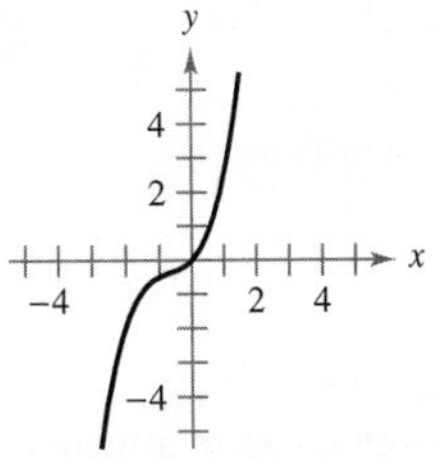

(b)

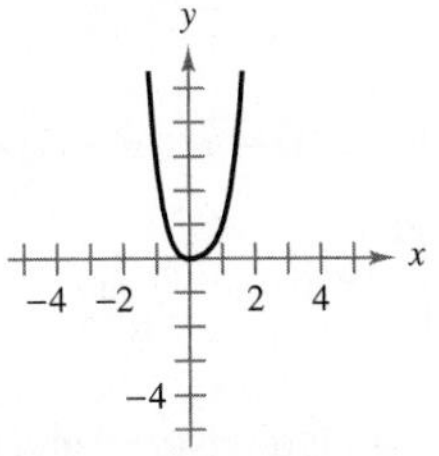

(c)

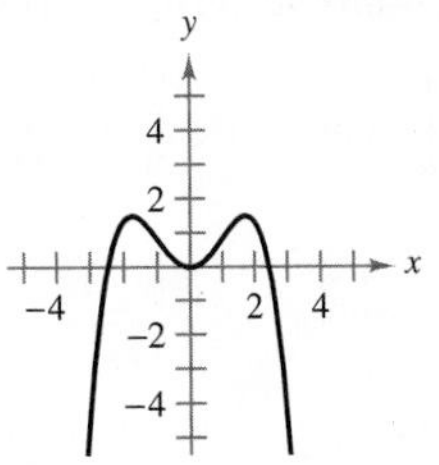

(d)

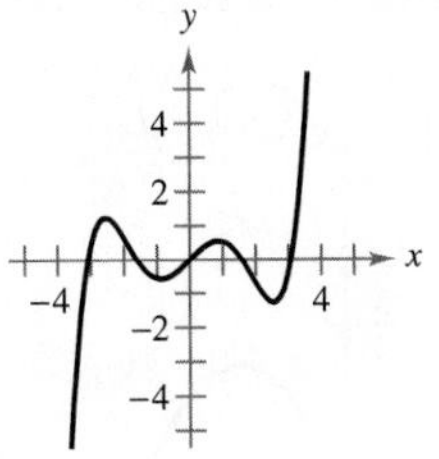

43. $y = x^2 - \dfrac{x^4}{3!}$
44. $y = x - \dfrac{x^3}{2!} + \dfrac{x^5}{4!}$
45. $y = x + x^2 + \dfrac{x^3}{2!}$
46. $y = x^2 - x^3 + x^4$

In Exercises 47 and 48, find a Maclaurin series for $f(x)$.

47. $f(x) = \int_0^x (e^{-t^2} - 1)\, dt$

48. $f(x) = \int_0^x \sqrt{1 + t^3}\, dt$

In Exercises 49–52, verify the sum. Then use a graphing utility to approximate the sum with an error of less than 0.0001.

49. $\sum_{n=1}^{\infty} (-1)^{n+1} \frac{1}{n} = \ln 2$

50. $\sum_{n=0}^{\infty} (-1)^n \left[\frac{1}{(2n + 1)!}\right] = \sin 1$

51. $\sum_{n=0}^{\infty} \frac{2^n}{n!} = e^2$

52. $\sum_{n=1}^{\infty} (-1)^{n-1} \left(\frac{1}{n!}\right) = \frac{e - 1}{e}$

In Exercises 53 and 54, use the series representation of the function f to find $\lim_{x \to 0} f(x)$ (if it exists).

53. $f(x) = \frac{1 - \cos x}{x}$

54. $f(x) = \frac{\sin x}{x}$

In Exercises 55–58, use a power series to approximate the value of the integral with an error of less than 0.0001. (In Exercises 55 and 56, assume that the integrand is defined as 1 when $x = 0$.)

55. $\int_0^1 \frac{\sin x}{x}\, dx$

56. $\int_0^{1/2} \frac{\arctan x}{x}\, dx$

57. $\int_{0.1}^{0.3} \sqrt{1 + x^3}\, dx$

58. $\int_0^{1/4} x \ln(x + 1)\, dx$

Area **In Exercises 59 and 60, use a power series to approximate the area of the region. Use a graphing utility to verify the result.**

59. $\int_0^{\pi/2} \sqrt{x} \cos x\, dx$

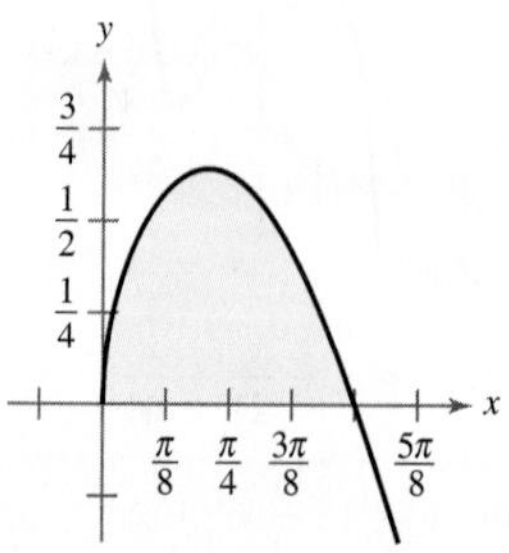

60. $\int_{0.5}^{1} \cos \sqrt{x}\, dx$

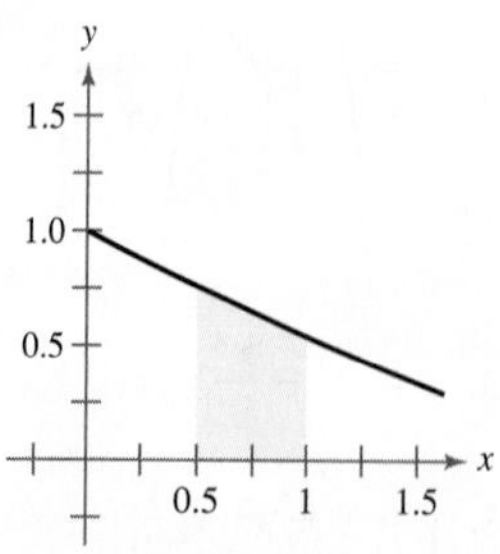

Probability **In Exercises 61 and 62, approximate the normal probability with an error of less than 0.0001, where the probability is given by**

$$P(a < x < b) = \frac{1}{\sqrt{2\pi}} \int_a^b e^{-x^2/2}\, dx.$$

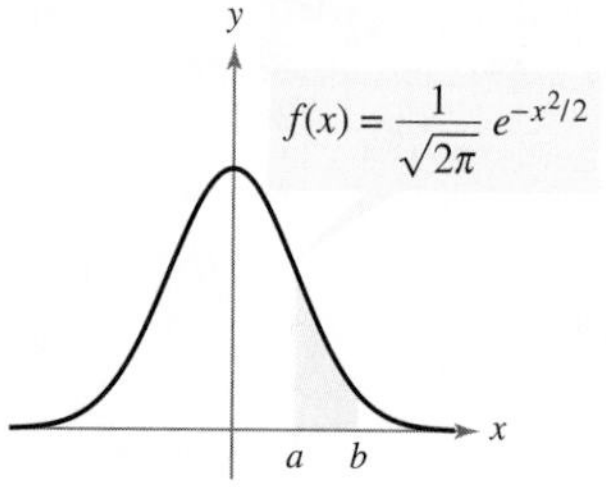

61. $P(0 < x < 1)$

62. $P(1 < x < 2)$

In Exercises 63–66, use a computer algebra system to find the fifth-degree Taylor polynomial (centered at c) for the function. Graph the function and the polynomial. Use the graph to determine the largest interval on which the polynomial is a reasonable approximation of the function.

63. $f(x) = x \cos 2x, \quad c = 0$

64. $f(x) = \sin \frac{x}{2} \ln(1 + x), \quad c = 0$

65. $g(x) = \sqrt{x} \ln x, \quad c = 1$

66. $h(x) = \sqrt[3]{x} \arctan x, \quad c = 1$

Writing About Concepts

67. State the guidelines for finding a Taylor series.

68. If f is an even function, what must be true about the coefficients a_n in the Maclaurin series

$$f(x) = \sum_{n=0}^{\infty} a_n x^n?$$

Explain your reasoning.

69. Explain how to use the series

$$g(x) = e^x = \sum_{n=0}^{\infty} \frac{x^n}{n!}$$

to find the series for each function. Do not find the series.

(a) $f(x) = e^{-x}$

(b) $f(x) = e^{3x}$

(c) $f(x) = xe^x$

(d) $f(x) = e^{2x} + e^{-2x}$

70. Define the binomial series. What is its radius of convergence?

71. ***Projectile Motion*** A projectile fired from the ground follows the trajectory given by

$$y = \left(\tan\theta - \frac{g}{kv_0\cos\theta}\right)x - \frac{g}{k^2}\ln\left(1 - \frac{kx}{v_0\cos\theta}\right)$$

where v_0 is the initial speed, θ is the angle of projection, g is the acceleration due to gravity, and k is the drag factor caused by air resistance. Using the power series representation

$$\ln(1+x) = x - \frac{x^2}{2} + \frac{x^3}{3} - \frac{x^4}{4} + \cdot\cdot\cdot, \quad -1 < x < 1$$

verify that the trajectory can be rewritten as

$$y = (\tan\theta)x + \frac{gx^2}{2v_0^2\cos^2\theta} + \frac{kgx^3}{3v_0^3\cos^3\theta} + \frac{k^2gx^4}{4v_0^4\cos^4\theta} + \cdot\cdot\cdot.$$

72. ***Projectile Motion*** Use the result of Exercise 71 to determine the series for the path of a projectile launched from ground level at an angle of $\theta = 60°$, with an initial speed of $v_0 = 64$ feet per second and a drag factor of $k = \frac{1}{16}$.

73. ***Investigation*** Consider the function f defined by

$$f(x) = \begin{cases} e^{-1/x^2}, & x \neq 0 \\ 0, & x = 0. \end{cases}$$

(a) Sketch a graph of the function.

(b) Use the alternative form of the definition of the derivative (Section 2.1) and L'Hôpital's Rule to show that $f'(0) = 0$. [By continuing this process, it can be shown that $f^{(n)}(0) = 0$ for $n > 1$.]

(c) Using the result in part (b), find the Maclaurin series for f. Does the series converge to f?

74. ***Investigation***

(a) Find the power series centered at 0 for the function

$$f(x) = \frac{\ln(x^2+1)}{x^2}.$$

(b) Use a graphing utility to graph f and the eighth-degree Taylor polynomial $P_8(x)$ for f.

(c) Complete the table, where

$$F(x) = \int_0^x \frac{\ln(t^2+1)}{t^2}\,dt \quad \text{and} \quad G(x) = \int_0^x P_8(t)\,dt.$$

x	0.25	0.50	0.75	1.00	1.50	2.00
$F(x)$						
$G(x)$						

(d) Describe the relationship between the graphs of f and P_8 and the results given in the table in part (c).

75. Prove that $\lim\limits_{n\to\infty} \dfrac{x^n}{n!} = 0$ for any real x.

76. Find the Maclaurin series for

$$f(x) = \ln\frac{1+x}{1-x}$$

and determine its radius of convergence. Use the first four terms of the series to approximate ln 3.

In Exercises 77–80, evaluate the binomial coefficient using the formula

$$\binom{k}{n} = \frac{k(k-1)(k-2)(k-3)\cdot\cdot\cdot(k-n+1)}{n!}$$

where k is a real number, n is a positive integer, and

$$\binom{k}{0} = 1.$$

77. $\binom{5}{3}$

78. $\binom{-2}{2}$

79. $\binom{0.5}{4}$

80. $\binom{-1/3}{5}$

81. Write the power series for $(1+x)^k$ in terms of binomial coefficients.

82. Prove that e is irrational. [Hint: Assume that $e = p/q$ is rational (p and q are integers) and consider

$$e = 1 + 1 + \frac{1}{2!} + \cdot\cdot\cdot + \frac{1}{n!} + \cdot\cdot\cdot.]$$

83. Show that the Maclaurin series of the function

$$g(x) = \frac{x}{1-x-x^2}$$

is

$$\sum_{n=1}^{\infty} F_n x^n$$

where F_n is the nth Fibonacci number with $F_1 = F_2 = 1$ and $F_n = F_{n-2} + F_{n-1}$, for $n \geq 3$.

(*Hint:* Write

$$\frac{x}{1-x-x^2} = a_0 + a_1x + a_2x^2 + \cdot\cdot\cdot$$

and multiply each side of this equation by $1 - x - x^2$.)

Putnam Exam Challenge

84. Assume that $|f(x)| \leq 1$ and $|f''(x)| \leq 1$ for all x on an interval of length at least 2. Show that $|f'(x)| \leq 2$ on the interval.

Review Exercises for Chapter 9

See www.CalcChat.com for worked-out solutions to odd-numbered exercises.

In Exercises 1 and 2, write an expression for the nth term of the sequence.

1. $1, \frac{1}{2}, \frac{1}{6}, \frac{1}{24}, \frac{1}{120}, \dots$

2. $\frac{1}{2}, \frac{2}{5}, \frac{3}{10}, \frac{4}{17}, \dots$

In Exercises 3–6, match the sequence with its graph. [The graphs are labeled (a), (b), (c), and (d).]

(a)

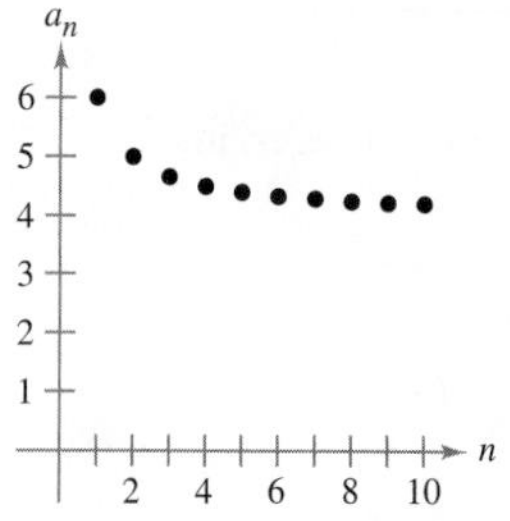

(b)

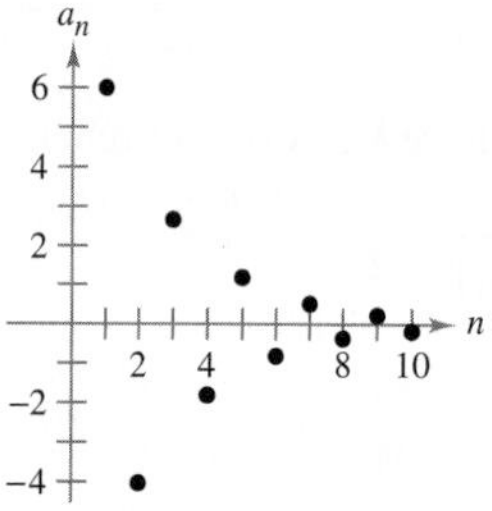

(c)

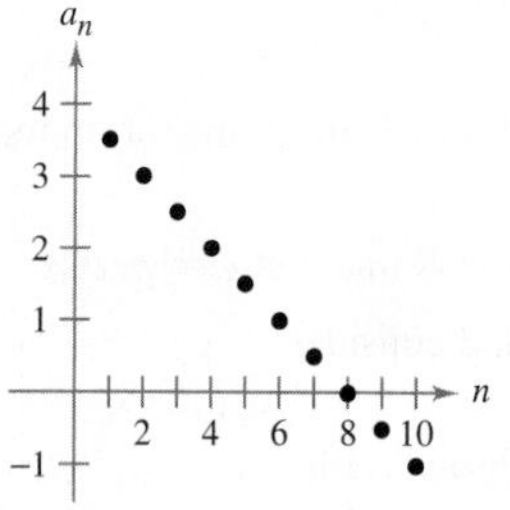

(d)

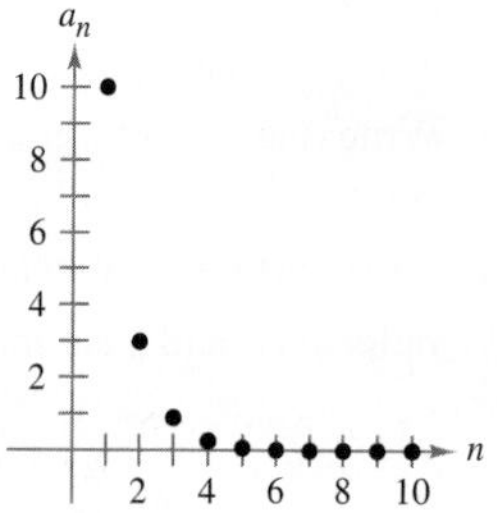

3. $a_n = 4 + \frac{2}{n}$

4. $a_n = 4 - \frac{1}{2}n$

5. $a_n = 10(0.3)^{n-1}$

6. $a_n = 6\left(-\frac{2}{3}\right)^{n-1}$

In Exercises 7 and 8, use a graphing utility to graph the first 10 terms of the sequence. Use the graph to make an inference about the convergence or divergence of the sequence. Verify your inference analytically and, if the sequence converges, find its limit.

7. $a_n = \frac{5n + 2}{n}$

8. $a_n = \sin \frac{n\pi}{2}$

In Exercises 9–16, determine the convergence or divergence of the sequence with the given nth term. If the sequence converges, find its limit. (b and c are positive real numbers.)

9. $a_n = \frac{n + 1}{n^2}$

10. $a_n = \frac{1}{\sqrt{n}}$

11. $a_n = \frac{n^3}{n^2 + 1}$

12. $a_n = \frac{n}{\ln n}$

13. $a_n = \sqrt{n + 1} - \sqrt{n}$

14. $a_n = \left(1 + \frac{1}{2n}\right)^n$

15. $a_n = \frac{\sin \sqrt{n}}{\sqrt{n}}$

16. $a_n = (b^n + c^n)^{1/n}$

17. ***Compound Interest*** A deposit of \$5000 is made in an account that earns 5% interest compounded quarterly. The balance in the account after n quarters is

$$A_n = 5000\left(1 + \frac{0.05}{4}\right)^n, \quad n = 1, 2, 3, \dots.$$

(a) Compute the first eight terms of the sequence $\{A_n\}$.

(b) Find the balance in the account after 10 years by computing the 40th term of the sequence.

18. ***Depreciation*** A company buys a machine for \$120,000. During the next 5 years the machine will depreciate at a rate of 30% per year. (That is, at the end of each year, the depreciated value will be 70% of what it was at the beginning of the year.)

(a) Find a formula for the nth term of the sequence that gives the value V of the machine t full years after it was purchased.

(b) Find the depreciated value of the machine at the end of 5 full years.

Numerical, Graphical, and Analytic Analysis **In Exercises 19–22, (a) use a graphing utility to find the indicated partial sum S_k and complete the table, and (b) use a graphing utility to graph the first 10 terms of the sequence of partial sums.**

k	5	10	15	20	25
S_k					

19. $\sum_{n=1}^{\infty} \left(\frac{3}{2}\right)^{n-1}$

20. $\sum_{n=1}^{\infty} \frac{(-1)^{n+1}}{2n}$

21. $\sum_{n=1}^{\infty} \frac{(-1)^{n+1}}{(2n)!}$

22. $\sum_{n=1}^{\infty} \frac{1}{n(n + 1)}$

In Exercises 23–26, determine the convergence or divergence of the series.

23. $\sum_{n=0}^{\infty} (0.82)^n$

24. $\sum_{n=0}^{\infty} (1.82)^n$

25. $\sum_{n=1}^{\infty} \frac{(-1)^n n}{\ln n}$

26. $\sum_{n=0}^{\infty} \frac{2n + 1}{3n + 2}$

In Exercises 27–30, find the sum of the convergent series.

27. $\sum_{n=0}^{\infty} \left(\frac{2}{3}\right)^n$

28. $\sum_{n=0}^{\infty} \frac{2^{n+2}}{3^n}$

29. $\sum_{n=0}^{\infty} \left(\frac{1}{2^n} - \frac{1}{3^n}\right)$

30. $\sum_{n=0}^{\infty} \left[\left(\frac{2}{3}\right)^n - \frac{1}{(n + 1)(n + 2)}\right]$

In Exercises 31 and 32, (a) write the repeating decimal as a geometric series and (b) write its sum as the ratio of two integers.

31. $0.\overline{09}$ **32.** $0.\overline{923076}$

33. *Distance* A ball is dropped from a height of 8 meters. Each time it drops h meters, it rebounds $0.7h$ meters. Find the total distance traveled by the ball.

34. *Salary* You accept a job that pays a salary of \$32,000 the first year. During the next 39 years, you will receive a 5.5% raise each year. What would be your total compensation over the 40-year period?

35. *Compound Interest* A deposit of \$200 is made at the end of each month for 2 years in an account that pays 6% interest, compounded continuously. Determine the balance in the account at the end of 2 years.

36. *Compound Interest* A deposit of \$100 is made at the end of each month for 10 years in an account that pays 3.5%, compounded monthly. Determine the balance in the account at the end of 10 years.

In Exercises 37–40, determine the convergence or divergence of the series.

37. $\displaystyle\sum_{n=1}^{\infty} \frac{\ln n}{n^4}$ **38.** $\displaystyle\sum_{n=1}^{\infty} \frac{1}{\sqrt[4]{n^3}}$

39. $\displaystyle\sum_{n=1}^{\infty} \left(\frac{1}{n^2} - \frac{1}{n}\right)$ **40.** $\displaystyle\sum_{n=1}^{\infty} \left(\frac{1}{n^2} - \frac{1}{2^n}\right)$

In Exercises 41–44, determine the convergence or divergence of the series.

41. $\displaystyle\sum_{n=1}^{\infty} \frac{1}{\sqrt{n^3 + 2n}}$ **42.** $\displaystyle\sum_{n=1}^{\infty} \frac{n + 1}{n(n + 2)}$

43. $\displaystyle\sum_{n=1}^{\infty} \frac{1 \cdot 3 \cdot 5 \cdots (2n - 1)}{2 \cdot 4 \cdot 6 \cdots (2n)}$ **44.** $\displaystyle\sum_{n=1}^{\infty} \frac{1}{3^n - 5}$

In Exercises 45–48, determine the convergence or divergence of the series.

45. $\displaystyle\sum_{n=2}^{\infty} \frac{(-1)^n n}{n^2 - 3}$ **46.** $\displaystyle\sum_{n=1}^{\infty} \frac{(-1)^n \sqrt{n}}{n + 1}$

47. $\displaystyle\sum_{n=4}^{\infty} \frac{(-1)^n n}{n - 3}$ **48.** $\displaystyle\sum_{n=2}^{\infty} \frac{(-1)^n \ln n^3}{n}$

In Exercises 49–52, determine the convergence or divergence of the series.

49. $\displaystyle\sum_{n=1}^{\infty} \frac{n}{e^{n^2}}$

50. $\displaystyle\sum_{n=1}^{\infty} \frac{n!}{e^n}$

51. $\displaystyle\sum_{n=1}^{\infty} \frac{2^n}{n^3}$

52. $\displaystyle\sum_{n=1}^{\infty} \frac{1 \cdot 3 \cdot 5 \cdots (2n - 1)}{2 \cdot 5 \cdot 8 \cdots (3n - 1)}$

Numerical, Graphical, and Analytic Analysis **In Exercises 53 and 54, (a) verify that the series converges, (b) use a graphing utility to find the indicated partial sum S_n and complete the table, (c) use a graphing utility to graph the first 10 terms of the sequence of partial sums, and (d) use the table to estimate the sum of the series.**

n	5	10	15	20	25
S_n					

53. $\displaystyle\sum_{n=1}^{\infty} n\left(\frac{3}{5}\right)^n$ **54.** $\displaystyle\sum_{n=1}^{\infty} \frac{(-1)^{n-1} n}{n^3 + 5}$

55. *Writing* Use a graphing utility to complete the table for (a) $p = 2$ and (b) $p = 5$. Write a short paragraph describing and comparing the entries in the table.

N	5	10	20	30	40
$\displaystyle\sum_{n=1}^{N} \frac{1}{n^p}$					
$\displaystyle\int_N^{\infty} \frac{1}{x^p}\,dx$					

56. *Writing* You are told that the terms of a positive series appear to approach zero very slowly as n approaches infinity. (In fact, $a_{75} = 0.7$.) If you are given no other information, can you conclude that the series diverges? Support your answer with an example.

In Exercises 57 and 58, find the third-degree Taylor polynomial centered at c.

57. $f(x) = e^{-x/2}, \quad c = 0$

58. $f(x) = \tan x, \quad c = -\dfrac{\pi}{4}$

In Exercises 59–62, use a Taylor polynomial to approximate the function with an error of less than 0.001.

59. $\sin 95°$ **60.** $\cos(0.75)$

61. $\ln(1.75)$ **62.** $e^{-0.25}$

63. A Taylor polynomial centered at 0 will be used to approximate the cosine function. Find the degree of the polynomial required to obtain the desired accuracy over each interval.

	Maximum Error	*Interval*
(a)	0.001	$[-0.5, 0.5]$
(b)	0.001	$[-1, 1]$
(c)	0.0001	$[-0.5, 0.5]$
(d)	0.0001	$[-2, 2]$

64. Use a graphing utility to graph the cosine function and the Taylor polynomials in Exercise 63.

In Exercises 65–70, find the interval of convergence of the power series. (Be sure to include a check for convergence at the endpoints of the interval.)

65. $\sum_{n=0}^{\infty} \left(\frac{x}{10}\right)^n$

66. $\sum_{n=0}^{\infty} (2x)^n$

67. $\sum_{n=0}^{\infty} \frac{(-1)^n (x-2)^n}{(n+1)^2}$

68. $\sum_{n=1}^{\infty} \frac{3^n (x-2)^n}{n}$

69. $\sum_{n=0}^{\infty} n!(x-2)^n$

70. $\sum_{n=0}^{\infty} \frac{(x-2)^n}{2^n}$

In Exercises 71 and 72, show that the function represented by the power series is a solution of the differential equation.

71. $y = \sum_{n=0}^{\infty} (-1)^n \frac{x^{2n}}{4^n (n!)^2}$

$x^2y'' + xy' + x^2y = 0$

72. $y = \sum_{n=0}^{\infty} \frac{(-3)^n x^{2n}}{2^n n!}$

$y'' + 3xy' + 3y = 0$

In Exercises 73 and 74, find a geometric power series centered at 0 for the function.

73. $g(x) = \frac{2}{3-x}$

74. $h(x) = \frac{3}{2+x}$

75. Find a power series for the derivative of the function in Exercise 73.

76. Find a power series for the integral of the function in Exercise 74.

In Exercises 77 and 78, find a function represented by the series and give the domain of the function.

77. $1 + \frac{2}{3}x + \frac{4}{9}x^2 + \frac{8}{27}x^3 + \cdots$

78. $8 - 2(x-3) + \frac{1}{2}(x-3)^2 - \frac{1}{8}(x-3)^3 + \cdots$

In Exercises 79–86, find a power series for the function centered at *c*.

79. $f(x) = \sin x, \quad c = \frac{3\pi}{4}$

80. $f(x) = \cos x, \quad c = -\frac{\pi}{4}$

81. $f(x) = 3^x, \quad c = 0$

82. $f(x) = \csc x, \quad c = \frac{\pi}{2}$

(first three terms)

83. $f(x) = \frac{1}{x}, \quad c = -1$

84. $f(x) = \sqrt{x}, \quad c = 4$

85. $g(x) = \sqrt[5]{1+x}, \quad c = 0$

86. $h(x) = \frac{1}{(1+x)^3}, \quad c = 0$

In Exercises 87–92, find the sum of the convergent series by using a well-known function. Identify the function and explain how you obtained the sum.

87. $\sum_{n=1}^{\infty} (-1)^{n+1} \frac{1}{4^n n}$

88. $\sum_{n=1}^{\infty} (-1)^{n+1} \frac{1}{5^n n}$

89. $\sum_{n=0}^{\infty} \frac{1}{2^n n!}$

90. $\sum_{n=0}^{\infty} \frac{2^n}{3^n n!}$

91. $\sum_{n=0}^{\infty} (-1)^n \frac{2^{2n}}{3^{2n} (2n)!}$

92. $\sum_{n=0}^{\infty} (-1)^n \frac{1}{3^{2n+1} (2n+1)!}$

93. ***Writing*** One of the series in Exercises 41 and 49 converges to its sum at a much lower rate than the other series. Which is it? Explain why this series converges so slowly. Use a graphing utility to illustrate the rate of convergence.

94. Use the binomial series to find the Maclaurin series for

$$f(x) = \frac{1}{\sqrt{1+x^3}}.$$

95. ***Forming Maclaurin Series*** Determine the first four terms of the Maclaurin series for e^{2x}

(a) by using the definition of the Maclaurin series and the formula for the coefficient of the nth term, $a_n = f^{(n)}(0)/n!$.

(b) by replacing x by $2x$ in the series for e^x.

(c) by multiplying the series for e^x by itself, because $e^{2x} = e^x \cdot e^x$.

96. ***Forming Maclaurin Series*** Follow the pattern of Exercise 95 to find the first four terms of the series for $\sin 2x$. (*Hint:* $\sin 2x = 2 \sin x \cos x$.)

In Exercises 97–100, find the series representation of the function defined by the integral.

97. $\int_0^x \frac{\sin t}{t}\, dt$

98. $\int_0^x \cos \frac{\sqrt{t}}{2}\, dt$

99. $\int_0^x \frac{\ln(t+1)}{t}\, dt$

100. $\int_0^x \frac{e^t - 1}{t}\, dt$

In Exercises 101 and 102, use a power series to find the limit (if it exists). Verify the result by using L'Hôpital's Rule.

101. $\lim_{x \to 0^+} \frac{\arctan x}{\sqrt{x}}$

102. $\lim_{x \to 0} \frac{\arcsin x}{x}$

P.S. Problem Solving

See www.CalcChat.com for worked-out solutions to odd-numbered exercises.

1. The Cantor set (Georg Cantor, 1845–1918) is a subset of the unit interval $[0, 1]$. To construct the Cantor set, first remove the middle third $\left(\frac{1}{3}, \frac{2}{3}\right)$ of the interval, leaving two line segments. For the second step, remove the middle third of each of the two remaining segments, leaving four line segments. Continue this procedure indefinitely, as shown in the figure. The Cantor set consists of all numbers in the unit interval $[0, 1]$ that still remain.

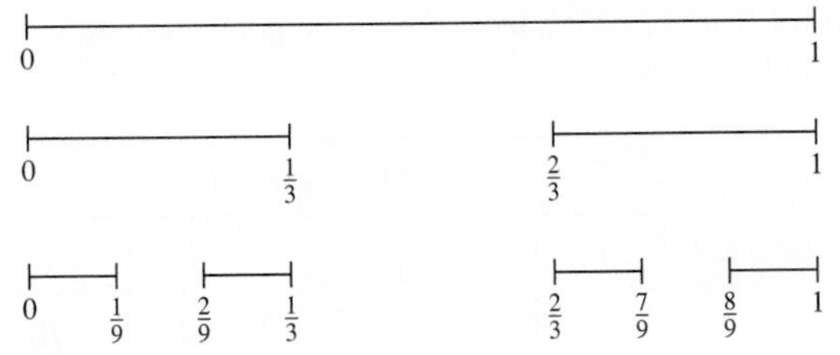

(a) Find the total length of all the line segments that are removed.

(b) Write down three numbers that are in the Cantor set.

(c) Let C_n denote the total length of the remaining line segments after n steps. Find $\lim_{n\to\infty} C_n$.

The Granger Collection

GEORG CANTOR (1845–1918)

Cantor was a German mathematician known for his work on the development of set theory, which is the basis of modern mathematical analysis. This theory extends to the concept of infinite (or transfinite) numbers.

2. It can be shown that

$$\sum_{n=1}^{\infty} \frac{1}{n^2} = \frac{\pi^2}{6} \text{ [see Example 3(b), Section 9.3].}$$

Use this fact to show that $\displaystyle\sum_{n=1}^{\infty} \frac{1}{(2n-1)^2} = \frac{\pi^2}{8}$.

3. Let T be an equilateral triangle with sides of length 1. Let a_n be the number of circles that can be packed tightly in n rows inside the triangle. For example, $a_1 = 1$, $a_2 = 3$, and $a_3 = 6$, as shown in the figure. Let A_n be the combined area of the a_n circles. Find $\lim_{n\to\infty} A_n$.

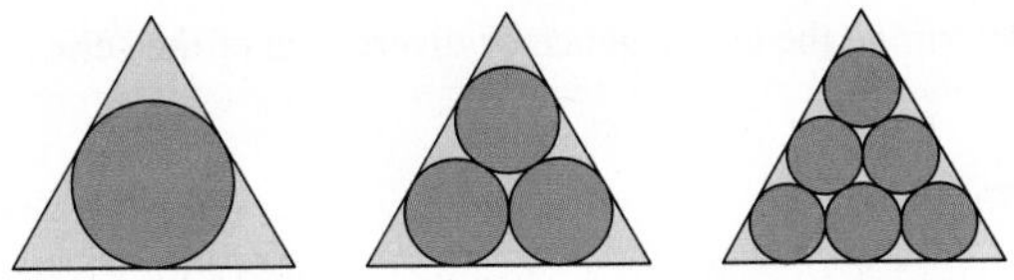

4. Identical blocks of unit length are stacked on top of each other at the edge of a table. The center of gravity of the top block must lie over the block below it, the center of gravity of the top two blocks must lie over the block below them, and so on (see figure).

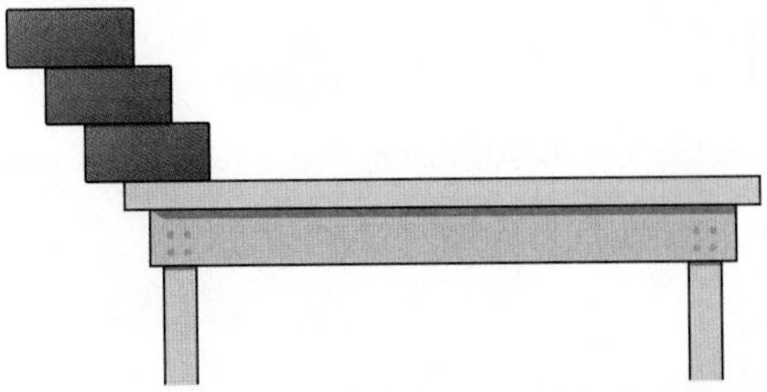

(a) If there are three blocks, show that it is possible to stack them so that the left edge of the top block extends $\frac{11}{12}$ unit beyond the edge of the table.

(b) Is it possible to stack the blocks so that the right edge of the top block extends beyond the edge of the table?

(c) How far beyond the table can the blocks be stacked?

5. (a) Consider the power series

$$\sum_{n=0}^{\infty} a_n x^n = 1 + 2x + 3x^2 + x^3 + 2x^4 + 3x^5 + x^6 + \cdots$$

in which the coefficients $a_n = 1, 2, 3, 1, 2, 3, 1, \ldots$ are periodic of period $p = 3$. Find the radius of convergence and the sum of this power series.

(b) Consider a power series

$$\sum_{n=0}^{\infty} a_n x^n$$

in which the coefficients are periodic, $(a_{n+p} = a_p)$ and $a_n > 0$. Find the radius of convergence and the sum of this power series.

6. For what values of the positive constants a and b does the following series converge absolutely? For what values does it converge conditionally?

$$a - \frac{b}{2} + \frac{a}{3} - \frac{b}{4} + \frac{a}{5} - \frac{b}{6} + \frac{a}{7} - \frac{b}{8} + \cdots$$

7. (a) Find a power series for the function

$$f(x) = xe^x$$

centered at 0. Use this representation to find the sum of the infinite series

$$\sum_{n=1}^{\infty} \frac{1}{n!(n+2)}.$$

(b) Differentiate the power series for $f(x) = xe^x$. Use the result to find the sum of the infinite series

$$\sum_{n=0}^{\infty} \frac{n+1}{n!}.$$

8. Find $f^{(12)}(0)$ if $f(x) = e^{x^2}$. (*Hint:* Do not calculate 12 derivatives.)

9. The graph of the function

$$f(x) = \begin{cases} 1, & x = 0 \\ \dfrac{\sin x}{x}, & x > 0 \end{cases}$$

is shown below. Use the Alternating Series Test to show that the improper integral $\int_1^{\infty} f(x)\,dx$ converges.

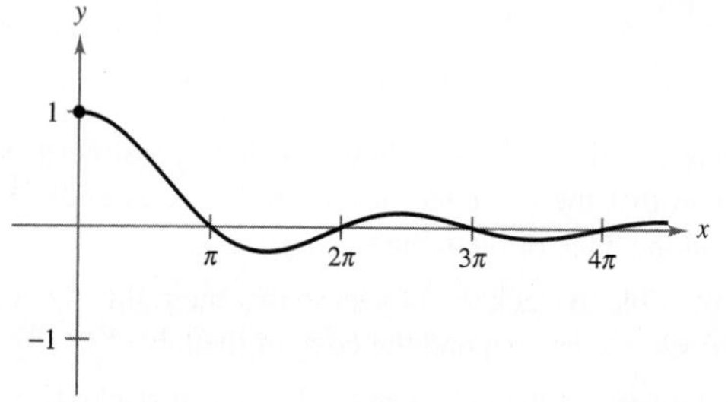

10. (a) Prove that $\displaystyle\int_2^{\infty} \frac{1}{x(\ln x)^p}\,dx$ converges if and only if $p > 1$.

(b) Determine the convergence or divergence of the series

$$\sum_{n=4}^{\infty} \frac{1}{n \ln(n^2)}.$$

11. (a) Consider the following sequence of numbers defined recursively.

$$a_1 = 3$$
$$a_2 = \sqrt{3}$$
$$a_3 = \sqrt{3 + \sqrt{3}}$$
$$\vdots$$
$$a_{n+1} = \sqrt{3 + a_n}$$

Write the decimal approximations for the first six terms of this sequence. Prove that the sequence converges and find its limit.

(b) Consider the following sequence defined recursively by $a_1 = \sqrt{a}$ and $a_{n+1} = \sqrt{a + a_n}$, where $a > 2$.

$$\sqrt{a},\ \sqrt{a + \sqrt{a}},\ \sqrt{a + \sqrt{a + \sqrt{a}}}, \ldots$$

Prove that this sequence converges and find its limit.

12. Let $\{a_n\}$ be a sequence of positive numbers satisfying $\lim_{n\to\infty} (a_n)^{1/n} = L < \dfrac{1}{r}$, $r > 0$. Prove that the series $\sum_{n=1}^{\infty} a_n r^n$ converges.

13. Consider the infinite series $\displaystyle\sum_{n=1}^{\infty} \frac{1}{2^{n+(-1)^n}}$.

(a) Find the first five terms of the sequence of partial sums.

(b) Show that the Ratio Test is inconclusive for this series.

(c) Use the Root Test to test for the convergence or divergence of this series.

14. Derive each identity using the appropriate geometric series.

(a) $\dfrac{1}{0.99} = 1.01010101\ldots$ (b) $\dfrac{1}{0.98} = 1.0204081632\ldots$

15. Consider an idealized population with the characteristic that each member of the population produces one offspring at the end of every time period. Each member has a life span of three time periods and the population begins with 10 newborn members. The following table shows the population during the first five time periods.

	Time Period				
Age Bracket	**1**	**2**	**3**	**4**	**5**
0–1	10	10	20	40	70
1–2		10	10	20	40
2–3			10	10	20
Total	10	20	40	70	130

The sequence for the total population has the property that

$$S_n = S_{n-1} + S_{n-2} + S_{n-3}, \qquad n > 3.$$

Find the total population during each of the next five time periods.

16. Imagine you are stacking an infinite number of spheres of decreasing radii on top of each other, as shown in the figure. The radii of the spheres are 1 meter, $1/\sqrt{2}$ meter, $1/\sqrt{3}$ meter, etc. The spheres are made of a material that weighs 1 newton per cubic meter.

(a) How high is this infinite stack of spheres?

(b) What is the total surface area of all the spheres in the stack?

(c) Show that the weight of the stack is finite.

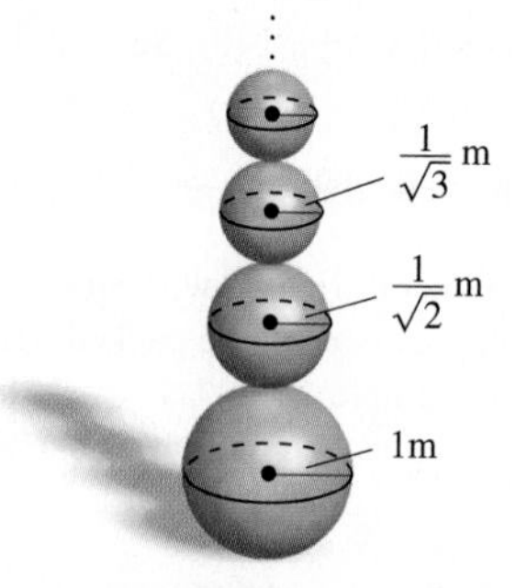

17. (a) Determine the convergence or divergence of the series

$$\sum_{n=1}^{\infty} \frac{1}{2n}.$$

(b) Determine the convergence or divergence of the series

$$\sum_{n=1}^{\infty} \left(\sin\frac{1}{2n} - \sin\frac{1}{2n+1}\right).$$

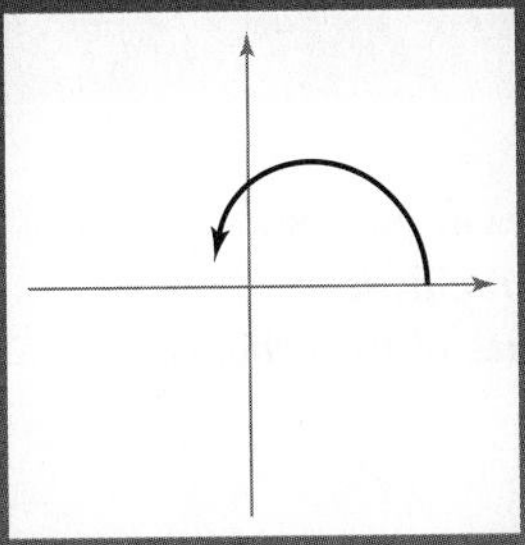

10 Conics, Parametric Equations, and Polar Coordinates

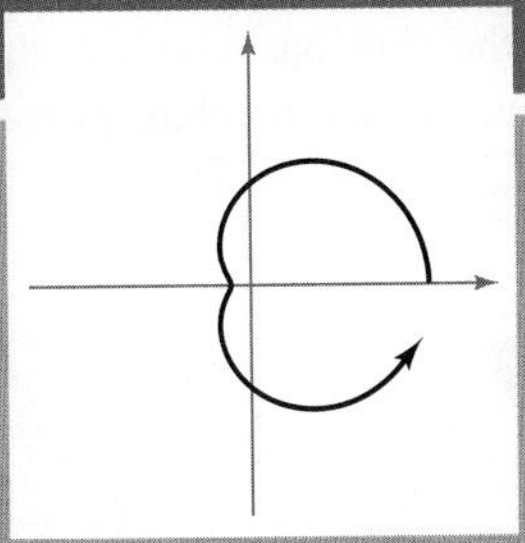

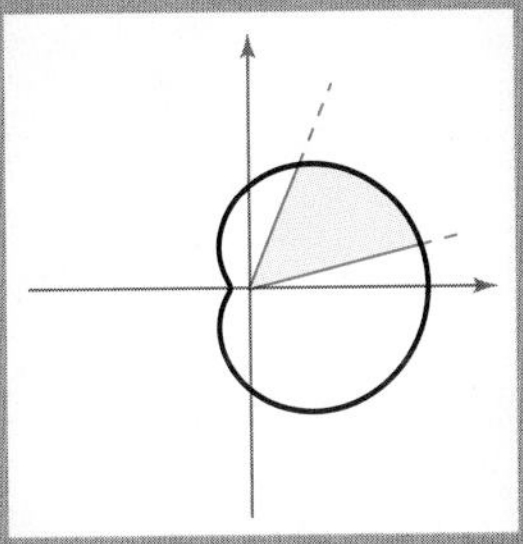

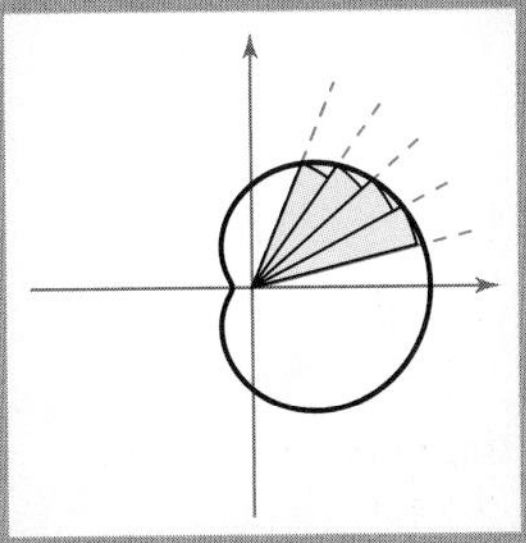

In the polar coordinate system, graphing an equation involves tracing a curve about a fixed point called the pole. Consider a region bounded by a curve and by the rays that contain the endpoints of an interval on the curve. You can use sectors of circles to approximate the area of such a region. In Chapter 10, you will see how the limit process can be used to find this area.

During the 2002 Winter Olympic Games, the Olympic rings were lighted high on a mountainside in Salt Lake City. The volunteers who installed the lights for the display took care to minimize the environmental impact of the project. How can you calculate the area enclosed by the display? Explain.

Harry Howe/Getty Images

Section 10.1 Conics and Calculus

- **Understand the definition of a conic section.**
- **Analyze and write equations of parabolas using properties of parabolas.**
- **Analyze and write equations of ellipses using properties of ellipses.**
- **Analyze and write equations of hyperbolas using properties of hyperbolas.**

Bettmann/Corbis

HYPATIA (370–415 A.D.)

The Greeks discovered conic sections sometime between 600 and 300 B.C. By the beginning of the Alexandrian period, enough was known about conics for Apollonius (262–190 B.C.) to produce an eight-volume work on the subject. Later, toward the end of the Alexandrian period, Hypatia wrote a textbook entitled *On the Conics of Apollonius*. Her death marked the end of major mathematical discoveries in Europe for several hundred years.

The early Greeks were largely concerned with the geometric properties of conics. It was not until 1900 years later, in the early seventeenth century, that the broader applicability of conics became apparent. Conics then played a prominent role in the development of calculus.

Conic Sections

Each **conic section** (or simply **conic**) can be described as the intersection of a plane and a double-napped cone. Notice in Figure 10.1 that for the four basic conics, the intersecting plane does not pass through the vertex of the cone. When the plane passes through the vertex, the resulting figure is a **degenerate conic,** as shown in Figure 10.2.

Circle Parabola Ellipse Hyperbola

Conic sections

Figure 10.1

Point Line Two intersecting lines

Degenerate conics

Figure 10.2

There are several ways to study conics. You could begin as the Greeks did by defining the conics in terms of the intersections of planes and cones, or you could define them algebraically in terms of the general second-degree equation

$$Ax^2 + Bxy + Cy^2 + Dx + Ey + F = 0. \qquad \text{General second-degree equation}$$

However, a third approach, in which each of the conics is defined as a **locus** (collection) of points satisfying a certain geometric property, works best. For example, a circle can be defined as the collection of all points (x, y) that are equidistant from a fixed point (h, k). This locus definition easily produces the standard equation of a circle

$$(x - h)^2 + (y - k)^2 = r^2. \qquad \text{Standard equation of a circle}$$

FOR FURTHER INFORMATION To learn more about the mathematical activities of Hypatia, see the article "Hypatia and Her Mathematics" by Michael A. B. Deakin in *The American Mathematical Monthly*. To view this article, go to the website *www.matharticles.com*.

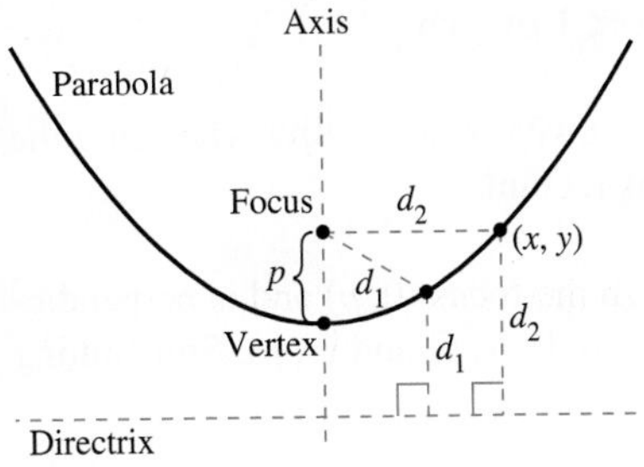

Figure 10.3

Parabolas

A **parabola** is the set of all points (x, y) that are equidistant from a fixed line called the **directrix** and a fixed point called the **focus** not on the line. The midpoint between the focus and the directrix is the **vertex,** and the line passing through the focus and the vertex is the **axis** of the parabola. Note in Figure 10.3 that a parabola is symmetric with respect to its axis.

THEOREM 10.1 Standard Equation of a Parabola

The **standard form** of the equation of a parabola with vertex (h, k) and directrix $y = k - p$ is

$$(x - h)^2 = 4p(y - k). \qquad \text{Vertical axis}$$

For directrix $x = h - p$, the equation is

$$(y - k)^2 = 4p(x - h). \qquad \text{Horizontal axis}$$

The focus lies on the axis p units (*directed distance*) from the vertex. The coordinates of the focus are as follows.

$$(h, k + p) \qquad \text{Vertical axis}$$
$$(h + p, k) \qquad \text{Horizontal axis}$$

EXAMPLE 1 Finding the Focus of a Parabola

Find the focus of the parabola given by $y = -\frac{1}{2}x^2 - x + \frac{1}{2}$.

Solution To find the focus, convert to standard form by completing the square.

$$y = \tfrac{1}{2} - x - \tfrac{1}{2}x^2 \qquad \text{Rewrite original equation.}$$
$$y = \tfrac{1}{2}(1 - 2x - x^2) \qquad \text{Factor out } \tfrac{1}{2}.$$
$$2y = 1 - 2x - x^2 \qquad \text{Multiply each side by 2.}$$
$$2y = 1 - (x^2 + 2x) \qquad \text{Group terms.}$$
$$2y = 2 - (x^2 + 2x + 1) \qquad \text{Add and subtract 1 on right side.}$$
$$x^2 + 2x + 1 = -2y + 2$$
$$(x + 1)^2 = -2(y - 1) \qquad \text{Write in standard form.}$$

Comparing this equation with $(x - h)^2 = 4p(y - k)$, you can conclude that

$$h = -1, \qquad k = 1, \qquad \text{and} \qquad p = -\tfrac{1}{2}.$$

Because p is negative, the parabola opens downward, as shown in Figure 10.4. So, the focus of the parabola is p units from the vertex, or

$$(h, k + p) = \left(-1, \tfrac{1}{2}\right). \qquad \text{Focus}$$

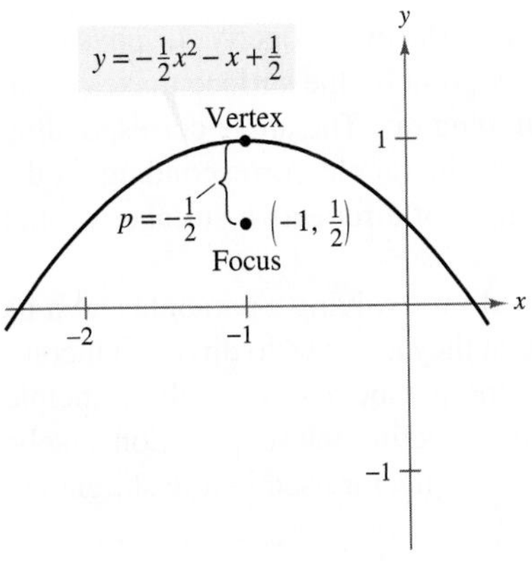

Parabola with a vertical axis, $p < 0$
Figure 10.4

A line segment that passes through the focus of a parabola and has endpoints on the parabola is called a **focal chord.** The specific focal chord perpendicular to the axis of the parabola is the **latus rectum.** The next example shows how to determine the length of the latus rectum and the length of the corresponding intercepted arc.

EXAMPLE 2 Focal Chord Length and Arc Length

Find the length of the latus rectum of the parabola given by $x^2 = 4py$. Then find the length of the parabolic arc intercepted by the latus rectum.

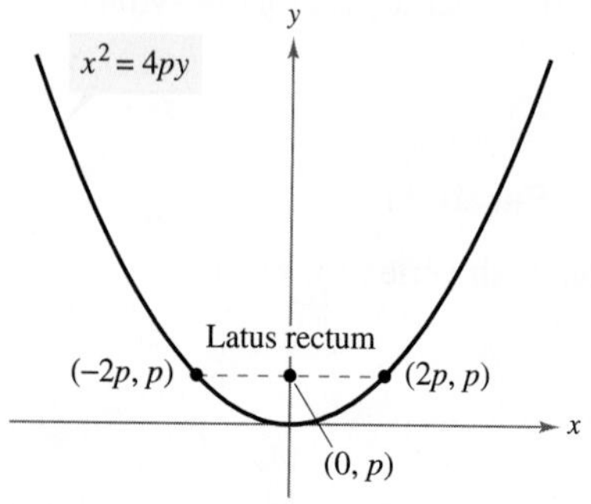

Length of latus rectum: $4p$
Figure 10.5

Solution Because the latus rectum passes through the focus $(0, p)$ and is perpendicular to the y-axis, the coordinates of its endpoints are $(-x, p)$ and (x, p). Substituting p for y in the equation of the parabola produces

$$x^2 = 4p(p) \Longrightarrow x = \pm 2p.$$

So, the endpoints of the latus rectum are $(-2p, p)$ and $(2p, p)$, and you can conclude that its length is $4p$, as shown in Figure 10.5. In contrast, the length of the intercepted arc is

$$s = \int_{-2p}^{2p} \sqrt{1 + (y')^2}\, dx \qquad \text{Use arc length formula.}$$

$$= 2\int_0^{2p} \sqrt{1 + \left(\frac{x}{2p}\right)^2}\, dx \qquad y = \frac{x^2}{4p} \Longrightarrow y' = \frac{x}{2p}$$

$$= \frac{1}{p}\int_0^{2p} \sqrt{4p^2 + x^2}\, dx \qquad \text{Simplify.}$$

$$= \frac{1}{2p}\left[x\sqrt{4p^2 + x^2} + 4p^2 \ln\left|x + \sqrt{4p^2 + x^2}\right|\right]_0^{2p} \qquad \text{Theorem 8.2}$$

$$= \frac{1}{2p}\left[2p\sqrt{8p^2} + 4p^2 \ln\left(2p + \sqrt{8p^2}\right) - 4p^2 \ln(2p)\right]$$

$$= 2p\left[\sqrt{2} + \ln\left(1 + \sqrt{2}\right)\right]$$

$$\approx 4.59p.$$

One widely used property of a parabola is its reflective property. In physics, a surface is called **reflective** if the tangent line at any point on the surface makes equal angles with an incoming ray and the resulting outgoing ray. The angle corresponding to the incoming ray is the **angle of incidence,** and the angle corresponding to the outgoing ray is the **angle of reflection.** One example of a reflective surface is a flat mirror.

Another type of reflective surface is that formed by revolving a parabola about its axis. A special property of parabolic reflectors is that they allow us to direct all incoming rays parallel to the axis through the focus of the parabola—this is the principle behind the design of the parabolic mirrors used in reflecting telescopes. Conversely, all light rays emanating from the focus of a parabolic reflector used in a flashlight are parallel, as shown in Figure 10.6.

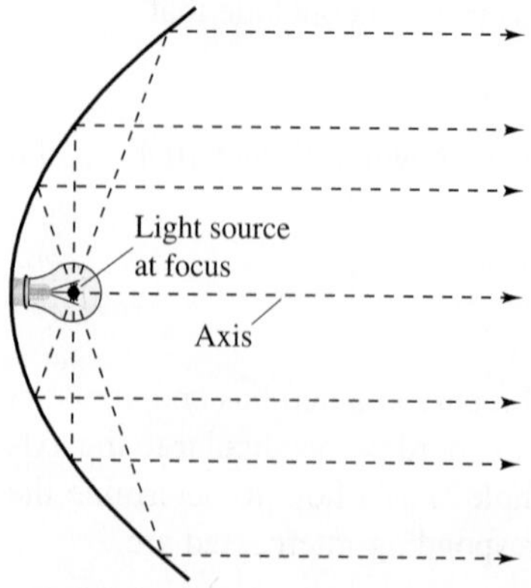

Parabolic reflector: light is reflected in parallel rays.
Figure 10.6

THEOREM 10.2 Reflective Property of a Parabola

Let P be a point on a parabola. The tangent line to the parabola at the point P makes equal angles with the following two lines.

1. The line passing through P and the focus
2. The line passing through P parallel to the axis of the parabola

indicates that in the HM mathSpace® *CD-ROM and the online* Eduspace® *system for this text, you will find an Open Exploration, which further explores this example using the computer algebra systems* Maple, Mathcad, Mathematica, *and* Derive.

NICOLAUS COPERNICUS (1473–1543)

Copernicus began to study planetary motion when asked to revise the calendar. At that time, the exact length of the year could not be accurately predicted using the theory that Earth was the center of the universe.

Ellipses

More than a thousand years after the close of the Alexandrian period of Greek mathematics, Western civilization finally began a Renaissance of mathematical and scientific discovery. One of the principal figures in this rebirth was the Polish astronomer Nicolaus Copernicus. In his work *On the Revolutions of the Heavenly Spheres*, Copernicus claimed that all of the planets, including Earth, revolved about the sun in circular orbits. Although some of Copernicus's claims were invalid, the controversy set off by his heliocentric theory motivated astronomers to search for a mathematical model to explain the observed movements of the sun and planets. The first to find an accurate model was the German astronomer Johannes Kepler (1571–1630). Kepler discovered that the planets move about the sun in elliptical orbits, with the sun not as the center but as a focal point of the orbit.

The use of ellipses to explain the movements of the planets is only one of many practical and aesthetic uses. As with parabolas, you will begin your study of this second type of conic by defining it as a locus of points. Now, however, *two* focal points are used rather than one.

An **ellipse** is the set of all points (x, y) the sum of whose distances from two distinct fixed points called **foci** is constant. (See Figure 10.7.) The line through the foci intersects the ellipse at two points, called the **vertices.** The chord joining the vertices is the **major axis,** and its midpoint is the **center** of the ellipse. The chord perpendicular to the major axis at the center is the **minor axis** of the ellipse. (See Figure 10.8.)

FOR FURTHER INFORMATION To learn about how an ellipse may be "exploded" into a parabola, see the article "Exploding the Ellipse" by Arnold Good in *Mathematics Teacher*. To view this article, go to the website *www.matharticles.com.*

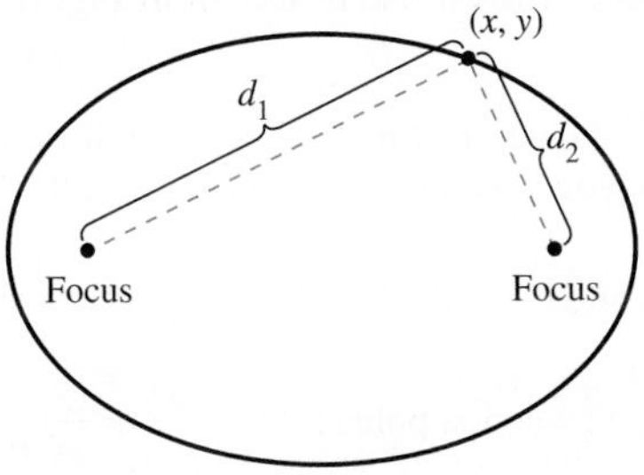

Figure 10.7

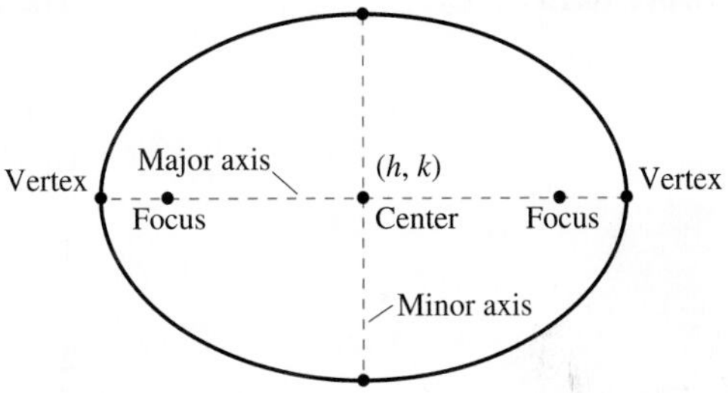

Figure 10.8

THEOREM 10.3 Standard Equation of an Ellipse

The standard form of the equation of an ellipse with center (h, k) and major and minor axes of lengths $2a$ and $2b$, where $a > b$, is

$$\frac{(x-h)^2}{a^2} + \frac{(y-k)^2}{b^2} = 1 \qquad \text{Major axis is horizontal.}$$

or

$$\frac{(x-h)^2}{b^2} + \frac{(y-k)^2}{a^2} = 1. \qquad \text{Major axis is vertical.}$$

The foci lie on the major axis, c units from the center, with $c^2 = a^2 - b^2$.

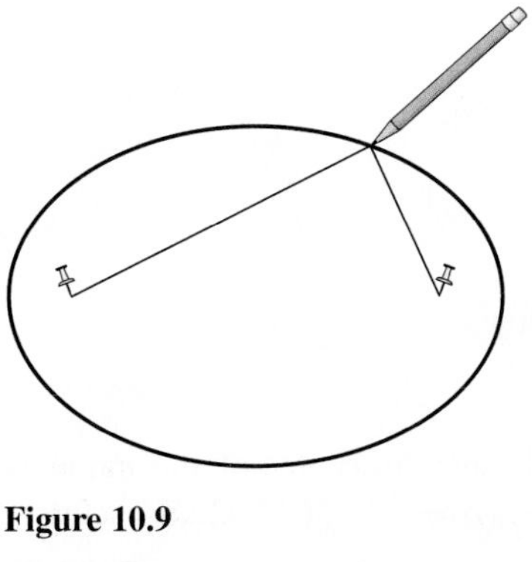

Figure 10.9

NOTE You can visualize the definition of an ellipse by imagining two thumbtacks placed at the foci, as shown in Figure 10.9. If the ends of a fixed length of string are fastened to the thumbtacks and the string is drawn taut with a pencil, the path traced by the pencil will be an ellipse.

EXAMPLE 3 Completing the Square

Find the center, vertices, and foci of the ellipse given by

$$4x^2 + y^2 - 8x + 4y - 8 = 0.$$

Solution By completing the square, you can write the original equation in standard form.

$$4x^2 + y^2 - 8x + 4y - 8 = 0 \qquad \text{Write original equation.}$$
$$4x^2 - 8x + y^2 + 4y = 8$$
$$4(x^2 - 2x + 1) + (y^2 + 4y + 4) = 8 + 4 + 4$$
$$4(x - 1)^2 + (y + 2)^2 = 16$$
$$\frac{(x-1)^2}{4} + \frac{(y+2)^2}{16} = 1 \qquad \text{Write in standard form.}$$

So, the major axis is parallel to the y-axis, where $h = 1$, $k = -2$, $a = 4$, $b = 2$, and $c = \sqrt{16 - 4} = 2\sqrt{3}$. So, you obtain the following.

Center: $(1, -2)$ $\qquad (h, k)$

Vertices: $(1, -6)$ and $(1, 2)$ $\qquad (h, k \pm a)$

Foci: $\left(1, -2 - 2\sqrt{3}\right)$ and $\left(1, -2 + 2\sqrt{3}\right)$ $\qquad (h, k \pm c)$

The graph of the ellipse is shown in Figure 10.10.

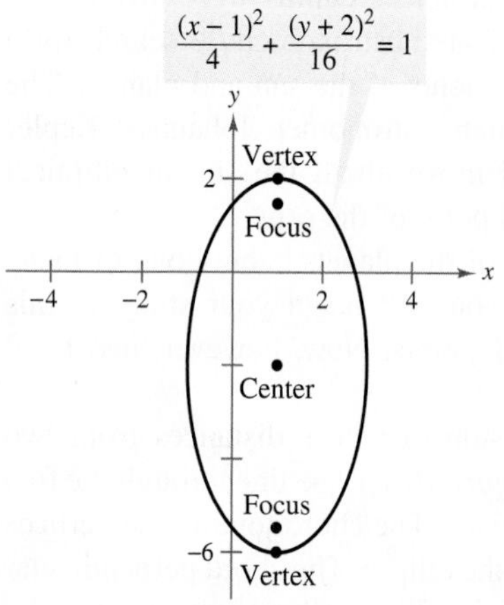

Ellipse with a vertical major axis
Figure 10.10

NOTE If the constant term $F = -8$ in the equation in Example 3 had been greater than or equal to 8, you would have obtained one of the following degenerate cases.

1. $F = 8$, single point, $(1, -2)$: $\dfrac{(x-1)^2}{4} + \dfrac{(y+2)^2}{16} = 0$

2. $F > 8$, no solution points: $\dfrac{(x-1)^2}{4} + \dfrac{(y+2)^2}{16} < 0$

EXAMPLE 4 The Orbit of the Moon

The moon orbits Earth in an elliptical path with the center of Earth at one focus, as shown in Figure 10.11. The major and minor axes of the orbit have lengths of 768,800 kilometers and 767,640 kilometers, respectively. Find the greatest and least distances (the apogee and perigee) from Earth's center to the moon's center.

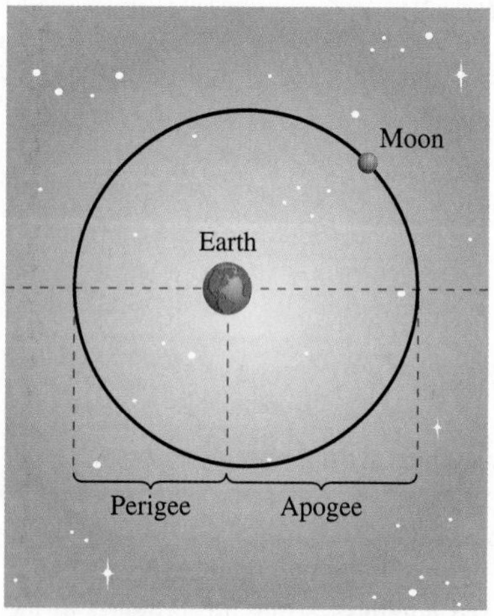

Figure 10.11

Solution Begin by solving for a and b.

$$2a = 768{,}800 \qquad \text{Length of major axis}$$
$$a = 384{,}400 \qquad \text{Solve for } a.$$
$$2b = 767{,}640 \qquad \text{Length of minor axis}$$
$$b = 383{,}820 \qquad \text{Solve for } b.$$

Now, using these values, you can solve for c as follows.

$$c = \sqrt{a^2 - b^2} \approx 21{,}108$$

The greatest distance between the center of Earth and the center of the moon is $a + c \approx 405{,}508$ kilometers, and the least distance is $a - c \approx 363{,}292$ kilometers.

FOR FURTHER INFORMATION For more information on some uses of the reflective properties of conics, see the article "Parabolic Mirrors, Elliptic and Hyperbolic Lenses" by Mohsen Maesumi in *The American Mathematical Monthly*. Also see the article "The Geometry of Microwave Antennas" by William R. Parzynski in *Mathematics Teacher*.

Theorem 10.2 presented a reflective property of parabolas. Ellipses have a similar reflective property. You are asked to prove the following theorem in Exercise 110.

THEOREM 10.4 Reflective Property of an Ellipse

Let P be a point on an ellipse. The tangent line to the ellipse at point P makes equal angles with the lines through P and the foci.

One of the reasons that astronomers had difficulty in detecting that the orbits of the planets are ellipses is that the foci of the planetary orbits are relatively close to the center of the sun, making the orbits nearly circular. To measure the ovalness of an ellipse, you can use the concept of **eccentricity.**

Definition of Eccentricity of an Ellipse

The **eccentricity** e of an ellipse is given by the ratio

$$e = \frac{c}{a}.$$

To see how this ratio is used to describe the shape of an ellipse, note that because the foci of an ellipse are located along the major axis between the vertices and the center, it follows that

$$0 < c < a.$$

For an ellipse that is nearly circular, the foci are close to the center and the ratio c/a is small, and for an elongated ellipse, the foci are close to the vertices and the ratio is close to 1, as shown in Figure 10.12. Note that $0 < e < 1$ for every ellipse.

The orbit of the moon has an eccentricity of $e = 0.0549$, and the eccentricities of the nine planetary orbits are as follows.

Mercury:	$e = 0.2056$	Saturn:	$e = 0.0542$
Venus:	$e = 0.0068$	Uranus:	$e = 0.0472$
Earth:	$e = 0.0167$	Neptune:	$e = 0.0086$
Mars:	$e = 0.0934$	Pluto:	$e = 0.2488$
Jupiter:	$e = 0.0484$		

You can use integration to show that the area of an ellipse is $A = \pi ab$. For instance, the area of the ellipse

$$\frac{x^2}{a^2} + \frac{y^2}{b^2} = 1$$

is given by

$$A = 4\int_0^a \frac{b}{a}\sqrt{a^2 - x^2}\, dx$$

$$= \frac{4b}{a}\int_0^{\pi/2} a^2 \cos^2 \theta \, d\theta. \quad \text{Trigonometric substitution } x = a \sin \theta.$$

However, it is not so simple to find the *circumference* of an ellipse. The next example shows how to use eccentricity to set up an "elliptic integral" for the circumference of an ellipse.

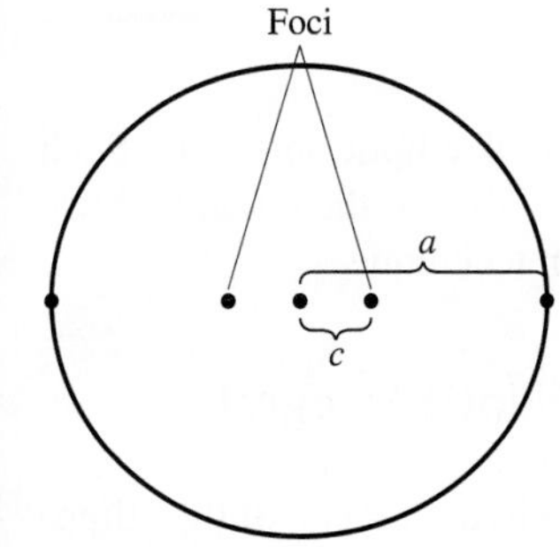

(a) $\frac{c}{a}$ is small.

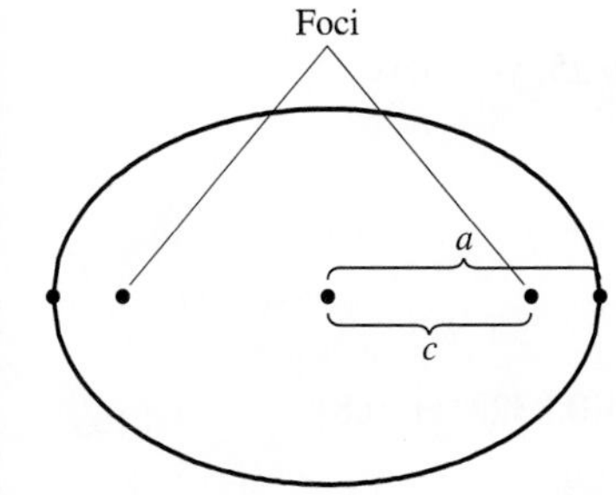

(b) $\frac{c}{a}$ is close to 1.

Eccentricity is the ratio $\frac{c}{a}$.
Figure 10.12

EXAMPLE 5 Finding the Circumference of an Ellipse

Show that the circumference of the ellipse $(x^2/a^2) + (y^2/b^2) = 1$ is

$$4a\int_0^{\pi/2} \sqrt{1 - e^2 \sin^2 \theta}\, d\theta. \qquad e = \frac{c}{a}$$

Solution Because the given ellipse is symmetric with respect to both the x-axis and the y-axis, you know that its circumference C is four times the arc length of $y = (b/a)\sqrt{a^2 - x^2}$ in the first quadrant. The function y is differentiable for all x in the interval $[0, a]$ except at $x = a$. So, the circumference is given by the improper integral

$$C = \lim_{d \to a} 4\int_0^d \sqrt{1 + (y')^2}\, dx = 4\int_0^a \sqrt{1 + (y')^2}\, dx = 4\int_0^a \sqrt{1 + \frac{b^2x^2}{a^2(a^2 - x^2)}}\, dx.$$

Using the trigonometric substitution $x = a \sin \theta$, you obtain

$$\begin{aligned} C &= 4\int_0^{\pi/2} \sqrt{1 + \frac{b^2 \sin^2 \theta}{a^2 \cos^2 \theta}}(a \cos \theta)\, d\theta \\ &= 4\int_0^{\pi/2} \sqrt{a^2 \cos^2 \theta + b^2 \sin^2 \theta}\, d\theta \\ &= 4\int_0^{\pi/2} \sqrt{a^2(1 - \sin^2 \theta) + b^2 \sin^2 \theta}\, d\theta \\ &= 4\int_0^{\pi/2} \sqrt{a^2 - (a^2 - b^2)\sin^2 \theta}\, d\theta. \end{aligned}$$

Because $e^2 = c^2/a^2 = (a^2 - b^2)/a^2$, you can rewrite this integral as

$$C = 4a\int_0^{\pi/2} \sqrt{1 - e^2 \sin^2 \theta}\, d\theta.$$

AREA AND CIRCUMFERENCE OF AN ELLIPSE

In his work with elliptic orbits in the early 1600's, Johannes Kepler successfully developed a formula for the area of an ellipse, $A = \pi ab$. He was less successful in developing a formula for the circumference of an ellipse, however; the best he could do was to give the approximate formula $C = \pi(a + b)$.

A great deal of time has been devoted to the study of elliptic integrals. Such integrals generally do not have elementary antiderivatives. To find the circumference of an ellipse, you must usually resort to an approximation technique.

EXAMPLE 6 Approximating the Value of an Elliptic Integral

Use the elliptic integral in Example 5 to approximate the circumference of the ellipse

$$\frac{x^2}{25} + \frac{y^2}{16} = 1.$$

Solution Because $e^2 = c^2/a^2 = (a^2 - b^2)/a^2 = 9/25$, you have

$$C = (4)(5)\int_0^{\pi/2} \sqrt{1 - \frac{9 \sin^2 \theta}{25}}\, d\theta.$$

Applying Simpson's Rule with $n = 4$ produces

$$\begin{aligned} C &\approx 20\left(\frac{\pi}{6}\right)\left(\frac{1}{4}\right)[1 + 4(0.9733) + 2(0.9055) + 4(0.8323) + 0.8] \\ &\approx 28.36. \end{aligned}$$

So, the ellipse has a circumference of about 28.36 units, as shown in Figure 10.13.

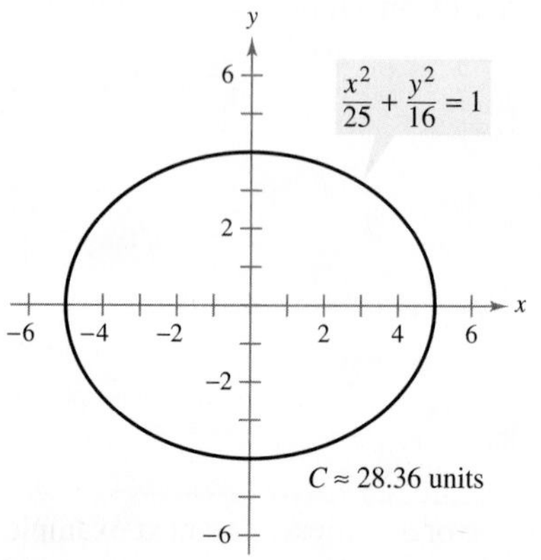

Figure 10.13

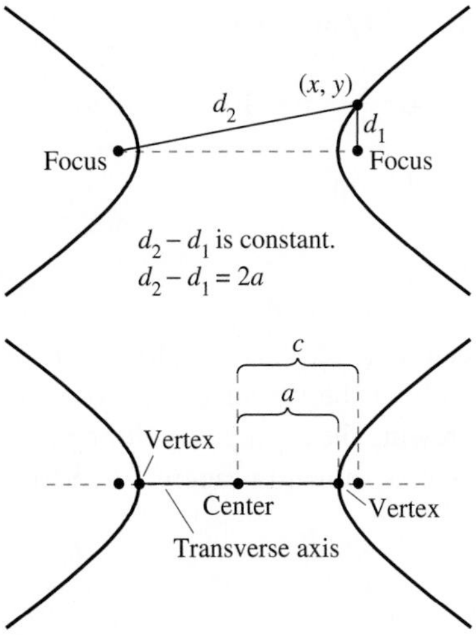

Figure 10.14

Hyperbolas

The definition of a hyperbola is similar to that of an ellipse. For an ellipse, the *sum* of the distances between the foci and a point on the ellipse is fixed, whereas for a hyperbola, the absolute value of the *difference* between these distances is fixed.

A **hyperbola** is the set of all points (x, y) for which the absolute value of the difference between the distances from two distinct fixed points called **foci** is constant. (See Figure 10.14.) The line through the two foci intersects a hyperbola at two points called the **vertices.** The line segment connecting the vertices is the **transverse axis,** and the midpoint of the transverse axis is the **center** of the hyperbola. One distinguishing feature of a hyperbola is that its graph has two separate *branches.*

THEOREM 10.5 Standard Equation of a Hyperbola

The standard form of the equation of a hyperbola with center at (h, k) is

$$\frac{(x-h)^2}{a^2} - \frac{(y-k)^2}{b^2} = 1$$ Transverse axis is horizontal.

or

$$\frac{(y-k)^2}{a^2} - \frac{(x-h)^2}{b^2} = 1.$$ Transverse axis is vertical.

The vertices are a units from the center, and the foci are c units from the center, where, $c^2 = a^2 + b^2$.

NOTE The constants a, b, and c do not have the same relationship for hyperbolas as they do for ellipses. For hyperbolas, $c^2 = a^2 + b^2$, but for ellipses, $c^2 = a^2 - b^2$.

An important aid in sketching the graph of a hyperbola is the determination of its **asymptotes,** as shown in Figure 10.15. Each hyperbola has two asymptotes that intersect at the center of the hyperbola. The asymptotes pass through the vertices of a rectangle of dimensions $2a$ by $2b$, with its center at (h, k). The line segment of length $2b$ joining $(h, k + b)$ and $(h, k - b)$ is referred to as the **conjugate axis** of the hyperbola.

THEOREM 10.6 Asymptotes of a Hyperbola

For a *horizontal* transverse axis, the equations of the asymptotes are

$$y = k + \frac{b}{a}(x - h) \qquad \text{and} \qquad y = k - \frac{b}{a}(x - h).$$

For a *vertical* transverse axis, the equations of the asymptotes are

$$y = k + \frac{a}{b}(x - h) \qquad \text{and} \qquad y = k - \frac{a}{b}(x - h).$$

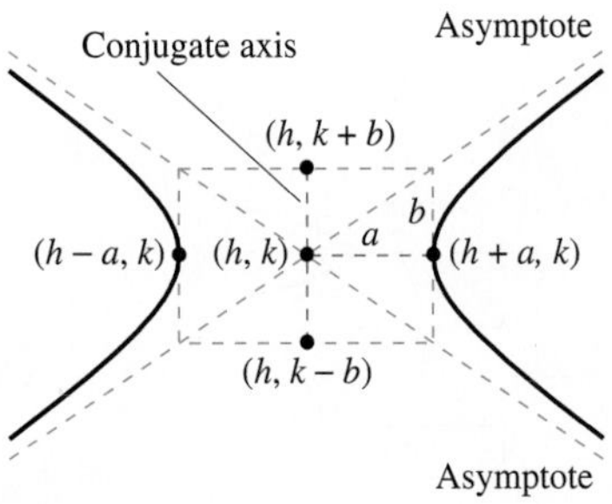

Figure 10.15

In Figure 10.15 you can see that the asymptotes coincide with the diagonals of the rectangle with dimensions $2a$ and $2b$, centered at (h, k). This provides you with a quick means of sketching the asymptotes, which in turn aids in sketching the hyperbola.

EXAMPLE 7 Using Asymptotes to Sketch a Hyperbola

Sketch the graph of the hyperbola whose equation is $4x^2 - y^2 = 16$.

Solution Begin by rewriting the equation in standard form.

$$\frac{x^2}{4} - \frac{y^2}{16} = 1$$

The transverse axis is horizontal and the vertices occur at $(-2, 0)$ and $(2, 0)$. The ends of the conjugate axis occur at $(0, -4)$ and $(0, 4)$. Using these four points, you can sketch the rectangle shown in Figure 10.16(a). By drawing the asymptotes through the corners of this rectangle, you can complete the sketch as shown in Figure 10.16(b).

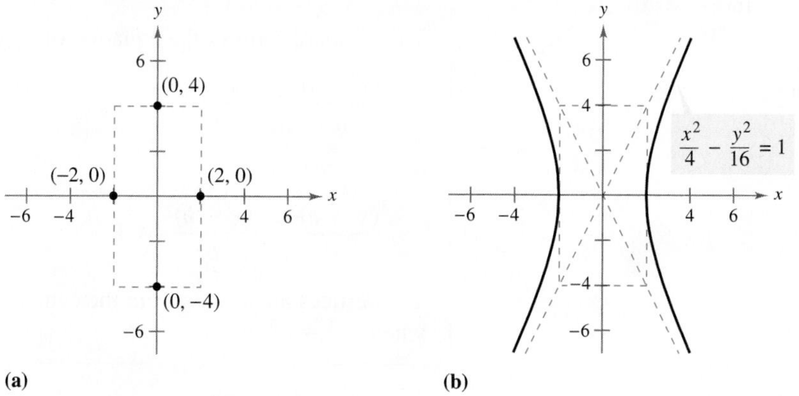

Figure 10.16

TECHNOLOGY You can use a graphing utility to verify the graph obtained in Example 7 by solving the original equation for y and graphing the following equations.

$$y_1 = \sqrt{4x^2 - 16}$$
$$y_2 = -\sqrt{4x^2 - 16}$$

Definition of Eccentricity of a Hyperbola

The **eccentricity** e of a hyperbola is given by the ratio

$$e = \frac{c}{a}.$$

As with an ellipse, the **eccentricity** of a hyperbola is $e = c/a$. Because $c > a$ for hyperbolas, it follows that $e > 1$ for hyperbolas. If the eccentricity is large, the branches of the hyperbola are nearly flat. If the eccentricity is close to 1, the branches of the hyperbola are more pointed, as shown in Figure 10.17.

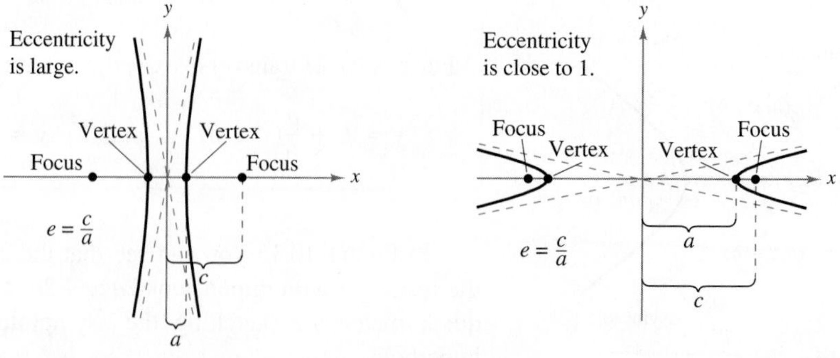

Figure 10.17

The following application was developed during World War II. It shows how the properties of hyperbolas can be used in radar and other detection systems.

EXAMPLE 8 A Hyperbolic Detection System

Two microphones, 1 mile apart, record an explosion. Microphone A receives the sound 2 seconds before microphone B. Where was the explosion?

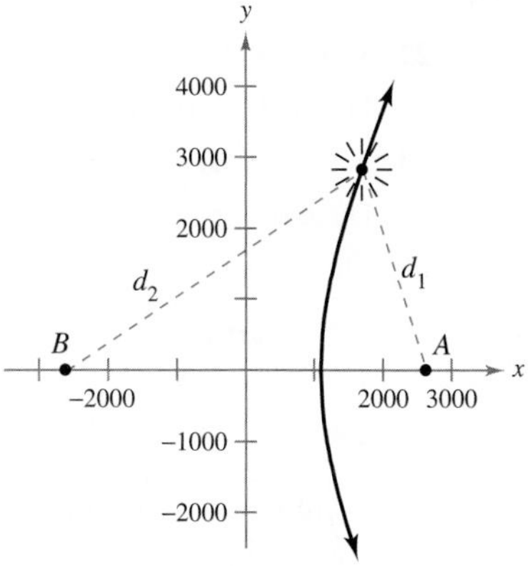

$2c = 5280$
$d_2 - d_1 = 2a = 2200$
Figure 10.18

Solution Assuming that sound travels at 1100 feet per second, you know that the explosion took place 2200 feet farther from B than from A, as shown in Figure 10.18. The locus of all points that are 2200 feet closer to A than to B is one branch of the hyperbola $(x^2/a^2) - (y^2/b^2) = 1$, where

$$c = \frac{1 \text{ mile}}{2} = \frac{5280 \text{ ft}}{2} = 2640 \text{ feet}$$

and

$$a = \frac{2200 \text{ ft}}{2} = 1100 \text{ feet.}$$

Because $c^2 = a^2 + b^2$, it follows that

$$\begin{aligned} b^2 &= c^2 - a^2 \\ &= 5{,}759{,}600 \end{aligned}$$

and you can conclude that the explosion occurred somewhere on the right branch of the hyperbola given by

$$\frac{x^2}{1{,}210{,}000} - \frac{y^2}{5{,}759{,}600} = 1.$$

In Example 8, you were able to determine only the hyperbola on which the explosion occurred, but not the exact location of the explosion. If, however, you had received the sound at a third position C, then two other hyperbolas would be determined. The exact location of the explosion would be the point at which these three hyperbolas intersect.

Another interesting application of conics involves the orbits of comets in our solar system. Of the 610 comets identified prior to 1970, 245 have elliptical orbits, 295 have parabolic orbits, and 70 have hyperbolic orbits. The center of the sun is a focus of each orbit, and each orbit has a vertex at the point at which the comet is closest to the sun. Undoubtedly, many comets with parabolic or hyperbolic orbits have not been identified—such comets pass through our solar system only once. Only comets with elliptical orbits such as Halley's comet remain in our solar system.

The type of orbit for a comet can be determined as follows.

1. Ellipse: $v < \sqrt{2GM/p}$
2. Parabola: $v = \sqrt{2GM/p}$
3. Hyperbola: $v > \sqrt{2GM/p}$

In these three formulas, p is the distance between one vertex and one focus of the comet's orbit (in meters), v is the velocity of the comet at the vertex (in meters per second), $M \approx 1.989 \times 10^{30}$ kilograms is the mass of the sun, and $G \approx 6.67 \times 10^{-8}$ cubic meters per kilogram-second squared is the gravitational constant.

CAROLINE HERSCHEL (1750–1848)

The first woman to be credited with detecting a new comet was the English astronomer Caroline Herschel. During her life, Caroline Herschel discovered a total of eight new comets.

Exercises for Section 10.1

See www.CalcChat.com for worked-out solutions to odd-numbered exercises.

In Exercises 1–8, match the equation with its graph. [The graphs are labeled (a), (b), (c), (d), (e), (f), (g), and (h).]

(a)

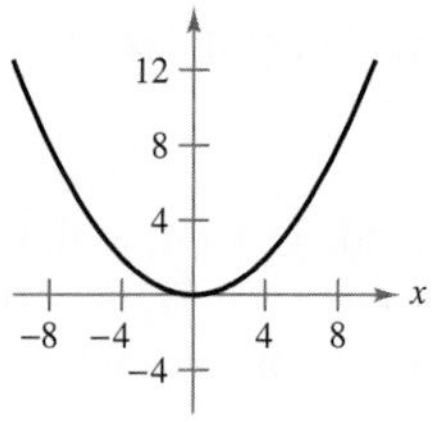

(b)

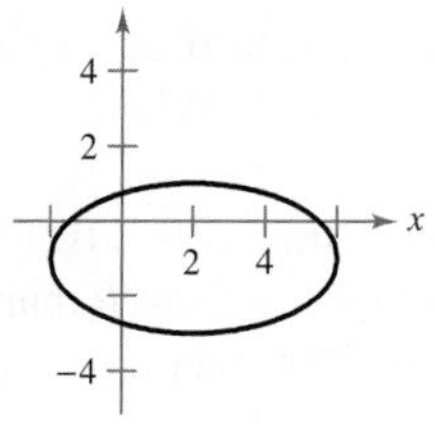

(c)

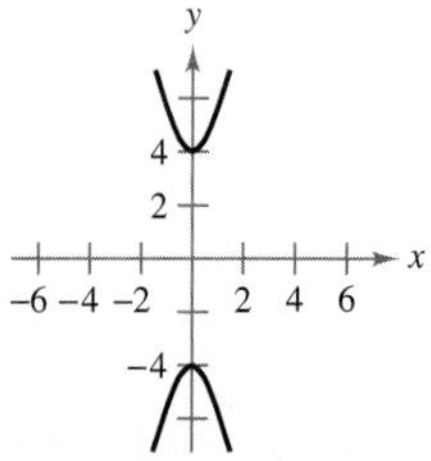

(d)

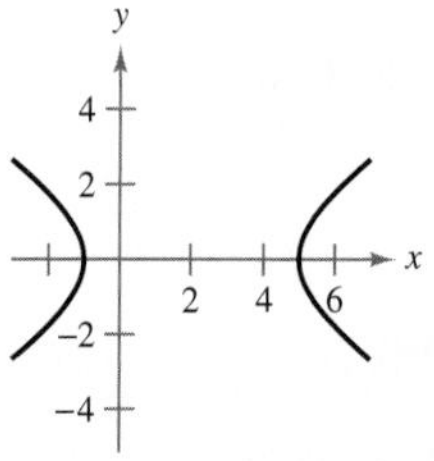

(e)

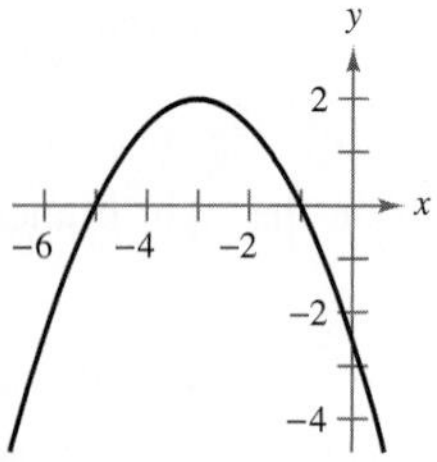

(f)

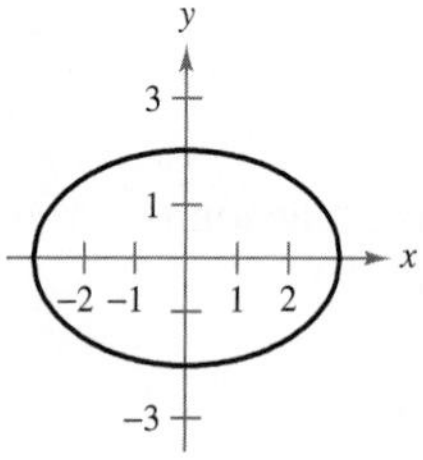

(g)

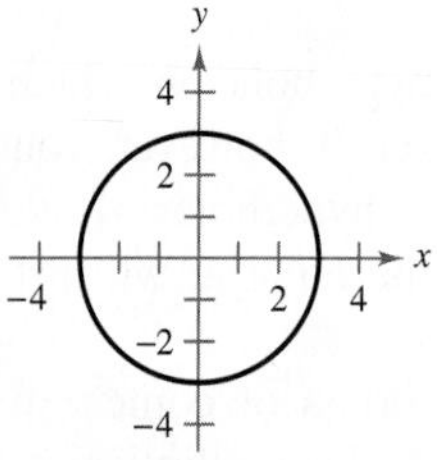

(h)

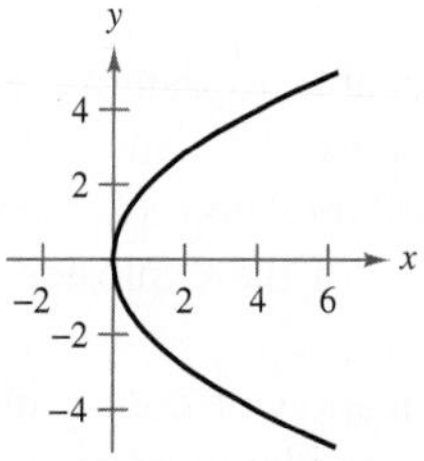

1. $y^2 = 4x$
2. $x^2 = 8y$
3. $(x + 3)^2 = -2(y - 2)$
4. $\dfrac{(x - 2)^2}{16} + \dfrac{(y + 1)^2}{4} = 1$
5. $\dfrac{x^2}{9} + \dfrac{y^2}{4} = 1$
6. $\dfrac{x^2}{9} + \dfrac{y^2}{9} = 1$
7. $\dfrac{y^2}{16} - \dfrac{x^2}{1} = 1$
8. $\dfrac{(x - 2)^2}{9} - \dfrac{y^2}{4} = 1$

In Exercises 9–16, find the vertex, focus, and directrix of the parabola, and sketch its graph.

9. $y^2 = -6x$
10. $x^2 + 8y = 0$
11. $(x + 3) + (y - 2)^2 = 0$
12. $(x - 1)^2 + 8(y + 2) = 0$
13. $y^2 - 4y - 4x = 0$
14. $y^2 + 6y + 8x + 25 = 0$
15. $x^2 + 4x + 4y - 4 = 0$
16. $y^2 + 4y + 8x - 12 = 0$

In Exercises 17–20, find the vertex, focus, and directrix of the parabola. Then use a graphing utility to graph the parabola.

17. $y^2 + x + y = 0$
18. $y = -\frac{1}{6}(x^2 - 8x + 6)$
19. $y^2 - 4x - 4 = 0$
20. $x^2 - 2x + 8y + 9 = 0$

In Exercises 21–28, find an equation of the parabola.

21. Vertex: $(3, 2)$
    Focus: $(1, 2)$
22. Vertex: $(-1, 2)$
    Focus: $(-1, 0)$
23. Vertex: $(0, 4)$
    Directrix: $y = -2$
24. Focus: $(2, 2)$
    Directrix: $x = -2$
25.

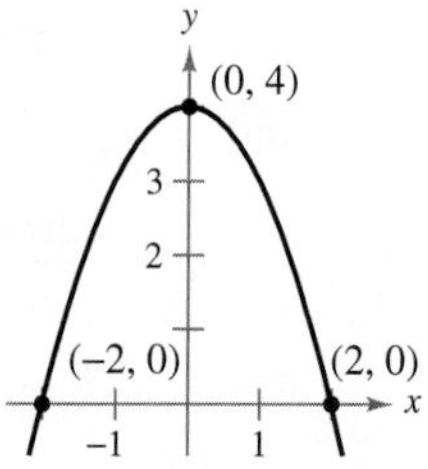

26.

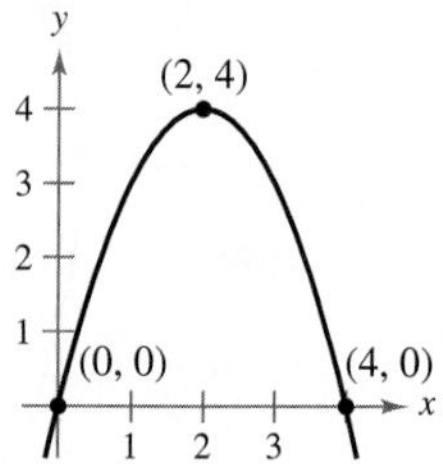

27. Axis is parallel to y-axis; graph passes through $(0, 3)$, $(3, 4)$, and $(4, 11)$.
28. Directrix: $y = -2$; endpoints of latus rectum are $(0, 2)$ and $(8, 2)$.

In Exercises 29–34, find the center, foci, vertices, and eccentricity of the ellipse, and sketch its graph.

29. $x^2 + 4y^2 = 4$
30. $5x^2 + 7y^2 = 70$
31. $\dfrac{(x - 1)^2}{9} + \dfrac{(y - 5)^2}{25} = 1$
32. $(x + 2)^2 + \dfrac{(y + 4)^2}{1/4} = 1$
33. $9x^2 + 4y^2 + 36x - 24y + 36 = 0$
34. $16x^2 + 25y^2 - 64x + 150y + 279 = 0$

In Exercises 35–38, find the center, foci, and vertices of the ellipse. Use a graphing utility to graph the ellipse.

35. $12x^2 + 20y^2 - 12x + 40y - 37 = 0$
36. $36x^2 + 9y^2 + 48x - 36y + 43 = 0$
37. $x^2 + 2y^2 - 3x + 4y + 0.25 = 0$
38. $2x^2 + y^2 + 4.8x - 6.4y + 3.12 = 0$

In Exercises 39–44, find an equation of the ellipse.

39. Center: $(0, 0)$
    Focus: $(2, 0)$
    Vertex: $(3, 0)$
40. Vertices: $(0, 2)$, $(4, 2)$
    Eccentricity: $\frac{1}{2}$
41. Vertices: $(3, 1)$, $(3, 9)$
    Minor axis length: 6
42. Foci: $(0, \pm 5)$
    Major axis length: 14

43. Center: $(0, 0)$
Major axis: horizontal
Points on the ellipse:
$(3, 1), (4, 0)$

44. Center: $(1, 2)$
Major axis: vertical
Points on the ellipse:
$(1, 6), (3, 2)$

In Exercises 45–52, find the center, foci, and vertices of the hyperbola, and sketch its graph using asymptotes as an aid.

45. $y^2 - \frac{x^2}{4} = 1$

46. $\frac{x^2}{25} - \frac{y^2}{9} = 1$

47. $\frac{(x-1)^2}{4} - \frac{(y+2)^2}{1} = 1$

48. $\frac{(y+1)^2}{144} - \frac{(x-4)^2}{25} = 1$

49. $9x^2 - y^2 - 36x - 6y + 18 = 0$

50. $y^2 - 9x^2 + 36x - 72 = 0$

51. $x^2 - 9y^2 + 2x - 54y - 80 = 0$

52. $9x^2 - 4y^2 + 54x + 8y + 78 = 0$

In Exercises 53–56, find the center, foci, and vertices of the hyperbola. Use a graphing utility to graph the hyperbola and its asymptotes.

53. $9y^2 - x^2 + 2x + 54y + 62 = 0$

54. $9x^2 - y^2 + 54x + 10y + 55 = 0$

55. $3x^2 - 2y^2 - 6x - 12y - 27 = 0$

56. $3y^2 - x^2 + 6x - 12y = 0$

In Exercises 57–64, find an equation of the hyperbola.

57. Vertices: $(\pm 1, 0)$
Asymptotes: $y = \pm 3x$

58. Vertices: $(0, \pm 3)$
Asymptotes: $y = \pm 3x$

59. Vertices: $(2, \pm 3)$
Point on graph: $(0, 5)$

60. Vertices: $(2, \pm 3)$
Foci: $(2, \pm 5)$

61. Center: $(0, 0)$
Vertex: $(0, 2)$
Focus: $(0, 4)$

62. Center: $(0, 0)$
Vertex: $(3, 0)$
Focus: $(5, 0)$

63. Vertices: $(0, 2), (6, 2)$
Asymptotes: $y = \frac{2}{3}x$
$y = 4 - \frac{2}{3}x$

64. Focus: $(10, 0)$
Asymptotes: $y = \pm\frac{3}{4}x$

In Exercises 65 and 66, find equations for (a) the tangent lines and (b) the normal lines to the hyperbola for the given value of x.

65. $\frac{x^2}{9} - y^2 = 1, \quad x = 6$

66. $\frac{y^2}{4} - \frac{x^2}{2} = 1, \quad x = 4$

In Exercises 67–76, classify the graph of the equation as a circle, a parabola, an ellipse, or a hyperbola.

67. $x^2 + 4y^2 - 6x + 16y + 21 = 0$

68. $4x^2 - y^2 - 4x - 3 = 0$

69. $y^2 - 4y - 4x = 0$

70. $25x^2 - 10x - 200y - 119 = 0$

71. $4x^2 + 4y^2 - 16y + 15 = 0$

72. $y^2 - 4y = x + 5$

73. $9x^2 + 9y^2 - 36x + 6y + 34 = 0$

74. $2x(x - y) = y(3 - y - 2x)$

75. $3(x - 1)^2 = 6 + 2(y + 1)^2$

76. $9(x + 3)^2 = 36 - 4(y - 2)^2$

Writing About Concepts

77. (a) Give the definition of a parabola.
(b) Give the standard forms of a parabola with vertex at (h, k).
(c) In your own words, state the reflective property of a parabola.

78. (a) Give the definition of an ellipse.
(b) Give the standard forms of an ellipse with center at (h, k).

79. (a) Give the definition of a hyperbola.
(b) Give the standard forms of a hyperbola with center at (h, k).
(c) Write equations for the asymptotes of a hyperbola.

80. Define the eccentricity of an ellipse. In your own words, describe how changes in the eccentricity affect the ellipse.

81. ***Solar Collector*** A solar collector for heating water is constructed with a sheet of stainless steel that is formed into the shape of a parabola (see figure). The water will flow through a pipe that is located at the focus of the parabola. At what distance from the vertex is the pipe?

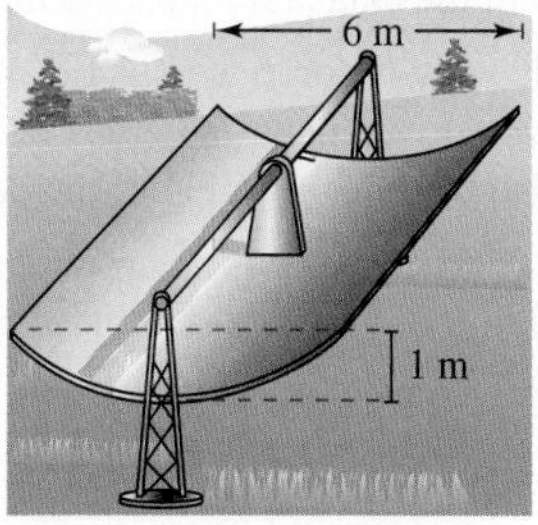

Figure for 81

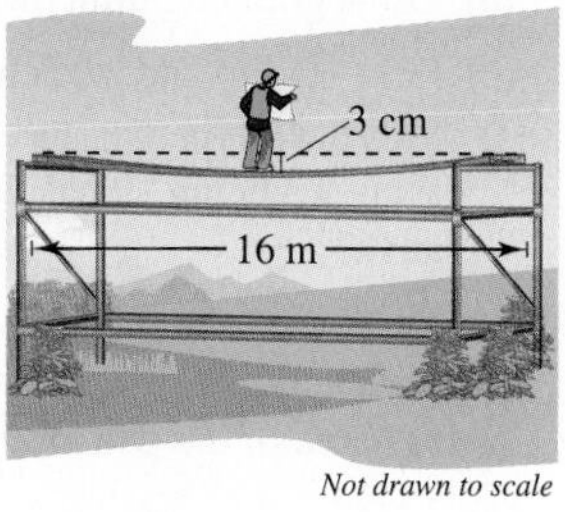

Figure for 82

82. ***Beam Deflection*** A simply supported beam that is 16 meters long has a load concentrated at the center (see figure). The deflection of the beam at its center is 3 centimeters. Assume that the shape of the deflected beam is parabolic.

(a) Find an equation of the parabola. (Assume that the origin is at the center of the beam.)

(b) How far from the center of the beam is the deflection 1 centimeter?

83. Find an equation of the tangent line to the parabola $y = ax^2$ at $x = x_0$. Prove that the x-intercept of this tangent line is $(x_0/2, 0)$.

84. (a) Prove that any two distinct tangent lines to a parabola intersect.

(b) Demonstrate the result of part (a) by finding the point of intersection of the tangent lines to the parabola $x^2 - 4x - 4y = 0$ at the points $(0, 0)$ and $(6, 3)$.

85. (a) Prove that if any two tangent lines to a parabola intersect at right angles, their point of intersection must lie on the directrix.

(b) Demonstrate the result of part (a) by proving that the tangent lines to the parabola $x^2 - 4x - 4y + 8 = 0$ at the points $(-2, 5)$ and $\left(3, \frac{5}{4}\right)$ intersect at right angles, and that the point of intersection lies on the directrix.

86. Find the point on the graph of $x^2 = 8y$ that is closest to the focus of the parabola.

87. ***Radio and Television Reception*** In mountainous areas, reception of radio and television is sometimes poor. Consider an idealized case where a hill is represented by the graph of the parabola $y = x - x^2$, a transmitter is located at the point $(-1, 1)$, and a receiver is located on the other side of the hill at the point $(x_0, 0)$. What is the closest the receiver can be to the hill so that the reception is unobstructed?

88. ***Modeling Data*** The table shows the average amounts of time A (in minutes) women spent watching television each day for the years 1996 to 2002. *(Source: Nielsen Media Research)*

Year	1996	1997	1998	1999	2000	2001	2002
A	274	273	273	280	286	291	298

(a) Use the regression capabilities of a graphing utility to find a model of the form $A = at^2 + bt + c$ for the data. Let t represent the year, with $t = 6$ corresponding to 1996.

(b) Use a graphing utility to plot the data and graph the model.

(c) Find dA/dt and sketch its graph for $6 \le t \le 12$. What information about the average amount of time women spent watching television is given by the graph of the derivative?

89. ***Architecture*** A church window is bounded above by a parabola and below by the arc of a circle (see figure). Find the surface area of the window.

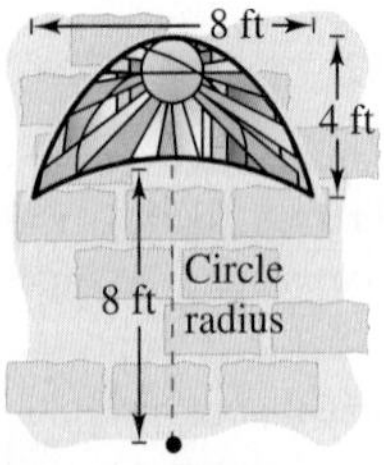

Figure for 89

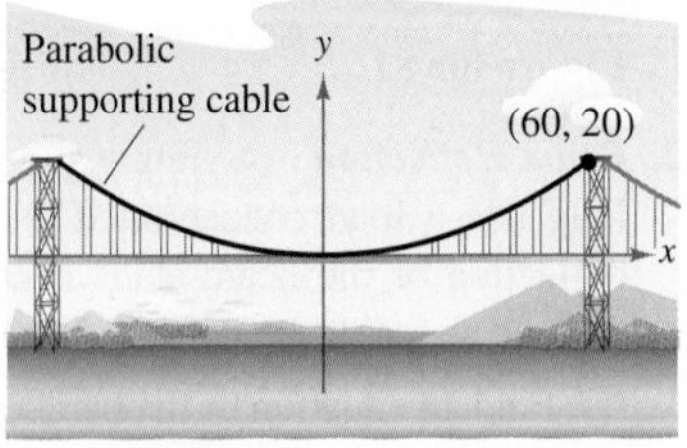

Figure for 91

90. ***Arc Length*** Find the arc length of the parabola $4x - y^2 = 0$ over the interval $0 \le y \le 4$.

91. ***Bridge Design*** A cable of a suspension bridge is suspended (in the shape of a parabola) between two towers that are 120 meters apart and 20 meters above the roadway (see figure). The cables touch the roadway midway between the towers.

(a) Find an equation for the parabolic shape of each cable.

(b) Find the length of the parabolic supporting cable.

92. ***Surface Area*** A satellite-signal receiving dish is formed by revolving the parabola given by $x^2 = 20y$ about the y-axis. The radius of the dish is r feet. Verify that the surface area of the dish is given by

$$2\pi \int_0^r x\sqrt{1 + \left(\frac{x}{10}\right)^2}\, dx = \frac{\pi}{15}[(100 + r^2)^{3/2} - 1000].$$

93. ***Investigation*** Sketch the graphs of $x^2 = 4py$ for $p = \frac{1}{4}, \frac{1}{2}, 1, \frac{3}{2}$, and 2 on the same coordinate axes. Discuss the change in the graphs as p increases.

94. ***Area*** Find a formula for the area of the shaded region in the figure.

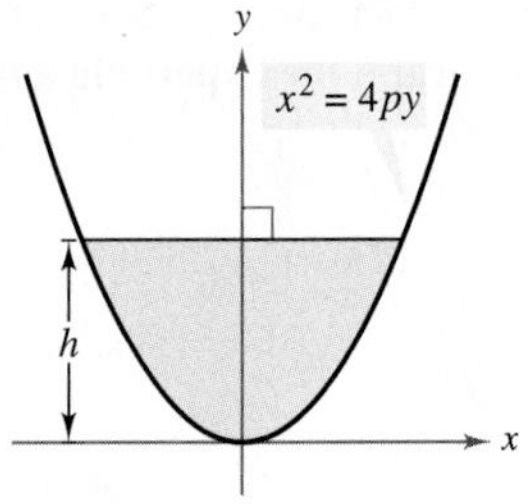

Figure for 94

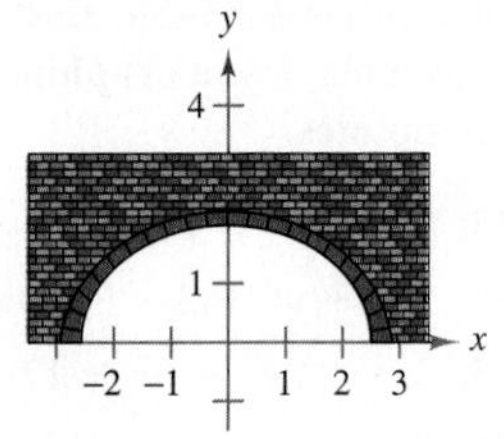

Figure for 96

95. ***Writing*** On page 697, it was noted that an ellipse can be drawn using two thumbtacks, a string of fixed length (greater than the distance between the tacks), and a pencil. If the ends of the string are fastened at the tacks and the string is drawn taut with a pencil, the path traced by the pencil will be an ellipse.

(a) What is the length of the string in terms of a?

(b) Explain why the path is an ellipse.

96. ***Construction of a Semielliptical Arch*** A fireplace arch is to be constructed in the shape of a semiellipse. The opening is to have a height of 2 feet at the center and a width of 5 feet along the base (see figure). The contractor draws the outline of the ellipse by the method shown in Exercise 95. Where should the tacks be placed and what should be the length of the piece of string?

97. Sketch the ellipse that consists of all points (x, y) such that the sum of the distances between (x, y) and two fixed points is 16 units, and the foci are located at the centers of the two sets of concentric circles in the figure. To print an enlarged copy of the graph, go to the website *www.mathgraphs.com*.

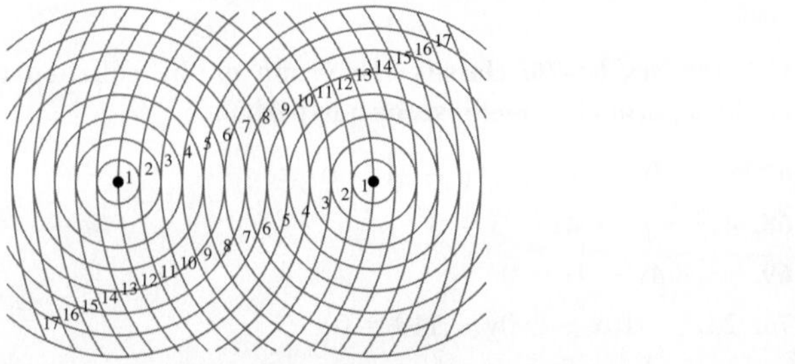

98. ***Orbit of Earth*** Earth moves in an elliptical orbit with the sun at one of the foci. The length of half of the major axis is 149,598,000 kilometers, and the eccentricity is 0.0167. Find the minimum distance (*perihelion*) and the maximum distance (*aphelion*) of Earth from the sun.

99. ***Satellite Orbit*** The *apogee* (the point in orbit farthest from Earth) and the *perigee* (the point in orbit closest to Earth) of an elliptical orbit of an Earth satellite are given by A and P. Show that the eccentricity of the orbit is

$$e = \frac{A - P}{A + P}.$$

100. ***Explorer 18*** On November 27, 1963, the United States launched Explorer 18. Its low and high points above the surface of Earth were 119 miles and 123,000 miles. Find the eccentricity of its elliptical orbit.

101. ***Halley's Comet*** Probably the most famous of all comets, Halley's comet, has an elliptical orbit with the sun at the focus. Its maximum distance from the sun is approximately 35.29 AU (astronomical unit $\approx 92.956 \times 10^6$ miles), and its minimum distance is approximately 0.59 AU. Find the eccentricity of the orbit.

102. The equation of an ellipse with its center at the origin can be written as

$$\frac{x^2}{a^2} + \frac{y^2}{a^2(1 - e^2)} = 1.$$

Show that as $e \to 0$, with a remaining fixed, the ellipse approaches a circle.

103. Consider a particle traveling clockwise on the elliptical path

$$\frac{x^2}{100} + \frac{y^2}{25} = 1.$$

The particle leaves the orbit at the point $(-8, 3)$ and travels in a straight line tangent to the ellipse. At what point will the particle cross the y-axis?

104. ***Volume*** The water tank on a fire truck is 16 feet long, and its cross sections are ellipses. Find the volume of water in the partially filled tank as shown in the figure.

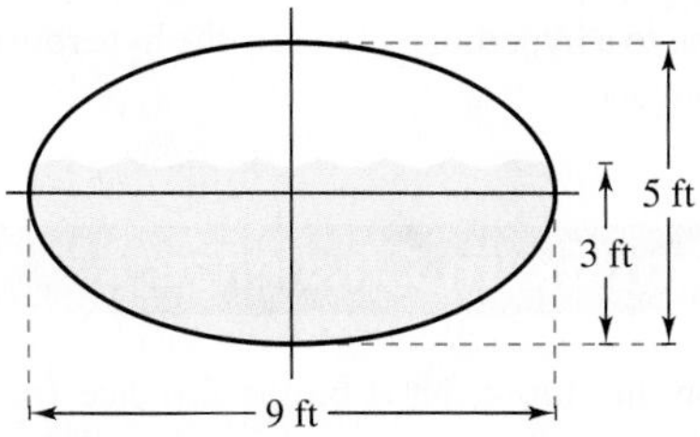

In Exercises 105 and 106, determine the points at which dy/dx is zero or does not exist to locate the endpoints of the major and minor axes of the ellipse.

105. $16x^2 + 9y^2 + 96x + 36y + 36 = 0$

106. $9x^2 + 4y^2 + 36x - 24y + 36 = 0$

Area and Volume **In Exercises 107 and 108, find (a) the area of the region bounded by the ellipse, (b) the volume and surface area of the solid generated by revolving the region about its major axis (prolate spheroid), and (c) the volume and surface area of the solid generated by revolving the region about its minor axis (oblate spheroid).**

107. $\dfrac{x^2}{4} + \dfrac{y^2}{1} = 1$

108. $\dfrac{x^2}{16} + \dfrac{y^2}{9} = 1$

109. ***Arc Length*** Use the integration capabilities of a graphing utility to approximate to two-decimal-place accuracy the elliptical integral representing the circumference of the ellipse

$$\frac{x^2}{25} + \frac{y^2}{49} = 1.$$

110. Prove that the tangent line to an ellipse at a point P makes equal angles with lines through P and the foci (see figure). [*Hint:* (1) Find the slope of the tangent line at P, (2) find the slopes of the lines through P and each focus, and (3) use the formula for the tangent of the angle between two lines.]

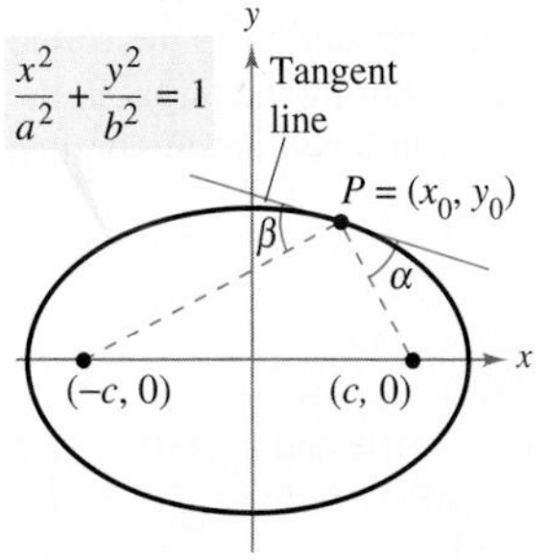

Figure for 110

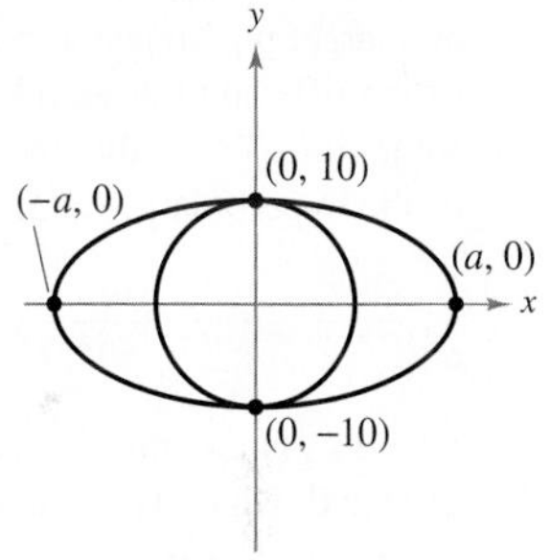

Figure for 111

111. ***Geometry*** The area of the ellipse in the figure is twice the area of the circle. What is the length of the major axis?

112. ***Conjecture***

(a) Show that the equation of an ellipse can be written as

$$\frac{(x - h)^2}{a^2} + \frac{(y - k)^2}{a^2(1 - e^2)} = 1.$$

(b) Use a graphing utility to graph the ellipse

$$\frac{(x - 2)^2}{4} + \frac{(y - 3)^2}{4(1 - e^2)} = 1$$

for $e = 0.95$, $e = 0.75$, $e = 0.5$, $e = 0.25$, and $e = 0$.

(c) Use the results of part (b) to make a conjecture about the change in the shape of the ellipse as e approaches 0.

113. Find an equation of the hyperbola such that for any point on the hyperbola, the difference between its distances from the points (2, 2) and (10, 2) is 6.

114. Find an equation of the hyperbola such that for any point on the hyperbola, the difference between its distances from the points $(-3, 0)$ and $(-3, 3)$ is 2.

115. Sketch the hyperbola that consists of all points (x, y) such that the difference of the distances between (x, y) and two fixed points is 10 units, and the foci are located at the centers of the two sets of concentric circles in the figure. To print an enlarged copy of the graph, go to the website *www.mathgraphs.com.*

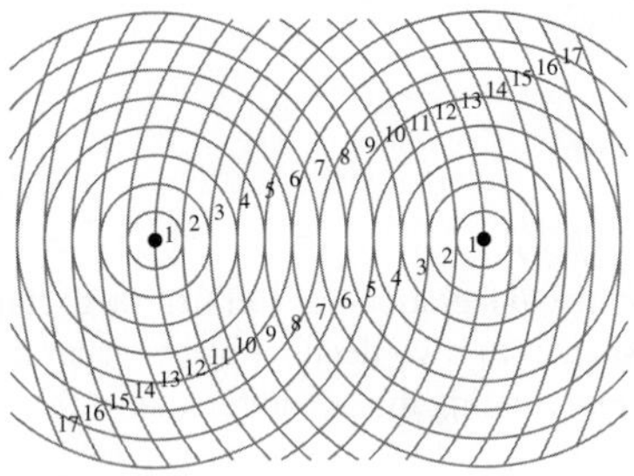

116. Consider a hyperbola centered at the origin with a horizontal transverse axis. Use the definition of a hyperbola to derive its standard form:

$$\frac{x^2}{a^2} - \frac{y^2}{b^2} = 1.$$

117. ***Sound Location*** A rifle positioned at point $(-c, 0)$ is fired at a target positioned at point $(c, 0)$. A person hears the sound of the rifle and the sound of the bullet hitting the target at the same time. Prove that the person is positioned on one branch of the hyperbola given by

$$\frac{x^2}{c^2 v_s^2/v_m^2} - \frac{y^2}{c^2(v_m^2 - v_s^2)/v_m^2} = 1$$

where v_m is the muzzle velocity of the rifle and v_s is the speed of sound, which is about 1100 feet per second.

118. ***Navigation*** LORAN (long distance radio navigation) for aircraft and ships uses synchronized pulses transmitted by widely separated transmitting stations. These pulses travel at the speed of light (186,000 miles per second). The difference in the times of arrival of these pulses at an aircraft or ship is constant on a hyperbola having the transmitting stations as foci. Assume that two stations, 300 miles apart, are positioned on the rectangular coordinate system at $(-150, 0)$ and $(150, 0)$ and that a ship is traveling on a path with coordinates $(x, 75)$ (see figure). Find the x-coordinate of the position of the ship if the time difference between the pulses from the transmitting stations is 1000 microseconds (0.001 second).

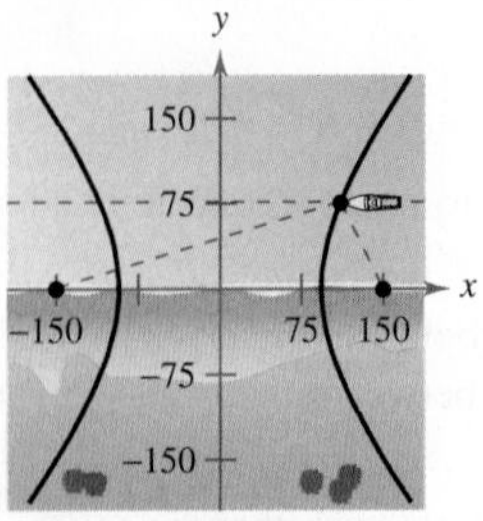

Figure for 118

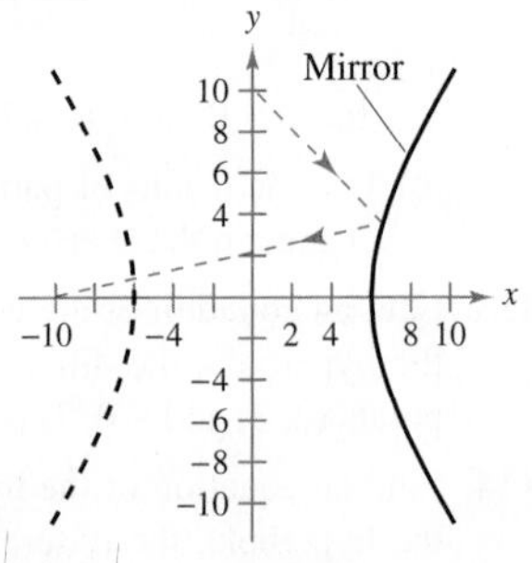

Figure for 119

119. ***Hyperbolic Mirror*** A hyperbolic mirror (used in some telescopes) has the property that a light ray directed at the focus will be reflected to the other focus. The mirror in the figure has the equation $(x^2/36) - (y^2/64) = 1$. At which point on the mirror will light from the point $(0, 10)$ be reflected to the other focus?

120. Show that the equation of the tangent line to $\dfrac{x^2}{a^2} - \dfrac{y^2}{b^2} = 1$ at the point (x_0, y_0) is $(x_0/a^2)x - (y_0/b^2)y = 1$.

121. Show that the graphs of the equations intersect at right angles:

$$\frac{x^2}{a^2} + \frac{2y^2}{b^2} = 1 \quad \text{and} \quad \frac{x^2}{a^2 - b^2} - \frac{2y^2}{b^2} = 1.$$

122. Prove that the graph of the equation

$$Ax^2 + Cy^2 + Dx + Ey + F = 0$$

is one of the following (except in degenerate cases).

Conic	*Condition*
(a) Circle	$A = C$
(b) Parabola	$A = 0$ or $C = 0$ (but not both)
(c) Ellipse	$AC > 0$
(d) Hyperbola	$AC < 0$

True or False? **In Exercises 123–128, determine whether the statement is true or false. If it is false, explain why or give an example that shows it is false.**

123. It is possible for a parabola to intersect its directrix.

124. The point on a parabola closest to its focus is its vertex.

125. If C is the circumference of the ellipse

$$\frac{x^2}{a^2} + \frac{y^2}{b^2} = 1, \quad b < a$$

then $2\pi b \le C \le 2\pi a$.

126. If $D \ne 0$ or $E \ne 0$, then the graph of $y^2 - x^2 + Dx + Ey = 0$ is a hyperbola.

127. If the asymptotes of the hyperbola $(x^2/a^2) - (y^2/b^2) = 1$ intersect at right angles, then $a = b$.

128. Every tangent line to a hyperbola intersects the hyperbola only at the point of tangency.

Putnam Exam Challenge

129. For a point P on an ellipse, let d be the distance from the center of the ellipse to the line tangent to the ellipse at P. Prove that $(PF_1)(PF_2)d^2$ is constant as P varies on the ellipse, where PF_1 and PF_2 are the distances from P to the foci F_1 and F_2 of the ellipse.

130. Find the minimum value of $(u - v)^2 + \left(\sqrt{2 - u^2} - \dfrac{9}{v}\right)^2$ for $0 < u < \sqrt{2}$ and $v > 0$.

These problems were composed by the Committee on the Putnam Prize Competition.

Section 10.2 Plane Curves and Parametric Equations

- **Sketch the graph of a curve given by a set of parametric equations.**
- **Eliminate the parameter in a set of parametric equations.**
- **Find a set of parametric equations to represent a curve.**
- **Understand two classic calculus problems, the tautochrone and brachistochrone problems.**

Plane Curves and Parametric Equations

Until now, you have been representing a graph by a single equation involving *two* variables. In this section you will study situations in which *three* variables are used to represent a curve in the plane.

Consider the path followed by an object that is propelled into the air at an angle of 45°. If the initial velocity of the object is 48 feet per second, the object travels the parabolic path given by

$$y = -\frac{x^2}{72} + x \qquad \text{Rectangular equation}$$

as shown in Figure 10.19. However, this equation does not tell the whole story. Although it does tell you *where* the object has been, it doesn't tell you *when* the object was at a given point (x, y). To determine this time, you can introduce a third variable t, called a **parameter.** By writing both x and y as functions of t, you obtain the **parametric equations**

$$x = 24\sqrt{2}\,t \qquad \text{Parametric equation for } x$$

and

$$y = -16t^2 + 24\sqrt{2}\,t. \qquad \text{Parametric equation for } y$$

From this set of equations, you can determine that at time $t = 0$, the object is at the point (0, 0). Similarly, at time $t = 1$, the object is at the point $\left(24\sqrt{2}, 24\sqrt{2} - 16\right)$, and so on. (You will learn a method for determining this particular set of parametric equations—the equations of motion—later, in Section 12.3.)

For this particular motion problem, x and y are continuous functions of t, and the resulting path is called a **plane curve.**

Rectangular equation:

$y = -\frac{x^2}{72} + x$

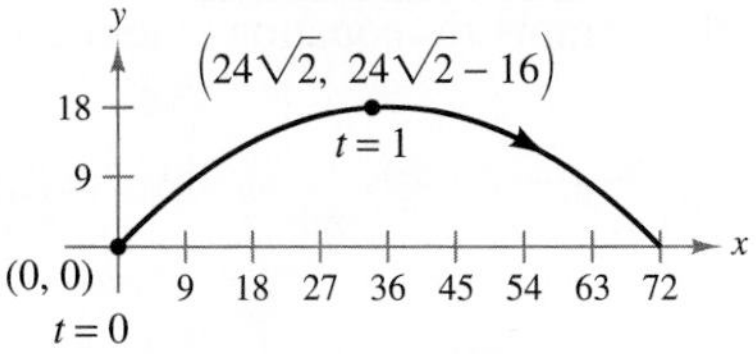

Parametric equations:

$x = 24\sqrt{2}t$

$y = -16t^2 + 24\sqrt{2}t$

Curvilinear motion: two variables for position, one variable for time
Figure 10.19

Definition of a Plane Curve

If f and g are continuous functions of t on an interval I, then the equations

$$x = f(t) \quad \text{and} \quad y = g(t)$$

are called **parametric equations** and t is called the **parameter.** The set of points (x, y) obtained as t varies over the interval I is called the **graph** of the parametric equations. Taken together, the parametric equations and the graph are called a **plane curve,** denoted by C.

NOTE At times it is important to distinguish between a graph (the set of points) and a curve (the points together with their defining parametric equations). When it is important, we will make the distinction explicit. When it is not important, we will use C to represent the graph or the curve.

When sketching (by hand) a curve represented by a set of parametric equations, you can plot points in the xy-plane. Each set of coordinates (x, y) is determined from a value chosen for the parameter t. By plotting the resulting points in order of increasing values of t, the curve is traced out in a specific direction. This is called the **orientation** of the curve.

EXAMPLE 1 Sketching a Curve

Sketch the curve described by the parametric equations

$$x = t^2 - 4 \quad \text{and} \quad y = \frac{t}{2}, \quad -2 \le t \le 3.$$

Solution For values of t on the given interval, the parametric equations yield the points (x, y) shown in the table.

t	-2	-1	0	1	2	3
x	0	-3	-4	-3	0	5
y	-1	$-\frac{1}{2}$	0	$\frac{1}{2}$	1	$\frac{3}{2}$

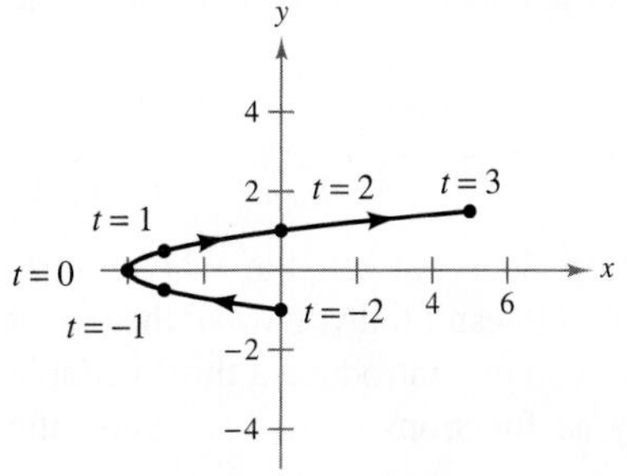

Parametric equations:
$x = t^2 - 4$ and $y = \frac{t}{2}, -2 \le t \le 3$

Figure 10.20

By plotting these points in order of increasing t and using the continuity of f and g, you obtain the curve C shown in Figure 10.20. Note that the arrows on the curve indicate its orientation as t increases from -2 to 3.

NOTE From the Vertical Line Test, you can see that the graph shown in Figure 10.20 does not define y as a function of x. This points out one benefit of parametric equations—they can be used to represent graphs that are more general than graphs of functions.

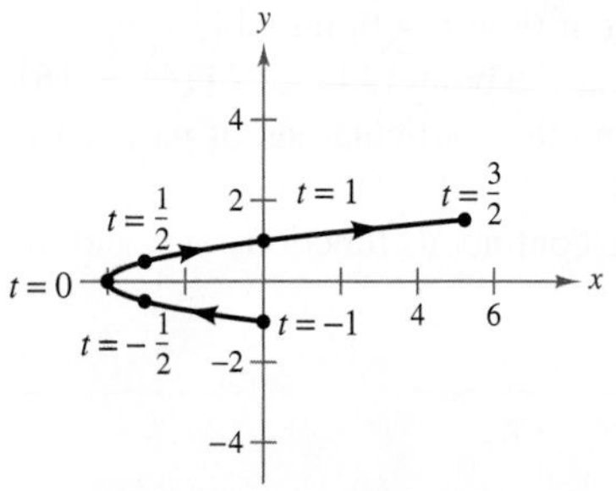

Parametric equations:
$x = 4t^2 - 4$ and $y = t, -1 \le t \le \frac{3}{2}$

Figure 10.21

It often happens that two different sets of parametric equations have the same graph. For example, the set of parametric equations

$$x = 4t^2 - 4 \quad \text{and} \quad y = t, \quad -1 \le t \le \frac{3}{2}$$

has the same graph as the set given in Example 1. However, comparing the values of t in Figures 10.20 and 10.21, you can see that the second graph is traced out more *rapidly* (considering t as time) than the first graph. So, in applications, different parametric representations can be used to represent various *speeds* at which objects travel along a given path.

TECHNOLOGY Most graphing utilities have a *parametric* graphing mode. If you have access to such a utility, use it to confirm the graphs shown in Figures 10.20 and 10.21. Does the curve given by

$$x = 4t^2 - 8t \quad \text{and} \quad y = 1 - t, \quad -\tfrac{1}{2} \le t \le 2$$

represent the same graph as that shown in Figures 10.20 and 10.21? What do you notice about the *orientation* of this curve?

Eliminating the Parameter

Finding a rectangular equation that represents the graph of a set of parametric equations is called **eliminating the parameter.** For instance, you can eliminate the parameter from the set of parametric equations in Example 1 as follows.

Parametric equations	⇒	Solve for t in one equation.	⇒	Substitute into second equation.	⇒	Rectangular equation
$x = t^2 - 4$ $y = t/2$		$t = 2y$		$x = (2y)^2 - 4$		$x = 4y^2 - 4$

Once you have eliminated the parameter, you can recognize that the equation $x = 4y^2 - 4$ represents a parabola with a horizontal axis and vertex at $(-4, 0)$, as shown in Figure 10.20.

The range of x and y implied by the parametric equations may be altered by the change to rectangular form. In such instances the domain of the rectangular equation must be adjusted so that its graph matches the graph of the parametric equations. Such a situation is demonstrated in the next example.

EXAMPLE 2 Adjusting the Domain After Eliminating the Parameter

Sketch the curve represented by the equations

$$x = \frac{1}{\sqrt{t + 1}} \quad \text{and} \quad y = \frac{t}{t + 1}, \quad t > -1$$

by eliminating the parameter and adjusting the domain of the resulting rectangular equation.

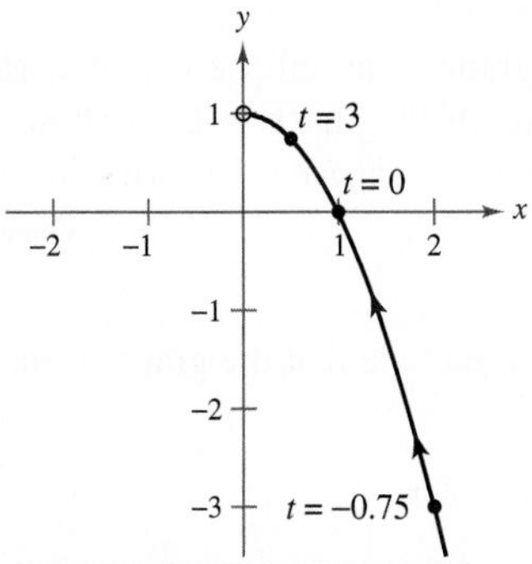

Parametric equations:
$x = \dfrac{1}{\sqrt{t+1}}, \; y = \dfrac{t}{t+1}, \; t > -1$

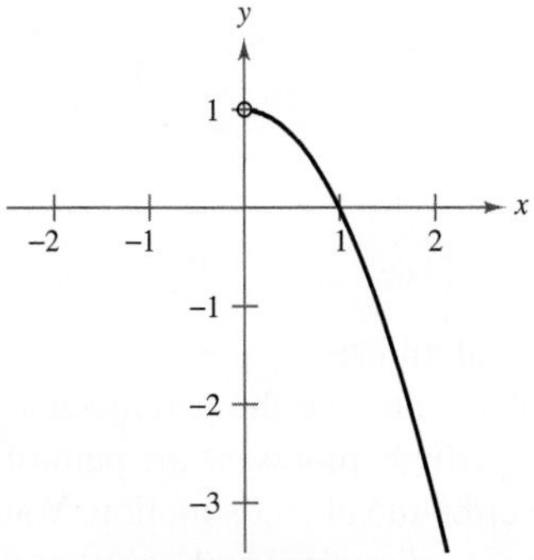

Rectangular equation:
$y = 1 - x^2, \; x > 0$

Figure 10.22

Solution Begin by solving one of the parametric equations for t. For instance, you can solve the first equation for t as follows.

$$x = \frac{1}{\sqrt{t + 1}} \qquad \text{Parametric equation for } x$$

$$x^2 = \frac{1}{t + 1} \qquad \text{Square each side.}$$

$$t + 1 = \frac{1}{x^2}$$

$$t = \frac{1}{x^2} - 1 = \frac{1 - x^2}{x^2} \qquad \text{Solve for } t.$$

Now, substituting into the parametric equation for y produces

$$y = \frac{t}{t + 1} \qquad \text{Parametric equation for } y$$

$$y = \frac{(1 - x^2)/x^2}{[(1 - x^2)/x^2] + 1} \qquad \text{Substitute } (1 - x^2)/x^2 \text{ for } t.$$

$$y = 1 - x^2. \qquad \text{Simplify.}$$

The rectangular equation, $y = 1 - x^2$, is defined for all values of x, but from the parametric equation for x you can see that the curve is defined only when $t > -1$. This implies that you should restrict the domain of x to positive values, as shown in Figure 10.22.

It is not necessary for the parameter in a set of parametric equations to represent time. The next example uses an *angle* as the parameter.

EXAMPLE 3 Using Trigonometry to Eliminate a Parameter

Sketch the curve represented by

$$x = 3\cos\theta \quad \text{and} \quad y = 4\sin\theta, \quad 0 \le \theta \le 2\pi$$

by eliminating the parameter and finding the corresponding rectangular equation.

Solution Begin by solving for $\cos\theta$ and $\sin\theta$ in the given equations.

$$\cos\theta = \frac{x}{3} \quad \text{and} \quad \sin\theta = \frac{y}{4} \qquad \text{Solve for } \cos\theta \text{ and } \sin\theta.$$

Next, make use of the identity $\sin^2\theta + \cos^2\theta = 1$ to form an equation involving only x and y.

$$\cos^2\theta + \sin^2\theta = 1 \qquad \text{Trigonometric identity}$$

$$\left(\frac{x}{3}\right)^2 + \left(\frac{y}{4}\right)^2 = 1 \qquad \text{Substitute.}$$

$$\frac{x^2}{9} + \frac{y^2}{16} = 1 \qquad \text{Rectangular equation}$$

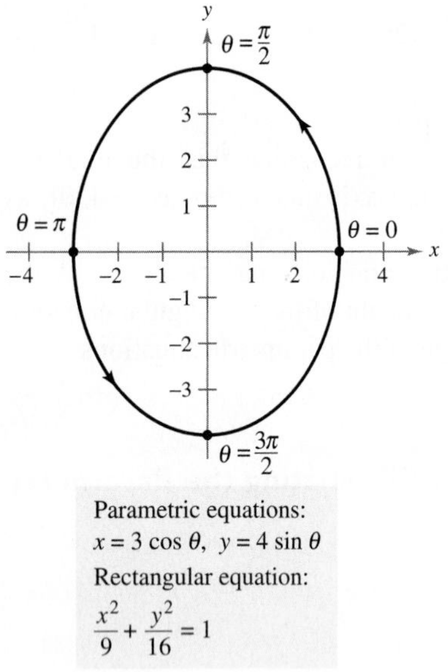

Figure 10.23

From this rectangular equation you can see that the graph is an ellipse centered at (0, 0), with vertices at (0, 4) and (0, −4) and minor axis of length $2b = 6$, as shown in Figure 10.23. Note that the ellipse is traced out *counterclockwise* as θ varies from 0 to 2π.

Using the technique shown in Example 3, you can conclude that the graph of the parametric equations

$$x = h + a\cos\theta \quad \text{and} \quad y = k + b\sin\theta, \quad 0 \le \theta \le 2\pi$$

is the ellipse (traced counterclockwise) given by

$$\frac{(x-h)^2}{a^2} + \frac{(y-k)^2}{b^2} = 1.$$

The graph of the parametric equations

$$x = h + a\sin\theta \quad \text{and} \quad y = k + b\cos\theta, \quad 0 \le \theta \le 2\pi$$

is also the ellipse (traced clockwise) given by

$$\frac{(x-h)^2}{a^2} + \frac{(y-k)^2}{b^2} = 1.$$

Use a graphing utility in *parametric* mode to graph several ellipses.

In Examples 2 and 3, it is important to realize that eliminating the parameter is primarily an *aid to curve sketching*. If the parametric equations represent the path of a moving object, the graph alone is not sufficient to describe the object's motion. You still need the parametric equations to tell you the *position*, *direction*, and *speed* at a given time.

Finding Parametric Equations

The first three examples in this section illustrate techniques for sketching the graph represented by a set of parametric equations. You will now investigate the reverse problem. How can you determine a set of parametric equations for a given graph or a given physical description? From the discussion following Example 1, you know that such a representation is not unique. This is demonstrated further in the following example, which finds two different parametric representations for a given graph.

EXAMPLE 4 Finding Parametric Equations for a Given Graph

Find a set of parametric equations to represent the graph of $y = 1 - x^2$, using each of the following parameters.

a. $t = x$ **b.** The slope $m = \dfrac{dy}{dx}$ at the point (x, y)

Solution

a. Letting $x = t$ produces the parametric equations

$$x = t \quad \text{and} \quad y = 1 - x^2 = 1 - t^2.$$

b. To write x and y in terms of the parameter m, you can proceed as follows.

$$m = \frac{dy}{dx} = -2x \qquad \text{Differentiate } y = 1 - x^2.$$

$$x = -\frac{m}{2} \qquad \text{Solve for } x.$$

This produces a parametric equation for x. To obtain a parametric equation for y, substitute $-m/2$ for x in the original equation.

$$y = 1 - x^2 \qquad \text{Write original rectangular equation.}$$

$$y = 1 - \left(-\frac{m}{2}\right)^2 \qquad \text{Substitute } -m/2 \text{ for } x.$$

$$y = 1 - \frac{m^2}{4} \qquad \text{Simplify.}$$

So, the parametric equations are

$$x = -\frac{m}{2} \quad \text{and} \quad y = 1 - \frac{m^2}{4}.$$

In Figure 10.24, note that the resulting curve has a right-to-left orientation as determined by the direction of increasing values of slope m. For part (a), the curve would have the opposite orientation.

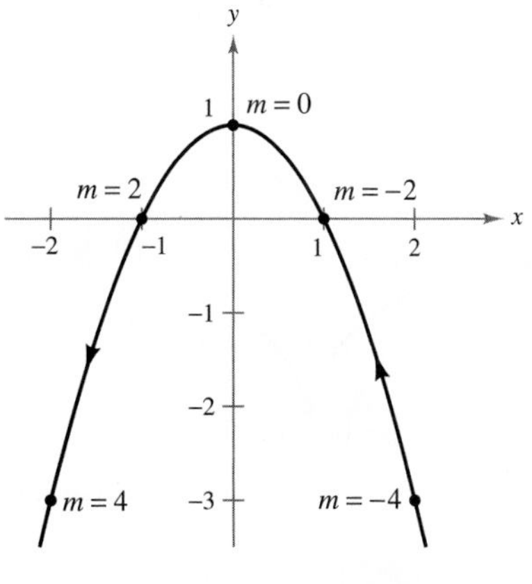

Rectangular equation: $y = 1 - x^2$
Parametric equations:
$x = -\dfrac{m}{2}, y = 1 - \dfrac{m^2}{4}$

Figure 10.24

TECHNOLOGY To be efficient at using a graphing utility, it is important that you develop skill in representing a graph by a set of parametric equations. The reason for this is that many graphing utilities have only three graphing modes—(1) functions, (2) parametric equations, and (3) polar equations. Most graphing utilities are not programmed to graph a general equation. For instance, suppose you want to graph the hyperbola $x^2 - y^2 = 1$. To graph the hyperbola in *function* mode, you need two equations: $y = \sqrt{x^2 - 1}$ and $y = -\sqrt{x^2 - 1}$. In *parametric* mode, you can represent the graph by $x = \sec t$ and $y = \tan t$.

CYCLOIDS

Galileo first called attention to the cycloid, once recommending that it be used for the arches of bridges. Pascal once spent 8 days attempting to solve many of the problems of cycloids, such as finding the area under one arch, and the volume of the solid of revolution formed by revolving the curve about a line. The cycloid has so many interesting properties and has caused so many quarrels among mathematicians that it has been called "the Helen of geometry" and "the apple of discord."

EXAMPLE 5 Parametric Equations for a Cycloid

Determine the curve traced by a point P on the circumference of a circle of radius a rolling along a straight line in a plane. Such a curve is called a **cycloid.**

Solution Let the parameter θ be the measure of the circle's rotation, and let the point $P = (x, y)$ begin at the origin. When $\theta = 0$, P is at the origin. When $\theta = \pi$, P is at a maximum point $(\pi a, 2a)$. When $\theta = 2\pi$, P is back on the x-axis at $(2\pi a, 0)$. From Figure 10.25, you can see that $\angle APC = 180° - \theta$. So,

$$\sin\theta = \sin(180° - \theta) = \sin(\angle APC) = \frac{AC}{a} = \frac{BD}{a}$$

$$\cos\theta = -\cos(180° - \theta) = -\cos(\angle APC) = \frac{AP}{-a}$$

which implies that

$$AP = -a\cos\theta \quad \text{and} \quad BD = a\sin\theta.$$

Because the circle rolls along the x-axis, you know that $OD = \widehat{PD} = a\theta$. Furthermore, because $BA = DC = a$, you have

$$x = OD - BD = a\theta - a\sin\theta$$
$$y = BA + AP = a - a\cos\theta.$$

So, the parametric equations are

$$x = a(\theta - \sin\theta) \quad \text{and} \quad y = a(1 - \cos\theta).$$

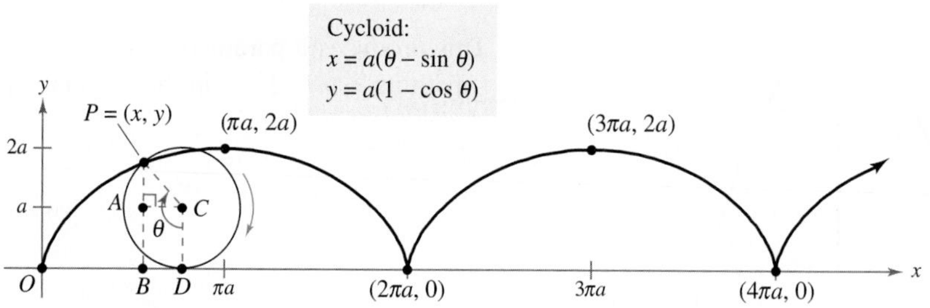

Figure 10.25

FOR FURTHER INFORMATION For more information on cycloids, see the article "The Geometry of Rolling Curves" by John Bloom and Lee Whitt in *The American Mathematical Monthly*. To view this article, go to the website *www.matharticles.com*.

The cycloid in Figure 10.25 has sharp corners at the values $x = 2n\pi a$. Notice that the derivatives $x'(\theta)$ and $y'(\theta)$ are both zero at the points for which $\theta = 2n\pi$.

$$x(\theta) = a(\theta - \sin\theta) \qquad y(\theta) = a(1 - \cos\theta)$$
$$x'(\theta) = a - a\cos\theta \qquad y'(\theta) = a\sin\theta$$
$$x'(2n\pi) = 0 \qquad y'(2n\pi) = 0$$

Between these points, the cycloid is called **smooth.**

TECHNOLOGY Some graphing utilities allow you to simulate the motion of an object that is moving in the plane or in space. If you have access to such a utility, use it to trace out the path of the cycloid shown in Figure 10.25.

Definition of a Smooth Curve

A curve C represented by $x = f(t)$ and $y = g(t)$ on an interval I is called **smooth** if f' and g' are continuous on I and not simultaneously 0, except possibly at the endpoints of I. The curve C is called **piecewise smooth** if it is smooth on each subinterval of some partition of I.

The Tautochrone and Brachistochrone Problems

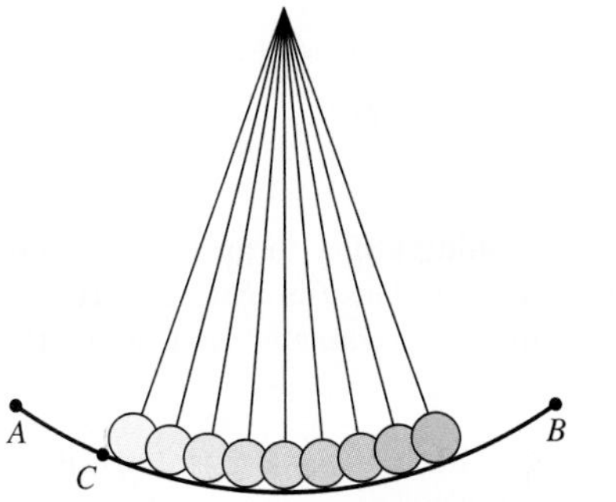

The time required to complete a full swing of the pendulum when starting from point C is only approximately the same as when starting from point A.
Figure 10.26

The type of curve described in Example 5 is related to one of the most famous pairs of problems in the history of calculus. The first problem (called the **tautochrone problem**) began with Galileo's discovery that the time required to complete a full swing of a given pendulum is *approximately* the same whether it makes a large movement at high speed or a small movement at lower speed (see Figure 10.26). Late in his life, Galileo (1564–1642) realized that he could use this principle to construct a clock. However, he was not able to conquer the mechanics of actual construction. Christian Huygens (1629–1695) was the first to design and construct a working model. In his work with pendulums, Huygens realized that a pendulum does not take exactly the same time to complete swings of varying lengths. (This doesn't affect a pendulum clock, because the length of the circular arc is kept constant by giving the pendulum a slight boost each time it passes its lowest point.) But, in studying the problem, Huygens discovered that a ball rolling back and forth on an inverted cycloid does complete each cycle in exactly the same time.

James Bernoulli (1654–1705)

James Bernoulli, also called Jacques, was the older brother of John. He was one of several accomplished mathematicians of the Swiss Bernoulli family. James's mathematical accomplishments have given him a prominent place in the early development of calculus.

An inverted cycloid is the path down which a ball will roll in the shortest time.
Figure 10.27

The second problem, which was posed by John Bernoulli in 1696, is called the **brachistochrone problem**—in Greek, *brachys* means short and chronos means time. The problem was to determine the path down which a particle will slide from point A to point B in the *shortest time*. Several mathematicians took up the challenge, and the following year the problem was solved by Newton, Leibniz, L'Hôpital, John Bernoulli, and James Bernoulli. As it turns out, the solution is not a straight line from A to B, but an inverted cycloid passing through the points A and B, as shown in Figure 10.27. The amazing part of the solution is that a particle starting at rest at *any* other point C of the cycloid between A and B will take exactly the same time to reach B, as shown in Figure 10.28.

A ball starting at point C takes the same time to reach point B as one that starts at point A.
Figure 10.28

FOR FURTHER INFORMATION To see a proof of the famous brachistochrone problem, see the article "A New Minimization Proof for the Brachistochrone" by Gary Lawlor in *The American Mathematical Monthly*. To view this article, go to the website *www.matharticles.com*.

Exercises for Section 10.2

See www.CalcChat.com for worked-out solutions to odd-numbered exercises.

1. Consider the parametric equations $x = \sqrt{t}$ and $y = 1 - t$.

(a) Complete the table.

t	0	1	2	3	4
x					
y					

(b) Plot the points (x, y) generated in the table, and sketch a graph of the parametric equations. Indicate the orientation of the graph.

(c) Use a graphing utility to confirm your graph in part (b).

(d) Find the rectangular equation by eliminating the parameter, and sketch its graph. Compare the graph in part (b) with the graph of the rectangular equation.

2. Consider the parametric equations $x = 4\cos^2\theta$ and $y = 2\sin\theta$.

(a) Complete the table.

θ	$-\frac{\pi}{2}$	$-\frac{\pi}{4}$	0	$\frac{\pi}{4}$	$\frac{\pi}{2}$
x					
y					

(b) Plot the points (x, y) generated in the table, and sketch a graph of the parametric equations. Indicate the orientation of the graph.

(c) Use a graphing utility to confirm your graph in part (b).

(d) Find the rectangular equation by eliminating the parameter, and sketch its graph. Compare the graph in part (b) with the graph of the rectangular equation.

(e) If values of θ were selected from the interval $[\pi/2, 3\pi/2]$ for the table in part (a), would the graph in part (b) be different? Explain.

In Exercises 3–20, sketch the curve represented by the parametric equations (indicate the orientation of the curve), and write the corresponding rectangular equation by eliminating the parameter.

3. $x = 3t - 1, \quad y = 2t + 1$

4. $x = 3 - 2t, \quad y = 2 + 3t$

5. $x = t + 1, \quad y = t^2$

6. $x = 2t^2, \quad y = t^4 + 1$

7. $x = t^3, \quad y = \dfrac{t^2}{2}$

8. $x = t^2 + t, \quad y = t^2 - t$

9. $x = \sqrt{t}, \quad y = t - 2$

10. $x = \sqrt[4]{t}, \quad y = 3 - t$

11. $x = t - 1, \quad y = \dfrac{t}{t - 1}$

12. $x = 1 + \dfrac{1}{t}, \quad y = t - 1$

13. $x = 2t, \quad y = |t - 2|$

14. $x = |t - 1|, \quad y = t + 2$

15. $x = e^t, \quad y = e^{3t} + 1$

16. $x = e^{-t}, \quad y = e^{2t} - 1$

17. $x = \sec\theta, \quad y = \cos\theta, \quad 0 \le \theta < \pi/2, \quad \pi/2 < \theta \le \pi$

18. $x = \tan^2\theta, \quad y = \sec^2\theta$

19. $x = 3\cos\theta, \quad y = 3\sin\theta$

20. $x = 2\cos\theta, \quad y = 6\sin\theta$

In Exercises 21–32, use a graphing utility to graph the curve represented by the parametric equations (indicate the orientation of the curve). Eliminate the parameter and write the corresponding rectangular equation.

21. $x = 4\sin 2\theta, y = 2\cos 2\theta$

22. $x = \cos\theta, y = 2\sin 2\theta$

23. $x = 4 + 2\cos\theta$
$y = -1 + \sin\theta$

24. $x = 4 + 2\cos\theta$
$y = -1 + 2\sin\theta$

25. $x = 4 + 2\cos\theta$
$y = -1 + 4\sin\theta$

26. $x = \sec\theta$
$y = \tan\theta$

27. $x = 4\sec\theta, \quad y = 3\tan\theta$

28. $x = \cos^3\theta, \quad y = \sin^3\theta$

29. $x = t^3, \quad y = 3\ln t$

30. $x = \ln 2t, \quad y = t^2$

31. $x = e^{-t}, \quad y = e^{3t}$

32. $x = e^{2t}, \quad y = e^t$

Comparing Plane Curves **In Exercises 33–36, determine any differences between the curves of the parametric equations. Are the graphs the same? Are the orientations the same? Are the curves smooth?**

33. (a) $x = t$
$y = 2t + 1$

(b) $x = \cos\theta$
$y = 2\cos\theta + 1$

(c) $x = e^{-t}$
$y = 2e^{-t} + 1$

(d) $x = e^t$
$y = 2e^t + 1$

34. (a) $x = 2\cos\theta$
$y = 2\sin\theta$

(b) $x = \sqrt{4t^2 - 1}/|t|$
$y = 1/t$

(c) $x = \sqrt{t}$
$y = \sqrt{4 - t}$

(d) $x = -\sqrt{4 - e^{2t}}$
$y = e^t$

35. (a) $x = \cos\theta$
$y = 2\sin^2\theta$
$0 < \theta < \pi$

(b) $x = \cos(-\theta)$
$y = 2\sin^2(-\theta)$
$0 < \theta < \pi$

36. (a) $x = t + 1, y = t^3$

(b) $x = -t + 1, y = (-t)^3$

37. ***Conjecture***

(a) Use a graphing utility to graph the curves represented by the two sets of parametric equations.

$x = 4\cos t \qquad x = 4\cos(-t)$
$y = 3\sin t \qquad y = 3\sin(-t)$

(b) Describe the change in the graph when the sign of the parameter is changed.

(c) Make a conjecture about the change in the graph of parametric equations when the sign of the parameter is changed.

(d) Test your conjecture with another set of parametric equations.

38. ***Writing*** Review Exercises 33–36 and write a short paragraph describing how the graphs of curves represented by different sets of parametric equations can differ even though eliminating the parameter from each yields the same rectangular equation.

In Exercises 39–42, eliminate the parameter and obtain the standard form of the rectangular equation.

39. Line through (x_1, y_1) and (x_2, y_2):

$x = x_1 + t(x_2 - x_1), \quad y = y_1 + t(y_2 - y_1)$

40. Circle: $x = h + r\cos\theta, \quad y = k + r\sin\theta$

41. Ellipse: $x = h + a\cos\theta, \quad y = k + b\sin\theta$

42. Hyperbola: $x = h + a\sec\theta, \quad y = k + b\tan\theta$

In Exercises 43–50, use the results of Exercises 39–42 to find a set of parametric equations for the line or conic.

43. Line: passes through $(0, 0)$ and $(5, -2)$

44. Line: passes through $(1, 4)$ and $(5, -2)$

45. Circle: center: $(2, 1)$; radius: 4

46. Circle: center: $(-3, 1)$; radius: 3

47. Ellipse: vertices: $(\pm 5, 0)$; foci: $(\pm 4, 0)$

48. Ellipse: vertices: $(4, 7)$, $(4, -3)$; foci: $(4, 5)$, $(4, -1)$

49. Hyperbola: vertices: $(\pm 4, 0)$; foci: $(\pm 5, 0)$

50. Hyperbola: vertices: $(0, \pm 1)$; foci: $(0, \pm 2)$

In Exercises 51–54, find two different sets of parametric equations for the rectangular equation.

51. $y = 3x - 2$

52. $y = \dfrac{2}{x - 1}$

53. $y = x^3$

54. $y = x^2$

In Exercises 55–62, use a graphing utility to graph the curve represented by the parametric equations. Indicate the direction of the curve. Identify any points at which the curve is not smooth.

55. Cycloid: $x = 2(\theta - \sin\theta), \quad y = 2(1 - \cos\theta)$

56. Cycloid: $x = \theta + \sin\theta, \quad y = 1 - \cos\theta$

57. Prolate cycloid: $x = \theta - \frac{3}{2}\sin\theta, \quad y = 1 - \frac{3}{2}\cos\theta$

58. Prolate cycloid: $x = 2\theta - 4\sin\theta, \quad y = 2 - 4\cos\theta$

59. Hypocycloid: $x = 3\cos^3\theta, \quad y = 3\sin^3\theta$

60. Curtate cycloid: $x = 2\theta - \sin\theta, \quad y = 2 - \cos\theta$

61. Witch of Agnesi: $x = 2\cot\theta, \quad y = 2\sin^2\theta$

62. Folium of Descartes: $x = \dfrac{3t}{1 + t^3}, \quad y = \dfrac{3t^2}{1 + t^3}$

Writing About Concepts

63. State the definition of a plane curve given by parametric equations.

64. Explain the process of sketching a plane curve given by parametric equations. What is meant by the orientation of the curve?

65. State the definition of a smooth curve.

Writing About Concepts *(continued)*

66. Match each set of parametric equations with the correct graph. [The graphs are labeled (a), (b), (c), (d), (e), and (f).] Explain your reasoning.

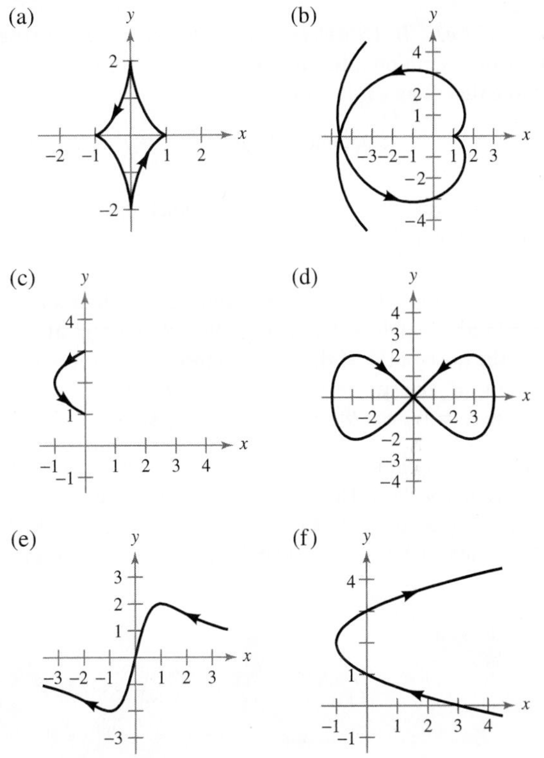

(i) $x = t^2 - 1, \quad y = t + 2$

(ii) $x = \sin^2\theta - 1, \quad y = \sin\theta + 2$

(iii) Lissajous curve: $x = 4\cos\theta, \quad y = 2\sin 2\theta$

(iv) Evolute of ellipse: $x = \cos^3\theta, \quad y = 2\sin^3\theta$

(v) Involute of circle: $x = \cos\theta + \theta\sin\theta$,
$y = \sin\theta - \theta\cos\theta$

(vi) Serpentine curve: $x = \cot\theta, \quad y = 4\sin\theta\cos\theta$

67. ***Curtate Cycloid*** A wheel of radius a rolls along a line without slipping. The curve traced by a point P that is b units from the center $(b < a)$ is called a **curtate cycloid** (see figure). Use the angle θ to find a set of parametric equations for this curve.

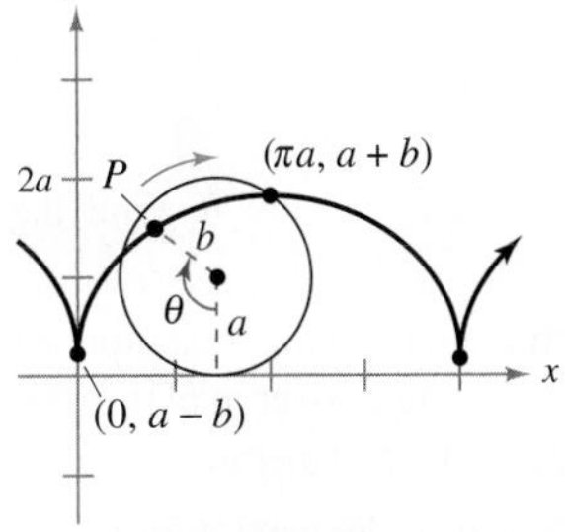

Figure for 67

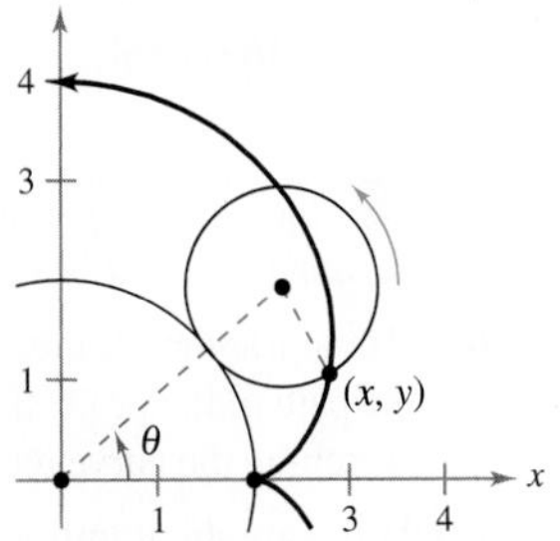

Figure for 68

68. ***Epicycloid*** A circle of radius 1 rolls around the outside of a circle of radius 2 without slipping. The curve traced by a point on the circumference of the smaller circle is called an epicycloid (see figure on previous page). Use the angle θ to find a set of parametric equations for this curve.

True or False? **In Exercises 69 and 70, determine whether the statement is true or false. If it is false, explain why or give an example that shows it is false.**

69. The graph of the parametric equations $x = t^2$ and $y = t^2$ is the line $y = x$.

70. If y is a function of t and x is a function of t, then y is a function of x.

Projectile Motion **In Exercises 71 and 72, consider a projectile launched at a height h feet above the ground and at an angle θ with the horizontal. If the initial velocity is v_0 feet per second, the path of the projectile is modeled by the parametric equations $x = (v_0 \cos \theta)t$ and $y = h + (v_0 \sin \theta)t - 16t^2$.**

71. The center field fence in a ballpark is 10 feet high and 400 feet from home plate. The ball is hit 3 feet above the ground. It leaves the bat at an angle of θ degrees with the horizontal at a speed of 100 miles per hour (see figure).

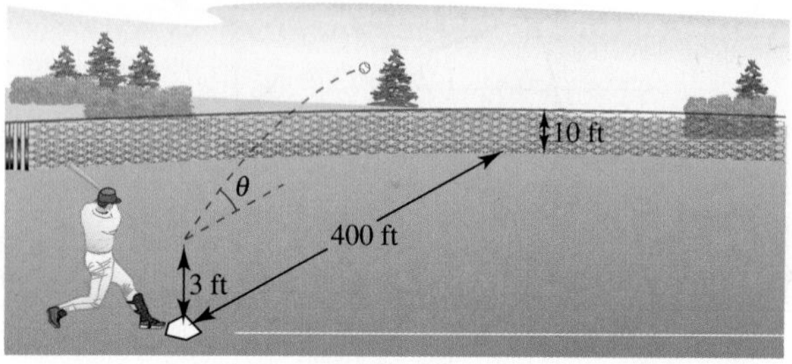

(a) Write a set of parametric equations for the path of the ball.

(b) Use a graphing utility to graph the path of the ball when $\theta = 15°$. Is the hit a home run?

(c) Use a graphing utility to graph the path of the ball when $\theta = 23°$. Is the hit a home run?

(d) Find the minimum angle at which the ball must leave the bat in order for the hit to be a home run.

72. A rectangular equation for the path of a projectile is $y = 5 + x - 0.005x^2$.

(a) Eliminate the parameter t from the position function for the motion of a projectile to show that the rectangular equation is

$$y = -\frac{16 \sec^2 \theta}{v_0^2} x^2 + (\tan \theta)\, x + h.$$

(b) Use the result of part (a) to find h, v_0, and θ. Find the parametric equations of the path.

(c) Use a graphing utility to graph the rectangular equation for the path of the projectile. Confirm your answer in part (b) by sketching the curve represented by the parametric equations.

(d) Use a graphing utility to approximate the maximum height of the projectile and its range.

Section Project: Cycloids

In Greek, the word *cycloid* means *wheel*, the word *hypocycloid* means *under the wheel*, and the word *epicycloid* means *upon the wheel*. Match the hypocycloid or epicycloid with its graph. [The graphs are labeled (a), (b), (c), (d), (e), and (f).]

Hypocycloid, H(A, B)

Path traced by a fixed point on a circle of radius B as it rolls around the inside of a circle of radius A

$$x = (A - B)\cos t + B\cos\left(\frac{A - B}{B}\right)t$$

$$y = (A - B)\sin t - B\sin\left(\frac{A - B}{B}\right)t$$

Epicycloid, E(A, B)

Path traced by a fixed point on a circle of radius B as it rolls around the *outside* of a circle of radius A

$$x = (A + B)\cos t - B\cos\left(\frac{A + B}{B}\right)t$$

$$y = (A + B)\sin t - B\sin\left(\frac{A + B}{B}\right)t$$

I. $H(8, 3)$ II. $E(8,3)$

III. $H(8, 7)$ IV. $E(24, 3)$

V. $H(24, 7)$ VI. $E(24, 7)$

(a)

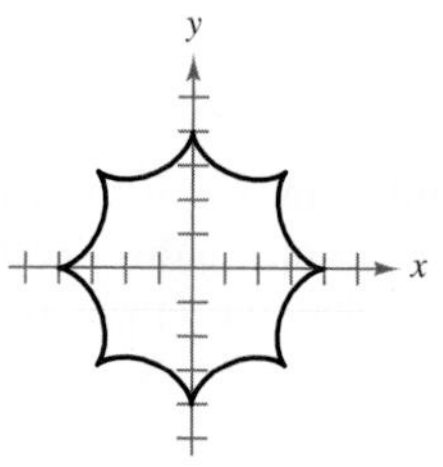

(b)

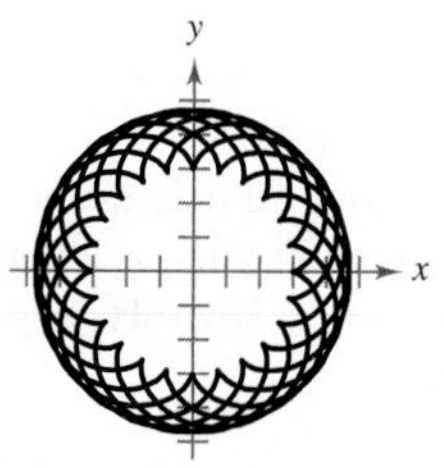

(c)

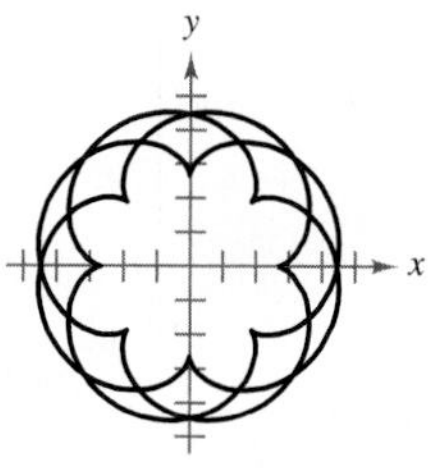

(d)

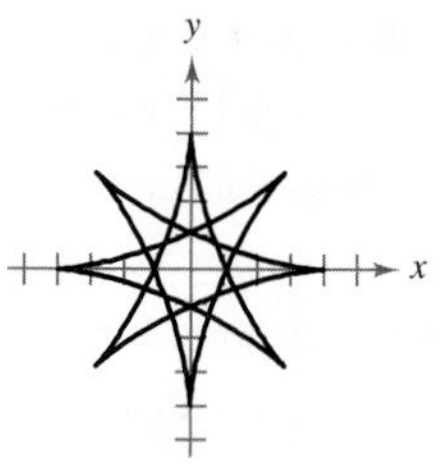

(e)

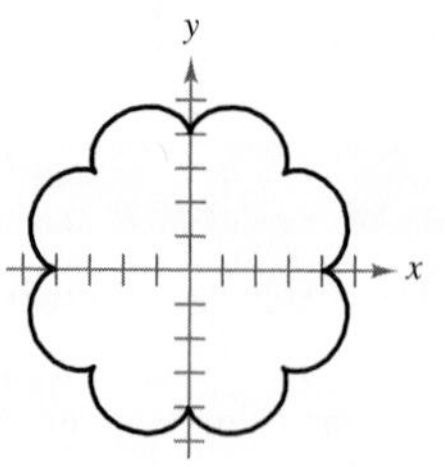

(f)

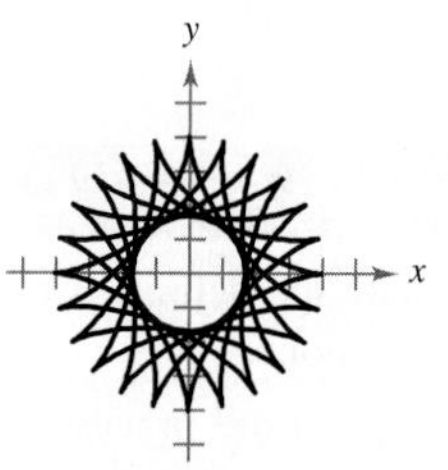

Exercises based on "Mathematical Discovery via Computer Graphics: Hypocycloids and Epicycloids" by Florence S. Gordon and Sheldon P. Gordon, *College Mathematics Journal*, November 1984, p.441. Used by permission of the authors.

Section 10.3 Parametric Equations and Calculus

- **Find the slope of a tangent line to a curve given by a set of parametric equations.**
- **Find the arc length of a curve given by a set of parametric equations.**
- **Find the area of a surface of revolution (parametric form).**

Slope and Tangent Lines

Now that you can represent a graph in the plane by a set of parametric equations, it is natural to ask how to use calculus to study plane curves. To begin, let's take another look at the projectile represented by the parametric equations

$$x = 24\sqrt{2}t \qquad \text{and} \qquad y = -16t^2 + 24\sqrt{2}t$$

as shown in Figure 10.29. From Section 10.2, you know that these equations enable you to locate the position of the projectile at a given time. You also know that the object is initially projected at an angle of 45°. But how can you find the angle θ representing the object's direction at some other time t? The following theorem answers this question by giving a formula for the slope of the tangent line as a function of t.

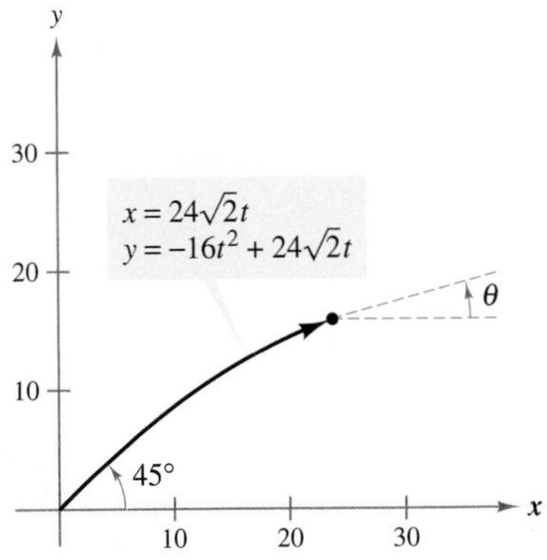

At time t, the angle of elevation of the projectile is θ, the slope of the tangent line at that point.
Figure 10.29

THEOREM 10.7 Parametric Form of the Derivative

If a smooth curve C is given by the equations $x = f(t)$ and $y = g(t)$, then the slope of C at (x, y) is

$$\frac{dy}{dx} = \frac{dy/dt}{dx/dt}, \qquad \frac{dx}{dt} \neq 0.$$

Proof In Figure 10.30, consider $\Delta t > 0$ and let

$$\Delta y = g(t + \Delta t) - g(t) \qquad \text{and} \qquad \Delta x = f(t + \Delta t) - f(t).$$

Because $\Delta x \to 0$ as $\Delta t \to 0$, you can write

$$\frac{dy}{dx} = \lim_{\Delta x \to 0} \frac{\Delta y}{\Delta x}$$
$$= \lim_{\Delta t \to 0} \frac{g(t + \Delta t) - g(t)}{f(t + \Delta t) - f(t)}.$$

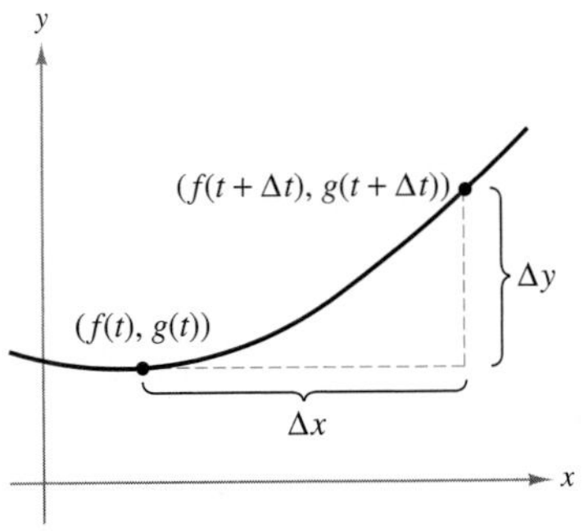

The slope of the secant line through the points $(f(t), g(t))$ and $(f(t + \Delta t), g(t + \Delta t))$ is $\Delta y / \Delta x$.
Figure 10.30

Dividing both the numerator and denominator by Δt, you can use the differentiability of f and g to conclude that

$$\frac{dy}{dx} = \lim_{\Delta t \to 0} \frac{[g(t + \Delta t) - g(t)]/\Delta t}{[f(t + \Delta t) - f(t)]/\Delta t}$$
$$= \frac{\displaystyle\lim_{\Delta t \to 0} \frac{g(t + \Delta t) - g(t)}{\Delta t}}{\displaystyle\lim_{\Delta t \to 0} \frac{f(t + \Delta t) - f(t)}{\Delta t}}$$
$$= \frac{g'(t)}{f'(t)}$$
$$= \frac{dy/dt}{dx/dt}.$$

EXAMPLE 1 Differentiation and Parametric Form

Find dy/dx for the curve given by $x = \sin t$ and $y = \cos t$.

STUDY TIP The curve traced out in Example 1 is a circle. Use the formula

$$\frac{dy}{dx} = -\tan t$$

to find the slopes at the points (1, 0) and (0, 1).

Solution

$$\frac{dy}{dx} = \frac{dy/dt}{dx/dt} = \frac{-\sin t}{\cos t} = -\tan t$$

Because dy/dx is a function of t, you can use Theorem 10.7 repeatedly to find *higher-order* derivatives. For instance,

$$\frac{d^2y}{dx^2} = \frac{d}{dx}\left[\frac{dy}{dx}\right] = \frac{\frac{d}{dt}\left[\frac{dy}{dx}\right]}{dx/dt}$$ Second derivative

$$\frac{d^3y}{dx^3} = \frac{d}{dx}\left[\frac{d^2y}{dx^2}\right] = \frac{\frac{d}{dt}\left[\frac{d^2y}{dx^2}\right]}{dx/dt}.$$ Third derivative

EXAMPLE 2 Finding Slope and Concavity

For the curve given by

$$x = \sqrt{t} \quad \text{and} \quad y = \frac{1}{4}(t^2 - 4), \quad t \geq 0$$

find the slope and concavity at the point (2, 3).

Solution Because

$$\frac{dy}{dx} = \frac{dy/dt}{dx/dt} = \frac{(1/2)t}{(1/2)t^{-1/2}} = t^{3/2}$$ Parametric form of first derivative

you can find the second derivative to be

$$\frac{d^2y}{dx^2} = \frac{\frac{d}{dt}[dy/dx]}{dx/dt} = \frac{\frac{d}{dt}[t^{3/2}]}{dx/dt} = \frac{(3/2)t^{1/2}}{(1/2)t^{-1/2}} = 3t.$$ Parametric form of second derivative

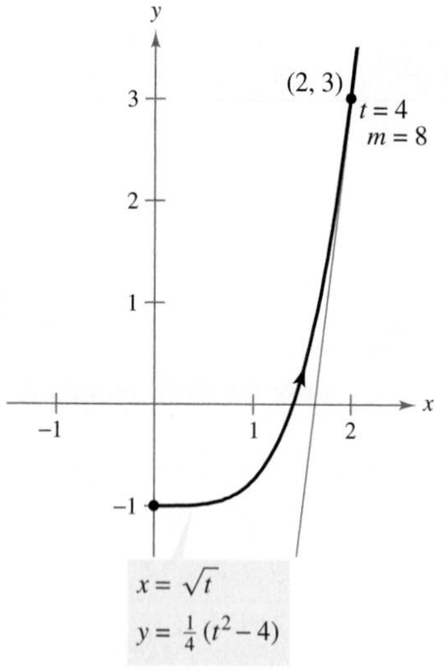

The graph is concave upward at (2, 3), when $t = 4$.
Figure 10.31

At $(x, y) = (2, 3)$, it follows that $t = 4$, and the slope is

$$\frac{dy}{dx} = (4)^{3/2} = 8.$$

Moreover, when $t = 4$, the second derivative is

$$\frac{d^2y}{dx^2} = 3(4) = 12 > 0$$

and you can conclude that the graph is concave upward at (2, 3), as shown in Figure 10.31.

Because the parametric equations $x = f(t)$ and $y = g(t)$ need not define y as a function of x, it is possible for a plane curve to loop around and cross itself. At such points the curve may have more than one tangent line, as shown in the next example.

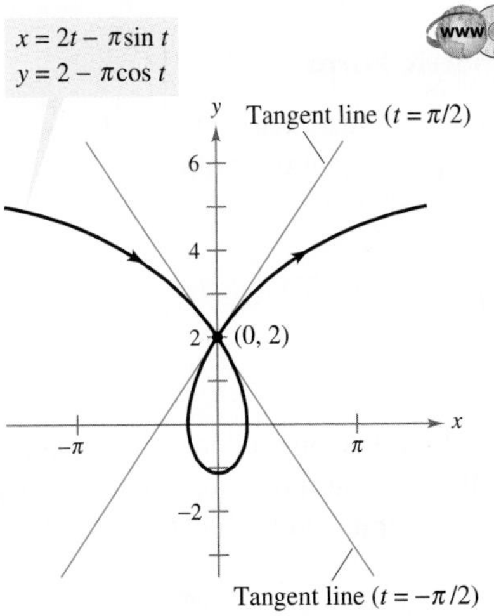

This prolate cycloid has two tangent lines at the point $(0, 2)$.
Figure 10.32

EXAMPLE 3 A Curve with Two Tangent Lines at a Point

The **prolate cycloid** given by

$$x = 2t - \pi \sin t \quad \text{and} \quad y = 2 - \pi \cos t$$

crosses itself at the point (0, 2), as shown in Figure 10.32. Find the equations of both tangent lines at this point.

Solution Because $x = 0$ and $y = 2$ when $t = \pm\pi/2$, and

$$\frac{dy}{dx} = \frac{dy/dt}{dx/dt} = \frac{\pi \sin t}{2 - \pi \cos t}$$

you have $dy/dx = -\pi/2$ when $t = -\pi/2$ and $dy/dx = \pi/2$ when $t = \pi/2$. So, the two tangent lines at (0, 2) are

$$y - 2 = -\left(\frac{\pi}{2}\right)x \qquad \text{Tangent line when } t = -\frac{\pi}{2}$$

$$y - 2 = \left(\frac{\pi}{2}\right)x. \qquad \text{Tangent line when } t = \frac{\pi}{2}$$

If $dy/dt = 0$ and $dx/dt \neq 0$ when $t = t_0$, the curve represented by $x = f(t)$ and $y = g(t)$ has a horizontal tangent at $(f(t_0), g(t_0))$. For instance, in Example 3, the given curve has a horizontal tangent at the point $(0, 2 - \pi)$ (when $t = 0$). Similarly, if $dx/dt = 0$ and $dy/dt \neq 0$ when $t = t_0$, the curve represented by $x = f(t)$ and $y = g(t)$ has a vertical tangent at $(f(t_0), g(t_0))$.

Arc Length

You have seen how parametric equations can be used to describe the path of a particle moving in the plane. You will now develop a formula for determining the *distance* traveled by the particle along its path.

Recall from Section 7.4 that the formula for the arc length of a curve C given by $y = h(x)$ over the interval $[x_0, x_1]$ is

$$\begin{aligned} s &= \int_{x_0}^{x_1} \sqrt{1 + [h'(x)]^2}\, dx \\ &= \int_{x_0}^{x_1} \sqrt{1 + \left(\frac{dy}{dx}\right)^2}\, dx. \end{aligned}$$

If C is represented by the parametric equations $x = f(t)$ and $y = g(t)$, $a \le t \le b$, and if $dx/dt = f'(t) > 0$, you can write

$$\begin{aligned} s = \int_{x_0}^{x_1} \sqrt{1 + \left(\frac{dy}{dx}\right)^2}\, dx &= \int_{x_0}^{x_1} \sqrt{1 + \left(\frac{dy/dt}{dx/dt}\right)^2}\, dx \\ &= \int_a^b \sqrt{\frac{(dx/dt)^2 + (dy/dt)^2}{(dx/dt)^2}}\, \frac{dx}{dt}\, dt \\ &= \int_a^b \sqrt{\left(\frac{dx}{dt}\right)^2 + \left(\frac{dy}{dt}\right)^2}\, dt \\ &= \int_a^b \sqrt{[f'(t)]^2 + [g'(t)]^2}\, dt. \end{aligned}$$

NOTE When applying the arc length formula to a curve, be sure that the curve is traced out only once on the interval of integration. For instance, the circle given by $x = \cos t$ and $y = \sin t$ is traced out once on the interval $0 \le t \le 2\pi$, but is traced out twice on the interval $0 \le t \le 4\pi$.

THEOREM 10.8 Arc Length in Parametric Form

If a smooth curve C is given by $x = f(t)$ and $y = g(t)$ such that C does not intersect itself on the interval $a \le t \le b$ (except possibly at the endpoints), then the arc length of C over the interval is given by

$$s = \int_a^b \sqrt{\left(\frac{dx}{dt}\right)^2 + \left(\frac{dy}{dt}\right)^2}\, dt = \int_a^b \sqrt{[f'(t)]^2 + [g'(t)]^2}\, dt.$$

In the preceding section you saw that if a circle rolls along a line, a point on its circumference will trace a path called a cycloid. If the circle rolls around the circumference of another circle, the path of the point is an **epicycloid.** The next example shows how to find the arc length of an epicycloid.

ARCH OF A CYCLOID

The arc length of an arch of a cycloid was first calculated in 1658 by British architect and mathematician Christopher Wren, famous for rebuilding many buildings and churches in London, including St. Paul's Cathedral.

EXAMPLE 4 Finding Arc Length

A circle of radius 1 rolls around the circumference of a larger circle of radius 4, as shown in Figure 10.33. The epicycloid traced by a point on the circumference of the smaller circle is given by

$$x = 5\cos t - \cos 5t$$

and

$$y = 5\sin t - \sin 5t.$$

Find the distance traveled by the point in one complete trip about the larger circle.

Solution Before applying Theorem 10.8, note in Figure 10.33 that the curve has sharp points when $t = 0$ and $t = \pi/2$. Between these two points, dx/dt and dy/dt are not simultaneously 0. So, the portion of the curve generated from $t = 0$ to $t = \pi/2$ is smooth. To find the total distance traveled by the point, you can find the arc length of that portion lying in the first quadrant and multiply by 4.

$$s = 4\int_0^{\pi/2} \sqrt{\left(\frac{dx}{dt}\right)^2 + \left(\frac{dy}{dt}\right)^2}\, dt \qquad \text{Parametric form for arc length}$$

$$= 4\int_0^{\pi/2} \sqrt{(-5\sin t + 5\sin 5t)^2 + (5\cos t - 5\cos 5t)^2}\, dt$$

$$= 20\int_0^{\pi/2} \sqrt{2 - 2\sin t \sin 5t - 2\cos t \cos 5t}\, dt$$

$$= 20\int_0^{\pi/2} \sqrt{2 - 2\cos 4t}\, dt$$

$$= 20\int_0^{\pi/2} \sqrt{4\sin^2 2t}\, dt \qquad \text{Trigonometric identity}$$

$$= 40\int_0^{\pi/2} \sin 2t\, dt$$

$$= -20\Big[\cos 2t\Big]_0^{\pi/2}$$

$$= 40$$

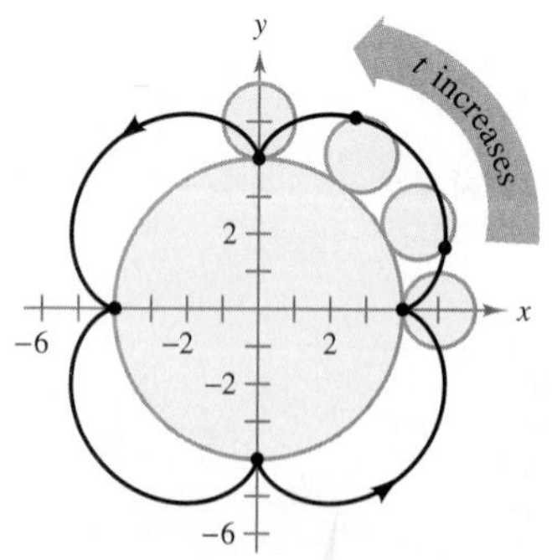

An epicycloid is traced by a point on the smaller circle as it rolls around the larger circle.
Figure 10.33

For the epicycloid shown in Figure 10.33, an arc length of 40 seems about right because the circumference of a circle of radius 6 is $2\pi r = 12\pi \approx 37.7$.

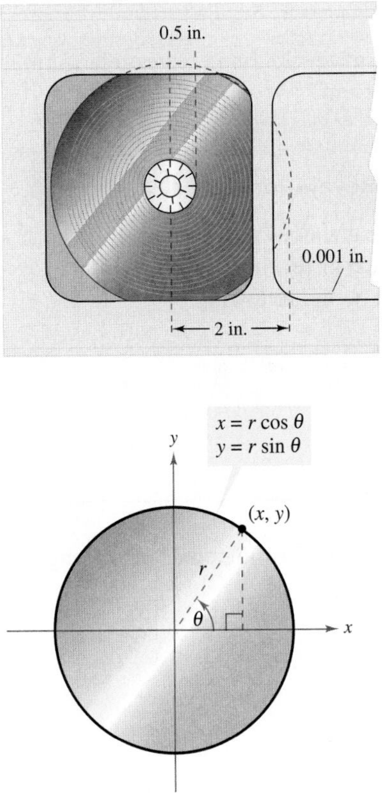

Figure 10.34

EXAMPLE 5 Length of a Recording Tape

A recording tape 0.001 inch thick is wound around a reel whose inner radius is 0.5 inch and whose outer radius is 2 inches, as shown in Figure 10.34. How much tape is required to fill the reel?

Solution To create a model for this problem, assume that as the tape is wound around the reel its distance r from the center increases linearly at a rate of 0.001 inch per revolution, or

$$r = (0.001)\frac{\theta}{2\pi} = \frac{\theta}{2000\pi}, \qquad 1000\pi \le \theta \le 4000\pi$$

where θ is measured in radians. You can determine the coordinates of the point (x, y) corresponding to a given radius to be

$$x = r\cos\theta$$

and

$$y = r\sin\theta.$$

Substituting for r, you obtain the parametric equations

$$x = \left(\frac{\theta}{2000\pi}\right)\cos\theta \quad \text{and} \quad y = \left(\frac{\theta}{2000\pi}\right)\sin\theta.$$

You can use the arc length formula to determine the total length of the tape to be

$$\begin{aligned} s &= \int_{1000\pi}^{4000\pi} \sqrt{\left(\frac{dx}{d\theta}\right)^2 + \left(\frac{dy}{d\theta}\right)^2}\, d\theta \\ &= \frac{1}{2000\pi}\int_{1000\pi}^{4000\pi} \sqrt{(-\theta\sin\theta + \cos\theta)^2 + (\theta\cos\theta + \sin\theta)^2}\, d\theta \\ &= \frac{1}{2000\pi}\int_{1000\pi}^{4000\pi} \sqrt{\theta^2 + 1}\, d\theta \\ &= \frac{1}{2000\pi}\left(\frac{1}{2}\right)\Big[\theta\sqrt{\theta^2+1} + \ln\left|\theta + \sqrt{\theta^2+1}\right|\Big]_{1000\pi}^{4000\pi} \qquad \text{Integration tables (Appendix B), Formula 26} \\ &\approx 11{,}781 \text{ inches} \\ &\approx 982 \text{ feet} \end{aligned}$$

FOR FURTHER INFORMATION For more information on the mathematics of recording tape, see "Tape Counters" by Richard L. Roth in *The American Mathematical Monthly*. To view this article, go to the website *www.matharticles.com*.

NOTE The graph of $r = a\theta$ is called the **spiral of Archimedes**. The graph of $r = \theta/2000\pi$ (in Example 5) is of this form.

The length of the tape in Example 5 can be approximated by adding the circumferences of circular pieces of tape. The smallest circle has a radius of 0.501 and the largest has a radius of 2.

$$\begin{aligned} s &\approx 2\pi(0.501) + 2\pi(0.502) + 2\pi(0.503) + \cdots + 2\pi(2.000) \\ &= \sum_{i=1}^{1500} 2\pi(0.5 + 0.001i) \\ &= 2\pi[1500(0.5) + 0.001(1500)(1501)/2] \\ &\approx 11{,}786 \text{ inches} \end{aligned}$$

Area of a Surface of Revolution

You can use the formula for the area of a surface of revolution in rectangular form to develop a formula for surface area in parametric form.

THEOREM 10.9 Area of a Surface of Revolution

If a smooth curve C given by $x = f(t)$ and $y = g(t)$ does not cross itself on an interval $a \le t \le b$, then the area S of the surface of revolution formed by revolving C about the coordinate axes is given by the following.

1. $S = 2\pi \int_a^b g(t) \sqrt{\left(\frac{dx}{dt}\right)^2 + \left(\frac{dy}{dt}\right)^2}\, dt$ Revolution about the x-axis: $g(t) \ge 0$

2. $S = 2\pi \int_a^b f(t) \sqrt{\left(\frac{dx}{dt}\right)^2 + \left(\frac{dy}{dt}\right)^2}\, dt$ Revolution about the y-axis: $f(t) \ge 0$

These formulas are easy to remember if you think of the differential of arc length as

$$ds = \sqrt{\left(\frac{dx}{dt}\right)^2 + \left(\frac{dy}{dt}\right)^2}\, dt.$$

Then the formulas are written as follows.

1. $S = 2\pi \int_a^b g(t)\, ds$ **2.** $S = 2\pi \int_a^b f(t)\, ds$

EXAMPLE 6 Finding the Area of a Surface of Revolution

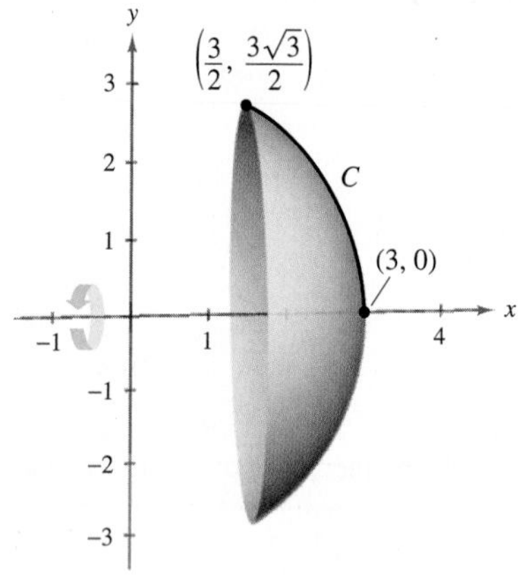

This surface of revolution has a surface area of 9π.
Figure 10.35

Let C be the arc of the circle

$$x^2 + y^2 = 9$$

from $(3, 0)$ to $(3/2, 3\sqrt{3}/2)$, as shown in Figure 10.35. Find the area of the surface formed by revolving C about the x-axis.

Solution You can represent C parametrically by the equations

$$x = 3\cos t \quad \text{and} \quad y = 3\sin t, \quad 0 \le t \le \pi/3.$$

(Note that you can determine the interval for t by observing that $t = 0$ when $x = 3$ and $t = \pi/3$ when $x = 3/2$.) On this interval, C is smooth and y is nonnegative, and you can apply Theorem 10.9 to obtain a surface area of

$$S = 2\pi \int_0^{\pi/3} (3\sin t)\sqrt{(-3\sin t)^2 + (3\cos t)^2}\, dt \quad \text{Formula for area of a surface of revolution}$$

$$= 6\pi \int_0^{\pi/3} \sin t \sqrt{9(\sin^2 t + \cos^2 t)}\, dt$$

$$= 6\pi \int_0^{\pi/3} 3\sin t\, dt \quad \text{Trigonometric identity}$$

$$= -18\pi \Big[\cos t\Big]_0^{\pi/3}$$

$$= -18\pi\left(\frac{1}{2} - 1\right)$$

$$= 9\pi.$$

Exercises for Section 10.3

See www.CalcChat.com for worked-out solutions to odd-numbered exercises.

In Exercises 1–4, find dy/dx.

1. $x = t^2,\ y = 5 - 4t$
2. $x = \sqrt[3]{t},\ y = 4 - t$
3. $x = \sin^2 \theta,\ y = \cos^2 \theta$
4. $x = 2e^{\theta},\ y = e^{-\theta/2}$

In Exercises 5–14, find dy/dx and d^2y/dx^2, and find the slope and concavity (if possible) at the given value of the parameter.

Parametric Equations	*Point*
5. $x = 2t,\ y = 3t - 1$	$t = 3$
6. $x = \sqrt{t},\ y = 3t - 1$	$t = 1$
7. $x = t + 1,\ y = t^2 + 3t$	$t = -1$
8. $x = t^2 + 3t + 2,\ y = 2t$	$t = 0$
9. $x = 2\cos\theta,\ y = 2\sin\theta$	$\theta = \dfrac{\pi}{4}$
10. $x = \cos\theta,\ y = 3\sin\theta$	$\theta = 0$
11. $x = 2 + \sec\theta,\ y = 1 + 2\tan\theta$	$\theta = \dfrac{\pi}{6}$
12. $x = \sqrt{t},\ y = \sqrt{t - 1}$	$t = 2$
13. $x = \cos^3\theta,\ y = \sin^3\theta$	$\theta = \dfrac{\pi}{4}$
14. $x = \theta - \sin\theta,\ y = 1 - \cos\theta$	$\theta = \pi$

In Exercises 15 and 16, find an equation of the tangent line at each given point on the curve.

15. $x = 2\cot\theta$
 $y = 2\sin^2\theta$

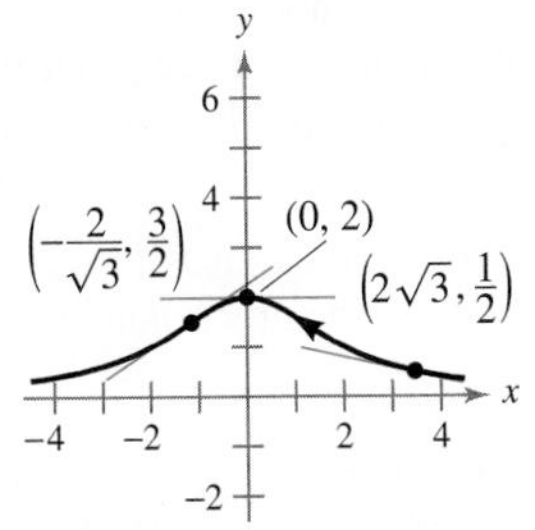

16. $x = 2 - 3\cos\theta$
 $y = 3 + 2\sin\theta$

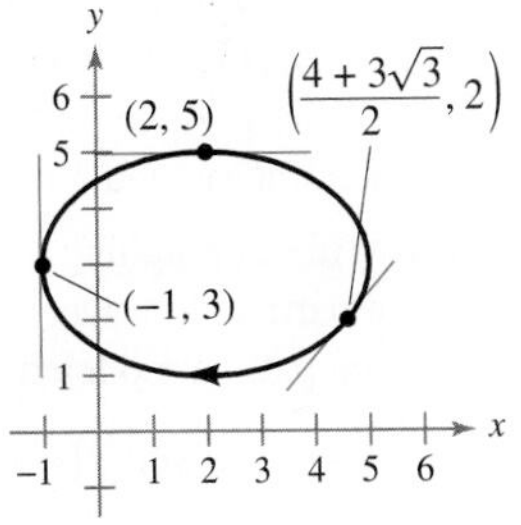

In Exercises 17–20, (a) use a graphing utility to graph the curve represented by the parametric equations, (b) use a graphing utility to find dx/dt, dy/dt, and dy/dx at the given value of the parameter, (c) find an equation of the tangent line to the curve at the given value of the parameter, and (d) use a graphing utility to graph the curve and the tangent line from part (c).

Parametric Equations	*Parameter*
17. $x = 2t,\ y = t^2 - 1$	$t = 2$
18. $x = t - 1,\ y = \dfrac{1}{t} + 1$	$t = 1$
19. $x = t^2 - t + 2,\ y = t^3 - 3t$	$t = -1$
20. $x = 4\cos\theta,\ y = 3\sin\theta$	$\theta = \dfrac{3\pi}{4}$

In Exercises 21–24, find the equations of the tangent lines at the point where the curve crosses itself.

21. $x = 2\sin 2t,\quad y = 3\sin t$
22. $x = 2 - \pi\cos t,\quad y = 2t - \pi\sin t$
23. $x = t^2 - t,\quad y = t^3 - 3t - 1$
24. $x = t^3 - 6t,\quad y = t^2$

In Exercises 25 and 26, find all points (if any) of horizontal and vertical tangency to the portion of the curve shown.

25. Involute of a circle:
 $x = \cos\theta + \theta\sin\theta$
 $y = \sin\theta - \theta\cos\theta$

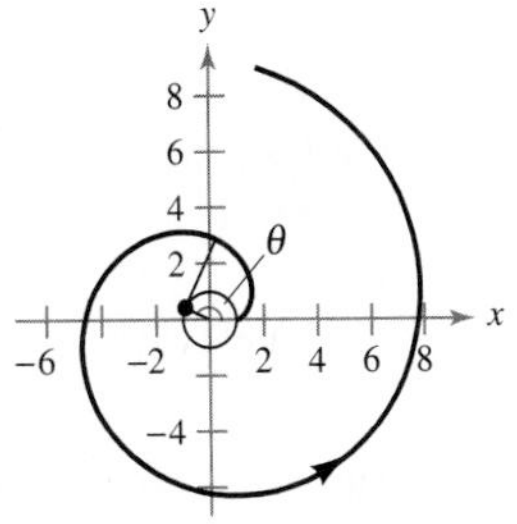

26. $x = 2\theta$
 $y = 2(1 - \cos\theta)$

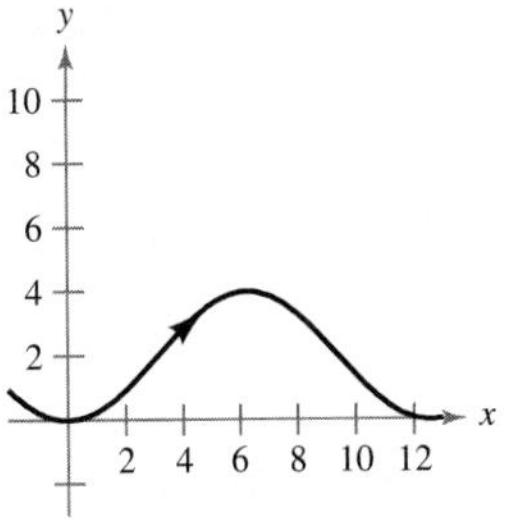

In Exercises 27–36, find all points (if any) of horizontal and vertical tangency to the curve. Use a graphing utility to confirm your results.

27. $x = 1 - t,\quad y = t^2$
28. $x = t + 1,\quad y = t^2 + 3t$
29. $x = 1 - t,\quad y = t^3 - 3t$
30. $x = t^2 - t + 2,\quad y = t^3 - 3t$
31. $x = 3\cos\theta,\quad y = 3\sin\theta$
32. $x = \cos\theta,\quad y = 2\sin 2\theta$
33. $x = 4 + 2\cos\theta,\quad y = -1 + \sin\theta$
34. $x = 4\cos^2\theta,\quad y = 2\sin\theta$
35. $x = \sec\theta,\quad y = \tan\theta$
36. $x = \cos^2\theta,\quad y = \cos\theta$

In Exercises 37–42, determine the t intervals on which the curve is concave downward or concave upward.

37. $x = t^2,\quad y = t^3 - t$
38. $x = 2 + t^2,\quad y = t^2 + t^3$
39. $x = 2t + \ln t,\quad y = 2t - \ln t$
40. $x = t^2,\quad y = \ln t$
41. $x = \sin t,\quad y = \cos t,\quad 0 < t < \pi$
42. $x = 2\cos t,\quad y = \sin t,\quad 0 < t < 2\pi$

Arc Length **In Exercises 43–46, write an integral that represents the arc length of the curve on the given interval. Do not evaluate the integral.**

Parametric Equations	*Interval*
43. $x = 2t - t^2, \quad y = 2t^{3/2}$	$1 \le t \le 2$
44. $x = \ln t, \quad y = t + 1$	$1 \le t \le 6$
45. $x = e^t + 2, \quad y = 2t + 1$	$-2 \le t \le 2$
46. $x = t + \sin t, \quad y = t - \cos t$	$0 \le t \le \pi$

Arc Length **In Exercises 47–52, find the arc length of the curve on the given interval.**

Parametric Equations	*Interval*
47. $x = t^2, \quad y = 2t$	$0 \le t \le 2$
48. $x = t^2 + 1, \quad y = 4t^3 + 3$	$-1 \le t \le 0$
49. $x = e^{-t}\cos t, \quad y = e^{-t}\sin t$	$0 \le t \le \frac{\pi}{2}$
50. $x = \arcsin t, \quad y = \ln\sqrt{1 - t^2}$	$0 \le t \le \frac{1}{2}$
51. $x = \sqrt{t}, \quad y = 3t - 1$	$0 \le t \le 1$
52. $x = t, \quad y = \frac{t^5}{10} + \frac{1}{6t^3}$	$1 \le t \le 2$

Arc Length **In Exercises 53–56, find the arc length of the curve on the interval $[0, 2\pi]$.**

53. Hypocycloid perimeter: $x = a\cos^3\theta, y = a\sin^3\theta$

54. Circle circumference: $x = a\cos\theta, y = a\sin\theta$

55. Cycloid arch: $x = a(\theta - \sin\theta), y = a(1 - \cos\theta)$

56. Involute of a circle: $x = \cos\theta + \theta\sin\theta, y = \sin\theta - \theta\cos\theta$

57. ***Path of a Projectile*** The path of a projectile is modeled by the parametric equations

$$x = (90\cos 30°)t \quad \text{and} \quad y = (90\sin 30°)t - 16t^2$$

where x and y are measured in feet.

(a) Use a graphing utility to graph the path of the projectile.

(b) Use a graphing utility to approximate the range of the projectile.

(c) Use the integration capabilities of a graphing utility to approximate the arc length of the path. Compare this result with the range of the projectile.

58. ***Path of a Projectile*** If the projectile in Exercise 57 is launched at an angle θ with the horizontal, its parametric equations are

$$x = (90\cos\theta)t \quad \text{and} \quad y = (90\sin\theta)t - 16t^2.$$

Use a graphing utility to find the angle that maximizes the range of the projectile. What angle maximizes the arc length of the trajectory?

59. ***Folium of Descartes*** Consider the parametric equations

$$x = \frac{4t}{1 + t^3} \quad \text{and} \quad y = \frac{4t^2}{1 + t^3}.$$

(a) Use a graphing utility to graph the curve represented by the parametric equations.

(b) Use a graphing utility to find the points of horizontal tangency to the curve.

(c) Use the integration capabilities of a graphing utility to approximate the arc length of the closed loop. (*Hint:* Use symmetry and integrate over the interval $0 \le t \le 1$.)

60. ***Witch of Agnesi*** Consider the parametric equations

$$x = 4\cot\theta \quad \text{and} \quad y = 4\sin^2\theta, \quad -\frac{\pi}{2} \le \theta \le \frac{\pi}{2}.$$

(a) Use a graphing utility to graph the curve represented by the parametric equations.

(b) Use a graphing utility to find the points of horizontal tangency to the curve.

(c) Use the integration capabilities of a graphing utility to approximate the arc length over the interval $\pi/4 \le \theta \le \pi/2$.

61. ***Writing***

(a) Use a graphing utility to graph each set of parametric equations.

$x = t - \sin t$	$x = 2t - \sin(2t)$
$y = 1 - \cos t$	$y = 1 - \cos(2t)$
$0 \le t \le 2\pi$	$0 \le t \le \pi$

(b) Compare the graphs of the two sets of parametric equations in part (a). If the curve represents the motion of a particle and t is time, what can you infer about the average speeds of the particle on the paths represented by the two sets of parametric equations?

(c) Without graphing the curve, determine the time required for a particle to traverse the same path as in parts (a) and (b) if the path is modeled by

$$x = \tfrac{1}{2}t - \sin\left(\tfrac{1}{2}t\right) \quad \text{and} \quad y = 1 - \cos\left(\tfrac{1}{2}t\right).$$

62. ***Writing***

(a) Each set of parametric equations represents the motion of a particle. Use a graphing utility to graph each set.

First Particle	*Second Particle*
$x = 3\cos t$	$x = 4\sin t$
$y = 4\sin t$	$y = 3\cos t$
$0 \le t \le 2\pi$	$0 \le t \le 2\pi$

(b) Determine the number of points of intersection.

(c) Will the particles ever be at the same place at the same time? If so, identify the points.

(d) Explain what happens if the motion of the second particle is represented by

$$x = 2 + 3\sin t, \quad y = 2 - 4\cos t, \quad 0 \le t \le 2\pi.$$

Surface Area **In Exercises 63–66, write an integral that represents the area of the surface generated by revolving the curve about the x-axis. Use a graphing utility to approximate the integral.**

Parametric Equations	*Interval*
63. $x = 4t, \quad y = t + 1$	$0 \le t \le 2$
64. $x = \frac{1}{4}t^2, \quad y = t + 2$	$0 \le t \le 4$
65. $x = \cos^2 \theta, \quad y = \cos \theta$	$0 \le \theta \le \frac{\pi}{2}$
66. $x = \theta + \sin \theta, \quad y = \theta + \cos \theta$	$0 \le \theta \le \frac{\pi}{2}$

Surface Area **In Exercises 67–72, find the area of the surface generated by revolving the curve about each given axis.**

67. $x = t, y = 2t, \quad 0 \le t \le 4,$ (a) x-axis (b) y-axis

68. $x = t, y = 4 - 2t, \quad 0 \le t \le 2,$ (a) x-axis (b) y-axis

69. $x = 4 \cos \theta, y = 4 \sin \theta, \quad 0 \le \theta \le \frac{\pi}{2}, \quad$ y-axis

70. $x = \frac{1}{3}t^3, y = t + 1, \quad 1 \le t \le 2, \quad$ y-axis

71. $x = a \cos^3 \theta, y = a \sin^3 \theta, \quad 0 \le \theta \le \pi, \quad$ x-axis

72. $x = a \cos \theta, y = b \sin \theta, \quad 0 \le \theta \le 2\pi,$

(a) x-axis (b) y-axis

Writing About Concepts

73. Give the parametric form of the derivative.

74. Mentally determine dy/dx.

(a) $x = t, \quad y = 4$ (b) $x = t, \quad y = 4t - 3$

75. Sketch a graph of a curve defined by the parametric equations $x = g(t)$ and $y = f(t)$ such that $dx/dt > 0$ and $dy/dt < 0$ for all real numbers t.

76. Sketch a graph of a curve defined by the parametric equations $x = g(t)$ and $y = f(t)$ such that $dx/dt < 0$ and $dy/dt < 0$ for all real numbers t.

77. Give the integral formula for arc length in parametric form.

78. Give the integral formulas for the areas of the surfaces of revolution formed when a smooth curve C is revolved about (a) the x-axis and (b) the y-axis.

79. Use integration by substitution to show that if y is a continuous function of x on the interval $a \le x \le b$, where $x = f(t)$ and $y = g(t)$, then

$$\int_a^b y\,dx = \int_{t_1}^{t_2} g(t)f'(t)\,dt$$

where $f(t_1) = a$, $f(t_2) = b$, and both g and f' are continuous on $[t_1, t_2]$.

80. ***Surface Area*** A portion of a sphere of radius r is removed by cutting out a circular cone with its vertex at the center of the sphere. The vertex of the cone forms an angle of 2θ. Find the surface area removed from the sphere.

Area **In Exercises 81 and 82, find the area of the region. (Use the result of Exercise 79.)**

81. $x = 2 \sin^2 \theta$

$y = 2 \sin^2 \theta \tan \theta$

$0 \le \theta < \frac{\pi}{2}$

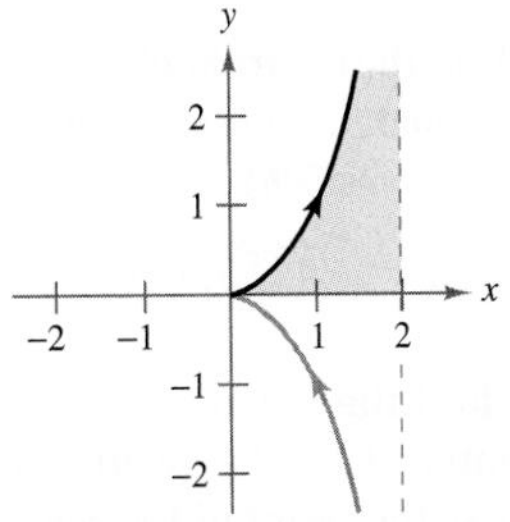

82. $x = 2 \cot \theta$

$y = 2 \sin^2 \theta$

$0 < \theta < \pi$

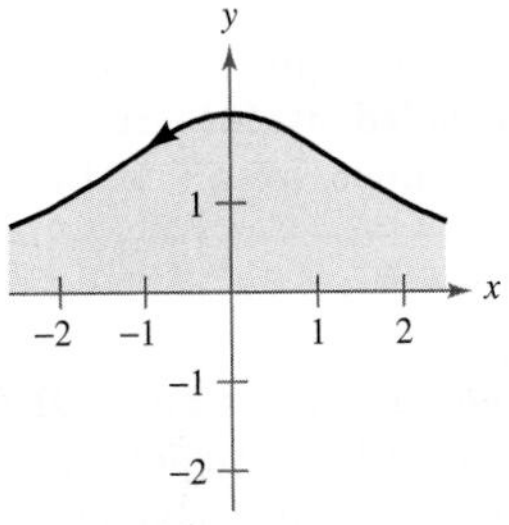

Areas of Simple Closed Curves **In Exercises 83–88, use a computer algebra system and the result of Exercise 79 to match the closed curve with its area. (These exercises were adapted from the article "The Surveyor's Area Formula" by Bart Braden in the September 1986 issue of the *College Mathematics Journal*, by permission of the author.)**

(a) $\frac{8}{3}ab$ **(b)** $\frac{3}{8}\pi a^2$ **(c)** $2\pi a^2$

(d) πab **(e)** $2\pi ab$ **(f)** $6\pi a^2$

83. Ellipse: $(0 \le t \le 2\pi)$

$x = b \cos t$

$y = a \sin t$

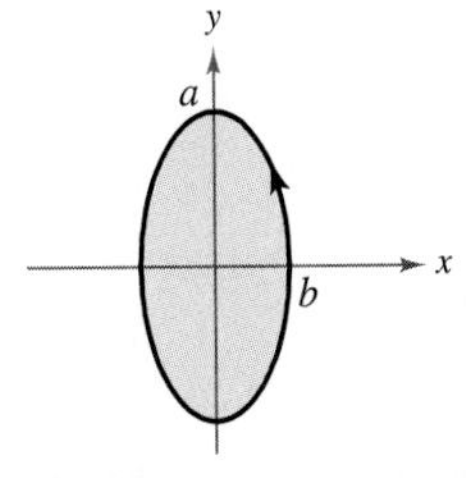

84. Astroid: $(0 \le t \le 2\pi)$

$x = a \cos^3 t$

$y = a \sin^3 t$

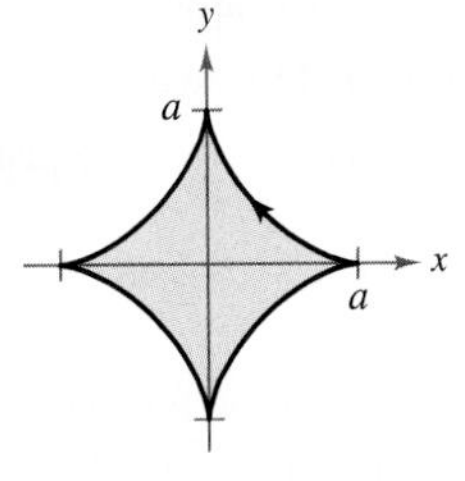

85. Cardioid: $(0 \le t \le 2\pi)$

$x = 2a \cos t - a \cos 2t$

$y = 2a \sin t - a \sin 2t$

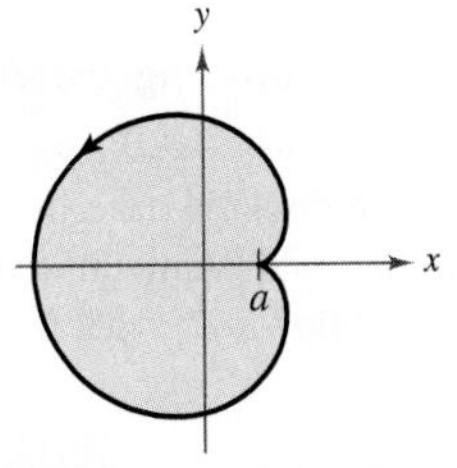

86. Deltoid: $(0 \le t \le 2\pi)$

$x = 2a \cos t + a \cos 2t$

$y = 2a \sin t - a \sin 2t$

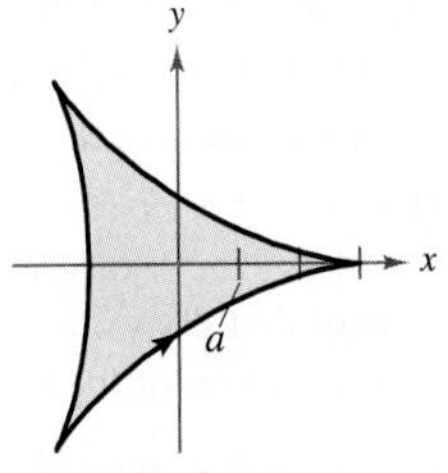

87. Hourglass: $(0 \le t \le 2\pi)$

$x = a \sin 2t$

$y = b \sin t$

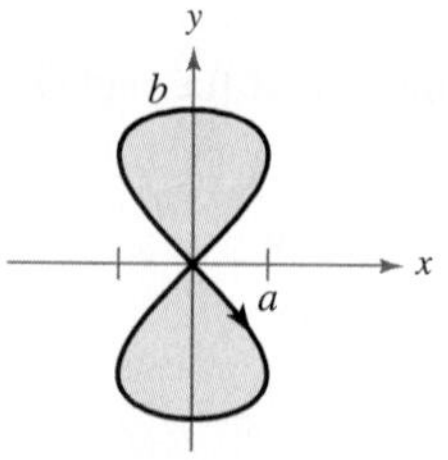

88. Teardrop: $(0 \le t \le 2\pi)$

$x = 2a \cos t - a \sin 2t$

$y = b \sin t$

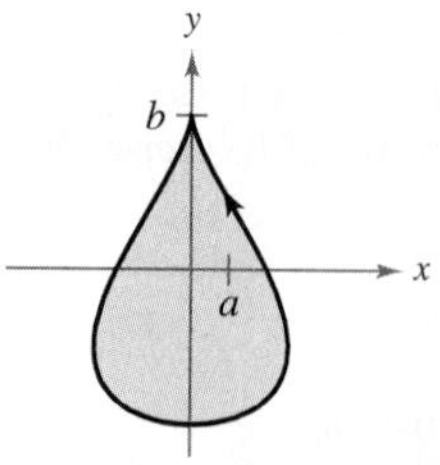

Centroid **In Exercises 89 and 90, find the centroid of the region bounded by the graph of the parametric equations and the coordinate axes. (Use the result of Exercise 79.)**

89. $x = \sqrt{t},\ y = 4 - t$

90. $x = \sqrt{4 - t},\ y = \sqrt{t}$

Volume **In Exercises 91 and 92, find the volume of the solid formed by revolving the region bounded by the graphs of the given equations about the x-axis. (Use the result of Exercise 79.)**

91. $x = 3 \cos \theta, \quad y = 3 \sin \theta$

92. $x = \cos \theta, \quad y = 3 \sin \theta, \quad a > 0$

93. ***Cycloid*** Use the parametric equations

$x = a(\theta - \sin \theta)$ and $y = a(1 - \cos \theta),\ a > 0$

to answer the following.

(a) Find dy/dx and d^2y/dx^2.

(b) Find the equations of the tangent line at the point where $\theta = \pi/6$.

(c) Find all points (if any) of horizontal tangency.

(d) Determine where the curve is concave upward or concave downward.

(e) Find the length of one arc of the curve.

94. Use the parametric equations

$x = t^2\sqrt{3}$ and $y = 3t - \frac{1}{3}t^3$

to answer the following.

(a) Use a graphing utility to graph the curve on the interval $-3 \le t \le 3$.

(b) Find dy/dx and d^2y/dx^2.

(c) Find the equation of the tangent line at the point $\left(\sqrt{3}, \frac{8}{3}\right)$.

(d) Find the length of the curve.

(e) Find the surface area generated by revolving the curve about the x-axis.

95. ***Involute of a Circle*** The involute of a circle is described by the endpoint P of a string that is held taut as it is unwound from a spool that does not turn (see figure). Show that a parametric representation of the involute is

$x = r(\cos \theta + \theta \sin \theta)$ and $y = r(\sin \theta - \theta \cos \theta)$.

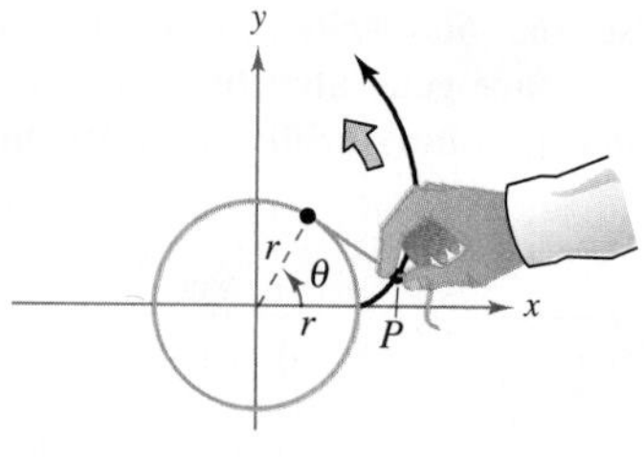

Figure for 95

96. ***Involute of a Circle*** The figure shows a piece of string tied to a circle with a radius of one unit. The string is just long enough to reach the opposite side of the circle. Find the area that is covered when the string is unwound counterclockwise.

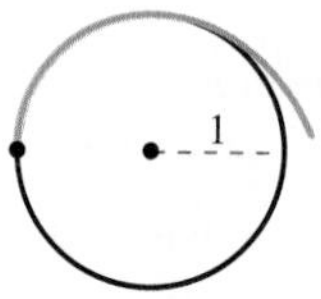

97. (a) Use a graphing utility to graph the curve given by

$$x = \frac{1 - t^2}{1 + t^2},\quad y = \frac{2t}{1 + t^2},\qquad -20 \le t \le 20.$$

(b) Describe the graph and confirm your result analytically.

(c) Discuss the speed at which the curve is traced as t increases from -20 to 20.

98. ***Tractrix*** A person moves from the origin along the positive y-axis pulling a weight at the end of a 12-meter rope. Initially, the weight is located at the point (12, 0).

(a) In Exercise 86 of Section 8.7, it was shown that the path of the weight is modeled by the rectangular equation

$$y = -12 \ln\left(\frac{12 - \sqrt{144 - x^2}}{x}\right) - \sqrt{144 - x^2}$$

where $0 < x \le 12$. Use a graphing utility to graph the rectangular equation.

(b) Use a graphing utility to graph the parametric equations

$$x = 12 \operatorname{sech} \frac{t}{12} \quad \text{and} \quad y = t - 12 \tanh \frac{t}{12}$$

where $t \ge 0$. How does this graph compare with the graph in part (a)? Which graph (if either) do you think is a better representation of the path?

(c) Use the parametric equations for the tractrix to verify that the distance from the y-intercept of the tangent line to the point of tangency is independent of the location of the point of tangency.

True or False? **In Exercises 99 and 100, determine whether the statement is true or false. If it is false, explain why or give an example that shows it is false.**

99. If $x = f(t)$ and $y = g(t)$, then $d^2y/dx^2 = g''(t)/f''(t)$.

100. The curve given by $x = t^3$, $y = t^2$ has a horizontal tangent at the origin because $dy/dt = 0$ when $t = 0$.

Section 10.4 Polar Coordinates and Polar Graphs

- **Understand the polar coordinate system.**
- **Rewrite rectangular coordinates and equations in polar form and vice versa.**
- **Sketch the graph of an equation given in polar form.**
- **Find the slope of a tangent line to a polar graph.**
- **Identify several types of special polar graphs.**

Polar Coordinates

So far, you have been representing graphs as collections of points (x, y) on the rectangular coordinate system. The corresponding equations for these graphs have been in either rectangular or parametric form. In this section you will study a coordinate system called the **polar coordinate system.**

To form the polar coordinate system in the plane, fix a point O, called the **pole** (or **origin**), and construct from O an initial ray called the **polar axis,** as shown in Figure 10.36. Then each point P in the plane can be assigned **polar coordinates** (r, θ), as follows.

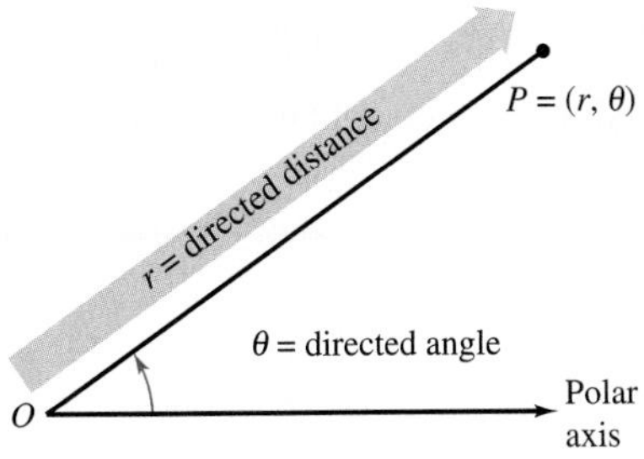

Polar coordinates
Figure 10.36

r = *directed distance* from O to P

θ = *directed angle*, counterclockwise from polar axis to segment $\overline{OP}$

Figure 10.37 shows three points on the polar coordinate system. Notice that in this system, it is convenient to locate points with respect to a grid of concentric circles intersected by **radial lines** through the pole.

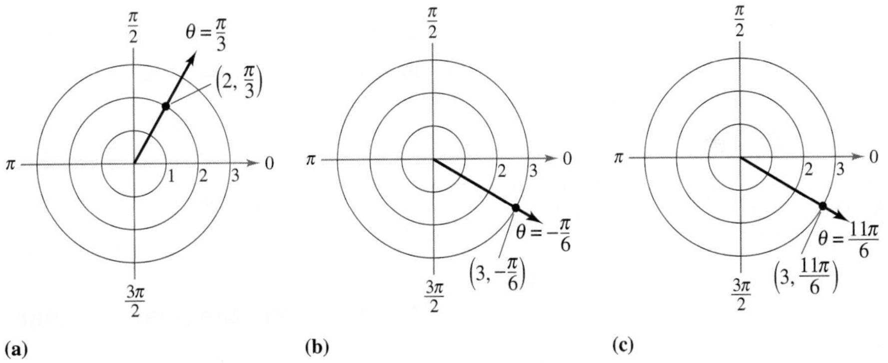

Figure 10.37

With rectangular coordinates, each point (x, y) has a unique representation. This is not true with polar coordinates. For instance, the coordinates (r, θ) and $(r, 2\pi + \theta)$ represent the same point [see parts (b) and (c) in Figure 10.37]. Also, because r is a *directed distance*, the coordinates (r, θ) and $(-r, \theta + \pi)$ represent the same point. In general, the point (r, θ) can be written as

$$(r, \theta) = (r, \theta + 2n\pi)$$

or

$$(r, \theta) = (-r, \theta + (2n + 1)\pi)$$

where n is any integer. Moreover, the pole is represented by $(0, \theta)$, where θ is any angle.

POLAR COORDINATES

The mathematician credited with first using polar coordinates was James Bernoulli, who introduced them in 1691. However, there is some evidence that it may have been Isaac Newton who first used them.

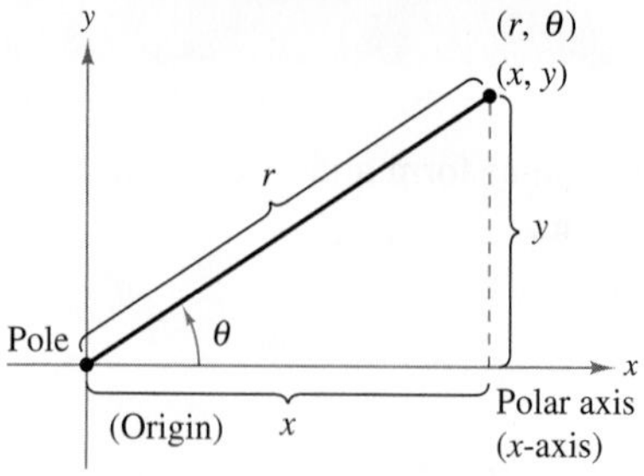

Relating polar and rectangular coordinates
Figure 10.38

Coordinate Conversion

To establish the relationship between polar and rectangular coordinates, let the polar axis coincide with the positive x-axis and the pole with the origin, as shown in Figure 10.38. Because (x, y) lies on a circle of radius r, it follows that $r^2 = x^2 + y^2$. Moreover, for $r > 0$, the definition of the trigonometric functions implies that

$$\tan \theta = \frac{y}{x}, \qquad \cos \theta = \frac{x}{r}, \qquad \text{and} \qquad \sin \theta = \frac{y}{r}.$$

If $r < 0$, you can show that the same relationships hold.

THEOREM 10.10 Coordinate Conversion

The polar coordinates (r, θ) of a point are related to the rectangular coordinates (x, y) of the point as follows.

1. $x = r \cos \theta$ **2.** $\tan \theta = \dfrac{y}{x}$

$y = r \sin \theta$ $r^2 = x^2 + y^2$

EXAMPLE 1 Polar-to-Rectangular Conversion

a. For the point $(r, \theta) = (2, \pi)$,

$$x = r \cos \theta = 2 \cos \pi = -2 \qquad \text{and} \qquad y = r \sin \theta = 2 \sin \pi = 0.$$

So, the rectangular coordinates are $(x, y) = (-2, 0)$.

b. For the point $(r, \theta) = (\sqrt{3}, \pi/6)$,

$$x = \sqrt{3} \cos \frac{\pi}{6} = \frac{3}{2} \qquad \text{and} \qquad y = \sqrt{3} \sin \frac{\pi}{6} = \frac{\sqrt{3}}{2}.$$

So, the rectangular coordinates are $(x, y) = (3/2, \sqrt{3}/2)$.
See Figure 10.39.

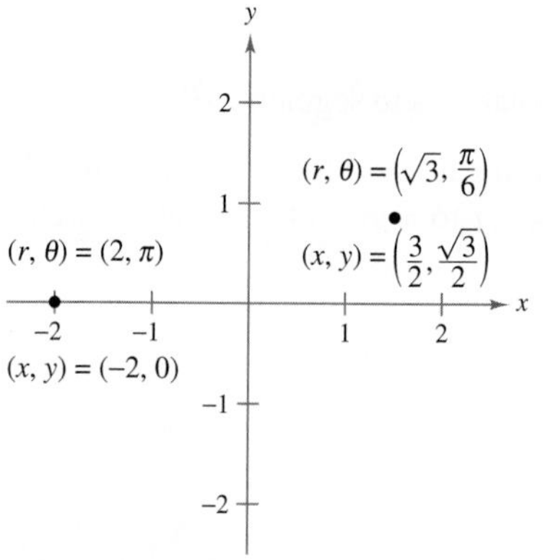

To convert from polar to rectangular coordinates, let $x = r \cos \theta$ and $y = r \sin \theta$.
Figure 10.39

EXAMPLE 2 Rectangular-to-Polar Conversion

a. For the second quadrant point $(x, y) = (-1, 1)$,

$$\tan \theta = \frac{y}{x} = -1 \quad \Longrightarrow \quad \theta = \frac{3\pi}{4}.$$

Because θ was chosen to be in the same quadrant as (x, y), you should use a positive value of r.

$$\begin{aligned} r &= \sqrt{x^2 + y^2} \\ &= \sqrt{(-1)^2 + (1)^2} \\ &= \sqrt{2} \end{aligned}$$

This implies that *one* set of polar coordinates is $(r, \theta) = (\sqrt{2}, 3\pi/4)$.

b. Because the point $(x, y) = (0, 2)$ lies on the positive y-axis, choose $\theta = \pi/2$ and $r = 2$, and one set of polar coordinates is $(r, \theta) = (2, \pi/2)$.

See Figure 10.40.

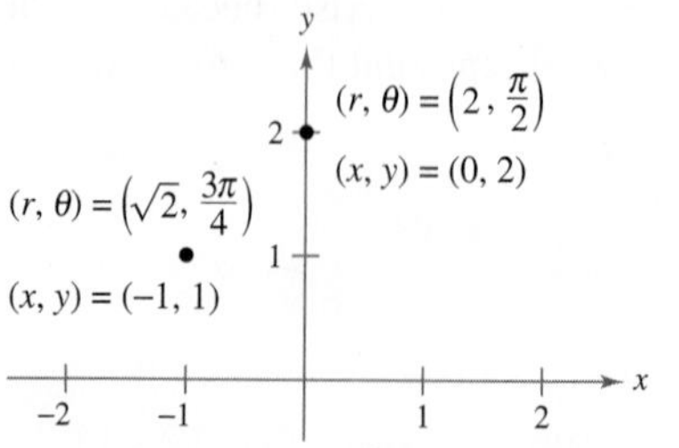

To convert from rectangular to polar coordinates, let $\tan \theta = y/x$ and $r = \sqrt{x^2 + y^2}$.
Figure 10.40

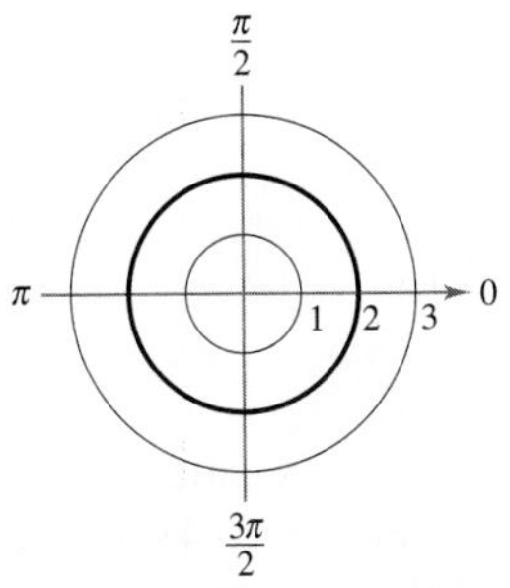

(a) Circle: $r = 2$

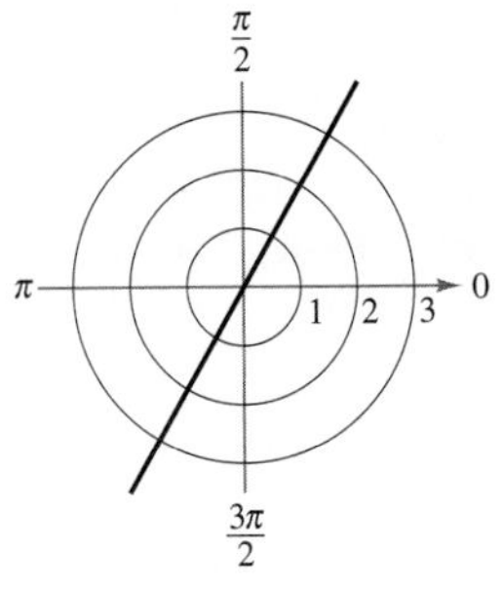

(b) Radial line: $\theta = \frac{\pi}{3}$

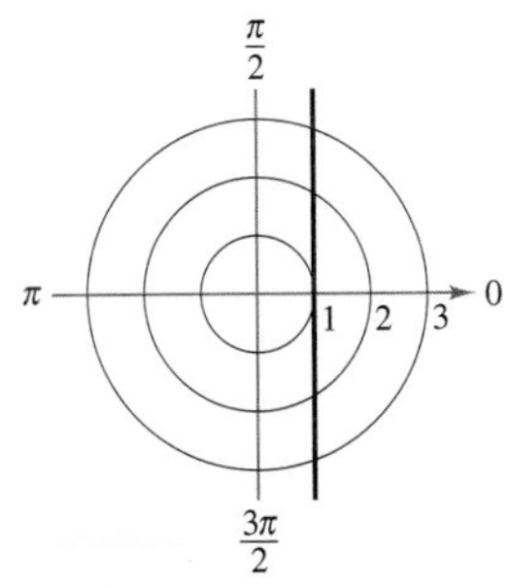

(c) Vertical line: $r = \sec \theta$

Figure 10.41

Polar Graphs

One way to sketch the graph of a polar equation is to convert to rectangular coordinates and then sketch the graph of the rectangular equation.

EXAMPLE 3 Graphing Polar Equations

Describe the graph of each polar equation. Confirm each description by converting to a rectangular equation.

a. $r = 2$ **b.** $\theta = \dfrac{\pi}{3}$ **c.** $r = \sec \theta$

Solution

a. The graph of the polar equation $r = 2$ consists of all points that are two units from the pole. In other words, this graph is a circle centered at the origin with a radius of 2. [See Figure 10.41(a).] You can confirm this by using the relationship $r^2 = x^2 + y^2$ to obtain the rectangular equation

$$x^2 + y^2 = 2^2. \qquad \text{Rectangular equation}$$

b. The graph of the polar equation $\theta = \pi/3$ consists of all points on the line that makes an angle of $\pi/3$ with the positive x-axis. [See Figure 10.41(b).] You can confirm this by using the relationship $\tan \theta = y/x$ to obtain the rectangular equation

$$y = \sqrt{3}\,x. \qquad \text{Rectangular equation}$$

c. The graph of the polar equation $r = \sec \theta$ is not evident by simple inspection, so you can begin by converting to rectangular form using the relationship $r \cos \theta = x$.

$$r = \sec \theta \qquad \text{Polar equation}$$
$$r \cos \theta = 1$$
$$x = 1 \qquad \text{Rectangular equation}$$

From the rectangular equation, you can see that the graph is a vertical line. [See Figure 10.41(c.)]

TECHNOLOGY Sketching the graphs of complicated polar equations *by hand* can be tedious. With technology, however, the task is not difficult. If your graphing utility has a *polar* mode, use it to graph the equations in the exercise set. If your graphing utility doesn't have a *polar* mode, but does have a *parametric* mode, you can graph $r = f(\theta)$ by writing the equation as

$$x = f(\theta) \cos \theta$$
$$y = f(\theta) \sin \theta.$$

For instance, the graph of $r = \frac{1}{2}\theta$ shown in Figure 10.42 was produced with a graphing calculator in *parametric* mode. This equation was graphed using the parametric equations

$$x = \frac{1}{2}\theta \cos \theta$$

$$y = \frac{1}{2}\theta \sin \theta$$

with the values of θ varying from -4π to 4π. This curve is of the form $r = a\theta$ and is called a **spiral of Archimedes.**

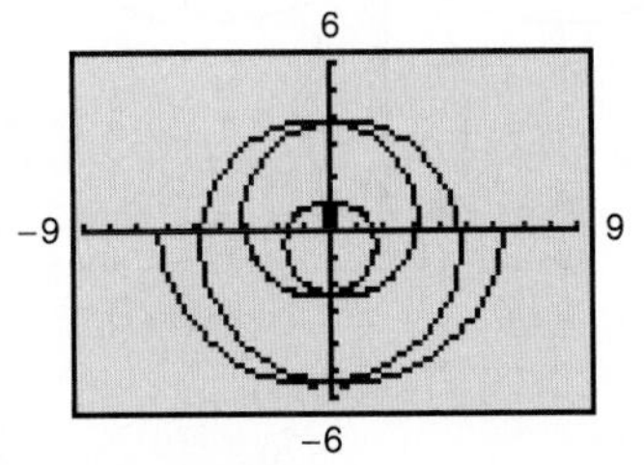

Spiral of Archimedes

Figure 10.42

EXAMPLE 4 Sketching a Polar Graph

NOTE One way to sketch the graph of $r = 2\cos 3\theta$ by hand is to make a table of values.

θ	0	$\frac{\pi}{6}$	$\frac{\pi}{3}$	$\frac{\pi}{2}$	$\frac{2\pi}{3}$
r	2	0	-2	0	2

By extending the table and plotting the points, you will obtain the curve shown in Example 4.

Sketch the graph of $r = 2\cos 3\theta$.

Solution Begin by writing the polar equation in parametric form.

$$x = 2\cos 3\theta \cos\theta \qquad \text{and} \qquad y = 2\cos 3\theta \sin\theta$$

After some experimentation, you will find that the entire curve, which is called a **rose curve,** can be sketched by letting θ vary from 0 to π, as shown in Figure 10.43. If you try duplicating this graph with a graphing utility, you will find that by letting θ vary from 0 to 2π, you will actually trace the entire curve *twice*.

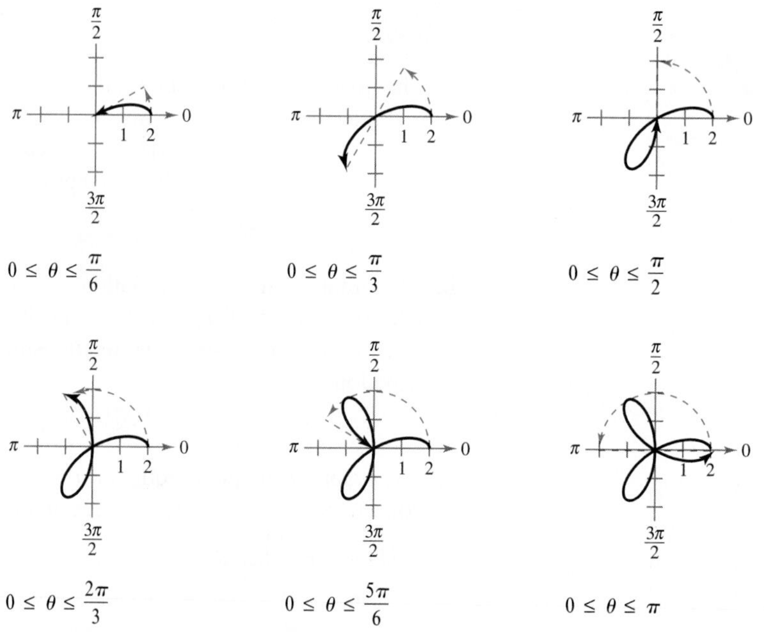

Figure 10.43

Use a graphing utility to experiment with other rose curves (they are of the form $r = a\cos n\theta$ or $r = a\sin n\theta$). For instance, Figure 10.44 shows the graphs of two other rose curves.

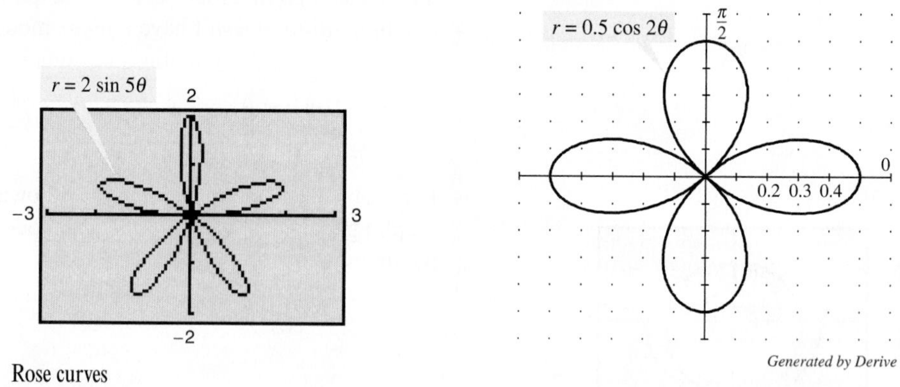

Rose curves
Figure 10.44

Slope and Tangent Lines

To find the slope of a tangent line to a polar graph, consider a differentiable function given by $r = f(\theta)$. To find the slope in polar form, use the parametric equations

$$x = r\cos\theta = f(\theta)\cos\theta \quad \text{and} \quad y = r\sin\theta = f(\theta)\sin\theta.$$

Using the parametric form of dy/dx given in Theorem 10.7, you have

$$\frac{dy}{dx} = \frac{dy/d\theta}{dx/d\theta} = \frac{f(\theta)\cos\theta + f'(\theta)\sin\theta}{-f(\theta)\sin\theta + f'(\theta)\cos\theta}$$

which establishes the following theorem.

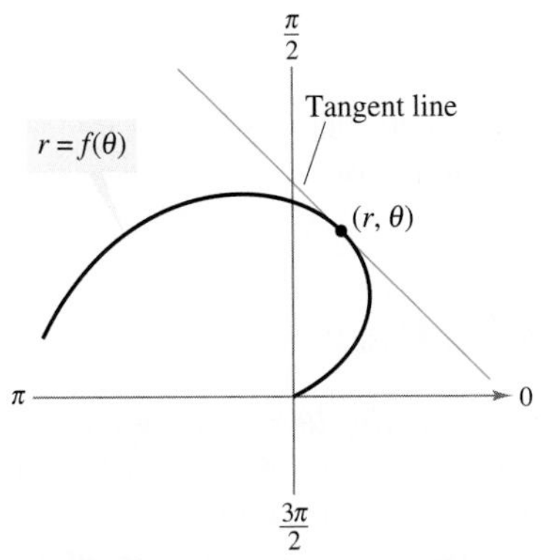

Tangent line to polar curve
Figure 10.45

THEOREM 10.11 Slope in Polar Form

If f is a differentiable function of θ, then the *slope* of the tangent line to the graph of $r = f(\theta)$ at the point (r, θ) is

$$\frac{dy}{dx} = \frac{dy/d\theta}{dx/d\theta} = \frac{f(\theta)\cos\theta + f'(\theta)\sin\theta}{-f(\theta)\sin\theta + f'(\theta)\cos\theta}$$

provided that $dx/d\theta \neq 0$ at (r, θ). (See Figure 10.45.)

From Theorem 10.11, you can make the following observations.

1. Solutions to $\dfrac{dy}{d\theta} = 0$ yield horizontal tangents, provided that $\dfrac{dx}{d\theta} \neq 0$.
2. Solutions to $\dfrac{dx}{d\theta} = 0$ yield vertical tangents, provided that $\dfrac{dy}{d\theta} \neq 0$.

If $dy/d\theta$ and $dx/d\theta$ are *simultaneously* 0, no conclusion can be drawn about tangent lines.

EXAMPLE 5 Finding Horizontal and Vertical Tangent Lines

Find the horizontal and vertical tangent lines of $r = \sin\theta$, $0 \le \theta \le \pi$.

Solution Begin by writing the equation in parametric form.

$$x = r\cos\theta = \sin\theta\cos\theta$$

and

$$y = r\sin\theta = \sin\theta\sin\theta = \sin^2\theta$$

Next, differentiate x and y with respect to θ and set each derivative equal to 0.

$$\frac{dx}{d\theta} = \cos^2\theta - \sin^2\theta = \cos 2\theta = 0 \quad \Longrightarrow \quad \theta = \frac{\pi}{4}, \frac{3\pi}{4}$$

$$\frac{dy}{d\theta} = 2\sin\theta\cos\theta = \sin 2\theta = 0 \quad \Longrightarrow \quad \theta = 0, \frac{\pi}{2}$$

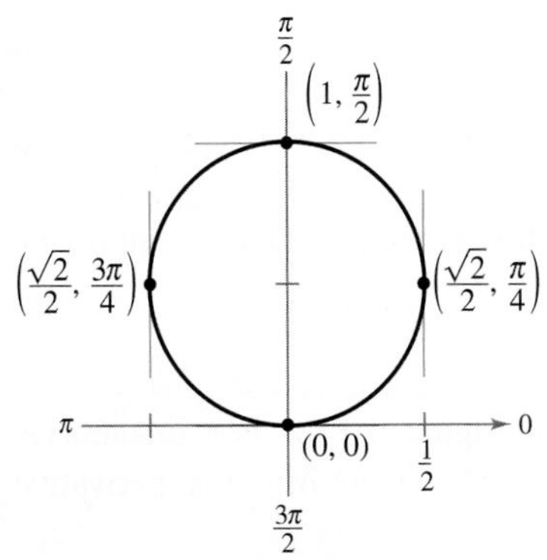

Horizontal and vertical tangent lines of $r = \sin\theta$
Figure 10.46

So, the graph has vertical tangent lines at $\left(\sqrt{2}/2, \pi/4\right)$ and $\left(\sqrt{2}/2, 3\pi/4\right)$, and it has horizontal tangent lines at $(0, 0)$ and $(1, \pi/2)$, as shown in Figure 10.46.

EXAMPLE 6 Finding Horizontal and Vertical Tangent Lines

Find the horizontal and vertical tangents to the graph of $r = 2(1 - \cos\theta)$.

Solution Using $y = r\sin\theta$, differentiate and set $dy/d\theta$ equal to 0.

$$y = r\sin\theta = 2(1 - \cos\theta)\sin\theta$$

$$\frac{dy}{d\theta} = 2[(1 - \cos\theta)(\cos\theta) + \sin\theta(\sin\theta)]$$

$$= -2(2\cos\theta + 1)(\cos\theta - 1) = 0$$

So, $\cos\theta = -\frac{1}{2}$ and $\cos\theta = 1$, and you can conclude that $dy/d\theta = 0$ when $\theta = 2\pi/3, 4\pi/3$, and 0. Similarly, using $x = r\cos\theta$, you have

$$x = r\cos\theta = 2\cos\theta - 2\cos^2\theta$$

$$\frac{dx}{d\theta} = -2\sin\theta + 4\cos\theta\sin\theta = 2\sin\theta(2\cos\theta - 1) = 0.$$

So, $\sin\theta = 0$ or $\cos\theta = \frac{1}{2}$, and you can conclude that $dx/d\theta = 0$ when $\theta = 0, \pi, \pi/3$, and $5\pi/3$. From these results, and from the graph shown in Figure 10.47, you can conclude that the graph has horizontal tangents at $(3, 2\pi/3)$ and $(3, 4\pi/3)$, and has vertical tangents at $(1, \pi/3)$, $(1, 5\pi/3)$, and $(4, \pi)$. This graph is called a **cardioid**. Note that both derivatives ($dy/d\theta$ and $dx/d\theta$) are 0 when $\theta = 0$. Using this information alone, you don't know whether the graph has a horizontal or vertical tangent line at the pole. From Figure 10.47, however, you can see that the graph has a cusp at the pole.

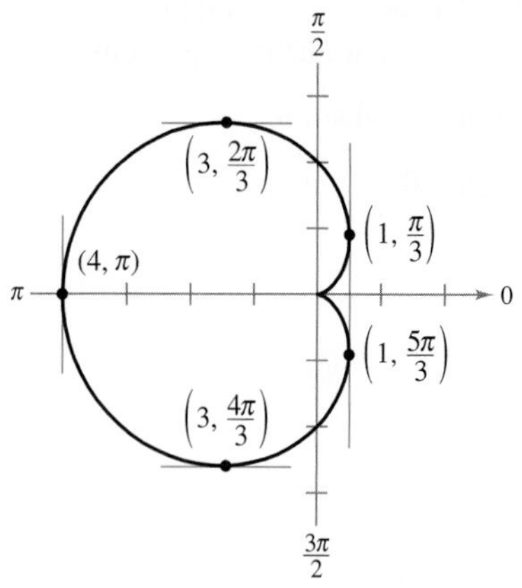

Horizontal and vertical tangent lines of $r = 2(1 - \cos\theta)$
Figure 10.47

Theorem 10.11 has an important consequence. Suppose the graph of $r = f(\theta)$ passes through the pole when $\theta = \alpha$ and $f'(\alpha) \neq 0$. Then the formula for dy/dx simplifies as follows.

$$\frac{dy}{dx} = \frac{f'(\alpha)\sin\alpha + f(\alpha)\cos\alpha}{f'(\alpha)\cos\alpha - f(\alpha)\sin\alpha} = \frac{f'(\alpha)\sin\alpha + 0}{f'(\alpha)\cos\alpha - 0} = \frac{\sin\alpha}{\cos\alpha} = \tan\alpha$$

So, the line $\theta = \alpha$ is tangent to the graph at the pole, $(0, \alpha)$.

THEOREM 10.12 Tangent Lines at the Pole

If $f(\alpha) = 0$ and $f'(\alpha) \neq 0$, then the line $\theta = \alpha$ is tangent at the pole to the graph of $r = f(\theta)$.

Theorem 10.12 is useful because it states that the zeros of $r = f(\theta)$ can be used to find the tangent lines at the pole. Note that because a polar curve can cross the pole more than once, it can have more than one tangent line at the pole. For example, the rose curve

$$f(\theta) = 2\cos 3\theta$$

has three tangent lines at the pole, as shown in Figure 10.48. For this curve, $f(\theta) = 2\cos 3\theta$ is 0 when θ is $\pi/6$, $\pi/2$, and $5\pi/6$. Moreover, the derivative $f'(\theta) = -6\sin 3\theta$ is not 0 for these values of θ.

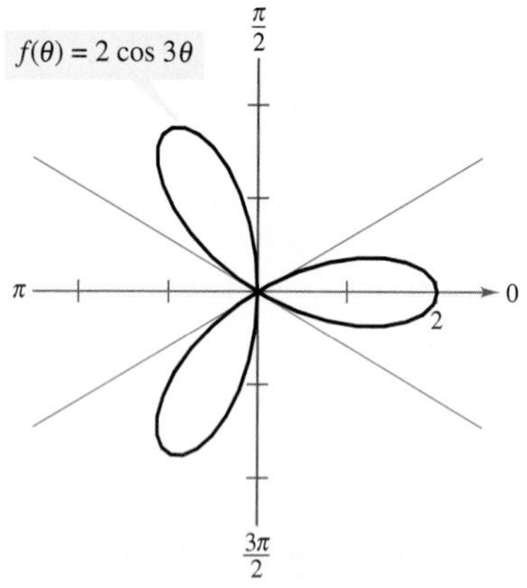

This rose curve has three tangent lines ($\theta = \pi/6$, $\theta = \pi/2$, and $\theta = 5\pi/6$) at the pole.
Figure 10.48

Special Polar Graphs

Several important types of graphs have equations that are simpler in polar form than in rectangular form. For example, the polar equation of a circle having a radius of a and centered at the origin is simply $r = a$. Later in the text you will come to appreciate this benefit. For now, several other types of graphs that have simpler equations in polar form are shown below. (Conics are considered in Section 10.6.)

Limaçons

$r = a \pm b \cos \theta$

$r = a \pm b \sin \theta$

$(a > 0, b > 0)$

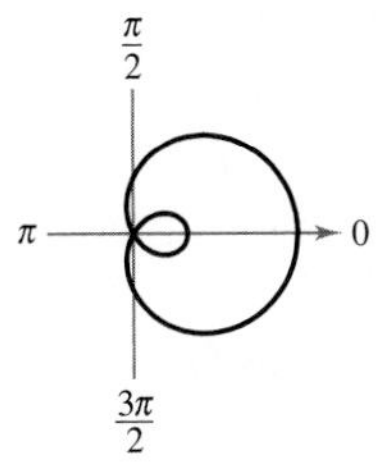

$\frac{a}{b} < 1$

Limaçon with inner loop

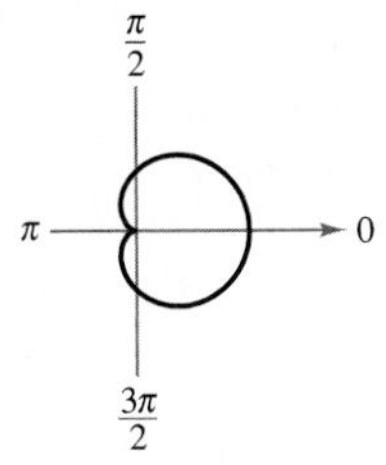

$\frac{a}{b} = 1$

Cardioid (heart-shaped)

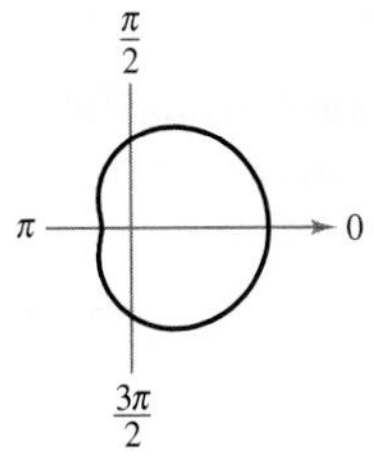

$1 < \frac{a}{b} < 2$

Dimpled limaçon

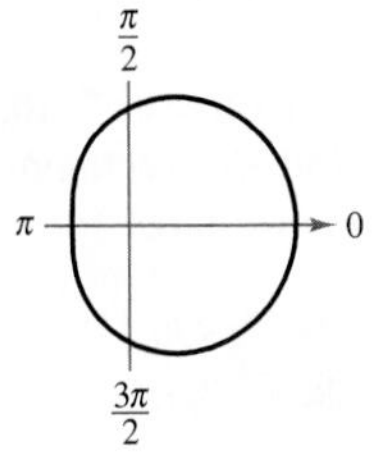

$\frac{a}{b} \geq 2$

Convex limaçon

Rose Curves

n petals if n is odd

$2n$ petals if n is even

$(n \geq 2)$

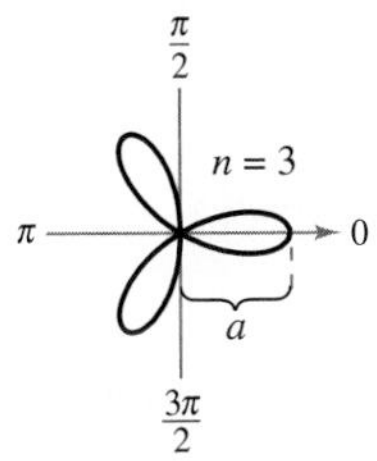

$r = a \cos n\theta$

Rose curve

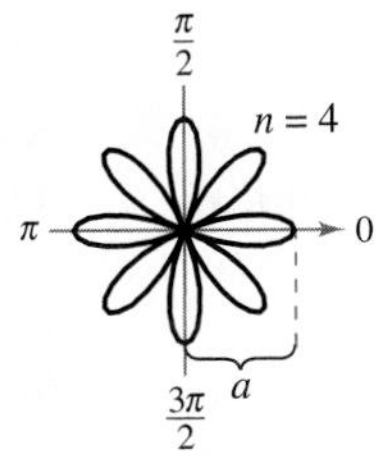

$r = a \cos n\theta$

Rose curve

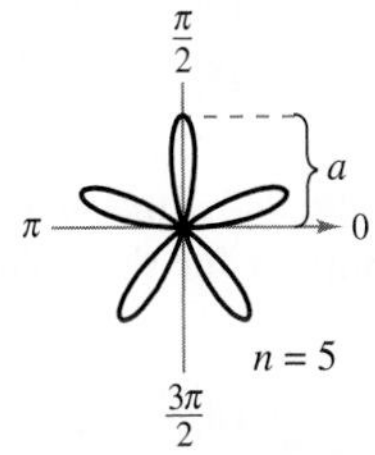

$r = a \sin n\theta$

Rose curve

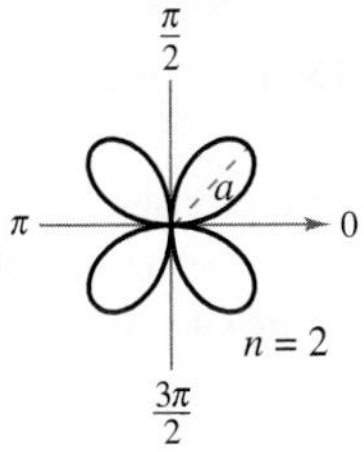

$r = a \sin n\theta$

Rose curve

Circles and Lemniscates

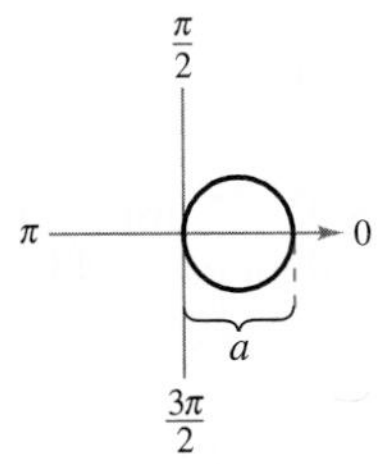

$r = a \cos \theta$

Circle

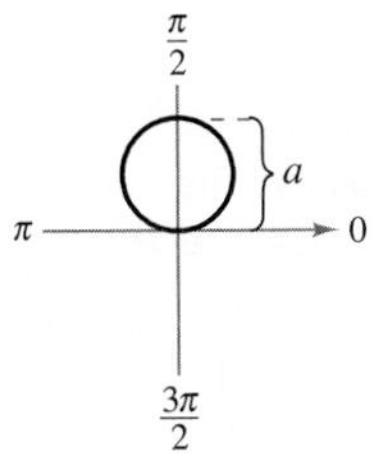

$r = a \sin \theta$

Circle

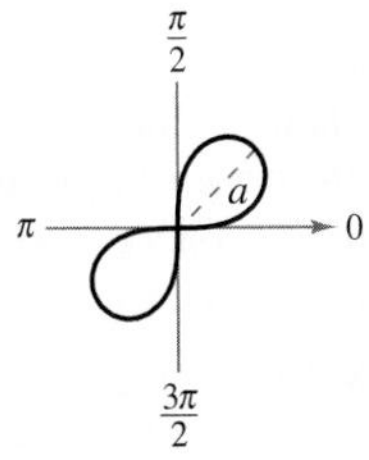

$r^2 = a^2 \sin 2\theta$

Lemniscate

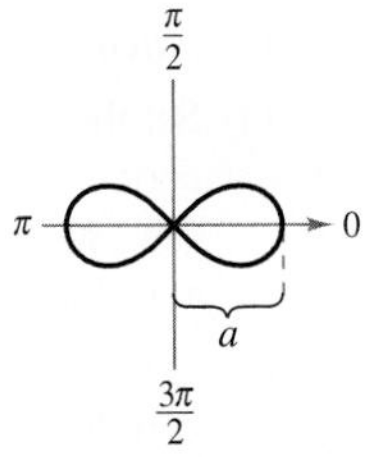

$r^2 = a^2 \cos 2\theta$

Lemniscate

TECHNOLOGY The rose curves described above are of the form $r = a \cos n\theta$ or $r = a \sin n\theta$, where n is a positive integer that is greater than or equal to 2. Use a graphing utility to graph $r = a \cos n\theta$ or $r = a \sin n\theta$ for some noninteger values of n. Are these graphs also rose curves? For example, try sketching the graph of $r = \cos \frac{2}{3}\theta$, $0 \leq \theta \leq 6\pi$.

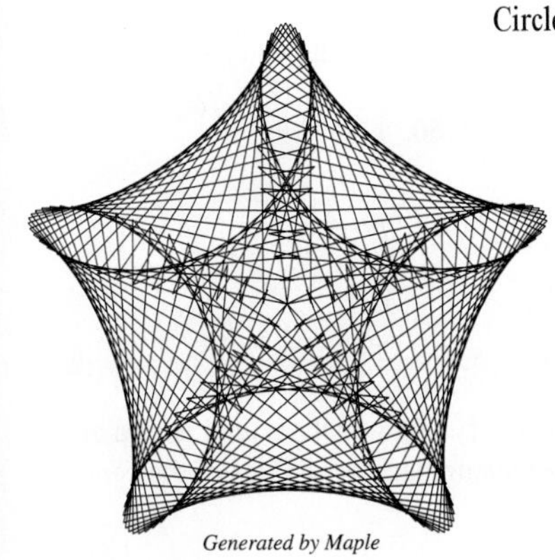

Generated by Maple

FOR FURTHER INFORMATION For more information on rose curves and related curves, see the article "A Rose is a Rose . . ." by Peter M. Maurer in *The American Mathematical Monthly*. The computer-generated graph at the left is the result of an algorithm that Maurer calls "The Rose." To view this article, go to the website *www.matharticles.com*.

Exercises for Section 10.4

See www.CalcChat.com for worked-out solutions to odd-numbered exercises.

In Exercises 1–6, plot the point in polar coordinates and find the corresponding rectangular coordinates for the point.

1. $(4, 3\pi/6)$
2. $(-2, 7\pi/4)$
3. $(-4, -\pi/3)$
4. $(0, -7\pi/6)$
5. $(\sqrt{2}, 2.36)$
6. $(-3, -1.57)$

In Exercises 7–10, use the *angle* feature of a graphing utility to find the rectangular coordinates for the point given in polar coordinates. Plot the point.

7. $(5, 3\pi/4)$
8. $(-2, 11\pi/6)$
9. $(-3.5, 2.5)$
10. $(8.25, 1.3)$

In Exercises 11–16, the rectangular coordinates of a point are given. Plot the point and find *two* sets of polar coordinates for the point for $0 \le \theta < 2\pi$.

11. $(1, 1)$
12. $(0, -5)$
13. $(-3, 4)$
14. $(4, -2)$
15. $(\sqrt{3}, -1)$
16. $(3, -\sqrt{3})$

In Exercises 17–20, use the *angle* feature of a graphing utility to find one set of polar coordinates for the point given in rectangular coordinates.

17. $(3, -2)$
18. $(3\sqrt{2}, 3\sqrt{2})$
19. $\left(\frac{5}{2}, \frac{4}{3}\right)$
20. $(0, -5)$

21. Plot the point $(4, 3.5)$ if the point is given in (a) rectangular coordinates and (b) polar coordinates.

22. ***Graphical Reasoning***

(a) Set the window format of a graphing utility to rectangular coordinates and locate the cursor at any position off the axes. Move the cursor horizontally and vertically. Describe any changes in the displayed coordinates of the points.

(b) Set the window format of a graphing utility to polar coordinates and locate the cursor at any position off the axes. Move the cursor horizontally and vertically. Describe any changes in the displayed coordinates of the points.

(c) Why are the results in parts (a) and (b) different?

In Exercises 23–26, match the graph with its polar equation. [The graphs are labeled (a), (b), (c), and (d).]

(a)
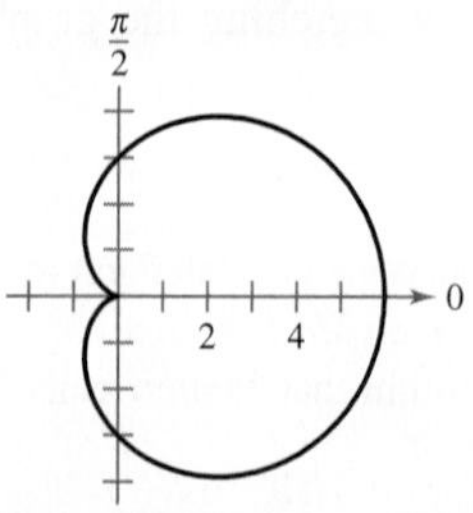

(b)
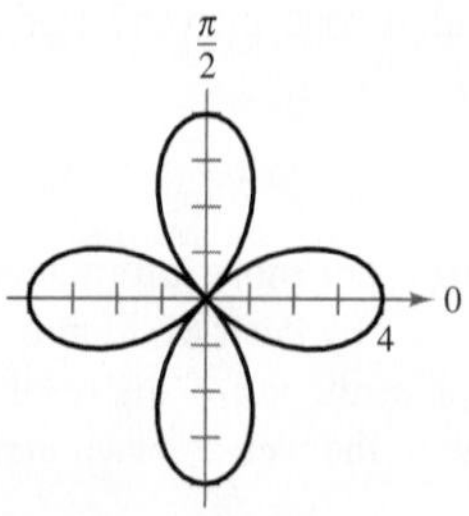

(c)
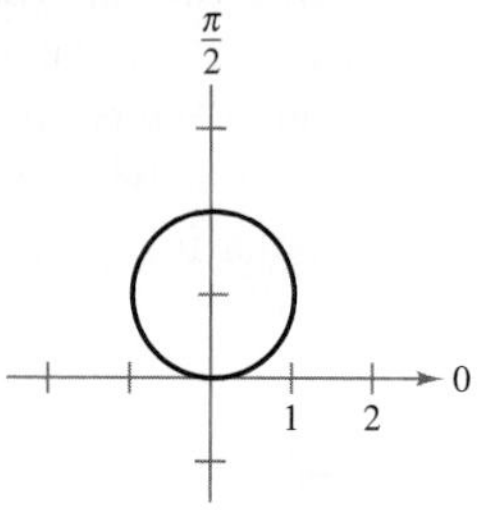

(d)
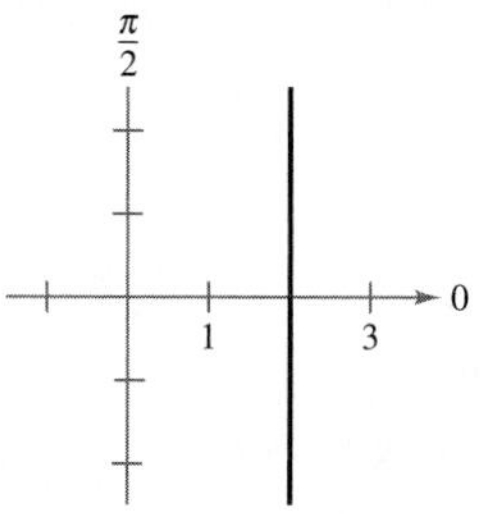

23. $r = 2\sin\theta$
24. $r = 4\cos 2\theta$
25. $r = 3(1 + \cos\theta)$
26. $r = 2\sec\theta$

In Exercises 27–34, convert the rectangular equation to polar form and sketch its graph.

27. $x^2 + y^2 = a^2$
28. $x^2 + y^2 - 2ax = 0$
29. $y = 4$
30. $x = 10$
31. $3x - y + 2 = 0$
32. $xy = 4$
33. $y^2 = 9x$
34. $(x^2 + y^2)^2 - 9(x^2 - y^2) = 0$

In Exercises 35–42, convert the polar equation to rectangular form and sketch its graph.

35. $r = 3$
36. $r = -2$
37. $r = \sin\theta$
38. $r = 5\cos\theta$
39. $r = \theta$
40. $\theta = \dfrac{5\pi}{6}$
41. $r = 3\sec\theta$
42. $r = 2\csc\theta$

In Exercises 43–52, use a graphing utility to graph the polar equation. Find an interval for θ over which the graph is traced *only once*.

43. $r = 3 - 4\cos\theta$
44. $r = 5(1 - 2\sin\theta)$
45. $r = 2 + \sin\theta$
46. $r = 4 + 3\cos\theta$
47. $r = \dfrac{2}{1 + \cos\theta}$
48. $r = \dfrac{2}{4 - 3\sin\theta}$
49. $r = 2\cos\left(\dfrac{3\theta}{2}\right)$
50. $r = 3\sin\left(\dfrac{5\theta}{2}\right)$
51. $r^2 = 4\sin 2\theta$
52. $r^2 = \dfrac{1}{\theta}$

53. Convert the equation

$$r = 2(h\cos\theta + k\sin\theta)$$

to rectangular form and verify that it is the equation of a circle. Find the radius and the rectangular coordinates of the center of the circle.

54. *Distance Formula*

(a) Verify that the Distance Formula for the distance between the two points (r_1, θ_1) and (r_2, θ_2) in polar coordinates is

$$d = \sqrt{r_1^2 + r_2^2 - 2r_1r_2\cos(\theta_1 - \theta_2)}.$$

(b) Describe the positions of the points relative to each other if $\theta_1 = \theta_2$. Simplify the Distance Formula for this case. Is the simplification what you expected? Explain.

(c) Simplify the Distance Formula if $\theta_1 - \theta_2 = 90°$. Is the simplification what you expected? Explain.

(d) Choose two points on the polar coordinate system and find the distance between them. Then choose different polar representations of the same two points and apply the Distance Formula again. Discuss the result.

In Exercises 55–58, use the result of Exercise 54 to approximate the distance between the two points in polar coordinates.

55. $\left(4, \frac{2\pi}{3}\right)$, $\left(2, \frac{\pi}{6}\right)$ **56.** $\left(10, \frac{7\pi}{6}\right)$, $(3, \pi)$

57. $(2, 0.5)$, $(7, 1.2)$ **58.** $(4, 2.5)$, $(12, 1)$

In Exercises 59 and 60, find dy/dx and the slopes of the tangent lines shown on the graph of the polar equation.

59. $r = 2 + 3\sin\theta$ **60.** $r = 2(1 - \sin\theta)$

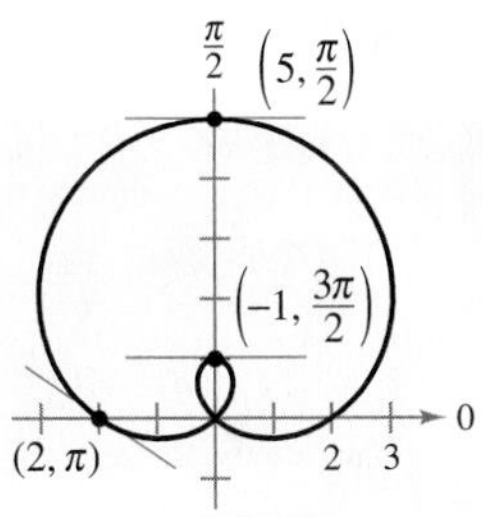

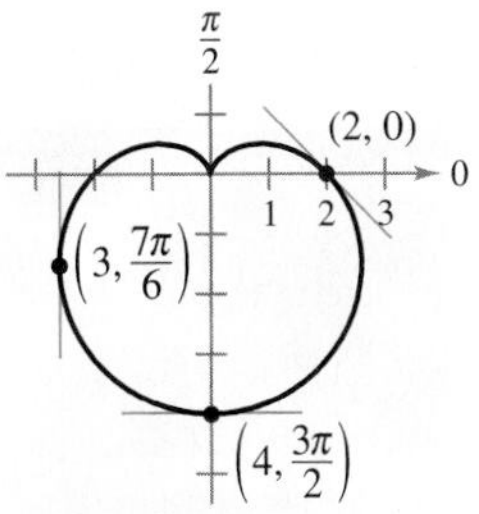

In Exercises 61–64, use a graphing utility to (a) graph the polar equation, (b) draw the tangent line at the given value of θ, and (c) find dy/dx at the given value of θ. (*Hint:* Let the increment between the values of θ equal $\pi/24$.)

61. $r = 3(1 - \cos\theta)$, $\theta = \frac{\pi}{2}$ **62.** $r = 3 - 2\cos\theta$, $\theta = 0$

63. $r = 3\sin\theta$, $\theta = \frac{\pi}{3}$ **64.** $r = 4$, $\theta = \frac{\pi}{4}$

In Exercises 65 and 66, find the points of horizontal and vertical tangency (if any) to the polar curve.

65. $r = 1 - \sin\theta$ **66.** $r = a\sin\theta$

In Exercises 67 and 68, find the points of horizontal tangency (if any) to the polar curve.

67. $r = 2\csc\theta + 3$ **68.** $r = a\sin\theta\cos^2\theta$

In Exercises 69–72, use a graphing utility to graph the polar equation and find all points of horizontal tangency.

69. $r = 4\sin\theta\cos^2\theta$ **70.** $r = 3\cos 2\theta\sec\theta$

71. $r = 2\csc\theta + 5$ **72.** $r = 2\cos(3\theta - 2)$

In Exercises 73–80, sketch a graph of the polar equation and find the tangents at the pole.

73. $r = 3\sin\theta$ **74.** $r = 3\cos\theta$

75. $r = 2(1 - \sin\theta)$ **76.** $r = 3(1 - \cos\theta)$

77. $r = 2\cos 3\theta$ **78.** $r = -\sin 5\theta$

79. $r = 3\sin 2\theta$ **80.** $r = 3\cos 2\theta$

In Exercises 81–92, sketch a graph of the polar equation.

81. $r = 5$ **82.** $r = 2$

83. $r = 4(1 + \cos\theta)$ **84.** $r = 1 + \sin\theta$

85. $r = 3 - 2\cos\theta$ **86.** $r = 5 - 4\sin\theta$

87. $r = 3\csc\theta$ **88.** $r = \frac{6}{2\sin\theta - 3\cos\theta}$

89. $r = 2\theta$ **90.** $r = \frac{1}{\theta}$

91. $r^2 = 4\cos 2\theta$ **92.** $r^2 = 4\sin\theta$

In Exercises 93–96, use a graphing utility to graph the equation and show that the given line is an asymptote of the graph.

	Name of Graph	*Polar Equation*	*Asymptote*
93.	Conchoid	$r = 2 - \sec\theta$	$x = -1$
94.	Conchoid	$r = 2 + \csc\theta$	$y = 1$
95.	Hyperbolic spiral	$r = 2/\theta$	$y = 2$
96.	Strophoid	$r = 2\cos 2\theta\sec\theta$	$x = -2$

Writing About Concepts

97. Describe the differences between the rectangular coordinate system and the polar coordinate system.

98. Give the equations for the coordinate conversion from rectangular to polar coordinates and vice versa.

99. For constants a and b, describe the graphs of the equations $r = a$ and $\theta = b$ in polar coordinates.

100. How are the slopes of tangent lines determined in polar coordinates? What are tangent lines at the pole and how are they determined?

101. Sketch the graph of $r = 4\sin\theta$ over each interval.

(a) $0 \le \theta \le \frac{\pi}{2}$ (b) $\frac{\pi}{2} \le \theta \le \pi$ (c) $-\frac{\pi}{2} \le \theta \le \frac{\pi}{2}$

102. *Think About It* Use a graphing utility to graph the polar equation $r = 6[1 + \cos(\theta - \phi)]$ for (a) $\phi = 0$, (b) $\phi = \pi/4$, and (c) $\phi = \pi/2$. Use the graphs to describe the effect of the angle ϕ. Write the equation as a function of $\sin\theta$ for part (c).

103. Verify that if the curve whose polar equation is $r = f(\theta)$ is rotated about the pole through an angle ϕ, then an equation for the rotated curve is $r = f(\theta - \phi)$.

104. The polar form of an equation for a curve is $r = f(\sin\theta)$. Show that the form becomes

(a) $r = f(-\cos\theta)$ if the curve is rotated counterclockwise $\pi/2$ radians about the pole.

(b) $r = f(-\sin\theta)$ if the curve is rotated counterclockwise π radians about the pole.

(c) $r = f(\cos\theta)$ if the curve is rotated counterclockwise $3\pi/2$ radians about the pole.

In Exercises 105–108, use the results of Exercises 103 and 104.

105. Write an equation for the limaçon $r = 2 - \sin\theta$ after it has been rotated by the given amount. Use a graphing utility to graph the rotated limaçon.

(a) $\dfrac{\pi}{4}$ (b) $\dfrac{\pi}{2}$ (c) π (d) $\dfrac{3\pi}{2}$

106. Write an equation for the rose curve $r = 2\sin 2\theta$ after it has been rotated by the given amount. Verify the results by using a graphing utility to graph the rotated rose curve.

(a) $\dfrac{\pi}{6}$ (b) $\dfrac{\pi}{2}$ (c) $\dfrac{2\pi}{3}$ (d) π

107. Sketch the graph of each equation.

(a) $r = 1 - \sin\theta$ (b) $r = 1 - \sin\left(\theta - \dfrac{\pi}{4}\right)$

108. Prove that the tangent of the angle $\psi\ (0 \le \psi \le \pi/2)$ between the radial line and the tangent line at the point (r, θ) on the graph of $r = f(\theta)$ (see figure) is given by $\tan\psi = |r/(dr/d\theta)|$.

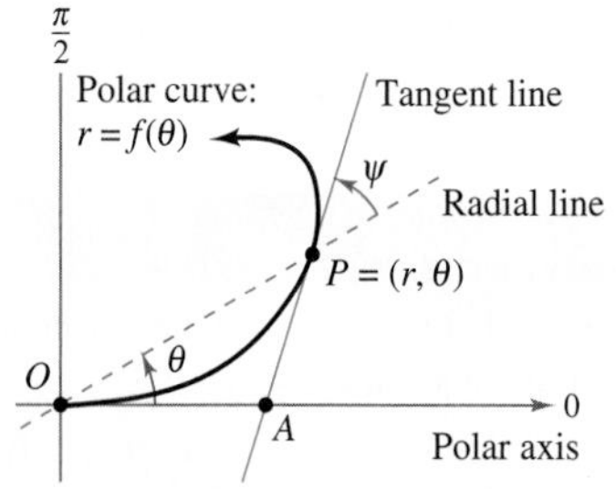

In Exercises 109–114, use the result of Exercise 108 to find the angle ψ between the radial and tangent lines to the graph for the indicated value of θ. Use a graphing utility to graph the polar equation, the radial line, and the tangent line for the indicated value of θ. Identify the angle ψ.

	Polar Equation	*Value of θ*
109.	$r = 2(1 - \cos\theta)$	$\theta = \pi$
110.	$r = 3(1 - \cos\theta)$	$\theta = 3\pi/4$
111.	$r = 2\cos 3\theta$	$\theta = \pi/4$
112.	$r = 4\sin 2\theta$	$\theta = \pi/6$
113.	$r = \dfrac{6}{1 - \cos\theta}$	$\theta = 2\pi/3$
114.	$r = 5$	$\theta = \pi/6$

True or False? **In Exercises 115–118, determine whether the statement is true or false. If it is false, explain why or give an example that shows it is false.**

115. If (r_1, θ_1) and (r_2, θ_2) represent the same point on the polar coordinate system, then $|r_1| = |r_2|$.

116. If (r, θ_1) and (r, θ_2) represent the same point on the polar coordinate system, then $\theta_1 = \theta_2 + 2\pi n$ for some integer n.

117. If $x > 0$, then the point (x, y) on the rectangular coordinate system can be represented by (r, θ) on the polar coordinate system, where $r = \sqrt{x^2 + y^2}$ and $\theta = \arctan(y/x)$.

118. The polar equations $r = \sin 2\theta$ and $r = -\sin 2\theta$ have the same graph.

Section Project: Anamorphic Art

Use the anamorphic transformations

$$r = y + 16 \quad \text{and} \quad \theta = -\frac{\pi}{8}x, \quad -\frac{3\pi}{4} \le \theta \le \frac{3\pi}{4}$$

to sketch the transformed polar image of the rectangular graph. When the reflection (in a cylindrical mirror centered at the pole) of each polar image is viewed from the polar axis, the viewer will see the original rectangular image.

(a) $y = 3$ (b) $x = 2$

(c) $y = x + 5$ (d) $x^2 + (y - 5)^2 = 5^2$

Museum of Science and Industry in Manchester, England

This example of anamorphic art is from the Museum of Science and Industry in Manchester, England. When the reflection of the transformed "polar painting" is viewed in the mirror, the viewer sees faces.

FOR FURTHER INFORMATION For more information on anamorphic art, see the article "Anamorphisms" by Philip Hickin in the *Mathematical Gazette*.

Section 10.5 Area and Arc Length in Polar Coordinates

- **Find the area of a region bounded by a polar graph.**
- **Find the points of intersection of two polar graphs.**
- **Find the arc length of a polar graph.**
- **Find the area of a surface of revolution (polar form).**

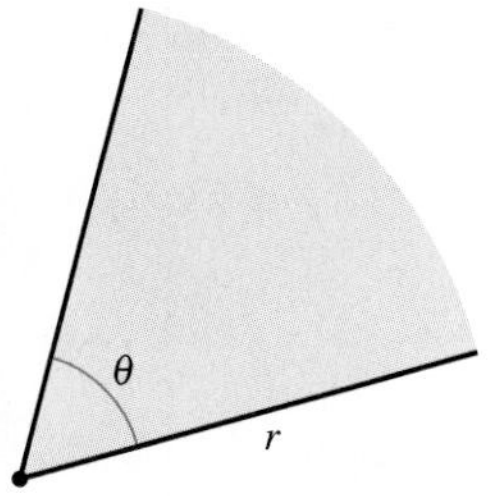

The area of a sector of a circle is $A = \frac{1}{2}\theta r^2$.
Figure 10.49

Area of a Polar Region

The development of a formula for the area of a polar region parallels that for the area of a region on the rectangular coordinate system, but uses sectors of a circle instead of rectangles as the basic element of area. In Figure 10.49, note that the area of a circular sector of radius r is given by $\frac{1}{2}\theta r^2$, provided θ is measured in radians.

Consider the function given by $r = f(\theta)$, where f is continuous and nonnegative in the interval given by $\alpha \le \theta \le \beta$. The region bounded by the graph of f and the radial lines $\theta = \alpha$ and $\theta = \beta$ is shown in Figure 10.50(a). To find the area of this region, partition the interval $[\alpha, \beta]$ into n equal subintervals

$$\alpha = \theta_0 < \theta_1 < \theta_2 < \cdot\cdot\cdot < \theta_{n-1} < \theta_n = \beta.$$

Then, approximate the area of the region by the sum of the areas of the n sectors, as shown in Figure 10.50(b).

$$\text{Radius of } i\text{th sector} = f(\theta_i)$$

$$\text{Central angle of } i\text{th sector} = \frac{\beta - \alpha}{n} = \Delta\theta$$

$$A \approx \sum_{i=1}^{n} \left(\frac{1}{2}\right) \Delta\theta [f(\theta_i)]^2$$

Taking the limit as $n \to \infty$ produces

$$\begin{aligned} A &= \lim_{n\to\infty} \frac{1}{2} \sum_{i=1}^{n} [f(\theta_i)]^2 \Delta\theta \\ &= \frac{1}{2}\int_{\alpha}^{\beta} [f(\theta)]^2 \, d\theta \end{aligned}$$

which leads to the following theorem.

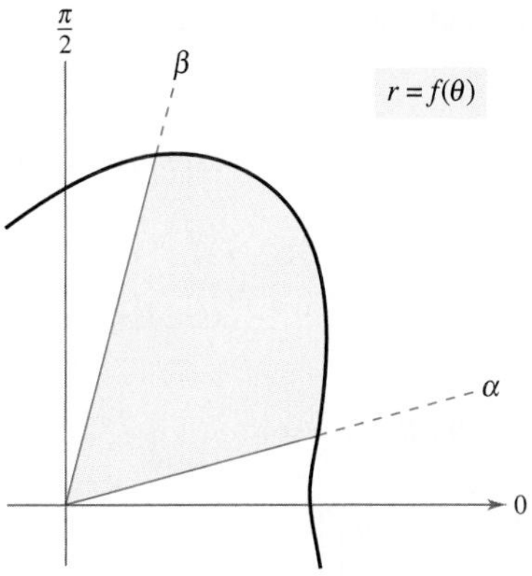

(a)

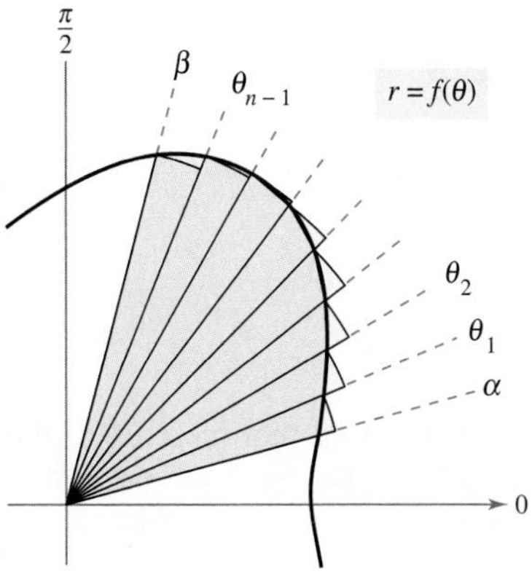

(b)
Figure 10.50

> **THEOREM 10.13 Area in Polar Coordinates**
>
> If f is continuous and nonnegative on the interval $[\alpha, \beta]$, $0 < \beta - \alpha \le 2\pi$, then the area of the region bounded by the graph of $r = f(\theta)$ between the radial lines $\theta = \alpha$ and $\theta = \beta$ is given by
>
> $$\begin{aligned} A &= \frac{1}{2}\int_{\alpha}^{\beta} [f(\theta)]^2 \, d\theta \\ &= \frac{1}{2}\int_{\alpha}^{\beta} r^2 \, d\theta. \qquad 0 < \beta - \alpha \le 2\pi \end{aligned}$$

NOTE You can use the same formula to find the area of a region bounded by the graph of a continuous *nonpositive* function. However, the formula is not necessarily valid if f takes on both positive *and* negative values in the interval $[\alpha, \beta]$.

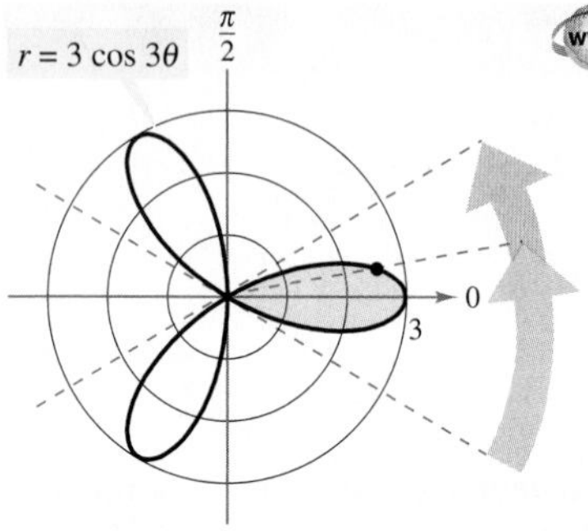

The area of one petal of the rose curve that lies between the radial lines $\theta = -\pi/6$ and $\theta = \pi/6$ is $3\pi/4$.
Figure 10.51

EXAMPLE 1 Finding the Area of a Polar Region

Find the area of one petal of the rose curve given by $r = 3\cos 3\theta$.

Solution In Figure 10.51, you can see that the right petal is traced as θ increases from $-\pi/6$ to $\pi/6$. So, the area is

$$\begin{aligned}
A &= \frac{1}{2}\int_{\alpha}^{\beta} r^2\,d\theta = \frac{1}{2}\int_{-\pi/6}^{\pi/6} (3\cos 3\theta)^2\,d\theta && \text{Formula for area in polar coordinates}\\
&= \frac{9}{2}\int_{-\pi/6}^{\pi/6} \frac{1+\cos 6\theta}{2}\,d\theta && \text{Trigonometric identity}\\
&= \frac{9}{4}\left[\theta + \frac{\sin 6\theta}{6}\right]_{-\pi/6}^{\pi/6}\\
&= \frac{9}{4}\left(\frac{\pi}{6}+\frac{\pi}{6}\right)\\
&= \frac{3\pi}{4}.
\end{aligned}$$

NOTE To find the area of the region lying inside all three petals of the rose curve in Example 1, you could not simply integrate between 0 and 2π. In doing this you would obtain $9\pi/2$, which is twice the area of the three petals. The duplication occurs because the rose curve is traced twice as θ increases from 0 to 2π.

EXAMPLE 2 Finding the Area Bounded by a Single Curve

Find the area of the region lying between the inner and outer loops of the limaçon $r = 1 - 2\sin\theta$.

Solution In Figure 10.52, note that the inner loop is traced as θ increases from $\pi/6$ to $5\pi/6$. So, the area inside the *inner loop* is

$$\begin{aligned}
A_1 &= \frac{1}{2}\int_{\alpha}^{\beta} r^2\,d\theta = \frac{1}{2}\int_{\pi/6}^{5\pi/6} (1-2\sin\theta)^2\,d\theta && \text{Formula for area in polar coordinates}\\
&= \frac{1}{2}\int_{\pi/6}^{5\pi/6} (1 - 4\sin\theta + 4\sin^2\theta)\,d\theta\\
&= \frac{1}{2}\int_{\pi/6}^{5\pi/6} \left[1 - 4\sin\theta + 4\left(\frac{1-\cos 2\theta}{2}\right)\right]d\theta && \text{Trigonometric identity}\\
&= \frac{1}{2}\int_{\pi/6}^{5\pi/6} (3 - 4\sin\theta - 2\cos 2\theta)\,d\theta && \text{Simplify.}\\
&= \frac{1}{2}\Big[3\theta + 4\cos\theta - \sin 2\theta\Big]_{\pi/6}^{5\pi/6}\\
&= \frac{1}{2}\left(2\pi - 3\sqrt{3}\right)\\
&= \pi - \frac{3\sqrt{3}}{2}.
\end{aligned}$$

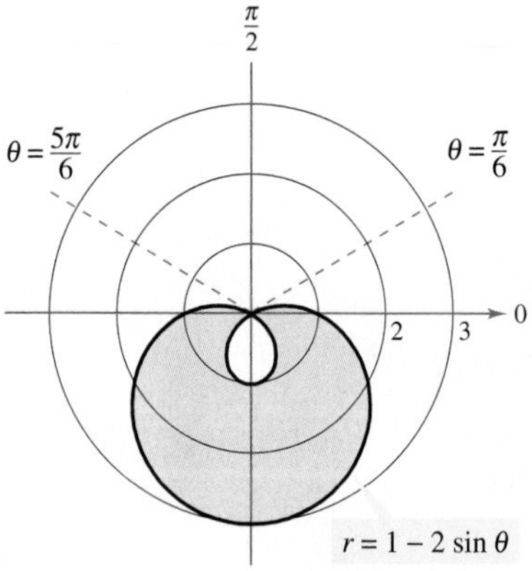

The area between the inner and outer loops is approximately 8.34.
Figure 10.52

In a similar way, you can integrate from $5\pi/6$ to $13\pi/6$ to find that the area of the region lying inside the outer loop is $A_2 = 2\pi + \left(3\sqrt{3}/2\right)$. The area of the region lying between the two loops is the difference of A_2 and A_1.

$$A = A_2 - A_1 = \left(2\pi + \frac{3\sqrt{3}}{2}\right) - \left(\pi - \frac{3\sqrt{3}}{2}\right) = \pi + 3\sqrt{3} \approx 8.34$$

Points of Intersection of Polar Graphs

Because a point may be represented in different ways in polar coordinates, care must be taken in determining the points of intersection of two polar graphs. For example, consider the points of intersection of the graphs of

$$r = 1 - 2\cos\theta \quad \text{and} \quad r = 1$$

as shown in Figure 10.53. If, as with rectangular equations, you attempted to find the points of intersection by solving the two equations simultaneously, you would obtain

$$r = 1 - 2\cos\theta \qquad \text{First equation}$$

$$1 = 1 - 2\cos\theta \qquad \text{Substitute } r = 1 \text{ from 2nd equation into 1st equation.}$$

$$\cos\theta = 0 \qquad \text{Simplify.}$$

$$\theta = \frac{\pi}{2}, \frac{3\pi}{2}. \qquad \text{Solve for } \theta.$$

FOR FURTHER INFORMATION For more information on using technology to find points of intersection, see the article "Finding Points of Intersection of Polar-Coordinate Graphs" by Warren W. Esty in *Mathematics Teacher*. To view this article, go to the website *www.matharticles.com*.

The corresponding points of intersection are $(1, \pi/2)$ and $(1, 3\pi/2)$. However, from Figure 10.53 you can see that there is a *third* point of intersection that did not show up when the two polar equations were solved simultaneously. (This is one reason why you should sketch a graph when finding the area of a polar region.) The reason the third point was not found is that it does not occur with the same coordinates in the two graphs. On the graph of $r = 1$, the point occurs with coordinates $(1, \pi)$, but on the graph of $r = 1 - 2\cos\theta$, the point occurs with coordinates $(-1, 0)$.

You can compare the problem of finding points of intersection of two polar graphs with that of finding collision points of two satellites in intersecting orbits about Earth, as shown in Figure 10.54. The satellites will not collide as long as they reach the points of intersection at different times (θ-values). Collisions will occur only at the points of intersection that are "simultaneous points"—those reached at the same time (θ-value).

NOTE Because the pole can be represented by $(0, \theta)$, where θ is *any* angle, you should check separately for the pole when finding points of intersection.

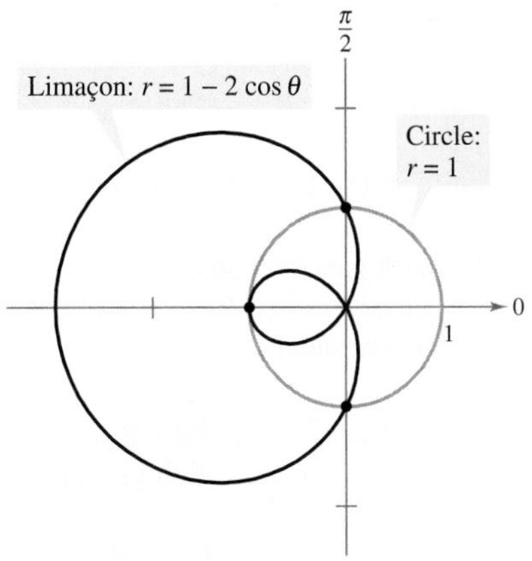

Three points of intersection: $(1, \pi/2)$, $(-1, 0)$, $(1, 3\pi/2)$
Figure 10.53

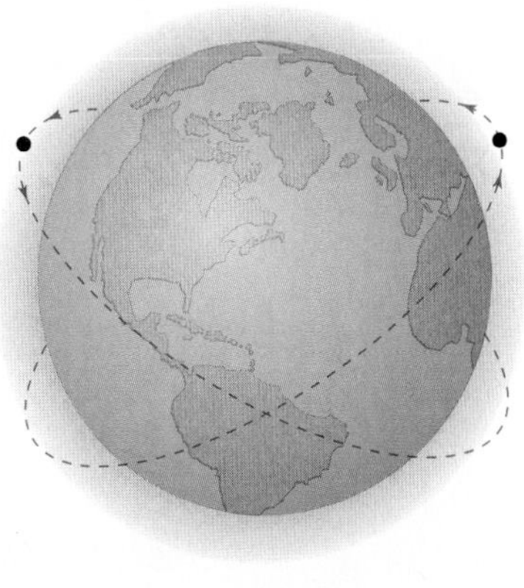

The paths of satellites can cross without causing a collision.
Figure 10.54

EXAMPLE 3 Finding the Area of a Region Between Two Curves

Find the area of the region common to the two regions bounded by the following curves.

$r = -6 \cos \theta$ Circle

$r = 2 - 2 \cos \theta$ Cardioid

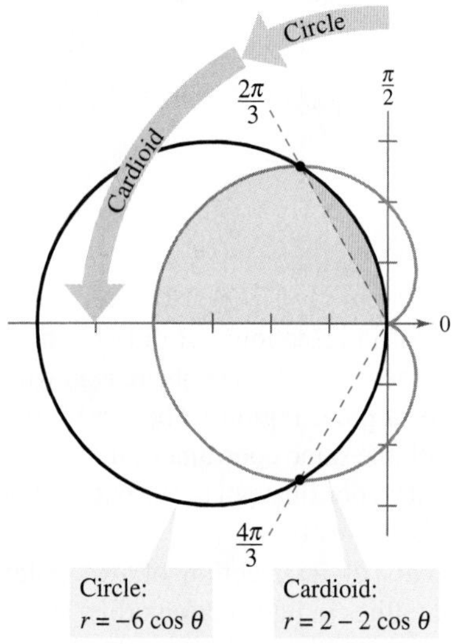

Figure 10.55

Solution Because both curves are symmetric with respect to the x-axis, you can work with the upper half-plane, as shown in Figure 10.55. The gray shaded region lies between the circle and the radial line $\theta = 2\pi/3$. Because the circle has coordinates $(0, \pi/2)$ at the pole, you can integrate between $\pi/2$ and $2\pi/3$ to obtain the area of this region. The region that is shaded red is bounded by the radial lines $\theta = 2\pi/3$ and $\theta = \pi$ and the cardioid. So, you can find the area of this second region by integrating between $2\pi/3$ and π. The sum of these two integrals gives the area of the common region lying *above* the radial line $\theta = \pi$.

$$\frac{A}{2} = \underbrace{\frac{1}{2}\int_{\pi/2}^{2\pi/3} (-6 \cos \theta)^2 \, d\theta}_{\text{Region between circle and radial line } \theta = 2\pi/3} + \underbrace{\frac{1}{2}\int_{2\pi/3}^{\pi} (2 - 2 \cos \theta)^2 \, d\theta}_{\text{Region between cardioid and radial lines } \theta = 2\pi/3 \text{ and } \theta = \pi}$$

$$= 18 \int_{\pi/2}^{2\pi/3} \cos^2 \theta \, d\theta + \frac{1}{2}\int_{2\pi/3}^{\pi} (4 - 8 \cos \theta + 4 \cos^2 \theta) \, d\theta$$

$$= 9 \int_{\pi/2}^{2\pi/3} (1 + \cos 2\theta) \, d\theta + \int_{2\pi/3}^{\pi} (3 - 4 \cos \theta + \cos 2\theta) \, d\theta$$

$$= 9\left[\theta + \frac{\sin 2\theta}{2}\right]_{\pi/2}^{2\pi/3} + \left[3\theta - 4 \sin \theta + \frac{\sin 2\theta}{2}\right]_{2\pi/3}^{\pi}$$

$$= 9\left(\frac{2\pi}{3} - \frac{\sqrt{3}}{4} - \frac{\pi}{2}\right) + \left(3\pi - 2\pi + 2\sqrt{3} + \frac{\sqrt{3}}{4}\right)$$

$$= \frac{5\pi}{2}$$

$$\approx 7.85$$

Finally, multiplying by 2, you can conclude that the total area is 5π.

NOTE To check the reasonableness of the result obtained in Example 3, note that the area of the circular region is $\pi r^2 = 9\pi$. So, it seems reasonable that the area of the region lying inside the circle and the cardioid is 5π.

To see the benefit of polar coordinates for finding the area in Example 3, consider the following integral, which gives the comparable area in rectangular coordinates.

$$\frac{A}{2} = \int_{-4}^{-3/2} \sqrt{2\sqrt{1 - 2x} - x^2 - 2x + 2} \, dx + \int_{-3/2}^{0} \sqrt{-x^2 - 6x} \, dx$$

Use the integration capabilities of a graphing utility to show that you obtain the same area as that found in Example 3.

Arc Length in Polar Form

NOTE When applying the arc length formula to a polar curve, be sure that the curve is traced out only once on the interval of integration. For instance, the rose curve given by $r = \cos 3\theta$ is traced out once on the interval $0 \le \theta \le \pi$, but is traced out twice on the interval $0 \le \theta \le 2\pi$.

The formula for the length of a polar arc can be obtained from the arc length formula for a curve described by parametric equations. (See Exercise 77.)

THEOREM 10.14 Arc Length of a Polar Curve

Let f be a function whose derivative is continuous on an interval $\alpha \le \theta \le \beta$. The length of the graph of $r = f(\theta)$ from $\theta = \alpha$ to $\theta = \beta$ is

$$s = \int_{\alpha}^{\beta} \sqrt{[f(\theta)]^2 + [f'(\theta)]^2}\, d\theta = \int_{\alpha}^{\beta} \sqrt{r^2 + \left(\frac{dr}{d\theta}\right)^2}\, d\theta.$$

EXAMPLE 4 Finding the Length of a Polar Curve

Find the length of the arc from $\theta = 0$ to $\theta = 2\pi$ for the cardioid

$$r = f(\theta) = 2 - 2\cos\theta$$

as shown in Figure 10.56.

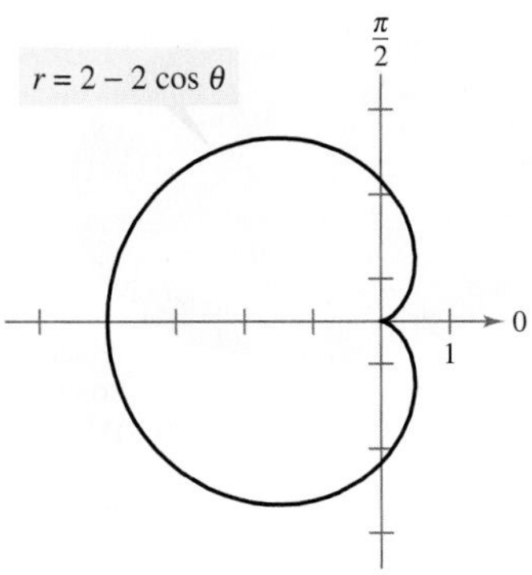

Figure 10.56

Solution Because $f'(\theta) = 2\sin\theta$, you can find the arc length as follows.

$$s = \int_{\alpha}^{\beta} \sqrt{[f(\theta)]^2 + [f'(\theta)]^2}\, d\theta \qquad \text{Formula for arc length of a polar curve}$$

$$= \int_0^{2\pi} \sqrt{(2 - 2\cos\theta)^2 + (2\sin\theta)^2}\, d\theta$$

$$= 2\sqrt{2}\int_0^{2\pi} \sqrt{1 - \cos\theta}\, d\theta \qquad \text{Simplify.}$$

$$= 2\sqrt{2}\int_0^{2\pi} \sqrt{2\sin^2\frac{\theta}{2}}\, d\theta \qquad \text{Trigonometric identity}$$

$$= 4\int_0^{2\pi} \sin\frac{\theta}{2}\, d\theta \qquad \sin\frac{\theta}{2} \ge 0 \text{ for } 0 \le \theta \le 2\pi$$

$$= 8\left[-\cos\frac{\theta}{2}\right]_0^{2\pi}$$

$$= 8(1 + 1)$$

$$= 16$$

In the fifth step of the solution, it is legitimate to write

$$\sqrt{2\sin^2(\theta/2)} = \sqrt{2}\sin(\theta/2)$$

rather than

$$\sqrt{2\sin^2(\theta/2)} = \sqrt{2}\,|\sin(\theta/2)|$$

because $\sin(\theta/2) \ge 0$ for $0 \le \theta \le 2\pi$.

NOTE Using Figure 10.56, you can determine the reasonableness of this answer by comparing it with the circumference of a circle. For example, a circle of radius $\frac{5}{2}$ has a circumference of $5\pi \approx 15.7$.

Area of a Surface of Revolution

The polar coordinate versions of the formulas for the area of a surface of revolution can be obtained from the parametric versions given in Theorem 10.9, using the equations $x = r\cos\theta$ and $y = r\sin\theta$.

NOTE When using Theorem 10.15, check to see that the graph of $r = f(\theta)$ is traced only once on the interval $\alpha \le \theta \le \beta$. For example, the circle given by $r = \cos\theta$ is traced only once on the interval $0 \le \theta \le \pi$.

THEOREM 10.15 Area of a Surface of Revolution

Let f be a function whose derivative is continuous on an interval $\alpha \le \theta \le \beta$. The area of the surface formed by revolving the graph of $r = f(\theta)$ from $\theta = \alpha$ to $\theta = \beta$ about the indicated line is as follows.

1. $S = 2\pi \displaystyle\int_{\alpha}^{\beta} f(\theta)\sin\theta\sqrt{[f(\theta)]^2 + [f'(\theta)]^2}\,d\theta$ About the polar axis

2. $S = 2\pi \displaystyle\int_{\alpha}^{\beta} f(\theta)\cos\theta\sqrt{[f(\theta)]^2 + [f'(\theta)]^2}\,d\theta$ About the line $\theta = \dfrac{\pi}{2}$

EXAMPLE 5 Finding the Area of a Surface of Revolution

Find the area of the surface formed by revolving the circle $r = f(\theta) = \cos\theta$ about the line $\theta = \pi/2$, as shown in Figure 10.57.

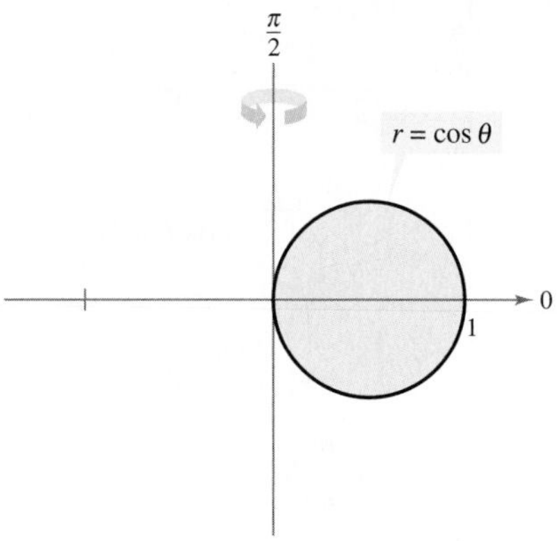

(a)

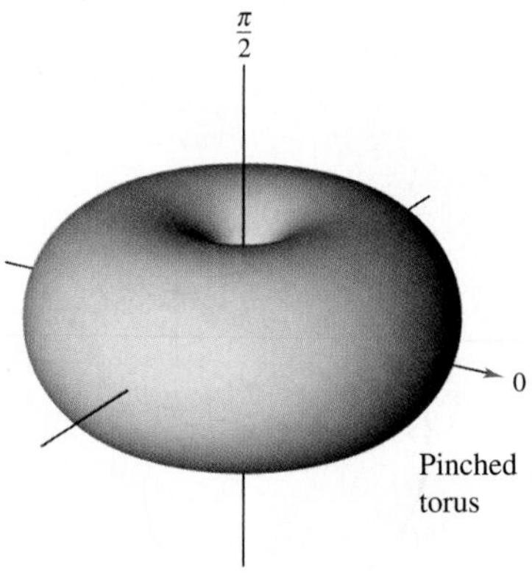

(b)

Figure 10.57

Solution You can use the second formula given in Theorem 10.15 with $f'(\theta) = -\sin\theta$. Because the circle is traced once as θ increases from 0 to π, you have

$$S = 2\pi \int_{\alpha}^{\beta} f(\theta)\cos\theta\sqrt{[f(\theta)]^2 + [f'(\theta)]^2}\,d\theta$$ Formula for area of a surface of revolution

$$= 2\pi \int_{0}^{\pi} \cos\theta(\cos\theta)\sqrt{\cos^2\theta + \sin^2\theta}\,d\theta$$

$$= 2\pi \int_{0}^{\pi} \cos^2\theta\,d\theta$$ Trigonometric identity

$$= \pi \int_{0}^{\pi} (1 + \cos 2\theta)\,d\theta$$ Trigonometric identity

$$= \pi\left[\theta + \frac{\sin 2\theta}{2}\right]_0^{\pi} = \pi^2.$$

Exercises for Section 10.5

See www.CalcChat.com for worked-out solutions to odd-numbered exercises.

In Exercises 1–4, write an integral that represents the area of the shaded region shown in the figure. Do not evaluate the integral.

1. $r = 2 \sin \theta$

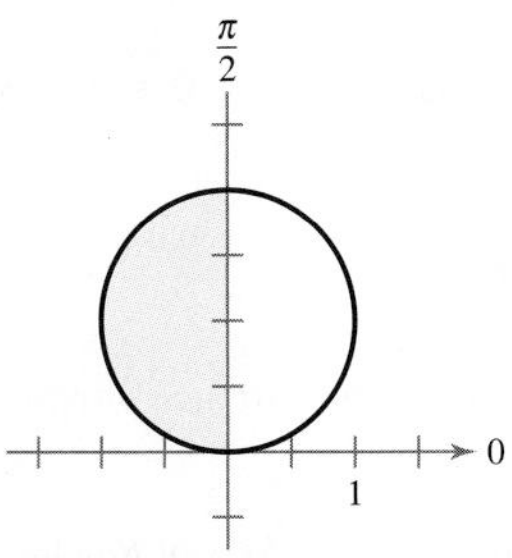

2. $r = \cos 2\theta$

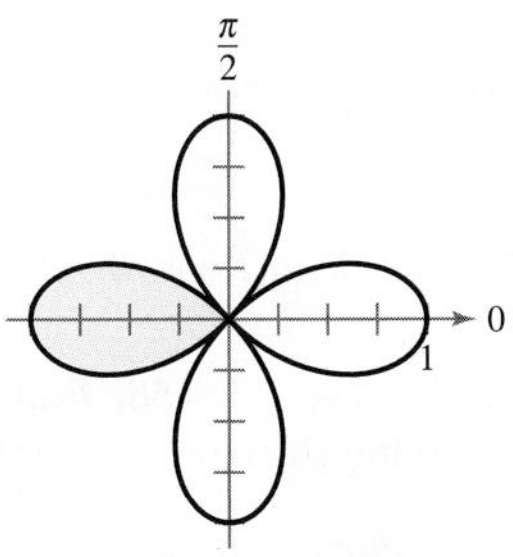

3. $r = 1 - \sin \theta$

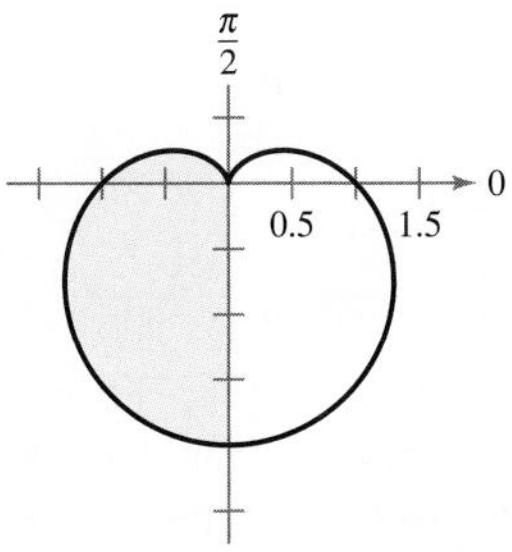

4. $r = 1 - \cos 2\theta$

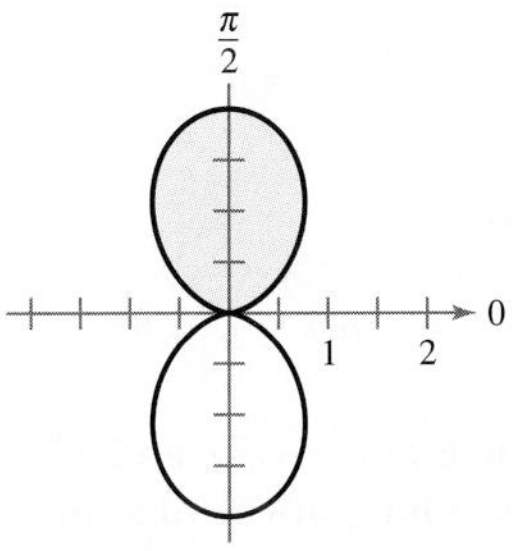

In Exercises 5 and 6, find the area of the region bounded by the graph of the polar equation using (a) a geometric formula and (b) integration.

5. $r = 8 \sin \theta$

6. $r = 3 \cos \theta$

In Exercises 7–12, find the area of the region.

7. One petal of $r = 2 \cos 3\theta$

8. One petal of $r = 6 \sin 2\theta$

9. One petal of $r = \cos 2\theta$

10. One petal of $r = \cos 5\theta$

11. Interior of $r = 1 - \sin \theta$

12. Interior of $r = 1 - \sin \theta$ (above the polar axis)

In Exercises 13–16, use a graphing utility to graph the polar equation and find the area of the given region.

13. Inner loop of $r = 1 + 2 \cos \theta$

14. Inner loop of $r = 4 - 6 \sin \theta$

15. Between the loops of $r = 1 + 2 \cos \theta$

16. Between the loops of $r = 2(1 + 2 \sin \theta)$

In Exercises 17–26, find the points of intersection of the graphs of the equations.

17. $r = 1 + \cos \theta$
$r = 1 - \cos \theta$

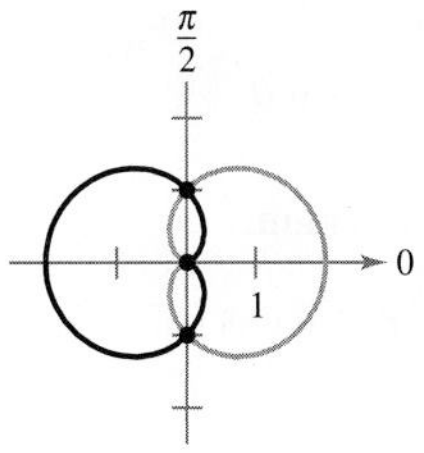

18. $r = 3(1 + \sin \theta)$
$r = 3(1 - \sin \theta)$

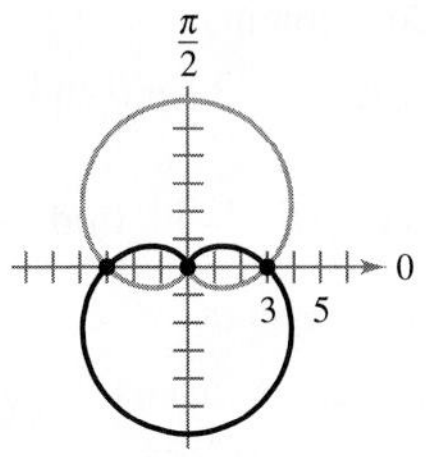

19. $r = 1 + \cos \theta$
$r = 1 - \sin \theta$

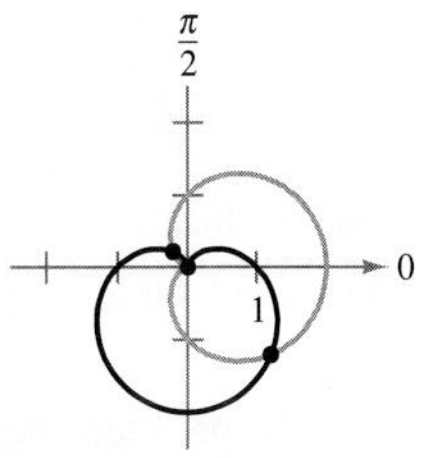

20. $r = 2 - 3 \cos \theta$
$r = \cos \theta$

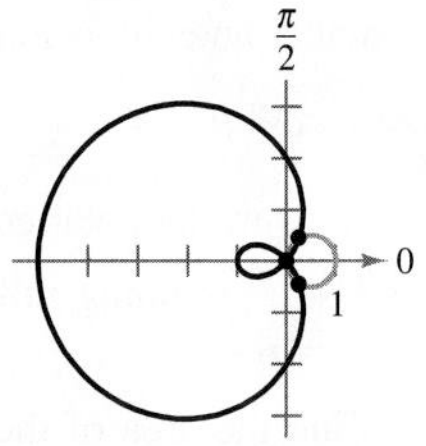

21. $r = 4 - 5 \sin \theta$
$r = 3 \sin \theta$

22. $r = 1 + \cos \theta$
$r = 3 \cos \theta$

23. $r = \dfrac{\theta}{2}$
$r = 2$

24. $\theta = \dfrac{\pi}{4}$
$r = 2$

25. $r = 4 \sin 2\theta$
$r = 2$

26. $r = 3 + \sin \theta$
$r = 2 \csc \theta$

In Exercises 27 and 28, use a graphing utility to approximate the points of intersection of the graphs of the polar equations. Confirm your results analytically.

27. $r = 2 + 3 \cos \theta$
$r = \dfrac{\sec \theta}{2}$

28. $r = 3(1 - \cos \theta)$
$r = \dfrac{6}{1 - \cos \theta}$

Writing **In Exercises 29 and 30, use a graphing utility to find the points of intersection of the graphs of the polar equations. Watch the graphs as they are traced in the viewing window. Explain why the pole is not a point of intersection obtained by solving the equations simultaneously.**

29. $r = \cos \theta$
$r = 2 - 3 \sin \theta$

30. $r = 4 \sin \theta$
$r = 2(1 + \sin \theta)$

In Exercises 31–36, use a graphing utility to graph the polar equations and find the area of the given region.

31. Common interior of $r = 4 \sin 2\theta$ and $r = 2$

32. Common interior of $r = 3(1 + \sin \theta)$ and $r = 3(1 - \sin \theta)$

33. Common interior of $r = 3 - 2 \sin \theta$ and $r = -3 + 2 \sin \theta$

34. Common interior of $r = 5 - 3 \sin \theta$ and $r = 5 - 3 \cos \theta$

35. Common interior of $r = 4 \sin \theta$ and $r = 2$

36. Inside $r = 3 \sin \theta$ and outside $r = 2 - \sin \theta$

In Exercises 37–40, find the area of the region.

37. Inside $r = a(1 + \cos \theta)$ and outside $r = a \cos \theta$

38. Inside $r = 2a \cos \theta$ and outside $r = a$

39. Common interior of $r = a(1 + \cos \theta)$ and $r = a \sin \theta$

40. Common interior of $r = a \cos \theta$ and $r = a \sin \theta$ where $a > 0$

41. ***Antenna Radiation*** The radiation from a transmitting antenna is not uniform in all directions. The intensity from a particular antenna is modeled by

$$r = a \cos^2 \theta.$$

(a) Convert the polar equation to rectangular form.

(b) Use a graphing utility to graph the model for $a = 4$ and $a = 6$.

(c) Find the area of the geographical region between the two curves in part (b).

42. ***Area*** The area inside one or more of the three interlocking circles

$$r = 2a \cos \theta, \quad r = 2a \sin \theta, \quad \text{and} \quad r = a$$

is divided into seven regions. Find the area of each region.

43. ***Conjecture*** Find the area of the region enclosed by

$$r = a \cos(n\theta)$$

for $n = 1, 2, 3, \ldots$. Use the results to make a conjecture about the area enclosed by the function if n is even and if n is odd.

44. ***Area*** Sketch the strophoid

$$r = \sec \theta - 2 \cos \theta, \quad -\frac{\pi}{2} < \theta < \frac{\pi}{2}.$$

Convert this equation to rectangular coordinates. Find the area enclosed by the loop.

In Exercises 45–48, find the length of the curve over the given interval.

Polar Equation	Interval
45. $r = a$	$0 \le \theta \le 2\pi$
46. $r = 2a \cos \theta$	$-\frac{\pi}{2} \le \theta \le \frac{\pi}{2}$
47. $r = 1 + \sin \theta$	$0 \le \theta \le 2\pi$
48. $r = 8(1 + \cos \theta)$	$0 \le \theta \le 2\pi$

In Exercises 49–54, use a graphing utility to graph the polar equation over the given interval. Use the integration capabilities of the graphing utility to approximate the length of the curve accurate to two decimal places.

49. $r = 2\theta, \quad 0 \le \theta \le \frac{\pi}{2}$

50. $r = \sec \theta, \quad 0 \le \theta \le \frac{\pi}{3}$

51. $r = \frac{1}{\theta}, \quad \pi \le \theta \le 2\pi$

52. $r = e^{\theta}, \quad 0 \le \theta \le \pi$

53. $r = \sin(3 \cos \theta), \quad 0 \le \theta \le \pi$

54. $r = 2 \sin(2 \cos \theta), \quad 0 \le \theta \le \pi$

In Exercises 55–58, find the area of the surface formed by revolving the curve about the given line.

Polar Equation	Interval	Axis of Revolution
55. $r = 6 \cos \theta$	$0 \le \theta \le \frac{\pi}{2}$	Polar axis
56. $r = a \cos \theta$	$0 \le \theta \le \frac{\pi}{2}$	$\theta = \frac{\pi}{2}$
57. $r = e^{a\theta}$	$0 \le \theta \le \frac{\pi}{2}$	$\theta = \frac{\pi}{2}$
58. $r = a(1 + \cos \theta)$	$0 \le \theta \le \pi$	Polar axis

In Exercises 59 and 60, use the integration capabilities of a graphing utility to approximate to two decimal places the area of the surface formed by revolving the curve about the polar axis.

59. $r = 4 \cos 2\theta, \quad 0 \le \theta \le \frac{\pi}{4}$

60. $r = \theta, \quad 0 \le \theta \le \pi$

Writing About Concepts

61. Give the integral formulas for area and arc length in polar coordinates.

62. Explain why finding points of intersection of polar graphs may require further analysis beyond solving two equations simultaneously.

63. Which integral yields the arc length of $r = 3(1 - \cos 2\theta)$? State why the other integrals are incorrect.

(a) $3\int_0^{2\pi} \sqrt{(1 - \cos 2\theta)^2 + 4 \sin^2 2\theta}\, d\theta$

(b) $12\int_0^{\pi/4} \sqrt{(1 - \cos 2\theta)^2 + 4 \sin^2 2\theta}\, d\theta$

(c) $3\int_0^{\pi} \sqrt{(1 - \cos 2\theta)^2 + 4 \sin^2 2\theta}\, d\theta$

(d) $6\int_0^{\pi/2} \sqrt{(1 - \cos 2\theta)^2 + 4 \sin^2 2\theta}\, d\theta$

64. Give the integral formulas for the area of the surface of revolution formed when the graph of $r = f(\theta)$ is revolved about (a) the x-axis and (b) the y-axis.

65. ***Surface Area of a Torus*** Find the surface area of the torus generated by revolving the circle given by $r = 2$ about the line $r = 5 \sec \theta$.

66. ***Surface Area of a Torus*** Find the surface area of the torus generated by revolving the circle given by $r = a$ about the line $r = b \sec \theta$, where $0 < a < b$.

67. ***Approximating Area*** Consider the circle $r = 8 \cos \theta$.

(a) Find the area of the circle.

(b) Complete the table giving the areas A of the sectors of the circle between $\theta = 0$ and the values of θ in the table.

θ	0.2	0.4	0.6	0.8	1.0	1.2	1.4
A							

(c) Use the table in part (b) to approximate the values of θ for which the sector of the circle composes $\frac{1}{4}$, $\frac{1}{2}$, and $\frac{3}{4}$ of the total area of the circle.

(d) Use a graphing utility to approximate, to two decimal places, the angles θ for which the sector of the circle composes $\frac{1}{4}$, $\frac{1}{2}$, and $\frac{3}{4}$ of the total area of the circle.

(e) Do the results of part (d) depend on the radius of the circle? Explain.

68. ***Approximate Area*** Consider the circle $r = 3 \sin \theta$.

(a) Find the area of the circle.

(b) Complete the table giving the areas A of the sectors of the circle between $\theta = 0$ and the values of θ in the table.

θ	0.2	0.4	0.6	0.8	1.0	1.2	1.4
A							

(c) Use the table in part (b) to approximate the values of θ for which the sector of the circle composes $\frac{1}{8}$, $\frac{1}{4}$, and $\frac{1}{2}$ of the total area of the circle.

(d) Use a graphing utility to approximate, to two decimal places, the angles θ for which the sector of the circle composes $\frac{1}{8}$, $\frac{1}{4}$, and $\frac{1}{2}$ of the total area of the circle.

69. What conic section does the following polar equation represent?

$$r = a \sin \theta + b \cos \theta$$

70. ***Area*** Find the area of the circle given by $r = \sin \theta + \cos \theta$. Check your result by converting the polar equation to rectangular form, then using the formula for the area of a circle.

71. ***Spiral of Archimedes*** The curve represented by the equation $r = a\theta$, where a is a constant, is called the spiral of Archimedes.

(a) Use a graphing utility to graph $r = \theta$, where $\theta \geq 0$. What happens to the graph of $r = a\theta$ as a increases? What happens if $\theta \leq 0$?

(b) Determine the points on the spiral $r = a\theta$ $(a > 0, \theta \geq 0)$, where the curve crosses the polar axis.

(c) Find the length of $r = \theta$ over the interval $0 \leq \theta \leq 2\pi$.

(d) Find the area under the curve $r = \theta$ for $0 \leq \theta \leq 2\pi$.

72. ***Logarithmic Spiral*** The curve represented by the equation $r = ae^{b\theta}$, where a and b are constants, is called a **logarithmic spiral.** The figure below shows the graph of $r = e^{\theta/6}$, $-2\pi \leq \theta \leq 2\pi$. Find the area of the shaded region.

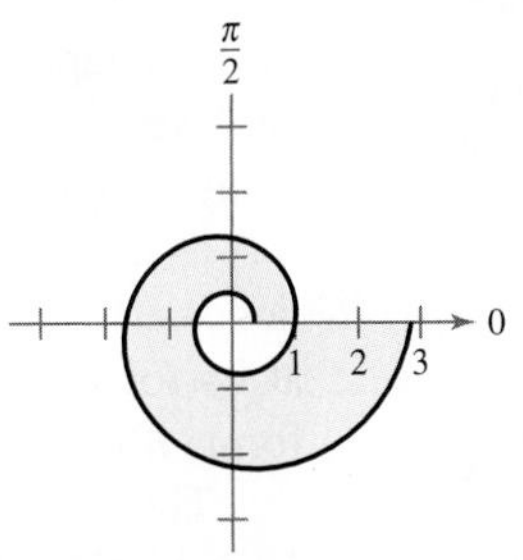

73. The larger circle in the figure below is the graph of $r = 1$. Find the polar equation of the smaller circle such that the shaded regions are equal.

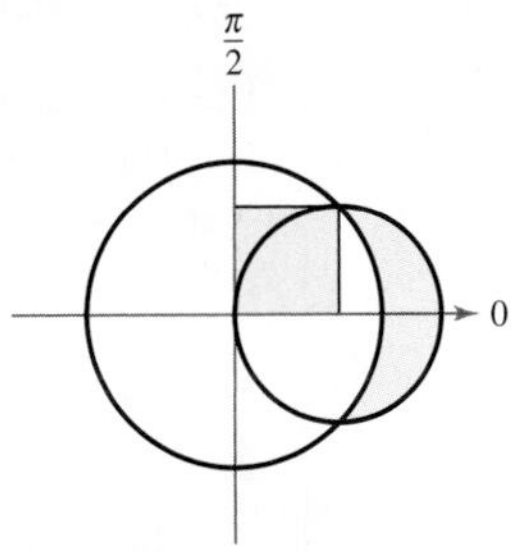

74. ***Folium of Descartes*** A curve called the **folium of Descartes** can be represented by the parametric equations

$$x = \frac{3t}{1 + t^3} \quad \text{and} \quad y = \frac{3t^2}{1 + t^3}.$$

(a) Convert the parametric equations to polar form.

(b) Sketch the graph of the polar equation from part (a).

(c) Use a graphing utility to approximate the area enclosed by the loop of the curve.

True or False? **In Exercises 75 and 76, determine whether the statement is true or false. If it is false, explain why or give an example that shows it is false.**

75. If $f(\theta) > 0$ for all θ and $g(\theta) < 0$ for all θ, then the graphs of $r = f(\theta)$ and $r = g(\theta)$ do not intersect.

76. If $f(\theta) = g(\theta)$ for $\theta = 0$, $\pi/2$, and $3\pi/2$, then the graphs of $r = f(\theta)$ and $r = g(\theta)$ have at least four points of intersection.

77. Use the formula for the arc length of a curve in parametric form to derive the formula for the arc length of a polar curve.

Section 10.6 Polar Equations of Conics and Kepler's Laws

- **Analyze and write polar equations of conics.**
- **Understand and use Kepler's Laws of planetary motion.**

EXPLORATION

Graphing Conics Set a graphing utility to *polar* mode and enter polar equations of the form

$$r = \frac{a}{1 \pm b \cos \theta}$$

or

$$r = \frac{a}{1 \pm b \sin \theta}.$$

As long as $a \neq 0$, the graph should be a conic. Describe the values of a and b that produce parabolas. What values produce ellipses? What values produce hyperbolas?

Polar Equations of Conics

In this chapter you have seen that the rectangular equations of ellipses and hyperbolas take simple forms when the origin lies at their *centers*. As it happens, there are many important applications of conics in which it is more convenient to use one of the foci as the reference point (the origin) for the coordinate system. For example, the sun lies at a focus of Earth's orbit. Similarly, the light source of a parabolic reflector lies at its focus. In this section you will see that polar equations of conics take simple forms if one of the foci lies at the pole.

The following theorem uses the concept of *eccentricity*, as defined in Section 10.1, to classify the three basic types of conics. A proof of this theorem is given in Appendix A.

THEOREM 10.16 Classification of Conics by Eccentricity

Let F be a fixed point (*focus*) and D be a fixed line (*directrix*) in the plane. Let P be another point in the plane and let e (*eccentricity*) be the ratio of the distance between P and F to the distance between P and D. The collection of all points P with a given eccentricity is a conic.

1. The conic is an ellipse if $0 < e < 1$.
2. The conic is a parabola if $e = 1$.
3. The conic is a hyperbola if $e > 1$.

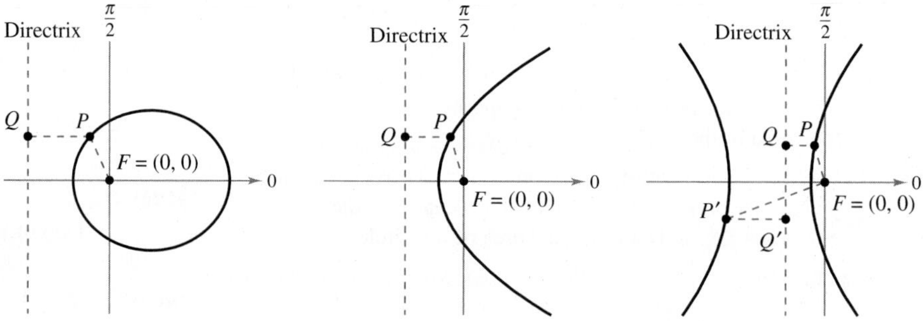

Ellipse: $0 < e < 1$
$\frac{PF}{PQ} < 1$

Parabola: $e = 1$
$PF = PQ$

Hyperbola: $e > 1$
$\frac{PF}{PQ} = \frac{P'F}{P'Q'} > 1$

Figure 10.58

In Figure 10.58, note that for each type of conic the pole corresponds to the fixed point (focus) given in the definition. The benefit of this location can be seen in the proof of the following theorem.

THEOREM 10.17 Polar Equations of Conics

The graph of a polar equation of the form

$$r = \frac{ed}{1 \pm e \cos \theta} \quad \text{or} \quad r = \frac{ed}{1 \pm e \sin \theta}$$

is a conic, where $e > 0$ is the eccentricity and $|d|$ is the distance between the focus at the pole and its corresponding directrix.

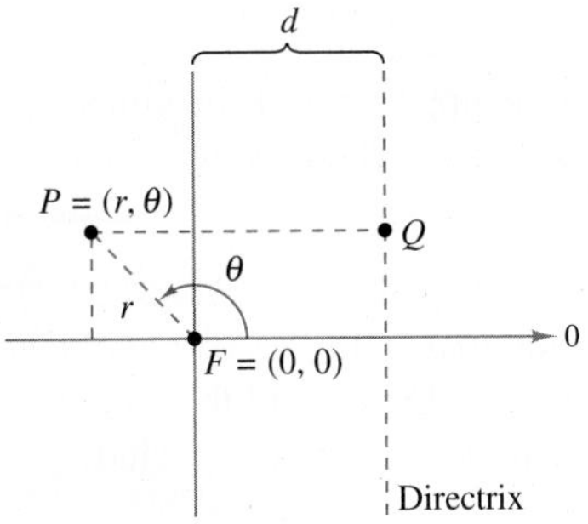

Figure 10.59

Proof The following is a proof for $r = ed/(1 + e \cos \theta)$ with $d > 0$. In Figure 10.59, consider a vertical directrix d units to the right of the focus $F = (0, 0)$. If $P = (r, \theta)$ is a point on the graph of $r = ed/(1 + e \cos \theta)$, the distance between P and the directrix can be shown to be

$$PQ = |d - x| = |d - r\cos\theta| = \left|\frac{r(1 + e\cos\theta)}{e} - r\cos\theta\right| = \left|\frac{r}{e}\right|.$$

Because the distance between P and the pole is simply $PF = |r|$, the ratio of PF to PQ is $PF/PQ = |r|/|r/e| = |e| = e$ and, by Theorem 10.16, the graph of the equation must be a conic. The proofs of the other cases are similar.

The four types of equations indicated in Theorem 10.17 can be classified as follows, where $d > 0$.

a. Horizontal directrix above the pole: $r = \dfrac{ed}{1 + e\sin\theta}$

b. Horizontal directrix below the pole: $r = \dfrac{ed}{1 - e\sin\theta}$

c. Vertical directrix to the right of the pole: $r = \dfrac{ed}{1 + e\cos\theta}$

d. Vertical directrix to the left of the pole: $r = \dfrac{ed}{1 - e\cos\theta}$

Figure 10.60 illustrates these four possibilities for a parabola.

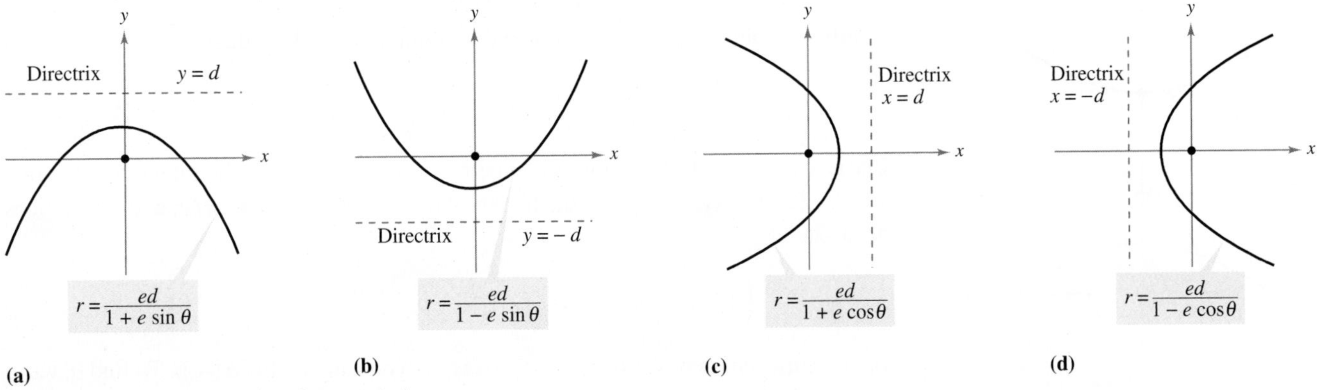

The four types of polar equations for a parabola
Figure 10.60

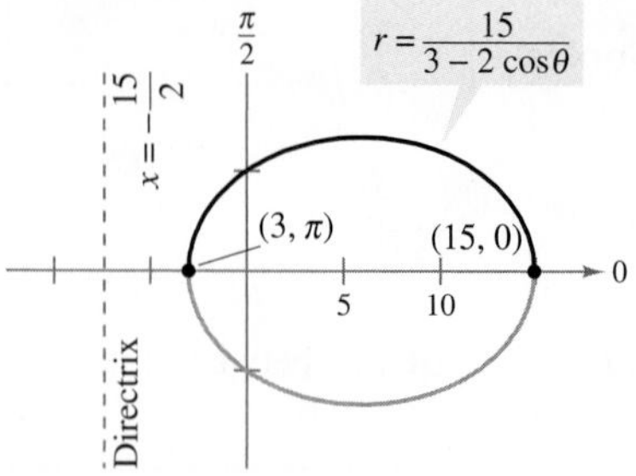

The graph of the conic is an ellipse with $e = \frac{2}{3}$.
Figure 10.61

EXAMPLE 1 Determining a Conic from Its Equation

Sketch the graph of the conic given by $r = \dfrac{15}{3 - 2\cos\theta}$.

Solution To determine the type of conic, rewrite the equation as

$$r = \frac{15}{3 - 2\cos\theta} \qquad \text{Write original equation.}$$

$$= \frac{5}{1 - (2/3)\cos\theta}. \qquad \text{Divide numerator and denominator by 3.}$$

So, the graph is an ellipse with $e = \frac{2}{3}$. You can sketch the upper half of the ellipse by plotting points from $\theta = 0$ to $\theta = \pi$, as shown in Figure 10.61. Then, using symmetry with respect to the polar axis, you can sketch the lower half.

For the ellipse in Figure 10.61, the major axis is horizontal and the vertices lie at $(15, 0)$ and $(3, \pi)$. So, the length of the *major* axis is $2a = 18$. To find the length of the minor axis, you can use the equations $e = c/a$ and $b^2 = a^2 - c^2$ to conclude

$$b^2 = a^2 - c^2 = a^2 - (ea)^2 = a^2(1 - e^2). \qquad \text{Ellipse}$$

Because $e = \frac{2}{3}$, you have

$$b^2 = 9^2\left[1 - \left(\tfrac{2}{3}\right)^2\right] = 45$$

which implies that $b = \sqrt{45} = 3\sqrt{5}$. So, the length of the minor axis is $2b = 6\sqrt{5}$. A similar analysis for hyperbolas yields

$$b^2 = c^2 - a^2 = (ea)^2 - a^2 = a^2(e^2 - 1). \qquad \text{Hyperbola}$$

EXAMPLE 2 Sketching a Conic from Its Polar Equation

Sketch the graph of the polar equation $r = \dfrac{32}{3 + 5\sin\theta}$.

Solution Dividing the numerator and denominator by 3 produces

$$r = \frac{32/3}{1 + (5/3)\sin\theta}.$$

Because $e = \frac{5}{3} > 1$, the graph is a hyperbola. Because $d = \frac{32}{5}$, the directrix is the line $y = \frac{32}{5}$. The transverse axis of the hyperbola lies on the line $\theta = \pi/2$, and the vertices occur at

$$(r, \theta) = \left(4, \frac{\pi}{2}\right) \quad \text{and} \quad (r, \theta) = \left(-16, \frac{3\pi}{2}\right).$$

Because the length of the transverse axis is 12, you can see that $a = 6$. To find b, write

$$b^2 = a^2(e^2 - 1) = 6^2\left[\left(\frac{5}{3}\right)^2 - 1\right] = 64.$$

Therefore, $b = 8$. Finally, you can use a and b to determine the asymptotes of the hyperbola and obtain the sketch shown in Figure 10.62.

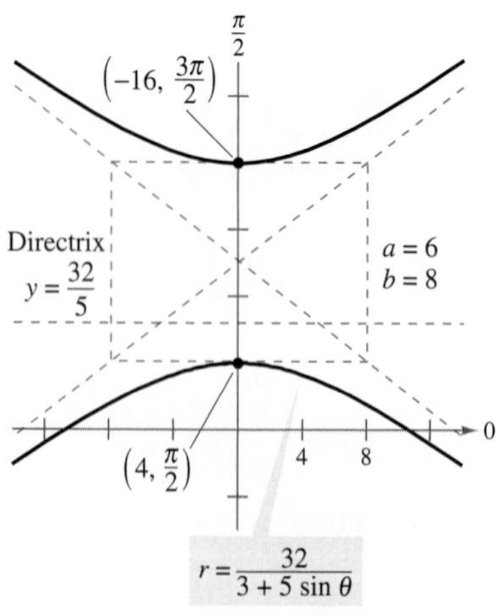

The graph of the conic is a hyperbola with $e = \frac{5}{3}$.
Figure 10.62

JOHANNES KEPLER (1571–1630)

Kepler formulated his three laws from the extensive data recorded by Danish astronomer Tycho Brahe, and from direct observation of the orbit of Mars.

Kepler's Laws

Kepler's Laws, named after the German astronomer Johannes Kepler, can be used to describe the orbits of the planets about the sun.

1. Each planet moves in an elliptical orbit with the sun as a focus.
2. A ray from the sun to the planet sweeps out equal areas of the ellipse in equal times.
3. The square of the period is proportional to the cube of the mean distance between the planet and the sun.*

Although Kepler derived these laws empirically, they were later validated by Newton. In fact, Newton was able to show that each law can be deduced from a set of universal laws of motion and gravitation that govern the movement of all heavenly bodies, including comets and satellites. This is shown in the next example, involving the comet named after the English mathematician and physicist Edmund Halley (1656–1742).

EXAMPLE 3 Halley's Comet

Halley's comet has an elliptical orbit with the sun at one focus and has an eccentricity of $e \approx 0.967$. The length of the major axis of the orbit is approximately 35.88 astronomical units. (An astronomical unit is defined to be the mean distance between Earth and the sun, 93 million miles.) Find a polar equation for the orbit. How close does Halley's comet come to the sun?

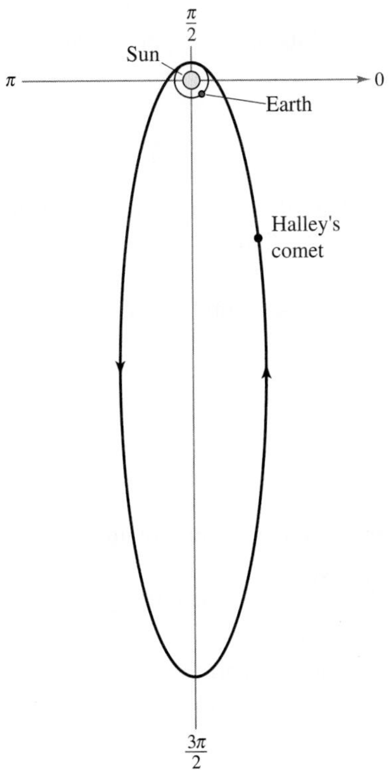

Figure 10.63

Solution Using a vertical axis, you can choose an equation of the form

$$r = \frac{ed}{(1 + e\sin\theta)}.$$

Because the vertices of the ellipse occur when $\theta = \pi/2$ and $\theta = 3\pi/2$, you can determine the length of the major axis to be the sum of the r-values of the vertices, as shown in Figure 10.63. That is,

$$2a = \frac{0.967d}{1 + 0.967} + \frac{0.967d}{1 - 0.967}$$

$$35.88 \approx 27.79d. \qquad 2a \approx 35.88$$

So, $d \approx 1.204$ and $ed \approx (0.967)(1.204) \approx 1.164$. Using this value in the equation produces

$$r = \frac{1.164}{1 + 0.967\sin\theta}$$

where r is measured in astronomical units. To find the closest point to the sun (the focus), you can write $c = ea \approx (0.967)(17.94) \approx 17.35$. Because c is the distance between the focus and the center, the closest point is

$$\begin{aligned} a - c &\approx 17.94 - 17.35 \\ &\approx 0.59 \text{ AU} \\ &\approx 55{,}000{,}000 \text{ miles} \end{aligned}$$

* *If Earth is used as a reference with a period of 1 year and a distance of 1 astronomical unit, the proportionality constant is 1. For example, because Mars has a mean distance to the sun of $D = 1.524$ AU, its period P is given by $D^3 = P^2$. So, the period for Mars is $P = 1.88$.*

Kepler's Second Law states that as a planet moves about the sun, a ray from the sun to the planet sweeps out equal areas in equal times. This law can also be applied to comets or asteroids with elliptical orbits. For example, Figure 10.64 shows the orbit of the asteroid Apollo about the sun. Applying Kepler's Second Law to this asteroid, you know that the closer it is to the sun, the greater its velocity, because a short ray must be moving quickly to sweep out as much area as a long ray.

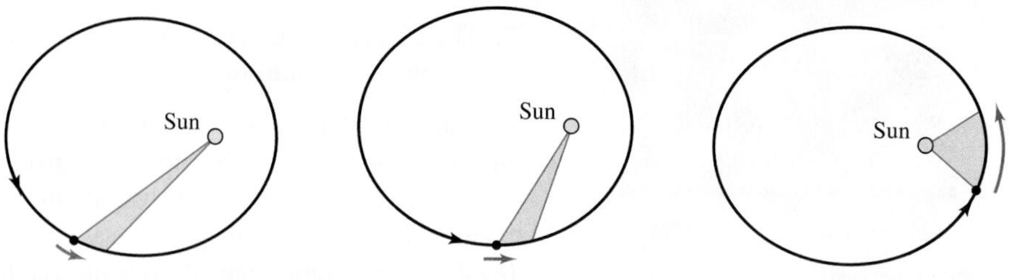

A ray from the sun to the asteroid sweeps out equal areas in equal times.
Figure 10.64

EXAMPLE 4 The Asteroid Apollo

The asteroid Apollo has a period of 661 Earth days, and its orbit is approximated by the ellipse

$$r = \frac{1}{1 + (5/9)\cos\theta} = \frac{9}{9 + 5\cos\theta}$$

where r is measured in astronomical units. How long does it take Apollo to move from the position given by $\theta = -\pi/2$ to $\theta = \pi/2$, as shown in Figure 10.65?

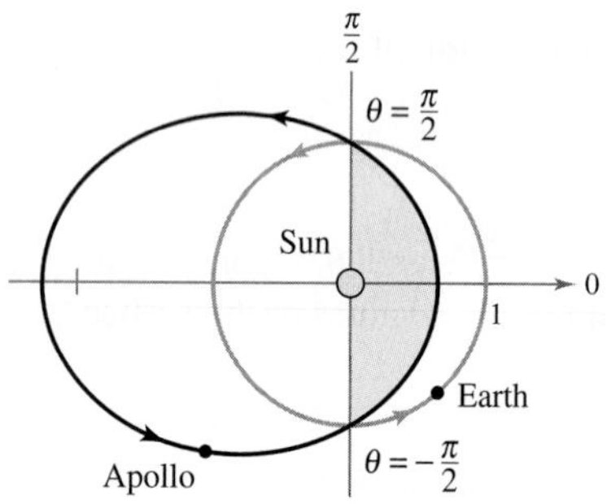

Figure 10.65

Solution Begin by finding the area swept out as θ increases from $-\pi/2$ to $\pi/2$.

$$A = \frac{1}{2}\int_{\alpha}^{\beta} r^2\,d\theta \qquad \text{Formula for area of a polar graph}$$

$$= \frac{1}{2}\int_{-\pi/2}^{\pi/2}\left(\frac{9}{9 + 5\cos\theta}\right)^2 d\theta$$

Using the substitution $u = \tan(\theta/2)$, as discussed in Section 8.6, you obtain

$$A = \frac{81}{112}\left[\frac{-5\sin\theta}{9 + 5\cos\theta} + \frac{18}{\sqrt{56}}\arctan\frac{\sqrt{56}\tan(\theta/2)}{14}\right]_{-\pi/2}^{\pi/2} \approx 0.90429.$$

Because the major axis of the ellipse has length $2a = 81/28$ and the eccentricity is $e = 5/9$, you can determine that $b = a\sqrt{1 - e^2} = 9/\sqrt{56}$. So, the area of the ellipse is

$$\text{Area of ellipse} = \pi ab = \pi\left(\frac{81}{56}\right)\left(\frac{9}{\sqrt{56}}\right) \approx 5.46507.$$

Because the time required to complete the orbit is 661 days, you can apply Kepler's Second Law to conclude that the time t required to move from the position $\theta = -\pi/2$ to $\theta = \pi/2$ is given by

$$\frac{t}{661} = \frac{\text{area of elliptical segment}}{\text{area of ellipse}} \approx \frac{0.90429}{5.46507}$$

which implies that $t \approx 109$ days.

Exercises for Section 10.6

See www.CalcChat.com for worked-out solutions to odd-numbered exercises.

Graphical Reasoning **In Exercises 1–4, use a graphing utility to graph the polar equation when (a) $e = 1$, (b) $e = 0.5$, and (c) $e = 1.5$. Identify the conic.**

1. $r = \dfrac{2e}{1 + e\cos\theta}$

2. $r = \dfrac{2e}{1 - e\cos\theta}$

3. $r = \dfrac{2e}{1 - e\sin\theta}$

4. $r = \dfrac{2e}{1 + e\sin\theta}$

5. ***Writing*** Consider the polar equation

$$r = \frac{4}{1 + e\sin\theta}.$$

(a) Use a graphing utility to graph the equation for $e = 0.1$, $e = 0.25$, $e = 0.5$, $e = 0.75$, and $e = 0.9$. Identify the conic and discuss the change in its shape as $e \to 1^-$ and $e \to 0^+$.

(b) Use a graphing utility to graph the equation for $e = 1$. Identify the conic.

(c) Use a graphing utility to graph the equation for $e = 1.1$, $e = 1.5$, and $e = 2$. Identify the conic and discuss the change in its shape as $e \to 1^+$ and $e \to \infty$.

6. Consider the polar equation

$$r = \frac{4}{1 - 0.4\cos\theta}.$$

(a) Identify the conic without graphing the equation.

(b) Without graphing the following polar equations, describe how each differs from the polar equation above.

$$r = \frac{4}{1 + 0.4\cos\theta}, \quad r = \frac{4}{1 - 0.4\sin\theta}$$

(c) Verify the results of part (b) graphically.

In Exercises 7–12, match the polar equation with the correct graph. [The graphs are labeled (a), (b), (c), (d), (e), and (f).]

(a)

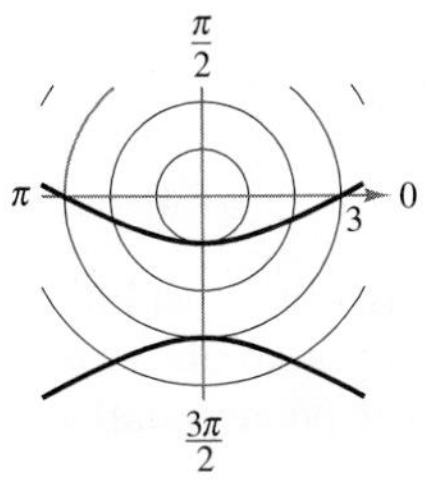

(b)

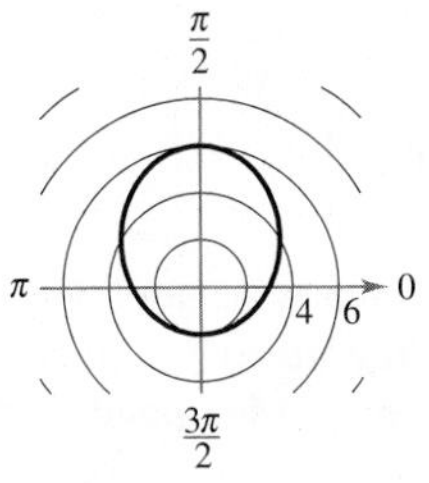

(c)

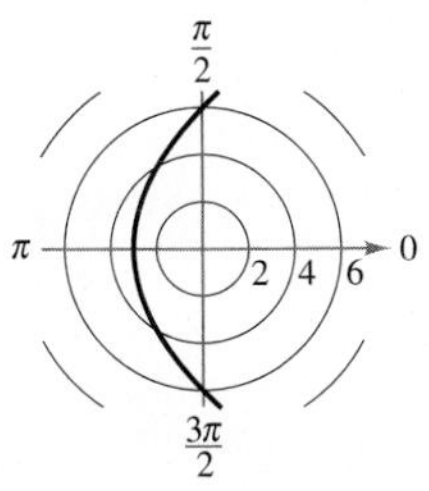

(d)

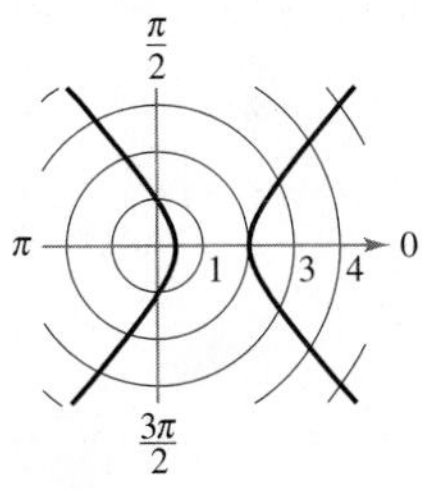

(e)

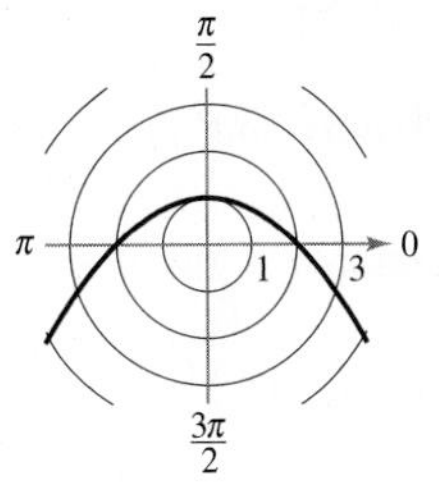

(f)

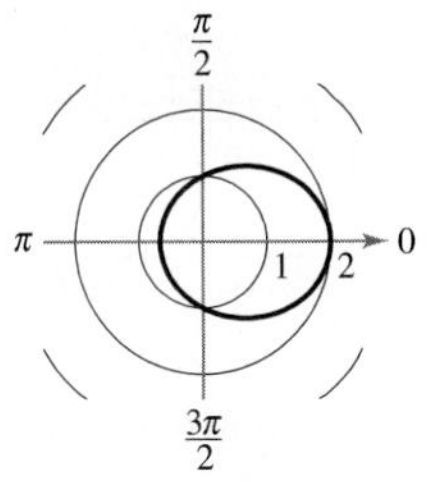

7. $r = \dfrac{6}{1 - \cos\theta}$

8. $r = \dfrac{2}{2 - \cos\theta}$

9. $r = \dfrac{3}{1 - 2\sin\theta}$

10. $r = \dfrac{2}{1 + \sin\theta}$

11. $r = \dfrac{6}{2 - \sin\theta}$

12. $r = \dfrac{2}{2 + 3\cos\theta}$

In Exercises 13–22, find the eccentricity and the distance from the pole to the directrix of the conic. Then sketch and identify the graph. Use a graphing utility to confirm your results.

13. $r = \dfrac{-1}{1 - \sin\theta}$

14. $r = \dfrac{6}{1 + \cos\theta}$

15. $r = \dfrac{6}{2 + \cos\theta}$

16. $r = \dfrac{5}{5 + 3\sin\theta}$

17. $r(2 + \sin\theta) = 4$

18. $r(3 - 2\cos\theta) = 6$

19. $r = \dfrac{5}{-1 + 2\cos\theta}$

20. $r = \dfrac{-6}{3 + 7\sin\theta}$

21. $r = \dfrac{3}{2 + 6\sin\theta}$

22. $r = \dfrac{4}{1 + 2\cos\theta}$

In Exercises 23–26, use a graphing utility to graph the polar equation. Identify the graph.

23. $r = \dfrac{3}{-4 + 2\sin\theta}$

24. $r = \dfrac{-3}{2 + 4\sin\theta}$

25. $r = \dfrac{-1}{1 - \cos\theta}$

26. $r = \dfrac{2}{2 + 3\sin\theta}$

In Exercises 27–30, use a graphing utility to graph the conic. Describe how the graph differs from that in the indicated exercise.

27. $r = \dfrac{-1}{1 - \sin(\theta - \pi/4)}$ (See Exercise 13.)

28. $r = \dfrac{6}{1 + \cos(\theta - \pi/3)}$ (See Exercise 14.)

29. $r = \dfrac{6}{2 + \cos(\theta + \pi/6)}$ (See Exercise 15.)

30. $r = \dfrac{-6}{3 + 7\sin(\theta + 2\pi/3)}$ (See Exercise 20.)

31. Write the equation for the ellipse rotated $\pi/4$ radian clockwise from the ellipse

$$r = \frac{5}{5 + 3\cos\theta}.$$

32. Write the equation for the parabola rotated $\pi/6$ radian counterclockwise from the parabola

$$r = \frac{2}{1 + \sin\theta}.$$

In Exercises 33–44, find a polar equation for the conic with its focus at the pole. (For convenience, the equation for the directrix is given in rectangular form.)

Conic	*Eccentricity*	*Directrix*
33. Parabola	$e = 1$	$x = -1$
34. Parabola	$e = 1$	$y = 1$
35. Ellipse	$e = \frac{1}{2}$	$y = 1$
36. Ellipse	$e = \frac{3}{4}$	$y = -2$
37. Hyperbola	$e = 2$	$x = 1$
38. Hyperbola	$e = \frac{3}{2}$	$x = -1$

Conic	*Vertex or Vertices*
39. Parabola	$\left(1, -\frac{\pi}{2}\right)$
40. Parabola	$(5, \pi)$
41. Ellipse	$(2, 0),\ (8, \pi)$
42. Ellipse	$\left(2, \frac{\pi}{2}\right),\ \left(4, \frac{3\pi}{2}\right)$
43. Hyperbola	$\left(1, \frac{3\pi}{2}\right),\ \left(9, \frac{3\pi}{2}\right)$
44. Hyperbola	$(2, 0),\ (10, 0)$

Writing About Concepts

45. Classify the conics by their eccentricities.

46. Explain how the graph of each conic differs from the graph of $r = \dfrac{4}{1 + \sin\theta}$.

(a) $r = \dfrac{4}{1 - \cos\theta}$ (b) $r = \dfrac{4}{1 - \sin\theta}$

(c) $r = \dfrac{4}{1 + \cos\theta}$ (d) $r = \dfrac{4}{1 - \sin(\theta - \pi/4)}$

47. Identify each conic.

(a) $r = \dfrac{5}{1 - 2\cos\theta}$ (b) $r = \dfrac{5}{10 - \sin\theta}$

(c) $r = \dfrac{5}{3 - 3\cos\theta}$ (d) $r = \dfrac{5}{1 - 3\sin(\theta - \pi/4)}$

48. Describe what happens to the distance between the directrix and the center of an ellipse if the foci remain fixed and e approaches 0.

49. Show that the polar equation for $\dfrac{x^2}{a^2} + \dfrac{y^2}{b^2} = 1$ is

$$r^2 = \frac{b^2}{1 - e^2\cos^2\theta}. \quad \text{Ellipse}$$

50. Show that the polar equation for $\dfrac{x^2}{a^2} - \dfrac{y^2}{b^2} = 1$ is

$$r^2 = \frac{-b^2}{1 - e^2\cos^2\theta}. \quad \text{Hyperbola}$$

In Exercises 51–54, use the results of Exercises 49 and 50 to write the polar form of the equation of the conic.

51. Ellipse: focus at $(4, 0)$; vertices at $(5, 0)$, $(5, \pi)$

52. Hyperbola: focus at $(5, 0)$; vertices at $(4, 0)$, $(4, \pi)$

53. $\dfrac{x^2}{9} - \dfrac{y^2}{16} = 1$

54. $\dfrac{x^2}{4} + y^2 = 1$

In Exercises 55 and 56, use the integration capabilities of a graphing utility to approximate to two decimal places the area of the region bounded by the graph of the polar equation.

55. $r = \dfrac{3}{2 - \cos\theta}$

56. $r = \dfrac{2}{3 - 2\sin\theta}$

57. ***Explorer 18*** On November 27, 1963, the United States launched Explorer 18. Its low and high points above the surface of Earth were approximately 119 miles and 123,000 miles (see figure). The center of Earth is the focus of the orbit. Find the polar equation for the orbit and find the distance between the surface of Earth and the satellite when $\theta = 60°$. (Assume that the radius of Earth is 4000 miles.)

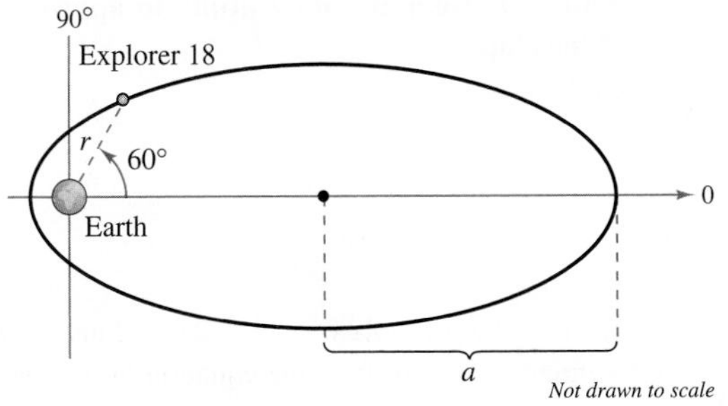

58. ***Planetary Motion*** The planets travel in elliptical orbits with the sun as a focus, as shown in the figure.

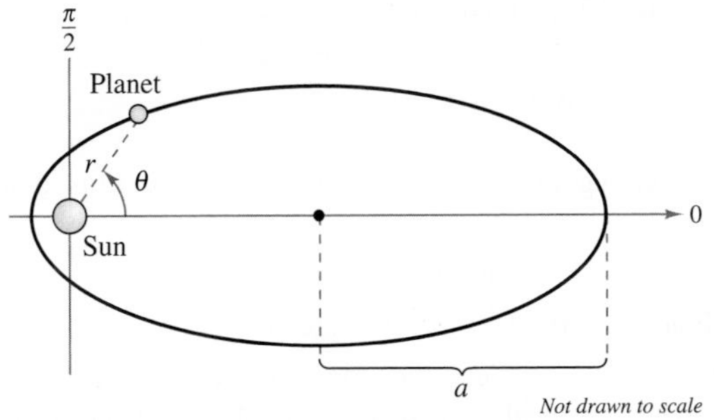

(a) Show that the polar equation of the orbit is given by

$$r = \frac{(1 - e^2)a}{1 - e\cos\theta}$$

where e is the eccentricity.

(b) Show that the minimum distance (*perihelion*) from the sun to the planet is $r = a(1 - e)$ and the maximum distance (*aphelion*) is $r = a(1 + e)$.

In Exercises 59–62, use Exercise 58 to find the polar equation of the elliptical orbit of the planet, and the perihelion and aphelion distances.

59. Earth $a = 1.496 \times 10^8$ kilometers
$e = 0.0167$

60. Saturn $a = 1.427 \times 10^9$ kilometers
$e = 0.0542$

61. Pluto $a = 5.906 \times 10^9$ kilometers
$e = 0.2488$

62. Mercury $a = 5.791 \times 10^7$ kilometers
$e = 0.2056$

63. ***Planetary Motion*** In Exercise 61, the polar equation for the elliptical orbit of Pluto was found. Use the equation and a computer algebra system to perform each of the following.

(a) Approximate the area swept out by a ray from the sun to the planet as θ increases from 0 to $\pi/9$. Use this result to determine the number of years for the planet to move through this arc if the period of one revolution around the sun is 248 years.

(b) By trial and error, approximate the angle α such that the area swept out by a ray from the sun to the planet as θ increases from π to α equals the area found in part (a) (see figure). Does the ray sweep through a larger or smaller angle than in part (a) to generate the same area? Why is this the case?

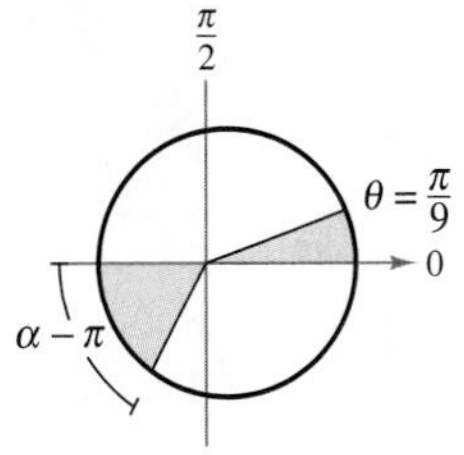

(c) Approximate the distances the planet traveled in parts (a) and (b). Use these distances to approximate the average number of kilometers per year the planet traveled in the two cases.

64. ***Comet Hale-Bopp*** The comet Hale-Bopp has an elliptical orbit with the sun at one focus and has an eccentricity of $e \approx 0.995$. The length of the major axis of the orbit is approximately 250 astronomical units.

(a) Find the length of its minor axis.

(b) Find a polar equation for the orbit.

(c) Find the perihelion and aphelion distances.

In Exercises 65 and 66, let r_0 represent the distance from the focus to the nearest vertex, and let r_1 represent the distance from the focus to the farthest vertex.

65. Show that the eccentricity of an ellipse can be written as

$$e = \frac{r_1 - r_0}{r_1 + r_0}. \text{ Then show that } \frac{r_1}{r_0} = \frac{1 + e}{1 - e}.$$

66. Show that the eccentricity of a hyperbola can be written as

$$e = \frac{r_1 + r_0}{r_1 - r_0}. \text{ Then show that } \frac{r_1}{r_0} = \frac{e + 1}{e - 1}.$$

In Exercises 67 and 68, show that the graphs of the given equations intersect at right angles.

67. $r = \dfrac{ed}{1 + \sin\theta}$ and $r = \dfrac{ed}{1 - \sin\theta}$

68. $r = \dfrac{c}{1 + \cos\theta}$ and $r = \dfrac{d}{1 - \cos\theta}$

Review Exercises for Chapter 10

See www.CalcChat.com for worked-out solutions to odd-numbered exercises.

In Exercises 1–6, match the equation with the correct graph. [The graphs are labeled (a), (b), (c), (d), (e), and (f).]

(a)

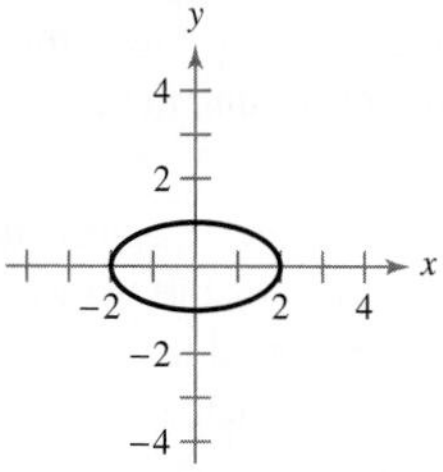

(b)

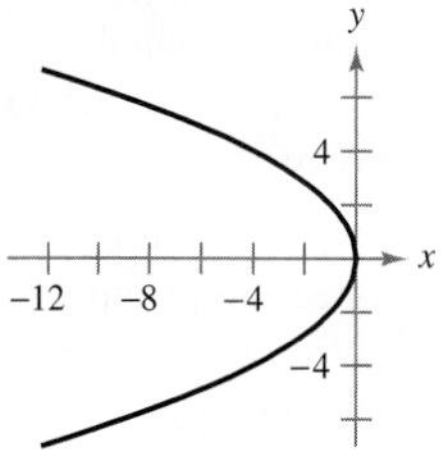

(c)

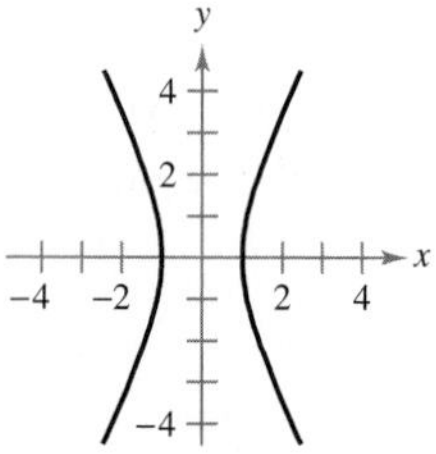

(d)

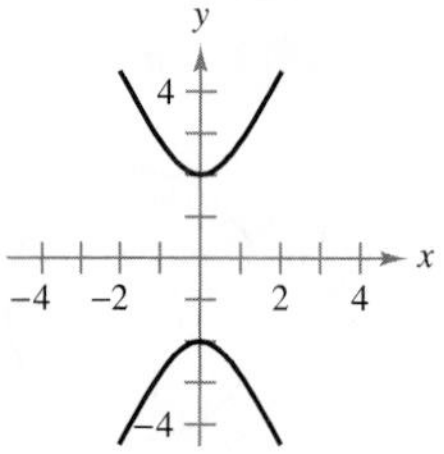

(e)

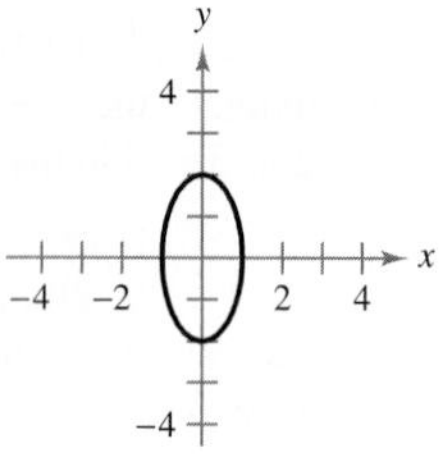

(f)

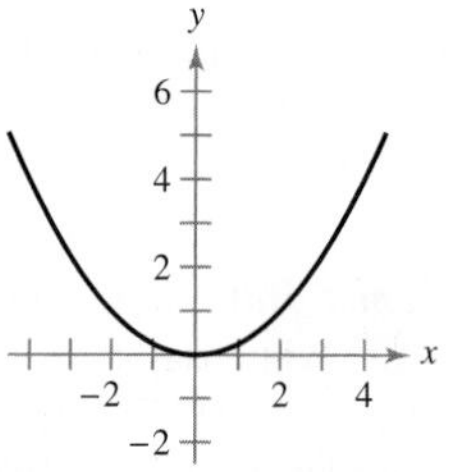

1. $4x^2 + y^2 = 4$

2. $4x^2 - y^2 = 4$

3. $y^2 = -4x$

4. $y^2 - 4x^2 = 4$

5. $x^2 + 4y^2 = 4$

6. $x^2 = 4y$

In Exercises 7–12, analyze the equation and sketch its graph. Use a graphing utility to confirm your results.

7. $16x^2 + 16y^2 - 16x + 24y - 3 = 0$

8. $y^2 - 12y - 8x + 20 = 0$

9. $3x^2 - 2y^2 + 24x + 12y + 24 = 0$

10. $4x^2 + y^2 - 16x + 15 = 0$

11. $3x^2 + 2y^2 - 12x + 12y + 29 = 0$

12. $4x^2 - 4y^2 - 4x + 8y - 11 = 0$

In Exercises 13 and 14, find an equation of the parabola.

13. Vertex: $(0, 2)$; directrix: $x = -3$

14. Vertex: $(4, 2)$; focus: $(4, 0)$

In Exercises 15 and 16, find an equation of the ellipse.

15. Vertices: $(-3, 0), (7, 0)$; foci: $(0, 0), (4, 0)$

16. Center: $(0, 0)$; solution points: $(1, 2), (2, 0)$

In Exercises 17 and 18, find an equation of the hyperbola.

17. Vertices: $(\pm 4, 0)$; foci: $(\pm 6, 0)$

18. Foci: $(0, \pm 8)$; asymptotes: $y = \pm 4x$

In Exercises 19 and 20, use a graphing utility to approximate the perimeter of the ellipse.

19. $\dfrac{x^2}{9} + \dfrac{y^2}{4} = 1$

20. $\dfrac{x^2}{4} + \dfrac{y^2}{25} = 1$

21. A line is tangent to the parabola $y = x^2 - 2x + 2$ and perpendicular to the line $y = x - 2$. Find the equation of the line.

22. A line is tangent to the parabola $3x^2 + y = x - 6$ and perpendicular to the line $2x + y = 5$. Find the equation of the line.

23. ***Satellite Antenna*** A cross section of a large parabolic antenna is modeled by the graph of

$$y = \frac{x^2}{200}, \quad -100 \le x \le 100.$$

The receiving and transmitting equipment is positioned at the focus.

(a) Find the coordinates of the focus.

(b) Find the surface area of the antenna.

24. ***Fire Truck*** Consider a fire truck with a water tank 16 feet long whose vertical cross sections are ellipses modeled by the equation

$$\frac{x^2}{16} + \frac{y^2}{9} = 1.$$

(a) Find the volume of the tank.

(b) Find the force on the end of the tank when it is full of water. (The density of water is 62.4 pounds per cubic foot.)

(c) Find the depth of the water in the tank if it is $\frac{3}{4}$ full (by volume) and the truck is on level ground.

(d) Approximate the tank's surface area.

In Exercises 25–30, sketch the curve represented by the parametric equations (indicate the orientation of the curve), and write the corresponding rectangular equation by eliminating the parameter.

25. $x = 1 + 4t, \ y = 2 - 3t$

26. $x = t + 4, \ y = t^2$

27. $x = 6 \cos \theta, \ y = 6 \sin \theta$

28. $x = 3 + 3 \cos \theta, \ y = 2 + 5 \sin \theta$

29. $x = 2 + \sec \theta, \ y = 3 + \tan \theta$

30. $x = 5 \sin^3 \theta, \ y = 5 \cos^3 \theta$

In Exercises 31–34, find a parametric representation of the line or conic.

31. Line: passes through $(-2, 6)$ and $(3, 2)$

32. Circle: center at $(5, 3)$; radius 2

33. Ellipse: center at $(-3, 4)$; horizontal major axis of length 8 and minor axis of length 6

34. Hyperbola: vertices at $(0, \pm 4)$; foci at $(0, \pm 5)$

35. ***Rotary Engine*** The rotary engine was developed by Felix Wankel in the 1950s. It features a rotor, which is a modified equilateral triangle. The rotor moves in a chamber that, in two dimensions, is an epitrochoid. Use a graphing utility to graph the chamber modeled by the parametric equations.

$x = \cos 3\theta + 5 \cos \theta$

and

$y = \sin 3\theta + 5 \sin \theta.$

36. ***Serpentine Curve*** Consider the parametric equations $x = 2 \cot \theta$ and $y = 4 \sin \theta \cos \theta$, $0 < \theta < \pi$.

(a) Use a graphing utility to graph the curve.

(b) Eliminate the parameter to show that the rectangular equation of the serpentine curve is $(4 + x^2)y = 8x$.

In Exercises 37–46, (a) find dy/dx and all points of horizontal tangency, (b) eliminate the parameter where possible, and (c) sketch the curve represented by the parametric equations.

37. $x = 1 + 4t, \quad y = 2 - 3t$

38. $x = t + 4, \quad y = t^2$

39. $x = \dfrac{1}{t}, \quad y = 2t + 3$

40. $x = \dfrac{1}{t}, \quad y = t^2$

41. $x = \dfrac{1}{2t + 1}$, $\quad y = \dfrac{1}{t^2 - 2t}$

42. $x = 2t - 1$, $\quad y = \dfrac{1}{t^2 - 2t}$

43. $x = 3 + 2 \cos \theta$, $\quad y = 2 + 5 \sin \theta$

44. $x = 6 \cos \theta$, $\quad y = 6 \sin \theta$

45. $x = \cos^3 \theta$, $\quad y = 4 \sin^3 \theta$

46. $x = e^t$, $\quad y = e^{-t}$

In Exercises 47–50, find all points (if any) of horizontal and vertical tangency to the curve. Use a graphing utility to confirm your results.

47. $x = 4 - t, \quad y = t^2$

48. $x = t + 2, \quad y = t^3 - 2t$

49. $x = 2 + 2 \sin \theta, \quad y = 1 + \cos \theta$

50. $x = 2 - 2 \cos \theta, \quad y = 2 \sin 2\theta$

In Exercises 51 and 52, (a) use a graphing utility to graph the curve represented by the parametric equations, (b) use a graphing utility to find $dx/d\theta$, $dy/d\theta$, and dy/dx for $\theta = \pi/6$, and (c) use a graphing utility to graph the tangent line to the curve when $\theta = \pi/6$.

51. $x = \cot \theta$, $\quad y = \sin 2\theta$

52. $x = 2\theta - \sin \theta$, $\quad y = 2 - \cos \theta$

Arc Length **In Exercises 53 and 54, find the arc length of the curve on the given interval.**

53. $x = r(\cos \theta + \theta \sin \theta)$, $\quad y = r(\sin \theta - \theta \cos \theta)$, $\quad 0 \le \theta \le \pi$

54. $x = 6 \cos \theta$, $\quad y = 6 \sin \theta$, $\quad 0 \le \theta \le \pi$

Surface Area **In Exercises 55 and 56, find the area of the surface generated by revolving the curve about (a) the x-axis and (b) the y-axis.**

55. $x = t, \quad y = 3t, \quad 0 \le t \le 2$

56. $x = 2 \cos \theta, \quad y = 2 \sin \theta, \quad 0 \le \theta \le \dfrac{\pi}{2}$

Area **In Exercises 57 and 58, find the area of the region.**

57. $x = 3 \sin \theta$, $\quad y = 2 \cos \theta$, $\quad -\dfrac{\pi}{2} \le \theta \le \dfrac{\pi}{2}$

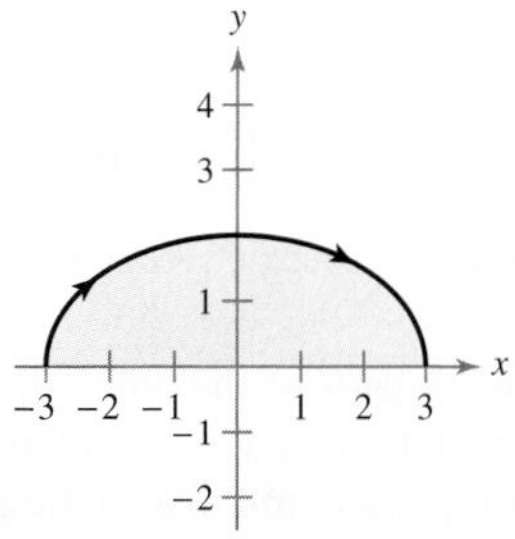

58. $x = 2 \cos \theta$, $\quad y = \sin \theta$, $\quad 0 \le \theta \le \pi$

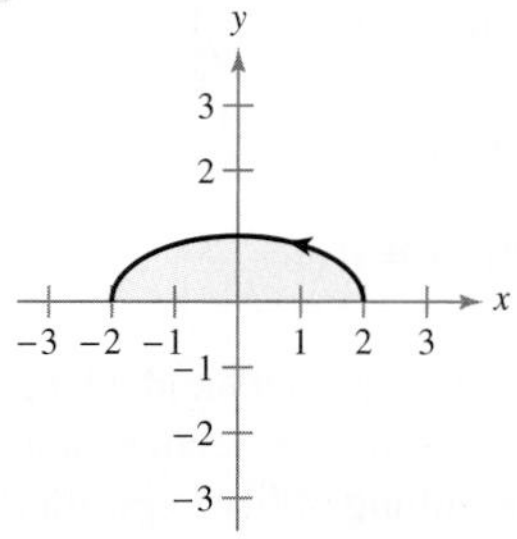

In Exercises 59–62, plot the point in polar coordinates and find the corresponding rectangular coordinates of the point.

59. $\left(3, \dfrac{\pi}{2}\right)$

60. $\left(-4, \dfrac{11\pi}{6}\right)$

61. $\left(\sqrt{3}, 1.56\right)$

62. $(-2, -2.45)$

In Exercises 63 and 64, the rectangular coordinates of a point are given. Plot the point and find *two* sets of polar coordinates of the point for $0 \le \theta < 2\pi$.

63. $(4, -4)$

64. $(-1, 3)$

In Exercises 65–72, convert the polar equation to rectangular form.

65. $r = 3\cos\theta$

66. $r = 10$

67. $r = -2(1 + \cos\theta)$

68. $r = \dfrac{1}{2 - \cos\theta}$

69. $r^2 = \cos 2\theta$

70. $r = 4\sec\left(\theta - \dfrac{\pi}{3}\right)$

71. $r = 4\cos 2\theta \sec\theta$

72. $\theta = \dfrac{3\pi}{4}$

In Exercises 73–76, convert the rectangular equation to polar form.

73. $(x^2 + y^2)^2 = ax^2y$

74. $x^2 + y^2 - 4x = 0$

75. $x^2 + y^2 = a^2\left(\arctan\dfrac{y}{x}\right)^2$

76. $(x^2 + y^2)\left(\arctan\dfrac{y}{x}\right)^2 = a^2$

In Exercises 77–88, sketch a graph of the polar equation.

77. $r = 4$

78. $\theta = \dfrac{\pi}{12}$

79. $r = -\sec\theta$

80. $r = 3\csc\theta$

81. $r = -2(1 + \cos\theta)$

82. $r = 3 - 4\cos\theta$

83. $r = 4 - 3\cos\theta$

84. $r = 2\theta$

85. $r = -3\cos 2\theta$

86. $r = \cos 5\theta$

87. $r^2 = 4\sin^2 2\theta$

88. $r^2 = \cos 2\theta$

In Exercises 89–92, use a graphing utility to graph the polar equation.

89. $r = \dfrac{3}{\cos(\theta - \pi/4)}$

90. $r = 2\sin\theta\cos^2\theta$

91. $r = 4\cos 2\theta\sec\theta$

92. $r = 4(\sec\theta - \cos\theta)$

In Exercises 93 and 94, (a) find the tangents at the pole, (b) find all points of vertical and horizontal tangency, and (c) use a graphing utility to graph the polar equation and draw a tangent line to the graph for $\theta = \pi/6$.

93. $r = 1 - 2\cos\theta$

94. $r^2 = 4\sin 2\theta$

95. Find the angle between the circle $r = 3\sin\theta$ and the limaçon $r = 4 - 5\sin\theta$ at the point of intersection $(3/2, \pi/6)$.

96. ***True or False?*** There is a unique polar coordinate representation for each point in the plane. Explain.

In Exercises 97 and 98, show that the graphs of the polar equations are orthogonal at the points of intersection. Use a graphing utility to confirm your results graphically.

97. $r = 1 + \cos\theta$
$r = 1 - \cos\theta$

98. $r = a\sin\theta$
$r = a\cos\theta$

In Exercises 99–102, find the area of the region.

99. Interior of $r = 2 + \cos\theta$

100. Interior of $r = 5(1 - \sin\theta)$

101. Interior of $r^2 = 4\sin 2\theta$

102. Common interior of $r = 4\cos\theta$ and $r = 2$

In Exercises 103–106, use a graphing utility to graph the polar equation. Set up an integral for finding the area of the given region and use the integration capabilities of a graphing utility to approximate the integral accurate to two decimal places.

103. Interior of $r = \sin\theta\cos^2\theta$

104. Interior of $r = 4\sin 3\theta$

105. Common interior of $r = 3$ and $r^2 = 18\sin 2\theta$

106. Region bounded by the polar axis and $r = e^{\theta}$ for $0 \le \theta \le \pi$

In Exercises 107 and 108, find the length of the curve over the given interval.

	Polar Equation	*Interval*
107.	$r = a(1 - \cos\theta)$	$0 \le \theta \le \pi$
108.	$r = a\cos 2\theta$	$-\dfrac{\pi}{2} \le \theta \le \dfrac{\pi}{2}$

In Exercises 109 and 110, write an integral that represents the area of the surface formed by revolving the curve about the given line. Use a graphing utility to approximate the integral.

	Polar Equation	*Interval*	*Axis of Revolution*
109.	$r = 1 + 4\cos\theta$	$0 \le \theta \le \dfrac{\pi}{2}$	Polar axis
110.	$r = 2\sin\theta$	$0 \le \theta \le \dfrac{\pi}{2}$	$\theta = \dfrac{\pi}{2}$

In Exercises 111–116, sketch and identify the graph. Use a graphing utility to confirm your results.

111. $r = \dfrac{2}{1 - \sin\theta}$

112. $r = \dfrac{2}{1 + \cos\theta}$

113. $r = \dfrac{6}{3 + 2\cos\theta}$

114. $r = \dfrac{4}{5 - 3\sin\theta}$

115. $r = \dfrac{4}{2 - 3\sin\theta}$

116. $r = \dfrac{8}{2 - 5\cos\theta}$

In Exercises 117–122, find a polar equation for the line or conic with its focus at the pole.

117. Circle
Center: $(5, \pi/2)$
Solution point: $(0, 0)$

118. Line
Solution point: $(0, 0)$
Slope: $\sqrt{3}$

119. Parabola
Vertex: $(2, \pi)$

120. Parabola
Vertex: $(2, \pi/2)$

121. Ellipse
Vertices: $(5, 0)$, $(1, \pi)$

122. Hyperbola
Vertices: $(1, 0)$, $(7, 0)$

P.S. Problem Solving

See www.CalcChat.com for worked-out solutions to odd-numbered exercises.

1. Consider the parabola $x^2 = 4y$ and the focal chord $y = \frac{3}{4}x + 1$.

(a) Sketch the graph of the parabola and the focal chord.

(b) Show that the tangent lines to the parabola at the endpoints of the focal chord intersect at right angles.

(c) Show that the tangent lines to the parabola at the endpoints of the focal chord intersect on the directrix of the parabola.

2. Consider the parabola $x^2 = 4py$ and one of its focal chords.

(a) Show that the tangent lines to the parabola at the endpoints of the focal chord intersect at right angles.

(b) Show that the tangent lines to the parabola at the endpoints of the focal chord intersect on the directrix of the parabola.

3. Prove Theorem 10.2, Reflective Property of a Parabola, as shown in the figure.

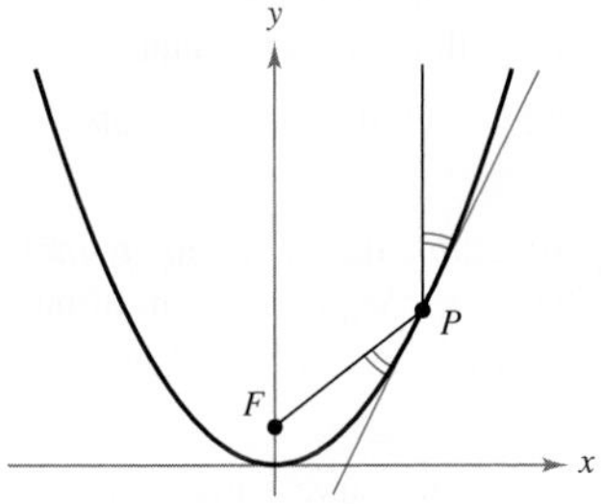

4. Consider the hyperbola

$$\frac{x^2}{a^2} - \frac{y^2}{b^2} = 1$$

with foci F_1 and F_2, as shown in the figure. Let T be the tangent line at a point M on the hyperbola. Show that incoming rays of light aimed at one focus are reflected by a hyperbolic mirror toward the other focus.

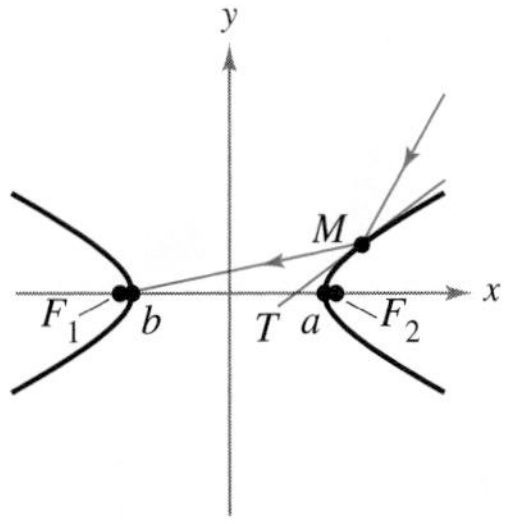

Figure for 4

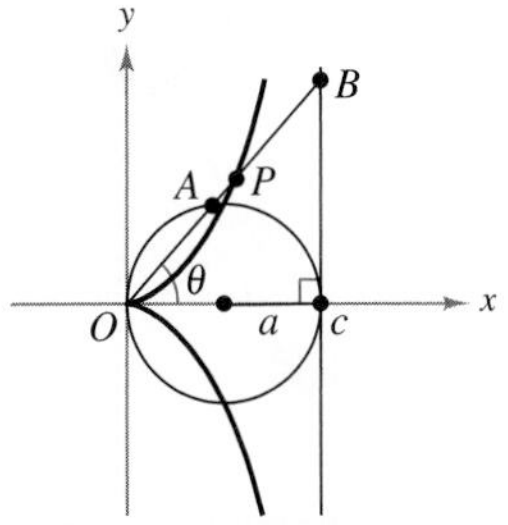

Figure for 5

5. Consider a circle of radius a tangent to the y-axis and the line $x = 2a$, as shown in the figure. Let A be the point where the segment OB intersects the circle. The **cissoid of Diocles** consists of all points P such that $OP = AB$.

(a) Find a polar equation of the cissoid.

(b) Find a set of parametric equations for the cissoid that does not contain trigonometric functions.

(c) Find a rectangular equation of the cissoid.

6. Consider the region bounded by the ellipse $x^2/a^2 + y^2/b^2 = 1$, with eccentricity $e = c/a$.

(a) Show that the area of the region is πab.

(b) Show that the solid (oblate spheroid) generated by revolving the region about the minor axis of the ellipse has a volume $V = 4\pi^2 b/3$ and a surface area of

$$S = 2\pi a^2 + \pi\left(\frac{b^2}{e}\right)\ln\left(\frac{1+e}{1-e}\right).$$

(c) Show that the solid (prolate spheroid) generated by revolving the region about the major axis of the ellipse has a volume of $V = 4\pi ab^2/3$ and a surface area of

$$S = 2\pi b^2 + 2\pi\left(\frac{ab}{e}\right)\arcsin e.$$

7. The curve given by the parametric equations

$$x(t) = \frac{1-t^2}{1+t^2} \quad \text{and} \quad y(t) = \frac{t(1-t^2)}{1+t^2}$$

is called a **strophoid.**

(a) Find a rectangular equation of the strophoid.

(b) Find a polar equation of the strophoid.

(c) Sketch a graph of the strophoid.

(d) Find the equations of the two tangent lines at the origin.

(e) Find the points on the graph where the tangent lines are horizontal.

8. Find a rectangular equation of the portion of the cycloid given by the parametric equations $x = a(\theta - \sin\theta)$ and $y = a(1 - \cos\theta)$, $0 \le \theta \le \pi$, as shown in the figure.

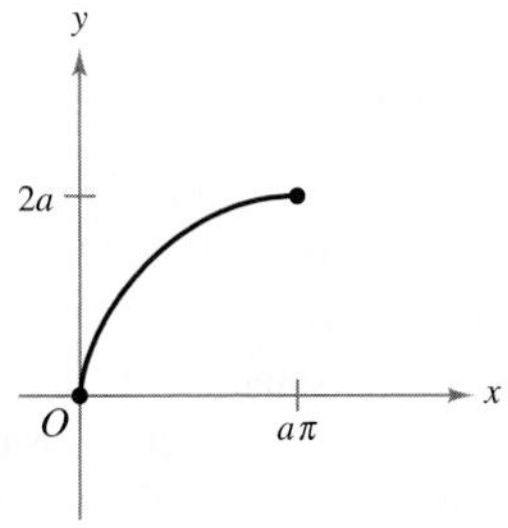

9. Consider the **cornu spiral** given by

$$x(t) = \int_0^t \cos\left(\frac{\pi u^2}{2}\right) du \quad \text{and} \quad y(t) = \int_0^t \sin\left(\frac{\pi u^2}{2}\right) du.$$

(a) Use a graphing utility to graph the spiral over the interval $-\pi \le t \le \pi$.

(b) Show that the cornu spiral is symmetric with respect to the origin.

(c) Find the length of the cornu spiral from $t = 0$ to $t = a$. What is the length of the spiral from $t = -\pi$ to $t = \pi$?

10. A particle is moving along the path described by the parametric equations $x = 1/t$ and $y = \sin t/t$, for $1 \le t < \infty$, as shown in the figure. Find the length of this path.

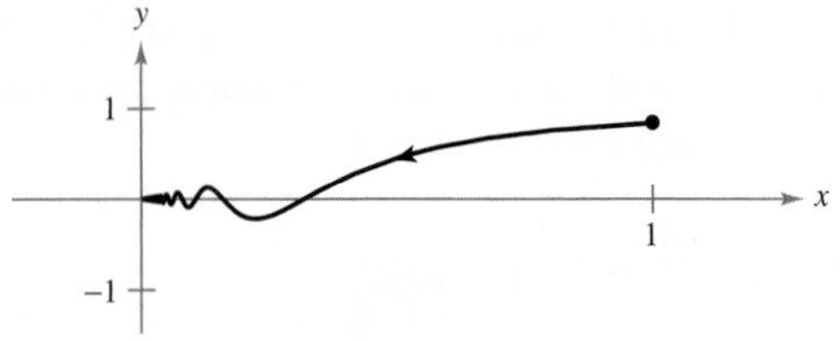

11. Let a and b be positive constants. Find the area of the region in the first quadrant bounded by the graph of the polar equation

$$r = \frac{ab}{(a \sin \theta + b \cos \theta)}, \quad 0 \le \theta \le \frac{\pi}{2}.$$

12. Consider the right triangle shown in the figure.

(a) Show that the area of the triangle is $A(\alpha) = \dfrac{1}{2}\displaystyle\int_0^{\alpha} \sec^2 \theta \, d\theta.$

(b) Show that $\tan \alpha = \displaystyle\int_0^{\alpha} \sec^2 \theta \, d\theta.$

(c) Use part (b) to derive the formula for the derivative of the tangent function.

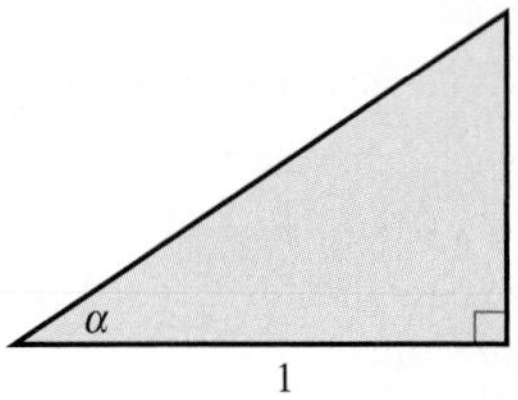

Figure for 12

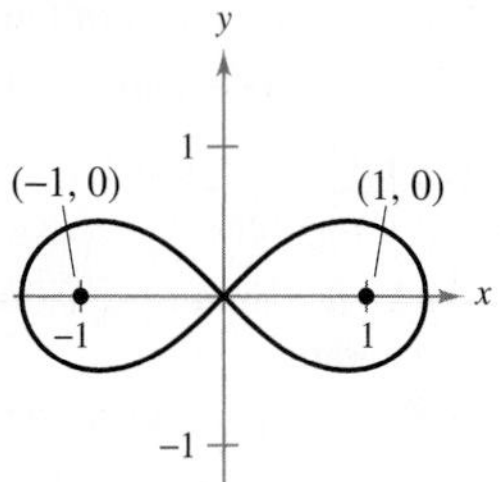

Figure for 13

13. Determine the polar equation of the set of all points (r, θ), the product of whose distances from the points $(1, 0)$ and $(-1, 0)$ is equal to 1, as shown in the figure.

14. Four dogs are located at the corners of a square with sides of length d. The dogs all move counterclockwise at the same speed directly toward the next dog, as shown in the figure. Find the polar equation of a dog's path as it spirals toward the center of the square.

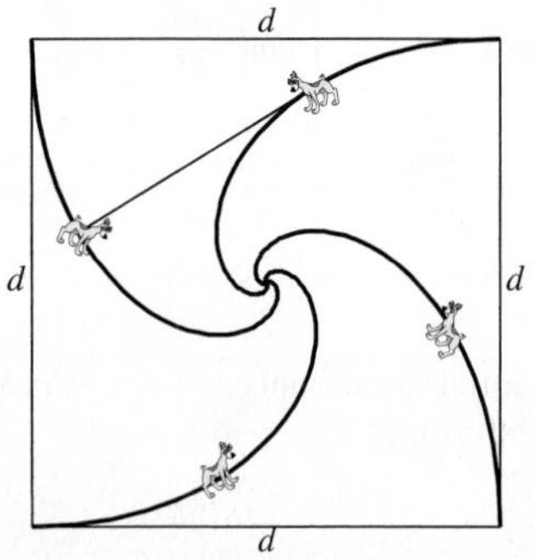

15. An air traffic controller spots two planes at the same altitude flying toward each other (see figure). Their flight paths are 20° and 315°. One plane is 150 miles from point P with a speed of 375 miles per hour. The other is 190 miles from point P with a speed of 450 miles per hour.

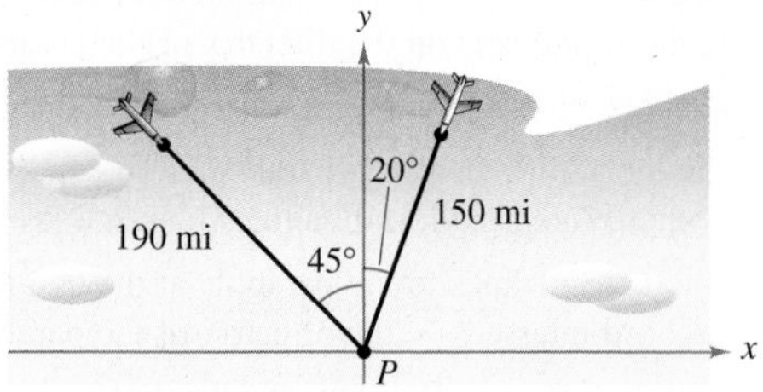

(a) Find parametric equations for the path of each plane where t is the time in hours, with $t = 0$ corresponding to the time at which the air traffic controller spots the planes.

(b) Use the result of part (a) to write the distance between the planes as a function of t.

(c) Use a graphing utility to graph the function in part (b). When will the distance between the planes be minimum? If the planes must keep a separation of at least 3 miles, is the requirement met?

16. Use a graphing utility to graph the curve shown below. The curve is given by

$$r = e^{\cos \theta} - 2 \cos 4\theta + \sin^5 \frac{\theta}{12}.$$

Over what interval must θ vary to produce the curve?

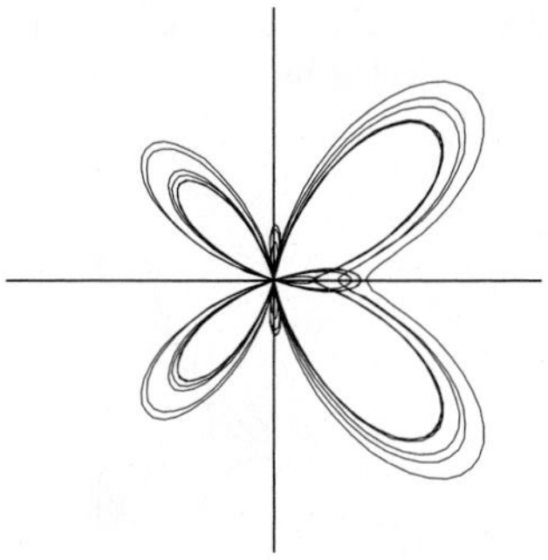

FOR FURTHER INFORMATION For more information on this curve, see the article "A Study in Step Size" by Temple H. Fay in *Mathematics Magazine*. To view this article, go to the website *www.matharticles.com*.

17. Use a graphing utility to graph the polar equation $r = \cos 5\theta + n \cos \theta$, for $0 \le \theta < \pi$ and for the integers $n = -5$ to $n = 5$. What values of n produce the "heart" portion of the curve? What values of n produce the "bell" portion? (This curve, created by Michael W. Chamberlin, appeared in *The College Mathematics Journal*.)

11 Vectors and the Geometry of Space

To make a hot air balloon rise, a pilot increases the flow of propane gas to a burner, which in turn increases the temperature of the air inside the balloon. To slow the balloon's ascent and eventually force it to descend, the pilot decreases the temperature of the air inside the balloon by opening the parachute valve at the top of the balloon. How do you think pilots control the horizontal speed and direction of the hot air balloon? Explain.

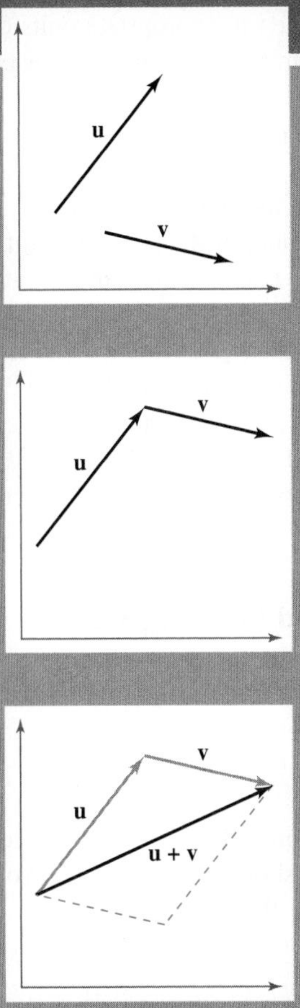

Vectors indicate quantities that involve both magnitude and direction. In Chapter 11, you will study operations of vectors in the plane and in space. You will also learn how to represent vector operations geometrically. For example, the graphs shown above represent vector addition in the plane.

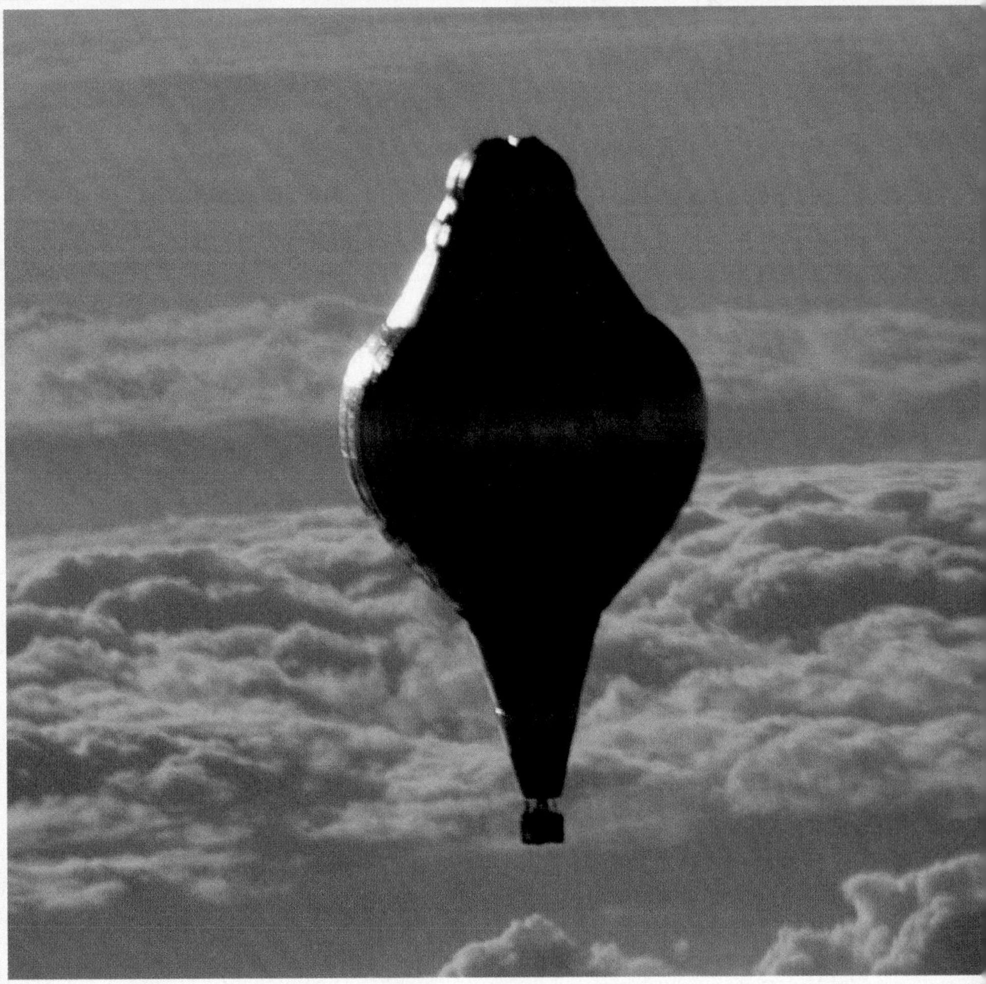

Reuters/Corbis

Section 11.1 Vectors in the Plane

- **Write the component form of a vector.**
- **Perform vector operations and interpret the results geometrically.**
- **Write a vector as a linear combination of standard unit vectors.**
- **Use vectors to solve problems involving force or velocity.**

Component Form of a Vector

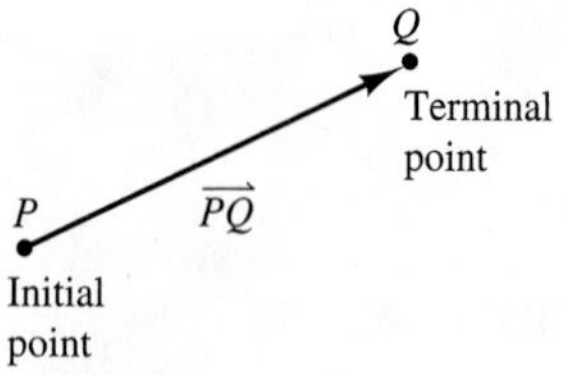

A directed line segment
Figure 11.1

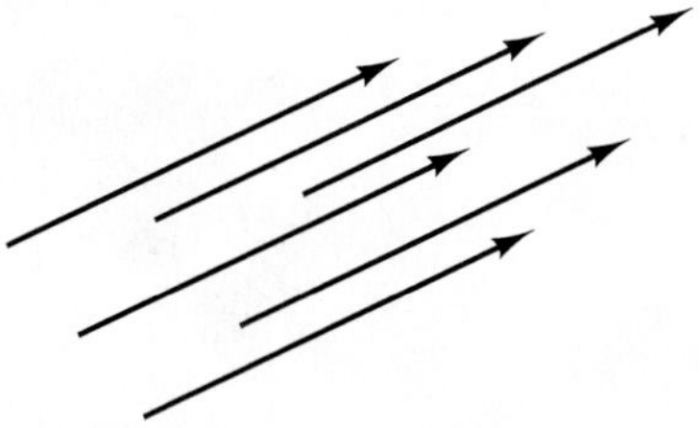
Equivalent directed line segments
Figure 11.2

Many quantities in geometry and physics, such as area, volume, temperature, mass, and time, can be characterized by a single real number scaled to appropriate units of measure. These are called **scalar quantities,** and the real number associated with each is called a **scalar.**

Other quantities, such as force, velocity, and acceleration, involve both magnitude and direction and cannot be characterized completely by a single real number. A **directed line segment** is used to represent such a quantity, as shown in Figure 11.1. The directed line segment $\overrightarrow{PQ}$ has **initial point** P and **terminal point** Q, and its **length** (or **magnitude**) is denoted by $\|\overrightarrow{PQ}\|$. Directed line segments that have the same length and direction are **equivalent,** as shown in Figure 11.2. The set of all directed line segments that are equivalent to a given directed line segment $\overrightarrow{PQ}$ is a **vector in the plane** and is denoted by $\mathbf{v} = \overrightarrow{PQ}$. In typeset material, vectors are usually denoted by lowercase, boldface letters such as **u**, **v**, and **w**. When written by hand, however, vectors are often denoted by letters with arrows above them, such as $\vec{u}$, $\vec{v}$, and $\vec{w}$.

Be sure you see that a vector in the plane can be represented by many different directed line segments—all pointing in the same direction and all of the same length.

EXAMPLE 1 Vector Representation by Directed Line Segments

Let **v** be represented by the directed line segment from $(0, 0)$ to $(3, 2)$, and let **u** be represented by the directed line segment from $(1, 2)$ to $(4, 4)$. Show that **v** and **u** are equivalent.

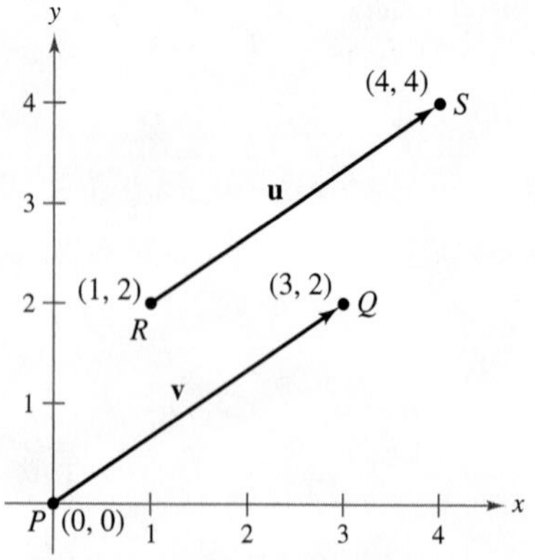

The vectors **u** and **v** are equivalent.
Figure 11.3

Solution Let $P(0, 0)$ and $Q(3, 2)$ be the initial and terminal points of **v**, and let $R(1, 2)$ and $S(4, 4)$ be the initial and terminal points of **u**, as shown in Figure 11.3. You can use the Distance Formula to show that $\overrightarrow{PQ}$ and $\overrightarrow{RS}$ have the *same length.*

$$\|\overrightarrow{PQ}\| = \sqrt{(3-0)^2 + (2-0)^2} = \sqrt{13} \qquad \text{Length of } \overrightarrow{PQ}$$
$$\|\overrightarrow{RS}\| = \sqrt{(4-1)^2 + (4-2)^2} = \sqrt{13} \qquad \text{Length of } \overrightarrow{RS}$$

Both line segments have the *same direction*, because they both are directed toward the upper right on lines having the same slope.

$$\text{Slope of } \overrightarrow{PQ} = \frac{2-0}{3-0} = \frac{2}{3}$$

and

$$\text{Slope of } \overrightarrow{RS} = \frac{4-2}{4-1} = \frac{2}{3}$$

Because $\overrightarrow{PQ}$ and $\overrightarrow{RS}$ have the same length and direction, you can conclude that the two vectors are equivalent. That is, **v** and **u** are equivalent.

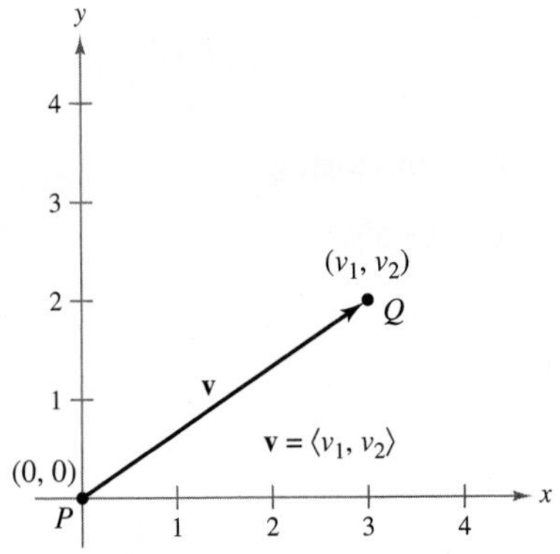

The standard position of a vector
Figure 11.4

The directed line segment whose initial point is the origin is often the most convenient representative of a set of equivalent directed line segments such as those shown in Figure 11.3. This representation of $\mathbf{v}$ is said to be in **standard position.** A directed line segment whose initial point is the origin can be uniquely represented by the coordinates of its terminal point $Q(v_1, v_2)$, as shown in Figure 11.4.

Definition of Component Form of a Vector in the Plane

If $\mathbf{v}$ is a vector in the plane whose initial point is the origin and whose terminal point is (v_1, v_2), then the **component form of v** is given by

$$\mathbf{v} = \langle v_1, v_2 \rangle.$$

The coordinates v_1 and v_2 are called the **components of v.** If both the initial point and the terminal point lie at the origin, then $\mathbf{v}$ is called the **zero vector** and is denoted by $\mathbf{0} = \langle 0, 0 \rangle$.

This definition implies that two vectors $\mathbf{u} = \langle u_1, u_2 \rangle$ and $\mathbf{v} = \langle v_1, v_2 \rangle$ are **equal** if and only if $u_1 = v_1$ and $u_2 = v_2$.

The following procedures can be used to convert directed line segments to component form or vice versa.

NOTE It is important to understand that a vector represents a *set* of directed line segments (each having the same length and direction). In practice, however, it is common not to distinguish between a vector and one of its representatives.

1. If $P(p_1, p_2)$ and $Q(q_1, q_2)$ are the initial and terminal points of a directed line segment, the component form of the vector $\mathbf{v}$ represented by $\overrightarrow{PQ}$ is $\langle v_1, v_2 \rangle = \langle q_1 - p_1, q_2 - p_2 \rangle$. Moreover, the **length** (or **magnitude**) **of v** is

$$\begin{aligned}\|\mathbf{v}\| &= \sqrt{(q_1 - p_1)^2 + (q_2 - p_2)^2} \\ &= \sqrt{v_1^2 + v_2^2}.\end{aligned}$$
Length of a vector

2. If $\mathbf{v} = \langle v_1, v_2 \rangle$, $\mathbf{v}$ can be represented by the directed line segment, in standard position, from $P(0, 0)$ to $Q(v_1, v_2)$.

The length of $\mathbf{v}$ is also called the **norm of v.** If $\|\mathbf{v}\| = 1$, $\mathbf{v}$ is a **unit vector.** Moreover, $\|\mathbf{v}\| = 0$ if and only if $\mathbf{v}$ is the zero vector $\mathbf{0}$.

EXAMPLE 2 Finding the Component Form and Length of a Vector

Find the component form and length of the vector $\mathbf{v}$ that has initial point $(3, -7)$ and terminal point $(-2, 5)$.

Solution Let $P(3, -7) = (p_1, p_2)$ and $Q(-2, 5) = (q_1, q_2)$. Then the components of $\mathbf{v} = \langle v_1, v_2 \rangle$ are

$$v_1 = q_1 - p_1 = -2 - 3 = -5$$
$$v_2 = q_2 - p_2 = 5 - (-7) = 12.$$

So, as shown in Figure 11.5, $\mathbf{v} = \langle -5, 12 \rangle$, and the length of $\mathbf{v}$ is

$$\begin{aligned}\|\mathbf{v}\| &= \sqrt{(-5)^2 + 12^2} \\ &= \sqrt{169} \\ &= 13.\end{aligned}$$

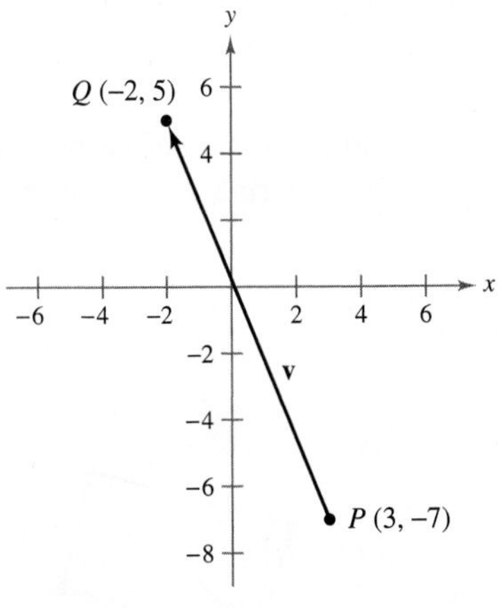

Component form of $\mathbf{v}$: $\mathbf{v} = \langle -5, 12 \rangle$
Figure 11.5

Vector Operations

Definitions of Vector Addition and Scalar Multiplication

Let $\mathbf{u} = \langle u_1, u_2 \rangle$ and $\mathbf{v} = \langle v_1, v_2 \rangle$ be vectors and let c be a scalar.

1. The **vector sum** of $\mathbf{u}$ and $\mathbf{v}$ is the vector $\mathbf{u} + \mathbf{v} = \langle u_1 + v_1, u_2 + v_2 \rangle$.
2. The **scalar multiple** of c and $\mathbf{u}$ is the vector $c\mathbf{u} = \langle cu_1, cu_2 \rangle$.
3. The **negative** of $\mathbf{v}$ is the vector

$$-\mathbf{v} = (-1)\mathbf{v} = \langle -v_1, -v_2 \rangle.$$

4. The **difference** of $\mathbf{u}$ and $\mathbf{v}$ is

$$\mathbf{u} - \mathbf{v} = \mathbf{u} + (-\mathbf{v}) = \langle u_1 - v_1, u_2 - v_2 \rangle.$$

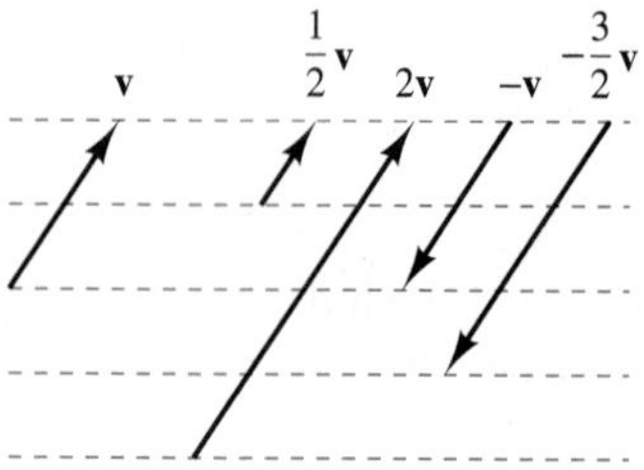

The scalar multiplication of **v**
Figure 11.6

Geometrically, the scalar multiple of a vector $\mathbf{v}$ and a scalar c is the vector that is $|c|$ times as long as $\mathbf{v}$, as shown in Figure 11.6. If c is positive, $c\mathbf{v}$ has the same direction as $\mathbf{v}$. If c is negative, $c\mathbf{v}$ has the opposite direction.

The sum of two vectors can be represented geometrically by positioning the vectors (without changing their magnitudes or directions) so that the initial point of one coincides with the terminal point of the other, as shown in Figure 11.7. The vector $\mathbf{u} + \mathbf{v}$, called the **resultant vector,** is the diagonal of a parallelogram having $\mathbf{u}$ and $\mathbf{v}$ as its adjacent sides.

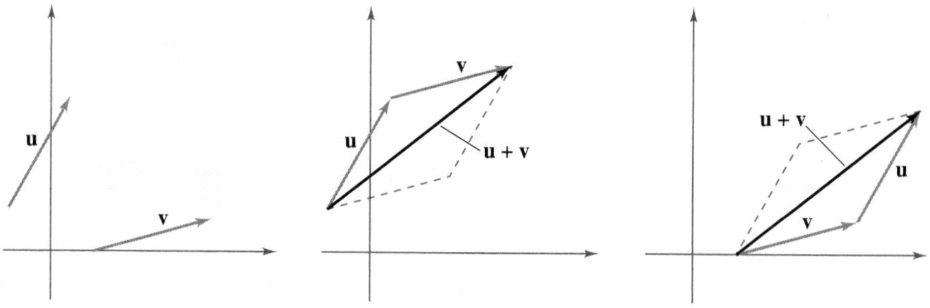

To find **u** + **v**, (1) move the initial point of **v** to the terminal point of **u**, or (2) move the initial point of **u** to the terminal point of **v**.

Figure 11.7

Isaac William Rowan Hamilton (1805–1865)

Some of the earliest work with vectors was done by the Irish mathematician William Rowan Hamilton. Hamilton spent many years developing a system of vector-like quantities called *quaternions*. Although Hamilton was convinced of the benefits of quaternions, the operations he defined did not produce good models for physical phenomena. It wasn't until the latter half of the nineteenth century that the Scottish physicist James Maxwell (1831–1879) restructured Hamilton's quaternions in a form useful for representing physical quantities such as force, velocity, and acceleration.

Figure 11.8 shows the equivalence of the geometric and algebraic definitions of vector addition and scalar multiplication, and presents (at far right) a geometric interpretation of $\mathbf{u} - \mathbf{v}$.

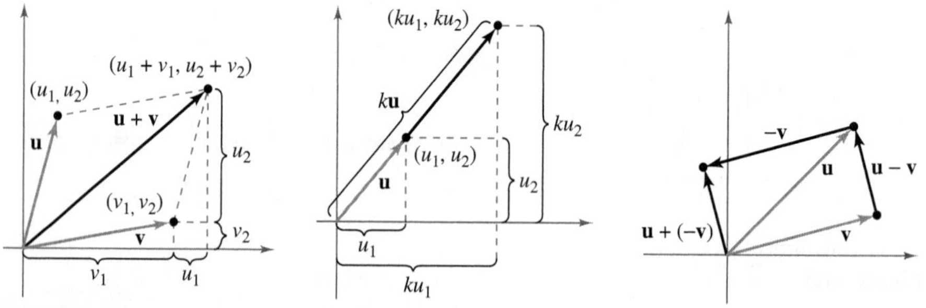

Vector addition　　Scalar multiplication　　Vector subtraction
Figure 11.8

EXAMPLE 3 Vector Operations

Given $\mathbf{v} = \langle -2, 5 \rangle$ and $\mathbf{w} = \langle 3, 4 \rangle$, find each of the vectors.

a. $\frac{1}{2}\mathbf{v}$ **b.** $\mathbf{w} - \mathbf{v}$ **c.** $\mathbf{v} + 2\mathbf{w}$

Solution

a. $\frac{1}{2}\mathbf{v} = \left\langle \frac{1}{2}(-2), \frac{1}{2}(5) \right\rangle = \left\langle -1, \frac{5}{2} \right\rangle$

b. $\mathbf{w} - \mathbf{v} = \langle w_1 - v_1, w_2 - v_2 \rangle = \langle 3 - (-2), 4 - 5 \rangle = \langle 5, -1 \rangle$

c. Using $2\mathbf{w} = \langle 6, 8 \rangle$, you have

$$\begin{aligned} \mathbf{v} + 2\mathbf{w} &= \langle -2, 5 \rangle + \langle 6, 8 \rangle \\ &= \langle -2 + 6, 5 + 8 \rangle \\ &= \langle 4, 13 \rangle. \end{aligned}$$

Vector addition and scalar multiplication share many properties of ordinary arithmetic, as shown in the following theorem.

THEOREM 11.1 Properties of Vector Operations

Let $\mathbf{u}$, $\mathbf{v}$, and $\mathbf{w}$ be vectors in the plane, and let c and d be scalars.

1. $\mathbf{u} + \mathbf{v} = \mathbf{v} + \mathbf{u}$ — Commutative Property
2. $(\mathbf{u} + \mathbf{v}) + \mathbf{w} = \mathbf{u} + (\mathbf{v} + \mathbf{w})$ — Associative Property
3. $\mathbf{u} + \mathbf{0} = \mathbf{u}$ — Additive Identity Property
4. $\mathbf{u} + (-\mathbf{u}) = \mathbf{0}$ — Additive Inverse Property
5. $c(d\mathbf{u}) = (cd)\mathbf{u}$
6. $(c + d)\mathbf{u} = c\mathbf{u} + d\mathbf{u}$ — Distributive Property
7. $c(\mathbf{u} + \mathbf{v}) = c\mathbf{u} + c\mathbf{v}$ — Distributive Property
8. $1(\mathbf{u}) = \mathbf{u}, 0(\mathbf{u}) = \mathbf{0}$

Proof The proof of the *Associative Property* of vector addition uses the Associative Property of addition of real numbers.

$$\begin{aligned} (\mathbf{u} + \mathbf{v}) + \mathbf{w} &= [\langle u_1, u_2 \rangle + \langle v_1, v_2 \rangle] + \langle w_1, w_2 \rangle \\ &= \langle u_1 + v_1, u_2 + v_2 \rangle + \langle w_1, w_2 \rangle \\ &= \langle (u_1 + v_1) + w_1, (u_2 + v_2) + w_2 \rangle \\ &= \langle u_1 + (v_1 + w_1), u_2 + (v_2 + w_2) \rangle \\ &= \langle u_1, u_2 \rangle + \langle v_1 + w_1, v_2 + w_2 \rangle = \mathbf{u} + (\mathbf{v} + \mathbf{w}) \end{aligned}$$

Similarly, the proof of the *Distributive Property* of vectors depends on the Distributive Property of real numbers.

$$\begin{aligned} (c + d)\mathbf{u} &= (c + d)\langle u_1, u_2 \rangle \\ &= \langle (c + d)u_1, (c + d)u_2 \rangle \\ &= \langle cu_1 + du_1, cu_2 + du_2 \rangle \\ &= \langle cu_1, cu_2 \rangle + \langle du_1, du_2 \rangle = c\mathbf{u} + d\mathbf{u} \end{aligned}$$

The other properties can be proved in a similar manner.

The Granger Collection

Emmy Noether (1882–1935)

One person who contributed to our knowledge of axiomatic systems was the German mathematician Emmy Noether. Noether is generally recognized as the leading woman mathematician in recent history.

FOR FURTHER INFORMATION For more information on Emmy Noether, see the article "Emmy Noether, Greatest Woman Mathematician" by Clark Kimberling in *The Mathematics Teacher*. To view this article, go to the website *www.matharticles.com*.

Any set of vectors (with an accompanying set of scalars) that satisfies the eight properties given in Theorem 11.1 is a **vector space.*** The eight properties are the *vector space axioms*. So, this theorem states that the set of vectors in the plane (with the set of real numbers) forms a vector space.

THEOREM 11.2 Length of a Scalar Multiple

Let $\mathbf{v}$ be a vector and let c be a scalar. Then

$$\|c\mathbf{v}\| = |c|\,\|\mathbf{v}\|. \qquad |c| \text{ is the absolute value of } c.$$

Proof Because $c\mathbf{v} = \langle cv_1, cv_2\rangle$, it follows that

$$\begin{aligned}\|c\mathbf{v}\| = \|\langle cv_1, cv_2\rangle\| &= \sqrt{(cv_1)^2 + (cv_2)^2}\\ &= \sqrt{c^2v_1^2 + c^2v_2^2}\\ &= \sqrt{c^2(v_1^2 + v_2^2)}\\ &= |c|\sqrt{v_1^2 + v_2^2}\\ &= |c|\,\|\mathbf{v}\|.\end{aligned}$$

In many applications of vectors, it is useful to find a unit vector that has the same direction as a given vector. The following theorem gives a procedure for doing this.

THEOREM 11.3 Unit Vector in the Direction of v

If $\mathbf{v}$ is a nonzero vector in the plane, then the vector

$$\mathbf{u} = \frac{\mathbf{v}}{\|\mathbf{v}\|} = \frac{1}{\|\mathbf{v}\|}\mathbf{v}$$

has length 1 and the same direction as $\mathbf{v}$.

Proof Because $1/\|\mathbf{v}\|$ is positive and $\mathbf{u} = (1/\|\mathbf{v}\|)\mathbf{v}$, you can conclude that $\mathbf{u}$ has the same direction as $\mathbf{v}$. To see that $\|\mathbf{u}\| = 1$, note that

$$\begin{aligned}\|\mathbf{u}\| &= \left\|\left(\frac{1}{\|\mathbf{v}\|}\right)\mathbf{v}\right\|\\ &= \left|\frac{1}{\|\mathbf{v}\|}\right|\,\|\mathbf{v}\|\\ &= \frac{1}{\|\mathbf{v}\|}\,\|\mathbf{v}\|\\ &= 1.\end{aligned}$$

So, $\mathbf{u}$ has length 1 and the same direction as $\mathbf{v}$.

In Theorem 11.3, $\mathbf{u}$ is called a **unit vector in the direction of v.** The process of multiplying $\mathbf{v}$ by $1/\|\mathbf{v}\|$ to get a unit vector is called **normalization of v.**

---

* *For more information about vector spaces, see* Elementary Linear Algebra, *Fifth Edition, by Larson, Edwards, and Falvo (Boston: Houghton Mifflin Company, 2004).*

EXAMPLE 4 Finding a Unit Vector

Find a unit vector in the direction of $\mathbf{v} = \langle -2, 5\rangle$ and verify that it has length 1.

Solution From Theorem 11.3, the unit vector in the direction of $\mathbf{v}$ is

$$\frac{\mathbf{v}}{\|\mathbf{v}\|} = \frac{\langle -2, 5\rangle}{\sqrt{(-2)^2 + (5)^2}} = \frac{1}{\sqrt{29}}\langle -2, 5\rangle = \left\langle \frac{-2}{\sqrt{29}}, \frac{5}{\sqrt{29}} \right\rangle.$$

This vector has length 1, because

$$\sqrt{\left(\frac{-2}{\sqrt{29}}\right)^2 + \left(\frac{5}{\sqrt{29}}\right)^2} = \sqrt{\frac{4}{29} + \frac{25}{29}} = \sqrt{\frac{29}{29}} = 1.$$

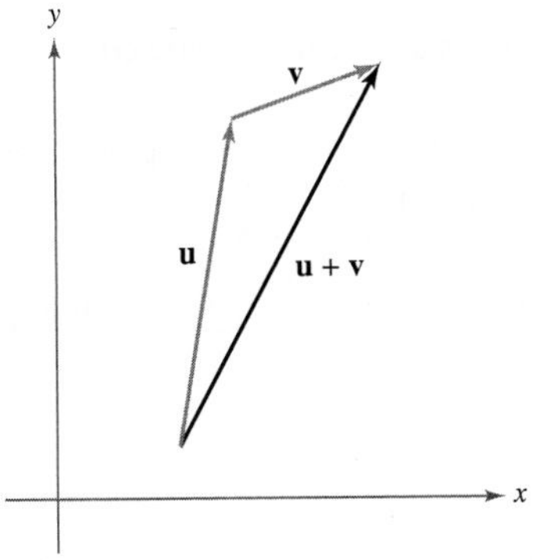

Triangle inequality
Figure 11.9

Generally, the length of the sum of two vectors is not equal to the sum of their lengths. To see this, consider the vectors $\mathbf{u}$ and $\mathbf{v}$ as shown in Figure 11.9. By considering $\mathbf{u}$ and $\mathbf{v}$ as two sides of a triangle, you can see that the length of the third side is $\|\mathbf{u} + \mathbf{v}\|$, and you have

$$\|\mathbf{u} + \mathbf{v}\| \le \|\mathbf{u}\| + \|\mathbf{v}\|.$$

Equality occurs only if the vectors $\mathbf{u}$ and $\mathbf{v}$ have the *same direction*. This result is called the **triangle inequality** for vectors. (You are asked to prove this in Exercise 89, Section 11.3.)

Standard Unit Vectors

The unit vectors $\langle 1, 0\rangle$ and $\langle 0, 1\rangle$ are called the **standard unit vectors** in the plane and are denoted by

$$\mathbf{i} = \langle 1, 0\rangle \quad \text{and} \quad \mathbf{j} = \langle 0, 1\rangle \qquad \text{Standard unit vectors}$$

as shown in Figure 11.10. These vectors can be used to represent any vector uniquely, as follows.

$$\mathbf{v} = \langle v_1, v_2\rangle = \langle v_1, 0\rangle + \langle 0, v_2\rangle = v_1\langle 1, 0\rangle + v_2\langle 0, 1\rangle = v_1\mathbf{i} + v_2\mathbf{j}$$

The vector $\mathbf{v} = v_1\mathbf{i} + v_2\mathbf{j}$ is called a **linear combination** of $\mathbf{i}$ and $\mathbf{j}$. The scalars v_1 and v_2 are called the **horizontal** and **vertical components of v.**

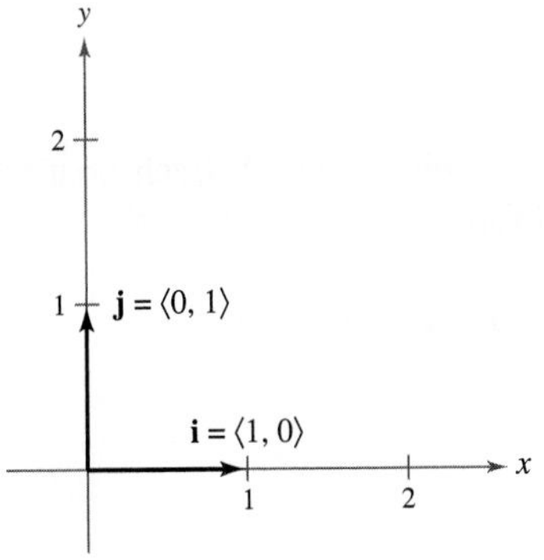

Standard unit vectors **i** and **j**
Figure 11.10

EXAMPLE 5 Writing a Linear Combination of Unit Vectors

Let $\mathbf{u}$ be the vector with initial point $(2, -5)$ and terminal point $(-1, 3)$, and let $\mathbf{v} = 2\mathbf{i} - \mathbf{j}$. Write each vector as a linear combination of $\mathbf{i}$ and $\mathbf{j}$.

a. $\mathbf{u}$ **b.** $\mathbf{w} = 2\mathbf{u} - 3\mathbf{v}$

Solution

a.
$$\begin{aligned}\mathbf{u} &= \langle q_1 - p_1, q_2 - p_2\rangle \\ &= \langle -1 - 2, 3 - (-5)\rangle \\ &= \langle -3, 8\rangle = -3\mathbf{i} + 8\mathbf{j}\end{aligned}$$

b.
$$\begin{aligned}\mathbf{w} = 2\mathbf{u} - 3\mathbf{v} &= 2(-3\mathbf{i} + 8\mathbf{j}) - 3(2\mathbf{i} - \mathbf{j}) \\ &= -6\mathbf{i} + 16\mathbf{j} - 6\mathbf{i} + 3\mathbf{j} \\ &= -12\mathbf{i} + 19\mathbf{j}\end{aligned}$$

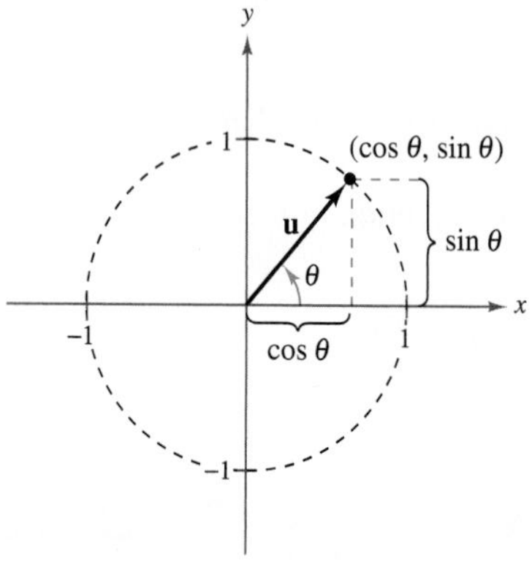

The angle θ from the positive x-axis to the vector **u**
Figure 11.11

If **u** is a unit vector and θ is the angle (measured counterclockwise) from the positive x-axis to **u**, then the terminal point of **u** lies on the unit circle, and you have

$$\mathbf{u} = \langle \cos\theta, \sin\theta \rangle = \cos\theta\mathbf{i} + \sin\theta\mathbf{j} \qquad \text{Unit vector}$$

as shown in Figure 11.11. Moreover, it follows that any other nonzero vector **v** making an angle θ with the positive x-axis has the same direction as **u**, and you can write

$$\mathbf{v} = \|\mathbf{v}\|\langle \cos\theta, \sin\theta \rangle = \|\mathbf{v}\|\cos\theta\mathbf{i} + \|\mathbf{v}\|\sin\theta\mathbf{j}.$$

EXAMPLE 6 Writing a Vector of Given Magnitude and Direction

The vector **v** has a magnitude of 3 and makes an angle of $30° = \pi/6$ with the positive x-axis. Write **v** as a linear combination of the unit vectors **i** and **j**.

Solution Because the angle between **v** and the positive x-axis is $\theta = \pi/6$, you can write the following.

$$\begin{aligned}\mathbf{v} &= \|\mathbf{v}\|\cos\theta\mathbf{i} + \|\mathbf{v}\|\sin\theta\mathbf{j}\\ &= 3\cos\frac{\pi}{6}\mathbf{i} + 3\sin\frac{\pi}{6}\mathbf{j}\\ &= \frac{3\sqrt{3}}{2}\mathbf{i} + \frac{3}{2}\mathbf{j}\end{aligned}$$

Applications of Vectors

Vectors have many applications in physics and engineering. One example is force. A vector can be used to represent force because force has both magnitude and direction. If two or more forces are acting on an object, then the **resultant force** on the object is the vector sum of the vector forces.

EXAMPLE 7 Finding the Resultant Force

Two tugboats are pushing an ocean liner, as shown in Figure 11.12. Each boat is exerting a force of 400 pounds. What is the resultant force on the ocean liner?

Solution Using Figure 11.12, you can represent the forces exerted by the first and second tugboats as

$$\begin{aligned}\mathbf{F}_1 &= 400\langle \cos 20°, \sin 20° \rangle\\ &= 400\cos(20°)\mathbf{i} + 400\sin(20°)\mathbf{j}\\ \mathbf{F}_2 &= 400\langle \cos(-20°), \sin(-20°) \rangle\\ &= 400\cos(20°)\mathbf{i} - 400\sin(20°)\mathbf{j}.\end{aligned}$$

The resultant force on the ocean liner is

$$\begin{aligned}\mathbf{F} &= \mathbf{F}_1 + \mathbf{F}_2\\ &= [400\cos(20°)\mathbf{i} + 400\sin(20°)\mathbf{j}] + [400\cos(20°)\mathbf{i} - 400\sin(20°)\mathbf{j}]\\ &= 800\cos(20°)\mathbf{i}\\ &\approx 752\mathbf{i}.\end{aligned}$$

So, the resultant force on the ocean liner is approximately 752 pounds in the direction of the positive x-axis.

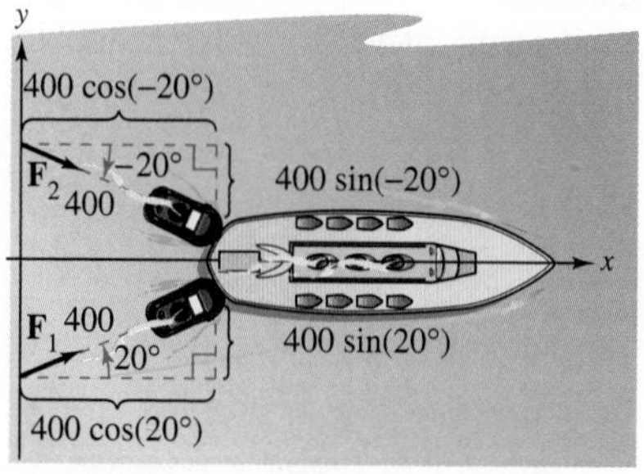

The resultant force on the ocean liner that is exerted by the two tugboats.
Figure 11.12

In surveying and navigation, a bearing is a direction that measures the acute angle that a path or line of sight makes with a fixed north-south line. In air navigation, bearings are measured in degrees clockwise from north.

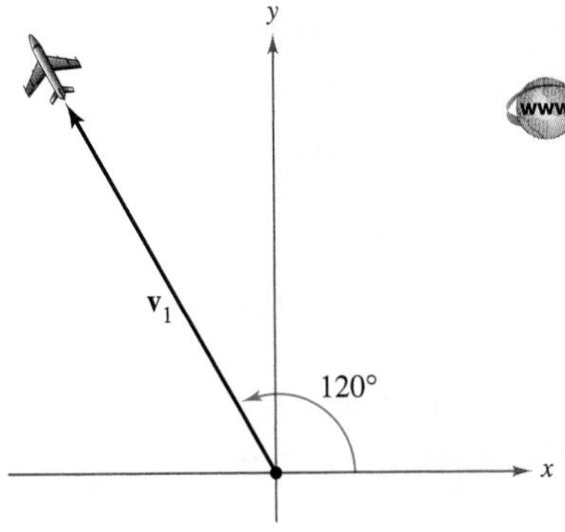

(a) Direction without wind

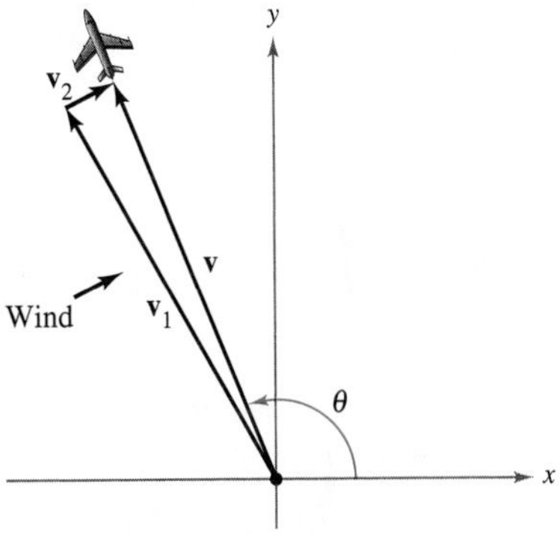

(b) Direction with wind
Figure 11.13

EXAMPLE 8 Finding a Velocity

An airplane is traveling at a fixed altitude with a negligible wind factor. The airplane is traveling at a speed of 500 miles per hour with a bearing of 330°, as shown in Figure 11.13(a). As the airplane reaches a certain point, it encounters wind with a velocity of 70 miles per hour in the direction N 45° E (45° east of north), as shown in Figure 11.13(b). What are the resultant speed and direction of the airplane?

Solution Using Figure 11.13(a), represent the velocity of the airplane (alone) as

$$\mathbf{v}_1 = 500\cos(120°)\mathbf{i} + 500\sin(120°)\mathbf{j}.$$

The velocity of the wind is represented by the vector

$$\mathbf{v}_2 = 70\cos(45°)\mathbf{i} + 70\sin(45°)\mathbf{j}.$$

The resultant velocity of the airplane (in the wind) is

$$\begin{aligned}\mathbf{v} = \mathbf{v}_1 + \mathbf{v}_2 &= 500\cos(120°)\mathbf{i} + 500\sin(120°)\mathbf{j} + 70\cos(45°)\mathbf{i} + 70\sin(45°)\mathbf{j}\\ &\approx -200.5\mathbf{i} + 482.5\mathbf{j}.\end{aligned}$$

To find the resultant speed and direction, write $\mathbf{v} = \|\mathbf{v}\|(\cos\theta\,\mathbf{i} + \sin\theta\,\mathbf{j})$. Because $\|\mathbf{v}\| \approx \sqrt{(-200.5)^2 + (482.5)^2} \approx 522.5$, you can write

$$\mathbf{v} \approx 522.5\left(\frac{-200.5}{522.5}\mathbf{i} + \frac{482.5}{522.5}\mathbf{j}\right) \approx 522.5,[\cos(112.6°)\mathbf{i} + \sin(112.6°)\mathbf{j}].$$

The new speed of the airplane, as altered by the wind, is approximately 522.5 miles per hour in a path that makes an angle of 112.6° with the positive x-axis.

Exercises for Section 11.1

See www.CalcChat.com for worked-out solutions to odd-numbered exercises.

In Exercises 1–4, (a) find the component form of the vector v and (b) sketch the vector with its initial point at the origin.

1.

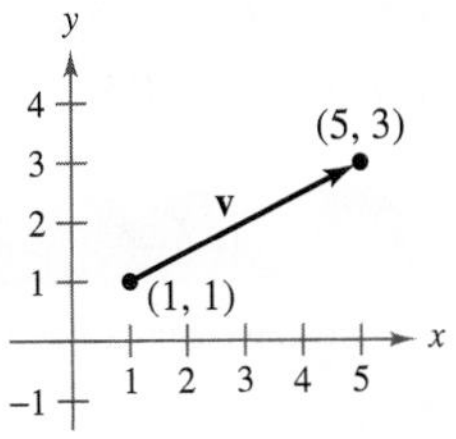

2.

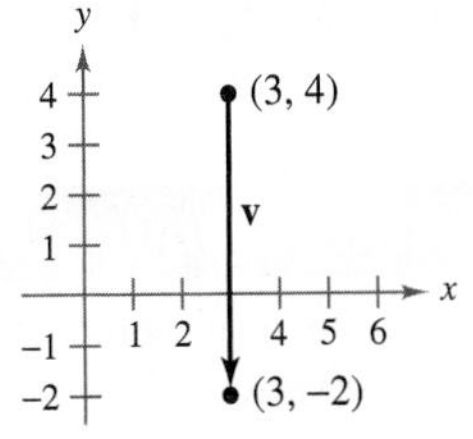

3.

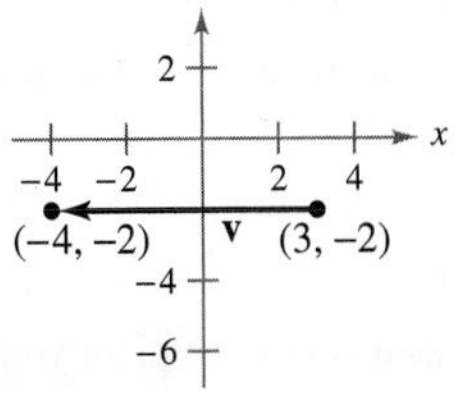

4.

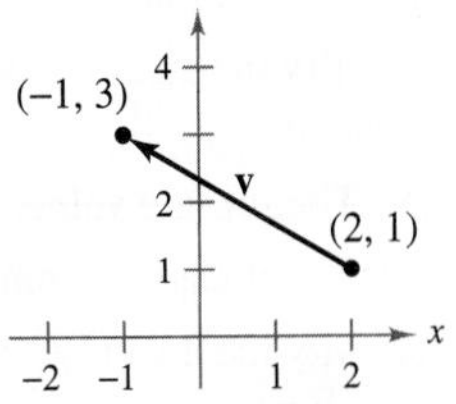

In Exercises 5–8, find the vectors u and v whose initial and terminal points are given. Show that u and v are equivalent.

5. **u**: (3, 2), (5, 6)
   v: (−1, 4), (1, 8)
6. **u**: (−4, 0), (1, 8)
   v: (2, −1), (7, 7)
7. **u**: (0, 3), (6, −2)
   v: (3, 10), (9, 5)
8. **u**: (−4, −1), (11, −4)
   v: (10, 13), (25, 10)

In Exercises 9–16, the initial and terminal points of a vector v are given. (a) Sketch the given directed line segment, (b) write the vector in component form, and (c) sketch the vector with its initial point at the origin.

	Initial Point	*Terminal Point*		*Initial Point*	*Terminal Point*
9.	(1, 2)	(5, 5)	**10.**	(2, −6)	(3, 6)
11.	(10, 2)	(6, −1)	**12.**	(0, −4)	(−5, −1)

indicates that in the HM mathSpace® *CD-ROM and the online* Eduspace® *system for this text, you will find an Open Exploration, which further explores this example using the computer algebra systems* Maple, Mathcad, Mathematica, *and* Derive.

	Initial Point	*Terminal Point*		*Initial Point*	*Terminal Point*
13.	$(6, 2)$	$(6, 6)$	**14.**	$(7, -1)$	$(-3, -1)$
15.	$\left(\frac{3}{2}, \frac{4}{3}\right)$	$\left(\frac{1}{2}, 3\right)$	**16.**	$(0.12, 0.60)$	$(0.84, 1.25)$

In Exercises 17 and 18, sketch each scalar multiple of v.

17. $\mathbf{v} = \langle 2, 3\rangle$

(a) $2\mathbf{v}$ (b) $-3\mathbf{v}$ (c) $\frac{7}{2}\mathbf{v}$ (d) $\frac{2}{3}\mathbf{v}$

18. $\mathbf{v} = \langle -1, 5\rangle$

(a) $4\mathbf{v}$ (b) $-\frac{1}{2}\mathbf{v}$ (c) $0\mathbf{v}$ (d) $-6\mathbf{v}$

In Exercises 19–22, use the figure to sketch a graph of the vector. To print an enlarged copy of the graph, go to the website *www.mathgraphs.com*.

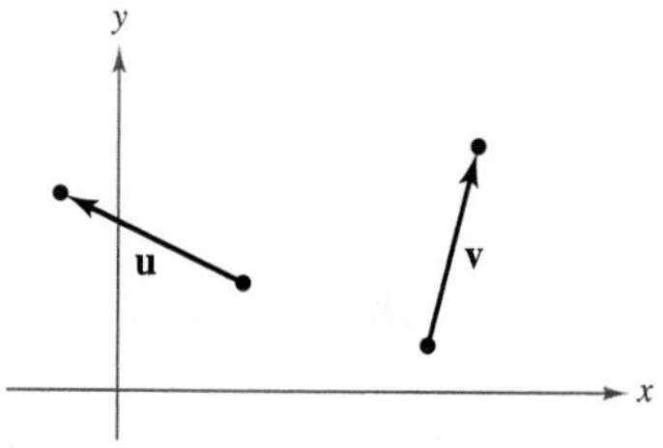

19. $-\mathbf{u}$

20. $2\mathbf{u}$

21. $\mathbf{u} - \mathbf{v}$

22. $\mathbf{u} + 2\mathbf{v}$

In Exercises 23 and 24, find (a) $\frac{2}{3}\mathbf{u}$, (b) $\mathbf{v} - \mathbf{u}$, and (c) $2\mathbf{u} + 5\mathbf{v}$.

23. $\mathbf{u} = \langle 4, 9\rangle$, $\mathbf{v} = \langle 2, -5\rangle$

24. $\mathbf{u} = \langle -3, -8\rangle$, $\mathbf{v} = \langle 8, 25\rangle$

In Exercises 25–28, find the vector v where $\mathbf{u} = \langle 2, -1\rangle$ and $\mathbf{w} = \langle 1, 2\rangle$. Illustrate the vector operations geometrically.

25. $\mathbf{v} = \frac{3}{2}\mathbf{u}$

26. $\mathbf{v} = \mathbf{u} + \mathbf{w}$

27. $\mathbf{v} = \mathbf{u} + 2\mathbf{w}$

28. $\mathbf{v} = 5\mathbf{u} - 3\mathbf{w}$

In Exercises 29 and 30, the vector v and its initial point are given. Find the terminal point.

29. $\mathbf{v} = \langle -1, 3\rangle$; Initial point: $(4, 2)$

30. $\mathbf{v} = \langle 4, -9\rangle$; Initial point: $(3, 2)$

In Exercises 31–36, find the magnitude of v.

31. $\mathbf{v} = \langle 4, 3\rangle$

32. $\mathbf{v} = \langle 12, -5\rangle$

33. $\mathbf{v} = 6\mathbf{i} - 5\mathbf{j}$

34. $\mathbf{v} = -10\mathbf{i} + 3\mathbf{j}$

35. $\mathbf{v} = 4\mathbf{j}$

36. $\mathbf{v} = \mathbf{i} - \mathbf{j}$

In Exercises 37–40, find the unit vector in the direction of u and verify that it has length 1.

37. $\mathbf{u} = \langle 3, 12\rangle$

38. $\mathbf{u} = \langle 5, 15\rangle$

39. $\mathbf{u} = \left\langle \frac{3}{2}, \frac{5}{2}\right\rangle$

40. $\mathbf{u} = \langle -6.2, 3.4\rangle$

In Exercises 41–44, find the following.

(a) $\|\mathbf{u}\|$ (b) $\|\mathbf{v}\|$ (c) $\|\mathbf{u} + \mathbf{v}\|$

(d) $\left\|\dfrac{\mathbf{u}}{\|\mathbf{u}\|}\right\|$ (e) $\left\|\dfrac{\mathbf{v}}{\|\mathbf{v}\|}\right\|$ (f) $\left\|\dfrac{\mathbf{u} + \mathbf{v}}{\|\mathbf{u} + \mathbf{v}\|}\right\|$

41. $\mathbf{u} = \langle 1, -1\rangle$, $\mathbf{v} = \langle -1, 2\rangle$

42. $\mathbf{u} = \langle 0, 1\rangle$, $\mathbf{v} = \langle 3, -3\rangle$

43. $\mathbf{u} = \left\langle 1, \frac{1}{2}\right\rangle$, $\mathbf{v} = \langle 2, 3\rangle$

44. $\mathbf{u} = \langle 2, -4\rangle$, $\mathbf{v} = \langle 5, 5\rangle$

In Exercises 45 and 46, sketch a graph of u, v, and u + v. Then demonstrate the triangle inequality using the vectors u and v.

45. $\mathbf{u} = \langle 2, 1\rangle$, $\mathbf{v} = \langle 5, 4\rangle$

46. $\mathbf{u} = \langle -3, 2\rangle$, $\mathbf{v} = \langle 1, -2\rangle$

In Exercises 47–50, find the vector v with the given magnitude and the same direction as u.

	Magnitude	*Direction*
47.	$\|\mathbf{v}\| = 4$	$\mathbf{u} = \langle 1, 1\rangle$
48.	$\|\mathbf{v}\| = 4$	$\mathbf{u} = \langle -1, 1\rangle$
49.	$\|\mathbf{v}\| = 2$	$\mathbf{u} = \langle \sqrt{3}, 3\rangle$
50.	$\|\mathbf{v}\| = 3$	$\mathbf{u} = \langle 0, 3\rangle$

In Exercises 51–54, find the component form of v given its magnitude and the angle it makes with the positive x-axis.

51. $\|\mathbf{v}\| = 3$, $\theta = 0°$

52. $\|\mathbf{v}\| = 5$, $\theta = 120°$

53. $\|\mathbf{v}\| = 2$, $\theta = 150°$

54. $\|\mathbf{v}\| = 1$, $\theta = 3.5°$

In Exercises 55–58, find the component form of u + v given the lengths of u and v and the angles that u and v make with the positive x-axis.

55. $\|\mathbf{u}\| = 1$, $\theta_{\mathbf{u}} = 0°$; $\|\mathbf{v}\| = 3$, $\theta_{\mathbf{v}} = 45°$

56. $\|\mathbf{u}\| = 4$, $\theta_{\mathbf{u}} = 0°$; $\|\mathbf{v}\| = 2$, $\theta_{\mathbf{v}} = 60°$

57. $\|\mathbf{u}\| = 2$, $\theta_{\mathbf{u}} = 4$; $\|\mathbf{v}\| = 1$, $\theta_{\mathbf{v}} = 2$

58. $\|\mathbf{u}\| = 5$, $\theta_{\mathbf{u}} = -0.5$; $\|\mathbf{v}\| = 5$, $\theta_{\mathbf{v}} = 0.5$

Writing About Concepts

59. In your own words, state the difference between a scalar and a vector. Give examples of each.

60. Give geometric descriptions of the operations of addition of vectors and multiplication of a vector by a scalar.

61. Identify the quantity as a scalar or as a vector. Explain your reasoning.

(a) The muzzle velocity of a gun

(b) The price of a company's stock

62. Identify the quantity as a scalar or as a vector. Explain your reasoning.

(a) The air temperature in a room

(b) The weight of a car

In Exercises 63–68, find a and b such that $\mathbf{v} = a\mathbf{u} + b\mathbf{w}$, where $\mathbf{u} = \langle 1, 2\rangle$ and $\mathbf{w} = \langle 1, -1\rangle$.

63. $\mathbf{v} = \langle 2, 1\rangle$

64. $\mathbf{v} = \langle 0, 3\rangle$

65. $\mathbf{v} = \langle 3, 0\rangle$

66. $\mathbf{v} = \langle 3, 3\rangle$

67. $\mathbf{v} = \langle 1, 1\rangle$

68. $\mathbf{v} = \langle -1, 7\rangle$

In Exercises 69–74, find a unit vector (a) parallel to and (b) normal to the graph of $f(x)$ at the given point. Then sketch a graph of the vectors and the function.

	Function	*Point*
69.	$f(x) = x^2$	$(3, 9)$
70.	$f(x) = -x^2 + 5$	$(1, 4)$
71.	$f(x) = x^3$	$(1, 1)$
72.	$f(x) = x^3$	$(-2, -8)$
73.	$f(x) = \sqrt{25 - x^2}$	$(3, 4)$
74.	$f(x) = \tan x$	$\left(\frac{\pi}{4}, 1\right)$

In Exercises 75 and 76, find the component form of v given the magnitudes of u and u + v and the angles that u and u + v make with the positive x-axis.

75. $\|\mathbf{u}\| = 1,\ \theta = 45°$
$\|\mathbf{u} + \mathbf{v}\| = \sqrt{2},\ \theta = 90°$

76. $\|\mathbf{u}\| = 4,\ \theta = 30°$
$\|\mathbf{u} + \mathbf{v}\| = 6,\ \theta = 120°$

77. ***Programming*** You are given the magnitudes of **u** and **v** and the angles **u** and **v** make with the positive x-axis. Write a program for a graphing utility in which the output is the following.

(a) $\mathbf{u} + \mathbf{v}$ (b) $\|\mathbf{u} + \mathbf{v}\|$

(c) The angle $\mathbf{u} + \mathbf{v}$ makes with the positive x-axis

78. ***Programming*** Use the program you wrote in Exercise 77 to find the magnitude and direction of the resultant of the vectors shown.

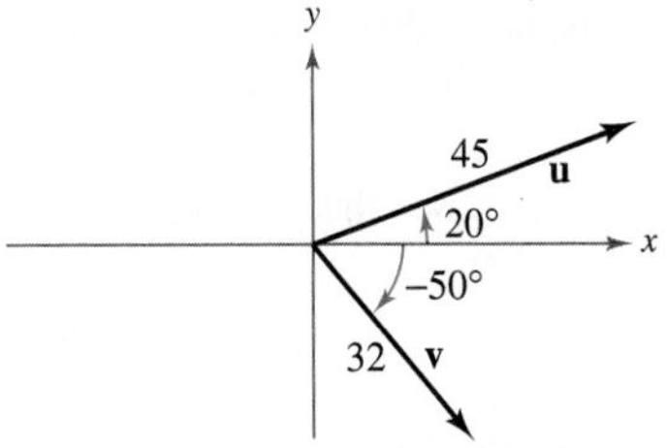

In Exercises 79 and 80, use a graphing utility to find the magnitude and direction of the resultant of the vectors.

79.

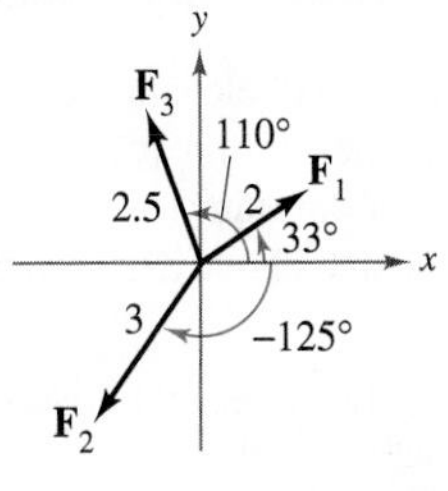

80.

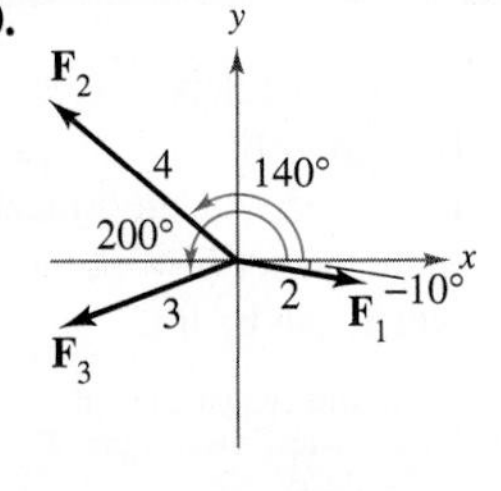

81. ***Numerical and Graphical Analysis*** Forces with magnitudes of 180 newtons and 275 newtons act on a hook (see figure). The angle between the two forces is θ degrees.

(a) If $\theta = 30°$, find the direction and magnitude of the resultant force.

(b) Write the magnitude M and direction α of the resultant force as functions of θ, where $0° \le \theta \le 180°$.

(c) Use a graphing utility to complete the table.

θ	0°	30°	60°	90°	120°	150°	180°
M							
α							

(d) Use a graphing utility to graph the two functions M and α.

(e) Explain why one of the functions decreases for increasing values of θ whereas the other does not.

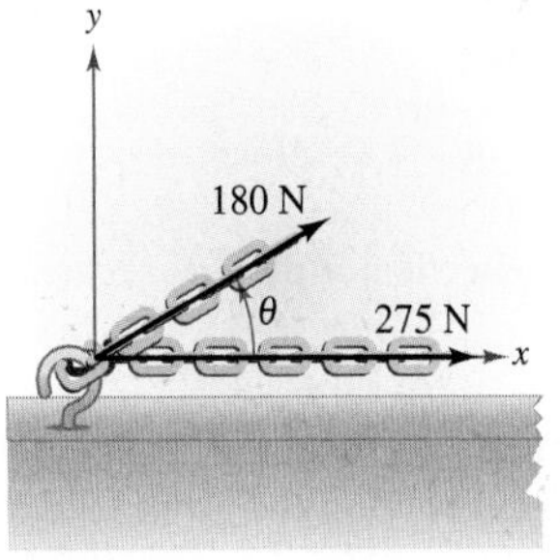

Figure for 81

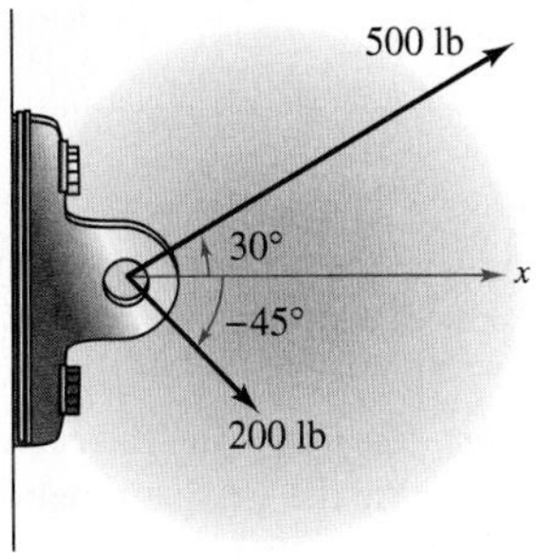

Figure for 82

82. ***Resultant Force*** Forces with magnitudes of 500 pounds and 200 pounds act on a machine part at angles of 30° and −45°, respectively, with the x-axis (see figure). Find the direction and magnitude of the resultant force.

83. ***Resultant Force*** Three forces with magnitudes of 75 pounds, 100 pounds, and 125 pounds act on an object at angles of 30°, 45°, and 120°, respectively, with the positive x-axis. Find the direction and magnitude of the resultant force.

84. ***Resultant Force*** Three forces with magnitudes of 400 newtons, 280 newtons, and 350 newtons act on an object at angles of −30°, 45°, and 135°, respectively, with the positive x-axis. Find the direction and magnitude of the resultant force.

85. ***Think About It*** Consider two forces of equal magnitude acting on a point.

(a) If the magnitude of the resultant is the sum of the magnitudes of the two forces, make a conjecture about the angle between the forces.

(b) If the resultant of the forces is **0**, make a conjecture about the angle between the forces.

(c) Can the magnitude of the resultant be greater than the sum of the magnitudes of the two forces? Explain.

86. ***Graphical Reasoning*** Consider two forces $\mathbf{F}_1 = \langle 20, 0 \rangle$ and $\mathbf{F}_2 = 10\langle \cos\theta, \sin\theta \rangle$.

(a) Find $\|\mathbf{F}_1 + \mathbf{F}_2\|$.

(b) Determine the magnitude of the resultant as a function of θ. Use a graphing utility to graph the function for $0 \le \theta < 2\pi$.

(c) Use the graph in part (b) to determine the range of the function. What is its maximum and for what value of θ does it occur? What is its minimum and for what value of θ does it occur?

(d) Explain why the magnitude of the resultant is never 0.

87. Three vertices of a parallelogram are $(1, 2)$, $(3, 1)$, and $(8, 4)$. Find the three possible fourth vertices (see figure).

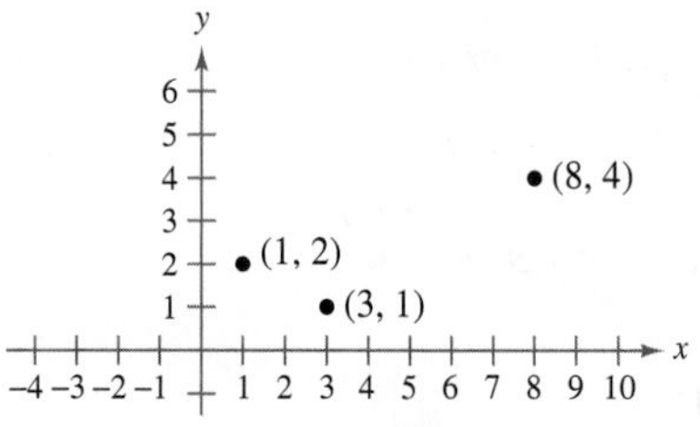

88. Use vectors to find the points of trisection of the line segment with endpoints $(1, 2)$ and $(7, 5)$.

Cable Tension **In Exercises 89 and 90, use the figure to determine the tension in each cable supporting the given load.**

89.

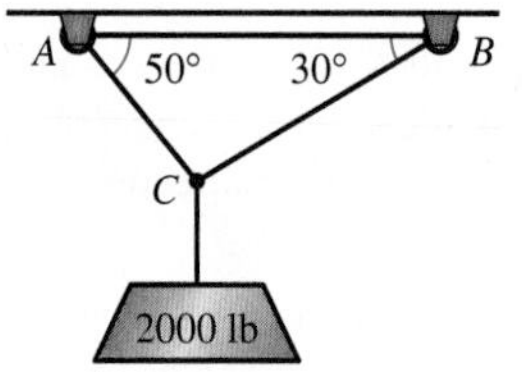

90.

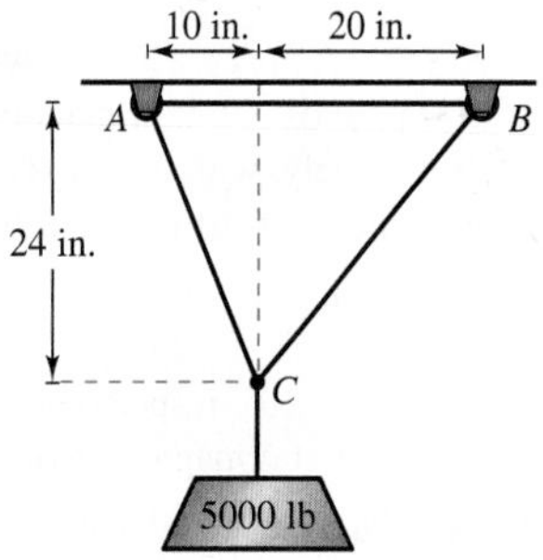

91. ***Projectile Motion*** A gun with a muzzle velocity of 1200 feet per second is fired at an angle of $6°$ above the horizontal. Find the vertical and horizontal components of the velocity.

92. ***Shared Load*** To carry a 100-pound cylindrical weight, two workers lift on the ends of short ropes tied to an eyelet on the top center of the cylinder. One rope makes a $20°$ angle away from the vertical and the other makes a $30°$ angle (see figure).

(a) Find each rope's tension if the resultant force is vertical.

(b) Find the vertical component of each worker's force.

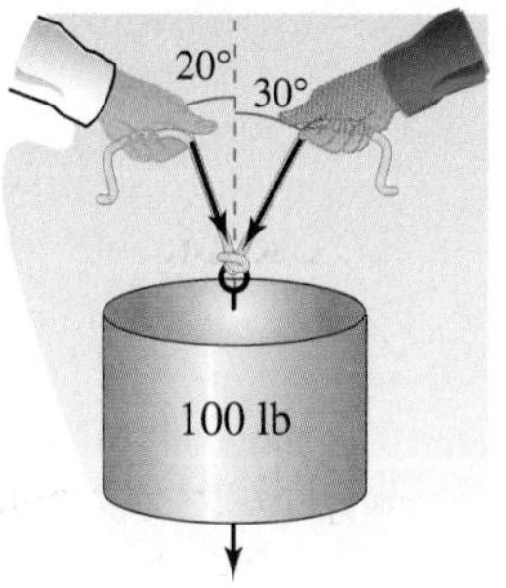

Figure for 92

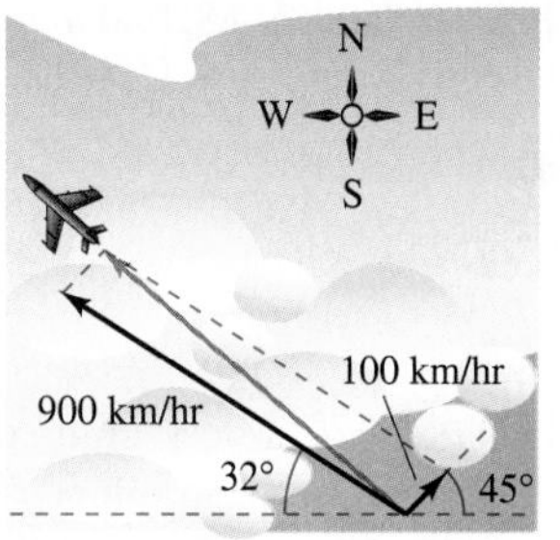

Figure for 93

93. ***Navigation*** A plane is flying in the direction $302°$. Its speed with respect to the air is 900 kilometers per hour. The wind at the plane's altitude is from the southwest at 100 kilometers per hour (see figure). What is the true direction of the plane, and what is its speed with respect to the ground?

94. ***Navigation*** A plane flies at a constant groundspeed of 400 miles per hour due east and encounters a 50-mile-per-hour wind from the northwest. Find the airspeed and compass direction that will allow the plane to maintain its groundspeed and eastward direction.

True or False? **In Exercises 95–100, determine whether the statement is true or false. If it is false, explain why or give an example that shows it is false.**

95. If $\mathbf{u}$ and $\mathbf{v}$ have the same magnitude and direction, then $\mathbf{u}$ and $\mathbf{v}$ are equivalent.

96. If $\mathbf{u}$ is a unit vector in the direction of $\mathbf{v}$, then $\mathbf{v} = \|\mathbf{v}\|\,\mathbf{u}$.

97. If $\mathbf{u} = a\mathbf{i} + b\mathbf{j}$ is a unit vector, then $a^2 + b^2 = 1$.

98. If $\mathbf{v} = a\mathbf{i} + b\mathbf{j} = \mathbf{0}$, then $a = -b$.

99. If $a = b$, then $\|a\mathbf{i} + b\mathbf{j}\| = \sqrt{2}a$.

100. If $\mathbf{u}$ and $\mathbf{v}$ have the same magnitude but opposite directions, then $\mathbf{u} + \mathbf{v} = \mathbf{0}$.

101. Prove that $\mathbf{u} = (\cos\theta)\mathbf{i} - (\sin\theta)\mathbf{j}$ and $\mathbf{v} = (\sin\theta)\mathbf{i} + (\cos\theta)\mathbf{j}$ are unit vectors for any angle θ.

102. ***Geometry*** Using vectors, prove that the line segment joining the midpoints of two sides of a triangle is parallel to, and one-half the length of, the third side.

103. ***Geometry*** Using vectors, prove that the diagonals of a parallelogram bisect each other.

104. Prove that the vector $\mathbf{w} = \|\mathbf{u}\|\mathbf{v} + \|\mathbf{v}\|\mathbf{u}$ bisects the angle between $\mathbf{u}$ and $\mathbf{v}$.

105. Consider the vector $\mathbf{u} = \langle x, y \rangle$. Describe the set of all points (x, y) such that $\|\mathbf{u}\| = 5$.

Putnam Exam Challenge

106. A coast artillery gun can fire at any angle of elevation between $0°$ and $90°$ in a fixed vertical plane. If air resistance is neglected and the muzzle velocity is constant $(= v_0)$, determine the set H of points in the plane and above the horizontal which can be hit.

Section 11.2 Space Coordinates and Vectors in Space

- **Understand the three-dimensional rectangular coordinate system.**
- **Analyze vectors in space.**
- **Use three-dimensional vectors to solve real-life problems.**

Coordinates in Space

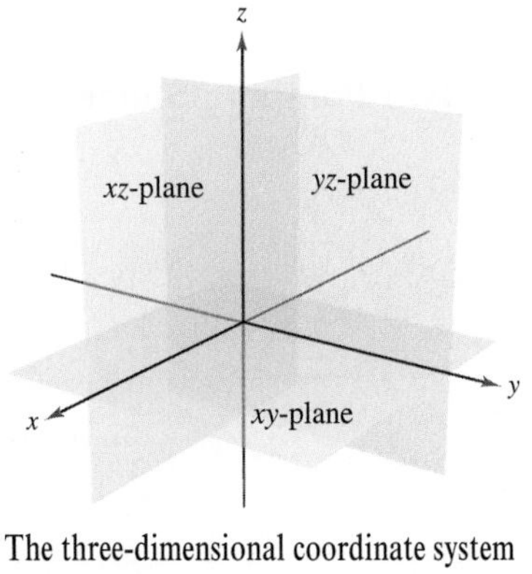

The three-dimensional coordinate system
Figure 11.14

Up to this point in the text, you have been primarily concerned with the two-dimensional coordinate system. Much of the remaining part of your study of calculus will involve the three-dimensional coordinate system.

Before extending the concept of a vector to three dimensions, you must be able to identify points in the **three-dimensional coordinate system.** You can construct this system by passing a z-axis perpendicular to both the x- and y-axes at the origin. Figure 11.14 shows the positive portion of each coordinate axis. Taken as pairs, the axes determine three **coordinate planes:** the ***xy*-plane,** the ***xz*-plane,** and the ***yz*-plane.** These three coordinate planes separate three-space into eight **octants.** The first octant is the one for which all three coordinates are positive. In this three-dimensional system, a point P in space is determined by an ordered triple (x, y, z) where x, y, and z are as follows.

$x =$ directed distance from yz-plane to P

$y =$ directed distance from xz-plane to P

$z =$ directed distance from xy-plane to P

Several points are shown in Figure 11.15.

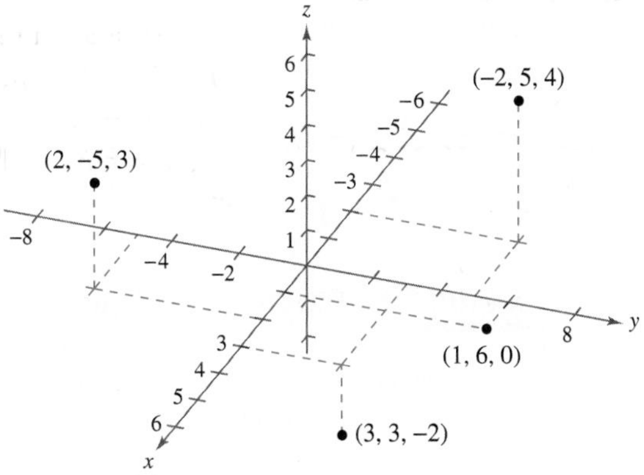

Points in the three-dimensional coordinate system are represented by ordered triples.
Figure 11.15

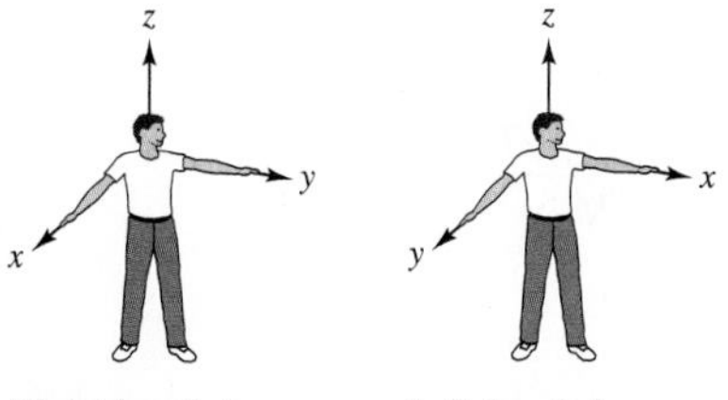

Right-handed system
Left-handed system
Figure 11.16

A three-dimensional coordinate system can have either a **left-handed** or a **right-handed** orientation. To determine the orientation of a system, imagine that you are standing at the origin, with your arms pointing in the direction of the positive x- and y-axes, and with the z-axis pointing up, as shown in Figure 11.16. The system is right-handed or left-handed depending on which hand points along the x-axis. In this text, you will work exclusively with the right-handed system.

NOTE The three-dimensional rotatable graphs that are available in the *HM mathSpace®* CD-ROM and the online *Eduspace®* system for this text will help you visualize points or objects in a three-dimensional coordinate system.

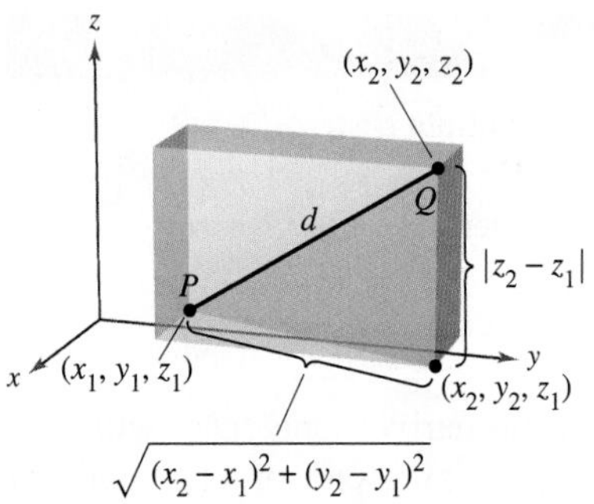

The distance between two points in space
Figure 11.17

Many of the formulas established for the two-dimensional coordinate system can be extended to three dimensions. For example, to find the distance between two points in space, you can use the Pythagorean Theorem twice, as shown in Figure 11.17. By doing this, you will obtain the formula for the distance between the points (x_1, y_1, z_1) and (x_2, y_2, z_2).

$$d = \sqrt{(x_2 - x_1)^2 + (y_2 - y_1)^2 + (z_2 - z_1)^2} \qquad \text{Distance Formula}$$

EXAMPLE 1 Finding the Distance Between Two Points in Space

The distance between the points $(2, -1, 3)$ and $(1, 0, -2)$ is

$$\begin{aligned} d &= \sqrt{(1-2)^2 + (0+1)^2 + (-2-3)^2} && \text{Distance Formula} \\ &= \sqrt{1+1+25} \\ &= \sqrt{27} \\ &= 3\sqrt{3}. \end{aligned}$$

A **sphere** with center at (x_0, y_0, z_0) and radius r is defined to be the set of all points (x, y, z) such that the distance between (x, y, z) and (x_0, y_0, z_0) is r. You can use the Distance Formula to find the **standard equation of a sphere** of radius r, centered at (x_0, y_0, z_0). If (x, y, z) is an arbitrary point on the sphere, the equation of the sphere is

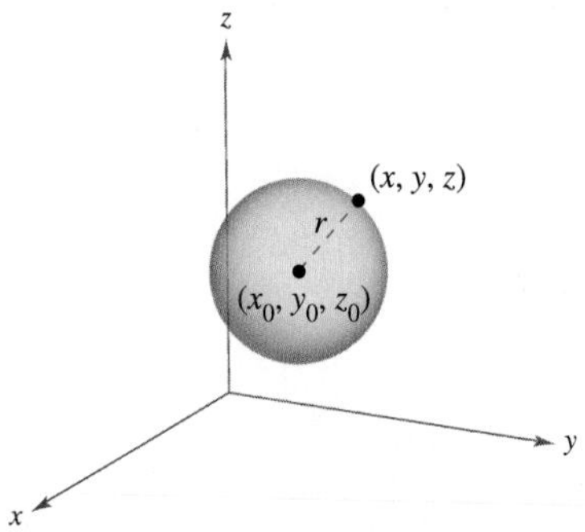

Figure 11.18

$$(x - x_0)^2 + (y - y_0)^2 + (z - z_0)^2 = r^2 \qquad \text{Equation of sphere}$$

as shown in Figure 11.18. Moreover, the midpoint of the line segment joining the points (x_1, y_1, z_1) and (x_2, y_2, z_2) has coordinates

$$\left(\frac{x_1 + x_2}{2}, \frac{y_1 + y_2}{2}, \frac{z_1 + z_2}{2}\right). \qquad \text{Midpoint Rule}$$

EXAMPLE 2 Finding the Equation of a Sphere

Find the standard equation of the sphere that has the points $(5, -2, 3)$ and $(0, 4, -3)$ as endpoints of a diameter.

Solution By the Midpoint Rule, the center of the sphere is

$$\left(\frac{5+0}{2}, \frac{-2+4}{2}, \frac{3-3}{2}\right) = \left(\frac{5}{2}, 1, 0\right). \qquad \text{Midpoint Rule}$$

By the Distance Formula, the radius is

$$r = \sqrt{\left(0 - \frac{5}{2}\right)^2 + (4-1)^2 + (-3-0)^2} = \sqrt{\frac{97}{4}} = \frac{\sqrt{97}}{2}.$$

Therefore, the standard equation of the sphere is

$$\left(x - \frac{5}{2}\right)^2 + (y-1)^2 + z^2 = \frac{97}{4}. \qquad \text{Equation of sphere}$$

Vectors in Space

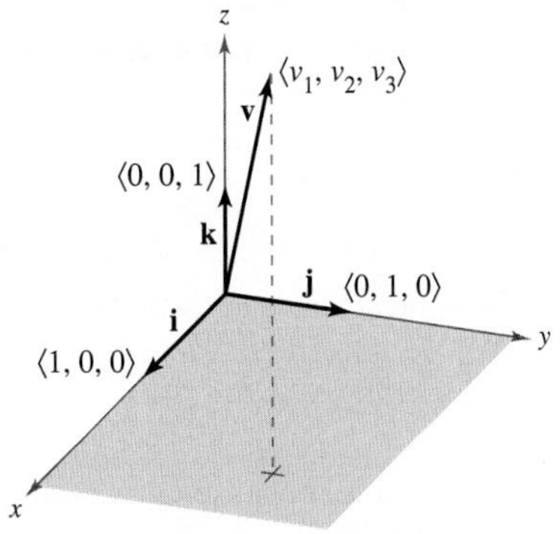

The standard unit vectors in space
Figure 11.19

In space, vectors are denoted by ordered triples $\mathbf{v} = \langle v_1, v_2, v_3 \rangle$. The **zero vector** is denoted by $\mathbf{0} = \langle 0, 0, 0 \rangle$. Using the unit vectors $\mathbf{i} = \langle 1, 0, 0 \rangle$, $\mathbf{j} = \langle 0, 1, 0 \rangle$, and $\mathbf{k} = \langle 0, 0, 1 \rangle$ in the direction of the positive z-axis, the **standard unit vector notation** for $\mathbf{v}$ is

$$\mathbf{v} = v_1\mathbf{i} + v_2\mathbf{j} + v_3\mathbf{k}$$

as shown in Figure 11.19. If $\mathbf{v}$ is represented by the directed line segment from $P(p_1, p_2, p_3)$ to $Q(q_1, q_2, q_3)$, as shown in Figure 11.20, the component form of $\mathbf{v}$ is given by subtracting the coordinates of the initial point from the coordinates of the terminal point, as follows.

$$\mathbf{v} = \langle v_1, v_2, v_3 \rangle = \langle q_1 - p_1, q_2 - p_2, q_3 - p_3 \rangle$$

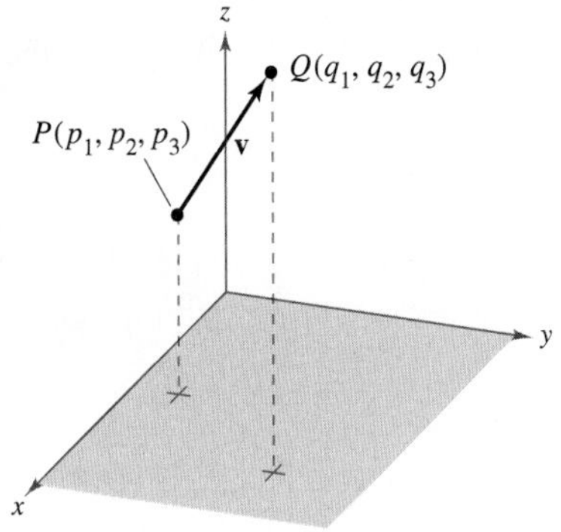

$\mathbf{v} = \langle q_1 - p_1, q_2 - p_2, q_3 - p_3 \rangle$

Figure 11.20

Vectors in Space

Let $\mathbf{u} = \langle u_1, u_2, u_3 \rangle$ and $\mathbf{v} = \langle v_1, v_2, v_3 \rangle$ be vectors in space and let c be a scalar.

1. *Equality of Vectors:* $\mathbf{u} = \mathbf{v}$ if and only if $u_1 = v_1$, $u_2 = v_2$, and $u_3 = v_3$.
2. *Component Form:* If $\mathbf{v}$ is represented by the directed line segment from $P(p_1, p_2, p_3)$ to $Q(q_1, q_2, q_3)$, then
$$\mathbf{v} = \langle v_1, v_2, v_3 \rangle = \langle q_1 - p_1, q_2 - p_2, q_3 - p_3 \rangle.$$
3. *Length:* $\|\mathbf{v}\| = \sqrt{v_1^2 + v_2^2 + v_3^2}$
4. *Unit Vector in the Direction of* $\mathbf{v}$*:* $\dfrac{\mathbf{v}}{\|\mathbf{v}\|} = \left(\dfrac{1}{\|\mathbf{v}\|}\right)\langle v_1, v_2, v_3 \rangle, \quad \mathbf{v} \neq \mathbf{0}$
5. *Vector Addition:* $\mathbf{v} + \mathbf{u} = \langle v_1 + u_1, v_2 + u_2, v_3 + u_3 \rangle$
6. *Scalar Multiplication:* $c\mathbf{v} = \langle cv_1, cv_2, cv_3 \rangle$

NOTE The properties of vector addition and scalar multiplication given in Theorem 11.1 are also valid for vectors in space.

EXAMPLE 3 Finding the Component Form of a Vector in Space

Find the component form and magnitude of the vector $\mathbf{v}$ having initial point $(-2, 3, 1)$ and terminal point $(0, -4, 4)$. Then find a unit vector in the direction of $\mathbf{v}$.

Solution The component form of $\mathbf{v}$ is

$$\begin{aligned}\mathbf{v} = \langle q_1 - p_1, q_2 - p_2, q_3 - p_3 \rangle &= \langle 0 - (-2), -4 - 3, 4 - 1 \rangle \\ &= \langle 2, -7, 3 \rangle\end{aligned}$$

which implies that its magnitude is

$$\|\mathbf{v}\| = \sqrt{2^2 + (-7)^2 + 3^2} = \sqrt{62}.$$

The unit vector in the direction of $\mathbf{v}$ is

$$\mathbf{u} = \frac{\mathbf{v}}{\|\mathbf{v}\|} = \frac{1}{\sqrt{62}}\langle 2, -7, 3 \rangle.$$

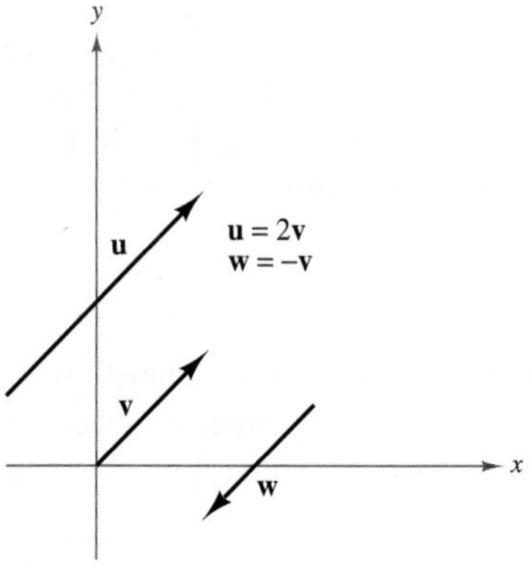

Parallel vectors
Figure 11.21

Recall from the definition of scalar multiplication that positive scalar multiples of a nonzero vector **v** have the same direction as **v**, whereas negative multiples have the direction opposite of **v**. In general, two nonzero vectors **u** and **v** are **parallel** if there is some scalar c such that $\mathbf{u} = c\mathbf{v}$.

Definition of Parallel Vectors

Two nonzero vectors **u** and **v** are **parallel** if there is some scalar c such that $\mathbf{u} = c\mathbf{v}$.

For example, in Figure 11.21, the vectors **u**, **v**, and **w** are parallel because $\mathbf{u} = 2\mathbf{v}$ and $\mathbf{w} = -\mathbf{v}$.

EXAMPLE 4 Parallel Vectors

Vector **w** has initial point $(2, -1, 3)$ and terminal point $(-4, 7, 5)$. Which of the following vectors is parallel to **w**?

a. $\mathbf{u} = \langle 3, -4, -1 \rangle$

b. $\mathbf{v} = \langle 12, -16, 4 \rangle$

Solution Begin by writing **w** in component form.

$$\mathbf{w} = \langle -4 - 2, 7 - (-1), 5 - 3 \rangle = \langle -6, 8, 2 \rangle$$

a. Because $\mathbf{u} = \langle 3, -4, -1 \rangle = -\frac{1}{2}\langle -6, 8, 2 \rangle = -\frac{1}{2}\mathbf{w}$, you can conclude that **u** is parallel to **w**.

b. In this case, you want to find a scalar c such that

$$\langle 12, -16, 4 \rangle = c\langle -6, 8, 2 \rangle.$$

$$\begin{aligned} 12 &= -6c \rightarrow c = -2 \\ -16 &= 8c \rightarrow c = -2 \\ 4 &= 2c \rightarrow c = 2 \end{aligned}$$

Because there is no c for which the equation has a solution, the vectors are not parallel.

EXAMPLE 5 Using Vectors to Determine Collinear Points

Determine whether the points $P(1, -2, 3)$, $Q(2, 1, 0)$, and $R(4, 7, -6)$ are collinear.

Solution The component forms of $\overrightarrow{PQ}$ and $\overrightarrow{PR}$ are

$$\overrightarrow{PQ} = \langle 2 - 1, 1 - (-2), 0 - 3 \rangle = \langle 1, 3, -3 \rangle$$

and

$$\overrightarrow{PR} = \langle 4 - 1, 7 - (-2), -6 - 3 \rangle = \langle 3, 9, -9 \rangle.$$

These two vectors have a common initial point. So, P, Q, and R lie on the same line if and only if $\overrightarrow{PQ}$ and $\overrightarrow{PR}$ are parallel—which they are because $\overrightarrow{PR} = 3\,\overrightarrow{PQ}$, as shown in Figure 11.22.

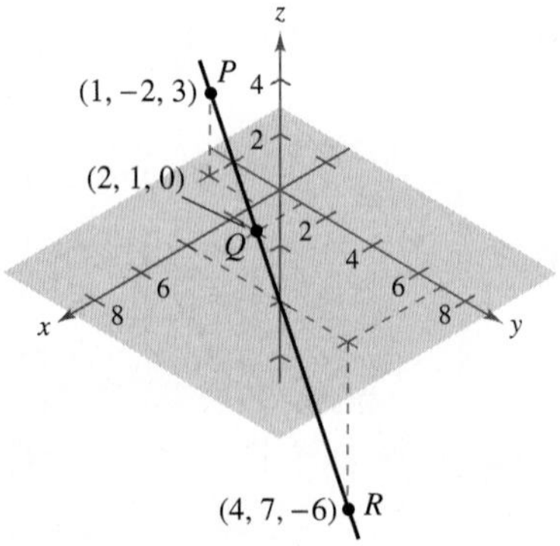

The points P, Q, and R lie on the same line.
Figure 11.22

EXAMPLE 6 Standard Unit Vector Notation

a. Write the vector $\mathbf{v} = 4\mathbf{i} - 5\mathbf{k}$ in component form.

b. Find the terminal point of the vector $\mathbf{v} = 7\mathbf{i} - \mathbf{j} + 3\mathbf{k}$, given that the initial point is $P(-2, 3, 5)$.

Solution

a. Because $\mathbf{j}$ is missing, its component is 0 and

$$\mathbf{v} = 4\mathbf{i} - 5\mathbf{k} = \langle 4, 0, -5\rangle.$$

b. You need to find $Q(q_1, q_2, q_3)$ such that $\mathbf{v} = \overrightarrow{PQ} = 7\mathbf{i} - \mathbf{j} + 3\mathbf{k}$. This implies that $q_1 - (-2) = 7$, $q_2 - 3 = -1$, and $q_3 - 5 = 3$. The solution of these three equations is $q_1 = 5$, $q_2 = 2$, and $q_3 = 8$. Therefore, Q is $(5, 2, 8)$.

Application

EXAMPLE 7 Measuring Force

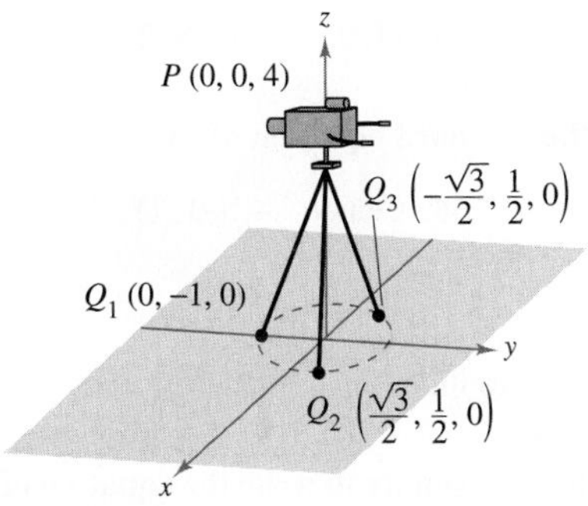

Figure 11.23

A television camera weighing 120 pounds is supported by a tripod, as shown in Figure 11.23. Represent the force exerted on each leg of the tripod as a vector.

Solution Let the vectors $\mathbf{F}_1$, $\mathbf{F}_2$, and $\mathbf{F}_3$ represent the forces exerted on the three legs. From Figure 11.23, you can determine the directions of $\mathbf{F}_1$, $\mathbf{F}_2$, and $\mathbf{F}_3$ to be as follows.

$$\overrightarrow{PQ_1} = \langle 0 - 0, -1 - 0, 0 - 4\rangle = \langle 0, -1, -4\rangle$$

$$\overrightarrow{PQ_2} = \left\langle \frac{\sqrt{3}}{2} - 0, \frac{1}{2} - 0, 0 - 4\right\rangle = \left\langle \frac{\sqrt{3}}{2}, \frac{1}{2}, -4\right\rangle$$

$$\overrightarrow{PQ_3} = \left\langle -\frac{\sqrt{3}}{2} - 0, \frac{1}{2} - 0, 0 - 4\right\rangle = \left\langle -\frac{\sqrt{3}}{2}, \frac{1}{2}, -4\right\rangle$$

Because each leg has the same length, and the total force is distributed equally among the three legs, you know that $\|\mathbf{F}_1\| = \|\mathbf{F}_2\| = \|\mathbf{F}_3\|$. So, there exists a constant c such that

$$\mathbf{F}_1 = c\langle 0, -1, -4\rangle, \quad \mathbf{F}_2 = c\left\langle \frac{\sqrt{3}}{2}, \frac{1}{2}, -4\right\rangle, \quad \text{and} \quad \mathbf{F}_3 = c\left\langle -\frac{\sqrt{3}}{2}, \frac{1}{2}, -4\right\rangle.$$

Let the total force exerted by the object be given by $\mathbf{F} = -120\mathbf{k}$. Then, using the fact that

$$\mathbf{F} = \mathbf{F}_1 + \mathbf{F}_2 + \mathbf{F}_3$$

you can conclude that $\mathbf{F}_1$, $\mathbf{F}_2$, and $\mathbf{F}_3$ all have a vertical component of -40. This implies that $c(-4) = -40$ and $c = 10$. Therefore, the forces exerted on the legs can be represented by

$$\mathbf{F}_1 = \langle 0, -10, -40\rangle$$
$$\mathbf{F}_2 = \langle 5\sqrt{3}, 5, -40\rangle$$
$$\mathbf{F}_3 = \langle -5\sqrt{3}, 5, -40\rangle.$$

Exercises for Section 11.2

See www.CalcChat.com for worked-out solutions to odd-numbered exercises.

In Exercises 1–4, plot the points on the same three-dimensional coordinate system.

1. (a) $(2, 1, 3)$ (b) $(-1, 2, 1)$
2. (a) $(3, -2, 5)$ (b) $\left(\frac{3}{2}, 4, -2\right)$
3. (a) $(5, -2, 2)$ (b) $(5, -2, -2)$
4. (a) $(0, 4, -5)$ (b) $(4, 0, 5)$

In Exercises 5 and 6, approximate the coordinates of the points.

5.

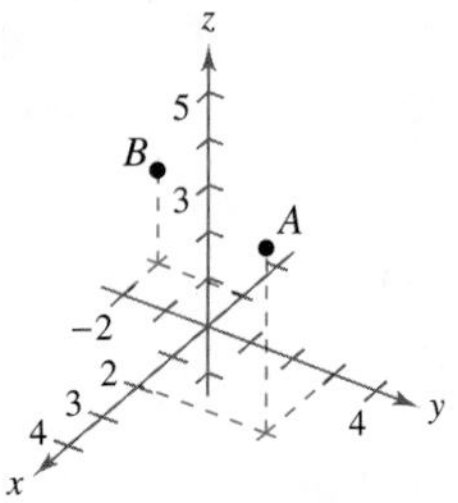

6.

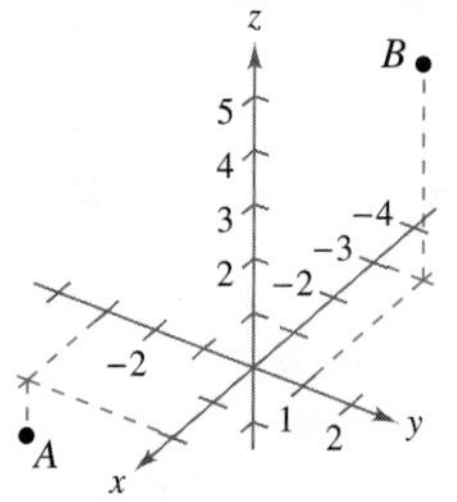

In Exercises 7–10, find the coordinates of the point.

7. The point is located three units behind the yz-plane, four units to the right of the xz-plane, and five units above the xy-plane.
8. The point is located seven units in front of the yz-plane, two units to the left of the xz-plane, and one unit below the xy-plane.
9. The point is located on the x-axis, 10 units in front of the yz-plane.
10. The point is located in the yz-plane, three units to the right of the xz-plane, and two units above the xy-plane.

11. ***Think About It*** What is the z-coordinate of any point in the xy-plane?
12. ***Think About It*** What is the x-coordinate of any point in the yz-plane?

In Exercises 13–24, determine the location of a point (x, y, z) that satisfies the condition(s).

13. $z = 6$
14. $y = 2$
15. $x = 4$
16. $z = -3$
17. $y < 0$
18. $x < 0$
19. $|y| \le 3$
20. $|x| > 4$
21. $xy > 0, \quad z = -3$
22. $xy < 0, \quad z = 4$
23. $xyz < 0$
24. $xyz > 0$

In Exercises 25–28, find the distance between the points.

25. $(0, 0, 0), \quad (5, 2, 6)$
26. $(-2, 3, 2), \quad (2, -5, -2)$
27. $(1, -2, 4), \quad (6, -2, -2)$
28. $(2, 2, 3), \quad (4, -5, 6)$

In Exercises 29–32, find the lengths of the sides of the triangle with the indicated vertices, and determine whether the triangle is a right triangle, an isosceles triangle, or neither.

29. $(0, 0, 0), (2, 2, 1), (2, -4, 4)$
30. $(5, 3, 4), (7, 1, 3), (3, 5, 3)$
31. $(1, -3, -2), (5, -1, 2), (-1, 1, 2)$
32. $(5, 0, 0), (0, 2, 0), (0, 0, -3)$

33. ***Think About It*** The triangle in Exercise 29 is translated five units upward along the z-axis. Determine the coordinates of the translated triangle.
34. ***Think About It*** The triangle in Exercise 30 is translated three units to the right along the y-axis. Determine the coordinates of the translated triangle.

In Exercises 35 and 36, find the coordinates of the midpoint of the line segment joining the points.

35. $(5, -9, 7), (-2, 3, 3)$
36. $(4, 0, -6), (8, 8, 20)$

In Exercises 37–40, find the standard equation of the sphere.

37. Center: $(0, 2, 5)$
    Radius: 2
38. Center: $(4, -1, 1)$
    Radius: 5
39. Endpoints of a diameter: $(2, 0, 0), (0, 6, 0)$
40. Center: $(-3, 2, 4)$, tangent to the yz-plane

In Exercises 41–44, complete the square to write the equation of the sphere in standard form. Find the center and radius.

41. $x^2 + y^2 + z^2 - 2x + 6y + 8z + 1 = 0$
42. $x^2 + y^2 + z^2 + 9x - 2y + 10z + 19 = 0$
43. $9x^2 + 9y^2 + 9z^2 - 6x + 18y + 1 = 0$
44. $4x^2 + 4y^2 + 4z^2 - 4x - 32y + 8z + 33 = 0$

In Exercises 45–48, describe the solid satisfying the condition.

45. $x^2 + y^2 + z^2 \le 36$
46. $x^2 + y^2 + z^2 > 4$
47. $x^2 + y^2 + z^2 < 4x - 6y + 8z - 13$
48. $x^2 + y^2 + z^2 > -4x + 6y - 8z - 13$

In Exercises 49–52, (a) find the component form of the vector v and (b) sketch the vector with its initial point at the origin.

49.

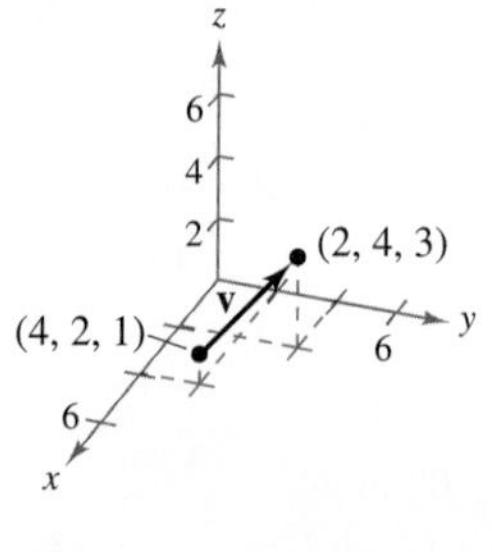

50.

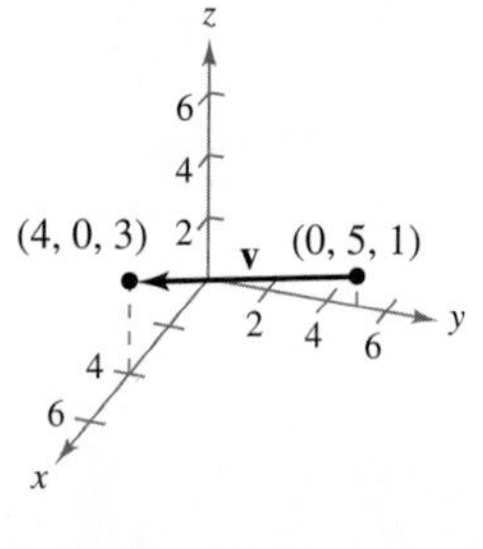

51.

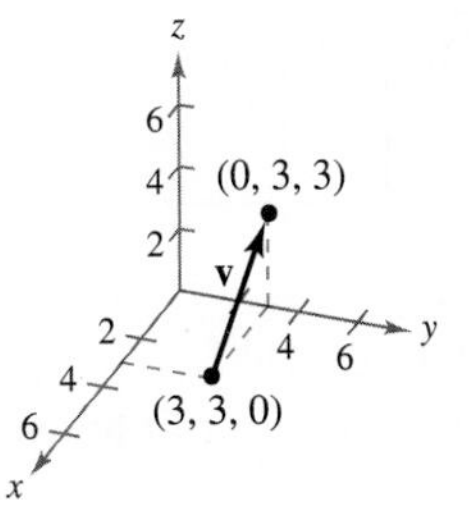

52.

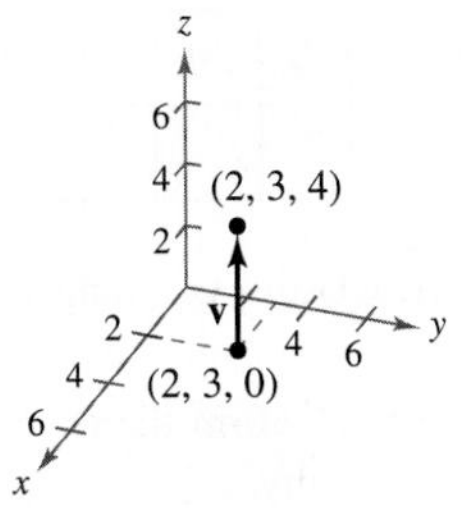

In Exercises 53–56, find the component form and magnitude of the vector u with the given initial and terminal points. Then find a unit vector in the direction of u.

	Initial Point	*Terminal Point*
53.	$(3, 2, 0)$	$(4, 1, 6)$
54.	$(4, -5, 2)$	$(-1, 7, -3)$
55.	$(-4, 3, 1)$	$(-5, 3, 0)$
56.	$(1, -2, 4)$	$(2, 4, -2)$

In Exercises 57 and 58, the initial and terminal points of a vector v are given. (a) Sketch the directed line segment, (b) find the component form of the vector, and (c) sketch the vector with its initial point at the origin.

57. Initial point: $(-1, 2, 3)$
Terminal point: $(3, 3, 4)$

58. Initial point: $(2, -1, -2)$
Terminal point: $(-4, 3, 7)$

In Exercises 59 and 60, the vector v and its initial point are given. Find the terminal point.

59. $\mathbf{v} = \langle 3, -5, 6 \rangle$
Initial point: $(0, 6, 2)$

60. $\mathbf{v} = \left\langle 1, -\frac{2}{3}, \frac{1}{2} \right\rangle$
Initial point: $\left(0, 2, \frac{5}{2}\right)$

In Exercises 61 and 62, find each scalar multiple of v and sketch its graph.

61. $\mathbf{v} = \langle 1, 2, 2 \rangle$
(a) $2\mathbf{v}$ (b) $-\mathbf{v}$
(c) $\frac{3}{2}\mathbf{v}$ (d) $0\mathbf{v}$

62. $\mathbf{v} = \langle 2, -2, 1 \rangle$
(a) $-\mathbf{v}$ (b) $2\mathbf{v}$
(c) $\frac{1}{2}\mathbf{v}$ (d) $\frac{5}{2}\mathbf{v}$

In Exercises 63–68, find the vector z, given that $\mathbf{u} = \langle 1, 2, 3 \rangle$, $\mathbf{v} = \langle 2, 2, -1 \rangle$, and $\mathbf{w} = \langle 4, 0, -4 \rangle$.

63. $\mathbf{z} = \mathbf{u} - \mathbf{v}$

64. $\mathbf{z} = \mathbf{u} - \mathbf{v} + 2\mathbf{w}$

65. $\mathbf{z} = 2\mathbf{u} + 4\mathbf{v} - \mathbf{w}$

66. $\mathbf{z} = 5\mathbf{u} - 3\mathbf{v} - \frac{1}{2}\mathbf{w}$

67. $2\mathbf{z} - 3\mathbf{u} = \mathbf{w}$

68. $2\mathbf{u} + \mathbf{v} - \mathbf{w} + 3\mathbf{z} = \mathbf{0}$

In Exercises 69–72, determine which of the vectors is (are) parallel to z. Use a graphing utility to confirm your results.

69. $\mathbf{z} = \langle 3, 2, -5 \rangle$
(a) $\langle -6, -4, 10 \rangle$
(b) $\left\langle 2, \frac{4}{3}, -\frac{10}{3} \right\rangle$
(c) $\langle 6, 4, 10 \rangle$
(d) $\langle 1, -4, 2 \rangle$

70. $\mathbf{z} = \frac{1}{2}\mathbf{i} - \frac{2}{3}\mathbf{j} + \frac{3}{4}\mathbf{k}$
(a) $6\mathbf{i} - 4\mathbf{j} + 9\mathbf{k}$
(b) $-\mathbf{i} + \frac{4}{3}\mathbf{j} - \frac{3}{2}\mathbf{k}$
(c) $12\mathbf{i} + 9\mathbf{k}$
(d) $\frac{3}{4}\mathbf{i} - \mathbf{j} + \frac{9}{8}\mathbf{k}$

71. $\mathbf{z}$ has initial point $(1, -1, 3)$ and terminal point $(-2, 3, 5)$.
(a) $-6\mathbf{i} + 8\mathbf{j} + 4\mathbf{k}$ (b) $4\mathbf{j} + 2\mathbf{k}$

72. $\mathbf{z}$ has initial point $(5, 4, 1)$ and terminal point $(-2, -4, 4)$.
(a) $\langle 7, 6, 2 \rangle$ (b) $\langle 14, 16, -6 \rangle$

In Exercises 73–76, use vectors to determine whether the points are collinear.

73. $(0, -2, -5), (3, 4, 4), (2, 2, 1)$

74. $(4, -2, 7), (-2, 0, 3), (7, -3, 9)$

75. $(1, 2, 4), (2, 5, 0), (0, 1, 5)$

76. $(0, 0, 0), (1, 3, -2), (2, -6, 4)$

In Exercises 77 and 78, use vectors to show that the points form the vertices of a parallelogram.

77. $(2, 9, 1), (3, 11, 4), (0, 10, 2), (1, 12, 5)$

78. $(1, 1, 3), (9, -1, -2), (11, 2, -9), (3, 4, -4)$

In Exercises 79–84, find the magnitude of v.

79. $\mathbf{v} = \langle 0, 0, 0 \rangle$

80. $\mathbf{v} = \langle 1, 0, 3 \rangle$

81. $\mathbf{v} = \mathbf{i} - 2\mathbf{j} - 3\mathbf{k}$

82. $\mathbf{v} = -4\mathbf{i} + 3\mathbf{j} + 7\mathbf{k}$

83. Initial point of $\mathbf{v}$: $(1, -3, 4)$
Terminal point of $\mathbf{v}$: $(1, 0, -1)$

84. Initial point of $\mathbf{v}$: $(0, -1, 0)$
Terminal point of $\mathbf{v}$: $(1, 2, -2)$

In Exercises 85–88, find a unit vector (a) in the direction of u and (b) in the direction opposite of u.

85. $\mathbf{u} = \langle 2, -1, 2 \rangle$

86. $\mathbf{u} = \langle 6, 0, 8 \rangle$

87. $\mathbf{u} = \langle 3, 2, -5 \rangle$

88. $\mathbf{u} = \langle 8, 0, 0 \rangle$

89. ***Programming*** You are given the component forms of the vectors **u** and **v**. Write a program for a graphing utility in which the output is (a) the component form of $\mathbf{u} + \mathbf{v}$, (b) $\|\mathbf{u} + \mathbf{v}\|$, (c) $\|\mathbf{u}\|$, and (d) $\|\mathbf{v}\|$.

90. ***Programming*** Run the program you wrote in Exercise 89 for the vectors $\mathbf{u} = \langle -1, 3, 4 \rangle$ and $\mathbf{v} = \langle 5, 4.5, -6 \rangle$.

In Exercises 91 and 92, determine the values of c that satisfy the equation. Let $\mathbf{u} = \mathbf{i} + 2\mathbf{j} + 3\mathbf{k}$ and $\mathbf{v} = 2\mathbf{i} + 2\mathbf{j} - \mathbf{k}$.

91. $\|c\mathbf{v}\| = 5$

92. $\|c\mathbf{u}\| = 3$

In Exercises 93–96, find the vector v with the given magnitude and direction u.

	Magnitude	*Direction*
93.	10	$\mathbf{u} = \langle 0, 3, 3 \rangle$
94.	3	$\mathbf{u} = \langle 1, 1, 1 \rangle$
95.	$\frac{3}{2}$	$\mathbf{u} = \langle 2, -2, 1 \rangle$
96.	$\sqrt{5}$	$\mathbf{u} = \langle -4, 6, 2 \rangle$

In Exercises 97 and 98, sketch the vector v and write its component form.

97. $\mathbf{v}$ lies in the yz-plane, has magnitude 2, and makes an angle of 30° with the positive y-axis.

98. $\mathbf{v}$ lies in the xz-plane, has magnitude 5, and makes an angle of 45° with the positive z-axis.

In Exercises 99 and 100, use vectors to find the point that lies two-thirds of the way from *P* to *Q*.

99. $P(4, 3, 0)$, $Q(1, -3, 3)$ **100.** $P(1, 2, 5)$, $Q(6, 8, 2)$

101. Let $\mathbf{u} = \mathbf{i} + \mathbf{j}$, $\mathbf{v} = \mathbf{j} + \mathbf{k}$, and $\mathbf{w} = a\mathbf{u} + b\mathbf{v}$.

(a) Sketch $\mathbf{u}$ and $\mathbf{v}$.

(b) If $\mathbf{w} = \mathbf{0}$, show that a and b must both be zero.

(c) Find a and b such that $\mathbf{w} = \mathbf{i} + 2\mathbf{j} + \mathbf{k}$.

(d) Show that no choice of a and b yields $\mathbf{w} = \mathbf{i} + 2\mathbf{j} + 3\mathbf{k}$.

102. ***Writing*** The initial and terminal points of the vector $\mathbf{v}$ are (x_1, y_1, z_1) and (x, y, z). Describe the set of all points (x, y, z) such that $\|\mathbf{v}\| = 4$.

Writing About Concepts

103. A point in the three-dimensional coordinate system has coordinates (x_0, y_0, z_0). Describe what each coordinate measures.

104. Give the formula for the distance between the points (x_1, y_1, z_1) and (x_2, y_2, z_2).

105. Give the standard equation of a sphere of radius r, centered at (x_0, y_0, z_0).

106. State the definition of parallel vectors.

107. Let A, B, and C be vertices of a triangle. Find $\overrightarrow{AB} + \overrightarrow{BC} + \overrightarrow{CA}$.

108. Let $\mathbf{r} = \langle x, y, z\rangle$ and $\mathbf{r}_0 = \langle 1, 1, 1\rangle$. Describe the set of all points (x, y, z) such that $\|\mathbf{r} - \mathbf{r}_0\| = 2$.

109. ***Numerical, Graphical, and Analytic Analysis*** The lights in an auditorium are 24-pound discs of radius 18 inches. Each disc is supported by three equally spaced cables that are L inches long (see figure).

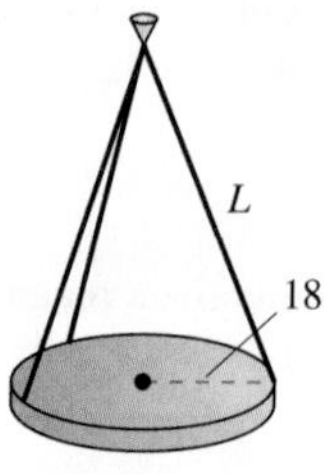

(a) Write the tension T in each cable as a function of L. Determine the domain of the function.

(b) Use a graphing utility and the function in part (a) to complete the table.

L	20	25	30	35	40	45	50
T							

(c) Use a graphing utility to graph the function in part (a). Determine the asymptotes of the graph.

(d) Confirm the asymptotes of the graph in part (c) analytically.

(e) Determine the minimum length of each cable if a cable is designed to carry a maximum load of 10 pounds.

110. ***Think About It*** Suppose the length of each cable in Exercise 109 has a fixed length $L = a$, and the radius of each disc is r_0 inches. Make a conjecture about the limit $\lim_{r_0 \to a^-} T$ and give a reason for your answer.

111. ***Diagonal of a Cube*** Find the component form of the unit vector $\mathbf{v}$ in the direction of the diagonal of the cube shown in the figure.

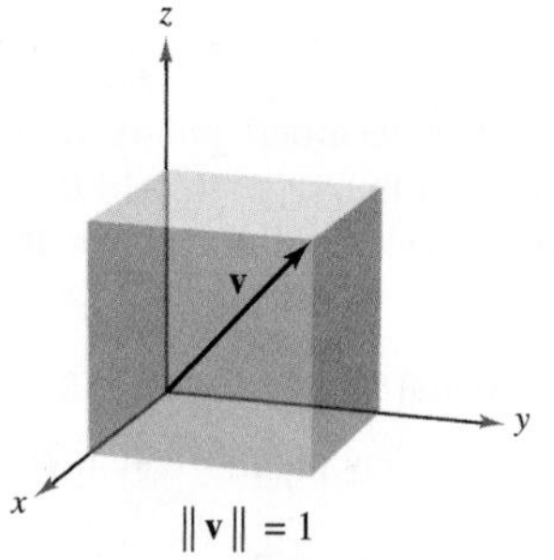

Figure for 111

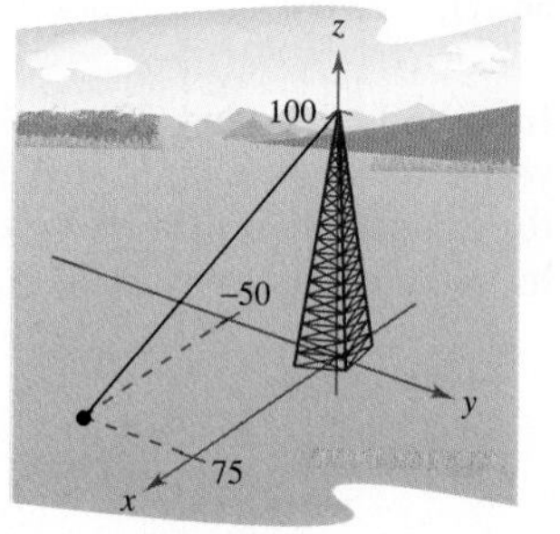

Figure for 112

112. ***Tower Guy Wire*** The guy wire to a 100-foot tower has a tension of 550 pounds. Using the distances shown in the figure, write the component form of the vector $\mathbf{F}$ representing the tension in the wire.

113. ***Load Supports*** Find the tension in each of the supporting cables in the figure if the weight of the crate is 500 newtons.

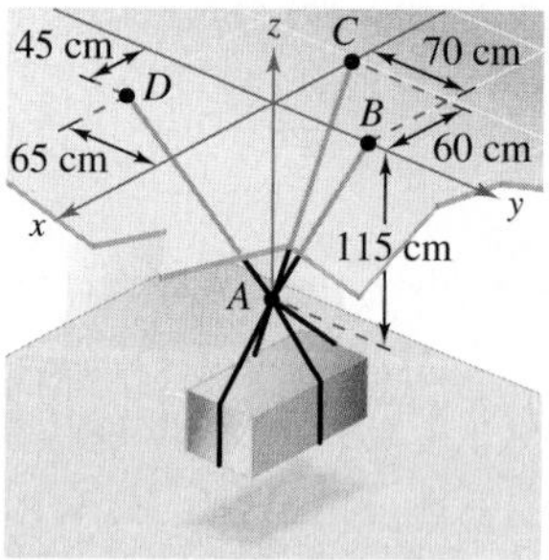

Figure for 113

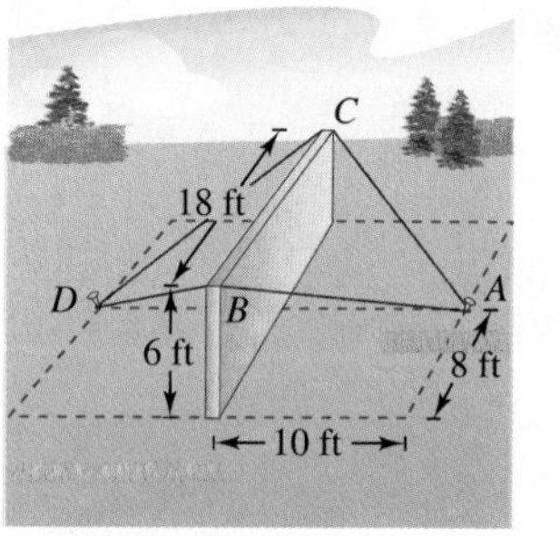

Figure for 114

114. ***Construction*** A precast concrete wall is temporarily kept in its vertical position by ropes (see figure). Find the total force exerted on the pin at position A. The tensions in AB and AC are 420 pounds and 650 pounds.

115. Write an equation whose graph consists of the set of points $P(x, y, z)$ that are twice as far from $A(0, -1, 1)$ as from $B(1, 2, 0)$.

Section 11.3 The Dot Product of Two Vectors

- **Use properties of the dot product of two vectors.**
- **Find the angle between two vectors using the dot product.**
- **Find the direction cosines of a vector in space.**
- **Find the projection of a vector onto another vector.**
- **Use vectors to find the work done by a constant force.**

EXPLORATION

Interpreting a Dot Product Several vectors are shown below on the unit circle. Find the dot products of several pairs of vectors. Then find the angle between each pair that you used. Make a conjecture about the relationship between the dot product of two vectors and the angle between the vectors.

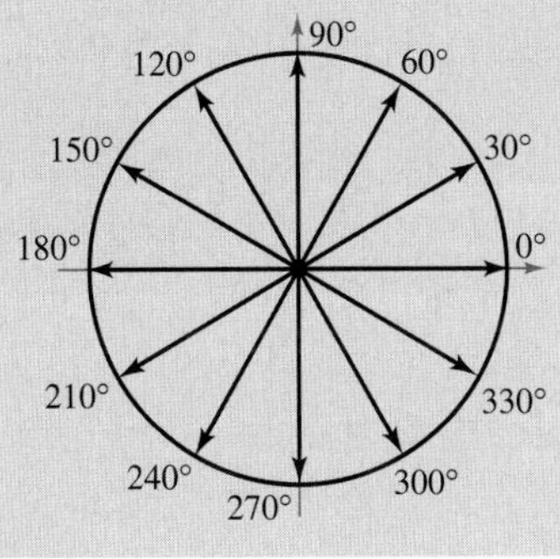

The Dot Product

So far you have studied two operations with vectors—vector addition and multiplication by a scalar—each of which yields another vector. In this section you will study a third vector operation, called the **dot product.** This product yields a scalar, rather than a vector.

Definition of Dot Product

The **dot product** of $\mathbf{u} = \langle u_1, u_2 \rangle$ and $\mathbf{v} = \langle v_1, v_2 \rangle$ is

$$\mathbf{u} \cdot \mathbf{v} = u_1v_1 + u_2v_2.$$

The **dot product** of $\mathbf{u} = \langle u_1, u_2, u_3 \rangle$ and $\mathbf{v} = \langle v_1, v_2, v_3 \rangle$ is

$$\mathbf{u} \cdot \mathbf{v} = u_1v_1 + u_2v_2 + u_3v_3.$$

NOTE Because the dot product of two vectors yields a scalar, it is also called the **inner product** (or **scalar product**) of the two vectors.

THEOREM 11.4 Properties of the Dot Product

Let $\mathbf{u}$, $\mathbf{v}$, and $\mathbf{w}$ be vectors in the plane or in space and let c be a scalar.

1. $\mathbf{u} \cdot \mathbf{v} = \mathbf{v} \cdot \mathbf{u}$ Commutative Property
2. $\mathbf{u} \cdot (\mathbf{v} + \mathbf{w}) = \mathbf{u} \cdot \mathbf{v} + \mathbf{u} \cdot \mathbf{w}$ Distributive Property
3. $c(\mathbf{u} \cdot \mathbf{v}) = c\mathbf{u} \cdot \mathbf{v} = \mathbf{u} \cdot c\mathbf{v}$
4. $\mathbf{0} \cdot \mathbf{v} = 0$
5. $\mathbf{v} \cdot \mathbf{v} = \|\mathbf{v}\|^2$

Proof To prove the first property, let $\mathbf{u} = \langle u_1, u_2, u_3 \rangle$ and $\mathbf{v} = \langle v_1, v_2, v_3 \rangle$. Then

$$\begin{aligned}\mathbf{u} \cdot \mathbf{v} &= u_1v_1 + u_2v_2 + u_3v_3 \\ &= v_1u_1 + v_2u_2 + v_3u_3 \\ &= \mathbf{v} \cdot \mathbf{u}.\end{aligned}$$

For the fifth property, let $\mathbf{v} = \langle v_1, v_2, v_3 \rangle$. Then

$$\begin{aligned}\mathbf{v} \cdot \mathbf{v} &= v_1^2 + v_2^2 + v_3^2 \\ &= \left(\sqrt{v_1^2 + v_2^2 + v_3^2}\right)^2 \\ &= \|\mathbf{v}\|^2.\end{aligned}$$

Proofs of the other properties are left to you.

EXAMPLE 1 Finding Dot Products

Given $\mathbf{u} = \langle 2, -2 \rangle$, $\mathbf{v} = \langle 5, 8 \rangle$, and $\mathbf{w} = \langle -4, 3 \rangle$, find each of the following.

a. $\mathbf{u} \cdot \mathbf{v}$ **b.** $(\mathbf{u} \cdot \mathbf{v})\mathbf{w}$

c. $\mathbf{u} \cdot (2\mathbf{v})$ **d.** $\|\mathbf{w}\|^2$

Solution

a. $\mathbf{u} \cdot \mathbf{v} = \langle 2, -2 \rangle \cdot \langle 5, 8 \rangle = 2(5) + (-2)(8) = -6$

b. $(\mathbf{u} \cdot \mathbf{v})\mathbf{w} = -6\langle -4, 3 \rangle = \langle 24, -18 \rangle$

c. $\mathbf{u} \cdot (2\mathbf{v}) = 2(\mathbf{u} \cdot \mathbf{v}) = 2(-6) = -12$ Theorem 11.4

d. $\|\mathbf{w}\|^2 = \mathbf{w} \cdot \mathbf{w}$ Theorem 11.4

$= \langle -4, 3 \rangle \cdot \langle -4, 3 \rangle$ Substitute $\langle -4, 3 \rangle$ for $\mathbf{w}$.

$= (-4)(-4) + (3)(3)$ Definition of dot product

$= 25$ Simplify.

Notice that the result of part (b) is a *vector* quantity, whereas the results of the other three parts are *scalar* quantities.

Angle Between Two Vectors

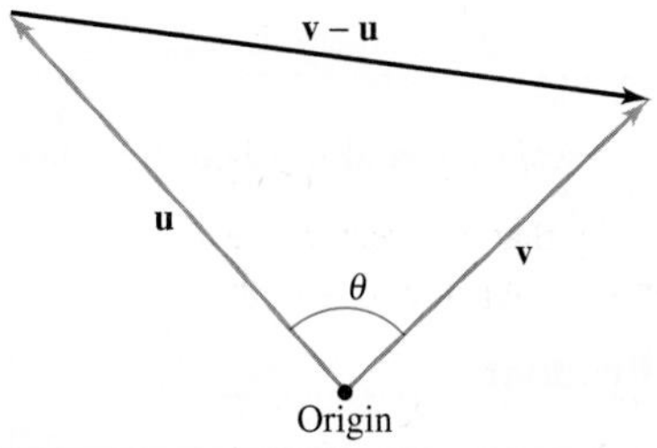

The angle between two vectors
Figure 11.24

The **angle between two nonzero vectors** is the angle θ, $0 \le \theta \le \pi$, between their respective standard position vectors, as shown in Figure 11.24. The next theorem shows how to find this angle using the dot product. (Note that the angle between the zero vector and another vector is not defined here.)

THEOREM 11.5 Angle Between Two Vectors

If θ is the angle between two nonzero vectors $\mathbf{u}$ and $\mathbf{v}$, then

$$\cos\theta = \frac{\mathbf{u} \cdot \mathbf{v}}{\|\mathbf{u}\|\,\|\mathbf{v}\|}.$$

Proof Consider the triangle determined by vectors $\mathbf{u}$, $\mathbf{v}$, and $\mathbf{v} - \mathbf{u}$, as shown in Figure 11.24. By the Law of Cosines, you can write

$$\|\mathbf{v} - \mathbf{u}\|^2 = \|\mathbf{u}\|^2 + \|\mathbf{v}\|^2 - 2\|\mathbf{u}\|\,\|\mathbf{v}\|\cos\theta.$$

Using the properties of the dot product, the left side can be rewritten as

$$\begin{aligned}
\|\mathbf{v} - \mathbf{u}\|^2 &= (\mathbf{v} - \mathbf{u}) \cdot (\mathbf{v} - \mathbf{u}) \\
&= (\mathbf{v} - \mathbf{u}) \cdot \mathbf{v} - (\mathbf{v} - \mathbf{u}) \cdot \mathbf{u} \\
&= \mathbf{v} \cdot \mathbf{v} - \mathbf{u} \cdot \mathbf{v} - \mathbf{v} \cdot \mathbf{u} + \mathbf{u} \cdot \mathbf{u} \\
&= \|\mathbf{v}\|^2 - 2\mathbf{u} \cdot \mathbf{v} + \|\mathbf{u}\|^2
\end{aligned}$$

and substitution back into the Law of Cosines yields

$$\begin{aligned}
\|\mathbf{v}\|^2 - 2\mathbf{u} \cdot \mathbf{v} + \|\mathbf{u}\|^2 &= \|\mathbf{u}\|^2 + \|\mathbf{v}\|^2 - 2\|\mathbf{u}\|\,\|\mathbf{v}\|\cos\theta \\
-2\mathbf{u} \cdot \mathbf{v} &= -2\|\mathbf{u}\|\,\|\mathbf{v}\|\cos\theta \\
\cos\theta &= \frac{\mathbf{u} \cdot \mathbf{v}}{\|\mathbf{u}\|\,\|\mathbf{v}\|}.
\end{aligned}$$

If the angle between two vectors is known, rewriting Theorem 11.5 in the form

$$\mathbf{u} \cdot \mathbf{v} = \|\mathbf{u}\| \, \|\mathbf{v}\| \cos \theta \qquad \text{Alternative form of dot product}$$

produces an alternative way to calculate the dot product. From this form, you can see that because $\|\mathbf{u}\|$ and $\|\mathbf{v}\|$ are always positive, $\mathbf{u} \cdot \mathbf{v}$ and $\cos \theta$ will always have the same sign. Figure 11.25 shows the possible orientations of two vectors.

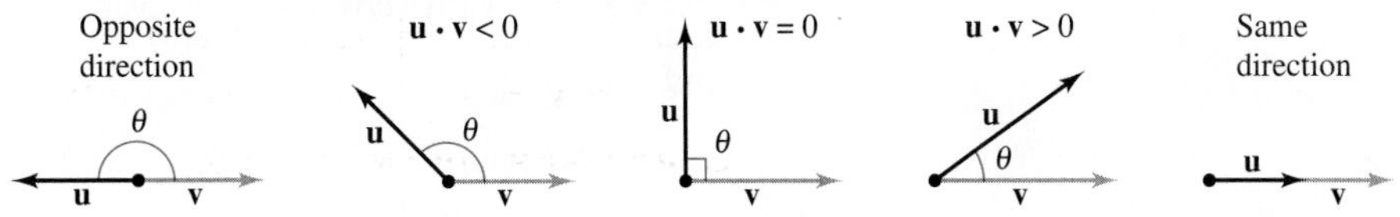

$\theta = \pi$, $\cos \theta = -1$ | $\pi/2 < \theta < \pi$, $-1 < \cos \theta < 0$ | $\theta = \pi/2$, $\cos \theta = 0$ | $0 < \theta < \pi/2$, $0 < \cos \theta < 1$ | $\theta = 0$, $\cos \theta = 1$

Figure 11.25

From Theorem 11.5, you can see that two nonzero vectors meet at a right angle if and only if their dot product is zero. Two such vectors are said to be **orthogonal.**

Definition of Orthogonal Vectors

The vectors $\mathbf{u}$ and $\mathbf{v}$ are orthogonal if $\mathbf{u} \cdot \mathbf{v} = 0$.

NOTE The terms "perpendicular," "orthogonal," and "normal" all mean essentially the same thing—meeting at right angles. However, it is common to say that two vectors are *orthogonal*, two lines or planes are *perpendicular*, and a vector is *normal* to a given line or plane.

From this definition, it follows that the zero vector is orthogonal to every vector $\mathbf{u}$, because $\mathbf{0} \cdot \mathbf{u} = 0$. Moreover, for $0 \le \theta \le \pi$, you know that $\cos \theta = 0$ if and only if $\theta = \pi/2$. So, you can use Theorem 11.5 to conclude that two *nonzero* vectors are orthogonal if and only if the angle between them is $\pi/2$.

EXAMPLE 2 Finding the Angle Between Two Vectors

For $\mathbf{u} = \langle 3, -1, 2 \rangle$, $\mathbf{v} = \langle -4, 0, 2 \rangle$, $\mathbf{w} = \langle 1, -1, -2 \rangle$, and $\mathbf{z} = \langle 2, 0, -1 \rangle$, find the angle between each pair of vectors.

a. $\mathbf{u}$ and $\mathbf{v}$ **b.** $\mathbf{u}$ and $\mathbf{w}$ **c.** $\mathbf{v}$ and $\mathbf{z}$

Solution

a. $\cos \theta = \dfrac{\mathbf{u} \cdot \mathbf{v}}{\|\mathbf{u}\| \, \|\mathbf{v}\|} = \dfrac{-12 + 0 + 4}{\sqrt{14}\sqrt{20}} = \dfrac{-8}{2\sqrt{14}\sqrt{5}} = \dfrac{-4}{\sqrt{70}}$

Because $\mathbf{u} \cdot \mathbf{v} < 0$, $\theta = \arccos \dfrac{-4}{\sqrt{70}} \approx 2.069$ radians.

b. $\cos \theta = \dfrac{\mathbf{u} \cdot \mathbf{w}}{\|\mathbf{u}\| \, \|\mathbf{w}\|} = \dfrac{3 + 1 - 4}{\sqrt{14}\sqrt{6}} = \dfrac{0}{\sqrt{84}} = 0$

Because $\mathbf{u} \cdot \mathbf{w} = 0$, $\mathbf{u}$ and $\mathbf{w}$ are *orthogonal*. So, $\theta = \pi/2$.

c. $\cos \theta = \dfrac{\mathbf{v} \cdot \mathbf{z}}{\|\mathbf{v}\| \, \|\mathbf{z}\|} = \dfrac{-8 + 0 - 2}{\sqrt{20}\sqrt{5}} = \dfrac{-10}{\sqrt{100}} = -1$

Consequently, $\theta = \pi$. Note that $\mathbf{v}$ and $\mathbf{z}$ are parallel, with $\mathbf{v} = -2\mathbf{z}$.

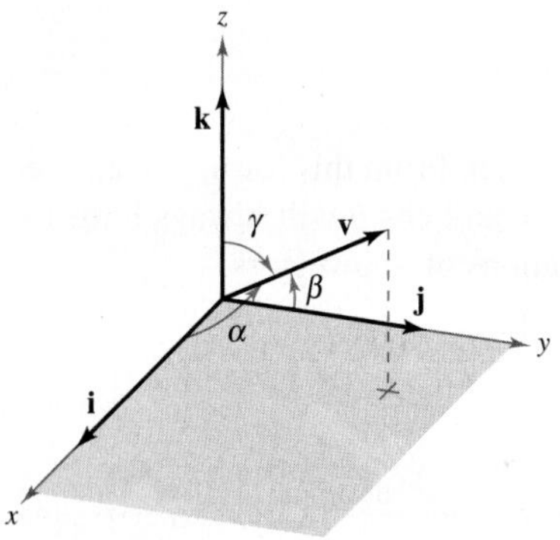

Direction angles
Figure 11.26

Direction Cosines

For a vector in the plane, you have seen that it is convenient to measure direction in terms of the angle, measured counterclockwise, *from* the positive x-axis *to* the vector. In space it is more convenient to measure direction in terms of the angles *between* the nonzero vector **v** and the three unit vectors **i**, **j**, and **k**, as shown in Figure 11.26. The angles α, β, and γ are the **direction angles of v,** and $\cos \alpha$, $\cos \beta$, and $\cos \gamma$ are the **direction cosines of v.** Because

$$\mathbf{v} \cdot \mathbf{i} = \|\mathbf{v}\| \|\mathbf{i}\| \cos \alpha = \|\mathbf{v}\| \cos \alpha$$

and

$$\mathbf{v} \cdot \mathbf{i} = \langle v_1, v_2, v_3 \rangle \cdot \langle 1, 0, 0 \rangle = v_1$$

it follows that $\cos \alpha = v_1/\|\mathbf{v}\|$. By similar reasoning with the unit vectors **j** and **k**, you have

$$\cos \alpha = \frac{v_1}{\|\mathbf{v}\|} \qquad \alpha \text{ is the angle between } \mathbf{v} \text{ and } \mathbf{i}.$$

$$\cos \beta = \frac{v_2}{\|\mathbf{v}\|} \qquad \beta \text{ is the angle between } \mathbf{v} \text{ and } \mathbf{j}.$$

$$\cos \gamma = \frac{v_3}{\|\mathbf{v}\|}. \qquad \gamma \text{ is the angle between } \mathbf{v} \text{ and } \mathbf{k}.$$

Consequently, any nonzero vector **v** in space has the normalized form

$$\frac{\mathbf{v}}{\|\mathbf{v}\|} = \frac{v_1}{\|\mathbf{v}\|}\mathbf{i} + \frac{v_2}{\|\mathbf{v}\|}\mathbf{j} + \frac{v_3}{\|\mathbf{v}\|}\mathbf{k} = \cos \alpha \mathbf{i} + \cos \beta \mathbf{j} + \cos \gamma \mathbf{k}$$

and because $\mathbf{v}/\|\mathbf{v}\|$ is a unit vector, it follows that

$$\cos^2 \alpha + \cos^2 \beta + \cos^2 \gamma = 1.$$

EXAMPLE 3 Finding Direction Angles

Find the direction cosines and angles for the vector $\mathbf{v} = 2\mathbf{i} + 3\mathbf{j} + 4\mathbf{k}$, and show that $\cos^2 \alpha + \cos^2 \beta + \cos^2 \gamma = 1$.

Solution Because $\|\mathbf{v}\| = \sqrt{2^2 + 3^2 + 4^2} = \sqrt{29}$, you can write the following.

$$\cos \alpha = \frac{v_1}{\|\mathbf{v}\|} = \frac{2}{\sqrt{29}} \Rightarrow \alpha \approx 68.2° \qquad \text{Angle between } \mathbf{v} \text{ and } \mathbf{i}$$

$$\cos \beta = \frac{v_2}{\|\mathbf{v}\|} = \frac{3}{\sqrt{29}} \Rightarrow \beta \approx 56.1° \qquad \text{Angle between } \mathbf{v} \text{ and } \mathbf{j}$$

$$\cos \gamma = \frac{v_3}{\|\mathbf{v}\|} = \frac{4}{\sqrt{29}} \Rightarrow \gamma \approx 42.0° \qquad \text{Angle between } \mathbf{v} \text{ and } \mathbf{k}$$

Furthermore, the sum of the squares of the direction cosines is

$$\begin{aligned}\cos^2 \alpha + \cos^2 \beta + \cos^2 \gamma &= \frac{4}{29} + \frac{9}{29} + \frac{16}{29} \\ &= \frac{29}{29} \\ &= 1.\end{aligned}$$

See Figure 11.27.

α = angle between **v** and **i**
β = angle between **v** and **j**
γ = angle between **v** and **k**

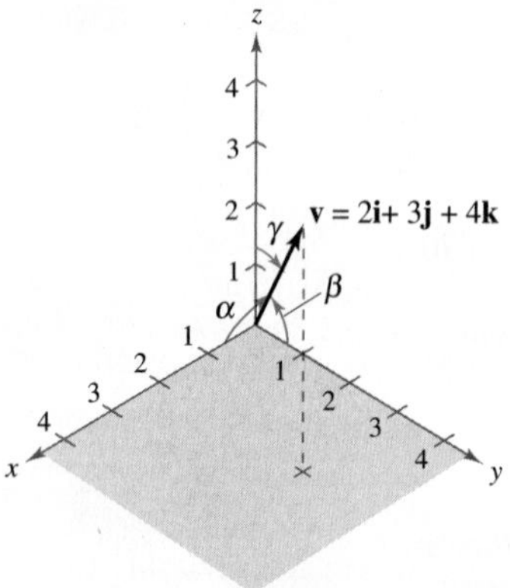

The direction angles of **v**
Figure 11.27

Projections and Vector Components

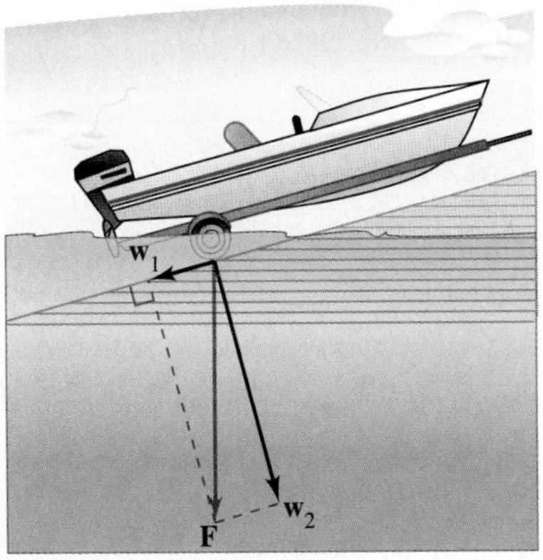

The force due to gravity pulls the boat against the ramp and down the ramp.
Figure 11.28

You have already seen applications in which two vectors are added to produce a resultant vector. Many applications in physics and engineering pose the reverse problem—decomposing a given vector into the sum of two **vector components.** The following physical example enables you to see the usefulness of this procedure.

Consider a boat on an inclined ramp, as shown in Figure 11.28. The force $\mathbf{F}$ due to gravity pulls the boat *down* the ramp and *against* the ramp. These two forces, $\mathbf{w}_1$ and $\mathbf{w}_2$, are orthogonal—they are called the vector components of $\mathbf{F}$.

$$\mathbf{F} = \mathbf{w}_1 + \mathbf{w}_2 \qquad \text{Vector components of } \mathbf{F}$$

The forces $\mathbf{w}_1$ and $\mathbf{w}_2$ help you analyze the effect of gravity on the boat. For example, $\mathbf{w}_1$ indicates the force necessary to keep the boat from rolling down the ramp, whereas $\mathbf{w}_2$ indicates the force that the tires must withstand.

Definition of Projection and Vector Components

Let $\mathbf{u}$ and $\mathbf{v}$ be nonzero vectors. Moreover, let $\mathbf{u} = \mathbf{w}_1 + \mathbf{w}_2$, where $\mathbf{w}_1$ is parallel to $\mathbf{v}$ and $\mathbf{w}_2$ is orthogonal to $\mathbf{v}$, as shown in Figure 11.29.

1. $\mathbf{w}_1$ is called the **projection of u onto v** or the **vector component of u along v,** and is denoted by $\mathbf{w}_1 = \text{proj}_{\mathbf{v}}\mathbf{u}$.
2. $\mathbf{w}_2 = \mathbf{u} - \mathbf{w}_1$ is called the **vector component of u orthogonal to v.**

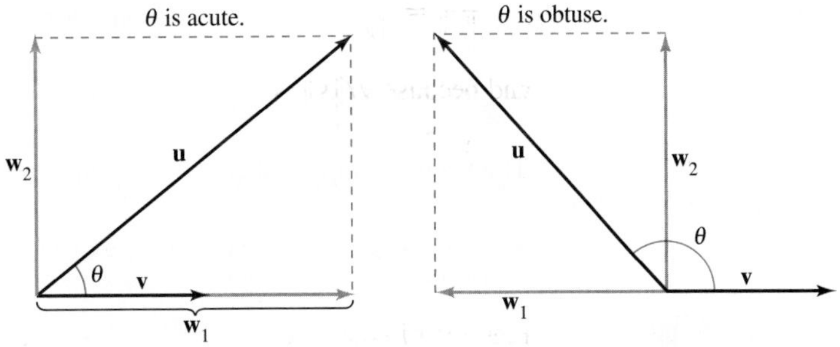

$\mathbf{w}_1 = \text{proj}_{\mathbf{v}}\mathbf{u}$ = projection of $\mathbf{u}$ onto $\mathbf{v}$ = vector component of $\mathbf{u}$ along $\mathbf{v}$
$\mathbf{w}_2$ = vector component of $\mathbf{u}$ orthogonal to $\mathbf{v}$
Figure 11.29

EXAMPLE 4 Finding a Vector Component of u Orthogonal to v

Find the vector component of $\mathbf{u} = \langle 7, 4\rangle$ that is orthogonal to $\mathbf{v} = \langle 2, 3\rangle$, given that $\mathbf{w}_1 = \text{proj}_{\mathbf{v}}\mathbf{u} = \langle 4, 6\rangle$ and

$$\mathbf{u} = \langle 7, 4\rangle = \mathbf{w}_1 + \mathbf{w}_2.$$

Solution Because $\mathbf{u} = \mathbf{w}_1 + \mathbf{w}_2$, where $\mathbf{w}_1$ is parallel to $\mathbf{v}$, it follows that $\mathbf{w}_2$ is the vector component of $\mathbf{u}$ orthogonal to $\mathbf{v}$. So, you have

$$\begin{aligned}\mathbf{w}_2 &= \mathbf{u} - \mathbf{w}_1\\ &= \langle 7, 4\rangle - \langle 4, 6\rangle\\ &= \langle 3, -2\rangle.\end{aligned}$$

Check to see that $\mathbf{w}_2$ is orthogonal to $\mathbf{v}$, as shown in Figure 11.30.

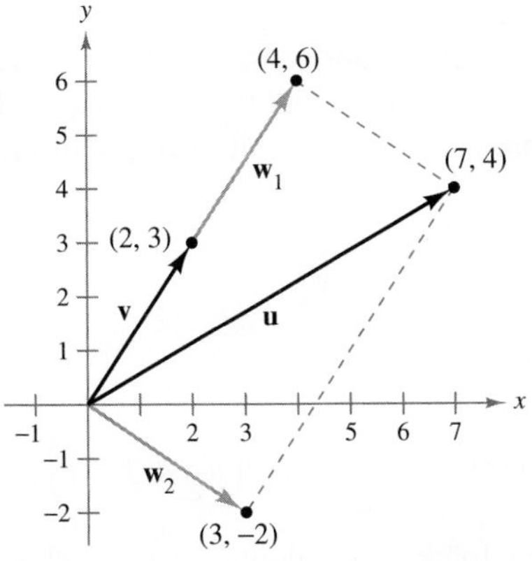

$\mathbf{u} = \mathbf{w}_1 + \mathbf{w}_2$
Figure 11.30

From Example 4, you can see that it is easy to find the vector component $\mathbf{w}_2$ once you have found the projection, $\mathbf{w}_1$, of $\mathbf{u}$ onto $\mathbf{v}$. To find this projection, use the dot product given in the theorem below, which you will prove in Exercise 90.

NOTE Note the distinction between the terms "component" and "vector component." For example, using the standard unit vectors with $\mathbf{u} = u_1\mathbf{i} + u_2\mathbf{j}$, u_1 is the *component* of $\mathbf{u}$ in the direction of $\mathbf{i}$ and $u_1\mathbf{i}$ is the *vector component* in the direction of $\mathbf{i}$.

THEOREM 11.6 Projection Using the Dot Product

If $\mathbf{u}$ and $\mathbf{v}$ are nonzero vectors, then the projection of $\mathbf{u}$ onto $\mathbf{v}$ is given by

$$\text{proj}_{\mathbf{v}}\mathbf{u} = \left(\frac{\mathbf{u} \cdot \mathbf{v}}{\|\mathbf{v}\|^2}\right)\mathbf{v}.$$

The projection of $\mathbf{u}$ onto $\mathbf{v}$ can be written as a scalar multiple of a unit vector in the direction of $\mathbf{v}$. That is,

$$\left(\frac{\mathbf{u} \cdot \mathbf{v}}{\|\mathbf{v}\|^2}\right)\mathbf{v} = \left(\frac{\mathbf{u} \cdot \mathbf{v}}{\|\mathbf{v}\|}\right)\frac{\mathbf{v}}{\|\mathbf{v}\|} = (k)\frac{\mathbf{v}}{\|\mathbf{v}\|} \quad \Longrightarrow \quad k = \frac{\mathbf{u} \cdot \mathbf{v}}{\|\mathbf{v}\|} = \|\mathbf{u}\| \cos \theta.$$

The scalar k is called the **component of u in the direction of v.**

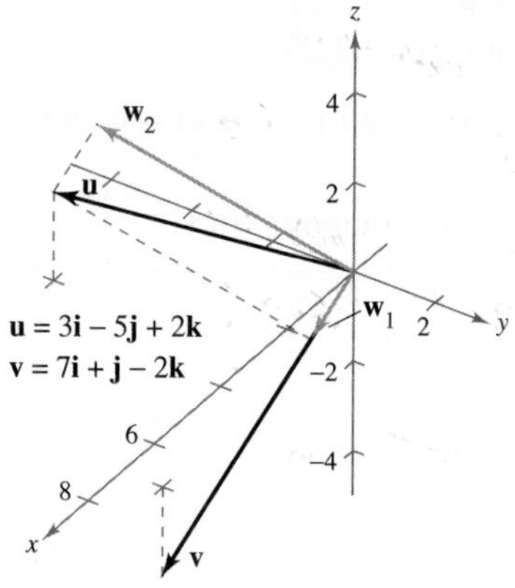

$\mathbf{u} = \mathbf{w}_1 + \mathbf{w}_2$

Figure 11.31

EXAMPLE 5 Decomposing a Vector into Vector Components

Find the projection of $\mathbf{u}$ onto $\mathbf{v}$ and the vector component of $\mathbf{u}$ orthogonal to $\mathbf{v}$ for the vectors $\mathbf{u} = 3\mathbf{i} - 5\mathbf{j} + 2\mathbf{k}$ and $\mathbf{v} = 7\mathbf{i} + \mathbf{j} - 2\mathbf{k}$ shown in Figure 11.31.

Solution The projection of $\mathbf{u}$ onto $\mathbf{v}$ is

$$\mathbf{w}_1 = \left(\frac{\mathbf{u} \cdot \mathbf{v}}{\|\mathbf{v}\|^2}\right)\mathbf{v} = \left(\frac{12}{54}\right)(7\mathbf{i} + \mathbf{j} - 2\mathbf{k}) = \frac{14}{9}\mathbf{i} + \frac{2}{9}\mathbf{j} - \frac{4}{9}\mathbf{k}.$$

The vector component of $\mathbf{u}$ orthogonal to $\mathbf{v}$ is the vector

$$\mathbf{w}_2 = \mathbf{u} - \mathbf{w}_1 = (3\mathbf{i} - 5\mathbf{j} + 2\mathbf{k}) - \left(\frac{14}{9}\mathbf{i} + \frac{2}{9}\mathbf{j} - \frac{4}{9}\mathbf{k}\right) = \frac{13}{9}\mathbf{i} - \frac{47}{9}\mathbf{j} + \frac{22}{9}\mathbf{k}.$$

EXAMPLE 6 Finding a Force

A 600-pound boat sits on a ramp inclined at 30°, as shown in Figure 11.32. What force is required to keep the boat from rolling down the ramp?

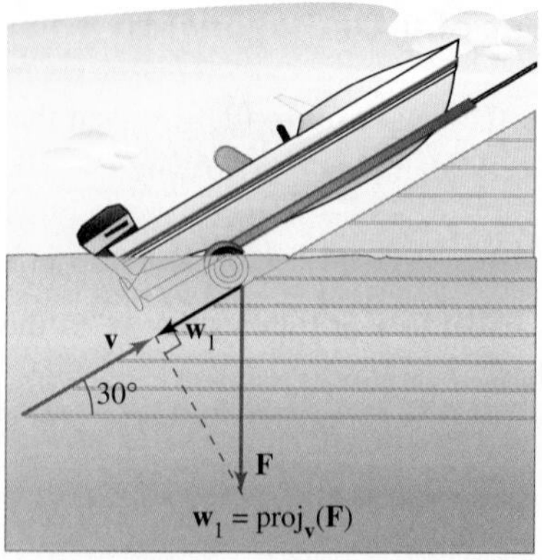

Figure 11.32

Solution Because the force due to gravity is vertical and downward, you can represent the gravitational force by the vector $\mathbf{F} = -600\mathbf{j}$. To find the force required to keep the boat from rolling down the ramp, project $\mathbf{F}$ onto a unit vector $\mathbf{v}$ in the direction of the ramp, as follows.

$$\mathbf{v} = \cos 30°\mathbf{i} + \sin 30°\mathbf{j} = \frac{\sqrt{3}}{2}\mathbf{i} + \frac{1}{2}\mathbf{j} \qquad \text{Unit vector along ramp}$$

Therefore, the projection of $\mathbf{F}$ onto $\mathbf{v}$ is given by

$$\mathbf{w}_1 = \text{proj}_{\mathbf{v}}\mathbf{F} = \left(\frac{\mathbf{F} \cdot \mathbf{v}}{\|\mathbf{v}\|^2}\right)\mathbf{v} = (\mathbf{F} \cdot \mathbf{v})\mathbf{v} = (-600)\left(\frac{1}{2}\right)\mathbf{v} = -300\left(\frac{\sqrt{3}}{2}\mathbf{i} + \frac{1}{2}\mathbf{j}\right).$$

The magnitude of this force is 300, and therefore a force of 300 pounds is required to keep the boat from rolling down the ramp.

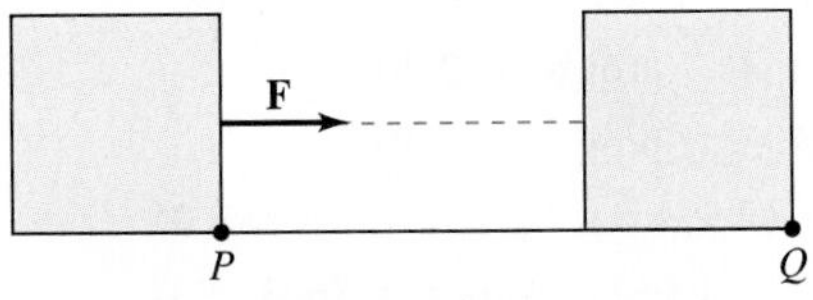

(a) Force acts along the line of motion.

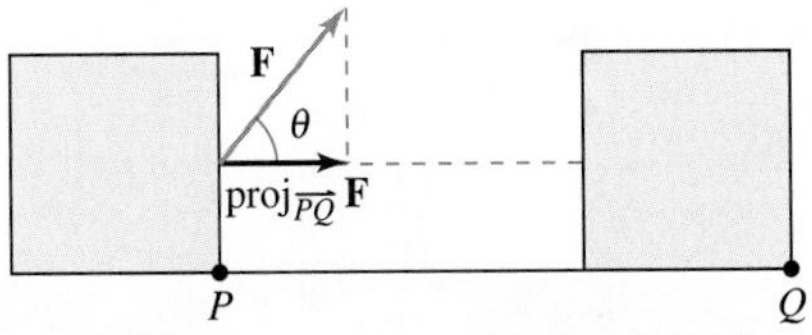

(b) Force acts at angle θ with the line of motion.

Figure 11.33

Work

The work W done by the constant force **F** acting along the line of motion of an object is given by

$$W = (\text{magnitude of force})(\text{distance}) = \|\mathbf{F}\|\,\|\overrightarrow{PQ}\|$$

as shown in Figure 11.33(a). If the constant force **F** is not directed along the line of motion, you can see from Figure 11.33(b) that the work W done by the force is

$$W = \|\text{proj}_{\overrightarrow{PQ}}\mathbf{F}\|\,\|\overrightarrow{PQ}\| = (\cos\theta)\|\mathbf{F}\|\,\|\overrightarrow{PQ}\| = \mathbf{F}\cdot\overrightarrow{PQ}.$$

This notion of work is summarized in the following definition.

Definition of Work

The work W done by a constant force **F** as its point of application moves along the vector $\overrightarrow{PQ}$ is given by either of the following.

1. $W = \|\text{proj}_{\overrightarrow{PQ}}\mathbf{F}\|\,\|\overrightarrow{PQ}\|$ Projection form
2. $W = \mathbf{F}\cdot\overrightarrow{PQ}$ Dot product form

EXAMPLE 7 Finding Work

Figure 11.34

To close a sliding door, a person pulls on a rope with a constant force of 50 pounds at a constant angle of 60°, as shown in Figure 11.34. Find the work done in moving the door 12 feet to its closed position.

Solution Using a projection, you can calculate the work as follows.

$$\begin{aligned} W &= \|\text{proj}_{\overrightarrow{PQ}}\mathbf{F}\|\,\|\overrightarrow{PQ}\| && \text{Projection form for work} \\ &= \cos(60^\circ)\,\|\mathbf{F}\|\,\|\overrightarrow{PQ}\| \\ &= \frac{1}{2}(50)(12) \\ &= 300 \text{ foot-pounds} \end{aligned}$$

Exercises for Section 11.3

See www.CalcChat.com for worked-out solutions to odd-numbered exercises.

In Exercises 1–8, find (a) $\mathbf{u}\cdot\mathbf{v}$, (b) $\mathbf{u}\cdot\mathbf{u}$, (c) $\|\mathbf{u}\|^2$, (d) $(\mathbf{u}\cdot\mathbf{v})\mathbf{v}$, and (e) $\mathbf{u}\cdot(2\mathbf{v})$.

1. $\mathbf{u} = \langle 3, 4\rangle$, $\mathbf{v} = \langle 2, -3\rangle$
2. $\mathbf{u} = \langle 4, 10\rangle$, $\mathbf{v} = \langle -2, 3\rangle$
3. $\mathbf{u} = \langle 5, -1\rangle$, $\mathbf{v} = \langle -3, 2\rangle$
4. $\mathbf{u} = \langle -4, 8\rangle$, $\mathbf{v} = \langle 6, 3\rangle$
5. $\mathbf{u} = \langle 2, -3, 4\rangle$, $\mathbf{v} = \langle 0, 6, 5\rangle$
6. $\mathbf{u} = \mathbf{i}$, $\mathbf{v} = \mathbf{i}$
7. $\mathbf{u} = 2\mathbf{i} - \mathbf{j} + \mathbf{k}$, $\mathbf{v} = \mathbf{i} - \mathbf{k}$
8. $\mathbf{u} = 2\mathbf{i} + \mathbf{j} - 2\mathbf{k}$, $\mathbf{v} = \mathbf{i} - 3\mathbf{j} + 2\mathbf{k}$

In Exercises 9 and 10, find $\mathbf{u}\cdot\mathbf{v}$.

9. $\|\mathbf{u}\| = 8$, $\|\mathbf{v}\| = 5$, and the angle between **u** and **v** is $\pi/3$.
10. $\|\mathbf{u}\| = 40$, $\|\mathbf{v}\| = 25$, and the angle between **u** and **v** is $5\pi/6$.

In Exercises 11–18, find the angle θ between the vectors.

11. $\mathbf{u} = \langle 1, 1\rangle$, $\mathbf{v} = \langle 2, -2\rangle$
12. $\mathbf{u} = \langle 3, 1\rangle$, $\mathbf{v} = \langle 2, -1\rangle$
13. $\mathbf{u} = 3\mathbf{i} + \mathbf{j}$, $\mathbf{v} = -2\mathbf{i} + 4\mathbf{j}$
14. $\mathbf{u} = \cos\left(\frac{\pi}{6}\right)\mathbf{i} + \sin\left(\frac{\pi}{6}\right)\mathbf{j}$, $\mathbf{v} = \cos\left(\frac{3\pi}{4}\right)\mathbf{i} + \sin\left(\frac{3\pi}{4}\right)\mathbf{j}$
15. $\mathbf{u} = \langle 1, 1, 1\rangle$, $\mathbf{v} = \langle 2, 1, -1\rangle$
16. $\mathbf{u} = 3\mathbf{i} + 2\mathbf{j} + \mathbf{k}$, $\mathbf{v} = 2\mathbf{i} - 3\mathbf{j}$
17. $\mathbf{u} = 3\mathbf{i} + 4\mathbf{j}$, $\mathbf{v} = -2\mathbf{j} + 3\mathbf{k}$
18. $\mathbf{u} = 2\mathbf{i} - 3\mathbf{j} + \mathbf{k}$, $\mathbf{v} = \mathbf{i} - 2\mathbf{j} + \mathbf{k}$

In Exercises 19–26, determine whether u and v are orthogonal, parallel, or neither.

19. $\mathbf{u} = \langle 4, 0\rangle$, $\mathbf{v} = \langle 1, 1\rangle$
20. $\mathbf{u} = \langle 2, 18\rangle$, $\mathbf{v} = \left\langle \frac{3}{2}, -\frac{1}{6}\right\rangle$

21. $\mathbf{u} = \langle 4, 3 \rangle$
$\mathbf{v} = \left\langle \frac{1}{2}, -\frac{2}{3} \right\rangle$

22. $\mathbf{u} = -\frac{1}{3}(\mathbf{i} - 2\mathbf{j})$
$\mathbf{v} = 2\mathbf{i} - 4\mathbf{j}$

23. $\mathbf{u} = \mathbf{j} + 6\mathbf{k}$
$\mathbf{v} = \mathbf{i} - 2\mathbf{j} - \mathbf{k}$

24. $\mathbf{u} = -2\mathbf{i} + 3\mathbf{j} - \mathbf{k}$
$\mathbf{v} = 2\mathbf{i} + \mathbf{j} - \mathbf{k}$

25. $\mathbf{u} = \langle 2, -3, 1 \rangle$
$\mathbf{v} = \langle -1, -1, -1 \rangle$

26. $\mathbf{u} = \langle \cos\theta, \sin\theta, -1 \rangle$
$\mathbf{v} = \langle \sin\theta, -\cos\theta, 0 \rangle$

In Exercises 27–30, the vertices of a triangle are given. Determine whether the triangle is an acute triangle, an obtuse triangle, or a right triangle. Explain your reasoning.

27. $(1, 2\ 0), (0, 0, 0), (-2, 1, 0)$

28. $(-3, 0, 0), (0, 0, 0), (1, 2, 3)$

29. $(2, -3, 4), (0, 1, 2), (-1, 2, 0)$

30. $(2, -7, 3), (-1, 5, 8), (4, 6, -1)$

In Exercises 31–34, find the direction cosines of u and demonstrate that the sum of the squares of the direction cosines is 1.

31. $\mathbf{u} = \mathbf{i} + 2\mathbf{j} + 2\mathbf{k}$

32. $\mathbf{u} = 5\mathbf{i} + 3\mathbf{j} - \mathbf{k}$

33. $\mathbf{u} = \langle 0, 6, -4 \rangle$

34. $\mathbf{u} = \langle a, b, c \rangle$

In Exercises 35–38, find the direction angles of the vector.

35. $\mathbf{u} = 3\mathbf{i} + 2\mathbf{j} - 2\mathbf{k}$

36. $\mathbf{u} = -4\mathbf{i} + 3\mathbf{j} + 5\mathbf{k}$

37. $\mathbf{u} = \langle -1, 5, 2 \rangle$

38. $\mathbf{u} = \langle -2, 6, 1 \rangle$

In Exercises 39 and 40, use a graphing utility to find the magnitude and direction angles of the resultant of forces $\mathbf{F}_1$ and $\mathbf{F}_2$ with initial points at the origin. The magnitude and terminal point of each vector are given.

	Vector	*Magnitude*	*Terminal Point*
39.	$\mathbf{F}_1$	50 lb	$(10, 5, 3)$
	$\mathbf{F}_2$	80 lb	$(12, 7, -5)$
40.	$\mathbf{F}_1$	300 N	$(-20, -10, 5)$
	$\mathbf{F}_2$	100 N	$(5, 15, 0)$

41. ***Load-Supporting Cables*** A load is supported by three cables, as shown in the figure. Find the direction angles of the load-supporting cable OA.

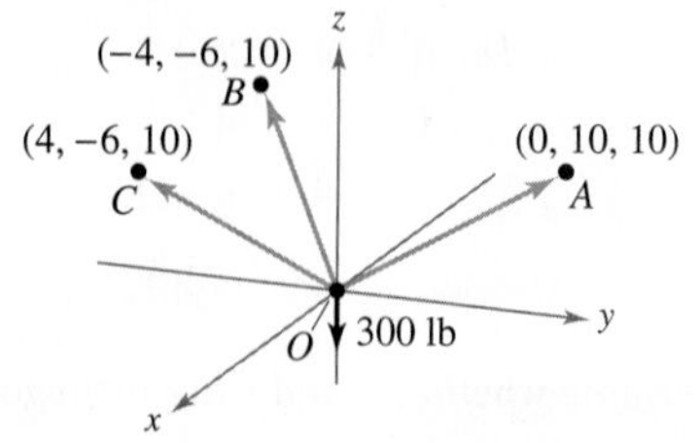

42. ***Load-Supporting Cables*** The tension in the cable OA in Exercise 41 is 200 newtons. Determine the weight of the load.

In Exercises 43–46, find the component of u that is orthogonal to v, given $\mathbf{w}_1 = \text{proj}_{\mathbf{v}}\mathbf{u}$.

43. $\mathbf{u} = \langle 6, 7 \rangle,\ \mathbf{v} = \langle 1, 4 \rangle,\ \text{proj}_{\mathbf{v}}\mathbf{u} = \langle 2, 8 \rangle$

44. $\mathbf{u} = \langle 9, 7 \rangle,\ \mathbf{v} = \langle 1, 3 \rangle,\ \text{proj}_{\mathbf{v}}\mathbf{u} = \langle 3, 9 \rangle$

45. $\mathbf{u} = \langle 0, 3, 3 \rangle,\ \mathbf{v} = \langle -1, 1, 1 \rangle,\ \text{proj}_{\mathbf{v}}\mathbf{u} = \langle -2, 2, 2 \rangle$

46. $\mathbf{u} = \langle 8, 2, 0 \rangle,\ \mathbf{v} = \langle 2, 1, -1 \rangle,\ \text{proj}_{\mathbf{v}}\mathbf{u} = \langle 6, 3, -3 \rangle$

In Exercises 47–50, (a) find the projection of u onto v, and (b) find the vector component of u orthogonal to v.

47. $\mathbf{u} = \langle 2, 3 \rangle,\ \mathbf{v} = \langle 5, 1 \rangle$

48. $\mathbf{u} = \langle 2, -3 \rangle,\ \mathbf{v} = \langle 3, 2 \rangle$

49. $\mathbf{u} = \langle 2, 1, 2 \rangle$
$\mathbf{v} = \langle 0, 3, 4 \rangle$

50. $\mathbf{u} = \langle 1, 0, 4 \rangle$
$\mathbf{v} = \langle 3, 0, 2 \rangle$

Writing About Concepts

51. Define the dot product of vectors **u** and **v**.

52. State the definition of orthogonal vectors. If vectors are neither parallel nor orthogonal, how do you find the angle between them? Explain.

53. What is known about θ, the angle between two nonzero vectors **u** and **v**, if

(a) $\mathbf{u} \cdot \mathbf{v} = 0$? (b) $\mathbf{u} \cdot \mathbf{v} > 0$? (c) $\mathbf{u} \cdot \mathbf{v} < 0$?

54. Determine which of the following are defined for nonzero vectors **u**, **v**, and **w**. Explain your reasoning.

(a) $\mathbf{u} \cdot (\mathbf{v} + \mathbf{w})$ (b) $(\mathbf{u} \cdot \mathbf{v})\mathbf{w}$

(c) $\mathbf{u} \cdot \mathbf{v} + \mathbf{w}$ (d) $\|\mathbf{u}\| \cdot (\mathbf{v} + \mathbf{w})$

55. Describe direction cosines and direction angles of a vector **v**.

56. Give a geometric description of the projection of **u** onto **v**.

57. What can be said about the vectors **u** and **v** if (a) the projection of **u** onto **v** equals **u** and (b) the projection of **u** onto **v** equals **0**?

58. If the projection of **u** onto **v** has the same magnitude as the projection of **v** onto **u**, can you conclude that $\|\mathbf{u}\| = \|\mathbf{v}\|$? Explain.

59. ***Revenue*** The vector $\mathbf{u} = \langle 3240, 1450, 2235 \rangle$ gives the numbers of hamburgers, chicken sandwiches, and cheeseburgers, respectively, sold at a fast-food restaurant in one week. The vector $\mathbf{v} = \langle 1.35, 2.65, 1.85 \rangle$ gives the prices (in dollars) per unit for the three food items. Find the dot product $\mathbf{u} \cdot \mathbf{v}$, and explain what information it gives.

60. ***Revenue*** Repeat Exercise 59 after increasing prices by 4%. Identify the vector operation used to increase prices by 4%.

61. ***Programming*** Given vectors **u** and **v** in component form, write a program for a graphing utility in which the output is (a) $\|\mathbf{u}\|$, (b) $\|\mathbf{v}\|$, and (c) the angle between **u** and **v**.

62. ***Programming*** Use the program you wrote in Exercise 61 to find the angle between the vectors $\mathbf{u} = \langle 8, -4, 2 \rangle$ and $\mathbf{v} = \langle 2, 5, 2 \rangle$.

63. ***Programming*** Given vectors **u** and **v** in component form, write a program for a graphing utility in which the output is the component form of the projection of **u** onto **v**.

64. ***Programming*** Use the program you wrote in Exercise 63 to find the projection of **u** onto **v** for $\mathbf{u} = \langle 5, 6, 2\rangle$ and $\mathbf{v} = \langle -1, 3, 4\rangle$.

Think About It **In Exercises 65 and 66, use the figure to determine mentally the projection of u onto v. (The coordinates of the terminal points of the vectors in standard position are given.) Verify your results analytically.**

65.

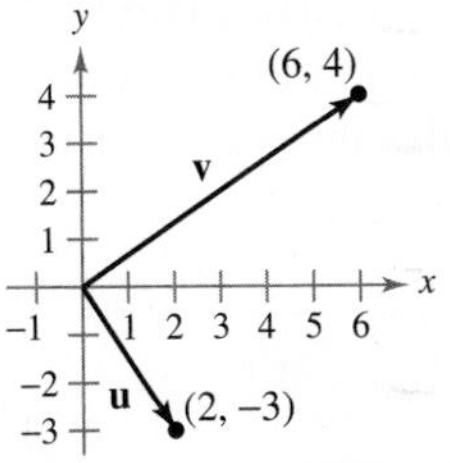

66.

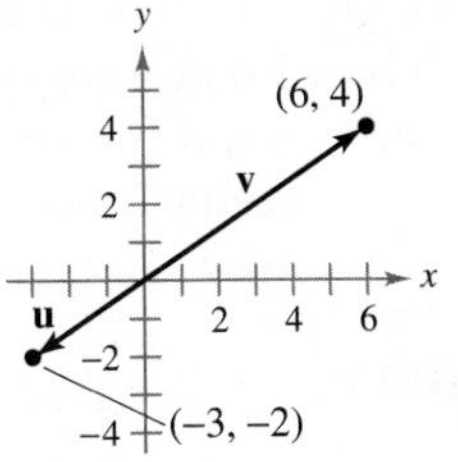

In Exercises 67–70, find two vectors in opposite directions that are orthogonal to the vector u. (The answers are not unique.)

67. $\mathbf{u} = \frac{1}{2}\mathbf{i} - \frac{2}{3}\mathbf{j}$

68. $\mathbf{u} = -8\mathbf{i} + 3\mathbf{j}$

69. $\mathbf{u} = \langle 3, 1, -2\rangle$

70. $\mathbf{u} = \langle 0, -3, 6\rangle$

71. ***Braking Load*** A 48,000-pound truck is parked on a 10° slope (see figure). Assume the only force to overcome is that due to gravity. Find (a) the force required to keep the truck from rolling down the hill and (b) the force perpendicular to the hill.

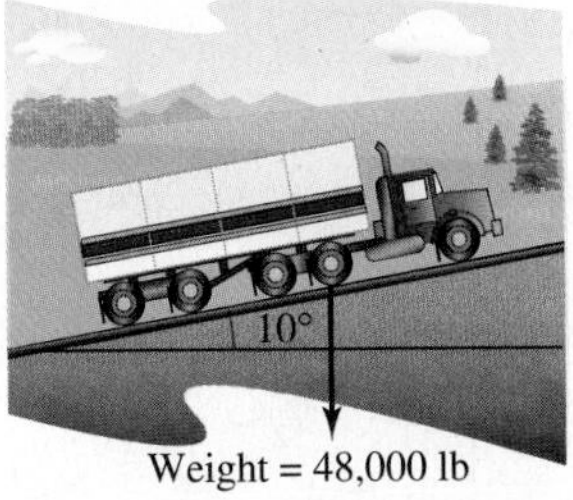

Figure for 71

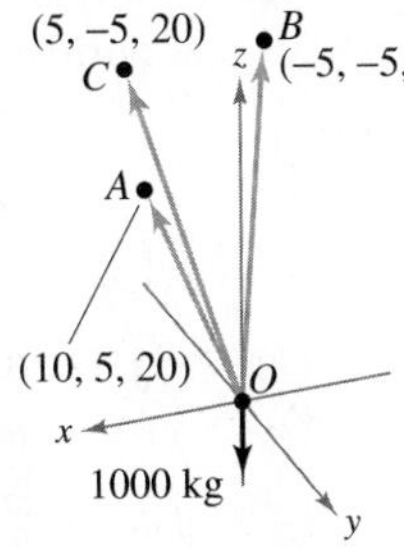

Figure for 72

72. ***Load-Supporting Cables*** Find the magnitude of the projection of the load-supporting cable *OA* onto the positive *z*-axis as shown in the figure.

73. ***Work*** An object is pulled 10 feet across a floor, using a force of 85 pounds. The direction of the force is 60° above the horizontal (see figure). Find the work done.

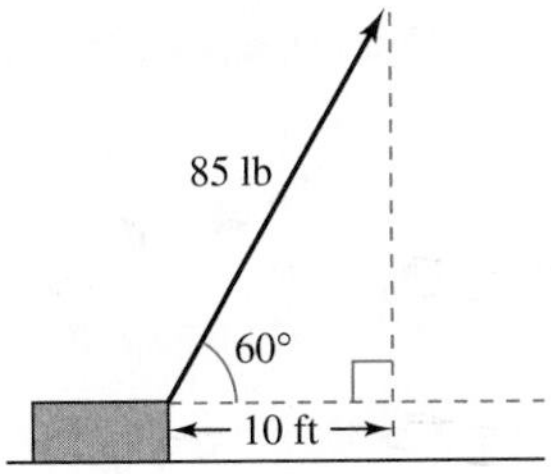

Not drawn to scale

Figure for 73

Figure for 74

74. ***Work*** A toy wagon is pulled by exerting a force of 25 pounds on a handle that makes a 20° angle with the horizontal (see figure in left column). Find the work done in pulling the wagon 50 feet.

True or False? **In Exercises 75 and 76, determine whether the statement is true or false. If it is false, explain why or give an example that shows it is false.**

75. If $\mathbf{u} \cdot \mathbf{v} = \mathbf{u} \cdot \mathbf{w}$ and $\mathbf{u} \neq \mathbf{0}$, then $\mathbf{v} = \mathbf{w}$.

76. If **u** and **v** are orthogonal to **w**, then $\mathbf{u} + \mathbf{v}$ is orthogonal to **w**.

77. Find the angle between a cube's diagonal and one of its edges.

78. Find the angle between the diagonal of a cube and the diagonal of one of its sides.

In Exercises 79–82, (a) find the unit tangent vectors to each curve at their points of intersection and (b) find the angles ($0 \le \theta \le 90°$) between the curves at their points of intersection.

79. $y = x^2, \quad y = x^{1/3}$

80. $y = x^3, \quad y = x^{1/3}$

81. $y = 1 - x^2, \quad y = x^2 - 1$

82. $(y + 1)^2 = x, \quad y = x^3 - 1$

83. Use vectors to prove that the diagonals of a rhombus are perpendicular.

84. Use vectors to prove that a parallelogram is a rectangle if and only if its diagonals are equal in length.

85. ***Bond Angle*** Consider a regular tetrahedron with vertices $(0, 0, 0)$, $(k, k, 0)$, $(k, 0, k)$, and $(0, k, k)$, where k is a positive real number.

(a) Sketch the graph of the tetrahedron.

(b) Find the length of each edge.

(c) Find the angle between any two edges.

(d) Find the angle between the line segments from the centroid $(k/2, k/2, k/2)$ to two vertices. This is the bond angle for a molecule such as CH_4 or $PbCl_4$, where the structure of the molecule is a tetrahedron.

86. Consider the vectors

$$\mathbf{u} = \langle \cos\alpha, \sin\alpha, 0\rangle$$

and

$$\mathbf{v} = \langle \cos\beta, \sin\beta, 0\rangle$$

where $\alpha > \beta$. Find the dot product of the vectors and use the result to prove the identity

$$\cos(\alpha - \beta) = \cos\alpha\cos\beta + \sin\alpha\sin\beta.$$

87. Prove that $\|\mathbf{u} - \mathbf{v}\|^2 = \|\mathbf{u}\|^2 + \|\mathbf{v}\|^2 - 2\mathbf{u} \cdot \mathbf{v}$.

88. Prove the **Cauchy-Schwarz Inequality** $|\mathbf{u} \cdot \mathbf{v}| \le \|\mathbf{u}\|\,\|\mathbf{v}\|$.

89. Prove the triangle inequality $\|\mathbf{u} + \mathbf{v}\| \le \|\mathbf{u}\| + \|\mathbf{v}\|$.

90. Prove Theorem 11.6.

Section 11.4 The Cross Product of Two Vectors in Space

- **Find the cross product of two vectors in space.**
- **Use the triple scalar product of three vectors in space.**

EXPLORATION

Geometric Property of the Cross Product Three pairs of vectors are shown below. Use the definition to find the cross product of each pair. Sketch all three vectors in a three-dimensional system. Describe any relationships among the three vectors. Use your description to write a conjecture about **u**, **v**, and **u** × **v**.

a. $\mathbf{u} = \langle 3, 0, 3 \rangle$, $\mathbf{v} = \langle 3, 0, -3 \rangle$

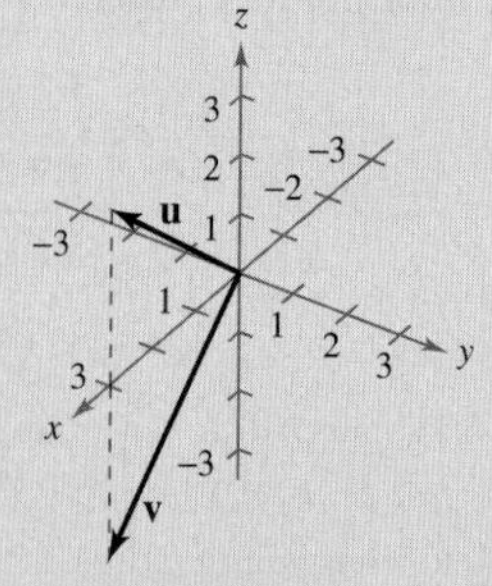

b. $\mathbf{u} = \langle 0, 3, 3 \rangle$, $\mathbf{v} = \langle 0, -3, 3 \rangle$

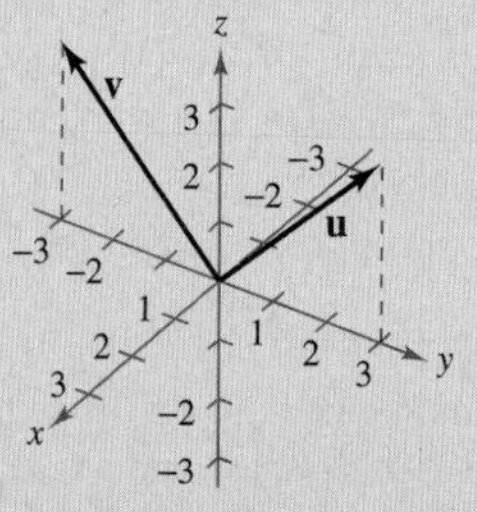

c. $\mathbf{u} = \langle 3, 3, 0 \rangle$, $\mathbf{v} = \langle 3, -3, 0 \rangle$

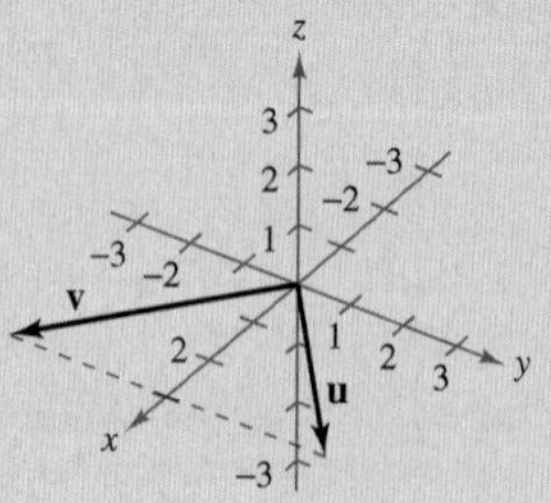

The Cross Product

Many applications in physics, engineering, and geometry involve finding a vector in space that is orthogonal to two given vectors. In this section you will study a product that will yield such a vector. It is called the **cross product,** and it is most conveniently defined and calculated using the standard unit vector form. Because the cross product yields a vector, it is also called the **vector product.**

Definition of Cross Product of Two Vectors in Space

Let $\mathbf{u} = u_1\mathbf{i} + u_2\mathbf{j} + u_3\mathbf{k}$ and $\mathbf{v} = v_1\mathbf{i} + v_2\mathbf{j} + v_3\mathbf{k}$ be vectors in space. The **cross product** of **u** and **v** is the vector

$$\mathbf{u} \times \mathbf{v} = (u_2v_3 - u_3v_2)\mathbf{i} - (u_1v_3 - u_3v_1)\mathbf{j} + (u_1v_2 - u_2v_1)\mathbf{k}.$$

NOTE Be sure you see that this definition applies only to three-dimensional vectors. The cross product is not defined for two-dimensional vectors.

A convenient way to calculate $\mathbf{u} \times \mathbf{v}$ is to use the following *determinant form* with cofactor expansion. (This 3 × 3 determinant form is used simply to help remember the formula for the cross product—it is technically not a determinant because the entries of the corresponding matrix are not all real numbers.)

$$\mathbf{u} \times \mathbf{v} = \begin{vmatrix} \mathbf{i} & \mathbf{j} & \mathbf{k} \\ u_1 & u_2 & u_3 \\ v_1 & v_2 & v_3 \end{vmatrix} \begin{array}{l} \\ \longleftarrow \text{Put "u" in Row 2.} \\ \longleftarrow \text{Put "v" in Row 3.} \end{array}$$

$$= \begin{vmatrix} \mathbf{i} & \mathbf{j} & \mathbf{k} \\ u_1 & u_2 & u_3 \\ v_1 & v_2 & v_3 \end{vmatrix}\mathbf{i} - \begin{vmatrix} \mathbf{i} & \mathbf{j} & \mathbf{k} \\ u_1 & u_2 & u_3 \\ v_1 & v_2 & v_3 \end{vmatrix}\mathbf{j} + \begin{vmatrix} \mathbf{i} & \mathbf{j} & \mathbf{k} \\ u_1 & u_2 & u_3 \\ v_1 & v_2 & v_3 \end{vmatrix}\mathbf{k}$$

$$= \begin{vmatrix} u_2 & u_3 \\ v_2 & v_3 \end{vmatrix}\mathbf{i} - \begin{vmatrix} u_1 & u_3 \\ v_1 & v_3 \end{vmatrix}\mathbf{j} + \begin{vmatrix} u_1 & u_2 \\ v_1 & v_2 \end{vmatrix}\mathbf{k}$$

$$= (u_2v_3 - u_3v_2)\mathbf{i} - (u_1v_3 - u_3v_1)\mathbf{j} + (u_1v_2 - u_2v_1)\mathbf{k}$$

Note the minus sign in front of the **j**-component. Each of the three 2 × 2 determinants can be evaluated by using the following diagonal pattern.

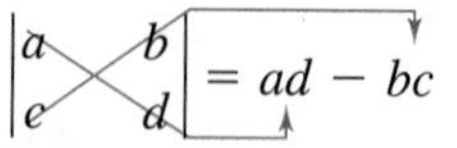

Here are a couple of examples.

$$\begin{vmatrix} 2 & 4 \\ 3 & -1 \end{vmatrix} = (2)(-1) - (4)(3) = -2 - 12 = -14$$

$$\begin{vmatrix} 4 & 0 \\ -6 & 3 \end{vmatrix} = (4)(3) - (0)(-6) = 12$$

Notation for Dot and Cross Products

The notation for the dot product and cross product of vectors was first introduced by the American physicist Josiah Willard Gibbs (1839–1903). In the early 1880s, Gibbs built a system to represent physical quantities called "vector analysis." The system was a departure from Hamilton's theory of quaternions.

EXAMPLE 1 Finding the Cross Product

Given $\mathbf{u} = \mathbf{i} - 2\mathbf{j} + \mathbf{k}$ and $\mathbf{v} = 3\mathbf{i} + \mathbf{j} - 2\mathbf{k}$, find each of the following.

a. $\mathbf{u} \times \mathbf{v}$ **b.** $\mathbf{v} \times \mathbf{u}$ **c.** $\mathbf{v} \times \mathbf{v}$

Solution

$$\textbf{a. } \mathbf{u} \times \mathbf{v} = \begin{vmatrix} \mathbf{i} & \mathbf{j} & \mathbf{k} \\ 1 & -2 & 1 \\ 3 & 1 & -2 \end{vmatrix} = \begin{vmatrix} -2 & 1 \\ 1 & -2 \end{vmatrix}\mathbf{i} - \begin{vmatrix} 1 & 1 \\ 3 & -2 \end{vmatrix}\mathbf{j} + \begin{vmatrix} 1 & -2 \\ 3 & 1 \end{vmatrix}\mathbf{k}$$

$$= (4 - 1)\mathbf{i} - (-2 - 3)\mathbf{j} + (1 + 6)\mathbf{k}$$

$$= 3\mathbf{i} + 5\mathbf{j} + 7\mathbf{k}$$

$$\textbf{b. } \mathbf{v} \times \mathbf{u} = \begin{vmatrix} \mathbf{i} & \mathbf{j} & \mathbf{k} \\ 3 & 1 & -2 \\ 1 & -2 & 1 \end{vmatrix} = \begin{vmatrix} 1 & -2 \\ -2 & 1 \end{vmatrix}\mathbf{i} - \begin{vmatrix} 3 & -2 \\ 1 & 1 \end{vmatrix}\mathbf{j} + \begin{vmatrix} 3 & 1 \\ 1 & -2 \end{vmatrix}\mathbf{k}$$

$$= (1 - 4)\mathbf{i} - (3 + 2)\mathbf{j} + (-6 - 1)\mathbf{k}$$

$$= -3\mathbf{i} - 5\mathbf{j} - 7\mathbf{k}$$

Note that this result is the negative of that in part (a).

$$\textbf{c. } \mathbf{v} \times \mathbf{v} = \begin{vmatrix} \mathbf{i} & \mathbf{j} & \mathbf{k} \\ 3 & 1 & -2 \\ 3 & 1 & -2 \end{vmatrix} = \mathbf{0}$$

The results obtained in Example 1 suggest some interesting *algebraic* properties of the cross product. For instance, $\mathbf{u} \times \mathbf{v} = -(\mathbf{v} \times \mathbf{u})$, and $\mathbf{v} \times \mathbf{v} = \mathbf{0}$. These properties, and several others, are summarized in the following theorem.

THEOREM 11.7 Algebraic Properties of the Cross Product

Let $\mathbf{u}$, $\mathbf{v}$, and $\mathbf{w}$ be vectors in space, and let c be a scalar.

1. $\mathbf{u} \times \mathbf{v} = -(\mathbf{v} \times \mathbf{u})$
2. $\mathbf{u} \times (\mathbf{v} + \mathbf{w}) = (\mathbf{u} \times \mathbf{v}) + (\mathbf{u} \times \mathbf{w})$
3. $c(\mathbf{u} \times \mathbf{v}) = (c\mathbf{u}) \times \mathbf{v} = \mathbf{u} \times (c\mathbf{v})$
4. $\mathbf{u} \times \mathbf{0} = \mathbf{0} \times \mathbf{u} = \mathbf{0}$
5. $\mathbf{u} \times \mathbf{u} = \mathbf{0}$
6. $\mathbf{u} \cdot (\mathbf{v} \times \mathbf{w}) = (\mathbf{u} \times \mathbf{v}) \cdot \mathbf{w}$

Proof To prove Property 1, let $\mathbf{u} = u_1\mathbf{i} + u_2\mathbf{j} + u_3\mathbf{k}$ and $\mathbf{v} = v_1\mathbf{i} + v_2\mathbf{j} + v_3\mathbf{k}$. Then,

$$\mathbf{u} \times \mathbf{v} = (u_2v_3 - u_3v_2)\mathbf{i} - (u_1v_3 - u_3v_1)\mathbf{j} + (u_1v_2 - u_2v_1)\mathbf{k}$$

and

$$\mathbf{v} \times \mathbf{u} = (v_2u_3 - v_3u_2)\mathbf{i} - (v_1u_3 - v_3u_1)\mathbf{j} + (v_1u_2 - v_2u_1)\mathbf{k}$$

which implies that $\mathbf{u} \times \mathbf{v} = -(\mathbf{v} \times \mathbf{u})$. Proofs of Properties 2, 3, 5, and 6 are left as exercises (see Exercises 57–60).

Note that Property 1 of Theorem 11.7 indicates that the cross product is *not commutative*. In particular, this property indicates that the vectors $\mathbf{u} \times \mathbf{v}$ and $\mathbf{v} \times \mathbf{u}$ have equal lengths but opposite directions. The following theorem lists some other *geometric* properties of the cross product of two vectors.

NOTE It follows from Properties 1 and 2 in Theorem 11.8 that if $\mathbf{n}$ is a unit vector orthogonal to both $\mathbf{u}$ and $\mathbf{v}$, then

$$\mathbf{u} \times \mathbf{v} = \pm(\|\mathbf{u}\|\ \|\mathbf{v}\| \sin\theta)\mathbf{n}.$$

THEOREM 11.8 Geometric Properties of the Cross Product

Let $\mathbf{u}$ and $\mathbf{v}$ be nonzero vectors in space, and let θ be the angle between $\mathbf{u}$ and $\mathbf{v}$.

1. $\mathbf{u} \times \mathbf{v}$ is orthogonal to both $\mathbf{u}$ and $\mathbf{v}$.
2. $\|\mathbf{u} \times \mathbf{v}\| = \|\mathbf{u}\|\ \|\mathbf{v}\| \sin\theta$
3. $\mathbf{u} \times \mathbf{v} = \mathbf{0}$ if and only if $\mathbf{u}$ and $\mathbf{v}$ are scalar multiples of each other.
4. $\|\mathbf{u} \times \mathbf{v}\|$ = area of parallelogram having $\mathbf{u}$ and $\mathbf{v}$ as adjacent sides.

Proof To prove Property 2, note because $\cos\theta = (\mathbf{u} \cdot \mathbf{v})/(\|\mathbf{u}\|\,\|\mathbf{v}\|)$, it follows that

$$\begin{aligned}
\|\mathbf{u}\|\ \|\mathbf{v}\| \sin\theta &= \|\mathbf{u}\|\ \|\mathbf{v}\|\sqrt{1-\cos^2\theta} \\
&= \|\mathbf{u}\|\ \|\mathbf{v}\|\sqrt{1-\frac{(\mathbf{u}\cdot\mathbf{v})^2}{\|\mathbf{u}\|^2\|\mathbf{v}\|^2}} \\
&= \sqrt{\|\mathbf{u}\|^2\|\mathbf{v}\|^2-(\mathbf{u}\cdot\mathbf{v})^2} \\
&= \sqrt{(u_1^2+u_2^2+u_3^2)(v_1^2+v_2^2+v_3^2)-(u_1v_1+u_2v_2+u_3v_3)^2} \\
&= \sqrt{(u_2v_3-u_3v_2)^2+(u_1v_3-u_3v_1)^2+(u_1v_2-u_2v_1)^2} \\
&= \|\mathbf{u}\times\mathbf{v}\|.
\end{aligned}$$

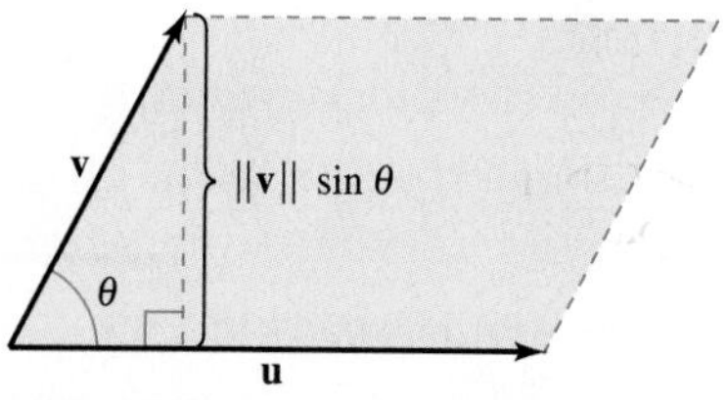

The vectors **u** and **v** form adjacent sides of a parallelogram.
Figure 11.35

To prove Property 4, refer to Figure 11.35, which is a parallelogram having $\mathbf{v}$ and $\mathbf{u}$ as adjacent sides. Because the height of the parallelogram is $\|\mathbf{v}\| \sin\theta$, the area is

$$\begin{aligned}
\text{Area} &= (\text{base})(\text{height}) \\
&= \|\mathbf{u}\|\ \|\mathbf{v}\| \sin\theta \\
&= \|\mathbf{u}\times\mathbf{v}\|.
\end{aligned}$$

Proofs of Properties 1 and 3 are left as exercises (see Exercises 61 and 62).

Both $\mathbf{u} \times \mathbf{v}$ and $\mathbf{v} \times \mathbf{u}$ are perpendicular to the plane determined by $\mathbf{u}$ and $\mathbf{v}$. One way to remember the orientations of the vectors $\mathbf{u}$, $\mathbf{v}$, and $\mathbf{u} \times \mathbf{v}$ is to compare them with the unit vectors $\mathbf{i}$, $\mathbf{j}$, and $\mathbf{k} = \mathbf{i} \times \mathbf{j}$, as shown in Figure 11.36. The three vectors $\mathbf{u}$, $\mathbf{v}$, and $\mathbf{u} \times \mathbf{v}$ form a *right-handed system*, whereas the three vectors $\mathbf{u}$, $\mathbf{v}$, and $\mathbf{v} \times \mathbf{u}$ form a *left-handed system*.

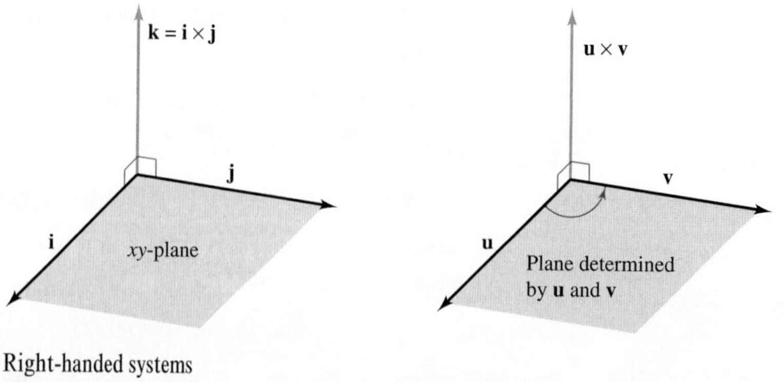

Right-handed systems
Figure 11.36

EXAMPLE 2 Using the Cross Product

Find a unit vector that is orthogonal to both

$$\mathbf{u} = \mathbf{i} - 4\mathbf{j} + \mathbf{k} \quad \text{and} \quad \mathbf{v} = 2\mathbf{i} + 3\mathbf{j}.$$

Solution The cross product $\mathbf{u} \times \mathbf{v}$, as shown in Figure 11.37, is orthogonal to both $\mathbf{u}$ and $\mathbf{v}$.

$$\mathbf{u} \times \mathbf{v} = \begin{vmatrix} \mathbf{i} & \mathbf{j} & \mathbf{k} \\ 1 & -4 & 1 \\ 2 & 3 & 0 \end{vmatrix} \qquad \text{Cross product}$$

$$= -3\mathbf{i} + 2\mathbf{j} + 11\mathbf{k}$$

Because

$$\|\mathbf{u} \times \mathbf{v}\| = \sqrt{(-3)^2 + 2^2 + 11^2} = \sqrt{134}$$

a unit vector orthogonal to both $\mathbf{u}$ and $\mathbf{v}$ is

$$\frac{\mathbf{u} \times \mathbf{v}}{\|\mathbf{u} \times \mathbf{v}\|} = -\frac{3}{\sqrt{134}}\mathbf{i} + \frac{2}{\sqrt{134}}\mathbf{j} + \frac{11}{\sqrt{134}}\mathbf{k}.$$

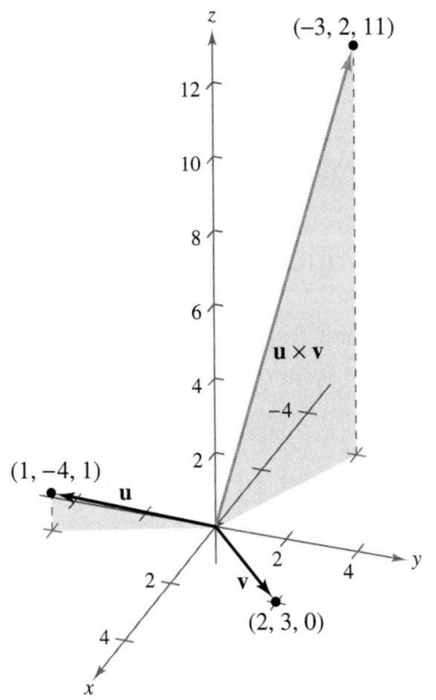

The vector $\mathbf{u} \times \mathbf{v}$ is orthogonal to both $\mathbf{u}$ and $\mathbf{v}$.
Figure 11.37

NOTE In Example 2, note that you could have used the cross product $\mathbf{v} \times \mathbf{u}$ to form a unit vector that is orthogonal to both $\mathbf{u}$ and $\mathbf{v}$. With that choice, you would have obtained the negative of the unit vector found in the example.

EXAMPLE 3 Geometric Application of the Cross Product

Show that the quadrilateral with vertices at the following points is a parallelogram, and find its area.

$$A = (5, 2, 0) \qquad B = (2, 6, 1)$$
$$C = (2, 4, 7) \qquad D = (5, 0, 6)$$

Solution From Figure 11.38 you can see that the sides of the quadrilateral correspond to the following four vectors.

$$\overrightarrow{AB} = -3\mathbf{i} + 4\mathbf{j} + \mathbf{k} \qquad \overrightarrow{CD} = 3\mathbf{i} - 4\mathbf{j} - \mathbf{k} = -\overrightarrow{AB}$$
$$\overrightarrow{AD} = 0\mathbf{i} - 2\mathbf{j} + 6\mathbf{k} \qquad \overrightarrow{CB} = 0\mathbf{i} + 2\mathbf{j} - 6\mathbf{k} = -\overrightarrow{AD}$$

So, $\overrightarrow{AB}$ is parallel to $\overrightarrow{CD}$ and $\overrightarrow{AD}$ is parallel to $\overrightarrow{CB}$, and you can conclude that the quadrilateral is a parallelogram with $\overrightarrow{AB}$ and $\overrightarrow{AD}$ as adjacent sides. Moreover, because

$$\overrightarrow{AB} \times \overrightarrow{AD} = \begin{vmatrix} \mathbf{i} & \mathbf{j} & \mathbf{k} \\ -3 & 4 & 1 \\ 0 & -2 & 6 \end{vmatrix} \qquad \text{Cross product}$$

$$= 26\mathbf{i} + 18\mathbf{j} + 6\mathbf{k}$$

the area of the parallelogram is

$$\|\overrightarrow{AB} \times \overrightarrow{AD}\| = \sqrt{1036} \approx 32.19.$$

Is the parallelogram a rectangle? You can determine whether it is by finding the angle between the vectors $\overrightarrow{AB}$ and $\overrightarrow{AD}$.

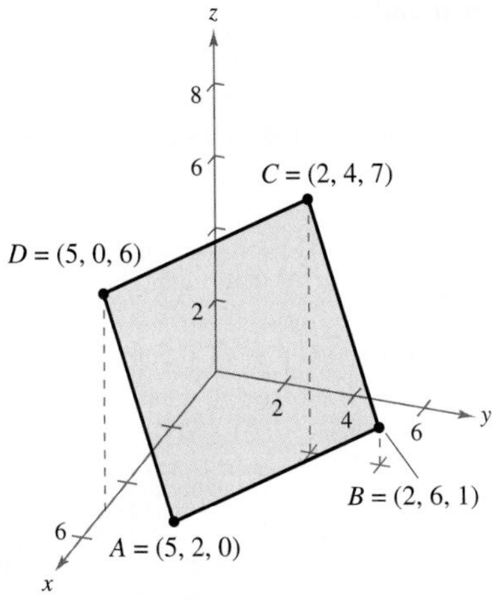

The area of the parallelogram is approximately 32.19.
Figure 11.38

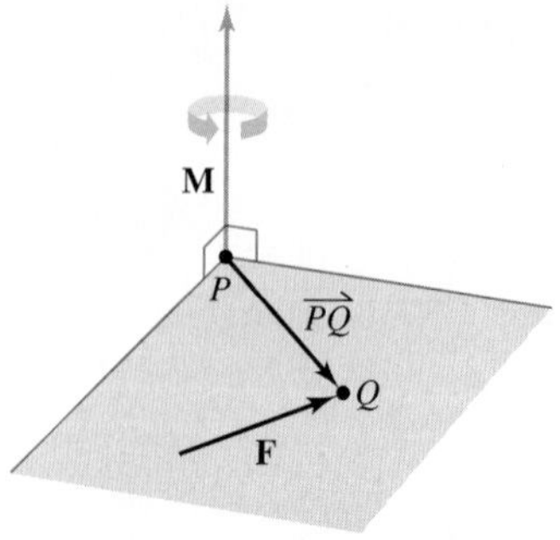

The moment of **F** about P
Figure 11.39

In physics, the cross product can be used to measure **torque**—the **moment M of a force F about a point *P*,** as shown in Figure 11.39. If the point of application of the force is Q, the moment of **F** about P is given by

$$\mathbf{M} = \overrightarrow{PQ} \times \mathbf{F}. \qquad \text{Moment of } \mathbf{F} \text{ about } P$$

The magnitude of the moment **M** measures the tendency of the vector $\overrightarrow{PQ}$ to rotate counterclockwise (using the right-hand rule) about an axis directed along the vector **M**.

EXAMPLE 4 An Application of the Cross Product

A vertical force of 50 pounds is applied to the end of a one-foot lever that is attached to an axle at point P, as shown in Figure 11.40. Find the moment of this force about the point P when $\theta = 60°$.

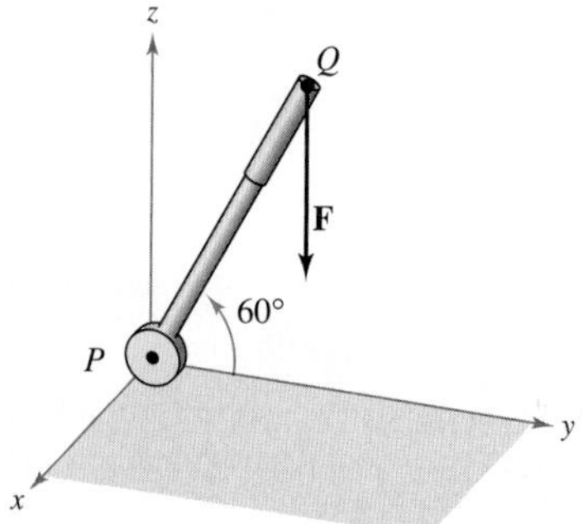

A vertical force of 50 pounds is applied at point Q.
Figure 11.40

Solution If you represent the 50-pound force as $\mathbf{F} = -50\mathbf{k}$ and the lever as

$$\overrightarrow{PQ} = \cos(60°)\mathbf{j} + \sin(60°)\mathbf{k} = \frac{1}{2}\mathbf{j} + \frac{\sqrt{3}}{2}\mathbf{k}$$

the moment of **F** about P is given by

$$\mathbf{M} = \overrightarrow{PQ} \times \mathbf{F} = \begin{vmatrix} \mathbf{i} & \mathbf{j} & \mathbf{k} \\ 0 & \frac{1}{2} & \frac{\sqrt{3}}{2} \\ 0 & 0 & -50 \end{vmatrix} = -25\mathbf{i}. \qquad \text{Moment of } \mathbf{F} \text{ about } P$$

The magnitude of this moment is 25 foot-pounds.

NOTE In Example 4, note that the moment (the tendency of the lever to rotate about its axle) is dependent on the angle θ. When $\theta = \pi/2$, the moment is **0**. The moment is greatest when $\theta = 0$.

The Triple Scalar Product

For vectors **u**, **v**, and **w** in space, the dot product of **u** and $\mathbf{v} \times \mathbf{w}$

$$\mathbf{u} \cdot (\mathbf{v} \times \mathbf{w})$$

is called the **triple scalar product,** as defined in Theorem 11.9. The proof of this theorem is left as an exercise (see Exercise 56).

FOR FURTHER INFORMATION To see how the cross product is used to model the torque of the robot arm of a space shuttle, see the article "The Long Arm of Calculus" by Ethan Berkove and Rich Marchand in *The College Mathematics Journal*. To view this article, go to the website *www.matharticles.com*.

THEOREM 11.9 The Triple Scalar Product

For $\mathbf{u} = u_1\mathbf{i} + u_2\mathbf{j} + u_3\mathbf{k}$, $\mathbf{v} = v_1\mathbf{i} + v_2\mathbf{j} + v_3\mathbf{k}$, and $\mathbf{w} = w_1\mathbf{i} + w_2\mathbf{j} + w_3\mathbf{k}$, the triple scalar product is given by

$$\mathbf{u} \cdot (\mathbf{v} \times \mathbf{w}) = \begin{vmatrix} u_1 & u_2 & u_3 \\ v_1 & v_2 & v_3 \\ w_1 & w_2 & w_3 \end{vmatrix}.$$

NOTE The value of a determinant is multiplied by -1 if two rows are interchanged. After two such interchanges, the value of the determinant will be unchanged. So, the following triple scalar products are equivalent.

$$\mathbf{u} \cdot (\mathbf{v} \times \mathbf{w}) = \mathbf{v} \cdot (\mathbf{w} \times \mathbf{u}) = \mathbf{w} \cdot (\mathbf{u} \times \mathbf{v})$$

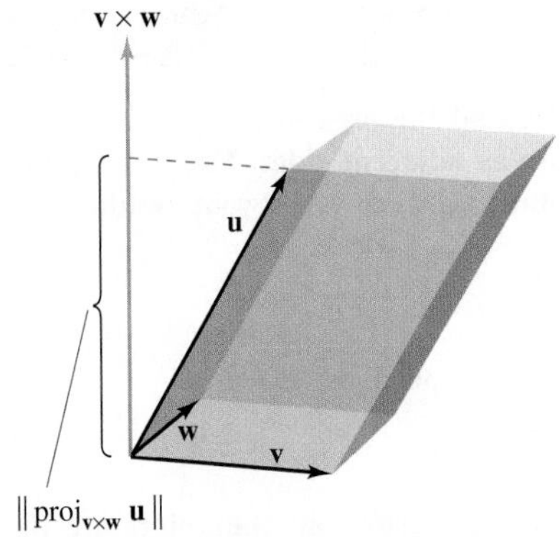

Area of base $= \|\mathbf{v} \times \mathbf{w}\|$
Volume of parallelepiped $= |\mathbf{u} \cdot (\mathbf{v} \times \mathbf{w})|$
Figure 11.41

If the vectors **u**, **v**, and **w** do not lie in the same plane, the triple scalar product $\mathbf{u} \cdot (\mathbf{v} \times \mathbf{w})$ can be used to determine the volume of the parallelepiped (a polyhedron, all of whose faces are parallelograms) with **u**, **v**, and **w** as adjacent edges, as shown in Figure 11.41. This is established in the following theorem.

THEOREM 11.10 Geometric Property of Triple Scalar Product

The volume V of a parallelepiped with vectors **u**, **v**, and **w** as adjacent edges is given by

$$V = |\mathbf{u} \cdot (\mathbf{v} \times \mathbf{w})|.$$

Proof In Figure 11.41, note that

$$\|\mathbf{v} \times \mathbf{w}\| = \text{area of base}$$

and

$$\|\text{proj}_{\mathbf{v}\times\mathbf{w}}\mathbf{u}\| = \text{height of parallelepiped.}$$

Therefore, the volume is

$$\begin{aligned} V = (\text{height})(\text{area of base}) &= \|\text{proj}_{\mathbf{v}\times\mathbf{w}}\mathbf{u}\|\,\|\mathbf{v} \times \mathbf{w}\| \\ &= \left|\frac{\mathbf{u} \cdot (\mathbf{v} \times \mathbf{w})}{\|\mathbf{v} \times \mathbf{w}\|}\right| \|\mathbf{v} \times \mathbf{w}\| \\ &= |\mathbf{u} \cdot (\mathbf{v} \times \mathbf{w})|. \end{aligned}$$

EXAMPLE 5 Volume by the Triple Scalar Product

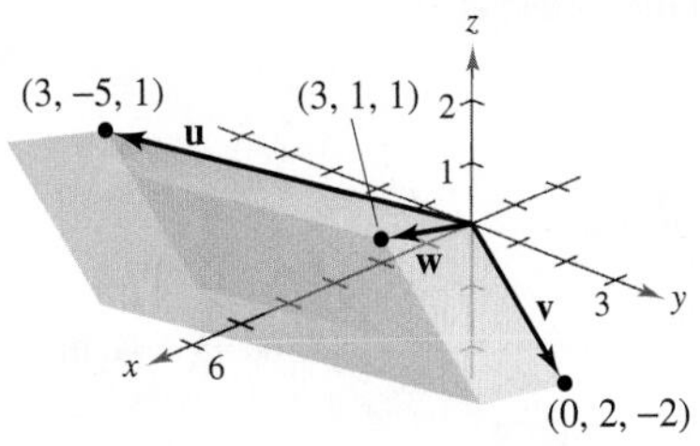

The parallelepiped has a volume of 36.
Figure 11.42

Find the volume of the parallelepiped shown in Figure 11.42 having $\mathbf{u} = 3\mathbf{i} - 5\mathbf{j} + \mathbf{k}$, $\mathbf{v} = 2\mathbf{j} - 2\mathbf{k}$, and $\mathbf{w} = 3\mathbf{i} + \mathbf{j} + \mathbf{k}$ as adjacent edges.

Solution By Theorem 11.10, you have

$$\begin{aligned} V &= |\mathbf{u} \cdot (\mathbf{v} \times \mathbf{w})| \qquad \text{Triple scalar product} \\ &= \begin{vmatrix} 3 & -5 & 1 \\ 0 & 2 & -2 \\ 3 & 1 & 1 \end{vmatrix} \\ &= 3\begin{vmatrix} 2 & -2 \\ 1 & 1 \end{vmatrix} - (-5)\begin{vmatrix} 0 & -2 \\ 3 & 1 \end{vmatrix} + (1)\begin{vmatrix} 0 & 2 \\ 3 & 1 \end{vmatrix} \\ &= 3(4) + 5(6) + 1(-6) \\ &= 36. \end{aligned}$$

A natural consequence of Theorem 11.10 is that the volume of the parallelepiped is 0 if and only if the three vectors are coplanar. That is, if the vectors $\mathbf{u} = \langle u_1, u_2, u_3 \rangle$, $\mathbf{v} = \langle v_1, v_2, v_3 \rangle$, and $\mathbf{w} = \langle w_1, w_2, w_3 \rangle$ have the same initial point, they lie in the same plane if and only if

$$\mathbf{u} \cdot (\mathbf{v} \times \mathbf{w}) = \begin{vmatrix} u_1 & u_2 & u_3 \\ v_1 & v_2 & v_3 \\ w_1 & w_2 & w_3 \end{vmatrix} = 0.$$

Exercises for Section 11.4

See www.CalcChat.com for worked-out solutions to odd-numbered exercises.

In Exercises 1–6, find the cross product of the unit vectors and sketch your result.

1. $\mathbf{j} \times \mathbf{i}$
2. $\mathbf{i} \times \mathbf{j}$
3. $\mathbf{j} \times \mathbf{k}$
4. $\mathbf{k} \times \mathbf{j}$
5. $\mathbf{i} \times \mathbf{k}$
6. $\mathbf{k} \times \mathbf{i}$

In Exercises 7–10, find (a) u × v, (b) v × u, and (c) v × v.

7. $\mathbf{u} = -2\mathbf{i} + 3\mathbf{j} + 4\mathbf{k}$, $\mathbf{v} = 3\mathbf{i} + 7\mathbf{j} + 2\mathbf{k}$
8. $\mathbf{u} = 3\mathbf{i} + 5\mathbf{k}$, $\mathbf{v} = 2\mathbf{i} + 3\mathbf{j} - 2\mathbf{k}$
9. $\mathbf{u} = \langle 7, 3, 2\rangle$, $\mathbf{v} = \langle 1, -1, 5\rangle$
10. $\mathbf{u} = \langle 3, -2, -2\rangle$, $\mathbf{v} = \langle 1, 5, 1\rangle$

In Exercises 11–16, find u × v and show that it is orthogonal to both u and v.

11. $\mathbf{u} = \langle 2, -3, 1\rangle$, $\mathbf{v} = \langle 1, -2, 1\rangle$
12. $\mathbf{u} = \langle -1, 1, 2\rangle$, $\mathbf{v} = \langle 0, 1, 0\rangle$
13. $\mathbf{u} = \langle 12, -3, 0\rangle$, $\mathbf{v} = \langle -2, 5, 0\rangle$
14. $\mathbf{u} = \langle -10, 0, 6\rangle$, $\mathbf{v} = \langle 7, 0, 0\rangle$
15. $\mathbf{u} = \mathbf{i} + \mathbf{j} + \mathbf{k}$, $\mathbf{v} = 2\mathbf{i} + \mathbf{j} - \mathbf{k}$
16. $\mathbf{u} = \mathbf{i} + 6\mathbf{j}$, $\mathbf{v} = -2\mathbf{i} + \mathbf{j} + \mathbf{k}$

Think About It **In Exercises 17–20, use the vectors u and v shown in the figure to sketch a vector in the direction of the indicated cross product in a right-handed system.**

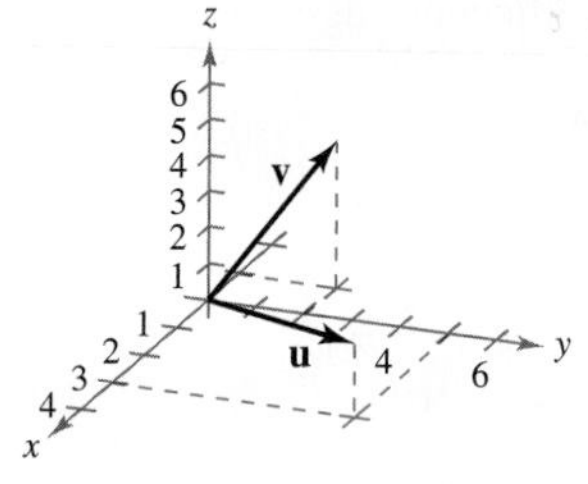

17. $\mathbf{u} \times \mathbf{v}$
18. $\mathbf{v} \times \mathbf{u}$
19. $(-\mathbf{v}) \times \mathbf{u}$
20. $\mathbf{u} \times (\mathbf{u} \times \mathbf{v})$

In Exercises 21–24, use a computer algebra system to find u × v and a unit vector orthogonal to u and v.

21. $\mathbf{u} = \langle 4, -3.5, 7\rangle$, $\mathbf{v} = \langle -1, 8, 4\rangle$
22. $\mathbf{u} = \langle -8, -6, 4\rangle$, $\mathbf{v} = \langle 10, -12, -2\rangle$
23. $\mathbf{u} = -3\mathbf{i} + 2\mathbf{j} - 5\mathbf{k}$, $\mathbf{v} = \frac{1}{2}\mathbf{i} - \frac{3}{4}\mathbf{j} + \frac{1}{10}\mathbf{k}$
24. $\mathbf{u} = \frac{2}{3}\mathbf{k}$, $\mathbf{v} = \frac{1}{2}\mathbf{i} + 6\mathbf{k}$

25. ***Programming*** Given the vectors **u** and **v** in component form, write a program for a graphing utility in which the output is $\mathbf{u} \times \mathbf{v}$ and $\|\mathbf{u} \times \mathbf{v}\|$.

26. ***Programming*** Use the program you wrote in Exercise 25 to find $\mathbf{u} \times \mathbf{v}$ and $\|\mathbf{u} \times \mathbf{v}\|$ for $\mathbf{u} = \langle -2, 6, 10\rangle$ and $\mathbf{v} = \langle 3, 8, 5\rangle$.

Area **In Exercises 27–30, find the area of the parallelogram that has the given vectors as adjacent sides. Use a computer algebra system or a graphing utility to verify your result.**

27. $\mathbf{u} = \mathbf{j}$, $\mathbf{v} = \mathbf{j} + \mathbf{k}$
28. $\mathbf{u} = \mathbf{i} + \mathbf{j} + \mathbf{k}$, $\mathbf{v} = \mathbf{j} + \mathbf{k}$
29. $\mathbf{u} = \langle 3, 2, -1\rangle$, $\mathbf{v} = \langle 1, 2, 3\rangle$
30. $\mathbf{u} = \langle 2, -1, 0\rangle$, $\mathbf{v} = \langle -1, 2, 0\rangle$

Area **In Exercises 31 and 32, verify that the points are the vertices of a parallelogram, and find its area.**

31. $(1, 1, 1)$, $(2, 3, 4)$, $(6, 5, 2)$, $(7, 7, 5)$
32. $(2, -3, 1)$, $(6, 5, -1)$, $(3, -6, 4)$, $(7, 2, 2)$

Area **In Exercises 33–36, find the area of the triangle with the given vertices. (*Hint:* $\frac{1}{2}\|\mathbf{u} \times \mathbf{v}\|$ is the area of the triangle having u and v as adjacent sides.)**

33. $(0, 0, 0)$, $(1, 2, 3)$, $(-3, 0, 0)$
34. $(2, -3, 4)$, $(0, 1, 2)$, $(-1, 2, 0)$
35. $(2, -7, 3)$, $(-1, 5, 8)$, $(4, 6, -1)$
36. $(1, 2, 0)$, $(-2, 1, 0)$, $(0, 0, 0)$

37. ***Torque*** A child applies the brakes on a bicycle by applying a downward force of 20 pounds on the pedal when the crank makes a 40° angle with the horizontal (see figure). The crank is 6 inches in length. Find the torque at P.

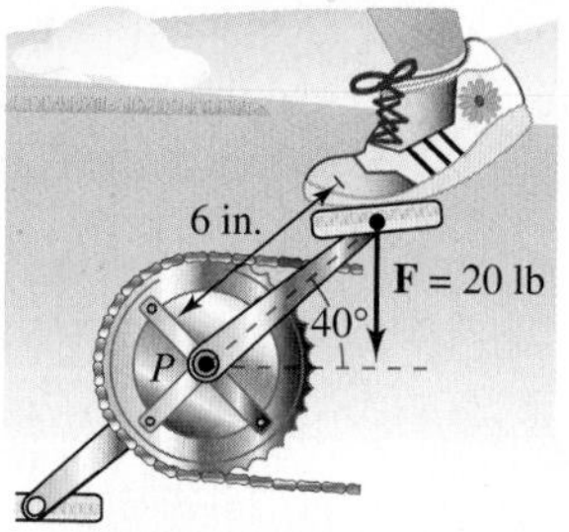

Figure for 37

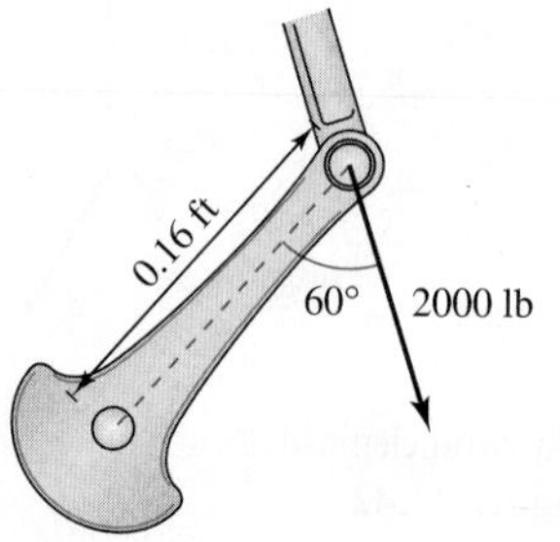

Figure for 38

38. ***Torque*** Both the magnitude and the direction of the force on a crankshaft change as the crankshaft rotates. Find the torque on the crankshaft using the position and data shown in the figure.

39. ***Optimization*** A force of 60 pounds acts on the pipe wrench shown in the figure on the next page.
    (a) Find the magnitude of the moment about O by evaluating $\|\overrightarrow{OA} \times \mathbf{F}\|$. Use a graphing utility to graph the resulting function of θ.
    (b) Use the result of part (a) to determine the magnitude of the moment when $\theta = 45°$.
    (c) Use the result of part (a) to determine the angle θ when the magnitude of the moment is maximum. Is the answer what you expected? Why or why not?

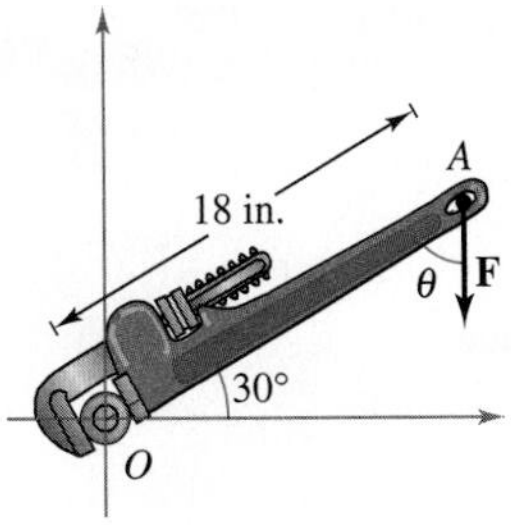

Figure for 39

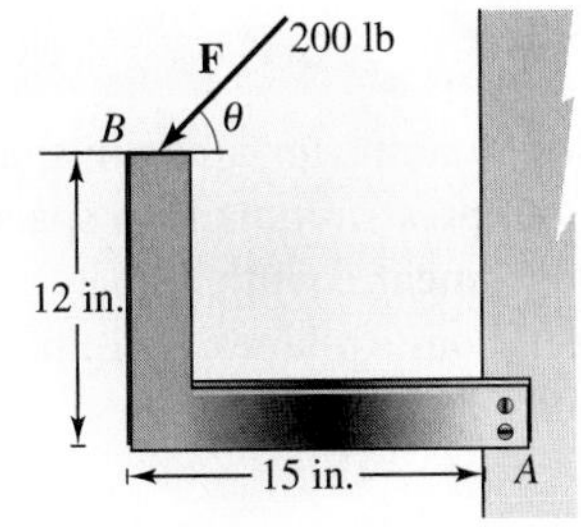

Figure for 40

40. *Optimization* A force of 200 pounds acts on the bracket shown in the figure.

(a) Determine the vector $\overrightarrow{AB}$ and the vector $\mathbf{F}$ representing the force. ($\mathbf{F}$ will be in terms of θ.)

(b) Find the magnitude of the moment about A by evaluating $\|\overrightarrow{AB} \times \mathbf{F}\|$.

(c) Use the result of part (b) to determine the magnitude of the moment when $\theta = 30°$.

(d) Use the result of part (b) to determine the angle θ when the magnitude of the moment is maximum. At that angle, what is the relationship between the vectors $\mathbf{F}$ and $\overrightarrow{AB}$? Is it what you expected? Why or why not?

(e) Use a graphing utility to graph the function for the magnitude of the moment about A for $0° \leq \theta \leq 180°$. Find the zero of the function in the given domain. Interpret the meaning of the zero in the context of the problem.

In Exercises 41–44, find $\mathbf{u} \cdot (\mathbf{v} \times \mathbf{w})$.

41. $\mathbf{u} = \mathbf{i}$
$\mathbf{v} = \mathbf{j}$
$\mathbf{w} = \mathbf{k}$

42. $\mathbf{u} = \langle 1, 1, 1 \rangle$
$\mathbf{v} = \langle 2, 1, 0 \rangle$
$\mathbf{w} = \langle 0, 0, 1 \rangle$

43. $\mathbf{u} = \langle 2, 0, 1 \rangle$
$\mathbf{v} = \langle 0, 3, 0 \rangle$
$\mathbf{w} = \langle 0, 0, 1 \rangle$

44. $\mathbf{u} = \langle 2, 0, 0 \rangle$
$\mathbf{v} = \langle 1, 1, 1 \rangle$
$\mathbf{w} = \langle 0, 2, 2 \rangle$

Volume **In Exercises 45 and 46, use the triple scalar product to find the volume of the parallelepiped having adjacent edges u, v, and w.**

45. $\mathbf{u} = \mathbf{i} + \mathbf{j}$
$\mathbf{v} = \mathbf{j} + \mathbf{k}$
$\mathbf{w} = \mathbf{i} + \mathbf{k}$

46. $\mathbf{u} = \langle 1, 3, 1 \rangle$
$\mathbf{v} = \langle 0, 6, 6 \rangle$
$\mathbf{w} = \langle -4, 0, -4 \rangle$

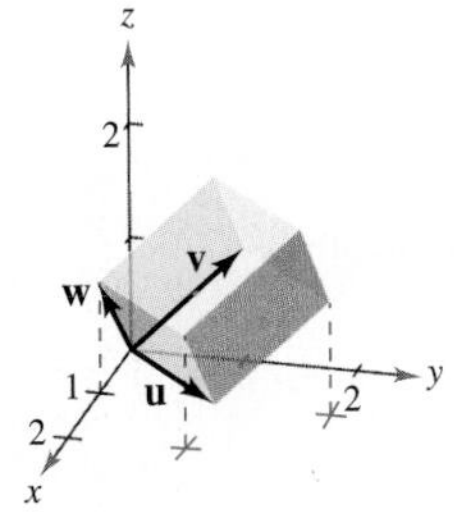

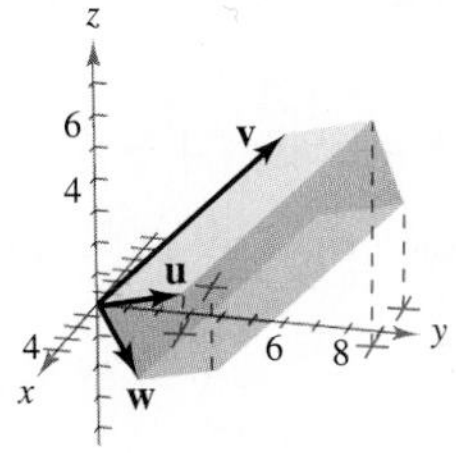

Volume **In Exercises 47 and 48, find the volume of the parallelepiped with the given vertices (see figures).**

47. $(0, 0, 0), (3, 0, 0), (0, 5, 1), (3, 5, 1)$
$(2, 0, 5), (5, 0, 5), (2, 5, 6), (5, 5, 6)$

48. $(0, 0, 0), (1, 1, 0), (1, 0, 2), (0, 1, 1)$
$(2, 1, 2), (1, 1, 3), (1, 2, 1), (2, 2, 3)$

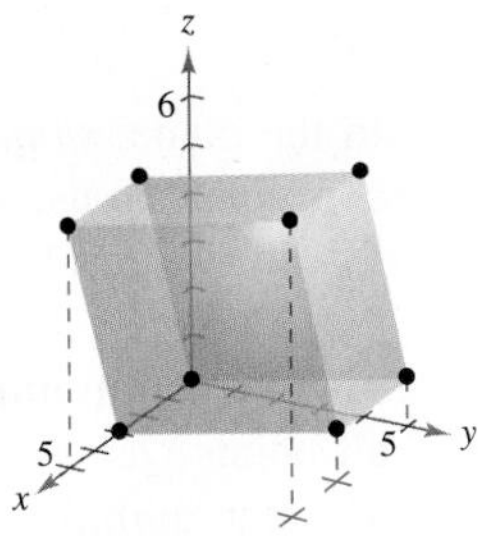

Figure for 47

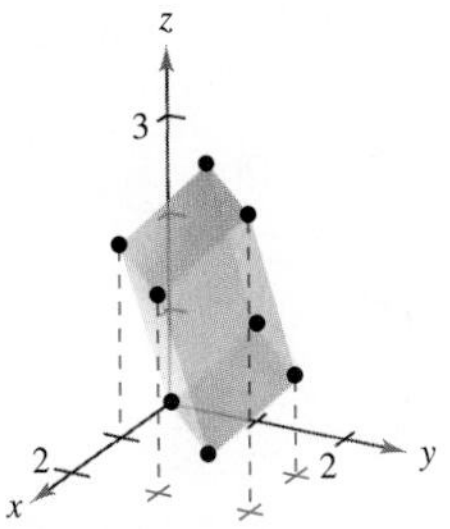

Figure for 48

Writing About Concepts

49. Define the cross product of vectors **u** and **v**.

50. State the geometric properties of the cross product.

51. If the magnitudes of two vectors are doubled, how will the magnitude of the cross product of the vectors change? Explain.

52. The vertices of a triangle in space are (x_1, y_1, z_1), (x_2, y_2, z_2), and (x_3, y_3, z_3). Explain how to find a vector perpendicular to the triangle.

True or False? **In Exercises 53–55, determine whether the statement is true or false. If it is false, explain why or give an example that shows it is false.**

53. It is possible to find the cross product of two vectors in a two-dimensional coordinate system.

54. If $\mathbf{u} \neq \mathbf{0}$ and $\mathbf{u} \times \mathbf{v} = \mathbf{u} \times \mathbf{w}$, then $\mathbf{v} = \mathbf{w}$.

55. If $\mathbf{u} \neq \mathbf{0}$, $\mathbf{u} \cdot \mathbf{v} = \mathbf{u} \cdot \mathbf{w}$, and $\mathbf{u} \times \mathbf{v} = \mathbf{u} \times \mathbf{w}$, then $\mathbf{v} = \mathbf{w}$.

56. Prove Theorem 11.9.

In Exercises 57–62, prove the property of the cross product.

57. $\mathbf{u} \times (\mathbf{v} + \mathbf{w}) = (\mathbf{u} \times \mathbf{v}) + (\mathbf{u} \times \mathbf{w})$

58. $c(\mathbf{u} \times \mathbf{v}) = (c\mathbf{u}) \times \mathbf{v} = \mathbf{u} \times (c\mathbf{v})$

59. $\mathbf{u} \times \mathbf{u} = \mathbf{0}$

60. $\mathbf{u} \cdot (\mathbf{v} \times \mathbf{w}) = (\mathbf{u} \times \mathbf{v}) \cdot \mathbf{w}$

61. $\mathbf{u} \times \mathbf{v}$ is orthogonal to both **u** and **v**.

62. $\mathbf{u} \times \mathbf{v} = \mathbf{0}$ if and only if **u** and **v** are scalar multiples of each other.

63. Prove $\|\mathbf{u} \times \mathbf{v}\| = \|\mathbf{u}\| \, \|\mathbf{v}\|$ if **u** and **v** are orthogonal.

64. Prove $\mathbf{u} \times (\mathbf{v} \times \mathbf{w}) = (\mathbf{u} \cdot \mathbf{w})\mathbf{v} - (\mathbf{u} \cdot \mathbf{v})\mathbf{w}$.

Section 11.5 Lines and Planes in Space

- Write a set of parametric equations for a line in space.
- Write a linear equation to represent a plane in space.
- Sketch the plane given by a linear equation.
- Find the distances between points, planes, and lines in space.

Lines in Space

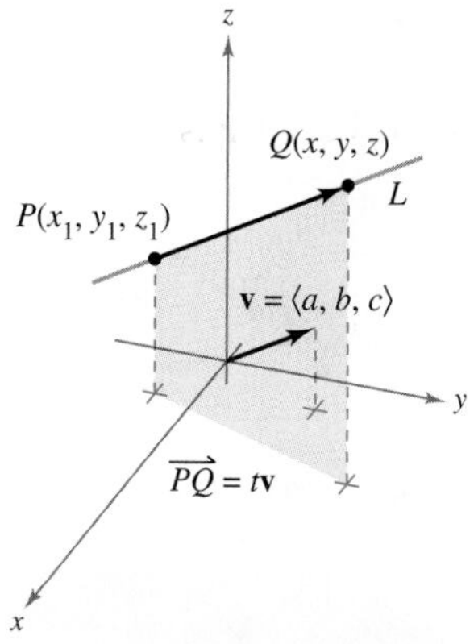

Line L and its direction vector $\mathbf{v}$
Figure 11.43

In the plane, *slope* is used to determine an equation of a line. In space, it is more convenient to use *vectors* to determine the equation of a line.

In Figure 11.43, consider the line L through the point $P(x_1, y_1, z_1)$ and parallel to the vector $\mathbf{v} = \langle a, b, c\rangle$. The vector $\mathbf{v}$ is a **direction vector** for the line L, and a, b, and c are **direction numbers.** One way of describing the line L is to say that it consists of all points $Q(x, y, z)$ for which the vector $\overrightarrow{PQ}$ is parallel to $\mathbf{v}$. This means that $\overrightarrow{PQ}$ is a scalar multiple of $\mathbf{v}$, and you can write $\overrightarrow{PQ} = t\mathbf{v}$, where t is a scalar (a real number).

$$\overrightarrow{PQ} = \langle x - x_1, y - y_1, z - z_1\rangle = \langle at, bt, ct\rangle = t\mathbf{v}$$

By equating corresponding components, you can obtain **parametric equations** of a line in space.

THEOREM 11.11 Parametric Equations of a Line in Space

A line L parallel to the vector $\mathbf{v} = \langle a, b, c\rangle$ and passing through the point $P(x_1, y_1, z_1)$ is represented by the **parametric equations**

$$x = x_1 + at, \quad y = y_1 + bt, \quad \text{and} \quad z = z_1 + ct.$$

If the direction numbers a, b, and c are all nonzero, you can eliminate the parameter t to obtain **symmetric equations** of the line.

$$\frac{x - x_1}{a} = \frac{y - y_1}{b} = \frac{z - z_1}{c} \qquad \text{Symmetric equations}$$

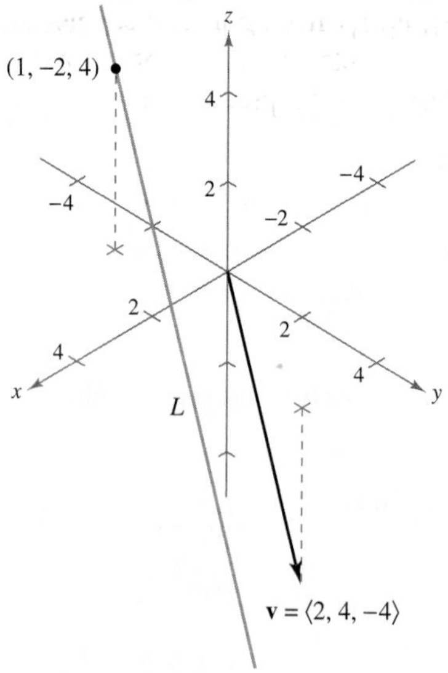

The vector $\mathbf{v}$ is parallel to the line L.
Figure 11.44

EXAMPLE 1 Finding Parametric and Symmetric Equations

Find parametric and symmetric equations of the line L that passes through the point $(1, -2, 4)$ and is parallel to $\mathbf{v} = \langle 2, 4, -4\rangle$.

Solution To find a set of parametric equations of the line, use the coordinates $x_1 = 1$, $y_1 = -2$, and $z_1 = 4$ and direction numbers $a = 2$, $b = 4$, and $c = -4$ (see Figure 11.44).

$$x = 1 + 2t, \quad y = -2 + 4t, \quad z = 4 - 4t \qquad \text{Parametric equations}$$

Because a, b, and c are all nonzero, a set of symmetric equations is

$$\frac{x - 1}{2} = \frac{y + 2}{4} = \frac{z - 4}{-4}. \qquad \text{Symmetric equations}$$

Neither parametric equations nor symmetric equations of a given line are unique. For instance, in Example 1, by letting $t = 1$ in the parametric equations you would obtain the point $(3, 2, 0)$. Using this point with the direction numbers $a = 2$, $b = 4$, and $c = -4$ would produce a different set of parametric equations

$$x = 3 + 2t, \quad y = 2 + 4t, \quad \text{and} \quad z = -4t.$$

EXAMPLE 2 Parametric Equations of a Line Through Two Points

Find a set of parametric equations of the line that passes through the points $(-2, 1, 0)$ and $(1, 3, 5)$.

Solution Begin by using the points $P(-2, 1, 0)$ and $Q(1, 3, 5)$ to find a direction vector for the line passing through P and Q, given by

$$\mathbf{v} = \overrightarrow{PQ} = \langle 1 - (-2), 3 - 1, 5 - 0 \rangle = \langle 3, 2, 5 \rangle = \langle a, b, c \rangle.$$

Using the direction numbers $a = 3$, $b = 2$, and $c = 5$ with the point $P(-2, 1, 0)$, you can obtain the parametric equations

$$x = -2 + 3t, \quad y = 1 + 2t, \quad \text{and} \quad z = 5t.$$

NOTE As t varies over all real numbers, the parametric equations in Example 2 determine the points (x, y, z) on the line. In particular, note that $t = 0$ and $t = 1$ give the original points $(-2, 1, 0)$ and $(1, 3, 5)$.

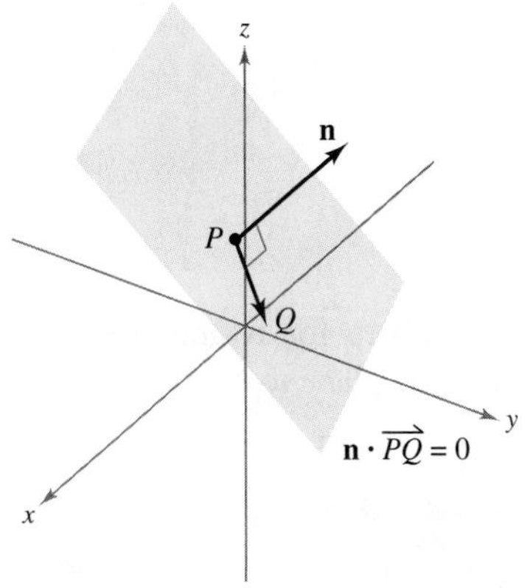

The normal vector $\mathbf{n}$ is orthogonal to each vector $\overrightarrow{PQ}$ in the plane.
Figure 11.45

Planes in Space

You have seen how an equation of a line in space can be obtained from a point on the line and a vector *parallel* to it. You will now see that an equation of a plane in space can be obtained from a point in the plane and a vector *normal* (perpendicular) to the plane.

Consider the plane containing the point $P(x_1, y_1, z_1)$ having a nonzero normal vector $\mathbf{n} = \langle a, b, c \rangle$, as shown in Figure 11.45. This plane consists of all points $Q(x, y, z)$ for which vector $\overrightarrow{PQ}$ is orthogonal to $\mathbf{n}$. Using the dot product, you can write the following.

$$\mathbf{n} \cdot \overrightarrow{PQ} = 0$$
$$\langle a, b, c \rangle \cdot \langle x - x_1, y - y_1, z - z_1 \rangle = 0$$
$$a(x - x_1) + b(y - y_1) + c(z - z_1) = 0$$

The third equation of the plane is said to be in **standard form.**

THEOREM 11.12 Standard Equation of a Plane in Space

The plane containing the point (x_1, y_1, z_1) and having a normal vector $\mathbf{n} = \langle a, b, c \rangle$ can be represented, in **standard form,** by the equation

$$a(x - x_1) + b(y - y_1) + c(z - z_1) = 0.$$

By regrouping terms, you obtain the **general form** of the equation of a plane in space.

$$ax + by + cz + d = 0 \qquad \text{General form of equation of plane}$$

Given the general form of the equation of a plane, it is easy to find a normal vector to the plane. Simply use the coefficients of x, y, and z and write $\mathbf{n} = \langle a, b, c \rangle$.

EXAMPLE 3 Finding an Equation of a Plane in Three-Space

Find the general equation of the plane containing the points $(2, 1, 1)$, $(0, 4, 1)$, and $(-2, 1, 4)$.

Solution To apply Theorem 11.12 you need a point in the plane and a vector that is normal to the plane. There are three choices for the point, but no normal vector is given. To obtain a normal vector, use the cross product of vectors $\mathbf{u}$ and $\mathbf{v}$ extending from the point $(2, 1, 1)$ to the points $(0, 4, 1)$ and $(-2, 1, 4)$, as shown in Figure 11.46. The component forms of $\mathbf{u}$ and $\mathbf{v}$ are

$$\mathbf{u} = \langle 0 - 2, 4 - 1, 1 - 1 \rangle = \langle -2, 3, 0 \rangle$$
$$\mathbf{v} = \langle -2 - 2, 1 - 1, 4 - 1 \rangle = \langle -4, 0, 3 \rangle$$

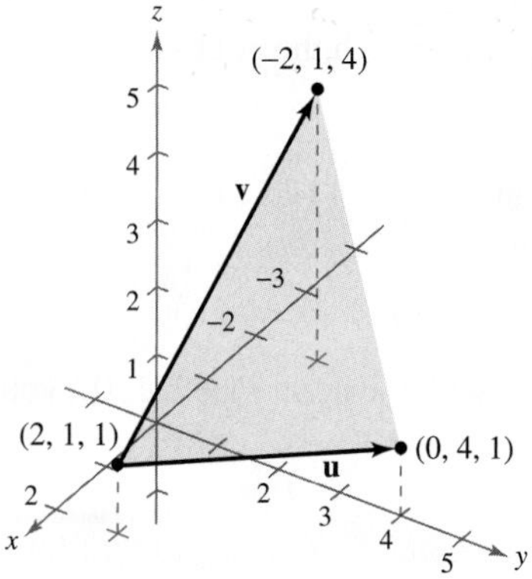

A plane determined by **u** and **v**
Figure 11.46

and it follows that

$$\begin{aligned}\mathbf{n} &= \mathbf{u} \times \mathbf{v}\\ &= \begin{vmatrix} \mathbf{i} & \mathbf{j} & \mathbf{k} \\ -2 & 3 & 0 \\ -4 & 0 & 3 \end{vmatrix}\\ &= 9\mathbf{i} + 6\mathbf{j} + 12\mathbf{k}\\ &= \langle a, b, c \rangle\end{aligned}$$

is normal to the given plane. Using the direction numbers for $\mathbf{n}$ and the point $(x_1, y_1, z_1) = (2, 1, 1)$, you can determine an equation of the plane to be

$$a(x - x_1) + b(y - y_1) + c(z - z_1) = 0$$
$$9(x - 2) + 6(y - 1) + 12(z - 1) = 0 \qquad \text{Standard form}$$
$$9x + 6y + 12z - 36 = 0 \qquad \text{General form}$$
$$3x + 2y + 4z - 12 = 0. \qquad \text{Simplified general form}$$

NOTE In Example 3, check to see that each of the three original points satisfies the equation $3x + 2y + 4z - 12 = 0$.

Two distinct planes in three-space either are parallel or intersect in a line. If they intersect, you can determine the angle $(0 \le \theta \le \pi/2)$ between them from the angle between their normal vectors, as shown in Figure 11.47. Specifically, if vectors $\mathbf{n}_1$ and $\mathbf{n}_2$ are normal to two intersecting planes, the angle θ between the normal vectors is equal to the angle between the two planes and is given by

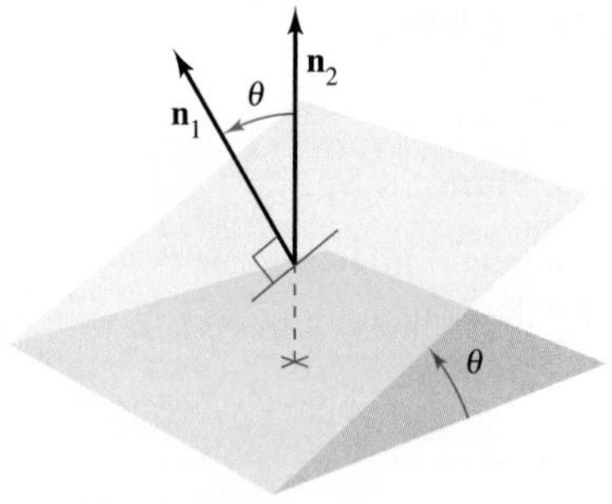

The angle θ between two planes
Figure 11.47

$$\cos \theta = \frac{|\mathbf{n}_1 \cdot \mathbf{n}_2|}{\|\mathbf{n}_1\| \, \|\mathbf{n}_2\|}. \qquad \text{Angle between two planes}$$

Consequently, two planes with normal vectors $\mathbf{n}_1$ and $\mathbf{n}_2$ are

1. *perpendicular* if $\mathbf{n}_1 \cdot \mathbf{n}_2 = 0$.
2. *parallel* if $\mathbf{n}_1$ is a scalar multiple of $\mathbf{n}_2$.

EXAMPLE 4 Finding the Line of Intersection of Two Planes

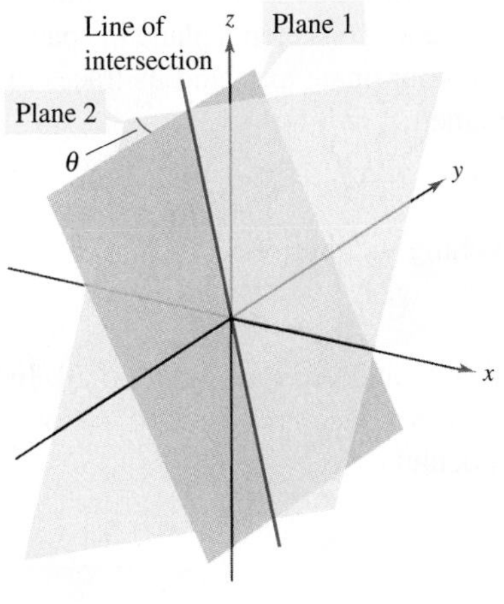

Figure 11.48

Find the angle between the two planes given by

$$x - 2y + z = 0 \qquad \text{Equation of plane 1}$$
$$2x + 3y - 2z = 0 \qquad \text{Equation of plane 2}$$

and find parametric equations of their line of intersection (see Figure 11.48).

Solution Normal vectors for the planes are $\mathbf{n}_1 = \langle 1, -2, 1\rangle$ and $\mathbf{n}_2 = \langle 2, 3, -2\rangle$. Consequently, the angle between the two planes is determined as follows.

$$\cos\theta = \frac{|\mathbf{n}_1 \cdot \mathbf{n}_2|}{\|\mathbf{n}_1\|\,\|\mathbf{n}_2\|} \qquad \text{Cosine of angle between } \mathbf{n}_1 \text{ and } \mathbf{n}_2$$
$$= \frac{|-6|}{\sqrt{6}\sqrt{17}}$$
$$= \frac{6}{\sqrt{102}}$$
$$\approx 0.59409$$

This implies that the angle between the two planes is $\theta \approx 53.55°$. You can find the line of intersection of the two planes by simultaneously solving the two linear equations representing the planes. One way to do this is to multiply the first equation by -2 and add the result to the second equation.

$$\begin{array}{rcr} x - 2y + z = 0 & \Longrightarrow & -2x + 4y - 2z = 0 \\ 2x + 3y - 2z = 0 & & 2x + 3y - 2z = 0 \\ \hline & & 7y - 4z = 0 \quad \Longrightarrow \quad y = \dfrac{4z}{7} \end{array}$$

Substituting $y = 4z/7$ back into one of the original equations, you can determine that $x = z/7$. Finally, by letting $t = z/7$, you obtain the parametric equations

$$x = t, \quad y = 4t, \quad \text{and} \quad z = 7t \qquad \text{Line of intersection}$$

which indicate that 1, 4, and 7 are direction numbers for the line of intersection.

Note that the direction numbers in Example 4 can be obtained from the cross product of the two normal vectors as follows.

$$\mathbf{n}_1 \times \mathbf{n}_2 = \begin{vmatrix} \mathbf{i} & \mathbf{j} & \mathbf{k} \\ 1 & -2 & 1 \\ 2 & 3 & -2 \end{vmatrix}$$
$$= \begin{vmatrix} -2 & 1 \\ 3 & -2 \end{vmatrix}\mathbf{i} - \begin{vmatrix} 1 & 1 \\ 2 & -2 \end{vmatrix}\mathbf{j} + \begin{vmatrix} 1 & -2 \\ 2 & 3 \end{vmatrix}\mathbf{k}$$
$$= \mathbf{i} + 4\mathbf{j} + 7\mathbf{k}$$

This means that the line of intersection of the two planes is parallel to the cross product of their normal vectors.

NOTE The three-dimensional rotatable graphs that are available in the *HM mathSpace®* CD-ROM and the online *Eduspace®* system for this text can help you visualize surfaces such as those shown in Figure 11.48. If you have access to these graphs, you should use them to help your spatial intuition when studying this section and other sections in the text that deal with vectors, curves, or surfaces in space.

Sketching Planes in Space

If a plane in space intersects one of the coordinate planes, the line of intersection is called the **trace** of the given plane in the coordinate plane. To sketch a plane in space, it is helpful to find its points of intersection with the coordinate axes and its traces in the coordinate planes. For example, consider the plane given by

$$3x + 2y + 4z = 12. \qquad \text{Equation of plane}$$

You can find the xy-trace by letting $z = 0$ and sketching the line

$$3x + 2y = 12 \qquad xy\text{-trace}$$

in the xy-plane. This line intersects the x-axis at $(4, 0, 0)$ and the y-axis at $(0, 6, 0)$. In Figure 11.49, this process is continued by finding the yz-trace and the xz-trace, and then shading the triangular region lying in the first octant.

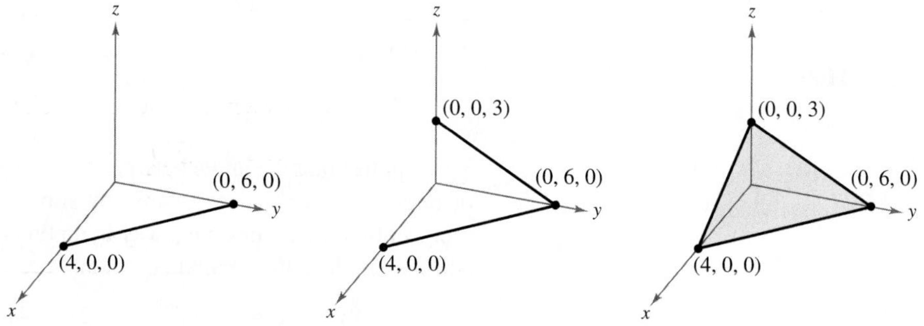

xy-trace ($z = 0$): $3x + 2y = 12$ | yz-trace ($x = 0$): $2y + 4z = 12$ | xz-trace ($y = 0$): $3x + 4z = 12$

Traces of the plane $3x + 2y + 4z = 12$

Figure 11.49

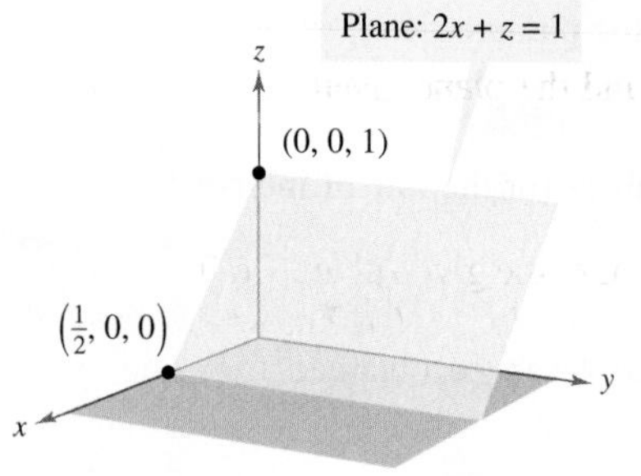

Plane $2x + z = 1$ is parallel to the y-axis.

Figure 11.50

If an equation of a plane has a missing variable, such as $2x + z = 1$, the plane must be *parallel to the axis* represented by the missing variable, as shown in Figure 11.50. If two variables are missing from an equation of a plane, it is *parallel to the coordinate plane* represented by the missing variables, as shown in Figure 11.51.

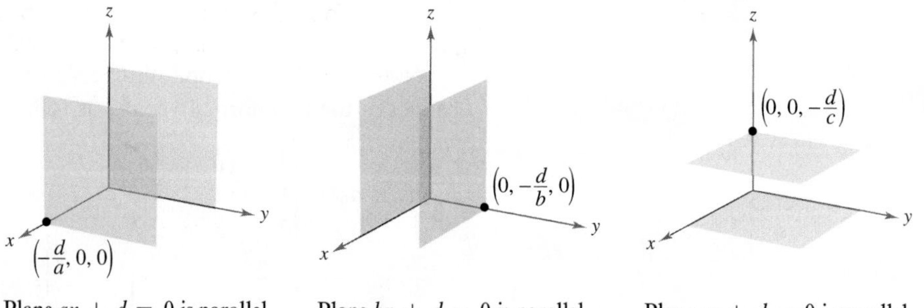

Plane $ax + d = 0$ is parallel to the yz-plane. | Plane $by + d = 0$ is parallel to the xz-plane. | Plane $cz + d = 0$ is parallel to the xy-plane.

Figure 11.51

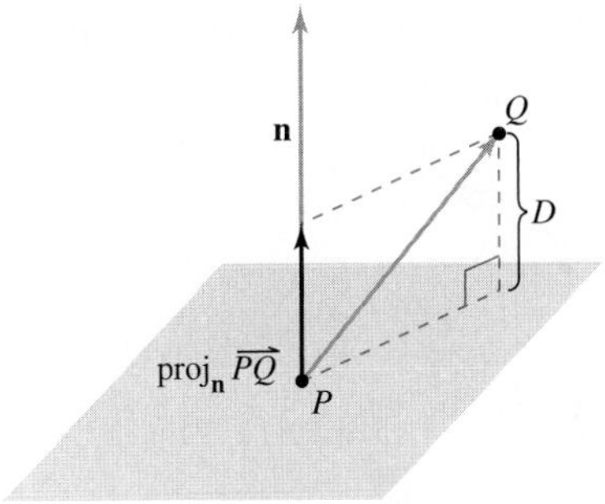

$D = \|\text{proj}_{\mathbf{n}}\overrightarrow{PQ}\|$

The distance between a point and a plane
Figure 11.52

Distances Between Points, Planes, and Lines

This section is concluded with the following discussion of two basic types of problems involving distance in space.

1. Finding the distance between a point and a plane
2. Finding the distance between a point and a line

The solutions of these problems illustrate the versatility and usefulness of vectors in coordinate geometry: the first problem uses the *dot product* of two vectors, and the second problem uses the *cross product.*

The distance D between a point Q and a plane is the length of the shortest line segment connecting Q to the plane, as shown in Figure 11.52. If P is *any* point in the plane, you can find this distance by projecting the vector $\overrightarrow{PQ}$ onto the normal vector $\mathbf{n}$. The length of this projection is the desired distance.

THEOREM 11.13 Distance Between a Point and a Plane

The distance between a plane and a point Q (not in the plane) is

$$D = \|\text{proj}_{\mathbf{n}}\overrightarrow{PQ}\| = \frac{|\overrightarrow{PQ} \cdot \mathbf{n}|}{\|\mathbf{n}\|}$$

where P is a point in the plane and $\mathbf{n}$ is normal to the plane.

To find a point in the plane given by $ax + by + cz + d = 0$ $(a \neq 0)$, let $y = 0$ and $z = 0$. Then, from the equation $ax + d = 0$, you can conclude that the point $(-d/a, 0, 0)$ lies in the plane.

EXAMPLE 5 Finding the Distance Between a Point and a Plane

Find the distance between the point $Q(1, 5, -4)$ and the plane given by

$$3x - y + 2z = 6.$$

Solution You know that $\mathbf{n} = \langle 3, -1, 2\rangle$ is normal to the given plane. To find a point in the plane, let $y = 0$ and $z = 0$, and obtain the point $P(2, 0, 0)$. The vector from P to Q is given by

$$\begin{aligned}\overrightarrow{PQ} &= \langle 1 - 2, 5 - 0, -4 - 0\rangle \\ &= \langle -1, 5, -4\rangle.\end{aligned}$$

Using the Distance Formula given in Theorem 11.13 produces

$$\begin{aligned}D = \frac{|\overrightarrow{PQ} \cdot \mathbf{n}|}{\|\mathbf{n}\|} &= \frac{|\langle -1, 5, -4\rangle \cdot \langle 3, -1, 2\rangle|}{\sqrt{9 + 1 + 4}} \qquad \text{Distance between a point and a plane} \\ &= \frac{|-3 - 5 - 8|}{\sqrt{14}} \\ &= \frac{16}{\sqrt{14}}.\end{aligned}$$

NOTE The choice of the point P in Example 5 is arbitrary. Try choosing a different point in the plane to verify that you obtain the same distance.

From Theorem 11.13, you can determine that the distance between the point $Q(x_0, y_0, z_0)$ and the plane given by $ax + by + cz + d = 0$ is

$$D = \frac{|a(x_0 - x_1) + b(y_0 - y_1) + c(z_0 - z_1)|}{\sqrt{a^2 + b^2 + c^2}}$$

or

$$D = \frac{|ax_0 + by_0 + cz_0 + d|}{\sqrt{a^2 + b^2 + c^2}}$$ Distance between a point and a plane

where $P(x_1, y_1, z_1)$ is a point in the plane and $d = -(ax_1 + by_1 + cz_1)$.

EXAMPLE 6 Finding the Distance Between Two Parallel Planes

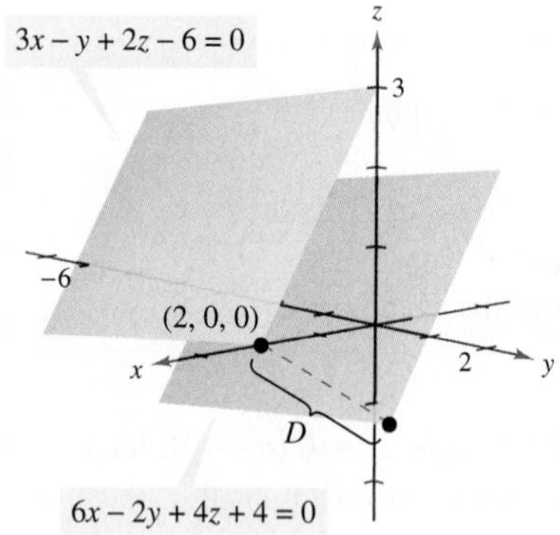

The distance between the parallel planes is approximately 2.14.
Figure 11.53

Find the distance between the two parallel planes given by

$$3x - y + 2z - 6 = 0 \quad \text{and} \quad 6x - 2y + 4z + 4 = 0.$$

Solution The two planes are shown in Figure 11.53. To find the distance between the planes, choose a point in the first plane, say $(x_0, y_0, z_0) = (2, 0, 0)$. Then, from the second plane, you can determine that $a = 6$, $b = -2$, $c = 4$, and $d = 4$, and conclude that the distance is

$$D = \frac{|ax_0 + by_0 + cz_0 + d|}{\sqrt{a^2 + b^2 + c^2}}$$ Distance between a point and a plane

$$= \frac{|6(2) + (-2)(0) + (4)(0) + 4|}{\sqrt{6^2 + (-2)^2 + 4^2}}$$

$$= \frac{16}{\sqrt{56}} = \frac{8}{\sqrt{14}} \approx 2.14.$$

The formula for the distance between a point and a line in space resembles that for the distance between a point and a plane—except that you replace the dot product with the length of the cross product and the normal vector **n** with a direction vector for the line.

THEOREM 11.14 Distance Between a Point and a Line in Space

The distance between a point Q and a line in space is given by

$$D = \frac{\|\overrightarrow{PQ} \times \mathbf{u}\|}{\|\mathbf{u}\|}$$

where **u** is a direction vector for the line and P is a point on the line.

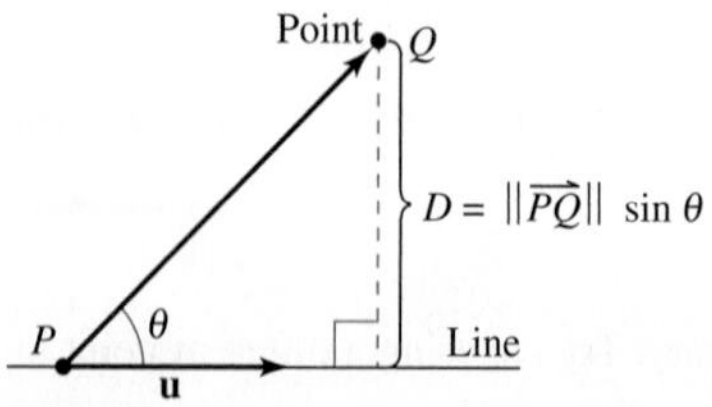

The distance between a point and a line
Figure 11.54

Proof In Figure 11.54, let D be the distance between the point Q and the given line. Then $D = \|\overrightarrow{PQ}\| \sin \theta$, where θ is the angle between **u** and $\overrightarrow{PQ}$. By Theorem 11.8, you have

$$\|\mathbf{u}\| \|\overrightarrow{PQ}\| \sin \theta = \|\mathbf{u} \times \overrightarrow{PQ}\| = \|\overrightarrow{PQ} \times \mathbf{u}\|.$$

Consequently,

$$D = \|\overrightarrow{PQ}\| \sin \theta = \frac{\|\overrightarrow{PQ} \times \mathbf{u}\|}{\|\mathbf{u}\|}.$$

EXAMPLE 7 Finding the Distance Between a Point and a Line

Find the distance between the point $Q(3, -1, 4)$ and the line given by

$$x = -2 + 3t, \quad y = -2t, \quad \text{and} \quad z = 1 + 4t.$$

Solution Using the direction numbers 3, -2, and 4, you know that a direction vector for the line is

$$\mathbf{u} = \langle 3, -2, 4 \rangle. \qquad \text{Direction vector for line}$$

To find a point on the line, let $t = 0$ and obtain

$$P = (-2, 0, 1). \qquad \text{Point on the line}$$

So,

$$\overrightarrow{PQ} = \langle 3 - (-2), -1 - 0, 4 - 1 \rangle = \langle 5, -1, 3 \rangle$$

and you can form the cross product

$$\overrightarrow{PQ} \times \mathbf{u} = \begin{vmatrix} \mathbf{i} & \mathbf{j} & \mathbf{k} \\ 5 & -1 & 3 \\ 3 & -2 & 4 \end{vmatrix} = 2\mathbf{i} - 11\mathbf{j} - 7\mathbf{k} = \langle 2, -11, -7 \rangle.$$

Finally, using Theorem 11.14, you can find the distance to be

$$D = \frac{\|\overrightarrow{PQ} \times \mathbf{u}\|}{\|\mathbf{u}\|} = \frac{\sqrt{174}}{\sqrt{29}} = \sqrt{6} \approx 2.45. \qquad \text{See Figure 11.55.}$$

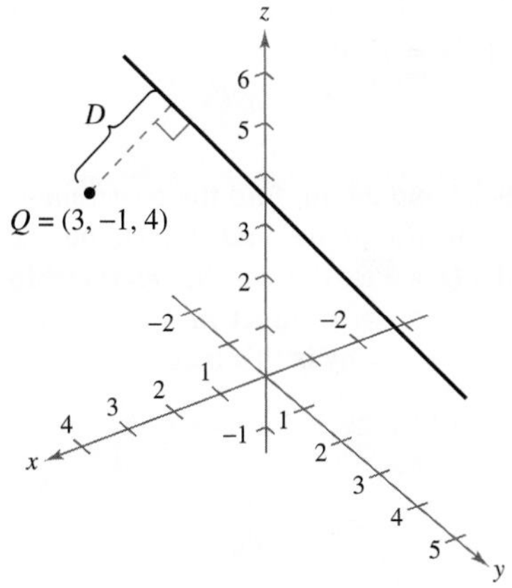

The distance between the point Q and the line is $\sqrt{6} \approx 2.45$.
Figure 11.55

Exercises for Section 11.5

See www.CalcChat.com for worked-out solutions to odd-numbered exercises.

In Exercises 1 and 2, the figure shows the graph of a line given by the parametric equations. (a) Draw an arrow on the line to indicate its orientation. To print an enlarged copy of the graph, go to the website *www.mathgraphs.com*. (b) Find the coordinates of two points, P and Q, on the line. Determine the vector $\overrightarrow{PQ}$. What is the relationship between the components of the vector and the coefficients of t in the parametric equations? Why is this true? (c) Determine the coordinates of any points of intersection with the coordinate planes. If the line does not intersect a coordinate plane, explain why.

1. $x = 1 + 3t$
$y = 2 - t$
$z = 2 + 5t$

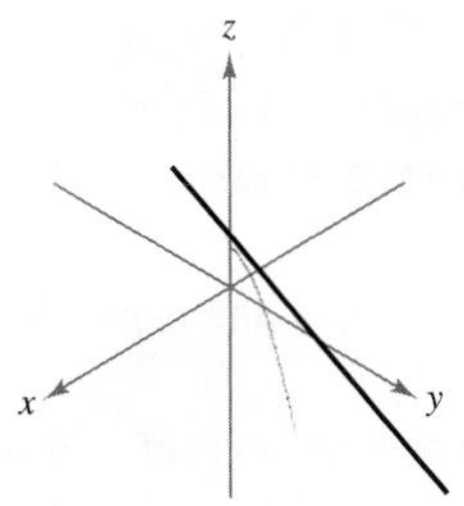

2. $x = 2 - 3t$
$y = 2$
$z = 1 - t$

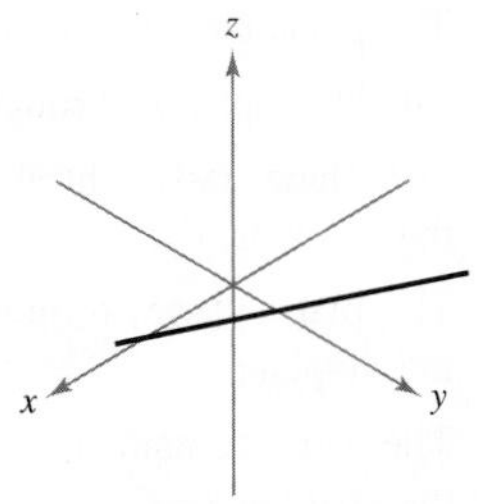

In Exercises 3–8, find sets of (a) parametric equations and (b) symmetric equations of the line through the point parallel to the given vector or line. (For each line, write the direction numbers as integers.)

	Point	*Parallel to*
3.	$(0, 0, 0)$	$\mathbf{v} = \langle 1, 2, 3 \rangle$
4.	$(0, 0, 0)$	$\mathbf{v} = \left\langle -2, \frac{5}{2}, 1 \right\rangle$
5.	$(-2, 0, 3)$	$\mathbf{v} = 2\mathbf{i} + 4\mathbf{j} - 2\mathbf{k}$
6.	$(-3, 0, 2)$	$\mathbf{v} = 6\mathbf{j} + 3\mathbf{k}$
7.	$(1, 0, 1)$	$x = 3 + 3t, y = 5 - 2t, z = -7 + t$
8.	$(-3, 5, 4)$	$\dfrac{x - 1}{3} = \dfrac{y + 1}{-2} = z - 3$

In Exercises 9–12, find sets of (a) parametric equations and (b) symmetric equations of the line through the two points. (For each line, write the direction numbers as integers.)

9. $(5, -3, -2), \left(-\frac{2}{3}, \frac{2}{3}, 1\right)$

10. $(2, 0, 2), (1, 4, -3)$

11. $(2, 3, 0), (10, 8, 12)$

12. $(0, 0, 25), (10, 10, 0)$

In Exercises 13–20, find a set of parametric equations of the line.

13. The line passes through the point $(2, 3, 4)$ and is parallel to the xz-plane and the yz-plane.

14. The line passes through the point $(-4, 5, 2)$ and is parallel to the xy-plane and the yz-plane.

15. The line passes through the point $(2, 3, 4)$ and is perpendicular to the plane given by $3x + 2y - z = 6$.

16. The line passes through the point $(-4, 5, 2)$ and is perpendicular to the plane given by $-x + 2y + z = 5$.

17. The line passes through the point $(5, -3, -4)$ and is parallel to $\mathbf{v} = \langle 2, -1, 3 \rangle$.

18. The line passes through the point $(-1, 4, -3)$ and is parallel to $\mathbf{v} = 5\mathbf{i} - \mathbf{j}$.

19. The line passes through the point $(2, 1, 2)$ and is parallel to the line $x = -t,\ y = 1 + t,\ z = -2 + t$.

20. The line passes through the point $(-6, 0, 8)$ and is parallel to the line $x = 5 - 2t,\ y = -4 + 2t,\ z = 0$.

In Exercises 21–24, find the coordinates of a point *P* on the line and a vector v parallel to the line.

21. $x = 3 - t, \quad y = -1 + 2t, \quad z = -2$

22. $x = 4t, \quad y = 5 - t, \quad z = 4 + 3t$

23. $\dfrac{x - 7}{4} = \dfrac{y + 6}{2} = z + 2$

24. $\dfrac{x + 3}{5} = \dfrac{y}{8} = \dfrac{z - 3}{6}$

In Exercises 25 and 26, determine if any of the lines are parallel or identical.

25. L_1: $x = 6 - 3t, \quad y = -2 + 2t, \quad z = 5 + 4t$

L_2: $x = 6t, \quad y = 2 - 4t, \quad z = 13 - 8t$

L_3: $x = 10 - 6t, \quad y = 3 + 4t, \quad z = 7 + 8t$

L_4: $x = -4 + 6t, \quad y = 3 + 4t, \quad z = 5 - 6t$

26. L_1: $\dfrac{x - 8}{4} = \dfrac{y + 5}{-2} = \dfrac{z + 9}{3}$

L_2: $\dfrac{x + 7}{2} = \dfrac{y - 4}{1} = \dfrac{z + 6}{5}$

L_3: $\dfrac{x + 4}{-8} = \dfrac{y - 1}{4} = \dfrac{z + 18}{-6}$

L_4: $\dfrac{x - 2}{-2} = \dfrac{y + 3}{1} = \dfrac{z - 4}{1.5}$

In Exercises 27–30, determine whether the lines intersect, and if so, find the point of intersection and the cosine of the angle of intersection.

27. $x = 4t + 2, \quad y = 3, \quad z = -t + 1$

$x = 2s + 2, \quad y = 2s + 3, \quad z = s + 1$

28. $x = -3t + 1,\ y = 4t + 1,\ z = 2t + 4$

$x = 3s + 1,\ y = 2s + 4,\ z = -s + 1$

29. $\dfrac{x}{3} = \dfrac{y - 2}{-1} = z + 1, \quad \dfrac{x - 1}{4} = y + 2 = \dfrac{z + 3}{-3}$

30. $\dfrac{x - 2}{-3} = \dfrac{y - 2}{6} = z - 3, \quad \dfrac{x - 3}{2} = y + 5 = \dfrac{z + 2}{4}$

In Exercises 31 and 32, use a computer algebra system to graph the pair of intersecting lines and find the point of intersection.

31. $x = 2t + 3,\ y = 5t - 2,\ z = -t + 1$

$x = -2s + 7,\ y = s + 8,\ z = 2s - 1$

32. $x = 2t - 1,\ y = -4t + 10,\ z = t$

$x = -5s - 12,\ y = 3s + 11,\ z = -2s - 4$

Cross Product **In Exercises 33 and 34, (a) find the coordinates of three points *P*, *Q*, and *R* in the plane, and determine the vectors $\overrightarrow{PQ}$ and $\overrightarrow{PR}$. (b) Find $\overrightarrow{PQ} \times \overrightarrow{PR}$. What is the relationship between the components of the cross product and the coefficients of the equation of the plane? Why is this true?**

33. $4x - 3y - 6z = 6$

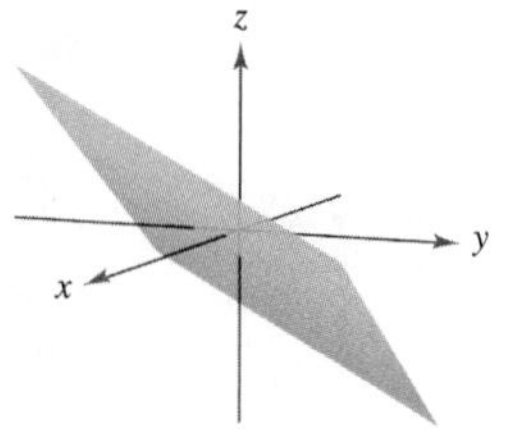

34. $2x + 3y + 4z = 4$

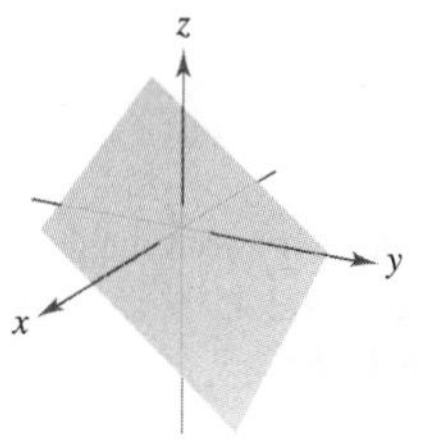

In Exercises 35–40, find an equation of the plane passing through the point perpendicular to the given vector or line.

	Point	*Perpendicular to*
35.	$(2, 1, 2)$	$\mathbf{n} = \mathbf{i}$
36.	$(1, 0, -3)$	$\mathbf{n} = \mathbf{k}$
37.	$(3, 2, 2)$	$\mathbf{n} = 2\mathbf{i} + 3\mathbf{j} - \mathbf{k}$
38.	$(0, 0, 0)$	$\mathbf{n} = -3\mathbf{i} + 2\mathbf{k}$
39.	$(0, 0, 6)$	$x = 1 - t,\ y = 2 + t,\ z = 4 - 2t$
40.	$(3, 2, 2)$	$\dfrac{x - 1}{4} = y + 2 = \dfrac{z + 3}{-3}$

In Exercises 41–52, find an equation of the plane.

41. The plane passes through $(0, 0, 0)$, $(1, 2, 3)$, and $(-2, 3, 3)$.

42. The plane passes through $(2, 3, -2)$, $(3, 4, 2)$, and $(1, -1, 0)$.

43. The plane passes through $(1, 2, 3)$, $(3, 2, 1)$, and $(-1, -2, 2)$.

44. The plane passes through the point $(1, 2, 3)$ and is parallel to the yz-plane.

45. The plane passes through the point $(1, 2, 3)$ and is parallel to the xy-plane.

46. The plane contains the y-axis and makes an angle of $\pi/6$ with the positive x-axis.

47. The plane contains the lines given by

$$\frac{x-1}{-2} = y - 4 = z \quad \text{and} \quad \frac{x-2}{-3} = \frac{y-1}{4} = \frac{z-2}{-1}.$$

48. The plane passes through the point $(2, 2, 1)$ and contains the line given by

$$\frac{x}{2} = \frac{y-4}{-1} = z.$$

49. The plane passes through the points $(2, 2, 1)$ and $(-1, 1, -1)$ and is perpendicular to the plane $2x - 3y + z = 3$.

50. The plane passes through the points $(3, 2, 1)$ and $(3, 1, -5)$ and is perpendicular to the plane $6x + 7y + 2z = 10$.

51. The plane passes through the points $(1, -2, -1)$ and $(2, 5, 6)$ and is parallel to the x-axis.

52. The plane passes through the points $(4, 2, 1)$ and $(-3, 5, 7)$ and is parallel to the z-axis.

In Exercises 53 and 54, sketch a graph of the line and find the points (if any) where the line intersects the *xy*-, *xz*-, and *yz*-planes.

53. $x = 1 - 2t, \quad y = -2 + 3t, \quad z = -4 + t$

54. $\dfrac{x-2}{3} = y + 1 = \dfrac{z-3}{2}$

In Exercises 55 and 56, find an equation of the plane that contains all the points that are equidistant from the given points.

55. $(2, 2, 0), \quad (0, 2, 2)$

56. $(-3, 1, 2), \quad (6, -2, 4)$

In Exercises 57–62, determine whether the planes are parallel, orthogonal, or neither. If they are neither parallel nor orthogonal, find the angle of intersection.

57. $5x - 3y + z = 4$
$x + 4y + 7z = 1$

58. $3x + y - 4z = 3$
$-9x - 3y + 12z = 4$

59. $x - 3y + 6z = 4$
$5x + y - z = 4$

60. $3x + 2y - z = 7$
$x - 4y + 2z = 0$

61. $x - 5y - z = 1$
$5x - 25y - 5z = -3$

62. $2x - z = 1$
$4x + y + 8z = 10$

In Exercises 63–70, label any intercepts and sketch a graph of the plane.

63. $4x + 2y + 6z = 12$

64. $3x + 6y + 2z = 6$

65. $2x - y + 3z = 4$

66. $2x - y + z = 4$

67. $y + z = 5$

68. $x + 2y = 4$

69. $x = 5$

70. $z = 8$

In Exercises 71–74, use a computer algebra system to graph the plane.

71. $2x + y - z = 6$

72. $x - 3z = 3$

73. $-5x + 4y - 6z = -8$

74. $2.1x - 4.7y - z = -3$

In Exercises 75 and 76, determine if any of the planes are parallel or identical.

75. P_1: $3x - 2y + 5z = 10$
P_2: $-6x + 4y - 10z = 5$
P_3: $-3x + 2y + 5z = 8$
P_4: $75x - 50y + 125z = 250$

76. P_1: $-60x + 90y + 30z = 27$
P_2: $6x - 9y - 3z = 2$
P_3: $-20x + 30y + 10z = 9$
P_4: $12x - 18y + 6z = 5$

In Exercises 77–80, describe the family of planes represented by the equation, where *c* is any real number.

77. $x + y + z = c$

78. $x + y = c$

79. $cy + z = 0$

80. $x + cz = 0$

In Exercises 81 and 82, find a set of parametric equations for the line of intersection of the planes.

81. $3x + 2y - z = 7$
$x - 4y + 2z = 0$

82. $6x - 3y + z = 5$
$-x + y + 5z = 5$

In Exercises 83–86, find the point(s) of intersection (if any) of the plane and the line. Also determine whether the line lies in the plane.

83. $2x - 2y + z = 12, \quad x - \dfrac{1}{2} = \dfrac{y + (3/2)}{-1} = \dfrac{z+1}{2}$

84. $2x + 3y = -5, \quad \dfrac{x-1}{4} = \dfrac{y}{2} = \dfrac{z-3}{6}$

85. $2x + 3y = 10, \quad \dfrac{x-1}{3} = \dfrac{y+1}{-2} = z - 3$

86. $5x + 3y = 17, \quad \dfrac{x-4}{2} = \dfrac{y+1}{-3} = \dfrac{z+2}{5}$

In Exercises 87–90, find the distance between the point and the plane.

87. $(0, 0, 0)$
$2x + 3y + z = 12$

88. $(0, 0, 0)$
$8x - 4y + z = 8$

89. $(2, 8, 4)$
$2x + y + z = 5$

90. $(3, 2, 1)$
$x - y + 2z = 4$

In Exercises 91–94, verify that the two planes are parallel, and find the distance between the planes.

91. $x - 3y + 4z = 10$
$x - 3y + 4z = 6$

92. $4x - 4y + 9z = 7$
$4x - 4y + 9z = 18$

93. $-3x + 6y + 7z = 1$
$6x - 12y - 14z = 25$

94. $2x - 4z = 4$
$2x - 4z = 10$

In Exercises 95–98, find the distance between the point and the line given by the set of parametric equations.

95. $(1, 5, -2)$; $x = 4t - 2$, $y = 3$, $z = -t + 1$

96. $(1, -2, 4)$; $x = 2t$, $y = t - 3$, $z = 2t + 2$

97. $(-2, 1, 3)$; $x = 1 - t$, $y = 2 + t$, $z = -2t$

98. $(4, -1, 5)$; $x = 3$, $y = 1 + 3t$, $z = 1 + t$

In Exercises 99 and 100, verify that the lines are parallel, and find the distance between them.

99. L_1: $x = 2 - t$, $y = 3 + 2t$, $z = 4 + t$
L_2: $x = 3t$, $y = 1 - 6t$, $z = 4 - 3t$

100. L_1: $x = 3 + 6t$, $y = -2 + 9t$, $z = 1 - 12t$
L_2: $x = -1 + 4t$, $y = 3 + 6t$, $z = -8t$

Writing About Concepts

101. Give the parametric equations and the symmetric equations of a line in space. Describe what is required to find these equations.

102. Give the standard equation of a plane in space. Describe what is required to find this equation.

103. Describe a method of finding the line of intersection of two planes.

104. Describe each surface given by the equations $x = a$, $y = b$, and $z = c$.

105. Describe a method for determining when two planes

$$a_1x + b_1y + c_1z + d_1 = 0$$

and

$$a_2x + b_2y + c_2z + d_2 = 0$$

are (a) parallel and (b) perpendicular. Explain your reasoning.

106. Let L_1 and L_2 be nonparallel lines that do not intersect. Is it possible to find a nonzero vector **v** such that **v** is perpendicular to both L_1 and L_2? Explain your reasoning.

107. Find an equation of the plane with x-intercept $(a, 0, 0)$, y-intercept $(0, b, 0)$, and z-intercept $(0, 0, c)$. (Assume a, b, and c are nonzero.)

108. (a) Describe and find an equation for the surface generated by all points (x, y, z) that are four units from the point $(3, -2, 5)$.

(b) Describe and find an equation for the surface generated by all points (x, y, z) that are four units from the plane

$$4x - 3y + z = 10.$$

109. ***Modeling Data*** Per capita consumptions (in gallons) of different types of plain milk in the United States from 1994 to 2000 are shown in the table. Consumption of light and skim milks, reduced-fat milk, and whole milk are represented by the variables x, y, and z, respectively. *(Source: U.S. Department of Agriculture)*

Year	1994	1995	1996	1997	1998	1999	2000
x	5.8	6.2	6.4	6.6	6.5	6.3	6.1
y	8.7	8.2	8.0	7.7	7.4	7.3	7.1
z	8.8	8.4	8.4	8.2	7.8	7.9	7.8

A model for the data is given by $0.04x - 0.64y + z = 3.4$.

(a) Complete a fourth row in the table using the model to approximate z for the given values of x and y. Compare the approximations with the actual values of z.

(b) According to this model, any increases in consumption of two types of milk will have what effect on the consumption of the third type?

110. ***Mechanical Design*** A chute at the top of a grain elevator of a combine funnels the grain into a bin (see figure). Find the angle between two adjacent sides.

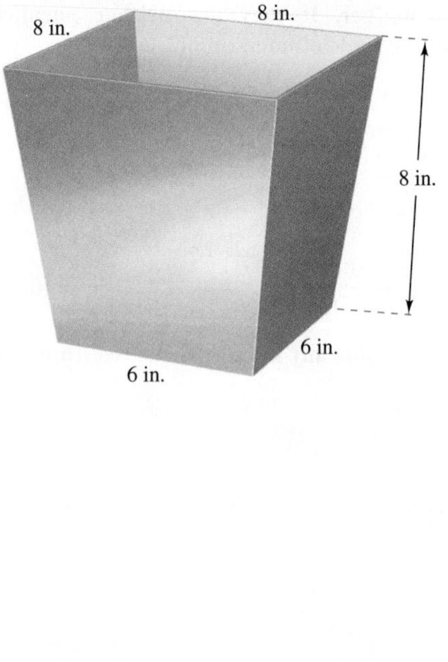

111. *Distance* Two insects are crawling along different lines in three-space. At time t (in minutes), the first insect is at the point (x, y, z) on the line

$$x = 6 + t, \quad y = 8 - t, \quad z = 3 + t.$$

Also, at time t, the second insect is at the point (x, y, z) on the line

$$x = 1 + t, \quad y = 2 + t, \quad z = 2t.$$

Assume distances are given in inches.

(a) Find the distance between the two insects at time $t = 0$.

(b) Use a graphing utility to graph the distance between the insects from $t = 0$ to $t = 10$.

(c) Using the graph from part (b), what can you conclude about the distance between the insects?

(d) How close do the insects get?

112. Find the standard equation of the sphere with center $(-3, 2, 4)$ that is tangent to the plane given by $2x + 4y - 3z = 8$.

113. Find the point of intersection of the plane $3x - y + 4z = 7$ and the line through $(5, 4, -3)$ that is perpendicular to this plane.

114. Show that the plane $2x - y - 3z = 4$ is parallel to the line $x = -2 + 2t$, $y = -1 + 4t$, $z = 4$, and find the distance between them.

115. Find the point of intersection of the line through $(1, -3, 1)$ and $(3, -4, 2)$, and the plane given by $x - y + z = 2$.

116. Find a set of parametric equations for the line passing through the point $(1, 0, 2)$ that is parallel to the plane given by $x + y + z = 5$, and perpendicular to the line $x = t$, $y = 1 + t$, $z = 1 + t$.

True or False? **In Exercises 117–120, determine whether the statement is true or false. If it is false, explain why or give an example that shows it is false.**

117. If $\mathbf{v} = a_1\mathbf{i} + b_1\mathbf{j} + c_1\mathbf{k}$ is any vector in the plane given by $a_2x + b_2y + c_2z + d_2 = 0$, then $a_1a_2 + b_1b_2 + c_1c_2 = 0$.

118. Every pair of lines in space are either intersecting or parallel.

119. Two planes in space are either intersecting or parallel.

120. If two lines L_1 and L_2 are parallel to a plane P, then L_1 and L_2 are parallel.

Section Project: Distances in Space

You have learned two distance formulas in this section—the distance between a point and a plane, and the distance between a point and a line. In this project you will study a third distance problem—the distance between two skew lines. Two lines in space are *skew* if they are neither parallel nor intersecting (see figure).

(a) Consider the following two lines in space.

L_1: $x = 4 + 5t$, $y = 5 + 5t$, $z = 1 - 4t$

L_2: $x = 4 + s$, $y = -6 + 8s$, $z = 7 - 3s$

(i) Show that these lines are not parallel.

(ii) Show that these lines do not intersect, and therefore are skew lines.

(iii) Show that the two lines lie in parallel planes.

(iv) Find the distance between the parallel planes from part (iii). This is the distance between the original skew lines.

(b) Use the procedure in part (a) to find the distance between the lines.

L_1: $x = 2t$, $y = 4t$, $z = 6t$

L_2: $x = 1 - s$, $y = 4 + s$, $z = -1 + s$

(c) Use the procedure in part (a) to find the distance between the lines.

L_1: $x = 3t$, $y = 2 - t$, $z = -1 + t$

L_2: $x = 1 + 4s$, $y = -2 + s$, $z = -3 - 3s$

(d) Develop a formula for finding the distance between the skew lines.

L_1: $x = x_1 + a_1t$, $y = y_1 + b_1t$, $z = z_1 + c_1t$

L_2: $x = x_2 + a_2s$, $y = y_2 + b_2s$, $z = z_2 + c_2s$

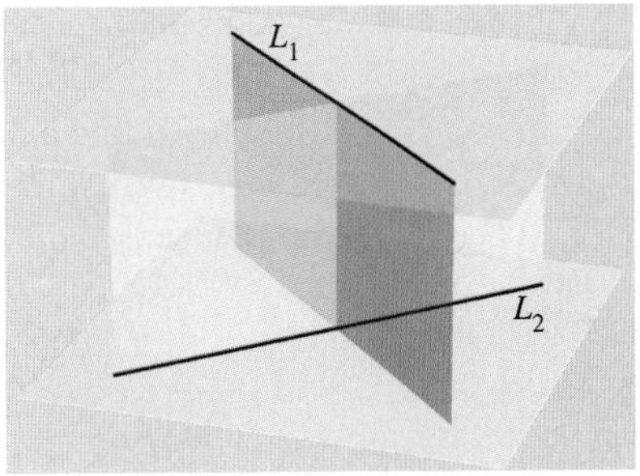

Section 11.6

Surfaces in Space

- **Recognize and write equations for cylindrical surfaces.**
- **Recognize and write equations for quadric surfaces.**
- **Recognize and write equations for surfaces of revolution.**

Cylindrical Surfaces

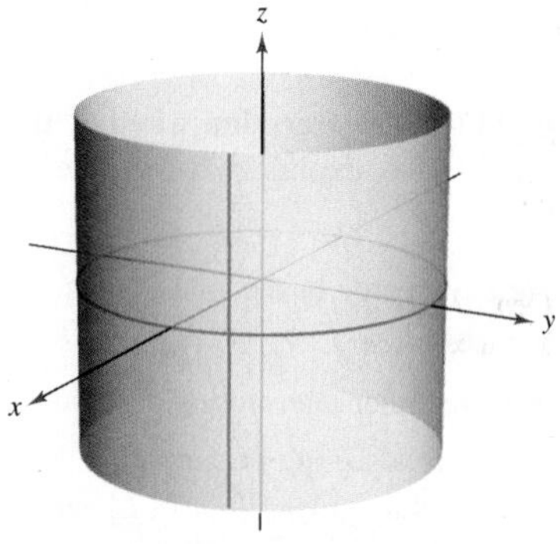

Rulings are parallel to z-axis.
Figure 11.56

The first five sections of this chapter contained the vector portion of the preliminary work necessary to study vector calculus and the calculus of space. In this and the next section, you will study surfaces in space and alternative coordinate systems for space. You have already studied two special types of surfaces.

1. Spheres: $(x - x_0)^2 + (y - y_0)^2 + (z - z_0)^2 = r^2$ — Section 11.2
2. Planes: $ax + by + cz + d = 0$ — Section 11.5

A third type of surface in space is called a **cylindrical surface,** or simply a **cylinder.** To define a cylinder, consider the familiar right circular cylinder shown in Figure 11.56. You can imagine that this cylinder is generated by a vertical line moving around the circle $x^2 + y^2 = a^2$ in the xy-plane. This circle is called a **generating curve** for the cylinder, as indicated in the following definition.

Definition of a Cylinder

Let C be a curve in a plane and let L be a line not in a parallel plane. The set of all lines parallel to L and intersecting C is called a **cylinder.** C is called the **generating curve** (or **directrix**) of the cylinder, and the parallel lines are called **rulings.**

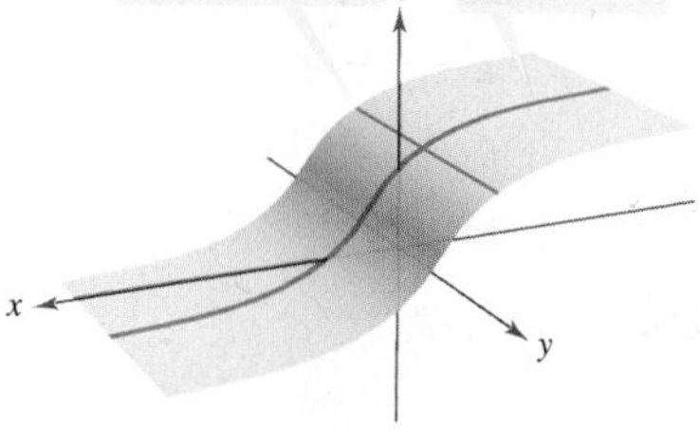

Cylinder: Rulings intersect C and are parallel to the given line.
Figure 11.57

NOTE Without loss of generality, you can assume that C lies in one of the three coordinate planes. Moreover, this text restricts the discussion to *right* cylinders—cylinders whose rulings are perpendicular to the coordinate plane containing C, as shown in Figure 11.57.

For the right circular cylinder shown in Figure 11.56, the equation of the generating curve is

$$x^2 + y^2 = a^2. \qquad \text{Equation of generating curve in } xy\text{-plane}$$

To find an equation for the cylinder, note that you can generate any one of the rulings by fixing the values of x and y and then allowing z to take on all real values. In this sense, the value of z is arbitrary and is, therefore, not included in the equation. In other words, the equation of this cylinder is simply the equation of its generating curve.

$$x^2 + y^2 = a^2 \qquad \text{Equation of cylinder in space}$$

Equations of Cylinders

The equation of a cylinder whose rulings are parallel to one of the coordinate axes contains only the variables corresponding to the other two axes.

EXAMPLE 1 Sketching a Cylinder

Sketch the surface represented by each equation.

a. $z = y^2$ **b.** $z = \sin x, \quad 0 \le x \le 2\pi$

Solution

a. The graph is a cylinder whose generating curve, $z = y^2$, is a parabola in the yz-plane. The rulings of the cylinder are parallel to the x-axis, as shown in Figure 11.58(a).

b. The graph is a cylinder generated by the sine curve in the xz-plane. The rulings are parallel to the y-axis, as shown in Figure 11.58(b).

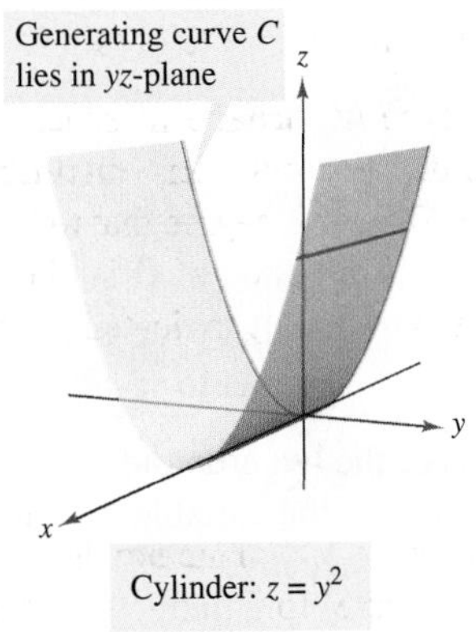

(a) Rulings are parallel to x-axis.

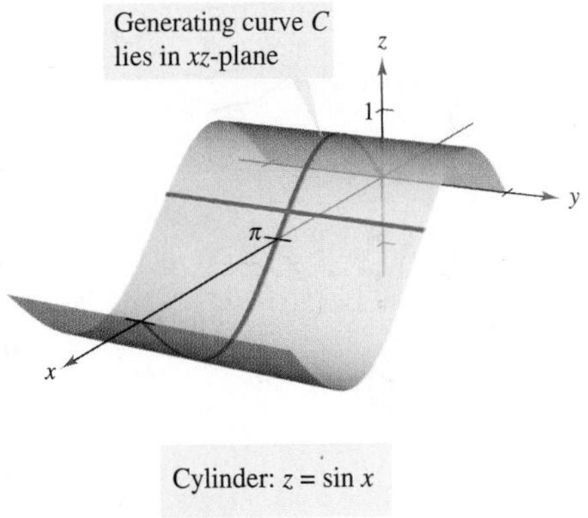

(b) Rulings are parallel to y-axis.

Figure 11.58

Quadric Surfaces

STUDY TIP In the table on pages 812 and 813, only one of several orientations of each quadric surface is shown. If the surface is oriented along a different axis, its standard equation will change accordingly, as illustrated in Examples 2 and 3. The fact that the two types of paraboloids have one variable raised to the first power can be helpful in classifying quadric surfaces. The other four types of basic quadric surfaces have equations that are of *second degree* in all three variables.

The fourth basic type of surface in space is a **quadric surface.** Quadric surfaces are the three-dimensional analogs of conic sections.

Quadric Surface

The equation of a **quadric surface** in space is a second-degree equation of the form

$$Ax^2 + By^2 + Cz^2 + Dxy + Exz + Fyz + Gx + Hy + Iz + J = 0.$$

There are six basic types of quadric surfaces: **ellipsoid, hyperboloid of one sheet, hyperboloid of two sheets, elliptic cone, elliptic paraboloid, and hyperbolic paraboloid.**

The intersection of a surface with a plane is called the **trace of the surface** in the plane. To visualize a surface in space, it is helpful to determine its traces in some well-chosen planes. The traces of quadric surfaces are conics. These traces, together with the **standard form** of the equation of each quadric surface, are shown in the table on pages 812 and 813.

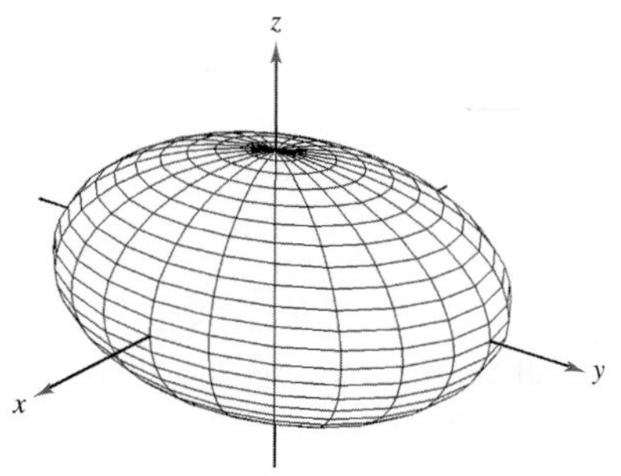

Ellipsoid

$$\frac{x^2}{a^2} + \frac{y^2}{b^2} + \frac{z^2}{c^2} = 1$$

Trace	*Plane*
Ellipse	Parallel to xy-plane
Ellipse	Parallel to xz-plane
Ellipse	Parallel to yz-plane

The surface is a sphere if $a = b = c \neq 0$.

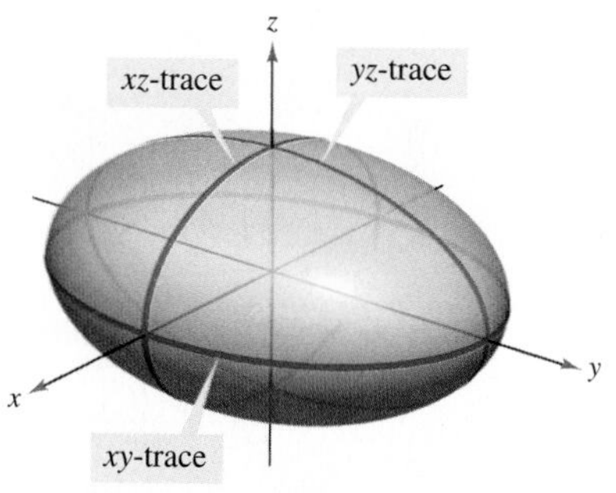

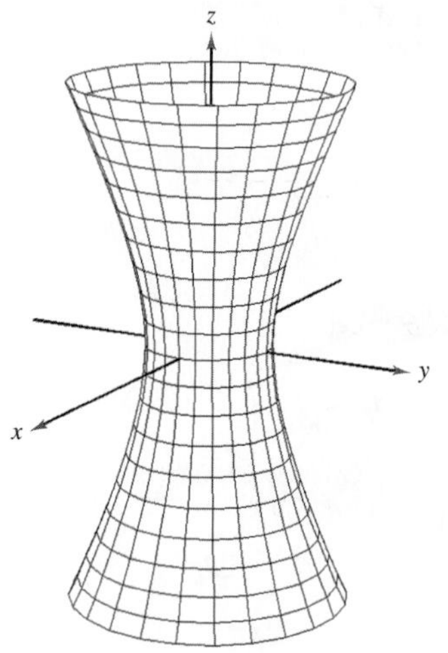

Hyperboloid of One Sheet

$$\frac{x^2}{a^2} + \frac{y^2}{b^2} - \frac{z^2}{c^2} = 1$$

Trace	*Plane*
Ellipse	Parallel to xy-plane
Hyperbola	Parallel to xz-plane
Hyperbola	Parallel to yz-plane

The axis of the hyperboloid corresponds to the variable whose coefficient is negative.

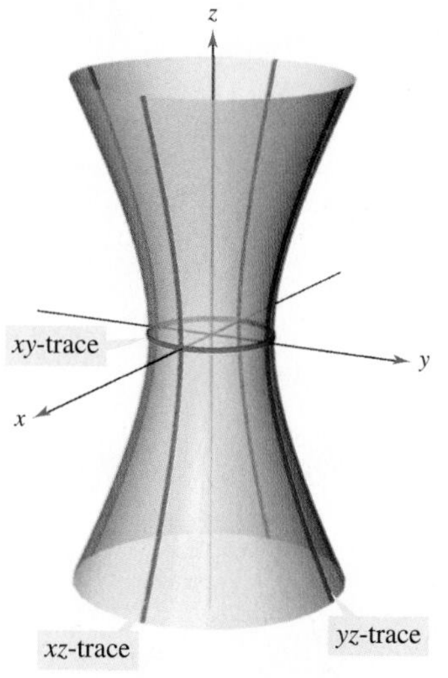

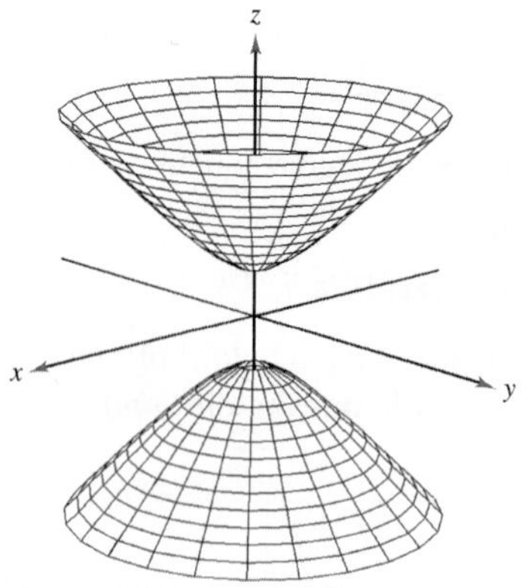

Hyperboloid of Two Sheets

$$\frac{z^2}{c^2} - \frac{x^2}{a^2} - \frac{y^2}{b^2} = 1$$

Trace	*Plane*
Ellipse	Parallel to xy-plane
Hyperbola	Parallel to xz-plane
Hyperbola	Parallel to yz-plane

The axis of the hyperboloid corresponds to the variable whose coefficient is positive. There is no trace in the coordinate plane perpendicular to this axis.

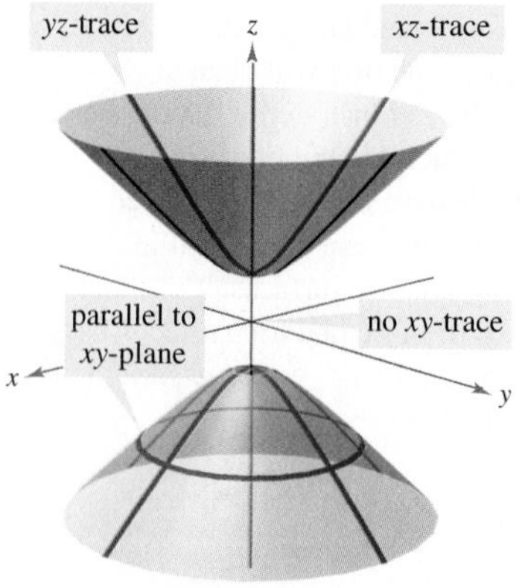

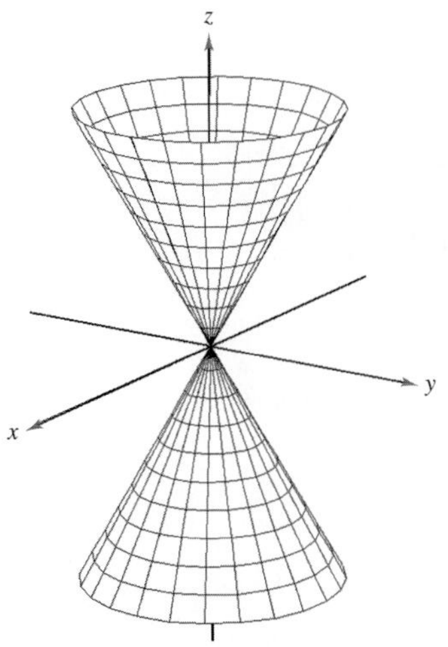

Elliptic Cone

$$\frac{x^2}{a^2} + \frac{y^2}{b^2} - \frac{z^2}{c^2} = 0$$

Trace	*Plane*
Ellipse	Parallel to *xy*-plane
Hyperbola	Parallel to *xz*-plane
Hyperbola	Parallel to *yz*-plane

The axis of the cone corresponds to the variable whose coefficient is negative. The traces in the coordinate planes parallel to this axis are intersecting lines.

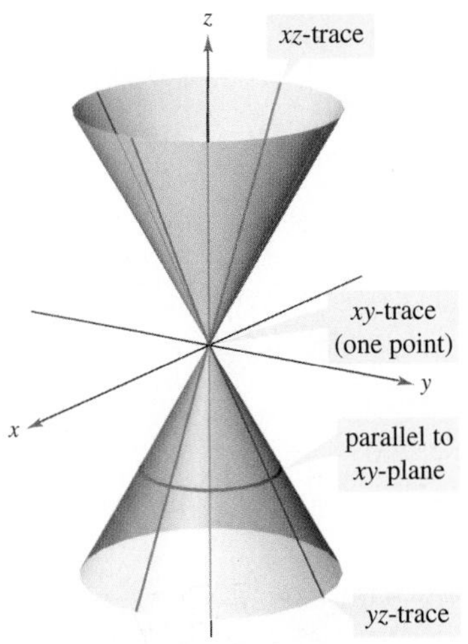

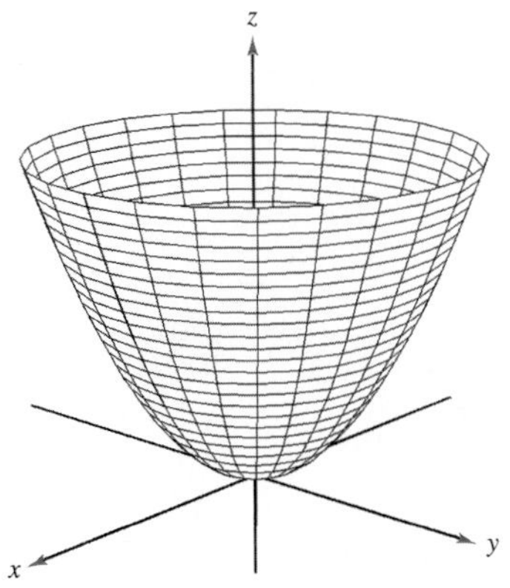

Elliptic Paraboloid

$$z = \frac{x^2}{a^2} + \frac{y^2}{b^2}$$

Trace	*Plane*
Ellipse	Parallel to *xy*-plane
Parabola	Parallel to *xz*-plane
Parabola	Parallel to *yz*-plane

The axis of the paraboloid corresponds to the variable raised to the first power.

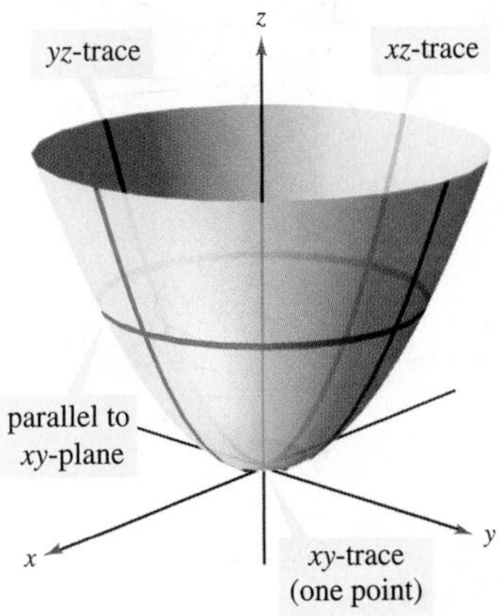

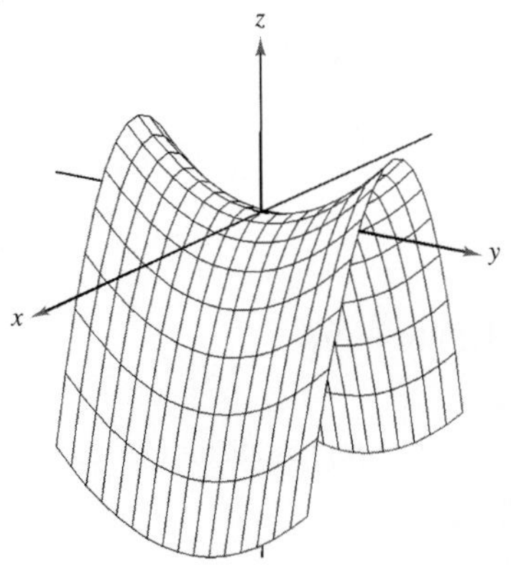

Hyperbolic Paraboloid

$$z = \frac{y^2}{b^2} - \frac{x^2}{a^2}$$

Trace	*Plane*
Hyperbola	Parallel to *xy*-plane
Parabola	Parallel to *xz*-plane
Parabola	Parallel to *yz*-plane

The axis of the paraboloid corresponds to the variable raised to the first power.

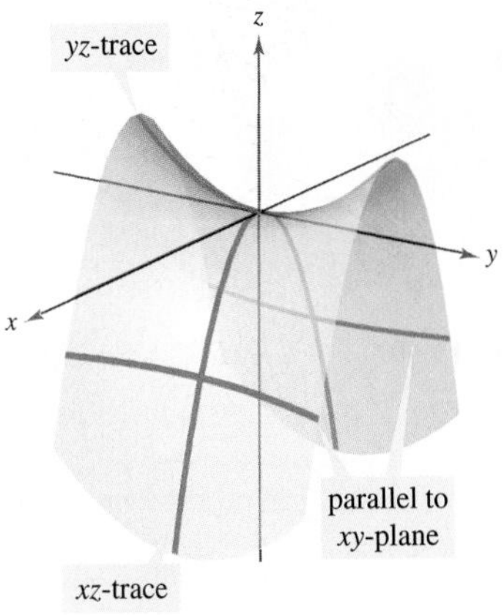

To classify a quadric surface, begin by writing the surface in standard form. Then, determine several traces taken in the coordinate planes *or* taken in planes that are parallel to the coordinate planes.

EXAMPLE 2 Sketching a Quadric Surface

Classify and sketch the surface given by $4x^2 - 3y^2 + 12z^2 + 12 = 0$.

Solution Begin by writing the equation in standard form.

$$4x^2 - 3y^2 + 12z^2 + 12 = 0 \qquad \text{Write original equation.}$$

$$\frac{x^2}{-3} + \frac{y^2}{4} - z^2 - 1 = 0 \qquad \text{Divide by } -12.$$

$$\frac{y^2}{4} - \frac{x^2}{3} - \frac{z^2}{1} = 1 \qquad \text{Standard form}$$

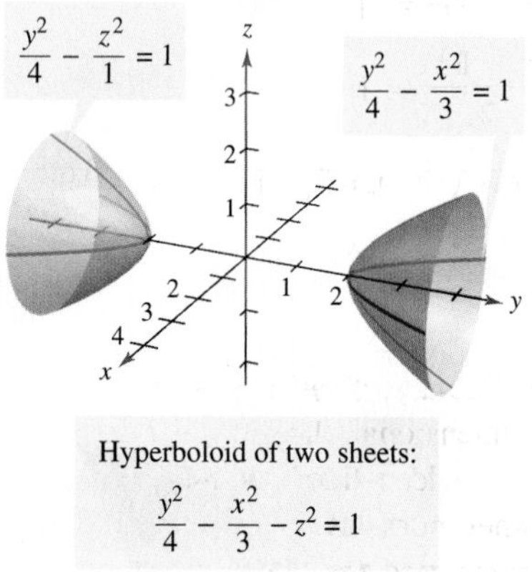

Figure 11.59

From the table on pages 812 and 813, you can conclude that the surface is a hyperboloid of two sheets with the y-axis as its axis. To sketch the graph of this surface, it helps to find the traces in the coordinate planes.

xy-trace ($z = 0$): $\quad \frac{y^2}{4} - \frac{x^2}{3} = 1 \qquad$ Hyperbola

xz-trace ($y = 0$): $\quad \frac{x^2}{3} + \frac{z^2}{1} = -1 \qquad$ No trace

yz-trace ($x = 0$): $\quad \frac{y^2}{4} - \frac{z^2}{1} = 1 \qquad$ Hyperbola

The graph is shown in Figure 11.59.

EXAMPLE 3 Sketching a Quadric Surface

Classify and sketch the surface given by $x - y^2 - 4z^2 = 0$.

Solution Because x is raised only to the first power, the surface is a paraboloid. The axis of the paraboloid is the x-axis. In the standard form, the equation is

$$x = y^2 + 4z^2. \qquad \text{Standard form}$$

Some convenient traces are as follows.

xy-trace ($z = 0$): $\quad x = y^2 \qquad$ Parabola

xz-trace ($y = 0$): $\quad x = 4z^2 \qquad$ Parabola

parallel to yz-plane ($x = 4$): $\quad \frac{y^2}{4} + \frac{z^2}{1} = 1 \qquad$ Ellipse

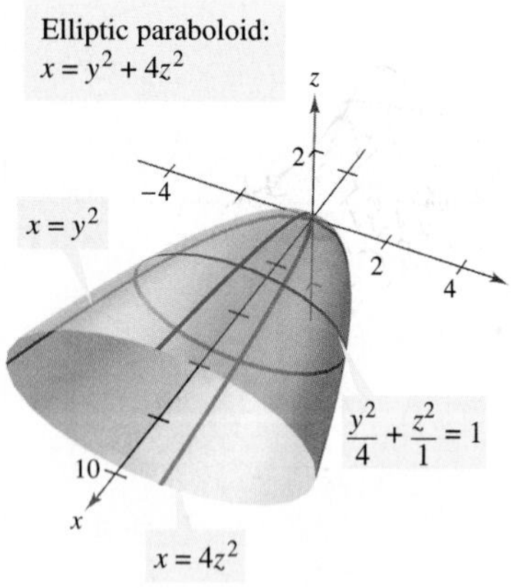

Figure 11.60

The surface is an *elliptic* paraboloid, as shown in Figure 11.60.

Some second-degree equations in x, y, and z do not represent any of the basic types of quadric surfaces. Here are two examples.

$$x^2 + y^2 + z^2 = 0 \qquad \text{Single point}$$

$$x^2 + y^2 = 1 \qquad \text{Right circular cylinder}$$

For a quadric surface not centered at the origin, you can form the standard equation by completing the square, as demonstrated in Example 4.

EXAMPLE 4 A Quadric Surface Not Centered at the Origin

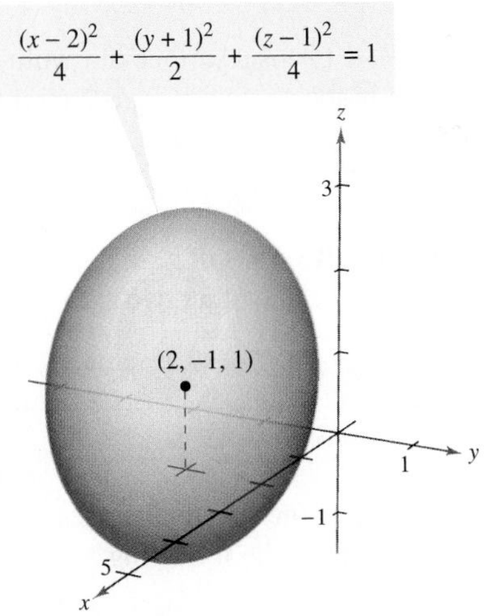

An ellipsoid centered at $(2, -1, 1)$
Figure 11.61

Classify and sketch the surface given by

$$x^2 + 2y^2 + z^2 - 4x + 4y - 2z + 3 = 0.$$

Solution Completing the square for each variable produces the following.

$$(x^2 - 4x + \quad) + 2(y^2 + 2y + \quad) + (z^2 - 2z + \quad) = -3$$
$$(x^2 - 4x + 4) + 2(y^2 + 2y + 1) + (z^2 - 2z + 1) = -3 + 4 + 2 + 1$$
$$(x - 2)^2 + 2(y + 1)^2 + (z - 1)^2 = 4$$
$$\frac{(x - 2)^2}{4} + \frac{(y + 1)^2}{2} + \frac{(z - 1)^2}{4} = 1$$

From this equation, you can see that the quadric surface is an ellipsoid that is centered at $(2, -1, 1)$. Its graph is shown in Figure 11.61.

TECHNOLOGY
A computer algebra system can help you visualize a surface in space.* Most of these computer algebra systems create three-dimensional illusions by sketching several traces of the surface and then applying a "hidden-line" routine that blocks out portions of the surface that lie behind other portions of the surface. Two examples of figures that were generated by *Mathematica* are shown below.

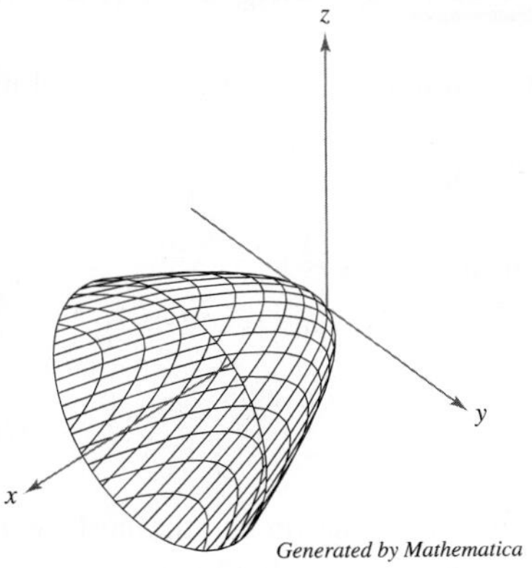

Elliptic paraboloid
$x = \frac{y^2}{2} + \frac{z^2}{2}$

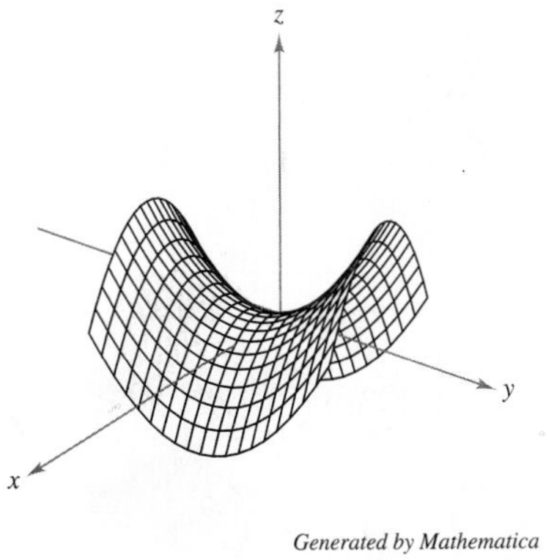

Hyperbolic paraboloid
$z = \frac{y^2}{16} - \frac{x^2}{16}$

Using a graphing utility to graph a surface in space requires practice. For one thing, you must know enough about the surface to be able to specify a *viewing window* that gives a representative view of the surface. Also, you can often improve the view of a surface by rotating the axes. For instance, note that the elliptic paraboloid in the figure is seen from a line of sight that is "higher" than the line of sight used to view the hyperbolic paraboloid.

Some 3-D graphing utilities require surfaces to be entered with parametric equations. For a discussion of this technique, see Section 15.5.

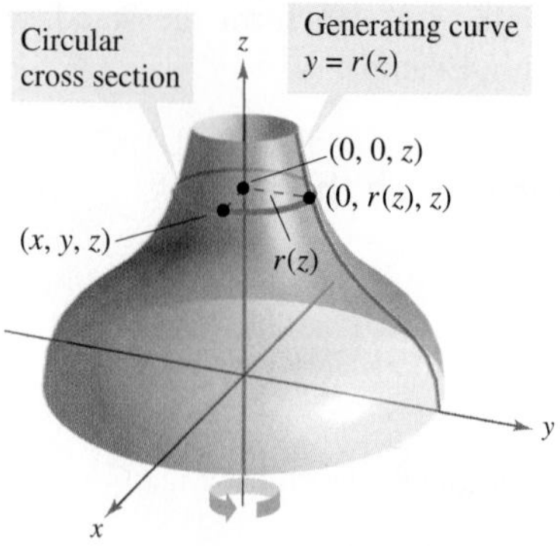

Figure 11.62

Surfaces of Revolution

The fifth special type of surface you will study is called a **surface of revolution.** In Section 7.4, you studied a method for finding the *area* of such a surface. You will now look at a procedure for finding its *equation*. Consider the graph of the **radius function**

$$y = r(z) \qquad \text{Generating curve}$$

in the yz-plane. If this graph is revolved about the z-axis, it forms a surface of revolution, as shown in Figure 11.62. The trace of the surface in the plane $z = z_0$ is a circle whose radius is $r(z_0)$ and whose equation is

$$x^2 + y^2 = [r(z_0)]^2. \qquad \text{Circular trace in plane: } z = z_0$$

Replacing z_0 with z produces an equation that is valid for all values of z. In a similar manner, you can obtain equations for surfaces of revolution for the other two axes, and the results are summarized as follows.

Surface of Revolution

If the graph of a radius function r is revolved about one of the coordinate axes, the equation of the resulting surface of revolution has one of the following forms.

1. Revolved about the x-axis: $y^2 + z^2 = [r(x)]^2$
2. Revolved about the y-axis: $x^2 + z^2 = [r(y)]^2$
3. Revolved about the z-axis: $x^2 + y^2 = [r(z)]^2$

EXAMPLE 5 Finding an Equation for a Surface of Revolution

a. An equation for the surface of revolution formed by revolving the graph of

$$y = \frac{1}{z} \qquad \text{Radius function}$$

about the z-axis is

$$x^2 + y^2 = [r(z)]^2 \qquad \text{Revolved about the } z\text{-axis}$$

$$x^2 + y^2 = \left(\frac{1}{z}\right)^2. \qquad \text{Substitute } 1/z \text{ for } r(z).$$

b. To find an equation for the surface formed by revolving the graph of $9x^2 = y^3$ about the y-axis, solve for x in terms of y to obtain

$$x = \tfrac{1}{3}y^{3/2} = r(y). \qquad \text{Radius function}$$

So, the equation for this surface is

$$x^2 + z^2 = [r(y)]^2 \qquad \text{Revolved about the } y\text{-axis}$$

$$x^2 + z^2 = \left(\tfrac{1}{3}y^{3/2}\right)^2 \qquad \text{Substitute } \tfrac{1}{3}y^{3/2} \text{ for } r(y).$$

$$x^2 + z^2 = \tfrac{1}{9}y^3. \qquad \text{Equation of surface}$$

The graph is shown in Figure 11.63.

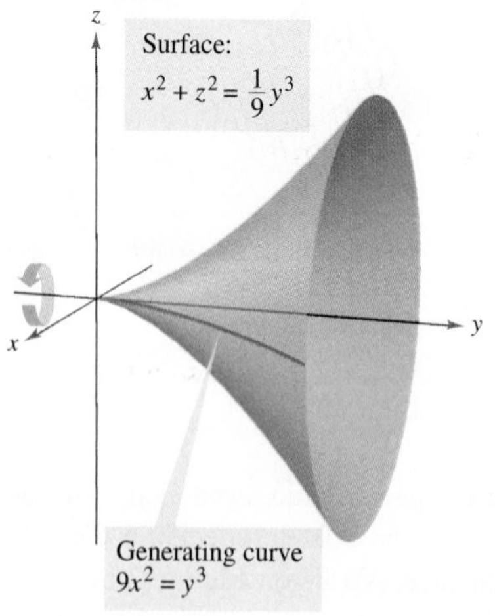

Figure 11.63

The generating curve for a surface of revolution is not unique. For instance, the surface

$$x^2 + z^2 = e^{-2y}$$

can be formed by revolving either the graph of $x = e^{-y}$ about the y-axis or the graph of $z = e^{-y}$ about the y-axis, as shown in Figure 11.64.

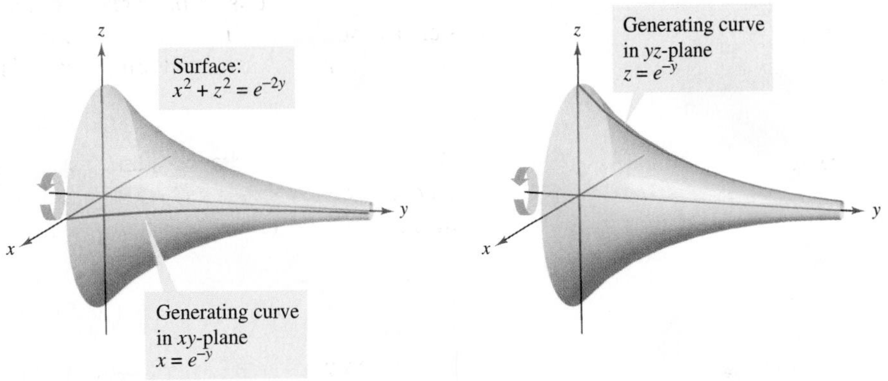

Figure 11.64

EXAMPLE 6 Finding a Generating Curve for a Surface of Revolution

Find a generating curve and the axis of revolution for the surface given by

$$x^2 + 3y^2 + z^2 = 9.$$

Solution You now know that the equation has one of the following forms.

$$x^2 + y^2 = [r(z)]^2 \quad \text{Revolved about } z\text{-axis}$$
$$y^2 + z^2 = [r(x)]^2 \quad \text{Revolved about } x\text{-axis}$$
$$x^2 + z^2 = [r(y)]^2 \quad \text{Revolved about } y\text{-axis}$$

Because the coefficients of x^2 and z^2 are equal, you should choose the third form and write

$$x^2 + z^2 = 9 - 3y^2.$$

The y-axis is the axis of revolution. You can choose a generating curve from either of the following traces.

$$x^2 = 9 - 3y^2 \quad \text{Trace in } xy\text{-plane}$$
$$z^2 = 9 - 3y^2 \quad \text{Trace in } yz\text{-plane}$$

For example, using the first trace, the generating curve is the semiellipse given by

$$x = \sqrt{9 - 3y^2}. \quad \text{Generating curve}$$

The graph of this surface is shown in Figure 11.65.

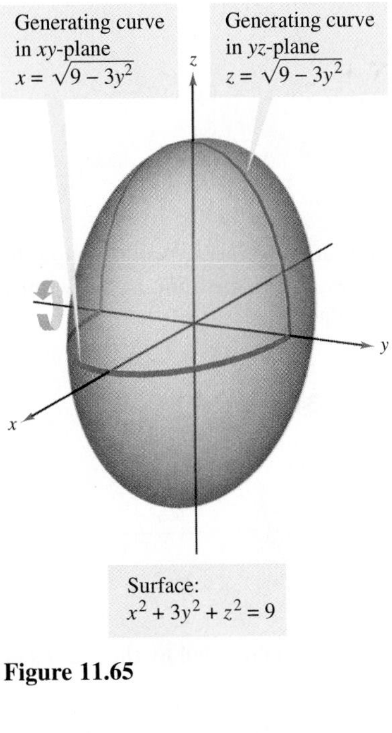

Figure 11.65

Exercises for Section 11.6

See www.CalcChat.com for worked-out solutions to odd-numbered exercises.

In Exercises 1–6, match the equation with its graph. [The graphs are labeled (a), (b), (c), (d), (e), and (f).]

(a)

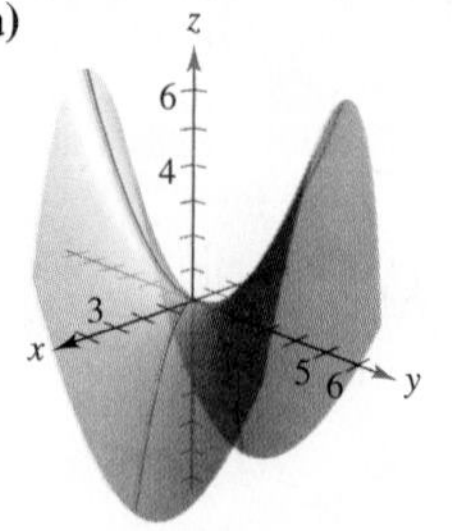

(b)

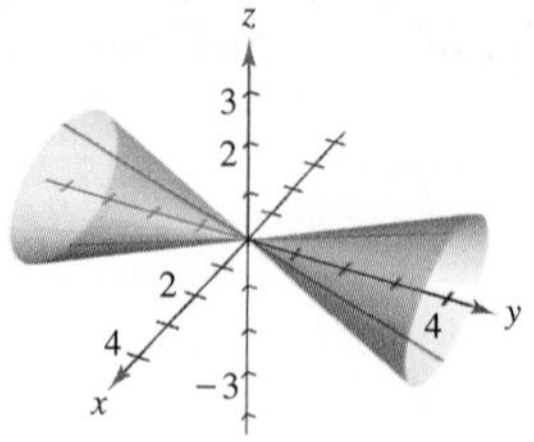

(c)

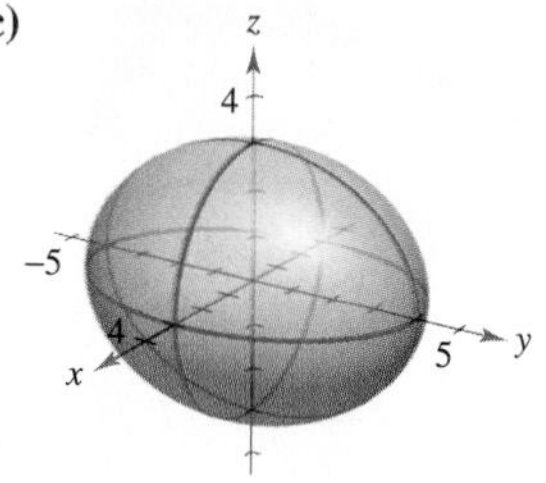

(d)

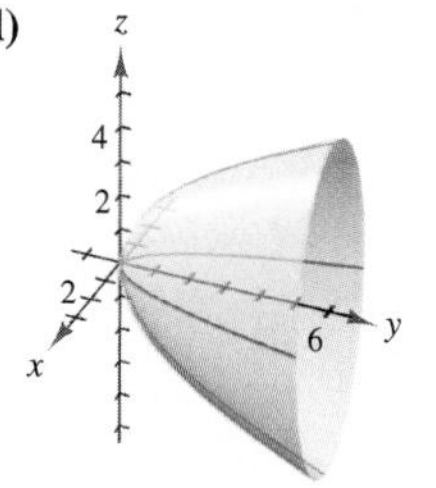

(e)

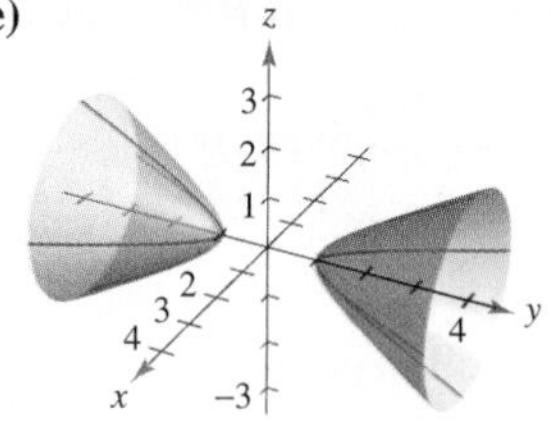

(f)

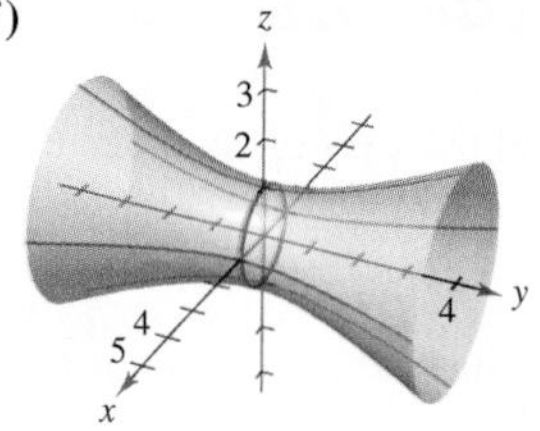

1. $\dfrac{x^2}{9} + \dfrac{y^2}{16} + \dfrac{z^2}{9} = 1$

2. $15x^2 - 4y^2 + 15z^2 = -4$

3. $4x^2 - y^2 + 4z^2 = 4$

4. $y^2 = 4x^2 + 9z^2$

5. $4x^2 - 4y + z^2 = 0$

6. $4x^2 - y^2 + 4z = 0$

In Exercises 7–16, describe and sketch the surface.

7. $z = 3$

8. $x = 4$

9. $y^2 + z^2 = 9$

10. $x^2 + z^2 = 25$

11. $x^2 - y = 0$

12. $y^2 + z = 4$

13. $4x^2 + y^2 = 4$

14. $y^2 - z^2 = 4$

15. $z - \sin y = 0$

16. $z - e^y = 0$

17. ***Think About It*** The four figures are graphs of the quadric surface $z = x^2 + y^2$. Match each of the four graphs with the point in space from which the paraboloid is viewed. The four points are $(0, 0, 20)$, $(0, 20, 0)$, $(20, 0, 0)$, and $(10, 10, 20)$.

(a)

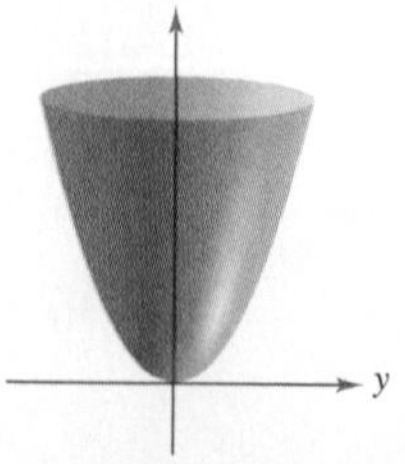

(b)

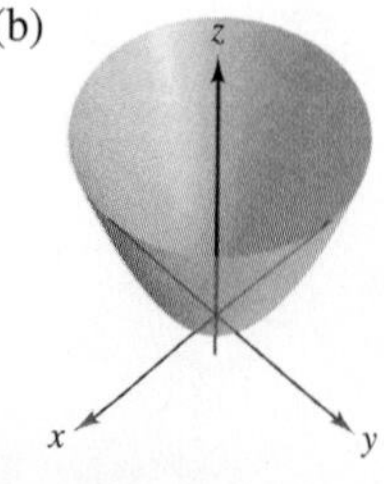

(c)

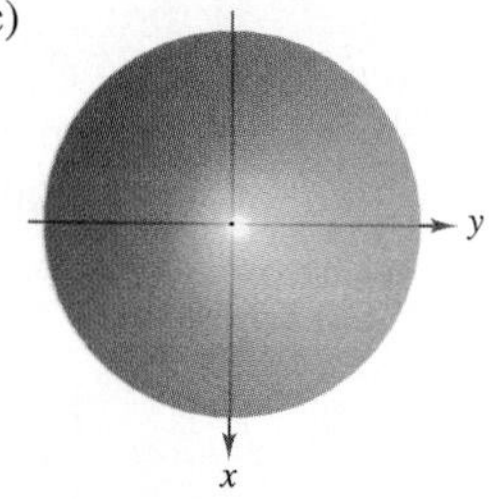

(d)

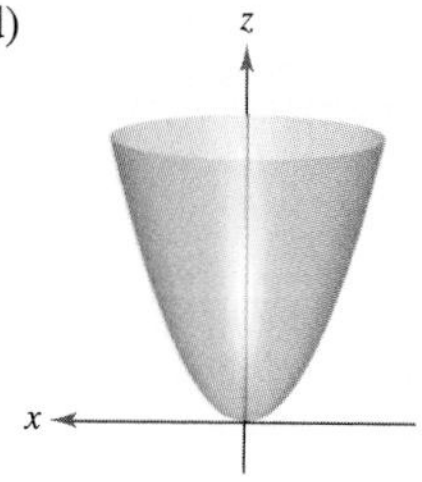

Figures for 17

18. Use a computer algebra system to graph a view of the cylinder $y^2 + z^2 = 4$ from each point.

(a) $(10, 0, 0)$

(b) $(0, 10, 0)$

(c) $(10, 10, 10)$

In Exercises 19–30, identify and sketch the quadric surface. Use a computer algebra system to confirm your sketch.

19. $x^2 + \dfrac{y^2}{4} + z^2 = 1$

20. $\dfrac{x^2}{16} + \dfrac{y^2}{25} + \dfrac{z^2}{25} = 1$

21. $16x^2 - y^2 + 16z^2 = 4$

22. $z^2 - x^2 - \dfrac{y^2}{4} = 1$

23. $x^2 - y + z^2 = 0$

24. $z = x^2 + 4y^2$

25. $x^2 - y^2 + z = 0$

26. $3z = -y^2 + x^2$

27. $z^2 = x^2 + \dfrac{y^2}{4}$

28. $x^2 = 2y^2 + 2z^2$

29. $16x^2 + 9y^2 + 16z^2 - 32x - 36y + 36 = 0$

30. $9x^2 + y^2 - 9z^2 - 54x - 4y - 54z + 4 = 0$

In Exercises 31–40, use a computer algebra system to graph the surface. (*Hint:* It may be necessary to solve for z and acquire two equations to graph the surface.)

31. $z = 2 \sin x$

32. $z = x^2 + 0.5y^2$

33. $z^2 = x^2 + 4y^2$

34. $4y = x^2 + z^2$

35. $x^2 + y^2 = \left(\dfrac{2}{z}\right)^2$

36. $x^2 + y^2 = e^{-z}$

37. $z = 4 - \sqrt{|xy|}$

38. $z = \dfrac{-x}{8 + x^2 + y^2}$

39. $4x^2 - y^2 + 4z^2 = -16$

40. $9x^2 + 4y^2 - 8z^2 = 72$

In Exercises 41–44, sketch the region bounded by the graphs of the equations.

41. $z = 2\sqrt{x^2 + y^2}$, $z = 2$

42. $z = \sqrt{4 - x^2}$, $y = \sqrt{4 - x^2}$, $x = 0$, $y = 0$, $z = 0$

43. $x^2 + y^2 = 1$, $x + z = 2$, $z = 0$

44. $z = \sqrt{4 - x^2 - y^2}$, $y = 2z$, $z = 0$

In Exercises 45–50, find an equation for the surface of revolution generated by revolving the curve in the indicated coordinate plane about the given axis.

	Equation of Curve	*Coordinate Plane*	*Axis of Revolution*
45.	$z^2 = 4y$	yz-plane	y-axis
46.	$z = 3y$	yz-plane	y-axis
47.	$z = 2y$	yz-plane	z-axis
48.	$2z = \sqrt{4 - x^2}$	xz-plane	x-axis
49.	$xy = 2$	xy-plane	x-axis
50.	$z = \ln y$	yz-plane	z-axis

In Exercises 51 and 52, find an equation of a generating curve given the equation of its surface of revolution.

51. $x^2 + y^2 - 2z = 0$ **52.** $x^2 + z^2 = \cos^2 y$

Writing About Concepts

53. State the definition of a cylinder.

54. What is meant by the trace of a surface? How do you find a trace?

55. Identify the six quadric surfaces and give the standard form of each.

56. What does the equation $z = x^2$ represent in the xz-plane? What does it represent in three-space?

In Exercises 57 and 58, use the shell method to find the volume of the solid below the surface of revolution and above the xy-plane.

57. The curve $z = 4x - x^2$ in the xz-plane is revolved about the z-axis.

58. The curve $z = \sin y$ $(0 \le y \le \pi)$ in the yz-plane is revolved about the z-axis.

In Exercises 59 and 60, analyze the trace when the surface

$$z = \frac{1}{2}x^2 + \frac{1}{4}y^2$$

is intersected by the indicated planes.

59. Find the lengths of the major and minor axes and the coordinates of the foci of the ellipse generated when the surface is intersected by the planes given by

(a) $z = 2$ and (b) $z = 8$.

60. Find the coordinates of the focus of the parabola formed when the surface is intersected by the planes given by

(a) $y = 4$ and (b) $x = 2$.

In Exercises 61 and 62, find an equation of the surface satisfying the conditions, and identify the surface.

61. The set of all points equidistant from the point $(0, 2, 0)$ and the plane $y = -2$

62. The set of all points equidistant from the point $(0, 0, 4)$ and the xy-plane

63. ***Geography*** Because of the forces caused by its rotation, Earth is an oblate ellipsoid rather than a sphere. The equatorial radius is 3963 miles and the polar radius is 3950 miles. Find an equation of the ellipsoid. (Assume that the center of Earth is at the origin and that the trace formed by the plane $z = 0$ corresponds to the equator.)

64. ***Machine Design*** The top of a rubber bushing designed to absorb vibrations in an automobile is the surface of revolution generated by revolving the curve $z = \frac{1}{2}y^2 + 1$ $(0 \le y \le 2)$ in the yz-plane about the z-axis.

(a) Find an equation for the surface of revolution.

(b) All measurements are in centimeters and the bushing is set on the xy-plane. Use the shell method to find its volume.

(c) The bushing has a hole of diameter 1 centimeter through its center and parallel to the axis of revolution. Find the volume of the rubber bushing.

65. Determine the intersection of the hyperbolic paraboloid $z = y^2/b^2 - x^2/a^2$ with the plane $bx + ay - z = 0$. (Assume $a, b > 0$.)

66. Explain why the curve of intersection of the surfaces $x^2 + 3y^2 - 2z^2 + 2y = 4$ and $2x^2 + 6y^2 - 4z^2 - 3x = 2$ lies in a plane.

True or False? **In Exercises 67 and 68, determine whether the statement is true or false. If it is false, explain why or give an example that shows it is false.**

67. A sphere is an ellipsoid.

68. The generating curve for a surface of revolution is unique.

69. ***Think About It*** Three types of classic "topological" surfaces are shown below. The sphere and torus have both an "inside" and an "outside." Does the Klein bottle have both an inside and an outside? Explain.

Sphere

Torus

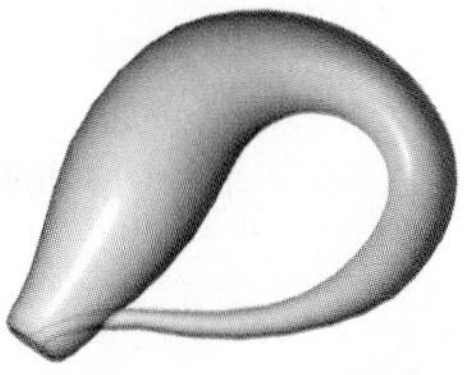

Klein bottle

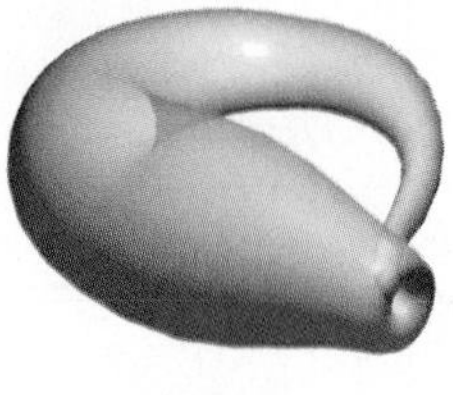

Klein bottle

Section 11.7 Cylindrical and Spherical Coordinates

- **Use cylindrical coordinates to represent surfaces in space.**
- **Use spherical coordinates to represent surfaces in space.**

Cylindrical Coordinates

You have already seen that some two-dimensional graphs are easier to represent in polar coordinates than in rectangular coordinates. A similar situation exists for surfaces in space. In this section, you will study two alternative space-coordinate systems. The first, the **cylindrical coordinate system,** is an extension of polar coordinates in the plane to three-dimensional space.

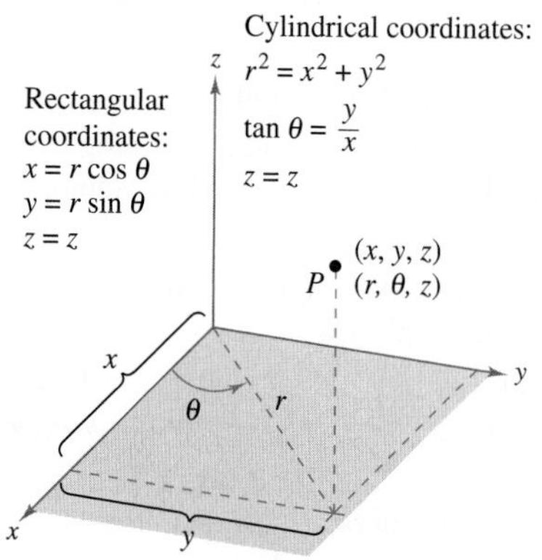

Figure 11.66

The Cylindrical Coordinate System

In a **cylindrical coordinate system,** a point P in space is represented by an ordered triple (r, θ, z).

1. (r, θ) is a polar representation of the projection of P in the xy-plane.
2. z is the directed distance from (r, θ) to P.

To convert from rectangular to cylindrical coordinates (or vice versa), use the following conversion guidelines for polar coordinates, as illustrated in Figure 11.66.

Cylindrical to rectangular:

$$x = r\cos\theta, \qquad y = r\sin\theta, \qquad z = z$$

Rectangular to cylindrical:

$$r^2 = x^2 + y^2, \qquad \tan\theta = \frac{y}{x}, \qquad z = z$$

The point $(0, 0, 0)$ is called the **pole.** Moreover, because the representation of a point in the polar coordinate system is not unique, it follows that the representation in the cylindrical coordinate system is also not unique.

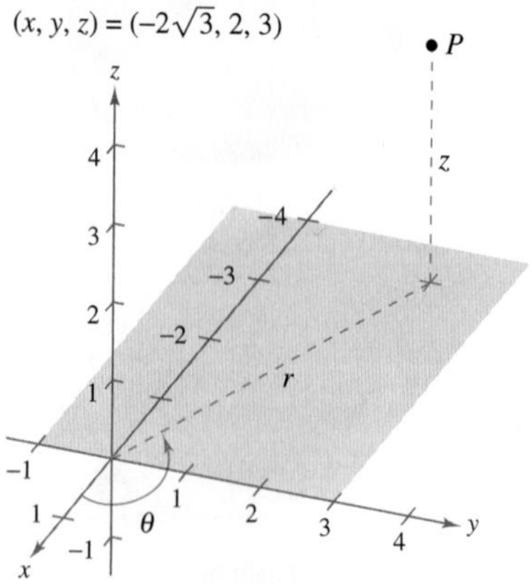

Figure 11.67

EXAMPLE 1 Converting from Cylindrical to Rectangular Coordinates

Convert the point $(r, \theta, z) = \left(4, \frac{5\pi}{6}, 3\right)$ to rectangular coordinates.

Solution Using the cylindrical-to-rectangular conversion equations produces

$$x = 4\cos\frac{5\pi}{6} = 4\left(-\frac{\sqrt{3}}{2}\right) = -2\sqrt{3}$$

$$y = 4\sin\frac{5\pi}{6} = 4\left(\frac{1}{2}\right) = 2$$

$$z = 3.$$

So, in rectangular coordinates, the point is $(x, y, z) = \left(-2\sqrt{3}, 2, 3\right)$, as shown in Figure 11.67.

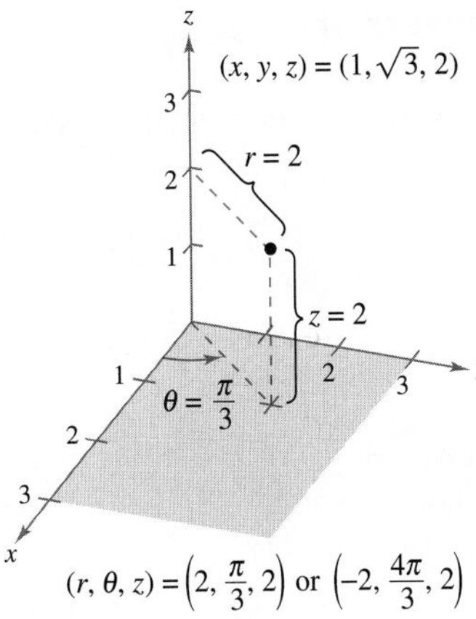

Figure 11.68

EXAMPLE 2 Converting from Rectangular to Cylindrical Coordinates

Convert the point $(x, y, z) = \left(1, \sqrt{3}, 2\right)$ to cylindrical coordinates.

Solution Use the rectangular-to-cylindrical conversion equations.

$$r = \pm\sqrt{1 + 3} = \pm 2$$

$$\tan\theta = \sqrt{3} \quad \Longrightarrow \quad \theta = \arctan\left(\sqrt{3}\right) + n\pi = \frac{\pi}{3} + n\pi$$

$$z = 2$$

You have two choices for r and infinitely many choices for θ. As shown in Figure 11.68, two convenient representations of the point are

$\left(2, \frac{\pi}{3}, 2\right)$ — $r > 0$ and θ in Quadrant I

$\left(-2, \frac{4\pi}{3}, 2\right).$ — $r < 0$ and θ in Quadrant III

Cylindrical coordinates are especially convenient for representing cylindrical surfaces and surfaces of revolution with the z-axis as the axis of symmetry, as shown in Figure 11.69.

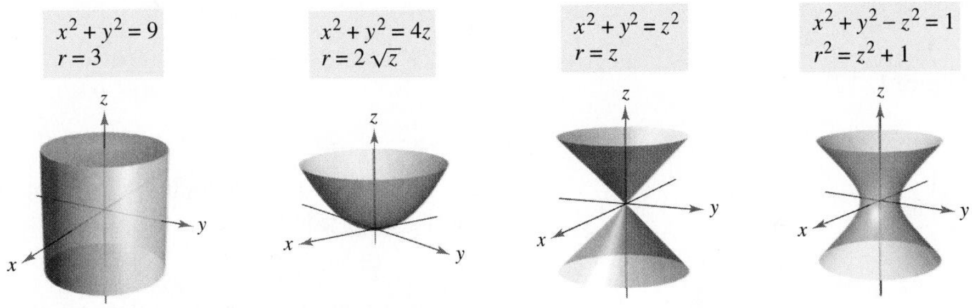

Cylinder Paraboloid Cone Hyperboloid

Figure 11.69

Vertical planes containing the z-axis and horizontal planes also have simple cylindrical coordinate equations, as shown in Figure 11.70.

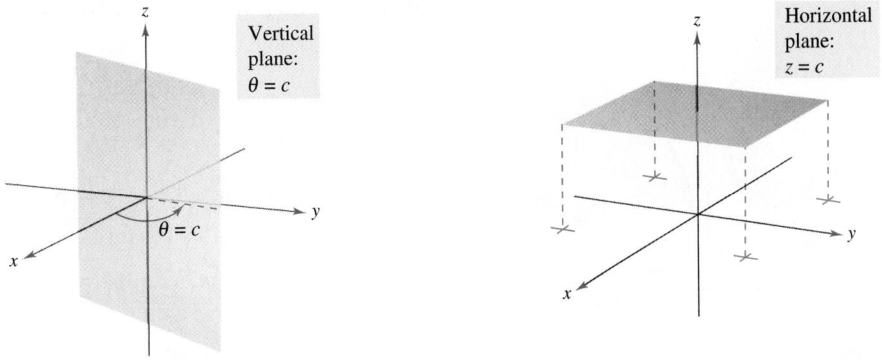

Figure 11.70

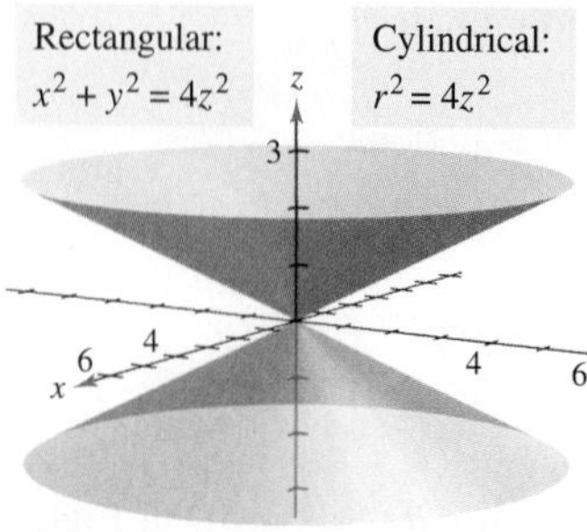

Figure 11.71

EXAMPLE 3 Rectangular-to-Cylindrical Conversion

Find an equation in cylindrical coordinates for the surface represented by each rectangular equation.

a. $x^2 + y^2 = 4z^2$

b. $y^2 = x$

Solution

a. From the preceding section, you know that the graph $x^2 + y^2 = 4z^2$ is a "double-napped" cone with its axis along the z-axis, as shown in Figure 11.71. If you replace $x^2 + y^2$ with r^2, the equation in cylindrical coordinates is

$$x^2 + y^2 = 4z^2 \qquad \text{Rectangular equation}$$
$$r^2 = 4z^2. \qquad \text{Cylindrical equation}$$

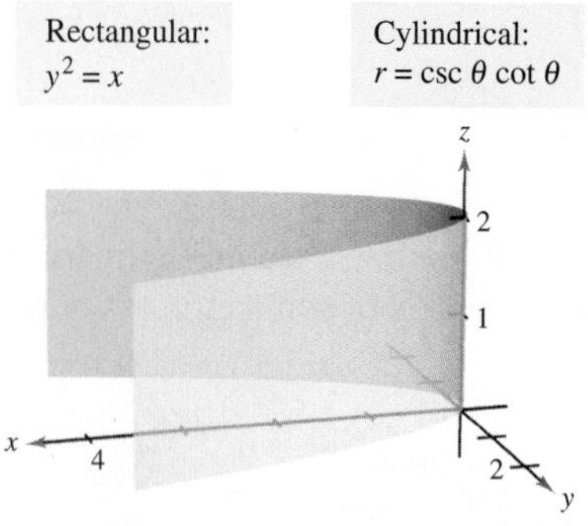

Figure 11.72

b. The graph of the surface $y^2 = x$ is a parabolic cylinder with rulings parallel to the z-axis, as shown in Figure 11.72. By replacing y^2 with $r^2 \sin^2\theta$ and x with $r\cos\theta$, you obtain the following equation in cylindrical coordinates.

$$y^2 = x \qquad \text{Rectangular equation}$$
$$r^2 \sin^2\theta = r\cos\theta \qquad \text{Substitute } r\sin\theta \text{ for } y \text{ and } r\cos\theta \text{ for } x.$$
$$r(r\sin^2\theta - \cos\theta) = 0 \qquad \text{Collect terms and factor.}$$
$$r\sin^2\theta - \cos\theta = 0 \qquad \text{Divide each side by } r.$$
$$r = \frac{\cos\theta}{\sin^2\theta} \qquad \text{Solve for } r.$$
$$r = \csc\theta\cot\theta \qquad \text{Cylindrical equation}$$

Note that this equation includes a point for which $r = 0$, so nothing was lost by dividing each side by the factor r.

Converting from rectangular coordinates to cylindrical coordinates is more straightforward than converting from cylindrical coordinates to rectangular coordinates, as demonstrated in Example 4.

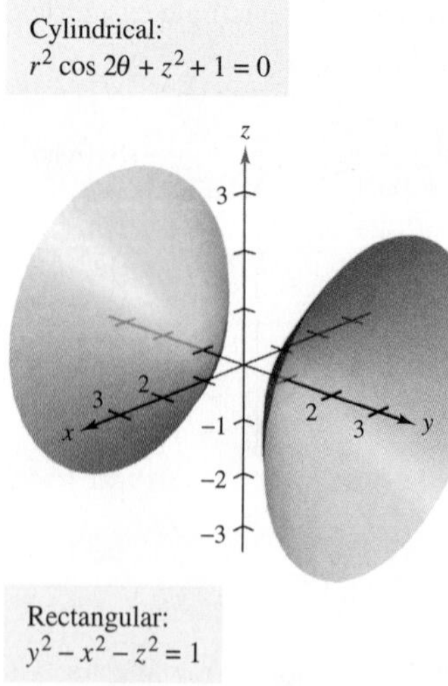

Figure 11.73

EXAMPLE 4 Cylindrical-to-Rectangular Conversion

Find an equation in rectangular coordinates for the surface represented by the cylindrical equation

$$r^2\cos 2\theta + z^2 + 1 = 0.$$

Solution

$$r^2\cos 2\theta + z^2 + 1 = 0 \qquad \text{Cylindrical equation}$$
$$r^2(\cos^2\theta - \sin^2\theta) + z^2 + 1 = 0 \qquad \text{Trigonometric identity}$$
$$r^2\cos^2\theta - r^2\sin^2\theta + z^2 = -1$$
$$x^2 - y^2 + z^2 = -1 \qquad \text{Replace } r\cos\theta \text{ with } x \text{ and } r\sin\theta \text{ with } y.$$
$$y^2 - x^2 - z^2 = 1 \qquad \text{Rectangular equation}$$

This is a hyperboloid of two sheets whose axis lies along the y-axis, as shown in Figure 11.73.

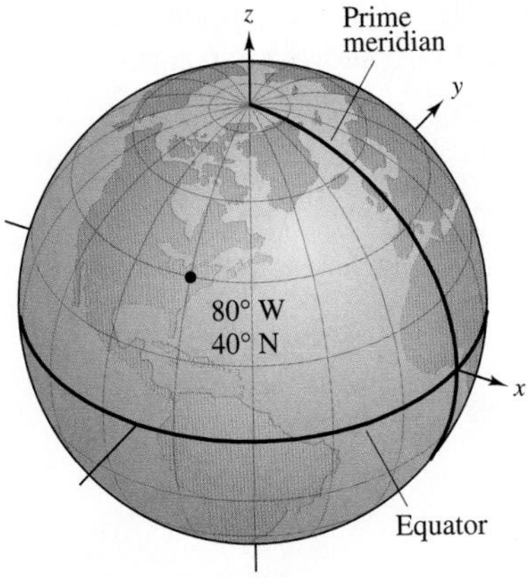

Figure 11.74

Spherical Coordinates

In the **spherical coordinate system,** each point is represented by an ordered triple: the first coordinate is a distance, and the second and third coordinates are angles. This system is similar to the latitude-longitude system used to identify points on the surface of Earth. For example, the point on the surface of Earth whose latitude is 40° North (of the equator) and whose longitude is 80° West (of the prime meridian) is shown in Figure 11.74. Assuming that the Earth is spherical and has a radius of 4000 miles, you would label this point as

$$(4000, -80°, 50°).$$

Radius — 80° clockwise from prime meridian — 50° down from North Pole

The Spherical Coordinate System

In a **spherical coordinate system,** a point P in space is represented by an ordered triple (ρ, θ, ϕ).

1. ρ is the distance between P and the origin, $\rho \geq 0$.
2. θ is the same angle used in cylindrical coordinates for $r \geq 0$.
3. ϕ is the angle *between* the positive z-axis and the line segment $\overrightarrow{OP}$, $0 \leq \phi \leq \pi$.

Note that the first and third coordinates, ρ and ϕ, are nonnegative. ρ is the lowercase Greek letter *rho*, and ϕ is the lowercase Greek letter *phi*.

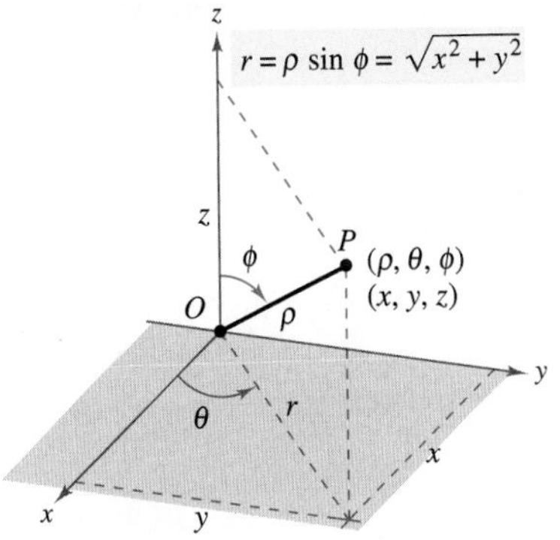

Spherical coordinates
Figure 11.75

The relationship between rectangular and spherical coordinates is illustrated in Figure 11.75. To convert from one system to the other, use the following.

Spherical to rectangular:

$$x = \rho \sin \phi \cos \theta, \qquad y = \rho \sin \phi \sin \theta, \qquad z = \rho \cos \phi$$

Rectangular to spherical:

$$\rho^2 = x^2 + y^2 + z^2, \qquad \tan \theta = \frac{y}{x}, \qquad \phi = \arccos\left(\frac{z}{\sqrt{x^2 + y^2 + z^2}}\right)$$

To change coordinates between the cylindrical and spherical systems, use the following.

Spherical to cylindrical* ($r \geq 0$)*:

$$r^2 = \rho^2 \sin^2 \phi, \qquad \theta = \theta, \qquad z = \rho \cos \phi$$

Cylindrical to spherical* ($r \geq 0$)*:

$$\rho = \sqrt{r^2 + z^2}, \qquad \theta = \theta, \qquad \phi = \arccos\left(\frac{z}{\sqrt{r^2 + z^2}}\right)$$

The spherical coordinate system is useful primarily for surfaces in space that have a *point* or *center* of symmetry. For example, Figure 11.76 shows three surfaces with simple spherical equations.

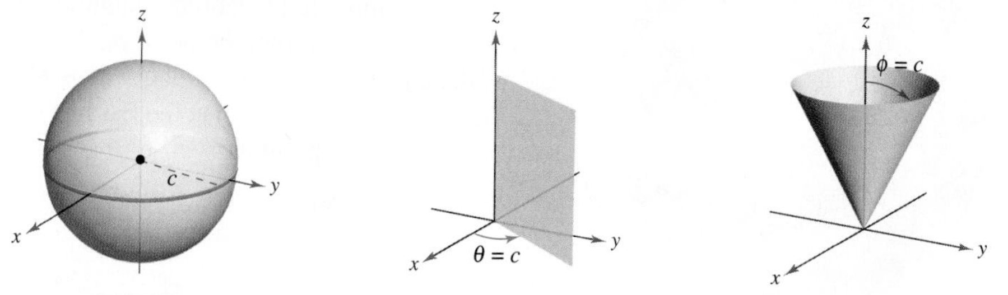

Sphere: $\rho = c$

Vertical half-plane: $\theta = c$

Half-cone: $\phi = c$ $\left(0 < c < \frac{\pi}{2}\right)$

Figure 11.76

EXAMPLE 5 Rectangular-to-Spherical Conversion

Find an equation in spherical coordinates for the surface represented by each rectangular equation.

a. Cone: $x^2 + y^2 = z^2$

b. Sphere: $x^2 + y^2 + z^2 - 4z = 0$

Solution

a. Making the appropriate replacements for x, y, and z in the given equation yields the following.

$$x^2 + y^2 = z^2$$
$$\rho^2 \sin^2 \phi \cos^2 \theta + \rho^2 \sin^2 \phi \sin^2 \theta = \rho^2 \cos^2 \phi$$
$$\rho^2 \sin^2 \phi (\cos^2 \theta + \sin^2 \theta) = \rho^2 \cos^2 \phi$$
$$\rho^2 \sin^2 \phi = \rho^2 \cos^2 \phi$$
$$\frac{\sin^2 \phi}{\cos^2 \phi} = 1 \qquad \rho \geq 0$$
$$\tan^2 \phi = 1 \qquad \phi = \pi/4 \text{ or } \phi = 3\pi/4$$

The equation $\phi = \pi/4$ represents the *upper* half-cone, and the equation $\phi = 3\pi/4$ represents the *lower* half-cone.

b. Because $\rho^2 = x^2 + y^2 + z^2$ and $z = \rho \cos \phi$, the given equation has the following spherical form.

$$\rho^2 - 4\rho \cos \phi = 0 \quad \Longrightarrow \quad \rho(\rho - 4 \cos \phi) = 0$$

Temporarily discarding the possibility that $\rho = 0$, you have the spherical equation

$$\rho - 4 \cos \phi = 0 \qquad \text{or} \qquad \rho = 4 \cos \phi.$$

Note that the solution set for this equation includes a point for which $\rho = 0$, so nothing is lost by discarding the factor ρ. The sphere represented by the equation $\rho = 4 \cos \phi$ is shown in Figure 11.77.

Rectangular: $x^2 + y^2 + z^2 - 4z = 0$

Spherical: $\rho = 4 \cos \phi$

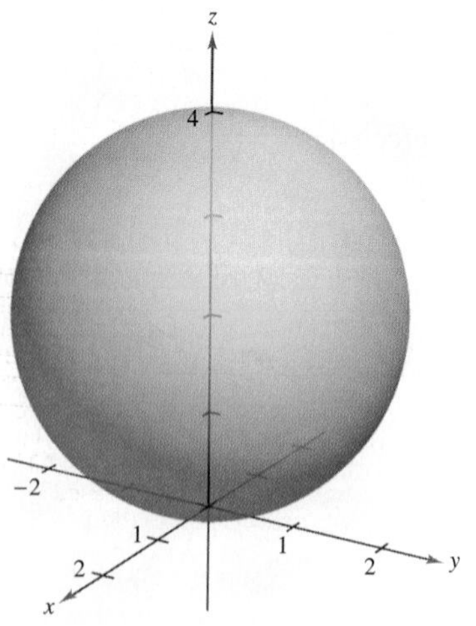

Figure 11.77

Exercises for Section 11.7

See www.CalcChat.com for worked-out solutions to odd-numbered exercises.

In Exercises 1–6, convert the point from cylindrical coordinates to rectangular coordinates.

1. $(5, 0, 2)$
2. $(4, \pi/2, -2)$
3. $(2, \pi/3, 2)$
4. $(6, -\pi/4, 2)$
5. $(4, 7\pi/6, 3)$
6. $(1, 3\pi/2, 1)$

In Exercises 7–12, convert the point from rectangular coordinates to cylindrical coordinates.

7. $(0, 5, 1)$
8. $(2\sqrt{2}, -2\sqrt{2}, 4)$
9. $(1, \sqrt{3}, 4)$
10. $(2\sqrt{3}, -2, 6)$
11. $(2, -2, -4)$
12. $(-3, 2, -1)$

In Exercises 13–20, find an equation in cylindrical coordinates for the equation given in rectangular coordinates.

13. $z = 5$
14. $x = 4$
15. $x^2 + y^2 + z^2 = 10$
16. $z = x^2 + y^2 - 2$
17. $y = x^2$
18. $x^2 + y^2 = 8x$
19. $y^2 = 10 - z^2$
20. $x^2 + y^2 + z^2 - 3z = 0$

In Exercises 21–28, find an equation in rectangular coordinates for the equation given in cylindrical coordinates, and sketch its graph.

21. $r = 2$
22. $z = 2$
23. $\theta = \pi/6$
24. $r = \frac{1}{2}z$
25. $r = 2 \sin \theta$
26. $r = 2 \cos \theta$
27. $r^2 + z^2 = 4$
28. $z = r^2 \cos^2 \theta$

In Exercises 29–34, convert the point from rectangular coordinates to spherical coordinates.

29. $(4, 0, 0)$
30. $(1, 1, 1)$
31. $(-2, 2\sqrt{3}, 4)$
32. $(2, 2, 4\sqrt{2})$
33. $(\sqrt{3}, 1, 2\sqrt{3})$
34. $(-4, 0, 0)$

In Exercises 35–40, convert the point from spherical coordinates to rectangular coordinates.

35. $(4, \pi/6, \pi/4)$
36. $(12, 3\pi/4, \pi/9)$
37. $(12, -\pi/4, 0)$
38. $(9, \pi/4, \pi)$
39. $(5, \pi/4, 3\pi/4)$
40. $(6, \pi, \pi/2)$

In Exercises 41–48, find an equation in spherical coordinates for the equation given in rectangular coordinates.

41. $y = 3$
42. $z = 2$
43. $x^2 + y^2 + z^2 = 36$
44. $x^2 + y^2 - 3z^2 = 0$
45. $x^2 + y^2 = 9$
46. $x = 10$
47. $x^2 + y^2 = 2z^2$
48. $x^2 + y^2 + z^2 - 9z = 0$

In Exercises 49–56, find an equation in rectangular coordinates for the equation given in spherical coordinates, and sketch its graph.

49. $\rho = 2$
50. $\theta = \dfrac{3\pi}{4}$
51. $\phi = \dfrac{\pi}{6}$
52. $\phi = \dfrac{\pi}{2}$
53. $\rho = 4 \cos \phi$
54. $\rho = 2 \sec \phi$
55. $\rho = \csc \phi$
56. $\rho = 4 \csc \phi \sec \theta$

In Exercises 57–64, convert the point from cylindrical coordinates to spherical coordinates.

57. $(4, \pi/4, 0)$
58. $(3, -\pi/4, 0)$
59. $(4, \pi/2, 4)$
60. $(2, 2\pi/3, -2)$
61. $(4, -\pi/6, 6)$
62. $(-4, \pi/3, 4)$
63. $(12, \pi, 5)$
64. $(4, \pi/2, 3)$

In Exercises 65–72, convert the point from spherical coordinates to cylindrical coordinates.

65. $(10, \pi/6, \pi/2)$
66. $(4, \pi/18, \pi/2)$
67. $(36, \pi, \pi/2)$
68. $(18, \pi/3, \pi/3)$
69. $(6, -\pi/6, \pi/3)$
70. $(5, -5\pi/6, \pi)$
71. $(8, 7\pi/6, \pi/6)$
72. $(7, \pi/4, 3\pi/4)$

In Exercises 73–86, use a computer algebra system or graphing utility to convert the point from one system to another among the rectangular, cylindrical, and spherical coordinate systems.

	Rectangular	*Cylindrical*	*Spherical*
73.	$(4, 6, 3)$		
74.	$(6, -2, -3)$		
75.		$(5, \pi/9, 8)$	
76.		$(10, -0.75, 6)$	
77.			$(20, 2\pi/3, \pi/4)$
78.			$(7.5, 0.25, 1)$
79.	$(3, -2, 2)$		
80.	$(3\sqrt{2}, 3\sqrt{2}, -3)$		
81.	$(5/2, 4/3, -3/2)$		
82.	$(0, -5, 4)$		
83.		$(5, 3\pi/4, -5)$	
84.		$(-2, 11\pi/6, 3)$	
85.		$(-3.5, 2.5, 6)$	
86.		$(8.25, 1.3, -4)$	

In Exercises 87–92, match the equation (written in terms of cylindrical or spherical coordinates) with its graph. [The graphs are labeled (a), (b), (c), (d), (e), and (f).]

(a)

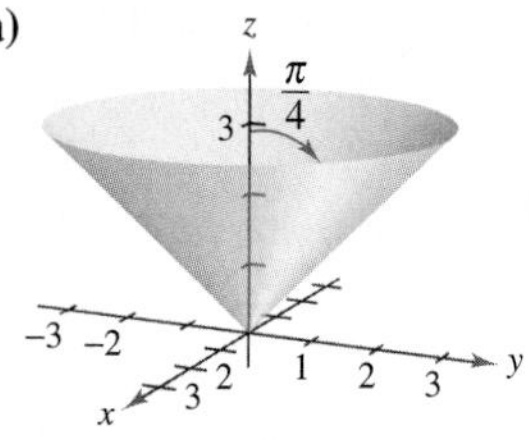

(b)

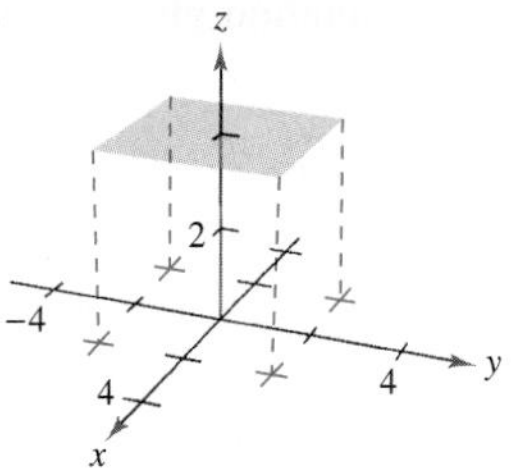

(c)

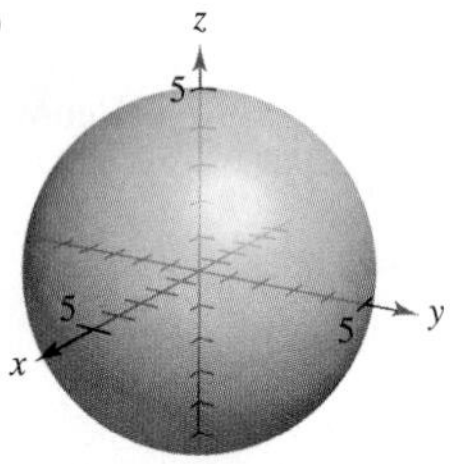

(d)

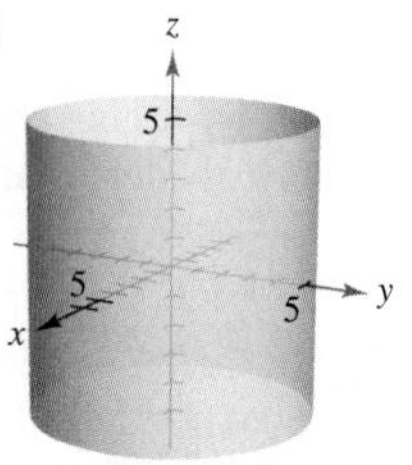

(e)

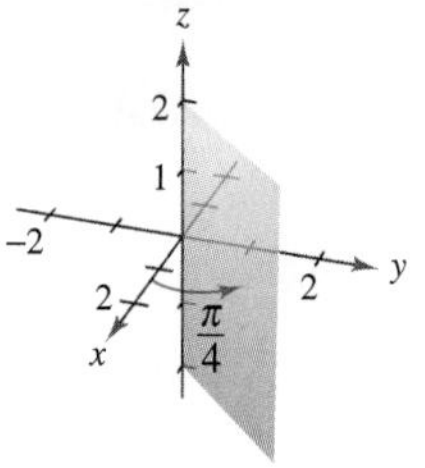

(f)

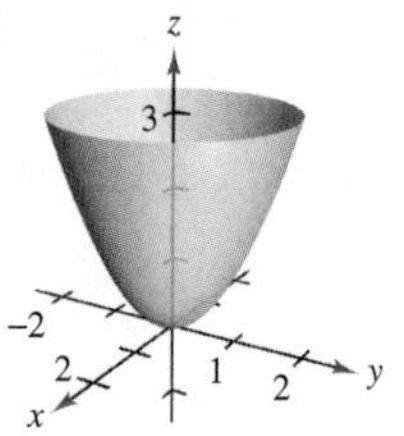

87. $r = 5$

88. $\theta = \frac{\pi}{4}$

89. $\rho = 5$

90. $\phi = \frac{\pi}{4}$

91. $r^2 = z$

92. $\rho = 4 \sec \phi$

Writing About Concepts

93. Give the equations for the coordinate conversion from rectangular to cylindrical coordinates and vice versa.

94. For constants a, b, and c, describe the graphs of the equations $r = a$, $\theta = b$, and $z = c$ in cylindrical coordinates.

95. Give the equations for the coordinate conversion from rectangular to spherical coordinates and vice versa.

96. For constants a, b, and c, describe the graphs of the equations $\rho = a$, $\theta = b$, and $\phi = c$ in spherical coordinates.

In Exercises 97–104, convert the rectangular equation to an equation in (a) cylindrical coordinates and (b) spherical coordinates.

97. $x^2 + y^2 + z^2 = 16$

98. $4(x^2 + y^2) = z^2$

99. $x^2 + y^2 + z^2 - 2z = 0$

100. $x^2 + y^2 = z$

101. $x^2 + y^2 = 4y$

102. $x^2 + y^2 = 16$

103. $x^2 - y^2 = 9$

104. $y = 4$

In Exercises 105–108, sketch the solid that has the given description in cylindrical coordinates.

105. $0 \le \theta \le \pi/2, 0 \le r \le 2, 0 \le z \le 4$

106. $-\pi/2 \le \theta \le \pi/2, 0 \le r \le 3, 0 \le z \le r \cos \theta$

107. $0 \le \theta \le 2\pi, 0 \le r \le a, r \le z \le a$

108. $0 \le \theta \le 2\pi, 2 \le r \le 4, z^2 \le -r^2 + 6r - 8$

In Exercises 109–112, sketch the solid that has the given description in spherical coordinates.

109. $0 \le \theta \le 2\pi, 0 \le \phi \le \pi/6, 0 \le \rho \le a \sec \phi$

110. $0 \le \theta \le 2\pi, \pi/4 \le \phi \le \pi/2, 0 \le \rho \le 1$

111. $0 \le \theta \le \pi/2, 0 \le \phi \le \pi/2, 0 \le \rho \le 2$

112. $0 \le \theta \le \pi, 0 \le \phi \le \pi/2, 1 \le \rho \le 3$

Think About It **In Exercises 113–118, find inequalities that describe the solid, and state the coordinate system used. Position the solid on the coordinate system such that the inequalities are as simple as possible.**

113. A cube with each edge 10 centimeters long

114. A cylindrical shell 8 meters long with an inside diameter of 0.75 meter and an outside diameter of 1.25 meters

115. A spherical shell with inside and outside radii of 4 inches and 6 inches, respectively

116. The solid that remains after a hole 1 inch in diameter is drilled through the center of a sphere 6 inches in diameter

117. The solid inside both $x^2 + y^2 + z^2 = 9$ and $\left(x - \frac{3}{2}\right)^2 + y^2 = \frac{9}{4}$

118. The solid between the spheres $x^2 + y^2 + z^2 = 4$ and $x^2 + y^2 + z^2 = 9$, and inside the cone $z^2 = x^2 + y^2$

True or False? **In Exercises 119–122, determine whether the statement is true or false. If it is false, explain why or give an example that shows it is false.**

119. In spherical coordinates, the equation $\theta = c$ represents an entire plane.

120. The equations $\rho = 2$ and $x^2 + y^2 + z^2 = 4$ represent the same surface.

121. The cylindrical coordinates of a point (x, y, z) are unique.

122. The spherical coordinates of a point (x, y, z) are unique.

123. Identify the curve of intersection of the surfaces (in cylindrical coordinates) $z = \sin \theta$ and $r = 1$.

124. Identify the curve of intersection of the surfaces (in spherical coordinates) $\rho = 2 \sec \phi$ and $\rho = 4$.

Review Exercises for Chapter 11

See www.CalcChat.com for worked-out solutions to odd-numbered exercises.

In Exercises 1 and 2, let $\mathbf{u} = \overrightarrow{PQ}$ and $\mathbf{v} = \overrightarrow{PR}$, and find (a) the component forms of u and v, (b) the magnitude of v, and (c) 2u + v.

1. $P = (1, 2), Q = (4, 1), R = (5, 4)$
2. $P = (-2, -1), Q = (5, -1), R = (2, 4)$

In Exercises 3 and 4, find the component form of v given its magnitude and the angle it makes with the positive x-axis.

3. $\|\mathbf{v}\| = 8,\ \theta = 120°$
4. $\|\mathbf{v}\| = \frac{1}{2},\ \theta = 225°$

5. Find the coordinates of the point in the xy-plane four units to the right of the xz-plane and five units behind the yz-plane.
6. Find the coordinates of the point located on the y-axis and seven units to the left of the xz-plane.

In Exercises 7 and 8, determine the location of a point (x, y, z) that satisfies the condition.

7. $yz > 0$
8. $xy < 0$

In Exercises 9 and 10, find the standard equation of the sphere.

9. Center: $(3, -2, 6)$; Diameter: 15
10. Endpoints of a diameter: $(0, 0, 4), (4, 6, 0)$

In Exercises 11 and 12, complete the square to write the equation of the sphere in standard form. Find the center and radius.

11. $x^2 + y^2 + z^2 - 4x - 6y + 4 = 0$
12. $x^2 + y^2 + z^2 - 10x + 6y - 4z + 34 = 0$

In Exercises 13 and 14, the initial and terminal points of a vector are given. Sketch the directed line segment and find the component form of the vector.

13. Initial point: $(2, -1, 3)$
    Terminal point: $(4, 4, -7)$
14. Initial point: $(6, 2, 0)$
    Terminal point: $(3, -3, 8)$

In Exercises 15 and 16, use vectors to determine whether the points are collinear.

15. $(3, 4, -1),\ (-1, 6, 9),\ (5, 3, -6)$
16. $(5, -4, 7),\ (8, -5, 5),\ (11, 6, 3)$

17. Find a unit vector in the direction of $\mathbf{u} = \langle 2, 3, 5\rangle$.
18. Find the vector $\mathbf{v}$ of magnitude 8 in the direction $\langle 6, -3, 2\rangle$.

In Exercises 19 and 20, let $\mathbf{u} = \overrightarrow{PQ}$ and $\mathbf{v} = \overrightarrow{PR}$, and find (a) the component forms of u and v, (b) u · v, and (c) v · v.

19. $P = (5, 0, 0),\ Q = (4, 4, 0),\ R = (2, 0, 6)$
20. $P = (2, -1, 3),\ Q = (0, 5, 1),\ R = (5, 5, 0)$

In Exercises 21 and 22, determine whether u and v are orthogonal, parallel, or neither.

21. $\mathbf{u} = \langle 7, -2, 3\rangle$
    $\mathbf{v} = \langle -1, 4, 5\rangle$
22. $\mathbf{u} = \langle -4, 3, -6\rangle$
    $\mathbf{v} = \langle 16, -12, 24\rangle$

In Exercises 23–26, find the angle θ between the vectors.

23. $\mathbf{u} = 5[\cos(3\pi/4)\mathbf{i} + \sin(3\pi/4)\mathbf{j}]$
    $\mathbf{v} = 2[\cos(2\pi/3)\mathbf{i} + \sin(2\pi/3)\mathbf{j}]$
24. $\mathbf{u} = \langle 4, -1, 5\rangle,\quad \mathbf{v} = \langle 3, 2, -2\rangle$
25. $\mathbf{u} = \langle 10, -5, 15\rangle,\quad \mathbf{v} = \langle -2, 1, -3\rangle$
26. $\mathbf{u} = \langle 1, 0, -3\rangle,\quad \mathbf{v} = \langle 2, -2, 1\rangle$

27. Find two vectors in opposite directions that are orthogonal to the vector $\mathbf{u} = \langle 5, 6, -3\rangle$.
28. ***Work*** An object is pulled 8 feet across a floor using a force of 75 pounds. The direction of the force is 30° above the horizontal. Find the work done.

In Exercises 29–32, let $\mathbf{u} = \langle 3, -2, 1\rangle$, $\mathbf{v} = \langle 2, -4, -3\rangle$, and $\mathbf{w} = \langle -1, 2, 2\rangle$.

29. Show that $\mathbf{u} \cdot \mathbf{u} = \|\mathbf{u}\|^2$.
30. Find the angle between $\mathbf{u}$ and $\mathbf{v}$.
31. Determine the projection of $\mathbf{w}$ onto $\mathbf{u}$.
32. Find the work done in moving an object along the vector $\mathbf{u}$ if the applied force is $\mathbf{w}$.

In Exercises 33–38, let $\mathbf{u} = \langle 3, -2, 1\rangle$, $\mathbf{v} = \langle 2, -4, -3\rangle$, and $\mathbf{w} = \langle -1, 2, 2\rangle$.

33. Determine a unit vector perpendicular to the plane containing $\mathbf{v}$ and $\mathbf{w}$.
34. Show that $\mathbf{u} \times \mathbf{v} = -(\mathbf{v} \times \mathbf{u})$.
35. Find the volume of the solid whose edges are $\mathbf{u}$, $\mathbf{v}$, and $\mathbf{w}$.
36. Show that $\mathbf{u} \times (\mathbf{v} + \mathbf{w}) = (\mathbf{u} \times \mathbf{v}) + (\mathbf{u} \times \mathbf{w})$.
37. Find the area of the parallelogram with adjacent sides $\mathbf{u}$ and $\mathbf{v}$.
38. Find the area of the triangle with adjacent sides $\mathbf{v}$ and $\mathbf{w}$.

39. ***Torque*** The specifications for a tractor state that the torque on a bolt with head size $\frac{7}{8}$ inch cannot exceed 200 foot-pounds. Determine the maximum force $\|\mathbf{F}\|$ that can be applied to the wrench in the figure.

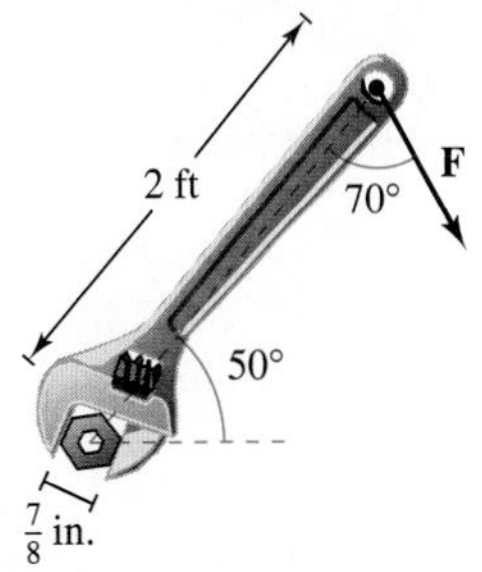

40. ***Volume*** Use the triple scalar product to find the volume of the parallelepiped having adjacent edges $\mathbf{u} = 2\mathbf{i} + \mathbf{j}$, $\mathbf{v} = 2\mathbf{j} + \mathbf{k}$, and $\mathbf{w} = -\mathbf{j} + 2\mathbf{k}$.

In Exercises 41 and 42, find sets of (a) parametric equations and (b) symmetric equations of the line through the two points. (For each line, write the direction numbers as integers.)

41. $(3, 0, 2)$, $(9, 11, 6)$

42. $(-1, 4, 3)$, $(8, 10, 5)$

In Exercises 43–46, find (a) a set of parametric equations and (b) a set of symmetric equations for the line.

43. The line passes through the point $(1, 2, 3)$ and is perpendicular to the xz-plane.

44. The line passes through the point $(1, 2, 3)$ and is parallel to the line given by $x = y = z$.

45. The intersection of the planes $3x - 3y - 7z = -4$ and $x - y + 2z = 3$

46. The line passes through the point $(0, 1, 4)$ and is perpendicular to $\mathbf{u} = \langle 2, -5, 1\rangle$ and $\mathbf{v} = \langle -3, 1, 4\rangle$.

In Exercises 47–50, find an equation of the plane.

47. The plane passes through

$(-3, -4, 2)$, $(-3, 4, 1)$, and $(1, 1, -2)$.

48. The plane passes through the point $(-2, 3, 1)$ and is perpendicular to $\mathbf{n} = 3\mathbf{i} - \mathbf{j} + \mathbf{k}$.

49. The plane contains the lines given by

$$\frac{x-1}{-2} = y = z + 1$$

and

$$\frac{x+1}{-2} = y - 1 = z - 2.$$

50. The plane passes through the points $(5, 1, 3)$ and $(2, -2, 1)$ and is perpendicular to the plane $2x + y - z = 4$.

51. Find the distance between the point $(1, 0, 2)$ and the plane $2x - 3y + 6z = 6$.

52. Find the distance between the point $(3, -2, 4)$ and the plane $2x - 5y + z = 10$.

53. Find the distance between the planes $5x - 3y + z = 2$ and $5x - 3y + z = -3$.

54. Find the distance between the point $(-5, 1, 3)$ and the line given by $x = 1 + t$, $y = 3 - 2t$, and $z = 5 - t$.

In Exercises 55–64, describe and sketch the surface.

55. $x + 2y + 3z = 6$

56. $y = z^2$

57. $y = \frac{1}{2}z$

58. $y = \cos z$

59. $\dfrac{x^2}{16} + \dfrac{y^2}{9} + z^2 = 1$

60. $16x^2 + 16y^2 - 9z^2 = 0$

61. $\dfrac{x^2}{16} - \dfrac{y^2}{9} + z^2 = -1$

62. $\dfrac{x^2}{25} + \dfrac{y^2}{4} - \dfrac{z^2}{100} = 1$

63. $x^2 + z^2 = 4$

64. $y^2 + z^2 = 16$

65. Find an equation of a generating curve of the surface of revolution $y^2 + z^2 - 4x = 0$.

66. Find an equation for the surface of revolution generated by revolving the curve $z^2 = 2y$ in the yz-plane about the y-axis.

In Exercises 67 and 68, convert the point from rectangular coordinates to (a) cylindrical coordinates and (b) spherical coordinates.

67. $\left(-2\sqrt{2}, 2\sqrt{2}, 2\right)$

68. $\left(\dfrac{\sqrt{3}}{4}, \dfrac{3}{4}, \dfrac{3\sqrt{3}}{2}\right)$

In Exercises 69 and 70, convert the point from cylindrical coordinates to spherical coordinates.

69. $\left(100, -\dfrac{\pi}{6}, 50\right)$

70. $\left(81, -\dfrac{5\pi}{6}, 27\sqrt{3}\right)$

In Exercises 71 and 72, convert the point from spherical coordinates to cylindrical coordinates.

71. $\left(25, -\dfrac{\pi}{4}, \dfrac{3\pi}{4}\right)$

72. $\left(12, -\dfrac{\pi}{2}, \dfrac{2\pi}{3}\right)$

In Exercises 73 and 74, convert the rectangular equation to an equation in (a) cylindrical coordinates and (b) spherical coordinates.

73. $x^2 - y^2 = 2z$

74. $x^2 + y^2 + z^2 = 16$

In Exercises 75 and 76, find an equation in rectangular coordinates for the equation given in cylindrical coordinates, and sketch its graph.

75. $r = 4 \sin \theta$

76. $z = 4$

In Exercises 77 and 78, find an equation in rectangular coordinates for the equation given in spherical coordinates, and sketch its graph.

77. $\theta = \dfrac{\pi}{4}$

78. $\rho = 2 \cos \theta$

P.S. Problem Solving

See www.CalcChat.com for worked-out solutions to odd-numbered exercises.

1. Using vectors, prove the Law of Sines: If **a**, **b**, and **c** are the three sides of the triangle shown in the figure, then

$$\frac{\sin A}{\|\mathbf{a}\|} = \frac{\sin B}{\|\mathbf{b}\|} = \frac{\sin C}{\|\mathbf{c}\|}.$$

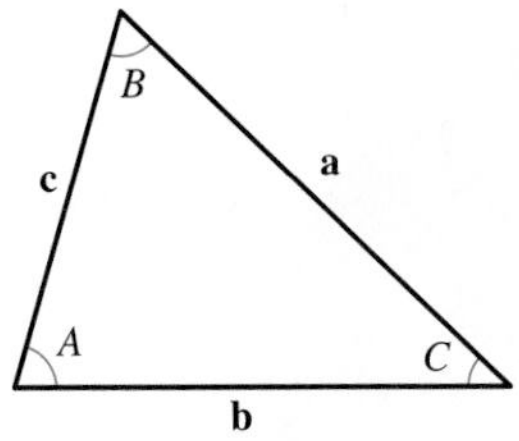

2. Consider the function $f(x) = \int_0^x \sqrt{t^4 + 1}\, dt$.

(a) Use a graphing utility to graph the function on the interval $-2 \le x \le 2$.

(b) Find a unit vector parallel to the graph of f at the point $(0, 0)$.

(c) Find a unit vector perpendicular to the graph of f at the point $(0, 0)$.

(d) Find the parametric equations of the tangent line to the graph of f at the point $(0, 0)$.

3. Using vectors, prove that the line segments joining the midpoints of the sides of a parallelogram form a parallelogram (see figure).

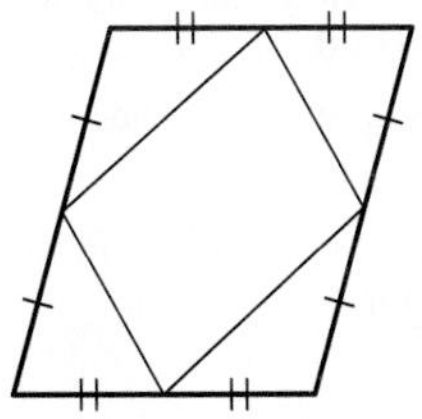

4. Using vectors, prove that the diagonals of a rhombus are perpendicular (see figure).

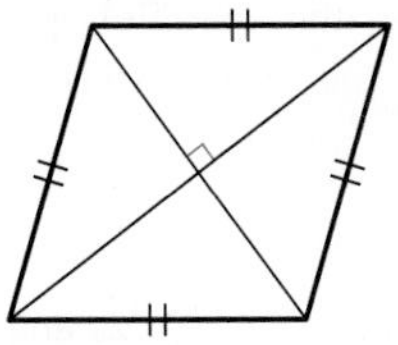

5. (a) Find the shortest distance between the point $Q(2, 0, 0)$ and the line determined by the points $P_1(0, 0, 1)$ and $P_2(0, 1, 2)$.

(b) Find the shortest distance between the point $Q(2, 0, 0)$ and the line segment joining the points $P_1(0, 0, 1)$ and $P_2(0, 1, 2)$.

6. Let P_0 be a point in the plane with normal vector **n**. Describe the set of points P in the plane for which $(\mathbf{n} + \overrightarrow{PP_0})$ is orthogonal to $(\mathbf{n} - \overrightarrow{PP_0})$.

7. (a) Find the volume of the solid bounded below by the paraboloid $z = x^2 + y^2$ and above by the plane $z = 1$.

(b) Find the volume of the solid bounded below by the elliptic paraboloid $z = \dfrac{x^2}{a^2} + \dfrac{y^2}{b^2}$ and above by the plane $z = k$, where $k > 0$.

(c) Show that the volume of the solid in part (b) is equal to one-half the product of the area of the base times the altitude, as shown in the figure.

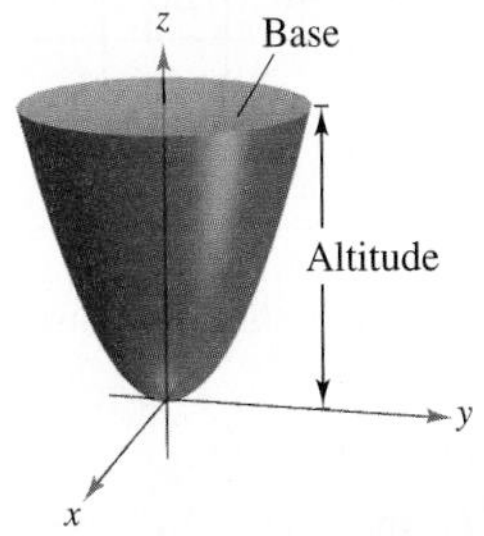

8. (a) Use the disk method to find the volume of the sphere $x^2 + y^2 + z^2 = r^2$.

(b) Find the volume of the ellipsoid $\dfrac{x^2}{a^2} + \dfrac{y^2}{b^2} + \dfrac{z^2}{c^2} = 1$.

9. Sketch the graph of each equation given in spherical coordinates.

(a) $\rho = 2 \sin \phi$

(b) $\rho = 2 \cos \phi$

10. Sketch the graph of each equation given in cylindrical coordinates.

(a) $r = 2 \cos \theta$

(b) $z = r^2 \cos 2\theta$

11. Prove the following property of the cross product.

$$(\mathbf{u} \times \mathbf{v}) \times (\mathbf{w} \times \mathbf{z}) = (\mathbf{u} \times \mathbf{v} \cdot \mathbf{z})\mathbf{w} - (\mathbf{u} \times \mathbf{v} \cdot \mathbf{w})\mathbf{z}$$

12. Consider the line given by the parametric equations

$$x = -t + 3, \quad y = \tfrac{1}{2}t + 1, \quad z = 2t - 1$$

and the point $(4, 3, s)$ for any real number s.

(a) Write the distance between the point and the line as a function of s.

(b) Use a graphing utility to graph the function in part (a). Use the graph to find the value of s such that the distance between the point and the line is minimum.

(c) Use the *zoom* feature of a graphing utility to zoom out several times on the graph in part (b). Does it appear that the graph has slant asymptotes? Explain. If it appears to have slant asymptotes, find them.

13. A tetherball weighing 1 pound is pulled outward from the pole by a horizontal force **u** until the rope makes an angle of θ degrees with the pole (see figure).

(a) Determine the resulting tension in the rope and the magnitude of **u** when $\theta = 30°$.

(b) Write the tension T in the rope and the magnitude of **u** as functions of θ. Determine the domains of the functions.

(c) Use a graphing utility to complete the table.

θ	0°	10°	20°	30°	40°	50°	60°
T							
$\lVert \mathbf{u} \rVert$							

(d) Use a graphing utility to graph the two functions for $0° \le \theta \le 60°$.

(e) Compare T and $\lVert \mathbf{u} \rVert$ as θ increases.

(f) Find (if possible) $\lim_{\theta \to \pi/2^-} T$ and $\lim_{\theta \to \pi/2^-} \lVert \mathbf{u} \rVert$. Are the results what you expected? Explain.

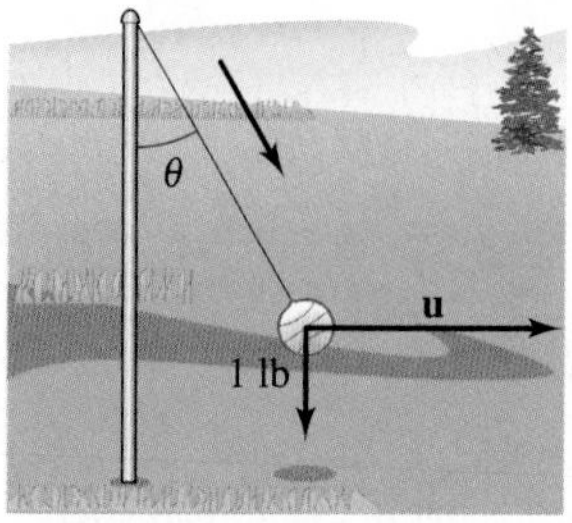

Figure for 13

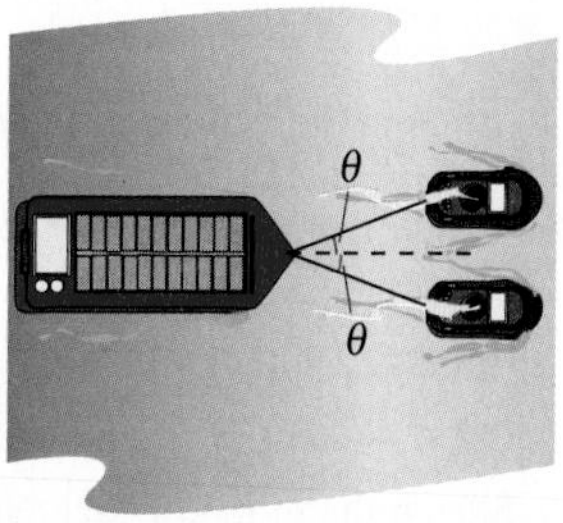

Figure for 14

14. A loaded barge is being towed by two tugboats, and the magnitude of the resultant is 6000 pounds directed along the axis of the barge (see figure). Each towline makes an angle of θ degrees with the axis of the barge.

(a) Find the tension in the towlines if $\theta = 20°$.

(b) Write the tension T of each line as a function of θ. Determine the domain of the function.

(c) Use a graphing utility to complete the table.

θ	10°	20°	30°	40°	50°	60°
T						

(d) Use a graphing utility to graph the tension function.

(e) Explain why the tension increases as θ increases.

15. Consider the vectors $\mathbf{u} = \langle \cos \alpha, \sin \alpha, 0 \rangle$ and $\mathbf{v} = \langle \cos \beta, \sin \beta, 0 \rangle$, where $\alpha > \beta$. Find the cross product of the vectors and use the result to prove the identity

$$\sin(\alpha - \beta) = \sin \alpha \cos \beta - \cos \alpha \sin \beta.$$

16. Los Angeles is located at 34.05° North latitude and 118.24° West longitude, and Rio de Janeiro, Brazil is located at 22.90° South latitude and 43.23° West longitude (see figure). Assume that Earth is spherical and has a radius of 4000 miles.

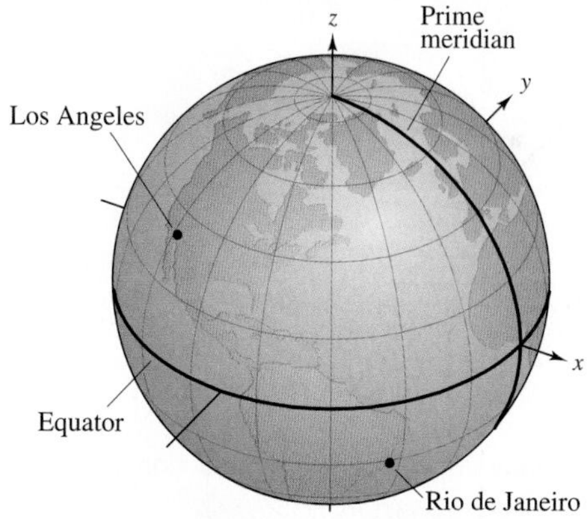

(a) Find the spherical coordinates for the location of each city.

(b) Find the rectangular coordinates for the location of each city.

(c) Find the angle (in radians) between the vectors from the center of Earth to each city.

(d) Find the great-circle distance s between the cities. (*Hint:* $s = r\theta$)

(e) Repeat parts (a)–(d) for the cities of Boston, located at 42.36° North latitude and 71.06° West longitude, and Honolulu, located at 21.31° North latitude and 157.86° West longitude.

17. Consider the plane that passes through the points P, R, and S. Show that the distance from a point Q to this plane is

$$\text{Distance} = \frac{|\mathbf{u} \cdot (\mathbf{v} \times \mathbf{w})|}{\lVert \mathbf{u} \times \mathbf{v} \rVert}$$

where $\mathbf{u} = \overrightarrow{PR}$, $\mathbf{v} = \overrightarrow{PS}$, and $\mathbf{w} = \overrightarrow{PQ}$.

18. Show that the distance between the parallel planes $ax + by + cz + d_1 = 0$ and $ax + by + cz + d_2 = 0$ is

$$\text{Distance} = \frac{|d_1 - d_2|}{\sqrt{a^2 + b^2 + c^2}}.$$

19. Show that the curve of intersection of the plane $z = 2y$ and the cylinder $x^2 + y^2 = 1$ is an ellipse.

20. Read the article "Tooth Tables: Solution of a Dental Problem by Vector Algebra" by Gary Hosler Meisters in *Mathematics Magazine*. (To view this article, go to the website *www.matharticles.com.*) Then write a paragraph explaining how vectors and vector algebra can be used in the construction of dental inlays.

12 Vector-Valued Functions

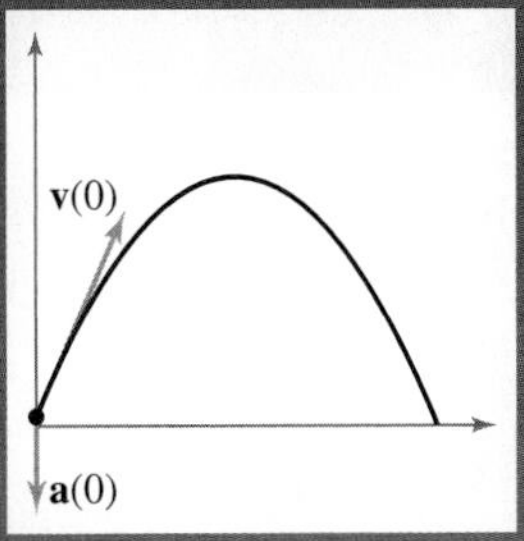

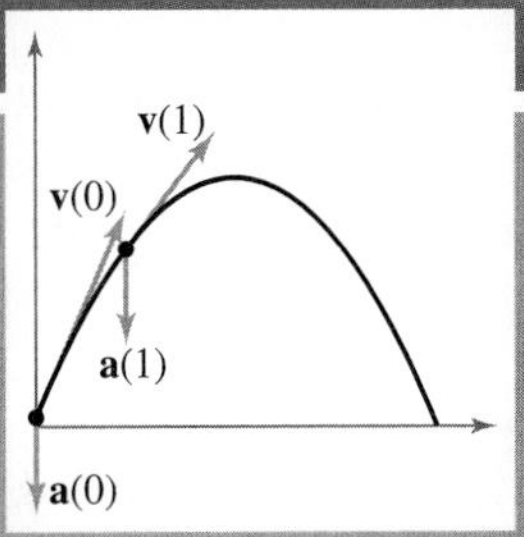

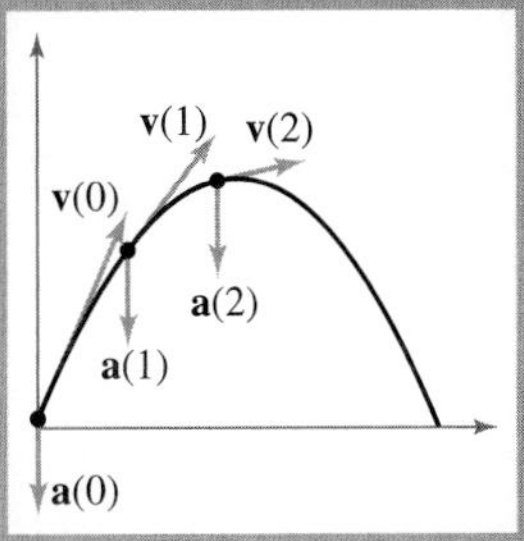

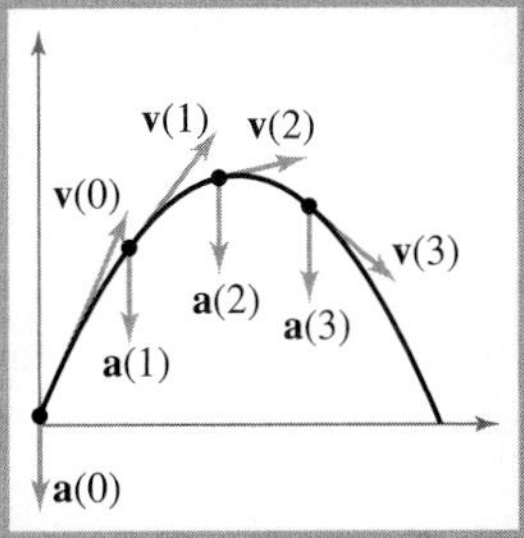

A *vector-valued function* maps real numbers to vectors. You can use a vector-valued function to represent the motion of a particle along a curve. In Section 12.3, you will use the first and second derivatives of a position vector to find a particle's velocity and acceleration.

A Ferris wheel is constructed using the basic principles of a bicycle wheel. When you are near the bottom of a moving Ferris wheel, the forces of rotation and weight combine, resulting in greater acceleration. Why do you think that the acceleration at the top of a moving Ferris wheel is not as great as at the bottom? Explain.

Paul Hardy/Corbis

Section 12.1 Vector-Valued Functions

- **Analyze and sketch a space curve given by a vector-valued function.**
- **Extend the concepts of limits and continuity to vector-valued functions.**

Space Curves and Vector-Valued Functions

In Section 10.2, a *plane curve* was defined as the set of ordered pairs $(f(t), g(t))$ together with their defining parametric equations

$$x = f(t) \qquad \text{and} \qquad y = g(t)$$

where f and g are continuous functions of t on an interval I. This definition can be extended naturally to three-dimensional space as follows. A **space curve** C is the set of all ordered triples $(f(t), g(t), h(t))$ together with their defining parametric equations

$$x = f(t), \qquad y = g(t), \qquad \text{and} \qquad z = h(t)$$

where f, g, and h are continuous functions of t on an interval I.

Before looking at examples of space curves, a new type of function, called a **vector-valued function,** is introduced. This type of function maps real numbers to vectors.

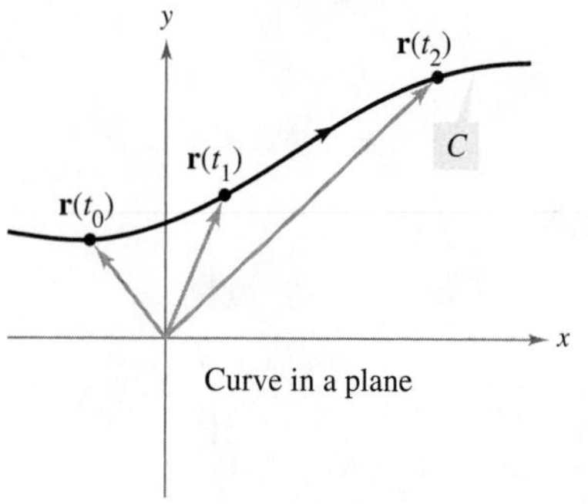

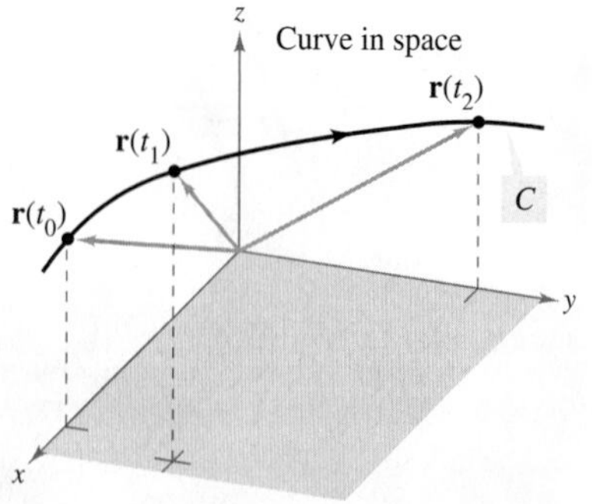

Curve C is traced out by the terminal point of position vector $\mathbf{r}(t)$.
Figure 12.1

Definition of Vector-Valued Function

A function of the form

$$\mathbf{r}(t) = f(t)\mathbf{i} + g(t)\mathbf{j} \qquad \text{Plane}$$

or

$$\mathbf{r}(t) = f(t)\mathbf{i} + g(t)\mathbf{j} + h(t)\mathbf{k} \qquad \text{Space}$$

is a **vector-valued function,** where the **component functions** f, g, and h are real-valued functions of the parameter t. Vector-valued functions are sometimes denoted as $\mathbf{r}(t) = \langle f(t), g(t)\rangle$ or $\mathbf{r}(t) = \langle f(t), g(t), h(t)\rangle$.

Technically, a curve in the plane or in space consists of a collection of points and the defining parametric equations. Two different curves can have the same graph. For instance, each of the curves given by

$$\mathbf{r} = \sin t\,\mathbf{i} + \cos t\,\mathbf{j} \qquad \text{and} \qquad \mathbf{r} = \sin t^2\,\mathbf{i} + \cos t^2\,\mathbf{j}$$

has the unit circle as its graph, but these equations do not represent the same curve—because the circle is traced out in different ways on the graphs.

Be sure you see the distinction between the vector-valued function $\mathbf{r}$ and the real-valued functions f, g, and h. All are functions of the real variable t, but $\mathbf{r}(t)$ is a vector, whereas $f(t)$, $g(t)$, and $h(t)$ are real numbers (for each specific value of t).

Vector-valued functions serve dual roles in the representation of curves. By letting the parameter t represent time, you can use a vector-valued function to represent *motion* along a curve. Or, in the more general case, you can use a vector-valued function to *trace the graph* of a curve. In either case, the terminal point of the position vector $\mathbf{r}(t)$ coincides with the point (x, y) or (x, y, z) on the curve given by the parametric equations, as shown in Figure 12.1. The arrowhead on the curve indicates the curve's *orientation* by pointing in the direction of increasing values of t.

Unless stated otherwise, the **domain** of a vector-valued function $\mathbf{r}$ is considered to be the intersection of the domains of the component functions f, g, and h. For instance, the domain of $\mathbf{r}(t) = (\ln t)\mathbf{i} + \sqrt{1-t}\,\mathbf{j} + t\mathbf{k}$ is the interval $(0, 1]$.

EXAMPLE 1 Sketching a Plane Curve

Sketch the plane curve represented by the vector-valued function

$$\mathbf{r}(t) = 2\cos t\mathbf{i} - 3\sin t\mathbf{j}, \quad 0 \le t \le 2\pi. \qquad \text{Vector-valued function}$$

Solution From the position vector $\mathbf{r}(t)$, you can write the parametric equations $x = 2\cos t$ and $y = -3\sin t$. Solving for $\cos t$ and $\sin t$ and using the identity $\cos^2 t + \sin^2 t = 1$ produces the rectangular equation

$$\frac{x^2}{2^2} + \frac{y^2}{3^2} = 1. \qquad \text{Rectangular equation}$$

The graph of this rectangular equation is the ellipse shown in Figure 12.2. The curve has a *clockwise* orientation. That is, as t increases from 0 to 2π, the position vector $\mathbf{r}(t)$ moves clockwise, and its terminal point traces the ellipse.

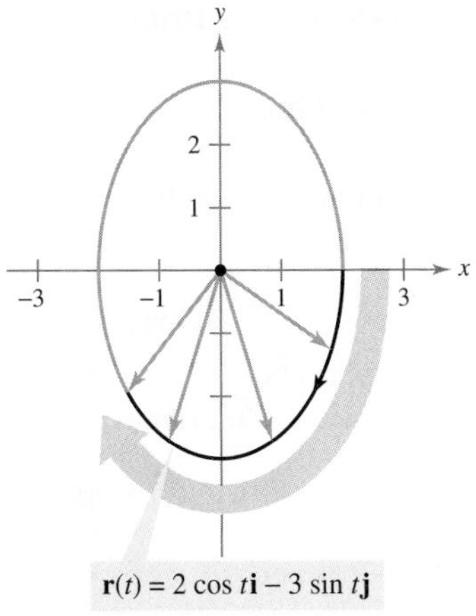

The ellipse is traced clockwise as t increases from 0 to 2π.
Figure 12.2

EXAMPLE 2 Sketching a Space Curve

Sketch the space curve represented by the vector-valued function

$$\mathbf{r}(t) = 4\cos t\mathbf{i} + 4\sin t\mathbf{j} + t\mathbf{k}, \quad 0 \le t \le 4\pi. \qquad \text{Vector-valued function}$$

Solution From the first two parametric equations $x = 4\cos t$ and $y = 4\sin t$, you can obtain

$$x^2 + y^2 = 16. \qquad \text{Rectangular equation}$$

This means that the curve lies on a right circular cylinder of radius 4, centered about the z-axis. To locate the curve on this cylinder, you can use the third parametric equation $z = t$. In Figure 12.3, note that as t increases from 0 to 4π, the point (x, y, z) spirals up the cylinder to produce a **helix.** A real-life example of a helix is shown in the drawing at the lower left.

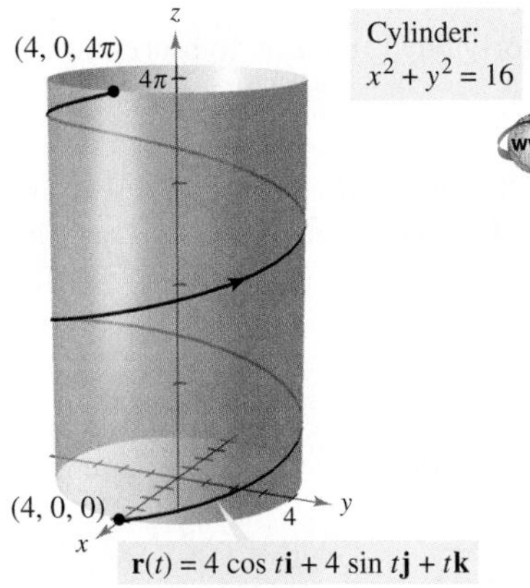

As t increases from 0 to 4π, two spirals on the helix are traced out.
Figure 12.3

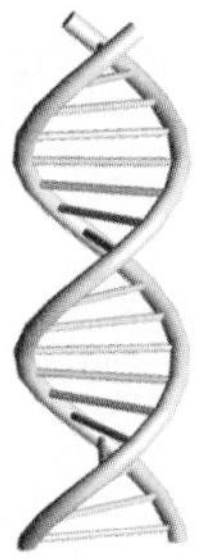

In 1953 Francis Crick and James D. Watson discovered the double helix structure of DNA, which led to the $30 billion per year biotechnology industry.

In Examples 1 and 2, you were given a vector-valued function and were asked to sketch the corresponding curve. The next two examples address the reverse problem—finding a vector-valued function to represent a given graph. Of course, if the graph is described parametrically, representation by a vector-valued function is straightforward. For instance, to represent the line in space given by

$$x = 2 + t, \quad y = 3t, \qquad \text{and} \qquad z = 4 - t$$

you can simply use the vector-valued function given by

$$\mathbf{r}(t) = (2 + t)\mathbf{i} + 3t\mathbf{j} + (4 - t)\mathbf{k}.$$

If a set of parametric equations for the graph is not given, the problem of representing the graph by a vector-valued function boils down to finding a set of parametric equations.

WWW *indicates that in the* HM mathSpace® *CD-ROM and the online* Eduspace® *system for this text, you will find an Open Exploration, which further explores this example using the computer algebra systems* Maple, Mathcad, Mathematica, *and* Derive.

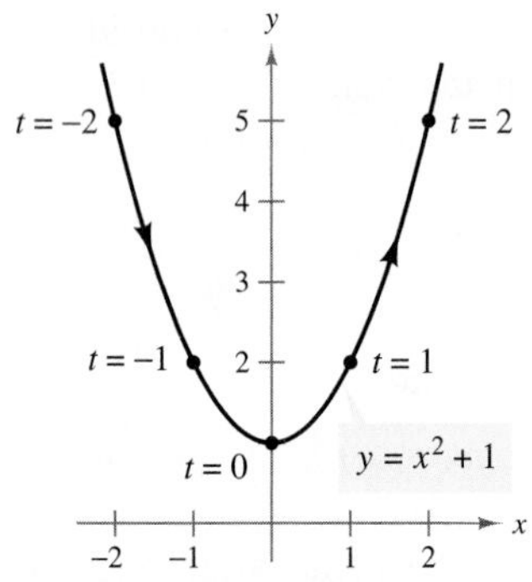

There are many ways to parametrize this graph. One way is to let $x = t$.
Figure 12.4

EXAMPLE 3 Representing a Graph by a Vector-Valued Function

Represent the parabola given by $y = x^2 + 1$ by a vector-valued function.

Solution Although there are many ways to choose the parameter t, a natural choice is to let $x = t$. Then $y = t^2 + 1$ and you have

$$\mathbf{r}(t) = t\mathbf{i} + (t^2 + 1)\mathbf{j}.$$ Vector-valued function

Note in Figure 12.4 the orientation produced by this particular choice of parameter. Had you chosen $x = -t$ as the parameter, the curve would have been oriented in the opposite direction.

EXAMPLE 4 Representing a Graph by a Vector-Valued Function

Sketch the graph C represented by the intersection of the semiellipsoid

$$\frac{x^2}{12} + \frac{y^2}{24} + \frac{z^2}{4} = 1, \quad z \geq 0$$

and the parabolic cylinder $y = x^2$. Then, find a vector-valued function to represent the graph.

Solution The intersection of the two surfaces is shown in Figure 12.5. As in Example 3, a natural choice of parameter is $x = t$. For this choice, you can use the given equation $y = x^2$ to obtain $y = t^2$. Then, it follows that

$$\frac{z^2}{4} = 1 - \frac{x^2}{12} - \frac{y^2}{24} = 1 - \frac{t^2}{12} - \frac{t^4}{24} = \frac{24 - 2t^2 - t^4}{24}.$$

NOTE Curves in space can be specified in various ways. For instance, the curve in Example 4 is described as the intersection of two surfaces in space.

Because the curve lies above the xy-plane, you should choose the positive square root for z and obtain the following parametric equations.

$$x = t, \quad y = t^2, \quad \text{and} \quad z = \sqrt{\frac{24 - 2t^2 - t^4}{6}}$$

The resulting vector-valued function is

$$\mathbf{r}(t) = t\mathbf{i} + t^2\mathbf{j} + \sqrt{\frac{24 - 2t^2 - t^4}{6}}\mathbf{k}, \quad -2 \leq t \leq 2.$$ Vector-valued function

From the points $(-2, 4, 0)$ and $(2, 4, 0)$ shown in Figure 12.5, you can see that the curve is traced as t increases from -2 to 2.

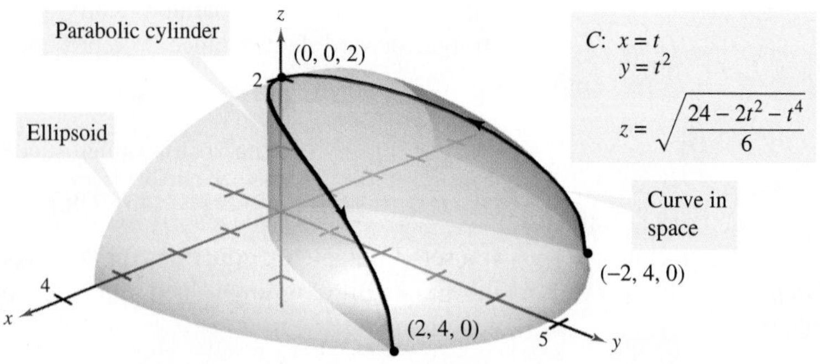

The curve C is the intersection of the semiellipsoid and the parabolic cylinder.
Figure 12.5

Limits and Continuity

Many techniques and definitions used in the calculus of real-valued functions can be applied to vector-valued functions. For instance, you can add and subtract vector-valued functions, multiply a vector-valued function by a scalar, take the limit of a vector-valued function, differentiate a vector-valued function, and so on. The basic approach is to capitalize on the linearity of vector operations by extending the definitions on a component-by-component basis. For example, to add or subtract two vector-valued functions (in the plane), you can write

$$\begin{aligned}\mathbf{r}_1(t) + \mathbf{r}_2(t) &= [f_1(t)\mathbf{i} + g_1(t)\mathbf{j}] + [f_2(t)\mathbf{i} + g_2(t)\mathbf{j}] && \text{Sum}\\ &= [f_1(t) + f_2(t)]\mathbf{i} + [g_1(t) + g_2(t)]\mathbf{j}\\ \mathbf{r}_1(t) - \mathbf{r}_2(t) &= [f_1(t)\mathbf{i} + g_1(t)\mathbf{j}] - [f_2(t)\mathbf{i} + g_2(t)\mathbf{j}] && \text{Difference}\\ &= [f_1(t) - f_2(t)]\mathbf{i} + [g_1(t) - g_2(t)]\mathbf{j}.\end{aligned}$$

Similarly, to multiply and divide a vector-valued function by a scalar, you can write

$$\begin{aligned}c\mathbf{r}(t) &= c[f_1(t)\mathbf{i} + g_1(t)\mathbf{j}] && \text{Scalar multiplication}\\ &= cf_1(t)\mathbf{i} + cg_1(t)\mathbf{j}\\ \frac{\mathbf{r}(t)}{c} &= \frac{[f_1(t)\mathbf{i} + g_1(t)\mathbf{j}]}{c}, \quad c \neq 0 && \text{Scalar division}\\ &= \frac{f_1(t)}{c}\mathbf{i} + \frac{g_1(t)}{c}\mathbf{j}.\end{aligned}$$

This component-by-component extension of operations with real-valued functions to vector-valued functions is further illustrated in the following definition of the limit of a vector-valued function.

Definition of the Limit of a Vector-Valued Function

1. If $\mathbf{r}$ is a vector-valued function such that $\mathbf{r}(t) = f(t)\mathbf{i} + g(t)\mathbf{j}$, then

$$\lim_{t\to a} \mathbf{r}(t) = \left[\lim_{t\to a} f(t)\right]\mathbf{i} + \left[\lim_{t\to a} g(t)\right]\mathbf{j} \qquad \text{Plane}$$

provided f and g have limits as $t \to a$.

2. If $\mathbf{r}$ is a vector-valued function such that $\mathbf{r}(t) = f(t)\mathbf{i} + g(t)\mathbf{j} + h(t)\mathbf{k}$, then

$$\lim_{t\to a} \mathbf{r}(t) = \left[\lim_{t\to a} f(t)\right]\mathbf{i} + \left[\lim_{t\to a} g(t)\right]\mathbf{j} + \left[\lim_{t\to a} h(t)\right]\mathbf{k} \qquad \text{Space}$$

provided f, g, and h have limits as $t \to a$.

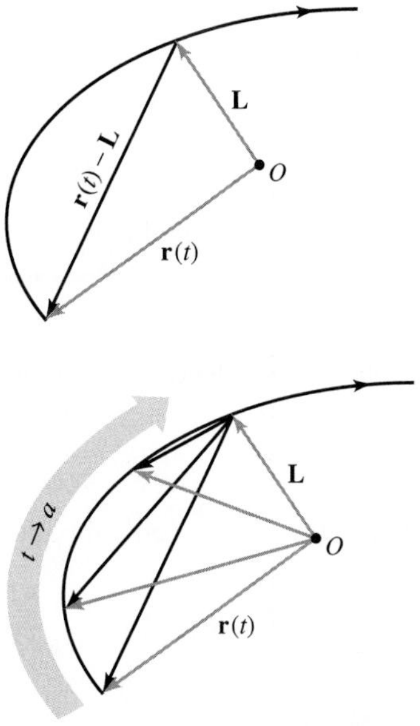

As t approaches a, $\mathbf{r}(t)$ approaches the limit $\mathbf{L}$. For the limit $\mathbf{L}$ to exist, it is not necessary that $\mathbf{r}(a)$ be defined or that $\mathbf{r}(a)$ be equal to $\mathbf{L}$.
Figure 12.6

If $\mathbf{r}(t)$ approaches the vector $\mathbf{L}$ as $t \to a$, the length of the vector $\mathbf{r}(t) - \mathbf{L}$ approaches 0. That is,

$$\|\mathbf{r}(t) - \mathbf{L}\| \to 0 \qquad \text{as} \qquad t \to a.$$

This is illustrated graphically in Figure 12.6. With this definition of the limit of a vector-valued function, you can develop vector versions of most of the limit theorems given in Chapter 1. For example, the limit of the sum of two vector-valued functions is the sum of their individual limits. Also, you can use the orientation of the curve $\mathbf{r}(t)$ to define one-sided limits of vector-valued functions. The next definition extends the notion of continuity to vector-valued functions.

Definition of Continuity of a Vector-Valued Function

A vector-valued function $\mathbf{r}$ is **continuous at the point** given by $t = a$ if the limit of $\mathbf{r}(t)$ exists as $t \to a$ and

$$\lim_{t\to a} \mathbf{r}(t) = \mathbf{r}(a).$$

A vector-valued function $\mathbf{r}$ is **continuous on an interval** I if it is continuous at every point in the interval.

From this definition, it follows that a vector-valued function is continuous at $t = a$ if and only if each of its component functions is continuous at $t = a$.

EXAMPLE 5 Continuity of Vector-Valued Functions

Discuss the continuity of the vector-valued function given by

$$\mathbf{r}(t) = t\mathbf{i} + a\mathbf{j} + (a^2 - t^2)\mathbf{k} \qquad a \text{ is a constant.}$$

at $t = 0$.

Solution As t approaches 0, the limit is

$$\begin{aligned}\lim_{t\to 0} \mathbf{r}(t) &= \left[\lim_{t\to 0} t\right]\mathbf{i} + \left[\lim_{t\to 0} a\right]\mathbf{j} + \left[\lim_{t\to 0} (a^2 - t^2)\right]\mathbf{k}\\ &= 0\mathbf{i} + a\mathbf{j} + a^2\mathbf{k}\\ &= a\mathbf{j} + a^2\mathbf{k}.\end{aligned}$$

Because

$$\begin{aligned}\mathbf{r}(0) &= (0)\mathbf{i} + (a)\mathbf{j} + (a^2)\mathbf{k}\\ &= a\mathbf{j} + a^2\mathbf{k}\end{aligned}$$

you can conclude that $\mathbf{r}$ is continuous at $t = 0$. By similar reasoning, you can conclude that the vector-valued function $\mathbf{r}$ is continuous at all real-number values of t.

For each value of a, the curve represented by the vector-valued function in Example 5,

$$\mathbf{r}(t) = t\mathbf{i} + a\mathbf{j} + (a^2 - t^2)\mathbf{k} \qquad a \text{ is a constant.}$$

is a parabola. You can think of each parabola as the intersection of the vertical plane $y = a$ and the hyperbolic paraboloid

$$y^2 - x^2 = z$$

as shown in Figure 12.7.

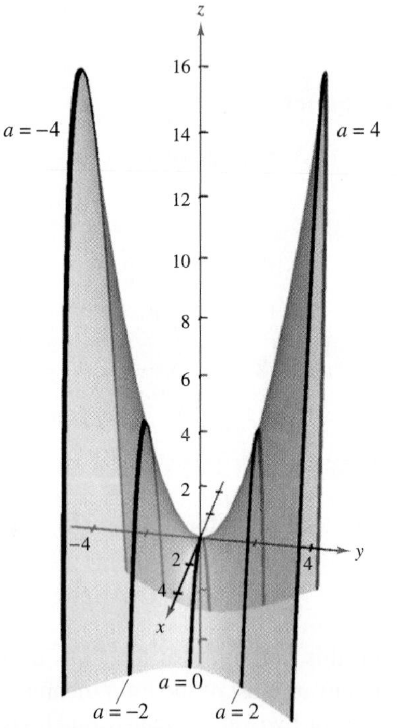

For each value of a, the curve represented by the vector-valued function $\mathbf{r}(t) = t\mathbf{i} + a\mathbf{j} + (a^2 - t^2)\mathbf{k}$ is a parabola.
Figure 12.7

TECHNOLOGY Almost any type of three-dimensional sketch is difficult to do by hand, but sketching curves in space is especially difficult. The problem is in trying to create the illusion of three dimensions. Graphing utilities use a variety of techniques to add "three-dimensionality" to graphs of space curves: one way is to show the curve on a surface, as in Figure 12.7.

Exercises for Section 12.1

See www.CalcChat.com for worked-out solutions to odd-numbered exercises.

In Exercises 1–8, find the domain of the vector-valued function.

1. $\mathbf{r}(t) = 5t\mathbf{i} - 4t\mathbf{j} - \dfrac{1}{t}\mathbf{k}$

2. $\mathbf{r}(t) = \sqrt{4 - t^2}\,\mathbf{i} + t^2\mathbf{j} - 6t\mathbf{k}$

3. $\mathbf{r}(t) = \ln t\,\mathbf{i} - e^t\,\mathbf{j} - t\mathbf{k}$

4. $\mathbf{r}(t) = \sin t\,\mathbf{i} + 4\cos t\,\mathbf{j} + t\mathbf{k}$

5. $\mathbf{r}(t) = \mathbf{F}(t) + \mathbf{G}(t)$ where

   $\mathbf{F}(t) = \cos t\,\mathbf{i} - \sin t\,\mathbf{j} + \sqrt{t}\,\mathbf{k}, \quad \mathbf{G}(t) = \cos t\,\mathbf{i} + \sin t\,\mathbf{j}$

6. $\mathbf{r}(t) = \mathbf{F}(t) - \mathbf{G}(t)$ where

   $\mathbf{F}(t) = \ln t\,\mathbf{i} + 5t\mathbf{j} - 3t^2\mathbf{k}, \quad \mathbf{G}(t) = \mathbf{i} + 4t\mathbf{j} - 3t^2\mathbf{k}$

7. $\mathbf{r}(t) = \mathbf{F}(t) \times \mathbf{G}(t)$ where

   $\mathbf{F}(t) = \sin t\,\mathbf{i} + \cos t\,\mathbf{j}, \quad \mathbf{G}(t) = \sin t\,\mathbf{j} + \cos t\,\mathbf{k}$

8. $\mathbf{r}(t) = \mathbf{F}(t) \times \mathbf{G}(t)$ where

   $\mathbf{F}(t) = t^3\mathbf{i} - t\mathbf{j} + t\mathbf{k}, \quad \mathbf{G}(t) = \sqrt[3]{t}\,\mathbf{i} + \dfrac{1}{t+1}\mathbf{j} + (t+2)\mathbf{k}$

In Exercises 9–12, evaluate (if possible) the vector-valued function at each given value of *t*.

9. $\mathbf{r}(t) = \frac{1}{2}t^2\mathbf{i} - (t-1)\mathbf{j}$

   (a) $\mathbf{r}(1)$ (b) $\mathbf{r}(0)$ (c) $\mathbf{r}(s+1)$

   (d) $\mathbf{r}(2 + \Delta t) - \mathbf{r}(2)$

10. $\mathbf{r}(t) = \cos t\,\mathbf{i} + 2\sin t\,\mathbf{j}$

    (a) $\mathbf{r}(0)$ (b) $\mathbf{r}(\pi/4)$ (c) $\mathbf{r}(\theta - \pi)$

    (d) $\mathbf{r}(\pi/6 + \Delta t) - \mathbf{r}(\pi/6)$

11. $\mathbf{r}(t) = \ln t\,\mathbf{i} + \dfrac{1}{t}\mathbf{j} + 3t\mathbf{k}$

    (a) $\mathbf{r}(2)$ (b) $\mathbf{r}(-3)$ (c) $\mathbf{r}(t-4)$

    (d) $\mathbf{r}(1 + \Delta t) - \mathbf{r}(1)$

12. $\mathbf{r}(t) = \sqrt{t}\,\mathbf{i} + t^{3/2}\,\mathbf{j} + e^{-t/4}\,\mathbf{k}$

    (a) $\mathbf{r}(0)$ (b) $\mathbf{r}(4)$ (c) $\mathbf{r}(c+2)$

    (d) $\mathbf{r}(9 + \Delta t) - \mathbf{r}(9)$

In Exercises 13 and 14, find $\|\mathbf{r}(t)\|$.

13. $\mathbf{r}(t) = \sin 3t\,\mathbf{i} + \cos 3t\,\mathbf{j} + t\mathbf{k}$

14. $\mathbf{r}(t) = \sqrt{t}\,\mathbf{i} + 3t\mathbf{j} - 4t\mathbf{k}$

Think About It **In Exercises 15 and 16, find $\mathbf{r}(t) \cdot \mathbf{u}(t)$. Is the result a vector-valued function? Explain.**

15. $\mathbf{r}(t) = (3t - 1)\mathbf{i} + \frac{1}{4}t^3\mathbf{j} + 4\mathbf{k}$

    $\mathbf{u}(t) = t^2\mathbf{i} - 8\mathbf{j} + t^3\mathbf{k}$

16. $\mathbf{r}(t) = \langle 3\cos t, 2\sin t, t - 2\rangle$

    $\mathbf{u}(t) = \langle 4\sin t, -6\cos t, t^2\rangle$

In Exercises 17–20, match the equation with its graph. [The graphs are labeled (a), (b), (c), and (d).]

(a)

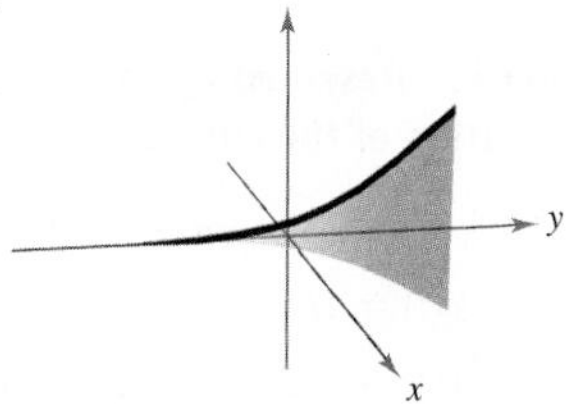

(b)

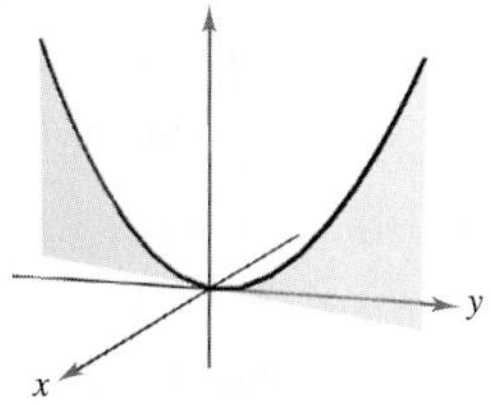

(c)

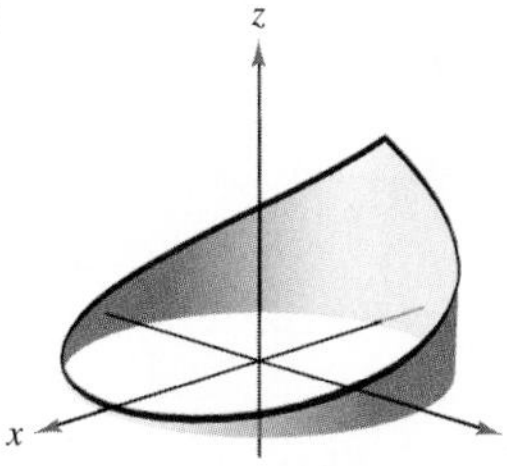

(d)

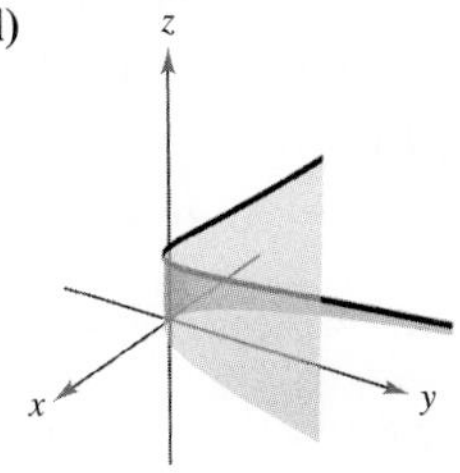

17. $\mathbf{r}(t) = t\mathbf{i} + 2t\mathbf{j} + t^2\mathbf{k}, \quad -2 \le t \le 2$

18. $\mathbf{r}(t) = \cos(\pi t)\mathbf{i} + \sin(\pi t)\mathbf{j} + t^2\mathbf{k}, \quad -1 \le t \le 1$

19. $\mathbf{r}(t) = t\mathbf{i} + t^2\mathbf{j} + e^{0.75t}\,\mathbf{k}, \quad -2 \le t \le 2$

20. $\mathbf{r}(t) = t\mathbf{i} + \ln t\,\mathbf{j} + \dfrac{2t}{3}\mathbf{k}, \quad 0.1 \le t \le 5$

21. ***Think About It*** The four figures below are graphs of the vector-valued function

    $$\mathbf{r}(t) = 4\cos t\,\mathbf{i} + 4\sin t\,\mathbf{j} + \frac{t}{4}\mathbf{k}.$$

    Match each of the four graphs with the point in space from which the helix is viewed. The four points are $(0, 0, 20)$, $(20, 0, 0)$, $(-20, 0, 0)$, and $(10, 20, 10)$.

(a)

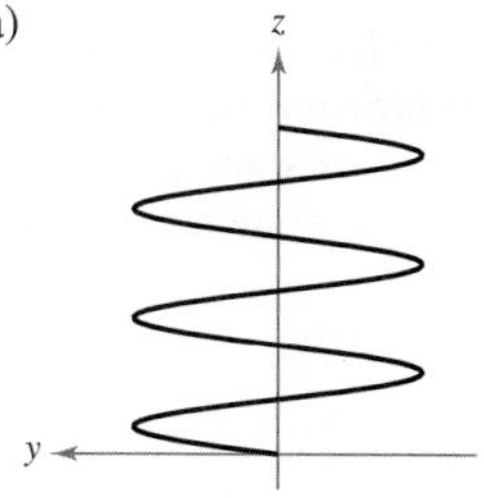

Generated by Mathematica

(b)

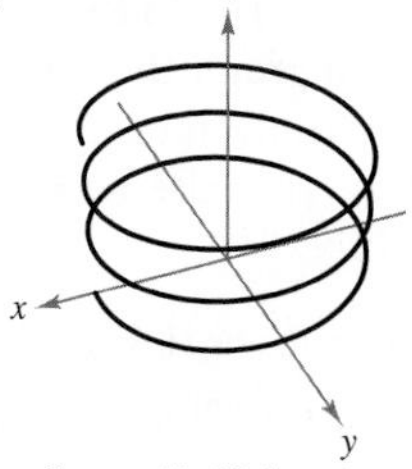

Generated by Mathematica

(c)

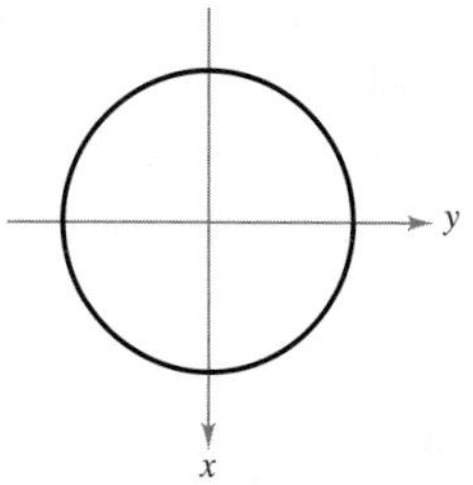

Generated by Mathematica

(d)

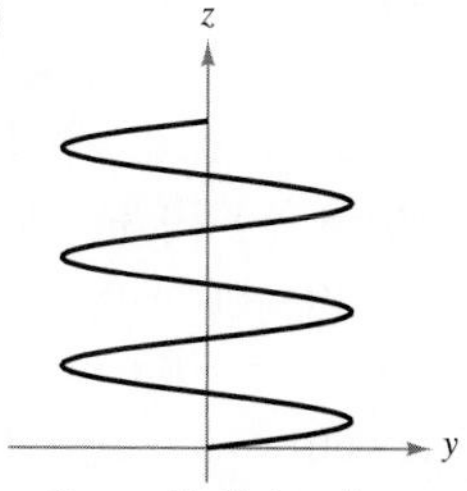

Generated by Mathematica

22. Sketch three graphs of the vector-valued function

$\mathbf{r}(t) = t\mathbf{i} + t\mathbf{j} + 2\mathbf{k}$

as viewed from each point.

(a) $(0, 0, 20)$ (b) $(10, 0, 0)$ (c) $(5, 5, 5)$

In Exercises 23–38, sketch the curve represented by the vector-valued function and give the orientation of the curve.

23. $\mathbf{r}(t) = 3t\mathbf{i} + (t - 1)\mathbf{j}$

24. $\mathbf{r}(t) = (1 - t)\mathbf{i} + \sqrt{t}\mathbf{j}$

25. $\mathbf{r}(t) = t^3\mathbf{i} + t^2\mathbf{j}$

26. $\mathbf{r}(t) = (t^2 + t)\mathbf{i} + (t^2 - t)\mathbf{j}$

27. $\mathbf{r}(\theta) = \cos\theta\mathbf{i} + 3\sin\theta\mathbf{j}$

28. $\mathbf{r}(t) = 2\cos t\mathbf{i} + 2\sin t\mathbf{j}$

29. $\mathbf{r}(\theta) = 3\sec\theta\mathbf{i} + 2\tan\theta\mathbf{j}$

30. $\mathbf{r}(t) = 2\cos^3 t\mathbf{i} + 2\sin^3 t\mathbf{j}$

31. $\mathbf{r}(t) = (-t + 1)\mathbf{i} + (4t + 2)\mathbf{j} + (2t + 3)\mathbf{k}$

32. $\mathbf{r}(t) = t\mathbf{i} + (2t - 5)\mathbf{j} + 3t\mathbf{k}$

33. $\mathbf{r}(t) = 2\cos t\mathbf{i} + 2\sin t\mathbf{j} + t\mathbf{k}$

34. $\mathbf{r}(t) = 3\cos t\mathbf{i} + 4\sin t\mathbf{j} + \frac{t}{2}\mathbf{k}$

35. $\mathbf{r}(t) = 2\sin t\mathbf{i} + 2\cos t\mathbf{j} + e^{-t}\mathbf{k}$

36. $\mathbf{r}(t) = t^2\mathbf{i} + 2t\mathbf{j} + \frac{3}{2}t\mathbf{k}$

37. $\mathbf{r}(t) = \langle t, t^2, \frac{2}{3}t^3 \rangle$

38. $\mathbf{r}(t) = \langle \cos t + t\sin t, \sin t - t\cos t, t \rangle$

In Exercises 39–42, use a computer algebra system to graph the vector-valued function and identify the common curve.

39. $\mathbf{r}(t) = -\frac{1}{2}t^2\mathbf{i} + t\mathbf{j} - \frac{\sqrt{3}}{2}t^2\mathbf{k}$

40. $\mathbf{r}(t) = t\mathbf{i} - \frac{\sqrt{3}}{2}t^2\mathbf{j} + \frac{1}{2}t^2\mathbf{k}$

41. $\mathbf{r}(t) = \sin t\mathbf{i} + \left(\frac{\sqrt{3}}{2}\cos t - \frac{1}{2}t\right)\mathbf{j} + \left(\frac{1}{2}\cos t + \frac{\sqrt{3}}{2}\right)\mathbf{k}$

42. $\mathbf{r}(t) = -\sqrt{2}\sin t\mathbf{i} + 2\cos t\mathbf{j} + \sqrt{2}\sin t\mathbf{k}$

Think About It **In Exercises 43 and 44, use a computer algebra system to graph the vector-valued function r(t). For each u(t), make a conjecture about the transformation (if any) of the graph of r(t). Use a computer algebra system to verify your conjecture.**

43. $\mathbf{r}(t) = 2\cos t\mathbf{i} + 2\sin t\mathbf{j} + \frac{1}{2}t\mathbf{k}$

(a) $\mathbf{u}(t) = 2(\cos t - 1)\mathbf{i} + 2\sin t\mathbf{j} + \frac{1}{2}t\mathbf{k}$

(b) $\mathbf{u}(t) = 2\cos t\mathbf{i} + 2\sin t\mathbf{j} + 2t\mathbf{k}$

(c) $\mathbf{u}(t) = 2\cos(-t)\mathbf{i} + 2\sin(-t)\mathbf{j} + \frac{1}{2}(-t)\mathbf{k}$

(d) $\mathbf{u}(t) = \frac{1}{2}t\mathbf{i} + 2\sin t\mathbf{j} + 2\cos t\mathbf{k}$

(e) $\mathbf{u}(t) = 6\cos t\mathbf{i} + 6\sin t\mathbf{j} + \frac{1}{2}t\mathbf{k}$

44. $\mathbf{r}(t) = t\mathbf{i} + t^2\mathbf{j} + \frac{1}{2}t^3\mathbf{k}$

(a) $\mathbf{u}(t) = t\mathbf{i} + (t^2 - 2)\mathbf{j} + \frac{1}{2}t^3\mathbf{k}$

(b) $\mathbf{u}(t) = t^2\mathbf{i} + t\mathbf{j} + \frac{1}{2}t^3\mathbf{k}$

(c) $\mathbf{u}(t) = t\mathbf{i} + t^2\mathbf{j} + (\frac{1}{2}t^3 + 4)\mathbf{k}$

(d) $\mathbf{u}(t) = t\mathbf{i} + t^2\mathbf{j} + \frac{1}{8}t^3\mathbf{k}$

(e) $\mathbf{u}(t) = (-t)\mathbf{i} + (-t)^2\mathbf{j} + \frac{1}{2}(-t)^3\mathbf{k}$

In Exercises 45–52, represent the plane curve by a vector-valued function. (There are many correct answers.)

45. $y = 4 - x$

46. $2x - 3y + 5 = 0$

47. $y = (x - 2)^2$

48. $y = 4 - x^2$

49. $x^2 + y^2 = 25$

50. $(x - 2)^2 + y^2 = 4$

51. $\frac{x^2}{16} - \frac{y^2}{4} = 1$

52. $\frac{x^2}{16} + \frac{y^2}{9} = 1$

53. A particle moves on a straight-line path that passes through the points $(2, 3, 0)$ and $(0, 8, 8)$. Find a vector-valued function for the path. Use a computer algebra system to graph your function. (There are many correct answers.)

54. The outer edge of a playground slide is in the shape of a helix of radius 1.5 meters. The slide has a height of 2 meters and makes one complete revolution from top to bottom. Find a vector-valued function for the helix. Use a computer algebra system to graph your function. (There are many correct answers.)

In Exercises 55–58, find vector-valued functions forming the boundaries of the region in the figure. State the interval for the parameter of each function.

55.

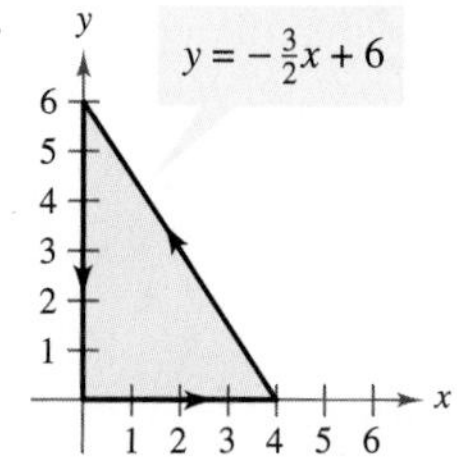

56.

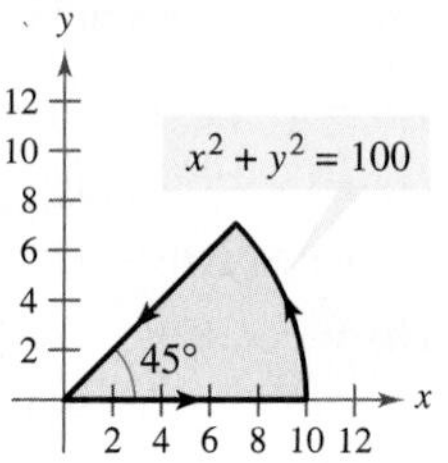

57.

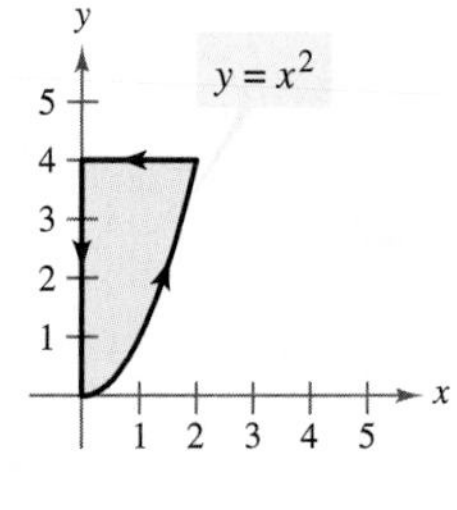

58.

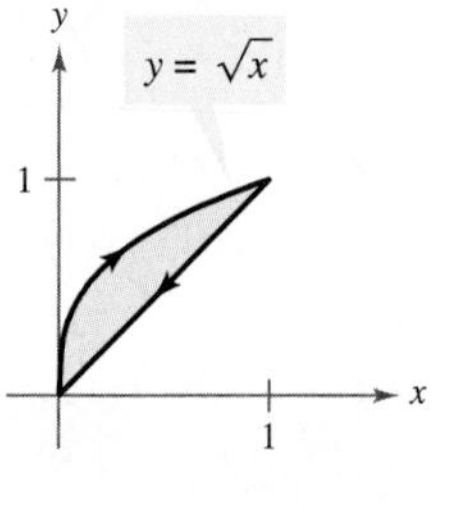

In Exercises 59–66, sketch the space curve represented by the intersection of the surfaces. Then represent the curve by a vector-valued function using the given parameter.

	Surfaces	*Parameter*
59.	$z = x^2 + y^2,\ x + y = 0$	$x = t$
60.	$z = x^2 + y^2,\ z = 4$	$x = 2\cos t$
61.	$x^2 + y^2 = 4,\ z = x^2$	$x = 2\sin t$
62.	$4x^2 + 4y^2 + z^2 = 16,\ x = z^2$	$z = t$
63.	$x^2 + y^2 + z^2 = 4,\ x + z = 2$	$x = 1 + \sin t$
64.	$x^2 + y^2 + z^2 = 10,\ x + y = 4$	$x = 2 + \sin t$
65.	$x^2 + z^2 = 4,\ y^2 + z^2 = 4$	$x = t$ (first octant)
66.	$x^2 + y^2 + z^2 = 16,\ xy = 4$	$x = t$ (first octant)

67. Show that the vector-valued function

$$\mathbf{r}(t) = t\mathbf{i} + 2t\cos t\mathbf{j} + 2t\sin t\mathbf{k}$$

lies on the cone $4x^2 = y^2 + z^2$. Sketch the curve.

68. Show that the vector-valued function

$$\mathbf{r}(t) = e^{-t}\cos t\mathbf{i} + e^{-t}\sin t\mathbf{j} + e^{-t}\mathbf{k}$$

lies on the cone $z^2 = x^2 + y^2$. Sketch the curve.

In Exercises 69–74, evaluate the limit.

69. $\lim_{t\to 2}\left(t\mathbf{i} + \frac{t^2-4}{t^2-2t}\mathbf{j} + \frac{1}{t}\mathbf{k}\right)$

70. $\lim_{t\to 0}\left(e^t\mathbf{i} + \frac{\sin t}{t}\mathbf{j} + e^{-t}\mathbf{k}\right)$

71. $\lim_{t\to 0}\left(t^2\mathbf{i} + 3t\mathbf{j} + \frac{1-\cos t}{t}\mathbf{k}\right)$

72. $\lim_{t\to 1}\left(\sqrt{t}\,\mathbf{i} + \frac{\ln t}{t^2-1}\mathbf{j} + 2t^2\,\mathbf{k}\right)$

73. $\lim_{t\to 0}\left(\frac{1}{t}\mathbf{i} + \cos t\mathbf{j} + \sin t\mathbf{k}\right)$

74. $\lim_{t\to\infty}\left(e^{-t}\mathbf{i} + \frac{1}{t}\mathbf{j} + \frac{t}{t^2+1}\mathbf{k}\right)$

In Exercises 75–80, determine the interval(s) on which the vector-valued function is continuous.

75. $\mathbf{r}(t) = t\mathbf{i} + \frac{1}{t}\mathbf{j}$

76. $\mathbf{r}(t) = \sqrt{t}\,\mathbf{i} + \sqrt{t-1}\,\mathbf{j}$

77. $\mathbf{r}(t) = t\mathbf{i} + \arcsin t\mathbf{j} + (t-1)\mathbf{k}$

78. $\mathbf{r}(t) = 2e^{-t}\mathbf{i} + e^{-t}\mathbf{j} + \ln(t-1)\mathbf{k}$

79. $\mathbf{r}(t) = \langle e^{-t}, t^2, \tan t\rangle$

80. $\mathbf{r}(t) = \left\langle 8, \sqrt{t}, \sqrt[3]{t}\right\rangle$

Writing About Concepts

81. State the definition of a vector-valued function in the plane and in space.

82. If $\mathbf{r}(t)$ is a vector-valued function, is the graph of the vector-valued function $\mathbf{u}(t) = \mathbf{r}(t-2)$ a horizontal translation of the graph of $\mathbf{r}(t)$? Explain your reasoning.

83. Consider the vector-valued function

$$\mathbf{r}(t) = t^2\mathbf{i} + (t-3)\mathbf{j} + t\mathbf{k}.$$

Write a vector-valued function $\mathbf{s}(t)$ that is the specified transformation of $\mathbf{r}$.

(a) A vertical translation three units upward

(b) A horizontal translation two units in the direction of the negative x-axis

(c) A horizontal translation five units in the direction of the positive y-axis

84. State the definition of continuity of a vector-valued function. Give an example of a vector-valued function that is defined but not continuous at $t = 2$.

85. Let $\mathbf{r}(t)$ and $\mathbf{u}(t)$ be vector-valued functions whose limits exist as $t \to c$. Prove that

$$\lim_{t\to c}[\mathbf{r}(t) \times \mathbf{u}(t)] = \lim_{t\to c}\mathbf{r}(t) \times \lim_{t\to c}\mathbf{u}(t).$$

86. Let $\mathbf{r}(t)$ and $\mathbf{u}(t)$ be vector-valued functions whose limits exist as $t \to c$. Prove that

$$\lim_{t\to c}[\mathbf{r}(t) \cdot \mathbf{u}(t)] = \lim_{t\to c}\mathbf{r}(t) \cdot \lim_{t\to c}\mathbf{u}(t).$$

87. Prove that if $\mathbf{r}$ is a vector-valued function that is continuous at c, then $\|\mathbf{r}\|$ is continuous at c.

88. Verify that the converse of Exercise 87 is not true by finding a vector-valued function $\mathbf{r}$ such that $\|\mathbf{r}\|$ is continuous at c but $\mathbf{r}$ is not continuous at c.

True or False? **In Exercises 89–92, determine whether the statement is true or false. If it is false, explain why or give an example that shows it is false.**

89. If f, g, and h are first-degree polynomial functions, then the curve given by $x = f(t)$, $y = g(t)$, and $z = h(t)$ is a line.

90. If the curve given by $x = f(t)$, $y = g(t)$, and $z = h(t)$ is a line, then f, g, and h are first-degree polynomial functions of t.

91. Two particles traveling along the curves $\mathbf{r}(t) = t\mathbf{i} + t^2\mathbf{j}$ and $\mathbf{u}(t) = (2+t)\mathbf{i} + 8t\mathbf{j}$ will collide.

92. The vector-valued function $\mathbf{r}(t) = t^2\mathbf{i} + t\sin t\,\mathbf{j} + t\cos t\,\mathbf{k}$ lies on the paraboloid $x = y^2 + z^2$.

Section Project: Witch of Agnesi

In Section 3.5, you studied a famous curve called the **Witch of Agnesi.** In this project you will take a closer look at this function.

Consider a circle of radius a centered on the y-axis at $(0, a)$. Let A be a point on the horizontal line $y = 2a$, let O be the origin, and let B be the point where the segment OA intersects the circle. A point P is on the Witch of Agnesi if P lies on the horizontal line through B and on the vertical line through A.

(a) Show that the point A is traced out by the vector-valued function

$$\mathbf{r}_A(\theta) = 2a\cot\theta\mathbf{i} + 2a\mathbf{j}, \quad 0 < \theta < \pi$$

where θ is the angle that OA makes with the positive x-axis.

(b) Show that the point B is traced out by the vector-valued function

$$\mathbf{r}_B(\theta) = a\sin 2\theta\mathbf{i} + a(1-\cos 2\theta)\mathbf{j}, \quad 0 < \theta < \pi.$$

(c) Combine the results in parts (a) and (b) to find the vector-valued function $\mathbf{r}(\theta)$ for the Witch of Agnesi. Use a graphing utility to graph this curve for $a = 1$.

(d) Describe the limits $\lim_{\theta\to 0^+}\mathbf{r}(\theta)$ and $\lim_{\theta\to\pi^-}\mathbf{r}(\theta)$.

(e) Eliminate the parameter θ and determine the rectangular equation of the Witch of Agnesi. Use a graphing utility to graph this function for $a = 1$ and compare your graph with that obtained in part (c).

Section 12.2 Differentiation and Integration of Vector-Valued Functions

- **Differentiate a vector-valued function.**
- **Integrate a vector-valued function.**

Differentiation of Vector-Valued Functions

In Sections 12.3–12.5, you will study several important applications involving the calculus of vector-valued functions. In preparation for that study, this section is devoted to the mechanics of differentiation and integration of vector-valued functions.

The definition of the derivative of a vector-valued function parallels that given for real-valued functions.

Definition of the Derivative of a Vector-Valued Function

The **derivative of a vector-valued function r** is defined by

$$\mathbf{r}'(t) = \lim_{\Delta t \to 0} \frac{\mathbf{r}(t + \Delta t) - \mathbf{r}(t)}{\Delta t}$$

for all t for which the limit exists. If $\mathbf{r}'(c)$ exists, then $\mathbf{r}$ is **differentiable at c.** If $\mathbf{r}'(c)$ exists for all c in an open interval I, then $\mathbf{r}$ is **differentiable on the interval I.** Differentiability of vector-valued functions can be extended to closed intervals by considering one-sided limits.

NOTE In addition to $\mathbf{r}'(t)$, other notations for the derivative of a vector-valued function are

$$D_t[\mathbf{r}(t)], \quad \frac{d}{dt}[\mathbf{r}(t)], \quad \text{and} \quad \frac{d\mathbf{r}}{dt}.$$

Differentiation of vector-valued functions can be done on a *component-by-component basis*. To see why this is true, consider the function given by

$$\mathbf{r}(t) = f(t)\mathbf{i} + g(t)\mathbf{j}.$$

Applying the definition of the derivative produces the following.

$$\begin{aligned}
\mathbf{r}'(t) &= \lim_{\Delta t \to 0} \frac{\mathbf{r}(t + \Delta t) - \mathbf{r}(t)}{\Delta t} \\
&= \lim_{\Delta t \to 0} \frac{f(t + \Delta t)\mathbf{i} + g(t + \Delta t)\mathbf{j} - f(t)\mathbf{i} - g(t)\mathbf{j}}{\Delta t} \\
&= \lim_{\Delta t \to 0} \left\{ \left[\frac{f(t + \Delta t) - f(t)}{\Delta t} \right]\mathbf{i} + \left[\frac{g(t + \Delta t) - g(t)}{\Delta t} \right]\mathbf{j} \right\} \\
&= \left\{ \lim_{\Delta t \to 0} \left[\frac{f(t + \Delta t) - f(t)}{\Delta t} \right] \right\}\mathbf{i} + \left\{ \lim_{\Delta t \to 0} \left[\frac{g(t + \Delta t) - g(t)}{\Delta t} \right] \right\}\mathbf{j} \\
&= f'(t)\mathbf{i} + g'(t)\mathbf{j}
\end{aligned}$$

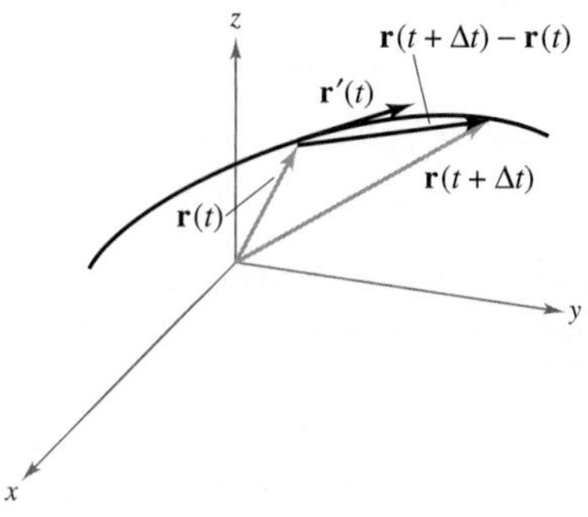

Figure 12.8

This important result is listed in the theorem on the next page. Note that the derivative of the vector-valued function $\mathbf{r}$ is itself a vector-valued function. You can see from Figure 12.8 that $\mathbf{r}'(t)$ is a vector tangent to the curve given by $\mathbf{r}(t)$ and pointing in the direction of increasing t-values.

THEOREM 12.1 Differentiation of Vector-Valued Functions

1. If $\mathbf{r}(t) = f(t)\mathbf{i} + g(t)\mathbf{j}$, where f and g are differentiable functions of t, then

$$\mathbf{r}'(t) = f'(t)\mathbf{i} + g'(t)\mathbf{j}. \qquad \text{Plane}$$

2. If $\mathbf{r}(t) = f(t)\mathbf{i} + g(t)\mathbf{j} + h(t)\mathbf{k}$, where f, g, and h are differentiable functions of t, then

$$\mathbf{r}'(t) = f'(t)\mathbf{i} + g'(t)\mathbf{j} + h'(t)\mathbf{k}. \qquad \text{Space}$$

EXAMPLE 1 Differentiation of Vector-Valued Functions

Find the derivative of each vector-valued function.

a. $\mathbf{r}(t) = t^2\mathbf{i} - 4\mathbf{j}$ **b.** $\mathbf{r}(t) = \dfrac{1}{t}\mathbf{i} + \ln t\mathbf{j} + e^{2t}\mathbf{k}$

Solution Differentiating on a component-by-component basis produces the following.

a. $\mathbf{r}'(t) = 2t\mathbf{i} - 0\mathbf{j}$

$= 2t\mathbf{i}$ — Derivative

b. $\mathbf{r}'(t) = -\dfrac{1}{t^2}\mathbf{i} + \dfrac{1}{t}\mathbf{j} + 2e^{2t}\mathbf{k}$ — Derivative

Higher-order derivatives of vector-valued functions are obtained by successive differentiation of each component function.

EXAMPLE 2 Higher-Order Differentiation

For the vector-valued function given by $\mathbf{r}(t) = \cos t\mathbf{i} + \sin t\mathbf{j} + 2t\mathbf{k}$, find each of the following.

a. $\mathbf{r}'(t)$ **b.** $\mathbf{r}''(t)$

c. $\mathbf{r}'(t) \cdot \mathbf{r}''(t)$ **d.** $\mathbf{r}'(t) \times \mathbf{r}''(t)$

Solution

a. $\mathbf{r}'(t) = -\sin t\mathbf{i} + \cos t\mathbf{j} + 2\mathbf{k}$ — First derivative

b. $\mathbf{r}''(t) = -\cos t\mathbf{i} - \sin t\mathbf{j} + 0\mathbf{k}$

$= -\cos t\mathbf{i} - \sin t\mathbf{j}$ — Second derivative

c. $\mathbf{r}'(t) \cdot \mathbf{r}''(t) = \sin t \cos t - \sin t \cos t = 0$ — Dot product

d. $\mathbf{r}'(t) \times \mathbf{r}''(t) = \begin{vmatrix} \mathbf{i} & \mathbf{j} & \mathbf{k} \\ -\sin t & \cos t & 2 \\ -\cos t & -\sin t & 0 \end{vmatrix}$ — Cross product

$$= \begin{vmatrix} \cos t & 2 \\ -\sin t & 0 \end{vmatrix}\mathbf{i} - \begin{vmatrix} -\sin t & 2 \\ -\cos t & 0 \end{vmatrix}\mathbf{j} + \begin{vmatrix} -\sin t & \cos t \\ -\cos t & -\sin t \end{vmatrix}\mathbf{k}$$

$$= 2\sin t\mathbf{i} - 2\cos t\mathbf{j} + \mathbf{k}$$

Note that the dot product in part (c) is a *real-valued* function, not a vector-valued function.

The parametrization of the curve represented by the vector-valued function

$$\mathbf{r}(t) = f(t)\mathbf{i} + g(t)\mathbf{j} + h(t)\mathbf{k}$$

is **smooth on an open interval** I if f', g', and h' are continuous on I and $\mathbf{r}'(t) \neq \mathbf{0}$ for any value of t in the interval I.

EXAMPLE 3 Finding Intervals on Which a Curve Is Smooth

Find the intervals on which the epicycloid C given by

$$\mathbf{r}(t) = (5\cos t - \cos 5t)\mathbf{i} + (5\sin t - \sin 5t)\mathbf{j}, \quad 0 \le t \le 2\pi$$

is smooth.

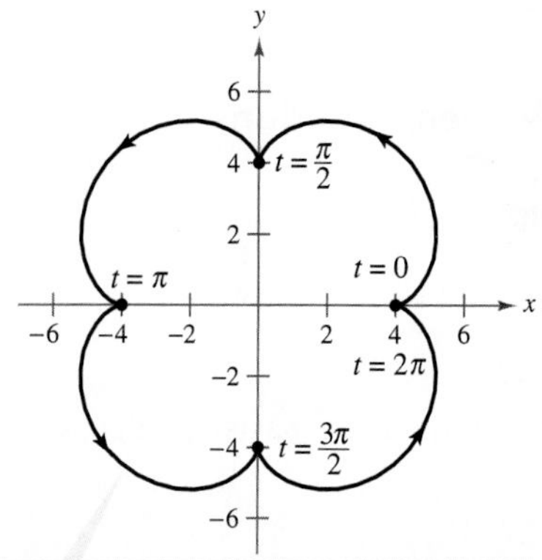

$\mathbf{r}(t) = (5\cos t - \cos 5t)\mathbf{i} + (5\sin t - \sin 5t)\mathbf{j}$

The epicycloid is not smooth at the points where it intersects the axes.
Figure 12.9

Solution The derivative of $\mathbf{r}$ is

$$\mathbf{r}'(t) = (-5\sin t + 5\sin 5t)\mathbf{i} + (5\cos t - 5\cos 5t)\mathbf{j}.$$

In the interval $[0, 2\pi]$, the only values of t for which

$$\mathbf{r}'(t) = 0\mathbf{i} + 0\mathbf{j}$$

are $t = 0, \pi/2, \pi, 3\pi/2$, and 2π. Therefore, you can conclude that C is smooth in the intervals

$$\left(0, \frac{\pi}{2}\right), \left(\frac{\pi}{2}, \pi\right), \left(\pi, \frac{3\pi}{2}\right), \quad \text{and} \quad \left(\frac{3\pi}{2}, 2\pi\right)$$

as shown in Figure 12.9.

NOTE In Figure 12.9, note that the curve is not smooth at points at which the curve makes abrupt changes in direction. Such points are called **cusps** or **nodes.**

Most of the differentiation rules in Chapter 2 have counterparts for vector-valued functions, and several are listed in the following theorem. Note that the theorem contains three versions of "product rules." Property 3 gives the derivative of the product of a real-valued function f and a vector-valued function $\mathbf{r}$, Property 4 gives the derivative of the dot product of two vector-valued functions, and Property 5 gives the derivative of the cross product of two vector-valued functions (in space). Note that Property 5 applies only to three-dimensional vector-valued functions, because the cross product is not defined for two-dimensional vectors.

THEOREM 12.2 Properties of the Derivative

Let $\mathbf{r}$ and $\mathbf{u}$ be differentiable vector-valued functions of t, let f be a differentiable real-valued function of t, and let c be a scalar.

1. $D_t[c\mathbf{r}(t)] = c\mathbf{r}'(t)$
2. $D_t[\mathbf{r}(t) \pm \mathbf{u}(t)] = \mathbf{r}'(t) \pm \mathbf{u}'(t)$
3. $D_t[f(t)\mathbf{r}(t)] = f(t)\mathbf{r}'(t) + f'(t)\mathbf{r}(t)$
4. $D_t[\mathbf{r}(t) \cdot \mathbf{u}(t)] = \mathbf{r}(t) \cdot \mathbf{u}'(t) + \mathbf{r}'(t) \cdot \mathbf{u}(t)$
5. $D_t[\mathbf{r}(t) \times \mathbf{u}(t)] = \mathbf{r}(t) \times \mathbf{u}'(t) + \mathbf{r}'(t) \times \mathbf{u}(t)$
6. $D_t[\mathbf{r}(f(t))] = \mathbf{r}'(f(t))f'(t)$
7. If $\mathbf{r}(t) \cdot \mathbf{r}(t) = c$, then $\mathbf{r}(t) \cdot \mathbf{r}'(t) = 0$.

Proof To prove Property 4, let

$$\mathbf{r}(t) = f_1(t)\mathbf{i} + g_1(t)\mathbf{j} \quad \text{and} \quad \mathbf{u}(t) = f_2(t)\mathbf{i} + g_2(t)\mathbf{j}$$

where f_1, f_2, g_1, and g_2 are differentiable functions of t. Then,

$$\mathbf{r}(t) \cdot \mathbf{u}(t) = f_1(t)f_2(t) + g_1(t)g_2(t)$$

and it follows that

$$\begin{aligned} D_t[\mathbf{r}(t) \cdot \mathbf{u}(t)] &= f_1(t)f_2'(t) + f_1'(t)f_2(t) + g_1(t)g_2'(t) + g_1'(t)g_2(t) \\ &= [f_1(t)f_2'(t) + g_1(t)g_2'(t)] + [f_1'(t)f_2(t) + g_1'(t)g_2(t)] \\ &= \mathbf{r}(t) \cdot \mathbf{u}'(t) + \mathbf{r}'(t) \cdot \mathbf{u}(t). \end{aligned}$$

Proofs of the other properties are left as exercises (see Exercises 73–77 and Exercise 80).

EXPLORATION

Let $\mathbf{r}(t) = \cos t\mathbf{i} + \sin t\mathbf{j}$. Sketch the graph of $\mathbf{r}(t)$. Explain why the graph is a circle of radius 1 centered at the origin. Calculate $\mathbf{r}(\pi/4)$ and $\mathbf{r}'(\pi/4)$. Position the vector $\mathbf{r}'(\pi/4)$ so that its initial point is at the terminal point of $\mathbf{r}(\pi/4)$. What do you observe? Show that $\mathbf{r}(t) \cdot \mathbf{r}(t)$ is constant and that $\mathbf{r}(t) \cdot \mathbf{r}'(t) = 0$ for all t. How does this example relate to Property 7 of Theorem 12.2?

EXAMPLE 4 Using Properties of the Derivative

For the vector-valued functions given by

$$\mathbf{r}(t) = \frac{1}{t}\mathbf{i} - \mathbf{j} + \ln t\mathbf{k} \quad \text{and} \quad \mathbf{u}(t) = t^2\mathbf{i} - 2t\mathbf{j} + \mathbf{k}$$

find

a. $D_t[\mathbf{r}(t) \cdot \mathbf{u}(t)]$ and **b.** $D_t[\mathbf{u}(t) \times \mathbf{u}'(t)]$.

Solution

a. Because $\mathbf{r}'(t) = -\frac{1}{t^2}\mathbf{i} + \frac{1}{t}\mathbf{k}$ and $\mathbf{u}'(t) = 2t\mathbf{i} - 2\mathbf{j}$, you have

$$\begin{aligned} D_t[\mathbf{r}(t) \cdot \mathbf{u}(t)] &= \mathbf{r}(t) \cdot \mathbf{u}'(t) + \mathbf{r}'(t) \cdot \mathbf{u}(t) \\ &= \left(\frac{1}{t}\mathbf{i} - \mathbf{j} + \ln t\mathbf{k}\right) \cdot (2t\mathbf{i} - 2\mathbf{j}) \\ &\quad + \left(-\frac{1}{t^2}\mathbf{i} + \frac{1}{t}\mathbf{k}\right) \cdot (t^2\mathbf{i} - 2t\mathbf{j} + \mathbf{k}) \\ &= 2 + 2 + (-1) + \frac{1}{t} \\ &= 3 + \frac{1}{t}. \end{aligned}$$

b. Because $\mathbf{u}'(t) = 2t\mathbf{i} - 2\mathbf{j}$ and $\mathbf{u}''(t) = 2\mathbf{i}$, you have

$$\begin{aligned} D_t[\mathbf{u}(t) \times \mathbf{u}'(t)] &= [\mathbf{u}(t) \times \mathbf{u}''(t)] + [\mathbf{u}'(t) \times \mathbf{u}'(t)] \\ &= \begin{vmatrix} \mathbf{i} & \mathbf{j} & \mathbf{k} \\ t^2 & -2t & 1 \\ 2 & 0 & 0 \end{vmatrix} + \mathbf{0} \\ &= \begin{vmatrix} -2t & 1 \\ 0 & 0 \end{vmatrix}\mathbf{i} - \begin{vmatrix} t^2 & 1 \\ 2 & 0 \end{vmatrix}\mathbf{j} + \begin{vmatrix} t^2 & -2t \\ 2 & 0 \end{vmatrix}\mathbf{k} \\ &= 0\mathbf{i} - (-2)\mathbf{j} + 4t\mathbf{k} \\ &= 2\mathbf{j} + 4t\mathbf{k}. \end{aligned}$$

NOTE Try reworking parts (a) and (b) in Example 4 by first forming the dot and cross products and then differentiating to see that you obtain the same results.

Integration of Vector-Valued Functions

The following definition is a rational consequence of the definition of the derivative of a vector-valued function.

Definition of Integration of Vector-Valued Functions

1. If $\mathbf{r}(t) = f(t)\mathbf{i} + g(t)\mathbf{j}$, where f and g are continuous on $[a, b]$, then the **indefinite integral (antiderivative)** of $\mathbf{r}$ is

$$\int \mathbf{r}(t)\,dt = \left[\int f(t)\,dt\right]\mathbf{i} + \left[\int g(t)\,dt\right]\mathbf{j}$$ Plane

and its **definite integral** over the interval $a \le t \le b$ is

$$\int_a^b \mathbf{r}(t)\,dt = \left[\int_a^b f(t)\,dt\right]\mathbf{i} + \left[\int_a^b g(t)\,dt\right]\mathbf{j}.$$

2. If $\mathbf{r}(t) = f(t)\mathbf{i} + g(t)\mathbf{j} + h(t)\mathbf{k}$, where f, g, and h are continuous on $[a, b]$, then the **indefinite integral (antiderivative)** of $\mathbf{r}$ is

$$\int \mathbf{r}(t)\,dt = \left[\int f(t)\,dt\right]\mathbf{i} + \left[\int g(t)\,dt\right]\mathbf{j} + \left[\int h(t)\,dt\right]\mathbf{k}$$ Space

and its **definite integral** over the interval $a \le t \le b$ is

$$\int_a^b \mathbf{r}(t)\,dt = \left[\int_a^b f(t)\,dt\right]\mathbf{i} + \left[\int_a^b g(t)\,dt\right]\mathbf{j} + \left[\int_a^b h(t)\,dt\right]\mathbf{k}.$$

The antiderivative of a vector-valued function is a family of vector-valued functions all differing by a constant vector $\mathbf{C}$. For instance, if $\mathbf{r}(t)$ is a three-dimensional vector-valued function, then for the indefinite integral $\int \mathbf{r}(t)\,dt$, you obtain three constants of integration

$$\int f(t)\,dt = F(t) + C_1, \qquad \int g(t)\,dt = G(t) + C_2, \qquad \int h(t)\,dt = H(t) + C_3$$

where $F'(t) = f(t)$, $G'(t) = g(t)$, and $H'(t) = h(t)$. These three *scalar* constants produce one *vector* constant of integration,

$$\begin{aligned}\int \mathbf{r}(t)\,dt &= [F(t) + C_1]\mathbf{i} + [G(t) + C_2]\mathbf{j} + [H(t) + C_3]\mathbf{k}\\ &= [F(t)\mathbf{i} + G(t)\mathbf{j} + H(t)\mathbf{k}] + [C_1\mathbf{i} + C_2\mathbf{j} + C_3\mathbf{k}]\\ &= \mathbf{R}(t) + \mathbf{C}\end{aligned}$$

where $\mathbf{R}'(t) = \mathbf{r}(t)$.

EXAMPLE 5 Integrating a Vector-Valued Function

Find the indefinite integral

$$\int (t\,\mathbf{i} + 3\mathbf{j})\,dt.$$

Solution Integrating on a component-by-component basis produces

$$\int (t\,\mathbf{i} + 3\mathbf{j})\,dt = \frac{t^2}{2}\mathbf{i} + 3t\mathbf{j} + \mathbf{C}.$$

Example 6 shows how to evaluate the definite integral of a vector-valued function.

EXAMPLE 6 Definite Integral of a Vector-Valued Function

Evaluate the integral

$$\int_0^1 \mathbf{r}(t)\,dt = \int_0^1 \left(\sqrt[3]{t}\,\mathbf{i} + \frac{1}{t+1}\mathbf{j} + e^{-t}\mathbf{k}\right) dt.$$

Solution

$$\begin{aligned}\int_0^1 \mathbf{r}(t)\,dt &= \left(\int_0^1 t^{1/3}\,dt\right)\mathbf{i} + \left(\int_0^1 \frac{1}{t+1}\,dt\right)\mathbf{j} + \left(\int_0^1 e^{-t}\,dt\right)\mathbf{k}\\ &= \left[\left(\frac{3}{4}\right)t^{4/3}\right]_0^1 \mathbf{i} + \Big[\ln|t+1|\Big]_0^1 \mathbf{j} + \Big[-e^{-t}\Big]_0^1 \mathbf{k}\\ &= \frac{3}{4}\mathbf{i} + (\ln 2)\mathbf{j} + \left(1 - \frac{1}{e}\right)\mathbf{k}\end{aligned}$$

As with real-valued functions, you can narrow the family of antiderivatives of a vector-valued function $\mathbf{r}'$ down to a single antiderivative by imposing an initial condition on the vector-valued function $\mathbf{r}$. This is demonstrated in the next example.

EXAMPLE 7 The Antiderivative of a Vector-Valued Function

Find the antiderivative of

$$\mathbf{r}'(t) = \cos 2t\mathbf{i} - 2\sin t\mathbf{j} + \frac{1}{1+t^2}\mathbf{k}$$

that satisfies the initial condition $\mathbf{r}(0) = 3\mathbf{i} - 2\mathbf{j} + \mathbf{k}$.

Solution

$$\begin{aligned}\mathbf{r}(t) &= \int \mathbf{r}'(t)\,dt\\ &= \left(\int \cos 2t\,dt\right)\mathbf{i} + \left(\int -2\sin t\,dt\right)\mathbf{j} + \left(\int \frac{1}{1+t^2}\,dt\right)\mathbf{k}\\ &= \left(\frac{1}{2}\sin 2t + C_1\right)\mathbf{i} + (2\cos t + C_2)\mathbf{j} + (\arctan t + C_3)\mathbf{k}\end{aligned}$$

Letting $t = 0$ and using the fact that $\mathbf{r}(0) = 3\mathbf{i} - 2\mathbf{j} + \mathbf{k}$, you have

$$\begin{aligned}\mathbf{r}(0) &= (0 + C_1)\mathbf{i} + (2 + C_2)\mathbf{j} + (0 + C_3)\mathbf{k}\\ &= 3\mathbf{i} + (-2)\mathbf{j} + \mathbf{k}.\end{aligned}$$

Equating corresponding components produces

$$C_1 = 3, \qquad 2 + C_2 = -2, \qquad \text{and} \qquad C_3 = 1.$$

So, the antiderivative that satisfies the given initial condition is

$$\mathbf{r}(t) = \left(\frac{1}{2}\sin 2t + 3\right)\mathbf{i} + (2\cos t - 4)\mathbf{j} + (\arctan t + 1)\mathbf{k}.$$

Exercises for Section 12.2

See www.CalcChat.com for worked-out solutions to odd-numbered exercises.

In Exercises 1–6, sketch the plane curve represented by the vector-valued function, and sketch the vectors $\mathbf{r}(t_0)$ and $\mathbf{r}'(t_0)$ for the given value of t_0. Position the vectors such that the initial point of $\mathbf{r}(t_0)$ is at the origin and the initial point of $\mathbf{r}'(t_0)$ is at the terminal point of $\mathbf{r}(t_0)$. What is the relationship between $\mathbf{r}'(t_0)$ and the curve?

1. $\mathbf{r}(t) = t^2\mathbf{i} + t\mathbf{j}, \quad t_0 = 2$

2. $\mathbf{r}(t) = t\mathbf{i} + t^3\mathbf{j}, \quad t_0 = 1$

3. $\mathbf{r}(t) = t^2\mathbf{i} + \dfrac{1}{t}\mathbf{j}, \quad t_0 = 2$

4. $\mathbf{r}(t) = (1 + t)\mathbf{i} + t^3\mathbf{j}, \quad t_0 = 1$

5. $\mathbf{r}(t) = \cos t\mathbf{i} + \sin t\mathbf{j}, \quad t_0 = \dfrac{\pi}{2}$

6. $\mathbf{r}(t) = e^t\mathbf{i} + e^{2t}\mathbf{j}, \quad t_0 = 0$

7. ***Investigation*** Consider the vector-valued function

$\mathbf{r}(t) = t\mathbf{i} + t^2\mathbf{j}$.

(a) Sketch the graph of $\mathbf{r}(t)$. Use a graphing utility to verify your graph.

(b) Sketch the vectors $\mathbf{r}(1/4)$, $\mathbf{r}(1/2)$, and $\mathbf{r}(1/2) - \mathbf{r}(1/4)$ on the graph in part (a).

(c) Compare the vector $\mathbf{r}'(1/4)$ with the vector

$$\frac{\mathbf{r}(1/2) - \mathbf{r}(1/4)}{1/2 - 1/4}.$$

8. ***Investigation*** Consider the vector-valued function

$\mathbf{r}(t) = t\mathbf{i} + (4 - t^2)\mathbf{j}$.

(a) Sketch the graph of $\mathbf{r}(t)$. Use a graphing utility to verify your graph.

(b) Sketch the vectors $\mathbf{r}(1)$, $\mathbf{r}(1.25)$, and $\mathbf{r}(1.25) - \mathbf{r}(1)$ on the graph in part (a).

(c) Compare the vector $\mathbf{r}'(1)$ with the vector $\dfrac{\mathbf{r}(1.25) - \mathbf{r}(1)}{1.25 - 1}$.

In Exercises 9 and 10, (a) sketch the space curve represented by the vector-valued function, and (b) sketch the vectors $\mathbf{r}(t_0)$ and $\mathbf{r}'(t_0)$ for the given value of t_0.

9. $\mathbf{r}(t) = 2\cos t\mathbf{i} + 2\sin t\mathbf{j} + t\mathbf{k}, \quad t_0 = \dfrac{3\pi}{2}$

10. $\mathbf{r}(t) = t\mathbf{i} + t^2\mathbf{j} + \frac{3}{2}\mathbf{k}, \quad t_0 = 2$

In Exercises 11–18, find $\mathbf{r}'(t)$.

11. $\mathbf{r}(t) = 6t\mathbf{i} - 7t^2\mathbf{j} + t^3\mathbf{k}$

12. $\mathbf{r}(t) = \dfrac{1}{t}\mathbf{i} + 16t\mathbf{j} + \dfrac{t^2}{2}\mathbf{k}$

13. $\mathbf{r}(t) = a\cos^3 t\mathbf{i} + a\sin^3 t\mathbf{j} + \mathbf{k}$

14. $\mathbf{r}(t) = 4\sqrt{t}\,\mathbf{i} + t^2\sqrt{t}\,\mathbf{j} + \ln t^2\mathbf{k}$

15. $\mathbf{r}(t) = e^{-t}\mathbf{i} + 4\mathbf{j}$

16. $\mathbf{r}(t) = \langle \sin t - t\cos t, \cos t + t\sin t, t^2\rangle$

17. $\mathbf{r}(t) = \langle t\sin t, t\cos t, t\rangle$

18. $\mathbf{r}(t) = \langle \arcsin t, \arccos t, 0\rangle$

In Exercises 19–26, find (a) $\mathbf{r}''(t)$ and (b) $\mathbf{r}'(t) \cdot \mathbf{r}''(t)$.

19. $\mathbf{r}(t) = t^3\mathbf{i} + \frac{1}{2}t^2\mathbf{j}$

20. $\mathbf{r}(t) = (t^2 + t)\mathbf{i} + (t^2 - t)\mathbf{j}$

21. $\mathbf{r}(t) = 4\cos t\mathbf{i} + 4\sin t\mathbf{j}$

22. $\mathbf{r}(t) = 8\cos t\mathbf{i} + 3\sin t\mathbf{j}$

23. $\mathbf{r}(t) = \frac{1}{2}t^2\mathbf{i} - t\mathbf{j} + \frac{1}{6}t^3\mathbf{k}$

24. $\mathbf{r}(t) = t\mathbf{i} + (2t + 3)\mathbf{j} + (3t - 5)\mathbf{k}$

25. $\mathbf{r}(t) = \langle \cos t + t\sin t, \sin t - t\cos t, t\rangle$

26. $\mathbf{r}(t) = \langle e^{-t}, t^2, \tan t\rangle$

In Exercises 27 and 28, a vector-valued function and its graph are given. The graph also shows the unit vectors $\mathbf{r}'(t_0)/\|\mathbf{r}'(t_0)\|$ and $\mathbf{r}''(t_0)/\|\mathbf{r}''(t_0)\|$. Find these two unit vectors and identify them on the graph.

27. $\mathbf{r}(t) = \cos(\pi t)\mathbf{i} + \sin(\pi t)\mathbf{j} + t^2\mathbf{k}, \quad t_0 = -\frac{1}{4}$

28. $\mathbf{r}(t) = t\mathbf{i} + t^2\mathbf{j} + e^{0.75t}\mathbf{k}, \quad t_0 = \frac{1}{4}$

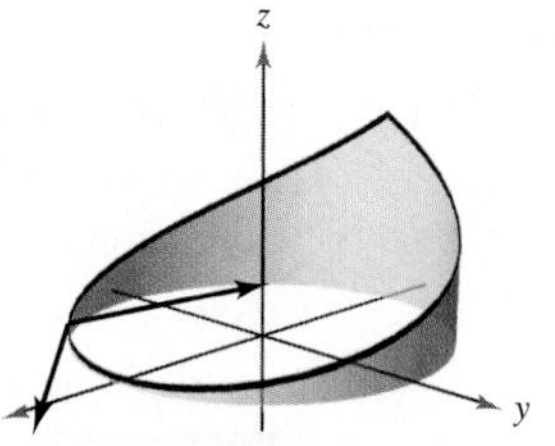

Figure for 27

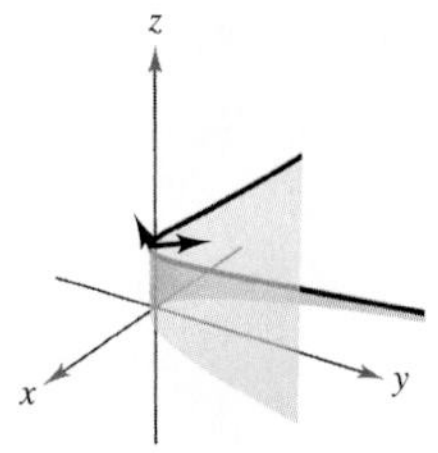

Figure for 28

In Exercises 29–38, find the open interval(s) on which the curve given by the vector-valued function is smooth.

29. $\mathbf{r}(t) = t^2\mathbf{i} + t^3\mathbf{j}$

30. $\mathbf{r}(t) = \dfrac{1}{t - 1}\mathbf{i} + 3t\mathbf{j}$

31. $\mathbf{r}(\theta) = 2\cos^3\theta\mathbf{i} + 3\sin^3\theta\mathbf{j}$

32. $\mathbf{r}(\theta) = (\theta + \sin\theta)\mathbf{i} + (1 - \cos\theta)\mathbf{j}$

33. $\mathbf{r}(\theta) = (\theta - 2\sin\theta)\mathbf{i} + (1 - 2\cos\theta)\mathbf{j}$

34. $\mathbf{r}(t) = \dfrac{2t}{8 + t^3}\mathbf{i} + \dfrac{2t^2}{8 + t^3}\mathbf{j}$

35. $\mathbf{r}(t) = (t - 1)\mathbf{i} + \dfrac{1}{t}\mathbf{j} - t^2\mathbf{k}$

36. $\mathbf{r}(t) = e^t\mathbf{i} - e^{-t}\mathbf{j} + 3t\mathbf{k}$

37. $\mathbf{r}(t) = t\mathbf{i} - 3t\mathbf{j} + \tan t\mathbf{k}$

38. $\mathbf{r}(t) = \sqrt{t}\,\mathbf{i} + (t^2 - 1)\mathbf{j} + \frac{1}{4}t\mathbf{k}$

In Exercises 39 and 40, use the properties of the derivative to find the following.

(a) $\mathbf{r}'(t)$ (b) $\mathbf{r}''(t)$ (c) $D_t[\mathbf{r}(t) \cdot \mathbf{u}(t)]$

(d) $D_t[3\mathbf{r}(t) - \mathbf{u}(t)]$ (e) $D_t[\mathbf{r}(t) \times \mathbf{u}(t)]$ (f) $D_t[\|\mathbf{r}(t)\|], \quad t > 0$

39. $\mathbf{r}(t) = t\mathbf{i} + 3t\mathbf{j} + t^2\mathbf{k}, \quad \mathbf{u}(t) = 4t\mathbf{i} + t^2\mathbf{j} + t^3\mathbf{k}$

40. $\mathbf{r}(t) = t\mathbf{i} + 2\sin t\mathbf{j} + 2\cos t\mathbf{k}$,

$\mathbf{u}(t) = \dfrac{1}{t}\mathbf{i} + 2\sin t\mathbf{j} + 2\cos t\mathbf{k}$

In Exercises 41 and 42, find (a) $D_t[\mathbf{r}(t) \cdot \mathbf{u}(t)]$ and (b) $D_t[\mathbf{r}(t) \times \mathbf{u}(t)]$ by differentiating the product, then applying the properties of Theorem 12.2.

41. $\mathbf{r}(t) = t\mathbf{i} + 2t^2\mathbf{j} + t^3\mathbf{k}$, $\mathbf{u}(t) = t^4\mathbf{k}$

42. $\mathbf{r}(t) = \cos t\mathbf{i} + \sin t\mathbf{j} + t\mathbf{k}$, $\mathbf{u}(t) = \mathbf{j} + t\mathbf{k}$

In Exercises 43 and 44, find the angle θ between $\mathbf{r}(t)$ and $\mathbf{r}'(t)$ as a function of t. Use a graphing utility to graph $\theta(t)$. Use the graph to find any extrema of the function. Find any values of t at which the vectors are orthogonal.

43. $\mathbf{r}(t) = 3\sin t\mathbf{i} + 4\cos t\mathbf{j}$

44. $\mathbf{r}(t) = t^2\mathbf{i} + t\mathbf{j}$

In Exercises 45–48, use the definition of the derivative to find $\mathbf{r}'(t)$.

45. $\mathbf{r}(t) = (3t + 2)\mathbf{i} + (1 - t^2)\mathbf{j}$

46. $\mathbf{r}(t) = \sqrt{t}\,\mathbf{i} + \dfrac{3}{t}\mathbf{j} - 2t\mathbf{k}$

47. $\mathbf{r}(t) = \langle t^2, 0, 2t\rangle$

48. $\mathbf{r}(t) = \langle 0, \sin t, 4t\rangle$

In Exercises 49–56, find the indefinite integral.

49. $\displaystyle\int (2t\mathbf{i} + \mathbf{j} + \mathbf{k})\,dt$

50. $\displaystyle\int \left(4t^3\,\mathbf{i} + 6t\mathbf{j} - 4\sqrt{t}\,\mathbf{k}\right)dt$

51. $\displaystyle\int \left(\frac{1}{t}\mathbf{i} + \mathbf{j} - t^{3/2}\,\mathbf{k}\right)dt$

52. $\displaystyle\int \left(\ln t\mathbf{i} + \frac{1}{t}\mathbf{j} + \mathbf{k}\right)dt$

53. $\displaystyle\int \left[(2t - 1)\mathbf{i} + 4t^3\mathbf{j} + 3\sqrt{t}\,\mathbf{k}\right]dt$

54. $\displaystyle\int (e^t\,\mathbf{i} + \sin t\mathbf{j} + \cos t\mathbf{k})\,dt$

55. $\displaystyle\int \left(\sec^2 t\mathbf{i} + \frac{1}{1 + t^2}\mathbf{j}\right)dt$

56. $\displaystyle\int (e^{-t}\sin t\mathbf{i} + e^{-t}\cos t\mathbf{j})\,dt$

In Exercises 57–62, evaluate the definite integral.

57. $\displaystyle\int_0^1 (8t\mathbf{i} + t\mathbf{j} - \mathbf{k})\,dt$

58. $\displaystyle\int_{-1}^1 \left(t\mathbf{i} + t^3\mathbf{j} + \sqrt[3]{t}\,\mathbf{k}\right)dt$

59. $\displaystyle\int_0^{\pi/2} [(a\cos t)\mathbf{i} + (a\sin t)\mathbf{j} + \mathbf{k}]\,dt$

60. $\displaystyle\int_0^{\pi/4} [(\sec t\tan t)\mathbf{i} + (\tan t)\mathbf{j} + (2\sin t\cos t)\mathbf{k}]\,dt$

61. $\displaystyle\int_0^2 (t\mathbf{i} + e^t\mathbf{j} - te^t\mathbf{k})\,dt$

62. $\displaystyle\int_0^3 \|t\mathbf{i} + t^2\mathbf{j}\|\,dt$

In Exercises 63–68, find $\mathbf{r}(t)$ for the given conditions.

63. $\mathbf{r}'(t) = 4e^{2t}\mathbf{i} + 3e^t\mathbf{j}$, $\mathbf{r}(0) = 2\mathbf{i}$

64. $\mathbf{r}'(t) = 3t^2\mathbf{j} + 6\sqrt{t}\,\mathbf{k}$, $\mathbf{r}(0) = \mathbf{i} + 2\mathbf{j}$

65. $\mathbf{r}''(t) = -32\mathbf{j}$, $\mathbf{r}'(0) = 600\sqrt{3}\mathbf{i} + 600\mathbf{j}$, $\mathbf{r}(0) = \mathbf{0}$

66. $\mathbf{r}''(t) = -4\cos t\mathbf{j} - 3\sin t\mathbf{k}$, $\mathbf{r}'(0) = 3\mathbf{k}$, $\mathbf{r}(0) = 4\mathbf{j}$

67. $\mathbf{r}'(t) = te^{-t^2}\mathbf{i} - e^{-t}\mathbf{j} + \mathbf{k}$, $\mathbf{r}(0) = \frac{1}{2}\mathbf{i} - \mathbf{j} + \mathbf{k}$

68. $\mathbf{r}'(t) = \dfrac{1}{1 + t^2}\mathbf{i} + \dfrac{1}{t^2}\mathbf{j} + \dfrac{1}{t}\mathbf{k}$, $\mathbf{r}(1) = 2\mathbf{i}$

Writing About Concepts

69. State the definition of the derivative of a vector-valued function. Describe how to find the derivative of a vector-valued function and give its geometric interpretation.

70. How do you find the integral of a vector-valued function?

71. The three components of the derivative of the vector-valued function $\mathbf{u}$ are positive at $t = t_0$. Describe the behavior of $\mathbf{u}$ at $t = t_0$.

72. The z-component of the derivative of the vector-valued function $\mathbf{u}$ is 0 for t in the domain of the function. What does this information imply about the graph of $\mathbf{u}$?

In Exercises 73–80, prove the property. In each case, assume r, u, and v are differentiable vector-valued functions of t, f is a differentiable real-valued function of t, and c is a scalar.

73. $D_t[c\mathbf{r}(t)] = c\mathbf{r}'(t)$

74. $D_t[\mathbf{r}(t) \pm \mathbf{u}(t)] = \mathbf{r}'(t) \pm \mathbf{u}'(t)$

75. $D_t[f(t)\mathbf{r}(t)] = f(t)\mathbf{r}'(t) + f'(t)\mathbf{r}(t)$

76. $D_t[\mathbf{r}(t) \times \mathbf{u}(t)] = \mathbf{r}(t) \times \mathbf{u}'(t) + \mathbf{r}'(t) \times \mathbf{u}(t)$

77. $D_t[\mathbf{r}(f(t))] = \mathbf{r}'(f(t))f'(t)$

78. $D_t[\mathbf{r}(t) \times \mathbf{r}'(t)] = \mathbf{r}(t) \times \mathbf{r}''(t)$

79. $D_t\{\mathbf{r}(t) \cdot [\mathbf{u}(t) \times \mathbf{v}(t)]\} = \mathbf{r}'(t) \cdot [\mathbf{u}(t) \times \mathbf{v}(t)] + \mathbf{r}(t) \cdot [\mathbf{u}'(t) \times \mathbf{v}(t)] + \mathbf{r}(t) \cdot [\mathbf{u}(t) \times \mathbf{v}'(t)]$

80. If $\mathbf{r}(t) \cdot \mathbf{r}(t)$ is a constant, then $\mathbf{r}(t) \cdot \mathbf{r}'(t) = 0$.

81. ***Particle Motion*** A particle moves in the xy-plane along the curve represented by the vector-valued function $\mathbf{r}(t) = (t - \sin t)\mathbf{i} + (1 - \cos t)\mathbf{j}$.

(a) Use a graphing utility to graph $\mathbf{r}$. Describe the curve.

(b) Find the minimum and maximum values of $\|\mathbf{r}'\|$ and $\|\mathbf{r}''\|$.

82. ***Particle Motion*** A particle moves in the yz-plane along the curve represented by the vector-valued function $\mathbf{r}(t) = (2\cos t)\mathbf{j} + (3\sin t)\mathbf{k}$.

(a) Describe the curve.

(b) Find the minimum and maximum values of $\|\mathbf{r}'\|$ and $\|\mathbf{r}''\|$.

True or False? **In Exercises 83–86, determine whether the statement is true or false. If it is false, explain why or give an example that shows it is false.**

83. If a particle moves along a sphere centered at the origin, then its derivative vector is always tangent to the sphere.

84. The definite integral of a vector-valued function is a real number.

85. $\dfrac{d}{dt}[\|\mathbf{r}(t)\|] = \|\mathbf{r}'(t)\|$

86. If $\mathbf{r}$ and $\mathbf{u}$ are differentiable vector-valued functions of t, then $D_t[\mathbf{r}(t) \cdot \mathbf{u}(t)] = \mathbf{r}'(t) \cdot \mathbf{u}'(t)$.

87. Consider the vector-valued function

$$\mathbf{r}(t) = (e^t\sin t)\mathbf{i} + (e^t\cos t)\mathbf{j}.$$

Show that $\mathbf{r}(t)$ and $\mathbf{r}''(t)$ are always perpendicular to each other.

Section 12.3 Velocity and Acceleration

- **Describe the velocity and acceleration associated with a vector-valued function.**
- **Use a vector-valued function to analyze projectile motion.**

EXPLORATION

Exploring Velocity Consider the circle given by

$$\mathbf{r}(t) = (\cos \omega t)\mathbf{i} + (\sin \omega t)\mathbf{j}.$$

Use a graphing utility in *parametric* mode to graph this circle for several values of ω. How does ω affect the velocity of the terminal point as it traces out the curve? For a given value of ω, does the speed appear constant? Does the acceleration appear constant? Explain your reasoning.

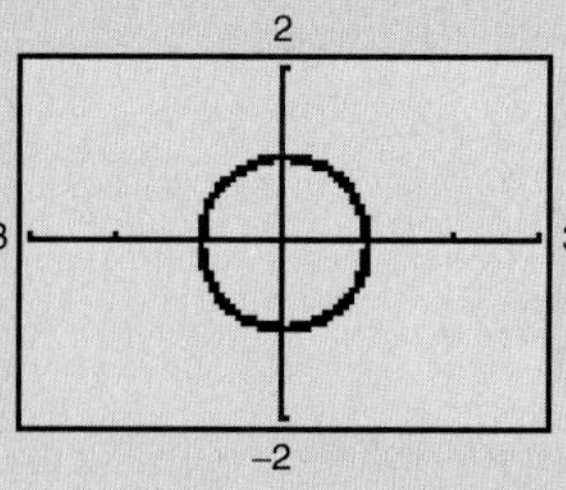

Velocity and Acceleration

You are now ready to combine your study of parametric equations, curves, vectors, and vector-valued functions to form a model for motion along a curve. You will begin by looking at the motion of an object in the plane. (The motion of an object in space can be developed similarly.)

As an object moves along a curve in the plane, the coordinates x and y of its center of mass are each functions of time t. Rather than using the letters f and g to represent these two functions, it is convenient to write $x = x(t)$ and $y = y(t)$. So, the position vector $\mathbf{r}(t)$ takes the form

$$\mathbf{r}(t) = x(t)\mathbf{i} + y(t)\mathbf{j}. \qquad \text{Position vector}$$

The beauty of this vector model for representing motion is that you can use the first and second derivatives of the vector-valued function $\mathbf{r}$ to find the object's velocity and acceleration. (Recall from the preceding chapter that velocity and acceleration are both vector quantities having magnitude and direction.) To find the velocity and acceleration vectors at a given time t, consider a point $Q(x(t + \Delta t), y(t + \Delta t))$ that is approaching the point $P(x(t), y(t))$ along the curve C given by $\mathbf{r}(t) = x(t)\mathbf{i} + y(t)\mathbf{j}$, as shown in Figure 12.10. As $\Delta t \to 0$, the direction of the vector $\overrightarrow{PQ}$ (denoted by $\Delta\mathbf{r}$) approaches the *direction of motion* at time t.

$$\Delta\mathbf{r} = \mathbf{r}(t + \Delta t) - \mathbf{r}(t)$$

$$\frac{\Delta\mathbf{r}}{\Delta t} = \frac{\mathbf{r}(t + \Delta t) - \mathbf{r}(t)}{\Delta t}$$

$$\lim_{\Delta t \to 0} \frac{\Delta\mathbf{r}}{\Delta t} = \lim_{\Delta t \to 0} \frac{\mathbf{r}(t + \Delta t) - \mathbf{r}(t)}{\Delta t}$$

If this limit exists, it is defined to be the **velocity vector** or **tangent vector** to the curve at point P. Note that this is the same limit used to define $\mathbf{r}'(t)$. So, the direction of $\mathbf{r}'(t)$ gives the direction of motion at time t. Moreover, the magnitude of the vector $\mathbf{r}'(t)$

$$\|\mathbf{r}'(t)\| = \|x'(t)\mathbf{i} + y'(t)\mathbf{j}\| = \sqrt{[x'(t)]^2 + [y'(t)]^2}$$

gives the **speed** of the object at time t. Similarly, you can use $\mathbf{r}''(t)$ to find acceleration, as indicated in the definitions at the top of page 849.

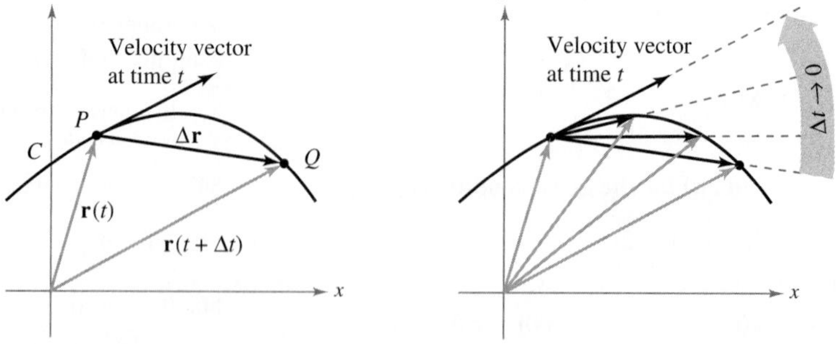

As $\Delta t \to 0$, $\dfrac{\Delta\mathbf{r}}{\Delta t}$ approaches the velocity vector.

Figure 12.10

Definitions of Velocity and Acceleration

If x and y are twice-differentiable functions of t, and $\mathbf{r}$ is a vector-valued function given by $\mathbf{r}(t) = x(t)\mathbf{i} + y(t)\mathbf{j}$, then the velocity vector, acceleration vector, and speed at time t are as follows.

$$\textbf{Velocity} = \mathbf{v}(t) = \mathbf{r}'(t) = x'(t)\mathbf{i} + y'(t)\mathbf{j}$$
$$\textbf{Acceleration} = \mathbf{a}(t) = \mathbf{r}''(t) = x''(t)\mathbf{i} + y''(t)\mathbf{j}$$
$$\textbf{Speed} = \|\mathbf{v}(t)\| = \|\mathbf{r}'(t)\| = \sqrt{[x'(t)]^2 + [y'(t)]^2}$$

For motion along a space curve, the definitions are similar. That is, if $\mathbf{r}(t) = x(t)\mathbf{i} + y(t)\mathbf{j} + z(t)\mathbf{k}$, you have

$$\textbf{Velocity} = \mathbf{v}(t) = \mathbf{r}'(t) = x'(t)\mathbf{i} + y'(t)\mathbf{j} + z'(t)\mathbf{k}$$
$$\textbf{Acceleration} = \mathbf{a}(t) = \mathbf{r}''(t) = x''(t)\mathbf{i} + y''(t)\mathbf{j} + z''(t)\mathbf{k}$$
$$\textbf{Speed} = \|\mathbf{v}(t)\| = \|\mathbf{r}'(t)\| = \sqrt{[x'(t)]^2 + [y'(t)]^2 + [z'(t)]^2}.$$

EXAMPLE 1 Finding Velocity and Acceleration Along a Plane Curve

NOTE In Example 1, note that the velocity and acceleration vectors are orthogonal at any point in time. This is characteristic of motion at a constant speed. (See Exercise 53.)

Find the velocity vector, speed, and acceleration vector of a particle that moves along the plane curve C described by

$$\mathbf{r}(t) = 2\sin\frac{t}{2}\mathbf{i} + 2\cos\frac{t}{2}\mathbf{j}. \qquad \text{Position vector}$$

Solution

The velocity vector is

$$\mathbf{v}(t) = \mathbf{r}'(t) = \cos\frac{t}{2}\mathbf{i} - \sin\frac{t}{2}\mathbf{j}. \qquad \text{Velocity vector}$$

The speed (at any time) is

$$\|\mathbf{r}'(t)\| = \sqrt{\cos^2\frac{t}{2} + \sin^2\frac{t}{2}} = 1. \qquad \text{Speed}$$

The acceleration vector is

$$\mathbf{a}(t) = \mathbf{r}''(t) = -\frac{1}{2}\sin\frac{t}{2}\mathbf{i} - \frac{1}{2}\cos\frac{t}{2}\mathbf{j}. \qquad \text{Acceleration vector}$$

The parametric equations for the curve in Example 1 are

$$x = 2\sin\frac{t}{2} \quad \text{and} \quad y = 2\cos\frac{t}{2}.$$

By eliminating the parameter t, you obtain the rectangular equation

$$x^2 + y^2 = 4. \qquad \text{Rectangular equation}$$

So, the curve is a circle of radius 2 centered at the origin, as shown in Figure 12.11. Because the velocity vector

$$\mathbf{v}(t) = \cos\frac{t}{2}\mathbf{i} - \sin\frac{t}{2}\mathbf{j}$$

has a constant magnitude but a changing direction as t increases, the particle moves around the circle at a constant speed.

Circle: $x^2 + y^2 = 4$

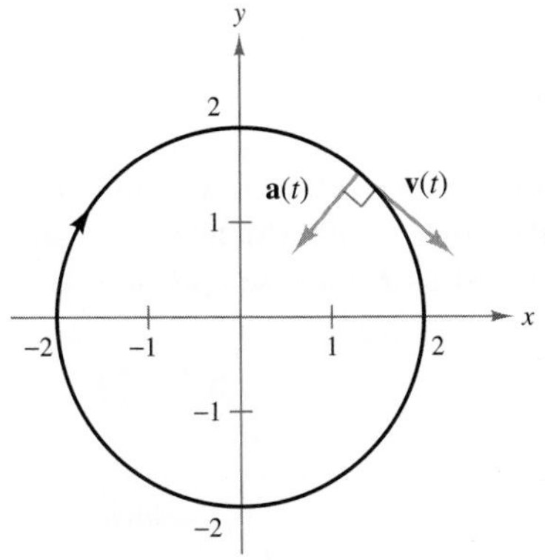

$\mathbf{r}(t) = 2\sin\frac{t}{2}\mathbf{i} + 2\cos\frac{t}{2}\mathbf{j}$

The particle moves around the circle at a constant speed.
Figure 12.11

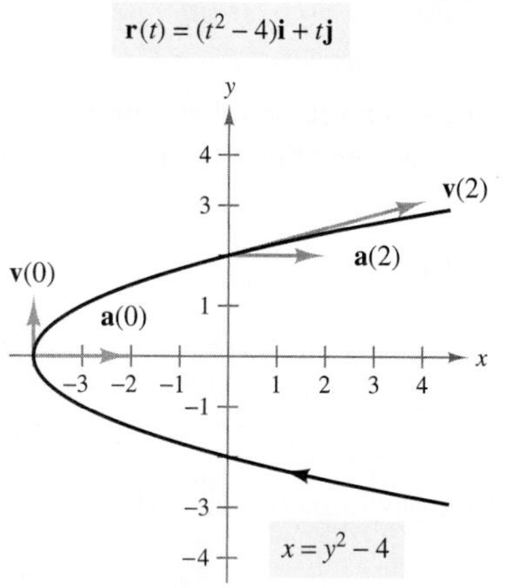

At each point on the curve, the acceleration vector points to the right.
Figure 12.12

EXAMPLE 2 Sketching Velocity and Acceleration Vectors in the Plane

Sketch the path of an object moving along the plane curve given by

$$\mathbf{r}(t) = (t^2 - 4)\mathbf{i} + t\mathbf{j} \qquad \text{Position vector}$$

and find the velocity and acceleration vectors when $t = 0$ and $t = 2$.

Solution Using the parametric equations $x = t^2 - 4$ and $y = t$, you can determine that the curve is a parabola given by $x = y^2 - 4$, as shown in Figure 12.12. The velocity vector (at any time) is

$$\mathbf{v}(t) = \mathbf{r}'(t) = 2t\mathbf{i} + \mathbf{j} \qquad \text{Velocity vector}$$

and the acceleration vector (at any time) is

$$\mathbf{a}(t) = \mathbf{r}''(t) = 2\mathbf{i}. \qquad \text{Acceleration vector}$$

When $t = 0$, the velocity and acceleration vectors are given by

$$\mathbf{v}(0) = 2(0)\mathbf{i} + \mathbf{j} = \mathbf{j} \quad \text{and} \quad \mathbf{a}(0) = 2\mathbf{i}.$$

When $t = 2$, the velocity and acceleration vectors are given by

$$\mathbf{v}(2) = 2(2)\mathbf{i} + \mathbf{j} = 4\mathbf{i} + \mathbf{j} \quad \text{and} \quad \mathbf{a}(2) = 2\mathbf{i}.$$

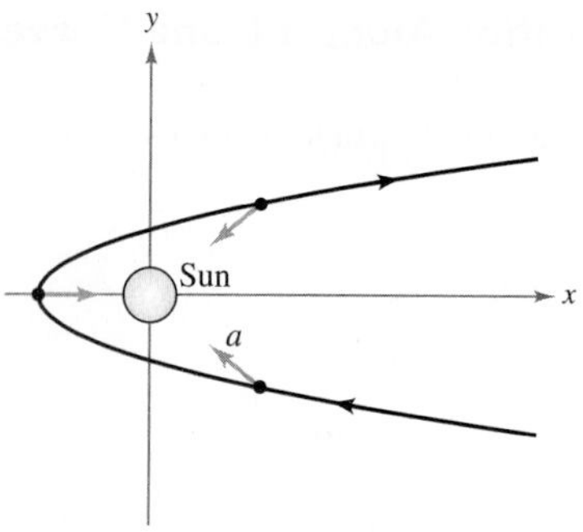

At each point in the comet's orbit, the acceleration vector points toward the sun.
Figure 12.13

For the object moving along the path shown in Figure 12.12, note that the acceleration vector is constant (it has a magnitude of 2 and points to the right). This implies that the speed of the object is decreasing as the object moves toward the vertex of the parabola, and the speed is increasing as the object moves away from the vertex of the parabola.

This type of motion is *not* characteristic of comets that travel on parabolic paths through our solar system. For such comets, the acceleration vector always points to the origin (the sun), which implies that the comet's speed increases as it approaches the vertex of the path and decreases as it moves away from the vertex. (See Figure 12.13.)

EXAMPLE 3 Sketching Velocity and Acceleration Vectors in Space

Sketch the path of an object moving along the space curve C given by

$$\mathbf{r}(t) = t\mathbf{i} + t^3\mathbf{j} + 3t\mathbf{k}, \quad t \geq 0 \qquad \text{Position vector}$$

and find the velocity and acceleration vectors when $t = 1$.

Solution Using the parametric equations $x = t$ and $y = t^3$, you can determine that the path of the object lies on the cubic cylinder given by $y = x^3$. Moreover, because $z = 3t$, the object starts at $(0, 0, 0)$ and moves upward as t increases, as shown in Figure 12.14. Because $\mathbf{r}(t) = t\mathbf{i} + t^3\mathbf{j} + 3t\mathbf{k}$, you have

$$\mathbf{v}(t) = \mathbf{r}'(t) = \mathbf{i} + 3t^2\mathbf{j} + 3\mathbf{k} \qquad \text{Velocity vector}$$

and

$$\mathbf{a}(t) = \mathbf{r}''(t) = 6t\mathbf{j}. \qquad \text{Acceleration vector}$$

When $t = 1$, the velocity and acceleration vectors are given by

$$\mathbf{v}(1) = \mathbf{r}'(1) = \mathbf{i} + 3\mathbf{j} + 3\mathbf{k} \quad \text{and} \quad \mathbf{a}(1) = \mathbf{r}''(1) = 6\mathbf{j}.$$

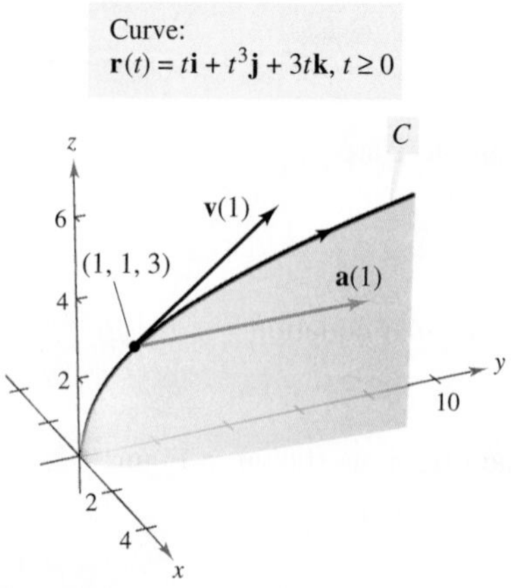

Figure 12.14

So far in this section, you have concentrated on finding the velocity and acceleration by differentiating the position function. Many practical applications involve the reverse problem—finding the position function for a given velocity or acceleration. This is demonstrated in the next example.

EXAMPLE 4 Finding a Position Function by Integration

An object starts from rest at the point $P(1, 2, 0)$ and moves with an acceleration of

$$\mathbf{a}(t) = \mathbf{j} + 2\mathbf{k} \qquad \text{Acceleration vector}$$

where $\|\mathbf{a}(t)\|$ is measured in feet per second per second. Find the location of the object after $t = 2$ seconds.

Solution From the description of the object's motion, you can deduce the following *initial conditions*. Because the object starts from rest, you have

$$\mathbf{v}(0) = \mathbf{0}.$$

Moreover, because the object starts at the point $(x, y, z) = (1, 2, 0)$, you have

$$\begin{aligned}\mathbf{r}(0) &= x(0)\mathbf{i} + y(0)\mathbf{j} + z(0)\mathbf{k}\\ &= 1\mathbf{i} + 2\mathbf{j} + 0\mathbf{k}\\ &= \mathbf{i} + 2\mathbf{j}.\end{aligned}$$

To find the position function, you should integrate twice, each time using one of the initial conditions to solve for the constant of integration. The velocity vector is

$$\begin{aligned}\mathbf{v}(t) = \int \mathbf{a}(t)\,dt &= \int (\mathbf{j} + 2\mathbf{k})\,dt\\ &= t\mathbf{j} + 2t\mathbf{k} + \mathbf{C}\end{aligned}$$

where $\mathbf{C} = C_1\mathbf{i} + C_2\mathbf{j} + C_3\mathbf{k}$. Letting $t = 0$ and applying the initial condition $\mathbf{v}(0) = \mathbf{0}$, you obtain

$$\mathbf{v}(0) = C_1\mathbf{i} + C_2\mathbf{j} + C_3\mathbf{k} = \mathbf{0} \quad \Longrightarrow \quad C_1 = C_2 = C_3 = 0.$$

So, the *velocity* at any time t is

$$\mathbf{v}(t) = t\mathbf{j} + 2t\mathbf{k}. \qquad \text{Velocity vector}$$

Integrating once more produces

$$\begin{aligned}\mathbf{r}(t) = \int \mathbf{v}(t)\,dt &= \int (t\mathbf{j} + 2t\mathbf{k})\,dt\\ &= \frac{t^2}{2}\mathbf{j} + t^2\mathbf{k} + \mathbf{C}\end{aligned}$$

where $\mathbf{C} = C_4\mathbf{i} + C_5\mathbf{j} + C_6\mathbf{k}$. Letting $t = 0$ and applying the initial condition $\mathbf{r}(0) = \mathbf{i} + 2\mathbf{j}$, you have

$$\mathbf{r}(0) = C_4\mathbf{i} + C_5\mathbf{j} + C_6\mathbf{k} = \mathbf{i} + 2\mathbf{j} \quad \Longrightarrow \quad C_4 = 1, C_5 = 2, C_6 = 0.$$

So, the *position* vector is

$$\mathbf{r}(t) = \mathbf{i} + \left(\frac{t^2}{2} + 2\right)\mathbf{j} + t^2\mathbf{k}. \qquad \text{Position vector}$$

The location of the object after $t = 2$ seconds is given by $\mathbf{r}(2) = \mathbf{i} + 4\mathbf{j} + 4\mathbf{k}$, as shown in Figure 12.15.

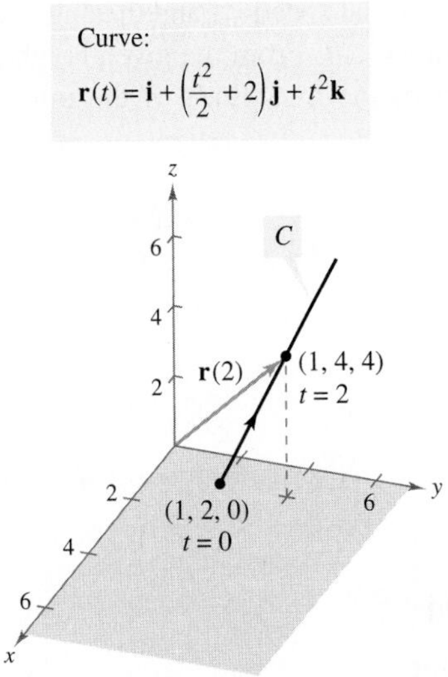

The object takes 2 seconds to move from point $(1, 2, 0)$ to point $(1, 4, 4)$ along C.
Figure 12.15

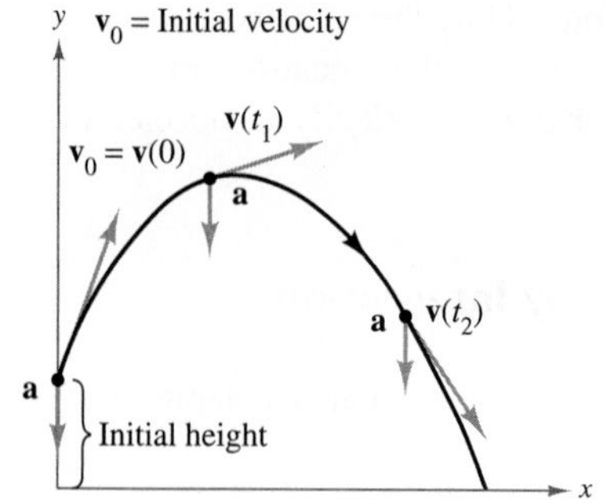

Figure 12.16

Projectile Motion

You now have the machinery to derive the parametric equations for the path of a projectile. Assume that gravity is the only force acting on the projectile after it is launched. So, the motion occurs in a vertical plane, which can be represented by the xy-coordinate system with the origin as a point on Earth's surface, as shown in Figure 12.16. For a projectile of mass m, the force due to gravity is

$$\mathbf{F} = -mg\mathbf{j}$$ Force due to gravity

where the gravitational constant is $g = 32$ feet per second per second, or 9.81 meters per second per second. By **Newton's Second Law of Motion,** this same force produces an acceleration $\mathbf{a} = \mathbf{a}(t)$, and satisfies the equation $\mathbf{F} = m\mathbf{a}$. Consequently, the acceleration of the projectile is given by $m\mathbf{a} = -mg\mathbf{j}$, which implies that

$$\mathbf{a} = -g\mathbf{j}.$$ Acceleration of projectile

EXAMPLE 5 Derivation of the Position Function for a Projectile

A projectile of mass m is launched from an initial position $\mathbf{r}_0$ with an initial velocity $\mathbf{v}_0$. Find its position vector as a function of time.

Solution Begin with the acceleration $\mathbf{a}(t) = -g\mathbf{j}$ and integrate twice.

$$\mathbf{v}(t) = \int \mathbf{a}(t)\,dt = \int -g\mathbf{j}\,dt = -gt\mathbf{j} + \mathbf{C}_1$$

$$\mathbf{r}(t) = \int \mathbf{v}(t)\,dt = \int (-gt\mathbf{j} + \mathbf{C}_1)\,dt = -\frac{1}{2}gt^2\mathbf{j} + \mathbf{C}_1 t + \mathbf{C}_2$$

You can use the facts that $\mathbf{v}(0) = \mathbf{v}_0$ and $\mathbf{r}(0) = \mathbf{r}_0$ to solve for the constant vectors $\mathbf{C}_1$ and $\mathbf{C}_2$. Doing this produces $\mathbf{C}_1 = \mathbf{v}_0$ and $\mathbf{C}_2 = \mathbf{r}_0$. Therefore, the position vector is

$$\mathbf{r}(t) = -\frac{1}{2}gt^2\mathbf{j} + t\mathbf{v}_0 + \mathbf{r}_0.$$ Position vector

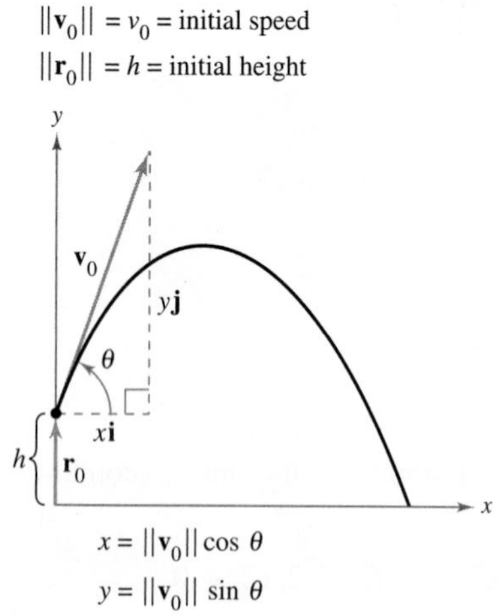

Figure 12.17

In many projectile problems, the constant vectors $\mathbf{r}_0$ and $\mathbf{v}_0$ are not given explicitly. Often you are given the initial height h, the initial speed v_0, and the angle θ at which the projectile is launched, as shown in Figure 12.17. From the given height, you can deduce that $\mathbf{r}_0 = h\mathbf{j}$. Because the speed gives the magnitude of the initial velocity, it follows that $v_0 = \|\mathbf{v}_0\|$ and you can write

$$\begin{aligned}\mathbf{v}_0 &= x\mathbf{i} + y\mathbf{j}\\ &= (\|\mathbf{v}_0\| \cos\theta)\mathbf{i} + (\|\mathbf{v}_0\| \sin\theta)\mathbf{j}\\ &= v_0 \cos\theta\mathbf{i} + v_0 \sin\theta\mathbf{j}.\end{aligned}$$

So, the position vector can be written in the form

$$\mathbf{r}(t) = -\frac{1}{2}gt^2\mathbf{j} + t\mathbf{v}_0 + \mathbf{r}_0$$ Position vector

$$= -\frac{1}{2}gt^2\mathbf{j} + tv_0\cos\theta\mathbf{i} + tv_0\sin\theta\mathbf{j} + h\mathbf{j}$$

$$= (v_0\cos\theta)t\mathbf{i} + \left[h + (v_0\sin\theta)t - \frac{1}{2}gt^2\right]\mathbf{j}.$$

THEOREM 12.3 Position Function for a Projectile

Neglecting air resistance, the path of a projectile launched from an initial height h with initial speed v_0 and angle of elevation θ is described by the vector function

$$\mathbf{r}(t) = (v_0 \cos \theta)t\mathbf{i} + \left[h + (v_0 \sin \theta)t - \frac{1}{2}gt^2\right]\mathbf{j}$$

where g is the gravitational constant.

EXAMPLE 6 Describing the Path of a Baseball

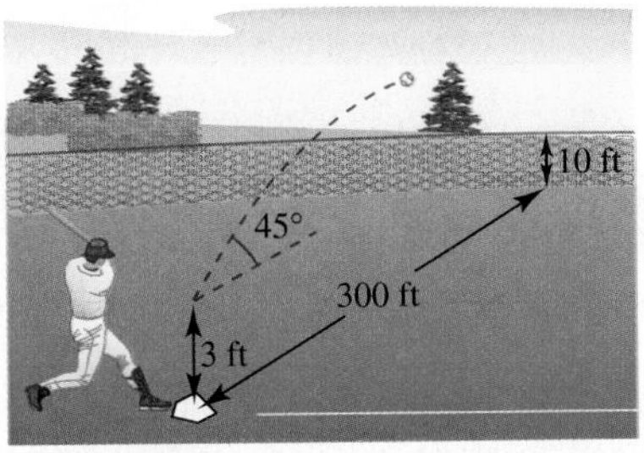

Figure 12.18

A baseball is hit 3 feet above ground level at 100 feet per second and at an angle of 45° with respect to the ground, as shown in Figure 12.18. Find the maximum height reached by the baseball. Will it clear a 10-foot-high fence located 300 feet from home plate?

Solution You are given $h = 3$, $v_0 = 100$, and $\theta = 45°$. So, using $g = 32$ feet per second per second produces

$$\begin{aligned}\mathbf{r}(t) &= \left(100 \cos \frac{\pi}{4}\right)t\mathbf{i} + \left[3 + \left(100 \sin \frac{\pi}{4}\right)t - 16t^2\right]\mathbf{j}\\ &= \left(50\sqrt{2}\,t\right)\mathbf{i} + \left(3 + 50\sqrt{2}\,t - 16t^2\right)\mathbf{j}\\ \mathbf{v}(t) &= \mathbf{r}'(t) = 50\sqrt{2}\,\mathbf{i} + \left(50\sqrt{2} - 32t\right)\mathbf{j}.\end{aligned}$$

The maximum height occurs when

$$y'(t) = 50\sqrt{2} - 32t = 0$$

which implies that

$$\begin{aligned}t &= \frac{25\sqrt{2}}{16}\\ &\approx 2.21 \text{ seconds.}\end{aligned}$$

So, the maximum height reached by the ball is

$$\begin{aligned}y &= 3 + 50\sqrt{2}\left(\frac{25\sqrt{2}}{16}\right) - 16\left(\frac{25\sqrt{2}}{16}\right)^2\\ &= \frac{649}{8}\\ &\approx 81 \text{ feet.}\end{aligned}$$

Maximum height when $t \approx 2.21$ seconds

The ball is 300 feet from where it was hit when

$$300 = x(t) = 50\sqrt{2}\,t.$$

Solving this equation for t produces $t = 3\sqrt{2} \approx 4.24$ seconds. At this time, the height of the ball is

$$\begin{aligned}y &= 3 + 50\sqrt{2}\left(3\sqrt{2}\right) - 16\left(3\sqrt{2}\right)^2\\ &= 303 - 288\\ &= 15 \text{ feet.}\end{aligned}$$

Height when $t \approx 4.24$ seconds

Therefore, the ball clears the 10-foot fence for a home run.

Exercises for Section 12.3

See www.CalcChat.com for worked-out solutions to odd-numbered exercises.

In Exercises 1–8, the position vector r describes the path of an object moving in the xy-plane. Sketch a graph of the path and sketch the velocity and acceleration vectors at the given point.

Position Function	*Point*
1. $\mathbf{r}(t) = 3t\mathbf{i} + (t - 1)\mathbf{j}$	$(3, 0)$
2. $\mathbf{r}(t) = (6 - t)\mathbf{i} + t\mathbf{j}$	$(3, 3)$
3. $\mathbf{r}(t) = t^2\mathbf{i} + t\mathbf{j}$	$(4, 2)$
4. $\mathbf{r}(t) = t^2\mathbf{i} + t^3\mathbf{j}$	$(1, 1)$
5. $\mathbf{r}(t) = 2\cos t\mathbf{i} + 2\sin t\mathbf{j}$	$(\sqrt{2}, \sqrt{2})$
6. $\mathbf{r}(t) = 3\cos t\mathbf{i} + 2\sin t\mathbf{j}$	$(3, 0)$
7. $\mathbf{r}(t) = \langle t - \sin t, 1 - \cos t\rangle$	$(\pi, 2)$
8. $\mathbf{r}(t) = \langle e^{-t}, e^t\rangle$	$(1, 1)$

In Exercises 9–16, the position vector r describes the path of an object moving in space. Find the velocity, speed, and acceleration of the object.

9. $\mathbf{r}(t) = t\mathbf{i} + (2t - 5)\mathbf{j} + 3t\mathbf{k}$

10. $\mathbf{r}(t) = 4t\mathbf{i} + 4t\mathbf{j} + 2t\mathbf{k}$

11. $\mathbf{r}(t) = t\mathbf{i} + t^2\mathbf{j} + \dfrac{t^2}{2}\mathbf{k}$

12. $\mathbf{r}(t) = 3t\mathbf{i} + t\mathbf{j} + \frac{1}{4}t^2\mathbf{k}$

13. $\mathbf{r}(t) = t\mathbf{i} + t\mathbf{j} + \sqrt{9 - t^2}\,\mathbf{k}$

14. $\mathbf{r}(t) = t^2\mathbf{i} + t\mathbf{j} + 2t^{3/2}\mathbf{k}$

15. $\mathbf{r}(t) = \langle 4t, 3\cos t, 3\sin t\rangle$

16. $\mathbf{r}(t) = \langle e^t\cos t, e^t\sin t, e^t\rangle$

Linear Approximation **In Exercises 17 and 18, the graph of the vector-valued function $\mathbf{r}(t)$ and a tangent vector to the graph at $t = t_0$ are given.**

(a) Find a set of parametric equations for the tangent line to the graph at $t = t_0$.

(b) Use the equations for the line to approximate $\mathbf{r}(t_0 + 0.1)$.

17. $\mathbf{r}(t) = \left\langle t, -t^2, \frac{1}{4}t^3\right\rangle, \quad t_0 = 1$

18. $\mathbf{r}(t) = \left\langle t, \sqrt{25 - t^2}, \sqrt{25 - t^2}\right\rangle, \quad t_0 = 3$

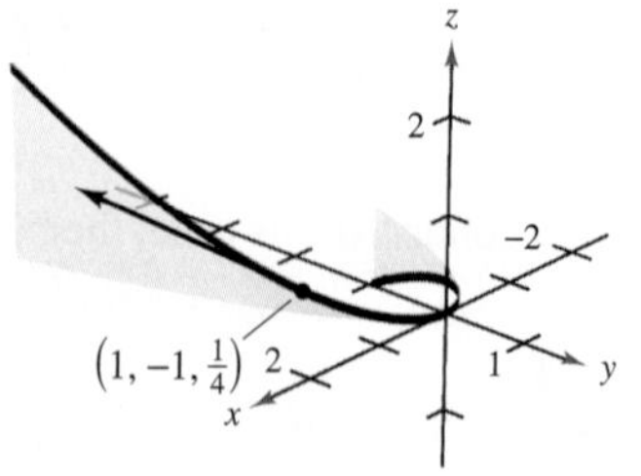

Figure for 17

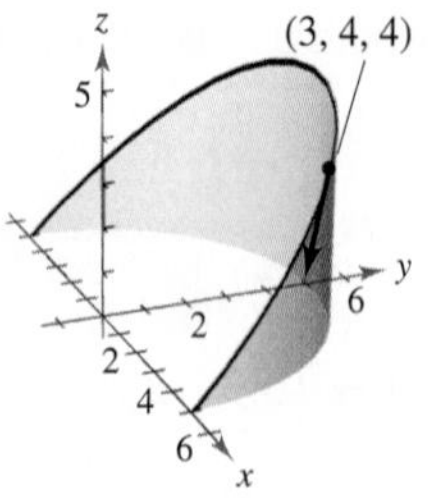

Figure for 18

In Exercises 19–22, use the given acceleration function to find the velocity and position vectors. Then find the position at time $t = 2$.

19. $\mathbf{a}(t) = \mathbf{i} + \mathbf{j} + \mathbf{k}$

$\mathbf{v}(0) = \mathbf{0}, \quad \mathbf{r}(0) = \mathbf{0}$

20. $\mathbf{a}(t) = 2\mathbf{i} + 3\mathbf{k}$

$\mathbf{v}(0) = 4\mathbf{j}, \quad \mathbf{r}(0) = \mathbf{0}$

21. $\mathbf{a}(t) = t\mathbf{j} + t\mathbf{k}$

$\mathbf{v}(1) = 5\mathbf{j}, \quad \mathbf{r}(1) = \mathbf{0}$

22. $\mathbf{a}(t) = -\cos t\mathbf{i} - \sin t\mathbf{j}$

$\mathbf{v}(0) = \mathbf{j} + \mathbf{k}, \quad \mathbf{r}(0) = \mathbf{i}$

Writing About Concepts

23. In your own words, explain the difference between the velocity of an object and its speed.

24. What is known about the speed of an object if the angle between the velocity and acceleration vectors is (a) acute and (b) obtuse?

Projectile Motion **In Exercises 25–40, use the model for projectile motion, assuming there is no air resistance.**

25. Find the vector-valued function for the path of a projectile launched at a height of 10 feet above the ground with an initial velocity of 88 feet per second and at an angle of 30° above the horizontal. Use a graphing utility to graph the path of the projectile.

26. Determine the maximum height and range of a projectile fired at a height of 3 feet above the ground with an initial velocity of 900 feet per second and at an angle of 45° above the horizontal.

27. A baseball, hit 3 feet above the ground, leaves the bat at an angle of 45° and is caught by an outfielder 3 feet above the ground and 300 feet from home plate. What is the initial speed of the ball, and how high does it rise?

28. A baseball player at second base throws a ball 90 feet to the player at first base. The ball is thrown 5 feet above the ground with an initial velocity of 50 miles per hour and at an angle of 15° above the horizontal. At what height does the player at first base catch the ball?

29. Eliminate the parameter t from the position function for the motion of a projectile to show that the rectangular equation is

$$y = -\frac{16\sec^2\theta}{v_0^2}x^2 + (\tan\theta)x + h.$$

30. The path of a ball is given by the rectangular equation

$$y = x - 0.005x^2.$$

Use the result of Exercise 29 to find the position function. Then find the speed and direction of the ball at the point at which it has traveled 60 feet horizontally.

31. *Modeling Data* After the path of a ball thrown by a baseball player is videotaped, it is analyzed on a television set with a grid covering the screen. The tape is paused three times and the positions of the ball are measured. The coordinates are approximately (0, 6.0), (15, 10.6), and (30, 13.4). (The x-coordinate measures the horizontal distance from the player in feet and the y-coordinate measures the height in feet.)

(a) Use a graphing utility to find a quadratic model for the data.

(b) Use a graphing utility to plot the data and graph the model.

(c) Determine the maximum height of the ball.

(d) Find the initial velocity of the ball and the angle at which it was thrown.

32. A baseball is hit from a height of 2.5 feet above the ground with an initial velocity of 140 feet per second and at an angle of 22° above the horizontal. Use a graphing utility to graph the path of the ball and determine whether it will clear a 10-foot-high fence located 375 feet from home plate.

33. The SkyDome in Toronto, Ontario has a center field fence that is 10 feet high and 400 feet from home plate. A ball is hit 3 feet above the ground and leaves the bat at a speed of 100 miles per hour.

(a) The ball leaves the bat at an angle of $\theta = \theta_0$ with the horizontal. Write the vector-valued function for the path of the ball.

(b) Use a graphing utility to graph the vector-valued function for $\theta_0 = 10°$, $\theta_0 = 15°$, $\theta_0 = 20°$, and $\theta_0 = 25°$. Use the graphs to approximate the minimum angle required for the hit to be a home run.

(c) Determine analytically the minimum angle required for the hit to be a home run.

34. The quarterback of a football team releases a pass at a height of 7 feet above the playing field, and the football is caught by a receiver 30 yards directly downfield at a height of 4 feet. The pass is released at an angle of 35° with the horizontal.

(a) Find the speed of the football when it is released.

(b) Find the maximum height of the football.

(c) Find the time the receiver has to reach the proper position after the quarterback releases the football.

35. A bale ejector consists of two variable-speed belts at the end of a baler. Its purpose is to toss bales into a trailing wagon. In loading the back of a wagon, a bale must be thrown to a position 8 feet above and 16 feet behind the ejector.

(a) Find the minimum initial speed of the bale and the corresponding angle at which it must be ejected from the baler.

(b) The ejector has a fixed angle of 45°. Find the initial speed required.

36. A bomber is flying at an altitude of 30,000 feet at a speed of 540 miles per hour (see figure). When should the bomb be released for it to hit the target? (Give your answer in terms of the angle of depression from the plane to the target.) What is the speed of the bomb at the time of impact?

Figure for 36

37. A shot fired from a gun with a muzzle velocity of 1200 feet per second is to hit a target 3000 feet away. Determine the minimum angle of elevation of the gun.

38. A projectile is fired from ground level at an angle of 12° with the horizontal. The projectile is to have a range of 150 feet. Find the minimum initial velocity necessary.

39. Use a graphing utility to graph the paths of a projectile for the given values of θ and v_0. For each case, use the graph to approximate the maximum height and range of the projectile. (Assume that the projectile is launched from ground level.)

(a) $\theta = 10°$, $v_0 = 66$ ft/sec

(b) $\theta = 10°$, $v_0 = 146$ ft/sec

(c) $\theta = 45°$, $v_0 = 66$ ft/sec

(d) $\theta = 45°$, $v_0 = 146$ ft/sec

(e) $\theta = 60°$, $v_0 = 66$ ft/sec

(f) $\theta = 60°$, $v_0 = 146$ ft/sec

40. Find the angle at which an object must be thrown to obtain (a) the maximum range and (b) the maximum height.

Projectile Motion **In Exercises 41 and 42, use the model for projectile motion, assuming there is no resistance. [$a(t) = -9.8$ meters per second per second]**

41. Determine the maximum height and range of a projectile fired at a height of 1.5 meters above the ground with an initial velocity of 100 meters per second and at an angle of 30° above the horizontal.

42. A projectile is fired from ground level at an angle of 8° with the horizontal. The projectile is to have a range of 50 meters. Find the minimum velocity necessary.

Cycloidal Motion **In Exercises 43 and 44, consider the motion of a point (or particle) on the circumference of a rolling circle. As the circle rolls, it generates the cycloid**

$$\mathbf{r}(t) = b(\omega t - \sin \omega t)\mathbf{i} + b(1 - \cos \omega t)\mathbf{j}$$

where ω is the constant angular velocity of the circle and b is the radius of the circle.

43. Find the velocity and acceleration vectors of the particle. Use the results to determine the times at which the speed of the particle will be (a) zero and (b) maximized.

44. Find the maximum speed of a point on the circumference of an automobile tire of radius 1 foot when the automobile is traveling at 55 miles per hour. Compare this speed with the speed of the automobile.

Circular Motion **In Exercises 45–48, consider a particle moving on a circular path of radius b described by**

$$\mathbf{r}(t) = b \cos \omega t\,\mathbf{i} + b \sin \omega t\,\mathbf{j}$$

where $\omega = d\theta/dt$ is the constant angular velocity.

45. Find the velocity vector and show that it is orthogonal to $\mathbf{r}(t)$.

46. (a) Show that the speed of the particle is $b\omega$.

(b) Use a graphing utility in *parametric* mode to graph the circle for $b = 6$. Try different values of ω. Does the graphing utility draw the circle faster for greater values of ω?

47. Find the acceleration vector and show that its direction is always toward the center of the circle.

48. Show that the magnitude of the acceleration vector is $b\omega^2$.

Circular Motion **In Exercises 49 and 50, use the results of Exercises 45–48.**

49. A stone weighing 1 pound is attached to a two-foot string and is whirled horizontally (see figure). The string will break under a force of 10 pounds. Find the maximum speed the stone can attain without breaking the string. (Use $\mathbf{F} = m\mathbf{a}$, where $m = \frac{1}{32}$.)

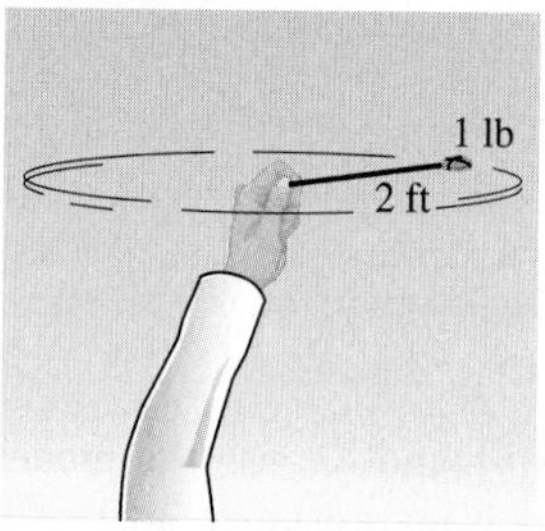

Figure for 49

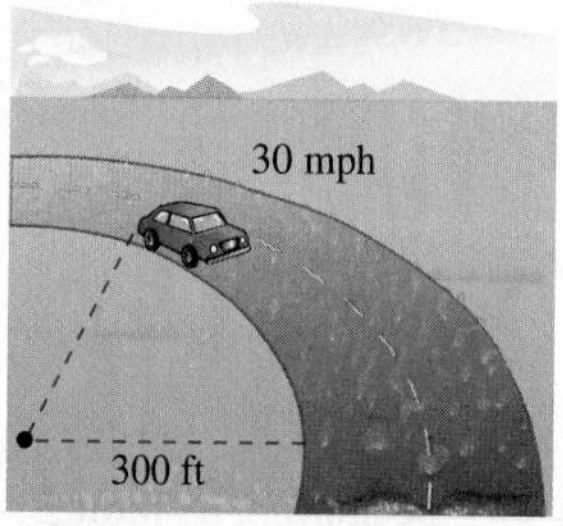

Figure for 50

50. A 3000-pound automobile is negotiating a circular interchange of radius 300 feet at 30 miles per hour (see figure). Assuming the roadway is level, find the force between the tires and the road such that the car stays on the circular path and does not skid. (Use $\mathbf{F} = m\mathbf{a}$, where $m = 3000/32$.) Find the angle at which the roadway should be banked so that no lateral frictional force is exerted on the tires of the automobile.

51. ***Shot-Put Throw*** The path of a shot thrown at an angle θ is

$$\mathbf{r}(t) = (v_0 \cos \theta)t\,\mathbf{i} + \left[h + (v_0 \sin \theta)t - \frac{1}{2}gt^2\right]\mathbf{j}$$

where v_0 is the initial speed, h is the initial height, t is the time in seconds, and g is the acceleration due to gravity. Verify that the shot will remain in the air for a total of

$$t = \frac{v_0 \sin \theta + \sqrt{v_0^2 \sin^2 \theta + 2gh}}{g} \text{ seconds}$$

and will travel a horizontal distance of

$$\frac{v_0^2 \cos \theta}{g}\left(\sin \theta + \sqrt{\sin^2 \theta + \frac{2gh}{v_0^2}}\right) \text{ feet.}$$

52. ***Shot-Put Throw*** A shot is thrown from a height of $h = 6$ feet with an initial speed of $v_0 = 45$ feet per second and at an angle of $\theta = 42.5°$ with the horizontal. Find the total time of travel and the total horizontal distance traveled.

53. Prove that if an object is traveling at a constant speed, its velocity and acceleration vectors are orthogonal.

54. Prove that an object moving in a straight line at a constant speed has an acceleration of 0.

55. ***Investigation*** An object moves on an elliptical path given by the vector-valued function

$$\mathbf{r}(t) = 6 \cos t\,\mathbf{i} + 3 \sin t\,\mathbf{j}.$$

(a) Find $\mathbf{v}(t)$, $\|\mathbf{v}(t)\|$, and $\mathbf{a}(t)$.

(b) Use a graphing utility to complete the table.

t	0	$\frac{\pi}{4}$	$\frac{\pi}{2}$	$\frac{2\pi}{3}$	π
Speed					

(c) Graph the elliptical path and the velocity and acceleration vectors at the values of t given in the table in part (b).

(d) Use the results in parts (b) and (c) to describe the geometric relationship between the velocity and acceleration vectors when the speed of the particle is increasing, and when it is decreasing.

56. ***Writing*** Consider a particle moving on the path

$$\mathbf{r}_1(t) = x(t)\mathbf{i} + y(t)\mathbf{j} + z(t)\mathbf{k}.$$

(a) Discuss any changes in the position, velocity, or acceleration of the particle if its position is given by the vector-valued function $\mathbf{r}_2(t) = \mathbf{r}_1(2t)$.

(b) Generalize the results for the position function $\mathbf{r}_3(t) = \mathbf{r}_1(\omega t)$.

True or False? **In Exercises 57 and 58, determine whether the statement is true or false. If it is false, explain why or give an example that shows it is false.**

57. The acceleration of an object is the derivative of the speed.

58. The velocity vector points in the direction of motion.

59. When $t = 0$, an object is at the point $(0, 1)$ and has a velocity vector $\mathbf{v}(0) = -\mathbf{i}$. It moves with an acceleration of

$$\mathbf{a}(t) = \sin t\,\mathbf{i} - \cos t\,\mathbf{j}.$$

Show that the path of the object is a circle.

Section 12.4 Tangent Vectors and Normal Vectors

- **Find a unit tangent vector at a point on a space curve.**
- **Find the tangential and normal components of acceleration.**

Tangent Vectors and Normal Vectors

In the preceding section, you learned that the velocity vector points in the direction of motion. This observation leads to the following definition, which applies to any smooth curve—not just to those for which the parameter represents time.

Definition of Unit Tangent Vector

Let C be a smooth curve represented by $\mathbf{r}$ on an open interval I. The **unit tangent vector** $\mathbf{T}(t)$ at t is defined to be

$$\mathbf{T}(t) = \frac{\mathbf{r}'(t)}{\|\mathbf{r}'(t)\|}, \quad \mathbf{r}'(t) \neq \mathbf{0}.$$

Recall that a curve is *smooth* on an interval if $\mathbf{r}'$ is continuous and nonzero on the interval. So, "smoothness" is sufficient to guarantee that a curve has a unit tangent vector.

EXAMPLE 1 **Finding the Unit Tangent Vector**

Find the unit tangent vector to the curve given by

$$\mathbf{r}(t) = t\mathbf{i} + t^2\mathbf{j}$$

when $t = 1$.

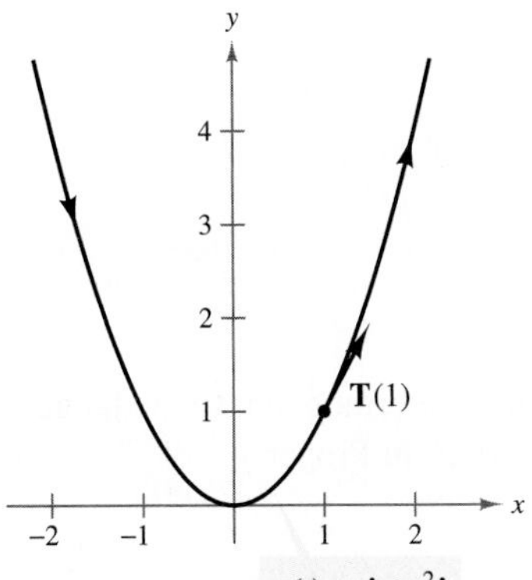

$\mathbf{r}(t) = t\mathbf{i} + t^2\mathbf{j}$

The direction of the unit tangent vector depends on the orientation of the curve.
Figure 12.19

Solution The derivative of $\mathbf{r}(t)$ is

$$\mathbf{r}'(t) = \mathbf{i} + 2t\mathbf{j}. \qquad \text{Derivative of } \mathbf{r}(t)$$

So, the unit tangent vector is

$$\mathbf{T}(t) = \frac{\mathbf{r}'(t)}{\|\mathbf{r}'(t)\|} \qquad \text{Definition of } \mathbf{T}(t)$$

$$= \frac{1}{\sqrt{1 + 4t^2}}(\mathbf{i} + 2t\mathbf{j}). \qquad \text{Substitute for } \mathbf{r}'(t).$$

When $t = 1$, the unit tangent vector is

$$\mathbf{T}(1) = \frac{1}{\sqrt{5}}(\mathbf{i} + 2\mathbf{j})$$

as shown in Figure 12.19.

NOTE In Example 1, note that the direction of the unit tangent vector depends on the orientation of the curve. For instance, if the parabola in Figure 12.19 were given by

$$\mathbf{r}(t) = -(t - 2)\mathbf{i} + (t - 2)^2\mathbf{j},$$

$\mathbf{T}(1)$ would still represent the unit tangent vector at the point $(1, 1)$, but it would point in the opposite direction. Try verifying this.

The **tangent line to a curve** at a point is the line passing through the point and parallel to the unit tangent vector. In Example 2, the unit tangent vector is used to find the tangent line at a point on a helix.

EXAMPLE 2 Finding the Tangent Line at a Point on a Curve

Find $\mathbf{T}(t)$ and then find a set of parametric equations for the tangent line to the helix given by

$$\mathbf{r}(t) = 2\cos t\mathbf{i} + 2\sin t\mathbf{j} + t\mathbf{k}$$

at the point corresponding to $t = \pi/4$.

Solution The derivative of $\mathbf{r}(t)$ is $\mathbf{r}'(t) = -2\sin t\mathbf{i} + 2\cos t\mathbf{j} + \mathbf{k}$, which implies that $\|\mathbf{r}'(t)\| = \sqrt{4\sin^2 t + 4\cos^2 t + 1} = \sqrt{5}$. Therefore, the unit tangent vector is

$$\mathbf{T}(t) = \frac{\mathbf{r}'(t)}{\|\mathbf{r}'(t)\|}$$

$$= \frac{1}{\sqrt{5}}(-2\sin t\mathbf{i} + 2\cos t\mathbf{j} + \mathbf{k}).$$ Unit tangent vector

When $t = \pi/4$, the unit tangent vector is

$$\mathbf{T}\left(\frac{\pi}{4}\right) = \frac{1}{\sqrt{5}}\left(-2\frac{\sqrt{2}}{2}\mathbf{i} + 2\frac{\sqrt{2}}{2}\mathbf{j} + \mathbf{k}\right)$$

$$= \frac{1}{\sqrt{5}}\left(-\sqrt{2}\,\mathbf{i} + \sqrt{2}\,\mathbf{j} + \mathbf{k}\right).$$

Curve:
$\mathbf{r}(t) = 2\cos t\mathbf{i} + 2\sin t\mathbf{j} + t\mathbf{k}$

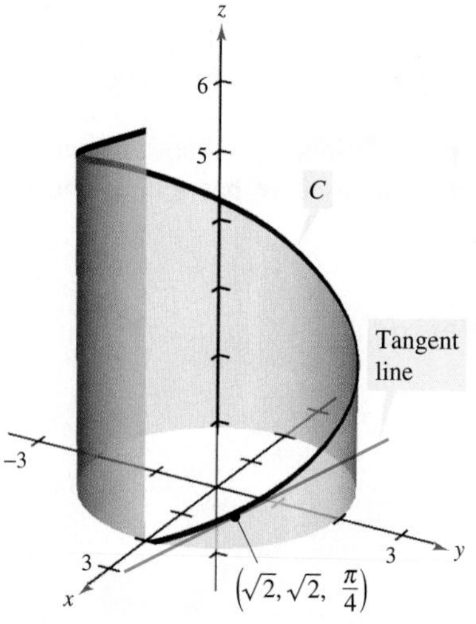

The tangent line to a curve at a point is determined by the unit tangent vector at the point.
Figure 12.20

Using the direction numbers $a = -\sqrt{2}$, $b = \sqrt{2}$, and $c = 1$, and the point $(x_1, y_1, z_1) = \left(\sqrt{2}, \sqrt{2}, \pi/4\right)$, you can obtain the following parametric equations (given with parameter s).

$$x = x_1 + as = \sqrt{2} - \sqrt{2}s$$

$$y = y_1 + bs = \sqrt{2} + \sqrt{2}s$$

$$z = z_1 + cs = \frac{\pi}{4} + s$$

This tangent line is shown in Figure 12.20.

In Example 2, there are infinitely many vectors that are orthogonal to the tangent vector $\mathbf{T}(t)$. One of these is the vector $\mathbf{T}'(t)$. This follows from Property 7 of Theorem 12.2. That is,

$$\mathbf{T}(t) \cdot \mathbf{T}(t) = \|\mathbf{T}(t)\|^2 = 1 \quad \Longrightarrow \quad \mathbf{T}(t) \cdot \mathbf{T}'(t) = 0.$$

By normalizing the vector $\mathbf{T}'(t)$, you obtain a special vector called the **principal unit normal vector,** as indicated in the following definition.

Definition of Principal Unit Normal Vector

Let C be a smooth curve represented by $\mathbf{r}$ on an open interval I. If $\mathbf{T}'(t) \neq \mathbf{0}$, then the **principal unit normal vector** at t is defined to be

$$\mathbf{N}(t) = \frac{\mathbf{T}'(t)}{\|\mathbf{T}'(t)\|}.$$

EXAMPLE 3 Finding the Principal Unit Normal Vector

Find $\mathbf{N}(t)$ and $\mathbf{N}(1)$ for the curve represented by

$$\mathbf{r}(t) = 3t\mathbf{i} + 2t^2\mathbf{j}.$$

Solution By differentiating, you obtain

$$\mathbf{r}'(t) = 3\mathbf{i} + 4t\mathbf{j} \quad \text{and} \quad \|\mathbf{r}'(t)\| = \sqrt{9 + 16t^2}$$

which implies that the unit tangent vector is

$$\mathbf{T}(t) = \frac{\mathbf{r}'(t)}{\|\mathbf{r}'(t)\|} = \frac{1}{\sqrt{9 + 16t^2}}(3\mathbf{i} + 4t\mathbf{j}). \qquad \text{Unit tangent vector}$$

Using Theorem 12.2, differentiate $\mathbf{T}(t)$ with respect to t to obtain

$$\mathbf{T}'(t) = \frac{1}{\sqrt{9 + 16t^2}}(4\mathbf{j}) - \frac{16t}{(9 + 16t^2)^{3/2}}(3\mathbf{i} + 4t\mathbf{j}) = \frac{12}{(9 + 16t^2)^{3/2}}(-4t\mathbf{i} + 3\mathbf{j})$$

$$\|\mathbf{T}'(t)\| = 12\sqrt{\frac{9 + 16t^2}{(9 + 16t^2)^3}} = \frac{12}{9 + 16t^2}.$$

Therefore, the principal unit normal vector is

$$\mathbf{N}(t) = \frac{\mathbf{T}'(t)}{\|\mathbf{T}'(t)\|} = \frac{1}{\sqrt{9 + 16t^2}}(-4t\mathbf{i} + 3\mathbf{j}). \qquad \text{Principal unit normal vector}$$

When $t = 1$, the principal unit normal vector is

$$\mathbf{N}(1) = \frac{1}{5}(-4\mathbf{i} + 3\mathbf{j})$$

as shown in Figure 12.21.

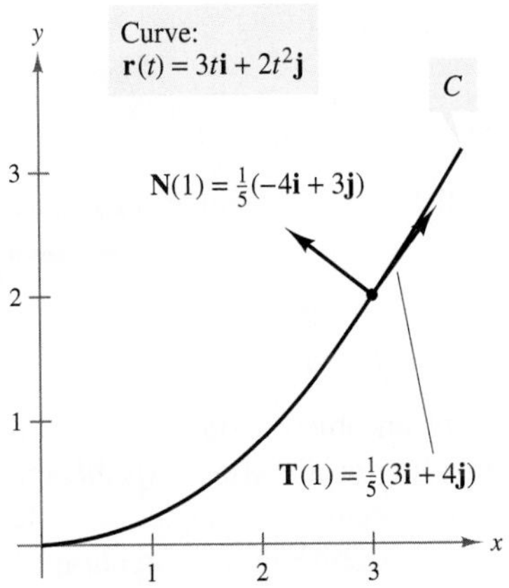

The principal unit normal vector points toward the concave side of the curve.
Figure 12.21

The principal unit normal vector can be difficult to evaluate algebraically. For plane curves, you can simplify the algebra by finding

$$\mathbf{T}(t) = x(t)\mathbf{i} + y(t)\mathbf{j} \qquad \text{Unit tangent vector}$$

and observing that $\mathbf{N}(t)$ must be either

$$\mathbf{N}_1(t) = y(t)\mathbf{i} - x(t)\mathbf{j} \quad \text{or} \quad \mathbf{N}_2(t) = -y(t)\mathbf{i} + x(t)\mathbf{j}.$$

Because $\sqrt{[x(t)]^2 + [y(t)]^2} = 1$, it follows that both $\mathbf{N}_1(t)$ and $\mathbf{N}_2(t)$ are unit normal vectors. The *principal* unit normal vector $\mathbf{N}$ is the one that points toward the concave side of the curve, as shown in Figure 12.21 (see Exercise 86). This also holds for curves in space. That is, for an object moving along a curve C in space, the vector $\mathbf{T}(t)$ points in the direction the object is moving, whereas the vector $\mathbf{N}(t)$ is orthogonal to $\mathbf{T}(t)$ and points in the direction in which the object is turning, as shown in Figure 12.22.

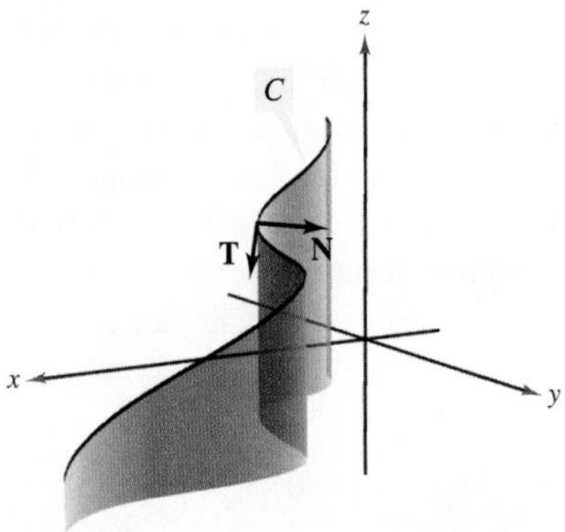

At any point on a curve, a unit normal vector is orthogonal to the unit tangent vector. The *principal* unit normal vector points in the direction in which the curve is turning.
Figure 12.22

Helix:
$\mathbf{r}(t) = 2\cos t\mathbf{i} + 2\sin t\mathbf{j} + t\mathbf{k}$

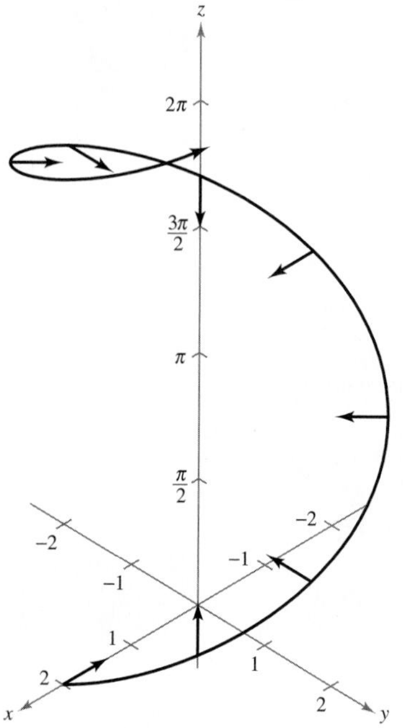

N(*t*) is horizontal and points toward the *z*-axis.
Figure 12.23

EXAMPLE 4 Finding the Principal Unit Normal Vector

Find the principal unit normal vector for the helix given by

$$\mathbf{r}(t) = 2\cos t\mathbf{i} + 2\sin t\mathbf{j} + t\mathbf{k}.$$

Solution From Example 2, you know that the unit tangent vector is

$$\mathbf{T}(t) = \frac{1}{\sqrt{5}}(-2\sin t\mathbf{i} + 2\cos t\mathbf{j} + \mathbf{k}). \qquad \text{Unit tangent vector}$$

So, $\mathbf{T}'(t)$ is given by

$$\mathbf{T}'(t) = \frac{1}{\sqrt{5}}(-2\cos t\mathbf{i} - 2\sin t\mathbf{j}).$$

Because $\|\mathbf{T}'(t)\| = 2/\sqrt{5}$, it follows that the principal unit normal vector is

$$\begin{aligned}\mathbf{N}(t) &= \frac{\mathbf{T}'(t)}{\|\mathbf{T}'(t)\|}\\ &= \frac{1}{2}(-2\cos t\mathbf{i} - 2\sin t\mathbf{j})\\ &= -\cos t\mathbf{i} - \sin t\mathbf{j}. \qquad \text{Principal unit normal vector}\end{aligned}$$

Note that this vector is horizontal and points toward the z-axis, as shown in Figure 12.23.

Tangential and Normal Components of Acceleration

Let's return to the problem of describing the motion of an object along a curve. In the preceding section, you saw that for an object traveling at a *constant speed*, the velocity and acceleration vectors are perpendicular. This seems reasonable, because the speed would not be constant if any acceleration were acting in the direction of motion. You can verify this observation by noting that

$$\mathbf{r}''(t) \cdot \mathbf{r}'(t) = 0$$

if $\|\mathbf{r}'(t)\|$ is a constant. (See Property 7 of Theorem 12.2.)

However, for an object traveling at a *variable speed*, the velocity and acceleration vectors are not necessarily perpendicular. For instance, you saw that the acceleration vector for a projectile always points down, regardless of the direction of motion.

In general, part of the acceleration (the tangential component) acts in the line of motion, and part (the normal component) acts perpendicular to the line of motion. In order to determine these two components, you can use the unit vectors $\mathbf{T}(t)$ and $\mathbf{N}(t)$, which serve in much the same way as do **i** and **j** in representing vectors in the plane. The following theorem states that the acceleration vector lies in the plane determined by $\mathbf{T}(t)$ and $\mathbf{N}(t)$.

THEOREM 12.4 Acceleration Vector

If $\mathbf{r}(t)$ is the position vector for a smooth curve C and $\mathbf{N}(t)$ exists, then the acceleration vector $\mathbf{a}(t)$ lies in the plane determined by $\mathbf{T}(t)$ and $\mathbf{N}(t)$.

Proof To simplify the notation, write **T** for **T**(t), **T**$'$ for **T**$'(t)$, and so on. Because $\mathbf{T} = \mathbf{r}'/\|\mathbf{r}'\| = \mathbf{v}/\|\mathbf{v}\|$, it follows that

$$\mathbf{v} = \|\mathbf{v}\|\mathbf{T}.$$

By differentiating, you obtain

$$\begin{aligned} \mathbf{a} = \mathbf{v}' &= D_t[\|\mathbf{v}\|]\mathbf{T} + \|\mathbf{v}\|\mathbf{T}' && \text{Product Rule} \\ &= D_t[\|\mathbf{v}\|]\mathbf{T} + \|\mathbf{v}\|\mathbf{T}'\left(\frac{\|\mathbf{T}'\|}{\|\mathbf{T}'\|}\right) \\ &= D_t[\|\mathbf{v}\|]\mathbf{T} + \|\mathbf{v}\|\,\|\mathbf{T}'\|\,\mathbf{N}. && \mathbf{N} = \mathbf{T}'/\|\mathbf{T}'\| \end{aligned}$$

Because **a** is written as a linear combination of **T** and **N**, it follows that **a** lies in the plane determined by **T** and **N**.

The coefficients of **T** and **N** in the proof of Theorem 12.4 are called the **tangential and normal components of acceleration** and are denoted by $a_{\mathbf{T}} = D_t[\|\mathbf{v}\|]$ and $a_{\mathbf{N}} = \|\mathbf{v}\|\,\|\mathbf{T}'\|$. So, you can write

$$\mathbf{a}(t) = a_{\mathbf{T}}\mathbf{T}(t) + a_{\mathbf{N}}\mathbf{N}(t).$$

The following theorem gives some convenient formulas for $a_{\mathbf{N}}$ and $a_{\mathbf{T}}$.

THEOREM 12.5 Tangential and Normal Components of Acceleration

If $\mathbf{r}(t)$ is the position vector for a smooth curve C [for which $\mathbf{N}(t)$ exists], then the tangential and normal components of acceleration are as follows.

$$a_{\mathbf{T}} = D_t[\|\mathbf{v}\|] = \mathbf{a}\cdot\mathbf{T} = \frac{\mathbf{v}\cdot\mathbf{a}}{\|\mathbf{v}\|}$$

$$a_{\mathbf{N}} = \|\mathbf{v}\|\,\|\mathbf{T}'\| = \mathbf{a}\cdot\mathbf{N} = \frac{\|\mathbf{v}\times\mathbf{a}\|}{\|\mathbf{v}\|} = \sqrt{\|\mathbf{a}\|^2 - a_{\mathbf{T}}^2}$$

Note that $a_{\mathbf{N}} \geq 0$. The normal component of acceleration is also called the **centripetal component of acceleration.**

Proof Note that a lies in the plane of **T** and **N**. So, you can use Figure 12.24 to conclude that, for any time t, the component of the projection of the acceleration vector onto **T** is given by $a_{\mathbf{T}} = \mathbf{a}\cdot\mathbf{T}$, and onto **N** is given by $a_{\mathbf{N}} = \mathbf{a}\cdot\mathbf{N}$. Moreover, because $\mathbf{a} = \mathbf{v}'$ and $\mathbf{T} = \mathbf{v}/\|\mathbf{v}\|$, you have

$$\begin{aligned} a_{\mathbf{T}} &= \mathbf{a}\cdot\mathbf{T} \\ &= \mathbf{T}\cdot\mathbf{a} \\ &= \frac{\mathbf{v}}{\|\mathbf{v}\|}\cdot\mathbf{a} \\ &= \frac{\mathbf{v}\cdot\mathbf{a}}{\|\mathbf{v}\|}. \end{aligned}$$

In Exercises 88 and 89, you are asked to prove the other parts of the theorem.

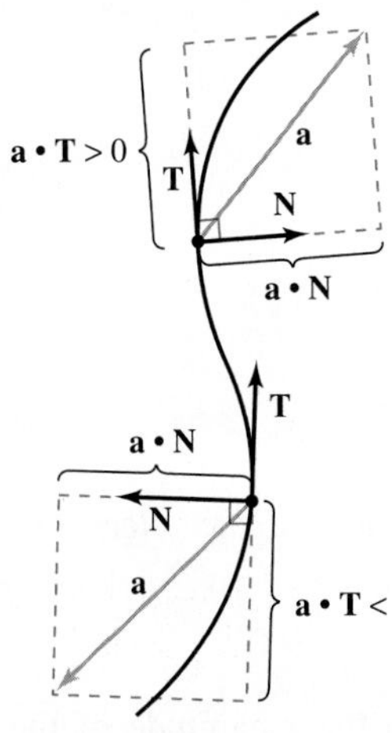

The tangential and normal components of acceleration are obtained by projecting **a** onto **T** and **N**.
Figure 12.24

NOTE The formulas from Theorem 12.5, together with several other formulas from this chapter, are summarized on page 875.

EXAMPLE 5 Tangential and Normal Components of Acceleration

Find the tangential and normal components of acceleration for the position vector given by $\mathbf{r}(t) = 3t\mathbf{i} - t\mathbf{j} + t^2\mathbf{k}$.

Solution Begin by finding the velocity, speed, and acceleration.

$$\mathbf{v}(t) = \mathbf{r}'(t) = 3\mathbf{i} - \mathbf{j} + 2t\mathbf{k}$$
$$\|\mathbf{v}(t)\| = \sqrt{9 + 1 + 4t^2} = \sqrt{10 + 4t^2}$$
$$\mathbf{a}(t) = \mathbf{r}''(t) = 2\mathbf{k}$$

By Theorem 12.5, the tangential component of acceleration is

$$a_{\mathbf{T}} = \frac{\mathbf{v} \cdot \mathbf{a}}{\|\mathbf{v}\|} = \frac{4t}{\sqrt{10 + 4t^2}}$$

Tangential component of acceleration

and because

$$\mathbf{v} \times \mathbf{a} = \begin{vmatrix} \mathbf{i} & \mathbf{j} & \mathbf{k} \\ 3 & -1 & 2t \\ 0 & 0 & 2 \end{vmatrix} = -2\mathbf{i} - 6\mathbf{j}$$

the normal component of acceleration is

$$a_{\mathbf{N}} = \frac{\|\mathbf{v} \times \mathbf{a}\|}{\|\mathbf{v}\|} = \frac{\sqrt{4 + 36}}{\sqrt{10 + 4t^2}} = \frac{2\sqrt{10}}{\sqrt{10 + 4t^2}}.$$

Normal component of acceleration

NOTE In Example 5, you could have used the alternative formula for $a_{\mathbf{N}}$ as follows.

$$a_{\mathbf{N}} = \sqrt{\|\mathbf{a}\|^2 - a_{\mathbf{T}}^2} = \sqrt{(2)^2 - \frac{16t^2}{10 + 4t^2}} = \frac{2\sqrt{10}}{\sqrt{10 + 4t^2}}$$

EXAMPLE 6 Finding $a_{\mathbf{T}}$ and $a_{\mathbf{N}}$ for a Circular Helix

Find the tangential and normal components of acceleration for the helix given by $\mathbf{r}(t) = b \cos t\mathbf{i} + b \sin t\mathbf{j} + ct\mathbf{k}$, $b > 0$.

Solution

$$\mathbf{v}(t) = \mathbf{r}'(t) = -b \sin t\mathbf{i} + b \cos t\mathbf{j} + c\mathbf{k}$$
$$\|\mathbf{v}(t)\| = \sqrt{b^2 \sin^2 t + b^2 \cos^2 t + c^2} = \sqrt{b^2 + c^2}$$
$$\mathbf{a}(t) = \mathbf{r}''(t) = -b \cos t\mathbf{i} - b \sin t\mathbf{j}$$

By Theorem 12.5, the tangential component of acceleration is

$$a_{\mathbf{T}} = \frac{\mathbf{v} \cdot \mathbf{a}}{\|\mathbf{v}\|} = \frac{b^2 \sin t \cos t - b^2 \sin t \cos t + 0}{\sqrt{b^2 + c^2}} = 0.$$

Tangential component of acceleration

Moreover, because $\|\mathbf{a}\| = \sqrt{b^2 \cos^2 t + b^2 \sin^2 t} = b$, you can use the alternative formula for the normal component of acceleration to obtain

$$a_{\mathbf{N}} = \sqrt{\|\mathbf{a}\|^2 - a_{\mathbf{T}}^2} = \sqrt{b^2 - 0^2} = b.$$

Normal component of acceleration

Note that the normal component of acceleration is equal to the magnitude of the acceleration. In other words, because the speed is constant, the acceleration is perpendicular to the velocity. See Figure 12.25.

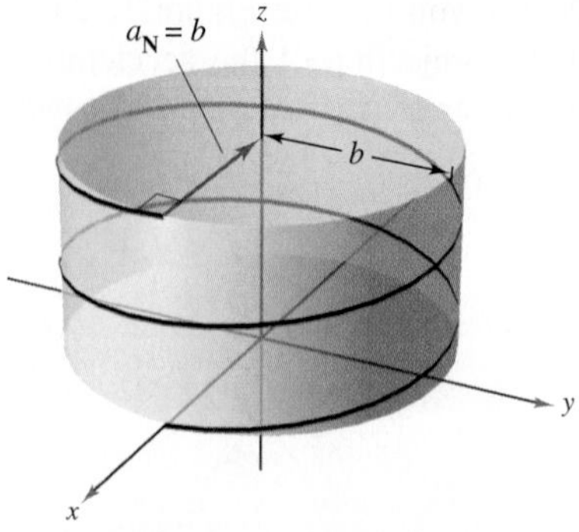

The normal component of acceleration is equal to the radius of the cylinder around which the helix is spiraling.
Figure 12.25

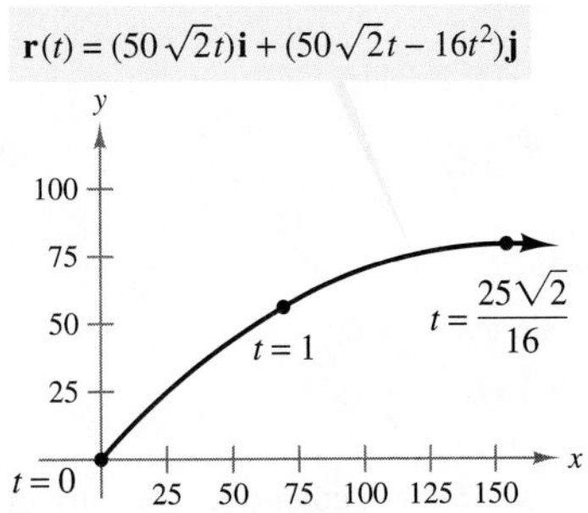

The path of a projectile
Figure 12.26

EXAMPLE 7 Projectile Motion

The position vector for the projectile shown in Figure 12.26 is given by

$$\mathbf{r}(t) = \left(50\sqrt{2}\,t\right)\mathbf{i} + \left(50\sqrt{2}\,t - 16t^2\right)\mathbf{j}. \qquad \text{Position vector}$$

Find the tangential component of acceleration when $t = 0$, 1, and $25\sqrt{2}/16$.

Solution

$$\mathbf{v}(t) = 50\sqrt{2}\mathbf{i} + \left(50\sqrt{2} - 32t\right)\mathbf{j} \qquad \text{Velocity vector}$$

$$\|\mathbf{v}(t)\| = 2\sqrt{50^2 - 16(50)\sqrt{2}t + 16^2t^2} \qquad \text{Speed}$$

$$\mathbf{a}(t) = -32\mathbf{j} \qquad \text{Acceleration vector}$$

The tangential component of acceleration is

$$a_{\mathbf{T}}(t) = \frac{\mathbf{v}(t) \cdot \mathbf{a}(t)}{\|\mathbf{v}(t)\|} = \frac{-32(50\sqrt{2} - 32t)}{2\sqrt{50^2 - 16(50)\sqrt{2}t + 16^2t^2}}. \qquad \text{Tangential component of acceleration}$$

At the specified times, you have

$$a_{\mathbf{T}}(0) = \frac{-32\left(50\sqrt{2}\right)}{100} = -16\sqrt{2} \approx -22.6$$

$$a_{\mathbf{T}}(1) = \frac{-32\left(50\sqrt{2} - 32\right)}{2\sqrt{50^2 - 16(50)\sqrt{2} + 16^2}} \approx -15.4$$

$$a_{\mathbf{T}}\left(\frac{25\sqrt{2}}{16}\right) = \frac{-32\left(50\sqrt{2} - 50\sqrt{2}\right)}{50\sqrt{2}} = 0.$$

You can see from Figure 12.26 that, at the maximum height, when $t = 25\sqrt{2}/16$, the tangential component is 0. This is reasonable because the direction of motion is horizontal at the point and the tangential component of the acceleration is equal to the horizontal component of the acceleration.

Exercises for Section 12.4

See www.CalcChat.com for worked-out solutions to odd-numbered exercises.

In Exercises 1–4, sketch the unit tangent and normal vectors at the given points. To print an enlarged copy of the graph, go to the website *www.mathgraphs.com*.

1.
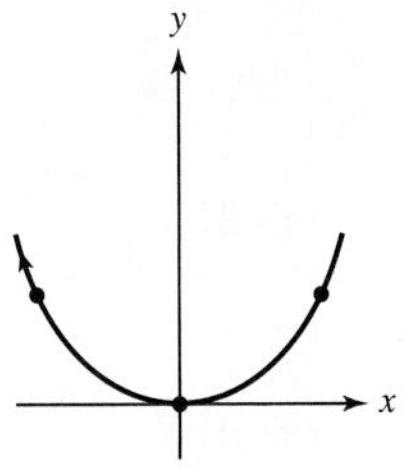

2.
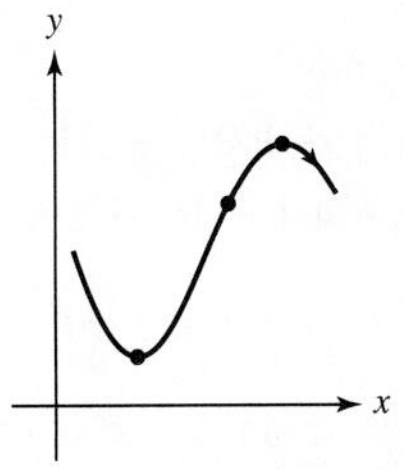

3.
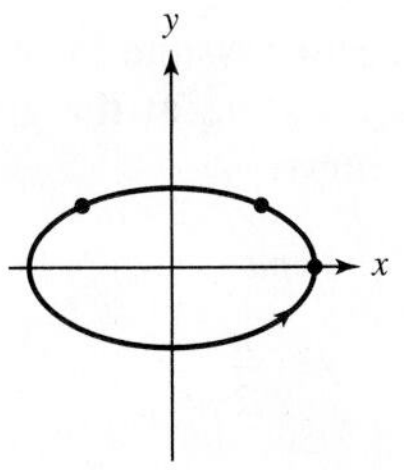

4.
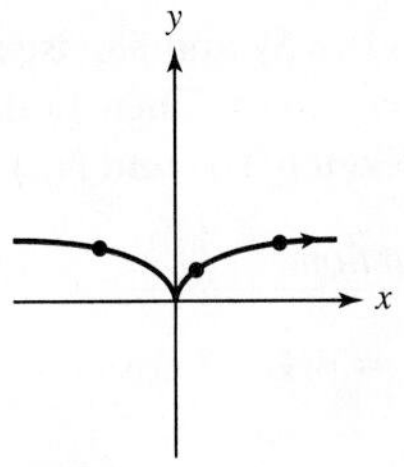

In Exercises 5–10, find the unit tangent vector to the curve at the specified value of the parameter.

5. $\mathbf{r}(t) = t^2\mathbf{i} + 2t\mathbf{j}, \quad t = 1$

6. $\mathbf{r}(t) = t^3\mathbf{i} + 2t^2\mathbf{j}, \quad t = 1$

7. $\mathbf{r}(t) = 4\cos t\mathbf{i} + 4\sin t\mathbf{j}, \quad t = \frac{\pi}{4}$

8. $\mathbf{r}(t) = 6\cos t\mathbf{i} + 2\sin t\mathbf{j}, \quad t = \frac{\pi}{3}$

9. $\mathbf{r}(t) = \ln t\mathbf{i} + 2t\mathbf{j}, \quad t = e$

10. $\mathbf{r}(t) = e^t\cos t\mathbf{i} + e^t\mathbf{j}, \quad t = 0$

In Exercises 11–16, find the unit tangent vector T(t) and find a set of parametric equations for the line tangent to the space curve at point *P*.

11. $\mathbf{r}(t) = t\mathbf{i} + t^2\mathbf{j} + t\mathbf{k}, \quad P(0, 0, 0)$

12. $\mathbf{r}(t) = t^2\mathbf{i} + t\mathbf{j} + \frac{4}{3}\mathbf{k}, \quad P\left(1, 1, \frac{4}{3}\right)$

13. $\mathbf{r}(t) = 2\cos t\mathbf{i} + 2\sin t\mathbf{j} + t\mathbf{k}, \quad P(2, 0, 0)$

14. $\mathbf{r}(t) = \left\langle t, t, \sqrt{4 - t^2}\right\rangle, \quad P\left(1, 1, \sqrt{3}\right)$

15. $\mathbf{r}(t) = \langle 2\cos t, 2\sin t, 4\rangle, \quad P\left(\sqrt{2}, \sqrt{2}, 4\right)$

16. $\mathbf{r}(t) = \langle 2\sin t, 2\cos t, 4\sin^2 t\rangle, \quad P\left(1, \sqrt{3}, 1\right)$

In Exercises 17 and 18, use a computer algebra system to graph the space curve. Then find $\mathbf{T}(t)$ and find a set of parametric equations for the line tangent to the space curve at point P. Graph the tangent line.

17. $\mathbf{r}(t) = \langle t, t^2, 2t^3/3\rangle, \quad P(3, 9, 18)$

18. $\mathbf{r}(t) = 3\cos t\mathbf{i} + 4\sin t\mathbf{j} + \frac{1}{2}t\mathbf{k}, \quad P(0, 4, \pi/4)$

Linear Approximation **In Exercises 19 and 20, find a set of parametric equations for the tangent line to the graph at $t = t_0$ and use the equations for the line to approximate $\mathbf{r}(t_0 + 0.1)$.**

19. $\mathbf{r}(t) = \langle t, \ln t, \sqrt{t}\rangle, \quad t_0 = 1$

20. $\mathbf{r}(t) = \langle e^{-t}, 2\cos t, 2\sin t\rangle, \quad t_0 = 0$

In Exercises 21 and 22, verify that the space curves intersect at the given values of the parameters. Find the angle between the tangent vectors to the curves at the point of intersection.

21. $\mathbf{r}(t) = \langle t - 2, t^2, \frac{1}{2}t\rangle, \quad t = 4$
$\mathbf{u}(s) = \langle \frac{1}{4}s, 2s, \sqrt[3]{s}\rangle, \quad s = 8$

22. $\mathbf{r}(t) = \langle t, \cos t, \sin t\rangle, \quad t = 0$
$\mathbf{u}(s) = \langle -\frac{1}{2}\sin^2 s - \sin s, 1 - \frac{1}{2}\sin^2 s - \sin s,$
$\frac{1}{2}\sin s\cos s + \frac{1}{2}s\rangle, \quad s = 0$

In Exercises 23–30, find the principal unit normal vector to the curve at the specified value of the parameter.

23. $\mathbf{r}(t) = t\mathbf{i} + \frac{1}{2}t^2\mathbf{j}, \quad t = 2$

24. $\mathbf{r}(t) = t\mathbf{i} + \dfrac{6}{t}\mathbf{j}, \quad t = 3$

25. $\mathbf{r}(t) = \ln t\mathbf{i} + (t + 1)\mathbf{j}, \quad t = 2$

26. $\mathbf{r}(t) = 3\cos t\mathbf{i} + 3\sin t\mathbf{j}, \quad t = \dfrac{\pi}{4}$

27. $\mathbf{r}(t) = t\mathbf{i} + t^2\mathbf{j} + \ln t\mathbf{k}, \quad t = 1$

28. $\mathbf{r}(t) = \sqrt{2}t\mathbf{i} + e^t\mathbf{j} + e^{-t}\mathbf{k}, \quad t = 0$

29. $\mathbf{r}(t) = 6\cos t\mathbf{i} + 6\sin t\mathbf{j} + \mathbf{k}, \quad t = \dfrac{3\pi}{4}$

30. $\mathbf{r}(t) = \cos t\mathbf{i} + 2\sin t\mathbf{j} + \mathbf{k}, \quad t = -\dfrac{\pi}{4}$

In Exercises 31–34, find $\mathbf{v}(t)$, $\mathbf{a}(t)$, $\mathbf{T}(t)$, and $\mathbf{N}(t)$ (if it exists) for an object moving along the path given by the vector-valued function $\mathbf{r}(t)$. Use the results to determine the form of the path. Is the speed of the object constant or changing?

31. $\mathbf{r}(t) = 4t\mathbf{i}$

32. $\mathbf{r}(t) = 4t\mathbf{i} - 2t\mathbf{j}$

33. $\mathbf{r}(t) = 4t^2\mathbf{i}$

34. $\mathbf{r}(t) = t^2\mathbf{j} + \mathbf{k}$

In Exercises 35–44, find $\mathbf{T}(t)$, $\mathbf{N}(t)$, $a_{\mathbf{T}}$, and $a_{\mathbf{N}}$ at the given time t for the plane curve $\mathbf{r}(t)$.

35. $\mathbf{r}(t) = t\mathbf{i} + \dfrac{1}{t}\mathbf{j}, \quad t = 1$

36. $\mathbf{r}(t) = t^2\mathbf{i} + 2t\mathbf{j}, \quad t = 1$

37. $\mathbf{r}(t) = (t - t^3)\mathbf{i} + 2t^2\mathbf{j}, \quad t = 1$

38. $\mathbf{r}(t) = (t^3 - 4t)\mathbf{i} + (t^2 - 1)\mathbf{j}, \quad t = 0$

39. $\mathbf{r}(t) = e^t\mathbf{i} + e^{-2t}\mathbf{j}, \quad t = 0$

40. $\mathbf{r}(t) = e^t\mathbf{i} + e^{-t}\mathbf{j} + t\mathbf{k}, \quad t = 0$

41. $\mathbf{r}(t) = e^t\cos t\mathbf{i} + e^t\sin t\mathbf{j}, \quad t = \dfrac{\pi}{2}$

42. $\mathbf{r}(t) = a\cos\omega t\mathbf{i} + b\sin\omega t\mathbf{j}, \quad t = 0$

43. $\mathbf{r}(t) = \langle\cos\omega t + \omega t\sin\omega t, \sin\omega t - \omega t\cos\omega t\rangle, \quad t = t_0$

44. $\mathbf{r}(t) = \langle\omega t - \sin\omega t, 1 - \cos\omega t\rangle, \quad t = t_0$

Circular Motion **In Exercises 45–48, consider an object moving according to the position function**

$$\mathbf{r}(t) = a\cos\omega t\,\mathbf{i} + a\sin\omega t\,\mathbf{j}.$$

45. Find $\mathbf{T}(t)$, $\mathbf{N}(t)$, $a_{\mathbf{T}}$, and $a_{\mathbf{N}}$.

46. Determine the directions of $\mathbf{T}$ and $\mathbf{N}$ relative to the position function $\mathbf{r}$.

47. Determine the speed of the object at any time t and explain its value relative to the value of $a_{\mathbf{T}}$.

48. If the angular velocity ω is halved, by what factor is $a_{\mathbf{N}}$ changed?

In Exercises 49–52, sketch the graph of the plane curve given by the vector-valued function, and, at the point on the curve determined by $\mathbf{r}(t_0)$, sketch the vectors $\mathbf{T}$ and $\mathbf{N}$. Note that $\mathbf{N}$ points toward the concave side of the curve.

Function	*Time*
49. $\mathbf{r}(t) = t\mathbf{i} + \dfrac{1}{t}\mathbf{j}$	$t_0 = 2$
50. $\mathbf{r}(t) = t^3\mathbf{i} + t\mathbf{j}$	$t_0 = 1$
51. $\mathbf{r}(t) = 2\cos t\mathbf{i} + 2\sin t\mathbf{j}$	$t_0 = \dfrac{\pi}{4}$
52. $\mathbf{r}(t) = 3\cos t\mathbf{i} + 2\sin t\mathbf{j}$	$t_0 = \pi$

In Exercises 53–56, find $\mathbf{T}(t)$, $\mathbf{N}(t)$, $a_{\mathbf{T}}$, and $a_{\mathbf{N}}$ at the given time t for the space curve $\mathbf{r}(t)$. [*Hint:* Find $\mathbf{a}(t)$, $\mathbf{T}(t)$, and $a_{\mathbf{N}}$. Solve for $\mathbf{N}$ in the equation $\mathbf{a}(t) = a_{\mathbf{T}}\mathbf{T} + a_{\mathbf{N}}\mathbf{N}$.]

Function	*Time*
53. $\mathbf{r}(t) = t\mathbf{i} + 2t\mathbf{j} - 3t\mathbf{k}$	$t = 1$
54. $\mathbf{r}(t) = 4t\mathbf{i} - 4t\mathbf{j} + 2t\mathbf{k}$	$t = 2$
55. $\mathbf{r}(t) = t\mathbf{i} + t^2\mathbf{j} + \dfrac{t^2}{2}\mathbf{k}$	$t = 1$
56. $\mathbf{r}(t) = e^t\sin t\mathbf{i} + e^t\cos t\mathbf{j} + e^t\mathbf{k}$	$t = 0$

In Exercises 57 and 58, use a computer algebra system to graph the space curve. Then find $\mathbf{T}(t)$, $\mathbf{N}(t)$, $a_{\mathbf{T}}$, and $a_{\mathbf{N}}$ at the given time t. Sketch $\mathbf{T}(t)$ and $\mathbf{N}(t)$ on the space curve.

Function	*Time*
57. $\mathbf{r}(t) = 4t\mathbf{i} + 3\cos t\mathbf{j} + 3\sin t\mathbf{k}$	$t = \dfrac{\pi}{2}$
58. $\mathbf{r}(t) = t\mathbf{i} + 3t^2\mathbf{j} + \dfrac{t^2}{2}\mathbf{k}$	$t = 2$

Writing About Concepts

59. Define the unit tangent vector, the principal unit normal vector, and the tangential and normal components of acceleration.

60. How is the unit tangent vector related to the orientation of a curve? Explain.

61. Describe the motion of a particle if the normal component of acceleration is 0.

62. Describe the motion of a particle if the tangential component of acceleration is 0.

63. ***Cycloidal Motion*** The figure shows the path of a particle modeled by the vector-valued function

$$\mathbf{r}(t) = \langle \pi t - \sin \pi t, 1 - \cos \pi t \rangle.$$

The figure also shows the vectors $\mathbf{v}(t)/\|\mathbf{v}(t)\|$ and $\mathbf{a}(t)/\|\mathbf{a}(t)\|$ at the indicated values of t.

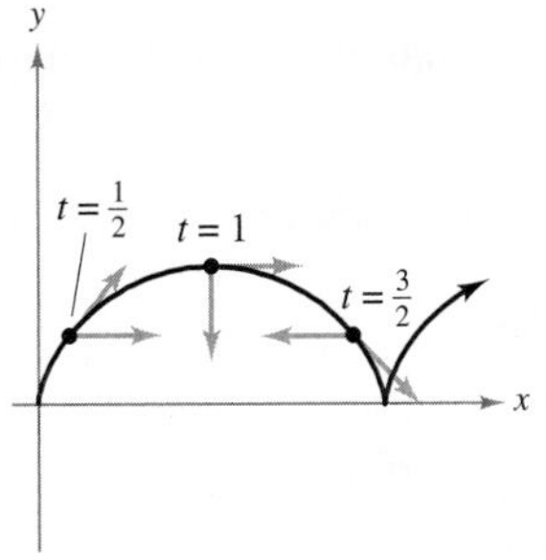

(a) Find $a_{\mathbf{T}}$ and $a_{\mathbf{N}}$ at $t = \frac{1}{2}$, $t = 1$, and $t = \frac{3}{2}$.

(b) Determine whether the speed of the particle is increasing or decreasing at each of the indicated values of t. Give reasons for your answers.

64. ***Motion Along an Involute of a Circle*** The figure shows a particle moving along a path modeled by

$$\mathbf{r}(t) = \langle \cos \pi t + \pi t \sin \pi t, \sin \pi t - \pi t \cos \pi t \rangle.$$

The figure also shows the vectors $\mathbf{v}(t)$ and $\mathbf{a}(t)$ for $t = 1$ and $t = 2$.

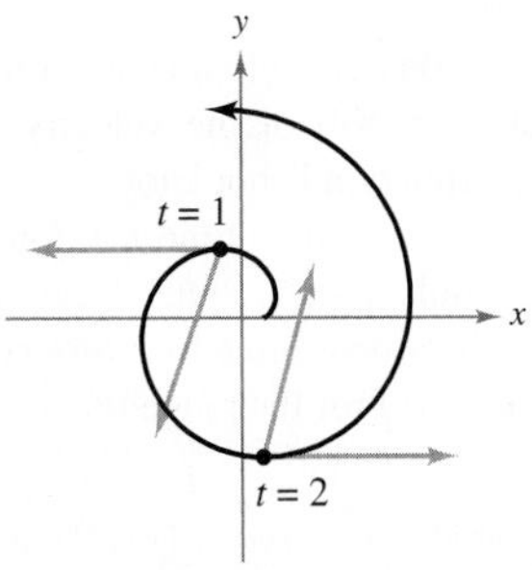

(a) Find $a_{\mathbf{T}}$ and $a_{\mathbf{N}}$ at $t = 1$ and $t = 2$.

(b) Determine whether the speed of the particle is increasing or decreasing at each of the indicated values of t. Give reasons for your answers.

In Exercises 65–70, find the vectors T and N, and the unit binormal vector B = T × N, for the vector-valued function r(*t*) at the given value of *t*.

65. $\mathbf{r}(t) = 2 \cos t\mathbf{i} + 2 \sin t\mathbf{j} + \dfrac{t}{2}\mathbf{k}$

$t_0 = \dfrac{\pi}{2}$

66. $\mathbf{r}(t) = t\mathbf{i} + t^2\mathbf{j} + \dfrac{t^3}{3}\mathbf{k}$

$t_0 = 1$

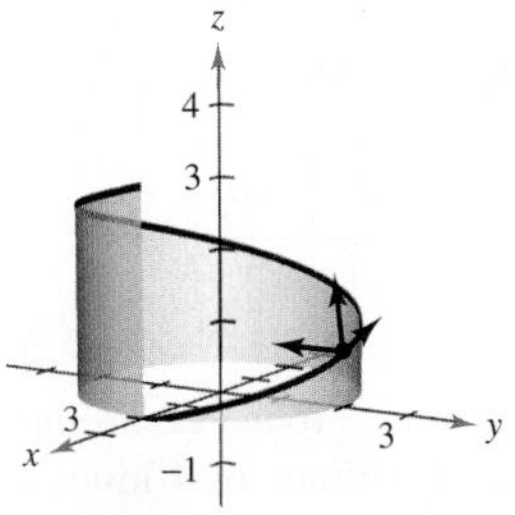

Figure for 65

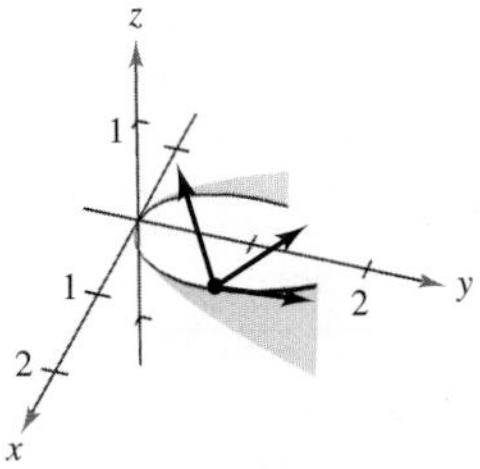

Figure for 66

67. $\mathbf{r}(t) = \mathbf{i} + \sin t\mathbf{j} + \cos t\mathbf{k}, \quad t_0 = \dfrac{\pi}{4}$

68. $\mathbf{r}(t) = 2e^t\mathbf{i} + e^t \cos t\mathbf{j} + e^t \sin t\mathbf{k}, \quad t_0 = 0$

69. $\mathbf{r}(t) = 4 \sin t\mathbf{i} + 4 \cos t\mathbf{j} + 2t\mathbf{k}, \quad t_0 = \dfrac{\pi}{3}$

70. $\mathbf{r}(t) = 2 \cos 2t\mathbf{i} + 2 \sin 2t\mathbf{j} + t\mathbf{k}, \quad t_0 = \dfrac{\pi}{4}$

71. ***Projectile Motion*** Find the tangential and normal components of acceleration for a projectile fired at an angle θ with the horizontal at an initial speed of v_0. What are the components when the projectile is at its maximum height?

72. ***Projectile Motion*** Use your results from Exercise 71 to find the tangential and normal components of acceleration for a projectile fired at an angle of 45° with the horizontal at an initial speed of 150 feet per second. What are the components when the projectile is at its maximum height?

73. ***Projectile Motion*** A projectile is launched with an initial velocity of 100 feet per second at a height of 5 feet and at an angle of 30° with the horizontal.

(a) Determine the vector-valued function for the path of the projectile.

(b) Use a graphing utility to graph the path and approximate the maximum height and range of the projectile.

(c) Find $\mathbf{v}(t)$, $\|\mathbf{v}(t)\|$, and $\mathbf{a}(t)$.

(d) Use a graphing utility to complete the table.

t	0.5	1.0	1.5	2.0	2.5	3.0
Speed						

(e) Use a graphing utility to graph the scalar functions $a_{\mathbf{T}}$ and $a_{\mathbf{N}}$. How is the speed of the projectile changing when $a_{\mathbf{T}}$ and $a_{\mathbf{N}}$ have opposite signs?

74. ***Projectile Motion*** A projectile is launched with an initial velocity of 200 feet per second at a height of 4 feet and at an angle of 45° with the horizontal.

(a) Determine the vector-valued function for the path of the projectile.

(b) Use a graphing utility to graph the path and approximate the maximum height and range of the projectile.

(c) Find $\mathbf{v}(t)$, $\|\mathbf{v}(t)\|$, and $\mathbf{a}(t)$.

(d) Use a graphing utility to complete the table.

t	0.5	1.0	1.5	2.0	2.5	3.0
Speed						

75. ***Air Traffic Control*** Because of a storm, ground controllers instruct the pilot of a plane flying at an altitude of 4 miles to make a 90° turn and climb to an altitude of 4.2 miles. The model for the path of the plane during this maneuver is

$$\mathbf{r}(t) = \langle 10 \cos 10\pi t, 10 \sin 10\pi t, 4 + 4t \rangle, \quad 0 \le t \le \tfrac{1}{20}$$

where t is the time in hours and $\mathbf{r}$ is the distance in miles.

(a) Determine the speed of the plane.

(b) Use a computer algebra system to calculate $a_{\mathbf{T}}$ and $a_{\mathbf{N}}$. Why is one of these equal to 0?

76. ***Projectile Motion*** A plane flying at an altitude of 36,000 feet at a speed of 600 miles per hour releases a bomb. Find the tangential and normal components of acceleration acting on the bomb.

77. ***Centripetal Acceleration*** An object is spinning at a constant speed on the end of a string, according to the position function given in Exercises 45–48.

(a) If the angular velocity ω is doubled, how is the centripetal component of acceleration changed?

(b) If the angular velocity is unchanged but the length of the string is halved, how is the centripetal component of acceleration changed?

78. ***Centripetal Force*** An object of mass m moves at a constant speed v in a circular path of radius r. The force required to produce the centripetal component of acceleration is called the *centripetal force* and is given by $F = mv^2/r$. Newton's Law of Universal Gravitation is given by $F = GMm/d^2$, where d is the distance between the centers of the two bodies of masses M and m, and G is a gravitational constant. Use this law to show that the speed required for circular motion is $v = \sqrt{GM/r}$.

Orbital Speed **In Exercises 79–82, use the result of Exercise 78 to find the speed necessary for the given circular orbit around Earth. Let $GM = 9.56 \times 10^4$ cubic miles per second per second, and assume the radius of Earth is 4000 miles.**

79. The orbit of a space shuttle 100 miles above the surface of Earth

80. The orbit of a space shuttle 200 miles above the surface of Earth

81. The orbit of a heat capacity mapping satellite 385 miles above the surface of Earth

82. The orbit of a SYNCOM satellite r miles above the surface of Earth that is in geosynchronous orbit [The satellite completes one orbit per sidereal day (approximately 23 hours, 56 minutes), and therefore appears to remain stationary above a point on Earth.]

True or False? **In Exercises 83 and 84, determine whether the statement is true or false. If it is false, explain why or give an example that shows it is false.**

83. If a car's speedometer is constant, then the car cannot be accelerating.

84. If $a_{\mathbf{N}} = 0$ for a moving object, then the object is moving in a straight line.

85. A particle moves along a path modeled by

$$\mathbf{r}(t) = \cosh(bt)\mathbf{i} + \sinh(bt)\mathbf{j}$$

where b is a positive constant.

(a) Show that the path of the particle is a hyperbola.

(b) Show that $\mathbf{a}(t) = b^2\,\mathbf{r}(t)$.

86. Prove that the principal unit normal vector $\mathbf{N}$ points toward the concave side of a plane curve.

87. Prove that the vector $\mathbf{T}'(t)$ is $\mathbf{0}$ for an object moving in a straight line.

88. Prove that $a_{\mathbf{N}} = \dfrac{\|\mathbf{v} \times \mathbf{a}\|}{\|\mathbf{v}\|}$.

89. Prove that $a_{\mathbf{N}} = \sqrt{\|\mathbf{a}\|^2 - a_{\mathbf{T}}^2}$.

Putnam Exam Challenge

90. A particle of unit mass moves on a straight line under the action of a force which is a function $f(v)$ of the velocity v of the particle, but the form of this function is not known. A motion is observed, and the distance x covered in time t is found to be connected with t by the formula $x = at + bt^2 + ct^3$, where a, b, and c have numerical values determined by observation of the motion. Find the function $f(v)$ for the range of v covered by the experiment.

This problem was composed by the Committee on the Putnam Prize Competition.

Section 12.5 Arc Length and Curvature

- **Find the arc length of a space curve.**
- **Use the arc length parameter to describe a plane curve or space curve.**
- **Find the curvature of a curve at a point on the curve.**
- **Use a vector-valued function to find frictional force.**

Arc Length

In Section 10.3, you saw that the arc length of a smooth *plane* curve C given by the parametric equations $x = x(t)$ and $y = y(t)$, $a \le t \le b$, is

$$s = \int_a^b \sqrt{[x'(t)]^2 + [y'(t)]^2}\, dt.$$

In vector form, where C is given by $\mathbf{r}(t) = x(t)\mathbf{i} + y(t)\mathbf{j}$, you can rewrite this equation for arc length as

$$s = \int_a^b \|\mathbf{r}'(t)\|\, dt.$$

The formula for the arc length of a plane curve has a natural extension to a smooth curve in *space*, as stated in the following theorem.

THEOREM 12.6 Arc Length of a Space Curve

If C is a smooth curve given by $\mathbf{r}(t) = x(t)\mathbf{i} + y(t)\mathbf{j} + z(t)\mathbf{k}$, on an interval $[a, b]$, then the arc length of C on the interval is

$$s = \int_a^b \sqrt{[x'(t)]^2 + [y'(t)]^2 + [z'(t)]^2}\, dt = \int_a^b \|\mathbf{r}'(t)\|\, dt.$$

EXPLORATION

Arc Length Formula The formula for the arc length of a space curve is given in terms of the parametric equations used to represent the curve. Does this mean that the arc length of the curve depends on the parameter being used? Would you want this to be true? Explain your reasoning.

Here is a different parametric representation of the curve in Example 1.

$$\mathbf{r}(t) = t^2\mathbf{i} + \frac{4}{3}t^3\mathbf{j} + \frac{1}{2}t^4\mathbf{k}$$

Find the arc length from $t = 0$ to $t = \sqrt{2}$ and compare the result with that found in Example 1.

EXAMPLE 1 Finding the Arc Length of a Curve in Space

Find the arc length of the curve given by

$$\mathbf{r}(t) = t\mathbf{i} + \frac{4}{3}t^{3/2}\mathbf{j} + \frac{1}{2}t^2\mathbf{k}$$

from $t = 0$ to $t = 2$, as shown in Figure 12.27.

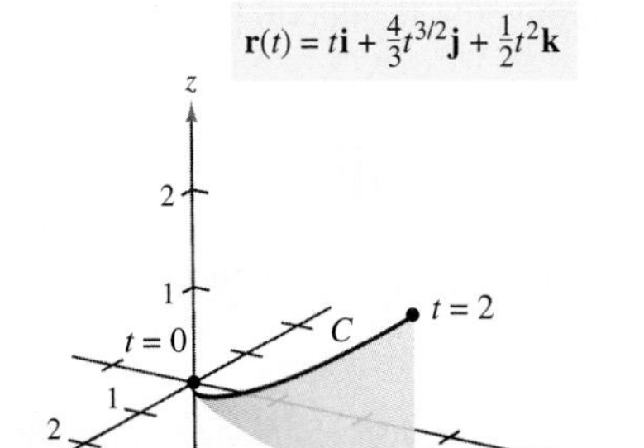

As t increases from 0 to 2, the vector $\mathbf{r}(t)$ traces out a curve.
Figure 12.27

Solution Using $x(t) = t$, $y(t) = \frac{4}{3}t^{3/2}$, and $z(t) = \frac{1}{2}t^2$, you obtain $x'(t) = 1$, $y'(t) = 2t^{1/2}$, and $z'(t) = t$. So, the arc length from $t = 0$ to $t = 2$ is given by

$$s = \int_0^2 \sqrt{[x'(t)]^2 + [y'(t)]^2 + [z'(t)]^2}\, dt \qquad \text{Formula for arc length}$$

$$= \int_0^2 \sqrt{1 + 4t + t^2}\, dt$$

$$= \int_0^2 \sqrt{(t+2)^2 - 3}\, dt \qquad \text{Integration tables (Appendix B), Formula 26}$$

$$= \left[\frac{t+2}{2}\sqrt{(t+2)^2 - 3} - \frac{3}{2}\ln\left|(t+2) + \sqrt{(t+2)^2 - 3}\right|\right]_0^2$$

$$= 2\sqrt{13} - \frac{3}{2}\ln\left(4 + \sqrt{13}\right) - 1 + \frac{3}{2}\ln 3 \approx 4.816.$$

Curve:
$\mathbf{r}(t) = b\cos t\mathbf{i} + b\sin t\mathbf{j} + \sqrt{1-b^2}\,t\mathbf{k}$

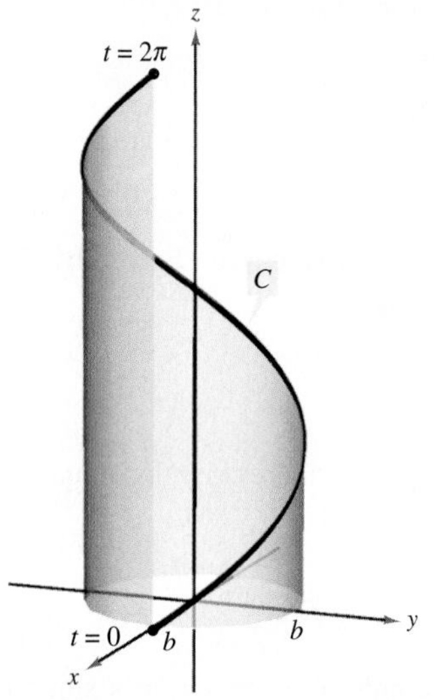

One turn of a helix
Figure 12.28

EXAMPLE 2 Finding the Arc Length of a Helix

Find the length of one turn of the helix given by

$$\mathbf{r}(t) = b\cos t\mathbf{i} + b\sin t\mathbf{j} + \sqrt{1-b^2}\,t\mathbf{k}$$

as shown in Figure 12.28.

Solution Begin by finding the derivative.

$$\mathbf{r}'(t) = -b\sin t\mathbf{i} + b\cos t\mathbf{j} + \sqrt{1-b^2}\,\mathbf{k} \qquad \text{Derivative}$$

Now, using the formula for arc length, you can find the length of one turn of the helix by integrating $\|\mathbf{r}'(t)\|$ from 0 to 2π.

$$\begin{aligned} s &= \int_0^{2\pi} \|\mathbf{r}'(t)\|\,dt && \text{Formula for arc length} \\ &= \int_0^{2\pi} \sqrt{b^2(\sin^2 t + \cos^2 t) + (1-b^2)}\,dt \\ &= \int_0^{2\pi} dt \\ &= t\Big]_0^{2\pi} = 2\pi. \end{aligned}$$

So, the length is 2π units.

Arc Length Parameter

You have seen that curves can be represented by vector-valued functions in different ways, depending on the choice of parameter. For *motion* along a curve, the convenient parameter is time t. However, for studying the *geometric properties* of a curve, the convenient parameter is often arc length s.

$s(t) = \int_a^t \sqrt{[x'(u)]^2 + [y'(u)]^2 + [z'(u)]^2}\,du$

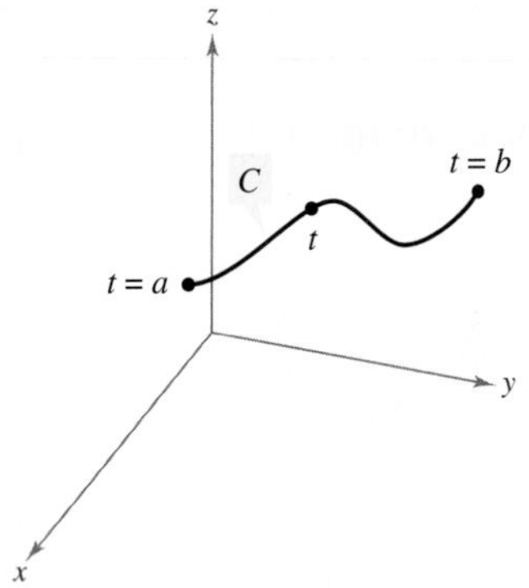

Figure 12.29

Definition of Arc Length Function

Let C be a smooth curve given by $\mathbf{r}(t)$ defined on the closed interval $[a, b]$. For $a \le t \le b$, the **arc length function** is given by

$$s(t) = \int_a^t \|\mathbf{r}'(u)\|\,du = \int_a^t \sqrt{[x'(u)]^2 + [y'(u)]^2 + [z'(u)]^2}\,du.$$

The arc length s is called the **arc length parameter.** (See Figure 12.29.)

NOTE The arc length function s is *nonnegative*. It measures the distance along C from the initial point $(x(a), y(a), z(a))$ to the point $(x(t), y(t), z(t))$.

Using the definition of the arc length function and the Second Fundamental Theorem of Calculus, you can conclude that

$$\frac{ds}{dt} = \|\mathbf{r}'(t)\|. \qquad \text{Derivative of arc length function}$$

In differential form, you can write

$$ds = \|\mathbf{r}'(t)\|\,dt.$$

EXAMPLE 3 Finding the Arc Length Function for a Line

Find the arc length function $s(t)$ for the line segment given by

$$\mathbf{r}(t) = (3 - 3t)\mathbf{i} + 4t\mathbf{j}, \quad 0 \le t \le 1$$

and write $\mathbf{r}$ as a function of the parameter s. (See Figure 12.30.)

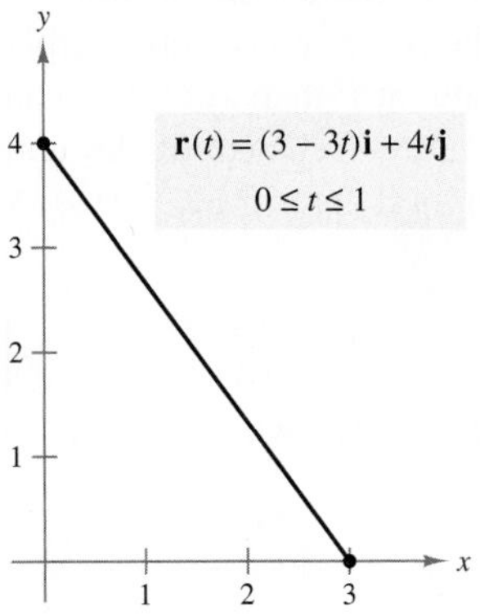

The line segment from $(3, 0)$ to $(0, 4)$ can be parametrized using the arc length parameter s.
Figure 12.30

Solution Because $\mathbf{r}'(t) = -3\mathbf{i} + 4\mathbf{j}$ and

$$\|\mathbf{r}'(t)\| = \sqrt{(-3)^2 + 4^2} = 5$$

you have

$$\begin{aligned} s(t) &= \int_0^t \|\mathbf{r}'(u)\|\,du \\ &= \int_0^t 5\,du \\ &= 5t. \end{aligned}$$

Using $s = 5t$ (or $t = s/5$), you can rewrite $\mathbf{r}$ using the arc length parameter as follows.

$$\mathbf{r}(s) = \left(3 - \tfrac{3}{5}s\right)\mathbf{i} + \tfrac{4}{5}s\,\mathbf{j}, \quad 0 \le s \le 5.$$

One of the advantages of writing a vector-valued function in terms of the arc length parameter is that $\|\mathbf{r}'(s)\| = 1$. For instance, in Example 3, you have

$$\|\mathbf{r}'(s)\| = \sqrt{\left(-\frac{3}{5}\right)^2 + \left(\frac{4}{5}\right)^2} = 1.$$

So, for a smooth curve C represented by $\mathbf{r}(s)$, where s is the arc length parameter, the arc length between a and b is

$$\begin{aligned} \text{Length of arc} &= \int_a^b \|\mathbf{r}'(s)\|\,ds \\ &= \int_a^b ds \\ &= b - a \\ &= \text{length of interval.} \end{aligned}$$

Furthermore, if t is *any* parameter such that $\|\mathbf{r}'(t)\| = 1$, then t must be the arc length parameter. These results are summarized in the following theorem, which is stated without proof.

THEOREM 12.7 Arc Length Parameter

If C is a smooth curve given by

$$\mathbf{r}(s) = x(s)\mathbf{i} + y(s)\mathbf{j} \quad \text{or} \quad \mathbf{r}(s) = x(s)\mathbf{i} + y(s)\mathbf{j} + z(s)\mathbf{k}$$

where s is the arc length parameter, then

$$\|\mathbf{r}'(s)\| = 1.$$

Moreover, if t is *any* parameter for the vector-valued function $\mathbf{r}$ such that $\|\mathbf{r}'(t)\| = 1$, then t must be the arc length parameter.

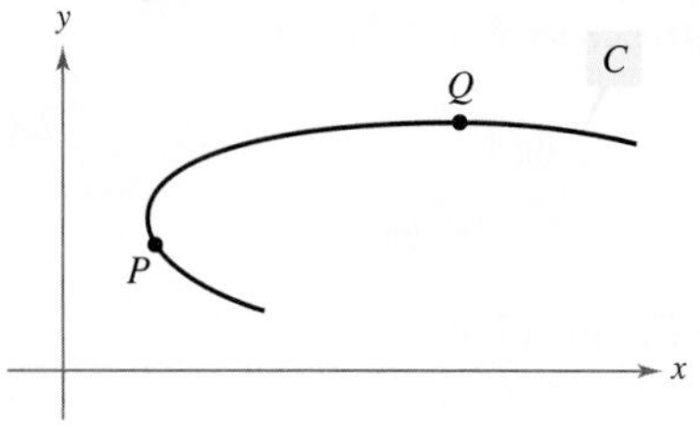

Curvature at P is greater than at Q.
Figure 12.31

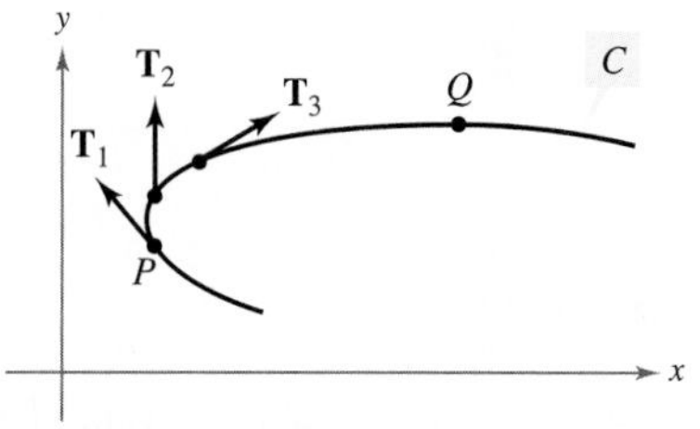

The magnitude of the rate of change of $\mathbf{T}$ with respect to the arc length is the curvature of a curve.
Figure 12.32

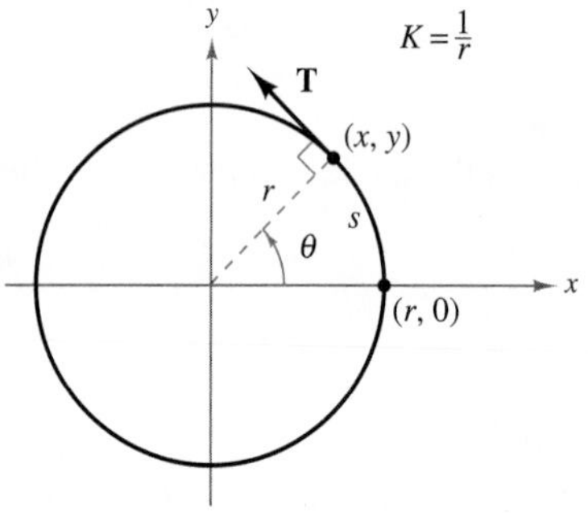

The curvature of a circle is constant.
Figure 12.33

Curvature

An important use of the arc length parameter is to find **curvature**—the measure of how sharply a curve bends. For instance, in Figure 12.31 the curve bends more sharply at P than at Q, and you can say that the curvature is greater at P than at Q. You can calculate curvature by calculating the magnitude of the rate of change of the unit tangent vector $\mathbf{T}$ with respect to the arc length s, as shown in Figure 12.32.

Definition of Curvature

Let C be a smooth curve (in the plane *or* in space) given by $\mathbf{r}(s)$, where s is the arc length parameter. The **curvature** K at s is given by

$$K = \left\| \frac{d\mathbf{T}}{ds} \right\| = \|\mathbf{T}'(s)\|.$$

A circle has the same curvature at any point. Moreover, the curvature and the radius of the circle are inversely related. That is, a circle with a large radius has a small curvature, and a circle with a small radius has a large curvature. This inverse relationship is made explicit in the following example.

EXAMPLE 4 Finding the Curvature of a Circle

Show that the curvature of a circle of radius r is $K = 1/r$.

Solution Without loss of generality you can consider the circle to be centered at the origin. Let (x, y) be any point on the circle and let s be the length of the arc from $(r, 0)$ to (x, y), as shown in Figure 12.33. By letting θ be the central angle of the circle, you can represent the circle by

$$\mathbf{r}(\theta) = r\cos\theta\mathbf{i} + r\sin\theta\mathbf{j}. \qquad \theta \text{ is the parameter.}$$

Using the formula for the length of a circular arc $s = r\theta$, you can rewrite $\mathbf{r}(\theta)$ in terms of the arc length parameter as follows.

$$\mathbf{r}(s) = r\cos\frac{s}{r}\mathbf{i} + r\sin\frac{s}{r}\mathbf{j} \qquad \text{Arc length } s \text{ is the parameter.}$$

So, $\mathbf{r}'(s) = -\sin\frac{s}{r}\mathbf{i} + \cos\frac{s}{r}\mathbf{j}$, and it follows that $\|\mathbf{r}'(s)\| = 1$, which implies that the unit tangent vector is

$$\mathbf{T}(s) = \frac{\mathbf{r}'(s)}{\|\mathbf{r}'(s)\|} = -\sin\frac{s}{r}\mathbf{i} + \cos\frac{s}{r}\mathbf{j}$$

and the curvature is given by

$$K = \|\mathbf{T}'(s)\| = \left\| -\frac{1}{r}\cos\frac{s}{r}\mathbf{i} - \frac{1}{r}\sin\frac{s}{r}\mathbf{j} \right\| = \frac{1}{r}$$

at every point on the circle.

NOTE Because a straight line doesn't curve, you would expect its curvature to be 0. Try checking this by finding the curvature of the line given by

$$\mathbf{r}(s) = \left(3 - \frac{3}{5}s\right)\mathbf{i} + \frac{4}{5}s\mathbf{j}.$$

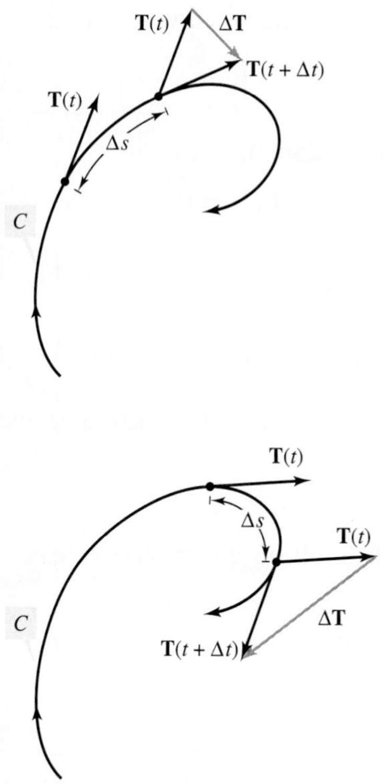

Figure 12.34

In Example 4, the curvature was found by applying the definition directly. This requires that the curve be written in terms of the arc length parameter s. The following theorem gives two other formulas for finding the curvature of a curve written in terms of an arbitrary parameter t. The proof of this theorem is left as an exercise [see Exercise 88, parts (a) and (b)].

THEOREM 12.8 Formulas for Curvature

If C is a smooth curve given by $\mathbf{r}(t)$, then the curvature K of C at t is given by

$$K = \frac{\|\mathbf{T}'(t)\|}{\|\mathbf{r}'(t)\|} = \frac{\|\mathbf{r}'(t) \times \mathbf{r}''(t)\|}{\|\mathbf{r}'(t)\|^3}.$$

Because $\|\mathbf{r}'(t)\| = ds/dt$, the first formula implies that curvature is the ratio of the rate of change in the tangent vector $\mathbf{T}$ to the rate of change in arc length. To see that this is reasonable, let Δt be a "small number." Then,

$$\frac{\mathbf{T}'(t)}{ds/dt} \approx \frac{[\mathbf{T}(t + \Delta t) - \mathbf{T}(t)]/\Delta t}{[s(t + \Delta t) - s(t)]/\Delta t} = \frac{\mathbf{T}(t + \Delta t) - \mathbf{T}(t)}{s(t + \Delta t) - s(t)} = \frac{\Delta \mathbf{T}}{\Delta s}.$$

In other words, for a given Δs, the greater the length of $\Delta \mathbf{T}$, the more the curve bends at t, as shown in Figure 12.34.

EXAMPLE 5 Finding the Curvature of a Space Curve

Find the curvature of the curve given by $\mathbf{r}(t) = 2t\mathbf{i} + t^2\mathbf{j} - \frac{1}{3}t^3\mathbf{k}$.

Solution It is not apparent whether this parameter represents arc length, so you should use the formula $K = \|\mathbf{T}'(t)\|/\|\mathbf{r}'(t)\|$.

$$\mathbf{r}'(t) = 2\mathbf{i} + 2t\mathbf{j} - t^2\mathbf{k}$$

$$\|\mathbf{r}'(t)\| = \sqrt{4 + 4t^2 + t^4} = t^2 + 2 \qquad \text{Length of } \mathbf{r}'(t)$$

$$\mathbf{T}(t) = \frac{\mathbf{r}'(t)}{\|\mathbf{r}'(t)\|} = \frac{2\mathbf{i} + 2t\mathbf{j} - t^2\mathbf{k}}{t^2 + 2}$$

$$\mathbf{T}'(t) = \frac{(t^2 + 2)(2\mathbf{j} - 2t\mathbf{k}) - (2t)(2\mathbf{i} + 2t\mathbf{j} - t^2\mathbf{k})}{(t^2 + 2)^2}$$

$$= \frac{-4t\mathbf{i} + (4 - 2t^2)\mathbf{j} - 4t\mathbf{k}}{(t^2 + 2)^2}$$

$$\|\mathbf{T}'(t)\| = \frac{\sqrt{16t^2 + 16 - 16t^2 + 4t^4 + 16t^2}}{(t^2 + 2)^2}$$

$$= \frac{2(t^2 + 2)}{(t^2 + 2)^2}$$

$$= \frac{2}{t^2 + 2} \qquad \text{Length of } \mathbf{T}'(t)$$

Therefore,

$$K = \frac{\|\mathbf{T}'(t)\|}{\|\mathbf{r}'(t)\|} = \frac{2}{(t^2 + 2)^2}. \qquad \text{Curvature}$$

The following theorem presents a formula for calculating the curvature of a plane curve given by $y = f(x)$.

THEOREM 12.9 Curvature in Rectangular Coordinates

If C is the graph of a twice-differentiable function given by $y = f(x)$, then the curvature K at the point (x, y) is given by

$$K = \frac{|y''|}{[1 + (y')^2]^{3/2}}.$$

Proof By representing the curve C by $\mathbf{r}(x) = x\mathbf{i} + f(x)\mathbf{j} + 0\mathbf{k}$ (where x is the parameter), you obtain $\mathbf{r}'(x) = \mathbf{i} + f'(x)\mathbf{j}$,

$$\|\mathbf{r}'(x)\| = \sqrt{1 + [f'(x)]^2}$$

and $\mathbf{r}''(x) = f''(x)\mathbf{j}$. Because $\mathbf{r}'(x) \times \mathbf{r}''(x) = f''(x)\mathbf{k}$, it follows that the curvature is

$$\begin{aligned} K &= \frac{\|\mathbf{r}'(x) \times \mathbf{r}''(x)\|}{\|\mathbf{r}'(x)\|^3} \\ &= \frac{|f''(x)|}{\{1 + [f'(x)]^2\}^{3/2}} \\ &= \frac{|y''|}{[1 + (y')^2]^{3/2}}. \end{aligned}$$

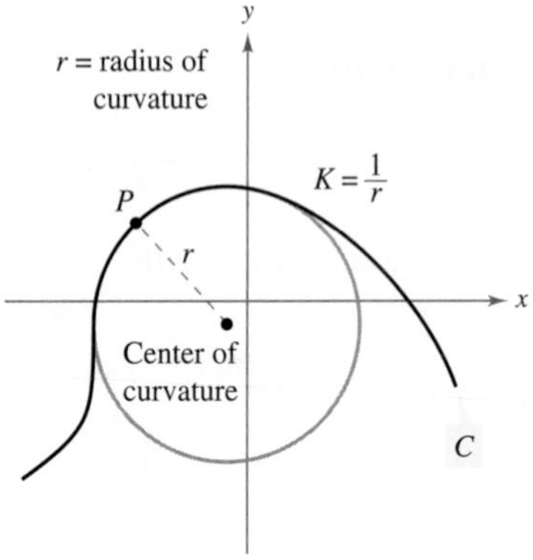

The circle of curvature
Figure 12.35

Let C be a curve with curvature K at point P. The circle passing through point P with radius $r = 1/K$ is called the **circle of curvature** if the circle lies on the concave side of the curve and shares a common tangent line with the curve at point P. The radius is called the **radius of curvature** at P, and the center of the circle is called the **center of curvature.**

The circle of curvature gives you a nice way to estimate graphically the curvature K at a point P on a curve. Using a compass, you can sketch a circle that lies against the concave side of the curve at point P, as shown in Figure 12.35. If the circle has a radius of r, you can estimate the curvature to be $K = 1/r$.

EXAMPLE 6 Finding Curvature in Rectangular Coordinates

Find the curvature of the parabola given by $y = x - \frac{1}{4}x^2$ at $x = 2$. Sketch the circle of curvature at $(2, 1)$.

Solution The curvature at $x = 2$ is as follows.

$$y' = 1 - \frac{x}{2} \qquad y' = 0$$

$$y'' = -\frac{1}{2} \qquad y'' = -\frac{1}{2}$$

$$K = \frac{|y''|}{[1 + (y')^2]^{3/2}} \qquad K = \frac{1}{2}$$

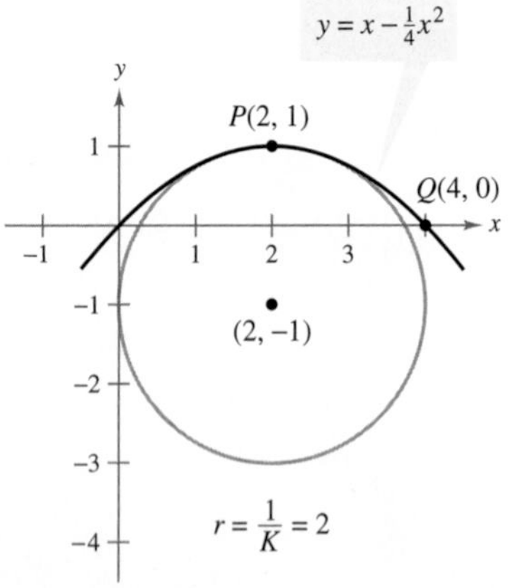

The circle of curvature
Figure 12.36

Because the curvature at $P(2, 1)$ is $\frac{1}{2}$, it follows that the radius of the circle of curvature at that point is 2. So, the center of curvature is $(2, -1)$, as shown in Figure 12.36. [In the figure, note that the curve has the greatest curvature at P. Try showing that the curvature at $Q(4, 0)$ is $1/2^{5/2} \approx 0.177$.]

The amount of thrust felt by passengers in a car that is turning depends on two things—the speed of the car and the sharpness of the turn.

Figure 12.37

Arc length and curvature are closely related to the tangential and normal components of acceleration. The tangential component of acceleration is the rate of change of the speed, which in turn is the rate of change of the arc length. This component is negative as a moving object slows down and positive as it speeds up—regardless of whether the object is turning or traveling in a straight line. So, the tangential component is solely a function of the arc length and is independent of the curvature.

On the other hand, the normal component of acceleration is a function of *both* speed and curvature. This component measures the acceleration acting perpendicular to the direction of motion. To see why the normal component is affected by both speed and curvature, imagine that you are driving a car around a turn, as shown in Figure 12.37. If your speed is high and the turn is sharp, you feel yourself thrown against the car door. By lowering your speed *or* taking a more gentle turn, you are able to lessen this sideways thrust.

The next theorem explicitly states the relationships among speed, curvature, and the components of acceleration.

NOTE Note that Theorem 12.10 gives additional formulas for $a_{\mathbf{T}}$ and $a_{\mathbf{N}}$.

THEOREM 12.10 Acceleration, Speed, and Curvature

If $\mathbf{r}(t)$ is the position vector for a smooth curve C, then the acceleration vector is given by

$$\mathbf{a}(t) = \frac{d^2s}{dt^2}\mathbf{T} + K\left(\frac{ds}{dt}\right)^2\mathbf{N}$$

where K is the curvature of C and ds/dt is the speed.

Proof For the position vector $\mathbf{r}(t)$, you have

$$\begin{aligned}\mathbf{a}(t) &= a_{\mathbf{T}}\mathbf{T} + a_{\mathbf{N}}\mathbf{N}\\ &= D_t[\|\mathbf{v}\|]\mathbf{T} + \|\mathbf{v}\|\,\|\mathbf{T}'\|\mathbf{N}\\ &= \frac{d^2s}{dt^2}\mathbf{T} + \frac{ds}{dt}(\|\mathbf{v}\|K)\mathbf{N}\\ &= \frac{d^2s}{dt^2}\mathbf{T} + K\left(\frac{ds}{dt}\right)^2\mathbf{N}.\end{aligned}$$

EXAMPLE 7 Tangential and Normal Components of Acceleration

Find $a_{\mathbf{T}}$ and $a_{\mathbf{N}}$ for the curve given by

$$\mathbf{r}(t) = 2t\mathbf{i} + t^2\mathbf{j} - \tfrac{1}{3}t^3\mathbf{k}.$$

Solution From Example 5, you know that

$$\frac{ds}{dt} = \|\mathbf{r}'(t)\| = t^2 + 2 \quad \text{and} \quad K = \frac{2}{(t^2+2)^2}.$$

Therefore,

$$a_{\mathbf{T}} = \frac{d^2s}{dt^2} = 2t \qquad \text{Tangential component}$$

and

$$a_{\mathbf{N}} = K\left(\frac{ds}{dt}\right)^2 = \frac{2}{(t^2+2)^2}(t^2+2)^2 = 2. \qquad \text{Normal component}$$

Application

There are many applications in physics and engineering dynamics that involve the relationships among speed, arc length, curvature, and acceleration. One such application concerns frictional force.

A moving object with mass m is in contact with a stationary object. The total force required to produce an acceleration **a** along a given path is

$$\begin{aligned}\mathbf{F} = m\mathbf{a} &= m\left(\frac{d^2s}{dt^2}\right)\mathbf{T} + mK\left(\frac{ds}{dt}\right)^2\mathbf{N} \\ &= ma_\mathbf{T}\mathbf{T} + ma_\mathbf{N}\mathbf{N}.\end{aligned}$$

The portion of this total force that is supplied by the stationary object is called the **force of friction.** For example, if a car moving with constant speed is rounding a turn, the roadway exerts a frictional force that keeps the car from sliding off the road. If the car is not sliding, the frictional force is perpendicular to the direction of motion and has magnitude equal to the normal component of acceleration, as shown in Figure 12.38. The potential frictional force of a road around a turn can be increased by banking the roadway.

The force of friction is perpendicular to the direction of the motion.
Figure 12.38

EXAMPLE 8 Frictional Force

A 360-kilogram go-cart is driven at a speed of 60 kilometers per hour around a circular racetrack of radius 12 meters, as shown in Figure 12.39. To keep the cart from skidding off course, what frictional force must the track surface exert on the tires?

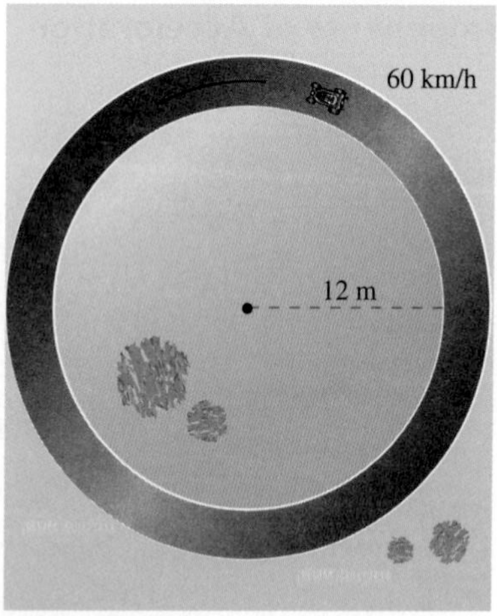

Figure 12.39

Solution The frictional force must equal the mass times the normal component of acceleration. For this circular path, you know that the curvature is

$$K = \frac{1}{12}. \qquad \text{Curvature of circular racetrack}$$

Therefore, the frictional force is

$$\begin{aligned}ma_\mathbf{N} &= mK\left(\frac{ds}{dt}\right)^2 \\ &= (360 \text{ kg})\left(\frac{1}{12 \text{ m}}\right)\left(\frac{60{,}000 \text{ m}}{3600 \text{ sec}}\right)^2 \\ &\approx 8333 \text{ (kg)(m)/sec}^2.\end{aligned}$$

Summary of Velocity, Acceleration, and Curvature

Let C be a curve (in the plane or in space) given by the position function

$\mathbf{r}(t) = x(t)\mathbf{i} + y(t)\mathbf{j}$ — Curve in the plane

$\mathbf{r}(t) = x(t)\mathbf{i} + y(t)\mathbf{j} + z(t)\mathbf{k}.$ — Curve in space

Velocity vector, speed, and acceleration vector:

$\mathbf{v}(t) = \mathbf{r}'(t)$ — Velocity vector

$\|\mathbf{v}(t)\| = \dfrac{ds}{dt} = \|\mathbf{r}'(t)\|$ — Speed

$\mathbf{a}(t) = \mathbf{r}''(t) = a_{\mathbf{T}}\mathbf{T}(t) + a_{\mathbf{N}}\mathbf{N}(t)$ — Acceleration vector

Unit tangent vector and principal unit normal vector:

$$\mathbf{T}(t) = \frac{\mathbf{r}'(t)}{\|\mathbf{r}'(t)\|} \quad \text{and} \quad \mathbf{N}(t) = \frac{\mathbf{T}'(t)}{\|\mathbf{T}'(t)\|}$$

Components of acceleration:

$$a_{\mathbf{T}} = \mathbf{a} \cdot \mathbf{T} = \frac{\mathbf{v} \cdot \mathbf{a}}{\|\mathbf{v}\|} = \frac{d^2s}{dt^2}$$

$$a_{\mathbf{N}} = \mathbf{a} \cdot \mathbf{N} = \frac{\|\mathbf{v} \times \mathbf{a}\|}{\|\mathbf{v}\|} = \sqrt{\|\mathbf{a}\|^2 - a_{\mathbf{T}}^2} = K\left(\frac{ds}{dt}\right)^2$$

Formulas for curvature in the plane:

$K = \dfrac{|y''|}{[1 + (y')^2]^{3/2}}$ — C given by $y = f(x)$

$K = \dfrac{|x'y'' - y'x''|}{[(x')^2 + (y')^2]^{3/2}}$ — C given by $x = x(t)$, $y = y(t)$

Formulas for curvature in the plane or in space:

$K = \|\mathbf{T}'(s)\| = \|\mathbf{r}''(s)\|$ — s is arc length parameter.

$K = \dfrac{\|\mathbf{T}'(t)\|}{\|\mathbf{r}'(t)\|} = \dfrac{\|\mathbf{r}'(t) \times \mathbf{r}''(t)\|}{\|\mathbf{r}'(t)\|^3}$ — t is general parameter.

$K = \dfrac{\mathbf{a}(t) \cdot \mathbf{N}(t)}{\|\mathbf{v}(t)\|^2}$

Cross product formulas apply only to curves in space.

Exercises for Section 12.5

See www.CalcChat.com for worked-out solutions to odd-numbered exercises.

In Exercises 1–6, sketch the plane curve and find its length over the given interval.

	Function	*Interval*
1.	$\mathbf{r}(t) = t\mathbf{i} + 3t\mathbf{j}$	$[0, 4]$
2.	$\mathbf{r}(t) = t\mathbf{i} + t^2\mathbf{k}$	$[0, 4]$
3.	$\mathbf{r}(t) = t^3\mathbf{i} + t^2\mathbf{j}$	$[0, 2]$
4.	$\mathbf{r}(t) = (t + 1)\mathbf{i} + t^2\mathbf{j}$	$[0, 6]$
5.	$\mathbf{r}(t) = a\cos^3 t\,\mathbf{i} + a\sin^3 t\,\mathbf{j}$	$[0, 2\pi]$
6.	$\mathbf{r}(t) = a\cos t\,\mathbf{i} + a\sin t\,\mathbf{j}$	$[0, 2\pi]$

7. ***Projectile Motion*** A baseball is hit 3 feet above the ground at 100 feet per second and at an angle of 45° with respect to the ground.
   (a) Find the vector-valued function for the path of the baseball.
   (b) Find the maximum height.
   (c) Find the range.
   (d) Find the arc length of the trajectory.

8. ***Projectile Motion*** An object is launched from ground level. Determine the angle of the launch to obtain (a) the maximum height, (b) the maximum range, and (c) the maximum length of the trajectory. For part (c), let $v_0 = 96$ feet per second.

In Exercises 9–14, sketch the space curve and find its length over the given interval.

	Function	*Interval*
9.	$\mathbf{r}(t) = 2t\mathbf{i} - 3t\mathbf{j} + t\mathbf{k}$	$[0, 2]$
10.	$\mathbf{r}(t) = \mathbf{i} + t^2\mathbf{j} + t^3\mathbf{k}$	$[0, 2]$
11.	$\mathbf{r}(t) = \langle 3t, 2\cos t, 2\sin t\rangle$	$\left[0, \frac{\pi}{2}\right]$
12.	$\mathbf{r}(t) = \langle 2\sin t, 5t, 2\cos t\rangle$	$[0, \pi]$
13.	$\mathbf{r}(t) = a\cos t\,\mathbf{i} + a\sin t\,\mathbf{j} + bt\,\mathbf{k}$	$[0, 2\pi]$
14.	$\mathbf{r}(t) = \langle \cos t + t\sin t, \sin t - t\cos t, t^2\rangle$	$\left[0, \frac{\pi}{2}\right]$

In Exercises 15 and 16, use the integration capabilities of a graphing utility to approximate the length of the space curve over the given interval.

Function	*Interval*
15. $\mathbf{r}(t) = t^2\mathbf{i} + t\mathbf{j} + \ln t\mathbf{k}$	$1 \leq t \leq 3$
16. $\mathbf{r}(t) = \sin \pi t\mathbf{i} + \cos \pi t\mathbf{j} + t^3\mathbf{k}$	$0 \leq t \leq 2$

17. ***Investigation*** Consider the graph of the vector-valued function

$$\mathbf{r}(t) = t\mathbf{i} + (4 - t^2)\mathbf{j} + t^3\mathbf{k}$$

on the interval $[0, 2]$.

(a) Approximate the length of the curve by finding the length of the line segment connecting its endpoints.

(b) Approximate the length of the curve by summing the lengths of the line segments connecting the terminal points of the vectors $\mathbf{r}(0)$, $\mathbf{r}(0.5)$, $\mathbf{r}(1)$, $\mathbf{r}(1.5)$, and $\mathbf{r}(2)$.

(c) Describe how you could obtain a more accurate approximation by continuing the processes in parts (a) and (b).

(d) Use the integration capabilities of a graphing utility to approximate the length of the curve. Compare this result with the answers in parts (a) and (b).

18. ***Investigation*** Repeat Exercise 17 for the vector-valued function $\mathbf{r}(t) = 6\cos(\pi t/4)\mathbf{i} + 2\sin(\pi t/4)\mathbf{j} + t\mathbf{k}$.

19. ***Investigation*** Consider the helix represented by the vector-valued function $\mathbf{r}(t) = \langle 2\cos t, 2\sin t, t\rangle$.

(a) Write the length of the arc s on the helix as a function of t by evaluating the integral

$$s = \int_0^t \sqrt{[x'(u)]^2 + [y'(u)]^2 + [z'(u)]^2}\, du.$$

(b) Solve for t in the relationship derived in part (a), and substitute the result into the original set of parametric equations. This yields a parametrization of the curve in terms of the arc length parameter s.

(c) Find the coordinates of the point on the helix for arc lengths $s = \sqrt{5}$ and $s = 4$.

(d) Verify that $\|\mathbf{r}'(s)\| = 1$.

20. ***Investigation*** Repeat Exercise 19 for the curve represented by the vector-valued function

$$\mathbf{r}(t) = \left\langle 4(\sin t - t\cos t), 4(\cos t + t\sin t), \tfrac{3}{2}t^2\right\rangle.$$

In Exercises 21–24, find the curvature K of the curve, where s is the arc length parameter.

21. $\mathbf{r}(s) = \left(1 + \dfrac{\sqrt{2}}{2}s\right)\mathbf{i} + \left(1 - \dfrac{\sqrt{2}}{2}s\right)\mathbf{j}$

22. $\mathbf{r}(s) = (3 + s)\mathbf{i} + \mathbf{j}$

23. Helix in Exercise 19: $\mathbf{r}(t) = \langle 2\cos t, 2\sin t, t\rangle$

24. Curve in Exercise 20:
$\mathbf{r}(t) = \left\langle 4(\sin t - t\cos t), 4(\cos t + t\sin t), \tfrac{3}{2}t^2\right\rangle$

In Exercises 25–30, find the curvature K of the plane curve at the given value of the parameter.

25. $\mathbf{r}(t) = 4t\mathbf{i} - 2t\mathbf{j}, \quad t = 1$

26. $\mathbf{r}(t) = t^2\mathbf{j} + \mathbf{k}, \quad t = 0$

27. $\mathbf{r}(t) = t\mathbf{i} + \dfrac{1}{t}\mathbf{j}, \quad t = 1$

28. $\mathbf{r}(t) = t\mathbf{i} + t^2\mathbf{j}, \quad t = 1$

29. $\mathbf{r}(t) = t\mathbf{i} + \cos t\mathbf{j}, \quad t = 0$

30. $\mathbf{r}(t) = 5\cos t\mathbf{i} + 4\sin t\mathbf{j}, \quad t = \dfrac{\pi}{3}$

In Exercises 31–40, find the curvature K of the curve.

31. $\mathbf{r}(t) = 4\cos 2\pi t\mathbf{i} + 4\sin 2\pi t\mathbf{j}$

32. $\mathbf{r}(t) = 2\cos \pi t\mathbf{i} + \sin \pi t\mathbf{j}$

33. $\mathbf{r}(t) = a\cos \omega t\mathbf{i} + a\sin \omega t\mathbf{j}$

34. $\mathbf{r}(t) = a\cos \omega t\mathbf{i} + b\sin \omega t\mathbf{j}$

35. $\mathbf{r}(t) = \langle a(\omega t - \sin \omega t), a(1 - \cos \omega t)\rangle$

36. $\mathbf{r}(t) = \langle \cos \omega t + \omega t\sin \omega t, \sin \omega t - \omega t\cos \omega t\rangle$

37. $\mathbf{r}(t) = t\mathbf{i} + t^2\mathbf{j} + \dfrac{t^2}{2}\mathbf{k}$

38. $\mathbf{r}(t) = 2t^2\mathbf{i} + t\mathbf{j} + \dfrac{1}{2}t^2\mathbf{k}$

39. $\mathbf{r}(t) = 4t\mathbf{i} + 3\cos t\mathbf{j} + 3\sin t\mathbf{k}$

40. $\mathbf{r}(t) = e^t\cos t\mathbf{i} + e^t\sin t\mathbf{j} + e^t\mathbf{k}$

In Exercises 41–46, find the curvature and radius of curvature of the plane curve at the given value of x.

41. $y = 3x - 2, \quad x = a$

42. $y = mx + b, \quad x = a$

43. $y = 2x^2 + 3, \quad x = -1$

44. $y = 2x + \dfrac{4}{x}, \quad x = 1$

45. $y = \sqrt{a^2 - x^2}, \quad x = 0$

46. $y = \tfrac{3}{4}\sqrt{16 - x^2}, \quad x = 0$

Writing **In Exercises 47 and 48, two circles of curvature to the graph of the function are given. (a) Find the equation of the smaller circle, and (b) write a short paragraph explaining why the circles have different radii.**

47. $f(x) = \sin x$

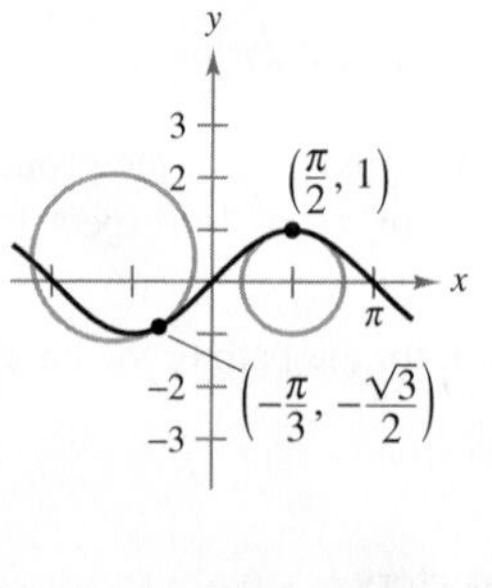

48. $f(x) = 4x^2/(x^2 + 3)$

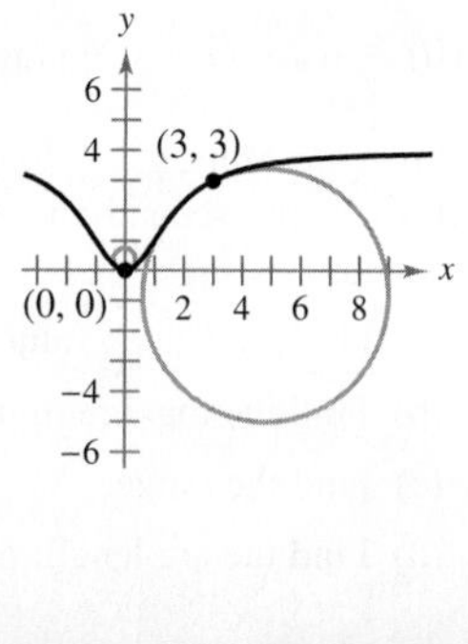

In Exercises 49–52, use a graphing utility to graph the function. In the same viewing window, graph the circle of curvature to the graph at the given value of x.

49. $y = x + \frac{1}{x}$, $x = 1$

50. $y = \ln x$, $x = 1$

51. $y = e^x$, $x = 0$

52. $y = \frac{1}{3}x^3$, $x = 1$

Evolute **An evolute is the curve formed by the set of centers of curvature of a curve. In Exercises 53 and 54, a curve and its evolute are given. Use a compass to sketch the circles of curvature with centers at points A and B. To print an enlarged copy of the graph, go to the website *www.mathgraphs.com*.**

53. Cycloid: $x = t - \sin t$
$y = 1 - \cos t$
Evolute: $x = \sin t + t$
$y = \cos t - 1$

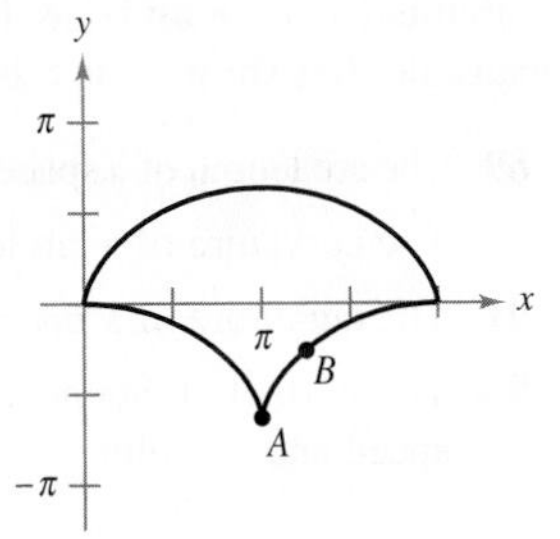

54. Ellipse: $x = 3\cos t$
$y = 2\sin t$
Evolute: $x = \frac{5}{3}\cos^3 t$
$y = \frac{5}{2}\sin^3 t$

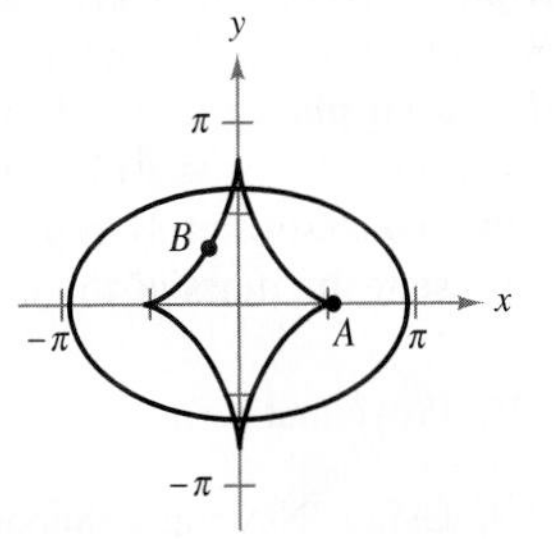

In Exercises 55–60, (a) find the point on the curve at which the curvature K is a maximum and (b) find the limit of K as $x \to \infty$.

55. $y = (x - 1)^2 + 3$

56. $y = x^3$

57. $y = x^{2/3}$

58. $y = \frac{1}{x}$

59. $y = \ln x$

60. $y = e^x$

In Exercises 61–64, find all points on the graph of the function such that the curvature is zero.

61. $y = 1 - x^3$

62. $y = (x - 1)^3 + 3$

63. $y = \cos x$

64. $y = \sin x$

Writing About Concepts

65. Describe the graph of a vector-valued function for which the curvature is 0 for all values of t in its domain.

66. Given a twice-differentiable function $y = f(x)$, determine its curvature at a relative extremum. Can the curvature ever be greater than it is at a relative extremum? Why or why not?

67. Show that the curvature is greatest at the endpoints of the major axis, and is least at the endpoints of the minor axis, for the ellipse given by $x^2 + 4y^2 = 4$.

68. ***Investigation*** Find all a and b such that the two curves given by

$$y_1 = ax(b - x) \quad \text{and} \quad y_2 = \frac{x}{x + 2}$$

intersect at only one point and have a common tangent line and equal curvature at that point. Sketch a graph for each set of values of a and b.

69. ***Investigation*** Consider the function $f(x) = x^4 - x^2$.

(a) Use a computer algebra system to find the curvature K of the curve as a function of x.

(b) Use the result of part (a) to find the circles of curvature to the graph of f when $x = 0$ and $x = 1$. Use a computer algebra system to graph the function and the two circles of curvature.

(c) Graph the function $K(x)$ and compare it with the graph of $f(x)$. For example, do the extrema of f and K occur at the same critical numbers? Explain your reasoning.

70. ***Investigation*** The surface of a goblet is formed by revolving the graph of the function

$$y = \frac{1}{4}x^{8/5}, \quad 0 \le x \le 5$$

about the y-axis. The measurements are given in centimeters.

(a) Use a computer algebra system to graph the surface.

(b) Find the volume of the goblet.

(c) Find the curvature K of the generating curve as a function of x. Use a graphing utility to graph K.

(d) If a spherical object is dropped into the goblet, is it possible for it to touch the bottom? Explain.

71. A sphere of radius 4 is dropped into the paraboloid given by $z = x^2 + y^2$.

(a) How close will the sphere come to the vertex of the paraboloid?

(b) What is the radius of the largest sphere that will touch the vertex?

72. ***Speed*** The smaller the curvature in a bend of a road, the faster a car can travel. Assume that the maximum speed around a turn is inversely proportional to the square root of the curvature. A car moving on the path $y = \frac{1}{3}x^3$ (x and y are measured in miles) can safely go 30 miles per hour at $\left(1, \frac{1}{3}\right)$. How fast can it go at $\left(\frac{3}{2}, \frac{9}{8}\right)$?

73. Let C be a curve given by $y = f(x)$. Let K be the curvature ($K \neq 0$) at the point $P(x_0, y_0)$ and let

$$z = \frac{1 + f'(x_0)^2}{f''(x_0)}.$$

Show that the coordinates (α, β) of the center of curvature at P are $(\alpha, \beta) = (x_0 - f'(x_0)z, y_0 + z)$.

74. Use the result of Exercise 73 to find the center of curvature for the curve at the given point.

(a) $y = e^x$, $(0, 1)$

(b) $y = \dfrac{x^2}{2}$, $\left(1, \dfrac{1}{2}\right)$

(c) $y = x^2$, $(0, 0)$

75. A curve C is given by the polar equation $r = f(\theta)$. Show that the curvature K at the point (r, θ) is

$$K = \frac{[2(r')^2 - rr'' + r^2]}{[(r')^2 + r^2]^{3/2}}.$$

[*Hint:* Represent the curve by $\mathbf{r}(\theta) = r\cos\theta\mathbf{i} + r\sin\theta\mathbf{j}$.]

76. Use the result of Exercise 75 to find the curvature of each polar curve.

(a) $r = 1 + \sin\theta$ (b) $r = \theta$

(c) $r = a\sin\theta$ (d) $r = e^\theta$

77. Given the polar curve $r = e^{a\theta}$, $a > 0$, find the curvature K and determine the limit of K as (a) $\theta \to \infty$ and (b) $a \to \infty$.

78. Show that the formula for the curvature of a polar curve $r = f(\theta)$ given in Exercise 75 reduces to $K = 2/|r'|$ for the curvature *at the pole*.

In Exercises 79 and 80, use the result of Exercise 78 to find the curvature of the rose curve at the pole.

79. $r = 4\sin 2\theta$ **80.** $r = 6\cos 3\theta$

81. For a smooth curve given by the parametric equations $x = f(t)$ and $y = g(t)$, prove that the curvature is given by

$$K = \frac{|f'(t)g''(t) - g'(t)f''(t)|}{\{[f'(t)]^2 + [g'(t)]^2\}^{3/2}}.$$

82. Use the result of Exercise 81 to find the curvature K of the curve represented by the parametric equations $x(t) = t^3$ and $y(t) = \frac{1}{2}t^2$. Use a graphing utility to graph K and determine any horizontal asymptotes. Interpret the asymptotes in the context of the problem.

83. Use the result of Exercise 81 to find the curvature K of the cycloid represented by the parametric equations

$$x(\theta) = a(\theta - \sin\theta) \quad \text{and} \quad y(\theta) = a(1 - \cos\theta).$$

What are the minimum and maximum values of K?

84. Use Theorem 12.10 to find $a_\mathbf{T}$ and $a_\mathbf{N}$ for each curve given by the vector-valued function.

(a) $\mathbf{r}(t) = 3t^2\mathbf{i} + (3t - t^3)\mathbf{j}$ (b) $\mathbf{r}(t) = t\mathbf{i} + t^2\mathbf{j} + \frac{1}{2}t^2\mathbf{k}$

85. ***Frictional Force*** A 5500-pound vehicle is driven at a speed of 30 miles per hour on a circular interchange of radius 100 feet. To keep the vehicle from skidding off course, what frictional force must the road surface exert on the tires?

86. ***Frictional Force*** A 6400-pound vehicle is driven at a speed of 35 miles per hour on a circular interchange of radius 250 feet. To keep the vehicle from skidding off course, what frictional force must the road surface exert on the tires?

87. Verify that the curvature at any point (x, y) on the graph of $y = \cosh x$ is $1/y^2$.

88. Use the definition of curvature in space, $K = \|\mathbf{T}'(s)\| = \|\mathbf{r}''(s)\|$, to verify each formula.

(a) $K = \dfrac{\|\mathbf{T}'(t)\|}{\|\mathbf{r}'(t)\|}$

(b) $K = \dfrac{\|\mathbf{r}'(t) \times \mathbf{r}''(t)\|}{\|\mathbf{r}'(t)\|^3}$

(c) $K = \dfrac{\mathbf{a}(t) \cdot \mathbf{N}(t)}{\|\mathbf{v}(t)\|^2}$

True or False? **In Exercises 89–92, determine whether the statement is true or false. If it is false, explain why or give an example that shows it is false.**

89. The arc length of a space curve depends on the parametrization.

90. The curvature of a circle is the same as its radius.

91. The curvature of a line is 0.

92. The normal component of acceleration is a function of both speed and curvature.

Kepler's Laws **In Exercises 93–100, you are asked to verify Kepler's Laws of Planetary Motion. For these exercises, assume that each planet moves in an orbit given by the vector-valued function r. Let $r = \|\mathbf{r}\|$, let G represent the universal gravitational constant, let M represent the mass of the sun, and let m represent the mass of the planet.**

93. Prove that $\mathbf{r} \cdot \mathbf{r}' = r\dfrac{dr}{dt}$.

94. Using Newton's Second Law of Motion, $\mathbf{F} = m\mathbf{a}$, and Newton's Second Law of Gravitation, $\mathbf{F} = -(GmM/r^3)\mathbf{r}$, show that $\mathbf{a}$ and $\mathbf{r}$ are parallel, and that $\mathbf{r}(t) \times \mathbf{r}'(t) = \mathbf{L}$ is a constant vector. So, $\mathbf{r}(t)$ moves in a *fixed plane*, orthogonal to $\mathbf{L}$.

95. Prove that $\dfrac{d}{dt}\left[\dfrac{\mathbf{r}}{r}\right] = \dfrac{1}{r^3}\{[\mathbf{r} \times \mathbf{r}'] \times \mathbf{r}\}$.

96. Show that $\dfrac{\mathbf{r}'}{GM} \times \mathbf{L} - \dfrac{\mathbf{r}}{r} = \mathbf{e}$ is a constant vector.

97. Prove Kepler's First Law: Each planet moves in an elliptical orbit with the sun as a focus.

98. Assume that the elliptical orbit

$$r = \frac{ed}{1 + e\cos\theta}$$

is in the xy-plane, with $\mathbf{L}$ along the z-axis. Prove that

$$\|\mathbf{L}\| = r^2\frac{d\theta}{dt}.$$

99. Prove Kepler's Second Law: Each ray from the sun to a planet sweeps out equal areas of the ellipse in equal times.

100. Prove Kepler's Third Law: The square of the period of a planet's orbit is proportional to the cube of the mean distance between the planet and the sun.

Review Exercises for Chapter 12

See www.CalcChat.com for worked-out solutions to odd-numbered exercises.

In Exercises 1–4, (a) find the domain of r and (b) determine the values (if any) of t for which the function is continuous.

1. $\mathbf{r}(t) = t\mathbf{i} + \csc t\mathbf{k}$

2. $\mathbf{r}(t) = \sqrt{t}\,\mathbf{i} + \dfrac{1}{t-4}\mathbf{j} + \mathbf{k}$

3. $\mathbf{r}(t) = \ln t\mathbf{i} + t\mathbf{j} + t\mathbf{k}$

4. $\mathbf{r}(t) = (2t+1)\mathbf{i} + t^2\mathbf{j} + t\mathbf{k}$

In Exercises 5 and 6, evaluate (if possible) the vector-valued function at each given value of t.

5. $\mathbf{r}(t) = (2t+1)\mathbf{i} + t^2\mathbf{j} - \frac{1}{3}t^3\mathbf{k}$
   (a) $\mathbf{r}(0)$ (b) $\mathbf{r}(-2)$ (c) $\mathbf{r}(c-1)$ (d) $\mathbf{r}(1+\Delta t) - \mathbf{r}(1)$

6. $\mathbf{r}(t) = 3\cos t\mathbf{i} + (1 - \sin t)\mathbf{j} - t\mathbf{k}$
   (a) $\mathbf{r}(0)$ (b) $\mathbf{r}\left(\dfrac{\pi}{2}\right)$ (c) $\mathbf{r}(s-\pi)$ (d) $\mathbf{r}(\pi + \Delta t) - \mathbf{r}(\pi)$

In Exercises 7 and 8, sketch the plane curve represented by the vector-valued function and give the orientation of the curve.

7. $\mathbf{r}(t) = \langle \cos t, 2\sin^2 t\rangle$

8. $\mathbf{r}(t) = \langle t, t/(t-1)\rangle$

In Exercises 9–14, use a computer algebra system to graph the space curve represented by the vector-valued function.

9. $\mathbf{r}(t) = \mathbf{i} + t\mathbf{j} + t^2\mathbf{k}$

10. $\mathbf{r}(t) = 2t\mathbf{i} + t\mathbf{j} + t^2\mathbf{k}$

11. $\mathbf{r}(t) = \langle 1, \sin t, 1\rangle$

12. $\mathbf{r}(t) = \langle 2\cos t, t, 2\sin t\rangle$

13. $\mathbf{r}(t) = \left\langle t, \ln t, \frac{1}{2}t^2\right\rangle$

14. $\mathbf{r}(t) = \left\langle \frac{1}{2}t, \sqrt{t}, \frac{1}{4}t^3\right\rangle$

In Exercises 15 and 16, find vector-valued functions forming the boundaries of the region in the figure.

15.

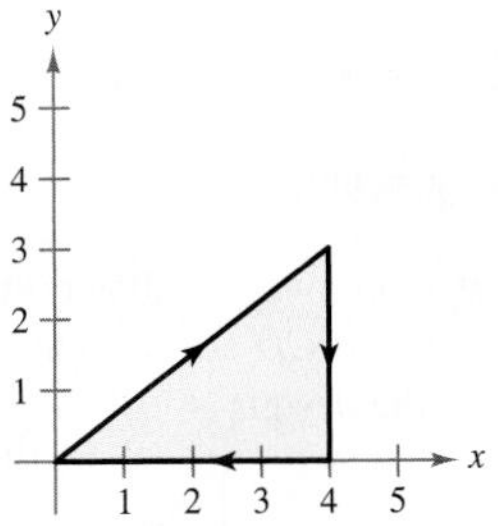

16.

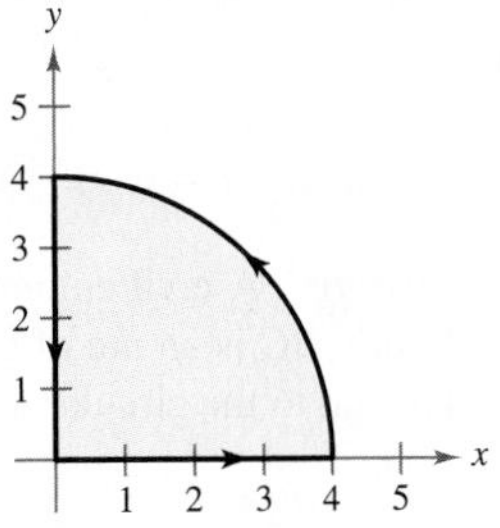

17. A particle moves on a straight-line path that passes through the points $(-2, -3, 8)$ and $(5, 1, -2)$. Find a vector-valued function for the path. (There are many correct answers.)

18. The outer edge of a spiral staircase is in the shape of a helix of radius 2 meters. The staircase has a height of 2 meters and is three-fourths of one complete revolution from bottom to top. Find a vector-valued function for the helix. (There are many correct answers.)

In Exercises 19 and 20, sketch the space curve represented by the intersection of the surfaces. Use the parameter $x = t$ to find a vector-valued function for the space curve.

19. $z = x^2 + y^2, \quad x + y = 0$

20. $x^2 + z^2 = 4, \quad x - y = 0$

In Exercises 21 and 22, evaluate the limit.

21. $\lim\limits_{t\to 2^-}\left(t^2\mathbf{i} + \sqrt{4-t^2}\,\mathbf{j} + \mathbf{k}\right)$

22. $\lim\limits_{t\to 0}\left(\dfrac{\sin 2t}{t}\mathbf{i} + e^{-t}\mathbf{j} + e^{t}\mathbf{k}\right)$

In Exercises 23 and 24, find the following.

(a) $\mathbf{r}'(t)$ (b) $\mathbf{r}''(t)$ (c) $D_t[\mathbf{r}(t)\cdot\mathbf{u}(t)]$

(d) $D_t[\mathbf{u}(t) - 2\mathbf{r}(t)]$ (e) $D_t[\|\mathbf{r}(t)\|],\ t > 0$ (f) $D_t[\mathbf{r}(t)\times\mathbf{u}(t)]$

23. $\mathbf{r}(t) = 3t\mathbf{i} + (t-1)\mathbf{j}, \quad \mathbf{u}(t) = t\mathbf{i} + t^2\mathbf{j} + \frac{2}{3}t^3\mathbf{k}$

24. $\mathbf{r}(t) = \sin t\mathbf{i} + \cos t\mathbf{j} + t\mathbf{k}, \quad \mathbf{u}(t) = \sin t\mathbf{i} + \cos t\mathbf{j} + \dfrac{1}{t}\mathbf{k}$

25. ***Writing*** The x- and y-components of the derivative of the vector-valued function **u** are positive at $t = t_0$, and the z-component is negative. Describe the behavior of **u** at $t = t_0$.

26. ***Writing*** The x-component of the derivative of the vector-valued function **u** is 0 for t in the domain of the function. What does this information imply about the graph of **u**?

In Exercises 27–30, find the indefinite integral.

27. $\displaystyle\int (\cos t\mathbf{i} + t\cos t\mathbf{j})\,dt$

28. $\displaystyle\int (\ln t\mathbf{i} + t\ln t\mathbf{j} + \mathbf{k})\,dt$

29. $\displaystyle\int \|\cos t\mathbf{i} + \sin t\mathbf{j} + t\mathbf{k}\|\,dt$

30. $\displaystyle\int (t\mathbf{j} + t^2\mathbf{k})\times(\mathbf{i} + t\mathbf{j} + t\mathbf{k})\,dt$

In Exercises 31 and 32, find r(t) for the given conditions.

31. $\mathbf{r}'(t) = 2t\mathbf{i} + e^t\mathbf{j} + e^{-t}\mathbf{k}, \quad \mathbf{r}(0) = \mathbf{i} + 3\mathbf{j} - 5\mathbf{k}$

32. $\mathbf{r}'(t) = \sec t\mathbf{i} + \tan t\mathbf{j} + t^2\mathbf{k}, \quad \mathbf{r}(0) = 3\mathbf{k}$

In Exercises 33–36, evaluate the definite integral.

33. $\displaystyle\int_{-2}^{2} (3t\mathbf{i} + 2t^2\mathbf{j} - t^3\mathbf{k})\,dt$

34. $\displaystyle\int_{0}^{1} \left(\sqrt{t}\,\mathbf{j} + t\sin t\mathbf{k}\right)dt$

35. $\displaystyle\int_{0}^{2} (e^{t/2}\mathbf{i} - 3t^2\mathbf{j} - \mathbf{k})\,dt$

36. $\displaystyle\int_{-1}^{1} (t^3\mathbf{i} + \arcsin t\mathbf{j} - t^2\mathbf{k})\,dt$

In Exercises 37 and 38, the position vector r describes the path of an object moving in space. Find the velocity, speed, and acceleration of the object.

37. $\mathbf{r}(t) = \langle \cos^3 t, \sin^3 t, 3t\rangle$

38. $\mathbf{r}(t) = \langle t, -\tan t, e^t\rangle$

Linear Approximation **In Exercises 39 and 40, find a set of parametric equations for the tangent line to the graph of the vector-valued function at $t = t_0$. Use the equations for the line to approximate r(t_0 + 0.1).**

39. $\mathbf{r}(t) = \ln(t-3)\mathbf{i} + t^2\mathbf{j} + \frac{1}{2}t\mathbf{k}, \quad t_0 = 4$

40. $\mathbf{r}(t) = 3\cosh t\mathbf{i} + \sinh t\mathbf{j} - 2t\mathbf{k}, \quad t_0 = 0$

Projectile Motion **In Exercises 41–44, use the model for projectile motion, assuming there is no air resistance. [$a(t) = -32$ feet per second per second or $a(t) = -9.8$ meters per second per second.]**

41. A projectile is fired from ground level with an initial velocity of 75 feet per second at an angle of 30° with the horizontal. Find the range of the projectile.

42. The center of a truck bed is 6 feet below and 4 feet horizontally from the end of a horizontal conveyor that is discharging gravel (see figure). Determine the speed ds/dt at which the conveyor belt should be moving so that the gravel falls onto the center of the truck bed.

43. A projectile is fired from ground level at an angle of 20° with the horizontal. The projectile has a range of 80 meters. Find the minimum initial velocity.

44. Use a graphing utility to graph the paths of a projectile if $v_0 = 20$ meters per second, $h = 0$ and (a) $\theta = 30°$, (b) $\theta = 45°$, and (c) $\theta = 60°$. Use the graphs to approximate the maximum height and range of the projectile for each case.

In Exercises 45–52, find the velocity, speed, and acceleration at time *t*. Then find a · T and a · N at time *t*.

45. $\mathbf{r}(t) = 5t\mathbf{i}$

46. $\mathbf{r}(t) = (1 + 4t)\mathbf{i} + (2 - 3t)\mathbf{j}$

47. $\mathbf{r}(t) = t\mathbf{i} + \sqrt{t}\mathbf{j}$

48. $\mathbf{r}(t) = 2(t + 1)\mathbf{i} + \dfrac{2}{t + 1}\mathbf{j}$

49. $\mathbf{r}(t) = e^t\mathbf{i} + e^{-t}\mathbf{j}$

50. $\mathbf{r}(t) = t\cos t\mathbf{i} + t\sin t\mathbf{j}$

51. $\mathbf{r}(t) = t\mathbf{i} + t^2\mathbf{j} + \dfrac{1}{2}t^2\mathbf{k}$

52. $\mathbf{r}(t) = (t - 1)\mathbf{i} + t\mathbf{j} + \dfrac{1}{t}\mathbf{k}$

In Exercises 53 and 54, find a set of parametric equations for the line tangent to the space curve at the given point.

53. $\mathbf{r}(t) = 2\cos t\mathbf{i} + 2\sin t\mathbf{j} + t\mathbf{k}, \quad t = \dfrac{3\pi}{4}$

54. $\mathbf{r}(t) = t\mathbf{i} + t^2\mathbf{j} + \frac{2}{3}t^3\mathbf{k}, \quad t = 2$

55. ***Satellite Orbit*** Find the speed necessary for a satellite to maintain a circular orbit 600 miles above the surface of Earth.

56. ***Centripetal Force*** An automobile in a circular traffic exchange is traveling at twice the posted speed. By what factor is the centripetal force increased over that which would occur at the posted speed?

In Exercises 57–60, sketch the plane curve and find its length over the given interval.

	Function	*Interval*
57.	$\mathbf{r}(t) = 2t\mathbf{i} - 3t\mathbf{j}$	$[0, 5]$
58.	$\mathbf{r}(t) = t^2\mathbf{i} + 2t\mathbf{k}$	$[0, 3]$
59.	$\mathbf{r}(t) = 10\cos^3 t\mathbf{i} + 10\sin^3 t\mathbf{j}$	$[0, 2\pi]$
60.	$\mathbf{r}(t) = 10\cos t\mathbf{i} + 10\sin t\mathbf{j}$	$[0, 2\pi]$

In Exercises 61–64, sketch the space curve and find its length over the given interval.

	Function	*Interval*
61.	$\mathbf{r}(t) = -3t\mathbf{i} + 2t\mathbf{j} + 4t\mathbf{k}$	$[0, 3]$
62.	$\mathbf{r}(t) = t\mathbf{i} + t^2\mathbf{j} + 2t\mathbf{k}$	$[0, 2]$
63.	$\mathbf{r}(t) = \langle 8\cos t, 8\sin t, t\rangle$	$[0, \pi/2]$
64.	$\mathbf{r}(t) = \langle 2(\sin t - t\cos t), 2(\cos t + t\sin t), t\rangle$	$[0, \pi/2]$

In Exercises 65 and 66, use a computer algebra system to find the length of the space curve over the given interval.

65. $\mathbf{r}(t) = \frac{1}{2}t\mathbf{i} + \sin t\mathbf{j} + \cos t\mathbf{k}, \quad 0 \le t \le \pi$

66. $\mathbf{r}(t) = e^t\sin t\mathbf{i} + e^t\cos t\mathbf{k}, \quad 0 \le t \le \pi$

In Exercises 67–70, find the curvature *K* of the curve.

67. $\mathbf{r}(t) = 3t\mathbf{i} + 2t\mathbf{j}$

68. $\mathbf{r}(t) = 2\sqrt{t}\mathbf{i} + 3t\mathbf{j}$

69. $\mathbf{r}(t) = 2t\mathbf{i} + \frac{1}{2}t^2\mathbf{j} + t^2\mathbf{k}$

70. $\mathbf{r}(t) = 2t\mathbf{i} + 5\cos t\mathbf{j} + 5\sin t\mathbf{k}$

In Exercises 71–74, find the curvature and radius of curvature of the plane curve at the given value of *x*.

71. $y = \frac{1}{2}x^2 + 2, \quad x = 4$

72. $y = e^{-x/2}, \quad x = 0$

73. $y = \ln x, \quad x = 1$

74. $y = \tan x, \quad x = \dfrac{\pi}{4}$

75. ***Writing*** A civil engineer designs a highway as shown in the figure. *BC* is an arc of the circle. *AB* and *CD* are straight lines tangent to the circular arc. Criticize the design.

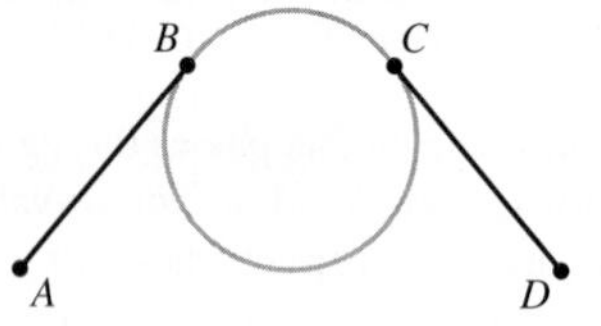

Figure for 75

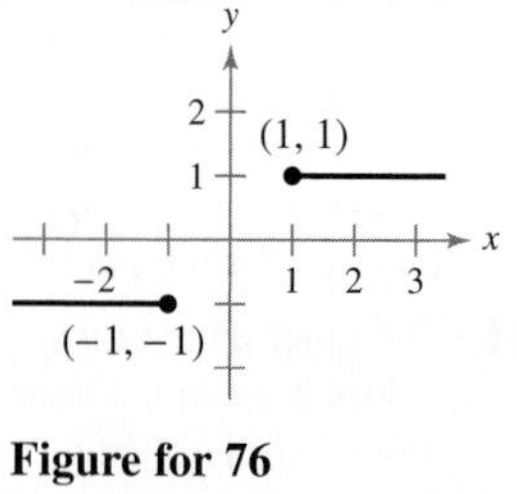

Figure for 76

76. A line segment extends horizontally to the left from the point $(-1, -1)$. Another line segment extends horizontally to the right from the point $(1, 1)$, as shown in the figure. Find a curve of the form

$$y = ax^5 + bx^3 + cx$$

that connects the points $(-1, -1)$ and $(1, 1)$ so that the slope and curvature of the curve are zero at the endpoints.

P.S. Problem Solving

See www.CalcChat.com for worked-out solutions to odd-numbered exercises.

1. The **cornu spiral** is given by

$$x(t) = \int_0^t \cos\left(\frac{\pi u^2}{2}\right) du \quad \text{and} \quad y(t) = \int_0^t \sin\left(\frac{\pi u^2}{2}\right) du.$$

The spiral shown in the figure was plotted over the interval $-\pi \le t \le \pi$.

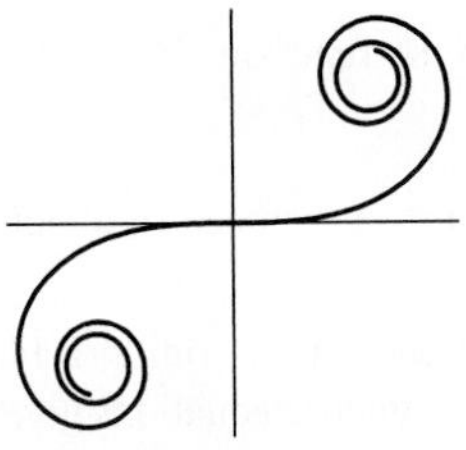

Generated by Mathematica

(a) Find the arc length of this curve from $t = 0$ to $t = a$.

(b) Find the curvature of the graph when $t = a$.

(c) The cornu spiral was discovered by James Bernoulli. He found that the spiral has an amazing relationship between curvature and arc length. What is this relationship?

2. Let T be the tangent line at the point $P(x, y)$ to the graph of the curve $x^{2/3} + y^{2/3} = a^{2/3}$, $a > 0$, as shown in the figure. Show that the radius of curvature at P is three times the distance from the origin to the tangent line T.

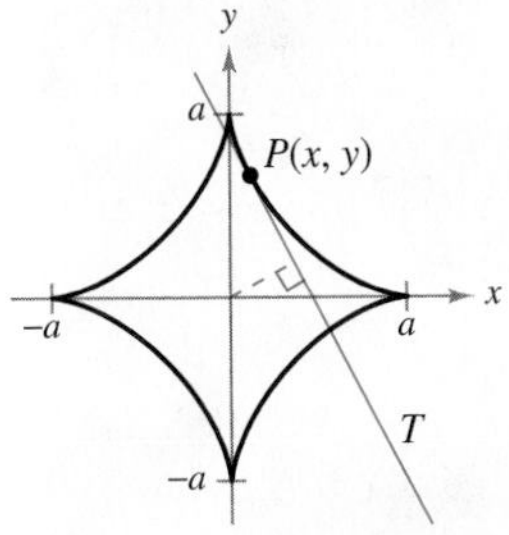

3. A bomber is flying horizontally at an altitude of 3200 feet with a velocity of 400 feet per second when it releases a bomb. A projectile is launched 5 seconds later from a cannon at a site facing the bomber and 5000 feet from the point beneath the original position of the bomber, as shown in the figure. The projectile is to intercept the bomb at an altitude of 1600 feet. Determine the initial speed and angle of inclination of the projectile. (Ignore air resistance.)

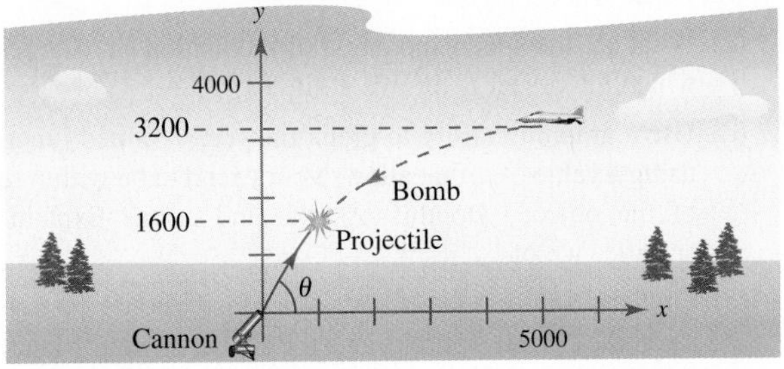

4. Repeat Exercise 3 if the bomber is facing *away* from the launch site, as shown in the figure.

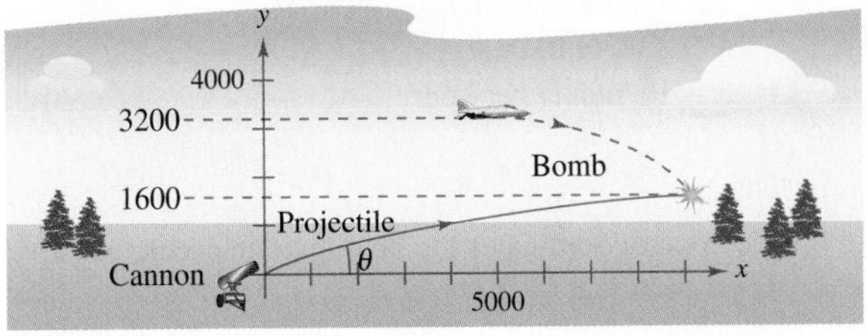

5. Consider one arch of the cycloid

$$\mathbf{r}(\theta) = (\theta - \sin\theta)\mathbf{i} + (1 - \cos\theta)\mathbf{j}, \quad 0 \le \theta \le 2\pi$$

as shown in the figure. Let $s(\theta)$ be the arc length from the highest point on the arch to the point $(x(\theta), y(\theta))$, and let $\rho(\theta) = \dfrac{1}{K}$ be the radius of curvature at the point $(x(\theta), y(\theta))$. Show that s and ρ are related by the equation $s^2 + \rho^2 = 16$. (This equation is called a *natural equation* for the curve.)

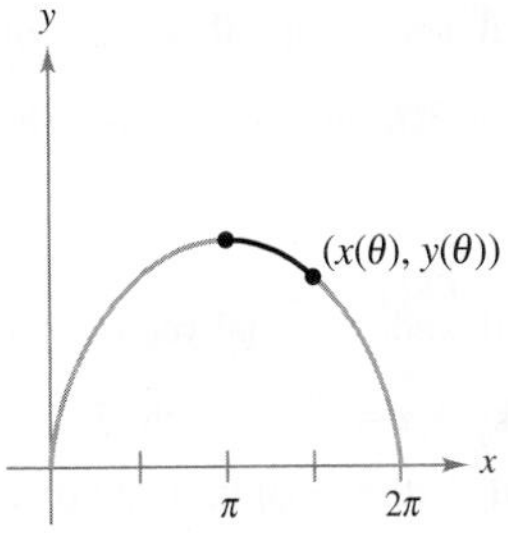

6. Consider the cardioid $r = 1 - \cos\theta$, $0 \le \theta \le 2\pi$, as shown in the figure. Let $s(\theta)$ be the arc length from the point $(2, \pi)$ on the cardioid to the point (r, θ), and let $\rho(\theta) = \dfrac{1}{K}$ be the radius of curvature at the point (r, θ). Show that s and ρ are related by the equation $s^2 + 9\rho^2 = 16$. (This equation is called a *natural equation* for the curve.)

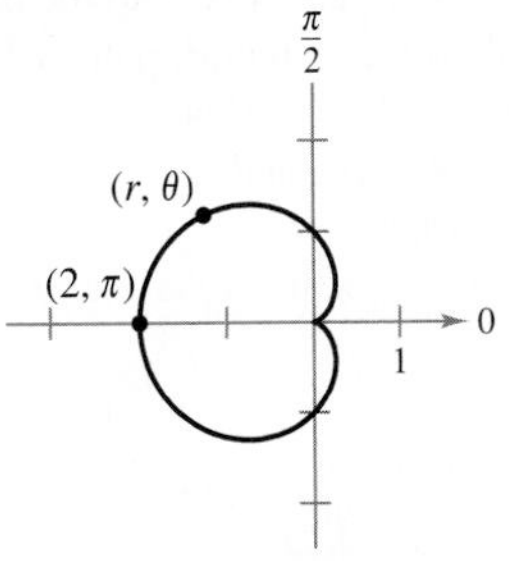

7. If $\mathbf{r}(t)$ is a nonzero differentiable function of t, prove that

$$\frac{d}{dt}(\|\mathbf{r}(t)\|) = \frac{1}{\|\mathbf{r}(t)\|}\mathbf{r}(t) \cdot \mathbf{r}'(t).$$

8. A communications satellite moves in a circular orbit around Earth at a distance of 42,000 kilometers from the center of Earth. The angular velocity

$$\frac{d\theta}{dt} = \omega = \frac{\pi}{12} \text{ radian per hour}$$

is constant.

(a) Use polar coordinates to show that the acceleration vector is given by

$$\mathbf{a} = \frac{d^2\mathbf{r}}{dt^2} = \left[\frac{d^2r}{dt^2} - r\left(\frac{d\theta}{dt}\right)^2\right]\mathbf{u_r} + \left[r\frac{d^2\theta}{dt^2} + 2\frac{dr}{dt}\frac{d\theta}{dt}\right]\mathbf{u}_\theta$$

where $\mathbf{u_r} = \cos\theta\mathbf{i} + \sin\theta\mathbf{j}$ is the unit vector in the radial direction and $\mathbf{u}_\theta = -\sin\theta\mathbf{i} + \cos\theta\mathbf{j}$.

(b) Find the radial and angular components of the acceleration for the satellite.

In Exercises 9–11, use the binormal vector defined by the equation $\mathbf{B} = \mathbf{T} \times \mathbf{N}$.

9. Find the unit tangent, unit normal, and binormal vectors for the helix $\mathbf{r}(t) = 4\cos t\mathbf{i} + 4\sin t\mathbf{j} + 3t\mathbf{k}$ at $t = \frac{\pi}{2}$. Sketch the helix together with these three mutually orthogonal unit vectors.

10. Find the unit tangent, unit normal, and binormal vectors for the curve $\mathbf{r}(t) = \cos t\mathbf{i} + \sin t\mathbf{j} - \mathbf{k}$ at $t = \frac{\pi}{4}$. Sketch the curve together with these three mutually orthogonal unit vectors.

11. (a) Prove that there exists a scalar τ, called the **torsion,** such that $d\mathbf{B}/ds = -\tau\mathbf{N}$.

(b) Prove that $\dfrac{d\mathbf{N}}{ds} = -K\mathbf{T} + \tau\mathbf{B}$.

(The three equations $d\mathbf{T}/ds = K\mathbf{N}$, $d\mathbf{N}/ds = -K\mathbf{T} + \tau\mathbf{B}$, and $d\mathbf{B}/ds = -\tau\mathbf{N}$ are called the *Frenet-Serret formulas.*)

12. A highway has an exit ramp that begins at the origin of a coordinate system and follows the curve $y = \frac{1}{32}x^{5/2}$ to the point (4, 1) (see figure). Then it follows a circular path whose curvature is that given by the curve at (4, 1). What is the radius of the circular arc? Explain why the curve and the circular arc should have the same curvature at (4, 1).

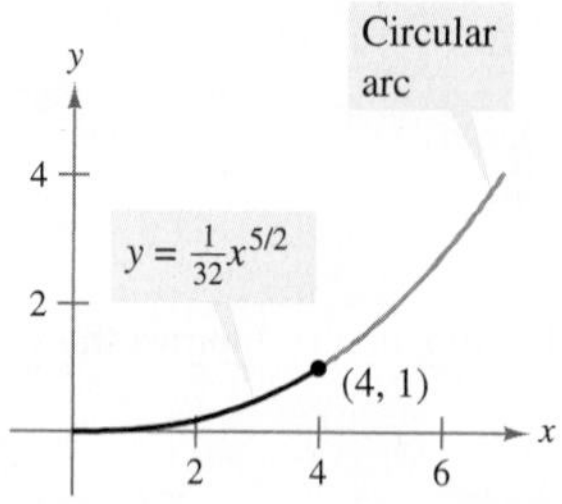

13. Consider the vector-valued function $\mathbf{r}(t) = \langle t\cos\pi t, t\sin\pi t\rangle$, $0 \le t \le 2$.

(a) Use a graphing utility to graph the function.

(b) Find the length of the arc in part (a).

(c) Find the curvature K as a function of t. Find the curvatures when t is 0, 1, and 2.

(d) Use a graphing utility to graph the function K.

(e) Find (if possible) $\lim_{t\to\infty} K$.

(f) Using the result of part (e), make a conjecture about the graph of $\mathbf{r}$ as $t \to \infty$.

14. You want to toss an object to a friend who is riding a Ferris wheel (see figure). The following parametric equations give the path of the friend $\mathbf{r}_1(t)$ and the path of the object $\mathbf{r}_2(t)$. Distance is measured in meters and time is measured in seconds.

$$\mathbf{r}_1(t) = 15\left(\sin\frac{\pi t}{10}\right)\mathbf{i} + \left(16 - 15\cos\frac{\pi t}{10}\right)\mathbf{j}$$

$$\mathbf{r}_2(t) = [22 - 8.03(t - t_0)]\mathbf{i} + [1 + 11.47(t - t_0) - 4.9(t - t_0)^2]\mathbf{j}$$

(a) Locate your friend's position on the Ferris wheel at time $t = 0$.

(b) Determine the number of revolutions per minute of the Ferris wheel.

(c) What are the speed and angle of inclination (in degrees) at which the object is thrown at time $t = t_0$?

(d) Use a graphing utility to graph the vector-valued functions using a value of t_0 that allows your friend to be within reach of the object. (Do this by trial and error.) Explain the significance of t_0.

(e) Find the approximate time your friend should be able to catch the object. Approximate the speeds of your friend and the object at that time.

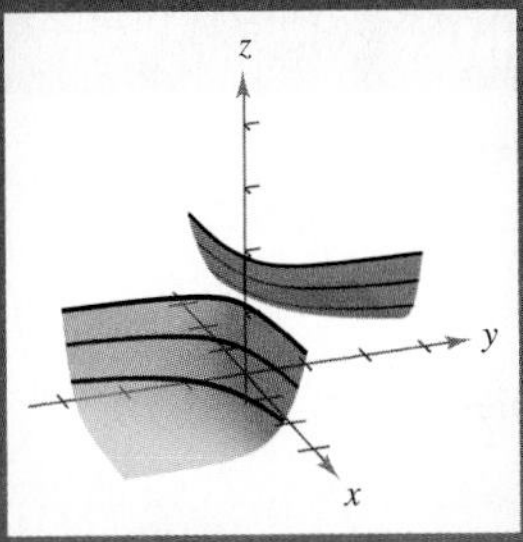

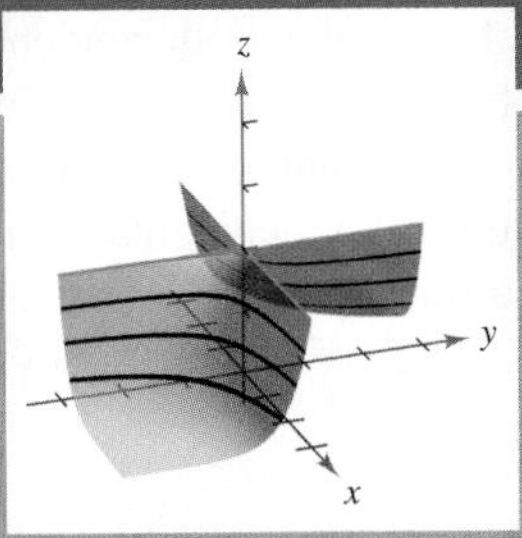

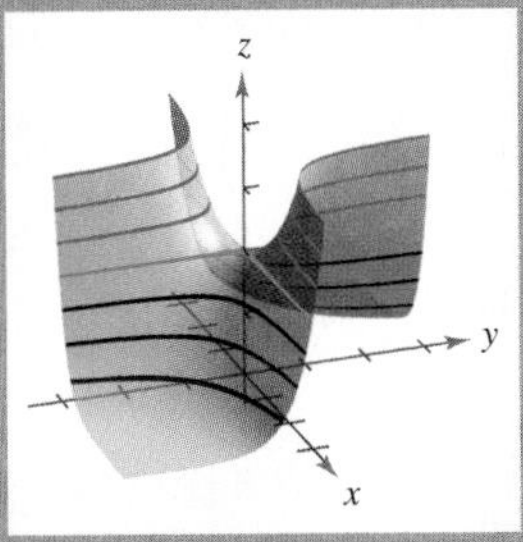

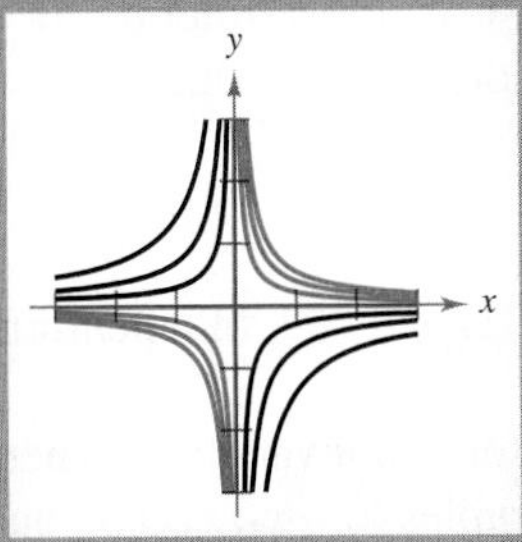

13 Functions of Several Variables

The colored areas shown on the weather map of the world depict the normal annual temperature range of regions on Earth. Which two geographic features have the greatest effect on temperature range? Why?

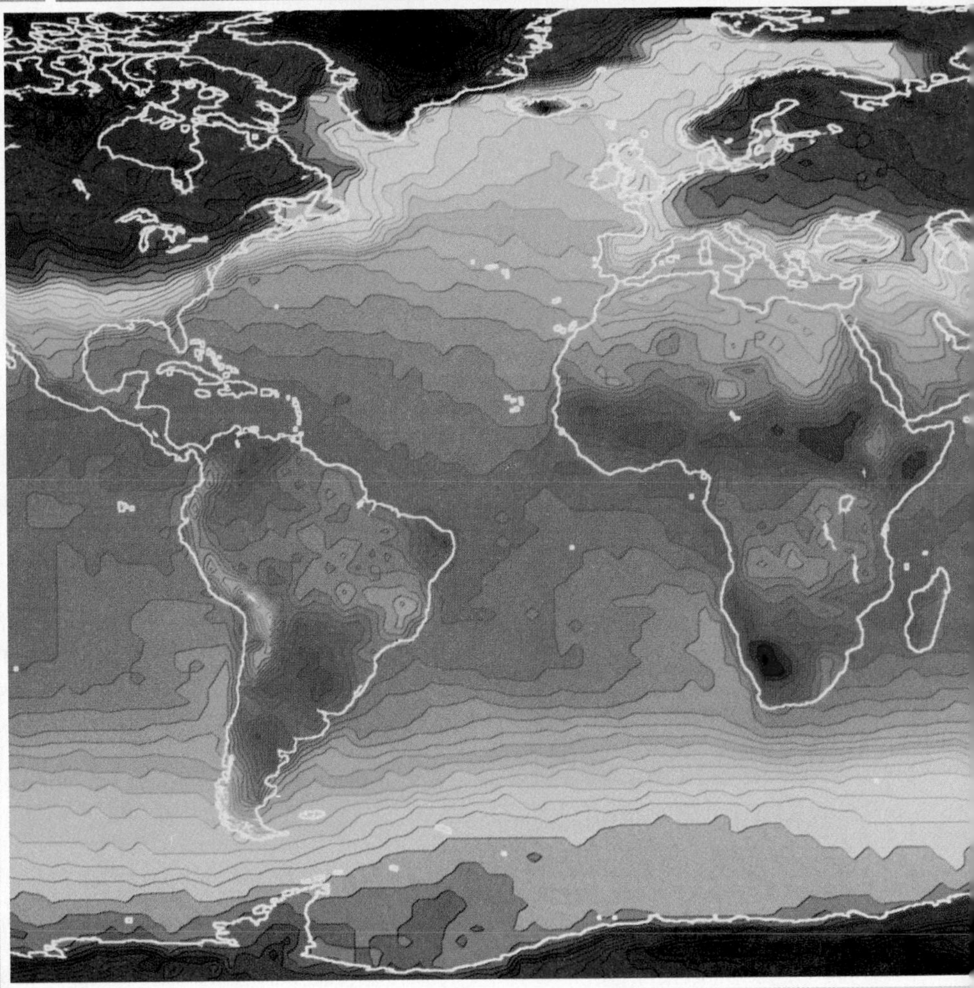

NASA/Science Photo Library/Photo Researchers, Inc.

Many real-life quantities are functions of two or more variables. In Section 13.1, you will learn how to graph a function of two variables, like the one shown above. The top three graphs show cut-away views of the surface at various traces. Another way to visualize this surface is to project the traces onto the *xy*-plane, as shown in the fourth graph.

Section 13.1 Introduction to Functions of Several Variables

- **Understand the notation for a function of several variables.**
- **Sketch the graph of a function of two variables.**
- **Sketch level curves for a function of two variables.**
- **Sketch level surfaces for a function of three variables.**
- **Use computer graphics to graph a function of two variables.**

EXPLORATION

Comparing Dimensions Without using a graphing utility, describe the graph of each function of two variables.

a. $z = x^2 + y^2$

b. $z = x + y$

c. $z = x^2 + y$

d. $z = \sqrt{x^2 + y^2}$

e. $z = \sqrt{1 - x^2 + y^2}$

Functions of Several Variables

So far in this text, you have dealt only with functions of a single (independent) variable. Many familiar quantities, however, are functions of two or more variables. For instance, the work done by a force $(W = FD)$ and the volume of a right circular cylinder $(V = \pi r^2 h)$ are both functions of two variables. The volume of a rectangular solid $(V = lwh)$ is a function of three variables. The notation for a function of two or more variables is similar to that for a function of a single variable. Here are two examples.

$$z = \underbrace{f(x, y)}_{\text{2 variables}} = x^2 + xy \qquad \text{Function of two variables}$$

and

$$w = \underbrace{f(x, y, z)}_{\text{3 variables}} = x + 2y - 3z \qquad \text{Function of three variables}$$

Definition of a Function of Two Variables

Let D be a set of ordered pairs of real numbers. If to each ordered pair (x, y) in D there corresponds a unique real number $f(x, y)$, then f is called a **function of x and y.** The set D is the **domain** of f, and the corresponding set of values for $f(x, y)$ is the **range** of f.

For the function given by $z = f(x, y)$, x and y are called the **independent variables** and z is called the **dependent variable.**

Similar definitions can be given for functions of three, four, or n variables, where the domains consist of ordered triples (x_1, x_2, x_3), quadruples (x_1, x_2, x_3, x_4), and n-tuples $(x_1, x_2, . . . , x_n)$. In all cases, the range is a set of real numbers. In this chapter, you will study only functions of two or three variables.

As with functions of one variable, the most common way to describe a function of several variables is with an *equation*, and unless otherwise restricted, you can assume that the domain is the set of all points for which the equation is defined. For instance, the domain of the function given by

$$f(x, y) = x^2 + y^2$$

is assumed to be the entire xy-plane. Similarly, the domain of

$$f(x, y) = \ln xy$$

is the set of all points (x, y) in the plane for which $xy > 0$. This consists of all points in the first and third quadrants.

Archive Photos

MARY FAIRFAX SOMERVILLE (1780–1872)

Somerville was interested in the problem of creating geometric models for functions of several variables. Her most well-known book, *The Mechanics of the Heavens*, was published in 1831.

EXAMPLE 1 Domains of Functions of Several Variables

Find the domain of each function.

a. $f(x, y) = \dfrac{\sqrt{x^2 + y^2 - 9}}{x}$ **b.** $g(x, y, z) = \dfrac{x}{\sqrt{9 - x^2 - y^2 - z^2}}$

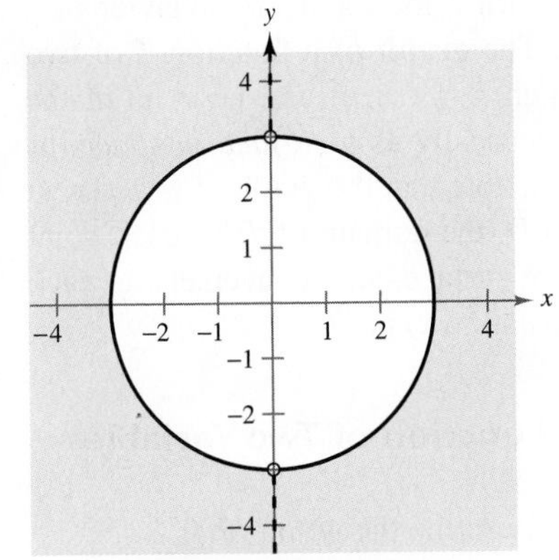

Domain of $f(x, y) = \dfrac{\sqrt{x^2 + y^2 - 9}}{x}$

Figure 13.1

Solution

a. The function f is defined for all points (x, y) such that $x \neq 0$ and

$$x^2 + y^2 \geq 9.$$

So, the domain is the set of all points lying on or outside the circle $x^2 + y^2 = 9$, *except* those points on the y-axis, as shown in Figure 13.1.

b. The function g is defined for all points (x, y, z) such that

$$x^2 + y^2 + z^2 < 9.$$

Consequently, the domain is the set of all points (x, y, z) lying inside a sphere of radius 3 that is centered at the origin.

Functions of several variables can be combined in the same ways as functions of single variables. For instance, you can form the sum, difference, product, and quotient of two functions of two variables as follows.

$$(f \pm g)(x, y) = f(x, y) \pm g(x, y) \qquad \text{Sum or difference}$$

$$(fg)(x, y) = f(x, y)g(x, y) \qquad \text{Product}$$

$$\frac{f}{g}(x, y) = \frac{f(x, y)}{g(x, y)} \quad g(x, y) \neq 0 \qquad \text{Quotient}$$

You cannot form the composite of two functions of several variables. However, if h is a function of several variables and g is a function of a single variable, you can form the **composite** function $(g \circ h)(x, y)$ as follows.

$$(g \circ h)(x, y) = g(h(x, y)) \qquad \text{Composition}$$

The domain of this composite function consists of all (x, y) in the domain of h such that $h(x, y)$ is in the domain of g. For example, the function given by

$$f(x, y) = \sqrt{16 - 4x^2 - y^2}$$

can be viewed as the composite of the function of two variables given by $h(x, y) = 16 - 4x^2 - y^2$ and the function of a single variable given by $g(u) = \sqrt{u}$. The domain of this function is the set of all points lying on or inside the ellipse given by $4x^2 + y^2 = 16$.

A function that can be written as a sum of functions of the form $cx^m y^n$ (where c is a real number and m and n are nonnegative integers) is called a **polynomial function** of two variables. For instance, the functions given by

$$f(x, y) = x^2 + y^2 - 2xy + x + 2 \quad \text{and} \quad g(x, y) = 3xy^2 + x - 2$$

are polynomial functions of two variables. A **rational function** is the quotient of two polynomial functions. Similar terminology is used for functions of more than two variables.

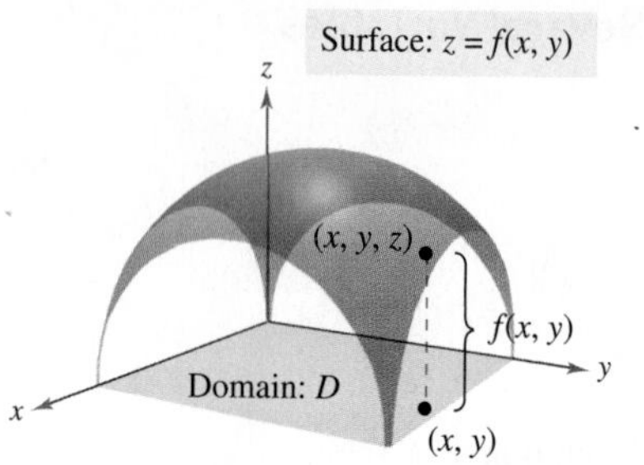

Figure 13.2

The Graph of a Function of Two Variables

As with functions of a single variable, you can learn a lot about the behavior of a function of two variables by sketching its graph. The **graph** of a function f of two variables is the set of all points (x, y, z) for which $z = f(x, y)$ and (x, y) is in the domain of f. This graph can be interpreted geometrically as a *surface in space*, as discussed in Sections 11.5 and 11.6. In Figure 13.2, note that the graph of $z = f(x, y)$ is a surface whose projection onto the xy-plane is D, the domain of f. To each point (x, y) in D there corresponds a point (x, y, z) on the surface, and, conversely, to each point (x, y, z) on the surface there corresponds a point (x, y) in D.

EXAMPLE 2 Describing the Graph of a Function of Two Variables

What is the range of $f(x, y) = \sqrt{16 - 4x^2 - y^2}$? Describe the graph of f.

Solution The domain D implied by the equation for f is the set of all points (x, y) such that $16 - 4x^2 - y^2 \geq 0$. So, D is the set of all points lying on or inside the ellipse given by

$$\frac{x^2}{4} + \frac{y^2}{16} = 1. \qquad \text{Ellipse in the } xy\text{-plane}$$

The range of f is all values $z = f(x, y)$ such that $0 \leq z \leq \sqrt{16}$ or

$$0 \leq z \leq 4. \qquad \text{Range of } f$$

A point (x, y, z) is on the graph of f if and only if

$$\begin{aligned} z &= \sqrt{16 - 4x^2 - y^2} \\ z^2 &= 16 - 4x^2 - y^2 \\ 4x^2 + y^2 + z^2 &= 16 \\ \frac{x^2}{4} + \frac{y^2}{16} + \frac{z^2}{16} &= 1, \qquad 0 \leq z \leq 4. \end{aligned}$$

From Section 11.6, you know that the graph of f is the upper half of an ellipsoid, as shown in Figure 13.3.

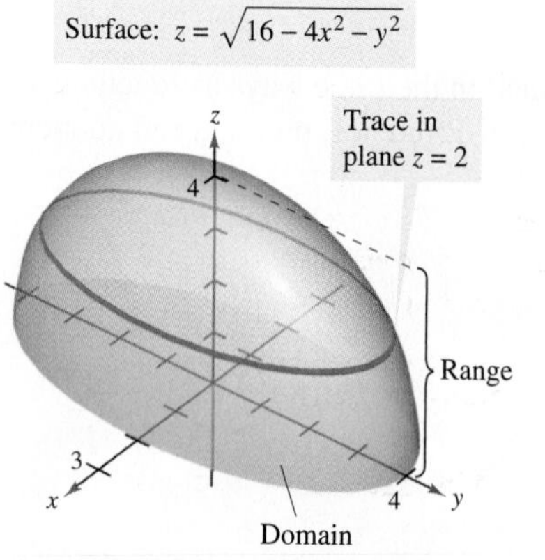

The graph of $f(x, y) = \sqrt{16 - 4x^2 - y^2}$ is the upper half of an ellipsoid.
Figure 13.3

To sketch a surface in space *by hand,* it helps to use traces in planes parallel to the coordinate planes, as shown in Figure 13.3. For example, to find the trace of the surface in the plane $z = 2$, substitute $z = 2$ in the equation $z = \sqrt{16 - 4x^2 - y^2}$ and obtain

$$2 = \sqrt{16 - 4x^2 - y^2} \quad \Longrightarrow \quad \frac{x^2}{3} + \frac{y^2}{12} = 1.$$

So, the trace is an ellipse centered at the point $(0, 0, 2)$ with major and minor axes of lengths $4\sqrt{3}$ and $2\sqrt{3}$.

Traces are also used with most three-dimensional graphing utilities. For instance, Figure 13.4 shows a computer-generated version of the surface given in Example 2. For this graph, the computer took 25 traces parallel to the xy-plane and 12 traces in vertical planes.

If you have access to a three-dimensional graphing utility, use it to graph several surfaces.

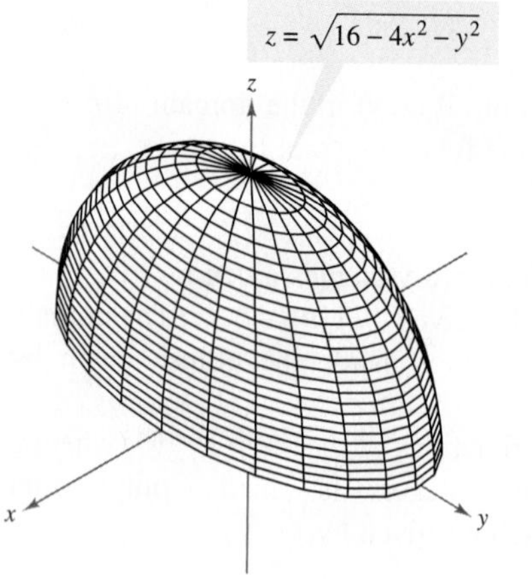

Figure 13.4

Level Curves

A second way to visualize a function of two variables is to use a **scalar field** in which the scalar $z = f(x, y)$ is assigned to the point (x, y). A scalar field can be characterized by **level curves** (or **contour lines**) along which the value of $f(x, y)$ is constant. For instance, the weather map in Figure 13.5 shows level curves of equal pressure called **isobars.** In weather maps for which the level curves represent points of equal temperature, the level curves are called **isotherms,** as shown in Figure 13.6. Another common use of level curves is in representing electric potential fields. In this type of map, the level curves are called **equipotential lines.**

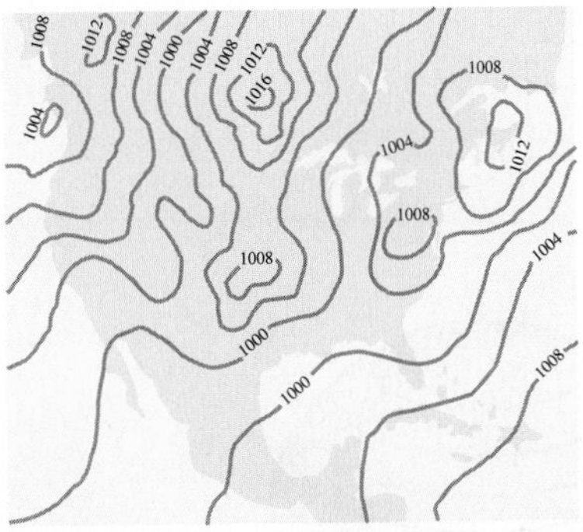

Level curves show the lines of equal pressure (isobars) measured in millibars.
Figure 13.5

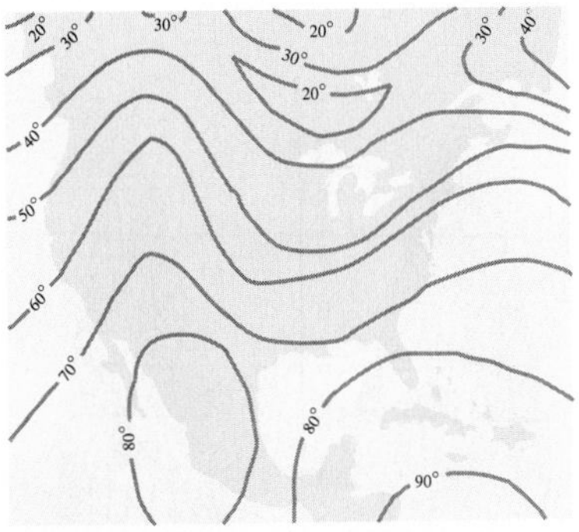

Level curves show the lines of equal temperature (isotherms) measured in degrees Fahrenheit.
Figure 13.6

Contour maps are commonly used to show regions on Earth's surface, with the level curves representing the height above sea level. This type of map is called a **topographic map.** For example, the mountain shown in Figure 13.7 is represented by the topographic map in Figure 13.8.

A contour map depicts the variation of z with respect to x and y by the spacing between level curves. Much space between level curves indicates that z is changing slowly, whereas little space indicates a rapid change in z. Furthermore, to give a good three-dimensional illusion in a contour map, it is important to choose c-values that are *evenly spaced.*

Alfred B. Thomas/Earth Scenes

Figure 13.7

USGS

Figure 13.8

EXAMPLE 3 Sketching a Contour Map

The hemisphere given by $f(x, y) = \sqrt{64 - x^2 - y^2}$ is shown in Figure 13.9. Sketch a contour map for this surface using level curves corresponding to $c = 0, 1, 2, \ldots, 8$.

Solution For each value of c, the equation given by $f(x, y) = c$ is a circle (or point) in the xy-plane. For example, when $c_1 = 0$, the level curve is

$$x^2 + y^2 = 64 \qquad \text{Circle of radius 8}$$

which is a circle of radius 8. Figure 13.10 shows the nine level curves for the hemisphere.

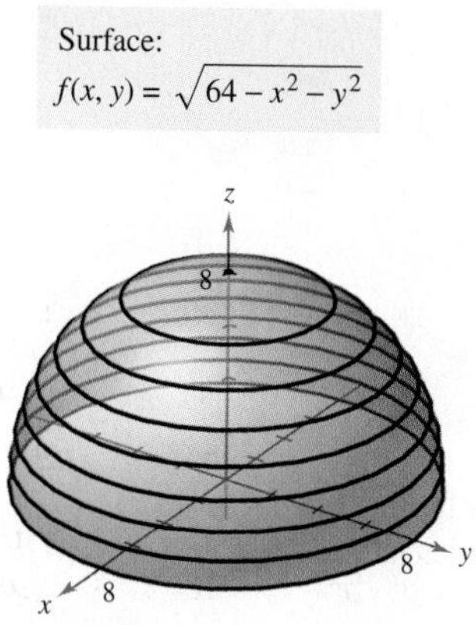

Hemisphere
Figure 13.9

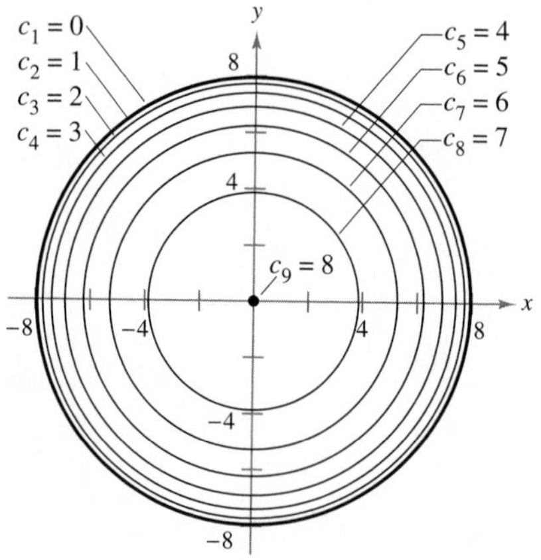

Contour map
Figure 13.10

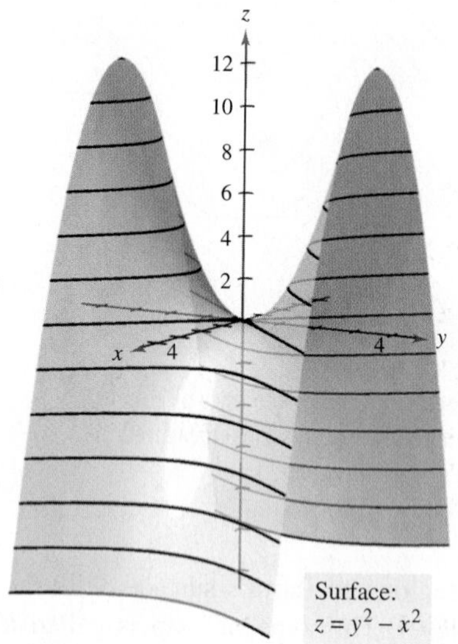

Hyperbolic paraboloid
Figure 13.11

EXAMPLE 4 Sketching a Contour Map

The hyperbolic paraboloid given by

$$z = y^2 - x^2$$

is shown in Figure 13.11. Sketch a contour map for this surface.

Solution For each value of c, let $f(x, y) = c$ and sketch the resulting level curve in the xy-plane. For this function, each of the level curves $(c \neq 0)$ is a hyperbola whose asymptotes are the lines $y = \pm x$. If $c < 0$, the transverse axis is horizontal. For instance, the level curve for $c = -4$ is given by

$$\frac{x^2}{2^2} - \frac{y^2}{2^2} = 1. \qquad \text{Hyperbola with horizontal transverse axis}$$

If $c > 0$, the transverse axis is vertical. For instance, the level curve for $c = 4$ is given by

$$\frac{y^2}{2^2} - \frac{x^2}{2^2} = 1. \qquad \text{Hyperbola with vertical transverse axis}$$

If $c = 0$, the level curve is the degenerate conic representing the intersecting asymptotes, as shown in Figure 13.12.

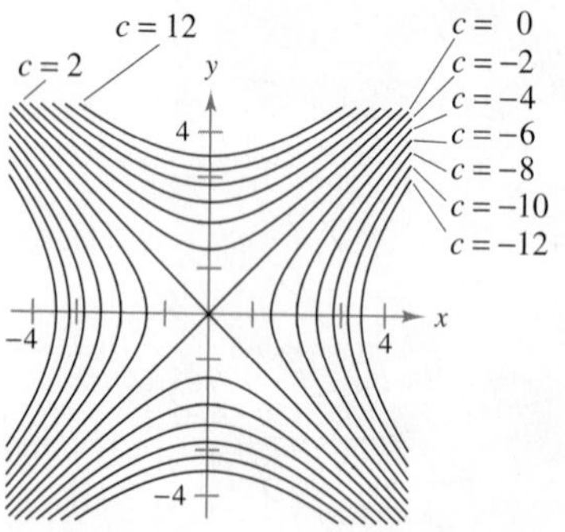

Hyperbolic level curves (at increments of 2)
Figure 13.12

---

indicates that in the HM mathSpace® *CD-ROM and the online* Eduspace® *system for this text, you will find an Open Exploration, which further explores this example using the computer algebra systems* Maple, Mathcad, Mathematica, *and* Derive.

One example of a function of two variables used in economics is the **Cobb-Douglas production function.** This function is used as a model to represent the number of units produced by varying amounts of labor and capital. If x measures the units of labor and y measures the units of capital, the number of units produced is given by

$$f(x, y) = Cx^a y^{1-a}$$

where C and a are constants with $0 < a < 1$.

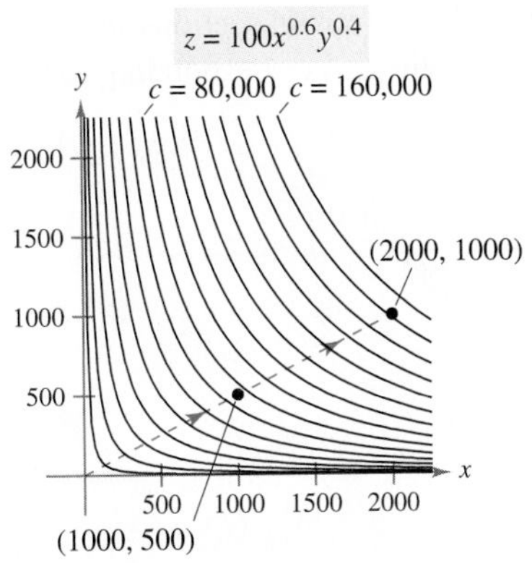

Level curves (at increments of 10,000)
Figure 13.13

EXAMPLE 5 The Cobb-Douglas Production Function

A toy manufacturer estimates a production function to be $f(x, y) = 100x^{0.6}y^{0.4}$, where x is the number of units of labor and y is the number of units of capital. Compare the production level when $x = 1000$ and $y = 500$ with the production level when $x = 2000$ and $y = 1000$.

Solution When $x = 1000$ and $y = 500$, the production level is

$$f(1000, 500) = 100(1000^{0.6})(500^{0.4}) \approx 75{,}786.$$

When $x = 2000$ and $y = 1000$, the production level is

$$f(2000, 1000) = 100(2000^{0.6})(1000^{0.4}) = 151{,}572.$$

The level curves of $z = f(x, y)$ are shown in Figure 13.13. Note that by doubling both x and y, you double the production level (see Exercise 79).

Level Surfaces

The concept of a level curve can be extended by one dimension to define a **level surface.** If f is a function of three variables and c is a constant, the graph of the equation $f(x, y, z) = c$ is a **level surface** of the function f, as shown in Figure 13.14.

With computers, engineers and scientists have developed other ways to view functions of three variables. For instance, Figure 13.15 shows a computer simulation that uses color to represent the optimal strain distribution of a car door.

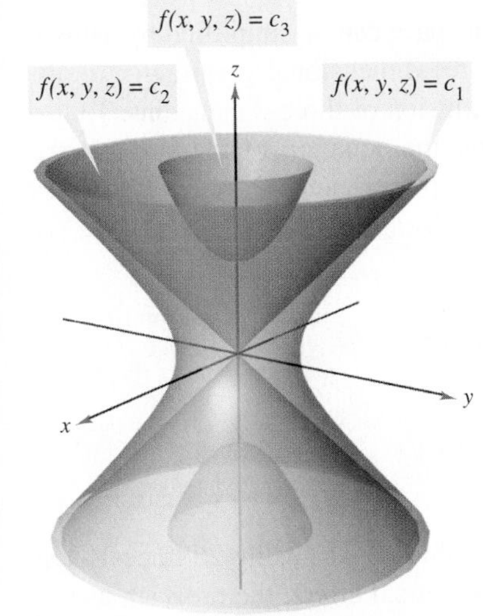

Level surfaces of f
Figure 13.14

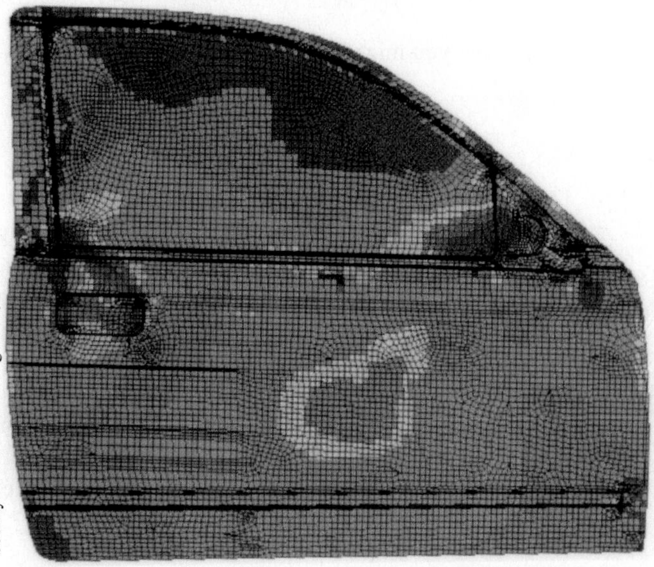

Figure 13.15

EXAMPLE 6 Level Surfaces

Describe the level surfaces of the function

$$f(x, y, z) = 4x^2 + y^2 + z^2.$$

Solution Each level surface has an equation of the form

$$4x^2 + y^2 + z^2 = c. \qquad \text{Equation of level surface}$$

So, the level surfaces are ellipsoids (whose cross sections parallel to the yz-plane are circles). As c increases, the radii of the circular cross sections increase according to the square root of c. For example, the level surfaces corresponding to the values $c = 0$, $c = 4$, and $c = 16$ are as follows.

$$4x^2 + y^2 + z^2 = 0 \qquad \text{Level surface for } c = 0 \text{ (single point)}$$

$$\frac{x^2}{1} + \frac{y^2}{4} + \frac{z^2}{4} = 1 \qquad \text{Level surface for } c = 4 \text{ (ellipsoid)}$$

$$\frac{x^2}{4} + \frac{y^2}{16} + \frac{z^2}{16} = 1 \qquad \text{Level surface for } c = 16 \text{ (ellipsoid)}$$

These level surfaces are shown in Figure 13.16.

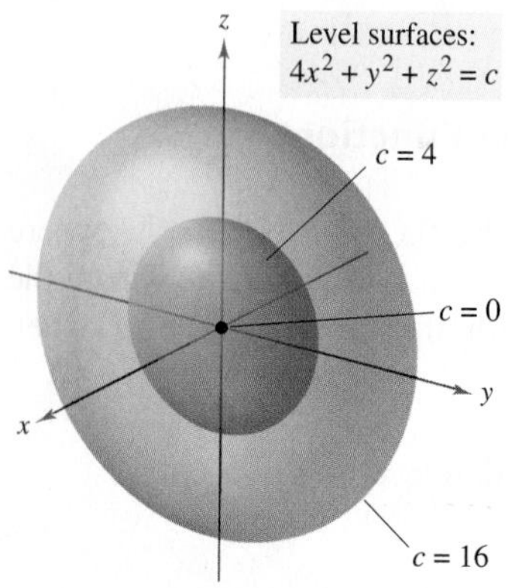

Figure 13.16

NOTE If the function in Example 6 represented the *temperature* at the point (x, y, z), the level surfaces shown in Figure 13.16 would be called **isothermal surfaces.**

Computer Graphics

The problem of sketching the graph of a surface in space can be simplified by using a computer. Although there are several types of three-dimensional graphing utilities, most use some form of trace analysis to give the illusion of three dimensions. To use such a graphing utility, you usually need to enter the equation of the surface, the region in the xy-plane over which the surface is to be plotted, and the number of traces to be taken. For instance, to graph the surface given by

$$f(x, y) = (x^2 + y^2)e^{1-x^2-y^2}$$

you might choose the following bounds for x, y, and z.

$$-3 \le x \le 3 \qquad \text{Bounds for } x$$

$$-3 \le y \le 3 \qquad \text{Bounds for } y$$

$$0 \le z \le 3 \qquad \text{Bounds for } z$$

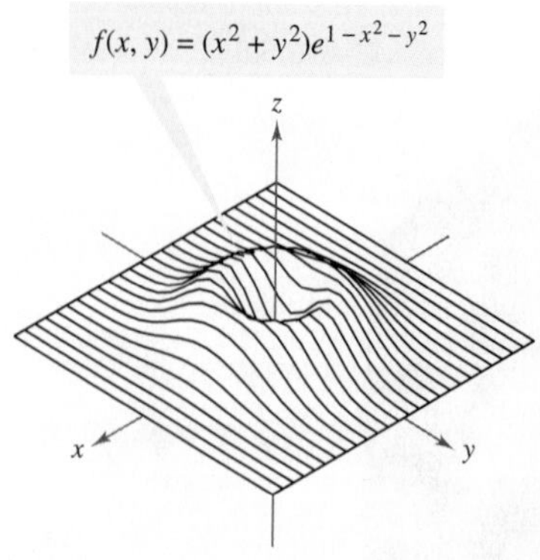

Figure 13.17

Figure 13.17 shows a computer-generated graph of this surface using 26 traces taken parallel to the yz-plane. To heighten the three-dimensional effect, the program uses a "hidden line" routine. That is, it begins by plotting the traces in the foreground (those corresponding to the largest x-values), and then, as each new trace is plotted, the program determines whether all or only part of the next trace should be shown.

The graphs on page 891 show a variety of surfaces that were plotted by computer. If you have access to a computer drawing program, use it to reproduce these surfaces. Remember also that the three-dimensional graphics in this text can be viewed and rotated. These rotatable graphs are available in the *HM mathSpace®* CD-Rom and the online *Eduspace®* system for this text.

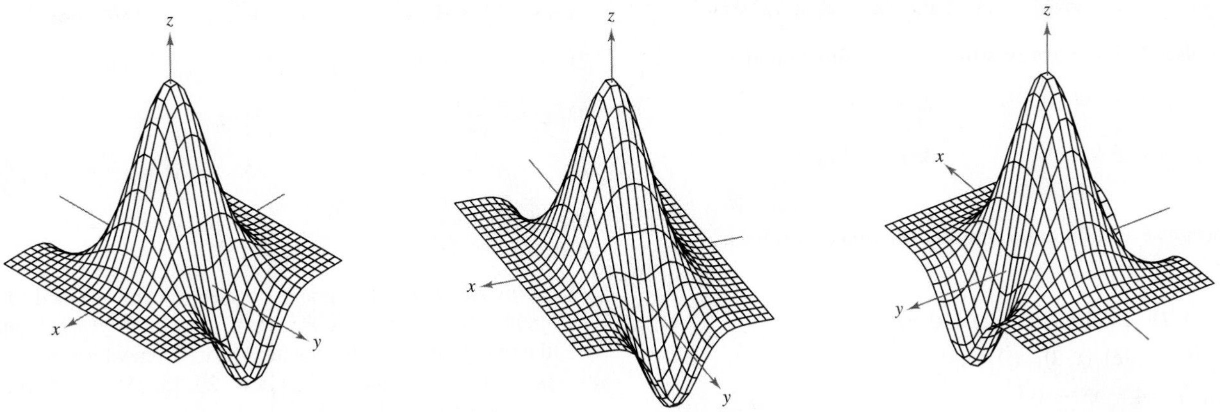

Three different views of the graph of $f(x, y) = (2 - y^2 + x^2)\,e^{1 - x^2 - (y^2/4)}$

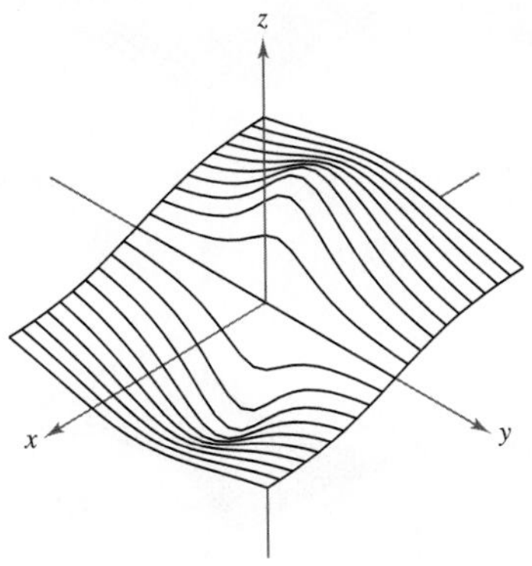

Single traces

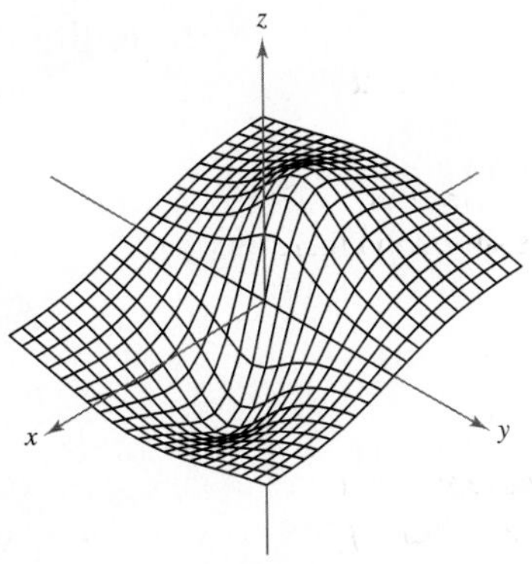

Double traces

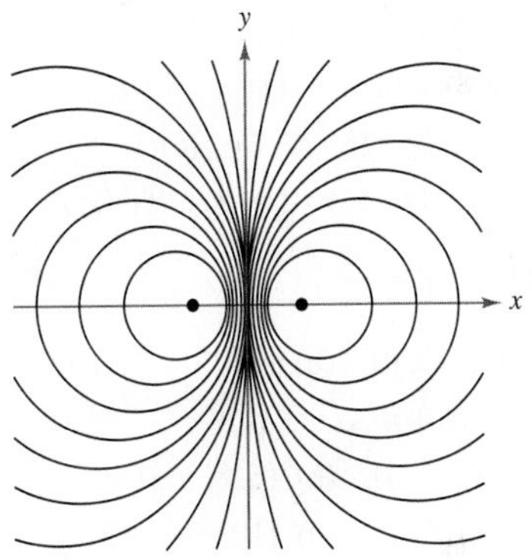

Level curves

Traces and level curves of the graph of $f(x, y) = \dfrac{-4x}{x^2 + y^2 + 1}$

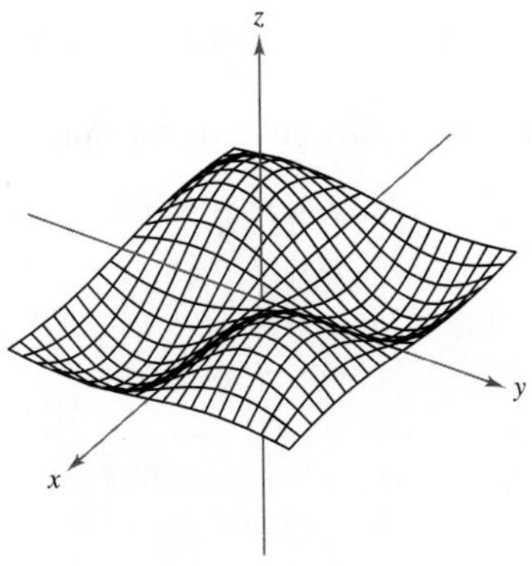

$f(x, y) = \sin x \sin y$

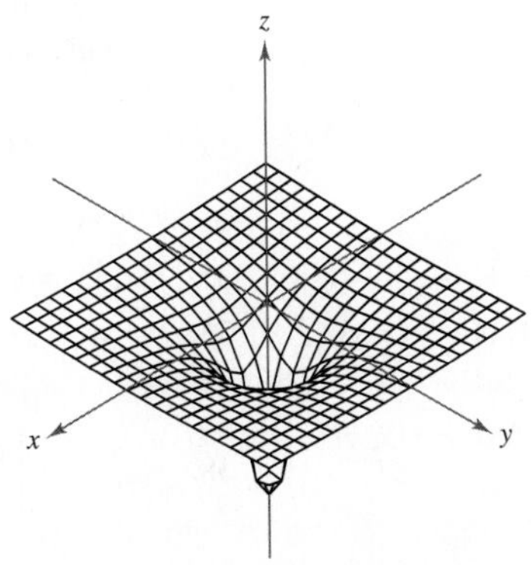

$f(x, y) = -\dfrac{1}{\sqrt{x^2 + y^2}}$

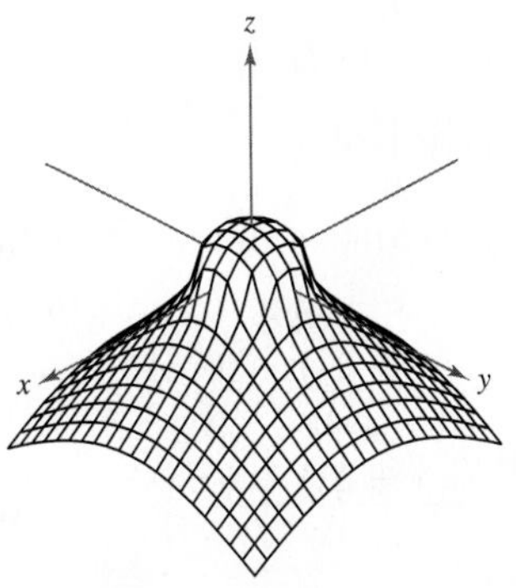

$f(x, y) = \dfrac{1 - x^2 - y^2}{\sqrt{|1 - x^2 - y^2|}}$

Exercises for Section 13.1

See www.CalcChat.com for worked-out solutions to odd-numbered exercises.

In Exercises 1–4, determine whether z is a function of x and y.

1. $x^2z + yz - xy = 10$
2. $xz^2 + 2xy - y^2 = 4$
3. $\frac{x^2}{4} + \frac{y^2}{9} + z^2 = 1$
4. $z + x \ln y - 8 = 0$

In Exercises 5–16, find and simplify the function values.

5. $f(x, y) = x/y$
   (a) $(3, 2)$ (b) $(-1, 4)$ (c) $(30, 5)$
   (d) $(5, y)$ (e) $(x, 2)$ (f) $(5, t)$
6. $f(x, y) = 4 - x^2 - 4y^2$
   (a) $(0, 0)$ (b) $(0, 1)$ (c) $(2, 3)$
   (d) $(1, y)$ (e) $(x, 0)$ (f) $(t, 1)$
7. $f(x, y) = xe^y$
   (a) $(5, 0)$ (b) $(3, 2)$ (c) $(2, -1)$
   (d) $(5, y)$ (e) $(x, 2)$ (f) (t, t)
8. $g(x, y) = \ln|x + y|$
   (a) $(2, 3)$ (b) $(5, 6)$ (c) $(e, 0)$
   (d) $(0, 1)$ (e) $(2, -3)$ (f) (e, e)
9. $h(x, y, z) = \frac{xy}{z}$
   (a) $(2, 3, 9)$ (b) $(1, 0, 1)$ (c) $(-2, 3, 4)$ (d) $(5, 4, -6)$
10. $f(x, y, z) = \sqrt{x + y + z}$
    (a) $(0, 5, 4)$ (b) $(6, 8, -3)$
    (c) $(4, 6, 2)$ (d) $(10, -4, -3)$
11. $f(x, y) = x \sin y$
    (a) $\left(2, \frac{\pi}{4}\right)$ (b) $(3, 1)$ (c) $\left(-3, \frac{\pi}{3}\right)$ (d) $\left(4, \frac{\pi}{2}\right)$
12. $V(r, h) = \pi r^2 h$
    (a) $(3, 10)$ (b) $(5, 2)$ (c) $(4, 8)$ (d) $(6, 4)$
13. $g(x, y) = \int_x^y (2t - 3)\, dt$
    (a) $(0, 4)$ (b) $(1, 4)$ (c) $\left(\frac{3}{2}, 4\right)$ (d) $\left(0, \frac{3}{2}\right)$
14. $g(x, y) = \int_x^y \frac{1}{t}\, dt$
    (a) $(4, 1)$ (b) $(6, 3)$ (c) $(2, 5)$ (d) $\left(\frac{1}{2}, 7\right)$
15. $f(x, y) = x^2 - 2y$
    (a) $\frac{f(x + \Delta x, y) - f(x, y)}{\Delta x}$
    (b) $\frac{f(x, y + \Delta y) - f(x, y)}{\Delta y}$
16. $f(x, y) = 3xy + y^2$
    (a) $\frac{f(x + \Delta x, y) - f(x, y)}{\Delta x}$
    (b) $\frac{f(x, y + \Delta y) - f(x, y)}{\Delta y}$

In Exercises 17–28, describe the domain and range of the function.

17. $f(x, y) = \sqrt{4 - x^2 - y^2}$
18. $f(x, y) = \sqrt{4 - x^2 - 4y^2}$
19. $f(x, y) = \arcsin(x + y)$
20. $f(x, y) = \arccos(y/x)$
21. $f(x, y) = \ln(4 - x - y)$
22. $f(x, y) = \ln(xy - 6)$
23. $z = \frac{x + y}{xy}$
24. $z = \frac{xy}{x - y}$
25. $f(x, y) = e^{x/y}$
26. $f(x, y) = x^2 + y^2$
27. $g(x, y) = \frac{1}{xy}$
28. $g(x, y) = x\sqrt{y}$

29. ***Think About It*** The graphs labeled (a), (b), (c), and (d) are graphs of the function $f(x, y) = -4x/(x^2 + y^2 + 1)$. Match the four graphs with the points in space from which the surface is viewed. The four points are $(20, 15, 25)$, $(-15, 10, 20)$, $(20, 20, 0)$, and $(20, 0, 0)$.

(a)

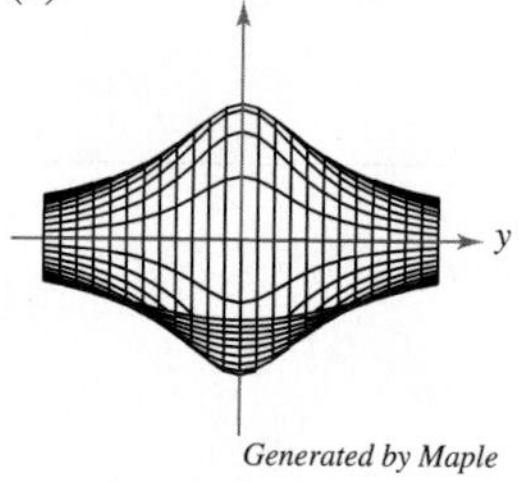

(b)

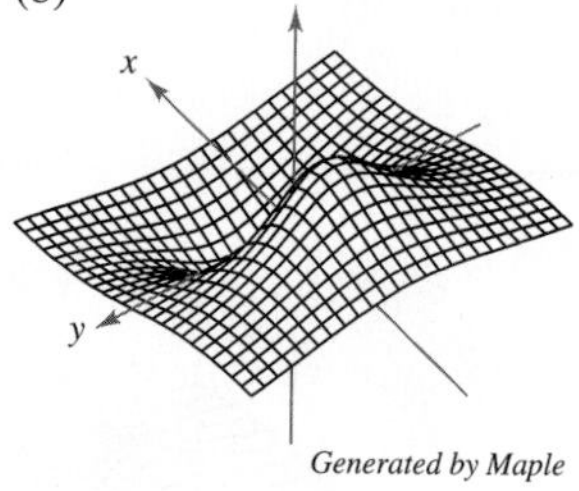

(c)

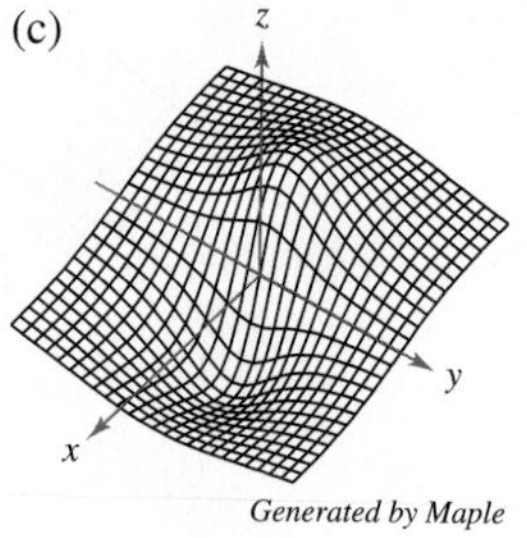

(d)

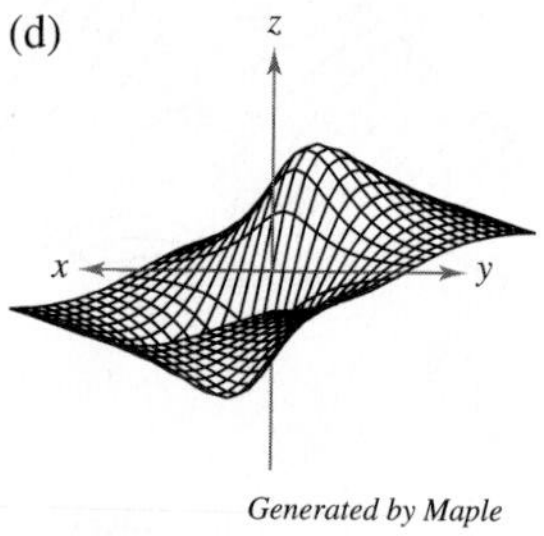

30. ***Think About It*** Use the function given in Exercise 29.
    (a) Find the domain and range of the function.
    (b) Identify the points in the xy-plane where the function value is 0.
    (c) Does the surface pass through all the octants of the rectangular coordinate system? Give reasons for your answer.

In Exercises 31–38, sketch the surface given by the function.

31. $f(x, y) = 5$
32. $f(x, y) = 6 - 2x - 3y$
33. $f(x, y) = y^2$
34. $g(x, y) = \frac{1}{2}x$
35. $z = 4 - x^2 - y^2$
36. $z = \frac{1}{2}\sqrt{x^2 + y^2}$
37. $f(x, y) = e^{-x}$
38. $f(x, y) = \begin{cases} xy, & x \geq 0, y \geq 0 \\ 0, & x < 0 \text{ or } y < 0 \end{cases}$

In Exercises 39–42, use a computer algebra system to graph the function.

39. $z = y^2 - x^2 + 1$
40. $z = \frac{1}{12}\sqrt{144 - 16x^2 - 9y^2}$
41. $f(x, y) = x^2 e^{(-xy/2)}$
42. $f(x, y) = x \sin y$

43. ***Conjecture*** Consider the function $f(x, y) = x^2 + y^2$.

(a) Sketch the graph of the surface given by f.

(b) Make a conjecture about the relationship between the graphs of f and $g(x, y) = f(x, y) + 2$. Use a computer algebra system to confirm your answer.

(c) Make a conjecture about the relationship between the graphs of f and $g(x, y) = f(x, y - 2)$. Use a computer algebra system to confirm your answer.

(d) Make a conjecture about the relationship between the graphs of f and $g(x, y) = 4 - f(x, y)$. Use a computer algebra system to confirm your answer.

(e) On the surface in part (a), sketch the graphs of $z = f(1, y)$ and $z = f(x, 1)$.

44. ***Conjecture*** Consider the function $f(x, y) = xy$, for $x \geq 0$ and $y \geq 0$.

(a) Sketch the graph of the surface given by f.

(b) Make a conjecture about the relationship between the graphs of f and $g(x, y) = f(x, y) - 3$. Use a computer algebra system to confirm your answer.

(c) Make a conjecture about the relationship between the graphs of f and $g(x, y) = -f(x, y)$. Use a computer algebra system to confirm your answer.

(d) Make a conjecture about the relationship between the graphs of f and $g(x, y) = \frac{1}{2}f(x, y)$. Use a computer algebra system to confirm your answer.

(e) On the surface in part (a), sketch the graph of $z = f(x, x)$.

In Exercises 45–48, match the graph of the surface with one of the contour maps. [The contour maps are labeled (a), (b), (c), and (d).]

(a)

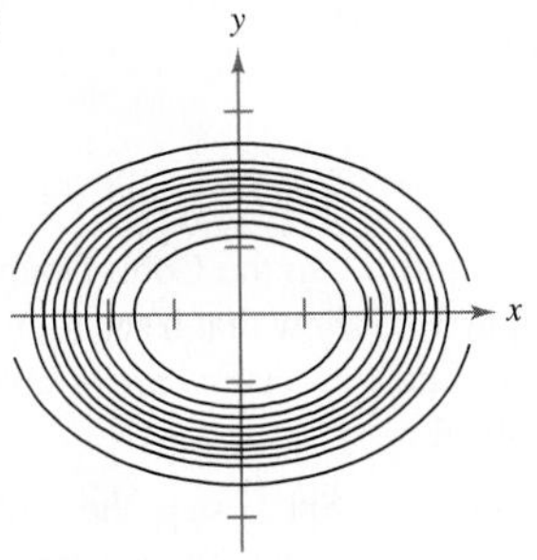

(b)

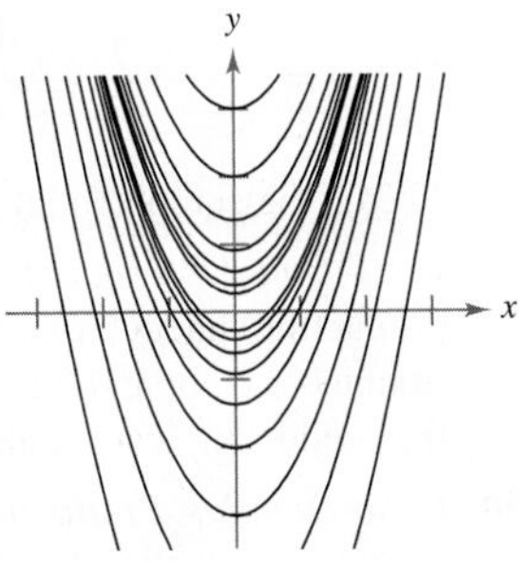

(c)

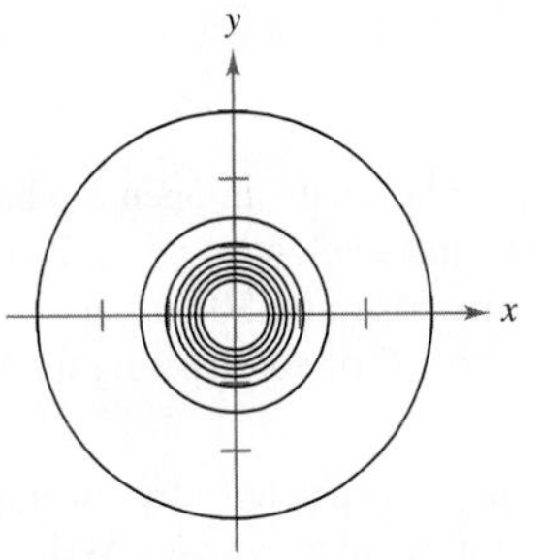

(d)

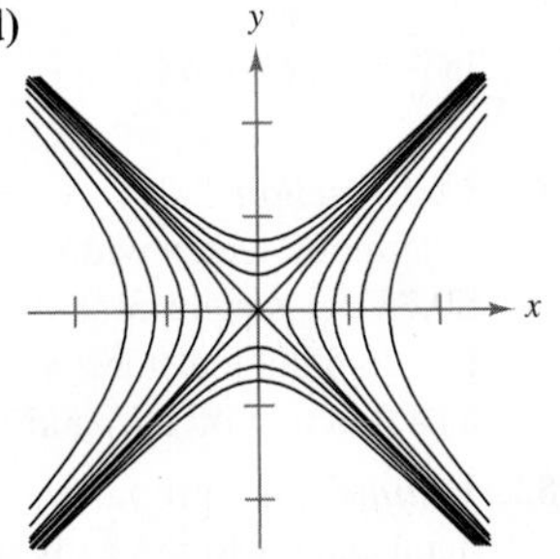

45. $f(x, y) = e^{1-x^2-y^2}$

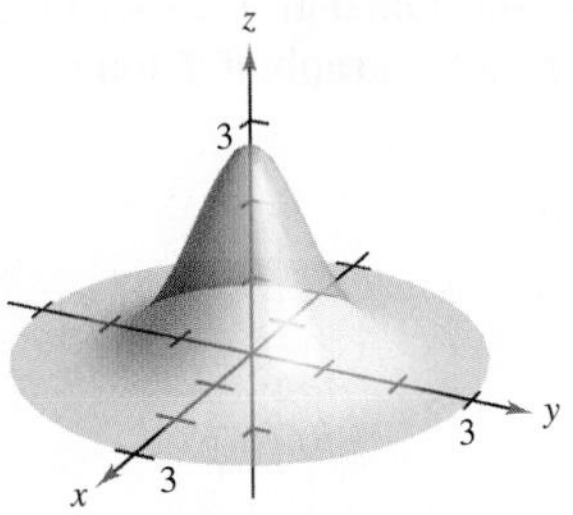

46. $f(x, y) = e^{1-x^2+y^2}$

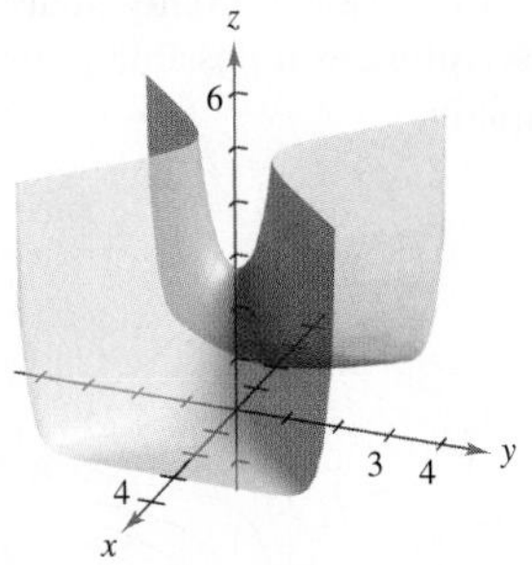

47. $f(x, y) = \ln|y - x^2|$

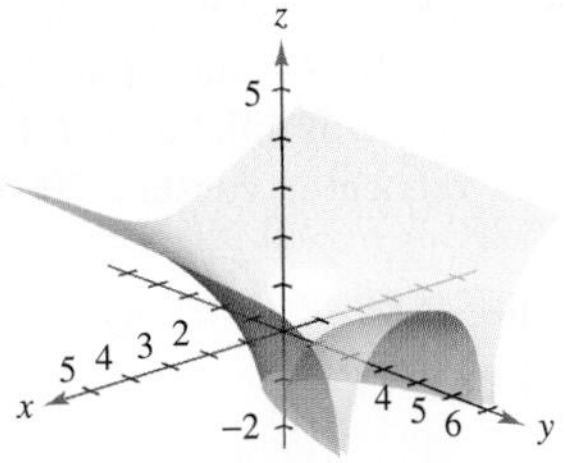

48. $f(x, y) = \cos\left(\frac{x^2 + 2y^2}{4}\right)$

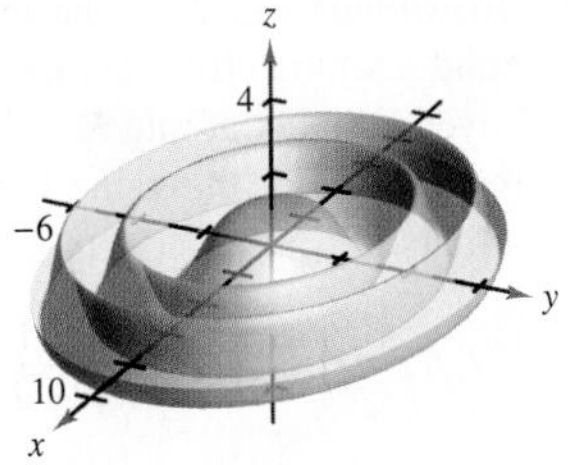

In Exercises 49–56, describe the level curves of the function. Sketch the level curves for the given c-values.

49. $z = x + y, \quad c = -1, 0, 2, 4$

50. $z = 6 - 2x - 3y, \quad c = 0, 2, 4, 6, 8, 10$

51. $z = \sqrt{25 - x^2 - y^2}, \quad c = 0, 1, 2, 3, 4, 5$

52. $f(x, y) = x^2 + 2y^2, \quad c = 0, 2, 4, 6, 8$

53. $f(x, y) = xy, \quad c = \pm 1, \pm 2, . . . , \pm 6$

54. $f(x, y) = e^{xy/2}, \quad c = 2, 3, 4, \frac{1}{2}, \frac{1}{3}, \frac{1}{4}$

55. $f(x, y) = \dfrac{x}{x^2 + y^2}, \quad c = \pm\frac{1}{2}, \pm 1, \pm\frac{3}{2}, \pm 2$

56. $f(x, y) = \ln(x - y), \quad c = 0, \pm\frac{1}{2}, \pm 1, \pm\frac{3}{2}, \pm 2$

In Exercises 57–60, use a graphing utility to graph six level curves of the function.

57. $f(x, y) = x^2 - y^2 + 2$

58. $f(x, y) = |xy|$

59. $g(x, y) = \dfrac{8}{1 + x^2 + y^2}$

60. $h(x, y) = 3\sin(|x| + |y|)$

Writing About Concepts

61. Define a function of two variables.

62. What is a graph of a function of two variables? How is it interpreted geometrically? Describe level curves.

63. All of the level curves of the surface given by $z = f(x, y)$ are concentric circles. Does this imply that the graph of f is a hemisphere? Illustrate your answer with an example.

64. Construct a function whose level curves are lines passing through the origin.

Writing **In Exercises 65 and 66, use the graphs of the level curves (c-values evenly spaced) of the function f to write a description of a possible graph of f. Is the graph of f unique? Explain.**

65.

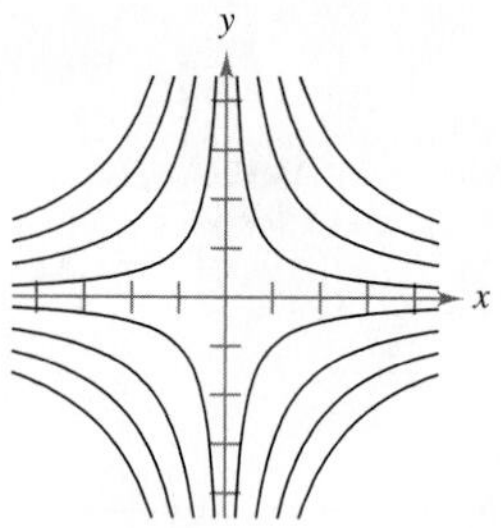

66.

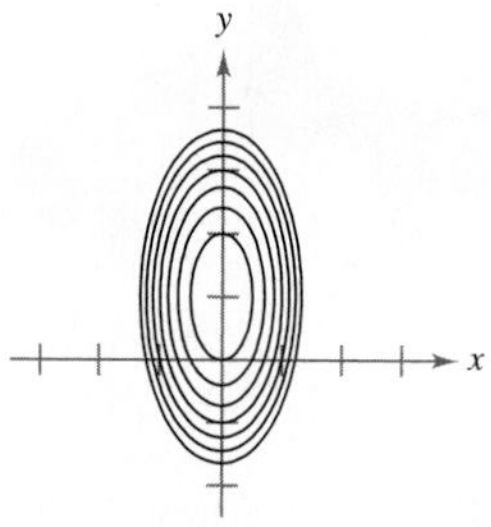

67. ***Investment*** In 2005, an investment of \$1000 was made in a bond earning 10% compounded annually. Assume that the buyer pays tax at rate R and the annual rate of inflation is I. In the year 2015, the value V of the investment in constant 2005 dollars is

$$V(I, R) = 1000\left[\frac{1 + 0.10(1 - R)}{1 + I}\right]^{10}.$$

Use this function of two variables to complete the table.

	Inflation Rate		
Tax Rate	**0**	**0.03**	**0.05**
0			
0.28			
0.35			

68. ***Investment*** A principal of \$1000 is deposited in a savings account that earns an interest rate of r (written as a decimal), compounded continuously. The amount $A(r, t)$ after t years is

$$A(r, t) = 1000e^{rt}.$$

Use this function of two variables to complete the table.

	Number of Years			
Rate	**5**	**10**	**15**	**20**
0.02				
0.04				
0.06				
0.08				

In Exercises 69–74, sketch the graph of the level surface $f(x, y, z) = c$ at the given value of c.

69. $f(x, y, z) = x - 2y + 3z, \quad c = 6$

70. $f(x, y, z) = 4x + y + 2z, \quad c = 4$

71. $f(x, y, z) = x^2 + y^2 + z^2, \quad c = 9$

72. $f(x, y, z) = x^2 + \frac{1}{4}y^2 - z, \quad c = 1$

73. $f(x, y, z) = 4x^2 + 4y^2 - z^2, \quad c = 0$

74. $f(x, y, z) = \sin x - z, \quad c = 0$

75. ***Forestry*** The **Doyle Log Rule** is one of several methods used to determine the lumber yield of a log (in board-feet) in terms of its diameter d (in inches) and its length L (in feet). The number of board-feet is

$$N(d, L) = \left(\frac{d - 4}{4}\right)^2 L.$$

(a) Find the number of board-feet of lumber in a log 22 inches in diameter and 12 feet in length.

(b) Find $N(30, 12)$.

76. ***Queuing Model*** The average length of time that a customer waits in line for service is

$$W(x, y) = \frac{1}{x - y}, \quad x > y$$

where y is the average arrival rate, written as the number of customers per unit of time, and x is the average service rate, written in the same units. Evaluate each of the following.

(a) $W(15, 10)$ (b) $W(12, 9)$ (c) $W(12, 6)$ (d) $W(4, 2)$

77. ***Temperature Distribution*** The temperature T (in degrees Celsius) at any point (x, y) in a circular steel plate of radius 10 meters is

$$T = 600 - 0.75x^2 - 0.75y^2$$

where x and y are measured in meters. Sketch some of the isothermal curves.

78. ***Electric Potential*** The electric potential V at any point (x, y) is

$$V(x, y) = \frac{5}{\sqrt{25 + x^2 + y^2}}.$$

Sketch the equipotential curves for $V = \frac{1}{2}$, $V = \frac{1}{3}$, and $V = \frac{1}{4}$.

79. ***Cobb-Douglas Production Function*** Use the Cobb-Douglas production function (see Example 5) to show that if the number of units of labor and the number of units of capital are doubled, the production level is also doubled.

80. ***Cobb-Douglas Production Function*** Show that the Cobb-Douglas production function $z = Cx^a y^{1-a}$ can be rewritten as

$$\ln\frac{z}{y} = \ln C + a\ln\frac{x}{y}.$$

81. ***Construction Cost*** A rectangular box with an open top has a length of x feet, a width of y feet, and a height of z feet. It costs \$0.75 per square foot to build the base and \$0.40 per square foot to build the sides. Write the cost C of constructing the box as a function of x, y, and z.

82. ***Volume*** A propane tank is constructed by welding hemispheres to the ends of a right circular cylinder. Write the volume V of the tank as a function of r and l, where r is the radius of the cylinder and hemispheres, and l is the length of the cylinder.

83. ***Ideal Gas Law*** According to the Ideal Gas Law, $PV = kT$, where P is pressure, V is volume, T is temperature (in Kelvins), and k is a constant of proportionality. A tank contains 2600 cubic inches of nitrogen at a pressure of 20 pounds per square inch and a temperature of 300 K.

(a) Determine k.

(b) Write P as a function of V and T and describe the level curves.

84. ***Modeling Data*** The table shows the net sales x (in billions of dollars), the total assets y (in billions of dollars), and the shareholder's equity z (in billions of dollars) for Wal-Mart for the years 1998 through 2003. *(Source: 2003 Annual Report for Wal-Mart)*

Year	1998	1999	2000	2001	2002	2003
x	118.0	137.6	165.0	191.3	217.8	244.5
y	45.4	50.0	70.3	78.1	83.5	94.7
z	18.5	21.1	25.8	31.3	35.1	39.3

A model for these data is

$$z = f(x, y) = 0.156x + 0.031y - 1.66.$$

(a) Use a graphing utility and the model to approximate z for the given values of x and y.

(b) Which of the two variables in this model has the greater influence on shareholder's equity?

(c) Simplify the expression for $f(x, 55)$ and interpret its meaning in the context of the problem.

85. ***Meteorology*** Meteorologists measure the atmospheric pressure in millibars. From these observations they create weather maps on which the curves of equal atmospheric pressure (isobars) are drawn (see figure). On the map, the closer the isobars the higher the wind speed. Match points A, B, and C with (a) highest pressure, (b) lowest pressure, and (c) highest wind velocity.

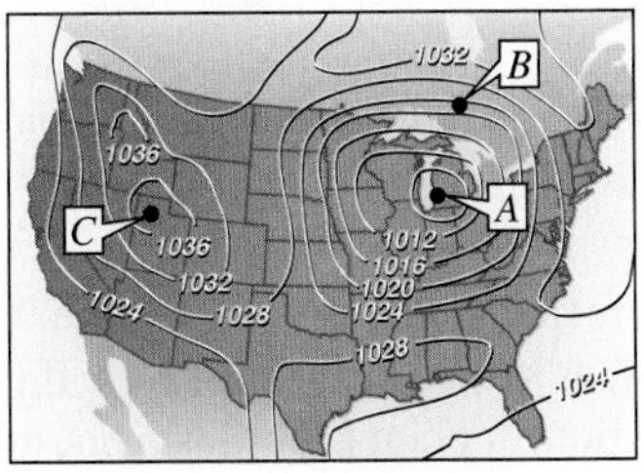

Figure for 85

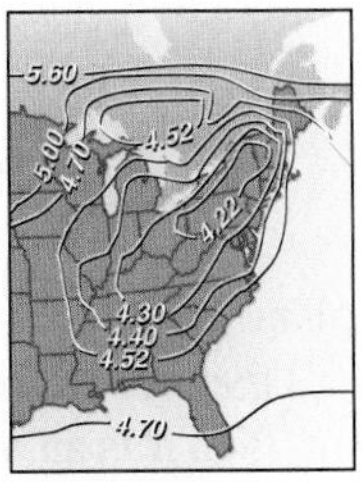

Figure for 86

86. ***Acid Rain*** The acidity of rainwater is measured in units called pH. A pH of 7 is neutral, smaller values are increasingly acidic, and larger values are increasingly alkaline. The map shows the curves of equal pH and gives evidence that downwind of heavily industrialized areas the acidity has been increasing. Using the level curves on the map, determine the direction of the prevailing winds in the northeastern United States.

87. ***Air Conditioner Use*** The contour map shown in the figure represents the estimated annual hours of air conditioner use for an average household. *(Source: Association of Home Appliance Manufacturers)*

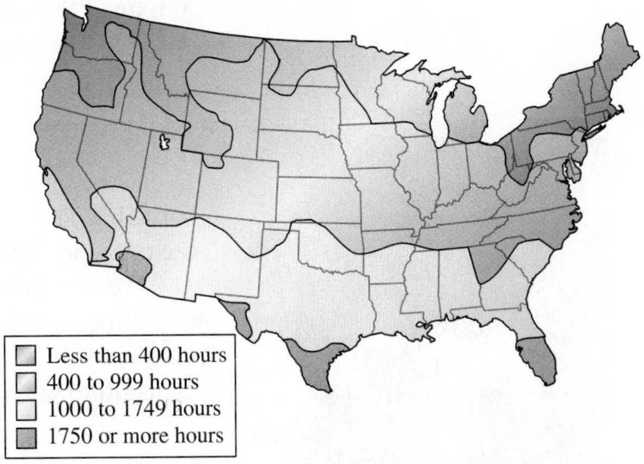

(a) Discuss the use of color to represent the level curves.

(b) Do the level curves correspond to equally spaced annual usage hours? Explain.

(c) Describe how to obtain a more detailed contour map.

88. ***Geology*** The contour map in the figure represents color-coded seismic amplitudes of a fault horizon and a projected contour map, which is used in earthquake studies. *(Source: Adapted from Shipman/Wilson/Todd,* An Introduction to Physical Science, *Eighth Edition)*

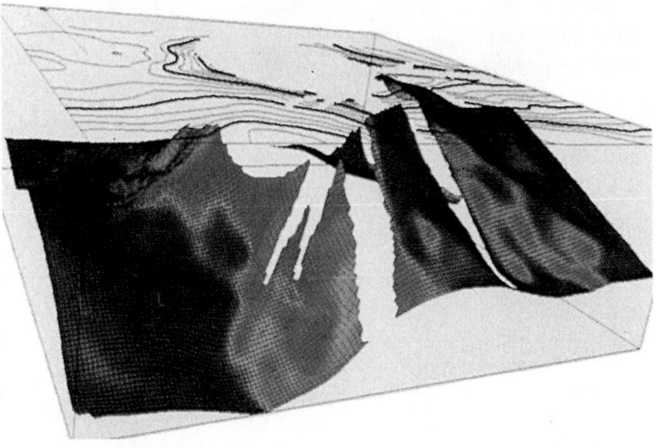

GeoQuest Systems, Inc.

(a) Discuss the use of color to represent the level curves.

(b) Do the level curves correspond to equally spaced amplitudes? Explain.

True or False? **In Exercises 89–92, determine whether the statement is true or false. If it is false, explain why or give an example that shows it is false.**

89. If $f(x_0, y_0) = f(x_1, y_1)$, then $x_0 = x_1$ and $y_0 = y_1$.

90. A vertical line can intersect the graph of $z = f(x, y)$ at most once.

91. If f is a function, then $f(ax, ay) = a^2 f(x, y)$.

92. The graph of $f(x, y) = x^2 - y^2$ is a hyperbolic paraboloid.

Section 13.2 Limits and Continuity

- **Understand the definition of a neighborhood in the plane.**
- **Understand and use the definition of the limit of a function of two variables.**
- **Extend the concept of continuity to a function of two variables.**
- **Extend the concept of continuity to a function of three variables.**

SONYA KOVALEVSKY (1850–1891)

Much of the terminology used to define limits and continuity of a function of two or three variables was introduced by the German mathematician Karl Weierstrass (1815–1897). Weierstrass's rigorous approach to limits and other topics in calculus gained him the reputation as the "father of modern analysis." Weierstrass was a gifted teacher. One of his best-known students was the Russian mathematician Sonya Kovalevsky, who applied many of Weierstrass's techniques to problems in mathematical physics and became one of the first women to gain acceptance as a research mathematician.

Neighborhoods in the Plane

In this section, you will study limits and continuity involving functions of two or three variables. The section begins with functions of two variables. At the end of the section, the concepts are extended to functions of three variables.

We begin our discussion of the limit of a function of two variables by defining a two-dimensional analog to an interval on the real line. Using the formula for the distance between two points (x, y) and (x_0, y_0) in the plane, you can define the **δ-neighborhood** about (x_0, y_0) to be the **disk** centered at (x_0, y_0) with radius $\delta > 0$

$$\{(x, y): \sqrt{(x - x_0)^2 + (y - y_0)^2} < \delta\}$$ Open disk

as shown in Figure 13.18. When this formula contains the *less than* inequality, <, the disk is called **open,** and when it contains the *less than or equal to* inequality, ≤, the disk is called **closed.** This corresponds to the use of < and ≤ to define open and closed intervals.

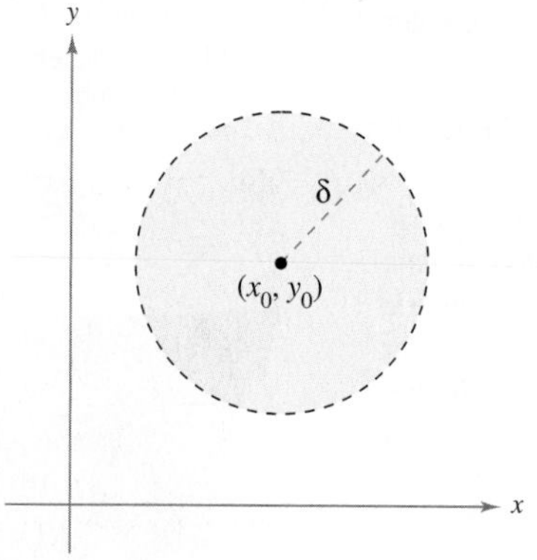

An open disk
Figure 13.18

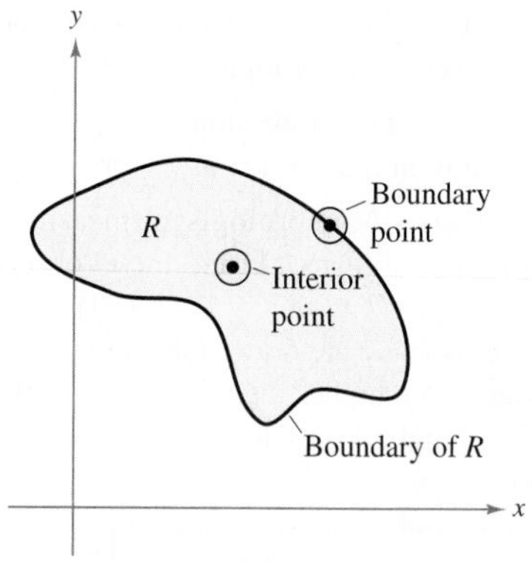

The boundary and interior points of a region R
Figure 13.19

A point (x_0, y_0) in a plane region R is an **interior point** of R if there exists a δ-neighborhood about (x_0, y_0) that lies entirely in R, as shown in Figure 13.19. If every point in R is an interior point, then R is an **open region.** A point (x_0, y_0) is a **boundary point** of R if every open disk centered at (x_0, y_0) contains points inside R *and* points outside R. By definition, a region must contain its interior points, but it need not contain its boundary points. If a region contains all its boundary points, the region is **closed.** A region that contains some but not all of its boundary points is neither open nor closed.

FOR FURTHER INFORMATION For more information on Sonya Kovalevsky, see the article "S. Kovalevsky: A Mathematical Lesson" by Karen D. Rappaport in *The American Mathematical Monthly*. To view this article, go to the website *www.matharticles.com.*

Limit of a Function of Two Variables

> **Definition of the Limit of a Function of Two Variables**
>
> Let f be a function of two variables defined, except possibly at (x_0, y_0), on an open disk centered at (x_0, y_0), and let L be a real number. Then
>
> $$\lim_{(x, y)\to(x_0, y_0)} f(x, y) = L$$
>
> if for each $\varepsilon > 0$ there corresponds a $\delta > 0$ such that
>
> $$|f(x, y) - L| < \varepsilon \quad \text{whenever} \quad 0 < \sqrt{(x - x_0)^2 + (y - y_0)^2} < \delta.$$

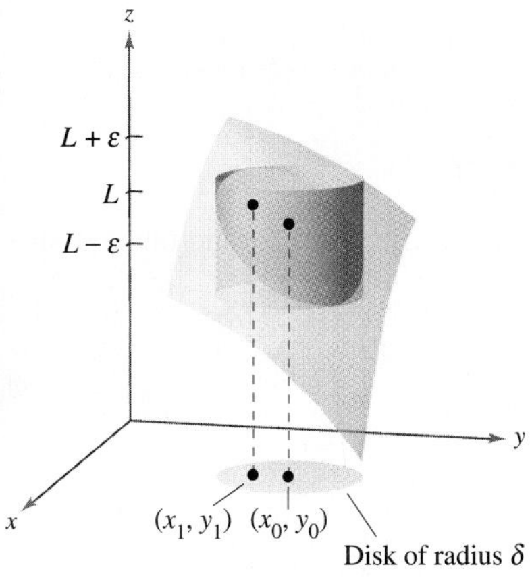

For any (x, y) in the circle of radius δ, the value $f(x, y)$ lies between $L + \varepsilon$ and $L - \varepsilon$.
Figure 13.20

NOTE Graphically, this definition of a limit implies that for any point $(x, y) \neq (x_0, y_0)$ in the disk of radius δ, the value $f(x, y)$ lies between $L + \varepsilon$ and $L - \varepsilon$, as shown in Figure 13.20.

The definition of the limit of a function of two variables is similar to the definition of the limit of a function of a single variable, yet there is a critical difference. To determine whether a function of a single variable has a limit, you need only test the approach from two directions—from the right and from the left. If the function approaches the same limit from the right and from the left, you can conclude that the limit exists. However, for a function of two variables, the statement

$$(x, y) \to (x_0, y_0)$$

means that the point (x, y) is allowed to approach (x_0, y_0) from any direction. If the value of

$$\lim_{(x, y)\to(x_0, y_0)} f(x, y)$$

is not the same for all possible approaches, or **paths,** to (x_0, y_0), the limit does not exist.

EXAMPLE 1 Verifying a Limit by the Definition

Show that

$$\lim_{(x, y)\to(a, b)} x = a.$$

Solution Let $f(x, y) = x$ and $L = a$. You need to show that for each $\varepsilon > 0$, there exists a δ-neighborhood about (a, b) such that

$$|f(x, y) - L| = |x - a| < \varepsilon$$

whenever $(x, y) \neq (a, b)$ lies in the neighborhood. You can first observe that from

$$0 < \sqrt{(x - a)^2 + (y - b)^2} < \delta$$

it follows that

$$\begin{aligned} |f(x, y) - a| &= |x - a| \\ &= \sqrt{(x - a)^2} \\ &\leq \sqrt{(x - a)^2 + (y - b)^2} \\ &< \delta. \end{aligned}$$

So, you can choose $\delta = \varepsilon$, and the limit is verified.

Limits of functions of several variables have the same properties regarding sums, differences, products, and quotients as do limits of functions of single variables. (See Theorem 1.2 in Section 1.3.) Some of these properties are used in the next example.

EXAMPLE 2 Verifying a Limit

Evaluate $\displaystyle\lim_{(x, y)\to(1, 2)} \frac{5x^2y}{x^2 + y^2}$.

Solution By using the properties of limits of products and sums, you obtain

$$\lim_{(x, y)\to(1, 2)} 5x^2y = 5(1^2)(2) = 10$$

and

$$\lim_{(x, y)\to(1, 2)} (x^2 + y^2) = (1^2 + 2^2) = 5.$$

Because the limit of a quotient is equal to the quotient of the limits (and the denominator is not 0), you have

$$\lim_{(x, y)\to(1, 2)} \frac{5x^2y}{x^2 + y^2} = \frac{10}{5} = 2.$$

EXAMPLE 3 Verifying a Limit

Evaluate $\displaystyle\lim_{(x, y)\to(0, 0)} \frac{5x^2y}{x^2 + y^2}$.

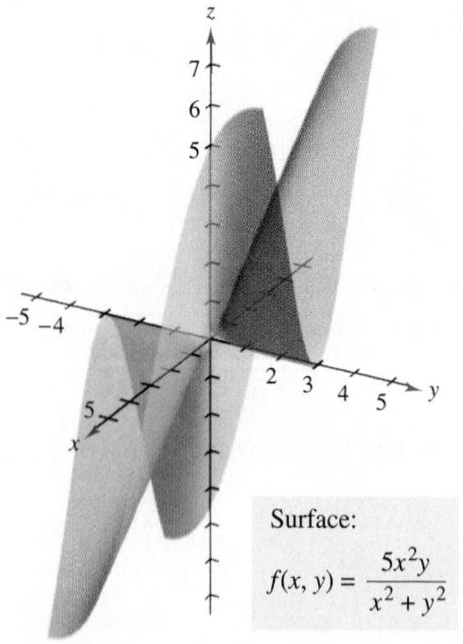

Figure 13.21

Solution In this case, the limits of the numerator and of the denominator are both 0, and so you cannot determine the existence (or nonexistence) of a limit by taking the limits of the numerator and denominator separately and then dividing. However, from the graph of f in Figure 13.21, it seems reasonable that the limit might be 0. So, you can try applying the definition to $L = 0$. First, note that

$$|y| \le \sqrt{x^2 + y^2} \qquad \text{and} \qquad \frac{x^2}{x^2 + y^2} \le 1.$$

Then, in a δ-neighborhood about (0, 0), you have $0 < \sqrt{x^2 + y^2} < \delta$, and it follows that, for $(x, y) \neq (0, 0)$,

$$\begin{aligned} |f(x, y) - 0| &= \left|\frac{5x^2y}{x^2 + y^2}\right| \\ &= 5|y|\left(\frac{x^2}{x^2 + y^2}\right) \\ &\le 5|y| \\ &\le 5\sqrt{x^2 + y^2} \\ &< 5\delta. \end{aligned}$$

So, you can choose $\delta = \varepsilon/5$ and conclude that

$$\lim_{(x, y)\to(0, 0)} \frac{5x^2y}{x^2 + y^2} = 0.$$

For some functions, it is easy to recognize that a limit does not exist. For instance, it is clear that the limit

$$\lim_{(x, y)\to(0, 0)} \frac{1}{x^2 + y^2}$$

does not exist because the values of $f(x, y)$ increase without bound as (x, y) approaches $(0, 0)$ along *any* path (see Figure 13.22).

For other functions, it is not so easy to recognize that a limit does not exist. For instance, the next example describes a limit that does not exist because the function approaches different values along different paths.

$\displaystyle\lim_{(x, y)\to(0, 0)} \frac{1}{x^2 + y^2}$ does not exist.

Figure 13.22

EXAMPLE 4 A Limit That Does Not Exist

Show that the following limit does not exist.

$$\lim_{(x, y)\to(0, 0)} \left(\frac{x^2 - y^2}{x^2 + y^2}\right)^2$$

Solution The domain of the function given by

$$f(x, y) = \left(\frac{x^2 - y^2}{x^2 + y^2}\right)^2$$

consists of all points in the xy-plane except for the point $(0, 0)$. To show that the limit as (x, y) approaches $(0, 0)$ does not exist, consider approaching $(0, 0)$ along two different "paths," as shown in Figure 13.23. Along the x-axis, every point is of the form $(x, 0)$, and the limit along this approach is

$$\lim_{(x, 0)\to(0, 0)} \left(\frac{x^2 - 0^2}{x^2 + 0^2}\right)^2 = \lim_{(x, 0)\to(0, 0)} 1^2 = 1. \qquad \text{Limit along } x\text{-axis}$$

However, if (x, y) approaches $(0, 0)$ along the line $y = x$, you obtain

$$\lim_{(x, x)\to(0, 0)} \left(\frac{x^2 - x^2}{x^2 + x^2}\right)^2 = \lim_{(x, x)\to(0, 0)} \left(\frac{0}{2x^2}\right)^2 = 0. \qquad \text{Limit along line } y = x$$

This means that in any open disk centered at $(0, 0)$ there are points (x, y) at which f takes on the value 1, and other points at which f takes on the value 0. For instance, $f(x, y) = 1$ at the points $(1, 0)$, $(0.1, 0)$, $(0.01, 0)$, and $(0.001, 0)$ and $f(x, y) = 0$ at the points $(1, 1)$, $(0.1, 0.1)$, $(0.01, 0.01)$, and $(0.001, 0.001)$. So, f does not have a limit as $(x, y) \to (0, 0)$.

NOTE In Example 4, you could conclude that the limit does not exist because you found two approaches that produced different limits. If two approaches had produced the same limit, you still could not have concluded that the limit exists. To form such a conclusion, you must show that the limit is the same along *all* possible approaches.

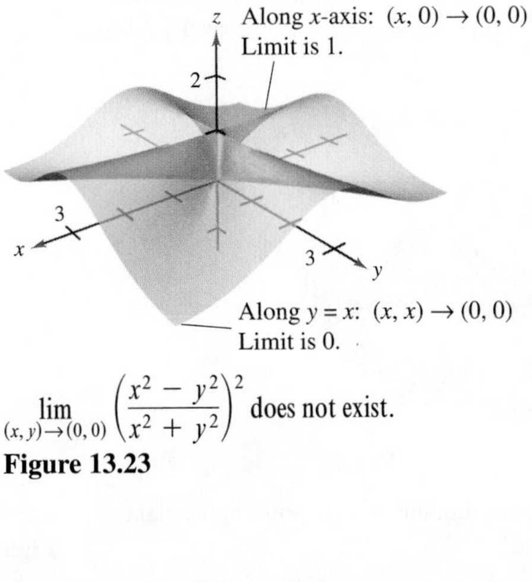

$\displaystyle\lim_{(x, y)\to(0, 0)} \left(\frac{x^2 - y^2}{x^2 + y^2}\right)^2$ does not exist.

Figure 13.23

Continuity of a Function of Two Variables

Notice in Example 2 that the limit of $f(x, y) = 5x^2y/(x^2 + y^2)$ as $(x, y) \to (1, 2)$ can be evaluated by direct substitution. That is, the limit is $f(1, 2) = 2$. In such cases the function f is said to be **continuous** at the point $(1, 2)$.

NOTE This definition of continuity can be extended to *boundary points* of the open region R by considering a special type of limit in which (x, y) is allowed to approach (x_0, y_0) along paths lying in the region R. This notion is similar to that of one-sided limits, as discussed in Chapter 1.

Definition of Continuity of a Function of Two Variables

A function f of two variables is **continuous at a point (x_0, y_0)** in an open region R if $f(x_0, y_0)$ is equal to the limit of $f(x, y)$ as (x, y) approaches (x_0, y_0). That is,

$$\lim_{(x, y)\to(x_0, y_0)} f(x, y) = f(x_0, y_0).$$

The function f is **continuous in the open region R** if it is continuous at every point in R.

In Example 3, it was shown that the function

$$f(x, y) = \frac{5x^2y}{x^2 + y^2}$$

is not continuous at $(0, 0)$. However, because the limit at this point exists, you can remove the discontinuity by defining f at $(0, 0)$ as being equal to its limit there. Such a discontinuity is called **removable.** In Example 4, the function

$$f(x, y) = \left(\frac{x^2 - y^2}{x^2 + y^2}\right)^2$$

was also shown not to be continuous at $(0, 0)$, but this discontinuity is **nonremovable.**

THEOREM 13.1 Continuous Functions of Two Variables

If k is a real number and f and g are continuous at (x_0, y_0), then the following functions are continuous at (x_0, y_0).

1. Scalar multiple: kf

2. Sum and difference: $f \pm g$

3. Product: fg

4. Quotient: f/g, if $g(x_0, y_0) \neq 0$

Theorem 13.1 establishes the continuity of *polynomial* and *rational* functions at every point in their domains. Furthermore, the continuity of other types of functions can be extended naturally from one to two variables. For instance, the functions whose graphs are shown in Figures 13.24 and 13.25 are continuous at every point in the plane.

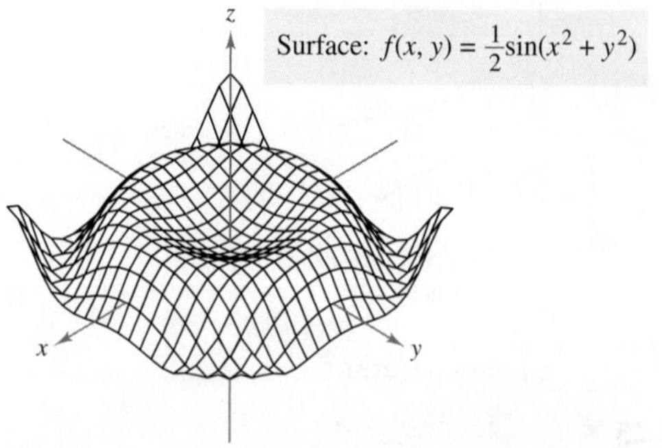

The function f is continuous at every point in the plane.
Figure 13.24

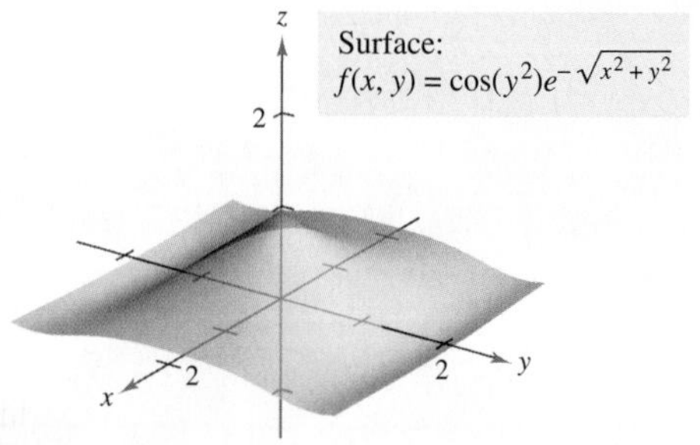

The function f is continuous at every point in the plane.
Figure 13.25

EXPLORATION

Hold a spoon a foot or so from your eyes. Look at your image in the spoon. It should be upside down. Now, move the spoon closer and closer to one eye. At some point, your image will be right side up. Could it be that your image is being continuously deformed? Talk about this question and the general meaning of continuity with other members of your class. (This exploration was suggested by Irvin Roy Hentzel, Iowa State University.)

The next theorem states conditions under which a composite function is continuous.

THEOREM 13.2 Continuity of a Composite Function

If h is continuous at (x_0, y_0) and g is continuous at $h(x_0, y_0)$, then the composite function given by $(g \circ h)(x, y) = g(h(x, y))$ is continuous at (x_0, y_0). That is,

$$\lim_{(x, y)\to(x_0, y_0)} g(h(x, y)) = g(h(x_0, y_0)).$$

NOTE Note in Theorem 13.2 that h is a function of two variables and g is a function of one variable.

EXAMPLE 5 Testing for Continuity

Discuss the continuity of each function.

a. $f(x, y) = \dfrac{x - 2y}{x^2 + y^2}$ **b.** $g(x, y) = \dfrac{2}{y - x^2}$

Solution

a. Because a rational function is continuous at every point in its domain, you can conclude that f is continuous at each point in the xy-plane except at $(0, 0)$, as shown in Figure 13.26.

b. The function given by $g(x, y) = 2/(y - x^2)$ is continuous except at the points at which the denominator is 0, $y - x^2 = 0$. So, you can conclude that the function is continuous at all points except those lying on the parabola $y = x^2$. Inside this parabola, you have $y > x^2$, and the surface represented by the function lies above the xy-plane, as shown in Figure 13.27. Outside the parabola, $y < x^2$, and the surface lies below the xy-plane.

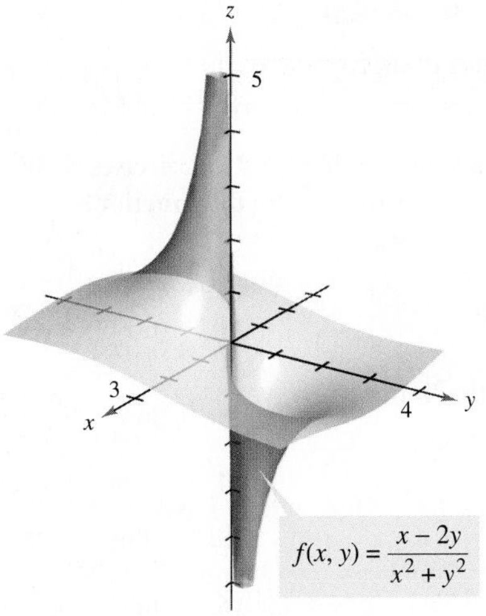

The function f is not continuous at $(0, 0)$.
Figure 13.26

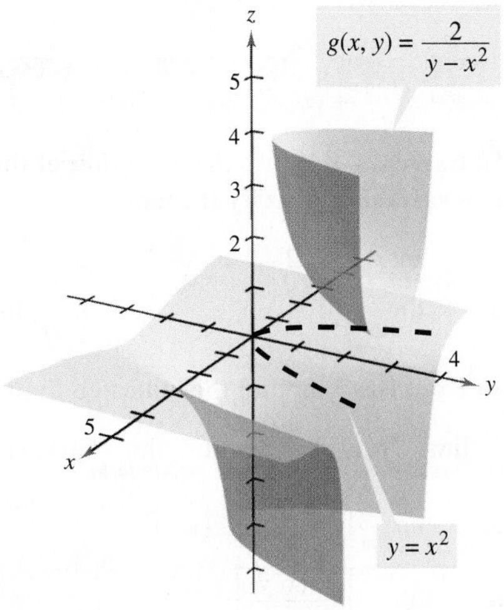

The function g is not continuous on the parabola $y = x^2$.
Figure 13.27

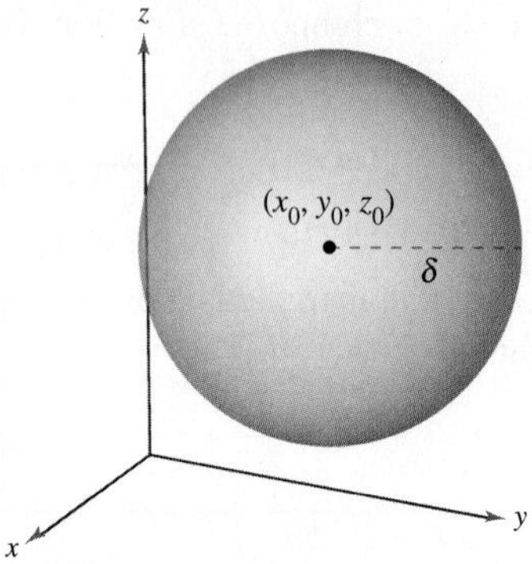

Open sphere in space
Figure 13.28

Continuity of a Function of Three Variables

The preceding definitions of limits and continuity can be extended to functions of three variables by considering points (x, y, z) within the *open sphere*

$$(x - x_0)^2 + (y - y_0)^2 + (z - z_0)^2 < \delta^2. \quad \text{Open sphere}$$

The radius of this sphere is δ, and the sphere is centered at (x_0, y_0, z_0), as shown in Figure 13.28. A point (x_0, y_0, z_0) in a region R in space is an **interior point** of R if there exists a δ-sphere about (x_0, y_0, z_0) that lies entirely in R. If every point in R is an interior point, then R is called **open.**

Definition of Continuity of a Function of Three Variables

A function f of three variables is **continuous at a point (x_0, y_0, z_0)** in an open region R if $f(x_0, y_0, z_0)$ is defined and is equal to the limit of $f(x, y, z)$ as (x, y, z) approaches (x_0, y_0, z_0). That is,

$$\lim_{(x, y, z)\to(x_0, y_0, z_0)} f(x, y, z) = f(x_0, y_0, z_0).$$

The function f is **continuous in the open region R** if it is continuous at every point in R.

EXAMPLE 6 Testing Continuity of a Function of Three Variables

The function

$$f(x, y, z) = \frac{1}{x^2 + y^2 - z}$$

is continuous at each point in space except at the points on the paraboloid given by $z = x^2 + y^2$.

Exercises for Section 13.2

See www.CalcChat.com for worked-out solutions to odd-numbered exercises.

In Exercises 1–4, use the definition of the limit of a function of two variables to verify the limit.

1. $\lim_{(x, y)\to(2, 3)} x = 2$
2. $\lim_{(x, y)\to(4, -1)} x = 4$
3. $\lim_{(x, y)\to(1, -3)} y = -3$
4. $\lim_{(x, y)\to(a, b)} y = b$

In Exercises 5–8, find the indicated limit by using the limits

$$\lim_{(x, y)\to(a, b)} f(x, y) = 5 \text{ and } \lim_{(x, y)\to(a, b)} g(x, y) = 3.$$

5. $\lim_{(x, y)\to(a, b)} [f(x, y) - g(x, y)]$
6. $\lim_{(x, y)\to(a, b)} \left[\frac{4f(x, y)}{g(x, y)}\right]$
7. $\lim_{(x, y)\to(a, b)} [f(x, y)g(x, y)]$
8. $\lim_{(x, y)\to(a, b)} \left[\frac{f(x, y) - g(x, y)}{f(x, y)}\right]$

In Exercises 9–18, find the limit and discuss the continuity of the function.

9. $\lim_{(x, y)\to(2, 1)} (x + 3y^2)$
10. $\lim_{(x, y)\to(0, 0)} (5x + y + 1)$
11. $\lim_{(x, y)\to(2, 4)} \frac{x + y}{x - y}$
12. $\lim_{(x, y)\to(1, 1)} \frac{x}{\sqrt{x + y}}$
13. $\lim_{(x, y)\to(0, 1)} \frac{\arcsin(x/y)}{1 + xy}$
14. $\lim_{(x, y)\to(\pi/4, 2)} y \cos xy$
15. $\lim_{(x, y)\to(-1, 2)} e^{xy}$
16. $\lim_{(x, y)\to(1, 1)} \frac{xy}{x^2 + y^2}$
17. $\lim_{(x, y, z)\to(1, 2, 5)} \sqrt{x + y + z}$
18. $\lim_{(x, y, z)\to(2, 0, 1)} xe^{yz}$

In Exercises 19–24, find the limit (if it exists). If the limit does not exist, explain why.

19. $\lim_{(x, y)\to(0, 0)} \dfrac{x + y}{x^2 + y}$

20. $\lim_{(x, y)\to(0, 0)} \dfrac{x}{x^2 - y^2}$

21. $\lim_{(x, y)\to(1, 1)} \dfrac{xy - 1}{1 + xy}$

22. $\lim_{(x, y)\to(0, 0)} \dfrac{x + y}{x + y^3}$

23. $\lim_{(x, y, z)\to(0, 0, 0)} \dfrac{xy + yz + xz}{x^2 + y^2 + z^2}$

24. $\lim_{(x, y, z)\to(0, 0, 0)} \dfrac{xy + yz^2 + xz^2}{x^2 + y^2 + z^2}$

In Exercises 25–28, discuss the continuity of the function and evaluate the limit of $f(x, y)$ (if it exists) as $(x, y) \to (0, 0)$.

25. $f(x, y) = e^{xy}$

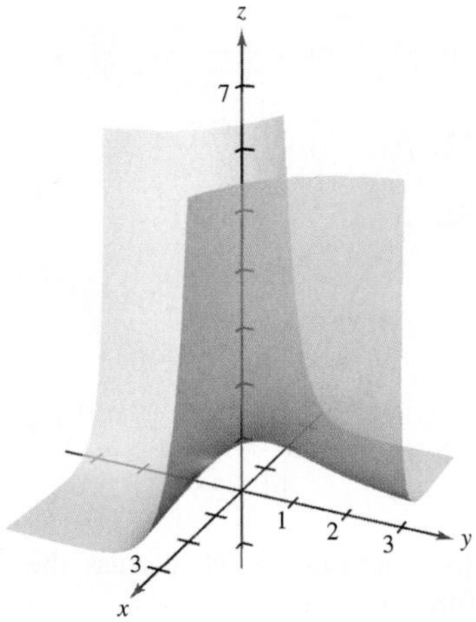

26. $f(x, y) = \dfrac{x^2}{(x^2 + 1)(y^2 + 1)}$

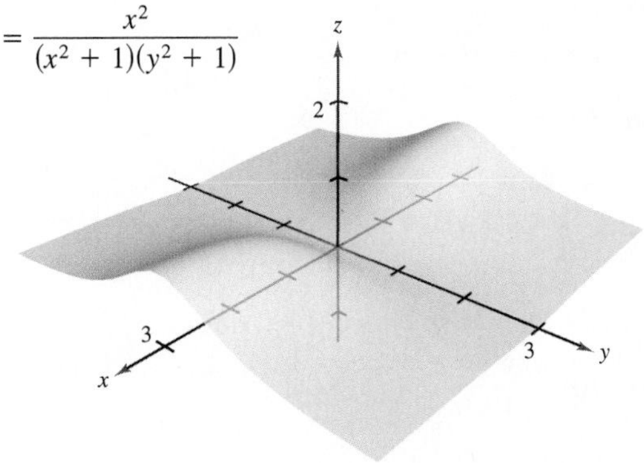

27. $f(x, y) = \ln(x^2 + y^2)$

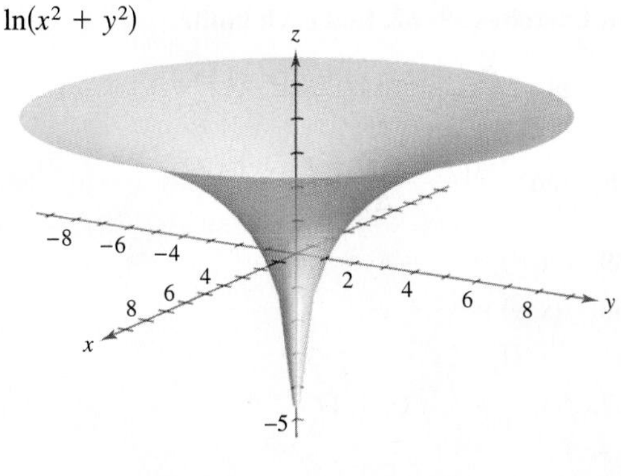

28. $f(x, y) = 1 - \dfrac{\cos(x^2 + y^2)}{x^2 + y^2}$

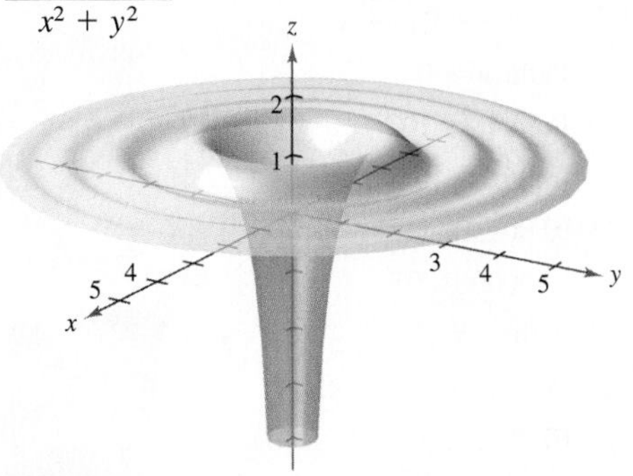

In Exercises 29–32, use a graphing utility to make a table showing the values of $f(x, y)$ at the given points. Use the result to make a conjecture about the limit of $f(x, y)$ as $(x, y) \to (0, 0)$. Determine whether the limit exists analytically and discuss the continuity of the function.

29. $f(x, y) = \dfrac{xy}{x^2 + y^2}$

Path: $y = 0$

Points: $(1, 0)$, $(0.5, 0)$, $(0.1, 0)$, $(0.01, 0)$, $(0.001, 0)$

Path: $y = x$

Points: $(1, 1)$, $(0.5, 0.5)$, $(0.1, 0.1)$, $(0.01, 0.01)$, $(0.001, 0.001)$

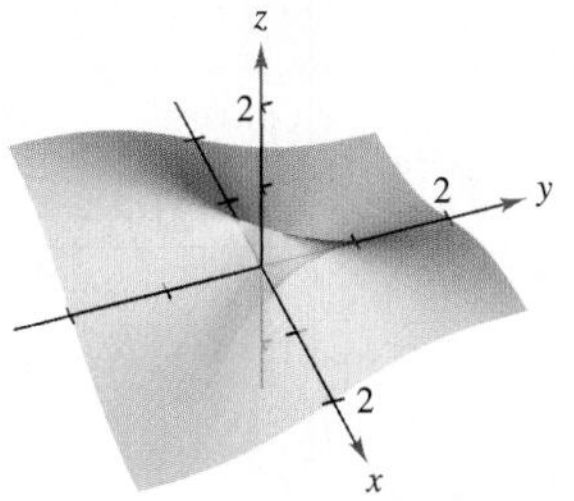

30. $f(x, y) = \dfrac{y}{x^2 + y^2}$

Path: $y = 0$

Points: $(1, 0)$, $(0.5, 0)$, $(0.1, 0)$, $(0.01, 0)$, $(0.001, 0)$

Path: $y = x$

Points: $(1, 1)$, $(0.5, 0.5)$, $(0.1, 0.1)$, $(0.01, 0.01)$, $(0.001, 0.001)$

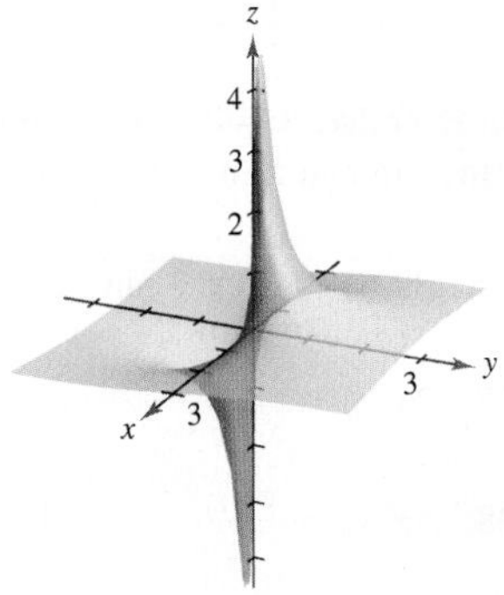

31. $f(x, y) = -\dfrac{xy^2}{x^2 + y^4}$

Path: $x = y^2$

Points: $(1, 1)$, $(0.25, 0.5)$, $(0.01, 0.1)$, $(0.0001, 0.01)$, $(0.000001, 0.001)$

Path: $x = -y^2$

Points: $(-1, 1)$, $(-0.25, 0.5)$, $(-0.01, 0.1)$, $(-0.0001, 0.01)$, $(-0.000001, 0.001)$

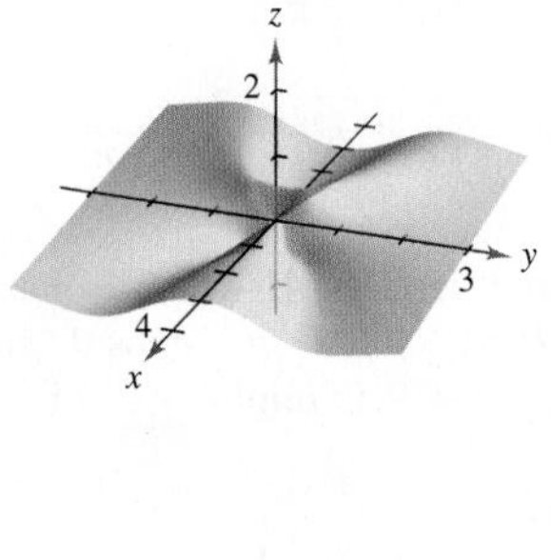

32. $f(x, y) = \dfrac{2x - y^2}{2x^2 + y}$

Path: $y = 0$

Points: $(1, 0)$, $(0.25, 0)$, $(0.01, 0)$, $(0.001, 0)$, $(0.000001, 0)$

Path: $y = x$

Points: $(1, 1)$, $(0.25, 0.25)$, $(0.01, 0.01)$, $(0.001, 0.001)$, $(0.0001, 0.0001)$

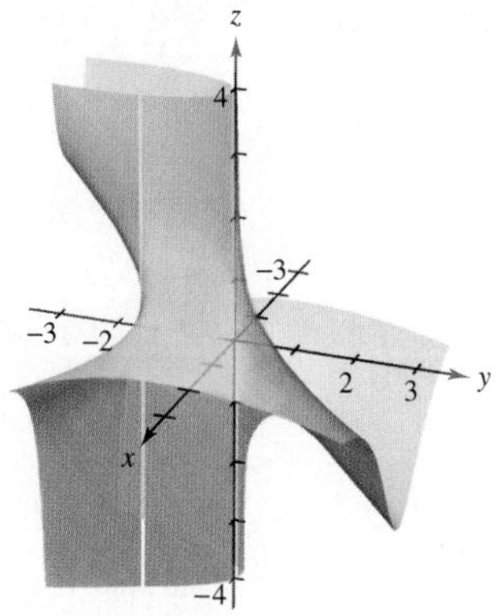

In Exercises 33 and 34, discuss the continuity of the functions f and g. Explain any differences.

33. $f(x, y) = \begin{cases} \dfrac{x^2 + 2xy^2 + y^2}{x^2 + y^2}, & (x, y) \neq (0, 0) \\ 0, & (x, y) = (0, 0) \end{cases}$

$g(x, y) = \begin{cases} \dfrac{x^2 + 2xy^2 + y^2}{x^2 + y^2}, & (x, y) \neq (0, 0) \\ 1, & (x, y) = (0, 0) \end{cases}$

34. $f(x, y) = \begin{cases} \dfrac{4x^2y^2}{x^2 + y^2}, & (x, y) \neq (0, 0) \\ 0, & (x, y) = (0, 0) \end{cases}$

$g(x, y) = \begin{cases} \dfrac{4x^2y^2}{x^2 + y^2}, & (x, y) \neq (0, 0) \\ 2, & (x, y) = (0, 0) \end{cases}$

In Exercises 35–40, use a computer algebra system to graph the function and find $\lim\limits_{(x, y)\to(0, 0)} f(x, y)$ (if it exists).

35. $f(x, y) = \sin x + \sin y$

36. $f(x, y) = \sin \dfrac{1}{x} + \cos \dfrac{1}{x}$

37. $f(x, y) = \dfrac{x^2y}{x^4 + 4y^2}$

38. $f(x, y) = \dfrac{x^2 + y^2}{x^2y}$

39. $f(x, y) = \dfrac{10xy}{2x^2 + 3y^2}$

40. $f(x, y) = \dfrac{2xy}{x^2 + y^2 + 1}$

In Exercises 41–48, use polar coordinates to find the limit. [*Hint:* Let $x = r\cos\theta$ and $y = r\sin\theta$, and note that $(x, y)\to(0, 0)$ implies $r \to 0$.]

41. $\lim\limits_{(x, y)\to(0, 0)} \dfrac{\sin(x^2 + y^2)}{x^2 + y^2}$

42. $\lim\limits_{(x, y)\to(0, 0)} \dfrac{xy^2}{x^2 + y^2}$

43. $\lim\limits_{(x, y)\to(0, 0)} \dfrac{x^3 + y^3}{x^2 + y^2}$

44. $\lim\limits_{(x, y)\to(0, 0)} \dfrac{x^2y^2}{x^2 + y^2}$

45. $\lim\limits_{(x, y)\to(0, 0)} \dfrac{x^2 - y^2}{\sqrt{x^2 + y^2}}$

46. $\lim\limits_{(x, y)\to(0, 0)} \dfrac{\sin\sqrt{x^2 + y^2}}{\sqrt{x^2 + y^2}}$

47. $\lim\limits_{(x, y)\to(0, 0)} (x^2 + y^2)\ln(x^2 + y^2)$

48. $\lim\limits_{(x, y)\to(0, 0)} \dfrac{1 - \cos(x^2 + y^2)}{x^2 + y^2}$

In Exercises 49–54, discuss the continuity of the function.

49. $f(x, y, z) = \dfrac{1}{\sqrt{x^2 + y^2 + z^2}}$

50. $f(x, y, z) = \dfrac{z}{x^2 + y^2 - 9}$

51. $f(x, y, z) = \dfrac{\sin z}{e^x + e^y}$

52. $f(x, y, z) = xy \sin z$

53. $f(x, y) = \begin{cases} \dfrac{\sin xy}{xy}, & xy \neq 0 \\ 1, & xy = 0 \end{cases}$

54. $f(x, y) = \begin{cases} \dfrac{\sin(x^2 - y^2)}{x^2 - y^2}, & x^2 \neq y^2 \\ 1, & x^2 = y^2 \end{cases}$

In Exercises 55–58, discuss the continuity of the composite function $f \circ g$.

55. $f(t) = t^2$

$g(x, y) = 3x - 2y$

56. $f(t) = \dfrac{1}{t}$

$g(x, y) = x^2 + y^2$

57. $f(t) = \dfrac{1}{t}$

$g(x, y) = 3x - 2y$

58. $f(t) = \dfrac{1}{4 - t}$

$g(x, y) = x^2 + y^2$

In Exercises 59–62, find each limit.

(a) $\lim\limits_{\Delta x\to 0} \dfrac{f(x + \Delta x, y) - f(x, y)}{\Delta x}$

(b) $\lim\limits_{\Delta y\to 0} \dfrac{f(x, y + \Delta y) - f(x, y)}{\Delta y}$

59. $f(x, y) = x^2 - 4y$

60. $f(x, y) = x^2 + y^2$

61. $f(x, y) = 2x + xy - 3y$

62. $f(x, y) = \sqrt{y}\,(y + 1)$

True or False? **In Exercises 63–66, determine whether the statement is true or false. If it is false, explain why or give an example that shows it is false.**

63. If $\lim_{(x, y)\to(0, 0)} f(x, y) = 0$, then $\lim_{x\to 0} f(x, 0) = 0$.

64. If $\lim_{(x, y)\to(0, 0)} f(0, y) = 0$, then $\lim_{(x, y)\to(0, 0)} f(x, y) = 0$.

65. If f is continuous for all nonzero x and y, and $f(0, 0) = 0$, then $\lim_{(x, y)\to(0, 0)} f(x, y) = 0$.

66. If g and h are continuous functions of x and y, and $f(x, y) = g(x) + h(y)$, then f is continuous.

Writing About Concepts

67. Define the limit of a function of two variables. Describe a method for showing that

$$\lim_{(x, y)\to(x_0, y_0)} f(x, y)$$

does not exist.

68. State the definition of continuity of a function of two variables.

69. If $f(2, 3) = 4$, can you conclude anything about

$$\lim_{(x, y)\to(2, 3)} f(x, y)?$$

Give reasons for your answer.

70. If $\lim_{(x, y)\to(2, 3)} f(x, y) = 4$, can you conclude anything about $f(2, 3)$? Give reasons for your answer.

71. Consider $\displaystyle\lim_{(x, y)\to(0, 0)} \frac{x^2 + y^2}{xy}$ (see figure).

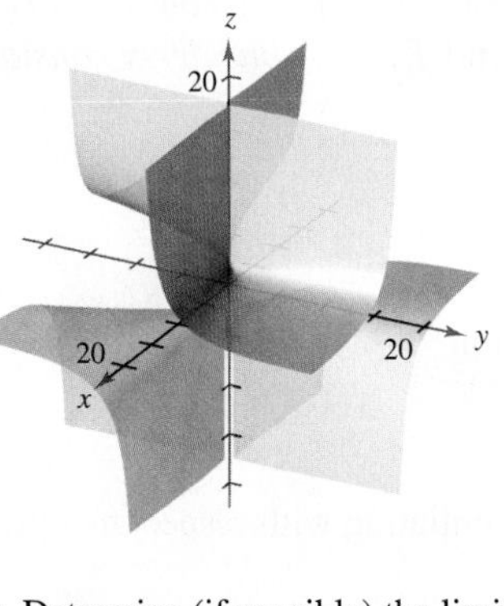

(a) Determine (if possible) the limit along any line of the form $y = ax$.

(b) Determine (if possible) the limit along the parabola $y = x^2$.

(c) Does the limit exist? Explain.

72. Consider $\displaystyle\lim_{(x, y)\to(0, 0)} \frac{x^2y}{x^4 + y^2}$ (see figure).

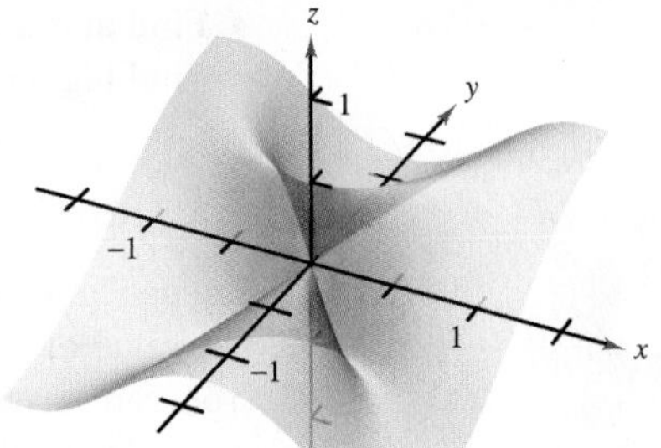

(a) Determine (if possible) the limit along any line of the form $y = ax$.

(b) Determine (if possible) the limit along the parabola $y = x^2$.

(c) Does the limit exist? Explain.

In Exercises 73 and 74, use spherical coordinates to find the limit. [*Hint:* Let $x = \rho \sin\phi \cos\theta$, $y = \rho \sin\phi \sin\theta$, and $z = \rho \cos\phi$, and note that $(x, y, z)\to(0, 0, 0)$ implies $\rho\to 0^+$.]

73. $\displaystyle\lim_{(x, y, z)\to(0, 0, 0)} \frac{xyz}{x^2 + y^2 + z^2}$

74. $\displaystyle\lim_{(x, y, z)\to(0, 0, 0)} \tan^{-1}\left[\frac{1}{x^2 + y^2 + z^2}\right]$

75. Find the following limit.

$$\lim_{(x, y)\to(0, 1)} \tan^{-1}\left[\frac{x^2 + 1}{x^2 + (y - 1)^2}\right]$$

76. For the function

$$f(x, y) = xy\left(\frac{x^2 - y^2}{x^2 + y^2}\right)$$

define $f(0, 0)$ such that f is continuous at the origin.

77. Prove that

$$\lim_{(x, y)\to(a, b)} [f(x, y) + g(x, y)] = L_1 + L_2$$

where $f(x, y)$ approaches L_1 and $g(x, y)$ approaches L_2 as $(x, y) \to (a, b)$.

78. Prove that if f is continuous and $f(a, b) < 0$, there exists a δ-neighborhood about (a, b) such that $f(x, y) < 0$ for every point (x, y) in the neighborhood.

Section 13.3 Partial Derivatives

- **Find and use partial derivatives of a function of two variables.**
- **Find and use partial derivatives of a function of three or more variables.**
- **Find higher-order partial derivatives of a function of two or three variables.**

Mary Evans Picture Library

JEAN LE ROND D'ALEMBERT (1717–1783)

The introduction of partial derivatives followed Newton's and Leibniz's work in calculus by several years. Between 1730 and 1760, Leonhard Euler and Jean Le Rond d'Alembert separately published several papers on dynamics, in which they established much of the theory of partial derivatives. These papers used functions of two or more variables to study problems involving equilibrium, fluid motion, and vibrating strings.

Partial Derivatives of a Function of Two Variables

In applications of functions of several variables, the question often arises, "How will the value of a function be affected by a change in one of its independent variables?" You can answer this by considering the independent variables one at a time. For example, to determine the effect of a catalyst in an experiment, a chemist could conduct the experiment several times using varying amounts of the catalyst, while keeping constant other variables such as temperature and pressure. You can use a similar procedure to determine the rate of change of a function f with respect to one of its several independent variables. This process is called **partial differentiation,** and the result is referred to as the **partial derivative** of f with respect to the chosen independent variable.

Definition of Partial Derivatives of a Function of Two Variables

If $z = f(x, y)$, then the **first partial derivatives** of f with respect to x and y are the functions f_x and f_y defined by

$$f_x(x, y) = \lim_{\Delta x \to 0} \frac{f(x + \Delta x, y) - f(x, y)}{\Delta x}$$

$$f_y(x, y) = \lim_{\Delta y \to 0} \frac{f(x, y + \Delta y) - f(x, y)}{\Delta y}$$

provided the limits exist.

This definition indicates that if $z = f(x, y)$, then to find f_x you *consider y constant* and differentiate with respect to x. Similarly, to find f_y, you *consider x constant* and differentiate with respect to y.

EXAMPLE 1 Finding Partial Derivatives

Find the partial derivatives f_x and f_y for the function

$$f(x, y) = 3x - x^2y^2 + 2x^3y. \qquad \text{Original function}$$

Solution Considering y to be constant and differentiating with respect to x produces

$$f(x, y) = 3x - x^2y^2 + 2x^3y \qquad \text{Write original function.}$$

$$f_x(x, y) = 3 - 2xy^2 + 6x^2y. \qquad \text{Partial derivative with respect to } x$$

Considering x to be constant and differentiating with respect to y produces

$$f(x, y) = 3x - x^2y^2 + 2x^3y \qquad \text{Write original function.}$$

$$f_y(x, y) = -2x^2y + 2x^3. \qquad \text{Partial derivative with respect to } y$$

Notation for First Partial Derivatives

For $z = f(x, y)$, the partial derivatives f_x and f_y are denoted by

$$\frac{\partial}{\partial x} f(x, y) = f_x(x, y) = z_x = \frac{\partial z}{\partial x}$$

and

$$\frac{\partial}{\partial y} f(x, y) = f_y(x, y) = z_y = \frac{\partial z}{\partial y}.$$

The first partials evaluated at the point (a, b) are denoted by

$$\left.\frac{\partial z}{\partial x}\right|_{(a,\,b)} = f_x(a, b) \qquad \text{and} \qquad \left.\frac{\partial z}{\partial y}\right|_{(a,\,b)} = f_y(a, b).$$

EXAMPLE 2 Finding and Evaluating Partial Derivatives

For $f(x, y) = xe^{x^2y}$, find f_x and f_y, and evaluate each at the point $(1, \ln 2)$.

Solution Because

$$f_x(x, y) = xe^{x^2y}(2xy) + e^{x^2y} \qquad \text{Partial derivative with respect to } x$$

the partial derivative of f with respect to x at $(1, \ln 2)$ is

$$\begin{aligned} f_x(1, \ln 2) &= e^{\ln 2}(2 \ln 2) + e^{\ln 2} \\ &= 4 \ln 2 + 2. \end{aligned}$$

Because

$$\begin{aligned} f_y(x, y) &= xe^{x^2y}(x^2) \\ &= x^3e^{x^2y} \qquad \text{Partial derivative with respect to } y \end{aligned}$$

the partial derivative of f with respect to y at $(1, \ln 2)$ is

$$\begin{aligned} f_y(1, \ln 2) &= e^{\ln 2} \\ &= 2. \end{aligned}$$

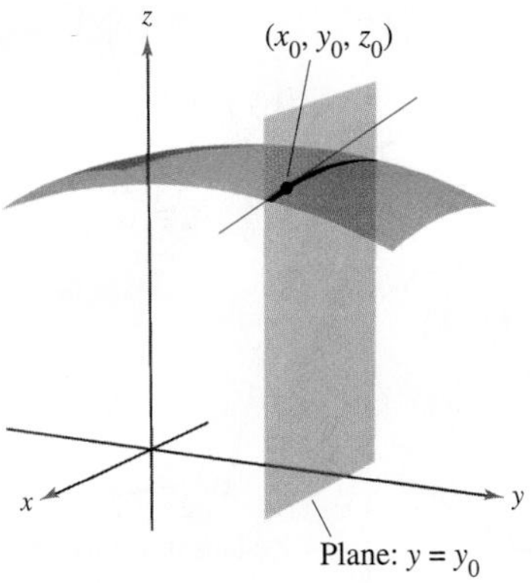

$\frac{\partial f}{\partial x}$ = slope in x-direction
Figure 13.29

The partial derivatives of a function of two variables, $z = f(x, y)$, have a useful geometric interpretation. If $y = y_0$, then $z = f(x, y_0)$ represents the curve formed by intersecting the surface $z = f(x, y)$ with the plane $y = y_0$, as shown in Figure 13.29. Therefore,

$$f_x(x_0, y_0) = \lim_{\Delta x \to 0} \frac{f(x_0 + \Delta x, y_0) - f(x_0, y_0)}{\Delta x}$$

represents the slope of this curve at the point $(x_0, y_0, f(x_0, y_0))$. Note that both the curve and the tangent line lie in the plane $y = y_0$. Similarly,

$$f_y(x_0, y_0) = \lim_{\Delta y \to 0} \frac{f(x_0, y_0 + \Delta y) - f(x_0, y_0)}{\Delta y}$$

represents the slope of the curve given by the intersection of $z = f(x, y)$ and the plane $x = x_0$ at $(x_0, y_0, f(x_0, y_0))$, as shown in Figure 13.30.

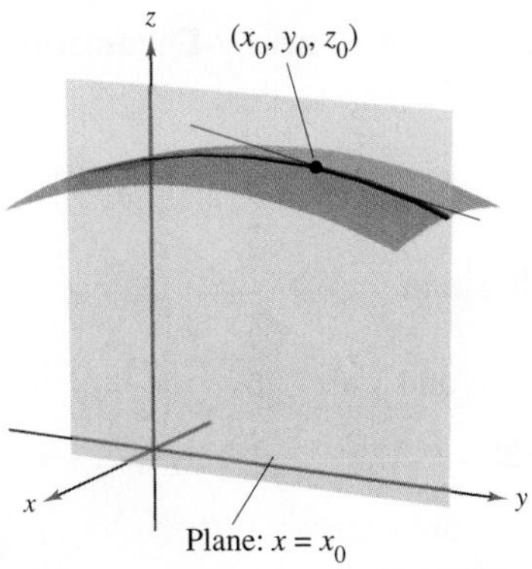

$\frac{\partial f}{\partial y}$ = slope in y-direction
Figure 13.30

Informally, the values of $\partial f/\partial x$ and $\partial f/\partial y$ at the point (x_0, y_0, z_0) denote the **slopes of the surface in the x- and y-directions,** respectively.

EXAMPLE 3 Finding the Slopes of a Surface in the x- and y-Directions

Find the slopes in the x-direction and in the y-direction of the surface given by

$$f(x, y) = -\frac{x^2}{2} - y^2 + \frac{25}{8}$$

at the point $\left(\frac{1}{2}, 1, 2\right)$.

Solution The partial derivatives of f with respect to x and y are

$$f_x(x, y) = -x \quad \text{and} \quad f_y(x, y) = -2y. \qquad \text{Partial derivatives}$$

So, in the x-direction, the slope is

$$f_x\left(\frac{1}{2}, 1\right) = -\frac{1}{2} \qquad \text{Figure 13.31(a)}$$

and in the y-direction, the slope is

$$f_y\left(\frac{1}{2}, 1\right) = -2. \qquad \text{Figure 13.31(b)}$$

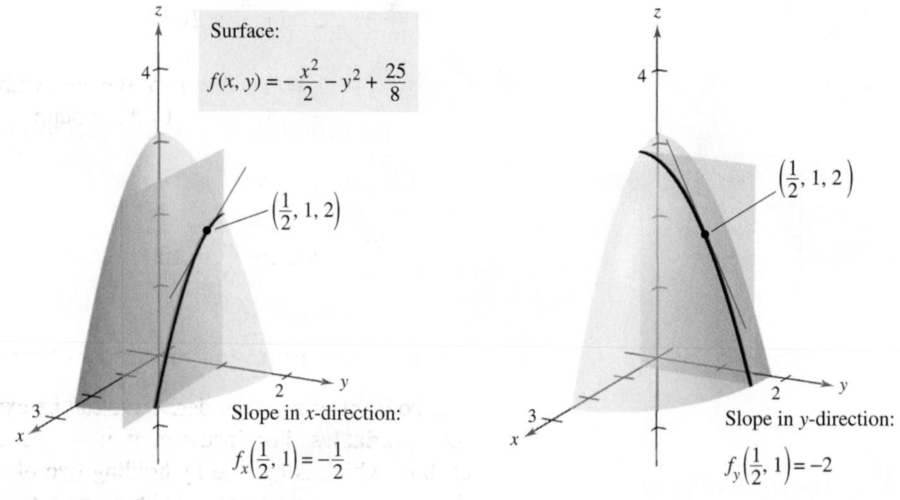

(a) **(b)**

Figure 13.31

EXAMPLE 4 Finding the Slopes of a Surface in the x- and y-Directions

Find the slopes of the surface given by

$$f(x, y) = 1 - (x - 1)^2 - (y - 2)^2$$

at the point $(1, 2, 1)$ in the x-direction and in the y-direction.

Solution The partial derivatives of f with respect to x and y are

$$f_x(x, y) = -2(x - 1) \quad \text{and} \quad f_y(x, y) = -2(y - 2). \qquad \text{Partial derivatives}$$

So, at the point $(1, 2, 1)$, the slopes in the x- and y-directions are

$$f_x(1, 2) = -2(1 - 1) = 0 \quad \text{and} \quad f_y(1, 2) = -2(2 - 2) = 0$$

as shown in Figure 13.32.

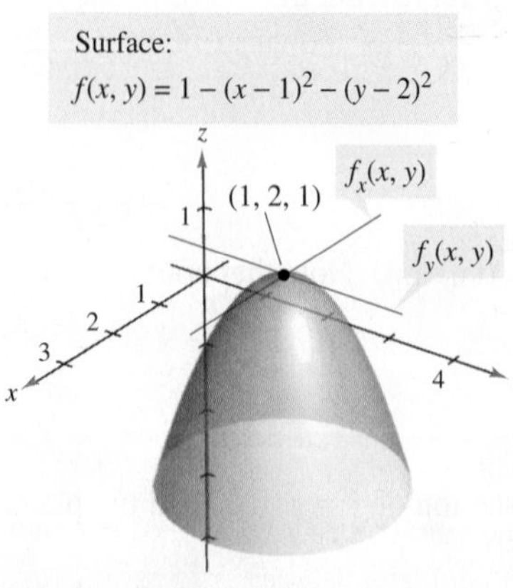

Figure 13.32

No matter how many variables are involved, partial derivatives can be interpreted as *rates of change*.

EXAMPLE 5 Using Partial Derivatives to Find Rates of Change

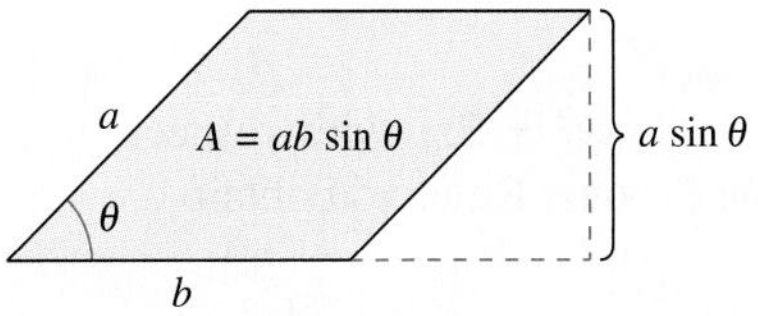

The area of the parallelogram is $ab \sin \theta$.
Figure 13.33

The area of a parallelogram with adjacent sides a and b and included angle θ is given by $A = ab \sin \theta$, as shown in Figure 13.33.

a. Find the rate of change of A with respect to a for $a = 10$, $b = 20$, and $\theta = \frac{\pi}{6}$.

b. Find the rate of change of A with respect to θ for $a = 10$, $b = 20$, and $\theta = \frac{\pi}{6}$.

Solution

a. To find the rate of change of the area with respect to a, hold b and θ constant and differentiate with respect to a to obtain

$$\frac{\partial A}{\partial a} = b \sin \theta \qquad \text{Find partial with respect to } a.$$

$$\frac{\partial A}{\partial a} = 20 \sin \frac{\pi}{6} = 10. \qquad \text{Substitute for } b \text{ and } \theta.$$

b. To find the rate of change of the area with respect to θ, hold a and b constant and differentiate with respect to θ to obtain

$$\frac{\partial A}{\partial \theta} = ab \cos \theta \qquad \text{Find partial with respect to } \theta.$$

$$\frac{\partial A}{\partial \theta} = 200 \cos \frac{\pi}{6} = 100\sqrt{3}. \qquad \text{Substitute for } a, b, \text{ and } \theta.$$

Partial Derivatives of a Function of Three or More Variables

The concept of a partial derivative can be extended naturally to functions of three or more variables. For instance, if $w = f(x, y, z)$, there are three partial derivatives, each of which is formed by holding two of the variables constant. That is, to define the partial derivative of w with respect to x, consider y and z to be constant and differentiate with respect to x. A similar process is used to find the derivatives of w with respect to y and with respect to z.

$$\frac{\partial w}{\partial x} = f_x(x, y, z) = \lim_{\Delta x \to 0} \frac{f(x + \Delta x, y, z) - f(x, y, z)}{\Delta x}$$

$$\frac{\partial w}{\partial y} = f_y(x, y, z) = \lim_{\Delta y \to 0} \frac{f(x, y + \Delta y, z) - f(x, y, z)}{\Delta y}$$

$$\frac{\partial w}{\partial z} = f_z(x, y, z) = \lim_{\Delta z \to 0} \frac{f(x, y, z + \Delta z) - f(x, y, z)}{\Delta z}$$

In general, if $w = f(x_1, x_2, \ldots, x_n)$, there are n partial derivatives denoted by

$$\frac{\partial w}{\partial x_k} = f_{x_k}(x_1, x_2, \ldots, x_n), \quad k = 1, 2, \ldots, n.$$

To find the partial derivative with respect to one of the variables, hold the other variables constant and differentiate with respect to the given variable.

EXAMPLE 6 Finding Partial Derivatives

a. To find the partial derivative of $f(x, y, z) = xy + yz^2 + xz$ with respect to z, consider x and y to be constant and obtain

$$\frac{\partial}{\partial z}[xy + yz^2 + xz] = 2yz + x.$$

b. To find the partial derivative of $f(x, y, z) = z \sin(xy^2 + 2z)$ with respect to z, consider x and y to be constant. Then, using the Product Rule, you obtain

$$\begin{aligned}\frac{\partial}{\partial z}[z \sin(xy^2 + 2z)] &= (z)\frac{\partial}{\partial z}[\sin(xy^2 + 2z)] + \sin(xy^2 + 2z)\frac{\partial}{\partial z}[z] \\ &= (z)[\cos(xy^2 + 2z)](2) + \sin(xy^2 + 2z) \\ &= 2z \cos(xy^2 + 2z) + \sin(xy^2 + 2z).\end{aligned}$$

c. To find the partial derivative of $f(x, y, z, w) = (x + y + z)/w$ with respect to w, consider x, y, and z to be constant and obtain

$$\frac{\partial}{\partial w}\left[\frac{x + y + z}{w}\right] = -\frac{x + y + z}{w^2}.$$

Higher-Order Partial Derivatives

As is true for ordinary derivatives, it is possible to take second, third, and higher partial derivatives of a function of several variables, provided such derivatives exist. Higher-order derivatives are denoted by the order in which the differentiation occurs. For instance, the function $z = f(x, y)$ has the following second partial derivatives.

1. Differentiate twice with respect to x:

$$\frac{\partial}{\partial x}\left(\frac{\partial f}{\partial x}\right) = \frac{\partial^2 f}{\partial x^2} = f_{xx}.$$

2. Differentiate twice with respect to y:

$$\frac{\partial}{\partial y}\left(\frac{\partial f}{\partial y}\right) = \frac{\partial^2 f}{\partial y^2} = f_{yy}.$$

3. Differentiate first with respect to x and then with respect to y:

$$\frac{\partial}{\partial y}\left(\frac{\partial f}{\partial x}\right) = \frac{\partial^2 f}{\partial y \partial x} = f_{xy}.$$

4. Differentiate first with respect to y and then with respect to x:

$$\frac{\partial}{\partial x}\left(\frac{\partial f}{\partial y}\right) = \frac{\partial^2 f}{\partial x \partial y} = f_{yx}.$$

The third and fourth cases are called **mixed partial derivatives.**

NOTE Note that the two types of notation for mixed partials have different conventions for indicating the order of differentiation.

$$\frac{\partial}{\partial y}\left(\frac{\partial f}{\partial x}\right) = \frac{\partial^2 f}{\partial y \partial x} \quad \text{Right-to-left order}$$

$$(f_x)_y = f_{xy} \quad \text{Left-to-right order}$$

You can remember the order by observing that in both notations, you differentiate first with respect to the variable "nearest" f.

EXAMPLE 7 Finding Second Partial Derivatives

Find the second partial derivatives of $f(x, y) = 3xy^2 - 2y + 5x^2y^2$, and determine the value of $f_{xy}(-1, 2)$.

Solution Begin by finding the first partial derivatives with respect to x and y.

$$f_x(x, y) = 3y^2 + 10xy^2 \qquad \text{and} \qquad f_y(x, y) = 6xy - 2 + 10x^2y$$

Then, differentiate each of these with respect to x and y.

$$f_{xx}(x, y) = 10y^2 \qquad \text{and} \qquad f_{yy}(x, y) = 6x + 10x^2$$

$$f_{xy}(x, y) = 6y + 20xy \qquad \text{and} \qquad f_{yx}(x, y) = 6y + 20xy$$

At $(-1, 2)$, the value of f_{xy} is $f_{xy}(-1, 2) = 12 - 40 = -28$.

NOTE Notice in Example 7 that the two mixed partials are equal. Sufficient conditions for this occurrence are given in Theorem 13.3.

THEOREM 13.3 Equality of Mixed Partial Derivatives

If f is a function of x and y such that f_{xy} and f_{yx} are continuous on an open disk R, then, for every (x, y) in R,

$$f_{xy}(x, y) = f_{yx}(x, y).$$

Theorem 13.3 also applies to a function f of *three or more variables* so long as all second partial derivatives are continuous. For example, if $w = f(x, y, z)$ and all the second partial derivatives are continuous in an open region R, then at each point in R the order of differentiation in the mixed second partial derivatives is irrelevant. If the third partial derivatives of f are also continuous, the order of differentiation of the mixed third partial derivatives is irrelevant.

EXAMPLE 8 Finding Higher-Order Partial Derivatives

Show that $f_{xz} = f_{zx}$ and $f_{xzz} = f_{zxz} = f_{zzx}$ for the function given by

$$f(x, y, z) = ye^x + x \ln z.$$

Solution

First partials:

$$f_x(x, y, z) = ye^x + \ln z, \qquad f_z(x, y, z) = \frac{x}{z}$$

Second partials (note that the first two are equal):

$$f_{xz}(x, y, z) = \frac{1}{z}, \qquad f_{zx}(x, y, z) = \frac{1}{z}, \qquad f_{zz}(x, y, z) = -\frac{x}{z^2}$$

Third partials (note that all three are equal):

$$f_{xzz}(x, y, z) = -\frac{1}{z^2}, \qquad f_{zxz}(x, y, z) = -\frac{1}{z^2}, \qquad f_{zzx}(x, y, z) = -\frac{1}{z^2}$$

Exercises for Section 13.3

See www.CalcChat.com for worked-out solutions to odd-numbered exercises.

Think About It **In Exercises 1–4, use the graph of the surface to determine the sign of the indicated partial derivative.**

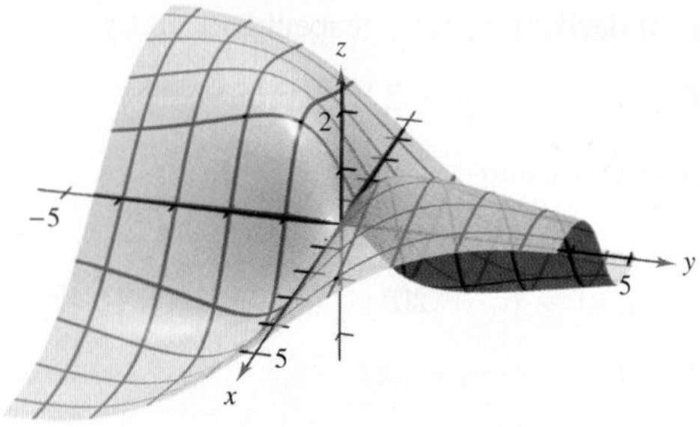

1. $f_x(4, 1)$
2. $f_y(-1, -2)$
3. $f_y(4, 1)$
4. $f_x(-1, -1)$

In Exercises 5–28, find both first partial derivatives.

5. $f(x, y) = 2x - 3y + 5$
6. $f(x, y) = x^2 - 3y^2 + 7$
7. $z = x\sqrt{y}$
8. $z = 2y^2\sqrt{x}$
9. $z = x^2 - 5xy + 3y^2$
10. $z = y^3 - 4xy^2 - 1$
11. $z = x^2e^{2y}$
12. $z = xe^{x/y}$
13. $z = \ln(x^2 + y^2)$
14. $z = \ln\sqrt{xy}$
15. $z = \ln\dfrac{x + y}{x - y}$
16. $z = \ln(x^2 - y^2)$
17. $z = \dfrac{x^2}{2y} + \dfrac{4y^2}{x}$
18. $z = \dfrac{xy}{x^2 + y^2}$
19. $h(x, y) = e^{-(x^2+y^2)}$
20. $g(x, y) = \ln\sqrt{x^2 + y^2}$
21. $f(x, y) = \sqrt{x^2 + y^2}$
22. $f(x, y) = \sqrt{2x + y^3}$
23. $z = \tan(2x - y)$
24. $z = \sin 3x \cos 3y$
25. $z = e^y \sin xy$
26. $z = \cos(x^2 + y^2)$
27. $f(x, y) = \displaystyle\int_x^y (t^2 - 1)\, dt$
28. $f(x, y) = \displaystyle\int_x^y (2t + 1)\, dt + \int_y^x (2t - 1)\, dt$

In Exercises 29–32, use the limit definition of partial derivatives to find $f_x(x, y)$ and $f_y(x, y)$.

29. $f(x, y) = 2x + 3y$
30. $f(x, y) = x^2 - 2xy + y^2$
31. $f(x, y) = \sqrt{x + y}$
32. $f(x, y) = \dfrac{1}{x + y}$

In Exercises 33–36, evaluate f_x and f_y at the given point.

33. $f(x, y) = \arctan\dfrac{y}{x}$, $(2, -2)$
34. $f(x, y) = \arccos xy$, $(1, 1)$
35. $f(x, y) = \dfrac{xy}{x - y}$, $(2, -2)$
36. $f(x, y) = \dfrac{6xy}{\sqrt{4x^2 + 5y^2}}$, $(1, 1)$

In Exercises 37–40, find the slopes of the surface in the x- and y-directions at the given point.

37. $g(x, y) = 4 - x^2 - y^2$
$(1, 1, 2)$

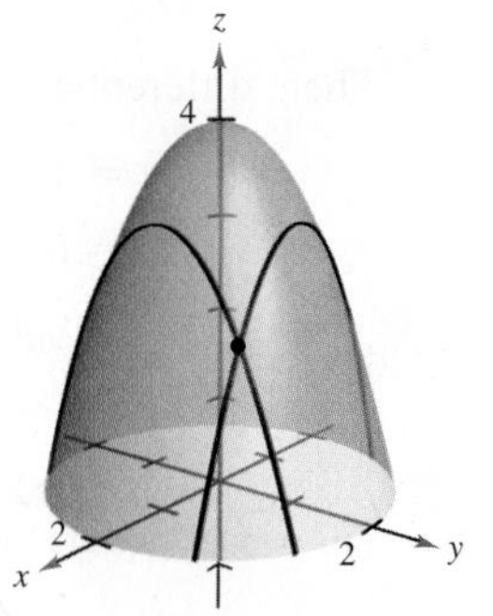

38. $h(x, y) = x^2 - y^2$
$(-2, 1, 3)$

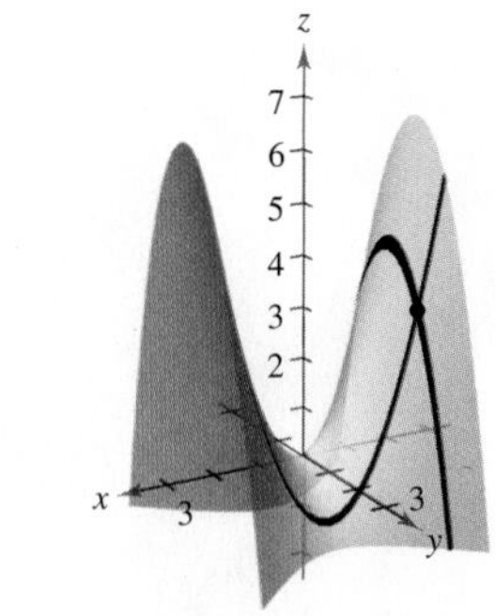

39. $z = e^{-x} \cos y$
$(0, 0, 1)$

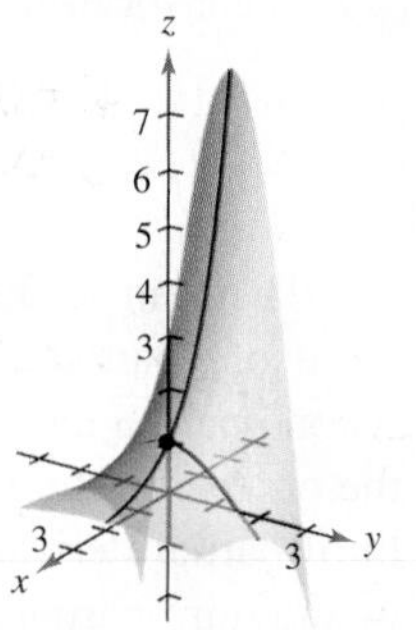

40. $z = \cos(2x - y)$
$\left(\dfrac{\pi}{4}, \dfrac{\pi}{3}, \dfrac{\sqrt{3}}{2}\right)$

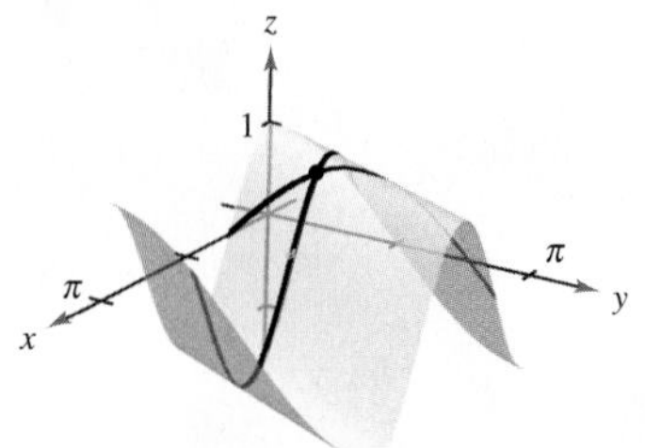

In Exercises 41–44, use a computer algebra system to graph the curve formed by the intersection of the surface and the plane. Find the slope of the curve at the given point.

	Surface	*Plane*	*Point*
41.	$z = \sqrt{49 - x^2 - y^2}$	$x = 2$	$(2, 3, 6)$
42.	$z = x^2 + 4y^2$	$y = 1$	$(2, 1, 8)$
43.	$z = 9x^2 - y^2$	$y = 3$	$(1, 3, 0)$
44.	$z = 9x^2 - y^2$	$x = 1$	$(1, 3, 0)$

In Exercises 45–48, for $f(x, y)$, find all values of x and y such that $f_x(x, y) = 0$ and $f_y(x, y) = 0$ simultaneously.

45. $f(x, y) = x^2 + 4xy + y^2 - 4x + 16y + 3$
46. $f(x, y) = 3x^3 - 12xy + y^3$
47. $f(x, y) = \dfrac{1}{x} + \dfrac{1}{y} + xy$
48. $f(x, y) = \ln(x^2 + y^2 + 1)$

Think About It **In Exercises 49 and 50, the graph of a function f and its two partial derivatives f_x and f_y are given. Identify f_x and f_y and give reasons for your answers.**

49.

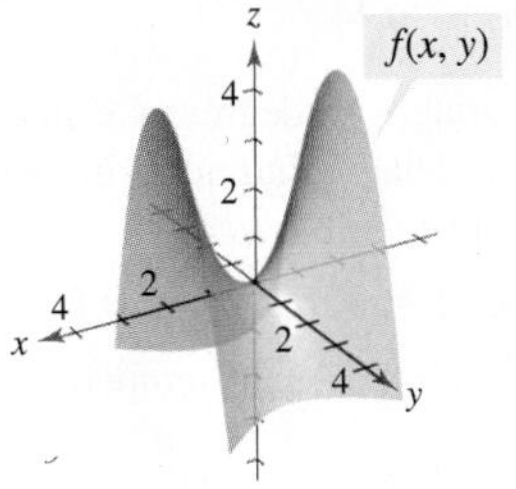

(a)

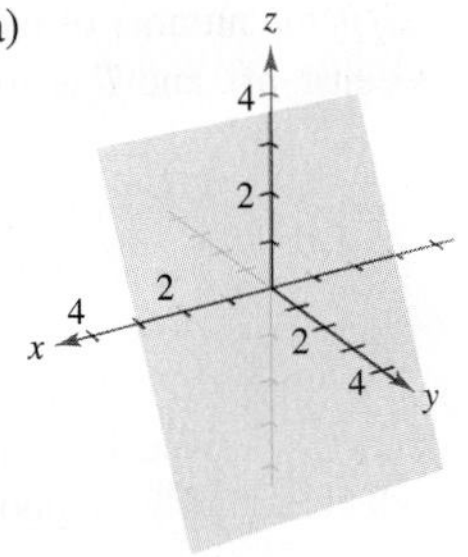

(b)

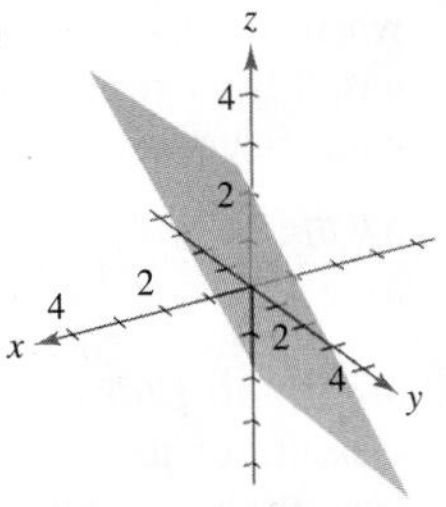

50.

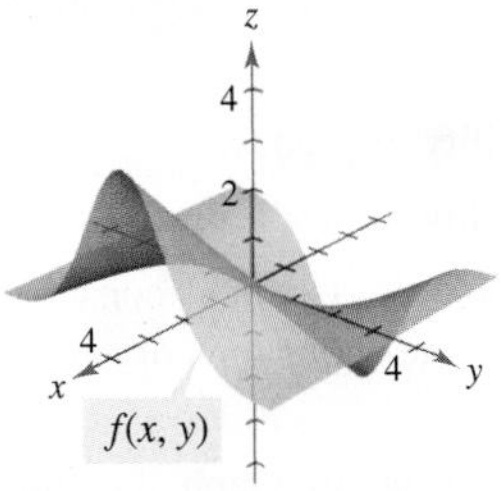

(a)

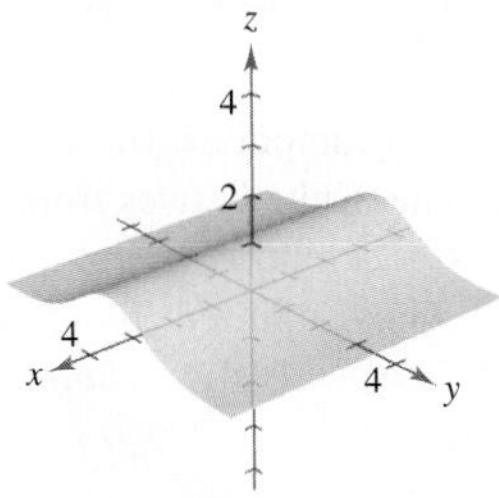

(b)

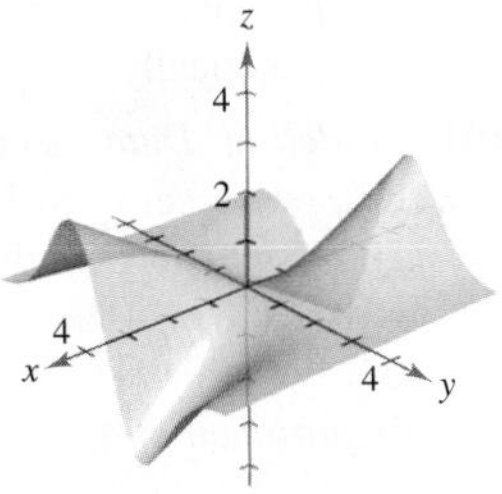

In Exercises 51–56, find the first partial derivatives with respect to x, y, and z.

51. $w = \sqrt{x^2 + y^2 + z^2}$

52. $w = \dfrac{3xz}{x + y}$

53. $F(x, y, z) = \ln\sqrt{x^2 + y^2 + z^2}$

54. $G(x, y, z) = \dfrac{1}{\sqrt{1 - x^2 - y^2 - z^2}}$

55. $H(x, y, z) = \sin(x + 2y + 3z)$

56. $f(x, y, z) = 3x^2y - 5xyz + 10yz^2$

In Exercises 57–60, evaluate f_x, f_y, and f_z at the given point.

57. $f(x, y, z) = \sqrt{3x^2 + y^2 - 2z^2}$, $(1, -2, 1)$

58. $f(x, y, z) = \dfrac{xy}{x + y + z}$, $(3, 1, -1)$

59. $f(x, y, z) = z\sin(x + y)$, $\left(0, \dfrac{\pi}{2}, -4\right)$

60. $f(x, y, z) = x^2y^3 + 2xyz - 3yz$, $(-2, 1, 2)$

In Exercises 61–68, find the four second partial derivatives. Observe that the second mixed partials are equal.

61. $z = x^2 - 2xy + 3y^2$

62. $z = x^4 - 3x^2y^2 + y^4$

63. $z = \sqrt{x^2 + y^2}$

64. $z = \ln(x - y)$

65. $z = e^x \tan y$

66. $z = 2xe^y - 3ye^{-x}$

67. $z = \arctan \dfrac{y}{x}$

68. $z = \sin(x - 2y)$

In Exercises 69–72, use a computer algebra system to find the first and second partial derivatives of the function. Determine whether there exist values of x and y such that $f_x(x, y) = 0$ and $f_y(x, y) = 0$ simultaneously.

69. $f(x, y) = x\sec y$

70. $f(x, y) = \sqrt{9 - x^2 - y^2}$

71. $f(x, y) = \ln \dfrac{x}{x^2 + y^2}$

72. $f(x, y) = \dfrac{xy}{x - y}$

In Exercises 73–76, show that the mixed partial derivatives f_{xyy}, f_{yxy}, and f_{yyx} are equal.

73. $f(x, y, z) = xyz$

74. $f(x, y, z) = x^2 - 3xy + 4yz + z^3$

75. $f(x, y, z) = e^{-x}\sin yz$

76. $f(x, y, z) = \dfrac{2z}{x + y}$

Laplace's Equation **In Exercises 77–80, show that the function satisfies Laplace's equation $\partial^2 z/\partial x^2 + \partial^2 z/\partial y^2 = 0$.**

77. $z = 5xy$

78. $z = \frac{1}{2}(e^y - e^{-y})\sin x$

79. $z = e^x \sin y$

80. $z = \arctan \dfrac{y}{x}$

Wave Equation **In Exercises 81–84, show that the function satisfies the wave equation $\partial^2 z/\partial t^2 = c^2(\partial^2 z/\partial x^2)$.**

81. $z = \sin(x - ct)$

82. $z = \cos(4x + 4ct)$

83. $z = \ln(x + ct)$

84. $z = \sin \omega ct \sin \omega x$

Heat Equation **In Exercises 85 and 86, show that the function satisfies the heat equation $\partial z/\partial t = c^2(\partial^2 z/\partial x^2)$.**

85. $z = e^{-t}\cos \dfrac{x}{c}$

86. $z = e^{-t}\sin \dfrac{x}{c}$

Writing About Concepts

87. Define the first partial derivatives of a function f of two variables x and y.

88. Let f be a function of two variables x and y. Describe the procedure for finding the first partial derivatives.

89. Sketch a surface representing a function f of two variables x and y. Use the sketch to give geometric interpretations of $\partial f/\partial x$ and $\partial f/\partial y$.

90. Sketch the graph of a function $z = f(x, y)$ whose derivative f_x is always negative and whose derivative f_y is always positive.

91. Sketch the graph of a function $z = f(x, y)$ whose derivatives f_x and f_y are always positive.

92. If f is a function of x and y such that f_{xy} and f_{yx} are continuous, what is the relationship between the mixed partial derivatives? Explain.

93. ***Marginal Costs*** A company manufactures two types of wood-burning stoves: a freestanding model and a fireplace-insert model. The cost function for producing x freestanding and y fireplace-insert stoves is

$$C = 32\sqrt{xy} + 175x + 205y + 1050.$$

(a) Find the marginal costs ($\partial C/\partial x$ and $\partial C/\partial y$) when $x = 80$ and $y = 20$.

(b) When additional production is required, which model of stove results in the cost increasing at a higher rate? How can this be determined from the cost model?

94. ***Marginal Productivity*** Consider the Cobb-Douglas production function $f(x, y) = 200x^{0.7}y^{0.3}$. When $x = 1000$ and $y = 500$, find

(a) the marginal productivity of labor, $\partial f/\partial x$.

(b) the marginal productivity of capital, $\partial f/\partial y$.

95. ***Think About It*** Let N be the number of applicants to a university, p the charge for food and housing at the university, and t the tuition. N is a function of p and t such that $\partial N/\partial p < 0$ and $\partial N/\partial t < 0$. What information is gained by noticing that both partials are negative?

96. ***Investment*** The value of an investment of \$1000 earning 10% compounded annually is

$$V(I, R) = 1000\left[\frac{1 + 0.10(1 - R)}{1 + I}\right]^{10}$$

where I is the annual rate of inflation and R is the tax rate for the person making the investment. Calculate $V_I(0.03, 0.28)$ and $V_R(0.03, 0.28)$. Determine whether the tax rate or the rate of inflation is the greater "negative" factor on the growth of the investment.

97. ***Temperature Distribution*** The temperature at any point (x, y) in a steel plate is $T = 500 - 0.6x^2 - 1.5y^2$, where x and y are measured in meters. At the point $(2, 3)$, find the rate of change of the temperature with respect to the distance moved along the plate in the directions of the x- and y-axes.

98. ***Apparent Temperature*** A measure of what hot weather feels like to two average persons is the Apparent Temperature Index. A model for this index is

$$A = 0.885t - 22.4h + 1.20th - 0.544$$

where A is the apparent temperature in degrees Celsius, t is the air temperature, and h is the relative humidity in decimal form. *(Source: The UMAP Journal, Fall 1984)*

(a) Find $\partial A/\partial t$ and $\partial A/\partial h$ when $t = 30°$ and $h = 0.80$.

(b) Which has a greater effect on A, air temperature or humidity? Explain.

99. ***Ideal Gas Law*** The Ideal Gas Law states that $PV = nRT$, where P is pressure, V is volume, n is the number of moles of gas, R is a fixed constant (the gas constant), and T is absolute temperature. Show that

$$\frac{\partial T}{\partial P}\frac{\partial P}{\partial V}\frac{\partial V}{\partial T} = -1.$$

100. ***Marginal Utility*** The utility function $U = f(x, y)$ is a measure of the utility (or satisfaction) derived by a person from the consumption of two products x and y. Suppose the utility function is

$$U = -5x^2 + xy - 3y^2.$$

(a) Determine the marginal utility of product x.

(b) Determine the marginal utility of product y.

(c) When $x = 2$ and $y = 3$, should a person consume one more unit of product x or one more unit of product y? Explain your reasoning.

(d) Use a computer algebra system to graph the function. Interpret the marginal utilities of products x and y graphically.

101. ***Modeling Data*** Per capita consumptions (in gallons) of different types of plain milk in the United States from 1994 to 2000 are shown in the table. Consumption of light and skim milks, reduced-fat milk, and whole milk are represented by the variables x, y, and z, respectively. *(Source: U.S. Department of Agriculture)*

Year	1994	1995	1996	1997	1998	1999	2000
x	5.8	6.2	6.4	6.6	6.5	6.3	6.1
y	8.7	8.2	8.0	7.7	7.4	7.3	7.1
z	8.8	8.4	8.4	8.2	7.8	7.9	7.8

A model for the data is given by

$$z = -0.04x + 0.64y + 3.4.$$

(a) Find $\dfrac{\partial z}{\partial x}$ and $\dfrac{\partial z}{\partial y}$.

(b) Interpret the partial derivatives in the context of the problem.

102. ***Modeling Data*** The table shows the amount of public medical expenditures (in billions of dollars) for worker's compensation x, public assistance y, and Medicare z for selected years. *(Source: Centers for Medicare and Medicaid Services)*

Year	1990	1996	1997	1998	1999	2000
x	17.5	21.9	20.5	20.8	22.5	23.3
y	78.7	157.6	164.8	176.6	191.8	208.5
z	110.2	197.5	208.2	209.5	212.6	224.4

A model for the data is given by

$$z = -1.3520x^2 - 0.0025y^2 + 56.080x + 1.537y - 562.23.$$

(a) Find $\dfrac{\partial^2 z}{\partial x^2}$ and $\dfrac{\partial^2 z}{\partial y^2}$.

(b) Determine the concavity of traces parallel to the xz-plane. Interpret the result in the context of the problem.

(c) Determine the concavity of traces parallel to the yz-plane. Interpret the result in the context of the problem.

True or False? **In Exercises 103–106, determine whether the statement is true or false. If it is false, explain why or give an example that shows it is false.**

103. If $z = f(x, y)$ and $\partial z/\partial x = \partial z/\partial y$, then $z = c(x + y)$.

104. If $z = f(x)g(y)$, then $(\partial z/\partial x) + (\partial z/\partial y) = f'(x)g(y) + f(x)g'(y)$.

105. If $z = e^{xy}$, then $\dfrac{\partial^2 z}{\partial y \partial x} = (xy + 1)e^{xy}$.

106. If a cylindrical surface $z = f(x, y)$ has rulings parallel to the y-axis, then $\partial z/\partial y = 0$.

107. Consider the function defined by

$$f(x, y) = \begin{cases} \dfrac{xy(x^2 - y^2)}{x^2 + y^2}, & (x, y) \neq (0, 0) \\ 0, & (x, y) = (0, 0) \end{cases}.$$

(a) Find $f_x(x, y)$ and $f_y(x, y)$ for $(x, y) \neq (0, 0)$.

(b) Use the definition of partial derivatives to find $f_x(0, 0)$ and $f_y(0, 0)$.

$$\left[\textit{Hint: } f_x(0, 0) = \lim_{\Delta x \to 0} \frac{f(\Delta x, 0) - f(0, 0)}{\Delta x}.\right]$$

(c) Use the definition of partial derivatives to find $f_{xy}(0, 0)$ and $f_{yx}(0, 0)$.

(d) Using Theorem 13.3 and the result of part (c), what can be said about f_{xy} or f_{yx}?

108. Let $f(x, y) = \displaystyle\int_x^y \sqrt{1 + t^3}\, dt$. Find $f_x(x, y)$ and $f_y(x, y)$.

109. Consider the function $f(x, y) = (x^3 + y^3)^{1/3}$.

(a) Show that $f_y(0, 0) = 1$.

(b) Determine the points (if any) at which $f_y(x, y)$ fails to exist.

110. Consider the function $f(x, y) = (x^2 + y^2)^{2/3}$. Show that

$$f_x(x, y) = \begin{cases} \dfrac{4x}{3(x^2 + y^2)^{1/3}}, & (x, y) \neq (0, 0) \\ 0, & (x, y) = (0, 0) \end{cases}.$$

FOR FURTHER INFORMATION For more information about this problem, see the article "A Classroom Note on a Naturally Occurring Piecewise Defined Function" by Don Cohen in *Mathematics and Computer Education.*

Section Project: Moiré Fringes

Read the article "Moiré Fringes and the Conic Sections" by Mike Cullen in *The College Mathematics Journal*. The article describes how two families of level curves given by

$$f(x, y) = a \qquad \text{and} \qquad g(x, y) = b$$

can form Moiré patterns. After reading the article, write a paper explaining how the expression

$$\frac{\partial f}{\partial x} \cdot \frac{\partial g}{\partial x} + \frac{\partial f}{\partial y} \cdot \frac{\partial g}{\partial y}$$

is related to the Moiré patterns formed by intersecting the two families of level curves. Use one of the following patterns as an example in your paper.

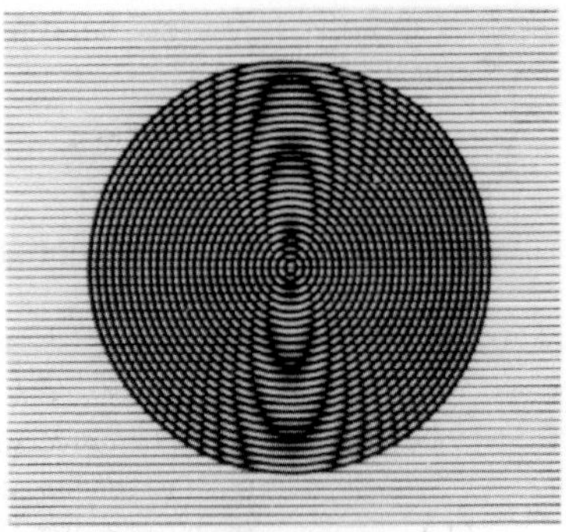

Mike Cullen

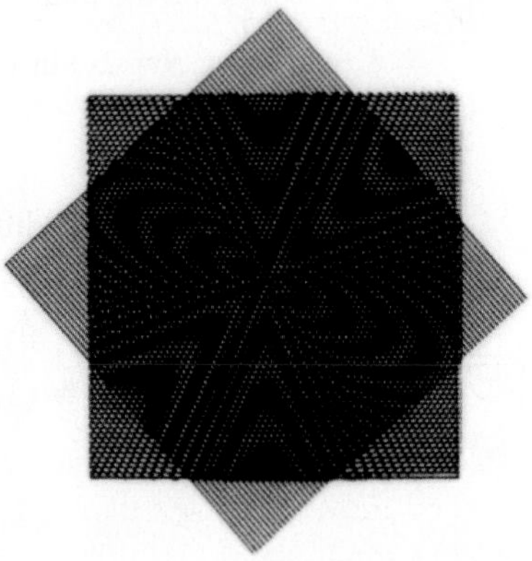

Mike Cullen

Section 13.4 Differentials

- **Understand the concepts of increments and differentials.**
- **Extend the concept of differentiability to a function of two variables.**
- **Use a differential as an approximation.**

Increments and Differentials

In this section, the concepts of increments and differentials are generalized to functions of two or more variables. Recall from Section 3.9 that for $y = f(x)$, the differential of y was defined as

$$dy = f'(x)\,dx.$$

Similar terminology is used for a function of two variables, $z = f(x, y)$. That is, Δx and Δy are the **increments of x and y,** and the **increment of z** is given by

$$\Delta z = f(x + \Delta x, y + \Delta y) - f(x, y).$$

Increment of z

Definition of Total Differential

If $z = f(x, y)$ and Δx and Δy are increments of x and y, then the **differentials** of the independent variables x and y are

$$dx = \Delta x \quad \text{and} \quad dy = \Delta y$$

and the **total differential** of the dependent variable z is

$$dz = \frac{\partial z}{\partial x}\,dx + \frac{\partial z}{\partial y}\,dy = f_x(x, y)\,dx + f_y(x, y)\,dy.$$

This definition can be extended to a function of three or more variables. For instance, if $w = f(x, y, z, u)$, then $dx = \Delta x$, $dy = \Delta y$, $dz = \Delta z$, $du = \Delta u$, and the total differential of w is

$$dw = \frac{\partial w}{\partial x}\,dx + \frac{\partial w}{\partial y}\,dy + \frac{\partial w}{\partial z}\,dz + \frac{\partial w}{\partial u}\,du.$$

EXAMPLE 1 Finding the Total Differential

Find the total differential for each function.

a. $z = 2x \sin y - 3x^2y^2$ **b.** $w = x^2 + y^2 + z^2$

Solution

a. The total differential dz for $z = 2x \sin y - 3x^2y^2$ is

$$dz = \frac{\partial z}{\partial x}\,dx + \frac{\partial z}{\partial y}\,dy$$

Total differential dz

$$= (2 \sin y - 6xy^2)\,dx + (2x \cos y - 6x^2y)\,dy.$$

b. The total differential dw for $w = x^2 + y^2 + z^2$ is

$$dw = \frac{\partial w}{\partial x}\,dx + \frac{\partial w}{\partial y}\,dy + \frac{\partial w}{\partial z}\,dz$$

Total differential dw

$$= 2x\,dx + 2y\,dy + 2z\,dz.$$

Differentiability

In Section 3.9, you learned that for a *differentiable* function given by $y = f(x)$, you can use the differential $dy = f'(x)\,dx$ as an approximation (for small Δx) to the value $\Delta y = f(x + \Delta x) - f(x)$. When a similar approximation is possible for a function of two variables, the function is said to be **differentiable.** This is stated explicitly in the following definition.

Definition of Differentiability

A function f given by $z = f(x, y)$ is **differentiable** at (x_0, y_0) if Δz can be written in the form

$$\Delta z = f_x(x_0, y_0)\,\Delta x + f_y(x_0, y_0)\,\Delta y + \varepsilon_1\Delta x + \varepsilon_2\Delta y$$

where both ε_1 and $\varepsilon_2 \to 0$ as $(\Delta x, \Delta y) \to (0, 0)$. The function f is **differentiable in a region R** if it is differentiable at each point in R.

EXAMPLE 2 Showing That a Function Is Differentiable

Show that the function given by

$$f(x, y) = x^2 + 3y$$

is differentiable at every point in the plane.

Solution Letting $z = f(x, y)$, the increment of z at an arbitrary point (x, y) in the plane is

$$\begin{aligned}\Delta z &= f(x + \Delta x, y + \Delta y) - f(x, y) && \text{Increment of } z\\ &= (x^2 + 2x\Delta x + \Delta x^2) + 3(y + \Delta y) - (x^2 + 3y)\\ &= 2x\Delta x + \Delta x^2 + 3\Delta y\\ &= 2x(\Delta x) + 3(\Delta y) + \Delta x(\Delta x) + 0(\Delta y)\\ &= f_x(x, y)\,\Delta x + f_y(x, y)\,\Delta y + \varepsilon_1\Delta x + \varepsilon_2\Delta y\end{aligned}$$

where $\varepsilon_1 = \Delta x$ and $\varepsilon_2 = 0$. Because $\varepsilon_1 \to 0$ and $\varepsilon_2 \to 0$ as $(\Delta x, \Delta y) \to (0, 0)$, it follows that f is differentiable at every point in the plane. The graph of f is shown in Figure 13.34.

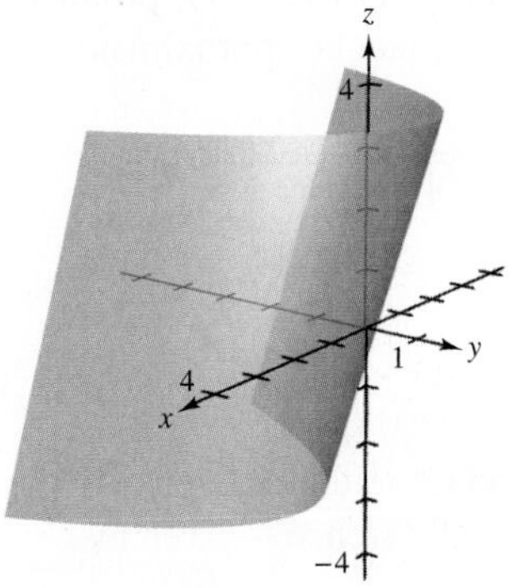

Figure 13.34

Be sure you see that the term "differentiable" is used differently for functions of two variables than for functions of one variable. A function of one variable is differentiable at a point if its derivative exists at the point. However, for a function of two variables, the existence of the partial derivatives f_x and f_y does not guarantee that the function is differentiable (see Example 5). The following theorem gives a *sufficient* condition for differentiability of a function of two variables. A proof of Theroem 13.4 is given in Appendix A.

THEOREM 13.4 Sufficient Condition for Differentiability

If f is a function of x and y, where f_x and f_y are continuous in an open region R, then f is differentiable on R.

Approximation by Differentials

Theorem 13.4 tells you that you can choose $(x + \Delta x, y + \Delta y)$ close enough to (x, y) to make $\varepsilon_1 \Delta x$ and $\varepsilon_2 \Delta y$ insignificant. In other words, for small Δx and Δy, you can use the approximation

$$\Delta z \approx dz.$$

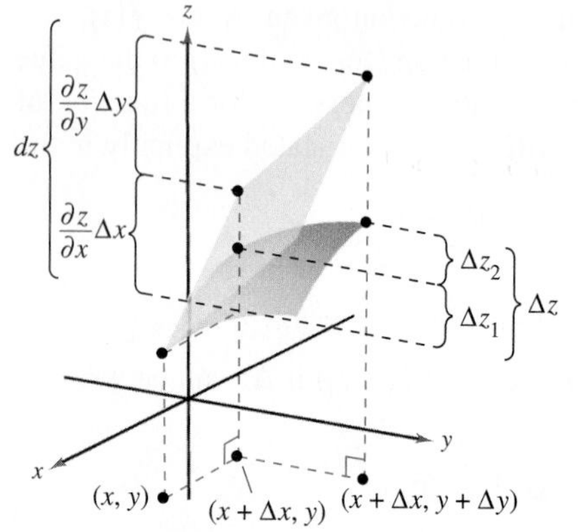

The exact change in z is Δz. This change can be approximated by the differential dz.
Figure 13.35

This approximation is illustrated graphically in Figure 13.35. Recall that the partial derivatives $\partial z/\partial x$ and $\partial z/\partial y$ can be interpreted as the slopes of the surface in the x- and y-directions. This means that

$$dz = \frac{\partial z}{\partial x}\Delta x + \frac{\partial z}{\partial y}\Delta y$$

represents the change in height of a plane that is tangent to the surface at the point $(x, y, f(x, y))$. Because a plane in space is represented by a linear equation in the variables x, y, and z, the approximation of Δz by dz is called a **linear approximation.** You will learn more about this geometric interpretation in Section 13.7.

EXAMPLE 3 Using a Differential as an Approximation

Use the differential dz to approximate the change in $z = \sqrt{4 - x^2 - y^2}$ as (x, y) moves from the point $(1, 1)$ to the point $(1.01, 0.97)$. Compare this approximation with the exact change in z.

Solution Letting $(x, y) = (1, 1)$ and $(x + \Delta x, y + \Delta y) = (1.01, 0.97)$ produces $dx = \Delta x = 0.01$ and $dy = \Delta y = -0.03$. So, the change in z can be approximated by

$$\Delta z \approx dz = \frac{\partial z}{\partial x}dx + \frac{\partial z}{\partial y}dy = \frac{-x}{\sqrt{4 - x^2 - y^2}}\Delta x + \frac{-y}{\sqrt{4 - x^2 - y^2}}\Delta y.$$

When $x = 1$ and $y = 1$, you have

$$\Delta z \approx -\frac{1}{\sqrt{2}}(0.01) - \frac{1}{\sqrt{2}}(-0.03) = \frac{0.02}{\sqrt{2}} = \sqrt{2}\,(0.01) \approx 0.0141.$$

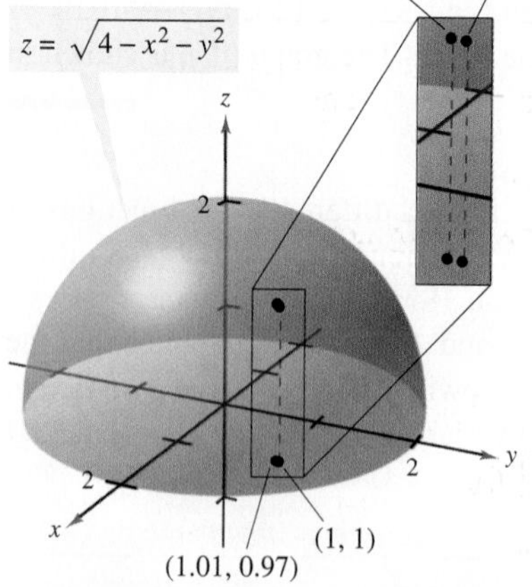

As (x, y) moves from $(1, 1)$ to the point $(1.01, 0.97)$, the value of $f(x, y)$ changes by about 0.0137.
Figure 13.36

In Figure 13.36 you can see that the exact change corresponds to the difference in the heights of two points on the surface of a hemisphere. This difference is given by

$$\begin{aligned}\Delta z &= f(1.01, 0.97) - f(1, 1)\\ &= \sqrt{4 - (1.01)^2 - (0.97)^2} - \sqrt{4 - 1^2 - 1^2} \approx 0.0137.\end{aligned}$$

A function of three variables $w = f(x, y, z)$ is called **differentiable** at (x, y, z) provided that

$$\Delta w = f(x + \Delta x, y + \Delta y, z + \Delta z) - f(x, y, z)$$

can be written in the form

$$\Delta w = f_x \Delta x + f_y \Delta y + f_z \Delta z + \varepsilon_1 \Delta x + \varepsilon_2 \Delta y + \varepsilon_3 \Delta z$$

where ε_1, ε_2, and $\varepsilon_3 \to 0$ as $(\Delta x, \Delta y, \Delta z) \to (0, 0, 0)$. With this definition of differentiability, Theorem 13.4 has the following extension for functions of three variables: If f is a function of x, y, and z, where f, f_x, f_y, and f_z are continuous in an open region R, then f is differentiable on R.

In Section 3.9, you used differentials to approximate the propagated error introduced by an error in measurement. This application of differentials is further illustrated in Example 4.

EXAMPLE 4 Error Analysis

The possible error involved in measuring each dimension of a rectangular box is ± 0.1 millimeter. The dimensions of the box are $x = 50$ centimeters, $y = 20$ centimeters, and $z = 15$ centimeters, as shown in Figure 13.37. Use dV to estimate the propagated error and the relative error in the calculated volume of the box.

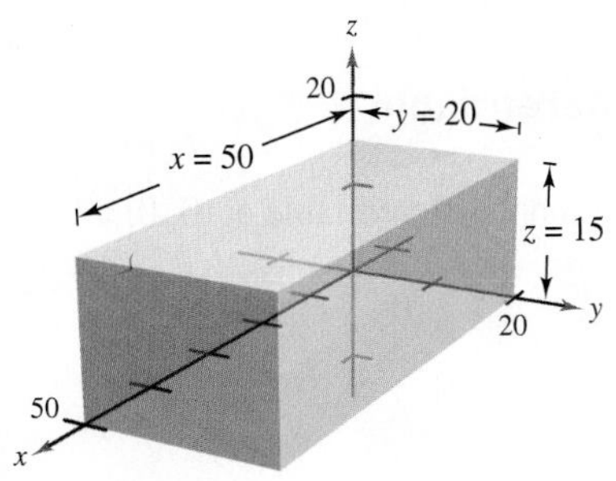

Volume $= xyz$
Figure 13.37

Solution The volume of the box is given by $V = xyz$, and so

$$\begin{aligned} dV &= \frac{\partial V}{\partial x}dx + \frac{\partial V}{\partial y}dy + \frac{\partial V}{\partial z}dz \\ &= yz\,dx + xz\,dy + xy\,dz. \end{aligned}$$

Using 0.1 millimeter $= 0.01$ centimeter, you have $dx = dy = dz = \pm 0.01$, and the propagated error is approximately

$$\begin{aligned} dV &= (20)(15)(\pm 0.01) + (50)(15)(\pm 0.01) + (50)(20)(\pm 0.01) \\ &= 300(\pm 0.01) + 750(\pm 0.01) + 1000(\pm 0.01) \\ &= 2050(\pm 0.01) = \pm 20.5 \text{ cubic centimeters.} \end{aligned}$$

Because the measured volume is

$$V = (50)(20)(15) = 15{,}000 \text{ cubic centimeters,}$$

the relative error, $\Delta V/V$, is approximately

$$\frac{\Delta V}{V} \approx \frac{dV}{V} = \frac{20.5}{15{,}000} \approx 0.14\%.$$

As is true for a function of a single variable, if a function in two or more variables is differentiable at a point, it is also continuous there.

THEOREM 13.5 Differentiability Implies Continuity

If a function of x and y is differentiable at (x_0, y_0), then it is continuous at (x_0, y_0).

Proof Let f be differentiable at (x_0, y_0), where $z = f(x, y)$. Then

$$\Delta z = [f_x(x_0, y_0) + \varepsilon_1]\,\Delta x + [f_y(x_0, y_0) + \varepsilon_2]\,\Delta y$$

where both ε_1 and $\varepsilon_2 \to 0$ as $(\Delta x, \Delta y) \to (0, 0)$. However, by definition, you know that Δz is given by

$$\Delta z = f(x_0 + \Delta x, y_0 + \Delta y) - f(x_0, y_0).$$

Letting $x = x_0 + \Delta x$ and $y = y_0 + \Delta y$ produces

$$\begin{aligned} f(x, y) - f(x_0, y_0) &= [f_x(x_0, y_0) + \varepsilon_1]\,\Delta x + [f_y(x_0, y_0) + \varepsilon_2]\,\Delta y \\ &= [f_x(x_0, y_0) + \varepsilon_1](x - x_0) + [f_y(x_0, y_0) + \varepsilon_2](y - y_0). \end{aligned}$$

Taking the limit as $(x, y) \to (x_0, y_0)$, you have

$$\lim_{(x, y) \to (x_0, y_0)} f(x, y) = f(x_0, y_0)$$

which means that f is continuous at (x_0, y_0).

Remember that the existence of f_x and f_y is not sufficient to guarantee differentiability, as illustrated in the next example.

EXAMPLE 5 A Function That Is Not Differentiable

Show that $f_x(0, 0)$ and $f_y(0, 0)$ both exist, but that f is not differentiable at $(0, 0)$ where f is defined as

$$f(x, y) = \begin{cases} \dfrac{-3xy}{x^2 + y^2}, & \text{if } (x, y) \neq (0, 0) \\ 0, & \text{if } (x, y) = (0, 0) \end{cases}.$$

TECHNOLOGY Use a graphing utility to graph the function given in Example 5. For instance, the graph shown below was generated by *Mathematica.*

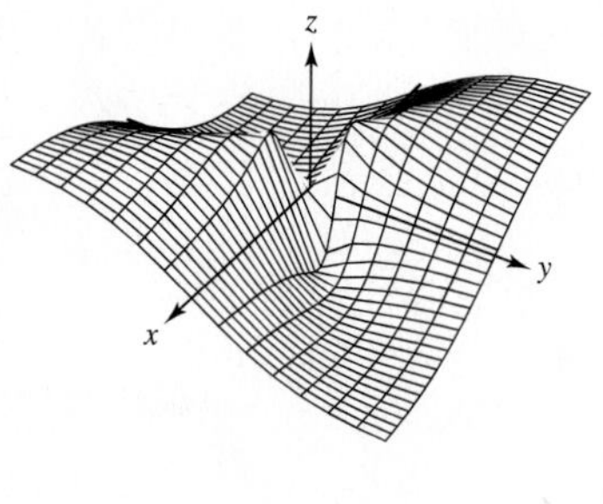

Generated by Mathematica

Solution You can show that f is not differentiable at $(0, 0)$ by showing that it is not continuous at this point. To see that f is not continuous at $(0, 0)$, look at the values of $f(x, y)$ along two different approaches to $(0, 0)$, as shown in Figure 13.38. Along the line $y = x$, the limit is

$$\lim_{(x, x)\to(0, 0)} f(x, y) = \lim_{(x, x)\to(0, 0)} \frac{-3x^2}{2x^2} = -\frac{3}{2}$$

whereas along $y = -x$ you have

$$\lim_{(x, -x)\to(0, 0)} f(x, y) = \lim_{(x, -x)\to(0, 0)} \frac{3x^2}{2x^2} = \frac{3}{2}.$$

So, the limit of $f(x, y)$ as $(x, y) \to (0, 0)$ does not exist, and you can conclude that f is not continuous at $(0, 0)$. Therefore, by Theorem 13.5, you know that f is not differentiable at $(0, 0)$. On the other hand, by the definition of the partial derivatives f_x and f_y, you have

$$f_x(0, 0) = \lim_{\Delta x\to 0} \frac{f(\Delta x, 0) - f(0, 0)}{\Delta x} = \lim_{\Delta x\to 0} \frac{0 - 0}{\Delta x} = 0$$

and

$$f_y(0, 0) = \lim_{\Delta y\to 0} \frac{f(0, \Delta y) - f(0, 0)}{\Delta y} = \lim_{\Delta y\to 0} \frac{0 - 0}{\Delta y} = 0.$$

So, the partial derivatives at $(0, 0)$ exist.

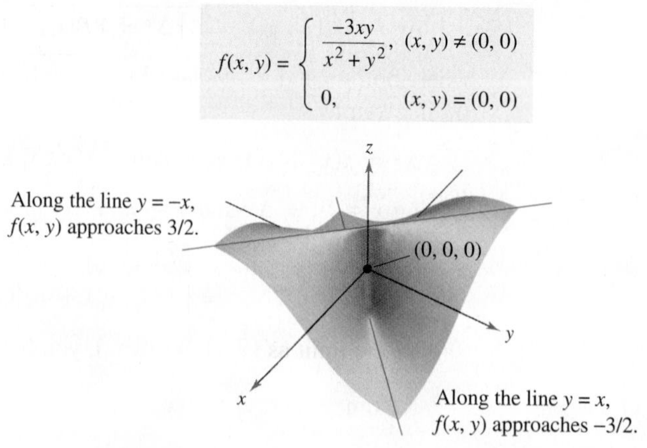

Figure 13.38

Exercises for Section 13.4

See www.CalcChat.com for worked-out solutions to odd-numbered exercises.

In Exercises 1–10, find the total differential.

1. $z = 3x^2y^3$
2. $z = \dfrac{x^2}{y}$
3. $z = \dfrac{-1}{x^2 + y^2}$
4. $w = \dfrac{x + y}{z - 2y}$
5. $z = x \cos y - y \cos x$
6. $z = \frac{1}{2}(e^{x^2+y^2} - e^{-x^2-y^2})$
7. $z = e^x \sin y$
8. $w = e^y \cos x + z^2$
9. $w = 2z^3y \sin x$
10. $w = x^2yz^2 + \sin yz$

In Exercises 11–16, (a) evaluate $f(1, 2)$ and $f(1.05, 2.1)$ and calculate Δz, and (b) use the total differential dz to approximate Δz.

11. $f(x, y) = 9 - x^2 - y^2$
12. $f(x, y) = \sqrt{x^2 + y^2}$
13. $f(x, y) = x \sin y$
14. $f(x, y) = xe^y$
15. $f(x, y) = 3x - 4y$
16. $f(x, y) = \dfrac{x}{y}$

In Exercises 17–20, find $z = f(x, y)$ and use the total differential to approximate the quantity.

17. $\sqrt{(5.05)^2 + (3.1)^2} - \sqrt{5^2 + 3^2}$
18. $(2.03)^2(1 + 8.9)^3 - 2^2(1 + 9)^3$
19. $\dfrac{1 - (3.05)^2}{(5.95)^2} - \dfrac{1 - 3^2}{6^2}$
20. $\sin[(1.05)^2 + (0.95)^2] - \sin(1^2 + 1^2)$

Writing About Concepts

21. Define the total differential of a function of two variables.
22. Describe the change in accuracy of dz as an approximation of Δz as Δx and Δy increase.
23. What is meant by a linear approximation of $z = f(x, y)$ at the point $P(x_0, y_0)$?
24. When using differentials, what is meant by the terms *propagated error* and *relative error*?

25. ***Area*** The area of the shaded rectangle in the figure is $A = lh$. The possible errors in the length and height are Δl and Δh, respectively. Find dA and identify the regions in the figure whose areas are given by the terms of dA. What region represents the difference between ΔA and dA?

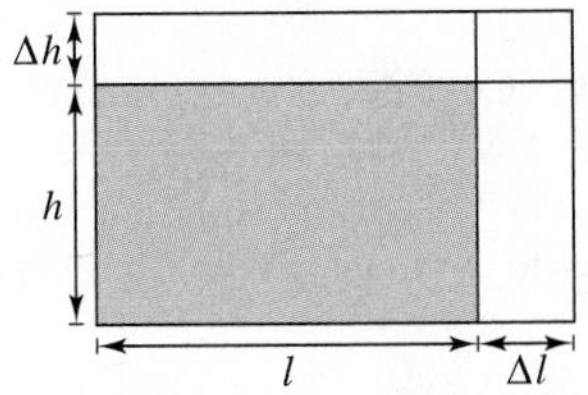

Figure for 25

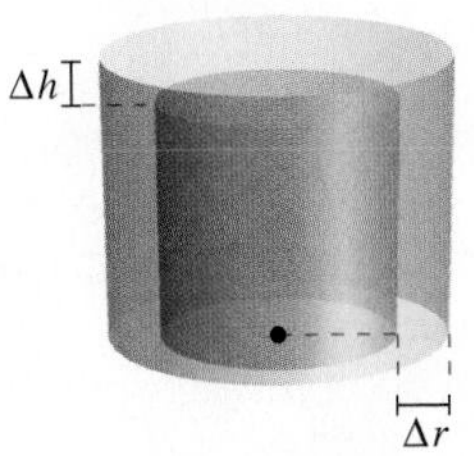

Figure for 26

26. ***Volume*** The volume of the red right circular cylinder in the figure is $V = \pi r^2 h$. The possible errors in the radius and the height are Δr and Δh, respectively. Find dV and identify the solids in the figure whose volumes are given by the terms of dV. What solid represents the difference between ΔV and dV?

27. ***Numerical Analysis*** A right circular cone of height $h = 6$ and radius $r = 3$ is constructed, and in the process errors Δr and Δh are made in the radius and height, respectively. Complete the table to show the relationship between ΔV and dV for the indicated errors.

Δr	Δh	dV or dS	ΔV or ΔS	$\Delta V - dV$ or $\Delta S - dS$
0.1	0.1			
0.1	−0.1			
0.001	0.002			
−0.0001	0.0002			

28. ***Numerical Analysis*** The height and radius of a right circular cone are measured as $h = 20$ meters and $r = 8$ meters. In the process of measuring, errors Δr and Δh are made. S is the lateral surface area of a cone. Complete the table above to show the relationship between ΔS and dS for the indicated errors.

29. ***Modeling Data*** Per capita consumptions (in gallons) of different types of plain milk in the United States from 1994 to 2000 are shown in the table. Consumption of light and skim milks, reduced-fat milk, and whole milk are represented by the variables x, y, and z, respectively. *(Source: U.S. Department of Agriculture)*

Year	1994	1995	1996	1997	1998	1999	2000
x	5.8	6.2	6.4	6.6	6.5	6.3	6.1
y	8.7	8.2	8.0	7.7	7.4	7.3	7.1
z	8.8	8.4	8.4	8.2	7.8	7.9	7.8

A model for the data is given by $z = -0.04x + 0.64y + 3.4$.

(a) Find the total differential of the model.

(b) A dairy industry forecast for a future year is that per capita consumption of light and skim milks will be 6.2 ± 0.25 gallons and that per capita consumption of reduced-fat milk will be 7.5 ± 0.25 gallons. Use dz to estimate the maximum possible propagated error and relative error in the prediction for the consumption of whole milk.

30. ***Rectangular to Polar Coordinates*** A rectangular coordinate system is placed over a map and the coordinates of a point of interest are $(8.5, 3.2)$. There is a possible error of 0.05 in each coordinate. Approximate the maximum possible error in measuring the polar coordinates for the point.

31. ***Volume*** The radius r and height h of a right circular cylinder are measured with possible errors of 4% and 2%, respectively. Approximate the maximum possible percent error in measuring the volume.

32. ***Area*** A triangle is measured and two adjacent sides are found to be 3 inches and 4 inches long, with an included angle of $\pi/4$. The possible errors in measurement are $\frac{1}{16}$ inch for the sides and 0.02 radian for the angle. Approximate the maximum possible error in the computation of the area.

33. ***Wind Chill*** The formula for wind chill C (in degrees Fahrenheit) is given by

$$C = 35.74 + 0.6215T - 35.75v^{0.16} + 0.4275Tv^{0.16}$$

where v is the wind speed in miles per hour and T is the temperature in degrees Fahrenheit. The wind speed is 23 ± 3 miles per hour and the temperature is $8° \pm 1°$. Use dC to estimate the maximum possible propagated error and relative error in calculating the wind chill.

34. ***Acceleration*** The centripetal acceleration of a particle moving in a circle is $a = v^2/r$, where v is the velocity and r is the radius of the circle. Approximate the maximum percent error in measuring the acceleration due to errors of 3% in v and 2% in r.

35. ***Volume*** A trough is 16 feet long (see figure). Its cross sections are isosceles triangles with each of the two equal sides measuring 18 inches. The angle between the two equal sides is θ.

(a) Write the volume of the trough as a function of θ and determine the value of θ such that the volume is a maximum.

(b) The maximum error in the linear measurements is one-half inch and the maximum error in the angle measure is 2°. Approximate the change from the maximum volume.

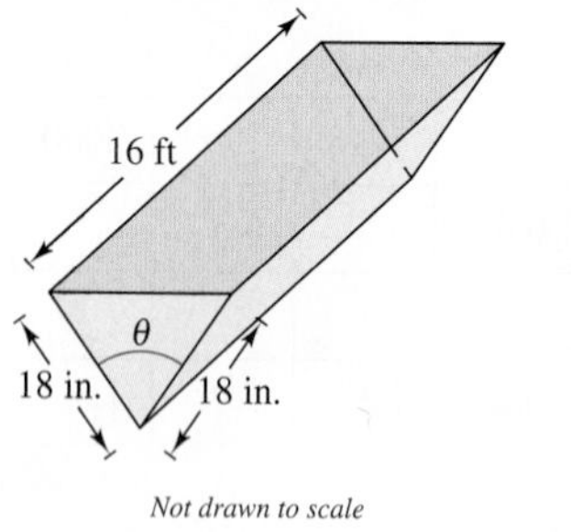

Figure for 35

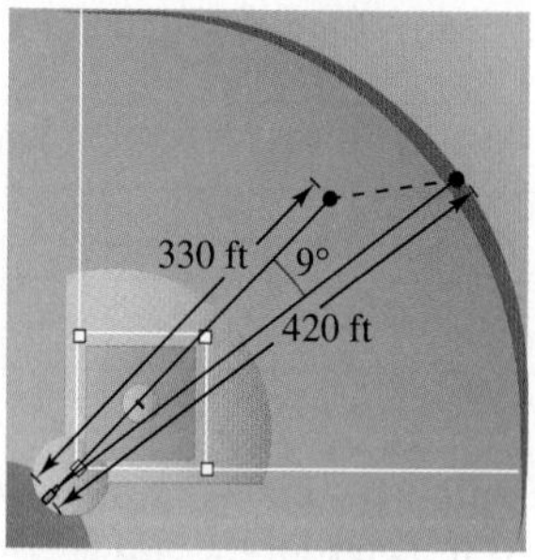

Figure for 36

36. ***Sports*** A baseball player in center field is playing approximately 330 feet from a television camera that is behind home plate. A batter hits a fly ball that goes to a wall 420 feet from the camera (see figure).

(a) The camera turns 9° to follow the play. Approximate the number of feet that the center fielder has to run to make the catch.

(b) The position of the center fielder could be in error by as much as 6 feet and the maximum error in measuring the rotation of the camera is 1°. Approximate the maximum possible error in the result of part (a).

37. ***Power*** Electrical power P is given by

$$P = \frac{E^2}{R}$$

where E is voltage and R is resistance. Approximate the maximum percent error in calculating power if 200 volts is applied to a 4000-ohm resistor and the possible percent errors in measuring E and R are 2% and 3%, respectively.

38. ***Resistance*** The total resistance R of two resistors connected in parallel is

$$\frac{1}{R} = \frac{1}{R_1} + \frac{1}{R_2}.$$

Approximate the change in R as R_1 is increased from 10 ohms to 10.5 ohms and R_2 is decreased from 15 ohms to 13 ohms.

39. ***Inductance*** The inductance L (in microhenrys) of a straight nonmagnetic wire in free space is

$$L = 0.00021\left(\ln\frac{2h}{r} - 0.75\right)$$

where h is the length of the wire in millimeters and r is the radius of a circular cross section. Approximate L when $r = 2 \pm \frac{1}{16}$ millimeters and $h = 100 \pm \frac{1}{100}$ millimeters.

40. ***Pendulum*** The period T of a pendulum of length L is $T = 2\pi\sqrt{L/g}$, where g is the acceleration due to gravity. A pendulum is moved from the Canal Zone, where $g = 32.09$ feet per second per second, to Greenland, where $g = 32.23$ feet per second per second. Because of the change in temperature, the length of the pendulum changes from 2.5 feet to 2.48 feet. Approximate the change in the period of the pendulum.

In Exercises 41–44, show that the function is differentiable by finding values for ε_1 and ε_2 as designated in the definition of differentiability, and verify that both ε_1 and $\varepsilon_2 \to 0$ as $(\Delta x, \Delta y) \to (0, 0)$.

41. $f(x, y) = x^2 - 2x + y$

42. $f(x, y) = x^2 + y^2$

43. $f(x, y) = x^2y$

44. $f(x, y) = 5x - 10y + y^3$

In Exercises 45 and 46, use the function to prove that (a) $f_x(0, 0)$ and $f_y(0, 0)$ exist, and (b) f is not differentiable at (0, 0).

45. $f(x, y) = \begin{cases} \dfrac{3x^2y}{x^4 + y^2}, & (x, y) \neq (0, 0) \\ 0, & (x, y) = (0, 0) \end{cases}$

46. $f(x, y) = \begin{cases} \dfrac{5x^2y}{x^3 + y^3}, & (x, y) \neq (0, 0) \\ 0, & (x, y) = (0, 0) \end{cases}$

47. ***Interdisciplinary Problem*** Consider measurements and formulas you are using, or have used, in other science or engineering courses. Show how to apply differentials to these measurements and formulas to estimate possible propagated errors.

Section 13.5 Chain Rules for Functions of Several Variables

- **Use the Chain Rules for functions of several variables.**
- **Find partial derivatives implicitly.**

Chain Rules for Functions of Several Variables

Your work with differentials in the preceding section provides the basis for the extension of the Chain Rule to functions of two variables. There are two cases—the first case involves w as a function of x and y, where x and y are functions of a single independent variable t. (A proof of this theorem is given in Appendix A.)

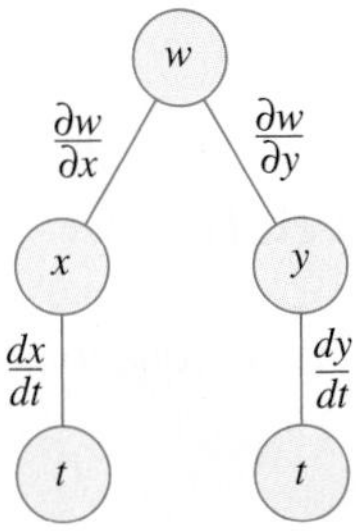

Chain Rule: one independent variable w is a function of x and y, which are each functions of t. This diagram represents the derivative of w with respect to t.
Figure 13.39

THEOREM 13.6 Chain Rule: One Independent Variable

Let $w = f(x, y)$, where f is a differentiable function of x and y. If $x = g(t)$ and $y = h(t)$, where g and h are differentiable functions of t, then w is a differentiable function of t, and

$$\frac{dw}{dt} = \frac{\partial w}{\partial x}\frac{dx}{dt} + \frac{\partial w}{\partial y}\frac{dy}{dt}.$$

See Figure 13.39.

EXAMPLE 1 Using the Chain Rule with One Independent Variable

Let $w = x^2y - y^2$, where $x = \sin t$ and $y = e^t$. Find dw/dt when $t = 0$.

Solution By the Chain Rule for one independent variable, you have

$$\begin{aligned}\frac{dw}{dt} &= \frac{\partial w}{\partial x}\frac{dx}{dt} + \frac{\partial w}{\partial y}\frac{dy}{dt}\\ &= 2xy(\cos t) + (x^2 - 2y)e^t\\ &= 2(\sin t)(e^t)(\cos t) + (\sin^2 t - 2e^t)e^t\\ &= 2e^t \sin t \cos t + e^t \sin^2 t - 2e^{2t}.\end{aligned}$$

When $t = 0$, it follows that

$$\frac{dw}{dt} = -2.$$

The Chain Rules presented in this section provide alternative techniques for solving many problems in single-variable calculus. For instance, in Example 1, you could have used single-variable techniques to find dw/dt by first writing w as a function of t,

$$\begin{aligned}w &= x^2y - y^2\\ &= (\sin t)^2(e^t) - (e^t)^2\\ &= e^t \sin^2 t - e^{2t}\end{aligned}$$

and then differentiating as usual.

$$\frac{dw}{dt} = 2e^t \sin t \cos t + e^t \sin^2 t - 2e^{2t}$$

The Chain Rule in Theorem 13.6 can be extended to any number of variables. For example, if each x_i is a differentiable function of a single variable t, then for

$$w = f(x_1, x_2, \ldots, x_n)$$

you have

$$\frac{dw}{dt} = \frac{\partial w}{\partial x_1}\frac{dx_1}{dt} + \frac{\partial w}{\partial x_2}\frac{dx_2}{dt} + \cdots + \frac{\partial w}{\partial x_n}\frac{dx_n}{dt}.$$

EXAMPLE 2 An Application of a Chain Rule to Related Rates

Two objects are traveling in elliptical paths given by the following parametric equations.

$$x_1 = 4\cos t \quad \text{and} \quad y_1 = 2\sin t \qquad \text{First object}$$

$$x_2 = 2\sin 2t \quad \text{and} \quad y_2 = 3\cos 2t \qquad \text{Second object}$$

At what rate is the distance between the two objects changing when $t = \pi$?

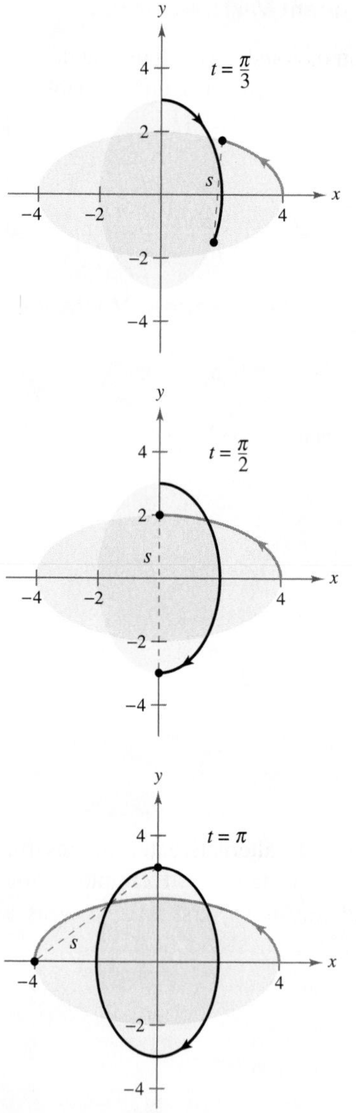

Paths of two objects traveling in elliptical orbits

Figure 13.40

Solution From Figure 13.40, you can see that the distance s between the two objects is given by

$$s = \sqrt{(x_2 - x_1)^2 + (y_2 - y_1)^2}$$

and that when $t = \pi$, you have $x_1 = -4$, $y_1 = 0$, $x_2 = 0$, $y_2 = 3$, and

$$s = \sqrt{(0 + 4)^2 + (3 - 0)^2} = 5.$$

When $t = \pi$, the partial derivatives of s are as follows.

$$\frac{\partial s}{\partial x_1} = \frac{-(x_2 - x_1)}{\sqrt{(x_2 - x_1)^2 + (y_2 - y_1)^2}} = -\frac{1}{5}(0 + 4) = -\frac{4}{5}$$

$$\frac{\partial s}{\partial y_1} = \frac{-(y_2 - y_1)}{\sqrt{(x_2 - x_1)^2 + (y_2 - y_1)^2}} = -\frac{1}{5}(3 - 0) = -\frac{3}{5}$$

$$\frac{\partial s}{\partial x_2} = \frac{(x_2 - x_1)}{\sqrt{(x_2 - x_1)^2 + (y_2 - y_1)^2}} = \frac{1}{5}(0 + 4) = \frac{4}{5}$$

$$\frac{\partial s}{\partial y_2} = \frac{(y_2 - y_1)}{\sqrt{(x_2 - x_1)^2 + (y_2 - y_1)^2}} = \frac{1}{5}(3 - 0) = \frac{3}{5}$$

When $t = \pi$, the derivatives of x_1, y_1, x_2, and y_2 are

$$\frac{dx_1}{dt} = -4\sin t = 0 \qquad \frac{dy_1}{dt} = 2\cos t = -2$$

$$\frac{dx_2}{dt} = 4\cos 2t = 4 \qquad \frac{dy_2}{dt} = -6\sin 2t = 0.$$

So, using the appropriate Chain Rule, you know that the distance is changing at a rate of

$$\frac{ds}{dt} = \frac{\partial s}{\partial x_1}\frac{dx_1}{dt} + \frac{\partial s}{\partial y_1}\frac{dy_1}{dt} + \frac{\partial s}{\partial x_2}\frac{dx_2}{dt} + \frac{\partial s}{\partial y_2}\frac{dy_2}{dt}$$

$$= \left(-\frac{4}{5}\right)(0) + \left(-\frac{3}{5}\right)(-2) + \left(\frac{4}{5}\right)(4) + \left(\frac{3}{5}\right)(0)$$

$$= \frac{22}{5}.$$

In Example 2, note that s is the function of four *intermediate* variables, x_1, y_1, x_2, and y_2, each of which is a function of a single variable t. Another type of composite function is one in which the intermediate variables are themselves functions of more than one variable. For instance, if $w = f(x, y)$, where $x = g(s, t)$ and $y = h(s, t)$, it follows that w is a function of s and t, and you can consider the partial derivatives of w with respect to s and t. One way to find these partial derivatives is to write w as a function of s and t explicitly by substituting the equations $x = g(s, t)$ and $y = h(s, t)$ into the equation $w = f(x, y)$. Then you can find the partial derivatives in the usual way, as demonstrated in the next example.

EXAMPLE 3 Finding Partial Derivatives by Substitution

Find $\partial w/\partial s$ and $\partial w/\partial t$ for $w = 2xy$, where $x = s^2 + t^2$ and $y = s/t$.

Solution Begin by substituting $x = s^2 + t^2$ and $y = s/t$ into the equation $w = 2xy$ to obtain

$$w = 2xy = 2(s^2 + t^2)\left(\frac{s}{t}\right) = 2\left(\frac{s^3}{t} + st\right).$$

Then, to find $\partial w/\partial s$, hold t constant and differentiate with respect to s.

$$\frac{\partial w}{\partial s} = 2\left(\frac{3s^2}{t} + t\right)$$
$$= \frac{6s^2 + 2t^2}{t}$$

Similarly, to find $\partial w/\partial t$, hold s constant and differentiate with respect to t to obtain

$$\frac{\partial w}{\partial t} = 2\left(-\frac{s^3}{t^2} + s\right)$$
$$= 2\left(\frac{-s^3 + st^2}{t^2}\right)$$
$$= \frac{2st^2 - 2s^3}{t^2}.$$

Theorem 13.7 gives an alternative method for finding the partial derivatives in Example 3, without explicitly writing w as a function of s and t.

THEOREM 13.7 Chain Rule: Two Independent Variables

Let $w = f(x, y)$, where f is a differentiable function of x and y. If $x = g(s, t)$ and $y = h(s, t)$ such that the first partials $\partial x/\partial s$, $\partial x/\partial t$, $\partial y/\partial s$, and $\partial y/\partial t$ all exist, then $\partial w/\partial s$ and $\partial w/\partial t$ exist and are given by

$$\frac{\partial w}{\partial s} = \frac{\partial w}{\partial x}\frac{\partial x}{\partial s} + \frac{\partial w}{\partial y}\frac{\partial y}{\partial s} \quad \text{and} \quad \frac{\partial w}{\partial t} = \frac{\partial w}{\partial x}\frac{\partial x}{\partial t} + \frac{\partial w}{\partial y}\frac{\partial y}{\partial t}.$$

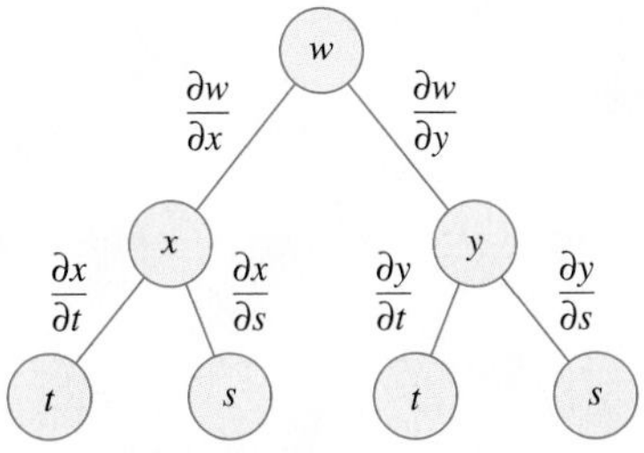

Chain Rule: two independent variables
Figure 13.41

Proof To obtain $\partial w/\partial s$, hold t constant and apply Theorem 13.6 to obtain the desired result. Similarly, for $\partial w/\partial t$ hold s constant and apply Theorem 13.6.

NOTE The Chain Rule in this theorem is shown schematically in Figure 13.41.

EXAMPLE 4 The Chain Rule with Two Independent Variables

Use the Chain Rule to find $\partial w/\partial s$ and $\partial w/\partial t$ for

$$w = 2xy$$

where $x = s^2 + t^2$ and $y = s/t$.

Solution Note that these same partials were found in Example 3. This time, using Theorem 13.7, you can hold t constant and differentiate with respect to s to obtain

$$\frac{\partial w}{\partial s} = \frac{\partial w}{\partial x}\frac{\partial x}{\partial s} + \frac{\partial w}{\partial y}\frac{\partial y}{\partial s}$$

$$= 2y(2s) + 2x\left(\frac{1}{t}\right)$$

$$= 2\left(\frac{s}{t}\right)(2s) + 2(s^2 + t^2)\left(\frac{1}{t}\right)$$

Substitute (s/t) for y and $s^2 + t^2$ for x.

$$= \frac{4s^2}{t} + \frac{2s^2 + 2t^2}{t}$$

$$= \frac{6s^2 + 2t^2}{t}.$$

Similarly, holding s constant gives

$$\frac{\partial w}{\partial t} = \frac{\partial w}{\partial x}\frac{\partial x}{\partial t} + \frac{\partial w}{\partial y}\frac{\partial y}{\partial t}$$

$$= 2y(2t) + 2x\left(\frac{-s}{t^2}\right)$$

$$= 2\left(\frac{s}{t}\right)(2t) + 2(s^2 + t^2)\left(\frac{-s}{t^2}\right)$$

Substitute (s/t) for y and $s^2 + t^2$ for x.

$$= 4s - \frac{2s^3 + 2st^2}{t^2}$$

$$= \frac{4st^2 - 2s^3 - 2st^2}{t^2}$$

$$= \frac{2st^2 - 2s^3}{t^2}.$$

The Chain Rule in Theorem 13.7 can also be extended to any number of variables. For example, if w is a differentiable function of the n variables $x_1, x_2, \ldots, x_n$, where each x_i is a differentiable function of the m variables $t_1, t_2, \ldots, t_m$, then for

$$w = f(x_1, x_2, \ldots, x_n)$$

you obtain the following.

$$\frac{\partial w}{\partial t_1} = \frac{\partial w}{\partial x_1}\frac{\partial x_1}{\partial t_1} + \frac{\partial w}{\partial x_2}\frac{\partial x_2}{\partial t_1} + \cdots + \frac{\partial w}{\partial x_n}\frac{\partial x_n}{\partial t_1}$$

$$\frac{\partial w}{\partial t_2} = \frac{\partial w}{\partial x_1}\frac{\partial x_1}{\partial t_2} + \frac{\partial w}{\partial x_2}\frac{\partial x_2}{\partial t_2} + \cdots + \frac{\partial w}{\partial x_n}\frac{\partial x_n}{\partial t_2}$$

$$\vdots$$

$$\frac{\partial w}{\partial t_m} = \frac{\partial w}{\partial x_1}\frac{\partial x_1}{\partial t_m} + \frac{\partial w}{\partial x_2}\frac{\partial x_2}{\partial t_m} + \cdots + \frac{\partial w}{\partial x_n}\frac{\partial x_n}{\partial t_m}$$

EXAMPLE 5 The Chain Rule for a Function of Three Variables

Find $\partial w/\partial s$ and $\partial w/\partial t$ when $s = 1$ and $t = 2\pi$ for the function given by

$$w = xy + yz + xz$$

where $x = s\cos t$, $y = s\sin t$, and $z = t$.

Solution By extending the result of Theorem 13.7, you have

$$\begin{aligned}\frac{\partial w}{\partial s} &= \frac{\partial w}{\partial x}\frac{\partial x}{\partial s} + \frac{\partial w}{\partial y}\frac{\partial y}{\partial s} + \frac{\partial w}{\partial z}\frac{\partial z}{\partial s}\\ &= (y + z)(\cos t) + (x + z)(\sin t) + (y + x)(0)\\ &= (y + z)(\cos t) + (x + z)(\sin t).\end{aligned}$$

When $s = 1$ and $t = 2\pi$, you have $x = 1$, $y = 0$, and $z = 2\pi$. So, $\partial w/\partial s = (0 + 2\pi)(1) + (1 + 2\pi)(0) = 2\pi$. Furthermore,

$$\begin{aligned}\frac{\partial w}{\partial t} &= \frac{\partial w}{\partial x}\frac{\partial x}{\partial t} + \frac{\partial w}{\partial y}\frac{\partial y}{\partial t} + \frac{\partial w}{\partial z}\frac{\partial z}{\partial t}\\ &= (y + z)(-s\sin t) + (x + z)(s\cos t) + (y + x)(1)\end{aligned}$$

and for $s = 1$ and $t = 2\pi$, it follows that

$$\begin{aligned}\frac{\partial w}{\partial t} &= (0 + 2\pi)(0) + (1 + 2\pi)(1) + (0 + 1)(1)\\ &= 2 + 2\pi.\end{aligned}$$

Implicit Partial Differentiation

This section concludes with an application of the Chain Rule to determine the derivative of a function defined *implicitly*. Suppose that x and y are related by the equation $F(x, y) = 0$, where it is assumed that $y = f(x)$ is a differentiable function of x. To find dy/dx, you could use the techniques discussed in Section 2.5. However, you will see that the Chain Rule provides a convenient alternative. If you consider the function given by

$$w = F(x, y) = F(x, f(x))$$

you can apply Theorem 13.6 to obtain

$$\frac{dw}{dx} = F_x(x, y)\frac{dx}{dx} + F_y(x, y)\frac{dy}{dx}.$$

Because $w = F(x, y) = 0$ for all x in the domain of f, you know that $dw/dx = 0$ and you have

$$F_x(x, y)\frac{dx}{dx} + F_y(x, y)\frac{dy}{dx} = 0.$$

Now, if $F_y(x, y) \neq 0$, you can use the fact that $dx/dx = 1$ to conclude that

$$\frac{dy}{dx} = -\frac{F_x(x, y)}{F_y(x, y)}.$$

A similar procedure can be used to find the partial derivatives of functions of several variables that are defined implicitly.

THEOREM 13.8 Chain Rule: Implicit Differentiation

If the equation $F(x, y) = 0$ defines y implicitly as a differentiable function of x, then

$$\frac{dy}{dx} = -\frac{F_x(x, y)}{F_y(x, y)}, \qquad F_y(x, y) \neq 0.$$

If the equation $F(x, y, z) = 0$ defines z implicitly as a differentiable function of x and y, then

$$\frac{\partial z}{\partial x} = -\frac{F_x(x, y, z)}{F_z(x, y, z)} \quad \text{and} \quad \frac{\partial z}{\partial y} = -\frac{F_y(x, y, z)}{F_z(x, y, z)}, \quad F_z(x, y, z) \neq 0.$$

This theorem can be extended to differentiable functions defined implicitly with any number of variables.

EXAMPLE 6 Finding a Derivative Implicitly

Find dy/dx, given $y^3 + y^2 - 5y - x^2 + 4 = 0$.

Solution Begin by defining a function F as

$$F(x, y) = y^3 + y^2 - 5y - x^2 + 4.$$

Then, using Theorem 13.8, you have

$$F_x(x, y) = -2x \quad \text{and} \quad F_y(x, y) = 3y^2 + 2y - 5$$

and it follows that

$$\frac{dy}{dx} = -\frac{F_x(x, y)}{F_y(x, y)} = \frac{-(-2x)}{3y^2 + 2y - 5} = \frac{2x}{3y^2 + 2y - 5}.$$

NOTE Compare the solution of Example 6 with the solution of Example 2 in Section 2.5.

EXAMPLE 7 Finding Partial Derivatives Implicitly

Find $\partial z/\partial x$ and $\partial z/\partial y$, given $3x^2z - x^2y^2 + 2z^3 + 3yz - 5 = 0$.

Solution To apply Theorem 13.8, let

$$F(x, y, z) = 3x^2z - x^2y^2 + 2z^3 + 3yz - 5.$$

Then

$$F_x(x, y, z) = 6xz - 2xy^2$$
$$F_y(x, y, z) = -2x^2y + 3z$$
$$F_z(x, y, z) = 3x^2 + 6z^2 + 3y$$

and you obtain

$$\frac{\partial z}{\partial x} = -\frac{F_x(x, y, z)}{F_z(x, y, z)} = \frac{2xy^2 - 6xz}{3x^2 + 6z^2 + 3y}$$

$$\frac{\partial z}{\partial y} = -\frac{F_y(x, y, z)}{F_z(x, y, z)} = \frac{2x^2y - 3z}{3x^2 + 6z^2 + 3y}.$$

Exercises for Section 13.5

See www.CalcChat.com for worked-out solutions to odd-numbered exercises.

In Exercises 1–4, find dw/dt using the appropriate Chain Rule.

1. $w = x^2 + y^2$, $x = e^t$, $y = e^{-t}$
2. $w = \sqrt{x^2 + y^2}$, $x = \cos t$, $y = e^t$
3. $w = x \sec y$, $x = e^t$, $y = \pi - t$
4. $w = \ln \dfrac{y}{x}$, $x = \cos t$, $y = \sin t$

In Exercises 5–10, find dw/dt (a) using the appropriate Chain Rule and (b) by converting w to a function of t before differentiating.

5. $w = xy$, $x = 2 \sin t$, $y = \cos t$
6. $w = \cos(x - y)$, $x = t^2$, $y = 1$
7. $w = x^2 + y^2 + z^2$, $x = e^t \cos t$, $y = e^t \sin t$, $z = e^t$
8. $w = xy \cos z$, $x = t$, $y = t^2$, $z = \arccos t$
9. $w = xy + xz + yz$, $x = t - 1$, $y = t^2 - 1$, $z = t$
10. $w = xyz$, $x = t^2$, $y = 2t$, $z = e^{-t}$

Projectile Motion **In Exercises 11 and 12, the parametric equations for the paths of two projectiles are given. At what rate is the distance between the two objects changing at the given value of t?**

11. $x_1 = 10 \cos 2t$, $y_1 = 6 \sin 2t$ — First object
$x_2 = 7 \cos t$, $y_2 = 4 \sin t$ — Second object
$t = \pi/2$
12. $x_1 = 48\sqrt{2}\,t$, $y_1 = 48\sqrt{2}\,t - 16t^2$ — First object
$x_2 = 48\sqrt{3}\,t$, $y_2 = 48t - 16t^2$ — Second object
$t = 1$

In Exercises 13 and 14, find d^2w/dt^2 using the appropriate Chain Rule. Evaluate d^2w/dt^2 at the given value of t.

13. $w = \arctan(2xy)$, $x = \cos t$, $y = \sin t$, $t = 0$
14. $w = \dfrac{x^2}{y}$, $x = t^2$, $y = t + 1$, $t = 1$

In Exercises 15–18, find $\partial w/\partial s$ and $\partial w/\partial t$ using the appropriate Chain Rule, and evaluate each partial derivative at the given values of s and t.

	Function	*Point*
15.	$w = x^2 + y^2$; $x = s + t$, $y = s - t$	$s = 2$, $t = -1$
16.	$w = y^3 - 3x^2y$; $x = e^s$, $y = e^t$	$s = 0$, $t = 1$
17.	$w = x^2 - y^2$; $x = s \cos t$, $y = s \sin t$	$s = 3$, $t = \dfrac{\pi}{4}$
18.	$w = \sin(2x + 3y)$; $x = s + t$, $y = s - t$	$s = 0$, $t = \dfrac{\pi}{2}$

In Exercises 19–22, find $\partial w/\partial r$ and $\partial w/\partial \theta$ (a) using the appropriate Chain Rule and (b) by converting w to a function of r and θ before differentiating.

19. $w = x^2 - 2xy + y^2$, $x = r + \theta$, $y = r - \theta$
20. $w = \sqrt{25 - 5x^2 - 5y^2}$, $x = r \cos \theta$, $y = r \sin \theta$
21. $w = \arctan \dfrac{y}{x}$, $x = r \cos \theta$, $y = r \sin \theta$
22. $w = \dfrac{yz}{x}$, $x = \theta^2$, $y = r + \theta$, $z = r - \theta$

In Exercises 23–26, find $\partial w/\partial s$ and $\partial w/\partial t$ by using the appropriate Chain Rule.

23. $w = xyz$, $x = s + t$, $y = s - t$, $z = st^2$
24. $w = x \cos yz$, $x = s^2$, $y = t^2$, $z = s - 2t$
25. $w = ze^{x/y}$, $x = s - t$, $y = s + t$, $z = st$
26. $w = x^2 + y^2 + z^2$, $x = t \sin s$, $y = t \cos s$, $z = st^2$

In Exercises 27–30, differentiate implicitly to find dy/dx.

27. $x^2 - 3xy + y^2 - 2x + y - 5 = 0$
28. $\cos x + \tan xy + 5 = 0$
29. $\ln\sqrt{x^2 + y^2} + xy = 4$
30. $\dfrac{x}{x^2 + y^2} - y^2 = 6$

In Exercises 31–38, differentiate implicitly to find the first partial derivatives of z.

31. $x^2 + y^2 + z^2 = 25$
32. $xz + yz + xy = 0$
33. $\tan(x + y) + \tan(y + z) = 1$
34. $z = e^x \sin(y + z)$
35. $x^2 + 2yz + z^2 = 1$
36. $x + \sin(y + z) = 0$
37. $e^{xz} + xy = 0$
38. $x \ln y + y^2z + z^2 = 8$

In Exercises 39–42, differentiate implicitly to find the first partial derivatives of w.

39. $xyz + xzw - yzw + w^2 = 5$
40. $x^2 + y^2 + z^2 - 5yw + 10w^2 = 2$
41. $\cos xy + \sin yz + wz = 20$
42. $w - \sqrt{x - y} - \sqrt{y - z} = 0$

Homogeneous Functions **In Exercises 43–46, the function f is homogeneous of degree n if $f(tx, ty) = t^nf(x, y)$. Determine the degree of the homogeneous function, and show that**

$$xf_x(x, y) + yf_y(x, y) = nf(x, y).$$

43. $f(x, y) = \dfrac{xy}{\sqrt{x^2 + y^2}}$
44. $f(x, y) = x^3 - 3xy^2 + y^3$
45. $f(x, y) = e^{x/y}$
46. $f(x, y) = \dfrac{x^2}{\sqrt{x^2 + y^2}}$

Writing About Concepts

47. Let $w = f(x, y)$ be a function where x and y are functions of a single variable t. Give the Chain Rule for finding dw/dt.

48. Let $w = f(x, y)$ be a function where x and y are functions of two variables s and t. Give the Chain Rule for finding $\partial w/\partial s$ and $\partial w/\partial t$.

49. Describe the difference between the explicit form of a function of two variables x and y and the implicit form. Give an example of each.

50. If $f(x, y) = 0$, give the rule for finding dy/dx implicitly. If $f(x, y, z) = 0$, give the rule for finding $\partial z/\partial x$ and $\partial z/\partial y$ implicitly.

51. ***Volume and Surface Area*** The radius of a right circular cylinder is increasing at a rate of 6 inches per minute, and the height is decreasing at a rate of 4 inches per minute. What are the rates of change of the volume and surface area when the radius is 12 inches and the height is 36 inches?

52. ***Volume and Surface Area*** Repeat Exercise 51 for a right circular cone.

53. ***Area*** Let θ be the angle between equal sides of an isosceles triangle and let x be the length of these sides. x is increasing at $\frac{1}{2}$ meter per hour and θ is increasing at $\pi/90$ radian per hour. Find the rate of increase of the area when $x = 6$ and $\theta = \pi/4$.

54. ***Volume and Surface Area*** The two radii of the frustum of a right circular cone are increasing at a rate of 4 centimeters per minute, and the height is increasing at a rate of 12 centimeters per minute (see figure). Find the rates at which the volume and surface area are changing when the two radii are 15 centimeters and 25 centimeters, and the height is 10 centimeters.

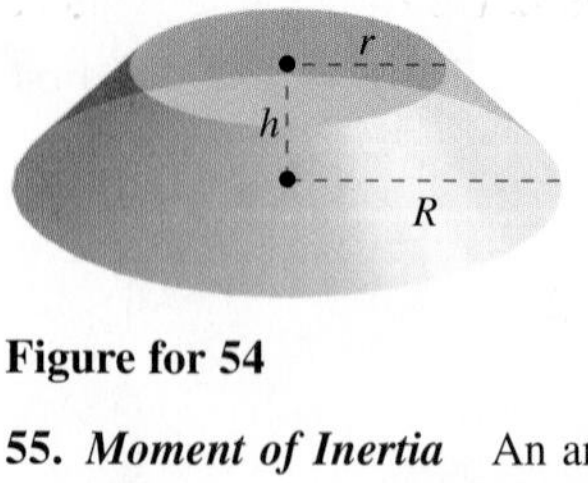

Figure for 54

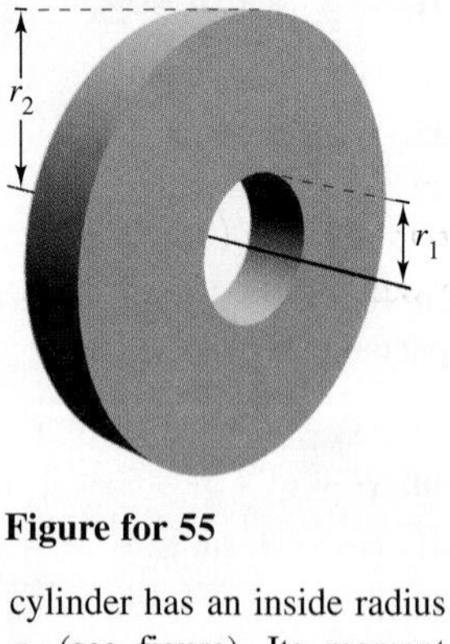

Figure for 55

55. ***Moment of Inertia*** An annular cylinder has an inside radius of r_1 and an outside radius of r_2 (see figure). Its moment of inertia is $I = \frac{1}{2}m(r_1^2 + r_2^2)$ where m is the mass. The two radii are increasing at a rate of 2 centimeters per second. Find the rate at which I is changing at the instant the radii are 6 centimeters and 8 centimeters. (Assume mass is a constant.)

56. ***Ideal Gas Law*** The Ideal Gas Law is $pV = mRT$, where R is a constant, m is a constant mass, and p and V are functions of time. Find dT/dt, the rate at which the temperature changes with respect to time.

57. ***Maximum Angle*** A two-foot-tall painting hangs on a wall such that the bottom is 6 feet from the floor. A child whose eyes are 4 feet above the floor stands x feet from the wall (see figure).

(a) Show that $x^2 \tan\theta - 2x + 8\tan\theta = 0$.

(b) Use implicit differentiation to find $d\theta/dx$.

(c) Find x such that θ is maximum.

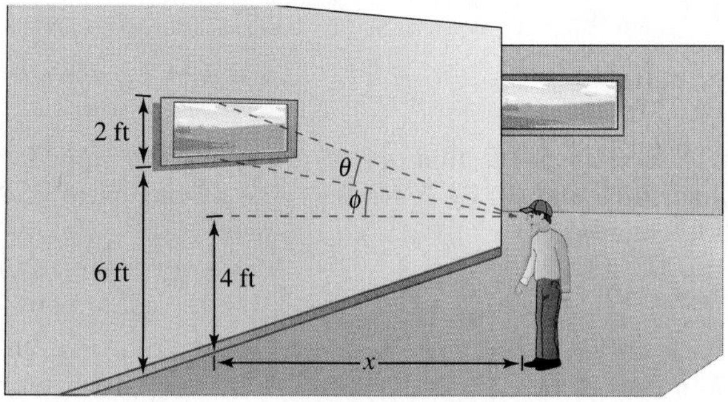

58. Show that if $f(x,y)$ is homogeneous of degree n, then

$$xf_x(x, y) + yf_y(x, y) = nf(x, y).$$

[*Hint:* Let $g(t) = f(tx, ty) = t^n f(x, y)$. Find $g'(t)$ and then let $t = 1$.]

59. Show that

$$\frac{\partial w}{\partial u} + \frac{\partial w}{\partial v} = 0$$

for $w = f(x, y)$, $x = u - v$, and $y = v - u$.

60. Demonstrate the result of Exercise 59 for

$$w = (x - y)\sin(y - x).$$

61. Consider the function $w = f(x, y)$, where $x = r\cos\theta$ and $y = r\sin\theta$. Prove each of the following.

(a) $$\frac{\partial w}{\partial x} = \frac{\partial w}{\partial r}\cos\theta - \frac{\partial w}{\partial \theta}\frac{\sin\theta}{r}$$

$$\frac{\partial w}{\partial y} = \frac{\partial w}{\partial r}\sin\theta + \frac{\partial w}{\partial \theta}\frac{\cos\theta}{r}$$

(b) $$\left(\frac{\partial w}{\partial x}\right)^2 + \left(\frac{\partial w}{\partial y}\right)^2 = \left(\frac{\partial w}{\partial r}\right)^2 + \left(\frac{1}{r^2}\right)\left(\frac{\partial w}{\partial \theta}\right)^2$$

62. Demonstrate the result of Exercise 61(b) for $w = \arctan(y/x)$.

63. ***Cauchy-Riemann Equations*** Given the functions $u(x, y)$ and $v(x, y)$, verify that the **Cauchy-Riemann differential equations**

$$\frac{\partial u}{\partial x} = \frac{\partial v}{\partial y} \quad \text{and} \quad \frac{\partial u}{\partial y} = -\frac{\partial v}{\partial x}$$

can be written in polar coordinate form as

$$\frac{\partial u}{\partial r} = \frac{1}{r}\frac{\partial v}{\partial \theta} \quad \text{and} \quad \frac{\partial v}{\partial r} = -\frac{1}{r}\frac{\partial u}{\partial \theta}.$$

64. Demonstrate the result of Exercise 63 for the functions

$$u = \ln\sqrt{x^2 + y^2} \quad \text{and} \quad v = \arctan\frac{y}{x}.$$

Section 13.6 Directional Derivatives and Gradients

- **Find and use directional derivatives of a function of two variables.**
- **Find the gradient of a function of two variables.**
- **Use the gradient of a function of two variables in applications.**
- **Find directional derivatives and gradients of functions of three variables.**

Directional Derivative

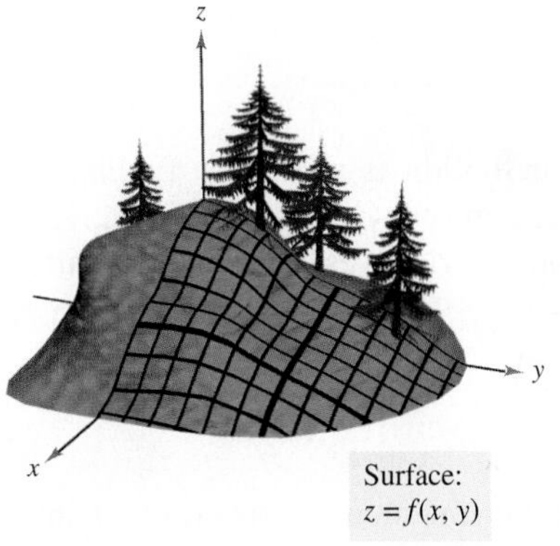

Figure 13.42

You are standing on the hillside pictured in Figure 13.42 and want to determine the hill's incline toward the z-axis. If the hill were represented by $z = f(x, y)$, you would already know how to determine the slopes in two different directions—the slope in the y-direction would be given by the partial derivative $f_y(x, y)$, and the slope in the x-direction would be given by the partial derivative $f_x(x, y)$. In this section, you will see that these two partial derivatives can be used to find the slope in *any* direction.

To determine the slope at a point on a surface, you will define a new type of derivative called a **directional derivative.** Begin by letting $z = f(x, y)$ be a *surface* and $P(x_0, y_0)$ a *point* in the domain of f, as shown in Figure 13.43. The "direction" of the directional derivative is given by a unit vector

$$\mathbf{u} = \cos\theta\mathbf{i} + \sin\theta\mathbf{j}$$

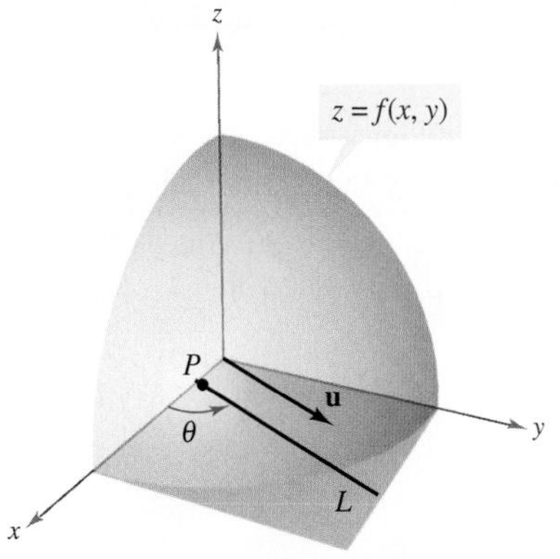

Figure 13.43

where θ is the angle the vector makes with the positive x-axis. To find the desired slope, reduce the problem to two dimensions by intersecting the surface with a vertical plane passing through the point P and parallel to $\mathbf{u}$, as shown in Figure 13.44. This vertical plane intersects the surface to form a curve C. The slope of the surface at $(x_0, y_0, f(x_0, y_0))$ in the direction of $\mathbf{u}$ is defined as the slope of the curve C at that point.

Informally, you can write the slope of the curve C as a limit that looks much like those used in single-variable calculus. The vertical plane used to form C intersects the xy-plane in a line L, represented by the parametric equations

$$x = x_0 + t\cos\theta$$

and

$$y = y_0 + t\sin\theta$$

so that for any value of t, the point $Q(x, y)$ lies on the line L. For each of the points P and Q, there is a corresponding point on the surface.

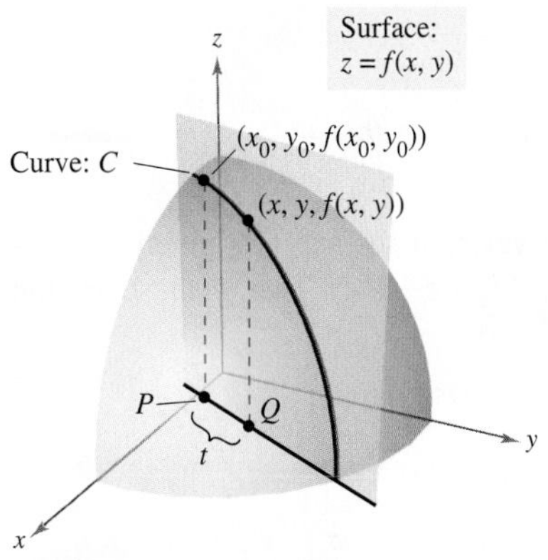

Figure 13.44

$(x_0, y_0, f(x_0, y_0))$ Point above P

$(x, y, f(x, y))$ Point above Q

Moreover, because the distance between P and Q is

$$\sqrt{(x - x_0)^2 + (y - y_0)^2} = \sqrt{(t\cos\theta)^2 + (t\sin\theta)^2} = |t|$$

you can write the slope of the secant line through $(x_0, y_0, f(x_0, y_0))$ and $(x, y, f(x, y))$ as

$$\frac{f(x, y) - f(x_0, y_0)}{t} = \frac{f(x_0 + t\cos\theta, y_0 + t\sin\theta) - f(x_0, y_0)}{t}.$$

Finally, by letting t approach 0, you arrive at the following definition.

Definition of Directional Derivative

Let f be a function of two variables x and y and let $\mathbf{u} = \cos\theta\,\mathbf{i} + \sin\theta\,\mathbf{j}$ be a unit vector. Then the **directional derivative of f in the direction of u,** denoted by $D_{\mathbf{u}}f$, is

$$D_{\mathbf{u}}f(x, y) = \lim_{t\to 0} \frac{f(x + t\cos\theta, y + t\sin\theta) - f(x, y)}{t}$$

provided this limit exists.

Calculating directional derivatives by this definition is similar to finding the derivative of a function of one variable by the limit process (given in Section 2.1). A simpler "working" formula for finding directional derivatives involves the partial derivatives f_x and f_y.

THEOREM 13.9 Directional Derivative

If f is a differentiable function of x and y, then the directional derivative of f in the direction of the unit vector $\mathbf{u} = \cos\theta\,\mathbf{i} + \sin\theta\,\mathbf{j}$ is

$$D_{\mathbf{u}}f(x, y) = f_x(x, y)\cos\theta + f_y(x, y)\sin\theta.$$

Proof For a fixed point (x_0, y_0), let $x = x_0 + t\cos\theta$ and let $y = y_0 + t\sin\theta$. Then, let $g(t) = f(x, y)$. Because f is differentiable, you can apply the Chain Rule given in Theorem 13.7 to obtain

$$g'(t) = f_x(x, y)x'(t) + f_y(x, y)y'(t) = f_x(x, y)\cos\theta + f_y(x, y)\sin\theta.$$

If $t = 0$, then $x = x_0$ and $y = y_0$, so

$$g'(0) = f_x(x_0, y_0)\cos\theta + f_y(x_0, y_0)\sin\theta.$$

By the definition of $g'(t)$, it is also true that

$$\begin{aligned} g'(0) &= \lim_{t\to 0}\frac{g(t) - g(0)}{t} \\ &= \lim_{t\to 0}\frac{f(x_0 + t\cos\theta, y_0 + t\sin\theta) - f(x_0, y_0)}{t}. \end{aligned}$$

Consequently, $D_{\mathbf{u}}f(x_0, y_0) = f_x(x_0, y_0)\cos\theta + f_y(x_0, y_0)\sin\theta$.

There are infinitely many directional derivatives to a surface at a given point—one for each direction specified by $\mathbf{u}$, as shown in Figure 13.45. Two of these are the partial derivatives f_x and f_y.

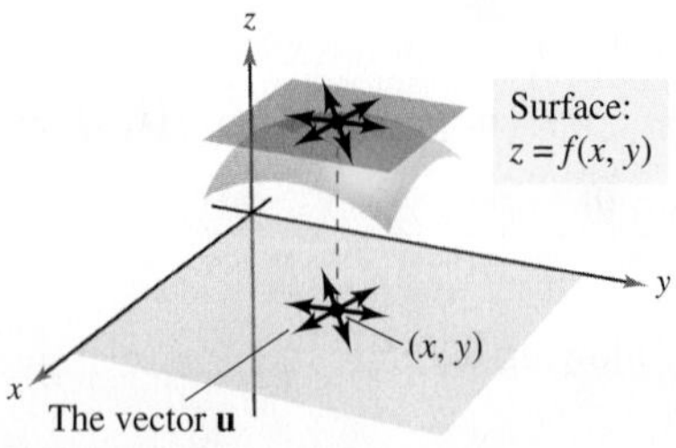

Figure 13.45

1. Direction of positive x-axis ($\theta = 0$): $\mathbf{u} = \cos 0\,\mathbf{i} + \sin 0\,\mathbf{j} = \mathbf{i}$

$$D_{\mathbf{i}}f(x, y) = f_x(x, y)\cos 0 + f_y(x, y)\sin 0 = f_x(x, y)$$

2. Direction of positive y-axis ($\theta = \pi/2$): $\mathbf{u} = \cos\frac{\pi}{2}\mathbf{i} + \sin\frac{\pi}{2}\mathbf{j} = \mathbf{j}$

$$D_{\mathbf{j}}f(x, y) = f_x(x, y)\cos\frac{\pi}{2} + f_y(x, y)\sin\frac{\pi}{2} = f_y(x, y)$$

EXAMPLE 1 Finding a Directional Derivative

Find the directional derivative of

$$f(x, y) = 4 - x^2 - \tfrac{1}{4}y^2 \qquad \text{Surface}$$

at $(1, 2)$ in the direction of

$$\mathbf{u} = \left(\cos \frac{\pi}{3}\right)\mathbf{i} + \left(\sin \frac{\pi}{3}\right)\mathbf{j}. \qquad \text{Direction}$$

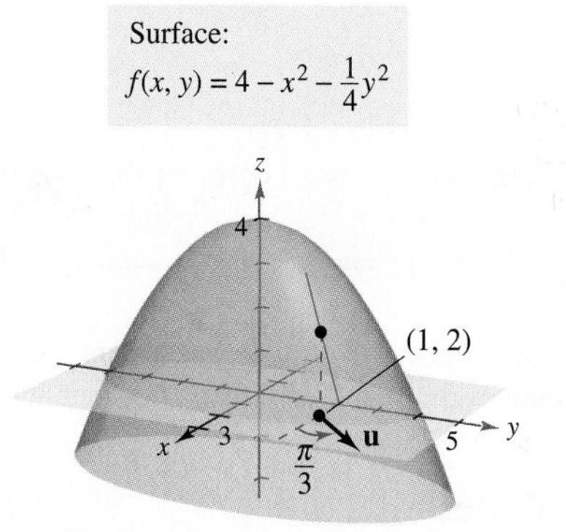

Figure 13.46

NOTE Note in Figure 13.46 that you can interpret the directional derivative as giving the slope of the surface at the point $(1, 2, 2)$ in the direction of the unit vector $\mathbf{u}$.

Solution Because f_x and f_y are continuous, f is differentiable, and you can apply Theorem 13.9.

$$\begin{aligned} D_{\mathbf{u}} f(x, y) &= f_x(x, y) \cos \theta + f_y(x, y) \sin \theta \\ &= (-2x) \cos \theta + \left(-\frac{y}{2}\right) \sin \theta \end{aligned}$$

Evaluating at $\theta = \pi/3$, $x = 1$, and $y = 2$ produces

$$\begin{aligned} D_{\mathbf{u}} f(1, 2) &= (-2)\left(\frac{1}{2}\right) + (-1)\left(\frac{\sqrt{3}}{2}\right) \\ &= -1 - \frac{\sqrt{3}}{2} \\ &\approx -1.866. \end{aligned}$$

See Figure 13.46.

You have been specifying direction by a unit vector $\mathbf{u}$. If the direction is given by a vector whose length is not 1, you must normalize the vector before applying the formula in Theorem 13.9.

EXAMPLE 2 Finding a Directional Derivative

Find the directional derivative of

$$f(x, y) = x^2 \sin 2y \qquad \text{Surface}$$

at $(1, \pi/2)$ in the direction of

$$\mathbf{v} = 3\mathbf{i} - 4\mathbf{j}. \qquad \text{Direction}$$

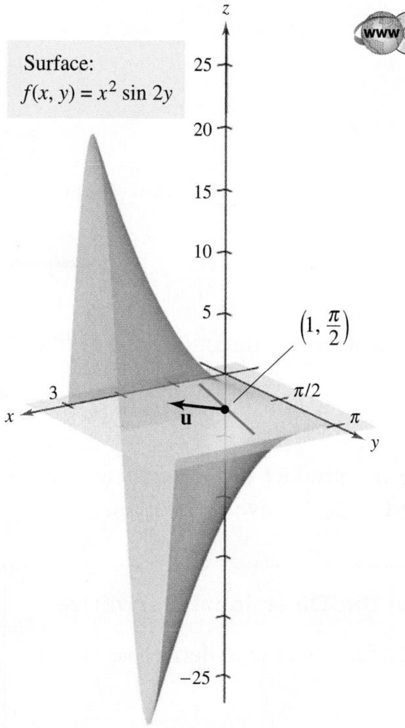

Figure 13.47

Solution Because f_x and f_y are continuous, f is differentiable, and you can apply Theorem 13.9. Begin by finding a unit vector in the direction of $\mathbf{v}$.

$$\mathbf{u} = \frac{\mathbf{v}}{\|\mathbf{v}\|} = \frac{3}{5}\mathbf{i} - \frac{4}{5}\mathbf{j} = \cos \theta \mathbf{i} + \sin \theta \mathbf{j}$$

Using this unit vector, you have

$$\begin{aligned} D_{\mathbf{u}} f(x, y) &= (2x \sin 2y)(\cos \theta) + (2x^2 \cos 2y)(\sin \theta) \\ D_{\mathbf{u}} f\left(1, \frac{\pi}{2}\right) &= (2 \sin \pi)\left(\frac{3}{5}\right) + (2 \cos \pi)\left(-\frac{4}{5}\right) \\ &= (0)\left(\frac{3}{5}\right) + (-2)\left(-\frac{4}{5}\right) \\ &= \frac{8}{5}. \end{aligned}$$

See Figure 13.47.

The Gradient of a Function of Two Variables

The **gradient** of a function of two variables is a vector-valued function of two variables. This function has many important uses, some of which are described later in this section.

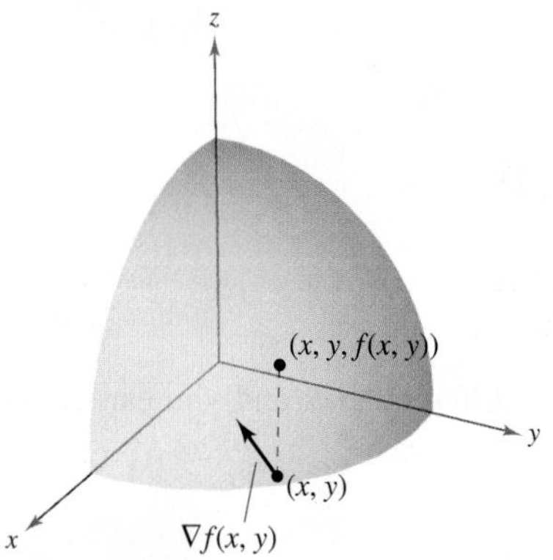

The gradient of f is a vector in the xy-plane
Figure 13.48

Definition of Gradient of a Function of Two Variables

Let $z = f(x, y)$ be a function of x and y such that f_x and f_y exist. Then the **gradient of *f*,** denoted by $\nabla f(x, y)$, is the vector

$$\nabla f(x, y) = f_x(x, y)\mathbf{i} + f_y(x, y)\mathbf{j}.$$

∇f is read as "del f." Another notation for the gradient is **grad** $f(x, y)$. In Figure 13.48, note that for each (x, y), the gradient $\nabla f(x, y)$ is a vector in the plane (not a vector in space).

NOTE No value is assigned to the symbol ∇ by itself. It is an operator in the same sense that d/dx is an operator. When ∇ operates on $f(x, y)$, it produces the vector $\nabla f(x, y)$.

EXAMPLE 3 Finding the Gradient of a Function

Find the gradient of $f(x, y) = y \ln x + xy^2$ at the point $(1, 2)$.

Solution Using

$$f_x(x, y) = \frac{y}{x} + y^2 \qquad \text{and} \qquad f_y(x, y) = \ln x + 2xy$$

you have

$$\nabla f(x, y) = \left(\frac{y}{x} + y^2\right)\mathbf{i} + (\ln x + 2xy)\mathbf{j}.$$

At the point $(1, 2)$, the gradient is

$$\begin{aligned}\nabla f(1, 2) &= \left(\frac{2}{1} + 2^2\right)\mathbf{i} + [\ln 1 + 2(1)(2)]\mathbf{j}\\ &= 6\mathbf{i} + 4\mathbf{j}.\end{aligned}$$

Because the gradient of f is a vector, you can write the directional derivative of f in the direction of $\mathbf{u}$ as

$$D_{\mathbf{u}}f(x, y) = [f_x(x, y)\mathbf{i} + f_y(x, y)\mathbf{j}] \cdot [\cos\theta\mathbf{i} + \sin\theta\mathbf{j}].$$

In other words, the directional derivative is the dot product of the gradient and the direction vector. This useful result is summarized in the following theorem.

THEOREM 13.10 Alternative Form of the Directional Derivative

If f is a differentiable function of x and y, then the directional derivative of f in the direction of the unit vector $\mathbf{u}$ is

$$D_{\mathbf{u}}f(x, y) = \nabla f(x, y) \cdot \mathbf{u}.$$

EXAMPLE 4 Using $\nabla f(x, y)$ to Find a Directional Derivative

Find the directional derivative of

$$f(x, y) = 3x^2 - 2y^2$$

at $\left(-\frac{3}{4}, 0\right)$ in the direction from $P\left(-\frac{3}{4}, 0\right)$ to $Q(0, 1)$.

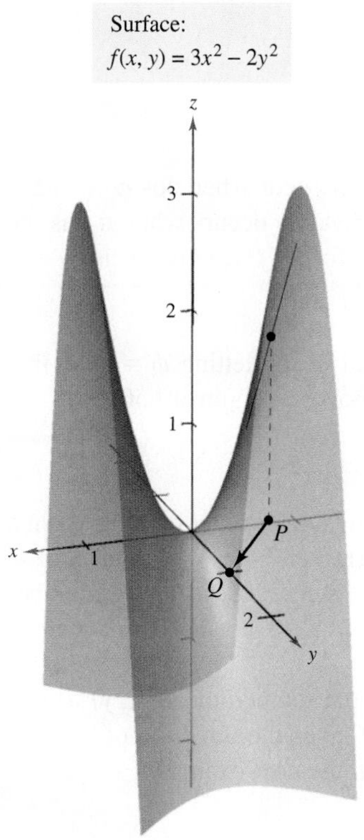

Figure 13.49

Solution Because the partials of f are continuous, f is differentiable and you can apply Theorem 13.10. A vector in the specified direction is

$$\overrightarrow{PQ} = \mathbf{v} = \left(0 + \frac{3}{4}\right)\mathbf{i} + (1 - 0)\mathbf{j}$$

$$= \frac{3}{4}\mathbf{i} + \mathbf{j}$$

and a unit vector in this direction is

$$\mathbf{u} = \frac{\mathbf{v}}{\|\mathbf{v}\|} = \frac{3}{5}\mathbf{i} + \frac{4}{5}\mathbf{j}. \qquad \text{Unit vector in direction of } \overrightarrow{PQ}$$

Because $\nabla f(x, y) = f_x(x, y)\mathbf{i} + f_y(x, y)\mathbf{j} = 6x\mathbf{i} - 4y\mathbf{j}$, the gradient at $\left(-\frac{3}{4}, 0\right)$ is

$$\nabla f\left(-\frac{3}{4}, 0\right) = -\frac{9}{2}\mathbf{i} + 0\mathbf{j}. \qquad \text{Gradient at } \left(-\tfrac{3}{4}, 0\right)$$

Consequently, at $\left(-\frac{3}{4}, 0\right)$ the directional derivative is

$$D_{\mathbf{u}} f\left(-\frac{3}{4}, 0\right) = \nabla f\left(-\frac{3}{4}, 0\right) \cdot \mathbf{u}$$

$$= \left(-\frac{9}{2}\mathbf{i} + 0\mathbf{j}\right) \cdot \left(\frac{3}{5}\mathbf{i} + \frac{4}{5}\mathbf{j}\right)$$

$$= -\frac{27}{10}. \qquad \text{Directional derivative at } \left(-\tfrac{3}{4}, 0\right)$$

See Figure 13.49.

Applications of the Gradient

You have already seen that there are many directional derivatives at the point (x, y) on a surface. In many applications, you may want to know in which direction to move so that $f(x, y)$ increases most rapidly. This direction is called the direction of steepest ascent, and it is given by the gradient, as stated in the following theorem.

NOTE Part 2 of Theorem 13.11 says that at the point (x, y), f increases most rapidly in the direction of the gradient, $\nabla f(x, y)$.

THEOREM 13.11 Properties of the Gradient

Let f be differentiable at the point (x, y).

1. If $\nabla f(x, y) = \mathbf{0}$, then $D_{\mathbf{u}} f(x, y) = 0$ for all $\mathbf{u}$.
2. The direction of *maximum* increase of f is given by $\nabla f(x, y)$. The maximum value of $D_{\mathbf{u}} f(x, y)$ is $\|\nabla f(x, y)\|$.
3. The direction of *minimum* increase of f is given by $-\nabla f(x, y)$. The minimum value of $D_{\mathbf{u}} f(x, y)$ is $-\|\nabla f(x, y)\|$.

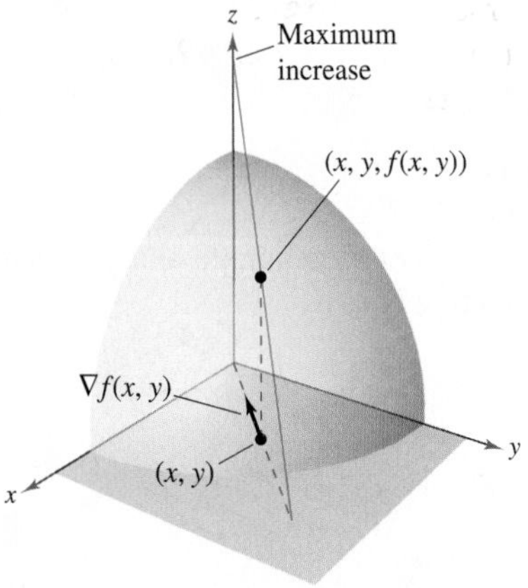

The gradient of f is a vector in the xy-plane that points in the direction of maximum increase on the surface given by $z = f(x, y)$.
Figure 13.50

Proof If $\nabla f(x, y) = \mathbf{0}$, then for any direction (any $\mathbf{u}$), you have

$$\begin{aligned} D_{\mathbf{u}} f(x, y) &= \nabla f(x, y) \cdot \mathbf{u} \\ &= (0\mathbf{i} + 0\mathbf{j}) \cdot (\cos\theta\mathbf{i} + \sin\theta\mathbf{j}) \\ &= 0. \end{aligned}$$

If $\nabla f(x, y) \neq \mathbf{0}$, then let ϕ be the angle between $\nabla f(x, y)$ and a unit vector $\mathbf{u}$. Using the dot product, you can apply Theorem 11.5 to conclude that

$$\begin{aligned} D_{\mathbf{u}} f(x, y) &= \nabla f(x, y) \cdot \mathbf{u} \\ &= \|\nabla f(x, y)\| \, \|\mathbf{u}\| \cos\phi \\ &= \|\nabla f(x, y)\| \cos\phi \end{aligned}$$

and it follows that the maximum value of $D_{\mathbf{u}} f(x, y)$ will occur when $\cos\phi = 1$. So, $\phi = 0$, and the maximum value for the directional derivative occurs when $\mathbf{u}$ has the same direction as $\nabla f(x, y)$. Moreover, this largest value for $D_{\mathbf{u}} f(x, y)$ is precisely

$$\|\nabla f(x, y)\| \cos\phi = \|\nabla f(x, y)\|.$$

Similarly, the minimum value of $D_{\mathbf{u}} f(x, y)$ can be obtained by letting $\phi = \pi$ so that $\mathbf{u}$ points in the direction opposite that of $\nabla f(x, y)$, as shown in Figure 13.50.

To visualize one of the properties of the gradient, imagine a skier coming down a mountainside. If $f(x, y)$ denotes the altitude of the skier, then $-\nabla f(x, y)$ indicates the *compass direction* the skier should take to ski the path of steepest descent. (Remember that the gradient indicates direction in the xy-plane and does not itself point up or down the mountainside.)

As another illustration of the gradient, consider the temperature $T(x, y)$ at any point (x, y) on a flat metal plate. In this case, $\nabla T(x, y)$ gives the direction of greatest temperature increase at the point (x, y), as illustrated in the next example.

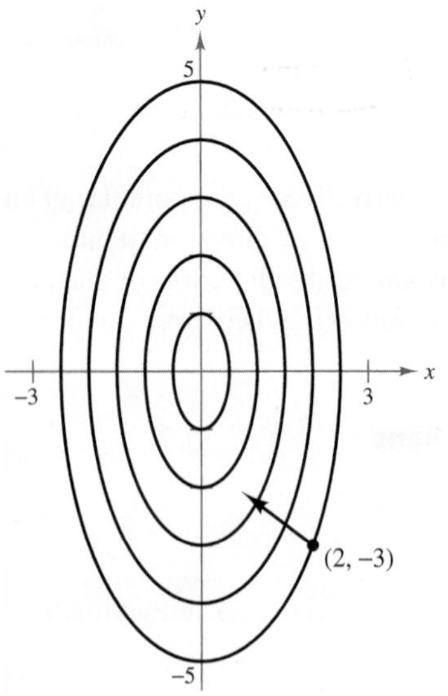

The direction of most rapid increase in temperature at $(2, -3)$ is given by $-16\mathbf{i} + 6\mathbf{j}$.
Figure 13.51

EXAMPLE 5 Finding the Direction of Maximum Increase

The temperature in degrees Celsius on the surface of a metal plate is

$$T(x, y) = 20 - 4x^2 - y^2$$

where x and y are measured in centimeters. In what direction from $(2, -3)$ does the temperature increase most rapidly? What is this rate of increase?

Solution The gradient is

$$\begin{aligned} \nabla T(x, y) &= T_x(x, y)\mathbf{i} + T_y(x, y)\mathbf{j} \\ &= -8x\mathbf{i} - 2y\mathbf{j}. \end{aligned}$$

It follows that the direction of maximum increase is given by

$$\nabla T(2, -3) = -16\mathbf{i} + 6\mathbf{j}$$

as shown in Figure 13.51, and the rate of increase is

$$\begin{aligned} \|\nabla T(2, -3)\| &= \sqrt{256 + 36} \\ &= \sqrt{292} \\ &\approx 17.09^\circ \text{ per centimeter.} \end{aligned}$$

The solution presented in Example 5 can be misleading. Although the gradient points in the direction of maximum temperature increase, it does not necessarily point toward the hottest spot on the plate. In other words, the gradient provides a local solution to finding an increase relative to the temperature at the point $(2, -3)$. *Once you leave that position, the direction of maximum increase may change.*

EXAMPLE 6 Finding the Path of a Heat-Seeking Particle

A heat-seeking particle is located at the point $(2, -3)$ on a metal plate whose temperature at (x, y) is

$$T(x, y) = 20 - 4x^2 - y^2.$$

Find the path of the particle as it continuously moves in the direction of maximum temperature increase.

Solution Let the path be represented by the position function

$$\mathbf{r}(t) = x(t)\mathbf{i} + y(t)\mathbf{j}.$$

A tangent vector at each point $(x(t), y(t))$ is given by

$$\mathbf{r}'(t) = \frac{dx}{dt}\mathbf{i} + \frac{dy}{dt}\mathbf{j}.$$

Because the particle seeks maximum temperature increase, the directions of $\mathbf{r}'(t)$ and $\nabla T(x, y) = -8x\mathbf{i} - 2y\mathbf{j}$ are the same at each point on the path. So,

$$-8x = k\frac{dx}{dt} \quad \text{and} \quad -2y = k\frac{dy}{dt}$$

where k depends on t. By solving each equation for dt/k and equating the results, you obtain

$$\frac{dx}{-8x} = \frac{dy}{-2y}.$$

The solution of this differential equation is $x = Cy^4$. Because the particle starts at the point $(2, -3)$, you can determine that $C = 2/81$. So, the path of the heat-seeking particle is

$$x = \frac{2}{81}y^4.$$

The path is shown in Figure 13.52.

Level curves:
$T(x, y) = 20 - 4x^2 - y^2$

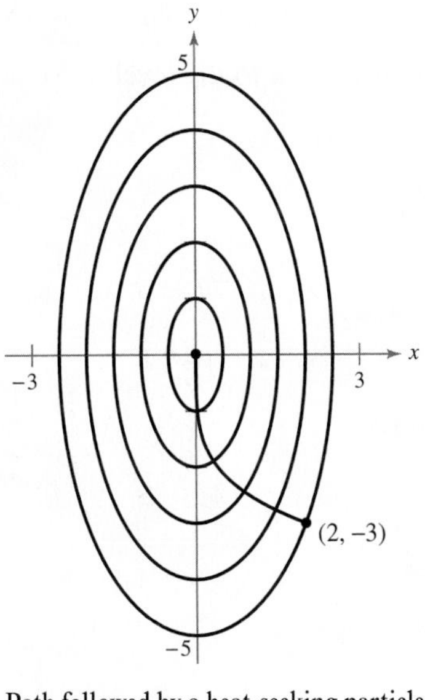

Path followed by a heat-seeking particle
Figure 13.52

In Figure 13.52, the path of the particle (determined by the gradient at each point) appears to be orthogonal to each of the level curves. This becomes clear when you consider that the temperature $T(x, y)$ is constant along a given level curve. So, at any point (x, y) on the curve, the rate of change of T in the direction of a unit tangent vector $\mathbf{u}$ is 0, and you can write

$$\nabla f(x, y) \cdot \mathbf{u} = D_{\mathbf{u}}T(x, y) = 0. \qquad \mathbf{u} \text{ is a unit tangent vector.}$$

Because the dot product of $\nabla f(x, y)$ and $\mathbf{u}$ is 0, you can conclude that they must be orthogonal. This result is stated in the following theorem.

THEOREM 13.12 Gradient Is Normal to Level Curves

If f is differentiable at (x_0, y_0) and $\nabla f(x_0, y_0) \neq \mathbf{0}$, then $\nabla f(x_0, y_0)$ is normal to the level curve through (x_0, y_0).

EXAMPLE 7 Finding a Normal Vector to a Level Curve

Sketch the level curve corresponding to $c = 0$ for the function given by

$$f(x, y) = y - \sin x$$

and find a normal vector at several points on the curve.

Solution The level curve for $c = 0$ is given by

$$0 = y - \sin x$$
$$y = \sin x$$

as shown in Figure 13.53(a). Because the gradient vector of f at (x, y) is

$$\nabla f(x, y) = f_x(x, y)\mathbf{i} + f_y(x, y)\mathbf{j}$$
$$= -\cos x\mathbf{i} + \mathbf{j}$$

you can use Theorem 13.12 to conclude that $\nabla f(x, y)$ is normal to the level curve at the point (x, y). Some gradient vectors are

$$\nabla f(-\pi, 0) = \mathbf{i} + \mathbf{j}$$
$$\nabla f\left(-\frac{2\pi}{3}, -\frac{\sqrt{3}}{2}\right) = \frac{1}{2}\mathbf{i} + \mathbf{j}$$
$$\nabla f\left(-\frac{\pi}{2}, -1\right) = \mathbf{j}$$
$$\nabla f\left(-\frac{\pi}{3}, -\frac{\sqrt{3}}{2}\right) = -\frac{1}{2}\mathbf{i} + \mathbf{j}$$
$$\nabla f(0, 0) = -\mathbf{i} + \mathbf{j}$$
$$\nabla f\left(\frac{\pi}{3}, \frac{\sqrt{3}}{2}\right) = -\frac{1}{2}\mathbf{i} + \mathbf{j}$$
$$\nabla f\left(\frac{\pi}{2}, 1\right) = \mathbf{j}.$$

These are shown in Figure 13.53(b).

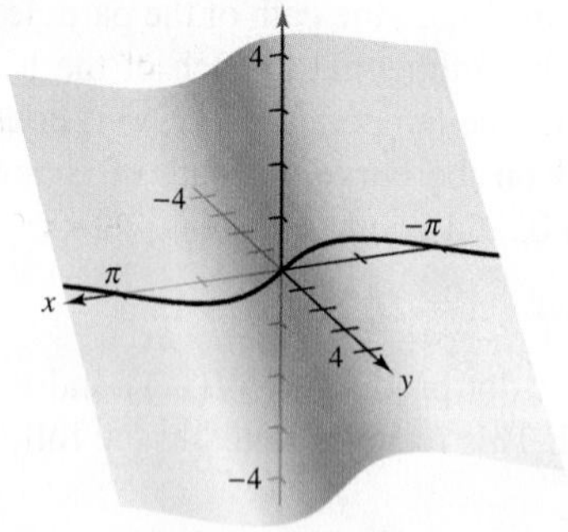

(a) The surface is given by $f(x, y) = y - \sin x$.

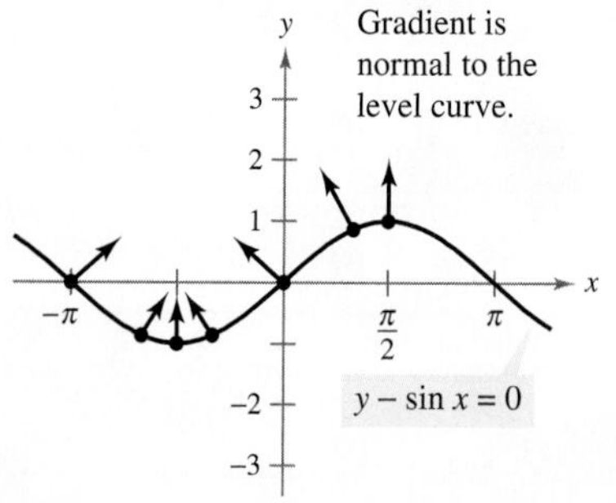

(b) The level curve is given by $f(x, y) = 0$.

Figure 13.53

Functions of Three Variables

The definitions of the directional derivative and the gradient can be extended naturally to functions of three or more variables. As often happens, some of the geometric interpretation is lost in the generalization from functions of two variables to those of three variables. For example, you cannot interpret the directional derivative of a function of three variables to represent slope.

The definitions and properties of the directional derivative and the gradient of a function of three variables are given in the following summary.

Directional Derivative and Gradient for Three Variables

Let f be a function of x, y, and z, with continuous first partial derivatives. The **directional derivative of f** in the direction of a unit vector $\mathbf{u} = a\mathbf{i} + b\mathbf{j} + c\mathbf{k}$ is given by

$$D_{\mathbf{u}} f(x, y, z) = af_x(x, y, z) + bf_y(x, y, z) + cf_z(x, y, z).$$

The **gradient of f** is defined to be

$$\nabla f(x, y, z) = f_x(x, y, z)\mathbf{i} + f_y(x, y, z)\mathbf{j} + f_z(x, y, z)\mathbf{k}.$$

Properties of the gradient are as follows.

1. $D_{\mathbf{u}} f(x, y, z) = \nabla f(x, y, z) \cdot \mathbf{u}$
2. If $\nabla f(x, y, z) = \mathbf{0}$, then $D_{\mathbf{u}} f(x, y, z) = 0$ for all $\mathbf{u}$.
3. The direction of *maximum* increase of f is given by $\nabla f(x, y, z)$. The maximum value of $D_{\mathbf{u}} f(x, y, z)$ is

   $\|\nabla f(x, y, z)\|$. Maximum value of $D_{\mathbf{u}} f(x, y, z)$

4. The direction of *minimum* increase of f is given by $-\nabla f(x, y, z)$. The minimum value of $D_{\mathbf{u}} f(x, y, z)$ is

   $-\|\nabla f(x, y, z)\|$. Minimum value of $D_{\mathbf{u}} f(x, y, z)$

NOTE You can generalize Theorem 13.12 to functions of three variables. Under suitable hypotheses,

$$\nabla f(x_0, y_0, z_0)$$

is normal to the level surface through (x_0, y_0, z_0).

EXAMPLE 8 Finding the Gradient for a Function of Three Variables

Find $\nabla f(x, y, z)$ for the function given by

$$f(x, y, z) = x^2 + y^2 - 4z$$

and find the direction of maximum increase of f at the point $(2, -1, 1)$.

Solution The gradient vector is given by

$$\begin{aligned}\nabla f(x, y, z) &= f_x(x, y, z)\mathbf{i} + f_y(x, y, z)\mathbf{j} + f_z(x, y, z)\mathbf{k}\\ &= 2x\mathbf{i} + 2y\mathbf{j} - 4\mathbf{k}.\end{aligned}$$

So, it follows that the direction of maximum increase at $(2, -1, 1)$ is

$$\nabla f(2, -1, 1) = 4\mathbf{i} - 2\mathbf{j} - 4\mathbf{k}.$$

Exercises for Section 13.6

See www.CalcChat.com for worked-out solutions to odd-numbered exercises.

In Exercises 1–12, find the directional derivative of the function at P in the direction of $\mathbf{v}$.

1. $f(x, y) = 3x - 4xy + 5y,\quad P(1, 2),\ \mathbf{v} = \frac{1}{2}(\mathbf{i} + \sqrt{3}\mathbf{j})$
2. $f(x, y) = x^3 - y^3,\quad P(4, 3),\ \mathbf{v} = \frac{\sqrt{2}}{2}(\mathbf{i} + \mathbf{j})$
3. $f(x, y) = xy,\quad P(2, 3),\ \mathbf{v} = \mathbf{i} + \mathbf{j}$
4. $f(x, y) = \frac{x}{y},\quad P(1, 1),\ \mathbf{v} = -\mathbf{j}$
5. $g(x, y) = \sqrt{x^2 + y^2},\quad P(3, 4),\ \mathbf{v} = 3\mathbf{i} - 4\mathbf{j}$
6. $g(x, y) = \arccos xy,\quad P(1, 0),\ \mathbf{v} = \mathbf{i} + 5\mathbf{j}$
7. $h(x, y) = e^x \sin y,\quad P\left(1, \frac{\pi}{2}\right),\ \mathbf{v} = -\mathbf{i}$
8. $h(x, y) = e^{-(x^2+y^2)},\quad P(0, 0),\ \mathbf{v} = \mathbf{i} + \mathbf{j}$
9. $f(x, y, z) = xy + yz + xz,\quad P(1, 1, 1),\ \mathbf{v} = 2\mathbf{i} + \mathbf{j} - \mathbf{k}$
10. $f(x, y, z), = x^2 + y^2 + z^2,\quad P(1, 2, -1),\ \mathbf{v} = \mathbf{i} - 2\mathbf{j} + 3\mathbf{k}$
11. $h(x, y, z) = x \arctan yz,\quad P(4, 1, 1),\ \mathbf{v} = \langle 1, 2, -1\rangle$
12. $h(x, y, z) = xyz,\quad P(2, 1, 1),\ \mathbf{v} = \langle 2, 1, 2\rangle$

In Exercises 13–16, find the directional derivative of the function in the direction of $\mathbf{u} = \cos\theta\mathbf{i} + \sin\theta\mathbf{j}$.

13. $f(x, y) = x^2 + y^2,\quad \theta = \frac{\pi}{4}$
14. $f(x, y) = \frac{y}{x + y},\quad \theta = -\frac{\pi}{6}$
15. $f(x, y) = \sin(2x - y),\quad \theta = -\frac{\pi}{3}$
16. $g(x, y) = xe^y,\quad \theta = \frac{2\pi}{3}$

In Exercises 17–20, find the directional derivative of the function at P in the direction of Q.

17. $f(x, y) = x^2 + 4y^2,\quad P(3, 1),\ Q(1, -1)$
18. $f(x, y) = \cos(x + y),\quad P(0, \pi),\ Q\left(\frac{\pi}{2}, 0\right)$
19. $h(x, y, z) = \ln(x + y + z),\quad P(1, 0, 0),\ Q(4, 3, 1)$
20. $g(x, y, z) = xye^z,\quad P(2, 4, 0),\ Q(0, 0, 0)$

In Exercises 21–26, find the gradient of the function at the given point.

21. $f(x, y) = 3x - 5y^2 + 10,\quad (2, 1)$
22. $g(x, y) = 2xe^{y/x},\quad (2, 0)$
23. $z = \cos(x^2 + y^2),\quad (3, -4)$
24. $z = \ln(x^2 - y),\quad (2, 3)$
25. $w = 3x^2y - 5yz + z^2,\quad (1, 1, -2)$
26. $w = x\tan(y + z),\quad (4, 3, -1)$

In Exercises 27–30, use the gradient to find the directional derivative of the function at P in the direction of Q.

27. $g(x, y) = x^2 + y^2 + 1,\quad P(1, 2),\ Q(3, 6)$
28. $f(x, y) = 3x^2 - y^2 + 4,\quad P(3, 1),\ Q(1, 8)$
29. $f(x, y) = e^{-x}\cos y,\quad P(0, 0),\ Q(2, 1)$
30. $f(x, y) = \sin 2x \cos y,\quad P(0, 0),\ Q\left(\frac{\pi}{2}, \pi\right)$

In Exercises 31–38, find the gradient of the function and the maximum value of the directional derivative at the given point.

	Function	*Point*
31.	$h(x, y) = x\tan y$	$\left(2, \frac{\pi}{4}\right)$
32.	$h(x, y) = y\cos(x - y)$	$\left(0, \frac{\pi}{3}\right)$
33.	$g(x, y) = \ln\sqrt[3]{x^2 + y^2}$	$(1, 2)$
34.	$g(x, y) = ye^{-x^2}$	$(0, 5)$
35.	$f(x, y, z) = \sqrt{x^2 + y^2 + z^2}$	$(1, 4, 2)$
36.	$w = \frac{1}{\sqrt{1 - x^2 - y^2 - z^2}}$	$(0, 0, 0)$
37.	$f(x, y, z) = xe^{yz}$	$(2, 0, -4)$
38.	$w = xy^2z^2$	$(2, 1, 1)$

In Exercises 39–46, use the function

$$f(x, y) = 3 - \frac{x}{3} - \frac{y}{2}.$$

39. Sketch the graph of f in the first octant and plot the point $(3, 2, 1)$ on the surface.
40. Find $D_{\mathbf{u}}f(3, 2)$, where $\mathbf{u} = \cos\theta\mathbf{i} + \sin\theta\mathbf{j}$.
    (a) $\theta = \frac{\pi}{4}$ (b) $\theta = \frac{2\pi}{3}$
41. Find $D_{\mathbf{u}}f(3, 2)$, where $\mathbf{u} = \cos\theta\mathbf{i} + \sin\theta\mathbf{j}$.
    (a) $\theta = \frac{4\pi}{3}$ (b) $\theta = -\frac{\pi}{6}$
42. Find $D_{\mathbf{u}}f(3, 2)$, where $\mathbf{u} = \frac{\mathbf{v}}{\|\mathbf{v}\|}$.
    (a) $\mathbf{v} = \mathbf{i} + \mathbf{j}$ (b) $\mathbf{v} = -3\mathbf{i} - 4\mathbf{j}$
43. Find $D_{\mathbf{u}}f(3, 2)$, where $\mathbf{u} = \frac{\mathbf{v}}{\|\mathbf{v}\|}$.
    (a) $\mathbf{v}$ is the vector from $(1, 2)$ to $(-2, 6)$.
    (b) $\mathbf{v}$ is the vector from $(3, 2)$ to $(4, 5)$.
44. Find $\nabla f(x, y)$.
45. Find the maximum value of the directional derivative at $(3, 2)$.
46. Find a unit vector $\mathbf{u}$ orthogonal to $\nabla f(3, 2)$ and calculate $D_{\mathbf{u}}f(3, 2)$. Discuss the geometric meaning of the result.

In Exercises 47–50, use the function

$$f(x, y) = 9 - x^2 - y^2.$$

47. Sketch the graph of f in the first octant and plot the point $(1, 2, 4)$ on the surface.

48. Find $D_{\mathbf{u}}f(1, 2)$, where $\mathbf{u} = \cos\theta\,\mathbf{i} + \sin\theta\,\mathbf{j}$.

(a) $\theta = -\dfrac{\pi}{4}$ (b) $\theta = \dfrac{\pi}{3}$

49. Find $\nabla f(1, 2)$ and $\|\nabla f(1, 2)\|$.

50. Find a unit vector $\mathbf{u}$ orthogonal to $\nabla f(1, 2)$ and calculate $D_{\mathbf{u}}f(1, 2)$. Discuss the geometric meaning of the result.

Investigation **In Exercises 51 and 52, (a) use the graph to estimate the components of the vector in the direction of the maximum rate of increase in the function at the given point. (b) Find the gradient at the point and compare it with your estimate in part (a). (c) In what direction would the function be decreasing at the greatest rate? Explain.**

51. $f(x, y) = \frac{1}{10}(x^2 - 3xy + y^2)$, $(1, 2)$

52. $f(x, y) = \frac{1}{2}y\sqrt{x}$, $(1, 2)$

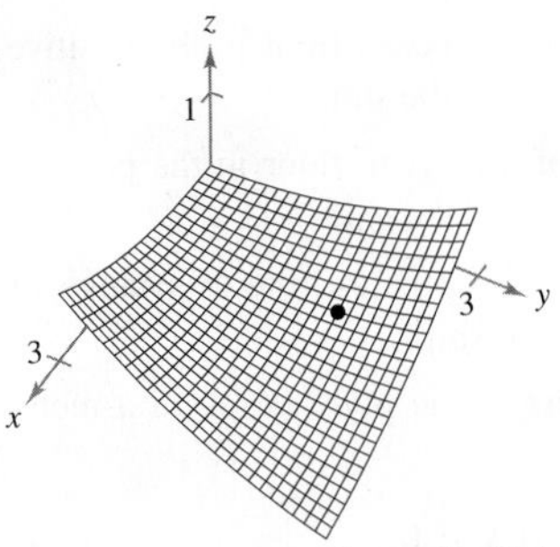

Generated by Maple

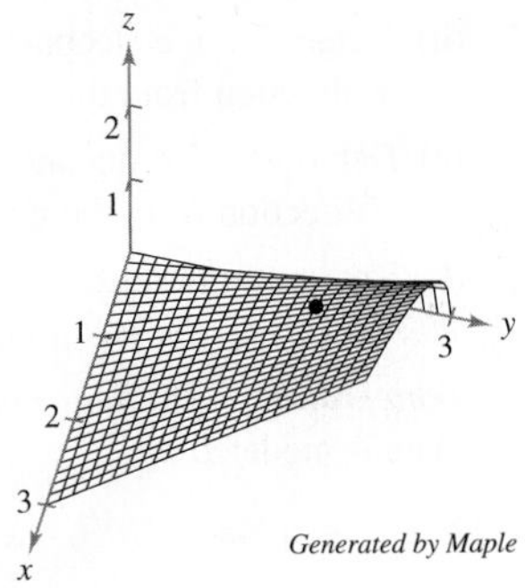

Generated by Maple

53. ***Investigation*** Consider the function

$$f(x, y) = x^2 - y^2$$

at the point $(4, -3, 7)$.

(a) Use a computer algebra system to graph the surface represented by the function.

(b) Determine the directional derivative $D_{\mathbf{u}}f(4, -3)$ as a function of θ, where $\mathbf{u} = \cos\theta\,\mathbf{i} + \sin\theta\,\mathbf{j}$. Use a computer algebra system to graph the function on the interval $[0, 2\pi)$.

(c) Approximate the zeros of the function in part (b) and interpret each in the context of the problem.

(d) Approximate the critical numbers of the function in part (b) and interpret each in the context of the problem.

(e) Find $\|\nabla f(4, -3)\|$ and explain its relationship to your answers in part (d).

(f) Use a computer algebra system to graph the level curve of the function f at the level $c = 7$. On this curve, graph the vector in the direction of $\nabla f(4, -3)$, and state its relationship to the level curve.

54. ***Investigation*** The figure below shows the level curve of the function

$$f(x, y) = \frac{8y}{1 + x^2 + y^2}$$

at the level $c = 2$.

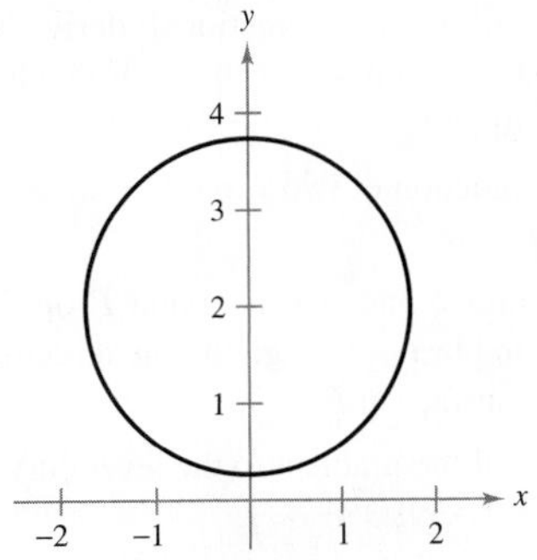

(a) Analytically verify that the curve is a circle.

(b) At the point $(\sqrt{3}, 2)$ on the level curve, sketch the vector showing the direction of the greatest rate of increase of the function. (To print an enlarged copy of the graph, go to the website *www.mathgraphs.com.*)

(c) At the point $(\sqrt{3}, 2)$ on the level curve, sketch the vector such that the directional derivative is 0.

(d) Use a computer algebra system to graph the surface to verify your answers in parts (a)–(c).

In Exercises 55–58, find a normal vector to the level curve $f(x, y) = c$ at P.

55. $f(x, y) = x^2 + y^2$, $c = 25$, $P(3, 4)$

56. $f(x, y) = 6 - 2x - 3y$, $c = 6$, $P(0, 0)$

57. $f(x, y) = \dfrac{x}{x^2 + y^2}$, $c = \frac{1}{2}$, $P(1, 1)$

58. $f(x, y) = xy$, $c = -3$, $P(-1, 3)$

In Exercises 59–62, use the gradient to find a unit normal vector to the graph of the equation at the given point. Sketch your results.

59. $4x^2 - y = 6$, $(2, 10)$

60. $3x^2 - 2y^2 = 1$, $(1, 1)$

61. $9x^2 + 4y^2 = 40$, $(2, -1)$

62. $xe^y - y = 5$, $(5, 0)$

63. ***Temperature Distribution*** The temperature at the point (x, y) on a metal plate is

$$T = \frac{x}{x^2 + y^2}.$$

Find the direction of greatest increase in heat from the point $(3, 4)$.

64. ***Topography*** The surface of a mountain is modeled by the equation $h(x, y) = 5000 - 0.001x^2 - 0.004y^2$. A mountain climber is at the point $(500, 300, 4390)$. In what direction should the climber move in order to ascend at the greatest rate?

Writing About Concepts

65. Define the derivative of the function $z = f(x, y)$ in the direction $\mathbf{u} = \cos\theta\mathbf{i} + \sin\theta\mathbf{j}$.

66. In your own words, give a geometric description of the directional derivative of $z = f(x, y)$.

67. Write a paragraph describing the directional derivative of the function f in the direction $\mathbf{u} = \cos\theta\mathbf{i} + \sin\theta\mathbf{j}$ when (a) $\theta = 0°$ and (b) $\theta = 90°$.

68. Define the gradient of a function of two variables. State the properties of the gradient.

69. Sketch the graph of a surface and select a point P on the surface. Sketch a vector in the xy-plane giving the direction of steepest ascent on the surface at P.

70. Describe the relationship of the gradient to the level curves of a surface given by $z = f(x, y)$.

71. ***Topography*** The figure shows a topographic map carried by a group of hikers. Sketch the paths of steepest descent if the hikers start at point A and if they start at point B. (To print an enlarged copy of the graph, go to the website *www.mathgraphs.com*.)

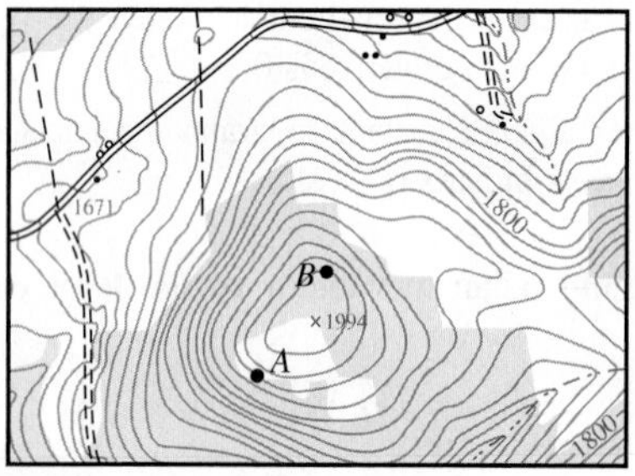

72. ***Meteorology*** Meteorologists measure the atmospheric pressure in units called millibars. From these observations they create weather maps on which the curves of equal atmospheric pressure (isobars) are drawn (see figure). These are level curves to the function $P(x, y)$ yielding the pressure at any point. Sketch the gradients to the isobars at the points A, B, and C. Although the magnitudes of the gradients are unknown, their lengths relative to each other can be estimated. At which of the three points is the wind speed greatest if the speed increases as the pressure gradient increases? (To print an enlarged copy of the graph, go to the website *www.mathgraphs.com*.)

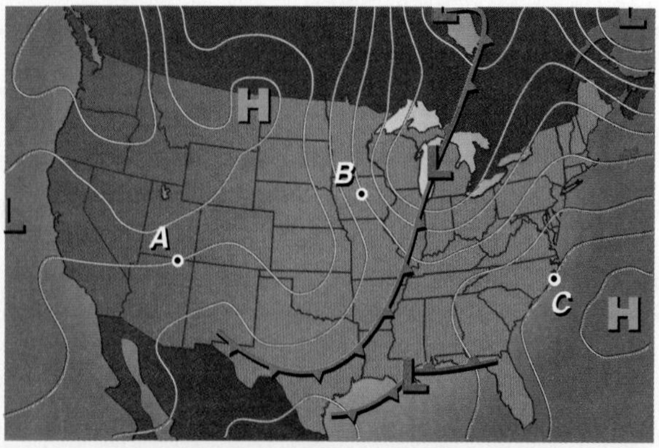

Heat-Seeking Path **In Exercises 73 and 74, find the path of a heat-seeking particle placed at point P on a metal plate with a temperature field $T(x, y)$.**

	Temperature Field	*Point*
73.	$T(x, y) = 400 - 2x^2 - y^2$	$P(10, 10)$
74.	$T(x, y) = 100 - x^2 - 2y^2$	$P(4, 3)$

75. ***Investigation*** A team of oceanographers is mapping the ocean floor to assist in the recovery of a sunken ship. Using sonar, they develop the model

$$D = 250 + 30x^2 + 50\sin\frac{\pi y}{2}, \quad 0 \le x \le 2, 0 \le y \le 2$$

where D is the depth in meters, and x and y are the distances in kilometers.

(a) Use a computer algebra system to graph the surface.

(b) Because the graph in part (a) is showing depth, it is not a map of the ocean floor. How could the model be changed so that the graph of the ocean floor could be obtained?

(c) What is the depth of the ship if it is located at the coordinates $x = 1$ and $y = 0.5$?

(d) Determine the steepness of the ocean floor in the positive x-direction from the position of the ship.

(e) Determine the steepness of the ocean floor in the positive y-direction from the position of the ship.

(f) Determine the direction of the greatest rate of change of depth from the position of the ship.

76. ***Temperature*** The temperature at the point (x, y) on a metal plate is modeled by

$$T(x, y) = 400e^{-(x^2+y)/2}, \quad x \ge 0, y \ge 0.$$

(a) Use a computer algebra system to graph the temperature distribution function.

(b) Find the directions of no change in heat on the plate from the point $(3, 5)$.

(c) Find the direction of greatest increase in heat from the point $(3, 5)$.

True or False? **In Exercises 77–80, determine whether the statement is true or false. If it is false, explain why or give an example that shows it is false.**

77. If $f(x, y) = \sqrt{1 - x^2 - y^2}$, then $D_{\mathbf{u}}f(0, 0) = 0$ for any unit vector $\mathbf{u}$.

78. If $f(x, y) = x + y$, then $-1 \le D_{\mathbf{u}}f(x, y) \le 1$.

79. If $D_{\mathbf{u}}f(x, y)$ exists, then $D_{\mathbf{u}}f(x, y) = -D_{-\mathbf{u}}f(x, y)$.

80. If $D_{\mathbf{u}}f(x_0, y_0) = c$ for any unit vector $\mathbf{u}$, then $c = 0$.

81. Find a function f such that

$$\nabla f = e^x \cos y\,\mathbf{i} - e^x \sin y\,\mathbf{j} + z\,\mathbf{k}.$$

Section 13.7 Tangent Planes and Normal Lines

- **Find equations of tangent planes and normal lines to surfaces.**
- **Find the angle of inclination of a plane in space.**
- **Compare the gradients $\nabla f(x, y)$ and $\nabla F(x, y, z)$.**

EXPLORATION

Billiard Balls and Normal Lines
In each of the three figures below, the cue ball is about to strike a stationary ball at point P. Explain how you can use the normal line to the stationary ball at point P to describe the resulting motion of each of the two balls. Assuming that each cue ball has the same speed, which stationary ball will acquire the greatest speed? Which will acquire the least? Explain your reasoning.

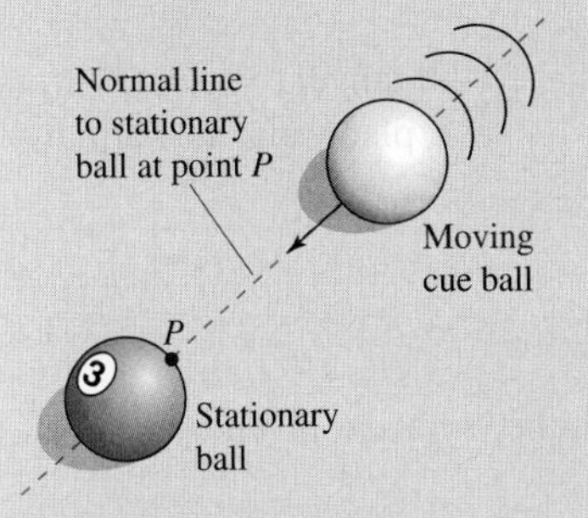

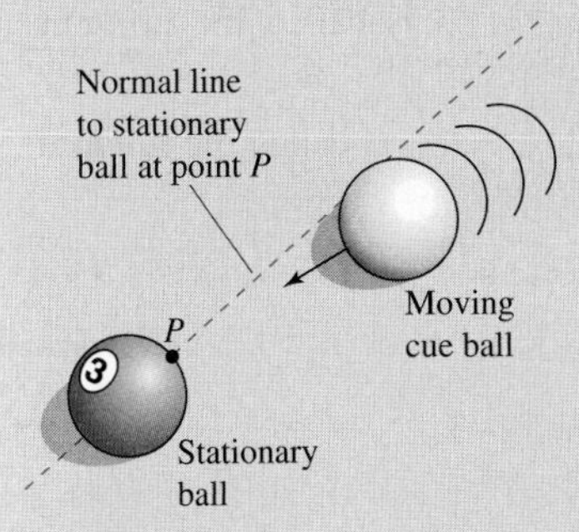

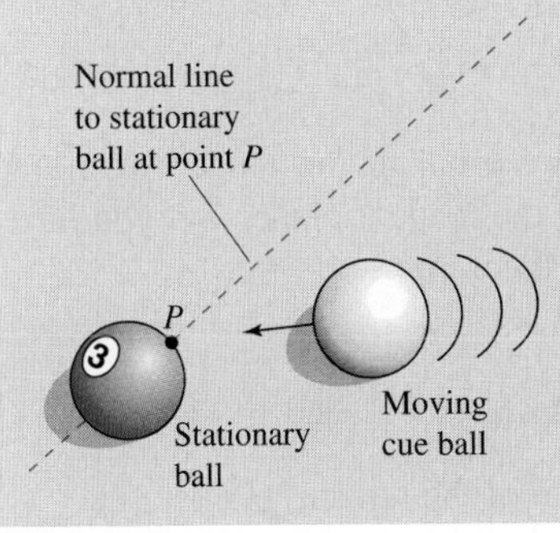

Tangent Plane and Normal Line to a Surface

So far you have represented surfaces in space primarily by equations of the form

$$z = f(x, y).$$ Equation of a surface S

In the development to follow, however, it is convenient to use the more general representation $F(x, y, z) = 0$. For a surface S given by $z = f(x, y)$, you can convert to the general form by defining F as

$$F(x, y, z) = f(x, y) - z.$$

Because $f(x, y) - z = 0$, you can consider S to be the level surface of F given by

$$F(x, y, z) = 0.$$ Alternative equation of surface S

EXAMPLE 1 Writing an Equation of a Surface

For the function given by

$$F(x, y, z) = x^2 + y^2 + z^2 - 4$$

describe the level surface given by $F(x, y, z) = 0$.

Solution The level surface given by $F(x, y, z) = 0$ can be written as

$$x^2 + y^2 + z^2 = 4$$

which is a sphere of radius 2 whose center is at the origin.

You have seen many examples of the usefulness of normal lines in applications involving curves. Normal lines are equally important in analyzing surfaces and solids. For example, consider the collision of two billiard balls. When a stationary ball is struck at a point P on its surface, it moves along the **line of impact** determined by P and the center of the ball. The impact can occur in *two* ways. If the cue ball is moving along the line of impact, it stops dead and imparts all of its momentum to the stationary ball, as shown in Figure 13.54. If the cue ball is not moving along the line of impact, it is deflected to one side or the other and retains part of its momentum. That part of the momentum that is transferred to the stationary ball occurs along the line of impact, *regardless* of the direction of the cue ball, as shown in Figure 13.55. This line of impact is called the **normal line** to the surface of the ball at the point P.

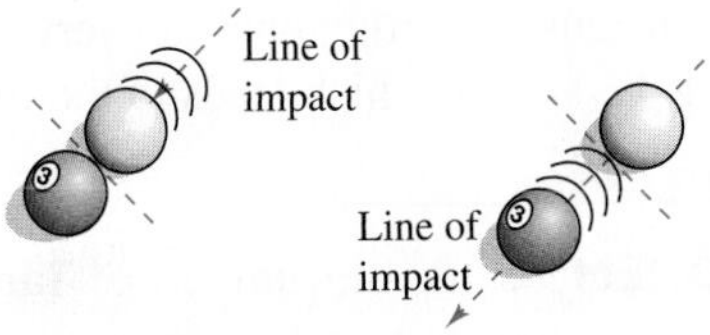

Figure 13.54

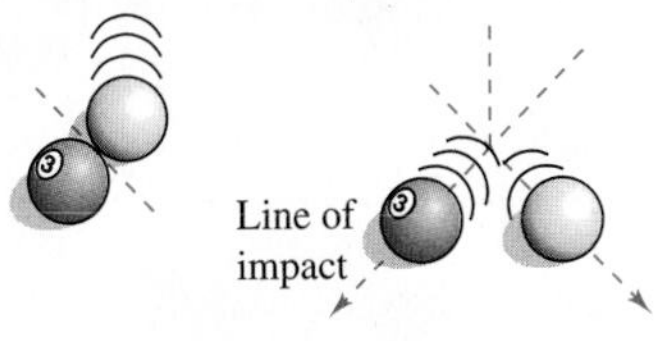

Figure 13.55

In the process of finding a normal line to a surface, you are also able to solve the problem of finding a **tangent plane** to the surface. Let S be a surface given by

$$F(x, y, z) = 0$$

and let $P(x_0, y_0, z_0)$ be a point on S. Let C be a curve on S through P that is defined by the vector-valued function

$$\mathbf{r}(t) = x(t)\mathbf{i} + y(t)\mathbf{j} + z(t)\mathbf{k}.$$

Then, for all t,

$$F(x(t),\ y(t),\ z(t)) = 0.$$

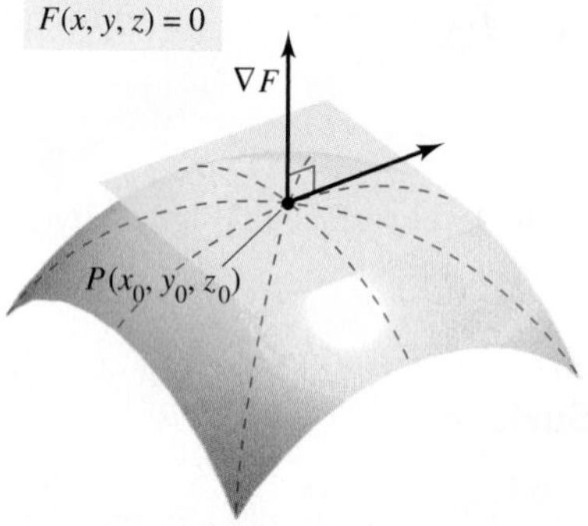

Tangent plane to surface S at P
Figure 13.56

If F is differentiable and $x'(t)$, $y'(t)$, and $z'(t)$ all exist, it follows from the Chain Rule that

$$\begin{aligned} 0 &= F'(t) \\ &= F_x(x, y, z)x'(t) + F_y(x, y, z)y'(t) + F_z(x, y, z)z'(t). \end{aligned}$$

At (x_0, y_0, z_0), the equivalent vector form is

$$0 = \underbrace{\nabla F(x_0, y_0, z_0)}_{\text{Gradient}} \cdot \underbrace{\mathbf{r}'(t_0)}_{\text{Tangent vector}}.$$

This result means that the gradient at P is orthogonal to the tangent vector of every curve on S through P. So, all tangent lines on S lie in a plane that is normal to $\nabla F(x_0, y_0, z_0)$ and contains P, as shown in Figure 13.56.

Definition of Tangent Plane and Normal Line

Let F be differentiable at the point $P(x_0, y_0, z_0)$ on the surface S given by $F(x, y, z) = 0$ such that $\nabla F(x_0, y_0, z_0) \neq \mathbf{0}$.

1. The plane through P that is normal to $\nabla F(x_0, y_0, z_0)$ is called the **tangent plane to S at P.**
2. The line through P having the direction of $\nabla F(x_0, y_0, z_0)$ is called the **normal line to S at P.**

NOTE In the remainder of this section, assume $\nabla F(x_0, y_0, z_0)$ to be nonzero unless stated otherwise.

To find an equation for the tangent plane to S at (x_0, y_0, z_0), let (x, y, z) be an arbitrary point in the tangent plane. Then the vector

$$\mathbf{v} = (x - x_0)\mathbf{i} + (y - y_0)\mathbf{j} + (z - z_0)\mathbf{k}$$

lies in the tangent plane. Because $\nabla F(x_0, y_0, z_0)$ is normal to the tangent plane at (x_0, y_0, z_0), it must be orthogonal to every vector in the tangent plane, and you have $\nabla F(x_0, y_0, z_0) \cdot \mathbf{v} = 0$, which leads to the following theorem.

THEOREM 13.13 Equation of Tangent Plane

If F is differentiable at (x_0, y_0, z_0), then an equation of the tangent plane to the surface given by $F(x, y, z) = 0$ at (x_0, y_0, z_0) is

$$F_x(x_0, y_0, z_0)(x - x_0) + F_y(x_0, y_0, z_0)(y - y_0) + F_z(x_0, y_0, z_0)(z - z_0) = 0.$$

EXAMPLE 2 Finding an Equation of a Tangent Plane

Find an equation of the tangent plane to the hyperboloid given by

$$z^2 - 2x^2 - 2y^2 = 12$$

at the point $(1, -1, 4)$.

Solution Begin by writing the equation of the surface as

$$z^2 - 2x^2 - 2y^2 - 12 = 0.$$

Then, considering

$$F(x, y, z) = z^2 - 2x^2 - 2y^2 - 12$$

you have

$$F_x(x, y, z) = -4x, \quad F_y(x, y, z) = -4y, \quad \text{and} \quad F_z(x, y, z) = 2z.$$

At the point $(1, -1, 4)$ the partial derivatives are

$$F_x(1, -1, 4) = -4, \quad F_y(1, -1, 4) = 4, \quad \text{and} \quad F_z(1, -1, 4) = 8.$$

So, an equation of the tangent plane at $(1, -1, 4)$ is

$$\begin{aligned} -4(x - 1) + 4(y + 1) + 8(z - 4) &= 0 \\ -4x + 4 + 4y + 4 + 8z - 32 &= 0 \\ -4x + 4y + 8z - 24 &= 0 \\ x - y - 2z + 6 &= 0. \end{aligned}$$

Figure 13.57 shows a portion of the hyperboloid and tangent plane.

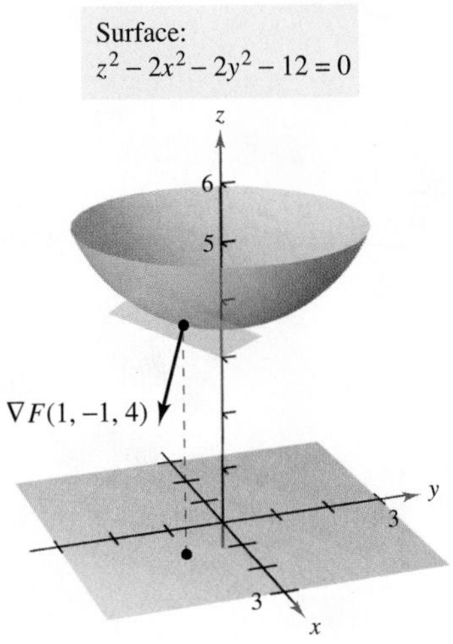

Tangent plane to surface
Figure 13.57

TECHNOLOGY Some three-dimensional graphing utilities are capable of graphing tangent planes to surfaces. Two examples are shown below.

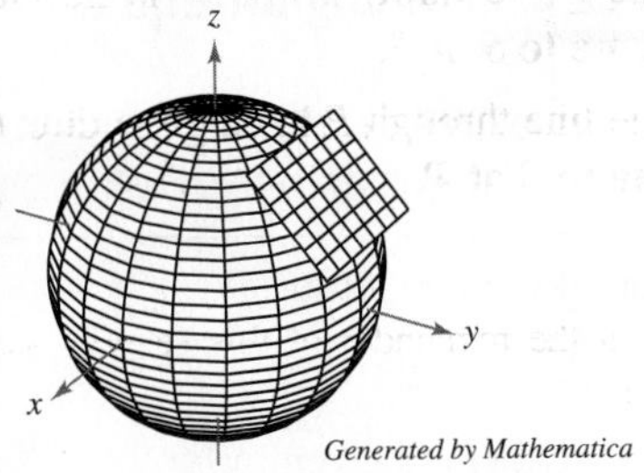

Sphere: $x^2 + y^2 + z^2 = 1$

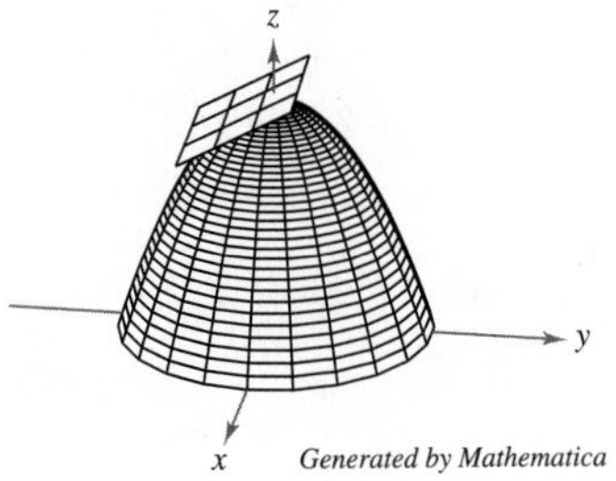

Paraboloid: $z = 2 - x^2 - y^2$

To find the equation of the tangent plane at a point on a surface given by $z = f(x, y)$, you can define the function F by

$$F(x, y, z) = f(x, y) - z.$$

Then S is given by the level surface $F(x, y, z) = 0$, and by Theorem 13.13 an equation of the tangent plane to S at the point (x_0, y_0, z_0) is

$$f_x(x_0, y_0)(x - x_0) + f_y(x_0, y_0)(y - y_0) - (z - z_0) = 0.$$

EXAMPLE 3 Finding an Equation of the Tangent Plane

Find the equation of the tangent plane to the paraboloid

$$z = 1 - \frac{1}{10}(x^2 + 4y^2)$$

at the point $\left(1, 1, \frac{1}{2}\right)$.

Solution From $z = f(x, y) = 1 - \frac{1}{10}(x^2 + 4y^2)$, you obtain

$$f_x(x, y) = -\frac{x}{5} \quad \Longrightarrow \quad f_x(1, 1) = -\frac{1}{5}$$

and

$$f_y(x, y) = -\frac{4y}{5} \quad \Longrightarrow \quad f_y(1, 1) = -\frac{4}{5}.$$

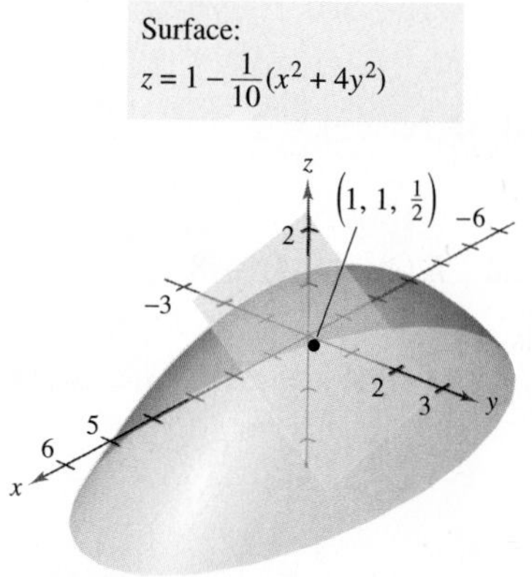

Figure 13.58

So, an equation of the tangent plane at $\left(1, 1, \frac{1}{2}\right)$ is

$$f_x(1, 1)(x - 1) + f_y(1, 1)(y - 1) - \left(z - \frac{1}{2}\right) = 0$$
$$-\frac{1}{5}(x - 1) - \frac{4}{5}(y - 1) - \left(z - \frac{1}{2}\right) = 0$$
$$-\frac{1}{5}x - \frac{4}{5}y - z + \frac{3}{2} = 0.$$

This tangent plane is shown in Figure 13.58.

The gradient $\nabla F(x, y, z)$ gives a convenient way to find equations of normal lines, as shown in Example 4.

EXAMPLE 4 Finding an Equation of a Normal Line to a Surface

Find a set of symmetric equations for the normal line to the surface given by $xyz = 12$ at the point $(2, -2, -3)$.

Solution Begin by letting

$$F(x, y, z) = xyz - 12.$$

Then, the gradient is given by

$$\nabla F(x, y, z) = F_x(x, y, z)\mathbf{i} + F_y(x, y, z)\mathbf{j} + F_z(x, y, z)\mathbf{k}$$
$$= yz\mathbf{i} + xz\mathbf{j} + xy\mathbf{k}$$

and at the point $(2, -2, -3)$ you have

$$\nabla F(2, -2, -3) = (-2)(-3)\mathbf{i} + (2)(-3)\mathbf{j} + (2)(-2)\mathbf{k}$$
$$= 6\mathbf{i} - 6\mathbf{j} - 4\mathbf{k}.$$

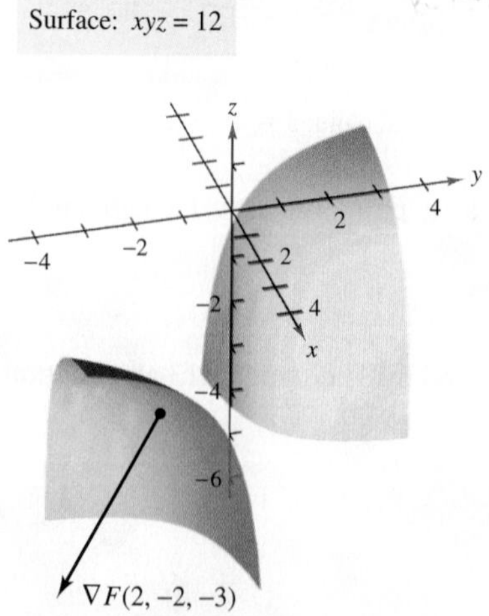

Figure 13.59

The normal line at $(2, -2, -3)$ has direction numbers 6, -6, and -4, and the corresponding set of symmetric equations is

$$\frac{x - 2}{6} = \frac{y + 2}{-6} = \frac{z + 3}{-4}.$$

See Figure 13.59.

Knowing that the gradient $\nabla F(x, y, z)$ is normal to the surface given by $F(x, y, z) = 0$ allows you to solve a variety of problems dealing with surfaces and curves in space.

EXAMPLE 5 Finding the Equation of a Tangent Line to a Curve

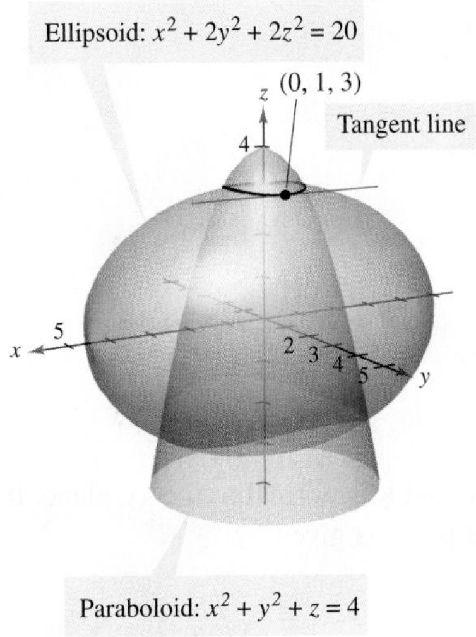

Figure 13.60

Describe the tangent line to the curve of intersection of the surfaces

$$x^2 + 2y^2 + 2z^2 = 20 \quad \text{Ellipsoid}$$
$$x^2 + y^2 + z = 4 \quad \text{Paraboloid}$$

at the point (0, 1, 3), as shown in Figure 13.60.

Solution Begin by finding the gradients to both surfaces at the point (0, 1, 3).

Ellipsoid	*Paraboloid*
$F(x, y, z) = x^2 + 2y^2 + 2z^2 - 20$	$G(x, y, z) = x^2 + y^2 + z - 4$
$\nabla F(x, y, z) = 2x\mathbf{i} + 4y\mathbf{j} + 4z\mathbf{k}$	$\nabla G(x, y, z) = 2x\mathbf{i} + 2y\mathbf{j} + \mathbf{k}$
$\nabla F(0, 1, 3) = 4\mathbf{j} + 12\mathbf{k}$	$\nabla G(0, 1, 3) = 2\mathbf{j} + \mathbf{k}$

The cross product of these two gradients is a vector that is tangent to both surfaces at the point (0, 1, 3).

$$\nabla F(0, 1, 3) \times \nabla G(0, 1, 3) = \begin{vmatrix} \mathbf{i} & \mathbf{j} & \mathbf{k} \\ 0 & 4 & 12 \\ 0 & 2 & 1 \end{vmatrix} = -20\mathbf{i}.$$

So, the tangent line to the curve of intersection of the two surfaces at the point (0, 1, 3) is a line that is parallel to the x-axis and passes through the point (0, 1, 3).

The Angle of Inclination of a Plane

Another use of the gradient $\nabla F(x, y, z)$ is to determine the angle of inclination of the tangent plane to a surface. The **angle of inclination** of a plane is defined to be the angle θ $(0 \le \theta \le \pi/2)$ between the given plane and the xy-plane, as shown in Figure 13.61. (The angle of inclination of a horizontal plane is defined to be zero.) Because the vector $\mathbf{k}$ is normal to the xy-plane, you can use the formula for the cosine of the angle between two planes (given in Section 11.5) to conclude that the angle of inclination of a plane with normal vector $\mathbf{n}$ is given by

$$\cos\theta = \frac{|\mathbf{n} \cdot \mathbf{k}|}{\|\mathbf{n}\|\,\|\mathbf{k}\|} = \frac{|\mathbf{n} \cdot \mathbf{k}|}{\|\mathbf{n}\|}. \quad \text{Angle of inclination of a plane}$$

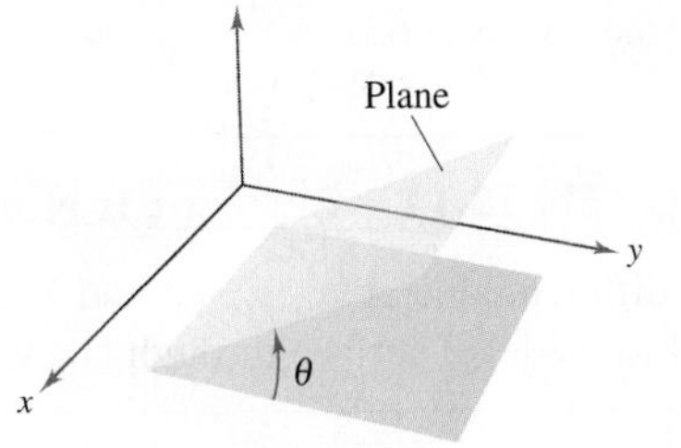

The angle of inclination
Figure 13.61

EXAMPLE 6 Finding the Angle of Inclination of a Tangent Plane

Find the angle of inclination of the tangent plane to the ellipsoid given by

$$\frac{x^2}{12} + \frac{y^2}{12} + \frac{z^2}{3} = 1$$

at the point (2, 2, 1).

Solution If you let

$$F(x, y, z) = \frac{x^2}{12} + \frac{y^2}{12} + \frac{z^2}{3} - 1$$

the gradient of F at the point (2, 2, 1) is given by

$$\nabla F(x, y, z) = \frac{x}{6}\mathbf{i} + \frac{y}{6}\mathbf{j} + \frac{2z}{3}\mathbf{k}$$

$$\nabla F(2, 2, 1) = \frac{1}{3}\mathbf{i} + \frac{1}{3}\mathbf{j} + \frac{2}{3}\mathbf{k}.$$

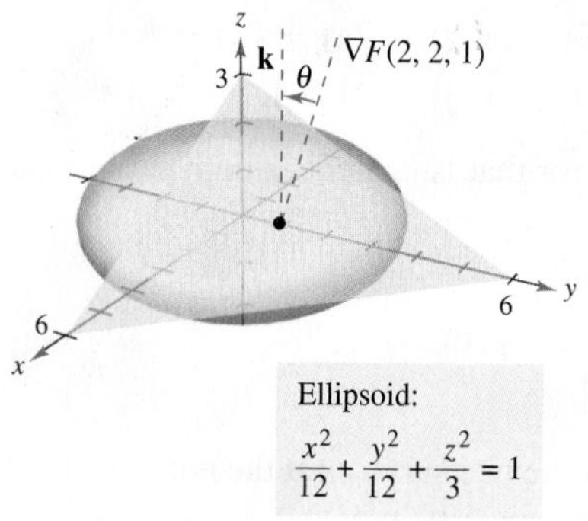

Figure 13.62

Because $\nabla F(2, 2, 1)$ is normal to the tangent plane and $\mathbf{k}$ is normal to the xy-plane, it follows that the angle of inclination of the tangent plane is given by

$$\cos\theta = \frac{|\nabla F(2, 2, 1) \cdot \mathbf{k}|}{\|\nabla F(2, 2, 1)\|} = \frac{2/3}{\sqrt{(1/3)^2 + (1/3)^2 + (2/3)^2}} = \sqrt{\frac{2}{3}}$$

which implies that

$$\theta = \arccos\sqrt{\frac{2}{3}} \approx 35.3°,$$

as shown in Figure 13.62.

NOTE A special case of the procedure shown in Example 6 is worth noting. The angle of inclination θ of the tangent plane to the surface $z = f(x, y)$ at (x_0, y_0, z_0) is given by

$$\cos\theta = \frac{1}{\sqrt{[f_x(x_0, y_0)]^2 + [f_y(x_0, y_0)]^2 + 1}}.$$

Alternative formula for angle of inclination (See Exercise 64.)

A Comparison of the Gradients $\nabla f(x, y)$ and $\nabla F(x, y, z)$

This section concludes with a comparison of the gradients $\nabla f(x, y)$ and $\nabla F(x, y, z)$. In the preceding section, you saw that the gradient of a function f of two variables is normal to the level curves of f. Specifically, Theorem 13.12 states that if f is differentiable at (x_0, y_0) and $\nabla f(x_0, y_0) \neq \mathbf{0}$, then $\nabla f(x_0, y_0)$ is normal to the level curve through (x_0, y_0). Having developed normal lines to surfaces, you can now extend this result to a function of three variables. The proof of Theorem 13.14 is left as an exercise (see Exercise 63).

THEOREM 13.14 Gradient Is Normal to Level Surfaces

If F is differentiable at (x_0, y_0, z_0) and $\nabla F(x_0, y_0, z_0) \neq \mathbf{0}$, then $\nabla F(x_0, y_0, z_0)$ is normal to the level surface through (x_0, y_0, z_0).

When working with the gradients $\nabla f(x, y)$ and $\nabla F(x, y, z)$, be sure you remember that $\nabla f(x, y)$ is a vector in the xy-plane and $\nabla F(x, y, z)$ is a vector in space.

Exercises for Section 13.7

See www.CalcChat.com for worked-out solutions to odd-numbered exercises.

In Exercises 1–4, describe the level surface $F(x, y, z) = 0$.

1. $F(x, y, z) = 3x - 5y + 3z - 15$
2. $F(x, y, z) = x^2 + y^2 + z^2 - 25$
3. $F(x, y, z) = 4x^2 + 9y^2 - 4z^2$
4. $F(x, y, z) = 16x^2 - 9y^2 + 144z$

In Exercises 5–14, find a unit normal vector to the surface at the given point. [*Hint:* Normalize the gradient vector $\nabla F(x, y, z)$.]

	Surface	*Point*
5.	$x + y + z = 4$	$(2, 0, 2)$
6.	$x^2 + y^2 + z^2 = 11$	$(3, 1, 1)$
7.	$z = \sqrt{x^2 + y^2}$	$(3, 4, 5)$
8.	$z = x^3$	$(2, 1, 8)$
9.	$x^2y^4 - z = 0$	$(1, 2, 16)$
10.	$x^2 + 3y + z^3 = 9$	$(2, -1, 2)$
11.	$\ln\left(\dfrac{x}{y - z}\right) = 0$	$(1, 4, 3)$
12.	$ze^{x^2-y^2} - 3 = 0$	$(2, 2, 3)$
13.	$z - x \sin y = 4$	$\left(6, \dfrac{\pi}{6}, 7\right)$
14.	$\sin(x - y) - z = 2$	$\left(\dfrac{\pi}{3}, \dfrac{\pi}{6}, -\dfrac{3}{2}\right)$

In Exercises 15–18, find an equation of the tangent plane to the surface at the given point.

15. $z = 25 - x^2 - y^2$

$(3, 1, 15)$

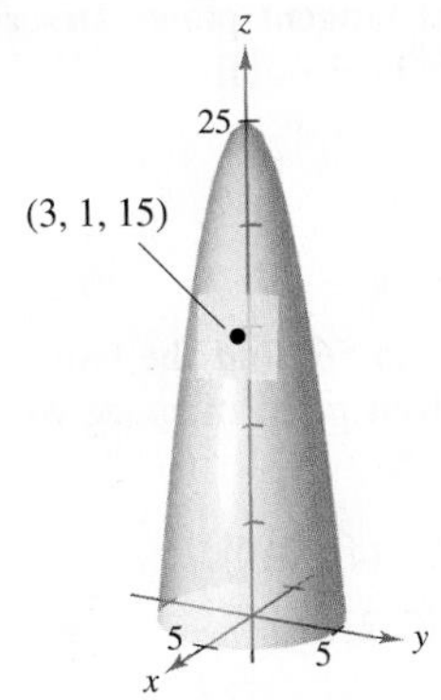

16. $f(x, y) = \dfrac{y}{x}$

$(1, 2, 2)$

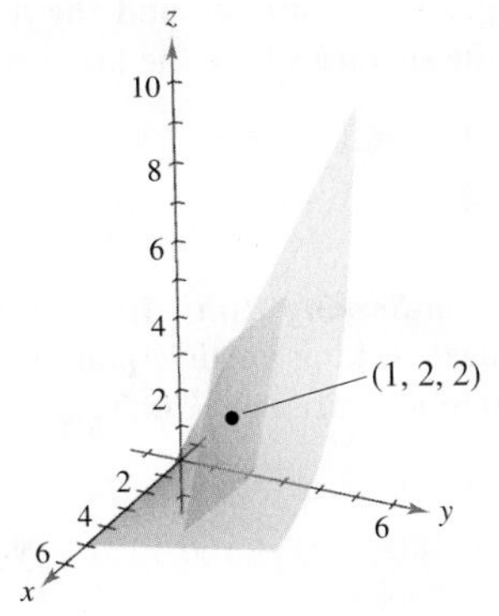

17. $z = \sqrt{x^2 + y^2}$

$(3, 4, 5)$

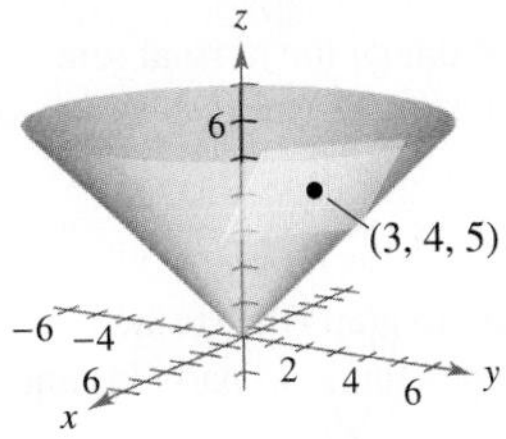

18. $g(x, y) = \arctan \dfrac{y}{x}$

$(1, 0, 0)$

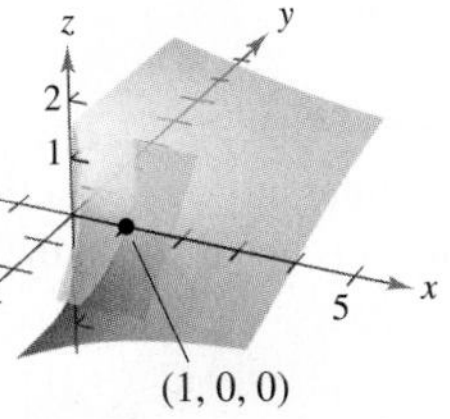

In Exercises 19–28, find an equation of the tangent plane to the surface at the given point.

19. $g(x, y) = x^2 - y^2, \quad (5, 4, 9)$
20. $f(x, y) = 2 - \frac{2}{3}x - y, \quad (3, -1, 1)$
21. $z = e^x(\sin y + 1), \quad \left(0, \dfrac{\pi}{2}, 2\right)$
22. $z = x^2 - 2xy + y^2, \quad (1, 2, 1)$
23. $h(x, y) = \ln \sqrt{x^2 + y^2}, \quad (3, 4, \ln 5)$
24. $h(x, y) = \cos y, \quad \left(5, \dfrac{\pi}{4}, \dfrac{\sqrt{2}}{2}\right)$
25. $x^2 + 4y^2 + z^2 = 36, \quad (2, -2, 4)$
26. $x^2 + 2z^2 = y^2, \quad (1, 3, -2)$
27. $xy^2 + 3x - z^2 = 4, \quad (2, 1, -2)$
28. $x = y(2z - 3), \quad (4, 4, 2)$

In Exercises 29–34, find an equation of the tangent plane and find symmetric equations of the normal line to the surface at the given point.

29. $x^2 + y^2 + z = 9, \quad (1, 2, 4)$
30. $x^2 + y^2 + z^2 = 9, \quad (1, 2, 2)$
31. $xy - z = 0, \quad (-2, -3, 6)$
32. $x^2 - y^2 + z^2 = 0, \quad (5, 13, -12)$
33. $z = \arctan \dfrac{y}{x}, \quad \left(1, 1, \dfrac{\pi}{4}\right)$
34. $xyz = 10, \quad (1, 2, 5)$

35. *Investigation* Consider the function

$$f(x, y) = \frac{4xy}{(x^2 + 1)(y^2 + 1)}$$

on the intervals $-2 \le x \le 2$ and $0 \le y \le 3$.

(a) Find a set of parametric equations of the normal line and an equation of the tangent plane to the surface at the point $(1, 1, 1)$.

(b) Repeat part (a) for the point $\left(-1, 2, -\frac{4}{5}\right)$.

(c) Use a computer algebra system to graph the surface, the normal lines, and the tangent planes found in parts (a) and (b).

(d) Use analytic and graphical analysis to write a brief description of the surface at the two indicated points.

36. *Investigation* Consider the function

$$f(x, y) = \frac{\sin y}{x}$$

on the intervals $-3 \le x \le 3$ and $0 \le y \le 2\pi$.

(a) Find a set of parametric equations of the normal line and an equation of the tangent plane to the surface at the point $\left(2, \frac{\pi}{2}, \frac{1}{2}\right)$.

(b) Repeat part (a) for the point $\left(-\frac{2}{3}, \frac{3\pi}{2}, \frac{3}{2}\right)$.

(c) Use a computer algebra system to graph the surface, the normal lines, and the tangent planes found in parts (a) and (b).

(d) Use analytic and graphical analysis to write a brief description of the surface at the two indicated points.

Writing About Concepts

37. Consider the function $F(x, y, z) = 0$, which is differentiable at $P(x_0, y_0, z_0)$. Give the definition of the tangent plane at P and the normal line at P.

38. Give the standard form of the equation of the tangent plane to a surface given by $F(x, y, z) = 0$ at (x_0, y_0, z_0).

39. For some surfaces, the normal lines at any point pass through the same geometric object. What is the common geometric object for a sphere? What is the common geometric object for a right circular cylinder? Explain.

40. Discuss the relationship between the tangent plane to a surface and approximation by differentials.

In Exercises 41–46, (a) find symmetric equations of the tangent line to the curve of intersection of the surfaces at the given point, and (b) find the cosine of the angle between the gradient vectors at this point. State whether or not the surfaces are orthogonal at the point of intersection.

41. $x^2 + y^2 = 5,\quad z = x,\quad (2, 1, 2)$

42. $z = x^2 + y^2,\quad z = 4 - y,\quad (2, -1, 5)$

43. $x^2 + z^2 = 25,\quad y^2 + z^2 = 25,\quad (3, 3, 4)$

44. $z = \sqrt{x^2 + y^2},\quad 5x - 2y + 3z = 22,\quad (3, 4, 5)$

45. $x^2 + y^2 + z^2 = 6,\quad x - y - z = 0,\quad (2, 1, 1)$

46. $z = x^2 + y^2,\quad x + y + 6z = 33,\quad (1, 2, 5)$

47. Consider the functions

$$f(x, y) = 6 - x^2 - y^2/4 \quad \text{and} \quad g(x, y) = 2x + y.$$

(a) Find a set of parametric equations of the tangent line to the curve of intersection of the surfaces at the point $(1, 2, 4)$, and find the angle between the gradient vectors.

(b) Use a computer algebra system to graph the surfaces. Graph the tangent line found in part (a).

48. Consider the functions

$$f(x, y) = \sqrt{16 - x^2 - y^2 + 2x - 4y}$$

and

$$g(x, y) = \frac{\sqrt{2}}{2}\sqrt{1 - 3x^2 + y^2 + 6x + 4y}.$$

(a) Use a computer algebra system to graph the first-octant portion of the surfaces represented by f and g.

(b) Find one first-octant point on the curve of intersection and show that the surfaces are orthogonal at this point.

(c) These surfaces are orthogonal along the curve of intersection. Does part (b) prove this fact? Explain.

In Exercises 49–52, find the angle of inclination θ of the tangent plane to the surface at the given point.

49. $3x^2 + 2y^2 - z = 15,\quad (2, 2, 5)$

50. $2xy - z^3 = 0,\quad (2, 2, 2)$

51. $x^2 - y^2 + z = 0,\quad (1, 2, 3)$

52. $x^2 + y^2 = 5,\quad (2, 1, 3)$

In Exercises 53 and 54, find the point on the surface where the tangent plane is horizontal. Use a computer algebra system to graph the surface and the horizontal tangent plane. Describe the surface where the tangent plane is horizontal.

53. $z = 3 - x^2 - y^2 + 6y$

54. $z = 3x^2 + 2y^2 - 3x + 4y - 5$

Heat-Seeking Path **In Exercises 55 and 56, find the path of a heat-seeking particle placed at the given point in space with a temperature field $T(x, y, z)$.**

55. $T(x, y, z) = 400 - 2x^2 - y^2 - 4z^2,\quad (4, 3, 10)$

56. $T(x, y, z) = 100 - 3x - y - z^2,\quad (2, 2, 5)$

In Exercises 57 and 58, show that the tangent plane to the quadric surface at the point (x_0, y_0, z_0) can be written in the given form.

57. Ellipsoid: $\dfrac{x^2}{a^2} + \dfrac{y^2}{b^2} + \dfrac{z^2}{c^2} = 1$

Plane: $\dfrac{x_0x}{a^2} + \dfrac{y_0y}{b^2} + \dfrac{z_0z}{c^2} = 1$

58. Hyperboloid: $\dfrac{x^2}{a^2} + \dfrac{y^2}{b^2} - \dfrac{z^2}{c^2} = 1$

Plane: $\dfrac{x_0x}{a^2} + \dfrac{y_0y}{b^2} - \dfrac{z_0z}{c^2} = 1$

59. Show that any tangent plane to the cone

$$z^2 = a^2x^2 + b^2y^2$$

passes through the origin.

60. Let f be a differentiable function and consider the surface $z = xf(y/x)$. Show that the tangent plane at any point $P(x_0, y_0, z_0)$ on the surface passes through the origin.

61. ***Approximation*** Consider the following approximations for a function $f(x, y)$ centered at $(0, 0)$.

Linear approximation:

$$P_1(x, y) = f(0, 0) + f_x(0, 0)x + f_y(0, 0)y$$

Quadratic approximation:

$$P_2(x, y) = f(0, 0) + f_x(0, 0)x + f_y(0, 0)y + \tfrac{1}{2}f_{xx}(0, 0)x^2 + f_{xy}(0, 0)xy + \tfrac{1}{2}f_{yy}(0, 0)y^2$$

[Note that the linear approximation is the tangent plane to the surface at $(0, 0, f(0, 0))$.]

(a) Find the linear approximation of $f(x, y) = e^{(x-y)}$ centered at $(0, 0)$.

(b) Find the quadratic approximation of $f(x, y) = e^{(x-y)}$ centered at $(0, 0)$.

(c) If $x = 0$ in the quadratic approximation, you obtain the second-degree Taylor polynomial for what function? Answer the same question for $y = 0$.

(d) Complete the table.

x	y	$f(x, y)$	$P_1(x, y)$	$P_2(x, y)$
0	0			
0	0.1			
0.2	0.1			
0.2	0.5			
1	0.5			

(e) Use a computer algebra system to graph the surfaces $z = f(x, y)$, $z = P_1(x, y)$, and $z = P_2(x, y)$.

62. ***Approximation*** Repeat Exercise 61 for the function $f(x, y) = \cos(x + y)$.

63. Prove Theorem 13.14.

64. Prove that the angle of inclination θ of the tangent plane to the surface $z = f(x, y)$ at the point (x_0, y_0, z_0) is given by

$$\cos \theta = \frac{1}{\sqrt{[f_x(x_0, y_0)]^2 + [f_y(x_0, y_0)]^2 + 1}}.$$

Section Project: Wildflowers

The diversity of wildflowers in a meadow can be measured by counting the number of daisies, buttercups, shooting stars, and so on. If there are n types of wildflowers, each with a proportion p_i of the total population, it follows that $p_1 + p_2 + \cdot \cdot \cdot + p_n = 1$. The measure of diversity of the population is defined as

$$H = -\sum_{i=1}^{n} p_i \log_2 p_i.$$

In this definition, it is understood that $p_i \log_2 p_i = 0$ when $p_i = 0$. The tables show proportions of wildflowers in a meadow in May, June, August, and September.

May

Flower type	1	2	3	4
Proportion	$\frac{5}{16}$	$\frac{5}{16}$	$\frac{5}{16}$	$\frac{1}{16}$

June

Flower type	1	2	3	4
Proportion	$\frac{1}{4}$	$\frac{1}{4}$	$\frac{1}{4}$	$\frac{1}{4}$

August

Flower type	1	2	3	4
Proportion	$\frac{1}{4}$	0	$\frac{1}{4}$	$\frac{1}{2}$

September

Flower type	1	2	3	4
Proportion	0	0	0	1

(a) Determine the wildflower diversity for each month. How would you interpret September's diversity? Which month had the greatest diversity?

(b) If the meadow contains 10 types of wildflowers in roughly equal proportions, is the diversity of the population greater than or less than the diversity of a similar distribution of 4 types of flowers? What type of distribution (of 10 types of wildflowers) would produce maximum diversity?

(c) Let H_n represent the maximum diversity of n types of wildflowers. Does H_n approach a limit as $n \to \infty$?

FOR FURTHER INFORMATION Biologists use the concept of diversity to measure the proportions of different types of organisms within an environment. For more information on this technique, see the article "Information Theory and Biological Diversity" by Steven Kolmes and Kevin Mitchell in the *UMAP Modules*.

Section 13.8 Extrema of Functions of Two Variables

- **Find absolute and relative extrema of a function of two variables.**
- **Use the Second Partials Test to find relative extrema of a function of two variables.**

Absolute Extrema and Relative Extrema

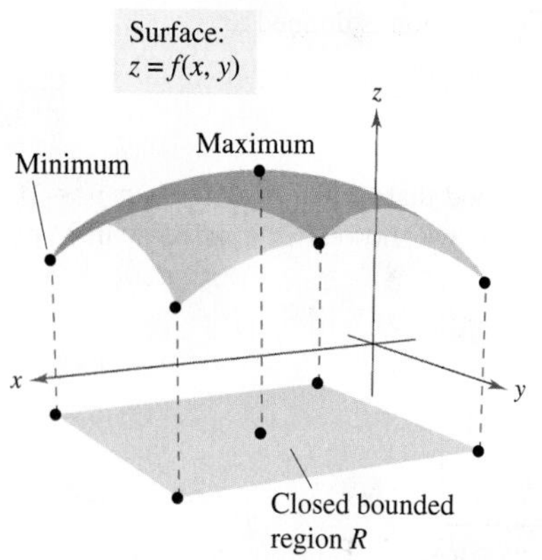

R contains point(s) at which $f(x, y)$ is a minimum and point(s) at which $f(x, y)$ is a maximum.
Figure 13.63

In Chapter 3, you studied techniques for finding the extreme values of a function of a single variable. In this section, you will extend these techniques to functions of two variables. For example, in Theorem 13.15 the Extreme Value Theorem for a function of a single variable is extended to a function of two variables.

Consider the continuous function f of two variables, defined on a closed bounded region R. The values $f(a, b)$ and $f(c, d)$ such that

$$f(a, b) \leq f(x, y) \leq f(c, d) \qquad (a, b) \text{ and } (c, d) \text{ are in } R.$$

for all (x, y) in R are called the **minimum** and **maximum** of f in the region R, as shown in Figure 13.63. Recall from Section 13.2 that a region in the plane is *closed* if it contains all of its boundary points. The Extreme Value Theorem deals with a region in the plane that is both closed and *bounded*. A region in the plane is called **bounded** if it is a subregion of a closed disk in the plane.

THEOREM 13.15 Extreme Value Theorem

Let f be a continuous function of two variables x and y defined on a closed bounded region R in the xy-plane.

1. There is at least one point in R where f takes on a minimum value.
2. There is at least one point in R where f takes on a maximum value.

A minimum is also called an **absolute minimum** and a maximum is also called an **absolute maximum.** As in single-variable calculus, there is a distinction made between absolute extrema and **relative extrema.**

Definition of Relative Extrema

Let f be a function defined on a region R containing (x_0, y_0).

1. The function f has a **relative minimum** at (x_0, y_0) if
$$f(x, y) \geq f(x_0, y_0)$$
for all (x, y) in an *open* disk containing (x_0, y_0).
2. The function f has a **relative maximum** at (x_0, y_0) if
$$f(x, y) \leq f(x_0, y_0)$$
for all (x, y) in an *open* disk containing (x_0, y_0).

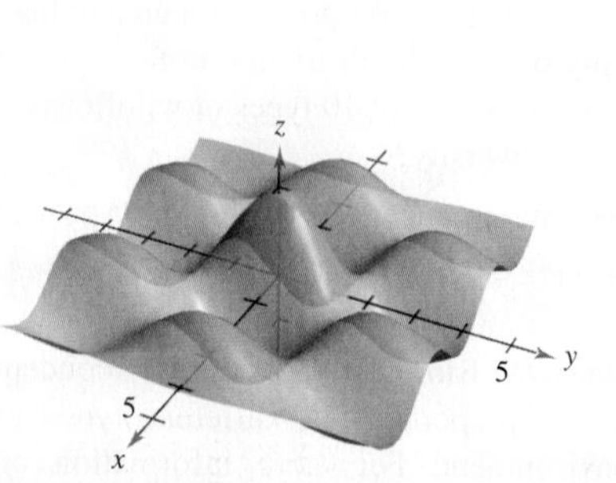

Relative extrema
Figure 13.64

To say that f has a relative maximum at (x_0, y_0) means that the point (x_0, y_0, z_0) is at least as high as all nearby points on the graph of $z = f(x, y)$. Similarly, f has a relative minimum at (x_0, y_0) if (x_0, y_0, z_0) is at least as low as all nearby points on the graph. (See Figure 13.64.)

The Granger Collection

KARL WEIERSTRASS (1815–1897)

Although the Extreme Value Theorem had been used by earlier mathematicians, the first to provide a rigorous proof was the German mathematician Karl Weierstrass. Weierstrass also provided rigorous justifications for many other mathematical results already in common use. We are indebted to him for much of the logical foundation on which modern calculus is built.

To locate relative extrema of f, you can investigate the points at which the gradient of f is $\mathbf{0}$ or the points at which one of the partial derivatives does not exist. Such points are called **critical points** of f.

Definition of Critical Point

Let f be defined on an open region R containing (x_0, y_0). The point (x_0, y_0) is a **critical point** of f if one of the following is true.

1. $f_x(x_0, y_0) = 0$ and $f_y(x_0, y_0) = 0$

2. $f_x(x_0, y_0)$ or $f_y(x_0, y_0)$ does not exist.

Recall from Theorem 13.11 that if f is differentiable and

$$\begin{aligned}\nabla f(x_0, y_0) &= f_x(x_0, y_0)\mathbf{i} + f_y(x_0, y_0)\mathbf{j} \\ &= 0\mathbf{i} + 0\mathbf{j}\end{aligned}$$

then every directional derivative at (x_0, y_0) must be 0. This implies that the function has a horizontal tangent plane at the point (x_0, y_0), as shown in Figure 13.65. It appears that such a point is a likely location of a relative extremum. This is confirmed by Theorem 13.16.

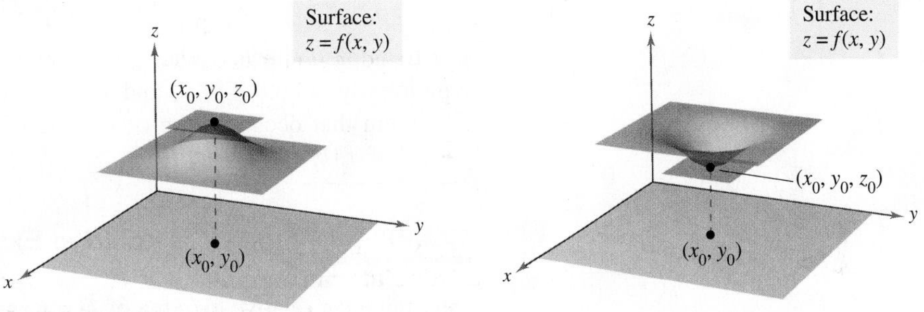

Relative maximum
Figure 13.65

Relative minimum

THEOREM 13.16 Relative Extrema Occur Only at Critical Points

If f has a relative extremum at (x_0, y_0) on an open region R, then (x_0, y_0) is a critical point of f.

EXPLORATION

Use a graphing utility to graph

$$z = x^3 - 3xy + y^3$$

using the bounds $0 \le x \le 3$, $0 \le y \le 3$, and $-3 \le z \le 3$. This view makes it appear as though the surface has an absolute minimum. But does it?

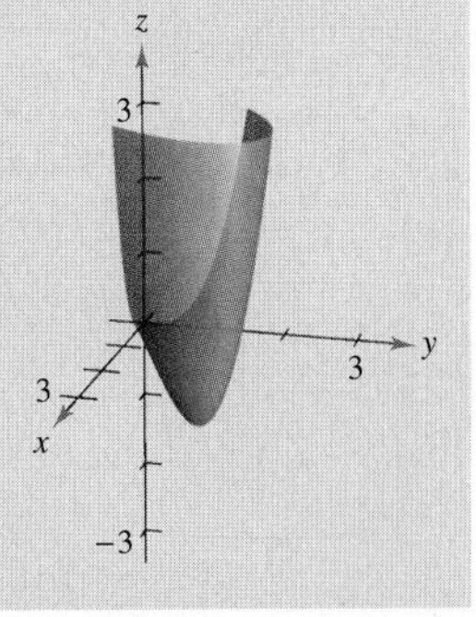

EXAMPLE 1 Finding a Relative Extremum

Determine the relative extrema of

$$f(x, y) = 2x^2 + y^2 + 8x - 6y + 20.$$

Solution Begin by finding the critical points of f. Because

$$f_x(x, y) = 4x + 8 \quad \text{Partial with respect to } x$$

and

$$f_y(x, y) = 2y - 6 \quad \text{Partial with respect to } y$$

are defined for all x and y, the only critical points are those for which both first partial derivatives are 0. To locate these points, let $f_x(x, y)$ and $f_y(x, y)$ be 0, and solve the equations

$$4x + 8 = 0 \quad \text{and} \quad 2y - 6 = 0$$

to obtain the critical point $(-2, 3)$. By completing the square, you can conclude that for all $(x, y) \neq (-2, 3)$

$$f(x, y) = 2(x + 2)^2 + (y - 3)^2 + 3 > 3.$$

So, a relative *minimum* of f occurs at $(-2, 3)$. The value of the relative minimum is $f(-2, 3) = 3$, as shown in Figure 13.66.

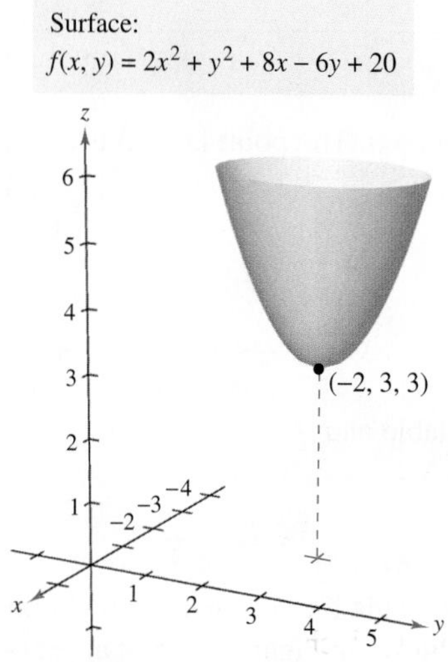

The function $z = f(x, y)$ has a relative minimum at $(-2, 3)$.
Figure 13.66

Example 1 shows a relative minimum occurring at one type of critical point—the type for which both $f_x(x, y)$ and $f_y(x, y)$ are 0. The next example concerns a relative maximum that occurs at the other type of critical point—the type for which either $f_x(x, y)$ or $f_y(x, y)$ does not exist.

EXAMPLE 2 Finding a Relative Extremum

Determine the relative extrema of $f(x, y) = 1 - (x^2 + y^2)^{1/3}$.

Solution Because

$$f_x(x, y) = -\frac{2x}{3(x^2 + y^2)^{2/3}} \quad \text{Partial with respect to } x$$

and

$$f_y(x, y) = -\frac{2y}{3(x^2 + y^2)^{2/3}} \quad \text{Partial with respect to } y$$

it follows that both partial derivatives exist for all points in the xy-plane except for $(0, 0)$. Moreover, because the partial derivatives cannot both be 0 unless both x and y are 0, you can conclude that $(0, 0)$ is the only critical point. In Figure 13.67, note that $f(0, 0)$ is 1. For all other (x, y) it is clear that

$$f(x, y) = 1 - (x^2 + y^2)^{1/3} < 1.$$

So, f has a relative *maximum* at $(0, 0)$.

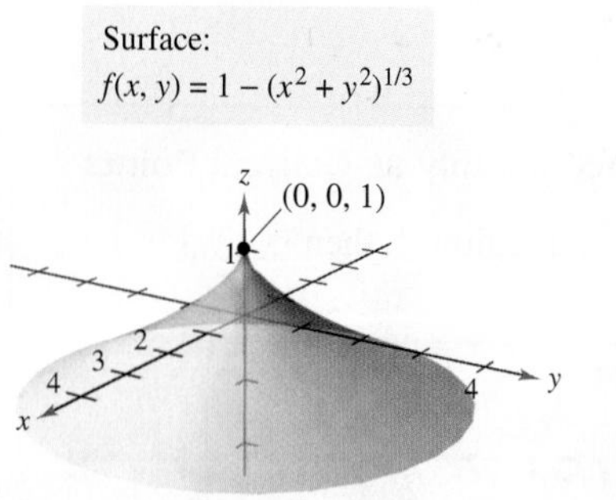

$f_x(x, y)$ and $f_y(x, y)$ are undefined at $(0, 0)$.
Figure 13.67

NOTE In Example 2, $f_x(x, y) = 0$ for every point on the y-axis other than $(0, 0)$. However, because $f_y(x, y)$ is nonzero, these are not critical points. Remember that *one* of the partials must not exist or *both* must be 0 in order to yield a critical point.

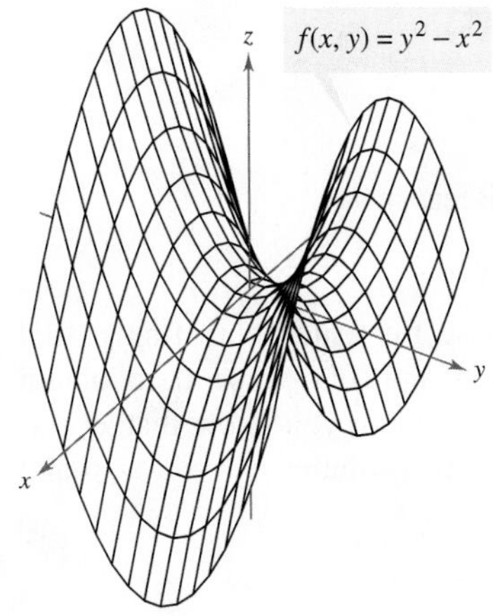

Saddle point at $(0, 0, 0)$:
$f_x(0, 0) = f_y(0, 0) = 0$
Figure 13.68

The Second Partials Test

Theorem 13.16 tells you that to find relative extrema you need only examine values of $f(x, y)$ at critical points. However, as is true for a function of one variable, the critical points of a function of two variables do not always yield relative maxima or minima. Some critical points yield **saddle points,** which are neither relative maxima nor relative minima.

As an example of a critical point that does not yield a relative extremum, consider the surface given by

$$f(x, y) = y^2 - x^2 \qquad \text{Hyperbolic paraboloid}$$

as shown in Figure 13.68. At the point $(0, 0)$, both partial derivatives are 0. The function f does not, however, have a relative extremum at this point because in any open disk centered at $(0, 0)$ the function takes on both negative values (along the x-axis) *and* positive values (along the y-axis). So, the point $(0, 0, 0)$ is a saddle point of the surface. (The term "saddle point" comes from the fact that the surface shown in Figure 13.68 resembles a saddle.)

For the functions in Examples 1 and 2, it was relatively easy to determine the relative extrema, because each function was either given, or able to be written, in completed square form. For more complicated functions, algebraic arguments are less convenient and it is better to rely on the analytic means presented in the following Second Partials Test. This is the two-variable counterpart of the Second Derivative Test for functions of one variable. The proof of this theorem is best left to a course in advanced calculus.

THEOREM 13.17 Second Partials Test

Let f have continuous second partial derivatives on an open region containing a point (a, b) for which

$$f_x(a, b) = 0 \quad \text{and} \quad f_y(a, b) = 0.$$

To test for relative extrema of f, consider the quantity

$$d = f_{xx}(a, b)f_{yy}(a, b) - [f_{xy}(a, b)]^2.$$

1. If $d > 0$ and $f_{xx}(a, b) > 0$, then f has a **relative minimum** at (a, b).
2. If $d > 0$ and $f_{xx}(a, b) < 0$, then f has a **relative maximum** at (a, b).
3. If $d < 0$, then $(a, b, f(a, b))$ is a **saddle point.**
4. The test is inconclusive if $d = 0$.

NOTE If $d > 0$, then $f_{xx}(a, b)$ and $f_{yy}(a, b)$ must have the same sign. This means that $f_{xx}(a, b)$ can be replaced by $f_{yy}(a, b)$ in the first two parts of the test.

A convenient device for remembering the formula for d in the Second Partials Test is given by the 2×2 determinant

$$d = \begin{vmatrix} f_{xx}(a, b) & f_{xy}(a, b) \\ f_{yx}(a, b) & f_{yy}(a, b) \end{vmatrix}$$

where $f_{xy}(a, b) = f_{yx}(a, b)$ by Theorem 13.3.

EXAMPLE 3 Using the Second Partials Test

Find the relative extrema of $f(x, y) = -x^3 + 4xy - 2y^2 + 1$.

Solution Begin by finding the critical points of f. Because

$$f_x(x, y) = -3x^2 + 4y \quad \text{and} \quad f_y(x, y) = 4x - 4y$$

exist for all x and y, the only critical points are those for which both first partial derivatives are 0. To locate these points, let $f_x(x, y)$ and $f_y(x, y)$ be 0 to obtain $-3x^2 + 4y = 0$ and $4x - 4y = 0$. From the second equation you know that $x = y$, and, by substitution into the first equation, you obtain two solutions: $y = x = 0$ and $y = x = \frac{4}{3}$. Because

$$f_{xx}(x, y) = -6x, \quad f_{yy}(x, y) = -4, \quad \text{and} \quad f_{xy}(x, y) = 4$$

it follows that, for the critical point $(0, 0)$,

$$d = f_{xx}(0, 0)f_{yy}(0, 0) - [f_{xy}(0, 0)]^2 = 0 - 16 < 0$$

and, by the Second Partials Test, you can conclude that $(0, 0, 1)$ is a saddle point of f. Furthermore, for the critical point $\left(\frac{4}{3}, \frac{4}{3}\right)$,

$$\begin{aligned} d &= f_{xx}\left(\tfrac{4}{3}, \tfrac{4}{3}\right) f_{yy}\left(\tfrac{4}{3}, \tfrac{4}{3}\right) - \left[f_{xy}\left(\tfrac{4}{3}, \tfrac{4}{3}\right)\right]^2 \\ &= -8(-4) - 16 \\ &= 16 \\ &> 0 \end{aligned}$$

and because $f_{xx}\left(\frac{4}{3}, \frac{4}{3}\right) = -8 < 0$ you can conclude that f has a relative maximum at $\left(\frac{4}{3}, \frac{4}{3}\right)$, as shown in Figure 13.69.

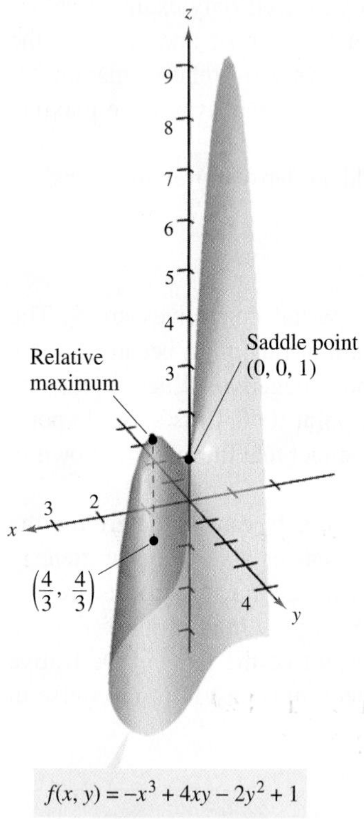

Figure 13.69

The Second Partials Test can fail to find relative extrema in two ways. If either of the first partial derivatives does not exist, you cannot use the test. Also, if

$$d = f_{xx}(a, b)f_{yy}(a, b) - [f_{xy}(a, b)]^2 = 0$$

the test fails. In such cases, you can try a sketch or some other approach, as demonstrated in the next example.

EXAMPLE 4 Failure of the Second Partials Test

Find the relative extrema of $f(x, y) = x^2y^2$.

Solution Because $f_x(x, y) = 2xy^2$ and $f_y(x, y) = 2x^2y$, you know that both partial derivatives are 0 if $x = 0$ or $y = 0$. That is, every point along the x- or y-axis is a critical point. Moreover, because

$$f_{xx}(x, y) = 2y^2, \quad f_{yy}(x, y) = 2x^2, \quad \text{and} \quad f_{xy}(x, y) = 4xy$$

you know that if either $x = 0$ or $y = 0$, then

$$\begin{aligned} d &= f_{xx}(x, y)f_{yy}(x, y) - [f_{xy}(x, y)]^2 \\ &= 4x^2y^2 - 16x^2y^2 = -12x^2y^2 = 0. \end{aligned}$$

So, the Second Partials Test fails. However, because $f(x, y) = 0$ for every point along the x- or y-axis and $f(x, y) = x^2y^2 > 0$ for all other points, you can conclude that each of these critical points yields an absolute minimum, as shown in Figure 13.70.

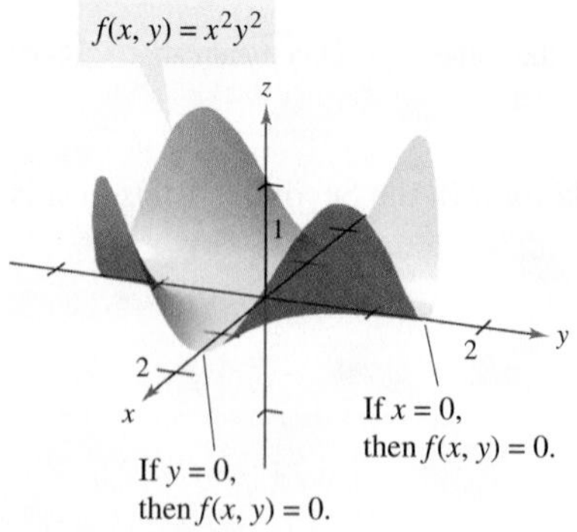

Figure 13.70

Absolute extrema of a function can occur in two ways. First, some relative extrema also happen to be absolute extrema. For instance, in Example 1, $f(-2, 3)$ is an absolute minimum of the function. (On the other hand, the relative maximum found in Example 3 is not an absolute maximum of the function.) Second, absolute extrema can occur at a boundary point of the domain. This is illustrated in Example 5.

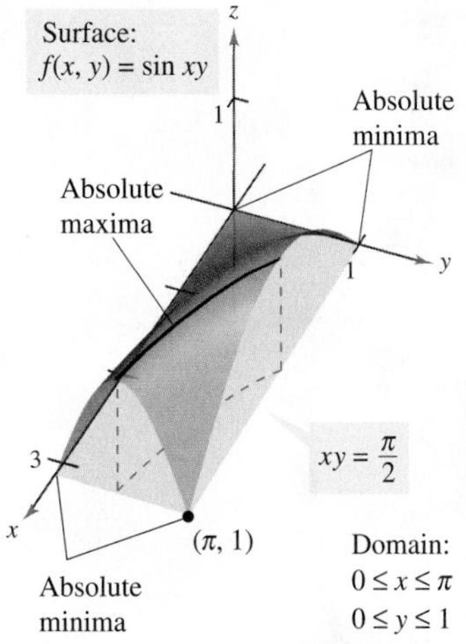

Figure 13.71

EXAMPLE 5 Finding Absolute Extrema

Find the absolute extrema of the function

$$f(x, y) = \sin xy$$

on the closed region given by $0 \le x \le \pi$ and $0 \le y \le 1$.

Solution From the partial derivatives

$$f_x(x, y) = y \cos xy \quad \text{and} \quad f_y(x, y) = x \cos xy$$

you can see that each point lying on the hyperbola given by $xy = \pi/2$ is a critical point. These points each yield the value

$$f(x, y) = \sin \frac{\pi}{2} = 1$$

which you know is the absolute maximum, as shown in Figure 13.71. The only other critical point of f *lying in the given region* is $(0, 0)$. It yields an absolute minimum of 0, because

$$0 \le xy \le \pi$$

implies that

$$0 \le \sin xy \le 1.$$

To locate other absolute extrema, you should consider the four boundaries of the region formed by taking traces with the vertical planes $x = 0$, $x = \pi$, $y = 0$, and $y = 1$. In doing this, you will find that $\sin xy = 0$ at all points on the x-axis, at all points on the y-axis, and at the point $(\pi, 1)$. Each of these points yields an absolute minimum for the surface, as shown in Figure 13.71.

The concepts of relative extrema and critical points can be extended to functions of three or more variables. If all first partial derivatives of

$$w = f(x_1, x_2, x_3, \ldots, x_n)$$

exist, it can be shown that a relative maximum or minimum can occur at $(x_1, x_2, x_3, \ldots, x_n)$ only if every first partial derivative is 0 at that point. This means that the critical points are obtained by solving the following system of equations.

$$\begin{aligned} f_{x_1}(x_1, x_2, x_3, \ldots, x_n) &= 0 \\ f_{x_2}(x_1, x_2, x_3, \ldots, x_n) &= 0 \\ &\vdots \\ f_{x_n}(x_1, x_2, x_3, \ldots, x_n) &= 0 \end{aligned}$$

The extension of Theorem 13.17 to three or more variables is also possible, although you will not consider such an extension in this text.

Exercises for Section 13.8

See www.CalcChat.com for worked-out solutions to odd-numbered exercises.

In Exercises 1–6, identify any extrema of the function by recognizing its given form or its form after completing the square. Verify your results by using the partial derivatives to locate any critical points and test for relative extrema. Use a computer algebra system to graph the function and label any extrema.

1. $g(x, y) = (x - 1)^2 + (y - 3)^2$
2. $g(x, y) = 9 - (x - 3)^2 - (y + 2)^2$
3. $f(x, y) = \sqrt{x^2 + y^2 + 1}$
4. $f(x, y) = \sqrt{25 - (x - 2)^2 - y^2}$
5. $f(x, y) = x^2 + y^2 + 2x - 6y + 6$
6. $f(x, y) = -x^2 - y^2 + 4x + 8y - 11$

In Exercises 7–16, examine the function for relative extrema.

7. $f(x, y) = 2x^2 + 2xy + y^2 + 2x - 3$
8. $f(x, y) = -x^2 - 5y^2 + 10x - 30y - 62$
9. $f(x, y) = -5x^2 + 4xy - y^2 + 16x + 10$
10. $f(x, y) = x^2 + 6xy + 10y^2 - 4y + 4$
11. $z = 2x^2 + 3y^2 - 4x - 12y + 13$
12. $z = -3x^2 - 2y^2 + 3x - 4y + 5$
13. $f(x, y) = 2\sqrt{x^2 + y^2} + 3$
14. $h(x, y) = (x^2 + y^2)^{1/3} + 2$
15. $g(x, y) = 4 - |x| - |y|$
16. $f(x, y) = |x + y| - 2$

In Exercises 17–20, use a computer algebra system to graph the surface and locate any relative extrema and saddle points.

17. $z = \dfrac{-4x}{x^2 + y^2 + 1}$
18. $f(x, y) = y^3 - 3yx^2 - 3y^2 - 3x^2 + 1$
19. $z = (x^2 + 4y^2)e^{1 - x^2 - y^2}$
20. $z = e^{xy}$

In Exercises 21–28, examine the function for relative extrema and saddle points.

21. $h(x, y) = x^2 - y^2 - 2x - 4y - 4$
22. $g(x, y) = 120x + 120y - xy - x^2 - y^2$
23. $h(x, y) = x^2 - 3xy - y^2$
24. $g(x, y) = xy$
25. $f(x, y) = x^3 - 3xy + y^3$

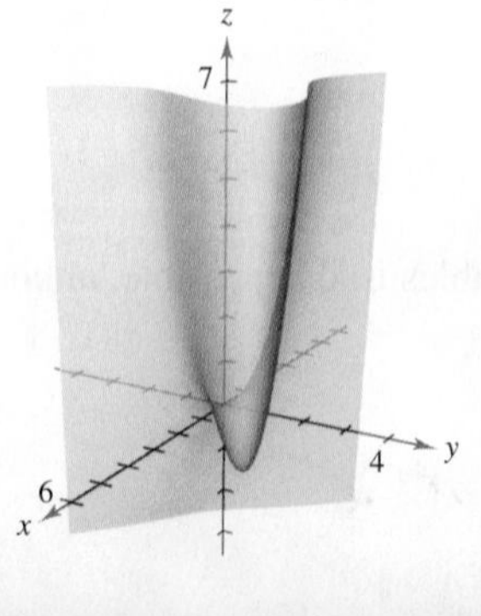

26. $f(x, y) = 2xy - \frac{1}{2}(x^4 + y^4) + 1$

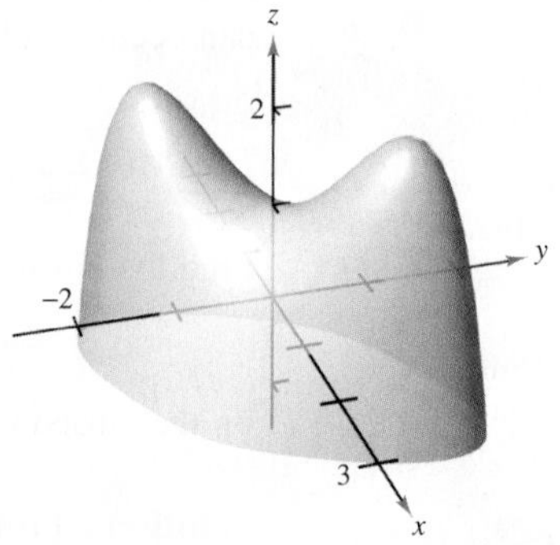

27. $z = e^{-x} \sin y$

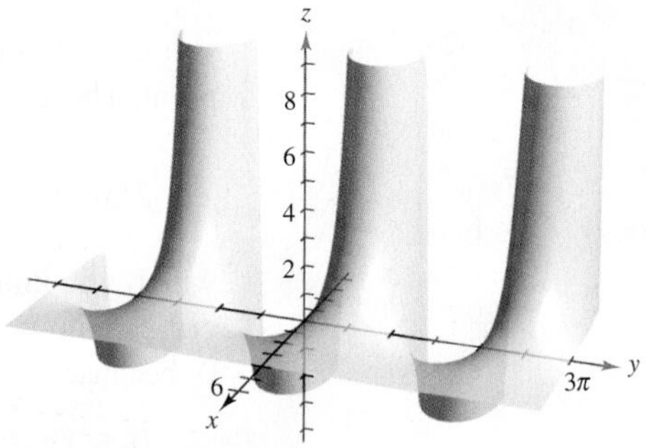

28. $z = \left(\dfrac{1}{2} - x^2 + y^2\right)e^{1 - x^2 - y^2}$

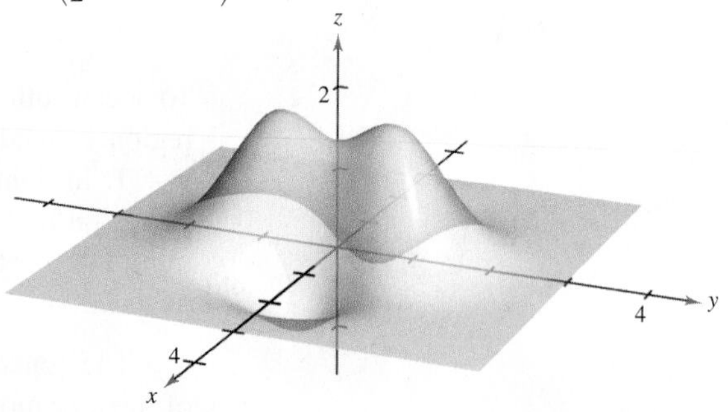

In Exercises 29 and 30, examine the function for extrema without using the derivative tests and use a computer algebra system to graph the surface. (*Hint:* By observation, determine if it is possible for z to be negative. When is z equal to 0?)

29. $z = \dfrac{(x - y)^4}{x^2 + y^2}$
30. $z = \dfrac{(x^2 - y^2)^2}{x^2 + y^2}$

Think About It **In Exercises 31–34, determine whether there is a relative maximum, a relative minimum, a saddle point, or insufficient information to determine the nature of the function $f(x, y)$ at the critical point (x_0, y_0).**

31. $f_{xx}(x_0, y_0) = 9,\ f_{yy}(x_0, y_0) = 4,\ f_{xy}(x_0, y_0) = 6$
32. $f_{xx}(x_0, y_0) = -3,\ f_{yy}(x_0, y_0) = -8,\ f_{xy}(x_0, y_0) = 2$
33. $f_{xx}(x_0, y_0) = -9,\ f_{yy}(x_0, y_0) = 6,\ f_{xy}(x_0, y_0) = 10$
34. $f_{xx}(x_0, y_0) = 25,\ f_{yy}(x_0, y_0) = 8,\ f_{xy}(x_0, y_0) = 10$

Writing About Concepts

35. Define each of the following for a function of two variables.

(a) Relative minimum

(b) Relative maximum

(c) Saddle point

(d) Critical point

36. State the Second Partials Test for relative extrema and saddle points.

In Exercises 37–40, sketch the graph of an arbitrary function f satisfying the given conditions. State whether the function has any extrema or saddle points. (There are many correct answers.)

37. $f_x(x, y) > 0$ and $f_y(x, y) < 0$ for all (x, y).

38. All of the first and second partial derivatives of f are 0.

39. $f_x(0, 0) = 0,\ f_y(0, 0) = 0$

$$f_x(x, y)\begin{cases} < 0, & x < 0 \\ > 0, & x > 0 \end{cases}, \quad f_y(x, y)\begin{cases} > 0, & y < 0 \\ < 0, & y > 0 \end{cases}$$

$f_{xx}(x, y) > 0$, $f_{yy}(x, y) < 0$, and $f_{xy}(x, y) = 0$ for all (x, y).

40. $f_x(2, 1) = 0, \quad f_y(2, 1) = 0$

$$f_x(x, y)\begin{cases} > 0, & x < 2 \\ < 0, & x > 2 \end{cases}, \quad f_y(x, y)\begin{cases} > 0, & y < 1 \\ < 0, & y > 1 \end{cases}$$

$f_{xx}(x, y) < 0$, $f_{yy}(x, y) < 0$, and $f_{xy}(x, y) = 0$ for all (x, y).

41. The figure shows the level curves for an unknown function $f(x, y)$. What, if any, information can be given about f at the point A? Explain your reasoning.

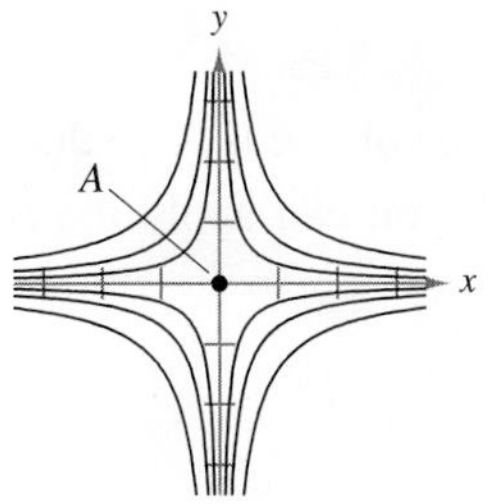

Figure for 41

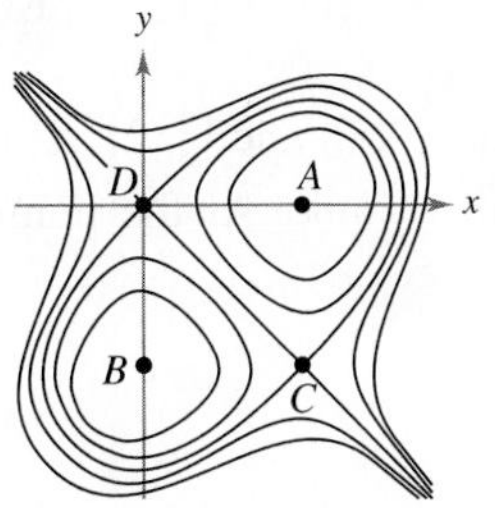

Figure for 42

42. The figure shows the level curves for an unknown function $f(x, y)$. What, if any, information can be given about f at the points A, B, C, and D? Explain your reasoning.

43. A function f has continuous second partial derivatives on an open region containing the critical point $(3, 7)$. The function has a minimum at $(3, 7)$ and $d > 0$ for the Second Partials Test. Determine the interval for $f_{xy}(3, 7)$ if $f_{xx}(3, 7) = 2$ and $f_{yy}(3, 7) = 8$.

44. A function f has continuous second partial derivatives on an open region containing the critical point (a, b). If $f_{xx}(a, b)$ and $f_{yy}(a, b)$ have opposite signs, what is implied? Explain.

In Exercises 45–50, find the critical points and test for relative extrema. List the critical points for which the Second Partials Test fails.

45. $f(x, y) = x^3 + y^3$

46. $f(x, y) = x^3 + y^3 - 6x^2 + 9y^2 + 12x + 27y + 19$

47. $f(x, y) = (x - 1)^2(y + 4)^2$

48. $f(x, y) = \sqrt{(x - 1)^2 + (y + 2)^2}$

49. $f(x, y) = x^{2/3} + y^{2/3}$

50. $f(x, y) = (x^2 + y^2)^{2/3}$

In Exercises 51 and 52, find the critical points of the function and, from the form of the function, determine whether a relative maximum or a relative minimum occurs at each point.

51. $f(x, y, z) = x^2 + (y - 3)^2 + (z + 1)^2$

52. $f(x, y, z) = 4 - [x(y - 1)(z + 2)]^2$

In Exercises 53–62, find the absolute extrema of the function over the region R. (In each case, R contains the boundaries.) Use a computer algebra system to confirm your results.

53. $f(x, y) = 12 - 3x - 2y$

R: The triangular region in the xy-plane with vertices $(2, 0)$, $(0, 1)$, and $(1, 2)$

54. $f(x, y) = (2x - y)^2$

R: The triangular region in the xy-plane with vertices $(2, 0)$, $(0, 1)$, and $(1, 2)$

55. $f(x, y) = 3x^2 + 2y^2 - 4y$

R: The region in the xy-plane bounded by the graphs of $y = x^2$ and $y = 4$

56. $f(x, y) = 2x - 2xy + y^2$

R: The region in the xy-plane bounded by the graphs of $y = x^2$ and $y = 1$

57. $f(x, y) = x^2 + xy, \quad R = \{(x, y) : |x| \le 2,\ |y| \le 1\}$

58. $f(x, y) = x^2 + 2xy + y^2, \quad R = \{(x, y) : |x| \le 2,\ |y| \le 1\}$

59. $f(x, y) = x^2 + 2xy + y^2, \quad R = \{(x, y) : x^2 + y^2 \le 8\}$

60. $f(x, y) = x^2 - 4xy + 5$

$R = \{(x, y) : 0 \le x \le 4,\ 0 \le y \le \sqrt{x}\}$

61. $f(x, y) = \dfrac{4xy}{(x^2 + 1)(y^2 + 1)}$

$R = \{(x, y) : 0 \le x \le 1,\ 0 \le y \le 1\}$

62. $f(x, y) = \dfrac{4xy}{(x^2 + 1)(y^2 + 1)}$

$R = \{(x, y) : x \ge 0,\ y \ge 0,\ x^2 + y^2 \le 1\}$

True or False? **In Exercises 63 and 64, determine whether the statement is true or false. If it is false, explain why or give an example that shows it is false.**

63. If f has a relative maximum at (x_0, y_0, z_0), then $f_x(x_0, y_0) = f_y(x_0, y_0) = 0$.

64. If f is continuous for all x and y and has two relative minima, then f must have at least one relative maximum.

Section 13.9 Applications of Extrema of Functions of Two Variables

- **Solve optimization problems involving functions of several variables.**
- **Use the method of least squares.**

Applied Optimization Problems

In this section, you will survey a few of the many applications of extrema of functions of two (or more) variables.

EXAMPLE 1 Finding Maximum Volume

A rectangular box is resting on the xy-plane with one vertex at the origin. The opposite vertex lies in the plane

$$6x + 4y + 3z = 24$$

as shown in Figure 13.72. Find the maximum volume of such a box.

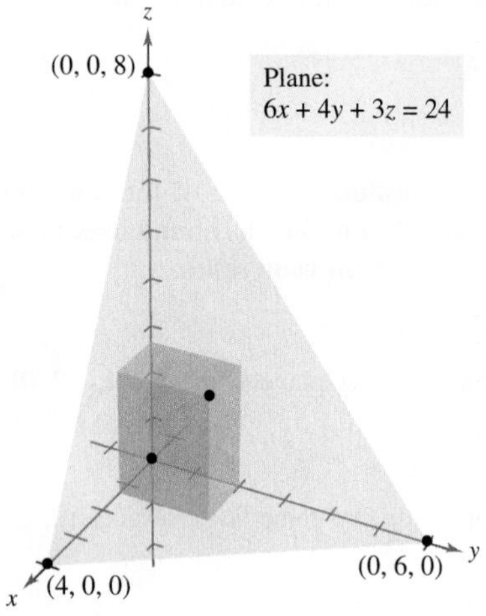

Figure 13.72

Solution Let x, y, and z represent the length, width, and height of the box. Because one vertex of the box lies in the plane $6x + 4y + 3z = 24$, you know that $z = \frac{1}{3}(24 - 6x - 4y)$, and you can write the volume xyz of the box as a function of two variables.

$$V(x, y) = (x)(y)\left[\tfrac{1}{3}(24 - 6x - 4y)\right]$$
$$= \tfrac{1}{3}(24xy - 6x^2y - 4xy^2)$$

By setting the first partial derivatives equal to 0

$$V_x(x, y) = \tfrac{1}{3}(24y - 12xy - 4y^2) = \frac{y}{3}(24 - 12x - 4y) = 0$$

$$V_y(x, y) = \tfrac{1}{3}(24x - 6x^2 - 8xy) = \frac{x}{3}(24 - 6x - 8y) = 0$$

you obtain the critical points $(0, 0)$ and $\left(\frac{4}{3}, 2\right)$. At $(0, 0)$ the volume is 0, so that point does not yield a maximum volume. At the point $\left(\frac{4}{3}, 2\right)$, you can apply the Second Partials Test.

$$V_{xx}(x, y) = -4y$$
$$V_{yy}(x, y) = \frac{-8x}{3}$$
$$V_{xy}(x, y) = \tfrac{1}{3}(24 - 12x - 8y)$$

Because

$$V_{xx}\left(\tfrac{4}{3}, 2\right)V_{yy}\left(\tfrac{4}{3}, 2\right) - \left[V_{xy}\left(\tfrac{4}{3}, 2\right)\right]^2 = (-8)\left(-\tfrac{32}{9}\right) - \left(-\tfrac{8}{3}\right)^2 = \tfrac{64}{3} > 0$$

and

$$V_{xx}\left(\tfrac{4}{3}, 2\right) = -8 < 0$$

you can conclude from the Second Partials Test that the maximum volume is

$$V\left(\tfrac{4}{3}, 2\right) = \tfrac{1}{3}\left[24\left(\tfrac{4}{3}\right)(2) - 6\left(\tfrac{4}{3}\right)^2(2) - 4\left(\tfrac{4}{3}\right)(2^2)\right]$$
$$= \tfrac{64}{9} \text{ cubic units.}$$

Note that the volume is 0 at the boundary points of the triangular domain of V.

NOTE In many applied problems, the domain of the function to be optimized is a closed bounded region. To find minimum or maximum points, you must not only test critical points, but also consider the values of the function at points on the boundary.

Applications of extrema in economics and business often involve more than one independent variable. For instance, a company may produce several models of one type of product. The price per unit and profit per unit are usually different for each model. Moreover, the demand for each model is often a function of the prices of the other models (as well as its own price). The next example illustrates an application involving two products.

EXAMPLE 2 Finding the Maximum Profit

An electronics manufacturer determines that the profit P (in dollars) obtained by producing x units of a DVD player and y units of a DVD recorder is approximated by the model

$$P(x, y) = 8x + 10y - (0.001)(x^2 + xy + y^2) - 10{,}000.$$

Find the production level that produces a maximum profit. What is the maximum profit?

Solution

The partial derivatives of the profit function are

$$P_x(x, y) = 8 - (0.001)(2x + y) \quad \text{and} \quad P_y(x, y) = 10 - (0.001)(x + 2y).$$

By setting these partial derivatives equal to 0, you obtain the following system of equations.

$$\begin{aligned} 8 - (0.001)(2x + y) &= 0 \\ 10 - (0.001)(x + 2y) &= 0 \end{aligned}$$

After simplifying, this system of linear equations can be written as

$$\begin{aligned} 2x + y &= 8000 \\ x + 2y &= 10{,}000. \end{aligned}$$

Solving this system produces $x = 2000$ and $y = 4000$. The second partial derivatives of P are

$$\begin{aligned} P_{xx}(2000, 4000) &= -0.002 \\ P_{yy}(2000, 4000) &= -0.002 \\ P_{xy}(2000, 4000) &= -0.001. \end{aligned}$$

Because $P_{xx} < 0$ and

$$P_{xx}(2000, 4000)P_{yy}(2000, 4000) - [P_{xy}(2000, 4000)]^2 = (-0.002)^2 - (-0.001)^2 > 0$$

you can conclude that the production level of $x = 2000$ units and $y = 4000$ units yields a *maximum* profit. The maximum profit is

$$\begin{aligned} P(2000, 4000) &= 8(2000) + 10(4000) - (0.001)[2000^2 + 2000(4000) + 4000^2)] - 10{,}000 \\ &= \$18{,}000. \end{aligned}$$

FOR FURTHER INFORMATION For more information on the use of mathematics in economics, see the article "Mathematical Methods of Economics" by Joel Franklin in *The American Mathematical Monthly.* To view this article, go to the website *www.matharticles.com.*

NOTE In Example 2, it was assumed that the manufacturing plant is able to produce the required number of units to yield a maximum profit. In actual practice, the production would be bounded by physical constraints. You will study such constrained optimization problems in the next section.

The Method of Least Squares

Many of the examples in this text have involved **mathematical models.** For instance, Example 2 involves a quadratic model for profit. There are several ways to develop such models; one is called the **method of least squares.**

In constructing a model to represent a particular phenomenon, the goals are simplicity and accuracy. Of course, these goals often conflict. For instance, a simple linear model for the points in Figure 13.73 is

$$y = 1.8566x - 5.0246.$$

However, Figure 13.74 shows that by choosing the slightly more complicated quadratic model*

$$y = 0.1996x^2 - 0.7281x + 1.3749$$

you can achieve greater accuracy.

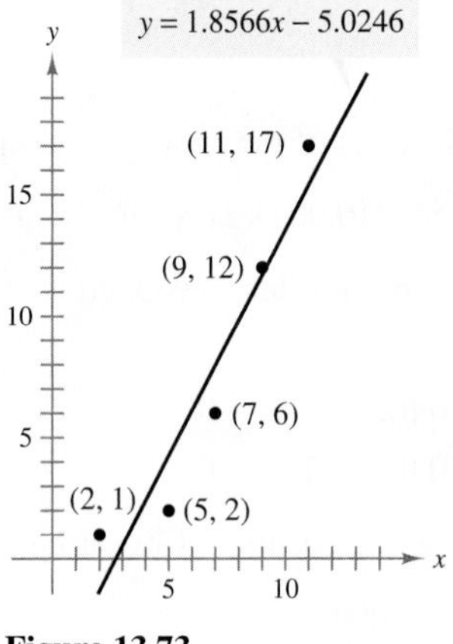

Figure 13.73

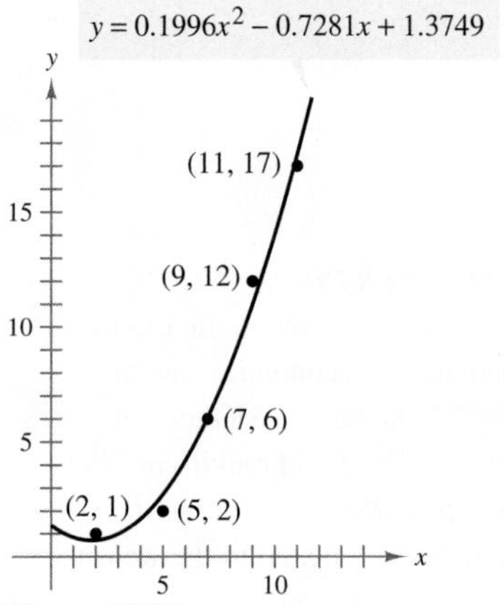

Figure 13.74

As a measure of how well the model $y = f(x)$ fits the collection of points

$$\{(x_1, y_1), (x_2, y_2), (x_3, y_3), \ldots, (x_n, y_n)\}$$

you can add the squares of the differences between the actual y-values and the values given by the model to obtain the **sum of the squared errors**

$$S = \sum_{i=1}^{n} [f(x_i) - y_i]^2. \qquad \text{Sum of the squared errors}$$

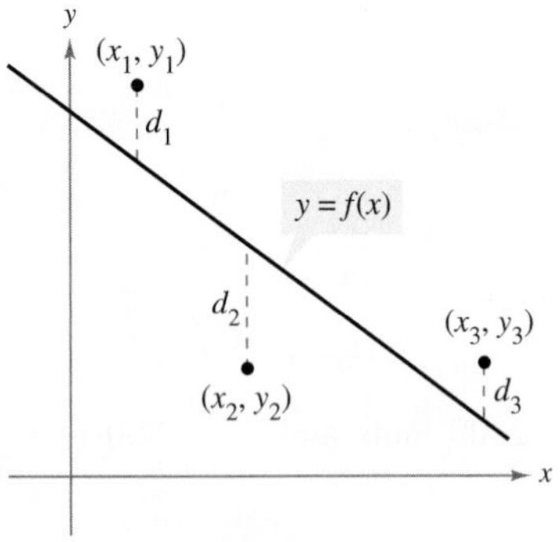

Sum of the squared errors:
$S = d_1^2 + d_2^2 + d_3^2$
Figure 13.75

Graphically, S can be interpreted as the sum of the squares of the vertical distances between the graph of f and the given points in the plane, as shown in Figure 13.75. If the model is perfect, then $S = 0$. However, when perfection is not feasible, you can settle for a model that minimizes S. For instance, the sum of the squared errors for the linear model in Figure 13.73 is $S \approx 17$. Statisticians call the *linear model* that minimizes S the **least squares regression line.** The proof that this line actually minimizes S involves the minimizing of a function of two variables.

---

* *A method for finding the least squares quadratic model for a collection of data is described in Exercise 39.*

ADRIEN-MARIE LEGENDRE (1752–1833)

The method of least squares was introduced by the French mathematician Adrien-Marie Legendre. Legendre is best known for his work in geometry. In fact, his text Elements of Geometry was so popular in the United States that it continued to be used for 33 editions, spanning a period of more than 100 years.

THEOREM 13.18 Least Squares Regression Line

The **least squares regression line** for $\{(x_1, y_1), (x_2, y_2), \ldots, (x_n, y_n)\}$ is given by $f(x) = ax + b$, where

$$a = \frac{n\sum_{i=1}^{n} x_i y_i - \sum_{i=1}^{n} x_i \sum_{i=1}^{n} y_i}{n\sum_{i=1}^{n} x_i^2 - \left(\sum_{i=1}^{n} x_i\right)^2} \quad \text{and} \quad b = \frac{1}{n}\left(\sum_{i=1}^{n} y_i - a\sum_{i=1}^{n} x_i\right).$$

Proof Let $S(a, b)$ represent the sum of the squared errors for the model $f(x) = ax + b$ and the given set of points. That is,

$$\begin{aligned} S(a, b) &= \sum_{i=1}^{n} [f(x_i) - y_i]^2 \\ &= \sum_{i=1}^{n} (ax_i + b - y_i)^2 \end{aligned}$$

where the points (x_i, y_i) represent constants. Because S is a function of a and b, you can use the methods discussed in the preceding section to find the minimum value of S. Specifically, the first partial derivatives of S are

$$\begin{aligned} S_a(a, b) &= \sum_{i=1}^{n} 2x_i(ax_i + b - y_i) \\ &= 2a\sum_{i=1}^{n} x_i^2 + 2b\sum_{i=1}^{n} x_i - 2\sum_{i=1}^{n} x_i y_i \\ S_b(a, b) &= \sum_{i=1}^{n} 2(ax_i + b - y_i) \\ &= 2a\sum_{i=1}^{n} x_i + 2nb - 2\sum_{i=1}^{n} y_i. \end{aligned}$$

By setting these two partial derivatives equal to 0, you obtain the values for a and b that are listed in the theorem. It is left to you to apply the Second Partials Test (see Exercise 40) to verify that these values of a and b yield a minimum.

If the x-values are symmetrically spaced about the y-axis, then $\Sigma\, x_i = 0$ and the formulas for a and b simplify to

$$a = \frac{\sum_{i=1}^{n} x_i y_i}{\sum_{i=1}^{n} x_i^2}$$

and

$$b = \frac{1}{n}\sum_{i=1}^{n} y_i.$$

This simplification is often possible with a translation of the x-values. For instance, if the x-values in a data collection consist of the years 2003, 2004, 2005, 2006, and 2007, you could let 2005 be represented by 0.

EXAMPLE 3 Finding the Least Squares Regression Line

Find the least squares regression line for the points $(-3, 0)$, $(-1, 1)$, $(0, 2)$, and $(2, 3)$.

Solution The table shows the calculations involved in finding the least squares regression line using $n = 4$.

x	y	xy	x^2
-3	0	0	9
-1	1	-1	1
0	2	0	0
2	3	6	4
$\sum_{i=1}^{n} x_i = -2$	$\sum_{i=1}^{n} y_i = 6$	$\sum_{i=1}^{n} x_i y_i = 5$	$\sum_{i=1}^{n} x_i^2 = 14$

TECHNOLOGY Many calculators have "built-in" least squares regression programs. If your calculator has such a program, use it to duplicate the results of Example 3.

Applying Theorem 13.18 produces

$$a = \frac{n\sum_{i=1}^{n} x_i y_i - \sum_{i=1}^{n} x_i \sum_{i=1}^{n} y_i}{n\sum_{i=1}^{n} x_i^2 - \left(\sum_{i=1}^{n} x_i\right)^2} = \frac{4(5) - (-2)(6)}{4(14) - (-2)^2} = \frac{8}{13}$$

and

$$b = \frac{1}{n}\left(\sum_{i=1}^{n} y_i - a\sum_{i=1}^{n} x_i\right) = \frac{1}{4}\left[6 - \frac{8}{13}(-2)\right] = \frac{47}{26}.$$

The least squares regression line is $f(x) = \frac{8}{13}x + \frac{47}{26}$, as shown in Figure 13.76.

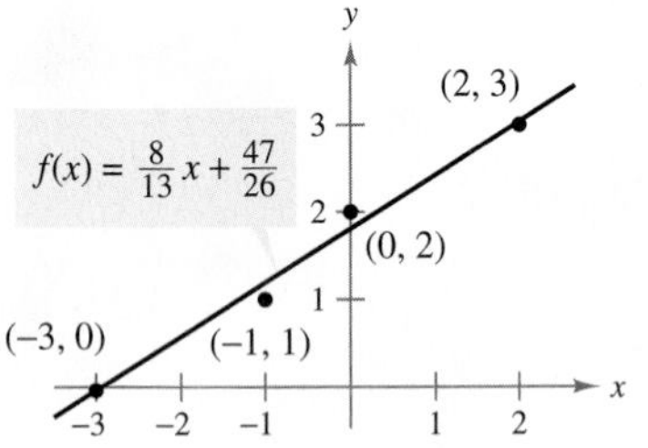

Least squares regression line
Figure 13.76

Exercises for Section 13.9

See www.CalcChat.com for worked-out solutions to odd-numbered exercises.

In Exercises 1 and 2, find the minimum distance from the point to the plane $2x + 3y + z = 12$. (*Hint:* To simplify the computations, minimize the square of the distance.)

1. $(0, 0, 0)$
2. $(1, 2, 3)$

In Exercises 3 and 4, find the minimum distance from the point to the paraboloid $z = x^2 + y^2$.

3. $(5, 5, 0)$
4. $(5, 0, 0)$

In Exercises 5–8, find three positive numbers x, y, and z that satisfy the given conditions.

5. The sum is 30 and the product is a maximum.
6. The sum is 32 and $P = xy^2z$ is a maximum.
7. The sum is 30 and the sum of the squares is a minimum.
8. The sum is 1 and the sum of the squares is a minimum.

9. ***Maximum Volume*** The sum of the length and the girth (perimeter of a cross section) of a package carried by a delivery service cannot exceed 108 inches. Find the dimensions of the rectangular package of largest volume that may be sent.

10. ***Maximum Volume*** The material for constructing the base of an open box costs 1.5 times as much per unit area as the material for constructing the sides. For a fixed amount of money C, find the dimensions of the box of largest volume that can be made.

11. ***Maximum Volume*** The volume of an ellipsoid

$$\frac{x^2}{a^2} + \frac{y^2}{b^2} + \frac{z^2}{c^2} = 1$$

is $4\pi abc/3$. For a fixed sum $a + b + c$, show that the ellipsoid of maximum volume is a sphere.

12. ***Maximum Volume*** Show that the rectangular box of maximum volume inscribed in a sphere of radius r is a cube.

13. ***Volume and Surface Area*** Show that a rectangular box of given volume and minimum surface area is a cube.

14. ***Maximum Volume*** Repeat Exercise 9 under the condition that the sum of the perimeters of the two cross sections shown in the figure cannot exceed 144 inches.

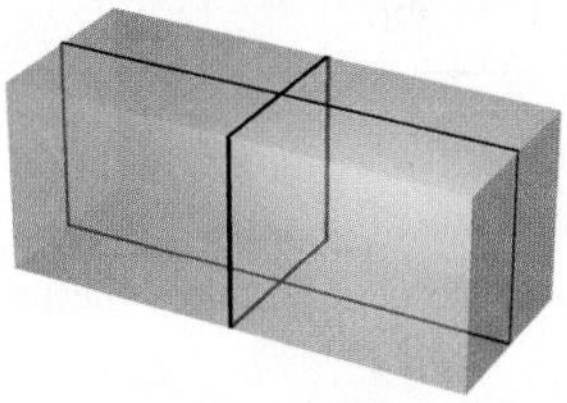

15. ***Area*** A trough with trapezoidal cross sections is formed by turning up the edges of a 30-inch-wide sheet of aluminum (see figure). Find the cross section of maximum area.

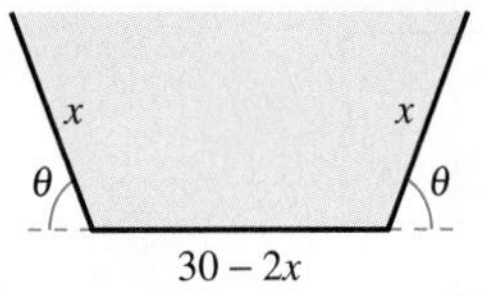

16. ***Area*** Repeat Exercise 15 for a sheet that is w inches wide.

17. ***Maximum Revenue*** A company manufactures two types of sneakers, running shoes and basketball shoes. The total revenue from x_1 units of running shoes and x_2 units of basketball shoes is $R = -5x_1^2 - 8x_2^2 - 2x_1x_2 + 42x_1 + 102x_2$, where x_1 and x_2 are in thousands of units. Find x_1 and x_2 so as to maximize the revenue.

18. ***Maximum Revenue*** A retail outlet sells two types of riding lawn mowers, the prices of which are p_1 and p_2. Find p_1 and p_2 so as to maximize total revenue, where $R = 515p_1 + 805p_2 + 1.5p_1p_2 - 1.5p_1^2 - p_2^2$.

19. ***Maximum Profit*** A corporation manufactures candles at two locations. The cost of producing x_1 units at location 1 is

$$C_1 = 0.02x_1^2 + 4x_1 + 500$$

and the cost of producing x_2 units at location 2 is

$$C_2 = 0.05x_2^2 + 4x_2 + 275.$$

The candles sell for \$15 per unit. Find the quantity that should be produced at each location to maximize the profit $P = 15(x_1 + x_2) - C_1 - C_2$.

20. ***Hardy-Weinberg Law*** Common blood types are determined genetically by three alleles A, B, and O. (An allele is any of a group of possible mutational forms of a gene.) A person whose blood type is AA, BB, or OO is homozygous. A person whose blood type is AB, AO, or BO is heterozygous. The Hardy-Weinberg Law states that the proportion P of heterozygous individuals in any given population is

$$P(p, q, r) = 2pq + 2pr + 2qr$$

where p represents the percent of allele A in the population, q represents the percent of allele B in the population, and r represents the percent of allele O in the population. Use the fact that $p + q + r = 1$ to show that the maximum proportion of heterozygous individuals in any population is $\frac{2}{3}$.

21. ***Minimum Cost*** A water line is to be built from point P to point S and must pass through regions where construction costs differ (see figure). The cost per kilometer in dollars is $3k$ from P to Q, $2k$ from Q to R, and k from R to S. Find x and y such that the total cost C will be minimized.

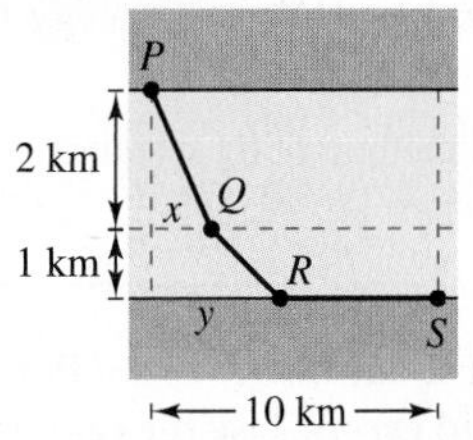

22. ***Distance*** A company has retail outlets located at the points $(0, 0)$, $(2, 2)$, and $(-2, 2)$ (see figure). Management plans to build a distribution center located such that the sum of the distances S from the center to the outlets is minimum. From the symmetry of the problem it is clear that the distribution center will be located on the y-axis, and therefore S is a function of the single variable y. Using techniques presented in Chapter 3, find the required value of y.

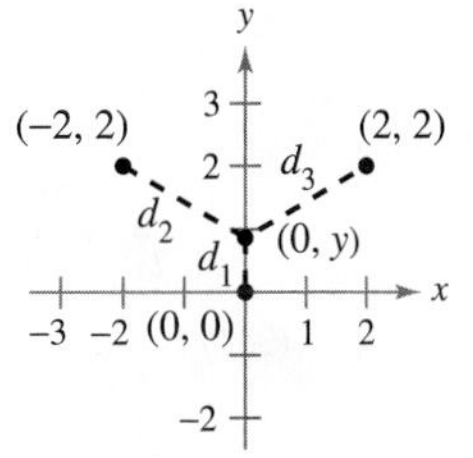

Figure for 22

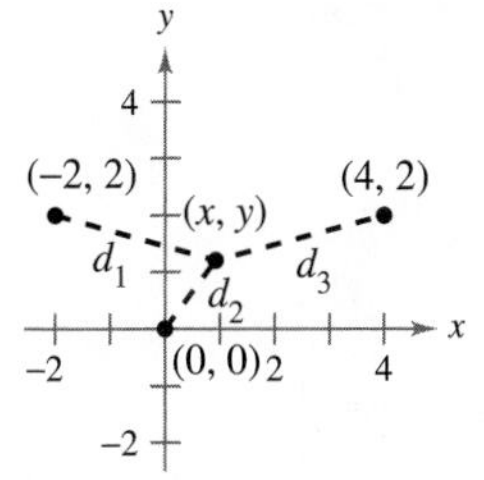

Figure for 23

23. ***Investigation*** The retail outlets described in Exercise 22 are located at $(0, 0)$, $(4, 2)$, and $(-2, 2)$ (see figure). The location of the distribution center is (x, y), and therefore the sum of the distances S is a function of x and y.

(a) Write the expression giving the sum of the distances S. Use a computer algebra system to graph S. Does the surface have a minimum?

(b) Use a computer algebra system to obtain S_x and S_y. Observe that solving the system $S_x = 0$ and $S_y = 0$ is very difficult. So, approximate the location of the distribution center.

(c) An initial estimate of the critical point is $(x_1, y_1) = (1, 1)$. Calculate $-\nabla S(1, 1)$ with components $-S_x(1, 1)$ and $-S_y(1, 1)$. What direction is given by the vector $-\nabla S(1, 1)$?

(d) The second estimate of the critical point is

$$(x_2, y_2) = (x_1 - S_x(x_1, y_1)t, y_1 - S_y(x_1, y_1)t).$$

If these coordinates are substituted into $S(x, y)$, then S becomes a function of the single variable t. Find the value of t that minimizes S. Use this value of t to estimate (x_2, y_2).

(e) Complete two more iterations of the process in part (d) to obtain (x_4, y_4). For this location of the distribution center, what is the sum of the distances to the retail outlets?

(f) Explain why $-\nabla S(x, y)$ was used to approximate the minimum value of S. In what types of problems would you use $\nabla S(x, y)$?

24. ***Investigation*** Repeat Exercise 23 for retail outlets located at the points $(-4, 0)$, $(1, 6)$, and $(12, 2)$.

Writing About Concepts

25. In your own words, state the problem-solving strategy for applied minimum and maximum problems.

26. In your own words, describe the method of least squares for finding mathematical models.

In Exercises 27–30, (a) find the least squares regression line and (b) calculate S, the sum of the squared errors. Use the regression capabilities of a graphing utility to verify your results.

27.

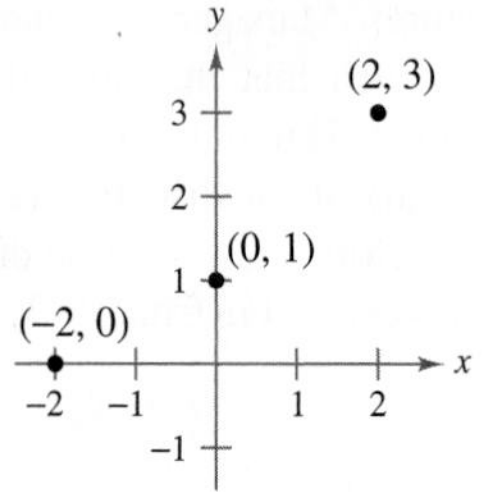

28.

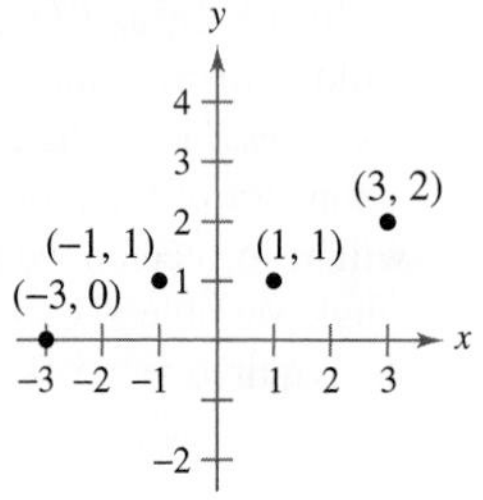

29.

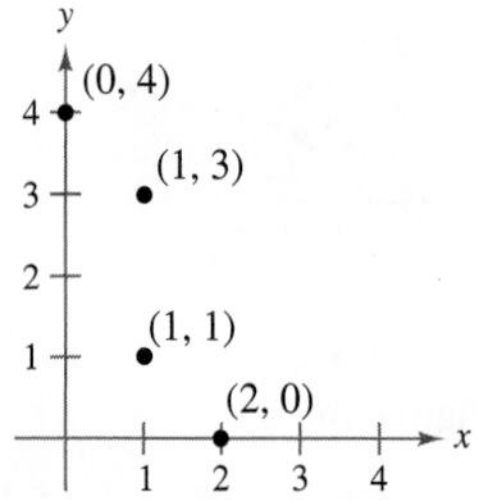

30.

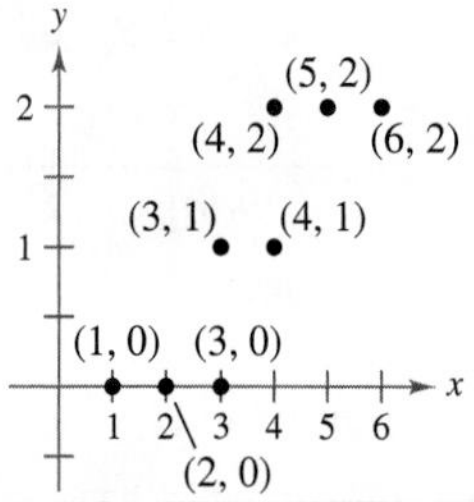

In Exercises 31–34, find the least squares regression line for the points. Use the regression capabilities of a graphing utility to verify your results. Use the graphing utility to plot the points and graph the regression line.

31. $(0, 0), (1, 1), (3, 4), (4, 2), (5, 5)$

32. $(1, 0), (3, 3), (5, 6)$

33. $(0, 6), (4, 3), (5, 0), (8, -4), (10, -5)$

34. $(6, 4), (1, 2), (3, 3), (8, 6), (11, 8), (13, 8)$

35. ***Modeling Data*** The ages x (in years) and systolic blood pressures y of seven men are shown in the table.

Age, x	16	25	39	45	49	64	70
Systolic Blood Pressure, y	109	122	143	132	199	185	199

(a) Use the regression capabilities of a graphing utility to find the least squares regression line for the data.

(b) Use a graphing utility to plot the data and graph the model.

(c) Use the model to approximate the change in systolic blood pressure for each one-year increase in age.

36. ***Modeling Data*** A store manager wants to know the demand y for an energy bar as a function of price x. The daily sales for three different prices of the energy bar are shown in the table.

Price, x	\$1.00	\$1.25	\$1.50
Demand, y	450	375	330

(a) Use the regression capabilities of a graphing utility to find the least squares regression line for the data.

(b) Use the model to estimate the demand when the price is \$1.40.

37. ***Modeling Data*** An agronomist used four test plots to determine the relationship between the wheat yield y (in bushels per acre) and the amount of fertilizer x (in hundreds of pounds per acre). The results are shown in the table.

Fertilizer, x	1.0	1.5	2.0	2.5
Yield, y	32	41	48	53

Use the regression capabilities of a graphing utility to find the least squares regression line for the data, and estimate the yield for a fertilizer application of 160 pounds per acre.

38. ***Modeling Data*** The table shows the percents x and numbers y (in millions) of women in the work force for selected years. *(Source: U.S. Bureau of Labor Statistics)*

Year	1965	1970	1975	1980
Percent, x	39.3	43.3	46.3	51.5
Number, y	26.2	31.5	37.5	45.5

Year	1985	1990	1995	2000
Percent, x	54.5	57.5	58.9	59.9
Number, y	51.1	56.8	60.9	66.3

(a) Use the regression capabilities of a graphing utility to find the least squares regression line for the data.

(b) According to this model, approximately how many women enter the labor force for each one-point increase in the percent of women in the labor force?

39. Find a system of equations whose solution yields the coefficients a, b, and c for the least squares regression quadratic $y = ax^2 + bx + c$ for the points $(x_1, y_1), (x_2, y_2), \ldots, (x_n, y_n)$ by minimizing the sum

$$S(a, b, c) = \sum_{i=1}^{n} (y_i - ax_i^2 - bx_i - c)^2.$$

40. Use the Second Partials Test to verify that the formulas for a and b given in Theorem 13.18 yield a minimum.

$\left[\textit{Hint:} \text{ Use the fact that } n\sum_{i=1}^{n} x_i^2 \geq \left(\sum_{i=1}^{n} x_i\right)^2.\right]$

In Exercises 41–44, use the result of Exercise 39 to find the least squares regression quadratic for the given points. Use the regression capabilities of a graphing utility to confirm your results. Use the graphing utility to plot the points and graph the least squares regression quadratic.

41. $(-2, 0), (-1, 0), (0, 1), (1, 2), (2, 5)$

42. $(-4, 5), (-2, 6), (2, 6), (4, 2)$

43. $(0, 0), (2, 2), (3, 6), (4, 12)$

44. $(0, 10), (1, 9), (2, 6), (3, 0)$

45. ***Modeling Data*** After a new turbocharger for an automobile engine was developed, the following experimental data were obtained for speed y in miles per hour at two-second time intervals x.

Time, x	0	2	4	6	8	10
Speed, y	0	15	30	50	65	70

(a) Find a least squares regression quadratic for the data. Use a graphing utility to confirm your results.

(b) Use a graphing utility to plot the points and graph the model.

46. ***Modeling Data*** The table shows the world populations y (in billions) for five different years. *(Source: U.S. Bureau of the Census, International Data Base)*

Year	1994	1996	1998	2000	2002
Population, y	5.6	5.8	5.9	6.1	6.2

Let $x = 4$ represent the year 1994.

(a) Use the regression capabilities of a graphing utility to find the least squares regression line for the data.

(b) Use the regression capabilities of a graphing utility to find the least squares regression quadratic for the data.

(c) Use a graphing utility to plot the data and graph the models.

(d) Use both models to forecast the world population for the year 2010. How do the two models differ as you extrapolate into the future?

47. ***Modeling Data*** A meteorologist measures the atmospheric pressure P (in kilograms per square meter) at altitude h (in kilometers). The data are shown below.

Altitude, h	0	5	10	15	20
Pressure, P	10,332	5583	2376	1240	517

(a) Use the regression capabilities of a graphing utility to find a least squares regression line for the points $(h, \ln P)$.

(b) The result in part (a) is an equation of the form $\ln P = ah + b$. Write this logarithmic form in exponential form.

(c) Use a graphing utility to plot the original data and graph the exponential model in part (b).

(d) If your graphing utility can fit logarithmic models to data, use it to verify the result in part (b).

48. ***Modeling Data*** The endpoints of the interval over which distinct vision is possible are called the near point and far point of the eye. With increasing age, these points normally change. The table shows the approximate near points y in inches for various ages x (in years).

Age, x	16	32	44	50	60
Near Point, y	3.0	4.7	9.8	19.7	39.4

(a) Find a rational model for the data by taking the reciprocal of the near points to generate the points $(x, 1/y)$. Use the regression capabilities of a graphing utility to find a least squares regression line for the revised data. The resulting line has the form

$$\frac{1}{y} = ax + b.$$

Solve for y.

(b) Use a graphing utility to plot the data and graph the model.

(c) Do you think the model can be used to predict the near point for a person who is 70 years old? Explain.

Section Project: Building a Pipeline

An oil company wishes to construct a pipeline from its offshore facility A to its refinery B. The offshore facility is 2 miles from shore, and the refinery is 1 mile inland. Furthermore, A and B are 5 miles apart, as shown in the figure.

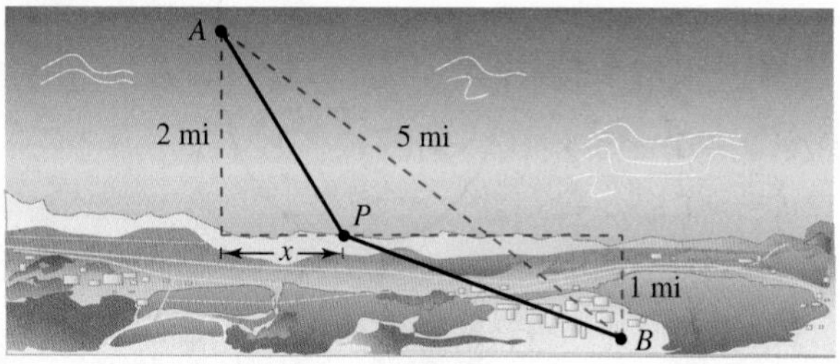

The cost of building the pipeline is \$3 million per mile in the water, and \$4 million per mile on land. So, the cost of the pipeline depends on the location of point P, where it meets the shore. What would be the most economical route of the pipeline?

Imagine that you are to write a report to the oil company about this problem. Let x be the distance shown in the figure. Determine the cost of building the pipeline from A to P, and the cost from P to B. Analyze some sample pipeline routes and their corresponding costs. For instance, what is the cost of the most direct route? Then use calculus to determine the route of the pipeline that minimizes the cost. Explain all steps of your development and include any relevant graphs.

Section 13.10 Lagrange Multipliers

- **Understand the Method of Lagrange Multipliers.**
- **Use Lagrange multipliers to solve constrained optimization problems.**
- **Use the Method of Lagrange Multipliers with two constraints.**

Lagrange Multipliers

Many optimization problems have restrictions, or **constraints,** on the values that can be used to produce the optimal solution. Such constraints tend to complicate optimization problems because the optimal solution can occur at a boundary point of the domain. In this section, you will study an ingenious technique for solving such problems. It is called the **Method of Lagrange Multipliers.**

To see how this technique works, suppose you want to find the rectangle of maximum area that can be inscribed in the ellipse given by

$$\frac{x^2}{3^2} + \frac{y^2}{4^2} = 1.$$

Let (x, y) be the vertex of the rectangle in the first quadrant, as shown in Figure 13.77. Because the rectangle has sides of lengths $2x$ and $2y$, its area is given by

$$f(x, y) = 4xy. \qquad \text{Objective function}$$

You want to find x and y such that $f(x, y)$ is a maximum. Your choice of (x, y) is restricted to first-quadrant points that lie on the ellipse

$$\frac{x^2}{3^2} + \frac{y^2}{4^2} = 1. \qquad \text{Constraint}$$

Now, consider the constraint equation to be a fixed level curve of

$$g(x, y) = \frac{x^2}{3^2} + \frac{y^2}{4^2}.$$

The level curves of f represent a family of hyperbolas

$$f(x, y) = 4xy = k.$$

In this family, the level curves that meet the given constraint correspond to the hyperbolas that intersect the ellipse. Moreover, to maximize $f(x, y)$, you want to find the hyperbola that just barely satisfies the constraint. The level curve that does this is the one that is *tangent* to the ellipse, as shown in Figure 13.78.

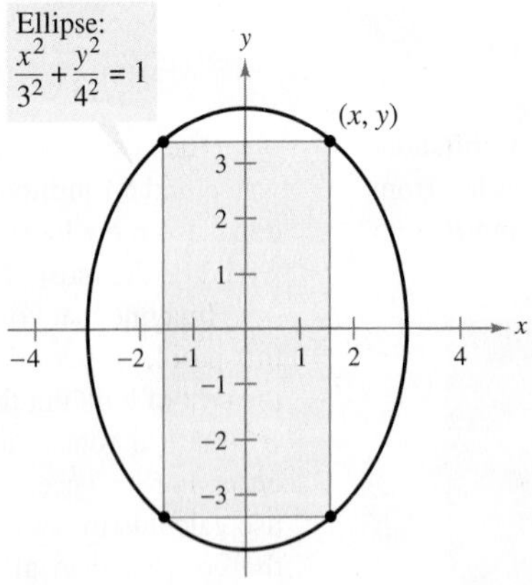

Objective function: $f(x, y) = 4xy$
Figure 13.77

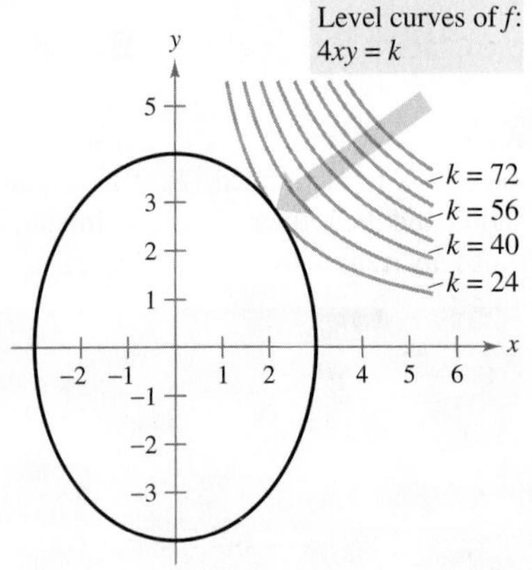

Constraint: $g(x, y) = \frac{x^2}{3^2} + \frac{y^2}{4^2} = 1$
Figure 13.78

To find the appropriate hyperbola, use the fact that two curves are tangent at a point if and only if their gradient vectors are parallel. This means that $\nabla f(x, y)$ must be a scalar multiple of $\nabla g(x, y)$ at the point of tangency. In the context of constrained optimization problems, this scalar is denoted by λ (the lowercase Greek letter lambda).

$$\nabla f(x, y) = \lambda \nabla g(x, y)$$

The scalar λ is called a **Lagrange multiplier.** Theorem 13.19 gives the necessary conditions for the existence of such multipliers.

THEOREM 13.19 Lagrange's Theorem

Let f and g have continuous first partial derivatives such that f has an extremum at a point (x_0, y_0) on the smooth constraint curve $g(x, y) = c$. If $\nabla g(x_0, y_0) \neq \mathbf{0}$, then there is a real number λ such that

$$\nabla f(x_0, y_0) = \lambda \nabla g(x_0, y_0).$$

JOSEPH-LOUIS LAGRANGE (1736–1813)

The Method of Lagrange Multipliers is named after the French mathematician Joseph-Louis Lagrange. Lagrange first introduced the method in his famous paper on mechanics, written when he was just 19 years old.

Proof To begin, represent the smooth curve given by $g(x, y) = c$ by the vector-valued function

$$\mathbf{r}(t) = x(t)\mathbf{i} + y(t)\mathbf{j}, \quad \mathbf{r}'(t) \neq \mathbf{0}$$

where x' and y' are continuous on an open interval I. Define the function h as $h(t) = f(x(t), y(t))$. Then, because $f(x_0, y_0)$ is an extreme value of f, you know that

$$h(t_0) = f(x(t_0), y(t_0)) = f(x_0, y_0)$$

is an extreme value of h. This implies that $h'(t_0) = 0$, and, by the Chain Rule,

$$h'(t_0) = f_x(x_0, y_0)x'(t_0) + f_y(x_0, y_0)y'(t_0) = \nabla f(x_0, y_0) \cdot \mathbf{r}'(t_0) = 0.$$

So, $\nabla f(x_0, y_0)$ is orthogonal to $\mathbf{r}'(t_0)$. Moreover, by Theorem 13.12, $\nabla g(x_0, y_0)$ is also orthogonal to $\mathbf{r}'(t_0)$. Consequently, the gradients $\nabla f(x_0, y_0)$ and $\nabla g(x_0, y_0)$ are parallel, and there must exist a scalar λ such that

$$\nabla f(x_0, y_0) = \lambda \nabla g(x_0, y_0).$$

NOTE Lagrange's Theorem can be shown to be true for functions of three variables, using a similar argument with level surfaces and Theorem 13.14.

The Method of Lagrange Multipliers uses Theorem 13.19 to find the extreme values of a function f subject to a constraint.

Method of Lagrange Multipliers

Let f and g satisfy the hypothesis of Lagrange's Theorem, and let f have a minimum or maximum subject to the constraint $g(x, y) = c$. To find the minimum or maximum of f, use the following steps.

1. Simultaneously solve the equations $\nabla f(x, y) = \lambda \nabla g(x, y)$ and $g(x, y) = c$ by solving the following system of equations.

$$f_x(x, y) = \lambda g_x(x, y)$$
$$f_y(x, y) = \lambda g_y(x, y)$$
$$g(x, y) = c$$

2. Evaluate f at each solution point obtained in the first step. The largest value yields the maximum of f subject to the constraint $g(x, y) = c$, and the smallest value yields the minimum of f subject to the constraint $g(x, y) = c$.

NOTE As you will see in Examples 1 and 2, the Method of Lagrange Multipliers requires solving systems of nonlinear equations. This often can require some tricky algebraic manipulation.

Constrained Optimization Problems

In the problem at the beginning of this section, you wanted to maximize the area of a rectangle that is inscribed in an ellipse. Example 1 shows how to use Lagrange multipliers to solve this problem.

EXAMPLE 1 Using a Lagrange Multiplier with One Constraint

Find the maximum value of $f(x, y) = 4xy$ where $x > 0$ and $y > 0$, subject to the constraint $(x^2/3^2) + (y^2/4^2) = 1$.

NOTE Example 1 can also be solved using the techniques you learned in Chapter 3. To see how, try to find the maximum value of $A = 4xy$ given that

$$\frac{x^2}{3^2} + \frac{y^2}{4^2} = 1.$$

To begin, solve the second equation for y to obtain

$$y = \tfrac{4}{3}\sqrt{9 - x^2}.$$

Then substitute into the first equation to obtain

$$A = 4x\left(\tfrac{4}{3}\sqrt{9 - x^2}\right).$$

Finally, use the techniques of Chapter 3 to maximize A.

Solution To begin, let

$$g(x, y) = \frac{x^2}{3^2} + \frac{y^2}{4^2} = 1.$$

By equating $\nabla f(x, y) = 4y\mathbf{i} + 4x\mathbf{j}$ and $\lambda \nabla g(x, y) = (2\lambda x/9)\mathbf{i} + (\lambda y/8)\mathbf{j}$, you can obtain the following system of equations.

$$4y = \frac{2}{9}\lambda x \qquad f_x(x, y) = \lambda g_x(x, y)$$

$$4x = \frac{1}{8}\lambda y \qquad f_y(x, y) = \lambda g_y(x, y)$$

$$\frac{x^2}{3^2} + \frac{y^2}{4^2} = 1 \qquad \text{Constraint}$$

From the first equation, you obtain $\lambda = 18y/x$, and substitution into the second equation produces

$$4x = \frac{1}{8}\left(\frac{18y}{x}\right)y \quad \Longrightarrow \quad x^2 = \frac{9}{16}y^2.$$

Substituting this value for x^2 into the third equation produces

$$\frac{1}{9}\left(\frac{9}{16}y^2\right) + \frac{1}{16}y^2 = 1 \quad \Longrightarrow \quad y^2 = 8.$$

So, $y = \pm 2\sqrt{2}$. Because it is required that $y > 0$, choose the positive value and find that

$$x^2 = \frac{9}{16}y^2$$

$$= \frac{9}{16}(8) = \frac{9}{2}$$

$$x = \frac{3}{\sqrt{2}}.$$

So, the maximum value of f is

$$f\left(\frac{3}{\sqrt{2}}, 2\sqrt{2}\right) = 4xy = 4\left(\frac{3}{\sqrt{2}}\right)\left(2\sqrt{2}\right) = 24.$$

Note that writing the constraint as

$$g(x, y) = \frac{x^2}{3^2} + \frac{y^2}{4^2} = 1 \qquad \text{or} \qquad g(x, y) = \frac{x^2}{3^2} + \frac{y^2}{4^2} - 1 = 0$$

does not affect the solution—the constant is eliminated when you form ∇g.

EXAMPLE 2 A Business Application

The Cobb-Douglas production function (see Example 5, Section 13.1) for a software manufacturer is given by

$$f(x, y) = 100x^{3/4}y^{1/4} \qquad \text{Objective function}$$

where x represents the units of labor (at \$150 per unit) and y represents the units of capital (at \$250 per unit). The total cost of labor and capital is limited to \$50,000. Find the maximum production level for this manufacturer.

FOR FURTHER INFORMATION For more information on the use of Lagrange multipliers in economics, see the article "Lagrange Multiplier Problems in Economics" by John V. Baxley and John C. Moorhouse in *The American Mathematical Monthly*. To view this article, go to the website *www.matharticles.com*.

Solution From the given function, you have

$$\nabla f(x, y) = 75x^{-1/4}y^{1/4}\mathbf{i} + 25x^{3/4}y^{-3/4}\mathbf{j}.$$

The limit on the cost of labor and capital produces the constraint

$$g(x, y) = 150x + 250y = 50{,}000. \qquad \text{Constraint}$$

So, $\lambda\nabla g(x, y) = 150\lambda\mathbf{i} + 250\lambda\mathbf{j}$. This gives rise to the following system of equations.

$$\begin{aligned} 75x^{-1/4}y^{1/4} &= 150\lambda && f_x(x, y) = \lambda g_x(x, y) \\ 25x^{3/4}y^{-3/4} &= 250\lambda && f_y(x, y) = \lambda g_y(x, y) \\ 150x + 250y &= 50{,}000 && \text{Constraint} \end{aligned}$$

By solving for λ in the first equation

$$\lambda = \frac{75x^{-1/4}y^{1/4}}{150} = \frac{x^{-1/4}y^{1/4}}{2}$$

and substituting into the second equation, you obtain

$$\begin{aligned} 25x^{3/4}y^{-3/4} &= 250\left(\frac{x^{-1/4}y^{1/4}}{2}\right) \\ 25x &= 125y. && \text{Multiply by } x^{1/4}y^{3/4}. \end{aligned}$$

So, $x = 5y$. By substituting into the third equation, you have

$$\begin{aligned} 150(5y) + 250y &= 50{,}000 \\ 1000y &= 50{,}000 \\ y &= 50 \text{ units of capital} \\ x &= 250 \text{ units of labor.} \end{aligned}$$

So, the maximum production level is

$$\begin{aligned} f(250, 50) &= 100(250)^{3/4}(50)^{1/4} \\ &\approx 16{,}719 \text{ product units.} \end{aligned}$$

Economists call the Lagrange multiplier obtained in a production function the **marginal productivity of money.** For instance, in Example 2 the marginal productivity of money at $x = 250$ and $y = 50$ is

$$\lambda = \frac{x^{-1/4}y^{1/4}}{2} = \frac{(250)^{-1/4}(50)^{1/4}}{2} \approx 0.334$$

which means that for each additional dollar spent on production, an additional 0.334 unit of the product can be produced.

EXAMPLE 3 Lagrange Multipliers and Three Variables

Find the minimum value of

$$f(x, y, z) = 2x^2 + y^2 + 3z^2 \qquad \text{Objective function}$$

subject to the constraint $2x - 3y - 4z = 49$.

Solution Let $g(x, y, z) = 2x - 3y - 4z = 49$. Then, because

$$\nabla f(x, y, z) = 4x\mathbf{i} + 2y\mathbf{j} + 6z\mathbf{k} \quad \text{and} \quad \lambda \nabla g(x, y, z) = 2\lambda\mathbf{i} - 3\lambda\mathbf{j} - 4\lambda\mathbf{k}$$

you obtain the following system of equations.

$$\begin{aligned} 4x &= 2\lambda & f_x(x, y, z) &= \lambda g_x(x, y, z) \\ 2y &= -3\lambda & f_y(x, y, z) &= \lambda g_y(x, y, z) \\ 6z &= -4\lambda & f_z(x, y, z) &= \lambda g_z(x, y, z) \\ 2x - 3y - 4z &= 49 & &\text{Constraint} \end{aligned}$$

The solution of this system is $x = 3$, $y = -9$, and $z = -4$. So, the optimum value of f is

$$\begin{aligned} f(3, -9, -4) &= 2(3)^2 + (-9)^2 + 3(-4)^2 \\ &= 147. \end{aligned}$$

From the original function and constraint, it is clear that $f(x, y, z)$ has no maximum. So, the optimum value of f determined above is a minimum.

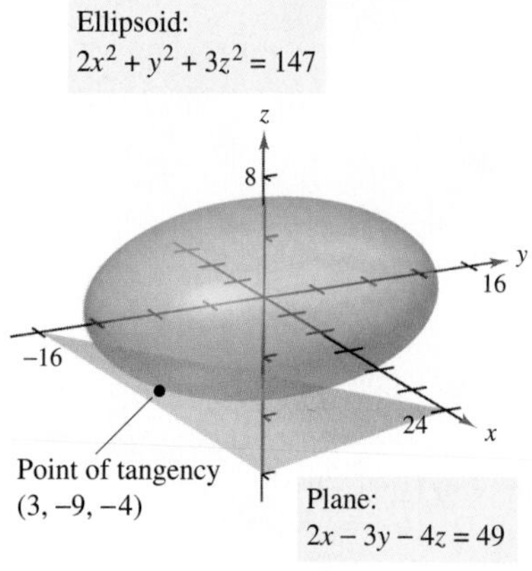

Figure 13.79

A graphical interpretation of constrained optimization problems in two variables was given at the beginning of this section. In three variables, the interpretation is similar, except that level surfaces are used instead of level curves. For instance, in Example 3, the level surfaces of f are ellipsoids centered at the origin, and the constraint

$$2x - 3y - 4z = 49$$

is a plane. The minimum value of f is represented by the ellipsoid that is tangent to the constraint plane, as shown in Figure 13.79.

EXAMPLE 4 Optimization Inside a Region

Find the extreme values of

$$f(x, y) = x^2 + 2y^2 - 2x + 3 \qquad \text{Objective function}$$

subject to the constraint $x^2 + y^2 \le 10$.

Solution To solve this problem, you can break the constraint into two cases.

a. For points *on the circle* $x^2 + y^2 = 10$, you can use Lagrange multipliers to find that the maximum value of $f(x, y)$ is 24—this value occurs at $(-1, 3)$ and at $(-1, -3)$. In a similar way, you can determine that the minimum value of $f(x, y)$ is approximately 6.675—this value occurs at $(\sqrt{10}, 0)$.

b. For points *inside the circle*, you can use the techniques discussed in Section 13.8 to conclude that the function has a relative minimum of 2 at the point $(1, 0)$.

By combining these two results, you can conclude that f has a maximum of 24 at $(-1, \pm 3)$ and a minimum of 2 at $(1, 0)$, as shown in Figure 13.80.

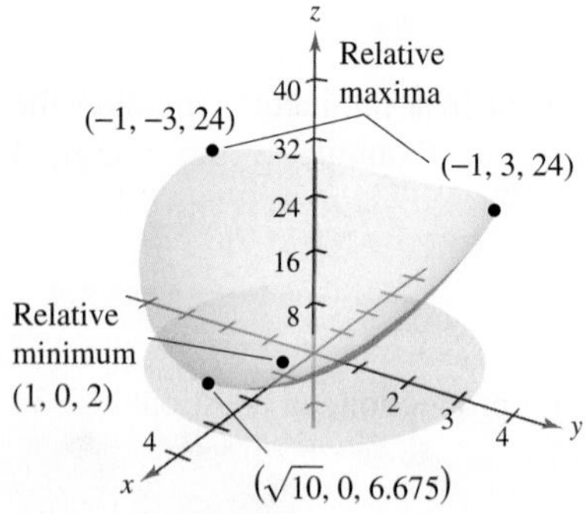

Figure 13.80

The Method of Lagrange Multipliers with Two Constraints

For optimization problems involving *two* constraint functions g and h, you can introduce a second Lagrange multiplier, μ (the lowercase Greek letter mu), and then solve the equation

$$\nabla f = \lambda \nabla g + \mu \nabla h$$

where the gradient vectors are not parallel, as illustrated in Example 5.

EXAMPLE 5 Optimization with Two Constraints

Let $T(x, y, z) = 20 + 2x + 2y + z^2$ represent the temperature at each point on the sphere $x^2 + y^2 + z^2 = 11$. Find the extreme temperatures on the curve formed by the intersection of the plane $x + y + z = 3$ and the sphere.

Solution The two constraints are

$$g(x, y, z) = x^2 + y^2 + z^2 = 11 \quad \text{and} \quad h(x, y, z) = x + y + z = 3.$$

Using

$$\nabla T(x, y, z) = 2\mathbf{i} + 2\mathbf{j} + 2z\mathbf{k}$$
$$\lambda \nabla g(x, y, z) = 2\lambda x\mathbf{i} + 2\lambda y\mathbf{j} + 2\lambda z\mathbf{k}$$

and

$$\mu \nabla h(x, y, z) = \mu\mathbf{i} + \mu\mathbf{j} + \mu\mathbf{k}$$

you can write the following system of equations.

$$2 = 2\lambda x + \mu \qquad T_x(x, y, z) = \lambda g_x(x, y, z) + \mu h_x(x, y, z)$$
$$2 = 2\lambda y + \mu \qquad T_y(x, y, z) = \lambda g_y(x, y, z) + \mu h_y(x, y, z)$$
$$2z = 2\lambda z + \mu \qquad T_z(x, y, z) = \lambda g_z(x, y, z) + \mu h_z(x, y, z)$$
$$x^2 + y^2 + z^2 = 11 \qquad \text{Constraint 1}$$
$$x + y + z = 3 \qquad \text{Constraint 2}$$

By subtracting the second equation from the first, you can obtain the following system.

STUDY TIP The system of equations that arises in the Method of Lagrange Multipliers is not, in general, a linear system, and finding the solution often requires ingenuity.

$$\lambda(x - y) = 0$$
$$2z(1 - \lambda) - \mu = 0$$
$$x^2 + y^2 + z^2 = 11$$
$$x + y + z = 3$$

From the first equation, you can conclude that $\lambda = 0$ or $x = y$. If $\lambda = 0$, you can show that the critical points are $(3, -1, 1)$ and $(-1, 3, 1)$. (Try doing this—it takes a little work.) If $\lambda \neq 0$, then $x = y$ and you can show that the critical points occur when $x = y = (3 \pm 2\sqrt{3})/3$ and $z = (3 \mp 4\sqrt{3})/3$. Finally, to find the optimal solutions, compare the temperatures at the four critical points.

$$T(3, -1, 1) = T(-1, 3, 1) = 25$$
$$T\left(\frac{3 - 2\sqrt{3}}{3}, \frac{3 - 2\sqrt{3}}{3}, \frac{3 + 4\sqrt{3}}{3}\right) = \frac{91}{3} \approx 30.33$$
$$T\left(\frac{3 + 2\sqrt{3}}{3}, \frac{3 + 2\sqrt{3}}{3}, \frac{3 - 4\sqrt{3}}{3}\right) = \frac{91}{3} \approx 30.33$$

So, $T = 25$ is the minimum temperature and $T = \frac{91}{3}$ is the maximum temperature on the curve.

Exercises for Section 13.10

See www.CalcChat.com for worked-out solutions to odd-numbered exercises.

In Exercises 1–4, identify the constraint and level curves of the objective function shown in the figure. Use the figure to approximate the indicated extrema, assuming that x and y are positive. Use Lagrange multipliers to verify your result.

1. Maximize $z = xy$

   Constraint: $x + y = 10$

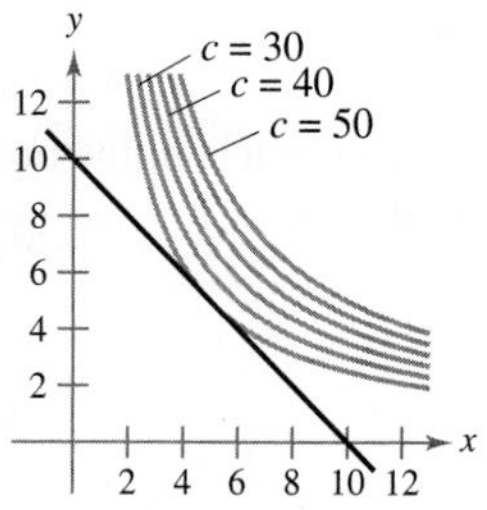

2. Maximize $z = xy$

   Constraint: $2x + y = 4$

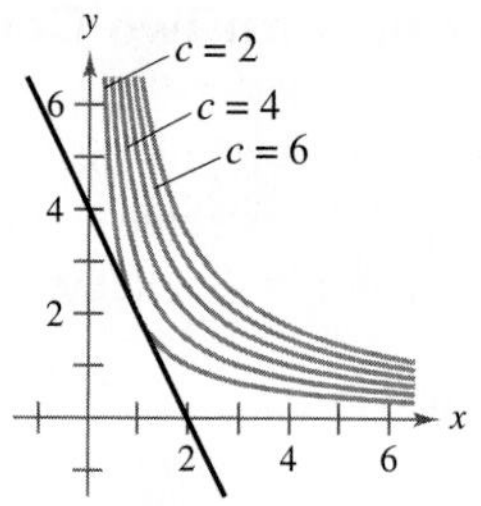

3. Minimize $z = x^2 + y^2$

   Constraint: $x + y - 4 = 0$

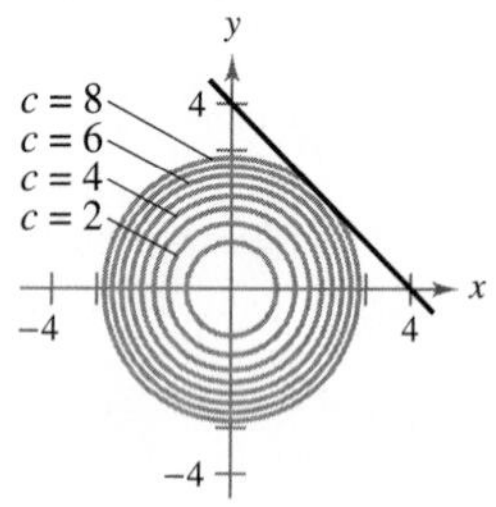

4. Minimize $z = x^2 + y^2$

   Constraint: $2x + 4y = 5$

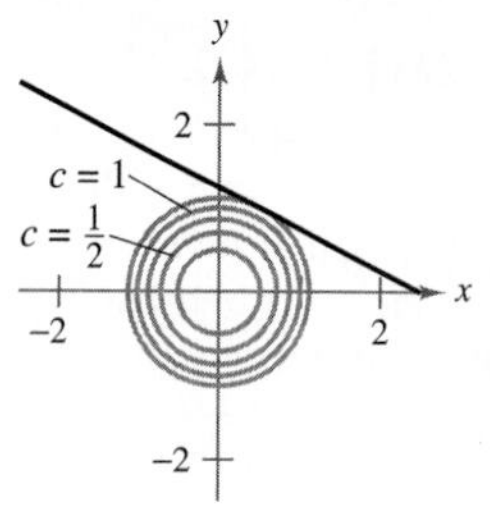

In Exercises 5–12, use Lagrange multipliers to find the indicated extrema, assuming that x and y are positive.

5. Minimize $f(x, y) = x^2 - y^2$

   Constraint: $x - 2y + 6 = 0$

6. Maximize $f(x, y) = x^2 - y^2$

   Constraint: $2y - x^2 = 0$

7. Maximize $f(x, y) = 2x + 2xy + y$

   Constraint: $2x + y = 100$

8. Minimize $f(x, y) = 3x + y + 10$

   Constraint: $x^2y = 6$

9. Maximize $f(x, y) = \sqrt{6 - x^2 - y^2}$

   Constraint: $x + y - 2 = 0$

10. Minimize $f(x, y) = \sqrt{x^2 + y^2}$

    Constraint: $2x + 4y - 15 = 0$

11. Maximize $f(x, y) = e^{xy}$

    Constraint: $x^2 + y^2 = 8$

12. Minimize $f(x, y) = 2x + y$

    Constraint: $xy = 32$

In Exercises 13 and 14, use Lagrange multipliers to find any extrema of the function subject to the constraint $x^2 + y^2 \le 1$.

13. $f(x, y) = x^2 + 3xy + y^2$

14. $f(x, y) = e^{-xy/4}$

In Exercises 15–18, use Lagrange multipliers to find the indicated extrema, assuming that x, y, and z are positive.

15. Minimize $f(x, y, z) = x^2 + y^2 + z^2$

    Constraint: $x + y + z - 6 = 0$

16. Maximize $f(x, y, z) = xyz$

    Constraint: $x + y + z - 6 = 0$

17. Minimize $f(x, y, z) = x^2 + y^2 + z^2$

    Constraint: $x + y + z = 1$

18. Minimize $f(x, y) = x^2 - 10x + y^2 - 14y + 70$

    Constraint: $x + y = 10$

In Exercises 19–22, use Lagrange multipliers to find the indicated extrema of f subject to two constraints. In each case, assume that x, y, and z are nonnegative.

19. Maximize $f(x, y, z) = xyz$

    Constraints: $x + y + z = 32$, $x - y + z = 0$

20. Minimize $f(x, y, z) = x^2 + y^2 + z^2$

    Constraints: $x + 2z = 6$, $x + y = 12$

21. Maximize $f(x, y, z) = xy + yz$

    Constraints: $x + 2y = 6$, $x - 3z = 0$

22. Maximize $f(x, y, z) = xyz$

    Constraints: $x^2 + z^2 = 5$, $x - 2y = 0$

In Exercises 23–26, use Lagrange multipliers to find the minimum distance from the curve or surface to the indicated point. [*Hint:* In Exercise 23, minimize $f(x, y) = x^2 + y^2$ subject to the constraint $2x + 3y = -1$.]

Curve	*Point*
23. Line: $2x + 3y = -1$	$(0, 0)$
24. Circle: $(x - 4)^2 + y^2 = 4$	$(0, 10)$
Surface	*Point*
25. Plane: $x + y + z = 1$	$(2, 1, 1)$
26. Cone: $z = \sqrt{x^2 + y^2}$	$(4, 0, 0)$

In Exercises 27 and 28, find the highest point on the curve of intersection of the surfaces.

27. Sphere: $x^2 + y^2 + z^2 = 36$, Plane: $2x + y - z = 2$

28. Cone: $x^2 + y^2 - z^2 = 0$, Plane: $x + 2z = 4$

Writing About Concepts

29. Explain what is meant by constrained optimization problems.

30. Explain the Method of Lagrange Multipliers for solving constrained optimization problems.

31. ***Maximum Volume*** Use Lagrange multipliers to find the dimensions of the rectangular package of largest volume subject to the constraint that the sum of the length and the girth cannot exceed 108 inches. Compare the answer with that obtained in Exercise 9, Section 13.9.

32. ***Maximum Volume*** The material for the base of an open box costs 1.5 times as much per unit area as the material for constructing the sides. Use Lagrange multipliers to find the dimensions of the box of largest volume that can be made for a fixed cost C. (Maximize $V = xyz$ subject to $1.5xy + 2xz + 2yz = C$.) Compare the answer to that obtained in Exercise 10, Section 13.9.

33. ***Minimum Cost*** A cargo container (in the shape of a rectangular solid) must have a volume of 480 cubic feet. The bottom will cost \$5 per square foot to construct and the sides and the top will cost \$3 per square foot to construct. Use Lagrange multipliers to find the dimensions of the container of this size that has minimum cost.

34. ***Minimum Surface Area*** Use Lagrange multipliers to find the dimensions of a right circular cylinder with volume V_0 cubic units and minimum surface area.

35. ***Maximum Volume*** Use Lagrange multipliers to find the dimensions of a rectangular box of maximum volume that can be inscribed (with edges parallel to the coordinate axes) in the ellipsoid

$$\frac{x^2}{a^2} + \frac{y^2}{b^2} + \frac{z^2}{c^2} = 1.$$

36. ***Geometric and Arithmetic Means***

(a) Use Lagrange multipliers to prove that the product of three positive numbers x, y, and z, whose sum has the constant value S, is a maximum when the three numbers are equal. Use this result to prove that

$$\sqrt[3]{xyz} \le \frac{x + y + z}{3}.$$

(b) Generalize the result of part (a) to prove that the product $x_1 x_2 x_3 \cdot \cdot \cdot x_n$ is a maximum when $x_1 = x_2 = x_3 = \cdot \cdot \cdot = x_n$, $\sum_{i=1}^{n} x_i = S$, and all $x_i \ge 0$. Then prove that

$$\sqrt[n]{x_1 x_2 x_3 \cdot \cdot \cdot x_n} \le \frac{x_1 + x_2 + x_3 + \cdot \cdot \cdot + x_n}{n}.$$

This shows that the geometric mean is never greater than the arithmetic mean.

37. ***Refraction of Light*** When light waves traveling in a transparent medium strike the surface of a second transparent medium, they tend to "bend" in order to follow the path of minimum time. This tendency is called *refraction* and is described by **Snell's Law of Refraction,**

$$\frac{\sin \theta_1}{v_1} = \frac{\sin \theta_2}{v_2}$$

where θ_1 and θ_2 are the magnitudes of the angles shown in the figure, and v_1 and v_2 are the velocities of light in the two media. Use Lagrange multipliers to derive this law using $x + y = a$.

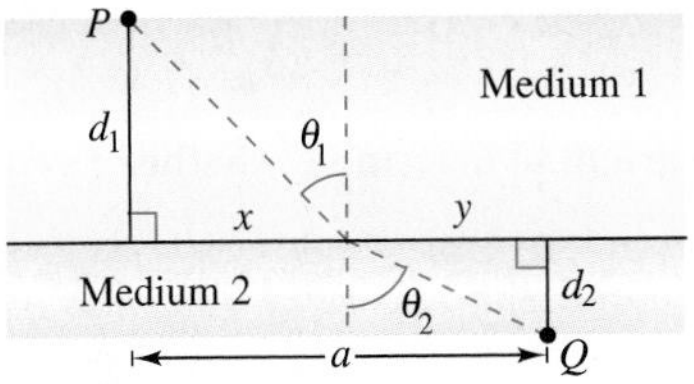

Figure for 37

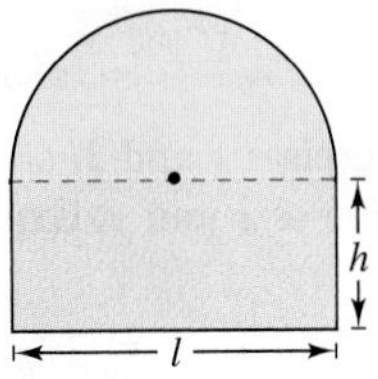

Figure for 38

38. ***Area and Perimeter*** A semicircle is on top of a rectangle (see figure). If the area is fixed and the perimeter is a minimum, or if the perimeter is fixed and the area is a maximum, use Lagrange multipliers to verify that the length of the rectangle is twice its height.

39. ***Hardy-Weinberg Law*** Use Lagrange multipliers to maximize $P(p, q, r) = 2pq + 2pr + 2qr$ subject to $p + q + r = 1$. (See Exercise 20 in Section 13.9.)

40. ***Temperature Distribution*** Let $T(x, y, z) = 100 + x^2 + y^2$ represent the temperature at each point on the sphere $x^2 + y^2 + z^2 = 50$. Find the maximum temperature on the curve formed by the intersection of the sphere and the plane $x - z = 0$.

Production Level **In Exercises 41 and 42, find the maximum production level P if the total cost of labor (at \$48 per unit) and capital (at \$36 per unit) is limited to \$100,000, where x is the number of units of labor and y is the number of units of capital.**

41. $P(x, y) = 100x^{0.25}y^{0.75}$

42. $P(x, y) = 100x^{0.4}y^{0.6}$

Cost **In Exercises 43 and 44, find the minimum cost of producing 20,000 units of a product, where x is the number of units of labor (at \$48 per unit) and y is the number of units of capital (at \$36 per unit).**

43. $P(x, y) = 100x^{0.25}y^{0.75}$

44. $P(x, y) = 100x^{0.6}y^{0.4}$

45. ***Investigation*** Consider the objective function $g(\alpha, \beta, \gamma) = \cos \alpha \cos \beta \cos \gamma$ subject to the constraint that α, β, and γ are the angles of a triangle.

(a) Use Lagrange multipliers to maximize g.

(b) Use the constraint to reduce the function g to a function of two independent variables. Use a computer algebra system to graph the surface represented by g. Identify the maximum values on the graph.

Putnam Exam Challenge

46. A can buoy is to be made of three pieces, namely, a cylinder and two equal cones, the altitude of each cone being equal to the altitude of the cylinder. For a given area of surface, what shape will have the greatest volume?

This problem was composed by the Committee on the Putnam Prize Competition.

Review Exercises for Chapter 13

See www.CalcChat.com for worked-out solutions to odd-numbered exercises.

In Exercises 1 and 2, use the graph to determine whether z is a function of x and y. Explain.

1.

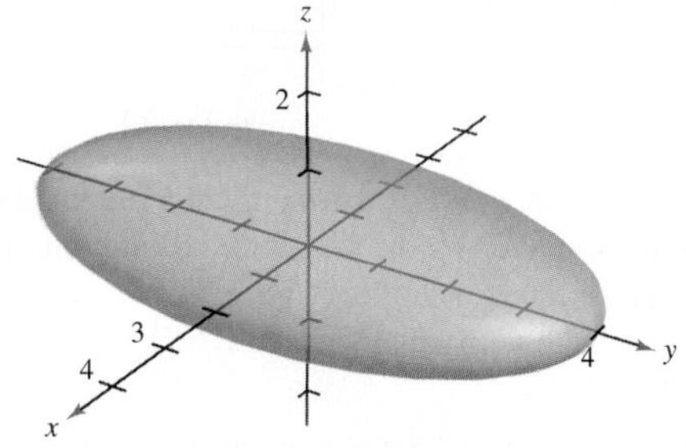

2.

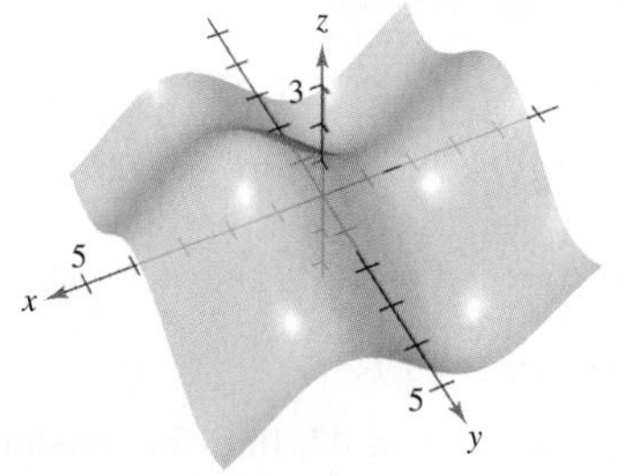

In Exercises 3–6, use a computer algebra system to graph several level curves of the function.

3. $f(x, y) = e^{x^2+y^2}$

4. $f(x, y) = \ln xy$

5. $f(x, y) = x^2 - y^2$

6. $f(x, y) = \dfrac{x}{x + y}$

In Exercises 7 and 8, use a computer algebra system to graph the function.

7. $f(x, y) = e^{-(x^2+y^2)}$

8. $g(x, y) = |y|^{1+|x|}$

In Exercises 9 and 10, sketch the graph of the level surface $f(x, y, z) = c$ at the given value of c.

9. $f(x, y, z) = x^2 - y + z^2, \quad c = 1$

10. $f(x, y, z) = 9x^2 - y^2 + 9z^2, \quad c = 0$

In Exercises 11–14, find the limit and discuss the continuity of the function (if it exists).

11. $\displaystyle\lim_{(x, y)\to(1, 1)} \frac{xy}{x^2 + y^2}$

12. $\displaystyle\lim_{(x, y)\to(1, 1)} \frac{xy}{x^2 - y^2}$

13. $\displaystyle\lim_{(x, y)\to(0, 0)} \frac{-4x^2y}{x^4 + y^2}$

14. $\displaystyle\lim_{(x, y)\to(0, 0)} \frac{y + xe^{-y^2}}{1 + x^2}$

In Exercises 15–24, find all first partial derivatives.

15. $f(x, y) = e^x \cos y$

16. $f(x, y) = \dfrac{xy}{x + y}$

17. $z = xe^y + ye^x$

18. $z = \ln(x^2 + y^2 + 1)$

19. $g(x, y) = \dfrac{xy}{x^2 + y^2}$

20. $w = \sqrt{x^2 + y^2 + z^2}$

21. $f(x, y, z) = z \arctan \dfrac{y}{x}$

22. $f(x, y, z) = \dfrac{1}{\sqrt{1 - x^2 - y^2 - z^2}}$

23. $u(x, t) = ce^{-n^2t} \sin nx$

24. $u(x, t) = c \sin(akx) \cos kt$

25. ***Think About It*** Sketch a graph of a function $z = f(x, y)$ whose derivative f_x is always negative and whose derivative f_y is always negative.

26. Find the slopes of the surface $z = x^2 \ln(y + 1)$ in the x- and y-directions at the point $(2, 0, 0)$.

In Exercises 27–30, find all second partial derivatives and verify that the second mixed partials are equal.

27. $f(x, y) = 3x^2 - xy + 2y^3$

28. $h(x, y) = \dfrac{x}{x + y}$

29. $h(x, y) = x \sin y + y \cos x$

30. $g(x, y) = \cos(x - 2y)$

Laplace's Equation **In Exercises 31–34, show that the function satisfies Laplace's equation**

$$\frac{\partial^2 z}{\partial x^2} + \frac{\partial^2 z}{\partial y^2} = 0.$$

31. $z = x^2 - y^2$

32. $z = x^3 - 3xy^2$

33. $z = \dfrac{y}{x^2 + y^2}$

34. $z = e^x \sin y$

In Exercises 35 and 36, find the total differential.

35. $z = x \sin \dfrac{y}{x}$

36. $z = \dfrac{xy}{\sqrt{x^2 + y^2}}$

37. ***Error Analysis*** The legs of a right triangle are measured to be 5 centimeters and 12 centimeters, with a possible error of $\frac{1}{2}$ centimeter. Approximate the maximum possible error in computing the length of the hypotenuse. Approximate the maximum percent error.

38. ***Error Analysis*** To determine the height of a tower, the angle of elevation to the top of the tower was measured from a point 100 feet $\pm\, \frac{1}{2}$ foot from the base. The angle is measured at 33°, with a possible error of 1°. Assuming that the ground is horizontal, approximate the maximum error in determining the height of the tower.

39. ***Volume*** A right circular cone is measured and the radius and height are found to be 2 inches and 5 inches, respectively. The possible error in measurement is $\frac{1}{8}$ inch. Approximate the maximum possible error in the computation of the volume.

40. ***Lateral Surface Area*** Approximate the error in the computation of the lateral surface area of the cone in Exercise 39. (The lateral surface area is given by $A = \pi r\sqrt{r^2 + h^2}$.)

In Exercises 41–44, find the indicated derivatives (a) using the appropriate Chain Rule and (b) by substitution before differentiating.

41. $w = \ln(x^2 + y^2)$, $\dfrac{dw}{dt}$

$x = 2t + 3$, $y = 4 - t$

42. $u = y^2 - x$, $\dfrac{du}{dt}$

$x = \cos t$, $y = \sin t$

43. $u = x^2 + y^2 + z^2$, $\dfrac{\partial u}{\partial r}, \dfrac{\partial u}{\partial t}$

$x = r\cos t$, $y = r\sin t$, $z = t$

44. $w = \dfrac{xy}{z}$, $\dfrac{\partial w}{\partial r}, \dfrac{\partial w}{\partial t}$

$x = 2r + t$, $y = rt$, $z = 2r - t$

In Exercises 45 and 46, differentiate implicitly to find the first partial derivatives of z.

45. $x^2y - 2yz - xz - z^2 = 0$

46. $xz^2 - y\sin z = 0$

In Exercises 47–50, find the directional derivative of the function at P in the direction of v.

47. $f(x, y) = x^2y$, $(2, 1)$, $\mathbf{v} = \mathbf{i} - \mathbf{j}$

48. $f(x, y) = \frac{1}{4}y^2 - x^2$, $(1, 4)$, $\mathbf{v} = 2\mathbf{i} + \mathbf{j}$

49. $w = y^2 + xz$, $(1, 2, 2)$, $\mathbf{v} = 2\mathbf{i} - \mathbf{j} + 2\mathbf{k}$

50. $w = 6x^2 + 3xy - 4y^2z$, $(1, 0, 1)$, $\mathbf{v} = \mathbf{i} + \mathbf{j} - \mathbf{k}$

In Exercises 51–54, find the gradient of the function and the maximum value of the directional derivative at the given point.

51. $z = \dfrac{y}{x^2 + y^2}$, $(1, 1)$

52. $z = \dfrac{x^2}{x - y}$, $(2, 1)$

53. $z = e^{-x}\cos y$, $\left(0, \dfrac{\pi}{4}\right)$

54. $z = x^2y$, $(2, 1)$

In Exercises 55 and 56, use the gradient to find a unit normal vector to the graph of the equation at the given point.

55. $9x^2 - 4y^2 = 65$, $(3, 2)$

56. $4y\sin x - y^2 = 3$, $\left(\dfrac{\pi}{2}, 1\right)$

In Exercises 57–60, find an equation of the tangent plane and parametric equations of the normal line to the surface at the given point.

	Surface	*Point*
57.	$f(x, y) = x^2y$	$(2, 1, 4)$
58.	$f(x, y) = \sqrt{25 - y^2}$	$(2, 3, 4)$
59.	$z = -9 + 4x - 6y - x^2 - y^2$	$(2, -3, 4)$
60.	$z = \sqrt{9 - x^2 - y^2}$	$(1, 2, 2)$

In Exercises 61 and 62, find symmetric equations of the tangent line to the curve of intersection of the surfaces at the given point.

	Surfaces	*Point*
61.	$z = x^2 - y^2$, $z = 3$	$(2, 1, 3)$
62.	$z = 25 - y^2$, $y = x$	$(4, 4, 9)$

63. Find the angle of inclination θ of the tangent plane to the surface $x^2 + y^2 + z^2 = 14$ at the point $(2, 1, 3)$.

64. ***Approximation*** Consider the following approximations for a function $f(x, y)$ centered at $(0, 0)$.

Linear approximation:

$$P_1(x, y) = f(0, 0) + f_x(0, 0)x + f_y(0, 0)y$$

Quadratic approximation:

$$P_2(x, y) = f(0, 0) + f_x(0, 0)x + f_y(0, 0)y + \tfrac{1}{2}f_{xx}(0, 0)x^2 + f_{xy}(0, 0)xy + \tfrac{1}{2}f_{yy}(0, 0)y^2$$

[Note that the linear approximation is the tangent plane to the surface at $(0, 0, f(0, 0))$.]

(a) Find the linear approximation of $f(x, y) = \cos x + \sin y$ centered at $(0, 0)$.

(b) Find the quadratic approximation of $f(x, y) = \cos x + \sin y$ centered at $(0, 0)$.

(c) If $y = 0$ in the quadratic approximation, you obtain the second-degree Taylor polynomial for what function?

(d) Complete the table.

x	y	$f(x, y)$	$P_1(x, y)$	$P_2(x, y)$
0	0			
0	0.1			
0.2	0.1			
0.5	0.3			
1	0.5			

(e) Use a computer algebra system to graph the surfaces $z = f(x, y)$, $z = P_1(x, y)$, and $z = P_2(x, y)$. How does the accuracy of the approximations change as the distance from $(0, 0)$ increases?

In Exercises 65–68, examine the function for relative extrema. Use a computer algebra system to graph the function and confirm your results.

65. $f(x, y) = x^3 - 3xy + y^2$

66. $f(x, y) = 2x^2 + 6xy + 9y^2 + 8x + 14$

67. $f(x, y) = xy + \dfrac{1}{x} + \dfrac{1}{y}$

68. $z = 50(x + y) - (0.1x^3 + 20x + 150) - (0.05y^3 + 20.6y + 125)$

Writing **In Exercises 69 and 70, write a short paragraph about the surface whose level curves (*c*-values evenly spaced) are shown. Comment on possible extrema, saddle points, the magnitude of the gradient, etc.**

69.

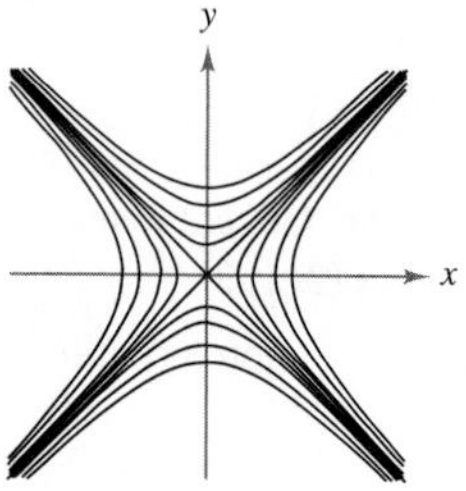

70.

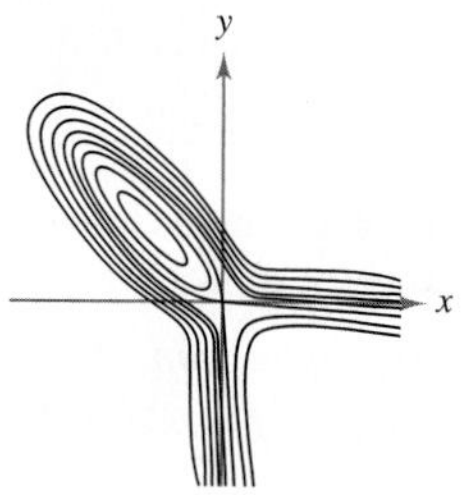

71. ***Maximum Profit*** A corporation manufactures digital cameras at two locations. The cost functions for producing x_1 units at location 1 and x_2 units at location 2 are

$$C_1 = 0.05x_1^2 + 15x_1 + 5400$$
$$C_2 = 0.03x_2^2 + 15x_2 + 6100$$

and the total revenue function is

$$R = [225 - 0.4(x_1 + x_2)](x_1 + x_2).$$

Find the production levels at the two locations that will maximize the profit $P(x_1, x_2) = R - C_1 - C_2$.

72. ***Minimum Cost*** A manufacturer has an order for 1000 units of wooden benches that can be produced at two locations. Let x_1 and x_2 be the numbers of units produced at the two locations. The cost function is

$$C = 0.25x_1^2 + 10x_1 + 0.15x_2^2 + 12x_2.$$

Find the number that should be produced at each location to meet the order and minimize cost.

73. ***Production Level*** The production function for a candy manufacturer is

$$f(x, y) = 4x + xy + 2y$$

where x is the number of units of labor and y is the number of units of capital. Assume that the total amount available for labor and capital is \$2000, and that units of labor and capital cost \$20 and \$4, respectively. Find the maximum production level for this manufacturer.

74. Find the minimum distance from the point $(2, 2, 0)$ to the surface $z = x^2 + y^2$.

75. ***Modeling Data*** The data in the table show the yield y (in milligrams) of a chemical reaction after t minutes.

Minutes, *t*	1	2	3	4
Yield, *y*	1.5	7.4	10.2	13.4

Minutes, *t*	5	6	7	8
Yield, *y*	15.8	16.3	18.2	18.3

(a) Use the regression capabilities of a graphing utility to find the least squares regression line for the data. Then use the graphing utility to plot the data and graph the model.

(b) Use a graphing utility to plot the points $(\ln t, y)$. Do these points appear to follow a linear pattern more closely than the plot of the given data in part (a)?

(c) Use the regression capabilities of a graphing utility to find the least squares regression line for the points $(\ln t, y)$ and obtain the logarithmic model $y = a + b \ln t$.

(d) Use a graphing utility to plot the data and graph the linear and logarithmic models. Which is a better model? Explain.

76. ***Modeling Data*** The table shows the drag force y in kilograms for a motor vehicle at indicated speeds x in kilometers per hour.

Speed, *x*	25	50	75	100	125
Drag, *y*	28	38	54	75	102

(a) Use the regression capabilities of a graphing utility to find the least squares regression quadratic for the data.

(b) Use the model to estimate the total drag when the vehicle is moving at 80 kilometers per hour.

In Exercises 77 and 78, use Lagrange multipliers to locate and classify any extrema of the function.

77. $w = xy + yz + xz$

Constraint: $x + y + z = 1$

78. $z = x^2y$

Constraint: $x + 2y = 2$

79. ***Minimum Cost*** A water line is to be built from point P to point S and must pass through regions where construction costs differ (see figure). The cost per kilometer in dollars is $3k$ from P to Q, $2k$ from Q to R, and k from R to S. For simplicity, let $k = 1$. Use Lagrange multipliers to find x, y, and z such that the total cost C will be minimized.

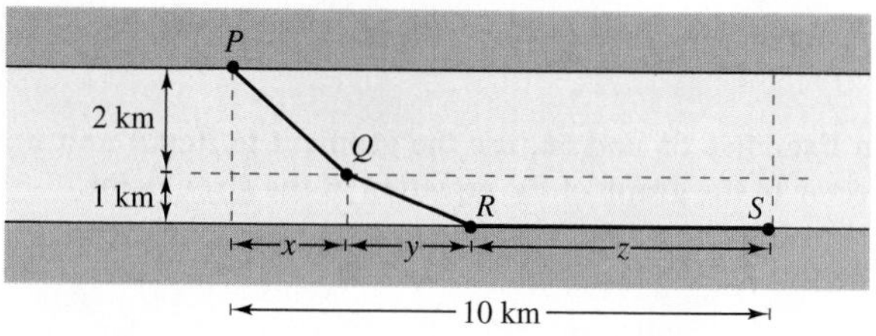

80. ***Investigation*** Consider the objective function $f(x, y) = ax + by$ subject to the constraint $x^2/64 + y^2/36 = 1$. Assume that x and y are positive.

(a) Use a computer algebra system to graph the constraint. If $a = 4$ and $b = 3$, use the computer algebra system to graph the level curves of the objective function. By trial and error, find the level curve that appears to be tangent to the ellipse. Use the result to approximate the maximum of f subject to the constraint.

(b) Repeat part (a) for $a = 4$ and $b = 9$.

P.S. Problem Solving

See www.CalcChat.com for worked-out solutions to odd-numbered exercises.

1. Heron's Formula states that the area of a triangle with sides of lengths a, b, and c is given by

$$A = \sqrt{s(s-a)(s-b)(s-c)}$$

where $s = \dfrac{a+b+c}{2}$, as shown in the figure.

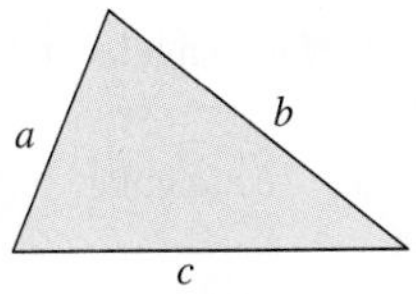

(a) Use Heron's Formula to find the area of the triangle with vertices $(0, 0)$, $(3, 4)$, and $(6, 0)$.

(b) Show that among all triangles having a fixed perimeter, the triangle with the largest area is an equilateral triangle.

(c) Show that among all triangles having a fixed area, the triangle with the smallest perimeter is an equilateral triangle.

2. An industrial container is in the shape of a cylinder with hemispherical ends, as shown in the figure. The container must hold 1000 liters of fluid. Determine the radius r and length h that minimize the amount of material used in the construction of the tank.

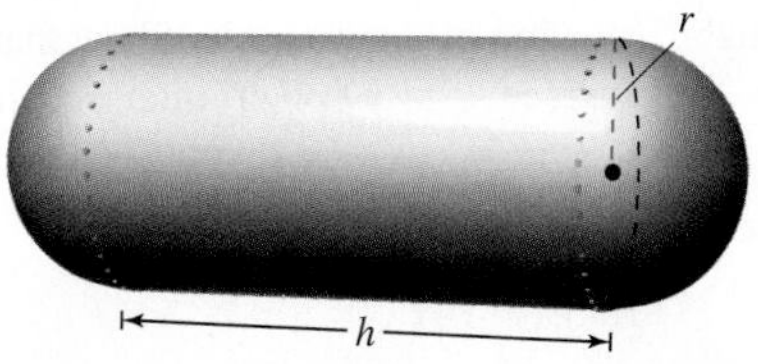

3. Let $P(x_0, y_0, z_0)$ be a point in the first octant on the surface $xyz = 1$.

(a) Find the equation of the tangent plane to the surface at the point P.

(b) Show that the volume of the tetrahedron formed by the three coordinate planes and the tangent plane is constant, independent of the point of tangency (see figure).

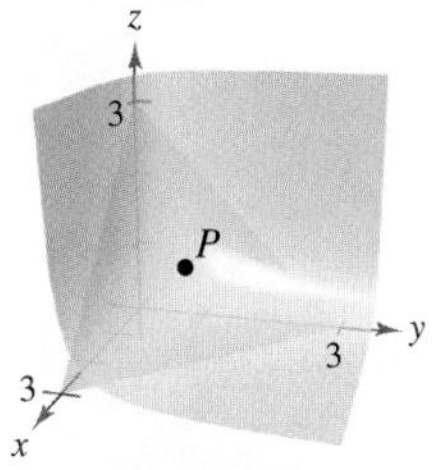

4. Use a graphing utility to graph the functions $f(x) = \sqrt[3]{x^3 - 1}$ and $g(x) = x$ in the same viewing window.

(a) Show that

$$\lim_{x\to\infty} [f(x) - g(x)] = 0 \text{ and } \lim_{x\to-\infty} [f(x) - g(x)] = 0.$$

(b) Find the point on the graph of f that is farthest from the graph of g.

5. Consider the function

$$f(x, y) = \begin{cases} \dfrac{4xy}{x^2 + y^2}, & (x, y) \neq (0, 0) \\ 0, & (x, y) = (0, 0) \end{cases}$$

and the unit vector $\mathbf{u} = \dfrac{1}{\sqrt{2}}(\mathbf{i} + \mathbf{j})$.

Does the directional derivative of f at $P(0, 0)$ in the direction of $\mathbf{u}$ exist? If $f(0, 0)$ were defined as 2 instead of 0, would the directional derivative exist?

6. A heated storage room is shaped like a rectangular box and has a volume of 1000 cubic feet, as shown in the figure. Because warm air rises, the heat loss per unit of area through the ceiling is five times as great as the heat loss through the floor. The heat loss through the four walls is three times as great as the heat loss through the floor. Determine the room dimensions that will minimize heat loss and therefore minimize heating costs.

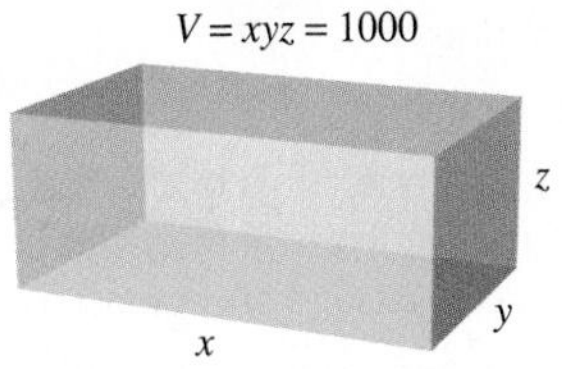

7. Repeat Exercise 6 assuming that the heat loss through the walls and ceiling remain the same, but the floor is insulated so that there is no heat loss through the floor.

8. Consider a circular plate of radius 1 given by $x^2 + y^2 \leq 1$, as shown in the figure. The temperature at any point $P(x, y)$ on the plate is $T(x, y) = 2x^2 + y^2 - y + 10$.

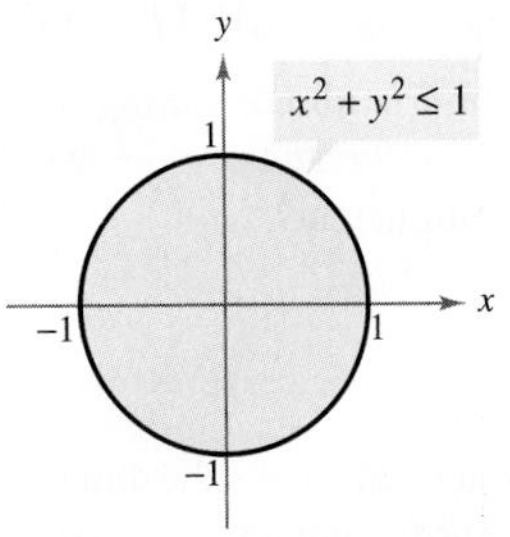

(a) Sketch the isotherm $T(x, y) = 10$. To print an enlarged copy of the graph, go to the website *www.mathgraphs.com*.

(b) Find the hottest and coldest points on the plate.

9. Consider the Cobb-Douglas production function

$$f(x, y) = Cx^a y^{1-a}, \quad 0 < a < 1.$$

(a) Show that f satisfies the equation $x\dfrac{\partial f}{dx} + y\dfrac{\partial f}{dy} = f$.

(b) Show that $f(tx, ty) = tf(x, y)$.

10. Rewrite Laplace's equation $\dfrac{\partial^2 u}{\partial x^2} + \dfrac{\partial^2 u}{\partial y^2} + \dfrac{\partial^2 u}{\partial z^2} = 0$ in cylindrical coordinates.

11. A projectile is launched at an angle of 45° with the horizontal and with an initial velocity of 64 feet per second. A television camera is located in the plane of the path of the projectile 50 feet behind the launch site (see figure).

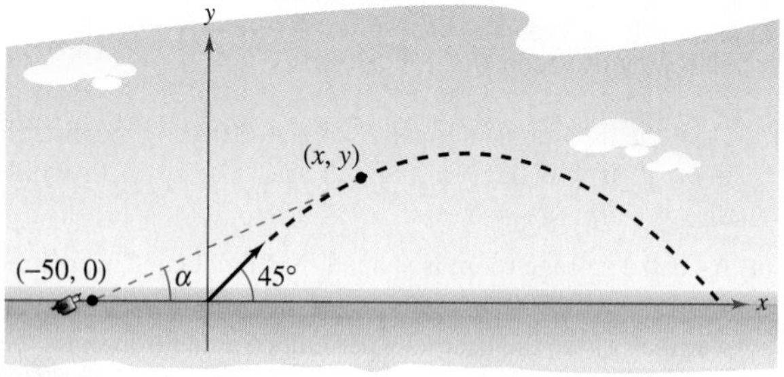

(a) Find parametric equations for the path of the projectile in terms of the parameter t representing time.

(b) Write the angle α that the camera makes with the horizontal in terms of x and y and in terms of t.

(c) Use the results of part (b) to find $d\alpha/dt$.

(d) Use a graphing utility to graph α in terms of t. Is the graph symmetric to the axis of the parabolic arch of the projectile? At what time is the rate of change of α greatest?

(e) At what time is the angle α maximum? Does this occur when the projectile is at its greatest height?

12. Consider the distance d between the launch site and the projectile in Exercise 11.

(a) Write the distance d in terms of x and y and in terms of the parameter t.

(b) Use the results of part (a) to find the rate of change of d.

(c) Find the rate of change of the distance when $t = 2$.

(d) When is the rate of change of d minimum during the flight of the projectile? Does this occur at the time when the projectile reaches its maximum height?

13. Consider the function

$$f(x, y) = (\alpha x^2 + \beta y^2)e^{-(x^2+y^2)}, \quad 0 < |\alpha| < \beta.$$

(a) Use a computer algebra system to graph the function for $\alpha = 1$ and $\beta = 2$, and identify any extrema or saddle points.

(b) Use a computer algebra system to graph the function for $\alpha = -1$ and $\beta = 2$, and identify any extrema or saddle points.

(c) Generalize the results in parts (a) and (b) for the function f.

14. Prove that if f is a differentiable function such that

$$\nabla f(x_0, y_0) = \mathbf{0}$$

then the tangent plane at (x_0, y_0) is horizontal.

15. The figure shows a rectangle that is approximately $l = 6$ centimeters long and $h = 1$ centimeter high.

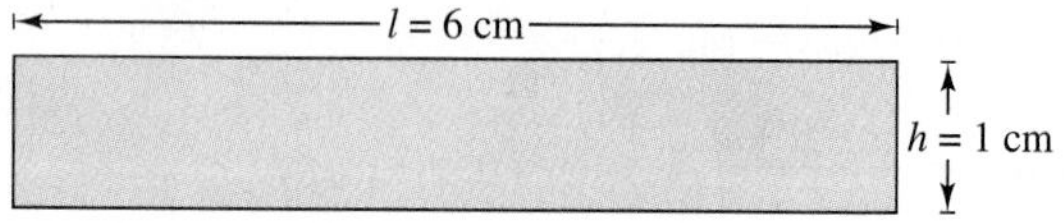

(a) Draw a rectangular strip along the rectangular region showing a small increase in length.

(b) Draw a rectangular strip along the rectangular region showing a small increase in height.

(c) Use the results in parts (a) and (b) to identify the measurement that has more effect on the area A of the rectangle.

(d) Verify your answer in part (c) analytically by comparing the value of dA when $dl = 0.01$ and when $dh = 0.01$.

16. Consider converting a point $(5 \pm 0.05, \pi/18 \pm 0.05)$ in polar coordinates to rectangular coordinates (x, y).

(a) Use a geometric argument to determine whether the accuracy in x is more dependent on the accuracy in r or on the accuracy in θ. Explain. Verify your answer analytically.

(b) Use a geometric argument to determine whether the accuracy in y is more dependent on the accuracy in r or on the accuracy in θ. Explain. Verify your answer analytically.

17. Let f be a differentiable function of one variable. Show that all tangent planes to the surface $z = yf(x/y)$ intersect in a common point.

18. Consider the ellipse

$$\frac{x^2}{a^2} + \frac{y^2}{b^2} = 1$$

that encloses the circle $x^2 + y^2 = 2x$. Find values of a and b that minimize the area of the ellipse.

19. Show that

$$u(x, t) = \frac{1}{2}[\sin(x - t) + \sin(x + t)]$$

is a solution to the one-dimensional wave equation

$$\frac{\partial^2 u}{\partial t^2} = \frac{\partial^2 u}{\partial x^2}.$$

20. Show that

$$u(x, t) = \frac{1}{2}[f(x - ct) + f(x + ct)]$$

is a solution to the one-dimensional wave equation

$$\frac{\partial^2 u}{\partial t^2} = c^2\frac{\partial^2 u}{\partial x^2}.$$

(This equation describes the small transverse vibration of an elastic string such as those on certain musical instruments.)

14 Multiple Integration

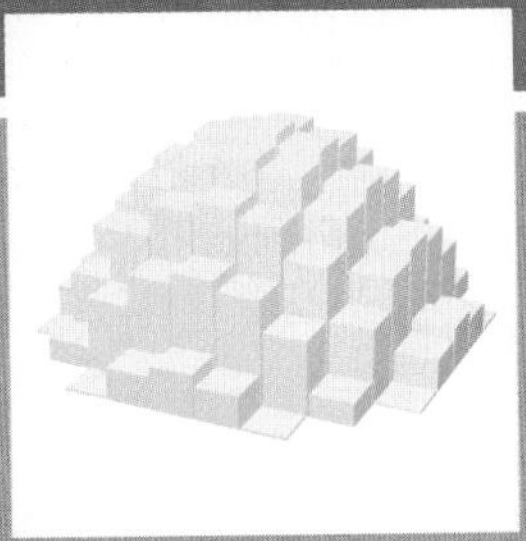

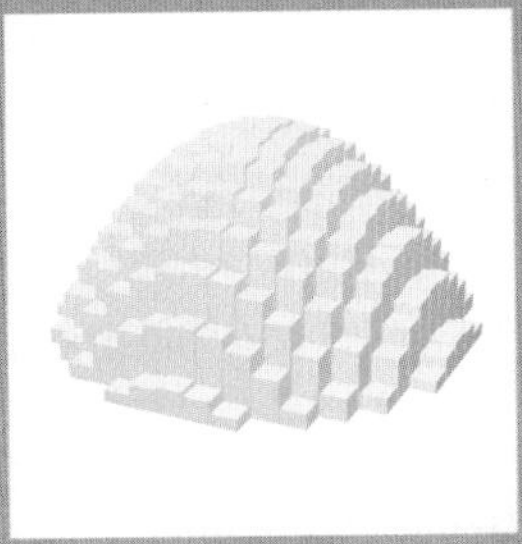

You can approximate the volume of a solid region by finding the sum of the volumes of representative rectangular prisms. As you increase the number of rectangular prisms, the approximation tends to become more and more accurate. In Chapter 14, you will learn how to use multiple integrals to find the volume of a solid region.

When sand is dry, you can pour it to form conical piles. When sand is wet, its physical properties change and you can use it to create a sand castle like the one shown in the photo. Why do you think wet sand is structurally stable?

Carl D. Jara, artcleveland.com, 10/1/02, artists: Carl D. Jara, Tom Morrison

Section 14.1 Iterated Integrals and Area in the Plane

- **Evaluate an iterated integral.**
- **Use an iterated integral to find the area of a plane region.**

Iterated Integrals

NOTE In Chapters 14 and 15, you will study several applications of integration involving functions of several variables. Chapter 14 is much like Chapter 7 in that it surveys the use of integration to find plane areas, volumes, surface areas, moments, and centers of mass.

In Chapter 13, you saw that it is meaningful to differentiate functions of several variables with respect to one variable while holding the other variables constant. You can *integrate* functions of several variables by a similar procedure. For example, if you are given the partial derivative

$$f_x(x, y) = 2xy$$

then, by considering y constant, you can integrate with respect to x to obtain

$$\begin{aligned} f(x, y) &= \int f_x(x, y)\, dx && \text{Integrate with respect to } x. \\ &= \int 2xy\, dx && \text{Hold } y \text{ constant.} \\ &= y\int 2x\, dx && \text{Factor out constant } y. \\ &= y(x^2) + C(y) && \text{Antiderivative of } 2x \text{ is } x^2. \\ &= x^2y + C(y). && C(y) \text{ is a function of } y. \end{aligned}$$

The "constant" of integration, $C(y)$, is a function of y. In other words, by integrating with respect to x, you are able to recover $f(x, y)$ only partially. The total recovery of a function of x and y from its partial derivatives is a topic you will study in Chapter 15. For now, we are more concerned with extending definite integrals to functions of several variables. For instance, by considering y constant, you can apply the Fundamental Theorem of Calculus to evaluate

$$\int_1^{2y} 2xy\, dx = x^2y\Big]_1^{2y} = (2y)^2y - (1)^2y = 4y^3 - y.$$

x is the variable of integration and y is fixed. — Replace x by the limits of integration. — The result is a function of y.

Similarly, you can integrate with respect to y by holding x fixed. Both procedures are summarized as follows.

$$\int_{h_1(y)}^{h_2(y)} f_x(x, y)\, dx = f(x, y)\Big]_{h_1(y)}^{h_2(y)} = f(h_2(y), y) - f(h_1(y), y) \qquad \text{With respect to } x$$

$$\int_{g_1(x)}^{g_2(x)} f_y(x, y)\, dy = f(x, y)\Big]_{g_1(x)}^{g_2(x)} = f(x, g_2(x)) - f(x, g_1(x)) \qquad \text{With respect to } y$$

Note that the variable of integration cannot appear in either limit of integration. For instance, it makes no sense to write

$$\int_0^x y\, dx.$$

EXAMPLE 1 Integrating with Respect to y

Evaluate $\int_1^x (2x^2y^{-2} + 2y)\, dy$.

Solution Considering x to be constant and integrating with respect to y produces

$$\int_1^x (2x^2y^{-2} + 2y)\, dy = \left[\frac{-2x^2}{y} + y^2\right]_1^x \qquad \text{Integrate with respect to } y.$$
$$= \left(\frac{-2x^2}{x} + x^2\right) - \left(\frac{-2x^2}{1} + 1\right)$$
$$= 3x^2 - 2x - 1.$$

Notice in Example 1 that the integral defines a function of x and can *itself* be integrated, as shown in the next example.

EXAMPLE 2 The Integral of an Integral

Evaluate $\int_1^2 \left[\int_1^x (2x^2y^{-2} + 2y)\, dy\right] dx$.

Solution Using the result of Example 1, you have

$$\int_1^2 \left[\int_1^x (2x^2y^{-2} + 2y)\, dy\right] dx = \int_1^2 (3x^2 - 2x - 1)\, dx$$
$$= \Big[x^3 - x^2 - x\Big]_1^2 \qquad \text{Integrate with respect to } x.$$
$$= 2 - (-1)$$
$$= 3.$$

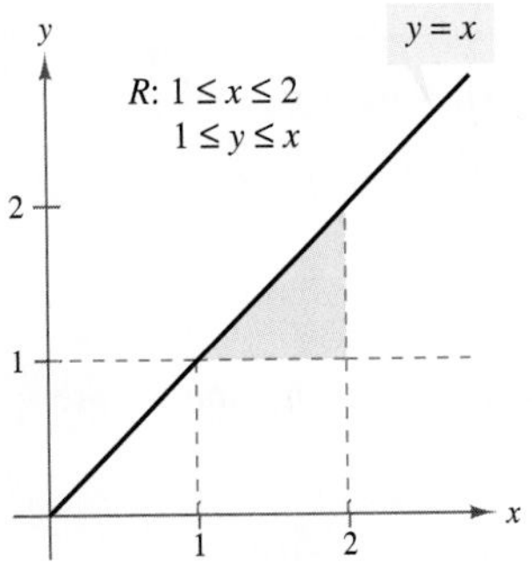

The region of integration for
$\int_1^2 \int_1^x f(x, y)\, dy\, dx$
Figure 14.1

The integral in Example 2 is an **iterated integral.** The brackets used in Example 2 are normally not written. Instead, iterated integrals are usually written simply as

$$\int_a^b \int_{g_1(x)}^{g_2(x)} f(x, y)\, dy\, dx \quad \text{and} \quad \int_c^d \int_{h_1(y)}^{h_2(y)} f(x, y)\, dx\, dy.$$

The **inside limits of integration** can be variable with respect to the outer variable of integration. However, the **outside limits of integration** *must be* constant with respect to both variables of integration. After performing the inside integration, you obtain a "standard" definite integral, and the second integration produces a real number. The limits of integration for an iterated integral identify two sets of boundary intervals for the variables. For instance, in Example 2, the outside limits indicate that x lies in the interval $1 \le x \le 2$ and the inside limits indicate that y lies in the interval $1 \le y \le x$. Together, these two intervals determine the **region of integration R** of the iterated integral, as shown in Figure 14.1.

Because an iterated integral is just a special type of definite integral—one in which the integrand is also an integral—you can use the properties of definite integrals to evaluate iterated integrals.

Area of a Plane Region

In the remainder of this section, you will take a new look at an old problem—that of finding the area of a plane region. Consider the plane region R bounded by $a \le x \le b$ and $g_1(x) \le y \le g_2(x)$, as shown in Figure 14.2. The area of R is given by the definite integral

$$\int_a^b [g_2(x) - g_1(x)]\, dx. \qquad \text{Area of } R$$

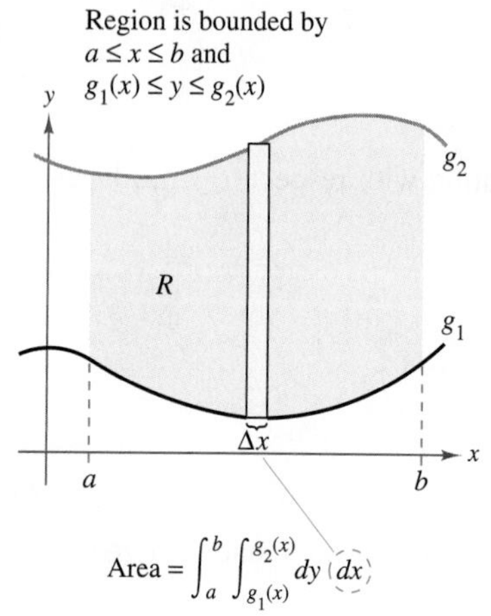

Vertically simple region
Figure 14.2

Using the Fundamental Theorem of Calculus, you can rewrite the integrand $g_2(x) - g_1(x)$ as a definite integral. Specifically, if you consider x to be fixed and let y vary from $g_1(x)$ to $g_2(x)$, you can write

$$\int_{g_1(x)}^{g_2(x)} dy = y\Big]_{g_1(x)}^{g_2(x)} = g_2(x) - g_1(x).$$

Combining these two integrals, you can write the area of the region R as an iterated integral

$$\int_a^b \int_{g_1(x)}^{g_2(x)} dy\, dx = \int_a^b y\Big]_{g_1(x)}^{g_2(x)} dx \qquad \text{Area of } R$$

$$= \int_a^b [g_2(x) - g_1(x)]\, dx.$$

Placing a representative rectangle in the region R helps determine both the order and the limits of integration. A vertical rectangle implies the order $dy\,dx$, with the inside limits corresponding to the upper and lower bounds of the rectangle, as shown in Figure 14.2. This type of region is called **vertically simple,** because the outside limits of integration represent the vertical lines $x = a$ and $x = b$.

Similarly, a horizontal rectangle implies the order $dx\,dy$, with the inside limits determined by the left and right bounds of the rectangle, as shown in Figure 14.3. This type of region is called **horizontally simple,** because the outside limits represent the horizontal lines $y = c$ and $y = d$. The iterated integrals used for these two types of simple regions are summarized as follows.

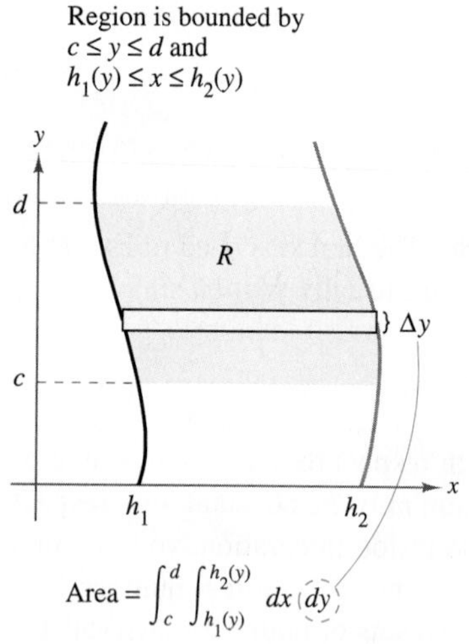

Horizontally simple region
Figure 14.3

Area of a Region in the Plane

1. If R is defined by $a \le x \le b$ and $g_1(x) \le y \le g_2(x)$, where g_1 and g_2 are continuous on $[a, b]$, then the area of R is given by

$$A = \int_a^b \int_{g_1(x)}^{g_2(x)} dy\, dx. \qquad \text{Figure 14.2 (vertically simple)}$$

2. If R is defined by $c \le y \le d$ and $h_1(y) \le x \le h_2(y)$, where h_1 and h_2 are continuous on $[c, d]$, then the area of R is given by

$$A = \int_c^d \int_{h_1(y)}^{h_2(y)} dx\, dy. \qquad \text{Figure 14.3 (horizontally simple)}$$

NOTE Be sure you see that the order of integration of these two integrals is different—the order $dy\,dx$ corresponds to a vertically simple region, and the order $dx\,dy$ corresponds to a horizontally simple region.

If all four limits of integration happen to be constants, the region of integration is rectangular, as shown in Example 3.

EXAMPLE 3 The Area of a Rectangular Region

Use an iterated integral to represent the area of the rectangle shown in Figure 14.4.

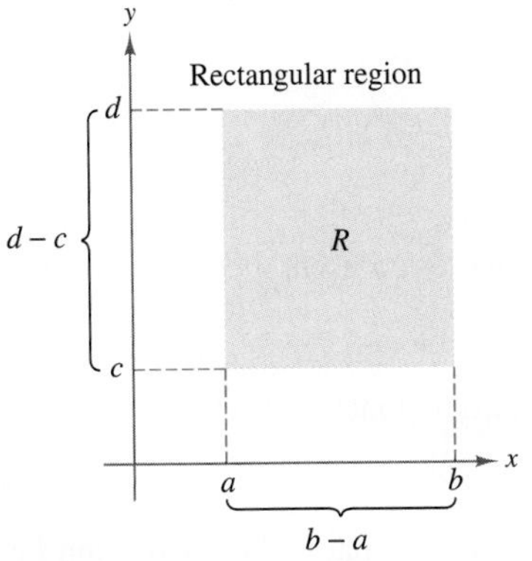

Figure 14.4

Solution The region shown in Figure 14.4 is both vertically simple and horizontally simple, so you can use either order of integration. By choosing the order *dy dx*, you obtain the following.

$$\int_a^b \int_c^d dy\,dx = \int_a^b \Big[y \Big]_c^d dx \qquad \text{Integrate with respect to } y.$$

$$= \int_a^b (d - c)\,dx$$

$$= \Big[(d - c)x \Big]_a^b \qquad \text{Integrate with respect to } x.$$

$$= (d - c)(b - a)$$

Notice that this answer is consistent with what you know from geometry.

EXAMPLE 4 Finding Area by an Iterated Integral

Use an iterated integral to find the area of the region bounded by the graphs of

$f(x) = \sin x$ Sine curve forms upper boundary.

$g(x) = \cos x$ Cosine curve forms lower boundary.

between $x = \pi/4$ and $x = 5\pi/4$.

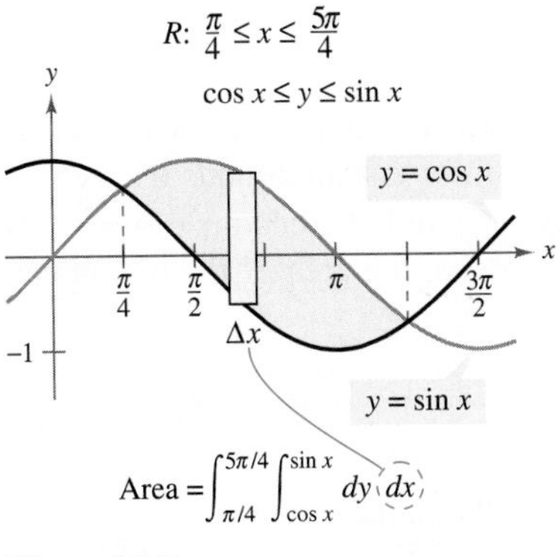

Figure 14.5

Solution Because f and g are given as functions of x, a vertical representative rectangle is convenient, and you can choose *dy dx* as the order of integration, as shown in Figure 14.5. The outside limits of integration are $\pi/4 \le x \le 5\pi/4$. Moreover, because the rectangle is bounded above by $f(x) = \sin x$ and below by $g(x) = \cos x$, you have

$$\text{Area of } R = \int_{\pi/4}^{5\pi/4} \int_{\cos x}^{\sin x} dy\,dx$$

$$= \int_{\pi/4}^{5\pi/4} \Big[y \Big]_{\cos x}^{\sin x} dx \qquad \text{Integrate with respect to } y.$$

$$= \int_{\pi/4}^{5\pi/4} (\sin x - \cos x)\,dx$$

$$= \Big[-\cos x - \sin x \Big]_{\pi/4}^{5\pi/4} \qquad \text{Integrate with respect to } x.$$

$$= 2\sqrt{2}.$$

NOTE The region of integration of an iterated integral need not have any straight lines as boundaries. For instance, the region of integration shown in Figure 14.5 is *vertically simple* even though it has no vertical lines as left and right boundaries. The quality that makes the region vertically simple is that it is bounded above and below by the graphs of *functions of x*.

One order of integration will often produce a simpler integration problem than the other order. For instance, try reworking Example 4 with the order $dx\,dy$—you may be surprised to see that the task is formidable. However, if you succeed, you will see that the answer is the same. In other words, the order of integration affects the ease of integration, but not the value of the integral.

EXAMPLE 5 Comparing Different Orders of Integration

Sketch the region whose area is represented by the integral

$$\int_0^2 \int_{y^2}^4 dx\,dy.$$

Then find another iterated integral using the order $dy\,dx$ to represent the same area and show that both integrals yield the same value.

Solution From the given limits of integration, you know that

$$y^2 \le x \le 4 \qquad \text{Inner limits of integration}$$

which means that the region R is bounded on the left by the parabola $x = y^2$ and on the right by the line $x = 4$. Furthermore, because

$$0 \le y \le 2 \qquad \text{Outer limits of integration}$$

you know that R is bounded below by the x-axis, as shown in Figure 14.6(a). The value of this integral is

$$\int_0^2 \int_{y^2}^4 dx\,dy = \int_0^2 x \Big]_{y^2}^4 dy \qquad \text{Integrate with respect to } x.$$

$$= \int_0^2 (4 - y^2)\,dy$$

$$= \left[4y - \frac{y^3}{3}\right]_0^2 = \frac{16}{3}. \qquad \text{Integrate with respect to } y.$$

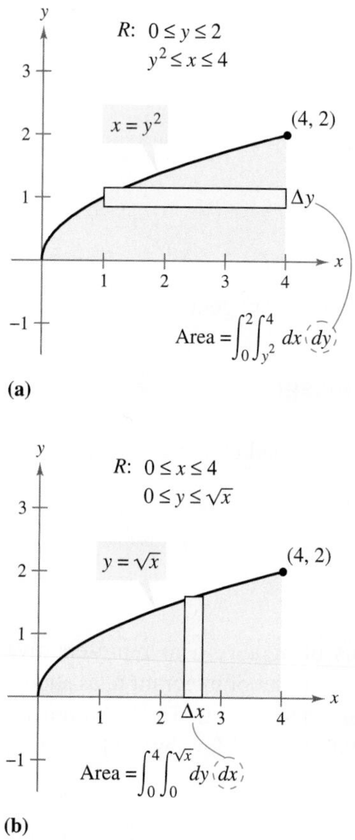

Figure 14.6

To change the order of integration to $dy\,dx$, place a vertical rectangle in the region, as shown in Figure 14.6(b). From this you can see that the constant bounds $0 \le x \le 4$ serve as the outer limits of integration. By solving for y in the equation $x = y^2$, you can conclude that the inner bounds are $0 \le y \le \sqrt{x}$. So, the area of the region can also be represented by

$$\int_0^4 \int_0^{\sqrt{x}} dy\,dx.$$

By evaluating this integral, you can see that it has the same value as the original integral.

$$\int_0^4 \int_0^{\sqrt{x}} dy\,dx = \int_0^4 y \Big]_0^{\sqrt{x}} dx \qquad \text{Integrate with respect to } y.$$

$$= \int_0^4 \sqrt{x}\,dx$$

$$= \frac{2}{3}x^{3/2}\Big]_0^4 = \frac{16}{3} \qquad \text{Integrate with respect to } x.$$

indicates that in the HM mathSpace® *CD-ROM and the online* Eduspace® *system for this text, you will find an Open Exploration, which further explores this example using the computer algebra systems* Maple, Mathcad, Mathematica, *and* Derive.

Sometimes it is not possible to calculate the area of a region with a single iterated integral. In these cases you can divide the region into subregions such that the area of each subregion can be calculated by an iterated integral. The total area is then the sum of the iterated integrals.

TECHNOLOGY Some computer software can perform symbolic integration for integrals such as those in Example 6. If you have access to such software, use it to evaluate the integrals in the exercises and examples given in this section.

EXAMPLE 6 An Area Represented by Two Iterated Integrals

Find the area of the region R that lies below the parabola

$y = 4x - x^2$ Parabola forms upper boundary.

above the x-axis, and above the line

$y = -3x + 6.$ Line and x-axis form lower boundary.

Solution Begin by dividing R into the two subregions R_1 and R_2 shown in Figure 14.7.

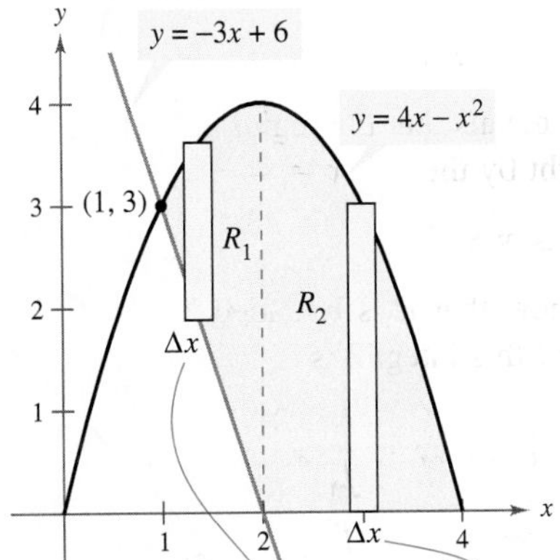

$$\text{Area} = \int_1^2 \int_{-3x+6}^{4x-x^2} dy\,dx + \int_2^4 \int_0^{4x-x^2} dy\,dx$$

Figure 14.7

In both regions, it is convenient to use vertical rectangles, and you have

$$\begin{aligned}
\text{Area} &= \int_1^2 \int_{-3x+6}^{4x-x^2} dy\,dx + \int_2^4 \int_0^{4x-x^2} dy\,dx \\
&= \int_1^2 (4x - x^2 + 3x - 6)\,dx + \int_2^4 (4x - x^2)\,dx \\
&= \left[\frac{7x^2}{2} - \frac{x^3}{3} - 6x\right]_1^2 + \left[2x^2 - \frac{x^3}{3}\right]_2^4 \\
&= \left(14 - \frac{8}{3} - 12 - \frac{7}{2} + \frac{1}{3} + 6\right) + \left(32 - \frac{64}{3} - 8 + \frac{8}{3}\right) = \frac{15}{2}.
\end{aligned}$$

The area of the region is $15/2$ square units. Try checking this using the procedure for finding the area between two curves, as presented in Section 7.1.

At this point you may be wondering why you would need iterated integrals. After all, you already know how to use conventional integration to find the area of a region in the plane. (For instance, compare the solution of Example 4 in this section with that given in Example 3 in Section 7.1.) The need for iterated integrals will become clear in the next section. In this section, primary attention is given to procedures for finding the limits of integration of the region of an iterated integral, and the following exercise set is designed to develop skill in this important procedure.

NOTE In Examples 3 to 6, be sure you see the benefit of sketching the region of integration. You should develop the habit of making sketches to help determine the limits of integration for all iterated integrals in this chapter.

Exercises for Section 14.1

See www.CalcChat.com for worked-out solutions to odd-numbered exercises.

In Exercises 1–10, evaluate the integral.

1. $\int_0^x (2x - y)\,dy$

2. $\int_x^{x^2} \frac{y}{x}\,dy$

3. $\int_1^{2y} \frac{y}{x}\,dx, \quad y > 0$

4. $\int_0^{\cos y} y\,dx$

5. $\int_0^{\sqrt{4-x^2}} x^2 y\,dy$

6. $\int_{x^3}^{\sqrt{x}} (x^2 + 3y^2)\,dy$

7. $\int_{e^y}^{y} \frac{y \ln x}{x}\,dx, \quad y > 0$

8. $\int_{-\sqrt{1-y^2}}^{\sqrt{1-y^2}} (x^2 + y^2)\,dx$

9. $\int_0^{x^3} ye^{-y/x}\,dy$

10. $\int_y^{\pi/2} \sin^3 x \cos y\,dx$

In Exercises 11–24, evaluate the iterated integral.

11. $\int_0^1 \int_0^2 (x + y)\,dy\,dx$

12. $\int_{-1}^1 \int_{-2}^2 (x^2 - y^2)\,dy\,dx$

13. $\int_0^{\pi} \int_0^{\sin x} (1 + \cos x)\,dy\,dx$

14. $\int_1^4 \int_1^{\sqrt{x}} 2ye^{-x}\,dy\,dx$

15. $\int_0^1 \int_0^x \sqrt{1 - x^2}\,dy\,dx$

16. $\int_{-4}^4 \int_0^{x^2} \sqrt{64 - x^3}\,dy\,dx$

17. $\int_1^2 \int_0^4 (x^2 - 2y^2 + 1)\,dx\,dy$

18. $\int_0^2 \int_y^{2y} (10 + 2x^2 + 2y^2)\,dx\,dy$

19. $\int_0^1 \int_0^{\sqrt{1-y^2}} (x + y)\,dx\,dy$

20. $\int_0^2 \int_{3y^2-6y}^{2y-y^2} 3y\,dx\,dy$

21. $\int_0^2 \int_0^{\sqrt{4-y^2}} \frac{2}{\sqrt{4 - y^2}}\,dx\,dy$

22. $\int_0^{\pi/2} \int_0^{2\cos\theta} r\,dr\,d\theta$

23. $\int_0^{\pi/2} \int_0^{\sin\theta} \theta r\,dr\,d\theta$

24. $\int_0^{\pi/4} \int_0^{\cos\theta} 3r^2 \sin\theta\,dr\,d\theta$

In Exercises 25–28, evaluate the improper iterated integral.

25. $\int_1^{\infty} \int_0^{1/x} y\,dy\,dx$

26. $\int_0^3 \int_0^{\infty} \frac{x^2}{1 + y^2}\,dy\,dx$

27. $\int_1^{\infty} \int_1^{\infty} \frac{1}{xy}\,dx\,dy$

28. $\int_0^{\infty} \int_0^{\infty} xye^{-(x^2+y^2)}\,dx\,dy$

In Exercises 29–34, use an iterated integral to find the area of the region.

29.
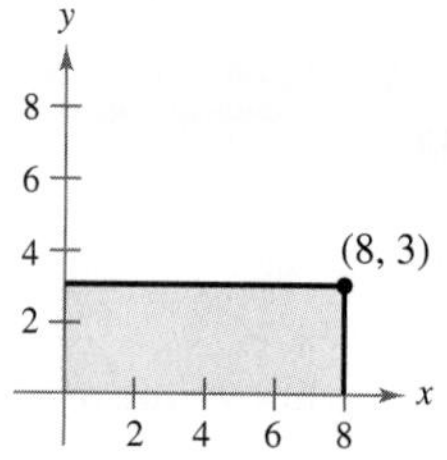

30.
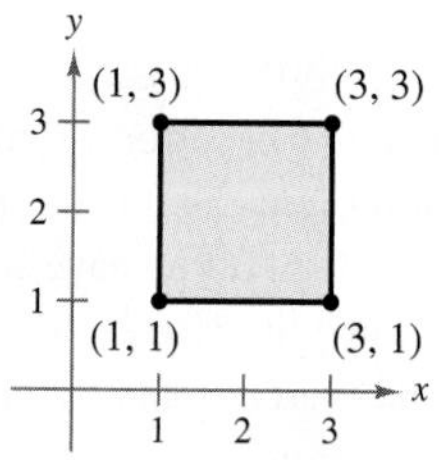

31.
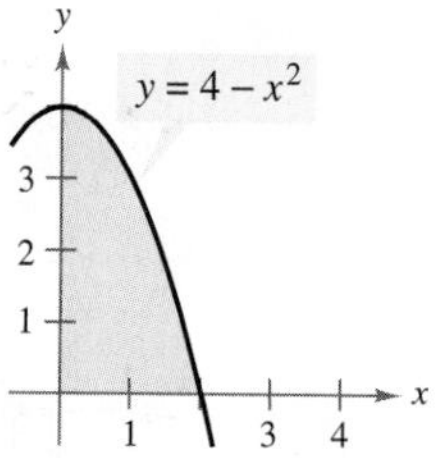

32.
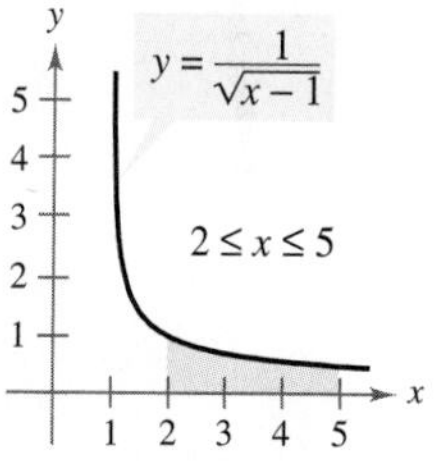

33.
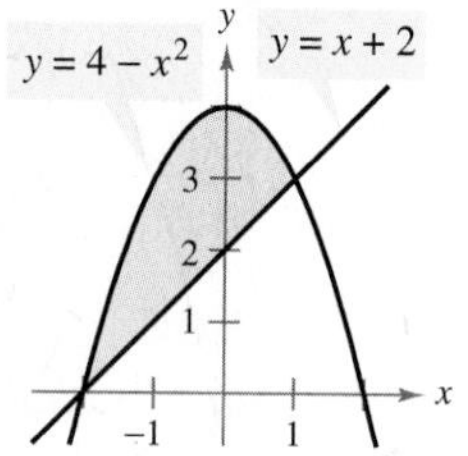

34.
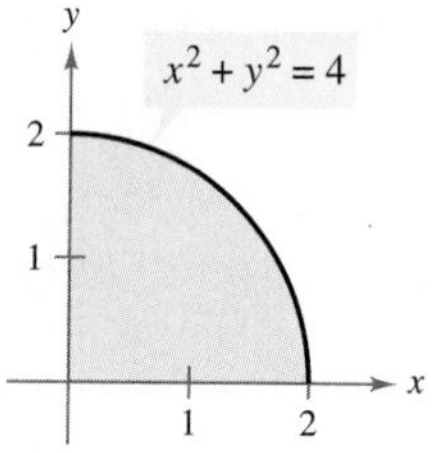

In Exercises 35–40, use an iterated integral to find the area of the region bounded by the graphs of the equations.

35. $\sqrt{x} + \sqrt{y} = 2, \quad x = 0, \quad y = 0$

36. $y = x^{3/2}, \quad y = 2x$

37. $2x - 3y = 0, \quad x + y = 5, \quad y = 0$

38. $xy = 9, \quad y = x, \quad y = 0, \quad x = 9$

39. $\frac{x^2}{a^2} + \frac{y^2}{b^2} = 1$

40. $y = x, \quad y = 2x, \quad x = 2$

In Exercises 41–48, sketch the region *R* of integration and switch the order of integration.

41. $\int_0^4 \int_0^y f(x, y)\,dx\,dy$

42. $\int_0^4 \int_{\sqrt{y}}^2 f(x, y)\,dx\,dy$

43. $\int_{-2}^2 \int_0^{\sqrt{4-x^2}} f(x, y)\,dy\,dx$

44. $\int_0^2 \int_0^{4-x^2} f(x, y)\,dy\,dx$

45. $\int_1^{10} \int_0^{\ln y} f(x, y)\,dx\,dy$

46. $\int_{-1}^2 \int_0^{e^{-x}} f(x, y)\,dy\,dx$

47. $\int_{-1}^1 \int_{x^2}^1 f(x, y)\,dy\,dx$

48. $\int_{-\pi/2}^{\pi/2} \int_0^{\cos x} f(x, y)\,dy\,dx$

In Exercises 49–58, sketch the region R whose area is given by the iterated integral. Then switch the order of integration and show that both orders yield the same area.

49. $\int_0^1 \int_0^2 dy\, dx$

50. $\int_1^2 \int_2^4 dx\, dy$

51. $\int_0^1 \int_{-\sqrt{1-y^2}}^{\sqrt{1-y^2}} dx\, dy$

52. $\int_{-2}^2 \int_{-\sqrt{4-x^2}}^{\sqrt{4-x^2}} dy\, dx$

53. $\int_0^2 \int_0^x dy\, dx + \int_2^4 \int_0^{4-x} dy\, dx$

54. $\int_0^4 \int_0^{x/2} dy\, dx + \int_4^6 \int_0^{6-x} dy\, dx$

55. $\int_0^2 \int_{x/2}^1 dy\, dx$

56. $\int_0^9 \int_{\sqrt{x}}^3 dy\, dx$

57. $\int_0^1 \int_{y^2}^{\sqrt[3]{y}} dx\, dy$

58. $\int_{-2}^2 \int_0^{4-y^2} dx\, dy$

Think About It **In Exercises 59 and 60, give a geometric argument for the given equality. Verify the equality analytically.**

59. $\int_0^5 \int_x^{\sqrt{50-x^2}} x^2y^2\, dy\, dx =$

$\int_0^5 \int_0^y x^2y^2\, dx\, dy + \int_5^{5\sqrt{2}} \int_0^{\sqrt{50-y^2}} x^2y^2\, dx\, dy$

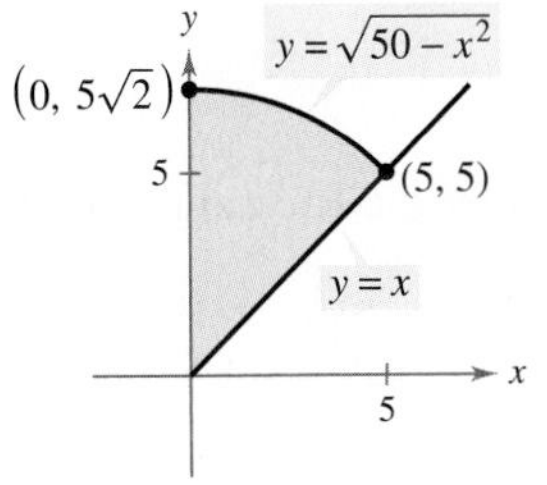

60. $\int_0^2 \int_{x^2}^{2x} x \sin y\, dy\, dx = \int_0^4 \int_{y/2}^{\sqrt{y}} x \sin y\, dx\, dy$

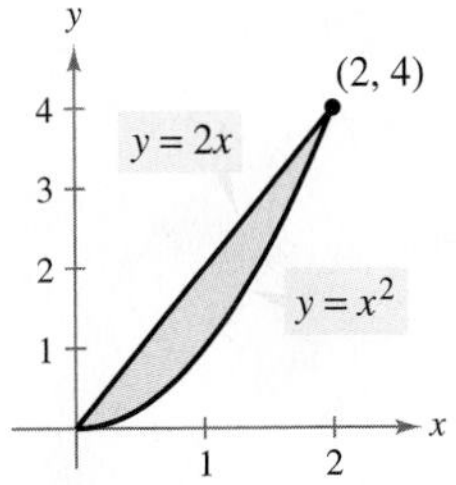

In Exercises 61–64, evaluate the iterated integral. (Note that it is necessary to switch the order of integration.)

61. $\int_0^2 \int_x^2 x\sqrt{1 + y^3}\, dy\, dx$

62. $\int_0^2 \int_x^2 e^{-y^2}\, dy\, dx$

63. $\int_0^1 \int_y^1 \sin x^2\, dx\, dy$

64. $\int_0^2 \int_{y^2}^4 \sqrt{x} \sin x\, dx\, dy$

In Exercises 65–68, use a computer algebra system to evaluate the iterated integral.

65. $\int_0^2 \int_{x^2}^{2x} (x^3 + 3y^2)\, dy\, dx$

66. $\int_0^1 \int_y^{2y} \sin(x + y)\, dx\, dy$

67. $\int_0^4 \int_0^y \dfrac{2}{(x + 1)(y + 1)}\, dx\, dy$

68. $\int_0^a \int_0^{a-x} (x^2 + y^2)\, dy\, dx$

In Exercises 69 and 70, (a) sketch the region of integration, (b) switch the order of integration, and (c) use a computer algebra system to show that both orders yield the same value.

69. $\int_0^2 \int_{y^3}^{4\sqrt{2y}} (x^2y - xy^2)\, dx\, dy$

70. $\int_0^2 \int_{\sqrt{4-x^2}}^{4-x^2/4} \dfrac{xy}{x^2 + y^2 + 1}\, dy\, dx$

In Exercises 71–74, use a computer algebra system to approximate the iterated integral.

71. $\int_0^2 \int_0^{4-x^2} e^{xy}\, dy\, dx$

72. $\int_0^2 \int_x^2 \sqrt{16 - x^3 - y^3}\, dy\, dx$

73. $\int_0^{2\pi} \int_0^{1+\cos\theta} 6r^2 \cos\theta\, dr\, d\theta$

74. $\int_0^{\pi/2} \int_0^{1+\sin\theta} 15\theta r\, dr\, d\theta$

Writing About Concepts

75. Explain what is meant by an iterated integral. How is it evaluated?

76. Describe regions that are vertically simple and regions that are horizontally simple.

77. Give a geometric description of the region of integration if the inside and outside limits of integration are constants.

78. Explain why it is sometimes an advantage to change the order of integration.

True or False? **In Exercises 79 and 80, determine whether the statement is true or false. If it is false, explain why or give an example that shows it is false.**

79. $\int_a^b \int_c^d f(x, y)\, dy\, dx = \int_c^d \int_a^b f(x, y)\, dx\, dy$

80. $\int_0^1 \int_0^x f(x, y)\, dy\, dx = \int_0^1 \int_0^y f(x, y)\, dx\, dy$

Section 14.2 Double Integrals and Volume

- **Use a double integral to represent the volume of a solid region.**
- **Use properties of double integrals.**
- **Evaluate a double integral as an iterated integral.**

Double Integrals and Volume of a Solid Region

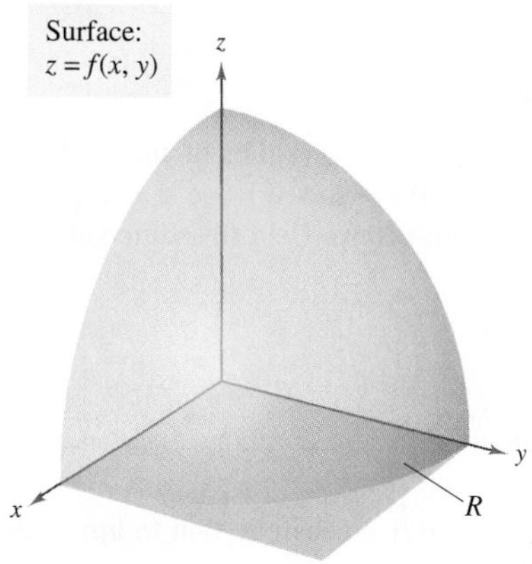

Figure 14.8

You already know that a definite integral over an *interval* uses a limit process to assign measure to quantities such as area, volume, arc length, and mass. In this section, you will use a similar process to define the **double integral** of a function of two variables over a *region in the plane*.

Consider a continuous function f such that $f(x, y) \geq 0$ for all (x, y) in a region R in the xy-plane. The goal is to find the volume of the solid region lying between the surface given by

$$z = f(x, y) \qquad \text{Surface lying above the } xy\text{-plane}$$

and the xy-plane, as shown in Figure 14.8. You can begin by superimposing a rectangular grid over the region, as shown in Figure 14.9. The rectangles lying entirely within R form an **inner partition** Δ, whose **norm** $\|\Delta\|$ is defined as the length of the longest diagonal of the n rectangles. Next, choose a point (x_i, y_i) in each rectangle and form the rectangular prism whose height is $f(x_i, y_i)$, as shown in Figure 14.10. Because the area of the ith rectangle is

$$\Delta A_i \qquad \text{Area of } i\text{th rectangle}$$

it follows that the volume of the ith prism is

$$f(x_i, y_i)\,\Delta A_i \qquad \text{Volume of } i\text{th prism}$$

and you can approximate the volume of the solid region by the Riemann sum of the volumes of all n prisms,

$$\sum_{i=1}^{n} f(x_i, y_i)\,\Delta A_i \qquad \text{Riemann sum}$$

as shown in Figure 14.11. This approximation can be improved by tightening the mesh of the grid to form smaller and smaller rectangles, as shown in Example 1.

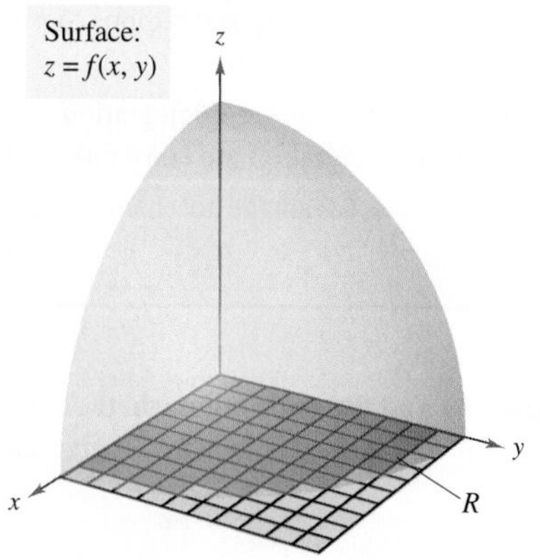

The rectangles lying within R form an inner partition of R.
Figure 14.9

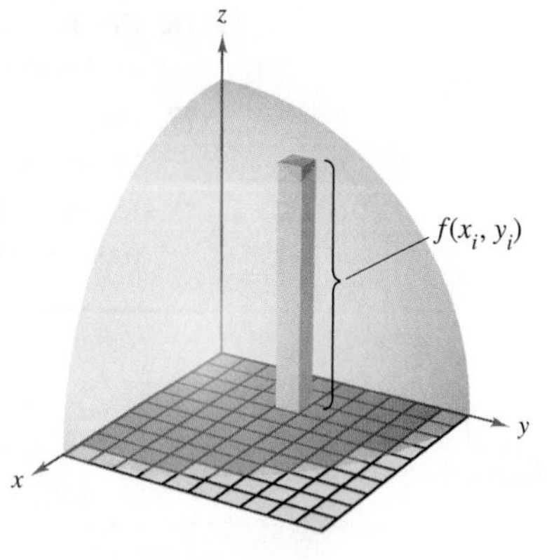

Rectangular prism whose base has an area of ΔA_i and whose height is $f(x_i, y_i)$
Figure 14.10

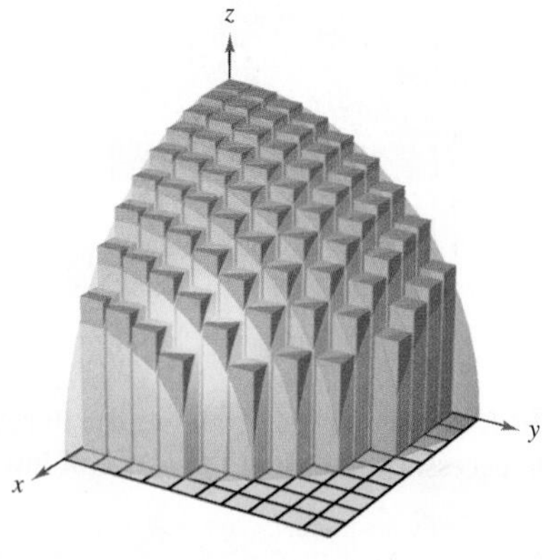

Volume approximated by rectangular prisms
Figure 14.11

EXAMPLE 1 Approximating the Volume of a Solid

Approximate the volume of the solid lying between the paraboloid

$$f(x, y) = 1 - \frac{1}{2}x^2 - \frac{1}{2}y^2$$

and the square region R given by $0 \le x \le 1, 0 \le y \le 1$. Use a partition made up of squares whose sides have a length of $\frac{1}{4}$.

Solution Begin by forming the specified partition of R. For this partition, it is convenient to choose the centers of the subregions as the points at which to evaluate $f(x, y)$.

$$\begin{array}{cccc} \left(\frac{1}{8}, \frac{1}{8}\right) & \left(\frac{1}{8}, \frac{3}{8}\right) & \left(\frac{1}{8}, \frac{5}{8}\right) & \left(\frac{1}{8}, \frac{7}{8}\right) \\ \left(\frac{3}{8}, \frac{1}{8}\right) & \left(\frac{3}{8}, \frac{3}{8}\right) & \left(\frac{3}{8}, \frac{5}{8}\right) & \left(\frac{3}{8}, \frac{7}{8}\right) \\ \left(\frac{5}{8}, \frac{1}{8}\right) & \left(\frac{5}{8}, \frac{3}{8}\right) & \left(\frac{5}{8}, \frac{5}{8}\right) & \left(\frac{5}{8}, \frac{7}{8}\right) \\ \left(\frac{7}{8}, \frac{1}{8}\right) & \left(\frac{7}{8}, \frac{3}{8}\right) & \left(\frac{7}{8}, \frac{5}{8}\right) & \left(\frac{7}{8}, \frac{7}{8}\right) \end{array}$$

Because the area of each square is $\Delta A_i = \frac{1}{16}$, you can approximate the volume by the sum

$$\sum_{i=1}^{16} f(x_i, y_i)\,\Delta A_i = \sum_{i=1}^{16} \left(1 - \frac{1}{2}x_i^2 - \frac{1}{2}y_i^2\right)\left(\frac{1}{16}\right)$$

$$\approx 0.672.$$

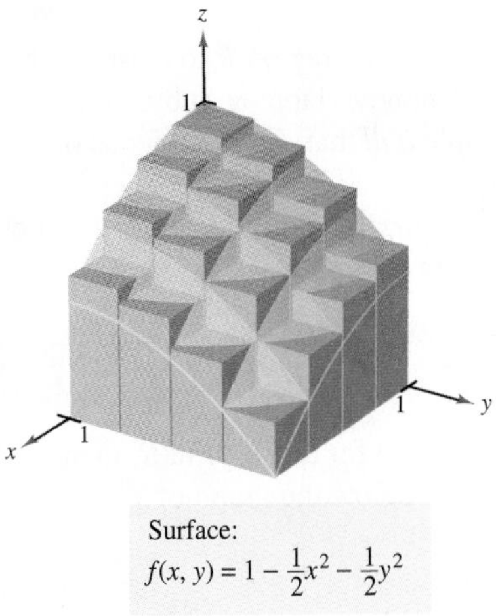

Figure 14.12

This approximation is shown graphically in Figure 14.12. The exact volume of the solid is $\frac{2}{3}$ (see Example 2). You can obtain a better approximation by using a finer partition. For example, with a partition of squares with sides of length $\frac{1}{10}$, the approximation is 0.668.

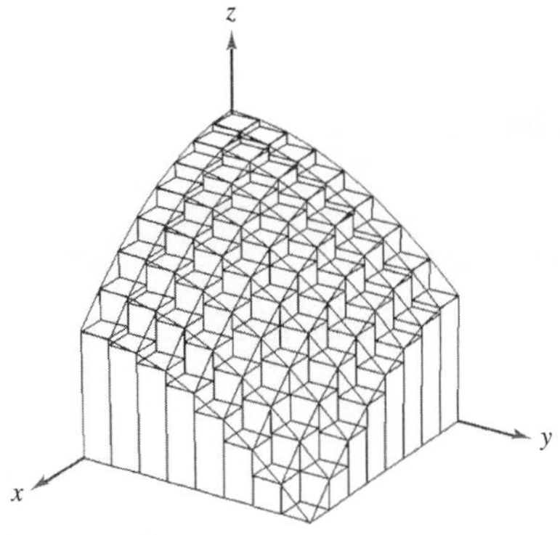

Figure 14.13

> **TECHNOLOGY** Some three-dimensional graphing utilities are capable of graphing figures such as that shown in Figure 14.12. For instance, the graph shown in Figure 14.13 was drawn with a computer program. In this graph, note that each of the rectangular prisms lies within the solid region.

In Example 1, note that by using finer partitions, you obtain better approximations of the volume. This observation suggests that you could obtain the exact volume by taking a limit. That is,

$$\text{Volume} = \lim_{\|\Delta\| \to 0} \sum_{i=1}^{n} f(x_i, y_i)\,\Delta A_i.$$

The precise meaning of this limit is that the limit is equal to L if for every $\varepsilon > 0$ there exists a $\delta > 0$ such that

$$\left| L - \sum_{i=1}^{n} f(x_i, y_i)\,\Delta A_i \right| < \varepsilon$$

for all partitions Δ of the plane region R (that satisfy $\|\Delta\| < \delta$) and for all possible choices of x_i and y_i in the ith region.

Using the limit of a Riemann sum to define volume is a special case of using the limit to define a **double integral.** The general case, however, does not require that the function be positive or continuous.

EXPLORATION

The entries in the table represent the depth (in 10-yard units) of earth at the center of each square in the figure below.

x \ y	1	2	3
1	10	9	7
2	7	7	4
3	5	5	4
4	4	5	3

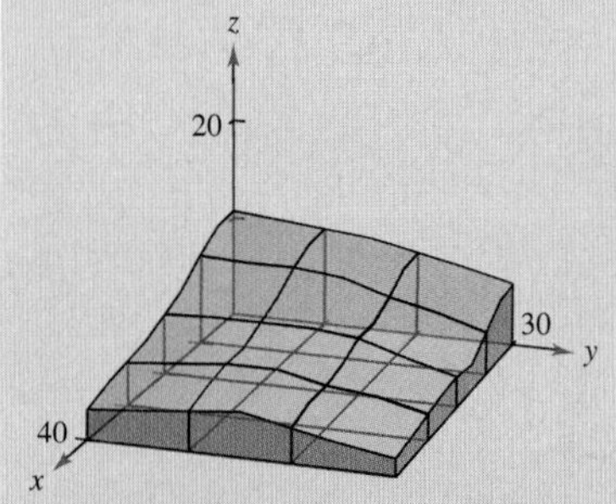

Approximate the number of cubic yards of earth in the first octant. (This exploration was submitted by Robert Vojack, Ridgewood High School, Ridgewood, NJ.)

Definition of Double Integral

If f is defined on a closed, bounded region R in the xy-plane, then the **double integral of f over R** is given by

$$\int_R\int f(x, y)\, dA = \lim_{\|\Delta\|\to 0} \sum_{i=1}^{n} f(x_i, y_i)\, \Delta A_i$$

provided the limit exists. If the limit exists, then f is **integrable** over R.

NOTE Having defined a double integral, you will see that a definite integral is occasionally referred to as a **single integral.**

Sufficient conditions for the double integral of f on the region R to exist are that R can be written as a union of a finite number of nonoverlapping subregions (see Figure 14.14) that are vertically or horizontally simple *and* that f is continuous on the region R.

A double integral can be used to find the volume of a solid region that lies between the xy-plane and the surface given by $z = f(x, y)$.

Volume of a Solid Region

If f is integrable over a plane region R and $f(x, y) \ge 0$ for all (x, y) in R, then the volume of the solid region that lies above R and below the graph of f is defined as

$$V = \int_R\int f(x, y)\, dA.$$

Properties of Double Integrals

Double integrals share many properties of single integrals.

THEOREM 14.1 Properties of Double Integrals

Let f and g be continuous over a closed, bounded plane region R, and let c be a constant.

1. $\displaystyle\int_R\int cf(x, y)\, dA = c\int_R\int f(x, y)\, dA$
2. $\displaystyle\int_R\int [f(x, y) \pm g(x, y)]\, dA = \int_R\int f(x, y)\, dA \pm \int_R\int g(x, y)\, dA$
3. $\displaystyle\int_R\int f(x, y)\, dA \ge 0, \quad \text{if } f(x, y) \ge 0$
4. $\displaystyle\int_R\int f(x, y)\, dA \ge \int_R\int g(x, y)\, dA, \quad \text{if } f(x, y) \ge g(x, y)$
5. $\displaystyle\int_R\int f(x, y)\, dA = \int_{R_1}\int f(x, y)\, dA + \int_{R_2}\int f(x, y)\, dA$, where R is the union of two nonoverlapping subregions R_1 and R_2.

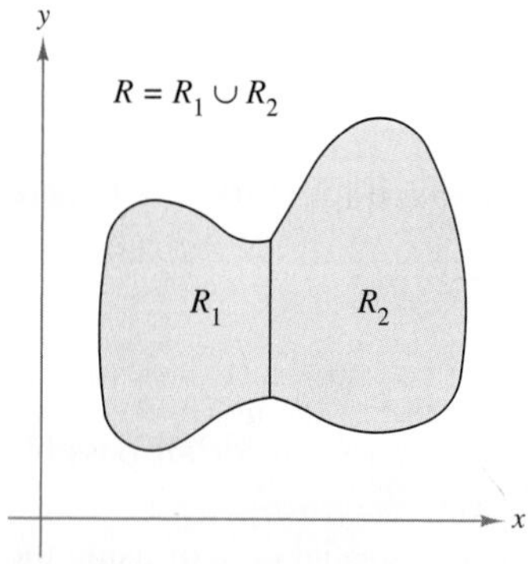

Two regions are nonoverlapping if their intersection is a set that has an area of 0. In this figure, the area of the line segment that is common to R_1 and R_2 is 0.
Figure 14.14

Evaluation of Double Integrals

Normally, the first step in evaluating a double integral is to rewrite it as an iterated integral. To show how this is done, a geometric model of a double integral is used as the volume of a solid.

Consider the solid region bounded by the plane $z = f(x, y) = 2 - x - 2y$ and the three coordinate planes, as shown in Figure 14.15. Each vertical cross section taken parallel to the yz-plane is a triangular region whose base has a length of $y = (2 - x)/2$ and whose height is $z = 2 - x$. This implies that for a fixed value of x, the area of the triangular cross section is

$$A(x) = \frac{1}{2}(\text{base})(\text{height}) = \frac{1}{2}\left(\frac{2 - x}{2}\right)(2 - x) = \frac{(2 - x)^2}{4}.$$

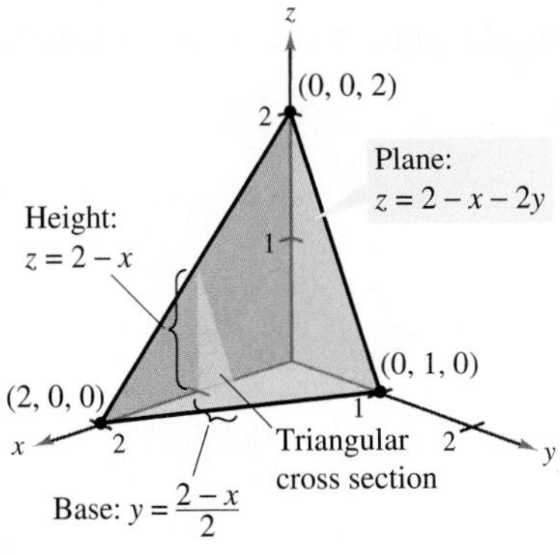

Figure 14.15

By the formula for the volume of a solid with known cross sections (Section 7.2), the volume of the solid is

$$\begin{aligned}\text{Volume} &= \int_a^b A(x)\,dx \\ &= \int_0^2 \frac{(2 - x)^2}{4}\,dx \\ &= -\frac{(2 - x)^3}{12}\Big]_0^2 = \frac{2}{3}.\end{aligned}$$

This procedure works no matter how $A(x)$ is obtained. In particular, you can find $A(x)$ by integration, as shown in Figure 14.16. That is, you consider x to be constant, and integrate $z = 2 - x - 2y$ from 0 to $(2 - x)/2$ to obtain

$$\begin{aligned}A(x) &= \int_0^{(2-x)/2} (2 - x - 2y)\,dy \\ &= \Big[(2 - x)y - y^2\Big]_0^{(2-x)/2} \\ &= \frac{(2 - x)^2}{4}.\end{aligned}$$

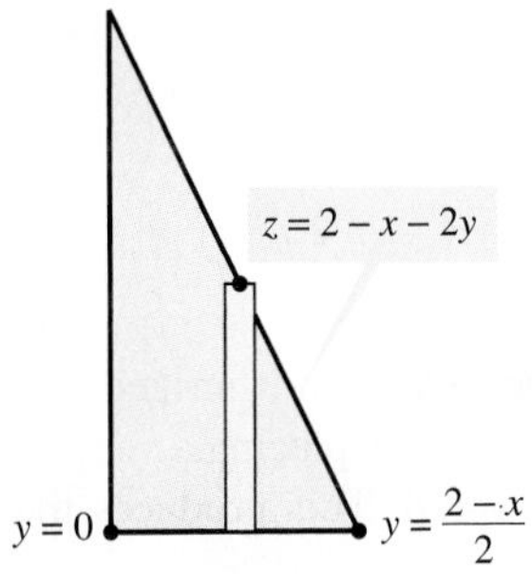

Triangular cross section
Figure 14.16

Combining these results, you have the *iterated integral*

$$\text{Volume} = \iint_R f(x, y)\,dA = \int_0^2 \int_0^{(2-x)/2} (2 - x - 2y)\,dy\,dx.$$

To understand this procedure better, it helps to imagine the integration as two sweeping motions. For the inner integration, a vertical line sweeps out the area of a cross section. For the outer integration, the triangular cross section sweeps out the volume, as shown in Figure 14.17.

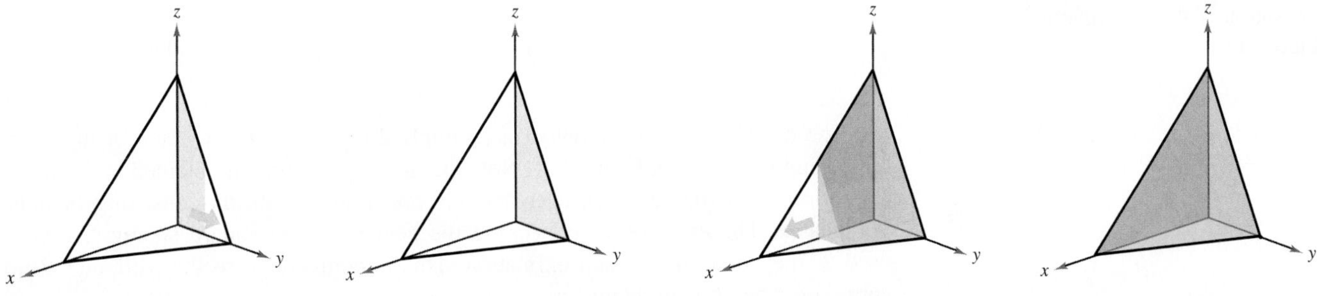

Integrate with respect to y to obtain the area of the cross section.
Integrate with respect to x to obtain the volume of the solid.
Figure 14.17

The following theorem was proved by the Italian mathematician Guido Fubini (1879–1943). The theorem states that if R is a vertically or horizontally simple region and f is continuous on R, the double integral of f on R is equal to an iterated integral.

THEOREM 14.2 Fubini's Theorem

Let f be continuous on a plane region R.

1. If R is defined by $a \le x \le b$ and $g_1(x) \le y \le g_2(x)$, where g_1 and g_2 are continuous on $[a, b]$, then
$$\int_R\int f(x, y)\, dA = \int_a^b \int_{g_1(x)}^{g_2(x)} f(x, y)\, dy\, dx.$$
2. If R is defined by $c \le y \le d$ and $h_1(y) \le x \le h_2(y)$, where h_1 and h_2 are continuous on $[c, d]$, then
$$\int_R\int f(x, y)\, dA = \int_c^d \int_{h_1(y)}^{h_2(y)} f(x, y)\, dx\, dy.$$

EXAMPLE 2 Evaluating a Double Integral as an Iterated Integral

Evaluate

$$\int_R\int \left(1 - \frac{1}{2}x^2 - \frac{1}{2}y^2\right) dA$$

where R is the region given by $0 \le x \le 1, 0 \le y \le 1$.

Solution Because the region R is a square, it is both vertically and horizontally simple, and you can use either order of integration. Choose $dy\, dx$ by placing a vertical representative rectangle in the region, as shown in Figure 14.18. This produces the following.

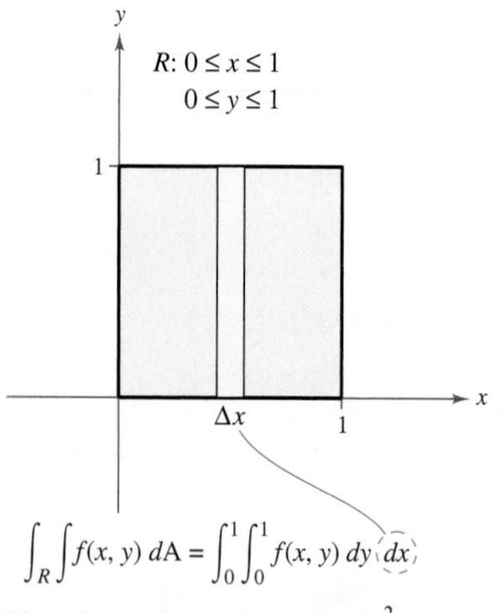

The volume of the solid region is $\frac{2}{3}$.
Figure 14.18

$$\begin{aligned}
\int_R\int \left(1 - \frac{1}{2}x^2 - \frac{1}{2}y^2\right) dA &= \int_0^1\int_0^1 \left(1 - \frac{1}{2}x^2 - \frac{1}{2}y^2\right) dy\, dx \\
&= \int_0^1 \left[\left(1 - \frac{1}{2}x^2\right)y - \frac{y^3}{6}\right]_0^1 dx \\
&= \int_0^1 \left(\frac{5}{6} - \frac{1}{2}x^2\right) dx \\
&= \left[\frac{5}{6}x - \frac{x^3}{6}\right]_0^1 \\
&= \frac{2}{3}
\end{aligned}$$

The double integral evaluated in Example 2 represents the volume of the solid region approximated in Example 1. Note that the approximation obtained in Example 1 is quite good $\left(0.672 \text{ vs. } \frac{2}{3}\right)$, even though you used a partition consisting of only 16 squares. The error resulted because the centers of the square subregions were used as the points in the approximation. This is comparable to the Midpoint Rule approximation of a single integral.

EXPLORATION

Volume of a Paraboloid Sector
The solid in Example 3 has an elliptical (not a circular) base. Consider the region bounded by the circular paraboloid

$$z = a^2 - x^2 - y^2, \quad a > 0$$

and the xy-plane. How many ways do you now know for finding the volume of this solid? For instance, you could use the disk method to find the volume as a solid of revolution. Does each method involve integration?

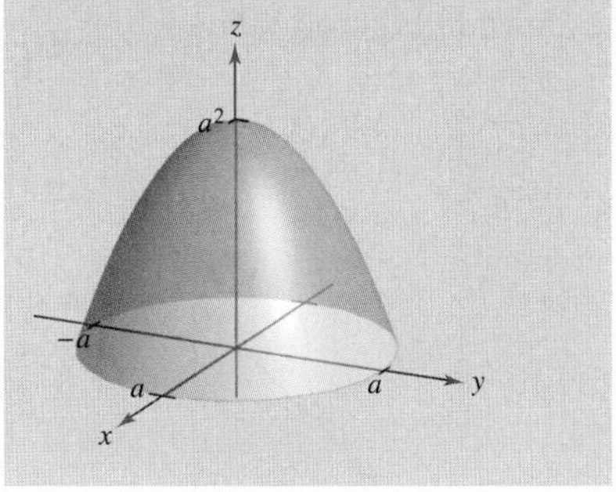

The difficulty of evaluating a single integral $\int_a^b f(x)\,dx$ usually depends on the function f, and not on the interval $[a, b]$. This is a major difference between single and double integrals. In the next example, you will integrate a function similar to that in Examples 1 and 2. Notice that a change in the region R produces a much more difficult integration problem.

EXAMPLE 3 Finding Volume by a Double Integral

Find the volume of the solid region bounded by the paraboloid $z = 4 - x^2 - 2y^2$ and the xy-plane.

Solution By letting $z = 0$, you can see that the base of the region in the xy-plane is the ellipse $x^2 + 2y^2 = 4$, as shown in Figure 14.19(a). This plane region is both vertically and horizontally simple, so the order $dy\,dx$ is appropriate.

Variable bounds for y: $-\sqrt{\dfrac{(4 - x^2)}{2}} \le y \le \sqrt{\dfrac{(4 - x^2)}{2}}$

Constant bounds for x: $-2 \le x \le 2$

The volume is given by

$$V = \int_{-2}^{2}\int_{-\sqrt{(4-x^2)/2}}^{\sqrt{(4-x^2)/2}} (4 - x^2 - 2y^2)\,dy\,dx \qquad \text{See Figure 14.19(b).}$$

$$= \int_{-2}^{2}\left[(4 - x^2)y - \frac{2y^3}{3}\right]_{-\sqrt{(4-x^2)/2}}^{\sqrt{(4-x^2)/2}} dx$$

$$= \frac{4}{3\sqrt{2}}\int_{-2}^{2}(4 - x^2)^{3/2}\,dx$$

$$= \frac{4}{3\sqrt{2}}\int_{-\pi/2}^{\pi/2} 16\cos^4\theta\,d\theta \qquad x = 2\sin\theta$$

$$= \frac{64}{3\sqrt{2}}(2)\int_{0}^{\pi/2}\cos^4\theta\,d\theta$$

$$= \frac{128}{3\sqrt{2}}\left(\frac{3\pi}{16}\right) \qquad \text{Wallis's Formula}$$

$$= 4\sqrt{2}\pi.$$

NOTE In Example 3, note the usefulness of Wallis's Formula to evaluate $\int_0^{\pi/2}\cos^n\theta\,d\theta$. You may want to review this formula in Section 8.3.

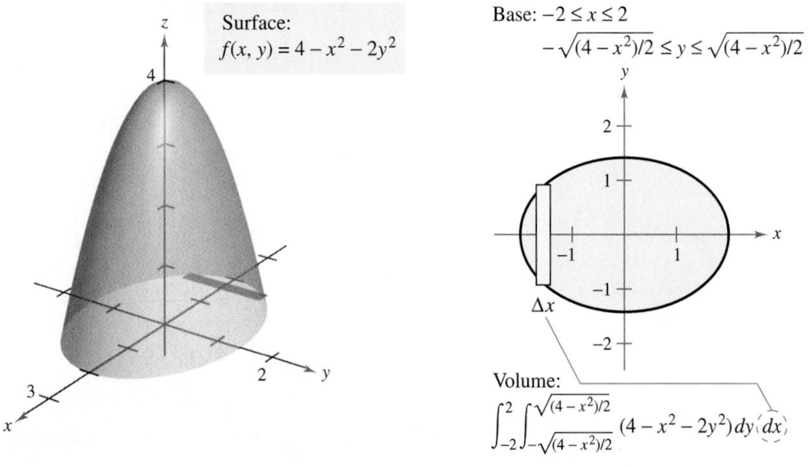

(a) **(b)**

Figure 14.19

In Examples 2 and 3, the problems could be solved with either order of integration because the regions were both vertically and horizontally simple. Moreover, had you used the order $dx\,dy$,you would have obtained integrals of comparable difficulty. There are, however, some occasions in which one order of integration is much more convenient than the other. Example 4 shows such a case.

EXAMPLE 4 Comparing Different Orders of Integration

Find the volume of the solid region R bounded by the surface

$$f(x, y) = e^{-x^2} \quad \text{Surface}$$

and the planes $z = 0$, $y = 0$, $y = x$, and $x = 1$, as shown in Figure 14.20.

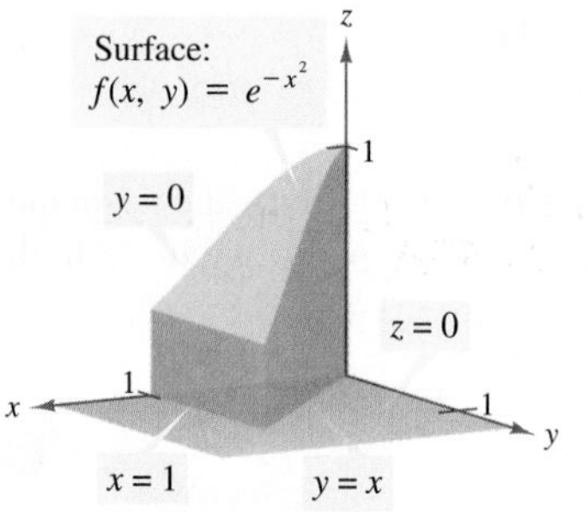

Base is bounded by $y = 0$, $y = x$, and $x = 1$.
Figure 14.20

Solution The base of R in the xy-plane is bounded by the lines $y = 0$, $x = 1$, and $y = x$. The two possible orders of integration are shown in Figure 14.21.

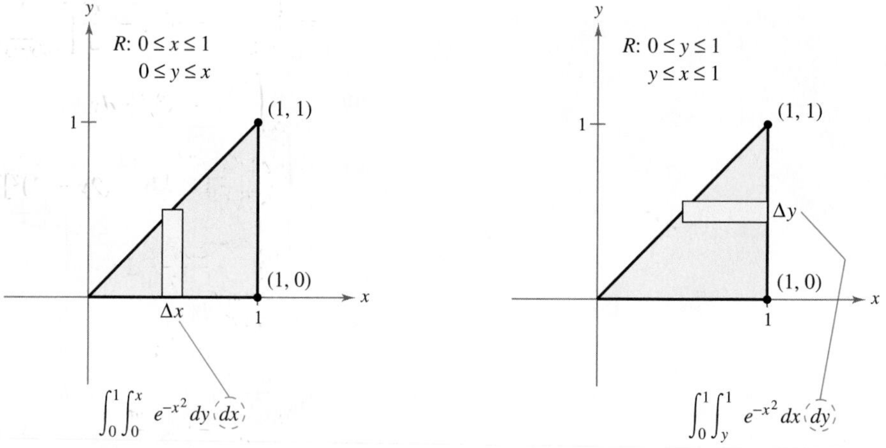

Figure 14.21

By setting up the corresponding iterated integrals, you can see that the order $dx\,dy$ requires the antiderivative $\int e^{-x^2}\,dx$, which is not an elementary function. On the other hand, the order $dy\,dx$ produces the integral

$$\begin{aligned}
\int_0^1 \int_0^x e^{-x^2}\,dy\,dx &= \int_0^1 e^{-x^2} y \Big]_0^x \,dx \\
&= \int_0^1 xe^{-x^2}\,dx \\
&= -\frac{1}{2}e^{-x^2}\Big]_0^1 \\
&= -\frac{1}{2}\left(\frac{1}{e} - 1\right) \\
&= \frac{e-1}{2e} \\
&\approx 0.316.
\end{aligned}$$

NOTE Try using a symbolic integration utility to evaluate the integral in Example 4.

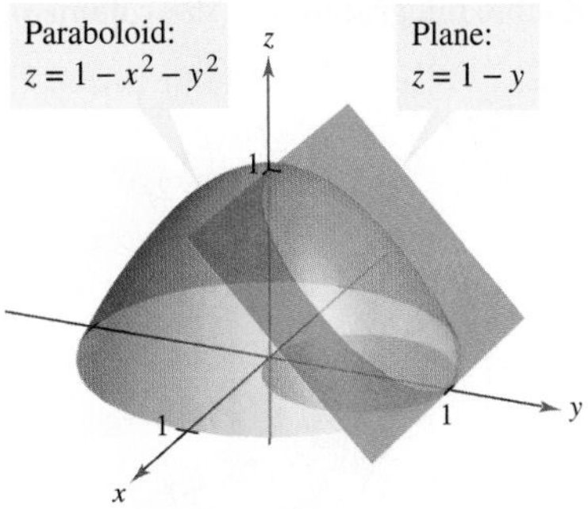

R: $0 \le y \le 1$

$-\sqrt{y - y^2} \le x \le \sqrt{y - y^2}$

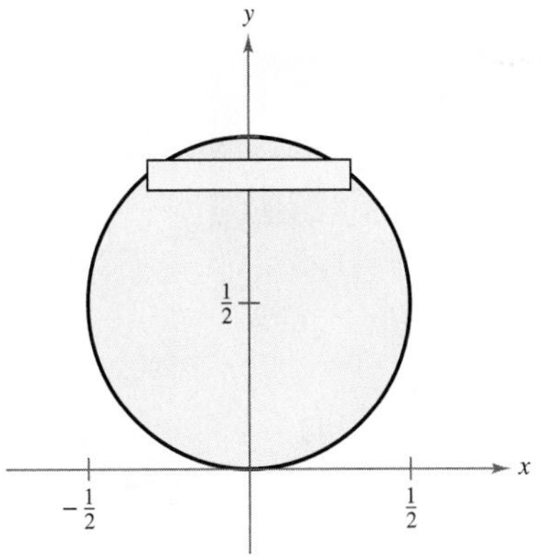

Figure 14.22

EXAMPLE 5 Volume of a Region Bounded by Two Surfaces

Find the volume of the solid region R bounded above by the paraboloid $z = 1 - x^2 - y^2$ and below by the plane $z = 1 - y$, as shown in Figure 14.22.

Solution Equating z-values, you can determine that the intersection of the two surfaces occurs on the right circular cylinder given by

$$1 - y = 1 - x^2 - y^2 \implies x^2 = y - y^2.$$

Because the volume of R is the difference between the volume under the paraboloid and the volume under the plane, you have

$$\begin{aligned}
\text{Volume} &= \int_0^1 \int_{-\sqrt{y-y^2}}^{\sqrt{y-y^2}} (1 - x^2 - y^2)\,dx\,dy - \int_0^1 \int_{-\sqrt{y-y^2}}^{\sqrt{y-y^2}} (1 - y)\,dx\,dy \\
&= \int_0^1 \int_{-\sqrt{y-y^2}}^{\sqrt{y-y^2}} (y - y^2 - x^2)\,dx\,dy \\
&= \int_0^1 \left[(y - y^2)x - \frac{x^3}{3} \right]_{-\sqrt{y-y^2}}^{\sqrt{y-y^2}} dy \\
&= \frac{4}{3}\int_0^1 (y - y^2)^{3/2}\,dy \\
&= \left(\frac{4}{3}\right)\left(\frac{1}{8}\right)\int_0^1 [1 - (2y - 1)^2]^{3/2}\,dy \\
&= \frac{1}{6}\int_{-\pi/2}^{\pi/2} \frac{\cos^4\theta}{2}\,d\theta \qquad 2y - 1 = \sin\theta \\
&= \frac{1}{6}\int_0^{\pi/2} \cos^4 d\theta \\
&= \left(\frac{1}{6}\right)\left(\frac{3\pi}{16}\right) = \frac{\pi}{32}. \qquad \text{Wallis's Formula}
\end{aligned}$$

Exercises for Section 14.2

See www.CalcChat.com for worked-out solutions to odd-numbered exercises.

Approximation **In Exercises 1–4, approximate the integral $\int_R\int f(x, y)\,dA$ by dividing the rectangle R with vertices $(0, 0)$, $(4, 0)$, $(4, 2)$, and $(0, 2)$ into eight equal squares and finding the sum**

$$\sum_{i=1}^{8} f(x_i, y_i)\,\Delta A_i$$

where (x_i, y_i) is the center of the ith square. Evaluate the iterated integral and compare it with the approximation.

1. $\displaystyle\int_0^4\int_0^2 (x + y)\,dy\,dx$

2. $\displaystyle\frac{1}{2}\int_0^4\int_0^2 x^2y\,dy\,dx$

3. $\displaystyle\int_0^4\int_0^2 (x^2 + y^2)\,dy\,dx$

4. $\displaystyle\int_0^4\int_0^2 \frac{1}{(x + 1)(y + 1)}\,dy\,dx$

5. ***Approximation*** The table shows values of a function f over a square region R. Divide the region into 16 equal squares and select (x_i, y_i) to be the point in the ith square closest to the origin. Compare this approximation with that obtained by using the point in the ith square farthest from the origin.

$$\int_0^4\int_0^4 f(x, y)\,dy\,dx$$

x \ y	0	1	2	3	4
0	32	31	28	23	16
1	31	30	27	22	15
2	28	27	24	19	12
3	23	22	19	14	7
4	16	15	12	7	0

6. ***Approximation*** The figure shows the level curves for a function f over a square region R. Approximate the integral using four squares, selecting the midpoint of each square as (x_i, y_i).

$$\int_0^2 \int_0^2 f(x, y)\, dy\, dx$$

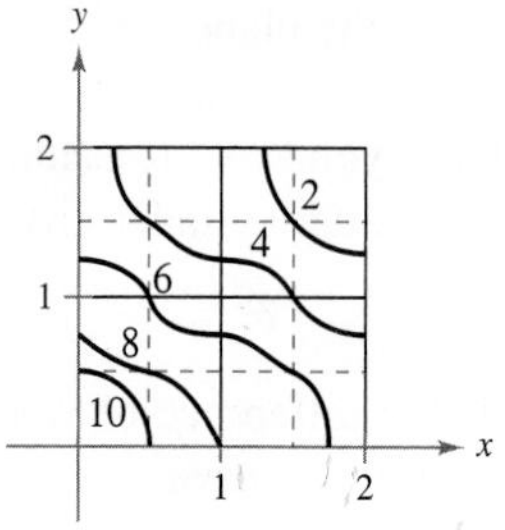

In Exercises 7–12, sketch the region R and evaluate the iterated integral $\int_R\int f(x, y)\, dA$.

7. $\displaystyle\int_0^2 \int_0^1 (1 + 2x + 2y)\, dy\, dx$

8. $\displaystyle\int_0^{\pi} \int_0^{\pi/2} \sin^2 x \cos^2 y\, dy\, dx$

9. $\displaystyle\int_0^6 \int_{y/2}^3 (x + y)\, dx\, dy$

10. $\displaystyle\int_0^4 \int_{\frac{1}{2}y}^{\sqrt{y}} x^2y^2\, dx\, dy$

11. $\displaystyle\int_{-a}^{a} \int_{-\sqrt{a^2-x^2}}^{\sqrt{a^2-x^2}} (x + y)\, dy\, dx$

12. $\displaystyle\int_0^1 \int_{y-1}^0 e^{x+y}\, dx\, dy + \int_0^1 \int_0^{1-y} e^{x+y}\, dx\, dy$

In Exercises 13–20, set up an integral for both orders of integration, and use the more convenient order to evaluate the integral over the region R.

13. $\displaystyle\int_R\int xy\, dA$

R: rectangle with vertices $(0, 0)$, $(0, 5)$, $(3, 5)$, $(3, 0)$

14. $\displaystyle\int_R\int \sin x \sin y\, dA$

R: rectangle with vertices $(-\pi, 0)$, $(\pi, 0)$, $(\pi, \pi/2)$, $(-\pi, \pi/2)$

15. $\displaystyle\int_R\int \frac{y}{x^2 + y^2}\, dA$

R: triangle bounded by $y = x$, $y = 2x$, $x = 2$

16. $\displaystyle\int_R\int xe^y\, dA$

R: triangle bounded by $y = 4 - x$, $y = 0$, $x = 0$

17. $\displaystyle\int_R\int -2y \ln x\, dA$

R: region bounded by $y = 4 - x^2$, $y = 4 - x$

18. $\displaystyle\int_R\int \frac{y}{1 + x^2}\, dA$

R: region bounded by $y = 0$, $y = \sqrt{x}$, $x = 4$

19. $\displaystyle\int_R\int x\, dA$

R: sector of a circle in the first quadrant bounded by $y = \sqrt{25 - x^2}$, $3x - 4y = 0$, $y = 0$

20. $\displaystyle\int_R\int (x^2 + y^2)\, dA$

R: semicircle bounded by $y = \sqrt{4 - x^2}$, $y = 0$

In Exercises 21–30, use a double integral to find the volume of the indicated solid.

21.

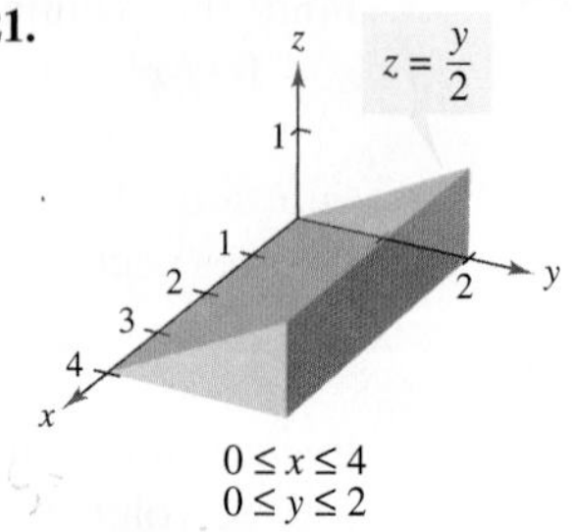

22.

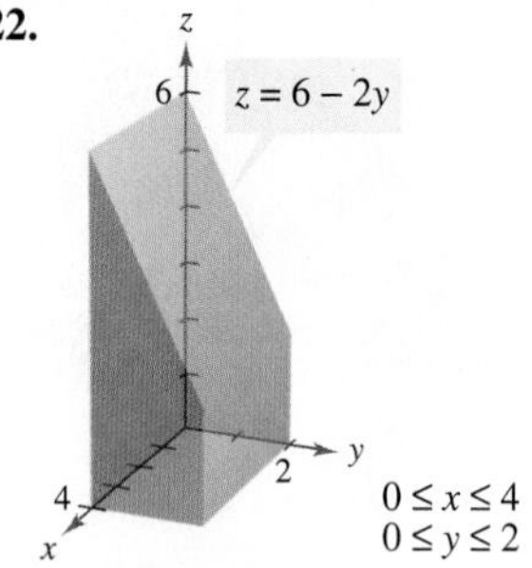

23.

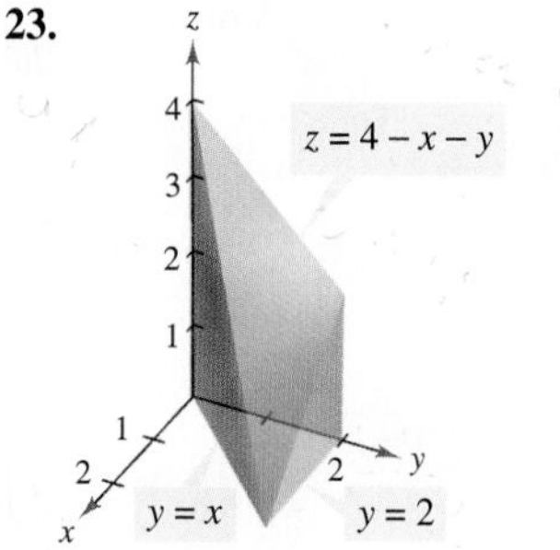

24.

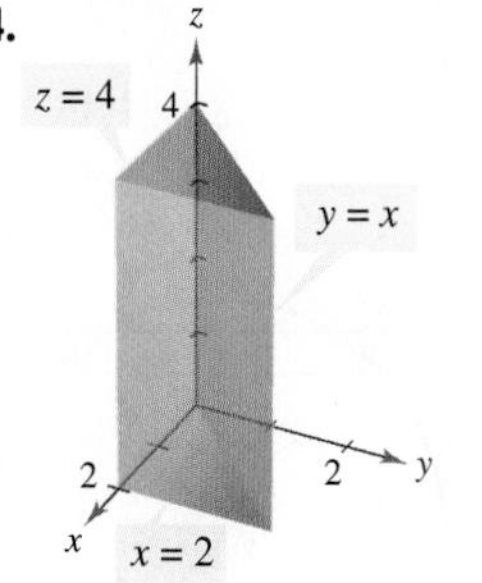

25.

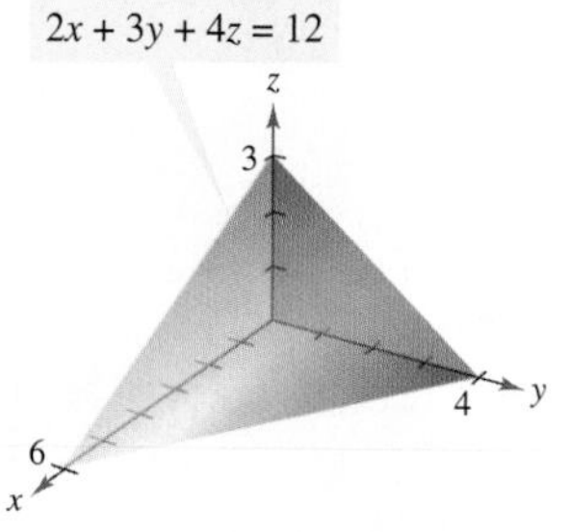

26.

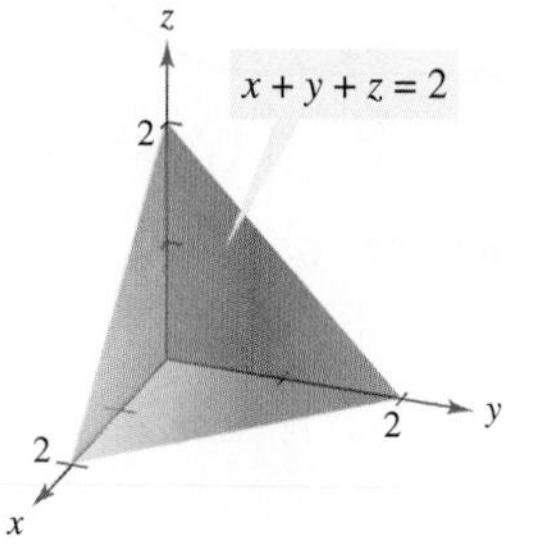

27.

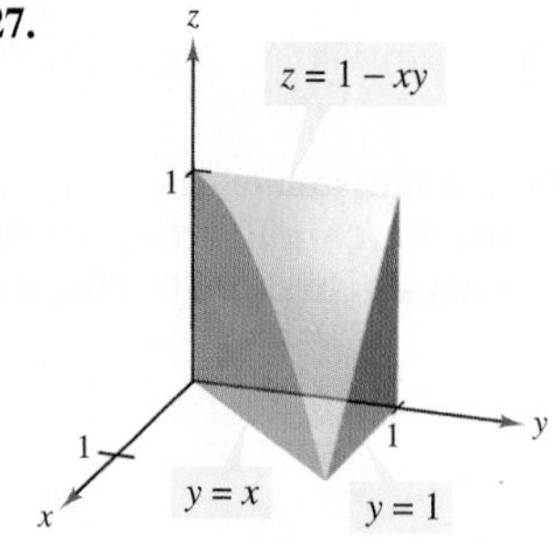

28.

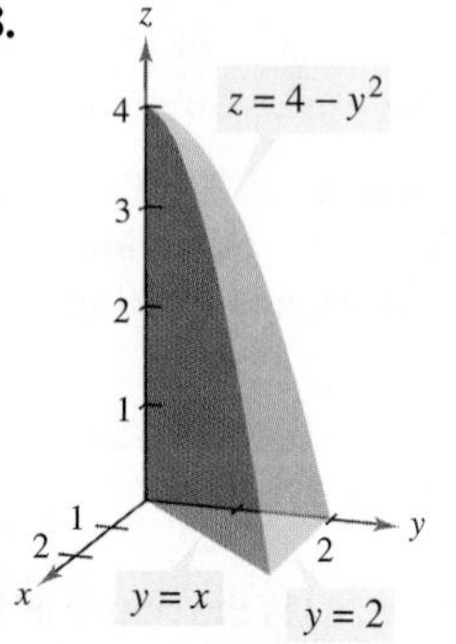

29. Improper integral

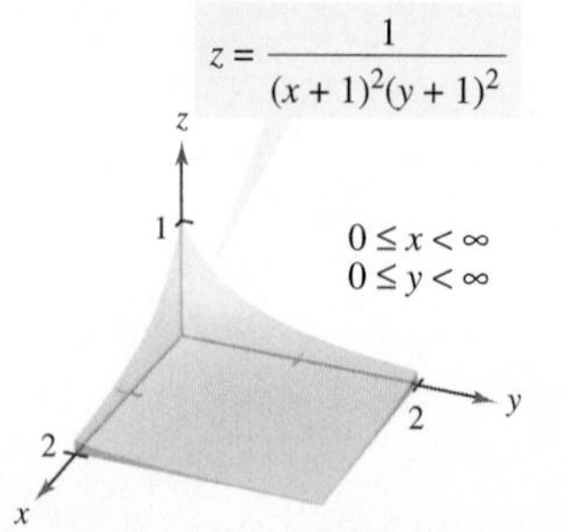

30. Improper integral

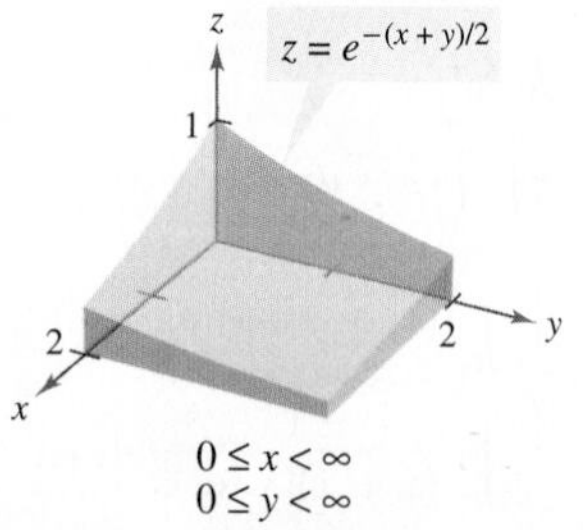

In Exercises 31 and 32, use a computer algebra system to find the volume of the solid.

31.

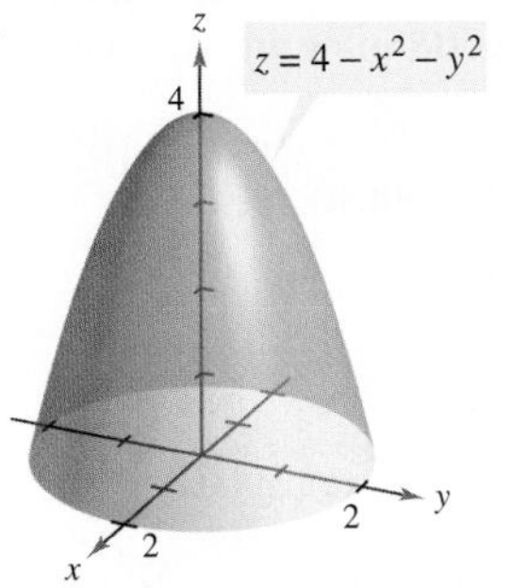

32.

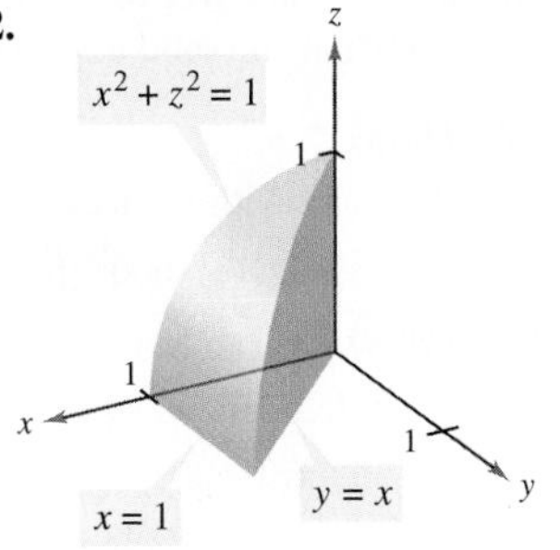

In Exercises 33–40, set up a double integral to find the volume of the solid bounded by the graphs of the equations.

33. $z = xy,\ z = 0,\ y = x,\ x = 1$, first octant

34. $y = 0,\ z = 0,\ y = x,\ z = x,\ x = 0,\ x = 5$

35. $z = 0,\ z = x^2,\ x = 0,\ x = 2,\ y = 0,\ y = 4$

36. $x^2 + y^2 + z^2 = r^2$

37. $x^2 + z^2 = 1,\ y^2 + z^2 = 1$, first octant

38. $y = 4 - x^2,\ z = 4 - x^2$, first octant

39. $z = x + y,\ x^2 + y^2 = 4$, first octant

40. $z = \dfrac{1}{1 + y^2},\ x = 0,\ x = 2,\ y \geq 0$

In Exercises 41 and 42, use Wallis's Formula to find the volume of the solid bounded by the graphs of the equations.

41. $z = x^2 + y^2,\ x^2 + y^2 = 4,\ z = 0$

42. $z = \sin^2 x,\ z = 0,\ 0 \leq x \leq \pi,\ 0 \leq y \leq 5$

In Exercises 43–46, use a computer algebra system to find the volume of the solid bounded by the graphs of the equations.

43. $z = 9 - x^2 - y^2,\ z = 0$

44. $x^2 = 9 - y,\ z^2 = 9 - y$, first octant

45. $z = \dfrac{2}{1 + x^2 + y^2},\ z = 0,\ y = 0,\ x = 0,\ y = -0.5x + 1$

46. $z = \ln(1 + x + y),\ z = 0,\ y = 0,\ x = 0,\ x = 4 - \sqrt{y}$

47. If f is a continuous function such that $0 \leq f(x, y) \leq 1$ over a region R of area 1, prove that $0 \leq \int_R\int f(x, y)\, dA \leq 1$.

48. Find the volume of the solid in the first octant bounded by the coordinate planes and the plane $(x/a) + (y/b) + (z/c) = 1$, where $a > 0$, $b > 0$, and $c > 0$.

In Exercises 49–52, evaluate the iterated integral. (Note that it is necessary to switch the order of integration.)

49. $\displaystyle\int_0^1 \int_{y/2}^{1/2} e^{-x^2}\, dx\, dy$

50. $\displaystyle\int_0^{\ln 10} \int_{e^x}^{10} \frac{1}{\ln y}\, dy\, dx$

51. $\displaystyle\int_0^1 \int_0^{\arccos y} \sin x\sqrt{1 + \sin^2 x}\, dx\, dy$

52. $\displaystyle\int_0^2 \int_{\frac{1}{2}x^2}^{2} \sqrt{y}\cos y\, dy\, dx$

In Exercises 53–56, find the average value of $f(x, y)$ over the region R where

$$\text{Average value} = \frac{1}{A}\int_R\int f(x, y)\, dA$$

and where A is the area of R.

53. $f(x, y) = x$

R: rectangle with vertices $(0, 0), (4, 0), (4, 2), (0, 2)$

54. $f(x, y) = xy$

R: rectangle with vertices $(0, 0), (4, 0), (4, 2), (0, 2)$

55. $f(x, y) = x^2 + y^2$

R: square with vertices $(0, 0), (2, 0), (2, 2), (0, 2)$

56. $f(x, y) = e^{x+y}$

R: triangle with vertices $(0, 0), (0, 1), (1, 1)$

Writing About Concepts

57. State the definition of a double integral. If the integrand is a nonnegative function over the region of integration, give the geometric interpretation of a double integral.

58. The following iterated integrals represent the solution to the same problem. Which iterated integral is easier to evaluate, and why?

$$\int_0^4 \int_{x/2}^{2} \sin y^2\, dy\, dx = \int_2^2 \int_0^{2y} \sin y^2\, dx\, dy$$

59. Let R be a region in the xy-plane whose area is B. If $f(x, y) = k$ for every point (x, y) in R, what is the value of $\int_R\int f(x, y)\, dA$? Explain.

60. Let R represent a county in the northern part of the United States, and let $f(x, y)$ represent the total annual snowfall at the point (x, y) in R. Interpret each of the following.

(a) $\displaystyle\int_R\int f(x, y)\, dA$ (b) $\dfrac{\displaystyle\int_R\int f(x, y)\, dA}{\displaystyle\int_R\int dA}$

61. ***Average Production*** The Cobb-Douglas production function for an automobile manufacturer is

$$f(x, y) = 100x^{0.6}y^{0.4}$$

where x is the number of units of labor and y is the number of units of capital. Estimate the average production level if the number of units of labor x varies between 200 and 250 and the number of units of capital y varies between 300 and 325.

62. ***Average Profit*** A firm's profit P in marketing two soft drinks is $P = 192x + 576y - x^2 - 5y^2 - 2xy - 5000$, where x and y represent the numbers of units of the two soft drinks. Use a computer algebra system to evaluate the double integral yielding the average weekly profit if x varies between 40 and 50 units and y varies between 45 and 60 units.

Probability **A joint density function of the continuous random variables x and y is a function $f(x, y)$ satisfying the following properties.**

(a) $f(x, y) \geq 0$ **for all** (x, y) **(b)** $\int_{-\infty}^{\infty}\int_{-\infty}^{\infty} f(x, y)\, dA = 1$

(c) $P[(x, y) \in R] = \int\int_R f(x, y)\, dA$

In Exercises 63–66, show that the function is a joint density function and find the required probability.

63. $f(x, y) = \begin{cases} \frac{1}{10}, & 0 \leq x \leq 5,\ 0 \leq y \leq 2 \\ 0, & \text{elsewhere} \end{cases}$

$P(0 \leq x \leq 2,\ 1 \leq y \leq 2)$

64. $f(x, y) = \begin{cases} \frac{1}{4}xy, & 0 \leq x \leq 2,\ 0 \leq y \leq 2 \\ 0, & \text{elsewhere} \end{cases}$

$P(0 \leq x \leq 1,\ 1 \leq y \leq 2)$

65. $f(x, y) = \begin{cases} \frac{1}{27}(9 - x - y), & 0 \leq x \leq 3,\ 3 \leq y \leq 6 \\ 0, & \text{elsewhere} \end{cases}$

$P(0 \leq x \leq 1,\ 4 \leq y \leq 6)$

66. $f(x, y) = \begin{cases} e^{-x-y}, & x \geq 0,\ y \geq 0 \\ 0, & \text{elsewhere} \end{cases}$

$P(0 \leq x \leq 1,\ x \leq y \leq 1)$

67. ***Approximation*** The base of a pile of sand at a cement plant is rectangular with approximate dimensions of 20 meters by 30 meters. If the base is placed on the xy-plane with one vertex at the origin, the coordinates on the surface of the pile are $(5, 5, 3)$, $(15, 5, 6)$, $(25, 5, 4)$, $(5, 15, 2)$, $(15, 15, 7)$, and $(25, 15, 3)$. Approximate the volume of sand in the pile.

68. ***Programming*** Consider a continuous function $f(x, y)$ over the rectangular region R with vertices (a, c), (b, c), (a, d), and (b, d), where $a < b$ and $c < d$. Partition the intervals $[a, b]$ and $[c, d]$ into m and n subintervals, so that the subintervals in a given direction are of equal length. Write a program for a graphing utility to compute the sum

$$\sum_{i=1}^{n}\sum_{j=1}^{m} f(x_i, y_j)\, \Delta A_i \approx \int_a^b\int_c^d f(x, y)\, dA$$

where (x_i, y_j) is the center of a representative rectangle in R.

Approximation **In Exercises 69–72, (a) use a computer algebra system to approximate the iterated integral, and (b) use the program in Exercise 68 to approximate the iterated integral for the given values of m and n.**

69. $\int_0^1\int_0^2 \sin\sqrt{x + y}\, dy\, dx$

$m = 4,\ n = 8$

70. $\int_0^2\int_0^4 20e^{-x^3/8} dy\, dx$

$m = 10,\ n = 20$

71. $\int_4^6\int_0^2 y\cos\sqrt{x}\, dx\, dy$

$m = 4,\ n = 8$

72. $\int_1^4\int_1^2 \sqrt{x^3 + y^3}\, dx\, dy$

$m = 6,\ n = 4$

Approximation **In Exercises 73 and 74, determine which value best approximates the volume of the solid between the xy-plane and the function over the region. (Make your selection on the basis of a sketch of the solid and *not* by performing any calculations.)**

73. $f(x, y) = 4x$

R: square with vertices $(0, 0)$, $(4, 0)$, $(4, 4)$, $(0, 4)$

(a) -200 (b) 600 (c) 50 (d) 125 (e) 1000

74. $f(x, y) = \sqrt{x^2 + y^2}$

R: circle bounded by $x^2 + y^2 = 9$

(a) 50 (b) 500 (c) -500 (d) 5 (e) 5000

True or False? **In Exercises 75 and 76, determine whether the statement is true or false. If it is false, explain why or give an example that shows it is false.**

75. The volume of the sphere $x^2 + y^2 + z^2 = 1$ is given by the integral

$$V = 8\int_0^1\int_0^1 \sqrt{1 - x^2 - y^2}\, dx\, dy.$$

76. If $f(x, y) \leq g(x, y)$ for all (x, y) in R, and both f and g are continuous over R, then

$$\int\int_R f(x, y)\, dA \leq \int\int_R g(x, y)\, dA.$$

77. Let $f(x) = \int_1^x e^{t^2}\, dt$. Find the average value of f on the interval $[0, 1]$.

78. Find $\displaystyle\int_0^{\infty} \frac{e^{-x} - e^{-2x}}{x}\, dx.$ $\left(\textit{Hint:}\ \text{Evaluate } \int_1^2 e^{-xy}\, dy.\right)$

79. Determine the region R in the xy-plane that maximizes the value of

$$\int\int_R (9 - x^2 - y^2)\, dA.$$

80. Determine the region R in the xy-plane that minimizes the value of

$$\int\int_R (x^2 + y^2 - 4)\, dA.$$

81. Find $\int_0^2 [\arctan(\pi x) - \arctan x]\, dx$. (*Hint:* Convert the integral to a double integral.)

82. Use a geometric argument to show that

$$\int_0^3\int_0^{\sqrt{9-y^2}} \sqrt{9 - x^2 - y^2}\, dx\, dy = \frac{9\pi}{2}.$$

Putnam Exam Challenge

83. Evaluate $\int_0^a\int_0^b e^{\max\{b^2x^2,\, a^2y^2\}}\, dy\, dx$, where a and b are positive.

84. Show that if $\lambda > \frac{1}{2}$ there does not exist a real-valued function u such that for all x in the closed interval $0 \leq x \leq 1$

$u(x) = 1 + \lambda\int_x^1 u(y)u(y - x)\, dy.$

These problems were composed by the Committee on the Putnam Prize Competition.

Section 14.3 Change of Variables: Polar Coordinates

- **Write and evaluate double integrals in polar coordinates.**

Double Integrals in Polar Coordinates

Some double integrals are *much* easier to evaluate in polar form than in rectangular form. This is especially true for regions such as circles, cardioids, and rose curves, and for integrands that involve $x^2 + y^2$.

In Section 10.4, you learned that the polar coordinates (r, θ) of a point are related to the rectangular coordinates (x, y) of the point as follows.

$$x = r\cos\theta \quad \text{and} \quad y = r\sin\theta$$

$$r^2 = x^2 + y^2 \quad \text{and} \quad \tan\theta = \frac{y}{x}$$

EXAMPLE 1 Using Polar Coordinates to Describe a Region

Use polar coordinates to describe each region shown in Figure 14.23.

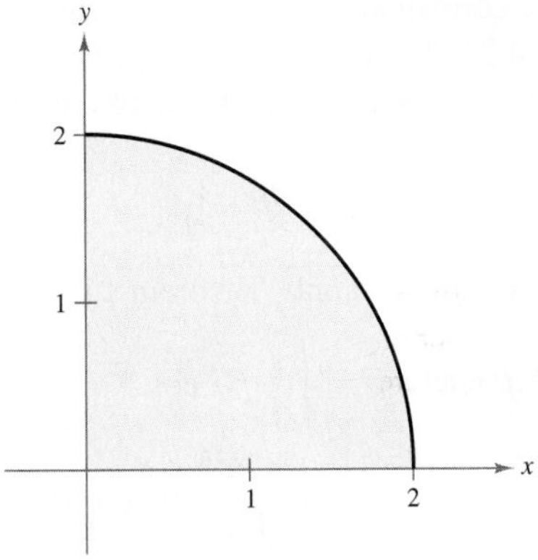

(a)

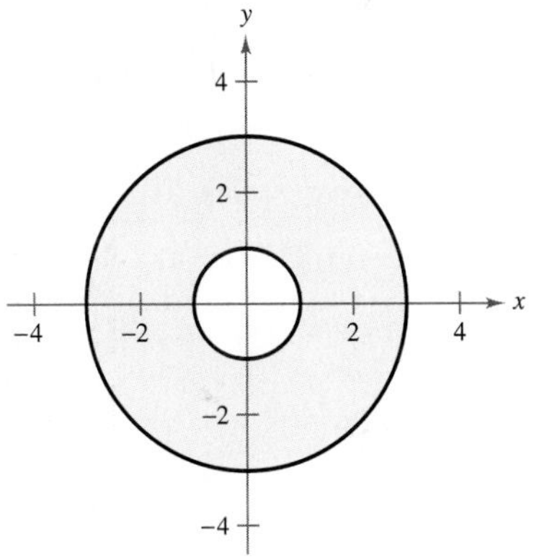

(b)

Figure 14.23

Solution

a. The region R is a quarter circle of radius 2. It can be described in polar coordinates as

$$R = \{(r, \theta): 0 \le r \le 2, \quad 0 \le \theta \le \pi/2\}.$$

b. The region R consists of all points between the concentric circles of radii 1 and 3. It can be described in polar coordinates as

$$R = \{(r, \theta): 1 \le r \le 3, \quad 0 \le \theta \le 2\pi\}.$$

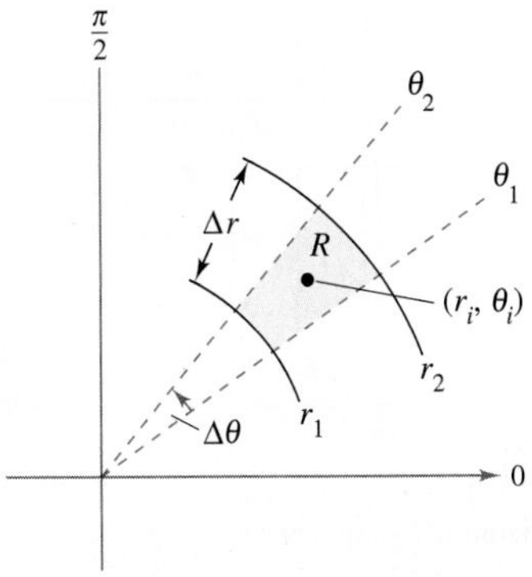

Polar sector
Figure 14.24

The regions in Example 1 are special cases of **polar sectors**

$$R = \{(r, \theta): r_1 \le r \le r_2, \quad \theta_1 \le \theta \le \theta_2\} \qquad \text{Polar sector}$$

as shown in Figure 14.24.

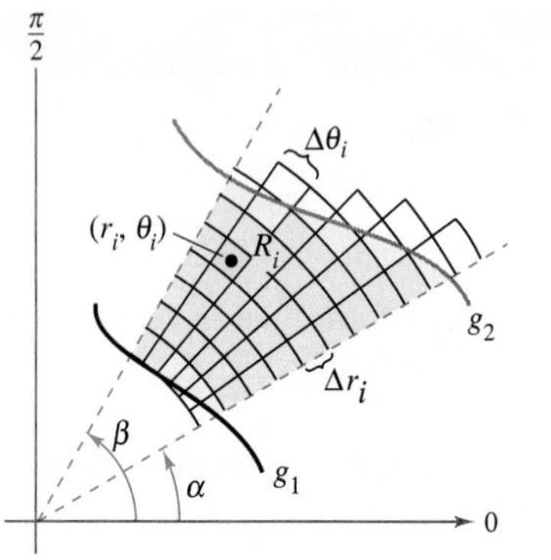

Polar grid is superimposed over region R.
Figure 14.25

To define a double integral of a continuous function $z = f(x, y)$ in polar coordinates, consider a region R bounded by the graphs of $r = g_1(\theta)$ and $r = g_2(\theta)$ and the lines $\theta = \alpha$ and $\theta = \beta$. Instead of partitioning R into small rectangles, use a partition of small polar sectors. On R, superimpose a polar grid made of rays and circular arcs, as shown in Figure 14.25. The polar sectors R_i lying entirely within R form an **inner polar partition** Δ, whose **norm** $\|\Delta\|$ is the length of the longest diagonal of the n polar sectors.

Consider a specific polar sector R_i, as shown in Figure 14.26. It can be shown (see Exercise 61) that the area of R_i is

$$\Delta A_i = r_i \Delta r_i \Delta \theta_i \qquad \text{Area of } R_i$$

where $\Delta r_i = r_2 - r_1$ and $\Delta \theta_i = \theta_2 - \theta_1$. This implies that the volume of the solid of height $f(r_i \cos \theta_i, r_i \sin \theta_i)$ above R_i is approximately

$$f(r_i \cos \theta_i, r_i \sin \theta_i) r_i \Delta r_i \Delta \theta_i$$

and you have

$$\int_R\int f(x, y)\, dA \approx \sum_{i=1}^{n} f(r_i \cos \theta_i, r_i \sin \theta_i) r_i \Delta r_i \Delta \theta_i.$$

The sum on the right can be interpreted as a Riemann sum for $f(r \cos \theta, r \sin \theta)r$. The region R corresponds to a *horizontally simple* region S in the $r\theta$-plane, as shown in Figure 14.27. The polar sectors R_i correspond to rectangles S_i, and the area ΔA_i of S_i is $\Delta r_i \Delta \theta_i$. So, the right-hand side of the equation corresponds to the double integral

$$\int_S\int f(r \cos \theta, r \sin \theta) r\, dA.$$

From this, you can apply Theorem 14.2 to write

$$\begin{aligned}\int_R\int f(x, y)\, dA &= \int_S\int f(r \cos \theta, r \sin \theta) r\, dA \\ &= \int_{\alpha}^{\beta}\int_{g_1(\theta)}^{g_2(\theta)} f(r \cos \theta, r \sin \theta) r\, dr\, d\theta.\end{aligned}$$

This suggests the following theorem, the proof of which is discussed in Section 14.8.

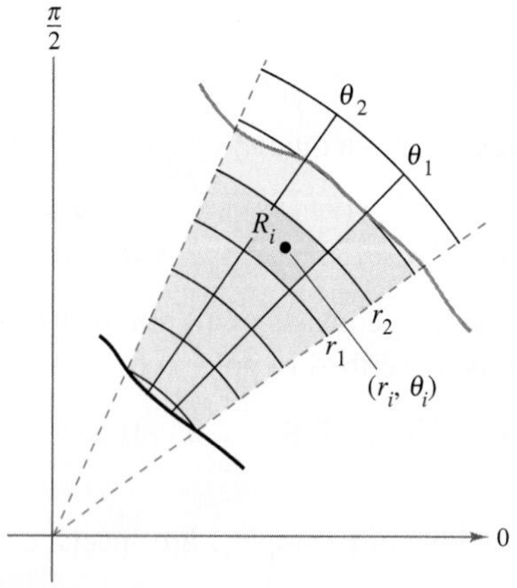

The polar sector R_i is the set of all points (r, θ) such that $r_1 \le r \le r_2$ and $\theta_1 \le \theta \le \theta_2$.
Figure 14.26

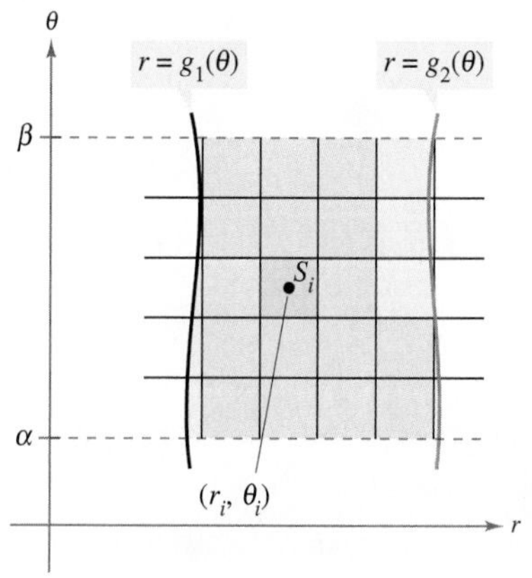

Horizontally simple region S
Figure 14.27

THEOREM 14.3 Change of Variables to Polar Form

Let R be a plane region consisting of all points $(x, y) = (r\cos\theta, r\sin\theta)$ satisfying the conditions $0 \le g_1(\theta) \le r \le g_2(\theta)$, $\alpha \le \theta \le \beta$, where $0 \le (\beta - \alpha) \le 2\pi$. If g_1 and g_2 are continuous on $[\alpha, \beta]$ and f is continuous on R, then

$$\int_R\int f(x, y)\, dA = \int_\alpha^\beta \int_{g_1(\theta)}^{g_2(\theta)} f(r\cos\theta, r\sin\theta) r\, dr\, d\theta.$$

EXPLORATION

Volume of a Paraboloid Sector
In the Exploration feature on page 995, you were asked to summarize the different ways you know for finding the volume of the solid bounded by the paraboloid

$$z = a^2 - x^2 - y^2, \quad a > 0$$

and the xy-plane. You now know another way. Use it to find the volume of the solid.

NOTE If $z = f(x, y)$ is nonnegative on R, then the integral in Theorem 14.3 can be interpreted as the volume of the solid region between the graph of f and the region R.

The region R is restricted to two basic types, ***r*-simple** regions and **θ-simple** regions, as shown in Figure 14.28.

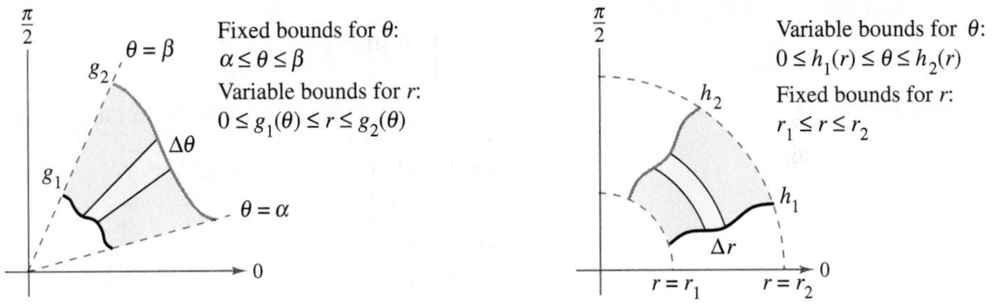

r-Simple region $\qquad$ θ-Simple region
Figure 14.28

EXAMPLE 2 Evaluating a Double Polar Integral

Let R be the annular region lying between the two circles $x^2 + y^2 = 1$ and $x^2 + y^2 = 5$. Evaluate the integral $\int_R\int (x^2 + y)\, dA$.

Solution The polar boundaries are $1 \le r \le \sqrt{5}$ and $0 \le \theta \le 2\pi$, as shown in Figure 14.29. Furthermore, $x^2 = (r\cos\theta)^2$ and $y = r\sin\theta$. So, you have

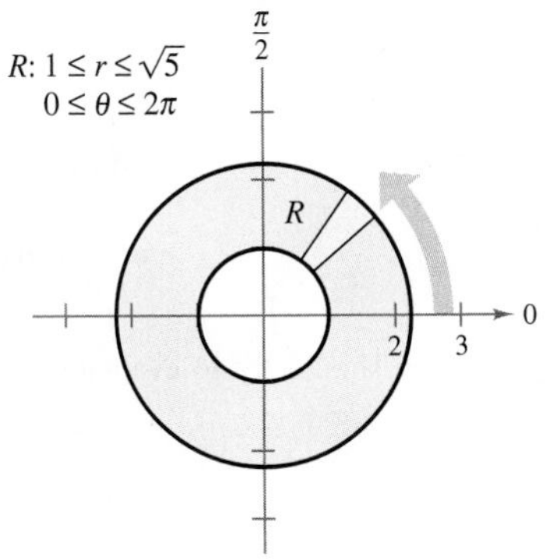

r-Simple region
Figure 14.29

$$\begin{aligned}
\int_R\int (x^2 + y)\, dA &= \int_0^{2\pi}\int_1^{\sqrt{5}} (r^2\cos^2\theta + r\sin\theta) r\, dr\, d\theta \\
&= \int_0^{2\pi}\int_1^{\sqrt{5}} (r^3\cos^2\theta + r^2\sin\theta)\, dr\, d\theta \\
&= \int_0^{2\pi} \left(\frac{r^4}{4}\cos^2\theta + \frac{r^3}{3}\sin\theta\right)\Bigg]_1^{\sqrt{5}} d\theta \\
&= \int_0^{2\pi} \left(6\cos^2\theta + \frac{5\sqrt{5} - 1}{3}\sin\theta\right) d\theta \\
&= \int_0^{2\pi} \left(3 + 3\cos 2\theta + \frac{5\sqrt{5} - 1}{3}\sin\theta\right) d\theta \\
&= \left(3\theta + \frac{3\sin 2\theta}{2} - \frac{5\sqrt{5} - 1}{3}\cos\theta\right)\Bigg]_0^{2\pi} \\
&= 6\pi.
\end{aligned}$$

In Example 2, be sure to notice the extra factor of r in the integrand. This comes from the formula for the area of a polar sector. In differential notation, you can write

$$dA = r\,dr\,d\theta$$

which indicates that the area of a polar sector increases as you move away from the origin.

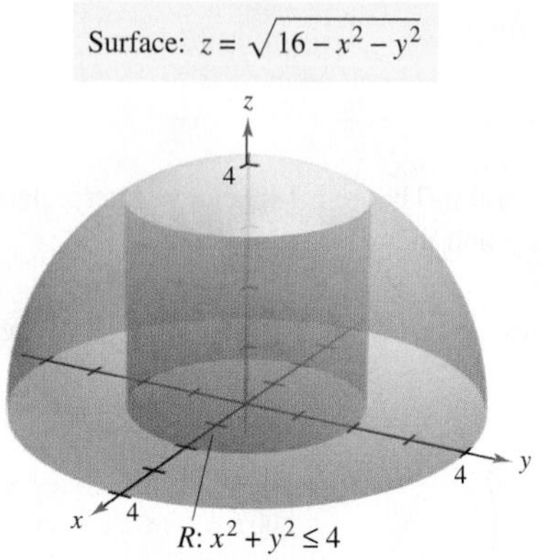

Figure 14.30

EXAMPLE 3 Change of Variables to Polar Coordinates

Use polar coordinates to find the volume of the solid region bounded above by the hemisphere

$$z = \sqrt{16 - x^2 - y^2}$$ Hemisphere forms upper surface.

and below by the circular region R given by

$$x^2 + y^2 \le 4$$ Circular region forms lower surface.

as shown in Figure 14.30.

Solution In Figure 14.30, you can see that R has the bounds

$$-\sqrt{4 - y^2} \le x \le \sqrt{4 - y^2}, \quad -2 \le y \le 2$$

and that $0 \le z \le \sqrt{16 - x^2 - y^2}$. In polar coordinates, the bounds are

$$0 \le r \le 2 \quad \text{and} \quad 0 \le \theta \le 2\pi$$

with height $z = \sqrt{16 - x^2 - y^2} = \sqrt{16 - r^2}$. Consequently, the volume V is given by

$$\begin{aligned} V = \int_R\int f(x, y)\,dA &= \int_0^{2\pi}\int_0^2 \sqrt{16 - r^2}\,r\,dr\,d\theta \\ &= -\frac{1}{3}\int_0^{2\pi} (16 - r^2)^{3/2}\Big]_0^2\,d\theta \\ &= -\frac{1}{3}\int_0^{2\pi} \left(24\sqrt{3} - 64\right)d\theta \\ &= -\frac{8}{3}\left(3\sqrt{3} - 8\right)\theta\Big]_0^{2\pi} \\ &= \frac{16\pi}{3}\left(8 - 3\sqrt{3}\right) \approx 46.979. \end{aligned}$$

NOTE To see the benefit of polar coordinates in Example 3, you should try to evaluate the corresponding rectangular iterated integral

$$\int_{-2}^{2}\int_{-\sqrt{4-y^2}}^{\sqrt{4-y^2}} \sqrt{16 - x^2 - y^2}\,dx\,dy.$$

TECHNOLOGY Any computer algebra system that can handle double integrals in rectangular coordinates can also handle double integrals in polar coordinates. The reason this is true is that once you have formed the iterated integral, its value is not changed by using different variables. In other words, if you use a computer algebra system to evaluate

$$\int_0^{2\pi}\int_0^2 \sqrt{16 - x^2}\,x\,dx\,dy$$

you should obtain the same value as that obtained in Example 3.

Just as with rectangular coordinates, the double integral

$$\int_R\int dA$$

can be used to find the area of a region in the plane.

EXAMPLE 4 Finding Areas of Polar Regions

Use a double integral to find the area enclosed by the graph of $r = 3\cos 3\theta$.

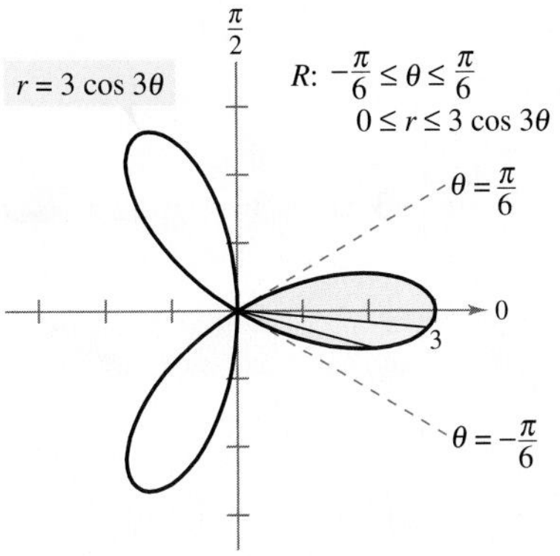

Figure 14.31

Solution Let R be one petal of the curve shown in Figure 14.31. This region is r-simple, and the boundaries are as follows.

$$-\frac{\pi}{6} \le \theta \le \frac{\pi}{6} \qquad \text{Fixed bounds on } \theta$$

$$0 \le r \le 3\cos 3\theta \qquad \text{Variable bounds on } r$$

So, the area of one petal is

$$\begin{aligned}
\frac{1}{3}A = \int_R\int dA &= \int_{-\pi/6}^{\pi/6}\int_0^{3\cos 3\theta} r\,dr\,d\theta \\
&= \int_{-\pi/6}^{\pi/6}\left[\frac{r^2}{2}\right]_0^{3\cos 3\theta} d\theta \\
&= \frac{9}{2}\int_{-\pi/6}^{\pi/6} \cos^2 3\theta\,d\theta \\
&= \frac{9}{4}\int_{-\pi/6}^{\pi/6} (1 + \cos 6\theta)\,d\theta \\
&= \frac{9}{4}\left[\theta + \frac{1}{6}\sin 6\theta\right]_{-\pi/6}^{\pi/6} \\
&= \frac{3\pi}{4}.
\end{aligned}$$

So, the total area is $A = 9\pi/4$.

As illustrated in Example 4, the area of a region in the plane can be represented by

$$A = \int_\alpha^\beta\int_{g_1(\theta)}^{g_2(\theta)} r\,dr\,d\theta.$$

If $g_1(\theta) = 0$, you obtain

$$A = \int_\alpha^\beta\int_0^{g_2(\theta)} r\,dr\,d\theta = \int_\alpha^\beta \left[\frac{r^2}{2}\right]_0^{g_2(\theta)} d\theta = \int_\alpha^\beta \frac{1}{2}(g_2(\theta))^2\,d\theta$$

which agrees with Theorem 10.13.

So far in this section, all of the examples of iterated integrals in polar form have been of the form

$$\int_\alpha^\beta\int_{g_1(\theta)}^{g_2(\theta)} f(r\cos\theta, r\sin\theta)r\,dr\,d\theta$$

in which the order of integration is with respect to r first. Sometimes you can obtain a simpler integration problem by switching the order of integration, as illustrated in the next example.

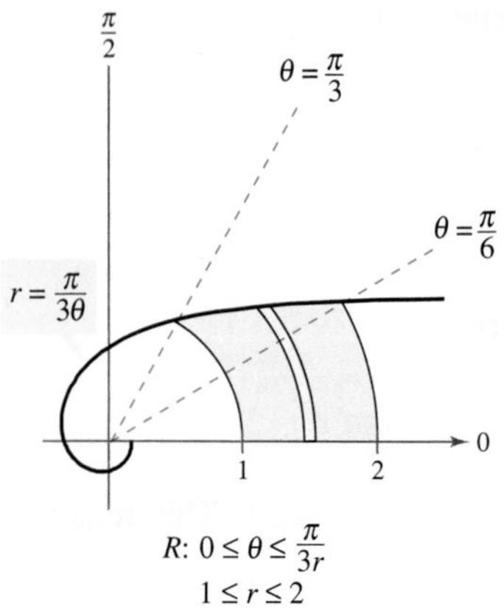

θ-Simple region
Figure 14.32

EXAMPLE 5 Changing the Order of Integration

Find the area of the region bounded above by the spiral

$$r = \frac{\pi}{3\theta}$$

and below by the polar axis, between $r = 1$ and $r = 2$.

Solution The region is shown in Figure 14.32. The polar boundaries for the region are

$$1 \le r \le 2 \quad \text{and} \quad 0 \le \theta \le \frac{\pi}{3r}.$$

So, the area of the region can be evaluated as follows.

$$A = \int_1^2 \int_0^{\pi/(3r)} r\, d\theta\, dr = \int_1^2 r\theta \Big]_0^{\pi/(3r)} dr = \int_1^2 \frac{\pi}{3}\, dr = \frac{\pi r}{3}\Big]_1^2 = \frac{\pi}{3}$$

Exercises for Section 14.3

See www.CalcChat.com for worked-out solutions to odd-numbered exercises.

In Exercises 1–4, the region R for the integral $\int_R\int f(x, y)\, dA$ is shown. State whether you would use rectangular or polar coordinates to evaluate the integral.

1. 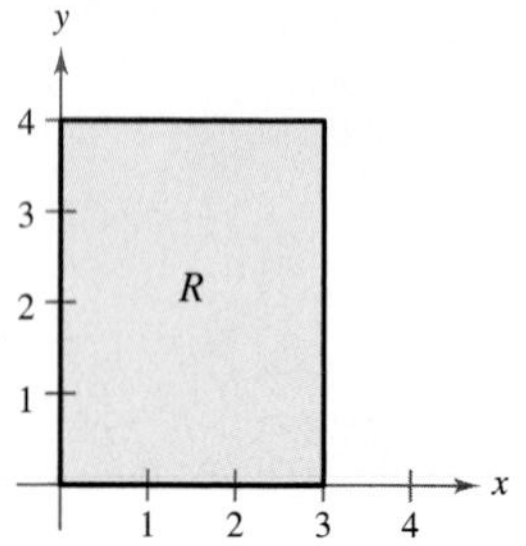

2. 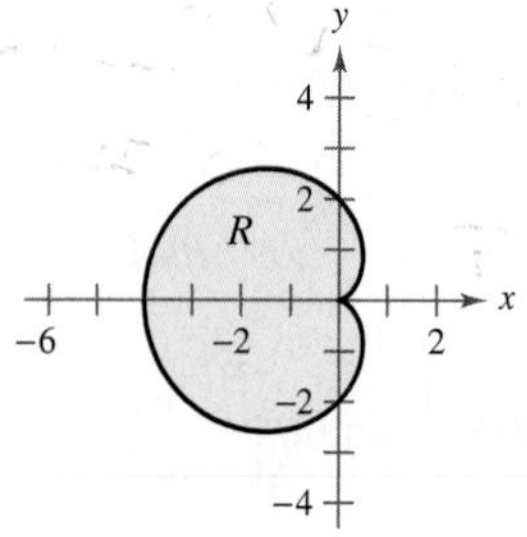

3. 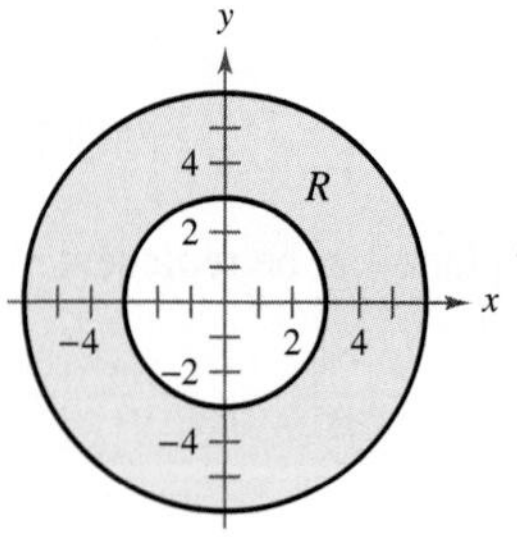

4. 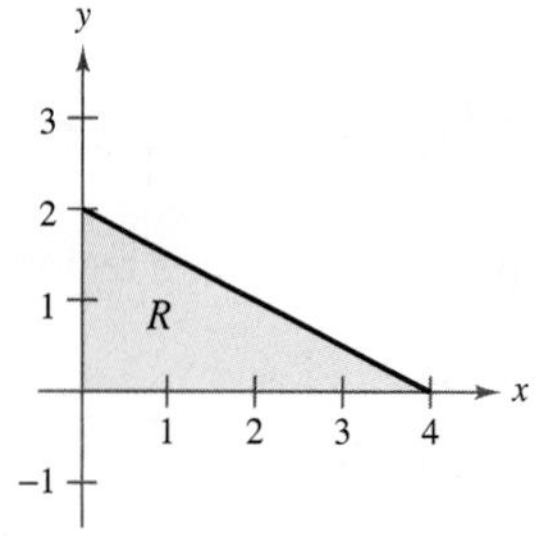

In Exercises 5–8, use polar coordinates to describe the region shown.

5. 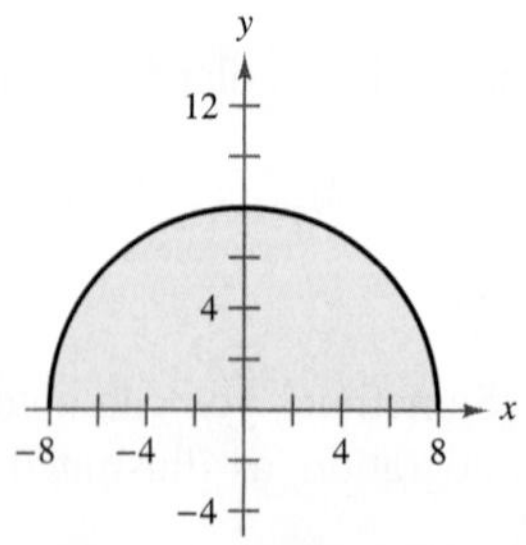

6. 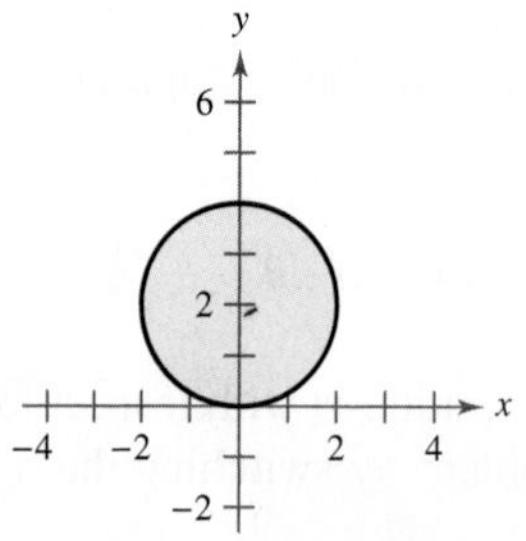

7. 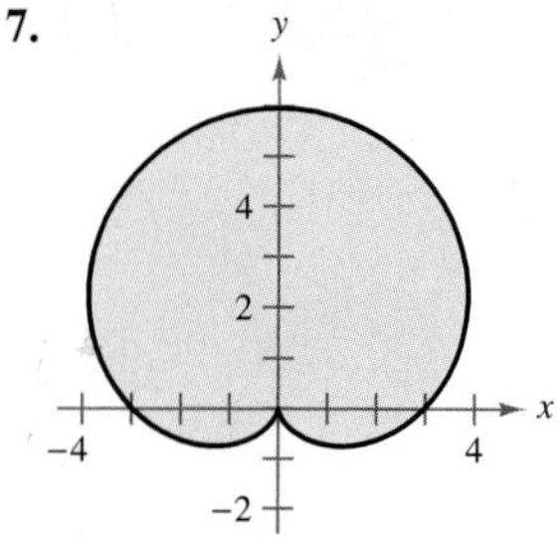

8. 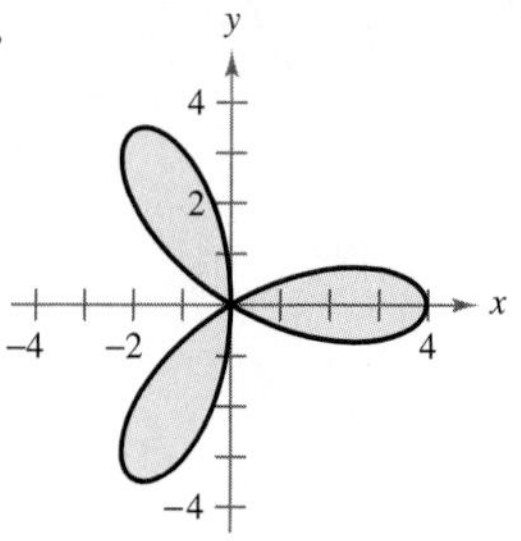

In Exercises 9–14, evaluate the double integral $\int_R\int f(r, \theta)\, dA$, and sketch the region R.

9. $\displaystyle\int_0^{2\pi}\int_0^6 3r^2 \sin\theta\, dr\, d\theta$

10. $\displaystyle\int_0^{\pi/4}\int_0^4 r^2 \sin\theta \cos\theta\, dr\, d\theta$

11. $\displaystyle\int_0^{\pi/2}\int_2^3 \sqrt{9 - r^2}\, r\, dr\, d\theta$

12. $\displaystyle\int_0^{\pi/2}\int_0^3 re^{-r^2}\, dr\, d\theta$

13. $\displaystyle\int_0^{\pi/2}\int_0^{1+\sin\theta} \theta r\, dr\, d\theta$

14. $\displaystyle\int_0^{\pi/2}\int_0^{1-\cos\theta} (\sin\theta) r\, dr\, d\theta$

In Exercises 15–20, evaluate the iterated integral by converting to polar coordinates.

15. $\displaystyle\int_0^a\int_0^{\sqrt{a^2-y^2}} y\, dx\, dy$

16. $\displaystyle\int_0^a\int_0^{\sqrt{a^2-x^2}} x\, dy\, dx$

17. $\displaystyle\int_0^3\int_0^{\sqrt{9-x^2}} (x^2 + y^2)^{3/2}\, dy\, dx$

18. $\displaystyle\int_0^2\int_y^{\sqrt{8-y^2}} \sqrt{x^2 + y^2}\, dx\, dy$

19. $\displaystyle\int_0^2\int_0^{\sqrt{2x-x^2}} xy\, dy\, dx$

20. $\displaystyle\int_0^4\int_0^{\sqrt{4y-y^2}} x^2\, dx\, dy$

In Exercises 21 and 22, combine the sum of the two iterated integrals into a single iterated integral by converting to polar coordinates. Evaluate the resulting iterated integral.

21. $\displaystyle\int_0^2\int_0^x \sqrt{x^2 + y^2}\, dy\, dx + \int_2^{2\sqrt{2}}\int_0^{\sqrt{8-x^2}} \sqrt{x^2 + y^2}\, dy\, dx$

22. $\displaystyle\int_0^{5\sqrt{2}/2}\int_0^x xy\,dy\,dx + \int_{5\sqrt{2}/2}^{5}\int_0^{\sqrt{25-x^2}} xy\,dy\,dx$

In Exercises 23–26, use polar coordinates to set up and evaluate the double integral $\int_R\int f(x, y)\,dA$.

23. $f(x, y) = x + y,\ R: x^2 + y^2 \le 4,\ x \ge 0,\ y \ge 0$

24. $f(x, y) = e^{-(x^2+y^2)/2},\ R: x^2 + y^2 \le 25,\ x \ge 0$

25. $f(x, y) = \arctan \dfrac{y}{x},\ R: x^2 + y^2 \ge 1,\ x^2 + y^2 \le 4,\ 0 \le y \le x$

26. $f(x, y) = 9 - x^2 - y^2,\ R: x^2 + y^2 \le 9,\ x \ge 0,\ y \ge 0$

Volume **In Exercises 27–32, use a double integral in polar coordinates to find the volume of the solid bounded by the graphs of the equations.**

27. $z = xy,\ x^2 + y^2 = 1$, first octant

28. $z = x^2 + y^2 + 3,\ z = 0,\ x^2 + y^2 = 1$

29. $z = \sqrt{x^2 + y^2},\ z = 0,\ x^2 + y^2 = 25$

30. $z = \ln(x^2 + y^2),\ z = 0,\ x^2 + y^2 \ge 1,\ x^2 + y^2 \le 4$

31. Inside the hemisphere $z = \sqrt{16 - x^2 - y^2}$ and inside the cylinder $x^2 + y^2 - 4x = 0$

32. Inside the hemisphere $z = \sqrt{16 - x^2 - y^2}$ and outside the cylinder $x^2 + y^2 = 1$

33. ***Volume*** Find a such that the volume inside the hemisphere $z = \sqrt{16 - x^2 - y^2}$ and outside the cylinder $x^2 + y^2 = a^2$ is one-half the volume of the hemisphere.

34. ***Volume*** Use a double integral in polar coordinates to find the volume of a sphere of radius a.

35. ***Volume*** Determine the diameter of a hole that is drilled vertically through the center of the solid bounded by the graphs of the equations

$$z = 25e^{-(x^2+y^2)/4}, \quad z = 0, \quad \text{and} \quad x^2 + y^2 = 16$$

if one-tenth of the volume of the solid is removed.

36. ***Machine Design*** The surfaces of a double-lobed cam are modeled by the inequalities $\frac{1}{4} \le r \le \frac{1}{2}(1 + \cos^2\theta)$ and

$$\frac{-9}{4(x^2 + y^2 + 9)} \le z \le \frac{9}{4(x^2 + y^2 + 9)}$$

where all measurements are in inches.

(a) Use a computer algebra system to graph the cam.

(b) Use a computer algebra system to approximate the perimeter of the polar curve

$$r = \tfrac{1}{2}(1 + \cos^2\theta).$$

This is the distance a roller must travel as it runs against the cam through one revolution of the cam.

(c) Use a computer algebra system to find the volume of steel in the cam.

In Exercises 37–42, use a double integral to find the area of the shaded region.

37.

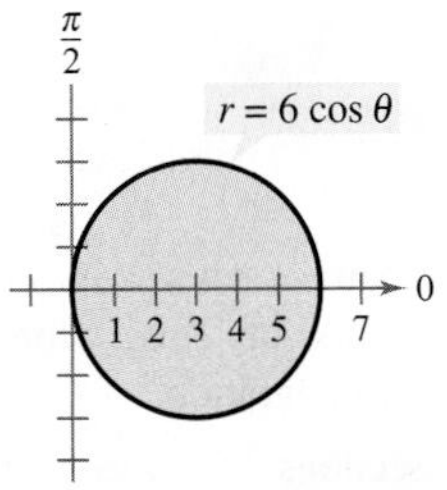

38.

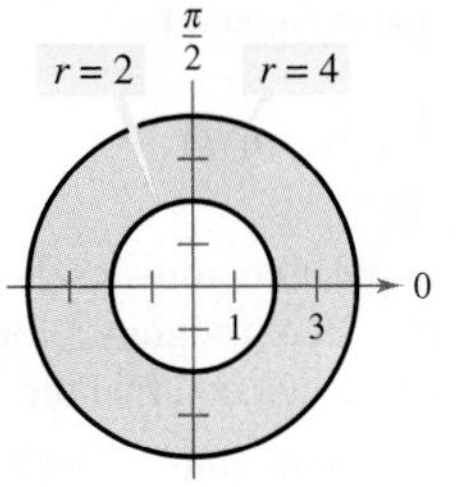

39.

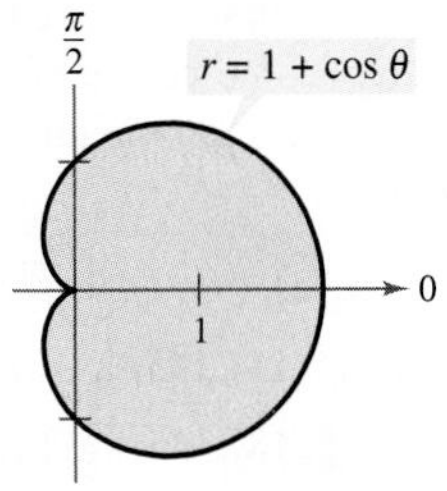

40.

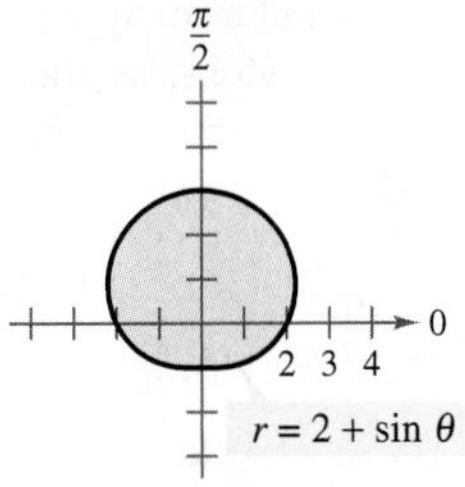

41.

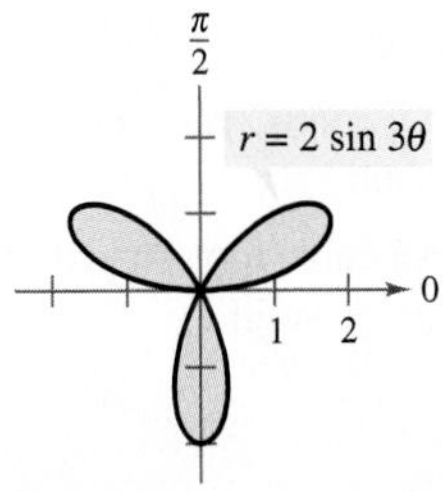

42.

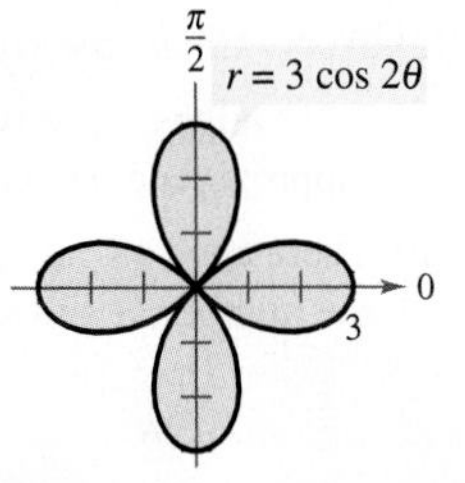

Writing About Concepts

43. Describe the partition of the region R of integration in the xy-plane when polar coordinates are used to evaluate a double integral.

44. Explain how to change from rectangular coordinates to polar coordinates in a double integral.

45. In your own words, describe r-simple regions and θ-simple regions.

46. Each figure shows a region of integration for the double integral $\int_R\int f(x, y)\,dA$. For each region, state whether horizontal representative elements, vertical representative elements, or polar sectors would yield the easiest method for obtaining the limits of integration. Explain your reasoning.

(a)

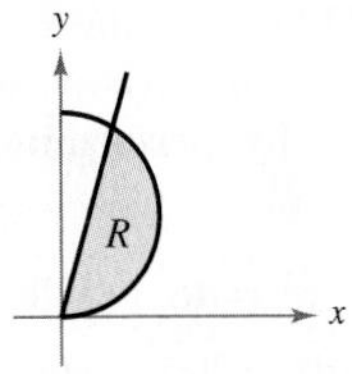

(b)

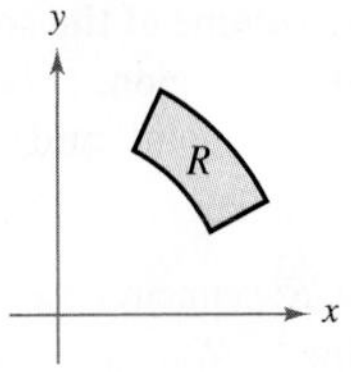

(c)

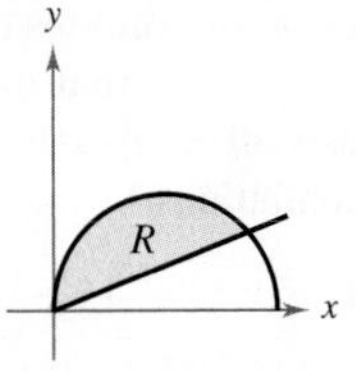

47. ***Think About It*** Consider the program you wrote to approximate double integrals in rectangular coordinates in Exercise 68, in Section 14.2. If the program is used to approximate the double integral

$$\int_R\int f(r, \theta)\, dA$$

in polar coordinates, how will you modify f when it is entered into the program? Because the limits of integration are constants, describe the plane region of integration.

48. ***Approximation*** Horizontal cross sections of a piece of ice that broke from a glacier are in the shape of a quarter of a circle with a radius of approximately 50 feet. The base is divided into 20 subregions, as shown in the figure. At the center of each subregion, the height of the ice is measured, yielding the following points in cylindrical coordinates.

$\left(5, \frac{\pi}{16}, 7\right), \left(15, \frac{\pi}{16}, 8\right), \left(25, \frac{\pi}{16}, 10\right), \left(35, \frac{\pi}{16}, 12\right), \left(45, \frac{\pi}{16}, 9\right),$
$\left(5, \frac{3\pi}{16}, 9\right), \left(15, \frac{3\pi}{16}, 10\right), \left(25, \frac{3\pi}{16}, 14\right), \left(35, \frac{3\pi}{16}, 15\right), \left(45, \frac{3\pi}{16}, 10\right),$
$\left(5, \frac{5\pi}{16}, 9\right), \left(15, \frac{5\pi}{16}, 11\right), \left(25, \frac{5\pi}{16}, 15\right), \left(35, \frac{5\pi}{16}, 18\right), \left(45, \frac{5\pi}{16}, 14\right),$
$\left(5, \frac{7\pi}{16}, 5\right), \left(15, \frac{7\pi}{16}, 8\right), \left(25, \frac{7\pi}{16}, 11\right), \left(35, \frac{7\pi}{16}, 16\right), \left(45, \frac{7\pi}{16}, 12\right)$

(a) Approximate the volume of the solid.

(b) Ice weighs approximately 57 pounds per cubic foot. Approximate the weight of the solid.

(c) There are 7.48 gallons of water per cubic foot. Approximate the number of gallons of water in the solid.

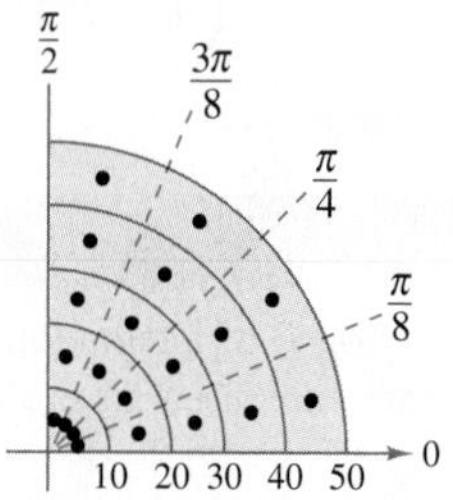

Approximation **In Exercises 49 and 50, use a computer algebra system to approximate the iterated integral.**

49. $\displaystyle\int_{\pi/4}^{\pi/2}\int_0^5 r\sqrt{1 + r^3}\sin\sqrt{\theta}\, dr\, d\theta$

50. $\displaystyle\int_0^{\pi/4}\int_0^4 5re^{\sqrt{r\theta}}\, dr\, d\theta$

Approximation **In Exercises 51 and 52, determine which value best approximates the volume of the solid between the** xy**-plane and the function over the region. (Make your selection on the basis of a sketch of the solid and** ***not*** **by performing any calculations.)**

51. $f(x, y) = 15 - 2y$; R: semicircle: $x^2 + y^2 = 16, y \geq 0$

(a) 100 (b) 200 (c) 300 (d) -200 (e) 800

52. $f(x, y) = xy + 2$; R: quarter circle: $x^2 + y^2 = 9, x \geq 0, y \geq 0$

(a) 25 (b) 8 (c) 100 (d) 50 (e) -30

True or False? **In Exercises 53 and 54, determine whether the statement is true or false. If it is false, explain why or give an example that shows it is false.**

53. If $\int_R\int f(r, \theta)\, dA > 0$, then $f(r, \theta) > 0$ for all (r, θ) in R.

54. If $f(r, \theta)$ is a constant function and the area of the region S is twice that of the region R, then $2\int_R\int f(r, \theta)\, dA = \int_S\int f(r, \theta)\, dA$.

55. ***Probability*** The value of the integral $I = \int_{-\infty}^{\infty} e^{-x^2/2}\, dx$ is required in the development of the normal probability density function.

(a) Use polar coordinates to evaluate the improper integral.

$$I^2 = \left(\int_{-\infty}^{\infty} e^{-x^2/2}\, dx\right)\left(\int_{-\infty}^{\infty} e^{-y^2/2}\, dy\right)$$
$$= \int_{-\infty}^{\infty}\int_{-\infty}^{\infty} e^{-(x^2+y^2)/2}\, dA$$

(b) Use the result of part (a) to determine I.

FOR FURTHER INFORMATION For more information on this problem, see the article "Integrating e^{-x^2} Without Polar Coordinates" by William Dunham in *Mathematics Teacher*. To view this article, go to the website *www.matharticles.com*.

56. Use the result of Exercise 55 and a change of variables to evaluate each integral. No integration is required.

(a) $\displaystyle\int_{-\infty}^{\infty} e^{-x^2}\, dx$ (b) $\displaystyle\int_{-\infty}^{\infty} e^{-4x^2}\, dx$

57. ***Population*** The population density of a city is approximated by the model $f(x, y) = 4000e^{-0.01(x^2+y^2)}$, $x^2 + y^2 \leq 49$, where x and y are measured in miles. Integrate the density function over the indicated circular region to approximate the population of the city.

58. ***Probability*** Find k such that the function

$$f(x, y) = \begin{cases} ke^{-(x^2+y^2)}, & x \geq 0, y \geq 0 \\ 0, & \text{elsewhere} \end{cases}$$

is a probability density function.

59. ***Think About It*** Consider the region bounded by the graphs of $y = 2$, $y = 4$, $y = x$, and $y = \sqrt{3}x$ and the double integral $\int_R\int f\, dA$. Determine the limits of integration if the region R is divided into (a) horizontal representative elements, (b) vertical representative elements, and (c) polar sectors.

60. Repeat Exercise 59 for a region R bounded by the graph of the equation $(x - 2)^2 + y^2 = 4$.

61. Show that the area A of the polar sector R (see figure) is $A = r\Delta r\Delta\theta$, where $r = (r_1 + r_2)/2$ is the average radius of R.

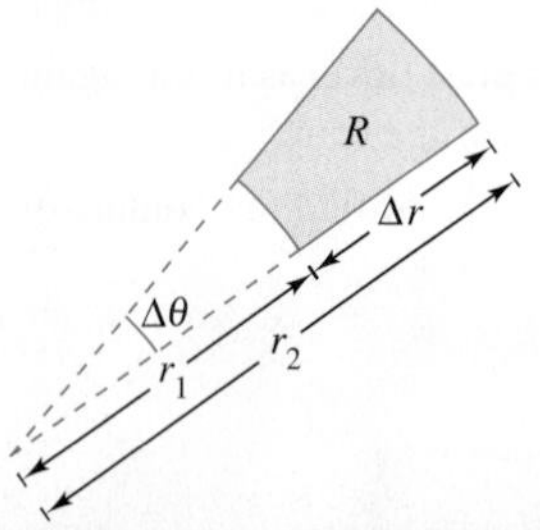

Section 14.4 Center of Mass and Moments of Inertia

- **Find the mass of a planar lamina using a double integral.**
- **Find the center of mass of a planar lamina using double integrals.**
- **Find moments of inertia using double integrals.**

Mass

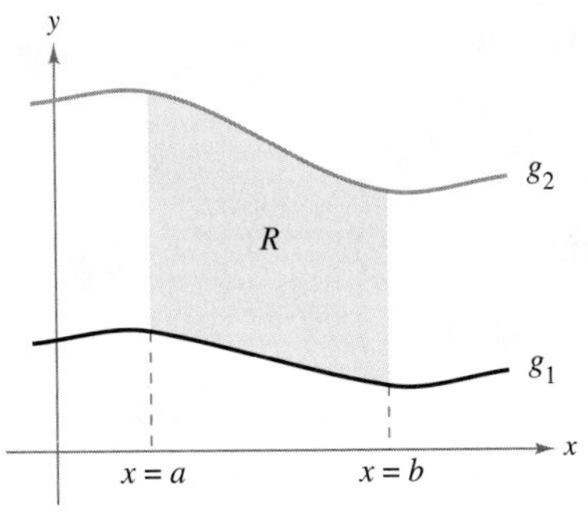

Lamina of constant density ρ
Figure 14.33

Section 7.6 discussed several applications of integration involving a lamina of *constant* density ρ. For example, if the lamina corresponding to the region R, as shown in Figure 14.33, has a constant density ρ, then the mass of the lamina is given by

$$\text{Mass} = \rho A = \rho \int_R\int dA = \int_R\int \rho\, dA. \qquad \text{Constant density}$$

If not otherwise stated, a lamina is assumed to have a constant density. In this section, however, you will extend the definition of the term *lamina* to include thin plates of *variable* density. Double integrals can be used to find the mass of a lamina of *variable* density, where the density at (x, y) is given by the **density function ρ.**

Definition of Mass of a Planar Lamina of Variable Density

If ρ is a continuous density function on the lamina corresponding to a plane region R, then the mass m of the lamina is given by

$$m = \int_R\int \rho(x, y)\, dA. \qquad \text{Variable density}$$

NOTE Density is normally expressed as mass per unit volume. For a planar lamina, however, density is mass per unit surface area.

EXAMPLE 1 Finding the Mass of a Planar Lamina

Find the mass of the triangular lamina with vertices $(0, 0)$, $(0, 3)$, and $(2, 3)$, given that the density at (x, y) is $\rho(x, y) = 2x + y$.

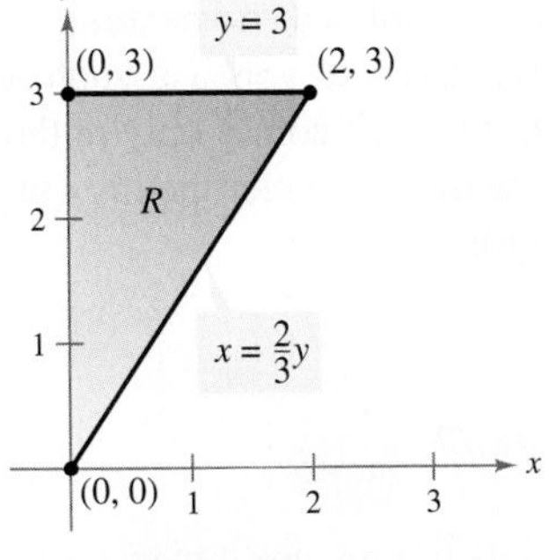

Lamina of variable density
$\rho(x, y) = 2x + y$
Figure 14.34

Solution As shown in Figure 14.34, region R has the boundaries $x = 0$, $y = 3$, and $y = 3x/2$ (or $x = 2y/3$). Therefore, the mass of the lamina is

$$\begin{aligned} m = \int_R\int (2x + y)\, dA &= \int_0^3\int_0^{2y/3} (2x + y)\, dx\, dy \\ &= \int_0^3 \Big[x^2 + xy\Big]_0^{2y/3} dy \\ &= \frac{10}{9}\int_0^3 y^2\, dy \\ &= \frac{10}{9}\left[\frac{y^3}{3}\right]_0^3 \\ &= 10. \end{aligned}$$

NOTE In Figure 14.34, note that the planar lamina is shaded so that the darkest shading corresponds to the densest part.

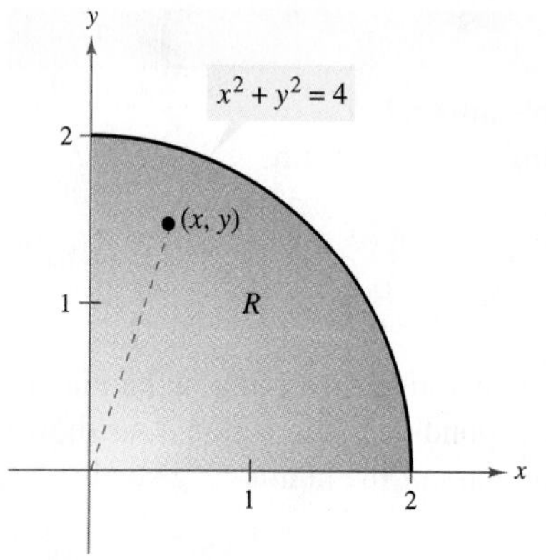

Density at (x, y): $\rho(x, y) = k\sqrt{x^2 + y^2}$
Figure 14.35

EXAMPLE 2 Finding Mass by Polar Coordinates

Find the mass of the lamina corresponding to the first-quadrant portion of the circle

$$x^2 + y^2 = 4$$

where the density at the point (x, y) is proportional to the distance between the point and the origin, as shown in Figure 14.35.

Solution At any point (x, y), the density of the lamina is

$$\begin{aligned}\rho(x, y) &= k\sqrt{(x - 0)^2 + (y - 0)^2}\\ &= k\sqrt{x^2 + y^2}.\end{aligned}$$

Because $0 \le x \le 2$ and $0 \le y \le \sqrt{4 - x^2}$, the mass is given by

$$\begin{aligned}m &= \iint_R k\sqrt{x^2 + y^2}\, dA\\ &= \int_0^2 \int_0^{\sqrt{4-x^2}} k\sqrt{x^2 + y^2}\, dy\, dx.\end{aligned}$$

To simplify the integration, you can convert to polar coordinates, using the bounds $0 \le \theta \le \pi/2$ and $0 \le r \le 2$. So, the mass is

$$\begin{aligned}m = \iint_R k\sqrt{x^2 + y^2}\, dA &= \int_0^{\pi/2} \int_0^2 k\sqrt{r^2}\, r\, dr\, d\theta\\ &= \int_0^{\pi/2} \int_0^2 kr^2\, dr\, d\theta\\ &= \int_0^{\pi/2} \frac{kr^3}{3}\Big]_0^2\, d\theta\\ &= \frac{8k}{3}\int_0^{\pi/2} d\theta\\ &= \frac{8k}{3}\Big[\theta\Big]_0^{\pi/2}\\ &= \frac{4\pi k}{3}.\end{aligned}$$

TECHNOLOGY On many occasions, this text has mentioned the benefits of computer programs that perform symbolic integration. Even if you use such a program regularly, you should remember that its greatest benefit comes only in the hands of a knowledgeable user. For instance, notice how much simpler the integral in Example 2 becomes when it is converted to polar form.

Rectangular Form	*Polar Form*
$\int_0^2 \int_0^{\sqrt{4-x^2}} k\sqrt{x^2 + y^2}\, dy\, dx$	$\int_0^{\pi/2} \int_0^2 kr^2\, dr\, d\theta$

If you have access to software that performs symbolic integration, use it to evaluate both integrals. Some software programs cannot handle the first integral, but any program that can handle double integrals can evaluate the second integral.

Moments and Center of Mass

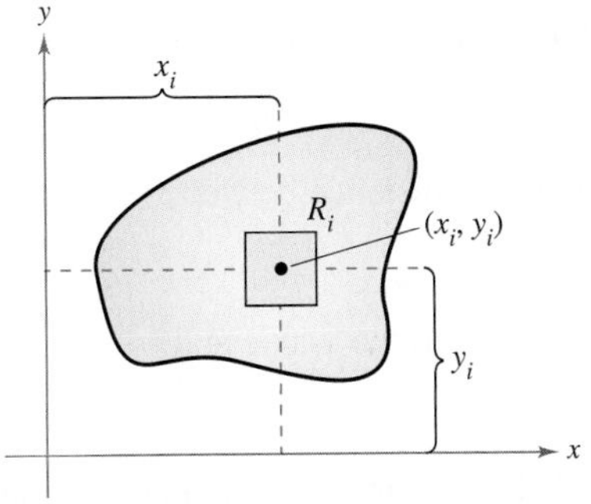

$M_x = (\text{mass})(y_i)$
$M_y = (\text{mass})(x_i)$
Figure 14.36

For a lamina of variable density, moments of mass are defined in a manner similar to that used for the uniform density case. For a partition Δ of a lamina corresponding to a plane region R, consider the ith rectangle R_i of one area ΔA_i, as shown in Figure 14.36. Assume that the mass of R_i is concentrated at one of its interior points (x_i, y_i). The moment of mass of R_i with respect to the x-axis can be approximated by

$$(\text{Mass})(y_i) \approx [\rho(x_i, y_i)\,\Delta A_i](y_i).$$

Similarly, the moment of mass with respect to the y-axis can be approximated by

$$(\text{Mass})(x_i) \approx [\rho(x_i, y_i)\,\Delta A_i](x_i).$$

By forming the Riemann sum of all such products and taking the limits as the norm of Δ approaches 0, you obtain the following definitions of moments of mass with respect to the x- and y-axes.

Moments and Center of Mass of a Variable Density Planar Lamina

Let ρ be a continuous density function on the planar lamina R. The **moments of mass** with respect to the x- and y-axes are

$$M_x = \int_R\int y\rho(x, y)\,dA \quad \text{and} \quad M_y = \int_R\int x\rho(x, y)\,dA.$$

If m is the mass of the lamina, then the **center of mass** is

$$(\bar{x}, \bar{y}) = \left(\frac{M_y}{m}, \frac{M_x}{m}\right).$$

If R represents a simple plane region rather than a lamina, the point $(\bar{x}, \bar{y})$ is called the **centroid** of the region.

For some planar laminas with a constant density ρ, you can determine the center of mass (or one of its coordinates) using symmetry rather than using integration. For instance, consider the laminas of constant density shown in Figure 14.37. Using symmetry, you can see that $\bar{y} = 0$ for the first lamina and $\bar{x} = 0$ for the second lamina.

$R: 0 \le x \le 1$
$-\sqrt{1-x^2} \le y \le \sqrt{1-x^2}$

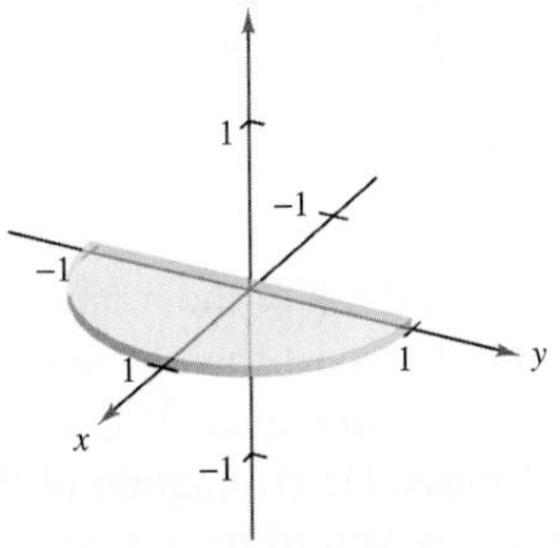

Lamina of constant density and symmetric with respect to the x-axis
Figure 14.37

$R: -\sqrt{1-y^2} \le x \le \sqrt{1-y^2}$
$0 \le y \le 1$

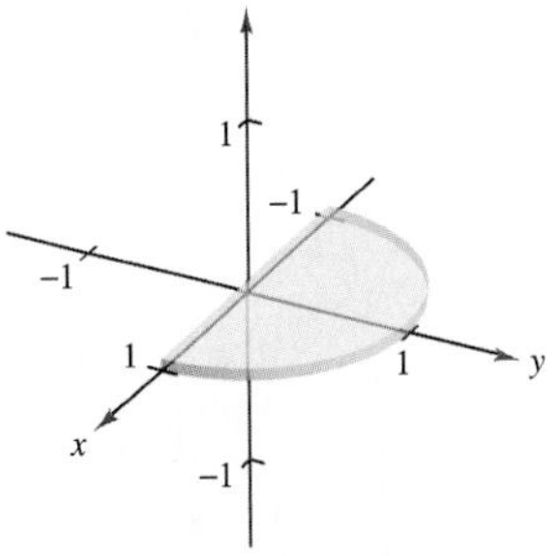

Lamina of constant density and symmetric with respect to the y-axis

EXAMPLE 3 Finding the Center of Mass

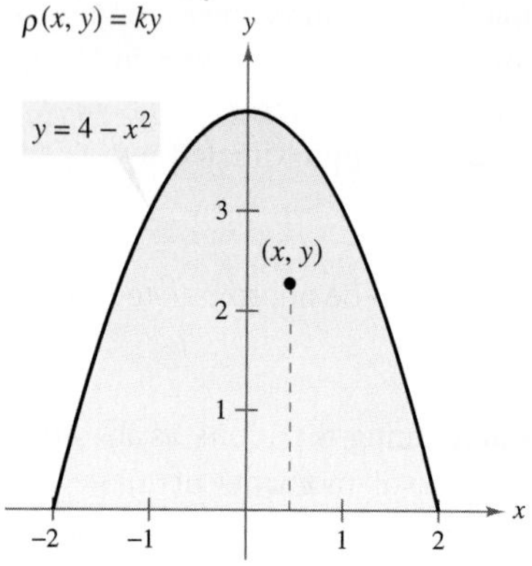

The parabolic region of variable density
Figure 14.38

Find the center of mass of the lamina corresponding to the parabolic region

$$0 \le y \le 4 - x^2 \qquad \text{Parabolic region}$$

where the density at the point (x, y) is proportional to the distance between (x, y) and the x-axis, as shown in Figure 14.38.

Solution Because the lamina is symmetric with respect to the y-axis and

$$\rho(x, y) = ky$$

the center of mass lies on the y-axis. So, $\bar{x} = 0$. To find $\bar{y}$, first find the mass of the lamina.

$$\begin{aligned}
\text{Mass} = \int_{-2}^{2}\int_{0}^{4-x^2} ky\,dy\,dx &= \frac{k}{2}\int_{-2}^{2} y^2\Big]_0^{4-x^2}\,dx \\
&= \frac{k}{2}\int_{-2}^{2} (16 - 8x^2 + x^4)\,dx \\
&= \frac{k}{2}\left[16x - \frac{8x^3}{3} + \frac{x^5}{5}\right]_{-2}^{2} \\
&= k\left(32 - \frac{64}{3} + \frac{32}{5}\right) \\
&= \frac{256k}{15}
\end{aligned}$$

Next, find the moment about the x-axis.

$$\begin{aligned}
M_x = \int_{-2}^{2}\int_{0}^{4-x^2} (y)(ky)\,dy\,dx &= \frac{k}{3}\int_{-2}^{2} y^3\Big]_0^{4-x^2}\,dx \\
&= \frac{k}{3}\int_{-2}^{2} (64 - 48x^2 + 12x^4 - x^6)\,dx \\
&= \frac{k}{3}\left[64x - 16x^3 + \frac{12x^5}{5} - \frac{x^7}{7}\right]_{-2}^{2} \\
&= \frac{4096k}{105}
\end{aligned}$$

So,

$$\bar{y} = \frac{M_x}{m} = \frac{4096k/105}{256k/15} = \frac{16}{7}$$

and the center of mass is $\left(0, \frac{16}{7}\right)$.

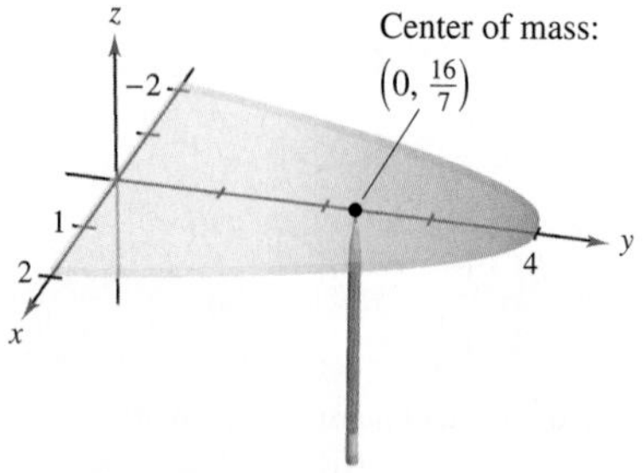

Figure 14.39

Although you can think of the moments M_x and M_y as measuring the tendency to rotate about the x- or y-axis, the calculation of moments is usually an intermediate step toward a more tangible goal. The use of the moments M_x and M_y is typical—to find the center of mass. Determination of the center of mass is useful in a variety of applications that allow you to treat a lamina as if its mass were concentrated at just one point. Intuitively, you can think of the center of mass as the balancing point of the lamina. For instance, the lamina in Example 3 should balance on the point of a pencil placed at $\left(0, \frac{16}{7}\right)$, as shown in Figure 14.39.

Moments of Inertia

The moments of M_x and M_y used in determining the center of mass of a lamina are sometimes called the **first moments** about the x- and y-axes. In each case, the moment is the product of a mass times a distance.

$$M_x = \int_R\int (y)\rho(x, y)\, dA \qquad M_y = \int_R\int (x)\rho(x, y)\, dA$$

Distance to x-axis; Mass — Distance to y-axis; Mass

You will now look at another type of moment—the **second moment,** or the **moment of inertia** of a lamina about a line. In the same way that mass is a measure of the tendency of matter to resist a change in straight-line motion, the moment of inertia about a line is a *measure of the tendency of matter to resist a change in rotational motion.* For example, if a particle of mass m is a distance d from a fixed line, its moment of inertia about the line is defined as

$$I = md^2 = (\text{mass})(\text{distance})^2.$$

As with moments of mass, you can generalize this concept to obtain the moments of inertia about the x- and y-axes of a lamina of variable density. These second moments are denoted by I_x and I_y, and in each case the moment is the product of a mass times the square of a distance.

$$I_x = \int_R\int (y^2)\rho(x, y)\, dA \qquad I_y = \int_R\int (x^2)\rho(x, y)\, dA$$

Square of distance to x-axis; Mass — Square of distance to y-axis; Mass

The sum of the moments I_x and I_y is called the **polar moment of inertia** and is denoted by I_0.

NOTE For a lamina in the xy-plane, I_0 represents the moment of inertia of the lamina about the z-axis. The term "polar moment of inertia" stems from the fact that the square of the polar distance r is used in the calculation.

$$I_0 = \int_R\int (x^2 + y^2)\rho(x, y)\, dA = \int_R\int r^2\rho(x, y)\, dA$$

EXAMPLE 4 Finding the Moment of Inertia

Find the moment of inertia about the x-axis of the lamina in Example 3.

Solution From the definition of moment of inertia, you have

$$\begin{aligned}
I_x &= \int_{-2}^{2}\int_{0}^{4-x^2} y^2(ky)\, dy\, dx \\
&= \frac{k}{4}\int_{-2}^{2} y^4\Big]_0^{4-x^2} dx \\
&= \frac{k}{4}\int_{-2}^{2} (256 - 256x^2 + 96x^4 - 16x^6 + x^8)\, dx \\
&= \frac{k}{4}\left[256x - \frac{256x^3}{3} + \frac{96x^5}{5} - \frac{16x^7}{7} + \frac{x^9}{9}\right]_{-2}^{2} \\
&= \frac{32{,}768k}{315}.
\end{aligned}$$

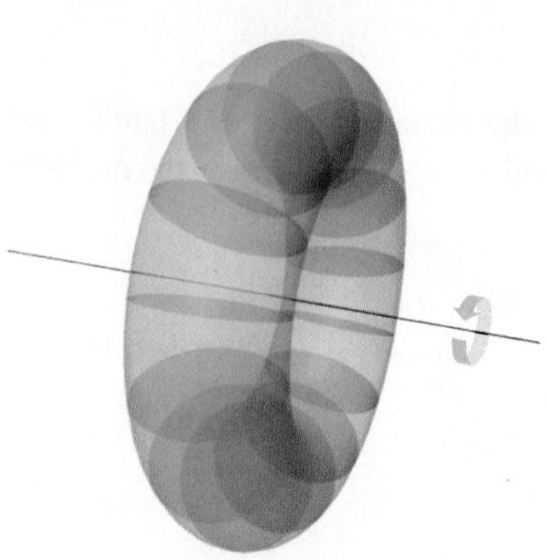

Planar lamina revolving at ω radians per second
Figure 14.40

The moment of inertia I of a revolving lamina can be used to measure its kinetic energy. For example, suppose a planar lamina is revolving about a line with an **angular speed** of ω radians per second, as shown in Figure 14.40. The kinetic energy E of the revolving lamina is

$$E = \frac{1}{2}I\omega^2. \qquad \text{Kinetic energy for rotational motion}$$

On the other hand, the kinetic energy E of a mass m moving in a straight line at a velocity v is

$$E = \frac{1}{2}mv^2. \qquad \text{Kinetic energy for linear motion}$$

So, the kinetic energy of a mass moving in a straight line is proportional to its mass, but the kinetic energy of a mass revolving about an axis is proportional to its moment of inertia.

The **radius of gyration** $\bar{\bar{r}}$ of a revolving mass m with moment of inertia I is defined to be

$$\bar{\bar{r}} = \sqrt{\frac{I}{m}}. \qquad \text{Radius of gyration}$$

If the entire mass were located at a distance $\bar{\bar{r}}$ from its axis of revolution, it would have the same moment of inertia and, consequently, the same kinetic energy. For instance, the radius of gyration of the lamina in Example 4 about the x-axis is given by

$$\bar{\bar{y}} = \sqrt{\frac{I_x}{m}} = \sqrt{\frac{32{,}768k/315}{256k/15}} = \sqrt{\frac{128}{21}} \approx 2.469.$$

EXAMPLE 5 Finding the Radius of Gyration

Find the radius of gyration about the y-axis for the lamina corresponding to the region R: $0 \le y \le \sin x$, $0 \le x \le \pi$, where the density at (x, y) is given by $\rho(x, y) = x$.

Solution The region R is shown in Figure 14.41. By integrating $\rho(x, y) = x$ over the region R, you can determine that the mass of the region is π. The moment of inertia about the y-axis is

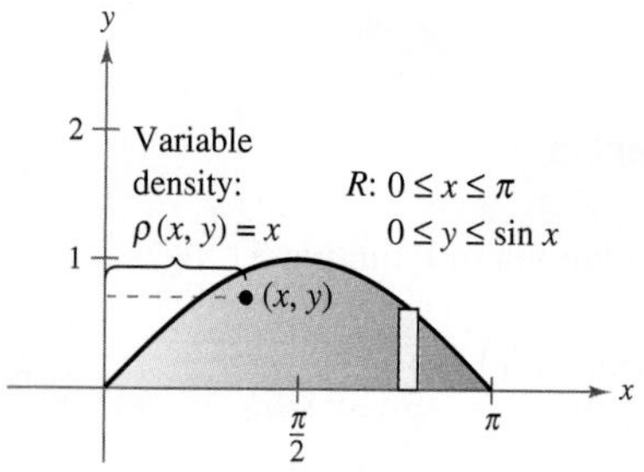

Figure 14.41

$$\begin{aligned}
I_y &= \int_0^{\pi}\int_0^{\sin x} x^3\,dy\,dx \\
&= \int_0^{\pi} x^3 y \Big]_0^{\sin x} dx \\
&= \int_0^{\pi} x^3 \sin x\,dx \\
&= \Big[(3x^2 - 6)(\sin x) - (x^3 - 6x)(\cos x)\Big]_0^{\pi} \\
&= \pi^3 - 6\pi.
\end{aligned}$$

So, the radius of gyration about the y-axis is

$$\begin{aligned}
\bar{\bar{x}} &= \sqrt{\frac{I_y}{m}} \\
&= \sqrt{\frac{\pi^3 - 6\pi}{\pi}} \\
&= \sqrt{\pi^2 - 6} \approx 1.967.
\end{aligned}$$

Exercises for Section 14.4

See www.CalcChat.com for worked-out solutions to odd-numbered exercises.

In Exercises 1–4, find the mass of the lamina described by the inequalities, given that its density is $\rho(x, y) = xy$. (*Hint:* Some of the integrals are simpler in polar coordinates.)

1. $0 \le x \le 4,\ 0 \le y \le 3$

2. $x \ge 0,\ 0 \le y \le 9 - x^2$

3. $x \ge 0,\ 0 \le y \le \sqrt{4 - x^2}$

4. $x \ge 0,\ 3 \le y \le 3 + \sqrt{9 - x^2}$

In Exercises 5–8, find the mass and center of mass of the lamina for each density.

5. R: rectangle with vertices $(0, 0),\ (a, 0),\ (0, b),\ (a, b)$

(a) $\rho = k$ (b) $\rho = ky$ (c) $\rho = kx$

6. R: rectangle with vertices $(0, 0),\ (a, 0),\ (0, b),\ (a, b)$

(a) $\rho = kxy$ (b) $\rho = k(x^2 + y^2)$

7. R: triangle with vertices $(0, 0),\ (b/2, h),\ (b, 0)$

(a) $\rho = k$ (b) $\rho = ky$ (c) $\rho = kx$

8. R: triangle with vertices $(0, 0),\ (0, a),\ (a, 0)$

(a) $\rho = k$ (b) $\rho = x^2 + y^2$

9. ***Translations in the Plane*** Translate the lamina in Exercise 5 to the right five units and determine the resulting center of mass.

10. ***Conjecture*** Use the result of Exercise 9 to make a conjecture about the change in the center of mass when a lamina of constant density is translated h units horizontally or k units vertically. Is the conjecture true if the density is not constant? Explain.

In Exercises 11–22, find the mass and center of mass of the lamina bounded by the graphs of the equations for the given density or densities. (*Hint:* Some of the integrals are simpler in polar coordinates.)

11. $y = \sqrt{a^2 - x^2},\ y = 0$

(a) $\rho = k$ (b) $\rho = k(a - y)y$

12. $x^2 + y^2 = a^2,\ 0 \le x,\ 0 \le y$

(a) $\rho = k$ (b) $\rho = k(x^2 + y^2)$

13. $y = \sqrt{x},\ y = 0,\ x = 4,\ \rho = kxy$

14. $y = x^3,\ y = 0,\ x = 2,\ \rho = kx$

15. $y = \dfrac{1}{1 + x^2},\ y = 0,\ x = -1,\ x = 1,\ \rho = k$

16. $xy = 4,\ x = 1,\ x = 4,\ \rho = kx^2$

17. $x = 16 - y^2,\ x = 0,\ \rho = kx$

18. $y = 9 - x^2,\ y = 0,\ \rho = ky^2$

19. $y = \sin \dfrac{\pi x}{L},\ y = 0,\ x = 0,\ x = L,\ \rho = ky$

20. $y = \cos \dfrac{\pi x}{L},\ y = 0,\ x = 0,\ x = \dfrac{L}{2},\ \rho = k$

21. $y = \sqrt{a^2 - x^2},\ 0 \le y \le x,\ \rho = k$

22. $y = \sqrt{a^2 - x^2},\ y = 0,\ y = x,\ \rho = k\sqrt{x^2 + y^2}$

In Exercises 23–26, use a computer algebra system to find the mass and center of mass of the lamina bounded by the graphs of the equations for the given density.

23. $y = e^{-x},\ y = 0,\ x = 0,\ x = 2,\ \rho = ky$

24. $y = \ln x,\ y = 0,\ x = 1,\ x = e,\ \rho = k/x$

25. $r = 2\cos 3\theta,\ -\dfrac{\pi}{6} \le \theta \le \dfrac{\pi}{6},\ \rho = k$

26. $r = 1 + \cos\theta,\ \rho = k$

In Exercises 27–32, verify the given moment(s) of inertia and find $\bar{x}$ and $\bar{y}$. Assume that each lamina has a density of $\rho = 1$. (These regions are common shapes used in engineering.)

27. Rectangle

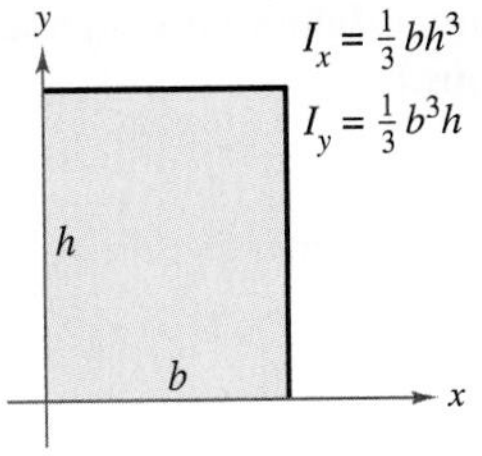

28. Right triangle

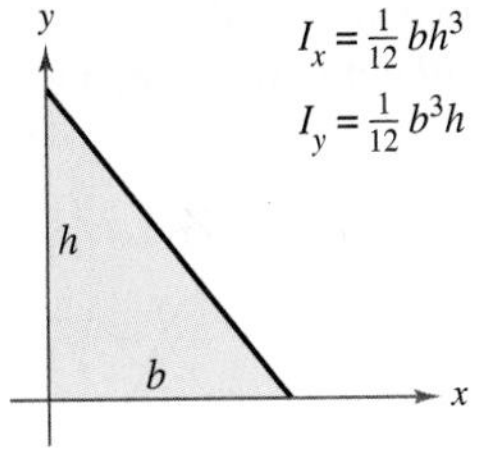

29. Circle

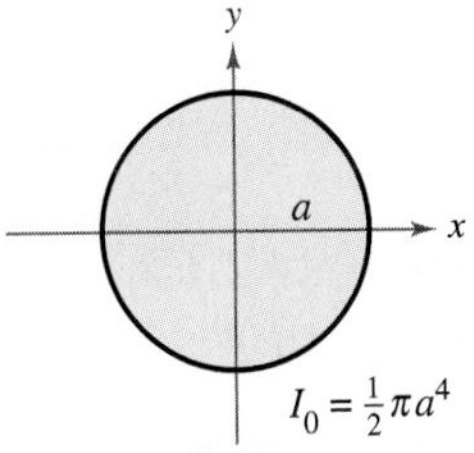

30. Semicircle

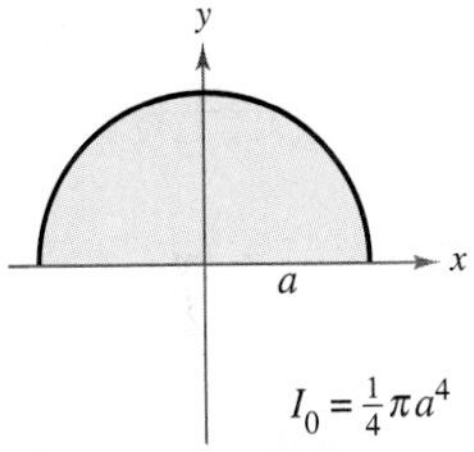

31. Quarter circle

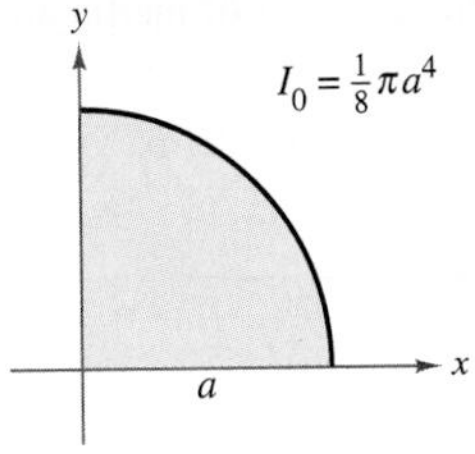

32. Ellipse

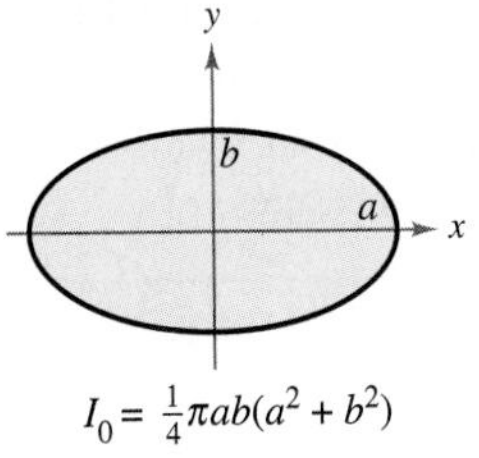

In Exercises 33–40, find I_x, I_y, I_0, $\bar{x}$, and $\bar{y}$ for the lamina bounded by the graphs of the equations. Use a computer algebra system to evaluate the double integrals.

33. $y = 0,\ y = b,\ x = 0,\ x = a,\ \rho = ky$

34. $y = \sqrt{a^2 - x^2},\ y = 0,\ \rho = ky$

35. $y = 4 - x^2,\ y = 0,\ x > 0,\ \rho = kx$

36. $y = x,\ y = x^2,\ \rho = kxy$

37. $y = \sqrt{x},\ y = 0,\ x = 4,\ \rho = kxy$

38. $y = x^2,\ y^2 = x,\ \rho = x^2 + y^2$

39. $y = x^2,\ y^2 = x,\ \rho = kx$

40. $y = x^3,\ y = 4x,\ \rho = k|y|$

In Exercises 41–46, set up the double integral required to find the moment of inertia I, about the given line, of the lamina bounded by the graphs of the equations. Use a computer algebra system to evaluate the double integral.

41. $x^2 + y^2 = b^2$, $\rho = k$, line: $x = a\ (a > b)$

42. $y = 0$, $y = 2$, $x = 0$, $x = 4$, $\rho = k$, line: $x = 6$

43. $y = \sqrt{x}$, $y = 0$, $x = 4$, $\rho = kx$, line: $x = 6$

44. $y = \sqrt{a^2 - x^2}$, $y = 0$, $\rho = ky$, line: $y = a$

45. $y = \sqrt{a^2 - x^2}$, $y = 0$, $x \geq 0$, $\rho = k(a - y)$, line: $y = a$

46. $y = 4 - x^2$, $y = 0$, $\rho = k$, line: $y = 2$

Writing About Concepts

The center of mass of the lamina of constant density shown in the figure is $\left(2, \frac{8}{5}\right)$. In Exercises 47–50, make a conjecture about how the center of mass $(\bar{x}, \bar{y})$ will change for the nonconstant density $\rho(x, y)$. Explain. (Make your conjecture *without* performing any calculations.)

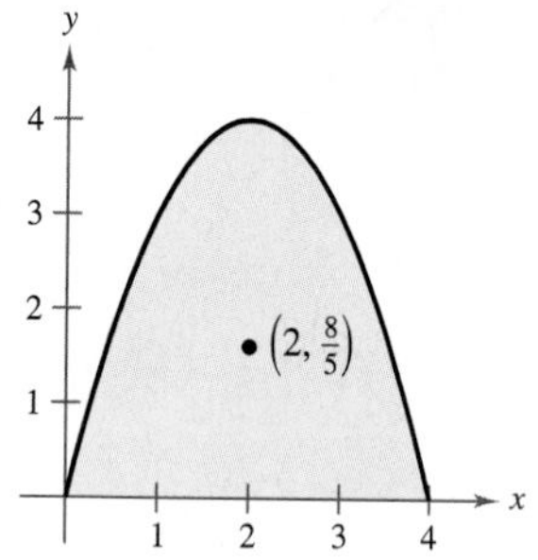

47. $\rho(x, y) = ky$

48. $\rho(x, y) = k|2 - x|$

49. $\rho(x, y) = kxy$

50. $\rho(x, y) = k(4 - x)(4 - y)$

51. Give the formulas for finding the moments and center of mass of a variable density planar lamina.

52. Give the formulas for finding the moments of inertia about the x- and y-axes for a variable density planar lamina.

53. In your own words, describe what the radius of gyration measures.

54. Prove the following Theorem of Pappus: Let R be a region in a plane and let L be a line in the same plane such that L does not intersect the interior of R. If r is the distance between the centroid of R and the line, then the volume V of the solid of revolution formed by revolving R about the line is given by $V = 2\pi rA$, where A is the area of R.

Hydraulics **In Exercises 55–58, determine the location of the horizontal axis y_a at which a vertical gate in a dam is to be hinged so that there is no moment causing rotation under the indicated loading (see figure). The model for y_a is**

$$y_a = \bar{y} - \frac{I_{\bar{y}}}{hA}$$

where $\bar{y}$ is the y-coordinate of the centroid of the gate, $I_{\bar{y}}$ is the moment of inertia of the gate about the line $y = \bar{y}$, h is the depth of the centroid below the surface, and A is the area of the gate.

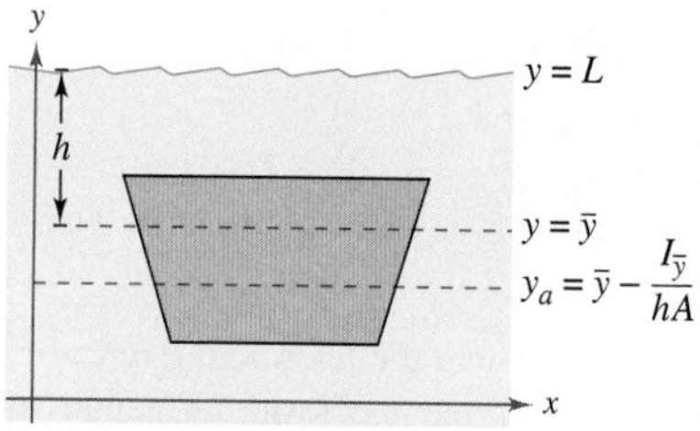

55.

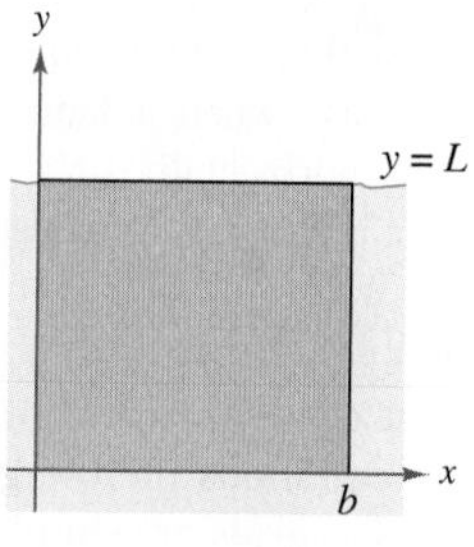

56.

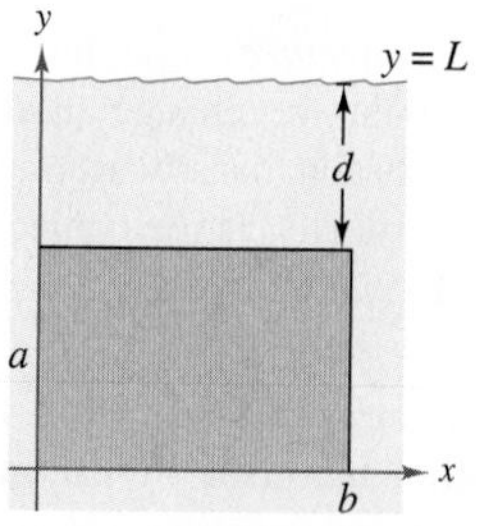

57.

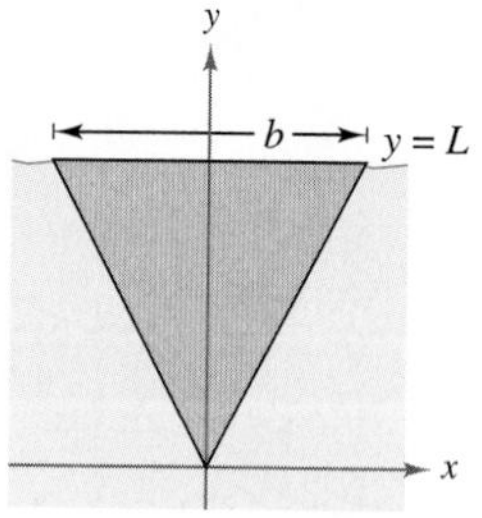

58.

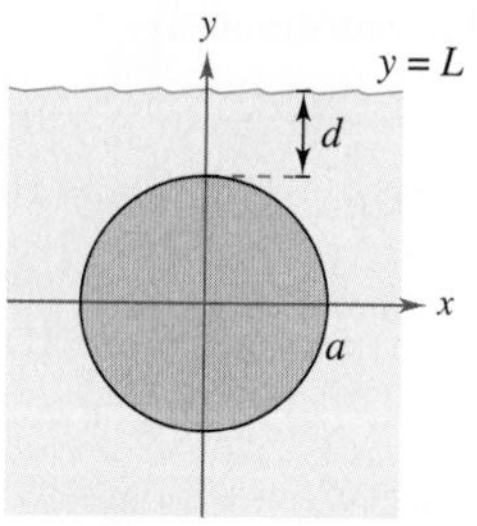

Section Project: Center of Pressure on a Sail

The center of pressure on a sail is that point (x_p, y_p) at which the total aerodynamic force may be assumed to act. If the sail is represented by a plane region R, the center of pressure is

$$x_p = \frac{\int_R\int xy\,dA}{\int_R\int y\,dA} \quad \text{and} \quad y_p = \frac{\int_R\int y^2\,dA}{\int_R\int y\,dA}.$$

Consider a triangular sail with vertices at $(0, 0)$, $(2, 1)$, and $(0, 5)$. Verify the values of each integral.

(a) $\displaystyle\int_R\int y\,dA = 10$ (b) $\displaystyle\int_R\int xy\,dA = \frac{35}{6}$ (c) $\displaystyle\int_R\int y^2\,dA = \frac{155}{6}$

Calculate the coordinates (x_p, y_p) of the center of pressure. Sketch a graph of the sail and indicate the location of the center of pressure.

Section 14.5 Surface Area

- **Use a double integral to find the area of a surface.**

Surface Area

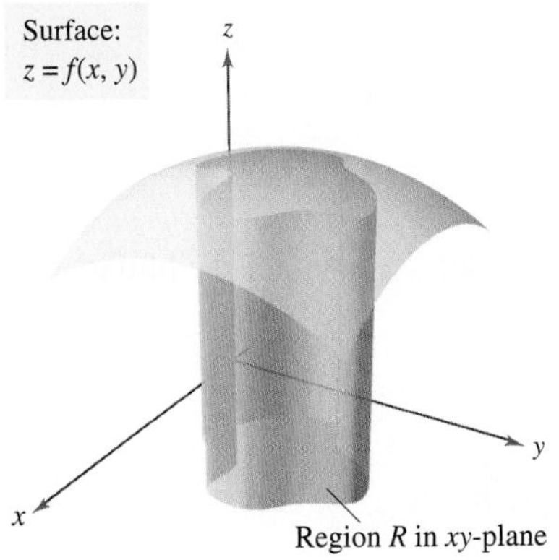

Figure 14.42

At this point you know a great deal about the solid region lying between a surface and a closed and bounded region R in the xy-plane, as shown in Figure 14.42. For example, you know how to find the extrema of f on R (Section 13.8), the area of the base R of the solid (Section 14.1), the volume of the solid (Section 14.2), and the centroid of the base R (Section 14.4).

In this section, you will learn how to find the upper **surface area** of the solid. Later, you will learn how to find the centroid of the solid (Section 14.6) and the lateral surface area (Section 15.2).

To begin, consider a surface S given by

$$z = f(x, y) \qquad \text{Surface defined over a region } R$$

defined over a region R. Assume that R is closed and bounded and that f has continuous first partial derivatives. To find the surface area, construct an inner partition of R consisting of n rectangles, where the area of the ith rectangle R_i is $\Delta A_i = \Delta x_i \Delta y_i$, as shown in Figure 14.43. In each R_i let (x_i, y_i) be the point that is closest to the origin. At the point $(x_i, y_i, z_i) = (x_i, y_i, f(x_i, y_i))$ on the surface S, construct a tangent plane T_i. The area of the portion of the tangent plane that lies directly above R_i is approximately equal to the area of the surface lying directly above R_i. That is, $\Delta T_i \approx \Delta S_i$. So, the surface area of S is given by

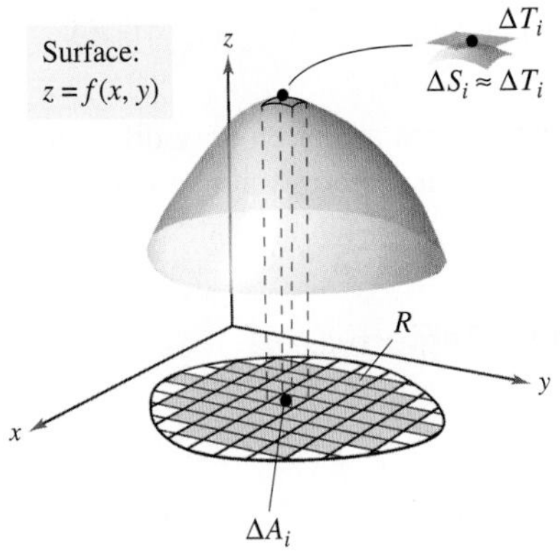

Figure 14.43

$$\sum_{i=1}^{n} \Delta S_i \approx \sum_{i=1}^{n} \Delta T_i.$$

To find the area of the parallelogram ΔT_i, note that its sides are given by the vectors

$$\mathbf{u} = \Delta x_i \mathbf{i} + f_x(x_i, y_i)\, \Delta x_i \mathbf{k}$$

and

$$\mathbf{v} = \Delta y_i \mathbf{j} + f_y(x_i, y_i)\, \Delta y_i \mathbf{k}.$$

From Theorem 11.8, the area of ΔT_i is given by $\|\mathbf{u} \times \mathbf{v}\|$, where

$$\begin{aligned}
\mathbf{u} \times \mathbf{v} &= \begin{vmatrix} \mathbf{i} & \mathbf{j} & \mathbf{k} \\ \Delta x_i & 0 & f_x(x_i, y_i)\, \Delta x_i \\ 0 & \Delta y_i & f_y(x_i, y_i)\, \Delta y_i \end{vmatrix} \\
&= -f_x(x_i, y_i)\, \Delta x_i \Delta y_i \mathbf{i} - f_y(x_i, y_i)\, \Delta x_i \Delta y_i \mathbf{j} + \Delta x_i \Delta y_i \mathbf{k} \\
&= (-f_x(x_i, y_i)\mathbf{i} - f_y(x_i, y_i)\mathbf{j} + \mathbf{k})\, \Delta A_i.
\end{aligned}$$

So, the area of ΔT_i is $\|\mathbf{u} \times \mathbf{v}\| = \sqrt{[f_x(x_i, y_i)]^2 + [f_y(x_i, y_i)]^2 + 1}\, \Delta A_i$, and

$$\begin{aligned}
\text{Surface area of } S &\approx \sum_{i=1}^{n} \Delta S_i \\
&\approx \sum_{i=1}^{n} \sqrt{1 + [f_x(x_i, y_i)]^2 + [f_y(x_i, y_i)]^2}\, \Delta A_i.
\end{aligned}$$

This suggests the following definition of surface area.

Definition of Surface Area

If f and its first partial derivatives are continuous on the closed region R in the xy-plane, then the **area of the surface S** given by $z = f(x, y)$ over R is given by

$$\text{Surface area} = \iint_R dS$$
$$= \iint_R \sqrt{1 + [f_x(x, y)]^2 + [f_y(x, y)]^2}\, dA.$$

As an aid to remembering the double integral for surface area, it is helpful to note its similarity to the integral for arc length.

Length on x-axis: $\displaystyle\int_a^b dx$

Arc length in xy-plane: $\displaystyle\int_a^b ds = \int_a^b \sqrt{1 + [f'(x)]^2}\, dx$

Area in xy-plane: $\displaystyle\iint_R dA$

Surface area in space: $\displaystyle\iint_R dS = \iint_R \sqrt{1 + [f_x(x, y)]^2 + [f_y(x, y)]^2}\, dA$

Like integrals for arc length, integrals for surface area are often very difficult to evaluate. However, one type that is easily evaluated is demonstrated in the next example.

EXAMPLE 1 The Surface Area of a Plane Region

Find the surface area of the portion of the plane

$$z = 2 - x - y$$

that lies above the circle $x^2 + y^2 \leq 1$ in the first quadrant, as shown in Figure 14.44.

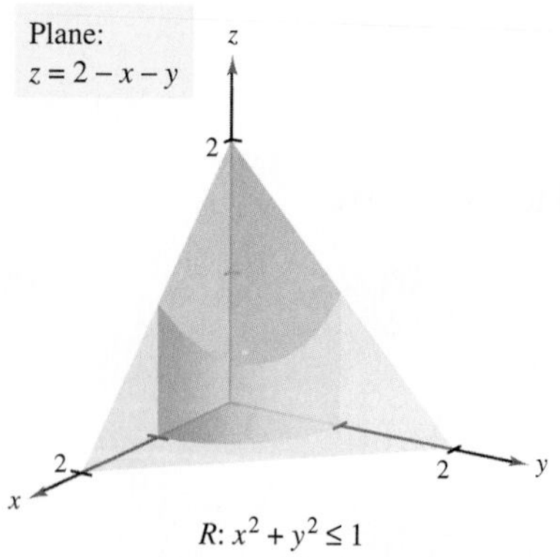

Figure 14.44

Solution Because $f_x(x, y) = -1$ and $f_y(x, y) = -1$, the surface area is given by

$$S = \iint_R \sqrt{1 + [f_x(x, y)]^2 + [f_y(x, y)]^2}\, dA \quad \text{Formula for surface area}$$
$$= \iint_R \sqrt{1 + (-1)^2 + (-1)^2}\, dA \quad \text{Substitute.}$$
$$= \iint_R \sqrt{3}\, dA$$
$$= \sqrt{3} \iint_R dA.$$

Note that the last integral is simply $\sqrt{3}$ times the area of the region R. R is a quarter circle of radius 1, with an area of $\frac{1}{4}\pi(1^2)$ or $\pi/4$. So, the area of S is

$$S = \sqrt{3}\,(\text{area of } R)$$
$$= \sqrt{3}\left(\frac{\pi}{4}\right)$$
$$= \frac{\sqrt{3}\,\pi}{4}.$$

EXAMPLE 2 Finding Surface Area

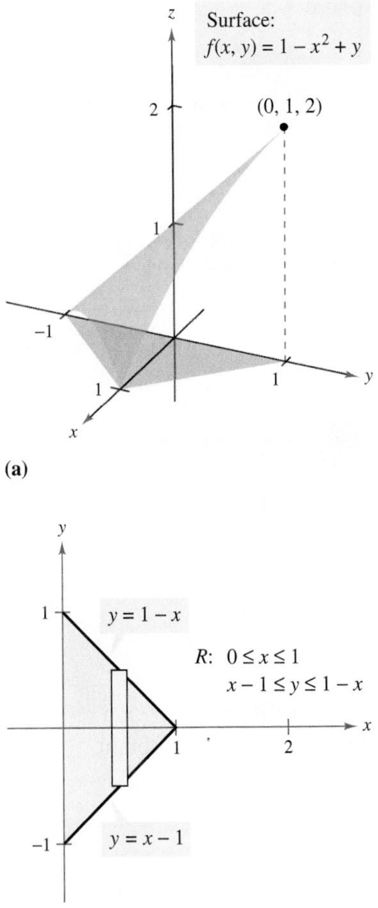

(a)

(b)

Figure 14.45

Find the area of the portion of the surface

$$f(x, y) = 1 - x^2 + y$$

that lies above the triangular region with vertices (1, 0, 0), (0, −1, 0), and (0, 1, 0), as shown in Figure 14.45(a).

Solution Because $f_x(x, y) = -2x$ and $f_y(x, y) = 1$, you have

$$S = \int\int_R \sqrt{1 + [f_x(x, y)]^2 + [f_y(x, y)]^2}\, dA = \int\int_R \sqrt{1 + 4x^2 + 1}\, dA.$$

In Figure 14.45(b), you can see that the bounds for R are $0 \le x \le 1$ and $x - 1 \le y \le 1 - x$. So, the integral becomes

$$\begin{aligned} S &= \int_0^1 \int_{x-1}^{1-x} \sqrt{2 + 4x^2}\, dy\, dx \\ &= \int_0^1 y\sqrt{2 + 4x^2}\Big]_{x-1}^{1-x} dx \\ &= \int_0^1 \left[(1 - x)\sqrt{2 + 4x^2} - (x - 1)\sqrt{2 + 4x^2}\right] dx \\ &= \int_0^1 \left(2\sqrt{2 + 4x^2} - 2x\sqrt{2 + 4x^2}\right) dx \\ &= \left[x\sqrt{2 + 4x^2} + \ln\left(2x + \sqrt{2 + 4x^2}\right) - \frac{(2 + 4x^2)^{3/2}}{6}\right]_0^1 \\ &= \sqrt{6} + \ln\left(2 + \sqrt{6}\right) - \sqrt{6} - \ln\sqrt{2} + \frac{1}{3}\sqrt{2} \approx 1.618. \end{aligned}$$

Integration tables (Appendix B), Formula 26 and Power Rule

EXAMPLE 3 Change of Variables to Polar Coordinates

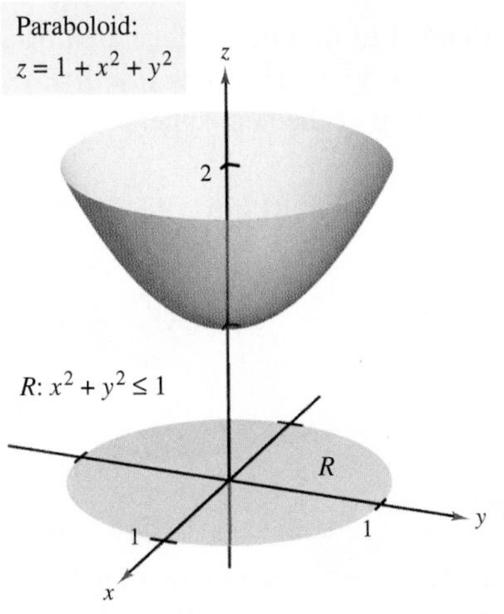

Figure 14.46

Find the surface area of the paraboloid $z = 1 + x^2 + y^2$ that lies above the unit circle, as shown in Figure 14.46.

Solution Because $f_x(x, y) = 2x$ and $f_y(x, y) = 2y$, you have

$$S = \int\int_R \sqrt{1 + [f_x(x, y)]^2 + [f_y(x, y)]^2}\, dA = \int\int_R \sqrt{1 + 4x^2 + 4y^2}\, dA.$$

You can convert to polar coordinates by letting $x = r\cos\theta$ and $y = r\sin\theta$. Then, because the region R is bounded by $0 \le r \le 1$ and $0 \le \theta \le 2\pi$, you have

$$\begin{aligned} S &= \int_0^{2\pi} \int_0^1 \sqrt{1 + 4r^2}\, r\, dr\, d\theta \\ &= \int_0^{2\pi} \frac{1}{12}(1 + 4r^2)^{3/2}\Big]_0^1 d\theta \\ &= \int_0^{2\pi} \frac{5\sqrt{5} - 1}{12}\, d\theta \\ &= \frac{5\sqrt{5} - 1}{12}\,\theta\Big]_0^{2\pi} \\ &= \frac{\pi\left(5\sqrt{5} - 1\right)}{6} \\ &\approx 5.33. \end{aligned}$$

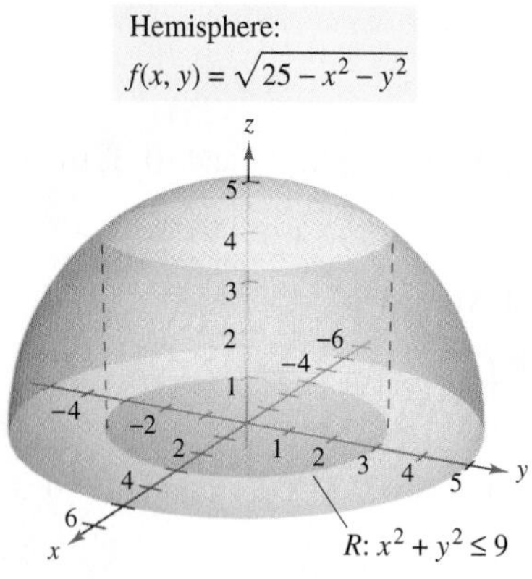

Figure 14.47

EXAMPLE 4 Finding Surface Area

Find the surface area S of the portion of the hemisphere

$$f(x, y) = \sqrt{25 - x^2 - y^2} \qquad \text{Hemisphere}$$

that lies above the region R bounded by the circle $x^2 + y^2 \le 9$, as shown in Figure 14.47.

Solution The first partial derivatives of f are

$$f_x(x, y) = \frac{-x}{\sqrt{25 - x^2 - y^2}} \quad \text{and} \quad f_y(x, y) = \frac{-y}{\sqrt{25 - x^2 - y^2}}$$

and, from the formula for surface area, you have

$$\begin{aligned} dS &= \sqrt{1 + [f_x(x, y)]^2 + [f_y(x, y)]^2}\, dA \\ &= \sqrt{1 + \left(\frac{-x}{\sqrt{25 - x^2 - y^2}}\right)^2 + \left(\frac{-y}{\sqrt{25 - x^2 - y^2}}\right)^2}\, dA \\ &= \frac{5}{\sqrt{25 - x^2 - y^2}}\, dA. \end{aligned}$$

So, the surface area is

$$S = \int_R\!\!\int \frac{5}{\sqrt{25 - x^2 - y^2}}\, dA.$$

You can convert to polar coordinates by letting $x = r\cos\theta$ and $y = r\sin\theta$. Then, because the region R is bounded by $0 \le r \le 3$ and $0 \le \theta \le 2\pi$, you obtain

$$\begin{aligned} S &= \int_0^{2\pi}\!\!\int_0^3 \frac{5}{\sqrt{25 - r^2}}\, r\, dr\, d\theta \\ &= 5\int_0^{2\pi} -\sqrt{25 - r^2}\,\Big]_0^3\, d\theta \\ &= 5\int_0^{2\pi} d\theta \\ &= 10\pi. \end{aligned}$$

The procedure used in Example 4 can be extended to find the surface area of a sphere by using the region R bounded by the circle $x^2 + y^2 \le a^2$, where $0 < a < 5$, as shown in Figure 14.48. The surface area of the portion of the hemisphere

$$f(x, y) = \sqrt{25 - x^2 - y^2}$$

lying above the circular region can be shown to be

$$\begin{aligned} S &= \int_R\!\!\int \frac{5}{\sqrt{25 - x^2 - y^2}}\, dA \\ &= \int_0^{2\pi}\!\!\int_0^a \frac{5}{\sqrt{25 - r^2}}\, r\, dr\, d\theta \\ &= 10\pi\left(5 - \sqrt{25 - a^2}\right). \end{aligned}$$

By taking the limit as a approaches 5 and doubling the result, you obtain a total area of 100π. (The surface area of a sphere of radius r is $S = 4\pi r^2$.)

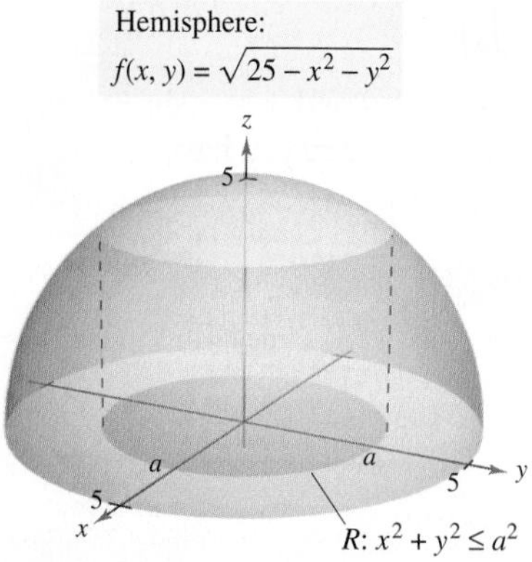

Figure 14.48

You can use Simpson's Rule or the Trapezoidal Rule to approximate the value of a double integral, *provided* you can get through the first integration. This is demonstrated in the next example.

EXAMPLE 5 Approximating Surface Area by Simpson's Rule

Find the area of the surface of the paraboloid

$$f(x, y) = 2 - x^2 - y^2 \qquad \text{Paraboloid}$$

that lies above the square region bounded by $-1 \le x \le 1$ and $-1 \le y \le 1$, as shown in Figure 14.49.

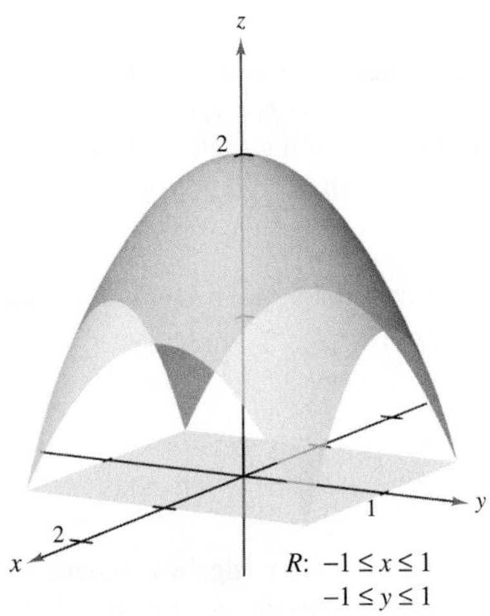

Figure 14.49

Solution Using the partial derivatives

$$f_x(x, y) = -2x \quad \text{and} \quad f_y(x, y) = -2y$$

you have a surface area of

$$\begin{aligned} S &= \iint_R \sqrt{1 + [f_x(x, y)]^2 + [f_y(x, y)]^2}\, dA \\ &= \iint_R \sqrt{1 + (-2x)^2 + (-2y)^2}\, dA \\ &= \iint_R \sqrt{1 + 4x^2 + 4y^2}\, dA. \end{aligned}$$

In polar coordinates, the line $x = 1$ is given by $r \cos \theta = 1$ or $r = \sec \theta$, and you can determine from Figure 14.50 that one-fourth of the region R is bounded by

$$0 \le r \le \sec \theta \quad \text{and} \quad -\frac{\pi}{4} \le \theta \le \frac{\pi}{4}.$$

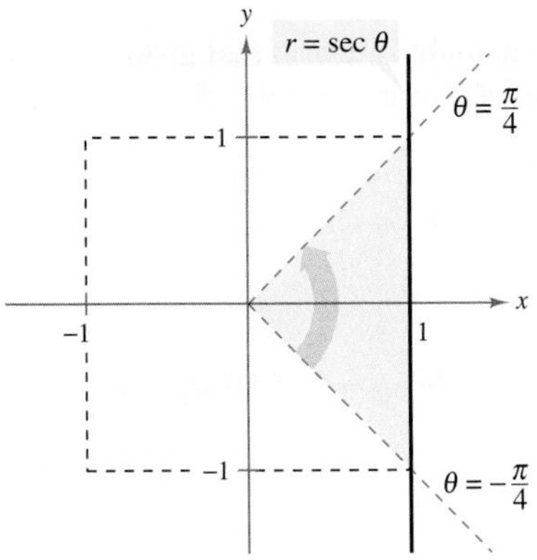

One-fourth of the region R is bounded by $0 \le r \le \sec \theta$ and $-\frac{\pi}{4} \le \theta \le \frac{\pi}{4}$.

Figure 14.50

Letting $x = r \cos \theta$ and $y = r \sin \theta$ produces

$$\begin{aligned} \frac{1}{4} S &= \frac{1}{4} \iint_R \sqrt{1 + 4x^2 + 4y^2}\, dA \\ &= \int_{-\pi/4}^{\pi/4} \int_0^{\sec \theta} \sqrt{1 + 4r^2}\, r\, dr\, d\theta \\ &= \int_{-\pi/4}^{\pi/4} \frac{1}{12} (1 + 4r^2)^{3/2} \Big]_0^{\sec \theta} d\theta \\ &= \frac{1}{12} \int_{-\pi/4}^{\pi/4} [(1 + 4 \sec^2 \theta)]^{3/2} - 1]\, d\theta. \end{aligned}$$

Finally, using Simpson's Rule with $n = 10$, you can approximate this single integral to be

$$\begin{aligned} S &= \frac{1}{3} \int_{-\pi/4}^{\pi/4} [(1 + 4 \sec^2 \theta)^{3/2} - 1]\, d\theta \\ &\approx 7.450. \end{aligned}$$

TECHNOLOGY Most computer programs that are capable of performing symbolic integration for multiple integrals are also capable of performing numerical approximation techniques. If you have access to such software, use it to approximate the value of the integral in Example 5.

Exercises for Section 14.5

See www.CalcChat.com for worked-out solutions to odd-numbered exercises.

In Exercises 1–14, find the area of the surface given by $z = f(x, y)$ over the region R. (*Hint:* Some of the integrals are simpler in polar coordinates.)

1. $f(x, y) = 2x + 2y$
   R: triangle with vertices $(0, 0)$, $(2, 0)$, $(0, 2)$
2. $f(x, y) = 15 + 2x - 3y$
   R: square with vertices $(0, 0)$, $(3, 0)$, $(0, 3)$, $(3, 3)$
3. $f(x, y) = 8 + 2x + 2y$
   $R = \{(x, y): x^2 + y^2 \le 4\}$
4. $f(x, y) = 10 + 2x - 3y$
   $R = \{(x, y): x^2 + y^2 \le 9\}$
5. $f(x, y) = 9 - x^2$
   R: square with vertices $(0, 0)$, $(3, 0)$, $(0, 3)$, $(3, 3)$
6. $f(x, y) = y^2$
   R: square with vertices $(0, 0)$, $(3, 0)$, $(0, 3)$, $(3, 3)$
7. $f(x, y) = 2 + x^{3/2}$
   R: rectangle with vertices $(0, 0)$, $(0, 4)$, $(3, 4)$, $(3, 0)$
8. $f(x, y) = 2 + \frac{2}{3}y^{3/2}$
   $R = \{(x, y): 0 \le x \le 2, 0 \le y \le 2 - x\}$
9. $f(x, y) = \ln|\sec x|$
   $R = \left\{(x, y): 0 \le x \le \dfrac{\pi}{4}, 0 \le y \le \tan x\right\}$
10. $f(x, y) = 9 + x^2 - y^2$
    $R = \{(x, y): x^2 + y^2 \le 4\}$
11. $f(x, y) = \sqrt{x^2 + y^2}$
    $R = \{(x, y): 0 \le f(x, y) \le 1\}$
12. $f(x, y) = xy$
    $R = \{(x, y): x^2 + y^2 \le 16\}$
13. $f(x, y) = \sqrt{a^2 - x^2 - y^2}$
    $R = \{(x, y): x^2 + y^2 \le b^2, 0 < b < a\}$
14. $f(x, y) = \sqrt{a^2 - x^2 - y^2}$
    $R = \{(x, y): x^2 + y^2 \le a^2\}$

In Exercises 15–18, find the area of the surface.

15. The portion of the plane $z = 24 - 3x - 2y$ in the first octant
16. The portion of the paraboloid $z = 16 - x^2 - y^2$ in the first octant
17. The portion of the sphere $x^2 + y^2 + z^2 = 25$ inside the cylinder $x^2 + y^2 = 9$
18. The portion of the cone $z = 2\sqrt{x^2 + y^2}$ inside the cylinder $x^2 + y^2 = 4$

In Exercises 19–24, write a double integral that represents the surface area of $z = f(x, y)$ over the region R. Use a computer algebra system to evaluate the double integral.

19. $f(x, y) = 2y + x^2$
    R: triangle with vertices $(0, 0)$, $(1, 0)$, $(1, 1)$
20. $f(x, y) = 2x + y^2$
    R: triangle with vertices $(0, 0)$, $(2, 0)$, $(2, 2)$
21. $f(x, y) = 4 - x^2 - y^2$
    $R = \{(x, y): 0 \le f(x, y)\}$
22. $f(x, y) = x^2 + y^2$
    $R = \{(x, y): 0 \le f(x, y) \le 16\}$
23. $f(x, y) = 4 - x^2 - y^2$
    $R = \{(x, y): 0 \le x \le 1, 0 \le y \le 1\}$
24. $f(x, y) = \frac{2}{3}x^{3/2} + \cos x$
    $R = \{(x, y): 0 \le x \le 1, 0 \le y \le 1\}$

Approximation **In Exercises 25 and 26, determine which value best approximates the surface area of $z = f(x, y)$ over the region R. (Make your selection on the basis of a sketch of the surface and *not* by performing any calculations.)**

25. $f(x, y) = 10 - \frac{1}{2}y^2$
    R: square with vertices $(0, 0)$, $(4, 0)$, $(4, 4)$, $(0, 4)$
    (a) 16 (b) 200 (c) -100 (d) 72 (e) 36
26. $f(x, y) = \frac{1}{4}\sqrt{x^2 + y^2}$
    R: circle bounded by $x^2 + y^2 = 9$
    (a) -100 (b) 150 (c) 9π (d) 55 (e) 500

In Exercises 27 and 28, use a computer algebra system to approximate the double integral that gives the surface area of the graph of f over the region $R = \{(x, y): 0 \le x \le 1, 0 \le y \le 1\}$.

27. $f(x, y) = e^x$
28. $f(x, y) = \frac{2}{5}y^{5/2}$

In Exercises 29–34, set up a double integral that gives the area of the surface on the graph of f over the region R.

29. $f(x, y) = x^3 - 3xy + y^3$
    R: square with vertices $(1, 1)$, $(-1, 1)$, $(-1, -1)$, $(1, -1)$
30. $f(x, y) = x^2 - 3xy - y^2$
    $R = \{(x, y): 0 \le x \le 4, 0 \le y \le x\}$
31. $f(x, y) = e^{-x} \sin y$
    $R = \{(x, y): x^2 + y^2 \le 4\}$
32. $f(x, y) = \cos(x^2 + y^2)$
    $R = \left\{(x, y): x^2 + y^2 \le \dfrac{\pi}{2}\right\}$
33. $f(x, y) = e^{xy}$
    $R = \{(x, y): 0 \le x \le 4, 0 \le y \le 10\}$
34. $f(x, y) = e^{-x} \sin y$
    $R = \{(x, y): 0 \le x \le 4, 0 \le y \le x\}$

Writing About Concepts

35. State the double integral definition of the area of a surface S given by $z = f(x, y)$ over the region R in the xy-plane.
36. Answer the following questions about the surface area S on a surface given by a positive function $z = f(x, y)$ over a region R in the xy-plane. Explain each answer.
    (a) Is it possible for S to equal the area of R?
    (b) Can S be greater than the area of R?
    (c) Can S be less than the area of R?

37. Find the surface area of the solid of intersection of the cylinders $x^2 + z^2 = 1$ and $y^2 + z^2 = 1$ (see figure).

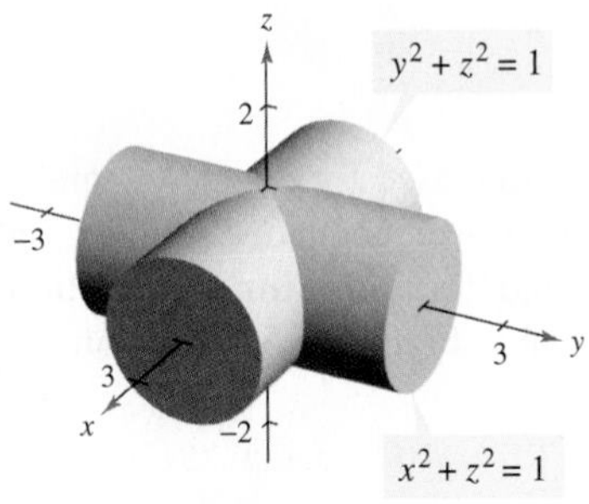

Figure for 37

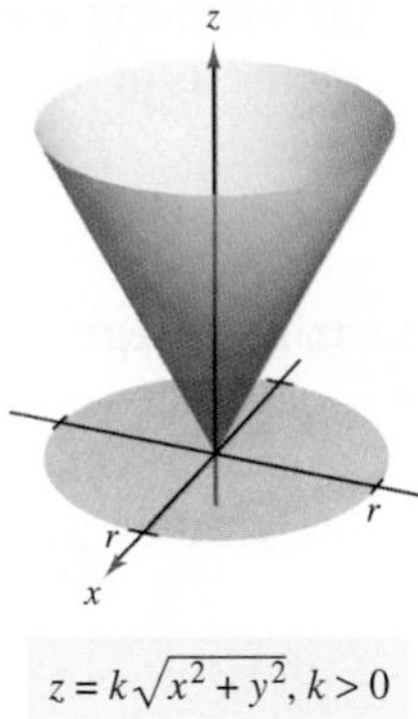

Figure for 38

38. Show that the surface area of the cone $z = k\sqrt{x^2 + y^2}$, $k > 0$ over the circular region $x^2 + y^2 \le r^2$ in the xy-plane is $\pi r^2 \sqrt{k^2 + 1}$ (see figure).

39. ***Building Design*** A new auditorium is built with a foundation in the shape of one-fourth of a circle of radius 50 feet. So, it forms a region R bounded by the graph of $x^2 + y^2 = 50^2$ with $x \ge 0$ and $y \ge 0$. The following equations are models for the floor and ceiling.

Floor: $z = \dfrac{x + y}{5}$

Ceiling: $z = 20 + \dfrac{xy}{100}$

(a) Calculate the volume of the room, which is needed to determine the heating and cooling requirements.

(b) Find the surface area of the ceiling.

40. ***Modeling Data*** A rancher builds a barn with dimensions 30 feet by 50 feet. The symmetrical shape and selected heights of the roof are shown in the figure.

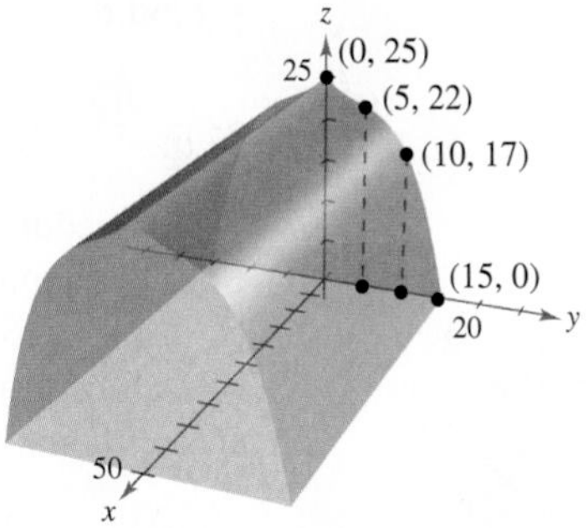

(a) Use the regression capabilities of a graphing utility to find a model of the form $z = ay^3 + by^2 + cy + d$ for the roof line.

(b) Use the numerical integration capabilities of a graphing utility and the model in part (a) to approximate the volume of storage space in the barn.

(c) Use the numerical integration capabilities of a graphing utility and the model in part (a) to approximate the surface area of the roof.

(d) Approximate the arc length of the roof line and find the surface area of the roof by multiplying the arc length by the length of the barn. Compare the results and the integrations with those found in part (c).

41. ***Product Design*** A company produces a spherical object of radius 25 centimeters. A hole of radius 4 centimeters is drilled through the center of the object. Find (a) the volume of the object and (b) the outer surface area of the object.

Section Project: Capillary Action

A well-known property of liquids is that they will rise in narrow vertical channels—this property is called "capillary action." The figure shows two plates, which form a narrow wedge, in a container of liquid. The upper surface of the liquid follows a hyperbolic shape given by

$$z = \frac{k}{\sqrt{x^2 + y^2}}$$

where x, y, and z are measured in inches. The constant k depends on the angle of the wedge, the type of liquid, and the material that comprises the flat plates.

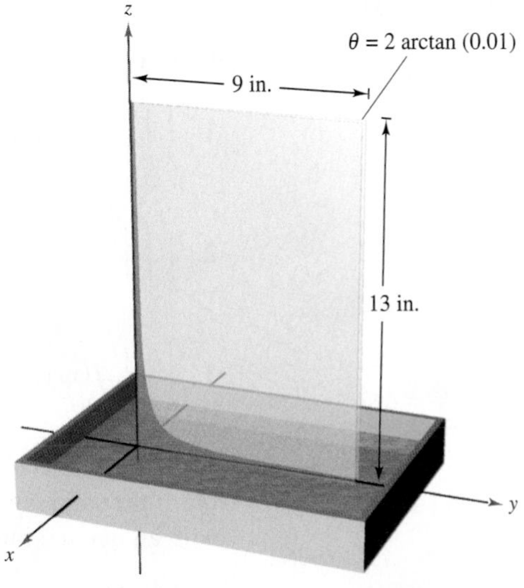

(a) Find the volume of the liquid that has risen in the wedge. (Assume $k = 1$.)

(b) Find the horizontal surface area of the liquid that has risen in the wedge.

Adaptation of Capillary Action problem from "Capillary Phenomena" by Thomas B. Greenslade, Jr., *Physics Teacher*, May 1992. By permission of the author.

Section 14.6 Triple Integrals and Applications

- **Use a triple integral to find the volume of a solid region.**
- **Find the center of mass and moments of inertia of a solid region.**

Triple Integrals

The procedure used to define a **triple integral** follows that used for double integrals. Consider a function f of three variables that is continuous over a bounded solid region Q. Then, encompass Q with a network of boxes and form the **inner partition** consisting of all boxes lying entirely within Q, as shown in Figure 14.51. The volume of the ith box is

$$\Delta V_i = \Delta x_i \Delta y_i \Delta z_i. \qquad \text{Volume of } i\text{th box}$$

The **norm** $\|\Delta\|$ of the partition is the length of the longest diagonal of the n boxes in the partition. Choose a point (x_i, y_i, z_i) in each box and form the Riemann sum

$$\sum_{i=1}^{n} f(x_i, y_i, z_i)\,\Delta V_i.$$

Taking the limit as $\|\Delta\| \to 0$ leads to the following definition.

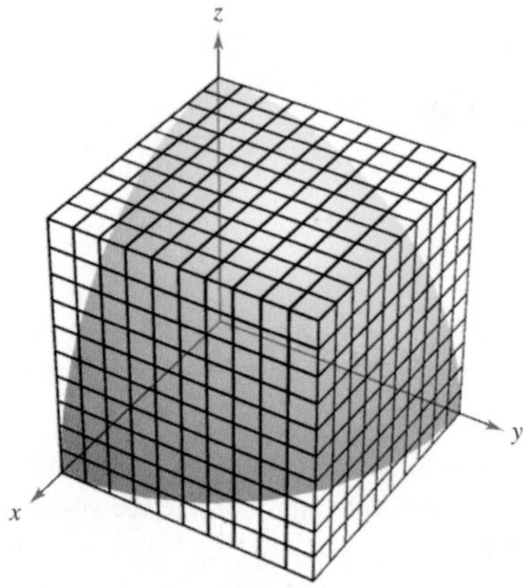

Solid region Q

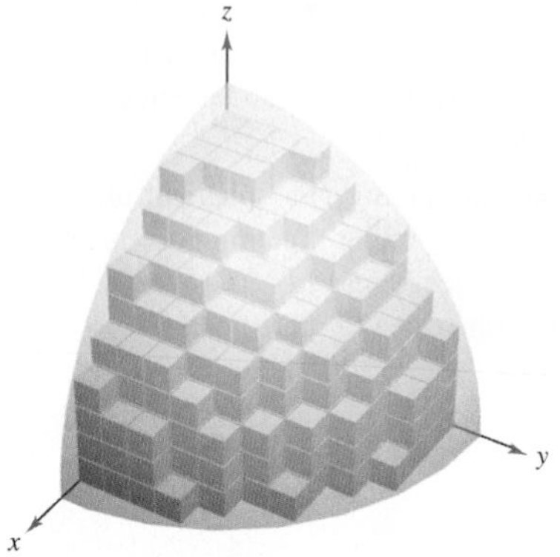

Volume of $Q \approx \sum_{i=1}^{n} f(x_i, y_i, z_i)\,\Delta V_i$

Figure 14.51

Definition of Triple Integral

If f is continuous over a bounded solid region Q, then the **triple integral of f over Q** is defined as

$$\iiint_Q f(x, y, z)\,dV = \lim_{\|\Delta\| \to 0} \sum_{i=1}^{n} f(x_i, y_i, z_i)\,\Delta V_i$$

provided the limit exists. The **volume** of the solid region Q is given by

$$\text{Volume of } Q = \iiint_Q dV.$$

Some of the properties of double integrals in Theorem 14.1 can be restated in terms of triple integrals.

1. $\displaystyle \iiint_Q cf(x, y, z)\,dV = c\iiint_Q f(x, y, z)\,dV$

2. $\displaystyle \iiint_Q [f(x, y, z) \pm g(x, y, z)]\,dV = \iiint_Q f(x, y, z)\,dV \pm \iiint_Q g(x, y, z)\,dV$

3. $\displaystyle \iiint_Q f(x, y, z)\,dV = \iiint_{Q_1} f(x, y, z)\,dV + \iiint_{Q_2} f(x, y, z)\,dV$

In the properties above, Q is the union of two nonoverlapping solid subregions Q_1 and Q_2. If the solid region Q is simple, the triple integral $\iiint f(x, y, z)\,dV$ can be evaluated with an iterated integral using one of the six possible orders of integration:

$$dx\,dy\,dz \quad dy\,dx\,dz \quad dz\,dx\,dy \quad dx\,dz\,dy \quad dy\,dz\,dx \quad dz\,dy\,dx.$$

EXPLORATION

Volume of a Paraboloid Sector On pages 995 and 1003, you were asked to summarize the different ways you know for finding the volume of the solid bounded by the paraboloid

$$z = a^2 - x^2 - y^2, \quad a > 0$$

and the xy-plane. You now know one more way. Use it to find the volume of the solid.

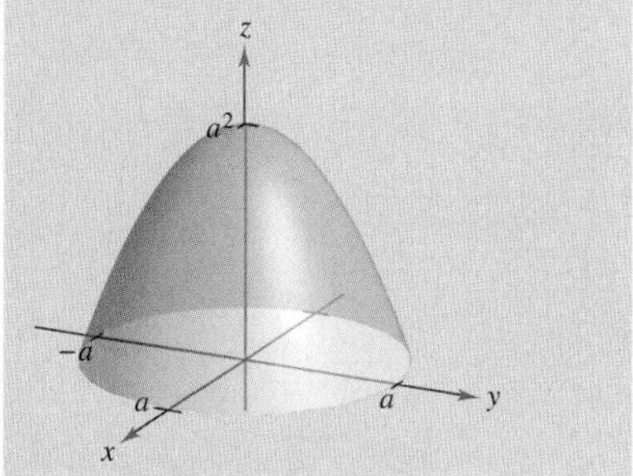

The following version of Fubini's Theorem describes a region that is considered simple with respect to the order $dz\,dy\,dx$. Similar descriptions can be given for the other five orders.

THEOREM 14.4 Evaluation by Iterated Integrals

Let f be continuous on a solid region Q defined by

$$a \le x \le b, \quad h_1(x) \le y \le h_2(x), \quad g_1(x, y) \le z \le g_2(x, y)$$

where h_1, h_2, g_1, and g_2 are continuous functions. Then,

$$\iiint\limits_Q f(x, y, z)\,dV = \int_a^b \int_{h_1(x)}^{h_2(x)} \int_{g_1(x,y)}^{g_2(x,y)} f(x, y, z)\,dz\,dy\,dx.$$

To evaluate a triple iterated integral in the order $dz\,dy\,dx$, hold *both* x and y constant for the innermost integration. Then, hold x constant for the second integration.

EXAMPLE 1 Evaluating a Triple Iterated Integral

Evaluate the triple iterated integral

$$\int_0^2 \int_0^x \int_0^{x+y} e^x(y + 2z)\,dz\,dy\,dx.$$

Solution For the first integration, hold x and y constant and integrate with respect to z.

$$\begin{aligned}\int_0^2 \int_0^x \int_0^{x+y} e^x(y + 2z)\,dz\,dy\,dx &= \int_0^2 \int_0^x e^x(yz + z^2)\Big]_0^{x+y} dy\,dx \\ &= \int_0^2 \int_0^x e^x(x^2 + 3xy + 2y^2)\,dy\,dx\end{aligned}$$

For the second integration, hold x constant and integrate with respect to y.

$$\begin{aligned}\int_0^2 \int_0^x e^x(x^2 + 3xy + 2y^2)\,dy\,dx &= \int_0^2 \left[e^x\left(x^2y + \frac{3xy^2}{2} + \frac{2y^3}{3}\right)\right]_0^x dx \\ &= \frac{19}{6}\int_0^2 x^3e^x\,dx\end{aligned}$$

Finally, integrate with respect to x.

$$\begin{aligned}\frac{19}{6}\int_0^2 x^3e^x\,dx &= \frac{19}{6}\Big[e^x(x^3 - 3x^2 + 6x - 6)\Big]_0^2 \\ &= 19\left(\frac{e^2}{3} + 1\right) \\ &\approx 65.797\end{aligned}$$

Example 1 demonstrates the integration order $dz\,dy\,dx$. For other orders, you can follow a similar procedure. For instance, to evaluate a triple iterated integral in the order $dx\,dy\,dz$, hold both y and z constant for the innermost integration and integrate with respect to x. Then, for the second integration, hold z constant and integrate with respect to y. Finally, for the third integration, integrate with respect to z.

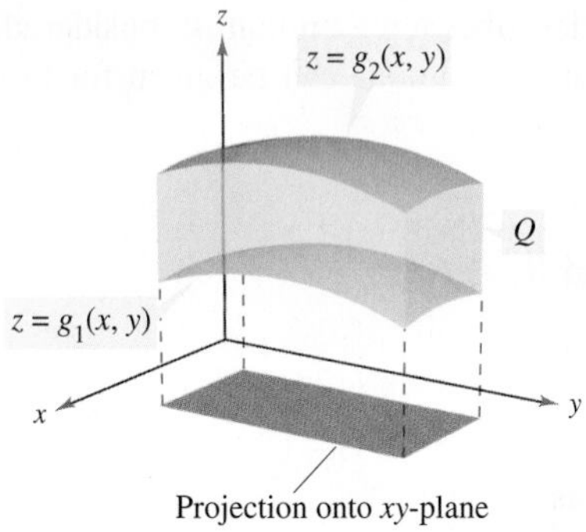

Solid region Q lies between two surfaces.
Figure 14.52

To find the limits for a particular order of integration, it is generally advisable first to determine the innermost limits, which may be functions of the outer two variables. Then, by projecting the solid Q onto the coordinate plane of the outer two variables, you can determine their limits of integration by the methods used for double integrals. For instance, to evaluate

$$\iiint_Q f(x, y, z)\, dz\, dy\, dx$$

first determine the limits for z, and then the integral has the form

$$\iint \left[\int_{g_1(x, y)}^{g_2(x, y)} f(x, y, z)\, dz \right] dy\, dx.$$

By projecting the solid Q onto the xy-plane, you can determine the limits for x and y as you did for double integrals, as shown in Figure 14.52.

EXAMPLE 2 Using a Triple Integral to Find Volume

Find the volume of the ellipsoid given by $4x^2 + 4y^2 + z^2 = 16$.

Solution Because x, y, and z play similar roles in the equation, the order of integration is probably immaterial, and you can arbitrarily choose $dz\, dy\, dx$. Moreover, you can simplify the calculation by considering only the portion of the ellipsoid lying in the first octant, as shown in Figure 14.53. From the order $dz\, dy\, dx$, you first determine the bounds for z.

$$0 \le z \le 2\sqrt{4 - x^2 - y^2}$$

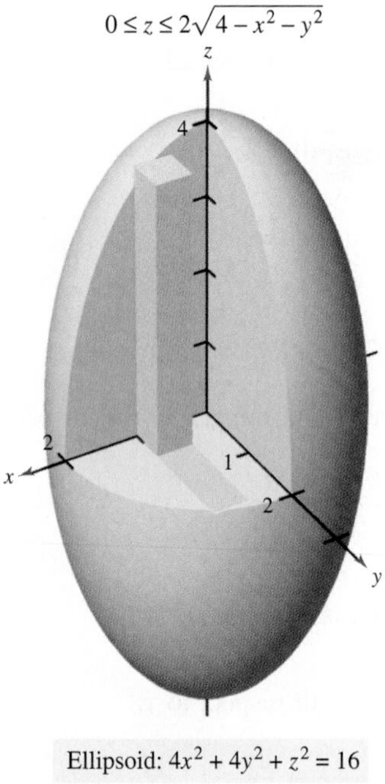

Figure 14.53

In Figure 14.54, you can see that the boundaries for x and y are $0 \le x \le 2$ and $0 \le y \le \sqrt{4 - x^2}$, so the volume of the ellipsoid is

$$\begin{aligned}
V &= \iiint_Q dV \\
&= 8 \int_0^2 \int_0^{\sqrt{4-x^2}} \int_0^{2\sqrt{4-x^2-y^2}} dz\, dy\, dx \\
&= 8 \int_0^2 \int_0^{\sqrt{4-x^2}} z \Big]_0^{2\sqrt{4-x^2-y^2}} dy\, dx \\
&= 16 \int_0^2 \int_0^{\sqrt{4-x^2}} \sqrt{(4 - x^2) - y^2}\, dy\, dx \\
&= 8 \int_0^2 \left[y\sqrt{4 - x^2 - y^2} + (4 - x^2) \arcsin\left(\frac{y}{\sqrt{4 - x^2}}\right) \right]_0^{\sqrt{4-x^2}} dx \\
&= 8 \int_0^2 [0 + (4 - x^2) \arcsin(1) - 0 - 0]\, dx \\
&= 8 \int_0^2 (4 - x^2)\left(\frac{\pi}{2}\right) dx \\
&= 4\pi \left[4x - \frac{x^3}{3} \right]_0^2 \\
&= \frac{64\pi}{3}.
\end{aligned}$$

Integration tables (Appendix B), Formula 37

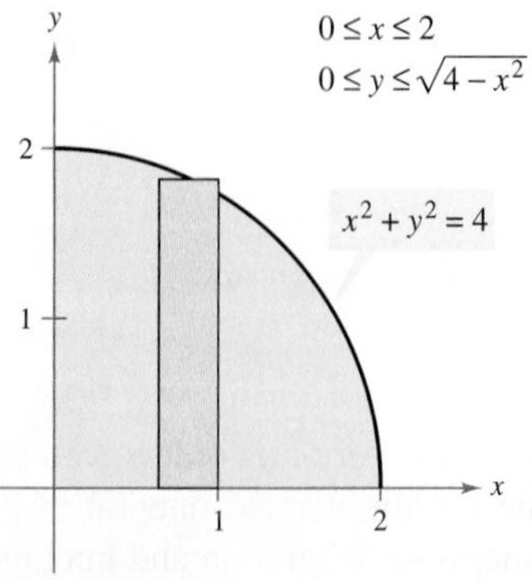

Figure 14.54

Example 2 is unusual in that all six possible orders of integration produce integrals of comparable difficulty. Try setting up some other possible orders of integration to find the volume of the ellipsoid. For instance, the order $dx\,dy\,dz$ yields the integral

$$V = 8\int_0^4 \int_0^{\sqrt{16-z^2}/2} \int_0^{\sqrt{16-4y^2-z^2}/2} dx\,dy\,dz.$$

If you solve this integral, you will obtain the same volume obtained in Example 2. This is always the case—the order of integration does not affect the value of the integral. However, the order of integration often does affect the complexity of the integral. In Example 3, the given order of integration is not convenient, so you can change the order to simplify the problem.

EXAMPLE 3 Changing the Order of Integration

Evaluate $\displaystyle\int_0^{\sqrt{\pi/2}} \int_x^{\sqrt{\pi/2}} \int_1^3 \sin(y^2)\,dz\,dy\,dx.$

Solution Note that after one integration in the given order, you would encounter the integral $2\int \sin(y^2)\,dy$, which is not an elementary function. To avoid this problem, change the order of integration to $dz\,dx\,dy$, so that y is the outer variable. The solid region Q is given by

$$0 \le x \le \sqrt{\frac{\pi}{2}}, \quad x \le y \le \sqrt{\frac{\pi}{2}}, \quad 1 \le z \le 3$$

and the projection of Q in the xy-plane yields the bounds

$$0 \le y \le \sqrt{\frac{\pi}{2}} \quad \text{and} \quad 0 \le x \le y.$$

So, you have

$$\begin{aligned}
V &= \iiint_Q dV \\
&= \int_0^{\sqrt{\pi/2}} \int_0^y \int_1^3 \sin(y^2)\,dz\,dx\,dy \\
&= \int_0^{\sqrt{\pi/2}} \int_0^y z\sin(y^2)\Big]_1^3\,dx\,dy \\
&= 2\int_0^{\sqrt{\pi/2}} \int_0^y \sin(y^2)\,dx\,dy \\
&= 2\int_0^{\sqrt{\pi/2}} x\sin(y^2)\Big]_0^y\,dy \\
&= 2\int_0^{\sqrt{\pi/2}} y\sin(y^2)\,dy \\
&= -\cos(y^2)\Big]_0^{\sqrt{\pi/2}} \\
&= 1.
\end{aligned}$$

See Figure 14.55.

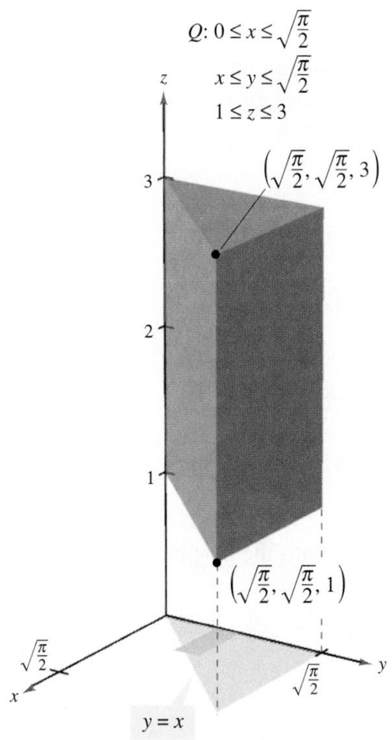

The volume of the solid region Q is 1.
Figure 14.55

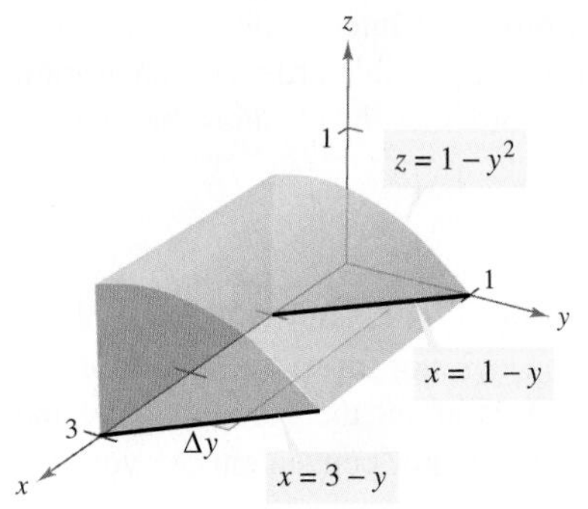

Q: $0 \le z \le 1 - y^2$
$1 - y \le x \le 3 - y$
$0 \le y \le 1$

Figure 14.56

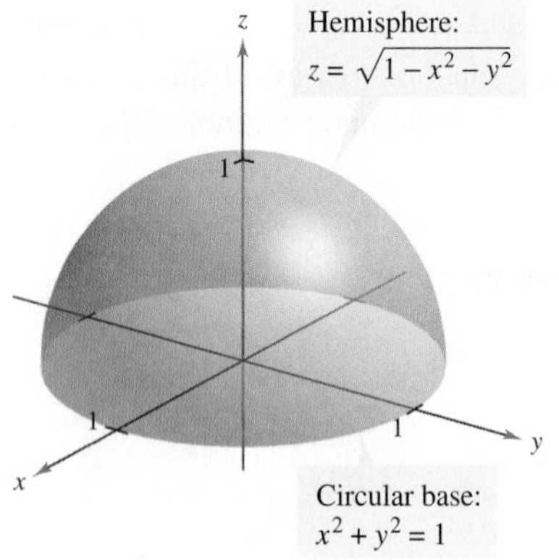

Q: $0 \le z \le \sqrt{1 - x^2 - y^2}$
$-\sqrt{1 - y^2} \le x \le \sqrt{1 - y^2}$
$-1 \le y \le 1$

Figure 14.57

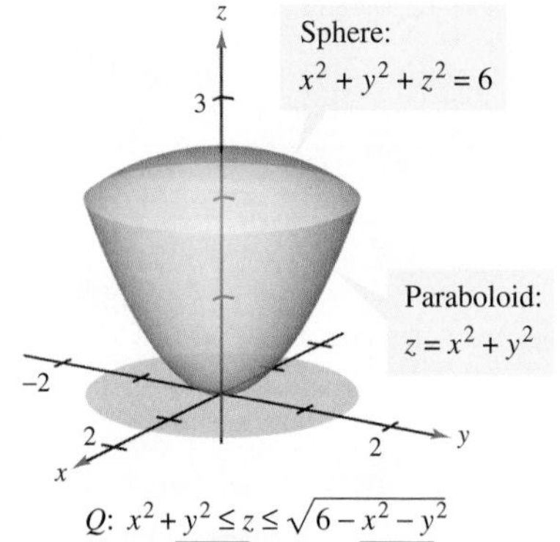

Q: $x^2 + y^2 \le z \le \sqrt{6 - x^2 - y^2}$
$-\sqrt{2 - x^2} \le y \le \sqrt{2 - x^2}$
$-\sqrt{2} \le x \le \sqrt{2}$

Figure 14.58

EXAMPLE 4 Determining the Limits of Integration

Set up a triple integral for the volume of each solid region.

a. The region in the first octant bounded above by the cylinder $z = 1 - y^2$ and lying between the vertical planes $x + y = 1$ and $x + y = 3$

b. The upper hemisphere given by $z = \sqrt{1 - x^2 - y^2}$

c. The region bounded below by the paraboloid $z = x^2 + y^2$ and above by the sphere $x^2 + y^2 + z^2 = 6$

Solution

a. In Figure 14.56, note that the solid is bounded below by the xy-plane ($z = 0$) and above by the cylinder $z = 1 - y^2$. So,

$$0 \le z \le 1 - y^2. \qquad \text{Bounds for } z$$

Projecting the region onto the xy-plane produces a parallelogram. Because two sides of the parallelogram are parallel to the x-axis, you have the following bounds:

$$1 - y \le x \le 3 - y \quad \text{and} \quad 0 \le y \le 1.$$

So, the volume of the region is given by

$$V = \iiint_Q dV = \int_0^1 \int_{1-y}^{3-y} \int_0^{1-y^2} dz\,dx\,dy.$$

b. For the upper hemisphere given by $z = \sqrt{1 - x^2 - y^2}$, you have

$$0 \le z \le \sqrt{1 - x^2 - y^2}. \qquad \text{Bounds for } z$$

In Figure 14.57, note that the projection of the hemisphere onto the xy-plane is the circle given by $x^2 + y^2 = 1$, and you can use either order $dx\,dy$ or $dy\,dx$. Choosing the first produces

$$-\sqrt{1 - y^2} \le x \le \sqrt{1 - y^2} \quad \text{and} \quad -1 \le y \le 1$$

which implies that the volume of the region is given by

$$V = \iiint_Q dV = \int_{-1}^1 \int_{-\sqrt{1-y^2}}^{\sqrt{1-y^2}} \int_0^{\sqrt{1-x^2-y^2}} dz\,dx\,dy.$$

c. For the region bounded below by the paraboloid $z = x^2 + y^2$ and above by the sphere $x^2 + y^2 + z^2 = 6$, you have

$$x^2 + y^2 \le z \le \sqrt{6 - x^2 - y^2}. \qquad \text{Bounds for } z$$

The sphere and the paraboloid intersect when $z = 2$. Moreover, you can see in Figure 14.58 that the projection of the solid region onto the xy-plane is the circle given by $x^2 + y^2 = 2$. Using the order $dy\,dx$ produces

$$-\sqrt{2 - x^2} \le y \le \sqrt{2 - x^2} \quad \text{and} \quad -\sqrt{2} \le x \le \sqrt{2}$$

which implies that the volume of the region is given by

$$V = \iiint_Q dV = \int_{-\sqrt{2}}^{\sqrt{2}} \int_{-\sqrt{2-x^2}}^{\sqrt{2-x^2}} \int_{x^2+y^2}^{\sqrt{6-x^2-y^2}} dz\,dy\,dx.$$

Center of Mass and Moments of Inertia

EXPLORATION

Sketch the solid (of uniform density) bounded by $z = 0$ and

$$z = \frac{1}{1 + x^2 + y^2}$$

where $x^2 + y^2 \le 1$. From your sketch, estimate the coordinates of the center of mass of the solid. Now use a computer algebra system to verify your estimate. What do you observe?

In the remainder of this section, two important engineering applications of triple integrals are discussed. Consider a solid region Q whose density is given by the **density function ρ.** The **center of mass** of a solid region Q of mass m is given by $(\bar{x}, \bar{y}, \bar{z})$, where

$$m = \iiint_Q \rho(x, y, z)\, dV \qquad \text{Mass of the solid}$$

$$M_{yz} = \iiint_Q x\rho(x, y, z)\, dV \qquad \text{First moment about } yz\text{-plane}$$

$$M_{xz} = \iiint_Q y\rho(x, y, z)\, dV \qquad \text{First moment about } xz\text{-plane}$$

$$M_{xy} = \iiint_Q z\rho(x, y, z)\, dV \qquad \text{First moment about } xy\text{-plane}$$

and

$$\bar{x} = \frac{M_{yz}}{m}, \quad \bar{y} = \frac{M_{xz}}{m}, \quad \bar{z} = \frac{M_{xy}}{m}.$$

The quantities M_{yz}, M_{xz}, and M_{xy} are called the **first moments** of the region Q about the yz-, xz-, and xy-planes, respectively.

The first moments for solid regions are taken about a plane, whereas the second moments for solids are taken about a line. The **second moments** (or **moments of inertia**) about the x-, y-, and z-axes are as follows.

NOTE In engineering and physics, the moment of inertia of a mass is used to find the time required for a mass to reach a given speed of rotation about an axis, as shown in Figure 14.59. The greater the moment of inertia, the longer a force must be applied for the mass to reach the given speed.

$$I_x = \iiint_Q (y^2 + z^2)\rho(x, y, z)\, dV \qquad \text{Moment of inertia about } x\text{-axis}$$

$$I_y = \iiint_Q (x^2 + z^2)\rho(x, y, z)\, dV \qquad \text{Moment of inertia about } y\text{-axis}$$

$$I_z = \iiint_Q (x^2 + y^2)\rho(x, y, z)\, dV \qquad \text{Moment of inertia about } z\text{-axis}$$

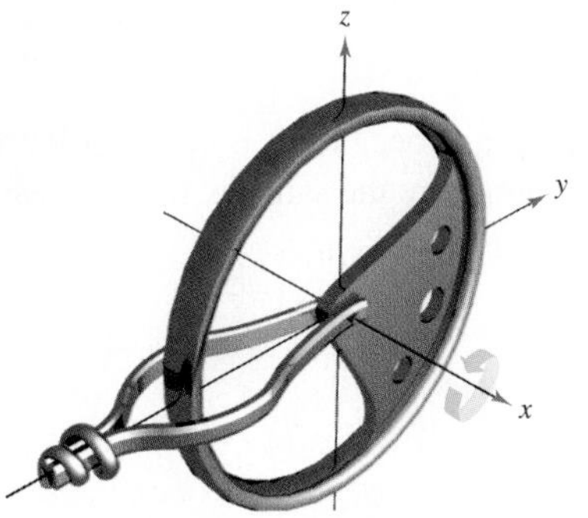

Figure 14.59

For problems requiring the calculation of all three moments, considerable effort can be saved by applying the additive property of triple integrals and writing

$$I_x = I_{xz} + I_{xy}, \quad I_y = I_{yz} + I_{xy}, \quad \text{and} \quad I_z = I_{yz} + I_{xz}$$

where I_{xy}, I_{xz}, and I_{yz} are as follows.

$$I_{xy} = \iiint_Q z^2\rho(x, y, z)\, dV$$

$$I_{xz} = \iiint_Q y^2\rho(x, y, z)\, dV$$

$$I_{yz} = \iiint_Q x^2\rho(x, y, z)\, dV$$

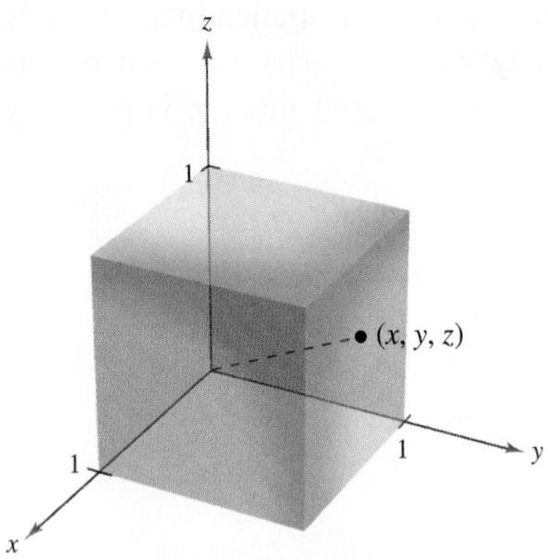

Variable density:
$\rho(x, y, z) = k(x^2 + y^2 + z^2)$
Figure 14.60

EXAMPLE 5 Finding the Center of Mass of a Solid Region

Find the center of mass of the unit cube shown in Figure 14.60, given that the density at the point (x, y, z) is proportional to the square of its distance from the origin.

Solution Because the density at (x, y, z) is proportional to the square of the distance between $(0, 0, 0)$ and (x, y, z), you have

$$\rho(x, y, z) = k(x^2 + y^2 + z^2).$$

You can use this density function to find the mass of the cube. Because of the symmetry of the region, any order of integration will produce an integral of comparable difficulty.

$$\begin{aligned}
m &= \int_0^1 \int_0^1 \int_0^1 k(x^2 + y^2 + z^2)\, dz\, dy\, dx \\
&= k \int_0^1 \int_0^1 \left[(x^2 + y^2)z + \frac{z^3}{3}\right]_0^1 dy\, dx \\
&= k \int_0^1 \int_0^1 \left(x^2 + y^2 + \frac{1}{3}\right) dy\, dx \\
&= k \int_0^1 \left[\left(x^2 + \frac{1}{3}\right)y + \frac{y^3}{3}\right]_0^1 dx \\
&= k \int_0^1 \left(x^2 + \frac{2}{3}\right) dx \\
&= k\left[\frac{x^3}{3} + \frac{2x}{3}\right]_0^1 = k
\end{aligned}$$

The first moment about the yz-plane is

$$\begin{aligned}
M_{yz} &= k \int_0^1 \int_0^1 \int_0^1 x(x^2 + y^2 + z^2)\, dz\, dy\, dx \\
&= k \int_0^1 x\left[\int_0^1 \int_0^1 (x^2 + y^2 + z^2)\, dz\, dy\right] dx.
\end{aligned}$$

Note that x can be factored out of the two inner integrals, because it is constant with respect to y and z. After factoring, the two inner integrals are the same as for the mass m. Therefore, you have

$$\begin{aligned}
M_{yz} &= k \int_0^1 x\left(x^2 + \frac{2}{3}\right) dx \\
&= k\left[\frac{x^4}{4} + \frac{x^2}{3}\right]_0^1 \\
&= \frac{7k}{12}.
\end{aligned}$$

So,

$$\bar{x} = \frac{M_{yz}}{m} = \frac{7k/12}{k} = \frac{7}{12}.$$

Finally, from the nature of ρ and the symmetry of x, y, and z in this solid region, you have $\bar{x} = \bar{y} = \bar{z}$, and the center of mass is $\left(\frac{7}{12}, \frac{7}{12}, \frac{7}{12}\right)$.

EXAMPLE 6 Moments of Inertia for a Solid Region

Find the moments of inertia about the x- and y-axes for the solid region lying between the hemisphere

$$z = \sqrt{4 - x^2 - y^2}$$

and the xy-plane, given that the density at (x, y, z) is proportional to the distance between (x, y, z) and the xy-plane.

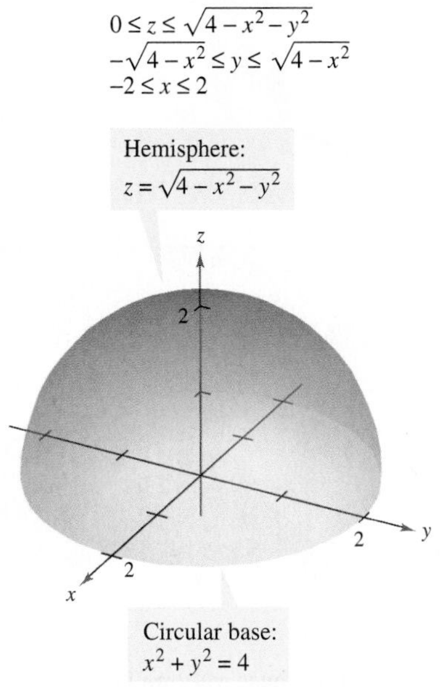

Variable density: $\rho(x, y, z) = kz$
Figure 14.61

Solution The density of the region is given by $\rho(x, y, z) = kz$. Considering the symmetry of this problem, you know that $I_x = I_y$, and you need to compute only one moment, say I_x. From Figure 14.61, choose the order $dz\,dy\,dx$ and write

$$\begin{aligned}
I_x &= \iiint_Q (y^2 + z^2)\rho(x, y, z)\,dV \\
&= \int_{-2}^{2}\int_{-\sqrt{4-x^2}}^{\sqrt{4-x^2}}\int_{0}^{\sqrt{4-x^2-y^2}} (y^2 + z^2)(kz)\,dz\,dy\,dx \\
&= k\int_{-2}^{2}\int_{-\sqrt{4-x^2}}^{\sqrt{4-x^2}} \left[\frac{y^2z^2}{2} + \frac{z^4}{4}\right]_0^{\sqrt{4-x^2-y^2}} dy\,dx \\
&= k\int_{-2}^{2}\int_{-\sqrt{4-x^2}}^{\sqrt{4-x^2}} \left[\frac{y^2(4 - x^2 - y^2)}{2} + \frac{(4 - x^2 - y^2)^2}{4}\right] dy\,dx \\
&= \frac{k}{4}\int_{-2}^{2}\int_{-\sqrt{4-x^2}}^{\sqrt{4-x^2}} [(4 - x^2)^2 - y^4]\,dy\,dx \\
&= \frac{k}{4}\int_{-2}^{2} \left[(4 - x^2)^2 y - \frac{y^5}{5}\right]_{-\sqrt{4-x^2}}^{\sqrt{4-x^2}} dx \\
&= \frac{k}{4}\int_{-2}^{2} \frac{8}{5}(4 - x^2)^{5/2}\,dx \\
&= \frac{4k}{5}\int_{0}^{2} (4 - x^2)^{5/2}\,dx && x = 2\sin\theta \\
&= \frac{4k}{5}\int_{0}^{\pi/2} 64\cos^6\theta\,d\theta \\
&= \left(\frac{256k}{5}\right)\left(\frac{5\pi}{32}\right) && \text{Wallis's Formula} \\
&= 8k\pi.
\end{aligned}$$

So, $I_x = 8k\pi = I_y$.

In Example 6, notice that the moments of inertia about the x- and y-axes are equal to each other. The moment about the z-axis, however, is different. Does it seem that the moment of inertia about the z-axis should be less than or greater than the moments calculated in Example 6? By performing the calculations, you can determine that

$$I_z = \frac{16}{3}k\pi.$$

This tells you that the solid shown in Figure 14.61 has a greater resistance to rotation about the x- or y-axis than about the z-axis.

Exercises for Section 14.6

See www.CalcChat.com for worked-out solutions to odd-numbered exercises.

In Exercises 1–8, evaluate the iterated integral.

1. $\displaystyle\int_0^3\int_0^2\int_0^1 (x+y+z)\,dx\,dy\,dz$

2. $\displaystyle\int_{-1}^1\int_{-1}^1\int_{-1}^1 x^2y^2z^2\,dx\,dy\,dz$

3. $\displaystyle\int_0^1\int_0^x\int_0^{xy} x\,dz\,dy\,dx$

4. $\displaystyle\int_0^9\int_0^{y/3}\int_0^{\sqrt{y^2-9x^2}} z\,dz\,dx\,dy$

5. $\displaystyle\int_1^4\int_0^1\int_0^x 2ze^{-x^2}\,dy\,dx\,dz$

6. $\displaystyle\int_1^4\int_1^{e^2}\int_0^{1/xz} \ln z\,dy\,dz\,dx$

7. $\displaystyle\int_0^4\int_0^{\pi/2}\int_0^{1-x} x\cos y\,dz\,dy\,dx$

8. $\displaystyle\int_0^{\pi/2}\int_0^{y/2}\int_0^{1/y} \sin y\,dz\,dx\,dy$

In Exercises 9 and 10, use a computer algebra system to evaluate the iterated integral.

9. $\displaystyle\int_0^2\int_{-\sqrt{4-x^2}}^{\sqrt{4-x^2}}\int_0^{x^2} x\,dz\,dy\,dx$

10. $\displaystyle\int_0^{\sqrt{2}}\int_0^{\sqrt{2-x^2}}\int_{2x^2+y^2}^{4-y^2} y\,dz\,dy\,dx$

In Exercises 11 and 12, use a computer algebra system to approximate the iterated integral.

11. $\displaystyle\int_0^2\int_0^{\sqrt{4-x^2}}\int_1^4 \frac{x^2\sin y}{z}\,dz\,dy\,dx$

12. $\displaystyle\int_0^3\int_0^{2-(2y/3)}\int_0^{6-2y-3z} ze^{-x^2y^2}\,dx\,dz\,dy$

In Exercises 13–16, set up a triple integral for the volume of the solid.

13. The solid in the first octant bounded by the coordinate planes and the plane $z = 4 - x - y$

14. The solid bounded by $z = 9 - x^2$, $z = 0$, $x = 0$, and $y = 2x$

15. The solid bounded by the paraboloid $z = 9 - x^2 - y^2$ and the plane $z = 0$

16. The solid that is the common interior below the sphere $x^2 + y^2 + z^2 = 80$ and above the paraboloid $z = \frac{1}{2}(x^2 + y^2)$

Volume **In Exercises 17–22, use a triple integral to find the volume of the solid shown in the figure.**

17.

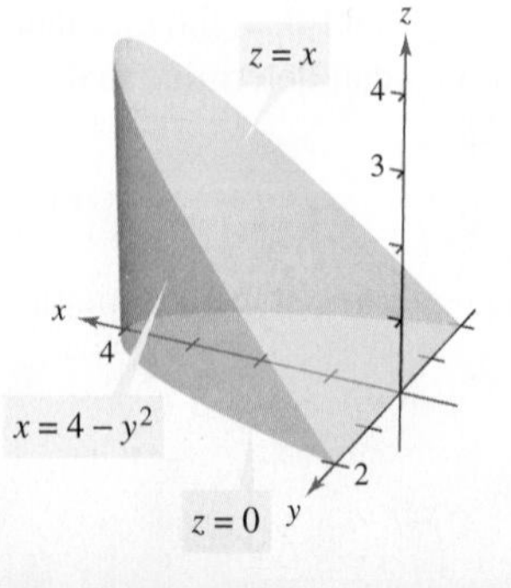

18.

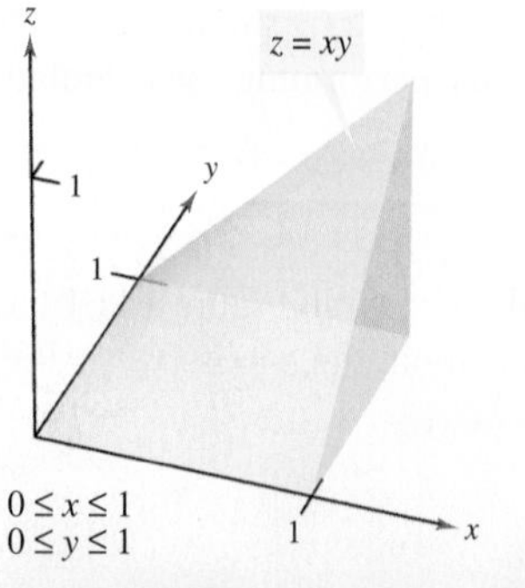

19.

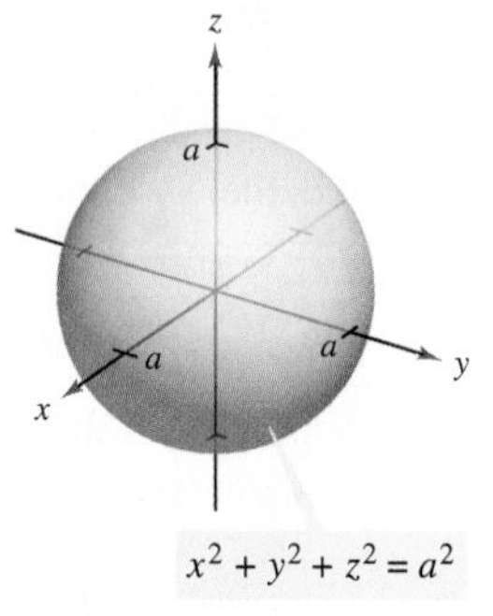

20.

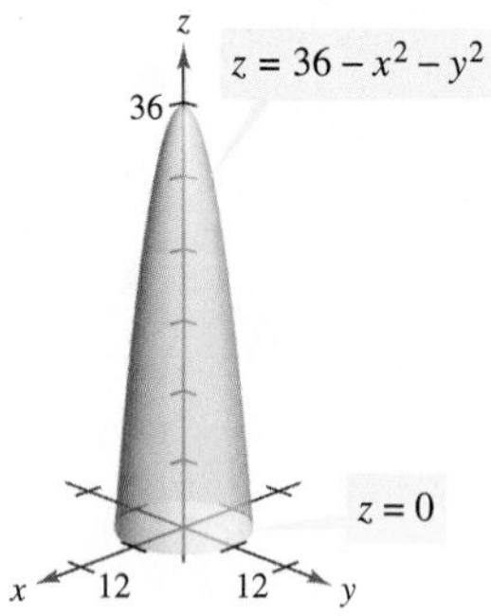

21.

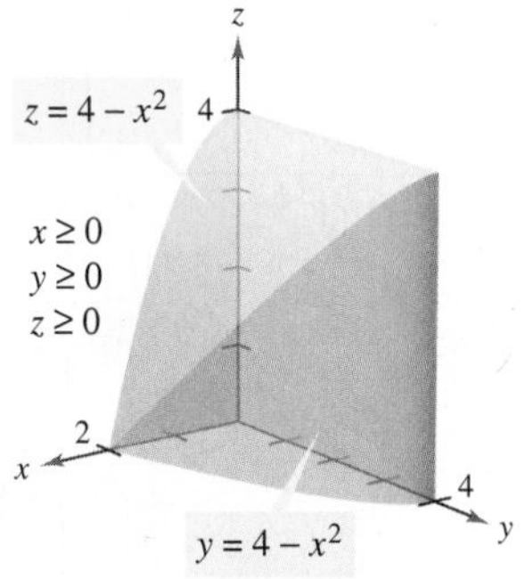

22.

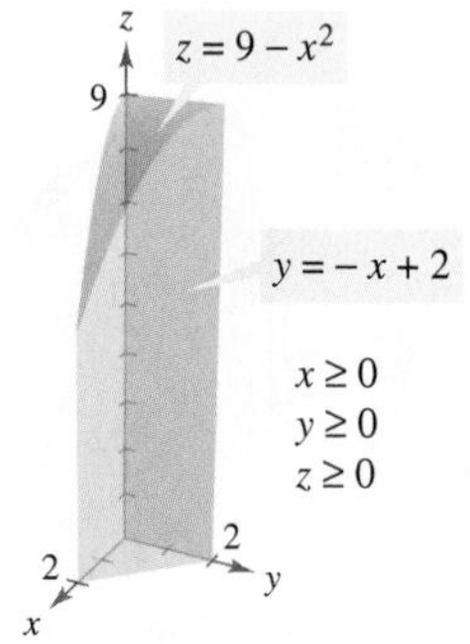

In Exercises 23–26, sketch the solid whose volume is given by the iterated integral and rewrite the integral using the indicated order of integration.

23. $\displaystyle\int_0^4\int_0^{(4-x)/2}\int_0^{(12-3x-6y)/4} dz\,dy\,dx$

Rewrite using the order $dy\,dx\,dz$.

24. $\displaystyle\int_0^3\int_0^{\sqrt{9-x^2}}\int_0^{6-x-y} dz\,dy\,dx$

Rewrite using the order $dz\,dx\,dy$.

25. $\displaystyle\int_0^1\int_y^1\int_0^{\sqrt{1-y^2}} dz\,dx\,dy$

Rewrite using the order $dz\,dy\,dx$.

26. $\displaystyle\int_0^2\int_{2x}^4\int_0^{\sqrt{y^2-4x^2}} dz\,dy\,dx$

Rewrite using the order $dx\,dy\,dz$.

In Exercises 27–30, list the six possible orders of integration for the triple integral over the solid region Q

$$\iiint_Q xyz\,dV.$$

27. $Q = \{(x, y, z): 0 \le x \le 1, 0 \le y \le x, 0 \le z \le 3\}$

28. $Q = \{(x, y, z): 0 \le x \le 2, x^2 \le y \le 4, 0 \le z \le 2 - x\}$

29. $Q = \{(x, y, z): x^2 + y^2 \le 9, 0 \le z \le 4\}$

30. $Q = \{(x, y, z): 0 \le x \le 1, y \le 1 - x^2, 0 \le z \le 6\}$

In Exercises 31 and 32, the figure shows the region of integration for the given integral. Rewrite the integral as an equivalent iterated integral in the five other orders.

31. $\displaystyle\int_0^1 \int_0^{1-y^2} \int_0^{1-y} dz\,dx\,dy$

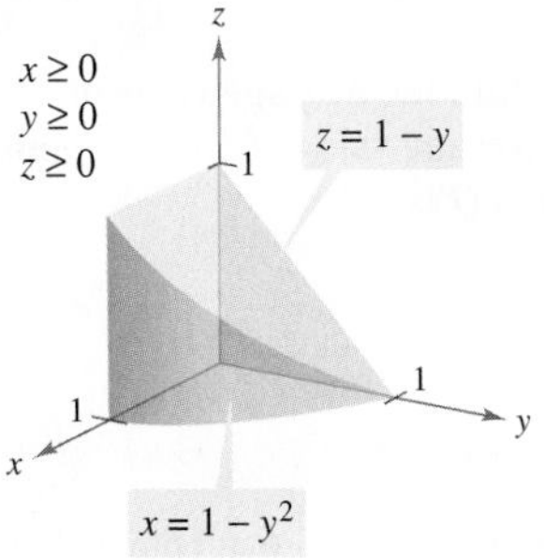

32. $\displaystyle\int_0^3 \int_0^x \int_0^{9-x^2} dz\,dy\,dx$

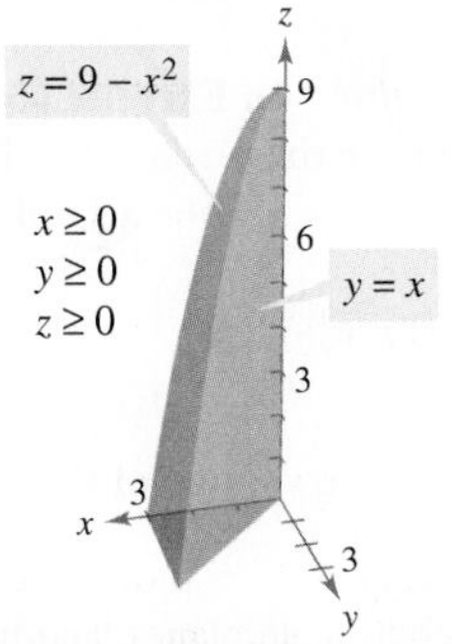

Mass and Center of Mass **In Exercises 33–36, find the mass and the indicated coordinates of the center of mass of the solid of given density bounded by the graphs of the equations.**

33. Find $\bar{x}$ using $\rho(x, y, z) = k$.

Q: $2x + 3y + 6z = 12,\ x = 0,\ y = 0,\ z = 0$

34. Find $\bar{y}$ using $\rho(x, y, z) = ky$.

Q: $3x + 3y + 5z = 15,\ x = 0,\ y = 0,\ z = 0$

35. Find $\bar{z}$ using $\rho(x, y, z) = kx$.

Q: $z = 4 - x,\ z = 0,\ y = 0,\ y = 4,\ x = 0$

36. Find $\bar{y}$ using $\rho(x, y, z) = k$.

Q: $\dfrac{x}{a} + \dfrac{y}{b} + \dfrac{z}{c} = 1\ (a, b, c > 0),\ x = 0,\ y = 0,\ z = 0$

Mass and Center of Mass **In Exercises 37 and 38, set up the triple integrals for finding the mass and the center of mass of the solid bounded by the graphs of the equations.**

37. $x = 0,\ x = b,\ y = 0,\ y = b,\ z = 0,\ z = b$

$\rho(x, y, z) = kxy$

38. $x = 0,\ x = a,\ y = 0,\ y = b,\ z = 0,\ z = c$

$\rho(x, y, z) = kz$

Think About It **The center of mass of a solid of constant density is shown in the figure. In Exercises 39–42, make a conjecture about how the center of mass $(\bar{x}, \bar{y}, \bar{z})$ will change for the nonconstant density $\rho(x, y, z)$. Explain.**

39. $\rho(x, y, z) = kx$

40. $\rho(x, y, z) = kz$

41. $\rho(x, y, z) = k(y + 2)$

42. $\rho(x, y, z) = kxz^2(y + 2)^2$

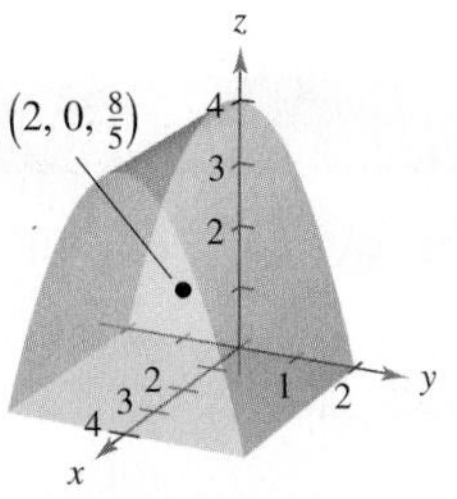

Centroid **In Exercises 43–48, find the centroid of the solid region bounded by the graphs of the equations or described by the figure. Use a computer algebra system to evaluate the triple integrals. (Assume uniform density and find the center of mass.)**

43. $z = \dfrac{h}{r}\sqrt{x^2 + y^2},\ z = h$

44. $y = \sqrt{4 - x^2},\ z = y,\ z = 0$

45. $z = \sqrt{4^2 - x^2 - y^2},\ z = 0$

46. $z = \dfrac{1}{y^2 + 1},\ z = 0,\ x = -2,\ x = 2,\ y = 0,\ y = 1$

47.

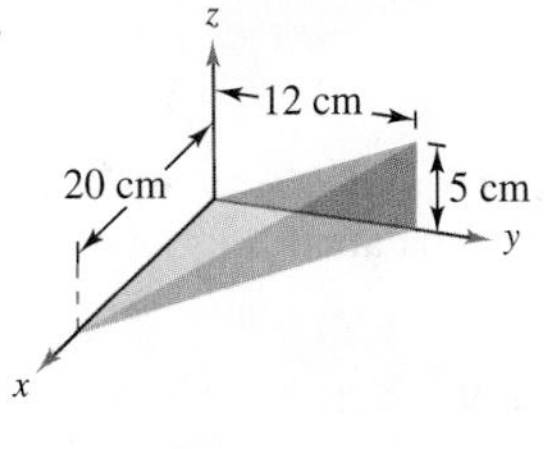

48.

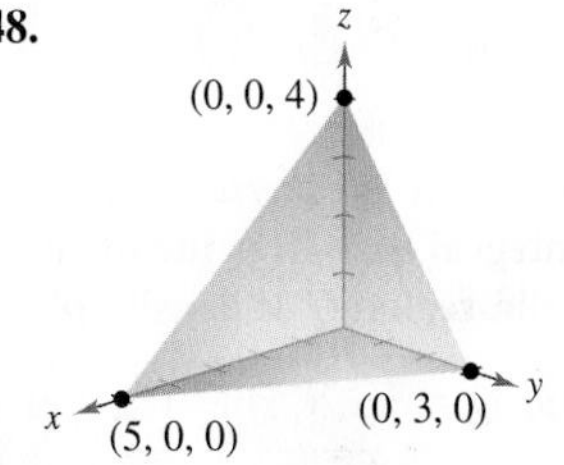

Moments of Inertia **In Exercises 49–52, find I_x, I_y, and I_z for the solid of given density. Use a computer algebra system to evaluate the triple integrals.**

49. (a) $\rho = k$

(b) $\rho = kxyz$

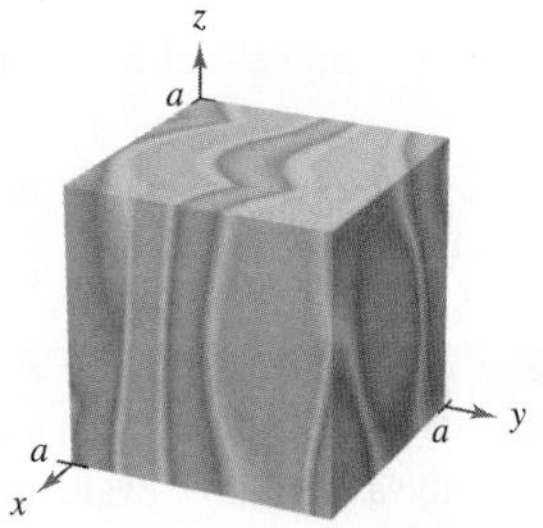

50. (a) $\rho(x, y, z) = k$

(b) $\rho(x, y, z) = k(x^2 + y^2)$

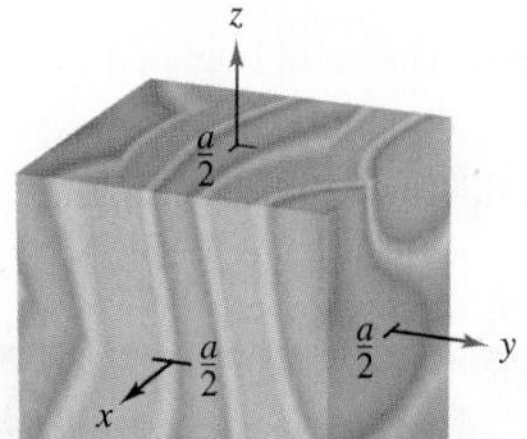

51. (a) $\rho(x, y, z) = k$

(b) $\rho = ky$

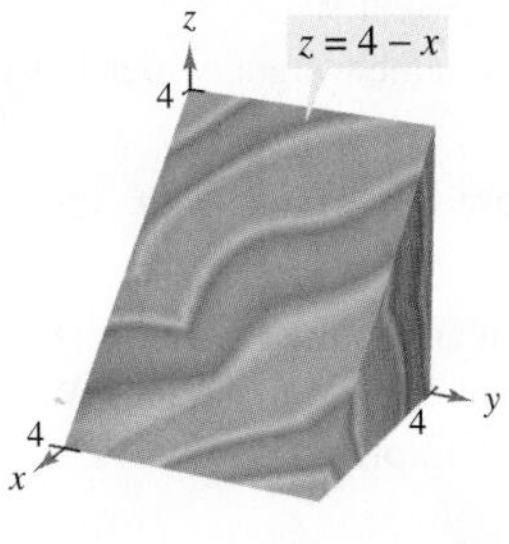

52. (a) $\rho = kz$

(b) $\rho = k(4 - z)$

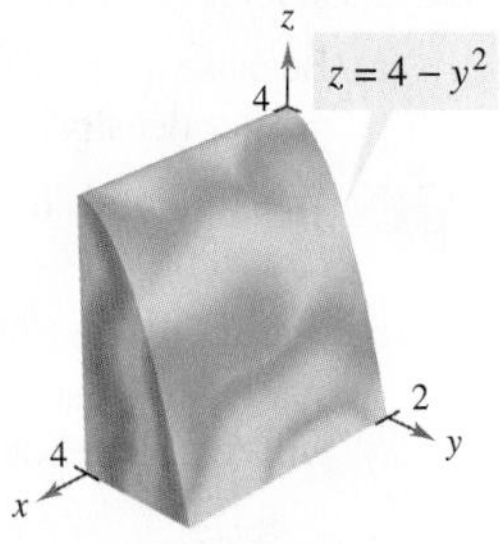

Moments of Inertia **In Exercises 53 and 54, verify the moments of inertia for the solid of uniform density. Use a computer algebra system to evaluate the triple integrals.**

53. $I_x = \frac{1}{12}m(3a^2 + L^2)$

$I_y = \frac{1}{2}ma^2$

$I_z = \frac{1}{12}m(3a^2 + L^2)$

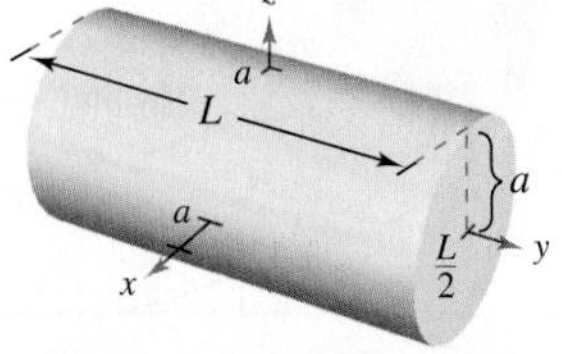

54. $I_x = \frac{1}{12}m(a^2 + b^2)$
$I_y = \frac{1}{12}m(b^2 + c^2)$
$I_z = \frac{1}{12}m(a^2 + c^2)$

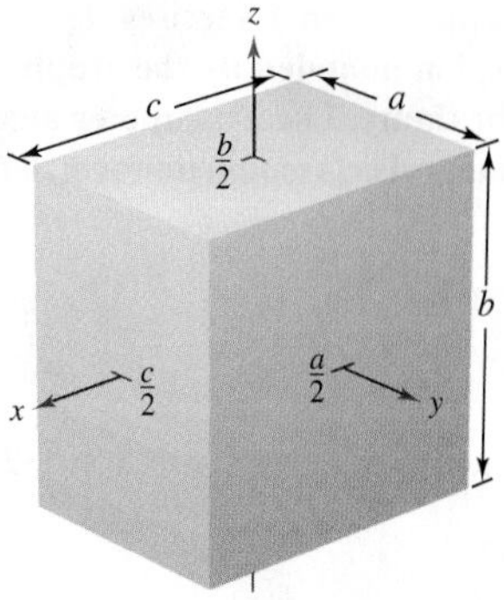

Moments of Inertia **In Exercises 55 and 56, set up a triple integral that gives the moment of inertia about the z-axis of the solid region Q of density ρ.**

55. $Q = \{(x, y, z): -1 \le x \le 1, -1 \le y \le 1, 0 \le z \le 1 - x\}$
$\rho = \sqrt{x^2 + y^2 + z^2}$

56. $Q = \{(x, y, z): x^2 + y^2 \le 1, 0 \le z \le 4 - x^2 - y^2\}$
$\rho = kx^2$

In Exercises 57 and 58, using the description of the solid region, set up the integral for (a) the mass, (b) the center of mass, and (c) the moment of inertia about the z-axis.

57. The solid bounded by $z = 4 - x^2 - y^2$ and $z = 0$ with density function $\rho = kz$

58. The solid in the first octant bounded by the coordinate planes and $x^2 + y^2 + z^2 = 25$ with density function $\rho = kxy$

Writing About Concepts

59. Define a triple integral and describe a method of evaluating a triple integral.

60. Give the number of possible orders of integration when evaluating a triple integral.

61. Consider solid A and solid B of equal weight shown below.

(a) Because the solids have the same weight, which has the greater density?

(b) Which solid has the greater moment of inertia? Explain.

(c) The solids are rolled down an inclined plane. They are started at the same time and at the same height. Which will reach the bottom first? Explain.

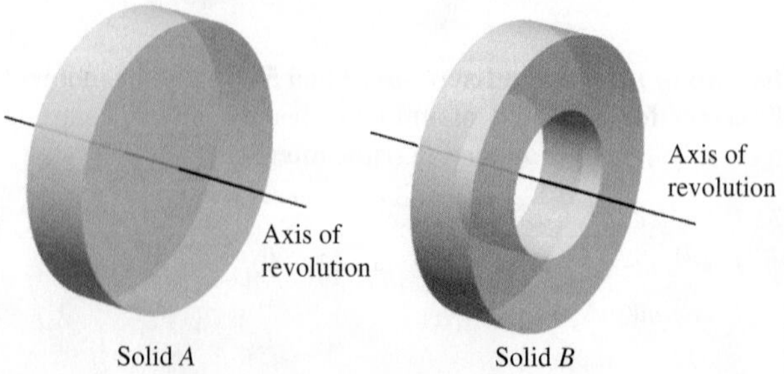

Writing About Concepts *(continued)*

62. Determine whether the moment of inertia about the y-axis of the cylinder in Exercise 53 will increase or decrease for the nonconstant density $\rho(x, y, z) = \sqrt{x^2 + z^2}$ and $a = 4$.

Average Value **In Exercise 63–66, find the average value of the function over the given solid. The average value of a continuous function $f(x, y, z)$ over a solid region Q is**

$$\frac{1}{V}\iiint_Q f(x, y, z)\, dV$$

where V is the volume of the solid region Q.

63. $f(x, y, z) = z^2 + 4$ over the cube in the first octant bounded by the coordinate planes, and the planes $x = 1$, $y = 1$, and $z = 1$

64. $f(x, y, z) = xyz$ over the cube in the first octant bounded by the coordinate planes, and the planes $x = 3$, $y = 3$, and $z = 3$

65. $f(x, y, z) = x + y + z$ over the tetrahedron in the first octant with vertices $(0, 0, 0)$, $(2, 0, 0)$, $(0, 2, 0)$ and $(0, 0, 2)$

66. $f(x, y, z) = x + y$ over the solid bounded by the sphere $x^2 + y^2 + z^2 = 2$

67. Find the solid region Q where the triple integral

$$\iiint_Q (1 - 2x^2 - y^2 - 3z^2)\, dV$$

is a maximum. Use a computer algebra system to approximate the maximum value. What is the exact maximum value?

68. Find the solid region Q where the triple integral

$$\iiint_Q (1 - x^2 - y^2 - z^2)\, dV$$

is a maximum. Use a computer algebra system to approximate the maximum value. What is the exact maximum value?

69. Solve for a in the triple integral.

$$\int_0^1 \int_0^{3-a-y^2} \int_a^{4-x-y^2} dz\, dx\, dy = \frac{14}{15}$$

70. Determine the value of b so that the volume of the ellipsoid

$$x^2 + \frac{y^2}{b^2} + \frac{z^2}{9} = 1$$

is 16π.

Putnam Exam Challenge

71. Evaluate

$$\lim_{n\to\infty} \int_0^1 \int_0^1 \cdots \int_0^1 \cos^2\left\{\frac{\pi}{2n}(x_1 + x_2 + \cdots + x_n)\right\} dx_1\, dx_2 \cdots dx_n.$$

This problem was composed by the Committee on the Putnam Prize Competition.

Section 14.7 Triple Integrals in Cylindrical and Spherical Coordinates

- **Write and evaluate a triple integral in cylindrical coordinates.**
- **Write and evaluate a triple integral in spherical coordinates.**

The Granger Collection

PIERRE SIMON DE LAPLACE (1749–1827)

One of the first to use a cylindrical coordinate system was the French mathematician Pierre Simon de Laplace. Laplace has been called the "Newton of France," and he published many important works in mechanics, differential equations, and probability.

Triple Integrals in Cylindrical Coordinates

Many common solid regions such as spheres, ellipsoids, cones, and paraboloids can yield difficult triple integrals in rectangular coordinates. In fact, it is precisely this difficulty that led to the introduction of nonrectangular coordinate systems. In this section, you will learn how to use *cylindrical* and *spherical* coordinates to evaluate triple integrals.

Recall from Section 11.7 that the rectangular conversion equations for cylindrical coordinates are

$$x = r\cos\theta$$
$$y = r\sin\theta$$
$$z = z.$$

STUDY TIP An easy way to remember these conversions is to note that the equations for x and y are the same as in polar coordinates and z is unchanged.

In this coordinate system, the simplest solid region is a cylindrical block determined by

$$r_1 \le r \le r_2, \quad \theta_1 \le \theta \le \theta_2, \quad z_1 \le z \le z_2$$

as shown in Figure 14.62. To obtain the cylindrical coordinate form of a triple integral, suppose that Q is a solid region whose projection R onto the xy-plane can be described in polar coordinates. That is,

$$Q = \{(x, y, z): (x, y) \text{ is in } R, \quad h_1(x, y) \le z \le h_2(x, y)\}$$

and

$$R = \{(r, \theta): \theta_1 \le \theta \le \theta_2, \quad g_1(\theta) \le r \le g_2(\theta)\}.$$

If f is a continuous function on the solid Q, you can write the triple integral of f over Q as

$$\iiint\limits_Q f(x, y, z)\,dV = \iint\limits_R \left[\int_{h_1(x, y)}^{h_2(x, y)} f(x, y, z)\,dz\right] dA$$

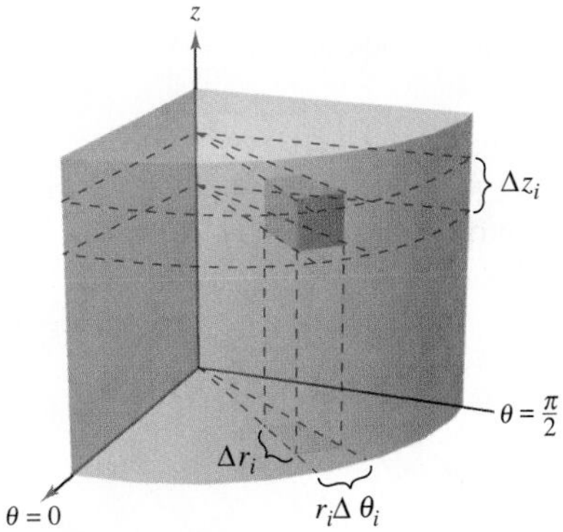

Volume of cylindrical block:
$\Delta V_i = r_i \Delta r_i \Delta\theta_i \Delta z_i$
Figure 14.62

where the double integral over R is evaluated in polar coordinates. That is, R is a plane region that is either r-simple or θ-simple. If R is r-simple, the iterated form of the triple integral in cylindrical form is

$$\iiint\limits_Q f(x, y, z)\,dV = \int_{\theta_1}^{\theta_2}\int_{g_1(\theta)}^{g_2(\theta)}\int_{h_1(r\cos\theta,\, r\sin\theta)}^{h_2(r\cos\theta,\, r\sin\theta)} f(r\cos\theta, r\sin\theta, z)r\,dz\,dr\,d\theta.$$

NOTE This is only one of six possible orders of integration. The other five are $dz\,d\theta\,dr$, $dr\,dz\,d\theta$, $dr\,d\theta\,dz$, $d\theta\,dz\,dr$, and $d\theta\,dr\,dz$.

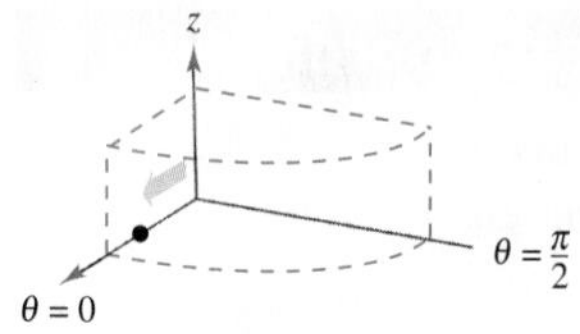

Integrate with respect to r.

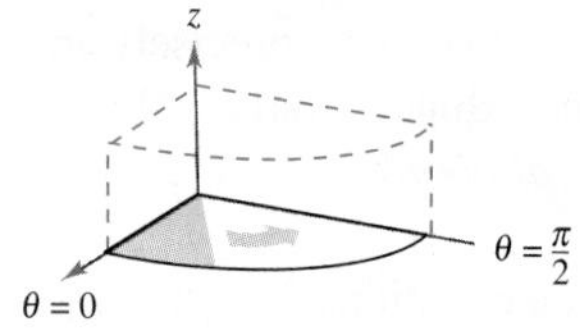

Integrate with respect to θ.

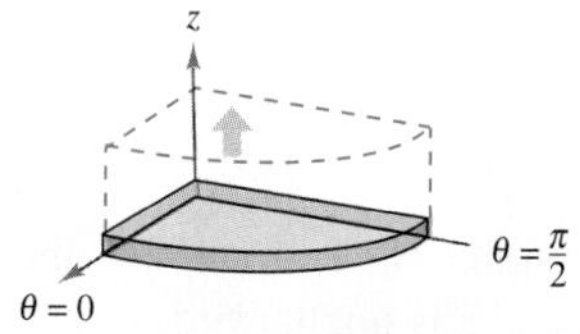

Integrate with respect to z.

Figure 14.63

To visualize a particular order of integration, it helps to view the iterated integral in terms of three sweeping motions—each adding another dimension to the solid. For instance, in the order $dr\,d\theta\,dz$, the first integration occurs in the r-direction as a point sweeps out a ray. Then, as θ increases, the line sweeps out a sector. Finally, as z increases, the sector sweeps out a solid wedge, as shown in Figure 14.63.

EXPLORATION

Volume of a Paraboloid Sector On pages 995, 1003, and 1025, you were asked to summarize the different ways you know for finding the volume of the solid bounded by the paraboloid

$$z = a^2 - x^2 - y^2, \quad a > 0$$

and the xy-plane. You now know one more way. Use it to find the volume of the solid. Compare the different methods. What are the advantages and disadvantages of each?

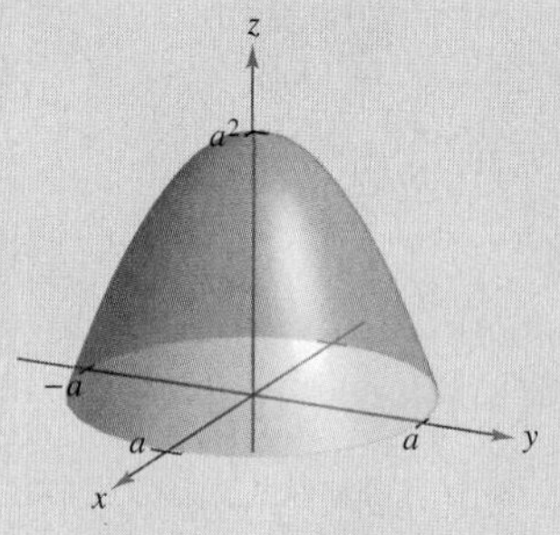

EXAMPLE 1 Finding Volume by Cylindrical Coordinates

Find the volume of the solid region Q cut from the sphere

$$x^2 + y^2 + z^2 = 4 \qquad \text{Sphere}$$

by the cylinder $r = 2 \sin \theta$, as shown in Figure 14.64.

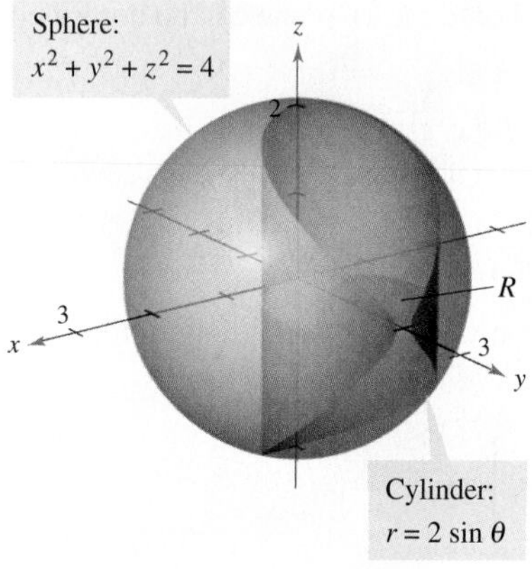

Figure 14.64

Solution Because $x^2 + y^2 + z^2 = r^2 + z^2 = 4$, the bounds on z are

$$-\sqrt{4 - r^2} \le z \le \sqrt{4 - r^2}.$$

Let R be the circular projection of the solid onto the $r\theta$-plane. Then the bounds on R are $0 \le r \le 2 \sin \theta$ and $0 \le \theta \le \pi$. So, the volume of Q is

$$\begin{aligned}
V &= \int_0^{\pi} \int_0^{2\sin\theta} \int_{-\sqrt{4-r^2}}^{\sqrt{4-r^2}} r\,dz\,dr\,d\theta \\
&= 2\int_0^{\pi/2} \int_0^{2\sin\theta} 2r\sqrt{4 - r^2}\,dr\,d\theta \\
&= 2\int_0^{\pi/2} -\frac{2}{3}(4 - r^2)^{3/2}\Big]_0^{2\sin\theta}\,d\theta \\
&= \frac{4}{3}\int_0^{\pi/2} (8 - 8\cos^3\theta)\,d\theta \\
&= \frac{32}{3}\int_0^{\pi/2} [1 - (\cos\theta)(1 - \sin^2\theta)]\,d\theta \\
&= \frac{32}{3}\left[\theta - \sin\theta + \frac{\sin^3\theta}{3}\right]_0^{\pi/2} \\
&= \frac{16}{9}(3\pi - 4) \\
&\approx 9.644.
\end{aligned}$$

EXAMPLE 2 Finding Mass by Cylindrical Coordinates

Find the mass of the ellipsoid Q given by $4x^2 + 4y^2 + z^2 = 16$, lying above the xy-plane. The density at a point in the solid is proportional to the distance between the point and the xy-plane.

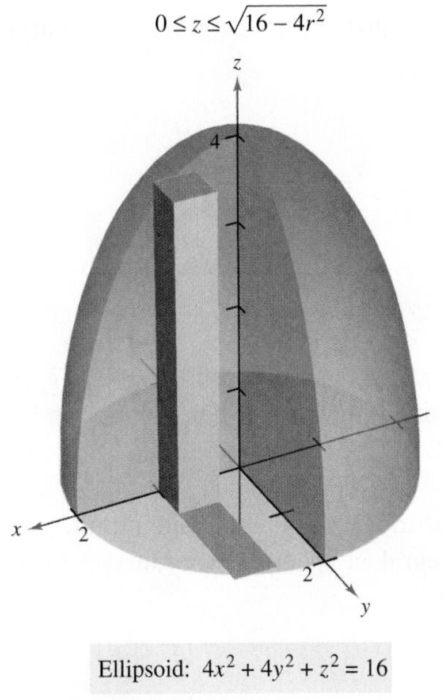

Figure 14.65

Solution The density function is $\rho(r, \theta, z) = kz$. The bounds on z are

$$0 \le z \le \sqrt{16 - 4x^2 - 4y^2} = \sqrt{16 - 4r^2}$$

where $0 \le r \le 2$ and $0 \le \theta \le 2\pi$, as shown in Figure 14.65. The mass of the solid is

$$\begin{aligned} m &= \int_0^{2\pi}\int_0^2\int_0^{\sqrt{16-4r^2}} kzr\,dz\,dr\,d\theta \\ &= \frac{k}{2}\int_0^{2\pi}\int_0^2 z^2r\Big]_0^{\sqrt{16-4r^2}} dr\,d\theta \\ &= \frac{k}{2}\int_0^{2\pi}\int_0^2 (16r - 4r^3)\,dr\,d\theta \\ &= \frac{k}{2}\int_0^{2\pi}\Big[8r^2 - r^4\Big]_0^2 d\theta \\ &= 8k\int_0^{2\pi} d\theta = 16\pi k. \end{aligned}$$

Integration in cylindrical coordinates is useful when factors involving $x^2 + y^2$ appear in the integrand, as illustrated in Example 3.

EXAMPLE 3 Finding a Moment of Inertia

Find the moment of inertia about the axis of symmetry of the solid Q bounded by the paraboloid $z = x^2 + y^2$ and the plane $z = 4$, as shown in Figure 14.66. The density at each point is proportional to the distance between the point and the z-axis.

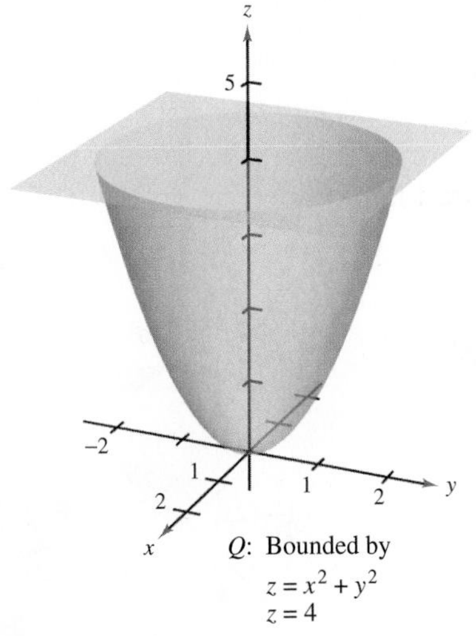

Figure 14.66

Solution Because the z-axis is the axis of symmetry, and $\rho(x, y, z) = k\sqrt{x^2 + y^2}$, it follows that

$$I_z = \iiint_Q k(x^2 + y^2)\sqrt{x^2 + y^2}\,dV.$$

In cylindrical coordinates, $0 \le r \le \sqrt{x^2 + y^2} = \sqrt{z}$. So, you have

$$\begin{aligned} I_z &= k\int_0^4\int_0^{2\pi}\int_0^{\sqrt{z}} r^2(r)r\,dr\,d\theta\,dz \\ &= k\int_0^4\int_0^{2\pi} \frac{r^5}{5}\Big]_0^{\sqrt{z}} d\theta\,dz \\ &= k\int_0^4\int_0^{2\pi} \frac{z^{5/2}}{5}\,d\theta\,dz \\ &= \frac{k}{5}\int_0^4 z^{5/2}(2\pi)\,dz \\ &= \frac{2\pi k}{5}\Big[\frac{2}{7}z^{7/2}\Big]_0^4 = \frac{512k\pi}{35}. \end{aligned}$$

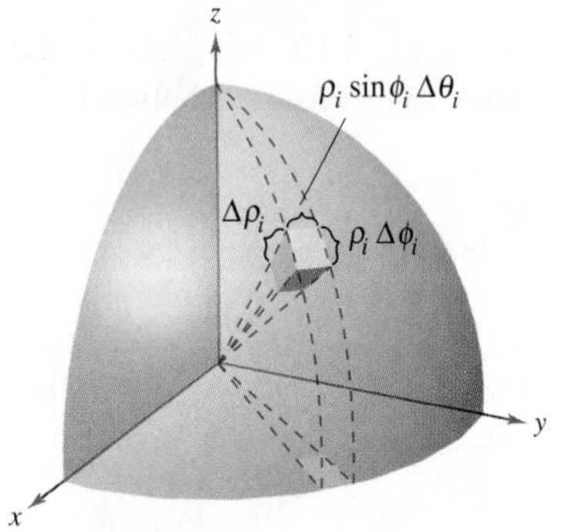

Spherical block:
$\Delta V_i \approx \rho_i^2 \sin\phi_i\,\Delta\rho_i\,\Delta\theta_i\,\Delta\phi_i$
Figure 14.67

Triple Integrals in Spherical Coordinates

Triple integrals involving spheres or cones are often easier to evaluate by converting to spherical coordinates. Recall from Section 11.7 that the rectangular conversion equations for spherical coordinates are

$$x = \rho \sin \phi \cos \theta$$
$$y = \rho \sin \phi \sin \theta$$
$$z = \rho \cos \phi.$$

In this coordinate system, the simplest region is a spherical block determined by

$$\{(\rho, \theta, \phi): \rho_1 \le \rho \le \rho_2, \quad \theta_1 \le \theta \le \theta_2, \quad \phi_1 \le \phi \le \phi_2\}$$

where $\rho_1 \ge 0$, $\theta_2 - \theta_1 \le 2\pi$, and $0 \le \phi_1 \le \phi_2 \le \pi$, as shown in Figure 14.67. If (ρ, θ, ϕ) is a point in the interior of such a block, then the volume of the block can be approximated by $\Delta V \approx \rho^2 \sin\phi\,\Delta\rho\,\Delta\phi\,\Delta\theta$ (see Exercise 17 in the Problem Solving exercises for this chapter).

Using the usual process involving an inner partition, summation, and a limit, you can develop the following version of a triple integral in spherical coordinates for a continuous function f defined on the solid region Q.

$$\iiint_Q f(x, y, z)\,dV = \int_{\theta_1}^{\theta_2}\int_{\phi_1}^{\phi_2}\int_{\rho_1}^{\rho_2} f(\rho\sin\phi\cos\theta, \rho\sin\phi\sin\theta, \rho\cos\phi)\rho^2\sin\phi\,d\rho\,d\phi\,d\theta.$$

This formula can be modified for different orders of integration and generalized to include regions with variable boundaries.

Like triple integrals in cylindrical coordinates, triple integrals in spherical coordinates are evaluated with iterated integrals. As with cylindrical coordinates, you can visualize a particular order of integration by viewing the iterated integral in terms of three sweeping motions—each adding another dimension to the solid. For instance, the iterated integral

$$\int_0^{2\pi}\int_0^{\pi/4}\int_0^3 \rho^2 \sin\phi\,d\rho\,d\phi\,d\theta$$

(which is used in Example 4) is illustrated in Figure 14.68.

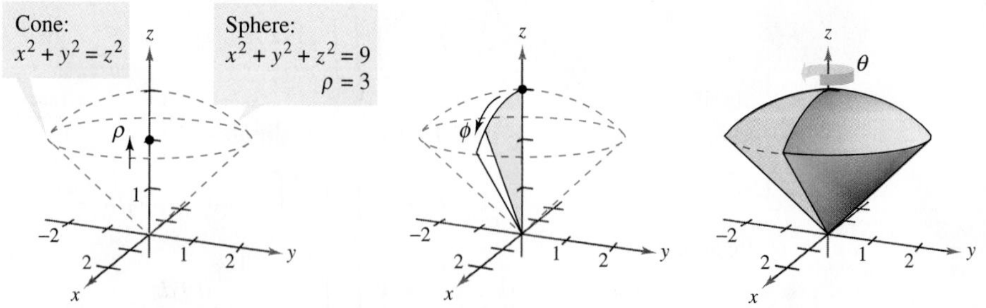

ρ varies from 0 to 3 with ϕ and θ held constant. ϕ varies from 0 to $\pi/4$ with θ held constant. θ varies from 0 to 2π.
Figure 14.68

NOTE The Greek letter ρ used in spherical coordinates is not related to density. Rather, it is the three-dimensional analog of the r used in polar coordinates. For problems involving spherical coordinates and a density function, this text uses a different symbol to denote density.

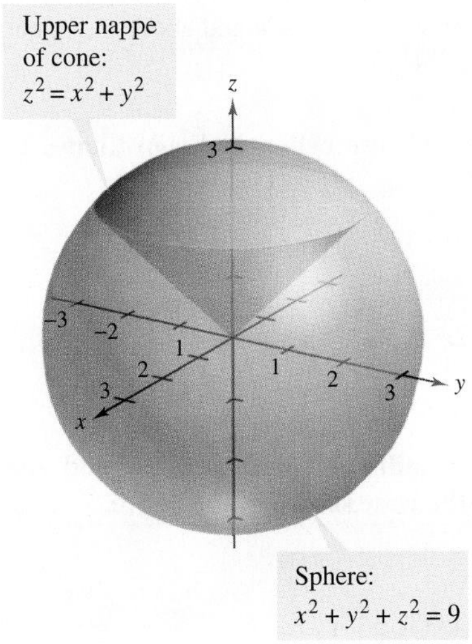

Figure 14.69

EXAMPLE 4 Finding Volume in Spherical Coordinates

Find the volume of the solid region Q bounded below by the upper nappe of the cone $z^2 = x^2 + y^2$ and above by the sphere $x^2 + y^2 + z^2 = 9$, as shown in Figure 14.69.

Solution In spherical coordinates, the equation of the sphere is

$$\rho^2 = x^2 + y^2 + z^2 = 9 \quad \Longrightarrow \quad \rho = 3.$$

Furthermore, the sphere and cone intersect when

$$(x^2 + y^2) + z^2 = (z^2) + z^2 = 9 \quad \Longrightarrow \quad z = \frac{3}{\sqrt{2}}$$

and, because $z = \rho \cos \phi$, it follows that

$$\left(\frac{3}{\sqrt{2}}\right)\left(\frac{1}{3}\right) = \cos \phi \quad \Longrightarrow \quad \phi = \frac{\pi}{4}.$$

Consequently, you can use the integration order $d\rho\, d\phi\, d\theta$, where $0 \le \rho \le 3$, $0 \le \phi \le \pi/4$, and $0 \le \theta \le 2\pi$. The volume is

$$\begin{aligned}
V = \iiint_Q dV &= \int_0^{2\pi} \int_0^{\pi/4} \int_0^3 \rho^2 \sin \phi \, d\rho \, d\phi \, d\theta \\
&= \int_0^{2\pi} \int_0^{\pi/4} 9 \sin \phi \, d\phi \, d\theta \\
&= 9 \int_0^{2\pi} -\cos \phi \Big]_0^{\pi/4} d\theta \\
&= 9 \int_0^{2\pi} \left(1 - \frac{\sqrt{2}}{2}\right) d\theta = 9\pi\left(2 - \sqrt{2}\right) \approx 16.563.
\end{aligned}$$

EXAMPLE 5 Finding the Center of Mass of a Solid Region

Find the center of mass of the solid region Q of uniform density, bounded below by the upper nappe of the cone $z^2 = x^2 + y^2$ and above by the sphere $x^2 + y^2 + z^2 = 9$.

Solution Because the density is uniform, you can consider the density at the point (x, y, z) to be k. By symmetry, the center of mass lies on the z-axis, and you need only calculate $\bar{z} = M_{xy}/m$, where $m = kV = 9k\pi\left(2 - \sqrt{2}\right)$ from Example 4. Because $z = \rho \cos \phi$, it follows that

$$\begin{aligned}
M_{xy} = \iiint_Q kz \, dV &= k \int_0^3 \int_0^{2\pi} \int_0^{\pi/4} (\rho \cos \phi)\rho^2 \sin \phi \, d\phi \, d\theta \, d\rho \\
&= k \int_0^3 \int_0^{2\pi} \rho^3 \frac{\sin^2 \phi}{2} \Big]_0^{\pi/4} d\theta \, d\rho \\
&= \frac{k}{4} \int_0^3 \int_0^{2\pi} \rho^3 \, d\theta \, d\rho = \frac{k\pi}{2} \int_0^3 \rho^3 \, d\rho = \frac{81k\pi}{8}.
\end{aligned}$$

So,

$$\bar{z} = \frac{M_{xy}}{m} = \frac{81k\pi/8}{9k\pi\left(2 - \sqrt{2}\right)} = \frac{9\left(2 + \sqrt{2}\right)}{16} \approx 1.920$$

and the center of mass is approximately $(0, 0, 1.92)$.

Exercises for Section 14.7

See www.CalcChat.com for worked-out solutions to odd-numbered exercises.

In Exercises 1–6, evaluate the iterated integral.

1. $\int_0^4 \int_0^{\pi/2} \int_0^2 r \cos \theta \, dr \, d\theta \, dz$

2. $\int_0^{\pi/4} \int_0^2 \int_0^{2-r} rz \, dz \, dr \, d\theta$

3. $\int_0^{\pi/2} \int_0^{2 \cos^2 \theta} \int_0^{4-r^2} r \sin\theta \, dz \, dr \, d\theta$

4. $\int_0^{\pi/2} \int_0^{\pi} \int_0^2 e^{-\rho^3} \rho^2 \, d\rho \, d\theta \, d\phi$

5. $\int_0^{2\pi} \int_0^{\pi/4} \int_0^{\cos \phi} \rho^2 \sin \phi \, d\rho \, d\phi \, d\theta$

6. $\int_0^{\pi/4} \int_0^{\pi/4} \int_0^{\cos \theta} \rho^2 \sin \phi \cos \phi \, d\rho \, d\theta \, d\phi$

In Exercises 7 and 8, use a computer algebra system to evaluate the iterated integral.

7. $\int_0^4 \int_0^z \int_0^{\pi/2} re^r \, d\theta \, dr \, dz$

8. $\int_0^{\pi/2} \int_0^{\pi} \int_0^{\sin \theta} (2 \cos \phi)\rho^2 \, d\rho \, d\theta \, d\phi$

In Exercises 9–12, sketch the solid region whose volume is given by the iterated integral, and evaluate the iterated integral.

9. $\int_0^{\pi/2} \int_0^3 \int_0^{e^{-r^2}} r \, dz \, dr \, d\theta$

10. $\int_0^{2\pi} \int_0^{\sqrt{3}} \int_0^{3-r^2} r \, dz \, dr \, d\theta$

11. $\int_0^{2\pi} \int_{\pi/6}^{\pi/2} \int_0^4 \rho^2 \sin \phi \, d\rho \, d\phi \, d\theta$

12. $\int_0^{2\pi} \int_0^{\pi} \int_2^5 \rho^2 \sin \phi \, d\rho \, d\phi \, d\theta$

In Exercises 13–16, convert the integral from rectangular coordinates to both cylindrical and spherical coordinates, and evaluate the simplest iterated integral.

13. $\int_{-2}^{2} \int_{-\sqrt{4-x^2}}^{\sqrt{4-x^2}} \int_{x^2+y^2}^{4} x \, dz \, dy \, dx$

14. $\int_0^2 \int_0^{\sqrt{4-x^2}} \int_0^{\sqrt{16-x^2-y^2}} \sqrt{x^2+y^2} \, dz \, dy \, dx$

15. $\int_{-a}^{a} \int_{-\sqrt{a^2-x^2}}^{\sqrt{a^2-x^2}} \int_a^{a+\sqrt{a^2-x^2-y^2}} x \, dz \, dy \, dx$

16. $\int_0^1 \int_0^{\sqrt{1-x^2}} \int_0^{\sqrt{1-x^2-y^2}} \sqrt{x^2+y^2+z^2} \, dz \, dy \, dx$

Volume **In Exercises 17–20, use cylindrical coordinates to find the volume of the solid.**

17. Solid inside both $x^2 + y^2 + z^2 = a^2$ and $(x - a/2)^2 + y^2 = (a/2)^2$

18. Solid inside $x^2 + y^2 + z^2 = 16$ and outside $z = \sqrt{x^2 + y^2}$

19. Solid bounded by the graphs of the sphere $r^2 + z^2 = a^2$ and the cylinder $r = a \cos \theta$

20. Solid inside the sphere $x^2 + y^2 + z^2 = 4$ and above the upper nappe of the cone $z^2 = x^2 + y^2$

Mass **In Exercises 21 and 22, use cylindrical coordinates to find the mass of the solid Q.**

21. $Q = \{(x, y, z): 0 \le z \le 9 - x - 2y, x^2 + y^2 \le 4\}$
$\rho(x, y, z) = k\sqrt{x^2 + y^2}$

22. $Q = \{(x, y, z): 0 \le z \le 12e^{-(x^2+y^2)}, x^2 + y^2 \le 4, x \ge 0, y \ge 0\}$
$\rho(x, y, z) = k$

In Exercises 23–28, use cylindrical coordinates to find the indicated characteristic of the cone shown in the figure.

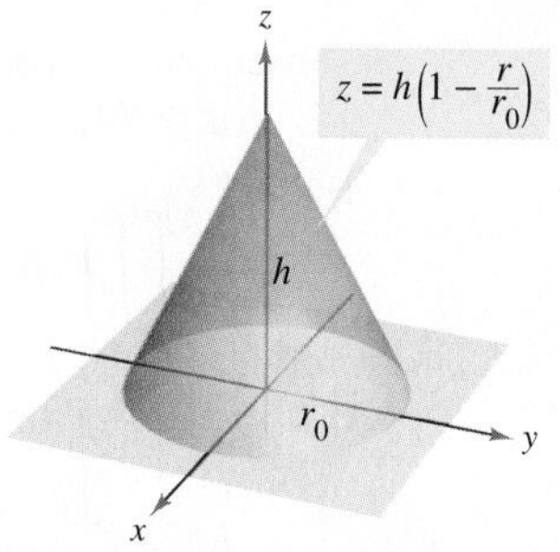

23. ***Volume*** Find the volume of the cone.

24. ***Centroid*** Find the centroid of the cone.

25. ***Center of Mass*** Find the center of mass of the cone assuming that its density at any point is proportional to the distance between the point and the axis of the cone. Use a computer algebra system to evaluate the triple integral.

26. ***Center of Mass*** Find the center of mass of the cone assuming that its density at any point is proportional to the distance between the point and the base. Use a computer algebra system to evaluate the triple integral.

27. ***Moment of Inertia*** Assume that the cone has uniform density and show that the moment of inertia about the z-axis is

$I_z = \frac{3}{10}mr_0^2.$

28. ***Moment of Inertia*** Assume that the density of the cone is

$\rho(x, y, z) = k\sqrt{x^2 + y^2}$

and find the moment of inertia about the z-axis.

Moment of Inertia **In Exercises 29 and 30, use cylindrical coordinates to verify the given formula for the moment of inertia of the solid of uniform density.**

29. Cylindrical shell: $I_z = \frac{1}{2}m(a^2 + b^2)$

$0 < a \le r \le b, \quad 0 \le z \le h$

30. Right circular cylinder: $I_z = \frac{3}{2}ma^2$

$r = 2a \sin \theta, \quad 0 \le z \le h$

Use a computer algebra system to evaluate the triple integral.

Volume **In Exercises 31 and 32, use spherical coordinates to find the volume of the solid.**

31. The torus given by $\rho = 4 \sin \phi$ (Use a computer algebra system to evaluate the triple integral.)

32. The solid between the spheres $x^2 + y^2 + z^2 = a^2$ and $x^2 + y^2 + z^2 = b^2$, $b > a$, and inside the cone $z^2 = x^2 + y^2$

Mass **In Exercises 33 and 34, use spherical coordinates to find the mass of the sphere $x^2 + y^2 + z^2 = a^2$ with the given density.**

33. The density at any point is proportional to the distance between the point and the origin.

34. The density at any point is proportional to the distance of the point from the z-axis.

Center of Mass **In Exercises 35 and 36, use spherical coordinates to find the center of mass of the solid of uniform density.**

35. Hemispherical solid of radius r

36. Solid lying between two concentric hemispheres of radii r and R, where $r < R$

Moment of Inertia **In Exercises 37 and 38, use spherical coordinates to find the moment of inertia about the z-axis of the solid of uniform density.**

37. Solid bounded by the hemisphere $\rho = \cos \phi$, $\pi/4 \le \phi \le \pi/2$, and the cone $\phi = \pi/4$

38. Solid lying between two concentric hemispheres of radii r and R, where $r < R$

Writing About Concepts

39. Give the equations for the coordinate conversion from rectangular to cylindrical coordinates and vice versa.

40. Give the equations for the coordinate conversion from rectangular to spherical coordinates and vice versa.

41. Give the iterated form of the triple integral $\iiint_Q f(x, y, z)\, dV$ in cylindrical form.

42. Give the iterated form of the triple integral $\iiint_Q f(x, y, z)\, dV$ in spherical form.

43. Describe the surface whose equation is a coordinate equal to a constant for each of the coordinates in (a) the cylindrical coordinate system and (b) the spherical coordinate system.

44. When evaluating a triple integral with constant limits of integration in the cylindrical coordinate system, you are integrating over a part of what solid? What is the solid when you are in spherical coordinates?

45. Find the "volume" of the "four-dimensional sphere"

$$x^2 + y^2 + z^2 + w^2 = a^2$$

by evaluating

$$16 \int_0^a \int_0^{\sqrt{a^2-x^2}} \int_0^{\sqrt{a^2-x^2-y^2}} \int_0^{\sqrt{a^2-x^2-y^2-z^2}} dw\, dz\, dy\, dx.$$

46. Use spherical coordinates to show that

$$\int_{-\infty}^{\infty} \int_{-\infty}^{\infty} \int_{-\infty}^{\infty} \sqrt{x^2 + y^2 + z^2}\, e^{-(x^2+y^2+z^2)}\, dx\, dy\, dz = 2\pi.$$

Section Project: Wrinkled and Bumpy Spheres

In parts (a) and (b), find the volume of the wrinkled sphere or bumpy sphere. These solids are used as models for tumors.

(a) Wrinkled sphere

$\rho = 1 + 0.2 \sin 8\theta \sin \phi$

$0 \le \theta \le 2\pi, 0 \le \phi \le \pi$

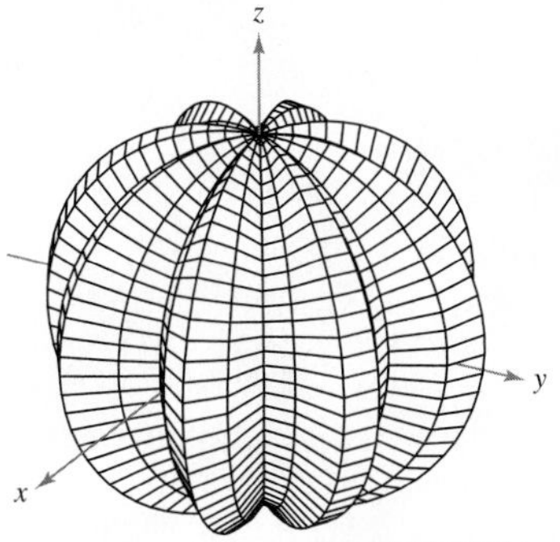

Generated by Maple

(b) Bumpy sphere

$\rho = 1 + 0.2 \sin 8\theta \sin 4\phi$

$0 \le \theta \le 2\pi, 0 \le \phi \le \pi$

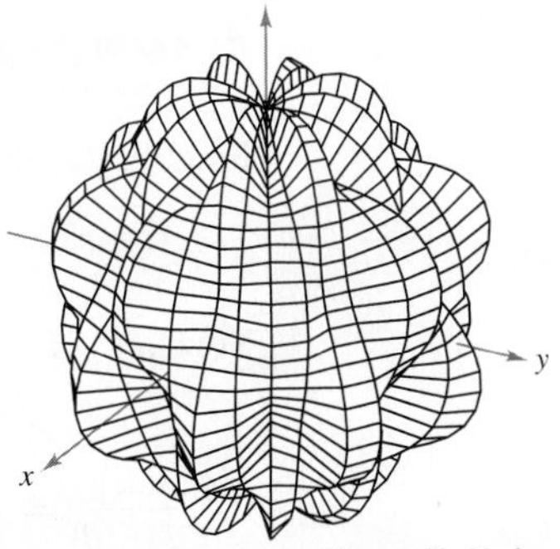

Generated by Maple

FOR FURTHER INFORMATION For more information on these types of spheres, see the article "Heat Therapy for Tumors" by Leah Edelstein-Keshet in *The UMAP Journal*.

Section 14.8 Change of Variables: Jacobians

- **Understand the concept of a Jacobian.**
- **Use a Jacobian to change variables in a double integral.**

CARL GUSTAV JACOBI (1804–1851)

The Jacobian is named after the German mathematician Carl Gustav Jacobi. Jacobi is known for his work in many areas of mathematics, but his interest in integration stemmed from the problem of finding the circumference of an ellipse.

Jacobians

For the single integral

$$\int_a^b f(x)\,dx$$

you can change variables by letting $x = g(u)$, so that $dx = g'(u)\,du$, and obtain

$$\int_a^b f(x)\,dx = \int_c^d f(g(u))g'(u)\,du$$

where $a = g(c)$ and $b = g(d)$. Note that the change-of-variables process introduces an additional factor $g'(u)$ into the integrand. This also occurs in the case of double integrals

$$\int_R\int f(x, y)\,dA = \int_S\int f(g(u, v), h(u, v))\underbrace{\left|\frac{\partial x}{\partial u}\frac{\partial y}{\partial v} - \frac{\partial y}{\partial u}\frac{\partial x}{\partial v}\right|}_{\text{Jacobian}}du\,dv$$

where the change of variables $x = g(u, v)$ and $y = h(u, v)$ introduces a factor called the **Jacobian** of x and y with respect to u and v. In defining the Jacobian, it is convenient to use the following determinant notation.

Definition of the Jacobian

If $x = g(u, v)$ and $y = h(u, v)$, then the **Jacobian** of x and y with respect to u and v, denoted by $\partial(x, y)/\partial(u, v)$, is

$$\frac{\partial(x, y)}{\partial(u, v)} = \begin{vmatrix} \frac{\partial x}{\partial u} & \frac{\partial x}{\partial v} \\ \frac{\partial y}{\partial u} & \frac{\partial y}{\partial v} \end{vmatrix} = \frac{\partial x}{\partial u}\frac{\partial y}{\partial v} - \frac{\partial y}{\partial u}\frac{\partial x}{\partial v}.$$

EXAMPLE 1 The Jacobian for Rectangular-to-Polar Conversion

Find the Jacobian for the change of variables defined by

$$x = r\cos\theta \qquad \text{and} \qquad y = r\sin\theta.$$

Solution From the definition of a Jacobian, you obtain

$$\begin{aligned}\frac{\partial(x, y)}{\partial(r, \theta)} &= \begin{vmatrix} \frac{\partial x}{\partial r} & \frac{\partial x}{\partial \theta} \\ \frac{\partial y}{\partial r} & \frac{\partial y}{\partial \theta} \end{vmatrix} \\ &= \begin{vmatrix} \cos\theta & -r\sin\theta \\ \sin\theta & r\cos\theta \end{vmatrix} \\ &= r\cos^2\theta + r\sin^2\theta \\ &= r.\end{aligned}$$

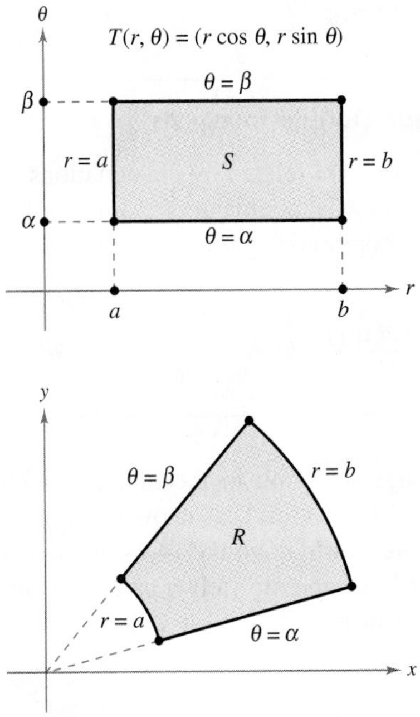

S is the region in the $r\theta$-plane that corresponds to R in the xy-plane.
Figure 14.70

Example 1 points out that the change of variables from rectangular to polar coordinates for a double integral can be written as

$$\int_R\int f(x, y)\, dA = \int_S\int f(r\cos\theta, r\sin\theta) r\, dr\, d\theta, \quad r > 0$$
$$= \int_S\int f(r\cos\theta, r\sin\theta)\left|\frac{\partial(x, y)}{\partial(r, \theta)}\right| dr\, d\theta$$

where S is the region in the $r\theta$-plane that corresponds to the region R in the xy-plane, as shown in Figure 14.70. This formula is similar to that found on page 1003.

In general, a change of variables is given by a one-to-one **transformation** T from a region S in the uv-plane to a region R in the xy-plane, to be given by

$$T(u, v) = (x, y) = (g(u, v), h(u, v))$$

where g and h have continuous first partial derivatives in the region S. Note that the point (u, v) lies in S and the point (x, y) lies in R. In most cases, you are hunting for a transformation in which the region S is simpler than the region R.

EXAMPLE 2 Finding a Change of Variables to Simplify a Region

Let R be the region bounded by the lines

$$x - 2y = 0, \qquad x - 2y = -4, \qquad x + y = 4, \qquad \text{and} \qquad x + y = 1$$

as shown in Figure 14.71. Find a transformation T from a region S to R such that S is a rectangular region (with sides parallel to the u- or v-axis).

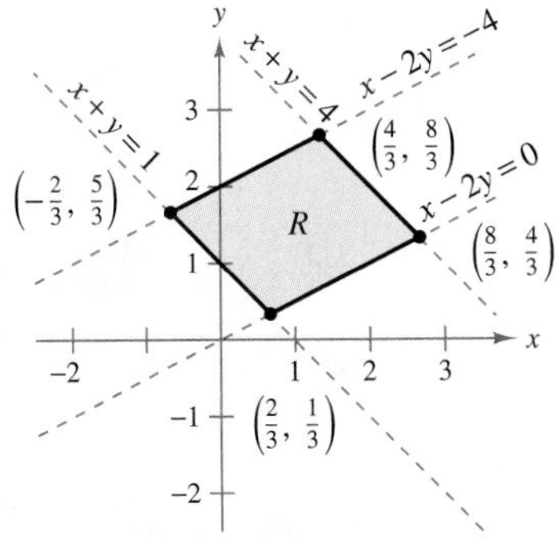

Region R in the xy-plane
Figure 14.71

Solution To begin, let $u = x + y$ and $v = x - 2y$. Solving this system of equations for x and y produces $T(u, v) = (x, y)$, where

$$x = \frac{1}{3}(2u + v) \qquad \text{and} \qquad y = \frac{1}{3}(u - v).$$

The four boundaries for R in the xy-plane give rise to the following bounds for S in the uv-plane.

Bounds in the xy-Plane		*Bounds in the uv-Plane*
$x + y = 1$	⇨	$u = 1$
$x + y = 4$	⇨	$u = 4$
$x - 2y = 0$	⇨	$v = 0$
$x - 2y = -4$	⇨	$v = -4$

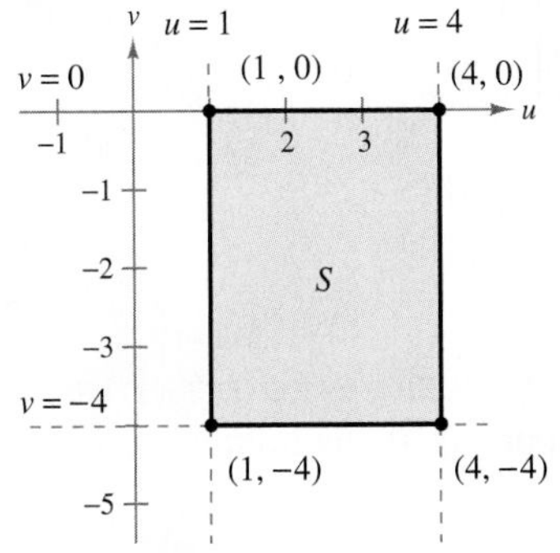

Region S in the uv-plane
Figure 14.72

The region S is shown in Figure 14.72. Note that the transformation T maps the vertices of the region S onto the vertices of the region R. For instance,

$$T(1, 0) = \left(\tfrac{1}{3}[2(1) + 0], \tfrac{1}{3}[1 - 0]\right)$$
$$= \left(\tfrac{2}{3}, \tfrac{1}{3}\right)$$
$$T(4, 0) = \left(\tfrac{1}{3}[2(4) + 0], \tfrac{1}{3}[4 - 0]\right)$$
$$= \left(\tfrac{8}{3}, \tfrac{4}{3}\right)$$
$$T(4, -4) = \left(\tfrac{1}{3}[2(4) - 4], \tfrac{1}{3}[4 - (-4)]\right)$$
$$= \left(\tfrac{4}{3}, \tfrac{8}{3}\right)$$
$$T(1, -4) = \left(\tfrac{1}{3}[2(1) - 4], \tfrac{1}{3}[1 - (-4)]\right)$$
$$= \left(-\tfrac{2}{3}, \tfrac{5}{3}\right).$$

Change of Variables for Double Integrals

THEOREM 14.5 Change of Variables for Double Integrals

Let R and S be regions in the xy- and uv-planes that are related by the equations $x = g(u, v)$ and $y = h(u, v)$ such that each point in R is the image of a unique point in S. If f is continuous on R, g and h have continuous partial derivatives on S, and $\partial(x, y)/\partial(u, v)$ is nonzero on S, then

$$\int_R\int f(x, y)\,dx\,dy = \int_S\int f(g(u, v), h(u, v))\left|\frac{\partial(x, y)}{\partial(u, v)}\right| du\,dv.$$

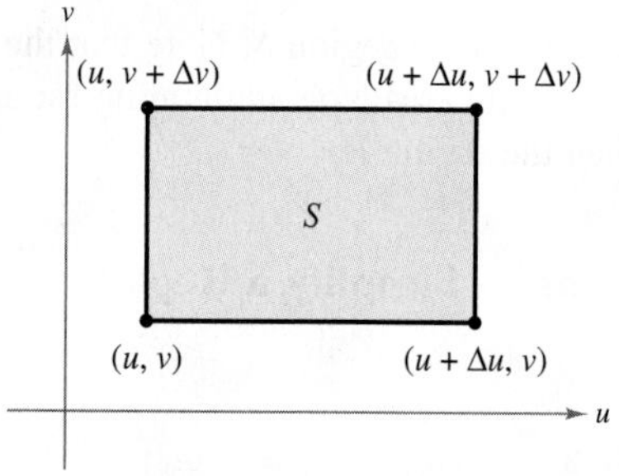

Area of $S = \Delta u\,\Delta v$
$\Delta u > 0,\ \Delta v > 0$
Figure 14.73

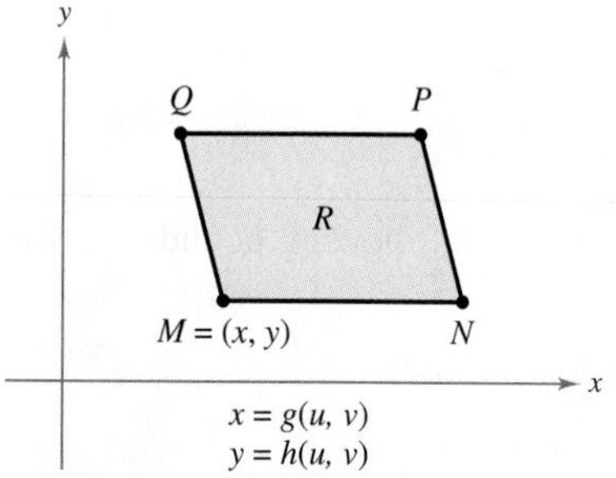

The vertices in the xy-plane are $M(g(u, v), h(u, v))$, $N(g(u + \Delta u, v), h(u + \Delta u, v))$, $P(g(u + \Delta u, v + \Delta v), h(u + \Delta u, v + \Delta v))$, and $Q(g(u, v + \Delta v), h(u, v + \Delta v))$.
Figure 14.74

Proof Consider the case in which S is a rectangular region in the uv-plane with vertices (u, v), $(u + \Delta u, v)$, $(u + \Delta u, v + \Delta v)$, and $(u, v + \Delta v)$, as shown in Figure 14.73. The images of these vertices in the xy-plane are shown in Figure 14.74. If Δu and Δv are small, the continuity of g and h implies that R is approximately a parallelogram determined by the vectors $\overrightarrow{MN}$ and $\overrightarrow{MQ}$. So, the area of R is

$$\Delta A \approx \|\overrightarrow{MN} \times \overrightarrow{MQ}\|.$$

Moreover, for small Δu and Δv, the partial derivatives of g and h with respect to u can be approximated by

$$g_u(u, v) \approx \frac{g(u + \Delta u, v) - g(u, v)}{\Delta u}$$

and

$$h_u(u, v) \approx \frac{h(u + \Delta u, v) - h(u, v)}{\Delta u}.$$

Consequently,

$$\begin{aligned}\overrightarrow{MN} &= [g(u + \Delta u, v) - g(u, v)]\mathbf{i} + [h(u + \Delta u, v) - h(u, v)]\mathbf{j}\\ &\approx [g_u(u, v)\,\Delta u]\mathbf{i} + [h_u(u, v)\,\Delta u]\mathbf{j}\\ &= \frac{\partial x}{\partial u}\Delta u\mathbf{i} + \frac{\partial y}{\partial u}\Delta u\mathbf{j}.\end{aligned}$$

Similarly, you can approximate $\overrightarrow{MQ}$ by $\dfrac{\partial x}{\partial v}\Delta v\mathbf{i} + \dfrac{\partial y}{\partial v}\Delta v\mathbf{j}$, which implies that

$$\overrightarrow{MN} \times \overrightarrow{MQ} \approx \begin{vmatrix} \mathbf{i} & \mathbf{j} & \mathbf{k} \\ \dfrac{\partial x}{\partial u}\Delta u & \dfrac{\partial y}{\partial u}\Delta u & 0 \\ \dfrac{\partial x}{\partial v}\Delta v & \dfrac{\partial y}{\partial v}\Delta v & 0 \end{vmatrix} = \begin{vmatrix} \dfrac{\partial x}{\partial u} & \dfrac{\partial y}{\partial u} \\ \dfrac{\partial x}{\partial v} & \dfrac{\partial y}{\partial v} \end{vmatrix} \Delta u\,\Delta v\mathbf{k}.$$

It follows that, in Jacobian notation,

$$\Delta A \approx \|\overrightarrow{MN} \times \overrightarrow{MQ}\| \approx \left|\frac{\partial(x, y)}{\partial(u, v)}\right| \Delta u\,\Delta v.$$

Because this approximation improves as Δu and Δv approach 0, the limiting case can be written as

$$dA \approx \|\overrightarrow{MN} \times \overrightarrow{MQ}\| \approx \left|\frac{\partial(x, y)}{\partial(u, v)}\right| du\,dv.$$

The next two examples show how a change of variables can simplify the integration process. The simplification can occur in various ways. You can make a change of variables to simplify either the *region R* or the *integrand* $f(x, y)$, or both.

EXAMPLE 3 Using a Change of Variables to Simplify a Region

Let R be the region bounded by the lines

$$x - 2y = 0, \quad x - 2y = -4, \quad x + y = 4, \quad \text{and} \quad x + y = 1$$

as shown in Figure 14.75. Evaluate the double integral

$$\int_R\int 3xy\, dA.$$

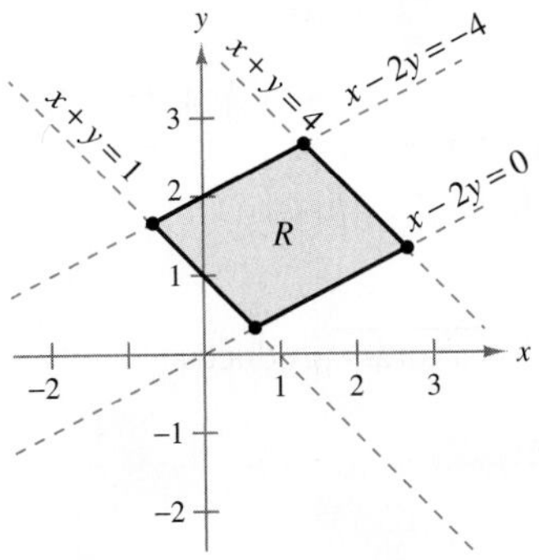

Figure 14.75

Solution From Example 2, you can use the following change of variables.

$$x = \frac{1}{3}(2u + v) \quad \text{and} \quad y = \frac{1}{3}(u - v)$$

The partial derivatives of x and y are

$$\frac{\partial x}{\partial u} = \frac{2}{3}, \quad \frac{\partial x}{\partial v} = \frac{1}{3}, \quad \frac{\partial y}{\partial u} = \frac{1}{3}, \quad \text{and} \quad \frac{\partial y}{\partial v} = -\frac{1}{3}$$

which implies that the Jacobian is

$$\begin{aligned}
\frac{\partial(x, y)}{\partial(u, v)} &= \begin{vmatrix} \frac{\partial x}{\partial u} & \frac{\partial x}{\partial v} \\ \frac{\partial y}{\partial u} & \frac{\partial y}{\partial v} \end{vmatrix} \\
&= \begin{vmatrix} \frac{2}{3} & \frac{1}{3} \\ \frac{1}{3} & -\frac{1}{3} \end{vmatrix} \\
&= -\frac{2}{9} - \frac{1}{9} \\
&= -\frac{1}{3}.
\end{aligned}$$

So, by Theorem 14.5, you obtain

$$\begin{aligned}
\int_R\int 3xy\, dA &= \int_S\int 3\left[\frac{1}{3}(2u + v)\frac{1}{3}(u - v)\right]\left|\frac{\partial(x, y)}{\partial(u, v)}\right| dv\, du \\
&= \int_1^4 \int_{-4}^0 \frac{1}{9}(2u^2 - uv - v^2)\, dv\, du \\
&= \frac{1}{9}\int_1^4 \left[2u^2v - \frac{uv^2}{2} - \frac{v^3}{3}\right]_{-4}^0 du \\
&= \frac{1}{9}\int_1^4 \left(8u^2 + 8u - \frac{64}{3}\right) du \\
&= \frac{1}{9}\left[\frac{8u^3}{3} + 4u^2 - \frac{64}{3}u\right]_1^4 \\
&= \frac{164}{9}.
\end{aligned}$$

EXAMPLE 4 Using a Change of Variables to Simplify an Integrand

Let R be the region bounded by the square with vertices $(0, 1)$, $(1, 2)$, $(2, 1)$, and $(1, 0)$. Evaluate the integral

$$\int_R\int (x + y)^2 \sin^2(x - y)\, dA.$$

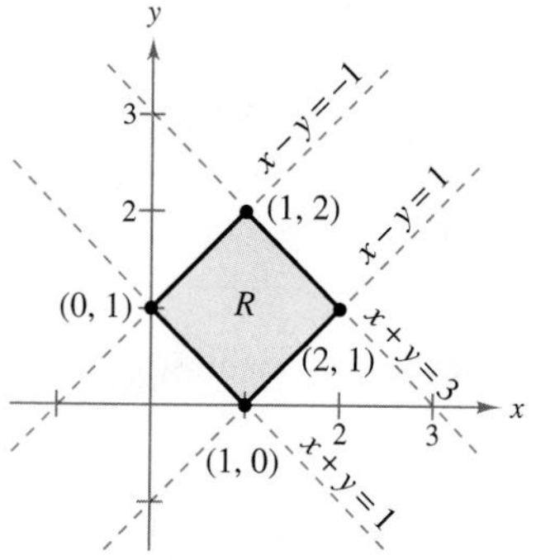

Region R in the xy-plane
Figure 14.76

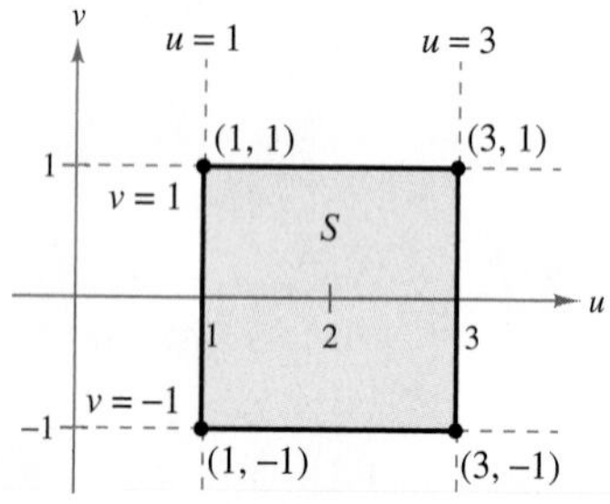

Region S in the uv-plane
Figure 14.77

Solution Note that the sides of R lie on the lines $x + y = 1$, $x - y = 1$, $x + y = 3$, and $x - y = -1$, as shown in Figure 14.76. Letting $u = x + y$ and $v = x - y$, you can determine the bounds for region S in the uv-plane to be

$$1 \le u \le 3 \quad \text{and} \quad -1 \le v \le 1$$

as shown in Figure 14.77. Solving for x and y in terms of u and v produces

$$x = \frac{1}{2}(u + v) \quad \text{and} \quad y = \frac{1}{2}(u - v).$$

The partial derivatives of x and y are

$$\frac{\partial x}{\partial u} = \frac{1}{2}, \quad \frac{\partial x}{\partial v} = \frac{1}{2}, \quad \frac{\partial y}{\partial u} = \frac{1}{2}, \quad \text{and} \quad \frac{\partial y}{\partial v} = -\frac{1}{2}$$

which implies that the Jacobian is

$$\frac{\partial(x, y)}{\partial(u, v)} = \begin{vmatrix} \frac{\partial x}{\partial u} & \frac{\partial x}{\partial v} \\ \frac{\partial y}{\partial u} & \frac{\partial y}{\partial v} \end{vmatrix} = \begin{vmatrix} \frac{1}{2} & \frac{1}{2} \\ \frac{1}{2} & -\frac{1}{2} \end{vmatrix} = -\frac{1}{4} - \frac{1}{4} = -\frac{1}{2}.$$

By Theorem 14.5, it follows that

$$\begin{aligned}
\int_R\int (x + y)^2 \sin^2(x - y)\, dA &= \int_{-1}^{1}\int_{1}^{3} u^2 \sin^2 v \left(\frac{1}{2}\right) du\, dv \\
&= \frac{1}{2}\int_{-1}^{1} (\sin^2 v) \frac{u^3}{3}\Big]_1^3\, dv \\
&= \frac{13}{3}\int_{-1}^{1} \sin^2 v\, dv \\
&= \frac{13}{6}\int_{-1}^{1} (1 - \cos 2v)\, dv \\
&= \frac{13}{6}\left[v - \frac{1}{2}\sin 2v\right]_{-1}^{1} \\
&= \frac{13}{6}\left[2 - \frac{1}{2}\sin 2 + \frac{1}{2}\sin(-2)\right] \\
&= \frac{13}{6}(2 - \sin 2) \\
&\approx 2.363.
\end{aligned}$$

In each of the change-of-variables examples in this section, the region S has been a rectangle with sides parallel to the u- or v-axis. Occasionally, a change of variables can be used for other types of regions. For instance, letting $T(u, v) = \left(x, \frac{1}{2}y\right)$ changes the circular region $u^2 + v^2 = 1$ to the elliptical region $x^2 + (y^2/4) = 1$.

Exercises for Section 14.8

See www.CalcChat.com for worked-out solutions to odd-numbered exercises.

In Exercises 1–8, find the Jacobian $\partial(x, y)/\partial(u, v)$ for the indicated change of variables.

1. $x = -\frac{1}{2}(u - v),\ y = \frac{1}{2}(u + v)$
2. $x = au + bv,\ y = cu + dv$
3. $x = u - v^2,\ y = u + v$
4. $x = uv - 2u,\ y = uv$
5. $x = u \cos\theta - v \sin\theta,\ y = u \sin\theta + v \cos\theta$
6. $x = u + a,\ y = v + a$
7. $x = e^u \sin v,\ y = e^u \cos v$
8. $x = \dfrac{u}{v},\ y = u + v$

In Exercises 9 and 10, sketch the image S in the uv-plane of the region R in the xy-plane using the given transformations.

9. $x = 3u + 2v$
   $y = 3v$

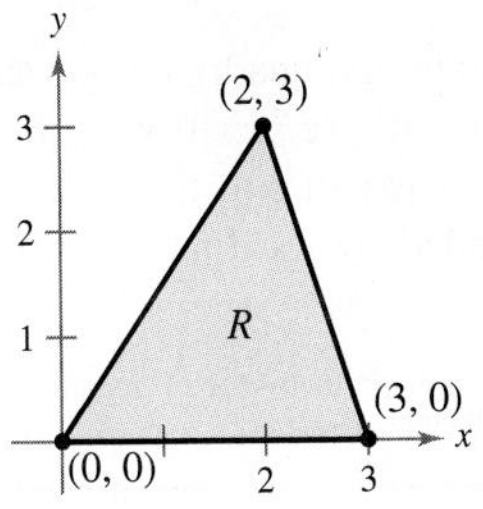

10. $x = \frac{1}{3}(4u - v)$
    $y = \frac{1}{3}(u - v)$

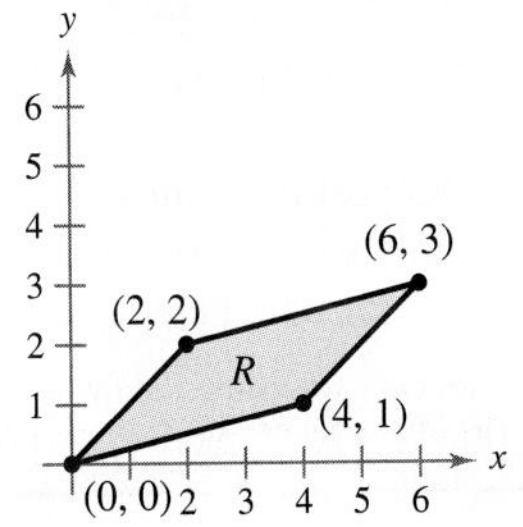

In Exercises 11–16, use the indicated change of variables to evaluate the double integral.

11. $\displaystyle\int_R\int 4(x^2 + y^2)\, dA$
    $x = \frac{1}{2}(u + v)$
    $y = \frac{1}{2}(u - v)$

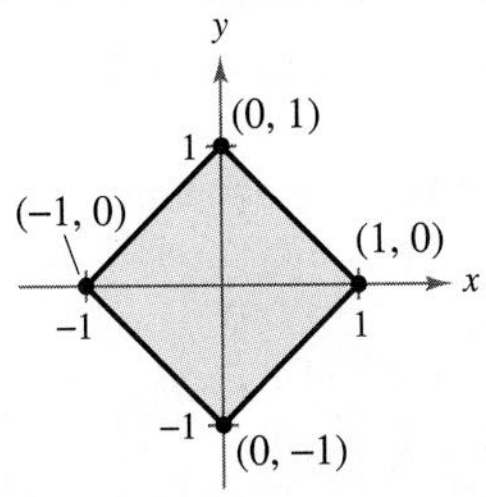

12. $\displaystyle\int_R\int 60xy\, dA$
    $x = \frac{1}{2}(u + v)$
    $y = -\frac{1}{2}(u - v)$

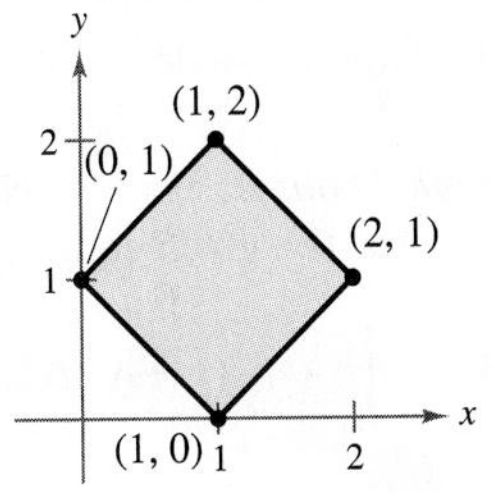

13. $\displaystyle\int_R\int y(x - y)\, dA$
    $x = u + v$
    $y = u$

14. $\displaystyle\int_R\int 4(x + y)e^{x-y}\, dA$
    $x = \frac{1}{2}(u + v)$
    $y = \frac{1}{2}(u - v)$

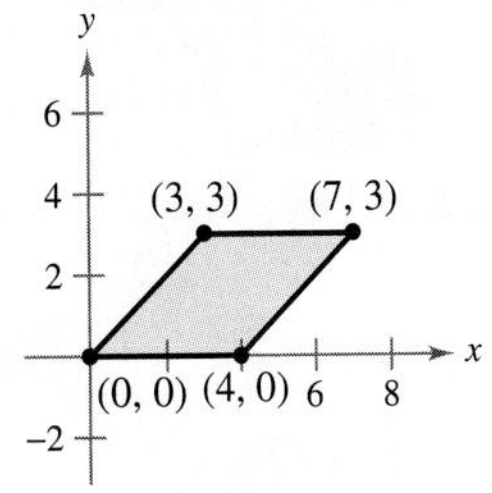

Figure for 13

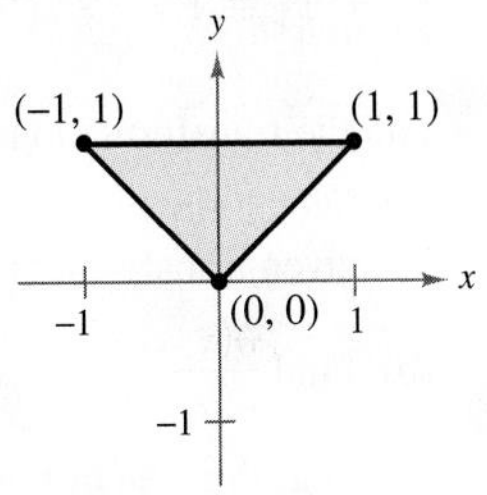

Figure for 14

15. $\displaystyle\int_R\int e^{-xy/2}\, dA$
    $x = \sqrt{\dfrac{v}{u}},\ y = \sqrt{uv}$
    R: first-quadrant region lying between the graphs of
    $y = \dfrac{1}{4}x,\quad y = 2x,\quad y = \dfrac{1}{x},\quad y = \dfrac{4}{x}$

16. $\displaystyle\int_R\int y \sin xy\, dA$
    $x = \dfrac{u}{v},\quad y = v$
    R: region lying between the graphs of $xy = 1$, $xy = 4$, $y = 1$, $y = 4$

In Exercises 17–22, use a change of variables to find the volume of the solid region lying below the surface $z = f(x, y)$ and above the plane region R.

17. $f(x, y) = (x + y)e^{x-y}$
    R: region bounded by the square with vertices $(4, 0)$, $(6, 2)$, $(4, 4)$, $(2, 2)$
18. $f(x, y) = (x + y)^2 \sin^2(x - y)$
    R: region bounded by the square with vertices $(\pi, 0)$, $(3\pi/2, \pi/2)$, (π, π), $(\pi/2, \pi/2)$
19. $f(x, y) = \sqrt{(x - y)(x + 4y)}$
    R: region bounded by the parallelogram with vertices $(0, 0)$, $(1, 1)$, $(5, 0)$, $(4, -1)$
20. $f(x, y) = (3x + 2y)(2y - x)^{3/2}$
    R: region bounded by the parallelogram with vertices $(0, 0)$, $(-2, 3)$, $(2, 5)$, $(4, 2)$
21. $f(x, y) = \sqrt{x + y}$
    R: region bounded by the triangle with vertices $(0, 0)$, $(a, 0)$, $(0, a)$, where $a > 0$
22. $f(x, y) = \dfrac{xy}{1 + x^2y^2}$
    R: region bounded by the graphs of $xy = 1$, $xy = 4$, $x = 1$, $x = 4$ (*Hint:* Let $x = u$, $y = v/u$.)

23. Consider the region R in the xy-plane bounded by the ellipse

$$\frac{x^2}{a^2} + \frac{y^2}{b^2} = 1$$

and the transformations $x = au$ and $y = bv$.

(a) Sketch the graph of the region R and its image S under the given transformation.

(b) Find $\dfrac{\partial(x, y)}{\partial(u, v)}$.

(c) Find the area of the ellipse.

24. Use the result of Exercise 23 to find the volume of each dome-shaped solid lying below the surface $z = f(x, y)$ and above the elliptical region R. (*Hint:* After making the change of variables given by the results in Exercise 23, make a second change of variables to polar coordinates.)

(a) $f(x, y) = 16 - x^2 - y^2$; R: $\dfrac{x^2}{16} + \dfrac{y^2}{9} \leq 1$

(b) $f(x, y) = A\cos\left(\dfrac{\pi}{2}\sqrt{\dfrac{x^2}{a^2} + \dfrac{y^2}{b^2}}\right)$; R: $\dfrac{x^2}{a^2} + \dfrac{y^2}{b^2} \leq 1$

Writing About Concepts

25. State the definition of the Jacobian.

26. Describe how to use the Jacobian to change variables in double integrals.

In Exercises 27–30, find the Jacobian $\partial(x, y, z)/\partial(u, v, w)$ for the indicated change of variables. If $x = f(u, v, w)$, $y = g(u, v, w)$, and $z = h(u, v, w)$, then the Jacobian of x, y, and z with respect to u, v, and w is

$$\frac{\partial(x, y, z)}{\partial(u, v, w)} = \begin{vmatrix} \frac{\partial x}{\partial u} & \frac{\partial x}{\partial v} & \frac{\partial x}{\partial w} \\ \frac{\partial y}{\partial u} & \frac{\partial y}{\partial v} & \frac{\partial y}{\partial w} \\ \frac{\partial z}{\partial u} & \frac{\partial z}{\partial v} & \frac{\partial z}{\partial w} \end{vmatrix}.$$

27. $x = u(1 - v)$, $y = uv(1 - w)$, $z = uvw$

28. $x = 4u - v$, $y = 4v - w$, $z = u + w$

29. ***Spherical Coordinates***

$x = \rho \sin \phi \cos \theta$, $y = \rho \sin \phi \sin \theta$, $z = \rho \cos \phi$

30. ***Cylindrical Coordinates***

$x = r \cos \theta$, $y = r \sin \theta$, $z = z$

Putnam Exam Challenge

31. Let A be the area of the region in the first quadrant bounded by the line $y = \frac{1}{2}x$, the x-axis, and the ellipse $\frac{1}{9}x^2 + y^2 = 1$. Find the positive number m such that A is equal to the area of the region in the first quadrant bounded by the line $y = mx$, the y-axis, and the ellipse $\frac{1}{9}x^2 + y^2 = 1$.

Review Exercises for Chapter 14

See www.CalcChat.com for worked-out solutions to odd-numbered exercises.

In Exercises 1 and 2, evaluate the integral.

1. $\displaystyle\int_1^{x^2} x \ln y \, dy$ **2.** $\displaystyle\int_y^{2y} (x^2 + y^2) \, dx$

In Exercises 3–6, evaluate the iterated integral. Change the coordinate system when convenient.

3. $\displaystyle\int_0^1 \int_0^{1+x} (3x + 2y) \, dy \, dx$ **4.** $\displaystyle\int_0^2 \int_{x^2}^{2x} (x^2 + 2y) \, dy \, dx$

5. $\displaystyle\int_0^3 \int_0^{\sqrt{9-x^2}} 4x \, dy \, dx$ **6.** $\displaystyle\int_0^{\sqrt{3}} \int_{2-\sqrt{4-y^2}}^{2+\sqrt{4-y^2}} dx \, dy$

Area **In Exercises 7–14, write the limits for the double integral**

$$\int_R \int f(x, y) \, dA$$

for both orders of integration. Compute the area of R by letting $f(x, y) = 1$ and integrating.

7. Triangle: vertices $(0, 0)$, $(3, 0)$, $(0, 1)$

8. Triangle: vertices $(0, 0)$, $(3, 0)$, $(2, 2)$

9. The larger area between the graphs of $x^2 + y^2 = 25$ and $x = 3$

10. Region bounded by the graphs of $y = 6x - x^2$ and $y = x^2 - 2x$

11. Region enclosed by the graph of $y^2 = x^2 - x^4$

12. Region bounded by the graphs of $x = y^2 + 1$, $x = 0$, $y = 0$, and $y = 2$

13. Region bounded by the graphs of $x = y + 3$ and $x = y^2 + 1$

14. Region bounded by the graphs of $x = -y$ and $x = 2y - y^2$

Think About It **In Exercises 15 and 16, give a geometric argument for the given equality. Verify the equality analytically.**

15. $\displaystyle\int_0^1 \int_{2y}^{2\sqrt{2-y^2}} (x + y) \, dx \, dy = \int_0^2 \int_0^{x/2} (x + y) \, dy \, dx + \int_2^{2\sqrt{2}} \int_0^{\sqrt{8-x^2}/2} (x + y) \, dy \, dx$

16. $\displaystyle\int_0^2 \int_{3y/2}^{5-y} e^{x+y} \, dx \, dy = \int_0^3 \int_0^{2x/3} e^{x+y} \, dy \, dx + \int_3^5 \int_0^{5-x} e^{x+y} \, dy \, dx$

Volume **In Exercises 17 and 18, use a multiple integral and a convenient coordinate system to find the volume of the solid.**

17. Solid bounded by the graphs of $z = x^2 - y + 4$, $z = 0$, $y = 0$, $x = 0$, and $x = 4$

18. Solid bounded by the graphs of $z = x + y$, $z = 0$, $x = 0$, $x = 3$, and $y = x$

Approximation **In Exercises 19 and 20, determine which value best approximates the volume of the solid between the xy-plane and the function over the region. (Make your selection on the basis of a sketch of the solid and *not* by performing any calculations.)**

19. $f(x, y) = x + y$

R: triangle with vertices $(0, 0)$, $(3, 0)$, $(3, 3)$

(a) $\frac{9}{2}$ (b) 5 (c) 13 (d) 100 (e) -100

20. $f(x, y) = 10x^2y^2$

R: circle bounded by $x^2 + y^2 = 1$

(a) π (b) -15 (c) $\frac{2}{3}$ (d) 3 (e) 15

Probability **In Exercises 21 and 22, find k such that the function is a joint density function and find the required probability, where**

$$P(a \le x \le b, c \le y \le d) = \int_c^d \int_a^b f(x, y)\,dx\,dy.$$

21. $f(x, y) = \begin{cases} kxye^{-(x+y)}, & x \ge 0, y \ge 0 \\ 0, & \text{elsewhere} \end{cases}$

$P(0 \le x \le 1, 0 \le y \le 1)$

22. $f(x, y) = \begin{cases} kxy, & 0 \le x \le 1, 0 \le y \le x \\ 0, & \text{elsewhere} \end{cases}$

$P(0 \le x \le 0.5, 0 \le y \le 0.25)$

True or False? **In Exercises 23–26, determine whether the statement is true or false. If it is false, explain why or give an example that shows it is false.**

23. $\displaystyle\int_a^b \int_c^d f(x)g(y)\,dy\,dx = \left[\int_a^b f(x)\,dx\right]\left[\int_c^d g(y)\,dy\right]$

24. If f is continuous over R_1 and R_2, and

$$\int_{R_1}\int dA = \int_{R_2}\int dA$$

then

$$\int_{R_1}\int f(x, y)\,dA = \int_{R_2}\int f(x, y)\,dA.$$

25. $\displaystyle\int_{-1}^1 \int_{-1}^1 \cos(x^2 + y^2)\,dx\,dy = 4\int_0^1 \int_0^1 \cos(x^2 + y^2)\,dx\,dy$

26. $\displaystyle\int_0^1 \int_0^1 \frac{1}{1 + x^2 + y^2}\,dx\,dy < \frac{\pi}{4}$

In Exercises 27 and 28, evaluate the iterated integral by converting to polar coordinates.

27. $\displaystyle\int_0^h \int_0^x \sqrt{x^2 + y^2}\,dy\,dx$

28. $\displaystyle\int_0^4 \int_0^{\sqrt{16-y^2}} (x^2 + y^2)\,dx\,dy$

Volume **In Exercises 29 and 30, use a multiple integral and a convenient coordinate system to find the volume of the solid.**

29. Solid bounded by the graphs of $z = 0$ and $z = h$, outside the cylinder $x^2 + y^2 = 1$ and inside the hyperboloid $x^2 + y^2 - z^2 = 1$

30. Solid that remains after drilling a hole of radius b through the center of a sphere of radius R $(b < R)$

31. Consider the region R in the xy-plane bounded by the graph of the equation

$$(x^2 + y^2)^2 = 9(x^2 - y^2).$$

(a) Convert the equation to polar coordinates. Use a graphing utility to graph the equation.

(b) Use a double integral to find the area of the region R.

(c) Use a computer algebra system to determine the volume of the solid over the region R and beneath the hemisphere $z = \sqrt{9 - x^2 - y^2}$.

32. Combine the sum of the two iterated integrals into a single iterated integral by converting to polar coordinates. Evaluate the resulting iterated integral.

$$\int_0^{8/\sqrt{13}} \int_0^{3x/2} xy\,dy\,dx + \int_{8/\sqrt{13}}^4 \int_0^{\sqrt{16-x^2}} xy\,dy\,dx$$

Mass and Center of Mass **In Exercises 33 and 34, find the mass and center of mass of the lamina bounded by the graphs of the equations for the given density or densities. Use a computer algebra system to evaluate the multiple integrals.**

33. $y = 2x$, $y = 2x^3$, first quadrant

(a) $\rho = kxy$ (b) $\rho = k(x^2 + y^2)$

34. $y = \dfrac{h}{2}\left(2 - \dfrac{x}{L} - \dfrac{x^2}{L^2}\right)$, $\rho = k$, first quadrant

In Exercises 35 and 36, find I_x, I_y, I_0, $\bar{x}$, and $\bar{y}$ for the lamina bounded by the graphs of the equations. Use a computer algebra system to evaluate the double integrals.

35. $y = 0$, $y = b$, $x = 0$, $x = a$, $\rho = kx$

36. $y = 4 - x^2$, $y = 0$, $x > 0$, $\rho = ky$

Surface Area **In Exercises 37 and 38, find the area of the surface given by $z = f(x, y)$ over the region R.**

37. $f(x, y) = 16 - x^2 - y^2$

$R = \{(x, y): x^2 + y^2 \le 16\}$

38. $f(x, y) = 16 - x - y^2$

$R = \{(x, y): 0 \le x \le 2, 0 \le y \le x\}$

Use a computer algebra system to evaluate the integral.

39. ***Surface Area*** Find the area of the surface of the cylinder $f(x, y) = 9 - y^2$ that lies above the triangle bounded by the graphs of the equations $y = x$, $y = -x$, and $y = 3$.

40. ***Surface Area*** The roof over the stage of an open air theater at a theme park is modeled by

$$f(x, y) = 25\left[1 + e^{-(x^2+y^2)/1000}\cos^2\left(\frac{x^2 + y^2}{1000}\right)\right]$$

where the stage is a semicircle bounded by the graphs of $y = \sqrt{50^2 - x^2}$ and $y = 0$.

(a) Use a computer algebra system to graph the surface.

(b) Use a computer algebra system to approximate the number of square feet of roofing required to cover the surface.

In Exercises 41–44, evaluate the iterated integral.

41. $\displaystyle\int_{-3}^{3}\int_{-\sqrt{9-x^2}}^{\sqrt{9-x^2}}\int_{x^2+y^2}^{9}\sqrt{x^2 + y^2}\,dz\,dy\,dx$

42. $\displaystyle\int_{-2}^{2}\int_{-\sqrt{4-x^2}}^{\sqrt{4-x^2}}\int_{0}^{(x^2+y^2)/2}(x^2 + y^2)\,dz\,dy\,dx$

43. $\displaystyle\int_{0}^{a}\int_{0}^{b}\int_{0}^{c}(x^2 + y^2 + z^2)\,dx\,dy\,dz$

44. $\displaystyle\int_{0}^{5}\int_{0}^{\sqrt{25-x^2}}\int_{0}^{\sqrt{25-x^2-y^2}}\frac{1}{1 + x^2 + y^2 + z^2}\,dz\,dy\,dx$

In Exercises 45 and 46, use a computer algebra system to evaluate the iterated integral.

45. $\displaystyle\int_{-1}^{1}\int_{-\sqrt{1-x^2}}^{\sqrt{1-x^2}}\int_{-\sqrt{1-x^2-y^2}}^{\sqrt{1-x^2-y^2}}(x^2 + y^2)\,dz\,dy\,dx$

46. $\displaystyle\int_{0}^{2}\int_{0}^{\sqrt{4-x^2}}\int_{0}^{\sqrt{4-x^2-y^2}}xyz\,dz\,dy\,dx$

Volume **In Exercises 47 and 48, use a multiple integral to find the volume of the solid.**

47. Solid inside the graphs of $r = 2\cos\theta$ and $r^2 + z^2 = 4$

48. Solid inside the graphs of $r^2 + z = 16$, $z = 0$, and $r = 2\sin\theta$

Center of Mass **In Exercises 49–52, find the center of mass of the solid of uniform density bounded by the graphs of the equations.**

49. Solid inside the hemisphere $\rho = \cos\phi$, $\pi/4 \le \phi \le \pi/2$, and outside the cone $\phi = \pi/4$

50. Wedge: $x^2 + y^2 = a^2$, $z = cy\ (c > 0)$, $y \ge 0$, $z \ge 0$

51. $x^2 + y^2 + z^2 = a^2$, first octant

52. $x^2 + y^2 + z^2 = 25$, $z = 4$ (the larger solid)

Moment of Inertia **In Exercises 53 and 54, find the moment of inertia I_z of the solid of given density.**

53. The solid of uniform density inside the paraboloid $z = 16 - x^2 - y^2$, and outside the cylinder $x^2 + y^2 = 9$, $z \ge 0$.

54. $x^2 + y^2 + z^2 = a^2$, density proportional to the distance from the center

55. ***Investigation*** Consider a spherical segment of height h from a sphere of radius a, where $h \le a$, and constant density $\rho(x, y, z) = k$ (see figure).

(a) Find the volume of the solid.

(b) Find the centroid of the solid.

(c) Use the result of part (b) to find the centroid of a hemisphere of radius a.

(d) Find $\lim_{h\to 0}\bar{z}$.

(e) Find I_z.

(f) Use the result of part (e) to find I_z for a hemisphere.

56. ***Moment of Inertia*** Find the moment of inertia about the z-axis of the ellipsoid $x^2 + y^2 + \dfrac{z^2}{a^2} = 1$, where $a > 0$.

In Exercises 57 and 58, give a geometric interpretation of the iterated integral.

57. $\displaystyle\int_{0}^{2\pi}\int_{0}^{\pi}\int_{0}^{6\sin\phi}\rho^2\sin\phi\,d\rho\,d\phi\,d\theta$

58. $\displaystyle\int_{0}^{\pi}\int_{0}^{2}\int_{0}^{1+r^2}r\,dz\,dr\,d\theta$

In Exercises 59 and 60, find the Jacobian $\partial(x, y)/\partial(u, v)$ for the indicated change of variables.

59. $x = u + 3v$, $y = 2u - 3v$

60. $x = u^2 + v^2$, $y = u^2 - v^2$

In Exercises 61 and 62, use the indicated change of variables to evaluate the double integral.

61. $\displaystyle\int_R\int \ln(x + y)\,dA$

$x = \frac{1}{2}(u + v)$, $y = \frac{1}{2}(u - v)$

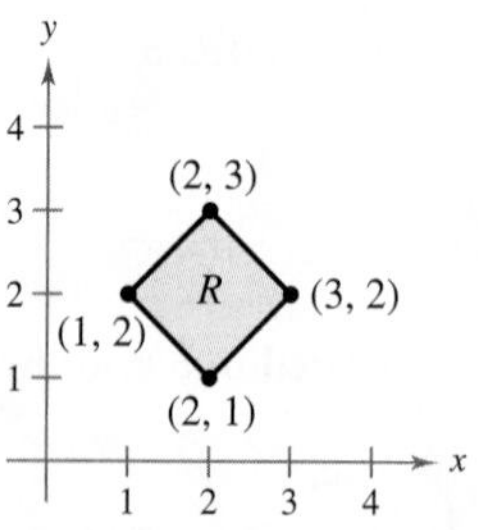

62. $\displaystyle\int_R\int \frac{x}{1 + x^2y^2}\,dA$

$x = u$, $y = \dfrac{v}{u}$

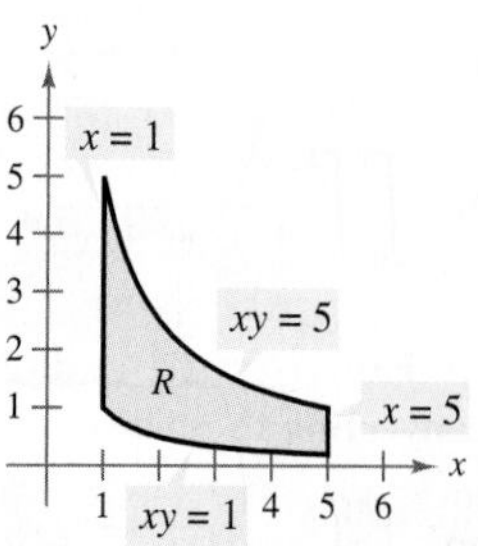

P.S. Problem Solving

See www.CalcChat.com for worked-out solutions to odd-numbered exercises.

1. (a) Find the volume of the solid of intersection of the three cylinders $x^2 + z^2 = 1$, $y^2 + z^2 = 1$, and $x^2 + y^2 = 1$ (see figure).

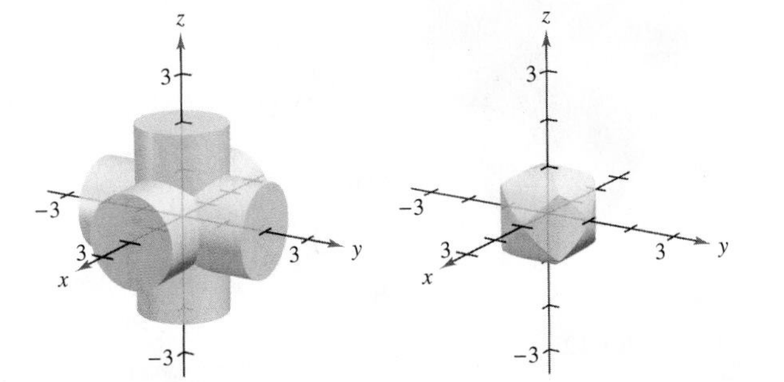

(b) Use the Monte Carlo Method (see Section 4.2 exercises) to confirm the answer in part (a). (*Hint:* Generate random points inside the cube of volume 8 centered at the origin.)

2. Let a, b, c, and d be positive real numbers. The first octant of the plane $ax + by + cz = d$ is shown in the figure. Show that the surface area of this portion of the plane is equal to

$$\frac{A(R)}{c}\sqrt{a^2 + b^2 + c^2}$$

where $A(R)$ is the area of the triangular region R in the xy-plane, as shown in the figure.

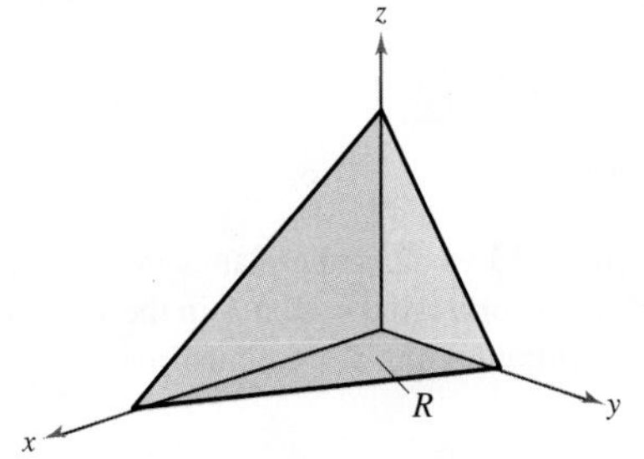

3. Derive Euler's famous result that was mentioned in Section 9.3, $\sum_{n=1}^{\infty} \frac{1}{n^2} = \frac{\pi^2}{6}$, by completing each step.

(a) Prove that $\displaystyle\int \frac{dv}{2 - u^2 + v^2} = \frac{1}{\sqrt{2 - u^2}} \arctan \frac{v}{\sqrt{2 - u^2}} + C.$

(b) Prove that $\displaystyle I_1 = \int_0^{\sqrt{2}/2}\int_{-u}^{u} \frac{2}{2 - u^2 + v^2}\, dv\, du = \frac{\pi^2}{18}$ by using the substitution $u = \sqrt{2}\sin\theta$.

(c) Prove that

$$I_2 = \int_{\sqrt{2}/2}^{\sqrt{2}}\int_{u-\sqrt{2}}^{-u+\sqrt{2}} \frac{2}{2 - u^2 + v^2}\, dv\, du$$

$$= 4\int_{\pi/6}^{\pi/2} \arctan \frac{1 - \sin\theta}{\cos\theta}\, d\theta$$

by using the substitution $u = \sqrt{2}\sin\theta$.

(d) Prove the trigonometric identity $\displaystyle\frac{1 - \sin\theta}{\cos\theta} = \tan\left(\frac{(\pi/2) - \theta}{2}\right)$.

(e) Prove that $\displaystyle I_2 = \int_{\sqrt{2}/2}^{\sqrt{2}}\int_{u-\sqrt{2}}^{-u+\sqrt{2}} \frac{2}{2 - u^2 + v^2}\, dv\, du = \frac{\pi^2}{9}$.

(f) Use the formula for the sum of an infinite geometric series to verify that $\displaystyle\sum_{n=1}^{\infty} \frac{1}{n^2} = \int_0^1\int_0^1 \frac{1}{1 - xy}\, dx\, dy$.

(g) Use the change of variables $u = \dfrac{x + y}{\sqrt{2}}$ and $v = \dfrac{y - x}{\sqrt{2}}$ to prove that $\displaystyle\sum_{n=1}^{\infty} \frac{1}{n^2} = \int_0^1\int_0^1 \frac{1}{1 - xy}\, dx\, dy = I_1 + I_2 = \frac{\pi^2}{6}$.

4. Consider a circular lawn with a radius of 10 feet, as shown in the figure. Assume that a sprinkler distributes water in a radial fashion according to the formula

$$f(r) = \frac{r}{16} - \frac{r^2}{160}$$

(measured in cubic feet of water per hour per square foot of lawn), where r is the distance in feet from the sprinkler. Find the amount of water that is distributed in 1 hour in the following two annular regions.

$A = \{(r, \theta): 4 \le r \le 5, 0 \le \theta \le 2\pi\}$

$B = \{(r, \theta): 9 \le r \le 10, 0 \le \theta \le 2\pi\}$

Is the distribution of water uniform? Determine the amount of water the entire lawn receives in 1 hour.

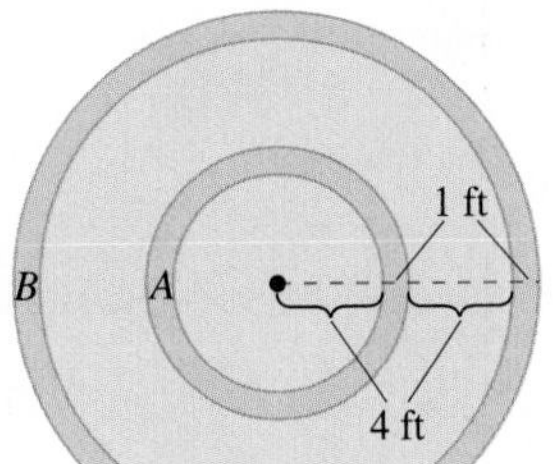

5. The figure shows the region R bounded by the curves $y = \sqrt{x}$, $y = \sqrt{2x}$, $y = \dfrac{x^2}{3}$, and $y = \dfrac{x^2}{4}$. Use the change of variables $x = u^{1/3}v^{2/3}$ and $y = u^{2/3}v^{1/3}$ to find the area of the region R.

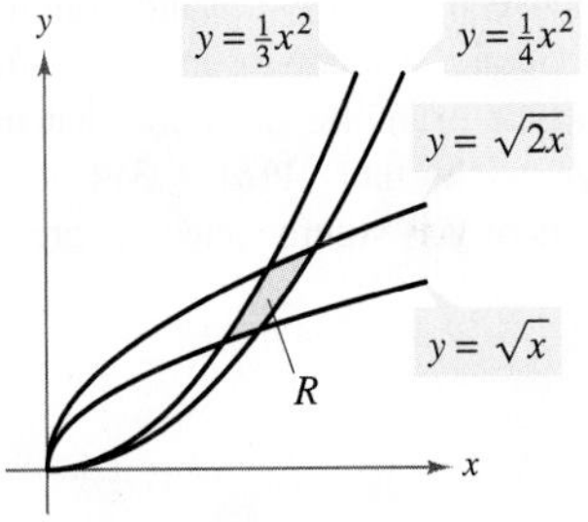

6. The figure shows a solid bounded below by the plane $z = 2$ and above by the sphere $x^2 + y^2 + z^2 = 8$.

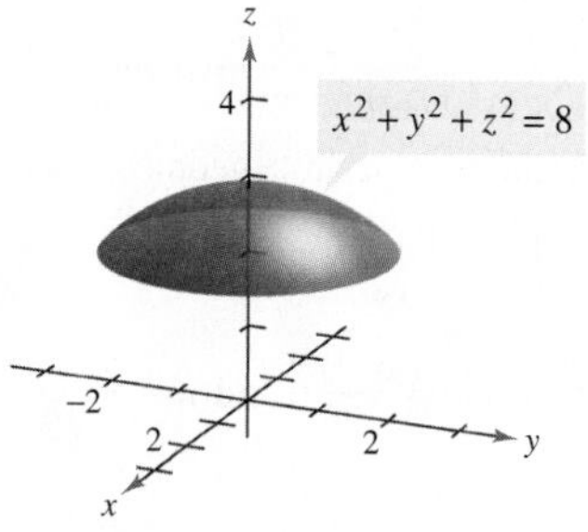

(a) Find the volume of the solid using cylindrical coordinates.

(b) Find the volume of the solid using spherical coordinates.

7. Sketch the solid whose volume is given by the sum of the iterated integrals

$$\int_0^6\int_{z/2}^3\int_{z/2}^y dx\,dy\,dz + \int_0^6\int_3^{(12-z)/2}\int_{z/2}^{6-y} dx\,dy\,dz.$$

Then write the volume as a single iterated integral in the order $dy\,dz\,dx$.

8. Prove that $\displaystyle\lim_{n\to\infty}\int_0^1\int_0^1 x^n y^n\,dx\,dy = 0.$

In Exercises 9 and 10, evaluate the integral. (*Hint:* See Exercise 55 in Section 14.3.)

9. $\displaystyle\int_0^\infty x^2 e^{-x^2}\,dx$

10. $\displaystyle\int_0^1 \sqrt{\ln\frac{1}{x}}\,dx$

11. Consider the function

$$f(x, y) = \begin{cases} ke^{-(x+y)/a}, & x \ge 0,\ y \ge 0 \\ 0, & \text{elsewhere.} \end{cases}$$

Find the relationship between the positive constants a and k such that f is a joint density function of the continuous random variables x and y.

12. From 1963 to 1986, the volume of the Great Salt Lake approximately tripled while its top surface area approximately doubled. Read the article "Relations between Surface Area and Volume in Lakes" by Daniel Cass and Gerald Wildenberg in *The College Mathematics Journal.* Then give examples of solids that have "water levels" a and b such that $V(b) = 3V(a)$ and $A(b) = 2A(a)$ (see figure), where V is volume and A is area.

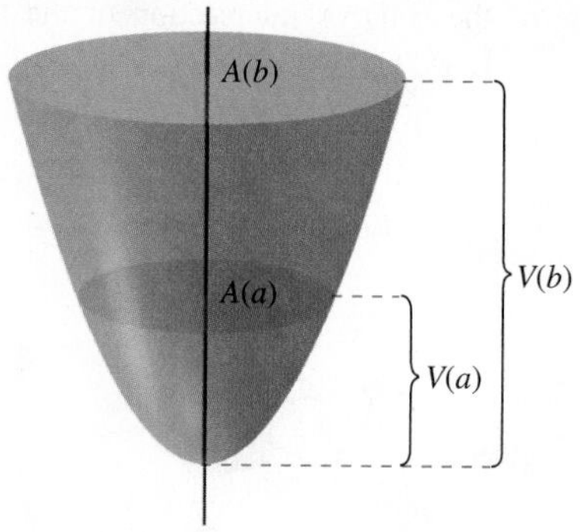

Figure for 12

13. The angle between a plane P and the xy-plane is θ, where $0 \le \theta < \pi/2$. The projection of a rectangular region in P onto the xy-plane is a rectangle whose sides have lengths Δx and Δy, as shown in the figure. Prove that the area of the rectangular region in P is $\sec\theta\,\Delta x\,\Delta y$.

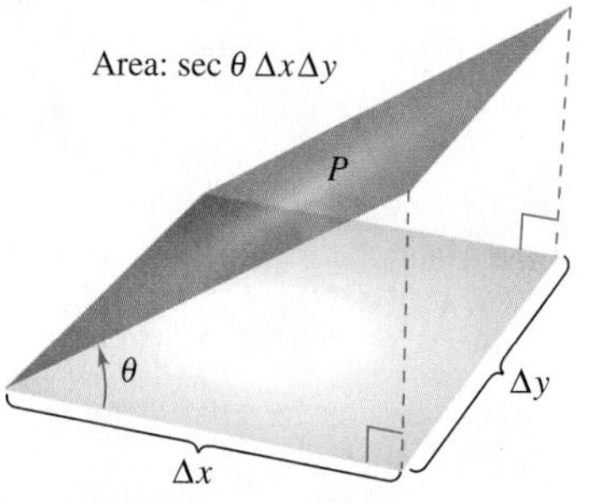

14. Use the result of Exercise 13 to order the planes in ascending order of their surface areas for a fixed region R in the xy-plane. Explain your ordering without doing any calculations.

(a) $z_1 = 2 + x$

(b) $z_2 = 5$

(c) $z_3 = 10 - 5x + 9y$

(d) $z_4 = 3 + x - 2y$

15. Evaluate the integral $\displaystyle\int_0^\infty\int_0^\infty \frac{1}{(1 + x^2 + y^2)^2}\,dx\,dy.$

16. Evaluate the integrals

$$\int_0^1\int_0^1 \frac{x-y}{(x+y)^3}\,dx\,dy \quad\text{and}\quad \int_0^1\int_0^1 \frac{x-y}{(x+y)^3}\,dy\,dx.$$

Are the results the same? Why or why not?

17. Show that the volume of a spherical block can be approximated by

$\Delta V \approx \rho^2 \sin\phi\,\Delta\rho\,\Delta\phi\,\Delta\theta.$

15

Vector Analysis

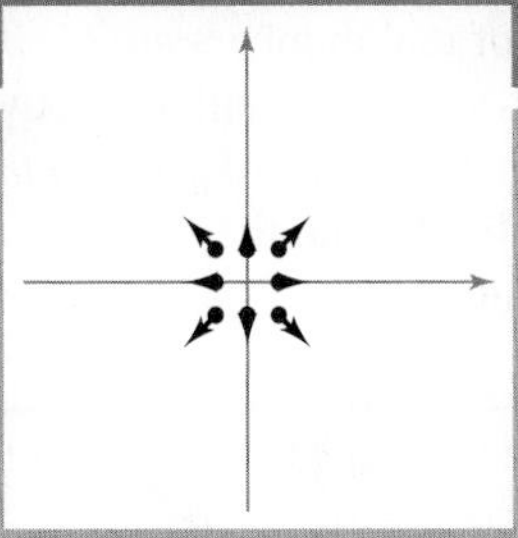

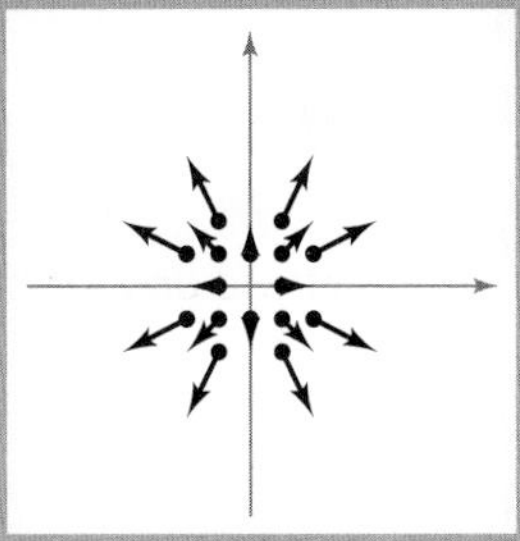

In Chapter 15, you will combine your knowledge of vectors with your knowledge of integral calculus. The chapter begins by introducing *vector fields*, like those shown above. Examples of vector fields are velocity fields, electromagnetic fields, and gravitational fields.

The rotation of a hurricane is an example of a vector field. The photo shows the position of Hurricane Andrew over a three-day period as it moved westward across Florida in August of 1992. Hurricanes in the northern hemisphere rotate counterclockwise and move through the ocean clockwise. Hurricanes in the southern hemisphere rotate clockwise and move counterclockwise. Why do you think this is so?

NASA

Section 15.1 Vector Fields

- **Understand the concept of a vector field.**
- **Determine whether a vector field is conservative.**
- **Find the curl of a vector field.**
- **Find the divergence of a vector field.**

Vector Fields

In Chapter 12, you studied vector-valued functions—functions that assign a vector to a *real number*. There you saw that vector-valued functions of real numbers are useful in representing curves and motion along a curve. In this chapter, you will study two other types of vector-valued functions—functions that assign a vector to a *point in the plane* or a *point in space*. Such functions are called **vector fields,** and they are useful in representing various types of **force fields** and **velocity fields.**

Definition of Vector Field

Let M and N be functions of two variables x and y, defined on a plane region R. The function $\mathbf{F}$ defined by

$$\mathbf{F}(x, y) = M\mathbf{i} + N\mathbf{j} \qquad \text{Plane}$$

is called a **vector field over *R*.**

Let M, N, and P be functions of three variables x, y, and z, defined on a solid region Q in space. The function $\mathbf{F}$ defined by

$$\mathbf{F}(x, y, z) = M\mathbf{i} + N\mathbf{j} + P\mathbf{k} \qquad \text{Space}$$

is called a **vector field over *Q*.**

NOTE Although a vector field consists of infinitely many vectors, you can get a good idea of what the vector field looks like by sketching several representative vectors $\mathbf{F}(x, y)$ whose initial points are (x, y).

From this definition you can see that the *gradient* is one example of a vector field. For example, if

$$f(x, y) = x^2 + y^2$$

then the gradient of f

$$\nabla f(x, y) = f_x(x, y)\mathbf{i} + f_y(x, y)\mathbf{j} = 2x\mathbf{i} + 2y\mathbf{j} \qquad \text{Vector field in the plane}$$

is a vector field in the plane. From Chapter 13, the graphical interpretation of this field is a family of vectors, each of which points in the direction of maximum increase along the surface given by $z = f(x, y)$. For this particular function, the surface is a paraboloid and the gradient tells you that the direction of maximum increase along the surface is the direction given by the ray from the origin through the point (x, y).

Similarly, if

$$f(x, y, z) = x^2 + y^2 + z^2$$

then the gradient of f

$$\begin{aligned}\nabla f(x, y, z) &= f_x(x, y, z)\mathbf{i} + f_y(x, y, z)\mathbf{j} + f_z(x, y, z)\mathbf{k}\\ &= 2x\mathbf{i} + 2y\mathbf{j} + 2z\mathbf{k}\end{aligned} \qquad \text{Vector field in space}$$

is a vector field in space.

A vector field is **continuous** at a point if each of its component functions M, N, and P is continuous at that point.

Some common *physical* examples of vector fields are **velocity fields, gravitational fields,** and **electric force fields.**

1. *Velocity fields* describe the motions of systems of particles in the plane or in space. For instance, Figure 15.1 shows the vector field determined by a wheel rotating on an axle. Notice that the velocity vectors are determined by the locations of their initial points—the farther a point is from the axle, the greater its velocity. Velocity fields are also determined by the flow of liquids through a container or by the flow of air currents around a moving object, as shown in Figure 15.2.

Velocity field

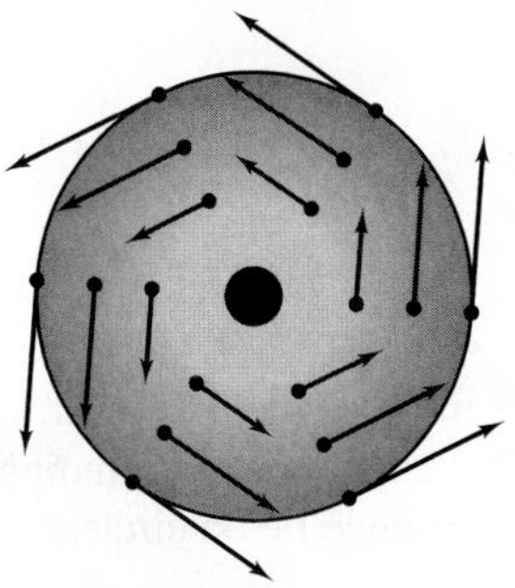

Rotating wheel
Figure 15.1

Air flow vector field
Figure 15.2

2. *Gravitational fields* are defined by **Newton's Law of Gravitation,** which states that the force of attraction exerted on a particle of mass m_1 located at (x, y, z) by a particle of mass m_2 located at $(0, 0, 0)$ is given by

$$\mathbf{F}(x, y, z) = \frac{-Gm_1m_2}{x^2 + y^2 + z^2}\mathbf{u}$$

where G is the gravitational constant and $\mathbf{u}$ is the unit vector in the direction from the origin to (x, y, z). In Figure 15.3, you can see that the gravitational field $\mathbf{F}$ has the properties that $\mathbf{F}(x, y, z)$ always points toward the origin, and that the magnitude of $\mathbf{F}(x, y, z)$ is the same at all points equidistant from the origin. A vector field with these two properties is called a **central force field.** Using the position vector

$$\mathbf{r} = x\mathbf{i} + y\mathbf{j} + z\mathbf{k}$$

for the point (x, y, z), you can write the gravitational field $\mathbf{F}$ as

$$\begin{aligned}\mathbf{F}(x, y, z) &= \frac{-Gm_1m_2}{\|\mathbf{r}\|^2}\left(\frac{\mathbf{r}}{\|\mathbf{r}\|}\right)\\ &= \frac{-Gm_1m_2}{\|\mathbf{r}\|^2}\mathbf{u}.\end{aligned}$$

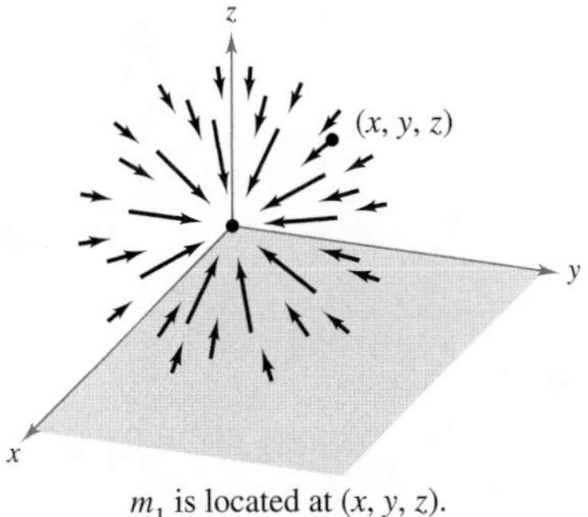

m_1 is located at (x, y, z).
m_2 is located at $(0, 0, 0)$.

Gravitational force field
Figure 15.3

3. *Electric force fields* are defined by **Coulomb's Law,** which states that the force exerted on a particle with electric charge q_1 located at (x, y, z) by a particle with electric charge q_2 located at $(0, 0, 0)$ is given by

$$\mathbf{F}(x, y, z) = \frac{cq_1q_2}{\|\mathbf{r}\|^2}\mathbf{u}$$

where $\mathbf{r} = x\mathbf{i} + y\mathbf{j} + z\mathbf{k}$, $\mathbf{u} = \mathbf{r}/\|\mathbf{r}\|$, and c is a constant that depends on the choice of units for $\|\mathbf{r}\|$, q_1, and q_2.

Note that an electric force field has the same form as a gravitational field. That is,

$$\mathbf{F}(x, y, z) = \frac{k}{\|\mathbf{r}\|^2}\mathbf{u}.$$

Such a force field is called an **inverse square field.**

Definition of Inverse Square Field

Let $\mathbf{r}(t) = x(t)\mathbf{i} + y(t)\mathbf{j} + z(t)\mathbf{k}$ be a position vector. The vector field $\mathbf{F}$ is an **inverse square field** if

$$\mathbf{F}(x, y, z) = \frac{k}{\|\mathbf{r}\|^2}\mathbf{u}$$

where k is a real number and $\mathbf{u} = \mathbf{r}/\|\mathbf{r}\|$ is a unit vector in the direction of $\mathbf{r}$.

Because vector fields consist of infinitely many vectors, it is not possible to create a sketch of the entire field. Instead, when you sketch a vector field, your goal is to sketch representative vectors that help you visualize the field.

EXAMPLE 1 Sketching a Vector Field

Sketch some vectors in the vector field given by

$$\mathbf{F}(x, y) = -y\mathbf{i} + x\mathbf{j}.$$

Solution You could plot vectors at several random points in the plane. However, it is more enlightening to plot vectors of equal magnitude. This corresponds to finding level curves in scalar fields. In this case, vectors of equal magnitude lie on circles.

$$\|\mathbf{F}\| = c \qquad \text{Vectors of length } c$$
$$\sqrt{x^2 + y^2} = c$$
$$x^2 + y^2 = c^2 \qquad \text{Equation of circle}$$

To begin making the sketch, choose a value for c and plot several vectors on the resulting circle. For instance, the following vectors occur on the unit circle.

Point	*Vector*
$(1, 0)$	$\mathbf{F}(1, 0) = \mathbf{j}$
$(0, 1)$	$\mathbf{F}(0, 1) = -\mathbf{i}$
$(-1, 0)$	$\mathbf{F}(-1, 0) = -\mathbf{j}$
$(0, -1)$	$\mathbf{F}(0, -1) = \mathbf{i}$

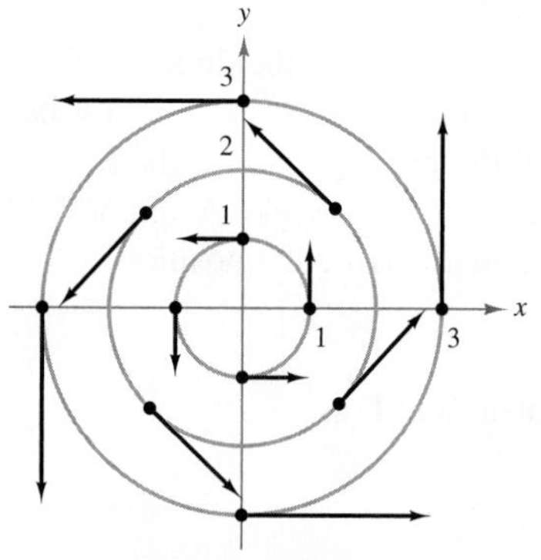

Vector field:
$\mathbf{F}(x, y) = -y\mathbf{i} + x\mathbf{j}$

Figure 15.4

These and several other vectors in the vector field are shown in Figure 15.4. Note in the figure that this vector field is similar to that given by the rotating wheel shown in Figure 15.1.

EXAMPLE 2 Sketching a Vector Field

Sketch some vectors in the vector field given by

$$\mathbf{F}(x, y) = 2x\mathbf{i} + y\mathbf{j}.$$

Solution For this vector field, vectors of equal length lie on ellipses given by

$$\|\mathbf{F}\| = \sqrt{(2x)^2 + (y)^2} = c$$

which implies that

$$4x^2 + y^2 = c^2.$$

For $c = 1$, sketch several vectors $2x\mathbf{i} + y\mathbf{j}$ of magnitude 1 at points on the ellipse given by

$$4x^2 + y^2 = 1.$$

For $c = 2$, sketch several vectors $2x\mathbf{i} + y\mathbf{j}$ of magnitude 2 at points on the ellipse given by

$$4x^2 + y^2 = 4.$$

These vectors are shown in Figure 15.5.

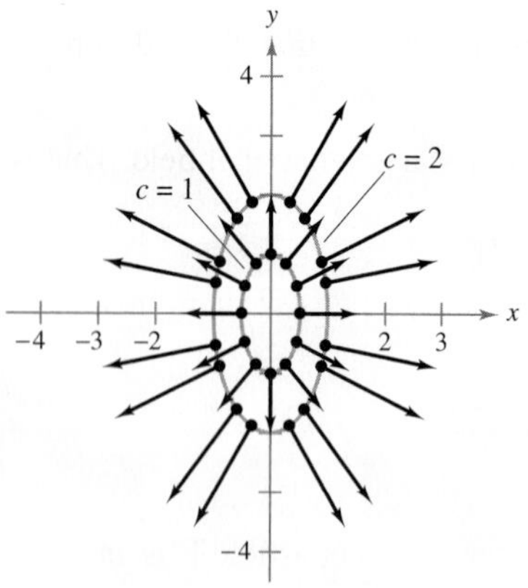

Vector field:
$\mathbf{F}(x, y) = 2x\mathbf{i} + y\mathbf{j}$

Figure 15.5

EXAMPLE 3 Sketching a Velocity Field

Sketch some vectors in the velocity field given by

$$\mathbf{v}(x, y, z) = (16 - x^2 - y^2)\mathbf{k}$$

where $x^2 + y^2 \leq 16$.

Solution You can imagine that $\mathbf{v}$ describes the velocity of a liquid flowing through a tube of radius 4. Vectors near the z-axis are longer than those near the edge of the tube. For instance, at the point $(0, 0, 0)$, the velocity vector is $\mathbf{v}(0, 0, 0) = 16\mathbf{k}$, whereas at the point $(0, 3, 0)$, the velocity vector is $\mathbf{v}(0, 3, 0) = 7\mathbf{k}$. Figure 15.6 shows these and several other vectors for the velocity field. From the figure, you can see that the speed of the liquid is greater near the center of the tube than near the edges of the tube.

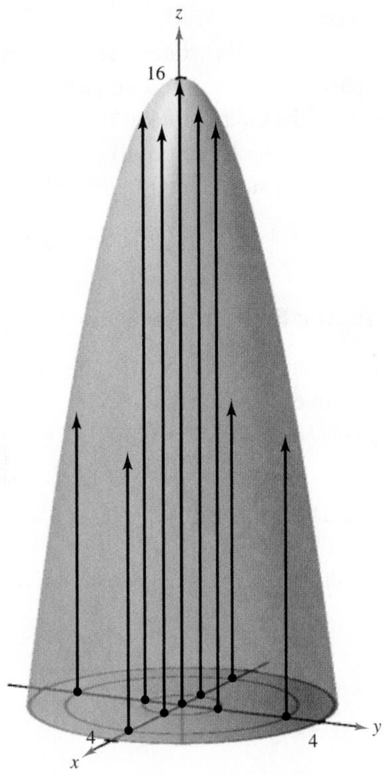

Velocity field:
$\mathbf{v}(x, y, z) = (16 - x^2 - y^2)\mathbf{k}$

Figure 15.6

Conservative Vector Fields

Notice in Figure 15.5 that all the vectors appear to be normal to the level curve from which they emanate. Because this is a property of gradients, it is natural to ask whether the vector field given by $\mathbf{F}(x, y) = 2x\mathbf{i} + y\mathbf{j}$ is the *gradient* for some differentiable function f. The answer is that some vector fields can be represented as the gradients of differentiable functions and some cannot—those that can are called **conservative** vector fields.

Definition of Conservative Vector Field

A vector field $\mathbf{F}$ is called **conservative** if there exists a differentiable function f such that $\mathbf{F} = \nabla f$. The function f is called the **potential function** for $\mathbf{F}$.

EXAMPLE 4 Conservative Vector Fields

a. The vector field given by $\mathbf{F}(x, y) = 2x\mathbf{i} + y\mathbf{j}$ is conservative. To see this, consider the potential function $f(x, y) = x^2 + \frac{1}{2}y^2$. Because

$$\nabla f = 2x\mathbf{i} + y\mathbf{j} = \mathbf{F}$$

it follows that $\mathbf{F}$ is conservative.

b. Every inverse square field is conservative. To see this, let

$$\mathbf{F}(x, y, z) = \frac{k}{\|\mathbf{r}\|^2}\mathbf{u} \quad \text{and} \quad f(x, y, z) = \frac{-k}{\sqrt{x^2 + y^2 + z^2}}$$

where $\mathbf{u} = \mathbf{r}/\|\mathbf{r}\|$. Because

$$\begin{aligned}
\nabla f &= \frac{kx}{(x^2 + y^2 + z^2)^{3/2}}\mathbf{i} + \frac{ky}{(x^2 + y^2 + z^2)^{3/2}}\mathbf{j} + \frac{kz}{(x^2 + y^2 + z^2)^{3/2}}\mathbf{k} \\
&= \frac{k}{x^2 + y^2 + z^2}\left(\frac{x\mathbf{i} + y\mathbf{j} + z\mathbf{k}}{\sqrt{x^2 + y^2 + z^2}}\right) \\
&= \frac{k}{\|\mathbf{r}\|^2}\frac{\mathbf{r}}{\|\mathbf{r}\|} \\
&= \frac{k}{\|\mathbf{r}\|^2}\mathbf{u}
\end{aligned}$$

it follows that $\mathbf{F}$ is conservative.

As can be seen in Example 4(b), many important vector fields, including gravitational fields and electric force fields, are conservative. Most of the terminology in this chapter comes from physics. For example, the term "conservative" is derived from the classic physical law regarding the conservation of energy. This law states that the sum of the kinetic energy and the potential energy of a particle moving in a conservative force field is constant. (The kinetic energy of a particle is the energy due to its motion, and the potential energy is the energy due to its position in the force field.)

The following important theorem gives a necessary and sufficient condition for a vector field *in the plane* to be conservative.

THEOREM 15.1 Test for Conservative Vector Field in the Plane

Let M and N have continuous first partial derivatives on an open disk R. The vector field given by $\mathbf{F}(x, y) = M\mathbf{i} + N\mathbf{j}$ is conservative if and only if

$$\frac{\partial N}{\partial x} = \frac{\partial M}{\partial y}.$$

Proof To prove that the given condition is necessary for $\mathbf{F}$ to be conservative, suppose there exists a potential function f such that

$$\mathbf{F}(x, y) = \nabla f(x, y) = M\mathbf{i} + N\mathbf{j}.$$

Then you have

$$f_x(x, y) = M \quad \Rightarrow \quad f_{xy}(x, y) = \frac{\partial M}{\partial y}$$

$$f_y(x, y) = N \quad \Rightarrow \quad f_{yx}(x, y) = \frac{\partial N}{\partial x}$$

and, by the equivalence of the mixed partials f_{xy} and f_{yx}, you can conclude that $\partial N/\partial x = \partial M/\partial y$ for all (x, y) in R. The sufficiency of the condition is proved in Section 15.4.

NOTE Theorem 15.1 requires that the domain of $\mathbf{F}$ be an open disk. If R is simply an open region, the given condition is necessary but not sufficient to produce a conservative vector field.

EXAMPLE 5 Testing for Conservative Vector Fields in the Plane

Decide whether the vector field given by $\mathbf{F}$ is conservative.

a. $\mathbf{F}(x, y) = x^2y\mathbf{i} + xy\mathbf{j}$ **b.** $\mathbf{F}(x, y) = 2x\mathbf{i} + y\mathbf{j}$

Solution

a. The vector field given by $\mathbf{F}(x, y) = x^2y\mathbf{i} + xy\mathbf{j}$ is not conservative because

$$\frac{\partial M}{\partial y} = \frac{\partial}{\partial y}[x^2y] = x^2 \quad \text{and} \quad \frac{\partial N}{\partial x} = \frac{\partial}{\partial x}[xy] = y.$$

b. The vector field given by $\mathbf{F}(x, y) = 2x\mathbf{i} + y\mathbf{j}$ is conservative because

$$\frac{\partial M}{\partial y} = \frac{\partial}{\partial y}[2x] = 0 \quad \text{and} \quad \frac{\partial N}{\partial x} = \frac{\partial}{\partial x}[y] = 0.$$

Theorem 15.1 tells you whether a vector field is conservative. It does not tell you how to find a potential function of **F**. The problem is comparable to antidifferentiation. Sometimes you will be able to find a potential function by simple inspection. For instance, in Example 4 you observed that

$$f(x, y) = x^2 + \frac{1}{2}y^2$$

has the property that $\nabla f(x, y) = 2x\mathbf{i} + y\mathbf{j}$.

EXAMPLE 6 Finding a Potential Function for F(x, y)

Find a potential function for

$$\mathbf{F}(x, y) = 2xy\mathbf{i} + (x^2 - y)\mathbf{j}.$$

Solution From Theorem 15.1 it follows that **F** is conservative because

$$\frac{\partial}{\partial y}[2xy] = 2x \quad \text{and} \quad \frac{\partial}{\partial x}[x^2 - y] = 2x.$$

If f is a function whose gradient is equal to $\mathbf{F}(x, y)$, then

$$\nabla f(x, y) = 2xy\mathbf{i} + (x^2 - y)\mathbf{j}$$

which implies that

$$f_x(x, y) = 2xy$$

and

$$f_y(x, y) = x^2 - y.$$

To reconstruct the function f from these two partial derivatives, integrate $f_x(x, y)$ with respect to x and $f_y(x, y)$ with respect to y, as follows.

$$f(x, y) = \int f_x(x, y)\, dx = \int 2xy\, dx = x^2y + g(y)$$

$$f(x, y) = \int f_y(x, y)\, dy = \int (x^2 - y)\, dy = x^2y - \frac{y^2}{2} + h(x)$$

Notice that $g(y)$ is constant with respect to x and $h(x)$ is constant with respect to y. To find a single expression that represents $f(x, y)$, let

$$g(y) = -\frac{y^2}{2} \quad \text{and} \quad h(x) = K.$$

Then, you can write

$$\begin{aligned} f(x, y) &= x^2y + g(y) + K \\ &= x^2y - \frac{y^2}{2} + K. \end{aligned}$$

You can check this result by forming the gradient of f. You will see that it is equal to the original function **F**.

NOTE Notice that the solution in Example 6 is comparable to that given by an indefinite integral. That is, the solution represents a family of potential functions, any two of which differ by a constant. To find a unique solution, you would have to be given an initial condition satisfied by the potential function.

Curl of a Vector Field

Theorem 15.1 has a counterpart for vector fields in space. Before stating that result, the definition of the **curl of a vector field** in space is given.

Definition of Curl of a Vector Field

The curl of $\mathbf{F}(x, y, z) = M\mathbf{i} + N\mathbf{j} + P\mathbf{k}$ is

$$\begin{aligned}\textbf{curl } \mathbf{F}(x, y, z) &= \nabla \times \mathbf{F}(x, y, z)\\ &= \left(\frac{\partial P}{\partial y} - \frac{\partial N}{\partial z}\right)\mathbf{i} - \left(\frac{\partial P}{\partial x} - \frac{\partial M}{\partial z}\right)\mathbf{j} + \left(\frac{\partial N}{\partial x} - \frac{\partial M}{\partial y}\right)\mathbf{k}.\end{aligned}$$

NOTE If $\textbf{curl } \mathbf{F} = \mathbf{0}$, then $\mathbf{F}$ is said to be **irrotational.**

The cross product notation used for curl comes from viewing the gradient ∇f as the result of the **differential operator** ∇ acting on the function f. In this context, you can use the following determinant form as an aid in remembering the formula for curl.

$$\begin{aligned}\textbf{curl } \mathbf{F}(x, y, z) &= \nabla \times \mathbf{F}(x, y, z)\\ &= \begin{vmatrix} \mathbf{i} & \mathbf{j} & \mathbf{k} \\ \dfrac{\partial}{\partial x} & \dfrac{\partial}{\partial y} & \dfrac{\partial}{\partial z} \\ M & N & P \end{vmatrix}\\ &= \left(\frac{\partial P}{\partial y} - \frac{\partial N}{\partial z}\right)\mathbf{i} - \left(\frac{\partial P}{\partial x} - \frac{\partial M}{\partial z}\right)\mathbf{j} + \left(\frac{\partial N}{\partial x} - \frac{\partial M}{\partial y}\right)\mathbf{k}\end{aligned}$$

EXAMPLE 7 Finding the Curl of a Vector Field

Find **curl F** for the vector field given by

$$\mathbf{F}(x, y, z) = 2xy\mathbf{i} + (x^2 + z^2)\mathbf{j} + 2yz\mathbf{k}.$$

Is **F** irrotational?

Solution The curl of **F** is given by

$$\begin{aligned}\textbf{curl } \mathbf{F}(x, y, z) &= \nabla \times \mathbf{F}(x, y, z)\\ &= \begin{vmatrix} \mathbf{i} & \mathbf{j} & \mathbf{k} \\ \dfrac{\partial}{\partial x} & \dfrac{\partial}{\partial y} & \dfrac{\partial}{\partial z} \\ 2xy & x^2 + z^2 & 2yz \end{vmatrix}\\ &= \begin{vmatrix} \dfrac{\partial}{\partial y} & \dfrac{\partial}{\partial z} \\ x^2 + z^2 & 2yz \end{vmatrix}\mathbf{i} - \begin{vmatrix} \dfrac{\partial}{\partial x} & \dfrac{\partial}{\partial z} \\ 2xy & 2yz \end{vmatrix}\mathbf{j} + \begin{vmatrix} \dfrac{\partial}{\partial x} & \dfrac{\partial}{\partial y} \\ 2xy & x^2 + z^2 \end{vmatrix}\mathbf{k}\\ &= (2z - 2z)\mathbf{i} - (0 - 0)\mathbf{j} + (2x - 2x)\mathbf{k}\\ &= \mathbf{0}.\end{aligned}$$

Because $\textbf{curl } \mathbf{F} = \mathbf{0}$, **F** is irrotational.

indicates that in the HM mathSpace® *CD-ROM and the online* Eduspace® *system for this text, you will find an Open Exploration, which further explores this example using the computer algebra systems* Maple, Mathcad, Mathematica, *and* Derive.

Later in this chapter, you will assign a physical interpretation to the curl of a vector field. But for now, the primary use of curl is shown in the following test for conservative vector fields in space. The test states that for a vector field whose domain is all of three-dimensional space (or an open sphere), the curl is **0** at every point in the domain if and only if **F** is conservative. The proof is similar to that given for Theorem 15.1.

> **THEOREM 15.2 Test for Conservative Vector Field in Space**
>
> Suppose that M, N, and P have continuous first partial derivatives in an open sphere Q in space. The vector field given by $\mathbf{F}(x, y, z) = M\mathbf{i} + N\mathbf{j} + P\mathbf{k}$ is conservative if and only if
>
> $$\mathbf{curl}\ \mathbf{F}(x, y, z) = \mathbf{0}.$$
>
> That is, **F** is conservative if and only if
>
> $$\frac{\partial P}{\partial y} = \frac{\partial N}{\partial z}, \quad \frac{\partial P}{\partial x} = \frac{\partial M}{\partial z}, \quad \text{and} \quad \frac{\partial N}{\partial x} = \frac{\partial M}{\partial y}.$$

From Theorem 15.2, you can see that the vector field given in Example 7 is conservative because $\mathbf{curl}\ \mathbf{F}(x, y, z) = \mathbf{0}$. Try showing that the vector field

$$\mathbf{F}(x, y, z) = x^3y^2z\mathbf{i} + x^2z\mathbf{j} + x^2y\mathbf{k}$$

is not conservative—you can do this by showing that its curl is

$$\mathbf{curl}\ \mathbf{F}(x, y, z) = (x^3y^2 - 2xy)\mathbf{j} + (2xz - 2x^3yz)\mathbf{k} \neq \mathbf{0}.$$

For vector fields in space that pass the test for being conservative, you can find a potential function by following the same pattern used in the plane (as demonstrated in Example 6).

EXAMPLE 8 Finding a Potential Function for F(x, y, z)

NOTE Examples 6 and 8 are illustrations of a type of problem called *recovering a function from its gradient*. If you go on to take a course in differential equations, you will study other methods for solving this type of problem. One popular method gives an interplay between successive "partial integrations" and partial differentiations.

Find a potential function for $\mathbf{F}(x, y, z) = 2xy\mathbf{i} + (x^2 + z^2)\mathbf{j} + 2yz\mathbf{k}$.

Solution From Example 7, you know that the vector field given by **F** is conservative. If f is a function such that $\mathbf{F}(x, y, z) = \nabla f(x, y, z)$, then

$$f_x(x, y, z) = 2xy, \quad f_y(x, y, z) = x^2 + z^2, \quad \text{and} \quad f_z(x, y, z) = 2yz$$

and integrating with respect to x, y, and z separately produces

$$f(x, y, z) = \int M\,dx = \int 2xy\,dx = x^2y + g(y, z)$$

$$f(x, y, z) = \int N\,dy = \int (x^2 + z^2)\,dy = x^2y + yz^2 + h(x, z)$$

$$f(x, y, z) = \int P\,dz = \int 2yz\,dz = yz^2 + k(x, y).$$

Comparing these three versions of $f(x, y, z)$, you can conclude that

$$g(y, z) = yz^2 + K, \quad h(x, z) = K, \quad \text{and} \quad k(x, y) = x^2y + K.$$

So, $f(x, y, z)$ is given by

$$f(x, y, z) = x^2y + yz^2 + K.$$

NOTE Divergence can be viewed as a type of derivative of **F** in that, for vector fields representing velocities of moving particles, the divergence measures the rate of particle flow per unit volume at a point. In hydrodynamics (the study of fluid motion), a velocity field that is divergence free is called **incompressible.** In the study of electricity and magnetism, a vector field that is divergence free is called **solenoidal.**

Divergence of a Vector Field

You have seen that the curl of a vector field **F** is itself a vector field. Another important function defined on a vector field is **divergence,** which is a scalar function.

Definition of Divergence of a Vector Field

The **divergence** of $\mathbf{F}(x, y) = M\mathbf{i} + N\mathbf{j}$ is

$$\text{div } \mathbf{F}(x, y) = \nabla \cdot \mathbf{F}(x, y) = \frac{\partial M}{\partial x} + \frac{\partial N}{\partial y}. \qquad \text{Plane}$$

The **divergence** of $\mathbf{F}(x, y, z) = M\mathbf{i} + N\mathbf{j} + P\mathbf{k}$ is

$$\text{div } \mathbf{F}(x, y, z) = \nabla \cdot \mathbf{F}(x, y, z) = \frac{\partial M}{\partial x} + \frac{\partial N}{\partial y} + \frac{\partial P}{\partial z}. \qquad \text{Space}$$

If div $\mathbf{F} = 0$, then **F** is said to be **divergence free.**

The dot product notation used for divergence comes from considering ∇ as a **differential operator,** as follows.

$$\nabla \cdot \mathbf{F}(x, y, z) = \left[\left(\frac{\partial}{\partial x}\right)\mathbf{i} + \left(\frac{\partial}{\partial y}\right)\mathbf{j} + \left(\frac{\partial}{\partial z}\right)\mathbf{k}\right] \cdot (M\mathbf{i} + N\mathbf{j} + P\mathbf{k})$$
$$= \frac{\partial M}{\partial x} + \frac{\partial N}{\partial y} + \frac{\partial P}{\partial z}$$

EXAMPLE 9 Finding the Divergence of a Vector Field

Find the divergence at $(2, 1, -1)$ for the vector field

$$\mathbf{F}(x, y, z) = x^3y^2z\mathbf{i} + x^2z\mathbf{j} + x^2y\mathbf{k}.$$

Solution The divergence of **F** is

$$\text{div } \mathbf{F}(x, y, z) = \frac{\partial}{\partial x}[x^3y^2z] + \frac{\partial}{\partial y}[x^2z] + \frac{\partial}{\partial z}[x^2y] = 3x^2y^2z.$$

At the point $(2, 1, -1)$, the divergence is

$$\text{div } \mathbf{F}(2, 1, -1) = 3(2^2)(1^2)(-1) = -12.$$

There are many important properties of the divergence and curl of a vector field **F** (see Exercises 77–83). One that is used often is described in Theorem 15.3. You are asked to prove this theorem in Exercise 84.

THEOREM 15.3 Relationship Between Divergence and Curl

If $\mathbf{F}(x, y, z) = M\mathbf{i} + N\mathbf{j} + P\mathbf{k}$ is a vector field and M, N, and P have continuous second partial derivatives, then

$$\text{div } (\mathbf{curl\ F}) = 0.$$

Exercises for Section 15.1

See www.CalcChat.com for worked-out solutions to odd-numbered exercises.

In Exercises 1–6, match the vector field with its graph. [The graphs are labeled (a), (b), (c), (d), (e), and (f).]

(a)

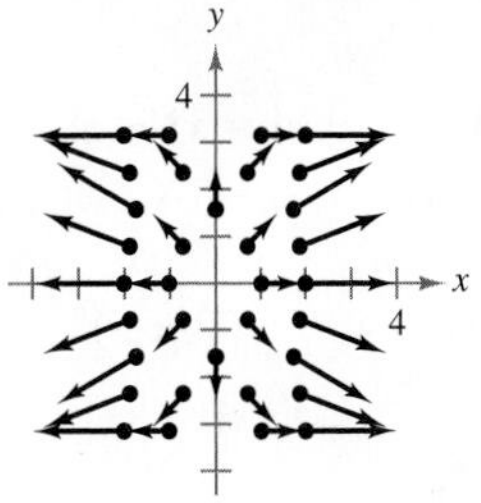

(b)

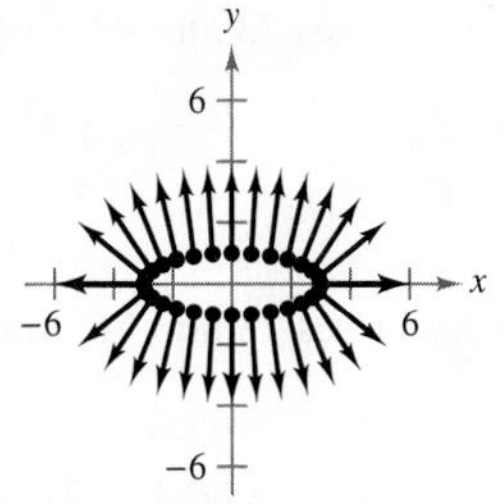

(c)

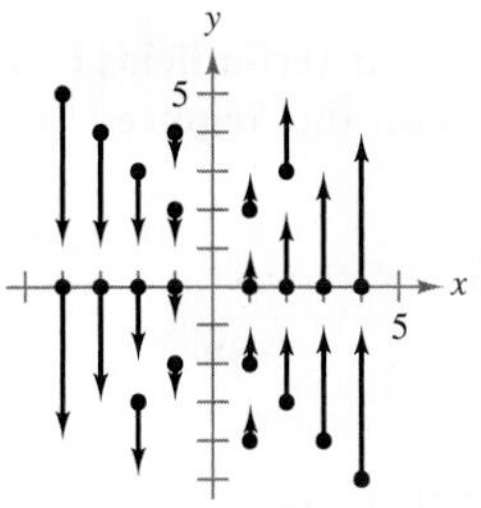

(d)

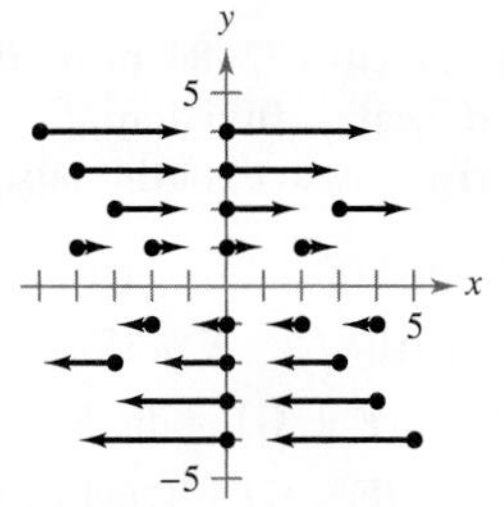

(e)

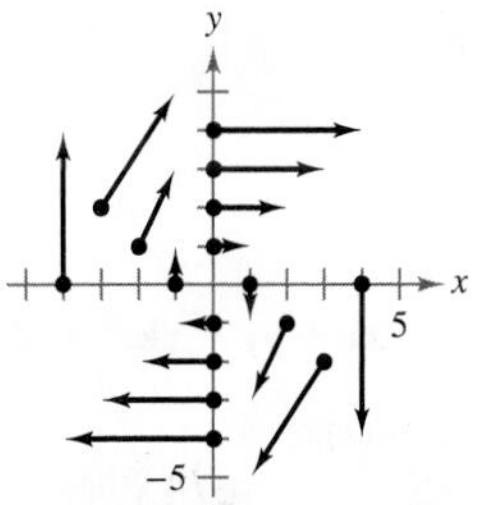

(f)

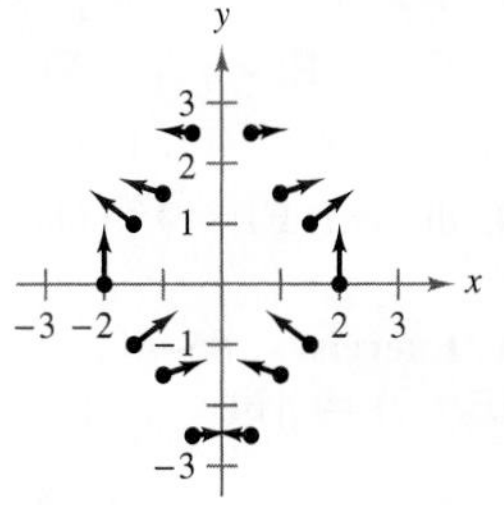

1. $\mathbf{F}(x, y) = x\mathbf{j}$
2. $\mathbf{F}(x, y) = y\mathbf{i}$
3. $\mathbf{F}(x, y) = x\mathbf{i} + 3y\mathbf{j}$
4. $\mathbf{F}(x, y) = y\mathbf{i} - x\mathbf{j}$
5. $\mathbf{F}(x, y) = \langle x, \sin y \rangle$
6. $\mathbf{F}(x, y) = \left\langle \frac{1}{2}xy, \frac{1}{4}x^2 \right\rangle$

In Exercises 7–16, sketch several representative vectors in the vector field.

7. $\mathbf{F}(x, y) = \mathbf{i} + \mathbf{j}$
8. $\mathbf{F}(x, y) = 2\mathbf{i}$
9. $\mathbf{F}(x, y) = x\mathbf{i} + y\mathbf{j}$
10. $\mathbf{F}(x, y) = x\mathbf{i} - y\mathbf{j}$
11. $\mathbf{F}(x, y, z) = 3y\mathbf{j}$
12. $\mathbf{F}(x, y) = x\mathbf{i}$
13. $\mathbf{F}(x, y) = 4x\mathbf{i} + y\mathbf{j}$
14. $\mathbf{F}(x, y) = (x^2 + y^2)\mathbf{i} + \mathbf{j}$
15. $\mathbf{F}(x, y, z) = \mathbf{i} + \mathbf{j} + \mathbf{k}$
16. $\mathbf{F}(x, y, z) = x\mathbf{i} + y\mathbf{j} + z\mathbf{k}$

In Exercises 17–20, use a computer algebra system to graph several representative vectors in the vector field.

17. $\mathbf{F}(x, y) = \frac{1}{8}(2xy\mathbf{i} + y^2\mathbf{j})$
18. $\mathbf{F}(x, y) = (2y - 3x)\mathbf{i} + (2y + 3x)\mathbf{j}$
19. $\mathbf{F}(x, y, z) = \dfrac{x\mathbf{i} + y\mathbf{j} + z\mathbf{k}}{\sqrt{x^2 + y^2 + z^2}}$
20. $\mathbf{F}(x, y, z) = x\mathbf{i} - y\mathbf{j} + z\mathbf{k}$

In Exercises 21–26, find the gradient vector field for the scalar function. (That is, find the conservative vector field for the potential function.)

21. $f(x, y) = 5x^2 + 3xy + 10y^2$
22. $f(x, y) = \sin 3x \cos 4y$
23. $f(x, y, z) = z - ye^{x^2}$
24. $f(x, y, z) = \dfrac{y}{z} + \dfrac{z}{x} - \dfrac{xz}{y}$
25. $g(x, y, z) = xy \ln(x + y)$
26. $g(x, y, z) = x \arcsin yz$

In Exercises 27–30, verify that the vector field is conservative.

27. $\mathbf{F}(x, y) = 12xy\mathbf{i} + 6(x^2 + y)\mathbf{j}$
28. $\mathbf{F}(x, y) = \dfrac{1}{x^2}(y\mathbf{i} - x\mathbf{j})$
29. $\mathbf{F}(x, y) = \sin y\mathbf{i} + x \cos y\mathbf{j}$
30. $\mathbf{F}(x, y) = \dfrac{1}{xy}(y\mathbf{i} - x\mathbf{j})$

In Exercises 31–34, determine if the vector field is conservative.

31. $\mathbf{F}(x, y) = 5y^2(3y\mathbf{i} - x\mathbf{j})$
32. $\mathbf{F}(x, y) = \dfrac{1}{\sqrt{x^2 + y^2}}(x\mathbf{i} + y\mathbf{j})$
33. $\mathbf{F}(x, y) = \dfrac{2}{y^2}e^{2x/y}(y\mathbf{i} - x\mathbf{j})$
34. $\mathbf{F}(x, y) = \dfrac{1}{\sqrt{1 - x^2y^2}}(y\mathbf{i} - x\mathbf{j})$

In Exercises 35–42, determine whether the vector field is conservative. If it is, find a potential function for the vector field.

35. $\mathbf{F}(x, y) = 2xy\mathbf{i} + x^2\mathbf{j}$
36. $\mathbf{F}(x, y) = \dfrac{1}{y^2}(y\mathbf{i} - 2x\mathbf{j})$
37. $\mathbf{F}(x, y) = xe^{x^2y}(2y\mathbf{i} + x\mathbf{j})$
38. $\mathbf{F}(x, y) = 3x^2y^2\mathbf{i} + 2x^3y\mathbf{j}$
39. $\mathbf{F}(x, y) = \dfrac{x\mathbf{i} + y\mathbf{j}}{x^2 + y^2}$
40. $\mathbf{F}(x, y) = \dfrac{2y}{x}\mathbf{i} - \dfrac{x^2}{y^2}\mathbf{j}$
41. $\mathbf{F}(x, y) = e^x(\cos y\mathbf{i} + \sin y\mathbf{j})$
42. $\mathbf{F}(x, y) = \dfrac{2x\mathbf{i} + 2y\mathbf{j}}{(x^2 + y^2)^2}$

In Exercises 43–46, find curl F for the vector field at the given point.

	Vector Field	*Point*
43.	$\mathbf{F}(x, y, z) = xyz\mathbf{i} + y\mathbf{j} + z\mathbf{k}$	$(1, 2, 1)$
44.	$\mathbf{F}(x, y, z) = x^2z\mathbf{i} - 2xz\mathbf{j} + yz\mathbf{k}$	$(2, -1, 3)$
45.	$\mathbf{F}(x, y, z) = e^x \sin y\mathbf{i} - e^x \cos y\mathbf{j}$	$(0, 0, 3)$
46.	$\mathbf{F}(x, y, z) = e^{-xyz}(\mathbf{i} + \mathbf{j} + \mathbf{k})$	$(3, 2, 0)$

In Exercises 47–50, use a computer algebra system to find the curl F for the vector field.

47. $\mathbf{F}(x, y, z) = \arctan\left(\frac{x}{y}\right)\mathbf{i} + \ln\sqrt{x^2 + y^2}\,\mathbf{j} + \mathbf{k}$

48. $\mathbf{F}(x, y, z) = \frac{yz}{y - z}\mathbf{i} + \frac{xz}{x - z}\mathbf{j} + \frac{xy}{x - y}\mathbf{k}$

49. $\mathbf{F}(x, y, z) = \sin(x - y)\mathbf{i} + \sin(y - z)\mathbf{j} + \sin(z - x)\mathbf{k}$

50. $\mathbf{F}(x, y, z) = \sqrt{x^2 + y^2 + z^2}\,(\mathbf{i} + \mathbf{j} + \mathbf{k})$

In Exercises 51–56, determine whether the vector field F is conservative. If it is, find a potential function for the vector field.

51. $\mathbf{F}(x, y, z) = \sin y\mathbf{i} - x\cos y\mathbf{j} + \mathbf{k}$

52. $\mathbf{F}(x, y, z) = e^z(y\mathbf{i} + x\mathbf{j} + \mathbf{k})$

53. $\mathbf{F}(x, y, z) = e^z(y\mathbf{i} + x\mathbf{j} + xy\mathbf{k})$

54. $\mathbf{F}(x, y, z) = y^2z^3\mathbf{i} + 2xyz^3\mathbf{j} + 3xy^2z^2\mathbf{k}$

55. $\mathbf{F}(x, y, z) = \frac{1}{y}\mathbf{i} - \frac{x}{y^2}\mathbf{j} + (2z - 1)\mathbf{k}$

56. $\mathbf{F}(x, y, z) = \frac{x}{x^2 + y^2}\mathbf{i} + \frac{y}{x^2 + y^2}\mathbf{j} + \mathbf{k}$

In Exercises 57–60, find the divergence of the vector field F.

57. $\mathbf{F}(x, y, z) = 6x^2\mathbf{i} - xy^2\mathbf{j}$

58. $\mathbf{F}(x, y, z) = xe^x\mathbf{i} + ye^y\mathbf{j}$

59. $\mathbf{F}(x, y, z) = \sin x\mathbf{i} + \cos y\mathbf{j} + z^2\mathbf{k}$

60. $\mathbf{F}(x, y, z) = \ln(x^2 + y^2)\mathbf{i} + xy\mathbf{j} + \ln(y^2 + z^2)\mathbf{k}$

In Exercises 61–64, find the divergence of the vector field F at the given point.

Vector Field	*Point*
61. $\mathbf{F}(x, y, z) = xyz\mathbf{i} + y\mathbf{j} + z\mathbf{k}$	$(1, 2, 1)$
62. $\mathbf{F}(x, y, z) = x^2z\mathbf{i} - 2xz\mathbf{j} + yz\mathbf{k}$	$(2, -1, 3)$
63. $\mathbf{F}(x, y, z) = e^x\sin y\mathbf{i} - e^x\cos y\mathbf{j}$	$(0, 0, 3)$
64. $\mathbf{F}(x, y, z) = \ln(xyz)(\mathbf{i} + \mathbf{j} + \mathbf{k})$	$(3, 2, 1)$

Writing About Concepts

65. Define a vector field in the plane and in space. Give some physical examples of vector fields.

66. What is a conservative vector field and how do you test for it in the plane and in space?

67. Define the curl of a vector field.

68. Define the divergence of a vector field in the plane and in space.

In Exercises 69 and 70, find curl (F × G).

69. $\mathbf{F}(x, y, z) = \mathbf{i} + 2x\mathbf{j} + 3y\mathbf{k}$
$\mathbf{G}(x, y, z) = x\mathbf{i} - y\mathbf{j} + z\mathbf{k}$

70. $\mathbf{F}(x, y, z) = x\mathbf{i} - z\mathbf{k}$
$\mathbf{G}(x, y, z) = x^2\mathbf{i} + y\mathbf{j} + z^2\mathbf{k}$

In Exercises 71 and 72, find curl (curl F) = ∇ × (∇ × F).

71. $\mathbf{F}(x, y, z) = xyz\mathbf{i} + y\mathbf{j} + z\mathbf{k}$

72. $\mathbf{F}(x, y, z) = x^2z\mathbf{i} - 2xz\mathbf{j} + yz\mathbf{k}$

In Exercises 73 and 74, find div (F × G).

73. $\mathbf{F}(x, y, z) = \mathbf{i} + 2x\mathbf{j} + 3y\mathbf{k}$
$\mathbf{G}(x, y, z) = x\mathbf{i} - y\mathbf{j} + z\mathbf{k}$

74. $\mathbf{F}(x, y, z) = x\mathbf{i} - z\mathbf{k}$
$\mathbf{G}(x, y, z) = x^2\mathbf{i} + y\mathbf{j} + z^2\mathbf{k}$

In Exercises 75 and 76, find div (curl F) = ∇ · (∇ × F).

75. $\mathbf{F}(x, y, z) = xyz\mathbf{i} + y\mathbf{j} + z\mathbf{k}$

76. $\mathbf{F}(x, y, z) = x^2z\mathbf{i} - 2xz\mathbf{j} + yz\mathbf{k}$

In Exercises 77–84, prove the property for vector fields F and G and scalar function *f*. (Assume that the required partial derivatives are continuous.)

77. $\textbf{curl}\,(\mathbf{F} + \mathbf{G}) = \textbf{curl}\,\mathbf{F} + \textbf{curl}\,\mathbf{G}$

78. $\textbf{curl}(\nabla f) = \nabla \times (\nabla f) = \mathbf{0}$

79. $\text{div}(\mathbf{F} + \mathbf{G}) = \text{div}\,\mathbf{F} + \text{div}\,\mathbf{G}$

80. $\text{div}(\mathbf{F} \times \mathbf{G}) = (\textbf{curl}\,\mathbf{F}) \cdot \mathbf{G} - \mathbf{F} \cdot (\textbf{curl}\,\mathbf{G})$

81. $\nabla \times [\nabla f + (\nabla \times \mathbf{F})] = \nabla \times (\nabla \times \mathbf{F})$

82. $\nabla \times (f\mathbf{F}) = f(\nabla \times \mathbf{F}) + (\nabla f) \times \mathbf{F}$

83. $\text{div}(f\mathbf{F}) = f\,\text{div}\,\mathbf{F} + \nabla f \cdot \mathbf{F}$

84. $\text{div}(\textbf{curl}\,\mathbf{F}) = 0$ (Theorem 15.3)

In Exercises 85–88, let $\mathbf{F}(x, y, z) = x\mathbf{i} + y\mathbf{j} + z\mathbf{k}$, and let $f(x, y, z) = \|\mathbf{F}(x, y, z)\|$.

85. Show that $\nabla(\ln f) = \frac{\mathbf{F}}{f^2}$.

86. Show that $\nabla\left(\frac{1}{f}\right) = -\frac{\mathbf{F}}{f^3}$.

87. Show that $\nabla f^n = nf^{n-2}\mathbf{F}$.

88. The **Laplacian** is the differential operator

$$\nabla^2 = \nabla \cdot \nabla = \frac{\partial^2}{\partial x^2} + \frac{\partial^2}{\partial y^2} + \frac{\partial^2}{\partial z^2}$$

and **Laplace's equation** is

$$\nabla^2 w = \frac{\partial^2 w}{\partial x^2} + \frac{\partial^2 w}{\partial y^2} + \frac{\partial^2 w}{\partial z^2} = 0.$$

Any function that satisfies this equation is called **harmonic.** Show that the function $1/f$ is harmonic.

True or False? **In Exercises 89–92, determine whether the statement is true or false. If it is false, explain why or give an example that shows it is false.**

89. If $\mathbf{F}(x, y) = 4x\mathbf{i} - y^2\mathbf{j}$, then $\|\mathbf{F}(x, y)\| \to 0$ as $(x, y) \to (0, 0)$.

90. If $\mathbf{F}(x, y) = 4x\mathbf{i} - y^2\mathbf{j}$ and (x, y) is on the positive y-axis, then the vector points in the negative y-direction.

91. If f is a scalar field, then curl f is a meaningful expression.

92. If $\mathbf{F}$ is a vector field and $\textbf{curl}\,\mathbf{F} = \mathbf{0}$, then $\mathbf{F}$ is irrotational but not conservative.

Section 15.2 Line Integrals

- **Understand and use the concept of a piecewise smooth curve.**
- **Write and evaluate a line integral.**
- **Write and evaluate a line integral of a vector field.**
- **Write and evaluate a line integral in differential form.**

JOSIAH WILLARD GIBBS (1839–1903)

Many physicists and mathematicians have contributed to the theory and applications described in this chapter—Newton, Gauss, Laplace, Hamilton, and Maxwell, among others. However, the use of vector analysis to describe these results is attributed primarily to the American mathematical physicist Josiah Willard Gibbs.

Piecewise Smooth Curves

A classic property of gravitational fields is that, subject to certain physical constraints, the work done by gravity on an object moving between two points in the field is independent of the path taken by the object. One of the constraints is that the **path** must be a piecewise smooth curve. Recall that a plane curve C given by

$$\mathbf{r}(t) = x(t)\mathbf{i} + y(t)\mathbf{j}, \quad a \le t \le b$$

is **smooth** if

$$\frac{dx}{dt} \quad \text{and} \quad \frac{dy}{dt}$$

are continuous on $[a, b]$ and not simultaneously 0 on (a, b). Similarly, a space curve C given by

$$\mathbf{r}(t) = x(t)\mathbf{i} + y(t)\mathbf{j} + z(t)\mathbf{k}, \quad a \le t \le b$$

is **smooth** if

$$\frac{dx}{dt}, \quad \frac{dy}{dt}, \quad \text{and} \quad \frac{dz}{dt}$$

are continuous on $[a, b]$ and not simultaneously 0 on (a, b). A curve C is **piecewise smooth** if the interval $[a, b]$ can be partitioned into a finite number of subintervals, on each of which C is smooth.

EXAMPLE 1 Finding a Piecewise Smooth Parametrization

Find a piecewise smooth parametrization of the graph of C shown in Figure 15.7.

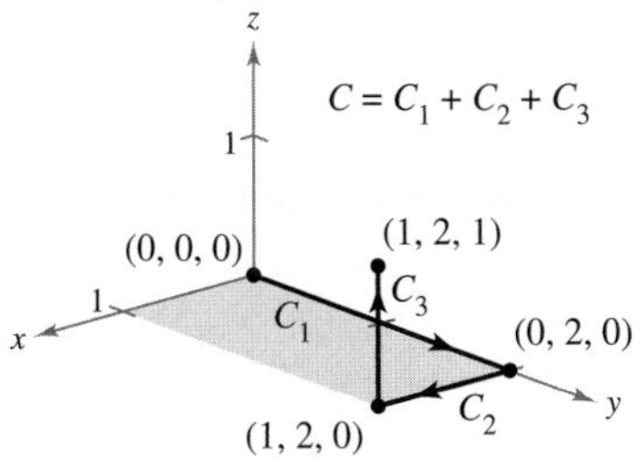

Figure 15.7

Solution Because C consists of three line segments C_1, C_2, and C_3, you can construct a smooth parametrization for each segment and piece them together by making the last t-value in C_i correspond to the first t-value in C_{i+1}, as follows.

$$\begin{array}{llll} C_1\text{: } x(t) = 0, & y(t) = 2t, & z(t) = 0, & 0 \le t \le 1 \\ C_2\text{: } x(t) = t - 1, & y(t) = 2, & z(t) = 0, & 1 \le t \le 2 \\ C_3\text{: } x(t) = 1, & y(t) = 2, & z(t) = t - 2, & 2 \le t \le 3 \end{array}$$

So, C is given by

$$\mathbf{r}(t) = \begin{cases} 2t\mathbf{j}, & 0 \le t \le 1 \\ (t - 1)\mathbf{i} + 2\mathbf{j}, & 1 \le t \le 2 \\ \mathbf{i} + 2\mathbf{j} + (t - 2)\mathbf{k}, & 2 \le t \le 3 \end{cases}.$$

Because C_1, C_2, and C_3 are smooth, it follows that C is piecewise smooth.

Recall that parametrization of a curve induces an **orientation** to the curve. For instance, in Example 1, the curve is oriented such that the positive direction is from $(0, 0, 0)$, following the curve to $(1, 2, 1)$. Try finding a parametrization that induces the opposite orientation.

Line Integrals

Up to this point in the text, you have studied various types of integrals. For a single integral

$$\int_a^b f(x)\,dx \qquad \text{Integrate over interval } [a, b].$$

you integrated over the interval $[a, b]$. Similarly, for a double integral

$$\int_R\int f(x, y)\,dA \qquad \text{Integrate over region } R.$$

you integrated over the region R in the plane. In this section, you will study a new type of integral called a **line integral**

$$\int_C f(x, y)\,ds \qquad \text{Integrate over curve } C.$$

for which you integrate over a piecewise smooth curve C. (The terminology is somewhat unfortunate—this type of integral might be better described as a "curve integral.")

To introduce the concept of a line integral, consider the mass of a wire of finite length, given by a curve C in space. The density (mass per unit length) of the wire at the point (x, y, z) is given by $f(x, y, z)$. Partition the curve C by the points

$$P_0, P_1, \ldots, P_n$$

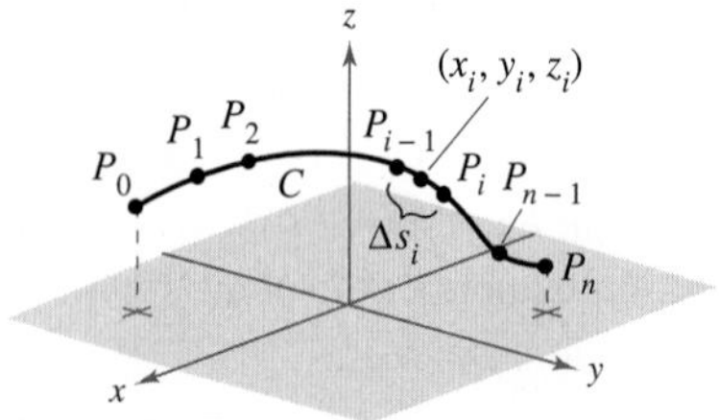

Partitioning of curve C
Figure 15.8

producing n subarcs, as shown in Figure 15.8. The length of the ith subarc is given by Δs_i. Next, choose a point (x_i, y_i, z_i) in each subarc. If the length of each subarc is small, the total mass of the wire can be approximated by the sum

$$\text{Mass of wire} \approx \sum_{i=1}^{n} f(x_i, y_i, z_i)\,\Delta s_i.$$

If you let $\|\Delta\|$ denote the length of the longest subarc and let $\|\Delta\|$ approach 0, it seems reasonable that the limit of this sum approaches the mass of the wire. This leads to the following definition.

Definition of Line Integral

If f is defined in a region containing a smooth curve C of finite length, then the **line integral of f along C** is given by

$$\int_C f(x, y)\,ds = \lim_{\|\Delta\|\to 0} \sum_{i=1}^{n} f(x_i, y_i)\,\Delta s_i \qquad \text{Plane}$$

or

$$\int_C f(x, y, z)\,ds = \lim_{\|\Delta\|\to 0} \sum_{i=1}^{n} f(x_i, y_i, z_i)\,\Delta s_i \qquad \text{Space}$$

provided this limit exists.

As with the integrals discussed in Chapter 14, evaluation of a line integral is best accomplished by converting to a definite integral. It can be shown that if f is *continuous*, the limit given above exists and is the same for all smooth parametrizations of C.

To evaluate a line integral over a plane curve C given by $\mathbf{r}(t) = x(t)\mathbf{i} + y(t)\mathbf{j}$, use the fact that

$$ds = \|\mathbf{r}'(t)\|\,dt = \sqrt{[x'(t)]^2 + [y'(t)]^2}\,dt.$$

A similar formula holds for a space curve, as indicated in Theorem 15.4.

THEOREM 15.4 Evaluation of a Line Integral as a Definite Integral

Let f be continuous in a region containing a smooth curve C. If C is given by $\mathbf{r}(t) = x(t)\mathbf{i} + y(t)\mathbf{j}$, where $a \le t \le b$, then

$$\int_C f(x, y)\,ds = \int_a^b f(x(t), y(t))\sqrt{[x'(t)]^2 + [y'(t)]^2}\,dt.$$

If C is given by $\mathbf{r}(t) = x(t)\mathbf{i} + y(t)\mathbf{j} + z(t)\mathbf{k}$, where $a \le t \le b$, then

$$\int_C f(x, y, z)\,ds = \int_a^b f(x(t), y(t), z(t))\sqrt{[x'(t)]^2 + [y'(t)]^2 + [z'(t)]^2}\,dt.$$

Note that if $f(x, y, z) = 1$, the line integral gives the arc length of the curve C, as defined in Section 12.5. That is,

$$\int_C 1\,ds = \int_a^b \|\mathbf{r}'(t)\|\,dt = \text{length of curve } C.$$

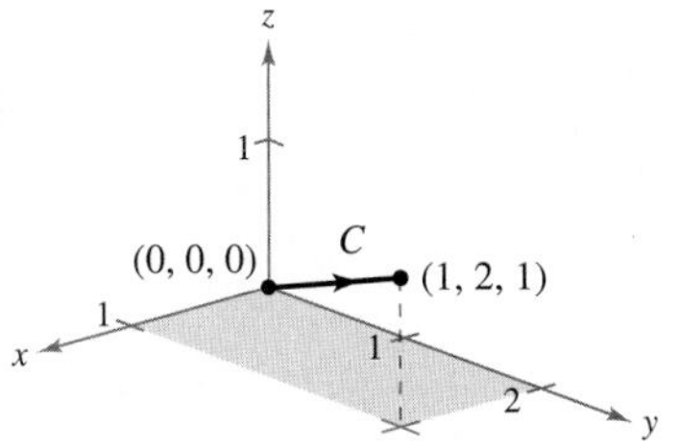

Figure 15.9

EXAMPLE 2 Evaluating a Line Integral

Evaluate

$$\int_C (x^2 - y + 3z)\,ds$$

where C is the line segment shown in Figure 15.9.

Solution Begin by writing a parametric form of the equation of a line:

$$x = t, \quad y = 2t, \quad \text{and} \quad z = t, \quad 0 \le t \le 1.$$

Therefore, $x'(t) = 1$, $y'(t) = 2$, and $z'(t) = 1$, which implies that

$$\sqrt{[x'(t)]^2 + [y'(t)]^2 + [z'(t)]^2} = \sqrt{1^2 + 2^2 + 1^2} = \sqrt{6}.$$

So, the line integral takes the following form.

$$\begin{aligned}\int_C (x^2 - y + 3z)\,ds &= \int_0^1 (t^2 - 2t + 3t)\sqrt{6}\,dt \\ &= \sqrt{6}\int_0^1 (t^2 + t)\,dt \\ &= \sqrt{6}\left[\frac{t^3}{3} + \frac{t^2}{2}\right]_0^1 \\ &= \frac{5\sqrt{6}}{6}\end{aligned}$$

NOTE The value of the line integral in Example 2 does not depend on the parametrization of the line segment C (any smooth parametrization will produce the same value). To convince yourself of this, try some other parametrizations, such as $x = 1 + 2t$, $y = 2 + 4t$, $z = 1 + 2t$, $-\frac{1}{2} \le t \le 0$, or $x = -t$, $y = -2t$, $z = -t$, $-1 \le t \le 0$.

Suppose C is a path composed of smooth curves $C_1, C_2, \ldots, C_n$. If f is continuous on C, it can be shown that

$$\int_C f(x, y)\, ds = \int_{C_1} f(x, y)\, ds + \int_{C_2} f(x, y)\, ds + \cdots + \int_{C_n} f(x, y)\, ds.$$

This property is used in Example 3.

EXAMPLE 3 Evaluating a Line Integral Over a Path

Evaluate $\int_C x\, ds$, where C is the piecewise smooth curve shown in Figure 15.10.

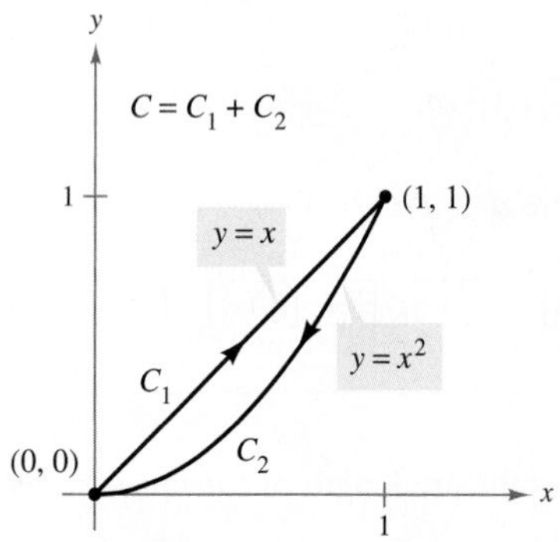

Figure 15.10

Solution Begin by integrating up the line $y = x$, using the following parametrization.

$$C_1\colon\ x = t,\ y = t,\quad 0 \le t \le 1$$

For this curve, $\mathbf{r}(t) = t\mathbf{i} + t\mathbf{j}$, which implies that $x'(t) = 1$ and $y'(t) = 1$. So,

$$\sqrt{[x'(t)]^2 + [y'(t)]^2} = \sqrt{2}$$

and you have

$$\int_{C_1} x\, ds = \int_0^1 t\sqrt{2}\, dt = \frac{\sqrt{2}}{2}t^2\Big]_0^1 = \frac{\sqrt{2}}{2}.$$

Next, integrate down the parabola $y = x^2$, using the parametrization

$$C_2\colon\ x = 1 - t,\quad y = (1 - t)^2,\quad 0 \le t \le 1.$$

For this curve, $\mathbf{r}(t) = (1 - t)\mathbf{i} + (1 - t)^2\mathbf{j}$, which implies that $x'(t) = -1$ and $y'(t) = -2(1 - t)$. So,

$$\sqrt{[x'(t)]^2 + [y'(t)]^2} = \sqrt{1 + 4(1 - t)^2}$$

and you have

$$\begin{aligned}\int_{C_2} x\, ds &= \int_0^1 (1 - t)\sqrt{1 + 4(1 - t)^2}\, dt \\ &= -\frac{1}{8}\left[\frac{2}{3}[1 + 4(1 - t)^2]^{3/2}\right]_0^1 \\ &= \frac{1}{12}(5^{3/2} - 1).\end{aligned}$$

Consequently,

$$\int_C x\, ds = \int_{C_1} x\, ds + \int_{C_2} x\, ds = \frac{\sqrt{2}}{2} + \frac{1}{12}(5^{3/2} - 1) \approx 1.56.$$

For parametrizations given by $\mathbf{r}(t) = x(t)\mathbf{i} + y(t)\mathbf{j} + z(t)\mathbf{k}$, it is helpful to remember the form of ds as

$$ds = \|\mathbf{r}'(t)\|\, dt = \sqrt{[x'(t)]^2 + [y'(t)]^2 + [z'(t)]^2}\, dt.$$

This is demonstrated in Example 4.

EXAMPLE 4 Evaluating a Line Integral

Evaluate $\int_C (x + 2)\, ds$, where C is the curve represented by

$$\mathbf{r}(t) = t\mathbf{i} + \frac{4}{3}t^{3/2}\mathbf{j} + \frac{1}{2}t^2\mathbf{k}, \quad 0 \le t \le 2.$$

Solution Because $\mathbf{r}'(t) = \mathbf{i} + 2t^{1/2}\mathbf{j} + t\mathbf{k}$, and

$$\|\mathbf{r}'(t)\| = \sqrt{[x'(t)]^2 + [y'(t)]^2 + [z'(t)]^2} = \sqrt{1 + 4t + t^2}$$

it follows that

$$\begin{aligned}\int_C (x + 2)\, ds &= \int_0^2 (t + 2)\sqrt{1 + 4t + t^2}\, dt \\ &= \frac{1}{2}\int_0^2 2(t + 2)(1 + 4t + t^2)^{1/2}\, dt \\ &= \frac{1}{3}\Big[(1 + 4t + t^2)^{3/2}\Big]_0^2 \\ &= \frac{1}{3}\left(13\sqrt{13} - 1\right) \\ &\approx 15.29.\end{aligned}$$

The next example shows how a line integral can be used to find the mass of a spring whose density varies. In Figure 15.11, note that the density of this spring increases as the spring spirals up the z-axis.

EXAMPLE 5 Finding the Mass of a Spring

Find the mass of a spring in the shape of the circular helix

$$\mathbf{r}(t) = \frac{1}{\sqrt{2}}(\cos t\mathbf{i} + \sin t\mathbf{j} + t\mathbf{k}), \quad 0 \le t \le 6\pi$$

where the density of the spring is $\rho(x, y, z) = 1 + z$, as shown in Figure 15.11.

Solution Because

$$\|\mathbf{r}'(t)\| = \frac{1}{\sqrt{2}}\sqrt{(-\sin t)^2 + (\cos t)^2 + (1)^2} = 1$$

it follows that the mass of the spring is

$$\begin{aligned}\text{Mass} = \int_C (1 + z)\, ds &= \int_0^{6\pi}\left(1 + \frac{t}{\sqrt{2}}\right) dt \\ &= \left[t + \frac{t^2}{2\sqrt{2}}\right]_0^{6\pi} \\ &= 6\pi\left(1 + \frac{3\pi}{\sqrt{2}}\right) \\ &\approx 144.47.\end{aligned}$$

The mass of the spring is approximately 144.47.

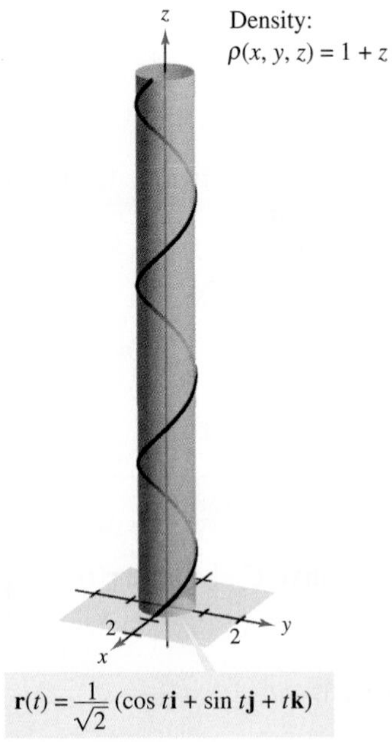

Figure 15.11

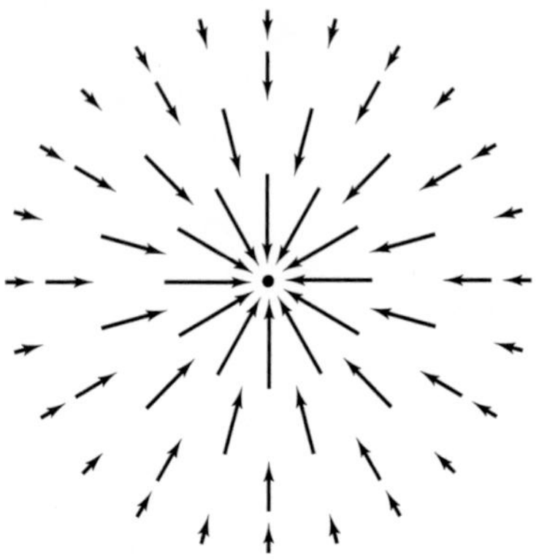

Inverse square force field **F**

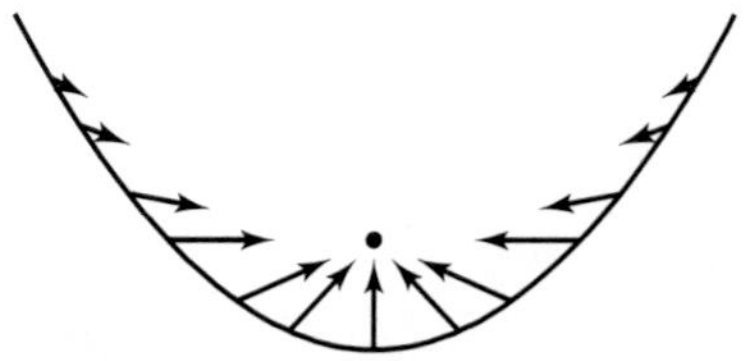

Vectors along a parabolic path in the force field **F**

Figure 15.12

Line Integrals of Vector Fields

One of the most important physical applications of line integrals is that of finding the **work** done on an object moving in a force field. For example, Figure 15.12 shows an inverse square force field similar to the gravitational field of the sun. Note that the magnitude of the force along a circular path about the center is constant, whereas the magnitude of the force along a parabolic path varies from point to point.

To see how a line integral can be used to find work done in a force field **F**, consider an object moving along a path C in the field, as shown in Figure 15.13. To determine the work done by the force, you need consider only that part of the force that is acting in the same direction as that in which the object is moving (or the opposite direction). This means that at each point on C, you can consider the projection $\mathbf{F} \cdot \mathbf{T}$ of the force vector **F** onto the unit tangent vector **T**. On a small subarc of length Δs_i, the increment of work is

$$\begin{aligned}\Delta W_i &= (\text{force})(\text{distance})\\ &\approx [\mathbf{F}(x_i, y_i, z_i) \cdot \mathbf{T}(x_i, y_i, z_i)]\,\Delta s_i\end{aligned}$$

where (x_i, y_i, z_i) is a point in the ith subarc. Consequently, the total work done is given by the following integral.

$$W = \int_C \mathbf{F}(x, y, z) \cdot \mathbf{T}(x, y, z)\, ds$$

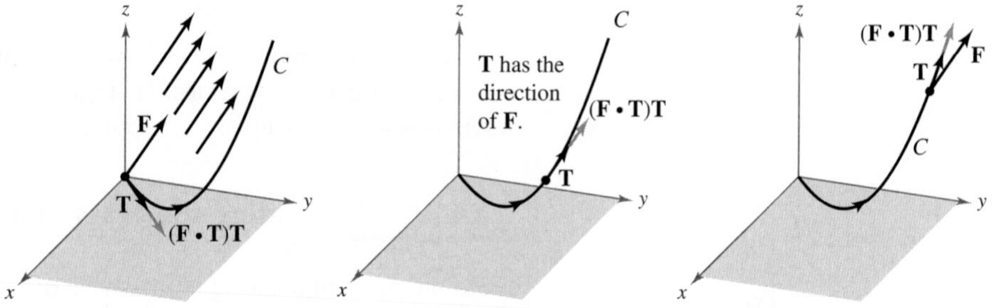

At each point on C, the force in the direction of motion is $(\mathbf{F} \cdot \mathbf{T})\mathbf{T}$.

Figure 15.13

This line integral appears in other contexts and is the basis of the following definition of the **line integral of a vector field.** Note in the definition that

$$\begin{aligned}\mathbf{F} \cdot \mathbf{T}\, ds &= \mathbf{F} \cdot \frac{\mathbf{r}'(t)}{\|\mathbf{r}'(t)\|}\|\mathbf{r}'(t)\|\, dt\\ &= \mathbf{F} \cdot \mathbf{r}'(t)\, dt\\ &= \mathbf{F} \cdot d\mathbf{r}.\end{aligned}$$

Definition of Line Integral of a Vector Field

Let **F** be a continuous vector field defined on a smooth curve C given by $\mathbf{r}(t)$, $a \le t \le b$. The **line integral** of **F** on C is given by

$$\int_C \mathbf{F} \cdot d\mathbf{r} = \int_C \mathbf{F} \cdot \mathbf{T}\, ds = \int_a^b \mathbf{F}(x(t),\ y(t), z(t)) \cdot \mathbf{r}'(t)\, dt.$$

EXAMPLE 6 Work Done by a Force

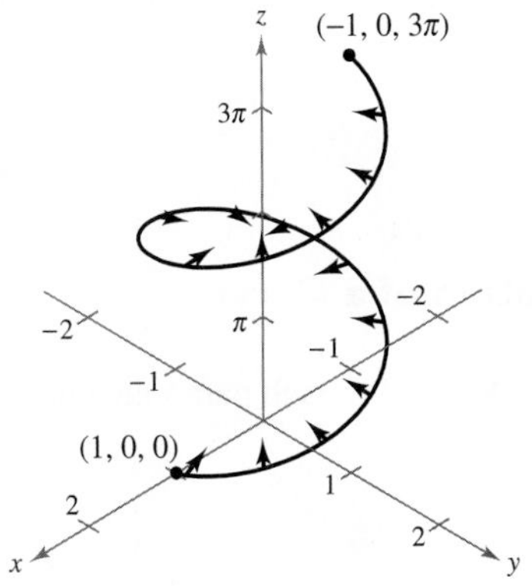

Figure 15.14

Find the work done by the force field

$$\mathbf{F}(x, y, z) = -\frac{1}{2}x\mathbf{i} - \frac{1}{2}y\mathbf{j} + \frac{1}{4}\mathbf{k} \qquad \text{Force field } \mathbf{F}$$

on a particle as it moves along the helix given by

$$\mathbf{r}(t) = \cos t\mathbf{i} + \sin t\mathbf{j} + t\mathbf{k} \qquad \text{Space curve } C$$

from the point $(1, 0, 0)$ to $(-1, 0, 3\pi)$, as shown in Figure 15.14.

Solution Because

$$\begin{aligned}\mathbf{r}(t) &= x(t)\mathbf{i} + y(t)\mathbf{j} + z(t)\mathbf{k}\\ &= \cos t\mathbf{i} + \sin t\mathbf{j} + t\mathbf{k}\end{aligned}$$

it follows that $x(t) = \cos t$, $y(t) = \sin t$, and $z(t) = t$. So, the force field can be written as

$$\mathbf{F}(x(t), y(t), z(t)) = -\frac{1}{2}\cos t\mathbf{i} - \frac{1}{2}\sin t\mathbf{j} + \frac{1}{4}\mathbf{k}.$$

To find the work done by the force field in moving a particle along the curve C, use the fact that

$$\mathbf{r}'(t) = -\sin t\mathbf{i} + \cos t\mathbf{j} + \mathbf{k}$$

and write the following.

$$\begin{aligned}W &= \int_C \mathbf{F} \cdot d\mathbf{r}\\ &= \int_a^b \mathbf{F}(x(t), y(t), z(t)) \cdot \mathbf{r}'(t)\,dt\\ &= \int_0^{3\pi} \left(-\frac{1}{2}\cos t\mathbf{i} - \frac{1}{2}\sin t\mathbf{j} + \frac{1}{4}\mathbf{k}\right) \cdot (-\sin t\mathbf{i} + \cos t\mathbf{j} + \mathbf{k})\,dt\\ &= \int_0^{3\pi} \left(\frac{1}{2}\sin t\cos t - \frac{1}{2}\sin t\cos t + \frac{1}{4}\right) dt\\ &= \int_0^{3\pi} \frac{1}{4}\,dt\\ &= \frac{1}{4}t\Big]_0^{3\pi}\\ &= \frac{3\pi}{4}\end{aligned}$$

NOTE In Example 6, note that the x- and y-components of the force field end up contributing nothing to the total work. This occurs because *in this particular example* the z-component of the force field is the only portion of the force that is acting in the same (or opposite) direction in which the particle is moving (see Figure 15.15).

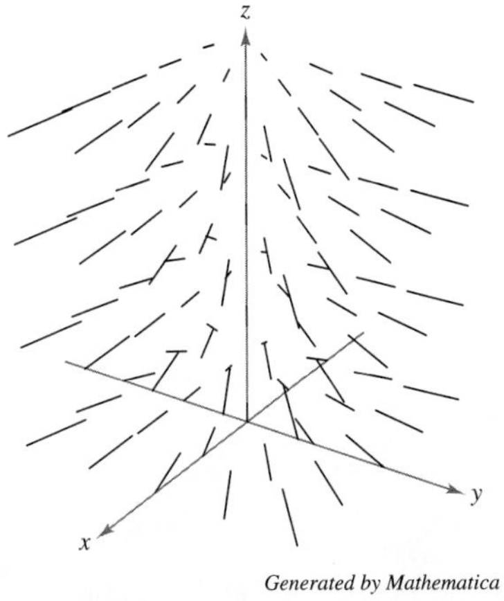

Figure 15.15

TECHNOLOGY The computer-generated view of the force field in Example 6 shown in Figure 15.15 indicates that each vector in the force field points toward the z-axis.

For line integrals of vector functions, the orientation of the curve C is important. If the orientation of the curve is reversed, the unit tangent vector $\mathbf{T}(t)$ is changed to $-\mathbf{T}(t)$, and you obtain

$$\int_{-C} \mathbf{F} \cdot d\mathbf{r} = -\int_{C} \mathbf{F} \cdot d\mathbf{r}.$$

EXAMPLE 7 Orientation and Parametrization of a Curve

Let $\mathbf{F}(x, y) = y\mathbf{i} + x^2\mathbf{j}$ and evaluate the line integral $\int_C \mathbf{F} \cdot d\mathbf{r}$ for each parabolic curve shown in Figure 15.16.

a. C_1: $\mathbf{r}_1(t) = (4 - t)\mathbf{i} + (4t - t^2)\mathbf{j}, \quad 0 \le t \le 3$

b. C_2: $\mathbf{r}_2(t) = t\mathbf{i} + (4t - t^2)\mathbf{j}, \quad 1 \le t \le 4$

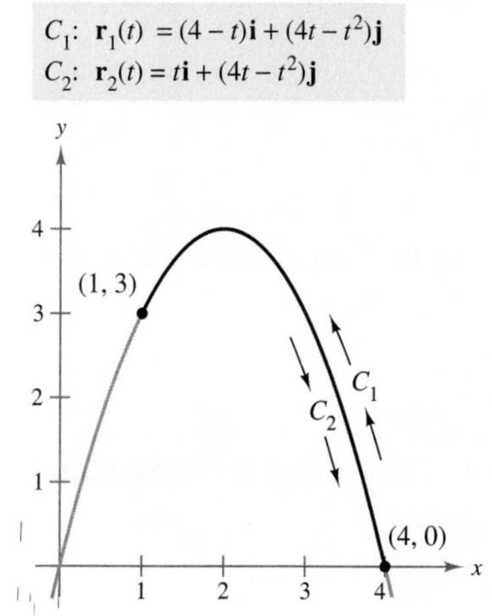

Figure 15.16

Solution

a. Because $\mathbf{r}_1'(t) = -\mathbf{i} + (4 - 2t)\mathbf{j}$ and

$$\mathbf{F}(x(t), y(t)) = (4t - t^2)\mathbf{i} + (4 - t)^2\mathbf{j}$$

the line integral is

$$\begin{aligned}
\int_{C_1} \mathbf{F} \cdot d\mathbf{r} &= \int_0^3 \left[(4t - t^2)\mathbf{i} + (4 - t)^2\mathbf{j}\right] \cdot \left[-\mathbf{i} + (4 - 2t)\mathbf{j}\right] dt \\
&= \int_0^3 (-4t + t^2 + 64 - 64t + 20t^2 - 2t^3)\, dt \\
&= \int_0^3 (-2t^3 + 21t^2 - 68t + 64)\, dt \\
&= \left[-\frac{t^4}{2} + 7t^3 - 34t^2 + 64t\right]_0^3 \\
&= \frac{69}{2}.
\end{aligned}$$

b. Because $\mathbf{r}_2'(t) = \mathbf{i} + (4 - 2t)\mathbf{j}$ and

$$\mathbf{F}(x(t), y(t)) = (4t - t^2)\mathbf{i} + t^2\mathbf{j}$$

the line integral is

$$\begin{aligned}
\int_{C_2} \mathbf{F} \cdot d\mathbf{r} &= \int_1^4 \left[(4t - t^2)\mathbf{i} + t^2\mathbf{j}\right] \cdot \left[\mathbf{i} + (4 - 2t)\mathbf{j}\right] dt \\
&= \int_1^4 (4t - t^2 + 4t^2 - 2t^3)\, dt \\
&= \int_1^4 (-2t^3 + 3t^2 + 4t)\, dt \\
&= \left[-\frac{t^4}{2} + t^3 + 2t^2\right]_1^4 \\
&= -\frac{69}{2}.
\end{aligned}$$

The answer in part (b) is the negative of that in part (a) because C_1 and C_2 represent opposite orientations of the same parabolic segment.

NOTE Although the value of the line integral in Example 7 depends on the orientation of C, it does not depend on the parametrization of C. To see this, let C_3 be represented by

$$\mathbf{r}_3 = (t + 2)\mathbf{i} + (4 - t^2)\mathbf{j}$$

where $-1 \le t \le 2$. The graph of this curve is the same parabolic segment shown in Figure 15.16. Does the value of the line integral over C_3 agree with the value over C_1 or C_2? Why or why not?

Line Integrals in Differential Form

A second commonly used form of line integrals is derived from the vector field notation used in the preceding section. If $\mathbf{F}$ is a vector field of the form $\mathbf{F}(x, y) = M\mathbf{i} + N\mathbf{j}$, and C is given by $\mathbf{r}(t) = x(t)\mathbf{i} + y(t)\mathbf{j}$, then $\mathbf{F} \cdot d\mathbf{r}$ is often written as $M\,dx + N\,dy$.

$$\begin{aligned}\int_C \mathbf{F} \cdot d\mathbf{r} &= \int_C \mathbf{F} \cdot \frac{d\mathbf{r}}{dt}\,dt \\ &= \int_a^b (M\mathbf{i} + N\mathbf{j}) \cdot (x'(t)\mathbf{i} + y'(t)\mathbf{j})\,dt \\ &= \int_a^b \left(M\frac{dx}{dt} + N\frac{dy}{dt}\right) dt \\ &= \int_C (M\,dx + N\,dy)\end{aligned}$$

This **differential form** can be extended to three variables. The parentheses are often omitted, as follows.

$$\int_C M\,dx + N\,dy \qquad \text{and} \qquad \int_C M\,dx + N\,dy + P\,dz$$

Notice how this differential notation is used in Example 8.

EXAMPLE 8 Evaluating a Line Integral in Differential Form

Let C be the circle of radius 3 given by

$$\mathbf{r}(t) = 3\cos t\mathbf{i} + 3\sin t\mathbf{j}, \quad 0 \le t \le 2\pi$$

as shown in Figure 15.17. Evaluate the line integral

$$\int_C y^3\,dx + (x^3 + 3xy^2)\,dy.$$

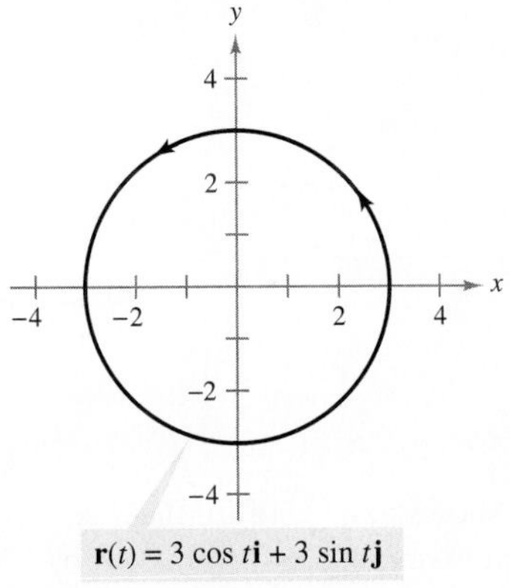

Figure 15.17

Solution Because $x = 3\cos t$ and $y = 3\sin t$, you have $dx = -3\sin t\,dt$ and $dy = 3\cos t\,dt$. So, the line integral is

$$\begin{aligned}&\int_C M\,dx + N\,dy \\ &= \int_C y^3\,dx + (x^3 + 3xy^2)\,dy \\ &= \int_0^{2\pi} \left[(27\sin^3 t)(-3\sin t) + (27\cos^3 t + 81\cos t\sin^2 t)(3\cos t)\right] dt \\ &= 81\int_0^{2\pi} (\cos^4 t - \sin^4 t + 3\cos^2 t\sin^2 t)\,dt \\ &= 81\int_0^{2\pi} \left(\cos^2 t - \sin^2 t + \frac{3}{4}\sin^2 2t\right) dt \\ &= 81\int_0^{2\pi} \left[\cos 2t + \frac{3}{4}\left(\frac{1 - \cos 4t}{2}\right)\right] dt \\ &= 81\left[\frac{\sin 2t}{2} + \frac{3}{8}t - \frac{3\sin 4t}{32}\right]_0^{2\pi} \\ &= \frac{243\pi}{4}.\end{aligned}$$

NOTE The orientation of C affects the value of the differential form of a line integral. Specifically, if $-C$ has the orientation opposite to that of C, then

$$\int_{-C} M\,dx + N\,dy = -\int_C M\,dx + N\,dy.$$

So, of the three line integral forms presented in this section, the orientation of C does not affect the form $\int_C f(x, y)\,ds$, but it does affect the vector form and the differential form.

For curves represented by $y = g(x)$, $a \le x \le b$, you can let $x = t$ and obtain the parametric form

$$x = t \quad \text{and} \quad y = g(t), \quad a \le t \le b.$$

Because $dx = dt$ for this form, you have the option of evaluating the line integral in the variable x or t. This is demonstrated in Example 9.

EXAMPLE 9 Evaluating a Line Integral in Differential Form

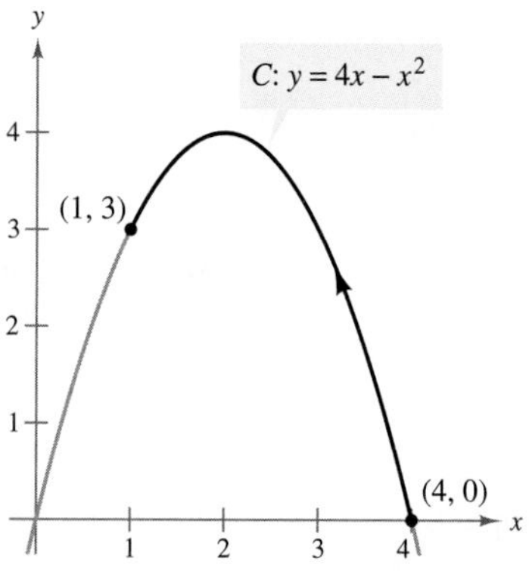

Figure 15.18

Evaluate

$$\int_C y\,dx + x^2\,dy$$

where C is the parabolic arc given by $y = 4x - x^2$ from $(4, 0)$ to $(1, 3)$, as shown in Figure 15.18.

Solution Rather than converting to the parameter t, you can simply retain the variable x and write

$$y = 4x - x^2 \quad \Longrightarrow \quad dy = (4 - 2x)\,dx.$$

Then, in the direction from $(4, 0)$ to $(1, 3)$, the line integral is

$$\begin{aligned}\int_C y\,dx + x^2\,dy &= \int_4^1 [(4x - x^2)\,dx + x^2(4 - 2x)\,dx] \\ &= \int_4^1 (4x + 3x^2 - 2x^3)\,dx \\ &= \left[2x^2 + x^3 - \frac{x^4}{2}\right]_4^1 = \frac{69}{2}.\end{aligned}$$

See Example 7.

EXPLORATION

Finding Lateral Surface Area The figure below shows a piece of tin that has been cut from a circular cylinder. The base of the circular cylinder is modeled by $x^2 + y^2 = 9$. At any point (x, y) on the base, the height of the object is given by

$$f(x, y) = 1 + \cos\frac{\pi x}{4}.$$

Explain how to use a line integral to find the surface area of the piece of tin.

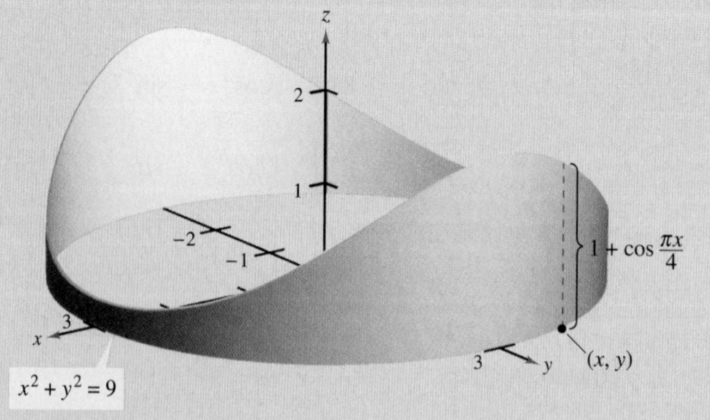

Exercises for Section 15.2

See www.CalcChat.com for worked-out solutions to odd-numbered exercises.

In Exercises 1–6, find a piecewise smooth parametrization of the path C.

1.

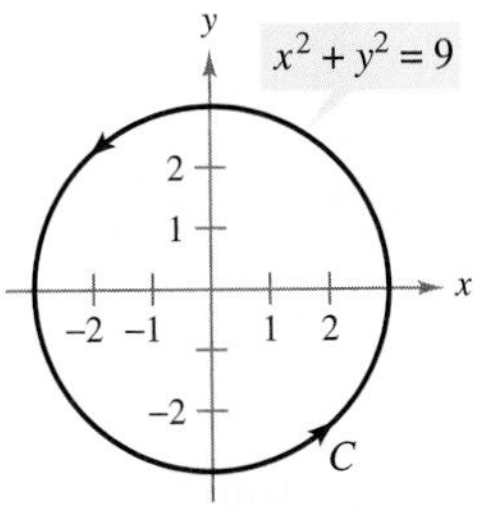

2.

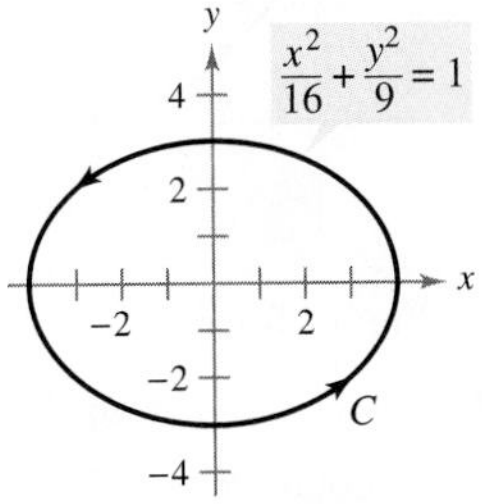

3.

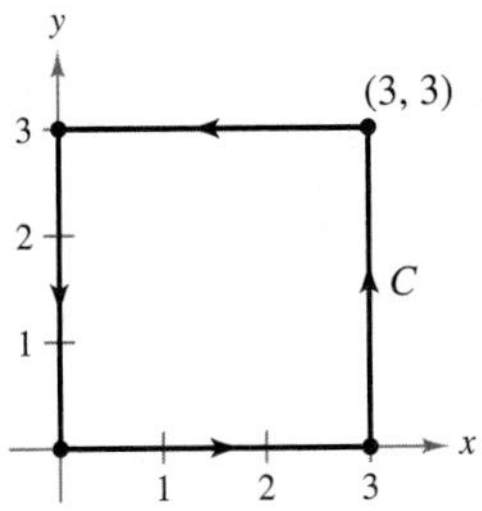

4.

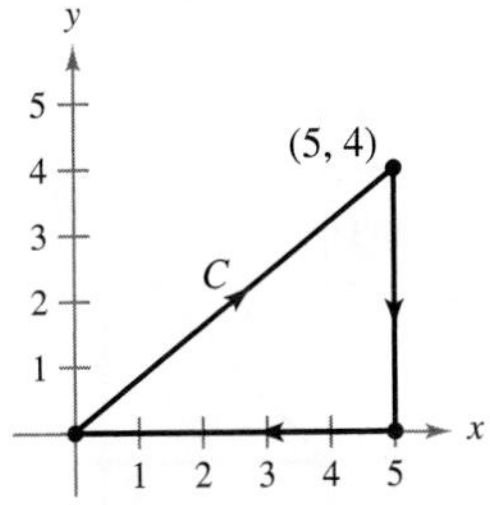

5.

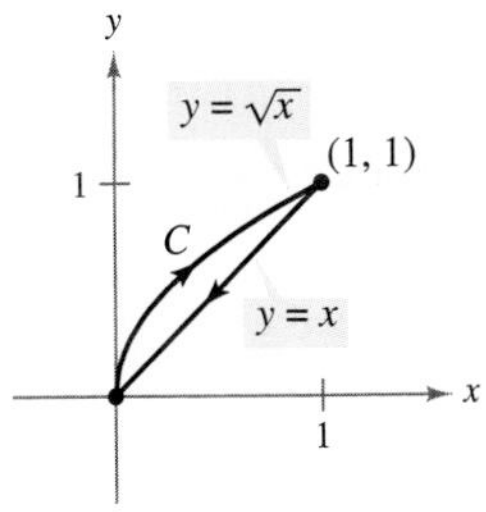

6.

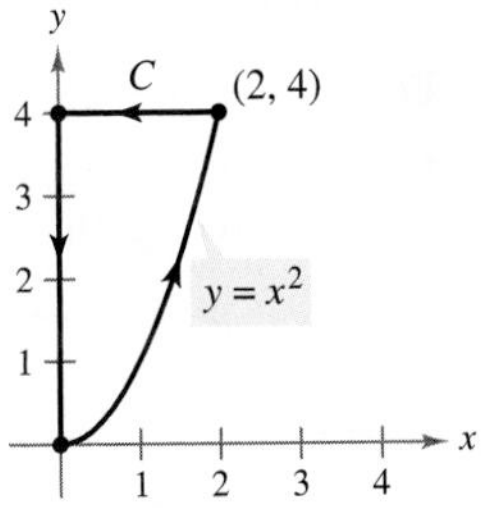

In Exercises 7–10, evaluate the line integral along the given path.

7. $\int_C (x - y)\, ds$

C: $\mathbf{r}(t) = 4t\mathbf{i} + 3t\mathbf{j}$

$0 \le t \le 2$

8. $\int_C 4xy\, ds$

C: $\mathbf{r}(t) = t\mathbf{i} + (2 - t)\mathbf{j}$

$0 \le t \le 2$

9. $\int_C (x^2 + y^2 + z^2)\, ds$

C: $\mathbf{r}(t) = \sin t\mathbf{i} + \cos t\mathbf{j} + 8t\mathbf{k}$

$0 \le t \le \pi/2$

10. $\int_C 8xyz\, ds$

C: $\mathbf{r}(t) = 12t\mathbf{i} + 5t\mathbf{j} + 3\mathbf{k}$

$0 \le t \le 2$

In Exercises 11–14, evaluate

$$\int_C (x^2 + y^2)\, ds$$

along the given path.

11. C: x-axis from $x = 0$ to $x = 3$
12. C: y-axis from $y = 1$ to $y = 10$
13. C: counterclockwise around the circle $x^2 + y^2 = 1$ from $(1, 0)$ to $(0, 1)$
14. C: counterclockwise around the circle $x^2 + y^2 = 4$ from $(2, 0)$ to $(0, 2)$

In Exercises 15–18, evaluate

$$\int_C (x + 4\sqrt{y})\, ds$$

along the given path.

15. C: line from $(0, 0)$ to $(1, 1)$
16. C: line from $(0, 0)$ to $(3, 9)$
17. C: counterclockwise around the triangle with vertices $(0, 0)$, $(1, 0)$, and $(0, 1)$
18. C: counterclockwise around the square with vertices $(0, 0)$, $(2, 0)$, $(2, 2)$, and $(0, 2)$

In Exercises 19 and 20, evaluate

$$\int_C (2x + y^2 - z)\, ds$$

along the path C shown in the figure.

19.

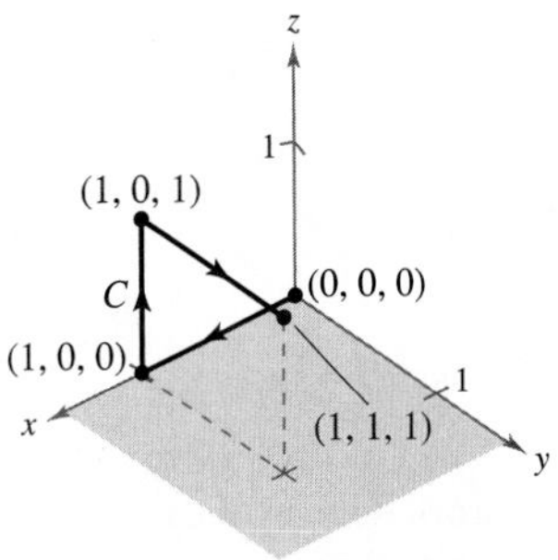

20.

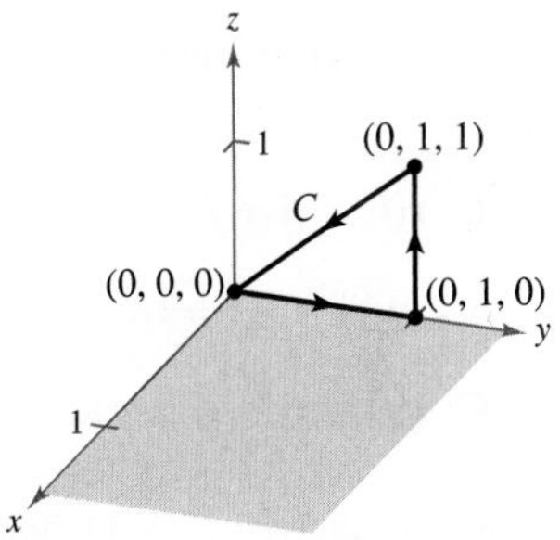

Mass **In Exercises 21 and 22, find the total mass of two turns of a spring with density ρ in the shape of the circular helix**

$$\mathbf{r}(t) = 3\cos t\mathbf{i} + 3\sin t\mathbf{j} + 2t\mathbf{k}.$$

21. $\rho(x, y, z) = \frac{1}{2}(x^2 + y^2 + z^2)$
22. $\rho(x, y, z) = z$

Mass **In Exercises 23–26, find the total mass of the wire with density ρ.**

23. $\mathbf{r}(t) = \cos t\mathbf{i} + \sin t\mathbf{j}$, $\rho(x, y) = x + y$, $0 \le t \le \pi$
24. $\mathbf{r}(t) = t^2\mathbf{i} + 2t\mathbf{j}$, $\rho(x, y) = \frac{3}{4}y$, $0 \le t \le 1$
25. $\mathbf{r}(t) = t^2\mathbf{i} + 2t\mathbf{j} + t\mathbf{k}$, $\rho(x, y, z) = kz$ $(k > 0)$, $1 \le t \le 3$
26. $\mathbf{r}(t) = 2\cos t\mathbf{i} + 2\sin t\mathbf{j} + 3t\mathbf{k}$, $\rho(x, y, z) = k + z$ $(k > 0)$, $0 \le t \le 2\pi$

In Exercises 27–32, evaluate

$$\int_C \mathbf{F} \cdot d\mathbf{r}$$

where C is represented by $\mathbf{r}(t)$.

27. $\mathbf{F}(x, y) = xy\mathbf{i} + y\mathbf{j}$

C: $\mathbf{r}(t) = 4t\mathbf{i} + t\mathbf{j}, \quad 0 \le t \le 1$

28. $\mathbf{F}(x, y) = xy\mathbf{i} + y\mathbf{j}$

C: $\mathbf{r}(t) = 4 \cos t\mathbf{i} + 4 \sin t\mathbf{j}, \quad 0 \le t \le \pi/2$

29. $\mathbf{F}(x, y) = 3x\mathbf{i} + 4y\mathbf{j}$

C: $\mathbf{r}(t) = 2 \cos t\mathbf{i} + 2 \sin t\mathbf{j}, \quad 0 \le t \le \pi/2$

30. $\mathbf{F}(x, y) = 3x\mathbf{i} + 4y\mathbf{j}$

C: $\mathbf{r}(t) = t\mathbf{i} + \sqrt{4 - t^2}\mathbf{j}, \quad -2 \le t \le 2$

31. $\mathbf{F}(x, y, z) = x^2y\mathbf{i} + (x - z)\mathbf{j} + xyz\mathbf{k}$

C: $\mathbf{r}(t) = t\mathbf{i} + t^2\mathbf{j} + 2\mathbf{k}, \quad 0 \le t \le 1$

32. $\mathbf{F}(x, y, z) = x^2\mathbf{i} + y^2\mathbf{j} + z^2\mathbf{k}$

C: $\mathbf{r}(t) = 2 \sin t\mathbf{i} + 2 \cos t\mathbf{j} + \frac{1}{2}t^2\mathbf{k}, \quad 0 \le t \le \pi$

In Exercises 33 and 34, use a computer algebra system to evaluate the integral

$$\int_C \mathbf{F} \cdot d\mathbf{r}$$

where C is represented by $\mathbf{r}(t)$.

33. $\mathbf{F}(x, y, z) = x^2z\mathbf{i} + 6y\mathbf{j} + yz^2\mathbf{k}$

C: $\mathbf{r}(t) = t\mathbf{i} + t^2\mathbf{j} + \ln t\mathbf{k}, \quad 1 \le t \le 3$

34. $\mathbf{F}(x, y, z) = \dfrac{x\mathbf{i} + y\mathbf{j} + z\mathbf{k}}{\sqrt{x^2 + y^2 + z^2}}$

C: $\mathbf{r}(t) = t\mathbf{i} + t\mathbf{j} + e^t\mathbf{k}, \quad 0 \le t \le 2$

Work **In Exercises 35–40, find the work done by the force field F on a particle moving along the given path.**

35. $\mathbf{F}(x, y) = -x\mathbf{i} - 2y\mathbf{j}$

C: $y = x^3$ from $(0, 0)$ to $(2, 8)$

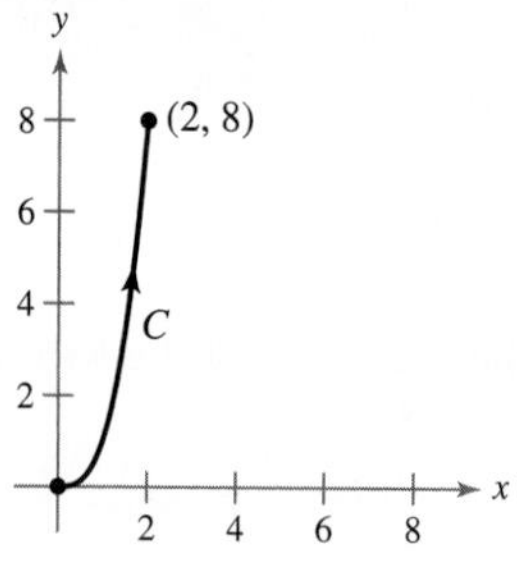

36. $\mathbf{F}(x, y) = x^2\mathbf{i} - xy\mathbf{j}$

C: $x = \cos^3 t$, $y = \sin^3 t$ from $(1, 0)$ to $(0, 1)$

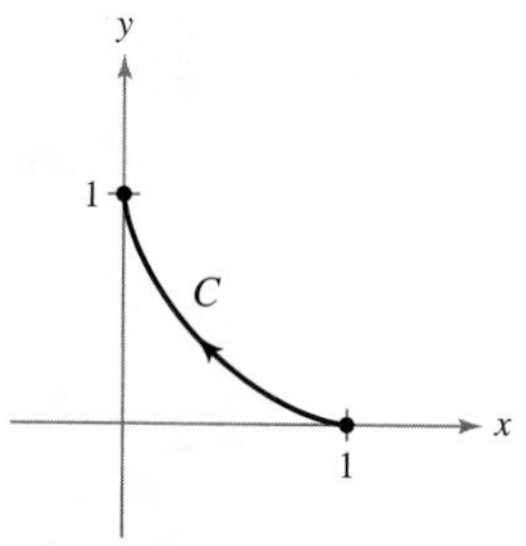

Figure for 36

37. $\mathbf{F}(x, y) = 2x\mathbf{i} + y\mathbf{j}$

C: counterclockwise around the triangle with vertices $(0, 0)$, $(1, 0)$, and $(1, 1)$

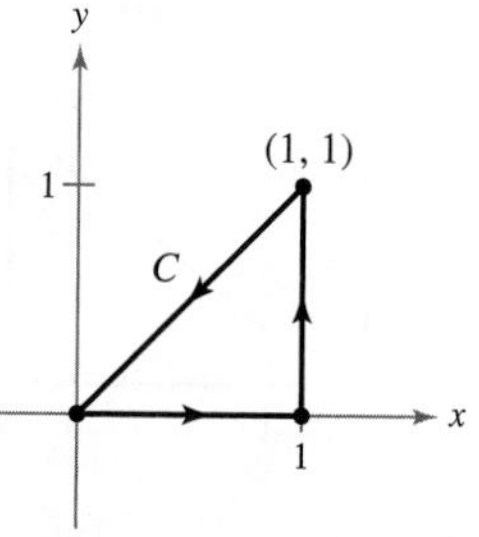

38. $\mathbf{F}(x, y) = -y\mathbf{i} - x\mathbf{j}$

C: counterclockwise along the semicircle $y = \sqrt{4 - x^2}$ from $(2, 0)$ to $(-2, 0)$

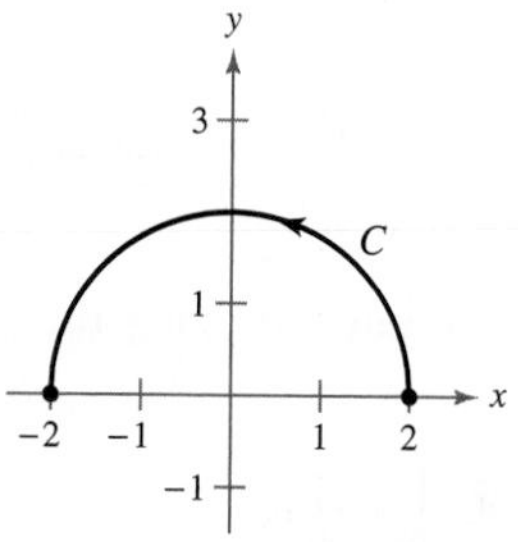

39. $\mathbf{F}(x, y, z) = x\mathbf{i} + y\mathbf{j} - 5z\mathbf{k}$

C: $\mathbf{r}(t) = 2 \cos t\mathbf{i} + 2 \sin t\mathbf{j} + t\mathbf{k}, \quad 0 \le t \le 2\pi$

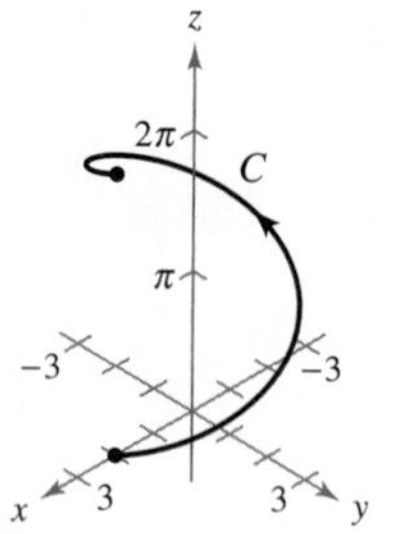

40. $\mathbf{F}(x, y, z) = yz\mathbf{i} + xz\mathbf{j} + xy\mathbf{k}$

C: line from $(0, 0, 0)$ to $(5, 3, 2)$

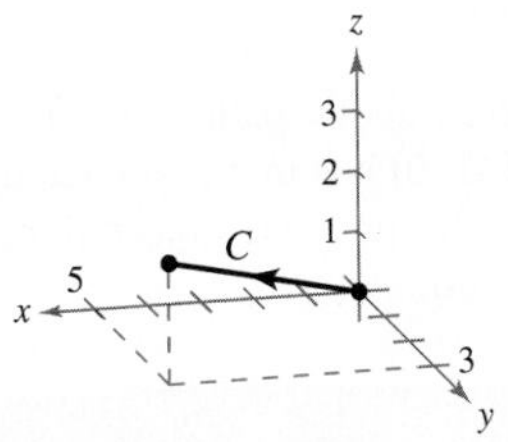

41. ***Work*** Find the work done by a person weighing 150 pounds walking exactly one revolution up a circular helical staircase of radius 3 feet if the person rises 10 feet.

42. ***Work*** A particle moves along the path $y = x^2$ from the point $(0, 0)$ to the point $(1, 1)$. The force field $\mathbf{F}$ is measured at five points along the path and the results are shown in the table. Use Simpson's Rule or a graphing utility to approximate the work done by the force field.

(x, y)	$(0, 0)$	$\left(\frac{1}{4}, \frac{1}{16}\right)$	$\left(\frac{1}{2}, \frac{1}{4}\right)$	$\left(\frac{3}{4}, \frac{9}{16}\right)$	$(1, 1)$
$\mathbf{F}(x, y)$	$\langle 5, 0\rangle$	$\langle 3.5, 1\rangle$	$\langle 2, 2\rangle$	$\langle 1.5, 3\rangle$	$\langle 1, 5\rangle$

In Exercises 43 and 44, evaluate $\int_C \mathbf{F} \cdot d\mathbf{r}$ for each curve. Discuss the orientation of the curve and its effect on the value of the integral.

43. $\mathbf{F}(x, y) = x^2\mathbf{i} + xy\mathbf{j}$

(a) $\mathbf{r}_1(t) = 2t\mathbf{i} + (t - 1)\mathbf{j}, \quad 1 \le t \le 3$

(b) $\mathbf{r}_2(t) = 2(3 - t)\mathbf{i} + (2 - t)\mathbf{j}, \quad 0 \le t \le 2$

44. $\mathbf{F}(x, y) = x^2y\mathbf{i} + xy^{3/2}\mathbf{j}$

(a) $\mathbf{r}_1(t) = (t + 1)\mathbf{i} + t^2\mathbf{j}, \quad 0 \le t \le 2$

(b) $\mathbf{r}_2(t) = (1 + 2\cos t)\mathbf{i} + (4\cos^2 t)\mathbf{j}, \quad 0 \le t \le \pi/2$

In Exercises 45–48, demonstrate the property that

$$\int_C \mathbf{F} \cdot d\mathbf{r} = 0$$

regardless of the initial and terminal points of C, if the tangent vector $\mathbf{r}'(t)$ is orthogonal to the force field $\mathbf{F}$.

45. $\mathbf{F}(x, y) = y\mathbf{i} - x\mathbf{j}$

C: $\mathbf{r}(t) = t\mathbf{i} - 2t\mathbf{j}$

46. $\mathbf{F}(x, y) = -3y\mathbf{i} + x\mathbf{j}$

C: $\mathbf{r}(t) = t\mathbf{i} - t^3\mathbf{j}$

47. $\mathbf{F}(x, y) = (x^3 - 2x^2)\mathbf{i} + \left(x - \frac{y}{2}\right)\mathbf{j}$

C: $\mathbf{r}(t) = t\mathbf{i} + t^2\mathbf{j}$

48. $\mathbf{F}(x, y) = x\mathbf{i} + y\mathbf{j}$

C: $\mathbf{r}(t) = 3\sin t\mathbf{i} + 3\cos t\mathbf{j}$

In Exercises 49–52, evaluate the line integral along the path C given by $x = 2t$, $y = 10t$, where $0 \le t \le 1$.

49. $\int_C (x + 3y^2)\, dy$

50. $\int_C (x + 3y^2)\, dx$

51. $\int_C xy\, dx + y\, dy$

52. $\int_C (3y - x)\, dx + y^2\, dy$

In Exercises 53–60, evaluate the integral

$$\int_C (2x - y)\, dx + (x + 3y)\, dy$$

along the path C.

53. C: x-axis from $x = 0$ to $x = 5$

54. C: y-axis from $y = 0$ to $y = 2$

55. C: line segments from $(0, 0)$ to $(3, 0)$ and $(3, 0)$ to $(3, 3)$

56. C: line segments from $(0, 0)$ to $(0, -3)$ and $(0, -3)$ to $(2, -3)$

57. C: arc on $y = 1 - x^2$ from $(0, 1)$ to $(1, 0)$

58. C: arc on $y = x^{3/2}$ from $(0, 0)$ to $(4, 8)$

59. C: parabolic path $x = t$, $y = 2t^2$, from $(0, 0)$ to $(2, 8)$

60. C: elliptic path $x = 4\sin t$, $y = 3\cos t$, from $(0, 3)$ to $(4, 0)$

Lateral Surface Area **In Exercises 61–68, find the area of the lateral surface (see figure) over the curve C in the xy-plane and under the surface $z = f(x, y)$, where**

$$\text{Lateral surface area} = \int_C f(x, y)\, ds.$$

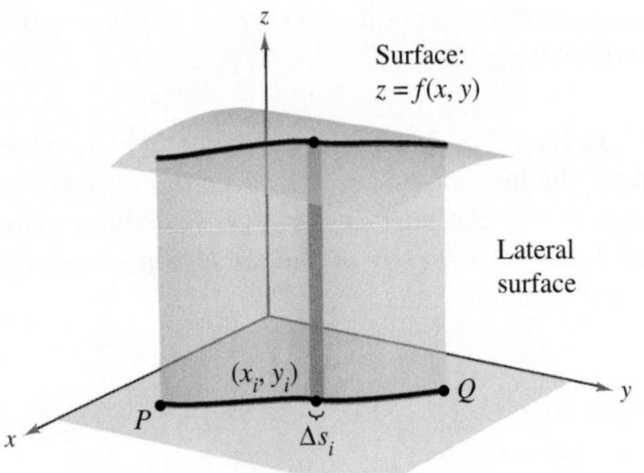

C: Curve in xy-plane

61. $f(x, y) = h$, C: line from $(0, 0)$ to $(3, 4)$

62. $f(x, y) = y$, C: line from $(0, 0)$ to $(4, 4)$

63. $f(x, y) = xy$, C: $x^2 + y^2 = 1$ from $(1, 0)$ to $(0, 1)$

64. $f(x, y) = x + y$, C: $x^2 + y^2 = 1$ from $(1, 0)$ to $(0, 1)$

65. $f(x, y) = h$, C: $y = 1 - x^2$ from $(1, 0)$ to $(0, 1)$

66. $f(x, y) = y + 1$, C: $y = 1 - x^2$ from $(1, 0)$ to $(0, 1)$

67. $f(x, y) = xy$, C: $y = 1 - x^2$ from $(1, 0)$ to $(0, 1)$

68. $f(x, y) = x^2 - y^2 + 4$, C: $x^2 + y^2 = 4$

69. ***Engine Design*** A tractor engine has a steel component with a circular base modeled by the vector-valued function $\mathbf{r}(t) = 2\cos t\mathbf{i} + 2\sin t\mathbf{j}$. Its height is given by $z = 1 + y^2$. (All measurements of the component are given in centimeters.)
    (a) Find the lateral surface area of the component.
    (b) The component is in the form of a shell of thickness 0.2 centimeter. Use the result of part (a) to approximate the amount of steel used in its manufacture.
    (c) Draw a sketch of the component.

70. ***Building Design*** The ceiling of a building has a height above the floor given by $z = 20 + \frac{1}{4}x$, and one of the walls follows a path modeled by $y = x^{3/2}$. Find the surface area of the wall if $0 \le x \le 40$. (All measurements are given in feet.)

Moments of Inertia **Consider a wire of density $\rho(x, y)$ given by the space curve**

$$C: \mathbf{r}(t) = x(t)\mathbf{i} + y(t)\mathbf{j}, \quad a \le t \le b.$$

The moments of inertia about the x- and y-axes are given by

$$I_x = \int_C y^2 \rho(x, y)\, ds$$

$$I_y = \int_C x^2 \rho(x, y)\, ds.$$

In Exercises 71 and 72, find the moments of inertia for the wire of density ρ.

71. A wire lies along $\mathbf{r}(t) = a\cos t\mathbf{i} + a\sin t\mathbf{j}$, $0 \le t \le 2\pi$ and $a > 0$, with density $\rho(x, y) = 1$.

72. A wire lies along $\mathbf{r}(t) = a\cos t\mathbf{i} + a\sin t\mathbf{j}$, $0 \le t \le 2\pi$ and $a > 0$, with density $\rho(x, y) = y$.

Approximation **In Exercises 73 and 74, determine which value best approximates the lateral surface area over the curve C in the xy-plane and under the surface $z = f(x, y)$. (Make your selection on the basis of a sketch of the surface and *not* by performing any calculations.)**

73. $f(x, y) = e^{xy}$
    C: line from $(0, 0)$ to $(2, 2)$
    (a) 54 (b) 25 (c) -250 (d) 75 (e) 100

74. $f(x, y) = y$
    C: $y = x^2$ from $(0, 0)$ to $(2, 4)$
    (a) 2 (b) 4 (c) 8 (d) 16

75. ***Investigation*** The top outer edge of a solid with vertical sides and resting on the xy-plane is modeled by $\mathbf{r}(t) = 3\cos t\mathbf{i} + 3\sin t\mathbf{j} + (1 + \sin^2 2t)\mathbf{k}$, where all measurements are in centimeters. The intersection of the plane $y = b\ (-3 < b < 3)$ with the top of the solid is a horizontal line.
    (a) Use a computer algebra system to graph the solid.
    (b) Use a computer algebra system to approximate the lateral surface area of the solid.
    (c) Find (if possible) the volume of the solid.

76. ***Investigation*** Determine the value of c such that the work done by the force field

$$\mathbf{F}(x, y) = 15[(4 - x^2y)\mathbf{i} - xy\mathbf{j}]$$

on an object moving along the parabolic path $y = c(1 - x^2)$ between the points $(-1, 0)$ and $(1, 0)$ is a minimum. Compare the result with the work required to move the object along the straight-line path connecting the points.

Writing About Concepts

77. Define a line integral of a function f along a smooth curve C in the plane and in space. How do you evaluate the line integral as a definite integral?

78. Define a line integral of a continuous vector field $\mathbf{F}$ on a smooth curve C. How do you evaluate the line integral as a definite integral?

79. Order the surfaces in ascending order of the lateral surface area under the surface and over the curve $y = \sqrt{x}$ from $(0, 0)$ to $(4, 2)$ in the xy-plane. Explain your ordering without doing any calculations.
    (a) $z_1 = 2 + x$ (b) $z_2 = 5 + x$
    (c) $z_3 = 2$ (d) $z_4 = 10 + x + 2y$

80. For each of the following, determine whether the work done in moving an object from the first to the second point through the force field shown in the figure is positive, negative, or zero. Explain your answer.
    (a) From $(-3, -3)$ to $(3, 3)$
    (b) From $(-3, 0)$ to $(0, 3)$
    (c) From $(5, 0)$ to $(0, 3)$

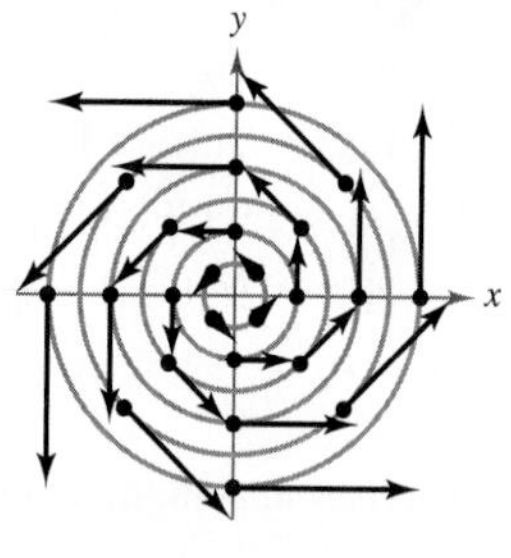

True or False? **In Exercises 81–84, determine whether the statement is true or false. If it is false, explain why or give an example that shows it is false.**

81. If C is given by $x(t) = t$, $y(t) = t$, $0 \le t \le 1$, then

$$\int_C xy\, ds = \int_0^1 t^2\, dt.$$

82. If $C_2 = -C_1$, then $\displaystyle\int_{C_1} f(x, y)\, ds + \int_{C_2} f(x, y)\, ds = 0$.

83. The vector functions $\mathbf{r}_1 = t\mathbf{i} + t^2\mathbf{j}$, $0 \le t \le 1$, and $\mathbf{r}_2 = (1 - t)\mathbf{i} + (1 - t)^2\mathbf{j}$, $0 \le t \le 1$, define the same curve.

84. If $\displaystyle\int_C \mathbf{F} \cdot \mathbf{T}\, ds = 0$, then $\mathbf{F}$ and $\mathbf{T}$ are orthogonal.

85. ***Work*** Consider a particle that moves through the force field $\mathbf{F}(x, y) = (y - x)\mathbf{i} + xy\mathbf{j}$ from the point $(0, 0)$ to the point $(0, 1)$ along the curve $x = kt(1 - t)$, $y = t$. Find the value of k such that the work done by the force field is 1.

Section 15.3 Conservative Vector Fields and Independence of Path

- **Understand and use the Fundamental Theorem of Line Integrals.**
- **Understand the concept of independence of path.**
- **Understand the concept of conservation of energy.**

Fundamental Theorem of Line Integrals

The discussion in the preceding section pointed out that in a gravitational field the work done by gravity on an object moving between two points in the field is independent of the path taken by the object. In this section, you will study an important generalization of this result—it is called the **Fundamental Theorem of Line Integrals.**

To begin, an example is presented in which the line integral of a *conservative vector field* is evaluated over three different paths.

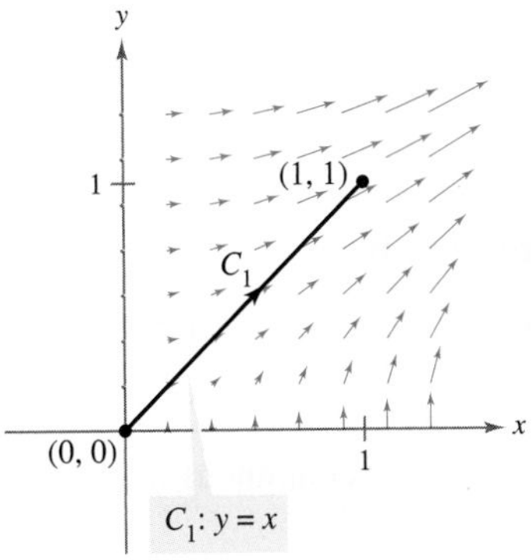

(a)

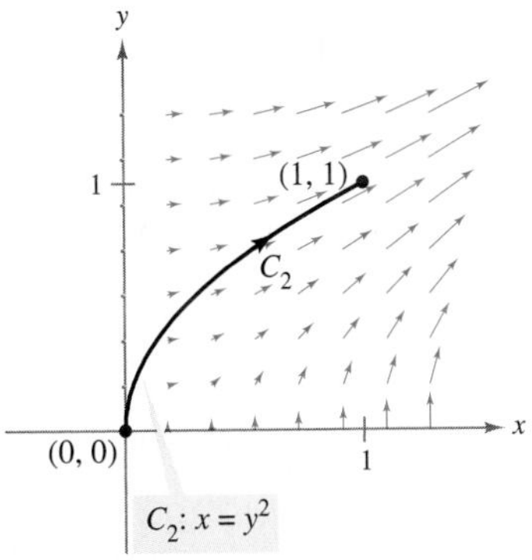

(b)

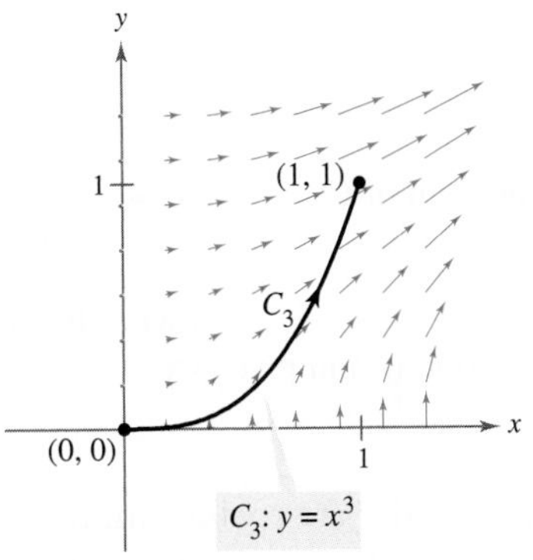

(c)

Figure 15.19

EXAMPLE 1 Line Integral of a Conservative Vector Field

Find the work done by the force field

$$\mathbf{F}(x, y) = \frac{1}{2}xy\mathbf{i} + \frac{1}{4}x^2\mathbf{j}$$

on a particle that moves from (0, 0) to (1, 1) along each path, as shown in Figure 15.19.

a. $C_1: y = x$ **b.** $C_2: x = y^2$ **c.** $C_3: y = x^3$

Solution

a. Let $\mathbf{r}(t) = t\mathbf{i} + t\mathbf{j}$ for $0 \leq t \leq 1$, so that

$$d\mathbf{r} = (\mathbf{i} + \mathbf{j})\,dt \quad \text{and} \quad \mathbf{F}(x, y) = \frac{1}{2}t^2\mathbf{i} + \frac{1}{4}t^2\mathbf{j}.$$

Then, the work done is

$$W = \int_{C_1} \mathbf{F} \cdot d\mathbf{r} = \int_0^1 \frac{3}{4}t^2\,dt = \frac{1}{4}t^3\Big]_0^1 = \frac{1}{4}.$$

b. Let $\mathbf{r}(t) = t\mathbf{i} + \sqrt{t}\mathbf{j}$ for $0 \leq t \leq 1$, so that

$$d\mathbf{r} = \left(\mathbf{i} + \frac{1}{2\sqrt{t}}\mathbf{j}\right)dt \quad \text{and} \quad \mathbf{F}(x, y) = \frac{1}{2}t^{3/2}\mathbf{i} + \frac{1}{4}t^2\mathbf{j}.$$

Then, the work done is

$$W = \int_{C_2} \mathbf{F} \cdot d\mathbf{r} = \int_0^1 \frac{5}{8}t^{3/2}\,dt = \frac{1}{4}t^{5/2}\Big]_0^1 = \frac{1}{4}.$$

c. Let $\mathbf{r}(t) = \frac{1}{2}t\mathbf{i} + \frac{1}{8}t^3\mathbf{j}$ for $0 \leq t \leq 2$, so that

$$d\mathbf{r} = \left(\frac{1}{2}\mathbf{i} + \frac{3}{8}t^2\mathbf{j}\right)dt \quad \text{and} \quad \mathbf{F}(x, y) = \frac{1}{32}t^4\mathbf{i} + \frac{1}{16}t^2\mathbf{j}.$$

Then, the work done is

$$W = \int_{C_3} \mathbf{F} \cdot d\mathbf{r} = \int_0^2 \frac{5}{128}t^4\,dt = \frac{1}{128}t^5\Big]_0^2 = \frac{1}{4}.$$

So, the work done by a conservative vector field is the same for all paths.

In Example 1, note that the vector field $\mathbf{F}(x, y) = \frac{1}{2}xy\mathbf{i} + \frac{1}{4}x^2\mathbf{j}$ is conservative because $\mathbf{F}(x, y) = \nabla f(x, y)$, where $f(x, y) = \frac{1}{4}x^2y$. In such cases, the following theorem states that the value of $\int_C \mathbf{F} \cdot d\mathbf{r}$ is given by

$$\begin{aligned}\int_C \mathbf{F} \cdot d\mathbf{r} &= f(x(1), y(1)) - f(x(0), y(0))\\ &= \frac{1}{4} - 0\\ &= \frac{1}{4}.\end{aligned}$$

THEOREM 15.5 Fundamental Theorem of Line Integrals

Let C be a piecewise smooth curve lying in an open region R and given by

$$\mathbf{r}(t) = x(t)\mathbf{i} + y(t)\mathbf{j}, \quad a \le t \le b.$$

If $\mathbf{F}(x, y) = M\mathbf{i} + N\mathbf{j}$ is conservative in R, and M and N are continuous in R, then

$$\int_C \mathbf{F} \cdot d\mathbf{r} = \int_C \nabla f \cdot d\mathbf{r} = f(x(b), y(b)) - f(x(a), y(a))$$

where f is a potential function of $\mathbf{F}$. That is, $\mathbf{F}(x, y) = \nabla f(x, y)$.

NOTE Notice how the Fundamental Theorem of Line Integrals is similar to the Fundamental Theorem of Calculus (Section 4.4), which states that

$$\int_a^b f(x)\,dx = F(b) - F(a)$$

where $F'(x) = f(x)$.

Proof A proof is provided only for a smooth curve. For piecewise smooth curves, the procedure is carried out separately on each smooth portion. Because $\mathbf{F}(x, y) = \nabla f(x, y) = f_x(x, y)\mathbf{i} + f_y(x, y)\mathbf{j}$, it follows that

$$\begin{aligned}\int_C \mathbf{F} \cdot d\mathbf{r} &= \int_a^b \mathbf{F} \cdot \frac{d\mathbf{r}}{dt}\,dt\\ &= \int_a^b \left[f_x(x, y)\frac{dx}{dt} + f_y(x, y)\frac{dy}{dt}\right] dt\end{aligned}$$

and, by the Chain Rule (Theorem 13.6), you have

$$\begin{aligned}\int_C \mathbf{F} \cdot d\mathbf{r} &= \int_a^b \frac{d}{dt}[f(x(t), y(t))]\,dt\\ &= f(x(b), y(b)) - f(x(a), y(a)).\end{aligned}$$

The last step is an application of the Fundamental Theorem of Calculus.

In space, the Fundamental Theorem of Line Integrals takes the following form. Let C be a piecewise smooth curve lying in an open region Q and given by

$$\mathbf{r}(t) = x(t)\mathbf{i} + y(t)\mathbf{j} + z(t)\mathbf{k}, \quad a \le t \le b.$$

If $\mathbf{F}(x, y, z) = M\mathbf{i} + N\mathbf{j} + P\mathbf{k}$ is conservative and M, N, and P are continuous, then

$$\begin{aligned}\int_C \mathbf{F} \cdot d\mathbf{r} &= \int_C \nabla f \cdot d\mathbf{r}\\ &= f(x(b), y(b), z(b)) - f(x(a), y(a), z(a))\end{aligned}$$

where $\mathbf{F}(x, y, z) = \nabla f(x, y, z)$.

The Fundamental Theorem of Line Integrals states that if the vector field $\mathbf{F}$ is conservative, then the line integral between any two points is simply the difference in the values of the *potential* function f at these points.

EXAMPLE 2 Using the Fundamental Theorem of Line Integrals

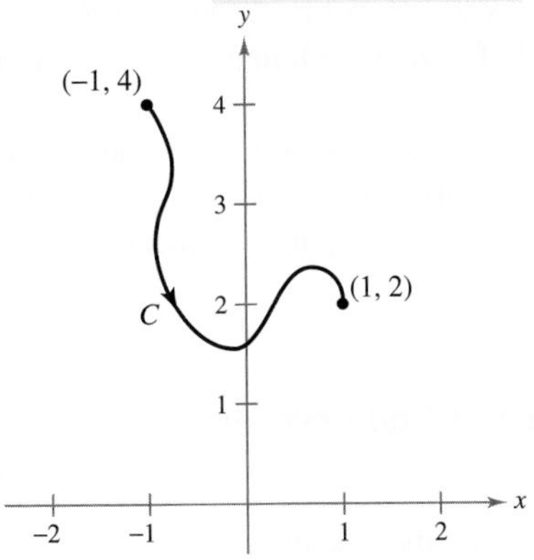

Using the Fundamental Theorem of Line Integrals, $\int_C \mathbf{F} \cdot d\mathbf{r}$.
Figure 15.20

Evaluate $\int_C \mathbf{F} \cdot d\mathbf{r}$, where C is a piecewise smooth curve from $(-1, 4)$ to $(1, 2)$ and

$$\mathbf{F}(x, y) = 2xy\mathbf{i} + (x^2 - y)\mathbf{j}$$

as shown in Figure 15.20.

Solution From Example 6 in Section 15.1, you know that $\mathbf{F}$ is the gradient of f where

$$f(x, y) = x^2y - \frac{y^2}{2} + K.$$

Consequently, $\mathbf{F}$ is conservative, and by the Fundamental Theorem of Line Integrals, it follows that

$$\begin{aligned}\int_C \mathbf{F} \cdot d\mathbf{r} &= f(1, 2) - f(-1, 4)\\ &= \left[1^2(2) - \frac{2^2}{2}\right] - \left[(-1)^2(4) - \frac{4^2}{2}\right]\\ &= 4.\end{aligned}$$

Note that it is unnecessary to include a constant K as part of f, because it is canceled by subtraction.

EXAMPLE 3 Using the Fundamental Theorem of Line Integrals

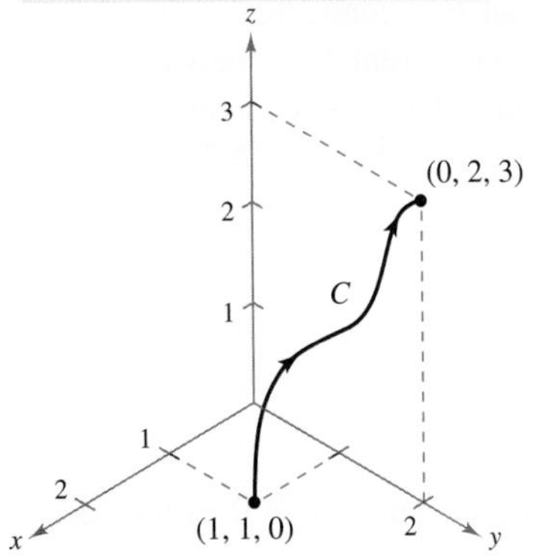

Using the Fundamental Theorem of Line Integrals, $\int_C \mathbf{F} \cdot d\mathbf{r}$.
Figure 15.21

Evaluate $\int_C \mathbf{F} \cdot d\mathbf{r}$, where C is a piecewise smooth curve from $(1, 1, 0)$ to $(0, 2, 3)$ and

$$\mathbf{F}(x, y, z) = 2xy\mathbf{i} + (x^2 + z^2)\mathbf{j} + 2yz\mathbf{k}$$

as shown in Figure 15.21.

Solution From Example 8 in Section 15.1, you know that $\mathbf{F}$ is the gradient of f where $f(x, y, z) = x^2y + yz^2 + K$. Consequently, $\mathbf{F}$ is conservative, and by the Fundamental Theorem of Line Integrals, it follows that

$$\begin{aligned}\int_C \mathbf{F} \cdot d\mathbf{r} &= f(0, 2, 3) - f(1, 1, 0)\\ &= [(0)^2(2) + (2)(3)^2] - [(1)^2(1) + (1)(0)^2]\\ &= 17.\end{aligned}$$

In Examples 2 and 3, be sure you see that the value of the line integral is the same for any smooth curve C that has the given initial and terminal points. For instance, in Example 3, try evaluating the line integral for the curve given by

$$\mathbf{r}(t) = (1 - t)\mathbf{i} + (1 + t)\mathbf{j} + 3t\mathbf{k}.$$

You should obtain

$$\begin{aligned}\int_C \mathbf{F} \cdot d\mathbf{r} &= \int_0^1 (30t^2 + 16t - 1)\, dt\\ &= 17.\end{aligned}$$

Independence of Path

From the Fundamental Theorem of Line Integrals it is clear that if **F** is continuous and conservative in an open region R, the value of $\int_C \mathbf{F} \cdot d\mathbf{r}$ is the same for every piecewise smooth curve C from one fixed point in R to another fixed point in R. This result is described by saying that the line integral $\int_C \mathbf{F} \cdot d\mathbf{r}$ is **independent of path** in the region R.

A region in the plane (or in space) is **connected** if any two points in the region can be joined by a piecewise smooth curve lying entirely within the region, as shown in Figure 15.22. In open regions that are *connected*, the path independence of $\int_C \mathbf{F} \cdot d\mathbf{r}$ is equivalent to the condition that **F** is conservative.

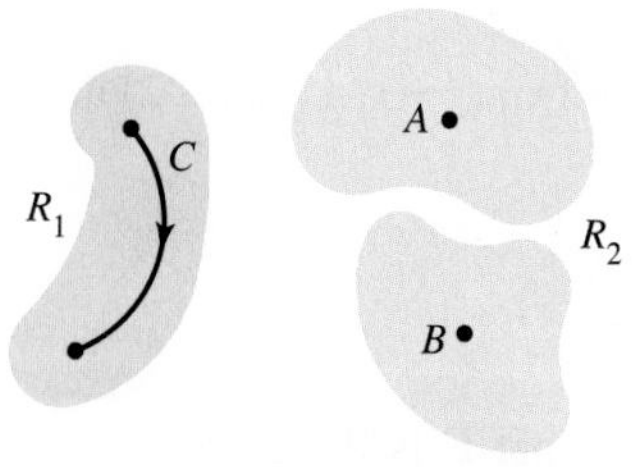

R_1 is connected. R_2 is not connected.

Figure 15.22

THEOREM 15.6 Independence of Path and Conservative Vector Fields

If **F** is continuous on an open connected region, then the line integral

$$\int_C \mathbf{F} \cdot d\mathbf{r}$$

is independent of path if and only if **F** is conservative.

Proof If **F** is conservative, then, by the Fundamental Theorem of Line Integrals, the line integral is independent of path. Now establish the converse for a plane region R. Let $\mathbf{F}(x, y) = M\mathbf{i} + N\mathbf{j}$, and let (x_0, y_0) be a fixed point in R. If (x, y) is any point in R, choose a piecewise smooth curve C running from (x_0, y_0) to (x, y), and define f by

$$f(x, y) = \int_C \mathbf{F} \cdot d\mathbf{r} = \int_C M\,dx + N\,dy.$$

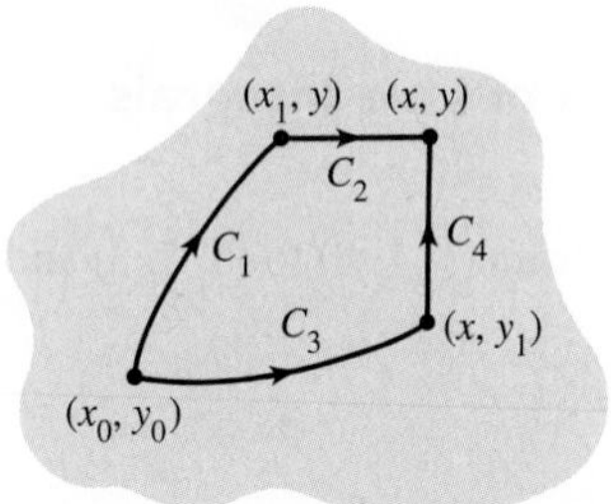

Figure 15.23

The existence of C in R is guaranteed by the fact that R is connected. You can show that f is a potential function of **F** by considering two different paths between (x_0, y_0) and (x, y). For the *first* path, choose (x_1, y) in R such that $x \neq x_1$. This is possible because R is open. Then choose C_1 and C_2, as shown in Figure 15.23. Using the independence of path, it follows that

$$\begin{aligned} f(x, y) &= \int_C M\,dx + N\,dy \\ &= \int_{C_1} M\,dx + N\,dy + \int_{C_2} M\,dx + N\,dy. \end{aligned}$$

Because the first integral does not depend on x, and because $dy = 0$ in the second integral, you have

$$f(x, y) = g(y) + \int_{C_2} M\,dx$$

and it follows that the partial derivative of f with respect to x is $f_x(x, y) = M$. For the *second* path, choose a point (x, y_1). Using reasoning similar to that used for the first path, you can conclude that $f_y(x, y) = N$. Therefore,

$$\begin{aligned} \nabla f(x, y) &= f_x(x, y)\mathbf{i} + f_y(x, y)\mathbf{j} \\ &= M\mathbf{i} + N\mathbf{j} \\ &= \mathbf{F}(x, y) \end{aligned}$$

and it follows that **F** is conservative.

EXAMPLE 4 Finding Work in a Conservative Force Field

For the force field given by

$$\mathbf{F}(x, y, z) = e^x \cos y\mathbf{i} - e^x \sin y\mathbf{j} + 2\mathbf{k}$$

show that $\int_C \mathbf{F} \cdot d\mathbf{r}$ is independent of path, and calculate the work done by $\mathbf{F}$ on an object moving along a curve C from $(0, \pi/2, 1)$ to $(1, \pi, 3)$.

Solution Writing the force field in the form $\mathbf{F}(x, y, z) = M\mathbf{i} + N\mathbf{j} + P\mathbf{k}$, you have $M = e^x \cos y$, $N = -e^x \sin y$, and $P = 2$, and it follows that

$$\frac{\partial P}{\partial y} = 0 = \frac{\partial N}{\partial z}$$
$$\frac{\partial P}{\partial x} = 0 = \frac{\partial M}{\partial z}$$
$$\frac{\partial N}{\partial x} = -e^x \sin y = \frac{\partial M}{\partial y}.$$

So, $\mathbf{F}$ is conservative. If f is a potential function of $\mathbf{F}$, then

$$f_x(x, y, z) = e^x \cos y$$
$$f_y(x, y, z) = -e^x \sin y$$
$$f_z(x, y, z) = 2.$$

By integrating with respect to x, y, and z separately, you obtain

$$f(x, y, z) = \int f_x(x, y, z)\, dx = \int e^x \cos y\, dx = e^x \cos y + g(y, z)$$
$$f(x, y, z) = \int f_y(x, y, z)\, dy = \int -e^x \sin y\, dy = e^x \cos y + h(x, z)$$
$$f(x, y, z) = \int f_z(x, y, z)\, dz = \int 2\, dz = 2z + k(x, y).$$

By comparing these three versions of $f(x, y, z)$, you can conclude that

$$f(x, y, z) = e^x \cos y + 2z + K.$$

Therefore, the work done by $\mathbf{F}$ along *any* curve C from $(0, \pi/2, 1)$ to $(1, \pi, 3)$ is

$$\begin{aligned} W &= \int_C \mathbf{F} \cdot d\mathbf{r} \\ &= \Big[e^x \cos y + 2z\Big]_{(0,\, \pi/2,\, 1)}^{(1,\, \pi,\, 3)} \\ &= (-e + 6) - (0 + 2) \\ &= 4 - e. \end{aligned}$$

How much work would be done if the object in Example 4 moved from the point $(0, \pi/2, 1)$ to $(1, \pi, 3)$ and then back to the starting point $(0, \pi/2, 1)$? The Fundamental Theorem of Line Integrals states that there is zero work done. Remember that, by definition, work can be negative. So, by the time the object gets back to its starting point, the amount of work that registers positively is canceled out by the amount of work that registers negatively.

A curve C given by $\mathbf{r}(t)$ for $a \le t \le b$ is **closed** if $\mathbf{r}(a) = \mathbf{r}(b)$. By the Fundamental Theorem of Line Integrals, you can conclude that if $\mathbf{F}$ is continuous and conservative on an open region R, then the line integral over every closed curve C is 0.

NOTE Theorem 15.7 gives you options for evaluating a line integral involving a conservative vector field. You can use a potential function, or it might be more convenient to choose a particularly simple path, such as a straight line.

THEOREM 15.7 Equivalent Conditions

Let $\mathbf{F}(x, y, z) = M\mathbf{i} + N\mathbf{j} + P\mathbf{k}$ have continuous first partial derivatives in an open connected region R, and let C be a piecewise smooth curve in R. The following conditions are equivalent.

1. $\mathbf{F}$ is conservative. That is, $\mathbf{F} = \nabla f$ for some function f.
2. $\displaystyle\int_C \mathbf{F} \cdot d\mathbf{r}$ is independent of path.
3. $\displaystyle\int_C \mathbf{F} \cdot d\mathbf{r} = 0$ for every *closed* curve C in R.

EXAMPLE 5 Evaluating a Line Integral

Evaluate $\displaystyle\int_{C_1} \mathbf{F} \cdot d\mathbf{r}$, where

$$\mathbf{F}(x, y) = (y^3 + 1)\mathbf{i} + (3xy^2 + 1)\mathbf{j}$$

and C_1 is the semicircular path from (0, 0) to (2, 0), as shown in Figure 15.24.

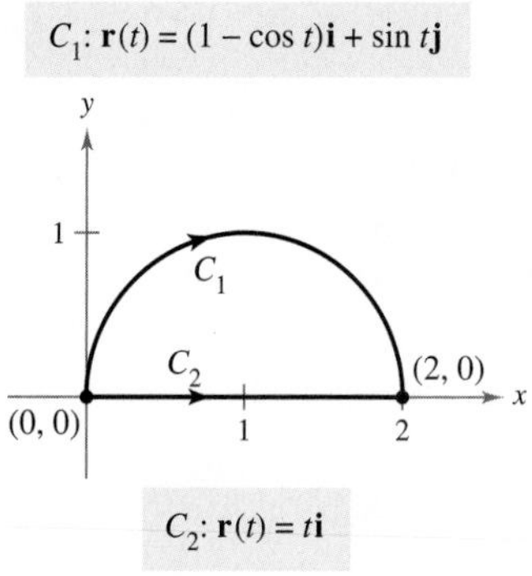

Figure 15.24

Solution You have the following three options.

a. You can use the method presented in the preceding section to evaluate the line integral along the *given curve*. To do this, you can use the parametrization $\mathbf{r}(t) = (1 - \cos t)\mathbf{i} + \sin t\mathbf{j}$, where $0 \le t \le \pi$. For this parametrization, it follows that $d\mathbf{r} = \mathbf{r}'(t)\,dt = (\sin t\mathbf{i} + \cos t\mathbf{j})\,dt$, and

$$\int_{C_1} \mathbf{F} \cdot d\mathbf{r} = \int_0^{\pi} (\sin t + \sin^4 t + \cos t + 3\sin^2 t\cos t - 3\sin^2 t\cos^2 t)\,dt.$$

This integral should dampen your enthusiasm for this option.

b. You can try to find a *potential function* and evaluate the line integral by the Fundamental Theorem of Line Integrals. Using the technique demonstrated in Example 4, you can find the potential function to be $f(x, y) = xy^3 + x + y + K$, and, by the Fundamental Theorem,

$$W = \int_{C_1} \mathbf{F} \cdot d\mathbf{r} = f(2, 0) - f(0, 0) = 2.$$

c. Knowing that $\mathbf{F}$ is conservative, you have a third option. Because the value of the line integral is independent of path, you can replace the semicircular path with a *simpler path*. Suppose you choose the straight-line path C_2 from (0, 0) to (2, 0). Then, $\mathbf{r}(t) = t\mathbf{i}$, where $0 \le t \le 2$. So, $d\mathbf{r} = \mathbf{i}\,dt$ and $\mathbf{F}(x, y) = (y^3 + 1)\mathbf{i} + (3xy^2 + 1)\mathbf{j} = \mathbf{i} + \mathbf{j}$, so that

$$\int_{C_1} \mathbf{F} \cdot d\mathbf{r} = \int_{C_2} \mathbf{F} \cdot d\mathbf{r} = \int_0^2 1\,dt = t\Big]_0^2 = 2.$$

Of the three options, obviously the third one is the easiest.

Conservation of Energy

In 1840, the English physicist Michael Faraday wrote, "Nowhere is there a pure creation or production of power without a corresponding exhaustion of something to supply it." This statement represents the first formulation of one of the most important laws of physics—**the Law of Conservation of Energy.** In modern terminology, the law is stated as follows: *In a conservative force field, the sum of the potential and kinetic energies of an object remains constant from point to point.*

You can use the Fundamental Theorem of Line Integrals to derive this law. From physics, the **kinetic energy** of a particle of mass m and speed v is $k = \frac{1}{2}mv^2$. The **potential energy** p of a particle at point (x, y, z) in a conservative vector field $\mathbf{F}$ is defined as $p(x, y, z) = -f(x, y, z)$, where f is the potential function for $\mathbf{F}$. Consequently, the work done by $\mathbf{F}$ along a smooth curve C from A to B is

$$\begin{aligned} W = \int_C \mathbf{F} \cdot d\mathbf{r} &= f(x, y, z)\Big]_A^B \\ &= -p(x, y, z)\Big]_A^B \\ &= p(A) - p(B) \end{aligned}$$

as shown in Figure 15.25. In other words, work W is equal to the difference in the potential energies of A and B. Now, suppose that $\mathbf{r}(t)$ is the position vector for a particle moving along C from $A = \mathbf{r}(a)$ to $B = \mathbf{r}(b)$. At any time t, the particle's velocity, acceleration, and speed are $\mathbf{v}(t) = \mathbf{r}'(t)$, $\mathbf{a}(t) = \mathbf{r}''(t)$, and $v(t) = \|\mathbf{v}(t)\|$, respectively. So, by Newton's Second Law of Motion, $\mathbf{F} = m\mathbf{a}(t) = m(\mathbf{v}'(t))$, and the work done by $\mathbf{F}$ is

$$\begin{aligned} W = \int_C \mathbf{F} \cdot d\mathbf{r} &= \int_a^b \mathbf{F} \cdot \mathbf{r}'(t)\,dt \\ &= \int_a^b \mathbf{F} \cdot \mathbf{v}(t)\,dt = \int_a^b [m\mathbf{v}'(t)] \cdot \mathbf{v}(t)\,dt \\ &= \int_a^b m[\mathbf{v}'(t) \cdot \mathbf{v}(t)]\,dt \\ &= \frac{m}{2}\int_a^b \frac{d}{dt}[\mathbf{v}(t) \cdot \mathbf{v}(t)]\,dt \\ &= \frac{m}{2}\int_a^b \frac{d}{dt}[\|\mathbf{v}(t)\|^2]\,dt \\ &= \frac{m}{2}\Big[\|\mathbf{v}(t)\|^2\Big]_a^b \\ &= \frac{m}{2}\Big[[v(t)]^2\Big]_a^b \\ &= \frac{1}{2}m[v(b)]^2 - \frac{1}{2}m[v(a)]^2 \\ &= k(B) - k(A). \end{aligned}$$

Equating these two results for W produces

$$\begin{aligned} p(A) - p(B) &= k(B) - k(A) \\ p(A) + k(A) &= p(B) + k(B) \end{aligned}$$

which implies that the sum of the potential and kinetic energies remains constant from point to point.

MICHAEL FARADAY (1791–1867)

Several philosophers of science have considered Faraday's Law of Conservation of Energy to be the greatest generalization ever conceived by humankind. Many physicists have contributed to our knowledge of this law. Two early and influential ones were James Prescott Joule (1818–1889) and Hermann Ludwig Helmholtz (1821–1894).

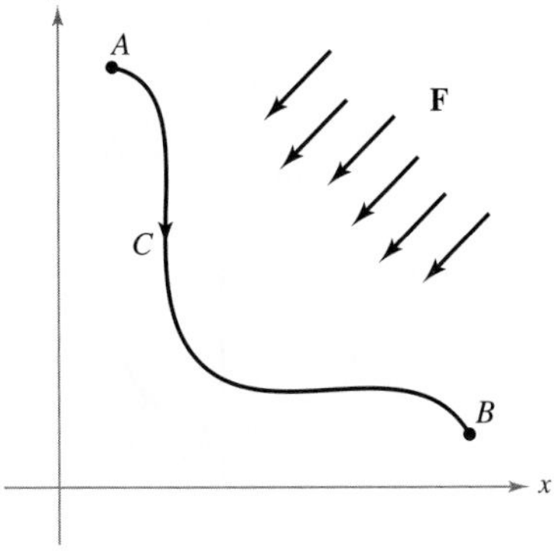

The work done by **F** along C is
$W = \int_C \mathbf{F} \cdot d\mathbf{r} = p(A) - p(B).$
Figure 15.25

Exercises for Section 15.3

See www.CalcChat.com for worked-out solutions to odd-numbered exercises.

In Exercises 1–4, show that the value of $\int_C \mathbf{F} \cdot d\mathbf{r}$ is the same for each parametric representation of C.

1. $\mathbf{F}(x, y) = x^2\mathbf{i} + xy\mathbf{j}$
   (a) $\mathbf{r}_1(t) = t\mathbf{i} + t^2\mathbf{j}, \quad 0 \le t \le 1$
   (b) $\mathbf{r}_2(\theta) = \sin\theta\mathbf{i} + \sin^2\theta\mathbf{j}, \quad 0 \le \theta \le \frac{\pi}{2}$
2. $\mathbf{F}(x, y) = (x^2 + y^2)\mathbf{i} - x\mathbf{j}$
   (a) $\mathbf{r}_1(t) = t\mathbf{i} + \sqrt{t}\mathbf{j}, \quad 0 \le t \le 4$
   (b) $\mathbf{r}_2(w) = w^2\mathbf{i} + w\mathbf{j}, \quad 0 \le w \le 2$
3. $\mathbf{F}(x, y) = y\mathbf{i} - x\mathbf{j}$
   (a) $\mathbf{r}_1(\theta) = \sec\theta\mathbf{i} + \tan\theta\mathbf{j}, \quad 0 \le \theta \le \frac{\pi}{3}$
   (b) $\mathbf{r}_2(t) = \sqrt{t+1}\mathbf{i} + \sqrt{t}\mathbf{j}, \quad 0 \le t \le 3$
4. $\mathbf{F}(x, y) = y\mathbf{i} + x^2\mathbf{j}$
   (a) $\mathbf{r}_1(t) = (2 + t)\mathbf{i} + (3 - t)\mathbf{j}, \quad 0 \le t \le 3$
   (b) $\mathbf{r}_2(w) = (2 + \ln w)\mathbf{i} + (3 - \ln w)\mathbf{j}, \quad 1 \le w \le e^3$

In Exercises 5–10, determine whether or not the vector field is conservative.

5. $\mathbf{F}(x, y) = e^x(\sin y\mathbf{i} + \cos y\mathbf{j})$
6. $\mathbf{F}(x, y) = 15x^2y^2\mathbf{i} + 10x^3y\mathbf{j}$
7. $\mathbf{F}(x, y) = \frac{1}{y^2}(y\mathbf{i} + x\mathbf{j})$
8. $\mathbf{F}(x, y, z) = y\ln z\mathbf{i} - x\ln z\mathbf{j} + \frac{xy}{z}\mathbf{k}$
9. $\mathbf{F}(x, y, z) = y^2z\mathbf{i} + 2xyz\mathbf{j} + xy^2\mathbf{k}$
10. $\mathbf{F}(x, y, z) = \sin yz\mathbf{i} + xz\cos yz\mathbf{j} + xy\sin yz\mathbf{k}$

In Exercises 11–24, find the value of the line integral

$$\int_C \mathbf{F} \cdot d\mathbf{r}.$$

(*Hint:* If F is conservative, the integration may be easier on an alternative path.)

11. $\mathbf{F}(x, y) = 2xy\mathbf{i} + x^2\mathbf{j}$
    (a) $\mathbf{r}_1(t) = t\mathbf{i} + t^2\mathbf{j}, \quad 0 \le t \le 1$
    (b) $\mathbf{r}_2(t) = t\mathbf{i} + t^3\mathbf{j}, \quad 0 \le t \le 1$
12. $\mathbf{F}(x, y) = ye^{xy}\mathbf{i} + xe^{xy}\mathbf{j}$
    (a) $\mathbf{r}_1(t) = t\mathbf{i} - (t - 3)\mathbf{j}, \quad 0 \le t \le 3$
    (b) The closed path consisting of line segments from $(0, 3)$ to $(0, 0)$, and then from $(0, 0)$ to $(3, 0)$
13. $\mathbf{F}(x, y) = y\mathbf{i} - x\mathbf{j}$
    (a) $\mathbf{r}_1(t) = t\mathbf{i} + t\mathbf{j}, \quad 0 \le t \le 1$
    (b) $\mathbf{r}_2(t) = t\mathbf{i} + t^2\mathbf{j}, \quad 0 \le t \le 1$
    (c) $\mathbf{r}_3(t) = t\mathbf{i} + t^3\mathbf{j}, \quad 0 \le t \le 1$
14. $\mathbf{F}(x, y) = xy^2\mathbf{i} + 2x^2y\mathbf{j}$
    (a) $\mathbf{r}_1(t) = t\mathbf{i} + \frac{1}{t}\mathbf{j}, \quad 1 \le t \le 3$
    (b) $\mathbf{r}_2(t) = (t + 1)\mathbf{i} - \frac{1}{3}(t - 3)\mathbf{j}, \quad 0 \le t \le 2$
15. $\int_C y^2\,dx + 2xy\,dy$

(a)
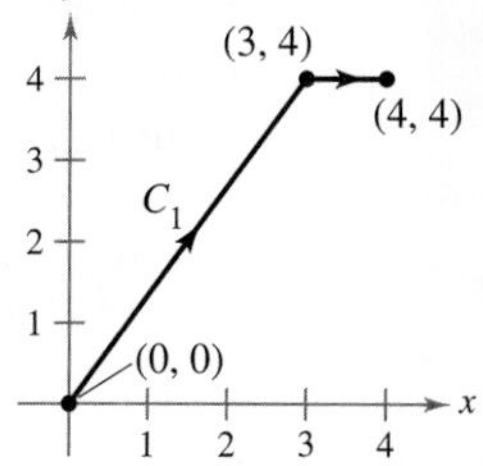

(b)
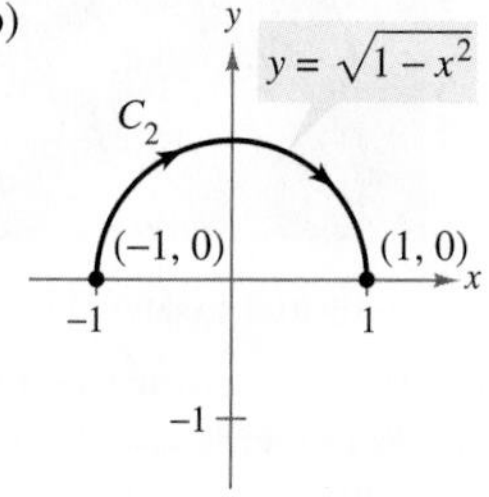

(c)
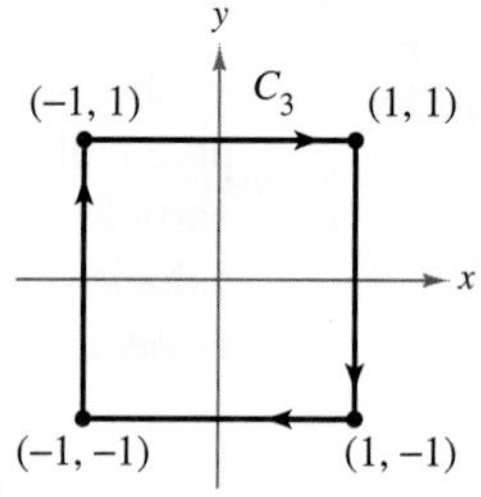

(d)
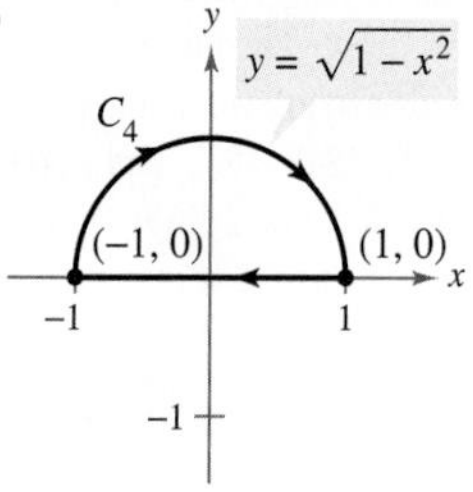

16. $\int_C (2x - 3y + 1)\,dx - (3x + y - 5)\,dy$

(a)
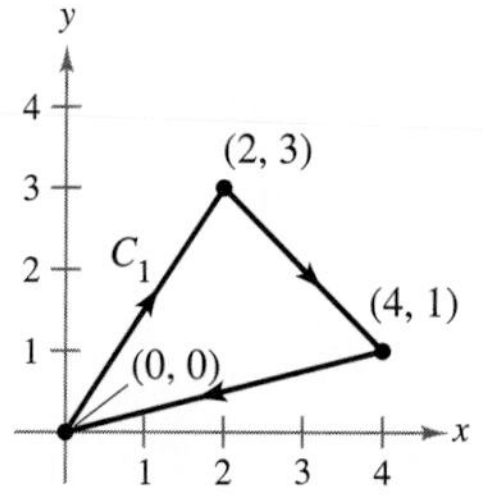

(b)
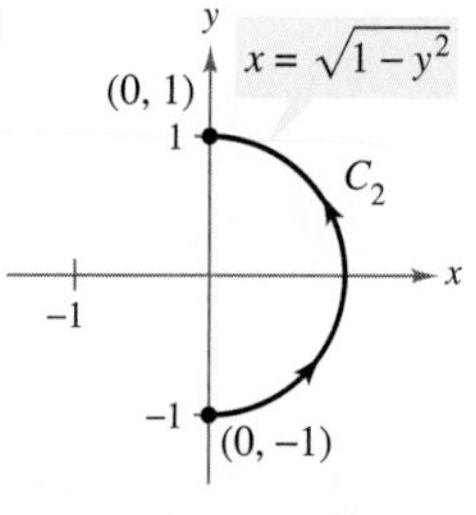

(c)
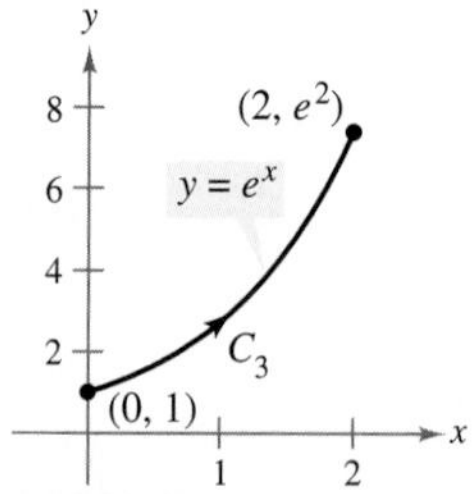

(d)
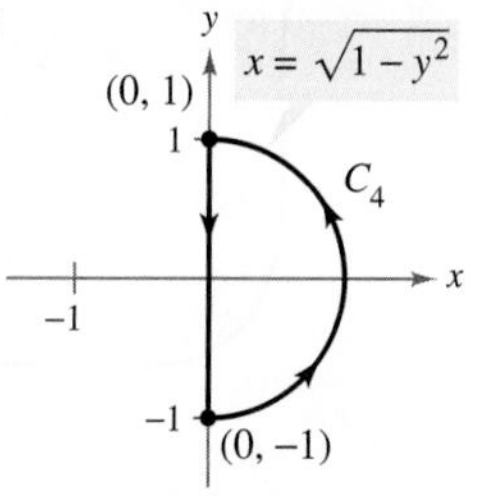

17. $\int_C 2xy\,dx + (x^2 + y^2)\,dy$
    (a) C: ellipse $\frac{x^2}{25} + \frac{y^2}{16} = 1$ from $(5, 0)$ to $(0, 4)$
    (b) C: parabola $y = 4 - x^2$ from $(2, 0)$ to $(0, 4)$

18. $\int_C (x^2 + y^2)\,dx + 2xy\,dy$

(a) $\mathbf{r}_1(t) = t^3\mathbf{i} + t^2\mathbf{j}, \quad 0 \le t \le 2$

(b) $\mathbf{r}_2(t) = 2\cos t\mathbf{i} + 2\sin t\mathbf{j}, \quad 0 \le t \le \frac{\pi}{2}$

19. $\mathbf{F}(x, y, z) = yz\mathbf{i} + xz\mathbf{j} + xy\mathbf{k}$

(a) $\mathbf{r}_1(t) = t\mathbf{i} + 2\mathbf{j} + t\mathbf{k}, \quad 0 \le t \le 4$

(b) $\mathbf{r}_2(t) = t^2\mathbf{i} + t\mathbf{j} + t^2\mathbf{k}, \quad 0 \le t \le 2$

20. $\mathbf{F}(x, y, z) = \mathbf{i} + z\mathbf{j} + y\mathbf{k}$

(a) $\mathbf{r}_1(t) = \cos t\mathbf{i} + \sin t\mathbf{j} + t^2\mathbf{k}, \quad 0 \le t \le \pi$

(b) $\mathbf{r}_2(t) = (1 - 2t)\mathbf{i} + \pi^2 t\mathbf{k}, \quad 0 \le t \le 1$

21. $\mathbf{F}(x, y, z) = (2y + x)\mathbf{i} + (x^2 - z)\mathbf{j} + (2y - 4z)\mathbf{k}$

(a) $\mathbf{r}_1(t) = t\mathbf{i} + t^2\mathbf{j} + \mathbf{k}, \quad 0 \le t \le 1$

(b) $\mathbf{r}_2(t) = t\mathbf{i} + t\mathbf{j} + (2t - 1)^2\mathbf{k}, \quad 0 \le t \le 1$

22. $\mathbf{F}(x, y, z) = -y\mathbf{i} + x\mathbf{j} + 3xz^2\mathbf{k}$

(a) $\mathbf{r}_1(t) = \cos t\mathbf{i} + \sin t\mathbf{j} + t\mathbf{k}, \quad 0 \le t \le \pi$

(b) $\mathbf{r}_2(t) = (1 - 2t)\mathbf{i} + \pi t\mathbf{k}, \quad 0 \le t \le 1$

23. $\mathbf{F}(x, y, z) = e^z(y\mathbf{i} + x\mathbf{j} + xy\mathbf{k})$

(a) $\mathbf{r}_1(t) = 4\cos t\mathbf{i} + 4\sin t\mathbf{j} + 3\mathbf{k}, \quad 0 \le t \le \pi$

(b) $\mathbf{r}_2(t) = (4 - 8t)\mathbf{i} + 3\mathbf{k}, \quad 0 \le t \le 1$

24. $\mathbf{F}(x, y, z) = y\sin z\mathbf{i} + x\sin z\mathbf{j} + xy\cos x\mathbf{k}$

(a) $\mathbf{r}_1(t) = t^2\mathbf{i} + t^2\mathbf{j}, \quad 0 \le t \le 2$

(b) $\mathbf{r}_2(t) = 4t\mathbf{i} + 4t\mathbf{j}, \quad 0 \le t \le 1$

In Exercises 25–34, evaluate the line integral using the Fundamental Theorem of Line Integrals. Use a computer algebra system to verify your results.

25. $\int_C (y\mathbf{i} + x\mathbf{j}) \cdot d\mathbf{r}$

C: smooth curve from $(0, 0)$ to $(3, 8)$

26. $\int_C [2(x + y)\mathbf{i} + 2(x + y)\mathbf{j}] \cdot d\mathbf{r}$

C: smooth curve from $(-2, 2)$ to $(4, 3)$

27. $\int_C \cos x \sin y\,dx + \sin x \cos y\,dy$

C: smooth curve from $(0, -\pi)$ to $\left(\frac{3\pi}{2}, \frac{\pi}{2}\right)$

28. $\int_C \frac{y\,dx - x\,dy}{x^2 + y^2}$

C: smooth curve from $(1, 1)$ to $(2\sqrt{3}, 2)$

29. $\int_C e^x \sin y\,dx + e^x \cos y\,dy$

C: cycloid $x = \theta - \sin\theta$, $y = 1 - \cos\theta$ from $(0, 0)$ to $(2\pi, 0)$

30. $\int_C \frac{2x}{(x^2 + y^2)^2}\,dx + \frac{2y}{(x^2 + y^2)^2}\,dy$

C: circle $(x - 4)^2 + (y - 5)^2 = 9$ clockwise from $(7, 5)$ to $(1, 5)$

31. $\int_C (y + 2z)\,dx + (x - 3z)\,dy + (2x - 3y)\,dz$

(a) C: line segment from $(0, 0, 0)$ to $(1, 1, 1)$

(b) C: line segments from $(0, 0, 0)$ to $(0, 0, 1)$ to $(1, 1, 1)$

(c) C: line segments from $(0, 0, 0)$ to $(1, 0, 0)$ to $(1, 1, 0)$ to $(1, 1, 1)$

32. Repeat Exercise 31 using the integral

$$\int_C zy\,dx + xz\,dy + xy\,dz.$$

33. $\int_C -\sin x\,dx + z\,dy + y\,dz$

C: smooth curve from $(0, 0, 0)$ to $\left(\frac{\pi}{2}, 3, 4\right)$

34. $\int_C 6x\,dx - 4z\,dy - (4y - 20z)\,dz$

C: smooth curve from $(0, 0, 0)$ to $(4, 3, 1)$

Work **In Exercises 35 and 36, find the work done by the force field F in moving an object from P to Q.**

35. $\mathbf{F}(x, y) = 9x^2y^2\mathbf{i} + (6x^3y - 1)\mathbf{j}$; $P(0, 0)$, $Q(5, 9)$

36. $\mathbf{F}(x, y) = \frac{2x}{y}\mathbf{i} - \frac{x^2}{y^2}\mathbf{j}$; $P(-3, 2)$, $Q(1, 4)$

37. ***Work*** A stone weighing 1 pound is attached to the end of a two-foot string and is whirled horizontally with one end held fixed. It makes 1 revolution per second. Find the work done by the force **F** that keeps the stone moving in a circular path. [*Hint:* Use Force = (mass)(centripetal acceleration).]

38. ***Work*** If $\mathbf{F}(x, y, z) = a_1\mathbf{i} + a_2\mathbf{j} + a_3\mathbf{k}$ is a constant force vector field, show that the work done in moving a particle along any path from P to Q is $W = \mathbf{F} \cdot \overrightarrow{PQ}$.

39. ***Work*** To allow a means of escape for workers in a hazardous job 50 meters above ground level, a slide wire is installed. It runs from their position to a point on the ground 50 meters from the base of the installation where they are located. Show that the work done by the gravitational force field for a 150-pound man moving the length of the slide wire is the same for each path.

(a) $\mathbf{r}(t) = t\mathbf{i} + (50 - t)\mathbf{j}$

(b) $\mathbf{r}(t) = t\mathbf{i} + \frac{1}{50}(50 - t)^2\mathbf{j}$

40. ***Work*** Can you find a path for the slide wire in Exercise 39 such that the work done by the gravitational force field would differ from the amounts of work done for the two paths given? Explain why or why not.

Writing About Concepts

41. State the Fundamental Theorem of Line Integrals.

42. What does it mean that a line integral is independent of path? State the method for determining if a line integral is independent of path.

Writing About Concepts *(continued)*

43. Consider the force field shown in the figure.

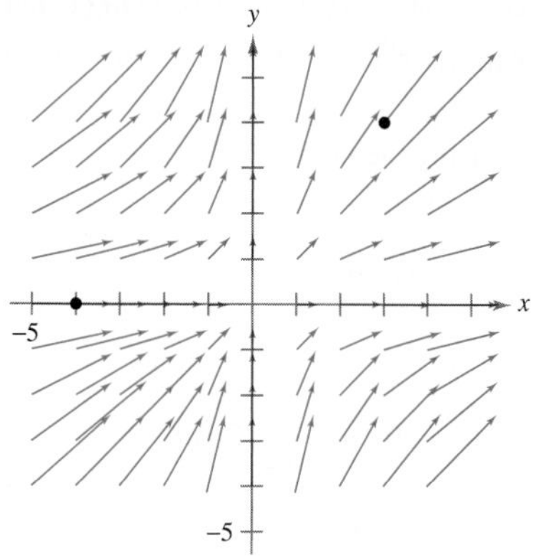

(a) Give a verbal argument that the force field is not conservative because you can identify two paths that require different amounts of work to move an object from $(-4, 0)$ to $(3, 4)$. Identify two paths and state which requires the greater amount of work. To print an enlarged copy of the graph, go to the website *www.mathgraphs.com*.

(b) Give a verbal argument that the force field is not conservative because you can find a closed curve C such that

$$\int_C \mathbf{F} \cdot d\mathbf{r} \neq 0.$$

44. ***Wind Speed and Direction*** The map shows the jet stream wind speed vectors over the United States for March 19, 2004. In planning a flight from Dallas to Atlanta in a small plane at an altitude of 5000 feet, is the amount of fuel required independent of the flight path? Is the vector field conservative? Explain.

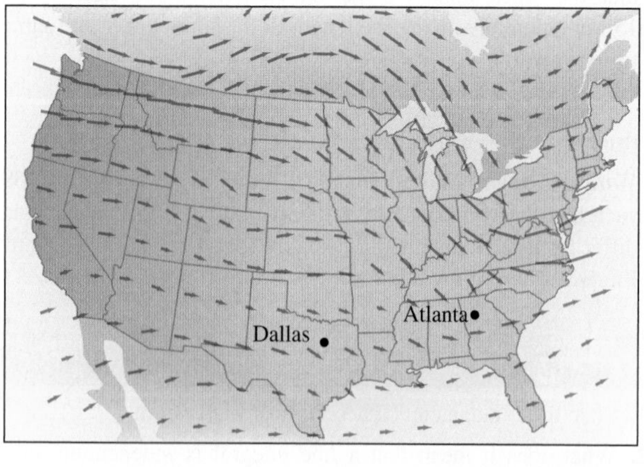

In Exercises 45 and 46, consider the force field shown in the figure. Is the force field conservative? Explain why or why not.

45.

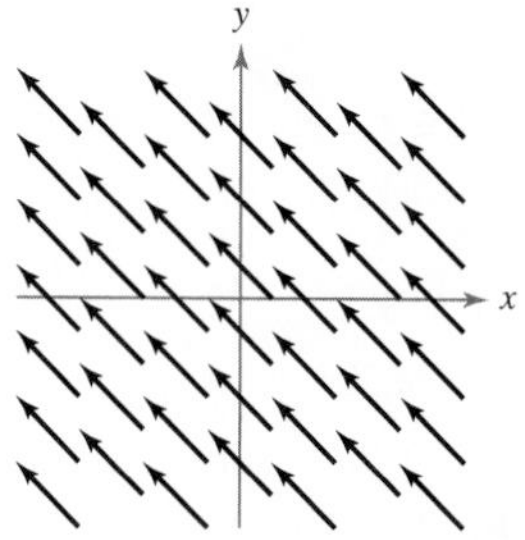

46.

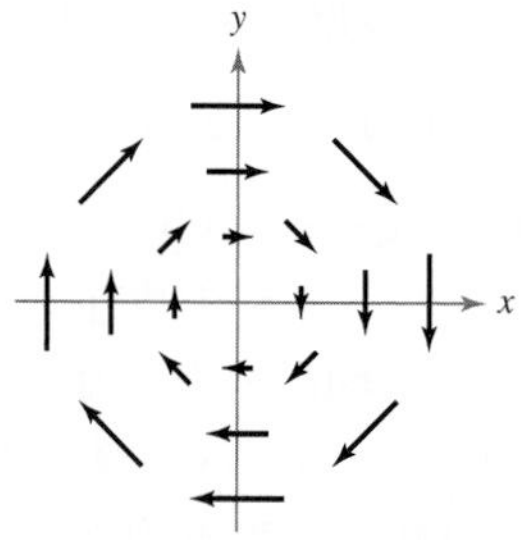

True or False? **In Exercises 47–50, determine whether the statement is true or false. If it is false, explain why or give an example that shows it is false.**

47. If C_1, C_2, and C_3 have the same initial and terminal points and $\int_{C_1} \mathbf{F} \cdot d\mathbf{r}_1 = \int_{C_2} \mathbf{F} \cdot d\mathbf{r}_2$, then $\int_{C_1} \mathbf{F} \cdot d\mathbf{r}_1 = \int_{C_3} \mathbf{F} \cdot d\mathbf{r}_3$.

48. If $\mathbf{F} = y\mathbf{i} + x\mathbf{j}$ and C is given by $\mathbf{r}(t) = (4 \sin t)\mathbf{i} + (3 \cos t)\mathbf{j}$, $0 \leq t \leq \pi$, then $\int_C \mathbf{F} \cdot d\mathbf{r} = 0$.

49. If $\mathbf{F}$ is conservative in a region R bounded by a simple closed path and C lies within R, then $\int_C \mathbf{F} \cdot d\mathbf{r}$ is independent of path.

50. If $\mathbf{F} = M\mathbf{i} + N\mathbf{j}$ and $\partial M/\partial x = \partial N/\partial y$, then $\mathbf{F}$ is conservative.

51. A function f is called harmonic if $\dfrac{\partial^2 f}{\partial x^2} + \dfrac{\partial^2 f}{\partial y^2} = 0$. Prove that if f is harmonic, then

$$\int_C \left(\frac{\partial f}{\partial y}\,dx - \frac{\partial f}{\partial x}\,dy\right) = 0$$

where C is a smooth closed curve in the plane.

52. ***Kinetic and Potential Energy*** The kinetic energy of an object moving through a conservative force field is decreasing at a rate of 10 units per minute. At what rate is the potential energy changing?

53. Let $\mathbf{F}(x, y) = \dfrac{y}{x^2 + y^2}\mathbf{i} - \dfrac{x}{x^2 + y^2}\mathbf{j}$.

(a) Show that

$$\frac{\partial N}{\partial x} = \frac{\partial M}{\partial y}$$

where

$$M = \frac{y}{x^2 + y^2} \text{ and } N = \frac{-x}{x^2 + y^2}.$$

(b) If $\mathbf{r}(t) = \cos t\mathbf{i} + \sin t\mathbf{j}$ for $0 \leq t \leq \pi$, find $\int_C \mathbf{F} \cdot d\mathbf{r}$.

(c) If $\mathbf{r}(t) = \cos t\mathbf{i} - \sin t\mathbf{j}$ for $0 \leq t \leq \pi$, find $\int_C \mathbf{F} \cdot d\mathbf{r}$.

(d) If $\mathbf{r}(t) = \cos t\mathbf{i} + \sin t\mathbf{j}$ for $0 \leq t \leq 2\pi$, find $\int_C \mathbf{F} \cdot d\mathbf{r}$. Why doesn't this contradict Theorem 15.7?

(e) Show that $\nabla\left(\arctan \dfrac{x}{y}\right) = \mathbf{F}$.

Section 15.4 Green's Theorem

- **Use Green's Theorem to evaluate a line integral.**
- **Use alternative forms of Green's Theorem.**

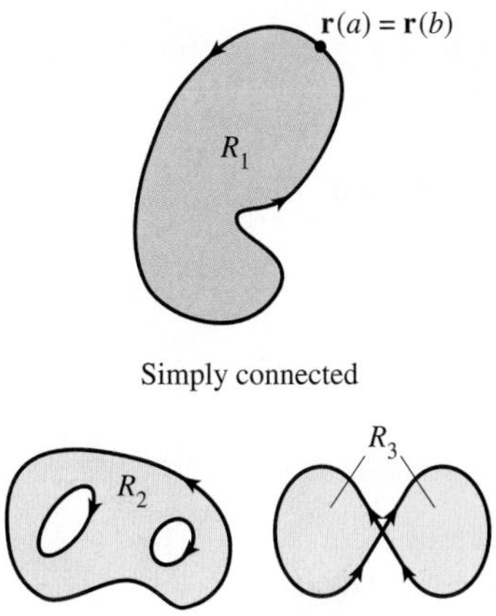

Simply connected

Not simply connected

Figure 15.26

Green's Theorem

In this section, you will study **Green's Theorem,** named after the English mathematician George Green (1793–1841). This theorem states that the value of a double integral over a *simply connected* plane region R is determined by the value of a line integral around the boundary of R.

A curve C given by $\mathbf{r}(t) = x(t)\mathbf{i} + y(t)\mathbf{j}$, where $a \le t \le b$, is **simple** if it does not cross itself—that is, $\mathbf{r}(c) \neq \mathbf{r}(d)$ for all c and d in the open interval (a, b). A plane region R is **simply connected** if its boundary consists of *one* simple closed curve, as shown in Figure 15.26.

> **THEOREM 15.8 Green's Theorem**
>
> Let R be a simply connected region with a piecewise smooth boundary C, oriented counterclockwise (that is, C is traversed *once* so that the region R always lies to the *left*). If M and N have continuous partial derivatives in an open region containing R, then
>
> $$\int_C M\,dx + N\,dy = \int_R\!\!\int \left(\frac{\partial N}{\partial x} - \frac{\partial M}{\partial y}\right) dA.$$

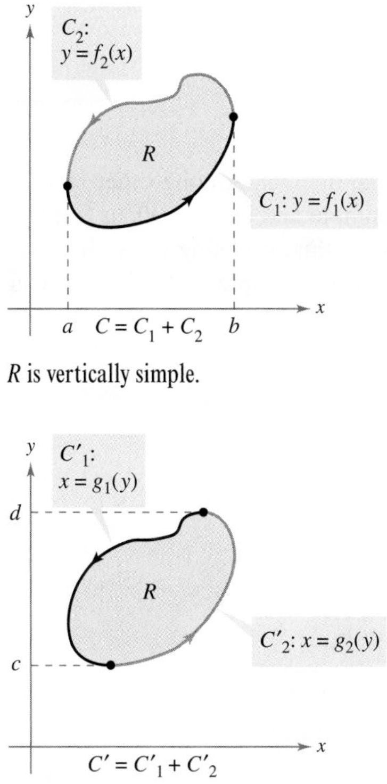

R is vertically simple.

R is horizontally simple.

Figure 15.27

Proof A proof is given only for a region that is both vertically simple and horizontally simple, as shown in Figure 15.27.

$$\begin{aligned}
\int_C M\,dx &= \int_{C_1} M\,dx + \int_{C_2} M\,dx \\
&= \int_a^b M(x, f_1(x))\,dx + \int_b^a M(x, f_2(x))\,dx \\
&= \int_a^b [M(x, f_1(x)) - M(x, f_2(x))]\,dx
\end{aligned}$$

On the other hand,

$$\begin{aligned}
\int_R\!\!\int \frac{\partial M}{\partial y}\,dA &= \int_a^b \int_{f_1(x)}^{f_2(x)} \frac{\partial M}{\partial y}\,dy\,dx \\
&= \int_a^b M(x, y)\Big]_{f_1(x)}^{f_2(x)}\,dx \\
&= \int_a^b [M(x, f_2(x)) - M(x, f_1(x))]\,dx.
\end{aligned}$$

Consequently,

$$\int_C M\,dx = -\int_R\!\!\int \frac{\partial M}{\partial y}\,dA.$$

Similarly, you can use $g_1(y)$ and $g_2(y)$ to show that $\int_C N\,dy = \int_R\int \partial N/\partial x\,dA$. By adding the integrals $\int_C M\,dx$ and $\int_C N\,dy$, you obtain the conclusion stated in the theorem.

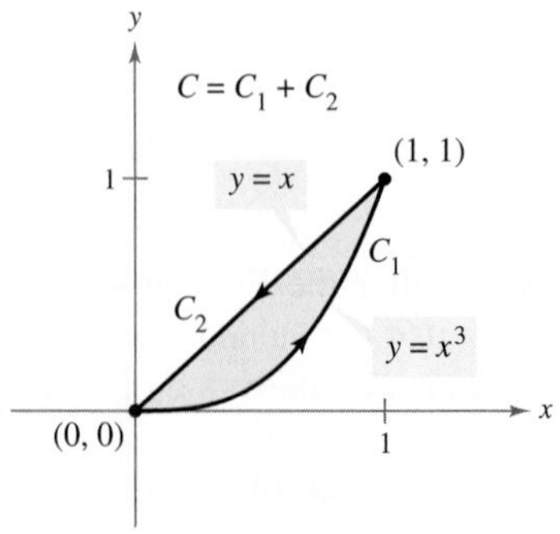

C is simple and closed, and the region R always lies to the left of C.
Figure 15.28

EXAMPLE 1 Using Green's Theorem

Use Green's Theorem to evaluate the line integral

$$\int_C y^3\,dx + (x^3 + 3xy^2)\,dy$$

where C is the path from $(0, 0)$ to $(1, 1)$ along the graph of $y = x^3$ and from $(1, 1)$ to $(0, 0)$ along the graph of $y = x$, as shown in Figure 15.28.

Solution Because $M = y^3$ and $N = x^3 + 3xy^2$, it follows that

$$\frac{\partial N}{\partial x} = 3x^2 + 3y^2 \quad \text{and} \quad \frac{\partial M}{\partial y} = 3y^2.$$

Applying Green's Theorem, you then have

$$\begin{aligned}
\int_C y^3\,dx + (x^3 + 3xy^2)\,dy &= \int_R\int \left(\frac{\partial N}{\partial x} - \frac{\partial M}{\partial y}\right) dA \\
&= \int_0^1 \int_{x^3}^{x} \left[(3x^2 + 3y^2) - 3y^2\right] dy\,dx \\
&= \int_0^1 \int_{x^3}^{x} 3x^2\,dy\,dx \\
&= \int_0^1 3x^2y \Big]_{x^3}^{x}\,dx \\
&= \int_0^1 (3x^3 - 3x^5)\,dx \\
&= \left[\frac{3x^4}{4} - \frac{x^6}{2}\right]_0^1 \\
&= \frac{1}{4}.
\end{aligned}$$

George Green (1793–1841)

Green, a self-educated miller's son, first published the theorem that bears his name in 1828 in an essay on electricity and magnetism. At that time there was almost no mathematical theory to explain electrical phenomena. "Considering how desirable it was that a power of universal agency, like electricity, should, as far as possible, be submitted to calculation, . . . I was induced to try whether it would be possible to discover any general relations existing between this function and the quantities of electricity in the bodies producing it."

Green's Theorem cannot be applied to every line integral. Among other restrictions stated in Theorem 15.8, the curve C must be simple and closed. When Green's Theorem does apply, however, it can save time. To see this, try using the techniques described in Section 15.2 to evaluate the line integral in Example 1. To do this, you would need to write the line integral as

$$\int_C y^3\,dx + (x^3 + 3xy^2)\,dy =$$
$$\int_{C_1} y^3\,dx + (x^3 + 3xy^2)\,dy + \int_{C_2} y^3\,dx + (x^3 + 3xy^2)\,dy$$

where C_1 is the cubic path given by

$$\mathbf{r}(t) = t\mathbf{i} + t^3\mathbf{j}$$

from $t = 0$ to $t = 1$, and C_2 is the line segment given by

$$\mathbf{r}(t) = (1 - t)\mathbf{i} + (1 - t)\mathbf{j}$$

from $t = 0$ to $t = 1$.

EXAMPLE 2 Using Green's Theorem to Calculate Work

While subject to the force

$$\mathbf{F}(x, y) = y^3\mathbf{i} + (x^3 + 3xy^2)\mathbf{j}$$

a particle travels once around the circle of radius 3 shown in Figure 15.29. Use Green's Theorem to find the work done by **F**.

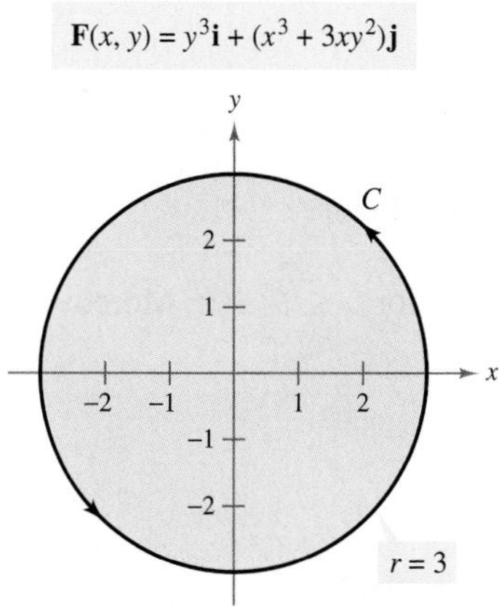

Figure 15.29

Solution From Example 1, you know by Green's Theorem that

$$\int_C y^3\,dx + (x^3 + 3xy^2)\,dy = \int_R\!\int 3x^2\,dA.$$

In polar coordinates, using $x = r\cos\theta$ and $dA = r\,dr\,d\theta$, the work done is

$$\begin{aligned} W = \int_R\!\int 3x^2\,dA &= \int_0^{2\pi}\!\int_0^3 3(r\cos\theta)^2\,r\,dr\,d\theta \\ &= 3\int_0^{2\pi}\!\int_0^3 r^3\cos^2\theta\,dr\,d\theta \\ &= 3\int_0^{2\pi} \frac{r^4}{4}\cos^2\theta\Big]_0^3\,d\theta \\ &= 3\int_0^{2\pi} \frac{81}{4}\cos^2\theta\,d\theta \\ &= \frac{243}{8}\int_0^{2\pi}(1 + \cos 2\theta)\,d\theta \\ &= \frac{243}{8}\left[\theta + \frac{\sin 2\theta}{2}\right]_0^{2\pi} \\ &= \frac{243\pi}{4}. \end{aligned}$$

When evaluating line integrals over closed curves, remember that for conservative vector fields (those for which $\partial N/\partial x = \partial M/\partial y$), the value of the line integral is 0. This is easily seen from the statement of Green's Theorem:

$$\int_C M\,dx + N\,dy = \int_R\!\int \left(\frac{\partial N}{\partial x} - \frac{\partial M}{\partial y}\right) dA = 0.$$

EXAMPLE 3 Green's Theorem and Conservative Vector Fields

Evaluate the line integral

$$\int_C y^3\,dx + 3xy^2\,dy$$

where C is the path shown in Figure 15.30.

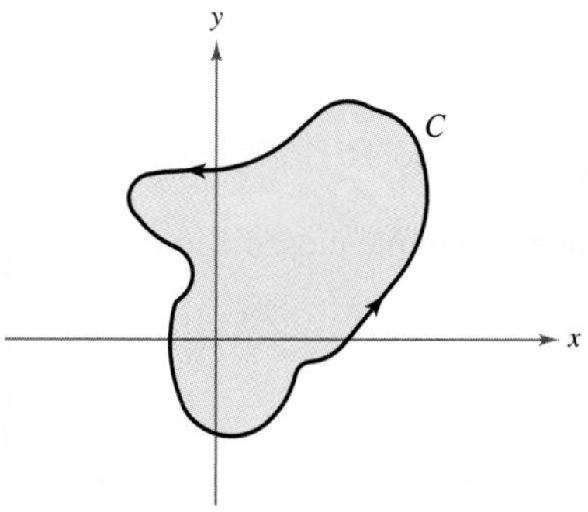

C is closed.
Figure 15.30

Solution From this line integral, $M = y^3$ and $N = 3xy^2$. So, $\partial N/\partial x = 3y^2$ and $\partial M/\partial y = 3y^2$. This implies that the vector field $\mathbf{F} = M\mathbf{i} + N\mathbf{j}$ is conservative, and because C is closed, you can conclude that

$$\int_C y^3\,dx + 3xy^2\,dy = 0.$$

EXAMPLE 4 Using Green's Theorem for a Piecewise Smooth Curve

Evaluate

$$\int_C (\arctan x + y^2)\,dx + (e^y - x^2)\,dy$$

where C is the path enclosing the annular region shown in Figure 15.31.

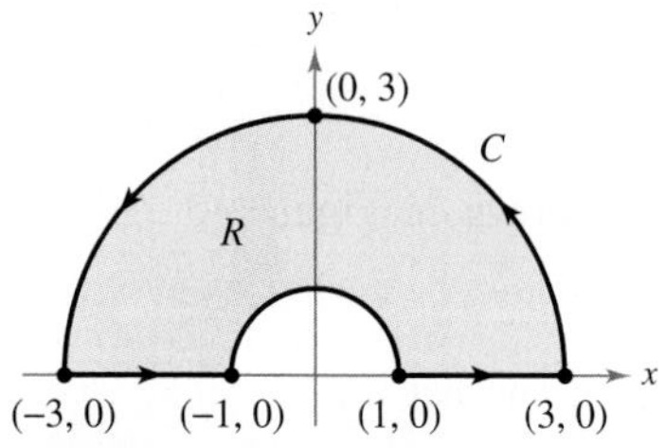

C is piecewise smooth.
Figure 15.31

Solution In polar coordinates, R is given by $1 \le r \le 3$ for $0 \le \theta \le \pi$. Moreover,

$$\frac{\partial N}{\partial x} - \frac{\partial M}{\partial y} = -2x - 2y = -2(r\cos\theta + r\sin\theta).$$

So, by Green's Theorem,

$$\begin{aligned}
\int_C (\arctan x + y^2)\,dx + (e^y - x^2)\,dy &= \int_R\int -2(x + y)\,dA \\
&= \int_0^{\pi}\int_1^3 -2r(\cos\theta + \sin\theta)r\,dr\,d\theta \\
&= \int_0^{\pi} -2(\cos\theta + \sin\theta)\frac{r^3}{3}\Big]_1^3\,d\theta \\
&= \int_0^{\pi}\left(-\frac{52}{3}\right)(\cos\theta + \sin\theta)\,d\theta \\
&= -\frac{52}{3}\Big[\sin\theta - \cos\theta\Big]_0^{\pi} \\
&= -\frac{104}{3}.
\end{aligned}$$

In Examples 1, 2, and 4, Green's Theorem was used to evaluate line integrals as double integrals. You can also use the theorem to evaluate double integrals as line integrals. One useful application occurs when $\partial N/\partial x - \partial M/\partial y = 1$.

$$\begin{aligned}
\int_C M\,dx + N\,dy &= \int_R\int\left(\frac{\partial N}{\partial x} - \frac{\partial M}{\partial y}\right)dA \\
&= \int_R\int 1\,dA \qquad \frac{\partial N}{\partial x} - \frac{\partial M}{\partial y} = 1 \\
&= \text{area of region } R
\end{aligned}$$

Among the many choices for M and N satisfying the stated condition, the choice of $M = -y/2$ and $N = x/2$ produces the following line integral for the area of region R.

THEOREM 15.9 Line Integral for Area

If R is a plane region bounded by a piecewise smooth simple closed curve C, oriented counterclockwise, then the area of R is given by

$$A = \frac{1}{2}\int_C x\,dy - y\,dx.$$

EXAMPLE 5 Finding Area by a Line Integral

Use a line integral to find the area of the ellipse

$$\frac{x^2}{a^2} + \frac{y^2}{b^2} = 1.$$

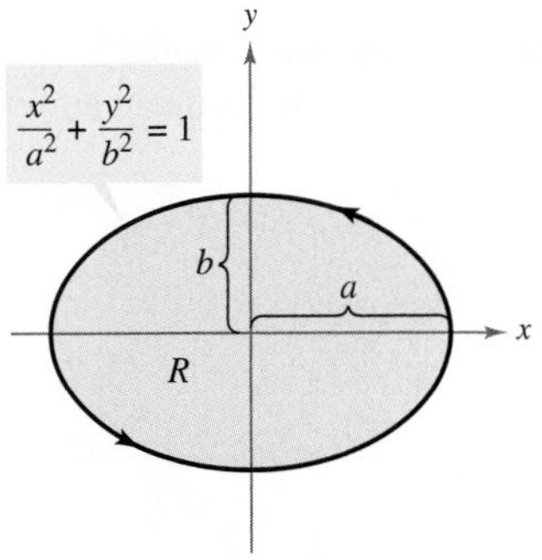

Figure 15.32

Solution Using Figure 15.32, you can induce a counterclockwise orientation to the elliptical path by letting

$$x = a\cos t \quad \text{and} \quad y = b\sin t, \quad 0 \le t \le 2\pi.$$

So, the area is

$$\begin{aligned} A = \frac{1}{2}\int_C x\,dy - y\,dx &= \frac{1}{2}\int_0^{2\pi} [(a\cos t)(b\cos t)\,dt - (b\sin t)(-a\sin t)\,dt] \\ &= \frac{ab}{2}\int_0^{2\pi} (\cos^2 t + \sin^2 t)\,dt \\ &= \frac{ab}{2}\Big[t\Big]_0^{2\pi} \\ &= \pi ab. \end{aligned}$$

Green's Theorem can be extended to cover some regions that are not simply connected. This is demonstrated in the next example.

EXAMPLE 6 Green's Theorem Extended to a Region with a Hole

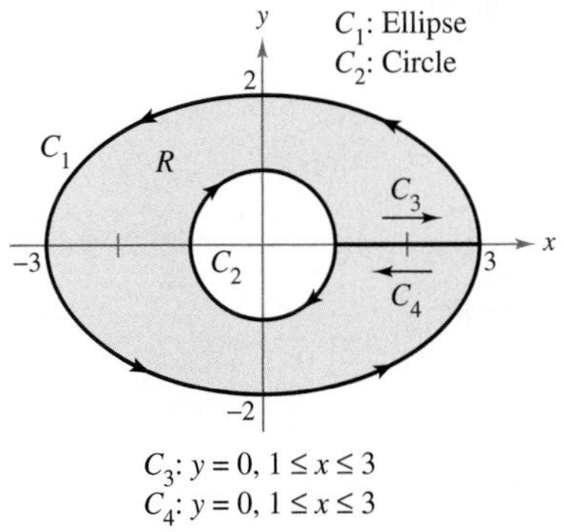

Figure 15.33

Let R be the region inside the ellipse $(x^2/9) + (y^2/4) = 1$ and outside the circle $x^2 + y^2 = 1$. Evaluate the line integral

$$\int_C 2xy\,dx + (x^2 + 2x)\,dy$$

where $C = C_1 + C_2$ is the boundary of R, as shown in Figure 15.33.

Solution To begin, you can introduce the line segments C_3 and C_4, as shown in Figure 15.33. Note that because the curves C_3 and C_4 have opposite orientations, the line integrals over them cancel. Furthermore, you can apply Green's Theorem to the region R using the boundary $C_1 + C_4 + C_2 + C_3$ to obtain

$$\begin{aligned} \int_C 2xy\,dx + (x^2 + 2x)\,dy &= \int_R\int \left(\frac{\partial N}{\partial x} - \frac{\partial M}{\partial y}\right) dA \\ &= \int_R\int (2x + 2 - 2x)\,dA \\ &= 2\int_R\int dA \\ &= 2(\text{area of } R) \\ &= 2(\pi ab - \pi r^2) \\ &= 2[\pi(3)(2) - \pi(1^2)] \\ &= 10\pi. \end{aligned}$$

In Section 15.1, a necessary and sufficient condition for conservative vector fields was listed. There, only one direction of the proof was shown. You can now outline the other direction, using Green's Theorem. Let $\mathbf{F}(x, y) = M\mathbf{i} + N\mathbf{j}$ be defined on an open disk R. You want to show that if M and N have continuous first partial derivatives and

$$\frac{\partial M}{\partial y} = \frac{\partial N}{\partial x}$$

then $\mathbf{F}$ is conservative. Suppose that C is a closed path forming the boundary of a connected region lying in R. Then, using the fact that $\partial M/\partial y = \partial N/\partial x$, you can apply Green's Theorem to conclude that

$$\begin{aligned}\int_C \mathbf{F} \cdot d\mathbf{r} &= \int_C M\,dx + N\,dy \\ &= \int_R\!\!\int \left(\frac{\partial N}{\partial x} - \frac{\partial M}{\partial y}\right) dA \\ &= 0.\end{aligned}$$

This, in turn, is equivalent to showing that $\mathbf{F}$ is conservative (see Theorem 15.7).

Alternative Forms of Green's Theorem

This section concludes with the derivation of two vector forms of Green's Theorem for regions in the plane. The extension of these vector forms to three dimensions is the basis for the discussion in the remaining sections of this chapter. If $\mathbf{F}$ is a vector field in the plane, you can write

$$\mathbf{F}(x, y, z) = M\mathbf{i} + N\mathbf{j} + 0\mathbf{k}$$

so that the curl of $\mathbf{F}$, as described in Section 15.1, is given by

$$\begin{aligned}\textbf{curl F} = \nabla \times \mathbf{F} &= \begin{vmatrix} \mathbf{i} & \mathbf{j} & \mathbf{k} \\ \dfrac{\partial}{\partial x} & \dfrac{\partial}{\partial y} & \dfrac{\partial}{\partial z} \\ M & N & 0 \end{vmatrix} \\ &= -\frac{\partial N}{\partial z}\mathbf{i} + \frac{\partial M}{\partial z}\mathbf{j} + \left(\frac{\partial N}{\partial x} - \frac{\partial M}{\partial y}\right)\mathbf{k}.\end{aligned}$$

Consequently,

$$\begin{aligned}(\textbf{curl F}) \cdot \mathbf{k} &= \left[-\frac{\partial N}{\partial z}\mathbf{i} + \frac{\partial M}{\partial z}\mathbf{j} + \left(\frac{\partial N}{\partial x} - \frac{\partial M}{\partial y}\right)\mathbf{k}\right] \cdot \mathbf{k} \\ &= \frac{\partial N}{\partial x} - \frac{\partial M}{\partial y}.\end{aligned}$$

With appropriate conditions on $\mathbf{F}$, C, and R, you can write Green's Theorem in the vector form

$$\begin{aligned}\int_C \mathbf{F} \cdot d\mathbf{r} &= \int_R\!\!\int \left(\frac{\partial N}{\partial x} - \frac{\partial M}{\partial y}\right) dA \\ &= \int_R\!\!\int (\textbf{curl F}) \cdot \mathbf{k}\, dA. \qquad \text{First alternative form}\end{aligned}$$

The extension of this vector form of Green's Theorem to surfaces in space produces **Stokes's Theorem,** discussed in Section 15.8.

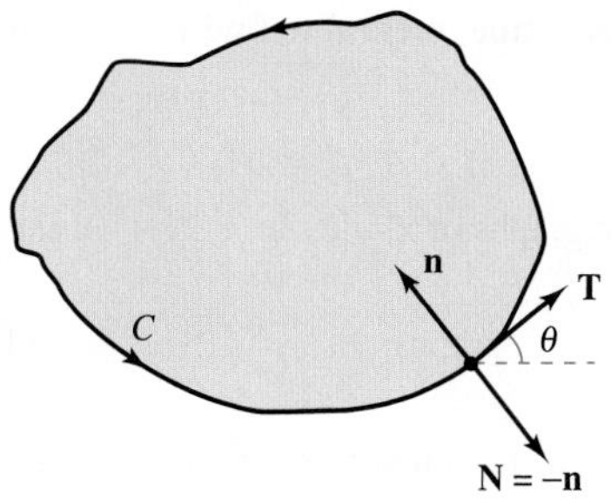

$\mathbf{T} = \cos\theta\mathbf{i} + \sin\theta\mathbf{j}$

$\mathbf{n} = \cos\left(\theta + \frac{\pi}{2}\right)\mathbf{i} + \sin\left(\theta + \frac{\pi}{2}\right)\mathbf{j}$

$= -\sin\theta\mathbf{i} + \cos\theta\mathbf{j}$

$\mathbf{N} = \sin\theta\mathbf{i} - \cos\theta\mathbf{j}$

Figure 15.34

For the second vector form of Green's Theorem, assume the same conditions for $\mathbf{F}$, C, and R. Using the arc length parameter s for C, you have $\mathbf{r}(s) = x(s)\mathbf{i} + y(s)\mathbf{j}$. So, a unit tangent vector $\mathbf{T}$ to curve C is given by

$$\mathbf{r}'(s) = \mathbf{T} = x'(s)\mathbf{i} + y'(s)\mathbf{j}.$$

From Figure 15.34 you can see that the *outward* unit normal vector $\mathbf{N}$ can then be written as

$$\mathbf{N} = y'(s)\mathbf{i} - x'(s)\mathbf{j}.$$

Consequently, for $\mathbf{F}(x, y) = M\mathbf{i} + N\mathbf{j}$, you can apply Green's Theorem to obtain

$$\begin{aligned}\int_C \mathbf{F}\cdot\mathbf{N}\,ds &= \int_a^b (M\mathbf{i} + N\mathbf{j})\cdot(y'(s)\mathbf{i} - x'(s)\mathbf{j})\,ds\\ &= \int_a^b \left(M\frac{dy}{ds} - N\frac{dx}{ds}\right)ds\\ &= \int_C M\,dy - N\,dx\\ &= \int_C -N\,dx + M\,dy\\ &= \int_R\int \left(\frac{\partial M}{\partial x} + \frac{\partial N}{\partial y}\right)dA \qquad \text{Green's Theorem}\\ &= \int_R\int \operatorname{div}\mathbf{F}\,dA.\end{aligned}$$

Therefore,

$$\int_C \mathbf{F}\cdot\mathbf{N}\,ds = \int_R\int \operatorname{div}\mathbf{F}\,dA. \qquad \text{Second alternative form}$$

The extension of this form to three dimensions is called the **Divergence Theorem,** discussed in Section 15.7. The physical interpretations of divergence and curl will be discussed in Sections 15.7 and 15.8.

Exercises for Section 15.4

See www.CalcChat.com for worked-out solutions to odd-numbered exercises.

In Exercises 1–4, verify Green's Theorem by evaluating both integrals

$$\int_C y^2\,dx + x^2\,dy = \int_R\int\left(\frac{\partial N}{\partial x} - \frac{\partial M}{\partial y}\right)dA$$

for the given path.

1. C: square with vertices $(0, 0), (4, 0), (4, 4), (0, 4)$
2. C: triangle with vertices $(0, 0), (4, 0), (4, 4)$
3. C: boundary of the region lying between the graphs of $y = x$ and $y = x^2/4$
4. C: circle given by $x^2 + y^2 = 1$

In Exercises 5 and 6, verify Green's Theorem by using a computer algebra system to evaluate both integrals

$$\int_C xe^y\,dx + e^x\,dy = \int_R\int\left(\frac{\partial N}{\partial x} - \frac{\partial M}{\partial y}\right)dA$$

for the given path.

5. C: circle given by $x^2 + y^2 = 4$
6. C: boundary of the region lying between the graphs of $y = x$ and $y = x^3$ in the first quadrant

In Exercises 7–10, use Green's Theorem to evaluate the integral

$$\int_C (y - x)\,dx + (2x - y)\,dy$$

for the given path.

7. C: boundary of the region lying between the graphs of $y = x$ and $y = x^2 - x$
8. C: $x = 2\cos\theta$, $y = \sin\theta$
9. C: boundary of the region lying inside the rectangle bounded by $x = -5$, $x = 5$, $y = -3$, and $y = 3$, and outside the square bounded by $x = -1$, $x = 1$, $y = -1$, and $y = 1$
10. C: boundary of the region lying inside the semicircle $y = \sqrt{25 - x^2}$ and outside the semicircle $y = \sqrt{9 - x^2}$

In Exercises 11–20, use Green's Theorem to evaluate the line integral.

11. $\int_C 2xy\,dx + (x + y)\,dy$

C: boundary of the region lying between the graphs of $y = 0$ and $y = 4 - x^2$

12. $\int_C y^2\,dx + xy\,dy$

C: boundary of the region lying between the graphs of $y = 0$, $y = \sqrt{x}$, and $x = 9$

13. $\int_C (x^2 - y^2)\,dx + 2xy\,dy$

C: $x^2 + y^2 = a^2$

14. $\int_C (x^2 - y^2)\,dx + 2xy\,dy$

C: $r = 1 + \cos\theta$

15. $\int_C 2\arctan\frac{y}{x}\,dx + \ln(x^2 + y^2)\,dy$

C: $x = 4 + 2\cos\theta$, $y = 4 + \sin\theta$

16. $\int_C e^x\cos 2y\,dx - 2e^x\sin 2y\,dy$

C: $x^2 + y^2 = a^2$

17. $\int_C \sin x\cos y\,dx + (xy + \cos x\sin y)\,dy$

C: boundary of the region lying between the graphs of $y = x$ and $y = \sqrt{x}$

18. $\int_C (e^{-x^2/2} - y)\,dx + (e^{-y^2/2} + x)\,dy$

C: boundary of the region lying between the graphs of the circle $x = 6\cos\theta$, $y = 6\sin\theta$ and the ellipse $x = 3\cos\theta$, $y = 2\sin\theta$

19. $\int_C xy\,dx + (x + y)\,dy$

C: boundary of the region lying between the graphs of $x^2 + y^2 = 1$ and $x^2 + y^2 = 9$

20. $\int_C 3x^2e^y\,dx + e^y\,dy$

C: boundary of the region lying between the squares with vertices $(1, 1)$, $(-1, 1)$, $(-1, -1)$, and $(1, -1)$, and $(2, 2)$, $(-2, 2)$, $(-2, -2)$, and $(2, -2)$

Work **In Exercises 21–24, use Green's Theorem to calculate the work done by the force F on a particle that is moving counterclockwise around the closed path *C*.**

21. $\mathbf{F}(x, y) = xy\mathbf{i} + (x + y)\mathbf{j}$

C: $x^2 + y^2 = 4$

22. $\mathbf{F}(x, y) = (e^x - 3y)\mathbf{i} + (e^y + 6x)\mathbf{j}$

C: $r = 2\cos\theta$

23. $\mathbf{F}(x, y) = (x^{3/2} - 3y)\mathbf{i} + \left(6x + 5\sqrt{y}\right)\mathbf{j}$

C: boundary of the triangle with vertices $(0, 0)$, $(5, 0)$, and $(0, 5)$

24. $\mathbf{F}(x, y) = (3x^2 + y)\mathbf{i} + 4xy^2\,\mathbf{j}$

C: boundary of the region lying between the graphs of $y = \sqrt{x}$, $y = 0$, and $x = 9$

Area **In Exercises 25–28, use a line integral to find the area of the region *R*.**

25. R: region bounded by the graph of $x^2 + y^2 = a^2$

26. R: triangle bounded by the graphs of $x = 0$, $3x - 2y = 0$, and $x + 2y = 8$

27. R: region bounded by the graphs of $y = 2x + 1$ and $y = 4 - x^2$

28. R: region inside the loop of the folium of Descartes bounded by the graph of

$$x = \frac{3t}{t^3 + 1}, \quad y = \frac{3t^2}{t^3 + 1}$$

Writing About Concepts

29. State Green's Theorem.

30. Give the line integral for the area of a region R bounded by a piecewise smooth simple curve C.

In Exercises 31 and 32, use Green's Theorem to verify the line integral formulas.

31. The centroid of the region having area A bounded by the simple closed path C is

$$\bar{x} = \frac{1}{2A}\int_C x^2\,dy, \quad \bar{y} = -\frac{1}{2A}\int_C y^2\,dx.$$

32. The area of a plane region bounded by the simple closed path C given in polar coordinates is $A = \frac{1}{2}\int_C r^2\,d\theta$.

Centroid **In Exercises 33–36, use a computer algebra system and the results of Exercise 31 to find the centroid of the region.**

33. R: region bounded by the graphs of $y = 0$ and $y = 4 - x^2$

34. R: region bounded by the graphs of $y = \sqrt{a^2 - x^2}$ and $y = 0$

35. R: region bounded by the graphs of $y = x^3$ and $y = x$, $0 \le x \le 1$

36. R: triangle with vertices $(-a, 0)$, $(a, 0)$, and (b, c), where $-a \le b \le a$

Area **In Exercises 37–40, use a computer algebra system and the results of Exercise 32 to find the area of the region bounded by the graph of the polar equation.**

37. $r = a(1 - \cos\theta)$

38. $r = a\cos 3\theta$

39. $r = 1 + 2\cos\theta$ (inner loop)

40. $r = \dfrac{3}{2 - \cos\theta}$

41. ***Think About It*** Let

$$I = \int_C \frac{y\,dx - x\,dy}{x^2 + y^2}$$

where C is a circle oriented counterclockwise. Show that $I = 0$ if C does not contain the origin. What is I if C contains the origin?

42. (a) Let C be the line segment joining (x_1, y_1) and (x_2, y_2). Show that

$$\int_C -y\,dx + x\,dy = x_1 y_2 - x_2 y_1.$$

(b) Let $(x_1, y_1), (x_2, y_2), \ldots, (x_n, y_n)$ be the vertices of a polygon. Prove that the area enclosed is

$$\tfrac{1}{2}[(x_1y_2 - x_2y_1) + (x_2y_3 - x_3y_2) + \cdots + (x_{n-1}y_n - x_ny_{n-1}) + (x_ny_1 - x_1y_n)].$$

Area **In Exercises 43 and 44, use the result of Exercise 42(b) to find the area enclosed by the polygon with the given vertices.**

43. Pentagon: $(0, 0), (2, 0), (3, 2), (1, 4), (-1, 1)$

44. Hexagon: $(0, 0), (2, 0), (3, 2), (2, 4), (0, 3), (-1, 1)$

45. ***Investigation*** Consider the line integral

$$\int_C y^n\,dx + x^n\,dy$$

where C is the boundary of the region lying between the graphs of $y = \sqrt{a^2 - x^2}$ $(a > 0)$ and $y = 0$.

(a) Use a computer algebra system to verify Green's Theorem for n, an odd integer from 1 through 7.

(b) Use a computer algebra system to verify Green's Theorem for n, an even integer from 2 through 8.

(c) For n an odd integer, make a conjecture about the value of the integral.

In Exercises 46 and 47, prove the identity where R is a simply connected region with boundary C. Assume that the required partial derivatives of the scalar functions f and g are continuous. The expressions $D_{\mathbf{N}}f$ and $D_{\mathbf{N}}g$ are the derivatives in the direction of the outward normal vector $\mathbf{N}$ of C, and are defined by $D_{\mathbf{N}}f = \nabla f \cdot \mathbf{N}$, and $D_{\mathbf{N}}g = \nabla g \cdot \mathbf{N}$.

46. Green's first identity:

$$\int_R\int (f\nabla^2 g + \nabla f \cdot \nabla g)\,dA = \int_C f D_{\mathbf{N}}g\,ds$$

[*Hint:* Use the second alternative form of Green's Theorem and the property $\operatorname{div}(f\mathbf{G}) = f \operatorname{div}\mathbf{G} + \nabla f \cdot \mathbf{G}$.]

47. Green's second identity:

$$\int_R\int (f\nabla^2 g - g\nabla^2 f)\,dA = \int_C (f D_{\mathbf{N}}g - g D_{\mathbf{N}}f)\,ds$$

(*Hint:* Use Exercise 46 twice.)

48. Use Green's Theorem to prove that

$$\int_C f(x)\,dx + g(y)\,dy = 0$$

if f and g are differentiable functions and C is a piecewise smooth simple closed path.

49. Let $\mathbf{F} = M\mathbf{i} + N\mathbf{j}$, where M and N have continuous first partial derivatives in a simply connected region R. Prove that if C is simple, smooth, and closed, and $N_x = M_y$, then

$$\int_C \mathbf{F} \cdot d\mathbf{r} = 0.$$

Section Project: Hyperbolic and Trigonometric Functions

(a) Sketch the plane curve represented by the vector-valued function $\mathbf{r}(t) = \cosh t\mathbf{i} + \sinh t\mathbf{j}$ on the interval $0 \le t \le 5$. Show that the rectangular equation corresponding to $\mathbf{r}(t)$ is the hyperbola $x^2 - y^2 = 1$. Verify your sketch by using a graphing utility to graph the hyperbola.

(b) Let $P = (\cosh \phi, \sinh \phi)$ be the point on the hyperbola corresponding to $\mathbf{r}(\phi)$ for $\phi > 0$. Use the formula for area

$$A = \frac{1}{2}\int_C x\,dy - y\,dx$$

to verify that the area of the region shown in the figure is $\frac{1}{2}\phi$.

(c) Show that the area of the indicated region is also given by the integral

$$A = \int_0^{\sinh \phi} \left[\sqrt{1 + y^2} - (\coth \phi)y\right] dy.$$

Confirm your answer in part (b) by numerically approximating this integral for $\phi = 1, 2, 4$, and 10.

(d) Consider the unit circle given by $x^2 + y^2 = 1$. Let θ be the angle formed by the x-axis and the radius to (x, y). The area of the corresponding sector is $\frac{1}{2}\theta$. That is, the trigonometric functions $f(\theta) = \cos\theta$ and $g(\theta) = \sin\theta$ could have been defined to be the coordinates of that point $(\cos\theta, \sin\theta)$ on the unit circle that determines a sector of area $\frac{1}{2}\theta$. Write a short paragraph explaining how you could define the hyperbolic functions in a similar manner, using the "unit hyperbola" $x^2 - y^2 = 1$.

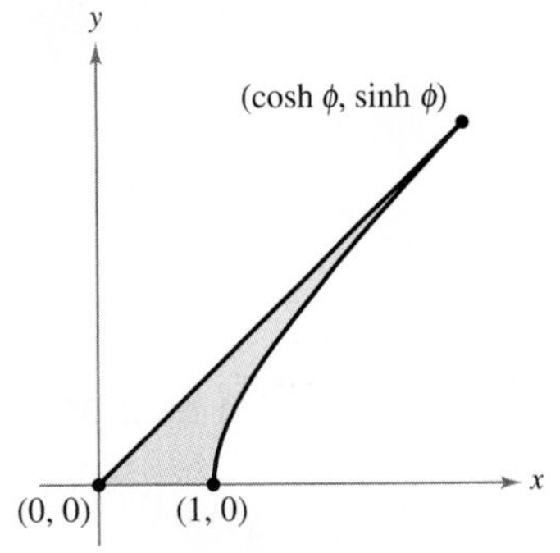

Section 15.5 Parametric Surfaces

- **Understand the definition of and sketch a parametric surface.**
- **Find a set of parametric equations to represent a surface.**
- **Find a normal vector and a tangent plane to a parametric surface.**
- **Find the area of a parametric surface.**

Parametric Surfaces

You already know how to represent a curve in the plane or in space by a set of parametric equations—or, equivalently, by a vector-valued function.

$$\mathbf{r}(t) = x(t)\mathbf{i} + y(t)\mathbf{j} \qquad \text{Plane curve}$$

$$\mathbf{r}(t) = x(t)\mathbf{i} + y(t)\mathbf{j} + z(t)\mathbf{k} \qquad \text{Space curve}$$

In this section, you will learn how to represent a surface in space by a set of parametric equations—or by a vector-valued function. For curves, note that the vector-valued function $\mathbf{r}$ is a function of a *single* parameter t. For surfaces, the vector-valued function is a function of *two* parameters u and v.

Definition of Parametric Surface

Let x, y, and z be functions of u and v that are continuous on a domain D in the uv-plane. The set of points (x, y, z) given by

$$\mathbf{r}(u, v) = x(u, v)\mathbf{i} + y(u, v)\mathbf{j} + z(u, v)\mathbf{k} \qquad \text{Parametric surface}$$

is called a **parametric surface.** The equations

$$x = x(u, v), \quad y = y(u, v), \quad \text{and} \quad z = z(u, v) \qquad \text{Parametric equations}$$

are the **parametric equations** for the surface.

If S is a parametric surface given by the vector-valued function $\mathbf{r}$, then S is traced out by the position vector $\mathbf{r}(u, v)$ as the point (u, v) moves throughout the domain D, as shown in Figure 15.35.

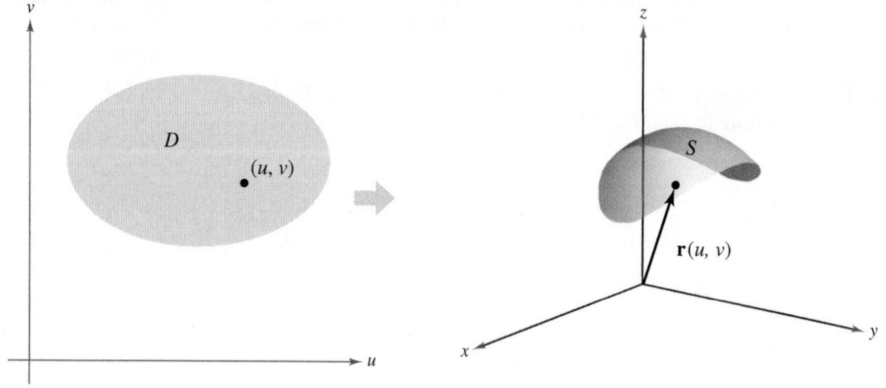

Figure 15.35

TECHNOLOGY Some computer algebra systems are capable of graphing surfaces that are represented parametrically. If you have access to such software, use it to graph some of the surfaces in the examples and exercises in this section.

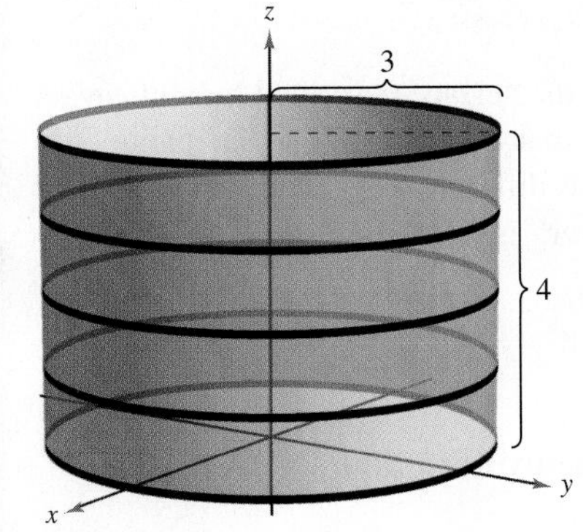

Figure 15.36

EXAMPLE 1 Sketching a Parametric Surface

Identify and sketch the parametric surface S given by

$$\mathbf{r}(u, v) = 3 \cos u\mathbf{i} + 3 \sin u\mathbf{j} + v\mathbf{k}$$

where $0 \le u \le 2\pi$ and $0 \le v \le 4$.

Solution Because $x = 3 \cos u$ and $y = 3 \sin u$, you know that for each point (x, y, z) on the surface, x and y are related by the equation $x^2 + y^2 = 3^2$. In other words, each cross section of S taken parallel to the xy-plane is a circle of radius 3, centered on the z-axis. Because $z = v$, where $0 \le v \le 4$, you can see that the surface is a right circular cylinder of height 4. The radius of the cylinder is 3, and the z-axis forms the axis of the cylinder, as shown in Figure 15.36.

As with parametric representations of curves, parametric representations of surfaces are not unique. That is, there are many other sets of parametric equations that could be used to represent the surface shown in Figure 15.36.

EXAMPLE 2 Sketching a Parametric Surface

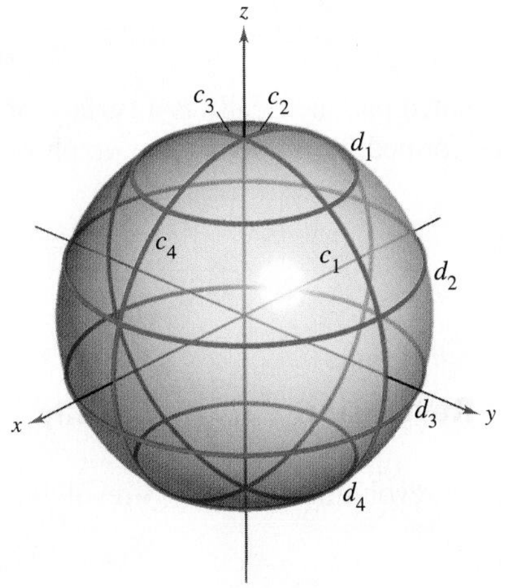

Figure 15.37

Identify and sketch the parametric surface S given by

$$\mathbf{r}(u, v) = \sin u \cos v\mathbf{i} + \sin u \sin v\mathbf{j} + \cos u\mathbf{k}$$

where $0 \le u \le \pi$ and $0 \le v \le 2\pi$.

Solution To identify the surface, you can try to use trigonometric identities to eliminate the parameters. After some experimentation, you can discover that

$$\begin{aligned} x^2 + y^2 + z^2 &= (\sin u \cos v)^2 + (\sin u \sin v)^2 + (\cos u)^2 \\ &= \sin^2 u \cos^2 v + \sin^2 u \sin^2 v + \cos^2 u \\ &= \sin^2 u(\cos^2 v + \sin^2 v) + \cos^2 u \\ &= \sin^2 u + \cos^2 u \\ &= 1. \end{aligned}$$

So, each point on S lies on the unit sphere, centered at the origin, as shown in Figure 15.37. For fixed $u = d_i$, $\mathbf{r}(u, v)$ traces out latitude circles

$$x^2 + y^2 = \sin^2 d_i, \quad 0 \le d_i \le \pi$$

that are parallel to the xy-plane, and for fixed $v = c_i$, $\mathbf{r}(u, v)$ traces out longitude (or meridian) half-circles.

NOTE To convince yourself further that the vector-valued function in Example 2 traces out the entire unit sphere, recall that the parametric equations

$$x = \rho \sin \phi \cos \theta, \quad y = \rho \sin \phi \sin \theta, \quad \text{and} \quad z = \rho \cos \phi$$

where $0 \le \theta \le 2\pi$ and $0 \le \phi \le \pi$, describe the conversion from spherical to rectangular coordinates, as discussed in Section 11.7.

Finding Parametric Equations for Surfaces

In Examples 1 and 2, you were asked to identify the surface described by a given set of parametric equations. The reverse problem—that of writing a set of parametric equations for a given surface—is generally more difficult. One type of surface for which this problem is straightforward, however, is a surface that is given by $z = f(x, y)$. You can parametrize such a surface as

$$\mathbf{r}(x, y) = x\mathbf{i} + y\mathbf{j} + f(x, y)\mathbf{k}.$$

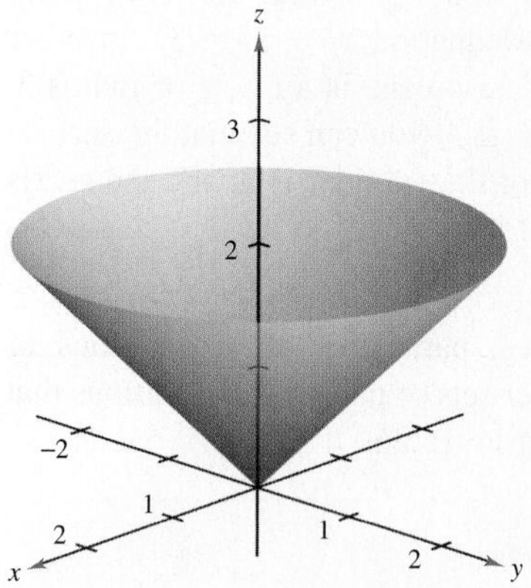

Figure 15.38

EXAMPLE 3 Representing a Surface Parametrically

Write a set of parametric equations for the cone given by

$$z = \sqrt{x^2 + y^2}$$

as shown in Figure 15.38.

Solution Because this surface is given in the form $z = f(x, y)$, you can let x and y be the parameters. Then the cone is represented by the vector-valued function

$$\mathbf{r}(x, y) = x\mathbf{i} + y\mathbf{j} + \sqrt{x^2 + y^2}\,\mathbf{k}$$

where (x, y) varies over the entire xy-plane.

A second type of surface that is easily represented parametrically is a surface of revolution. For instance, to represent the surface formed by revolving the graph of $y = f(x)$, $a \le x \le b$, about the x-axis, use

$$x = u, \quad y = f(u)\cos v, \quad \text{and} \quad z = f(u)\sin v$$

where $a \le u \le b$ and $0 \le v \le 2\pi$.

EXAMPLE 4 Representing a Surface of Revolution Parametrically

Write a set of parametric equations for the surface of revolution obtained by revolving

$$f(x) = \frac{1}{x}, \quad 1 \le x \le 10$$

about the x-axis.

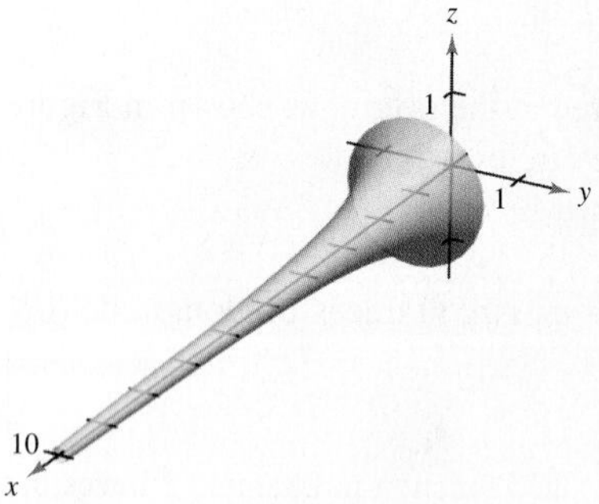

Figure 15.39

Solution Use the parameters u and v as described above to write

$$x = u, \quad y = f(u)\cos v = \frac{1}{u}\cos v, \quad \text{and} \quad z = f(u)\sin v = \frac{1}{u}\sin v$$

where $1 \le u \le 10$ and $0 \le v \le 2\pi$. The resulting surface is a portion of *Gabriel's Horn*, as shown in Figure 15.39.

The surface of revolution in Example 4 is formed by revolving the graph of $y = f(x)$ about the x-axis. For other types of surfaces of revolution, a similar parametrization can be used. For instance, to parametrize the surface formed by revolving the graph of $x = f(z)$ about the z-axis, you can use

$$z = u, \quad x = f(u)\cos v, \quad \text{and} \quad y = f(u)\sin v.$$

Normal Vectors and Tangent Planes

Let S be a parametric surface given by

$$\mathbf{r}(u, v) = x(u, v)\mathbf{i} + y(u, v)\mathbf{j} + z(u, v)\mathbf{k}$$

over an open region D such that x, y, and z have continuous partial derivatives on D. The **partial derivatives of r** with respect to u and v are defined as

$$\mathbf{r}_u = \frac{\partial x}{\partial u}(u, v)\mathbf{i} + \frac{\partial y}{\partial u}(u, v)\mathbf{j} + \frac{\partial z}{\partial u}(u, v)\mathbf{k}$$

and

$$\mathbf{r}_v = \frac{\partial x}{\partial v}(u, v)\mathbf{i} + \frac{\partial y}{\partial v}(u, v)\mathbf{j} + \frac{\partial z}{\partial v}(u, v)\mathbf{k}.$$

Each of these partial derivatives is a vector-valued function that can be interpreted geometrically in terms of tangent vectors. For instance, if $v = v_0$ is held constant, then $\mathbf{r}(u, v_0)$ is a vector-valued function of a single parameter and defines a curve C_1 that lies on the surface S. The tangent vector to C_1 at the point $(x(u_0, v_0), y(u_0, v_0), z(u_0, v_0))$ is given by

$$\mathbf{r}_u(u_0, v_0) = \frac{\partial x}{\partial u}(u_0, v_0)\mathbf{i} + \frac{\partial y}{\partial u}(u_0, v_0)\mathbf{j} + \frac{\partial z}{\partial u}(u_0, v_0)\mathbf{k}$$

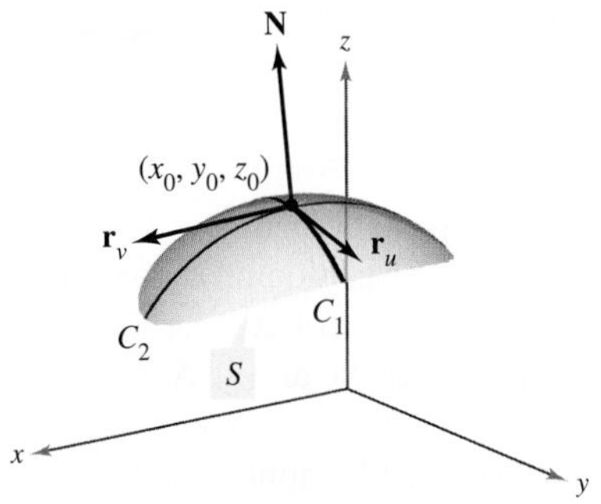

Figure 15.40

as shown in Figure 15.40. In a similar way, if $u = u_0$ is held constant, then $\mathbf{r}(u_0, v)$ is a vector-valued function of a single parameter and defines a curve C_2 that lies on the surface S. The tangent vector to C_2 at the point $(x(u_0, v_0), y(u_0, v_0), z(u_0, v_0))$ is given by

$$\mathbf{r}_v(u_0, v_0) = \frac{\partial x}{\partial v}(u_0, v_0)\mathbf{i} + \frac{\partial y}{\partial v}(u_0, v_0)\mathbf{j} + \frac{\partial z}{\partial v}(u_0, v_0)\mathbf{k}.$$

If the normal vector $\mathbf{r}_u \times \mathbf{r}_v$ is not $\mathbf{0}$ for any (u, v) in D, the surface S is called **smooth** and will have a tangent plane. Informally, a smooth surface is one that has no sharp points or cusps. For instance, spheres, ellipsoids, and paraboloids are smooth, whereas the cone given in Example 3 is not smooth.

Normal Vector to a Smooth Parametric Surface

Let S be a smooth parametric surface

$$\mathbf{r}(u, v) = x(u, v)\mathbf{i} + y(u, v)\mathbf{j} + z(u, v)\mathbf{k}$$

defined over an open region D in the uv-plane. Let (u_0, v_0) be a point in D. A normal vector at the point

$$(x_0, y_0, z_0) = (x(u_0, v_0), y(u_0, v_0), z(u_0, v_0))$$

is given by

$$\mathbf{N} = \mathbf{r}_u(u_0, v_0) \times \mathbf{r}_v(u_0, v_0) = \begin{vmatrix} \mathbf{i} & \mathbf{j} & \mathbf{k} \\ \frac{\partial x}{\partial u} & \frac{\partial y}{\partial u} & \frac{\partial z}{\partial u} \\ \frac{\partial x}{\partial v} & \frac{\partial y}{\partial v} & \frac{\partial z}{\partial v} \end{vmatrix}.$$

NOTE Figure 15.40 shows the normal vector $\mathbf{r}_u \times \mathbf{r}_v$. The vector $\mathbf{r}_v \times \mathbf{r}_u$ is also normal to S and points in the opposite direction.

EXAMPLE 5 Finding a Tangent Plane to a Parametric Surface

Find an equation of the tangent plane to the paraboloid given by

$$\mathbf{r}(u, v) = u\mathbf{i} + v\mathbf{j} + (u^2 + v^2)\mathbf{k}$$

at the point (1, 2, 5).

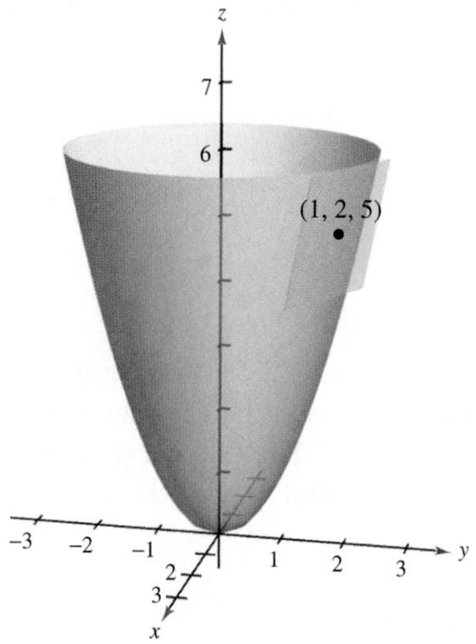

Figure 15.41

Solution The point in the uv-plane that is mapped to the point $(x, y, z) = (1, 2, 5)$ is $(u, v) = (1, 2)$. The partial derivatives of $\mathbf{r}$ are

$$\mathbf{r}_u = \mathbf{i} + 2u\mathbf{k} \quad \text{and} \quad \mathbf{r}_v = \mathbf{j} + 2v\mathbf{k}.$$

The normal vector is given by

$$\mathbf{r}_u \times \mathbf{r}_v = \begin{vmatrix} \mathbf{i} & \mathbf{j} & \mathbf{k} \\ 1 & 0 & 2u \\ 0 & 1 & 2v \end{vmatrix} = -2u\mathbf{i} - 2v\mathbf{j} + \mathbf{k}$$

which implies that the normal vector at (1, 2, 5) is $\mathbf{r}_u \times \mathbf{r}_v = -2\mathbf{i} - 4\mathbf{j} + \mathbf{k}$. So, an equation of the tangent plane at (1, 2, 5) is

$$-2(x - 1) - 4(y - 2) + (z - 5) = 0$$
$$-2x - 4y + z = -5.$$

The tangent plane is shown in Figure 15.41.

Area of a Parametric Surface

To define the area of a parametric surface, you can use a development that is similar to that given in Section 14.5. Begin by constructing an inner partition of D consisting of n rectangles, where the area of the ith rectangle D_i is $\Delta A_i = \Delta u_i \, \Delta v_i$, as shown in Figure 15.42. In each D_i let (u_i, v_i) be the point that is closest to the origin. At the point $(x_i, y_i, z_i) = (x(u_i, v_i), y(u_i, v_i), z(u_i, v_i))$ on the surface S, construct a tangent plane T_i. The area of the portion of S that corresponds to D_i, ΔT_i, can be approximated by a parallelogram in the tangent plane. That is, $\Delta T_i \approx \Delta S_i$. So, the surface of S is given by $\Sigma \, \Delta S_i \approx \Sigma \, \Delta T_i$. The area of the parallelogram in the tangent plane is

$$\|\Delta u_i \mathbf{r}_u \times \Delta v_i \mathbf{r}_v\| = \|\mathbf{r}_u \times \mathbf{r}_v\| \, \Delta u_i \, \Delta v_i$$

which leads to the following definition.

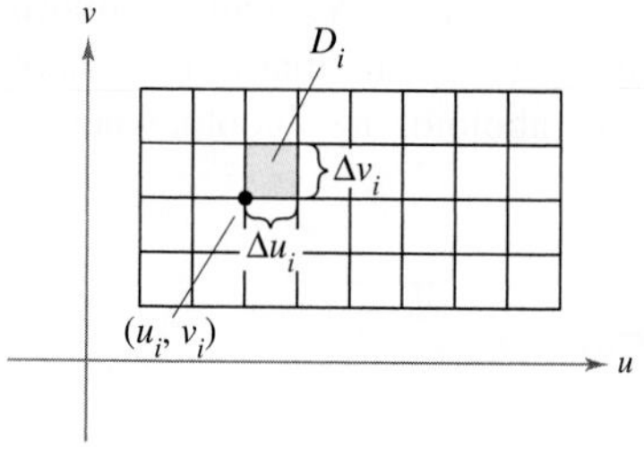

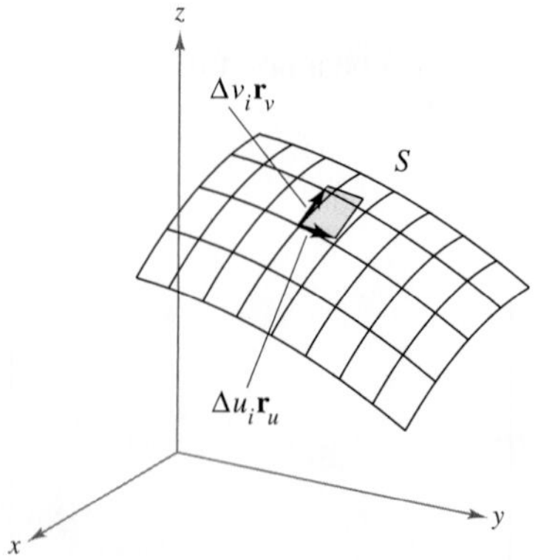

Figure 15.42

> **Area of a Parametric Surface**
>
> Let S be a smooth parametric surface
>
> $$\mathbf{r}(u, v) = x(u, v)\mathbf{i} + y(u, v)\mathbf{j} + z(u, v)\mathbf{k}$$
>
> defined over an open region D in the uv-plane. If each point on the surface S corresponds to exactly one point in the domain D, then the **surface area** of S is given by
>
> $$\text{Surface area} = \iint_S dS = \iint_D \|\mathbf{r}_u \times \mathbf{r}_v\| \, dA$$
>
> where $\mathbf{r}_u = \dfrac{\partial x}{\partial u}\mathbf{i} + \dfrac{\partial y}{\partial u}\mathbf{j} + \dfrac{\partial z}{\partial u}\mathbf{k}$ and $\mathbf{r}_v = \dfrac{\partial x}{\partial v}\mathbf{i} + \dfrac{\partial y}{\partial v}\mathbf{j} + \dfrac{\partial z}{\partial v}\mathbf{k}$.

For a surface S given by $z = f(x, y)$, this formula for surface area corresponds to that given in Section 14.5. To see this, you can parametrize the surface using the vector-valued function

$$\mathbf{r}(x, y) = x\mathbf{i} + y\mathbf{j} + f(x, y)\mathbf{k}$$

defined over the region R in the xy-plane. Using

$$\mathbf{r}_x = \mathbf{i} + f_x(x, y)\mathbf{k} \quad \text{and} \quad \mathbf{r}_y = \mathbf{j} + f_y(x, y)\mathbf{k}$$

you have

$$\mathbf{r}_x \times \mathbf{r}_y = \begin{vmatrix} \mathbf{i} & \mathbf{j} & \mathbf{k} \\ 1 & 0 & f_x(x, y) \\ 0 & 1 & f_y(x, y) \end{vmatrix} = -f_x(x, y)\mathbf{i} - f_y(x, y)\mathbf{j} + \mathbf{k}$$

and $\|\mathbf{r}_x \times \mathbf{r}_y\| = \sqrt{[f_x(x, y)]^2 + [f_y(x, y)]^2 + 1}$. This implies that the surface area of S is

$$\begin{aligned} \text{Surface area} &= \int\!\!\int_R \|\mathbf{r}_x \times \mathbf{r}_y\|\, dA \\ &= \int\!\!\int_R \sqrt{1 + [f_x(x, y)]^2 + [f_y(x, y)]^2}\, dA. \end{aligned}$$

EXAMPLE 6 Finding Surface Area

NOTE The surface in Example 6 does not quite fulfill the hypothesis that each point on the surface corresponds to exactly one point in D. For this surface, $\mathbf{r}(u, 0) = \mathbf{r}(u, 2\pi)$ for any fixed value of u. However, because the overlap consists of only a semicircle (which has no area), you can still apply the formula for the area of a parametric surface.

Find the surface area of the unit sphere given by

$$\mathbf{r}(u, v) = \sin u \cos v\mathbf{i} + \sin u \sin v\mathbf{j} + \cos u\mathbf{k}$$

where the domain D is given by $0 \le u \le \pi$ and $0 \le v \le 2\pi$.

Solution Begin by calculating $\mathbf{r}_u$ and $\mathbf{r}_v$.

$$\begin{aligned} \mathbf{r}_u &= \cos u \cos v\mathbf{i} + \cos u \sin v\mathbf{j} - \sin u\mathbf{k} \\ \mathbf{r}_v &= -\sin u \sin v\mathbf{i} + \sin u \cos v\mathbf{j} \end{aligned}$$

The cross product of these two vectors is

$$\begin{aligned} \mathbf{r}_u \times \mathbf{r}_v &= \begin{vmatrix} \mathbf{i} & \mathbf{j} & \mathbf{k} \\ \cos u \cos v & \cos u \sin v & -\sin u \\ -\sin u \sin v & \sin u \cos v & 0 \end{vmatrix} \\ &= \sin^2 u \cos v\mathbf{i} + \sin^2 u \sin v\mathbf{j} + \sin u \cos u\mathbf{k} \end{aligned}$$

which implies that

$$\begin{aligned} \|\mathbf{r}_u \times \mathbf{r}_v\| &= \sqrt{(\sin^2 u \cos v)^2 + (\sin^2 u \sin v)^2 + (\sin u \cos u)^2} \\ &= \sqrt{\sin^4 u + \sin^2 u \cos^2 u} \\ &= \sqrt{\sin^2 u} \\ &= \sin u. \qquad \sin u > 0 \text{ for } 0 \le u \le \pi \end{aligned}$$

Finally, the surface area of the sphere is

$$\begin{aligned} A = \int\!\!\int_D \|\mathbf{r}_u \times \mathbf{r}_v\|\, dA &= \int_0^{2\pi}\!\!\int_0^{\pi} \sin u\, du\, dv \\ &= \int_0^{2\pi} 2\, dv \\ &= 4\pi. \end{aligned}$$

EXPLORATION

For the torus in Example 7, describe the function $\mathbf{r}(u, v)$ for fixed u. Then describe the function $\mathbf{r}(u, v)$ for fixed v.

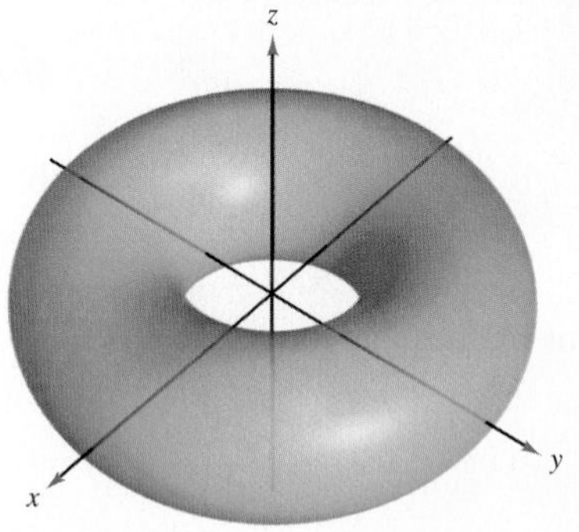

Figure 15.43

EXAMPLE 7 Finding Surface Area

Find the surface area of the torus given by

$$\mathbf{r}(u, v) = (2 + \cos u)\cos v\mathbf{i} + (2 + \cos u)\sin v\mathbf{j} + \sin u\mathbf{k}$$

where the domain D is given by $0 \le u \le 2\pi$ and $0 \le v \le 2\pi$. (See Figure 15.43.)

Solution Begin by calculating $\mathbf{r}_u$ and $\mathbf{r}_v$.

$$\mathbf{r}_u = -\sin u \cos v\mathbf{i} - \sin u \sin v\mathbf{j} + \cos u\mathbf{k}$$
$$\mathbf{r}_v = -(2 + \cos u)\sin v\mathbf{i} + (2 + \cos u)\cos v\mathbf{j}$$

The cross product of these two vectors is

$$\begin{aligned}\mathbf{r}_u \times \mathbf{r}_v &= \begin{vmatrix} \mathbf{i} & \mathbf{j} & \mathbf{k} \\ -\sin u \cos v & -\sin u \sin v & \cos u \\ -(2 + \cos u)\sin v & (2 + \cos u)\cos v & 0 \end{vmatrix} \\ &= -(2 + \cos u)(\cos v \cos u\mathbf{i} + \sin v \cos u\mathbf{j} + \sin u\mathbf{k})\end{aligned}$$

which implies that

$$\begin{aligned}\|\mathbf{r}_u \times \mathbf{r}_v\| &= (2 + \cos u)\sqrt{(\cos v \cos u)^2 + (\sin v \cos u)^2 + \sin^2 u} \\ &= (2 + \cos u)\sqrt{\cos^2 u(\cos^2 v + \sin^2 v) + \sin^2 u} \\ &= (2 + \cos u)\sqrt{\cos^2 u + \sin^2 u} \\ &= 2 + \cos u.\end{aligned}$$

Finally, the surface area of the torus is

$$\begin{aligned}A = \int\!\!\int_D \|\mathbf{r}_u \times \mathbf{r}_v\|\, dA &= \int_0^{2\pi}\!\!\int_0^{2\pi} (2 + \cos u)\, du\, dv \\ &= \int_0^{2\pi} 4\pi\, dv \\ &= 8\pi^2.\end{aligned}$$

If the surface S is a surface of revolution, you can show that the formula for surface area given in Section 7.4 is equivalent to the formula given in this section. For instance, suppose f is a nonnegative function such that f' is continuous over the interval $[a, b]$. Let S be the surface of revolution formed by revolving the graph of f, where $a \le x \le b$, about the x-axis. From Section 7.4, you know that the surface area is given by

$$\text{Surface area} = 2\pi \int_a^b f(x)\sqrt{1 + [f'(x)]^2}\, dx.$$

To represent S parametrically, let $x = u$, $y = f(u)\cos v$, and $z = f(u)\sin v$, where $a \le u \le b$ and $0 \le v \le 2\pi$. Then,

$$\mathbf{r}(u, v) = u\mathbf{i} + f(u)\cos v\mathbf{j} + f(u)\sin v\mathbf{k}.$$

Try showing that the formula

$$\text{Surface area} = \int\!\!\int_D \|\mathbf{r}_u \times \mathbf{r}_v\|\, dA$$

is equivalent to the formula given above (see Exercise 52).

Exercises for Section 15.5

In Exercises 1–4, match the vector-valued function with its graph. [The graphs are labeled (a), (b), (c), and (d).]

(a)

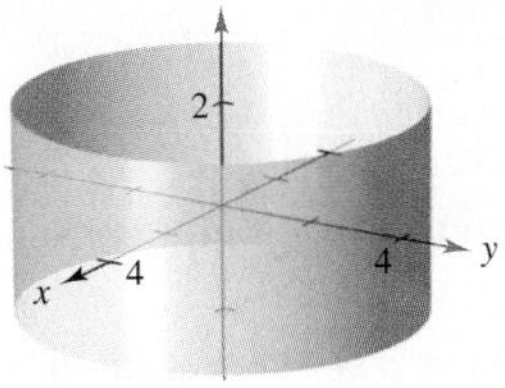

(b)

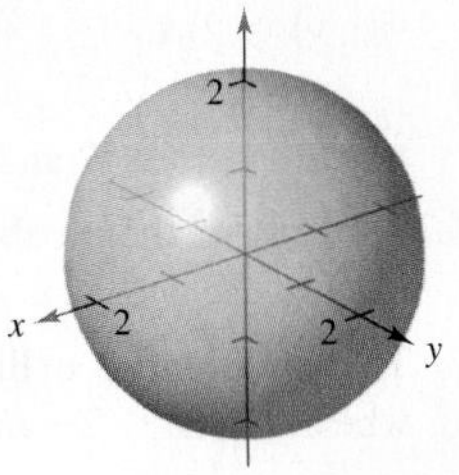

(c)

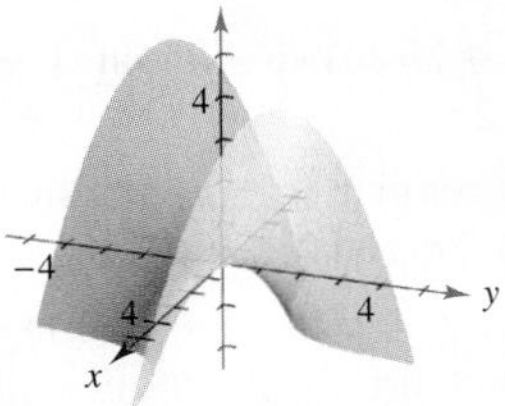

(d)

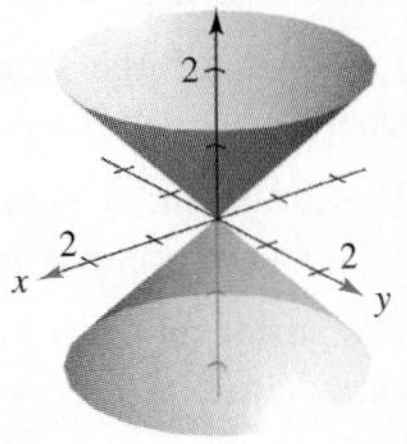

1. $\mathbf{r}(u, v) = u\mathbf{i} + v\mathbf{j} + uv\mathbf{k}$
2. $\mathbf{r}(u, v) = u \cos v\mathbf{i} + u \sin v\mathbf{j} + u\mathbf{k}$
3. $\mathbf{r}(u, v) = 2 \cos v \cos u\mathbf{i} + 2 \cos v \sin u\mathbf{j} + 2 \sin v\mathbf{k}$
4. $\mathbf{r}(u, v) = 4 \cos u\mathbf{i} + 4 \sin u\mathbf{j} + v\mathbf{k}$

In Exercises 5–8, find the rectangular equation for the surface by eliminating the parameters from the vector-valued function. Identify the surface and sketch its graph.

5. $\mathbf{r}(u, v) = u\mathbf{i} + v\mathbf{j} + \dfrac{v}{2}\mathbf{k}$
6. $\mathbf{r}(u, v) = 2u \cos v\mathbf{i} + 2u \sin v\mathbf{j} + \frac{1}{2}u^2\mathbf{k}$
7. $\mathbf{r}(u, v) = 2 \cos u\mathbf{i} + v\mathbf{j} + 2 \sin u\mathbf{k}$
8. $\mathbf{r}(u, v) = 3 \cos v \cos u\mathbf{i} + 3 \cos v \sin u\mathbf{j} + 5 \sin v\mathbf{k}$

Think About It **In Exercises 9–12, determine how the graph of the surface $\mathbf{s}(u, v)$ differs from the graph of $\mathbf{r}(u, v) = u \cos v\mathbf{i} + u \sin v\mathbf{j} + u^2\mathbf{k}$ (see figure) where $0 \le u \le 2$ and $0 \le v \le 2\pi$. (It is not necessary to graph s.)**

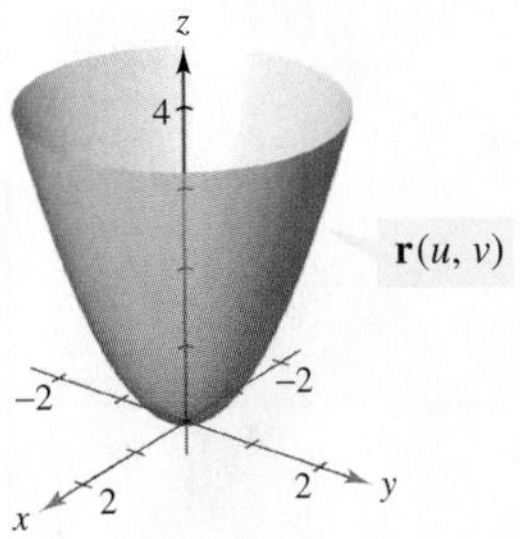

9. $\mathbf{s}(u, v) = u \cos v\mathbf{i} + u \sin v\mathbf{j} - u^2\mathbf{k}$
   $0 \le u \le 2,\quad 0 \le v \le 2\pi$
10. $\mathbf{s}(u, v) = u \cos v\mathbf{i} + u^2\mathbf{j} + u \sin v\mathbf{k}$
    $0 \le u \le 2,\quad 0 \le v \le 2\pi$
11. $\mathbf{s}(u, v) = u \cos v\mathbf{i} + u \sin v\mathbf{j} + u^2\mathbf{k}$
    $0 \le u \le 3,\quad 0 \le v \le 2\pi$
12. $\mathbf{s}(u, v) = 4u \cos v\mathbf{i} + 4u \sin v\mathbf{j} + u^2\mathbf{k}$
    $0 \le u \le 2,\quad 0 \le v \le 2\pi$

In Exercises 13–18, use a computer algebra system to graph the surface represented by the vector-valued function.

13. $\mathbf{r}(u, v) = 2u \cos v\mathbf{i} + 2u \sin v\mathbf{j} + u^4\mathbf{k}$
    $0 \le u \le 1,\quad 0 \le v \le 2\pi$
14. $\mathbf{r}(u, v) = 2 \cos v \cos u\mathbf{i} + 4 \cos v \sin u\mathbf{j} + \sin v\mathbf{k}$
    $0 \le u \le 2\pi,\quad 0 \le v \le 2\pi$
15. $\mathbf{r}(u, v) = 2 \sinh u \cos v\mathbf{i} + \sinh u \sin v\mathbf{j} + \cosh u\mathbf{k}$
    $0 \le u \le 2,\quad 0 \le v \le 2\pi$
16. $\mathbf{r}(u, v) = 2u \cos v\mathbf{i} + 2u \sin v\mathbf{j} + v\mathbf{k}$
    $0 \le u \le 1,\quad 0 \le v \le 3\pi$
17. $\mathbf{r}(u, v) = (u - \sin u)\cos v\mathbf{i} + (1 - \cos u)\sin v\mathbf{j} + u\mathbf{k}$
    $0 \le u \le \pi,\quad 0 \le v \le 2\pi$
18. $\mathbf{r}(u, v) = \cos^3 u \cos v\mathbf{i} + \sin^3 u \sin v\mathbf{j} + u\mathbf{k}$
    $0 \le u \le \dfrac{\pi}{2},\quad 0 \le v \le 2\pi$

In Exercises 19–26, find a vector-valued function whose graph is the indicated surface.

19. The plane $z = y$
20. The plane $x + y + z = 6$
21. The cylinder $x^2 + y^2 = 16$
22. The cylinder $4x^2 + y^2 = 16$
23. The cylinder $z = x^2$
24. The ellipsoid $\dfrac{x^2}{9} + \dfrac{y^2}{4} + \dfrac{z^2}{1} = 1$
25. The part of the plane $z = 4$ that lies inside the cylinder $x^2 + y^2 = 9$
26. The part of the paraboloid $z = x^2 + y^2$ that lies inside the cylinder $x^2 + y^2 = 9$

Surface of Revolution **In Exercises 27–30, write a set of parametric equations for the surface of revolution obtained by revolving the graph of the function about the given axis.**

	Function	*Axis of Revolution*
27.	$y = \dfrac{x}{2},\quad 0 \le x \le 6$	x-axis
28.	$y = x^{3/2},\quad 0 \le x \le 4$	x-axis
29.	$x = \sin z,\quad 0 \le z \le \pi$	z-axis
30.	$z = 4 - y^2,\quad 0 \le y \le 2$	y-axis

Tangent Plane **In Exercises 31–34, find an equation of the tangent plane to the surface represented by the vector-valued function at the given point.**

31. $\mathbf{r}(u, v) = (u + v)\mathbf{i} + (u - v)\mathbf{j} + v\mathbf{k}, \quad (1, -1, 1)$

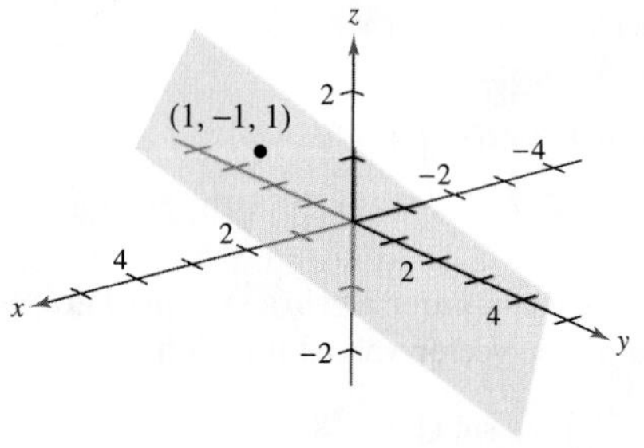

32. $\mathbf{r}(u, v) = u\mathbf{i} + v\mathbf{j} + \sqrt{uv}\,\mathbf{k}, \quad (1, 1, 1)$

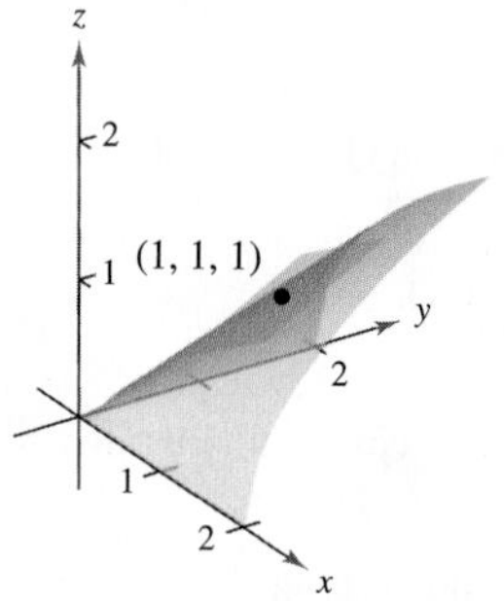

33. $\mathbf{r}(u, v) = 2u \cos v\mathbf{i} + 3u \sin v\mathbf{j} + u^2\mathbf{k}, \quad (0, 6, 4)$

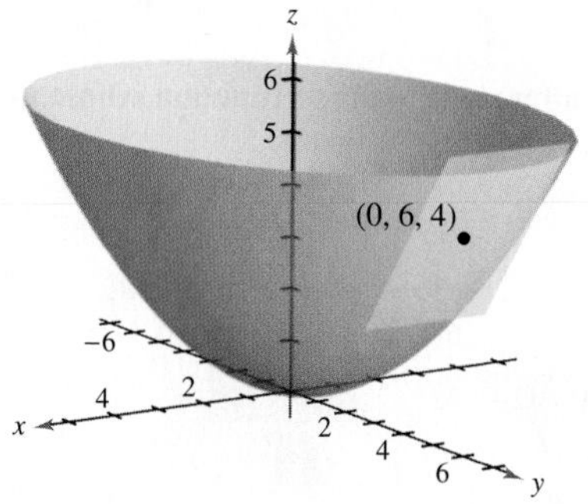

34. $\mathbf{r}(u, v) = 2u \cosh v\mathbf{i} + 2u \sinh v\mathbf{j} + \frac{1}{2}u^2\mathbf{k}, \quad (-4, 0, 2)$

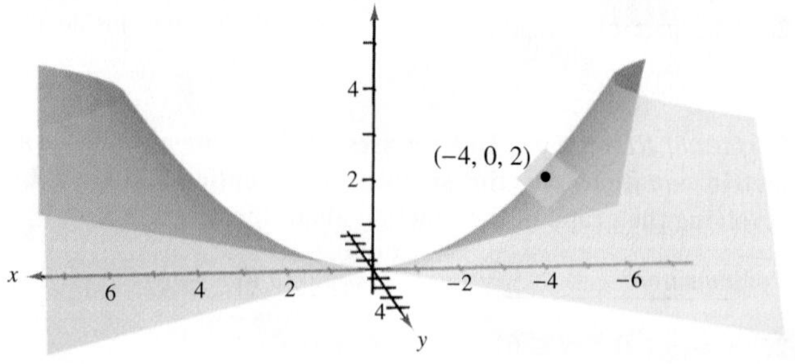

Area **In Exercises 35–42, find the area of the surface over the given region. Use a computer algebra system to verify your results.**

35. The part of the plane

$$\mathbf{r}(u, v) = 2u\mathbf{i} - \frac{v}{2}\mathbf{j} + \frac{v}{2}\mathbf{k}$$

where $0 \le u \le 2$ and $0 \le v \le 1$

36. The part of the paraboloid $\mathbf{r}(u, v) = 4u \cos v\mathbf{i} + 4u \sin v\mathbf{j} + u^2\mathbf{k}$, where $0 \le u \le 2$ and $0 \le v \le 2\pi$

37. The part of the cylinder $\mathbf{r}(u, v) = a \cos u\mathbf{i} + a \sin u\mathbf{j} + v\mathbf{k}$, where $0 \le u \le 2\pi$ and $0 \le v \le b$

38. The sphere $\mathbf{r}(u, v) = a \sin u \cos v\mathbf{i} + a \sin u \sin v\mathbf{j} + a \cos u\mathbf{k}$, where $0 \le u \le \pi$ and $0 \le v \le 2\pi$

39. The part of the cone $\mathbf{r}(u, v) = au \cos v\mathbf{i} + au \sin v\mathbf{j} + u\mathbf{k}$, where $0 \le u \le b$ and $0 \le v \le 2\pi$

40. The torus $\mathbf{r}(u, v) = (a + b \cos v)\cos u\mathbf{i} + (a + b \cos v)\sin u\mathbf{j} + b \sin v\mathbf{k}$, where $a > b$, $0 \le u \le 2\pi$, and $0 \le v \le 2\pi$

41. The surface of revolution $\mathbf{r}(u, v) = \sqrt{u} \cos v\mathbf{i} + \sqrt{u} \sin v\mathbf{j} + u\mathbf{k}$, where $0 \le u \le 4$ and $0 \le v \le 2\pi$

42. The surface of revolution $\mathbf{r}(u, v) = \sin u \cos v\mathbf{i} + u\mathbf{j} + \sin u \sin v\mathbf{k}$, where $0 \le u \le \pi$ and $0 \le v \le 2\pi$

Writing About Concepts

43. Define a parametric surface.

44. Give the double integral that yields the surface area of a parametric surface over an open region D.

45. The four figures are graphs of the surface $\mathbf{r}(u, v) = u\mathbf{i} + \sin u \cos v\mathbf{j} + \sin u \sin v\mathbf{k}$,

$$0 \le u \le \frac{\pi}{2}, \quad 0 \le v \le 2\pi.$$

Match each of the four graphs with the point in space from which the surface is viewed. The four points are $(10, 0, 0)$, $(-10, 10, 0)$, $(0, 10, 0)$, and $(10, 10, 10)$.

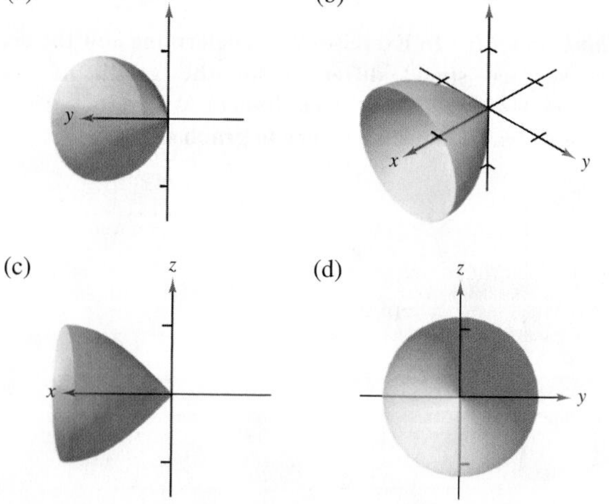

46. Use a computer algebra system to graph three views of the graph of the vector-valued function

$$\mathbf{r}(u, v) = u \cos v\mathbf{i} + u \sin v\mathbf{j} + v\mathbf{k}, \quad 0 \le u \le \pi, \quad 0 \le v \le \pi$$

from the points (10, 0, 0), (0, 0, 10), and (10, 10, 10).

47. ***Investigation*** Use a computer algebra system to graph the torus

$$\mathbf{r}(u, v) = (a + b \cos v) \cos u\mathbf{i} + (a + b \cos v) \sin u\mathbf{j} + b \sin v\mathbf{k}$$

for each set of values of a and b, where $0 \le u \le 2\pi$ and $0 \le v \le 2\pi$. Use the results to describe the effects of a and b on the shape of the torus.

(a) $a = 4, \quad b = 1$ (b) $a = 4, \quad b = 2$

(c) $a = 8, \quad b = 1$ (d) $a = 8, \quad b = 3$

48. ***Investigation*** Consider the function in Exercise 16.

(a) Sketch a graph of the function where u is held constant at $u = 1$. Identify the graph.

(b) Sketch a graph of the function where v is held constant at $v = 2\pi/3$. Identify the graph.

(c) Assume that a surface is represented by the vector-valued function $\mathbf{r} = \mathbf{r}(u, v)$. What generalization can you make about the graph of the function if one of the parameters is held constant?

49. ***Surface Area*** The surface of the dome on a new museum is given by

$$\mathbf{r}(u, v) = 20 \sin u \cos v\mathbf{i} + 20 \sin u \sin v\mathbf{j} + 20 \cos u\mathbf{k}$$

where $0 \le u \le \pi/3$ and $0 \le v \le 2\pi$ and $\mathbf{r}$ is in meters. Find the surface area of the dome.

50. Find a vector-valued function for the hyperboloid

$$x^2 + y^2 - z^2 = 1$$

and determine the tangent plane at (1, 0, 0).

51. Graph and find the area of one turn of the spiral ramp

$$\mathbf{r}(u, v) = u \cos v\mathbf{i} + u \sin v\mathbf{j} + 2v\mathbf{k}$$

where $0 \le u \le 3$, and $0 \le v \le 2\pi$.

52. Let f be a nonnegative function such that f' is continuous over the interval $[a, b]$. Let S be the surface of revolution formed by revolving the graph of f, where $a \le x \le b$, about the x-axis. Let $x = u$, $y = f(u) \cos v$, and $z = f(u)\sin v$, where $a \le u \le b$ and $0 \le v \le 2\pi$. Then, S is represented parametrically by

$$\mathbf{r}(u, v) = u\mathbf{i} + f(u) \cos v\mathbf{j} + f(u) \sin v\mathbf{k}.$$

Show that the following formulas are equivalent.

$$\text{Surface area} = 2\pi \int_a^b f(x)\sqrt{1 + [f'(x)]^2}\, dx$$

$$\text{Surface area} = \int\!\!\int_D \|\mathbf{r}_u \times \mathbf{r}_v\|\, dA$$

53. ***Open-Ended Project*** The parametric equations

$$x = 3 + \sin u[7 - \cos(3u - 2v) - 2 \cos(3u + v)]$$

$$y = 3 + \cos u[7 - \cos(3u - 2v) - 2 \cos(3u + v)]$$

$$z = \sin(3u - 2v) + 2 \sin(3u + v)$$

where $-\pi \le u \le \pi$ and $-\pi \le v \le \pi$, represent the surface shown below. Try to create your own parametric surface using a computer algebra system.

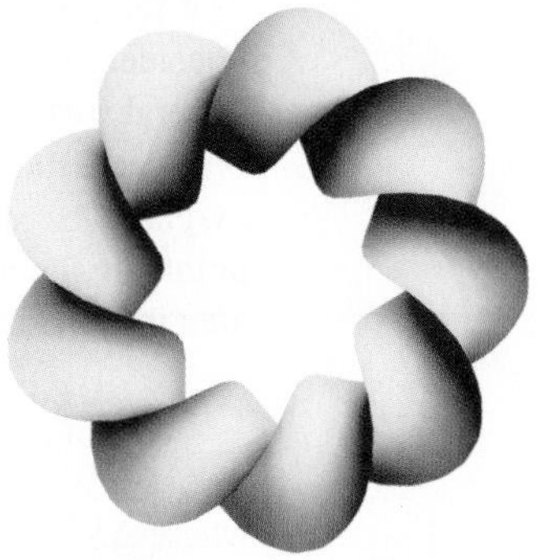

54. ***Möbius Strip*** The surface shown in the figure is called a **Möbius Strip** and can be represented by the parametric equations

$$x = \left(a + u \cos\frac{v}{2}\right) \cos v, \; y = \left(a + u \cos\frac{v}{2}\right) \sin v, \; z = u \sin\frac{v}{2}$$

where $-1 \le u \le 1$, $0 \le v \le 2\pi$, and $a = 3$. Try to graph other Möbius strips for different values of a using a computer algebra system.

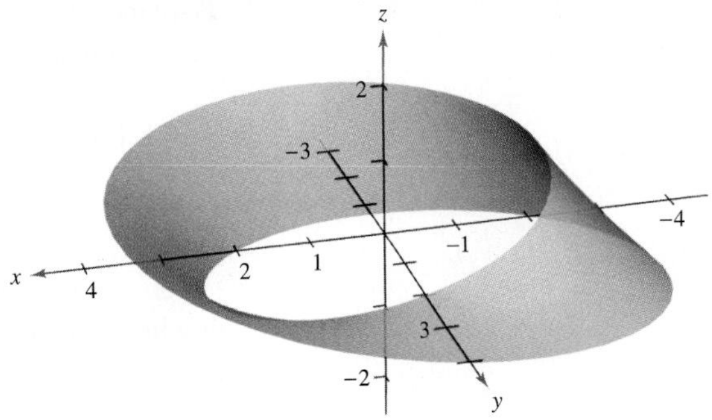

Section 15.6 Surface Integrals

- **Evaluate a surface integral as a double integral.**
- **Evaluate a surface integral for a parametric surface.**
- **Determine the orientation of a surface.**
- **Understand the concept of a flux integral.**

Surface Integrals

The remainder of this chapter deals primarily with **surface integrals.** You will first consider surfaces given by $z = g(x, y)$. Later in this section you will consider more general surfaces given in parametric form.

Let S be a surface given by $z = g(x, y)$ and let R be its projection onto the xy-plane, as shown in Figure 15.44. Suppose that g, g_x, and g_y are continuous at all points in R and that f is defined on S. Employing the procedure used to find surface area in Section 14.5, evaluate f at (x_i, y_i, z_i) and form the sum

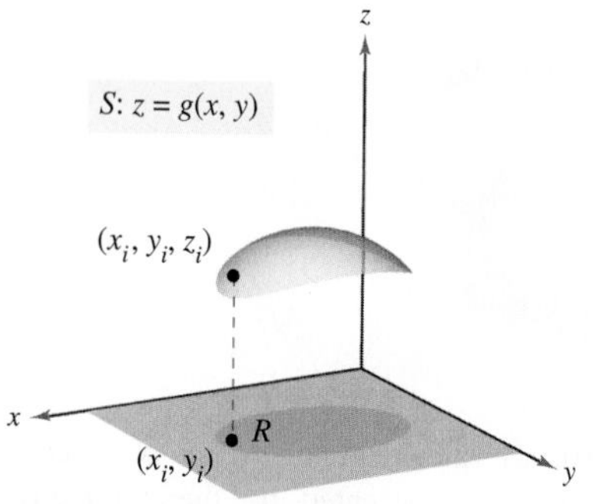

Scalar function f assigns a number to each point of S.
Figure 15.44

$$\sum_{i=1}^{n} f(x_i, y_i, z_i)\,\Delta S_i$$

where $\Delta S_i \approx \sqrt{1 + [g_x(x_i, y_i)]^2 + [g_y(x_i, y_i)]^2}\,\Delta A_i$. Provided the limit of the above sum as $\|\Delta\|$ approaches 0 exists, the **surface integral of f over S** is defined as

$$\int_S\!\!\int f(x, y, z)\,dS = \lim_{\|\Delta\| \to 0} \sum_{i=1}^{n} f(x_i, y_i, z_i)\,\Delta S_i.$$

This integral can be evaluated by a double integral.

THEOREM 15.10 Evaluating a Surface Integral

Let S be a surface with equation $z = g(x, y)$ and let R be its projection onto the xy-plane. If g, g_x, and g_y are continuous on R and f is continuous on S, then the surface integral of f over S is

$$\int_S\!\!\int f(x, y, z)\,dS = \int_R\!\!\int f(x, y, g(x, y))\sqrt{1 + [g_x(x, y)]^2 + [g_y(x, y)]^2}\,dA.$$

For surfaces described by functions of x and z (or y and z), you can make the following adjustments to Theorem 15.10. If S is the graph of $y = g(x, z)$ and R is its projection onto the xz-plane, then

$$\int_S\!\!\int f(x, y, z)\,dS = \int_R\!\!\int f(x, g(x, z), z)\sqrt{1 + [g_x(x, z)]^2 + [g_z(x, z)]^2}\,dA.$$

If S is the graph of $x = g(y, z)$ and R is its projection onto the yz-plane, then

$$\int_S\!\!\int f(x, y, z)\,dS = \int_R\!\!\int f(g(y, z), y, z)\sqrt{1 + [g_y(y, z)]^2 + [g_z(y, z)]^2}\,dA.$$

If $f(x, y, z) = 1$, the surface integral over S yields the surface area of S. For instance, suppose the surface S is the plane given by $z = x$, where $0 \le x \le 1$ and $0 \le y \le 1$. The surface area of S is $\sqrt{2}$ square units. Try verifying that $\int_S\!\int f(x, y, z)\,dS = \sqrt{2}$.

EXAMPLE 1 Evaluating a Surface Integral

Evaluate the surface integral

$$\int_S\int (y^2 + 2yz)\, dS$$

where S is the first-octant portion of the plane $2x + y + 2z = 6$.

Solution Begin by writing S as

$$z = \frac{1}{2}(6 - 2x - y)$$

$$g(x, y) = \frac{1}{2}(6 - 2x - y).$$

Using the partial derivatives $g_x(x, y) = -1$ and $g_y(x, y) = -\frac{1}{2}$, you can write

$$\sqrt{1 + [g_x(x, y)]^2 + [g_y(x, y)]^2} = \sqrt{1 + 1 + \frac{1}{4}} = \frac{3}{2}.$$

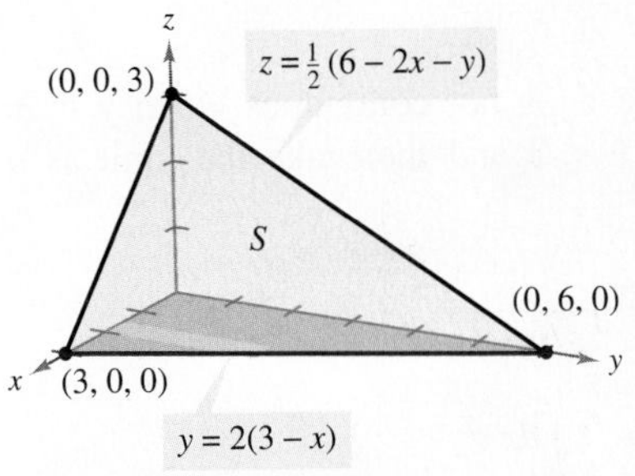

Figure 15.45

Using Figure 15.45 and Theorem 15.10, you obtain

$$\begin{aligned}\int_S\int (y^2 + 2yz)\, dS &= \int_R\int f(x, y, g(x, y))\sqrt{1 + [g_x(x, y)]^2 + [g_y(x, y)]^2}\, dA \\ &= \int_R\int \left[y^2 + 2y\left(\frac{1}{2}\right)(6 - 2x - y)\right]\left(\frac{3}{2}\right) dA \\ &= 3\int_0^3\int_0^{2(3-x)} y(3 - x)\, dy\, dx \\ &= 6\int_0^3 (3 - x)^3\, dx \\ &= -\frac{3}{2}(3 - x)^4\Big]_0^3 \\ &= \frac{243}{2}.\end{aligned}$$

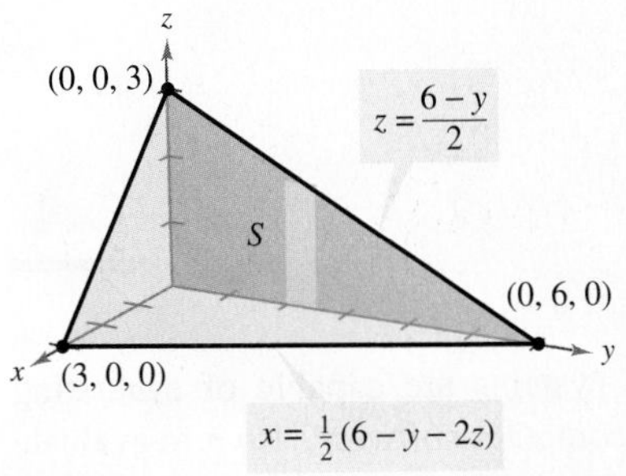

Figure 15.46

An alternative solution to Example 1 would be to project S onto the yz-plane, as shown in Figure 15.46. Then, $x = \frac{1}{2}(6 - y - 2z)$, and

$$\sqrt{1 + [g_y(y, z)]^2 + [g_z(y, z)]^2} = \sqrt{1 + \frac{1}{4} + 1} = \frac{3}{2}.$$

So, the surface integral is

$$\begin{aligned}\int_S\int (y^2 + 2yz)\, dS &= \int_R\int f(g(y, z), y, z)\sqrt{1 + [g_y(y, z)]^2 + [g_z(y, z)]^2}\, dA \\ &= \int_0^6\int_0^{(6-y)/2} (y^2 + 2yz)\left(\frac{3}{2}\right) dz\, dy \\ &= \frac{3}{8}\int_0^6 (36y - y^3)\, dy \\ &= \frac{243}{2}.\end{aligned}$$

Try reworking Example 1 by projecting S onto the xz-plane.

In Example 1, you could have projected the surface S onto any one of the three coordinate planes. In Example 2, S is a portion of a cylinder centered about the x-axis, and you can project it onto either the xz-plane or the xy-plane.

EXAMPLE 2 Evaluating a Surface Integral

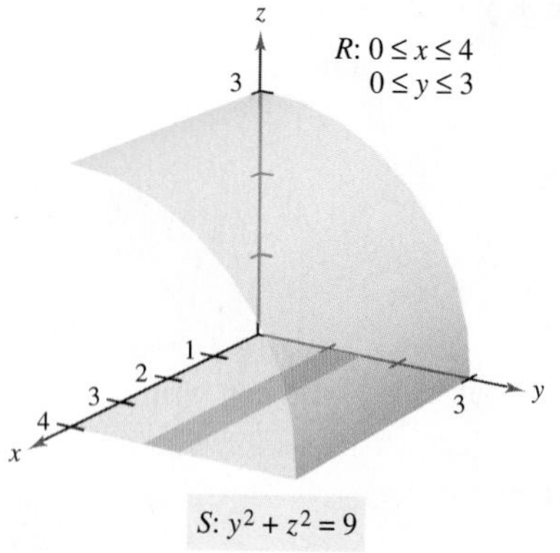

Figure 15.47

Evaluate the surface integral

$$\int_S\int (x+z)\,dS$$

where S is the first-octant portion of the cylinder $y^2 + z^2 = 9$ between $x = 0$ and $x = 4$, as shown in Figure 15.47.

Solution Project S onto the xy-plane, so that $z = g(x, y) = \sqrt{9 - y^2}$, and obtain

$$\sqrt{1 + [g_x(x,y)]^2 + [g_y(x,y)]^2} = \sqrt{1 + \left(\frac{-y}{\sqrt{9-y^2}}\right)^2}$$
$$= \frac{3}{\sqrt{9-y^2}}.$$

Theorem 15.10 does not apply directly because g_y is not continuous when $y = 3$. However, you can apply the theorem for $0 \le b < 3$ and then take the limit as b approaches 3, as follows.

$$\begin{aligned}
\int_S\int (x+z)\,dS &= \lim_{b\to 3^-} \int_0^b\int_0^4 \left(x + \sqrt{9-y^2}\right)\frac{3}{\sqrt{9-y^2}}\,dx\,dy \\
&= \lim_{b\to 3^-} 3\int_0^b\int_0^4 \left(\frac{x}{\sqrt{9-y^2}} + 1\right)dx\,dy \\
&= \lim_{b\to 3^-} 3\int_0^b \left[\frac{x^2}{2\sqrt{9-y^2}} + x\right]_0^4 dy \\
&= \lim_{b\to 3^-} 3\int_0^b \left(\frac{8}{\sqrt{9-y^2}} + 4\right)dy \\
&= \lim_{b\to 3^-} 3\left[4y + 8\arcsin\frac{y}{3}\right]_0^b \\
&= \lim_{b\to 3^-} 3\left(4b + 8\arcsin\frac{b}{3}\right) \\
&= 36 + 24\left(\frac{\pi}{2}\right) \\
&= 36 + 12\pi
\end{aligned}$$

TECHNOLOGY Some computer algebra systems are capable of evaluating improper integrals. If you have access to such computer software, use it to evaluate the improper integral

$$\int_0^3\int_0^4 \left(x + \sqrt{9-y^2}\right)\frac{3}{\sqrt{9-y^2}}\,dx\,dy.$$

Do you obtain the same result as in Example 2?

You have already seen that if the function f defined on the surface S is simply $f(x, y, z) = 1$, the surface integral yields the *surface area* of S.

$$\text{Area of surface} = \int_S\int 1\, dS$$

On the other hand, if S is a lamina of variable density and $\rho(x, y, z)$ is the density at the point (x, y, z), then the *mass* of the lamina is given by

$$\text{Mass of lamina} = \int_S\int \rho(x, y, z)\, dS.$$

EXAMPLE 3 Finding the Mass of a Surface Lamina

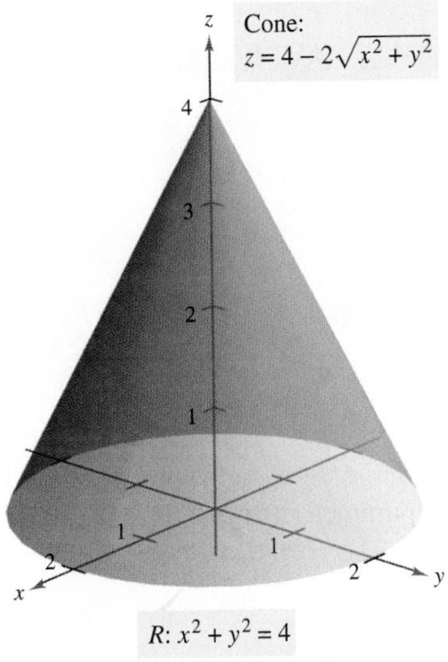

Figure 15.48

A cone-shaped surface lamina S is given by

$$z = 4 - 2\sqrt{x^2 + y^2}, \quad 0 \le z \le 4$$

as shown in Figure 15.48. At each point on S, the density is proportional to the distance between the point and the z-axis. Find the mass m of the lamina.

Solution Projecting S onto the xy-plane produces

$$S: z = 4 - 2\sqrt{x^2 + y^2} = g(x, y), \quad 0 \le z \le 4$$
$$R: x^2 + y^2 \le 4$$

with a density of $\rho(x, y, z) = k\sqrt{x^2 + y^2}$. Using a surface integral, you can find the mass to be

$$\begin{aligned}
m &= \int_S\int \rho(x, y, z)\, dS \\
&= \int_R\int k\sqrt{x^2 + y^2}\sqrt{1 + [g_x(x, y)]^2 + [g_y(x, y)]^2}\, dA \\
&= k\int_R\int \sqrt{x^2 + y^2}\sqrt{1 + \frac{4x^2}{x^2 + y^2} + \frac{4y^2}{x^2 + y^2}}\, dA \\
&= k\int_R\int \sqrt{5}\sqrt{x^2 + y^2}\, dA \\
&= k\int_0^{2\pi}\int_0^2 \left(\sqrt{5}r\right) r\, dr\, d\theta \qquad \text{Polar coordinates} \\
&= \frac{\sqrt{5}k}{3}\int_0^{2\pi} r^3\Big]_0^2\, d\theta \\
&= \frac{8\sqrt{5}k}{3}\int_0^{2\pi} d\theta \\
&= \frac{8\sqrt{5}k}{3}\Big[\theta\Big]_0^{2\pi} = \frac{16\sqrt{5}k\pi}{3}.
\end{aligned}$$

TECHNOLOGY Use a computer algebra system to confirm the result shown in Example 3. The computer algebra system *Derive* evaluated the integral as follows.

$$k\int_{-2}^{2}\int_{-\sqrt{4-y^2}}^{\sqrt{4-y^2}} \sqrt{5}\sqrt{x^2 + y^2}\, dx\, dy = \frac{16\sqrt{5}k\pi}{3}$$

Parametric Surfaces and Surface Integrals

For a surface S given by the vector-valued function

$$\mathbf{r}(u, v) = x(u, v)\mathbf{i} + y(u, v)\mathbf{j} + z(u, v)\mathbf{k} \qquad \text{Parametric surface}$$

defined over a region D in the uv-plane, you can show that the surface integral of $f(x, y, z)$ over S is given by

$$\int_S\int f(x, y, z)\, dS = \int_D\int f(x(u, v), y(u, v), z(u, v))\, \|\mathbf{r}_u(u, v) \times \mathbf{r}_v(u, v)\|\, dA.$$

Note the similarity to a line integral over a space curve C.

$$\int_C f(x, y, z)\, ds = \int_a^b f(x(t), y(t), z(t))\, \|\mathbf{r}'(t)\|\, dt \qquad \text{Line integral}$$

NOTE Notice that ds and dS can be written as $ds = \|\mathbf{r}'(t)\|\, dt$ and $dS = \|\mathbf{r}_u(u, v) \times \mathbf{r}_v(u, v)\|\, dA$.

EXAMPLE 4 Evaluating a Surface Integral

Example 2 demonstrated an evaluation of the surface integral

$$\int_S\int (x + z)\, dS$$

where S is the first-octant portion of the cylinder $y^2 + z^2 = 9$ between $x = 0$ and $x = 4$ (see Figure 15.49). Reevaluate this integral in parametric form.

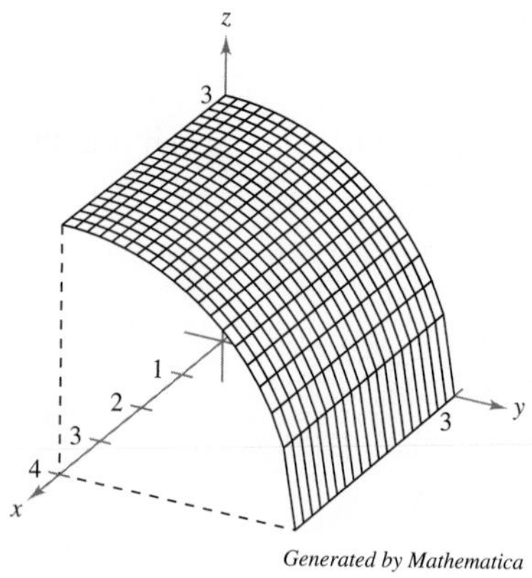

Figure 15.49

Solution In parametric form, the surface is given by

$$\mathbf{r}(x, \theta) = x\mathbf{i} + 3\cos\theta\mathbf{j} + 3\sin\theta\mathbf{k}$$

where $0 \le x \le 4$ and $0 \le \theta \le \pi/2$. To evaluate the surface integral in parametric form, begin by calculating the following.

$$\mathbf{r}_x = \mathbf{i}$$

$$\mathbf{r}_\theta = -3\sin\theta\mathbf{j} + 3\cos\theta\mathbf{k}$$

$$\mathbf{r}_x \times \mathbf{r}_\theta = \begin{vmatrix} \mathbf{i} & \mathbf{j} & \mathbf{k} \\ 1 & 0 & 0 \\ 0 & -3\sin\theta & 3\cos\theta \end{vmatrix} = -3\cos\theta\mathbf{j} - 3\sin\theta\mathbf{k}$$

$$\|\mathbf{r}_x \times \mathbf{r}_\theta\| = \sqrt{9\cos^2\theta + 9\sin^2\theta} = 3$$

So, the surface integral can be evaluated as follows.

$$\begin{aligned}\int_D\int (x + 3\sin\theta)3\, dA &= \int_0^4\int_0^{\pi/2} (3x + 9\sin\theta)\, d\theta\, dx \\ &= \int_0^4 \Big[3x\theta - 9\cos\theta\Big]_0^{\pi/2} dx \\ &= \int_0^4 \left(\frac{3\pi}{2}x + 9\right) dx \\ &= \left[\frac{3\pi}{4}x^2 + 9x\right]_0^4 \\ &= 12\pi + 36\end{aligned}$$

Orientation of a Surface

Unit normal vectors are used to induce an orientation to a surface S in space. A surface is called **orientable** if a unit normal vector $\mathbf{N}$ can be defined at every nonboundary point of S in such a way that the normal vectors vary continuously over the surface S. If this is possible, S is called an **oriented surface.**

An orientable surface S has two distinct sides. So, when you orient a surface, you are selecting one of the two possible unit normal vectors. If S is a closed surface such as a sphere, it is customary to choose the unit normal vector $\mathbf{N}$ to be the one that points outward from the sphere.

Most common surfaces, such as spheres, paraboloids, ellipses, and planes, are orientable. (See Exercise 43 for an example of a surface that is *not* orientable.) Moreover, for an orientable surface, the gradient vector provides a convenient way to find a unit normal vector. That is, for an orientable surface S given by

$$z = g(x, y) \qquad \text{Orientable surface}$$

let

$$G(x, y, z) = z - g(x, y).$$

Then, S can be oriented by either the unit normal vector

$$\mathbf{N} = \frac{\nabla G(x, y, z)}{\|\nabla G(x, y, z)\|} = \frac{-g_x(x, y)\mathbf{i} - g_y(x, y)\mathbf{j} + \mathbf{k}}{\sqrt{1 + [g_x(x, y)]^2 + [g_y(x, y)]^2}} \qquad \text{Upward unit normal}$$

or the unit normal vector

$$\mathbf{N} = \frac{-\nabla G(x, y, z)}{\|\nabla G(x, y, z)\|} = \frac{g_x(x, y)\mathbf{i} + g_y(x, y)\mathbf{j} - \mathbf{k}}{\sqrt{1 + [g_x(x, y)]^2 + [g_y(x, y)]^2}} \qquad \text{Downward unit normal}$$

as shown in Figure 15.50. If the smooth orientable surface S is given in parametric form by

$$\mathbf{r}(u, v) = x(u, v)\mathbf{i} + y(u, v)\mathbf{j} + z(u, v)\mathbf{k} \qquad \text{Parametric surface}$$

the unit normal vectors are given by

$$\mathbf{N} = \frac{\mathbf{r}_u \times \mathbf{r}_v}{\|\mathbf{r}_u \times \mathbf{r}_v\|}$$

and

$$\mathbf{N} = \frac{\mathbf{r}_v \times \mathbf{r}_u}{\|\mathbf{r}_v \times \mathbf{r}_u\|}.$$

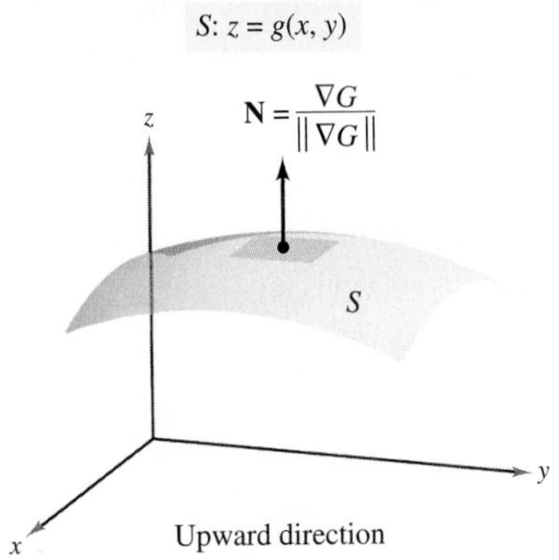

S is oriented in an upward direction.

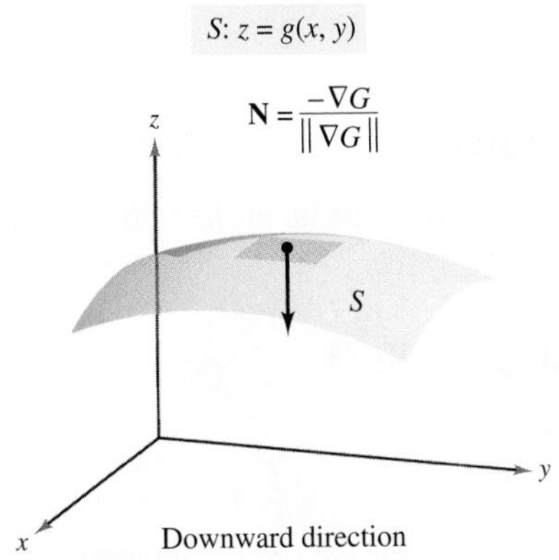

S is oriented in a downward direction.
Figure 15.50

NOTE Suppose that the orientable surface is given by $y = g(x, z)$ or $x = g(y, z)$. Then you can use the gradient vector

$$\nabla G(x, y, z) = -g_x(x, z)\mathbf{i} + \mathbf{j} - g_z(x, z)\mathbf{k} \qquad G(x, y, z) = y - g(x, z)$$

or

$$\nabla G(x, y, z) = \mathbf{i} - g_y(y, z)\mathbf{j} - g_z(y, z)\mathbf{k} \qquad G(x, y, z) = x - g(y, z)$$

to orient the surface.

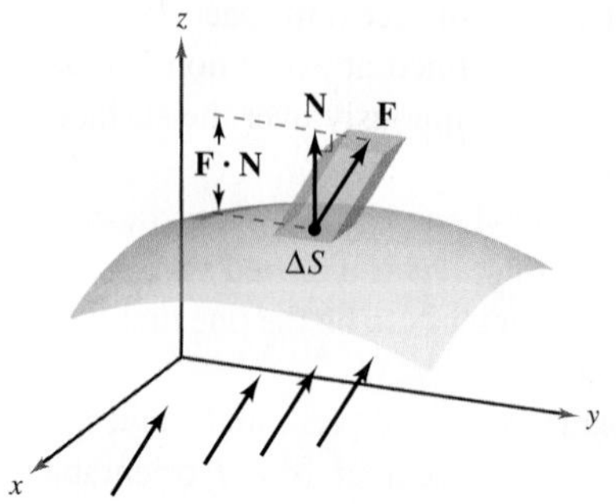

The velocity field **F** indicates the direction of the fluid flow.
Figure 15.51

Flux Integrals

One of the principal applications involving the vector form of a surface integral relates to the flow of a fluid through a surface S. Suppose an oriented surface S is submerged in a fluid having a continuous velocity field $\mathbf{F}$. Let ΔS be the area of a small patch of the surface S over which $\mathbf{F}$ is nearly constant. Then the amount of fluid crossing this region per unit of time is approximated by the volume of the column of height $\mathbf{F} \cdot \mathbf{N}$, as shown in Figure 15.51. That is,

$$\Delta V = (\text{height})(\text{area of base}) = (\mathbf{F} \cdot \mathbf{N})\Delta S.$$

Consequently, the volume of fluid crossing the surface S per unit of time (called the **flux of F across** S) is given by the surface integral in the following definition.

Definition of Flux Integral

Let $\mathbf{F}(x, y, z) = M\mathbf{i} + N\mathbf{j} + P\mathbf{k}$, where M, N, and P have continuous first partial derivatives on the surface S oriented by a unit normal vector $\mathbf{N}$. The **flux integral of F across** S is given by

$$\int\!\!\int_S \mathbf{F} \cdot \mathbf{N}\, dS.$$

Geometrically, a flux integral is the surface integral over S of the *normal component* of $\mathbf{F}$. If $\rho(x, y, z)$ is the density of the fluid at (x, y, z), the flux integral

$$\int\!\!\int_S \rho\,\mathbf{F} \cdot \mathbf{N}\, dS$$

represents the *mass* of the fluid flowing across S per unit of time.

To evaluate a flux integral for a surface given by $z = g(x, y)$, let

$$G(x, y, z) = z - g(x, y).$$

Then, $\mathbf{N}\, dS$ can be written as follows.

$$\begin{aligned}\mathbf{N}\, dS &= \frac{\nabla G(x, y, z)}{\|\nabla G(x, y, z)\|}\, dS\\ &= \frac{\nabla G(x, y, z)}{\sqrt{(g_x)^2 + (g_y)^2 + 1}}\sqrt{(g_x)^2 + (g_y)^2 + 1}\, dA\\ &= \nabla G(x, y, z)\, dA\end{aligned}$$

THEOREM 15.11 Evaluating a Flux Integral

Let S be an oriented surface given by $z = g(x, y)$ and let R be its projection onto the xy-plane.

$$\int\!\!\int_S \mathbf{F} \cdot \mathbf{N}\, dS = \int\!\!\int_R \mathbf{F} \cdot [-g_x(x, y)\mathbf{i} - g_y(x, y)\mathbf{j} + \mathbf{k}]\, dA \quad \text{Oriented upward}$$

$$\int\!\!\int_S \mathbf{F} \cdot \mathbf{N}\, dS = \int\!\!\int_R \mathbf{F} \cdot [g_x(x, y)\mathbf{i} + g_y(x, y)\mathbf{j} - \mathbf{k}]\, dA \quad \text{Oriented downward}$$

For the first integral, the surface is oriented upward, and for the second integral, the surface is oriented downward.

EXAMPLE 5 Using a Flux Integral to Find the Rate of Mass Flow

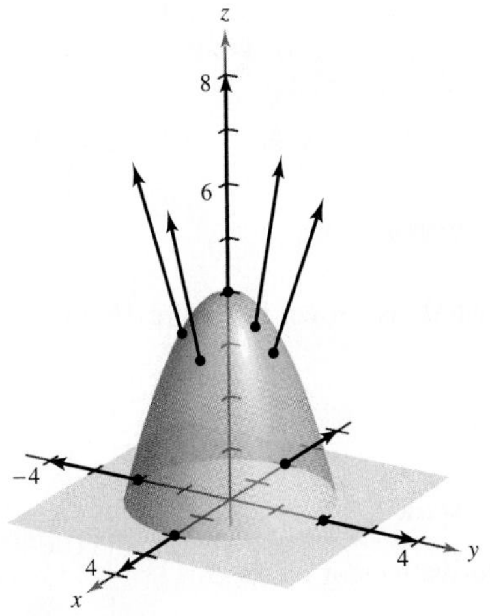

Figure 15.52

Let S be the portion of the paraboloid

$$z = g(x, y) = 4 - x^2 - y^2$$

lying above the xy-plane, oriented by an upward unit normal vector, as shown in Figure 15.52. A fluid of constant density ρ is flowing through the surface S according to the vector field

$$\mathbf{F}(x, y, z) = x\mathbf{i} + y\mathbf{j} + z\mathbf{k}.$$

Find the rate of mass flow through S.

Solution Begin by computing the partial derivatives of g.

$$g_x(x, y) = -2x$$

and

$$g_y(x, y) = -2y$$

The rate of mass flow through the surface S is

$$\begin{aligned}\int\!\!\int_S \rho\mathbf{F}\cdot\mathbf{N}\,dS &= \rho\int\!\!\int_R \mathbf{F}\cdot[-g_x(x, y)\mathbf{i} - g_y(x, y)\mathbf{j} + \mathbf{k}]\,dA \\ &= \rho\int\!\!\int_R [x\mathbf{i} + y\mathbf{j} + (4 - x^2 - y^2)\mathbf{k}]\cdot(2x\mathbf{i} + 2y\mathbf{j} + \mathbf{k})\,dA \\ &= \rho\int\!\!\int_R [2x^2 + 2y^2 + (4 - x^2 - y^2)]\,dA \\ &= \rho\int\!\!\int_R (4 + x^2 + y^2)\,dA \\ &= \rho\int_0^{2\pi}\int_0^2 (4 + r^2)r\,dr\,d\theta \qquad \text{Polar coordinates} \\ &= \rho\int_0^{2\pi} 12\,d\theta \\ &= 24\pi\rho.\end{aligned}$$

For an oriented surface S given by the vector-valued function

$$\mathbf{r}(u, v) = x(u, v)\mathbf{i} + y(u, v)\mathbf{j} + z(u, v)\mathbf{k} \qquad \text{Parametric surface}$$

defined over a region D in the uv-plane, you can define the flux integral of $\mathbf{F}$ across S as

$$\begin{aligned}\int\!\!\int_S \mathbf{F}\cdot\mathbf{N}\,dS &= \int\!\!\int_D \mathbf{F}\cdot\left(\frac{\mathbf{r}_u\times\mathbf{r}_v}{\|\mathbf{r}_u\times\mathbf{r}_v\|}\right)\|\mathbf{r}_u\times\mathbf{r}_v\|\,dA \\ &= \int\!\!\int_D \mathbf{F}\cdot(\mathbf{r}_u\times\mathbf{r}_v)\,dA.\end{aligned}$$

Note the similarity of this integral to the line integral

$$\int_C \mathbf{F}\cdot d\mathbf{r} = \int_C \mathbf{F}\cdot\mathbf{T}\,ds.$$

A summary of formulas for line and surface integrals is presented on page 1117.

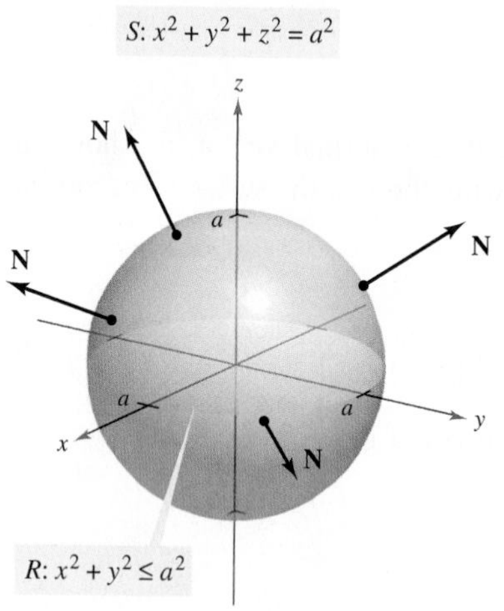

Figure 15.53

EXAMPLE 6 Finding the Flux of an Inverse Square Field

Find the flux over the sphere S given by

$$x^2 + y^2 + z^2 = a^2 \qquad \text{Sphere } S$$

where $\mathbf{F}$ is an inverse square field given by

$$\mathbf{F}(x, y, z) = \frac{kq}{\|\mathbf{r}\|^2}\frac{\mathbf{r}}{\|\mathbf{r}\|} = \frac{kq\mathbf{r}}{\|\mathbf{r}\|^3} \qquad \text{Inverse square field } \mathbf{F}$$

and $\mathbf{r} = x\mathbf{i} + y\mathbf{j} + z\mathbf{k}$. Assume S is oriented outward, as shown in Figure 15.53.

Solution The sphere is given by

$$\begin{aligned}\mathbf{r}(u, v) &= x(u, v)\mathbf{i} + y(u, v)\mathbf{j} + z(u, v)\mathbf{k}\\ &= a \sin u \cos v\mathbf{i} + a \sin u \sin v\mathbf{j} + a \cos u\mathbf{k}\end{aligned}$$

where $0 \le u \le \pi$ and $0 \le v \le 2\pi$. The partial derivatives of $\mathbf{r}$ are

$$\mathbf{r}_u(u, v) = a \cos u \cos v\mathbf{i} + a \cos u \sin v\mathbf{j} - a \sin u\mathbf{k}$$

and

$$\mathbf{r}_v(u, v) = -a \sin u \sin v\mathbf{i} + a \sin u \cos v\mathbf{j}$$

which implies that the normal vector $\mathbf{r}_u \times \mathbf{r}_v$ is

$$\begin{aligned}\mathbf{r}_u \times \mathbf{r}_v &= \begin{vmatrix} \mathbf{i} & \mathbf{j} & \mathbf{k} \\ a \cos u \cos v & a \cos u \sin v & -a \sin u \\ -a \sin u \sin v & a \sin u \cos v & 0 \end{vmatrix}\\ &= a^2(\sin^2 u \cos v\mathbf{i} + \sin^2 u \sin v\mathbf{j} + \sin u \cos u\mathbf{k}).\end{aligned}$$

Now, using

$$\begin{aligned}\mathbf{F}(x, y, z) &= \frac{kq\mathbf{r}}{\|\mathbf{r}\|^3}\\ &= kq\frac{x\mathbf{i} + y\mathbf{j} + z\mathbf{k}}{\|x\mathbf{i} + y\mathbf{j} + z\mathbf{k}\|^3}\\ &= \frac{kq}{a^3}(a \sin u \cos v\mathbf{i} + a \sin u \sin v\mathbf{j} + a \cos u\mathbf{k})\end{aligned}$$

it follows that

$$\begin{aligned}\mathbf{F} \cdot (\mathbf{r}_u \times \mathbf{r}_v) &= \frac{kq}{a^3}[(a \sin u \cos v\mathbf{i} + a \sin u \sin v\mathbf{j} + a \cos u\mathbf{k}) \cdot\\ &\qquad a^2(\sin^2 u \cos v\mathbf{i} + \sin^2 u \sin v\mathbf{j} + \sin u \cos u\mathbf{k})]\\ &= kq(\sin^3 u \cos^2 v + \sin^3 u \sin^2 v + \sin u \cos^2 u)\\ &= kq \sin u.\end{aligned}$$

Finally, the flux over the sphere S is given by

$$\begin{aligned}\int_S\!\!\int \mathbf{F} \cdot \mathbf{N}\, dS &= \int_D\!\!\int (kq \sin u)\, dA\\ &= \int_0^{2\pi}\!\!\int_0^{\pi} kq \sin u\, du\, dv\\ &= 4\pi kq.\end{aligned}$$

The result in Example 6 shows that the flux across a sphere S in an inverse square field is independent of the radius of S. In particular, if $\mathbf{E}$ is an electric field, the result in Example 6, along with Coulomb's Law, yields one of the basic laws of electrostatics, known as **Gauss's Law:**

$$\int\!\!\int_S \mathbf{E} \cdot \mathbf{N}\, dS = 4\pi kq \qquad \text{Gauss's Law}$$

where q is a point charge located at the center of the sphere and k is the Coulomb constant. Gauss's Law is valid for more general closed surfaces that enclose the origin, and relates the flux out of the surface to the total charge q inside the surface.

This section concludes with a summary of different forms of line integrals and surface integrals.

Summary of Line and Surface Integrals

Line Integrals

$$ds = \|\mathbf{r}'(t)\|\, dt$$
$$= \sqrt{[x'(t)]^2 + [y'(t)]^2 + [z'(t)]^2}\, dt$$

$$\int_C f(x, y, z)\, ds = \int_a^b f(x(t), y(t), z(t))\, ds \qquad \text{Scalar form}$$

$$\int_C \mathbf{F} \cdot d\mathbf{r} = \int_C \mathbf{F} \cdot \mathbf{T}\, ds$$
$$= \int_a^b \mathbf{F}(x(t), y(t), z(t)) \cdot \mathbf{r}'(t)\, dt \qquad \text{Vector form}$$

Surface Integrals $[z = g(x, y)]$

$$dS = \sqrt{1 + [g_x(x, y)]^2 + [g_y(x, y)]^2}\, dA$$

$$\int\!\!\int_S f(x, y, z)\, dS = \int\!\!\int_R f(x, y, g(x, y))\sqrt{1 + [g_x(x, y)]^2 + [g_y(x, y)]^2}\, dA \qquad \text{Scalar form}$$

$$\int\!\!\int_S \mathbf{F} \cdot \mathbf{N}\, dS = \int\!\!\int_R \mathbf{F} \cdot [-g_x(x, y)\mathbf{i} - g_y(x, y)\mathbf{j} + \mathbf{k}]\, dA \qquad \text{Vector form (upward normal)}$$

Surface Integrals (parametric form)

$$dS = \|\mathbf{r}_u(u, v) \times \mathbf{r}_v(u, v)\|\, dA$$

$$\int\!\!\int_S f(x, y, z)\, dS = \int\!\!\int_D f(x(u, v), y(u, v), z(u, v))\, dS \qquad \text{Scalar form}$$

$$\int\!\!\int_S \mathbf{F} \cdot \mathbf{N}\, dS = \int\!\!\int_D \mathbf{F} \cdot (\mathbf{r}_u \times \mathbf{r}_v)\, dA \qquad \text{Vector form}$$

Exercises for Section 15.6

See www.CalcChat.com for worked-out solutions to odd-numbered exercises.

In Exercises 1–4, evaluate $\int_S\int (x - 2y + z)\, dS$.

1. S: $z = 4 - x$, $0 \le x \le 4$, $0 \le y \le 4$
2. S: $z = 15 - 2x + 3y$, $0 \le x \le 2$, $0 \le y \le 4$
3. S: $z = 10$, $x^2 + y^2 \le 1$
4. S: $z = \frac{2}{3}x^{3/2}$, $0 \le x \le 1$, $0 \le y \le x$

In Exercises 5 and 6, evaluate $\int_S\int xy\, dS$.

5. S: $z = 6 - x - 2y$, first octant
6. S: $z = h$, $0 \le x \le 2$, $0 \le y \le \sqrt{4 - x^2}$

In Exercises 7 and 8, use a computer algebra system to evaluate

$$\int_S\int xy\, dS.$$

7. S: $z = 9 - x^2$, $0 \le x \le 2$, $0 \le y \le x$
8. S: $z = \frac{1}{2}xy$, $0 \le x \le 4$, $0 \le y \le 4$

In Exercises 9 and 10, use a computer algebra system to evaluate

$$\int_S\int (x^2 - 2xy)\, dS.$$

9. S: $z = 10 - x^2 - y^2$, $0 \le x \le 2$, $0 \le y \le 2$
10. S: $z = \cos x$, $0 \le x \le \dfrac{\pi}{2}$, $0 \le y \le \dfrac{1}{2}x$

Mass **In Exercises 11 and 12, find the mass of the surface lamina S of density ρ.**

11. S: $2x + 3y + 6z = 12$, first octant, $\rho(x, y, z) = x^2 + y^2$
12. S: $z = \sqrt{a^2 - x^2 - y^2}$, $\rho(x, y, z) = kz$

In Exercises 13–16, evaluate $\int_S\int f(x, y)\, dS$.

13. $f(x, y) = y + 5$

    S: $\mathbf{r}(u, v) = u\mathbf{i} + v\mathbf{j} + \dfrac{v}{2}\mathbf{k}$, $0 \le u \le 1$, $0 \le v \le 2$
14. $f(x, y) = x + y$

    S: $\mathbf{r}(u, v) = 2\cos u\,\mathbf{i} + 2\sin u\,\mathbf{j} + v\mathbf{k}$

    $0 \le u \le \dfrac{\pi}{2}$, $0 \le v \le 2$
15. $f(x, y) = xy$

    S: $\mathbf{r}(u, v) = 2\cos u\,\mathbf{i} + 2\sin u\,\mathbf{j} + v\mathbf{k}$

    $0 \le u \le \dfrac{\pi}{2}$, $0 \le v \le 2$
16. $f(x, y) = x + y$

    S: $\mathbf{r}(u, v) = 4u\cos v\,\mathbf{i} + 4u\sin v\,\mathbf{j} + 3u\mathbf{k}$

    $0 \le u \le 4$, $0 \le v \le \pi$

In Exercises 17–22, evaluate $\int_S\int f(x, y, z)\, dS$.

17. $f(x, y, z) = x^2 + y^2 + z^2$

    S: $z = x + 2$, $x^2 + y^2 \le 1$
18. $f(x, y, z) = \dfrac{xy}{z}$

    S: $z = x^2 + y^2$, $4 \le x^2 + y^2 \le 16$
19. $f(x, y, z) = \sqrt{x^2 + y^2 + z^2}$

    S: $z = \sqrt{x^2 + y^2}$, $x^2 + y^2 \le 4$
20. $f(x, y, z) = \sqrt{x^2 + y^2 + z^2}$

    S: $z = \sqrt{x^2 + y^2}$, $(x - 1)^2 + y^2 \le 1$
21. $f(x, y, z) = x^2 + y^2 + z^2$

    S: $x^2 + y^2 = 9$, $0 \le x \le 3$, $0 \le y \le 3$, $0 \le z \le 9$
22. $f(x, y, z) = x^2 + y^2 + z^2$

    S: $x^2 + y^2 = 9$, $0 \le x \le 3$, $0 \le z \le x$

In Exercises 23–28, find the flux of F through S,

$$\int_S\int \mathbf{F} \cdot \mathbf{N}\, dS$$

where N is the upward unit normal vector to S.

23. $\mathbf{F}(x, y, z) = 3z\mathbf{i} - 4\mathbf{j} + y\mathbf{k}$

    S: $x + y + z = 1$, first octant
24. $\mathbf{F}(x, y, z) = x\mathbf{i} + y\mathbf{j}$

    S: $2x + 3y + z = 6$, first octant
25. $\mathbf{F}(x, y, z) = x\mathbf{i} + y\mathbf{j} + z\mathbf{k}$

    S: $z = 9 - x^2 - y^2$, $z \ge 0$
26. $\mathbf{F}(x, y, z) = x\mathbf{i} + y\mathbf{j} + z\mathbf{k}$

    S: $x^2 + y^2 + z^2 = 36$, first octant
27. $\mathbf{F}(x, y, z) = 4\mathbf{i} - 3\mathbf{j} + 5\mathbf{k}$

    S: $z = x^2 + y^2$, $x^2 + y^2 \le 4$
28. $\mathbf{F}(x, y, z) = x\mathbf{i} + y\mathbf{j} - 2z\mathbf{k}$

    S: $z = \sqrt{a^2 - x^2 - y^2}$

In Exercises 29 and 30, find the flux of F over the closed surface. (Let N be the outward unit normal vector of the surface.)

29. $\mathbf{F}(x, y, z) = 4xy\mathbf{i} + z^2\mathbf{j} + yz\mathbf{k}$

    S: unit cube bounded by $x = 0$, $x = 1$, $y = 0$, $y = 1$, $z = 0$, $z = 1$
30. $\mathbf{F}(x, y, z) = (x + y)\mathbf{i} + y\mathbf{j} + z\mathbf{k}$

    S: $z = 1 - x^2 - y^2$, $z = 0$

Writing About Concepts

31. Define a surface integral of the scalar function f over a surface $z = g(x, y)$. Explain how to evaluate the surface integral.
32. Describe an orientable surface.
33. Define a flux integral and explain how it is evaluated.
34. Is the surface shown in the figure orientable? Explain.

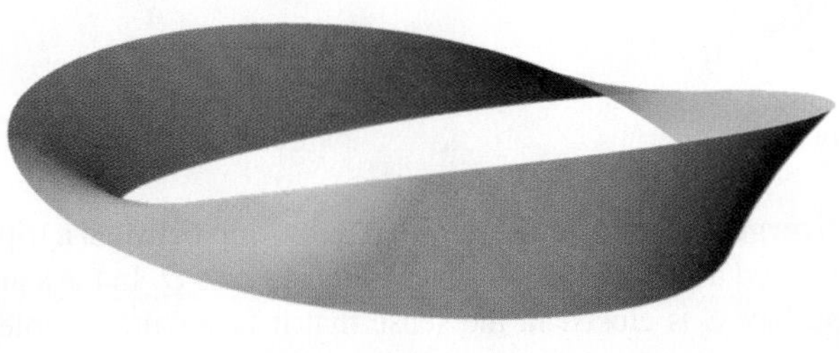

Double twist

35. ***Electrical Charge*** Let $\mathbf{E} = yz\mathbf{i} + xz\mathbf{j} + xy\mathbf{k}$ be an electrostatic field. Use Gauss's Law to find the total charge enclosed by the closed surface consisting of the hemisphere $z = \sqrt{1 - x^2 - y^2}$ and its circular base in the xy-plane.
36. ***Electrical Charge*** Let $\mathbf{E} = x\mathbf{i} + y\mathbf{j} + 2z\mathbf{k}$ be an electrostatic field. Use Gauss's Law to find the total charge enclosed by the closed surface consisting of the hemisphere $z = \sqrt{1 - x^2 - y^2}$ and its circular base in the xy-plane.

Moment of Inertia **In Exercises 37 and 38, use the following formulas for the moments of inertia about the coordinate axes of a surface lamina of density ρ.**

$$I_x = \iint_S (y^2 + z^2)\rho(x, y, z)\, dS$$

$$I_y = \iint_S (x^2 + z^2)\rho(x, y, z)\, dS$$

$$I_z = \iint_S (x^2 + y^2)\rho(x, y, z)\, dS$$

37. Verify that the moment of inertia of a conical shell of uniform density about its axis is $\frac{1}{2}ma^2$, where m is the mass and a is the radius and height.
38. Verify that the moment of inertia of a spherical shell of uniform density about its diameter is $\frac{2}{3}ma^2$, where m is the mass and a is the radius.

Moment of Inertia **In Exercises 39 and 40, find I_z for the given lamina with uniform density of 1. Use a computer algebra system to verify your results.**

39. $x^2 + y^2 = a^2, \quad 0 \le z \le h$
40. $z = x^2 + y^2, \quad 0 \le z \le h$

Flow Rate **In Exercises 41 and 42, use a computer algebra system to find the rate of mass flow of a fluid of density ρ through the surface S oriented upward if the velocity field is given by $\mathbf{F}(x, y, z) = 0.5z\mathbf{k}$.**

41. $S\colon z = 16 - x^2 - y^2, \quad z \ge 0$
42. $S\colon z = \sqrt{16 - x^2 - y^2}$

43. ***Investigation***
    (a) Use a computer algebra system to graph the vector-valued function

    $\mathbf{r}(u, v) = (4 - v \sin u)\cos(2u)\mathbf{i} + (4 - v \sin u)\sin(2u)\mathbf{j} + v \cos u\mathbf{k}, \quad 0 \le u \le \pi, \quad -1 \le v \le 1.$

    This surface is called a Möbius strip.
    (b) Explain why this surface is not orientable.
    (c) Use a computer algebra system to graph the space curve represented by $\mathbf{r}(u, 0)$. Identify the curve.
    (d) Construct a Möbius strip by cutting a strip of paper, making a single twist, and pasting the ends together.
    (e) Cut the Möbius strip along the space curve graphed in part (c), and describe the result.

Section Project: Hyperboloid of One Sheet

Consider the parametric surface given by the function

$\mathbf{r}(u, v) = a \cosh u \cos v\mathbf{i} + a \cosh u \sin v\mathbf{j} + b \sinh u\mathbf{k}.$

(a) Use a graphing utility to graph $\mathbf{r}$ for various values of the constants a and b. Describe the effect of the constants on the shape of the surface.

(b) Show that the surface is a hyperboloid of one sheet given by

$$\frac{x^2}{a^2} + \frac{y^2}{a^2} - \frac{z^2}{b^2} = 1.$$

(c) For fixed values $u = u_0$, describe the curves given by

$\mathbf{r}(u_0, v) = a \cosh u_0 \cos v\mathbf{i} + a \cosh u_0 \sin v\mathbf{j} + b \sinh u_0\mathbf{k}.$

(d) For fixed values $v = v_0$, describe the curves given by

$\mathbf{r}(u, v_0) = a \cosh u \cos v_0\mathbf{i} + a \cosh u \sin v_0\mathbf{j} + b \sinh u\mathbf{k}.$

(e) Find a normal vector to the surface at $(u, v) = (0, 0)$.

Section 15.7 Divergence Theorem

- **Understand and use the Divergence Theorem.**
- **Use the Divergence Theorem to calculate flux.**

CARL FRIEDRICH GAUSS (1777–1855)

The **Divergence Theorem** is also called **Gauss's Theorem,** after the famous German mathematician Carl Friedrich Gauss. Gauss is recognized, with Newton and Archimedes, as one of the three greatest mathematicians in history. One of his many contributions to mathematics was made at the age of 22, when, as part of his doctoral dissertation, he proved the *Fundamental Theorem of Algebra.*

Divergence Theorem

Recall from Section 15.4 that an alternative form of Green's Theorem is

$$\begin{aligned}\int_C \mathbf{F}\cdot\mathbf{N}\,ds &= \int\!\!\int_R \left(\frac{\partial M}{\partial x}+\frac{\partial N}{\partial y}\right) dA\\ &= \int\!\!\int_R \operatorname{div}\mathbf{F}\,dA.\end{aligned}$$

In an analogous way, the **Divergence Theorem** gives the relationship between a triple integral over a solid region Q and a surface integral over the surface of Q. In the statement of the theorem, the surface S is **closed** in the sense that it forms the complete boundary of the solid Q. Regions bounded by spheres, ellipsoids, cubes, tetrahedrons, or combinations of these surfaces are typical examples of closed surfaces. Assume that Q is a solid region on which a triple integral can be evaluated, and that the closed surface S is oriented by *outward* unit normal vectors, as shown in Figure 15.54. With these restrictions on S and Q, the Divergence Theorem is as follows.

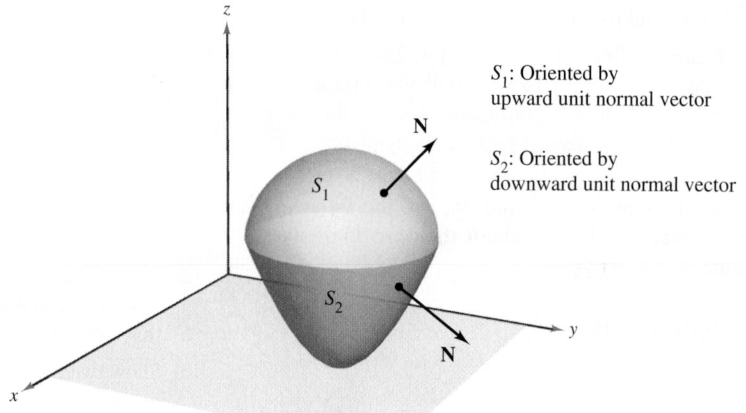

Figure 15.54

> **THEOREM 15.12 The Divergence Theorem**
>
> Let Q be a solid region bounded by a closed surface S oriented by a unit normal vector directed outward from Q. If $\mathbf{F}$ is a vector field whose component functions have continuous partial derivatives in Q, then
>
> $$\int\!\!\int_S \mathbf{F}\cdot\mathbf{N}\,dS = \int\!\!\int\!\!\int_Q \operatorname{div}\mathbf{F}\,dV.$$

NOTE As noted at the left above, the Divergence Theorem is sometimes called Gauss's Theorem. It is also sometimes called Ostrogradsky's Theorem, after the Russian mathematician Michel Ostrogradsky (1801–1861).

Proof If you let $\mathbf{F}(x, y, z) = M\mathbf{i} + N\mathbf{j} + P\mathbf{k}$, the theorem takes the form

$$\int_S\int \mathbf{F} \cdot \mathbf{N}\, dS = \int_S\int (M\mathbf{i} \cdot \mathbf{N} + N\mathbf{j} \cdot \mathbf{N} + P\mathbf{k} \cdot \mathbf{N})\, dS$$

$$= \iiint_Q \left(\frac{\partial M}{\partial x} + \frac{\partial N}{\partial y} + \frac{\partial P}{\partial z}\right) dV.$$

You can prove this by verifying that the following three equations are valid.

$$\int_S\int M\mathbf{i} \cdot \mathbf{N}\, dS = \iiint_Q \frac{\partial M}{\partial x}\, dV$$

$$\int_S\int N\mathbf{j} \cdot \mathbf{N}\, dS = \iiint_Q \frac{\partial N}{\partial y}\, dV$$

$$\int_S\int P\mathbf{k} \cdot \mathbf{N}\, dS = \iiint_Q \frac{\partial P}{\partial z}\, dV$$

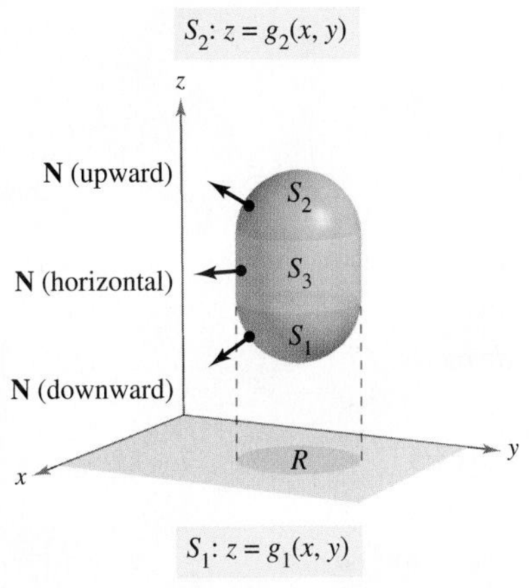

Figure 15.55

Because the verifications of the three equations are similar, only the third is discussed. Restrict the proof to a **simple solid** region with upper surface

$z = g_2(x, y)$ Upper surface

and lower surface

$z = g_1(x, y)$ Lower surface

whose projections onto the xy-plane coincide and form region R. If Q has a lateral surface like S_3 in Figure 15.55, then a normal vector is horizontal, which implies that $P\mathbf{k} \cdot \mathbf{N} = 0$. Consequently, you have

$$\int_S\int P\mathbf{k} \cdot \mathbf{N}\, dS = \int_{S_1}\int P\mathbf{k} \cdot \mathbf{N}\, dS + \int_{S_2}\int P\mathbf{k} \cdot \mathbf{N}\, dS + 0.$$

On the upper surface S_2, the outward normal vector is upward, whereas on the lower surface S_1, the outward normal vector is downward. So, by Theorem 15.11, you have the following.

$$\int_{S_1}\int P\mathbf{k} \cdot \mathbf{N}\, dS = \int_R\int P(x, y, g_1(x, y))\mathbf{k} \cdot \left(\frac{\partial g_1}{\partial x}\mathbf{i} + \frac{\partial g_1}{\partial y}\mathbf{j} - \mathbf{k}\right) dA$$

$$= -\int_R\int P(x, y, g_1(x, y))\, dA$$

$$\int_{S_2}\int P\mathbf{k} \cdot \mathbf{N}\, dS = \int_R\int P(x, y, g_2(x, y))\mathbf{k} \cdot \left(-\frac{\partial g_2}{\partial x}\mathbf{i} - \frac{\partial g_2}{\partial y}\mathbf{j} + \mathbf{k}\right) dA$$

$$= \int_R\int P(x, y, g_2(x, y))\, dA$$

Adding these results, you obtain

$$\int_S\int P\mathbf{k} \cdot \mathbf{N}\, dS = \int_R\int [P(x, y, g_2(x, y)) - P(x, y, g_1(x, y))]\, dA$$

$$= \int_R\int \left[\int_{g_1(x, y)}^{g_2(x, y)} \frac{\partial P}{\partial z}\, dz\right] dA$$

$$= \iiint_Q \frac{\partial P}{\partial z}\, dV.$$

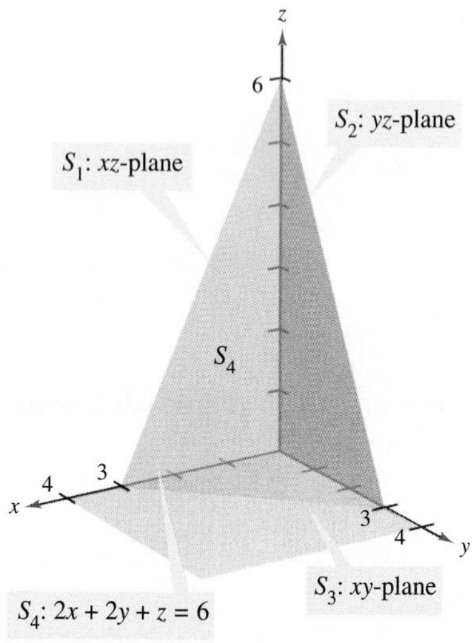

Figure 15.56

EXAMPLE 1 Using the Divergence Theorem

Let Q be the solid region bounded by the coordinate planes and the plane $2x + 2y + z = 6$, and let $\mathbf{F} = x\mathbf{i} + y^2\mathbf{j} + z\mathbf{k}$. Find

$$\int_S\int \mathbf{F} \cdot \mathbf{N}\, dS$$

where S is the surface of Q.

Solution From Figure 15.56, you can see that Q is bounded by four subsurfaces. So, you would need four *surface integrals* to evaluate

$$\int_S\int \mathbf{F} \cdot \mathbf{N}\, dS.$$

However, by the Divergence Theorem, you need only one triple integral. Because

$$\begin{aligned}\text{div } \mathbf{F} &= \frac{\partial M}{\partial x} + \frac{\partial N}{\partial y} + \frac{\partial P}{\partial z}\\ &= 1 + 2y + 1\\ &= 2 + 2y\end{aligned}$$

you have

$$\begin{aligned}\int_S\int \mathbf{F} \cdot \mathbf{N}\, dS &= \iiint_Q \text{div } \mathbf{F}\, dV\\ &= \int_0^3\int_0^{3-y}\int_0^{6-2x-2y} (2 + 2y)\, dz\, dx\, dy\\ &= \int_0^3\int_0^{3-y} (2z + 2yz)\Big]_0^{6-2x-2y} dx\, dy\\ &= \int_0^3\int_0^{3-y} (12 - 4x + 8y - 4xy - 4y^2)\, dx\, dy\\ &= \int_0^3 \Big[12x - 2x^2 + 8xy - 2x^2y - 4xy^2\Big]_0^{3-y} dy\\ &= \int_0^3 (18 + 6y - 10y^2 + 2y^3)\, dy\\ &= \Big[18y + 3y^2 - \frac{10y^3}{3} + \frac{y^4}{2}\Big]_0^3\\ &= \frac{63}{2}.\end{aligned}$$

TECHNOLOGY If you have access to a computer algebra system that can evaluate triple-iterated integrals, use it to verify the result in Example 1. When you are using such a utility, note that the first step is to convert the triple integral to an iterated integral—this step must be done by hand. To give yourself some practice with this important step, find the limits of integration for the following iterated integrals. Then use a computer to verify that the value is the same as that obtained in Example 1.

$$\int_?^?\int_?^?\int_?^? (2 + 2y)\, dy\, dz\, dx, \qquad \int_?^?\int_?^?\int_?^? (2 + 2y)\, dx\, dy\, dz$$

EXAMPLE 2 Verifying the Divergence Theorem

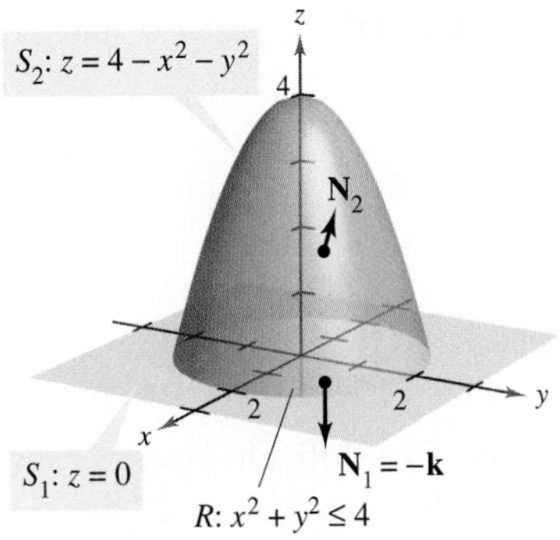

Figure 15.57

Let Q be the solid region between the paraboloid

$$z = 4 - x^2 - y^2$$

and the xy-plane. Verify the Divergence Theorem for

$$\mathbf{F}(x, y, z) = 2z\mathbf{i} + x\mathbf{j} + y^2\mathbf{k}.$$

Solution From Figure 15.57 you can see that the outward normal vector for the surface S_1 is $\mathbf{N}_1 = -\mathbf{k}$, whereas the outward normal vector for the surface S_2 is

$$\mathbf{N}_2 = \frac{2x\mathbf{i} + 2y\mathbf{j} + \mathbf{k}}{\sqrt{4x^2 + 4y^2 + 1}}.$$

So, by Theorem 15.11, you have

$$\begin{aligned}
\int_S\!\!\int \mathbf{F} \cdot \mathbf{N}\, dS
&= \int_{S_1}\!\!\int \mathbf{F} \cdot \mathbf{N}_1\, dS + \int_{S_2}\!\!\int \mathbf{F} \cdot \mathbf{N}_2\, dS \\
&= \int_{S_1}\!\!\int \mathbf{F} \cdot (-\mathbf{k})\, dS + \int_{S_2}\!\!\int \mathbf{F} \cdot (2x\mathbf{i} + 2y\mathbf{j} + \mathbf{k})\, dS \\
&= \int_R\!\!\int -y^2\, dA + \int_R\!\!\int (4xz + 2xy + y^2)\, dA \\
&= -\int_{-2}^{2}\int_{-\sqrt{4-y^2}}^{\sqrt{4-y^2}} y^2\, dx\, dy + \int_{-2}^{2}\int_{-\sqrt{4-y^2}}^{\sqrt{4-y^2}} (4xz + 2xy + y^2)\, dx\, dy \\
&= \int_{-2}^{2}\int_{-\sqrt{4-y^2}}^{\sqrt{4-y^2}} (4xz + 2xy)\, dx\, dy \\
&= \int_{-2}^{2}\int_{-\sqrt{4-y^2}}^{\sqrt{4-y^2}} [4x(4 - x^2 - y^2) + 2xy]\, dx\, dy \\
&= \int_{-2}^{2}\int_{-\sqrt{4-y^2}}^{\sqrt{4-y^2}} (16x - 4x^3 - 4xy^2 + 2xy)\, dx\, dy \\
&= \int_{-2}^{2} \Big[8x^2 - x^4 - 2x^2y^2 + x^2y\Big]_{-\sqrt{4-y^2}}^{\sqrt{4-y^2}}\, dy \\
&= \int_{-2}^{2} 0\, dy \\
&= 0.
\end{aligned}$$

On the other hand, because

$$\text{div}\,\mathbf{F} = \frac{\partial}{\partial x}[2z] + \frac{\partial}{\partial y}[x] + \frac{\partial}{\partial z}[y^2] = 0 + 0 + 0 = 0$$

you can apply the Divergence Theorem to obtain the equivalent result

$$\begin{aligned}
\int_S\!\!\int \mathbf{F} \cdot \mathbf{N}\, dS &= \iiint_Q \text{div}\,\mathbf{F}\, dV \\
&= \iiint_Q 0\, dV = 0.
\end{aligned}$$

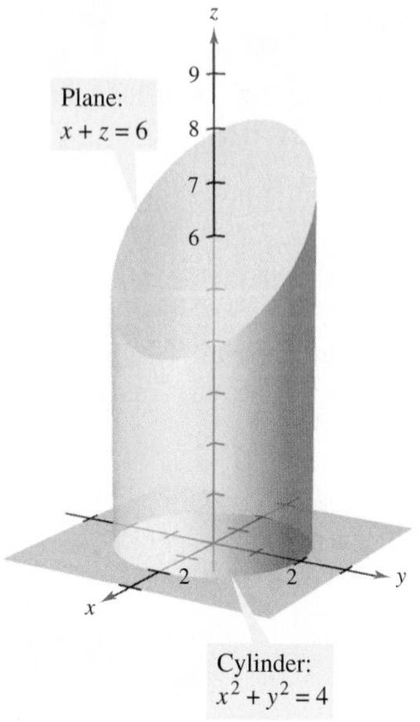

Figure 15.58

EXAMPLE 3 Using the Divergence Theorem

Let Q be the solid bounded by the cylinder $x^2 + y^2 = 4$, the plane $x + z = 6$, and the xy-plane, as shown in Figure 15.58. Find

$$\int_S\int \mathbf{F} \cdot \mathbf{N}\, dS$$

where S is the surface of Q and

$$\mathbf{F}(x, y, z) = (x^2 + \sin z)\mathbf{i} + (xy + \cos z)\mathbf{j} + e^y\mathbf{k}.$$

Solution Direct evaluation of this surface integral would be difficult. However, by the Divergence Theorem, you can evaluate the integral as follows.

$$\begin{aligned}
\int_S\int \mathbf{F} \cdot \mathbf{N}\, dS &= \iiint_Q \operatorname{div} \mathbf{F}\, dV \\
&= \iiint_Q (2x + x + 0)\, dV \\
&= \iiint_Q 3x\, dV \\
&= \int_0^{2\pi}\int_0^2\int_0^{6 - r\cos\theta} (3r\cos\theta) r\, dz\, dr\, d\theta \\
&= \int_0^{2\pi}\int_0^2 (18r^2\cos\theta - 3r^3\cos^2\theta)\, dr\, d\theta \\
&= \int_0^{2\pi} (48\cos\theta - 12\cos^2\theta) d\theta \\
&= \left[48\sin\theta - 6\left(\theta + \frac{1}{2}\sin 2\theta\right)\right]_0^{2\pi} \\
&= -12\pi
\end{aligned}$$

Notice that cylindrical coordinates with $x = r\cos\theta$ and $dV = r\, dz\, dr\, d\theta$ were used to evaluate the triple integral.

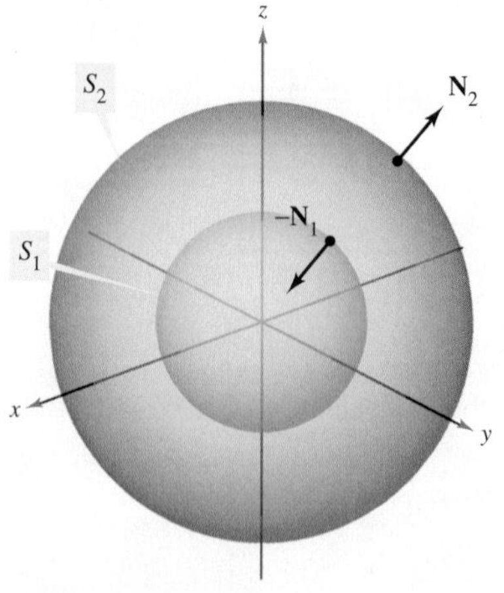

Figure 15.59

Even though the Divergence Theorem was stated for a simple solid region Q bounded by a closed surface, the theorem is also valid for regions that are the finite unions of simple solid regions. For example, let Q be the solid bounded by the closed surfaces S_1 and S_2, as shown in Figure 15.59. To apply the Divergence Theorem to this solid, let $S = S_1 \cup S_2$. The normal vector $\mathbf{N}$ to S is given by $-\mathbf{N}_1$ on S_1 and by $\mathbf{N}_2$ on S_2. So, you can write

$$\begin{aligned}
\iiint_Q \operatorname{div} \mathbf{F}\, dV &= \int_S\int \mathbf{F} \cdot \mathbf{N}\, dS \\
&= \int_{S_1}\int \mathbf{F} \cdot (-\mathbf{N}_1)\, dS + \int_{S_2}\int \mathbf{F} \cdot \mathbf{N}_2\, dS \\
&= -\int_{S_1}\int \mathbf{F} \cdot \mathbf{N}_1\, dS + \int_{S_2}\int \mathbf{F} \cdot \mathbf{N}_2\, dS.
\end{aligned}$$

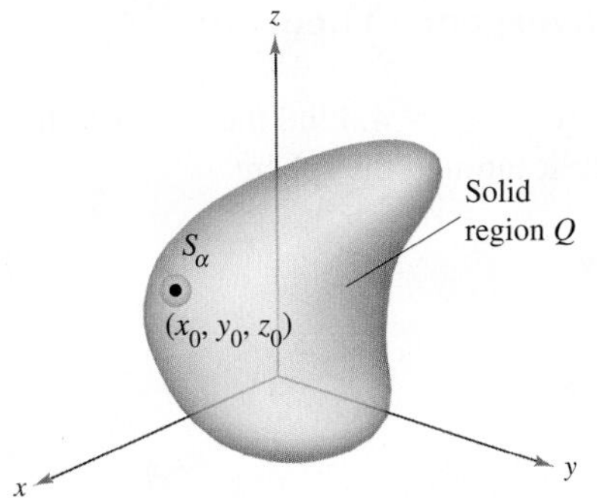

Figure 15.60

Flux and the Divergence Theorem

To help understand the Divergence Theorem, consider the two sides of the equation

$$\int_S\int \mathbf{F} \cdot \mathbf{N}\, dS = \iiint_Q \text{div}\, \mathbf{F}\, dV.$$

You know from Section 15.6 that the flux integral on the left determines the total fluid flow across the surface S per unit of time. This can be approximated by summing the fluid flow across small patches of the surface. The triple integral on the right measures this same fluid flow across S, but from a very different perspective—namely, by calculating the flow of fluid into (or out of) small *cubes* of volume ΔV_i. The flux of the ith cube is approximately

$$\text{Flux of } i\text{th cube} \approx \text{div}\, \mathbf{F}(x_i, y_i, z_i)\, \Delta V_i$$

for some point (x_i, y_i, z_i) in the ith cube. Note that for a cube in the interior of Q, the gain (or loss) of fluid through any one of its six sides is offset by a corresponding loss (or gain) through one of the sides of an adjacent cube. After summing over all the cubes in Q, the only fluid flow that is not canceled by adjoining cubes is that on the outside edges of the cubes on the boundary. So, the sum

$$\sum_{i=1}^{n} \text{div}\, \mathbf{F}(x_i, y_i, z_i)\, \Delta V_i$$

approximates the total flux into (or out of) Q, and therefore through the surface S.

To see what is meant by the divergence of $\mathbf{F}$ at a point, consider ΔV_α to be the volume of a small sphere S_α of radius α and center (x_0, y_0, z_0), contained in region Q, as shown in Figure 15.60. Applying the Divergence Theorem to S_α produces

$$\begin{aligned}\text{Flux of } \mathbf{F} \text{ across } S_\alpha &= \iiint_{Q_\alpha} \text{div}\, \mathbf{F}\, dV \\ &\approx \text{div}\, \mathbf{F}(x_0, y_0, z_0,)\, \Delta V_\alpha\end{aligned}$$

where Q_α is the interior of S_α. Consequently, you have

$$\text{div}\, \mathbf{F}(x_0, y_0, z_0) \approx \frac{\text{flux of } \mathbf{F} \text{ across } S_\alpha}{\Delta V_\alpha}$$

and, by taking the limit as $\alpha \to 0$, you obtain the divergence of $\mathbf{F}$ at the point (x_0, y_0, z_0).

$$\begin{aligned}\text{div}\, \mathbf{F}(x_0, y_0, z_0) &= \lim_{\alpha \to 0} \frac{\text{flux of } \mathbf{F} \text{ across } S_\alpha}{\Delta V_\alpha} \\ &= \text{flux per unit volume at } (x_0, y_0, z_0)\end{aligned}$$

The point (x_0, y_0, z_0) in a vector field is classified as a source, a sink, or incompressible, as follows.

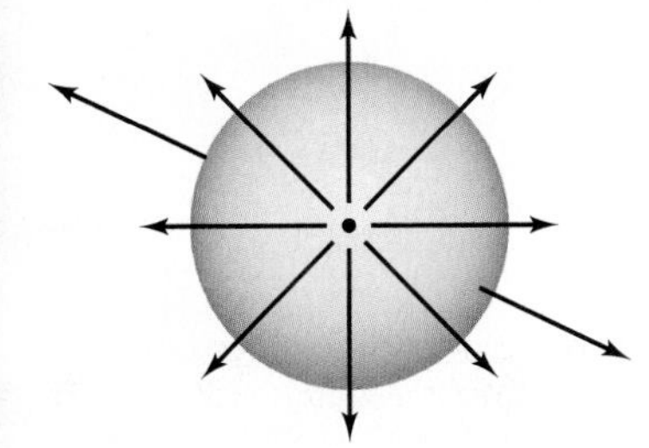

(a) Source: div $\mathbf{F} > 0$

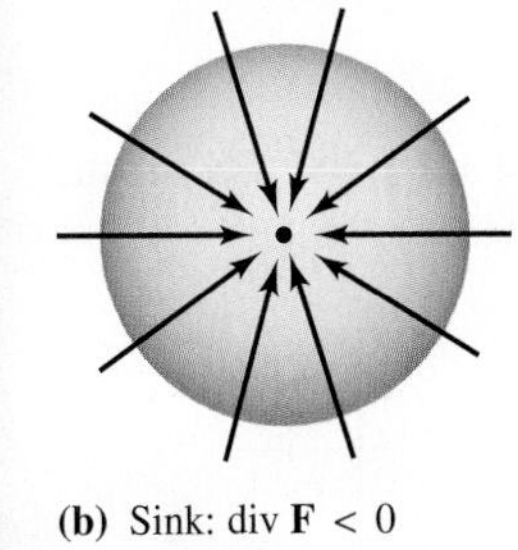

(b) Sink: div $\mathbf{F} < 0$

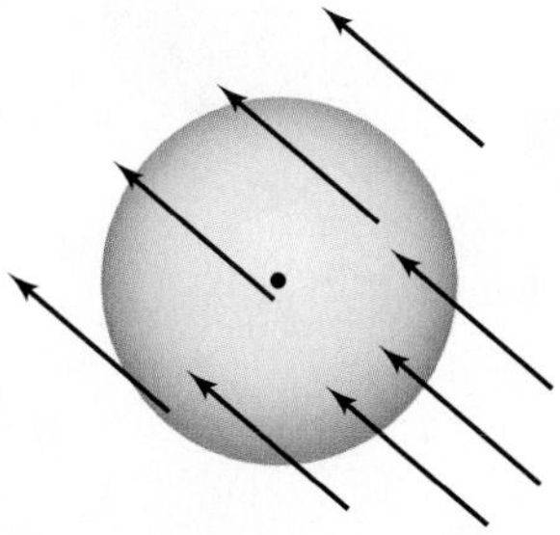

(c) Incompressible: div $\mathbf{F} = 0$

Figure 15.61

1. **Source,** if div $\mathbf{F} > 0$ — See Figure 15.61(a).
2. **Sink,** if div $\mathbf{F} < 0$ — See Figure 15.61(b).
3. **Incompressible,** if div $\mathbf{F} = 0$ — See Figure 15.61(c).

NOTE In hydrodynamics, a *source* is a point at which additional fluid is considered as being introduced to the region occupied by the fluid. A *sink* is a point at which fluid is considered as being removed.

EXAMPLE 4 Calculating Flux by the Divergence Theorem

Let Q be the region bounded by the sphere $x^2 + y^2 + z^2 = 4$. Find the outward flux of the vector field $\mathbf{F}(x, y, z) = 2x^3\mathbf{i} + 2y^3\mathbf{j} + 2z^3\mathbf{k}$ through the sphere.

Solution By the Divergence Theorem, you have

$$\begin{aligned}\text{Flux across } S &= \int_S\!\!\int \mathbf{F}\cdot\mathbf{N}\,dS = \iiint_Q \text{div}\,\mathbf{F}\,dV\\
&= \iiint_Q 6(x^2 + y^2 + z^2)\,dV\\
&= 6\int_0^2\!\int_0^{\pi}\!\int_0^{2\pi} \rho^4 \sin\phi\,d\theta\,d\phi\,d\rho \qquad \text{Spherical coordinates}\\
&= 6\int_0^2\!\int_0^{\pi} 2\pi\rho^4 \sin\phi\,d\phi\,d\rho\\
&= 12\pi\int_0^2 2\rho^4\,d\rho\\
&= 24\pi\left(\frac{32}{5}\right)\\
&= \frac{768\pi}{5}.\end{aligned}$$

Exercises for Section 15.7

See www.CalcChat.com for worked-out solutions to odd-numbered exercises.

In Exercises 1–4, verify the Divergence Theorem by evaluating

$$\int_S\!\!\int \mathbf{F}\cdot\mathbf{N}\,dS$$

as a surface integral and as a triple integral.

1. $\mathbf{F}(x, y, z) = 2x\mathbf{i} - 2y\mathbf{j} + z^2\mathbf{k}$
   S: cube bounded by the planes $x = 0,\ x = a,\ y = 0,\ y = a,\ z = 0,\ z = a$

2. $\mathbf{F}(x, y, z) = 2x\mathbf{i} - 2y\mathbf{j} + z^2\mathbf{k}$
   S: cylinder $x^2 + y^2 = 4, \quad 0 \le z \le h$

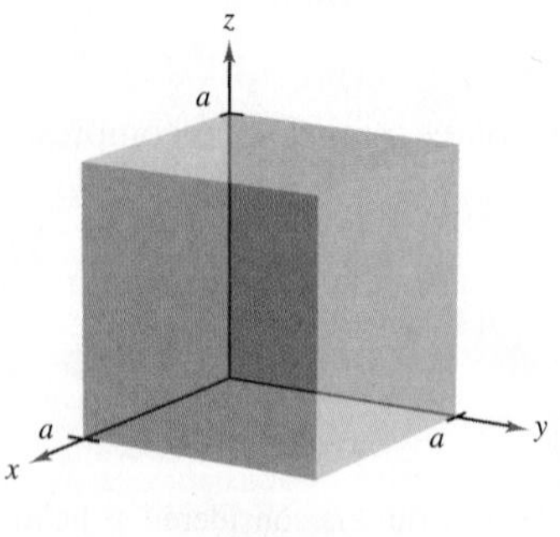

Figure for 1

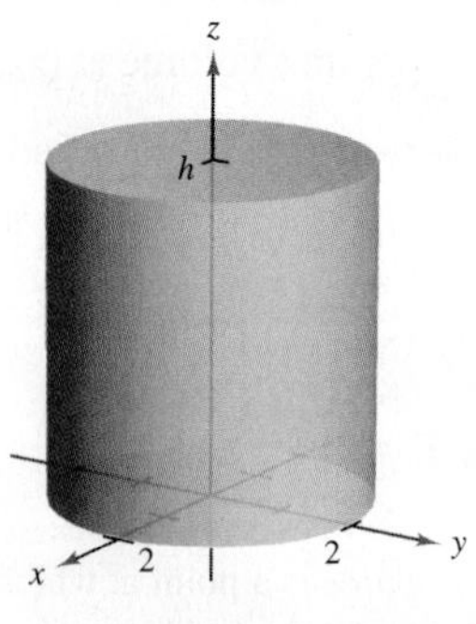

Figure for 2

3. $\mathbf{F}(x, y, z) = (2x - y)\mathbf{i} - (2y - z)\mathbf{j} + z\mathbf{k}$
   S: surface bounded by the plane $2x + 4y + 2z = 12$ and the coordinate planes

4. $\mathbf{F}(x, y, z) = xy\mathbf{i} + z\mathbf{j} + (x + y)\mathbf{k}$
   S: surface bounded by the planes $y = 4$ and $z = 4 - x$ and the coordinate planes

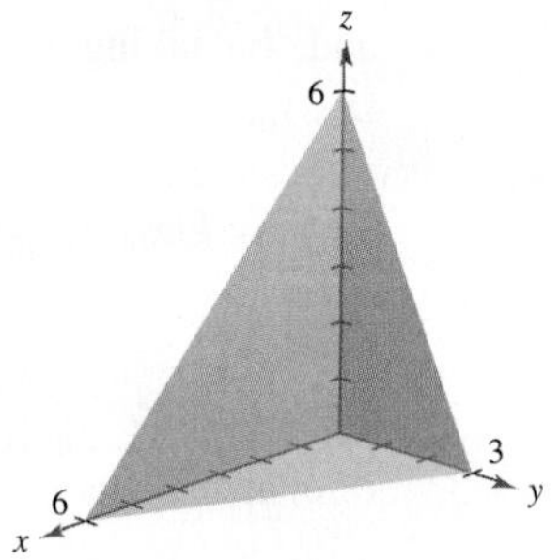

Figure for 3

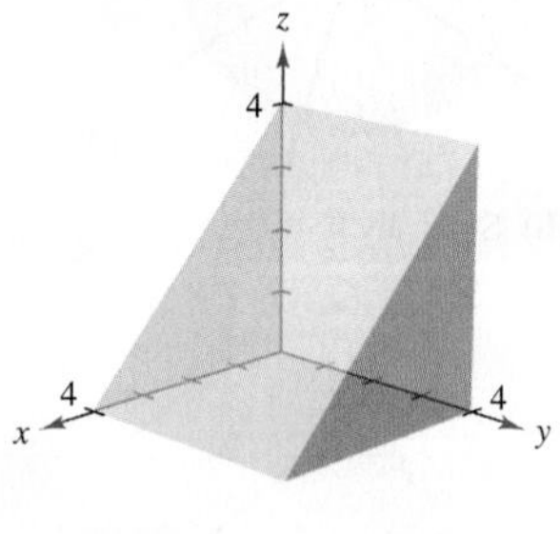

Figure for 4

In Exercises 5–16, use the Divergence Theorem to evaluate

$$\iint_S \mathbf{F}\cdot\mathbf{N}\,dS$$

and find the outward flux of F through the surface of the solid bounded by the graphs of the equations. Use a computer algebra system to verify your results.

5. $\mathbf{F}(x, y, z) = x^2\mathbf{i} + y^2\mathbf{j} + z^2\mathbf{k}$

S: $x = 0, x = a, y = 0, y = a, z = 0, z = a$

6. $\mathbf{F}(x, y, z) = x^2z^2\mathbf{i} - 2y\mathbf{j} + 3xyz\mathbf{k}$

S: $x = 0, x = a, y = 0, y = a, z = 0, z = a$

7. $\mathbf{F}(x, y, z) = x^2\mathbf{i} - 2xy\mathbf{j} + xyz^2\mathbf{k}$

S: $z = \sqrt{a^2 - x^2 - y^2}, z = 0$

8. $\mathbf{F}(x, y, z) = xy\mathbf{i} + yz\mathbf{j} - yz\mathbf{k}$

S: $z = \sqrt{a^2 - x^2 - y^2}, z = 0$

9. $\mathbf{F}(x, y, z) = x\mathbf{i} + y\mathbf{j} + z\mathbf{k}$

S: $x^2 + y^2 + z^2 = 4$

10. $\mathbf{F}(x, y, z) = xyz\mathbf{j}$

S: $x^2 + y^2 = 9, z = 0, z = 4$

11. $\mathbf{F}(x, y, z) = x\mathbf{i} + y^2\mathbf{j} - z\mathbf{k}$

S: $x^2 + y^2 = 9, z = 0, z = 4$

12. $\mathbf{F}(x, y, z) = (xy^2 + \cos z)\mathbf{i} + (x^2y + \sin z)\mathbf{j} + e^z\mathbf{k}$

S: $z = \frac{1}{2}\sqrt{x^2 + y^2}, z = 8$

13. $\mathbf{F}(x, y, z) = x^3\mathbf{i} + x^2y\mathbf{j} + x^2e^y\mathbf{k}$

S: $z = 4 - y, z = 0, x = 0, x = 6, y = 0$

14. $\mathbf{F}(x, y, z) = xe^z\mathbf{i} + ye^z\mathbf{j} + e^z\mathbf{k}$

S: $z = 4 - y, z = 0, x = 0, x = 6, y = 0$

15. $\mathbf{F}(x, y, z) = xy\mathbf{i} + 4y\mathbf{j} + xz\mathbf{k}$

S: $x^2 + y^2 + z^2 = 9$

16. $\mathbf{F}(x, y, z) = 2(x\mathbf{i} + y\mathbf{j} + z\mathbf{k})$

S: $z = \sqrt{4 - x^2 - y^2}, z = 0$

In Exercises 17 and 18, evaluate

$$\iint_S \text{curl } \mathbf{F}\cdot\mathbf{N}\,dS$$

where S is the closed surface of the solid bounded by the graphs of $x = 4$, and $z = 9 - y^2$, and the coordinate planes.

17. $\mathbf{F}(x, y, z) = (4xy + z^2)\mathbf{i} + (2x^2 + 6yz)\mathbf{j} + 2xz\mathbf{k}$

18. $\mathbf{F}(x, y, z) = xy\cos z\mathbf{i} + yz\sin x\mathbf{j} + xyz\mathbf{k}$

Writing About Concepts

19. State the Divergence Theorem.

20. How do you determine if a point (x_0, y_0, z_0) in a vector field is a source, a sink, or incompressible?

21. Use the Divergence Theorem to verify that the volume of the solid bounded by a surface S is

$$\iint_S x\,dy\,dz = \iint_S y\,dz\,dx = \iint_S z\,dx\,dy.$$

22. Verify the result of Exercise 21 for the cube bounded by $x = 0$, $x = a$, $y = 0$, $y = a$, $z = 0$, and $z = a$.

23. Verify that

$$\iint_S \text{curl } \mathbf{F}\cdot\mathbf{N}\,dS = 0$$

for any closed surface S.

24. For the constant vector field given by

$$\mathbf{F}(x, y, z) = a_1\mathbf{i} + a_2\mathbf{j} + a_3\mathbf{k}$$

verify that

$$\iint_S \mathbf{F}\cdot\mathbf{N}\,dS = 0$$

where V is the volume of the solid bounded by the closed surface S.

25. Given the vector field

$$\mathbf{F}(x, y, z) = x\mathbf{i} + y\mathbf{j} + z\mathbf{k}$$

verify that

$$\iint_S \mathbf{F}\cdot\mathbf{N}\,dS = 3V$$

where V is the volume of the solid bounded by the closed surface S.

26. Given the vector field

$$\mathbf{F}(x, y, z) = x\mathbf{i} + y\mathbf{j} + z\mathbf{k}$$

verify that

$$\frac{1}{\|\mathbf{F}\|}\iint_S \mathbf{F}\cdot\mathbf{N}\,dS = \frac{3}{\|\mathbf{F}\|}\iiint_Q dV.$$

In Exercises 27 and 28, prove the identity, assuming that Q, S, and N meet the conditions of the Divergence Theorem and that the required partial derivatives of the scalar functions f and g are continuous. The expressions $D_{\mathbf{N}}f$ and $D_{\mathbf{N}}g$ are the derivatives in the direction of the vector N and are defined by

$$D_{\mathbf{N}}f = \nabla f\cdot\mathbf{N},\quad D_{\mathbf{N}}g = \nabla g\cdot\mathbf{N}.$$

27. $\displaystyle\iiint_Q (f\nabla^2 g + \nabla f\cdot\nabla g)\,dV = \iint_S f D_{\mathbf{N}}g\,dS$

[*Hint:* Use div $(f\mathbf{G}) = f\,\text{div}\,\mathbf{G} + \nabla f\cdot\mathbf{G}$.]

28. $\displaystyle\iiint_Q (f\nabla^2 g - g\nabla^2 f)\,dV = \iint_S (f D_{\mathbf{N}}g - g D_{\mathbf{N}}f)\,dS$

(*Hint:* Use Exercise 27 twice.)

Section 15.8

Stokes's Theorem

- **Understand and use Stokes's Theorem.**
- **Use curl to analyze the motion of a rotating liquid.**

GEORGE GABRIEL STOKES (1819–1903)

Stokes became a Lucasian professor of mathematics at Cambridge in 1849. Five years later, he published the theorem that bears his name as a prize examination question there.

Stokes's Theorem

A second higher-dimension analog of Green's Theorem is called **Stokes's Theorem,** after the English mathematical physicist George Gabriel Stokes. Stokes was part of a group of English mathematical physicists referred to as the Cambridge School, which included William Thomson (Lord Kelvin) and James Clerk Maxwell. In addition to making contributions to physics, Stokes worked with infinite series and differential equations, as well as with the integration results presented in this section.

Stokes's Theorem gives the relationship between a surface integral over an oriented surface S and a line integral along a closed space curve C forming the boundary of S, as shown in Figure 15.62. The positive direction along C is counterclockwise relative to the normal vector **N**. That is, if you imagine grasping the normal vector **N** with your right hand, with your thumb pointing in the direction of **N**, your fingers will point in the positive direction C, as shown in Figure 15.63.

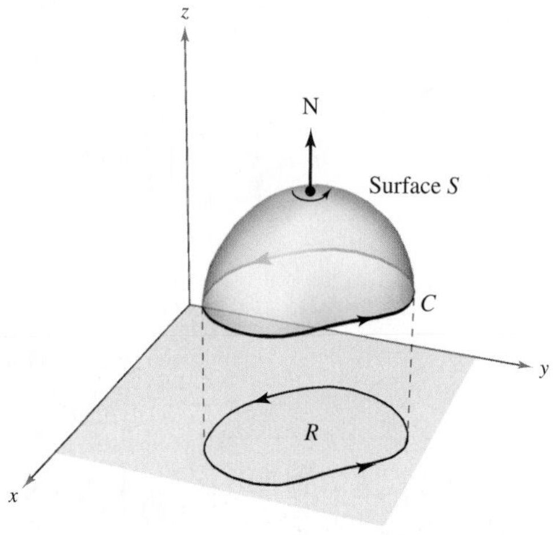

Figure 15.62

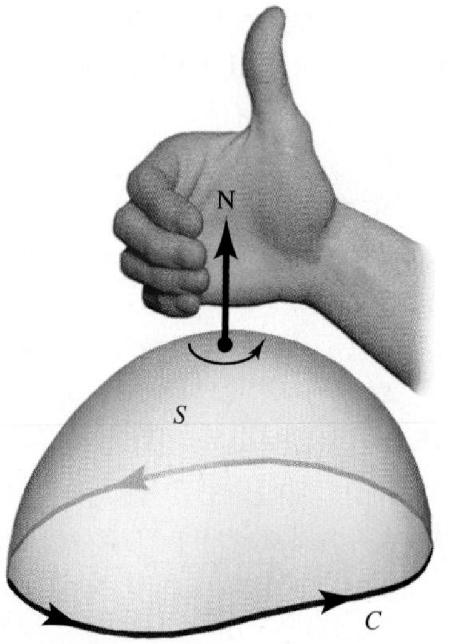

Direction along C is counterclockwise relative to **N**.
Figure 15.63

THEOREM 15.13 Stokes's Theorem

Let S be an oriented surface with unit normal vector **N**, bounded by a piecewise smooth simple closed curve C with a positive orientation. If **F** is a vector field whose component functions have continuous partial derivatives on an open region containing S and C, then

$$\int_C \mathbf{F} \cdot d\mathbf{r} = \int_S\int (\mathbf{curl\ F}) \cdot \mathbf{N}\, dS.$$

NOTE The line integral may be written in the differential form $\int_C M\,dx + N\,dy + P\,dz$ or in the vector form $\int_C \mathbf{F} \cdot \mathbf{T}\,ds$.

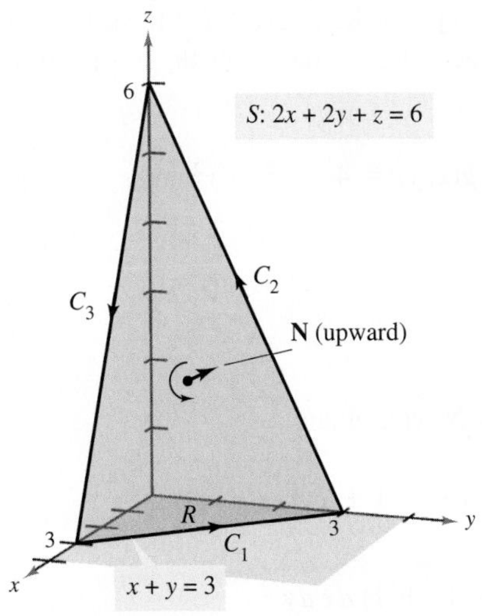

Figure 15.64

EXAMPLE 1 Using Stokes's Theorem

Let C be the oriented triangle lying in the plane $2x + 2y + z = 6$, as shown in Figure 15.64. Evaluate

$$\int_C \mathbf{F} \cdot d\mathbf{r}$$

where $\mathbf{F}(x, y, z) = -y^2\mathbf{i} + z\mathbf{j} + x\mathbf{k}$.

Solution Using Stokes's Theorem, begin by finding the curl of $\mathbf{F}$.

$$\mathbf{curl\ F} = \begin{vmatrix} \mathbf{i} & \mathbf{j} & \mathbf{k} \\ \dfrac{\partial}{\partial x} & \dfrac{\partial}{\partial y} & \dfrac{\partial}{\partial z} \\ -y^2 & z & x \end{vmatrix} = -\mathbf{i} - \mathbf{j} + 2y\mathbf{k}$$

Considering $z = 6 - 2x - 2y = g(x, y)$, you can use Theorem 15.11 for an upward normal vector to obtain

$$\begin{aligned}
\int_C \mathbf{F} \cdot d\mathbf{r} &= \int_S\int (\mathbf{curl\ F}) \cdot \mathbf{N}\, dS \\
&= \int_R\int (-\mathbf{i} - \mathbf{j} + 2y\mathbf{k}) \cdot [-g_x(x, y)\mathbf{i} - g_y(x, y)\mathbf{j} + \mathbf{k}]\, dA \\
&= \int_R\int (-\mathbf{i} - \mathbf{j} + 2y\mathbf{k}) \cdot (2\mathbf{i} + 2\mathbf{j} + \mathbf{k})\, dA \\
&= \int_0^3\int_0^{3-y} (2y - 4)\, dx\, dy \\
&= \int_0^3 (-2y^2 + 10y - 12)\, dy \\
&= \left[-\frac{2y^3}{3} + 5y^2 - 12y\right]_0^3 \\
&= -9.
\end{aligned}$$

Try evaluating the line integral in Example 1 directly, *without* using Stokes's Theorem. One way to do this would be to consider C as the union of C_1, C_2, and C_3, as follows.

$$\begin{aligned}
C_1&: \mathbf{r}_1(t) = (3 - t)\mathbf{i} + t\mathbf{j}, \quad 0 \le t \le 3 \\
C_2&: \mathbf{r}_2(t) = (6 - t)\mathbf{j} + (2t - 6)\mathbf{k}, \quad 3 \le t \le 6 \\
C_3&: \mathbf{r}_3(t) = (t - 6)\mathbf{i} + (18 - 2t)\mathbf{k}, \quad 6 \le t \le 9
\end{aligned}$$

The value of the line integral is

$$\begin{aligned}
\int_C \mathbf{F} \cdot d\mathbf{r} &= \int_{C_1} \mathbf{F} \cdot \mathbf{r}_1'(t)\, dt + \int_{C_2} \mathbf{F} \cdot \mathbf{r}_2'(t)\, dt + \int_{C_3} \mathbf{F} \cdot \mathbf{r}_3'(t)\, dt \\
&= \int_0^3 t^2\, dt + \int_3^6 (-2t + 6)\, dt + \int_6^9 (-2t + 12)\, dt \\
&= 9 - 9 - 9 \\
&= -9.
\end{aligned}$$

EXAMPLE 2 Verifying Stokes's Theorem

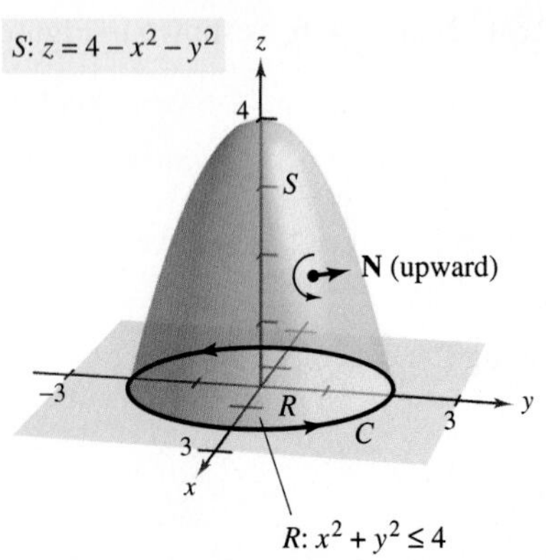

Figure 15.65

Verify Stokes's Theorem for $\mathbf{F}(x, y, z) = 2z\mathbf{i} + x\mathbf{j} + y^2\mathbf{k}$, where S is the surface of the paraboloid $z = 4 - x^2 - y^2$ and C is the trace of S in the xy-plane, as shown in Figure 15.65.

Solution As a *surface integral*, you have $z = g(x, y) = 4 - x^2 - y^2$ and

$$\textbf{curl F} = \begin{vmatrix} \mathbf{i} & \mathbf{j} & \mathbf{k} \\ \dfrac{\partial}{\partial x} & \dfrac{\partial}{\partial y} & \dfrac{\partial}{\partial z} \\ 2z & x & y^2 \end{vmatrix} = 2y\mathbf{i} + 2\mathbf{j} + \mathbf{k}.$$

By Theorem 15.11 for an upward normal vector $\mathbf{N}$, you obtain

$$\begin{aligned}\int_S\int (\textbf{curl F}) \cdot \mathbf{N}\,dS &= \int_R\int (2y\mathbf{i} + 2\mathbf{j} + \mathbf{k}) \cdot (2x\mathbf{i} + 2y\mathbf{j} + \mathbf{k})\,dA \\ &= \int_{-2}^{2}\int_{-\sqrt{4-y^2}}^{\sqrt{4-y^2}} (4xy + 4y + 1)\,dx\,dy \\ &= \int_{-2}^{2} \Big[2x^2y + (4y + 1)x\Big]_{-\sqrt{4-y^2}}^{\sqrt{4-y^2}} dy \\ &= \int_{-2}^{2} 2(4y + 1)\sqrt{4 - y^2}\,dy \\ &= \int_{-2}^{2} \left(8y\sqrt{4 - y^2} + 2\sqrt{4 - y^2}\right) dy \\ &= \left[-\frac{8}{3}(4 - y^2)^{3/2} + y\sqrt{4 - y^2} + 4\arcsin\frac{y}{2}\right]_{-2}^{2} \\ &= 4\pi.\end{aligned}$$

As a *line integral*, you can parametrize C by

$$\mathbf{r}(t) = 2\cos t\mathbf{i} + 2\sin t\mathbf{j} + 0\mathbf{k}, \quad 0 \le t \le 2\pi.$$

For $\mathbf{F}(x, y, z) = 2z\mathbf{i} + x\mathbf{j} + y^2\mathbf{k}$, you obtain

$$\begin{aligned}\int_C \mathbf{F} \cdot d\mathbf{r} &= \int_C M\,dx + N\,dy + P\,dz \\ &= \int_C 2z\,dx + x\,dy + y^2\,dz \\ &= \int_0^{2\pi} [0 + 2\cos t(2\cos t) + 0]\,dt \\ &= \int_0^{2\pi} 4\cos^2 t\,dt \\ &= 2\int_0^{2\pi} (1 + \cos 2t)\,dt \\ &= 2\left[t + \frac{1}{2}\sin 2t\right]_0^{2\pi} \\ &= 4\pi.\end{aligned}$$

Physical Interpretation of Curl

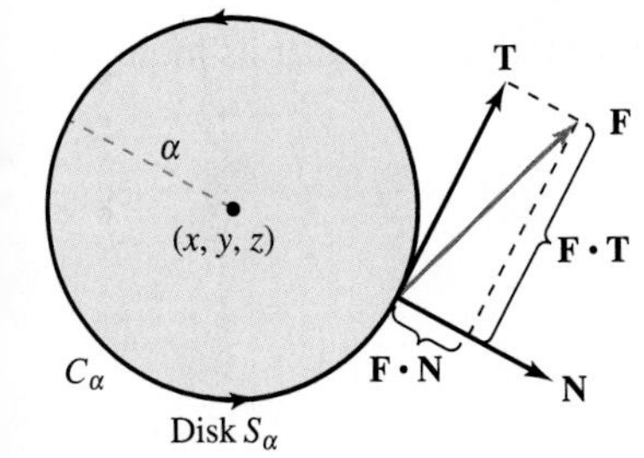

Figure 15.66

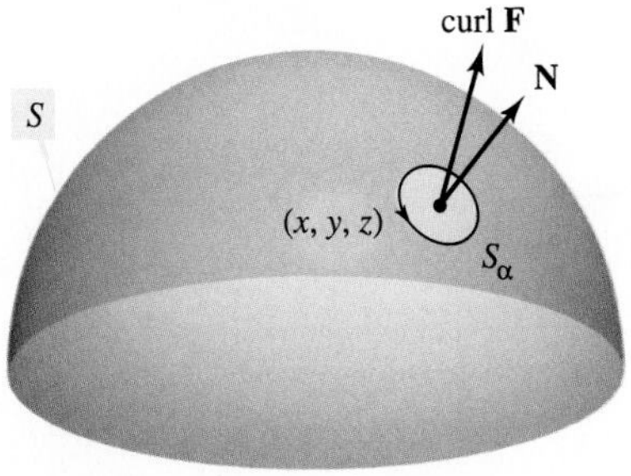

Figure 15.67

Stokes's Theorem provides insight into a physical interpretation of curl. In a vector field **F**, let S_α be a *small* circular disk of radius α, centered at (x, y, z) and with boundary C_α, as shown in Figure 15.66. At each point on the circle C_α, **F** has a normal component $\mathbf{F} \cdot \mathbf{N}$ and a tangential component $\mathbf{F} \cdot \mathbf{T}$. The more closely **F** and **T** are aligned, the greater the value of $\mathbf{F} \cdot \mathbf{T}$. So, a fluid tends to move along the circle rather than across it. Consequently, you say that the line integral around C_α measures the **circulation of F around C_α.** That is,

$$\int_{C_\alpha} \mathbf{F} \cdot \mathbf{T}\, ds = \text{circulation of } \mathbf{F} \text{ around } C_\alpha.$$

Now consider a small disk S_α to be centered at some point (x, y, z) on the surface S, as shown in Figure 15.67. On such a small disk, **curl F** is nearly constant, because it varies little from its value at (x, y, z). Moreover, $\mathbf{curl\,F} \cdot \mathbf{N}$ is also nearly constant on S_α, because all unit normals to S_α are about the same. Consequently, Stokes's Theorem yields

$$\begin{aligned}\int_{C_\alpha} \mathbf{F} \cdot \mathbf{T}\, ds &= \int\!\!\int_{S_\alpha} (\mathbf{curl\,F}) \cdot \mathbf{N}\, dS \\ &\approx (\mathbf{curl\,F}) \cdot \mathbf{N} \int\!\!\int_{S_\alpha} dS \\ &\approx (\mathbf{curl\,F}) \cdot \mathbf{N}(\pi\alpha^2).\end{aligned}$$

So,

$$\begin{aligned}(\mathbf{curl\,F}) \cdot \mathbf{N} &\approx \frac{\displaystyle\int_{C_\alpha} \mathbf{F} \cdot \mathbf{T}\, ds}{\pi\alpha^2} \\ &= \frac{\text{circulation of } \mathbf{F} \text{ around } C_\alpha}{\text{area of disk } S_\alpha} \\ &= \text{rate of circulation.}\end{aligned}$$

Assuming conditions are such that the approximation improves for smaller and smaller disks $(\alpha \to 0)$, it follows that

$$(\mathbf{curl\,F}) \cdot \mathbf{N} = \lim_{\alpha \to 0} \frac{1}{\pi\alpha^2} \int_{C_\alpha} \mathbf{F} \cdot \mathbf{T}\, ds$$

which is referred to as the **rotation of F about N.** That is,

$$\mathbf{curl\,F}(x, y, z) \cdot \mathbf{N} = \text{rotation of } \mathbf{F} \text{ about } \mathbf{N} \text{ at } (x, y, z).$$

In this case, the rotation of **F** is maximum when curl **F** and **N** have the same direction. Normally, this tendency to rotate will vary from point to point on the surface S, and Stokes's Theorem

$$\underbrace{\int\!\!\int_S (\mathbf{curl\,F}) \cdot \mathbf{N}\, dS}_{\text{Surface integral}} = \underbrace{\int_C \mathbf{F} \cdot d\mathbf{r}}_{\text{Line integral}}$$

says that the collective measure of this *rotational* tendency taken over the entire surface S (surface integral) is equal to the tendency of a fluid to *circulate* around the boundary C (line integral).

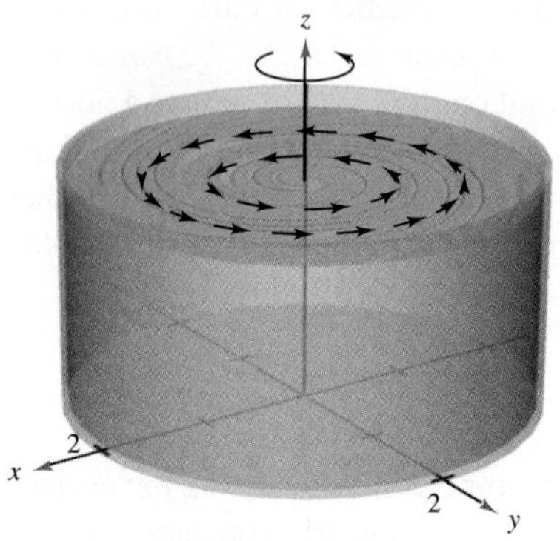

Figure 15.68

EXAMPLE 3 An Application of Curl

A liquid is swirling around in a cylindrical container of radius 2, so that its motion is described by the velocity field

$$\mathbf{F}(x, y, z) = -y\sqrt{x^2 + y^2}\,\mathbf{i} + x\sqrt{x^2 + y^2}\,\mathbf{j}$$

as shown in Figure 15.68. Find

$$\int_S\int (\textbf{curl F}) \cdot \mathbf{N}\, dS$$

where S is the upper surface of the cylindrical container.

Solution The curl of **F** is given by

$$\textbf{curl F} = \begin{vmatrix} \mathbf{i} & \mathbf{j} & \mathbf{k} \\ \dfrac{\partial}{\partial x} & \dfrac{\partial}{\partial y} & \dfrac{\partial}{\partial z} \\ -y\sqrt{x^2 + y^2} & x\sqrt{x^2 + y^2} & 0 \end{vmatrix} = 3\sqrt{x^2 + y^2}\,\mathbf{k}.$$

Letting $\mathbf{N} = \mathbf{k}$, you have

$$\begin{aligned} \int_S\int (\textbf{curl F}) \cdot \mathbf{N}\, dS &= \int_R\int 3\sqrt{x^2 + y^2}\, dA \\ &= \int_0^{2\pi}\int_0^2 (3r)r\, dr\, d\theta \\ &= \int_0^{2\pi} r^3\Big]_0^2\, d\theta \\ &= \int_0^{2\pi} 8\, d\theta \\ &= 16\pi. \end{aligned}$$

NOTE If $\textbf{curl F} = \mathbf{0}$ throughout region Q, the rotation of **F** about each unit normal **N** is 0. That is, **F** is irrotational. From earlier work, you know that this is a characteristic of conservative vector fields.

Summary of Integration Formulas

Fundamental Theorem of Calculus:

$$\int_a^b F'(x)\, dx = F(b) - F(a)$$

Fundamental Theorem of Line Integrals:

$$\int_C \mathbf{F} \cdot d\mathbf{r} = \int_C \nabla f \cdot d\mathbf{r} = f(x(b), y(b)) - f(x(a), y(a))$$

Green's Theorem:

$$\int_C M\, dx + N\, dy = \int_R\int \left(\frac{\partial N}{\partial x} - \frac{\partial M}{\partial y}\right) dA = \int_C \mathbf{F} \cdot \mathbf{T}\, ds = \int_C \mathbf{F} \cdot d\mathbf{r} = \int_R\int (\textbf{curl F}) \cdot \mathbf{k}\, dA$$

$$\int_C \mathbf{F} \cdot \mathbf{N}\, ds = \int_R\int \text{div}\,\mathbf{F}\, dA$$

Divergence Theorem:

$$\int_S\int \mathbf{F} \cdot \mathbf{N}\, dS = \int\int_Q\int \text{div}\,\mathbf{F}\, dV$$

Stokes's Theorem:

$$\int_C \mathbf{F} \cdot d\mathbf{r} = \int_S\int (\textbf{curl F}) \cdot \mathbf{N}\, dS$$

Exercises for Section 15.8

See www.CalcChat.com for worked-out solutions to odd-numbered exercises.

In Exercises 1–6, find the curl of the vector field F.

1. $\mathbf{F}(x, y, z) = (2y - z)\mathbf{i} + xyz\mathbf{j} + e^z\mathbf{k}$
2. $\mathbf{F}(x, y, z) = x^2\mathbf{i} + y^2\mathbf{j} + x^2\mathbf{k}$
3. $\mathbf{F}(x, y, z) = 2z\mathbf{i} - 4x^2\mathbf{j} + \arctan x\mathbf{k}$
4. $\mathbf{F}(x, y, z) = x \sin y\mathbf{i} - y \cos x\mathbf{j} + yz^2\mathbf{k}$
5. $\mathbf{F}(x, y, z) = e^{x^2+y^2}\mathbf{i} + e^{y^2+z^2}\mathbf{j} + xyz\mathbf{k}$
6. $\mathbf{F}(x, y, z) = \arcsin y\mathbf{i} + \sqrt{1 - x^2}\,\mathbf{j} + y^2\mathbf{k}$

In Exercises 7–10, verify Stokes's Theorem by evaluating

$$\int_C \mathbf{F} \cdot \mathbf{T}\, ds = \int_C \mathbf{F} \cdot d\mathbf{r}$$

as a line integral and as a double integral.

7. $\mathbf{F}(x, y, z) = (-y + z)\mathbf{i} + (x - z)\mathbf{j} + (x - y)\mathbf{k}$
   S: $z = \sqrt{1 - x^2 - y^2}$
8. $\mathbf{F}(x, y, z) = (-y + z)\mathbf{i} + (x - z)\mathbf{j} + (x - y)\mathbf{k}$
   S: $z = 4 - x^2 - y^2, \quad z \ge 0$
9. $\mathbf{F}(x, y, z) = xyz\mathbf{i} + y\mathbf{j} + z\mathbf{k}$
   S: $3x + 4y + 2z = 12$, first octant
10. $\mathbf{F}(x, y, z) = z^2\mathbf{i} + x^2\mathbf{j} + y^2\mathbf{k}$
    S: $z = y^2, \quad 0 \le x \le a, \quad 0 \le y \le a$

In Exercises 11–20, use Stokes's Theorem to evaluate $\int_C \mathbf{F} \cdot d\mathbf{r}$. Use a computer algebra system to verify your results. In each case, C is oriented counterclockwise as viewed from above.

11. $\mathbf{F}(x, y, z) = 2y\mathbf{i} + 3z\mathbf{j} + x\mathbf{k}$
    C: triangle with vertices $(0, 0, 0)$, $(0, 2, 0)$, $(1, 1, 1)$
12. $\mathbf{F}(x, y, z) = \arctan \dfrac{x}{y}\mathbf{i} + \ln\sqrt{x^2 + y^2}\,\mathbf{j} + \mathbf{k}$
    C: triangle with vertices $(0, 0, 0)$, $(1, 1, 1)$, $(0, 0, 2)$
13. $\mathbf{F}(x, y, z) = z^2\mathbf{i} + x^2\mathbf{j} + y^2\mathbf{k}$
    S: $z = 4 - x^2 - y^2, \quad z \ge 0$
14. $\mathbf{F}(x, y, z) = 4xz\mathbf{i} + y\mathbf{j} + 4xy\mathbf{k}$
    S: $z = 9 - x^2 - y^2, \quad z \ge 0$
15. $\mathbf{F}(x, y, z) = z^2\mathbf{i} + y\mathbf{j} + xz\mathbf{k}$
    S: $z = \sqrt{4 - x^2 - y^2}$
16. $\mathbf{F}(x, y, z) = x^2\mathbf{i} + z^2\mathbf{j} - xyz\mathbf{k}$
    S: $z = \sqrt{4 - x^2 - y^2}$
17. $\mathbf{F}(x, y, z) = -\ln\sqrt{x^2 + y^2}\,\mathbf{i} + \arctan \dfrac{x}{y}\mathbf{j} + \mathbf{k}$
    S: $z = 9 - 2x - 3y$ over one petal of $r = 2 \sin 2\theta$ in the first octant
18. $\mathbf{F}(x, y, z) = yz\mathbf{i} + (2 - 3y)\mathbf{j} + (x^2 + y^2)\mathbf{k}, \quad x^2 + y^2 \le 16$
    S: the first-octant portion of $x^2 + z^2 = 16$ over $x^2 + y^2 = 16$
19. $\mathbf{F}(x, y, z) = xyz\mathbf{i} + y\mathbf{j} + z\mathbf{k}$
    S: $z = x^2, \quad 0 \le x \le a, \quad 0 \le y \le a$
    $\mathbf{N}$ is the downward unit normal to the surface.
20. $\mathbf{F}(x, y, z) = xyz\mathbf{i} + y\mathbf{j} + z\mathbf{k}, \quad x^2 + y^2 \le a^2$
    S: the first-octant portion of $z = x^2$ over $x^2 + y^2 = a^2$

Motion of a Liquid **In Exercises 21 and 22, the motion of a liquid in a cylindrical container of radius 1 is described by the velocity field $\mathbf{F}(x, y, z)$. Find**

$$\int_S\int (\text{curl } \mathbf{F}) \cdot \mathbf{N}\, dS$$

where S is the upper surface of the cylindrical container.

21. $\mathbf{F}(x, y, z) = \mathbf{i} + \mathbf{j} - 2\mathbf{k}$
22. $\mathbf{F}(x, y, z) = -z\mathbf{i} + y\mathbf{k}$

Writing About Concepts

23. State Stokes's Theorem.
24. Give a physical interpretation of curl.
25. According to Stokes's Theorem, what can you conclude about the circulation in a field whose curl is $\mathbf{0}$? Explain your reasoning.

26. Let f and g be scalar functions with continuous partial derivatives, and let C and S satisfy the conditions of Stokes's Theorem. Verify each identity.
    (a) $\displaystyle\int_C (f\nabla g) \cdot d\mathbf{r} = \int_S\int (\nabla f \times \nabla g) \cdot \mathbf{N}\, dS$
    (b) $\displaystyle\int_C (f\nabla f) \cdot d\mathbf{r} = 0$
    (c) $\displaystyle\int_C (f\nabla g + g\nabla f) \cdot d\mathbf{r} = 0$
27. Demonstrate the results of Exercise 26 for the functions $f(x, y, z) = xyz$ and $g(x, y, z) = z$. Let S be the hemisphere $z = \sqrt{4 - x^2 - y^2}$.
28. Let $\mathbf{C}$ be a constant vector. Let S be an oriented surface with a unit normal vector $\mathbf{N}$, bounded by a smooth curve C. Prove that
    $$\int_S\int \mathbf{C} \cdot \mathbf{N}\, dS = \frac{1}{2}\int_C (\mathbf{C} \times \mathbf{r}) \cdot d\mathbf{r}.$$

Putnam Exam Challenge

29. Let $\mathbf{G}(x, y) = \left(\dfrac{-y}{x^2 + 4y^2}, \dfrac{x}{x^2 + 4y^2}, 0\right)$.
    Prove or disprove that there is a vector-valued function $\mathbf{F}(x, y, z) = (M(x, y, z), N(x, y, z), P(x, y, z))$ with the following properties.
    (i) M, N, P have continuous partial derivatives for all $(x, y, z) \ne (0, 0, 0)$;
    (ii) $\textbf{Curl F} = \mathbf{0}$ for all $(x, y, z) \ne (0, 0, 0)$;
    (iii) $\mathbf{F}(x, y, 0) = \mathbf{G}(x, y)$.

This problem was composed by the Committee on the Putnam Prize Competition.

Review Exercises for Chapter 15

See www.CalcChat.com for worked-out solutions to odd-numbered exercises.

In Exercises 1 and 2, sketch several representative vectors in the vector field. Use a computer algebra system to verify your results.

1. $\mathbf{F}(x, y, z) = x\mathbf{i} + \mathbf{j} + 2\mathbf{k}$
2. $\mathbf{F}(x, y) = \mathbf{i} - 2y\mathbf{j}$

In Exercises 3 and 4, find the gradient vector field for the scalar function.

3. $f(x, y, z) = 8x^2 + xy + z^2$
4. $f(x, y, z) = x^2e^{yz}$

In Exercises 5–12, determine if the vector field is conservative. If it is, find a potential function for the vector field.

5. $\mathbf{F}(x, y) = \dfrac{1}{y}\mathbf{i} - \dfrac{y}{x^2}\mathbf{j}$
6. $\mathbf{F}(x, y) = -\dfrac{y}{x^2}\mathbf{i} + \dfrac{1}{x}\mathbf{j}$
7. $\mathbf{F}(x, y) = (6xy^2 - 3x^2)\mathbf{i} + (6x^2y + 3y^2 - 7)\mathbf{j}$
8. $\mathbf{F}(x, y) = (-2y^3 \sin 2x)\mathbf{i} + 3y^2(1 + \cos 2x)\mathbf{j}$
9. $\mathbf{F}(x, y, z) = (4xy + z)\mathbf{i} + (2x^2 + 6y)\mathbf{j} + 2z\mathbf{k}$
10. $\mathbf{F}(x, y, z) = (4xy + z^2)\mathbf{i} + (2x^2 + 6yz)\mathbf{j} + 2xz\mathbf{k}$
11. $\mathbf{F}(x, y, z) = \dfrac{yz\mathbf{i} - xz\mathbf{j} - xy\mathbf{k}}{y^2z^2}$
12. $\mathbf{F}(x, y, z) = \sin z(y\mathbf{i} + x\mathbf{j} + \mathbf{k})$

In Exercises 13–20, find (a) the divergence of the vector field F and (b) the curl of the vector field F.

13. $\mathbf{F}(x, y, z) = x^2\mathbf{i} + y^2\mathbf{j} + z^2\mathbf{k}$
14. $\mathbf{F}(x, y, z) = xy^2\mathbf{j} - zx^2\mathbf{k}$
15. $\mathbf{F}(x, y, z) = (\cos y + y \cos x)\mathbf{i} + (\sin x - x \sin y)\mathbf{j} + xyz\mathbf{k}$
16. $\mathbf{F}(x, y, z) = (3x - y)\mathbf{i} + (y - 2z)\mathbf{j} + (z - 3x)\mathbf{k}$
17. $\mathbf{F}(x, y, z) = \arcsin x\mathbf{i} + xy^2\mathbf{j} + yz^2\mathbf{k}$
18. $\mathbf{F}(x, y, z) = (x^2 - y)\mathbf{i} - (x + \sin^2 y)\mathbf{j}$
19. $\mathbf{F}(x, y, z) = \ln(x^2 + y^2)\mathbf{i} + \ln(x^2 + y^2)\mathbf{j} + z\mathbf{k}$
20. $\mathbf{F}(x, y, z) = \dfrac{z}{x}\mathbf{i} + \dfrac{z}{y}\mathbf{j} + z^2\mathbf{k}$

In Exercises 21–26, evaluate the line integral along the given path(s).

21. $\displaystyle\int_C (x^2 + y^2)\, ds$
   (a) C: line segment from $(-1, -1)$ to $(2, 2)$
   (b) C: $x^2 + y^2 = 16$, one revolution counterclockwise, starting at $(4, 0)$
22. $\displaystyle\int_C xy\, ds$
   (a) C: line segment from $(0, 0)$ to $(5, 4)$
   (b) C: counterclockwise around the triangle with vertices $(0, 0)$, $(4, 0)$, $(0, 2)$
23. $\displaystyle\int_C (x^2 + y^2)\, ds$
   C: $\mathbf{r}(t) = (\cos t + t \sin t)\mathbf{i} + (\sin t - t \cos t)\mathbf{j}, \quad 0 \le t \le 2\pi$
24. $\displaystyle\int_C x\, ds$
   C: $\mathbf{r}(t) = (t - \sin t)\mathbf{i} + (1 - \cos t)\mathbf{j}, \quad 0 \le t \le 2\pi$
25. $\displaystyle\int_C (2x - y)\, dx + (x + 3y)\, dy$
   (a) C: line segment from $(0, 0)$ to $(2, -3)$
   (b) C: counterclockwise around the circle $x = 3 \cos t$, $y = 3 \sin t$
26. $\displaystyle\int_C (2x - y)\, dx + (x + 3y)\, dy$
   C: $\mathbf{r}(t) = (\cos t + t \sin t)\mathbf{i} + (\sin t - t \sin t)\mathbf{j}, \quad 0 \le t \le \pi/2$

In Exercises 27 and 28, use a computer algebra system to evaluate the line integral over the given path.

27. $\displaystyle\int_C (2x + y)\, ds$
   $\mathbf{r}(t) = a \cos^3 t\,\mathbf{i} + a \sin^3 t\,\mathbf{j}$, $0 \le t \le \pi/2$
28. $\displaystyle\int_C (x^2 + y^2 + z^2)\, ds$
   $\mathbf{r}(t) = t\mathbf{i} + t^2\mathbf{j} + t^{3/2}\mathbf{k}$, $0 \le t \le 4$

Lateral Surface Area **In Exercises 29 and 30, find the lateral surface area over the curve C in the xy-plane and under the surface $z = f(x, y)$.**

29. $f(x, y) = 5 + \sin(x + y)$
   C: $y = 3x$ from $(0, 0)$ to $(2, 6)$
30. $f(x, y) = 12 - x - y$
   C: $y = x^2$ from $(0, 0)$ to $(2, 4)$

In Exercises 31–36, evaluate $\displaystyle\int_C \mathbf{F} \cdot d\mathbf{r}$.

31. $\mathbf{F}(x, y) = xy\mathbf{i} + x^2\mathbf{j}$
   C: $\mathbf{r}(t) = t^2\mathbf{i} + t^3\mathbf{j}, \quad 0 \le t \le 1$
32. $\mathbf{F}(x, y) = (x - y)\mathbf{i} + (x + y)\mathbf{j}$
   C: $\mathbf{r}(t) = 4 \cos t\mathbf{i} + 3 \sin t\mathbf{j}, \quad 0 \le t \le 2\pi$
33. $\mathbf{F}(x, y, z) = x\mathbf{i} + y\mathbf{j} + z\mathbf{k}$
   C: $\mathbf{r}(t) = 2 \cos t\mathbf{i} + 2 \sin t\mathbf{j} + t\mathbf{k}, \quad 0 \le t \le 2\pi$
34. $\mathbf{F}(x, y, z) = (2y - z)\mathbf{i} + (z - x)\mathbf{j} + (x - y)\mathbf{k}$
   C: curve of intersection of $x^2 + z^2 = 4$ and $y^2 + z^2 = 4$ from $(2, 2, 0)$ to $(0, 0, 2)$
35. $\mathbf{F}(x, y, z) = (y - z)\mathbf{i} + (z - x)\mathbf{j} + (x - y)\mathbf{k}$
   C: curve of intersection of $z = x^2 + y^2$ and $x + y = 0$ from $(-2, 2, 8)$ to $(2, -2, 8)$
36. $\mathbf{F}(x, y, z) = (x^2 - z)\mathbf{i} + (y^2 + z)\mathbf{j} + x\mathbf{k}$
   C: curve of intersection of $z = x^2$ and $x^2 + y^2 = 4$ from $(0, -2, 0)$ to $(0, 2, 0)$

In Exercises 37 and 38, use a computer algebra system to evaluate the line integral.

37. $\int_C xy\,dx + (x^2 + y^2)\,dy$

C: $y = x^2$ from $(0, 0)$ to $(2, 4)$ and $y = 2x$ from $(2, 4)$ to $(0, 0)$

38. $\int_C \mathbf{F} \cdot d\mathbf{r}$

$\mathbf{F}(x, y) = (2x - y)\mathbf{i} + (2y - x)\mathbf{j}$

C: $\mathbf{r}(t) = (2\cos t + 2t\sin t)\mathbf{i} + (2\sin t - 2t\cos t)\mathbf{j}$, $0 \le t \le \pi$

39. ***Work*** Find the work done by the force field $\mathbf{F} = x\mathbf{i} - \sqrt{y}\mathbf{j}$ along the path $y = x^{3/2}$ from $(0, 0)$ to $(4, 8)$.

40. ***Work*** A 20-ton aircraft climbs 2000 feet while making a 90° turn in a circular arc of radius 10 miles. Find the work done by the engines.

In Exercises 41 and 42, evaluate the integral using the Fundamental Theorem of Line Integrals.

41. $\int_C 2xyz\,dx + x^2z\,dy + x^2y\,dz$

C: smooth curve from $(0, 0, 0)$ to $(1, 4, 3)$

42. $\int_C y\,dx + x\,dy + \frac{1}{z}\,dz$

C: smooth curve from $(0, 0, 1)$ to $(4, 4, 4)$

43. Evaluate the line integral $\int_C y^2\,dx + 2xy\,dy$.

(a) C: $\mathbf{r}(t) = (1 + 3t)\mathbf{i} + (1 + t)\mathbf{j}$, $0 \le t \le 1$

(b) C: $\mathbf{r}(t) = t\mathbf{i} + \sqrt{t}\mathbf{j}$, $1 \le t \le 4$

(c) Use the Fundamental Theorem of Line Integrals, where C is a smooth curve from $(1, 1)$ to $(4, 2)$.

44. ***Area and Centroid*** Consider the region bounded by the x-axis and one arch of the cycloid with parametric equations $x = a(\theta - \sin\theta)$ and $y = a(1 - \cos\theta)$. Use line integrals to find (a) the area of the region and (b) the centroid of the region.

In Exercises 45–50, use Green's Theorem to evaluate the line integral.

45. $\int_C y\,dx + 2x\,dy$

C: boundary of the square with vertices $(0, 0)$, $(0, 2)$, $(2, 0)$, $(2, 2)$

46. $\int_C xy\,dx + (x^2 + y^2)\,dy$

C: boundary of the square with vertices $(0, 0)$, $(0, 2)$, $(2, 0)$, $(2, 2)$

47. $\int_C xy^2\,dx + x^2y\,dy$

C: $x = 4\cos t$, $y = 2\sin t$

48. $\int_C (x^2 - y^2)\,dx + 2xy\,dy$

C: $x^2 + y^2 = a^2$

49. $\int_C xy\,dx + x^2\,dy$

C: boundary of the region between the graphs of $y = x^2$ and $y = x$

50. $\int_C y^2\,dx + x^{4/3}\,dy$

C: $x^{2/3} + y^{2/3} = 1$

In Exercises 51 and 52, use a computer algebra system to graph the surface represented by the vector-valued function.

51. $\mathbf{r}(u, v) = \sec u \cos v\,\mathbf{i} + (1 + 2\tan u)\sin v\,\mathbf{j} + 2u\,\mathbf{k}$

$0 \le u \le \frac{\pi}{3}$, $0 \le v \le 2\pi$

52. $\mathbf{r}(u, v) = e^{-u/4}\cos v\,\mathbf{i} + e^{-u/4}\sin v\,\mathbf{j} + \frac{u}{6}\mathbf{k}$

$0 \le u \le 4$, $0 \le v \le 2\pi$

53. ***Investigation*** Consider the surface represented by the vector-valued function

$\mathbf{r}(u, v) = 3\cos v \cos u\,\mathbf{i} + 3\cos v \sin u\,\mathbf{j} + \sin v\,\mathbf{k}$.

Use a computer algebra system to do the following.

(a) Graph the surface for $0 \le u \le 2\pi$ and $-\frac{\pi}{2} \le v \le \frac{\pi}{2}$.

(b) Graph the surface for $0 \le u \le 2\pi$ and $\frac{\pi}{4} \le v \le \frac{\pi}{2}$.

(c) Graph the surface for $0 \le u \le \frac{\pi}{4}$ and $0 \le v \le \frac{\pi}{2}$.

(d) Graph and identify the space curve for $0 \le u \le 2\pi$ and $v = \frac{\pi}{4}$.

(e) Approximate the area of the surface graphed in part (b).

(f) Approximate the area of the surface graphed in part (c).

54. Evaluate the surface integral $\int\int_S z\,dS$ over the surface S:

$\mathbf{r}(u, v) = (u + v)\mathbf{i} + (u - v)\mathbf{j} + \sin v\,\mathbf{k}$

where $0 \le u \le 2$ and $0 \le v \le \pi$.

55. Use a computer algebra system to graph the surface S and approximate the surface integral

$\int\int_S (x + y)\,dS$

where S is the surface

S: $\mathbf{r}(u, v) = u\cos v\,\mathbf{i} + u\sin v\,\mathbf{j} + (u - 1)(2 - u)\mathbf{k}$

over $0 \le u \le 2$ and $0 \le v \le 2\pi$.

56. *Mass* A cone-shaped surface lamina S is given by

$$z = a\left(a - \sqrt{x^2 + y^2}\right), \quad 0 \le z \le a^2.$$

At each point on S, the density is proportional to the distance between the point and the z-axis.

(a) Sketch the cone-shaped surface.

(b) Find the mass m of the lamina.

In Exercises 57 and 58, verify the Divergence Theorem by evaluating

$$\int_S\int \mathbf{F} \cdot \mathbf{N}\,dS$$

as a surface integral and as a triple integral.

57. $\mathbf{F}(x, y, z) = x^2\mathbf{i} + xy\mathbf{j} + z\mathbf{k}$

Q: solid region bounded by the coordinate planes and the plane $2x + 3y + 4z = 12$

58. $\mathbf{F}(x, y, z) = x\mathbf{i} + y\mathbf{j} + z\mathbf{k}$

Q: solid region bounded by the coordinate planes and the plane $2x + 3y + 4z = 12$

In Exercises 59 and 60, verify Stokes's Theorem by evaluating

$$\int_C \mathbf{F} \cdot d\mathbf{r}$$

as a line integral and as a double integral.

59. $\mathbf{F}(x, y, z) = (\cos y + y \cos x)\mathbf{i} + (\sin x - x \sin y)\mathbf{j} + xyz\mathbf{k}$

S: portion of $z = y^2$ over the square in the xy-plane with vertices $(0, 0)$, $(a, 0)$, (a, a), $(0, a)$

$\mathbf{N}$ is the upward unit normal vector to the surface.

60. $\mathbf{F}(x, y, z) = (x - z)\mathbf{i} + (y - z)\mathbf{j} + x^2\mathbf{k}$

S: first-octant portion of the plane $3x + y + 2z = 12$

61. Prove that it is not possible for a vector field with twice-differentiable components to have a curl of $x\mathbf{i} + y\mathbf{j} + z\mathbf{k}$.

Section Project: The Planimeter

You have learned many calculus techniques for finding the area of a planar region. Engineers use a mechanical device called a *planimeter* for measuring planar areas, which is based on the area formula given in Theorem 15.9 (page 1092). As you can see in the figure, the planimeter is fixed at point O (but free to pivot) and has a hinge at A. The end of the tracer arm AB moves counterclockwise around the region R. A small wheel at B is perpendicular to $\overline{AB}$ and is marked with a scale to measure how much it rolls as B traces out the boundary of region R. In this project you will show that the area of R is given by the length L of the tracer arm $\overline{AB}$ multiplied by the distance D that the wheel rolls.

Assume that point B traces out the boundary of R for $a \le t \le b$. Point A will move back and forth along a circular arc around the origin O. Let $\theta(t)$ denote the angle in the figure and let $(x(t), y(t))$ denote the coordinates of A.

(a) Show that the vector $\overrightarrow{OB}$ is given by the vector-valued function

$$\mathbf{r}(t) = [x(t) + L\cos\theta(t)]\mathbf{i} + [y(t) + L\sin\theta(t)]\mathbf{j}.$$

(b) Show that the following two integrals are equal to zero.

$$I_1 = \int_a^b \frac{1}{2}L^2\frac{d\theta}{dt}\,dt$$

$$I_2 = \int_a^b \frac{1}{2}\left(x\frac{dy}{dt} - y\frac{dx}{dt}\right)dt$$

(c) Use the integral $\int_a^b [x(t)\sin\theta(t) - y(t)\cos\theta(t)]'dt$ to show that the following two integrals are equal.

$$I_3 = \int_a^b \frac{1}{2}L\left(y\sin\theta\frac{d\theta}{dt} + x\cos\theta\frac{d\theta}{dt}\right)dt$$

$$I_4 = \int_a^b \frac{1}{2}L\left(-\sin\theta\frac{dx}{dt} + \cos\theta\frac{dy}{dt}\right)dt$$

(d) Let $\mathbf{N} = -\sin\theta\mathbf{i} + \cos\theta\mathbf{j}$. Explain why the distance D that the wheel rolls is given by

$$D = \int_C \mathbf{N} \cdot \mathbf{T}\,ds.$$

(e) Show that the area of region R is given by $I_1 + I_2 + I_3 + I_4 = DL$.

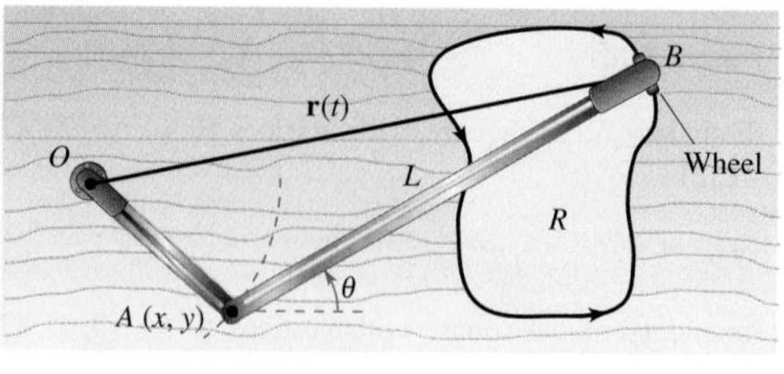

FOR FURTHER INFORMATION For more information about using calculus to find irregular areas, see "The Amateur Scientist" by C. L. Strong in the August 1958 issue of *Scientific American*.

P.S. Problem Solving

See www.CalcChat.com for worked-out solutions to odd-numbered exercises.

1. Heat flows from areas of higher temperature to areas of lower temperature in the direction of greatest change. As a result, measuring heat flux involves the gradient of the temperature. The flux depends on the area of the surface. It is the normal direction to the surface that is important, because heat that flows in directions tangential to the surface will give no heat loss. So, assume that the heat flux across a portion of the surface of area ΔS is given by $\Delta H \approx -k\nabla T \cdot \mathbf{N}\, dS$, where T is the temperature, $\mathbf{N}$ is the unit normal vector to the surface in the direction of the heat flow, and k is the thermal diffusivity of the material. The heat flux across the surface S is given by

$$H = \iint_S -k\nabla T \cdot \mathbf{N}\, dS.$$

Consider a single heat source located at the origin with temperature

$$T(x, y, z) = \frac{25}{\sqrt{x^2 + y^2 + z^2}}.$$

(a) Calculate the heat flux across the surface

$$S = \left\{(x, y, z): z = \sqrt{1 - x^2},\ -\frac{1}{2} \le x \le \frac{1}{2},\ 0 \le y \le 1\right\}$$

as shown in the figure.

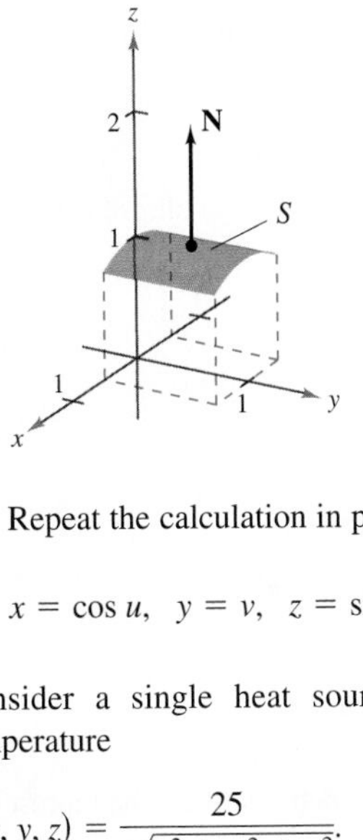

(b) Repeat the calculation in part (a) using the parametrization

$$x = \cos u,\quad y = v,\quad z = \sin u,\quad \frac{\pi}{3} \le u \le \frac{2\pi}{3},\quad 0 \le v \le 1.$$

2. Consider a single heat source located at the origin with temperature

$$T(x, y, z) = \frac{25}{\sqrt{x^2 + y^2 + z^2}}.$$

(a) Calculate the heat flux across the surface

$$S = \left\{(x, y, z): z = \sqrt{1 - x^2 - y^2},\ x^2 + y^2 \le 1\right\}$$

as shown in the figure.

(b) Repeat the calculation in part (a) using the parametrization

$x = \sin u \cos v,\quad y = \sin u \sin v,\quad z = \cos u,\quad 0 \le u \le \dfrac{\pi}{2},$
$0 \le v \le 2\pi.$

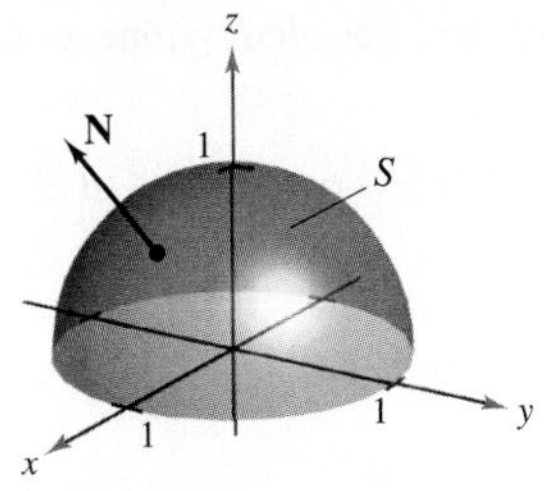

Figure for 2

3. Consider a wire of density $\rho(x, y, z)$ given by the space curve

$$C: \mathbf{r}(t) = x(t)\mathbf{i} + y(t)\mathbf{j} + z(t)\mathbf{k},\quad a \le t \le b.$$

The **moments of inertia** about the x-, y-, and z-axes are given by

$$I_x = \int_C (y^2 + z^2)\rho(x, y, z)\, ds$$

$$I_y = \int_C (x^2 + z^2)\rho(x, y, z)\, ds$$

$$I_z = \int_C (x^2 + y^2)\rho(x, y, z)\, ds.$$

Find the moments of inertia for a wire of uniform density $\rho = 1$ in the shape of the helix

$$\mathbf{r}(t) = 3\cos t\mathbf{i} + 3\sin t\mathbf{j} + 2t\mathbf{k},\quad 0 \le t \le 2\pi \text{ (see figure).}$$

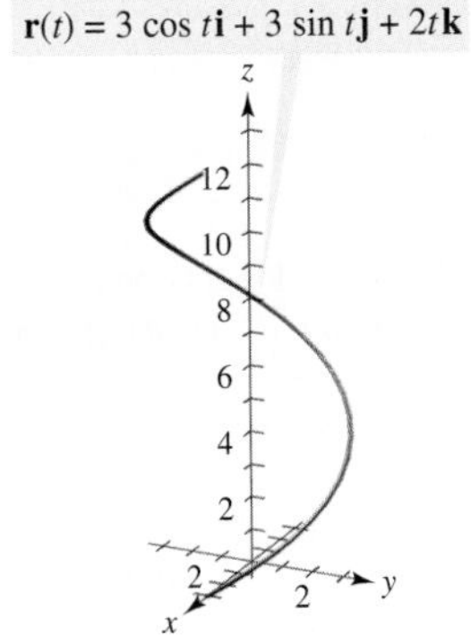

4. Find the moments of inertia for the wire of density $\rho = \dfrac{1}{1 + t}$ given by the curve

$$C: \mathbf{r}(t) = \frac{t^2}{2}\mathbf{i} + t\mathbf{j} + \frac{2\sqrt{2}\,t^{3/2}}{3}\mathbf{k},\quad 0 \le t \le 1 \text{ (see figure).}$$

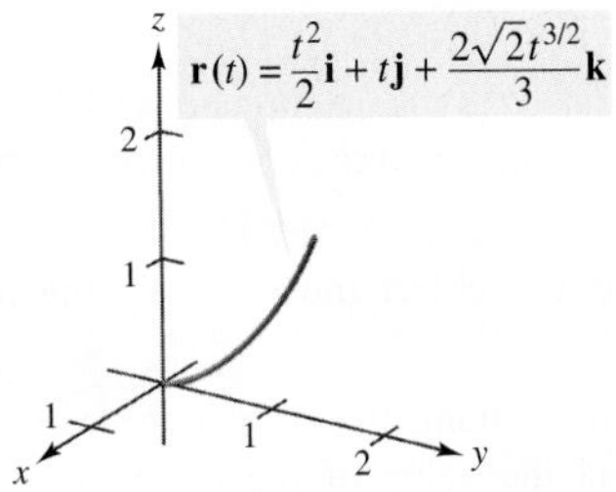

5. Use a line integral to find the area bounded by one arch of the cycloid

$$x(\theta) = a(\theta - \sin\theta), \quad y(\theta) = a(1 - \cos\theta), \quad 0 \le \theta \le 2\pi$$

as shown in the figure.

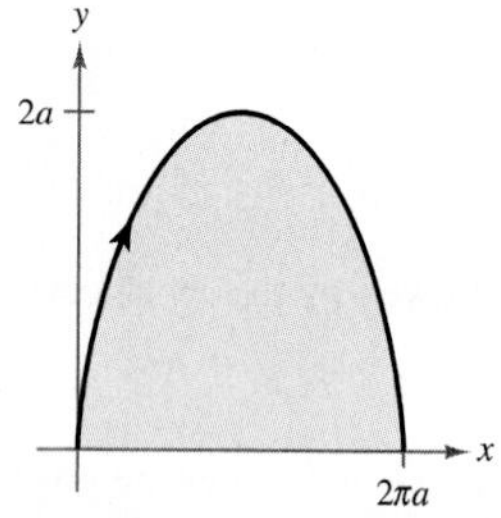

6. Use a line integral to find the area bounded by the two loops of the eight curve

$$x(t) = \frac{1}{2}\sin 2t, \quad y(t) = \sin t, \quad 0 \le t \le 2\pi$$

as shown in the figure.

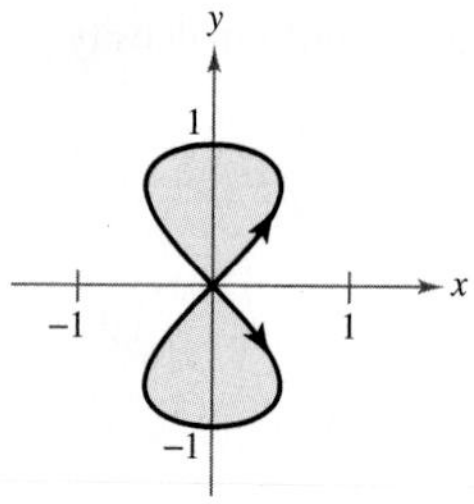

7. The force field $\mathbf{F}(x, y) = (x + y)\mathbf{i} + (x^2 + 1)\mathbf{j}$ acts on an object moving from the point $(0, 0)$ to the point $(0, 1)$, as shown in the figure.

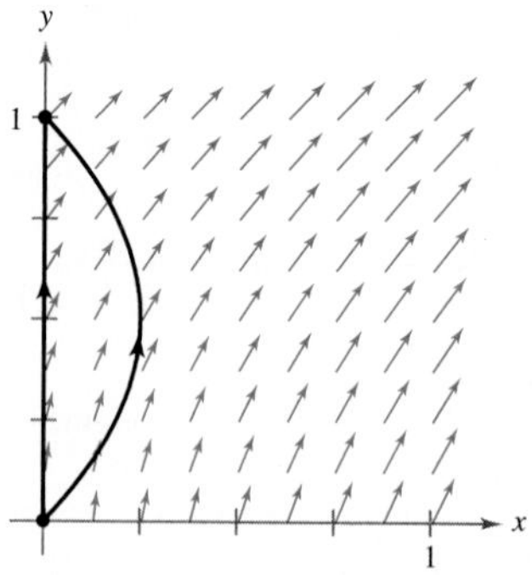

(a) Find the work done if the object moves along the path $x = 0$, $0 \le y \le 1$.

(b) Find the work done if the object moves along the path $x = y - y^2$, $0 \le y \le 1$.

(c) Suppose the object moves along the path $x = c(y - y^2)$, $0 \le y \le 1$, $c > 0$. Find the value of the constant c that minimizes the work.

8. The force field $\mathbf{F}(x, y) = (3x^2y^2)\mathbf{i} + (2x^3y)\mathbf{j}$ is shown in the figure below. Three particles move from the point $(1, 1)$ to the point $(2, 4)$ along different paths. Explain why the work done is the same for each particle, and find the value of the work.

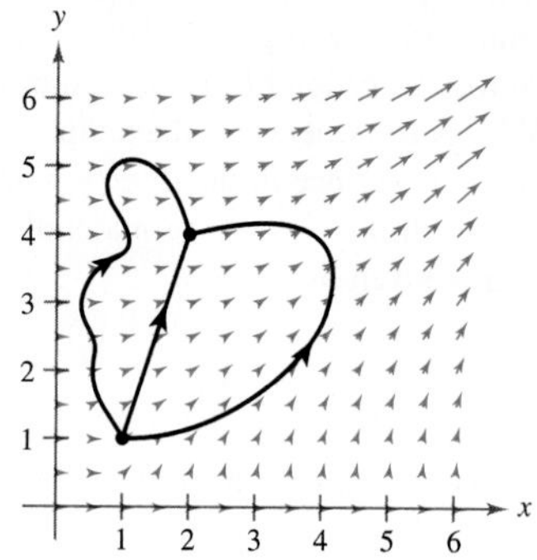

9. Let S be a smooth oriented surface with normal vector $\mathbf{N}$, bounded by a smooth simple closed curve C. Let $\mathbf{v}$ be a constant vector, and prove that

$$\int_S\int (2\mathbf{v} \cdot \mathbf{N})\, dS = \int_C (\mathbf{v} \times \mathbf{r}) \cdot d\mathbf{r}.$$

10. How does the area of the ellipse $\dfrac{x^2}{a^2} + \dfrac{y^2}{b^2} = 1$ compare with the magnitude of the work done by the force field

$$\mathbf{F}(x, y) = -\frac{1}{2}y\mathbf{i} + \frac{1}{2}x\mathbf{j}$$

on a particle that moves once around the ellipse (see figure)?

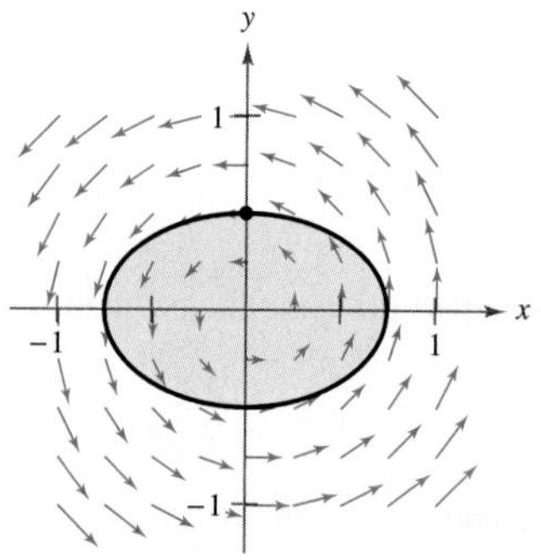

11. A cross section of Earth's magnetic field can be represented as a vector field in which the center of Earth is located at the origin and the positive y-axis points in the direction of the magnetic north pole. The equation for this field is

$$\mathbf{F}(x, y) = M(x, y)\mathbf{i} + N(x, y)\mathbf{j}$$
$$= \frac{m}{(x^2 + y^2)^{5/2}}[3xy\mathbf{i} + (2y^2 - x^2)\mathbf{j}]$$

where m is the magnetic moment of Earth. Show that this vector field is conservative.

Appendices

The remaining appendices are located on the website that accompanies this text at *college.hmco.com*.

A Proofs of Selected Theorems

THEOREM 1.2 Properties of Limits (Properties 2, 3, 4, and 5) (page 59)

Let b and c be real numbers, let n be a positive integer, and let f and g be functions with the following limits.

$$\lim_{x \to c} f(x) = L \quad \text{and} \quad \lim_{x \to c} g(x) = K$$

2. Sum or difference: $\lim_{x \to c} [f(x) \pm g(x)] = L \pm K$

3. Product: $\lim_{x \to c} [f(x)g(x)] = LK$

4. Quotient: $\lim_{x \to c} \frac{f(x)}{g(x)} = \frac{L}{K}$, provided $K \neq 0$

5. Power: $\lim_{x \to c} [f(x)]^n = L^n$

Proof To prove Property 2, choose $\varepsilon > 0$. Because $\varepsilon/2 > 0$, you know that there exists $\delta_1 > 0$ such that $0 < |x - c| < \delta_1$ implies $|f(x) - L| < \varepsilon/2$. You also know that there exists $\delta_2 > 0$ such that $0 < |x - c| < \delta_2$ implies $|g(x) - K| < \varepsilon/2$. Let δ be the smaller of δ_1 and δ_2; then $0 < |x - c| < \delta$ implies that

$$|f(x) - L| < \frac{\varepsilon}{2} \quad \text{and} \quad |g(x) - K| < \frac{\varepsilon}{2}.$$

So, you can apply the triangle inequality to conclude that

$$|[f(x) + g(x)] - (L + K)| \leq |f(x) - L| + |g(x) - K| < \frac{\varepsilon}{2} + \frac{\varepsilon}{2} = \varepsilon$$

which implies that

$$\lim_{x \to c} [f(x) + g(x)] = L + K = \lim_{x \to c} f(x) + \lim_{x \to c} g(x).$$

The proof that

$$\lim_{x \to c} [f(x) - g(x)] = L - K$$

is similar.

To prove Property 3, given that

$$\lim_{x \to c} f(x) = L \quad \text{and} \quad \lim_{x \to c} g(x) = K$$

you can write

$$f(x)g(x) = [f(x) - L][g(x) - K] + [Lg(x) + Kf(x)] - LK.$$

Because the limit of $f(x)$ is L, and the limit of $g(x)$ is K, you have

$$\lim_{x \to c} [f(x) - L] = 0 \quad \text{and} \quad \lim_{x \to c} [g(x) - K] = 0.$$

Let $0 < \varepsilon < 1$. Then there exists $\delta > 0$ such that if $0 < |x - c| < \delta$, then

$$|f(x) - L - 0| < \varepsilon \quad \text{and} \quad |g(x) - K - 0| < \varepsilon$$

which implies that

$$|[f(x) - L][g(x) - K] - 0| = |f(x) - L|\,|g(x) - K| < \varepsilon\varepsilon < \varepsilon.$$

So,

$$\lim_{x\to c} [f(x) - L][g(x) - K] = 0.$$

Furthermore, by Property 1, you have

$$\lim_{x\to c} Lg(x) = LK \quad \text{and} \quad \lim_{x\to c} Kf(x) = KL.$$

Finally, by Property 2, you obtain

$$\begin{aligned}\lim_{x\to c} f(x)g(x) &= \lim_{x\to c} [f(x) - L][g(x) - K] + \lim_{x\to c} Lg(x) + \lim_{x\to c} Kf(x) - \lim_{x\to c} LK\\ &= 0 + LK + KL - LK\\ &= LK.\end{aligned}$$

To prove Property 4, note that it is sufficient to prove that

$$\lim_{x\to c} \frac{1}{g(x)} = \frac{1}{K}.$$

Then you can use Property 3 to write

$$\lim_{x\to c} \frac{f(x)}{g(x)} = \lim_{x\to c} f(x)\frac{1}{g(x)} = \lim_{x\to c} f(x) \cdot \lim_{x\to c} \frac{1}{g(x)} = \frac{L}{K}.$$

Let $\varepsilon > 0$. Because $\lim_{x\to c} g(x) = K$, there exists $\delta_1 > 0$ such that if

$$0 < |x - c| < \delta_1, \text{ then } |g(x) - K| < \frac{|K|}{2}$$

which implies that

$$|K| = |g(x) + [|K| - g(x)]| \le |g(x)| + ||K| - g(x)| < |g(x)| + \frac{|K|}{2}.$$

That is, for $0 < |x - c| < \delta_1$,

$$\frac{|K|}{2} < |g(x)| \quad \text{or} \quad \frac{1}{|g(x)|} < \frac{2}{|K|}.$$

Similarly, there exists a $\delta_2 > 0$ such that if $0 < |x - c| < \delta_2$, then

$$|g(x) - K| < \frac{|K|^2}{2}\varepsilon.$$

Let δ be the smaller of δ_1 and δ_2. For $0 < |x - c| < \delta$, you have

$$\left|\frac{1}{g(x)} - \frac{1}{K}\right| = \left|\frac{K - g(x)}{g(x)K}\right| = \frac{1}{|K|} \cdot \frac{1}{|g(x)|}|K - g(x)| < \frac{1}{|K|} \cdot \frac{2}{|K|}\frac{|K|^2}{2}\varepsilon = \varepsilon.$$

So, $\lim_{x\to c} \frac{1}{g(x)} = \frac{1}{K}$.

Finally, the proof of Property 5 can be obtained by a straightforward application of mathematical induction coupled with Property 3. ▬

THEOREM 1.4 The Limit of a Function Involving a Radical (page 60)

Let n be a positive integer. The following limit is valid for all c if n is odd, and is valid for $c > 0$ if n is even.

$$\lim_{x \to c} \sqrt[n]{x} = \sqrt[n]{c}.$$

Proof Consider the case for which $c > 0$ and n is any positive integer. For a given $\varepsilon > 0$, you need to find $\delta > 0$ such that

$$\left|\sqrt[n]{x} - \sqrt[n]{c}\right| < \varepsilon \quad \text{whenever} \quad 0 < |x - c| < \delta$$

which is the same as saying

$$-\varepsilon < \sqrt[n]{x} - \sqrt[n]{c} < \varepsilon \quad \text{whenever} \quad -\delta < x - c < \delta.$$

Assume $\varepsilon < \sqrt[n]{c}$, which implies that $0 < \sqrt[n]{c} - \varepsilon < \sqrt[n]{c}$. Now, let δ be the smaller of the two numbers.

$$c - \left(\sqrt[n]{c} - \varepsilon\right)^n \quad \text{and} \quad \left(\sqrt[n]{c} + \varepsilon\right)^n - c$$

Then you have

$$\begin{aligned}
-\delta < x - c &< \delta \\
-\left[c - \left(\sqrt[n]{c} - \varepsilon\right)^n\right] < x - c &< \left(\sqrt[n]{c} + \varepsilon\right)^n - c \\
\left(\sqrt[n]{c} - \varepsilon\right)^n - c < x - c &< \left(\sqrt[n]{c} + \varepsilon\right)^n - c \\
\left(\sqrt[n]{c} - \varepsilon\right)^n < x &< \left(\sqrt[n]{c} + \varepsilon\right)^n \\
\sqrt[n]{c} - \varepsilon < \sqrt[n]{x} &< \sqrt[n]{c} + \varepsilon \\
-\varepsilon < \sqrt[n]{x} - \sqrt[n]{c} &< \varepsilon.
\end{aligned}$$

THEOREM 1.5 The Limit of a Composite Function (page 61)

If f and g are functions such that $\lim_{x \to c} g(x) = L$ and $\lim_{x \to L} f(x) = f(L)$, then

$$\lim_{x \to c} f(g(x)) = f\left(\lim_{x \to c} g(x)\right) = f(L).$$

Proof For a given $\varepsilon > 0$, you must find $\delta > 0$ such that

$$|f(g(x)) - f(L)| < \varepsilon \quad \text{whenever} \quad 0 < |x - c| < \delta.$$

Because the limit of $f(x)$ as $x \to L$ is $f(L)$, you know there exists $\delta_1 > 0$ such that

$$|f(u) - f(L)| < \varepsilon \quad \text{whenever} \quad |u - L| < \delta_1.$$

Moreover, because the limit of $g(x)$ as $x \to c$ is L, you know there exists $\delta > 0$ such that

$$|g(x) - L| < \delta_1 \quad \text{whenever} \quad 0 < |x - c| < \delta.$$

Finally, letting $u = g(x)$, you have

$$|f(g(x)) - f(L)| < \varepsilon \quad \text{whenever} \quad 0 < |x - c| < \delta.$$

THEOREM 1.7 Functions That Agree at All But One Point (page 62)

Let c be a real number and let $f(x) = g(x)$ for all $x \neq c$ in an open interval containing c. If the limit of $g(x)$ as x approaches c exists, then the limit of $f(x)$ also exists and

$$\lim_{x \to c} f(x) = \lim_{x \to c} g(x).$$

Proof Let L be the limit of $g(x)$ as $x \to c$. Then, for each $\varepsilon > 0$ there exists a $\delta > 0$ such that $f(x) = g(x)$ in the open intervals $(c - \delta, c)$ and $(c, c + \delta)$, and

$$|g(x) - L| < \varepsilon \quad \text{whenever} \quad 0 < |x - c| < \delta.$$

Because $f(x) = g(x)$ for all x in the open interval other than $x = c$, it follows that

$$|f(x) - L| < \varepsilon \quad \text{whenever} \quad 0 < |x - c| < \delta.$$

So, the limit of $f(x)$ as $x \to c$ is also L.

THEOREM 1.8 The Squeeze Theorem (page 65)

If $h(x) \leq f(x) \leq g(x)$ for all x in an open interval containing c, except possibly at c itself, and if

$$\lim_{x \to c} h(x) = L = \lim_{x \to c} g(x)$$

then $\lim_{x \to c} f(x)$ exists and is equal to L.

Proof For $\varepsilon > 0$ there exist $\delta_1 > 0$ and $\delta_2 > 0$ such that

$$|h(x) - L| < \varepsilon \quad \text{whenever} \quad 0 < |x - c| < \delta_1$$

and

$$|g(x) - L| < \varepsilon \quad \text{whenever} \quad 0 < |x - c| < \delta_2.$$

Because $h(x) \leq f(x) \leq g(x)$ for all x in an open interval containing c, except possibly at c itself, there exists $\delta_3 > 0$ such that $h(x) \leq f(x) \leq g(x)$ for $0 < |x - c| < \delta_3$. Let δ be the smallest of δ_1, δ_2, and δ_3. Then, if $0 < |x - c| < \delta$, it follows that $|h(x) - L| < \varepsilon$ and $|g(x) - L| < \varepsilon$, which implies that

$$-\varepsilon < h(x) - L < \varepsilon \quad \text{and} \quad -\varepsilon < g(x) - L < \varepsilon$$
$$L - \varepsilon < h(x) \quad \text{and} \quad g(x) < L + \varepsilon.$$

Now, because $h(x) \leq f(x) \leq g(x)$, it follows that $L - \varepsilon < f(x) < L + \varepsilon$, which implies that $|f(x) - L| < \varepsilon$. Therefore,

$$\lim_{x \to c} f(x) = L.$$

THEOREM 1.14 Vertical Asymptotes (page 85)

Let f and g be continuous on an open interval containing c. If $f(c) \neq 0$, $g(c) = 0$, and there exists an open interval containing c such that $g(x) \neq 0$ for all $x \neq c$ in the interval, then the graph of the function given by

$$h(x) = \frac{f(x)}{g(x)}$$

has a vertical asymptote at $x = c$.

Proof Consider the case for which $f(c) > 0$, and there exists $b > c$ such that $c < x < b$ implies $g(x) > 0$. Then for $M > 0$, choose δ_1 such that

$$0 < x - c < \delta_1 \quad \text{implies that} \quad \frac{f(c)}{2} < f(x) < \frac{3f(c)}{2}$$

and δ_2 such that

$$0 < x - c < \delta_2 \quad \text{implies that} \quad 0 < g(x) < \frac{f(c)}{2M}.$$

Now let δ be the smaller of δ_1 and δ_2. Then it follows that

$$0 < x - c < \delta \quad \text{implies that} \quad \frac{f(x)}{g(x)} > \frac{f(c)}{2}\left[\frac{2M}{f(c)}\right] = M.$$

So, it follows that

$$\lim_{x\to c^+} \frac{f(x)}{g(x)} = \infty$$

and the line $x = c$ is a vertical asymptote of the graph of h.

Alternative Form of the Derivative (page 101)

The derivative of f at c is given by

$$f'(c) = \lim_{x\to c} \frac{f(x) - f(c)}{x - c}$$

provided this limit exists.

Proof The derivative of f at c is given by

$$f'(c) = \lim_{\Delta x\to 0} \frac{f(c + \Delta x) - f(c)}{\Delta x}.$$

Let $x = c + \Delta x$. Then $x \to c$ as $\Delta x \to 0$. So, replacing $c + \Delta x$ by x, you have

$$f'(c) = \lim_{\Delta x\to 0} \frac{f(c + \Delta x) - f(c)}{\Delta x} = \lim_{x\to c} \frac{f(x) - f(c)}{x - c}.$$

THEOREM 2.10 The Chain Rule (page 131)

If $y = f(u)$ is a differentiable function of u, and $u = g(x)$ is a differentiable function of x, then $y = f(g(x))$ is a differentiable function of x and

$$\frac{dy}{dx} = \frac{dy}{du} \cdot \frac{du}{dx}$$

or, equivalently,

$$\frac{d}{dx}[f(g(x))] = f'(g(x))g'(x).$$

Proof In Section 2.4, you let $h(x) = f(g(x))$ and used the alternative form of the derivative to show that $h'(c) = f'(g(c))g'(c)$, provided $g(x) \neq g(c)$ for values of x other than c. Now consider a more general proof. Begin by considering the derivative of f.

$$f'(x) = \lim_{\Delta x \to 0} \frac{f(x + \Delta x) - f(x)}{\Delta x} = \lim_{\Delta x \to 0} \frac{\Delta y}{\Delta x}$$

For a fixed value of x, define a function η such that

$$\eta(\Delta x) = \begin{cases} 0, & \Delta x = 0 \\ \dfrac{\Delta y}{\Delta x} - f'(x), & \Delta x \neq 0. \end{cases}$$

Because the limit of $\eta(\Delta x)$ as $\Delta x \to 0$ doesn't depend on the value of $\eta(0)$, you have

$$\lim_{\Delta x \to 0} \eta(\Delta x) = \lim_{\Delta x \to 0} \left[\frac{\Delta y}{\Delta x} - f'(x)\right] = 0$$

and you can conclude that η is continuous at 0. Moreover, because $\Delta y = 0$ when $\Delta x = 0$, the equation

$$\Delta y = \Delta x \eta(\Delta x) + \Delta x f'(x)$$

is valid whether Δx is zero or not. Now, by letting $\Delta u = g(x + \Delta x) - g(x)$, you can use the continuity of g to conclude that

$$\lim_{\Delta x \to 0} \Delta u = \lim_{\Delta x \to 0} [g(x + \Delta x) - g(x)] = 0$$

which implies that

$$\lim_{\Delta x \to 0} \eta(\Delta u) = 0.$$

Finally,

$$\Delta y = \Delta u \eta(\Delta u) + \Delta u f'(u) \to \frac{\Delta y}{\Delta x} = \frac{\Delta u}{\Delta x} \eta(\Delta u) + \frac{\Delta u}{\Delta x} f'(u), \quad \Delta x \neq 0$$

and taking the limit as $\Delta x \to 0$, you have

$$\begin{aligned} \frac{dy}{dx} = \frac{du}{dx}\left[\lim_{\Delta x \to 0} \eta(\Delta u)\right] + \frac{du}{dx} f'(u) &= \frac{dy}{dx}(0) + \frac{du}{dx} f'(u) \\ &= \frac{du}{dx} f'(u) \\ &= \frac{du}{dx} \cdot \frac{dy}{du}. \end{aligned}$$

Concavity Interpretation (page 190)

1. Let f be differentiable on an open interval I. If the graph of f is concave *upward* on I, then the graph of f lies *above* all of its tangent lines on I.
2. Let f be differentiable on an open interval I. If the graph of f is concave *downward* on I, then the graph of f lies *below* all of its tangent lines on I.

Proof Assume that f is concave upward on $I = (a, b)$. Then, f' is increasing on (a, b). Let c be a point in the interval $I = (a, b)$. The equation of the tangent line to the graph of f at c is given by

$$g(x) = f(c) + f'(c)(x - c).$$

If x is in the open interval (c, b), then the directed distance from point $(x, f(x))$ (on the graph of f) to the point $(x, g(x))$ (on the tangent line) is given by

$$\begin{aligned} d &= f(x) - [f(c) + f'(c)(x - c)] \\ &= f(x) - f(c) - f'(c)(x - c). \end{aligned}$$

Moreover, by the Mean Value Theorem there exists a number z in (c, x) such that

$$f'(z) = \frac{f(x) - f(c)}{x - c}.$$

So, you have

$$\begin{aligned} d &= f(x) - f(c) - f'(c)(x - c) \\ &= f'(z)(x - c) - f'(c)(x - c) \\ &= [f'(z) - f'(c)](x - c). \end{aligned}$$

The second factor $(x - c)$ is positive because $c < x$. Moreover, because f' is increasing, it follows that the first factor $[f'(z) - f'(c)]$ is also positive. Therefore, $d > 0$ and you can conclude that the graph of f lies above the tangent line at x. If x is in the open interval (a, c), a similar argument can be given. This proves the first statement. The proof of the second statement is similar.

THEOREM 3.10 Limits at Infinity (page 199)

If r is a positive rational number and c is any real number, then

$$\lim_{x\to\infty} \frac{c}{x^r} = 0.$$

Furthermore, if x^r is defined when $x < 0$, then $\lim_{x\to-\infty} \frac{c}{x^r} = 0.$

Proof Begin by proving that

$$\lim_{x\to\infty} \frac{1}{x} = 0.$$

For $\varepsilon > 0$, let $M = 1/\varepsilon$. Then, for $x > M$, you have

$$x > M = \frac{1}{\varepsilon} \quad \Longrightarrow \quad \frac{1}{x} < \varepsilon \quad \Longrightarrow \quad \left|\frac{1}{x} - 0\right| < \varepsilon.$$

So, by the definition of a limit at infinity, you can conclude that the limit of $1/x$ as $x \to \infty$ is 0. Now, using this result, and letting $r = m/n$, you can write the following.

$$\begin{aligned}
\lim_{x\to\infty} \frac{c}{x^r} &= \lim_{x\to\infty} \frac{c}{x^{m/n}} \\
&= c\left[\lim_{x\to\infty}\left(\frac{1}{\sqrt[n]{x}}\right)^m\right] \\
&= c\left(\lim_{x\to\infty}\sqrt[n]{\frac{1}{x}}\right)^m \\
&= c\left(\sqrt[n]{\lim_{x\to\infty}\frac{1}{x}}\right)^m \\
&= c\left(\sqrt[n]{0}\right)^m \\
&= 0
\end{aligned}$$

The proof of the second part of the theorem is similar.

THEOREM 4.2 Summation Formulas (page 260)

1. $\displaystyle\sum_{i=1}^{n} c = cn$
2. $\displaystyle\sum_{i=1}^{n} i = \frac{n(n+1)}{2}$
3. $\displaystyle\sum_{i=1}^{n} i^2 = \frac{n(n+1)(2n+1)}{6}$
4. $\displaystyle\sum_{i=1}^{n} i^3 = \frac{n^2(n+1)^2}{4}$

Proof The proof of Property 1 is straightforward. By adding c to itself n times, you obtain a sum of cn.

To prove Property 2, write the sum in increasing and decreasing order and add corresponding terms, as follows.

$$\begin{aligned}
\sum_{i=1}^{n} i &= 1 + 2 + 3 + \cdots + (n-1) + n \\
\sum_{i=1}^{n} i &= n + (n-1) + (n-2) + \cdots + 2 + 1 \\
2\sum_{i=1}^{n} i &= \underbrace{(n+1) + (n+1) + (n+1) + \cdots + (n+1) + (n+1)}_{n \text{ terms}}
\end{aligned}$$

So,

$$\sum_{i=1}^{n} i = \frac{n(n+1)}{2}.$$

To prove Property 3, use mathematical induction. First, if $n = 1$, the result is true because

$$\sum_{i=1}^{1} i^2 = 1^2 = 1 = \frac{1(1+1)(2+1)}{6}.$$

Now, assuming the result is true for $n = k$, you can show that it is true for $n = k + 1$, as follows.

$$\begin{aligned}\sum_{i=1}^{k+1} i^2 &= \sum_{i=1}^{k} i^2 + (k+1)^2 \\ &= \frac{k(k+1)(2k+1)}{6} + (k+1)^2 \\ &= \frac{k+1}{6}(2k^2 + k + 6k + 6) \\ &= \frac{k+1}{6}[(2k+3)(k+2)] \\ &= \frac{(k+1)(k+2)[2(k+1)+1]}{6}\end{aligned}$$

Property 4 can be proved using a similar argument with mathematical induction.

THEOREM 4.8 Preservation of Inequality (page 278)

1. If f is integrable and nonnegative on the closed interval $[a, b]$, then

$$0 \le \int_a^b f(x)\, dx.$$

2. If f and g are integrable on the closed interval $[a, b]$ and $f(x) \le g(x)$ for every x in $[a, b]$, then

$$\int_a^b f(x)\, dx \le \int_a^b g(x)\, dx.$$

Proof To prove Property 1, suppose, on the contrary, that

$$\int_a^b f(x)\, dx = I < 0.$$

Then, let $a = x_0 < x_1 < x_2 < \cdot \cdot \cdot < x_n = b$ be a partition of $[a, b]$, and let

$$R = \sum_{i=1}^{n} f(c_i)\,\Delta x_i$$

be a Riemann sum. Because $f(x) \ge 0$, it follows that $R \ge 0$. Now, for $\|\Delta\|$ sufficiently small, you have $|R - I| < -I/2$, which implies that

$$\sum_{i=1}^{n} f(c_i)\,\Delta x_i = R < I - \frac{I}{2} < 0$$

which is not possible. From this contradiction, you can conclude that

$$0 \le \int_a^b f(x)\, dx.$$

To prove Property 2 of the theorem, note that $f(x) \leq g(x)$ implies that $g(x) - f(x) \geq 0$. So, you can apply the result of Property 1 to conclude that

$$0 \leq \int_a^b [g(x) - f(x)]\, dx$$
$$0 \leq \int_a^b g(x)\, dx - \int_a^b f(x)\, dx$$
$$\int_a^b f(x)\, dx \leq \int_a^b g(x)\, dx.$$

Properties of the Natural Logarithmic Function (page 323)

$$\lim_{x\to 0^+} \ln x = -\infty \quad \text{and} \quad \lim_{x\to\infty} \ln x = \infty$$

Proof To begin, show that $\ln 2 \geq \frac{1}{2}$. From the Mean Value Theorem for Integrals, you can write

$$\ln 2 = \int_1^2 \frac{1}{x}\, dx = \frac{1}{c}(2 - 1) = \frac{1}{c}$$

where c is in $[1, 2]$. This implies that

$$1 \leq c \leq 2$$
$$1 \geq \frac{1}{c} \geq \frac{1}{2}$$
$$1 \geq \ln 2 \geq \frac{1}{2}.$$

Now, let N be any positive (large) number. Because $\ln x$ is increasing, it follows that if $x > 2^{2N}$, then

$$\ln x > \ln 2^{2N} = 2N \ln 2.$$

However, because $\ln 2 \geq \frac{1}{2}$, it follows that

$$\ln x > 2N \ln 2 \geq 2N\left(\frac{1}{2}\right) = N.$$

This verifies the second limit. To verify the first limit, let $z = 1/x$. Then, $z \to \infty$ as $x \to 0^+$, and you can write

$$\begin{aligned}\lim_{x\to 0^+} \ln x &= \lim_{x\to 0^+}\left(-\ln \frac{1}{x}\right)\\ &= \lim_{z\to\infty} (-\ln z)\\ &= -\lim_{z\to\infty} \ln z\\ &= -\infty\end{aligned}$$

THEOREM 5.8 Continuity and Differentiability of Inverse Functions (page 345)

Let f be a function whose domain is an interval I. If f has an inverse function, then the following statements are true.

1. If f is continuous on its domain, then f^{-1} is continuous on its domain.
2. If f is increasing on its domain, then f^{-1} is increasing on its domain.
3. If f is decreasing on its domain, then f^{-1} is decreasing on its domain.
4. If f is differentiable at c and $f'(c) \neq 0$, then f^{-1} is differentiable at $f(c)$.

Proof To prove Property 1, first show that if f is continuous on I and has an inverse function, then f is strictly monotonic on I. Suppose that f were not strictly monotonic. Then there would exist numbers x_1, x_2, x_3 in I such that $x_1 < x_2 < x_3$, but $f(x_2)$ is not between $f(x_1)$ and $f(x_3)$. Without loss of generality, assume $f(x_1) < f(x_3) < f(x_2)$. By the Intermediate Value Theorem, there exists a number x_0 between x_1 and x_2 such that $f(x_0) = f(x_3)$. So, f is not one-to-one and cannot have an inverse function. So, f must be strictly monotonic.

Because f is continuous, the Intermediate Value Theorem implies that the set of values of f

$$\{f(x): x \in I\}$$

forms an interval J. Assume that a is an interior point of J. From the previous argument, $f^{-1}(a)$ is an interior point of I. Let $\varepsilon > 0$. There exists $0 < \varepsilon_1 < \varepsilon$ such that

$$I_1 = (f^{-1}(a) - \varepsilon_1, f^{-1}(a) + \varepsilon_1) \subseteq I.$$

Because f is strictly monotonic on I_1, the set of values $\{f(x): x \in I_1\}$ forms an interval $J_1 \subseteq J$. Let $\delta > 0$ such that $(a - \delta, a + \delta) \subseteq J_1$. Finally, if

$$|y - a| < \delta, \text{ then } |f^{-1}(y) - f^{-1}(a)| < \varepsilon_1 < \varepsilon.$$

So, f^{-1} is continuous at a. A similar proof can be given if a is an endpoint.

To prove Property 2, let y_1 and y_2 be in the domain of f^{-1}, with $y_1 < y_2$. Then, there exist x_1 and x_2 in the domain of f such that

$$f(x_1) = y_1 < y_2 = f(x_2).$$

Because f is increasing, $f(x_1) < f(x_2)$ holds precisely when $x_1 < x_2$. Therefore,

$$f^{-1}(y_1) = x_1 < x_2 = f^{-1}(y_2)$$

which implies that f^{-1} is increasing. (Property 3 can be proved in a similar way.)

Finally, to prove Property 4, consider the limit

$$(f^{-1})'(a) = \lim_{y \to a} \frac{f^{-1}(y) - f^{-1}(a)}{y - a}$$

where a is in the domain of f^{-1} and $f^{-1}(a) = c$. Because f is differentiable at c, f is continuous at c, and so is f^{-1} at a. So, $y \to a$ implies that $x \to c$, and you have

$$\begin{aligned}(f^{-1})'(a) &= \lim_{x \to c} \frac{x - c}{f(x) - f(c)} \\ &= \lim_{x \to c} \frac{1}{\left(\dfrac{f(x) - f(c)}{x - c}\right)} \\ &= \frac{1}{\displaystyle\lim_{x \to c} \frac{f(x) - f(c)}{x - c}} \\ &= \frac{1}{f'(c)}.\end{aligned}$$

So, $(f^{-1})'(a)$ exists, and f^{-1} is differentiable at $f(c)$.

THEOREM 5.9 The Derivative of an Inverse Function (page 345)

Let f be a function that is differentiable on an interval I. If f has an inverse function g, then g is differentiable at any x for which $f'(g(x)) \neq 0$. Moreover,

$$g'(x) = \frac{1}{f'(g(x))}, \quad f'(g(x)) \neq 0.$$

Proof From the proof of Theorem 5.8, letting $a = x$, you know that g is differentiable. Using the Chain Rule, differentiate both sides of the equation $x = f(g(x))$ to obtain

$$1 = f'(g(x)) \frac{d}{dx}[g(x)].$$

Because $f'(g(x)) \neq 0$, you can divide by this quantity to obtain

$$\frac{d}{dx}[g(x)] = \frac{1}{f'(g(x))}.$$

THEOREM 5.15 A Limit Involving e (page 364)

$$\lim_{x \to \infty} \left(1 + \frac{1}{x}\right)^x = \lim_{x \to \infty} \left(\frac{x + 1}{x}\right)^x = e$$

Proof Let $y = \lim_{x \to \infty} \left(1 + \frac{1}{x}\right)^x$. Taking the natural logarithm of each side, you have

$$\ln y = \ln\left[\lim_{x \to \infty} \left(1 + \frac{1}{x}\right)^x\right].$$

Because the natural logarithmic function is continuous, you can write

$$\ln y = \lim_{x\to\infty}\left[x\ln\left(1+\frac{1}{x}\right)\right] = \lim_{x\to\infty}\left\{\frac{\ln[1+(1/x)]}{1/x}\right\}.$$

Letting $x = \frac{1}{t}$, you have

$$\begin{aligned}\ln y &= \lim_{t\to 0^+}\frac{\ln(1+t)}{t}\\ &= \lim_{t\to 0^+}\frac{\ln(1+t)-\ln 1}{t}\\ &= \frac{d}{dx}\ln x \text{ at } x=1\\ &= \frac{1}{x} \text{ at } x = 1\\ &= 1.\end{aligned}$$

Finally, because $\ln y = 1$, you know that $y = e$, and you can conclude that

$$\lim_{x\to\infty}\left(1+\frac{1}{x}\right)^x = e.$$

THEOREM 8.3 The Extended Mean Value Theorem (page 568)

If f and g are differentiable on an open interval (a, b) and continuous on $[a, b]$ such that $g'(x) \neq 0$ for any x in (a, b), then there exists a point c in (a, b) such that

$$\frac{f'(c)}{g'(c)} = \frac{f(b)-f(a)}{g(b)-g(a)}.$$

Proof You can assume that $g(a) \neq g(b)$, because otherwise, by Rolle's Theorem, it would follow that $g'(x) = 0$ for some x in (a, b). Now, define $h(x)$ to be

$$h(x) = f(x) - \left[\frac{f(b)-f(a)}{g(b)-g(a)}\right]g(x).$$

Then

$$h(a) = f(a) - \left[\frac{f(b)-f(a)}{g(b)-g(a)}\right]g(a) = \frac{f(a)g(b)-f(b)g(a)}{g(b)-g(a)}$$

and

$$h(b) = f(b) - \left[\frac{f(b)-f(a)}{g(b)-g(a)}\right]g(b) = \frac{f(a)g(b)-f(b)g(a)}{g(b)-g(a)}$$

and by Rolle's Theorem there exists a point c in (a, b) such that

$$h'(c) = f'(c) - \frac{f(b)-f(a)}{g(b)-g(a)}g'(c) = 0$$

which implies that

$$\frac{f'(c)}{g'(c)} = \frac{f(b)-f(a)}{g(b)-g(a)}.$$

THEOREM 8.4 L'Hôpital's Rule (page 568)

Let f and g be functions that are differentiable on an open interval (a, b) containing c, except possibly at c itself. Assume that $g'(x) \neq 0$ for all x in (a, b), except possibly at c itself. If the limit of $f(x)/g(x)$ as x approaches c produces the indeterminate form $0/0$, then

$$\lim_{x \to c} \frac{f(x)}{g(x)} = \lim_{x \to c} \frac{f'(x)}{g'(x)}$$

provided the limit on the right exists (or is infinite). This result also applies if the limit of $f(x)/g(x)$ as x approaches c produces any one of the indeterminate forms ∞/∞, $(-\infty)/\infty$, $\infty/(-\infty)$, or $(-\infty)/(-\infty)$.

You can use the Extended Mean Value Theorem to prove L'Hôpital's Rule. Of the several different cases of this rule, the proof of only one case is illustrated. The remaining cases where $x \to c^-$ and $x \to c$ are left for you to prove.

Proof Consider the case for which

$$\lim_{x \to c^+} f(x) = 0 \quad \text{and} \quad \lim_{x \to c^+} g(x) = 0.$$

Define the following new functions:

$$F(x) = \begin{cases} f(x), & x \neq c \\ 0, & x = c \end{cases} \quad \text{and} \quad G(x) = \begin{cases} g(x), & x \neq c \\ 0, & x = c \end{cases}.$$

For any x, $c < x < b$, F and G are differentiable on $(c, x]$ and continuous on $[c, x]$. You can apply the Extended Mean Value Theorem to conclude that there exists a number z in (c, x) such that

$$\begin{aligned} \frac{F'(z)}{G'(z)} &= \frac{F(x) - F(c)}{G(x) - G(c)} \\ &= \frac{F(x)}{G(x)} \\ &= \frac{f'(z)}{g'(z)} \\ &= \frac{f(x)}{g(x)}. \end{aligned}$$

Finally, by letting x approach c from the right, $x \to c^+$, you have $z \to c^+$ because $c < z < x$, and

$$\begin{aligned} \lim_{x \to c^+} \frac{f(x)}{g(x)} &= \lim_{x \to c^+} \frac{f'(z)}{g'(z)} \\ &= \lim_{z \to c^+} \frac{f'(z)}{g'(z)} \\ &= \lim_{x \to c^+} \frac{f'(x)}{g'(x)}. \end{aligned}$$

THEOREM 9.19 Taylor's Theorem (page 654)

If a function f is differentiable through order $n + 1$ in an interval I containing c, then, for each x in I, there exists z between x and c such that

$$f(x) = f(c) + f'(c)(x - c) + \frac{f''(c)}{2!}(x - c)^2 + \cdot \cdot \cdot + \frac{f^{(n)}(c)}{n!}(x - c)^n + R_n(x)$$

where

$$R_n(x) = \frac{f^{(n+1)}(z)}{(n + 1)!}(x - c)^{n+1}.$$

Proof To find $R_n(x)$, fix x in I $(x \neq c)$ and write

$$R_n(x) = f(x) - P_n(x)$$

where $P_n(x)$ is the nth Taylor polynomial for $f(x)$. Then let g be a function of t defined by

$$g(t) = f(x) - f(t) - f'(t)(x - t) - \cdot \cdot \cdot - \frac{f^{(n)}(t)}{n!}(x - t)^n - R_n(x)\frac{(x - t)^{n+1}}{(x - c)^{n+1}}.$$

The reason for defining g in this way is that differentiation with respect to t has a telescoping effect. For example, you have

$$\begin{aligned} \frac{d}{dt}[-f(t) - f'(t)(x - t)] &= -f'(t) + f'(t) - f''(t)(x - t) \\ &= -f''(t)(x - t). \end{aligned}$$

The result is that the derivative $g'(t)$ simplifies to

$$g'(t) = -\frac{f^{(n+1)}(t)}{n!}(x - t)^n + (n + 1)R_n(x)\frac{(x - t)^n}{(x - c)^{n+1}}$$

for all t between c and x. Moreover, for a fixed x,

$$g(c) = f(x) - [P_n(x) + R_n(x)] = f(x) - f(x) = 0$$

and

$$g(x) = f(x) - f(x) - 0 - \cdot \cdot \cdot - 0 = f(x) - f(x) = 0.$$

Therefore, g satisfies the conditions of Rolle's Theorem, and it follows that there is a number z between c and x such that $g'(z) = 0$. Substituting z for t in the equation for $g'(t)$ and then solving for $R_n(x)$, you obtain

$$\begin{aligned} g'(z) &= -\frac{f^{(n+1)}(z)}{n!}(x - z)^n + (n + 1)R_n(x)\frac{(x - z)^n}{(x - c)^{n+1}} = 0 \\ R_n(x) &= \frac{f^{(n+1)}(z)}{(n + 1)!}(x - c)^{n+1}. \end{aligned}$$

Finally, because $g(c) = 0$, you have

$$\begin{aligned} 0 &= f(x) - f(c) - f'(c)(x - c) - \cdot \cdot \cdot - \frac{f^{(n)}(c)}{n!}(x - c)^n - R_n(x) \\ f(x) &= f(c) + f'(c)(x - c) + \cdot \cdot \cdot + \frac{f^{(n)}(c)}{n!}(x - c)^n + R_n(x). \end{aligned}$$

THEOREM 9.20 Convergence of a Power Series (page 660)

For a power series centered at c, precisely one of the following is true.

1. The series converges only at c.
2. There exists a real number $R > 0$ such that the series converges absolutely for $|x - c| < R$, and diverges for $|x - c| > R$.
3. The series converges absolutely for all x.

The number R is the **radius of convergence** of the power series. If the series converges only at c, the radius of convergence is $R = 0$, and if the series converges for all x, the radius of convergence is $R = \infty$. The set of all values of x for which the power series converges is the **interval of convergence** of the power series.

Proof In order to simplify the notation, the theorem for the power series $\Sigma\, a_n x^n$ centered at $x = 0$ will be proved. The proof for a power series centered at $x = c$ follows easily. A key step in this proof uses the completeness property of the set of real numbers: If a nonempty set S of real numbers has an upper bound, then it must have a least upper bound (see page 601).

It must be shown that if a power series $\Sigma\, a_n x^n$ converges at $x = d$, $d \neq 0$, then it converges for all b satisfying $|b| < |d|$. Because $\Sigma\, a_n x^n$ converges, $\lim\limits_{x \to \infty} a_n d^n = 0$. So, there exists $N > 0$ such that $a_n d^n < 1$ for all $n \geq N$. Then for $n \geq N$,

$$|a_n b^n| = \left|a_n b^n \frac{d^n}{d^n}\right| = |a_n d^n| \left|\frac{b^n}{d^n}\right| < \left|\frac{b^n}{d^n}\right|.$$

So, for $|b| < |d|$, $\left|\dfrac{b}{d}\right| < 1$, which implies that

$$\Sigma \left|\frac{b^n}{d^n}\right|$$

is a convergent geometric series. By the Comparison Test, the series $\Sigma\, a_n b^n$ converges.

Similarly, if the power series $\Sigma\, a_n x^n$ diverges at $x = b$, where $b \neq 0$, then it diverges for all d satisfying $|d| > |b|$. If $\Sigma\, a_n d^n$ converged, then the argument above would imply that $\Sigma\, a_n b^n$ converged as well.

Finally, to prove the theorem, suppose that neither case 1 nor case 3 is true. Then there exist points b and d such that $\Sigma\, a_n x^n$ converges at b and diverges at d. Let $S = \{x\colon \Sigma\, a_n x^n \text{ converges}\}$. S is nonempty because $b \in S$. If $x \in S$ then $|x| \leq |d|$, which shows that $|d|$ is an upper bound for the nonempty set S. By the completeness property, S has a least upper bound, R.

Now, if $|x| > R$, then $x \notin S$ so $\Sigma\, a_n x^n$ diverges. And if $|x| < R$, then $|x|$ is not an upper bound for S, so there exists b in S satisfying $|b| > |x|$. Since $b \in S$, $\Sigma\, a_n b^n$ converges, which implies that $\Sigma\, a_n x^n$ converges.

THEOREM 10.16 Classification of Conics by Eccentricity (page 748)

Let F be a fixed point (*focus*) and let D be a fixed line (*directrix*) in the plane. Let P be another point in the plane and let e (*eccentricity*) be the ratio of the distance between P and F to the distance between P and D. The collection of all points P with a given eccentricity is a conic.

1. The conic is an ellipse if $0 < e < 1$.
2. The conic is a parabola if $e = 1$.
3. The conic is a hyperbola if $e > 1$.

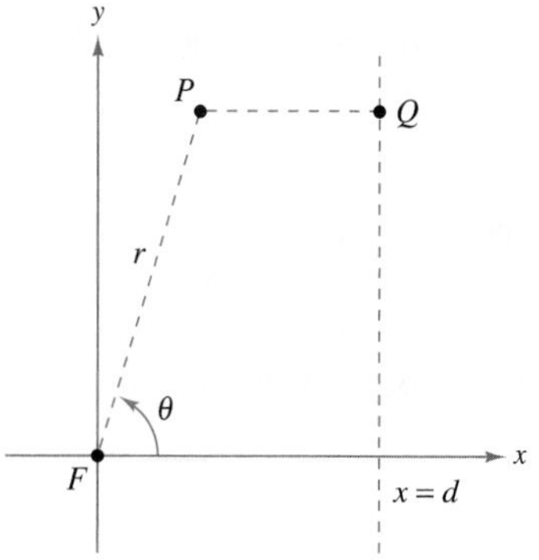

Figure A.1

Proof If $e = 1$, then, by definition, the conic must be a parabola. If $e \neq 1$, then you can consider the focus F to lie at the origin and the directrix $x = d$ to lie to the right of the origin, as shown in Figure A.1. For the point $P = (r, \theta) = (x, y)$, you have $|PF| = r$ and $|PQ| = d - r\cos\theta$. Given that $e = |PF|/|PQ|$, it follows that

$$|PF| = |PQ|e \implies r = e(d - r\cos\theta).$$

By converting to rectangular coordinates and squaring each side, you obtain

$$x^2 + y^2 = e^2(d - x)^2 = e^2(d^2 - 2dx + x^2).$$

Completing the square produces

$$\left(x + \frac{e^2 d}{1 - e^2}\right)^2 + \frac{y^2}{1 - e^2} = \frac{e^2 d^2}{(1 - e^2)^2}.$$

If $e < 1$, this equation represents an ellipse. If $e > 1$, then $1 - e^2 < 0$, and the equation represents a hyperbola.

THEOREM 13.4 Sufficient Condition for Differentiability (page 917)

If f is a function of x and y, where f_x and f_y are continuous in an open region R, then f is differentiable on R.

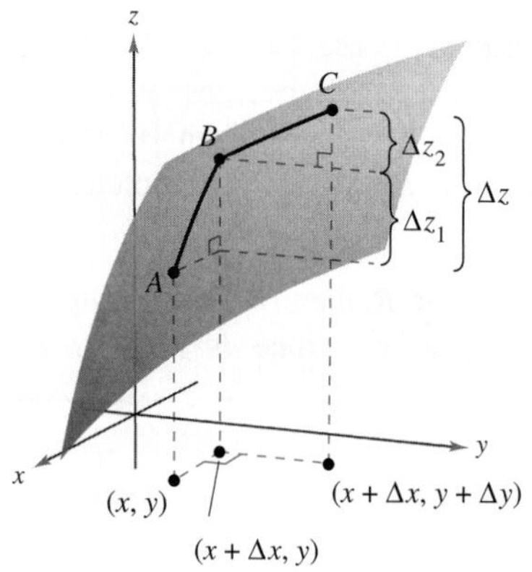

$\Delta z = f(x + \Delta x, y + \Delta y) - f(x, y)$
Figure A.2

Proof Let S be the surface defined by $z = f(x, y)$, where f, f_x, and f_y are continuous at (x, y). Let A, B, and C be points on surface S, as shown in Figure A.2. From this figure, you can see that the change in f from point A to point C is given by

$$\begin{aligned}\Delta z &= f(x + \Delta x, y + \Delta y) - f(x, y)\\ &= [f(x + \Delta x, y) - f(x, y)] + [f(x + \Delta x, y + \Delta y) - f(x + \Delta x, y)]\\ &= \Delta z_1 + \Delta z_2.\end{aligned}$$

Between A and B, y is fixed and x changes. So, by the Mean Value Theorem, there is a value x_1 between x and $x + \Delta x$ such that

$$\Delta z_1 = f(x + \Delta x, y) - f(x, y) = f_x(x_1, y)\,\Delta x.$$

Similarly, between B and C, x is fixed and y changes, and there is a value y_1 between y and $y + \Delta y$ such that

$$\Delta z_2 = f(x + \Delta x, y + \Delta y) - f(x + \Delta x, y) = f_y(x + \Delta x, y_1)\,\Delta y.$$

By combining these two results, you can write

$$\Delta z = \Delta z_1 + \Delta z_2 = f_x(x_1, y)\Delta x + f_y(x + \Delta x, y_1)\,\Delta y.$$

If you define ε_1 and ε_2 as

$$\varepsilon_1 = f_x(x_1, y) - f_x(x, y) \quad \text{and} \quad \varepsilon_2 = f_y(x + \Delta x, y_1) - f_y(x, y)$$

it follows that

$$\begin{aligned}\Delta z = \Delta z_1 + \Delta z_2 &= [\varepsilon_1 + f_x(x, y)]\,\Delta x + [\varepsilon_2 + f_y(x, y)]\,\Delta y\\ &= [f_x(x, y)\,\Delta x + f_y(x, y)\,\Delta y] + \varepsilon_1\Delta x + \varepsilon_2\Delta y.\end{aligned}$$

By the continuity of f_x and f_y and the fact that $x \le x_1 \le x + \Delta x$ and $y \le y_1 \le y + \Delta y$, it follows that $\varepsilon_1 \to 0$ and $\varepsilon_2 \to 0$ as $\Delta x \to 0$ and $\Delta y \to 0$. Therefore, by definition, f is differentiable.

THEOREM 13.6 Chain Rule: One Independent Variable (page 923)

Let $w = f(x, y)$, where f is a differentiable function of x and y. If $x = g(t)$ and $y = h(t)$, where g and h are differentiable functions of t, then w is a differentiable function of t, and

$$\frac{dw}{dt} = \frac{\partial w}{\partial x}\frac{dx}{dt} + \frac{\partial w}{\partial y}\frac{dy}{dt}.$$

Proof Because g and h are differentiable functions of t, you know that both Δx and Δy approach zero as Δt approaches zero. Moreover, because f is a differentiable function of x and y, you know that

$$\Delta w = \frac{\partial w}{\partial x}\Delta x + \frac{\partial w}{\partial y}\Delta y + \varepsilon_1\Delta x + \varepsilon_2\Delta y$$

where both ε_1 and $\varepsilon_2 \to 0$ as $(\Delta x, \Delta y) \to (0, 0)$. So, for $\Delta t \ne 0$

$$\frac{\Delta w}{\Delta t} = \frac{\partial w}{\partial x}\frac{\Delta x}{\Delta t} + \frac{\partial w}{\partial y}\frac{\Delta y}{\Delta t} + \varepsilon_1\frac{\Delta x}{\Delta t} + \varepsilon_2\frac{\Delta y}{\Delta t}$$

from which it follows that

$$\begin{aligned}\frac{dw}{dt} = \lim_{\Delta t \to 0}\frac{\Delta w}{\Delta t} &= \frac{\partial w}{\partial x}\frac{dx}{dt} + \frac{\partial w}{\partial y}\frac{dy}{dt} + 0\left(\frac{dx}{dt}\right) + 0\left(\frac{dy}{dt}\right)\\ &= \frac{\partial w}{\partial x}\frac{dx}{dt} + \frac{\partial w}{\partial y}\frac{dy}{dt}.\end{aligned}$$

B Integration Tables

Forms Involving u^n

1. $\displaystyle\int u^n\,du = \frac{u^{n+1}}{n+1} + C,\ n \neq -1$

2. $\displaystyle\int \frac{1}{u}\,du = \ln|u| + C$

Forms Involving $a + bu$

3. $\displaystyle\int \frac{u}{a+bu}\,du = \frac{1}{b^2}\big(bu - a\ln|a+bu|\big) + C$

4. $\displaystyle\int \frac{u}{(a+bu)^2}\,du = \frac{1}{b^2}\left(\frac{a}{a+bu} + \ln|a+bu|\right) + C$

5. $\displaystyle\int \frac{u}{(a+bu)^n}\,du = \frac{1}{b^2}\left[\frac{-1}{(n-2)(a+bu)^{n-2}} + \frac{a}{(n-1)(a+bu)^{n-1}}\right] + C,\quad n \neq 1, 2$

6. $\displaystyle\int \frac{u^2}{a+bu}\,du = \frac{1}{b^3}\left[-\frac{bu}{2}(2a - bu) + a^2\ln|a+bu|\right] + C$

7. $\displaystyle\int \frac{u^2}{(a+bu)^2}\,du = \frac{1}{b^3}\left(bu - \frac{a^2}{a+bu} - 2a\ln|a+bu|\right) + C$

8. $\displaystyle\int \frac{u^2}{(a+bu)^3}\,du = \frac{1}{b^3}\left[\frac{2a}{a+bu} - \frac{a^2}{2(a+bu)^2} + \ln|a+bu|\right] + C$

9. $\displaystyle\int \frac{u^2}{(a+bu)^n}\,du = \frac{1}{b^3}\left[\frac{-1}{(n-3)(a+bu)^{n-3}} + \frac{2a}{(n-2)(a+bu)^{n-2}} - \frac{a^2}{(n-1)(a+bu)^{n-1}}\right] + C,\quad n \neq 1, 2, 3$

10. $\displaystyle\int \frac{1}{u(a+bu)}\,du = \frac{1}{a}\ln\left|\frac{u}{a+bu}\right| + C$

11. $\displaystyle\int \frac{1}{u(a+bu)^2}\,du = \frac{1}{a}\left(\frac{1}{a+bu} + \frac{1}{a}\ln\left|\frac{u}{a+bu}\right|\right) + C$

12. $\displaystyle\int \frac{1}{u^2(a+bu)}\,du = -\frac{1}{a}\left(\frac{1}{u} + \frac{b}{a}\ln\left|\frac{u}{a+bu}\right|\right) + C$

13. $\displaystyle\int \frac{1}{u^2(a+bu)^2}\,du = -\frac{1}{a^2}\left[\frac{a+2bu}{u(a+bu)} + \frac{2b}{a}\ln\left|\frac{u}{a+bu}\right|\right] + C$

Forms Involving $a + bu + cu^2,\ b^2 \neq 4ac$

14. $$\int \frac{1}{a + bu + cu^2}\, du = \begin{cases} \dfrac{2}{\sqrt{4ac - b^2}} \arctan \dfrac{2cu + b}{\sqrt{4ac - b^2}} + C, & b^2 < 4ac \\ \dfrac{1}{\sqrt{b^2 - 4ac}} \ln\left|\dfrac{2cu + b - \sqrt{b^2 - 4ac}}{2cu + b + \sqrt{b^2 - 4ac}}\right| + C, & b^2 > 4ac \end{cases}$$

15. $$\int \frac{u}{a + bu + cu^2}\, du = \frac{1}{2c}\left(\ln|a + bu + cu^2| - b \int \frac{1}{a + bu + cu^2}\, du\right)$$

Forms Involving $\sqrt{a + bu}$

16. $$\int u^n \sqrt{a + bu}\, du = \frac{2}{b(2n + 3)}\left[u^n (a + bu)^{3/2} - na \int u^{n-1} \sqrt{a + bu}\, du\right]$$

17. $$\int \frac{1}{u\sqrt{a + bu}}\, du = \begin{cases} \dfrac{1}{\sqrt{a}} \ln\left|\dfrac{\sqrt{a + bu} - \sqrt{a}}{\sqrt{a + bu} + \sqrt{a}}\right| + C, & a > 0 \\ \dfrac{2}{\sqrt{-a}} \arctan \sqrt{\dfrac{a + bu}{-a}} + C, & a < 0 \end{cases}$$

18. $$\int \frac{1}{u^n \sqrt{a + bu}}\, du = \frac{-1}{a(n - 1)}\left[\frac{\sqrt{a + bu}}{u^{n-1}} + \frac{(2n - 3)b}{2} \int \frac{1}{u^{n-1}\sqrt{a + bu}}\, du\right],\quad n \neq 1$$

19. $$\int \frac{\sqrt{a + bu}}{u}\, du = 2\sqrt{a + bu} + a \int \frac{1}{u\sqrt{a + bu}}\, du$$

20. $$\int \frac{\sqrt{a + bu}}{u^n}\, du = \frac{-1}{a(n - 1)}\left[\frac{(a + bu)^{3/2}}{u^{n-1}} + \frac{(2n - 5)b}{2} \int \frac{\sqrt{a + bu}}{u^{n-1}}\, du\right],\quad n \neq 1$$

21. $$\int \frac{u}{\sqrt{a + bu}}\, du = \frac{-2(2a - bu)}{3b^2}\sqrt{a + bu} + C$$

22. $$\int \frac{u^n}{\sqrt{a + bu}}\, du = \frac{2}{(2n + 1)b}\left(u^n \sqrt{a + bu} - na \int \frac{u^{n-1}}{\sqrt{a + bu}}\, du\right)$$

Forms Involving $a^2 \pm u^2,\ a > 0$

23. $$\int \frac{1}{a^2 + u^2}\, du = \frac{1}{a} \arctan \frac{u}{a} + C$$

24. $$\int \frac{1}{u^2 - a^2}\, du = -\int \frac{1}{a^2 - u^2}\, du = \frac{1}{2a} \ln\left|\frac{u - a}{u + a}\right| + C$$

25. $$\int \frac{1}{(a^2 \pm u^2)^n}\, du = \frac{1}{2a^2(n - 1)}\left[\frac{u}{(a^2 \pm u^2)^{n-1}} + (2n - 3)\int \frac{1}{(a^2 \pm u^2)^{n-1}}\, du\right],\quad n \neq 1$$

Forms Involving $\sqrt{u^2 \pm a^2},\ a > 0$

26. $$\int \sqrt{u^2 \pm a^2}\, du = \frac{1}{2}\left(u\sqrt{u^2 \pm a^2} \pm a^2 \ln\left|u + \sqrt{u^2 \pm a^2}\right|\right) + C$$

27. $$\int u^2 \sqrt{u^2 \pm a^2}\, du = \frac{1}{8}\left[u(2u^2 \pm a^2)\sqrt{u^2 \pm a^2} - a^4 \ln\left|u + \sqrt{u^2 \pm a^2}\right|\right] + C$$

28. $$\int \frac{\sqrt{u^2 + a^2}}{u}\, du = \sqrt{u^2 + a^2} - a \ln\left|\frac{a + \sqrt{u^2 + a^2}}{u}\right| + C$$

29. $\displaystyle\int \frac{\sqrt{u^2 - a^2}}{u}\,du = \sqrt{u^2 - a^2} - a \operatorname{arcsec} \frac{|u|}{a} + C$

30. $\displaystyle\int \frac{\sqrt{u^2 \pm a^2}}{u^2}\,du = \frac{-\sqrt{u^2 \pm a^2}}{u} + \ln\left|u + \sqrt{u^2 \pm a^2}\right| + C$

31. $\displaystyle\int \frac{1}{\sqrt{u^2 \pm a^2}}\,du = \ln\left|u + \sqrt{u^2 \pm a^2}\right| + C$

32. $\displaystyle\int \frac{1}{u\sqrt{u^2 + a^2}}\,du = \frac{-1}{a} \ln\left|\frac{a + \sqrt{u^2 + a^2}}{u}\right| + C$

33. $\displaystyle\int \frac{1}{u\sqrt{u^2 - a^2}}\,du = \frac{1}{a} \operatorname{arcsec} \frac{|u|}{a} + C$

34. $\displaystyle\int \frac{u^2}{\sqrt{u^2 \pm a^2}}\,du = \frac{1}{2}\left(u\sqrt{u^2 \pm a^2} \mp a^2 \ln\left|u + \sqrt{u^2 \pm a^2}\right|\right) + C$

35. $\displaystyle\int \frac{1}{u^2\sqrt{u^2 \pm a^2}}\,du = \mp \frac{\sqrt{u^2 \pm a^2}}{a^2 u} + C$

36. $\displaystyle\int \frac{1}{(u^2 \pm a^2)^{3/2}}\,du = \frac{\pm u}{a^2\sqrt{u^2 \pm a^2}} + C$

Forms Involving $\sqrt{a^2 - u^2}$, $a > 0$

37. $\displaystyle\int \sqrt{a^2 - u^2}\,du = \frac{1}{2}\left(u\sqrt{a^2 - u^2} + a^2 \arcsin \frac{u}{a}\right) + C$

38. $\displaystyle\int u^2\sqrt{a^2 - u^2}\,du = \frac{1}{8}\left[u(2u^2 - a^2)\sqrt{a^2 - u^2} + a^4 \arcsin \frac{u}{a}\right] + C$

39. $\displaystyle\int \frac{\sqrt{a^2 - u^2}}{u}\,du = \sqrt{a^2 - u^2} - a \ln\left|\frac{a + \sqrt{a^2 - u^2}}{u}\right| + C$

40. $\displaystyle\int \frac{\sqrt{a^2 - u^2}}{u^2}\,du = \frac{-\sqrt{a^2 - u^2}}{u} - \arcsin \frac{u}{a} + C$

41. $\displaystyle\int \frac{1}{\sqrt{a^2 - u^2}}\,du = \arcsin \frac{u}{a} + C$

42. $\displaystyle\int \frac{1}{u\sqrt{a^2 - u^2}}\,du = \frac{-1}{a} \ln\left|\frac{a + \sqrt{a^2 - u^2}}{u}\right| + C$

43. $\displaystyle\int \frac{u^2}{\sqrt{a^2 - u^2}}\,du = \frac{1}{2}\left(-u\sqrt{a^2 - u^2} + a^2 \arcsin \frac{u}{a}\right) + C$

44. $\displaystyle\int \frac{1}{u^2\sqrt{a^2 - u^2}}\,du = \frac{-\sqrt{a^2 - u^2}}{a^2 u} + C$

45. $\displaystyle\int \frac{1}{(a^2 - u^2)^{3/2}}\,du = \frac{u}{a^2\sqrt{a^2 - u^2}} + C$

Forms Involving $\sin u$ or $\cos u$

46. $\displaystyle\int \sin u\, du = -\cos u + C$

47. $\displaystyle\int \cos u\, du = \sin u + C$

48. $\displaystyle\int \sin^2 u\, du = \frac{1}{2}(u - \sin u \cos u) + C$

49. $\displaystyle\int \cos^2 u\, du = \frac{1}{2}(u + \sin u \cos u) + C$

50. $\displaystyle\int \sin^n u\, du = -\frac{\sin^{n-1} u \cos u}{n} + \frac{n-1}{n}\int \sin^{n-2} u\, du$

51. $\displaystyle\int \cos^n u\, du = \frac{\cos^{n-1} u \sin u}{n} + \frac{n-1}{n}\int \cos^{n-2} u\, du$

52. $\displaystyle\int u \sin u\, du = \sin u - u \cos u + C$

53. $\displaystyle\int u \cos u\, du = \cos u + u \sin u + C$

54. $\displaystyle\int u^n \sin u\, du = -u^n \cos u + n\int u^{n-1} \cos u\, du$

55. $\displaystyle\int u^n \cos u\, du = u^n \sin u - n\int u^{n-1} \sin u\, du$

56. $\displaystyle\int \frac{1}{1 \pm \sin u}\, du = \tan u \mp \sec u + C$

57. $\displaystyle\int \frac{1}{1 \pm \cos u}\, du = -\cot u \pm \csc u + C$

58. $\displaystyle\int \frac{1}{\sin u \cos u}\, du = \ln|\tan u| + C$

Forms Involving $\tan u$, $\cot u$, $\sec u$, $\csc u$

59. $\displaystyle\int \tan u\, du = -\ln|\cos u| + C$

60. $\displaystyle\int \cot u\, du = \ln|\sin u| + C$

61. $\displaystyle\int \sec u\, du = \ln|\sec u + \tan u| + C$

62. $\displaystyle\int \csc u\, du = \ln|\csc u - \cot u| + C$ or $\displaystyle\int \csc u\, du = -\ln|\csc u + \cot u| + C$

63. $\displaystyle\int \tan^2 u\, du = -u + \tan u + C$

64. $\displaystyle\int \cot^2 u\, du = -u - \cot u + C$

65. $\displaystyle\int \sec^2 u\, du = \tan u + C$

66. $\displaystyle\int \csc^2 u\, du = -\cot u + C$

67. $\displaystyle\int \tan^n u\, du = \frac{\tan^{n-1} u}{n-1} - \int \tan^{n-2} u\, du,\ n \neq 1$

68. $\displaystyle\int \cot^n u\, du = -\frac{\cot^{n-1} u}{n-1} - \int (\cot^{n-2} u)\, du,\ n \neq 1$

69. $\displaystyle\int \sec^n u\, du = \frac{\sec^{n-2} u \tan u}{n-1} + \frac{n-2}{n-1}\int \sec^{n-2} u\, du,\ n \neq 1$

70. $\displaystyle\int \csc^n u\, du = -\frac{\csc^{n-2} u \cot u}{n-1} + \frac{n-2}{n-1}\int \csc^{n-2} u\, du,\ n \neq 1$

71. $\displaystyle\int \frac{1}{1 \pm \tan u}\, du = \frac{1}{2}(u \pm \ln|\cos u \pm \sin u|) + C$

72. $\displaystyle\int \frac{1}{1 \pm \cot u}\, du = \frac{1}{2}(u \mp \ln|\sin u \pm \cos u|) + C$

73. $\displaystyle\int \frac{1}{1 \pm \sec u}\, du = u + \cot u \mp \csc u + C$

74. $\displaystyle\int \frac{1}{1 \pm \csc u}\, du = u - \tan u \pm \sec u + C$

Forms Involving Inverse Trigonometric Functions

75. $\int \arcsin u\, du = u \arcsin u + \sqrt{1 - u^2} + C$

76. $\int \arccos u\, du = u \arccos u - \sqrt{1 - u^2} + C$

77. $\int \arctan u\, du = u \arctan u - \ln\sqrt{1 + u^2} + C$

78. $\int \operatorname{arccot} u\, du = u \operatorname{arccot} u + \ln\sqrt{1 + u^2} + C$

79. $\int \operatorname{arcsec} u\, du = u \operatorname{arcsec} u - \ln\left|u + \sqrt{u^2 - 1}\right| + C$

80. $\int \operatorname{arccsc} u\, du = u \operatorname{arccsc} u + \ln\left|u + \sqrt{u^2 - 1}\right| + C$

Forms Involving e^u

81. $\int e^u\, du = e^u + C$

82. $\int ue^u\, du = (u - 1)e^u + C$

83. $\int u^n e^u\, du = u^n e^u - n\int u^{n-1}e^u\, du$

84. $\int \frac{1}{1 + e^u}\, du = u - \ln(1 + e^u) + C$

85. $\int e^{au} \sin bu\, du = \frac{e^{au}}{a^2 + b^2}(a \sin bu - b \cos bu) + C$

86. $\int e^{au} \cos bu\, du = \frac{e^{au}}{a^2 + b^2}(a \cos bu + b \sin bu) + C$

Forms Involving $\ln u$

87. $\int \ln u\, du = u(-1 + \ln u) + C$

88. $\int u \ln u\, du = \frac{u^2}{4}(-1 + 2 \ln u) + C$

89. $\int u^n \ln u\, du = \frac{u^{n+1}}{(n + 1)^2}[-1 + (n + 1) \ln u] + C,\ n \neq -1$

90. $\int (\ln u)^2\, du = u\,[2 - 2 \ln u + (\ln u)^2] + C$

91. $\int (\ln u)^n\, du = u(\ln u)^n - n\int (\ln u)^{n-1}\, du$

Forms Involving Hyperbolic Functions

92. $\int \cosh u\, du = \sinh u + C$

93. $\int \sinh u\, du = \cosh u + C$

94. $\int \operatorname{sech}^2 u\, du = \tanh u + C$

95. $\int \operatorname{csch}^2 u\, du = -\coth u + C$

96. $\int \operatorname{sech} u \tanh u\, du = -\operatorname{sech} u + C$

97. $\int \operatorname{csch} u \coth u\, du = -\operatorname{csch} u + C$

Forms Involving Inverse Hyperbolic Functions (in logarithmic form)

98. $\int \frac{du}{\sqrt{u^2 \pm a^2}} = \ln\left(u + \sqrt{u^2 \pm a^2}\right) + C$

99. $\int \frac{du}{a^2 - u^2} = \frac{1}{2a} \ln\left|\frac{a + u}{a - u}\right| + C$

100. $\int \frac{du}{u\sqrt{a^2 \pm u^2}} = -\frac{1}{a} \ln \frac{a + \sqrt{a^2 \pm u^2}}{|u|} + C$

Answers to Odd-Numbered Exercises

Chapter P

Section P.1 (page 8)

1. b **2.** d **3.** a **4.** c

5. Answers will vary.

x	-4	-2	0	2	4
y	-5	-2	1	4	7

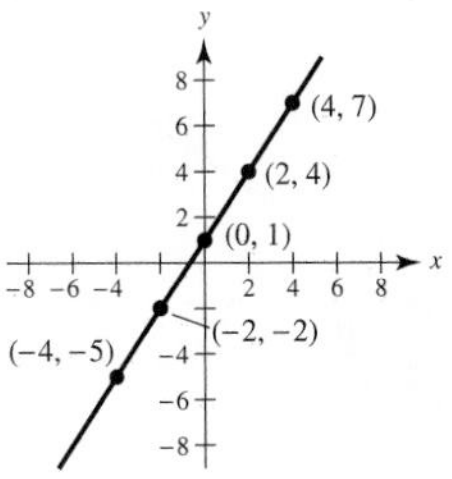

7. Answers will vary.

x	-3	-2	0	2	3
y	-5	0	4	0	-5

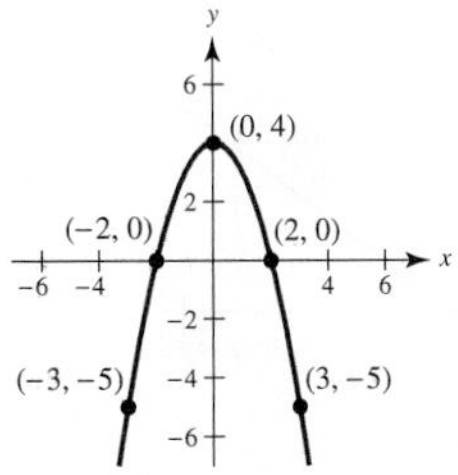

9. Answers will vary.

x	-5	-4	-3	-2	-1	0	1
y	3	2	1	0	1	2	3

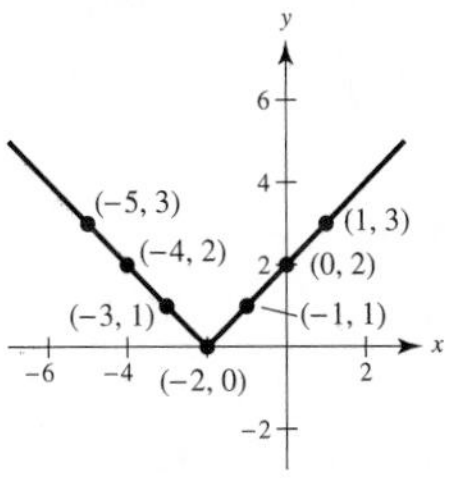

11. Answers will vary.

x	0	1	4	9	16
y	-4	-3	-2	-1	0

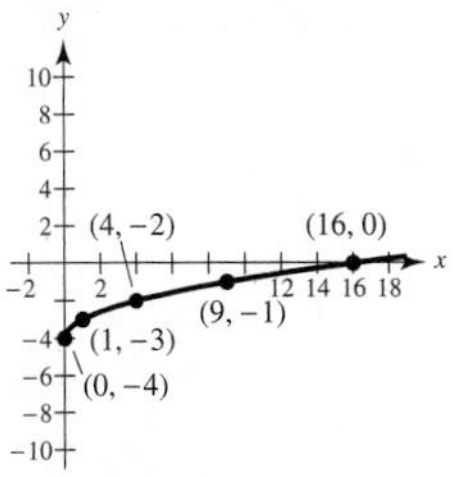

13. Answers will vary.

x	-3	-2	-1	0	1	2	3
y	$-\frac{2}{3}$	-1	-2	Undef.	2	1	$\frac{2}{3}$

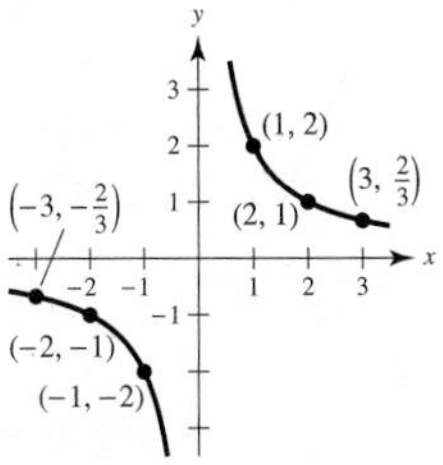

15.

Xmin = -3
Xmax = 5
Xscl = 1
Ymin = -3
Ymax = 5
Yscl = 1

17. $y = \sqrt{5 - x}$

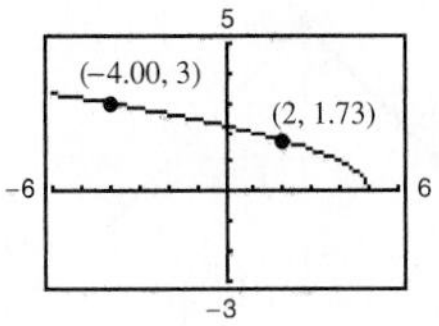

(a) $y \approx 1.73$ (b) $x = -4.00$

19. $(0, -2), (-2, 0), (1, 0)$ **21.** $(0, 0), (5, 0), (-5, 0)$

23. $(4, 0)$ **25.** $(0, 0)$ **27.** Symmetric with respect to the y-axis

29. Symmetric with respect to the x-axis

31. Symmetric with respect to the origin **33.** No symmetry

35. Symmetric with respect to the origin

37. Symmetric with respect to the y-axis

39. $y = -3x + 2$
Symmetry: none

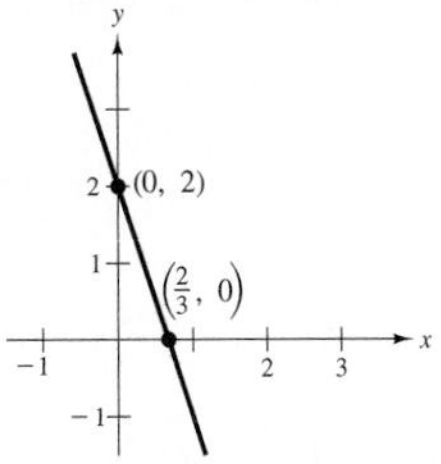

41. $y = \frac{1}{2}x - 4$
Symmetry: none

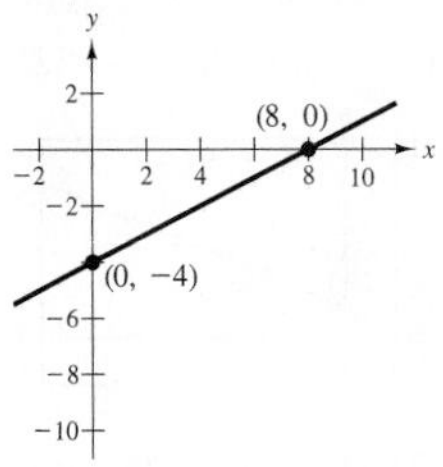

43. $y = 1 - x^2$
Symmetry: y-axis

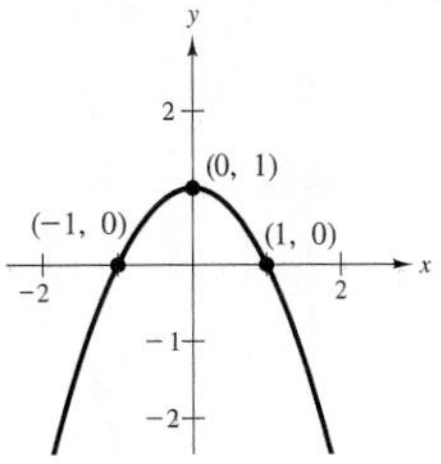

45. $y = (x + 3)^2$
Symmetry: none

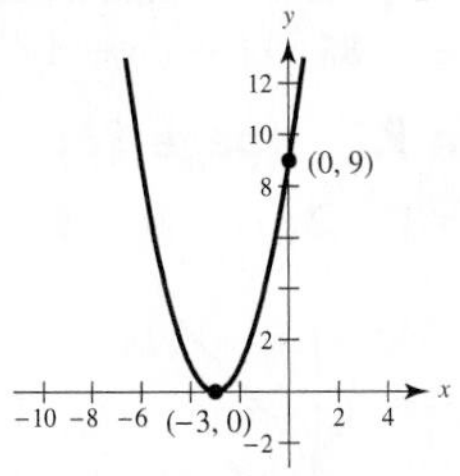

47. $y = x^3 + 2$
Symmetry: none

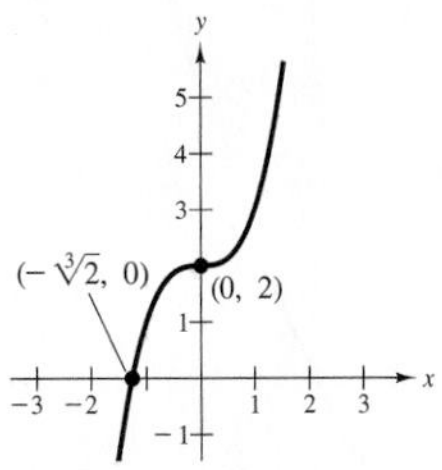

49. $y = x\sqrt{x + 2}$
Symmetry: none

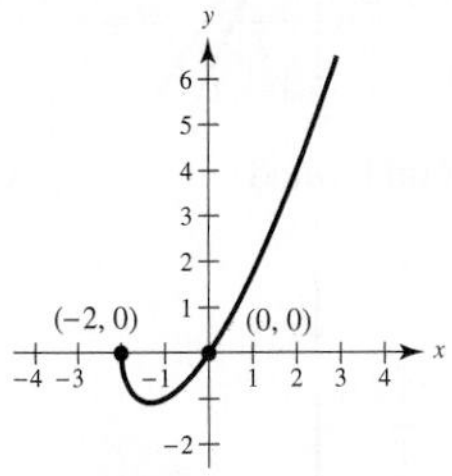

51. $x = y^3$
Symmetry: origin

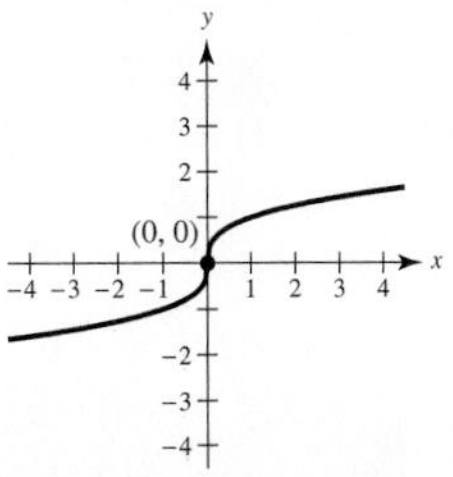

53. $y = 1/x$
Symmetry: origin

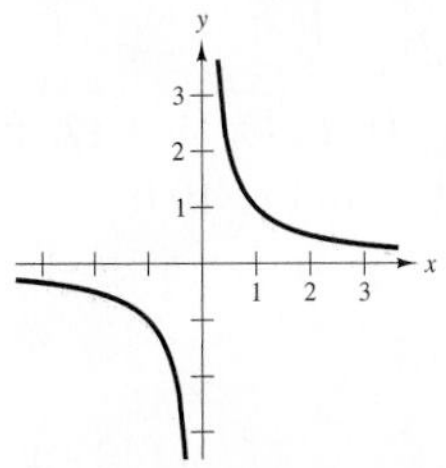

55. $y = 6 - |x|$
Symmetry: y-axis

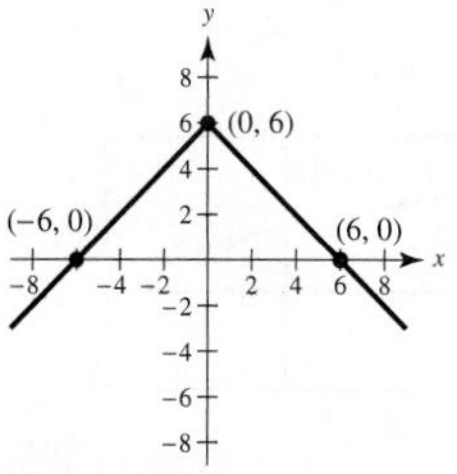

57. $y_1 = \sqrt{x + 9}$
$y_2 = -\sqrt{x + 9}$
Symmetry: x-axis

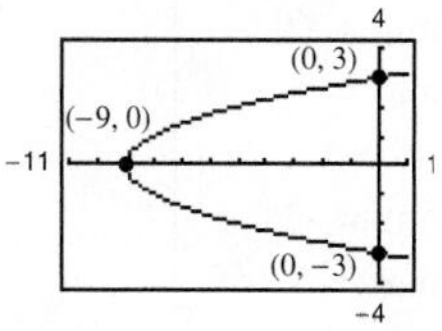

59. $y_1 = \sqrt{\dfrac{6 - x}{3}}$
$y_2 = -\sqrt{\dfrac{6 - x}{3}}$
Symmetry: x-axis

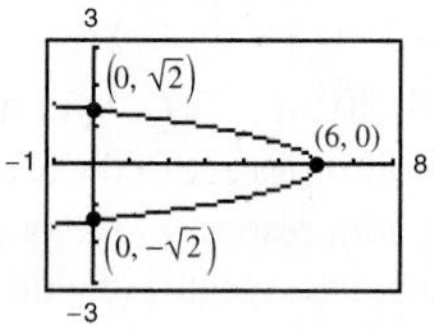

61. $(1, 1)$ **63.** $(-1, 5), (2, 2)$ **65.** $(-1, -2), (2, 1)$
67. $(-1, -1), (0, 0), (1, 1)$
69. $(-1, -5), (0, -1), (2, 1)$ **71.** $(-2, 2), \left(-3, \sqrt{3}\right)$
73. (a) $y = -0.007t^2 + 4.82t + 35.4$
(b)

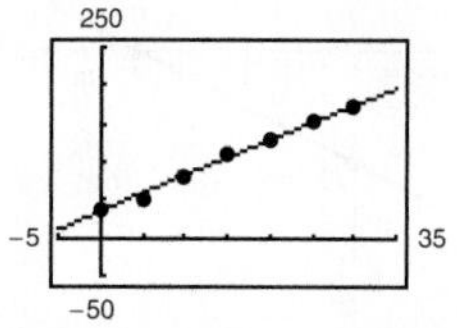

(c) 217

75. $x \approx 3133$ units **77.** $y = (x + 2)(x - 4)(x - 6)$
79. (i) b; $k = 2$ (ii) d; $k = -10$ (iii) a; $k = 3$ (iv) c; $k = 36$
81. False. $(-1, -2)$ is not a point on the graph of $x = \frac{1}{4}y^2$.
83. True **85.** $x^2 + (y - 4)^2 = 4$

Section P.2 (page 16)

1. $m = 1$ **3.** $m = 0$ **5.** $m = -12$
7.

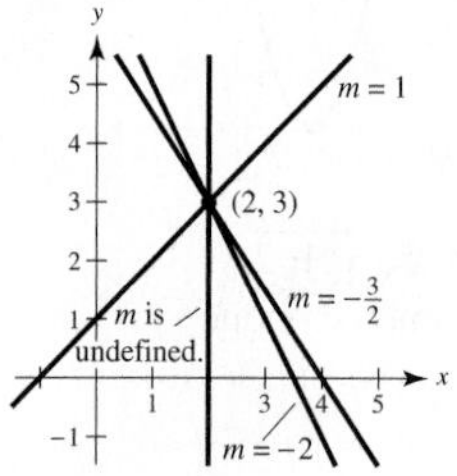

9. $m = 3$

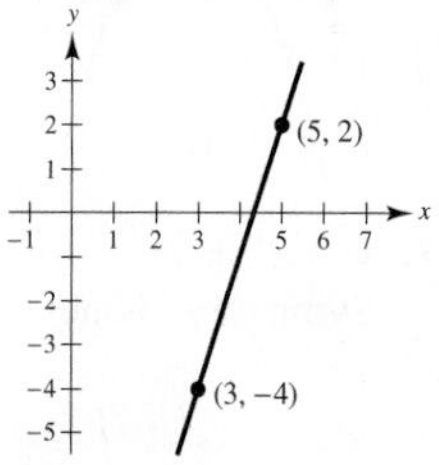

11. m is undefined.

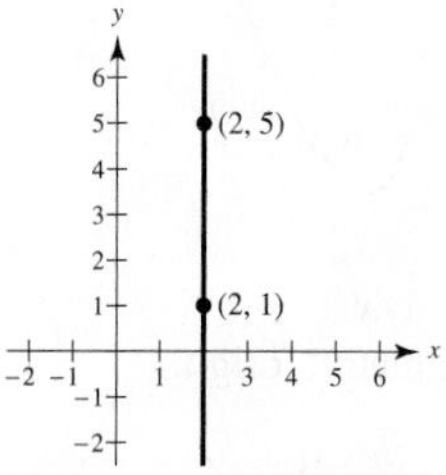

13. $m = 2$

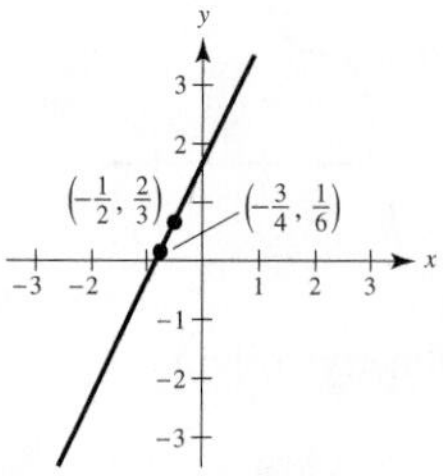

15. $(0, 1), (1, 1), (3, 1)$ **17.** $(0, 10), (2, 4), (3, 1)$
19. (a) $\frac{1}{3}$ (b) $10\sqrt{10}$ ft

21. (a)

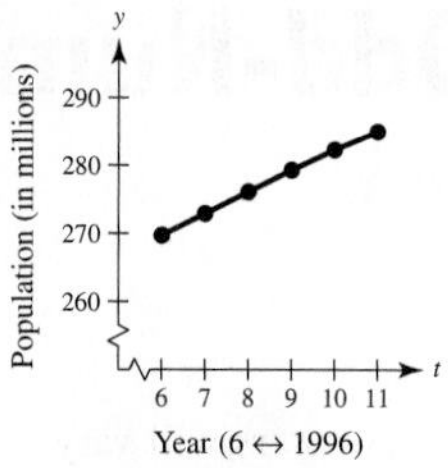

(b) Population increased least rapidly: 2000–2001

23. $m = -\frac{1}{5}, (0, 4)$ **25.** m is undefined, no y-intercept
27. $3x - 4y + 12 = 0$

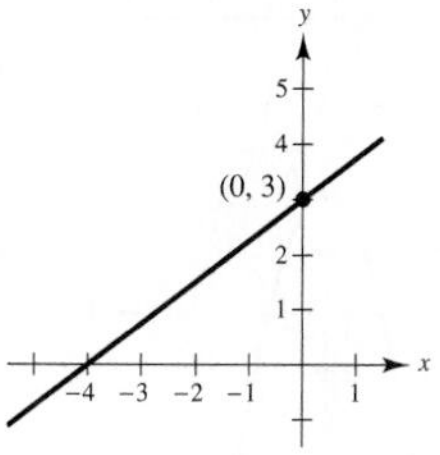

29. $2x - 3y = 0$

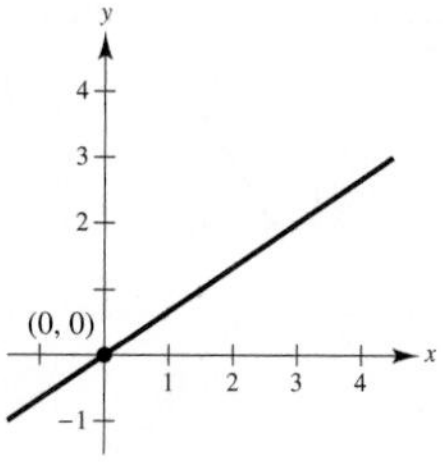

31. $3x - y - 11 = 0$

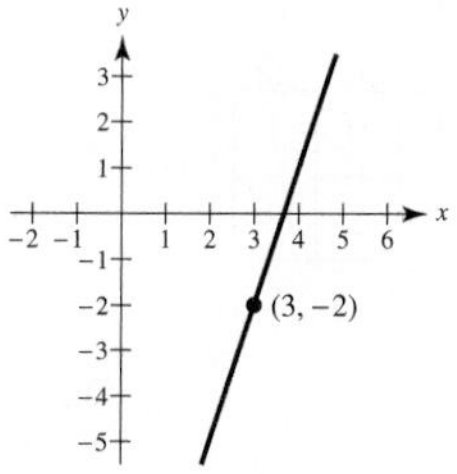

33. $3x - y = 0$

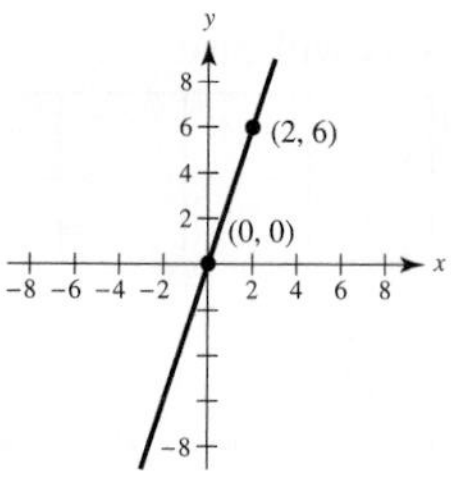

35. $2x - y - 3 = 0$

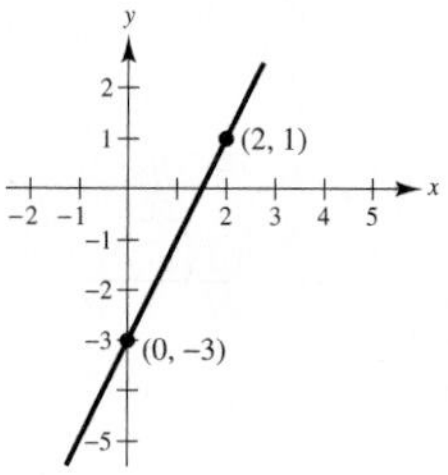

37. $8x + 3y - 40 = 0$

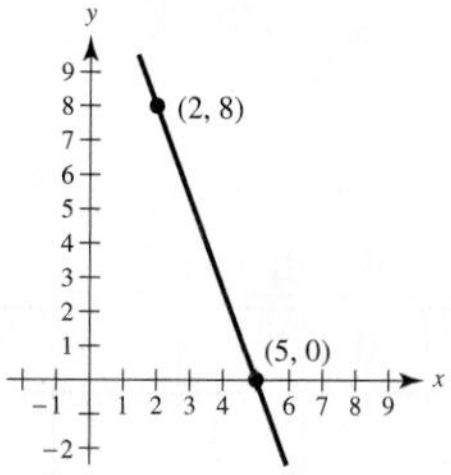

39. $x - 5 = 0$

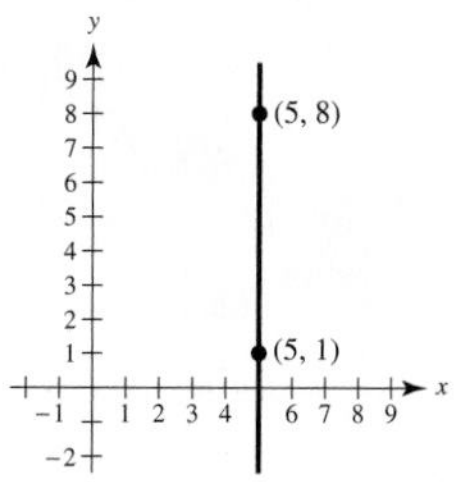

41. $22x - 4y + 3 = 0$

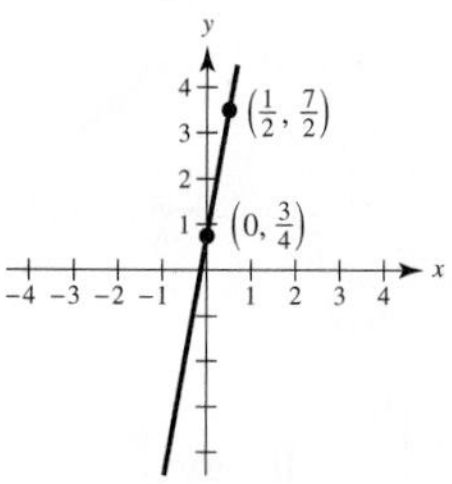

43. $x - 3 = 0$ **45.** $3x + 2y - 6 = 0$ **47.** $x + y - 3 = 0$
49.

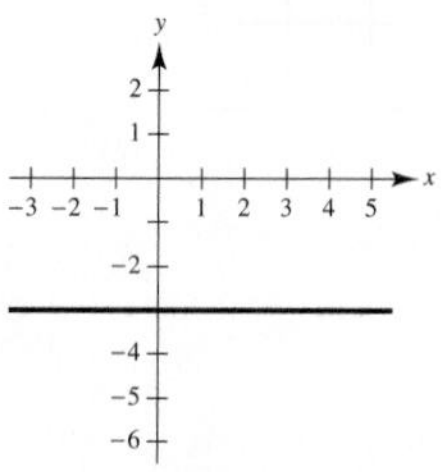

51.

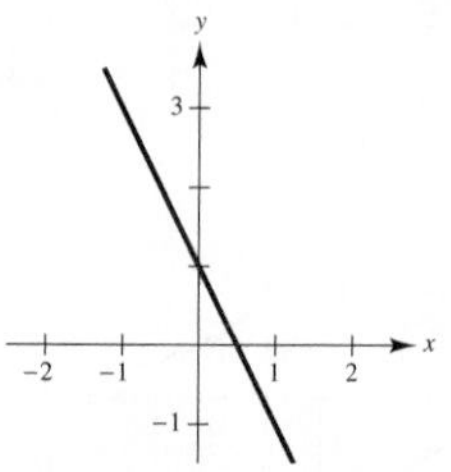

53.

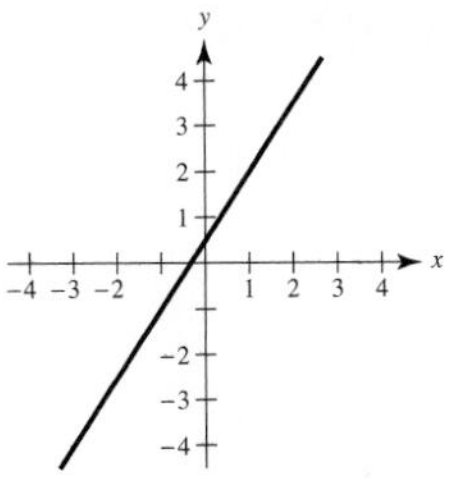

55.

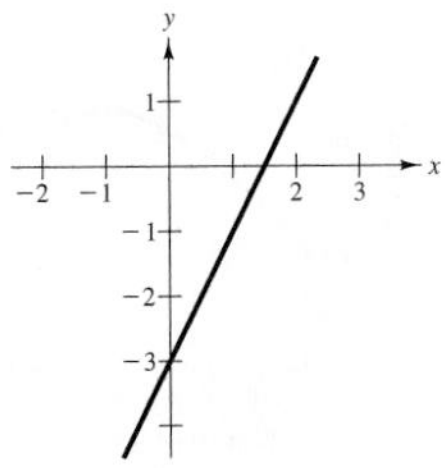

57. (a)

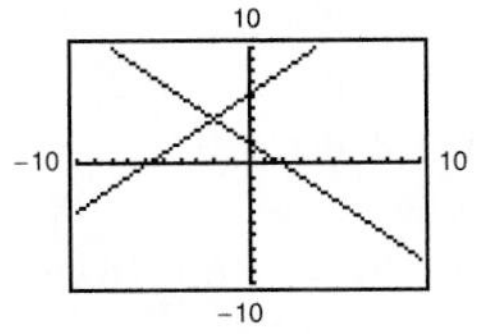

(b)

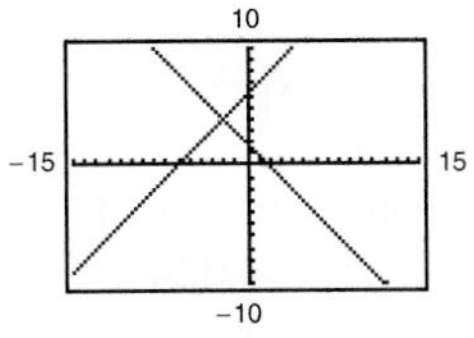

The lines in (a) do not appear perpendicular, but they do in (b) because a square setting is used. The lines are perpendicular.

59. (a) $2x - y - 3 = 0$ (b) $x + 2y - 4 = 0$

61. (a) $40x - 24y - 9 = 0$ (b) $24x + 40y - 53 = 0$

63. (a) $x - 2 = 0$ (b) $y - 5 = 0$

65. $V = 125t + 2040$ **67.** $V = -2000t + 28{,}400$

69. $y = 2x$ **71.** Not collinear, because $m_1 \neq m_2$

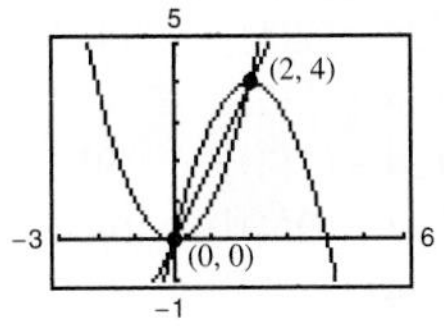

73. $\left(0, \dfrac{-a^2 + b^2 + c^2}{2c}\right)$ **75.** $\left(b, \dfrac{a^2 - b^2}{c}\right)$

77. $5F - 9C - 160 = 0$; $72°\text{F} \approx 22.2°\text{C}$

79. (a) $W_1 = 12.50 + 0.75x$; $W_2 = 9.20 + 1.30x$

(b)

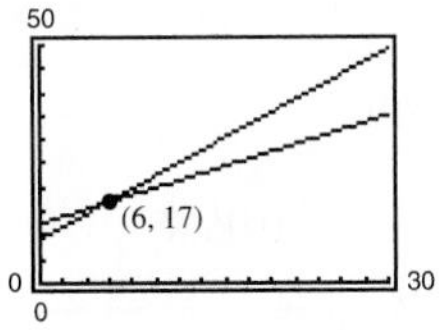

(c) When six units are produced, wages are \$17.00 per hour with either option. Choose position 1 when less than six units are produced and position 2 otherwise.

81. (a) $x = (1330 - p)/15$

(b)

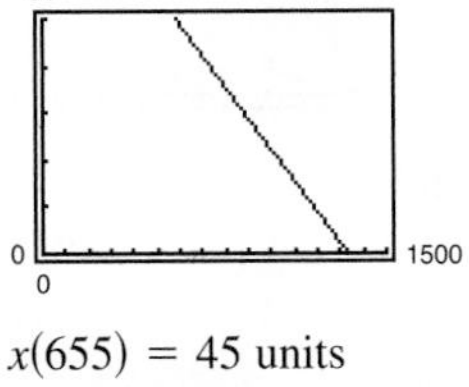

(c) 49 units

$x(655) = 45$ units

83. $12y + 5x - 169 = 0$ **85.** 2 **87.** $(5\sqrt{2})/2$ **89.** $2\sqrt{2}$

91. Proof **93.** Proof **95.** Proof **97.** True

Section P.3 (page 27)

1. (a) Domain of f: $[-4, 4]$; Range of f: $[-3, 5]$
Domain of g: $[-3, 3]$; Range of g: $[-4, 4]$
(b) $f(-2) = -1$; $g(3) = -4$
(c) $x = -1$ (d) $x \approx 1$ (e) $x \approx -1, x \approx 1$, and $x \approx 2$

3. (a) -3 (b) -9 (c) $2b - 3$ (d) $2x - 5$

5. (a) 3 (b) 0 (c) -1 (d) $2 + 2t - t^2$

7. (a) 1 (b) 0 (c) $-\frac{1}{2}$ **9.** $3x^2 + 3x\,\Delta x + (\Delta x)^2, \Delta x \neq 0$

11. $\left(\sqrt{x-1} - x + 1\right)/[(x - 2)(x - 1)]$
$= -1/\left[\sqrt{x-1}\left(1 + \sqrt{x-1}\right)\right], x \neq 2$

13. Domain: $[-3, \infty)$; Range: $(-\infty, 0]$

15. Domain: All real numbers t such that $t \neq 4n + 2$, where n is an integer; Range: $(-\infty, -1] \cup [1, \infty)$

17. Domain: $(-\infty, 0) \cup (0, \infty)$; Range: $(-\infty, 0) \cup (0, \infty)$

19. Domain: $[0, 1]$

21. Domain: All real numbers x such that $x \neq 2n\pi$, where n is an integer.

23. Domain: $(-\infty, -3) \cup (-3, \infty)$

25. (a) -1 (b) 2 (c) 6 (d) $2t^2 + 4$
Domain: $(-\infty, \infty)$; Range: $(-\infty, 1) \cup [2, \infty)$

27. (a) 4 (b) 0 (c) -2 (d) $-b^2$
Domain: $(-\infty, \infty)$; Range: $(-\infty, 0] \cup [1, \infty)$

29. $f(x) = 4 - x$
Domain: $(-\infty, \infty)$
Range: $(-\infty, \infty)$

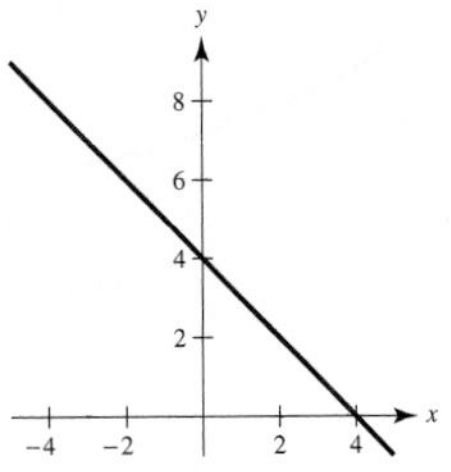

31. $h(x) = \sqrt{x - 1}$
Domain: $[1, \infty)$
Range: $[0, \infty)$

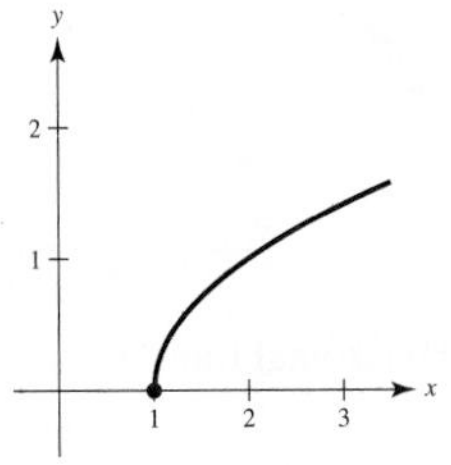

33. $f(x) = \sqrt{9 - x^2}$
Domain: $[-3, 3]$
Range: $[0, 3]$

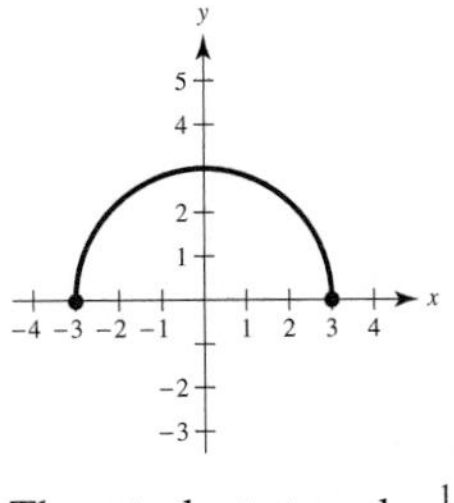

35. $g(t) = 2 \sin \pi t$
Domain: $(-\infty, \infty)$
Range: $[-2, 2]$

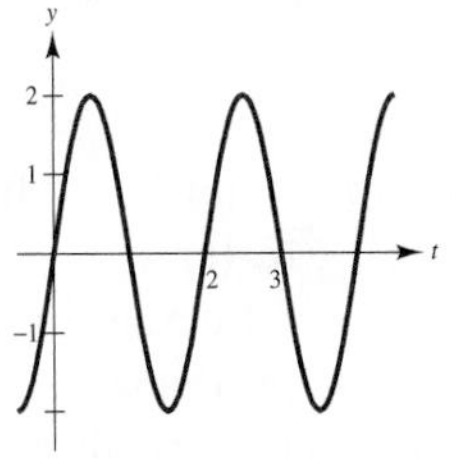

37. The student travels $\frac{1}{2}$ mi/min during the first 4 min, is stationary for the next 2 min, and travels 1 mi/min during the final 4 min.

39. y is not a function of x. **41.** y is a function of x.

43. y is not a function of x. **45.** y is not a function of x.

47. d **48.** b **49.** c **50.** a **51.** e **52.** g

53. (a)

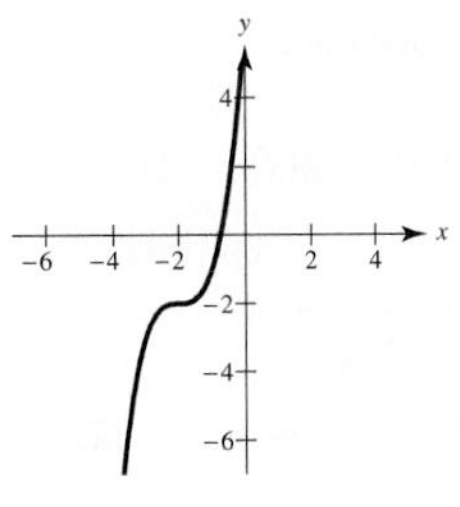

(b)

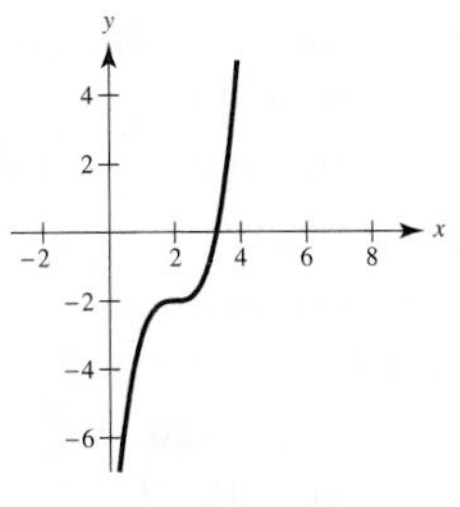

(c)

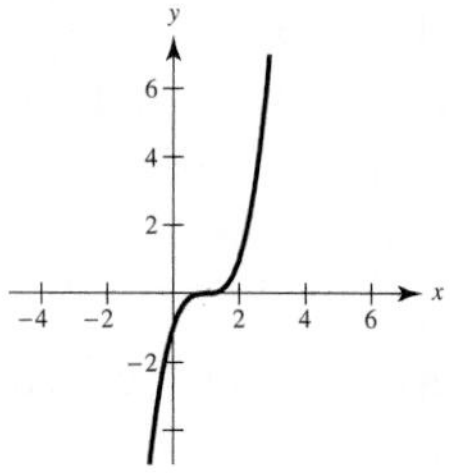

(d)

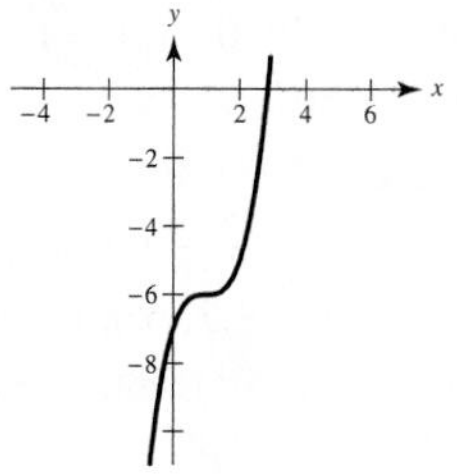

(e)

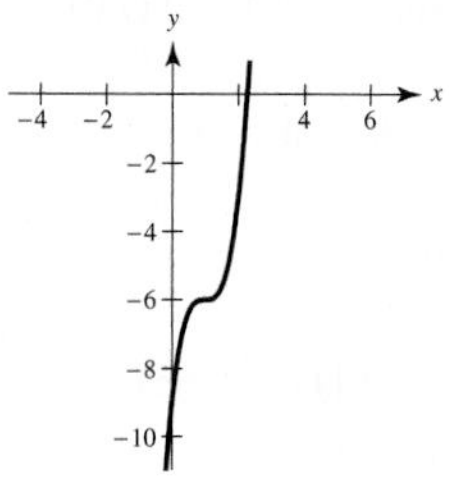

(f)

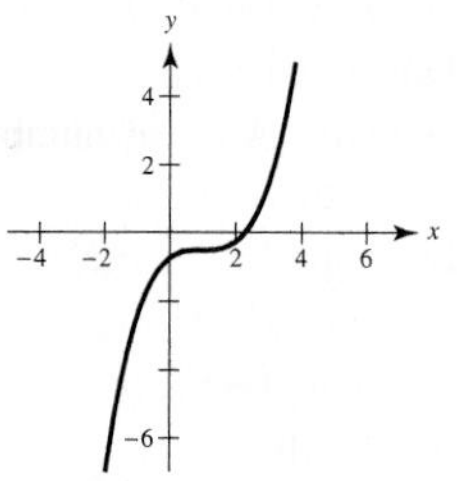

55. (a) Vertical translation

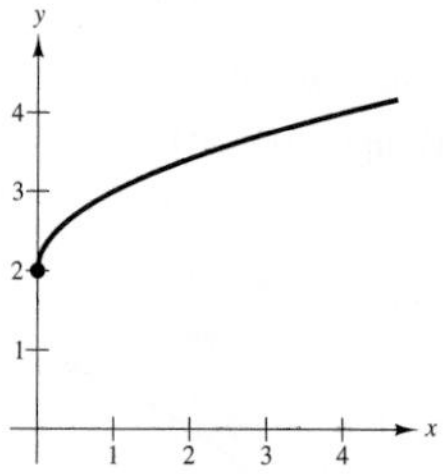

(b) Reflection about the x-axis

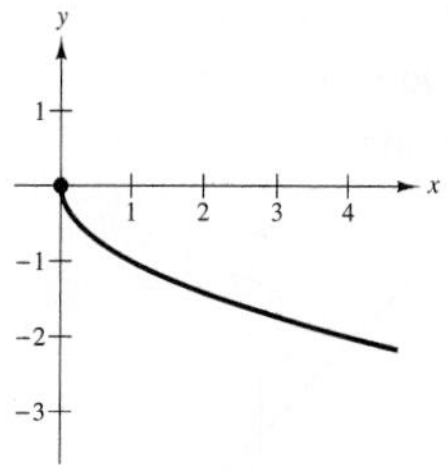

(c) Horizontal translation

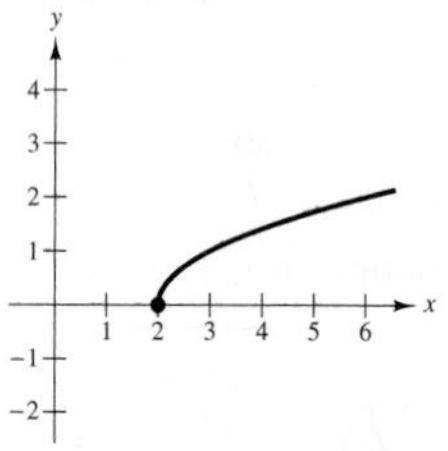

57. (a) 0 (b) 0 (c) -1 (d) $\sqrt{15}$
(e) $\sqrt{x^2 - 1}$ (f) $x - 1 \ (x \geq 0)$

59. $(f \circ g)(x) = x$; Domain: $[0, \infty)$
$(g \circ f)(x) = |x|$; Domain: $(-\infty, \infty)$
No, their domains are different.

61. $(f \circ g)(x) = 3/(x^2 - 1)$; Domain: $(-\infty, -1) \cup (-1, 1) \cup (1, \infty)$
$(g \circ f)(x) = 9/x^2 - 1$; Domain: $(-\infty, 0) \cup (0, \infty)$
No

63. (a) 4 (b) -2
(c) Undefined. The graph of g does not exist at $x = -5$.
(d) 3 (e) 2
(f) Underfined. The graph of f does not exist at $x = -4$.

65. Answers will vary.
Example: $f(x) = \sqrt{x}$; $g(x) = x - 2$; $h(x) = 2x$

67. Even **69.** Odd **71.** (a) $\left(\frac{3}{2}, 4\right)$ (b) $\left(\frac{3}{2}, -4\right)$

73. f is even. g is neither even nor odd. h is odd.

75. $f(x) = -2x - 5$ **77.** $y = -\sqrt{-x}$

79. $ii, c = -2$ **80.** $i, c = \frac{1}{4}$ **81.** $iv, c = 32$ **82.** $iii, c = 3$

83. (a) $T(4) = 16°$, $T(15) = 24°$
(b) The changes in temperature will occur 1 hr later.
(c) The temperatures are 1° lower.

85. (a)

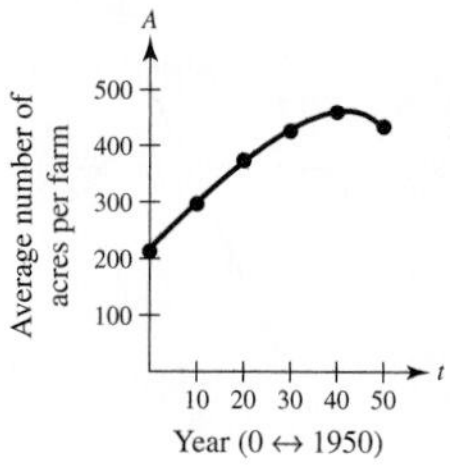

(b) $A(15) \approx 345$ acres/farm

87. $f(x) = |x| + |x - 2| = \begin{cases} 2x - 2, & \text{if } x \geq 2 \\ 2, & \text{if } 0 < x < 2 \\ -2x + 2, & \text{if } x \leq 0 \end{cases}$

89. Proof **91.** Proof

93. (a) $V(x) = x(24 - 2x)^2, x > 0$ (b) $4 \times 16 \times 16$ cm

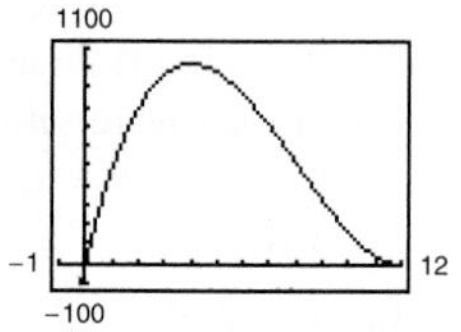

(c)

Height, x	Length and Width	Volume, V
1	$24 - 2(1)$	$1[24 - 2(1)]^2 = 484$
2	$24 - 2(2)$	$2[24 - 2(2)]^2 = 800$
3	$24 - 2(3)$	$3[24 - 2(3)]^2 = 972$
4	$24 - 2(4)$	$4[24 - 2(4)]^2 = 1024$
5	$24 - 2(5)$	$5[24 - 2(5)]^2 = 980$
6	$24 - 2(6)$	$6[24 - 2(6)]^2 = 864$

The dimensions of the box that yield a maximum volume are $4 \times 16 \times 16$ cm.

95. False. For example, if $f(x) = x^2$, then $f(-1) = f(1)$. **97.** True

99. Putnam Problem A1, 1988

Section P.4 (page 34)

1. Quadratic **3.** Linear

5. (a) and (b)

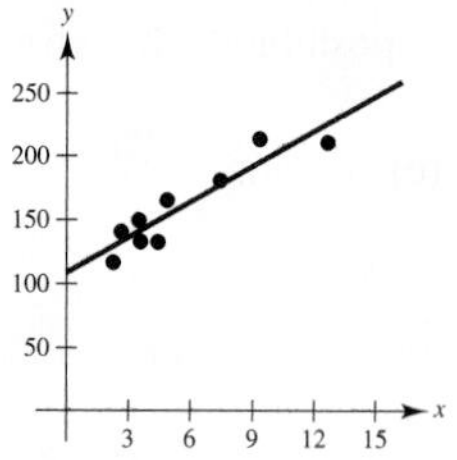

Approximately linear
(c) 136

7. (a) $d = 0.066F$
(b)

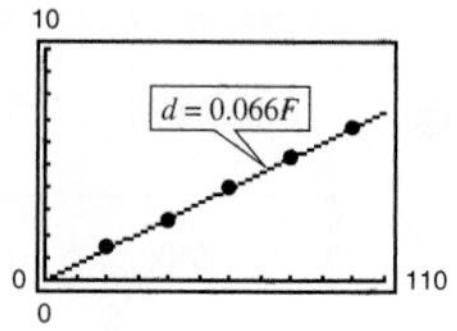

(c) 3.63 cm

9. (a) $y = 0.124x + 0.82$
$r \approx 0.838$
(b)

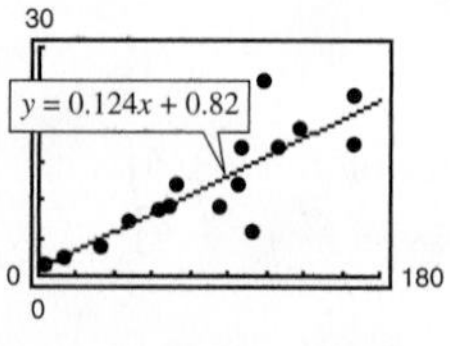

(c) Greater per capita electricity consumption by a country tends to relate to greater per capita gross national product of the country. Hong Kong, Venezuela, South Korea

(d) $y = 0.134x + 0.28$
$r \approx 0.968$

11. (a) $y_1 = 0.03434t^3 - 0.3451t^2 + 0.884t + 5.61$
$y_2 = 0.110t + 2.07$
$y_3 = 0.092t + 0.79$
(b) $y_1 + y_2 + y_3 = 0.03434t^3 - 0.3451t^2 + 1.086t + 8.47$

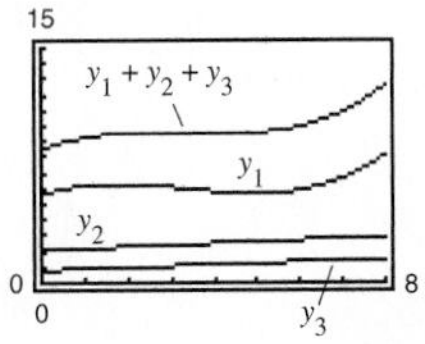

31.1 cents/mi

13. (a) Linear: $y_1 = 4.83t + 28.6$
Cubic: $y_2 = -0.1289t^3 + 2.235t^2 - 4.86t + 35.2$
(b)

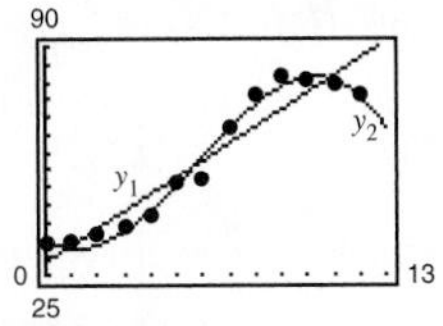

(c) Cubic

(d) $y = -0.084t^2 + 5.84t + 26.7$

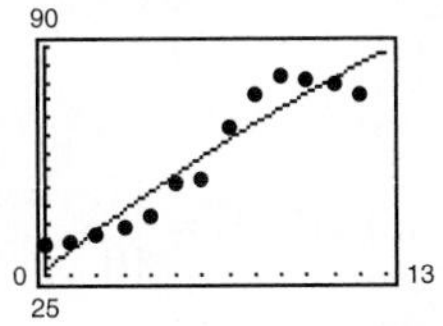

(e) Linear:
$N(14) \approx 96.2$ million people
Cubic:
$N(14) \approx 51.5$ million people
(f) Answers will vary.

15. (a) $y = -1.806x^3 + 14.58x^2 + 16.4x + 10$
(b)

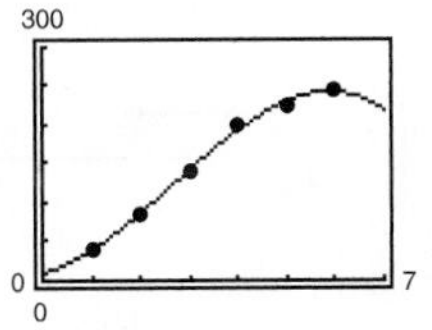

(c) 214

17. (a) Yes. At time t there is one and only one displacement y.
(b) Amplitude: 0.35; Period: 0.5 (c) $y = 0.35\sin(4\pi t) + 2$
(d)

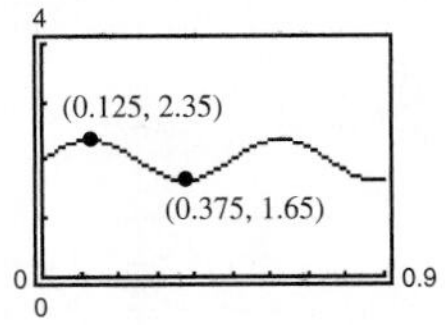

The model appears to fit the data.

19. Answers will vary.

Review Exercises for Chapter P (page 37)

1. $\left(\frac{3}{2}, 0\right), (0, -3)$ **3.** $(1, 0), \left(0, \frac{1}{2}\right)$ **5.** y-axis symmetry

7.

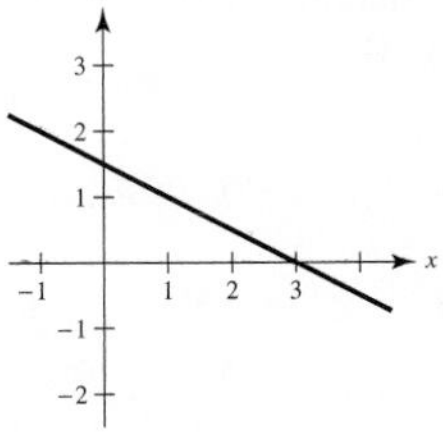

9.

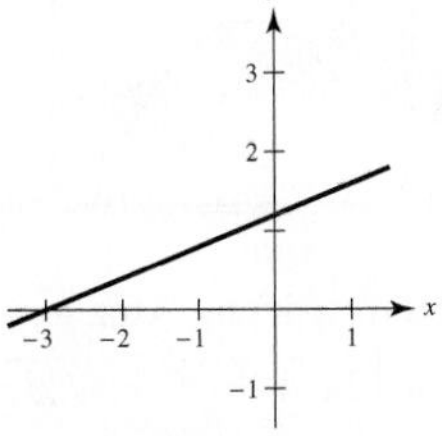

11.

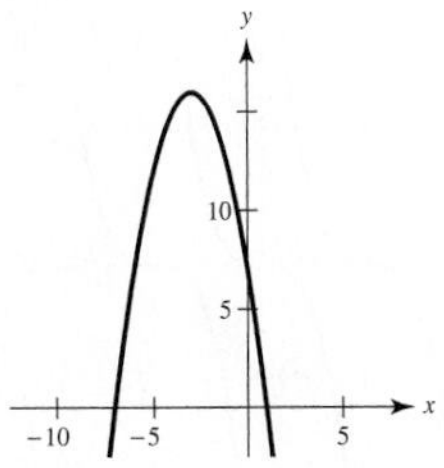

13.

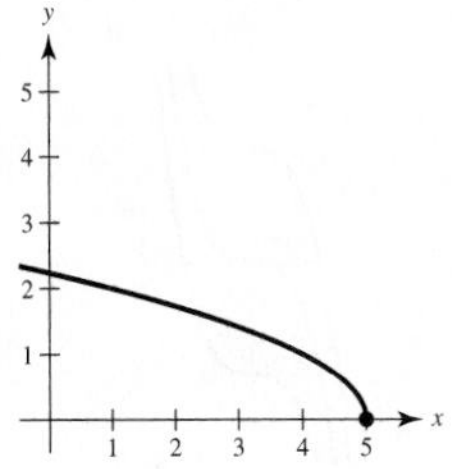

15.

Xmin = -5
Xmax = 5
Xscl = 1
Ymin = -30
Ymax = 10
Yscl = 5

17. $(4, 1)$ **19.** $y = x^3 - 4x$

21. $m = \frac{3}{7}$

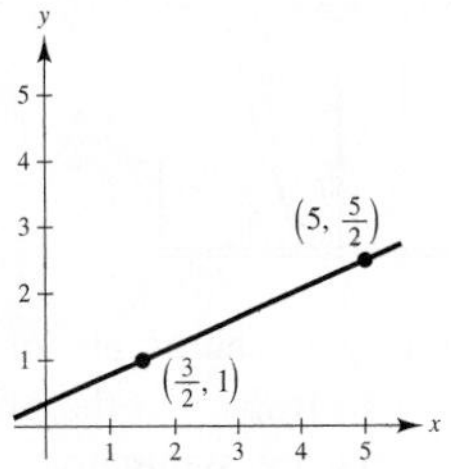

23. $t = \frac{7}{3}$

25. $y = \frac{3}{2}x - 5$ or $3x - 2y - 10 = 0$

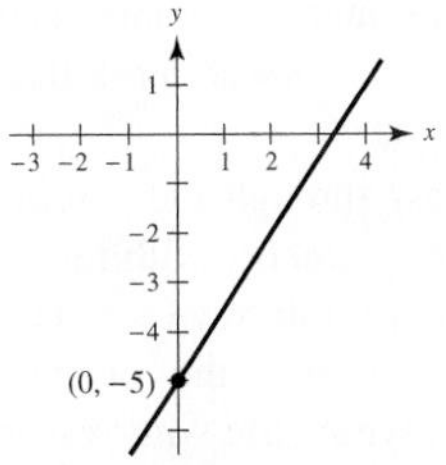

27. $y = -\frac{2}{3}x - 2$ or $2x + 3y + 6 = 0$

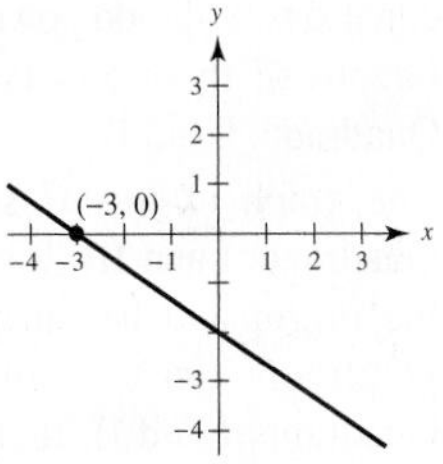

29. (a) $7x - 16y + 78 = 0$ (b) $5x - 3y + 22 = 0$
(c) $2x + y = 0$ (d) $x + 2 = 0$

31. $V = 12{,}500 - 850t$; \$9950

33. Not a function

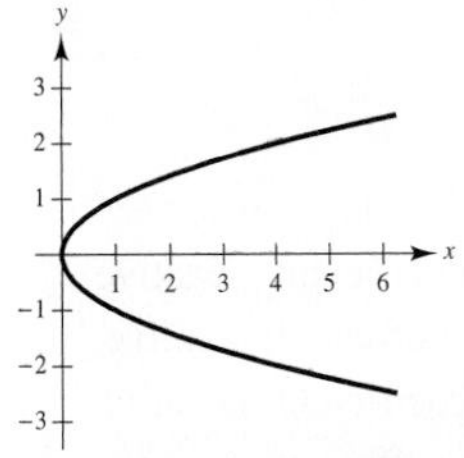

35. Function

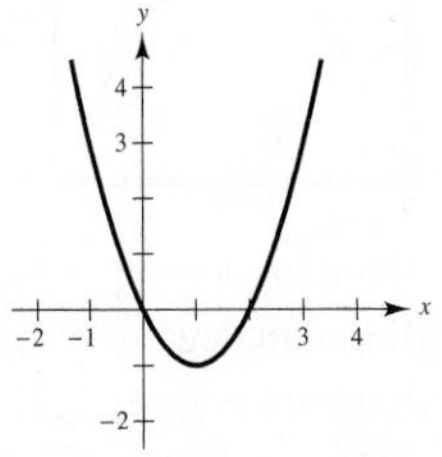

37. (a) Undefined (b) $-1/(1 + \Delta x), \Delta x \neq 0, -1$

39. (a) Domain: $[-6, 6]$; Range: $[0, 6]$
(b) Domain: $(-\infty, 5) \cup (5, \infty)$; Range: $(-\infty, 0) \cup (0, \infty)$
(c) Domain: $(-\infty, \infty)$; Range: $(-\infty, \infty)$

41. (a)

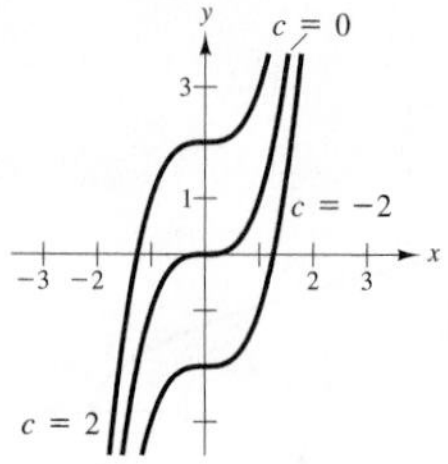

(b)

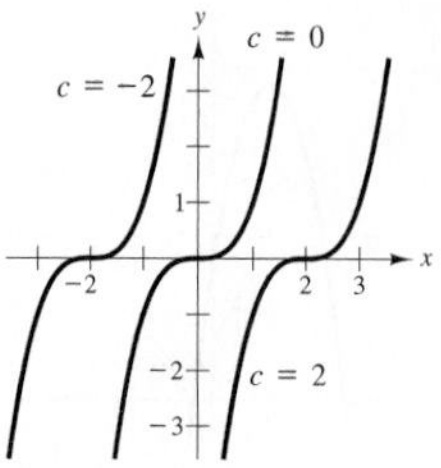

(c)

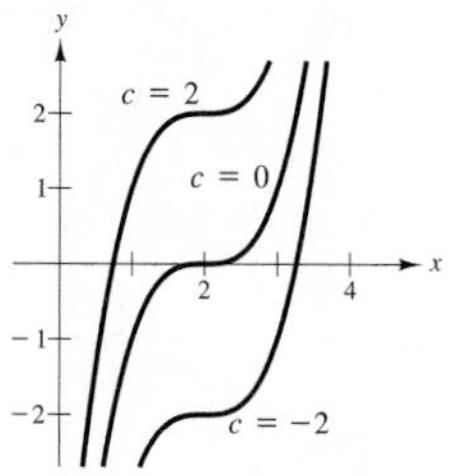

(d)

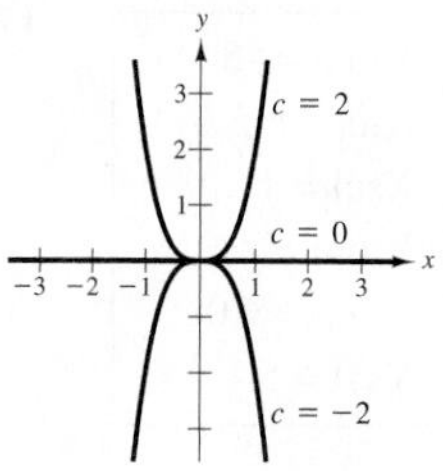

43. (a)

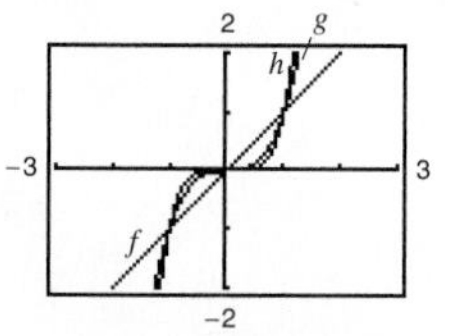

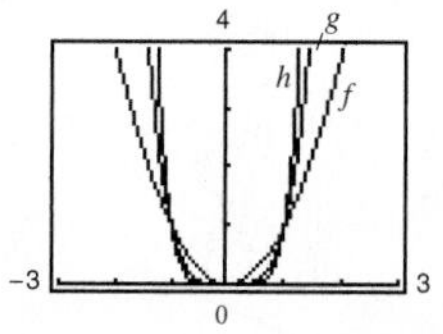

All the graphs pass through the origin. The graphs of the odd powers of x are symmetric with respect to the origin and the graphs of the even powers are symmetric with respect to the y-axis. As the powers increase, the graphs become flatter in the interval $-1 < x < 1$. Graphs of these equations with odd powers pass through Quadrants I and III. Graphs of these equations with even powers pass through Quadrants I and II.

(b) The graph of $y = x^7$ should pass through the origin and Quadrants I and III. It should be symmetric with respect to the origin and be fairly flat in the interval $(-1, 1)$. The graph of $y = x^8$ should pass through the origin and Quadrants I and II. It should be symmetric with respect to the y-axis and be fairly flat in the interval $(-1, 1)$.

45. (a) $A = x(12 - x)$

(b) Domain: $(0, 12)$

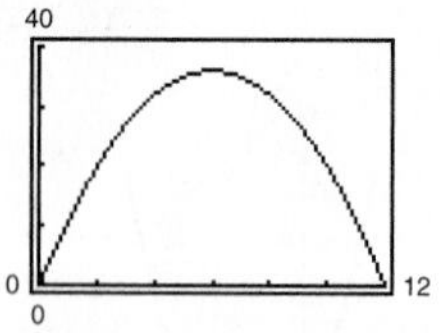

(c) Maximum area: 36 in.^{2}; 6×6 in.

47. (a) Minimum degree: 3; Leading coefficient: negative

(b) Minimum degree: 4; Leading coefficient: positive

(c) Minimum degree: 2; Leading coefficient: negative

(d) Minimum degree: 5; Leading coefficient: positive

49. (a) Yes. For each time t there corresponds one and only one displacement y.

(b) Amplitude: 0.25; Period: 1.1 (c) $y = \frac{1}{4}\cos(5.7t)$

(d)

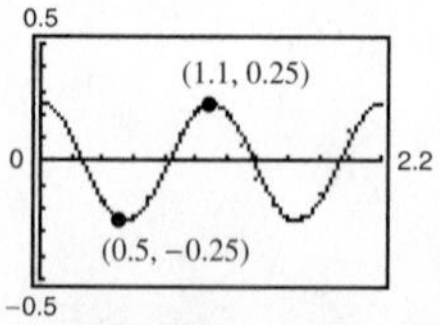

The model appears to fit the data.

P.S. Problem Solving (page 39)

1. (a) Center: $(3, 4)$; Radius: 5

(b) $y = -\frac{3}{4}x$ (c) $y = \frac{3}{4}x - \frac{9}{2}$ (d) $\left(3, -\frac{9}{4}\right)$

3.

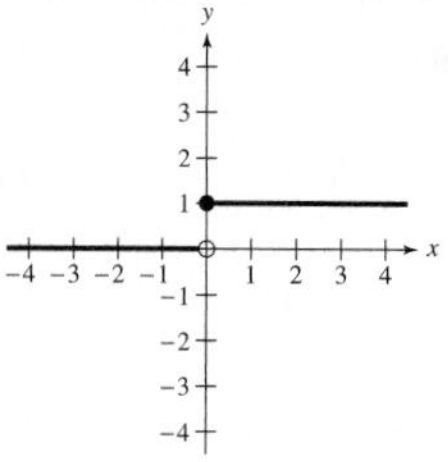

(a) $H(x) - 2 = \begin{cases} -1, & x \geq 0 \\ -2, & x < 0 \end{cases}$ (b) $H(x - 2) = \begin{cases} 1, & x \geq 2 \\ 0, & x < 2 \end{cases}$

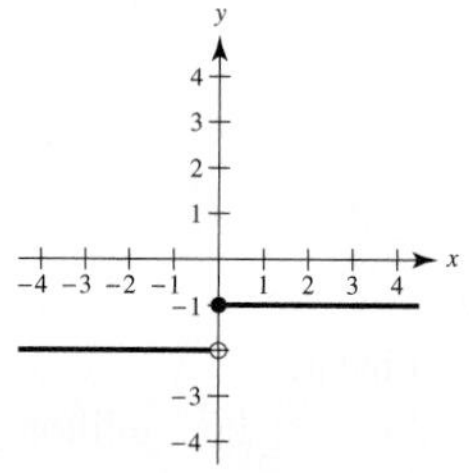

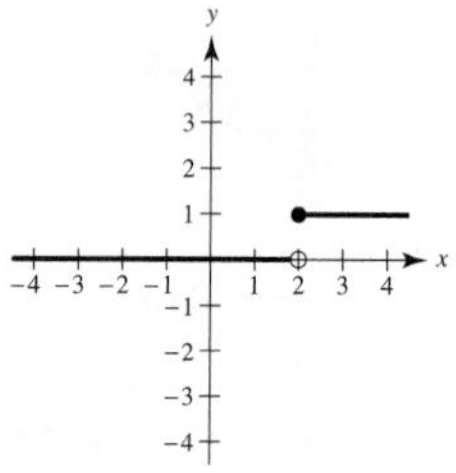

(c) $-H(x) = \begin{cases} -1, & x \geq 0 \\ 0, & x < 0 \end{cases}$ (d) $H(-x) = \begin{cases} 1, & x \leq 0 \\ 0, & x > 0 \end{cases}$

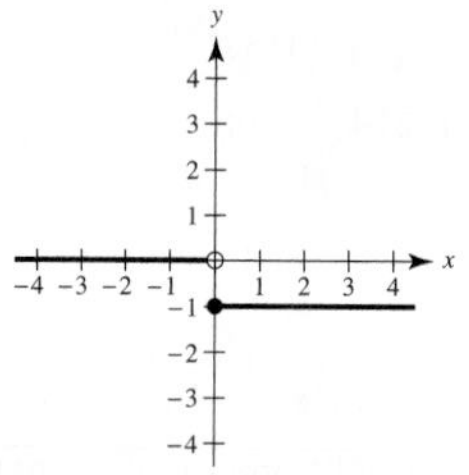

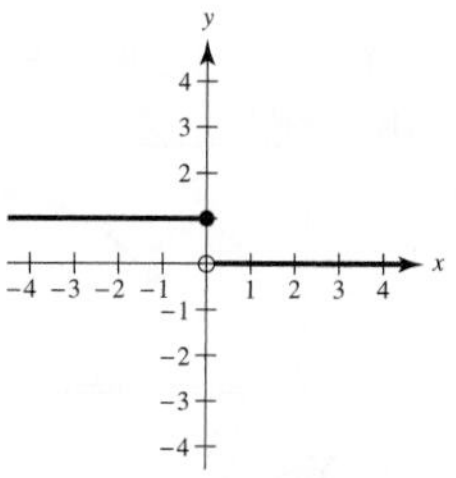

(e) $\frac{1}{2}H(x) = \begin{cases} \frac{1}{2}, & x \geq 0 \\ 0, & x < 0 \end{cases}$ (f) $-H(x - 2) + 2 = \begin{cases} 1, & x \geq 2 \\ 2, & x < 2 \end{cases}$

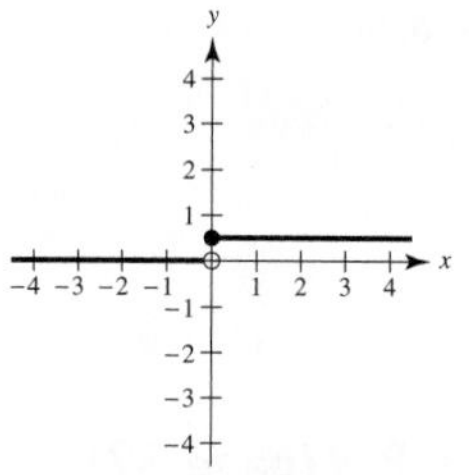

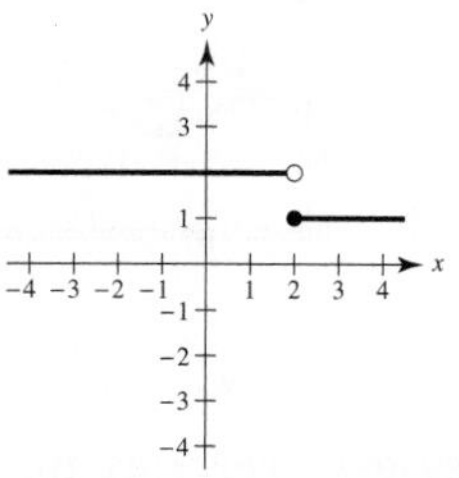

5. (a) $A(x) = x[(100 - x)/2]$; Domain: $(0, 100)$

(b)

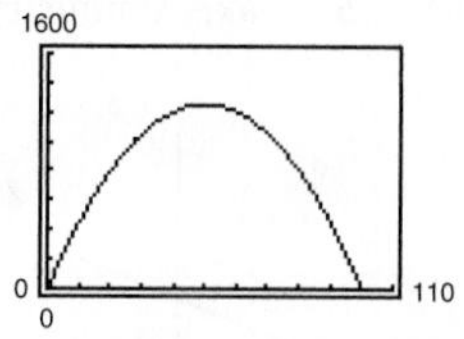

Dimensions 50 m × 25 m yield maximum area of 1250 m^2.

(c) 50 m × 25 m; Area = 1250 m^2

7. $T(x) = \left[2\sqrt{4 + x^2} + \sqrt{(3 - x)^2 + 1}\right]/4$

9. (a) 5, less (b) 3, greater (c) 4.1, less
(d) $4 + h$ (e) 4; Answers will vary.

11. (a) $x = -3 + \sqrt{18} \approx 1.2426,$
$x = -3 - \sqrt{18} \approx -7.2426$
(b) $(x + 3)^2 + y^2 = 18$

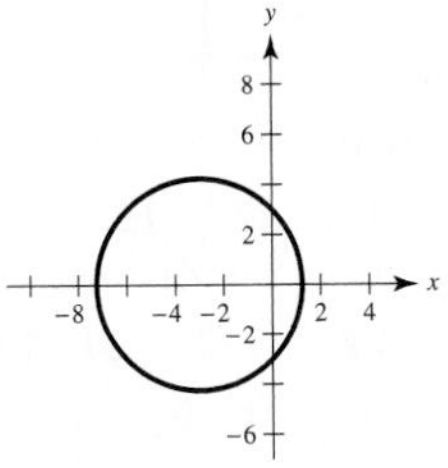

13. Answers will vary.

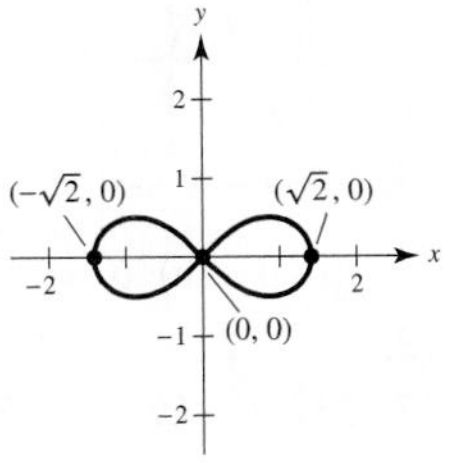

Chapter 1

Section 1.1 (page 47)

1. Precalculus: 300 ft

3. Calculus: Slope of the tangent line at $x = 2$ is 0.16.

5. Precalculus: $\frac{15}{2}$ square units

7. (a)

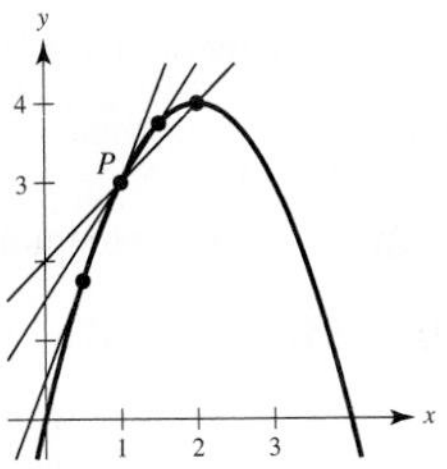

(b) $1; \frac{3}{2}; \frac{5}{2}$
(c) 2. Use points closer to P.

9. (a) Area ≈ 10.417; Area ≈ 9.145 (b) Use more rectangles.

11. (a) 5.66 (b) 6.11 (c) Increase the number of line segments.

Section 1.2 (page 54)

1.

x	1.9	1.99	1.999	2.001	2.01	2.1
$f(x)$	0.3448	0.3344	0.3334	0.3332	0.3322	0.3226

$$\lim_{x\to 2} \frac{x-2}{x^2-x-2} \approx 0.3333 \left(\text{Actual limit is } \frac{1}{3}.\right)$$

3.

x	−0.1	−0.01	−0.001	0.001	0.01	0.1
$f(x)$	0.2911	0.2889	0.2887	0.2887	0.2884	0.2863

$$\lim_{x\to 0} \frac{\sqrt{x+3}-\sqrt{3}}{x} \approx 0.2887 \left(\text{Actual limit is } \frac{1}{2\sqrt{3}}.\right)$$

5.

x	2.9	2.99	2.999
$f(x)$	−0.0641	−0.0627	−0.0625

x	3.001	3.01	3.1
$f(x)$	−0.0625	−0.0623	−0.0610

$$\lim_{x\to 3} \frac{[1/(x+1)]-(1/4)}{x-3} \approx -0.0625 \left(\text{Actual limit is } -\frac{1}{16}.\right)$$

7.

x	−0.1	−0.01	−0.001	0.001	0.01	0.1
$f(x)$	0.9983	0.99998	1.0000	1.0000	0.99998	0.9983

$$\lim_{x\to 0} \frac{\sin x}{x} \approx 1.0000 \text{ (Actual limit is 1.)}$$

9. 1 **11.** 2

13. Limit does not exist. The function approaches 1 from the right side of 5 but it approaches −1 from the left side of 5.

15. 0

17. Limit does not exist. The function oscillates between 1 and −1 as x approaches 0.

19. (a) 2
(b) Limit does not exist. The function approaches 1 from the right side of 1 but it approaches 3.5 from the left side of 1.
(c) Value does not exist. The function is undefined at $x = 4$.
(d) 2

21. $\lim_{x\to c} f(x)$ exists at all points on the graph except where $c = -3$.

23.

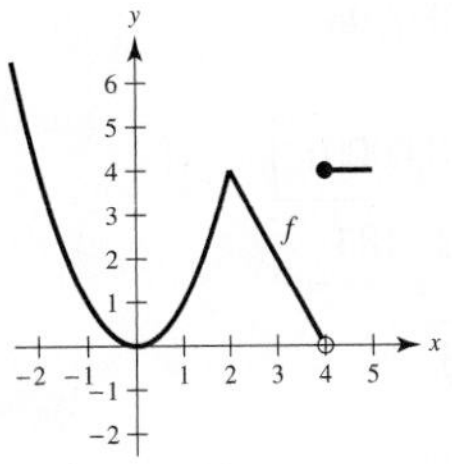

$\lim_{x\to c} f(x)$ exists at all points on the graph except where $c = 4$.

25.

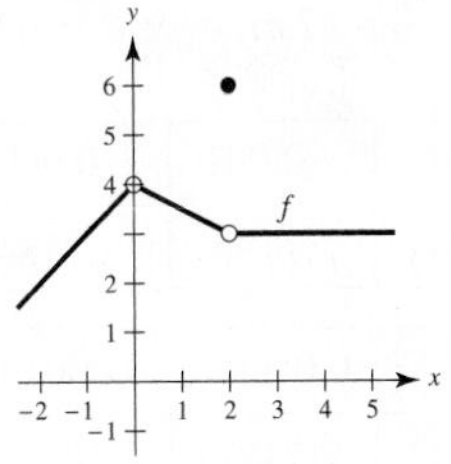

27. (a)

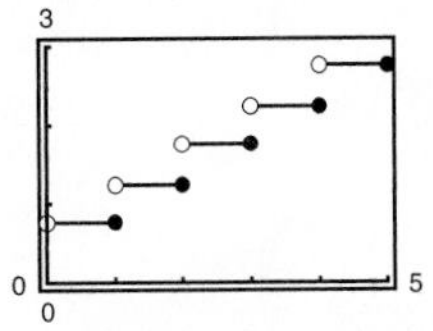

(b)

t	3	3.3	3.4	3.5	3.6	3.7	4
C	1.75	2.25	2.25	2.25	2.25	2.25	2.25

$\lim_{t\to 3.5} C(t) = 2.25$

(c)

t	2	2.5	2.9	3	3.1	3.5	4
C	1.25	1.75	1.75	1.75	2.25	2.25	2.25

The limit does not exist, because the limits from the right and left are not equal.

29. $\delta = 0.4$ **31.** $\delta = \frac{1}{11} \approx 0.091$

33. $L = 8$. Let $\delta = 0.01/3 \approx 0.0033$.

35. $L = 1$. Let $\delta = 0.01/5 = 0.002$.

37. 5 **39.** −3 **41.** 3 **43.** 0 **45.** 4 **47.** 2

49.

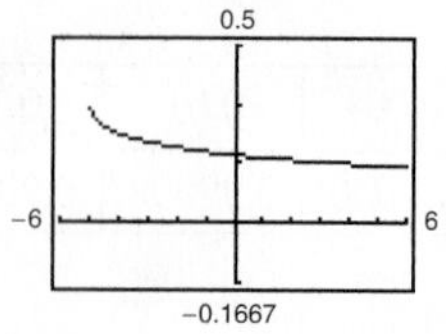

$\lim_{x\to4} f(x) = \frac{1}{6}$

Domain: $[-5, 4) \cup (4, \infty)$

The graph has a hole at $x = 4$.

51.

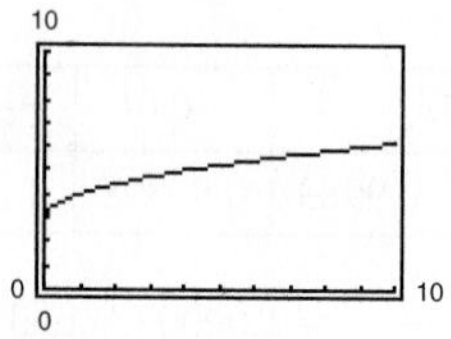

$\lim_{x\to9} f(x) = 6$

Domain: $[0, 9) \cup (9, \infty)$

The graph has a hole at $x = 9$.

53. Answers will vary. Sample answer: As x approaches 8 from either side, $f(x)$ becomes arbitrarily close to 25.

55. No. The fact that $\lim_{x\to2} f(x) = 4$ has no bearing on the value of f at 2.

57. (a) $r = \frac{3}{\pi} \approx 0.9549$ cm

(b) $\frac{5.5}{2\pi} \le r \le \frac{6.5}{2\pi}$, or approximately $0.8754 < r < 1.0345$

(c) $\lim_{r\to3/\pi} 2\pi r = 6$; $\varepsilon = 0.5$; $\delta \approx 0.0796$

59.

x	−0.001	−0.0001	−0.00001
$f(x)$	2.7196	2.7184	2.7183

x	0.00001	0.0001	0.001
$f(x)$	2.7183	2.7181	2.7169

$\lim_{x\to0} f(x) \approx 2.7183$

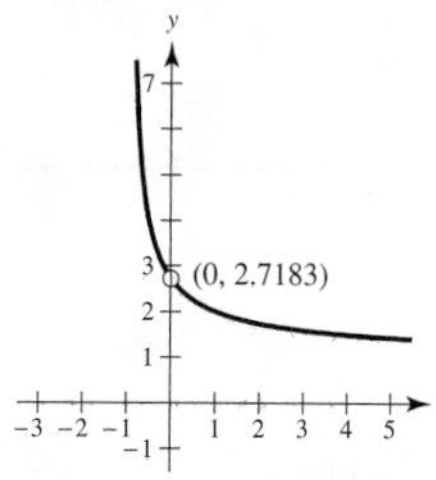

61.

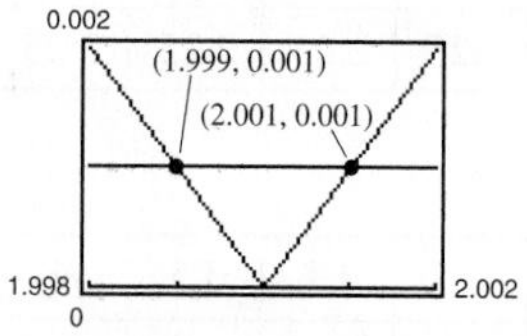

$\delta \approx 0.001$

(1.999, 2.001)

63. False. The existence or nonexistence of $f(x)$ at $x = c$ has no bearing on the existence of the limit of $f(x)$ as $x \to c$.

65. False. See Exercise 11.

67. (a) Yes. As x approaches 0.25 from either side, $\sqrt{x}$ becomes arbitrarily close to 0.5.

(b) No. $\lim_{x\to0} \sqrt{x}$ does not exist because for $x < 0$, $\sqrt{x}$ does not exist.

69–71. Proofs **73.** Answers will vary.

75. Putnam Problem B1, 1986

Section 1.3 (page 67)

1.

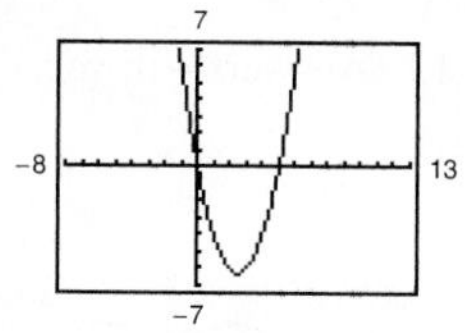

(a) 0 (b) 6

3.

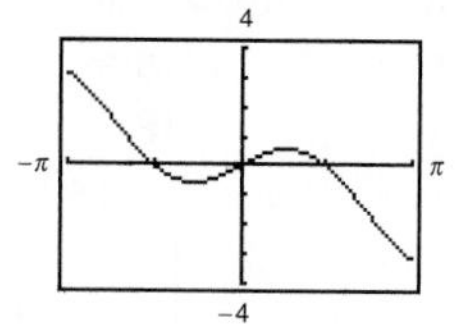

(a) 0 (b) ≈0.52 or $\pi/6$

5. 16 **7.** −1 **9.** 0 **11.** 7 **13.** 1/2 **15.** −2/5

17. 35/3 **19.** 2 **21.** 1 **23.** (a) 4 (b) 64 (c) 64

25. (a) 3 (b) 2 (c) 2 **27.** 1 **29.** −1/2 **31.** 1

33. 1/2 **35.** −1 **37.** (a) 15 (b) 5 (c) 6 (d) 2/3

39. (a) 64 (b) 2 (c) 12 (d) 8

41. (a) 1 (b) 3

$g(x) = \frac{-2x^2 + x}{x}$ and $f(x) = -2x + 1$ agree except at $x = 0$.

43. (a) 2 (b) 0

$g(x) = \frac{x^3 - x}{x - 1}$ and $f(x) = x^2 + x$ agree except at $x = 1$.

45. −2

$f(x) = \frac{x^2 - 1}{x + 1}$ and $g(x) = x - 1$ agree except at $x = -1$.

47. 12

$f(x) = \frac{x^3 - 8}{x - 2}$ and $g(x) = x^2 + 2x + 4$ agree except at $x = 2$.

49. 1/10 **51.** 5/6 **53.** $\sqrt{5}/10$ **55.** 1/6 **57.** −1/9

59. 2 **61.** $2x - 2$

63.

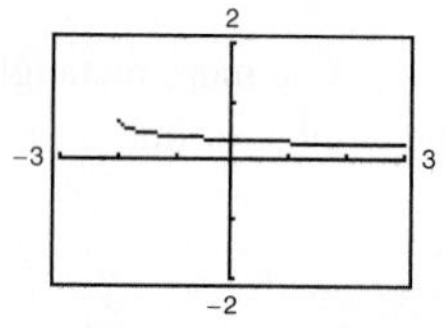

The graph has a hole at $x = 0$.

Answers will vary. Example:

x	−0.1	−0.01	−0.001	0.001	0.01	0.1
$f(x)$	0.358	0.354	0.354	0.354	0.353	0.349

$\lim_{x\to0} \frac{\sqrt{x+2} - \sqrt{2}}{x} \approx 0.354 \left(\text{Actual limit is } \frac{1}{2\sqrt{2}} = \frac{\sqrt{2}}{4}.\right)$

65.

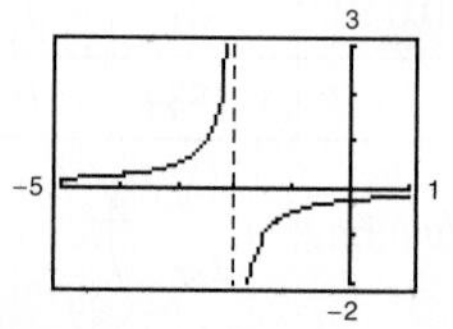

The graph has a hole at $x = 0$.

Answers will vary. Example:

x	−0.1	−0.01	−0.001
$f(x)$	−0.263	−0.251	−0.250

x	0.001	0.01	0.1
$f(x)$	−0.250	−0.249	−0.238

$\lim_{x\to0} \frac{[1/(2+x)] - (1/2)}{x} \approx -0.250 \left(\text{Actual limit is } -\frac{1}{4}.\right)$

67. 1/5 **69.** 0 **71.** 0 **73.** 0 **75.** 1 **77.** 3/2

79.

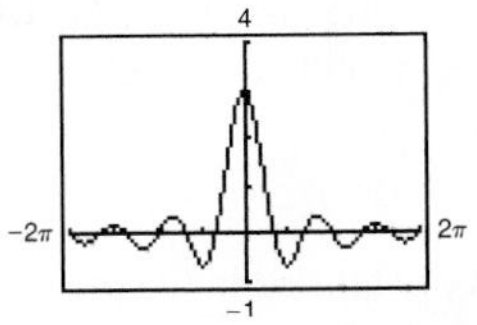

The graph has a hole at $t = 0$.

Answers will vary. Example:

t	-0.1	-0.01	0	0.01	0.1
$f(t)$	2.96	2.9996	?	2.9996	2.96

$\lim_{t \to 0} \frac{\sin 3t}{t} = 3$

81.

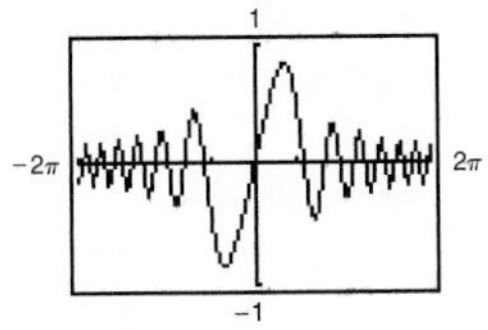

The graph has a hole at $x = 0$.

Answers will vary. Example:

x	-0.1	-0.01	-0.001	0	0.001	0.01	0.1
$f(x)$	-0.1	-0.01	-0.001	?	0.001	0.01	0.1

$\lim_{x \to 0} \frac{\sin x^2}{x} = 0$

83. 2 **85.** $-4/x^2$ **87.** 4

89. 0

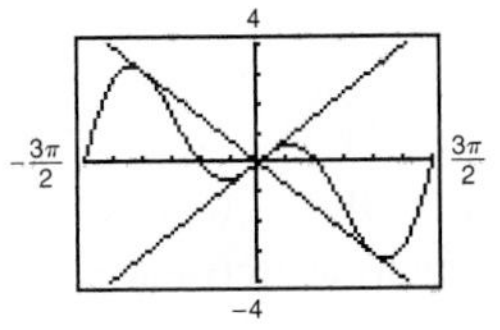

91. 0

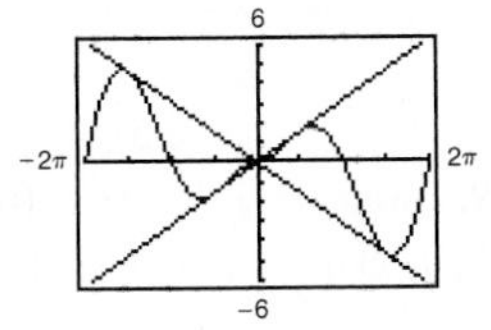

93. 0

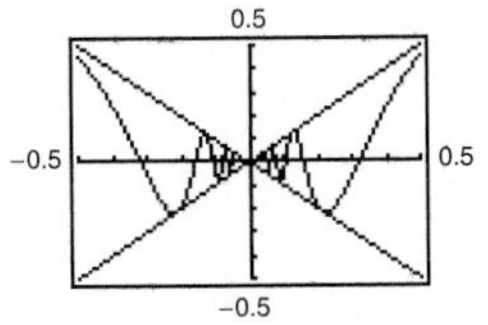

The graph has a hole at $x = 0$.

95. f and g agree at all but one point if c is a real number such that $f(x) = g(x)$ for all $x \neq c$.

97. An indeterminate form is obtained when evaluating a limit using direct substitution produces a meaningless fractional form, such as $\frac{0}{0}$.

99.

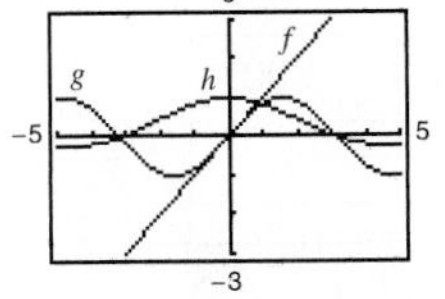

The magnitudes of $f(x)$ and $g(x)$ are approximately equal when x is close to 0. Therefore, their ratio is approximately 1.

101. 160 ft/sec **103.** -29.4 m/sec

105. Let $f(x) = 1/x$ and $g(x) = -1/x$.

$\lim_{x \to 0} f(x)$ and $\lim_{x \to 0} g(x)$ do not exist. However,

$$\lim_{x \to 0} [f(x) + g(x)] = \lim_{x \to 0} \left[\frac{1}{x} + \left(-\frac{1}{x}\right)\right] = \lim_{x \to 0} 0 = 0$$

and therefore does exist.

107–111. Proofs

113. False. The limit does not exist because the function approaches 1 from the right side of 0 and approaches -1 from the left side of 0. (See graph below.)

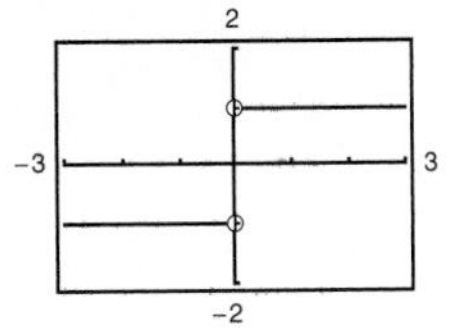

115. True.

117. False. The limit does not exist because $f(x)$ approaches 3 from the left side of 2 and approaches 0 from the right side of 2. (See graph below.)

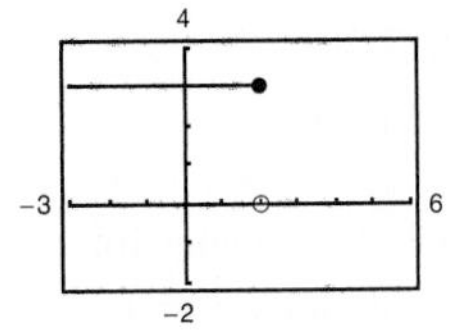

119. Let $f(x) = \begin{cases} 4, & \text{if } x \geq 0 \\ -4, & \text{if } x < 0 \end{cases}$

$\lim_{x \to 0} |f(x)| = \lim_{x \to 0} 4 = 4$

$\lim_{x \to 0} f(x)$ does not exist because for $x < 0$, $f(x) = -4$ and for $x \geq 0$, $f(x) = 4$.

121. $\lim_{x \to 0} f(x)$ does not exist because $f(x)$ oscillates between two fixed values as x approaches 0.

$\lim_{x \to 0} g(x) = 0$ because, as x gets increasingly closer to 0, the values of $g(x)$ become increasingly closer to 0.

123. (a) 1/2

(b) Because $\frac{1 - \cos x}{x^2} \approx \frac{1}{2}$, it follows that $1 - \cos x \approx \frac{1}{2}x^2$

$\cos x \approx 1 - \frac{1}{2}x^2$ when $x \approx 0$.

(c) 0.995 (d) Calculator: $\cos(0.1) \approx 0.9950$

Section 1.4 (page 78)

1. (a) 1 (b) 1 (c) 1; $f(x)$ is continuous on $(-\infty, \infty)$.

3. (a) 0 (b) 0 (c) 0; Discontinuity at $x = 3$

5. (a) 2 (b) -2 (c) Limit does not exist. Discontinuity at $x = 4$

7. $\frac{1}{10}$

9. Limit does not exist. The function decreases without bound as x approaches -3 from the left.

11. -1 **13.** $-1/x^2$ **15.** 5/2 **17.** 2

19. Limit does not exist. The function decreases without bound as x approaches π from the left and increases without bound as x approaches π from the right.

21. 4

23. Limit does not exist. The function approaches 5 from the left side of 3 but approaches 6 from the right side of 3.

25. Discontinuous at $x = -2$ and $x = 2$

27. Discontinuous at every integer

29. Continuous on $[-5, 5]$ **31.** Continuous on $[-1, 4]$

33. Continuous for all real x **35.** Continuous for all real x

37. Nonremovable discontinuity at $x = 1$
Removable discontinuity at $x = 0$

39. Continuous for all real x

41. Removable discontinuity at $x = -2$
Nonremovable discontinuity at $x = 5$

43. Nonremovable discontinuity at $x = -2$

45. Continuous for all real x

47. Nonremovable discontinuity at $x = 2$

49. Continuous for all real x

51. Nonremovable discontinuities at integer multiples of $\pi/2$

53. Nonremovable discontinuity at each integer

55.

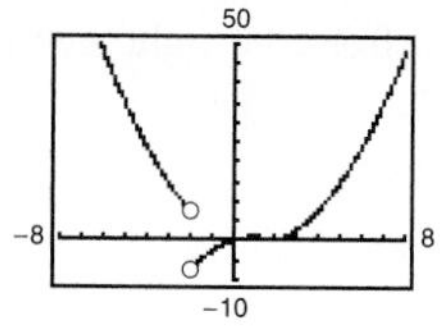

$\lim_{x\to 0^+} f(x) = 0$

$\lim_{x\to 0^-} f(x) = 0$

Discontinuity at $x = -2$

57. $a = 2$ **59.** $a = -1,\ b = 1$ **61.** Continuous for all real x

63. Nonremovable discontinuities at $x = 1$ and $x = -1$

65.

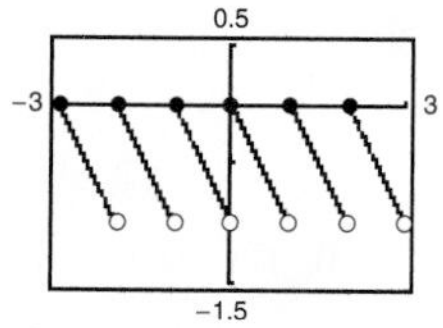

Nonremovable discontinuity at each integer

67.

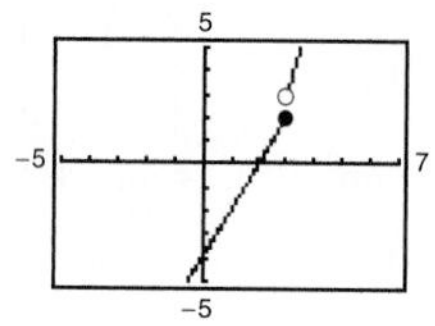

Discontinuous at $x = 3$

69. Continuous on $(-\infty, \infty)$

71. Continuous on the open intervals . . . $(-6, -2)$, $(-2, 2)$, $(2, 6)$, . . .

73.

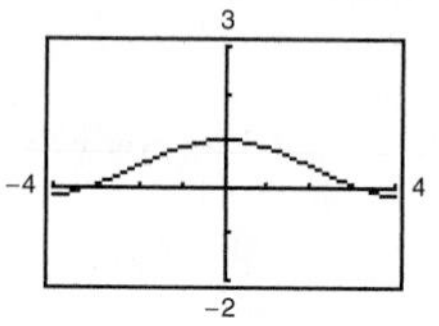

The graph has a hole at $x = 0$. The graph appears continuous but the function is not continuous on $[-4, 4]$. It is not obvious from the graph that the function has a discontinuity at $x = 0$.

75. Because $f(x)$ is continuous on the interval $[1, 2]$ and $f(1) = 2.0625$ and $f(2) = -4$, by the Intermediate Value Theorem there exists a real number c in $[1, 2]$ such that $f(c) = 0$.

77. Because $f(x)$ is continuous on the interval $[0, \pi]$ and $f(0) = -3$ and $f(\pi) \approx 8.87$, by the Intermediate Value Theorem there exists a real number c in $[0, \pi]$ such that $f(c) = 0$.

79. 0.68, 0.6823 **81.** 0.56, 0.5636

83. $f(3) = 11$ **85.** $f(2) = 4$

87. (a) The limit does not exist at $x = c$.
(b) The function is not defined at $x = c$.
(c) The limit exists, but it is not equal to the value of the function at $x = c$.
(d) The limit does not exist at $x = c$.

89.

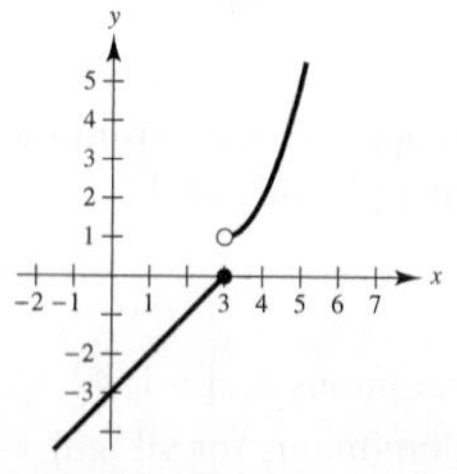

Not continuous because $\lim_{x\to 3} f(x)$ does not exist.

91. True

93. False. A rational function can be written as $P(x)/Q(x)$ where P and Q are polynomials of degree m and n, respectively. It can have, at most, n discontinuities.

95. $\lim_{t\to 4^-} f(t) \approx 28$; $\lim_{t\to 4^+} f(t) \approx 56$

At the end of day 3, the amount of chlorine in the pool is about 28 oz. At the beginning of day 4, the amount of chlorine in the pool is about 56 oz.

97. $C = \begin{cases} 1.04, & 0 < t \le 2 \\ 1.04 + 0.36[\![t - 1]\!], & t > 2,\ t \text{ is not an integer} \\ 1.04 + 0.36(t - 2), & t > 2,\ t \text{ is an integer} \end{cases}$

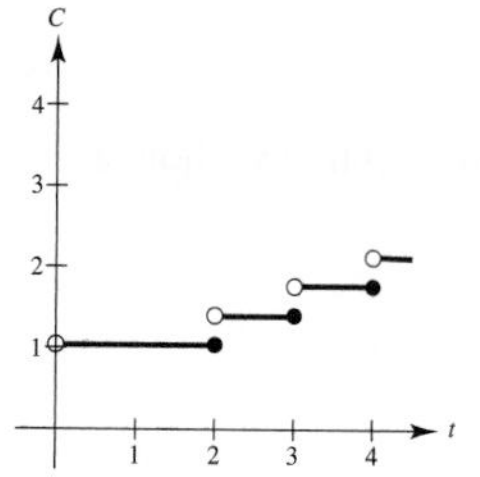

Nonremovable discontinuity at each integer greater than or equal to 2.

99–101. Proofs **103.** Answers will vary.

105. (a)

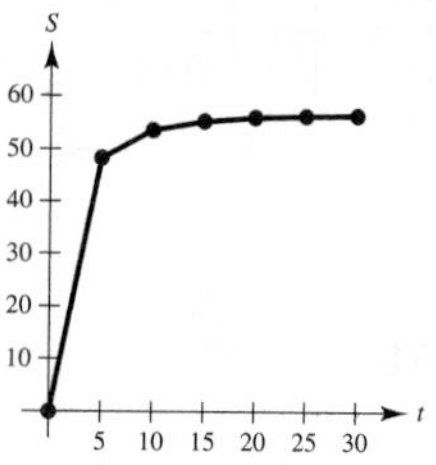

(b) There appears to be a limiting speed, and a possible cause is air resistance.

107. $c = \left(-1 \pm \sqrt{5}\right)/2$

109. Domain: $[-c^2, 0) \cup (0, \infty)$; Let $f(0) = 1/(2c)$

111. $h(x)$ has a nonremovable discontinuity at every integer except 0.

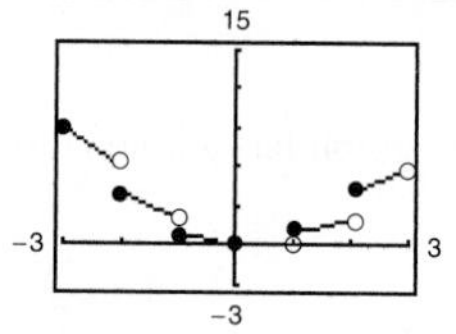

113. Putnam Problem B2, 1988

Section 1.5 (page 88)

1. $\lim_{x\to -2^+} 2\left|\dfrac{x}{x^2 - 4}\right| = \infty$ $\lim_{x\to -2^-} 2\left|\dfrac{x}{x^2 - 4}\right| = \infty$

3. $\lim_{x\to -2^+} \tan((\pi x)/4) = -\infty$ $\lim_{x\to -2^-} \tan((\pi x)/4) = \infty$

5.

x	−3.5	−3.1	−3.01	−3.001
$f(x)$	0.31	1.64	16.6	167

x	−2.999	−2.99	−2.9	−2.5
$f(x)$	−167	−16.7	−1.69	−0.36

$\lim_{x\to -3^+} f(x) = -\infty$ $\lim_{x\to -3^-} f(x) = \infty$

7.

x	-3.5	-3.1	-3.01	-3.001
$f(x)$	3.8	16	151	1501

x	-2.999	-2.99	-2.9	-2.5
$f(x)$	-1499	-149	-14	-2.3

$\lim_{x\to-3^+} f(x) = -\infty$ $\lim_{x\to-3^-} f(x) = \infty$

9. $x = 0$ **11.** $x = 2,\ x = -1$ **13.** $x = \pm 2$
15. No vertical asymptote **17.** $x = \pi/4 + (n\pi)/2$, n is an integer.
19. $t = 0$ **21.** $x = -2, x = 1$ **23.** No vertical asymptote
25. No vertical asymptote **27.** $t = n\pi$, n is a nonzero integer.
29. Removable discontinuity at $x = -1$
31. Vertical asymptote at $x = -1$ **33.** $-\infty$ **35.** ∞ **37.** $\frac{4}{5}$
39. $\frac{1}{2}$ **41.** $-\infty$ **43.** ∞ **45.** 0 **47.** Does not exist

49.

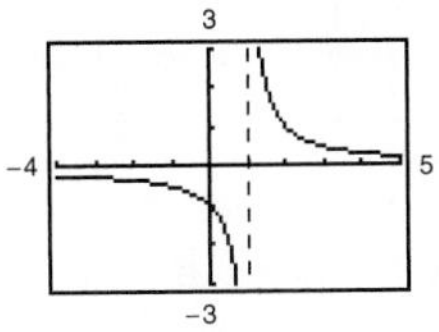

$\lim_{x\to1^+} f(x) = \infty$

51.

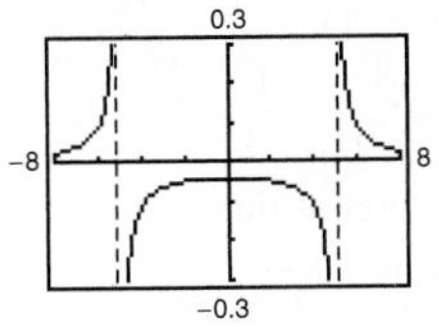

$\lim_{x\to5^-} f(x) = -\infty$

53. Answers will vary.

55. Answers will vary. Example: $f(x) = \dfrac{x-3}{x^2-4x-12}$

57.

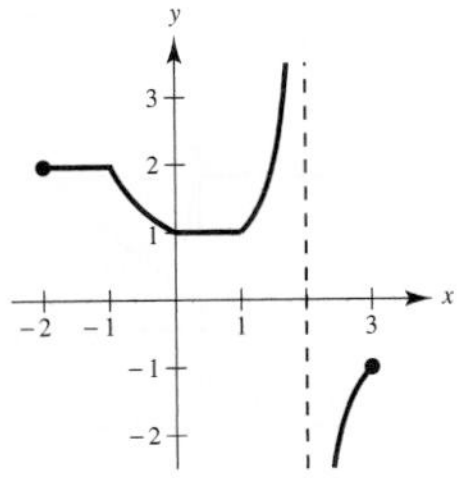

59. (a) $\frac{1}{3}(200\pi)$ ft/sec
(b) 200π ft/sec
(c) $\lim_{\theta\to(\pi/2)^-} [50\pi \sec^2 \theta] = \infty$

61. ∞

63. (a) Domain: $x > 25$
(b)

x	30	40	50	60
y	150	66.667	50	42.857

(c) $\lim_{x\to25^+} \dfrac{25x}{x-25} = \infty$

As x gets close to 25 mph, y becomes larger and larger.

65. (a) $A = 50 \tan\theta - 50\theta$; Domain: $(0, \pi/2)$
(b)

θ	0.3	0.6	0.9	1.2	1.5
$f(\theta)$	0.47	4.21	18.0	68.6	630.1

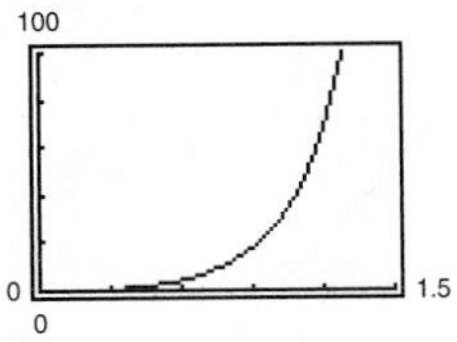

(c) $\lim_{\theta\to\pi/2^-} A = \infty$

67. False; let $f(x) = (x^2-1)/(x-1)$ **69.** True

71. Let $f(x) = \dfrac{1}{x^2}$ and $g(x) = \dfrac{1}{x^4}$, and $c = 0$. $\lim_{x\to0} \dfrac{1}{x^2} = \infty$ and

$$\lim_{x\to0} \frac{1}{x^4} = \infty, \text{ but } \lim_{x\to0}\left(\frac{1}{x^2} - \frac{1}{x^4}\right) = \lim_{x\to0}\left(\frac{x^2-1}{x^4}\right) = -\infty \neq 0.$$

73. Given $\lim_{x\to c} f(x) = \infty$, let $g(x) = 1$. Then $\lim_{x\to c} \dfrac{g(x)}{f(x)} = 0$ by Theorem 1.15.

75. Answers will vary.

Review Exercises for Chapter 1 (page 91)

1. Calculus

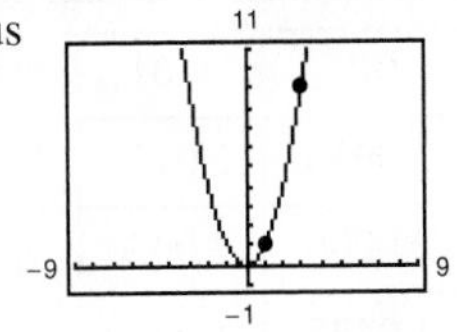

Estimate: 8.261

3.

x	-0.1	-0.01	-0.001
$f(x)$	-1.0526	-1.0050	-1.0005

x	0.001	0.01	0.1
$f(x)$	-0.9995	-0.9950	-0.9524

The estimate of the limit of $f(x)$, as x approaches zero, is -1.00.

5. (a) -2 (b) -3 **7.** 2; Proof **9.** 1; Proof
11. $\sqrt{6} \approx 2.45$ **13.** $-\frac{1}{4}$ **15.** $\frac{1}{4}$ **17.** -1 **19.** 75
21. 0 **23.** $\sqrt{3}/2$ **25.** $-\frac{1}{2}$

27. (a)

x	1.1	1.01	1.001	1.0001
$f(x)$	0.5680	0.5764	0.5773	0.5773

$\lim_{x\to1^+} f(x) \approx 0.5773$

(b)

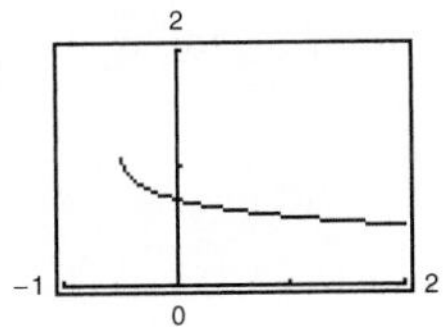

The graph has a hole at $x = 1$.
$\lim_{x\to1^+} f(x) \approx 0.5774$

(c) $\sqrt{3}/3$

29. -39.2 m/sec **31.** -1 **33.** 0
35. Limit does not exist. The limit as t approaches 1 from the left is 2 whereas the limit as t approaches 1 from the right is 1.
37. Nonremovable discontinuity at each integer
Continuous on $(k, k+1)$ for all integers k
39. Removable discontinuity at $x = 1$
Continuous on $(-\infty, 1) \cup (1, \infty)$
41. Nonremovable discontinuity at $x = 2$
Continuous on $(-\infty, 2) \cup (2, \infty)$
43. Nonremovable discontinuity at $x = -1$
Continuous on $(-\infty, -1) \cup (-1, \infty)$
45. Nonremovable discontinuity at each even integer
Continuous on $(2k, 2k+2)$ for all integers k
47. $c = -\frac{1}{2}$ **49.** Proof
51. (a) -4 (b) 4 (c) Limit does not exist.
53. $x = 0$ **55.** $x = 10$ **57.** $-\infty$ **59.** $\frac{1}{3}$
61. $-\infty$ **63.** $-\infty$ **65.** $\frac{4}{5}$ **67.** ∞
69. (a) \$14,117.65 (b) \$80,000.00 (c) \$720,000.00 (d) ∞

P.S. Problem Solving (page 93)

1. (a) Perimeter $\triangle PAO = 1 + \sqrt{(x^2-1)^2 + x^2} + \sqrt{x^4 + x^2}$
Perimeter $\triangle PBO = 1 + \sqrt{x^4 + (x-1)^2} + \sqrt{x^4 + x^2}$

(b)

x	4	2	1
Perimeter $\triangle PAO$	33.0166	9.0777	3.4142
Perimeter $\triangle PBO$	33.7712	9.5952	3.4142
$r(x)$	0.9777	0.9461	1.0000

x	0.1	0.01
Perimeter $\triangle PAO$	2.0955	2.0100
Perimeter $\triangle PBO$	2.0006	2.0000
$r(x)$	1.0475	1.0050

(c) 1

3. (a) Area (hexagon) $= (3\sqrt{3})/2 \approx 2.5981$
Area (circle) $= \pi \approx 3.1416$
Area (circle) $-$ Area (hexagon) ≈ 0.5435
(b) $A_n = (n/2)\sin(2\pi/n)$
(c)

n	6	12	24	48	96
A_n	2.5981	3.0000	3.1058	3.1326	3.1394

(d) 3.1416 or π

5. (a) $m = -\frac{12}{5}$ (b) $y = \frac{5}{12}x - \frac{169}{12}$
(c) $m_x = \dfrac{-\sqrt{169 - x^2} + 12}{x - 5}$
(d) $\frac{5}{12}$; It is the same as the slope of the tangent line found in (b).

7. (a) Domain: $[-27, 1) \cup (1, \infty)$
(b)
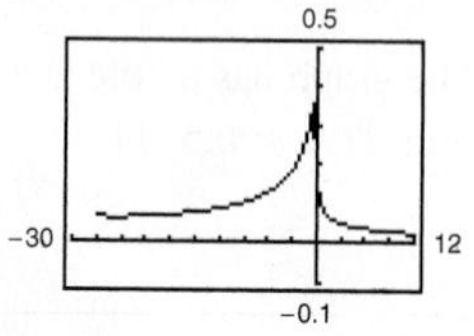

(c) $\frac{1}{14}$ (d) $\frac{1}{12}$

The graph has a hole at $x = 1$.

9. (a) g_1, g_4 (b) g_1 (c) g_1, g_3, g_4

11.
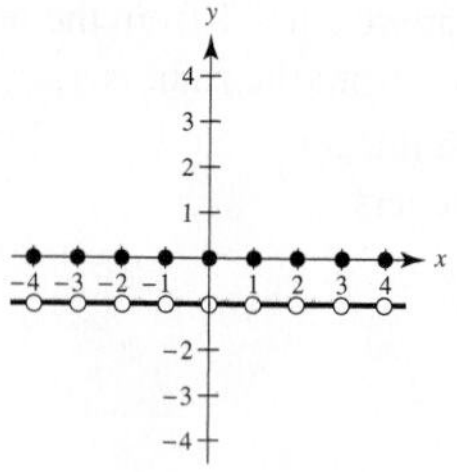

The graph jumps at every integer.

(a) $f(1) = 0,\ f(0) = 0,\ f\left(\frac{1}{2}\right) = -1,\ f(-2.7) = -1$
(b) $\lim_{x\to 1^-} f(x) = -1,\ \lim_{x\to 1^+} f(x) = -1,\ \lim_{x\to 1/2} f(x) = -1$
(c) There is a discontinuity at each integer.

13. (a)
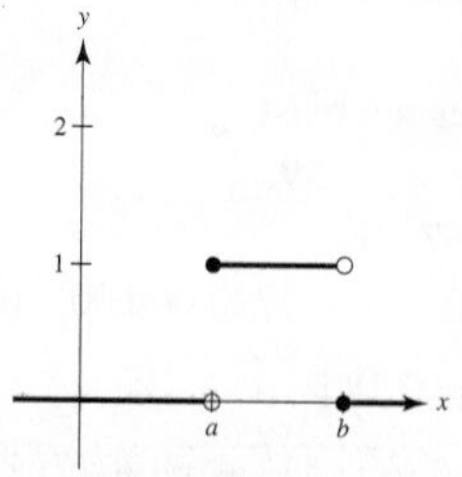

(b) (i) $\lim_{x\to a^+} P_{a,b}(x) = 1$
(ii) $\lim_{x\to a^-} P_{a,b}(x) = 0$
(iii) $\lim_{x\to b^+} P_{a,b}(x) = 0$
(iv) $\lim_{x\to b^-} P_{a,b}(x) = 1$

(c) Continuous for all positive real numbers except a and b
(d) The area under the curve gives a value of 1.

Chapter 2

Section 2.1 (page 103)

1. (a) $m_1 = 0, m_2 = 5/2$ (b) $m_1 = -5/2, m_2 = 2$

3.
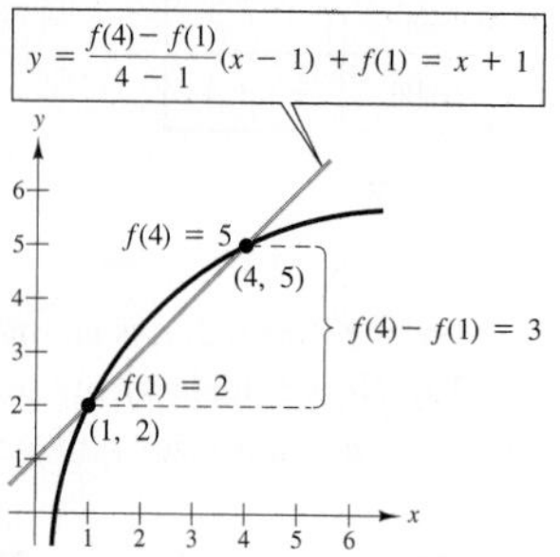

5. $m = -2$ **7.** $m = 2$

9. $m = 3$ **11.** $f'(x) = 0$ **13.** $f'(x) = -5$ **15.** $h'(s) = \frac{2}{3}$

17. $f'(x) = 4x + 1$ **19.** $f'(x) = 3x^2 - 12$

21. $f'(x) = \dfrac{-1}{(x-1)^2}$ **23.** $f'(x) = \dfrac{1}{2\sqrt{x+1}}$

25. (a) Tangent line: $y = 4x - 3$
(b)
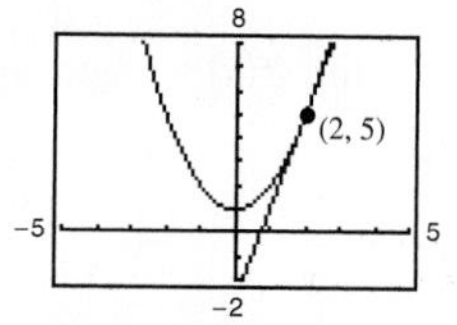

27. (a) Tangent line: $y = 12x - 16$
(b)
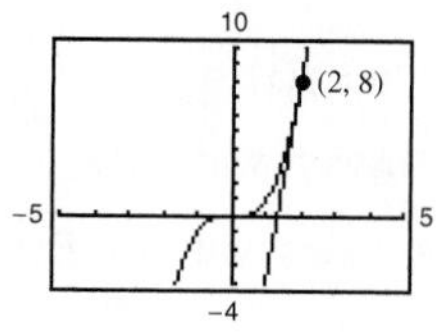

29. (a) Tangent line: $y = \frac{1}{2}x + \frac{1}{2}$
(b)
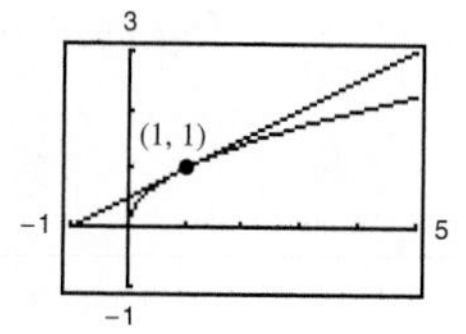

31. (a) Tangent line: $y = \frac{3}{4}x + 2$
(b)
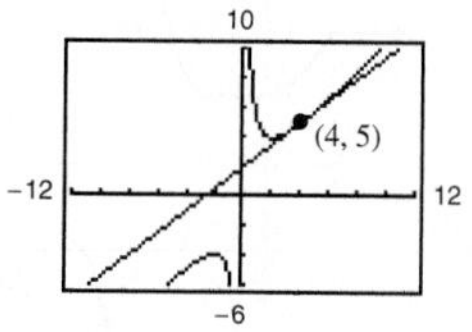

33. $y = 3x - 2;\ y = 3x + 2$ **35.** $y = -\frac{1}{2}x + \frac{3}{2}$

37. b **38.** d **39.** a **40.** c **41.** $g(5) = 2;\ g'(5) = -\frac{1}{2}$

43. $f'(x) = 1$
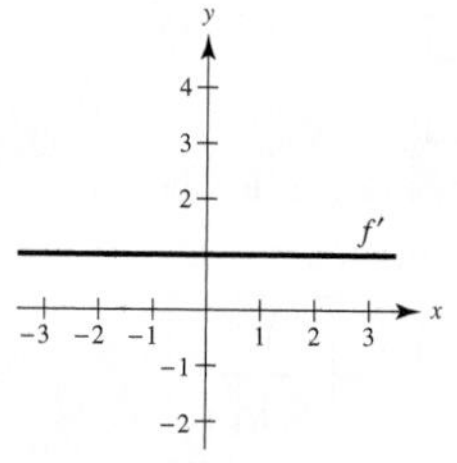

45. $f'(x) = 2x - 8$
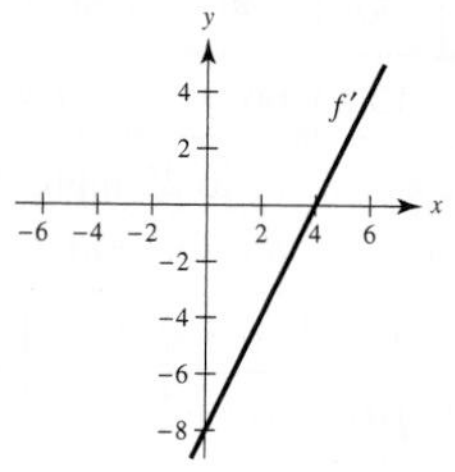

47. Answers will vary.
Sample answer: $y = -x$
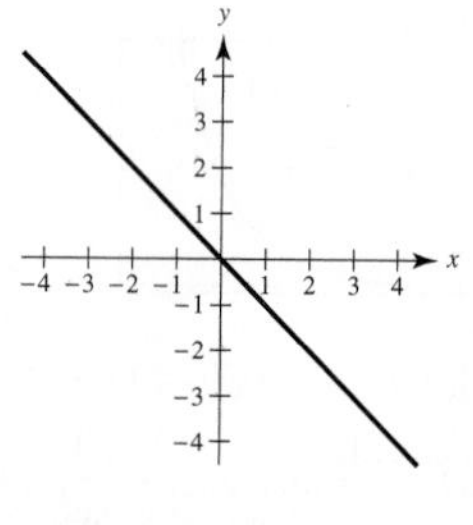

49. $f(x) = 5 - 3x$
$c = 1$

51. $f(x) = -x^2$
$c = 6$

53. $f(x) = -3x + 2$

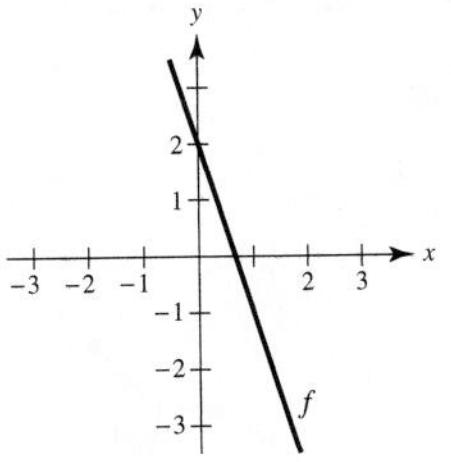

55. Answers will vary.
Sample answer: $f(x) = x^3$

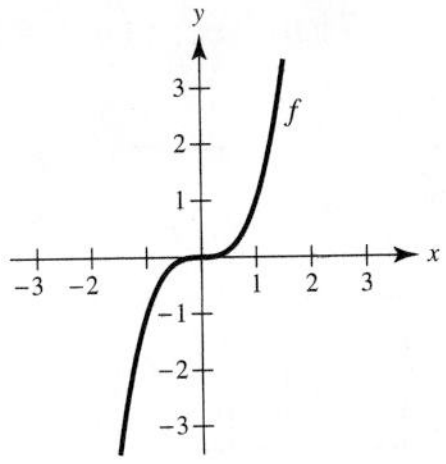

57. $y = 2x + 1;\ y = -2x + 9$

59. (a) -3 (b) 0
(c) The graph is moving downward to the right when $x = 1$.
(d) The graph is moving upward to the right when $x = -4$.
(e) Positive. Because $g'(x) > 0$ on $[3, 6]$, the graph of g is moving upward to the right.
(f) No. Knowing only $g'(2)$ is not sufficient information. $g'(2)$ remains the same for any vertical translation of g.

61.

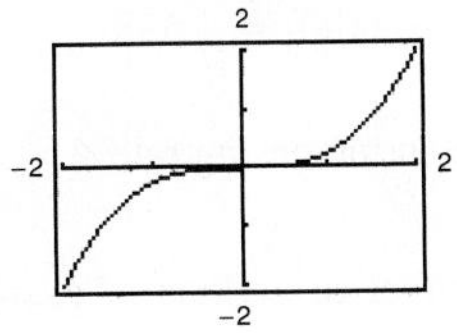

x	-2	-1.5	-1	-0.5	0	0.5	1	1.5	2
$f(x)$	-2	$-\frac{27}{32}$	$-\frac{1}{4}$	$-\frac{1}{32}$	0	$\frac{1}{32}$	$\frac{1}{4}$	$\frac{27}{32}$	2
$f'(x)$	3	$\frac{27}{16}$	$\frac{3}{4}$	$\frac{3}{16}$	0	$\frac{3}{16}$	$\frac{3}{4}$	$\frac{27}{16}$	3

63.

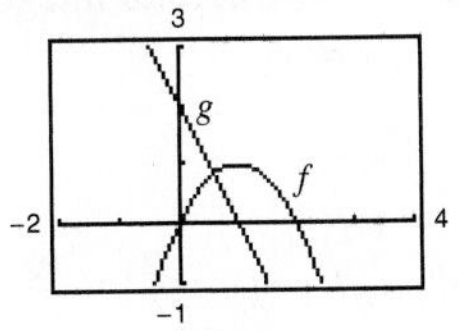

$g(x) \approx f'(x)$

65. $f(2) = 4;\ f(2.1) = 3.99;\ f'(2) \approx -0.1$

67.

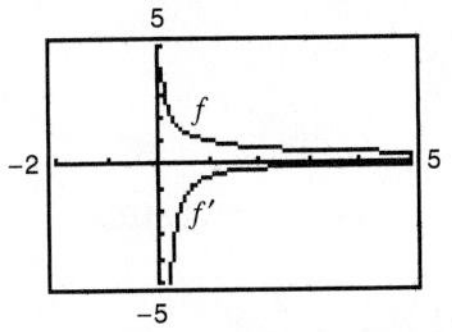

As x approaches infinity, the graph of f approaches a line of slope 0. Thus $f'(x)$ approaches 0.

69. (a)

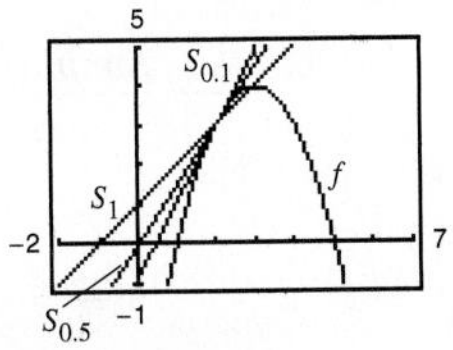

(b) The graphs of S for decreasing values of Δx are secant lines approaching the tangent line to the graph of f at the point $(2, f(2))$.

71. 4 **73.** 4 **75.** $g(x)$ is not differentiable at $x = 0$.

77. $f(x)$ is not differentiable at $x = 6$.

79. $h(x)$ is not differentiable at $x = -5$.

81. $(-\infty, -1) \cup (-1, \infty)$ **83.** $(-\infty, 3) \cup (3, \infty)$ **85.** $(1, \infty)$

87. $(-\infty, -3) \cup (-3, \infty)$

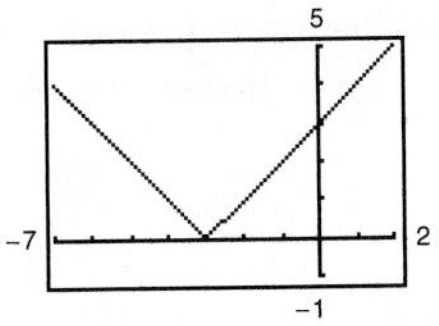

89. $(-\infty, 0) \cup (0, \infty)$

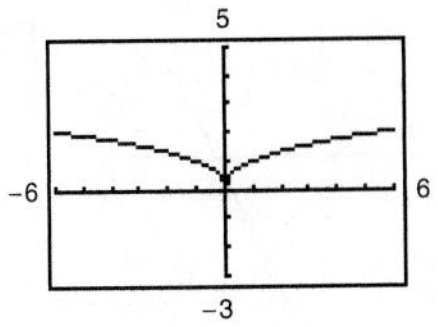

91. The derivative from the left is -1 and the derivative from the right is 1, so f is not differentiable at $x = 1$.

93. The derivatives from both the right and the left are 0, so $f'(1) = 0$.

95. f is differentiable at $x = 2$.

97. (a) $d = \left(3|m + 1|\right)/\sqrt{m^2 + 1}$
(b)

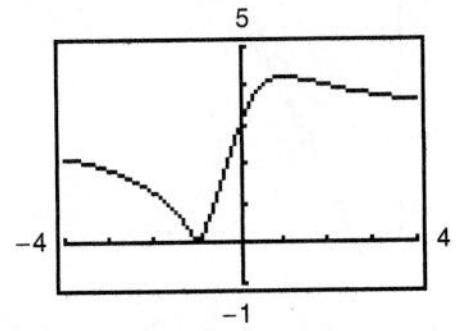

Not differentiable at $m = -1$

99. False. The slope is $\lim_{\Delta x \to 0} \dfrac{f(2 + \Delta x) - f(2)}{\Delta x}$.

101. False. For example: $f(x) = |x|$. The derivative from the left and the derivative from the right both exist but are not equal.

103. Proof

Section 2.2 (page 115)

1. (a) $\frac{1}{2}$ (b) 3 **3.** 0 **5.** $6x^5$ **7.** $-7/x^8$ **9.** $1/(5x^{4/5})$

11. 1 **13.** $-4t + 3$ **15.** $2x + 12x^2$ **17.** $3t^2 - 2$

19. $\dfrac{\pi}{2}\cos\theta + \sin\theta$ **21.** $2x + \dfrac{1}{2}\sin x$ **23.** $-\dfrac{1}{x^2} - 3\cos x$

	Function	*Rewrite*	*Derivative*	*Simplify*
25.	$y = \dfrac{5}{2x^2}$	$y = \dfrac{5}{2}x^{-2}$	$y' = -5x^{-3}$	$y' = -\dfrac{5}{x^3}$
27.	$y = \dfrac{3}{(2x)^3}$	$y = \dfrac{3}{8}x^{-3}$	$y' = -\dfrac{9}{8}x^{-4}$	$y' = -\dfrac{9}{8x^4}$
29.	$y = \dfrac{\sqrt{x}}{x}$	$y = x^{-1/2}$	$y' = -\dfrac{1}{2}x^{-3/2}$	$y' = -\dfrac{1}{2x^{3/2}}$

31. -6 **33.** 0 **35.** 4 **37.** 3 **39.** $2x + 6/x^3$

41. $2t + 12/t^4$ **43.** $(x^3 - 8)/x^3$ **45.** $3x^2 + 1$

47. $\dfrac{1}{2\sqrt{x}} - \dfrac{2}{x^{2/3}}$ **49.** $\dfrac{4}{5s^{1/5}} - \dfrac{2}{3s^{1/3}}$ **51.** $\dfrac{3}{\sqrt{x}} - 5\sin x$

53. (a) $2x + y - 2 = 0$
(b)

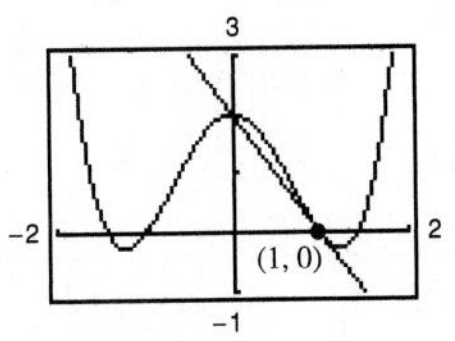

55. (a) $3x + 2y - 7 = 0$
(b)

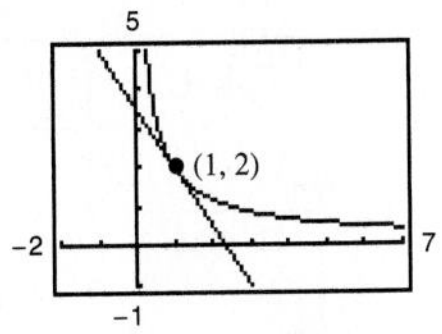

57. $(0, 2), (-2, -14), (2, -14)$ **59.** No horizontal tangents

61. (π, π) **63.** $k = 2,\ k = -10$ **65.** $k = 3$

67. (a) A and B (b) Greater (c)

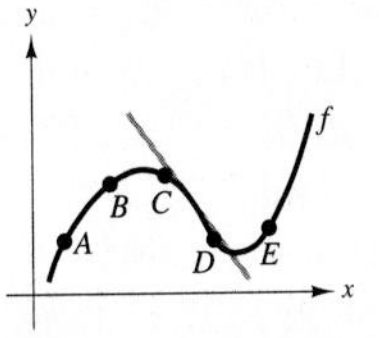

69. $g'(x) = f'(x)$

71.

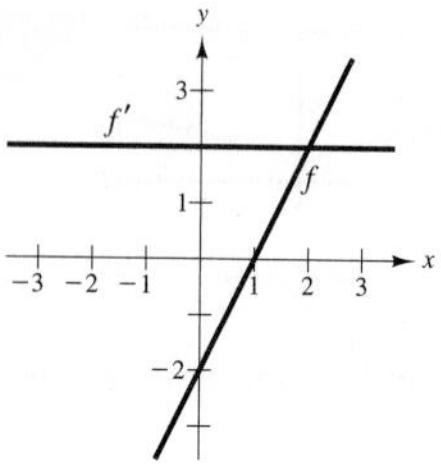

The rate of change of f is constant and therefore f' is a constant function.

73. $y = 2x - 1$ $\qquad$ $y = 4x - 4$

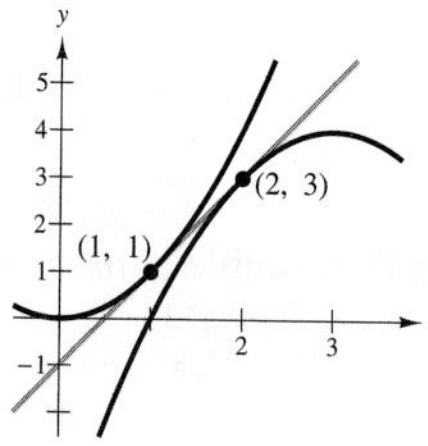

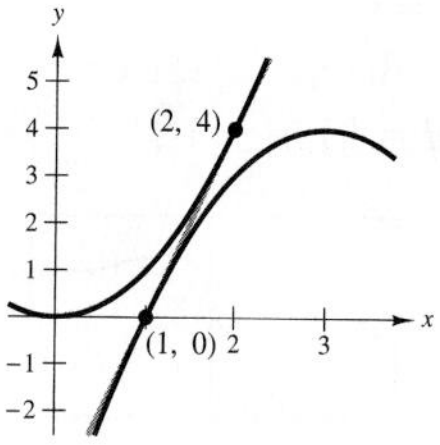

75. $f'(x) = 3 + \cos x \neq 0$ for all x. **77.** $x - 4y + 4 = 0$

79.

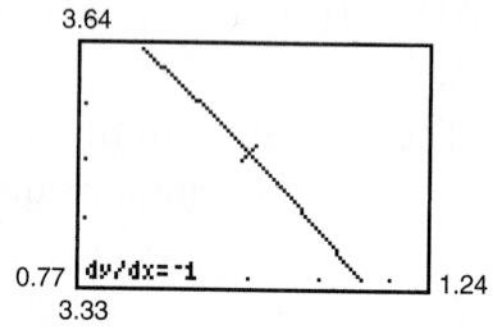

$f'(1)$ appears to be close to -1.

$f'(1) = -1$

81. (a)

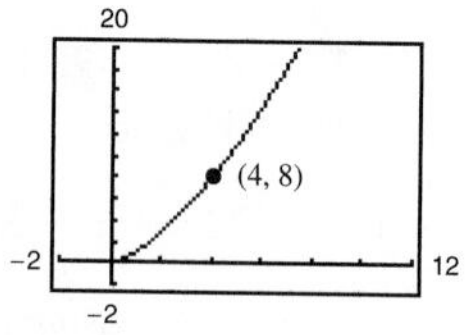

(3.9, 7.7019), $S(x) = 2.981x - 3.924$

(b) $T(x) = 3(x - 4) + 8 = 3x - 4$

The slope (and equation) of the secant line approaches that of the tangent line at (4, 8) as you choose points closer and closer to (4, 8).

(c)

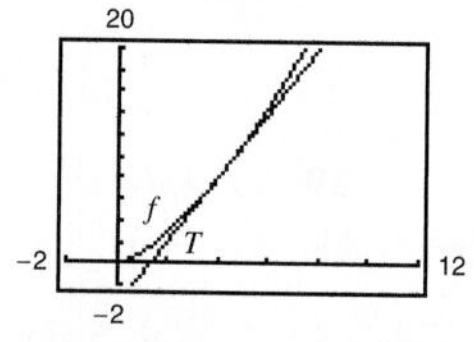

The approximation becomes less accurate.

(d)

Δx	-3	-2	-1	-0.5	-0.1	0
$f(4 + \Delta x)$	1	2.828	5.196	6.548	7.702	8
$T(4 + \Delta x)$	-1	2	5	6.5	7.7	8

Δx	0.1	0.5	1	2	3
$f(4 + \Delta x)$	8.302	9.546	11.180	14.697	18.520
$T(4 + \Delta x)$	8.3	9.5	11	14	17

83. False. Let $f(x) = x$ and $g(x) = x + 1$.

85. False. $dy/dx = 0$ **87.** True

89. Average rate: 2
Instantaneous rates:
$f'(1) = 2$; $f'(2) = 2$

91. Average rate: $\frac{1}{2}$
Instantaneous rates:
$f'(1) = 1$; $f'(2) = \frac{1}{4}$

93. (a) $s(t) = -16t^2 + 1362$; $v(t) = -32t$ (b) -48 ft/sec
(c) $s'(1) = -32$ ft/sec; $s'(2) = -64$ ft/sec
(d) $t = \dfrac{\sqrt{1362}}{4} \approx 9.226$ sec (e) -295.242 ft/sec

95. $v(5) = 71$ m/sec; $v(10) = 22$ m/sec

97.

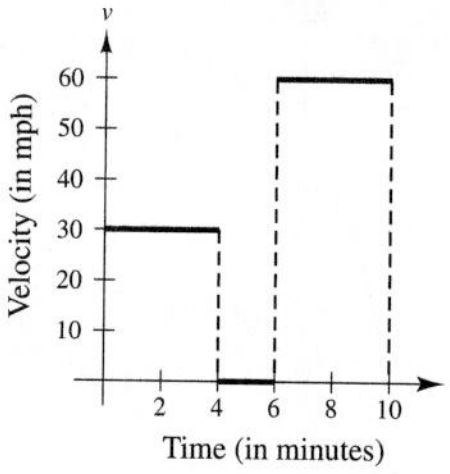

99.

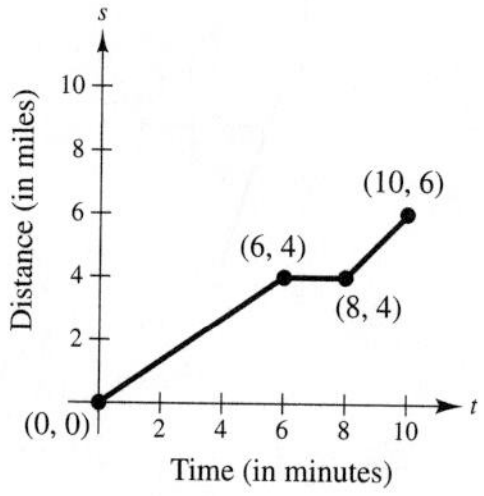

101. (a) $R(v) = 0.417v - 0.02$
(b) $B(v) = 0.0056v^2 + 0.001v + 0.04$
(c) $T(v) = 0.0056v^2 + 0.418v + 0.02$
(d)

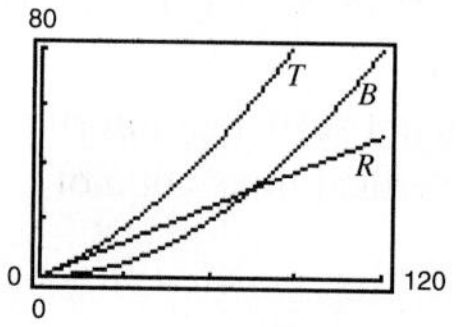

(e) $T'(v) = 0.0112v + 0.418$
$T'(40) = 0.866$
$T'(80) = 1.314$
$T'(100) = 1.538$

(f) Stopping distance increases at an increasing rate.

103. $V'(4) = 48\text{ cm}^2$ **105.** Proof

107. (a) The rate of change of the number of gallons of gasoline sold when the price is \$1.479
(b) In general, the rate of change when $p = 1.479$ should be negative. As prices go up, sales go down.

109. $y = 2x^2 - 3x + 1$ **111.** $y = -9x$, $y = -\frac{9}{4}x - \frac{27}{4}$

113. $a = \frac{1}{3}$, $b = -\frac{4}{3}$

115. $f_1(x) = |\sin x|$ is differentiable for all $x \neq n\pi$, n an integer.
$f_2(x) = \sin|x|$ is differentiable for all $x \neq 0$.

Section 2.3 (page 126)

1. $2(2x^3 - 3x^2 + x - 1)$ **3.** $(7t^2 + 4)/(3t^{2/3})$

5. $x^2(3\cos x - x\sin x)$ **7.** $(1 - x^2)/(x^2 + 1)^2$

9. $(1 - 8x^3)/[3x^{2/3}(x^3 + 1)^2]$ **11.** $(x\cos x - 2\sin x)/x^3$

13. $f'(x) = (x^3 - 3x)(4x + 3) + (2x^2 + 3x + 5)(3x^2 - 3)$
$= 10x^4 + 12x^3 - 3x^2 - 18x - 15$
$f'(0) = -15$

15. $f'(x) = \dfrac{x^2 - 6x + 4}{(x - 3)^2}$
$f'(1) = -\dfrac{1}{4}$

17. $f'(x) = \cos x - x\sin x$
$f'\left(\dfrac{\pi}{4}\right) = \dfrac{\sqrt{2}}{8}(4 - \pi)$

	Function	*Rewrite*	*Differentiate*	*Simplify*
19.	$y = \dfrac{x^2 + 2x}{3}$	$y = \dfrac{1}{3}(x^2 + 2x)$	$y' = \dfrac{1}{3}(2x + 2)$	$y' = \dfrac{2(x + 1)}{3}$
21.	$y = \dfrac{7}{3x^3}$	$y = \dfrac{7}{3}x^{-3}$	$y' = -7x^{-4}$	$y' = -\dfrac{7}{x^4}$
23.	$y = \dfrac{4x^{3/2}}{x}$	$y = 4x^{1/2}$, $x > 0$	$y' = 2x^{-1/2}$	$y' = \dfrac{2}{\sqrt{x}}$, $x > 0$

25. $\dfrac{(x^2 - 1)(-2 - 2x) - (3 - 2x - x^2)(2x)}{(x^2 - 1)^2} = \dfrac{2}{(x + 1)^2}$, $x \neq 1$

27. $1 - 12/(x+3)^2 = (x^2+6x-3)/(x+3)^2$

29. $\left[2\sqrt{x} - (2x+5)\dfrac{1}{2\sqrt{x}}\right]/x = (2x-5)/2x^{3/2}$

31. $6s^2(s^3-2)$ **33.** $-(2x^2-2x+3)/[x^2(x-3)^2]$

35. $(3x^3+4x)[(x-5)\cdot 1 + (x+1)\cdot 1]$
$+ [(x-5)(x+1)](9x^2+4)$
$= 15x^4 - 48x^3 - 33x^2 - 32x - 20$

37. $\dfrac{(x^2-c^2)(2x)-(x^2+c^2)(2x)}{(x^2-c^2)^2} = -\dfrac{4xc^2}{(x^2-c^2)^2}$

39. $t(t\cos t + 2\sin t)$ **41.** $-(t\sin t + \cos t)/t^2$

43. $-1 + \sec^2 x = \tan^2 x$ **45.** $\dfrac{1}{4t^{3/4}} + 8\sec t\tan t$

47. $\dfrac{-6\cos^2 x + 6\sin x - 6\sin^2 x}{4\cos^2 x} = \dfrac{3}{2}(-1+\tan x\sec x - \tan^2 x)$
$= \frac{3}{2}\sec x(\tan x - \sec x)$

49. $\csc x\cot x - \cos x = \cos x\cot^2 x$ **51.** $x(x\sec^2 x + 2\tan x)$

53. $2x\cos x + 2\sin x - x^2\sin x + 2x\cos x$
$= 4x\cos x + (2-x^2)\sin x$

55. $\left(\dfrac{x+1}{x+2}\right)(2) + (2x-5)\left[\dfrac{(x+2)(1)-(x+1)(1)}{(x+2)^2}\right]$
$= \dfrac{2x^2+8x-1}{(x+2)^2}$

57. $\dfrac{1-\sin\theta+\theta\cos\theta}{(1-\sin\theta)^2}$ **59.** $y' = \dfrac{-2\csc x\cot x}{(1-\csc x)^2}$, $-4\sqrt{3}$

61. $h'(t) = \sec t(t\tan t - 1)/t^2$, $1/\pi^2$

63. (a) $y = -x - 2$
(b)

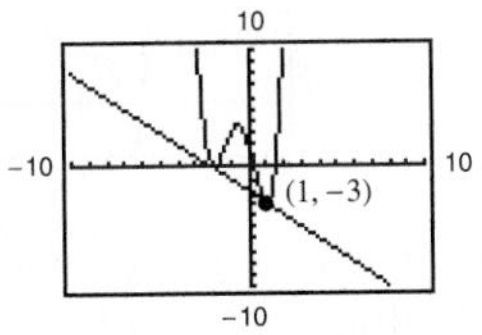

65. (a) $y = -x + 4$
(b)

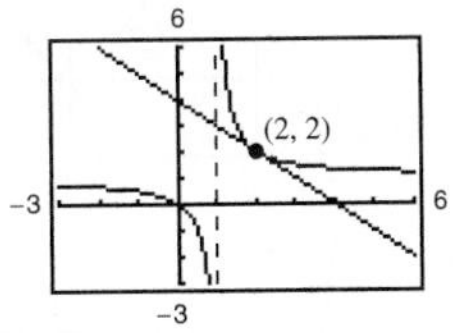

67. (a) $4x - 2y - \pi + 2 = 0$
(b)

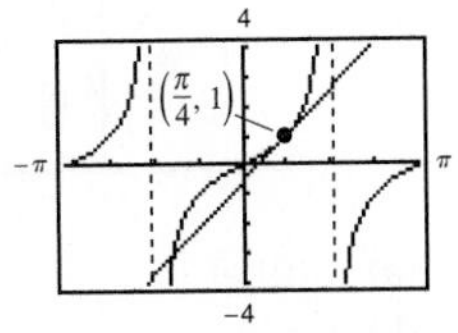

69. $2y + x - 4 = 0$

71. $25y - 12x + 16 = 0$ **73.** $(0, 0), (2, 4)$ **75.** $(1, 2)$

77. Tangent lines: $2y + x = 7$; $2y + x = -1$

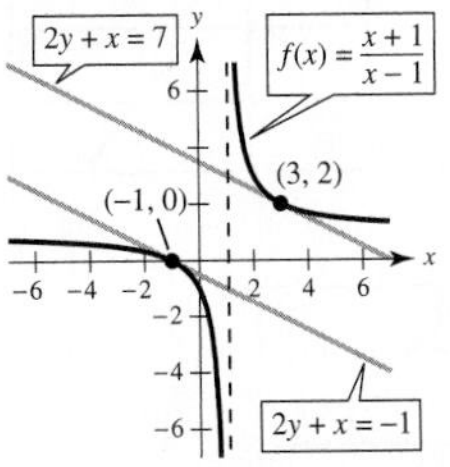

79. $f(x) + 2 = g(x)$ **81.** (a) $p'(1) = 1$ (b) $q'(4) = -1/3$

83. $(6t+1)/(2\sqrt{t})$ cm²/sec

85. (a) $-\$38.13$ (b) $-\$10.37$ (c) $-\$3.80$
The costs decrease with increasing order size.

87. 31.55 bacteria/hr **89.** Proof

91. (a) $n(t) = -3.5806t^3 + 82.577t^2 - 603.60t + 1667.5$
$v(t) = -0.1361t^3 + 3.165t^2 - 23.02t + 59.8$
(b)

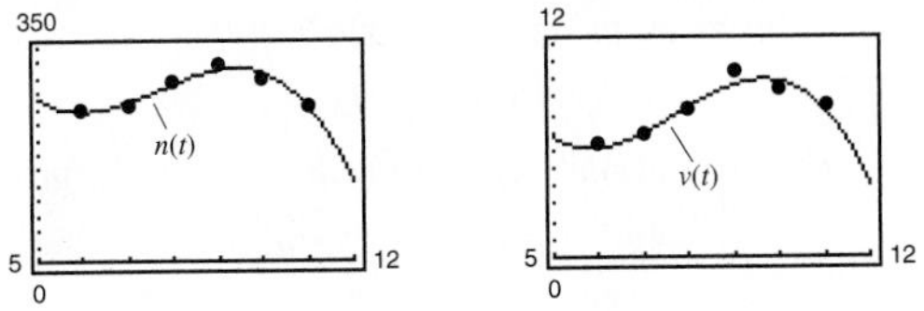

(c) $A = \dfrac{-0.1361t^3 + 3.165t^2 - 23.02t + 59.8}{-3.5806t^3 + 82.577t^2 - 603.60t + 1667.5}$

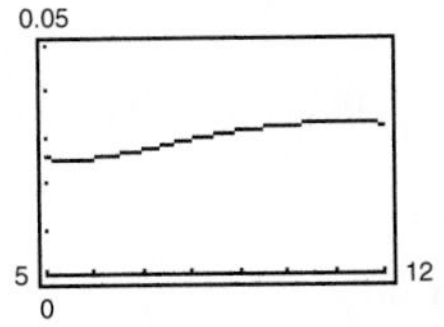

A represents the average retail value (in millions of dollars) per 1000 motor homes.

(d) $A'(t)$ represents the rate of change of the average retail value per 1000 motor homes for the given year.

93. $3/\sqrt{x}$ **95.** $2/(x-1)^3$ **97.** $-3\sin x$

99. $2x$ **101.** $1/\sqrt{x}$

103. Answers will vary. For example: $(x-2)^2$

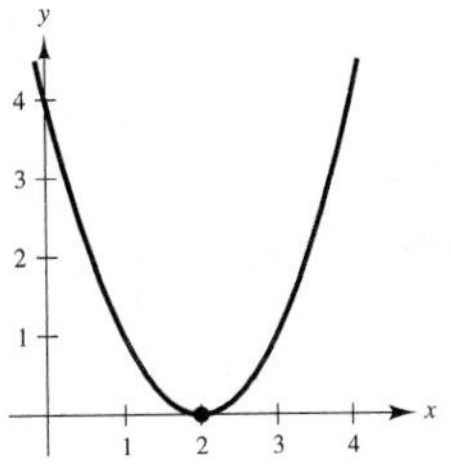

105. 0 **107.** -10

109.

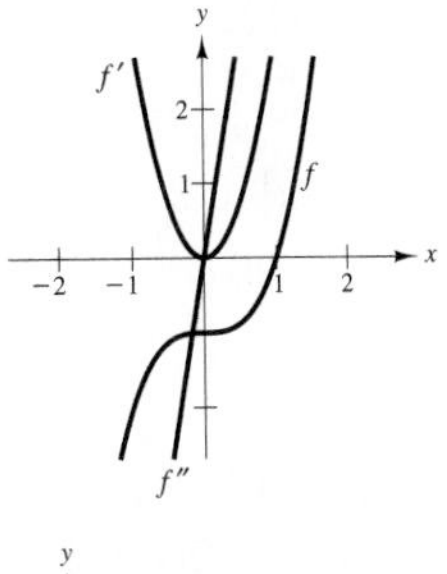

111.

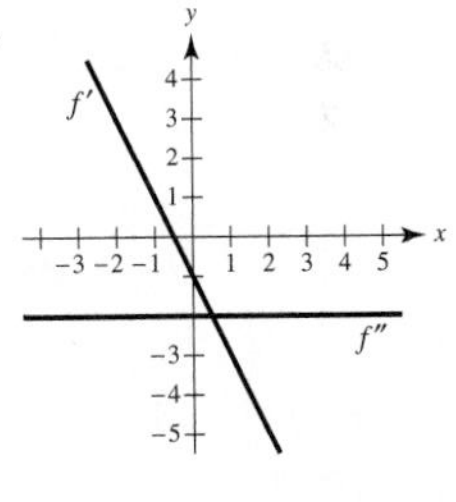

113.

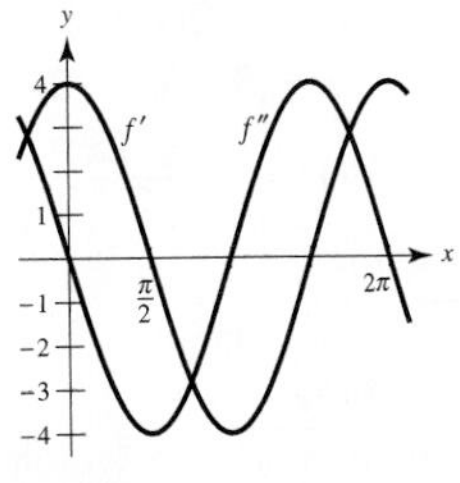

115. $v(3) = 27$ m/sec
$a(3) = -6$ m/sec²
The speed of the object is decreasing, but the rate of that decrease is increasing.

117.

t	0	1	2	3	4
$s(t)$	0	57.75	99	123.75	132
$v(t)$	66	49.5	33	16.5	0
$a(t)$	−16.5	−16.5	−16.5	−16.5	−16.5

The average velocity on $[0, 1]$ is 57.75, on $[1, 2]$ is 41.25, on $[2, 3]$ is 24.75, and on $[3, 4]$ is 8.25.

119. $f^{(n)}(x) = n(n-1)(n-2)\cdots(2)(1) = n!$

121. (a) $f''(x) = g(x)h''(x) + 2g'(x)h'(x) + g''(x)h(x)$

$f'''(x) = g(x)h'''(x) + 3g'(x)h''(x) + 3g''(x)h'(x) + g'''(x)h(x)$

$f^{(4)}(x) = g(x)h^{(4)}(x) + 4g'(x)h'''(x) + 6g''(x)h''(x) + 4g'''(x)h'(x) + g^{(4)}(x)h(x)$

(b) $f^{(n)}(x) = g(x)h^{(n)}(x) + \dfrac{n!}{1!(n-1)!}g'(x)h^{(n-1)}(x) + \dfrac{n!}{2!(n-2)!}g''(x)h^{(n-2)}(x) + \cdots + \dfrac{n!}{(n-1)!1!}g^{(n-1)}(x)h'(x) + g^{(n)}(x)h(x)$

123. $n = 1$: $f'(x) = x\cos x + \sin x$

$n = 2$: $f'(x) = x^2\cos x + 2x\sin x$

$n = 3$: $f'(x) = x^3\cos x + 3x^2\sin x$

$n = 4$: $f'(x) = x^4\cos x + 4x^3\sin x$

General rule: $f'(x) = x^n\cos x + nx^{(n-1)}\sin x$

125. $y' = -1/x^2$, $y'' = 2/x^3$,

$x^3y'' + 2x^2y' = x^3(2/x^3) + 2x^2(-1/x^2) = 2 - 2 = 0$

127. $y' = 2\cos x$, $y'' = -2\sin x$,

$y'' + y = -2\sin x + 2\sin x + 3 = 3$

129. False. $dy/dx = f(x)g'(x) + g(x)f'(x)$ **131.** True

133. True **135.** $f(x) = 3x^2 - 2x - 1$

137. $f'(x) = 2|x|$; $f''(0)$ does not exist.

Section 2.4 (page 137)

$y = f(g(x))$	$u = g(x)$	$y = f(u)$
1. $y = (6x-5)^4$	$u = 6x - 5$	$y = u^4$
3. $y = \sqrt{x^2-1}$	$u = x^2 - 1$	$y = \sqrt{u}$
5. $y = \csc^3 x$	$u = \csc x$	$y = u^3$

7. $6(2x-7)^2$ **9.** $-108(4-9x)^3$

11. $\frac{1}{2}(1-t)^{-1/2}(-1) = -1/\left(2\sqrt{1-t}\right)$

13. $\frac{1}{3}(9x^2+4)^{-2/3}(18x) = 6x/(9x^2+4)^{2/3}$

15. $\frac{1}{4}(4-x^2)^{-3/4}(-2x) = -x/\sqrt[4]{(4-x^2)^3}$ **17.** $-1/(x-2)^2$

19. $-2(t-3)^{-3}(1) = -2/(t-3)^3$ **21.** $-1/[2(x+2)^{3/2}]$

23. $x^2[4(x-2)^3(1)] + (x-2)^4(2x) = 2x(x-2)^3(3x-2)$

25. $x\left(\dfrac{1}{2}\right)(1-x^2)^{-1/2}(-2x) + (1-x^2)^{1/2}(1) = \dfrac{1-2x^2}{\sqrt{1-x^2}}$

27. $\dfrac{(x^2+1)^{1/2}(1) - x(1/2)(x^2+1)^{-1/2}(2x)}{x^2+1} = \dfrac{1}{(x^2+1)^{3/2}}$

29. $\dfrac{-2(x+5)(x^2+10x-2)}{(x^2+2)^3}$ **31.** $\dfrac{-9(2v-1)^2}{(v+1)^4}$

33. $(1 - 3x^2 - 4x^{3/2})/[2\sqrt{x}(x^2+1)^2]$

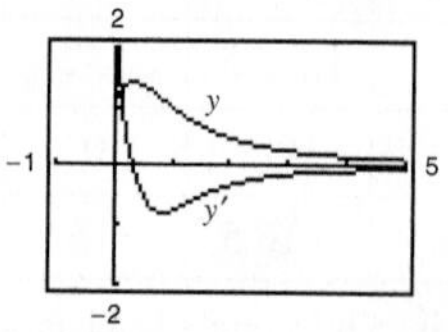

The zero of y' corresponds to the point on the graph of the function where the tangent line is horizontal.

35. $-\dfrac{\sqrt{\dfrac{x+1}{x}}}{2x(x+1)}$

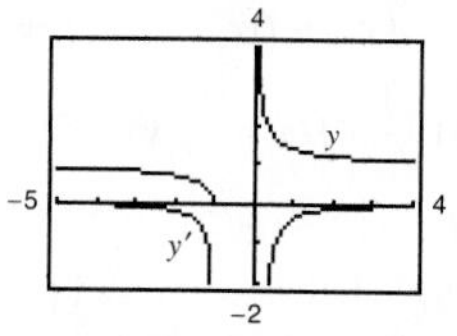

y' has no zeros.

37. $-[\pi x\sin(\pi x) + \cos(\pi x) + 1]/x^2$

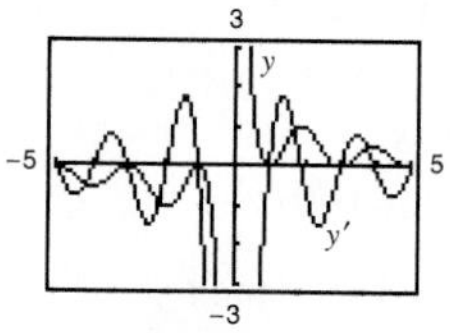

The zeros of y' correspond to the points on the graph of the function where the tangent lines are horizontal.

39. (a) 1 (b) 2; The slope of $\sin ax$ at the origin is a.

41. $-3\sin(3x)$ **43.** $12\sec^2(4x)$ **45.** $2\pi^2 x\cos(\pi x)^2$

47. $2\cos(4x)$ **49.** $(-1-\cos^2 x)/\sin^3 x$

51. $8\sec^2 x\tan x = \dfrac{8\sin x}{\cos^3 x}$ **53.** $\sin(2\theta)\cos(2\theta) = \dfrac{1}{2}\sin(4\theta)$

55. $\dfrac{6\pi\sin(\pi t-1)}{\cos^3(\pi t-1)}$ **57.** $\dfrac{1}{2\sqrt{x}} + 2x\cos(2x)^2$

59. $s'(t) = \dfrac{t+1}{\sqrt{t^2+2t+8}}, \dfrac{3}{4}$ **61.** $f'(x) = \dfrac{-9x^2}{(x^3-4)^2}, -\dfrac{9}{25}$

63. $f'(t) = \dfrac{-5}{(t-1)^2}, -5$ **65.** $y' = -6\sec^3(2x)\tan(2x), 0$

67. (a) $9x - 5y - 2 = 0$

(b)

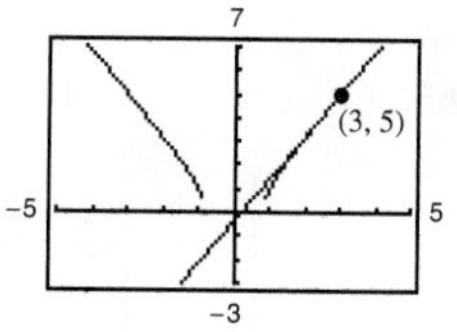

69. (a) $12x + y + 11 = 0$

(b)

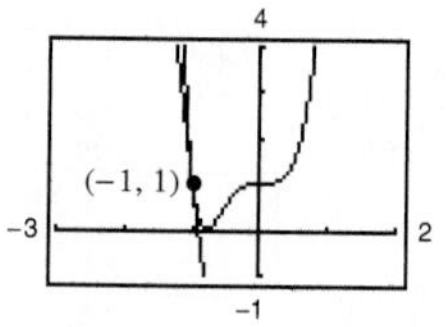

71. (a) $2x - y - 2\pi = 0$

(b)

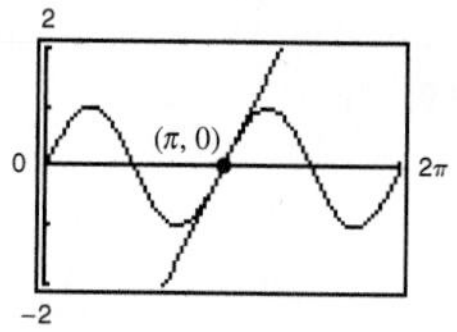

73. (a) $4x - y + (1 - \pi) = 0$

(b)

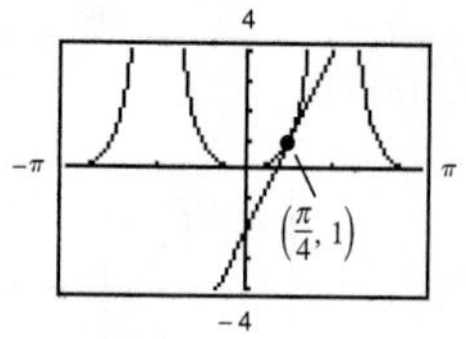

75. (a) $g'(1/2) = -3$

(b) $3x + y - 3 = 0$

(c)

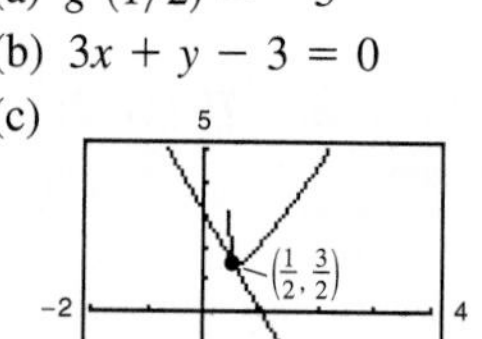

77. (a) $s'(0) = 0$

(b) $y = \frac{4}{3}$

(c)

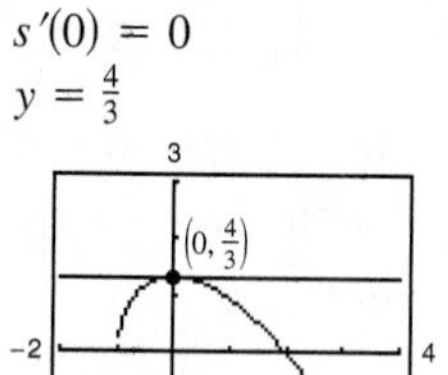

79. $3x + 4y - 25 = 0$

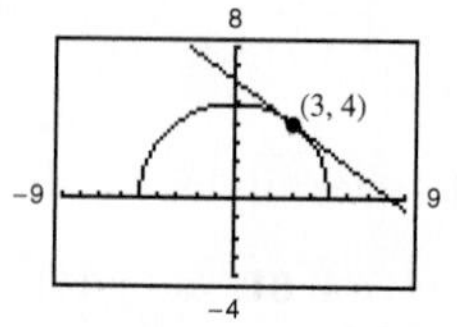

81. $\left(\dfrac{\pi}{6}, \dfrac{3\sqrt{3}}{2}\right), \left(\dfrac{5\pi}{6}, -\dfrac{3\sqrt{3}}{2}\right), \left(\dfrac{3\pi}{2}, 0\right)$ **83.** $12(5x^2-1)(x^2-1)$

85. $2(\cos x^2 - 2x^2\sin x^2)$ **87.** $h''(x) = 18x + 6, 24$

89. $f''(x) = -4x^2\cos(x^2) - 2\sin(x^2), 0$

91.

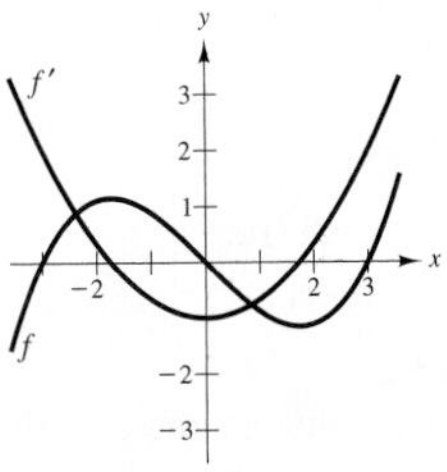

The zeros of f' correspond to the points where the graph of f has horizontal tangents.

93.

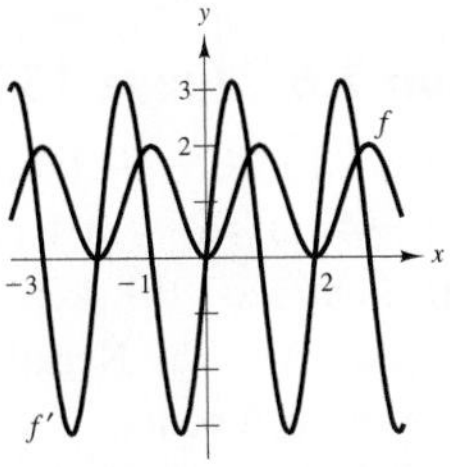

The zeros of f' correspond to the points where the graph of f has horizontal tangents.

95. The rate of change of g will be three times as fast as the rate of change of f.

97. (a) 24 (b) Not possible because $g'(h(5))$ is not known. (c) $\frac{4}{3}$ (d) 162

99. (a) $\frac{1}{2}$
(b) $s'(5)$ does not exist because g is not differentiable at 6.

101. (a) 1.461 (b) -1.016

103. 0.2 radian, 1.45 radians/sec **105.** 0.04224 cm/sec

107. (a) $x = -1.637t^3 + 19.31t^2 - 0.5t - 1$
(b) $\dfrac{dC}{dt} = -294.66t^2 + 2317.2t - 30$
(c) Because x, the number of units produced in t hours, is not a linear function, and therefore the cost with respect to time t is not linear.

109. $f(x + p) = f(x)$ for all x.
(a) Yes, $f'(x + p) = f'(x)$, which shows that f' is periodic as well.
(b) Yes, let $g(x) = f(2x)$, so $g'(x) = 2f'(2x)$. Because f' is periodic, so is g'.

111. (a) 0
(b) $f'(x) = 2\sec x \cdot \sec x\tan x = 2\sec^2 x\tan x$
$g'(x) = 2\tan x\sec^2 x = 2\sec^2 x\tan x$
$f'(x) = g'(x)$

113. Proof **115.** $f'(x) = 2x\left(\dfrac{x^2 - 4}{|x^2 - 4|}\right), x \neq \pm 2$

117. $f'(x) = \cos x\sin x/|\sin x|, x \neq k\pi$

119. (a) $P_1(x) = 4\sqrt{3}(x - \pi/6) + 2$
$P_2(x) = 1/2(56)(x - \pi/6)^2 + 4\sqrt{3}(x - \pi/6) + 2$
$= 28(x - \pi/6)^2 + 4\sqrt{3}(x - \pi/6) + 2$

(b)

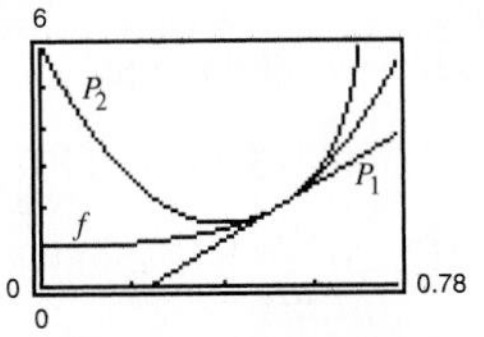

(c) P_2
(d) The accuracy worsens as you move away from $x = \pi/6$.

121. False. If $f(x) = \sin^2(2x)$, then $f'(x) = 2(\sin 2x)(2\cos 2x)$.

123. Putnam Problem A1, 1967

Section 2.5 (page 146)

1. $-x/y$ **3.** $-\sqrt{y/x}$ **5.** $(y - 3x^2)/(2y - x)$

7. $(1 - 3x^2y^3)/(3x^3y^2 - 1)$

9. $(6xy - 3x^2 - 2y^2)/(4xy - 3x^2)$

11. $\cos x/[4\sin(2y)]$ **13.** $(\cos x - \tan y - 1)/(x\sec^2 y)$

15. $[y\cos(xy)]/[1 - x\cos(xy)]$

17. (a) $y_1 = \sqrt{16 - x^2}$; $y_2 = -\sqrt{16 - x^2}$

(b)

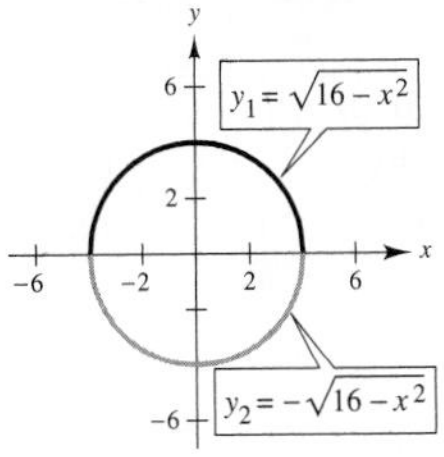

(c) $y' = \mp\dfrac{x}{\sqrt{16 - x^2}} = -\dfrac{x}{y}$
(d) $y' = -\dfrac{x}{y}$

19. (a) $y_1 = \frac{3}{4}\sqrt{16 - x^2}$; $y_2 = -\frac{3}{4}\sqrt{16 - x^2}$

(b)

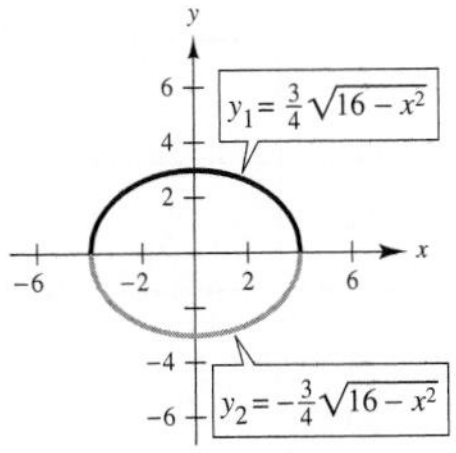

(c) $y' = \mp\dfrac{3x}{4\sqrt{16 - x^2}} = -\dfrac{9x}{16y}$
(d) $y' = -\dfrac{9x}{16y}$

21. $-\dfrac{y}{x}, -\dfrac{1}{4}$ **23.** $\dfrac{8x}{y(x^2 + 4)^2}$, Undefined **25.** $-\sqrt[3]{\dfrac{y}{x}}, -\dfrac{1}{2}$

27. $-\sin^2(x + y)$ or $-\dfrac{x^2}{x^2 + 1}, 0$ **29.** $-\dfrac{1}{2}$ **31.** 0

33. $y = -x + 4$ **35.** $y = -x + 2$

37. $y = \sqrt{3}x/6 + 4\sqrt{3}$ **39.** $y = -\frac{2}{11}x + \frac{30}{11}$

41. (a) $y = -2x + 4$ (b) Answers will vary.

43. $\cos^2 y, -\dfrac{\pi}{2} < y < \dfrac{\pi}{2}, \dfrac{1}{1 + x^2}$ **45.** $-36/y^3$

47. $-16/y^3$ **49.** $(3x)/(4y)$

51. $x + 3y - 12 = 0$

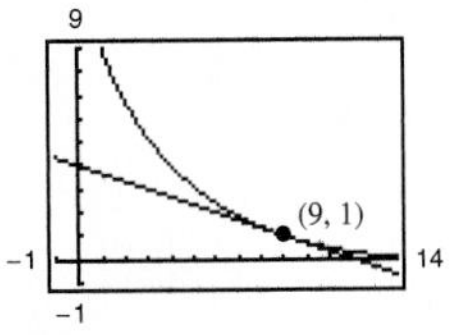

53. At (4, 3):
Tangent line: $4x + 3y - 25 = 0$
Normal line: $3x - 4y = 0$

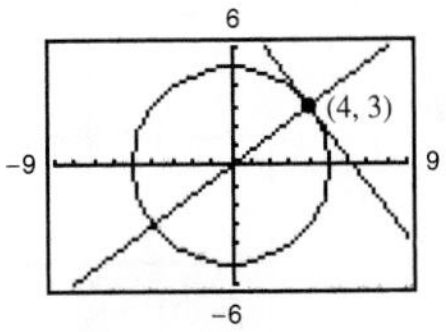

At $(-3, 4)$:
Tangent line: $3x - 4y + 25 = 0$
Normal line: $4x + 3y = 0$

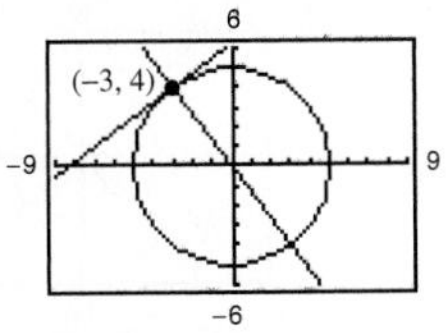

55. $x^2 + y^2 = r^2 \Rightarrow y' = -x/y \Rightarrow y/x =$ slope of normal line. Then for (x_0, y_0) on the circle, $x_0 \neq 0$, an equation of the normal line is $y = (y_0/x_0)x$, which passes through the origin. If $x_0 = 0$, the normal line is vertical and passes through the origin.

57. Horizontal tangents: $(-4, 0), (-4, 10)$
Vertical tangents: $(0, 5), (-8, 5)$

59.

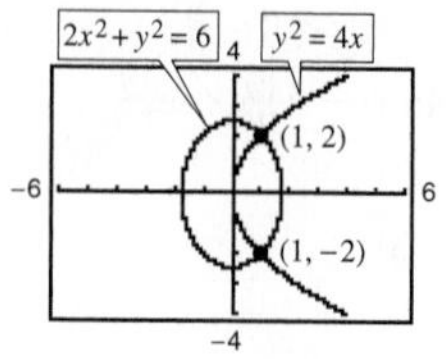

At $(1, 2)$:
Slope of ellipse: -1
Slope of parabola: 1
At $(1, -2)$:
Slope of ellipse: 1
Slope of parabola: -1

61.

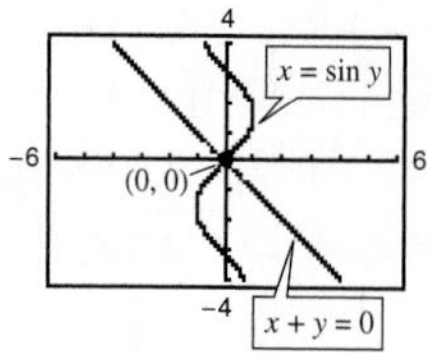

At $(0, 0)$:
Slope of line: -1
Slope of sine curve: 1

63. Derivatives: $\dfrac{dy}{dx} = -\dfrac{y}{x}, \dfrac{dy}{dx} = \dfrac{x}{y}$

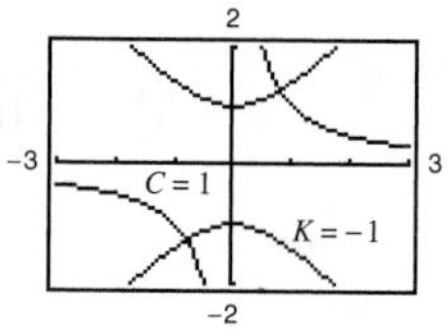

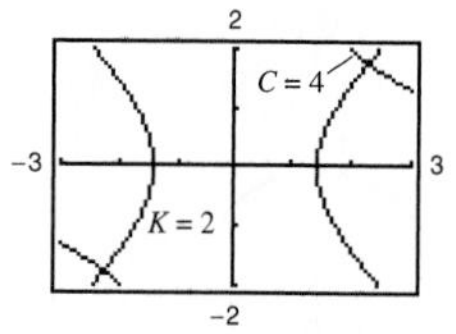

65. (a) $y\dfrac{dy}{dx} - 3x^3 = 0$ (b) $y\dfrac{dy}{dt} - 3x^3\dfrac{dx}{dt} = 0$

67. (a) $-\pi \sin \pi y\left(\dfrac{dy}{dx}\right) - 3\pi \cos \pi x = 0$

(b) $-\pi \sin \pi y\left(\dfrac{dy}{dt}\right) - 3\pi \cos \pi x\left(\dfrac{dx}{dt}\right) = 0$

69. Answers will vary. In the explicit form of a function, the variable is explicitly written as a function of x. In an implicit equation, the function is only implied by an equation. An example of an implicit function is $x^2 + xy = 5$. In explicit form it would be $y = (5 - x^2)/x$.

71.

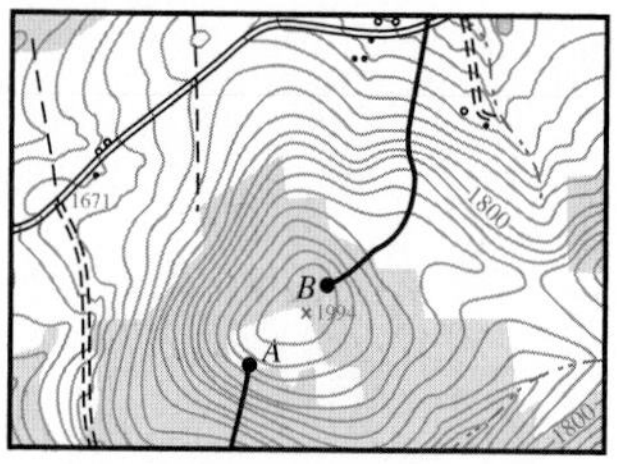

Use starting point B.

73. (a)

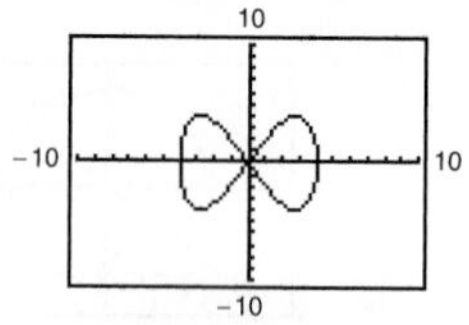

(b)

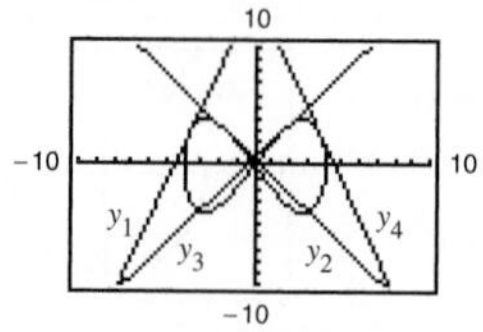

(c) $\left(\dfrac{8\sqrt{7}}{7}, 5\right)$

$y_1 = \frac{1}{3}\left[\left(\sqrt{7} + 7\right)x + \left(8\sqrt{7} + 23\right)\right]$
$y_2 = -\frac{1}{3}\left[\left(-\sqrt{7} + 7\right)x - \left(23 - 8\sqrt{7}\right)\right]$
$y_3 = -\frac{1}{3}\left[\left(\sqrt{7} - 7\right)x - \left(23 - 8\sqrt{7}\right)\right]$
$y_4 = -\frac{1}{3}\left[\left(\sqrt{7} + 7\right)x - \left(8\sqrt{7} + 23\right)\right]$

75. Proof **77.** $(0, \pm 1)$ **79.** (a) 1 (b) 1 (c) 3
$x_0 = 3/4$

Section 2.6 (page 154)

1. (a) $\frac{3}{4}$ (b) 20 **3.** (a) $-\frac{5}{8}$ (b) $\frac{3}{2}$
5. (a) -4 cm/sec (b) 0 cm/sec (c) 4 cm/sec
7. (a) 8 cm/sec (b) 4 cm/sec (c) 2 cm/sec
9. (a) Positive (b) Negative
11. In a linear function, if x changes at a constant rate, so does y. However, unless $a = 1$, y does not change at the same rate as x.
13. $\left[2(2x^3 + 3x)\right]/\sqrt{x^4 + 3x^2 + 1}$
15. (a) 36π cm²/min (b) 144π cm²/min
17. (a) Proof

(b) When $\theta = \dfrac{\pi}{6}, \dfrac{dA}{dt} = \dfrac{\sqrt{3}}{8}s^2$. When $\theta = \dfrac{\pi}{3}, \dfrac{dA}{dt} = \dfrac{1}{8}s^2$.

(c) If s and $d\theta/dt$ are constant, dA/dt is proportional to $\cos \theta$.
19. (a) $2/(9\pi)$ cm/min (b) $1/(18\pi)$ cm/min
21. (a) 36 cm²/sec (b) 360 cm²/sec **23.** $8/(405\pi)$ ft/min
25. (a) 12.5% (b) $\frac{1}{144}$ m/min
27. (a) $-\frac{7}{12}$ ft/sec; $-\frac{3}{2}$ ft/sec; $-\frac{48}{7}$ ft/sec
(b) $\frac{527}{24}$ ft²/sec (c) $\frac{1}{12}$ rad/sec
29. Rate of vertical change: $\frac{1}{5}$ m/sec
Rate of horizontal change: $-\sqrt{3}/15$ m/sec
31. (a) -750 mph (b) 20 min
33. $-28/\sqrt{10} \approx -8.85$ ft/sec **35.** (a) $\frac{25}{3}$ ft/sec (b) $\frac{10}{3}$ ft/sec
37. (a) 12 sec (b) $\frac{1}{2}\sqrt{3}$ m (c) $\left(\sqrt{5}\pi\right)/120$ m/sec
39. Evaporation rate proportional to $S \Rightarrow \dfrac{dV}{dt} = k(4\pi r^2)$

$V = \left(\dfrac{4}{3}\right)\pi r^3 \Rightarrow \dfrac{dV}{dt} = 4\pi r^2 \dfrac{dr}{dt}$. So $k = \dfrac{dr}{dt}$.

41. $-V\dfrac{dp}{dt} = 1.3p\dfrac{dV}{dt}$ **43.** $\dfrac{1}{20}$ rad/sec
45. (a) $\frac{1}{2}$ rad/min (b) $\frac{3}{2}$ rad/min
(c) 1.87 rad/min
47. (a) $\dfrac{dx}{dt} = -600\pi \sin \theta$

(b)

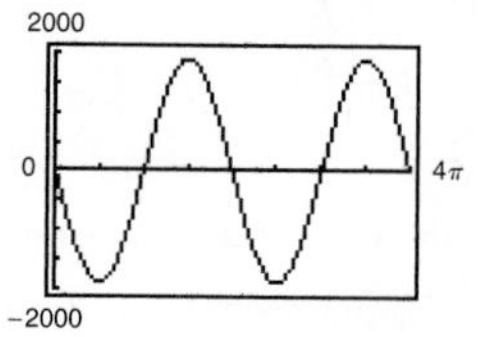

(c) $\theta = 90° + n \cdot 180°$;
$\theta = 0° + n \cdot 180°$
(d) -300π cm/sec;
$-300\sqrt{3}\pi$ cm/sec

49. $\dfrac{1}{25}\cos^2 \theta, -\dfrac{\pi}{4} \le \theta \le \dfrac{\pi}{4}$ **51.** -0.1808 ft/sec²
53. (a) $m(s) = 0.3754s^3 - 18.780s^2 + 313.23s - 1707.8$
(b) $(1.1262s^2 - 37.560s + 313.23)\dfrac{ds}{dt}$; 1.12 million

Review Exercises for Chapter 2 (page 158)

1. $f'(x) = 2x - 2$ **3.** $f'(x) = -2/(x - 1)^2$
5. f is differentiable at all $x \neq -1$.

7. 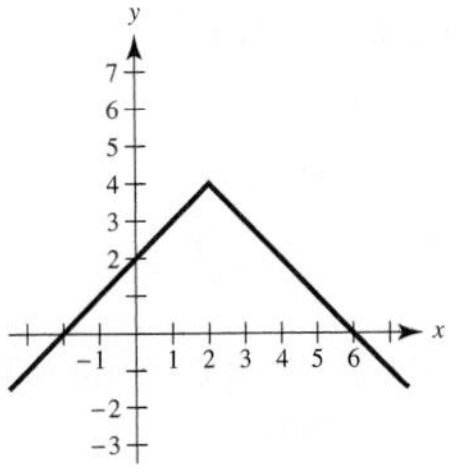

(a) Yes
(b) No, because the derivatives from the left and right are not equal.

9. $-\frac{3}{2}$

11. (a) $y = 3x + 1$
(b) 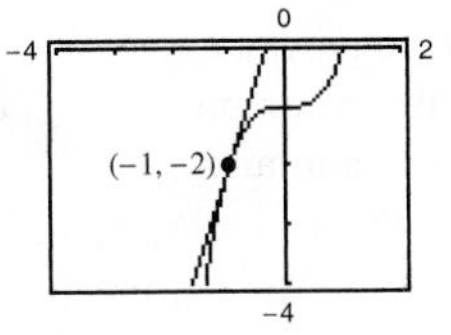

13. 8

15. 0 **17.** $8x^7$ **19.** $12t^3$ **21.** $3x(x - 2)$ **23.** $\dfrac{3}{\sqrt{x}} + \dfrac{1}{x^{2/3}}$

25. $-4/(3t^3)$ **27.** $2 - 3\cos\theta$ **29.** $-3\sin\theta - \cos\theta/4$

31. 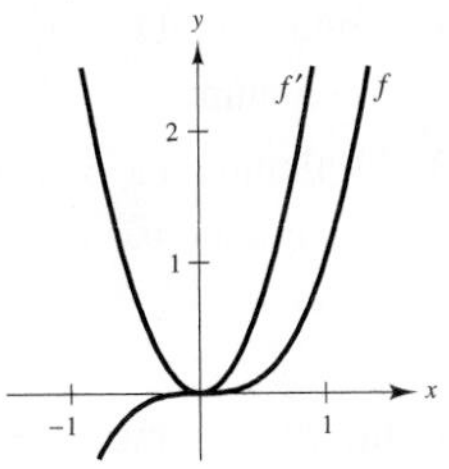

$f' > 0$ where the slopes of tangent lines to the graph of f are positive.

33. (a) 50 vibrations/sec/lb
(b) 33.33 vibrations/sec/lb

35. 414.74 m or 1354 ft

37. (a) 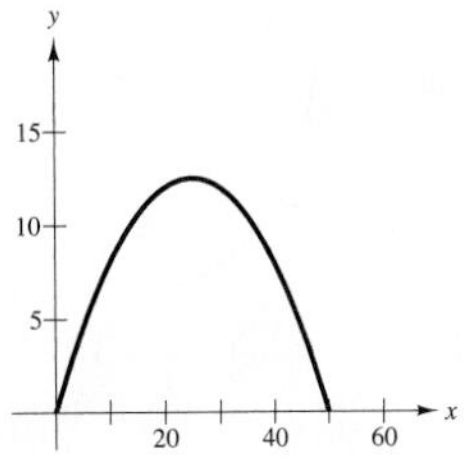

(b) 50
(c) $x = 25$
(d) $y' = 1 - 0.04x$

x	0	10	25	30	50
y'	1	0.6	0	−0.2	−1

(e) $y'(25) = 0$

39. (a) $x'(t) = 2t - 3$ (b) $(-\infty, 1.5)$ (c) $x = -\frac{1}{4}$ (d) 1

41. $2(6x^3 - 9x^2 + 16x - 7)$ **43.** $\sqrt{x}\cos x + \sin x/(2\sqrt{x})$

45. $-(x^2 + 1)/(x^2 - 1)^2$ **47.** $(6x)/(4 - 3x^2)^2$

49. $\dfrac{2x\cos x + x^2\sin x}{\cos^2 x}$ **51.** $3x^2\sec x\tan x + 6x\sec x$

53. $-x\sin x$ **55.** $y = 4x - 3$ **57.** $y = 0$

59. $v(4) = 20$ m/sec; $a(4) = -8$ m/sec^2

61. $6t$ **63.** $6\sec^2\theta\tan\theta$

65. $y'' + y = -(2\sin x + 3\cos x) + (2\sin x + 3\cos x) = 0$

67. $\dfrac{2(x - 3)(-x^2 + 6x + 1)}{(x^2 + 1)^3}$ **69.** $s(s^2 - 1)^{3/2}(8s^3 - 3s + 25)$

71. $-9\sin(3x + 1)$ **73.** $\frac{1}{2}(1 - \cos 2x) = \sin^2 x$

75. $\sin^{1/2}x\cos x - \sin^{5/2}x\cos x = \cos^3 x\sqrt{\sin x}$

77. $\dfrac{(x + 2)[\pi\cos(\pi x)] - \sin(\pi x)}{(x + 2)^2}$ **79.** -2 **81.** 0

83. $(x + 2)/(x + 1)^{3/2}$

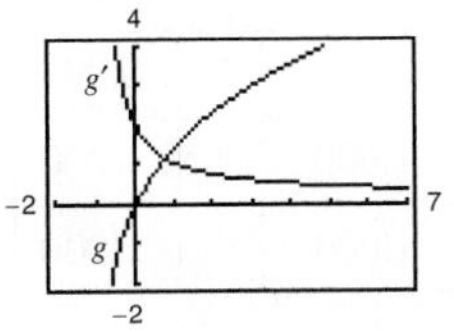

g' is not equal to zero for any x.

85. $5/[6(t + 1)^{1/6}]$

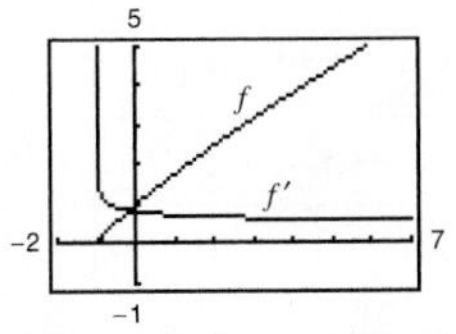

f' has no zeros.

87. (a) $f'(2) = 24$ (b) $g(t) = 24t - 44$
(c) 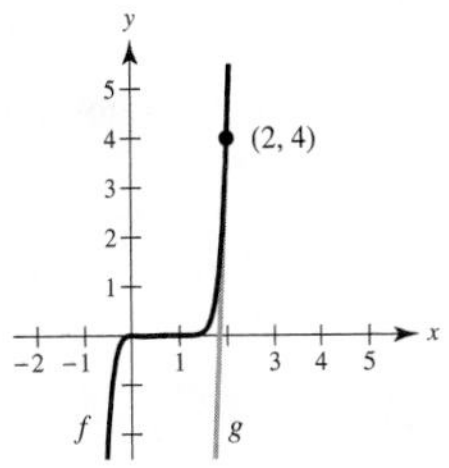

89. (a) $y'(-2) = -\dfrac{1}{2\sqrt{3}\cos^2\sqrt{3}} \approx -11.1983$
(b) $y_1 = -\dfrac{(x + 2)}{2\sqrt{3}\cos^2\sqrt{3}} + \tan\sqrt{3}$
(c) 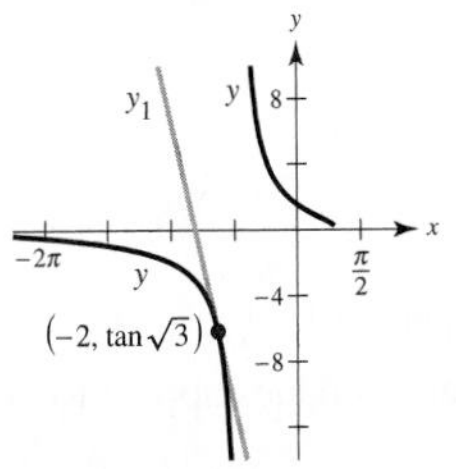

91. $4 - 4\sin 2x$ **93.** $2\csc^2 x\cot x$

95. $[2(t + 2)]/(1 - t)^4$ **97.** $18\sec^2(3\theta)\tan(3\theta) + \sin(\theta - 1)$

99. (a) −18.667 degrees per hour (b) −7.284 degrees per hour
(c) −3.240 degrees per hour (d) −0.747 degree per hour

101. $-\dfrac{2x + 3y}{3(x + y^2)}$ **103.** $\dfrac{2y\sqrt{x} - y\sqrt{y}}{2x\sqrt{y} - x\sqrt{x}}$ **105.** $\dfrac{y\sin x + \sin y}{\cos x - x\cos y}$

107. Tangent line: $x + 2y - 10 = 0$
Normal line: $2x - y = 0$

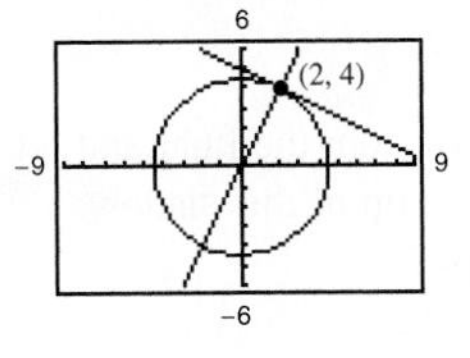

109. (a) $2\sqrt{2}$ units/sec (b) 4 units/sec (c) 8 units/sec

111. $\frac{2}{25}$ m/min **113.** −38.34 m/sec

P.S. Problem Solving (page 161)

1. (a) $r = \frac{1}{2}$; $x^2 + \left(y - \frac{1}{2}\right)^2 = \frac{1}{4}$
(b) Center: $\left(0, \frac{5}{4}\right)$; $x^2 + \left(y - \frac{5}{4}\right)^2 = 1$

3. (a) $P_1(x) = 1$ (b) $P_2(x) = 1 - \frac{1}{2}x^2$

(c)

x	-1.0	-0.1	-0.001	0	0.001
$\cos x$	0.5403	0.9950	1.000	1	1.000
$P_2(x)$	0.5	0.995	1.000	1	1.000

x	0.1	1.0
$\cos x$	0.9950	0.5403
$P_2(x)$	0.995	0.5

$P_2(x)$ is a good approximation of $f(x) = \cos x$ when x is very close to 0.

(d) $P_3(x) = x - \frac{1}{6}x^3$

5. $p(x) = 2x^3 + 4x^2 - 5$

7. (a) Graph $\begin{cases} y_1 = \dfrac{1}{a}\sqrt{x^2(a^2 - x^2)} \\ y_2 = -\dfrac{1}{a}\sqrt{x^2(a^2 - x^2)} \end{cases}$ as separate equations.

(b) Answers will vary. Sample answer:

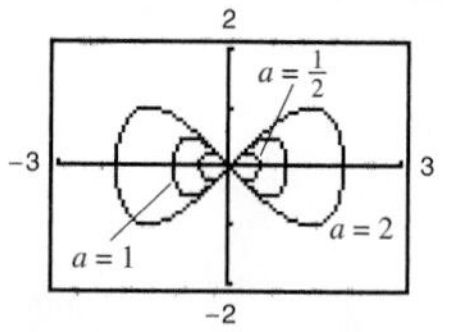

The intercepts will always be $(0, 0)$, $(a, 0)$, and $(-a, 0)$, and the maximum and minimum y-values appear to be $\pm\frac{1}{2}a$.

(c) $\left(\frac{a\sqrt{2}}{2}, \frac{a}{2}\right), \left(\frac{a\sqrt{2}}{2}, -\frac{a}{2}\right), \left(-\frac{a\sqrt{2}}{2}, \frac{a}{2}\right), \left(-\frac{a\sqrt{2}}{2}, -\frac{a}{2}\right)$

9. (a) When the man is 90 ft from the light, the tip of his shadow is $112\frac{1}{2}$ ft from the light. The tip of the child's shadow is $111\frac{1}{9}$ ft from the light, so the man's shadow extends $1\frac{7}{18}$ ft beyond the child's shadow.

(b) When the man is 60 ft from the light, the tip of his shadow is 75 ft from the light. The tip of the child's shadow is $77\frac{7}{9}$ ft from the light, so the child's shadow extends $2\frac{7}{9}$ ft beyond the man's shadow.

(c) $d = 80$ ft

(d) Let x be the distance of the man from the light and let s be the distance from the light to the tip of the shadow.
If $0 < x < 80$, $ds/dt = -50/9$.
If $x > 80$, $ds/dt = -25/4$
There is a discontinuity at $x = 80$.

11. Proof. The graph of L is a line passing through the origin $(0, 0)$.

13. (a)

$z°$	0.1	0.01	0.0001
$\dfrac{\sin z}{z}$	0.0174532837	0.0174532924	0.0174532925

(b) $\pi/180$ (c) $(\pi/180)\cos z$

(d) $S(90) = 1$, $C(180) = -1$; $(\pi/180)C(z)$

(e) Answers will vary.

15. (a) j would be the rate of change of acceleration.

(b) $j = 0$. Acceleration is constant, so there is no change in acceleration.

(c) a: position function, d: velocity function, b: acceleration function, c: jerk function

Chapter 3

Section 3.1 (page 169)

1. A: neither, B: absolute maximum (and relative maximum), C: neither, D: neither, E: relative maximum, F: relative minimum, G: neither

3. $f'(0) = 0$ **5.** $f'(3) = 0$ **7.** $f'(-2)$ is undefined.

9. 2, absolute maximum (and relative maximum)

11. 1, absolute maximum (and relative maximum); 2, absolute minimum (and relative maximum); 3, absolute maximum (and relative minimum)

13. $x = 0, x = 2$ **15.** $t = 8/3$ **17.** $x = \pi/3, \pi, 5\pi/3$

19. Minimum: $(2, 2)$
Maximum: $(-1, 8)$

21. Minima: $(0, 0)$ and $(3, 0)$
Maximum: $\left(\frac{3}{2}, \frac{9}{4}\right)$

23. Minimum: $\left(-1, -\frac{5}{2}\right)$
Maximum: $(2, 2)$

25. Minimum: $(0, 0)$
Maximum: $(-1, 5)$

27. Minimum: $(0, 0)$
Maxima: $\left(-1, \frac{1}{4}\right)$ and $\left(1, \frac{1}{4}\right)$

29. Minimum: $(1, -1)$
Maximum: $\left(0, -\frac{1}{2}\right)$

31. Minimum: $(-1, -1)$
Maximum: $(3, 3)$

33. Minimum: $\left(1/6, \sqrt{3}/2\right)$
Maximum: $(0, 1)$

35. Minimum: $(2, 3)$
Maximum: $\left(1, \sqrt{2} + 3\right)$

37. (a) Minimum: $(0, -3)$; Maximum: $(2, 1)$
(b) Minimum: $(0, -3)$
(c) Maximum: $(2, 1)$
(d) No extrema

39. (a) Minimum: $(1, -1)$; Maximum: $(-1, 3)$
(b) Maximum: $(3, 3)$
(c) Minimum: $(1, -1)$
(d) Minimum: $(1, -1)$

41.

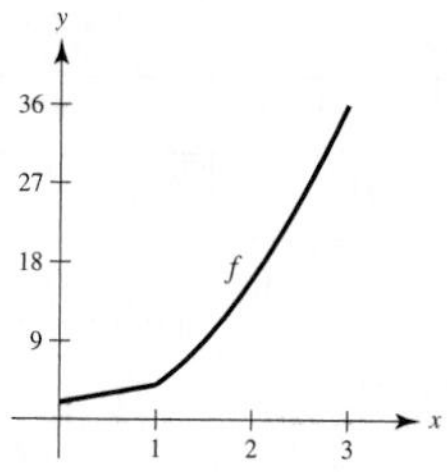

Minimum: $(0, 2)$
Maximum: $(3, 36)$

43.

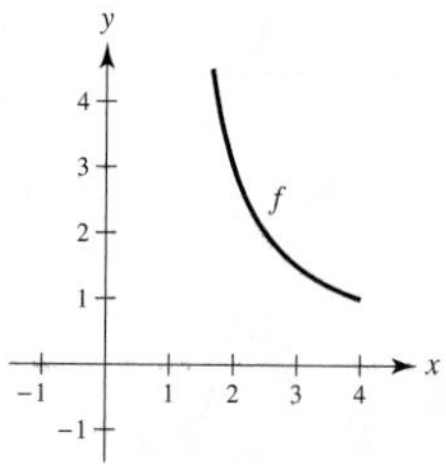

Minimum: $(4, 1)$

45.

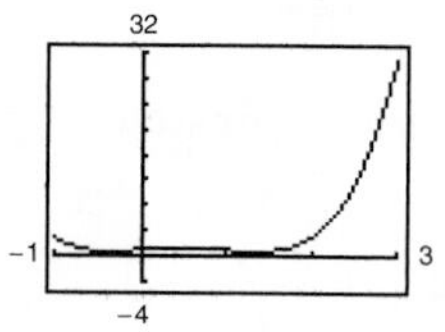

Minima: $\left(\frac{-\sqrt{3} + 1}{2}, \frac{3}{4}\right)$ and $\left(\frac{\sqrt{3} + 1}{2}, \frac{3}{4}\right)$
Maximum: $(3, 31)$

47. (a)

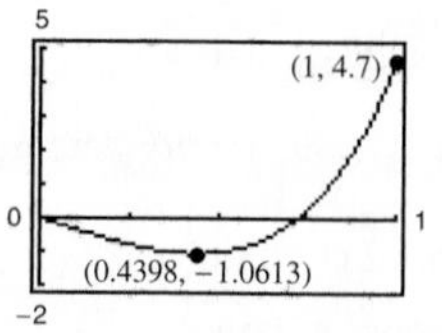

(b) Minimum: $(0.4398, -1.0613)$

49. Maximum: $\left|f''\left(\sqrt[3]{-10 + \sqrt{108}}\right)\right| = f''\left(\sqrt{3} - 1\right) \approx 1.47$

51. Maximum: $\left|f^{(4)}(0)\right| = \frac{56}{81}$

53. Answers will vary. Example:

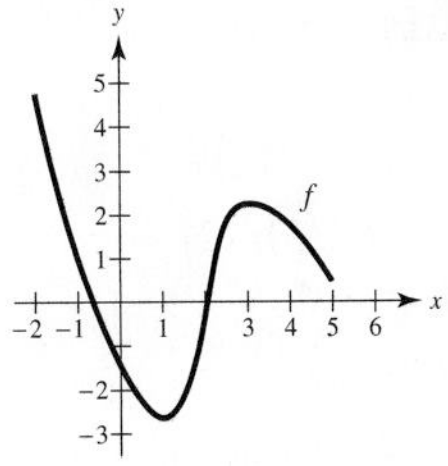

55. (a) Yes (b) No **57.** (a) No (b) Yes

59. Maximum: $P(12) = 72$ No. P is decreasing for $I \geq 12$.

61. $\dfrac{dx}{dt} = \dfrac{v^2 \cos 2\theta}{16} \dfrac{d\theta}{dt}$

In the interval $[\pi/4, 3\pi/4]$, $\theta = \pi/4, 3\pi/4$ indicate minima for dx/dt and $\theta = \pi/2$ indicates a maximum for dx/dt. This implies that the sprinkler waters longest when $\theta = \pi/4$ and $3\pi/4$. So, the lawn farthest from the sprinkler gets the most water.

63. True **65.** True **67.** Proof

69. (a) $y = (3/40{,}000)x^2 - (3/200)x + 75/4$

(b)

x	-500	-400	-300	-200	-100	0
d	0	0.75	3	6.75	12	18.75

x	100	200	300	400	500
d	12	6.75	3	0.75	0

(c) Lowest point $\approx (100, 18)$; No

Section 3.2 (page 176)

1. $f(0) = f(2) = 0$; f is not differentiable on $(0, 2)$.

3. $f(-1) = f(1) = 1$; f is not continuous on $(-1, 1)$.

5. $(2, 0), (-1, 0); f'(\frac{1}{2}) = 0$ **7.** $(0, 0), (-4, 0); f'(-\frac{8}{3}) = 0$

9. $f'(-\frac{3}{2}) = 0$ **11.** $f'(1) = 0$

13. $f'\left(\dfrac{6 - \sqrt{3}}{3}\right) = 0;\ f'\left(\dfrac{6 + \sqrt{3}}{3}\right) = 0$

15. Not differentiable at $x = 0$ **17.** $f'(-2 + \sqrt{5}) = 0$

19. $f'(\pi/2) = 0;\ f'(3\pi/2) = 0$ **21.** $f'(0.249) \approx 0$

23. Not continuous on $[0, \pi]$

25.

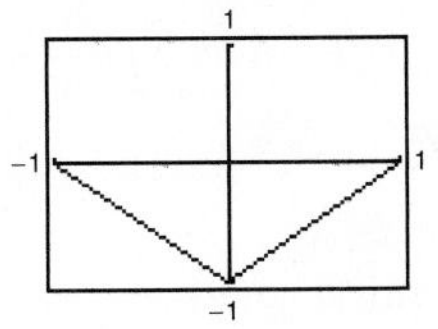

Rolle's Theorem does not apply.

27.

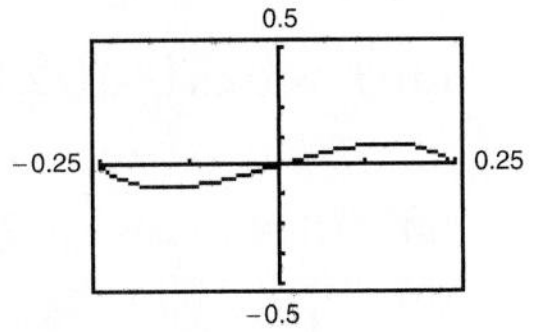

$f'(\pm 0.1533) = 0$

29. (a) $f(1) = f(2) = 64$

(b) Velocity $= 0$ for some t in $(1, 2)$; $t = \frac{3}{2}$ sec

31.

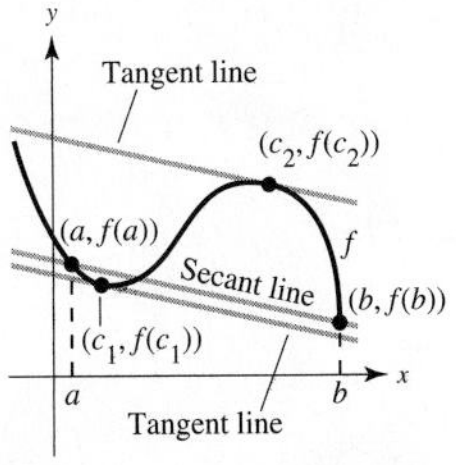

33. The function is not continuous on $[0, 6]$.

35. The function is not continuous on $[0, 6]$.

37. (a) Secant line: $x - y + 3 = 0$ (b) $c = \frac{1}{2}$

(c) Tangent line: $4x - 4y + 3 = 0$

(d)

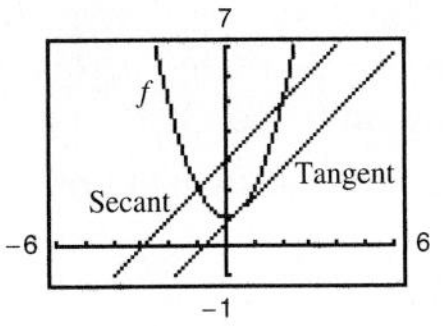

39. $f'(-1/2) = -1$ **41.** $f'(8/27) = 1$

43. $f'(-1/4) = -1/3$ **45.** $f'(\pi/2) = 0$

47. (a)–(c)

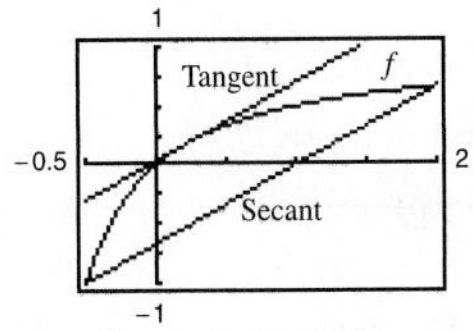

(b) $y = \frac{2}{3}(x - 1)$

(c) $y = \frac{1}{3}(2x + 5 - 2\sqrt{6})$

49. (a)–(c)

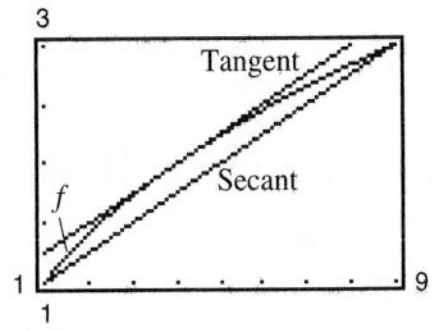

(b) $y = \frac{1}{4}x + \frac{3}{4}$

(c) $y = \frac{1}{4}x + 1$

51. (a) -14.7 m/sec (b) 1.5 sec

53. No. Let $f(x) = x^2$ on $[-1, 2]$.

55. No. $f(x)$ is not continuous on $[0, 1]$. So it does not satisfy the hypothesis of Rolle's Theorem.

57. By the Mean Value Theorem, there is a time when the speed of the plane must equal the average speed of 454.5 mph. The speed was 400 mph when the plane was accelerating to 454.5 mph and decelerating from 454.5 mph.

59. Proof

61. (a) f is continuous and changes signs in $[-10, 4]$ (Intermediate Value Theorem).

(b) There exist real numbers a and b such that $-10 < a < b < 4$ and $f(a) = f(b) = 2$. Therefore, f' has a zero in the interval by Rolle's Theorem.

(c)

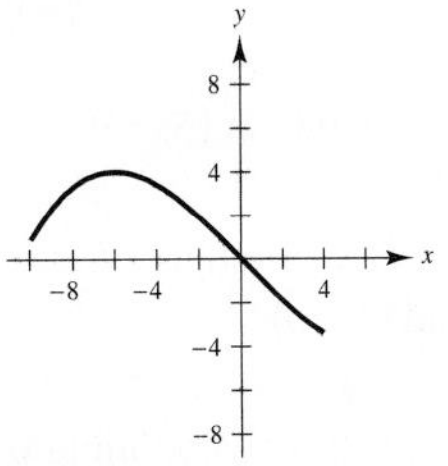

(d)

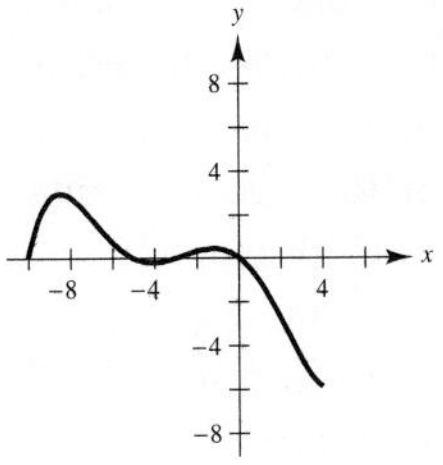

(e) No, by Theorem 2.1.

63.

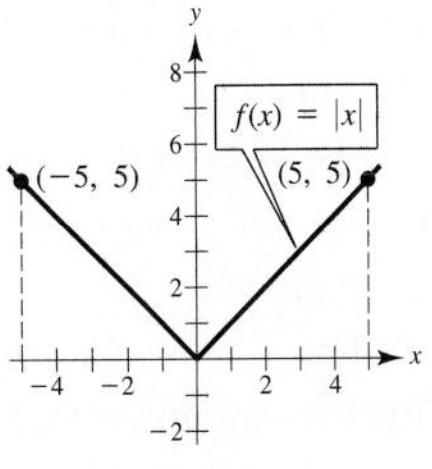

65. Proof

67. $a = 6, b = 1, c = 2$ **69.** $f(x) = 5$ **71.** $f(x) = x^2 - 1$
73. False. f is not continuous on $[-1, 1]$. **75.** True
77–85. Proofs

Section 3.3 (page 186)

1. (a) $(0, 6)$ (b) $(6, 8)$
3. Increasing on $(3, \infty)$; Decreasing on $(-\infty, 3)$
5. Increasing on $(-\infty, -2)$ and $(2, \infty)$; Decreasing on $(-2, 2)$
7. Increasing on $(0, \pi/2)$ and $(3\pi/2, 2\pi)$;
Decreasing on $(\pi/2, 3\pi/2)$
9. Increasing on $(-\infty, 0)$; Decreasing on $(0, \infty)$
11. Increasing on $(1, \infty)$; Decreasing on $(-\infty, 1)$
13. Increasing on $(-2\sqrt{2}, 2\sqrt{2})$
Decreasing on $(-4, -2\sqrt{2}), (2\sqrt{2}, 4)$
15. Increasing on $(0, 7\pi/6)$ and $(11\pi/6, 2\pi)$;
Decreasing on $(7\pi/6, 11\pi/6)$
17. (a) Critical number: $x = 3$
(b) Increasing on $(3, \infty)$; Decreasing on $(-\infty, 3)$
(c) Relative minimum: $(3, -9)$
19. (a) Critical number: $x = 1$
(b) Increasing on $(-\infty, 1)$; Decreasing on $(1, \infty)$
(c) Relative maximum: $(1, 5)$
21. (a) Critical numbers: $x = -2, 1$
(b) Increasing on $(-\infty, -2)$ and $(1, \infty)$; Decreasing on $(-2, 1)$
(c) Relative maximum: $(-2, 20)$; Relative minimum: $(1, -7)$
23. (a) Critical numbers: $x = 0, 2$
(b) Increasing on $(0, 2)$; Decreasing on $(-\infty, 0), (2, \infty)$
(c) Relative maximum: $(2, 4)$; Relative minimum: $(0, 0)$
25. (a) Critical numbers: $x = \pm 1$
(b) Increasing on $(-\infty, -1)$ and $(1, \infty)$; Decreasing on $(-1, 1)$
(c) Relative maximum: $\left(-1, \frac{4}{5}\right)$; Relative minimum: $\left(1, -\frac{4}{5}\right)$
27. (a) Critical number: $x = 0$
(b) Increasing on $(-\infty, \infty)$
(c) No relative extrema
29. (a) Critical number: $x = 1$
(b) Increasing on $(1, \infty)$; Decreasing on $(-\infty, 1)$
(c) Relative minimum: $(1, 0)$
31. (a) Critical number: $x = 5$
(b) Increasing on $(-\infty, 5)$; Decreasing on $(5, \infty)$
(c) Relative maximum: $(5, 5)$
33. (a) Critical numbers: $x = \pm 1$; Discontinuity: $x = 0$
(b) Increasing on $(-\infty, -1)$ and $(1, \infty)$
Decreasing on $(-1, 0)$ and $(0, 1)$
(c) Relative maximum: $(-1, -2)$; Relative minimum: $(1, 2)$
35. (a) Critical number: $x = 0$; Discontinuities: $x = \pm 3$
(b) Increasing on $(-\infty, -3)$ and $(-3, 0)$
Decreasing on $(0, 3)$ and $(3, \infty)$
(c) Relative maximum: $(0, 0)$
37. (a) Critical numbers: $x = -3, 1$; Discontinuity: $x = -1$
(b) Increasing on $(-\infty, -3)$ and $(1, \infty)$
Decreasing on $(-3, -1)$ and $(-1, 1)$
(c) Relative maximum: $(-3, -8)$; Relative minimum: $(1, 0)$
39. (a) Critical numbers: $x = \pi/6, 5\pi/6$
Increasing on $(0, \pi/6), (5\pi/6, 2\pi)$
Decreasing on $(\pi/6, 5\pi/6)$
(b) Relative maximum: $\left(\pi/6, (\pi + 6\sqrt{3})/12\right)$
Relative minimum: $\left(5\pi/6, (5\pi - 6\sqrt{3})/12\right)$
41. (a) Critical numbers: $x = \pi/4, 5\pi/4$
Increasing on $(0, \pi/4), (5\pi/4, 2\pi)$
Decreasing on $(\pi/4, 5\pi/4)$
(b) Relative maximum: $(\pi/4, \sqrt{2})$
Relative minimum: $(5\pi/4, -\sqrt{2})$
43. (a) Critical numbers:
$x = \pi/4, \pi/2, 3\pi/4, \pi, 5\pi/4, 3\pi/2, 7\pi/4$
Increasing on $(\pi/4, \pi/2), (3\pi/4, \pi), (5\pi/4, 3\pi/2), (7\pi/4, 2\pi)$
Decreasing on $(0, \pi/4), (\pi/2, 3\pi/4), (\pi, 5\pi/4), (3\pi/2, 7\pi/4)$
(b) Relative maxima: $(\pi/2, 1), (\pi, 1), (3\pi/2, 1)$
Relative minima: $(\pi/4, 0), (3\pi/4, 0), (5\pi/4, 0), (7\pi/4, 0)$
45. (a) Critical numbers: $\pi/2, 7\pi/6, 3\pi/2, 11\pi/6$
Increasing on $\left(0, \dfrac{\pi}{2}\right), \left(\dfrac{7\pi}{6}, \dfrac{3\pi}{2}\right), \left(\dfrac{11\pi}{6}, 2\pi\right)$
Decreasing on $\left(\dfrac{\pi}{2}, \dfrac{7\pi}{6}\right), \left(\dfrac{3\pi}{2}, \dfrac{11\pi}{6}\right)$
(b) Relative maxima: $\left(\dfrac{\pi}{2}, 2\right), \left(\dfrac{3\pi}{2}, 0\right)$
Relative minima: $\left(\dfrac{7\pi}{6}, -\dfrac{1}{4}\right), \left(\dfrac{11\pi}{6}, -\dfrac{1}{4}\right)$
47. (a) $f'(x) = 2(9 - 2x^2)/\sqrt{9 - x^2}$
(b)

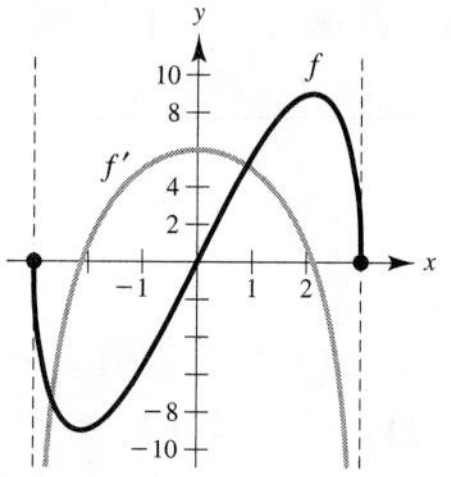

(c) Critical numbers: $x = \pm 3\sqrt{2}/2$
(d) $f' > 0$ on $(-3\sqrt{2}/2, 3\sqrt{2}/2)$
$f' < 0$ on $(-3, -3\sqrt{2}/2), (3\sqrt{2}/2, 3)$
49. (a) $f'(t) = t(t \cos t + 2 \sin t)$
(b)

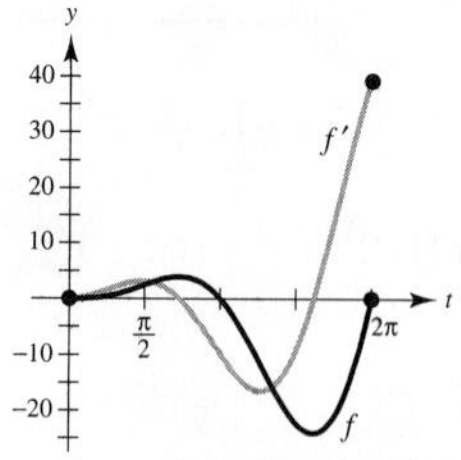

(c) Critical numbers: $t = 2.2889, 5.0870$
(d) $f' > 0$ on $(0, 2.2889), (5.0870, 2\pi)$
$f' < 0$ on $(2.2889, 5.0870)$

51. (a) $f'(x) = -\cos(x/3)$

(b)

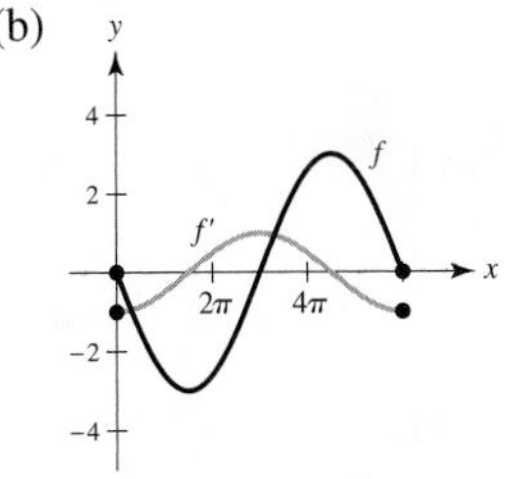

(c) Critical numbers: $x = 3\pi/2, 9\pi/2$

(d) $f' > 0$ on $\left(\frac{3\pi}{2}, \frac{9\pi}{2}\right)$

$f' < 0$ on $\left(0, \frac{3\pi}{2}\right), \left(\frac{9\pi}{2}, 6\pi\right)$

53. $f(x)$ is symmetric with respect to the origin.
Zeros: $(0, 0), (\pm\sqrt{3}, 0)$

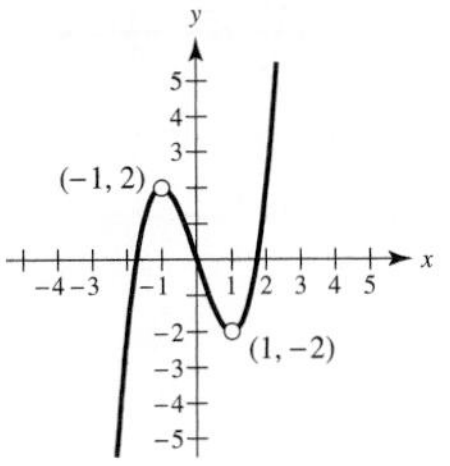

$g(x)$ is continuous on $(-\infty, \infty)$ and $f(x)$ has holes at $x = 1$ and $x = -1$.

55.

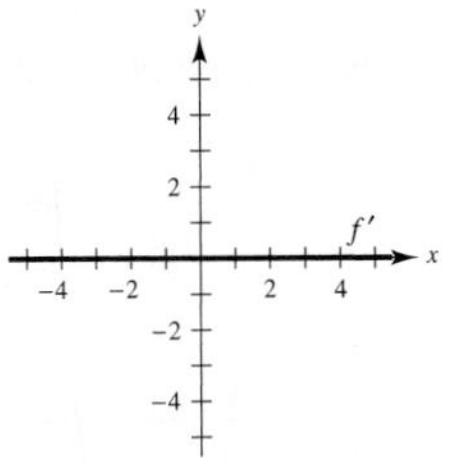

57.

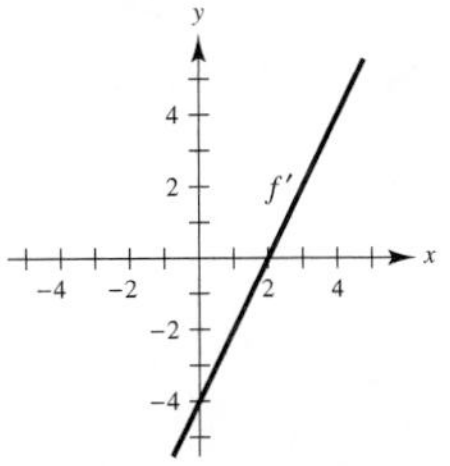

59.

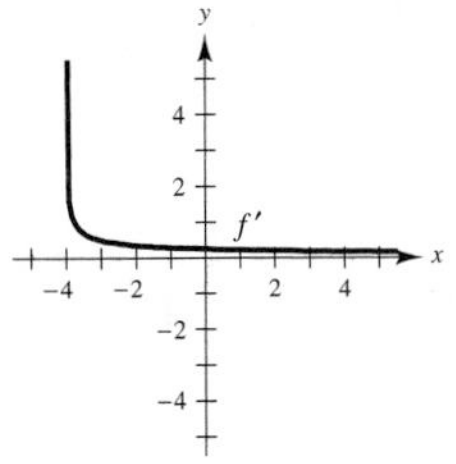

61. (a) Increasing on $(2, \infty)$; Decreasing on $(-\infty, 2)$
(b) Relative minimum: $x = 2$

63. (a) Increasing on $(-\infty, 0)$ and $(1, \infty)$; Decreasing on $(0, 1)$
(b) Relative maximum: $x = 0$; Relative minimum: $x = 1$

65. $g'(0) < 0$ **67.** $g'(-6) < 0$ **69.** $g'(0) > 0$

71.

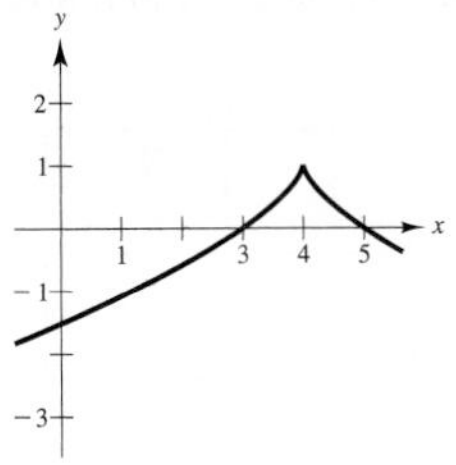

73.

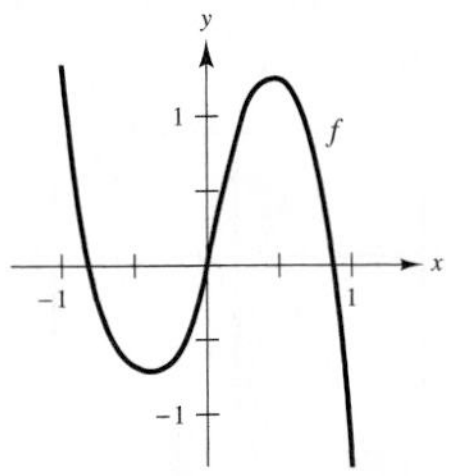

Relative minimum at the approximate critical number $x = -0.40$; Relative maximum at the approximate critical number $x = 0.48$

75. (a) $s'(t) = 9.8(\sin\theta)t$; speed $= |9.8(\sin\theta)t|$

(b)

θ	0	$\pi/4$	$\pi/3$	$\pi/2$	$2\pi/3$	$3\pi/4$	π
$s'(t)$	0	$4.9\sqrt{2}t$	$4.9\sqrt{3}t$	$9.8t$	$4.9\sqrt{3}t$	$4.9\sqrt{2}t$	0

The speed is maximum at $\theta = \pi/2$.

77. (a)

x	0.5	1	1.5	2	2.5	3
$f(x)$	0.5	1	1.5	2	2.5	3
$g(x)$	0.48	0.84	1.00	0.91	0.60	0.14

$f(x) > g(x)$

(b)

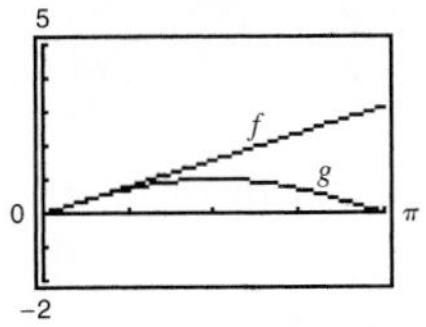

$f(x) > g(x)$

(c) Proof

79. $r = 2R/3$ **81.** Maximum when $R_2 = R_1$

83. (a) $M = 5.267t^2 - 71.19t + 356.9$

(b)

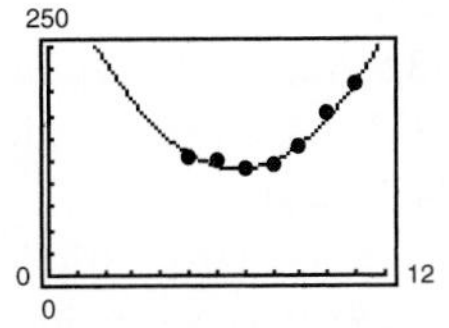

(c) $(6.8, 116.3)$

85. (a) $v(t) = 6 - 2t$ (b) $[0, 3)$ (c) $(3, \infty)$ (d) $t = 3$

87. (a) $v(t) = 3t^2 - 10t + 4$
(b) $[0, (5 - \sqrt{13})/3)$ and $(5 + \sqrt{13}/3, \infty)$
(c) $\left(\frac{5 - \sqrt{13}}{3}, \frac{5 + \sqrt{13}}{3}\right)$ (d) $t = \frac{5 \pm \sqrt{13}}{3}$

89. Answers will vary.

91. (a) 3
(b) $a_3(0)^3 + a_2(0)^2 + a_1(0) + a_0 = 0$
$a_3(2)^3 + a_2(2)^2 + a_1(2) + a_0 = 2$
$3a_3(0)^2 + 2a_2(0) + a_1 = 0$
$3a_3(2)^2 + 2a_2(2) + a_1 = 0$
(c) $f(x) = -\frac{1}{2}x^3 + \frac{3}{2}x^2$

93. (a) 4
(b) $a_4(0)^4 + a_3(0)^3 + a_2(0)^2 + a_1(0) + a_0 = 0$
$a_4(2)^4 + a_3(2)^3 + a_2(2)^2 + a_1(2) + a_0 = 4$
$a_4(4)^4 + a_3(4)^3 + a_2(4)^2 + a_1(4) + a_0 = 0$
$4a_4(0)^3 + 3a_3(0)^2 + 2a_2(0) + a_1 = 0$
$4a_4(2)^3 + 3a_3(2)^2 + 2a_2(2) + a_1 = 0$
(c) $f(x) = \frac{1}{4}x^4 - 2x^3 + 4x^2$

95. True **97.** False. Let $f(x) = x^3$.

99. False. Let $f(x) = x^3$. There is a critical number at $x = 0$, but not a relative extremum.

101–105. Proofs

Section 3.4 (page 195)

1. Concave upward: $(-\infty, \infty)$

3. Concave upward: $(-\infty, -2), (2, \infty)$
Concave downward: $(-2, 2)$

5. Concave upward: $(-\infty, -1), (1, \infty)$
Concave downward: $(-1, 1)$

7. Concave upward: $(-\infty, 1)$; Concave downward: $(1, \infty)$

9. Concave upward: $(-\pi/2, 0)$; Concave downward: $(0, \pi/2)$

11. Point of inflection: $(2, 8)$; Concave downward: $(-\infty, 2)$; Concave upward: $(2, \infty)$

13. Points of inflection: $(\pm 2\sqrt{3}/3, -20/9)$
Concave upward: $(-\infty, -2\sqrt{3}/3), (2\sqrt{3}/3, \infty)$
Concave downward: $(-2\sqrt{3}/3, 2\sqrt{3}/3)$

15. Points of inflection: $(2, -16), (4, 0)$;
Concave upward: $(-\infty, 2), (4, \infty)$; Concave downward: $(2, 4)$

17. Concave upward: $(-3, \infty)$

19. Points of inflection: $(-\sqrt{3}, -\sqrt{3}/4), (0, 0), (\sqrt{3}, \sqrt{3}/4)$
Concave upward: $(-\sqrt{3}, 0), (\sqrt{3}, \infty)$
Concave downward: $(-\infty, -\sqrt{3}), (0, \sqrt{3})$

21. Point of inflection: $(2\pi, 0)$
Concave upward: $(2\pi, 4\pi)$; Concave downward: $(0, 2\pi)$

23. Concave upward: $(0, \pi), (2\pi, 3\pi)$
Concave downward: $(\pi, 2\pi), (3\pi, 4\pi)$

25. Points of inflection: $(\pi, 0), (1.823, 1.452), (4.46, -1.452)$
Concave upward: $(1.823, \pi), (4.46, 2\pi)$
Concave downward: $(0, 1.823), (\pi, 4.46)$

27. Relative minimum: $(3, -25)$ **29.** Relative minimum: $(5, 0)$

31. Relative maximum: $(0, 3)$; Relative minimum: $(2, -1)$

33. Relative maximum: $(2.4, 268.74)$; Relative minimum: $(0, 0)$

35. Relative minimum: $(0, -3)$

37. Relative maximum: $(-2, -4)$; Relative minimum: $(2, 4)$

39. No relative extrema, because f is nonincreasing.

41. (a) $f'(x) = 0.2x(x-3)^2(5x-6)$
$f''(x) = 0.4(x-3)(10x^2 - 24x + 9)$
(b) Relative maximum: $(0, 0)$
Relative minimum: $(1.2, -1.6796)$
Points of inflection: $(0.4652, -0.7048)$, $(1.9348, -0.9048)$, $(3, 0)$
(c)

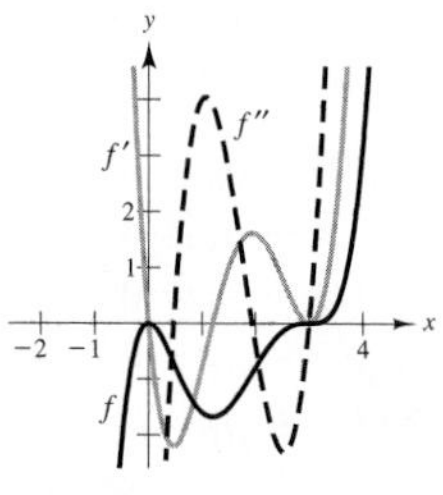

f is increasing when f' is positive, and decreasing when f' is negative. f is concave upward when f'' is positive, and concave downward when f'' is negative.

43. (a) $f'(x) = \cos x - \cos 3x + \cos 5x$
$f''(x) = -\sin x + 3\sin 3x - 5\sin 5x$
(b) Relative maximum: $(\pi/2, 1.53333)$
Points of inflection:
$(0.5236, 0.2667), (1.1731, 0.9637)$,
$(1.9685, 0.9637), (2.6180, 0.2667)$
(c)

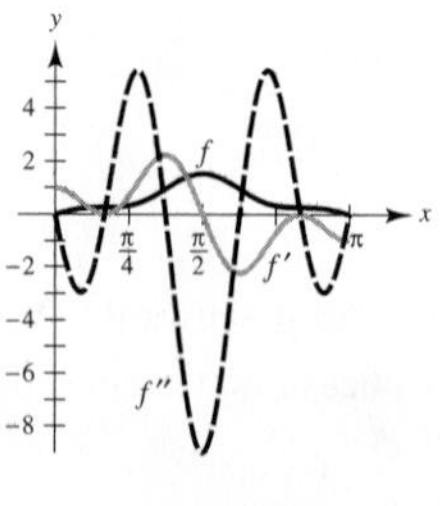

f is increasing when f' is positive, and decreasing when f' is negative. f is concave upward when f'' is positive, and concave downward when f'' is negative.

45. (a)

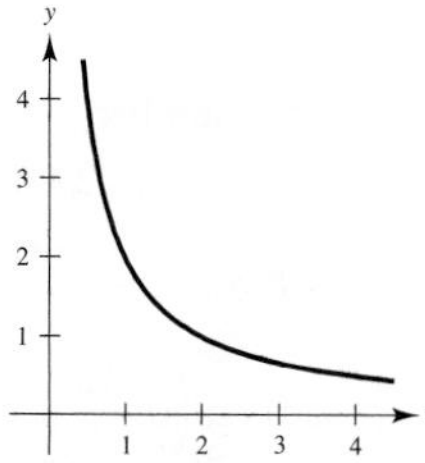

(b)

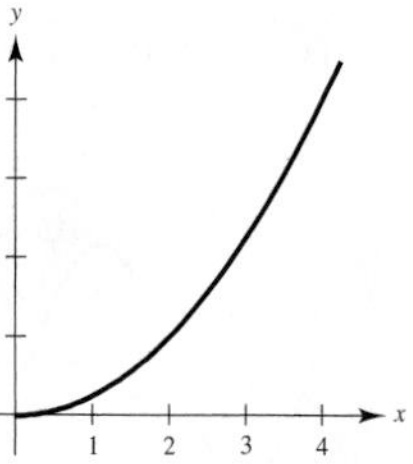

47. Answers will vary. Example: $f(x) = x^4$; $f''(0) = 0$, but $(0, 0)$ is not a point of inflection.

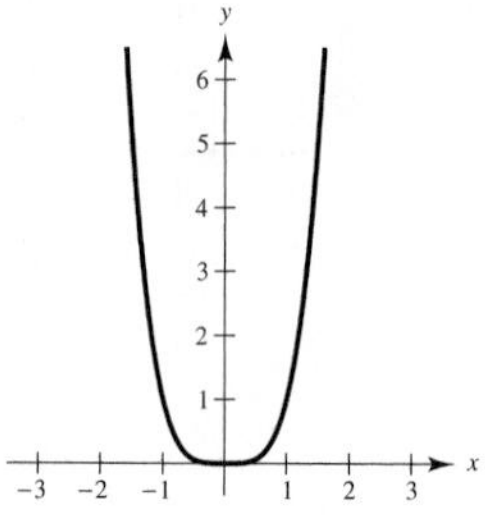

49.

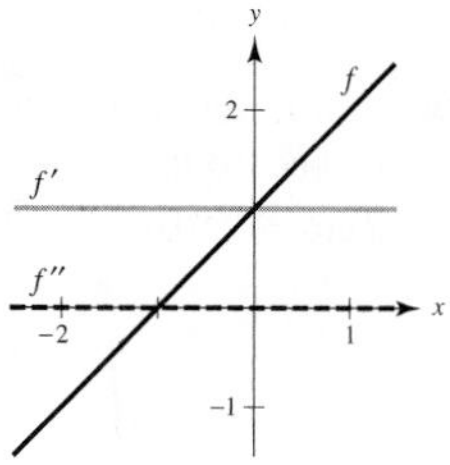

51.

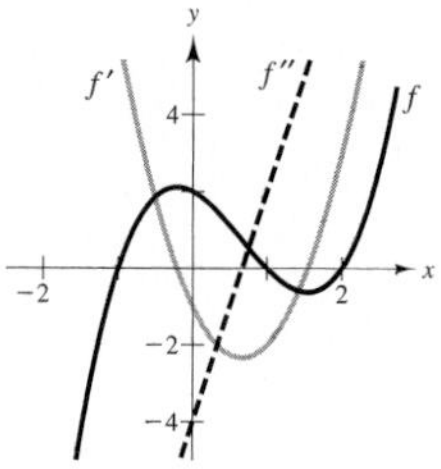

53.

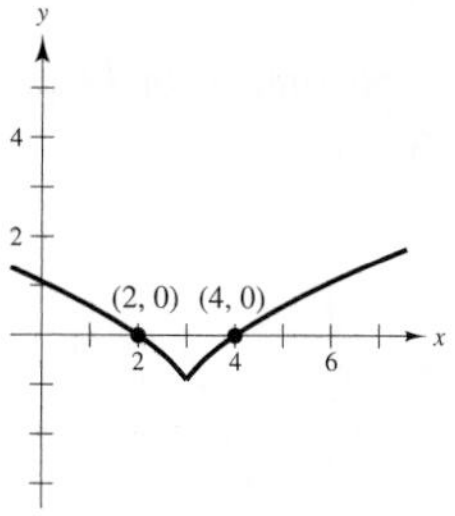

55.

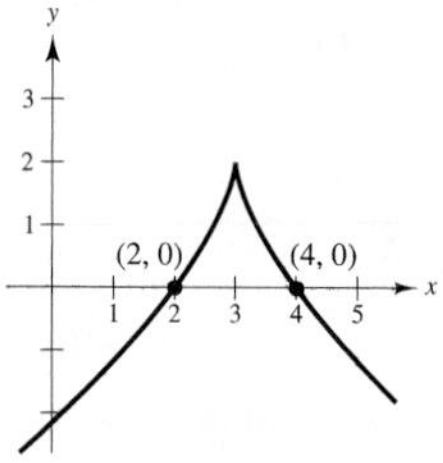

57. Example:

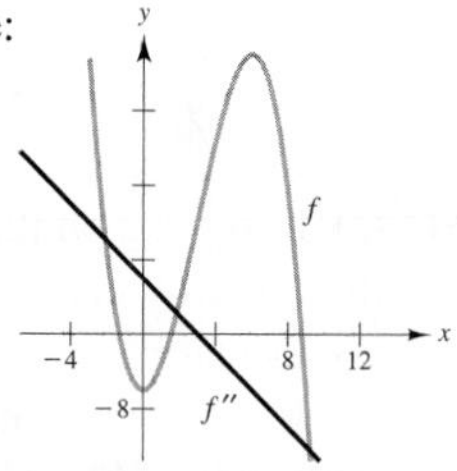

59. (a) $f(x) = (x-2)^n$ has a point of inflection at $(2, 0)$ if n is odd and $n \geq 3$.

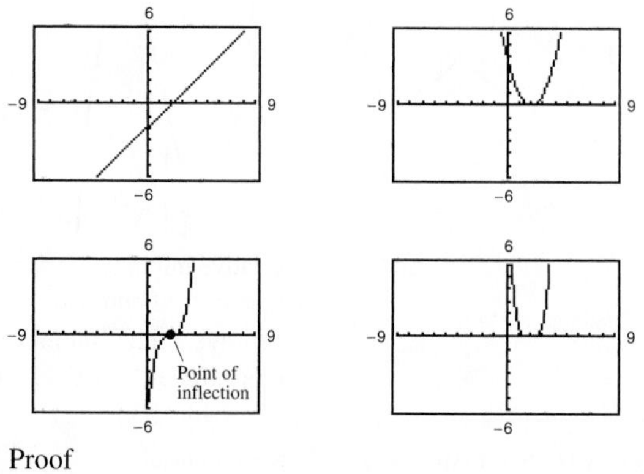

(b) Proof

61. $f(x) = \frac{1}{2}x^3 - 6x^2 + \frac{45}{2}x - 24$

63. (a) $f(x) = \frac{1}{32}x^3 + \frac{3}{16}x^2$ (b) Two miles from touchdown

65. $x = \left(\dfrac{15 - \sqrt{33}}{16}\right)L \approx 0.578L$ **67.** $x = 100$ units

69. (a)

t	0.5	1	1.5	2	2.5	3
S	151.5	555.6	1097.6	1666.7	2193.0	2647.1

$t \approx 1.6$

(b)

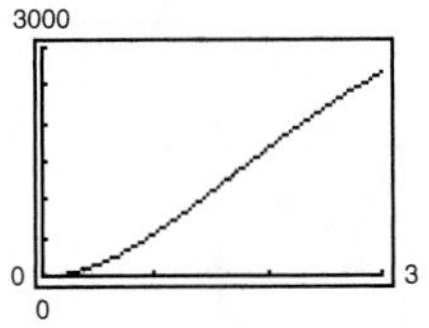

$1.5 < t < 2$

(c) ≈ 1.633 yrs

71. $P_1(x) = 2\sqrt{2}$

$P_2(x) = 2\sqrt{2} - \sqrt{2}(x - \pi/4)^2$

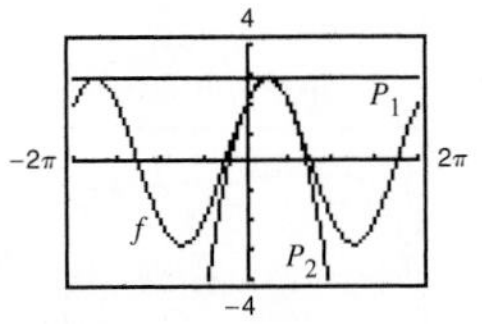

The values of f, P_1, and P_2 and their first derivatives are equal when $x = \pi/4$. The approximations worsen as you move away from $x = \pi/4$.

73. $P_1(x) = 1 - x/2$

$P_2(x) = 1 - x/2 - x^2/8$

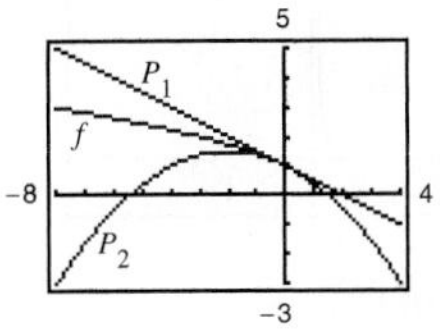

The values of f, P_1, and P_2 and their first derivatives are equal when $x = 0$. The approximations worsen as you move away from $x = 0$.

75.

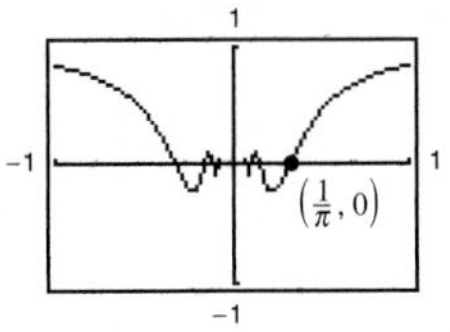

77. Proof **79.** True

81. False. f is concave upward at $x = c$ if $f''(c) > 0$. **83.** Proof

Section 3.5 (page 205)

1. As x becomes large, $f(x)$ approaches 4.

3. f **4.** c **5.** d **6.** a **7.** b **8.** e

9.

x	10^0	10^1	10^2	10^3
$f(x)$	7	2.2632	2.0251	2.0025

x	10^4	10^5	10^6
$f(x)$	2.0003	2.0000	2.0000

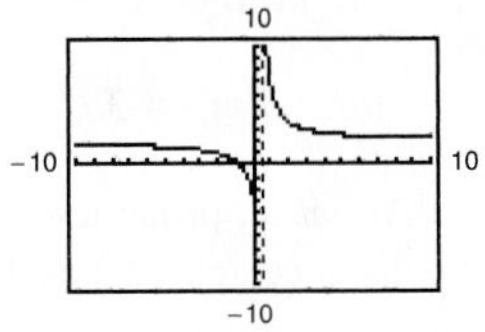

$$\lim_{x\to\infty} \frac{4x + 3}{2x - 1} = 2$$

11.

x	10^0	10^1	10^2	10^3
$f(x)$	-2	-2.9814	-2.9998	-3.0000

x	10^4	10^5	10^6
$f(x)$	-3.0000	-3.0000	-3.0000

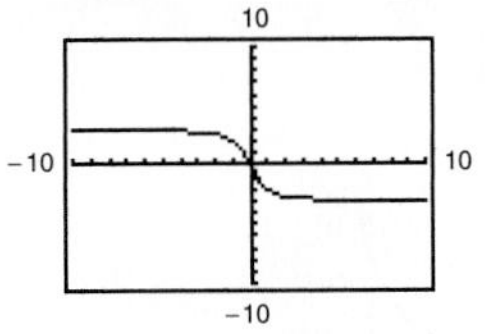

$$\lim_{x\to\infty} \frac{-6x}{\sqrt{4x^2 + 5}} = -3$$

13.

x	10^0	10^1	10^2	10^3
$f(x)$	4.5000	4.9901	4.9999	5.0000

x	10^4	10^5	10^6
$f(x)$	5.0000	5.0000	5.0000

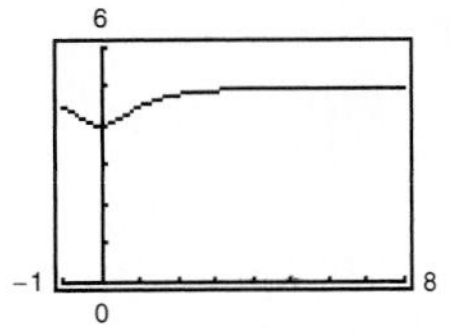

$$\lim_{x\to\infty} \left(5 - \frac{1}{x^2 + 1}\right) = 5$$

15. (a) ∞ (b) 5 (c) 0 **17.** (a) 0 (b) 1 (c) ∞

19. (a) 0 (b) $-\frac{2}{3}$ (c) $-\infty$ **21.** $\frac{2}{3}$ **23.** 0

25. $-\infty$ **27.** -1 **29.** -2 **31.** 0 **33.** 0

35.

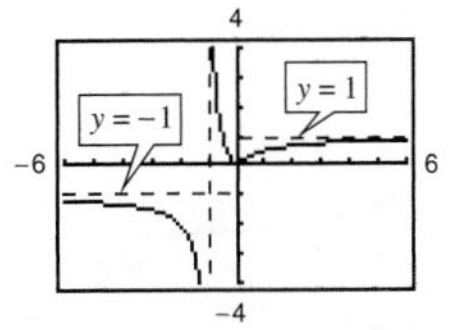

37.

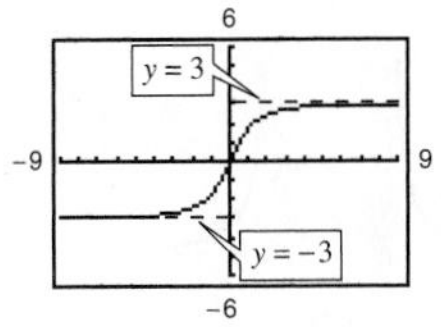

39. 1 **41.** 0 **43.** $-\frac{1}{2}$ **45.** $\frac{1}{8}$

47.

x	10^0	10^1	10^2	10^3	10^4	10^5	10^6
$f(x)$	1.000	0.513	0.501	0.500	0.500	0.500	0.500

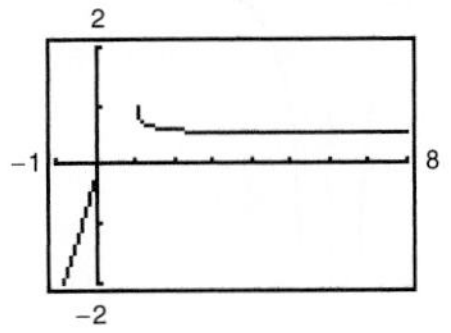

$$\lim_{x\to\infty} \left[x - \sqrt{x(x - 1)}\right] = \frac{1}{2}$$

49.

x	10^0	10^1	10^2	10^3	10^4	10^5	10^6
$f(x)$	0.479	0.500	0.500	0.500	0.500	0.500	0.500

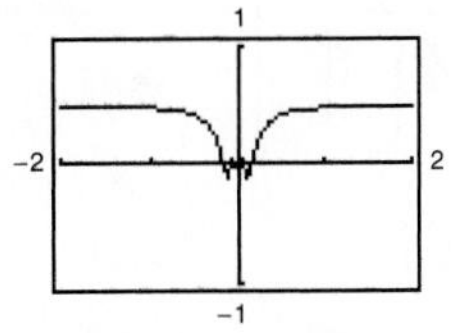

The graph has a hole at $x = 0$.

$$\lim_{x\to\infty} x \sin\frac{1}{2x} = \frac{1}{2}$$

51. (a)

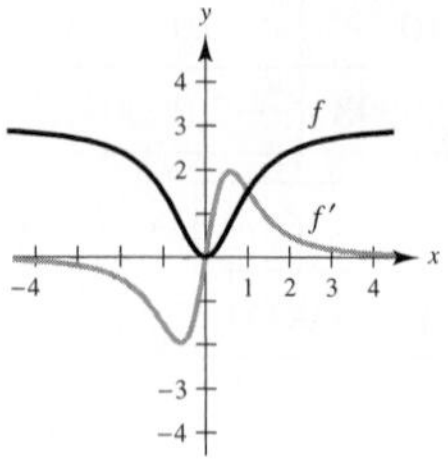

(b) $\lim_{x\to\infty} f(x) = 3, \ \lim_{x\to\infty} f'(x) = 0$

(c) $y = 3$ is a horizontal asymptote. The rate of increase of the function approaches 0 as the graph approaches $y = 3$.

53. Yes. For example, let $f(x) = \dfrac{6|x-2|}{\sqrt{(x-2)^2+1}}$.

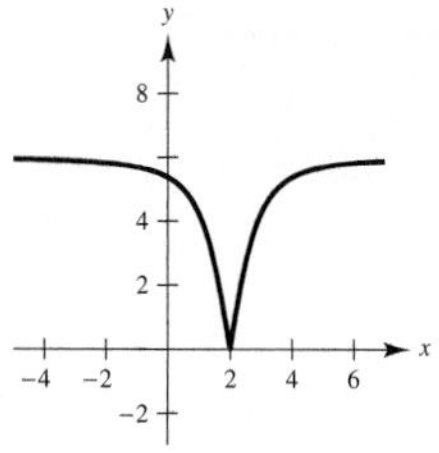

55.

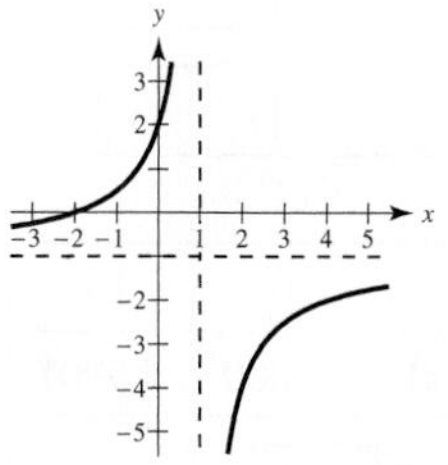

57.

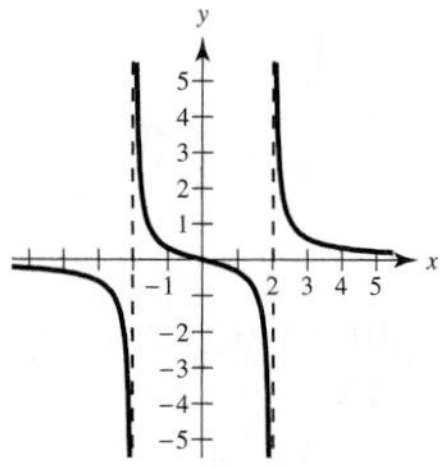

59.

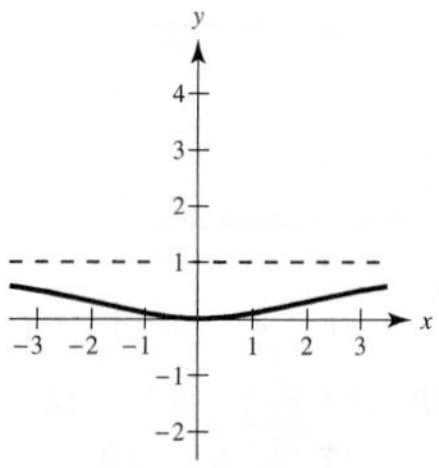

61.

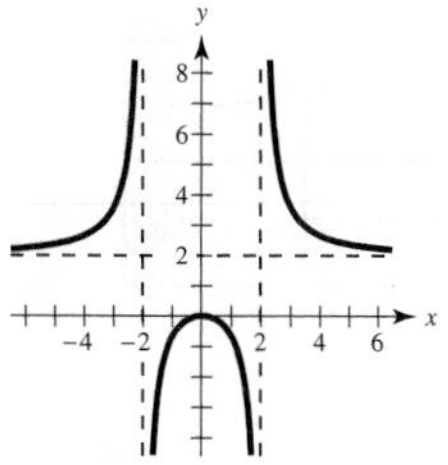

63.

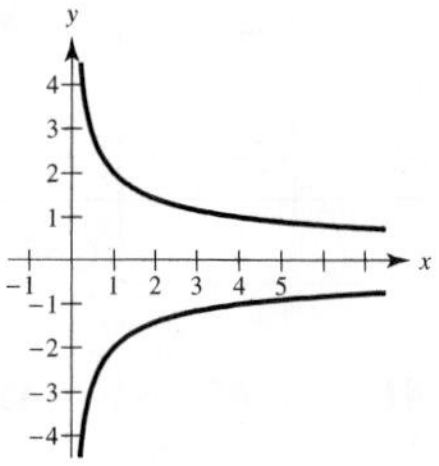

65.

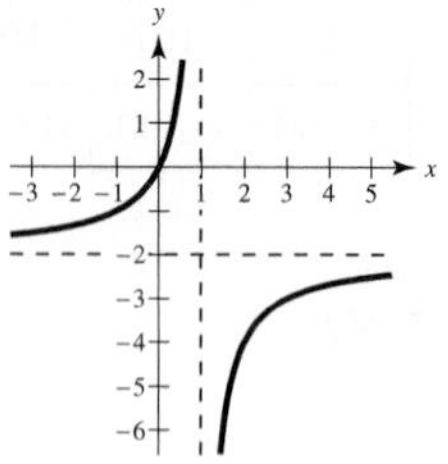

67.

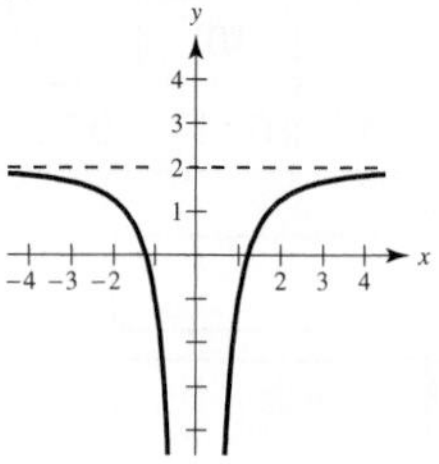

69.

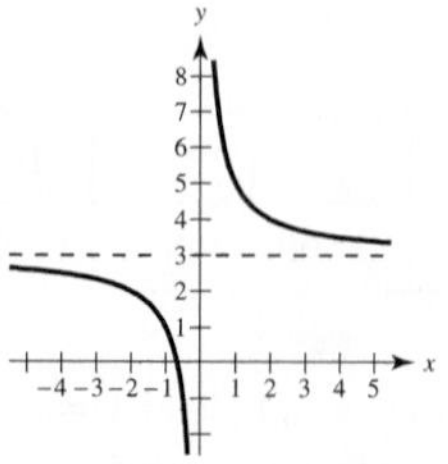

71.

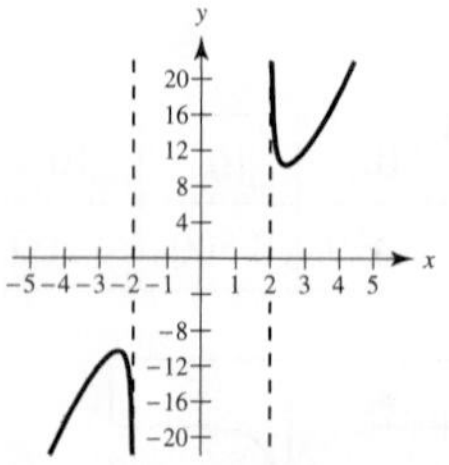

73.

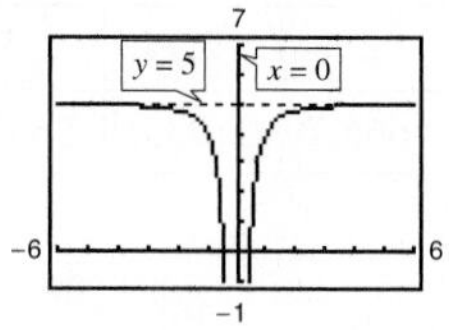

75.

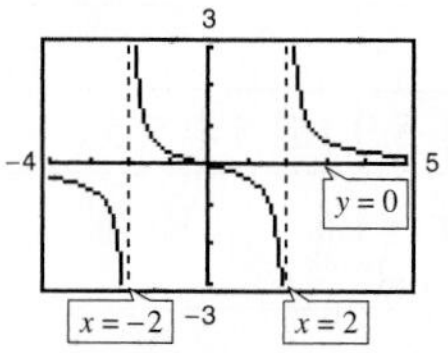

77.

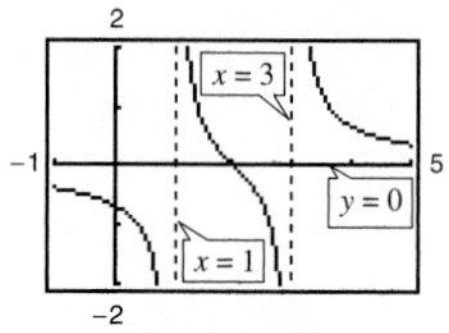

79.

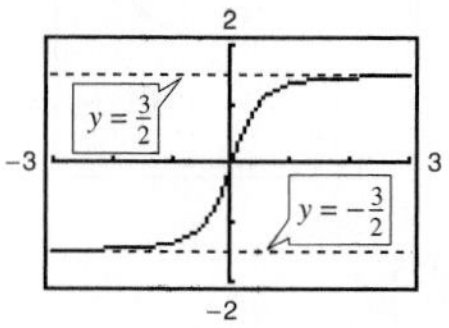

81.

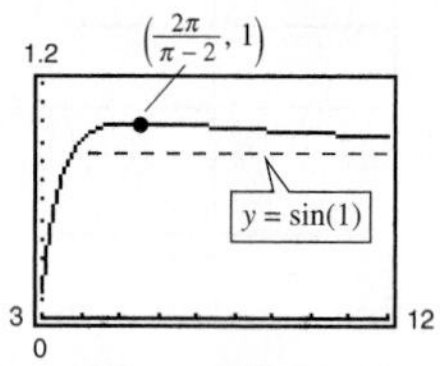

83. (a)

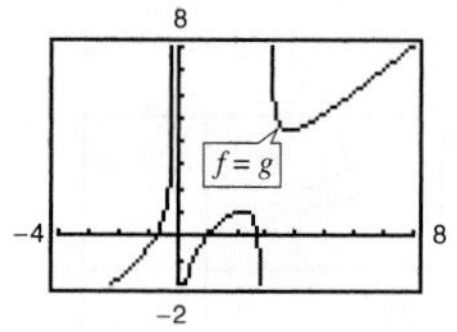

(b) Proof

(c)

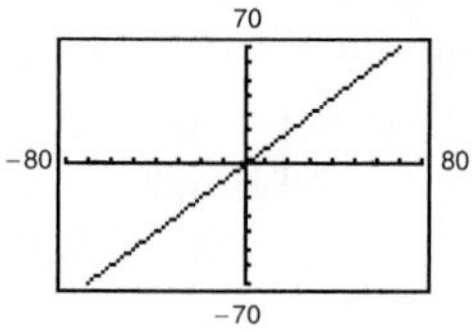

The slant asymptote $y = x$

85. $\frac{1}{2}$ **87.** $\lim_{t\to\infty} N(t) = +\infty;\ \lim_{t\to\infty} E(t) = c$

89. (a)

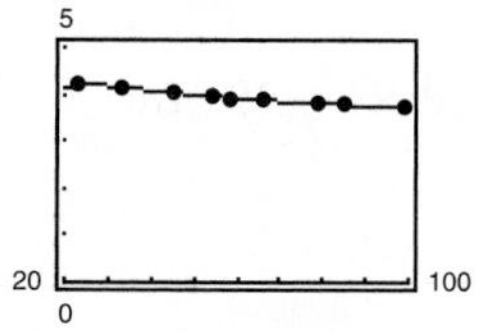

(b) Yes. $\lim_{t\to\infty} y = 3.351$

91. (a) $T_1 = -0.003t^2 + 0.68t + 26.6$

(b)

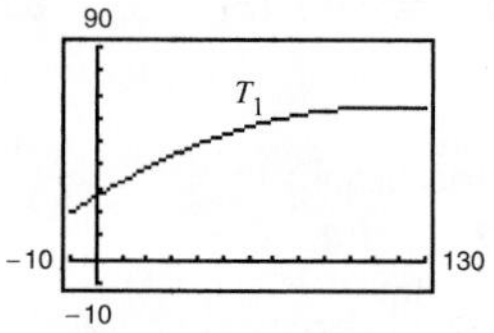

(c)

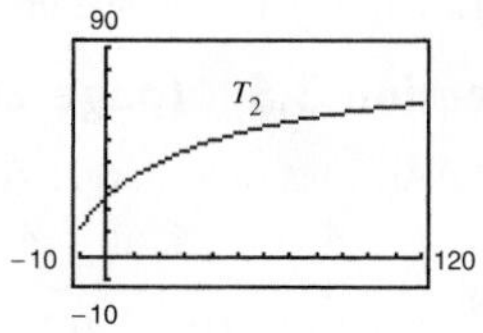

(d) $T_1(0) \approx 26.6°,\ T_2(0) \approx 25.0°$ (e) 86

(f) The limiting temperature is 86°.
No. T_1 has no horizontal asymptote.

93. (a) $d(m) = \dfrac{|3m+3|}{\sqrt{m^2+1}}$

(b)

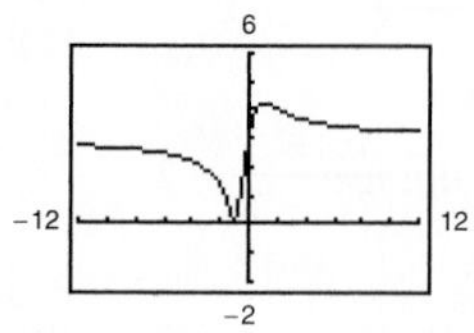

(c) $\lim_{m\to\infty} d(m) = 3$

$\lim_{m\to-\infty} d(m) = 3$

As m approaches $\pm\infty$, distance approaches 3.

95. (a) $\lim_{x\to\infty} f(x) = 2$ (b) $x_1 = \sqrt{\dfrac{4-2\varepsilon}{\varepsilon}},\ x_2 = -\sqrt{\dfrac{4-2\varepsilon}{\varepsilon}}$

(c) $\sqrt{\dfrac{4-2\varepsilon}{\varepsilon}}$ (d) $-\sqrt{\dfrac{4-2\varepsilon}{\varepsilon}}$

97. (a) Answers will vary. $M = \dfrac{5\sqrt{33}}{11}$

(b) Answers will vary. $M = \dfrac{29\sqrt{177}}{59}$

99–103. Proofs

105. False. Let $f(x) = \dfrac{2x}{\sqrt{x^2 + 2}}$. $f'(x) > 0$ for all real numbers.

Section 3.6 (page 215)

1. d **2.** c **3.** a **4.** b

5. (a) $f'(x) = 0$ for $x = \pm 2$; $f'(x) > 0$ for $(-\infty, -2), (2, \infty)$
$f'(x) < 0$ for $(-2, 2)$

(b) $f''(x) = 0$ for $x = 0$; $f''(x) > 0$ for $(0, \infty)$
$f''(x) < 0$ for $(-\infty, 0)$

(c) $(0, \infty)$

(d) f' is minimum for $x = 0$.
f is decreasing at the greatest rate.

7.

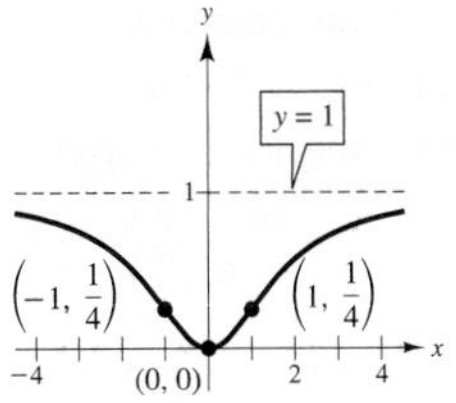

9.

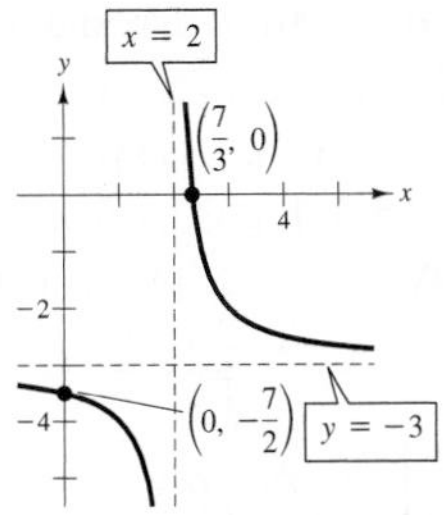

11.

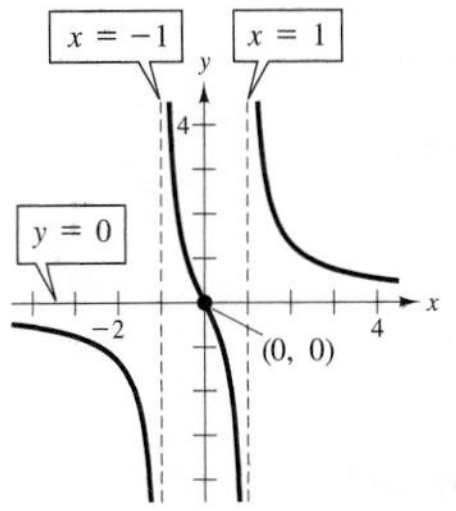

13.

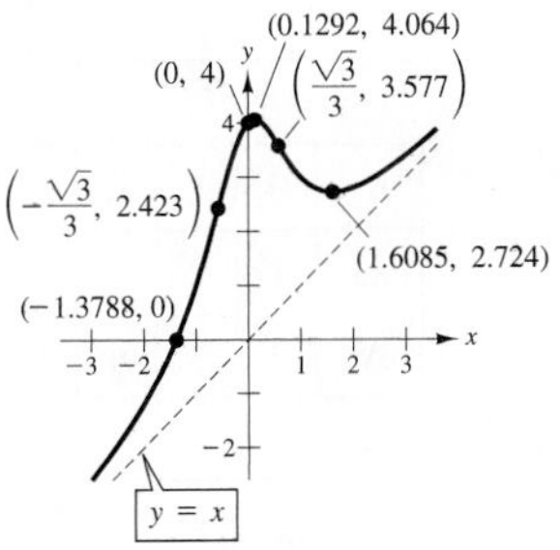

15.

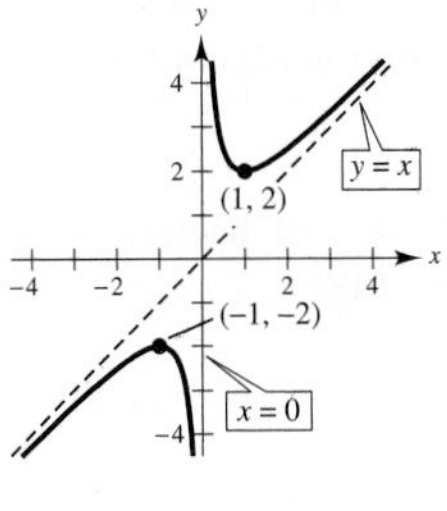

17.

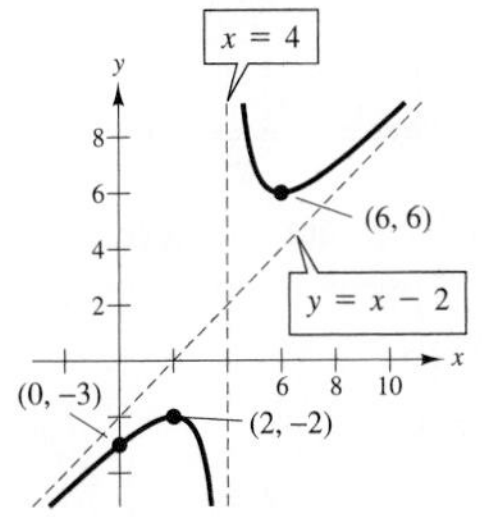

19.

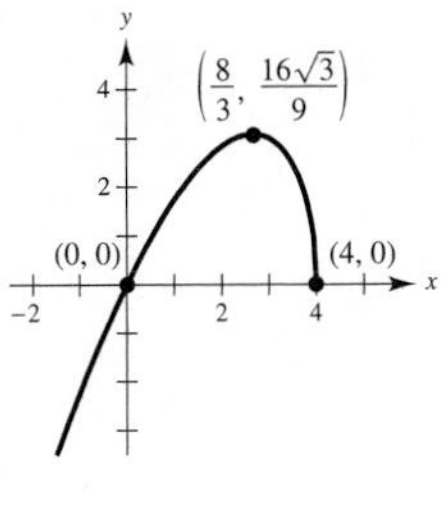

21.

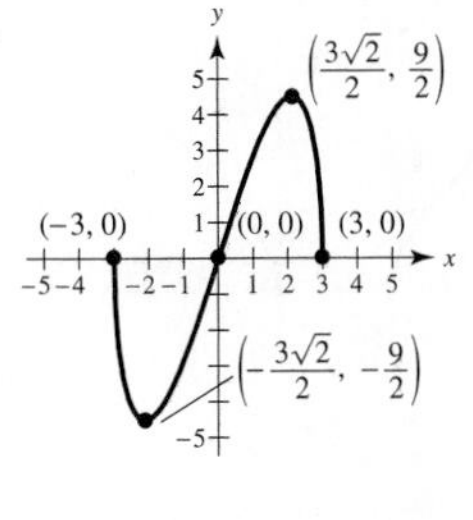

23.

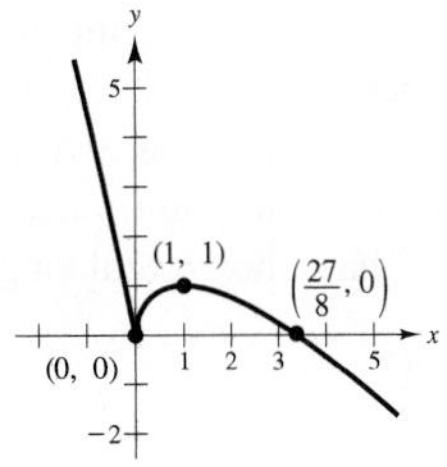

25.

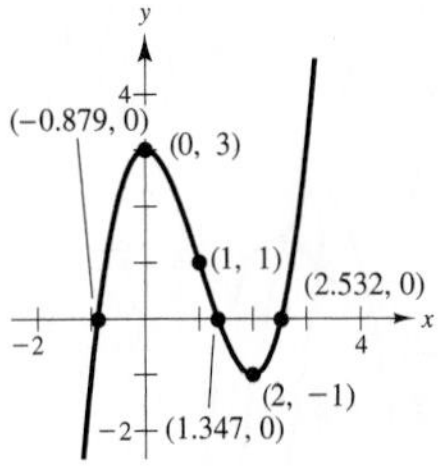

27.

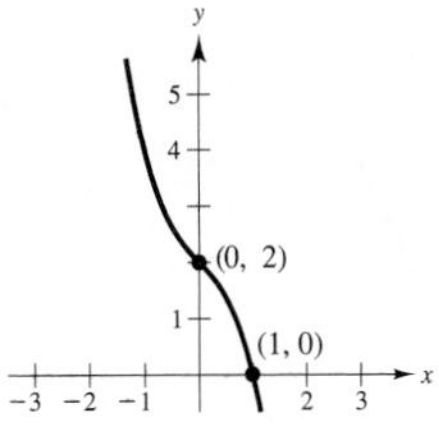

29.

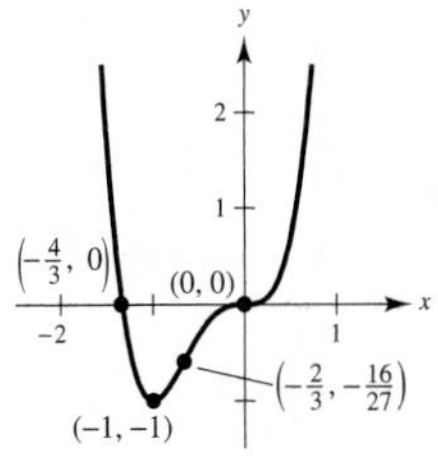

31.

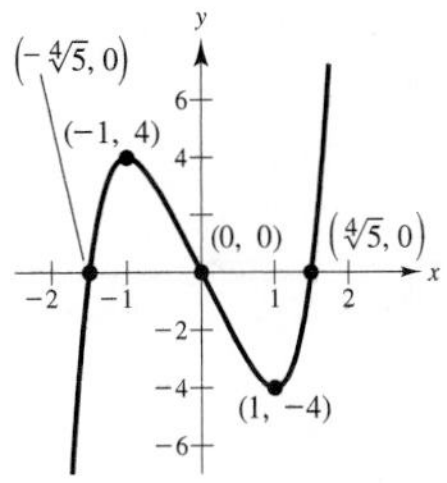

33.

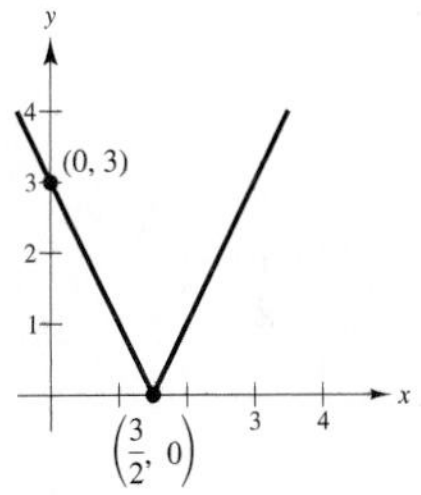

35.

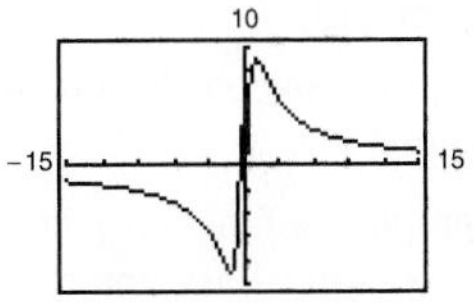

Minimum: $(-1.10, -9.05)$
Maximum: $(1.10, 9.05)$
Points of inflection:
$(-1.84, -7.86), (1.84, 7.86)$
Vertical asymptote: $x = 0$
Horizontal asymptote: $y = 0$

37.

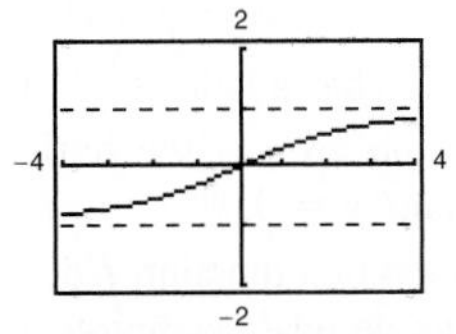

Point of inflection: $(0, 0)$
Horizontal asymptotes: $y = \pm 1$

39.

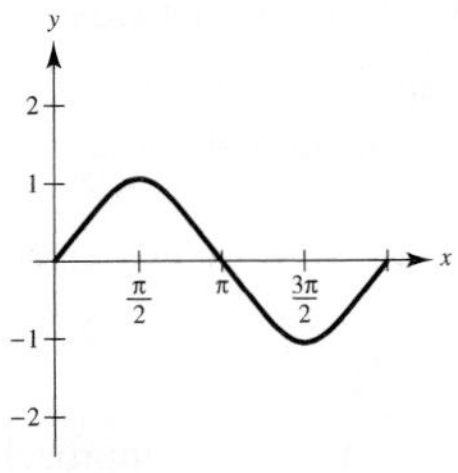

41.

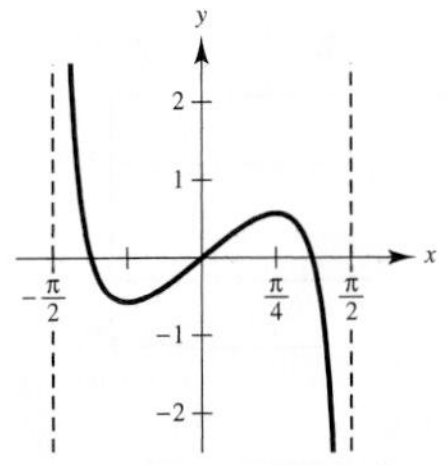

43.

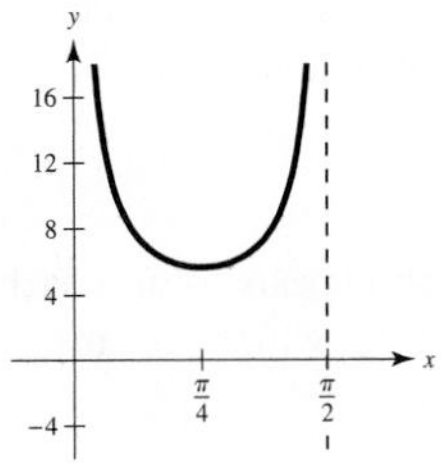

45.

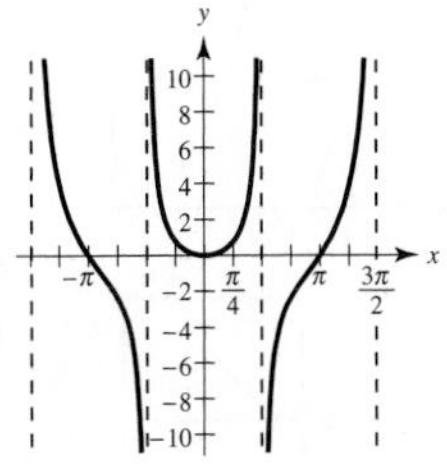

47.

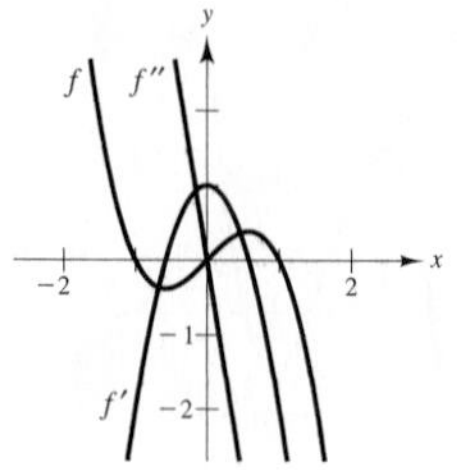

The zeros of f' correspond to the points where the graph of f has horizontal tangents. The zero of f'' corresponds to the point where the graph of f' has a horizontal tangent.

49.

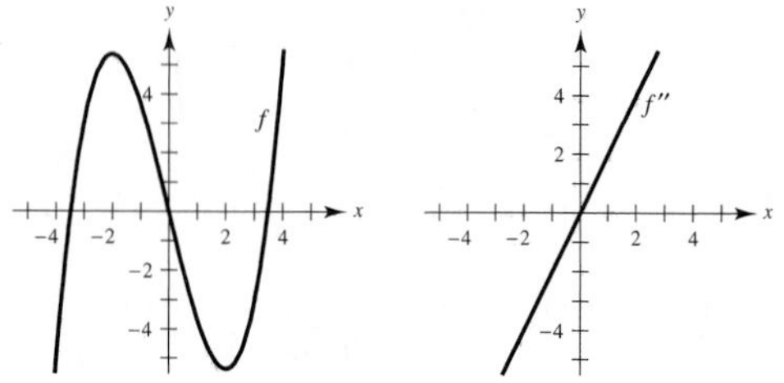

51.

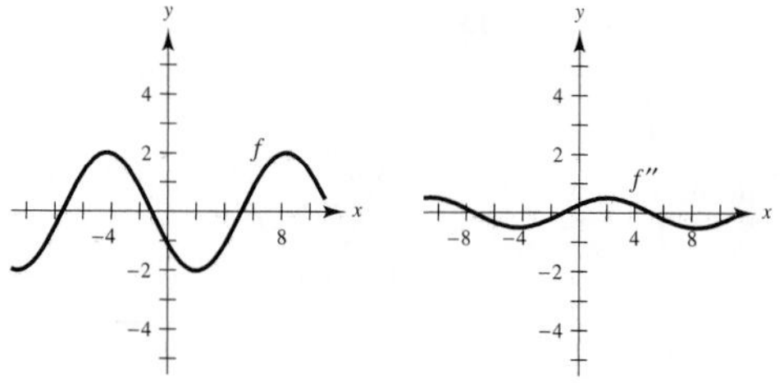

53. f is decreasing on $(2, 8)$ and therefore $f(3) > f(5)$.

55.

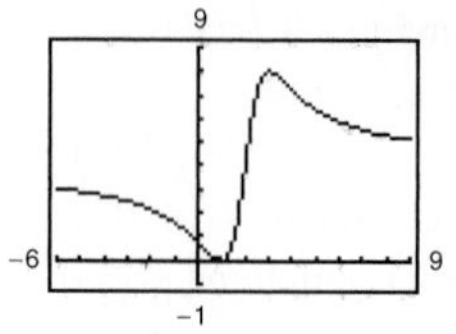

The graph crosses the horizontal asymptote $y = 4$.

The graph of a function f does not cross its vertical asymptote $x = c$ because $f(c)$ does not exist.

57.

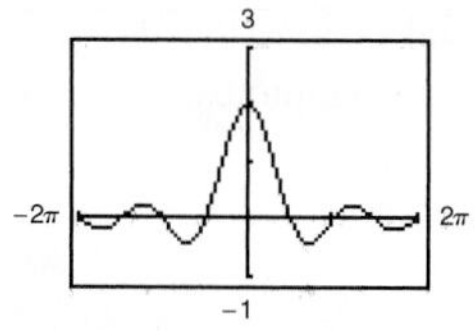

The graph has a hole at $x = 0$.
The graph crosses the horizontal asymptote $y = 0$.
The graph of a function f does not cross its vertical asymptote $x = c$ because $f(c)$ does not exist.

59.

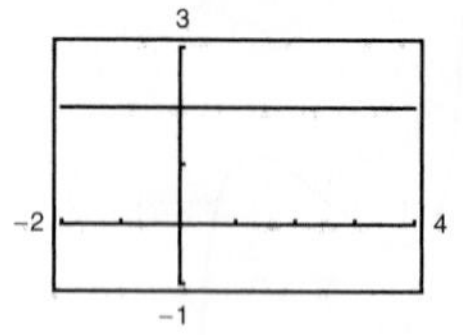

The graph has a hole at $x = 3$. The rational function is not reduced to lowest terms.

61.

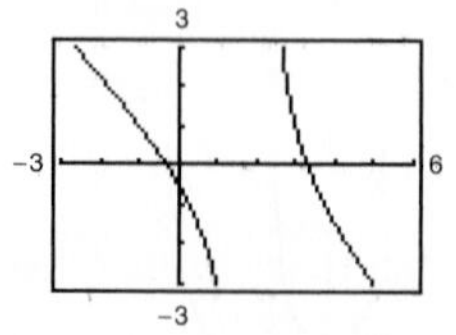

The graph appears to approach the line $y = -x + 1$, which is the slant asymptote.

63.

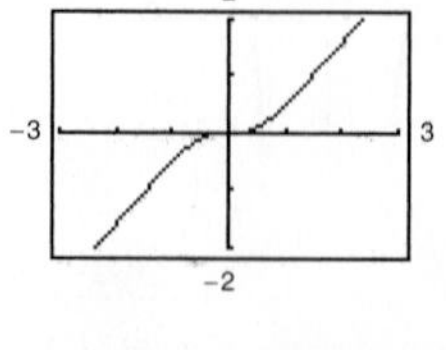

The graph appears to approach the line $y = x$, which is the slant asymptote.

65. (a)

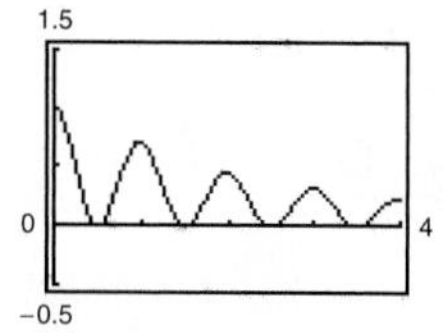

The graph has holes at $x = 0$ and at $x = 4$.

Visual approximate critical numbers: $\frac{1}{2}, 1, \frac{3}{2}, 2, \frac{5}{2}, 3, \frac{7}{2}$

(b) $f'(x) = \dfrac{-x\cos^2(\pi x)}{(x^2 + 1)^{3/2}} - \dfrac{2\pi \sin(\pi x)\cos(\pi x)}{\sqrt{x^2 + 1}}$

Approximate critical numbers: $\frac{1}{2}, 0.97, \frac{3}{2}, 1.98, \frac{5}{2}, 2.98, \frac{7}{2}$

The critical numbers where maxima occur appear to be integers in part (a), but approximating them using f' you see that they are not integers.

67. Answers will vary. Example: $y = 1/(x - 5)$

69. Answers will vary. Example: $y = (3x^2 - 13x - 9)/(x - 5)$

71. (a) Rate of change of f changes as a varies. If the sign of a is changed, the graph is reflected through the x-axis.

(b) The locations of the vertical asymptote and the minimum (if $a > 0$) or maximum (if $a < 0$) are changed.

73. (a) If n is even, f is symmetric with respect to the y-axis.
If n is odd, f is symmetric with respect to the origin.

(b) $n = 0, 1, 2, 3$ (c) $n = 4$

(d) When $n = 5$, the slant asymptote is $y = 3x$.

(e)

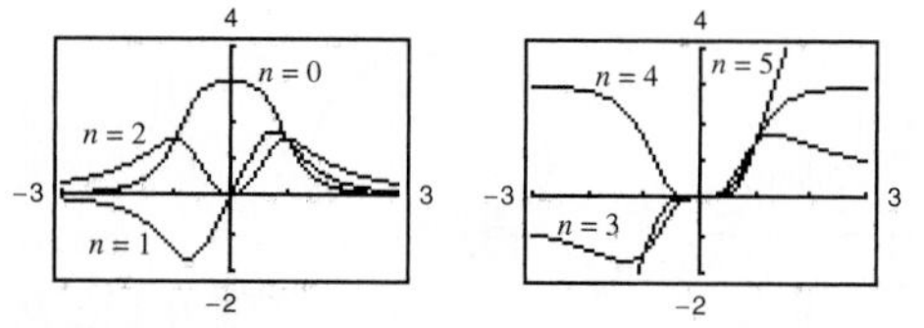

n	0	1	2	3	4	5
M	1	2	3	2	1	0
N	2	3	4	5	2	3

75. (a)

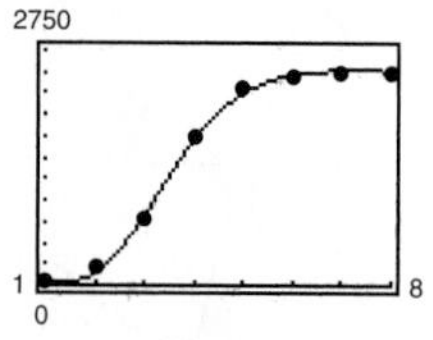

(b) 2434

(c) The number of bacteria reaches its maximum early on the seventh day.

(d) The rate of increase in the number of bacteria is greatest approximately in the early part of the third day.

(e) 13,250/7

77. $y = x + 3$, $y = -x - 3$

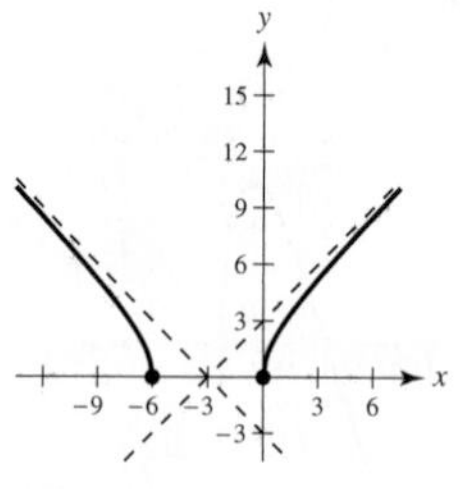

Section 3.7 (page 223)

1. (a) and (b)

First Number x	*Second Number*	*Product P*
10	110 − 10	10(110 − 10) = 1000
20	110 − 20	20(110 − 20) = 1800
30	110 − 30	30(110 − 30) = 2400
40	110 − 40	40(110 − 40) = 2800
50	110 − 50	50(110 − 50) = 3000
60	110 − 60	60(110 − 60) = 3000
70	110 − 70	70(110 − 70) = 2800
80	110 − 80	80(110 − 80) = 2400
90	110 − 90	90(110 − 90) = 1800
100	110 − 100	100(110 − 100) = 1000

(c) $P = x(110 - x)$

(d)

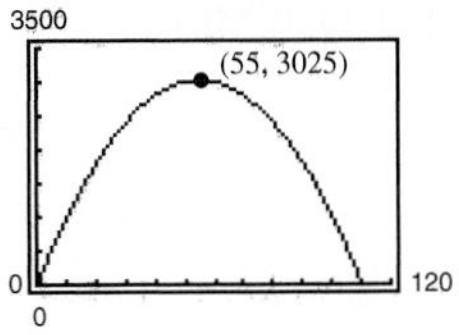

(e) 55 and 55

3. $S/2$ and $S/2$ **5.** 24 and 8 **7.** 50 and 25

9. $l = w = 25$ m **11.** $l = w = 8$ ft **13.** $\left(\frac{7}{2}, \sqrt{\frac{7}{2}}\right)$

15. (1, 1) **17.** $x = Q_0/2$ **19.** 600 × 300 m

21. (a) Proof (b) $V_1 = 99$ in.^{3}, $V_2 = 125$ in.^{3}, $V_3 = 117$ in.^{3}

(c) 5 × 5 × 5 in.

23. Rectangular portion: $16/(\pi + 4) \times 32/(\pi + 4)$ ft

25. (a) $L = \sqrt{x^2 + 4 + \dfrac{8}{x - 1} + \dfrac{4}{(x - 1)^2}}, \quad x > 1$

(b)

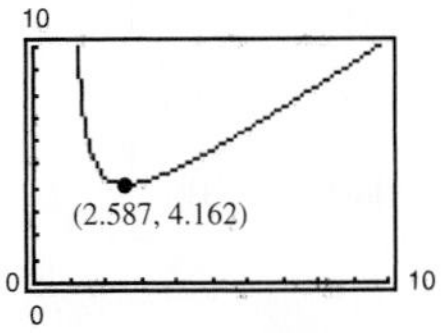

Minimum when $x \approx 2.587$

(c) (0, 0), (2, 0), (0, 4)

27. Width: $5\sqrt{2}/2$; Length: $5\sqrt{2}$

29. Dimensions of page: $(2 + \sqrt{30})$ in. × $(2 + \sqrt{30})$ in.

31. (a)

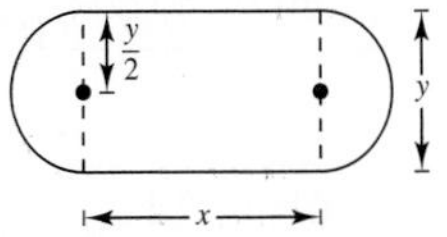

(b)

Length x	Width y	Area xy
10	$2/\pi(100 - 10)$	$(10)(2/\pi)(100 - 10) \approx 573$
20	$2/\pi(100 - 20)$	$(20)(2/\pi)(100 - 20) \approx 1019$
30	$2/\pi(100 - 30)$	$(30)(2/\pi)(100 - 30) \approx 1337$
40	$2/\pi(100 - 40)$	$(40)(2/\pi)(100 - 40) \approx 1528$
50	$2/\pi(100 - 50)$	$(50)(2/\pi)(100 - 50) \approx 1592$
60	$2/\pi(100 - 60)$	$(60)(2/\pi)(100 - 60) \approx 1528$

The maximum area of the rectangle is approximately 1592 m^2.

(c) $A = 2/\pi(100x - x^2), \quad 0 < x < 100$

(d) $\dfrac{dA}{dx} = \dfrac{2}{\pi}(100 - 2x)$

$= 0$ when $x = 50$

The maximum value is approximately 1592 when $x = 50$.

(e)

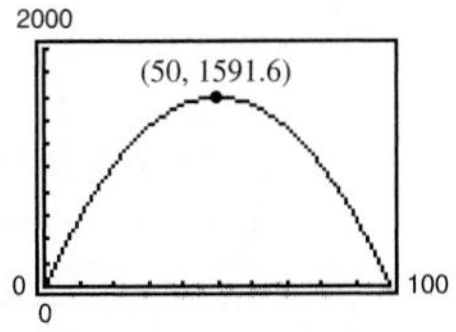

33. 18 × 18 × 36 in. **35.** $32\pi r^3/81$

37. Answers will vary. If area is expressed as a function of either length or width, the feasible domain is the interval (0, 10). No dimensions will yield a minimum area because the second derivative on this open interval is always negative.

39. $r = \sqrt[3]{9/\pi} \approx 1.42$ cm

41. Side of square: $\dfrac{10\sqrt{3}}{9 + 4\sqrt{3}}$; Side of triangle: $\dfrac{30}{9 + 4\sqrt{3}}$

43. $w = 8\sqrt{3}$ in., $h = 8\sqrt{6}$ in. **45.** $\theta = \pi/4$ **47.** $h = \sqrt{2}$ ft

49. One mile from the nearest point on the coast **51.** Proof

53.

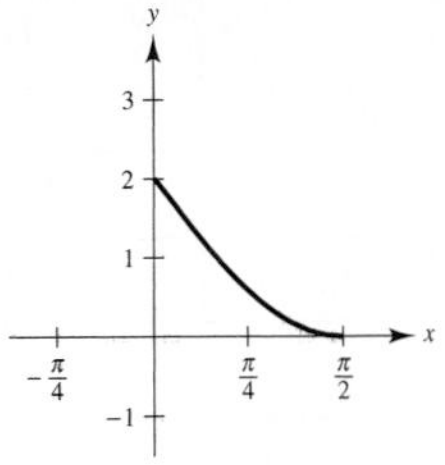

(a) Origin to y-intercept: 2

Origin to x-intercept: $\pi/2$

(b) $d = \sqrt{x^2 + (2 - 2\sin x)^2}$

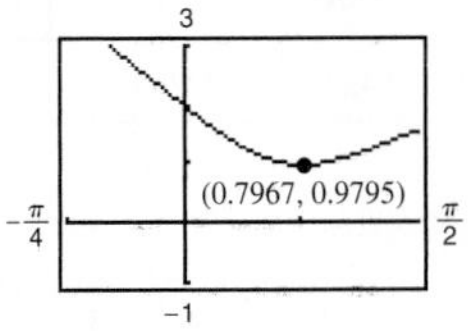

(c) Minimum distance is 0.9795 when $x \approx 0.7967$.

55. $F = kW/\sqrt{k^2 + 1}$; $\theta = \arctan k$

57. (a)

Base 1	Base 2	Altitude	Area
8	8 + 16 cos 10°	8 sin 10°	≈ 22.1
8	8 + 16 cos 20°	8 sin 20°	≈ 42.5
8	8 + 16 cos 30°	8 sin 30°	≈ 59.7
8	8 + 16 cos 40°	8 sin 40°	≈ 72.7
8	8 + 16 cos 50°	8 sin 50°	≈ 80.5
8	8 + 16 cos 60°	8 sin 60°	≈ 83.1

(b)

Base 1	Base 2	Altitude	Area
8	8 + 16 cos 10°	8 sin 10°	≈ 22.1
8	8 + 16 cos 20°	8 sin 20°	≈ 42.5
8	8 + 16 cos 30°	8 sin 30°	≈ 59.7
8	8 + 16 cos 40°	8 sin 40°	≈ 72.7
8	8 + 16 cos 50°	8 sin 50°	≈ 80.5
8	8 + 16 cos 60°	8 sin 60°	≈ 83.1
8	8 + 16 cos 70°	8 sin 70°	≈ 80.7
8	8 + 16 cos 80°	8 sin 80°	≈ 74.0
8	8 + 16 cos 90°	8 sin 90°	= 64.0

The maximum cross-sectional area is approximately 83.1 ft^2

(c) $A = 64(1 + \cos\theta)\sin\theta,\ 0° < \theta < 90°$

(d) $\dfrac{dA}{d\theta} = 64(2\cos\theta - 1)(\cos\theta + 1)$

$= 0$ when $\theta = 60°, 180°, 300°$

The maximum occurs when $\theta = 60°$.

(e) 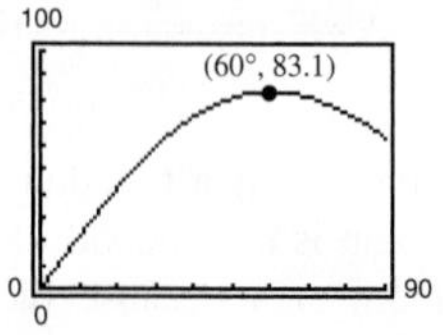

59. 4045 units **61.** $y = \frac{64}{141}x;\ S_1 \approx 6.1$ mi

63. $y = \frac{3}{10}x;\ S_3 \approx 4.50$ mi

65. Putnam Problem A1, 1986

Section 3.8 (page 233)

1.

n	x_n	$f(x_n)$	$f'(x_n)$	$\dfrac{f(x_n)}{f'(x_n)}$	$x_n - \dfrac{f(x_n)}{f'(x_n)}$
1	1.7000	−0.1100	3.4000	−0.0324	1.7324
2	1.7324	0.0012	3.4648	0.0003	1.7321

3.

n	x_n	$f(x_n)$	$f'(x_n)$	$\dfrac{f(x_n)}{f'(x_n)}$	$x_n - \dfrac{f(x_n)}{f'(x_n)}$
1	3	0.1411	−0.9900	−0.1425	3.1425
2	3.1425	−0.0009	−1.0000	0.0009	3.1416

5. 0.682 **7.** 1.146, 7.854 **9.** −1.442

11. 0.900, 1.100, 1.900 **13.** −0.489 **15.** 0.569

17. 4.493 **19.** (a) Proof (b) $\sqrt{5} \approx 2.236;\ \sqrt{7} \approx 2.646$

21. $f'(x_1) = 0$ **23.** $2 = x_1 = x_3 = \ldots;\ 1 = x_2 = x_4 = \ldots$

25. Answers will vary. Sample answer: If f is a function continuous on $[a, b]$ and differentiable on (a, b), where $c \in [a, b]$ and $f(c) = 0$, Newton's Method uses tangent lines to approximate c. First, estimate an initial x_1 close to c. (See graph.) Then determine x_2 by $x_2 = x_1 - f(x_1)/f'(x_1)$. Calculate a third estimate by $x_3 = x_2 - f(x_2)/f'(x_2)$. Continue this process until $|x_n - x_{n+1}|$ is within the desired accuracy and let x_{n+1} be the final approximation of c.

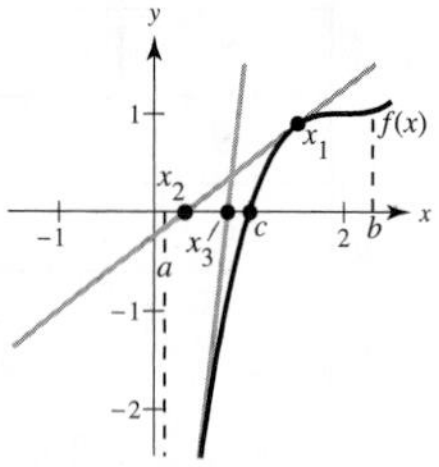

27. 0.74

29. (a) 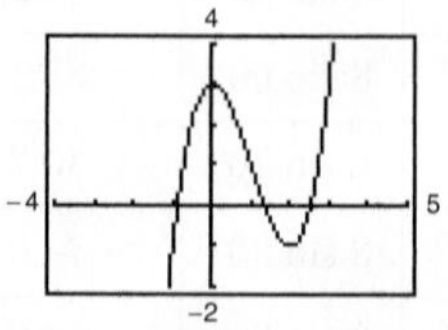

(b) 1.347
(c) 2.532

(d) 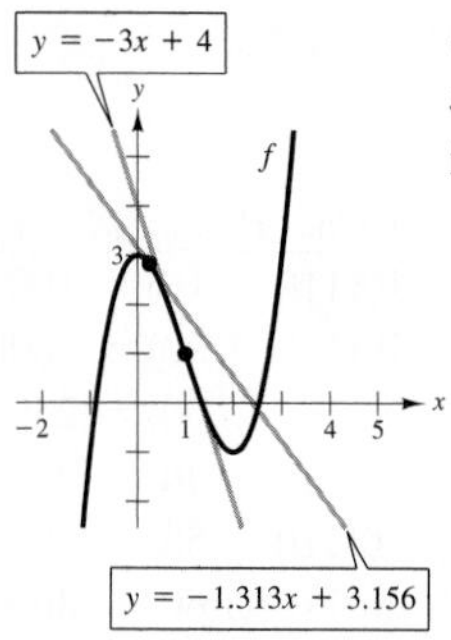

x-intercept of $y = -3x + 4$ is $\frac{4}{3}$. x-intercept of $y = -1.313x + 3.156$ is approximately 2.404.

(e) If the initial estimate $x = x_1$ is not sufficiently close to the desired zero of a function, the x-intercept of the corresponding tangent line to the function may approximate a second zero of the function.

31. Proof

33. 0.860 **35.** (1.939, 0.240)

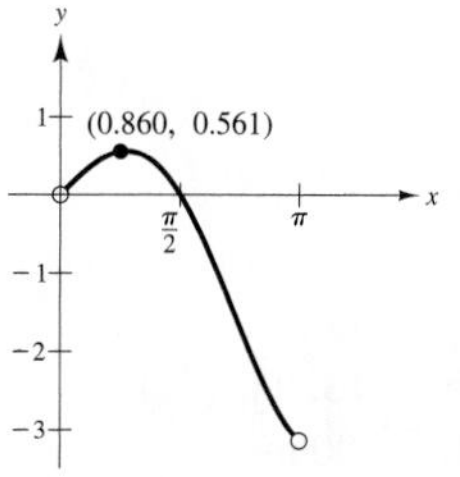

37. $x \approx 1.563$ mi **39.** \$384,356 **41.** False: let $f(x) = \dfrac{x^2 - 1}{x - 1}$.

43. True **45.** 0.217

Section 3.9 (page 240)

1. $T(x) = 4x - 4$

x	1.9	1.99	2	2.01	2.1
$f(x)$	3.610	3.960	4	4.040	4.410
$T(x)$	3.600	3.960	4	4.040	4.400

3. $T(x) = 80x - 128$

x	1.9	1.99	2	2.01	2.1
$f(x)$	24.761	31.208	32	32.808	40.841
$T(x)$	24.000	31.200	32	32.800	40.000

5. $T(x) = (\cos 2)(x - 2) + \sin 2$

x	1.9	1.99	2	2.01	2.1
$f(x)$	0.946	0.913	0.909	0.905	0.863
$T(x)$	0.951	0.913	0.909	0.905	0.868

7. $\Delta y = 0.6305;\ dy = 0.6000$ **9.** $\Delta y = -0.039;\ dy = -0.040$

11. $6x\,dx$ **13.** $-\dfrac{3}{(2x - 1)^2}\,dx$ **15.** $\dfrac{1 - 2x^2}{\sqrt{1 - x^2}}\,dx$

17. $(2 + 2\cot x + 2\cot^3 x)\,dx$ **19.** $-\pi\sin\left(\dfrac{6\pi x - 1}{2}\right)dx$

21. (a) 0.9 (b) 1.04 **23.** (a) 1.05 (b) 0.98

25. (a) 8.035 (b) 7.95 **27.** $\pm\frac{3}{8}$ in.^{2} **29.** $\pm 7\pi$ in.^{2}

31. (a) $\frac{2}{3}\%$ (b) 1.25%

33. (a) $\pm 2.88\pi$ in.^{3} (b) $\pm 0.96\pi$ in.^{2} (c) $1\%, \frac{2}{3}\%$

35. 80π cm^3 **37.** (a) $\frac{1}{4}\%$ (b) 216 sec = 3.6 min

39. (a) 0.87% (b) 2.16% **41.** 4961 ft

43. $f(x) = \sqrt{x}, dy = \dfrac{1}{2\sqrt{x}}\,dx$

$f(99.4) \approx \sqrt{100} + \dfrac{1}{2\sqrt{100}}(-0.6) = 9.97$

Calculator: 9.97

45. $f(x) = \sqrt[4]{x}, dy = \dfrac{1}{4x^{3/4}}\,dx$

$f(624) \approx \sqrt[4]{625} + \dfrac{1}{4(625)^{3/4}}(-1) = 4.998$

Calculator: 4.998

47. $f(x) = \sqrt{x};\ dy = \dfrac{1}{2\sqrt{x}}\,dx$

$f(4.02) \approx \sqrt{4} + \dfrac{1}{2\sqrt{4}}(0.02) = 2 + \dfrac{1}{4}(0.02)$

49. $y - f(0) = f'(0)(x - 0)$

$y - 2 = \frac{1}{4}x$

$y = 2 + x/4$

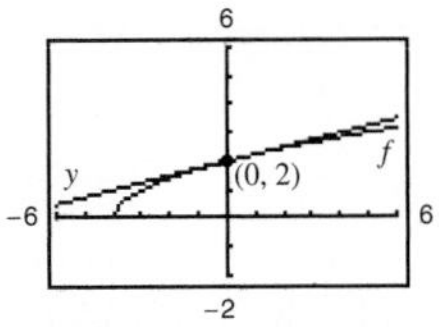

51. $y - f(0) = f'(0)(x - 0)$

$y - 0 = 1(x)$

$y = x$

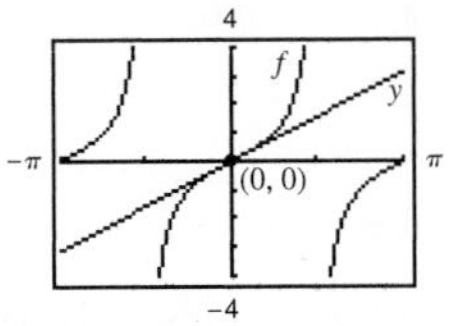

53. The value of dy becomes closer to the value of Δy as Δx decreases.

55. True **57.** True

Review Exercises for Chapter 3 (page 242)

1. Let f be defined at c. If $f'(c) = 0$ or if f' is undefined at c, then c is a critical number of f.

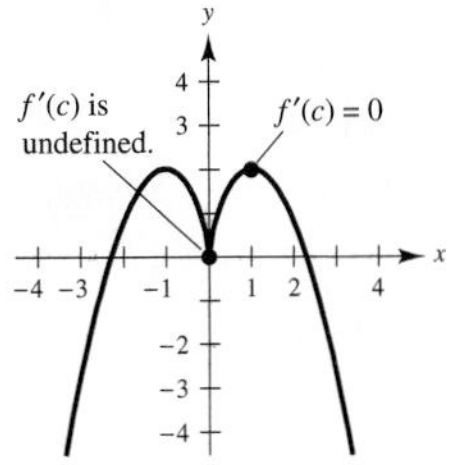

3. Maximum: $(2\pi, 17.57)$

Minimum: $(2.73, 0.88)$

5. $f'\left(\frac{1}{3}\right) = 0$

7. (a)

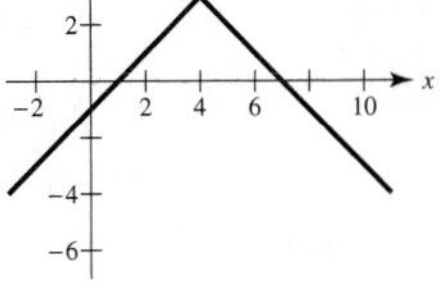

(b) f is not differentiable at $x = 4$.

9. $f'\left(\dfrac{2744}{729}\right) = \dfrac{3}{7}$ **11.** $f'(0) = 1$ **13.** $c = \dfrac{x_1 + x_2}{2}$

15. Critical numbers: $x = 1, \frac{7}{3}$

Increasing on $(-\infty, 1), \left(\frac{7}{3}, \infty\right)$; Decreasing on $\left(1, \frac{7}{3}\right)$

17. Critical number: $x = 1$

Increasing on $(1, \infty)$; Decreasing on $(0, 1)$

19. Relative minimum: $(2, -12)$

21. (a) $y = \frac{1}{4}$ in.; $v = 4$ in./sec (b) Proof

(c) Period: $\pi/6$; Frequency: $6/\pi$

23. $(\pi/2, \pi/2), (3\pi/2, 3\pi/2)$; Concave upward: $(\pi/2, 3\pi/2)$

Concave downward: $(0, \pi/2), (3\pi/2, 2\pi)$

25. Relative maxima: $\left(\sqrt{2}/2, 1/2\right), \left(-\sqrt{2}/2, 1/2\right)$

Relative minimum: $(0, 0)$

27.

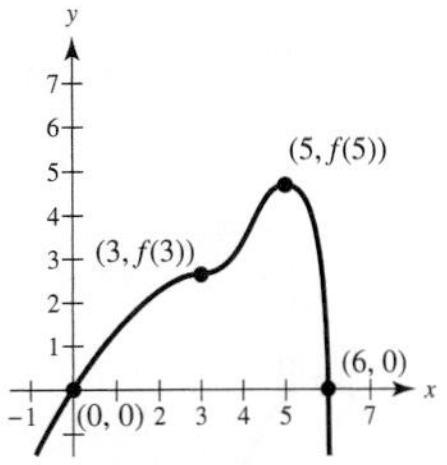

29. Increasing and concave down

31. (a) $D = 0.00340t^4 - 0.2352t^3 + 4.942t^2 - 20.86t + 94.4$

(b)

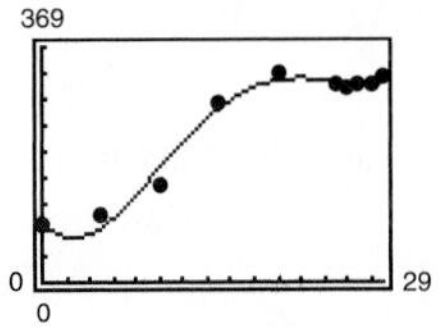

(c) Maximum occurs in 1991; Minimum occurs in 1972.

(d) 1979

33. $\frac{2}{3}$ **35.** $-\infty$ **37.** 0 **39.** 6

41. Vertical asymptote: $x = 4$; Horizontal asymptote: $y = 2$

43. Vertical asymptote: $x = 0$; Horizontal asymptote: $y = -2$

45.

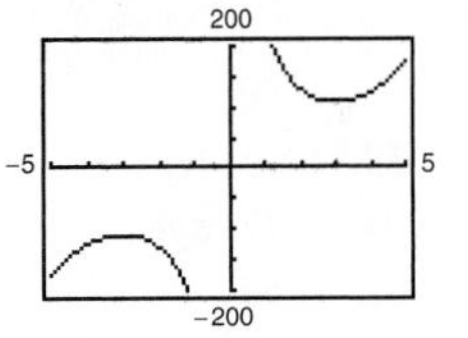

Vertical asymptote: $x = 0$

Relative minimum: $(3, 108)$

Relative maximum: $(-3, -108)$

47.

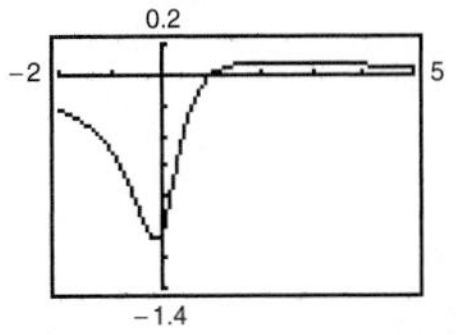

Horizontal asymptote: $y = 0$

Relative minimum: $(-0.155, -1.077)$

Relative maximum: $(2.155, 0.077)$

49.

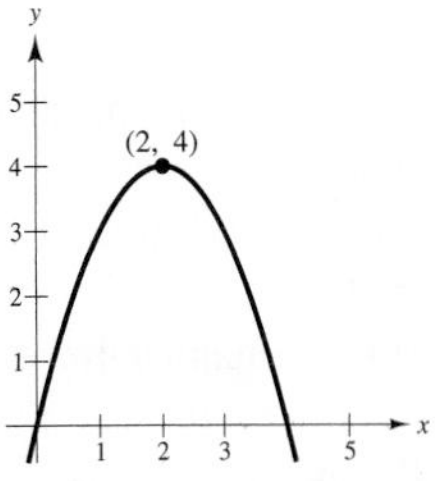

51.

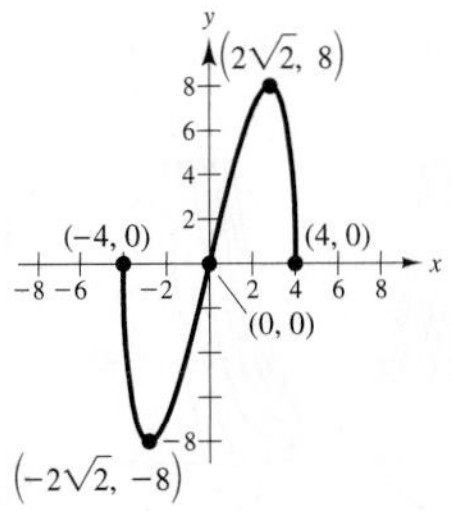

53.
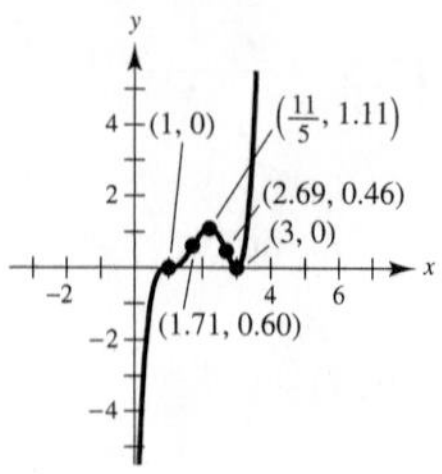

55.
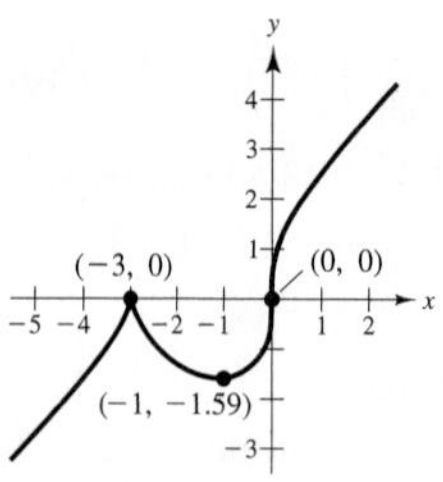

57.
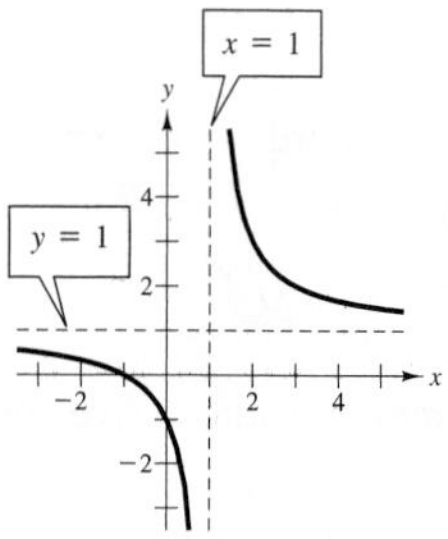

59.
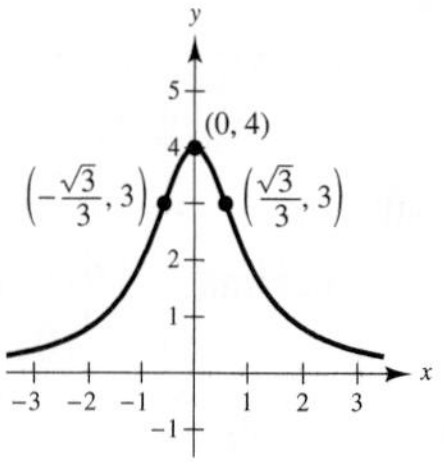

61.
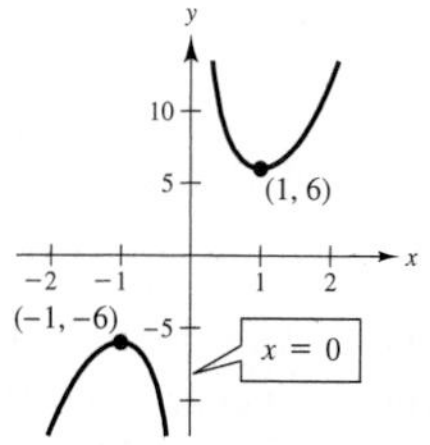

63.
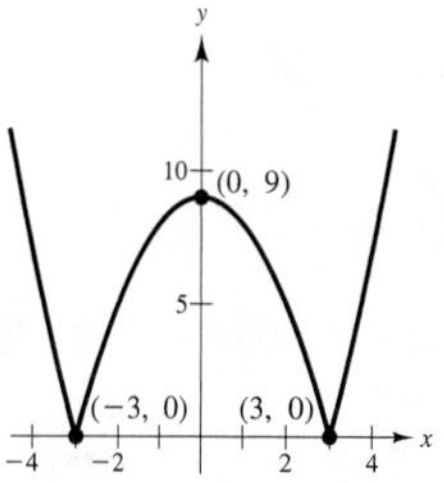

65.
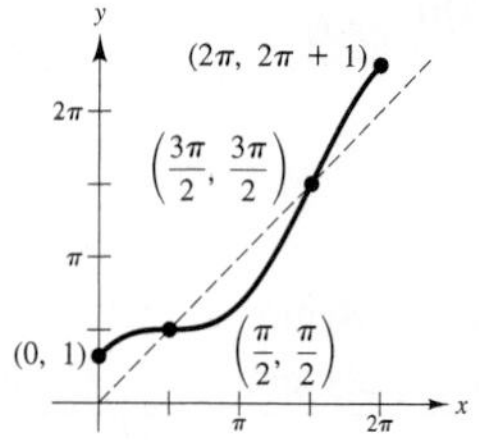

67. (a) and (b) Maximum: (1, 3)
Minimum: (1, 1)

69. $t \approx 4.92 \approx$ 4:55 P.M.; $d \approx 64$ km
71. (0, 0), (5, 0), (0, 10) **73.** Proof **75.** 14.05 ft
77. $3(3^{2/3} + 2^{2/3})^{3/2} \approx 21.07$ ft **79.** $v \approx 54.77$ mph
81. $-1.532, -0.347, 1.879$ **83.** $-1.164, 1.453$
85. $dy = (1 - \cos x + x \sin x)\, dx$
87. $dS = \pm 1.8\pi \text{ cm}^2, \dfrac{dS}{S} \times 100 \approx \pm 0.56\%$

$dV = \pm 8.1\pi \text{ cm}^3, \dfrac{dV}{V} \times 100 \approx \pm 0.83\%$

P.S. Problem Solving (page 245)

1. Choices of a may vary.

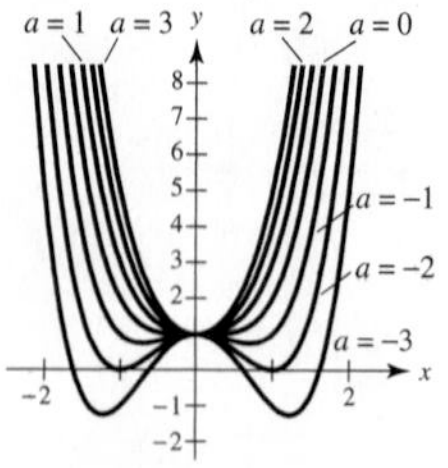

(a) One relative minimum at (0, 1) for $a \ge 0$
(b) One relative maximum at (0, 1) for $a < 0$
(c) Two relative minima for $a < 0$ when $x = \pm\sqrt{-a/2}$
(d) If $a < 0$, there are three critical points; if $a \ge 0$, there is only one critical point.

3. All c where c is a real number **5–7.** Proofs
9. About 9.19 ft
11. Minimum: $(\sqrt{2} - 1)d$; There is no maximum.
13. (a)–(c). Proofs
15. (a)

x	0	0.5	1	2
$\sqrt{1+x}$	1	1.2247	1.4142	1.7321
$\frac{1}{2}x + 1$	1	1.25	1.5	2

(b) Proof

17. (a)

v	20	40	60	80	100
s	5.56	11.11	16.67	22.22	27.78
d	5.1	13.7	27.2	44.2	66.4

$d(s) = 0.071s^2 + 0.389s + 0.727$

(b) The distance between the back of the first vehicle and the front of the second vehicle is $d(s)$, the safe stopping distance. The first vehicle passes the given point in $5.5/s$ seconds, and the second vehicle takes $d(s)/s$ more seconds. Hence,

$T = d(s)/s + 5.5/s.$

(c)
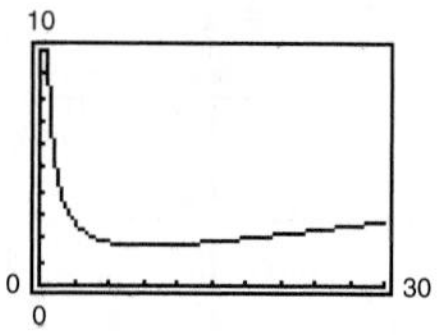

$s \approx 9.365$ m/sec

(d) $s \approx 9.365$ m/sec; 1.719 sec; 33.714 kph (e) 10.597 m

19. (a) $P(x) = x - x^2$
(b)
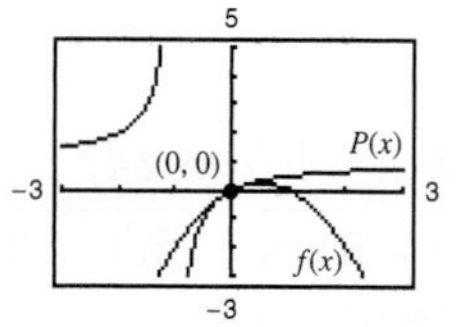

Chapter 4

Section 4.1 (page 255)

1–3. Proofs **5.** $y = t^3 + C$ **7.** $y = \frac{2}{5}x^{5/2} + C$

	Original Integral	Rewrite	Integrate	Simplify
9.	$\int \sqrt[3]{x}\, dx$	$\int x^{1/3}\, dx$	$\dfrac{x^{4/3}}{4/3} + C$	$\dfrac{3}{4}x^{4/3} + C$
11.	$\int \dfrac{1}{x\sqrt{x}}\, dx$	$\int x^{-3/2}\, dx$	$\dfrac{x^{-1/2}}{-1/2} + C$	$-\dfrac{2}{\sqrt{x}} + C$
13.	$\int \dfrac{1}{2x^3}\, dx$	$\dfrac{1}{2}\int x^{-3}\, dx$	$\dfrac{1}{2}\left(\dfrac{x^{-2}}{-2}\right) + C$	$-\dfrac{1}{4x^2} + C$

15. $\frac{1}{2}x^2 + 3x + C$ **17.** $x^2 - x^3 + C$ **19.** $\frac{1}{4}x^4 + 2x + C$
21. $\frac{2}{5}x^{5/2} + x^2 + x + C$ **23.** $\frac{3}{5}x^{5/3} + C$ **25.** $-1/(2x^2) + C$
27. $\frac{2}{15}x^{1/2}(3x^2 + 5x + 15) + C$ **29.** $x^3 + \frac{1}{2}x^2 - 2x + C$
31. $\frac{2}{7}y^{7/2} + C$ **33.** $x + C$ **35.** $-2\cos x + 3\sin x + C$
37. $t + \csc t + C$ **39.** $\tan\theta + \cos\theta + C$ **41.** $\tan y + C$

43. Answers will vary.
Example:

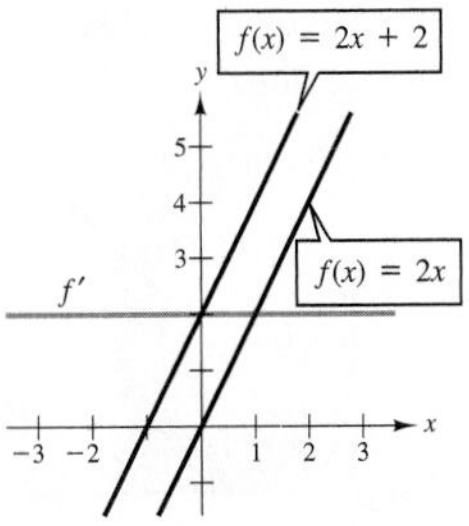

45. Answers will vary.
Example:

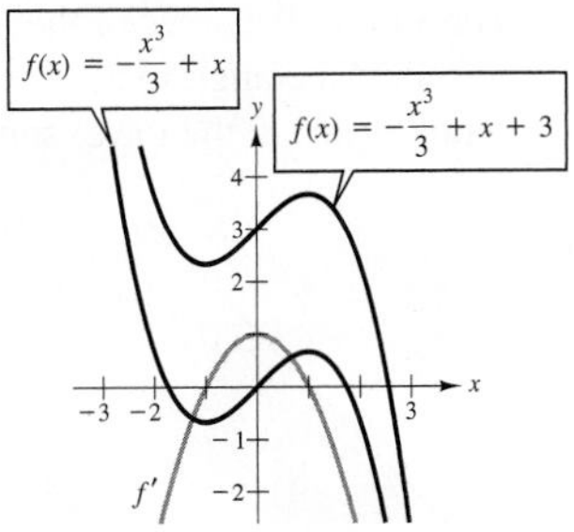

47. $y = x^2 - x + 1$

49. (a) Answers will vary.
Example:

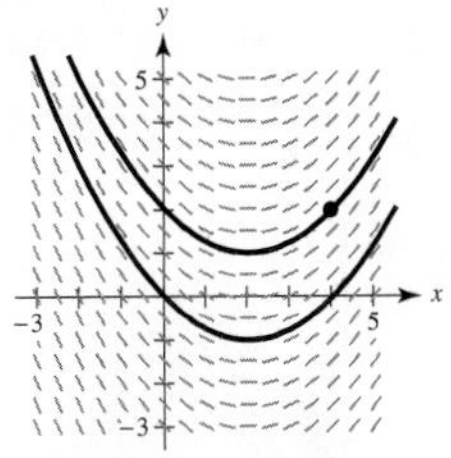

(b) $y = \frac{1}{4}x^2 - x + 2$

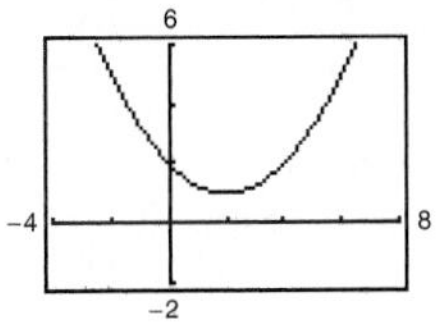

51. (a) Answers will vary.
Example:

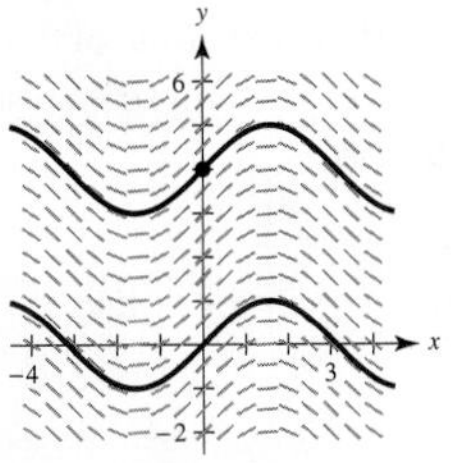

(b) $y = \sin x + 4$

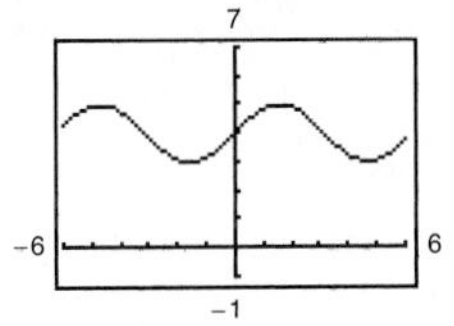

53. (a)

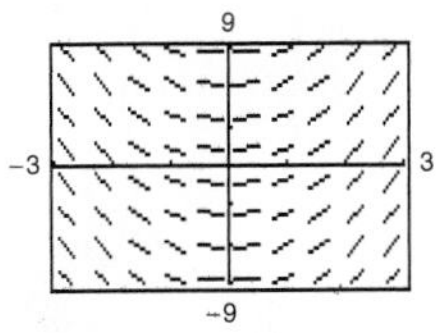

(b) $y = x^2 - 6$

(c)

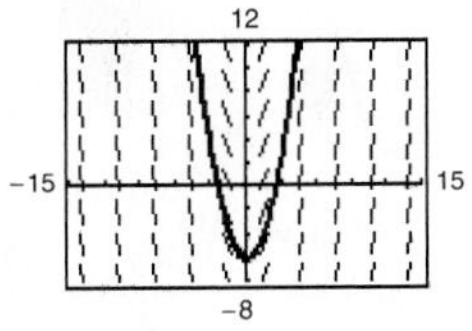

55. $f(x) = 2x^2 + 6$ **57.** $h(t) = 2t^4 + 5t - 11$

59. $f(x) = x^2 + x + 4$ **61.** $f(x) = -4\sqrt{x} + 3x$

63. (a) $h(t) = \frac{3}{4}t^2 + 5t + 12$ (b) 69 cm

65. (a) -1; $f'(4)$ represents the slope of f at $x = 4$.
(b) No. The slope of the tangent lines are greater than 2 on $[0, 2]$. Therefore, f must increase more than four units on $[0, 2]$.
(c) No. The function is decreasing on $[4, 5]$.
(d) 3.5; $f'(3.5) \approx 0$
(e) Concave upward: $(-\infty, 1)$, $(5, \infty)$
Concave downward: $(1, 5)$
Points of inflection at $x \approx 1$ and $x \approx 5$
(f) 3 (g)

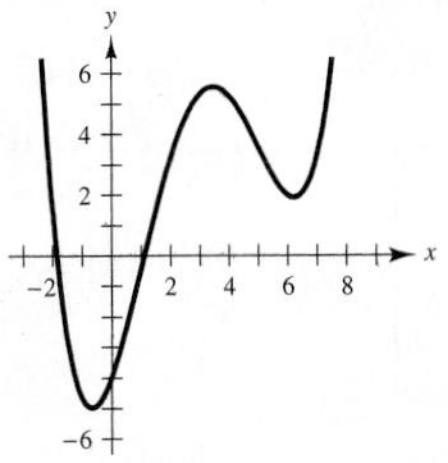

67. 62.25 ft **69.** $v_0 \approx 187.617$ ft/sec

71. $v(t) = -9.8t + C_1 = -9.8t + v_0$
$f(t) = -4.9t^2 + v_0 t + C_2 = -4.9t^2 + v_0 t + s_0$

73. 7.1 m **75.** 320 m; -32 m/sec

77. (a) $v(t) = 3t^2 - 12t + 9$; $a(t) = 6t - 12$
(b) $(0, 1)$, $(3, 5)$ (c) -3

79. $a(t) = -1/(2t^{3/2})$; $x(t) = 2\sqrt{t} + 2$

81. (a) 1.18 m/sec² (b) 190 m

83. (a) 300 ft (b) 60 ft/sec $\approx$ 41 mph

85. 7.45 ft/sec² **87.** True **89.** True

91. False. Let $f(x) = x$ and $g(x) = x + 1$.

93. $f(x) = \frac{1}{3}x^3 - 4x + \frac{16}{3}$

95. $f(x) = \begin{cases} x + 2, & 0 \le x < 2 \\ \frac{3}{2}x^2 - 2, & 2 \le x \le 5 \end{cases}$

f is not differentiable at $x = 2$ because the left- and right-hand derivatives at $x = 2$ do not agree.

97. Putnam Problem B2, 1991

Section 4.2 (page 267)

1. 35 **3.** $\frac{158}{85}$ **5.** $4c$ **7.** $\sum_{i=1}^{9} \frac{1}{3i}$ **9.** $\sum_{j=1}^{8} \left[5\left(\frac{j}{8}\right) + 3\right]$

11. $\frac{2}{n}\sum_{i=1}^{n}\left[\left(\frac{2i}{n}\right)^3 - \left(\frac{2i}{n}\right)\right]$ **13.** $\frac{3}{n}\sum_{i=1}^{n}\left[2\left(1 + \frac{3i}{n}\right)^2\right]$

15. 420 **17.** 2470 **19.** 12,040 **21.** 2930

23. The area of the shaded region falls between 12.5 units² and 16.5 units²

25. The area of the shaded region falls between 7 units² and 11 units².

27. $A \approx S \approx 0.768$
$A \approx s \approx 0.518$

29. $A \approx S \approx 0.746$
$A \approx s \approx 0.646$

31. $\frac{81}{4}$ **33.** 9

35. $(n + 2)/n$
$n = 10$; $S = 1.2$
$n = 100$; $S = 1.02$
$n = 1000$; $S = 1.002$
$n = 10{,}000$; $S = 1.0002$

37. $[2(n + 1)(n - 1)]/n^2$
$n = 10$; $S = 1.98$
$n = 100$; $S = 1.9998$
$n = 1000$; $S = 1.999998$
$n = 10{,}000$; $S = 1.99999998$

39. $\lim_{n\to\infty}\left[8\left(\frac{n^2 + n}{n^2}\right)\right] = 8$ **41.** $\lim_{n\to\infty}\frac{1}{6}\left(\frac{2n^3 - 3n^2 + n}{n^3}\right) = \frac{1}{3}$

43. $\lim_{n\to\infty}[(3n + 1)/n] = 3$

45. (a)

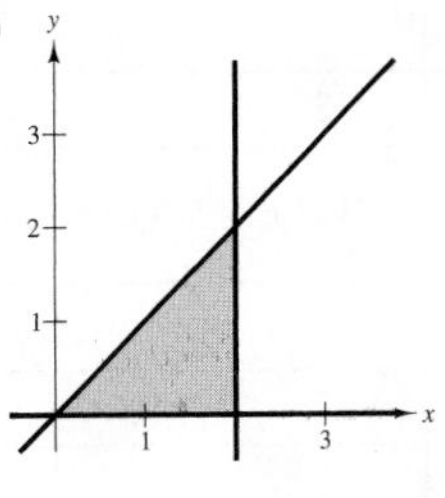

(b) $\Delta x = (2 - 0)/n = 2/n$

(c) $s(n) = \sum_{i=1}^{n} f(x_{i-1})\,\Delta x$
$= \sum_{i=1}^{n}[(i - 1)(2/n)](2/n)$

(d) $S(n) = \sum_{i=1}^{n} f(x_i)\,\Delta x$
$= \sum_{i=1}^{n}[i(2/n)](2/n)$

(e)

n	5	10	50	100
$s(n)$	1.6	1.8	1.96	1.98
$S(n)$	2.4	2.2	2.04	2.02

(f) $\lim_{n\to\infty}\sum_{i=1}^{n}[(i - 1)(2/n)](2/n) = 2$; $\lim_{n\to\infty}\sum_{i=1}^{n}[i(2/n)](2/n) = 2$

47. $A = 2$

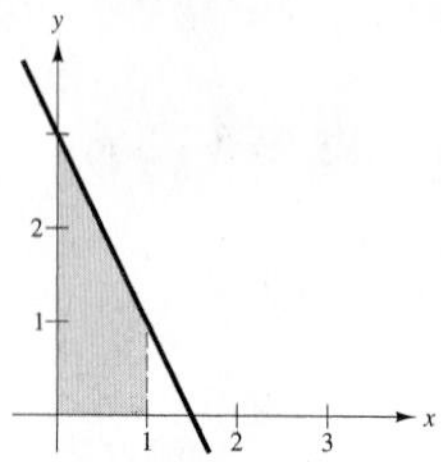

49. $A = \frac{7}{3}$

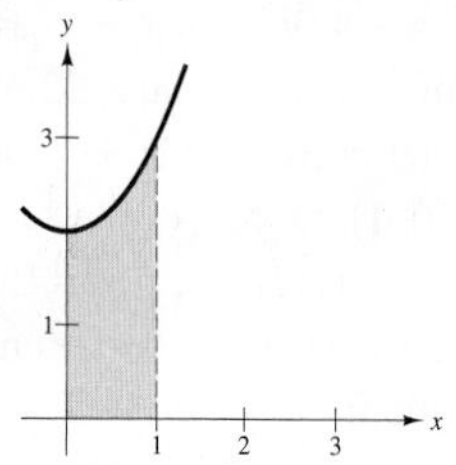

51. $A = \frac{70}{3}$

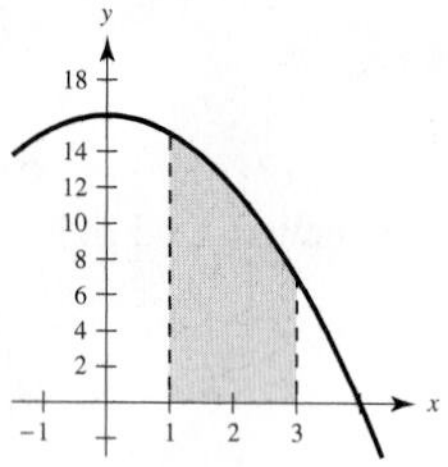

53. $A = \frac{513}{4}$

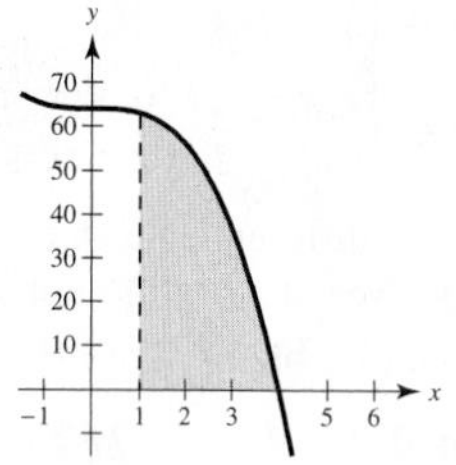

55. $A = \frac{2}{3}$

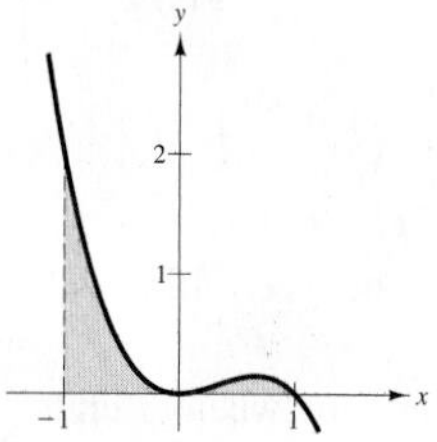

57. $A = 6$

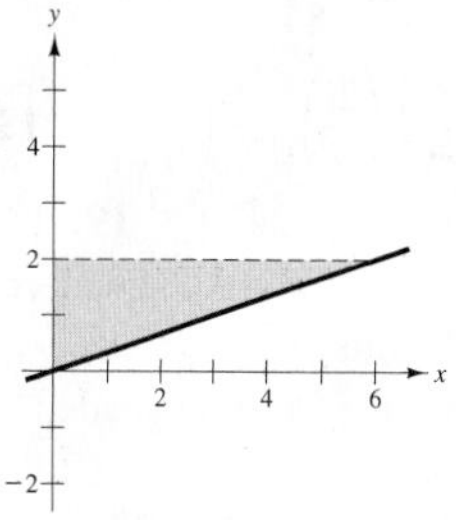

59. $A = 9$

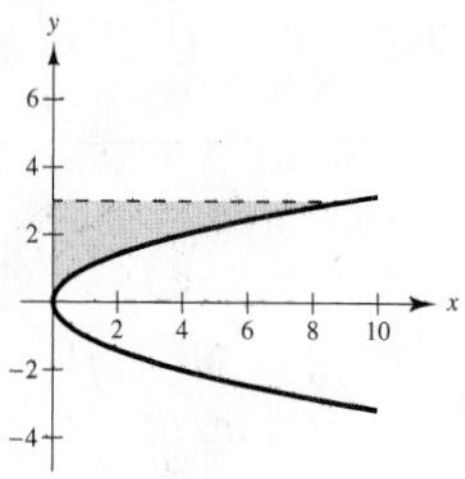

61. $A = \frac{44}{3}$

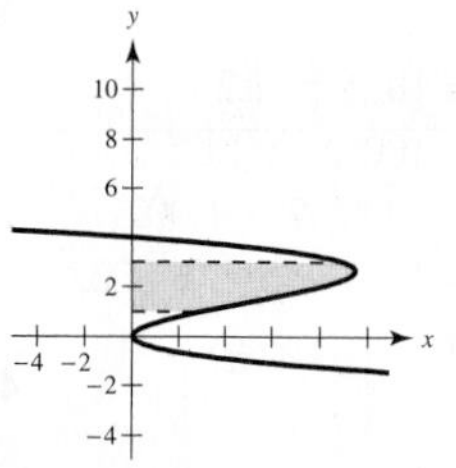

63. $\frac{69}{8}$ **65.** 0.345

67.

n	4	8	12	16	20
Approximate Area	5.3838	5.3523	5.3439	5.3403	5.3384

69.

n	4	8	12	16	20
Approximate Area	2.2223	2.2387	2.2418	2.2430	2.2435

71. b

73. You can use the line $y = x$ bounded by $x = a$ and $x = b$. The sum of the areas of the inscribed rectangles in the figure below is the lower sum.

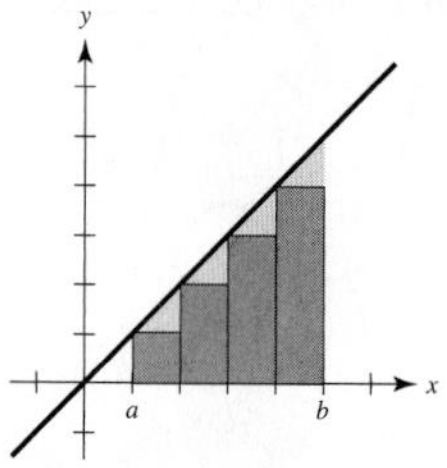

The sum of the areas of the circumscribed rectangles in the figure below is the upper sum.

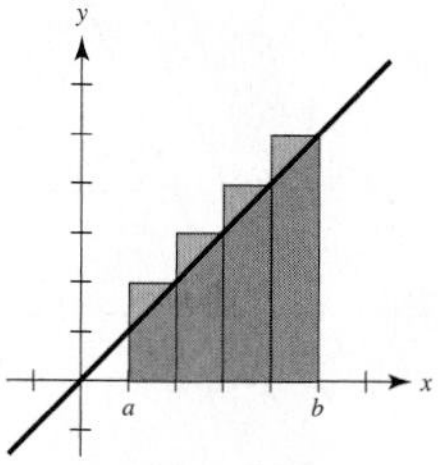

The rectangles in the first graph do not contain all of the area of the region, and the rectangles in the second graph cover more than the area of the region. The exact value of the area lies between these two sums.

75. (a)

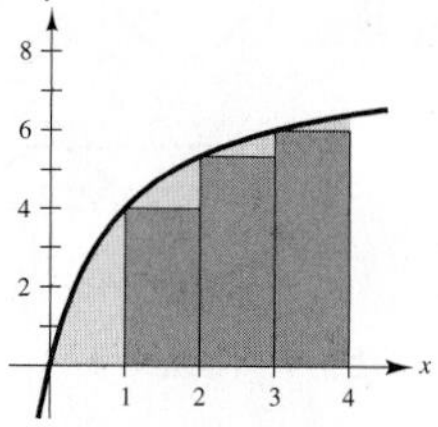

$s(4) = \frac{46}{3}$

(b)

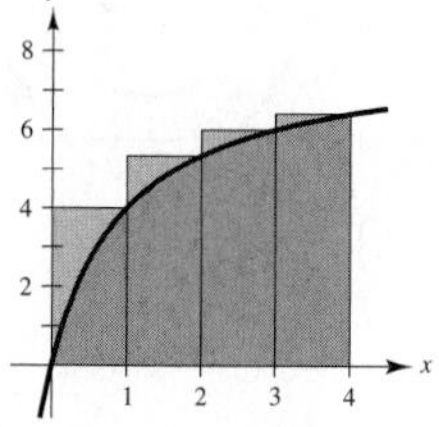

$S(4) = \frac{326}{15}$

(c)

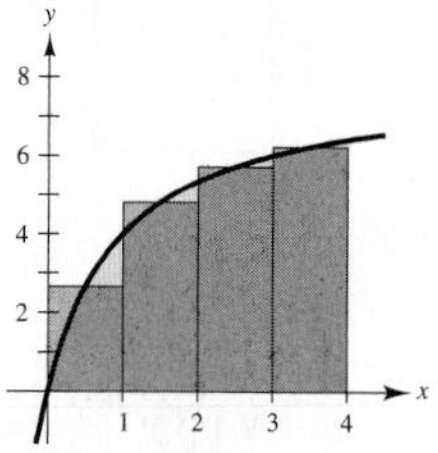

$M(4) = \frac{6112}{315}$

(d) Proof

(e)

n	4	8	20	100	200
$s(n)$	15.333	17.368	18.459	18.995	19.060
$S(n)$	21.733	20.568	19.739	19.251	19.188
$M(n)$	19.403	19.201	19.137	19.125	19.125

(f) Because f is an increasing function, $s(n)$ is always increasing and $S(n)$ is always decreasing.

77. True

79. Suppose there are n rows in the figure. The stars on the left total $1 + 2 + \cdots + n$, as do the stars on the right. There are $n(n + 1)$ stars in total. So, $2[1 + 2 + \cdots + n] = n(n + 1)$
$1 + 2 + \cdots + n = [n(n + 1)]/2$.

81. (a) $y = (-4.09 \times 10^{-5})x^3 + 0.016x^2 - 2.67x + 452.9$

(b)

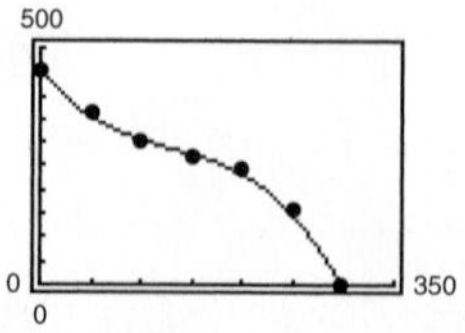

(c) $76{,}897.5 \text{ ft}^2$

83. Proof

Section 4.3 (page 278)

1. $2\sqrt{3} \approx 3.464$ **3.** 36 **5.** 0 **7.** $\frac{10}{3}$ **9.** $\int_{-1}^{5} (3x + 10)\,dx$

11. $\int_0^3 \sqrt{x^2 + 4}\,dx$ **13.** $\int_0^5 3\,dx$ **15.** $\int_{-4}^{4} (4 - |x|)\,dx$

17. $\int_{-2}^{2} (4 - x^2)\,dx$ **19.** $\int_0^{\pi} \sin x\,dx$ **21.** $\int_0^2 y^3\,dy$

23.

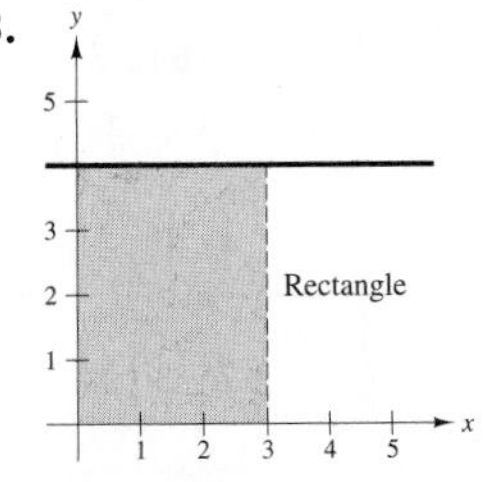

$A = 12$

25.

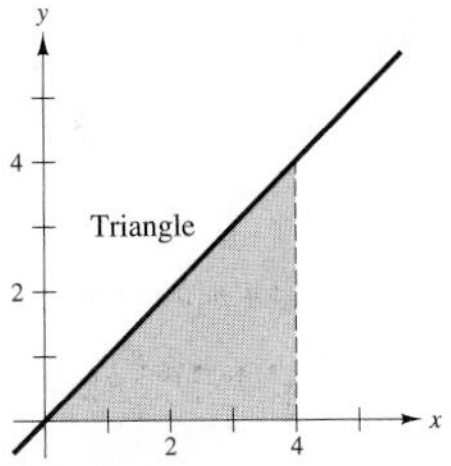

$A = 8$

27.

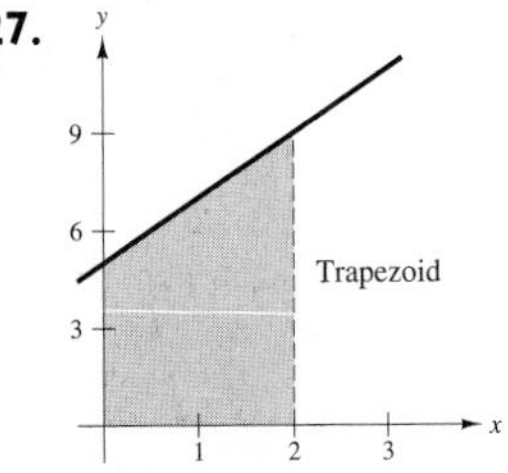

$A = 14$

29.

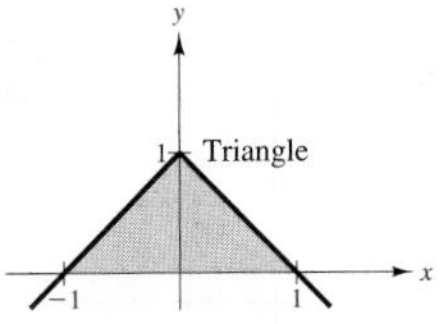

$A = 1$

31.

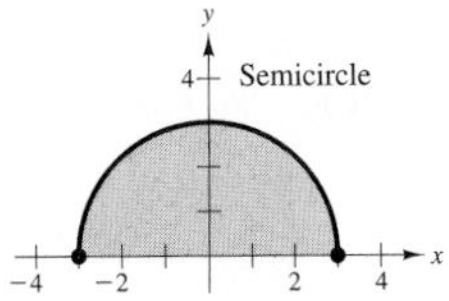

$A = 9\pi/2$

33. -6 **35.** 24 **37.** -10

39. 16 **41.** (a) 13 (b) -10 (c) 0 (d) 30

43. (a) 8 (b) -12 (c) -4 (d) 30 **45.** $-48, 88$

47. (a) $-\pi$ (b) 4 (c) $-(1 + 2\pi)$ (d) $3 - 2\pi$ (e) $5 + 2\pi$ (f) $23 - 2\pi$

49. (a) 14 (b) 4 (c) 8 (d) 0 **51.** $\sum_{i=1}^{n} f(x_i)\,\Delta x > \int_1^5 f(x)\,dx$

53. No. There is a discontinuity at $x = 4$. **55.** a **57.** d

59.

n	4	8	12	16	20
$L(n)$	3.6830	3.9956	4.0707	4.1016	4.1177
$M(n)$	4.3082	4.2076	4.1838	4.1740	4.1690
$R(n)$	3.6830	3.9956	4.0707	4.1016	4.1177

61.

n	4	8	12	16	20
$L(n)$	0.5890	0.6872	0.7199	0.7363	0.7461
$M(n)$	0.4076	0.5909	0.6551	0.7854	0.7854
$R(n)$	0.9817	0.8836	0.8508	0.8345	0.8247

63. True **65.** True

67. False: $\int_0^2 (-x)\,dx = -2$ **69.** 272 **71.** Proof

73. No. No matter how small the subintervals, the number of both rational and irrational numbers within each subinterval is infinite and $f(c_i) = 0$ or $f(c_i) = 1$.

75. $a = -1$ and $b = 1$ maximize the integral. **77.** $\frac{1}{3}$

Section 4.4 (page 291)

1.

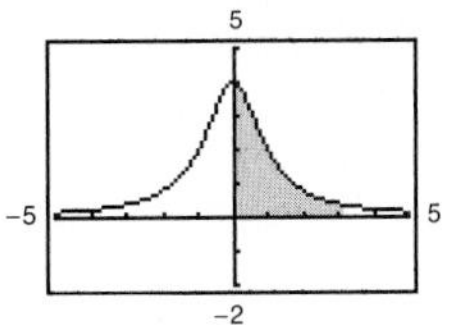

Positive

3.

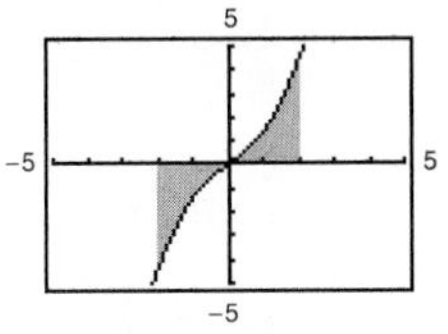

Zero

5. 1 **7.** $-\frac{5}{2}$ **9.** $-\frac{10}{3}$ **11.** $\frac{1}{3}$ **13.** $\frac{1}{2}$ **15.** $\frac{2}{3}$ **17.** -4

19. $-\frac{1}{18}$ **21.** $-\frac{27}{20}$ **23.** $\frac{9}{2}$ **25.** $\frac{23}{3}$ **27.** $\pi + 2$

29. $2\sqrt{3}/3$ **31.** 0 **33.** $\frac{1}{6}$ **35.** $12\sqrt{3}/5$ **37.** 1 **39.** 10

41. 6 **43.** 0.4380, 1.7908 **45.** $\pm\arccos\sqrt{\pi}/2 \approx \pm 0.4817$

47. Average value $= \frac{8}{3}$; $x = \pm 2\sqrt{3}/3 \approx \pm 1.155$

49. Average value $= 2/\pi$; $x \approx 0.690$, $x \approx 2.451$

51. about 540 ft

53. The Fundamental Theorem of Calculus states that if a function f is continuous on $[a, b]$ and F is an antiderivative of f on $[a, b]$, then $\int_a^b f(x)\,dx = F(b) - F(a)$.

55. -1.5 **57.** 6.5 **59.** 15.5

61. (a) $F(x) = 500 \sec^2 x$ (b) $1500\sqrt{3}/\pi \approx 827$ N

63. ≈ 0.5318 L

65. (a) $v = -0.00086t^3 + 0.0782t^2 - 0.208t + 0.10$

(b)

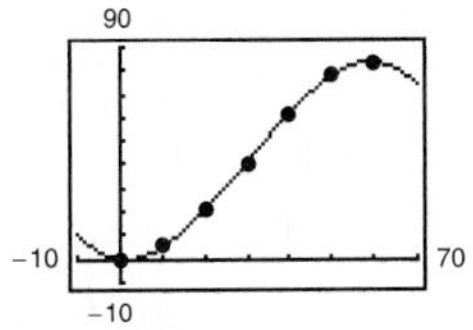

(c) 2475.6 m

67. $F(x) = \frac{1}{2}x^2 - 5x$
$F(2) = -8$
$F(5) = -12\frac{1}{2}$
$F(8) = -8$

69. $F(x) = -10/x + 10$
$F(2) = 5$
$F(5) = 8$
$F(8) = 8\frac{3}{4}$

71. $F(x) = \sin x - \sin 1$
$F(2) = \sin 2 - \sin 1 \approx 0.0678$
$F(5) = \sin 5 - \sin 1 \approx -1.8004$
$F(8) = \sin 8 - \sin 1 \approx 0.1479$

73. (a) $g(0) = 0$, $g(2) \approx 7$, $g(4) \approx 9$, $g(6) \approx 8$, $g(8) \approx 5$
(b) Increasing: $(0, 4)$; Decreasing: $(4, 8)$
(c) A maximum occurs at $x = 4$.
(d)

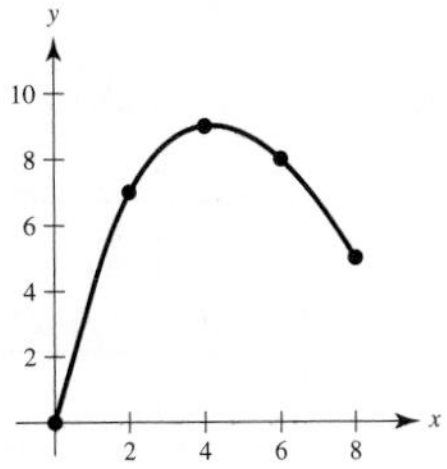

75. $\frac{1}{2}x^2 + 2x$ **77.** $\frac{3}{4}x^{4/3} - 12$ **79.** $\tan x - 1$

81. $x^2 - 2x$ **83.** $\sqrt{x^4 + 1}$ **85.** $x \cos x$ **87.** 8

89. $\cos x\sqrt{\sin x}$ **91.** $3x^2 \sin x^6$

93.

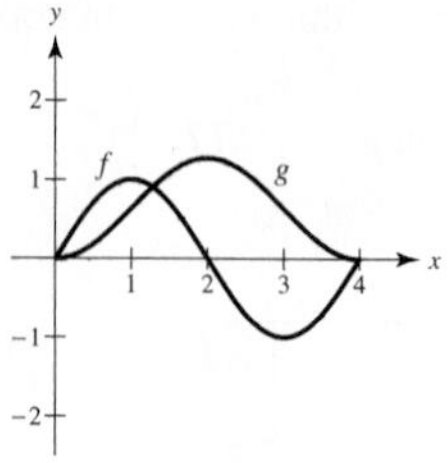

An extremum of g occurs at $x = 2$.

95. (a) $C(x) = 1000(12x^{5/4} + 125)$
(b) $C(1) = \$137{,}000$
$C(5) \approx \$214{,}721$
$C(10) \approx \$338{,}394$

97. 28 units **99.** 2 units **101.** True

103. $f(x) = x^{-2}$ has a nonremovable discontinuity at $x = 0$.

105. $f'(x) = \dfrac{1}{(1/x)^2 + 1}\left(-\dfrac{1}{x^2}\right) + \dfrac{1}{x^2 + 1} = 0$

Because $f'(x) = 0$, $f(x)$ is constant.

Section 4.5 (page 304)

$\int f(g(x))g'(x)\,dx$	$u = g(x)$	$du = g'(x)\,dx$
1. $\int (5x^2 + 1)^2(10x)\,dx$	$5x^2 + 1$	$10x\,dx$
3. $\int \dfrac{x}{\sqrt{x^2 + 1}}\,dx$	$x^2 + 1$	$2x\,dx$
5. $\int \tan^2 x \sec^2 x\,dx$	$\tan x$	$\sec^2 x\,dx$

7. $\frac{1}{5}(1 + 2x)^5 + C$ **9.** $\frac{2}{3}(9 - x^2)^{3/2} + C$

11. $\frac{1}{12}(x^4 + 3)^3 + C$ **13.** $\frac{1}{15}(x^3 - 1)^5 + C$

15. $\frac{1}{3}(t^2 + 2)^{3/2} + C$ **17.** $-\frac{15}{8}(1 - x^2)^{4/3} + C$

19. $1/[4(1 - x^2)^2] + C$ **21.** $-1/[3(1 + x^3)] + C$

23. $-\sqrt{1 - x^2} + C$ **25.** $-\frac{1}{4}(1 + 1/t)^4 + C$ **27.** $\sqrt{2x} + C$

29. $\frac{2}{5}x^{5/2} + 2x^{3/2} + 14x^{1/2} + C = \frac{2}{5}\sqrt{x}(x^2 + 5x + 35) + C$

31. $\frac{1}{4}t^4 - t^2 + C$ **33.** $6y^{3/2} - \frac{2}{5}y^{5/2} + C = \frac{2}{5}y^{3/2}(15 - y) + C$

35. $2x^2 - 4\sqrt{16 - x^2} + C$ **37.** $-1/[2(x^2 + 2x - 3)] + C$

39. (a) Answers will vary.
Example:

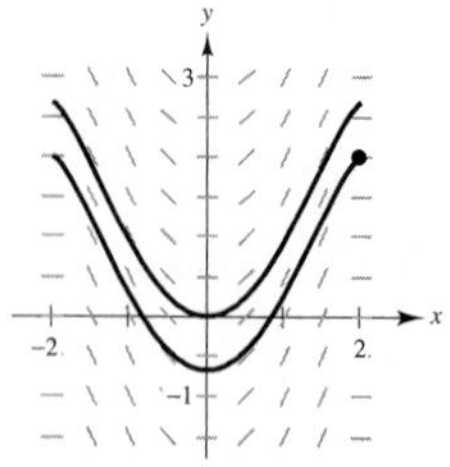

(b) $y = -\frac{1}{3}(4 - x^2)^{3/2} + 2$

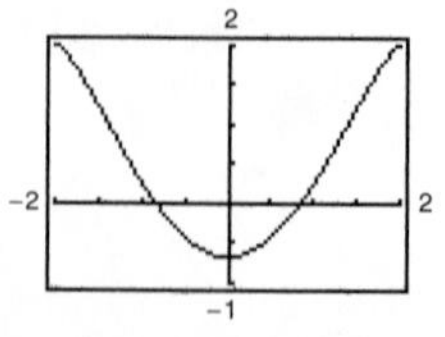

41. (a) Answers will vary.
Example:

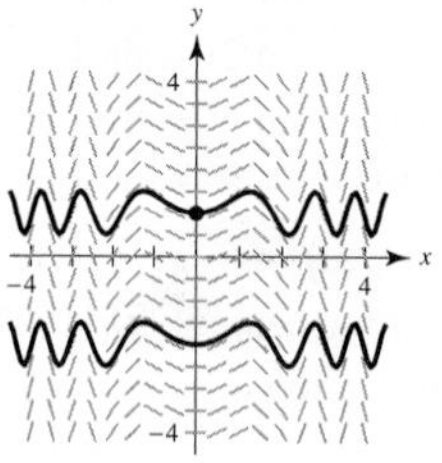

(b) $y = \frac{1}{2}\sin x^2 + 1$

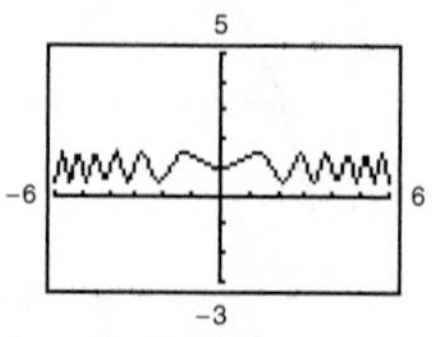

43. $-\cos(\pi x) + C$ **45.** $-\frac{1}{2}\cos 2x + C$ **47.** $-\sin(1/\theta) + C$

49. $\frac{1}{4}\sin^2 2x + C_1$ or $-\frac{1}{4}\cos^2 2x + C_2$ or $-\frac{1}{8}\cos 4x + C_3$

51. $\frac{1}{5}\tan^5 x + C$ **53.** $\frac{1}{2}\tan^2 x + C$ or $\frac{1}{2}\sec^2 x + C_1$

55. $-\cot x - x + C$ **57.** $f(x) = 2\sin(x/2) + 3$

59. $f(x) = -\frac{1}{4}\cos 4x - 1$ **61.** $f(x) = \frac{1}{12}(4x^2 - 10)^3 - 8$

63. $\frac{2}{15}(x + 2)^{3/2}(3x - 4) + C$

65. $-\frac{2}{105}(1 - x)^{3/2}(15x^2 + 12x + 8) + C$

67. $\left(\sqrt{2x - 1}/15\right)(3x^2 + 2x - 13) + C$

69. $-x - 1 - 2\sqrt{x + 1} + C$ or $-\left(x + 2\sqrt{x + 1}\right) + C_1$

71. 0 **73.** $12 - \frac{8}{9}\sqrt{2}$ **75.** 2 **77.** $\frac{1}{2}$ **79.** $\frac{4}{15}$ **81.** $3\sqrt{3}/4$

83. $f(x) = (2x^3 + 1)^3 + 3$ **85.** $f(x) = \sqrt{2x^2 - 1} - 3$

87. $1209/28$ **89.** 4 **91.** $2\left(\sqrt{3} - 1\right)$

93. $\frac{10}{3}$

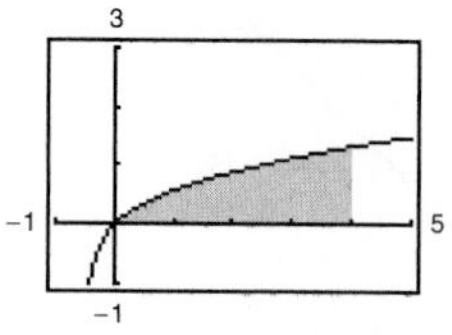

95. $\frac{144}{5}$

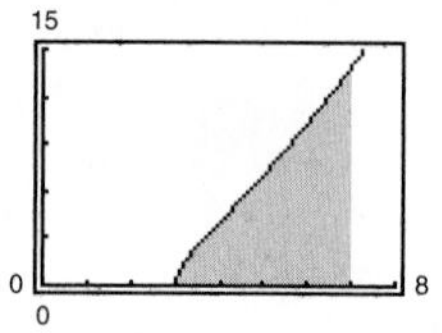

97. 7.38

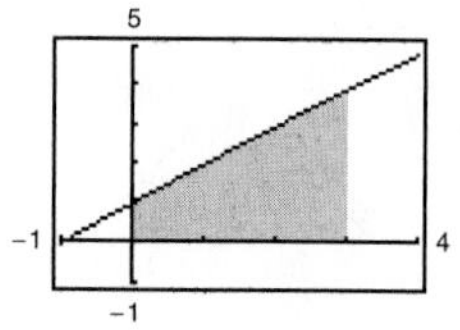

99. $\frac{1}{6}(2x - 1)^3 + C_1 = \frac{4}{3}x^3 - 2x^2 + x - \frac{1}{6} + C_1$
or $\frac{4}{3}x^3 - 2x^2 + x + C_2$
Answers differ by a constant: $C_2 = C_1 - \frac{1}{6}$

101. $\frac{272}{15}$ **103.** $\frac{2}{3}$ **105.** (a) $\frac{8}{3}$ (b) $\frac{16}{3}$ (c) $-\frac{8}{3}$ (d) 8

107. $2\int_0^4 (6x^2 - 3)\,dx = 232$

109. If $u = 5 - x^2$, then $du = -2x\,dx$ and
$\int x(5 - x^2)^3\,dx = -\frac{1}{2}\int(5 - x^2)^3(-2x)\,dx = -\frac{1}{2}\int u^3\,du$.

111. $250,000

113. (a) Relative minimum: (6.4, 0.7) or June
Relative maximum: (0.4, 5.5) or January
(b) 37.47 in. (c) 4.33 in.

115. (a)

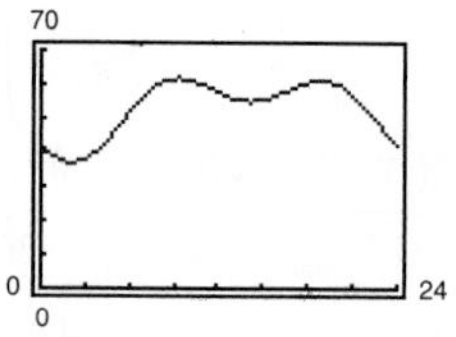

Maximum flow:
$R \approx 61.713$ at $t = 9.36$.

(b) 1272 thousand gallons

117. (a) $P_{50,\,75} \approx 35.3\%$ (b) $b \approx 58.6\%$

119. (a) $9.17 (b) $3.14

121. (a)

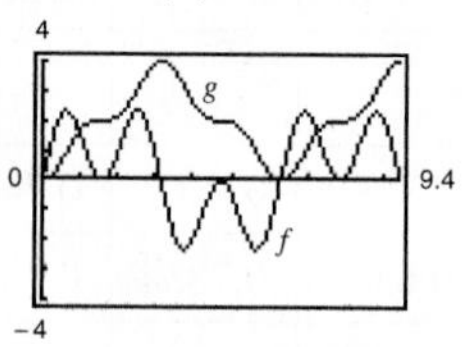

(b) g is nonnegative because the graph of f is positive at the beginning, and generally has more positive sections than negative ones.

(c) The points on g that correspond to the extrema of f are points of inflection of g.

(d) No, some zeros of f, such as $x = \pi/2$, do not correspond to extrema of g. The graph of g continues to increase after $x = \pi/2$ because f remains above the x-axis.

(e)

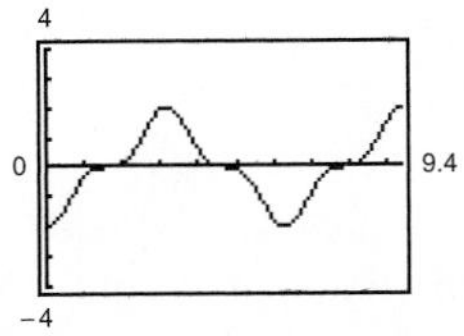

The graph of h is that of g shifted 2 units downward.

123. (a) Proof (b) Proof

125. False. $\int (2x+1)^2\,dx = \frac{1}{6}(2x+1)^3 + C$

127. True **129.** True **131–133.** Proofs

135. Putnam Problem A1, 1958

Section 4.6 (page 314)

	Trapezoidal	*Simpson's*	*Exact*
1.	2.7500	2.6667	2.6667
3.	4.2500	4.0000	4.0000
5.	4.0625	4.0000	4.0000
7.	12.6640	12.6667	12.6667
9.	0.1676	0.1667	0.1667

	Trapezoidal	*Simpson's*	*Graphing Utility*
11.	3.2833	3.2396	3.2413
13.	0.3415	0.3720	0.3927
15.	0.9567	0.9778	0.9775
17.	0.0891	0.0888	0.0891
19.	0.1940	0.1860	0.1858

21. The Trapezoidal Rule will yield a result greater than $\int_a^b f(x)\,dx$ if f is concave upward on $[a, b]$ because the graph of f will lie within the trapezoids.

23. (a) 0.500 (b) 0.000 **25.** (a) 0.01 (b) 0.0005

27. (a) 0.1615 (b) 0.0066 **29.** (a) $n = 366$ (b) $n = 26$

31. (a) $n = 77$ (b) $n = 8$ **33.** (a) $n = 287$ (b) $n = 16$

35. (a) $n = 130$ (b) $n = 12$ **37.** (a) $n = 643$ (b) $n = 48$

39. (a) 24.5 (b) 25.67 **41.** Answers will vary.

43.

n	$L(n)$	$M(n)$	$R(n)$	$T(n)$	$S(n)$
4	0.8739	0.7960	0.6239	0.7489	0.7709
8	0.8350	0.7892	0.7100	0.7725	0.7803
10	0.8261	0.7881	0.7261	0.7761	0.7818
12	0.8200	0.7875	0.7367	0.7783	0.7826
16	0.8121	0.7867	0.7496	0.7808	0.7836
20	0.8071	0.7864	0.7571	0.7821	0.7841

45. 0.701 **47.** 10,233.58 ft-lb **49.** 3.1416

51. 89,250 m^2 **53–55.** Proofs

Review Exercises for Chapter 4 (page 316)

1.

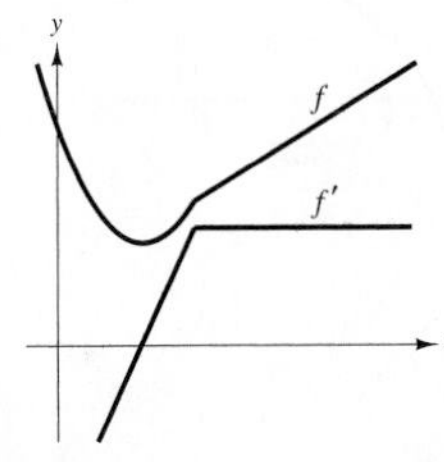

3. $\frac{2}{3}x^3 + \frac{1}{2}x^2 - x + C$

5. $x^2/2 - 1/x + C$ **7.** $2x^2 + 3\cos x + C$ **9.** $y = 2 - x^2$

11. (a) Answers will vary.
Example:

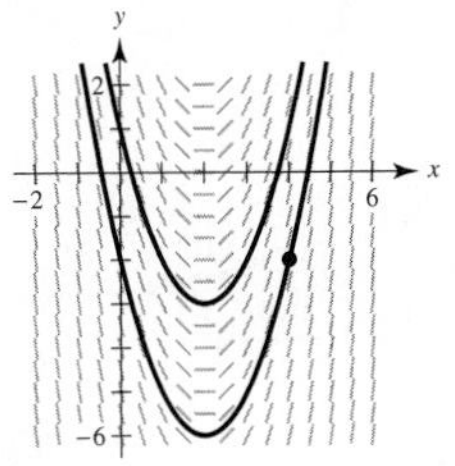

(b) $y = x^2 - 4x - 2$

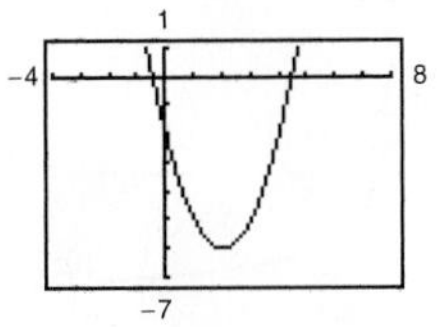

13. 240 ft/sec **15.** (a) 3 sec (b) 144 ft (c) $\frac{3}{2}$ sec (d) 108 ft

17. $\sum_{n=1}^{8} \frac{1}{4n}$ **19.** $\frac{3}{n}\sum_{i=1}^{n}\left(\frac{i+1}{n}\right)^2$ **21.** 165 **23.** 3310

25. (a) $\sum_{i=1}^{10}(2i-1)$ (b) $\sum_{i=1}^{n} i^3$ (c) $\sum_{i=1}^{10}(4i+2)$

27. $9.038 < (\text{Area of region}) < 13.038$

29. $A = 16$

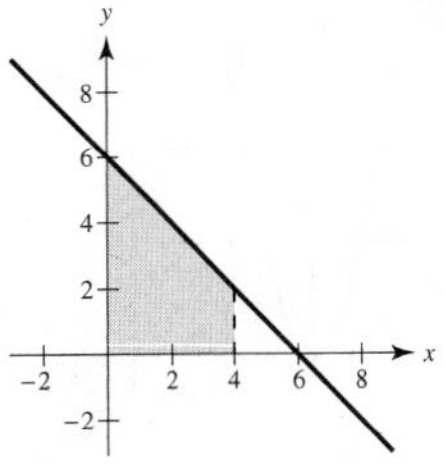

31. $A = 12$

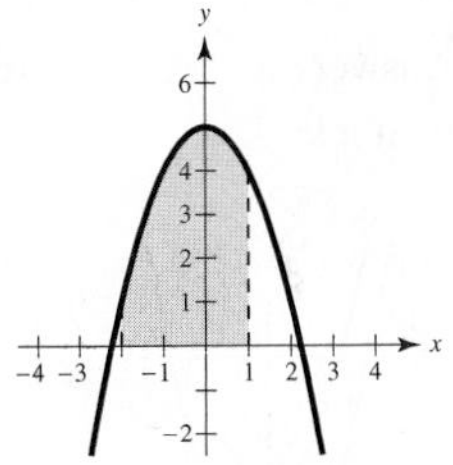

33. $\frac{27}{2}$ **35.** $\int_4^6 (2x-3)\,dx$ **37.** $\int_{-2}^{0} (3x+6)\,dx$

39.

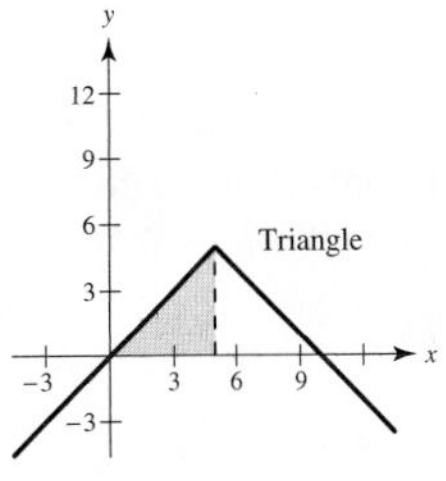

$A = \frac{25}{2}$

41. (a) 13 (b) 7 (c) 11 (d) 50

43. 16 **45.** 0 **47.** $\frac{422}{5}$ **49.** $(\sqrt{2} + 2)/2$

51.

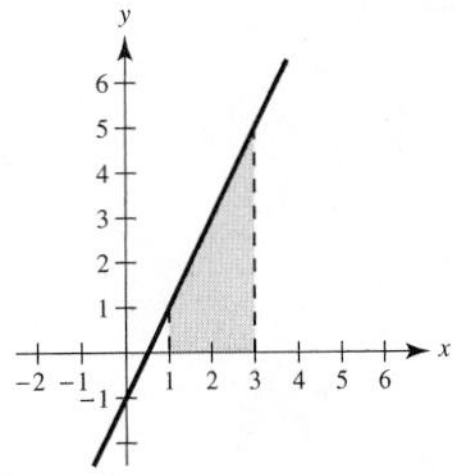

$A = 6$

53.

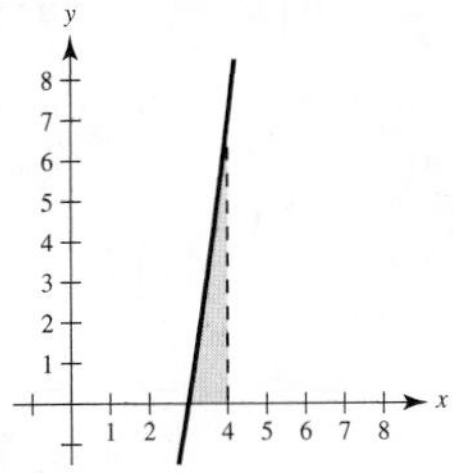

$A = \frac{10}{3}$

55.

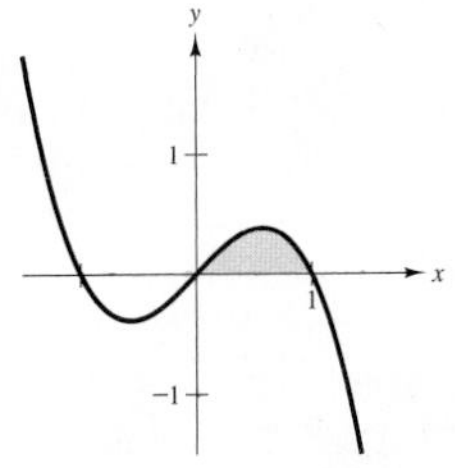

$A = \frac{1}{4}$

57. $-\cos 1 + 1 \approx 0.460$

59.

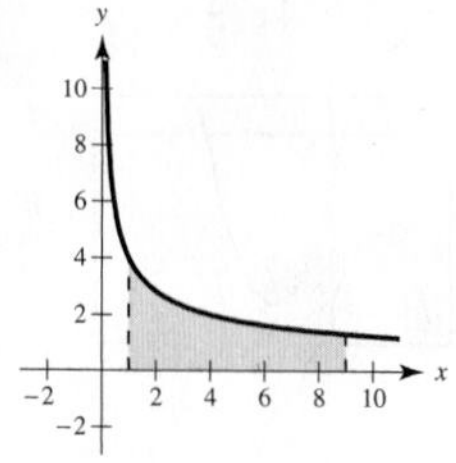

$A = 16$

61. Average value $= \frac{2}{5}$, $x = \frac{25}{4}$

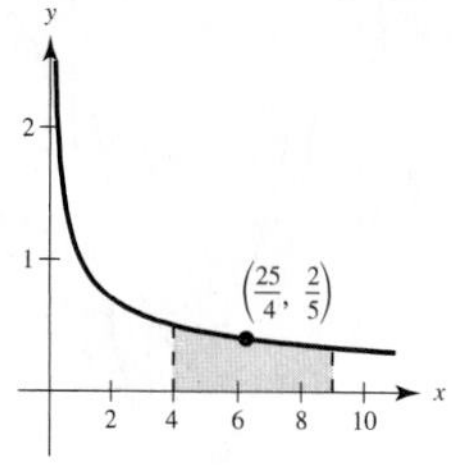

63. $x^2\sqrt{1 + x^3}$ **65.** $x^2 + 3x + 2$

67. $\frac{1}{7}x^7 + \frac{3}{5}x^5 + x^3 + x + C$ **69.** $\frac{2}{3}\sqrt{x^3 + 3} + C$

71. $-\frac{1}{30}(1 - 3x^2)^5 + C = \frac{1}{30}(3x^2 - 1)^5 + C$

73. $\frac{1}{4}\sin^4 x + C$ **75.** $2\sqrt{1 - \cos\theta} + C$

77. $\dfrac{\tan^{n+1} x}{n + 1} + C$ **79.** $\dfrac{1}{3\pi}(1 + \sec \pi x)^3 + C$

81. $-9/4$ **83.** 2 **85.** $28\pi/15$ **87.** 2

89. (a) Answers will vary. (b) $y = -\frac{1}{3}(9 - x^2)^{3/2} + 5$
Example:

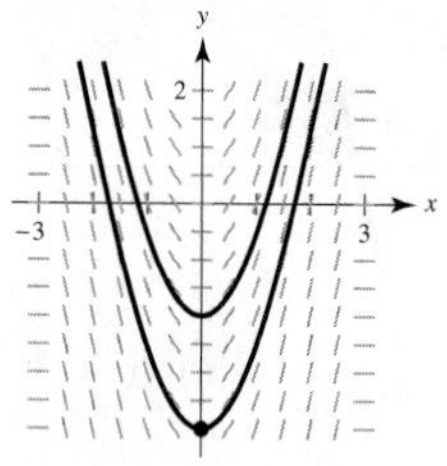

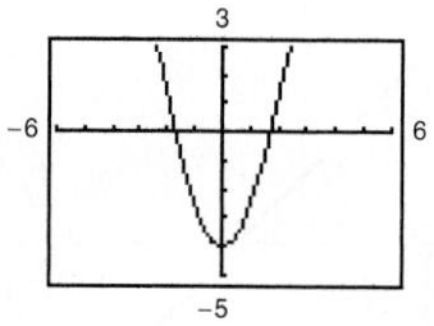

91. $\frac{468}{7}$ **93.** (a) $24{,}300/M$ (b) $27{,}300/M$

95. Trapezoidal Rule: 0.257
Simpson's Rule: 0.254
Graphing Utility: 0.254

97. Trapezoidal Rule: 0.637
Simpson's Rule: 0.685
Graphing Utility: 0.704

P.S. Problem Solving (page 319)

1. (a) $L(1) = 0$ (b) $L'(x) = 1/x$, $L'(1) = 1$
(c) $x \approx 2.718$ (d) Proof

3. (a) $\displaystyle\lim_{n\to\infty}\left[\frac{32}{n^5}\sum_{i=1}^{n} i^4 - \frac{64}{n^4}\sum_{i=1}^{n} i^3 + \frac{32}{n^3}\sum_{i=1}^{n} i^2\right]$
(b) $(16n^4 - 16)/(15n^4)$ (c) $16/15$

5. (a)

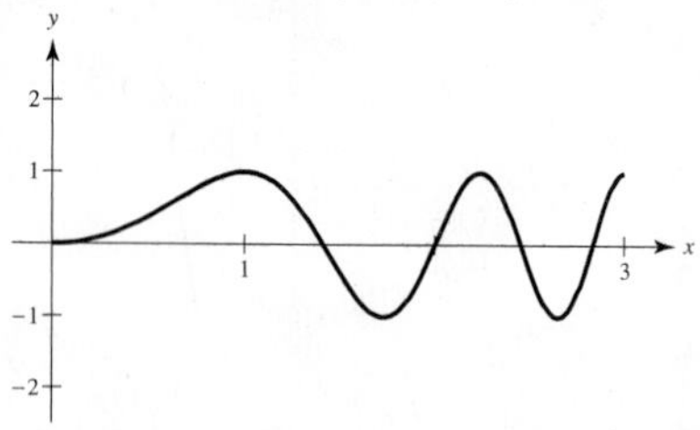

(b)

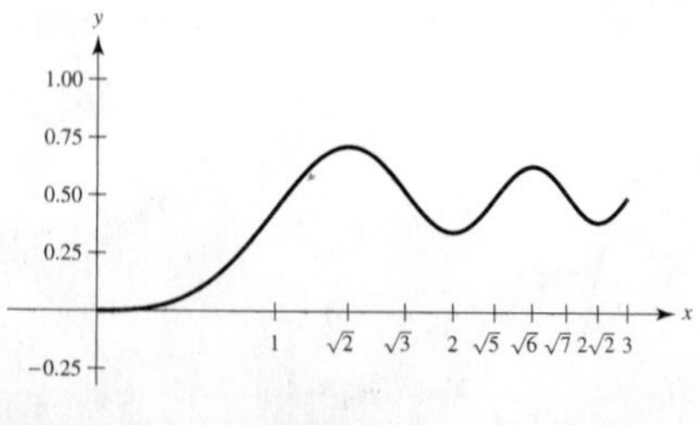

(c) Relative maxima at $x = \sqrt{2}, \sqrt{6}$
Relative minima at $x = 2, 2\sqrt{2}$
(d) Points of inflection at $x = 1, \sqrt{3}, \sqrt{5}, \sqrt{7}$

7. (a)

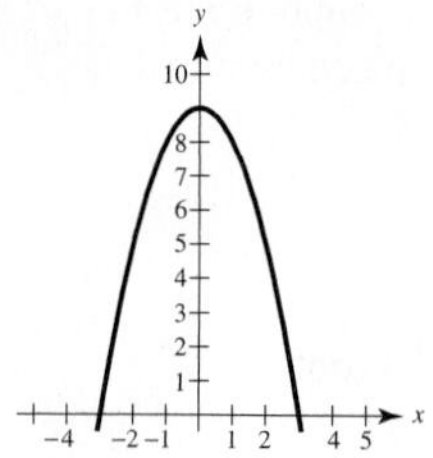

Area $= 36$

(b) Base $= 6$, height $= 9$
Area $= \frac{2}{3}bh = \frac{2}{3}(6)(9) = 36$
(c) Proof

9. (a)

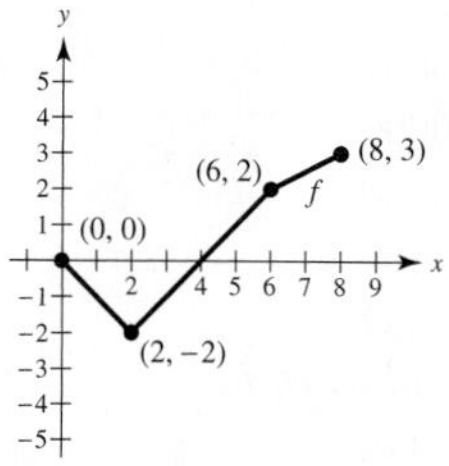

(b)

x	0	1	2	3
$F(x)$	0	$-\frac{1}{2}$	-2	$-\frac{7}{2}$

x	4	5	6	7	8
$F(x)$	-4	$-\frac{7}{2}$	-2	$\frac{1}{4}$	3

(c) $x = 4, 8$ (d) $x = 2$

11. Proof **13.** $\frac{2}{3}$ **15.** $1 \le \displaystyle\int_0^1 \sqrt{1 + x^4}\,dx \le \sqrt{2}$

17. (a) Proof (b) Proof (c) Proof

19. (a) $R(n), I, T(n), L(n)$
(b) $S(4) = \frac{1}{3}[f(0) + 4f(1) + 2f(2) + 4f(3) + f(4)] \approx 5.42$

Chapter 5

Section 5.1 (page 329)

1.

x	0.5	1.5	2	2.5
$\int_1^x (1/t)\,dt$	-0.6932	0.4055	0.6932	0.9163

x	3	3.5	4
$\int_1^x (1/t)\,dt$	1.0987	1.2529	1.3865

3. (a) 3.8067 (b) $\ln 45 = \displaystyle\int_1^{45} \frac{1}{t}\,dt \approx 3.8067$

5. (a) -0.2231 (b) $\ln 0.8 = \displaystyle\int_1^{0.8} \frac{1}{t}\,dt \approx -0.2231$

7. b **8.** d **9.** a **10.** c

11. Domain: $x > 0$

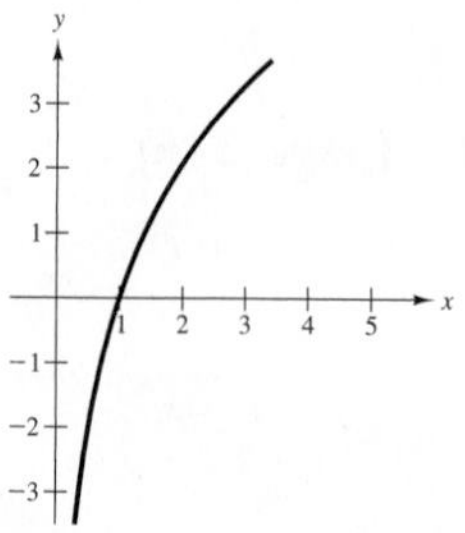

13. Domain: $x > 0$

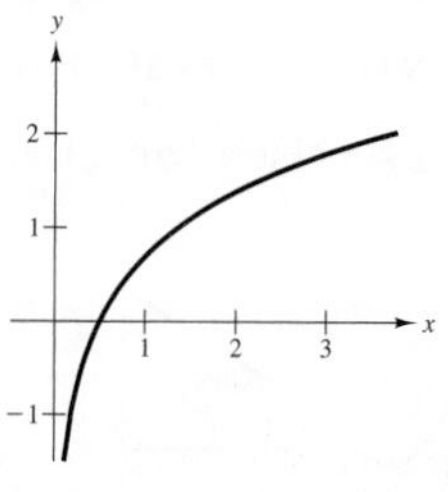

15. Domain: $x > 1$

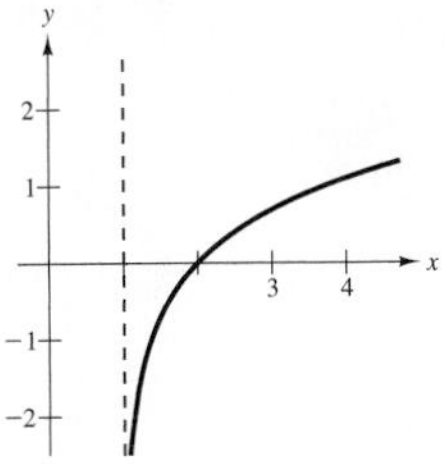

17. (a) 1.7917 (b) -0.4055 (c) 4.3944 (d) 0.5493

19. $\ln 2 - \ln 3$ **21.** $\ln x + \ln y - \ln z$ **23.** $\frac{1}{3}\ln(a^2 + 1)$

25. $3[\ln(x + 1) + \ln(x - 1) - 3\ln x]$ **27.** $\ln z + 2\ln(z - 1)$

29. $\ln \dfrac{x - 2}{x + 2}$ **31.** $\ln \sqrt[3]{\dfrac{x(x + 3)^2}{x^2 - 1}}$ **33.** $\ln\left(9/\sqrt{x^2 + 1}\right)$

35. (a)

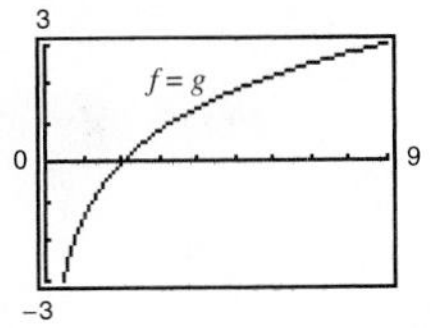

(b) $f(x) = \ln \dfrac{x^2}{4} = \ln x^2 - \ln 4$
$= 2\ln x - \ln 4$
$= g(x)$

37. $-\infty$ **39.** $\ln 4 \approx 1.3863$ **41.** $y = 3x - 3$

43. $y = 2x - 2$ **45.** $2/x$ **47.** $4(\ln x)^3/x$ **49.** $\dfrac{2x^2 - 1}{x(x^2 - 1)}$

51. $\dfrac{1 - x^2}{x(x^2 + 1)}$ **53.** $\dfrac{1 - 2\ln t}{t^3}$ **55.** $\dfrac{2}{x\ln x^2} = \dfrac{1}{x\ln x}$

57. $\dfrac{1}{1 - x^2}$ **59.** $\dfrac{-4}{x(x^2 + 4)}$ **61.** $\dfrac{\sqrt{x^2 + 1}}{x^2}$ **63.** $\cot x$

65. $-\tan x + \dfrac{\sin x}{\cos x - 1}$ **67.** $\dfrac{3\cos x}{(\sin x - 1)(\sin x + 2)}$

69. $[\ln(2x) + 1]/x$

71. (a) $5x - y - 2 = 0$
(b)

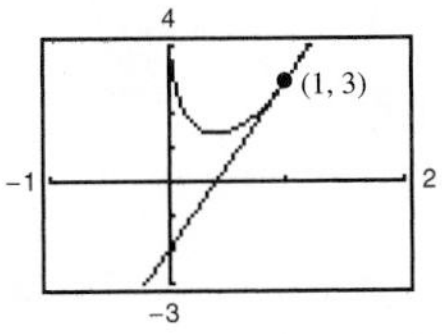

73. (a) $y = \frac{1}{3}x - \frac{1}{12}\pi + \frac{1}{2}\ln\left(\frac{3}{2}\right)$
(b)

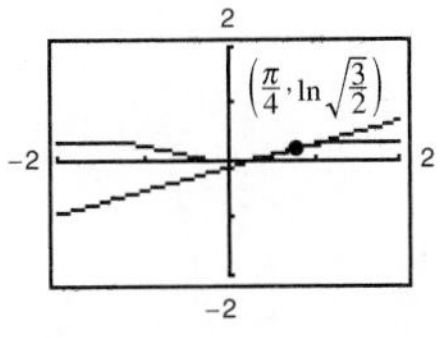

75. (a) $y = x - 1$
(b)

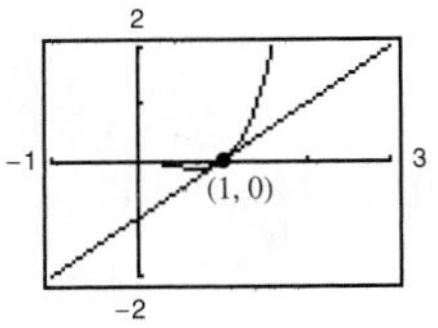

77. $2xy/(3 - 2y^2)$

79. $y = x - 1$ **81.** $xy'' + y' = x(-2/x^2) + 2/x = 0$

83. Relative minimum: $\left(1, \frac{1}{2}\right)$

85. Relative minimum: $(e^{-1}, -e^{-1})$

87. Relative minimum: (e, e); Point of inflection: $(e^2, e^2/2)$

89. $P_1(x) = x - 1$; $P_2(x) = x - 1 - \frac{1}{2}(x - 1)^2$

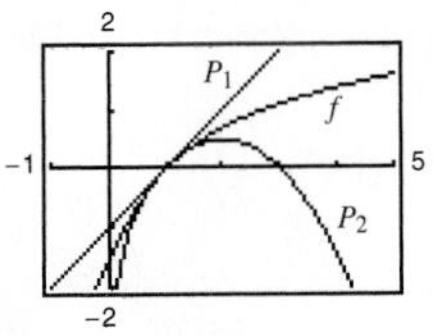

The values of f, P_1, and P_2 and their first derivatives agree at $x = 1$.

91. $x \approx 0.567$ **93.** $(2x^2 - 1)/\sqrt{x^2 - 1}$

95. $\dfrac{3x^3 - 15x^2 + 8x}{2(x - 1)^3\sqrt{3x - 2}}$ **97.** $\dfrac{(2x^2 + 2x - 1)\sqrt{x - 1}}{(x + 1)^{3/2}}$

99. The domain of the natural logarithmic function is $(0, \infty)$ and the range is $(-\infty, \infty)$. The function is continuous, increasing, and one-to-one and its graph is concave downward. In addition, if a and b are positive numbers and n is rational, then $\ln(1) = 0$, $\ln(a \cdot b) = \ln a + \ln b$, $\ln(a^n) = n\ln a$, and $\ln(a/b) = \ln a - \ln b$.

101. (a) Yes. If the graph of g is increasing , then $g'(x) > 0$. Since $f(x) > 0$, you know that $f'(x) = g'(x)f(x)$ and thus $f'(x) > 0$. Therefore, the graph of f is increasing.
(b) No. Let $f(x) = x^2 + 1$ (positive and concave up) and let $g(x) = \ln(x^2 + 1)$ (not concave up).

103. False: $\ln x + \ln 25 = \ln 25x$.

105. (a)

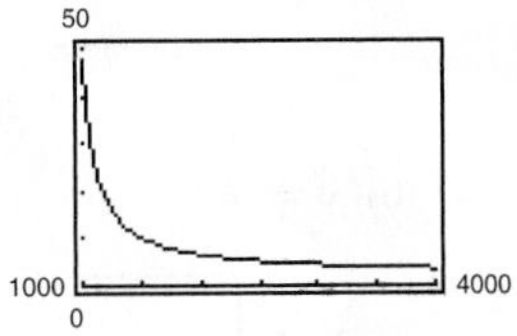

(b) 20 yrs; \$280,178.40
(c) 30 yrs; \$384,642.00

(d) When $x = 1167.41$, $dt/dx \approx -0.0645$.
When $x = 1068.45$, $dt/dx \approx -0.1585$.

(e) There are two obvious benefits of a higher monthly payment: 1. The term is lower.
2. The total amount paid is lower.

107. (a)

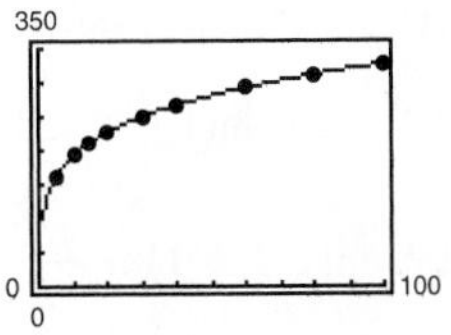

(c)

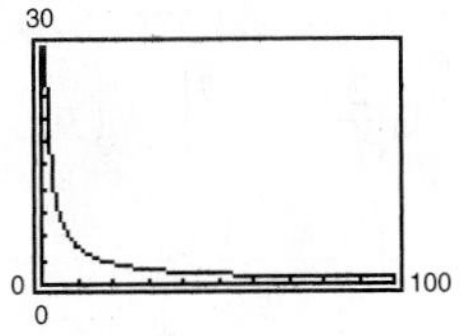

(b) $T'(10) \approx 4.75$ deg/lb/in.^{2} $\lim_{p\to\infty} T'(p) = 0$
$T'(70) \approx 0.97$ deg/lb/in.^{2} Answers will vary.

109. (a)

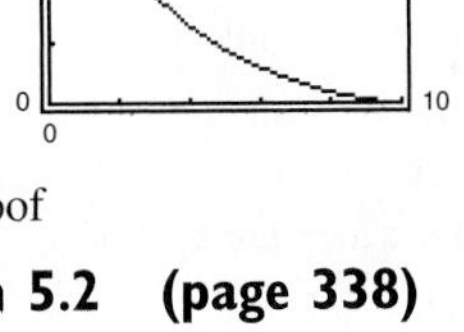

(b) When $x = 5$, $dy/dx = -\sqrt{3}$.
When $x = 9$, $dy/dx = -\sqrt{19}/9$.
(c) $\lim_{x\to 10^-} \dfrac{dy}{dx} = 0$

111. Proof

Section 5.2 (page 338)

1. $5\ln|x| + C$ **3.** $\ln|x + 1| + C$ **5.** $-\frac{1}{2}\ln|3 - 2x| + C$

7. $\ln\sqrt{x^2 + 1} + C$ **9.** $x^2/2 - \ln(x^4) + C$

11. $\frac{1}{3}\ln|x^3 + 3x^2 + 9x| + C$

13. $\frac{1}{2}x^2 - 4x + 6\ln|x + 1| + C$ **15.** $\frac{1}{3}x^3 + 5\ln|x - 3| + C$

17. $\frac{1}{3}x^3 - 2x + \ln\sqrt{x^2 + 2} + C$ **19.** $\frac{1}{3}(\ln x)^3 + C$

21. $2\sqrt{x + 1} + C$ **23.** $2\ln|x - 1| - 2/(x - 1) + C$

25. $\sqrt{2x} - \ln\left|1 + \sqrt{2x}\right| + C$

27. $x + 6\sqrt{x} + 18\ln\left|\sqrt{x} - 3\right| + C$ **29.** $\ln|\sin\theta| + C$

31. $-\frac{1}{2}\ln|\csc 2x + \cot 2x| + C$ **33.** $\ln|1 + \sin t| + C$

35. $\ln|\sec x - 1| + C$

37. $y = -3 \ln|2 - x| + C$

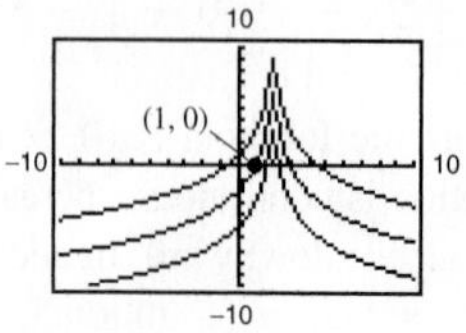

The graph has a hole at $x = 2$.

39. $s = -\frac{1}{2} \ln|\cos 2\theta| + C$

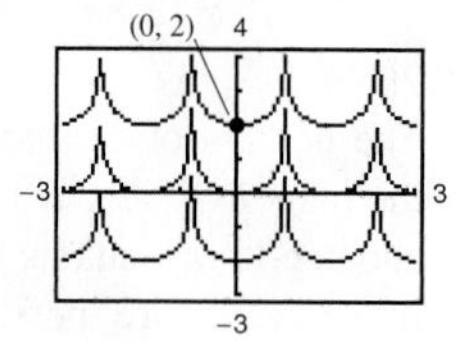

41. $f(x) = -2 \ln x + 3x - 2$

43. (a)

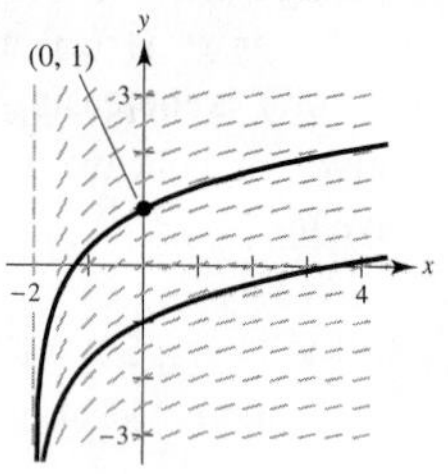

(b) $y = \ln\left|\dfrac{x+2}{2}\right| + 1$

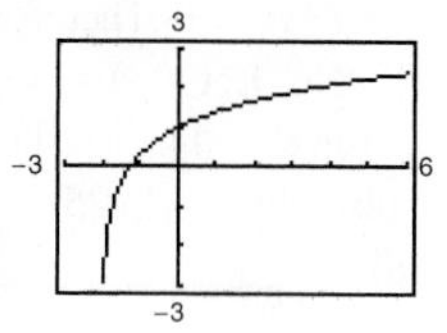

45. (a)

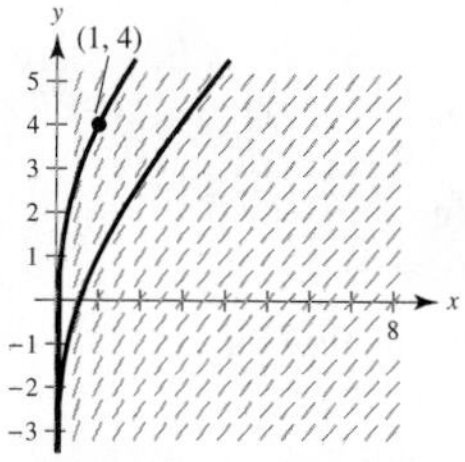

(b) $y = \ln x + x + 3$

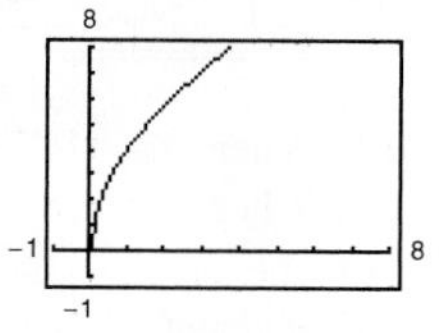

47. $\frac{5}{3} \ln 13 \approx 4.275$ **49.** $\frac{7}{3}$ **51.** $-\ln 3 \approx -1.099$

53. $\ln\left|\dfrac{2 - \sin 2}{1 - \sin 1}\right| \approx 1.929$ **55.** $2\left[\sqrt{x} - \ln\left(1 + \sqrt{x}\right)\right] + C$

57. $\ln\left(\dfrac{\sqrt{x} - 1}{\sqrt{x} + 1}\right) + 2\sqrt{x} + C$ **59.** $\ln\left(\sqrt{2} + 1\right) - \dfrac{\sqrt{2}}{2} \approx 0.174$

61. $1/x$ **63.** $1/x$ **65.** d **67.** $4 \ln 3$ **69.** $\frac{1}{2} \ln 2$

71. $\frac{15}{2} + 8 \ln 2 \approx 13.045$

73. $(12/\pi)\left[2 \ln\left(\sqrt{3} + 1\right) - \ln 2\right] \approx 5.03$

75. Trapezoidal Rule: 20.2
Simpson's Rule: 19.4667

77. Trapezoidal Rule: 5.3368
Simpson's Rule: 5.3632

79. Power Rule **81.** Log rule

83. $-\ln|\cos x| + C = \ln|1/\cos x| + C = \ln|\sec x| + C$

85. $\ln|\sec x + \tan x| + C = \ln\left|\dfrac{\sec^2 x - \tan^2 x}{\sec x - \tan x}\right| + C$
$= -\ln|\sec x - \tan x| + C$

87. 1 **89.** $1/[2(e - 1)] \approx 0.291$

91. $P(t) = 1000(12 \ln|1 + 0.25t| + 1)$; $P(3) \approx 7715$

93. \$168.27

95. (a)

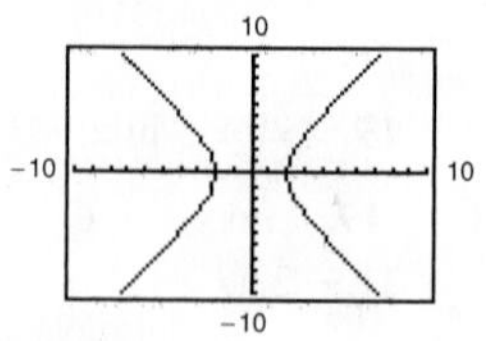

(b) Answers will vary.
Example:
$y^2 = e^{-\ln x + \ln 4} = 4/x$

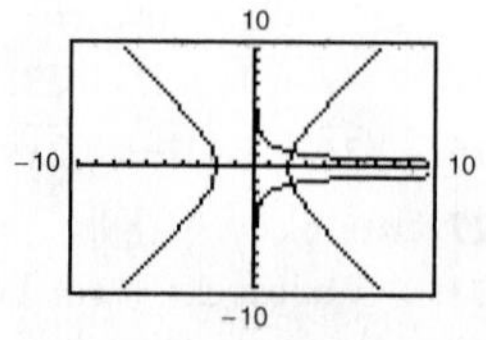

(c) Answers will vary.

97. False. $\frac{1}{2}(\ln x) = \ln x^{1/2}$ **99.** True

101.

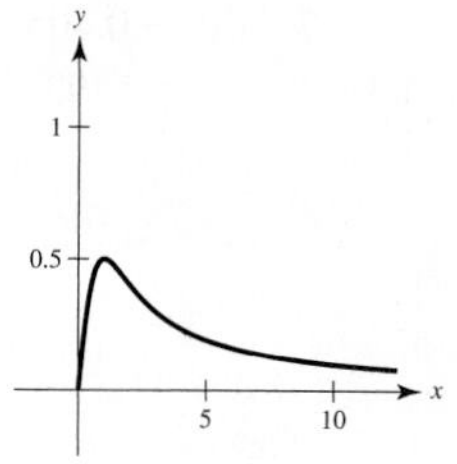

(a) $\frac{1}{2} \ln 2 - \frac{1}{4} \approx 0.0966$
(b) $0 < m < 1$
(c) $\frac{1}{2}(m - \ln m - 1)$

Section 5.3 (page 347)

1. (a) $f(g(x)) = 5[(x - 1)/5] + 1 = x$
$g(f(x)) = [(5x + 1) - 1]/5 = x$

(b)

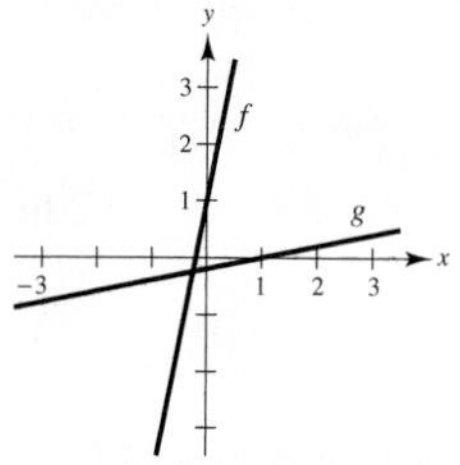

3. (a) $f(g(x)) = \left(\sqrt[3]{x}\right)^3 = x$; $g(f(x)) = \sqrt[3]{x^3} = x$

(b)

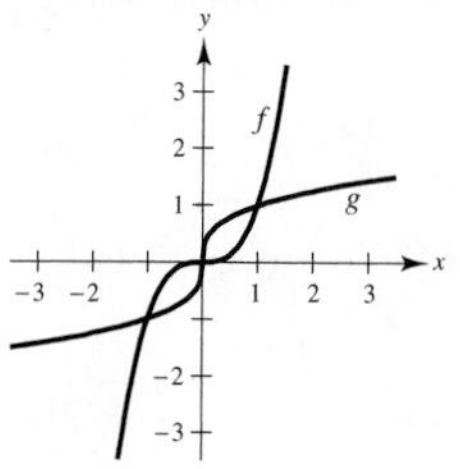

5. (a) $f(g(x)) = \sqrt{x^2 + 4 - 4} = x$;
$g(f(x)) = \left(\sqrt{x - 4}\right)^2 + 4 = x$

(b)

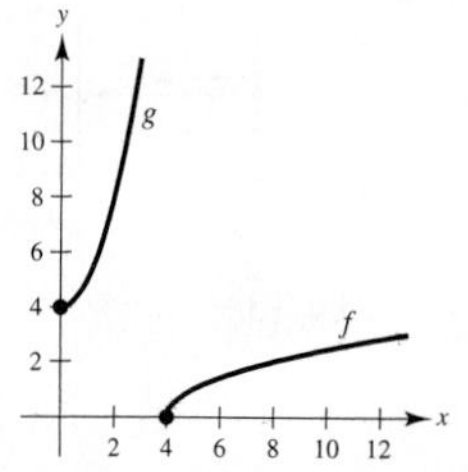

7. (a) $f(g(x)) = \dfrac{1}{1/x} = x$; $g(f(x)) = \dfrac{1}{1/x} = x$

(b)

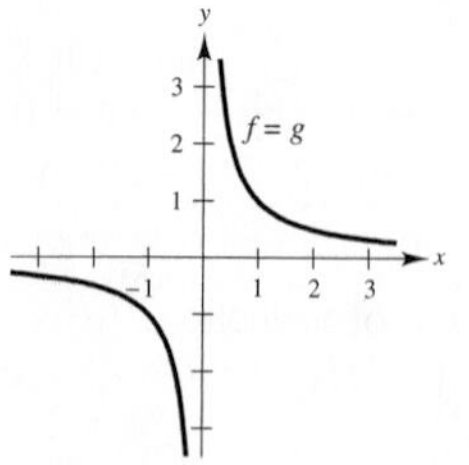

9. c **10.** b **11.** a **12.** d

13. Inverse exists. **15.** Inverse does not exist.

17. One-to-one

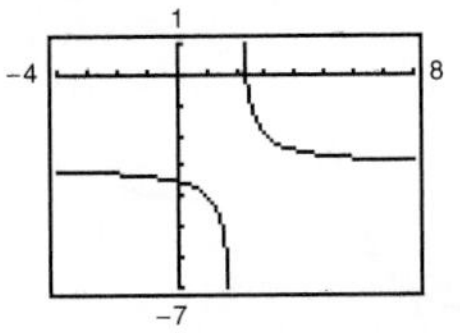

19. One-to-one

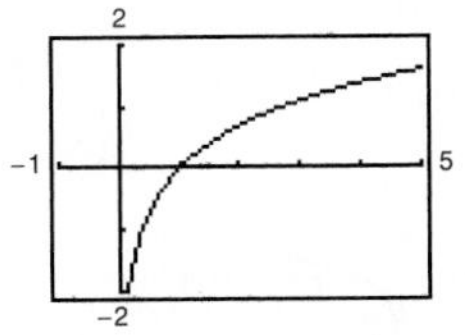

21. One-to-one

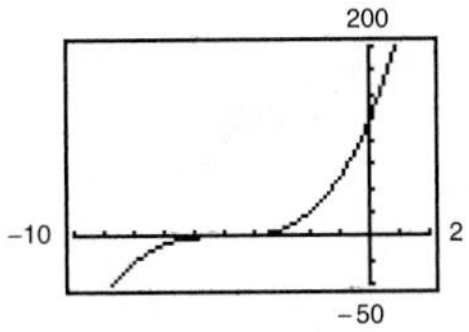

23. Inverse does not exist.

25. Inverse does not exist.

27. Inverse exists.

29. $f^{-1}(x) = (x + 3)/2$

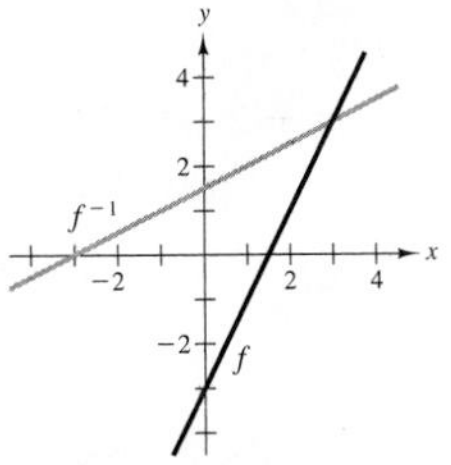

f and f^{-1} are symmetric about $y = x$.

31. $f^{-1}(x) = x^{1/5}$

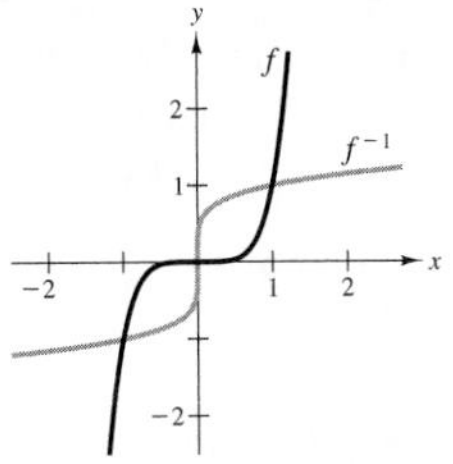

f and f^{-1} are symmetric about $y = x$.

33. $f^{-1}(x) = x^2,\ x \ge 0$

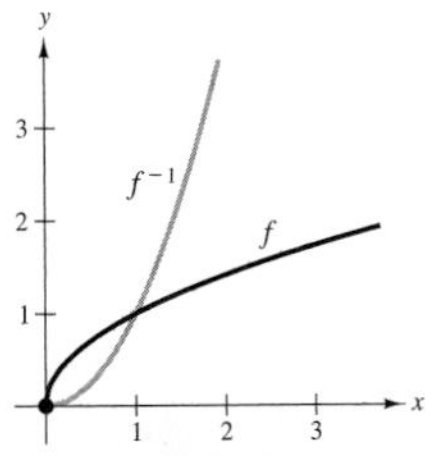

f and f^{-1} are symmetric about $y = x$.

35. $f^{-1}(x) = \sqrt{4 - x^2},\ 0 \le x \le 2$

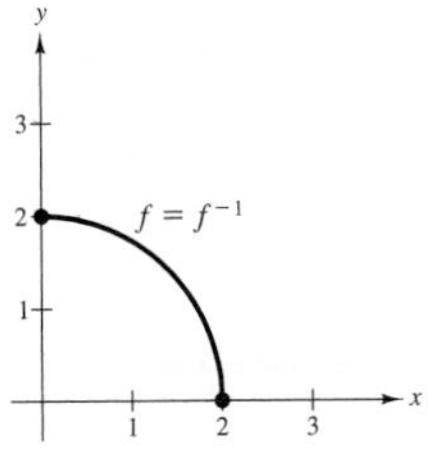

f and f^{-1} are symmetric about $y = x$.

37. $f^{-1}(x) = x^3 + 1$

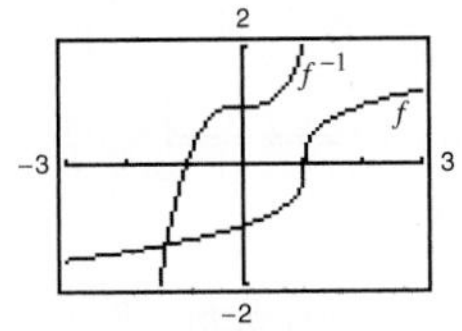

f and f^{-1} are symmetric about $y = x$.

39. $f^{-1}(x) = x^{3/2},\ x \ge 0$

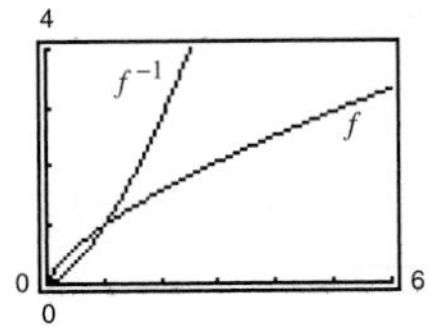

f and f^{-1} are symmetric about $y = x$.

41. $f^{-1}(x) = \sqrt{7}x/\sqrt{1 - x^2},\ -1 < x < 1$

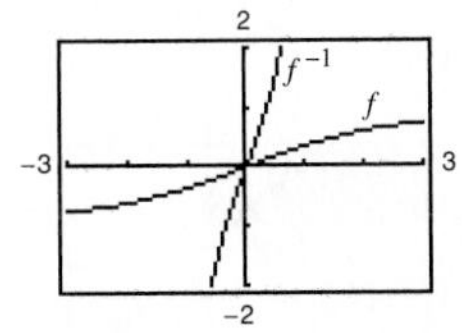

f and f^{-1} are symmetric about $y = x$.

43.

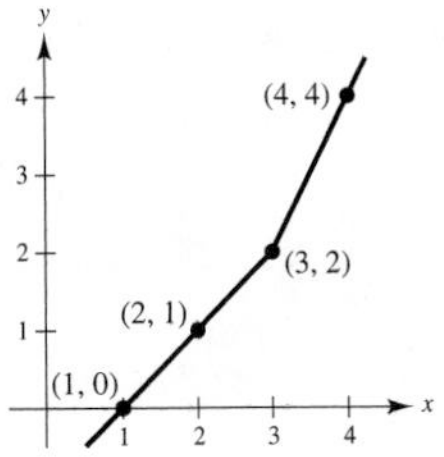

x	1	2	3	4
$f^{-1}(x)$	0	1	2	4

45. (a) Proof

(b) $y = \frac{20}{7}(80 - x)$

x: total cost

y: number of pounds of the less expensive commodity

(c) $[62.5, 80]$ (d) 20 pounds

47. $f'(x) = 2(x - 4) > 0$ on $(4, \infty)$

49. $f'(x) = -8/x^3 < 0$ on $(0, \infty)$

51. $f'(x) = -\sin x < 0$ on $(0, \pi)$

53. $f^{-1}(x) = \begin{cases} \left[1 - \sqrt{1 + 16x^2}\right]/(2x), & \text{if } x \ne 0 \\ 0, & \text{if } x = 0 \end{cases}$

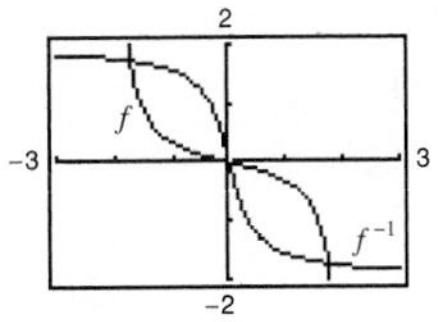

The graph of f^{-1} is a reflection of the graph of f in the line $y = x$.

55. (a) and (b)

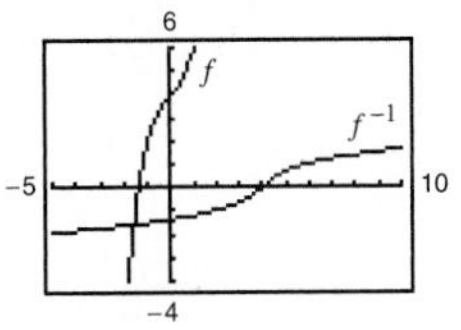

(c) f is one-to-one and has an inverse function.

57. (a) and (b)

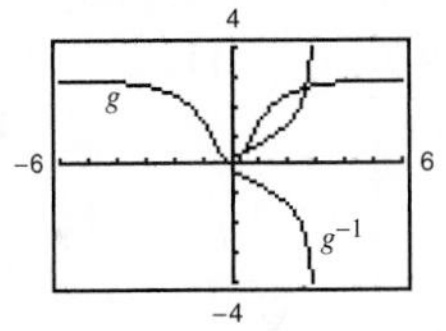

(c) g is not one-to-one and does not have an inverse function.

59. One-to-one

$f^{-1}(x) = x^2 + 2,\ x \ge 0$

61. One-to-one

$f^{-1}(x) = 2 - x,\ x \ge 0$

63. $f^{-1}(x) = \sqrt{x} + 3,\ x \ge 0$ (Answer is not unique.)

65. $f^{-1}(x) = x - 3,\ x \ge 0$ (Answer is not unique.)

67. Inverse exists. Volume is an increasing function, and therefore is one-to-one. The inverse function gives the time t corresponding to the volume V.

69. Inverse does not exist. **71.** $1/5$

73. $2\sqrt{3}/3$ **75.** $1/13$

77. (a) Domain of f: $(-\infty, \infty)$
Domain of f^{-1}: $(-\infty, \infty)$

(b) Range of f: $(-\infty, \infty)$
Range of f^{-1}: $(-\infty, \infty)$

(c)

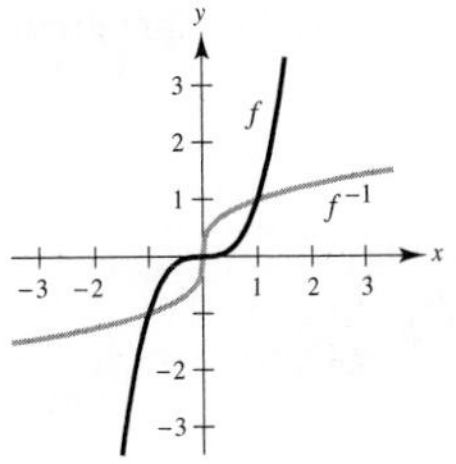

(d) $f'\left(\frac{1}{2}\right) = \frac{3}{4},\ (f^{-1})'\left(\frac{1}{8}\right) = \frac{4}{3}$

79. (a) Domain of f: $[4, \infty)$
Domain of f^{-1}: $[0, \infty)$

(b) Range of f: $[0, \infty)$
Range of f^{-1}: $[4, \infty)$

(c)

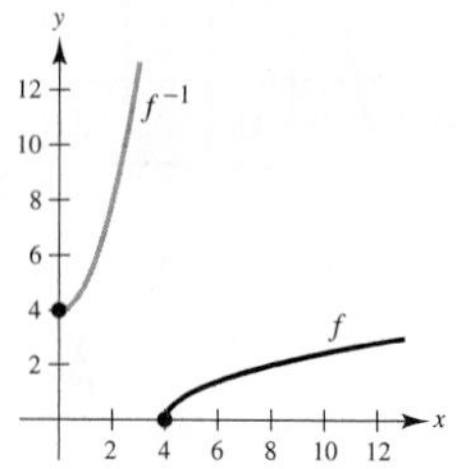

(d) $f'(5) = \frac{1}{2}$, $(f^{-1})'(1) = 2$

81. $-\frac{1}{11}$ **83.** 32 **85.** 600

87. $(g^{-1} \circ f^{-1})(x) = (x + 1)/2$ **89.** $(f \circ g)^{-1}(x) = (x + 1)/2$

91. Let $y = f(x)$ be one-to-one. Solve for x as a function of y. Interchange x and y to get $y = f^{-1}(x)$. Let the domain of f^{-1} be the range of f. Verify that $f(f^{-1}(x)) = x$ and $f^{-1}(f(x)) = x$.
Example: $f(x) = x^3$; $y = x^3$; $x = \sqrt[3]{y}$; $y = \sqrt[3]{x}$; $f^{-1}(x) = \sqrt[3]{x}$

93. Many x-values yield the same y-value. For example, $f(\pi) = 0 = f(0)$. The graph is not continuous at $[(2n - 1)\pi]/2$ where n is an integer.

95. $\frac{1}{4}$ **97.** (a) and (b) Proofs **99.** Proof

101. False. Let $f(x) = x^2$. **103.** True

105. No. Let $f(x) = \begin{cases} x, & 0 \le x \le 1 \\ 1 - x, & 1 < x \le 2 \end{cases}$.

107. $\sqrt{17}$ **109.** Proof

Section 5.4 (page 356)

1. $x = 4$ **3.** $x \approx 2.485$ **5.** $x = 0$ **7.** $x \approx 0.511$

9. $x \approx 7.389$ **11.** $x \approx 10.389$ **13.** $x \approx 5.389$

15.

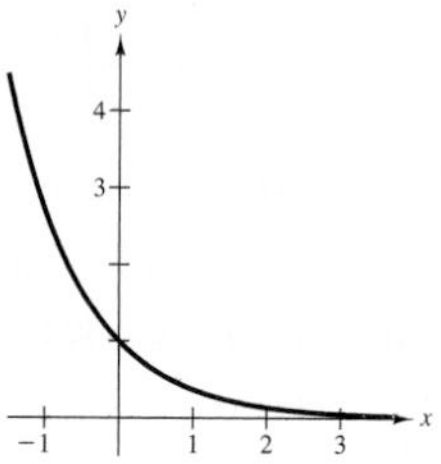

17.

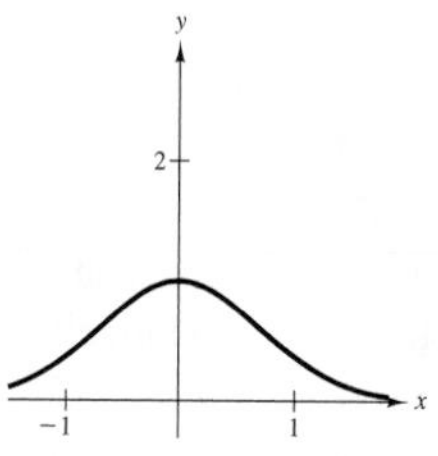

19. (a)

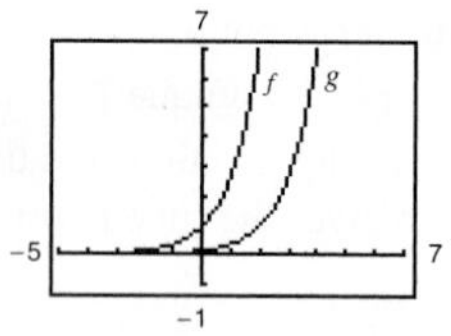

Translation two units to the right

(b)

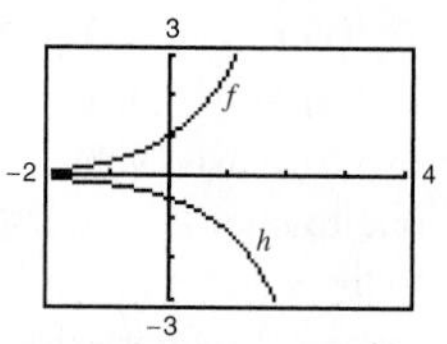

Reflection in the x-axis and a vertical shrink

(c)

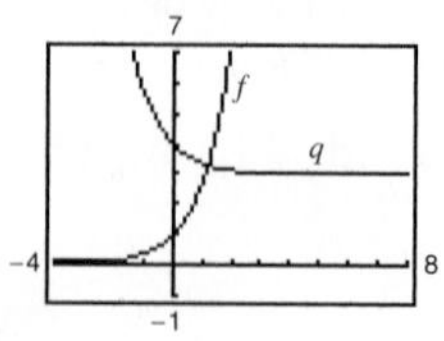

Reflection in the y-axis and a translation three units upward

21. c **22.** d **23.** a **24.** b

25.

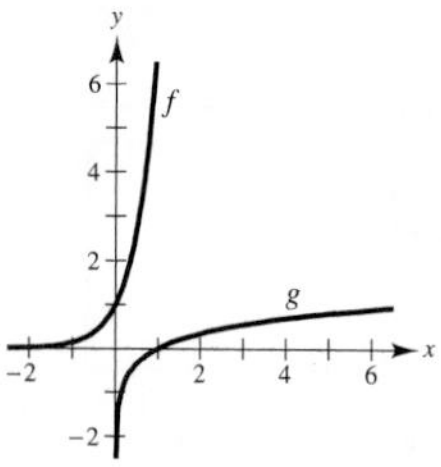

27.

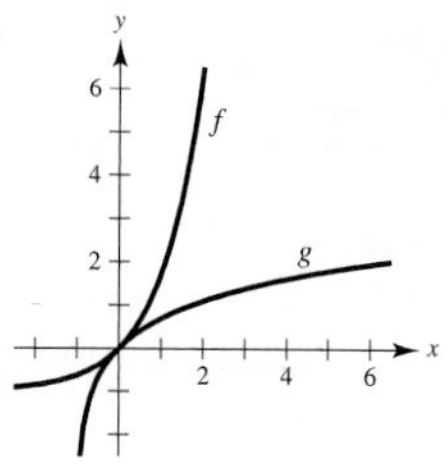

29.

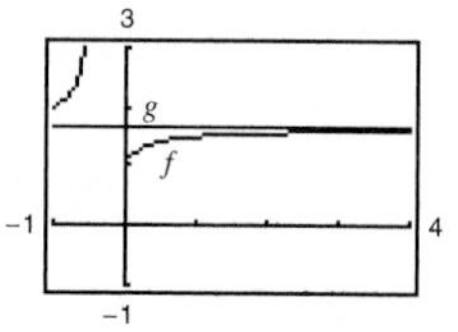

$\lim_{x\to\infty} f(x) = \lim_{x\to\infty} g(x) = e^{0.5}$

31. $2.7182805 < e$ **33.** (a) $y = 3x + 1$ (b) $y = -3x + 1$

35. $2e^{2x}$ **37.** $e^{\sqrt{x}}/(2\sqrt{x})$ **39.** $3(e^{-t} + e^{t})^2(e^{t} - e^{-t})$

41. $2e^{2x}/(1 + e^{2x})$ **43.** $[-2(e^x - e^{-x})]/(e^x + e^{-x})^2$

45. $2e^x \cos x$ **47.** $\cos(x)/x$ **49.** $y = -x + 2$

51. $y = -4(x + 1)$ **53.** $y = ex$ **55.** $y = (1/e)x - 1/e$

57. $\dfrac{10 - e^y}{xe^y + 3}$ **59.** $y = (-e - 1)x + 1$ **61.** $3(6x + 5)e^{-3x}$

63. $y'' - 2y' + 3y = 0$

$e^x[-\cos\sqrt{2}x - \sin\sqrt{2}x - 2\sqrt{2}\sin\sqrt{2}x + 2\sqrt{2}\cos\sqrt{2}x] -$
$2e^x[-\sqrt{2}\sin\sqrt{2}x + \sqrt{2}\cos\sqrt{2}x + \cos\sqrt{2}x + \sin\sqrt{2}x] +$
$3e^x[\cos\sqrt{2}x + \sin\sqrt{2}x] = 0$
$0 = 0$

65. Relative minimum: $(0, 1)$

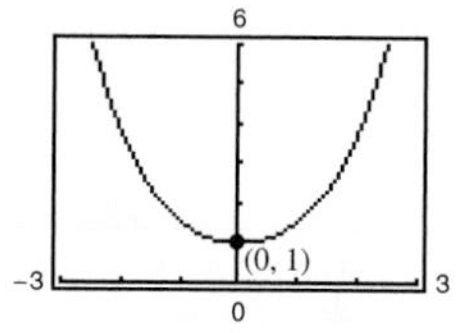

67. Relative maximum: $(2, 1/\sqrt{2\pi})$

Points of inflection: $\left(1, \dfrac{e^{-0.5}}{\sqrt{2\pi}}\right), \left(3, \dfrac{e^{-0.5}}{\sqrt{2\pi}}\right)$

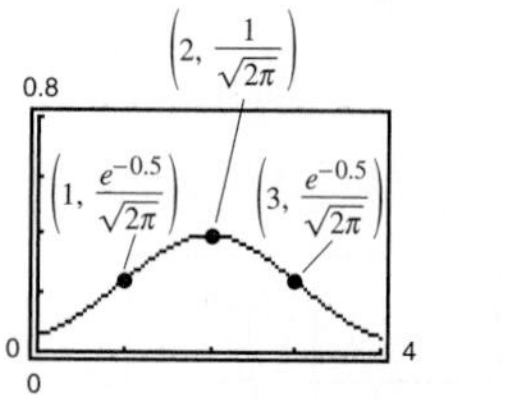

69. Relative minimum: $(0, 0)$

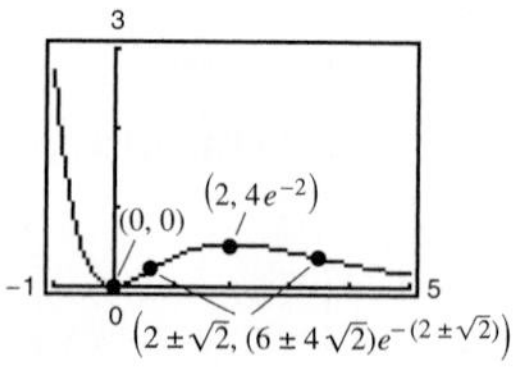

Relative maximum: $(2, 4e^{-2})$

Points of inflection: $(2 \pm \sqrt{2}, (6 \pm 4\sqrt{2})e^{-(2 \pm \sqrt{2})})$

71. Relative maximum: $(-1, 1 + e)$

Point of inflection: $(0, 3)$

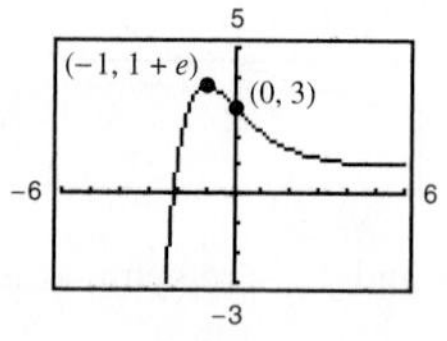

73. $A = \sqrt{2}e^{-1/2}$ **75.** Proof

77. $\left(\frac{1}{2}, e\right)$

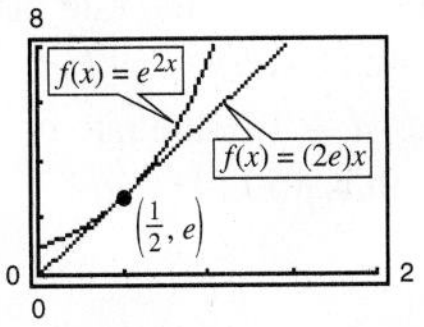

79. (a)

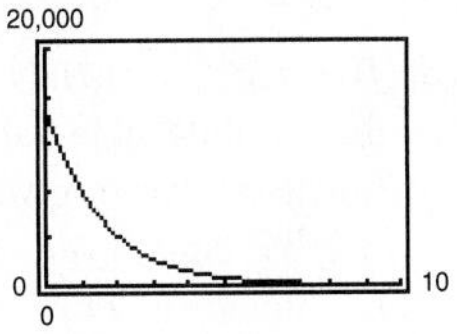

(c)

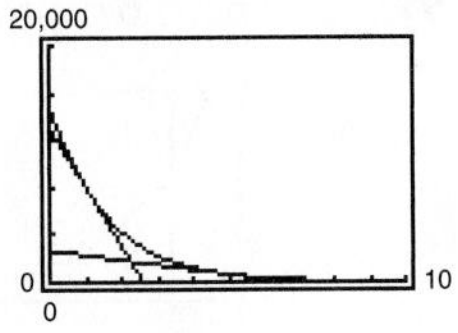

(b) When $t = 1$, $\frac{dV}{dt} \approx -5028.84$.

When $t = 5$, $\frac{dV}{dt} \approx -406.89$.

81. (a) $\ln P = -0.1499h + 9.3018$

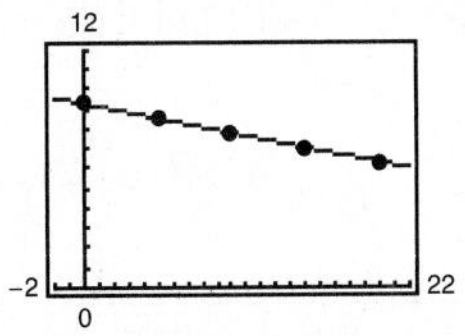

(b) $P = 10{,}957.7e^{-0.1499h}$

(c)

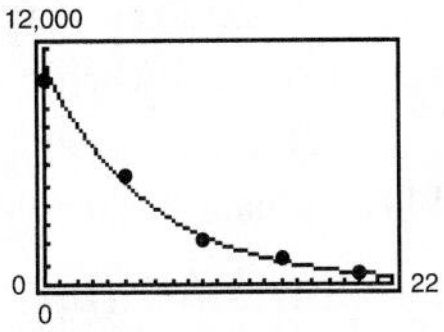

(d) $h = 5$: -776
$h = 18$: -111

83. $P_1 = 1 + \frac{1}{2}x$; $P_2 = 1 + \frac{1}{2}x + \frac{1}{8}x^2$

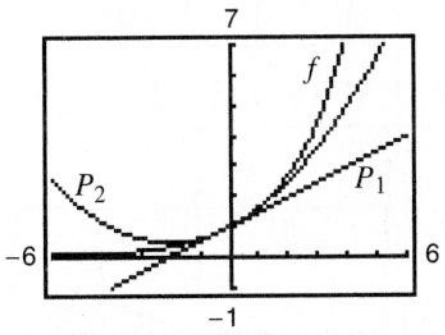

The values of f, P_1, and P_2 and their first derivatives agree at $x = 0$.

85. $e^{5x} + C$ **87.** $2e^{\sqrt{x}} + C$

89. $x - \ln(e^x + 1) + C_1$ or $-\ln(1 + e^{-x}) + C_2$

91. $-\frac{2}{3}(1 - e^x)^{3/2} + C$ **93.** $\ln|e^x - e^{-x}| + C$

95. $-\frac{5}{2}e^{-2x} + e^{-x} + C$ **97.** $\ln|\cos e^{-x}| + C$

99. $(e^2 - 1)/(2e^2)$ **101.** $(e - 1)/(2e)$ **103.** $(e/3)(e^2 - 1)$

105. $(1/\pi)[e^{(\sin \pi^2/2)} - 1]$ **107.** $[1/(2a)]e^{ax^2} + C$

109. $f(x) = \frac{1}{2}(e^x + e^{-x})$

111. (a)

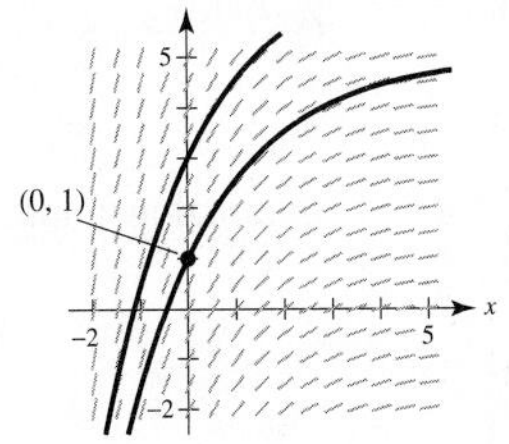

(b) $y = -4e^{-x/2} + 5$

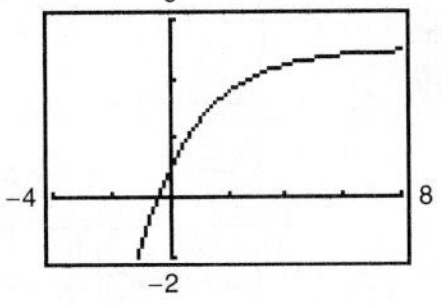

113. $e^5 - 1 \approx 147.413$

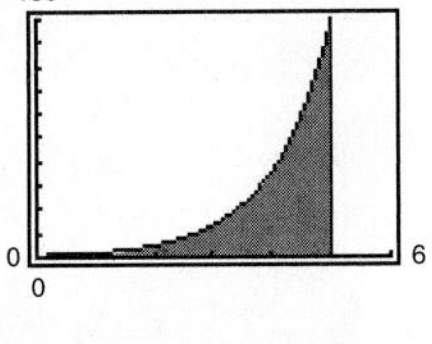

115. $2(1 - e^{-3/2}) \approx 1.554$

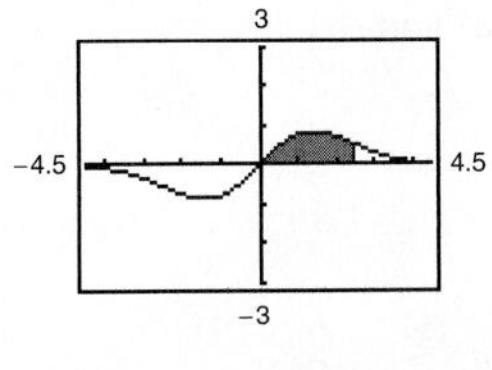

117. Midpoint Rule: 92.190; Trapezoidal Rule: 93.837;
Simpson's Rule: 92.7385

119. The probability that a given battery will last between 48 months and 60 months is approximately 47.72%.

121. $\int_0^x e^t\,dt \geq \int_0^x 1\,dt$; $e^x - 1 \geq x$; $e^x > x + 1$ for $x \geq 0$

123. $f(x) = e^x$
The domain of $f(x)$ is $(-\infty, \infty)$ and the range of $f(x)$ is $(0, \infty)$. $f(x)$ is continuous, increasing, one-to-one, and concave upward on its entire domain.
$\lim_{x\to-\infty} e^x = 0$ and $\lim_{x\to\infty} e^x = \infty$

125. $f(x) = e^x = f'(x)$ **127.** $x \approx 0.567$ **129.** Proof

Section 5.5 (page 366)

1. -3 **3.** 0 **5.** (a) $\log_2 8 = 3$ (b) $\log_3(1/3) = -1$

7. (a) $10^{-2} = 0.01$ (b) $\left(\frac{1}{2}\right)^{-3} = 8$

9.

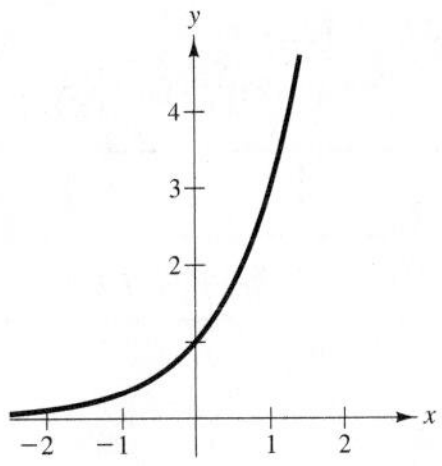

11.

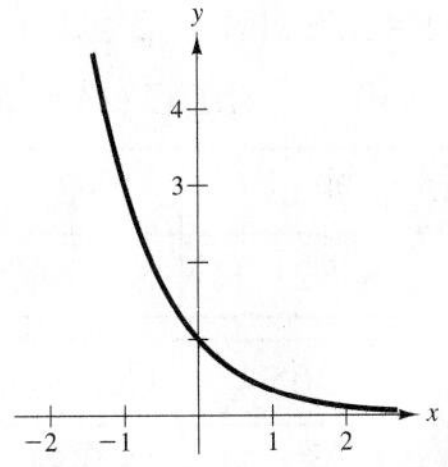

13.

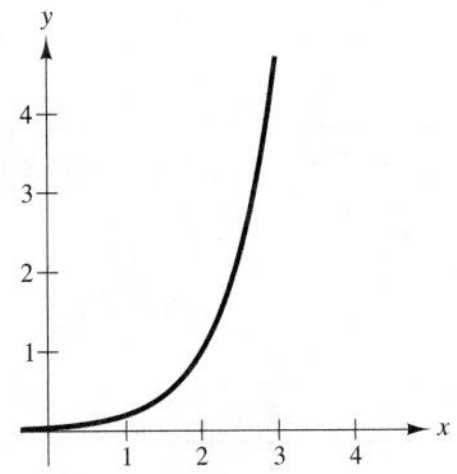

15. (a) $x = 3$ (b) $x = -1$

17. (a) $x = \frac{1}{3}$ (b) $x = \frac{1}{16}$ **19.** (a) $x = -1, 2$ (b) $x = \frac{1}{3}$

21. 1.965 **23.** -6.288 **25.** 12.253

27. 33.000 **29.** ± 11.845

31.

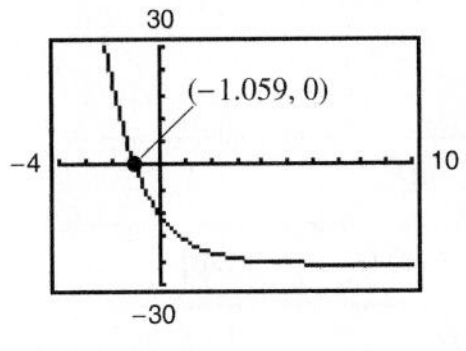

$(-1.059, 0)$

33.

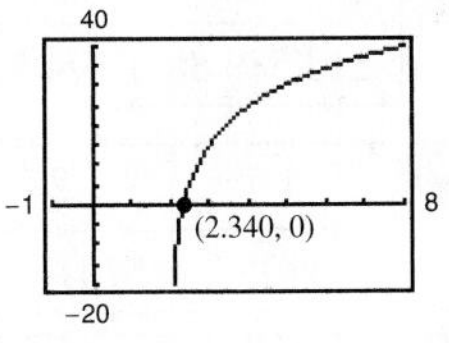

$(2.340, 0)$

35.

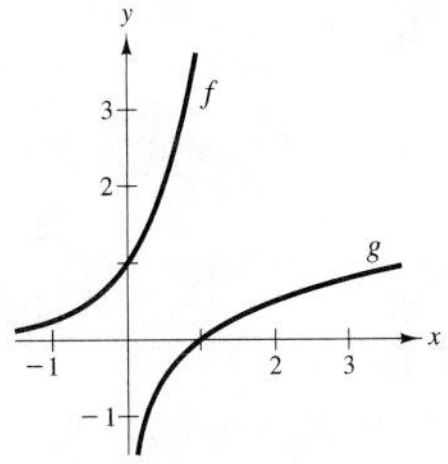

37. $(\ln 4)4^x$

39. $t2^t(t \ln 2 + 2)$ **41.** $-2^{-\theta}[(\ln 2)\cos \pi\theta + \pi \sin \pi\theta]$

43. $(x - 2)/[(\ln 2)x(x - 1)]$ **45.** $x/[(\ln 5)(x^2 - 1)]$

47. $5(1 - \ln t)/(t^2 \ln 2)$ **49.** $y = -2x \ln 2 - 2 \ln 2 + 2$

51. $y = [1/(27 \ln 3)]x + 3 - 1/\ln 3$ **53.** $2(1 - \ln x)x^{(2/x)-2}$

55. $(x-2)^{x+1}[(x+1)/(x-2)+\ln(x-2)]$

57. $y=x$ **59.** $y=\dfrac{\cos e}{e}x-\cos e+1$ **61.** $3^x/\ln 3+C$

63. $[-1/(2\ln 5)](5^{-x^2})+C$ **65.** $\ln(3^{2x}+1)/(2\ln 3)+C$

67. $7/(2\ln 2)$ **69.** $4/\ln 5-2/\ln 3$ **71.** $26/\ln 3$

73. (a)

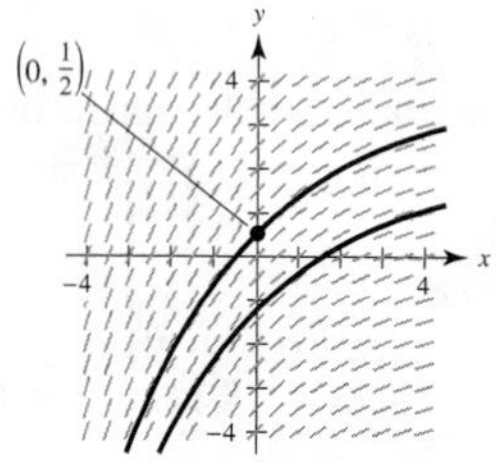

(b) $y=\dfrac{3(1-0.4^{x/3})}{\ln 2.5}+\dfrac{1}{2}$

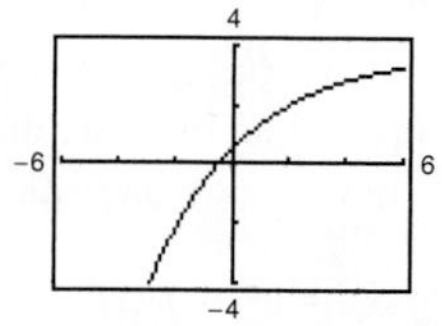

75. (a) False. $y=a^x \Rightarrow 0=a^1 \Rightarrow a=0$, but exponential functions are not defined for $a=0$.
(b) True: $y=\log_2 x$ (c) True: $2^y=x$
(d) False. (1, 0), (2, 1), and (8, 3) are not collinear.

77. $g(x)=x^x$, $k(x)=2^x$, $h(x)=x^2$, $f(x)=\log_2 x$

79. (a) \$40.64 (b) $C'(1)\approx 0.051P$, $C'(8)\approx 0.072P$ (c) ln 1.05

81.

n	1	2	4	12
A	\$4321.94	\$4399.79	\$4440.21	\$4467.74

n	365	Continuous
A	\$4481.23	\$4481.69

83.

t	1	10	20	30
P	\$95,122.94	\$60,653.07	\$36,787.94	\$22,313.02

t	40	50
P	\$13,533.53	\$8208.50

85.

t	1	10	20	30
P	\$95,132.82	\$60,716.10	\$36,864.45	\$22,382.66

t	40	50
P	\$13,589.88	\$8251.24

87.

n	1	2	4	12
A	\$1410.60	\$1414.78	\$1416.91	\$1418.34

n	365	Continuous
A	\$1419.04	\$1419.07

89. c

91. (a) 6.7 million ft^3/acre
(b) $t=20$: $\dfrac{dV}{dt}=0.073$; $t=60$: $\dfrac{dV}{dt}=0.040$

93. (a)

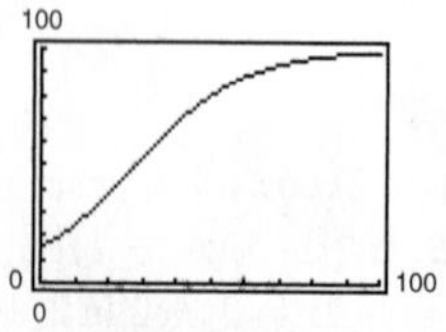

(b) 16.7%
(c) $x\approx 38.8$ or 38,800 egg masses
(d) $x\approx 27.75$ or 27,750 egg masses

95. (a) $B=4.75(6.774)^d$
(b)

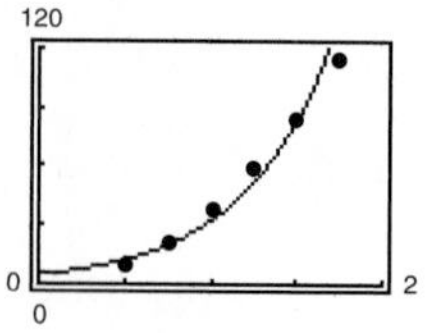

(c) When $d=0.8$, the rate of growth is 41.99.
When $d=1.5$, the rate of growth is 160.21.

97. (a) 5.67; 5.67; 5.67
(b)

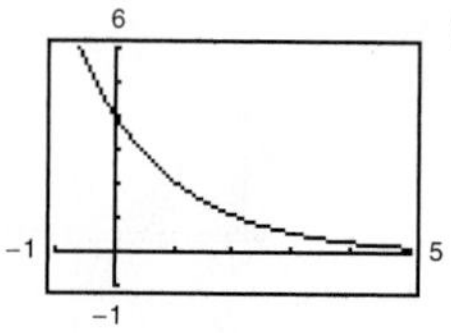

(c) $f(t)=g(t)=h(t)$. No, because the definite integrals of two functions over a given interval may be equal even though the functions are not equal.

99. $y=1200(0.6^t)$ **101.** False: e is an irrational number.

103. True **105.** True **107.** Proof

109. (a) $\dfrac{dy}{dx}=\dfrac{y^2-yx\ln y}{x^2-xy\ln x}$
(b) (i) 1 when $c\neq 0$, $c\neq e$ (ii) -3.1774 (iii) -0.3147
(c) (0, 0), (e, e)

111. Putnam Problem A15, 1940

Section 5.6 (page 377)

1. (a)

x	-1	-0.8	-0.6	-0.4	-0.2
y	-1.57	-0.93	-0.64	-0.41	-0.20

x	0	0.2	0.4	0.6	0.8	1
y	0	0.20	0.41	0.64	0.93	1.57

(b)

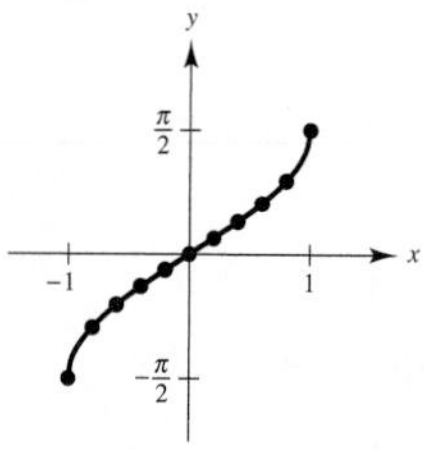

(c)

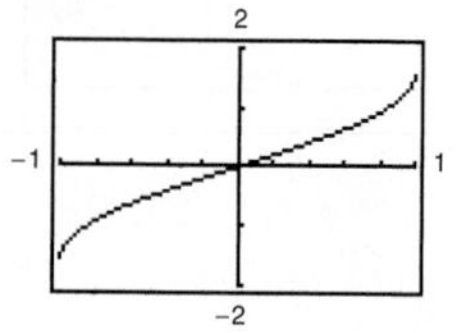

(d) Intercept: (0, 0); Symmetry: origin

3. $(-\sqrt{2}/2, 3\pi/4)$, $(1/2, \pi/3)$, $(\sqrt{3}/2, \pi/6)$

5. $\pi/6$ **7.** $\pi/3$ **9.** $\pi/6$ **11.** $-\pi/4$ **13.** 2.50

15. $\arccos(1/1.269)\approx 0.66$ **17.** (a) $3/5$ (b) $5/3$

19. (a) $-\sqrt{3}$ (b) $-\frac{13}{5}$ **21.** $\sqrt{1-4x^2}$ **23.** $\sqrt{x^2-1}/|x|$

25. $\sqrt{x^2-9}/3$ **27.** $\sqrt{x^2+2}/x$

29. (a)

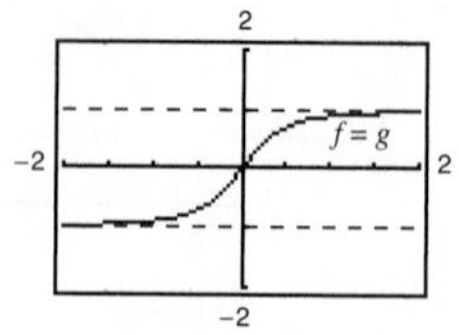

(b) Proof
(c) Horizontal asymptotes: $y=-1, y=1$

31. $x=\frac{1}{3}\left[\sin\left(\frac{1}{2}\right)+\pi\right]\approx 1.207$ **33.** $x=\frac{1}{3}$

35. (a) and (b) Proofs

37.

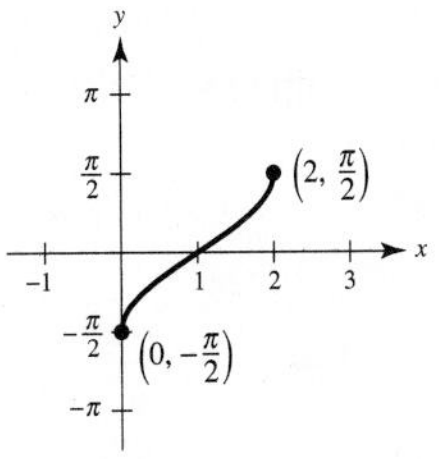

39.

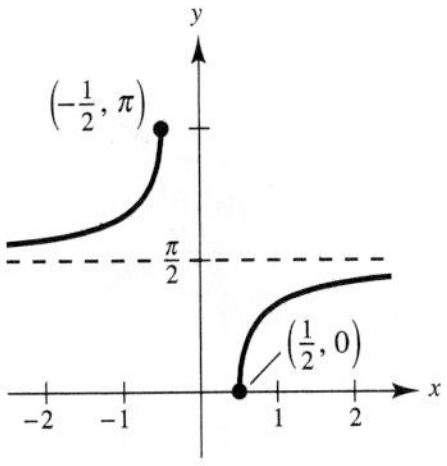

41. $2/\sqrt{2x - x^2}$ **43.** $-3/\sqrt{4 - x^2}$ **45.** $a/(a^2 + x^2)$

47. $\left(3x - \sqrt{1 - 9x^2} \arcsin 3x\right)/\left(x^2\sqrt{1 - 9x^2}\right)$

49. $-t/\sqrt{1 - t^2}$ **51.** $\arccos x$ **53.** $1/(1 - x^4)$

55. $\arcsin x$ **57.** $x^2/\sqrt{16 - x^2}$ **59.** $2/(1 + x^2)^2$

61. $y = \frac{1}{3}\left(4\sqrt{3}x - 2\sqrt{3} + \pi\right)$

63. $y = \frac{1}{4}x + (\pi - 2)/4$ **65.** $y = (2\pi - 4)x + 4$

67. $P_1(x) = x$; $P_2(x) = x$

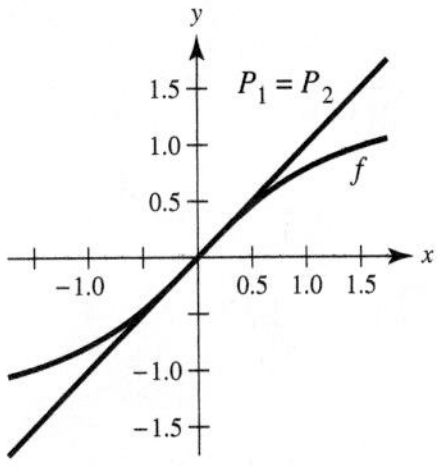

69. $P_1(x) = \dfrac{\pi}{6} + \dfrac{2\sqrt{3}}{3}\left(x - \dfrac{1}{2}\right)$

$P_2(x) = \dfrac{\pi}{6} + \dfrac{2\sqrt{3}}{3}\left(x - \dfrac{1}{2}\right) + \dfrac{2\sqrt{3}}{9}\left(x - \dfrac{1}{2}\right)^2$

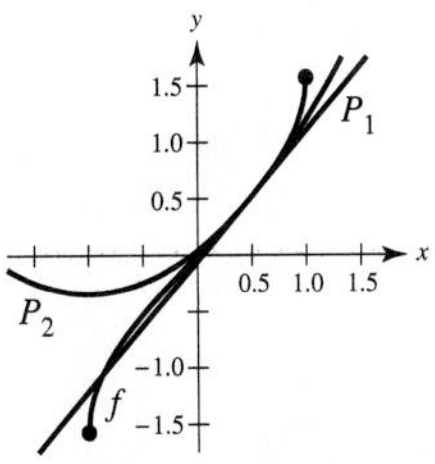

71. Relative maximum: $(1.272, -0.606)$
Relative minimum: $(-1.272, 3.747)$

73. Relative minimum: $(2, 2.214)$

75. $y = -2\pi x/(\pi + 8) + 1 - \pi^2/(2\pi + 16)$

77. $y = -x + \sqrt{2}$

79. If the domains were not restricted, then the trigonometric functions would not be one-to-one and hence would have no inverses.

81. If $x > 0$, $y = \operatorname{arccot} x = \arctan \dfrac{1}{x}$; If $x < 0$, $y = \arctan \dfrac{1}{x} + \pi$.

83. False. The range of arccos is $[0, \pi]$. **85.** True **87.** True

89. (a) $\theta = \operatorname{arccot}(x/5)$
(b) $x = 10$: 16 rad/h; $x = 3$: 58.824 rad/hr

91. (a) $h(t) = -16t^2 + 256$; $t = 4$ sec
(b) $t = 1$: -0.0520 rad/sec; $t = 2$: -0.1116 rad/sec

93. (a) and (b) Proofs **95.** $k \le -1$ or $k \ge 1$

97. (a)

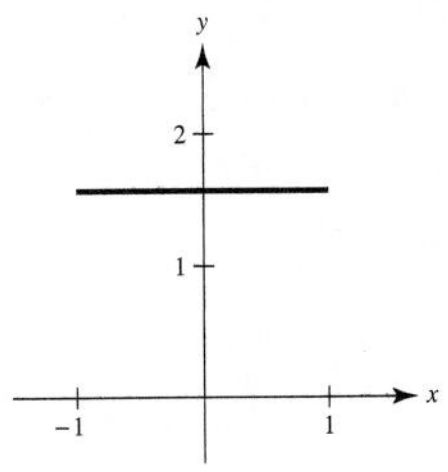

(b) The graph is a horizontal line at $\dfrac{\pi}{2}$.
(c) Proof

99. $c = 2$

101. (a)

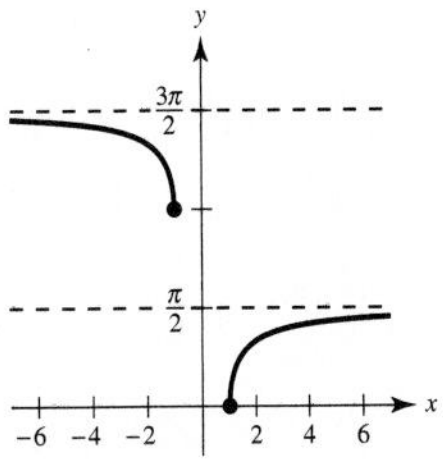

(b) Proof

Section 5.7 (page 385)

1. $5 \arcsin \dfrac{x}{3} + C$ **3.** $\dfrac{7}{4} \arctan \dfrac{x}{4} + C$ **5.** $\operatorname{arcsec}|2x| + C$

7. $\frac{1}{2}x^2 - \frac{1}{2}\ln(x^2 + 1) + C$ **9.** $\arcsin(x + 1) + C$

11. $\frac{1}{2}\arcsin t^2 + C$ **13.** $\frac{1}{4}\arctan(e^{2x}/2) + C$

15. $2\arcsin\sqrt{x} + C$ **17.** $\frac{1}{2}\ln(x^2 + 1) - 3\arctan x + C$

19. $8\arcsin[(x - 3)/3] - \sqrt{6x - x^2} + C$ **21.** $\pi/18$

23. $\pi/6$ **25.** $\frac{1}{32}\pi^2 \approx 0.308$ **27.** $\frac{1}{2}\left(\sqrt{3} - 2\right) \approx -0.134$

29. $\pi/4$ **31.** $\pi/2$

33. $\ln|x^2 + 6x + 13| - 3\arctan[(x + 3)/2] + C$

35. $\arcsin[(x + 2)/2] + C$ **37.** $-\sqrt{-x^2 - 4x} + C$

39. $4 - 2\sqrt{3} + \frac{1}{6}\pi \approx 1.059$ **41.** $\frac{1}{2}\arctan(x^2 + 1) + C$

43. $2\sqrt{e^t - 3} - 2\sqrt{3}\arctan\left(\sqrt{e^t - 3}/\sqrt{3}\right) + C$ **45.** $\pi/6$

47. a and b **49.** a, b, and c **51.** c **53.** $y = \arcsin(x/2) + \pi$

55. (a)

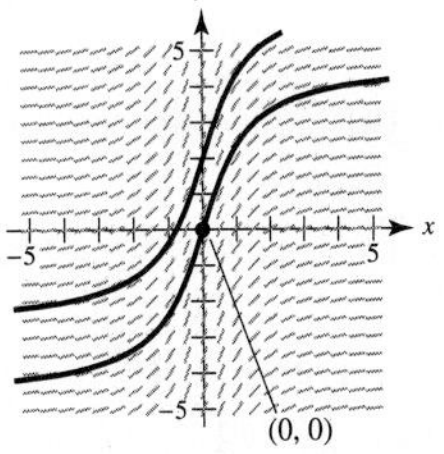

(b) $y = 3\arctan x$

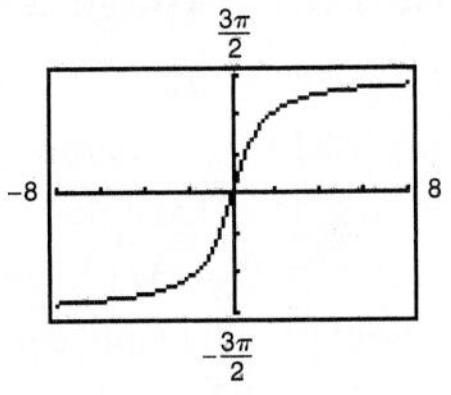

57. (a)

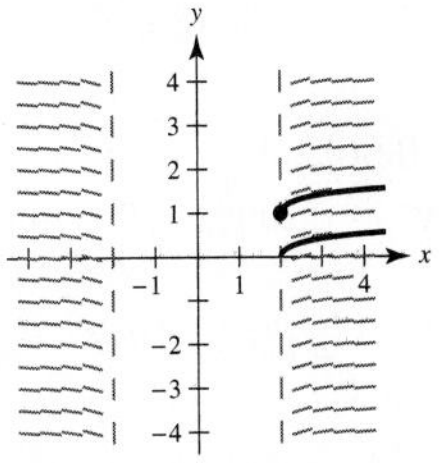

(b) $y = \frac{1}{2}\operatorname{arcsec}(x/2) + 1$, $x \ge 2$

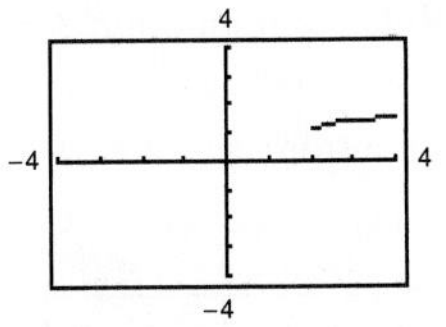

59.

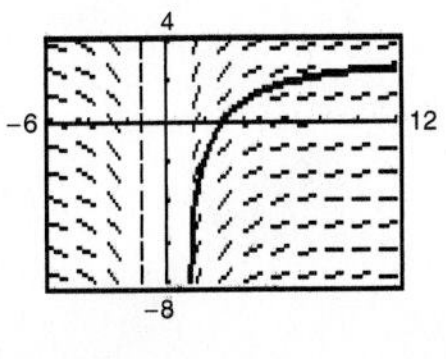

61.

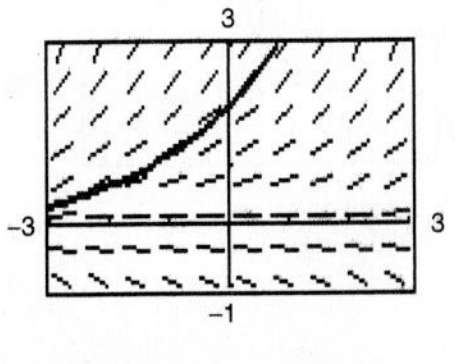

63. $\pi/8$ **65.** $\pi/6$ **67.** $3\pi/2$

69. (a) Proof (b) $\ln(\sqrt{6}/2) + (9\pi - 4\pi\sqrt{3})/36$

71. (a)

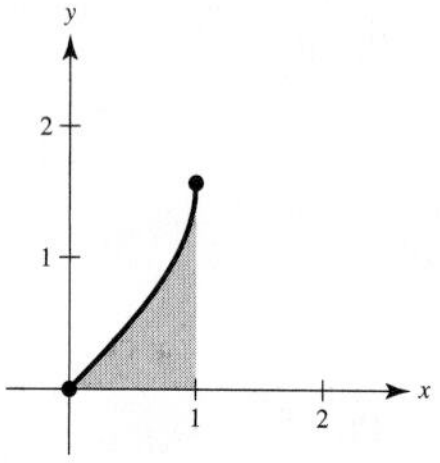

(b) 0.5708
(c) $(\pi - 2)/2$

73. (a) $F(x)$ represents the average value of $f(x)$ over the interval $[x, x + 2]$. Maximum at $x = -1$.
(b) $x = -1$.

75. False. $\displaystyle\int \frac{dx}{3x\sqrt{9x^2 - 16}} = \frac{1}{12}\operatorname{arcsec}\frac{|3x|}{4} + C$

77. True **79–81.** Proofs

83. (a) $v(t) = -32t + 500$

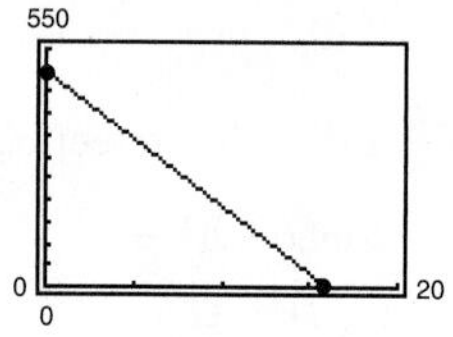

(b) $s(t) = -16t^2 + 500t$; 3906.25 ft

(c) $v(t) = \sqrt{\dfrac{32}{k}}\tan\left[\arctan\left(500\sqrt{\dfrac{k}{32}}\right) - \sqrt{32k}\,t\right]$

(d)

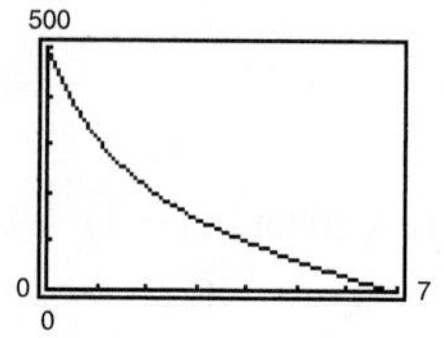

$t_0 = 6.86$ sec

(e) 1088 ft
(f) When air resistance is taken into account, the maximum height of the object is not as great.

Section 5.8 (page 396)

1. (a) 10.018 (b) -0.964 **3.** (a) $\frac{4}{3}$ (b) $\frac{13}{12}$

5. (a) 1.317 (b) 0.962 **7–11.** Proofs

13. $\cosh x = \sqrt{13}/2$; $\tanh x = 3\sqrt{13}/13$; $\operatorname{csch} x = 2/3$;
$\operatorname{sech} x = 2\sqrt{13}/13$; $\coth x = \sqrt{13}/3$

15. $-\operatorname{sech}(x + 1)\tanh(x + 1)$ **17.** $\coth x$ **19.** $\operatorname{csch} x$

21. $\sinh^2 x$ **23.** $\operatorname{sech} t$ **25.** $y = -2x + 2$ **27.** $y = 1 - 2x$

29. Relative maxima: $(\pm\pi, \cosh \pi)$; Relative minimum: $(0, -1)$

31. Relative maximum: $(1.20, 0.66)$
Relative minimum: $(-1.20, -0.66)$

33. $y = a \sinh x$; $y' = a \cosh x$; $y'' = a \sinh x$; $y''' = a \cosh x$;
Therefore, $y''' - y' = 0$.

35. $P_1(x) = x$; $P_2(x) = x$

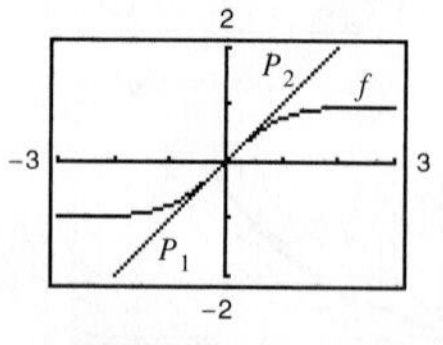

37. (a)

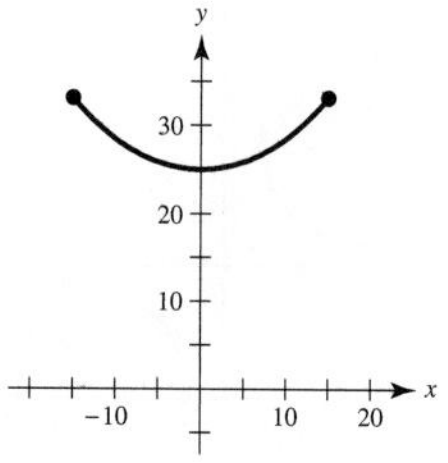

(b) 33.146 units; 25 units
(c) $m = \sinh(1) \approx 1.175$

39. $-\frac{1}{2}\cosh(1 - 2x) + C$ **41.** $\frac{1}{3}\cosh^3(x - 1) + C$

43. $\ln|\sinh x| + C$ **45.** $-\coth(x^2/2) + C$

47. $\operatorname{csch}(1/x) + C$ **49.** $\frac{1}{2}\arctan x^2 + C$ **51.** $\ln 5 - 2\ln 2$

53. $\frac{1}{5}\ln 3$ **55.** $\pi/4$ **57.** $3/\sqrt{9x^2 - 1}$ **59.** $|\sec x|$

61. $2\sec 2x$ **63.** $2\sinh^{-1}(2x)$ **65.** Answers will vary.

67. ∞ **69.** 0 **71.** 1 **73.** $\ln(\sqrt{e^{2x} + 1} - 1) - x + C$

75. $2\sinh^{-1}\sqrt{x} + C = 2\ln(\sqrt{x} + \sqrt{1 + x}) + C$

77. $\dfrac{1}{4}\ln\left|\dfrac{x - 4}{x}\right| + C$ **79.** $\dfrac{1}{2\sqrt{6}}\ln\left|\dfrac{\sqrt{2}(x + 1) + \sqrt{3}}{\sqrt{2}(x + 1) - \sqrt{3}}\right| + C$

81. $\dfrac{1}{4}\arcsin\left(\dfrac{4x - 1}{9}\right) + C$ **83.** $-\dfrac{x^2}{2} - 4x - \dfrac{10}{3}\ln\left|\dfrac{x - 5}{x + 1}\right| + C$

85. $8\arctan(e^2) - 2\pi \approx 5.207$ **87.** $\frac{5}{2}\ln(\sqrt{17} + 4) \approx 5.237$

89. (a) $\ln(\sqrt{3} + 2)$ (b) $\sinh^{-1}\sqrt{3}$

91. $\frac{52}{31}$ kg **93.** $-\sqrt{a^2 - x^2}/x$ **95–101.** Proofs

103. Putnam Problem 8, 1939

Review Exercises for Chapter 5 (page 399)

1.

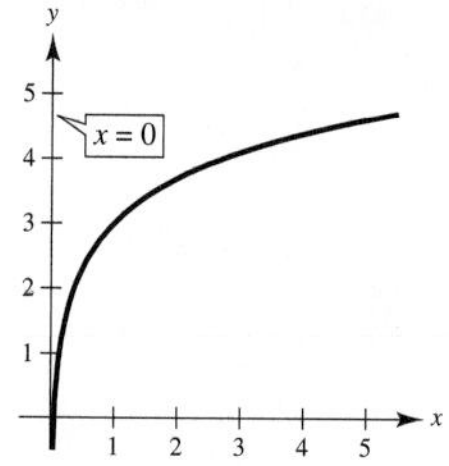

Vertical asymptote: $x = 0$

3. $\frac{1}{5}[\ln(2x + 1) + \ln(2x - 1) - \ln(4x^2 + 1)]$

5. $\ln(3\sqrt[3]{4 - x^2}/x)$ **7.** $e^4 - 1 \approx 53.598$

9. $1/(2x)$ **11.** $(1 + 2\ln x)/(2\sqrt{\ln x})$

13. $\dfrac{dy}{dx} = \dfrac{1}{b^2}\left(b - \dfrac{ab}{a + bx}\right) = \dfrac{x}{a + bx}$ **15.** $y = -x + 1$

17. $\frac{1}{7}\ln|7x - 2| + C$ **19.** $-\ln|1 + \cos x| + C$

21. $3 + \ln 4$ **23.** $\ln(2 + \sqrt{3})$

25. (a) $f^{-1}(x) = 2x + 6$
(b)

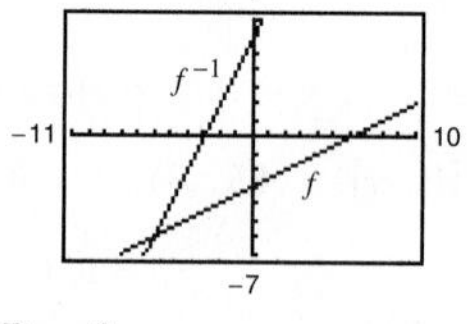

(c) Proof

27. (a) $f^{-1}(x) = x^2 - 1,\quad x \geq 0$
(b)

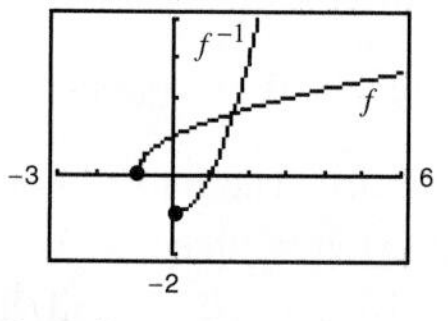

(c) Proof

29. (a) $f^{-1}(x) = x^3 - 1$
(b)

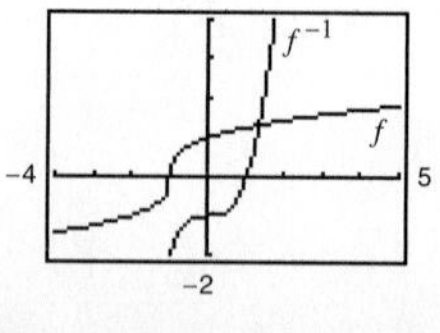

(c) Proof

31. $1/\left[3\left(\sqrt[3]{-3}\right)^2\right] \approx 0.160$ **33.** $3/4$

35. (a) $f^{-1}(x) = e^{2x}$

(b)

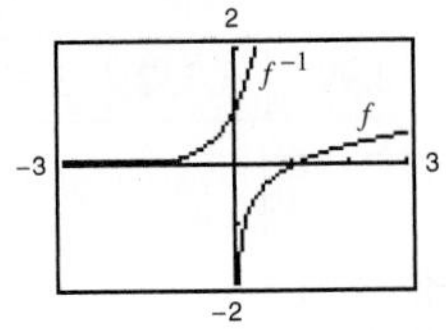

(c) Proof

37.

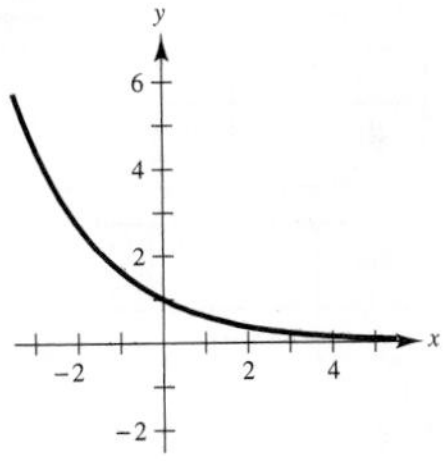

39. $te^t(t + 2)$ **41.** $(e^{2x} - e^{-2x})/\sqrt{e^{2x} + e^{-2x}}$

43. $x(2 - x)/e^x$ **45.** $y = -4x + 4$ **47.** $-y/[x(2y + \ln x)]$

49. $(1 - e^{-3})/6 \approx 0.158$ **51.** $(e^{4x} - 3e^{2x} - 3)/(3e^x) + C$

53. $-\frac{1}{2}e^{1-x^2} + C$ **55.** $\ln(e^2 + e + 1) \approx 2.408$

57. $y = e^x(a \cos 3x + b \sin 3x)$

$y' = e^x[(-3a + b) \sin 3x + (a + 3b) \cos 3x]$

$y'' = e^x[(-6a - 8b) \sin 3x + (-8a + 6b) \cos 3x]$

$y'' - 2y' + 10y$

$= e^x\{[(-6a - 8b) - 2(-3a + b) + 10b] \sin 3x + [(-8a + 6b) - 2(a + 3b) + 10a] \cos 3x\} = 0$

59. $-\frac{1}{2}(e^{-16} - 1) \approx 0.500$

61.

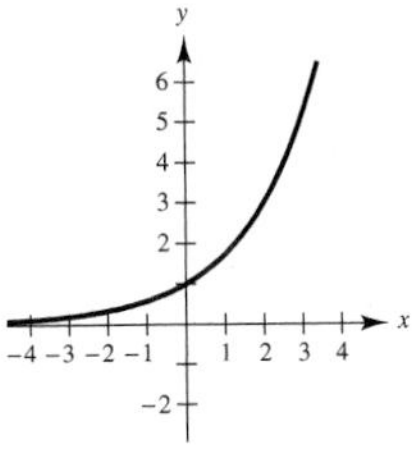

63.

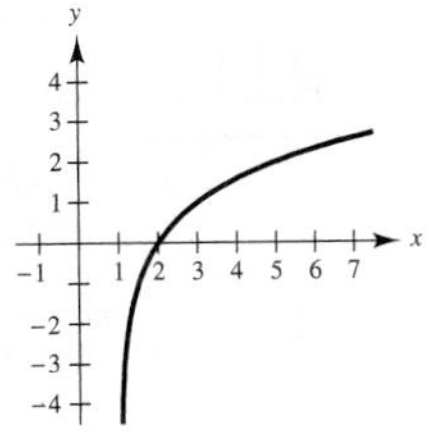

65. $3^{x-1} \ln 3$ **67.** $x^{2x+1}(2 \ln x + 2 + 1/x)$

69. $-1/[\ln 3(2 - 2x)]$ **71.** $5^{(x+1)^2}/(2 \ln 5) + C$

73. (a) Domain: $0 \le h < 18{,}000$

(b)

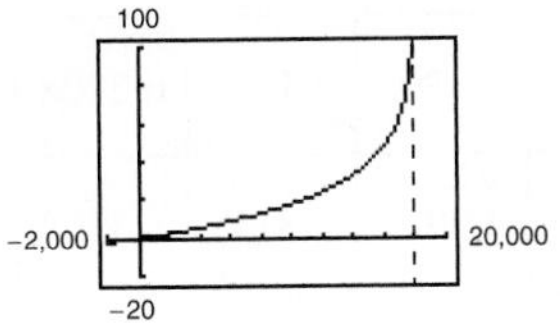

(c) $t = 0$

Vertical asymptote: $h = 18{,}000$

75.

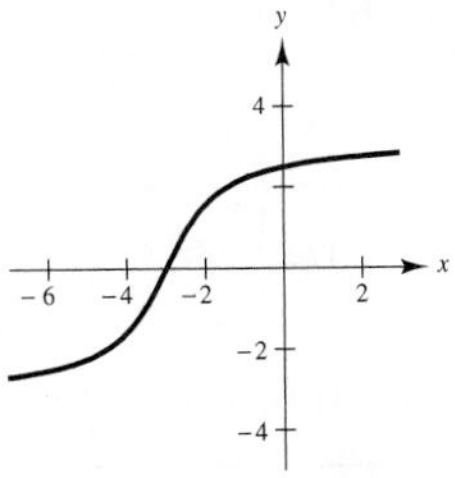

77. (a) $1/2$ (b) $\sqrt{3}/2$

79. $(1 - x^2)^{-3/2}$ **81.** $\dfrac{x}{|x|\sqrt{x^2 - 1}} + \operatorname{arcsec} x$ **83.** $(\arcsin x)^2$

85. $\frac{1}{2} \arctan(e^{2x}) + C$ **87.** $\frac{1}{2} \arcsin x^2 + C$

89. $\frac{1}{4}[\arctan (x/2)]^2 + C$ **91.** $\frac{2}{3}\pi + \sqrt{3} - 2 \approx 1.826$

93. $y = A \sin\left(t\sqrt{k/m}\right)$

95. $y' = x\left(\dfrac{2}{1 - 4x^2}\right) + \tanh^{-1} 2x = \dfrac{2x}{1 - 4x^2} + \tanh^{-1} 2x$

97. $\frac{1}{3} \tanh x^3 + C$

P.S. Problem Solving (page 401)

1. $a \approx 4.7648$; $\theta \approx 1.7263$ or $98.9°$

3. (a)

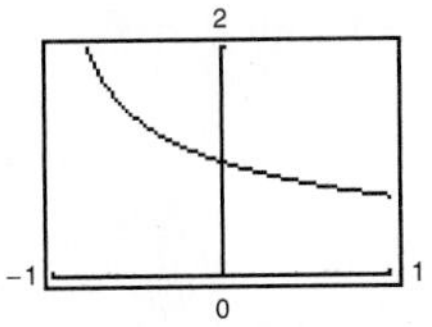

(b) 1 (c) Proof

5.

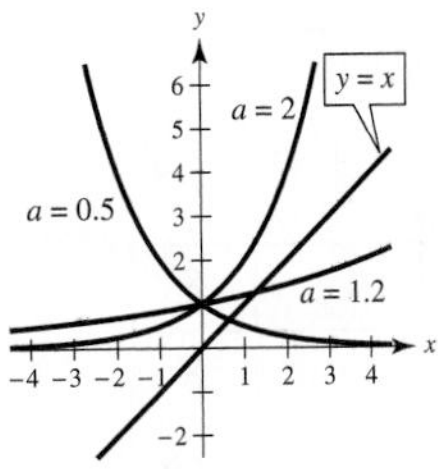

$y = 0.5^x$ and $y = 1.2^x$ intersect the line $y = x$; $0 < a < e^{1/e}$

7. (a) Area of region $A = \left(\sqrt{3} - \sqrt{2}\right)/2 \approx 0.1589$

Area of region $B = \pi/12 \approx 0.2618$

(b) $\frac{1}{24}\left[3\pi\sqrt{2} - 12\left(\sqrt{3} - \sqrt{2}\right) - 2\pi\right] \approx 0.1346$

(c) 1.2958 (d) 0.6818

9. Proof **11.** 0.2318

13. (a)

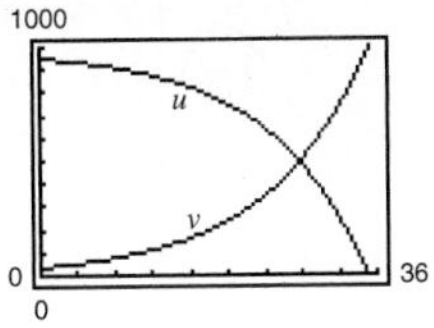

(b) Interest charged on unpaid balance; $t \approx 27.68$ yrs

(c) The slopes are negatives of each other; $u'(15) \approx -14.0572$; $v'(15) \approx 14.0572$

(d)

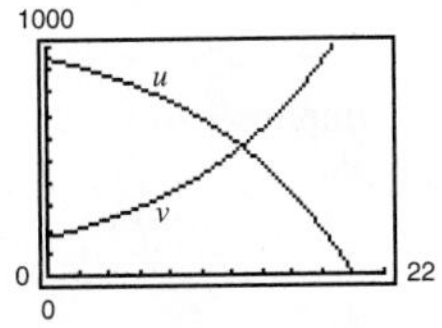

$t \approx 12.67$ yrs

Answers will vary.

Chapter 6

Section 6.1 (page 409)

1–11. Proofs **13.** Not a solution **15.** Solution

17. Solution **19.** Not a solution **21.** Solution

23. Not a solution **25.** $y = 3e^{-x/2}$ **27.** $4y^2 = x^3$

29.

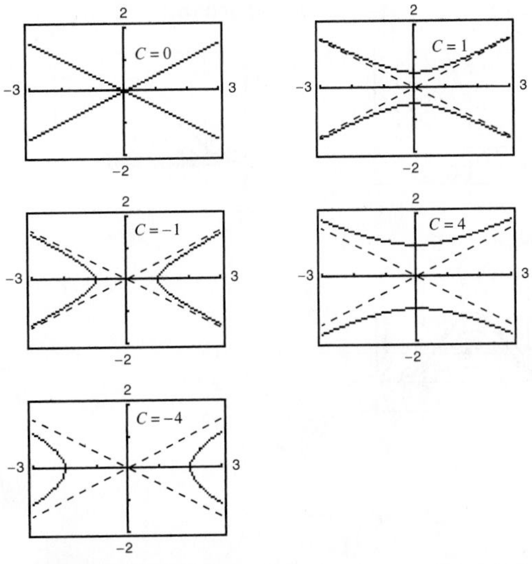

31. $y = 3e^{-2x}$ **33.** $y = 2 \sin 3x - \frac{1}{3} \cos 3x$
35. $y = -2x + \frac{1}{2}x^3$ **37.** $y = x^3 + C$
39. $y = \frac{1}{2} \ln(1 + x^2) + C$ **41.** $y = x - \ln x^2 + C$
43. $y = -\frac{1}{2} \cos 2x + C$
45. $y = \frac{2}{5}(x - 3)^{5/2} + 2(x - 3)^{3/2} + C$
47. $y = \frac{1}{2}e^{x^2} + C$

49.

x	-4	-2	0	2	4	8
y	2	0	4	4	6	8
dy/dx	-2	Undef.	0	$\frac{1}{2}$	$\frac{2}{3}$	1

51.

x	-4	-2	0	2	4	8
y	2	0	4	4	6	8
dy/dx	$-2\sqrt{2}$	-2	0	0	$-2\sqrt{2}$	-8

53. b **54.** c **55.** d **56.** a

57. (a) and (b)

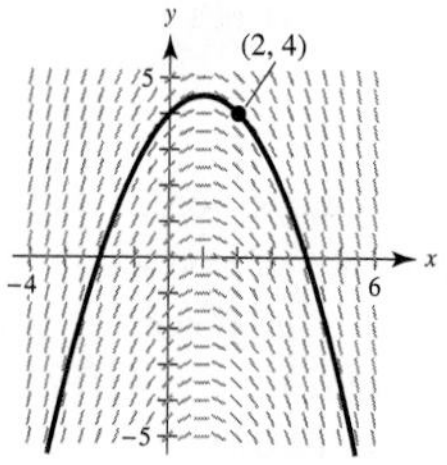

(c) $x \to \infty$ implies $y \to -\infty$; $x \to -\infty$ implies $y \to \infty$

59. (a) and (b)

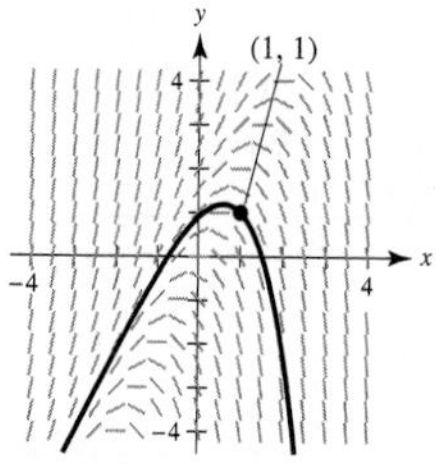

(c) $x \to \infty$ implies $y \to -\infty$; $x \to -\infty$ implies $y \to -\infty$

61. (a)

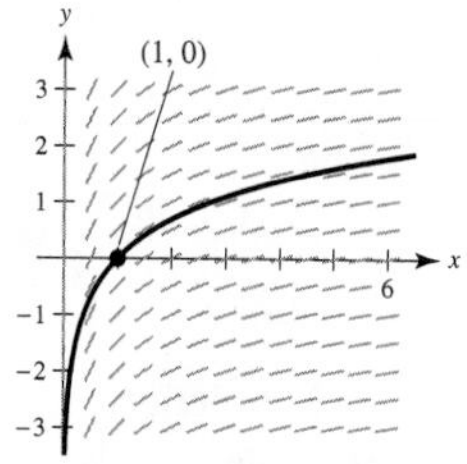

$x \to \infty, y \to \infty$

(b)

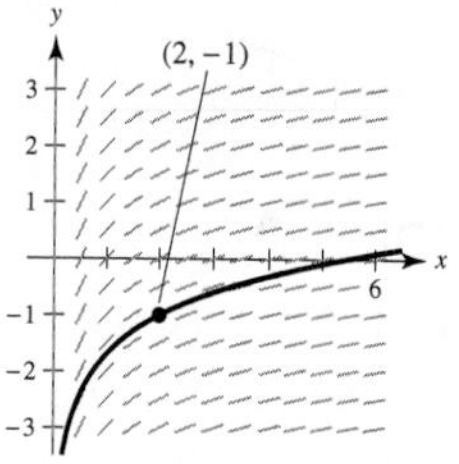

$x \to \infty, y \to \infty$

63. (a) and (b)

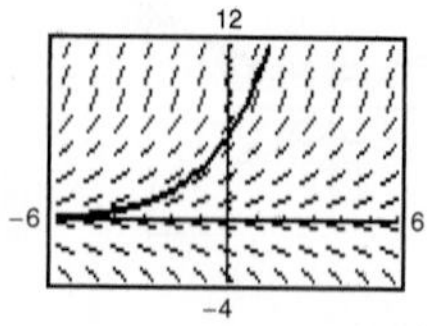

65. (a) and (b)

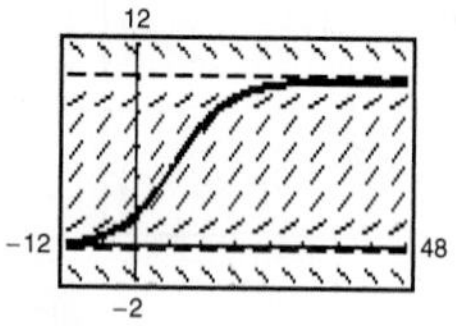

67. (a) and (b)

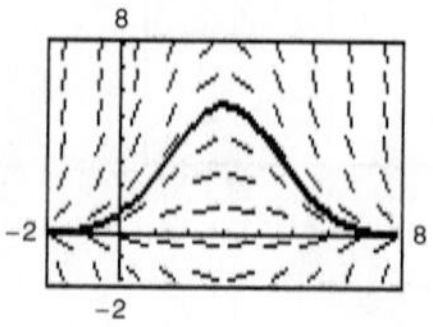

69.

n	0	1	2	3	4	5	6
x_n	0	0.1	0.2	0.3	0.4	0.5	0.6
y_n	2	2.2	2.43	2.693	2.992	3.332	3.715

n	7	8	9	10
x_n	0.7	0.8	0.9	1.0
y_n	4.146	4.631	5.174	5.781

71.

n	0	1	2	3	4	5	6
x_n	0	0.05	0.1	0.15	0.2	0.25	0.3
y_n	3	2.7	2.438	2.209	2.010	1.839	1.693

n	7	8	9	10
x_n	0.35	0.4	0.45	0.5
y_n	1.569	1.464	1.378	1.308

73.

n	0	1	2	3	4	5	6
x_n	0	0.1	0.2	0.3	0.4	0.5	0.6
y_n	1	1.1	1.212	1.339	1.488	1.670	1.900

n	7	8	9	10
x_n	0.7	0.8	0.9	1.0
y_n	2.213	2.684	3.540	5.958

75.

x	0	0.2	0.4	0.6	0.8	1
$y(x)$ (exact)	3.0000	3.6642	4.4755	5.4664	6.6766	8.1548
$y(x)$ ($h = 0.2$)	3.0000	3.6000	4.3200	5.1840	6.2208	7.4650
$y(x)$ ($h = 0.1$)	3.0000	3.6300	4.3923	5.3147	6.4308	7.7812

77.

x	0	0.2	0.4	0.6	0.8	1
$y(x)$ (exact)	0.0000	0.2200	0.4801	0.7807	1.1231	1.5097
$y(x)$ ($h = 0.2$)	0.0000	0.2000	0.4360	0.7074	1.0140	1.3561
$y(x)$ ($h = 0.1$)	0.0000	0.2095	0.4568	0.7418	1.0649	1.4273

79. (a) $y(1) = 112.7141°$; $y(2) = 96.3770°$; $y(3) = 86.5954°$
(b) $y(1) = 113.2441°$; $y(2) = 97.0158°$; $y(3) = 87.1729°$

81. The general solution is a family of curves that satisfies the differential equation. A particular solution is one member of the family that satisfies given conditions.

83. Begin with a point (x_0, y_0) that satisfies the initial condition $y(x_0) = y_0$. Then using a small step size h, calculate the point $(x_1, y_1) = (x_0 + h, y_0 + hF(x_0, y_0))$. Continue generating the sequence of points $(x_n + h, y_n + hF(x_n, y_n))$ or (x_{n+1}, y_{n+1}).

85. False: $y = x^3$ is a solution of $xy' - 3y = 0$, but $y = x^3 + 1$ is not a solution.

87. True

89. (a)

x	0	0.2	0.4	0.6	0.8	1
y	4	2.6813	1.7973	1.2048	0.8076	0.5413
y_1	4	2.56	1.6384	1.0486	0.6711	0.4295
y_2	4	2.4	1.44	0.864	0.5184	0.3110
e_1	0	0.1213	0.1589	0.1562	0.1365	0.1118
e_2	0	0.2813	0.3573	0.3408	0.2892	0.2303
r	0	0.4312	0.4447	0.4583	0.4720	0.4855

(b) If h is halved, then the error is approximately halved because r is approximately 0.5.

(c) The error will again be halved.

91. (a)

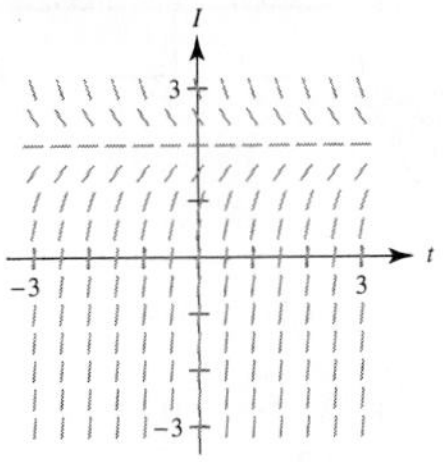

(b) $\lim_{t\to\infty} I(t) = 2$

93. $\omega = \pm 4$ rad/sec

95. Putnam Problem 3, Morning Session, 1954

Section 6.2 (page 418)

1. $y = \frac{1}{2}x^2 + 2x + C$ **3.** $y = Ce^x - 2$ **5.** $y^2 - 5x^2 = C$

7. $y = Ce^{(2x^{3/2})/3}$ **9.** $y = C(1 + x^2)$

11. $dQ/dt = k/t^2$
$Q = -k/t + C$

13. $dN/ds = k(250 - s)$
$N = -(k/2)(250 - s)^2 + C$

15. (a)

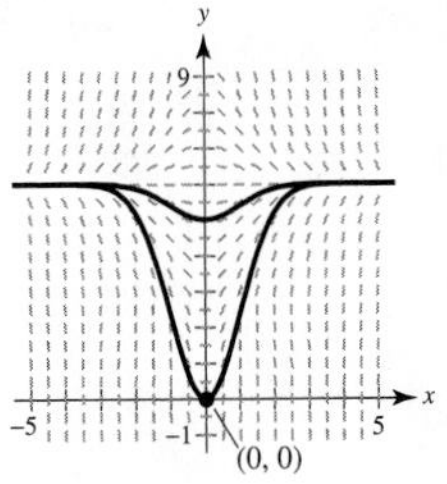

(b) $y = 6 - 6e^{-x^2/2}$

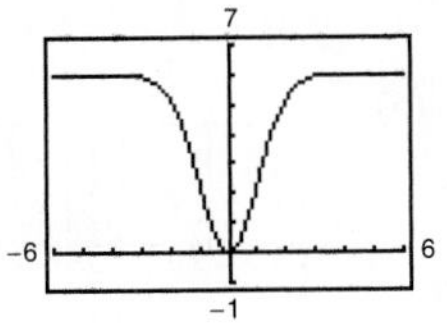

17. $y = \frac{1}{4}t^2 + 10$

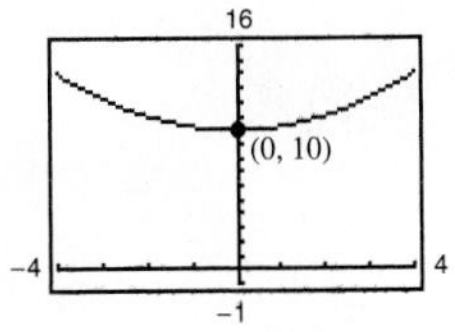

19. $y = 10e^{-t/2}$

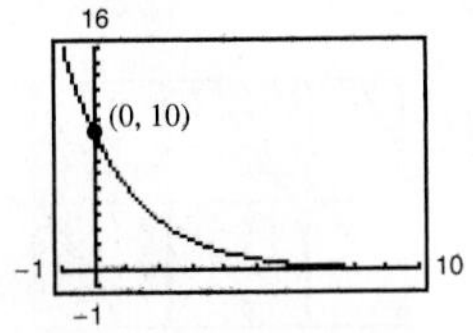

21. $dy/dx = ky$
$y = 4e^{0.3054x}$
$y(6) \approx 25$

23. $dV/dt = kV$
$V = 20{,}000e^{-0.1175t}$
$V(6) \approx 9882$

25. $y = \frac{1}{2}e^{0.4605t}$ **27.** $y = 0.6687e^{0.4024t}$

29. C is the initial value of y, and k is the proportionality constant.

31. Quadrants I and III; dy/dx is positive when both x and y are positive (Quadrant I) or when both x and y are negative (Quadrant III).

33. Amount after 1000 yrs: 6.48 g;
Amount after 10,000 yrs: 0.13 g

35. Initial quantity: 38.16 g;
Amount after 1000 yrs: 24.74 g

37. Amount after 1000 yrs: 4.43 g;
Amount after 10,000 yrs: 1.49 g

39. Initial quantity: 2.16 g;
Amount after 10,000 yrs: 1.62 g

41. 95.76%

43. Time to double: 11.55 yrs; Amount after 10 yrs: \$1822.12

45. Annual rate: 8.94%; Amount after 10 yrs: \$1833.67

47. Annual rate: 9.50%; Time to double: 7.30 yrs

49. \$112,087.09 **51.** \$30,688.87

53. (a) 10.24 yrs (b) 9.93 yrs (c) 9.90 yrs (d) 9.90 yrs

55. (a) 8.50 yrs (b) 8.18 yrs (c) 8.16 yrs (d) 8.15 yrs

57. (a) $P = 7.77e^{-0.009t}$ (b) 6.79 million
(c) Since $k < 0$, the population is decreasing.

59. (a) $P = 5.07e^{0.026t}$ (b) 7.48 million
(c) Since $k > 0$, the population is increasing.

61. (a) $100.1596(1.2455)^t$ (b) 6.3 hours

63. (a) $N \approx 30(1 - e^{-0.0502t})$ (b) 36 days

65. (a) $P_1 = 181e^{0.01245t}$ or $P_1 = 181(1.01253)^t$
(b) $P_2 = 182.3248(1.01091)^t$
(c)

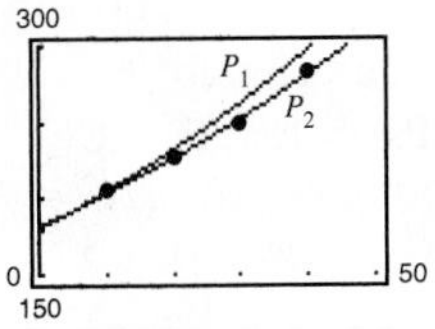

P_2 is a better approximation.
(d) 2011

67. (a) 20 decibels (b) 70 decibels
(c) 95 decibels (d) 120 decibels

69. 2014 ($t = 16$) **71.** 379.2°

73. False. The rate of growth dy/dx is proportional to y.

75. True

Section 6.3 (page 429)

1. $y^2 - x^2 = C$ **3.** $r = Ce^{0.05s}$ **5.** $y = C(x + 2)^3$

7. $y^2 = C - 2\cos x$ **9.** $y = -\frac{1}{4}\sqrt{1 - 4x^2} + C$

11. $y = Ce^{(\ln x)^2/2}$ **13.** $y^2 = 2e^x + 14$ **15.** $y = e^{-(x^2+2x)/2}$

17. $y^2 = 4x^2 + 3$ **19.** $u = e^{(1-\cos v^2)/2}$ **21.** $P = P_0e^{kt}$

23. $9x^2 + 16y^2 = 25$ **25.** $f(x) = Ce^{-x/2}$

27. Homogeneous of degree 3 **29.** Homogeneous of degree 3

31. Not homogeneous **33.** Homogeneous of degree 0

35. $|x| = C(x - y)^2$ **37.** $|y^2 + 2xy - x^2| = C$

39. $y = Ce^{-x^2/2y^2}$ **41.** $e^{y/x} = 1 + \ln x^2$ **43.** $x = e^{\sin(y/x)}$

45.

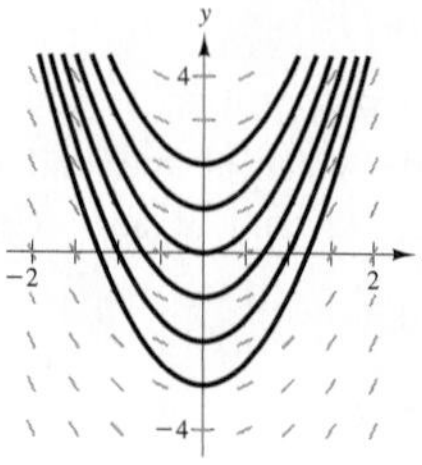

$y = \frac{1}{2}x^2 + C$

47.

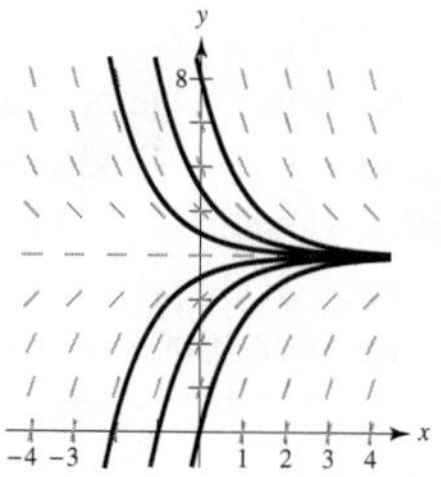

$y = 4 + Ce^{-x}$

49. (a) $y = 0.1602$ (b) $y = 5e^{-3x^2}$ (c) $y = 0.2489$

51. (a) $y = 3.0318$ (b) $y^3 - 4y = x^2 + 12x - 13$ (c) $y = 3$

53. 98.9% of the original amount

55. (a) $dy/dx = k(y - 4)$ (b) a (c) Proof

56. (a) $dy/dx = k(x - 4)$ (b) b (c) Proof

57. (a) $dy/dx = ky(y - 4)$ (b) c (c) Proof

58. (a) $dy/dx = ky^2$ (b) d (c) Proof

59. (a) $w = 1200 - 1140e^{-0.8t}$

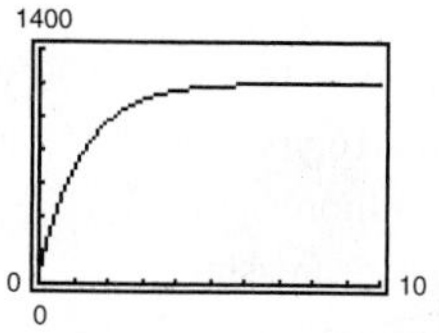

$w = 1200 - 1140e^{-0.9t}$

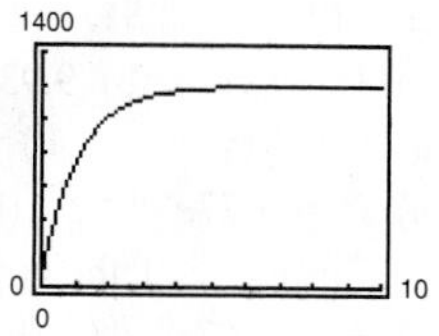

$w = 1200 - 1140e^{-t}$

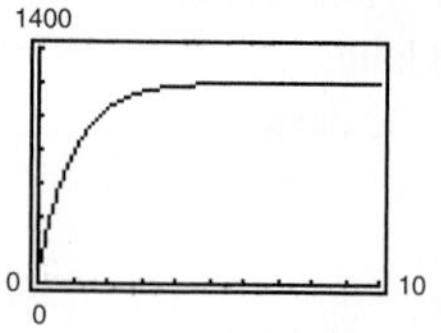

(b) 1.31 yrs; 1.16 yrs; 1.05 yrs (c) 1200 lb

61. Circles: $x^2 + y^2 = C$
Lines: $y = Kx$
Graphs will vary.

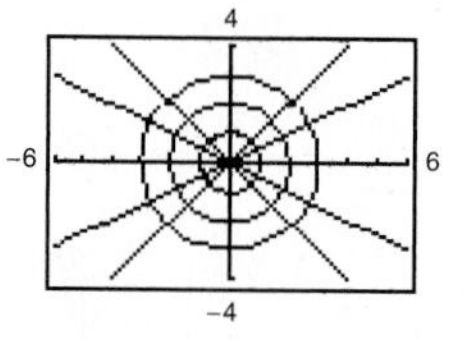

63. Parabolas: $x^2 = Cy$
Ellipses: $x^2 + 2y^2 = K$
Graphs will vary.

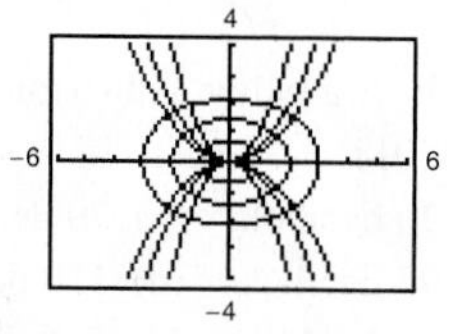

65. Curves: $y^2 = Cx^3$
Ellipses: $2x^2 + 3y^2 = K$
Graphs will vary.

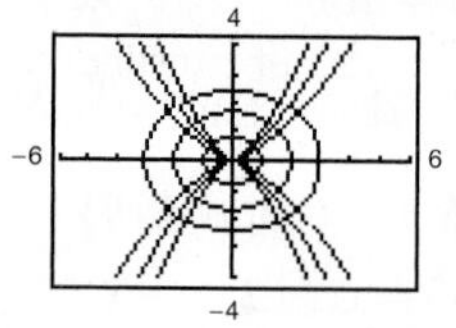

67. d **68.** a **69.** b **70.** c

71. (a) 0.75 (b) 1500 units (c) 60 units (d) 4.24 yrs
(e) $dP/dt = 0.75P(1 - P/1500)$

73. (a) 3 (b) 100 units
(c)

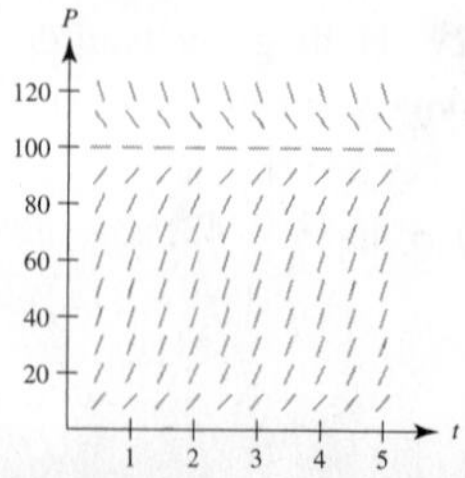

(d) 50 units

75. $y = 40/(1 + 4e^{-t})$ **77.** $y = 120/(1 + 14e^{-0.8t})$

79. (a) $P = \dfrac{200}{1 + 7e^{-0.2640t}}$ (b) 70 panthers (c) 7.37 years
(d) $dP/dt = 0.2640P(1 - P/200)$; 65.6 (e) 100 years

81. Answers will vary.

83. Two families of curves are mutually orthogonal if each curve in the first family intersects each curve in the second family at right angles.

85. False. $y' = x/y$ is separable, but $y = 0$ is not a solution.

87. False: $f(tx, ty) \neq t^n f(x, y)$. **89.** Proof

Section 6.4 (page 438)

1. Linear; can be written in the form $dy/dx + P(x)y = Q(x)$

3. Not linear; can't be written in the form $dy/dx + P(x)y = Q(x)$

5. $y = x^2 + 2x + C/x$ **7.** $y = -10 + Ce^x$

9. $y = -1 + Ce^{\sin x}$ **11.** $y = (x^3 - 3x + C)/[3(x - 1)]$

13. $y = e^{x^3}(x + C)$

15. (a) Answers will vary.

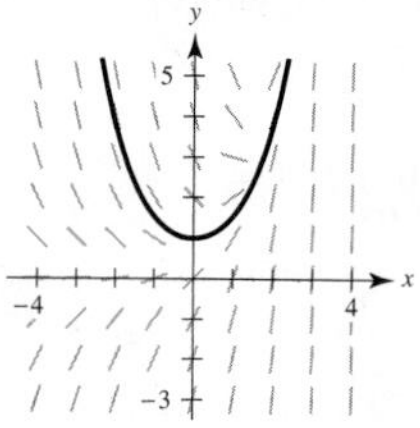

(b) $y = \frac{1}{2}(e^x + e^{-x})$

(c)

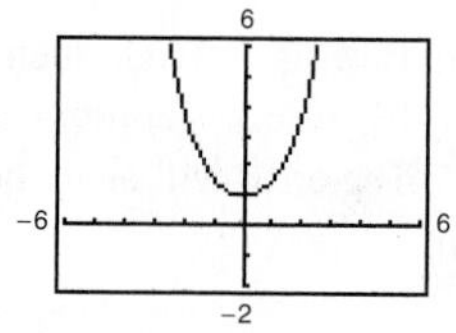

17. $y = 1 + 4/e^{\tan x}$ **19.** $y = \sin x + (x + 1)\cos x$

21. $xy = 4$ **23.** $y = -2 + x\ln|x| + 12x$

25. $1/y^2 = Ce^{2x^3} + \frac{1}{3}$ **27.** $y = 1/(Cx - x^2)$

29. $y^{2/3} = 2e^x + Ce^{2x/3}$

31. (a)

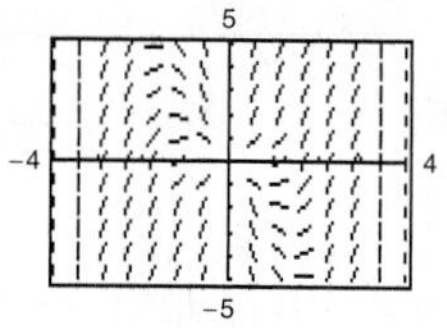

(c)

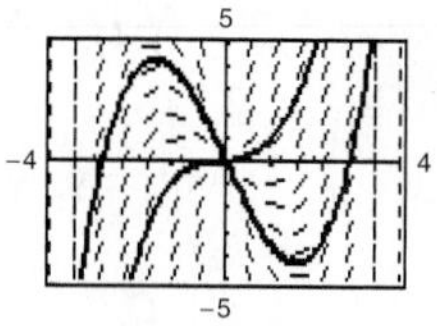

(b) $(-2, 4)$: $y = \frac{1}{2}x(x^2 - 8)$
$(2, 8)$: $y = \frac{1}{2}x(x^2 + 4)$

33. (a)

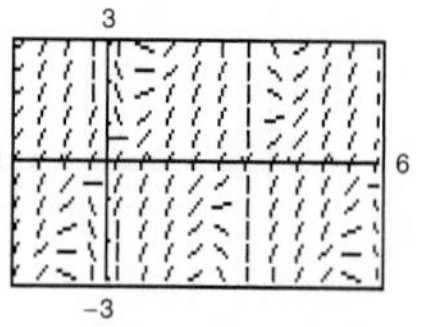

(b) $(1, 1)$: $y = (2\cos 1 + \sin 1)\csc x - 2\cot x$
$(3, -1)$: $y = (2\cos 3 - \sin 3)\csc x - 2\cot x$

(c)

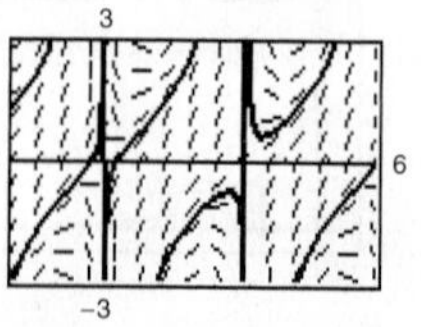

35. $P = -N/k + (N/k + P_0)e^{kt}$

37. (a) \$583,098.01 (b) \$3,243,606.35

39. (a) $\frac{dQ}{dt} = q - kQ$ (b) $Q = \frac{q}{k} + \left(Q_0 - \frac{q}{k}\right)e^{-kt}$ (c) $\frac{q}{k}$

41. Proof

43. (a) $Q = 25e^{-t/20}$ (b) $-20 \ln\left(\frac{3}{5}\right) \approx 10.2$ min (c) 0

45. (a) $t = 50$ min (b) $100 - \frac{25}{\sqrt{2}} \approx 82.32$ lb

47. $V(t) = -159.47(1 - e^{-0.2007t})$; -159.47 ft/sec

49. $I = \frac{E_0}{R} + Ce^{-Rt/L}$ **51.** $\frac{dy}{dx} + P(x)y = Q(x)$; $u(x) = e^{\int P(x)\,dx}$

53. c **54.** d **55.** a **56.** b **57.** $2e^x + e^{-2y} = C$

59. $y = Ce^{-\sin x} + 1$ **61.** $x^3y^2 + x^4y = C$

63. $y = [e^x(x - 1) + C]/x^2$ **65.** $x^4y^4 - 2x^2 = C$

67. $y = \frac{12}{5}x^2 + C/x^3$ **69.** False. $y' + xy = x^2$ is linear.

Review Exercises for Chapter 6 (page 441)

1. No **3.** $y = \frac{2}{3}x^3 + 5x + C$ **5.** $y = \frac{1}{2}\sin 2x + C$

7. $y = 4(x - 7)^{3/2}(3x + 14)/15 + C$

9.

x	-4	-2	0	2	4	8
y	2	0	4	4	6	8
dy/dx	-4	Undef.	0	1	$\frac{4}{3}$	2

11. (a) and (b)

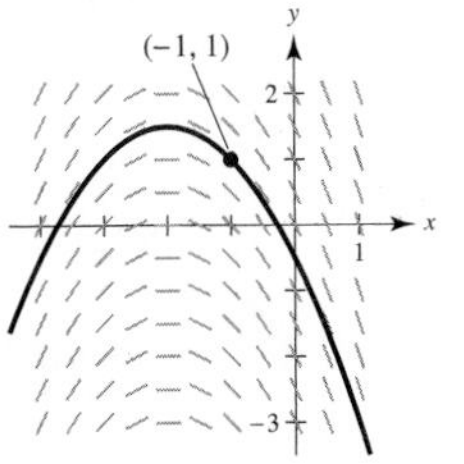

13. (a) and (b)

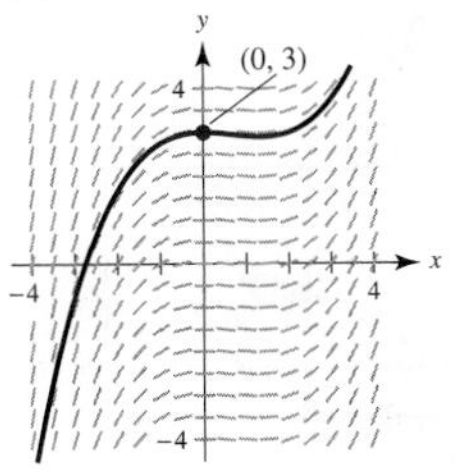

15. (a) and (b)

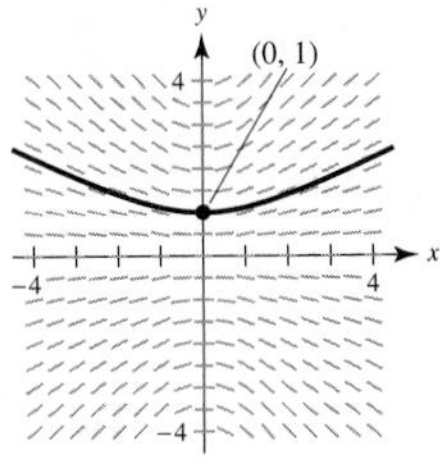

17. $y = 6x - \frac{1}{2}x^2 + C$

19. $y = -3 - 1/(x + c)$ **21.** $y = Ce^x/(2 + x)^2$

23. $y = \frac{3}{4}e^{0.379t}$ **25.** $y = 5e^{-0.680t}$ **27.** ≈ 7.79 in.

29. (a) $S \approx 30e^{-1.7918/t}$ (b) 20,965 units

(c)

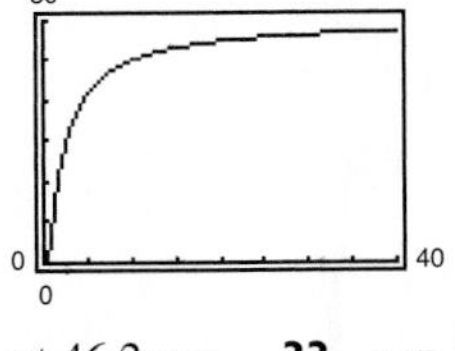

31. About 46.2 yrs **33.** $y = \frac{1}{2}x^2 + 3 \ln|x| + C$

35. $y = Ce^{x^2}$ **37.** $x/(x^2 - y^2) = C$

39. Proof; $y = -2x + \frac{1}{2}x^3$

41.

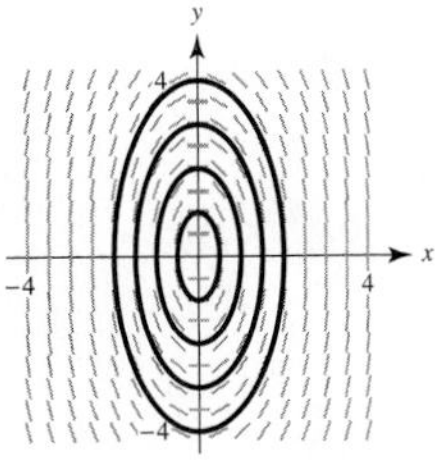

Graphs will vary.

$4x^2 + y^2 = C$

43. (a) 0.55 (b) 7200 (c) 160 (d) 6.88 yrs

(e) $\frac{dP}{dt} = 0.55P\left(1 - \frac{P}{7200}\right)$

45. (a) $P(t) = \frac{20{,}400}{1 + 16e^{-0.553t}}$ (b) 17,118 trout (c) 4.94 yrs

47. $y = -8 + Ce^x$ **49.** $y = e^{x/4}\left[\frac{1}{4}(x + C)\right]$

51. $y = (x + C)/(x - 2)$

53. $y = Ce^{3x} - \frac{1}{13}(2 \cos 2x + 3 \sin 2x)$

55. $y = e^{5x}/10 + Ce^{-5x}$ **57.** $y = 1/(1 + x + Ce^x)$

59. $y^{-2} = Cx^2 + 2/(3x)$

61. Answers will vary. Sample answer:
$(x^2 + 3y^2)\,dx - 2xy\,dy = 0$; $x^3 = C(x^2 + y^2)$

63. Answers will vary. Sample answer:
$x^3y' + 2x^2y = 1$; $x^2y = \ln|x| + C$

P.S. Problem Solving (page 443)

1. (a) $y = 1/(1 - 0.01t)^{100}$; $T = 100$

(b) $y = 1\Big/\left[\left(\frac{1}{y_0}\right)^{\varepsilon} - k\varepsilon t\right]^{1/\varepsilon}$; Answers will vary.

3. (a) $dS/dt = kS(L - S)$; $S = 100/(1 + 9e^{-0.8109t})$

(b) 2.7 months

(c)

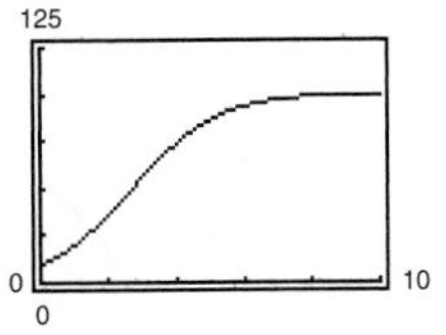

(d)

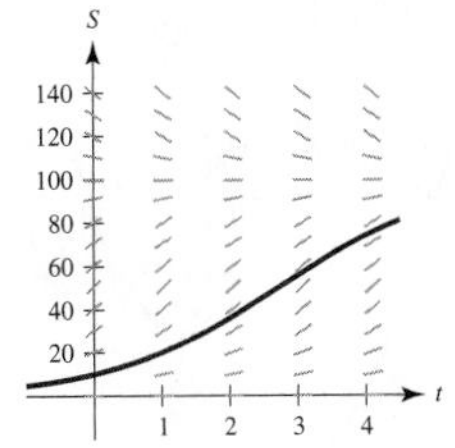

(e) Sales will decrease toward the line $S = L$.

5. Proof **7.** (a) 9809.1 sec (b) 7.21 ft

9. (a) $(1.155, 0)$ (b) $\left(\frac{h\sqrt{2 + h}}{2h + 4 - 3\sqrt{2 + h}} - 1, 0\right)$

(c)

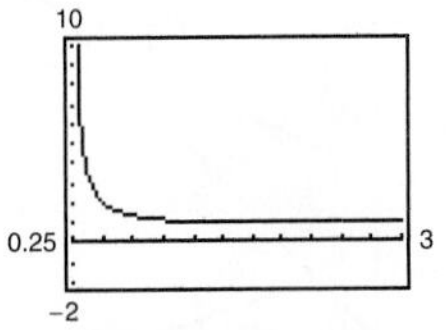

There is a vertical asymptote at $h = \frac{1}{4}$, which is the height of the mountain.

11. (a) $C = C_0e^{-Rt/V}$ (b) $\lim_{t\to\infty} C(t) = 0$

13. (a) $C = (Q/R)(1 - e^{-Rt/V})$ (b) $\lim_{t\to\infty} C(t) = Q/R$

Chapter 7

Section 7.1 (page 452)

1. $-\int_0^6 (x^2 - 6x)\,dx$ **3.** $\int_0^3 (-2x^2 + 6x)\,dx$

5. $-6\int_0^1 (x^3 - x)\,dx$

7.

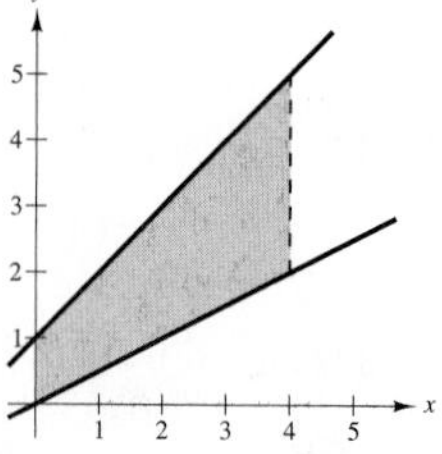

9.

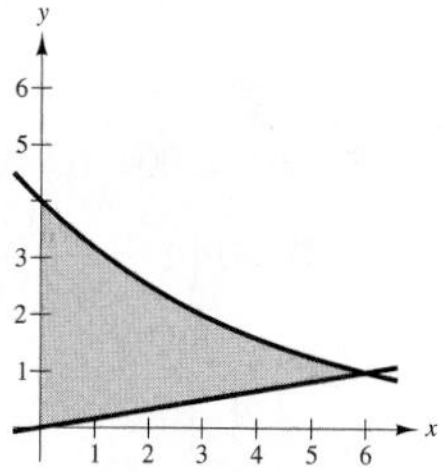

11.

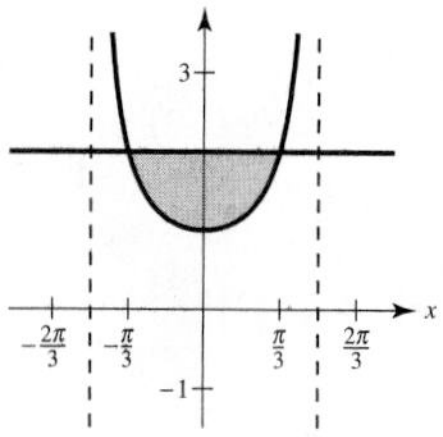

13. (a) $\frac{125}{6}$ (b) $\frac{125}{6}$ **15.** d

17.

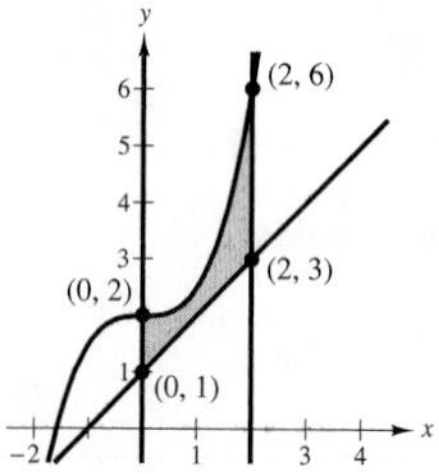

2

19.

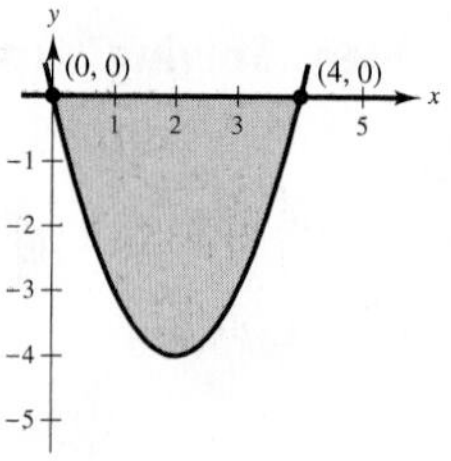

$\frac{32}{3}$

21.

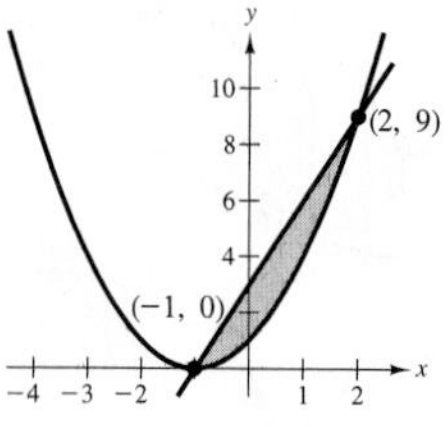

$\frac{9}{2}$

23.

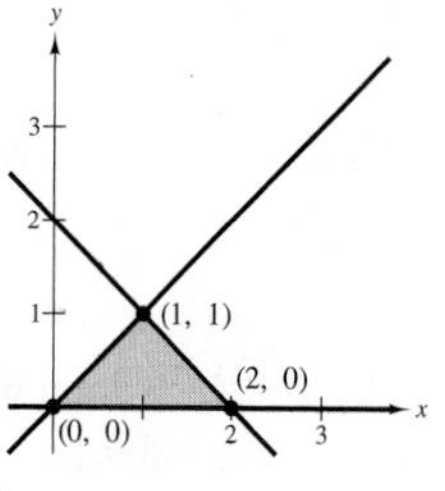

1

25.

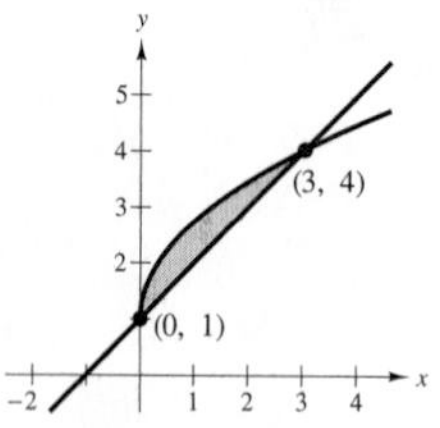

$\frac{3}{2}$

27.

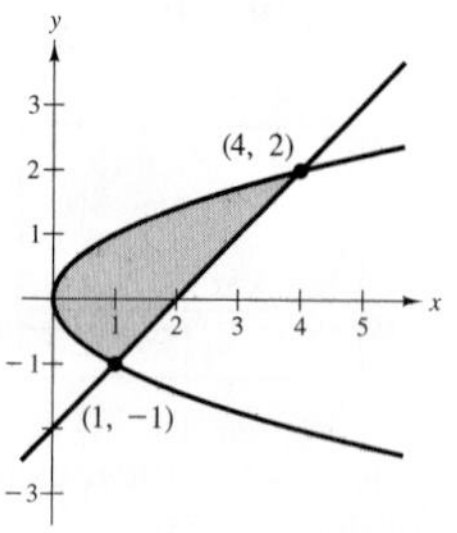

$\frac{9}{2}$

29.

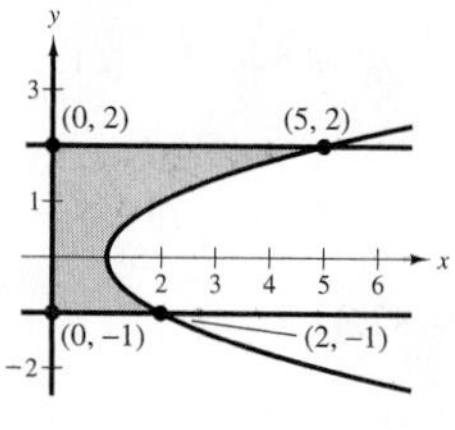

6

31.

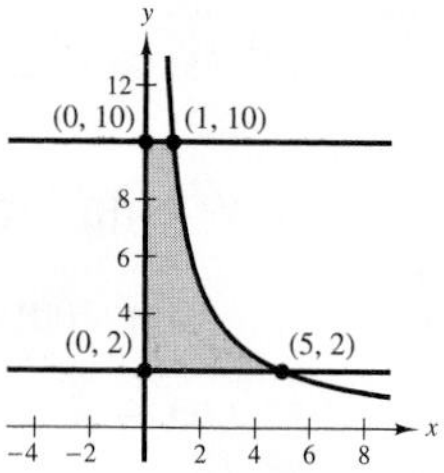

$10 \ln 5 \approx 16.094$

33. (a)

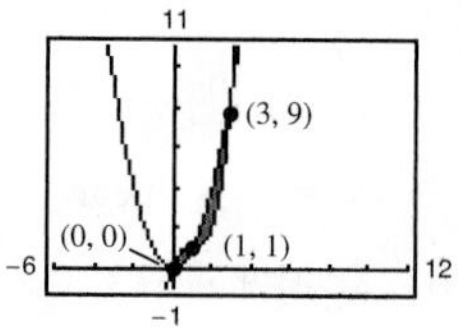

(b) $\frac{37}{12}$

35. (a)

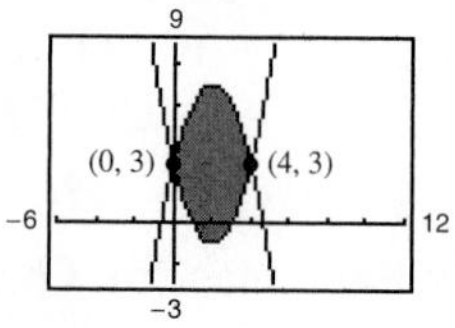

(b) $\frac{64}{3}$

37. (a)

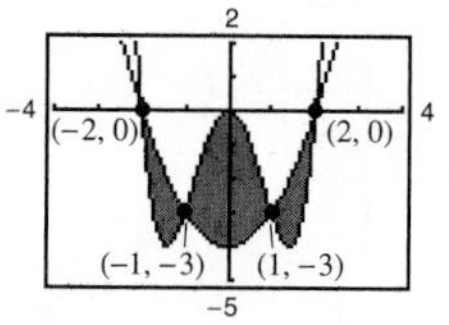

(b) 8

39. (a)

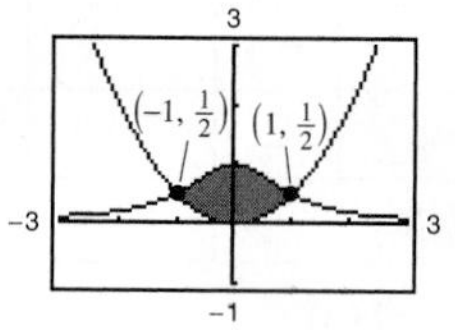

(b) $\pi/2 - 1/3 \approx 1.237$

41. (a)

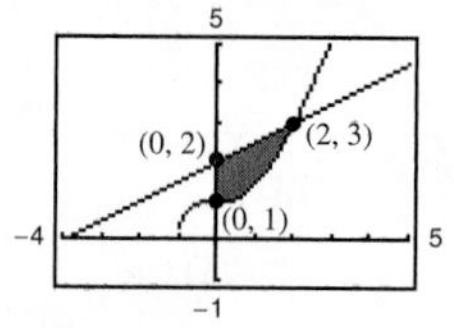

(b) ≈ 1.759

43.

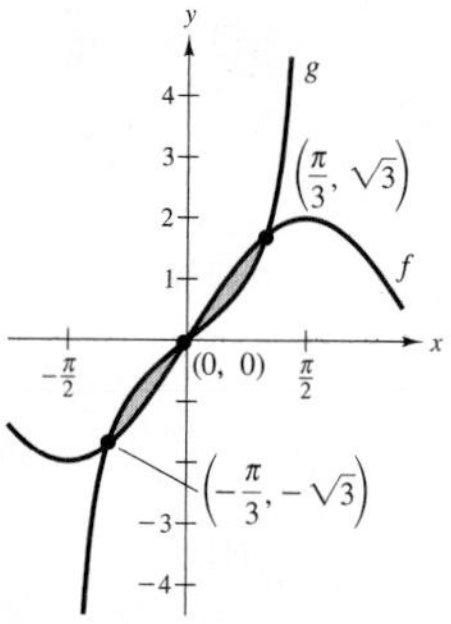

$2(1 - \ln 2) \approx 0.614$

45.

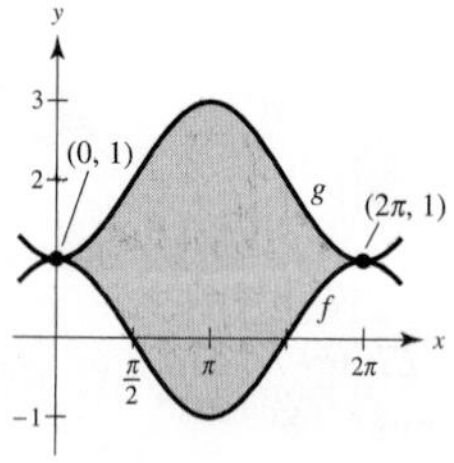

$4\pi \approx 12.566$

47.

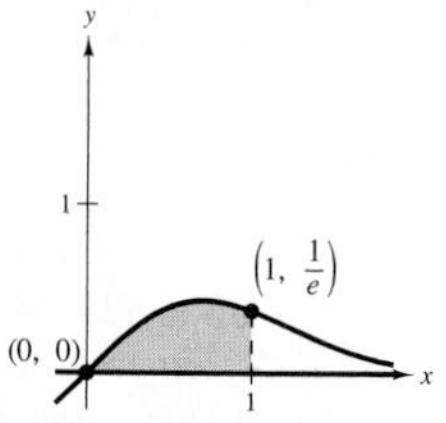

$(1/2)(1 - 1/e) \approx 0.316$

49. (a)

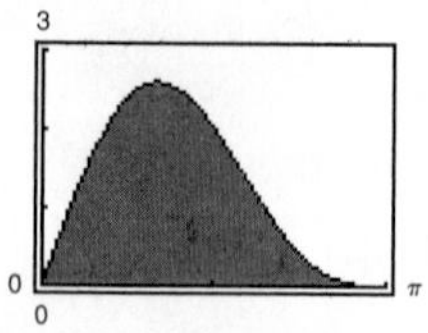

(b) 4

51. (a)

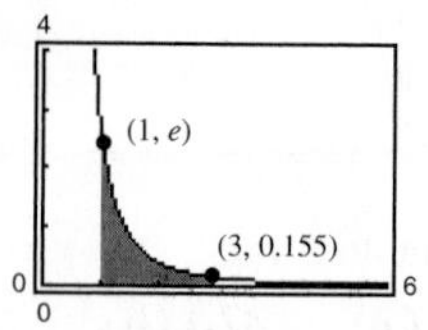

(b) ≈ 1.323

53. (a)

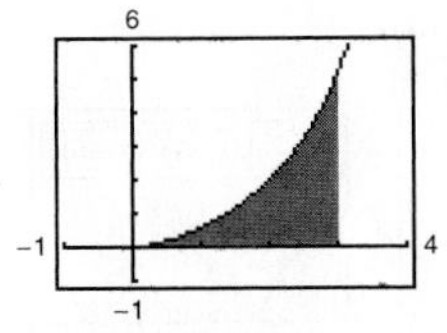

(b) Function is difficult to integrate.
(c) ≈ 4.772

55. (a)

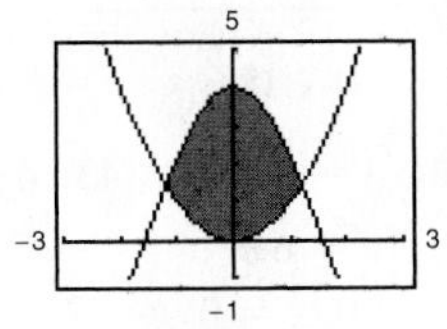

(b) Intersections are difficult to find.
(c) ≈ 6.304

57. $F(x) = \frac{1}{4}x^2 + x$

(a) $F(0) = 0$

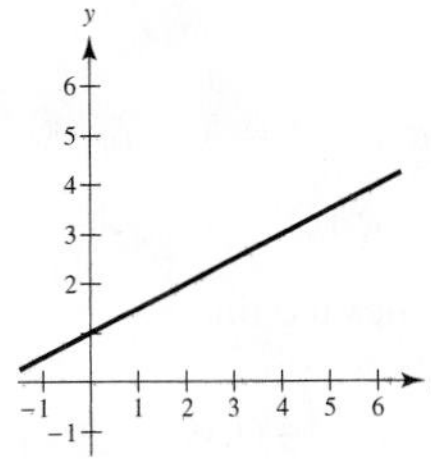

(b) $F(2) = 3$

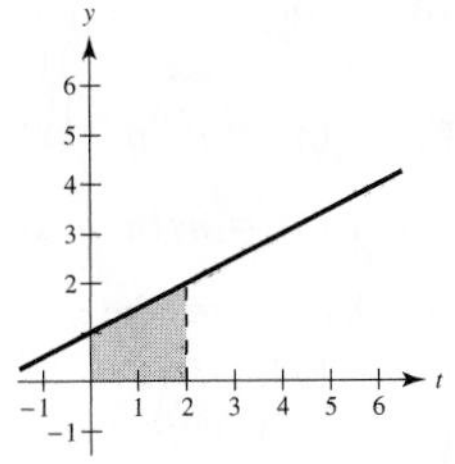

(c) $F(6) = 15$

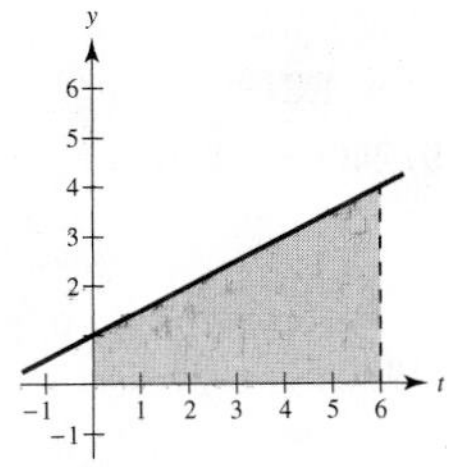

59. $F(\alpha) = (2/\pi)[\sin(\pi\alpha/2) + 1]$

(a) $F(-1) = 0$

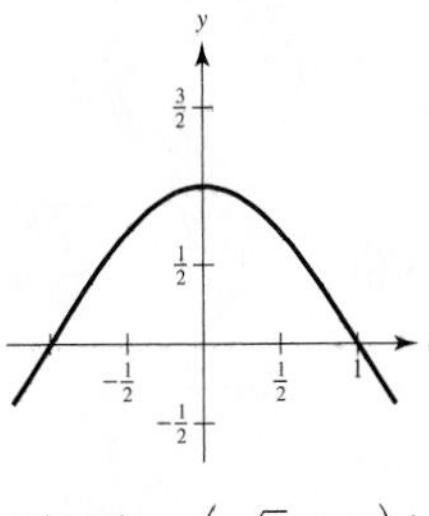

(b) $F(0) = 2/\pi \approx 0.6366$

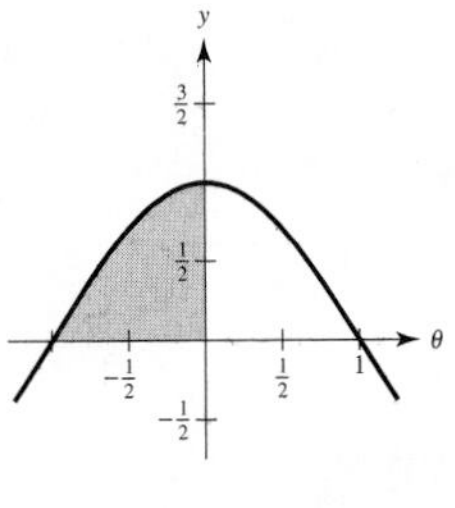

(c) $F(1/2) = (\sqrt{2} + 2)/\pi \approx 1.0868$

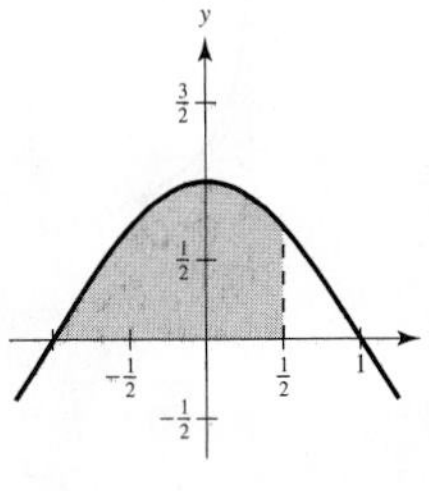

61. 14 **63.** 16

65. Answers will vary. Sample answers:
(a) ≈ 846 ft^2 (b) ≈ 848 ft^2

67. $\displaystyle\int_{-2}^{1} [x^3 - (3x - 2)]\,dx = \frac{27}{4}$

69. $\displaystyle\int_0^1 \left[\frac{1}{x^2 + 1} - \left(-\frac{1}{2}x + 1\right)\right] dx \approx 0.0354$

71. Answers will vary. Example: $x^4 - 2x^2 + 1 \le 1 - x^2$ on $[-1, 1]$

$$\int_{-1}^{1} [(1 - x^2) - (x^4 - 2x^2 + 1)]\,dx = \frac{4}{15}$$

73. Offer 2 is better because the cumulative salary (area under the curve) is greater.

75. $b = 9(1 - 1/\sqrt[3]{4}) \approx 3.330$ **77.** $a = 4 - 2\sqrt{2} \approx 1.172$

79. Answers will vary.
Sample answer: $\frac{1}{6}$

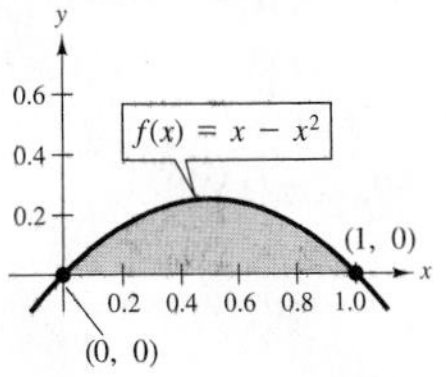

81. \$1.625 billion

83. (a) $R = 270.32e^{0.057t}$
or
$R = 270.32(1.058^t)$

(b) $E = 239.97e^{0.040t}$
or
$E = 239.97(1.042^t)$

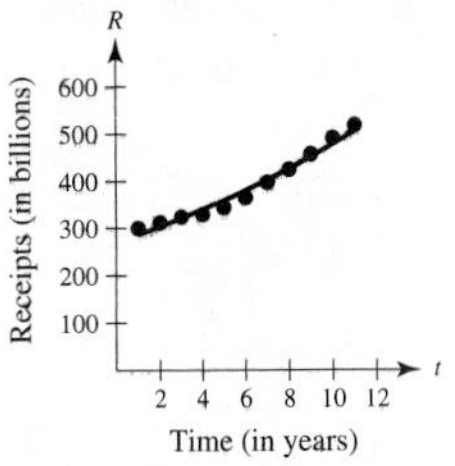

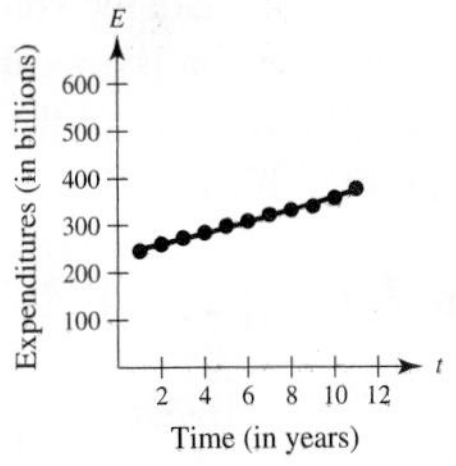

(c) Answer will vary. Sample answer: ≈ 931.6 billion
(d) No. The model for total receipts is increasing at a greater rate than the model for total expenditures. No.

85. Answers will vary. Sample answer: \$193,156

87. (a) $k = 3.125$ (b) 13.02083

89. (a) 5.908 m^2 (b) 11.816 m^3 (c) 59,082 lb

91. True **93.** $\sqrt{3}/2 + 7\pi/24 \approx 2.7823$

95. Putnam Problem A1, 1993

Section 7.2 (page 463)

1. $\pi\displaystyle\int_0^1 (-x + 1)^2\,dx = \frac{\pi}{3}$ **3.** $\pi\displaystyle\int_1^4 (\sqrt{x})^2\,dx = \frac{15\pi}{2}$

5. $\pi\displaystyle\int_0^1 [(x^2)^2 - (x^3)^2]\,dx = \frac{2\pi}{35}$ **7.** $\pi\displaystyle\int_0^4 (\sqrt{y})^2\,dy = 8\pi$

9. $\pi\displaystyle\int_0^1 (y^{3/2})^2\,dy = \frac{\pi}{4}$

11. (a) 8π (b) $128\pi/5$ (c) $256\pi/15$ (d) $192\pi/5$

13. (a) $32\pi/3$ (b) $64\pi/3$ **15.** 18π

17. $\pi(16 \ln 2 - \frac{3}{4}) \approx 32.485$ **19.** $208\pi/3$ **21.** $384\pi/5$

23. $\pi \ln 4$ **25.** $3\pi/4$ **27.** $(\pi/2)(1 - 1/e^2) \approx 1.358$

29. $277\pi/3$ **31.** 8π **33.** $\pi^2/2 \approx 4.935$

35. $(\pi/2)(e^2 - 1) \approx 10.036$ **37.** 1.969 **39.** 15.4115

41. A sine curve from $[0, \pi/2]$ revolved about the x-axis.

43. (a) The area appears to be close to 1 and therefore the volume (Area$^2 \times \pi$) is near 3.

45. The parabola $y = 4x - x^2$ is a horizontal translation of the parabola $y = 4 - x^2$. Therefore, their volumes are equal.

47. 18π **49.** Proof **51.** $\pi r^2 h[1 - h/H + h^2/(3H^2)]$

53. $\pi/30$ **55.** (a) 60π (b) 50π

57. One-fourth: 32.64 ft; Three-fourths: 67.36 ft

59. (a) ii; right circular cylinder of radius r and height h
(b) iv; ellipsoid whose underlying ellipse has the equation $(x/b)^2 + (y/a)^2 = 1$
(c) iii; sphere of radius r
(d) i; right circular cone of radius r and height h
(e) v; torus of cross-sectional radius r and other radius R

61. (a) $\frac{81}{10}$ (b) $\frac{9}{2}$

63. (a) $1/10$ (b) $\pi/80$ (c) $\sqrt{3}/40$ (d) $\pi/20$

65. $V = \frac{4}{3}\pi(R^2 - r^2)^{3/2}$ **67.** $\pi/3$ **69.** $2\pi/15$

71. $\pi/2$ **73.** $\pi/6$

75. (a) When $a = 1$: represents a square
When $a = 2$: represents a circle
(b) $A = 4\int_0^1 (1 - x^a)^{1/a}\,dx$
To approximate the volume of the solid, form n slices, each of whose area is approximated by the integral above. Then sum the volumes of these n slices.

77. (a) Proof (b) $V = 2\pi^2 r^2 R$

Section 7.3 (page 472)

1. $2\pi\int_0^2 x^2\,dx = \frac{16\pi}{3}$ **3.** $2\pi\int_0^4 x\sqrt{x}\,dx = \frac{128\pi}{5}$

5. $2\pi\int_0^2 x^3\,dx = 8\pi$ **7.** $2\pi\int_0^2 x(4x - 2x^2)\,dx = \frac{16\pi}{3}$

9. $2\pi\int_0^2 x(x^2 - 4x + 4)\,dx = \frac{8\pi}{3}$

11. $2\pi\int_0^1 x\left(\frac{1}{\sqrt{2\pi}}e^{-x^2/2}\right)dx = \sqrt{2\pi}\left(1 - \frac{1}{\sqrt{e}}\right) \approx 0.986$

13. $2\pi\int_0^2 y(2 - y)\,dy = \frac{8\pi}{3}$

15. $2\pi\left[\int_0^{1/2} y\,dy + \int_{1/2}^1 y\left(\frac{1}{y} - 1\right)dy\right] = \frac{\pi}{2}$

17. $2\pi\left[\int_0^8 y^{4/3}\,dy\right] = \frac{768\pi}{7}$ **19.** $2\pi\int_0^2 y(4 - 2y)\,dy = 16\pi/3$

21. 16π **23.** 64π

25. Shell method; it is much easier to put x in terms of y rather than vice versa.

27. (a) $128\pi/7$ (b) $64\pi/5$ (c) $96\pi/5$

29. (a) $\pi a^3/15$ (b) $\pi a^3/15$ (c) $4\pi a^3/15$

31. (a) The rectangles would be vertical.
(b) The rectangles would be horizontal.

33. Both integrals yield the volume of the solid generated by revolving the region bounded by the graphs of $y = \sqrt{x - 1}$, $y = 0$, and $x = 5$ about the x-axis.

35. (a)

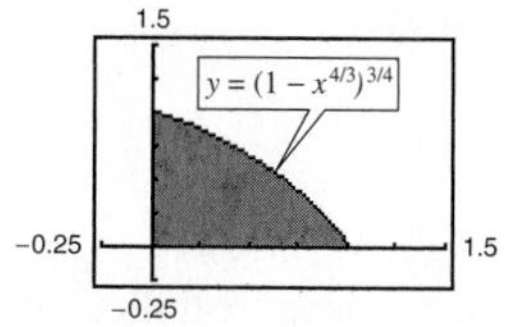

(b) 1.506

37. (a)

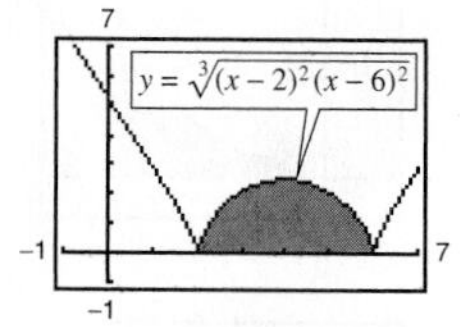

(b) 187.25

39. d **41.** Diameter $= 2\sqrt{4 - 2\sqrt{3}} \approx 1.464$ **43.** $4\pi^2$

45. (a) Proof (b) (i) $V = 2\pi$ (ii) $V = 6\pi^2$

47. (a) region bounded by $y = x^2$, $y = 0$, $x = 0$, $x = 2$
(b) revolved about the y-axis

49. (a) region bounded by $x = \sqrt{6 - y}$, $y = 0$, $x = 0$
(b) revolved about $y = -2$

51. Proof

53. (a) $R_1(n) = n/(n + 1)$ (b) $\lim_{n\to\infty} R_1(n) = 1$
(c) $R_2(n) = n/(n + 2)$ (d) $\lim_{n\to\infty} R_2(n) = 1$
(e) As $n \to \infty$, the graph approaches the line $x = 1$.

55. (a) $\approx 121{,}475$ ft^3 (b) $\approx 121{,}475$ ft^3

57. (a) $64\pi/3$ (b) $2048\pi/35$ (c) $8192\pi/105$ **59.** $c = 2$

Section 7.4 (page 483)

1. (a) and (b) 13 **3.** $\frac{2}{3}(2\sqrt{2} - 1) \approx 1.219$

5. $5\sqrt{5} - 2\sqrt{2} \approx 8.352$ **7.** $779/240 \approx 3.246$

9. $\ln\left[(\sqrt{2} + 1)/(\sqrt{2} - 1)\right] \approx 1.763$

11. $\frac{1}{2}(e^2 - 1/e^2) \approx 3.627$ **13.** $\frac{76}{3}$

15. (a)

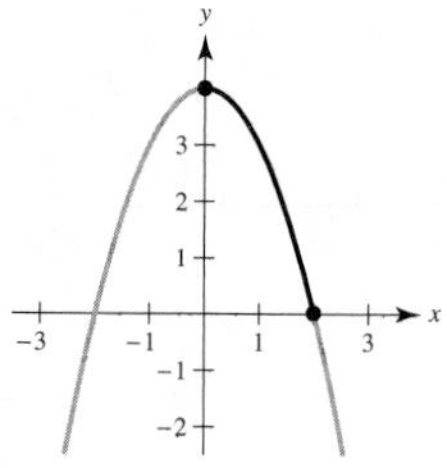

(b) $\int_0^2 \sqrt{1 + 4x^2}\,dx$
(c) ≈ 4.647

17. (a)

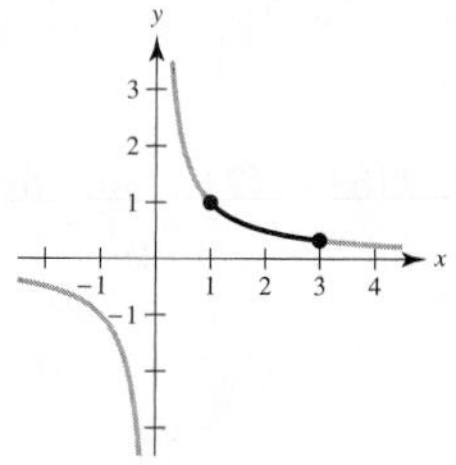

(b) $\int_1^3 \sqrt{1 + \frac{1}{x^4}}\,dx$
(c) ≈ 2.147

19. (a)

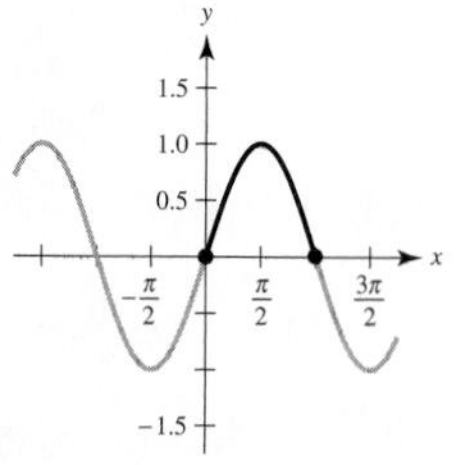

(b) $\int_0^\pi \sqrt{1 + \cos^2 x}\,dx$
(c) ≈ 3.820

21. (a)

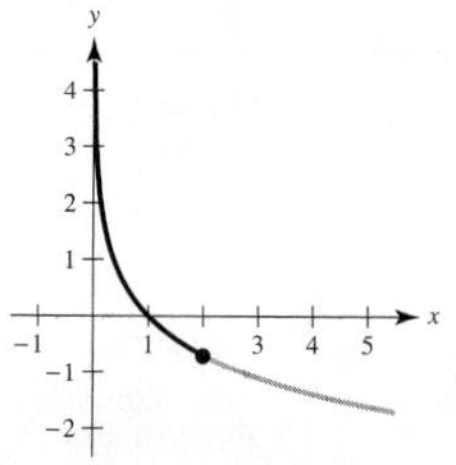

(b) $\int_0^2 \sqrt{1 + e^{-2y}}\,dy = \int_{e^{-2}}^1 \sqrt{1 + \frac{1}{x^2}}\,dx$
(c) ≈ 2.221

23. (a)

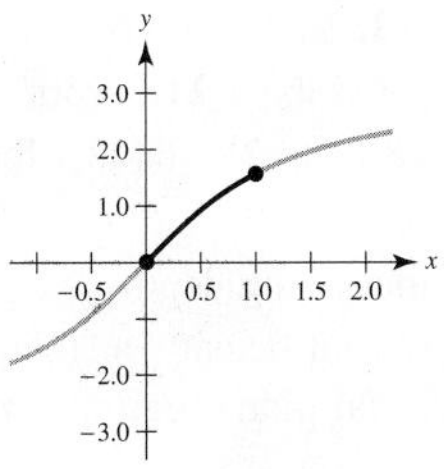

(b) $\int_0^1 \sqrt{1+\left(\frac{2}{1+x^2}\right)^2}\,dx$
(c) ≈ 1.871

25. b **27.** (a) 64.125 (b) 64.525 (c) 64.666 (d) 64.672

29. (a)

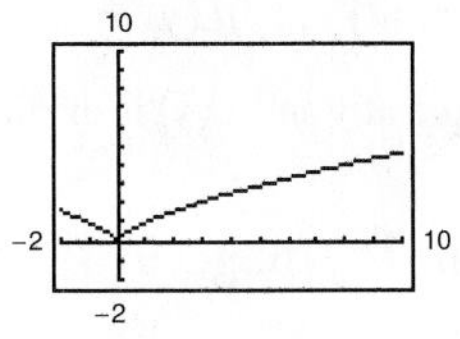

(b) No; $f'(0)$ is not defined.
(c) ≈ 10.5131

31. (a)

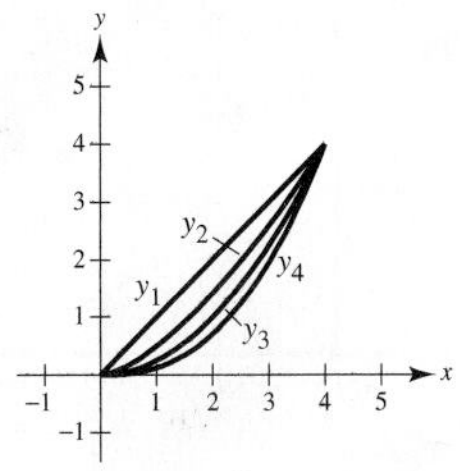

(b) y_1, y_2, y_3, y_4
(c) $s_1 \approx 5.657$; $s_2 \approx 5.759$; $s_3 \approx 5.916$; $s_4 \approx 6.063$

33. Fleeing object: $\frac{2}{3}$ unit

Pursuer: $\frac{1}{2}\int_0^1 \frac{x+1}{\sqrt{x}}\,dx = \frac{1}{2}\left[\frac{2}{3}x^{3/2} + 2x^{1/2}\right]_0^1 = \frac{4}{3} = 2\left(\frac{2}{3}\right)$

35. $20[\sinh 1 - \sinh(-1)] \approx 47.0$ m **37.** $3 \arcsin \frac{2}{3} \approx 2.1892$

39. $2\pi\int_0^3 \frac{1}{3}x^3\sqrt{1+x^4}\,dx = \frac{\pi}{9}\left(82\sqrt{82} - 1\right) \approx 258.85$

41. $2\pi\int_1^2 \left(\frac{x^3}{6} + \frac{1}{2x}\right)\left(\frac{x^2}{2} + \frac{1}{2x^2}\right)dx = \frac{47\pi}{16} \approx 9.23$

43. $2\pi\int_1^8 x\sqrt{1 + \frac{1}{9x^{4/3}}}\,dx = \frac{\pi}{27}\left(145\sqrt{145} - 10\sqrt{10}\right) \approx 199.48$

45. 14.424

47. A rectifiable curve is a curve with a finite arc length.

49. The integral formula for the area of a surface of revolution is derived from the formula for the lateral surface area of the frustum of a right circular cone. The formula is $S = 2\pi rL$, where $r = \frac{1}{2}(r_1 + r_2)$, which is the average radius of the frustum, and L is the length of a line segment on the frustum.

51. Proof **53.** $6\pi\left(3 - \sqrt{5}\right) \approx 14.40$

55. Surface area $= \pi/27$ ft$^2 \approx 16.8$ in.^{2}

Amount of glass $= (\pi/27)(0.015/12)$ ft^3
≈ 0.00015 ft^3
≈ 0.25 in.^{3}

57. (a) Answers will vary. Sample answers: 5207.62 in.^{3}
(b) Answers will vary. Sample answers: 1168.64 in.^{2}
(c) $r = 0.0040y^3 - 0.142y^2 + 1.23y + 7.9$

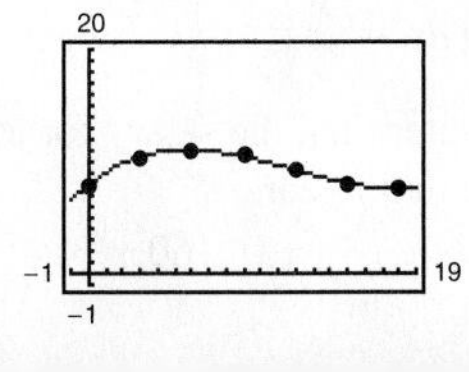

(d) 5279.64 in.^{3}; 1179.5 in.^{2}

59. (a) $\pi(1 - 1/b)$ (b) $2\pi\int_1^b \sqrt{x^4+1}/x^3\,dx$

(c) $\lim_{b\to\infty} V = \lim_{b\to\infty} \pi(1 - 1/b) = \pi$

(d) Since $\frac{\sqrt{x^4+1}}{x^3} > \frac{\sqrt{x^4}}{x^3} = \frac{1}{x} > 0$ on $[1, b]$,

we have $\int_1^b \frac{\sqrt{x^4+1}}{x^3}\,dx > \int_1^b \frac{1}{x}\,dx = \left[\ln x\right]_1^b = \ln b$

and $\lim_{b\to\infty} \ln b \to \infty$. Thus, $\lim_{b\to\infty} 2\pi\int_1^b \frac{\sqrt{x^4+1}}{x^3}\,dx = \infty$.

61. Answers will vary. **63.** $192\pi/5$ **65.** Proof

67. Putnam Problem A1, 1939

Section 7.5 (page 493)

1. 1000 ft-lb **3.** 448 N-m

5. If an object is moved a distance D in the direction of an applied constant force F, then the work W done by the force is defined as $W = FD$.

7. c, d, a, b; The area under the curves increases in this order.

9. 30.625 in.-lb $\approx$ 2.55 ft-lb **11.** 8750 N-cm = 87.5 N-m

13. 160 in.-lb $\approx$ 13.3 ft-lb **15.** 37.125 ft-lb

17. (a) 487.805 mile-tons $\approx 5.151 \times 10^9$ ft-lb
(b) 1395.349 mile-tons $\approx 1.473 \times 10^{10}$ ft-lb

19. (a) 2.93×10^4 mile-tons $\approx 3.10 \times 10^{11}$ ft-lb
(b) 3.38×10^4 mile-tons $\approx 3.57 \times 10^{11}$ ft-lb

21. (a) 2496 ft-lb (b) 9984 ft-lb **23.** $470{,}400\pi$ N-m

25. 2995.2π ft-lb **27.** $20{,}217.6\pi$ ft-lb **29.** 2457π ft-lb

31. 337.5 ft-lb **33.** 300 ft-lb **35.** 168.75 ft-lb

37. 7987.5 ft-lb **39.** $2000 \ln(3/2) \approx 810.93$ ft-lb **41.** $3k/4$

43. 3249.4 ft-lb **45.** 10,330.3 ft-lb

Section 7.6 (page 504)

1. $\bar{x} = -\frac{6}{7}$ **3.** $\bar{x} = 12$ **5.** (a) $\bar{x} = 17$ (b) $\bar{x} = -3$

7. $x = 6$ ft **9.** $(\bar{x}, \bar{y}) = \left(\frac{10}{9}, -\frac{1}{9}\right)$ **11.** $(\bar{x}, \bar{y}) = \left(\frac{5}{8}, \frac{13}{16}\right)$

13. $M_x = 4\rho$, $M_y = 64\rho/5$, $(\bar{x}, \bar{y}) = (12/5, 3/4)$

15. $M_x = \rho/35$, $M_y = \rho/20$, $(\bar{x}, \bar{y}) = (3/5, 12/35)$

17. $M_x = 99\rho/5$, $M_y = 27\rho/4$, $(\bar{x}, \bar{y}) = (3/2, 22/5)$

19. $M_x = 192\rho/7$, $M_y = 96\rho$, $(\bar{x}, \bar{y}) = (5, 10/7)$

21. $M_x = 0$, $M_y = 256\rho/15$, $(\bar{x}, \bar{y}) = (8/5, 0)$

23. $M_x = 27\rho/4$, $M_y = -27\rho/10$, $(\bar{x}, \bar{y}) = (-3/5, 3/2)$

25. $A = \int_0^1 (x - x^2)\,dx = \frac{1}{6}$

$M_x = \int_0^1 \left(\frac{x + x^2}{2}\right)(x - x^2)\,dx = \frac{1}{15}$

$M_y = \int_0^1 x(x - x^2)\,dx = \frac{1}{12}$

27. $A = \int_0^3 (2x + 4)\,dx = 21$

$M_x = \int_0^3 \left(\frac{2x + 4}{2}\right)(2x + 4)\,dx = 78$

$M_y = \int_0^3 x(2x + 4)\,dx = 36$

29.

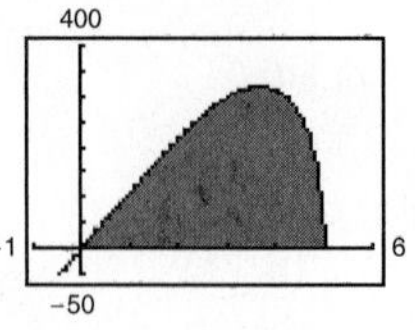

$(\bar{x}, \bar{y}) = (3.0, 126.0)$

31.

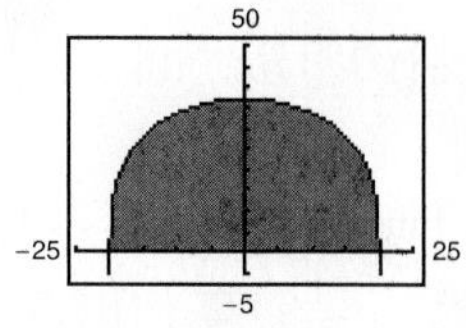

$(\bar{x}, \bar{y}) = (0, 16.2)$

33. $(\bar{x}, \bar{y}) = \left(\frac{b}{3}, \frac{c}{3}\right)$ **35.** $(\bar{x}, \bar{y}) = \left(\frac{(a + 2b)c}{3(a + b)}, \frac{a^2 + ab + b^2}{3(a + b)}\right)$

37. $(\bar{x}, \bar{y}) = (0, 4b/(3\pi))$

39. (a)

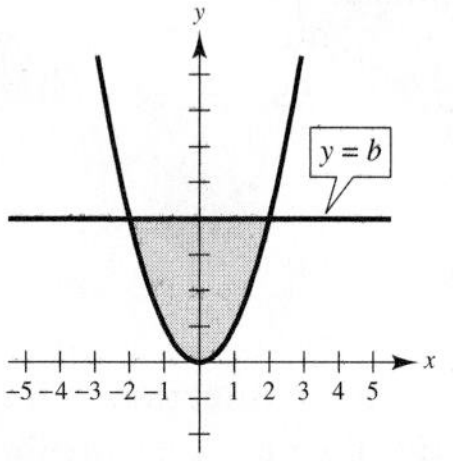

(b) $\bar{x} = 0$ by symmetry

(c) $M_y = \int_{-\sqrt{b}}^{\sqrt{b}} x(b - x^2)\,dx = 0$ because $x(b - x^2)$ is an odd function.

(d) $\bar{y} > b/2$ because the area is greater for $y > b/2$.

(e) $\bar{y} = (3/5)b$

41. (a) $(\bar{x}, \bar{y}) = (0, 12.98)$

(b) $y = (-1.02 \times 10^{-5})x^4 - 0.0019x^2 + 29.28$

(c) $(\bar{x}, \bar{y}) = (0, 12.85)$

43.

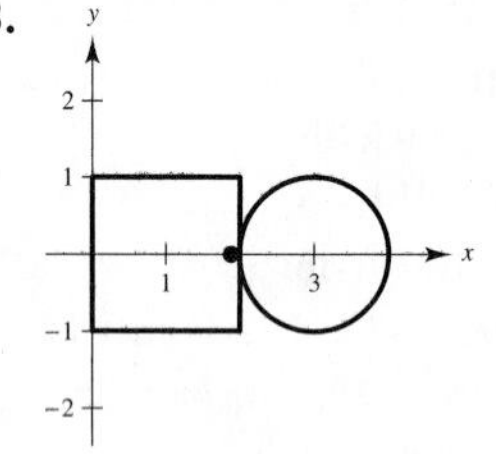

$(\bar{x}, \bar{y}) = \left(\frac{4 + 3\pi}{4 + \pi}, 0\right)$

45.

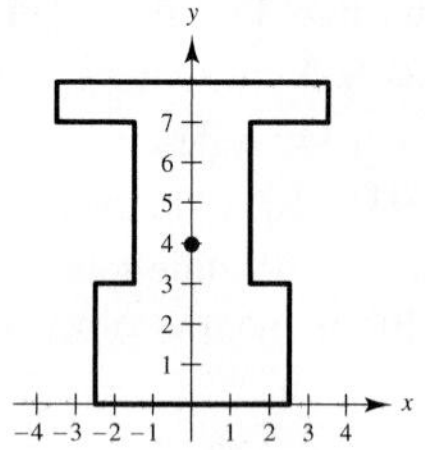

$(\bar{x}, \bar{y}) = \left(0, \frac{135}{34}\right)$

47. $(\bar{x}, \bar{y}) = \left(\frac{2 + 3\pi}{2 + \pi}, 0\right)$ **49.** $160\pi^2 \approx 1579.14$

51. $128\pi/3 \approx 134.04$

53. The center of mass $(\bar{x}, \bar{y})$ is $\bar{x} = M_y/m$ and $\bar{y} = M_x/m$, where:

1. $m = m_1 + m_2 + \cdots + m_n$ is the total mass of the system.
2. $M_y = m_1 x_1 + m_2 x_2 + \cdots + m_n x_n$ is the moment about the y-axis.
3. $M_x = m_1 y_1 + m_2 y_2 + \cdots + m_n y_n$ is the moment about the x-axis.

55. (a) $\left(\frac{5}{6}, 2\frac{5}{18}\right)$; The plane region has been translated 2 units up.

(b) $\left(2\frac{5}{6}, \frac{5}{18}\right)$; The plane region has been translated 2 units to the right.

(c) $\left(\frac{5}{6}, -\frac{5}{18}\right)$; The plane region has been reflected across the x-axis.

(d) Not possible

57. $(\bar{x}, \bar{y}) = (0, 2r/\pi)$

59. $(\bar{x}, \bar{y}) = \left(\frac{n + 1}{n + 2}, \frac{n + 1}{4n + 2}\right)$; As $n \to \infty$, the region shrinks toward the line segments $y = 0$ for $0 \le x \le 1$ and $x = 1$ for $0 \le y \le 1$; $(\bar{x}, \bar{y}) \to \left(1, \frac{1}{4}\right)$.

Section 7.7 (page 511)

1. 936 lb **3.** 748.8 lb **5.** 1123.2 lb **7.** 748.8 lb

9. 1064.96 lb **11.** 117,600 N **13.** 2,381,400 N

15. 2814 lb **17.** 6753.6 lb **19.** 94.5 lb **21.** Proof

23. Proof **25.** 960 lb **27.** 3010.8 lb **29.** 6448.7 lb

31. (a) $3\sqrt{2}/2 \approx 2.12$ ft

(b) The pressure increases with increasing depth.

33. The fluid force F of constant weight-density w (per unit of volume) against a submerged vertical plane region from $y = c$ to $y = d$ is

$$F = w \lim_{\|\Delta\| \to 0} \sum_{i=1}^{n} h(y_i)L(y_i)\,\Delta y = w\int_c^d h(y)L(y)\,dy$$

where $h(y)$ is the depth of the fluid at y and $L(y)$ is the horizontal length of the region at y.

Review Exercises for Chapter 7 (page 513)

1.

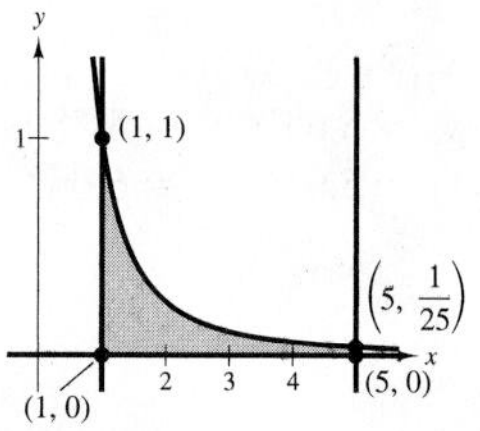

$4/5$

3.

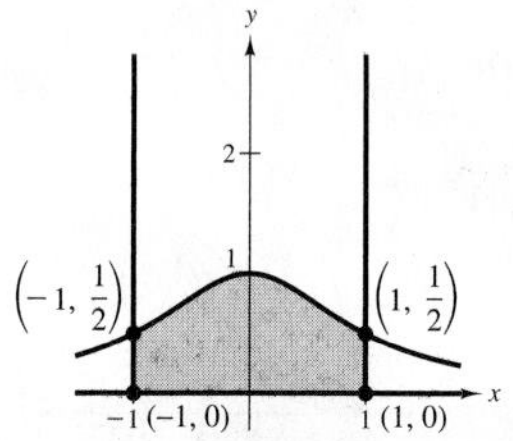

$\pi/2$

5.

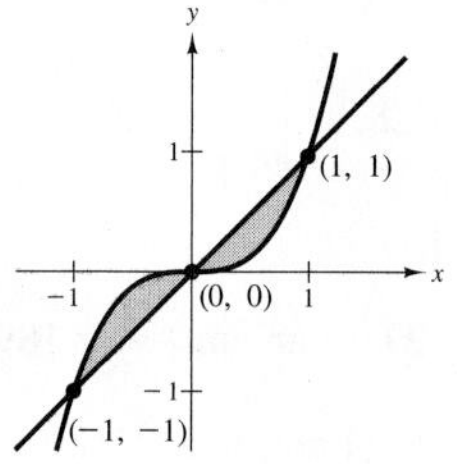

$\frac{1}{2}$

7.

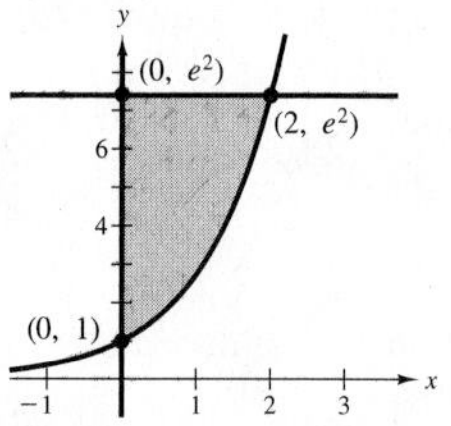

$e^2 + 1$

9.

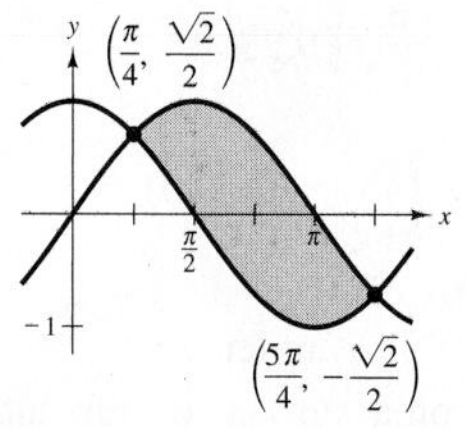

$2\sqrt{2}$

11.

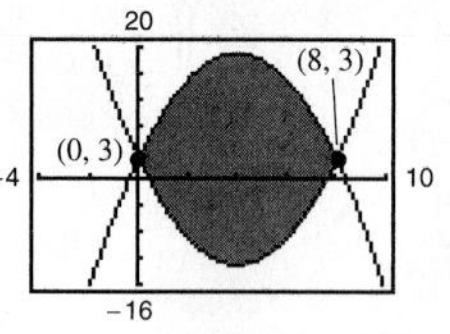

$\frac{512}{3}$

13.

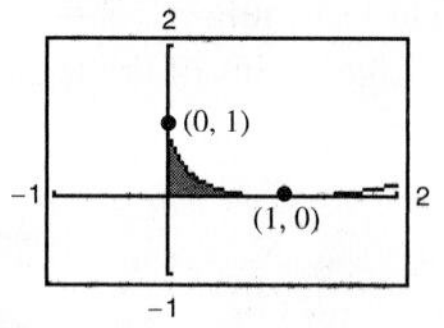

$\frac{1}{6}$

15. $\int_0^2 [0 - (y^2 - 2y)]\,dy = \int_{-1}^0 2\sqrt{x + 1}\,dx = \frac{4}{3}$

17. $\int_0^2 \left[1 - \left(1 - \frac{x}{2}\right)\right] dx + \int_2^3 [1 - (x - 2)]\,dx$

$= \int_0^1 [(y + 2) - (2 - 2y)]\,dy = \frac{3}{2}$

19. Job 1. The salary for job 1 is greater than the salary for job 2 for all the years except the first and tenth years.

21. (a) $64\pi/3$ (b) $128\pi/3$ (c) $64\pi/3$ (d) $160\pi/3$

23. (a) 64π (b) 48π **25.** $\pi^2/4$

27. $(4\pi/3)(20 - 9\ln 3) \approx 42.359$ **29.** $\frac{4}{15}$ **31.** 1.958 ft

33. $\frac{8}{15}(1 + 6\sqrt{3}) \approx 6.076$ **35.** 4018.2 ft

37. 15π **39.** 50 in.-lb $\approx$ 4.167 ft-lb

41. $104{,}000\pi$ ft-lb $\approx$ 163.4 foot-tons **43.** 250 ft-lb

45. $a = 15/4$ **47.** $(\bar{x}, \bar{y}) = (a/5, a/5)$ **49.** $(\bar{x}, \bar{y}) = (0, 2a^2/5)$

51. $(\bar{x}, \bar{y}) = \left(\dfrac{2(9\pi + 49)}{3(\pi + 9)}, 0\right)$

53. Let D = surface of liquid; ρ = weight per cubic volume.

$$F = \rho\int_c^d (D - y)[f(y) - g(y)]\,dy$$

$$= \rho\left[\int_c^d D[f(y) - g(y)]\,dy - \int_c^d y[f(y) - g(y)]\,dy\right]$$

$$= \rho\left[\int_c^d [f(y) - g(y)]\,dy\right]\left[D - \frac{\int_c^d y[f(y) - g(y)]\,dy}{\int_c^d [f(y) - g(y)]\,dy}\right]$$

$$= \rho(\text{area})(D - \bar{y})$$

$$= \rho(\text{area})(\text{depth of centroid})$$

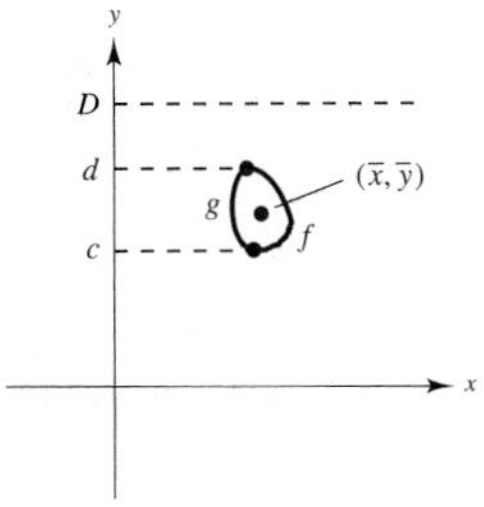

P.S. Problem Solving (page 515)

1. 3 **3.** (a) $4\pi^2$ (b) $2\pi^2 r^2 R$ **5.** $\pi h^3/6$

7. (a) Area S is 16 times area R.

(b) Let point A be (a, a^3). The equation of the tangent line to the curve $y = x^3$ at A is $y = 3a^2x - 2a^3$, and point B is $(-2a, -8a^3)$. Area R is

$$\int_{-2a}^{a} (x^3 - 3a^2x + 2a^3)\,dx = \frac{27a^4}{4}.$$

Then, the equation of the tangent line to the curve $y = x^3$ at B is $y = 12a^2x + 16a^3$, and point C is $(4a, 64a^3)$. Area S is

$$\int_{-2a}^{4a} (12a^2x + 16a^3 - x^3)\,dx = 108a^4.$$

Therefore, area S is 16 times area R.

9. (a) $\dfrac{ds}{dx} = \sqrt{1 + [f'(x)]^2}$

(b) $ds = \sqrt{1 + [f'(x)]^2}\,dx$; $(ds)^2 = (dx)^2 + (dy)^2$

(c) $\displaystyle\int_1^x \sqrt{1 + \tfrac{9}{4}t}\,dt$

(d) $s(2) \approx 2.0858$. This is the arc length of the curve.

11.

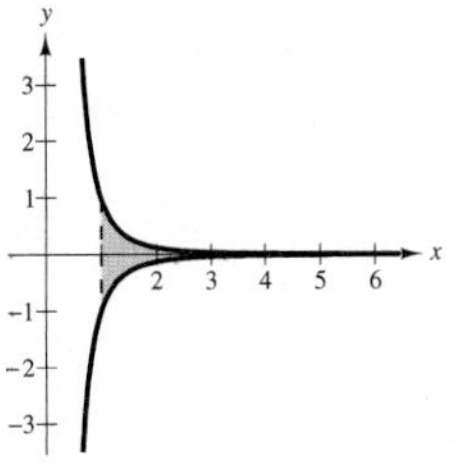

(a) $\left(\dfrac{12}{7}, 0\right)$

(b) $\left(\dfrac{2b}{b + 1}, 0\right)$

(c) $(2, 0)$

13. (a) 12 (b) 7.5

15. Consumer surplus: 1600; Producer surplus: 400

17. Wall at shallow end: 9984 pounds
Wall at deep end: 39,936 pounds
Side wall: 19,968 + 26,624 = 46,592 lb

Chapter 8

Section 8.1 (page 522)

1. b **3.** c

5. $\displaystyle\int u^n\,du$
$u = 3x - 2$, $n = 4$

7. $\displaystyle\int \frac{du}{u}$
$u = 1 - 2\sqrt{x}$

9. $\displaystyle\int \frac{du}{\sqrt{a^2 - u^2}}$
$u = t$, $a = 1$

11. $\displaystyle\int \sin u\,du$
$u = t^2$

13. $\displaystyle\int e^u\,du$
$u = \sin x$

15. $(x - 4)^6 + C$

17. $-5/[4(z - 4)^4] + C$ **19.** $\frac{1}{2}v^2 - 1/[6(3v - 1)^2] + C$

21. $-\frac{1}{3}\ln|-t^3 + 9t + 1| + C$ **23.** $\frac{1}{2}x^2 + x + \ln|x - 1| + C$

25. $\ln(1 + e^x) + C$ **27.** $\frac{1}{15}x(12x^4 + 20x^2 + 15) + C$

29. $\sin(2\pi x^2)/(4\pi) + C$ **31.** $-(1/\pi)\csc \pi x + C$

33. $\frac{1}{5}e^{5x} + C$ **35.** $2\ln(1 + e^x) + C$ **37.** $(\ln x)^2 + C$

39. $-\ln(1 - \sin x) + C = \ln|\sec x(\sec x + \tan x)| + C$

41. $\csc\theta + \cot\theta + C = (1 + \cos\theta)/\sin\theta + C$

43. $-\frac{1}{2}\arcsin(2t - 1) + C$ **45.** $\frac{1}{2}\ln|\cos(2/t)| + C$

47. $3\arcsin[(x - 3)/3] + C$ **49.** $\frac{1}{4}\arctan[(2x + 1)/8] + C$

51. (a)

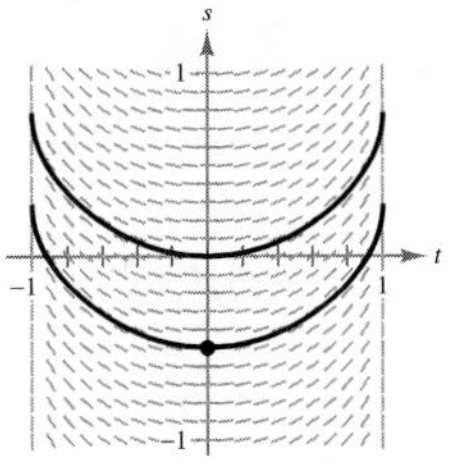

(b) $\frac{1}{2}\arcsin t^2 - \frac{1}{2}$

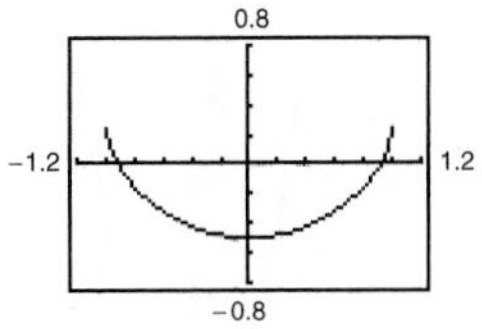

53.

(a)

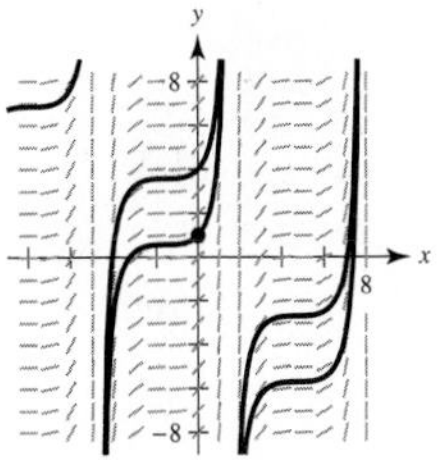

(b) $2\tan x + 2\sec x - x - 1 + C$

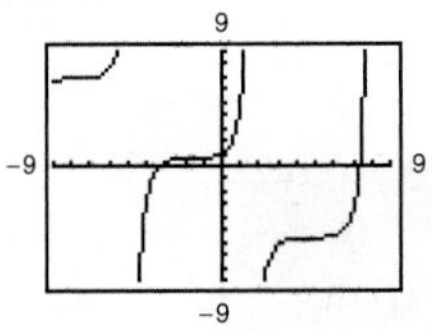

55. $y = 3e^{0.2x}$

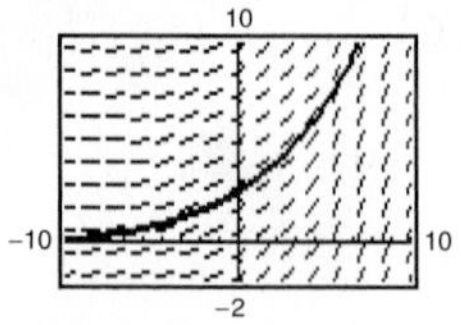

57. $y = \frac{1}{2}e^{2x} + 2e^x + x + C$

59. $y = \frac{1}{2}\arctan(\tan x/2) + C$ **61.** $\frac{1}{2}$

63. $\frac{1}{2}(1 - e^{-1}) \approx 0.316$ **65.** 4 **67.** $\pi/18$

69. $5\sqrt{5} \approx 11.18$ **71.** $\frac{3}{2}\ln\left(\frac{34}{9}\right) + \frac{2}{3}\arctan\left(\frac{5}{3}\right) \approx 2.68$

73. $\frac{4}{3} \approx 1.333$

75. $\frac{1}{3}\arctan\left[\frac{1}{3}(x + 2)\right] + C$
Graphs will vary.
Example:

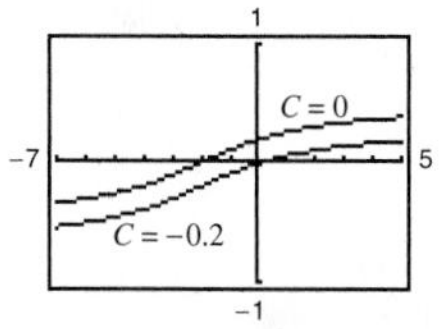

One graph is a vertical translation of the other.

77. $\tan\theta - \sec\theta + C$
Graphs will vary.
Example:

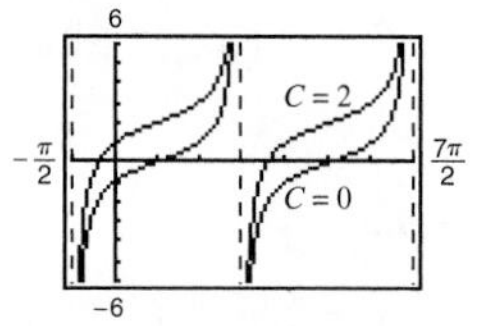

One graph is a vertical translation of the other.

79. Power rule: $\int u^n\,du = \dfrac{u^{n+1}}{n+1} + C;\ u = x^2 + 1,\ du = 2x,\ n = 3$

81. Log rule: $\int \dfrac{du}{u} = \ln|u| + C;\ u = x^2 + 1,\ du = 2x$

83. Using laws of logarithms, $y_1 = e^{x+C_1} = e^x \cdot e^{C_1}$, where e^{C_1} is a constant. Therefore, e^{C_1} can be replaced by C, resulting in $y_2 = Ce^x$.

85. $a = \sqrt{2},\ b = \dfrac{\pi}{4};\ -\dfrac{1}{\sqrt{2}}\ln\left|\csc\left(x + \dfrac{\pi}{4}\right) + \cot\left(x + \dfrac{\pi}{4}\right)\right| + C$

87.

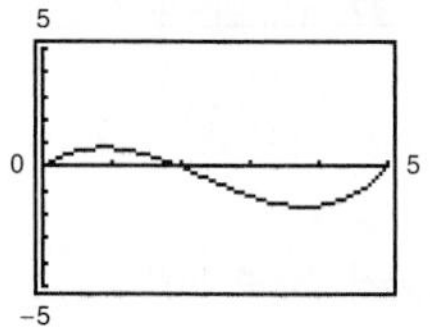

Negative; more area below the x-axis than above

89. a

91. (a)

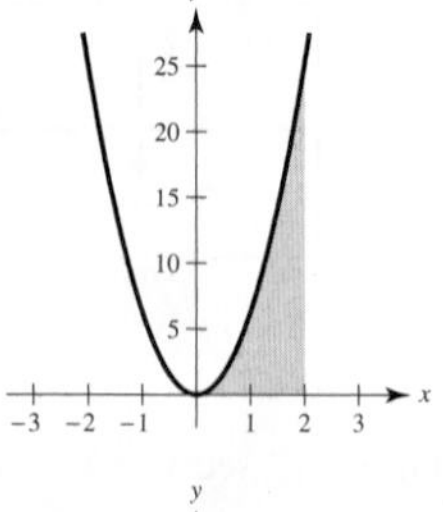

(b)

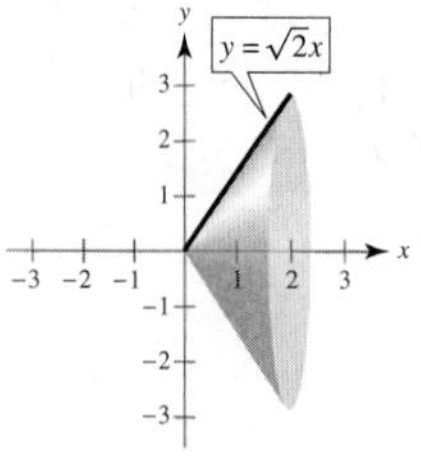

(c)

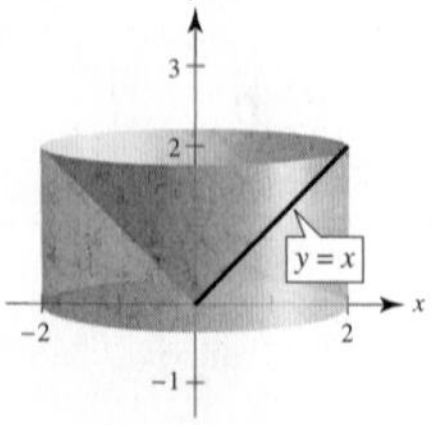

93. (a) $\pi(1 - e^{-1}) \approx 1.986$ (b) $b = \sqrt{\ln\left(\dfrac{3\pi}{3\pi - 4}\right)} \approx 0.743$

95. $(8\pi/3)(10\sqrt{10} - 1) \approx 256.545$

97. $\frac{1}{3}\arctan 3 \approx 0.416$ **99.** ≈ 1.0320

101. (a) $\frac{1}{3}\sin x(\cos^2 x + 2)$
(b) $\frac{1}{15}\sin x(3\cos^4 x + 4\cos^2 x + 8)$
(c) $\frac{1}{35}\sin x(5\cos^6 x + 6\cos^4 x + 8\cos^2 x + 16)$
(d) Use the power reducing formula, $\int \cos^n x\,dx = \dfrac{\cos^{n-1}\sin^n x}{n} + \dfrac{n-1}{n}\int \cos^{n-2} x\,dx$ until your integral is $\int \cos x\,dx$, which you can then integrate.

103. Proof

Section 8.2 (page 531)

1. b **2.** d **3.** c **4.** a **5.** $u = x,\ dv = e^{2x}\,dx$

7. $u = (\ln x)^2,\ dv = dx$ **9.** $u = x,\ dv = \sec^2 x\,dx$

11. $-(1/4e^{2x})(2x + 1) + C$ **13.** $e^x(x^3 - 3x^2 + 6x - 6) + C$

15. $\frac{1}{3}e^{x^3} + C$ **17.** $\frac{1}{4}[2(t^2 - 1)\ln|t + 1| - t^2 + 2t] + C$

19. $\frac{1}{3}(\ln x)^3 + C$ **21.** $e^{2x}/[4(2x + 1)] + C$

23. $(x - 1)^2 e^x + C$ **25.** $\frac{2}{15}(x - 1)^{3/2}(3x + 2) + C$

27. $x\sin x + \cos x + C$

29. $(6x - x^3)\cos x + (3x^2 - 6)\sin x + C$

31. $-t\csc t - \ln|\csc t + \cot t| + C$

33. $x\arctan x - \frac{1}{2}\ln(1 + x^2) + C$

35. $\frac{1}{5}e^{2x}(2\sin x - \cos x) + C$ **37.** $y = \frac{1}{2}e^{x^2} + C$

39. $y = \frac{2}{405}(27t^2 - 24t + 32)\sqrt{2 + 3t} + C$ **41.** $\sin y = x^2 + C$

43. (a)

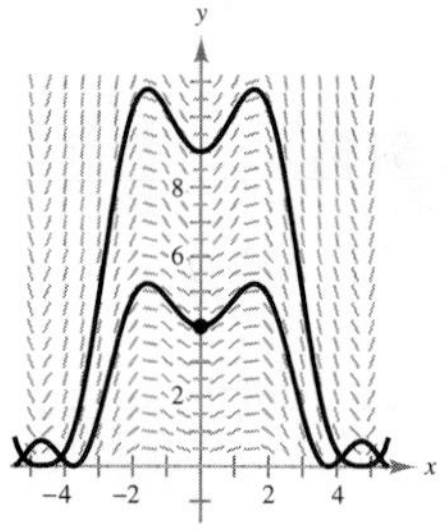

(b) $2\sqrt{y} - \cos x - x\sin x = 3$

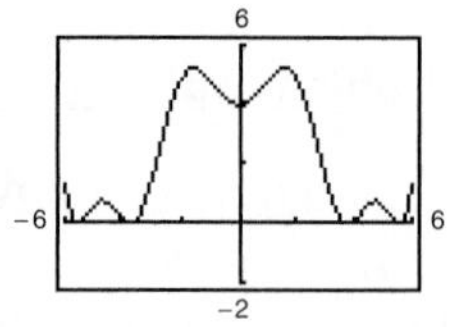

45.

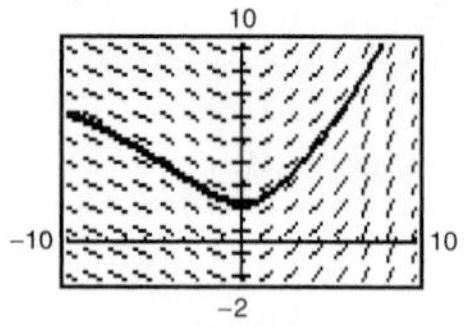

47. $4 - 12/e^2$ **49.** $\pi/2 - 1$

51. $(\pi - 3\sqrt{3} + 6)/6 \approx 0.658$

53. $\frac{1}{2}[e(\sin 1 - \cos 1) + 1] \approx 0.909$

55. $(24\ln 2 - 7)/9 \approx 1.071$

57. $8\,\text{arcsec}\,4 + \sqrt{3}/2 - \sqrt{15}/2 - 2\pi/3 \approx 7.380$

59. $(e^{2x}/4)(2x^2 - 2x + 1) + C$

61. $(3x^2 - 6)\sin x - (x^3 - 6x)\cos x + C$

63. $x\tan x + \ln|\cos x| + C$ **65.** $2(\sin\sqrt{x} - \sqrt{x}\cos\sqrt{x}) + C$

67. $\frac{128}{15}$ **69.** $\frac{1}{2}x[\cos(\ln x) + \sin(\ln x)] + C$ **71.** Product Rule

73. No **75.** Yes. Let $u = x^2$ and $dv = e^{2x}\,dx$.

77. Yes. Let $u = x,\ dv = (1/\sqrt{x + 1})\,dx$. (Substitution also works. Let $u = \sqrt{x + 1}$.)

79. (a) $-(e^{-4t}/128)(32t^3 + 24t^2 + 12t + 3) + C$
(b) Graphs will vary. Example:

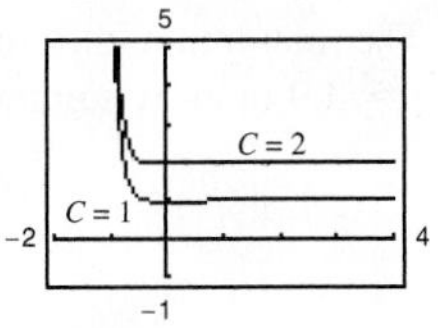

(c) One graph is a vertical translation of the other.

81. (a) $\frac{1}{13}(2e^{-\pi} + 3) \approx 0.2374$
(b) Graphs will vary. Example:

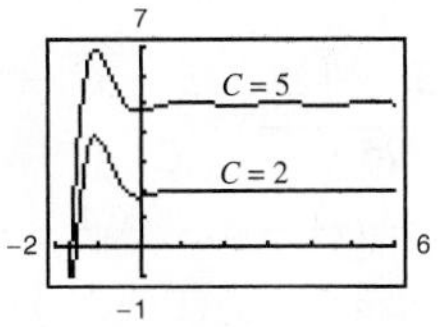

(c) One graph is a vertical translation of the other.

83. $\frac{2}{5}(2x - 3)^{3/2}(x + 1) + C$ **85.** $\frac{1}{3}\sqrt{4 + x^2}(x^2 - 8) + C$

87. $n = 0$: $x(\ln x - 1) + C$
$n = 1$: $\frac{1}{4}x^2(2 \ln x - 1) + C$
$n = 2$: $\frac{1}{9}x^3(3 \ln x - 1) + C$
$n = 3$: $\frac{1}{16}x^4(4 \ln x - 1) + C$
$n = 4$: $\frac{1}{25}x^5(5 \ln x - 1) + C$
$$\int x^n \ln x\, dx = \frac{x^{n+1}}{(n + 1)^2}[(n + 1)\ln x - 1] + C$$

89–93. Proofs

95. $\frac{1}{16}x^4(4 \ln x - 1) + C$ **97.** $\frac{1}{13}e^{2x}(2 \cos 3x + 3 \sin 3x) + C$

99.

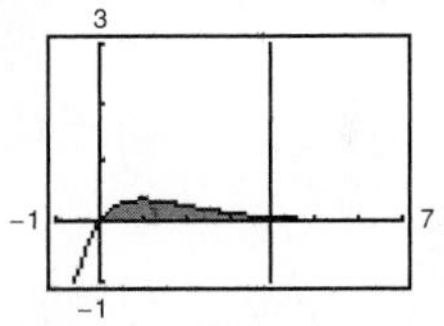

$1 - \dfrac{5}{e^4} \approx 0.908$

101.

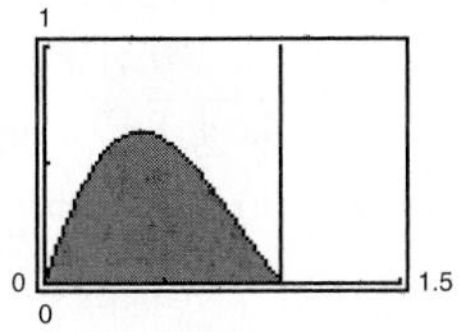

$\dfrac{\pi}{1 + \pi^2}\left(\dfrac{1}{e} + 1\right) \approx 0.395$

103. (a) 1 (b) $\pi(e - 2) \approx 2.257$ (c) $\frac{1}{2}\pi(e^2 + 1) \approx 13.177$
(d) $\left(\dfrac{e^2 + 1}{4}, \dfrac{e - 2}{2}\right) \approx (2.097, 0.359)$

105. In Example 6, we showed that the centroid of an equivalent region was $(1, \pi/8)$. By symmetry, the centroid of this region is $(1, \pi/8)$.

107. $[7/(10\pi)](1 - e^{-4\pi}) \approx 0.223$ **109.** $931,265

111. Proof **113.** $b_n = [8h/(n\pi)^2] \sin(n\pi/2)$

115. Shell: $V = \pi\left[b^2 f(b) - a^2 f(a) - \displaystyle\int_a^b x^2 f'(x)\, dx\right]$

Disk: $V = \pi\left[b^2 f(b) - a^2 f(a) - \displaystyle\int_{f(a)}^{f(b)} [f^{-1}(y)]^2\, dy\right]$

Both methods yield the same volume because $x = f^{-1}(y)$, $f'(x)\, dx = dy$, if $y = f(a)$ then $x = a$, and if $y = f(b)$ then $x = b$.

117. (a) $y = \frac{1}{4}(3 \sin 2x - 6x \cos 2x)$ (b)

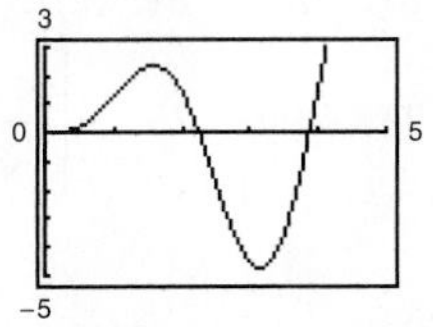

(c) You obtain the following points.

n	x^n	y^n
0	0	0
1	0.05	0.000250
2	0.10	0.001992
3	0.15	0.006689
4	0.20	0.015745
80	4.00	1.615019

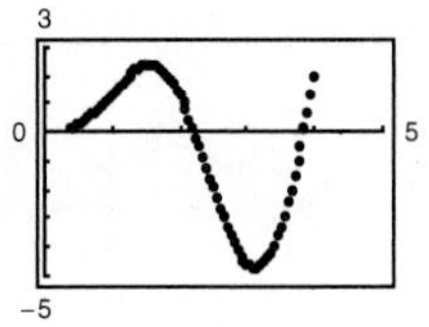

(d) You obtain the following points.

n	x^n	y^n
0	0	0
1	0.1	0.001992
2	0.2	0.015745
3	0.3	0.052081
4	0.4	0.119993
40	4.0	1.615019

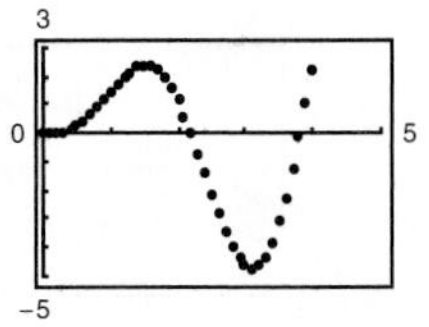

(e) Smaller step size, more data

119. The graph of $y = x \sin x$ is below the graph of $y = x$ on $[0, \pi/2]$.

Section 8.3 (page 540)

1. c **2.** a **3.** d **4.** b

5. $-\frac{1}{4}\cos^4 x + C$ **7.** $\frac{1}{12}\sin^6 2x + C$

9. $-\frac{1}{3}\cos^3 x + \frac{2}{5}\cos^5 x - \frac{1}{7}\cos^7 x + C$

11. $\frac{2}{3}\sin^{3/2}\theta - \frac{2}{7}\sin^{7/2}\theta + C$ **13.** $\frac{1}{12}(6x + \sin 6x) + C$

15. $\frac{1}{8}(\alpha - \frac{1}{4}\sin 4\alpha) + C$ or $\frac{1}{8}\alpha - \frac{1}{32}\sin 4\alpha + C$

17. $\frac{1}{8}(2x^2 - 2x \sin 2x - \cos 2x) + C$ **19.** $\frac{2}{3}$ **21.** $\frac{16}{35}$

23. $5\pi/32$ **25.** $\frac{1}{3}\ln|\sec 3x + \tan 3x| + C$

27. $\frac{1}{15}\tan 5x(3 + \tan^2 5x) + C$

29. $(\sec \pi x \tan \pi x + \ln|\sec \pi x + \tan \pi x|)/(2\pi) + C$

31. $\tan^4(x/4) - 2\tan^2(x/4) - 4\ln|\cos(x/4)| + C$

33. $\frac{1}{2}\tan^2 x + C$ **35.** $\frac{1}{3}\tan^3 x + C$ **37.** $\frac{1}{24}\sec^6 4x + C$

39. $\frac{1}{3}\sec^3 x + C$ **41.** $\ln|\sec x + \tan x| - \sin x + C$

43. $(12\pi\theta - 8 \sin 2\pi\theta + \sin 4\pi\theta)/(32\pi) + C$

45. $y = \frac{1}{9}\sec^3 3x - \frac{1}{3}\sec 3x + C$

47. (a)

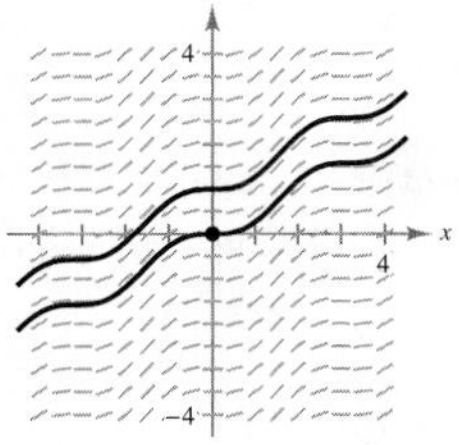

(b) $y = \frac{1}{2}x - \frac{1}{4}\sin 2x$

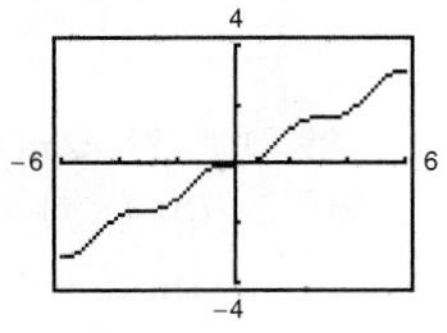

49.
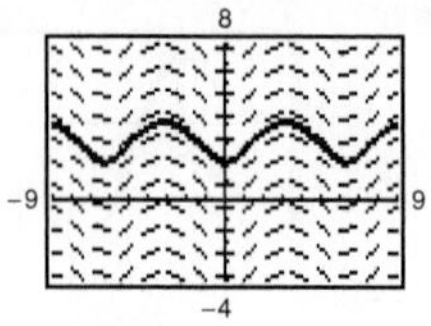

51. $-\frac{1}{10}(\cos 5x + 5\cos x) + C$

53. $\frac{1}{8}(2\sin 2\theta - \sin 4\theta) + C$ **55.** $\frac{1}{4}(\ln|\csc^2 2x| - \cot^2 2x) + C$

57. $-\cot\theta - \frac{1}{3}\cot^3\theta + C$ **59.** $\ln|\csc t - \cot t| + \cos t + C$

61. $\ln|\csc x - \cot x| + \cos x + C$ **63.** $t - 2\tan t + C$

65. π **67.** $\frac{1}{2}(1 - \ln 2)$ **69.** $\ln 2$ **71.** $\frac{4}{3}$

73. $\frac{1}{16}(6x + 8\sin x + \sin 2x) + C$
Graphs will vary. Example:

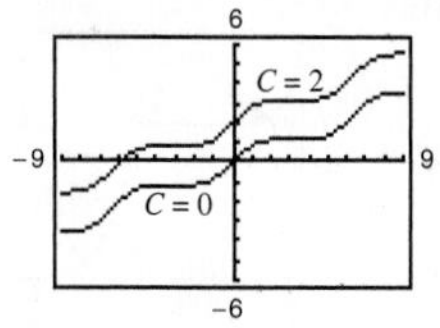

75. $\left[\sec^3 \pi x \tan \pi x + \frac{3}{2}\left(\sec \pi x \tan \pi x + \ln|\sec \pi x + \tan \pi x|\right)\right]/(4\pi) + C$
Graphs will vary. Example:

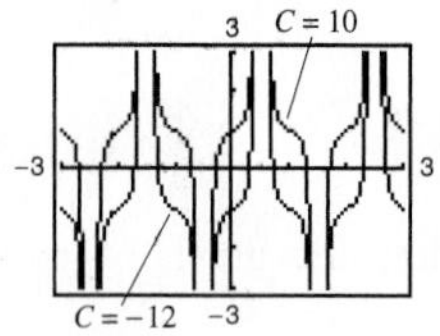

77. $(\sec^5 \pi x)/(5\pi) + C$ **79.** $3\sqrt{2}/10$ **81.** $3\pi/16$
Graphs will vary. Example:

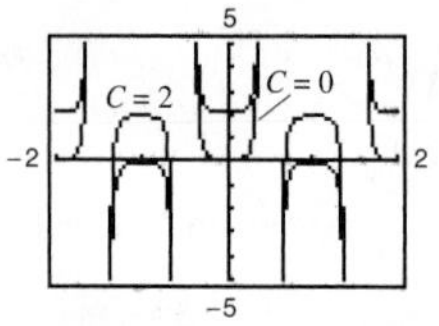

83. (a) Save one sine factor and convert the remaining factors to cosines. Then, expand and integrate.
(b) Save one cosine factor and convert the remaining factors to sines. Then, expand and integrate.
(c) Make repeated use of the power reducing formulas to convert the integrand to odd powers of the cosine. Then, proceed as in part (b).

85. (a) $\frac{1}{18}\tan^6 3x + \frac{1}{12}\tan^4 3x + C_1$, $\frac{1}{18}\sec^6 3x - \frac{1}{12}\sec^4 3x + C_2$
(b)

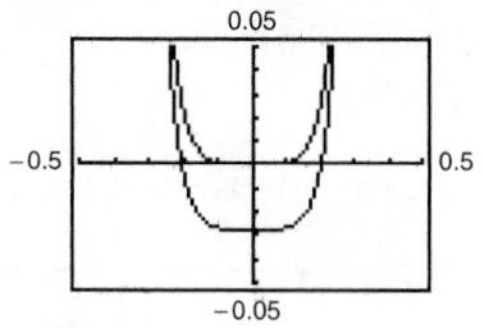

(c) Proof

87. $\frac{1}{3}$ **89.** 1 **91.** $2\pi(1 - \pi/4) \approx 1.348$

93. (a) $\pi^2/2$ (b) $(\bar{x}, \bar{y}) = (\pi/2, \pi/8)$ **95–97.** Proofs

99. $-\frac{1}{15}\cos x(3\sin^4 x + 4\sin^2 x + 8) + C$

101. $\dfrac{5}{6\pi}\tan\dfrac{2\pi x}{5}\left(\sec^2\dfrac{2\pi x}{5} + 2\right) + C$

103. (a) $H(t) \approx 57.72 - 23.36\cos(\pi t/6) - 2.75\sin(\pi t/6)$
(b) $L(t) \approx 42.04 - 20.91\cos(\pi t/6) - 4.33\sin(\pi t/6)$
(c)

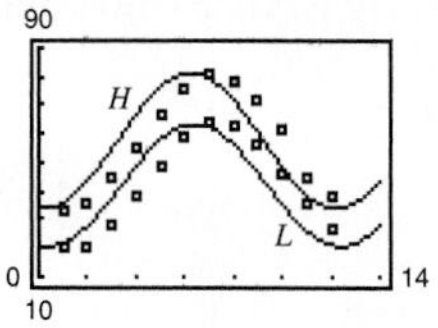

The maximum difference is at $t \approx 4.9$ or early summer.

105. Proof

Section 8.4 (page 549)

1. b **2.** d **3.** a **4.** c **5.** $x/\left(25\sqrt{25 - x^2}\right) + C$

7. $5\ln\left|\left(5 - \sqrt{25 - x^2}\right)/x\right| + \sqrt{25 - x^2} + C$

9. $\ln\left|x + \sqrt{x^2 - 4}\right| + C$ **11.** $\frac{1}{15}(x^2 - 4)^{3/2}(3x^2 + 8) + C$

13. $\frac{1}{3}(1 + x^2)^{3/2} + C$ **15.** $\frac{1}{2}[\arctan x + x/(1 + x^2)] + C$

17. $\frac{1}{2}x\sqrt{4 + 9x^2} + \frac{2}{3}\ln\left|3x + \sqrt{4 + 9x^2}\right| + C$

19. $\frac{25}{4}\arcsin(2x/5) + \frac{1}{2}x\sqrt{25 - 4x^2} + C$

21. $\sqrt{x^2 + 9} + C$ **23.** $\arcsin(x/4) + C$

25. $4\arcsin(x/2) + x\sqrt{4 - x^2} + C$ **27.** $\ln\left|x + \sqrt{x^2 - 9}\right| + C$

29. $-\dfrac{(1 - x^2)^{3/2}}{3x^3} + C$ **31.** $-\dfrac{1}{3}\ln\left|\dfrac{\sqrt{4x^2 + 9} + 3}{2x}\right| + C$

33. $5\sqrt{x^2 + 5}/(x + 5) + C$ **35.** $\frac{1}{3}(1 + e^{2x})^{3/2} + C$

37. $\frac{1}{2}\left(\arcsin e^x + e^x\sqrt{1 - e^{2x}}\right) + C$

39. $\frac{1}{4}\left(x/(x^2 + 2) + \left(1/\sqrt{2}\right)\arctan\left(x/\sqrt{2}\right)\right) + C$

41. $x\,\text{arcsec}\,2x - \frac{1}{2}\ln\left|2x + \sqrt{4x^2 - 1}\right| + C$

43. $\arcsin[(x - 2)/2] + C$

45. $\sqrt{x^2 + 4x + 8} - 2\ln\left|\sqrt{x^2 + 4x + 8} + (x + 2)\right| + C$

47. (a) and (b) $\sqrt{3} - \pi/3 \approx 0.685$

49. (a) and (b) $9\left(2 - \sqrt{2}\right) \approx 5.272$

51. (a) and (b) $-(9/2)\ln\left(2\sqrt{7}/3 - 4\sqrt{3}/3 - \sqrt{21}/3 + 8/3\right) + 9\sqrt{3} - 2\sqrt{7} \approx 12.644$

53. $\sqrt{x^2 - 9} - 3\arctan\left(\sqrt{x^2 - 9}/3\right) + 1$

55. $\frac{1}{2}(x - 15)\sqrt{x^2 + 10x + 9} + 33\ln\left|\sqrt{x^2 + 10x + 9} + (x + 5)\right| + C$

57. $\frac{1}{2}\left(x\sqrt{x^2 - 1} + \ln\left|x + \sqrt{x^2 - 1}\right|\right) + C$

59. (a) Let $u = a\sin\theta$, $\sqrt{a^2 - u^2} = a\cos\theta$, where $-\pi/2 \le \theta \le \pi/2$.
(b) Let $u = a\tan\theta$, $\sqrt{a^2 + u^2} = a\sec\theta$, where $-\pi/2 < \theta < \pi/2$.
(c) Let $u = a\sec\theta$, $\sqrt{u^2 - a^2} = \tan\theta$ if $u > a$ and $\sqrt{u^2 - a^2} = -\tan\theta$ if $u < -a$, where $0 \le \theta < \pi/2$ or $\pi/2 < \theta \le \pi$.

61. $\ln\sqrt{x^2 + 9} + C$ **63.** True

65. False: $\displaystyle\int_0^{\sqrt{3}} \frac{dx}{(1 + x^2)^{3/2}} = \int_0^{\pi/3} \cos\theta\, d\theta$

67. πab **69.** (a) $5\sqrt{2}$ (b) $25(1 - \pi/4)$ (c) $r^2(1 - \pi/4)$

71. $6\pi^2$ **73.** $\ln\left[\dfrac{5\left(\sqrt{2} + 1\right)}{\sqrt{26} + 1}\right] + \sqrt{26} - \sqrt{2} \approx 4.367$

75. Length of one arch of sine curve: $y = \sin x$, $y' = \cos x$

$$L_1 = \int_0^{\pi} \sqrt{1 + \cos^2 x}\, dx$$

Length of one arch of cosine curve: $y = \cos x$, $y' = -\sin x$

$$L_2 = \int_{-\pi/2}^{\pi/2} \sqrt{1 + \sin^2 x}\, dx$$
$$= \int_{-\pi/2}^{\pi/2} \sqrt{1 + \cos^2(x - \pi/2)}\, dx, \quad u = x - \pi/2,\ du = dx$$
$$= \int_{-\pi}^{0} \sqrt{1 + \cos^2 u}\, du = \int_0^{\pi} \sqrt{1 + \cos^2 u}\, du = L_1$$

77. (a)

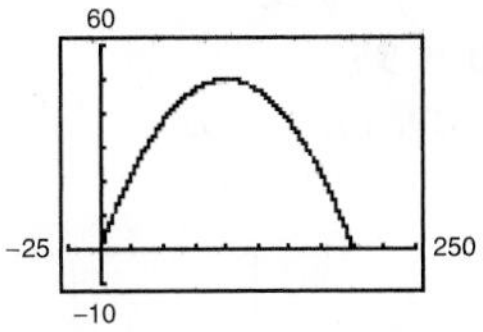

(b) 200

(c) $100\sqrt{2} + 50 \ln[(\sqrt{2} + 1)/(\sqrt{2} - 1)] \approx 229.559$

79. $(0, 0.422)$ **81.** $(\pi/32)[102\sqrt{2} - \ln(3 + 2\sqrt{2})] \approx 13.989$

83. (a) 187.2π lb (b) $62.4\pi d$ lb

85. Proof **87.** $12 + 9\pi/2 - 25 \arcsin(3/5) \approx 10.050$

Section 8.5 (page 559)

1. $\dfrac{A}{x} + \dfrac{B}{x - 10}$ **3.** $\dfrac{A}{x} + \dfrac{Bx + C}{x^2 + 10}$ **5.** $\dfrac{A}{x} + \dfrac{B}{x - 10}$

7. $\frac{1}{2} \ln|(x - 1)/(x + 1)| + C$ **9.** $\ln|(x - 1)/(x + 2)| + C$

11. $\frac{3}{2} \ln|2x - 1| - 2 \ln|x + 1| + C$

13. $5 \ln|x - 2| - \ln|x + 2| - 3 \ln|x| + C$

15. $x^2 + \frac{3}{2} \ln|x - 4| - \frac{1}{2} \ln|x + 2| + C$

17. $1/x + \ln|x^4 + x^3| + C$

19. $2 \ln|x - 2| - \ln|x| - 3/(x - 2) + C$

21. $\ln|(x^2 + 1)/x| + C$

23. $\frac{1}{6}[\ln|(x - 2)/(x + 2)| + \sqrt{2} \arctan(x/\sqrt{2})] + C$

25. $\frac{1}{16} \ln|(4x^2 - 1)/(4x^2 + 1)| + C$

27. $\ln|x + 1| + \sqrt{2} \arctan[(x - 1)/\sqrt{2}] + C$

29. $\ln 2$ **31.** $\frac{1}{2} \ln(8/5) - \pi/4 + \arctan 2 \approx 0.557$

33. $y = 3 \ln|x - 3| - 9/(x - 3) + 9$

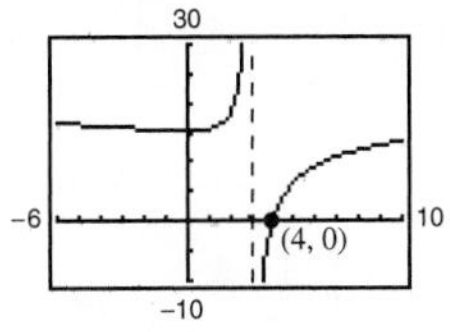

35. $y = (\sqrt{2}/2) \arctan(x/\sqrt{2}) - 1/[2(x^2 + 2)] + 5/4$

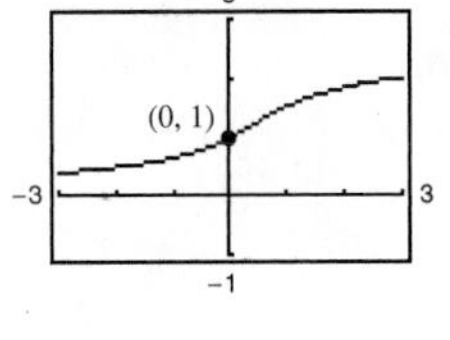

37. $y = \ln|x - 2| + \frac{1}{2} \ln|x^2 + x + 1|$
$- \sqrt{3} \arctan[(2x + 1)/\sqrt{3}] - \frac{1}{2} \ln 13$
$+ \sqrt{3} \arctan(7/\sqrt{3}) + 10$

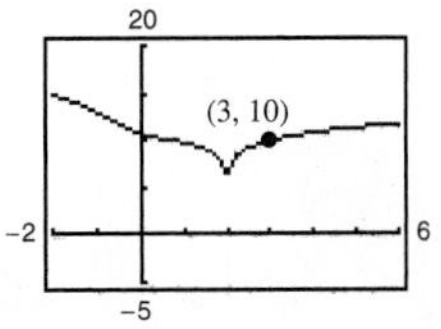

39. $y = \frac{1}{4} \ln|(x - 2)/(x + 2)| + \frac{1}{4} \ln 2 + 4$

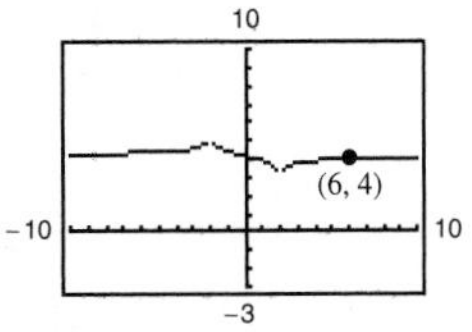

41. $\ln\left|\dfrac{\cos x}{\cos x - 1}\right| + C$ **43.** $\ln\left|\dfrac{-1 + \sin x}{2 + \sin x}\right| + C$

45. $\dfrac{1}{5} \ln\left|\dfrac{e^x - 1}{e^x + 4}\right| + C$ **47–49.** Proofs

51. $y = \dfrac{3}{2} \ln\left|\dfrac{2 + x}{2 - x}\right| + 3$ **53.** First divide x^3 by $(x - 5)$.

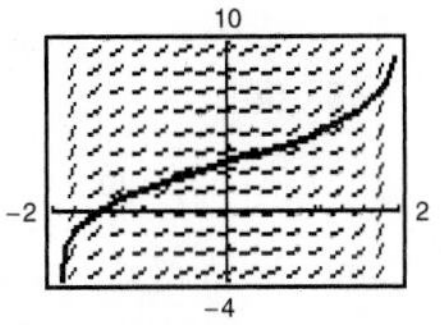

55. (a) Log Rule (b) Partial fractions (c) Inverse Tangent Rule

57. $12 \ln(\frac{9}{8}) \approx 1.4134$ **59.** 4.90 or \$490,000

61. $V = 2\pi(\arctan 3 - \frac{3}{10}) \approx 5.963$; $(\bar{x}, \bar{y}) \approx (1.521, 0.412)$

63. $x = n[e^{(n+1)kt} - 1]/[n + e^{(n+1)kt}]$ **65.** $\pi/8$

Section 8.6 (page 565)

1. $-\frac{1}{2}x(2 - x) + \ln|1 + x| + C$

3. $\frac{1}{2}[e^x\sqrt{e^{2x} + 1} + \ln(e^x + \sqrt{e^{2x} + 1})] + C$

5. $-\sqrt{1 - x^2}/x + C$

7. $\frac{1}{16}(6x - 3 \sin 2x \cos 2x - 2 \sin^3 2x \cos 2x) + C$

9. $-2(\cot\sqrt{x} + \csc\sqrt{x}) + C$ **11.** $x - \frac{1}{2}\ln(1 + e^{2x}) + C$

13. $\frac{1}{16}x^4(4 \ln x - 1) + C$

15. (a) and (b) $e^x(x^2 - 2x + 2) + C$

17. (a) and (b) $\ln|(x + 1)/x| - 1/x + C$

19. $\frac{1}{2}[(x^2 + 1) \operatorname{arcsec}(x^2 + 1) - \ln(x^2 + 1 + \sqrt{x^4 + 2x^2})] + C$

21. $\sqrt{x^2 - 4}/(4x) + C$ **23.** $\frac{2}{9}[\ln|1 - 3x| + 1/(1 - 3x)] + C$

25. $e^x \arccos(e^x) - \sqrt{1 - e^{2x}} + C$

27. $\frac{1}{2}(x^2 + \cot x^2 + \csc x^2) + C$

29. $(\sqrt{2}/2) \arctan[(1 + \sin\theta)/\sqrt{2}] + C$

31. $-\sqrt{2 + 9x^2}/(2x) + C$

33. $\frac{1}{4}(2 \ln|x| - 3 \ln|3 + 2 \ln|x||) + C$

35. $(3x - 10)/[2(x^2 - 6x + 10)] + \frac{3}{2} \arctan(x - 3) + C$

37. $\frac{1}{2} \ln|x^2 - 3 + \sqrt{x^4 - 6x^2 + 5}| + C$

39. $-\frac{1}{3}\sqrt{4 - x^2}(x^2 + 8) + C$

41. $2/(1 + e^x) - 1/[2(1 + e^x)^2] + \ln(1 + e^x) + C$

43. $\frac{1}{2}(e - 1) \approx 0.8591$ **45.** $9\ln(3) - \frac{26}{9} \approx 6.9986$

47. $\pi/2$ **49.** $\pi^3/8 - 3\pi + 6 \approx 0.4510$ **51–55.** Proofs

57. $y = -2\sqrt{1-x}/\sqrt{x} + 7$

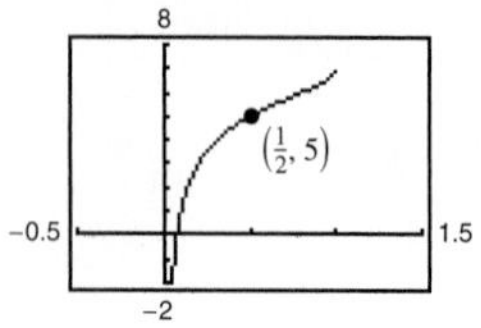

59. $y = \frac{1}{2}[(x-3)/(x^2 - 6x + 10) + \arctan(x-3)]$

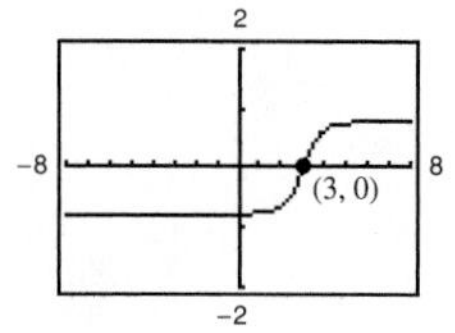

61. $y = -\csc\theta + \sqrt{2} + 2$

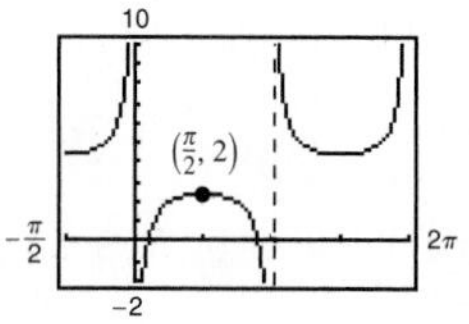

63. $\dfrac{1}{\sqrt{5}}\ln\left|\dfrac{2\tan(\theta/2) - 3 - \sqrt{5}}{2\tan(\theta/2) - 3 + \sqrt{5}}\right| + C$ **65.** $\ln 2$

67. $\frac{1}{2}\ln(3 - 2\cos\theta) + C$ **69.** $2\sin\sqrt{\theta} + C$ **71.** $\frac{40}{3}$

73. Use Formula 23 and let $a = 1$, $u = e^x$, and $du = e^x\,dx$, because it is in the form $\displaystyle\int \frac{1}{a^2 + u^2}\,du$.

75. Use Formula 81 and let $u = x^2$ and $du = 2x\,dx$, because it is in the form $\displaystyle\int e^u\,du$.

77. Impossible; there is no integration formula that fits.

79. (a) $\displaystyle\int x\ln x\,dx = \frac{1}{2}x^2\ln x - \frac{1}{4}x^2 + C$

$\displaystyle\int x^2\ln x\,dx = \frac{1}{3}x^3\ln x - \frac{1}{9}x^3 + C$

$\displaystyle\int x^3\ln x\,dx = \frac{1}{4}x^4\ln x - \frac{1}{16}x^4 + C$

(b) $\displaystyle\int x^n\ln x\,dx = x^{n+1}\ln x/(n+1) - x^{n+1}/(n+1)^2 + C$

81. False. Substitutions may first have to be made to rewrite the integral in a form that appears in the table.

83. 1919.145 ft-lb

85. (a) $V = 80\ln(\sqrt{10} + 3) \approx 145.5\ \text{ft}^3$
$W = 11{,}840\ln(\sqrt{10} + 3) \approx 21{,}530.4$ lb
(b) (0, 1.19)

87. (a) $k = 30/\ln 7 \approx 15.42$
(b)

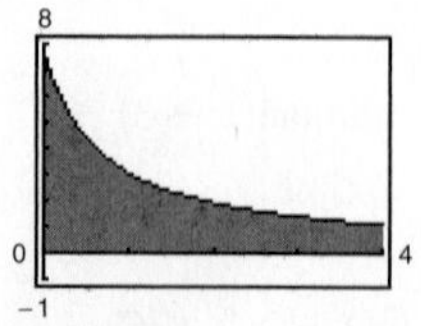

89. Putnam problem A3, 1980

Section 8.7 (page 574)

1.

x	−0.1	−0.01	−0.001	0.001	0.01	0.1
$f(x)$	2.4132	2.4991	2.500	2.500	2.4991	2.4132

2.5

3.

x	1	10	10^2	10^3	10^4	10^5
$f(x)$	0.9900	90,483.7	3.7×10^9	4.5×10^{10}	0	0

0

5. $\frac{1}{3}$ **7.** $\frac{1}{4}$ **9.** $\frac{5}{3}$ **11.** 3 **13.** 0 **15.** 2

17. ∞ **19.** $\frac{2}{3}$ **21.** 1 **23.** $\frac{3}{2}$ **25.** ∞

27. 0 **29.** 1 **31.** 0 **33.** 0 **35.** ∞

37. (a) Not indeterminate
(b) ∞
(c)

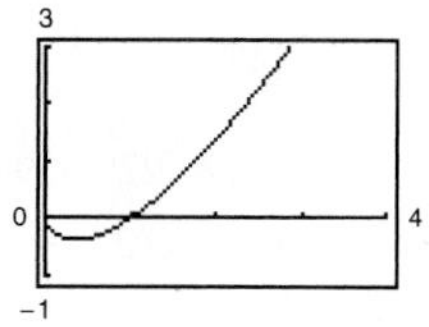

39. (a) $0 \cdot \infty$
(b) 1
(c)

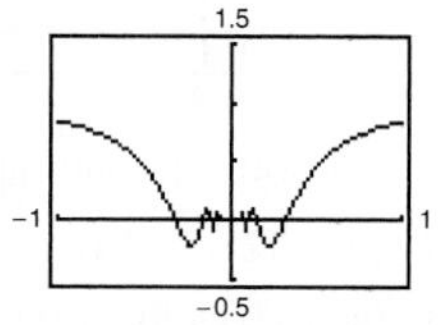

41. (a) Not indeterminate
(b) 0
(c)

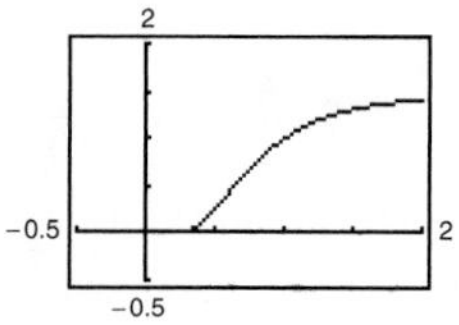

43. (a) ∞^0
(b) 1
(c)

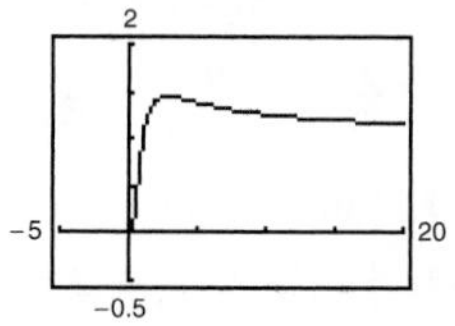

45. (a) 1^∞ (b) e
(c)

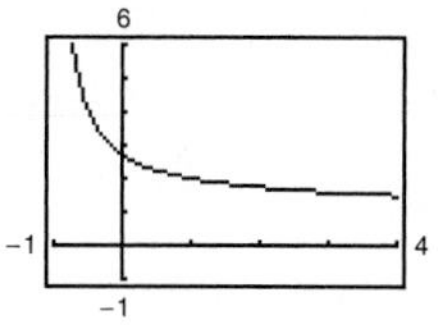

47. (a) 0^0 (b) 3
(c)

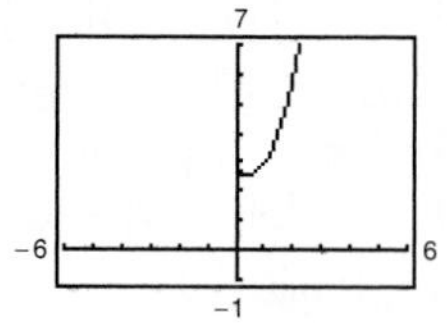

49. (a) 0^0 (b) 1
(c)

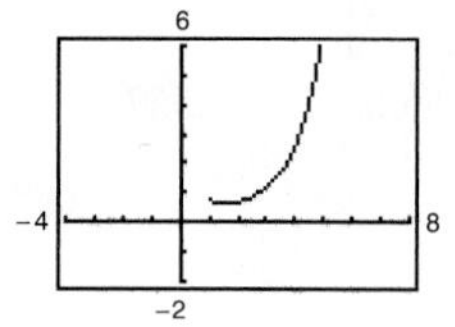

51. (a) $\infty - \infty$ (b) $-\frac{3}{2}$
(c)

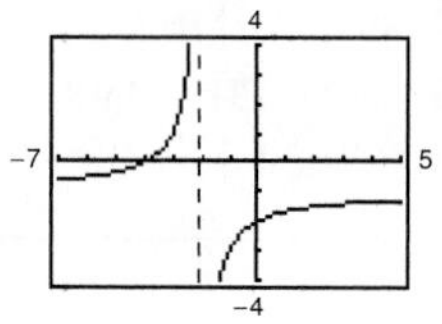

53. (a) $\infty - \infty$ (b) ∞
(c)

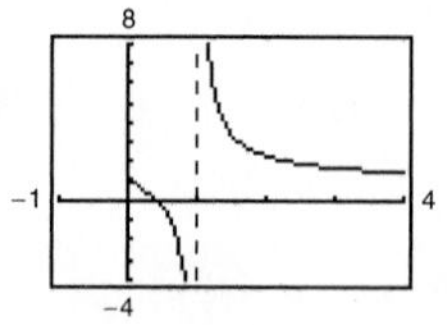

55. (a)

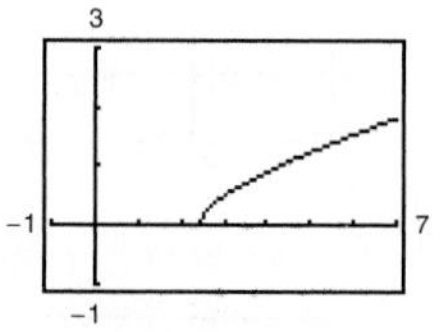

(b) $\frac{1}{2}$

57. (a)

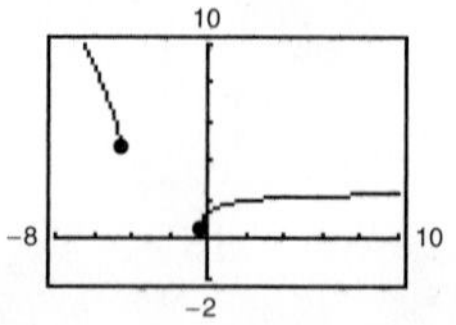

(b) $\frac{5}{2}$

59. $\frac{0}{0}, \frac{\infty}{\infty}, 0 \cdot \infty, 1^\infty, 0^0, \infty - \infty$

61. Answers will vary. Examples:
(a) $f(x) = x^2 - 25$, $g(x) = x - 5$
(b) $f(x) = (x - 5)^2$, $g(x) = x^2 - 25$
(c) $f(x) = x^2 - 25$, $g(x) = (x - 5)^3$

63.

x	10	10^2	10^4	10^6	10^8	10^{10}
$\frac{(\ln x)^4}{x}$	2.811	4.498	0.720	0.036	0.001	0.000

65. 0 **67.** 0 **69.** 0

71. Horizontal asymptote: $y = 1$
Relative maximum: $(e, e^{1/e})$

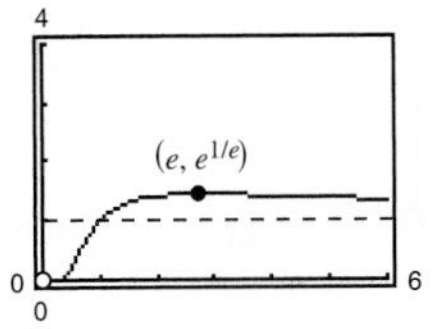

73. Horizontal asymptote: $y = 0$
Relative maximum: $(1, 2/e)$

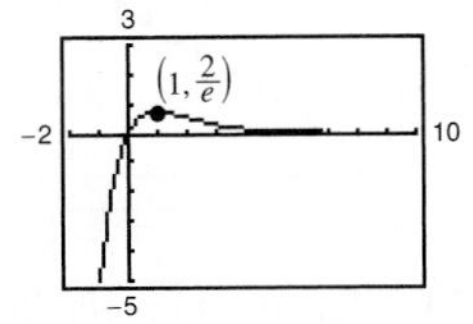

75. Limit is not of the form $0/0$ or ∞/∞.

77. Limit is not of the form $0/0$ or ∞/∞.

79. (a) $\lim\limits_{x\to\infty} \frac{x}{\sqrt{x^2+1}} = \lim\limits_{x\to\infty} \frac{\sqrt{x^2+1}}{x} = \lim\limits_{x\to\infty} \frac{x}{\sqrt{x^2+1}}$
Applying L'Hôpital's Rule twice results in the original limit, so L'Hôpital's Rule fails.
(b) 1
(c)

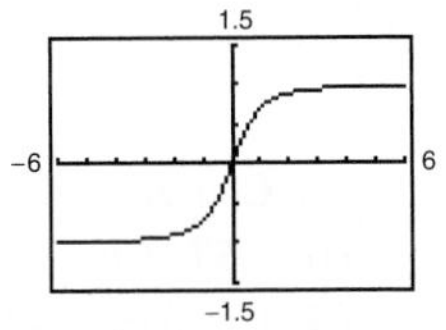

81.

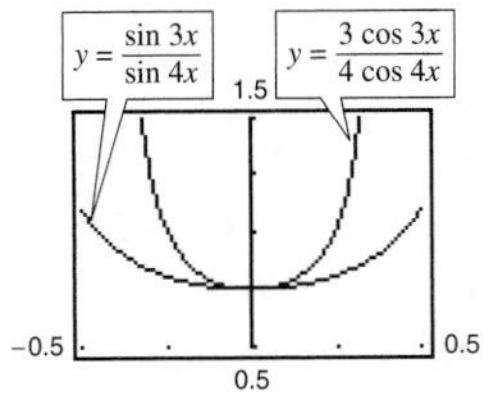

As $x \to 0$, the graphs get closer together.

83. $v = 32t + v_0$ **85.** Proof **87.** $c = \frac{2}{3}$ **89.** $c = \pi/4$

91. False: L'Hôpital's Rule does not apply, because $\lim\limits_{x\to 0}(x^2 + x + 1) \neq 0$. **93.** True

95. $\frac{3}{4}$ **97.** $\frac{4}{3}$ **99.** $a = 1, b = \pm 2$ **101.** Proof

103.

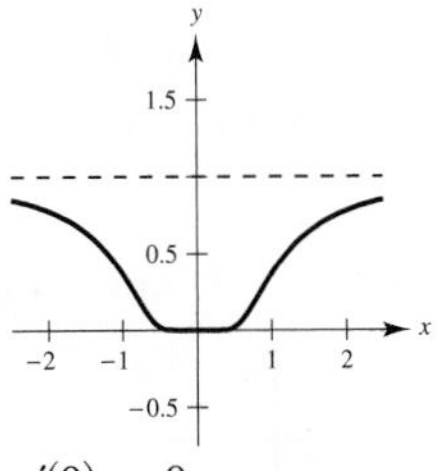

$g'(0) = 0$

105. (a) $0 \cdot \infty$ (b) 0

107–109. Proofs

111. (a)

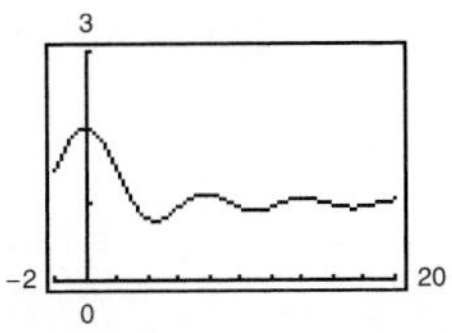

(b) $\lim\limits_{x\to\infty} h(x) = 1$
(c) No

Section 8.8 (page 585)

1. Improper; $0 < \frac{2}{3} < 1$ **3.** Not improper; continuous on $[0, 1]$

5. Infinite discontinuity at $x = 0$; 4

7. Infinite discontinuity at $x = 1$; diverges

9. Infinite limit of integration; 1

11. Infinite discontinuity at $x = 0$; diverges

13. Infinite limit of integration; converges to 1 **15.** 1

17. Diverges **19.** Diverges **21.** 2 **23.** $\frac{1}{2}$

25. $1/[2(\ln 4)^2]$ **27.** π **29.** $\pi/4$ **31.** Diverges

33. Diverges **35.** 6 **37.** $-\frac{1}{4}$ **39.** Diverges **41.** $\pi/3$

43. $\ln(2 + \sqrt{3})$ **45.** 0 **47.** $2\pi\sqrt{6}/3$ **49.** $p > 1$

51. Proof **53.** Diverges **55.** Converges **57.** Converges

59. Diverges **61.** Converges

63. An integral with infinite integration limits, an integral with an infinite discontinuity at or between the integration limits

65. The improper integral diverges. **67.** e **69.** π

71. (a) 1 (b) $\pi/2$ (c) 2π

73.

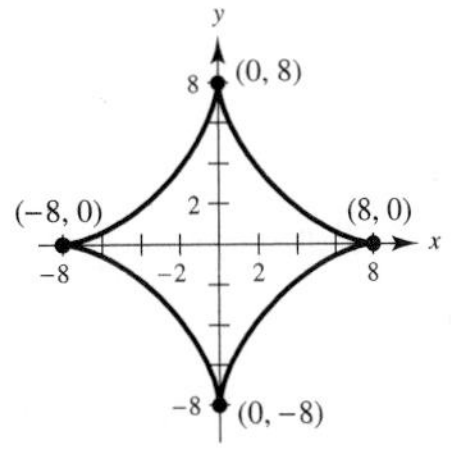

Perimeter = 48

75. $8\pi^2$ **77.** (a) $W = 20{,}000$ mile-tons (b) 4000 mi

79. (a) Proof (b) $P = 43.53\%$ (c) $E(x) = 7$

81. (a) $757,992.41 (b) $837,995.15 (c) $1,066,666.67

83. $P = \left[2\pi NI\left(\sqrt{r^2 + c^2} - c\right)\right]/\left(kr\sqrt{r^2 + c^2}\right)$

85. False. Let $f(x) = 1/(x + 1)$. **87.** True

89. (a) $\int_1^\infty \frac{1}{x^n}\,dx$ will converge if $n > 1$ and diverge if $n \leq 1$.
(b)

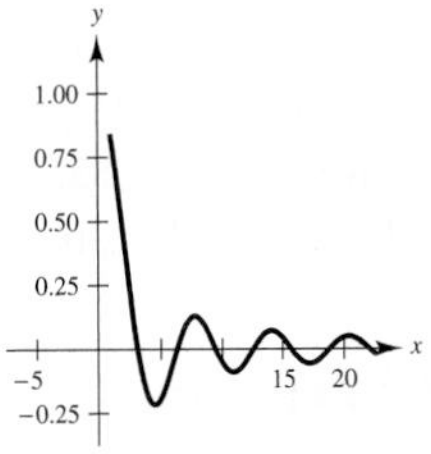

(c) Converges

91. (a) $\Gamma(1) = 1, \Gamma(2) = 1, \Gamma(3) = 2$ (b) Proof
(c) $\Gamma(n) = (n - 1)!$

93. $1/s,\ s > 0$ **95.** $2/s^3,\ s > 0$ **97.** $s/(s^2 + a^2),\ s > 0$

99. $s/(s^2 - a^2),\ s > |a|$

101. (a)

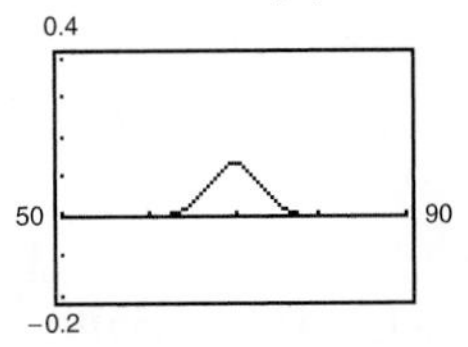

(b) ≈ 0.2525
(c) 0.2525; same by symmetry

103. $c = 1$; $\ln(2)$

105. $8\pi[(\ln 2)^{2/3} - (\ln 4)/9 + 2/27] \approx 2.01545$ **107.** 0.6278

109. (a)

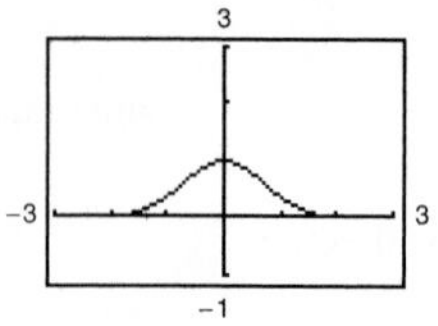

(b) Proof

Review Exercises for Chapter 8 (page 589)

1. $\frac{1}{3}(x^2-1)^{3/2}+C$ **3.** $\frac{1}{2}\ln|x^2-1|+C$
5. $\ln(2)+\frac{1}{2}\approx 1.1931$ **7.** $16\arcsin(x/4)+C$
9. $\frac{1}{13}e^{2x}(2\sin 3x-3\cos 3x)+C$
11. $\frac{2}{15}(x-5)^{3/2}(3x+10)+C$
13. $-\frac{1}{2}x^2\cos 2x+\frac{1}{2}x\sin 2x+\frac{1}{4}\cos 2x+C$
15. $\frac{1}{16}\left[(8x^2-1)\arcsin 2x+2x\sqrt{1-4x^2}\right]+C$
17. $\sin(\pi x-1)[\cos^2(\pi x-1)+2]/(3\pi)+C$
19. $\frac{2}{3}[\tan^3(x/2)+3\tan(x/2)]+C$ **21.** $\tan\theta+\sec\theta+C$
23. $3\pi/16+\frac{1}{2}\approx 1.0890$ **25.** $3\sqrt{4-x^2}/x+C$
27. $\frac{1}{3}(x^2+4)^{1/2}(x^2-8)+C$ **29.** π
31. (a), (b), and (c) $\frac{1}{3}\sqrt{4+x^2}(x^2-8)+C$
33. $6\ln|x+2|-5\ln|x-3|+C$
35. $\frac{1}{4}[6\ln|x-1|-\ln(x^2+1)+6\arctan x]+C$
37. $x+\frac{9}{8}\ln|x-3|-\frac{25}{8}\ln|x+5|+C$
39. $\frac{1}{9}\left[2/(2+3x)+\ln|2+3x|\right]+C$ **41.** $1-\sqrt{2}/2$
43. $\frac{1}{2}\ln|x^2+4x+8|-\arctan[(x+2)/2]+C$
45. $\ln|\tan\pi x|/\pi+C$ **47.** Proof
49. $\frac{1}{8}(\sin 2\theta-2\theta\cos 2\theta)+C$
51. $\frac{4}{3}[x^{3/4}-3x^{1/4}+3\arctan(x^{1/4})]+C$
53. $2\sqrt{1-\cos x}+C$ **55.** $\sin x\ln(\sin x)-\sin x+C$
57. $y=\frac{3}{2}\ln|(x-3)/(x+3)|+C$
59. $y=x\ln|x^2+x|-2x+\ln|x+1|+C$ **61.** $\frac{1}{5}$
63. $\frac{1}{2}(\ln 4)^2\approx 0.961$ **65.** π **67.** $\frac{128}{15}$
69. $(\bar{x},\bar{y})=(0,4/(3\pi))$ **71.** 3.82 **73.** 0 **75.** ∞ **77.** 1
79. $1000e^{0.09}\approx 1094.17$ **81.** Converges; $\frac{32}{3}$ **83.** Diverges
85. Converges; 1 **87.** (a) $6,321,205.59 (b) $10,000,000
89. (a) 0.4581 (b) 0.0135

P.S. Problem Solving (page 591)

1. (a) $\frac{4}{3}, \frac{16}{15}$ (b) Proof **3.** $\ln 3$
5. 2
7. (a)

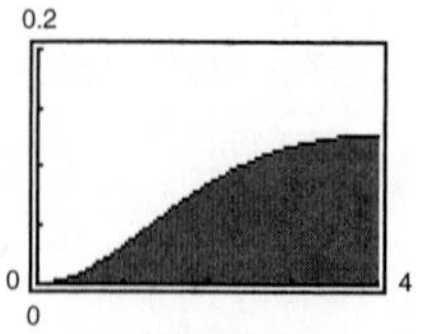

Area ≈ 0.2986

(b) $\ln 3-\frac{4}{5}$
(c) $\ln 3-\frac{4}{5}$

9. $\ln 3-\frac{1}{2}\approx 0.5986$ **11.** Proof **13.** ≈ 0.8670
15. (a) ∞ (b) 0 (c) $-\frac{2}{3}$

17. $\dfrac{1/12}{x}+\dfrac{1/42}{x-3}+\dfrac{1/10}{x-1}+\dfrac{111/140}{x+4}$
19. Proof **21.** ≈ 0.0158

Chapter 9

Section 9.1 (page 602)

1. 2, 4, 8, 16, 32 **3.** $-\frac{1}{2}, \frac{1}{4}, -\frac{1}{8}, \frac{1}{16}, -\frac{1}{32}$ **5.** 1, 0, −1, 0, 1
7. $-1, -\frac{1}{4}, \frac{1}{9}, \frac{1}{16}, -\frac{1}{25}$ **9.** $5, \frac{19}{4}, \frac{43}{9}, \frac{77}{16}, \frac{121}{25}$
11. 3, 4, 6, 10, 18 **13.** 32, 16, 8, 4, 2
15. f **16.** a **17.** e **18.** b **19.** d **20.** c
21.

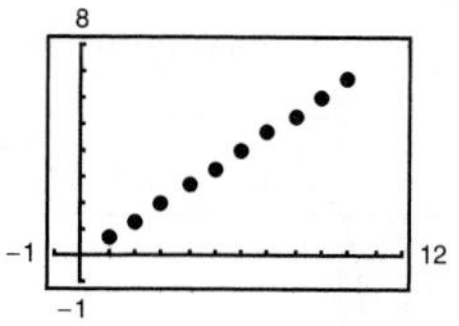

23.

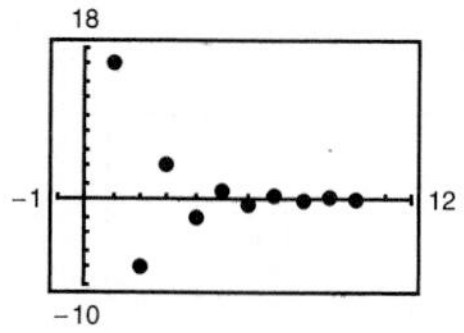

25. 14, 17; add 3 to preceding term
27. 80, 160; multiply preceding term by 2.
29. $\frac{3}{16}, -\frac{3}{32}$; multiply preceding term by $-\frac{1}{2}$ **31.** $10\cdot 9=90$
33. $n+1$ **35.** $1/[(2n+1)(2n)]$ **37.** 5 **39.** 2 **41.** 0
43.

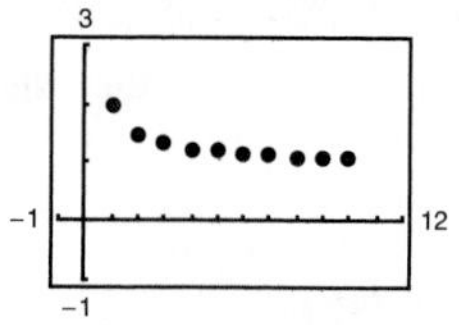

Converges to 1

45.

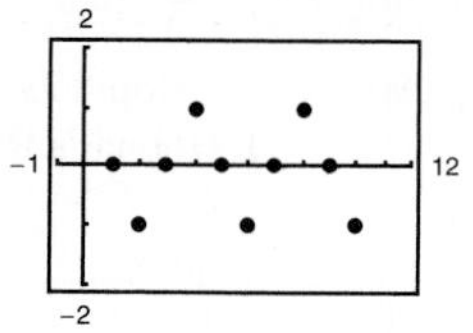

Diverges

47. Diverges **49.** Converges to $\frac{3}{2}$ **51.** Converges to 0
53. Converges to 0 **55.** Converges to 0 **57.** Converges to 0
59. Diverges **61.** Converges to 0 **63.** Converges to 0
65. Converges to e^k **67.** Converges to 0
69. Answers will vary. Sample answer: $3n-2$
71. Answers will vary. Sample answer: n^2-2
73. Answers will vary. Sample answer: $(n+1)/(n+2)$
75. Answers will vary. Sample answer: $(n+1)/n$
77. Answers will vary. Sample answer: $n/[(n+1)(n+2)]$
79. Answers will vary. Sample answer:
$$\frac{(-1)^{n-1}}{1\cdot 3\cdot 5\cdots(2n-1)}=\frac{(-1)^{n-1}2^n n!}{(2n)!}$$
81. Answers will vary. Sample answer: $(2n)!$
83. Monotonic, bounded **85.** Monotonic, bounded
87. Not monotonic, bounded **89.** Monotonic, bounded
91. Not monotonic, bounded **93.** Not monotonic, bounded
95. (a) $\left|5+\frac{1}{n}\right|\le 6 \Rightarrow$ bounded
$a_n>a_{n+1}\Rightarrow$ monotonic
So, $\{a_n\}$ converges.

(b)

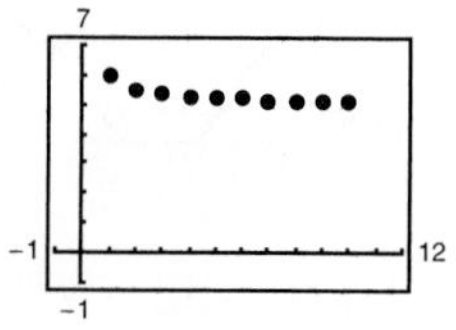

Limit = 5

97. (a) $\left|\frac{1}{3}\left(1 - \frac{1}{3^n}\right)\right| < \frac{1}{3} \Rightarrow$ bounded

$a_n < a_{n+1} \Rightarrow$ monotonic

So, $\{a_n\}$ converges.

(b)

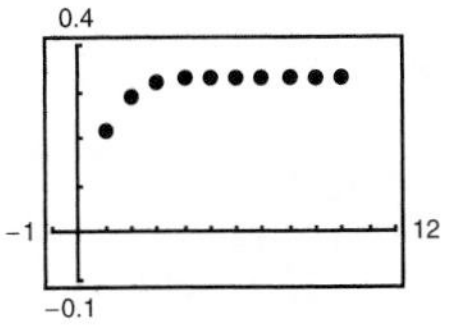

Limit $= \frac{1}{3}$

99. $\{a_n\}$ has a limit because it is bounded and monotonic; since $2 \le a_n \le 4$, $2 \le L \le 4$.

101. (a) No; $\lim_{n\to\infty} a_n$ does not exist.

(b)

n	1	2	3	4	5
A_n	\$9041.25	\$9082.69	\$9124.32	\$9166.14	\$9208.15

n	6	7	8	9	10
A_n	\$9250.35	\$9292.75	\$9335.34	\$9378.13	\$9421.11

103. (a) A sequence is a function whose domain is the set of positive integers.

(b) A sequence converges if it has a limit.

(c) A monotonic sequence is a sequence that has nonincreasing or nondecreasing terms.

(d) A bounded sequence is a sequence that has both an upper and a lower bound.

105. Answers will vary. Example: $a_n = 10n/(n + 1)$

107. Answers will vary. Example: $a_n = (3n^2 - n)/(4n^2 + 1)$

109. (a) $\$2{,}500{,}000{,}000(0.8)^n$

(b)

Year	1	2
Budget	\$2,000,000,000	\$1,600,000,000

Year	3	4
Budget	\$1,280,000,000	\$1,024,000,000

(c) Converges to 0

111. (a) $a_n = -6.60n^2 + 151.7n + 387$ (b) 979 species

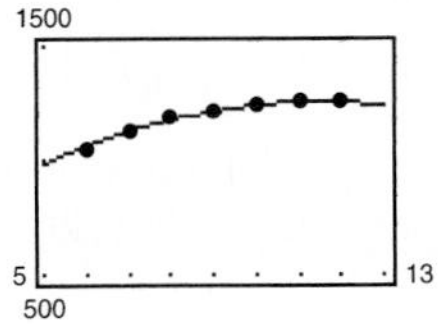

113. (a) $a_9 = a_{10} = 1{,}562{,}500/567$ (b) Decreasing

(c) Factorials increase more rapidly than exponentials.

115. 1, 1.4142, 1.4422, 1.4142, 1.3797, 1.3480; Converges to 1

117. True **119.** True

121. (a) 1, 1, 2, 3, 5, 8, 13, 21, 34, 55, 89, 144

(b) 1, 2, 1.5, 1.6667, 1.6, 1.6250, 1.6154, 1.6190, 1.6176, 1.6182

(c) Proof (d) $\rho = (1 + \sqrt{5})/2 \approx 1.6180$

123. (a) 1.4142, 1.8478, 1.9616, 1.9904, 1.9976

(b) $a_n = \sqrt{2 + a_{n-1}}$ (c) $\lim_{n\to\infty} a_n = 2$

125. (a) Proof (b) Proof (c) $\lim_{n\to\infty} a_n = \left(1 + \sqrt{1 + 4k}\right)/2$

127. (a) Proof (b) Proof

129. (a) Proof

(b)

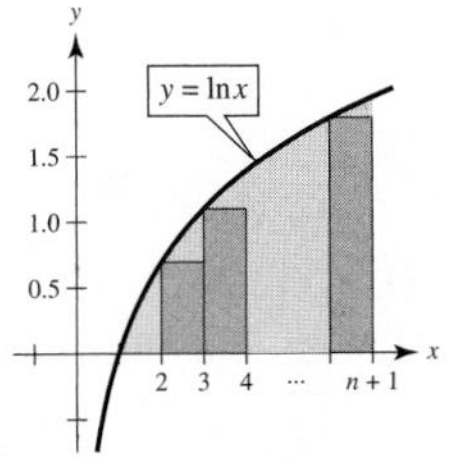

(c) Proof (d) Proof

(e) $\frac{\sqrt[20]{20!}}{20} \approx 0.4152$; $\frac{\sqrt[50]{50!}}{50} \approx 0.3897$; $\frac{\sqrt[100]{100!}}{100} \approx 0.3799$

131–133. Proofs **135.** Putnam Problem A1, 1990

Section 9.2 (page 612)

1. 1, 1.25, 1.361, 1.424, 1.464

3. 3, −1.5, 5.25, −4.875, 10.3125

5. 3, 4.5, 5.25, 5.625, 5.8125 **7.** Geometric series: $r = \frac{3}{2} > 1$

9. Geometric series: $r = 1.055 > 1$ **11.** $\lim_{n\to\infty} a_n = 1 \ne 0$

13. $\lim_{n\to\infty} a_n = 1 \ne 0$ **15.** $\lim_{n\to\infty} a_n = \frac{1}{2} \ne 0$ **17.** c; 3

18. b; 3 **19.** a; 3 **20.** d; 3 **21.** f; $\frac{34}{9}$ **22.** e; $\frac{5}{3}$

23. Telescoping series: $a_n = 1/n - 1/(n + 1)$; Converges to 1.

25. Geometric series: $r = \frac{3}{4} < 1$

27. Geometric series: $r = 0.9 < 1$

29. (a) $\frac{11}{3}$

(b)

n	5	10	20	50	100
S_n	2.7976	3.1643	3.3936	3.5513	3.6078

(c)

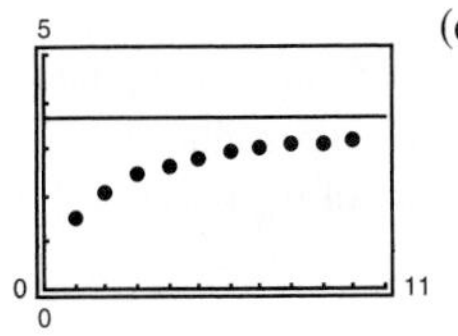

(d) The terms of the series decrease in magnitude relatively slowly, and the sequence of partial sums approaches the sum of the series relatively slowly.

31. (a) 20

(b)

n	5	10	20	50	100
S_n	8.1902	13.0264	17.5685	19.8969	19.9995

(c)

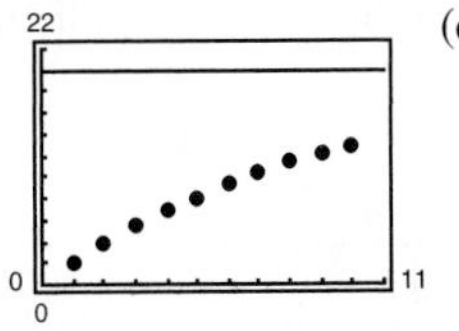

(d) The terms of the series decrease in magnitude relatively slowly, and the sequence of partial sums approaches the sum of the series relatively slowly.

33. (a) $\frac{40}{3}$

(b)

n	5	10	20	50	100
S_n	13.3203	13.3333	13.3333	13.3333	13.3333

(c)

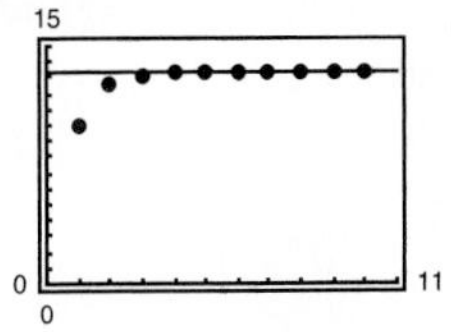

(d) The terms of the series decrease in magnitude relatively rapidly, and the sequence of partial sums approaches the sum of the series relatively rapidly.

35. $\frac{3}{4}$ **37.** 4 **39.** 2 **41.** $\frac{2}{3}$ **43.** $\frac{10}{9}$ **45.** $\frac{9}{4}$ **47.** $\frac{1}{2}$

49. $\dfrac{\sin(1)}{1 - \sin(1)}$ **51.** (a) $\displaystyle\sum_{n=0}^{\infty} \frac{4}{10}(0.1)^n$ (b) $\frac{4}{9}$

53. (a) $\displaystyle\sum_{n=0}^{\infty} \frac{81}{100}(0.01)^n$ (b) $\frac{9}{11}$

55. (a) $\displaystyle\sum_{n=0}^{\infty} \frac{3}{40}(0.01)^n$ (b) $\frac{5}{66}$ **57.** Diverges **59.** Converges

61. Diverges **63.** Converges **65.** Diverges **67.** Diverges

69. Diverges **71.** Diverges **73.** See definitions on page 606.

75. The series given by

$$\sum_{n=0}^{\infty} ar^n = a + ar + ar^2 + \cdot \cdot \cdot + ar^n + \cdot \cdot \cdot, a \neq 0$$

is a geometric series with ratio r. When $0 < |r| < 1$, the series converges to the sum $\displaystyle\sum_{n=0}^{\infty} ar^n = \frac{a}{1 - r}$.

77. $\{a_n\}$ converges to 1 and $\displaystyle\sum_{n=1}^{\infty} a_n$ diverges. This fits Theorem 9.9, which states that if $\displaystyle\lim_{n\to\infty} a_n \neq 0$, then $\displaystyle\sum_{n=1}^{\infty} a_n$ diverges.

79. $-2 < x < 2$; $x/(2 - x)$ **81.** $0 < x < 2$; $(x - 1)/(2 - x)$

83. $-1 < x < 1$; $1/(1 + x)$

85. x: $(-\infty, -1) \cup (1, \infty)$; $x/(x - 1)$

87. (a) Yes. Answers will vary. (b) Yes. Answers will vary.

89. (a) x (b) $f(x) = 1/(1 - x)$, $|x| < 1$

(c)

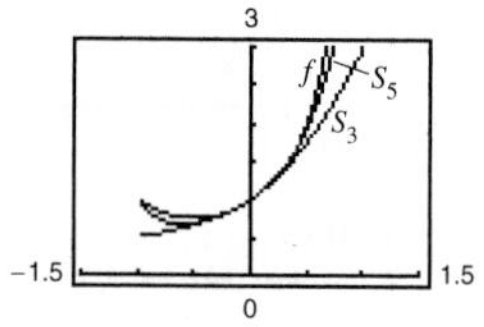

91.

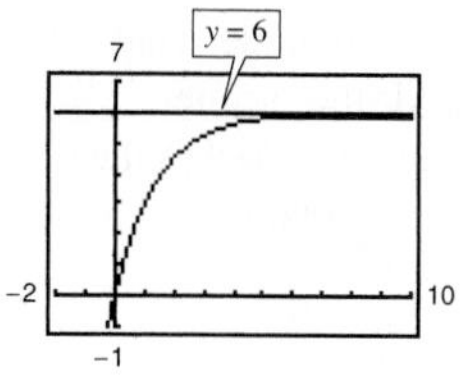

Horizontal asymptote: $y = 6$
The horizontal asymptote is the sum of the series.

93. The required terms for the two series are $n = 100$ and $n = 5$, respectively. The second series converges at a higher rate.

95. $80{,}000(1 - 0.9^n)$ units

97. $400(1 - 0.75^n)$ million dollars; Sum = \$400 million

99. 152.42 feet **101.** $\frac{1}{8}$; $\displaystyle\sum_{n=0}^{\infty} \frac{1}{2}\left(\frac{1}{2}\right)^n = \frac{1/2}{1 - 1/2} = 1$

103. (a) $-1 + \displaystyle\sum_{n=0}^{\infty}\left(\frac{1}{2}\right)^n = -1 + \frac{a}{1 - r} = -1 + \frac{1}{1 - 1/2} = 1$

(b) No (c) 2

105. (a) 126 in.^{2} (b) 128 in.^{2}

107. \$573,496.06; The \$1,000,000 sweepstakes has a present value of \$573,496.06. After accruing interest over the 20-year period, it attains its full value.

109. (a) \$5,368,709.11 (b) \$10,737,418.23 (c) \$21,474,836.47

111. (a) \$16,415.10 (b) \$16,421.83

113. (a) \$118,196.13 (b) \$118,393.43

115. (a) $a_n = 3484.1363e^{0.0490n}$

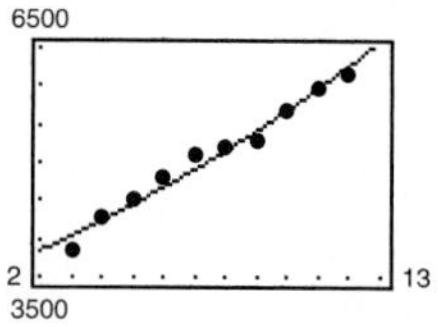

(b) \$50,809 million
(c) \$50,815 million

117. False. $\displaystyle\lim_{n\to\infty} \frac{1}{n} = 0$, but $\displaystyle\sum_{n=1}^{\infty} \frac{1}{n}$ diverges.

119. False. $\displaystyle\sum_{n=1}^{\infty} ar^n = \left(\frac{a}{1 - r}\right) - a$

The formula requires that the geometric series begins with $n = 0$.

121. True **123.** Proof

125. Answers will vary. Example: $\displaystyle\sum_{n=0}^{\infty} 1$, $\displaystyle\sum_{n=0}^{\infty} (-1)$

127–131. Proofs

133. H = half-life of the drug
n = number of equal doses
P = number of units of the drug
t = equal time intervals
The total amount of the drug in the patient's system at the time the last dose is given is
$T_n = P + Pe^{kt} + Pe^{2kt} + \cdot \cdot \cdot + Pe^{(n-1)kt}$
where $k = -(\ln 2)/H$. One time interval after the last dose is given is
$T_{n+1} = Pe^{kt} + Pe^{2kt} + Pe^{3kt} + \cdot \cdot \cdot + Pe^{nkt}$
and so on. Because $k < 0$, $T_{n+s} \to 0$ as $s \to \infty$.

135. Putnam Problem A1, 1966

Section 9.3 (page 620)

1. Diverges **3.** Converges **5.** Converges **7.** Diverges

9. Diverges **11.** Diverges **13.** Converges **15.** Converges

17. Diverges **19.** Diverges

21. $f(x)$ is not positive for $x \geq 1$.

23. $f(x)$ is not always decreasing. **25.** Converges **27.** Diverges

29. Diverges **31.** Diverges **33.** Converges **35.** Converges

37. c; diverges **38.** f; diverges **39.** b; converges

40. a; diverges **41.** d; converges **42.** e; converges

43. (a)

n	5	10	20	50	100
S_n	3.7488	3.75	3.75	3.75	3.75

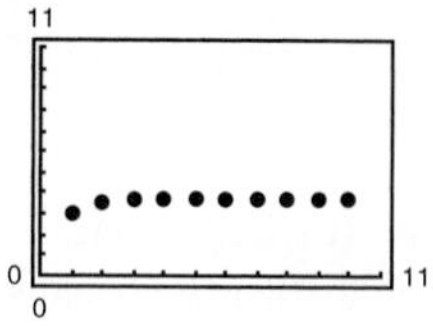

The partial sums approach the sum 3.75 very quickly.

(b)

n	5	10	20	50	100
S_n	1.4636	1.5498	1.5962	1.6251	1.635

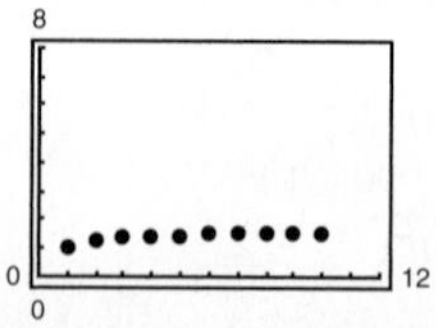

The partial sums approach the sum $\pi^2/6 \approx 1.6449$ more slowly than the series in part (a).

45. See Theorem 9.10 on page 617. Answers will vary. For example, convergence or divergence can be determined for the series $\sum_{n=1}^{\infty} \frac{1}{n^2+1}$.

47. No. Because $\sum_{n=1}^{\infty} \frac{1}{n}$ diverges, $\sum_{n=10{,}000}^{\infty} \frac{1}{n}$ also diverges. The convergence or divergence of a series is not determined by the first finite number of terms of the series.

49.

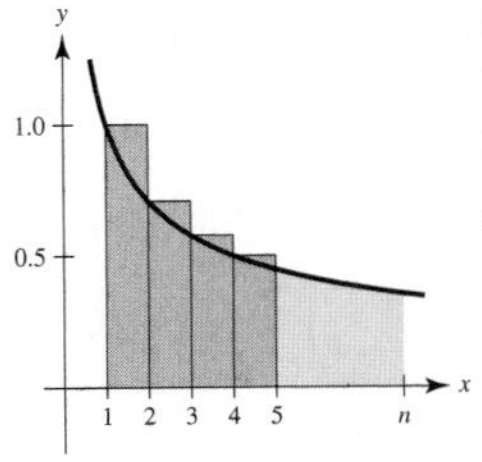

The series diverges because the area under the rectangles is greater than the infinite area under the curve $y = 1/\sqrt{x}$ for $x \geq 1$.

51. $p > 1$ **53.** $p > 1$ **55.** Diverges

57. Converges **59.** Proof

61. $S_6 \approx 1.0811$, $R_6 \approx 0.0015$ **63.** $S_{10} \approx 0.9818$, $R_{10} \approx 0.0997$ **65.** $S_4 \approx 0.4049$, $R_4 \approx 5.6 \times 10^{-8}$

67. $N \geq 7$ **69.** $N \geq 2$ **71.** $N \geq 1000$

73. (a) $\sum_{n=2}^{\infty} \frac{1}{n^{1.1}}$ converges by the p-Series Test since $1.1 > 1$.

$\sum_{n=2}^{\infty} \frac{1}{n \ln n}$ diverges by the Integral Test since $\int_2^{\infty} \frac{1}{x \ln x}\,dx$ diverges.

(b) $\sum_{n=2}^{\infty} \frac{1}{n^{1.1}} = 0.4665 + 0.2987 + 0.2176 + 0.1703 + 0.1393 + \cdots$

$\sum_{n=2}^{\infty} \frac{1}{n \ln n} = 0.7213 + 0.3034 + 0.1803 + 0.1243 + 0.0930 + \cdots$

(c) $n \geq 3.431 \times 10^{15}$

75. (a) Let $f(x) = 1/x$. f is positive, continuous, and decreasing on $[1, \infty)$.

$$S_n - 1 \leq \int_1^n \frac{1}{x}\,dx = \ln n$$

$$S_n \geq \int_1^{n+1} \frac{1}{x}\,dx = \ln(n+1)$$

So, $\ln(n+1) \leq S_n \leq 1 + \ln n$.

(b) $\ln(n+1) - \ln n \leq S_n - \ln n \leq 1$.
Also, $\ln(n+1) - \ln n > 0$ for $n \geq 1$. So, $0 \leq S_n - \ln n \leq 1$ and the sequence $\{a_n\}$ is bounded.

(c) $a_n - a_{n+1} = [S_n - \ln n] - [S_{n+1} - \ln(n+1)]$

$$= \int_n^{n+1} \frac{1}{x}\,dx - \frac{1}{n+1} \geq 0$$

So, $a_n \geq a_{n+1}$.

(d) Because the sequence is bounded and monotonic, it converges to a limit, γ.

(e) 0.5822

77. (a) Diverges (b) Diverges

(c) $\sum_{n=2}^{\infty} x^{\ln n}$ converges for $x < 1/e$.

79. Diverges **81.** Converges **83.** Converges **85.** Diverges

87. Diverges **89.** Converges

Section 9.4 (page 628)

1. (a)

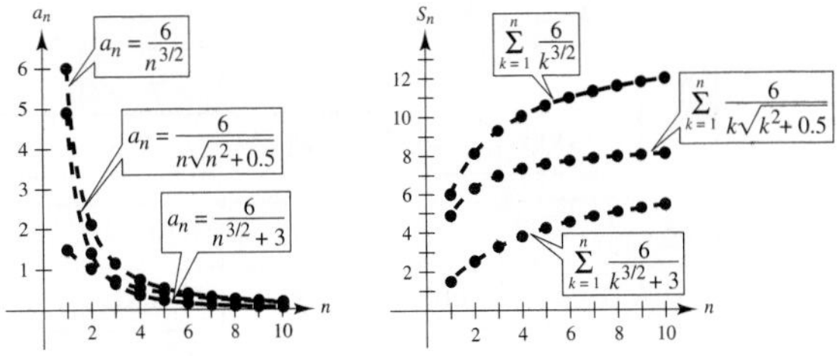

(b) $\sum_{n=1}^{\infty} \frac{6}{n^{3/2}}$; Converges

(c) The magnitudes of the terms are less than the magnitudes of the terms of the p-series. Therefore, the series converges.

(d) The smaller the magnitudes of the terms, the smaller the magnitudes of the terms of the sequence of partial sums.

3. Converges **5.** Diverges **7.** Converges **9.** Diverges

11. Converges **13.** Converges **15.** Diverges **17.** Diverges

19. Converges **21.** Diverges **23.** Converges **25.** Diverges

27. Diverges **29.** Diverges; p-Series Test

31. Converges; Direct Comparison Test with $\sum_{n=1}^{\infty} \left(\frac{1}{3}\right)^n$

33. Diverges; nth-Term Test **35.** Converges; Integral Test

37. $\lim_{n\to\infty} \frac{a_n}{1/n} = \lim_{n\to\infty} na_n$

$\lim_{n\to\infty} na_n \neq 0$, but is finite.

The series diverges by the Limit Comparison Test.

39. Diverges **41.** Converges

43. $\lim_{n\to\infty} n\left(\frac{n^3}{5n^4+3}\right) = \frac{1}{5} \neq 0$

So, $\sum_{n=1}^{\infty} \frac{n^3}{5n^4+3}$ diverges.

45. Diverges **47.** Converges

49. Convergence or divergence is dependent on the form of the general term for the series and not necessarily on the magnitude of the terms.

51. See Theorem 9.13 on page 626. Answers will vary. For example,

$\sum_{n=2}^{\infty} \frac{1}{\sqrt{n-1}}$ diverges because $\lim_{n\to\infty} \frac{1/\sqrt{n-1}}{1/\sqrt{n}} = 1$ and

$\sum_{n=2}^{\infty} \frac{1}{\sqrt{n}}$ diverges (p-series).

53.

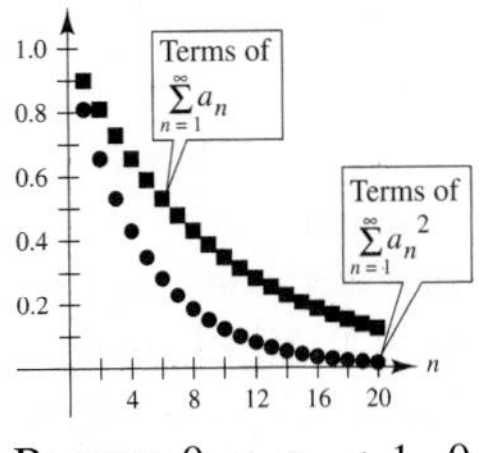

Because $0 < a_n < 1$, $0 < a_n^2 < a_n < 1$.

55. False. Let $a_n = 1/n^3$ and $b_n = 1/n^2$.

57. True **59.** True **61.** Proof **63.** $\sum_{n=1}^{\infty} \frac{1}{n^2}, \sum_{n=1}^{\infty} \frac{1}{n^3}$

65–69. Proofs **71.** Putnam Problem 1, afternoon session, 1953

Section 9.5 (page 636)

1. d **2.** f **3.** a **4.** b **5.** e **6.** c

7. (a)

n	1	2	3	4	5
S_n	1.0000	0.6667	0.8667	0.7238	0.8349

n	6	7	8	9	10
S_n	0.7440	0.8209	0.7543	0.8131	0.7605

(b)

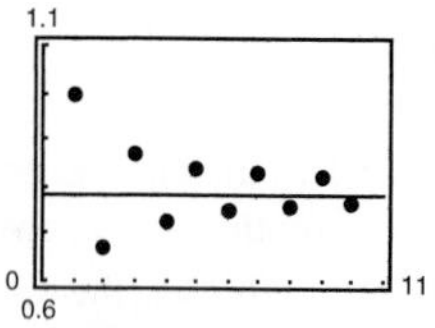

(c) The points alternate sides of the horizontal line $y = \pi/4$ that represents the sum of the series. The distances between the successive points and the line decrease.

(d) The distance in part (c) is always less than the magnitude of the next term of the series.

9. (a)

n	1	2	3	4	5
S_n	1.0000	0.7500	0.8611	0.7986	0.8386

n	6	7	8	9	10
S_n	0.8108	0.8312	0.8156	0.8280	0.8180

(b)

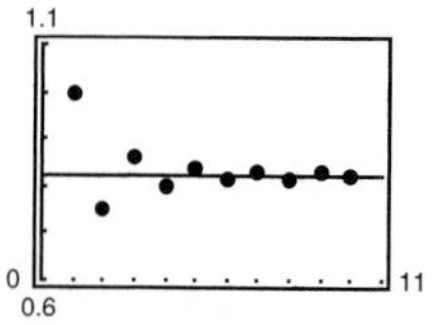

(c) The points alternate sides of the horizontal line $y = \pi^2/12$ that represents the sum of the series. The distances between the successive points and the line decrease.

(d) The distance in part (c) is always less than the magnitude of the next term of the series.

11. Converges **13.** Converges **15.** Diverges **17.** Converges
19. Diverges **21.** Diverges **23.** Diverges **25.** Converges
27. Converges **29.** Converges **31.** Converges
33. $2.3713 \le S \le 2.4937$ **35.** $0.7305 \le S \le 0.7361$
37. (a) 7 terms (Note that the sum begins with $N = 0$.) (b) 0.368
39. (a) 3 terms (Note that the sum begins with $N = 0$.) (b) 0.842
41. (a) 1000 terms (b) 0.693 **43.** 10 **45.** 7
47. Converges absolutely **49.** Converges conditionally
51. Diverges **53.** Converges conditionally
55. Converges absolutely **57.** Converges absolutely
59. Converges conditionally **61.** Converges absolutely
63. An alternating series is a series whose terms alternate in sign. See Theorem 9.14 on page 631 for the Alternating Series Test.
65. A series $\Sigma\, a_n$ is absolutely convergent if $\Sigma\, |a_n|$ converges. A series $\Sigma\, a_n$ is conditionally convergent if $\Sigma\, a_n$ converges and $\Sigma\, |a_n|$ diverges.
67. False. Let $a_n = \dfrac{(-1)^n}{n}$. **69.** True **71.** $p > 0$
73. (a) Proof
(b) The converse is false. For example: Let $a_n = 1/n$.
75. $\displaystyle\sum_{n=1}^{\infty} \frac{1}{n^2}$ converges, hence so does $\displaystyle\sum_{n=1}^{\infty} \frac{1}{n^4}$.
77. (a) No; $a_{n+1} \le a_n$ is not satisfied for all n. For example, $\frac{1}{9} < \frac{1}{8}$. (b) Yes; 0.5
79. Converges; p-Series Test **81.** Diverges; nth-Term Test
83. Converges; Geometric Series Test
85. Converges; Integral Test
87. Converges; Alternating Series Test
89. The first term of the series is 0, not 1. You cannot regroup series terms arbitrarily.
91. Putnam Problem 2, afternoon session, 1954

Section 9.6 (page 645)

1–3. Proofs **5.** d **6.** c **7.** f **8.** b **9.** a **10.** e
11. (a) Proof

(b)

n	5	10	15	20	25
S_n	9.2104	16.7598	18.8016	19.1878	19.2491

(c)

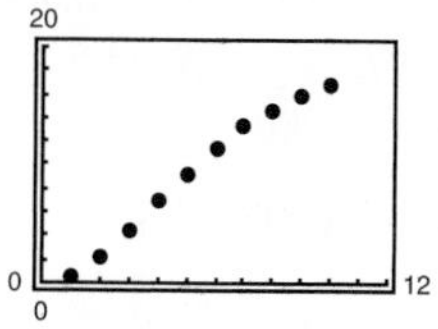

(d) 19.26

(e) The more rapidly the terms of the series approach 0, the more rapidly the sequence of partial sums approaches the sum of the series.

13. Diverges **15.** Converges **17.** Converges **19.** Diverges
21. Converges **23.** Diverges **25.** Converges
27. Converges **29.** Diverges **31.** Converges
33–35. Proofs **37.** Converges **39.** Diverges **41.** Converges
43. Diverges **45.** Converges **47.** Converges
49. Converges **51.** Converges; Alternating Series Test
53. Converges; p-Series Test **55.** Diverges; nth-Term Test
57. Diverges; Ratio Test
59. Converges; Limit Comparison Test with $b_n = 1/2^n$
61. Converges; Direct Comparison Test with $b_n = 1/2^n$
63. Converges; Ratio Test **65.** Converges; Ratio Test
67. Converges; Ratio Test **69.** a and c **71.** a and b
73. $\displaystyle\sum_{n=0}^{\infty} \frac{n+1}{4^{n+1}}$ **75.** (a) 9 (b) -0.7769
77. Diverges; $\displaystyle\lim_{n\to\infty} \left|\frac{a_{n+1}}{a_n}\right| > 1$
79. Converges; $\displaystyle\lim_{n\to\infty} \left|\frac{a_{n+1}}{a_n}\right| < 1$
81. Diverges; $\lim a_n \ne 0$ **83.** Converges **85.** Converges
87. $(-3, 3)$ **89.** $(-2, 0]$ **91.** $x = 0$
93. See Theorem 9.17 on page 639.
95. No; the series $\displaystyle\sum_{n=1}^{\infty} \frac{1}{n + 10{,}000}$ diverges.
97. Absolutely; by definition **99–103.** Proofs
105. Putnam Problem 3, morning session, 1942

Section 9.7 (page 656)

1. d **2.** c **3.** a **4.** b

5. $P_1 = 6 - 2x$

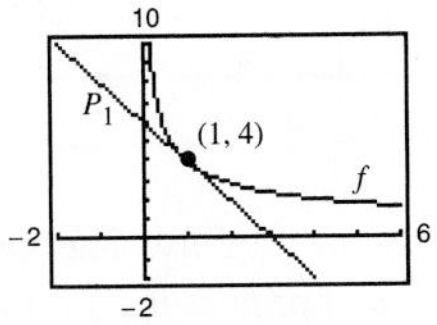

P_1 is the tangent line to the curve $f(x) = 4/\sqrt{x}$ at the point $(1, 4)$.

7. $P_1 = \sqrt{2}x + \sqrt{2}(4 - \pi)/4$

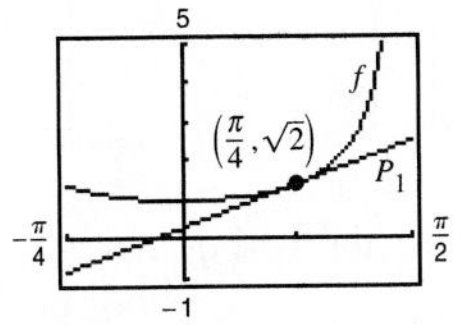

P_1 is the tangent line to the curve $f(x) = \sec x$ at the point $(\pi/4, \sqrt{2})$.

9.

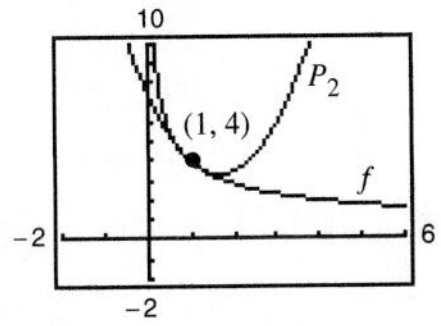

x	0	0.8	0.9	1	1.1
$f(x)$	Error	4.4721	4.2164	4.0000	3.8139
$P_2(x)$	7.5000	4.4600	4.2150	4.0000	3.8150

x	1.2	2
$f(x)$	3.6515	2.8284
$P_2(x)$	3.6600	3.5000

11. (a)

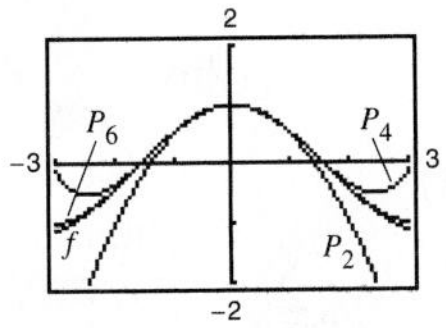

(b) $f^{(2)}(0) = -1$ $\quad P_2^{(2)}(0) = -1$
$f^{(4)}(0) = 1$ $\quad P_4^{(4)}(0) = 1$
$f^{(6)}(0) = -1$ $\quad P_6^{(6)}(0) = -1$

(c) $f^{(n)}(0) = P_n^{(n)}(0)$

13. $1 - x + \frac{1}{2}x^2 - \frac{1}{6}x^3$ **15.** $1 + 2x + 2x^2 + \frac{4}{3}x^3 + \frac{2}{3}x^4$

17. $x - \frac{1}{6}x^3 + \frac{1}{120}x^5$ **19.** $x + x^2 + \frac{1}{2}x^3 + \frac{1}{6}x^4$

21. $1 - x + x^2 - x^3 + x^4$ **23.** $1 + \frac{1}{2}x^2$

25. $1 - (x - 1) + (x - 1)^2 - (x - 1)^3 + (x - 1)^4$

27. $1 + \frac{1}{2}(x - 1) - \frac{1}{8}(x - 1)^2 + \frac{1}{16}(x - 1)^3 - \frac{5}{128}(x - 1)^4$

29. $(x - 1) - \frac{1}{2}(x - 1)^2 + \frac{1}{3}(x - 1)^3 - \frac{1}{4}(x - 1)^4$

31. (a) $P_3(x) = x + (1/3)x^3$

(b) $Q_3(x) = 1 + 2(x - \pi/4) + 2(x - \pi/4)^2 + \frac{8}{3}(x - \pi/4)^3$

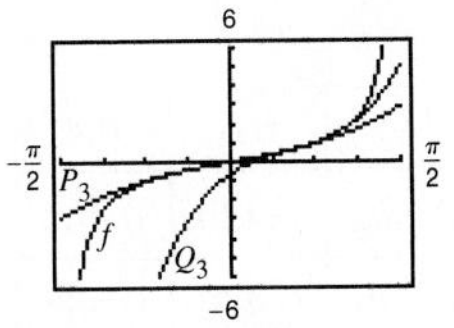

33. (a)

x	0	0.25	0.50	0.75	1.00
$\sin x$	0	0.2474	0.4794	0.6816	0.8415
$P_1(x)$	0	0.25	0.50	0.75	1.00
$P_3(x)$	0	0.2474	0.4792	0.6797	0.8333
$P_5(x)$	0	0.2474	0.4794	0.6817	0.8417

(b)

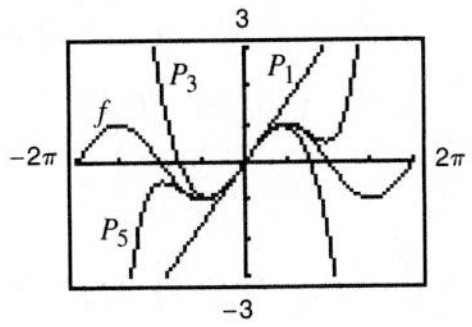

(c) As the distance increases, the polynomial approximation becomes less accurate.

35. (a) $P_3(x) = x + \frac{1}{6}x^3$

(b)

x	−0.75	−0.50	−0.25	0	0.25
$f(x)$	−0.848	−0.524	−0.253	0	0.253
$P_3(x)$	−0.820	−0.521	−0.253	0	0.253

x	0.50	0.75
$f(x)$	0.524	0.848
$P_3(x)$	0.521	0.820

(c)

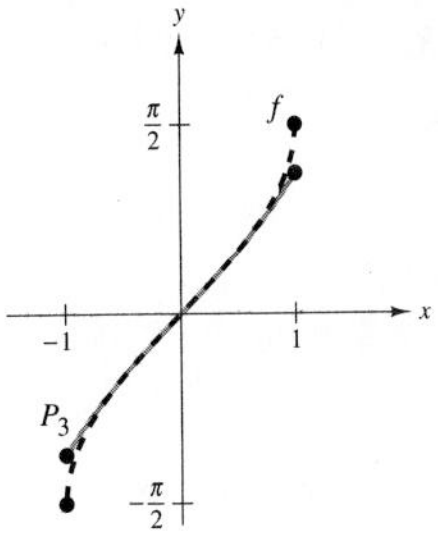

37.

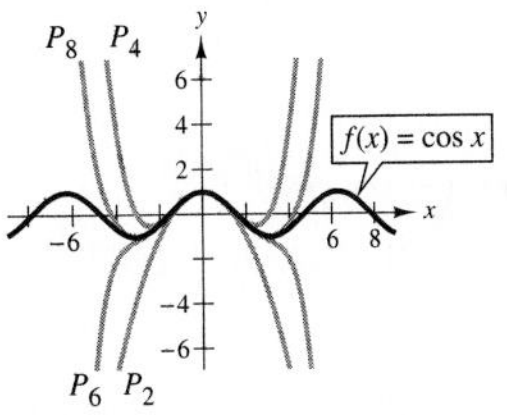

39.

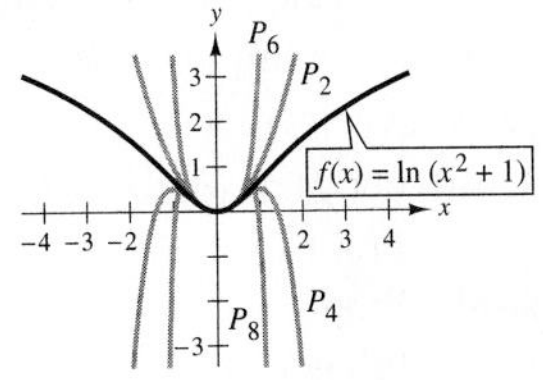

41. 0.6042 **43.** 0.1823 **45.** $R_4 \le 2.03 \times 10^{-5}$; 0.000001

47. $R_3 \le 7.82 \times 10^{-3}$ **49.** 3 **51.** 5

53. $n = 9$; $\ln(1.5) \approx 0.4055$ **55.** $n = 16$; $e^{-\pi(1.3)} \approx 0.16838$

57. $-0.3936 < x < 0$ **59.** $-0.9467 < x < 0.9467$

61. The graph of the approximating polynomial P and the elementary function f both pass through the point $(c, f(c))$, and the slope of P is the same as the slope of the graph of f at the point $(c, f(c))$. If P is of degree n, then the first n derivatives of f and P agree at c. This allows for the graph of P to resemble the graph of f near the point $(c, f(c))$.

63. See "Definitions of nth Taylor Polynomial and nth Maclaurin Polynomial" on page 650.

65. As the degree of the polynomial increases, the graph of the Taylor polynomial becomes a better and better approximation of the function within the interval of convergence. Therefore, the accuracy is increased.

67. (a) $f(x) \approx P_4(x) = 1 + x + (1/2)x^2 + (1/6)x^3 + (1/24)x^4$
$g(x) \approx Q_5(x) = x + x^2 + (1/2)x^3 + (1/6)x^4 + (1/24)x^5$
$Q_5(x) = xP_4(x)$
(b) $g(x) \approx P_6(x) = x^2 - x^4/3! + x^6/5!$
(c) $g(x) \approx P_4(x) = 1 - x^2/3! + x^4/5!$

69. (a) $Q_2(x) = -1 + (\pi^2/32)(x + 2)^2$
(b) $R_2(x) = -1 + (\pi^2/32)(x - 6)^2$
(c) No. Horizontal translations of the result in part (a) are possible only at $x = -2 + 8n$ (where n is an integer) because the period of f is 8.

71. Proof

73. As you move away from $x = c$, the Taylor polynomial becomes less and less accurate.

Section 9.8 (page 666)

1. 0 **3.** 2 **5.** $R = 1$ **7.** $R = \frac{1}{2}$ **9.** $R = \infty$
11. $(-2, 2)$ **13.** $(-1, 1]$ **15.** $(-\infty, \infty)$ **17.** $x = 0$
19. $(-4, 4)$ **21.** $(0, 10]$ **23.** $(0, 2]$ **25.** $(0, 6)$
27. $\left(-\frac{1}{2}, \frac{1}{2}\right)$ **29.** $(-\infty, \infty)$ **31.** $(-1, 1)$ **33.** $x = 3$
35. $R = c$ **37.** $(-k, k)$ **39.** $(-1, 1)$

41. $\displaystyle\sum_{n=1}^{\infty} \frac{x^{n-1}}{(n-1)!}$ **43.** $\displaystyle\sum_{n=1}^{\infty} \frac{x^{2n-1}}{(2n-1)!}$

45. (a) $(-2, 2)$ (b) $(-2, 2)$ (c) $(-2, 2)$ (d) $[-2, 2)$
47. (a) $(0, 2]$ (b) $(0, 2)$ (c) $(0, 2)$ (d) $[0, 2]$
49. c; $S_1 = 1, S_2 = 1.33$ **50.** a; $S_1 = 1, S_2 = 1.67$
51. b; diverges **52.** d; alternating
53. b **54.** c **55.** d **56.** a

57. A series of the form

$$\sum_{n=0}^{\infty} a_n(x - c)^n = a_0 + a_1(x - c) + a_2(x - c)^2 + \cdots + a_n(x - c)^n + \cdots$$

is called a power series centered at c, where c is a constant.

59. 1. A single point 2. An interval centered at c
3. The entire real line

61. Answers will vary.

63. (a) For $f(x)$: $(-\infty, \infty)$; For $g(x)$: $(-\infty, \infty)$
(b) Proof (c) Proof (d) $f(x) = \sin x$; $g(x) = \cos x$

65–69. Proofs

71. (a) Proof (b) Proof
(c)

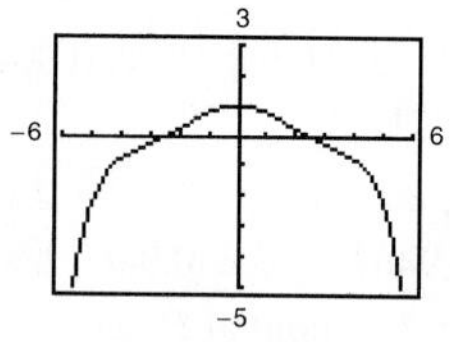

(d) 0.92

73. $f(x) = \cos x$

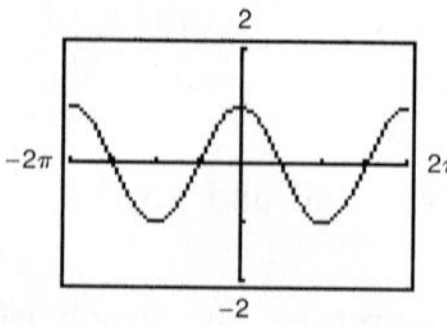

75. $f(x) = \dfrac{1}{1 + x}$

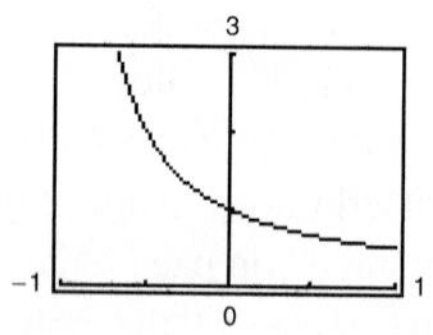

77. (a) $\frac{8}{5}$

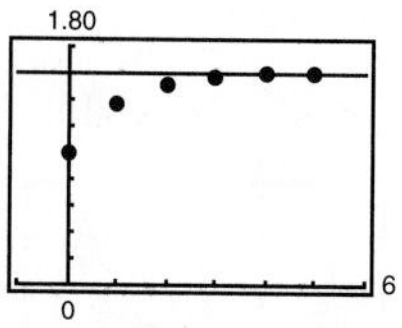

(b) $\frac{8}{11}$

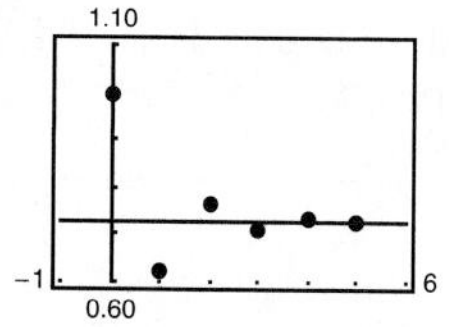

(c) The alternating series converges more rapidly. The partial sums of the series of positive terms approach the sum from below. The partial sums of the alternating series alternate sides of the horizontal line representing the sum.

(d)

M	10	100	1000	10,000
N	4	9	15	21

79. False. Let $a_n = (-1)^n/(n2^n)$ **81.** True **83.** Proof
85. (a) $(-1, 1)$ (b) $f(x) = (c_0 + c_1x + c_2x^2)/(1 - x^3)$
87. Proof

Section 9.9 (page 674)

1. $\displaystyle\sum_{n=0}^{\infty} \frac{x^n}{2^{n+1}}$ **3.** $\displaystyle\sum_{n=0}^{\infty} \frac{(-1)^n x^n}{2^{n+1}}$

5. $\displaystyle\sum_{n=0}^{\infty} \frac{(x - 5)^n}{(-3)^{n+1}}$ **7.** $\displaystyle -3\sum_{n=0}^{\infty} (2x)^n$
$(2, 8)$ $\left(-\frac{1}{2}, \frac{1}{2}\right)$

9. $\displaystyle -\frac{1}{11}\sum_{n=0}^{\infty} \left[\frac{2}{11}(x + 3)\right]^n$ **11.** $\displaystyle \frac{3}{2}\sum_{n=0}^{\infty} \left(-\frac{x}{2}\right)^n$
$\left(-\dfrac{17}{2}, \dfrac{5}{2}\right)$ $(-2, 2)$

13. $\displaystyle\sum_{n=0}^{\infty} \left[\left(-\frac{1}{2}\right)^n - 1\right]x^n$ **15.** $\displaystyle\sum_{n=0}^{\infty} x^n[1 + (-1)^n] = 2\sum_{n=0}^{\infty} x^{2n}$
$(-1, 1)$ $(-1, 1)$

17. $\displaystyle 2\sum_{n=0}^{\infty} x^{2n}$ **19.** $\displaystyle\sum_{n=1}^{\infty} n(-1)^n x^{n-1}$ **21.** $\displaystyle\sum_{n=0}^{\infty} \frac{(-1)^n x^{n+1}}{n + 1}$
$(-1, 1)$ $(-1, 1)$ $(-1, 1]$

23. $\displaystyle\sum_{n=0}^{\infty} (-1)^n x^{2n}$ **25.** $\displaystyle\sum_{n=0}^{\infty} (-1)^n (2x)^{2n}$
$(-1, 1)$ $\left(-\frac{1}{2}, \frac{1}{2}\right)$

27.

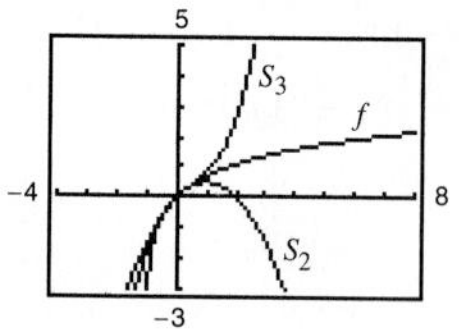

x	0.0	0.2	0.4	0.6	0.8	1.0
S_2	0.000	0.180	0.320	0.420	0.480	0.500
$\ln(x + 1)$	0.000	0.182	0.336	0.470	0.588	0.693
S_3	0.000	0.183	0.341	0.492	0.651	0.833

29. (a)

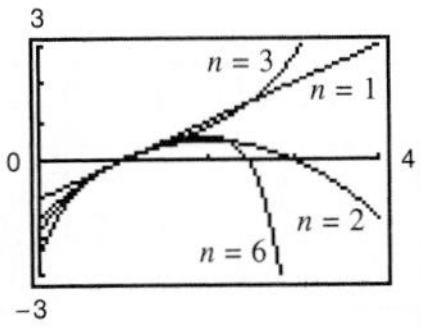

(b) $R = 1$
(c) -0.6931
(d) $\ln(0.5)$

31. c **32.** d **33.** a **34.** b **35.** 0.245 **37.** 0.125

39. $\sum_{n=1}^{\infty} nx^{n-1},\ -1 < x < 1$ **41.** $\sum_{n=0}^{\infty} (2n+1)x^n,\ -1 < x < 1$

43. $E(n) = 2$. Because the probability of obtaining a head on a single toss is $\frac{1}{2}$, it is expected that, on average, a head will be obtained in two tosses.

45. Since $\dfrac{1}{1+x} = \dfrac{1}{1-(-x)}$, substitute $(-x)$ into the geometric series.

47. Since $\dfrac{5}{1+x} = 5\left(\dfrac{1}{1-(-x)}\right)$, substitute $(-x)$ into the geometric series and then multiply the series by 5.

49. Proof **51.** (a) Proof (b) 3.14

53. $\ln\frac{3}{2} \approx 0.4055$; See Exercise 21.

55. $\ln\frac{7}{5} \approx 0.3365$; See Exercise 53.

57. $\arctan\frac{1}{2} \approx 0.4636$; See Exercise 56.

59. $f(x) = \arctan x$ is an odd function (symmetric to the origin).

61. The series in Exercise 56 converges to its sum at a lower rate because its terms approach 0 at a much lower rate.

63. The series converges on the interval $(-5, 3)$ and perhaps also at one or both endpoints.

65. $\sqrt{3}\pi/6$

Section 9.10 (page 685)

1. $\sum_{n=0}^{\infty} \dfrac{(2x)^n}{n!}$ **3.** $\dfrac{\sqrt{2}}{2}\sum_{n=0}^{\infty} \dfrac{(-1)^{n(n+1)/2}}{n!}\left(x - \dfrac{\pi}{4}\right)^n$

5. $\sum_{n=0}^{\infty} \dfrac{(-1)^n(x-1)^{n+1}}{n+1}$ **7.** $\sum_{n=0}^{\infty} \dfrac{(-1)^n(2x)^{2n+1}}{(2n+1)!}$

9. $1 + x^2/2! + 5x^4/4! + \cdots$ **11–13.** Proofs

15. $\sum_{n=0}^{\infty} (-1)^n(n+1)x^n$

17. $\dfrac{1}{2}\left[1 + \sum_{n=1}^{\infty} \dfrac{(-1)^n 1 \cdot 3 \cdot 5 \cdots (2n-1)x^{2n}}{2^{3n}n!}\right]$

19. $1 + \dfrac{x^2}{2} + \sum_{n=2}^{\infty} \dfrac{(-1)^{n+1} 1 \cdot 3 \cdot 5 \cdots (2n-3)x^{2n}}{2^n n!}$

21. $\sum_{n=0}^{\infty} \dfrac{x^{2n}}{2^n n!}$ **23.** $\sum_{n=0}^{\infty} \dfrac{(-1)^n(3x)^{2n+1}}{(2n+1)!}$

25. $\sum_{n=0}^{\infty} \dfrac{(-1)^n x^{3n}}{(2n)!}$ **27.** $\sum_{n=0}^{\infty} \dfrac{x^{2n+1}}{(2n+1)!}$

29. $\dfrac{1}{2}\left[1 + \sum_{n=0}^{\infty} \dfrac{(-1)^n(2x)^{2n}}{(2n)!}\right]$ **31.** $\sum_{n=0}^{\infty} \dfrac{(-1)^n x^{2n+2}}{(2n+1)!}$

33. $\begin{cases} \sum_{n=0}^{\infty} \dfrac{(-1)^n x^{2n}}{(2n+1)!}, & x \neq 0 \\ 1, & x = 0 \end{cases}$ **35.** Proof

37. $P_5(x) = x + x^2 + \frac{1}{3}x^3 - \frac{1}{30}x^5 + \cdots$

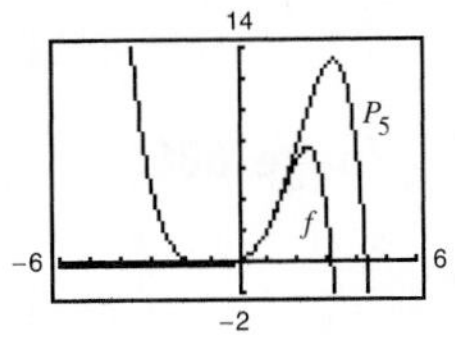

39. $P_5(x) = x - \frac{1}{2}x^2 - \frac{1}{6}x^3 + \frac{3}{40}x^5 + \cdots$

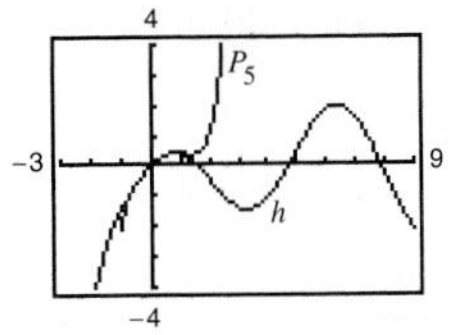

41. $P_4(x) = x - x^2 + \frac{5}{6}x^3 - \frac{5}{6}x^4 + \cdots$

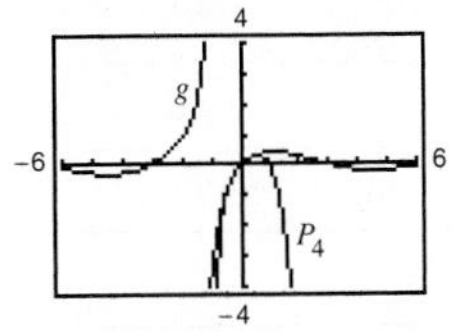

43. c; $f(x) = x \sin x$ **44.** d; $f(x) = x \cos x$ **45.** a; $f(x) = xe^x$

46. b; $f(x) = x^2\left(\dfrac{1}{1+x}\right)$ **47.** $\sum_{n=0}^{\infty} \dfrac{(-1)^{(n+1)}x^{2n+3}}{(2n+3)(n+1)!}$

49. 0.6931 **51.** 7.3891 **53.** 0 **55.** 0.9461

57. 0.2010 **59.** 0.7040 **61.** 0.3413

63. $P_5(x) = x - 2x^3 + \frac{2}{3}x^5$

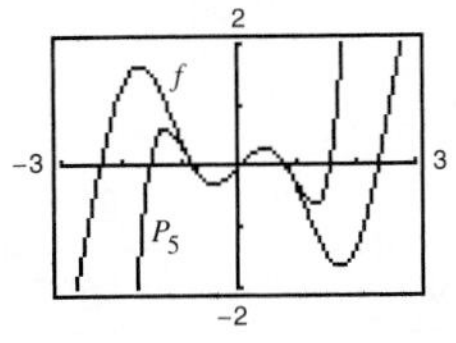

$\left[-\frac{3}{4}, \frac{3}{4}\right]$

65. $P_5(x) = (x-1) - \frac{1}{24}(x-1)^3 + \frac{1}{24}(x-1)^4 - \frac{71}{1920}(x-1)^5$

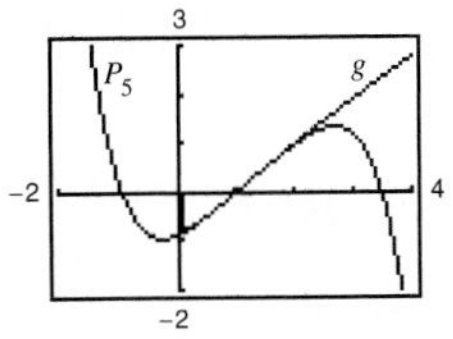

$\left[\frac{1}{4}, 2\right]$

67. See "Guidelines for Finding a Taylor Series" on page 680.

69. (a) Replace x with $-x$ in the series for e^x.
(b) Replace x with $3x$ in the series for e^x.
(c) Multiply the series for e^x by x.
(d) Replace x with $2x$ in the series for e^x. Then replace x with $-2x$ in the series for e^x. Then add the two together.

71. Proof

73. (a)

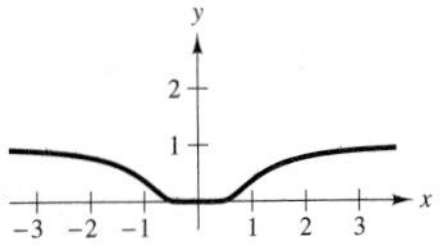

(b) Proof
(c) $\sum_{n=0}^{\infty} 0x^n = 0 \neq f(x)$

75. Proof **77.** 10 **79.** -0.0390625

81. $\sum_{n=0}^{\infty} \binom{k}{n} x^n$ **83.** Proof

Review Exercises for Chapter 9 (page 688)

1. $a_n = 1/n!$ **3.** a **4.** c **5.** d **6.** b

7.

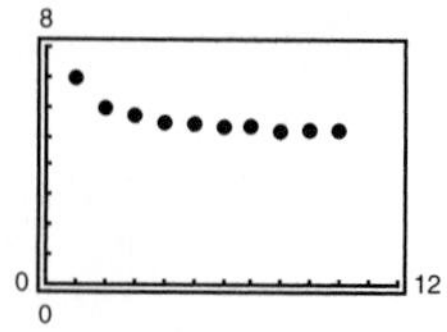

Converges to 5

9. Converges to 0 **11.** Diverges

13. Converges to 0 **15.** Converges to 0

17. (a)

n	1	2	3	4
A_n	\$5062.50	\$5125.78	\$5189.85	\$5254.73

n	5	6	7	8
A_n	\$5320.41	\$5386.92	\$5454.25	\$5522.43

(b) \$8218.10

19. (a)

k	5	10	15	20	25
S_k	13.2	113.3	873.8	6648.5	50,500.3

(b)

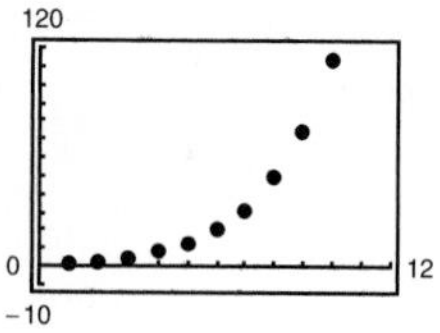

21. (a)

k	5	10	15	20	25
S_k	0.4597	0.4597	0.4597	0.4597	0.4597

(b)

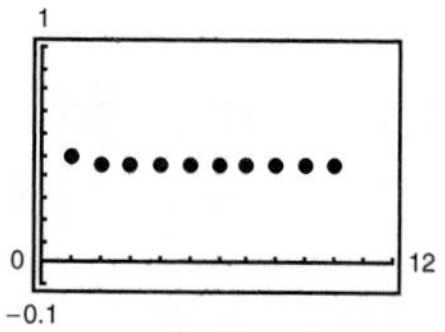

23. Converges **25.** Diverges **27.** 3 **29.** $\frac{1}{2}$

31. (a) $\sum_{n=0}^{\infty} (0.09)(0.01)^n$ (b) $\frac{1}{11}$ **33.** $45\frac{1}{3}$ m

35. \$5087.14 **37.** Converges **39.** Diverges **41.** Converges

43. Diverges **45.** Converges **47.** Diverges **49.** Converges

51. Diverges

53. (a) Proof

(b)

n	5	10	15	20	25
S_n	2.8752	3.6366	3.7377	3.7488	3.7499

(c)

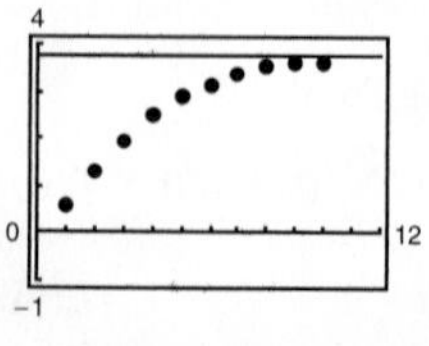

(d) 3.75

55. (a)

N	5	10	20	30	40
$\sum_{n=1}^{N} \frac{1}{n^p}$	1.4636	1.5498	1.5962	1.6122	1.6202
$\int_N^{\infty} \frac{1}{x^p}\,dx$	0.2000	0.1000	0.0500	0.0333	0.0250

(b)

N	5	10	20	30	40
$\sum_{n=1}^{N} \frac{1}{n^p}$	1.0367	1.0369	1.0369	1.0369	1.0369
$\int_N^{\infty} \frac{1}{x^p}\,dx$	0.0004	0.0000	0.0000	0.0000	0.0000

The series in part (b) converges more rapidly. This is evident from the integrals that give the remainders of the partial sums.

57. $P_3(x) = 1 - \frac{x}{2} + \frac{x^2}{8} - \frac{x^3}{48}$ **59.** 0.996 **61.** 0.560

63. (a) 4 (b) 6 (c) 5 (d) 10

65. $(-10, 10)$ **67.** $[1, 3]$ **69.** Converges only at $x = 2$

71. Proof **73.** $\sum_{n=0}^{\infty} \frac{2}{3}\left(\frac{x}{3}\right)^n$ **75.** $\sum_{n=0}^{\infty} \frac{2}{9}(n+1)\left(\frac{x}{3}\right)^n$

77. $f(x) = \frac{3}{3-2x}, \left(-\frac{3}{2}, \frac{3}{2}\right)$

79. $\frac{\sqrt{2}}{2} \sum_{n=0}^{\infty} \frac{(-1)^{n(n+1)/2}}{n!}\left(x - \frac{3\pi}{4}\right)^n$

81. $\sum_{n=0}^{\infty} \frac{(x \ln 3)^n}{n!}$ **83.** $-\sum_{n=0}^{\infty} (x+1)^n$

85. $1 + x/5 - 2x^2/25 + 6x^3/125 - 21x^4/625 + \cdots$

87. $\ln \frac{5}{4} \approx 0.2231$ **89.** $e^{1/2} \approx 1.6487$ **91.** $\cos \frac{2}{3} \approx 0.7859$

93. The series for Exercise 41 converges to its sum at a lower rate because its terms approach 0 at a lower rate.

95. $1 + 2x + 2x^2 + \frac{4}{3}x^3$ **97.** $\sum_{n=0}^{\infty} \frac{(-1)^n x^{2n+1}}{(2n+1)(2n+1)!}$

99. $\sum_{n=0}^{\infty} \frac{(-1)^n x^{n+1}}{(n+1)^2}$ **101.** 0

P.S. Problem Solving (page 691)

1. (a) 1 (b) Answers will vary. Example: $0, \frac{1}{3}, \frac{2}{3}$ (c) 0

3. $\pi/8$

5. (a) $R = 1$; Sum $= (3x^2 + 2x + 1)/(1 - x^3)$

(b) $R = 1$; Sum $= \frac{a_{p-1}x^{p-1} + a_{p-2}x^{p-2} + \cdots + a_1 x + a_0}{1 - x^p}$

7. (a) $\sum_{n=0}^{\infty} \frac{x^{n+1}}{n!}$; $\sum_{n=1}^{\infty} \frac{1}{n!(n+2)} = \frac{1}{2}$

(b) $\sum_{n=0}^{\infty} \frac{(n+1)x^n}{n!}$; $\sum_{n=0}^{\infty} \frac{n+1}{n!} = 2e \approx 5.4366$

9. Let $a_1 = \int_0^{\pi} \frac{\sin x}{x}\,dx = 1.8519$

$a_2 = \int_{\pi}^{2\pi} \frac{\sin x}{x}\,dx = -0.4338$

$a_3 = \int_{2\pi}^{3\pi} \frac{\sin x}{x}\,dx = 0.2566$

$a_4 = \int_{3\pi}^{4\pi} \frac{\sin x}{x}\,dx = -0.1826.$

It follows that the total area is

$\int_0^{\infty} \frac{\sin x}{x}\,dx = a_1 - a_2 + a_3 - a_4 + \cdots.$

Also, $\lim_{n\to\infty} a_n = 0$ and $0 < a_{n+1} \le a_n$. Therefore, it follows by the Alternating Series Test that $\int_0^{\infty} f(x)\,dx$ converges.

11. (a) $a_1 = 3, a_2 \approx 1.7321, a_3 \approx 2.1753, a_4 \approx 2.2749,$ $a_5 \approx 2.2967, a_6 \approx 2.3015$

Proof; $L = \left(1 + \sqrt{13}\right)/2$

(b) Proof; $L = \left(1 + \sqrt{1 + 4a}\right)/2$

13. (a) $1, \frac{9}{8}, \frac{11}{8}, \frac{45}{32}, \frac{47}{32}$ (b) Proof (c) Proof

15. $S_6 = 240; S_7 = 440; S_8 = 810; S_9 = 1490; S_{10} = 2740$

17. (a) Diverges (b) Converges

Chapter 10

Section 10.1 (page 704)

1. h **2.** a **3.** e **4.** b **5.** f **6.** g **7.** c **8.** d

9. Vertex: $(0, 0)$
Focus: $\left(-\frac{3}{2}, 0\right)$
Directrix: $x = \frac{3}{2}$

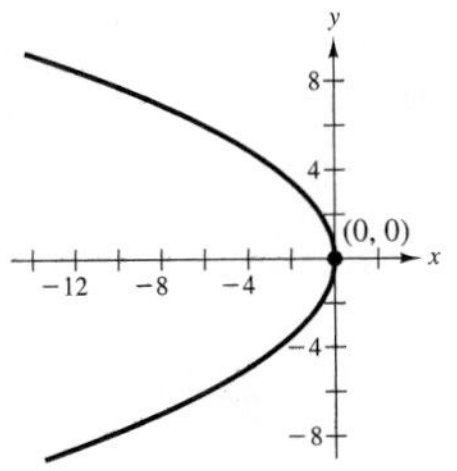

11. Vertex: $(-3, 2)$
Focus: $\left(-\frac{13}{4}, 2\right)$
Directrix: $x = -\frac{11}{4}$

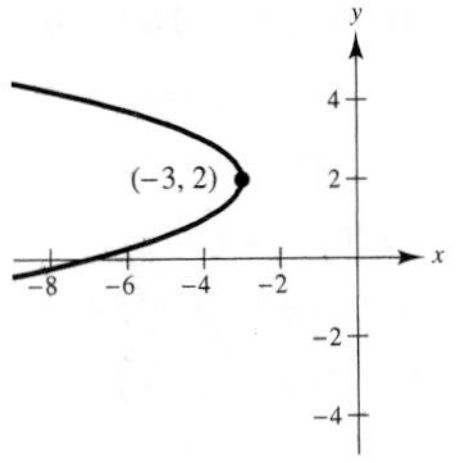

13. Vertex: $(-1, 2)$
Focus: $(0, 2)$
Directrix: $x = -2$

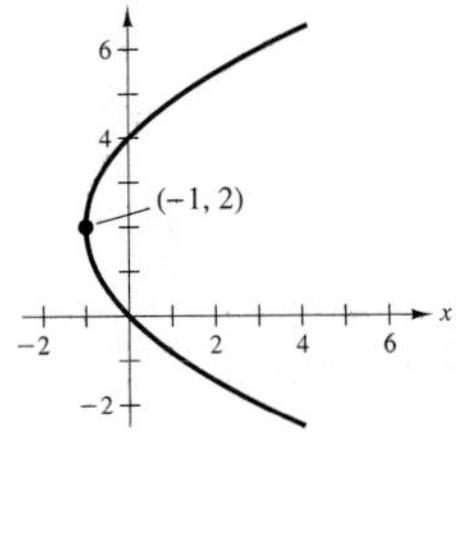

15. Vertex: $(-2, 2)$
Focus: $(-2, 1)$
Directrix: $y = 3$

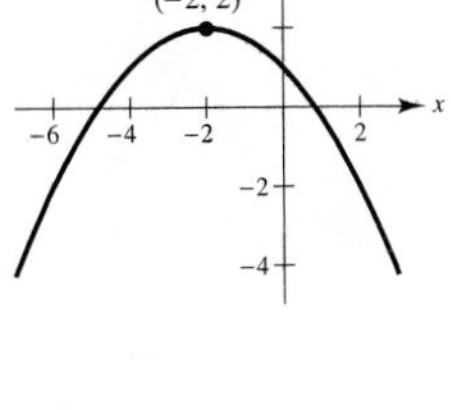

17. Vertex: $\left(\frac{1}{4}, -\frac{1}{2}\right)$
Focus: $\left(0, -\frac{1}{2}\right)$
Directrix: $x = \frac{1}{2}$

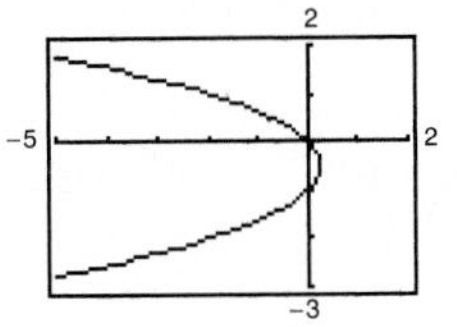

19. Vertex: $(-1, 0)$
Focus: $(0, 0)$
Directrix: $x = -2$

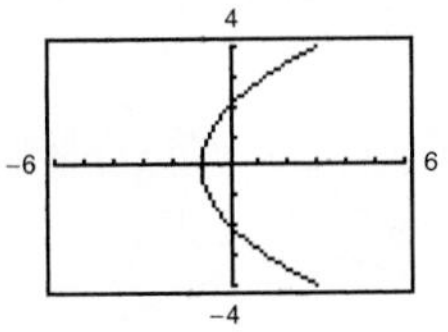

21. $y^2 - 4y + 8x - 20 = 0$ **23.** $x^2 - 24y + 96 = 0$

25. $x^2 + y - 4 = 0$ **27.** $5x^2 - 14x - 3y + 9 = 0$

29. Center: $(0, 0)$
Foci: $\left(\pm\sqrt{3}, 0\right)$
Vertices: $(\pm 2, 0)$
$e = \sqrt{3}/2$

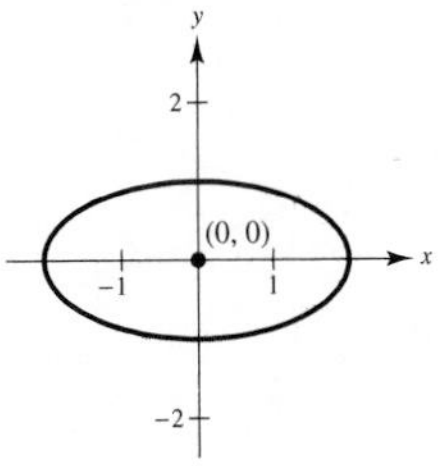

31. Center: $(1, 5)$
Foci: $(1, 9), (1, 1)$
Vertices: $(1, 10), (1, 0)$
$e = \frac{4}{5}$

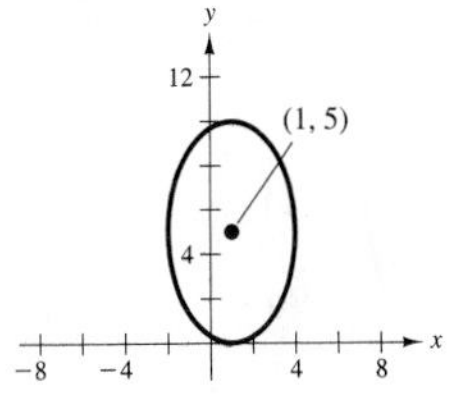

33. Center: $(-2, 3)$
Foci: $\left(-2, 3 \pm \sqrt{5}\right)$
Vertices: $(-2, 6), (-2, 0)$
$e = \sqrt{5}/3$

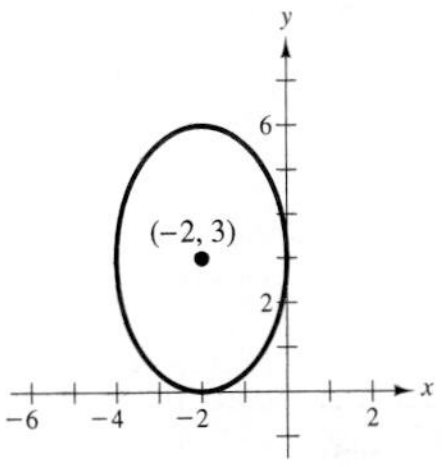

35. Center: $\left(\frac{1}{2}, -1\right)$
Foci: $\left(\frac{1}{2} \pm \sqrt{2}, -1\right)$
Vertices: $\left(\frac{1}{2} \pm \sqrt{5}, -1\right)$
To obtain the graph, solve for y and get

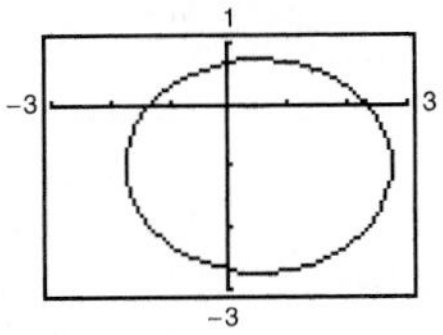

$y_1 = -1 + \sqrt{(57 + 12x - 12x^2)/20}$ and
$y_2 = -1 - \sqrt{(57 + 12x - 12x^2)/20}.$
Graph these equations in the same viewing window.

37. Center: $\left(\frac{3}{2}, -1\right)$
Foci: $\left(\frac{3}{2} - \sqrt{2}, -1\right), \left(\frac{3}{2} + \sqrt{2}, -1\right)$
Vertices: $\left(-\frac{1}{2}, -1\right), \left(\frac{7}{2}, -1\right)$
To obtain the graph, solve for y and get

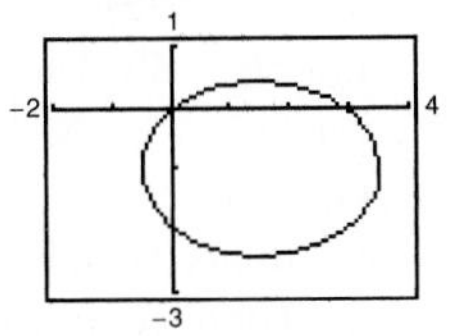

$y_1 = -1 + \sqrt{(7 + 12x - 4x^2)/8}$ and
$y_2 = -1 - \sqrt{(7 + 12x - 4x^2)/8}$
Graph these equations in the same viewing window.

39. $x^2/9 + y^2/5 = 1$ **41.** $(x - 3)^2/9 + (y - 5)^2/16 = 1$

43. $x^2/16 + 7y^2/16 = 1$

45. Center: $(0, 0)$
Foci: $\left(0, \pm\sqrt{5}\right)$
Vertices: $(0, \pm 1)$

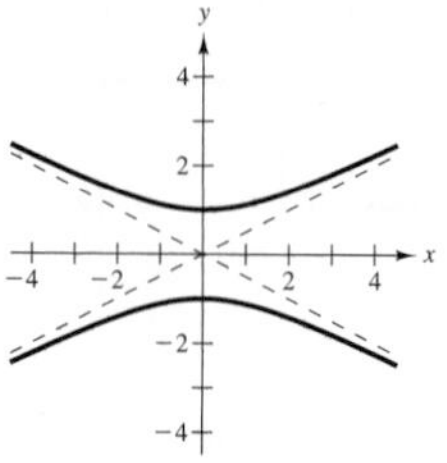

47. Center: $(1, -2)$
Foci: $\left(1 \pm \sqrt{5}, -2\right)$
Vertices: $(-1, -2), (3, -2)$

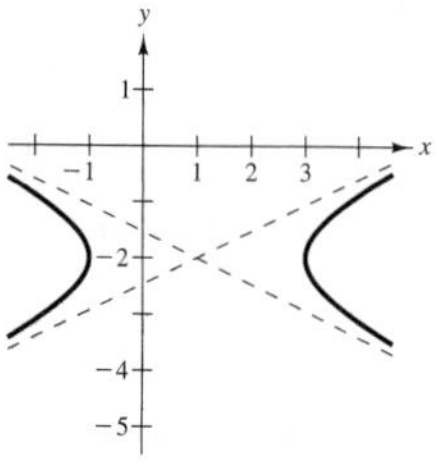

49. Center: $(2, -3)$
Foci: $\left(2 \pm \sqrt{10}, -3\right)$
Vertices: $(1, -3), (3, -3)$

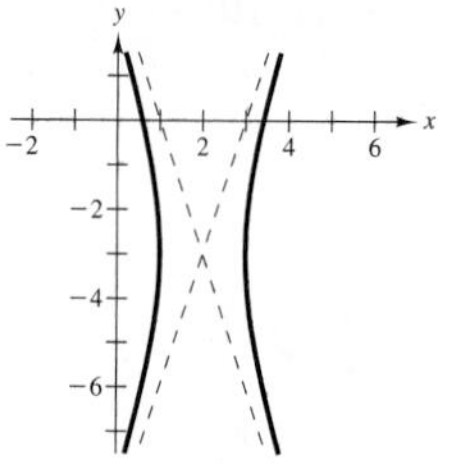

51. Degenerate hyperbola
Graph is two lines
$y = -3 \pm \frac{1}{3}(x + 1)$
intersecting at $(-1, -3)$.

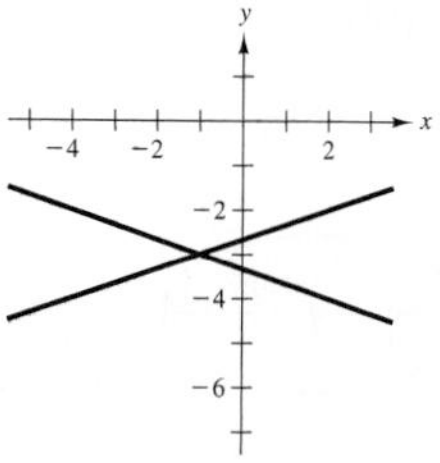

53. Center: $(1, -3)$
Foci: $\left(1, -3 \pm 2\sqrt{5}\right)$
Vertices: $\left(1, -3 \pm \sqrt{2}\right)$
Asymptotes:
$y = \frac{1}{3}x - \frac{1}{3} - 3$;
$y = -\frac{1}{3}x + \frac{1}{3} - 3$

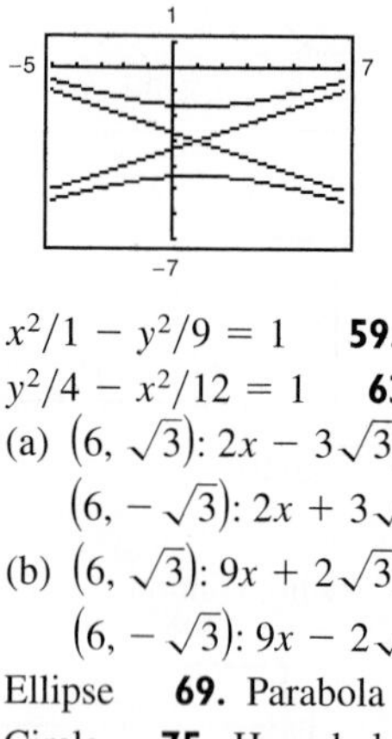

55. Center: $(1, -3)$
Foci: $\left(1 \pm \sqrt{10}, -3\right)$
Vertices: $(-1, -3), (3, -3)$
Asymptotes:
$y = \sqrt{6}x/2 - \sqrt{6}/2 - 3$;
$y = -\sqrt{6}x/2 + \sqrt{6}/2 - 3$

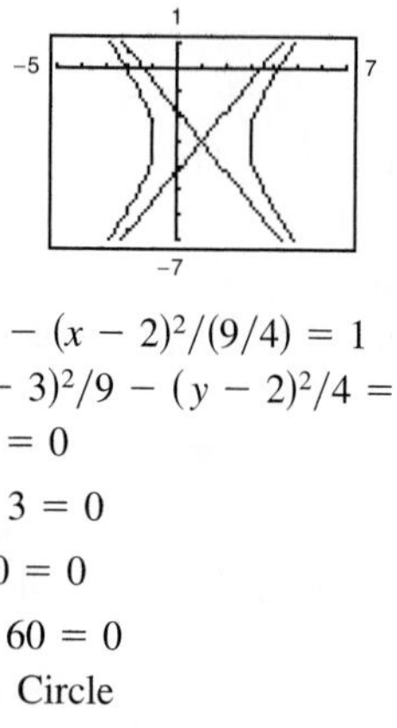

57. $x^2/1 - y^2/9 = 1$ **59.** $y^2/9 - (x - 2)^2/(9/4) = 1$
61. $y^2/4 - x^2/12 = 1$ **63.** $(x - 3)^2/9 - (y - 2)^2/4 = 1$
65. (a) $\left(6, \sqrt{3}\right)$: $2x - 3\sqrt{3}y - 3 = 0$
$\left(6, -\sqrt{3}\right)$: $2x + 3\sqrt{3}y - 3 = 0$
(b) $\left(6, \sqrt{3}\right)$: $9x + 2\sqrt{3}y - 60 = 0$
$\left(6, -\sqrt{3}\right)$: $9x - 2\sqrt{3}y - 60 = 0$
67. Ellipse **69.** Parabola **71.** Circle
73. Circle **75.** Hyperbola
77. (a) A parabola is the set of all points (x, y) that are equidistant from a fixed line and a fixed point not on the line.
(b) For directrix $y = k - p$: $(x - h)^2 = 4p(y - k)$
For directrix $x = h - p$: $(y - k)^2 = 4p(x - h)$
(c) If P is a point on a parabola, then the tangent line to the parabola at P makes equal angles with the line passing through P and the focus, and with the line passing through P parallel to the axis of the parabola.
79. (a) A hyperbola is the set of all points (x, y) for which the absolute value of the difference between the distances from two distinct fixed points is constant.
(b) Transverse axis is horizontal: $\dfrac{(x - h)^2}{a^2} - \dfrac{(y - k)^2}{b^2} = 1$
Transverse axis is vertical: $\dfrac{(y - k)^2}{a^2} - \dfrac{(x - h)^2}{b^2} = 1$
(c) Transverse axis is horizontal:
$y = k + (b/a)(x - h)$ and $y = k - (b/a)(x - h)$
Transverse axis is vertical:
$y = k + (a/b)(x - h)$ and $y = k - (a/b)(x - h)$
81. $\frac{9}{4}$ m **83.** $y = 2ax_0x - ax_0^2$ **85.** (a) Proof (b) Proof
87. $x_0 = 2\sqrt{3}/3$; Distance from hill: $2\sqrt{3}/3 - 1$
89. $[16(4 + 3\sqrt{3} - 2\pi)]/3 \approx 15.536$ ft²
91. (a) $y = (1/180)x^2$
(b) $10\left[2\sqrt{13} + 9\ln\left(\dfrac{2 + \sqrt{13}}{3}\right)\right] \approx 128.4$ m
93.

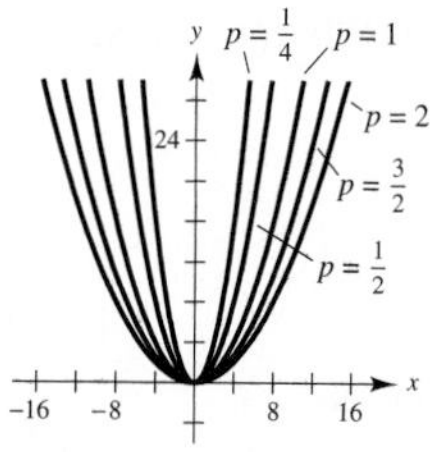

As p increases, the graph of $x^2 = 4py$ gets wider.

95. (a) $L = 2a$
(b) The thumbtacks are located at the foci and the length of string is the constant sum of distances from the foci.
97.

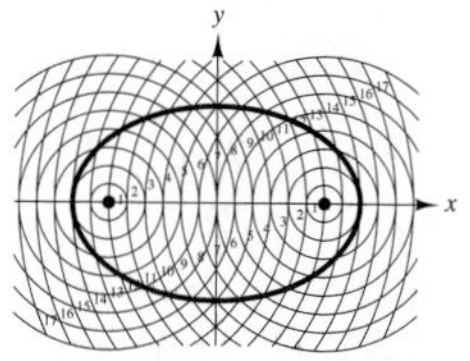

99. Proof
101. $e \approx 0.9672$ **103.** $\left(0, \frac{25}{3}\right)$
105. Minor-axis endpoints: $(-6, -2), (0, -2)$
Major-axis endpoints: $(-3, -6), (-3, 2)$
107. (a) Area $= 2\pi$
(b) Volume $= 8\pi/3$
Surface area $= [2\pi(9 + 4\sqrt{3}\pi)]/9 \approx 21.48$
(c) Volume $= 16\pi/3$
Surface area $= \dfrac{4\pi\left[6 + \sqrt{3}\ln\left(2 + \sqrt{3}\right)\right]}{3} \approx 34.69$
109. 37.96 **111.** 40 **113.** $(x - 6)^2/9 - (y - 2)^2/7 = 1$
115.

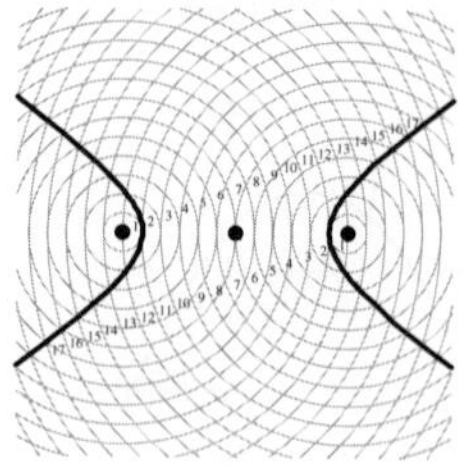

117. Proof
119. $x = \left(-90 + 96\sqrt{2}\right)/7 \approx 6.538$
$y = \left(160 - 96\sqrt{2}\right)/7 \approx 3.462$

121. There are four points of intersection.
At $\left(\dfrac{\sqrt{2}\,ac}{\sqrt{2a^2-b^2}}, \dfrac{b^2}{\sqrt{2}\sqrt{2a^2-b^2}}\right)$, the slopes of the tangent lines are $y'_e = -c/a$ and $y'_h = a/c$.
Since the slopes are negative reciprocals, the tangent lines are perpendicular. Similarly, the curves are perpendicular at the other three points of intersection.

123. False. See the definition of a parabola. **125.** True

127. True **129.** Putnam Problem B4, 1976

Section 10.2 (page 716)

1. (a)

t	0	1	2	3	4
x	0	1	$\sqrt{2}$	$\sqrt{3}$	2
y	1	0	-1	-2	-3

(b) and (c)

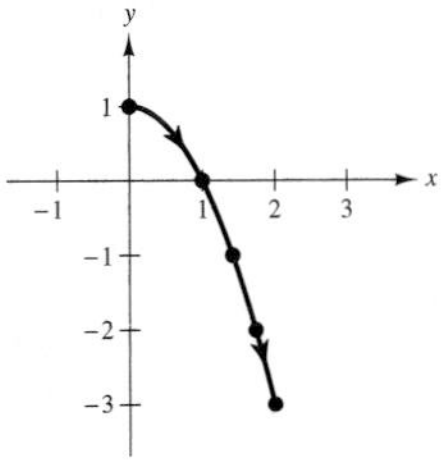

(d) $y = 1 - x^2, \quad x \geq 0$

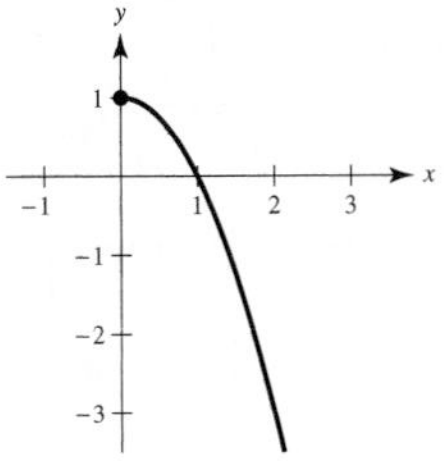

3.

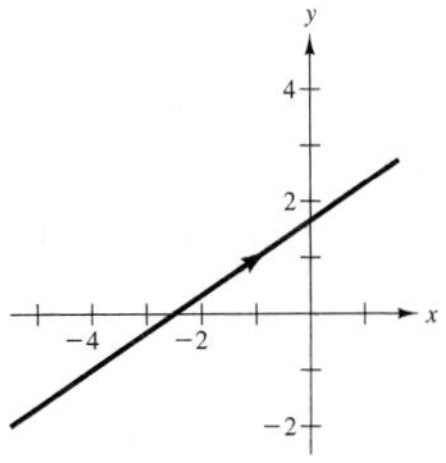

$2x - 3y + 5 = 0$

5.

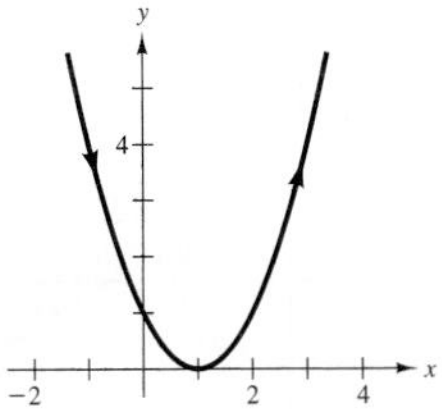

$y = (x - 1)^2$

7.

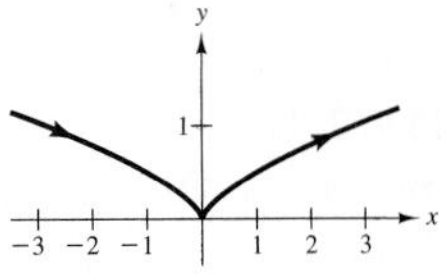

$y = \frac{1}{2}x^{2/3}$

9.

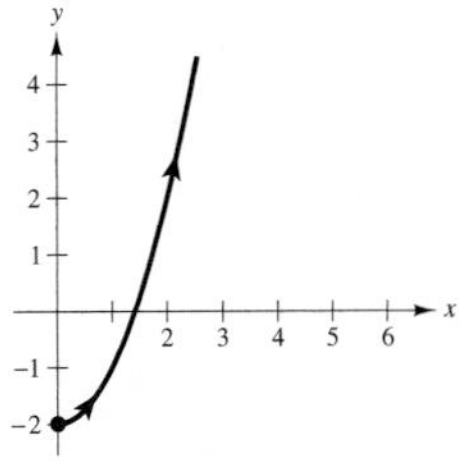

$y = x^2 - 2, \quad x \geq 0$

11.

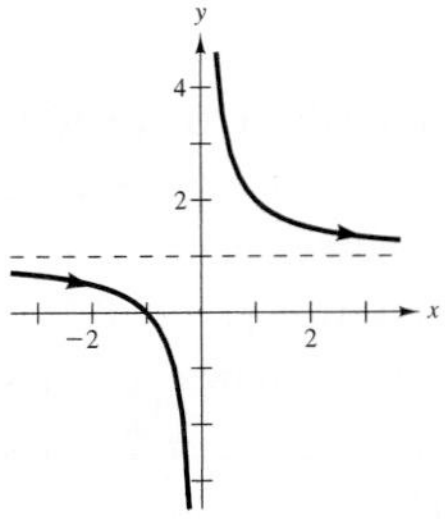

$y = (x + 1)/x$

13.

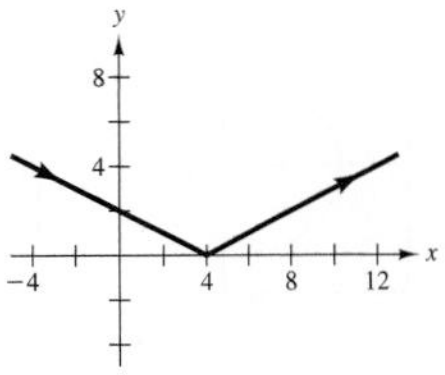

$y = |x - 4|/2$

15.

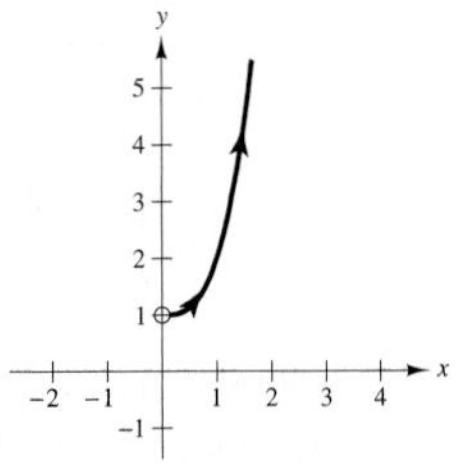

$y = x^3 + 1, \quad x > 0$

17.

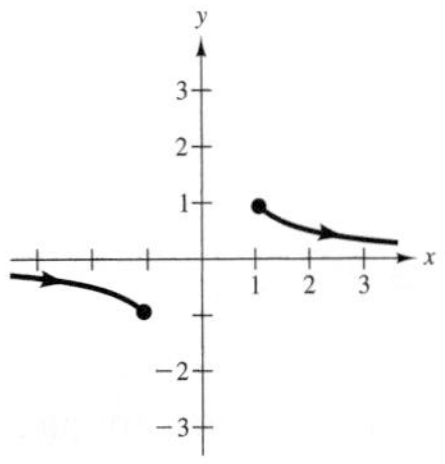

$y = 1/x, \quad |x| \geq 1$

19.

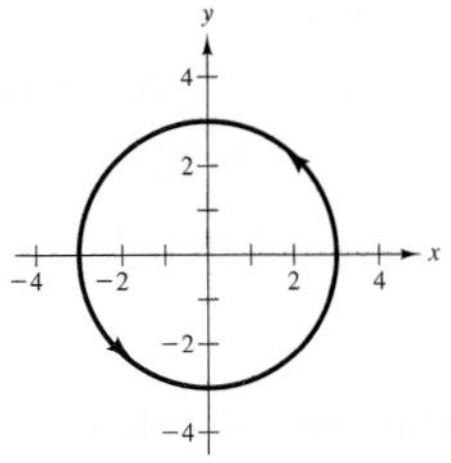

$x^2 + y^2 = 9$

21.

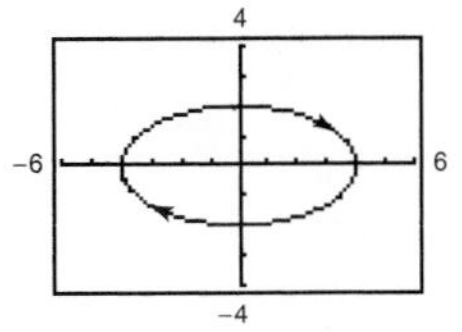

$\dfrac{x^2}{16} + \dfrac{y^2}{4} = 1$

23.

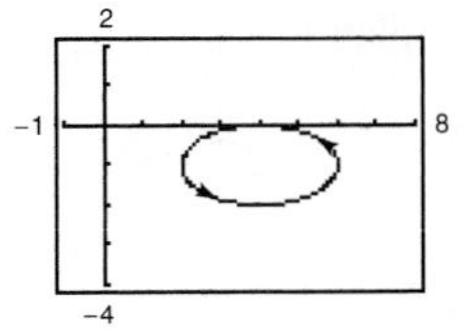

$\dfrac{(x-4)^2}{4} + \dfrac{(y+1)^2}{1} = 1$

25.

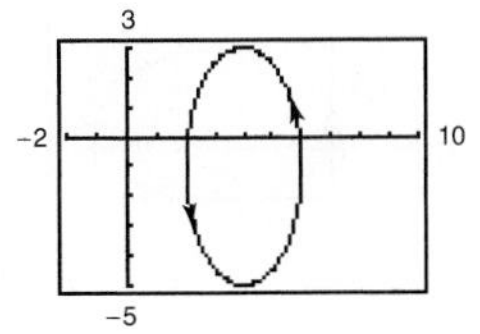

$\dfrac{(x-4)^2}{4} + \dfrac{(y+1)^2}{16} = 1$

27.

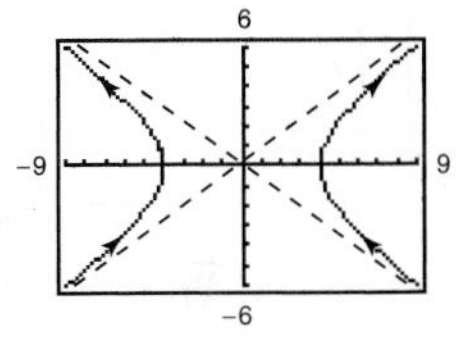

$\dfrac{x^2}{16} - \dfrac{y^2}{9} = 1$

29.

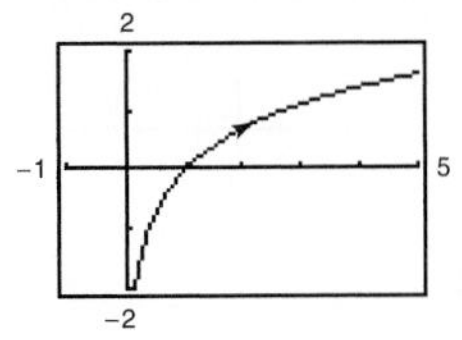

$y = \ln x$

31.

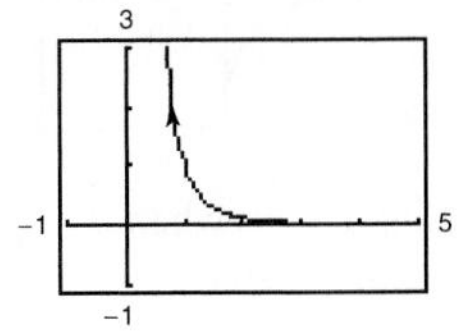

$y = 1/x^3, \quad x > 0$

33. Each curve represents a portion of the line $y = 2x + 1$.

	Domain	*Orientation*	*Smooth*
(a)	$-\infty < x < \infty$	Up	Yes
(b)	$-1 \leq x \leq 1$	Oscillates	No, $\dfrac{dx}{d\theta} = \dfrac{dy}{d\theta} = 0$ when $\theta = 0, \pm\pi, \pm 2\pi, \ldots$
(c)	$0 < x < \infty$	Down	Yes
(d)	$0 < x < \infty$	Up	Yes

35. (a) and (b) represent the parabola $y = 2(1 - x^2)$ for $-1 \leq x \leq 1$. The curve is smooth. The orientation is from right to left in part (a) and in part (b).

37. (a)

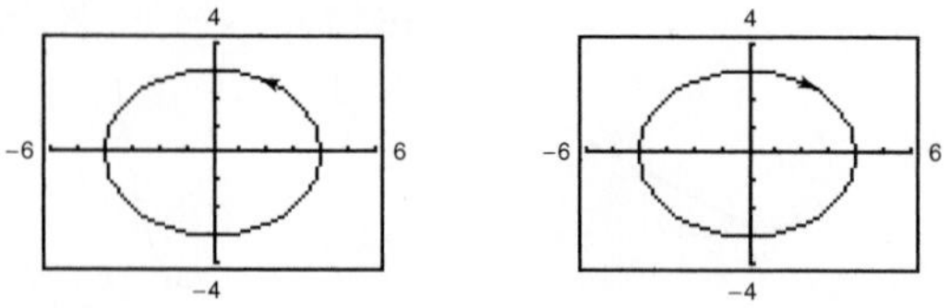

(b) The orientation is reversed. (c) The orientation is reversed.

(d) Answers will vary. For example,

$x = 2 \sec t$ $\quad x = 2 \sec(-t)$

$y = 5 \sin t$ $\quad y = 5 \sin(-t)$

have the same graphs, but their orientations are reversed.

39. $y - y_1 = \dfrac{y_2 - y_1}{x_2 - x_1}(x - x_1)$ **41.** $\dfrac{(x-h)^2}{a^2} + \dfrac{(y-k)^2}{b^2} = 1$

43. $x = 5t$
$y = -2t$
(Solution is not unique.)

45. $x = 2 + 4 \cos \theta$
$y = 1 + 4 \sin \theta$
(Solution is not unique.)

47. $x = 5 \cos \theta$
$y = 3 \sin \theta$
(Solution is not unique.)

49. $x = 4 \sec \theta$
$y = 3 \tan \theta$
(Solution is not unique.)

51. $x = t$
$y = 3t - 2;$
$x = t - 3$
$y = 3t - 11$
(Solution is not unique.)

53. $x = t$
$y = t^3;$
$x = \tan t$
$y = \tan^3 t$
(Solution is not unique.)

55.

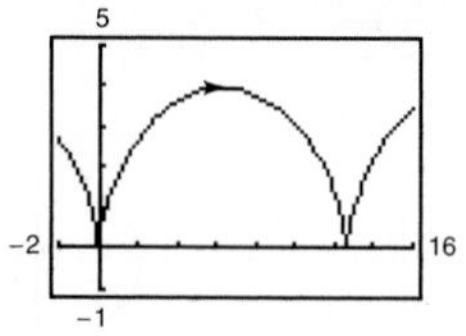

Not smooth when $\theta = 2n\pi$

57.

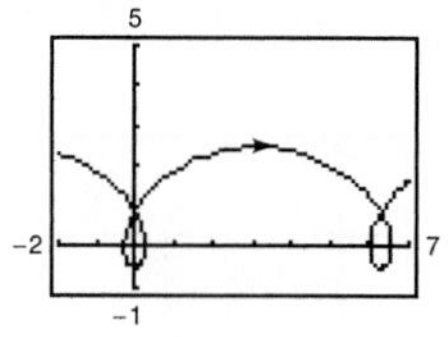

59.

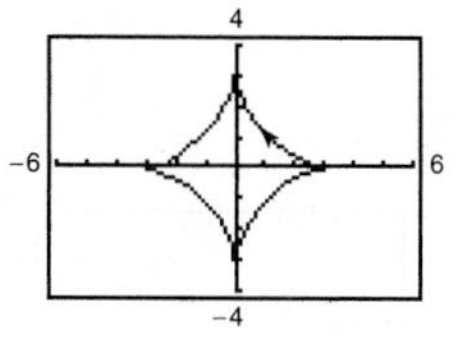

61.

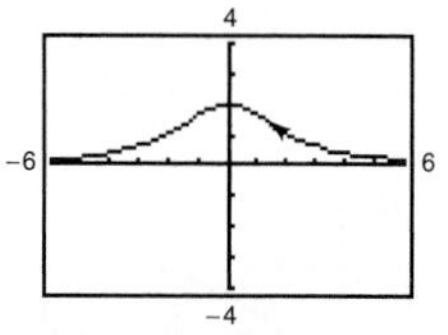

Not smooth when $\theta = \frac{1}{2}n\pi$

63. See page 709. **65.** See page 714.

67. $x = a\theta - b \sin \theta;\ y = a - b \cos \theta$

69. False. The graph of the parametric equations is the portion of the line $y = x$ when $x \geq 0$.

71. (a) $x = \left(\frac{440}{3} \cos \theta\right)t;\ y = 3 + \left(\frac{440}{3} \sin \theta\right)t - 16t^2$

(b)

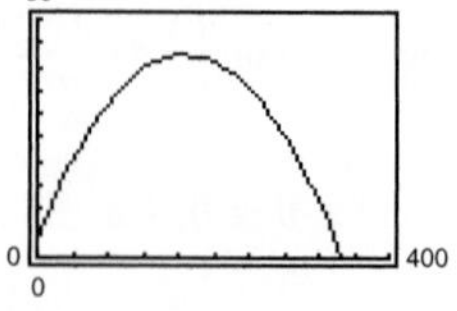

Not a home run

(c)

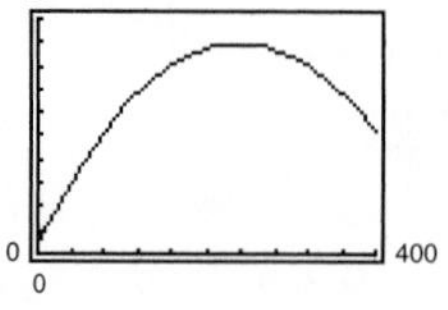

Home run

(d) 19.4°

Section 10.3 (page 725)

1. $-2/t$ **3.** -1

5. $\dfrac{dy}{dx} = \dfrac{3}{2}, \dfrac{d^2y}{dx^2} = 0$; Neither concave upward nor concave downward

7. $dy/dx = 2t + 3,\ d^2y/dx^2 = 2$
At $t = -1$, $dy/dx = 1$, $d^2y/dx^2 = 2$; Concave upward

9. $dy/dx = -\cot \theta,\ d^2y/dx^2 = -\csc^3 \theta/2$
At $\theta = \pi/4$, $dy/dx = -1$, $d^2y/dx^2 = -\sqrt{2}$;
Concave downward

11. $dy/dx = 2 \csc \theta,\ d^2y/dx^2 = -2 \cot^3 \theta$
At $\theta = \pi/6$, $dy/dx = 4$, $d^2y/dx^2 = -6\sqrt{3}$;
Concave downward

13. $dy/dx = -\tan \theta,\ d^2y/dx^2 = \sec^4 \theta \csc \theta/3$
At $\theta = \pi/4$, $dy/dx = -1$, $d^2y/dx^2 = 4\sqrt{2}/3$;
Concave upward

15. $\left(-2/\sqrt{3}, 3/2\right)$: $3\sqrt{3}x - 8y + 18 = 0$
$(0, 2)$: $y - 2 = 0$
$\left(2\sqrt{3}, 1/2\right)$: $\sqrt{3}x + 8y - 10 = 0$

17. (a) and (d)

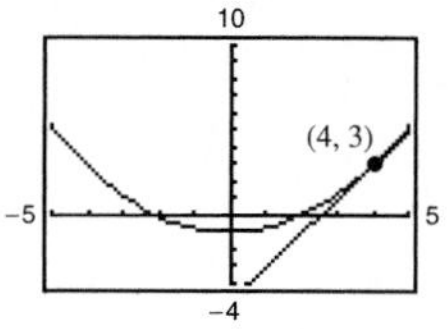

(b) At $t = 2$, $dx/dt = 2$, $dy/dt = 4$, and $dy/dx = 2$.
(c) $y = 2x - 5$

19. (a) and (d)

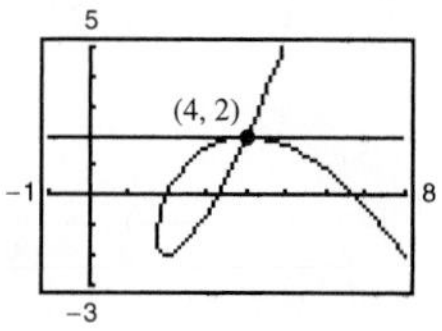

(b) At $t = -1$, $dx/dt = -3$, $dy/dt = 0$, and $dy/dx = 0$.
(c) $y = 2$

21. $y = \pm\frac{3}{4}x$ **23.** $y = 3x - 5$ and $y = 1$

25. Horizontal: $(1, 0), (-1, \pi), (1, -2\pi)$
Vertical: $(\pi/2, 1), (-3\pi/2, -1), (5\pi/2, 1)$

27. Horizontal: $(1, 0)$
Vertical: None

29. Horizontal: $(0, -2), (2, 2)$
Vertical: None

31. Horizontal: $(0, 3), (0, -3)$
Vertical: $(3, 0), (-3, 0)$

33. Horizontal: $(4, 0), (4, -2)$
Vertical: $(2, -1), (6, -1)$

35. Horizontal: None
Vertical: $(1, 0), (-1, 0)$

37. Concave down: $-\infty < t < 0$
Concave up: $0 < t < \infty$

39. Concave up: $t > 0$

41. Concave down: $0 < t < \pi/2$
Concave up: $\pi/2 < t < \pi$

43. $\displaystyle\int_1^2 \sqrt{4t^2 + t + 4}\, dt$ **45.** $\displaystyle\int_{-2}^2 \sqrt{e^{2t} + 4}\, dt$

47. $2\sqrt{5} + \ln\left(2 + \sqrt{5}\right) \approx 5.916$ **49.** $\sqrt{2}(1 - e^{-\pi/2}) \approx 1.12$

51. $\frac{1}{12}\left[\ln\left(\sqrt{37} + 6\right) + 6\sqrt{37}\right] \approx 3.249$ **53.** $6a$ **55.** $8a$

57. (a)
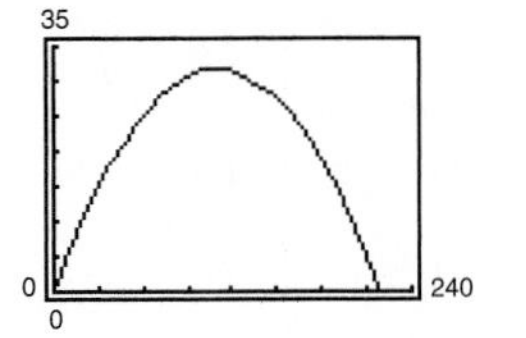

(b) 219.2 ft
(c) 230.8 ft

59. (a)
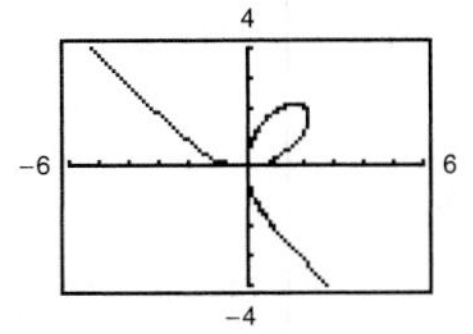

(b) $(0, 0), \left(4\sqrt[3]{2}/3, 4\sqrt[3]{2}/3\right)$
(c) ≈ 6.557

61. (a)
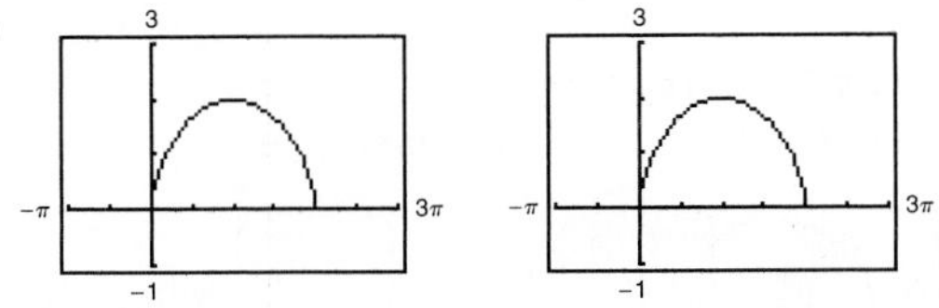

(b) The average speed of the particle on the second path is twice the average speed of the particle on the first path.
(c) 4π

63. $S = 2\pi \int_0^2 \sqrt{17}(t + 1)\, dt = 8\pi\sqrt{17} \approx 103.625$

65. $S = 2\pi \int_0^{\pi/2} \left(\sin\theta \cos\theta \sqrt{4\cos^2\theta + 1}\right) d\theta = \dfrac{(5\sqrt{5} - 1)\pi}{6}$
≈ 5.330

67. (a) $32\pi\sqrt{5}$ (b) $16\pi\sqrt{5}$ **69.** 32π **71.** $12\pi a^2/5$

73. See Theorem 10.7, Parametric Form of the Derivative, on page 719.

75. Answers will vary. Example:
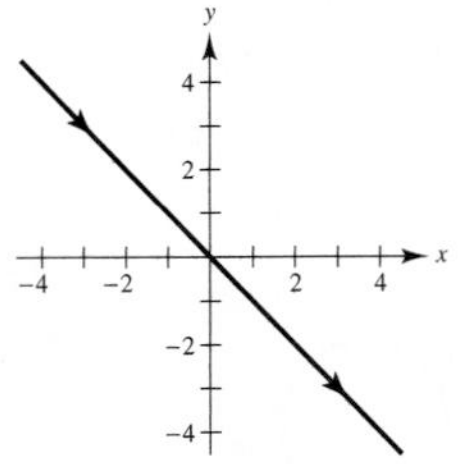

77. See Theorem 10.8, Arc Length in Parametric Form, on page 722.

79. Proof **81.** $3\pi/2$ **83.** d **84.** b **85.** f **86.** c

87. a **88.** e **89.** $\left(\frac{3}{4}, \frac{8}{5}\right)$ **91.** $V = 36\pi$

93. (a) $dy/dx = \sin\theta/(1 - \cos\theta)$; $d^2y/dx^2 = -1/[a(\cos\theta - 1)^2]$
(b) $y = \left(2 + \sqrt{3}\right)\left[x - a\left(\pi/6 - \frac{1}{2}\right)\right] + a\left(1 - \sqrt{3}/2\right)$
(c) $(a(2n + 1)\pi, 2a)$
(d) Concave down on $(0, 2\pi)$, $(2\pi, 4\pi)$, etc.
(e) $s = 8a$

95. Proof

97. (a)
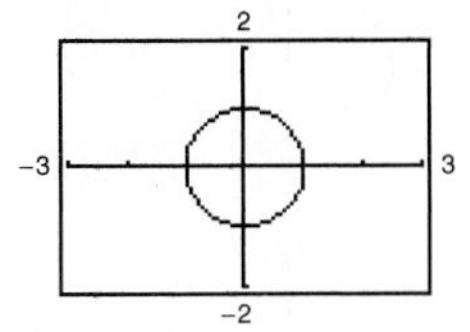

(b) Circle of radius 1 and center at $(0, 0)$ except the point $(-1, 0)$
(c) As t increases from -20 to 0, the speed increases, and as t increases from 0 to 20, the speed decreases.

99. False: $\dfrac{d^2y}{dx^2} = \dfrac{\dfrac{d}{dt}\left[\dfrac{g'(t)}{f'(t)}\right]}{f'(t)} = \dfrac{f'(t)g''(t) - g'(t)f''(t)}{[f'(t)]^3}$.

Section 10.4 (page 736)

1.
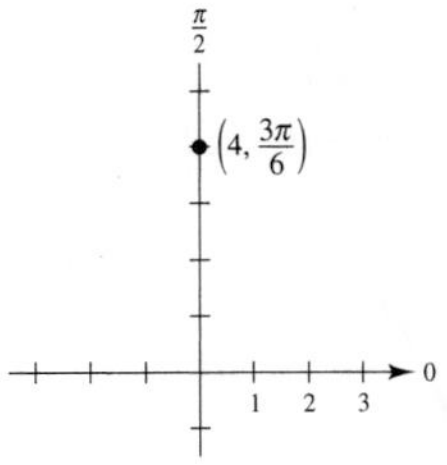

$(0, 4)$

3.
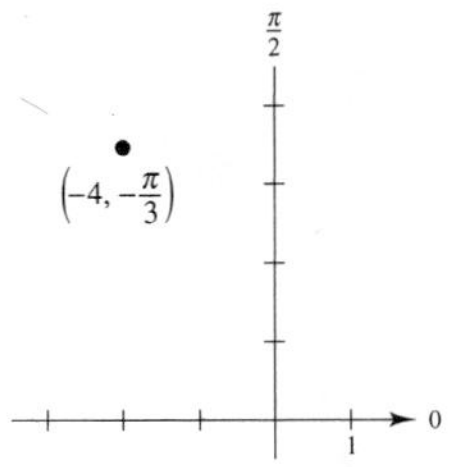

$\left(-2, 2\sqrt{3}\right) \approx (-2, 3.464)$

5.
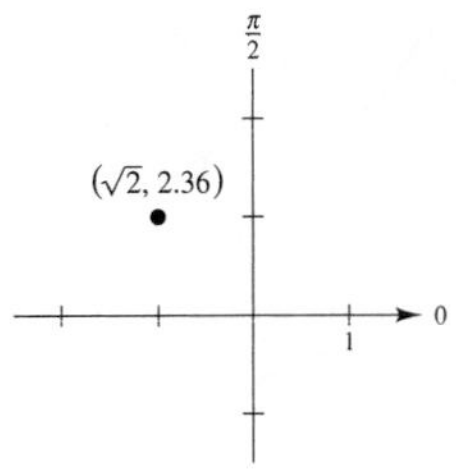

$(-1.004, 0.996)$

7.
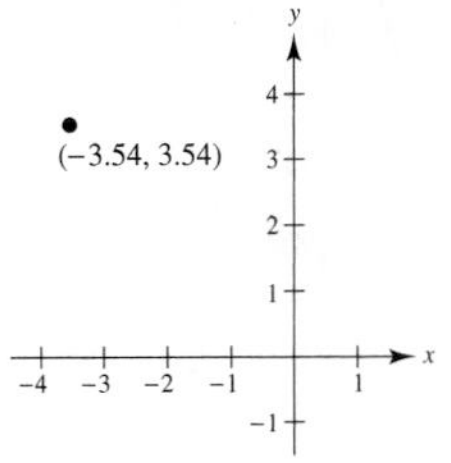

9.
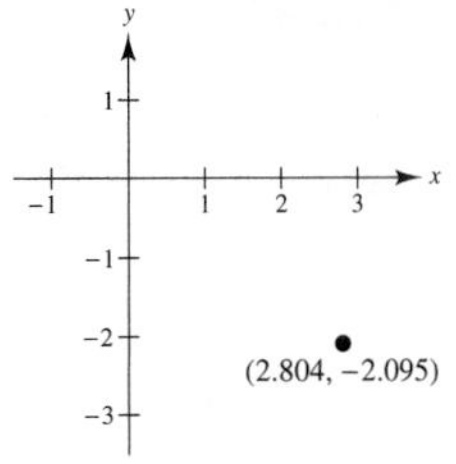

11.
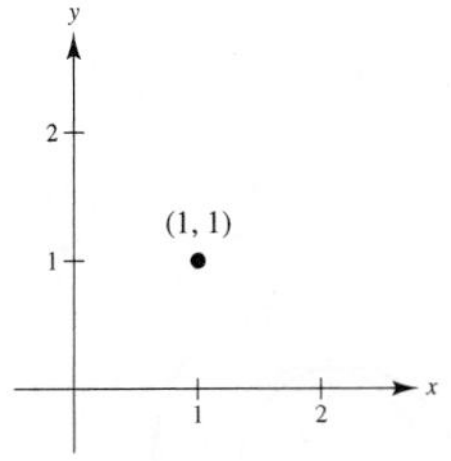

$\left(\sqrt{2}, \pi/4\right), \left(-\sqrt{2}, 5\pi/4\right)$

13.
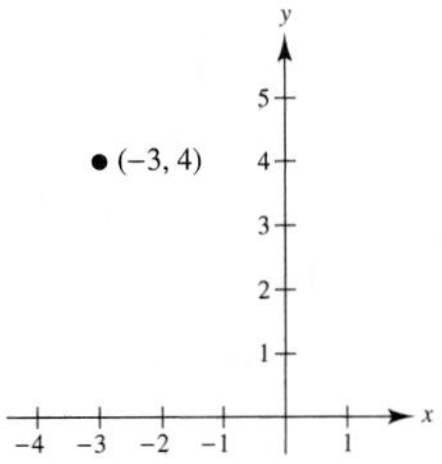

$(5, 2.214), (-5, 5.356)$

15.
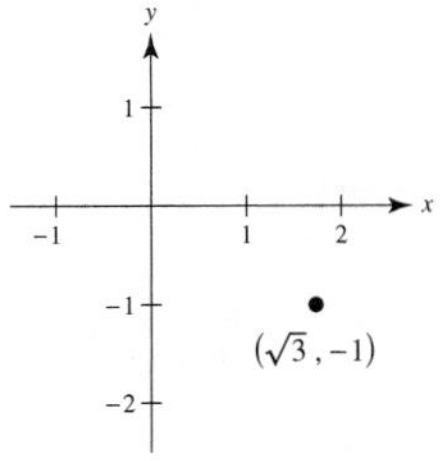

$\left(-2, 5\pi/6\right), \left(2, 11\pi/6\right)$

17. $(3.606, -0.588)$ **19.** $(2.833, 0.490)$

21. (a)
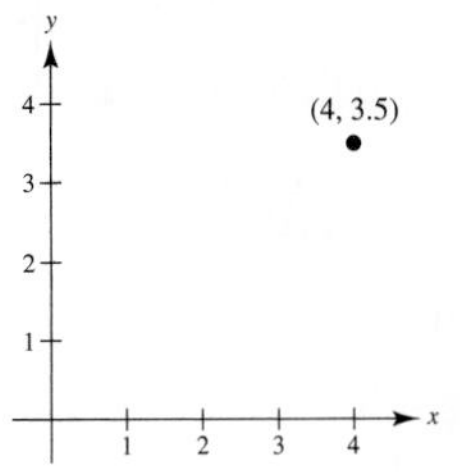

(b)
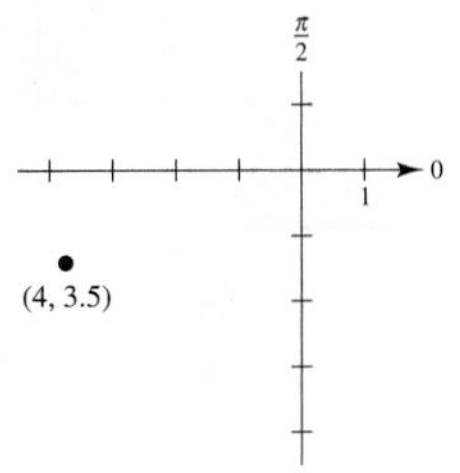

23. c **24.** b **25.** a **26.** d

27. $r = a$

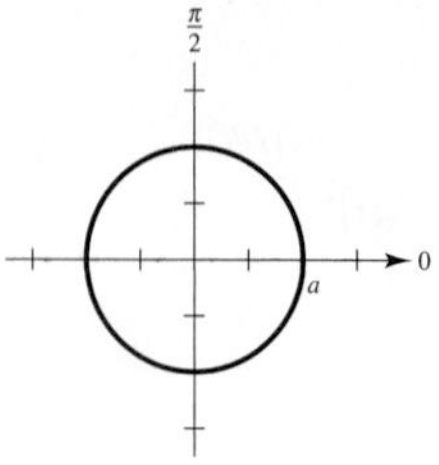

29. $r = 4 \csc \theta$

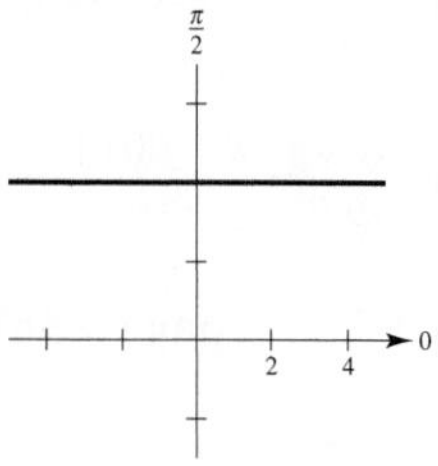

31. $r = \dfrac{-2}{3\cos\theta - \sin\theta}$

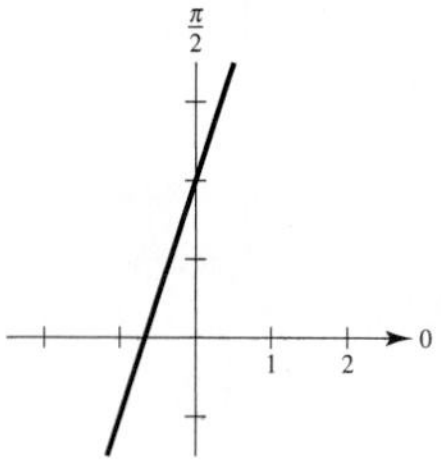

33. $r = 9 \csc^2 \theta \cos \theta$

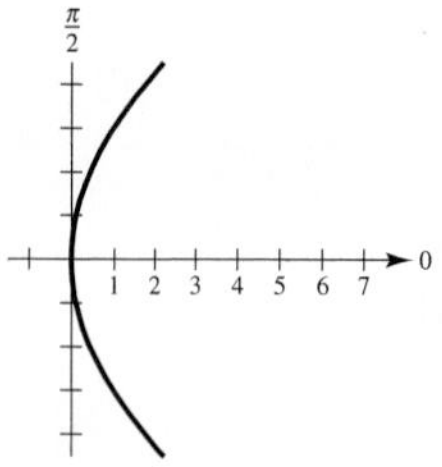

35. $x^2 + y^2 = 9$

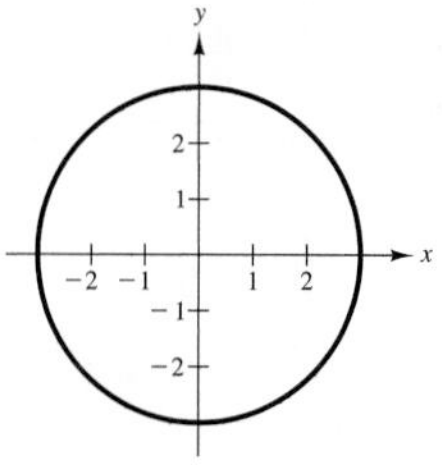

37. $x^2 + y^2 - y = 0$

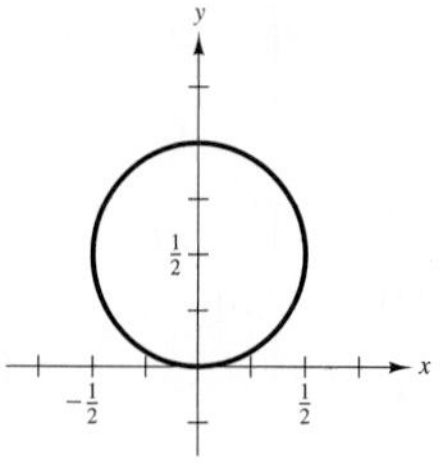

39. $\sqrt{x^2 + y^2} = \arctan(y/x)$

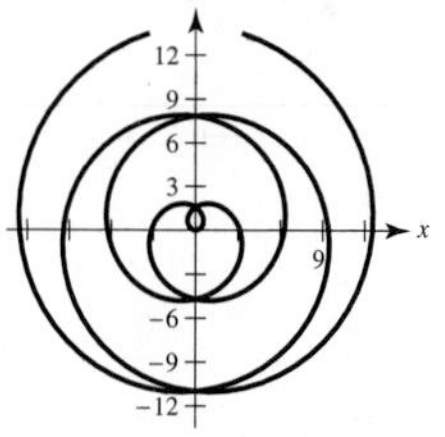

41. $x - 3 = 0$

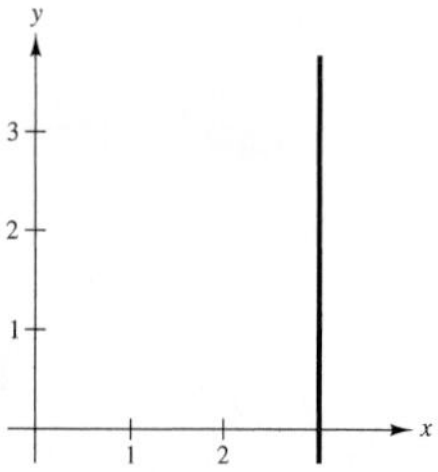

43.

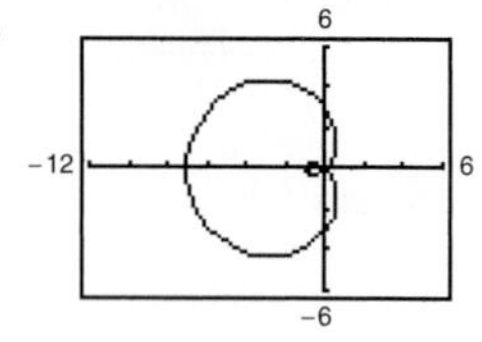

$0 \le \theta < 2\pi$

45.

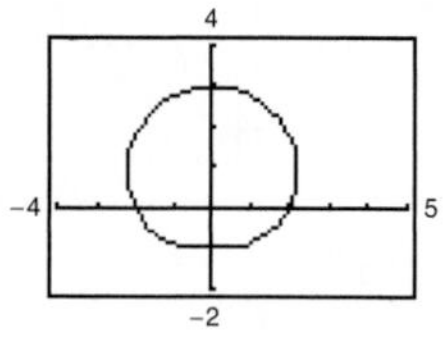

$0 \le \theta < 2\pi$

47.

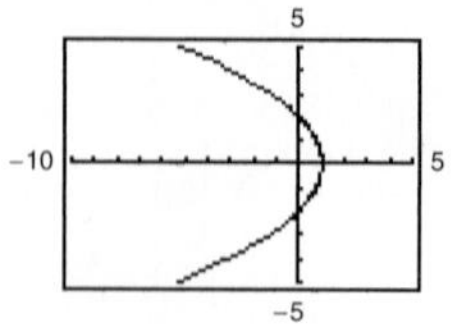

$-\pi < \theta < \pi$

49.

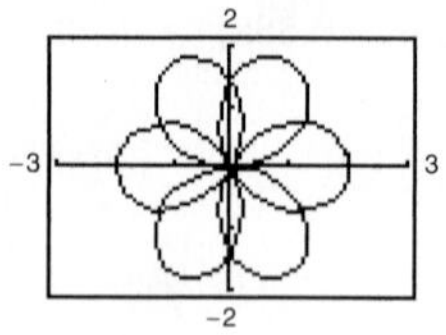

$0 \le \theta < 4\pi$

51.

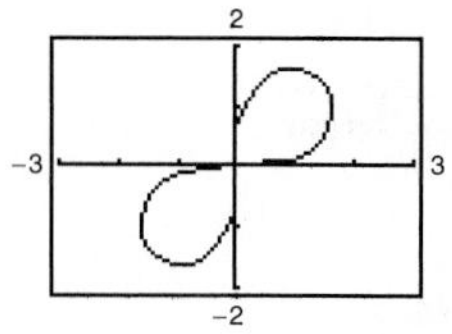

$0 \le \theta < \pi/2$

53. $(x - h)^2 + (y - k)^2 = h^2 + k^2$
Radius: $\sqrt{h^2 + k^2}$
Center: (h, k)

55. $2\sqrt{5}$ **57.** ≈ 5.6

59. $\dfrac{dy}{dx} = \dfrac{2\cos\theta(3\sin\theta + 1)}{6\cos^2\theta - 2\sin\theta - 3}$

$(5, \pi/2)$: $dy/dx = 0$
$(2, \pi)$: $dy/dx = -2/3$
$(-1, 3\pi/2)$: $dy/dx = 0$

61. (a) and (b)

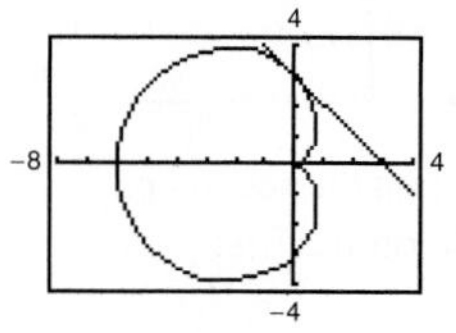

(c) $dy/dx = -1$

63. (a) and (b)

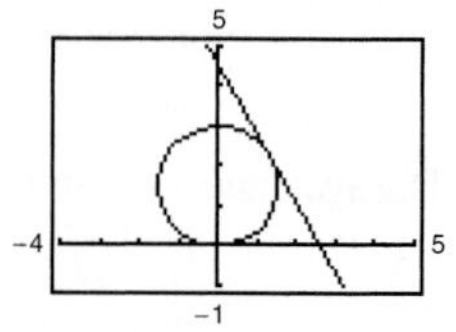

(c) $dy/dx = -\sqrt{3}$

65. Horizontal: $(2, 3\pi/2)$, $\left(\frac{1}{2}, \pi/6\right)$, $\left(\frac{1}{2}, 5\pi/6\right)$
Vertical: $\left(\frac{3}{2}, 7\pi/6\right)$, $\left(\frac{3}{2}, 11\pi/6\right)$

67. $(5, \pi/2)$, $(1, 3\pi/2)$

69.

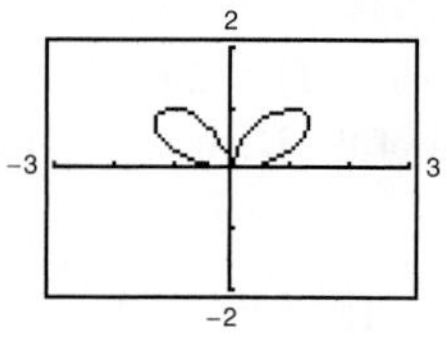

(0, 0), (1.4142, 0.7854), (1.4142, 2.3562)

71.

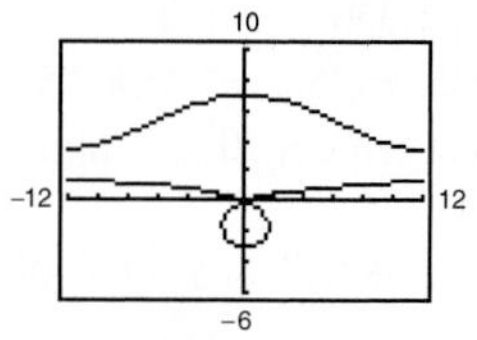

(7, 1.5708), (3, 4.7124)

73.

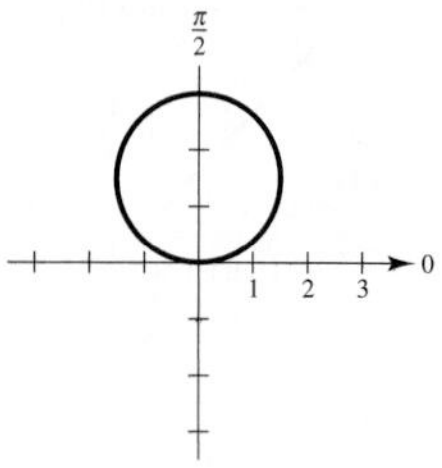

$\theta = 0$

75.

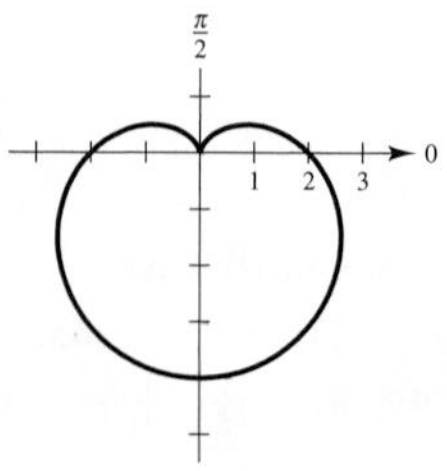

$\theta = \pi/2$

77.

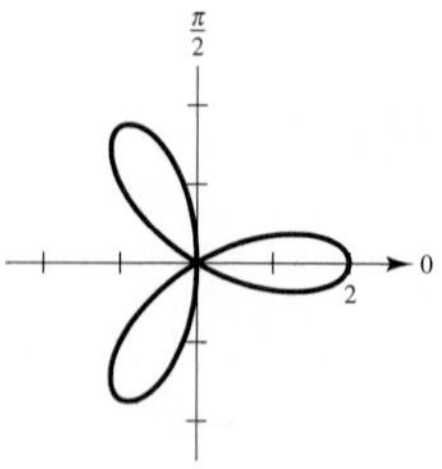

$\theta = \pi/6, \pi/2, 5\pi/6$

79.

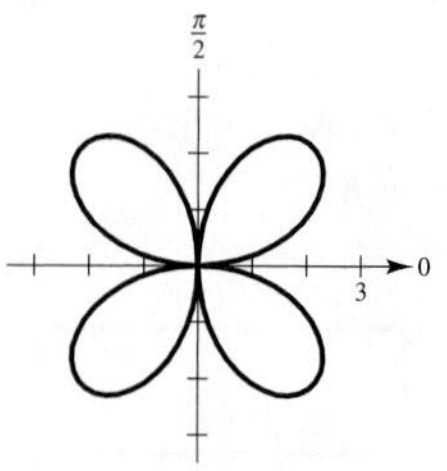

$\theta = 0, \pi/2$

81.

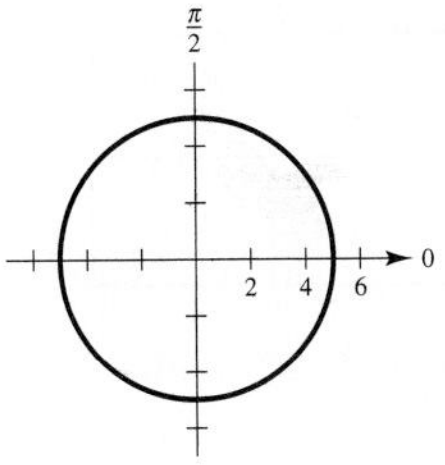

83.

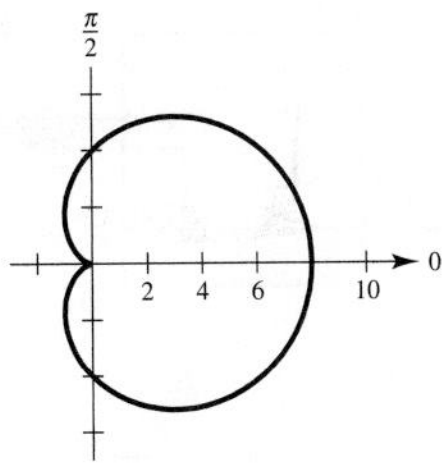

85.

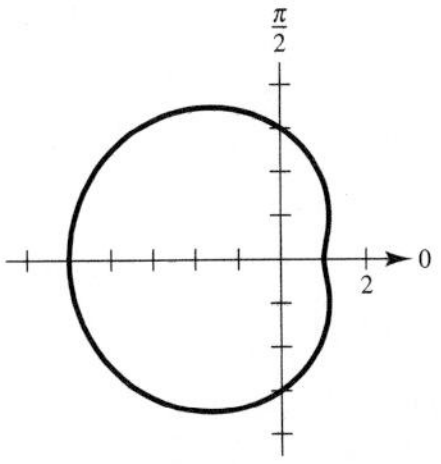

87.

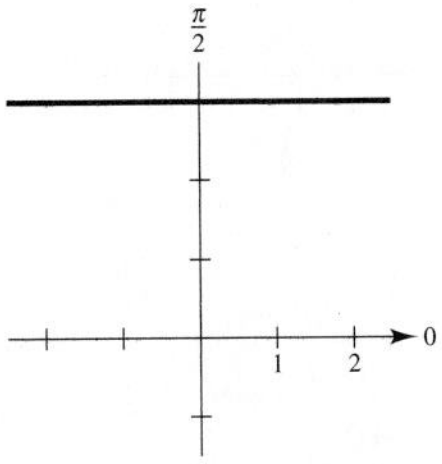

89.

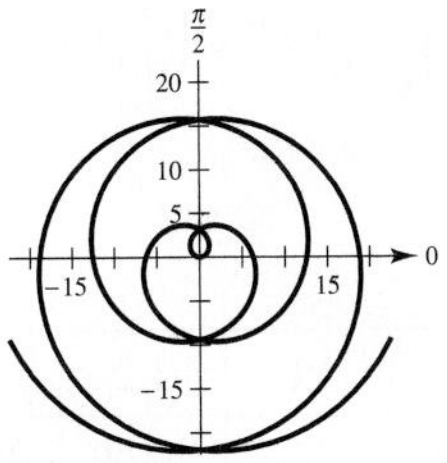

91.

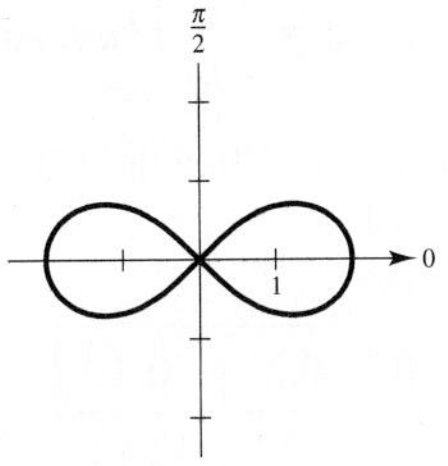

93.

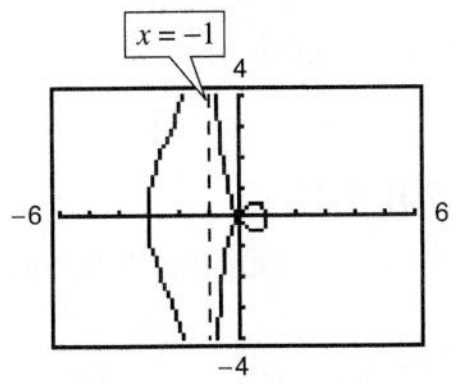

95.

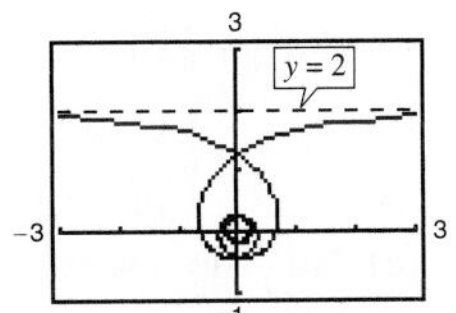

97. The rectangular coordinate system is a collection of points of the form (x, y), where x is the directed distance from the y-axis to the point and y is the directed distance from the x-axis to the point. Every point has a unique representation.
The polar coordinate system is a collection of points of the form (r, θ), where r is the directed distance from the origin O to a point P and θ is the directed angle, measured counterclockwise, from the polar axis to the segment $\overline{OP}$. Polar coordinates do not have unique representations.

99. $r = a$: Circle of radius a centered at the pole
$\theta = b$: Line passing through the pole

101. (a)

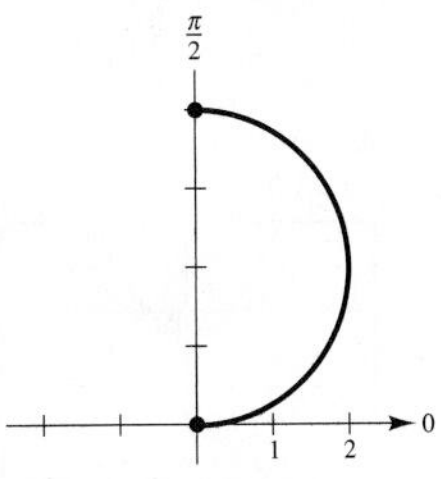

(b)

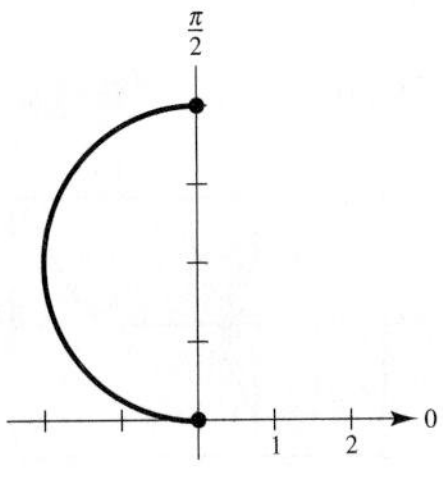

(c)

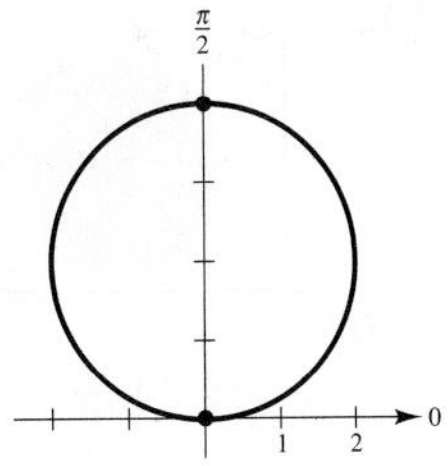

103. Proof

105. (a) $r = 2 - \sin(\theta - \pi/4)$
$$= 2 - \frac{\sqrt{2}(\sin\theta - \cos\theta)}{2}$$

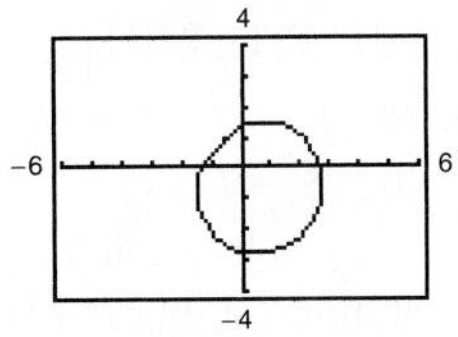

(b) $r = 2 + \cos\theta$

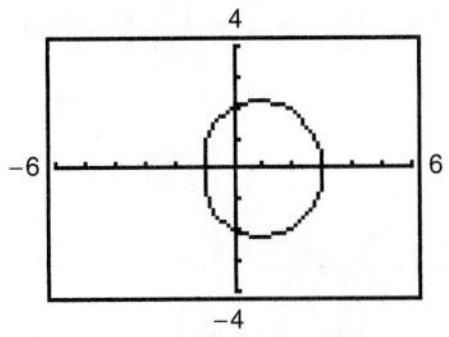

(c) $r = 2 + \sin\theta$

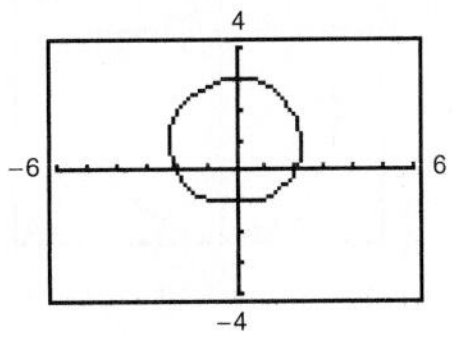

(d) $r = 2 - \cos\theta$

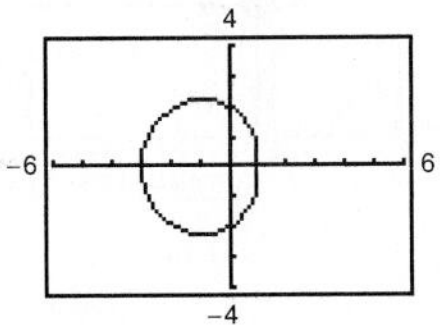

107. (a)

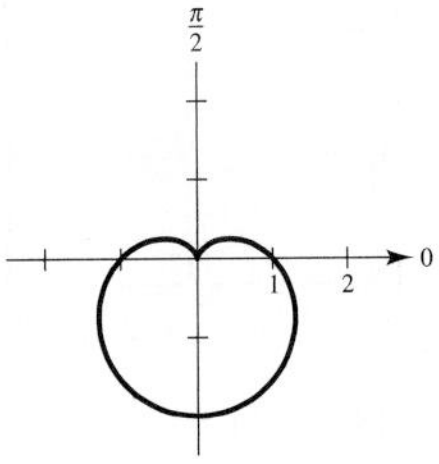

(b)

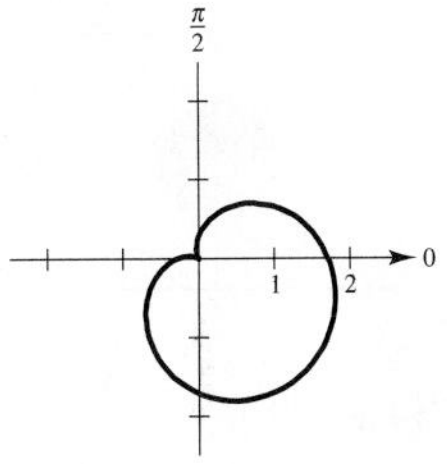

109.

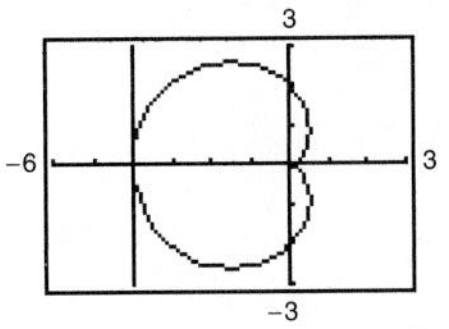

$\psi = \pi/2$

111.

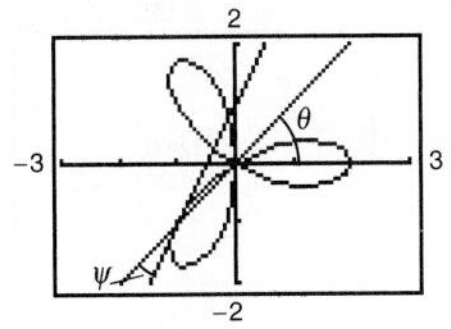

$\psi = \arctan \frac{1}{3} \approx 18.4°$

113.

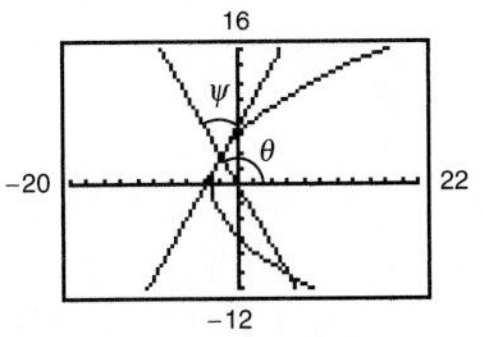

$\psi = \pi/3, 60°$

115. True **117.** True

Section 10.5 (page 745)

1. $2\int_{\pi/2}^{\pi} \sin^2\theta\, d\theta$ **3.** $\frac{1}{2}\int_{\pi/2}^{3\pi/2} (1 - \sin\theta)^2\, d\theta$

5. (a) and (b) 16π **7.** $\pi/3$ **9.** $\pi/8$ **11.** $3\pi/2$

13. 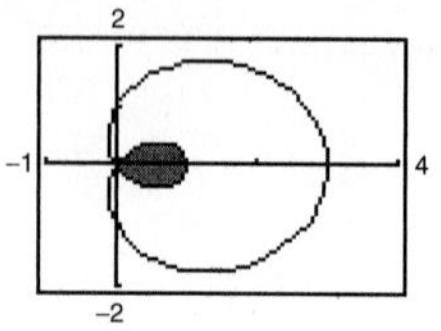

$(2\pi - 3\sqrt{3})/2$

15. 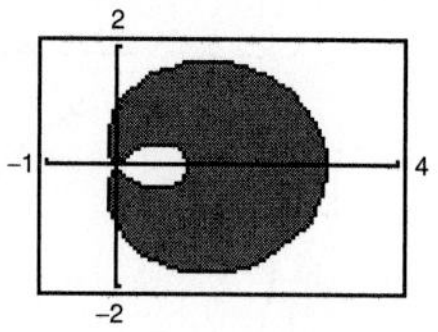

$\pi + 3\sqrt{3}$

17. $(1, \pi/2), (1, 3\pi/2), (0, 0)$

19. $\left(\dfrac{2-\sqrt{2}}{2}, \dfrac{3\pi}{4}\right), \left(\dfrac{2+\sqrt{2}}{2}, \dfrac{7\pi}{4}\right), (0, 0)$

21. $\left(\frac{3}{2}, \pi/6\right), \left(\frac{3}{2}, 5\pi/6\right), (0, 0)$ **23.** $(2, 4), (-2, -4)$

25. $(2, \pi/12), (2, 5\pi/12), (2, 7\pi/12), (2, 11\pi/12)$
$(2, 13\pi/12), (2, 17\pi/12), (2, 19\pi/12), (2, 23\pi/12)$

27. $(-0.581, \pm 2.607)$, $(2.581, \pm 1.376)$

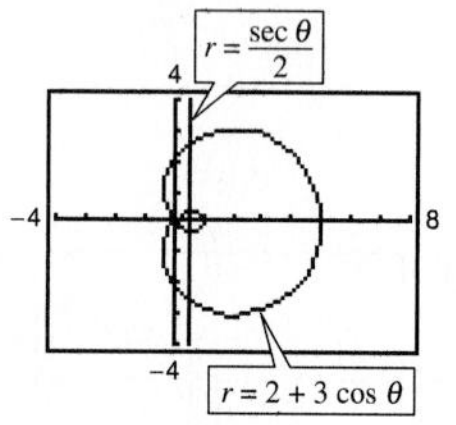

29. $(0, 0), (0.935, 0.363)$, $(0.535, -1.006)$
The graphs reach the pole at different times (θ-values).

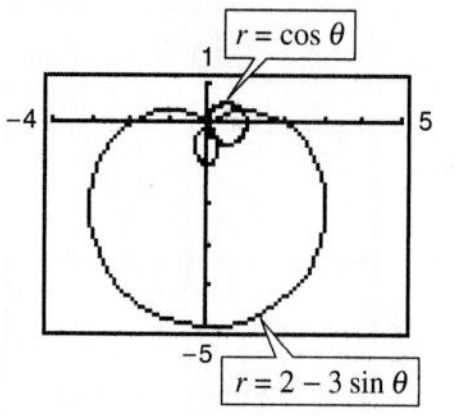

31. 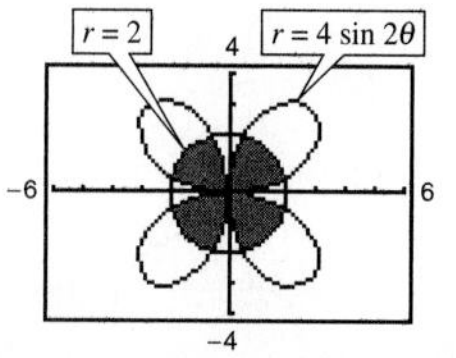

$\frac{4}{3}(4\pi - 3\sqrt{3})$

33. 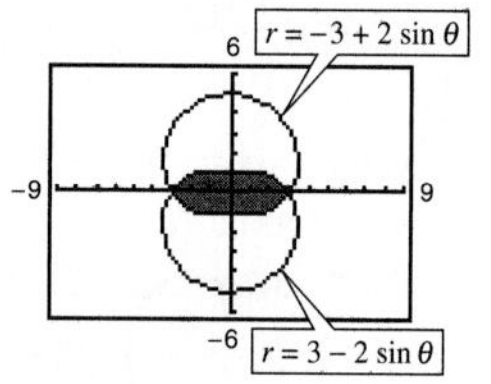

$11\pi - 24$

35. 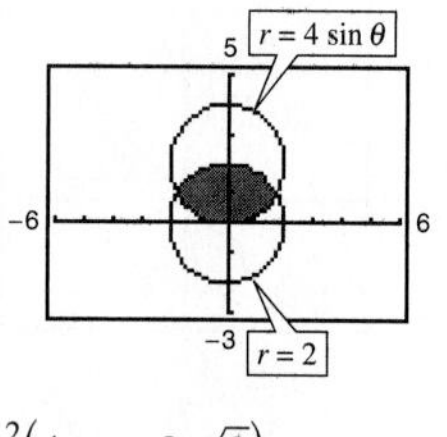

$\frac{2}{3}(4\pi - 3\sqrt{3})$

37. $5\pi a^2/4$ **39.** $(a^2/2)(\pi - 2)$

41. (a) $(x^2 + y^2)^{3/2} = ax^2$
(b) 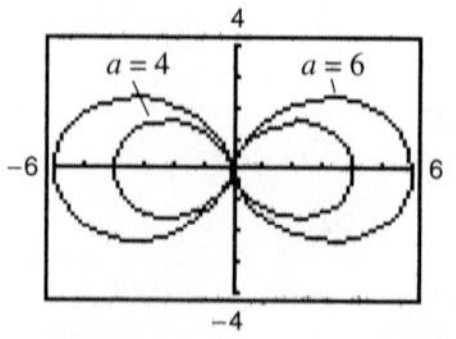

(c) $15\pi/2$

43. The area enclosed by the function is $\pi a^2/4$ if n is odd and is $\pi a^2/2$ if n is even.

45. $2\pi a$ **47.** 8

49. 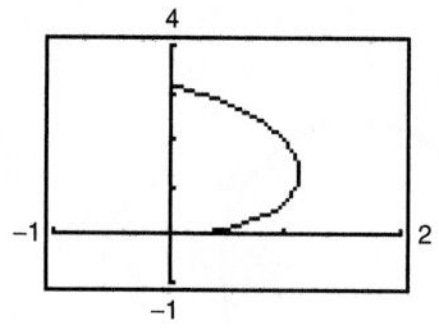

≈ 4.16

51. 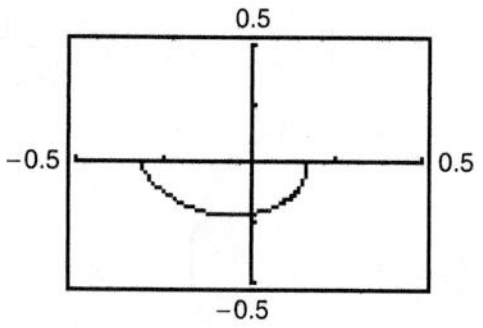

≈ 0.71

53. 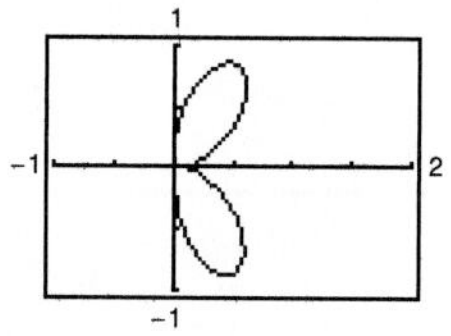

≈ 4.39

55. 36π

57. $\dfrac{2\pi\sqrt{1+a^2}}{1+4a^2}(e^{\pi a} - 2a)$ **59.** 21.87

61. Area $= \dfrac{1}{2}\displaystyle\int_\alpha^\beta r^2\,d\theta$; Arc length $= \displaystyle\int_\alpha^\beta \sqrt{r^2 + \left(\frac{dr}{d\theta}\right)^2}\,d\theta$

63. a; Answers will vary. **65.** $40\pi^2$

67. (a) 16π
(b)

θ	0.2	0.4	0.6	0.8	1.0	1.2	1.4
A	6.32	12.14	17.06	20.80	23.27	24.60	25.08

(c) and (d) $\frac{1}{4}$ of circle $(4\pi \approx 12.57) \approx 0.42$
$\frac{1}{2}$ of circle $(8\pi \approx 25.13) \approx 1.57(\pi/2)$
$\frac{3}{4}$ of circle $(12\pi \approx 37.70) \approx 2.73$
(e) No. The results do not depend on the radius. Answers will vary.

69. Circle

71. (a) 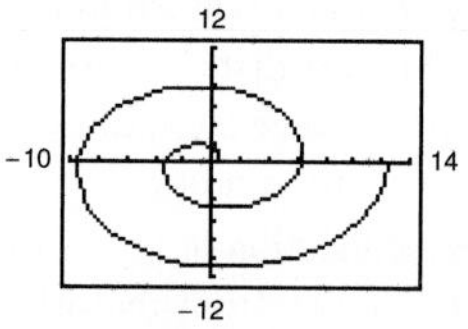

The graph becomes larger and more spread out. The graph is reflected over the y-axis.

(b) $(n\pi, an\pi)$ where $n = 1, 2, 3, \ldots$
(c) ≈ 21.26 (d) $4/3\pi^3$

73. $r = \sqrt{2}\cos\theta$

75. False. The graphs of $f(\theta) = 1$ and $g(\theta) = -1$ coincide.

77. Proof

Section 10.6 (page 753)

1. 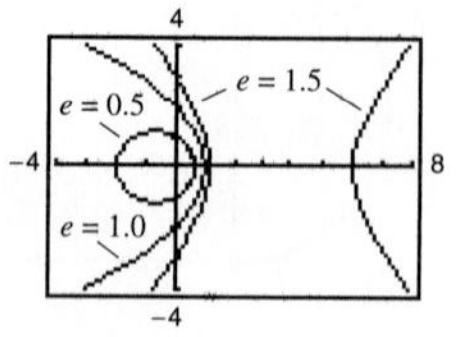

(a) Parabola (b) Ellipse
(c) Hyperbola

3. 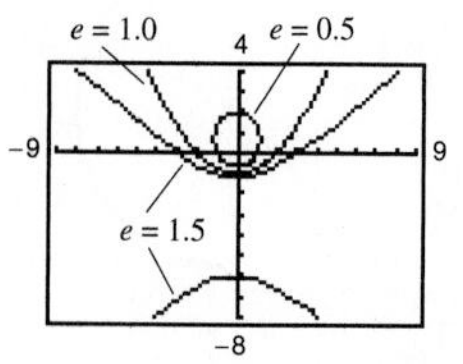

(a) Parabola (b) Ellipse
(c) Hyperbola

5. (a)

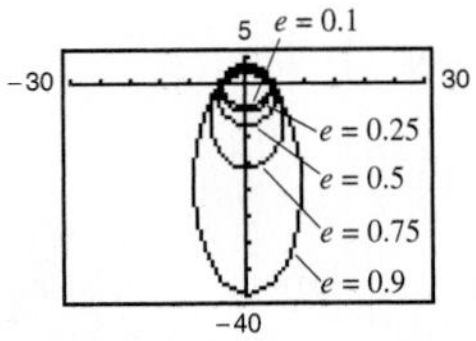

Ellipse

As $e \to 1^-$, the ellipse becomes more elliptical, and as $e \to 0^+$, it becomes more circular.

(b)

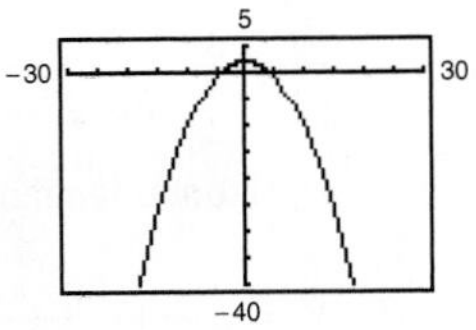

Parabola

(c)

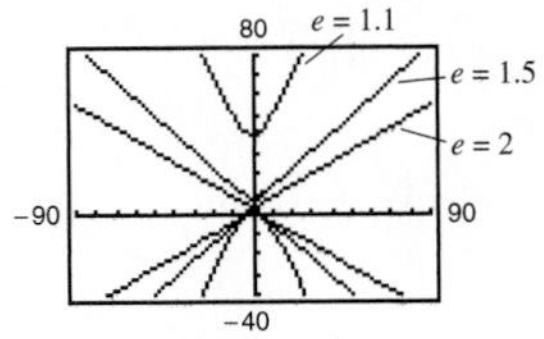

Hyperbola

As $e \to 1^+$, the hyperbola opens more slowly, and as $e \to \infty$, it opens more rapidly.

7. c **8.** f **9.** a **10.** e **11.** b **12.** d

13. $e = 1$

Distance $= 1$

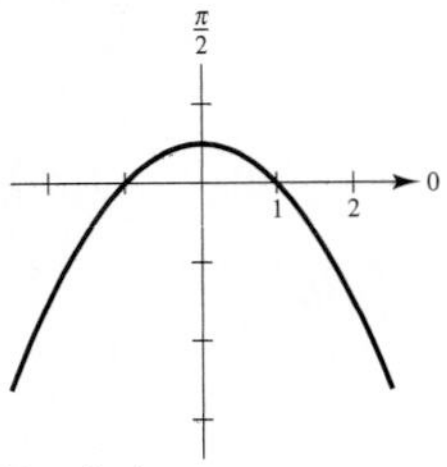

Parabola

15. $e = \frac{1}{2}$

Distance $= 6$

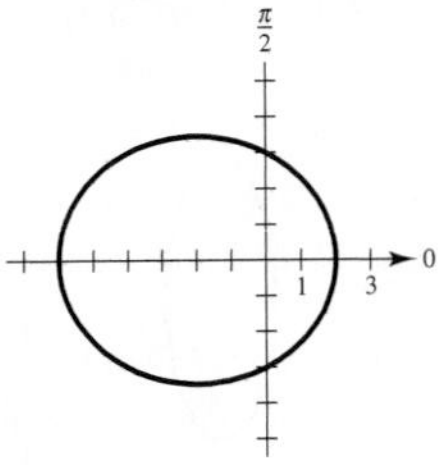

Ellipse

17. $e = \frac{1}{2}$

Distance $= 4$

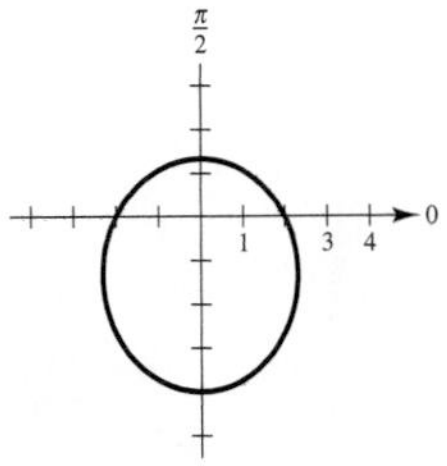

Ellipse

19. $e = 2$

Distance $= \frac{5}{2}$

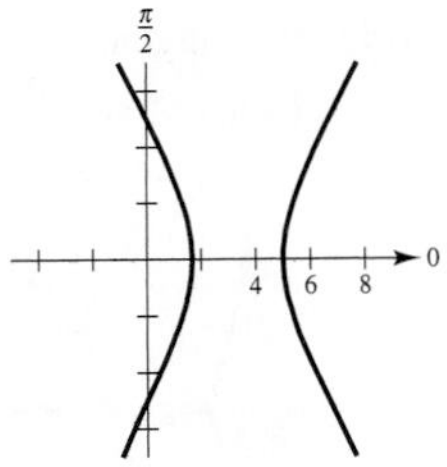

Hyperbola

21. $e = 3$

Distance $= \frac{1}{2}$

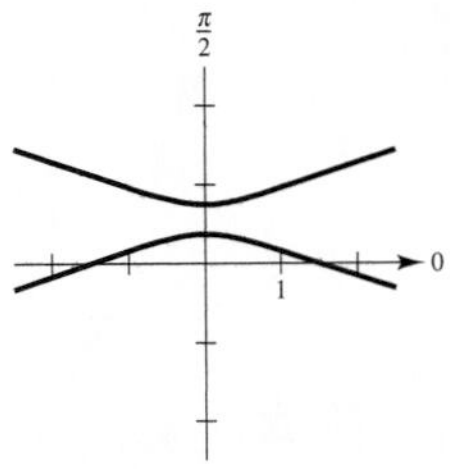

Hyperbola

23.

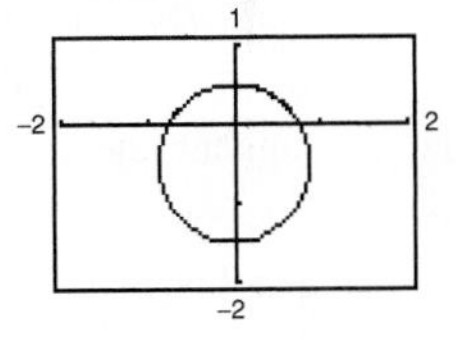

Ellipse

25.

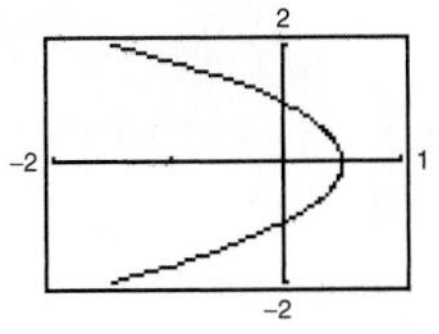

Parabola

27.

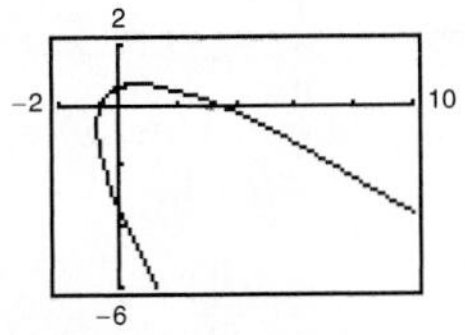

Rotated $\pi/4$ radian counterclockwise.

29.

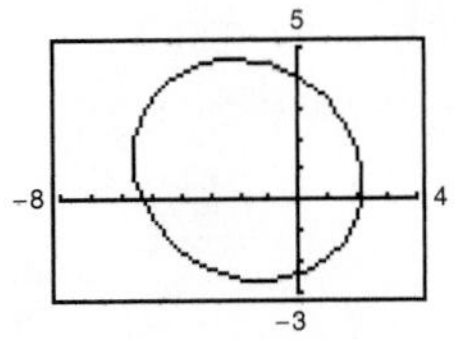

Rotated $\pi/6$ radian counterclockwise.

31. $r = \dfrac{5}{5 + 3\cos\left(\theta + \dfrac{\pi}{4}\right)}$

33. $r = 1/(1 - \cos\theta)$ **35.** $r = 1/(2 + \sin\theta)$

37. $r = 2/(1 + 2\cos\theta)$ **39.** $r = 2/(1 - \sin\theta)$

41. $r = 16/(5 + 3\cos\theta)$ **43.** $r = 9/(4 - 5\sin\theta)$

45. If $0 < e < 1$, the conic is an ellipse.
If $e = 1$, the conic is a parabola.
If $e > 1$, the conic is a hyperbola.

47. (a) Hyperbola (b) Ellipse (c) Parabola (d) Hyperbola

49. Proof

51. $r^2 = \dfrac{9}{1 - (16/25)\cos^2\theta}$ **53.** $r^2 = \dfrac{-16}{1 - (25/9)\cos^2\theta}$

55. ≈ 10.88 **57.** $\dfrac{7979.21}{1 - 0.9372\cos\theta}$; 11,015 mi

59. $r = \dfrac{149{,}558{,}278.0560}{1 - 0.0167\cos\theta}$

Perihelion: 147,101,680 km

Aphelion: 152,098,320 km

61. $r = \dfrac{5{,}540{,}410{,}095.36}{1 - 0.2488\cos\theta}$

Perihelion: 4,436,587,200 km

Aphelion: 7,375,412,800 km

63. Answers will vary. Sample answers:
(a) 9.341×10^{18} km^2; 21.867 yrs
(b) 0.8995 rad; Larger angle with the smaller ray to generate an equal area
(c) Part (a): 2.559×10^9 km; 1.17×10^8 km/yr
Part (b): 4.119×10^9 km; 1.88×10^8 km/yr

65. Proof

67. Let $r_1 = ed/(1 + \sin\theta)$ and $r_2 = ed/(1 - \sin\theta)$.
The points of intersection of r_1 and r_2 are $(ed, 0)$ and (ed, π). The slope of the tangent line to r_1 at $(ed, 0)$ is -1 and at (ed, π) is 1. The slope of the tangent line to r_2 at $(ed, 0)$ is 1 and at (ed, π) is -1. Therefore, at $(ed, 0)$, $m_1 m_2 = -1$ and at (ed, π), $m_1 m_2 = -1$ and the curves intersect at right angles.

Review Exercises for Chapter 10 (page 756)

1. e **2.** c **3.** b **4.** d **5.** a **6.** f

7. Circle

Center: $\left(\frac{1}{2}, -\frac{3}{4}\right)$

Radius: 1

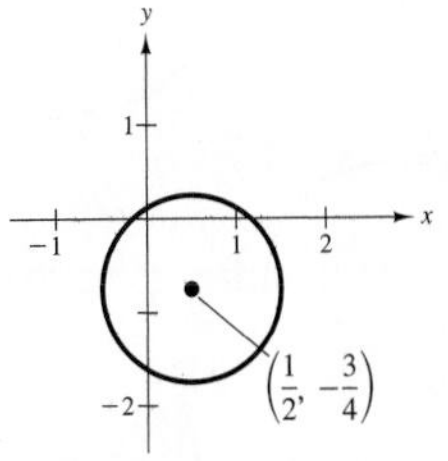

9. Hyperbola

Center: $(-4, 3)$

Vertices: $\left(-4 \pm \sqrt{2}, 3\right)$

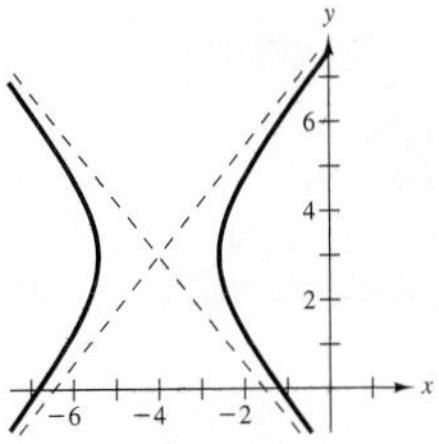

11. Ellipse
Center: $(2, -3)$
Vertices: $\left(2, -3 \pm \sqrt{2}/2\right)$

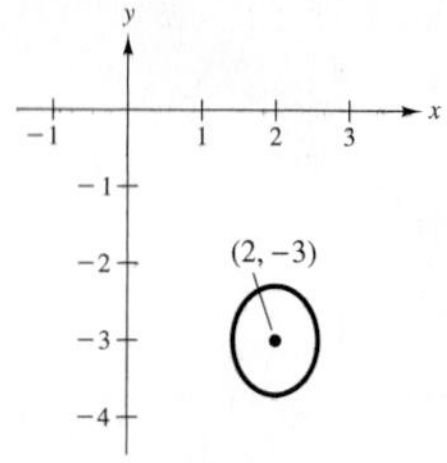

13. $y^2 - 4y - 12x + 4 = 0$ **15.** $(x-2)^2/25 + y^2/21 = 1$

17. $x^2/16 - y^2/20 = 1$ **19.** ≈ 15.87 **21.** $4x + 4y - 7 = 0$

23. (a) $(0, 50)$ (b) $\approx 38{,}294.49$

25.

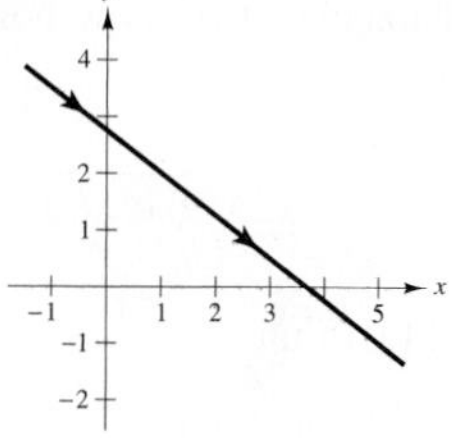

$4y + 3x - 11 = 0$

27.

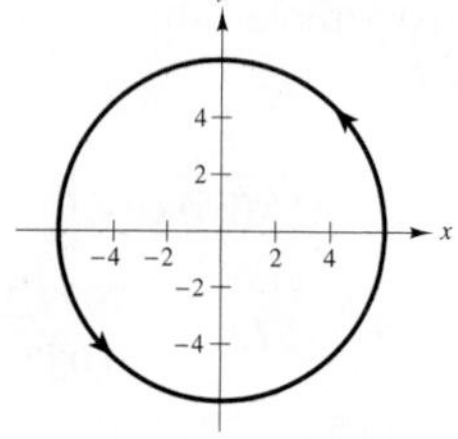

$x^2 + y^2 = 36$

29.

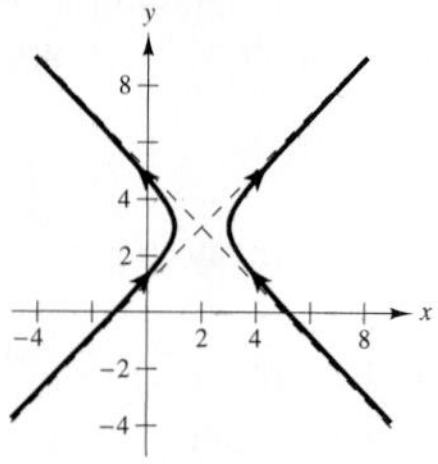

$(x-2)^2 - (y-3)^2 = 1$

31. $x = 5t - 2$
$y = 6 - 4t$

33. $x = 4\cos\theta - 3$
$y = 4 + 3\sin\theta$

35.

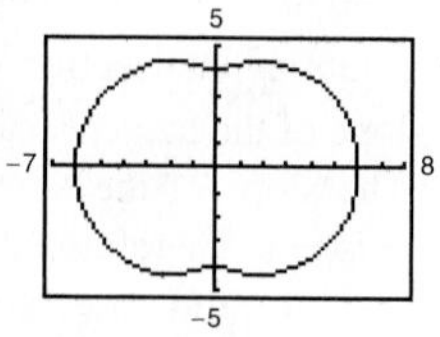

37. (a) $dy/dx = -\frac{3}{4}$;
Horizontal tangents: None
(b) $y = (-3x + 11)/4$
(c)

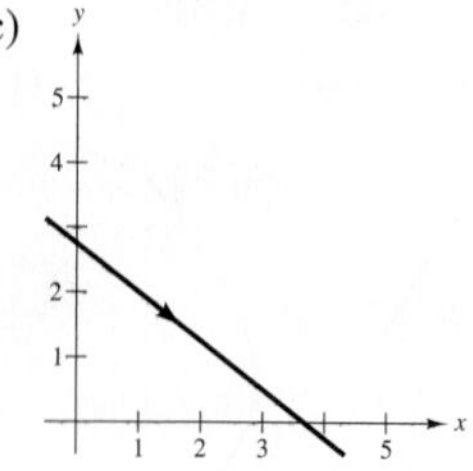

39. (a) $dy/dx = -2t^2$;
Horizontal tangents: None
(b) $y = 3 + 2/x$
(c)

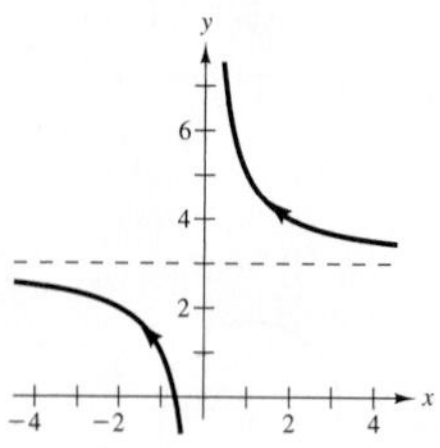

41. (a) $\dfrac{dy}{dx} = \dfrac{(t-1)(2t+1)^2}{t^2(t-2)^2}$;
Horizontal tangent: $\left(\frac{1}{3}, -1\right)$
(b) $y = \dfrac{4x^2}{(5x-1)(x-1)}$
(c)

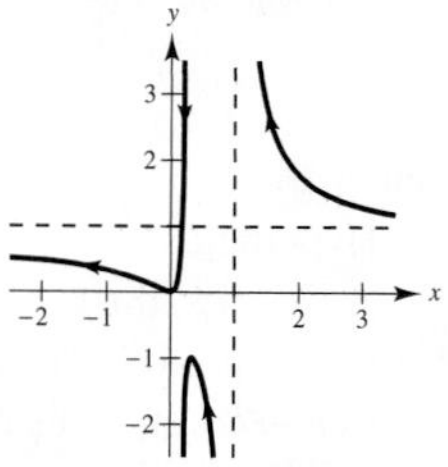

43. (a) $\dfrac{dy}{dx} = -\dfrac{5}{2}\cot\theta$;
Horizontal tangents: $(3, 7), (3, -3)$
(b) $\dfrac{(x-3)^2}{4} + \dfrac{(y-2)^2}{25} = 1$
(c)

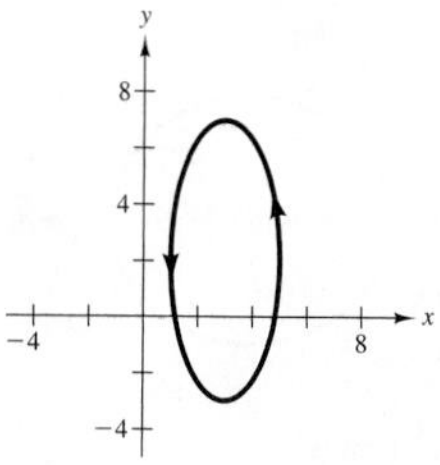

45. (a) $\dfrac{dy}{dx} = -4\tan\theta$;
Horizontal tangents: None
(b) $x^{2/3} + (y/4)^{2/3} = 1$
(c)

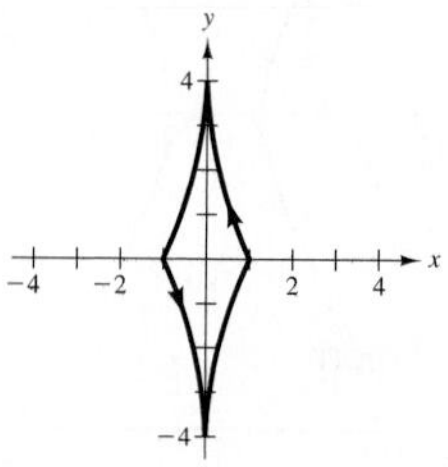

47. Horizontal tangent line: $t = 0$ $(4, 0)$

49. Horizontal tangent lines: $\theta = 0$ and $\theta = \pi$ ((2, 2) and (2, 0))
Vertical tangent lines: $\theta = \pi/2$ and $\theta = 3\pi/2$ ((4, 1) and (0, 1))

51. (a) and (c)

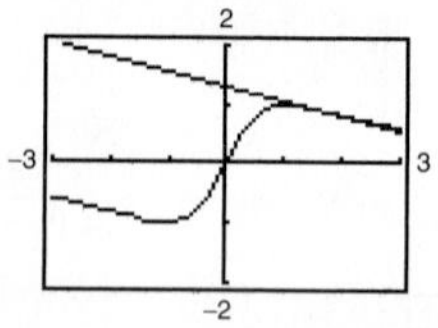

(b) $dx/d\theta = -4$, $dy/d\theta = 1$, $dy/dx = -\frac{1}{4}$

53. $\frac{1}{2}\pi^2 r$

55. (a) $s = 12\pi\sqrt{10} \approx 119.215$ **57.** $A = 3\pi$

(b) $s = 4\pi\sqrt{10} \approx 39.738$

59.

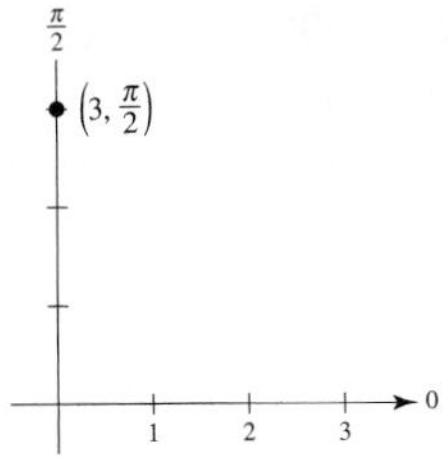

Rectangular: $(0, 3)$

61.

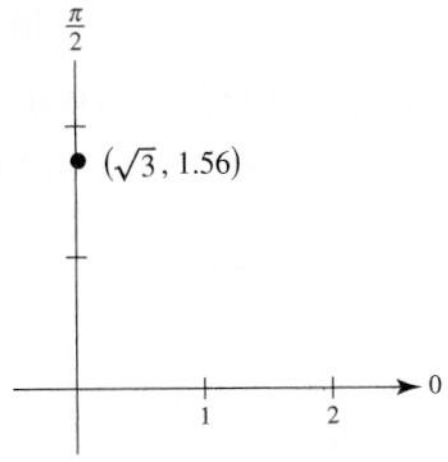

Rectangular: $(0.0187, 1.7320)$

63.

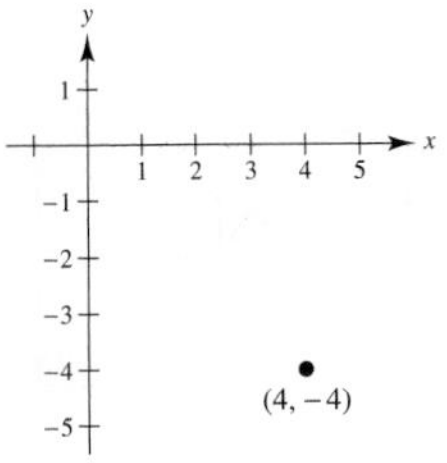

$\left(4\sqrt{2}, \dfrac{7\pi}{4}\right), \left(-4\sqrt{2}, \dfrac{3\pi}{4}\right)$

65. $x^2 + y^2 - 3x = 0$ **67.** $(x^2 + y^2 + 2x)^2 = 4(x^2 + y^2)$

69. $(x^2 + y^2)^2 = x^2 - y^2$ **71.** $y^2 = x^2[(4 - x)/(4 + x)]$

73. $r = a\cos^2\theta\sin\theta$ **75.** $r^2 = a^2\theta^2$

77. Circle

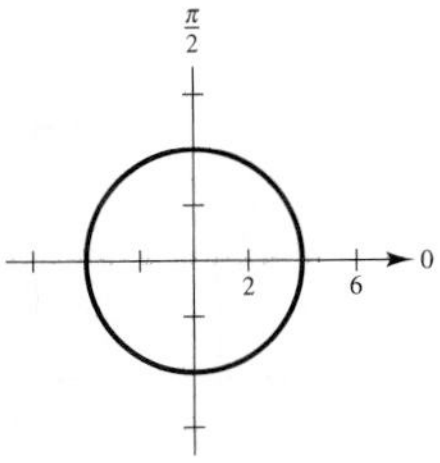

79. Line

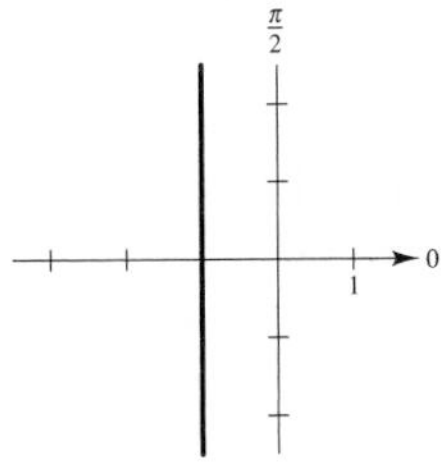

81. Cardioid

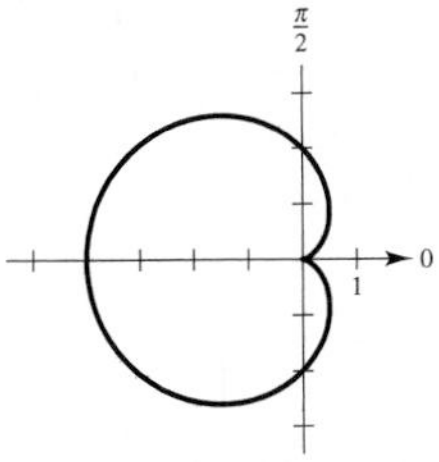

83. Limaçon

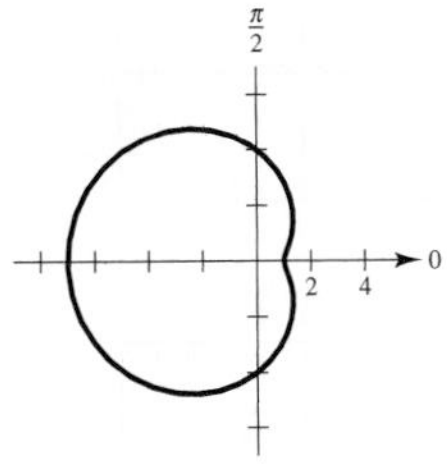

85. Rose curve

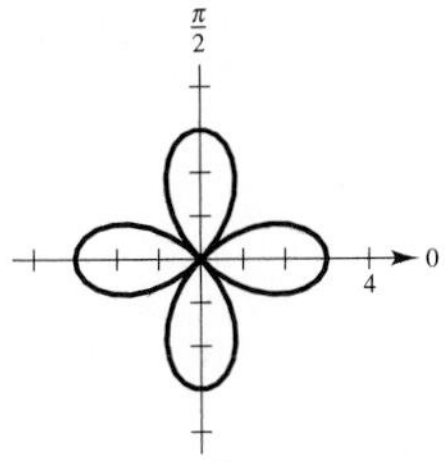

87. Rose curve

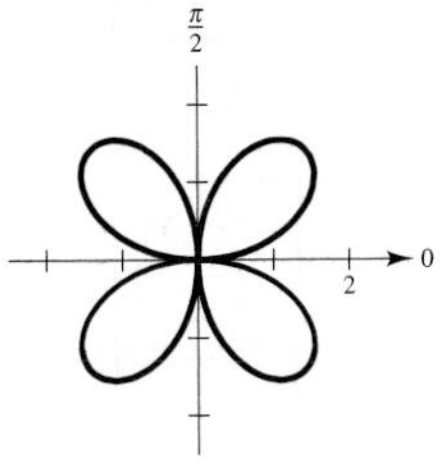

89.

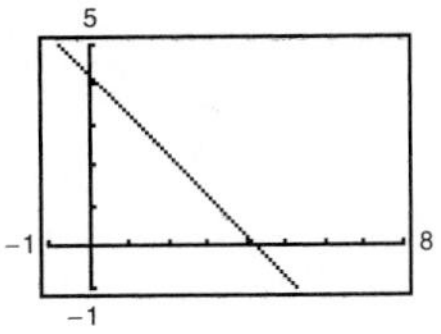

91.

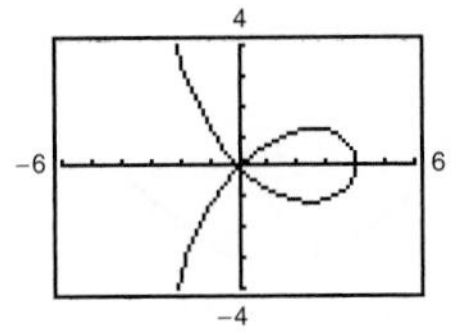

93. (a) $\pm\pi/3$

(b) Vertical: $(-1, 0), (3, \pi), \left(\frac{1}{2}, \pm 1.318\right)$

Horizontal: $(-0.686, \pm 0.568), (2.186, \pm 2.206)$

(c)

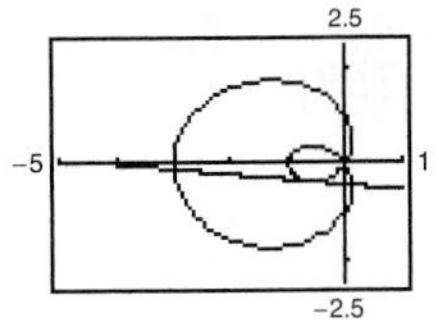

95. $\arctan(2\sqrt{3}/3) \approx 49.1°$ **97.** Proof

99. $A = 2\left(\frac{1}{2}\right)\int_0^{\pi} (2 + \cos\theta)^2\, d\theta \approx 14.14$

101. $A = 2\left(\frac{1}{2}\right)\int_0^{\pi/2} 4\sin 2\theta\, d\theta \approx 4.00$

103.

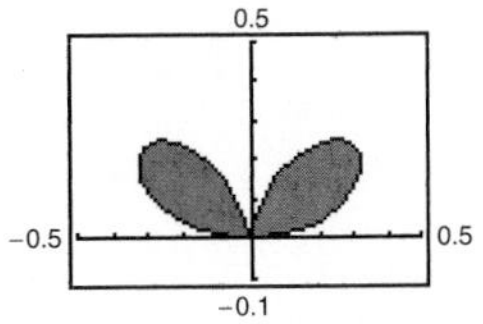

$A = 2\left(\frac{1}{2}\right)\int_0^{\pi/2} \sin^2\theta\cos^4\theta\, d\theta \approx 0.10$

105.

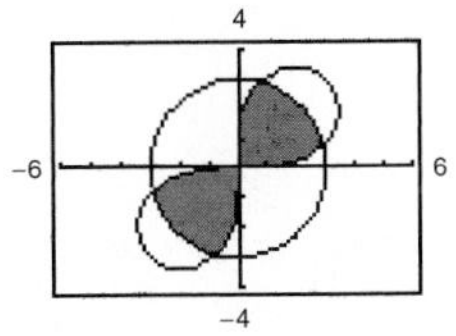

$$A = 2\left[\frac{1}{2}\int_0^{\pi/12} 18\sin 2\theta + \frac{1}{2}\int_{\pi/12}^{5\pi/12} 9\, d\theta + \frac{1}{2}\int_{5\pi/12}^{\pi/2} 18\sin 2\theta\, d\theta\right]$$

$\approx 1.2058 + 9.4248 + 1.2058 = 11.8364$

107. $4a$

109. $S = 2\pi\int_0^{\pi/2} (1 + 4\cos\theta)\sin\theta\sqrt{17 + 8\cos\theta}\, d\theta$

$= 34\pi\sqrt{17}/5 \approx 88.08$

111. Parabola

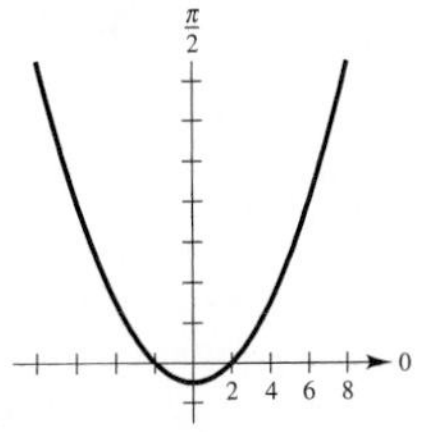

113. Ellipse

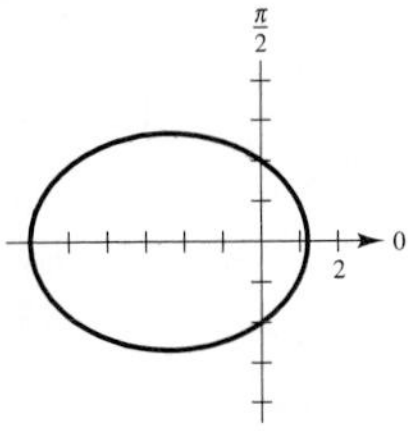

115. Hyperbola

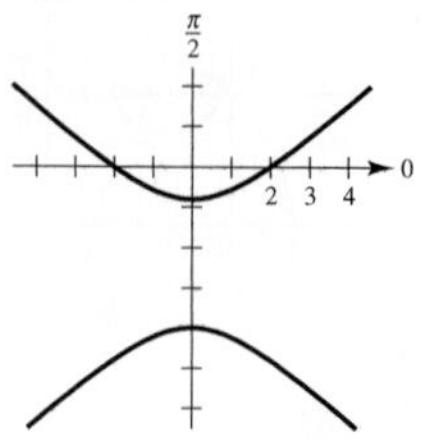

117. $r = 10 \sin \theta$

119. $r = 4/(1 - \cos \theta)$ **121.** $r = 5/(3 - 2 \cos \theta)$

P.S. Problem Solving (page 759)

1. (a)

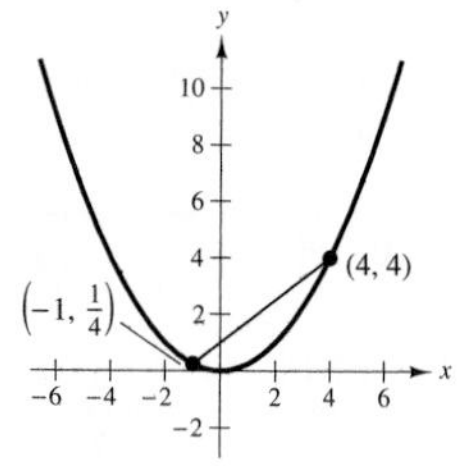

(b)–(c) Proofs

3. Proof

5. (a) $r = 2a \tan \theta \sin \theta$

(b) $x = 2at^2/(1 + t^2)$

$y = 2at^3/(1 + t^2)$

(c) $y^2 = x^3/(2a - x)$

7. (a) $y^2 = x^2[(1 - x)/(1 + x)]$

(b) $r = \cos 2\theta \cdot \sec \theta$

(c)

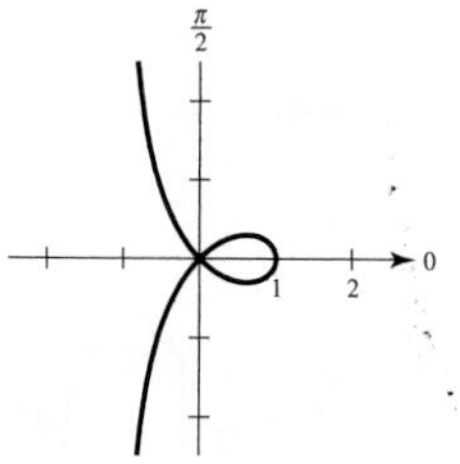

(d) $y = x, y = -x$

(e) $\left(\frac{\sqrt{5} - 1}{2}, \pm \frac{\sqrt{5} - 1}{2}\sqrt{-2 + \sqrt{5}}\right)$

9. (a)

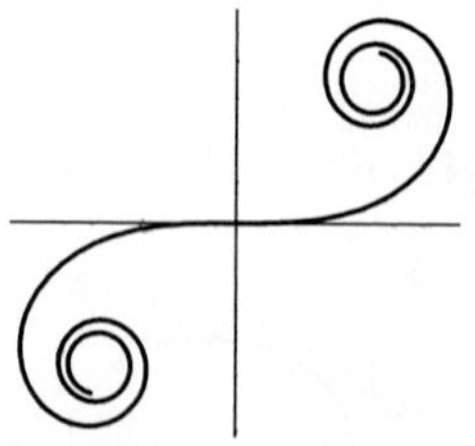

Generated by Mathematica

(b) Proof

(c) $a, 2\pi$

11. $A = \frac{1}{2}ab$ **13.** $r^2 = 2 \cos 2\theta$

15. (a) First plane: $x_1 = \cos 70°(150 - 375t)$

$y_1 = \sin 70°(150 - 375t)$

Second plane: $x_2 = \cos 45°(450t - 190)$

$y_2 = \sin 45°(190 - 450t)$

(b) $\{[\cos 45°(450t - 190) - \cos 70°(150 - 375t)]^2 + [\sin 45°(190 - 450t) - \sin 70°(150 - 375t)]^2\}^{1/2}$

(c)

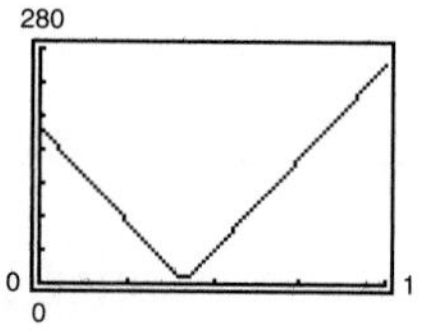

0.4145 hr; Yes

17.

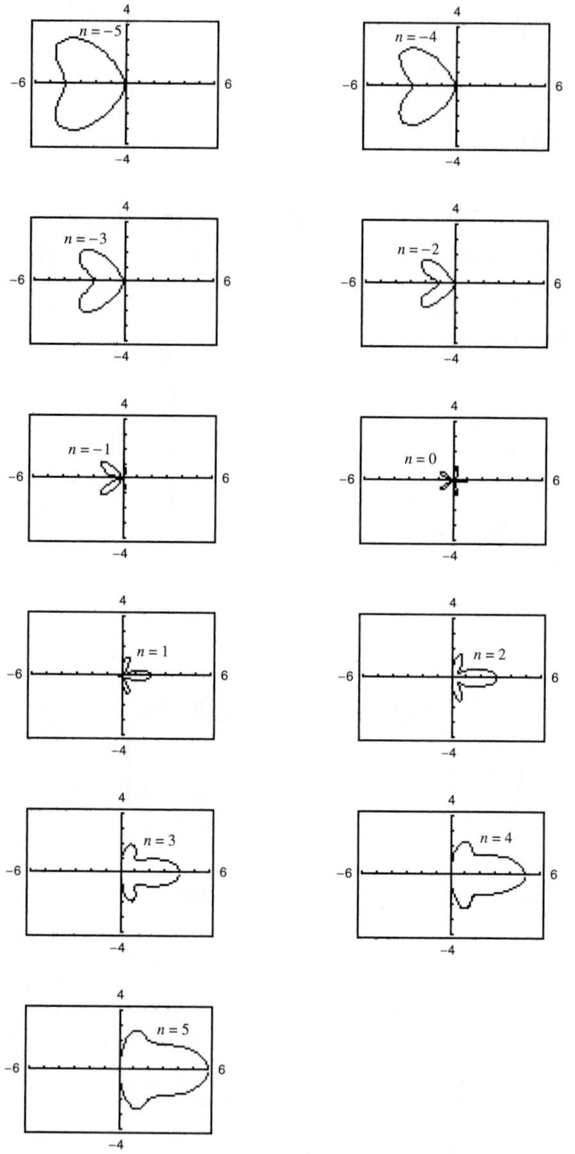

$n = 1, 2, 3, 4, 5$ produce "bells"; $n = -1, -2, -3, -4, -5$ produce "hearts."

Chapter 11

Section 11.1 (page 769)

1. (a) $\langle 4, 2\rangle$
(b)

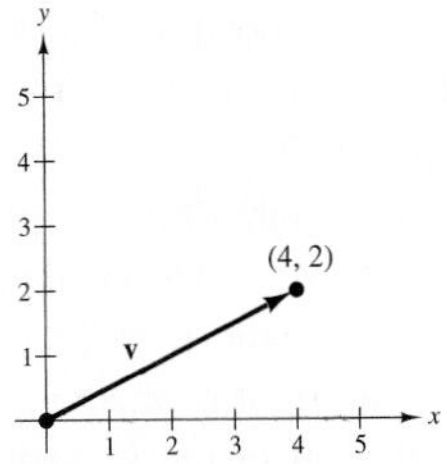

3. (a) $\langle -7, 0\rangle$
(b)

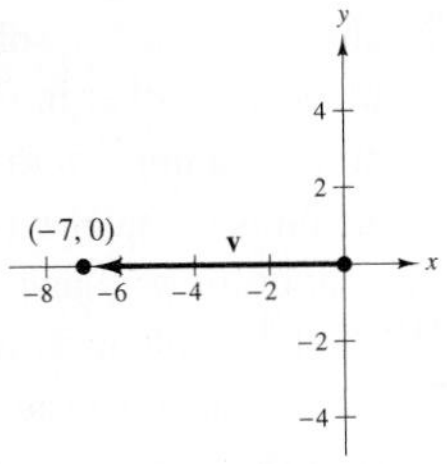

5. $\mathbf{u} = \mathbf{v} = \langle 2, 4\rangle$ **7.** $\mathbf{u} = \mathbf{v} = \langle 6, -5\rangle$

9. (a) and (c)

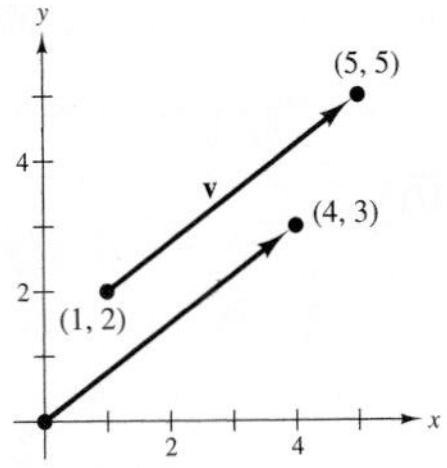

(b) $\langle 4, 3\rangle$

11. (a) and (c)

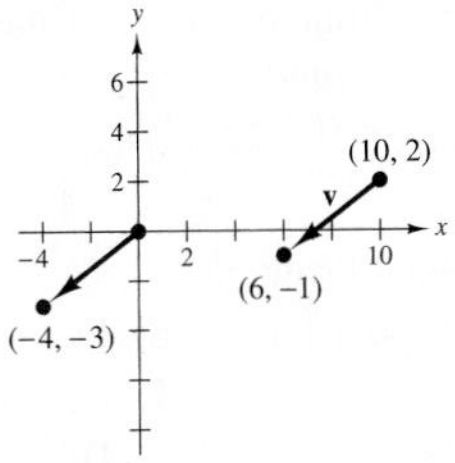

(b) $\langle -4, -3\rangle$

13. (a) and (c)

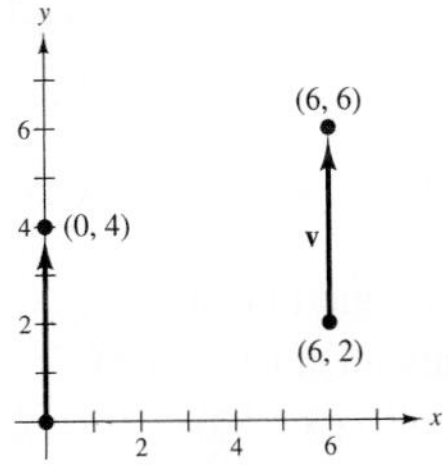

(b) $\langle 0, 4\rangle$

15. (a) and (c)

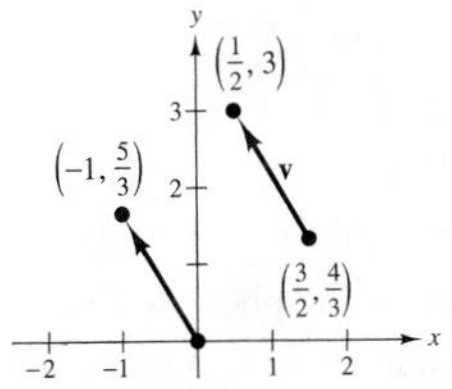

(b) $\left\langle -1, \frac{5}{3}\right\rangle$

17. (a) $\langle 4, 6\rangle$

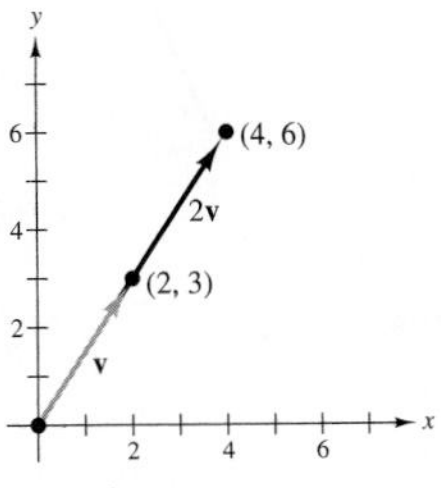

(b) $\langle -6, -9\rangle$

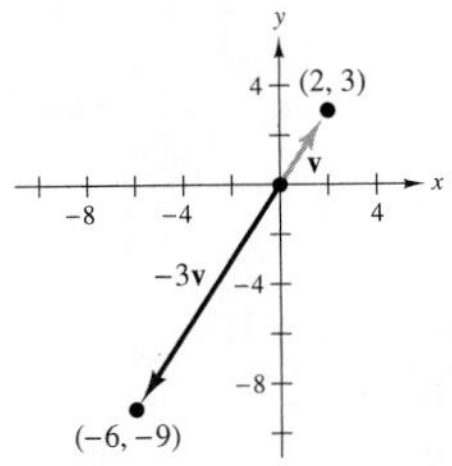

(c) $\left\langle 7, \frac{21}{2}\right\rangle$

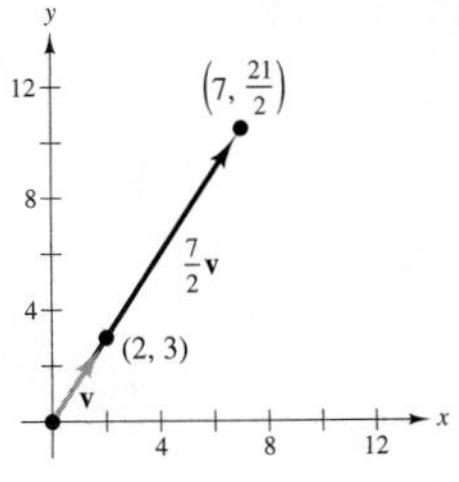

(d) $\left\langle \frac{4}{3}, 2\right\rangle$

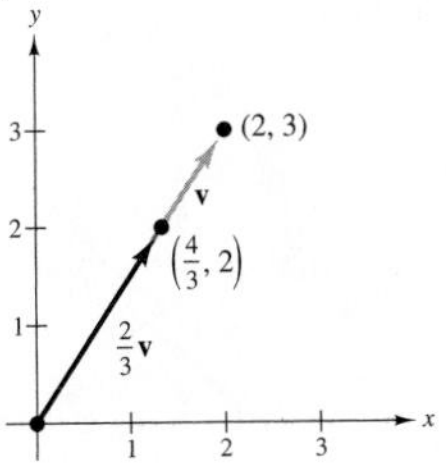

19.

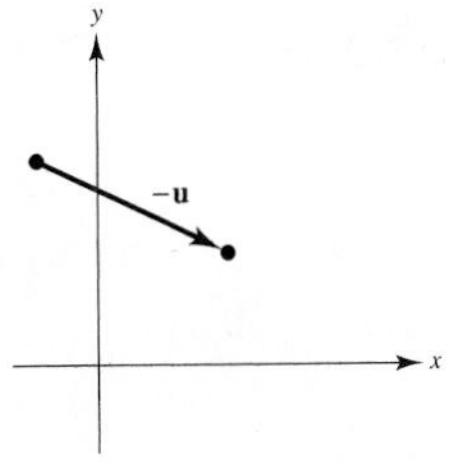

21.

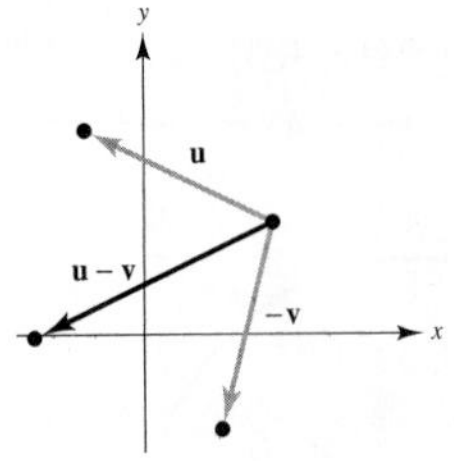

23. (a) $\left\langle \frac{8}{3}, 6\right\rangle$ (b) $\langle -2, -14\rangle$ (c) $\langle 18, -7\rangle$

25. $\left\langle 3, -\frac{3}{2}\right\rangle$

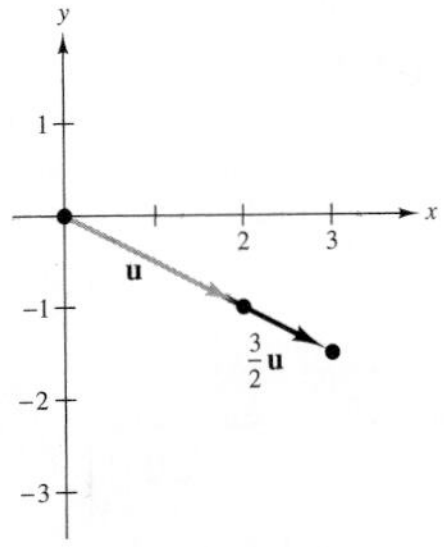

27. $\langle 4, 3\rangle$

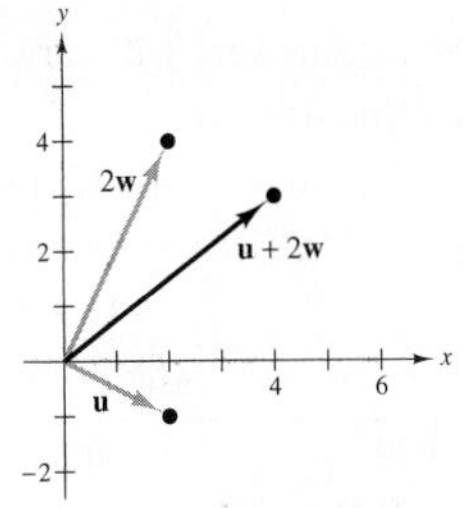

29. $(3, 5)$ **31.** 5 **33.** $\sqrt{61}$ **35.** 4

37. $\left\langle \sqrt{17}/17, 4\sqrt{17}/17\right\rangle$ **39.** $\left\langle 3\sqrt{34}/34, 5\sqrt{34}/34\right\rangle$

41. (a) $\sqrt{2}$ (b) $\sqrt{5}$ (c) 1 (d) 1 (e) 1 (f) 1

43. (a) $\sqrt{5}/2$ (b) $\sqrt{13}$ (c) $\sqrt{85}/2$ (d) 1 (e) 1 (f) 1

45.

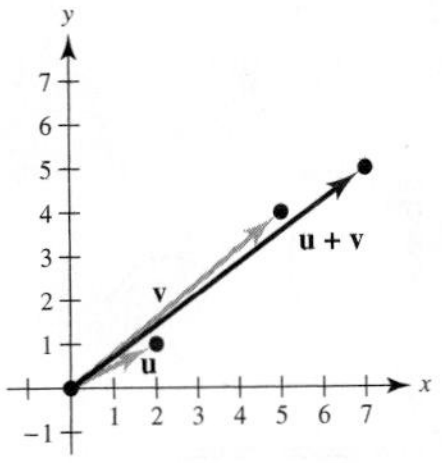

$\|\mathbf{u}\| + \|\mathbf{v}\| = \sqrt{5} + \sqrt{41}$ and $\|\mathbf{u} + \mathbf{v}\| = \sqrt{74}$

$\sqrt{74} < \sqrt{5} + \sqrt{41}$

47. $\left\langle 2\sqrt{2}, 2\sqrt{2}\right\rangle$ **49.** $\left\langle 1, \sqrt{3}\right\rangle$ **51.** $\langle 3, 0\rangle$ **53.** $\left\langle -\sqrt{3}, 1\right\rangle$

55. $\left\langle \dfrac{2 + 3\sqrt{2}}{2}, \dfrac{3\sqrt{2}}{2}\right\rangle$ **57.** $\langle 2\cos 4 + \cos 2, 2\sin 4 + \sin 2\rangle$

59. Answers will vary. Example: A scalar is a single real number such as 2. A vector is a line segment having both direction and magnitude. The vector $\left\langle \sqrt{3}, 1\right\rangle$, given in component form, has a direction of $\pi/6$ and a magnitude of 2.

61. (a) Vector; has magnitude and direction
(b) Scalar; has only magnitude

63. $a = 1, b = 1$ **65.** $a = 1, b = 2$ **67.** $a = \frac{2}{3}, b = \frac{1}{3}$

69. (a) $\pm\left(1/\sqrt{37}\right)\langle 1, 6\rangle$
(b) $\pm\left(1/\sqrt{37}\right)\langle 6, -1\rangle$

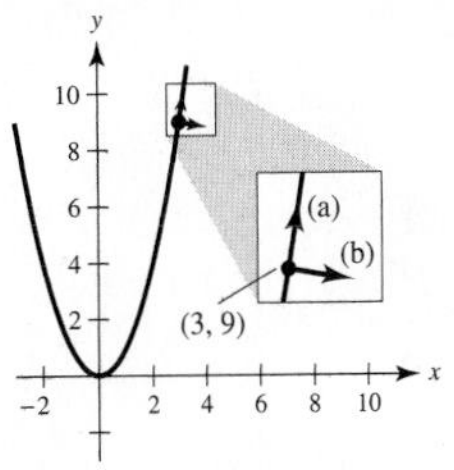

71. (a) $\pm\left(1/\sqrt{10}\right)\langle 1, 3\rangle$
(b) $\pm\left(1/\sqrt{10}\right)\langle 3, -1\rangle$

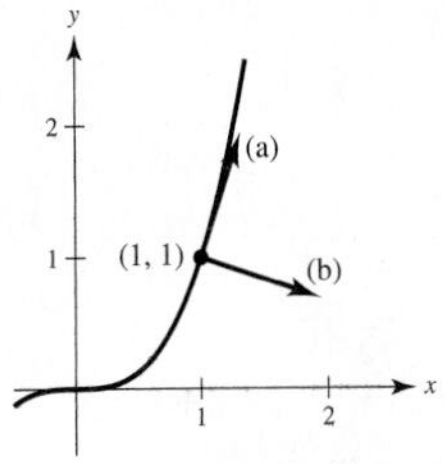

73. (a) $\pm\frac{1}{5}\langle -4, 3\rangle$
(b) $\pm\frac{1}{5}\langle 3, 4\rangle$

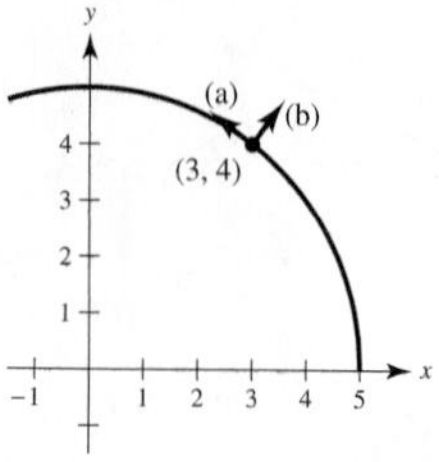

75. $\langle -\sqrt{2}/2, \sqrt{2}/2\rangle$

77. (a)–(c) Answers will vary. **79.** 1.33, 132.5°

81. (a) Direction: $\alpha = 11.8°$
Magnitude: 440.2 N
(b) $M = \sqrt{(275 + 180\cos\theta)^2 + (180\sin\theta)^2}$
$\alpha = \arctan\left(\dfrac{180\sin\theta}{275 + 180\cos\theta}\right)$
(c)

θ	0°	30°	60°	90°	120°
M	455.0	440.2	396.9	328.7	241.9
α	0°	11.8°	23.1°	33.2°	40.1°

θ	150°	180°
M	149.3	95.0
α	37.1°	0°

(d)
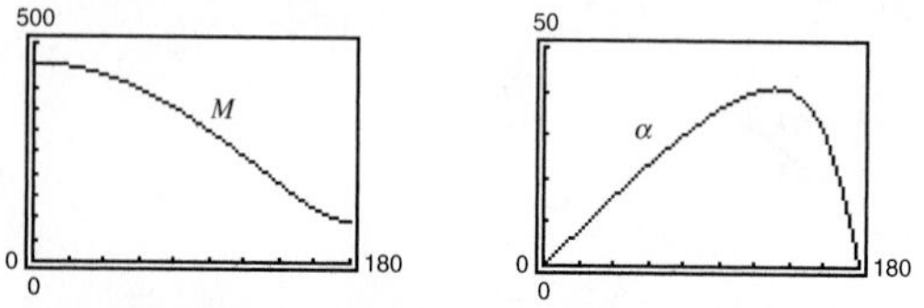

(e) M decreases because the forces change from acting in the same direction to acting in opposite directions as θ increases from 0° to 180°.

83. 71.3°, 228.5 lb

85. (a) $\theta = 0°$ (b) $\theta = 180°$
(c) No, the resultant can only be less than or equal to the sum.

87. $(-4, -1)$, $(6, 5)$, $(10, 3)$

89. Tension in cable $AC \approx 1758.8$ lb
Tension in cable $BC \approx 1305.4$ lb

91. Horizontal: 1193.43 ft/sec
Vertical: 125.43 ft/sec

93. 38.3° north of west
882.9 kph

95. True **97.** True **99.** False. $\|a\mathbf{i} + b\mathbf{j}\| = \sqrt{2}|a|$

101–103. Proofs **105.** $x^2 + y^2 = 25$

Section 11.2 (page 778)

1.

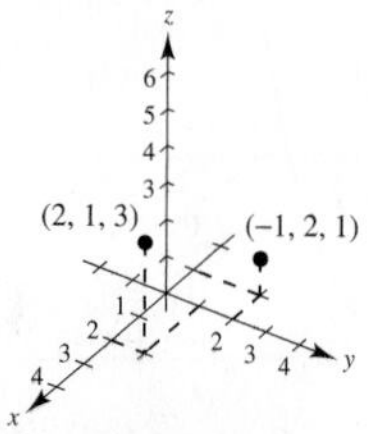

3.

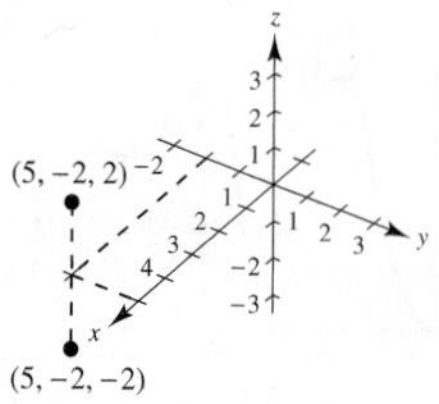

5. $A(2, 3, 4)$
$B(-1, -2, 2)$

7. $(-3, 4, 5)$ **9.** $(10, 0, 0)$ **11.** 0

13. Six units above the xy-plane

15. Four units in front of the yz-plane

17. To the left of the xz-plane and either above, below, or on the xy-plane and either in front of, behind, or on the yz-plane

19. Within three units of the xz-plane

21. Three units below the xy-plane, to the right of the xz-plane, and in front of the yz-plane, *or* three units below the xy-plane, to the left of the xz-plane, and behind the yz-plane

23. 1. Above the xy-plane and (a) to the right of the xz-plane and behind the yz-plane or (b) to the left of the xz-plane and in front of the yz-plane, *or*
2. Below the xy-plane and (a) to the right of the xz-plane and in front of the yz-plane or (b) to the left of the xz-plane and behind the yz-plane

25. $\sqrt{65}$ **27.** $\sqrt{61}$

29. $3, 3\sqrt{5}, 6$
Right triangle

31. $6, 6, 2\sqrt{10}$
Isosceles triangle

33. $(0, 0, 5)$, $(2, 2, 6)$, $(2, -4, 9)$

35. $\left(\frac{3}{2}, -3, 5\right)$ **37.** $(x - 0)^2 + (y - 2)^2 + (z - 5)^2 = 4$

39. $(x - 1)^2 + (y - 3)^2 + (z - 0)^2 = 10$

41. $(x - 1)^2 + (y + 3)^2 + (z + 4)^2 = 25$
Center: $(1, -3, -4)$
Radius: 5

43. $\left(x - \frac{1}{3}\right)^2 + (y + 1)^2 + z^2 = 1$
Center: $\left(\frac{1}{3}, -1, 0\right)$
Radius: 1

45. A solid sphere with center $(0, 0, 0)$ and radius 6

47. Interior of sphere of radius 4 centered at $(2, -3, 4)$

49. (a) $\langle -2, 2, 2\rangle$
(b)

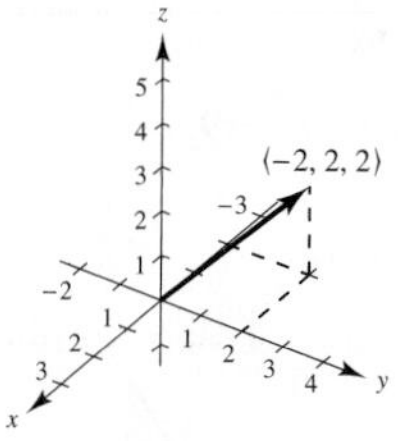

51. (a) $\langle -3, 0, 3\rangle$
(b)

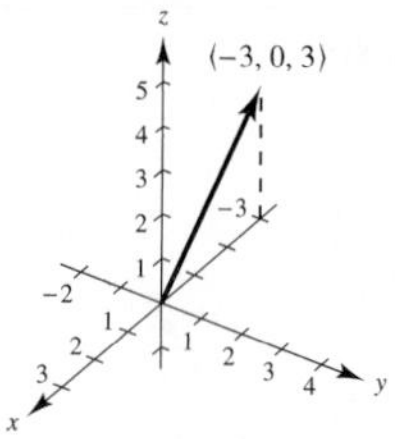

53. $\mathbf{u} = \langle 1, -1, 6\rangle$
$\|\mathbf{u}\| = \sqrt{38}$
$\dfrac{\mathbf{u}}{\|\mathbf{u}\|} = \dfrac{1}{\sqrt{38}}\langle 1, -1, 6\rangle$

55. $\mathbf{u} = \langle -1, 0, -1\rangle$
$\|\mathbf{u}\| = \sqrt{2}$
$\dfrac{\mathbf{u}}{\|\mathbf{u}\|} = \dfrac{1}{\sqrt{2}}\langle -1, 0, -1\rangle$

57. (a) and (c)

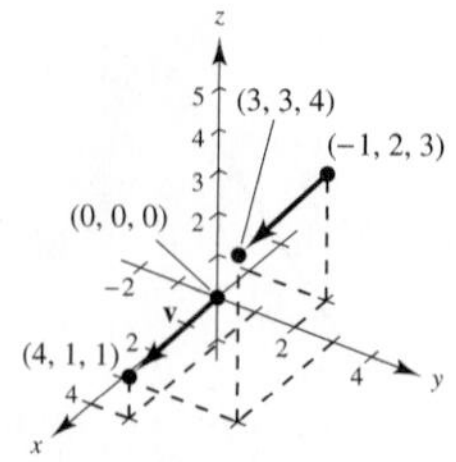

(b) $\langle 4, 1, 1\rangle$

59. $(3, 1, 8)$

61. (a)

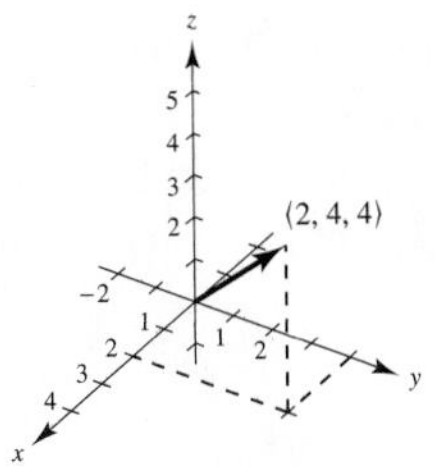

(b)

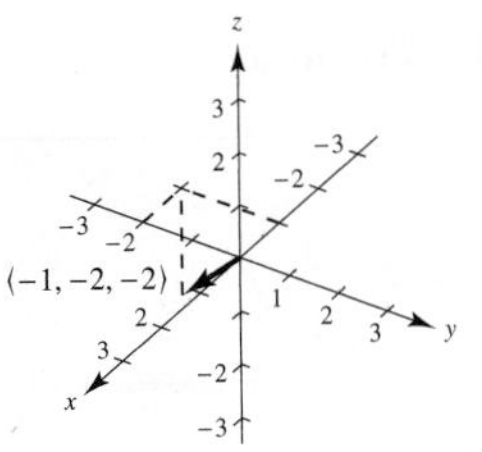

(c)

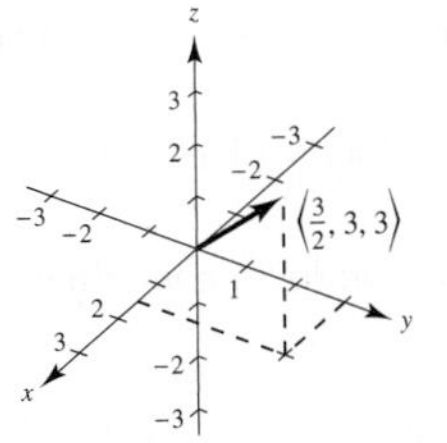

(d)

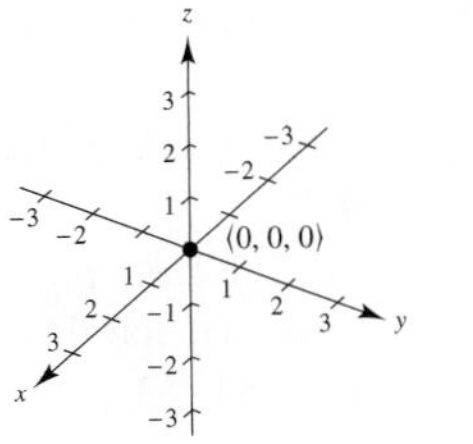

63. $\langle -1, 0, 4\rangle$ **65.** $\langle 6, 12, 6\rangle$ **67.** $\left\langle \frac{7}{2}, 3, \frac{5}{2}\right\rangle$

69. a and b **71.** a **73.** Collinear **75.** Not collinear

77. $\overrightarrow{AB} = \langle 1, 2, 3\rangle$
$\overrightarrow{CD} = \langle 1, 2, 3\rangle$
$\overrightarrow{BD} = \langle -2, 1, 1\rangle$
$\overrightarrow{AC} = \langle -2, 1, 1\rangle$
Since $\overrightarrow{AB} = \overrightarrow{CD}$ and $\overrightarrow{BD} = \overrightarrow{AC}$, the given points form the vertices of a parallelogram.

79. 0 **81.** $\sqrt{14}$ **83.** $\sqrt{34}$

85. (a) $\frac{1}{3}\langle 2, -1, 2\rangle$ (b) $-\frac{1}{3}\langle 2, -1, 2\rangle$

87. (a) $\left(1/\sqrt{38}\right)\langle 3, 2, -5\rangle$ (b) $-\left(1/\sqrt{38}\right)\langle 3, 2, -5\rangle$

89. (a)–(d) Answers will vary. **91.** $\pm\frac{5}{3}$

93. $\langle 0, 10/\sqrt{2}, 10/\sqrt{2}\,\rangle$ **95.** $\left\langle 1, -1, \frac{1}{2}\right\rangle$

97.

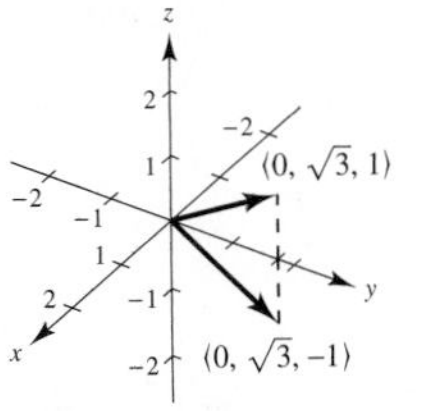

$\left\langle 0, \sqrt{3}, \pm 1\right\rangle$

99. $(2, -1, 2)$

101. (a)

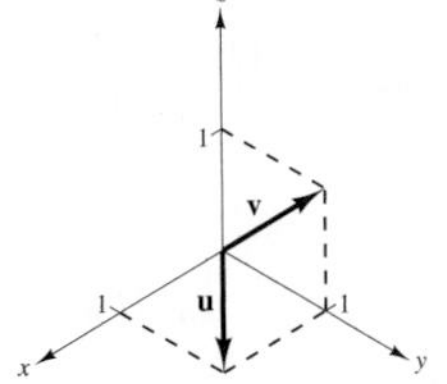

(b) $a = 0, a + b = 0, b = 0$
(c) $a = 1, a + b = 2, b = 1$
(d) Not possible

103. x_0 is directed distance to yz-plane.
y_0 is directed distance to xz-plane.
z_0 is directed distance to xy-plane.

105. $(x - x_0)^2 + (y - y_0)^2 + (z - z_0)^2 = r^2$ **107.** 0

109. (a) $T = 8L/\sqrt{L^2 - 18^2}, L > 18$

(b)

L	20	25	30	35	40	45	50
T	18.4	11.5	10	9.3	9.0	8.7	8.6

(c)

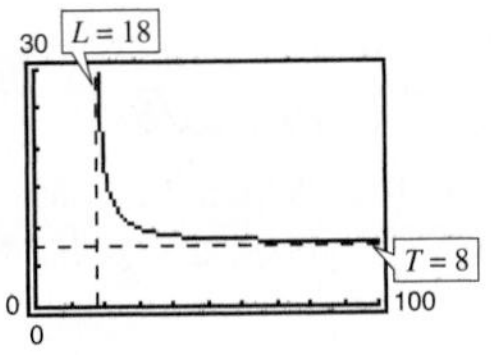

(d) Proof (e) 30 in.

111. $\left(\sqrt{3}/3\right)\langle 1, 1, 1\rangle$

113. Tension in cable AB: 202.919 N
Tension in cable AC: 157.909 N
Tension in cable AD: 226.521 N

115. $\left(x - \frac{4}{3}\right)^2 + (y - 3)^2 + \left(z + \frac{1}{3}\right)^2 = \frac{44}{9}$

Section 11.3 (page 787)

1. (a) -6 (b) 25 (c) 25 (d) $\langle -12, 18\rangle$ (e) -12

3. (a) -17 (b) 26 (c) 26 (d) $\langle 51, -34\rangle$ (e) -34

5. (a) 2 (b) 29 (c) 29 (d) $\langle 0, 12, 10\rangle$ (e) 4

7. (a) 1 (b) 6 (c) 6 (d) $\mathbf{i} - \mathbf{k}$ (e) 2

9. 20 **11.** $\pi/2$ **13.** $\arccos\left(-1/5\sqrt{2}\right) \approx 98.1°$

15. $\arccos\left(\sqrt{2}/3\right) \approx 61.9°$ **17.** $\arccos\left(-8\sqrt{13}/65\right) \approx 116.3°$

19. Neither **21.** Orthogonal **23.** Neither

25. Orthogonal **27.** Right triangle; answers will vary.

29. Acute triangle; answers will vary.

31. $\cos\alpha = \frac{1}{3}$
$\cos\beta = \frac{2}{3}$
$\cos\gamma = \frac{2}{3}$

33. $\cos\alpha = 0$
$\cos\beta = 3/\sqrt{13}$
$\cos\gamma = -2/\sqrt{13}$

35. $\alpha \approx 43.3°, \beta \approx 61.0°, \gamma \approx 119.0°$

37. $\alpha \approx 100.5°, \beta \approx 24.1°, \gamma \approx 68.6°$

39. Magnitude: 124.310 lb
$\alpha \approx 29.48°, \beta \approx 61.39°, \gamma \approx 96.53°$

41. $\alpha = 90°, \beta = 45°, \gamma = 45°$ **43.** $\langle 4, -1\rangle$ **45.** $\langle 2, 1, 1\rangle$

47. (a) $\left\langle \frac{5}{2}, \frac{1}{2}\right\rangle$ (b) $\left\langle -\frac{1}{2}, \frac{5}{2}\right\rangle$

49. (a) $\left\langle 0, \frac{33}{25}, \frac{44}{25}\right\rangle$ (b) $\left\langle 2, -\frac{8}{25}, \frac{6}{25}\right\rangle$

51. See "Definition of Dot Product," page 781.

53. (a) $\theta = \pi/2$ (b) $0 < \theta < \pi/2$ (c) $\pi/2 < \theta < \pi$

55. See the definitions of direction cosines and direction angles on page 784.

57. (a) The vectors are parallel. (b) The vectors are orthogonal.

59. \$12,351.25; Total revenue **61.** (a)–(c) Answers will vary.

63. Answers will vary. **65.** $\langle 0, 0\rangle$

67. Answers will vary. Example: $\langle 4, 3\rangle$ and $\langle -4, -3\rangle$

69. Answers will vary. Example: $\langle 2, 0, 3\rangle$ and $\langle -2, 0, -3\rangle$

71. (a) 8335.1 lb (b) 47,270.8 lb **73.** 425 ft-lb

75. False. For example, $\langle 1, 1\rangle \cdot \langle 2, 3\rangle = 5$ and $\langle 1, 1\rangle \cdot \langle 1, 4\rangle = 5$, but $\langle 2, 3\rangle \neq \langle 1, 4\rangle$.

77. $\arccos\left(1/\sqrt{3}\right) \approx 54.7°$

79. (a) To $y = x^2$ at (1, 1): $\left\langle \pm\sqrt{5}/5, \pm 2\sqrt{5}/5\right\rangle$
To $y = x^{1/3}$ at (1, 1): $\left\langle \pm 3\sqrt{10}/10, \pm\sqrt{10}/10\right\rangle$
To $y = x^2$ at (0, 0): $\langle \pm 1, 0\rangle$
To $y = x^{1/3}$ at (0, 0): $\langle 0, \pm 1\rangle$
(b) At (1, 1), $\theta = 45°$
At (0, 0), $\theta = 90°$

81. (a) To $y = 1 - x^2$ at $(1, 0)$: $\langle \pm\sqrt{5}/5, -2\sqrt{5}/5 \rangle$
To $y = x^2 - 1$ at $(1, 0)$: $\langle \pm\sqrt{5}/5, \pm 2\sqrt{5}/5 \rangle$
To $y = 1 - x^2$ at $(-1, 0)$: $\langle \pm\sqrt{5}/5, \pm 2\sqrt{5}/5 \rangle$
To $y = x^2 - 1$ at $(-1, 0)$: $\langle \pm\sqrt{5}/5, \mp -2\sqrt{5}/5 \rangle$
(b) At $(1, 0)$, $\theta = 53.13°$
At $(-1, 0)$, $\theta = 53.13°$

83. Proof

85. (a)

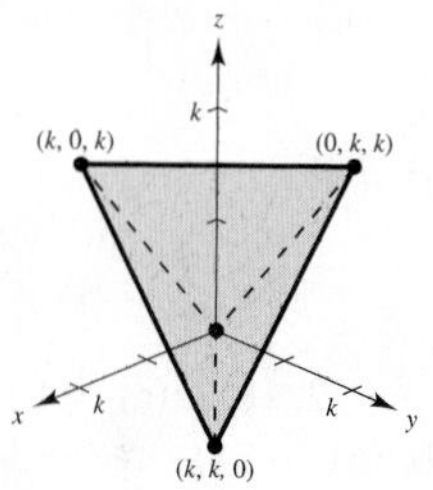

(b) $k\sqrt{2}$ (c) $60°$ (d) $109.5°$

87–89. Proofs

Section 11.4 (page 796)

1. $-\mathbf{k}$

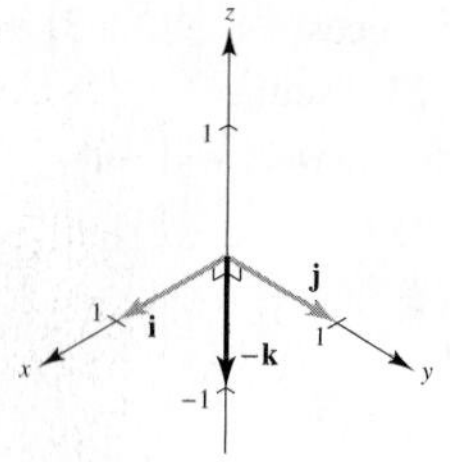

3. $\mathbf{i}$

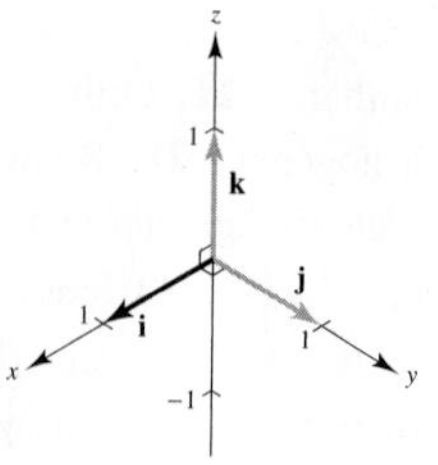

5. $-\mathbf{j}$

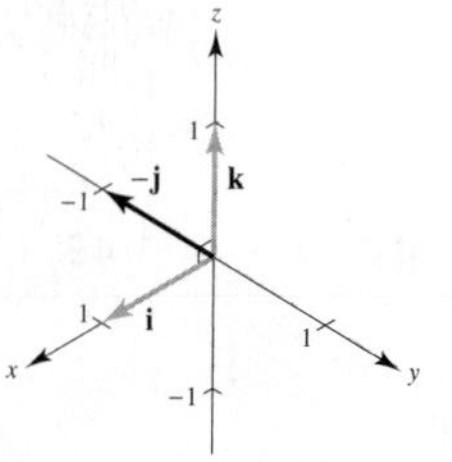

7. (a) $-22\mathbf{i} + 16\mathbf{j} - 23\mathbf{k}$
(b) $22\mathbf{i} - 16\mathbf{j} + 23\mathbf{k}$
(c) $\mathbf{0}$

9. (a) $17\mathbf{i} - 33\mathbf{j} - 10\mathbf{k}$
(b) $-17\mathbf{i} + 33\mathbf{j} + 10\mathbf{k}$
(c) $\mathbf{0}$

11. $\langle -1, -1, -1 \rangle$ **13.** $\langle 0, 0, 54 \rangle$ **15.** $\langle -2, 3, -1 \rangle$

17.

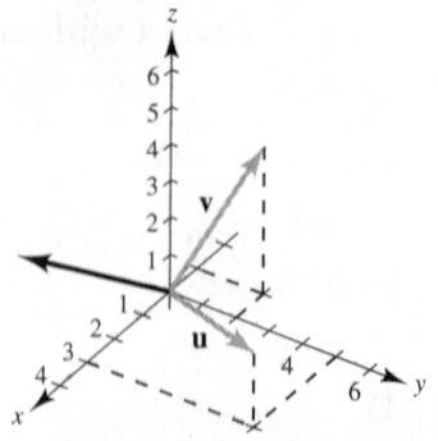

19.

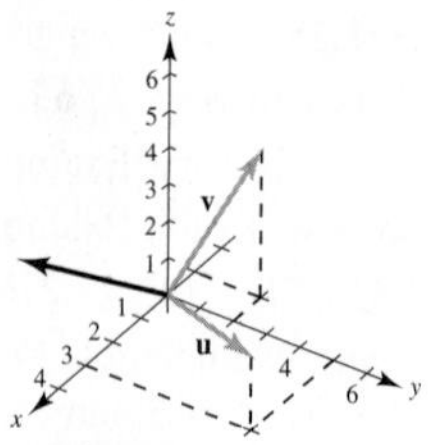

21. $\langle -70, -23, 57/2 \rangle$
$\langle -140/\sqrt{24{,}965}, -46/\sqrt{24{,}965}, 57/\sqrt{24{,}965} \rangle$

23. $\langle -\frac{71}{20}, -\frac{11}{5}, \frac{5}{4} \rangle$
$\langle -71/\sqrt{7602}, -44/\sqrt{7602}, 25/\sqrt{7602} \rangle$

25. Answers will vary. **27.** 1 **29.** $6\sqrt{5}$ **31.** $2\sqrt{83}$

33. $3\sqrt{13}/2$ **35.** $\sqrt{16{,}742}/2$ **37.** $10 \cos 40 \approx 7.66$ ft-lb

39. (a) $90 \sin \theta$

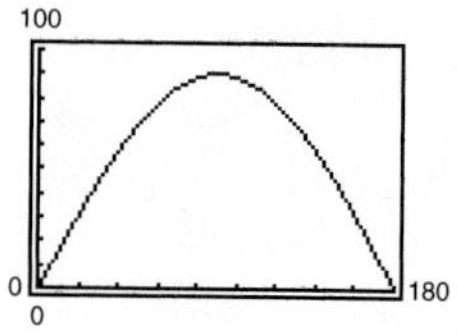

(b) $45\sqrt{2} \approx 63.64$
(c) $\theta = 90°$. This is what should be expected. When $\theta = 90°$, the pipe wrench is horizontal.

41. 1 **43.** 6 **45.** 2 **47.** 75

49. See "Definition of Cross Product of Two Vectors in Space," page 790.

51. The magnitude of the cross product will increase by a factor of 4.

53. False. The cross product of two vectors is not defined in a two-dimensional coordinate system.

55. True **57–63.** Proofs

Section 11.5 (page 805)

1. (a)

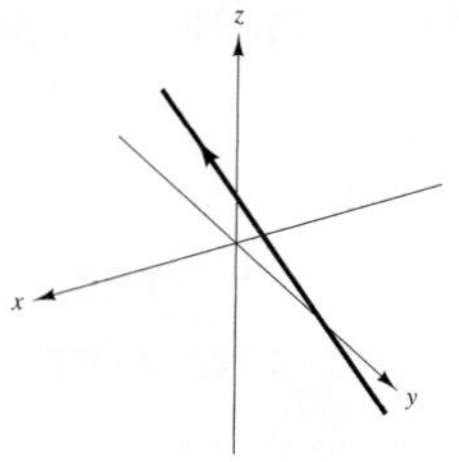

(b) $P = (1, 2, 2)$, $Q = (10, -1, 17)$, $\overrightarrow{PQ} = \langle 9, -3, 15 \rangle$
(There are many correct answers.) The components of the vector and the coefficients of t are proportional because the line is parallel to $\overrightarrow{PQ}$.
(c) $\left(-\frac{1}{5}, \frac{12}{5}, 0\right)$, $(7, 0, 12)$, $\left(0, \frac{7}{3}, \frac{1}{3}\right)$

	Parametric Equations	*Symmetric Equations*	*Direction Numbers*
3.	$x = t$, $y = 2t$, $z = 3t$	$x = \dfrac{y}{2} = \dfrac{z}{3}$	1, 2, 3
5.	$x = -2 + 2t$, $y = 4t$, $z = 3 - 2t$	$\dfrac{x+2}{2} = \dfrac{y}{4} = \dfrac{z-3}{-2}$	2, 4, −2
7.	$x = 1 + 3t$, $y = -2t$, $z = 1 + t$	$\dfrac{x-1}{3} = \dfrac{y}{-2} = \dfrac{z-1}{1}$	3, −2, 1
9.	$x = 5 + 17t$, $y = -3 - 11t$, $z = -2 - 9t$	$\dfrac{x-5}{17} = \dfrac{y+3}{-11} = \dfrac{z+2}{-9}$	17, −11, −9
11.	$x = 2 + 8t$, $y = 3 + 5t$, $z = 12t$	$\dfrac{x-2}{8} = \dfrac{y-3}{5} = \dfrac{z}{12}$	8, 5, 12

13. $x = 2$, $y = 3$, $z = 4 + t$

15. $x = 2 + 3t$, $y = 3 + 2t$, $z = 4 - t$

17. $x = 5 + 2t$, $y = -3 - t$, $z = -4 + 3t$

19. $x = 2 - t$
$y = 1 + t$
$z = 2 + t$

21. $P(3, -1, -2)$; $\mathbf{v} = \langle -1, 2, 0\rangle$

23. $P(7, -6, -2)$; $\mathbf{v} = \langle 4, 2, 1\rangle$

25. $L_1 = L_2$ and is parallel to L_3.

27. $(2, 3, 1)$; $\cos\theta = 7\sqrt{17}/51$ **29.** Not intersecting

31.

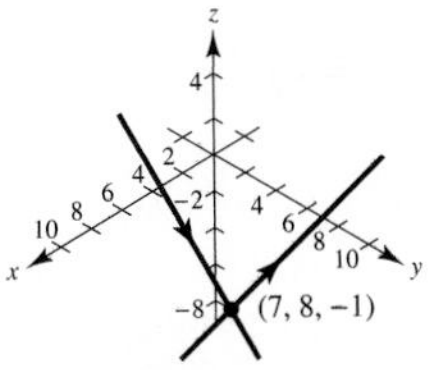

$(7, 8, -1)$

33. (a) $P = (0, 0, -1), Q = (0, -2, 0), R = (3, 4, -1)$
$\overrightarrow{PQ} = \langle 0, -2, 1\rangle, \overrightarrow{PR} = \langle 3, 4, 0\rangle$
(There are many correct answers.)
(b) $\overrightarrow{PQ} \times \overrightarrow{PR} = \langle -4, 3, 6\rangle$
The components of the cross product are proportional to the coefficients of the variables in the equation. The cross product is parallel to the normal vector.

35. $x - 2 = 0$ **37.** $2x + 3y - z = 10$

39. $x - y + 2z = 12$ **41.** $3x + 9y - 7z = 0$

43. $4x - 3y + 4z = 10$ **45.** $z = 3$ **47.** $x + y + z = 5$

49. $7x + y - 11z = 5$ **51.** $y - z = -1$

53.

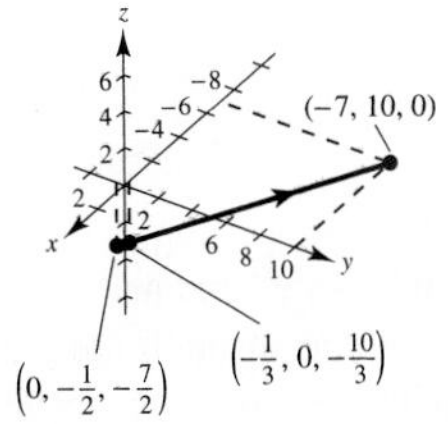

55. $x - z = 0$

57. Orthogonal **59.** Neither; 83.5° **61.** Parallel

63.

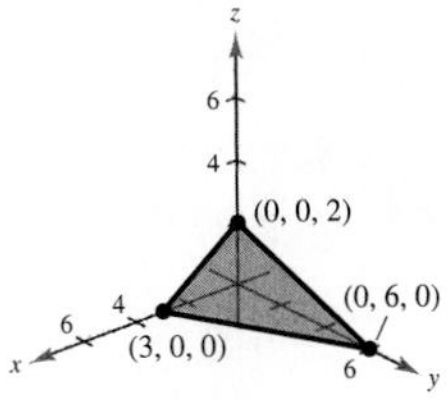

65.

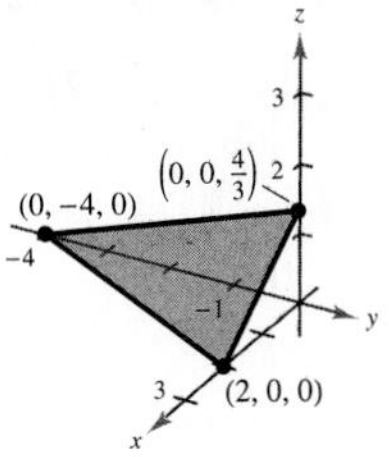

67.

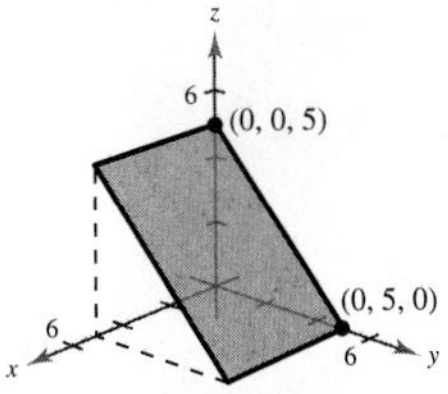

69.

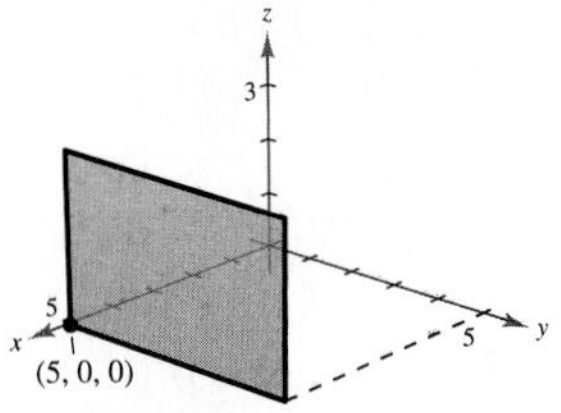

71.

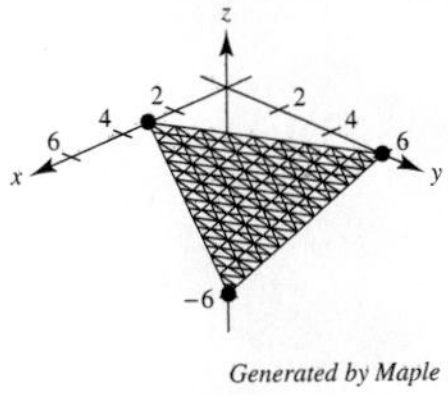

Generated by Maple

73.

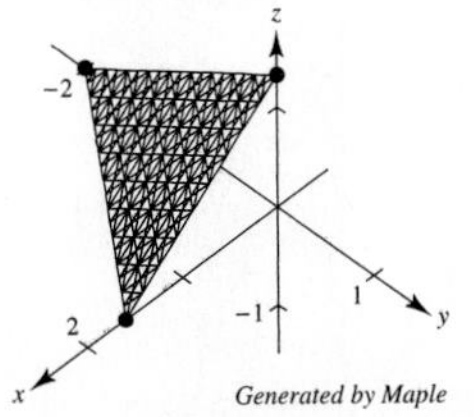

Generated by Maple

75. $P_1 = P_4$ and is parallel to P_2.

77. The planes have intercepts at $(c, 0, 0)$, $(0, c, 0)$, and $(0, 0, c)$ for each value of c.

79. If $c = 0$, $z = 0$ is xy-plane; If $c \neq 0$, plane is parallel to the x-axis and passes through $(0, 0, 0)$ and $(0, 1, -c)$.

81. $x = 2$
$y = 1 + t$
$z = 1 + 2t$

83. $(2, -3, 2)$ The line does not lie in the plane.

85. Not intersecting **87.** $6\sqrt{14}/7$ **89.** $11\sqrt{6}/6$

91. $2\sqrt{26}/13$ **93.** $27\sqrt{94}/188$ **95.** $\sqrt{2533}/17$

97. $7\sqrt{3}/3$ **99.** $\sqrt{66}/3$

101. Parametric equations: $x = x_1 + at$, $y = y_1 + bt$, and $z = z_1 + ct$
Symmetric equations: $\dfrac{x - x_1}{a} = \dfrac{y - y_1}{b} = \dfrac{z - z_1}{c}$
You need a vector $\mathbf{v} = \langle a, b, c\rangle$ parallel to the line and a point $P(x_1, y_1, z_1)$ on the line.

103. Simultaneously solve the two linear equations representing the planes and substitute the values back into one of the original equations. Then choose a value for t and form the corresponding parametric equations for the line of intersection.

105. (a) Parallel if vector $\langle a_1, b_1, c_1\rangle$ is a scalar multiple of $\langle a_2, b_2, c_2\rangle$; $\theta = 0$.
(b) Perpendicular if $a_1a_2 + b_1b_2 + c_1c_2 = 0$; $\theta = \pi/2$.

107. $cbx + acy + abz = abc$

109. (a)

Year	1994	1995	1996	1997	1998
z (approx.)	8.74	8.40	8.26	8.06	7.88

Year	1999	2000
z (approx.)	7.82	7.70

(b) Answers will vary.

111. (a) $\sqrt{70}$ in.
(b)

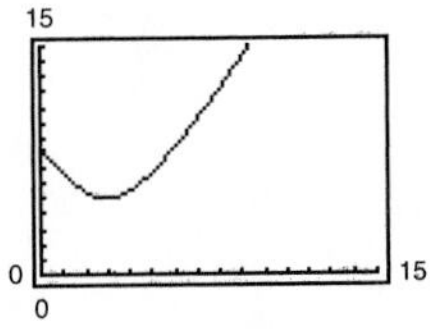

(c) The distance is never zero.
(d) 5 in.

113. $\left(\frac{77}{13}, \frac{48}{13}, -\frac{23}{13}\right)$ **115.** $\left(-\frac{1}{2}, -\frac{9}{4}, \frac{1}{4}\right)$ **117.** True **119.** True

Section 11.6 (page 818)

1. c **2.** e **3.** f **4.** b **5.** d **6.** a

7. Plane

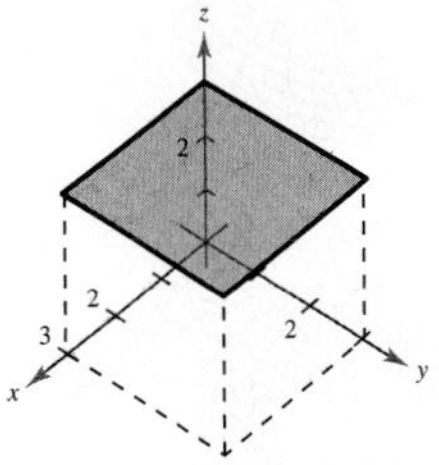

9. Right circular cylinder

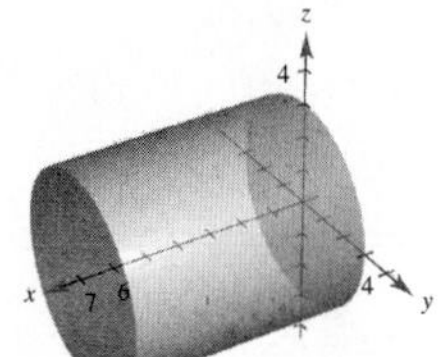

11. Parabolic cylinder

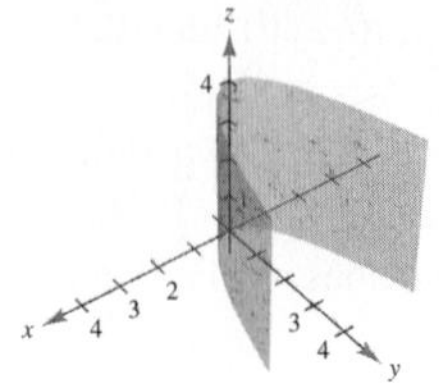

13. Elliptic cylinder

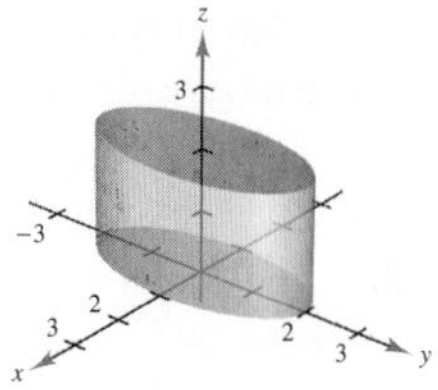

15. Cylinder

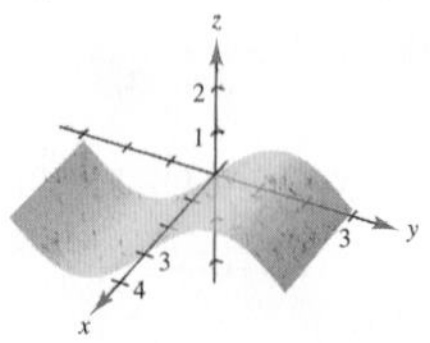

17. (a) $(20, 0, 0)$ (b) $(10, 10, 20)$
(c) $(0, 0, 20)$ (d) $(0, 20, 0)$

19. Ellipsoid

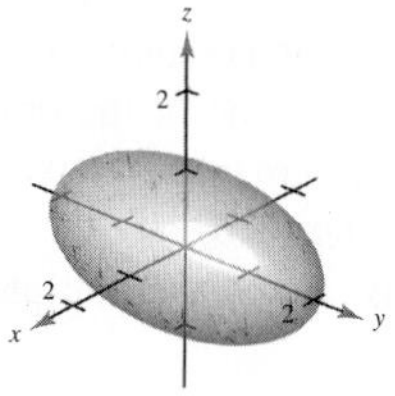

21. Hyperboloid of one sheet

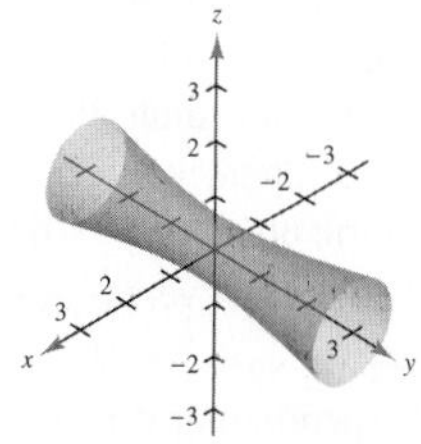

23. Elliptic paraboloid

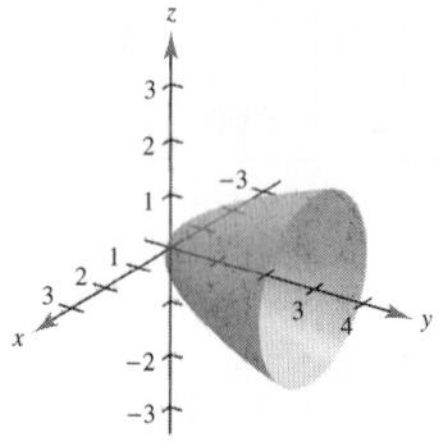

25. Hyperbolic paraboloid

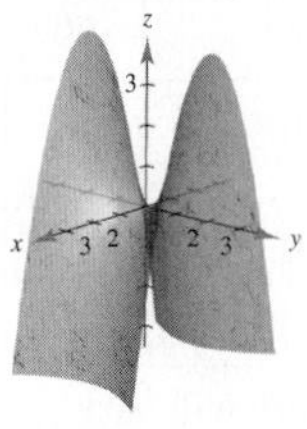

27. Elliptic cone

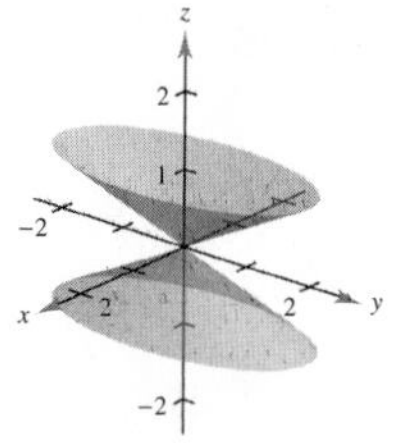

29. Ellipsoid

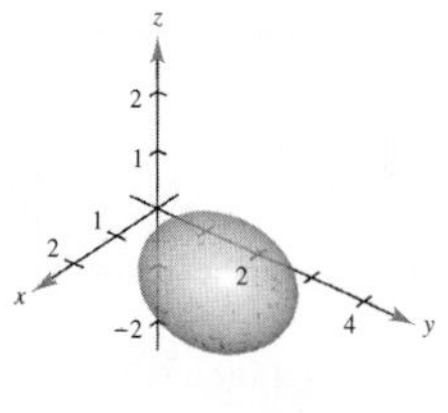

31.

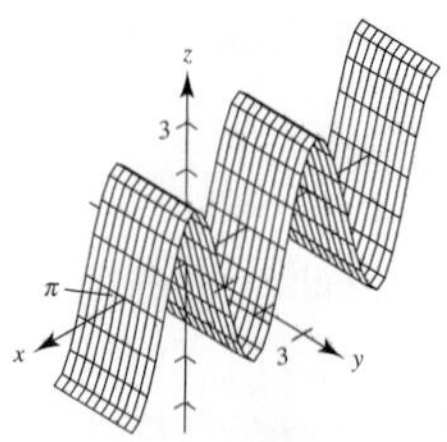

33.

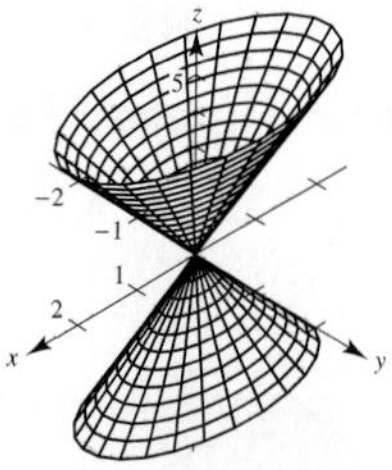

35.

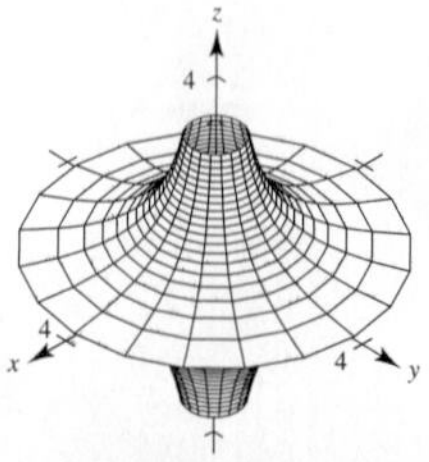

37.

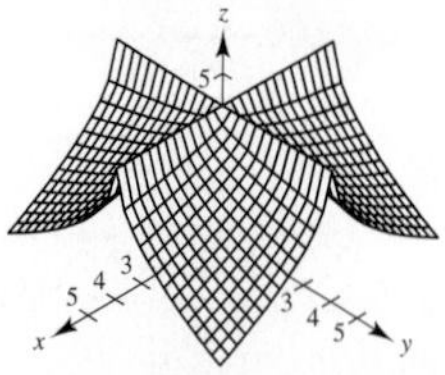

39.

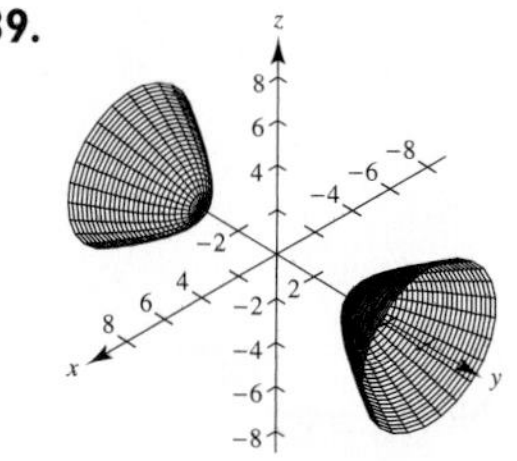

41.

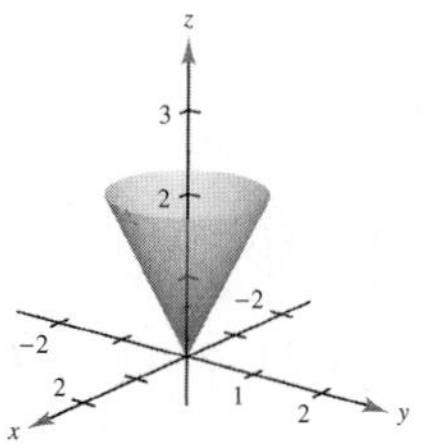

43.

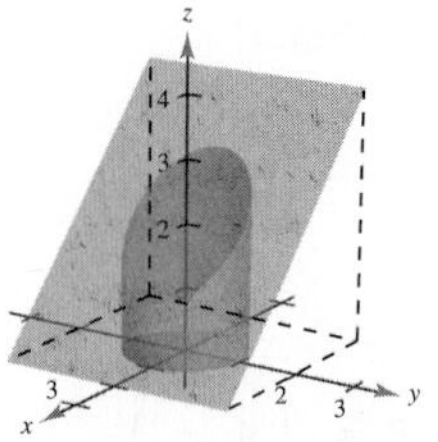

45. $x^2 + z^2 = 4y$

47. $4x^2 + 4y^2 = z^2$ **49.** $y^2 + z^2 = 4/x^2$

51. $y = \sqrt{2z}\ (\text{or } x = \sqrt{2z})$

53. Let C be a curve in a plane and let L be a line not in a parallel plane. The set of all lines parallel to L and intersecting C is called a cylinder. C is called the generating curve of the cylinder, and the parallel lines are called rulings.

55. See pages 812 and 813. **57.** $128\pi/3$

59. (a) Major axis: $4\sqrt{2}$ (b) Major axis: $8\sqrt{2}$
Minor axis: 4 Minor axis: 8
Foci: $(0, \pm 2, 2)$ Foci: $(0, \pm 4, 8)$

61. $x^2 + z^2 = 8y$; Elliptic paraboloid

63. $x^2/3963^2 + y^2/3963^2 + z^2/3950^2 = 1$

65. $x = at, y = -bt, z = 0$;
$x = at, y = bt + ab^2, z = 2abt + a^2b^2$

67. True

69. The Klein bottle does not have both an "inside" and an "outside." It is formed by inserting the small open end through the side of the bottle and making it contiguous with the top of the bottle.

Section 11.7 (page 825)

1. $(5, 0, 2)$ **3.** $(1, \sqrt{3}, 2)$ **5.** $(-2\sqrt{3}, -2, 3)$

7. $(5, \pi/2, 1)$ **9.** $(2, \pi/3, 4)$ **11.** $(2\sqrt{2}, -\pi/4, -4)$

13. $z = 5$ **15.** $r^2 + z^2 = 10$ **17.** $r = \sec\theta\tan\theta$

19. $r^2\sin^2\theta = 10 - z^2$

21. $x^2 + y^2 = 4$

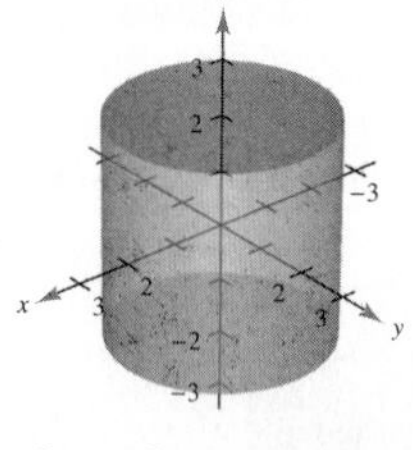

23. $x - \sqrt{3}y = 0$

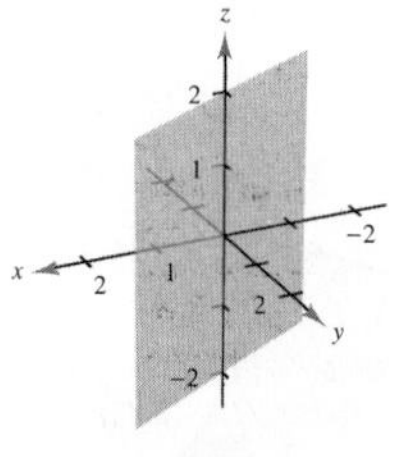

25. $x^2 + y^2 - 2y = 0$

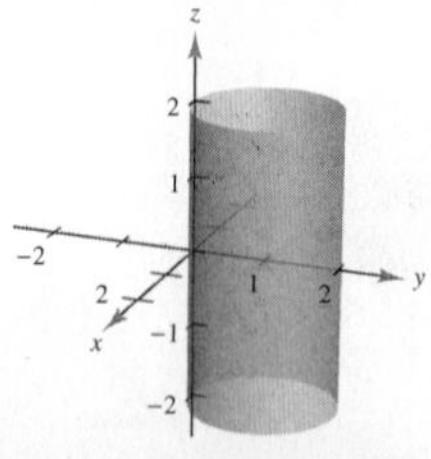

27. $x^2 + y^2 + z^2 = 4$

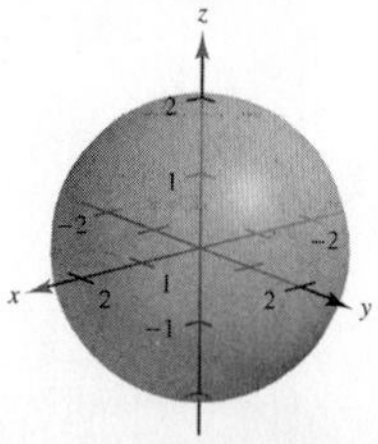

29. $(4, 0, \pi/2)$ **31.** $(4\sqrt{2}, 2\pi/3, \pi/4)$ **33.** $(4, \pi/6, \pi/6)$
35. $(\sqrt{6}, \sqrt{2}, 2\sqrt{2})$ **37.** $(0, 0, 12)$ **39.** $(\frac{5}{2}, \frac{5}{2}, -5\sqrt{2}/2)$
41. $\rho = 3 \csc\phi \csc\theta$ **43.** $\rho = 6$
45. $\rho = 3\csc\phi$ **47.** $\tan^2\phi = 2$
49. $x^2 + y^2 + z^2 = 4$ **51.** $3x^2 + 3y^2 - z^2 = 0$

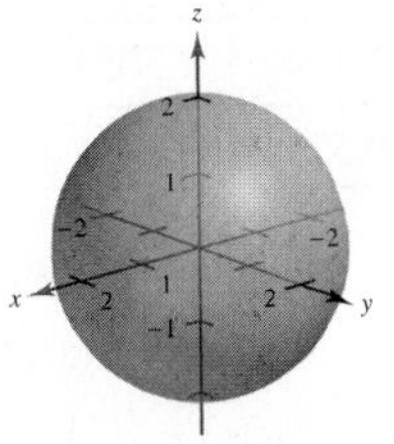

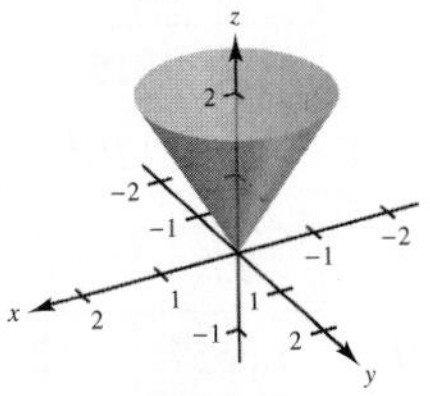

53. $x^2 + y^2 + (z - 2)^2 = 4$ **55.** $x^2 + y^2 = 1$

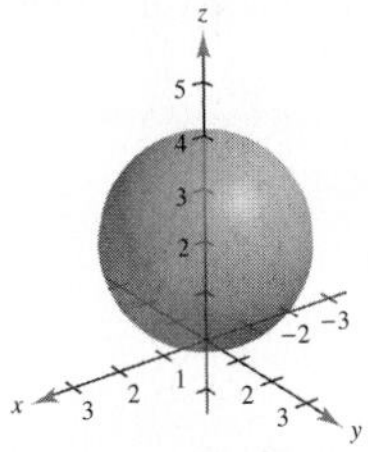

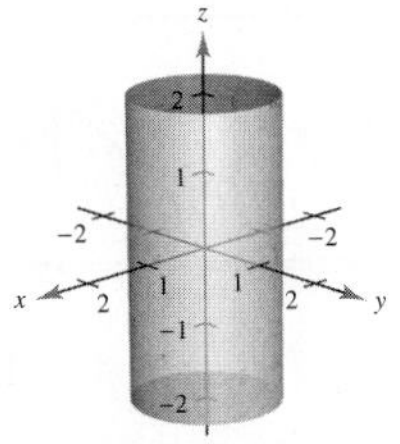

57. $(4, \pi/4, \pi/2)$ **59.** $(4\sqrt{2}, \pi/2, \pi/4)$
61. $(2\sqrt{13}, -\pi/6, \arccos[3/\sqrt{13}])$ **63.** $(13, \pi, \arccos[5/13])$
65. $(10, \pi/6, 0)$ **67.** $(36, \pi, 0)$
69. $(3\sqrt{3}, -\pi/6, 3)$ **71.** $(4, 7\pi/6, 4\sqrt{3})$

	Rectangular	*Cylindrical*	*Spherical*
73.	(4, 6, 3)	(7.211, 0.983, 3)	(7.810, 0.983, 1.177)
75.	(4.698, 1.710, 8)	$(5, \pi/9, 8)$	(9.434, 0.349, 0.559)
77.	(−7.071, 12.247, 14.142)	(14.142, 2.094, 14.142)	$(20, 2\pi/3, \pi/4)$
79.	(3, −2, 2)	(3.606, −0.588, 2)	(4.123, −0.588, 1.064)
81.	$(\frac{5}{2}, \frac{4}{3}, -\frac{3}{2})$	(2.833, 0.490, −1.5)	(3.206, 0.490, 2.058)
83.	(−3.536, 3.536, −5)	$(5, 3\pi/4, -5)$	(7.071, 2.356, 2.356)
85.	(2.804, −2.095, 6)	(−3.5, 2.5, 6)	(6.946, 5.642, 0.528)

87. d **88.** e **89.** c **90.** a **91.** f **92.** b

93. Rectangular to cylindrical:
$r^2 = x^2 + y^2$, $\tan\theta = y/x$, $z = z$
Cylindrical to rectangular:
$x = r\cos\theta$, $y = r\sin\theta$, $z = z$

95. Rectangular to spherical:
$\rho^2 = x^2 + y^2 + z^2$, $\tan\theta = y/x$, $\phi = \arccos(z/\sqrt{x^2 + y^2 + z^2})$
Spherical to rectangular:
$x = \rho\sin\phi\cos\theta$, $y = \rho\sin\phi\sin\theta$, $z = \rho\cos\phi$

97. (a) $r^2 + z^2 = 16$ (b) $\rho = 4$
99. (a) $r^2 + (z - 1)^2 = 1$ (b) $\rho = 2\cos\phi$
101. (a) $r = 4\sin\theta$ (b) $\rho = 4\sin\theta/\sin\phi = 4\sin\theta\csc\phi$
103. (a) $r^2 = 9/(\cos^2\theta - \sin^2\theta)$
(b) $\rho^2 = 9\csc^2\phi/(\cos^2\theta - \sin^2\theta)$

105.

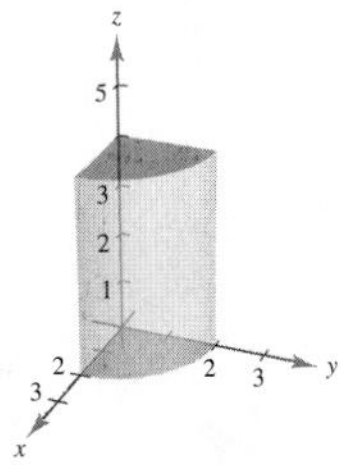

107.

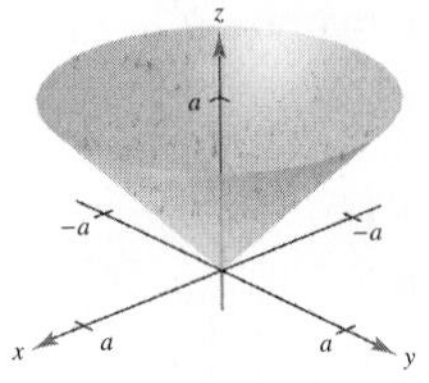

109.

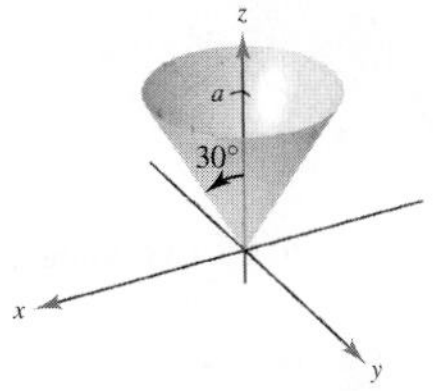

111.

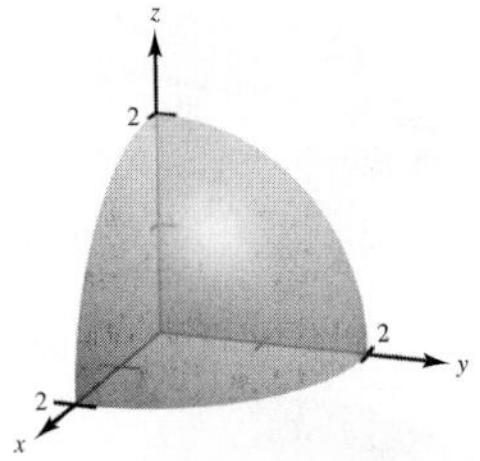

113. Rectangular: $0 \le x \le 10$
$0 \le y \le 10$
$0 \le z \le 10$

115. Spherical: $4 \le \rho \le 6$

117. Cylindrical: $r^2 + z^2 \le 9$, $r \le 3\cos\theta$, $0 \le \theta \le \pi$
119. False. $\theta = c$ represents a vertical half-plane.
121. False. See page 821. **123.** Ellipse

Review Exercises for Chapter 11 (page 827)

1. (a) $\mathbf{u} = 3\mathbf{i} - \mathbf{j}$
$\mathbf{v} = 4\mathbf{i} + 2\mathbf{j}$
(b) $2\sqrt{5}$ (c) $10\mathbf{i}$
3. $\mathbf{v} = \langle -4, 4\sqrt{3}\rangle$ **5.** $(-5, 4, 0)$
7. Above the xy-plane and to the right of the xz-plane *or* below the xy-plane and to the left of the xz-plane
9. $(x - 3)^2 + (y + 2)^2 + (z - 6)^2 = \frac{225}{4}$
11. $(x - 2)^2 + (y - 3)^2 + z^2 = 9$
Center: (2, 3, 0)
Radius: 3

13.

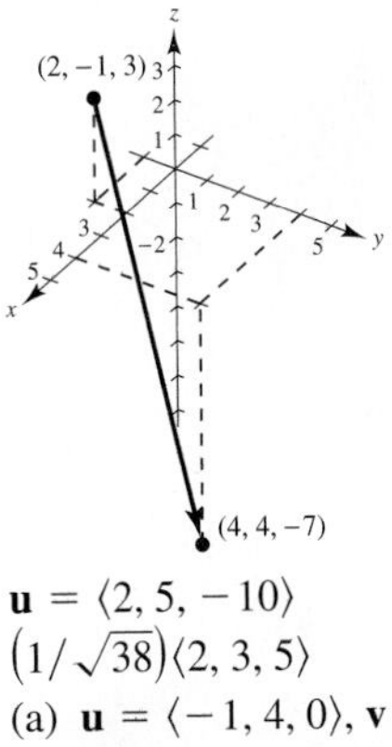

$\mathbf{u} = \langle 2, 5, -10\rangle$

15. Collinear

17. $(1/\sqrt{38})\langle 2, 3, 5\rangle$
19. (a) $\mathbf{u} = \langle -1, 4, 0\rangle$, $\mathbf{v} = \langle -3, 0, 6\rangle$ (b) 3 (c) 45
21. Orthogonal **23.** $\theta = \arccos\left(\dfrac{\sqrt{2} + \sqrt{6}}{4}\right) = 15°$ **25.** π
27. Answers will vary. Example: $\langle -6, 5, 0\rangle$, $\langle 6, -5, 0\rangle$
29. $\mathbf{u}\cdot\mathbf{u} = 14 = \|\mathbf{u}\|^2$ **31.** $\langle -\frac{15}{14}, \frac{5}{7}, -\frac{5}{14}\rangle$
33. $(1/\sqrt{5})(-2\mathbf{i} - \mathbf{j})$ or $(1/\sqrt{5})(2\mathbf{i} + \mathbf{j})$
35. 4 **37.** $\sqrt{285}$ **39.** $100 \sec 20° \approx 106.4$ lb
41. (a) $x = 3 + 6t$, $y = 11t$, $z = 2 + 4t$
(b) $(x - 3)/6 = y/11 = (z - 2)/4$

43. (a) $x = 1, y = 2 + t, z = 3$ (b) None
45. (a) $x = t, y = -1 + t, z = 1$ (b) $x = y + 1, z = 1$
47. $27x + 4y + 32z + 33 = 0$ **49.** $x + 2y = 1$
51. $\frac{8}{7}$ **53.** $\sqrt{35}/7$
55. Plane

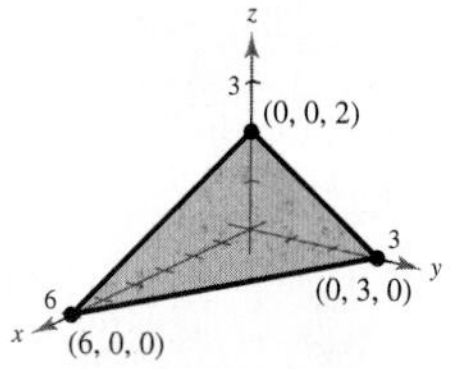

57. Plane

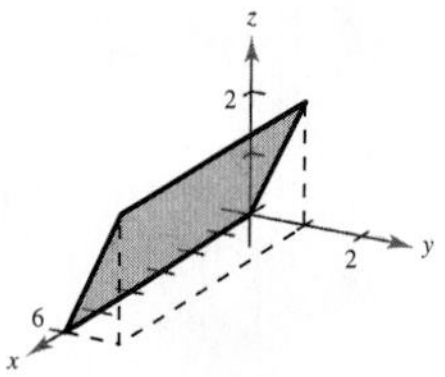

59. Ellipsoid

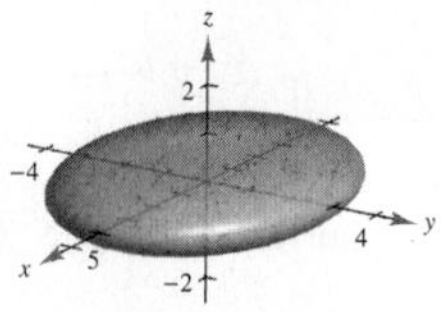

61. Hyperboloid of two sheets

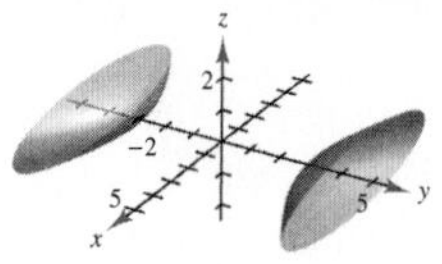

63. Cylinder

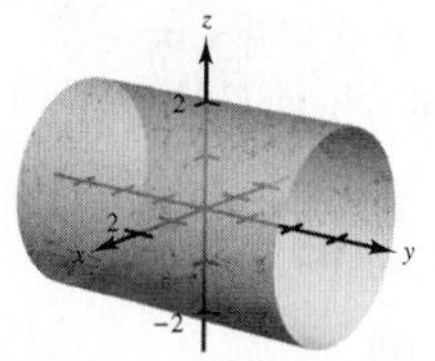

65. Let $y = 2\sqrt{x}$ and revolve around the x-axis.
67. (a) $(4, 3\pi/4, 2)$ (b) $(2\sqrt{5}, 3\pi/4, \arccos[\sqrt{5}/5])$
69. $(50\sqrt{5}, -\pi/6, \arccos[1/\sqrt{5}])$
71. $(25\sqrt{2}/2, -\pi/4, -25\sqrt{2}/2)$
73. (a) $r^2 \cos 2\theta = 2z$ (b) $\rho = 2 \sec 2\theta \cos \phi \csc^2 \phi$
75. $x^2 + (y - 2)^2 = 4$

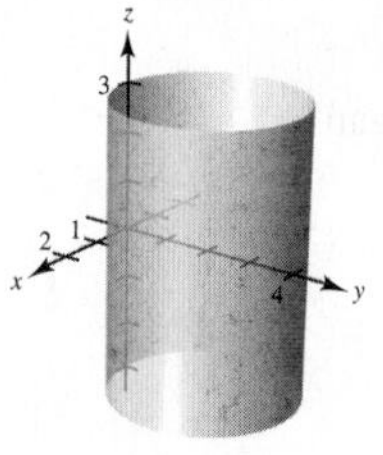

77. $x = y$

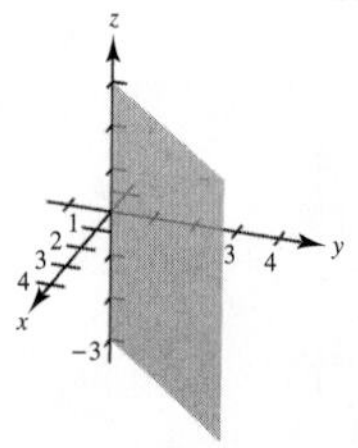

P.S. Problem Solving (page 829)

1–3. Proofs
5. (a) $3\sqrt{2}/2 \approx 2.12$ (b) $\sqrt{5} \approx 2.24$
7. (a) $\pi/2$ (b) $\frac{1}{2}(\pi ab)k^2$
(c) $V = \frac{1}{2}(\pi ab)k^2$
$V = \frac{1}{2}$(area of base)height
9. (a)

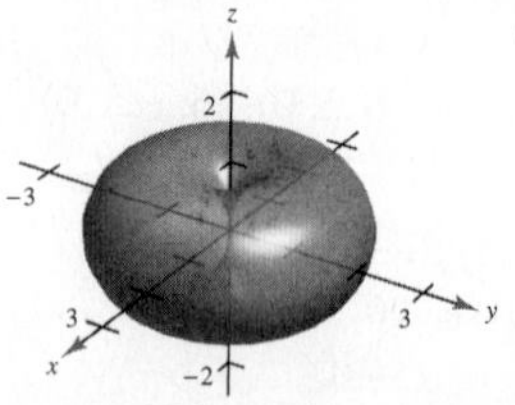

(b)

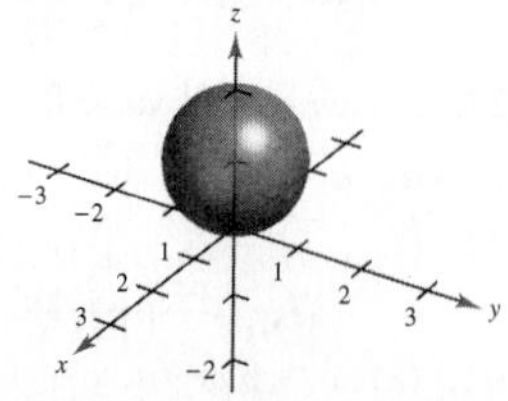

11. Proof
13. (a) Tension: $2\sqrt{3}/3 \approx 1.1547$ lb
Magnitude of **u**: $\sqrt{3}/3 \approx 0.5774$ lb
(b) $T = \sec \theta$; $\|\mathbf{u}\| = \tan \theta$; Domain: $0 \le \theta \le 90°0$
(c)

θ	0°	10°	20°	30°	40°	50°	60°
T	1	1.0154	1.0642	1.1547	1.3054	1.5557	2
$\|\mathbf{u}\|$	0	0.1763	0.3640	0.5774	0.8391	1.1918	1.7321

(d)

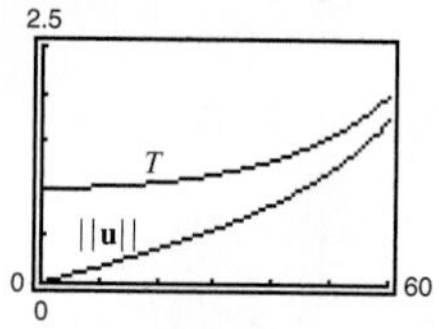

(e) Both are increasing functions.
(f) $\lim_{\theta \to \pi/2^-} T = \infty$ and $\lim_{\theta \to \pi/2^-} \|\mathbf{u}\| = \infty$
Yes. As θ increases, both T and $\|\mathbf{u}\|$ increase.
15. $\langle 0, 0, \cos \alpha \sin \beta - \cos \beta \sin \alpha \rangle$; Proof
17. $$D = \frac{|\overrightarrow{PQ} \cdot \mathbf{n}|}{\|\mathbf{n}\|} = \frac{|\mathbf{w} \cdot (\mathbf{u} \times \mathbf{v})|}{\|\mathbf{u} \times \mathbf{v}\|} = \frac{|(\mathbf{u} \times \mathbf{v}) \cdot \mathbf{w}|}{\|\mathbf{u} \times \mathbf{v}\|} = \frac{|\mathbf{u} \cdot (\mathbf{v} \times \mathbf{w})|}{\|\mathbf{u} \times \mathbf{v}\|}$$
19. Proof

Chapter 12

Section 12.1 (page 837)

1. $(-\infty, 0) \cup (0, \infty)$ **3.** $(0, \infty)$ **5.** $[0, \infty)$ **7.** $(-\infty, \infty)$
9. (a) $\frac{1}{2}\mathbf{i}$ (b) $\mathbf{j}$ (c) $\frac{1}{2}(s + 1)^2\mathbf{i} - s\mathbf{j}$ (d) $\frac{1}{2}\Delta t(\Delta t + 4)\mathbf{i} - \Delta t\mathbf{j}$
11. (a) $\ln 2\mathbf{i} + \frac{1}{2}\mathbf{j} + 6\mathbf{k}$ (b) Not possible
(c) $\ln(t - 4)\mathbf{i} + \dfrac{1}{t - 4}\mathbf{j} + 3(t - 4)\mathbf{k}$
(d) $\ln(1 + \Delta t)\mathbf{i} - \dfrac{\Delta t}{1 + \Delta t}\mathbf{j} + 3\Delta t\mathbf{k}$
13. $\sqrt{1 + t^2}$ **15.** $t^2(5t - 1)$; The dot product is a scalar.
17. b **18.** c **19.** d **20.** a
21. (a) $(-20, 0, 0)$ (b) $(10, 20, 10)$
(c) $(0, 0, 20)$ (d) $(20, 0, 0)$
23.

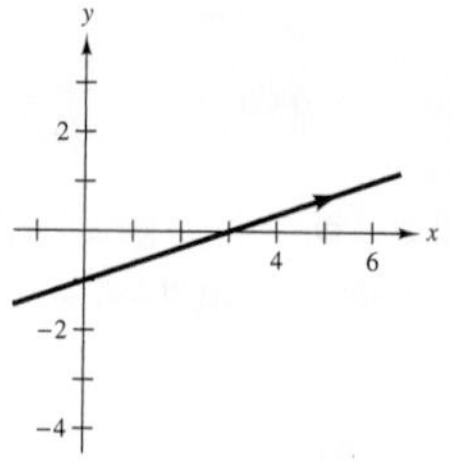

25.

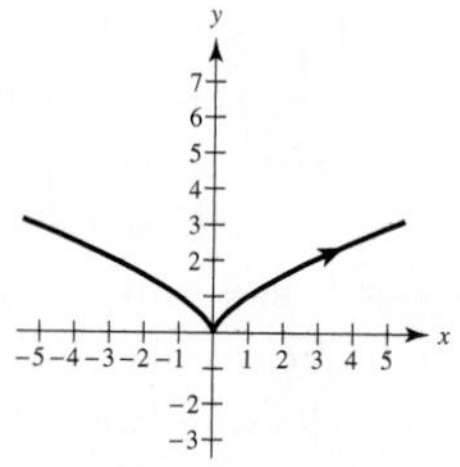

27.

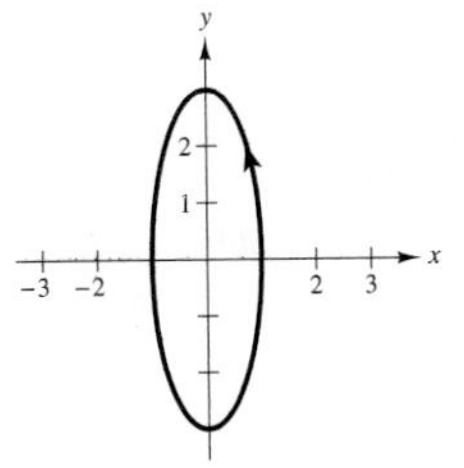

29.

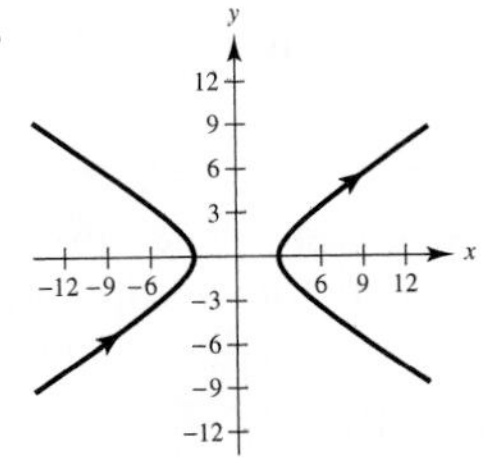

31.

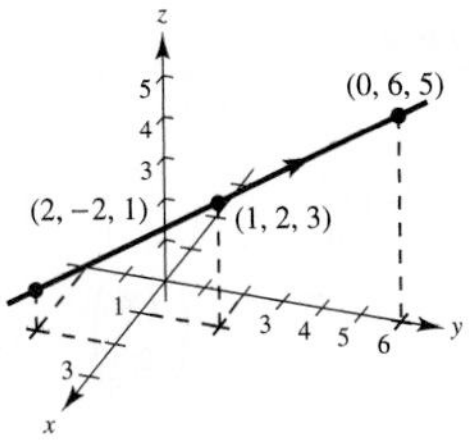

33.

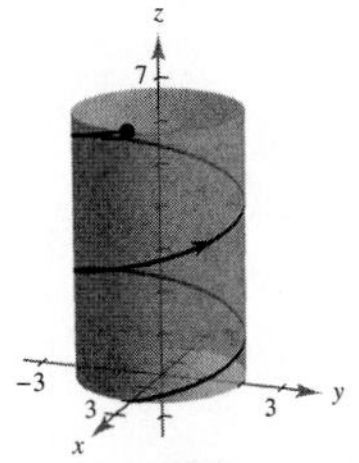

35.

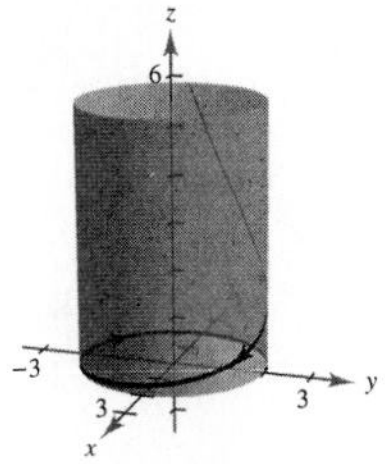

37.

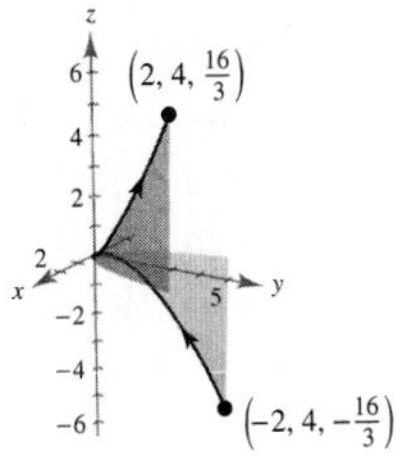

39.

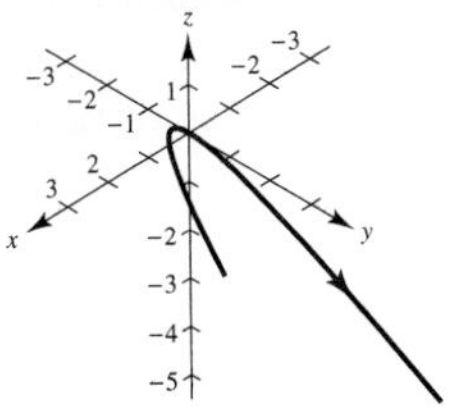

Parabola

41.

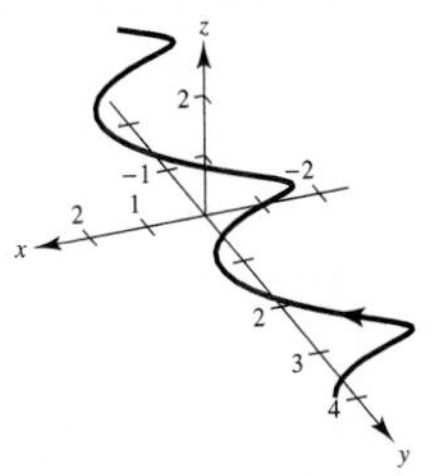

Helix

43.

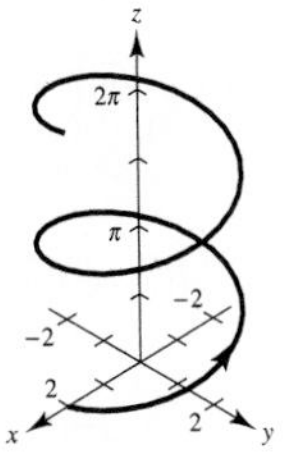

(a) The helix is translated two units back on the x-axis.
(b) The height of the helix increases at a greater rate.
(c) The orientation of the graph is reversed.
(d) The axis of the helix is the x-axis.
(e) The radius of the helix is increased from 2 to 6.

45–51. Answers will vary.

53. $\mathbf{r}(t) = \langle 2 - 2t, 3 + 5t, 8t \rangle$

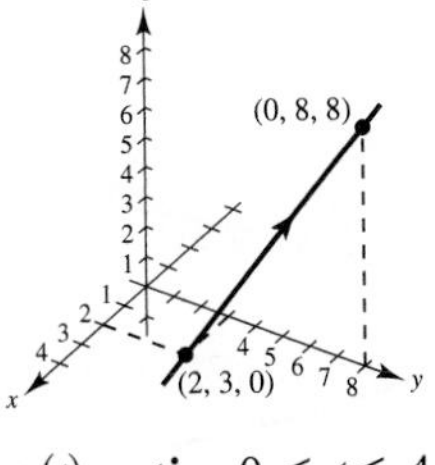

55. $\mathbf{r}_1(t) = t\mathbf{i}, \quad 0 \le t \le 4$
$\mathbf{r}_2(t) = (4 - 4t)\mathbf{i} + 6t\mathbf{j}, \quad 0 \le t \le 1$
$\mathbf{r}_3(t) = (6 - t)\mathbf{j}, \quad 0 \le t \le 6$

57. $\mathbf{r}_1(t) = t\mathbf{i} + t^2\mathbf{j}, \quad 0 \le t \le 2$
$\mathbf{r}_2(t) = (2 - t)\mathbf{i} + 4\mathbf{j}, \quad 0 \le t \le 2$
$\mathbf{r}_3(t) = (4 - t)\mathbf{j}, \quad 0 \le t \le 4$

59.

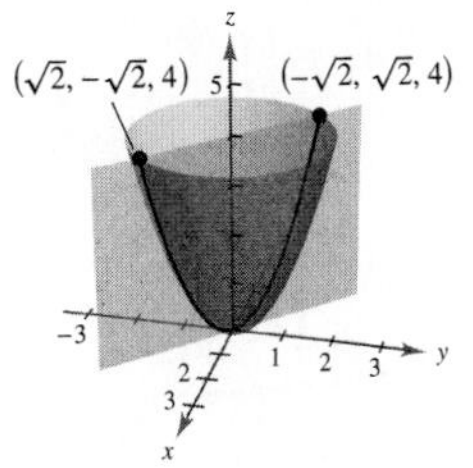

$\mathbf{r}(t) = t\mathbf{i} - t\mathbf{j} + 2t^2\mathbf{k}$

61.

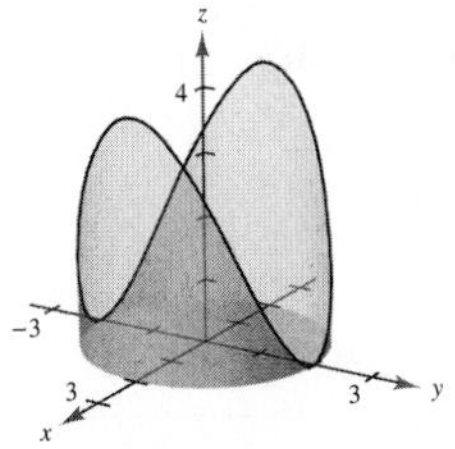

$\mathbf{r}(t) = 2 \sin t\mathbf{i} + 2 \cos t\mathbf{j} + 4 \sin^2 t\mathbf{k}$

63.

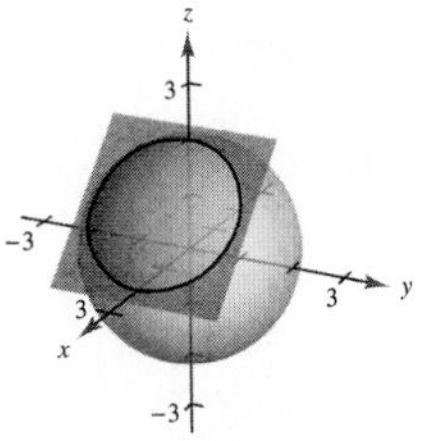

$\mathbf{r}(t) = (1 + \sin t)\mathbf{i} + \sqrt{2} \cos t\mathbf{j} + (1 - \sin t)\mathbf{k}$ and
$\mathbf{r}(t) = (1 + \sin t)\mathbf{i} - \sqrt{2} \cos t\mathbf{j} + (1 - \sin t)\mathbf{k}$

65.

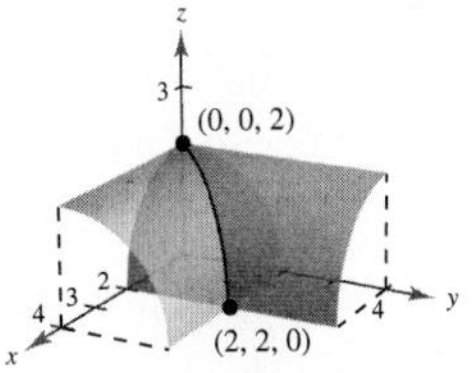

$\mathbf{r}(t) = t\mathbf{i} + t\mathbf{j} + \sqrt{4 - t^2}\mathbf{k}$

67. Let $x = t$, $y = 2t \cos t$, and $z = 2t \sin t$. Then
$$y^2 + z^2 = (2t \cos t)^2 + (2t \sin t)^2 = 4t^2 \cos^2 t + 4t^2 \sin^2 t = 4t^2(\cos^2 t + \sin^2 t) = 4t^2.$$
Since $x = t$, $y^2 + z^2 = 4x^2$.

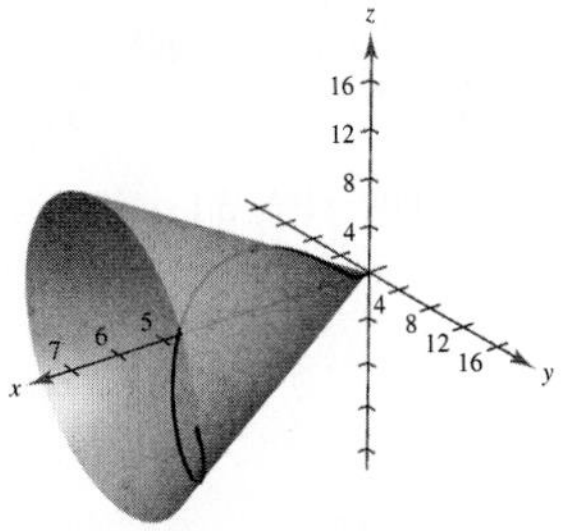

69. $2\mathbf{i} + 2\mathbf{j} + \frac{1}{2}\mathbf{k}$ **71.** $\mathbf{0}$ **73.** Limit does not exist.

75. $(-\infty, 0), (0, \infty)$ **77.** $[-1, 1]$

79. $(-\pi/2 + n\pi, \pi/2 + n\pi)$, n is an integer.

81. A function of the form $\mathbf{r}(t) = f(t)\mathbf{i} + g(t)\mathbf{j}$ (plane) or $\mathbf{r}(t) = f(t)\mathbf{i} + g(t)\mathbf{j} + h(t)\mathbf{k}$ (space) is a vector-valued function, where the component functions f, g, and h are real-valued functions of the parameter t.

83. (a) $\mathbf{s}(t) = t^2\mathbf{i} + (t - 3)\mathbf{j} + (t + 3)\mathbf{k}$
(b) $\mathbf{s}(t) = (t^2 - 2)\mathbf{i} + (t - 3)\mathbf{j} + t\mathbf{k}$
(c) $\mathbf{s}(t) = t^2\mathbf{i} + (t + 2)\mathbf{j} + t\mathbf{k}$

85–87. Proofs **89.** True

91. False; although $\mathbf{r}(4) = \mathbf{u}(2) = \langle 4, 16\rangle$, the particles do not collide because they reach this point at different times.

Section 12.2 (page 846)

1. $\mathbf{r}(2) = 4\mathbf{i} + 2\mathbf{j}$
$\mathbf{r}'(2) = 4\mathbf{i} + \mathbf{j}$

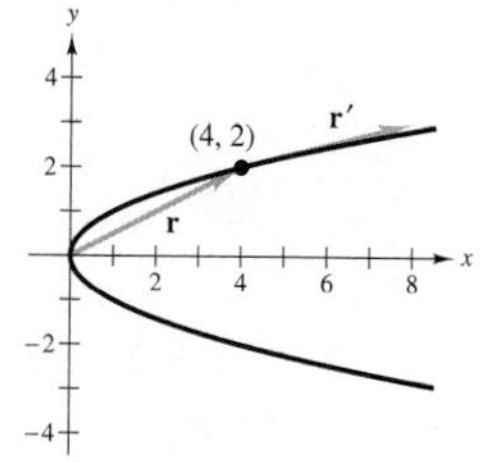

$\mathbf{r}'(t_0)$ is tangent to the curve at t_0.

3. $\mathbf{r}(2) = 4\mathbf{i} + \frac{1}{2}\mathbf{j}$
$\mathbf{r}'(2) = 4\mathbf{i} - \frac{1}{4}\mathbf{j}$

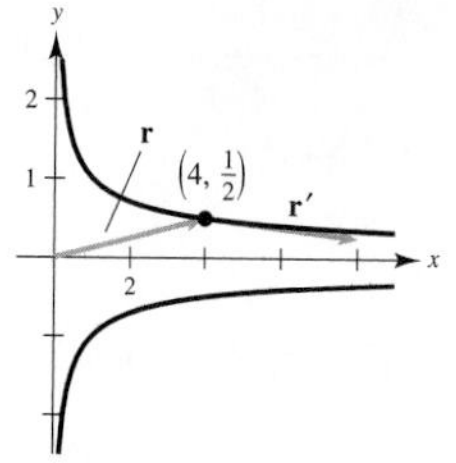

$\mathbf{r}'(t_0)$ is tangent to the curve at t_0.

5. $\mathbf{r}(\pi/2) = \mathbf{j}$
$\mathbf{r}'(\pi/2) = -\mathbf{i}$

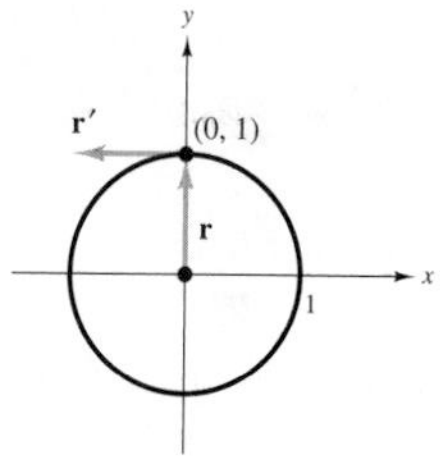

$\mathbf{r}'(t_0)$ is tangent to the curve at t_0.

7. (a) and (b)

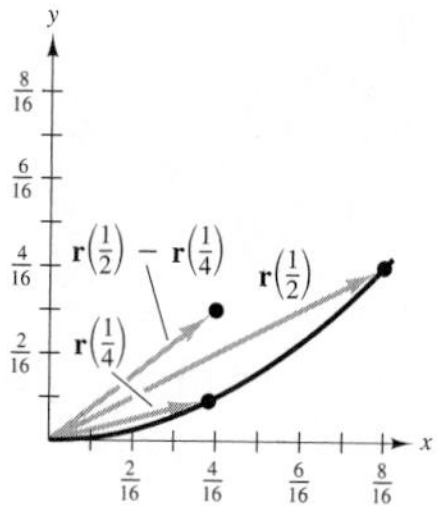

(c) The vector $\dfrac{\mathbf{r}(1/2) - \mathbf{r}(1/4)}{1/2 - 1/4}$ approximates the tangent vector $\mathbf{r}'(1/4)$.

9. $\mathbf{r}\left(\dfrac{3\pi}{2}\right) = -2\mathbf{j} + \left(\dfrac{3\pi}{2}\right)\mathbf{k}$
$\mathbf{r}'\left(\dfrac{3\pi}{2}\right) = 2\mathbf{i} + \mathbf{k}$

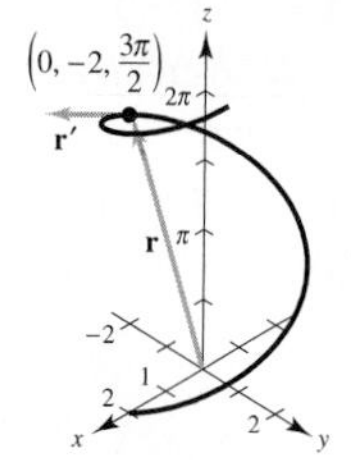

11. $6\mathbf{i} - 14t\mathbf{j} + 3t^2\mathbf{k}$ **13.** $-3a \sin t \cos^2 t\mathbf{i} + 3a \sin^2 t \cos t\mathbf{j}$

15. $-e^{-t}\mathbf{i}$ **17.** $\langle \sin t + t \cos t, \cos t - t \sin t, 1\rangle$

19. (a) $6t\mathbf{i} + \mathbf{j}$ (b) $18t^3 + t$

21. (a) $-4 \cos t\mathbf{i} - 4 \sin t\mathbf{j}$ (b) 0

23. (a) $\mathbf{i} + t\mathbf{k}$ (b) $t^3/2 + t$

25. (a) $\langle \cos t - t \sin t, \sin t + t \cos t, 0\rangle$ (b) t

27. $$\frac{\mathbf{r}'(-1/4)}{\|\mathbf{r}'(-1/4)\|} = \frac{1}{\sqrt{4\pi^2 + 1}}(\sqrt{2}\pi\mathbf{i} + \sqrt{2}\pi\mathbf{j} - \mathbf{k})$$
$$\frac{\mathbf{r}''(-1/4)}{\|\mathbf{r}''(-1/4)\|} = \frac{1}{2\sqrt{\pi^4 + 4}}(-\sqrt{2}\pi^2\mathbf{i} + \sqrt{2}\pi^2\mathbf{j} + 4\mathbf{k})$$

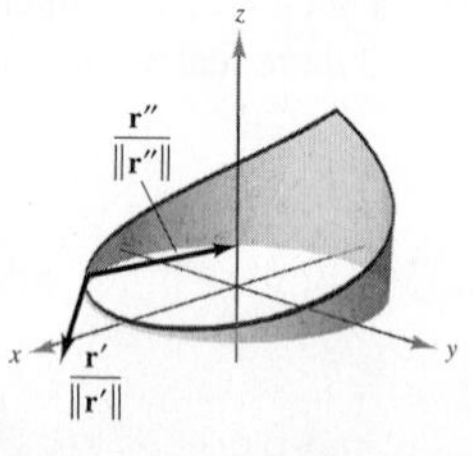

29. $(-\infty, 0), (0, \infty)$ **31.** $(n\pi/2, (n + 1)\pi/2)$

33. $(-\infty, \infty)$ **35.** $(-\infty, 0), (0, \infty)$

37. $(-\pi/2 + n\pi, \pi/2 + n\pi)$, n is an integer.

39. (a) $\mathbf{i} + 3\mathbf{j} + 2t\mathbf{k}$ (b) $2\mathbf{k}$ (c) $8t + 9t^2 + 5t^4$
(d) $-\mathbf{i} + (9 - 2t)\mathbf{j} + (6t - 3t^2)\mathbf{k}$
(e) $8t^3\mathbf{i} + (12t^2 - 4t^3)\mathbf{j} + (3t^2 - 24t)\mathbf{k}$
(f) $(10 + 2t^2)/\sqrt{10 + t^2}$

41. (a) $7t^6$ (b) $12t^5\mathbf{i} - 5t^4\mathbf{j}$

43. $$\theta(t) = \arccos\left(\frac{-7 \sin t \cos t}{\sqrt{9 \sin^2 t + 16 \cos^2 t}\sqrt{9 \cos^2 t + 16 \sin^2 t}}\right)$$

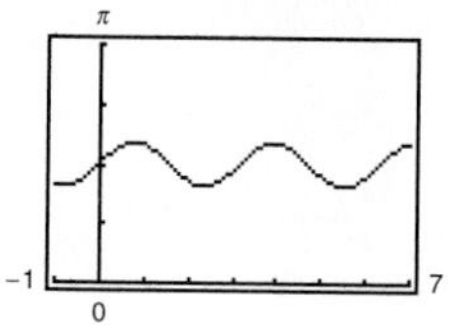

Maximum: $\theta\left(\dfrac{\pi}{4}\right) = \theta\left(\dfrac{5\pi}{4}\right) \approx 1.855$

Minimum: $\theta\left(\dfrac{3\pi}{4}\right) = \theta\left(\dfrac{7\pi}{4}\right) \approx 1.287$

Orthogonal: $\dfrac{n\pi}{2}$, n is an integer

45. $\mathbf{r}'(t) = 3\mathbf{i} - 2t\mathbf{j}$ **47.** $\mathbf{r}'(t) = 2t\mathbf{i} + 2\mathbf{k}$

49. $t^2\mathbf{i} + t\mathbf{j} + t\mathbf{k} + \mathbf{C}$ **51.** $\ln t\mathbf{i} + t\mathbf{j} - \frac{2}{5}t^{5/2}\mathbf{k} + \mathbf{C}$

53. $(t^2 - t)\mathbf{i} + t^4\mathbf{j} + 2t^{3/2}\mathbf{k} + \mathbf{C}$

55. $\tan t\mathbf{i} + \arctan t\mathbf{j} + \mathbf{C}$ **57.** $4\mathbf{i} + \frac{1}{2}\mathbf{j} - \mathbf{k}$

59. $a\mathbf{i} + a\mathbf{j} + (\pi/2)\mathbf{k}$ **61.** $2\mathbf{i} + (e^2 - 1)\mathbf{j} - (e^2 + 1)\mathbf{k}$

63. $2e^{2t}\mathbf{i} + 3(e^t - 1)\mathbf{j}$ **65.** $600\sqrt{3}t\mathbf{i} + (-16t^2 + 600t)\mathbf{j}$

67. $((2 - e^{-t^2})/2)\mathbf{i} + (e^{-t} - 2)\mathbf{j} + (t + 1)\mathbf{k}$

69. See "Definition of the Derivative of a Vector-Valued Function" and Figure 12.8 on page 840.

71. The three components of $\mathbf{u}$ are increasing functions of t at $t = t_0$.

73–79. Proofs

81. (a)

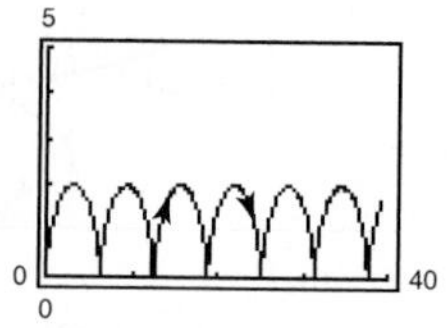

(b) The maximum of $\|\mathbf{r}'\|$ is 2; the minimum of $\|\mathbf{r}'\|$ is 0. The maximum and the minimum of $\|\mathbf{r}''\|$ is 1.

83. True

85. False: Let $\mathbf{r}(t) = \cos t\mathbf{i} + \sin t\mathbf{j} + \mathbf{k}$, then $d/dt[\|\mathbf{r}(t)\|] = 0$, but $\|\mathbf{r}'(t)\| = 1$.

87. Proof

Section 12.3 (page 854)

1. $\mathbf{v}(1) = 3\mathbf{i} + \mathbf{j}$
$\mathbf{a}(1) = \mathbf{0}$

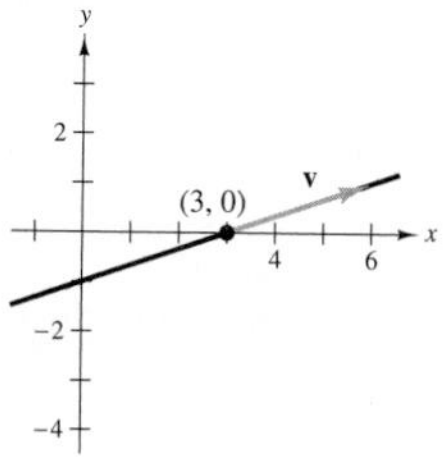

3. $\mathbf{v}(2) = 4\mathbf{i} + \mathbf{j}$
$\mathbf{a}(2) = 2\mathbf{i}$

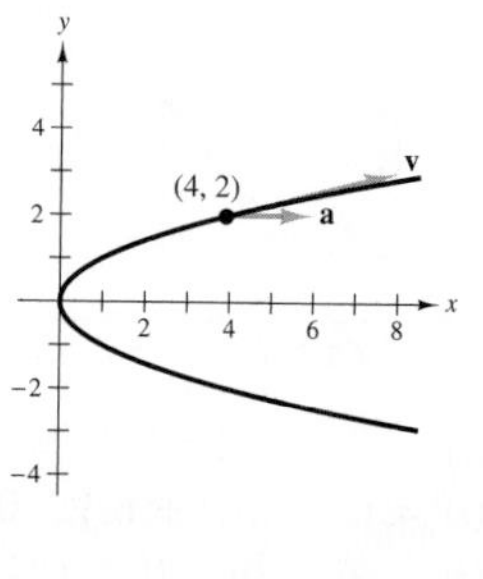

5. $\mathbf{v}(\pi/4) = -\sqrt{2}\mathbf{i} + \sqrt{2}\mathbf{j}$
$\mathbf{a}(\pi/4) = -\sqrt{2}\mathbf{i} - \sqrt{2}\mathbf{j}$

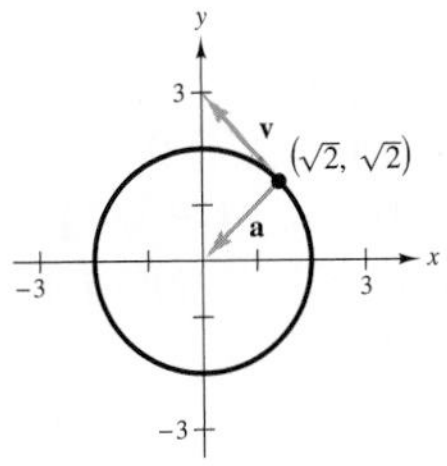

7. $\mathbf{v}(\pi) = 2\mathbf{i}$
$\mathbf{a}(\pi) = -\mathbf{j}$

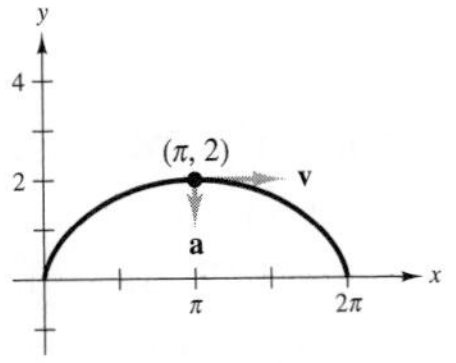

9. $\mathbf{v}(t) = \mathbf{i} + 2\mathbf{j} + 3\mathbf{k}$
$\|\mathbf{v}(t)\| = \sqrt{14}$
$\mathbf{a}(t) = \mathbf{0}$

11. $\mathbf{v}(t) = \mathbf{i} + 2t\mathbf{j} + t\mathbf{k}$
$\|\mathbf{v}(t)\| = \sqrt{1 + 5t^2}$
$\mathbf{a}(t) = 2\mathbf{j} + \mathbf{k}$

13. $\mathbf{v}(t) = \mathbf{i} + \mathbf{j} - \left(t/\sqrt{9 - t^2}\right)\mathbf{k}$
$\|\mathbf{v}(t)\| = \sqrt{(18 - t^2)/(9 - t^2)}$
$\mathbf{a}(t) = \left(-9/(9 - t^2)^{3/2}\right)\mathbf{k}$

15. $\mathbf{v}(t) = 4\mathbf{i} - 3\sin t\mathbf{j} + 3\cos t\mathbf{k}$
$\|\mathbf{v}(t)\| = 5$
$\mathbf{a}(t) = -3\cos t\mathbf{j} - 3\sin t\mathbf{k}$

17. (a) $x = 1 + t$
$y = -1 - 2t$
$z = \frac{1}{4} + \frac{3}{4}t$
(b) $(1.100, -1.200, 0.325)$

19. $\mathbf{v}(t) = t(\mathbf{i} + \mathbf{j} + \mathbf{k})$
$\mathbf{r}(t) = (t^2/2)(\mathbf{i} + \mathbf{j} + \mathbf{k})$
$\mathbf{r}(2) = 2(\mathbf{i} + \mathbf{j} + \mathbf{k})$

21. $\mathbf{v}(t) = \left(t^2/2 + \frac{9}{2}\right)\mathbf{j} + \left(t^2/2 - \frac{1}{2}\right)\mathbf{k}$
$\mathbf{r}(t) = \left(t^3/6 + \frac{9}{2}t - \frac{14}{3}\right)\mathbf{j} + \left(t^3/6 - \frac{1}{2}t + \frac{1}{3}\right)\mathbf{k}$
$\mathbf{r}(2) = \frac{17}{3}\mathbf{j} + \frac{2}{3}\mathbf{k}$

23. The velocity of an object involves both magnitude and direction of motion, whereas speed involves only magnitude.

25. $\mathbf{r}(t) = 44\sqrt{3}t\mathbf{i} + (10 + 44t - 16t^2)\mathbf{j}$

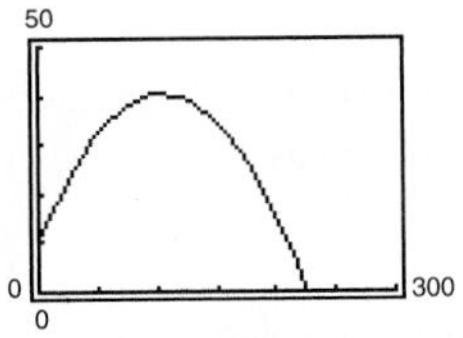

27. $v_0 = 40\sqrt{6}$ ft/sec; 78 ft

29. Proof

31. (a) $y = -0.004x^2 + 0.37x + 6$
$\mathbf{r}(t) = t\mathbf{i} + (-0.004t^2 + 0.37t + 6)\mathbf{j}$

(b)

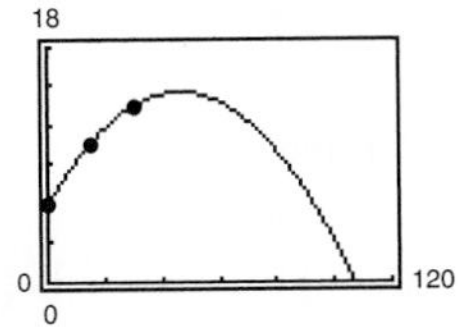

(c) 14.56 ft
(d) Initial velocity: 67.4 ft/sec; $\theta \approx 20.14°$

33. (a) $\mathbf{r}(t) = \left(\frac{440}{3}\cos\theta_0\right)t\mathbf{i} + \left[3 + \left(\frac{440}{3}\sin\theta_0\right)t - 16t^2\right]\mathbf{j}$

(b)

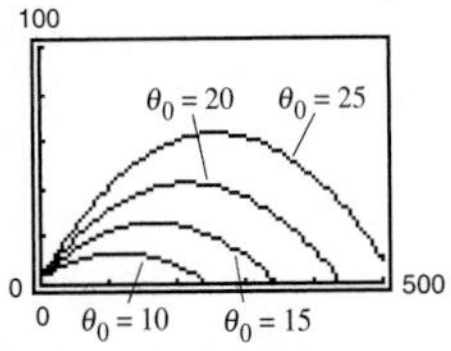

The minimum angle appears to be $\theta_0 = 20°$.

(c) $\theta_0 \approx 19.38°$

35. (a) $v_0 = 28.78$ ft/sec; $\theta = 58.28°$ (b) $v_0 \approx 32$ ft/sec

37. 1.91°

39. (a)

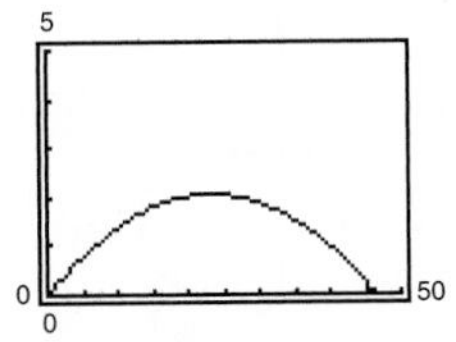

Maximum height: 2.1 ft
Range: 46.6 ft

(b)

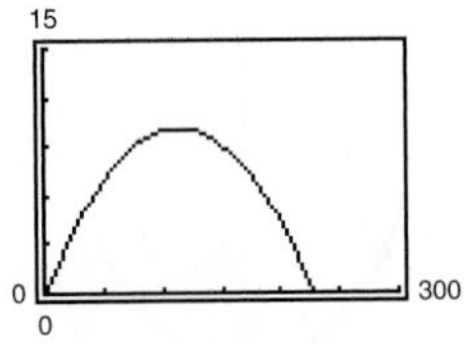

Maximum height: 10.0 ft
Range: 227.8 ft

(c)

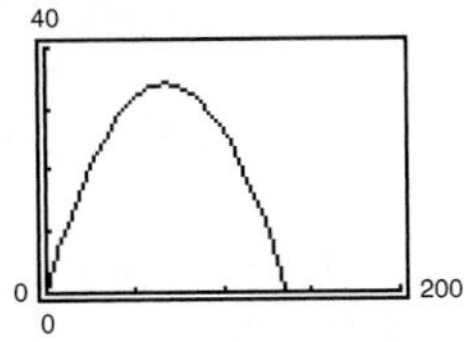

Maximum height: 34.0 ft
Range: 136.1 ft

(d)

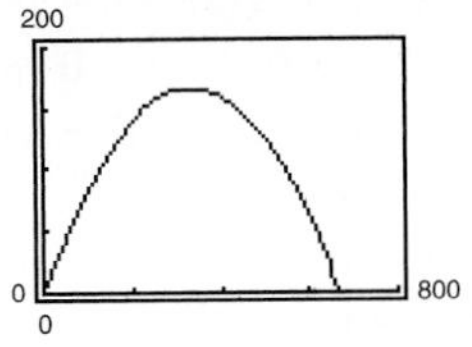

Maximum height: 166.5 ft
Range: 666.1 ft

(e)

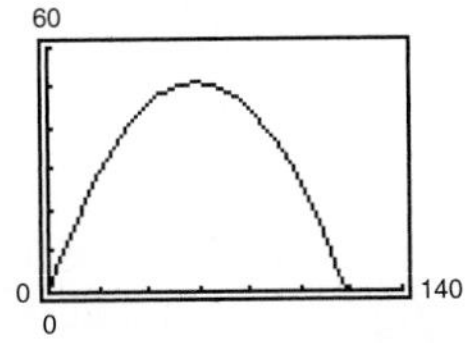

Maximum height: 51.0 ft
Range: 117.9 ft

(f)

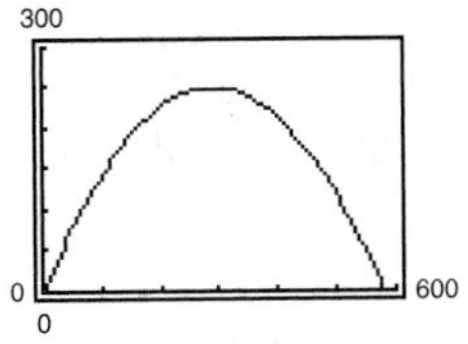

Maximum height: 249.8 ft
Range: 576.9 ft

41. Maximum height: 129.1 m
Range: 886.3 m

43. $\mathbf{v}(t) = b\omega[(1 - \cos\omega t)\mathbf{i} + \sin\omega t\mathbf{j}]$
$\mathbf{a}(t) = b\omega^2(\sin\omega t\mathbf{i} + \cos\omega t\mathbf{j})$
(a) $\|\mathbf{v}(t)\| = 0$ when $\omega t = 0, 2\pi, 4\pi, . . .$
(b) $\|\mathbf{v}(t)\|$ is maximum when $\omega t = \pi, 3\pi, . . .$

45. $\mathbf{v}(t) = -b\omega\sin\omega t\mathbf{i} + b\omega\cos\omega t\mathbf{j}$
$\mathbf{v}(t) \cdot \mathbf{r}(t) = 0$

47. $\mathbf{a}(t) = -b\omega^2(\cos\omega t\mathbf{i} + \sin\omega t\mathbf{j}) = -\omega^2\mathbf{r}(t)$

49. $8\sqrt{10}$ ft/sec

51–53. Proofs

55. (a) $\mathbf{v}(t) = -6\sin t\mathbf{i} + 3\cos t\mathbf{j}$
$\|\mathbf{v}(t)\| = 3\sqrt{3\sin^2 t + 1}$
$\mathbf{a}(t) = -6\cos t\mathbf{i} - 3\sin t\mathbf{j}$

(b)

t	0	$\pi/4$	$\pi/2$	$2\pi/3$	π
Speed	3	$3\sqrt{10}/2$	6	$3\sqrt{13}/2$	3

(c)

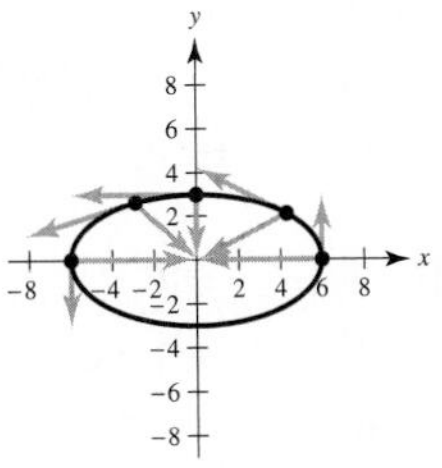

(d) The speed is increasing when the angle between **v** and **a** is in the interval $[0, \pi/2)$, and decreasing when the angle is in the interval $(\pi/2, \pi]$.

57. False; acceleration is the derivative of the velocity.

59. Proof

Section 12.4 (page 863)

1.

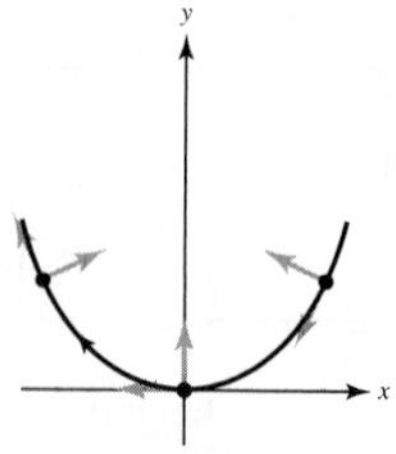

3.

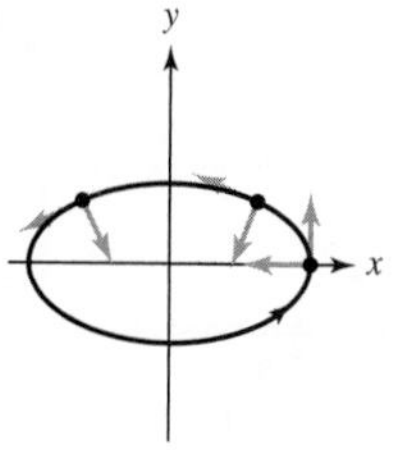

5. $\mathbf{T}(1) = (\sqrt{2}/2)(\mathbf{i} + \mathbf{j})$ **7.** $\mathbf{T}(\pi/4) = (\sqrt{2}/2)(-\mathbf{i} + \mathbf{j})$

9. $\mathbf{T}(e) \approx 0.1809\mathbf{i} + 0.9835\mathbf{j}$

11. $\mathbf{T}(0) = (\sqrt{2}/2)(\mathbf{i} + \mathbf{k})$
$x = t$
$y = 0$
$z = t$

13. $\mathbf{T}(0) = (\sqrt{5}/5)(2\mathbf{j} + \mathbf{k})$
$x = 2$
$y = 2t$
$z = t$

15. $\mathbf{T}(\pi/4) = \frac{1}{2}\langle -\sqrt{2}, \sqrt{2}, 0\rangle$
$x = \sqrt{2} - \sqrt{2}t$
$y = \sqrt{2} + \sqrt{2}t$
$z = 4$

17. $\mathbf{T}(3) = \frac{1}{19}\langle 1, 6, 18\rangle$
$x = 3 + t$
$y = 9 + 6t$
$z = 18 + 18t$

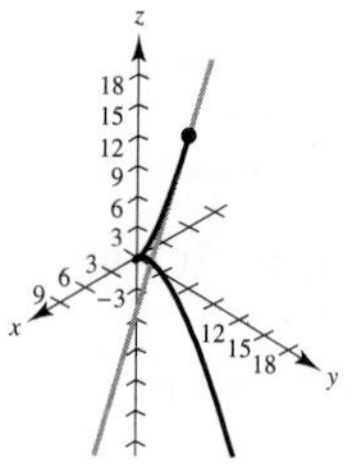

19. Tangent line: $x = 1 + t$
$y = t$
$z = 1 + \frac{1}{2}t$
$\mathbf{r}(1.1) \approx \langle 1.1, 0.1, 1.05\rangle$

21. $1.2°$ **23.** $\mathbf{N}(2) = (\sqrt{5}/5)(-2\mathbf{i} + \mathbf{j})$

25. $N(2) = (-\sqrt{5}/5)(2\mathbf{i} - \mathbf{j})$

27. $N(1) = (-\sqrt{14}/14)(\mathbf{i} - 2\mathbf{j} + 3\mathbf{k})$

29. $\mathbf{N}(3\pi/4) = (\sqrt{2}/2)(\mathbf{i} - \mathbf{j})$

31. $\mathbf{v}(t) = 4\mathbf{i}$
$\mathbf{a}(t) = \mathbf{0}$
$\mathbf{T}(t) = \mathbf{i}$
$\mathbf{N}(t)$ is undefined. The path is a line and the speed is constant.

33. $\mathbf{v}(t) = 8t\mathbf{i}$
$\mathbf{a}(t) = 8\mathbf{i}$
$\mathbf{T}(t) = \mathbf{i}$
$\mathbf{N}(t)$ is undefined. The path is a line and the speed is variable.

35. $\mathbf{T}(1) = (\sqrt{2}/2)(\mathbf{i} - \mathbf{j})$
$\mathbf{N}(1) = (\sqrt{2}/2)(\mathbf{i} + \mathbf{j})$
$a_{\mathbf{T}} = -\sqrt{2}$
$a_{\mathbf{N}} = \sqrt{2}$

37. $\mathbf{T}(1) = (-\sqrt{5}/5)(\mathbf{i} - 2\mathbf{j})$
$\mathbf{N}(1) = (-\sqrt{5}/5)(2\mathbf{i} + \mathbf{j})$
$a_{\mathbf{T}} = 14\sqrt{5}/5$
$a_{\mathbf{N}} = 8\sqrt{5}/5$

39. $\mathbf{T}(0) = (\sqrt{5}/5)(\mathbf{i} - 2\mathbf{j})$
$\mathbf{N}(0) = (\sqrt{5}/5)(2\mathbf{i} + \mathbf{j})$
$a_{\mathbf{T}} = -7\sqrt{5}/5$
$a_{\mathbf{N}} = 6\sqrt{5}/5$

41. $\mathbf{T}(\pi/2) = (\sqrt{2}/2)(-\mathbf{i} + \mathbf{j})$
$\mathbf{N}(\pi/2) = (-\sqrt{2}/2)(\mathbf{i} + \mathbf{j})$
$a_{\mathbf{T}} = \sqrt{2}e^{\pi/2}$
$a_{\mathbf{N}} = \sqrt{2}e^{\pi/2}$

43. $\mathbf{T}(t_0) = (\cos \omega t_0)\mathbf{i} + (\sin \omega t_0)\mathbf{j}$
$\mathbf{N}(t_0) = (-\sin \omega t_0)\mathbf{i} + (\cos \omega t_0)\mathbf{j}$
$a_{\mathbf{T}} = \omega^2$
$a_{\mathbf{N}} = \omega^3 t_0$

45. $\mathbf{T}(t) = -\sin(\omega t)\mathbf{i} + \cos(\omega t)\mathbf{j}$
$\mathbf{N}(t) = -\cos(\omega t)\mathbf{i} - \sin(\omega t)\mathbf{j}$
$a_{\mathbf{T}} = 0$
$a_{\mathbf{N}} = a\omega^2$

47. $\|\mathbf{v}(t)\| = a\omega$; The speed is constant because $a_{\mathbf{T}} = 0$.

49. $\mathbf{r}(2) = 2\mathbf{i} + \frac{1}{2}\mathbf{j}$
$\mathbf{T}(2) = (\sqrt{17}/17)(4\mathbf{i} - \mathbf{j})$
$\mathbf{N}(2) = (\sqrt{17}/17)(\mathbf{i} + 4\mathbf{j})$

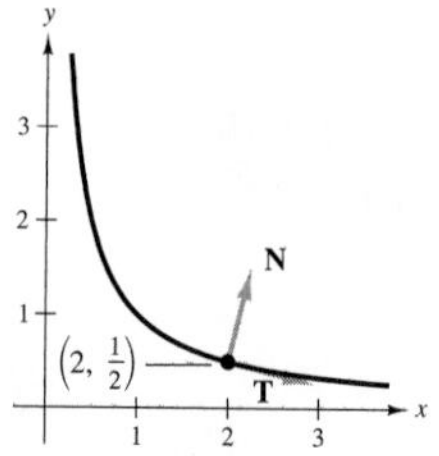

51. $\mathbf{r}(\pi/4) = \sqrt{2}\mathbf{i} + \sqrt{2}\mathbf{j}$
$\mathbf{T}(\pi/4) = (\sqrt{2}/2)(-\mathbf{i} + \mathbf{j})$
$\mathbf{N}(\pi/4) = (\sqrt{2}/2)(-\mathbf{i} - \mathbf{j})$

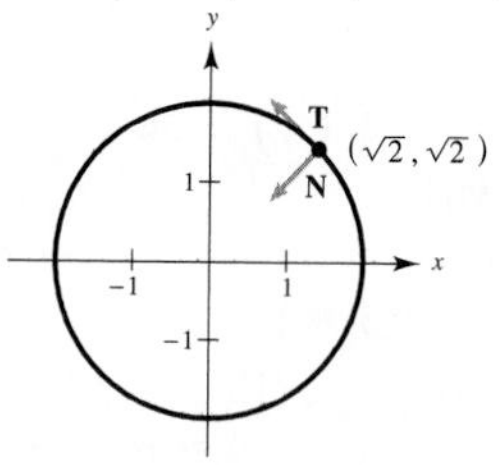

53. $\mathbf{T}(1) = (\sqrt{14}/14)(\mathbf{i} + 2\mathbf{j} - 3\mathbf{k})$
$\mathbf{N}(1)$ is undefined.
$a_{\mathbf{T}}$ is undefined.
$a_{\mathbf{N}}$ is undefined.

55. $\mathbf{T}(1) = (\sqrt{6}/6)(\mathbf{i} + 2\mathbf{j} + \mathbf{k})$
$\mathbf{N}(1) = (\sqrt{30}/30)(-5\mathbf{i} + 2\mathbf{j} + \mathbf{k})$
$a_{\mathbf{T}} = 5\sqrt{6}/6$
$a_{\mathbf{N}} = \sqrt{30}/6$

57. $\mathbf{T}(\pi/2) = \frac{1}{5}(4\mathbf{i} - 3\mathbf{j})$
$\mathbf{N}(\pi/2) = -\mathbf{k}$
$a_{\mathbf{T}} = 0$
$a_{\mathbf{N}} = 3$

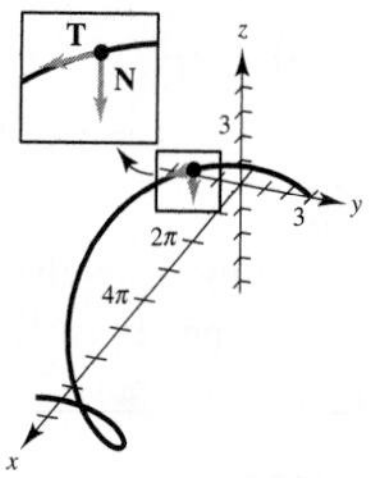

59. Let C be a smooth curve represented by $\mathbf{r}$ on an open interval I. The unit tangent vector $\mathbf{T}(t)$ at t is defined as

$$\mathbf{T}(t) = \frac{\mathbf{r}'(t)}{\|\mathbf{r}'(t)\|},\ \mathbf{r}'(t) \neq 0.$$

The principal unit normal vector $\mathbf{N}(t)$ at t is defined as

$$\mathbf{N}(t) = \frac{\mathbf{T}'(t)}{\|\mathbf{T}'(t)\|},\ \mathbf{T}'(t) \neq 0.$$

The tangential and normal components of acceleration are defined as follows

$\mathbf{a}(t) = a_{\mathbf{T}}\mathbf{T}(t) + a_{\mathbf{N}}\mathbf{N}(t).$

61. The particle's motion is in a straight line.

63. (a) $t = \frac{1}{2}$: $a_{\mathbf{T}} = \sqrt{2}\pi^2/2$, $a_{\mathbf{N}} = \sqrt{2}\pi^2/2$
$t = 1$: $a_{\mathbf{T}} = 0$, $a_{\mathbf{N}} = \pi^2$
$t = \frac{3}{2}$: $a_{\mathbf{T}} = -\sqrt{2}\pi^2/2$, $a_{\mathbf{N}} = \sqrt{2}\pi^2/2$

(b) $t = \frac{1}{2}$: Increasing since $a_{\mathbf{T}} > 0$.
$t = 1$: Maximum since $a_{\mathbf{T}} = 0$.
$t = \frac{3}{2}$: Decreasing since $a_{\mathbf{T}} < 0$.

65. $\mathbf{T}(\pi/2) = (\sqrt{17}/17)(-4\mathbf{i} + \mathbf{k})$
$\mathbf{N}(\pi/2) = -\mathbf{j}$
$\mathbf{B}(\pi/2) = (\sqrt{17}/17)(\mathbf{i} + 4\mathbf{k})$

67. $\mathbf{T}(\pi/4) = (\sqrt{2}/2)(\mathbf{j} - \mathbf{k})$
$\mathbf{N}(\pi/4) = -(\sqrt{2}/2)(\mathbf{j} + \mathbf{k})$
$\mathbf{B}(\pi/4) = -\mathbf{i}$

69. $\mathbf{T}(\pi/3) = (\sqrt{5}/5)(\mathbf{i} - \sqrt{3}\mathbf{j} + \mathbf{k})$
$\mathbf{N}(\pi/3) = -\frac{1}{2}(\sqrt{3}\mathbf{i} + \mathbf{j})$
$\mathbf{B}(\pi/3) = (\sqrt{5}/10)(\mathbf{i} - \sqrt{3}\mathbf{j} - 4\mathbf{k})$

71. $a_{\mathbf{T}} = \dfrac{-32(v_0 \sin\theta - 32t)}{\sqrt{v_0^2\cos^2\theta + (v_0\sin\theta - 32t)^2}}$

$a_{\mathbf{N}} = \dfrac{32v_0\cos\theta}{\sqrt{v_0^2\cos^2\theta + (v_0\sin\theta - 32t)^2}}$

At maximum height, $a_{\mathbf{T}} = 0$ and $a_{\mathbf{N}} = 32$.

73. (a) $\mathbf{r}(t) = 50\sqrt{3}t\mathbf{i} + (5 + 50t - 16t^2)\mathbf{j}$

(b)

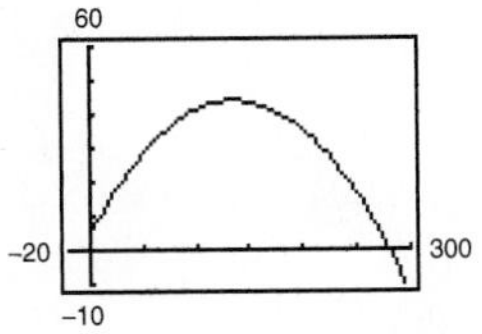

Maximum height $\approx$ 44.0625 ft
Range $\approx$ 279.0325 ft

(c) $\mathbf{v}(t) = 50\sqrt{3}\mathbf{i} + (50 - 32t)\mathbf{j}$
$\|\mathbf{v}(t)\| = 4\sqrt{64t^2 - 200t + 625}$
$\mathbf{a}(t) = -32\mathbf{j}$

(d)

T	0.5	1.0	1.5	2.0	2.5	3.0
Speed	93.04	88.45	86.63	87.73	91.65	98.06

(e)

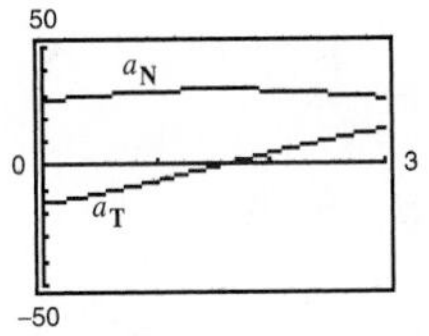

The speed is increasing when $a_{\mathbf{T}}$ and $a_{\mathbf{N}}$ have opposite signs.

75. (a) $4\sqrt{625\pi^2 + 1} \approx 314$ mph
(b) $a_{\mathbf{T}} = 0,\quad a_{\mathbf{N}} = 1000\pi^2$
$a_{\mathbf{T}} = 0$ because the speed is constant.

77. (a) The centripetal component is quadrupled.
(b) The centripetal component is halved.

79. 4.83 mi/sec **81.** 4.67 mi/sec

83. False; centripital acceleration may occur with constant speed.

85. (a) Proof (b) Proof **87–89.** Proofs

Section 12.5 (page 875)

1.

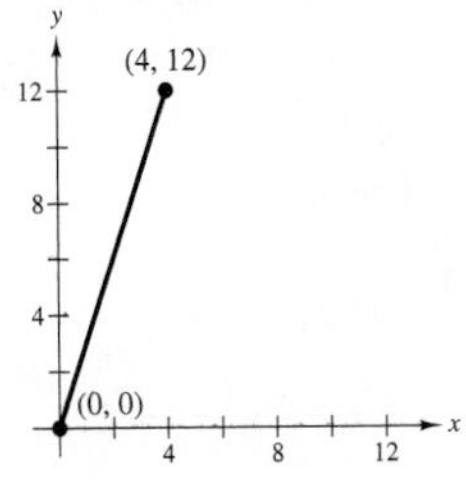

$4\sqrt{10}$

3.

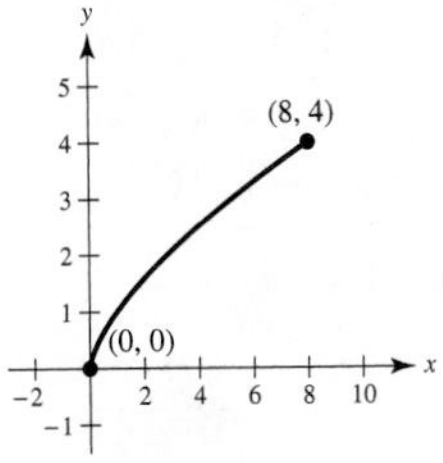

$8(10\sqrt{10} - 1)/27$

5.

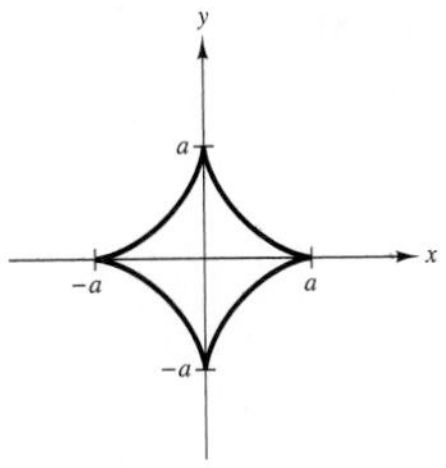

$6a$

7. (a) $\mathbf{r}(t) = (50t\sqrt{2})\mathbf{i} + (3 + 50t\sqrt{2} - 16t^2)\mathbf{j}$
(b) $\frac{649}{8} \approx 81$ ft (c) 315.5 ft (d) 362.9 ft

9.

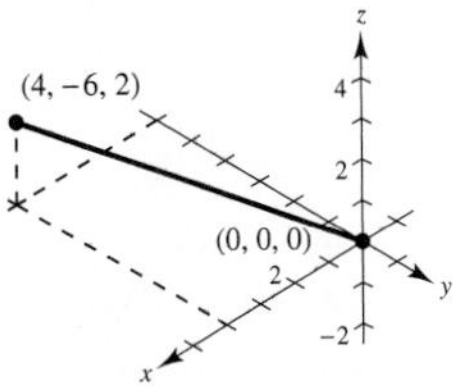

$2\sqrt{14}$

11.

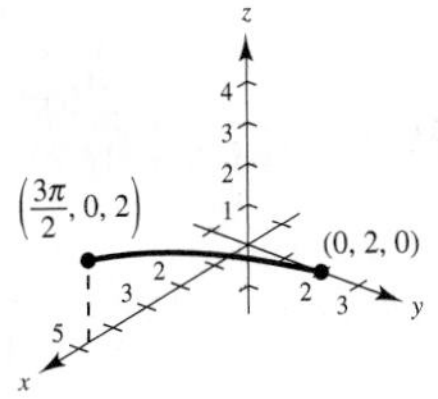

$\sqrt{13}\pi/2$

13.

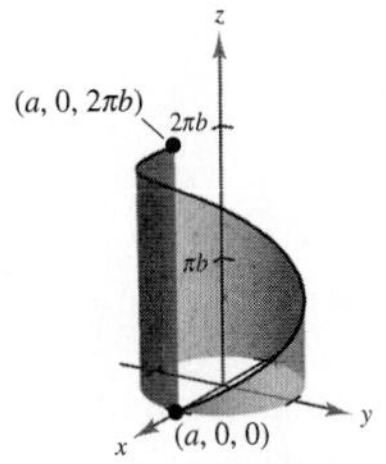

$2\pi\sqrt{a^2 + b^2}$

15. 8.37

17. (a) $2\sqrt{21} \approx 9.165$ (b) 9.529
(c) Increase the number of line segments. (d) 9.571

19. (a) $s = \sqrt{5}t$ (b) $\mathbf{r}(s) = 2\cos\dfrac{s}{\sqrt{5}}\mathbf{i} + 2\sin\dfrac{s}{\sqrt{5}}\mathbf{j} + \dfrac{s}{\sqrt{5}}\mathbf{k}$
(c) $s = \sqrt{5}$: (1.081, 1.683, 1.000)
$s = 4$: (−0.433, 1.953, 1.789)
(d) Proof

21. 0 **23.** $\frac{2}{5}$ **25.** 0 **27.** $\sqrt{2}/2$ **29.** 1

31. $\frac{1}{4}$ **33.** $1/a$ **35.** $\sqrt{2}/(4a\sqrt{1 - \cos\omega t})$

37. $\sqrt{5}/(1 + 5t^2)^{3/2}$ **39.** $\frac{3}{25}$ **41.** $K = 0$, $1/K$ is undefined.

43. $K = 4/17^{3/2}$, $1/K = 17^{3/2}/4$ **45.** $K = 1/a$, $1/K = a$

47. (a) $(x - \pi/2)^2 + y^2 = 1$
(b) Because the curvature is not as great, the radius of the curvature is greater.

49. $(x - 1)^2 + (y - \frac{5}{2})^2 = (\frac{1}{2})^2$

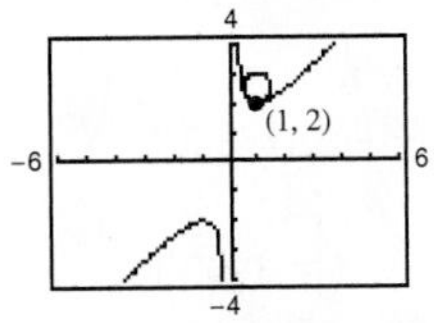

51. $(x + 2)^2 + (y - 3)^2 = 8$

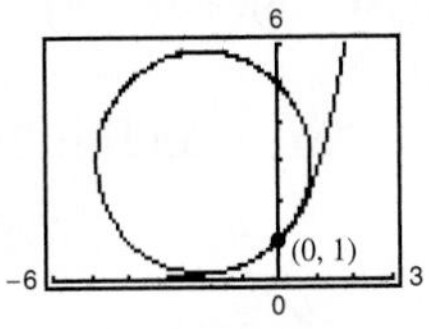

53.

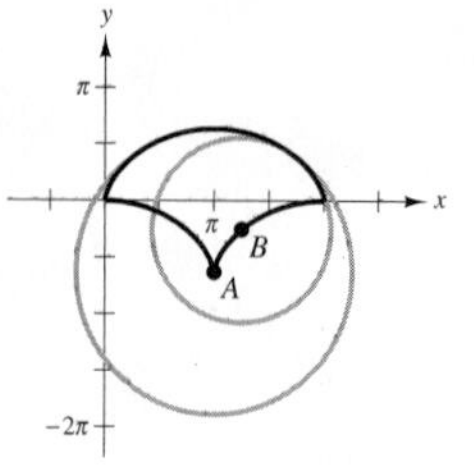

55. (a) $(1, 3)$
(b) 0

57. (a) $K \to \infty$ as $x \to 0$ (b) 0

59. (a) $\left(1/\sqrt{2}, -\ln 2/2\right)$ (b) 0 **61.** $(0, 1)$

63. $(\pi/2 + k\pi, 0)$ **65.** The graph is a line.

67. Proof

69. (a) $K = \dfrac{2|6x^2 - 1|}{(16x^6 - 16x^4 + 4x^2 + 1)^{3/2}}$

(b) $x = 0$: $x^2 + \left(y + \dfrac{1}{2}\right)^2 = \dfrac{1}{4}$

$x = 1$: $x^2 + \left(y - \dfrac{1}{2}\right)^2 = \dfrac{5}{4}$

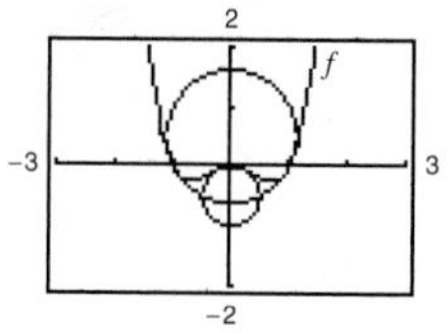

(c)

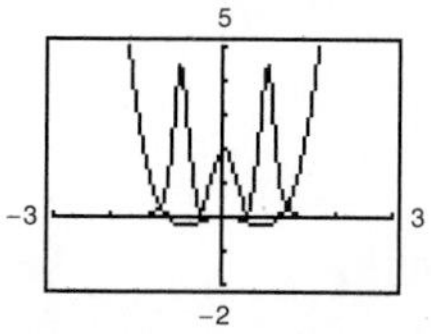

The curvature tends to be greatest near the extrema of the function and decreases as $x \to \pm\infty$. However, f and K do not have the same critical numbers.

Critical numbers of f: $x = 0, \pm\sqrt{2}/2 \approx \pm 0.7071$

Critical numbers of K: $x = 0, \pm 0.7647, \pm 0.4082$

71. (a) 12.25 units (b) $\frac{1}{2}$ **73–75.** Proofs

77. (a) 0 (b) 0 **79.** $\frac{1}{4}$ **81.** Proof

83. $K = [1/(4a)]|\csc(\theta/2)|$ **85.** 3327.5 lbs
Minimum: $K = 1/(4a)$
There is no maximum.

87. Proof

89. False. See Exploration on page 867. **91.** True

93–99. Proofs

Review Exercises for Chapter 12 (page 879)

1. (a) All reals except $n\pi$, n is an integer
(b) Continuous except at $t = n\pi$, n is an integer

3. (a) $(0, \infty)$ (b) Continuous for all $t > 0$

5. (a) $\mathbf{i}$ (b) $-3\mathbf{i} + 4\mathbf{j} + \frac{8}{3}\mathbf{k}$
(c) $(2c - 1)\mathbf{i} + (c - 1)^2\mathbf{j} + \frac{1}{3}(1 - c)^3\mathbf{k}$
(d) $2\Delta t\mathbf{i} + \Delta t(\Delta t + 2)\mathbf{j} - \frac{1}{3}\Delta t[(\Delta t)^2 + 3\Delta t + 3]\mathbf{k}$

7.

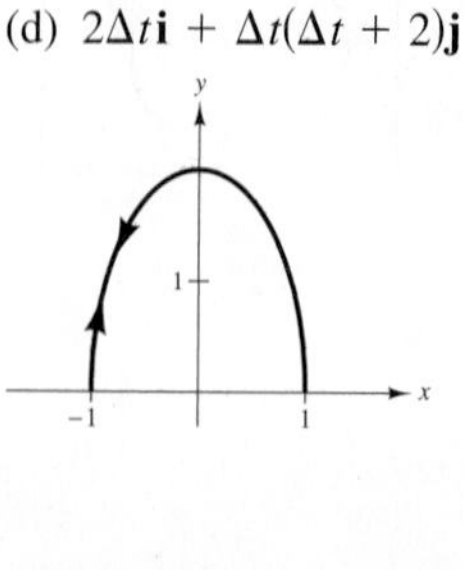

9.

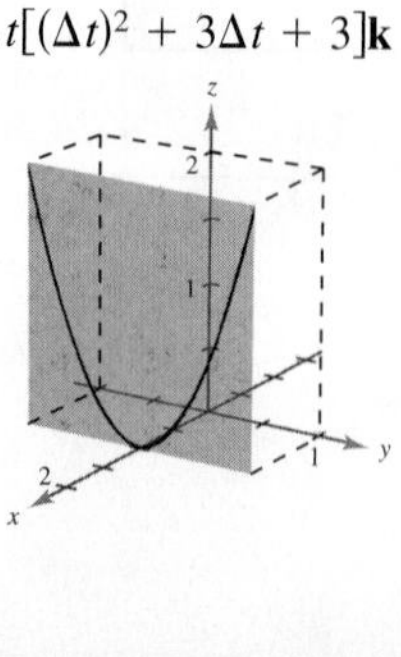

11.

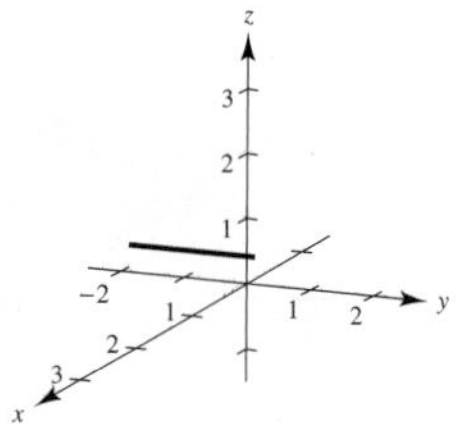

13.

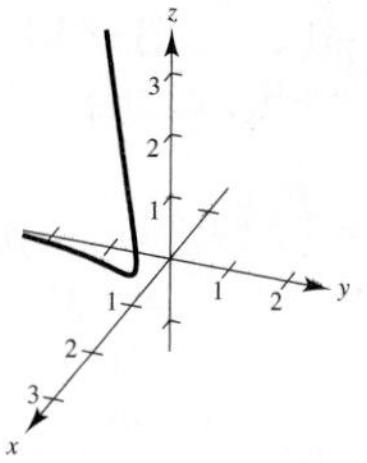

15. $\mathbf{r}_1(t) = 4t\mathbf{i} + 3t\mathbf{j}$, $0 \le t \le 1$
$\mathbf{r}_2(t) = 4\mathbf{i} + (3 - t)\mathbf{j}$, $0 \le t \le 3$
$\mathbf{r}_3(t) = (4 - t)\mathbf{i}$, $0 \le t \le 4$

17. $\mathbf{r}(t) = \langle -2 + 7t, -3 + 4t, 8 - 10t \rangle$
(Answer is not unique.)

19.

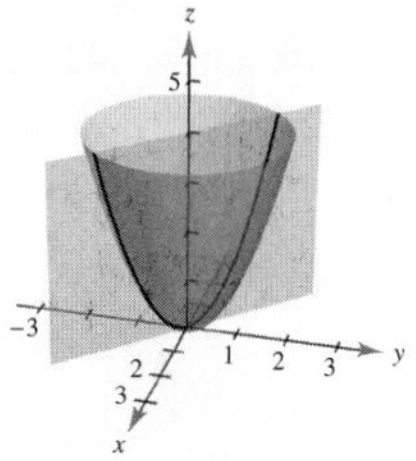

$x = t$, $y = -t$, $z = 2t^2$

21. $4\mathbf{i} + \mathbf{k}$

23. (a) $3\mathbf{i} + \mathbf{j}$ (b) $\mathbf{0}$ (c) $4t + 3t^2$
(d) $-5\mathbf{i} + (2t - 2)\mathbf{j} + 2t^2\mathbf{k}$ (e) $(10t - 1)/\sqrt{10t^2 - 2t + 1}$
(f) $\left(\frac{8}{3}t^3 - 2t^2\right)\mathbf{i} - 8t^3\mathbf{j} + (9t^2 - 2t + 1)\mathbf{k}$

25. $x(t)$ and $y(t)$ are increasing functions at $t = t_0$, and $z(t)$ is a decreasing function at $t = t_0$.

27. $\sin t\mathbf{i} + (t\sin t + \cos t)\mathbf{j} + \mathbf{C}$

29. $\frac{1}{2}\left(t\sqrt{1 + t^2} + \ln\left|t + \sqrt{1 + t^2}\right|\right) + \mathbf{C}$

31. $\mathbf{r}(t) = (t^2 + 1)\mathbf{i} + (e^t + 2)\mathbf{j} - (e^{-t} + 4)\mathbf{k}$

33. $\frac{32}{3}\mathbf{j}$ **35.** $2(e - 1)\mathbf{i} - 8\mathbf{j} - 2\mathbf{k}$

37. $\mathbf{v}(t) = \langle -3\cos^2 t \sin t, 3\sin^2 t \cos t, 3 \rangle$
$\|\mathbf{v}(t)\| = 3\sqrt{\sin^2 t \cos^2 t + 1}$
$\mathbf{a}(t) = \langle 3\cos t(3\sin^2 t - 1), 3\sin t(2\cos^2 t - \sin^2 t), 0 \rangle$

39. $x(t) = t$, $y(t) = 16 + 8t$, $z(t) = 2 + \frac{1}{2}t$
$\mathbf{r}(4.1) \approx \langle 0.1, 16.8, 2.05 \rangle$

41. 152 ft **43.** 34.9 m/sec

45. $\mathbf{v} = 5\mathbf{i}$
$\|\mathbf{v}\| = 5$
$\mathbf{a} = \mathbf{0}$
$\mathbf{a} \cdot \mathbf{T} = 0$
$\mathbf{a} \cdot \mathbf{N}$ does not exist.

47. $\mathbf{v} = \mathbf{i} + \left[1/\left(2\sqrt{t}\right)\right]\mathbf{j}$
$\|\mathbf{v}\| = \sqrt{4t + 1}/\left(2\sqrt{t}\right)$
$\mathbf{a} = \left[-1/\left(4t\sqrt{t}\right)\right]\mathbf{j}$
$\mathbf{a} \cdot \mathbf{T} = -1/\left(4t\sqrt{t}\sqrt{4t + 1}\right)$
$\mathbf{a} \cdot \mathbf{N} = 1/\left(2t\sqrt{4t + 1}\right)$

49. $\mathbf{v} = e^t\mathbf{i} - e^{-t}\mathbf{j}$
$\|\mathbf{v}\| = \sqrt{e^{2t} + e^{-2t}}$
$\mathbf{a} = e^t\mathbf{i} + e^{-t}\mathbf{j}$
$\mathbf{a} \cdot \mathbf{T} = \dfrac{e^{2t} - e^{-2t}}{\sqrt{e^{2t} + e^{-2t}}}$
$\mathbf{a} \cdot \mathbf{N} = \dfrac{2}{\sqrt{e^{2t} + e^{-2t}}}$

51. $\mathbf{v} = \mathbf{i} + 2t\mathbf{j} + t\mathbf{k}$
$\|\mathbf{v}\| = \sqrt{1 + 5t^2}$
$\mathbf{a} = 2\mathbf{j} + \mathbf{k}$
$\mathbf{a} \cdot \mathbf{T} = \dfrac{5t}{\sqrt{1 + 5t^2}}$
$\mathbf{a} \cdot \mathbf{N} = \dfrac{\sqrt{5}}{\sqrt{1 + 5t^2}}$

53. $x = -\sqrt{2} - \sqrt{2}t$
$y = \sqrt{2} - \sqrt{2}t$
$z = 3\pi/4 + t$

55. 4.56 mi/sec

57.

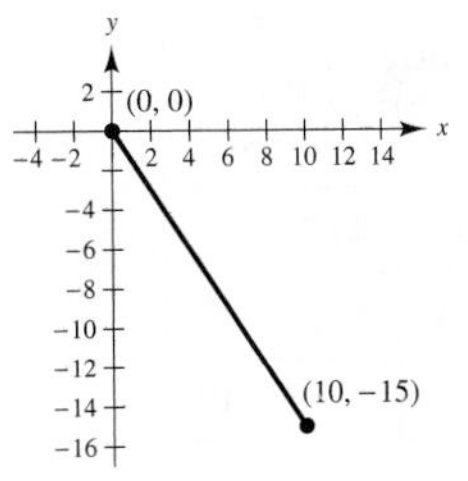

$5\sqrt{13}$

59.

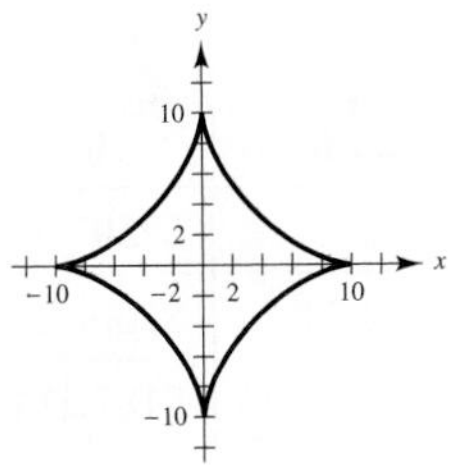

60

61.

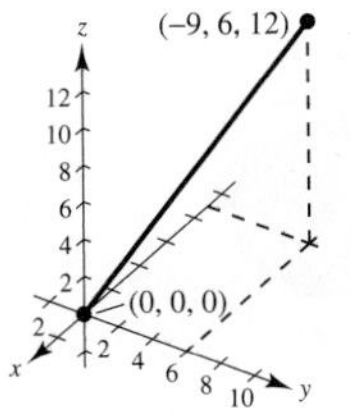

$3\sqrt{29}$

63.

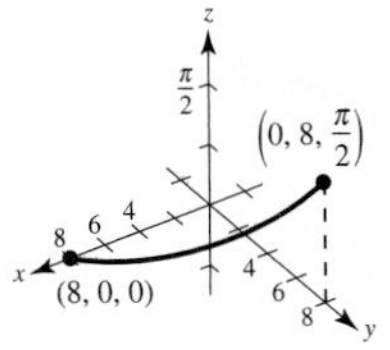

$\sqrt{65}\pi/2$

65. $\sqrt{5}\pi/2$ **67.** 0 **69.** $(2\sqrt{5})/(4 + 5t^2)^{3/2}$

71. $K = \sqrt{17}/289;\ r = 17\sqrt{17}$ **73.** $K = \sqrt{2}/4;\ r = 2\sqrt{2}$

75. The curvature changes abruptly from zero to a nonzero constant.

P.S. Problem Solving (page 881)

1. (a) a (b) πa (c) $K = \pi a$

3. Initial speed: 447.21 ft/sec; $\theta \approx 63.43°$

5–7. Proofs

9. Unit tangent: $\langle -\frac{4}{5}, 0, \frac{3}{5} \rangle$
Unit normal: $\langle 0, -1, 0 \rangle$
Binormal: $\langle \frac{3}{5}, 0, \frac{4}{5} \rangle$

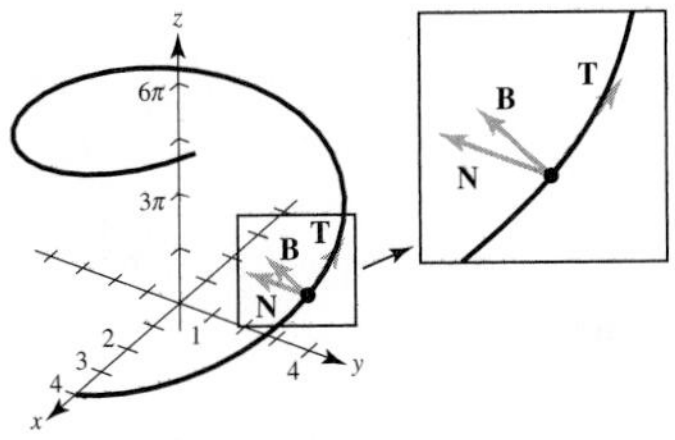

11. (a) Proof (b) Proof

13. (a)

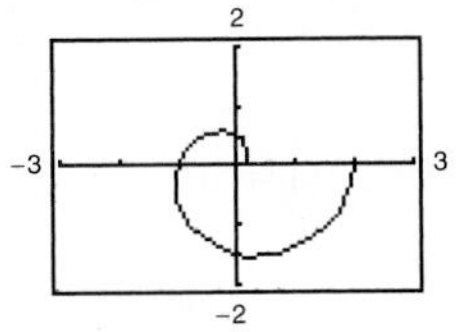

(b) 6.766

(c) $K = [\pi(\pi^2t^2 + 2)]/(\pi^2t^2 + 1)^{3/2}$
$K(0) = 2\pi$
$K(1) = [\pi(\pi^2 + 2)]/(\pi^2 + 1)^{3/2}$
$K(2) \approx 0.51$

(d)

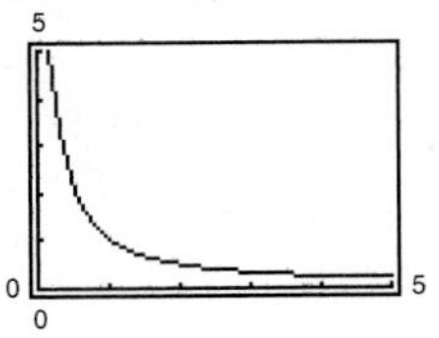

(e) $\lim_{t\to\infty} K = 0$

(f) As $t \to \infty$, the graph spirals outward and the curvature decreases.

Chapter 13

Section 13.1 (page 892)

1. z is a function of x and y. **3.** z is not a function of x and y.

5. (a) $\frac{3}{2}$ (b) $-\frac{1}{4}$ (c) 6 (d) $5/y$ (e) $x/2$ (f) $5/t$

7. (a) 5 (b) $3e^2$ (c) $2/e$ (d) $5e^y$ (e) xe^2 (f) te^t

9. (a) $\frac{2}{3}$ (b) 0 (c) $-\frac{3}{2}$ (d) $-\frac{10}{3}$

11. (a) $\sqrt{2}$ (b) $3\sin 1$ (c) $-3\sqrt{3}/2$ (d) 4

13. (a) 4 (b) 6 (c) $\frac{25}{4}$ (d) $-\frac{9}{4}$

15. (a) $2x + \Delta x,\ \Delta x \neq 0$ (b) $-2,\ \Delta y \neq 0$

17. Domain: $\{(x, y):\ x^2 + y^2 \leq 4\}$
Range: $0 \leq z \leq 2$

19. Domain: $\{(x, y):\ -1 \leq x + y \leq 1\}$
Range: $-\pi/2 \leq z \leq \pi/2$

21. Domain: $\{(x, y):\ y < -x + 4\}$
Range: all real numbers

23. Domain: $\{(x, y):\ x \neq 0, y \neq 0\}$
Range: all real numbers

25. Domain: $\{(x, y):\ y \neq 0\}$
Range: $z > 0$

27. Domain: $\{(x, y):\ x \neq 0, y \neq 0\}$
Range: $|z| > 0$

29. (a) (20, 0, 0) (b) (−15, 10, 20)
(c) (20, 15, 25) (d) (20, 20, 0)

31.

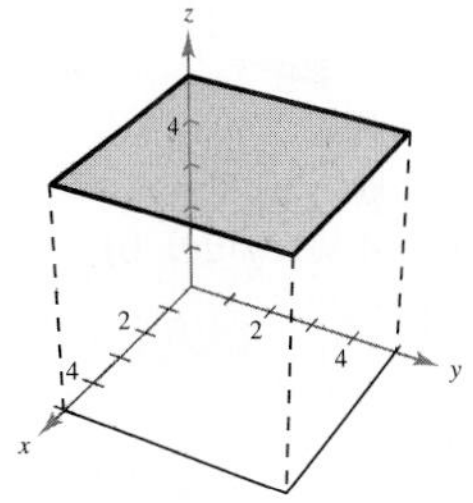

33.

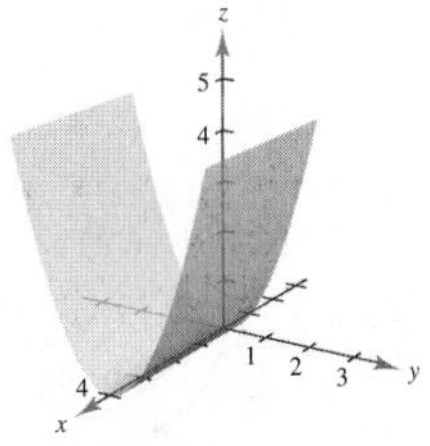

35.

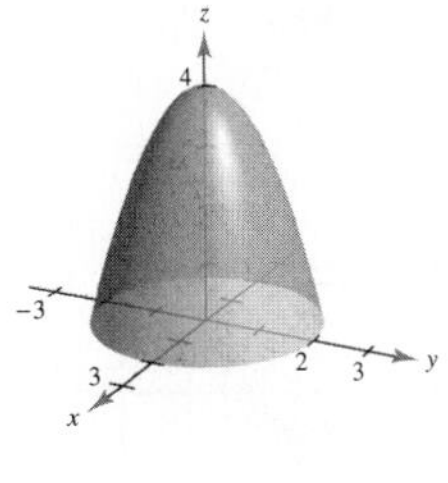

37.

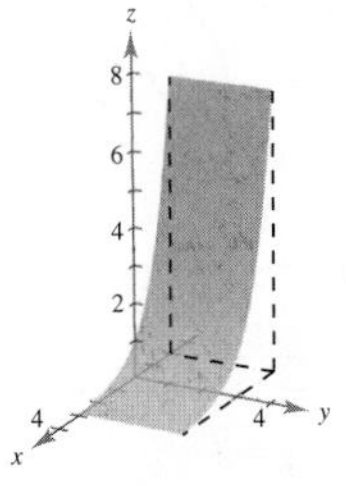

39.

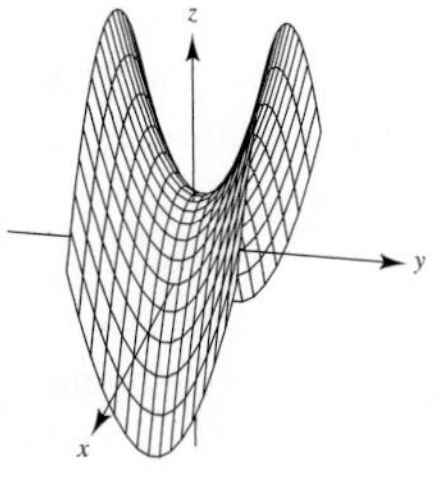

41.

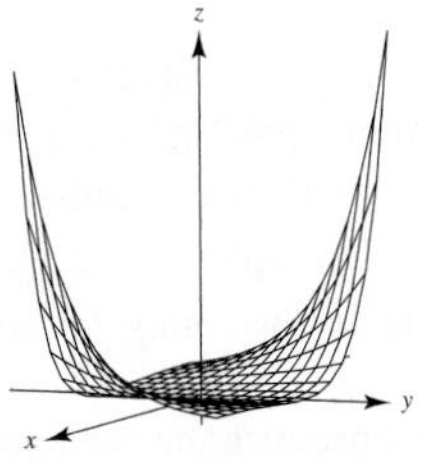

43. (a)

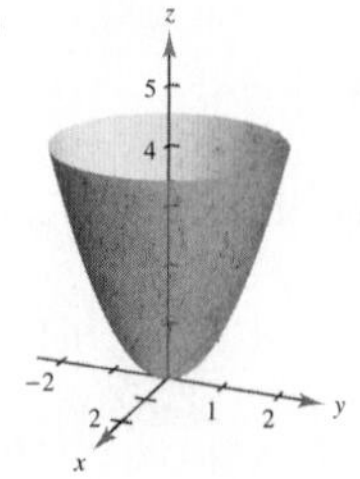

(b) g is a vertical translation of f two units upward.
(c) g is a horizontal translation of f two units to the right.
(d) g is a reflection of f in the xy-plane followed by a vertical translation four units upward.

(e)

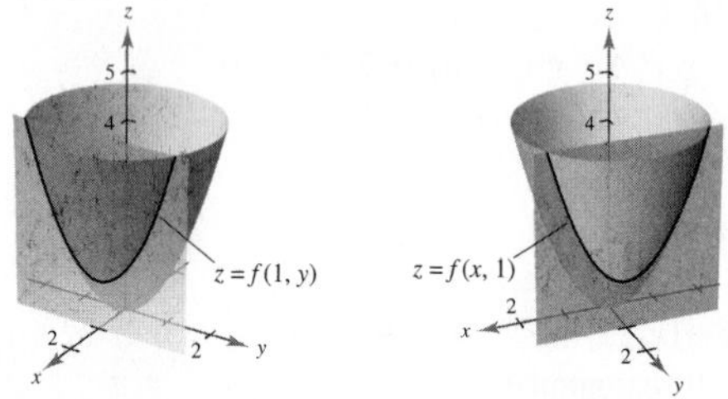

45. c **46.** d **47.** b **48.** a

49. Lines: $x + y = c$

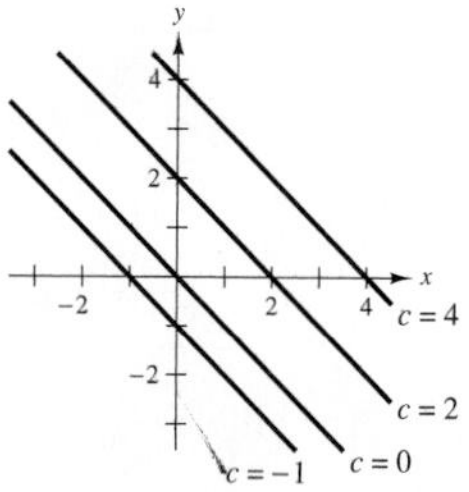

51. Circles centered at $(0, 0)$
Radius ≤ 5

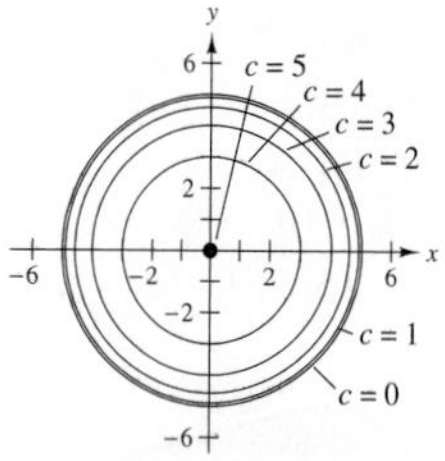

53. Hyperbolas: $xy = c$

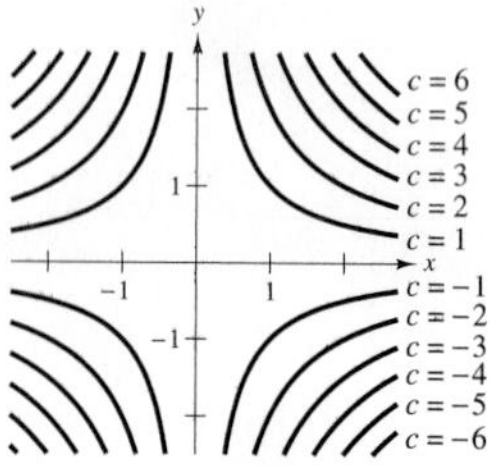

55. Circles passing through $(0, 0)$
Centered at $(1/(2c), 0)$

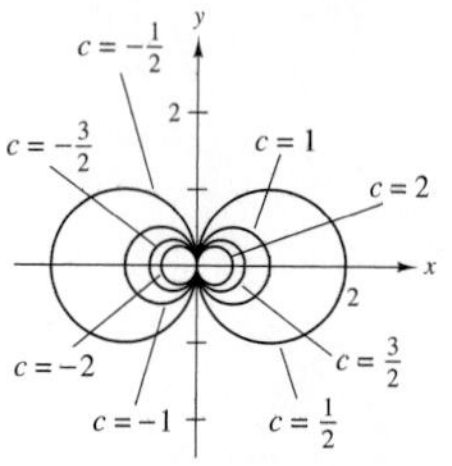

57.

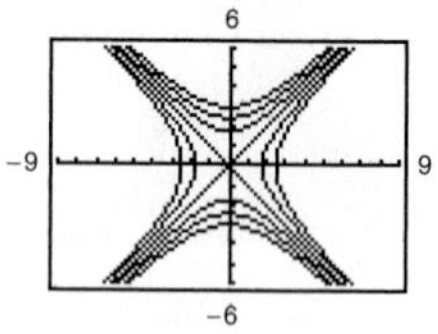

59.

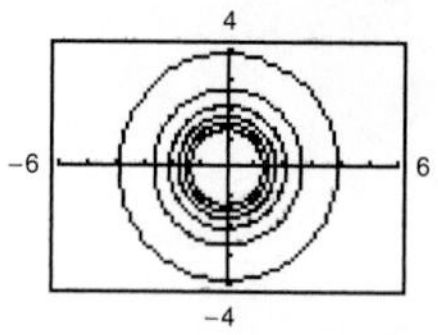

61. Let D be a set of ordered pairs of real numbers. If to each ordered pair (x, y) in D there corresponds a unique real number $f(x, y)$, then f is called a function of x and y.

63. No. Example: $z = e^{-(x^2+y^2)}$

65. The surface may be shaped like a saddle. For example, let $f(x, y) = xy$. The graph is not unique; any vertical translation will produce the same level curves.

67.

	Inflation Rate		
Tax Rate	**0**	**0.03**	**0.05**
0	\$2593.74	\$1929.99	\$1592.33
0.28	\$2004.23	\$1491.34	\$1230.42
0.35	\$1877.14	\$1396.77	\$1152.40

69.

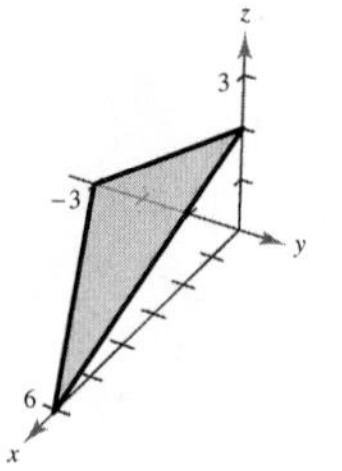

71.

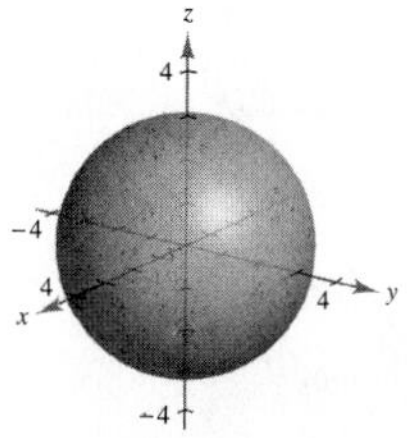

73.

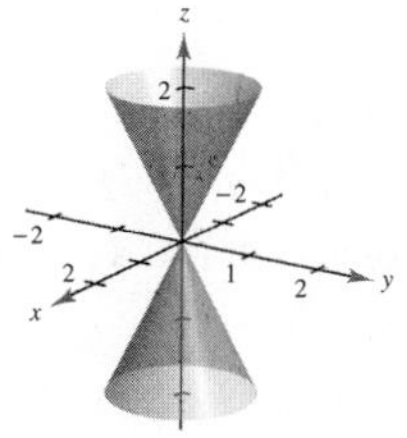

75. (a) 243 board-ft
(b) 507 board-ft

77.

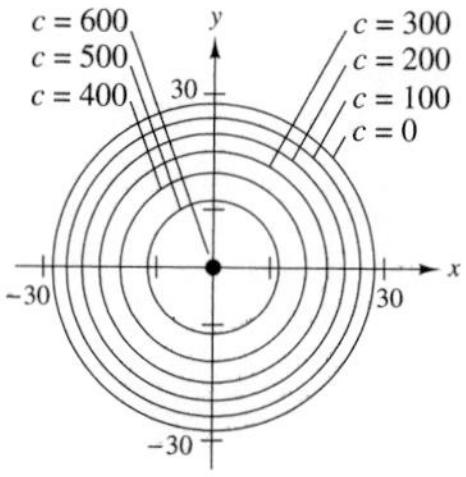

79. Proof

81. $C = 0.75xy + 0.80(xz + yz)$

83. (a) $k = \frac{520}{3}$
(b) $P = 520T/(3V)$
The level curves are lines.

85. (a) C (b) A (c) B

87. (a) The boundaries between colors represent level curves.
(b) No: the colors represent intervals of different lengths.
(c) Use more colors.

89. False: let $f(x, y) = 4$.

91. False: let $f(x, y) = xy^2$. Then $f(ax, ay) = a^3 f(x, y)$.

Section 13.2 (page 902)

1–3. Proofs **5.** 2 **7.** 15 **9.** 5, continuous

11. -3, continuous for $x \neq y$

13. 0, continuous for $xy \neq -1$, $y \neq 0$, $|x/y| \leq 1$

15. $1/e^2$, continuous **17.** $2\sqrt{2}$, continuous for $x + y + z \geq 0$

19. Limit does not exist. **21.** 0

23. Limit does not exist. **25.** Continuous, 1

27. Continuous except at $(0, 0)$; the limit does not exist.

29.

(x, y)	$(1, 0)$	$(0.5, 0)$	$(0.1, 0)$	$(0.01, 0)$	$(0.001, 0)$
$f(x, y)$	0	0	0	0	0

$y = 0$: 0

(x, y)	$(1, 1)$	$(0.5, 0.5)$	$(0.1, 0.1)$
$f(x, y)$	$\frac{1}{2}$	$\frac{1}{2}$	$\frac{1}{2}$

(x, y)	$(0.01, 0.01)$	$(0.001, 0.001)$
$f(x, y)$	$\frac{1}{2}$	$\frac{1}{2}$

$y = x$: $\frac{1}{2}$
Limit does not exist.
Continuous except at $(0, 0)$

31.

(x, y)	$(1, 1)$	$(0.25, 0.5)$	$(0.01, 0.1)$
$f(x, y)$	$-\frac{1}{2}$	$-\frac{1}{2}$	$-\frac{1}{2}$

(x, y)	$(0.0001, 0.01)$	$(0.000001, 0.001)$
$f(x, y)$	$-\frac{1}{2}$	$-\frac{1}{2}$

$x = y^2$: $-\frac{1}{2}$

(x, y)	$(-1, 1)$	$(-0.25, 0.5)$	$(-0.01, 0.1)$
$f(x, y)$	$\frac{1}{2}$	$\frac{1}{2}$	$\frac{1}{2}$

(x, y)	$(-0.0001, 0.01)$	$(-0.000001, 0.001)$
$f(x, y)$	$\frac{1}{2}$	$\frac{1}{2}$

$x = -y^2$: $\frac{1}{2}$
Limit does not exist.
Continuous except at $(0, 0)$

33. f is continuous except at $(0, 0)$.
g is continuous.
f has a removable discontinuity at $(0, 0)$.

35. 0

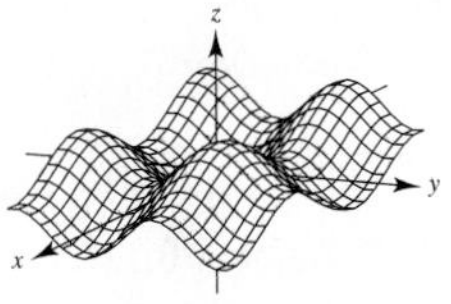

37. Limit does not exist.

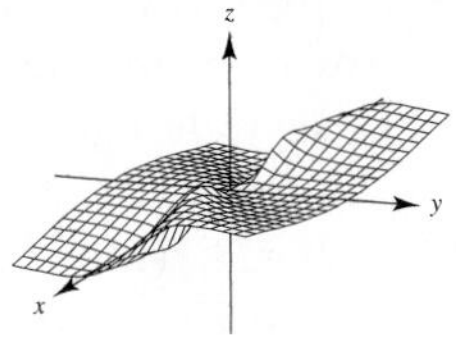

39. Limit does not exist.

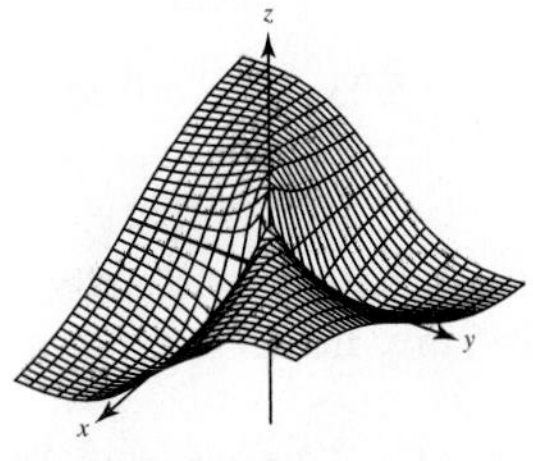

41. 1 **43.** 0 **45.** 0 **47.** 0

49. Continuous except at $(0, 0, 0)$

51. Continuous **53.** Continuous **55.** Continuous

57. Continuous for $y \neq 3x/2$ **59.** (a) $2x$ (b) -4

61. (a) $2 + y$ (b) $x - 3$ **63.** True

65. False: let $f(x, y) = \begin{cases} \ln(x^2 + y^2), & x \neq 0, y \neq 0 \\ 0, & x = 0, y = 0 \end{cases}$.

67. See "Definition of the Limit of a Function of Two Variables," on page 897; show that the value of $\lim_{(x, y) \to (x_0, y_0)} f(x, y)$ is not the same for two different paths to (x_0, y_0).

69. No: the existence of $f(2, 3)$ has no bearing on the existence of the limit as $(x, y) \to (2, 3)$.

71. (a) $(1 + a^2)/a$, $a \neq 0$ (b) Limit does not exist.
(c) No, the limit does not exist. Different paths result in different limits.

73. 0 **75.** $\pi/2$ **77.** Proof

Section 13.3 (page 912)

1. $f_x = (4, 1) < 0$

3. $f_y = (4, 1) > 0$

5. $f_x(x, y) = 2$
$f_y(x, y) = -3$

7. $\partial z/\partial x = \sqrt{y}$
$\partial z/\partial y = x/(2\sqrt{y})$

9. $\partial z/\partial x = 2x - 5y$
$\partial z/\partial y = -5x + 6y$

11. $\partial z/\partial x = 2xe^{2y}$
$\partial z/\partial y = 2x^2e^{2y}$

13. $\partial z/\partial x = 2x/(x^2 + y^2)$
$\partial z/\partial y = 2y/(x^2 + y^2)$

15. $\partial z/\partial x = -2y/(x^2 - y^2)$
$\partial z/\partial y = 2x/(x^2 - y^2)$

17. $\partial z/\partial x = (x^3 - 4y^3)/(x^2y)$
$\partial z/\partial y = (-x^3 + 16y^3)/(2xy^2)$

19. $h_x(x, y) = -2xe^{-(x^2+y^2)}$
$h_y(x, y) = -2ye^{-(x^2+y^2)}$

21. $f_x(x, y) = x/\sqrt{x^2 + y^2}$
$f_y(x, y) = y/\sqrt{x^2 + y^2}$

23. $\partial z/\partial x = 2\sec^2(2x - y)$
$\partial z/\partial y = -\sec^2(2x - y)$

25. $\partial z/\partial x = ye^y \cos xy$
$\partial z/\partial y = e^y(x \cos xy + \sin xy)$

27. $f_x(x, y) = 1 - x^2$
$f_y(x, y) = y^2 - 1$

29. $f_x(x, y) = 2$
$f_y(x, y) = 3$

31. $f_x(x, y) = 1/(2\sqrt{x + y})$
$f_y(x, y) = 1/(2\sqrt{x + y})$

33. $\partial z/\partial x = \frac{1}{4}$
$\partial z/\partial y = \frac{1}{4}$

35. $\partial z/\partial x = -\frac{1}{4}$
$\partial z/\partial y = \frac{1}{4}$

37. $g_x(1, 1) = -2$
$g_y(1, 1) = -2$

39. $\partial z/\partial x = -1$
$\partial z/\partial y = 0$

41.

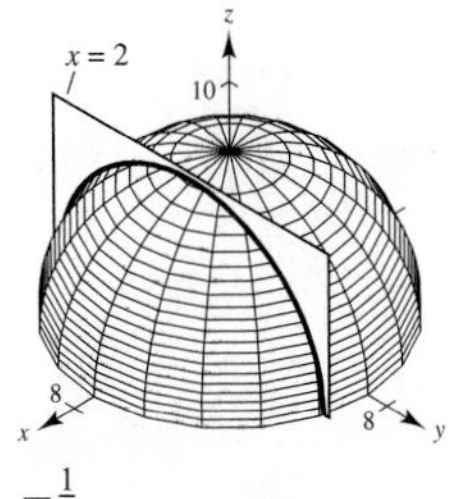

$-\frac{1}{2}$

43.

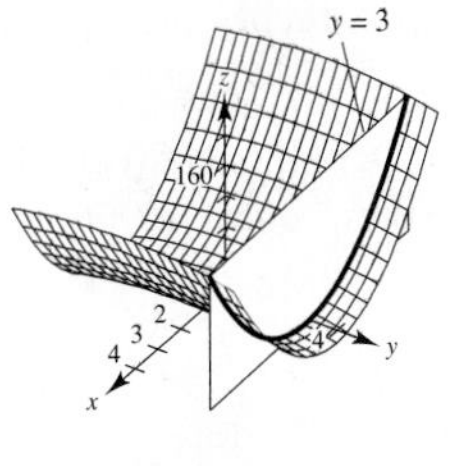

18

45. $x = -6, y = 4$ **47.** $x = 1, y = 1$

49. (a) f_y (b) f_x
f_x represents the slope in the x-direction, and f_y represents the slope in the y-direction.

51. $\dfrac{\partial w}{\partial x} = \dfrac{x}{\sqrt{x^2 + y^2 + z^2}}$
$\dfrac{\partial w}{\partial y} = \dfrac{y}{\sqrt{x^2 + y^2 + z^2}}$
$\dfrac{\partial w}{\partial z} = \dfrac{z}{\sqrt{x^2 + y^2 + z^2}}$

53. $F_x(x, y, z) = \dfrac{x}{x^2 + y^2 + z^2}$
$F_y(x, y, z) = \dfrac{y}{x^2 + y^2 + z^2}$
$F_z(x, y, z) = \dfrac{z}{x^2 + y^2 + z^2}$

55. $H_x(x, y, z) = \cos(x + 2y + 3z)$
$H_y(x, y, z) = 2\cos(x + 2y + 3z)$
$H_z(x, y, z) = 3\cos(x + 2y + 3z)$

57. $f_x = 3\sqrt{5}/5; f_y = -2\sqrt{5}/5; f_z = -2\sqrt{5}/5$

59. $f_x = 0; f_y = 0; f_z = 1$

61. $\dfrac{\partial^2 z}{\partial x^2} = 2$
$\dfrac{\partial^2 z}{\partial y^2} = 6$
$\dfrac{\partial^2 z}{\partial y \partial x} = \dfrac{\partial^2 z}{\partial x \partial y} = -2$

63. $\dfrac{\partial^2 z}{\partial x^2} = \dfrac{y^2}{(x^2 + y^2)^{3/2}}$
$\dfrac{\partial^2 z}{\partial y^2} = \dfrac{x^2}{(x^2 + y^2)^{3/2}}$
$\dfrac{\partial^2 z}{\partial y \partial x} = \dfrac{\partial^2 z}{\partial x \partial y} = \dfrac{-xy}{(x^2 + y^2)^{3/2}}$

65. $\dfrac{\partial^2 z}{\partial x^2} = e^x \tan y$
$\dfrac{\partial^2 z}{\partial y^2} = 2e^x \sec^2 y \tan y$
$\dfrac{\partial^2 z}{\partial y \partial x} = \dfrac{\partial^2 z}{\partial x \partial y} = e^x \sec^2 y$

67. $\dfrac{\partial^2 z}{\partial x^2} = \dfrac{2xy}{(x^2 + y^2)^2}$
$\dfrac{\partial^2 z}{\partial y^2} = \dfrac{-2xy}{(x^2 + y^2)^2}$
$\dfrac{\partial^2 z}{\partial y \partial x} = \dfrac{\partial^2 z}{\partial x \partial y} = \dfrac{y^2 - x^2}{(x^2 + y^2)^2}$

69. $\partial z/\partial x = \sec y$
$\partial z/\partial y = x \sec y \tan y$
$\partial^2 z/\partial x^2 = 0$
$\partial^2 z/\partial y^2 = x \sec y(\sec^2 y + \tan^2 y)$
$\partial^2 z/\partial y \partial x = \partial^2 z/\partial x \partial y = \sec y \tan y$
No values of x and y exist such that $f_x(x, y) = f_y(x, y) = 0$.

71. $\partial z/\partial x = (y^2 - x^2)/[x(x^2 + y^2)]$
$\partial z/\partial y = -2y/(x^2 + y^2)$
$\partial^2 z/\partial x^2 = (x^4 - 4x^2y^2 - y^4)/[x^2(x^2 + y^2)^2]$
$\partial^2 z/\partial y^2 = 2(y^2 - x^2)/(x^2 + y^2)^2$
$\partial^2 z/\partial y \partial x = \partial^2 z/\partial x \partial y = 4xy/(x^2 + y^2)^2$
No values of x and y exist such that $f_x(x, y) = f_y(x, y) = 0$.

73. $f_{xyy}(x, y, z) = f_{yxy}(x, y, z) = f_{yyx}(x, y, z) = 0$

75. $f_{xyy}(x, y, z) = f_{yxy}(x, y, z) = f_{yyx}(x, y, z) = z^2e^{-x} \sin yz$

77. $\partial^2 z/\partial x^2 + \partial^2 z/\partial y^2 = 0 + 0 = 0$

79. $\partial^2 z/\partial x^2 + \partial^2 z/\partial y^2 = e^x \sin y - e^x \sin y = 0$

81. $\partial^2 z/\partial t^2 = -c^2 \sin(x - ct) = c^2(\partial^2 z/\partial x^2)$

83. $\partial^2 z/\partial t^2 = -c^2/(x + ct)^2 = c^2(\partial^2 z/\partial x^2)$

85. $\partial z/\partial t = -e^{-t} \cos x/c = c^2(\partial^2 z/\partial x^2)$

87. See "Definition of Partial Derivatives of a Function of Two Variables," on page 906.

89.

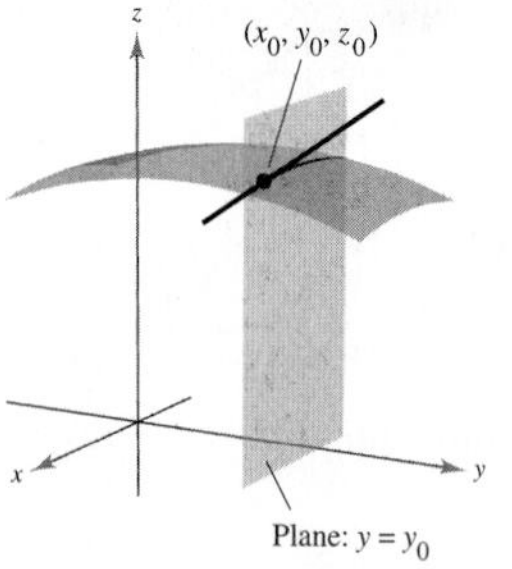

$\partial f/\partial x$ represents the slope of the curve formed by the intersection of the surface $z = f(x, y)$ and the plane $y = y_0$ at any point on the curve.

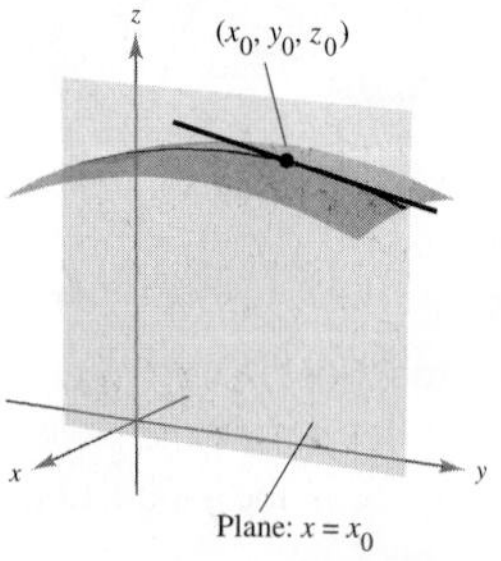

$\partial f/\partial y$ represents the slope of the curve formed by the intersection of the surface $z = f(x, y)$ and the plane $x = x_0$ at any point on the curve.

91.

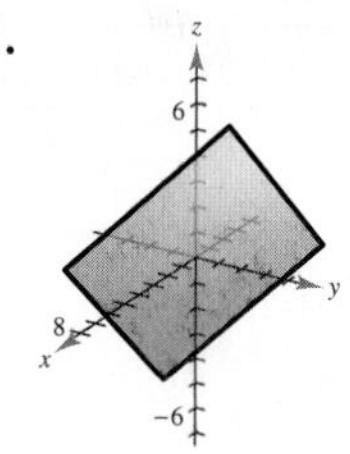

93. (a) $\partial C/\partial x = 183$, $\partial C/\partial y = 237$
(b) The fireplace-insert stove results in the cost increasing at a higher rate because the coefficient of y is greater in magnitude than the coefficient of x.

95. An increase in either the charge for food and housing or the tuition will cause a decrease in the number of applicants.

97. $\partial T/\partial x = -2.4°$ per m, $\partial T/\partial y = -9°$ per m

99. $T = PV/(nR) \Rightarrow \partial T/\partial P = V/(nR)$
$P = nRT/V \Rightarrow \partial P/\partial V = -nRT/V^2$
$V = nRT/P \Rightarrow \partial V/\partial T = nR/P$
$\partial T/\partial P \cdot \partial P/\partial V \cdot \partial V/\partial T = -nRT/(VP) = -nRT/(nRT) = -1$

101. (a) $\partial z/\partial x = -0.04$; $\partial z/\partial y = 0.64$
(b) For every decrease of 0.04 gallon of whole milk there is an increase of one gallon of skim milk. For every increase of 0.64 gallon of whole milk there is an decrease of one gallon of reduced-fat milk.

103. False; Let $z = x + y + 1$. **105.** True

107. (a) $f_x(x, y) = \dfrac{y(x^4 + 4x^2y^2 - y^4)}{(x^2 + y^2)^2}$
$f_y(x, y) = \dfrac{x(x^4 - 4x^2y^2 - y^4)}{(x^2 + y^2)^2}$
(b) $f_x(0, 0) = 0$, $f_y(0, 0) = 0$
(c) $f_{xy}(0, 0) = -1$, $f_{yx}(0, 0) = 1$
(d) f_{xy} or f_{yx} or both are not continuous at (0, 0).

109. (a) Proof (b) $f_y = (x, y)$ does not exist when $y = -x$.

Section 13.4 (page 921)

1. $dz = 6xy^3\,dx + 9x^2y^2\,dy$

3. $dz = 2(x\,dx + y\,dy)/(x^2 + y^2)^2$

5. $dz = (\cos y + y \sin x)\,dx - (x \sin y + \cos x)dy$

7. $dz = (e^x \sin y)\,dx + (e^x \cos y)\,dy$

9. $dw = 2z^3y \cos x\,dx + 2z^3 \sin x\,dy + 6z^2 y \sin x\,dz$

11. (a) $f(1, 2) = 4, f(1.05, 2.1) = 3.4875, \Delta z = -0.5125$
(b) $dz = -0.5$

13. (a) $f(1, 2) \approx 0.90930, f(1.05, 2.1) \approx 0.90637, \Delta z \approx -0.00293$
(b) $dz \approx 0.00385$

15. (a) $f(1, 2) = -5, f(1.05, 2.1) = -5.25, \Delta z = -0.25$
(b) $dz = -0.25$

17. 0.094 **19.** −0.012

21. If $z = f(x, y)$ and Δx and Δy are increments of x and y, and x and y are independent variables, then the total differential of the dependent variable z is
$dz = (\partial z/\partial x)\,dx + (\partial z/\partial y)\,dy = f_x(x, y)\,\Delta x + f_y(x, y)\,\Delta y$.

23. The approximation of Δz by dz is called a linear approximation, where dz represents the change in height of a plane that is tangent to the surface at the point $P(x_0, y_0)$.

25. $dA = h\,dl + l\,dh$

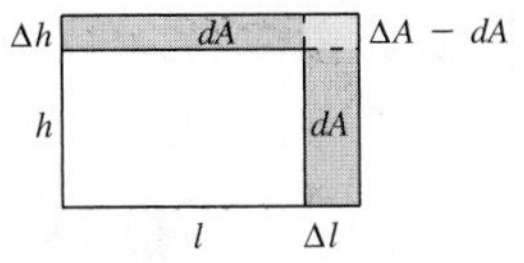

27.

Δr	Δh	dV	ΔV	$\Delta V - dV$
0.1	0.1	4.7124	4.8391	0.1267
0.1	−0.1	2.8274	2.8264	−0.0010
0.001	0.002	0.0565	0.0565	0.0001
−0.0001	0.0002	−0.0019	−0.0019	0.0000

29. (a) $dz = -0.04\,dx + 0.64\,dy$
(b) $dz = \pm 0.17$; $dz/z = 2.1\%$

31. 10% **33.** $dC = \pm 0.24418$; $dC/C = 19\%$

35. (a) $V = 18 \sin\theta \text{ ft}^3$; $\theta = \pi/2$
(b) 1.047 ft^3

37. 1% **39.** $L \approx 8.096 \times 10^{-4} \pm 6.6 \times 10^{-6}$ microhenrys

41. Answers will vary.
Example:
$\varepsilon_1 = \Delta x$
$\varepsilon_2 = 0$

43. Answers will vary.
Example:
$\varepsilon_1 = y\,\Delta x$
$\varepsilon_2 = 2x\,\Delta x + (\Delta x)^2$

45. Proof

47. Answers will vary. For example, we can use the equation $F = ma$.
Then $dF = (\partial F/\partial m)\,dm + (\partial F/\partial a)\,da = a\,dm + m\,da$.
We can estimate the possible propagated errors when given the error in measurement.

Section 13.5 (page 929)

1. $2(e^{2t} - e^{-2t})$ **3.** $e^t \sec(\pi - t)[1 - \tan(\pi - t)]$

5. $2\cos 2t$ **7.** $4e^{2t}$ **9.** $3(2t^2 - 1)$

11. $-11\sqrt{29}/29 \approx -2.04$

13. $\dfrac{8\sin t\cos t(4\sin^2 t\cos^2 t - 3)}{(4\sin^2 t\cos^2 t + 1)^2}$; 0

15. $\partial w/\partial s = 4s, 8$
$\partial w/\partial t = 4t, -4$

17. $\partial w/\partial s = 2s\cos 2t, 0$
$\partial w/\partial t = -2s^2 \sin 2t, -18$

19. $\partial w/\partial r = 0$
$\partial w/\partial\theta = 8\theta$

21. $\partial w/\partial r = 0$
$\partial w/\partial\theta = 1$

23. $\dfrac{\partial w}{\partial s} = t^2(3s^2 - t^2)$
$\dfrac{\partial w}{\partial t} = 2st(s^2 - 2t^2)$

25. $\dfrac{\partial w}{\partial s} = \dfrac{te^{(s-t)/(s+t)}(s^2 + 4st + t^2)}{(s+t)^2}$
$\dfrac{\partial w}{\partial t} = \dfrac{se^{(s-t)/(s+t)}(s^2 + t^2)}{(s+t)^2}$

27. $\dfrac{3y - 2x + 2}{2y - 3x + 1}$

29. $-\dfrac{x + y(x^2 + y^2)}{y + x(x^2 + y^2)}$

31. $\dfrac{\partial z}{\partial x} = \dfrac{-x}{z}$
$\dfrac{\partial z}{\partial y} = \dfrac{-y}{z}$

33. $\dfrac{\partial z}{\partial x} = \dfrac{-\sec^2(x + y)}{\sec^2(y + z)}$
$\dfrac{\partial z}{\partial y} = -1 - \dfrac{\sec^2(x + y)}{\sec^2(y + z)}$

35. $\partial z/\partial x = -x/(y + z)$
$\partial z/\partial y = -z/(y + z)$

37. $\partial z/\partial x = -(ze^{xz} + y)/xe^{xz}$
$\partial z/\partial y = -e^{-xz}$

39. $\dfrac{\partial w}{\partial x} = \dfrac{-yz - zw}{xz - yz + 2w}$
$\dfrac{\partial w}{\partial y} = \dfrac{-xz + zw}{xz - yz + 2w}$
$\dfrac{\partial w}{\partial z} = \dfrac{yw - xy - xw}{xz - yz + 2w}$

41. $\dfrac{\partial w}{\partial x} = \dfrac{y\sin xy}{z}$
$\dfrac{\partial w}{\partial y} = \dfrac{x\sin xy - z\cos yz}{z}$
$\dfrac{\partial w}{\partial z} = -\dfrac{y\cos yz + w}{z}$

43. 1; $xf_x(x, y) + yf_y(x, y) = \dfrac{xy}{\sqrt{x^2 + y^2}} = 1f(x, y)$

45. 0; $xf_x(x, y) + yf_y(x, y) = \dfrac{xe^{x/y}}{y} - \dfrac{xe^{x/y}}{y} = 0$

47. $dw/dt = \partial w/\partial x \cdot dx/dt + \partial w/\partial y \cdot dy/dt$

49. The explicit form of a function of two variables is of the form $z = f(x, y)$, as in $z = x^2 + y^2$. The implicit form of a function of two variables is of the form $F(x, y, z) = 0$, as in $z - x^2 - y^2 = 0$.

51. $4608\pi \text{ in.}^3/\text{min}$; $624\pi \text{ in.}^2/\text{min}$

53. $(\sqrt{2}/10)(15 + \pi)\text{m}^2/\text{hr}$ **55.** $28m \text{ cm}^2/\text{sec}$

57. (a) Proof
(b) $d\theta/dx = (2\cos^2\theta - 2x\cos\theta\sin\theta)/(x^2 + 8)$
(c) $x = 2\sqrt{2}$

59. Proof **61.** (a) Proof (b) Proof **63.** Proof

Section 13.6 (page 940)

1. $(\sqrt{3} - 5)/2$ **3.** $5\sqrt{2}/2$ **5.** $-\frac{7}{25}$ **7.** $-e$ **9.** $2\sqrt{6}/3$

11. $(8 + \pi)\sqrt{6}/24$ **13.** $\sqrt{2}(x + y)$

15. $\left[(2 + \sqrt{3})/2\right]\cos(2x - y)$

17. $-7\sqrt{2}$ **19.** $7\sqrt{19}/19$ **21.** $3\mathbf{i} - 10\mathbf{j}$

23. $-6\sin 25\mathbf{i} + 8\sin 25\mathbf{j} \approx 0.7941\mathbf{i} - 1.0588\mathbf{j}$

25. $6\mathbf{i} + 13\mathbf{j} - 9\mathbf{k}$ **27.** $2\sqrt{5}$ **29.** $-2\sqrt{5}/5$

31. $\tan y\mathbf{i} + x\sec^2 y\mathbf{j}$, $\sqrt{17}$

33. $\dfrac{2}{3(x^2 + y^2)}(x\mathbf{i} + y\mathbf{j}), \dfrac{2\sqrt{5}}{15}$ **35.** $\dfrac{x\mathbf{i} + y\mathbf{j} + z\mathbf{k}}{\sqrt{x^2 + y^2 + z^2}}$, 1

37. $e^{yz}\mathbf{i} + xze^{yz}\mathbf{j} + xye^{yz}\mathbf{k}$; $\sqrt{65}$

39.

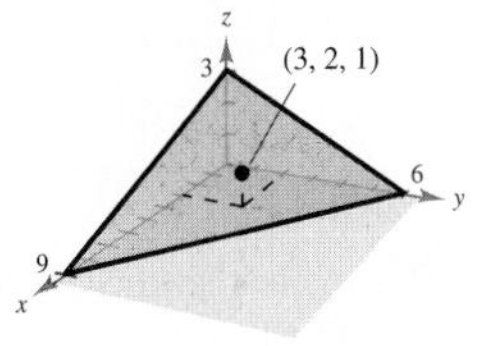

41. (a) $(2 + 3\sqrt{3})/12$
(b) $(3 - 2\sqrt{3})/12$

43. (a) $-\frac{1}{5}$ (b) $-11\sqrt{10}/60$ **45.** $\sqrt{13}/6$

47.

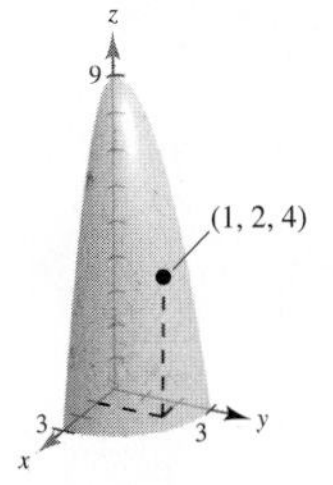

49. $-2\mathbf{i} - 4\mathbf{j}, 2\sqrt{5}$

51. (a) Answers will vary. Example: $-4\mathbf{i} + \mathbf{j}$
(b) $-\frac{2}{5}\mathbf{i} + \frac{1}{10}\mathbf{j}$ (c) $\frac{2}{5}\mathbf{i} - \frac{1}{10}\mathbf{j}$
The direction opposite that of the gradient

53. (a)

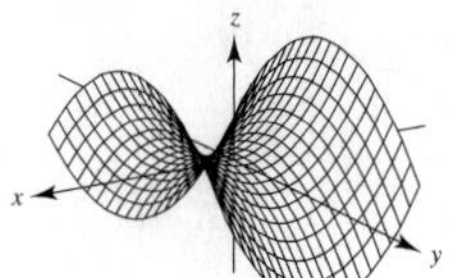

(b) $D_{\mathbf{u}}f(4, -3) = 8\cos\theta + 6\sin\theta$

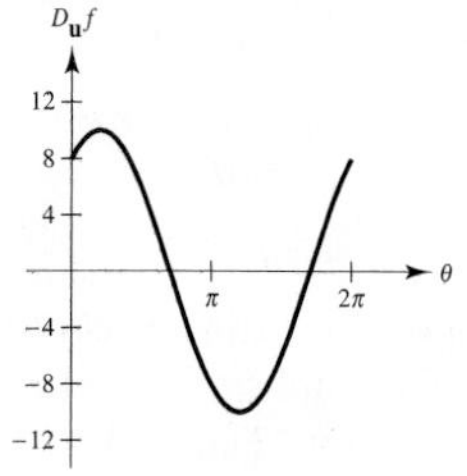

Generated by Mathematica

(c) $\theta \approx 2.21$, $\theta \approx 5.36$
Directions in which there is no change in f

(d) $\theta \approx 0.64$, $\theta \approx 3.79$
Directions of greatest rate of change in f

(e) 10; Magnitude of the greatest rate of change

(f)

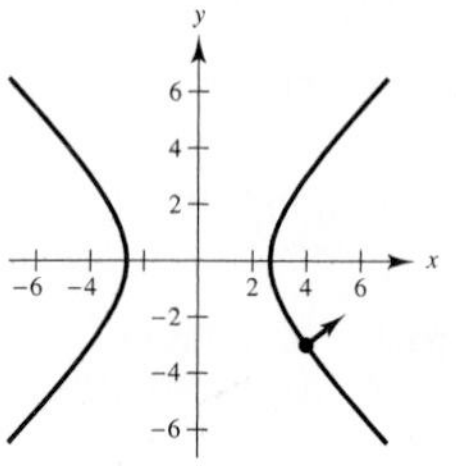

Generated by Mathematica

Orthogonal to the level curve

55. $6\mathbf{i} + 8\mathbf{j}$ **57.** $-\frac{1}{2}\mathbf{j}$

59. $(\sqrt{257}/257)(16\mathbf{i} - \mathbf{j})$ **61.** $(\sqrt{85}/85)(9\mathbf{i} - 2\mathbf{j})$

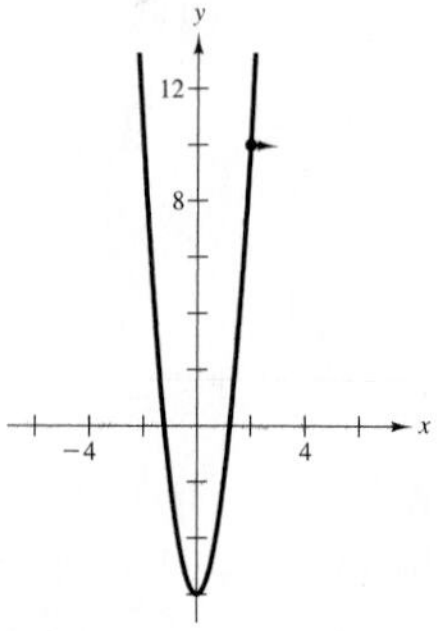

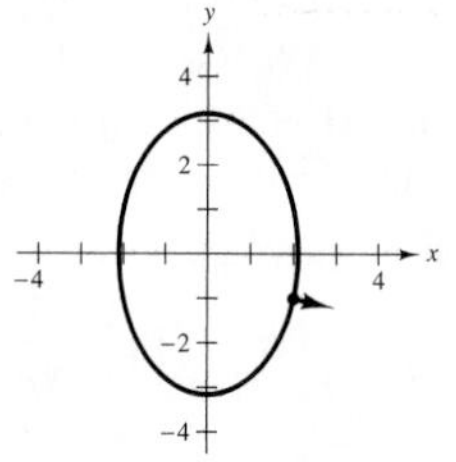

63. $\frac{1}{625}(7\mathbf{i} - 24\mathbf{j})$

65. The directional derivative of $z = f(x, y)$ in the direction of $\mathbf{u} = \cos t\,\mathbf{i} + \sin t\,\mathbf{j}$ is

$$D_{\mathbf{u}}f(x, y) = \lim_{t \to 0} \frac{f(x + t\cos\theta, y + t\sin\theta) - f(x, y)}{t}$$

if the limit exists.

67. Let $f(x, y)$ be a function of two variables and let $\mathbf{u} = \cos\theta\,\mathbf{i} + \sin\theta\,\mathbf{j}$ be a unit vector.
(a) If $\theta = 0°$, then $D_{\mathbf{u}}f = \partial f/\partial x$.
(b) If $\theta = 90°$, then $D_{\mathbf{u}}f = \partial f/\partial y$.

69. Answers will vary. Sample answer:

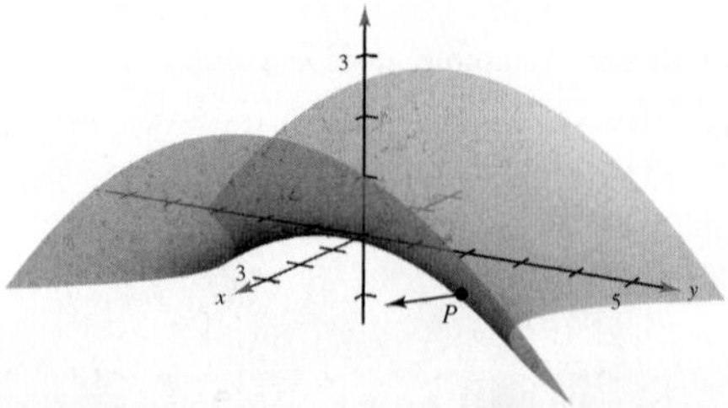

71.

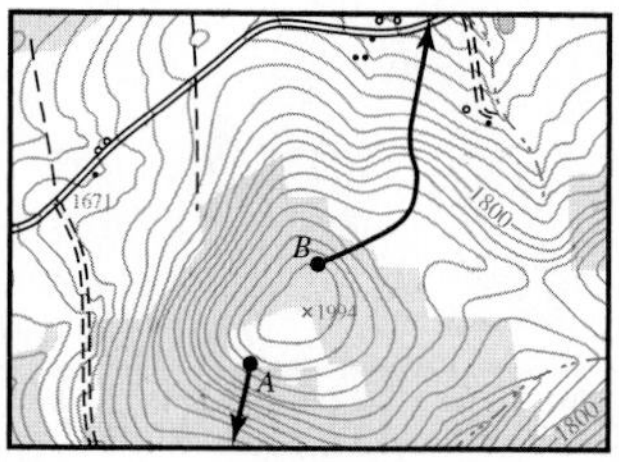

73. $y^2 = 10x$

75. (a)

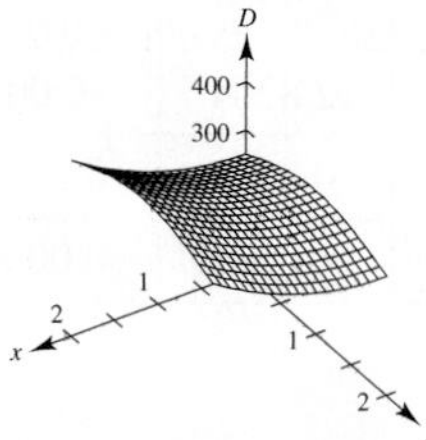

(b) Graph $-D$.
(c) 315 m
(d) 60.0
(e) 55.5
(f) $60.0\mathbf{i} + 55.5\mathbf{j}$

77. True **79.** True **81.** $f(x, y, z) = e^x \cos y + \frac{1}{2}z^2 + C$

Section 13.7 (page 949)

1. The level surface can be written as $3x - 5y + 3z = 15$, which is an equation of a plane in space.

3. The level surface can be written as $4x^2 + 9y^2 - 4z^2 = 0$, which is an elliptic cone that lies on the z-axis.

5. $(\sqrt{3}/3)(\mathbf{i} + \mathbf{j} + \mathbf{k})$ **7.** $(\sqrt{2}/10)(3\mathbf{i} + 4\mathbf{j} - 5\mathbf{k})$

9. $(\sqrt{2049}/2049)(32\mathbf{i} + 32\mathbf{j} - \mathbf{k})$ **11.** $(\sqrt{3}/3)(\mathbf{i} - \mathbf{j} + \mathbf{k})$

13. $(\sqrt{113}/113)(-\mathbf{i} - 6\sqrt{3}\mathbf{j} + 2\mathbf{k})$ **15.** $6x + 2y + z = 35$

17. $3x + 4y - 5z = 0$ **19.** $10x - 8y - z = 9$

21. $2x - z = -2$ **23.** $3x + 4y - 25z = 25(1 - \ln 5)$

25. $x - 4y + 2z = 18$ **27.** $x + y + z = 1$

29. $2x + 4y + z = 14$
$$\frac{x-1}{2} = \frac{y-2}{4} = \frac{z-4}{1}$$

31. $3x + 2y + z = -6$
$$\frac{x+2}{3} = \frac{y+3}{2} = \frac{z-6}{1}$$

33. $x - y + 2z = \pi/2$
$(x - 1)/1 = (y - 1)/-1 = (z - \pi/4)/2$

35. (a) Line: $x = 1, y = 1, z = 1 - t$
Plane: $z = 1$
(b) Line: $x = -1, y = 2 + \frac{6}{25}t, z = -\frac{4}{5} - t$
Plane: $6y - 25z - 32 = 0$
(c)

(d) At $(1, 1, 1)$, the tangent plane is parallel to the xy-plane, implying that the surface is level. At $\left(-1, 2, -\frac{4}{5}\right)$, the function does not change in the x-direction.

37. See "Definition of Tangent Plane and Normal Line" on page 944.

39. For a sphere, the common object is the center of the sphere. For a cylinder, the common object is its axis.

41. (a) $\dfrac{x-2}{1} = \dfrac{y-1}{-2} = \dfrac{z-2}{1}$ (b) $\dfrac{\sqrt{10}}{5}$, not orthogonal

43. (a) $\dfrac{x-3}{4} = \dfrac{y-3}{4} = \dfrac{z-4}{-3}$ (b) $\dfrac{16}{25}$, not orthogonal

45. (a) $\dfrac{y-1}{1} = \dfrac{z-1}{-1}$, $x = 2$ (b) 0, orthogonal

47. (a) $x = 1 + t$
$y = 2 - 2t$
$z = 4$
$\theta \approx 48.2°$

(b)
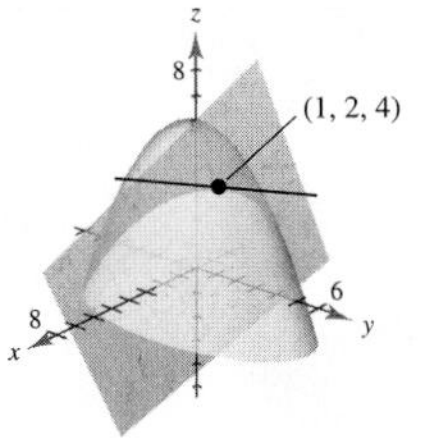

49. 86.0° **51.** 77.4°

53. (0, 3, 12)

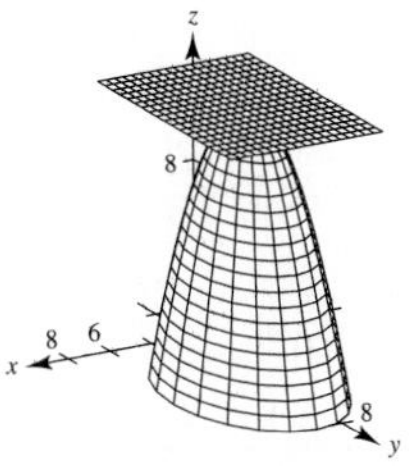

The function is maximum.

55. $x = 4e^{-4kt},\ y = 3e^{-2kt},\ z = 10e^{-8kt}$

57. $F(x, y, z) = \dfrac{x^2}{a^2} + \dfrac{y^2}{b^2} + \dfrac{z^2}{c^2} - 1$

$F_x(x, y, z) = 2x/a^2$

$F_y(x, y, z) = 2y/b^2$

$F_z(x, y, z) = 2z/c^2$

Plane: $\dfrac{2x_0}{a^2}(x - x_0) + \dfrac{2y_0}{b^2}(y - y_0) + \dfrac{2z_0}{c^2}(z - z_0) = 0$

$\dfrac{x_0x}{a^2} + \dfrac{y_0y}{b^2} + \dfrac{z_0z}{c^2} = 1$

59. $F(x, y, z) = a^2x^2 + b^2y^2 - z^2$

$F_x(x, y, z) = 2a^2x$

$F_y(x, y, z) = 2b^2y$

$F_z(x, y, z) = -2z$

Plane: $2a^2x_0(x - x_0) + 2b^2y_0(y - y_0) - 2z_0(z - z_0) = 0$

$a^2x_0x + b^2y_0y - z_0z = 0$

Hence, the plane passes through the origin.

61. (a) $P_1(x, y) = 1 + x - y$

(b) $P_2(x, y) = 1 + x - y + \frac{1}{2}x^2 - xy + \frac{1}{2}y^2$

(c) If $x = 0$, $P_2(0, y) = 1 - y + \frac{1}{2}y^2$.
This is the second-degree Taylor polynomial for e^{-y}.
If $y = 0$, $P_2(x, 0) = 1 + x + \frac{1}{2}x^2$.
This is the second-degree Taylor polynomial for e^x.

(d)

x	y	$f(x, y)$	$P_1(x, y)$	$P_2(x, y)$
0	0	1	1	1
0	0.1	0.9048	0.9000	0.9050
0.2	0.1	1.1052	1.1000	1.1050
0.2	0.5	0.7408	1.7000	0.7450
1	0.5	1.6487	1.5000	1.6250

(e)
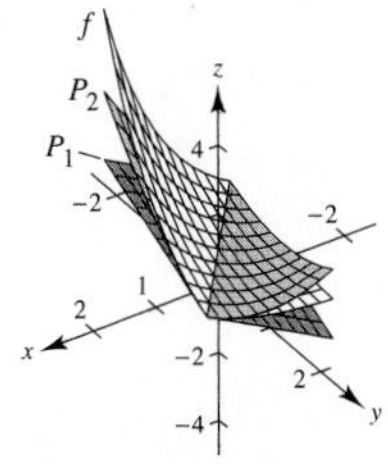

63. Proof

Section 13.8 (page 958)

1. Relative minimum: (1, 3, 0)

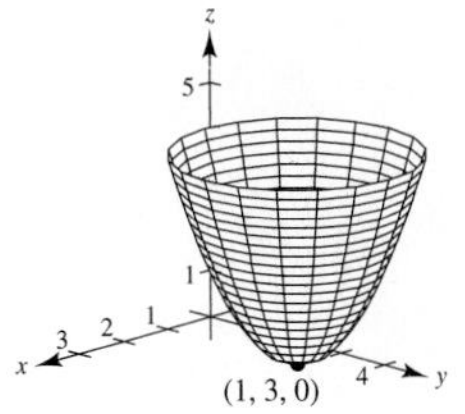

3. Relative minimum: (0, 0, 1)

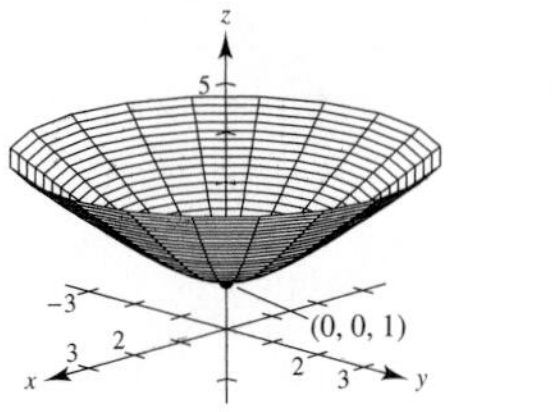

5. Relative minimum: $(-1, 3, -4)$

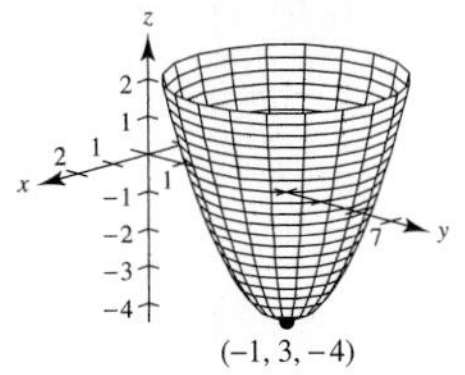

7. Relative minimum: $(-1, 1, -4)$

9. Relative maximum: (8, 16, 74)

11. Relative minimum: $(1, 2, -1)$

13. Relative minimum: (0, 0, 3) **15.** Relative maximum: (0, 0, 4)

17.

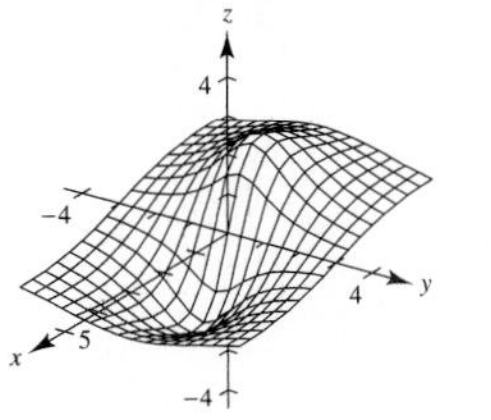

Relative maximum: $(-1, 0, 2)$
Relative minimum: $(1, 0, -2)$

19.

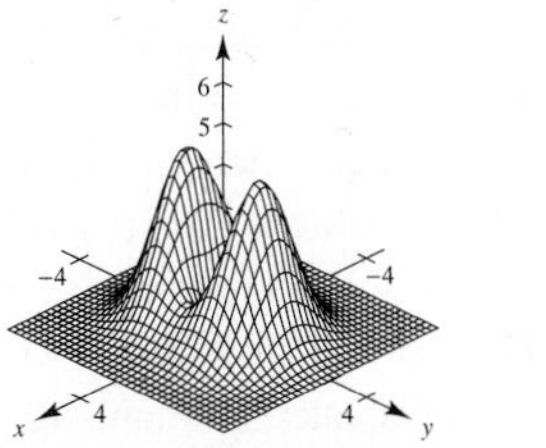

Relative minimum: (0, 0, 0)
Relative maxima: $(0, \pm 1, 4)$
Saddle points: $(\pm 1, 0, 1)$

21. Saddle point: $(1, -2, -1)$ **23.** Saddle point: (0, 0, 0)

25. Saddle point: (0, 0, 0); Relative minimum: $(1, 1, -1)$

27. There are no critical numbers.

29. z is never negative. Minimum: $z = 0$ when $x = y \neq 0$.

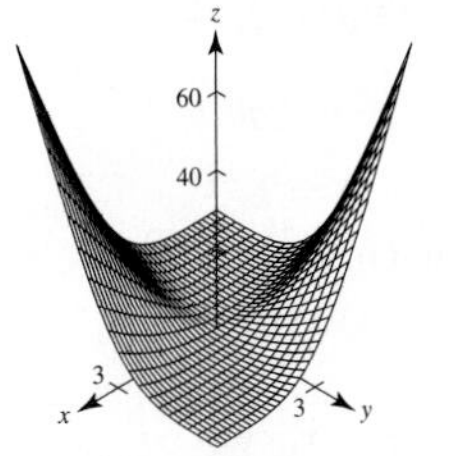

31. Insufficient information **33.** Saddle point

35. (a) The function f defined on a region R containing (x_0, y_0) has a relative minimum at (x_0, y_0) if $f(x, y) \geq f(x_0, y_0)$.

(b) The function f defined on a region R containing (x_0, y_0) has a relative maximum at (x_0, y_0) if $f(x, y) \leq f(x_0, y_0)$.

(c) A saddle point is a critical point (x_0, y_0) that is not an extremum.

(d) Let f be defined on an open region R containing (x_0, y_0). The point (x_0, y_0) is a critical point of f if one of the following is true:

1. $f_x(x_0, y_0) = 0$ and $f_y(x_0, y_0) = 0$
2. $f_x(x_0, y_0)$ or $f_y(x_0, y_0)$ does not exist.

37. Answers will vary.
Sample answer:

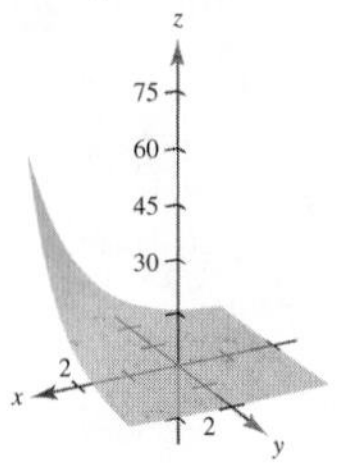

No extrema

39. Answers will vary.
Sample answer:

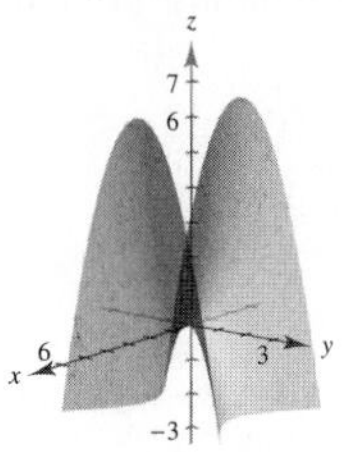

Saddle point

41. Point A is a saddle point.

43. $-4 < f_{xy}(3, 7) < 4$

45. Saddle point: $(0, 0, 0)$
Test fails

47. Absolute minima: $(1, a, 0)$, $(b, -4, 0)$
Test fails

49. Absolute minimum: $(0, 0, 0)$
Test fails

51. Relative minimum: $(0, 3, -1)$

53. Absolute maximum: $(0, 1, 10)$
Absolute minimum: $(1, 2, 5)$

55. Absolute maxima: $(\pm 2, 4, 28)$
Absolute minimum: $(0, 1, -2)$

57. Absolute maxima: $(2, 1, 6)$, $(-2, -1, 6)$
Absolute minima: $\left(-\frac{1}{2}, 1, -\frac{1}{4}\right)$, $\left(\frac{1}{2}, -1, -\frac{1}{4}\right)$

59. Absolute maxima: $(2, 2, 16)$, $(-2, -2, 16)$
Absolute minima: $(x, -x, 0)$, $|x| \le 2$

61. Absolute maximum: $(1, 1, 1)$
Absolute minimum: $(0, 0, 0)$

63. False. Let $f(x, y) = 1 - |x| - |y|$ at the point $(0, 0, 1)$.

Section 13.9 (page 964)

1. $6\sqrt{14}/7$ **3.** 6 **5.** 10, 10, 10

7. 10, 10, 10 **9.** $36 \times 18 \times 18$ in.

11. Let $a + b + c = k$.
$V = 4\pi abc/3 = \frac{4}{3}\pi ab(k - a - b) = \frac{4}{3}(kab - a^2b - ab^2)$

$$\left.\begin{aligned} V_a &= \tfrac{4}{3}\pi(kb - 2ab - b^2) = 0 \\ V_b &= \tfrac{4}{3}\pi(ka - a^2 - 2ab) = 0 \end{aligned}\right\} \begin{aligned} kb - 2ab - b^2 &= 0 \\ kb - a^2 - 2ab &= 0 \end{aligned}$$

So, $a = b$ and $b = k/3$. Thus, $a = b = c = k/3$.

13. Let x, y, and z be the length, width, and height, respectively, and let V_0 be the given volume. Then $V_0 = xyz$ and $z = V_0/xy$. The surface area is
$S = 2xy + 2yz + 2xz = 2(xy + V_0/x + V_0/y)$.

$$\left.\begin{aligned} S_x &= 2(y - V_0/x^2) = 0 \\ S_y &= 2(x - V_0/x^2) = 0 \end{aligned}\right\} \begin{aligned} x^2y - V_0 &= 0 \\ xy^2 - V_0 &= 0 \end{aligned}$$

So, $x = \sqrt[3]{V_0}$, $y = \sqrt[3]{V_0}$, and $z = \sqrt[3]{V_0}$.

15. $x = 10$ and $\theta = 60°$

17. $x_1 = 3$; $x_2 = 6$ **19.** $x_1 = 275$; $x_2 = 110$

21. $x = \sqrt{2}/2 \approx 0.707$ km
$y = \left(3\sqrt{2} + 2\sqrt{3}\right)/6 \approx 1.284$ km

23. (a) $S = \sqrt{x^2 + y^2} + \sqrt{(x + 2)^2 + (y - 2)^2} + \sqrt{(x - 4)^2 + (y - 2)^2}$

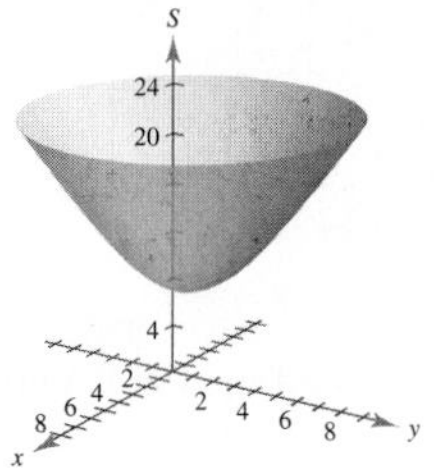

The surface has a minimum.

(b) $S_x = \dfrac{x}{\sqrt{x^2 + y^2}} + \dfrac{x + 2}{\sqrt{(x + 2)^2 + (y - 2)^2}} + \dfrac{x - 4}{\sqrt{(x - 4)^2 + (y - 2)^2}}$

$S_y = \dfrac{y}{\sqrt{x^2 + y^2}} + \dfrac{y - 2}{\sqrt{(x + 2)^2 + (y - 2)^2}} + \dfrac{y - 2}{\sqrt{(x - 4)^2 + (y - 2)^2}}$

(c) $-\dfrac{1}{\sqrt{2}}\mathbf{i} - \left(\dfrac{1}{\sqrt{2}} - \dfrac{2}{\sqrt{10}}\right)\mathbf{j}$
$\theta \approx 186.0°$

(d) $t = 1.344$; $(x_2, y_2) \approx (0.05, 0.90)$

(e) $(x_4, y_4) \approx (0.06, 0.45)$; $S = 7.266$

(f) $-\nabla S(x, y)$ gives the direction of greatest rate of decrease of S. Use $\nabla S(x, y)$ when finding a maximum.

25. Write the equation to be maximized or minimized as a function of two variables. Take the partial derivatives and set them equal to zero or undefined to obtain the critical points. Use the Second Partials Test to test for relative extrema using the critical points. Check the boundary points.

27. (a) $y = \frac{3}{4}x + \frac{4}{3}$ (b) $\frac{1}{6}$

29. (a) $y = -2x + 4$ (b) 2

31. $y = \frac{37}{43}x + \frac{7}{43}$

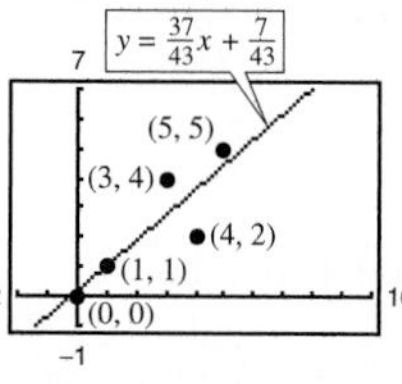

33. $y = -\frac{175}{148}x + \frac{945}{148}$

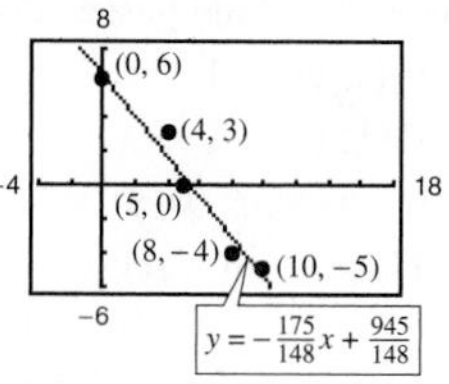

35. (a) $y = 1.724x + 79.733$
(b)

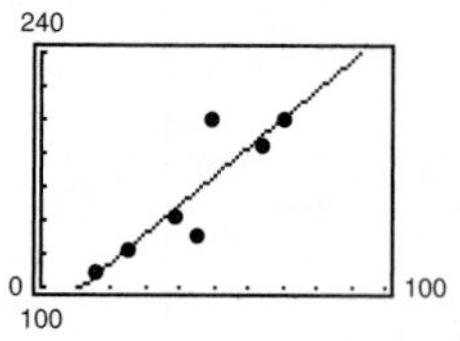

(c) 1.724

37. $y = 14x + 19$
41.4 bushels per acre

39.
$$a\sum_{i=1}^{n} x_i^4 + b\sum_{i=1}^{n} x_i^3 + c\sum_{i=1}^{n} x_i^2 = \sum_{i=1}^{n} x_i^2 y_i$$
$$a\sum_{i=1}^{n} x_i^3 + b\sum_{i=1}^{n} x_i^2 + c\sum_{i=1}^{n} x_i = \sum_{i=1}^{n} x_i y_i$$
$$a\sum_{i=1}^{n} x_i^2 + b\sum_{i=1}^{n} x_i + cn = \sum_{i=1}^{n} y_i$$

41. $y = \frac{3}{7}x^2 + \frac{6}{5}x + \frac{26}{35}$

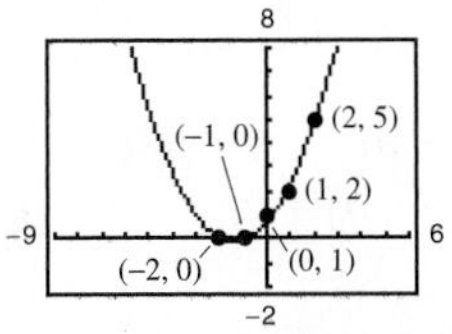

43. $y = x^2 - x$

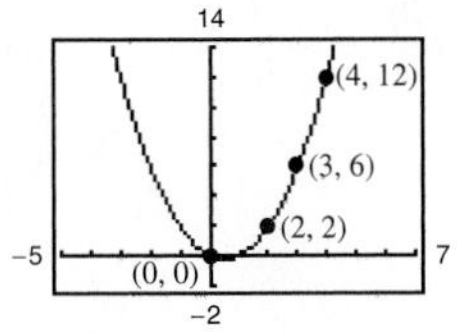

45. (a) $y = -0.22x^2 + 9.66x - 1.79$

(b)

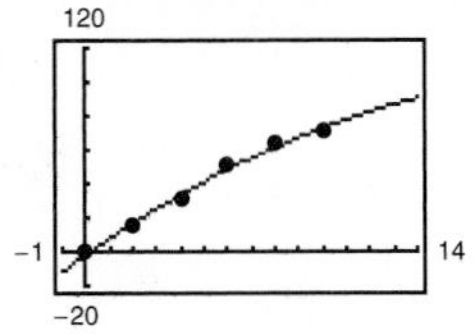

47. (a) $\ln P = -0.1499h + 9.3018$ (b) $P = 10{,}957.7e^{-0.1499h}$

(c)

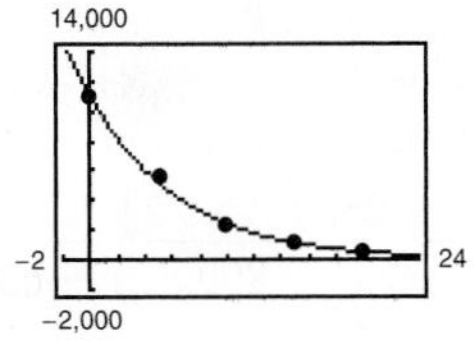

(d) Proof

Section 13.10 (page 974)

1.

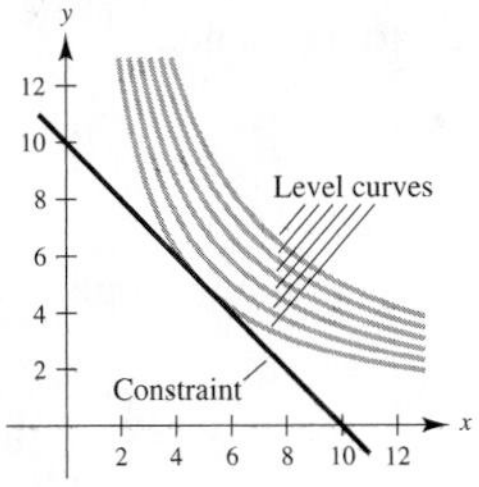

$f(5, 5) = 25$

3.

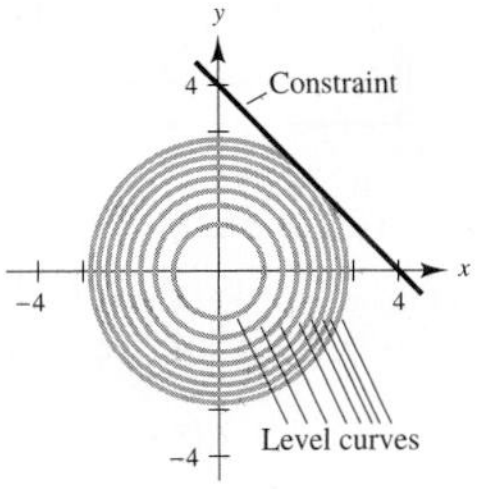

$f(2, 2) = 8$

5. $f(2, 4) = -12$ **7.** $f(25, 50) = 2600$

9. $f(1, 1) = 2$ **11.** $f(2, 2) = e^4$

13. Maxima: $f(\sqrt{2}/2, \sqrt{2}/2) = \frac{5}{2}$

$f(-\sqrt{2}/2, -\sqrt{2}/2) = \frac{5}{2}$

Minima: $f(-\sqrt{2}/2, \sqrt{2}/2) = -\frac{1}{2}$

$f(\sqrt{2}/2, -\sqrt{2}/2) = -\frac{1}{2}$

15. $f(2, 2, 2) = 12$ **17.** $f(\frac{1}{3}, \frac{1}{3}, \frac{1}{3}) = \frac{1}{3}$

19. $f(8, 16, 8) = 1024$ **21.** $f(3, \frac{3}{2}, 1) = 6$

23. $\sqrt{13}/13$ **25.** $\sqrt{3}$

27. $x = (10 + 2\sqrt{265})/15$

$y = (5 + \sqrt{265})/15$

$z = (-1 + \sqrt{265})/3$

29. Optimization problems that have restrictions or constraints on the values that can be used to produce the optimal solution are called constrained optimization problems.

31. $36 \times 18 \times 18$ in. **33.** $\sqrt[3]{360} \times \sqrt[3]{360} \times \frac{4}{3}\sqrt[3]{360}$ ft

35. $2\sqrt{3}a/3 \times 2\sqrt{3}b/3 \times 2\sqrt{3}c/3$ **37.** Proof **39.** $\frac{2}{3}$

41. $P(3125/6, 6250/3) = 147{,}314$

43. $x = 50\sqrt{2}$

$y = 200\sqrt{2}$

Cost = \$13,576.45

45. (a) $g(\pi/3, \pi/3, \pi/3) = \frac{1}{8}$

(b)

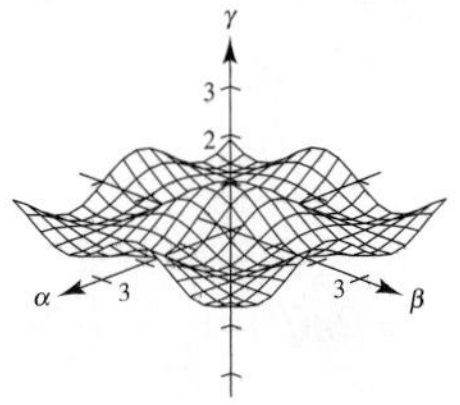

Maximum values occur when $\alpha = \beta$.

Review Exercises for Chapter 13 (page 976)

1. Not a function because not every $f(x, y)$ has a unique z-value.

3.

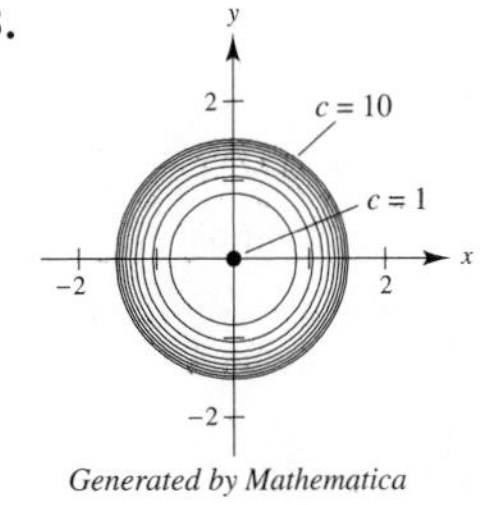

Generated by Mathematica

5.

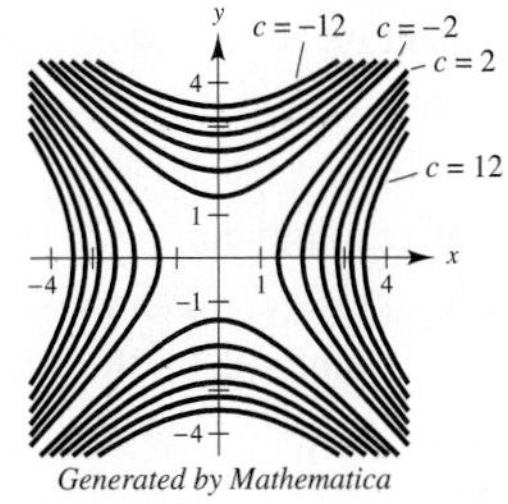

Generated by Mathematica

7.

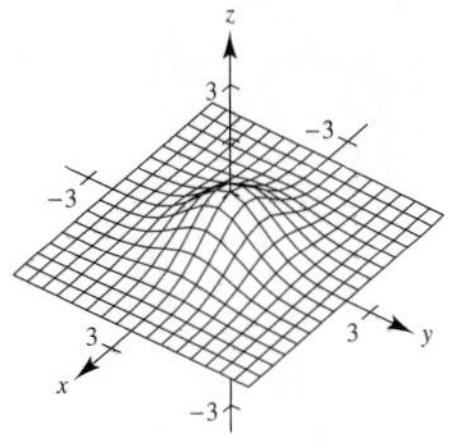

9.

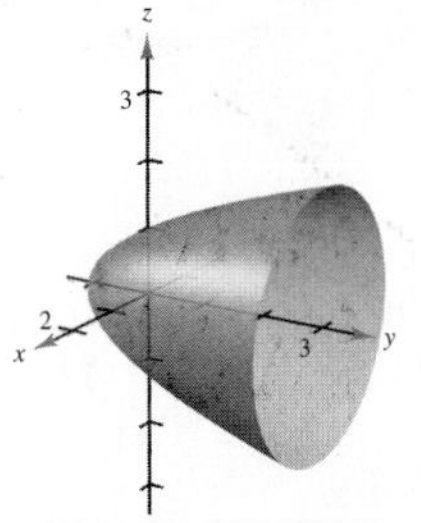

11. Continuous except at (0, 0)
Limit: $\frac{1}{2}$

13. Continuous except at (0, 0)
Limit does not exist.

15. $f_x(x, y) = e^x \cos y$
$f_y(x, y) = -e^x \sin y$

17. $\partial z/\partial x = e^y + ye^x$
$\partial z/\partial y = xe^y + e^x$

19. $g_x(x, y) = [y(y^2 - x^2)]/(x^2 + y^2)^2$
$g_y(x, y) = [x(x^2 - y^2)]/(x^2 + y^2)^2$

21. $f_x(x, y, z) = -yz/(x^2 + y^2)$
$f_y(x, y, z) = xz/(x^2 + y^2)$
$f_z(x, y, z) = \arctan y/x$

23. $u_x(x, t) = cne^{-n^2t} \cos nx$
$u_t(x, t) = -cn^2e^{-n^2t} \sin nx$

25. Answers will vary. Example:

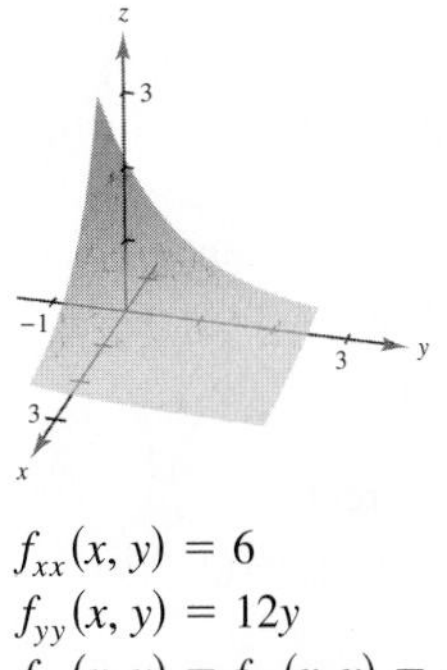

27. $f_{xx}(x, y) = 6$
$f_{yy}(x, y) = 12y$
$f_{xy}(x, y) = f_{yx}(x, y) = -1$

29. $h_{xx}(x, y) = -y \cos x$
$h_{yy}(x, y) = -x \sin y$
$h_{xy}(x, y) = h_{yx}(x, y) = \cos y - \sin x$

31. $\partial^2 z/\partial x^2 + \partial^2 z/\partial y^2 = 2 + (-2) = 0$

33. $\dfrac{\partial^2 z}{\partial x^2} + \dfrac{\partial^2 z}{\partial y^2} = \dfrac{6x^2y - 2y^3}{(x^2+y^2)^3} + \dfrac{-6x^2y + 2y^3}{(x^2+y^2)^3} = 0$

35. $(\sin y/x - y/x \cos y/x)\,dx + (\cos y/x)\,dy$

37. 0.6538 cm, 5.03% **39.** $\pm\pi$ in.^{3}

41. $dw/dt = (10t + 4)/(5t^2 + 4t + 25)$

43. $\partial u/\partial r = 2r$
$\partial u/\partial t = 2t$

45. $\partial z/\partial x = (2xy - z)/(x + 2y + 2z)$
$\partial z/\partial y = (x^2 - 2z)/(x + 2y + 2z)$

47. 0 **49.** $\frac{2}{3}$ **51.** $\langle -\frac{1}{2}, 0\rangle, \frac{1}{2}$

53. $\langle -\sqrt{2}/2, -\sqrt{2}/2\rangle, 1$ **55.** $27/\sqrt{793}\,\mathbf{i} - 8/\sqrt{793}\,\mathbf{j}$

57. Tangent plane: $4x + 4y - z = 8$
Normal line: $x = 2 + 4t,\ y = 1 + 4t,\ z = 4 - t$

59. Tangent plane: $z = 4$
Normal line: $x = 2,\ y = -3,\ z = 4 + t$

61. $(x - 2)/1 = (y - 1)/2,\ z = 3$ **63.** $\theta \approx 36.7°$

65. Relative minimum: $\left(\frac{3}{2}, \frac{9}{4}, -\frac{27}{16}\right)$
Saddle point: $(0, 0, 0)$

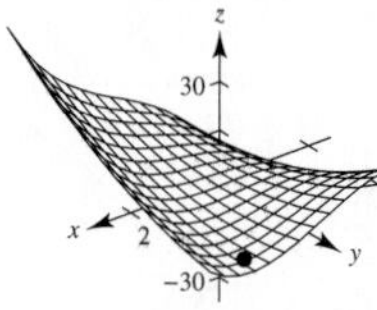

67. Relative minimum: $(1, 1, 3)$

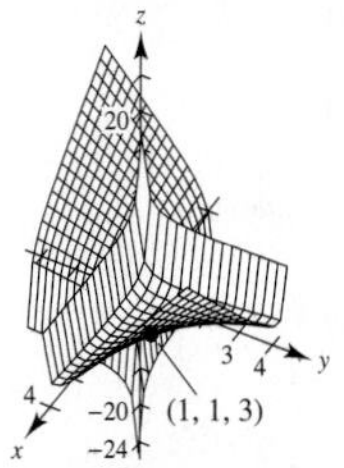

69. The level curves are hyperbolas. The critical point $(0, 0)$ may be a saddle point or an extremum.

71. $x_1 = 94,\ x_2 = 157$ **73.** $f(49.4, 253) = 13{,}201.8$

75. (a) $y = 2.29t + 2.34$

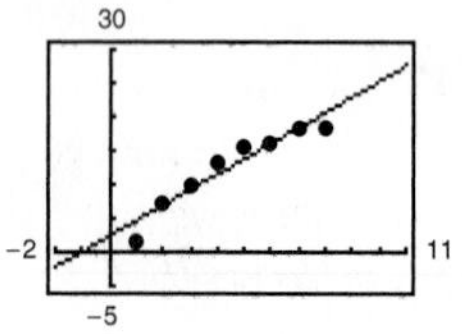

(b)

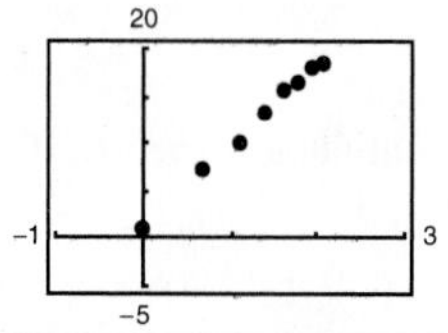

More closely linear

(c) $y = 1.54 + 8.37 \ln t$

(d)

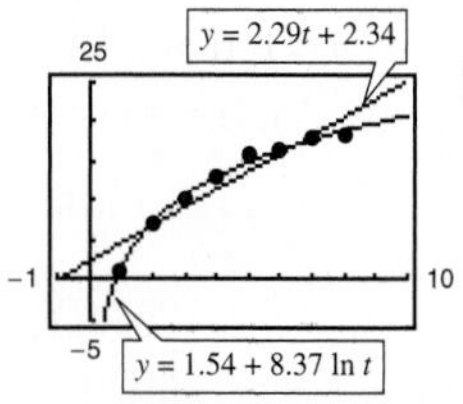

Logarithmic model is a better fit.

77. Maximum: $f\left(\frac{1}{3}, \frac{1}{3}, \frac{1}{3}\right) = \frac{1}{3}$

79. $x = \sqrt{2}/2 \approx 0.707$ km; $y = \sqrt{3}/3 \approx 0.577$ km;
$z = \left(60 - 3\sqrt{2} - 2\sqrt{3}\right)/6 \approx 8.716$ km

P.S. Problem Solving (page 979)

1. (a) 12 square units (b) Proof (c) Proof

3. (a) $y_0z_0(x - x_0) + x_0z_0(y - y_0) + x_0y_0(z - z_0) = 0$
(b) $x_0y_0z_0 = 1 \Rightarrow z_0 = 1/x_0y_0$
Then the tangent plane is

$$y_0\left(\frac{1}{x_0y_0}\right)(x - x_0) + x_0\left(\frac{1}{x_0y_0}\right)(y - y_0) + x_0y_0\left(z - \frac{1}{x_0y_0}\right) = 0.$$

Intercepts: $(3x_0, 0, 0),\ (0, 3y_0, 0),\ \left(0, 0, \dfrac{3}{x_0y_0}\right)$

$V = \frac{1}{3}bh = \frac{9}{2}$

5. No; Yes **7.** $2\sqrt[3]{150} \times 2\sqrt[3]{150} \times 5\sqrt[3]{150}/3$

9. (a)
$$\begin{aligned} x\frac{\partial f}{dx} + y\frac{\partial f}{dy} &= xCy^{1-a}ax^{a-1} + yCx^a(1-a)y^{1-a-1} \\ &= ax^aCy^{1-a} + (1-a)x^aC(y^{1-a}) \\ &= Cx^ay^{1-a}[a + (1-a)] \\ &= Cx^ay^{1-a} \\ &= f(x, y) \end{aligned}$$

(b)
$$\begin{aligned} f(tx, ty) &= C(tx)^a(ty)^{1-a} \\ &= Ctx^ay^{1-a} \\ &= tCx^ay^{1-a} \\ &= tf(x, y) \end{aligned}$$

11. (a) $x = 32\sqrt{2}t$
$y = 32\sqrt{2}t - 16t^2$

(b) $\alpha = \arctan\left(\dfrac{y}{x + 50}\right) = \arctan\left(\dfrac{32\sqrt{2}t - 16t^2}{32\sqrt{2}t + 50}\right)$

(c) $\dfrac{d\alpha}{dt} = \dfrac{-16\left(8\sqrt{2}t^2 + 25t - 25\sqrt{2}\right)}{64t^4 - 256\sqrt{2}t^3 + 1024t^2 + 800\sqrt{2}t + 625}$

(d)

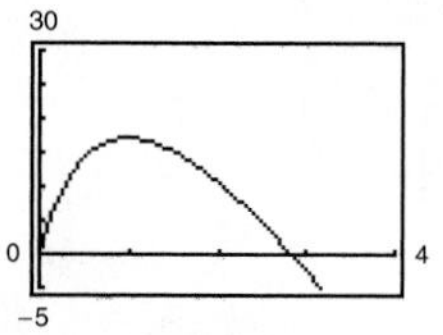

No; The rate of change of α is greatest when the projectile is closest to the camera.

(e) α is maximum when $t = 0.98$ second.
No; the projectile is at its maximum height when $t = \sqrt{2} \approx 1.41$ seconds.

13. (a)

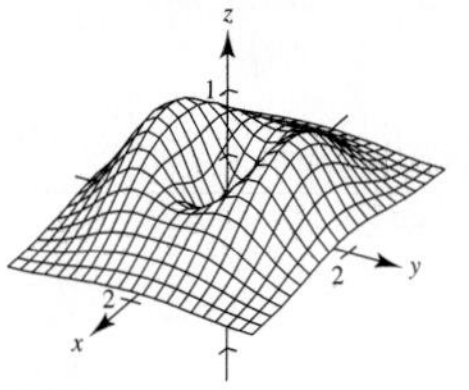

(b)

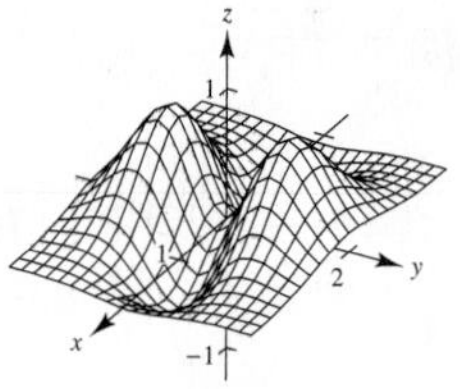

(a) Minimum: $(0, 0, 0)$
Maxima: $(0, \pm 1, 2e^{-1})$
Saddle points: $(\pm 1, 0, e^{-1})$

(b) Minima: $(\pm 1, 0, -e^{-1})$
Maxima: $(0, \pm 1, 2e^{-1})$
Saddle point: $(0, 0, 0)$

(c) $\alpha > 0$
Minimum: $(0, 0, 0)$
Maxima: $(0, \pm 1, \beta e^{-1})$
Saddle points: $(\pm 1, 0, \alpha e^{-1})$

$\alpha < 0$
Minima: $(\pm 1, 0, \alpha e^{-1})$
Maxima: $(0, \pm 1, \beta e^{-1})$
Saddle point: $(0, 0, 0)$

15. (a)

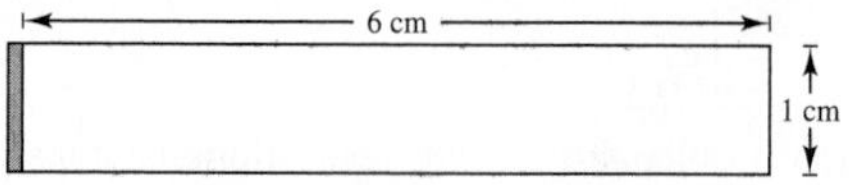

(b)

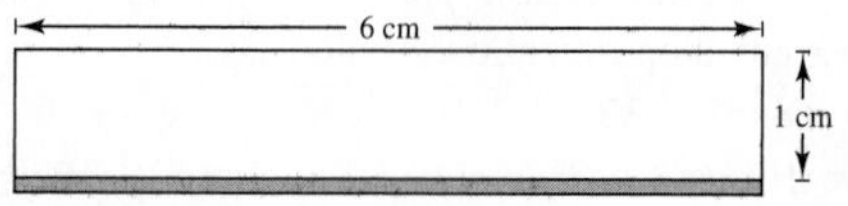

(c) Height
(d) $dl = 0.01,\ dh = 0{:}\ dA = 0.01$
$dl = 0,\ dh = 0.01{:}\ dA = 0.06$

17–19. Proofs

Chapter 14

Section 14.1 (page 988)

1. $3x^2/2$ **3.** $y \ln(2y)$ **5.** $(4x^2 - x^4)/2$
7. $(y/2)[(\ln y)^2 - y^2]$ **9.** $x^2(1 - e^{-x^2} - x^2e^{-x^2})$ **11.** 3
13. 2 **15.** $\frac{1}{3}$ **17.** $\frac{20}{3}$ **19.** $\frac{2}{3}$ **21.** 4 **23.** $\pi^2/32 + \frac{1}{8}$
25. $\frac{1}{2}$ **27.** Diverges **29.** 24 **31.** $\frac{16}{3}$ **33.** $\frac{9}{2}$ **35.** $\frac{8}{3}$
37. 5 **39.** πab

41.

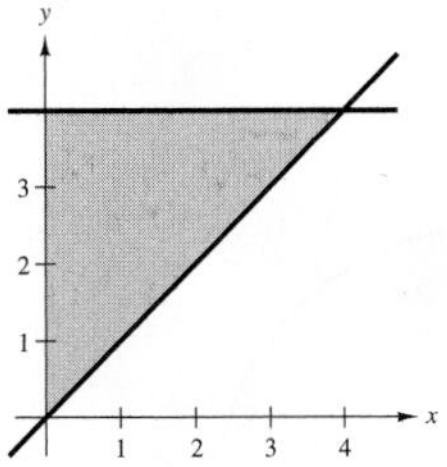

$$\int_0^4 \int_x^4 f(x, y)\, dy\, dx$$

43.

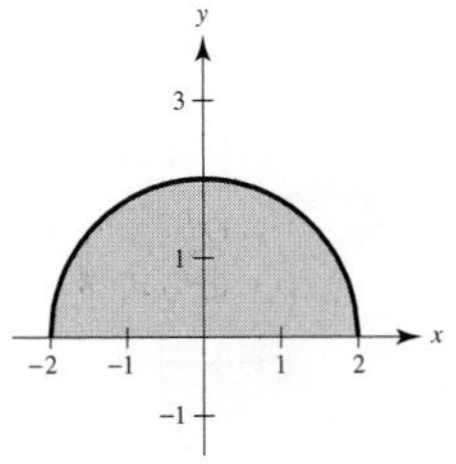

$$\int_0^2 \int_{-\sqrt{4-y^2}}^{\sqrt{4-y^2}} f(x, y)\, dx\, dy$$

45.

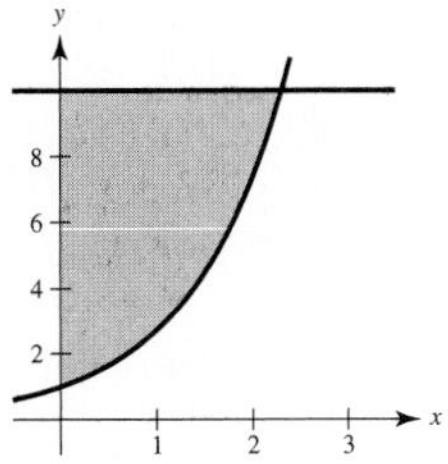

$$\int_0^{\ln 10} \int_{e^x}^{10} f(x, y)\, dy\, dx$$

47.

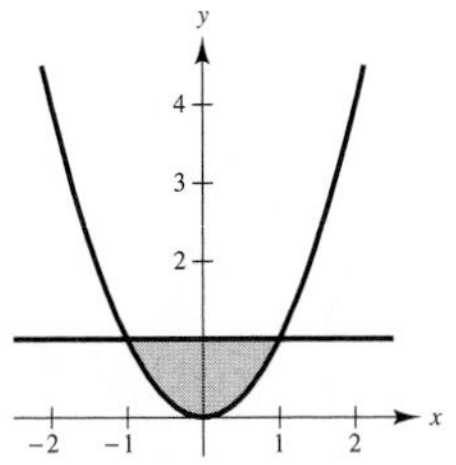

$$\int_0^1 \int_{-\sqrt{y}}^{\sqrt{y}} f(x, y)\, dx\, dy$$

49.

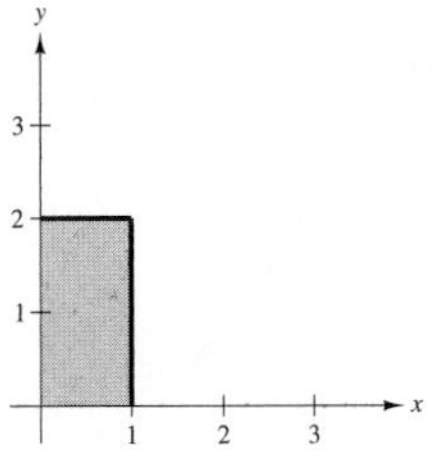

$$\int_0^1 \int_0^2 dy\, dx = \int_0^2 \int_0^1 dx\, dy = 2$$

51.

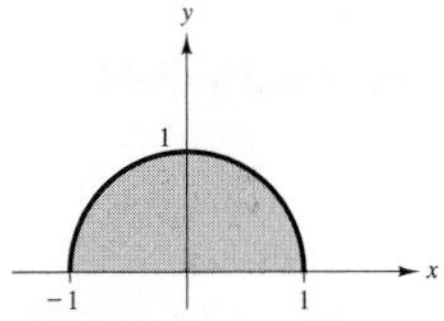

$$\int_0^1 \int_{-\sqrt{1-y^2}}^{\sqrt{1-y^2}} dx\, dy = \int_{-1}^1 \int_0^{\sqrt{1-x^2}} dy\, dx = \frac{\pi}{2}$$

53.

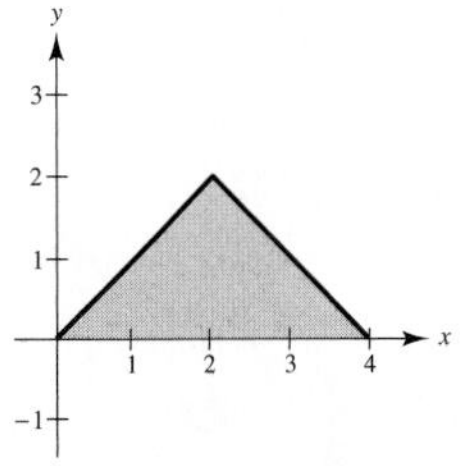

$$\int_0^2 \int_0^x dy\, dx + \int_2^4 \int_0^{4-x} dy\, dx = \int_0^2 \int_y^{4-y} dx\, dy = 4$$

55.

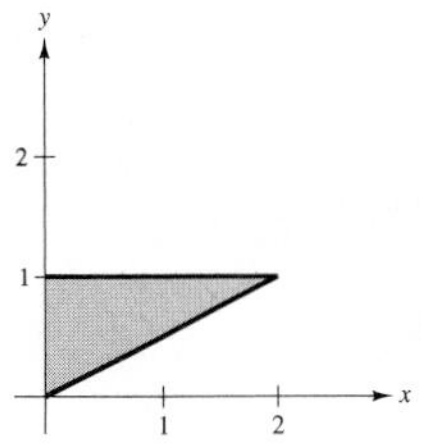

$$\int_0^2 \int_{x/2}^1 dy\, dx = \int_0^1 \int_0^{2y} dx\, dy = 1$$

57.

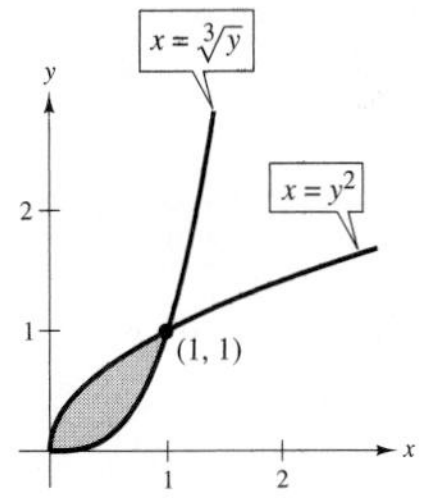

$$\int_0^1 \int_{y^2}^{\sqrt[3]{y}} dx\, dy = \int_0^1 \int_{x^3}^{\sqrt{x}} dy\, dx = \frac{5}{12}$$

59. The first integral arises using vertical representative rectangles. The second two integrals arise using horizontal representative rectangles.
Value of the integrals: $15{,}625\pi/24$

61. $\frac{26}{9}$ **63.** $\frac{1}{2}(1 - \cos 1) \approx 0.230$ **65.** $\frac{1664}{105}$ **67.** $(\ln 5)^2$

69. (a)

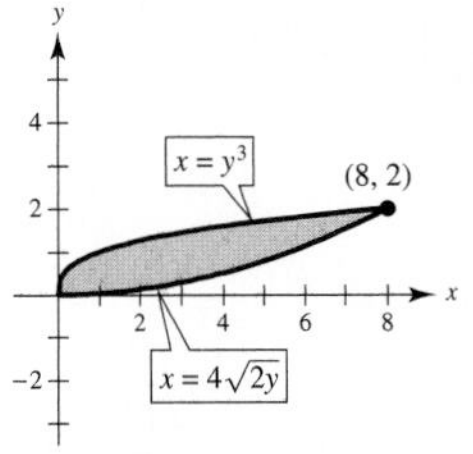

(b) $\displaystyle\int_0^8 \int_{x^2/32}^{\sqrt[3]{x}} (x^2y - xy^2)\, dy\, dx$ (c) 67,520/693

71. 20.5648 **73.** $15\pi/2$

75. An iterated integral is integration of a function of several variables. Integrate with respect to one variable while holding the other variables constant.

77. If all four limits of integration are constant, the region of integration is rectangular.

79. True

Section 14.2 (page 997)

1. 24 (approximation is exact)
3. Approximation: 52; Exact: $\frac{160}{3}$ **5.** 400; 272
7. 8 **9.** 36

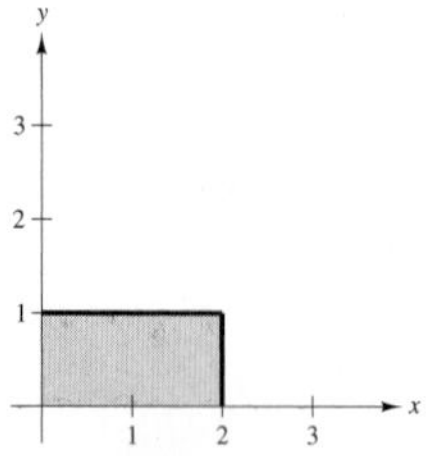

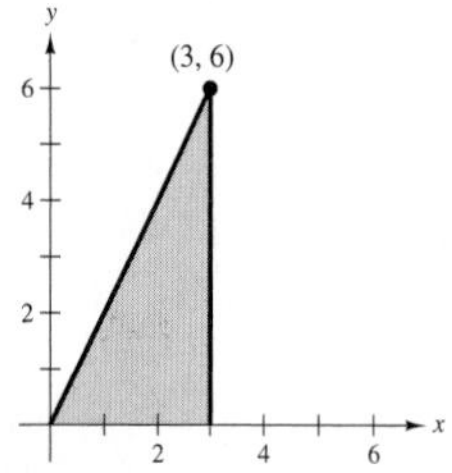

11. 0

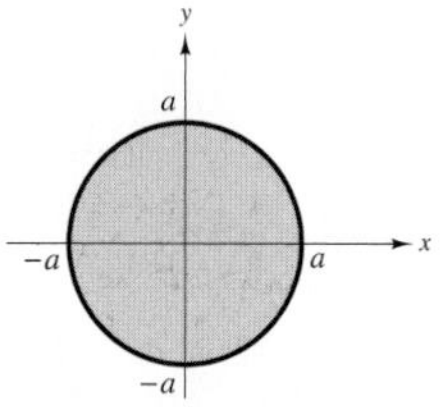

13. $\int_0^3\int_0^5 xy\,dy\,dx = \frac{225}{4}$

$\int_0^5\int_0^3 xy\,dx\,dy = \frac{225}{4}$

15. $\int_0^2\int_x^{2x} \frac{y}{x^2+y^2}\,dy\,dx = \ln\frac{5}{2}$

$\int_0^2\int_{y/2}^{y} \frac{y}{x^2+y^2}\,dx\,dy + \int_2^4\int_{y/2}^{2} \frac{y}{x^2+y^2}\,dx\,dy = \ln\frac{5}{2}$

17. $\int_0^1\int_{4-x}^{4-x^2} -2y\ln x\,dy\,dx = \frac{26}{25}$

$\int_3^4\int_{4-y}^{\sqrt{4-y}} -2y\ln x\,dx\,dy = \frac{26}{25}$

19. $\int_0^3\int_{4y/3}^{\sqrt{25-y^2}} x\,dx\,dy = 25$

$\int_0^4\int_0^{3x/4} x\,dy\,dx + \int_4^5\int_0^{\sqrt{25-x^2}} x\,dy\,dx = 25$

21. 4 **23.** 4 **25.** 12 **27.** $\frac{3}{8}$ **29.** 1 **31.** 8π

33. $\int_0^1\int_0^x xy\,dy\,dx = \frac{1}{8}$ **35.** $\int_0^2\int_0^4 x^2\,dy\,dx = \frac{32}{3}$

37. $2\int_0^1\int_0^x \sqrt{1-x^2}\,dy\,dx = \frac{2}{3}$

39. $\int_0^2\int_0^{\sqrt{4-x^2}} (x+y)\,dy\,dx = \frac{16}{3}$

41. 8π **43.** $81\pi/2$ **45.** 1.2315 **47.** Proof
49. $1 - e^{-1/4} \approx 0.221$ **51.** $\frac{1}{3}\left[2\sqrt{2}-1\right]$ **53.** 2 **55.** $\frac{8}{3}$
57. See "Definition of Double Integral" on page 992. The double integral of a function $f(x, y) \geq 0$ over the region of integration yields the volume of that region.
59. kB **61.** 25,645.24 **63.** Proof; $\frac{1}{5}$ **65.** Proof; $\frac{7}{27}$
67. 2500 m^3 **69.** (a) 1.784 (b) 1.788
71. (a) 11.057 (b) 11.041 **73.** d
75. False: $V = 8\int_0^1\int_0^{\sqrt{1-y^2}} \sqrt{1-x^2-y^2}\,dx\,dy$.
77. $\frac{1}{2}(1-e)$ **79.** R: $x^2+y^2 \leq 9$ **81.** ≈ 0.82736
83. Putnam problem A2, 1989

Section 14.3 (page 1006)

1. Rectangular **3.** Polar
5. The region R is a half-circle of radius 8. It can be described in polar coordinates as
$R = \{(r, \theta): 0 \leq r \leq 8, 0 \leq \theta \leq \pi\}$.
7. The region R is a cardioid with $a = b = 3$. It can be described in polar coordinates as
$R = \{(r, \theta): 0 \leq r \leq 3 + 3\sin\theta, 0 \leq \theta \leq 2\pi\}$.
9. 0 **11.** $5\sqrt{5}\pi/6$

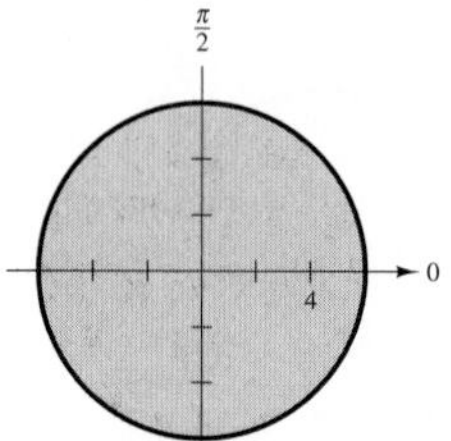

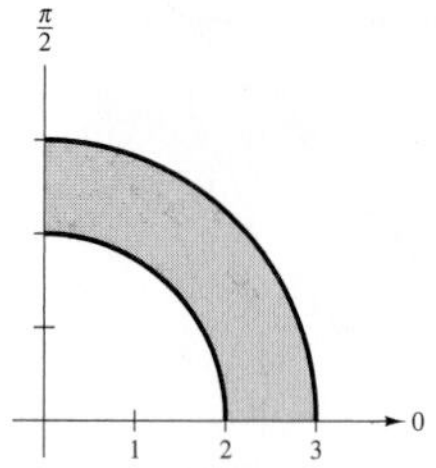

13. $\frac{9}{8} + 3\pi^2/32$ **15.** $a^3/3$ **17.** $243\pi/10$ **19.** $\frac{2}{3}$

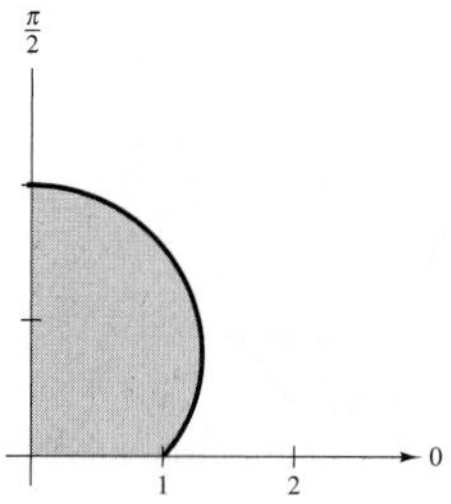

21. $\int_0^{\pi/4}\int_0^{2\sqrt{2}} r^2\,dr\,d\theta = \frac{4\sqrt{2}\pi}{3}$

23. $\int_0^{\pi/2}\int_0^2 r^2(\cos\theta + \sin\theta)\,dr\,d\theta = \frac{16}{3}$

25. $\int_0^{\pi/4}\int_1^2 r\theta\,dr\,d\theta = \frac{3\pi^2}{64}$ **27.** $\frac{1}{8}$ **29.** $\frac{250\pi}{3}$

31. $\frac{64}{9}(3\pi - 4)$ **33.** $2\sqrt{4 - 2\sqrt[3]{2}}$ **35.** 1.2858
37. 9π **39.** $3\pi/2$ **41.** π
43. Let R be a region bounded by the graphs of $r = g_1(\theta)$ and $r = g_2(\theta)$ and the lines $\theta = a$ and $\theta = b$. When using polar coordinates to evaluate a double integral over R, R can be partitioned into small polar sectors.
45. r-simple regions have fixed bounds for θ and variable bounds for r.
θ-simple regions have variable bounds for θ and fixed bounds for r.
47. Insert a factor of r; Sector of a circle **49.** 56.051 **51.** c
53. False: Let $f(r, \theta) = r - 1$ and let R be a sector where $0 \leq r \leq 6$ and $0 \leq \theta \leq \pi$.
55. (a) 2π (b) $\sqrt{2\pi}$ **57.** 486,788

59. (a) $\int_2^4\int_{y/\sqrt{3}}^{y} f\,dx\,dy$

(b) $\int_{2/\sqrt{3}}^{2}\int_2^{\sqrt{3}x} f\,dy\,dx + \int_2^{4/\sqrt{3}}\int_x^{\sqrt{3}x} f\,dy\,dx + \int_{4/\sqrt{3}}^{4}\int_x^4 f\,dy\,dx$

(c) $\int_{\pi/4}^{\pi/3}\int_{2\csc\theta}^{4\csc\theta} f\,r\,dr\,d\theta$

61. $A = \dfrac{\Delta\theta r_2^2}{2} - \dfrac{\Delta\theta r_1^2}{2} = \Delta\theta\left(\dfrac{r_1 + r_2}{2}\right)(r_2 - r_1) = r\Delta r\Delta\theta$

Section 14.4 (page 1015)

1. $m = 36$ **3.** $m = 2$

5. (a) $m = kab, (a/2, b/2)$ (b) $m = kab^2/2, (a/2, 2b/3)$
(c) $m = ka^2b/2, (2a/3, b/2)$

7. (a) $m = kbh/2, (b/2, h/3)$ (b) $m = kh^2b/6, (b/2, h/2)$
(c) $m = khb^2/4, (7b/12, h/3)$

9. (a) $(a/2 + 5, b/2)$ (b) $(a/2 + 5, 2b/3)$
(c) $\left(\dfrac{2(a^2 + 15a + 75)}{3(a + 10)}, \dfrac{b}{2}\right)$

11. (a) $m = k\pi a^2/2, (0, 4a/3\pi)$
(b) $m = \dfrac{ka^4}{24}(16 - 3\pi), \left(0, \dfrac{a}{5}\left[\dfrac{15\pi - 32}{16 - 3\pi}\right]\right)$

13. $m = 32k/3, \left(3, \frac{8}{7}\right)$ **15.** $m = k\pi/2, (0, (\pi + 2)/(4\pi))$

17. $m = 8192k/15, \left(\frac{64}{7}, 0\right)$ **19.** $m = kL/4, (L/2, 16/(9\pi))$

21. $m = \dfrac{k\pi a^2}{8}, \left(\dfrac{4\sqrt{2}a}{3\pi}, \dfrac{4a\left(2 - \sqrt{2}\right)}{3\pi}\right)$

23. $m = \dfrac{k}{4}(1 - e^{-4}), \left(\dfrac{e^4 - 5}{2(e^4 - 1)}, \dfrac{4}{9}\left[\dfrac{e^6 - 1}{e^6 - e^2}\right]\right)$

25. $m = k\pi/3, \left(81\sqrt{3}/(40\pi), 0\right)$

27. $\bar{\bar{x}} = \sqrt{3}b/3$, $\bar{\bar{y}} = \sqrt{3}h/3$ **29.** $\bar{\bar{x}} = a/2$, $\bar{\bar{y}} = a/2$ **31.** $\bar{\bar{x}} = a/2$, $\bar{\bar{y}} = a/2$

33. $I_x = kab^4/4$
$I_y = kb^2a^3/6$
$I_0 = (3kab^4 + 2ka^3b^2)/12$
$\bar{\bar{x}} = \sqrt{3}a/3$
$\bar{\bar{y}} = \sqrt{2}b/2$

35. $I_x = 32k/3$
$I_y = 16k/3$
$I_0 = 16k$
$\bar{\bar{x}} = 2\sqrt{3}/3$
$\bar{\bar{y}} = 2\sqrt{6}/3$

37. $I_x = 16k$
$I_y = 512k/5$
$I_0 = 592k/5$
$\bar{\bar{x}} = 4\sqrt{15}/5$
$\bar{\bar{y}} = \sqrt{6}/2$

39. $I_x = 3k/56$
$I_y = k/18$
$I_0 = 55k/504$
$\bar{\bar{x}} = \sqrt{30}/9$
$\bar{\bar{y}} = \sqrt{70}/14$

41. $2k\displaystyle\int_{-b}^{b}\int_0^{\sqrt{b^2-x^2}} (x - a)^2\, dy\, dx = \dfrac{k\pi b^2}{4}(b^2 + 4a^2)$

43. $\displaystyle\int_0^4\int_0^{\sqrt{x}} kx(x - 6)^2\, dy\, dx = \dfrac{42{,}752k}{315}$

45. $\displaystyle\int_0^a\int_0^{\sqrt{a^2-x^2}} k(a - y)(y - a)^2\, dy\, dx = ka^5\left(\dfrac{7\pi}{16} - \dfrac{17}{15}\right)$

47. $\bar{y}$ will increase. **49.** $\bar{x}$ and $\bar{y}$ will both increase.

51. See definition on page 1011. **53.** Answers will vary.

55. $L/3$ **57.** $L/2$

Section 14.5 (page 1022)

1. 6 **3.** 12π **5.** $\frac{3}{4}\left[6\sqrt{37} + \ln\left(\sqrt{37} + 6\right)\right]$

7. $\frac{4}{27}\left(31\sqrt{31} - 8\right)$ **9.** $\sqrt{2} - 1$ **11.** $\sqrt{2}\pi$

13. $2\pi a\left(a - \sqrt{a^2 - b^2}\right)$ **15.** $48\sqrt{14}$ **17.** 20π

19. $\displaystyle\int_0^1\int_0^x \sqrt{5 + 4x^2}\, dy\, dx = \dfrac{27 - 5\sqrt{5}}{12} \approx 1.3183$

21. $\displaystyle\int_{-2}^{2}\int_{-\sqrt{4-x^2}}^{\sqrt{4-x^2}} \sqrt{1 + 4x^2 + 4y^2}\, dy\, dx$
$= \dfrac{\pi}{6}\left(17\sqrt{17} - 1\right) \approx 36.1769$

23. $\displaystyle\int_0^1\int_0^1 \sqrt{1 + 4x^2 + 4y^2}\, dy\, dx \approx 1.8616$ **25.** e

27. 2.0035 **29.** $\displaystyle\int_{-1}^{1}\int_{-1}^{1} \sqrt{1 + 9(x^2 - y)^2 + 9(y^2 - x)^2}\, dy\, dx$

31. $\displaystyle\int_{-2}^{2}\int_{-\sqrt{4-x^2}}^{\sqrt{4-x^2}} \sqrt{1 + e^{-2x}}\, dy\, dx$

33. $\displaystyle\int_0^4\int_0^{10} \sqrt{1 + e^{2xy}(x^2 + y^2)}\, dy\, dx$

35. If f and its first partial derivatives are continuous on the closed region R in the xy-plane, then the area of the surface S given by $z = f(x, y)$ over R is
$$\int\!\!\int_R \sqrt{1 + [f_x(x, y)]^2 + [f_y(x, y)]^2}\, dA.$$

37. 16 **39.** (a) 30,415.74 ft^3 (b) 2081.53 ft^2

41. (a) $812\pi\sqrt{609}$ cm^3 (b) $100\pi\sqrt{609}$ cm^2

Section 14.6 (page 1032)

1. 18 **3.** $\frac{1}{10}$ **5.** $\frac{15}{2}(1 - 1/e)$ **7.** $-\frac{40}{3}$ **9.** $\frac{128}{15}$

11. 2.44167 **13.** $V = \displaystyle\int_0^4\int_0^{4-x}\int_0^{4-x-y} dz\, dy\, dx$

15. $V = \displaystyle\int_{-3}^{3}\int_{-\sqrt{9-y^2}}^{\sqrt{9-y^2}}\int_0^{9-x^2-y^2} dz\, dx\, dy$

17. $\frac{256}{15}$ **19.** $4\pi a^3/3$ **21.** $\frac{256}{15}$

23.

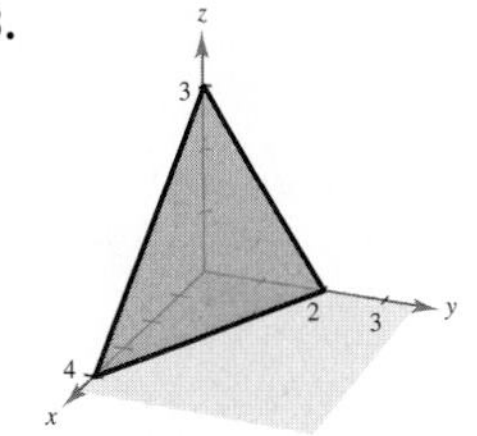

$\displaystyle\int_0^3\int_0^{(12-4z)/3}\int_0^{(12-4z-3x)/6} dy\, dx\, dz$

25.

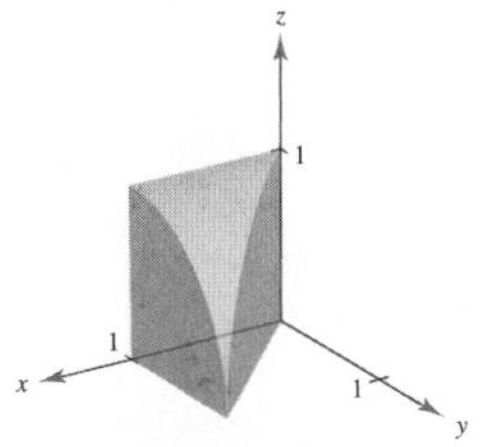

$\displaystyle\int_0^1\int_0^x\int_0^{\sqrt{1-y^2}} dz\, dy\, dx$

27. $\displaystyle\int_0^1\int_0^x\int_0^3 xyz\, dz\, dy\, dx$, $\displaystyle\int_0^1\int_y^1\int_0^3 xyz\, dz\, dx\, dy$,
$\displaystyle\int_0^1\int_0^3\int_0^x xyz\, dy\, dz\, dx$, $\displaystyle\int_0^3\int_0^1\int_0^x xyz\, dy\, dx\, dz$,
$\displaystyle\int_0^3\int_0^1\int_y^1 xyz\, dx\, dy\, dz$, $\displaystyle\int_0^1\int_0^3\int_y^1 xyz\, dx\, dz\, dy$

29. $\displaystyle\int_{-3}^{3}\int_{-\sqrt{9-x^2}}^{\sqrt{9-x^2}}\int_{0}^{4} xyz\,dz\,dy\,dx,\ \int_{-3}^{3}\int_{-\sqrt{9-y^2}}^{\sqrt{9-y^2}}\int_{0}^{4} xyz\,dz\,dx\,dy$

$\displaystyle\int_{-3}^{3}\int_{0}^{4}\int_{-\sqrt{9-x^2}}^{\sqrt{9-x^2}} xyz\,dy\,dz\,dx,\ \int_{0}^{4}\int_{-3}^{3}\int_{-\sqrt{9-x^2}}^{\sqrt{9-x^2}} xyz\,dy\,dx\,dz$

$\displaystyle\int_{0}^{4}\int_{-3}^{3}\int_{-\sqrt{9-y^2}}^{\sqrt{9-y^2}} xyz\,dx\,dy\,dz,\ \int_{-3}^{3}\int_{0}^{4}\int_{-\sqrt{9-y^2}}^{\sqrt{9-y^2}} xyz\,dx\,dz\,dy$

31. $\displaystyle\int_{0}^{1}\int_{0}^{\sqrt{1-x}}\int_{0}^{1-y} dz\,dy\,dx,\ \int_{0}^{1}\int_{0}^{1-z}\int_{0}^{1-y^2} dx\,dy\,dz$

$\displaystyle\int_{0}^{1}\int_{0}^{1-y}\int_{0}^{1-y^2} dx\,dz\,dy,\ \int_{0}^{1}\int_{0}^{1-y^2}\int_{0}^{1-y} dz\,dx\,dy$

$\displaystyle\int_{0}^{1}\int_{0}^{2z-z^2}\int_{0}^{1-z} 1\,dy\,dx\,dz + \int_{0}^{1}\int_{2z-z^2}^{1}\int_{0}^{\sqrt{1-x}} 1\,dy\,dx\,dz,$

$\displaystyle\int_{0}^{1}\int_{\sqrt{1-x}}^{1}\int_{0}^{1-z} 1\,dy\,dz\,dx + \int_{0}^{1}\int_{0}^{1-\sqrt{1-x}}\int_{0}^{\sqrt{1-x}} 1\,dy\,dz\,dx$

33. $m = 8k$ **35.** $m = 128k/3$
$\bar{x} = \frac{3}{2}$ $\quad\bar{z} = 1$

37. $\displaystyle m = k\int_0^b\int_0^b\int_0^b xy\,dz\,dy\,dx$

$\displaystyle M_{yz} = k\int_0^b\int_0^b\int_0^b x^2y\,dz\,dy\,dx$

$\displaystyle M_{xz} = k\int_0^b\int_0^b\int_0^b xy^2\,dz\,dy\,dx$

$\displaystyle M_{xy} = k\int_0^b\int_0^b\int_0^b xyz\,dz\,dy\,dx$

39. $\bar{x}$ will be greater than 2, and $\bar{y}$ and $\bar{z}$ will be unchanged.
41. $\bar{x}$ and $\bar{z}$ will be unchanged, and $\bar{y}$ will be greater than 0.
43. $(0, 0, 3h/4)$ **45.** $\left(0, 0, \frac{3}{2}\right)$ **47.** $\left(5, 6, \frac{5}{4}\right)$
49. (a) $I_x = 2ka^5/3$ (b) $I_x = ka^8/8$
$I_y = 2ka^5/3$ $\quad I_y = ka^8/8$
$I_z = 2ka^5/3$ $\quad I_z = ka^8/8$
51. (a) $I_x = 256k$ (b) $I_x = 2048k/3$
$I_y = 512k/3$ $\quad I_y = 1024k/3$
$I_z = 256k$ $\quad I_z = 2048k/3$

53. Proof **55.** $\displaystyle\int_{-1}^{1}\int_{-1}^{1}\int_{0}^{1-x}(x^2+y^2)\sqrt{x^2+y^2+z^2}\,dz\,dy\,dx$

57. (a) $\displaystyle m = \int_{-2}^{2}\int_{-\sqrt{4-x^2}}^{\sqrt{4-x^2}}\int_{0}^{4-x^2-y^2} kz\,dz\,dy\,dx$

(b) $\bar{x} = \bar{y} = 0$, by symmetry.

$\displaystyle \bar{z} = \frac{1}{m}\int_{-2}^{2}\int_{-\sqrt{4-x^2}}^{\sqrt{4-x^2}}\int_{0}^{4-x^2-y^2} kz^2\,dz\,dy\,dx$

(c) $\displaystyle I_z = \int_{-2}^{2}\int_{-\sqrt{4-x^2}}^{\sqrt{4-x^2}}\int_{0}^{4-x^2-y^2} kz(x^2+y^2)\,dz\,dy\,dx$

59. See "Definition of Triple Integral" on page 1024 and Theorem 14.4, "Evaluation by Iterated Integrals" on page 1025.
61. (a) Solid *B*
(b) Solid *B* has the greater moment of inertia because it is more dense.
(c) Solid *A* will reach the bottom first. Since Solid *B* has a greater moment of inertia, it has a greater resistance to rotational motion.

63. $\frac{13}{3}$ **65.** $\frac{3}{2}$
67. Q: $3z^2 + y^2 + 2x^2 \le 1$; $4\sqrt{6}\pi/45 \approx 0.684$
69. $a = 2, \frac{16}{3}$ **71.** Putnam problem B1, 1965

Section 14.7 (page 1040)

1. 8 **3.** $\frac{52}{45}$ **5.** $\pi/8$ **7.** $\pi(e^4 + 3)$

9.

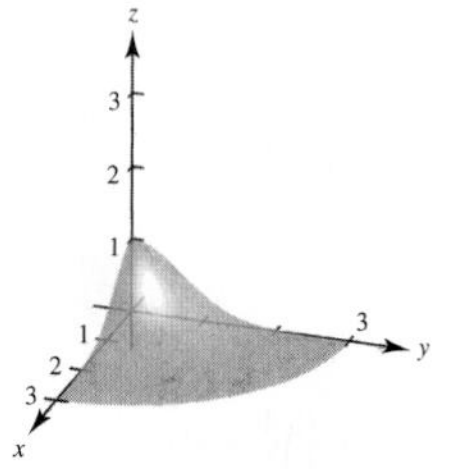

$(1 - e^{-9})\pi/4$

11.

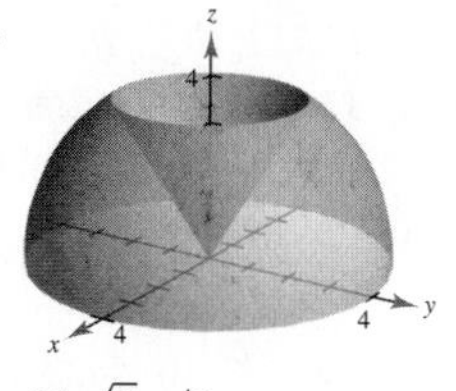

$64\sqrt{3}\pi/3$

13. Cylindrical: $\displaystyle\int_0^{2\pi}\int_0^2\int_{r^2}^4 r^2\cos\theta\,dz\,dr\,d\theta = 0$

Spherical: $\displaystyle\int_0^{2\pi}\int_0^{\arctan(1/2)}\int_0^{4\sec\phi} \rho^3\sin^2\phi\cos\theta\,d\rho\,d\phi\,d\theta$

$\displaystyle + \int_0^{2\pi}\int_{\arctan(1/2)}^{\pi/2}\int_0^{\cot\phi\csc\phi} \rho^3\sin^2\phi\cos\phi\,d\rho\,d\phi\,d\theta = 0$

15. Cylindrical: $\displaystyle\int_0^{2\pi}\int_0^{a}\int_{a}^{a+\sqrt{a^2-r^2}} r^2\cos\theta\,dz\,dr\,d\theta = 0$

Spherical: $\displaystyle\int_0^{\pi/4}\int_0^{2\pi}\int_{a\sec\phi}^{2a\cos\phi} \rho^3\sin^2\phi\cos\theta\,d\rho\,d\theta\,d\phi = 0$

17. $(2a^3/9)(3\pi - 4)$ **19.** $(2a^3/9)(3\pi - 4)$
21. $48k\pi$ **23.** $\pi r_0^2h/3$ **25.** $(0, 0, h/5)$

27. $\displaystyle I_z = 4k\int_0^{\pi/2}\int_0^{r_0}\int_0^{h(r_0-r)/r_0} r^3\,dz\,dr\,d\theta$
$= 3mr_0^2/10$

29. Proof **31.** $16\pi^2$ **33.** $k\pi a^4$
35. $(0, 0, 3r/8)$ **37.** $k\pi/192$
39. Rectangular to cylindrical: $r^2 = x^2 + y^2$
$\tan\theta = y/x$
$z = z$
Cylindrical to rectangular: $x = r\cos\theta$
$y = r\sin\theta$
$z = z$

41. $\displaystyle\int_{\theta_1}^{\theta_2}\int_{g_1(\theta)}^{g_2(\theta)}\int_{h_1(r\cos\theta,\,r\sin\theta)}^{h_2(r\cos\theta,\,r\sin\theta)} f(r\cos\theta, r\sin\theta, z)r\,dz\,dr\,d\theta$

43. (a) r constant: right circular cylinder about z-axis
θ constant: plane parallel to z-axis
z constant: plane parallel to xy-plane
(b) ρ constant: sphere
θ constant: plane parallel to z-axis
ϕ constant: cone

45. $\frac{1}{2}\pi^2a^4$

Section 14.8 (page 1047)

1. $-\frac{1}{2}$ **3.** $1 + 2v$ **5.** 1 **7.** $-e^{2u}$

9.

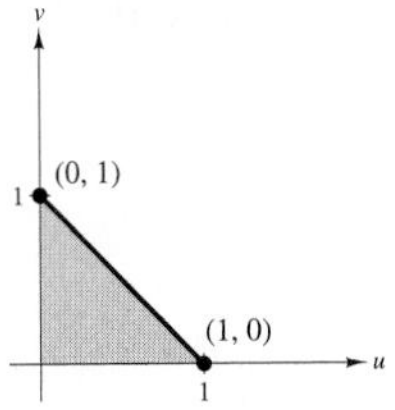

11. $\frac{8}{3}$ **13.** 36

15. $(e^{-1/2} - e^{-2}) \ln 8 \approx 0.9798$

17. $12(e^4 - 1)$ **19.** $\frac{100}{9}$ **21.** $\frac{2}{5}a^{5/2}$

23. (a)

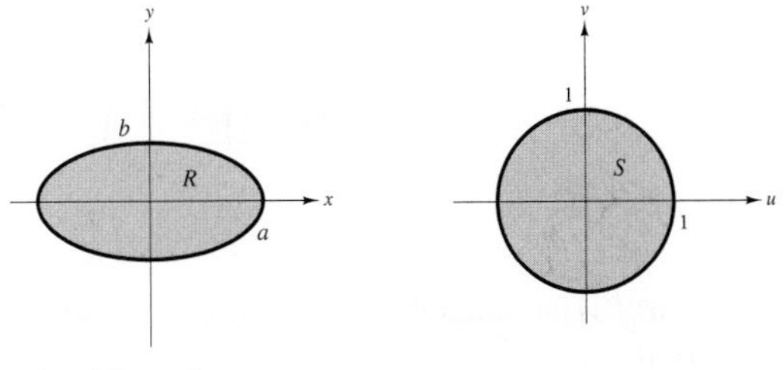

(b) ab (c) πab

25. See "Definition of the Jacobian" on page 1042.

27. u^2v **29.** $-\rho^2 \sin \phi$ **31.** Putnam problem A2, 1994

Review Exercises for Chapter 14 (page 1048)

1. $x - x^3 + x^3 \ln x^2$ **3.** $\frac{29}{6}$ **5.** 36

7. $\displaystyle\int_0^3 \int_0^{(3-x)/3} dy\, dx = \int_0^1 \int_0^{3-3y} dx\, dy = \frac{3}{2}$

9. $\displaystyle\int_{-5}^3 \int_{-\sqrt{25-x^2}}^{\sqrt{25-x^2}} dy\, dx$

$\displaystyle= \int_{-5}^{-4} \int_{-\sqrt{25-y^2}}^{\sqrt{25-y^2}} dx\, dy + \int_{-4}^{4} \int_{-\sqrt{25-y^2}}^{3} dx\, dy$

$\displaystyle+ \int_{4}^{5} \int_{-\sqrt{25-y^2}}^{\sqrt{25-y^2}} dx\, dy$

$= 25\pi/2 + 12 + 25 \arcsin \frac{3}{5} \approx 67.36$

11. $\displaystyle 4\int_0^1 \int_0^{x\sqrt{1-x^2}} dy\, dx = 4\int_0^{1/2} \int_{\sqrt{(1-\sqrt{1-4y^2})/2}}^{\sqrt{(1+\sqrt{1-4y^2})/2}} dx\, dy = \frac{4}{3}$

13. $\displaystyle\int_2^5 \int_{x-3}^{\sqrt{x-1}} dy\, dx + 2\int_1^2 \int_0^{\sqrt{x-1}} dy\, dx = \int_{-1}^2 \int_{y^2+1}^{y+3} dx\, dy = \frac{9}{2}$

15. Both integrations are over the common region R, as shown in the figure. Both integrals yield $\frac{4}{3} + \frac{4}{3}\sqrt{2}$.

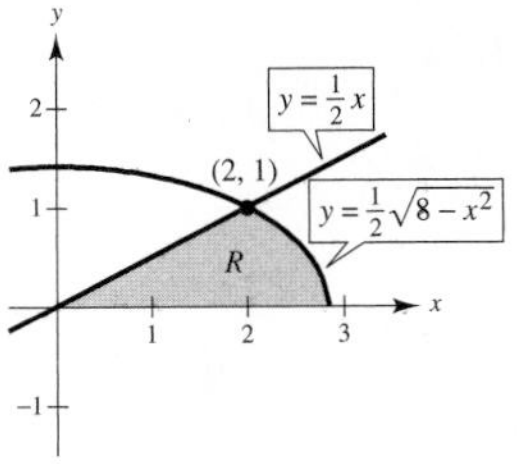

17. $\frac{3296}{15}$ **19.** c **21.** $k = 1, 0.070$ **23.** True

25. True **27.** $(h^3/6)\left[\ln(\sqrt{2} + 1) + \sqrt{2}\right]$ **29.** $\pi h^3/3$

31. (a) $r = 3\sqrt{\cos 2\theta}$

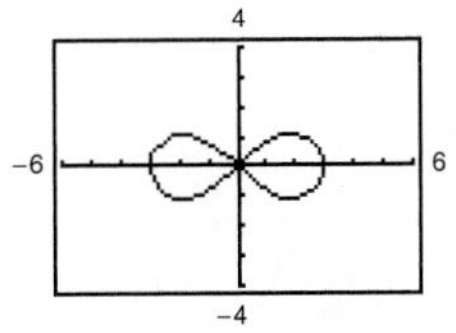

(b) 9

(c) $3(3\pi - 16\sqrt{2} + 20) \approx 20.392$

33. (a) $m = k/4$, $\left(\frac{32}{45}, \frac{64}{55}\right)$ (b) $m = 17k/30$, $\left(\frac{936}{1309}, \frac{784}{663}\right)$

35. $I_x = ka^2b^3/6$

$I_y = ka^4b/4$

$I_0 = (2ka^2b^3 + 3ka^4b)/12$

$\bar{\bar{x}} = a/\sqrt{2}$

$\bar{\bar{y}} = b/\sqrt{3}$

37. $(\pi/6)(65\sqrt{65} - 1)$ **39.** $\frac{1}{6}(37\sqrt{37} - 1)$ **41.** $324\pi/5$

43. $(abc/3)(a^2 + b^2 + c^2)$ **45.** $8\pi/15$ **47.** $\frac{32}{3}\left(\pi/2 - \frac{2}{3}\right)$

49. $\left(0, 0, \frac{1}{4}\right)$ **51.** $(3a/8, 3a/8, 3a/8)$ **53.** $833k\pi/3$

55. (a) $\dfrac{1}{3}\pi h^2(3a - h)$ (b) $\left(0, 0, \dfrac{3(2a - h)^2}{4(3a - h)}\right)$ (c) $\left(0, 0, \dfrac{3}{8}a\right)$

(d) a (e) $(\pi/30)h^3(20a^2 - 15ah + 3h^2)$ (f) $4\pi a^5/15$

57. Volume of a torus formed by a circle of radius 3, centered at $(0, 3, 0)$ and revolved about the z-axis

59. -9 **61.** $5 \ln 5 - 3 \ln 3 - 2 \approx 2.751$

P.S. Problem Solving (page 1051)

1. (a) $8(2 - \sqrt{2})$ (b) Programs will vary.

3. (a) Proof (b) Proof (c) Proof (d) Proof (e) Proof (f) Proof (g) Proof

5. $\frac{1}{3}$

7.

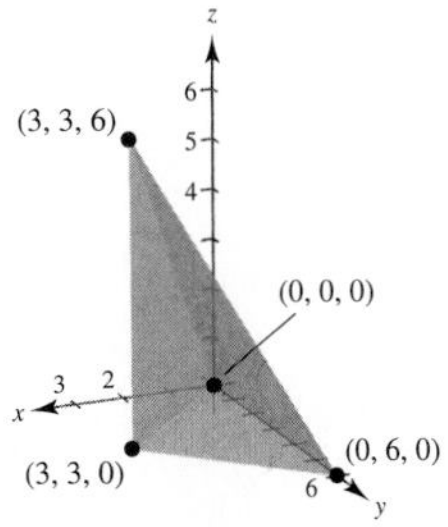

$\displaystyle\int_0^3 \int_0^{2x} \int_x^{6-x} dy\, dz\, dx = 18$

9. $\sqrt{\pi}/4$

11. If $a, k > 0$, then $1 = ka^2$ or $a = 1/\sqrt{k}$.

13. Proof **15.** $\pi/4$ **17.** Proof

Chapter 15

Section 15.1 (page 1063)

1. c **2.** d **3.** b **4.** e **5.** a **6.** f

7.

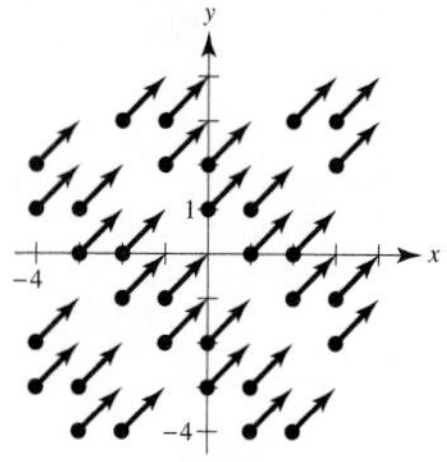

9.

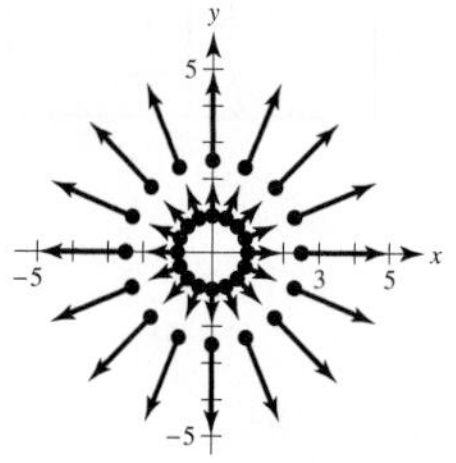

11.

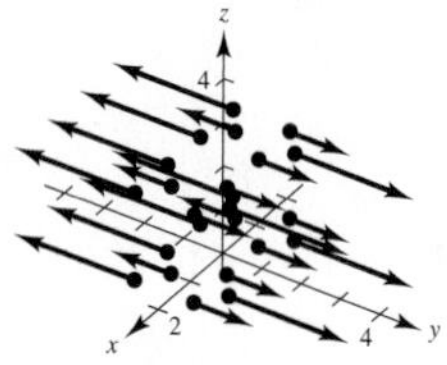

13.

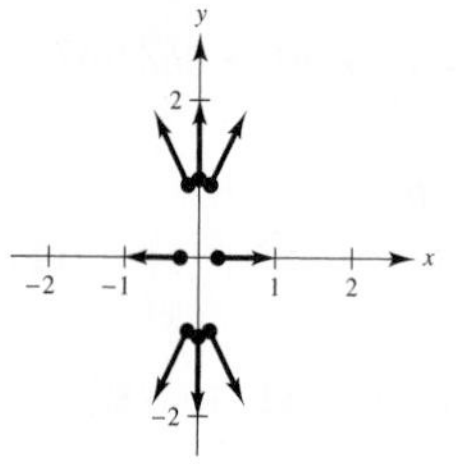

15.

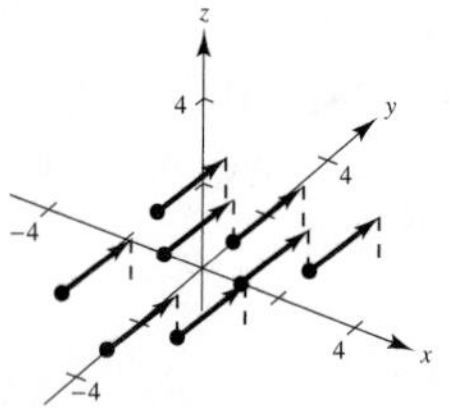

17.

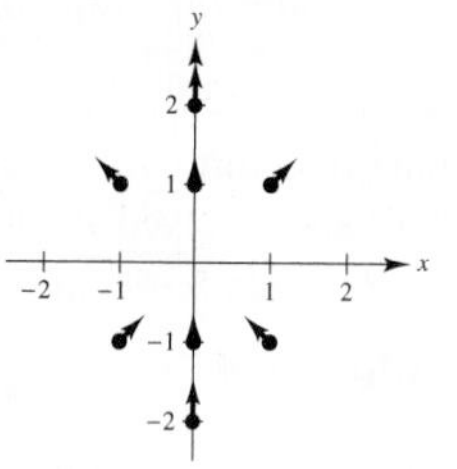

19.

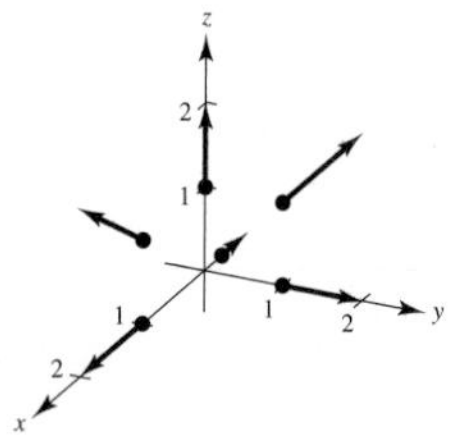

21. $(10x + 3y)\mathbf{i} + (3x + 20y)\mathbf{j}$

23. $-2xye^{x^2}\mathbf{i} - e^{x^2}\mathbf{j} + \mathbf{k}$

25. $[xy/(x + y) + y\ln(x + y)]\mathbf{i} + [xy/(x + y) + x\ln(x + y)]\mathbf{j}$

27–29. Proofs **31.** Not conservative

33. Conservative **35.** Conservative: $f(x, y) = x^2y + K$

37. Conservative: $f(x, y) = e^{x^2y} + K$

39. Conservative: $f(x, y) = \frac{1}{2}\ln(x^2 + y^2) + K$

41. Not conservative **43.** $2\mathbf{j} - \mathbf{k}$ **45.** $-2\mathbf{k}$

47. $2x/(x^2 + y^2)\mathbf{k}$

49. $\cos(y - z)\mathbf{i} + \cos(z - x)\mathbf{j} + \cos(x - y)\mathbf{k}$

51. Not conservative **53.** Conservative: $f(x, y, z) = xye^z + K$

55. Conservative: $f(x, y, z) = x/y + z^2 - z + K$

57. $12x - 2xy$ **59.** $\cos x - \sin y + 2z$ **61.** 4 **63.** 0

65. See "Definition of Vector Field" on page 1054. Some physical examples of vector fields include velocity fields, gravitational fields, and electric force fields.

67. See "Definition of Curl of a Vector Field" on page 1060.

69. $6x\mathbf{j} - 3y\mathbf{k}$ **71.** $z\mathbf{j} + y\mathbf{k}$ **73.** $2z + 3x$ **75.** 0

77–83. Proofs

85. $f(x, y, z) = \|\mathbf{F}(x, y, z)\| = \sqrt{x^2 + y^2 + z^2}$

$$\ln f = \frac{1}{2}\ln(x^2 + y^2 + z^2)$$

$$\nabla \ln f = \frac{x}{x^2 + y^2 + z^2}\mathbf{i} + \frac{y}{x^2 + y^2 + z^2}\mathbf{j} + \frac{z}{x^2 + y^2 + z^2}\mathbf{k}$$

$$= \frac{\mathbf{F}}{f^2}$$

87. $f^n = \|\mathbf{F}(x, y, z)\|^n = \left(\sqrt{x^2 + y^2 + z^2}\right)^n$

$$\nabla f^n = n\left(\sqrt{x^2 + y^2 + z^2}\right)^{n-1}\left(\frac{x\mathbf{i} + y\mathbf{j} + z\mathbf{k}}{\sqrt{x^2 + y^2 + z^2}}\right)$$

$$= nf^{n-2}\mathbf{F}$$

89. True

91. False. Curl f is meaningful only for vector fields, when direction is involved.

Section 15.2 (page 1075)

1. $\mathbf{r}(t) = 3\cos t\mathbf{i} + 3\sin t\mathbf{j}, \quad 0 \le t \le 2\pi$

3. $\mathbf{r}(t) = \begin{cases} t\mathbf{i}, & 0 \le t \le 3 \\ 3\mathbf{i} + (t - 3)\mathbf{j}, & 3 \le t \le 6 \\ (9 - t)\mathbf{i} + 3\mathbf{j}, & 6 \le t \le 9 \\ (12 - t)\mathbf{j}, & 9 \le t \le 12 \end{cases}$

5. $\mathbf{r}(t) = \begin{cases} t\mathbf{i} + \sqrt{t}\mathbf{j}, & 0 \le t \le 1 \\ (2 - t)\mathbf{i} + (2 - t)\mathbf{j}, & 1 \le t \le 2 \end{cases}$

7. 10 **9.** $(\sqrt{65}\pi/6)(3 + 16\pi^2)$ **11.** 9 **13.** $\pi/2$

15. $19\sqrt{2}/6$ **17.** $\frac{19}{6}(1 + \sqrt{2})$ **19.** $\frac{23}{6}$

21. $(2\sqrt{13}\pi/3)(27 + 64\pi^2) \approx 4973.8$ **23.** 2

25. $(k/12)(41\sqrt{41} - 27)$ **27.** $\frac{35}{6}$ **29.** 2 **31.** $-\frac{17}{15}$

33. ≈ 249.49 **35.** -66 **37.** 0 **39.** $-10\pi^2$ **41.** 1500 ft-lb

43. (a) $\frac{236}{3}$; Orientation is from left to right so the value is positive.

(b) $-\frac{236}{3}$; Orientation is from right to left so the value is negative.

45. $\mathbf{F}(t) = -2t\mathbf{i} - t\mathbf{j}$

$\mathbf{r}'(t) = \mathbf{i} - 2\mathbf{j}$

$\mathbf{F}(t) \cdot \mathbf{r}'(t) = -2t + 2t = 0$

$\int_C \mathbf{F} \cdot d\mathbf{r} = 0$

47. $\mathbf{F}(t) = (t^3 - 2t^2)\mathbf{i} + (t - t^2/2)\mathbf{j}$

$\mathbf{r}'(t) = \mathbf{i} + 2t\mathbf{j}$

$\mathbf{F}(t) \cdot \mathbf{r}'(t) = t^3 - 2t^2 + 2t^2 - t^3 = 0$

$\int_C \mathbf{F} \cdot d\mathbf{r} = 0$

49. 1010 **51.** $\frac{190}{3}$ **53.** 25 **55.** $\frac{63}{2}$

57. $-\frac{11}{6}$ **59.** $\frac{316}{3}$ **61.** $5h$ **63.** $\frac{1}{2}$

65. $(h/4)\left[2\sqrt{5} + \ln(2 + \sqrt{5})\right]$ **67.** $\frac{1}{120}(25\sqrt{5} - 11)$

69. (a) $12\pi \approx 37.70 \text{ cm}^2$ (b) $12\pi/5 \approx 7.54 \text{ cm}^3$
(c)

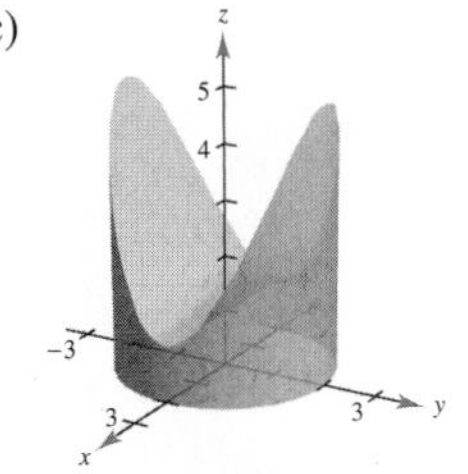

71. $I_x = I_y = a^3\pi$ **73.** b

75. (a)

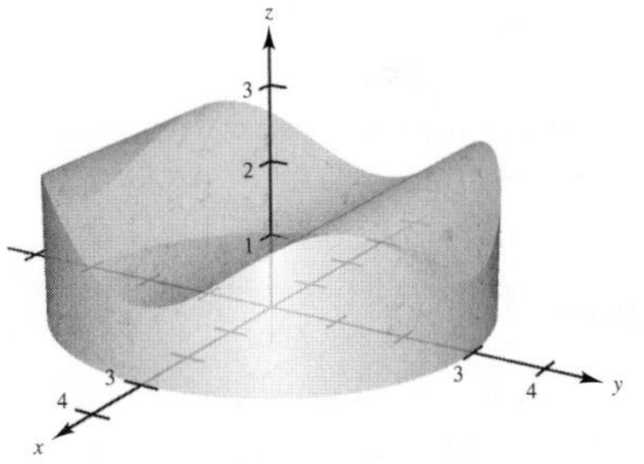

(b) $9\pi \text{ cm}^2 \approx 28.274 \text{ cm}^2$

(c) $\text{Volume} = 2\int_0^3 2\sqrt{9-y^2}\left[1 + 4\frac{y^2}{9}\left(1 - \frac{y^2}{9}\right)\right] dy$

$= 27\pi/2 \approx 42.412 \text{ cm}^3$

77. See "Definition of Line Integral" on page 1066 and Theorem 15.4, "Evaluation of a Line Integral as a Definite Integral" on page 1067.

79. z_3, z_1, z_2, z_4; The greater the height of the surface over the curve $y = \sqrt{x}$, the greater the lateral surface area.

81. False: $\int_C xy\, ds = \sqrt{2}\int_0^1 t^2\, dt.$

83. False: the orientations are different. **85.** -12

Section 15.3 (page 1086)

1. (a) $\int_0^1 (t^2 + 2t^4)\, dt = \frac{11}{15}$

(b) $\int_0^{\pi/2} (\sin^2\theta\cos\theta + 2\sin^4\theta\cos\theta)\, d\theta = \frac{11}{15}$

3. (a) $\int_0^{\pi/3} (\sec\theta\tan^2\theta - \sec^3\theta)\, d\theta \approx -1.317$

(b) $\int_0^3 \left[\frac{\sqrt{t}}{2\sqrt{t+1}} - \frac{\sqrt{t+1}}{2\sqrt{t}}\right] dt \approx -1.317$

5. Conservative **7.** Not conservative **9.** Conservative

11. (a) 1 (b) 1 **13.** (a) 0 (b) $-\frac{1}{3}$ (c) $-\frac{1}{2}$

15. (a) 64 (b) 0 (c) 0 (d) 0 **17.** (a) $\frac{64}{3}$ (b) $\frac{64}{3}$

19. (a) 32 (b) 32 **21.** (a) $\frac{2}{3}$ (b) $\frac{17}{6}$ **23.** (a) 0 (b) 0

25. 24 **27.** -1 **29.** 0 **31.** (a) 0 (b) 0 (c) 0

33. 11 **35.** 30,366 **37.** 0

39. (a) $d\mathbf{r} = (\mathbf{i} - \mathbf{j})\, dt \Rightarrow \int_0^{50} 150\, dt = 7500$ ft-lb

(b) $d\mathbf{r} = \left(\mathbf{i} - \frac{1}{25}(50 - t)\mathbf{j}\right) dt \Rightarrow 6\int_0^{50} (50 - t)\, dt$

$= 7500$ ft-lb

41. See Theorem 15.5, "Fundamental Theorem of Line Integrals," on page 1080.

43. (a) The direct path along the line segment joining $(-4, 0)$ to $(3, 4)$ requires less work than the path going from $(-4, 0)$ to $(-4, 4)$ and then to $(3, 4)$.

(b) The closed curve given by the line segments joining $(-4, 0)$, $(-4, 4)$, $(3, 4)$, and $(-4, 0)$ satisfies $\int_C \mathbf{F} \cdot d\mathbf{r} \neq 0$.

45. Yes, because the work required to get from point to point is irrespective of the path taken.

47. False. It would be true if **F** were conservative.

49. True **51.** Proof

53. (a) Proof (b) $-\pi$ (c) π

(d) -2π; does not contradict Theorem 15.7 because **F** is not continuous at $(0, 0)$ in R enclosed by C.

(e) $\nabla\left(\arctan\frac{x}{y}\right) = \frac{1/y}{1 + (x/y)^2}\mathbf{i} + \frac{-x/y^2}{1 + (x/y)^2}\mathbf{j}$

Section 15.4 (page 1095)

1. 0 **3.** $\frac{32}{15}$ **5.** ≈ 19.99 **7.** $\frac{4}{3}$ **9.** 56 **11.** $\frac{32}{3}$ **13.** 0

15. 0 **17.** $\frac{1}{12}$ **19.** 8π **21.** 4π **23.** $\frac{225}{2}$ **25.** πa^2

27. $\frac{32}{3}$ **29.** See Theorem 15.8 on page 1089. **31.** Proof

33. $\left(0, \frac{8}{5}\right)$ **35.** $\left(\frac{8}{15}, \frac{8}{21}\right)$ **37.** $3\pi a^2/2$ **39.** $\pi - 3\sqrt{3}/2$

41. $\int_C \mathbf{F} \cdot d\mathbf{r} = \int_C M\, dx + N\, dy = \int\int_R \left(\frac{\partial N}{\partial x} - \frac{\partial M}{\partial y}\right) dA = 0;$

$I = -2\pi$ when C is a circle that contains the origin.

43. $\frac{19}{2}$

45. (a) $n = 1$: 0 (b) $n = 2$: $-\frac{4}{3}a^3$ (c) 0
$n = 3$: 0 $n = 4$: $-\frac{16}{15}a^5$
$n = 5$: 0 $n = 6$: $-\frac{32}{35}a^7$
$n = 7$: 0 $n = 8$: $-\frac{256}{315}a^9$

47–49. Proofs

Section 15.5 (page 1105)

1. c **2.** d **3.** b **4.** a

5. $y - 2z = 0$
Plane

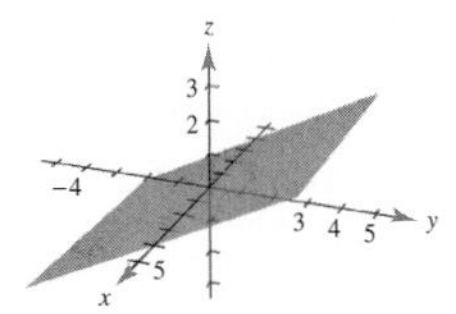

7. $x^2 + z^2 = 4$
Cylinder

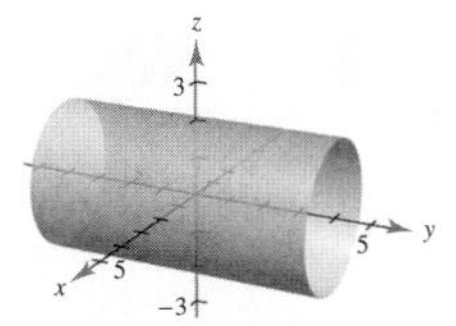

9. The paraboloid is reflected (inverted) through the xy-plane.

11. The height of the paraboloid is increased from 4 to 9.

13.

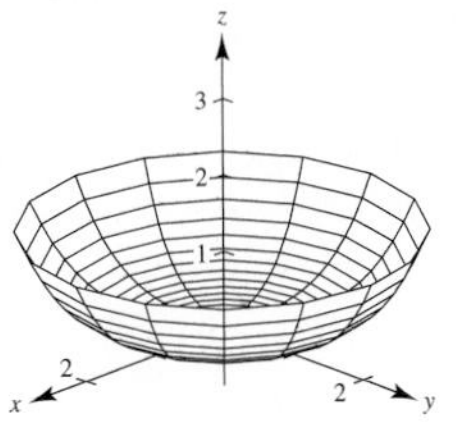

15.

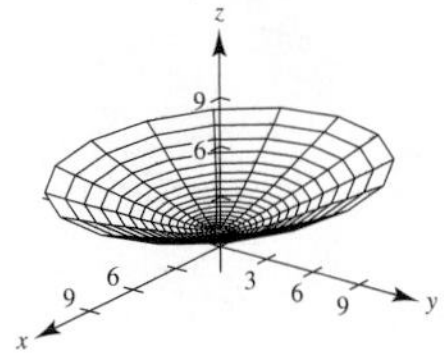

17.

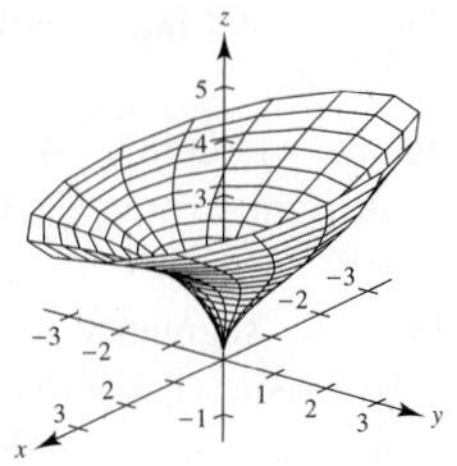

19. $\mathbf{r}(u, v) = u\mathbf{i} + v\mathbf{j} + v\mathbf{k}$

21. $\mathbf{r}(u, v) = 4 \cos u\mathbf{i} + 4 \sin u\mathbf{j} + v\mathbf{k}$

23. $\mathbf{r}(u, v) = u\mathbf{i} + v\mathbf{j} + u^2\mathbf{k}$

25. $\mathbf{r}(u, v) = v \cos u\mathbf{i} + v \sin u\mathbf{j} + 4\mathbf{k}, \quad 0 \le v \le 3$

27. $x = u, y = \dfrac{u}{2} \cos v, z = \dfrac{u}{2} \sin v, \quad 0 \le u \le 6, 0 \le v \le 2\pi$

29. $x = \sin u \cos v, y = \sin u \sin v, z = u$
$0 \le u \le \pi, 0 \le v \le 2\pi$

31. $x - y - 2z = 0$ **33.** $4y - 3z = 12$ **35.** $2\sqrt{2}$

37. $2\pi ab$ **39.** $\pi ab^2\sqrt{a^2 + 1}$

41. $(\pi/6)(17\sqrt{17} - 1) \approx 36.177$

43. See "Definition of Parametric Surface" on page 1098.

45. (a) $(-10, 10, 0)$ (b) $(10, 10, 10)$
(c) $(0, 10, 0)$ (d) $(10, 0, 0)$

47. (a)

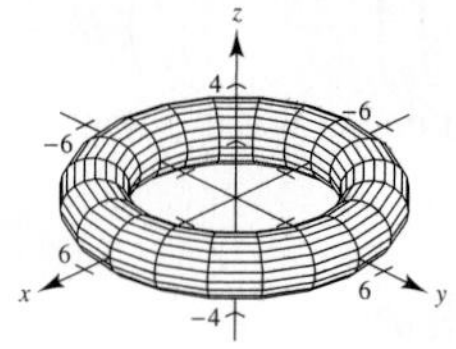

(b)

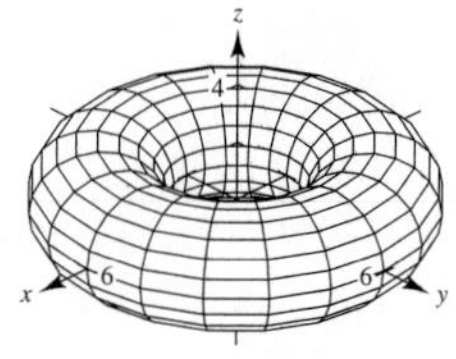

(c)

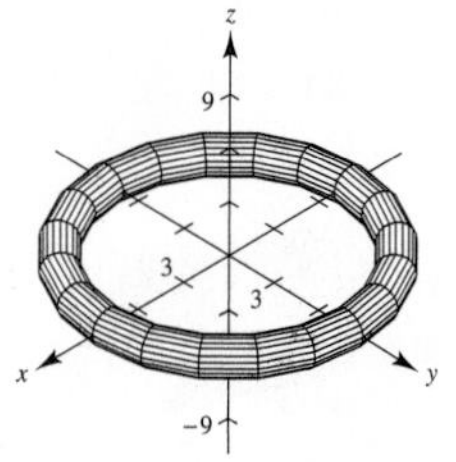

(d)

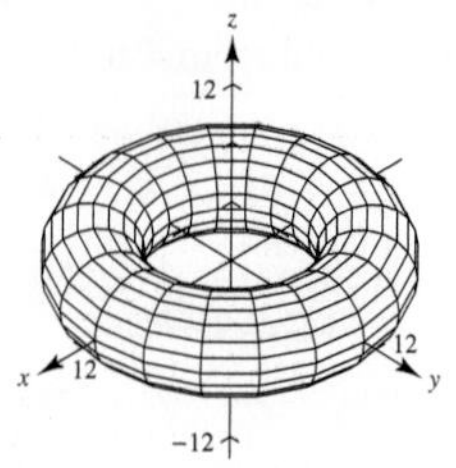

The radius of the generating circle that is revolved about the z-axis is b, and its center is a units from the axis of revolution.

49. $400\pi \text{ m}^2$

51. $2\pi\left[\frac{3}{2}\sqrt{13} + 2 \ln(3 + \sqrt{13}) - 2 \ln 2\right]$

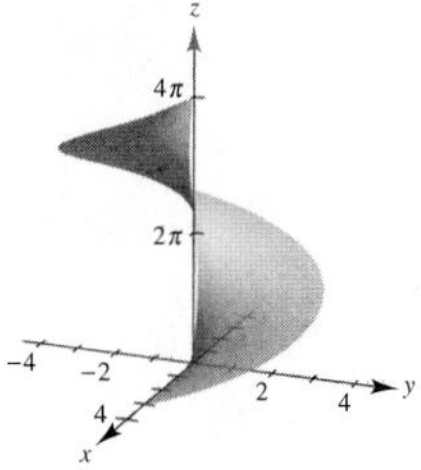

53. Answers will vary. Sample answer: Let
$x = (2 - u)(5 + \cos v) \cos 3\pi u$
$y = (2 - u)(5 + \cos v) \sin 3\pi u$
$z = 5u + (2 - u) \sin v$
where $-\pi \le u \le \pi$ and $-\pi \le v \le \pi$.

Section 15.6 (page 1118)

1. 0 **3.** 10π **5.** $27\sqrt{6}/2$ **7.** $(391\sqrt{17} + 1)/240$

9. ≈ -11.47 **11.** $\frac{364}{3}$ **13.** $6\sqrt{5}$ **15.** 8

17. $19\sqrt{2}\pi/4$ **19.** $32\pi/3$ **21.** 486π **23.** $-\frac{4}{3}$

25. $243\pi/2$ **27.** 20π **29.** $\frac{5}{2}$

31. See Theorem 15.10, "Evaluating a Surface Integral," on page 1108.

33. See "Definition of Flux Integral," on page 1114; see Theorem 15.11, "Evaluating a Flux Integral," on page 1114.

35. 0 **37.** Proof **39.** $2\pi a^3 h$ **41.** $64\pi\rho$

43. (a)

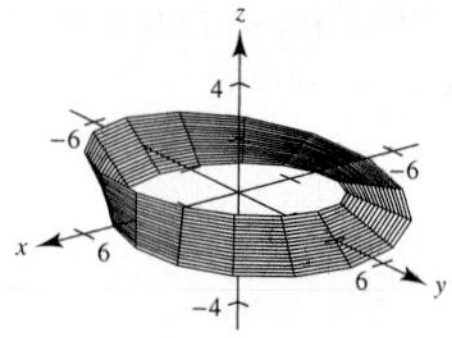

(b) If a normal vector at a point P on the surface is moved around the Möbius strip once, it will point in the opposite direction.

(c) Circle

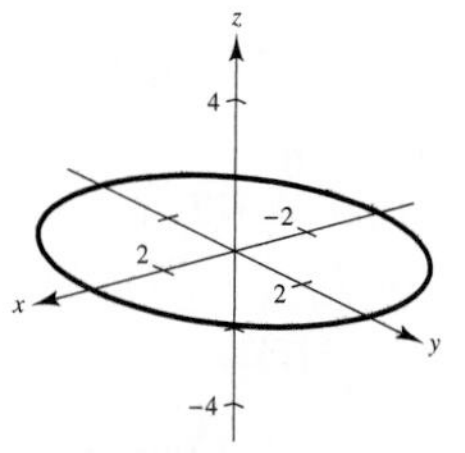

(d) Construction

(e) A strip with a double twist that is twice as long as the Möbius strip.

Section 15.7 (page 1126)

1. a^4 **3.** 18 **5.** $3a^4$ **7.** 0 **9.** 32π

11. 0 **13.** 2304 **15.** 144π **17.** 0

19. See Theorem 15.12, "The Divergence Theorem," on page 1120.

21–27. Proofs

Section 15.8 (page 1133)

1. $-xy\mathbf{i} - \mathbf{j} + (yz - 2)\mathbf{k}$ **3.** $[2 - 1/(1 + x^2)]\mathbf{j} - 8x\mathbf{k}$
5. $z(x - 2e^{y^2+z^2})\mathbf{i} - yz\mathbf{j} - 2ye^{x^2+y^2}\mathbf{k}$ **7.** 2π **9.** 0
11. 1 **13.** 0 **15.** 0 **17.** $\frac{8}{3}$ **19.** $a^5/4$ **21.** 0
23. See Theorem 15.13, "Stoke's Theorem," on page 1128.
25. Since circulation is determined by finding the dot product of **curl F** and **N**, the circulation would be 0.
27. Answers will vary. **29.** Putnam Problem A5, 1987

Review Exercises for Chapter 15 (page 1134)

1.

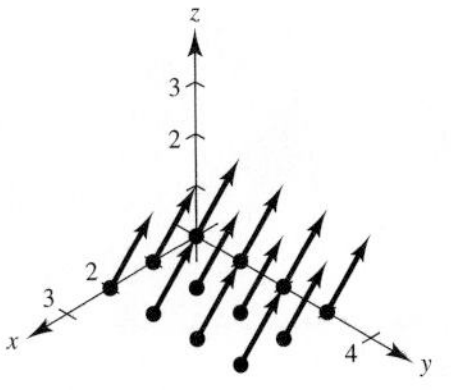

3. $(16x + y)\mathbf{i} + x\mathbf{j} + 2z\mathbf{k}$
5. Not conservative
7. Conservative: $f(x, y) = 3x^2y^2 - x^3 + y^3 - 7y + K$
9. Not conservative
11. Conservative: $f(x, y, z) = x/(yz) + K$
13. (a) div $\mathbf{F} = 2x + 2y + 2z$ (b) $\mathbf{curl\ F} = \mathbf{0}$
15. (a) div $\mathbf{F} = -y \sin x - x \cos y + xy$
(b) $\mathbf{curl\ F} = xz\mathbf{i} - yz\mathbf{j}$
17. (a) div $\mathbf{F} = 1/\sqrt{1 - x^2} + 2xy + 2yz$
(b) $\mathbf{curl\ F} = z^2\mathbf{i} + y^2\mathbf{k}$
19. (a) div $\mathbf{F} = \dfrac{2x + 2y}{x^2 + y^2} + 1$
(b) $\mathbf{curl\ F} = \dfrac{2x - 2y}{x^2 + y^2}\mathbf{k}$
21. (a) $6\sqrt{2}$ (b) 128π **23.** $2\pi^2(1 + 2\pi^2)$
25. (a) $\frac{35}{2}$ (b) 18π **27.** $9a^2/5$
29. $(\sqrt{10}/4)(41 - \cos 8) \approx 32.528$ **31.** $\frac{5}{7}$ **33.** $2\pi^2$
35. $\frac{64}{3}$ **37.** $\frac{4}{3}$ **39.** $\frac{8}{3}(3 - 4\sqrt{2}) \approx -7.085$
41. 12 **43.** (a) 15 (b) 15 (c) 15
45. 4 **47.** 0 **49.** $\frac{1}{12}$
51.

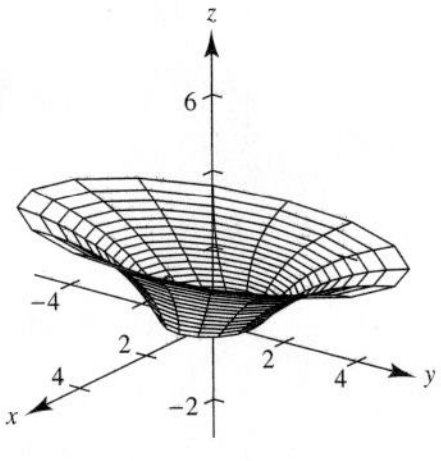

53. (a)

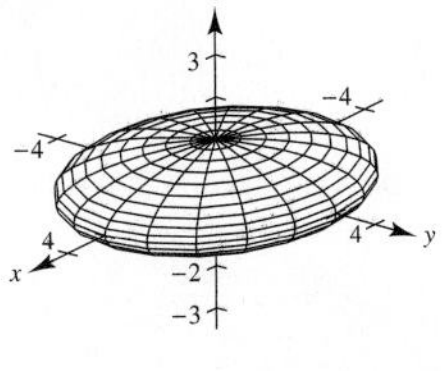

(b)

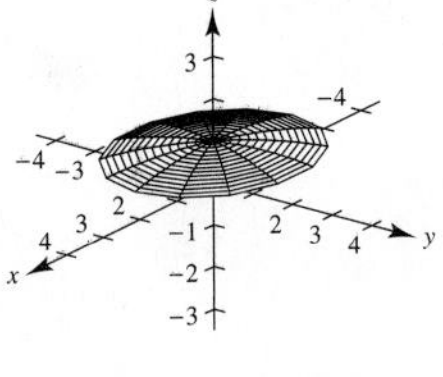

(c)

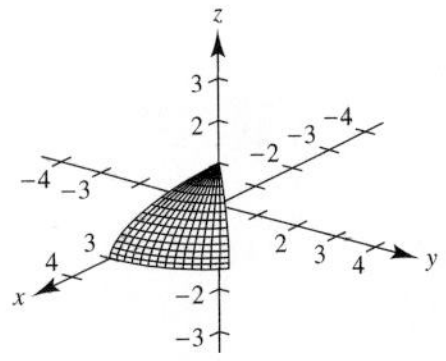

(d)

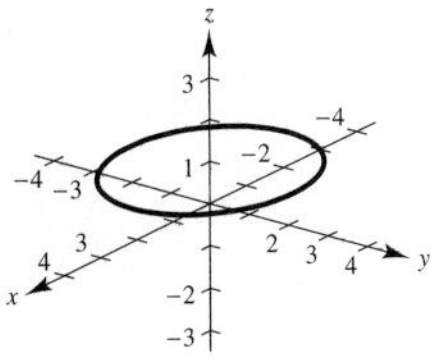

Circle

(e) ≈ 14.436 (f) ≈ 4.269

55.

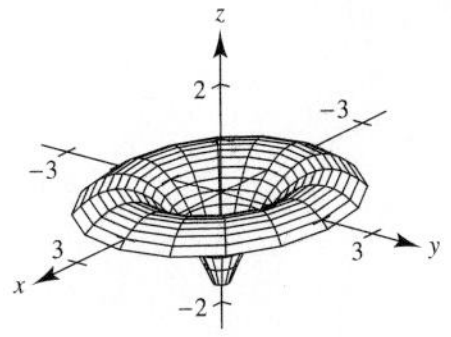

0

57. 66 **59.** $2a^6/5$ **61.** Proof

P.S. Problem Solving (page 1137)

1. (a) $(25\sqrt{2}/6)k\pi$ (b) $(25\sqrt{2}/6)k\pi$
3. $I_x = (\sqrt{13}\pi/3)(27 + 32\pi^2)$; $I_y = (\sqrt{13}\pi/3)(27 + 32\pi^2)$;
$I_z = 18\sqrt{13}\pi$
5. $3a^2\pi$ **7.** (a) 1 (b) $\frac{13}{15}$ (c) $\frac{5}{2}$ **9.** Proof
11. $M = 3mxy(x^2 + y^2)^{-5/2}$
$\partial M/\partial y = 3mx(x^2 - 4y^2)/(x^2 + y^2)^{7/2}$
$N = m(2y^2 - x^2)(x^2 + y^2)^{-5/2}$
$\partial N/\partial x = 3mx(x^2 - 4y^2)/(x^2 + y^2)^{7/2}$
Therefore, $\partial N/\partial x = \partial M/\partial y$ and **F** is conservative.

Index of Applications

Engineering and Physical Sciences

Business and Economics

Social and Behavioral Sciences

Life Sciences

General

Index

D

E

F

INDEX

G

H

I

J

K

L

M

Q

R

W

X

Y

Z

Additional Appendices C through G from Text Website

APPENDIX C

Additional Topics in Differential Equations

APPENDIX C.1 Exact First-Order Equations

Exact Differential Equations • Integrating Factors

Exact Differential Equations

In Chapter 6, you studied applications of differential equations to growth and decay problems. You also learned more about the basic ideas of differential equations and studied the solution technique known as separation of variables. In this appendix, you will learn more about solving differential equations and using them in real-life applications. This section introduces you to a method for solving the first-order differential equation

$$M(x, y)\,dx + N(x, y)\,dy = 0$$

for the special case in which this equation represents the exact differential of a function $z = f(x, y)$.

Definition of an Exact Differential Equation

The equation $M(x, y)\,dx + N(x, y)\,dy = 0$ is an **exact differential equation** if there exists a function f of two variables x and y having continuous partial derivatives such that

$$f_x(x, y) = M(x, y) \quad \text{and} \quad f_y(x, y) = N(x, y).$$

The general solution of the equation is $f(x, y) = C$.

If f has continuous second partials, then

$$\frac{\partial M}{\partial y} = \frac{\partial^2 f}{\partial y\,\partial x} = \frac{\partial^2 f}{\partial x\,\partial y} = \frac{\partial N}{\partial x}.$$

This suggests the following test for exactness.

THEOREM C.1 Test for Exactness

Let M and N have continuous partial derivatives on an open disk R. The differential equation $M(x, y)\,dx + N(x, y)\,dy = 0$ is exact if and only if

$$\frac{\partial M}{\partial y} = \frac{\partial N}{\partial x}.$$

Exactness is a fragile condition in the sense that seemingly minor alterations in an exact equation can destroy its exactness. This is demonstrated in the following example.

NOTE Every differential equation of the form

$$M(x)\,dx + N(y)\,dy = 0$$

is exact. In other words, a separable differential equation is actually a special type of an exact equation.

EXAMPLE 1 Testing for Exactness

a. The differential equation $(xy^2 + x)\,dx + yx^2\,dy = 0$ is exact because

$$\frac{\partial M}{\partial y} = \frac{\partial}{\partial y}[xy^2 + x] = 2xy \quad \text{and} \quad \frac{\partial N}{\partial x} = \frac{\partial}{\partial x}[yx^2] = 2xy.$$

But the equation $(y^2 + 1)\,dx + xy\,dy = 0$ is not exact, even though it is obtained by dividing each side of the first equation by x.

b. The differential equation $\cos y\,dx + (y^2 - x\sin y)\,dy = 0$ is exact because

$$\frac{\partial M}{\partial y} = \frac{\partial}{\partial y}[\cos y] = -\sin y \quad \text{and} \quad \frac{\partial N}{\partial x} = \frac{\partial}{\partial x}[y^2 - x\sin y] = -\sin y.$$

But the equation $\cos y\,dx + (y^2 + x\sin y)\,dy = 0$ is not exact, even though it differs from the first equation only by a single sign.

Note that the test for exactness of $M(x, y)\,dx + N(x, y)\,dy = 0$ is the same as the test for determining whether $\mathbf{F}(x, y) = M(x, y)\mathbf{i} + N(x, y)\mathbf{j}$ is the gradient of a potential function. This means that a general solution $f(x, y) = C$ to an exact differential equation can be found by the method used to find a potential function for a conservative vector field.

EXAMPLE 2 Solving an Exact Differential Equation

Solve the differential equation $(2xy - 3x^2)\,dx + (x^2 - 2y)\,dy = 0$.

Solution The given differential equation is exact because

$$\frac{\partial M}{\partial y} = \frac{\partial}{\partial y}[2xy - 3x^2] = 2x = \frac{\partial N}{\partial x} = \frac{\partial}{\partial x}[x^2 - 2y].$$

The general solution, $f(x, y) = C$, is given by

$$\begin{aligned} f(x, y) &= \int M(x, y)\,dx \\ &= \int (2xy - 3x^2)\,dx = x^2y - x^3 + g(y). \end{aligned}$$

You can determine $g(y)$ by integrating $N(x, y)$ with respect to y and reconciling the two expressions for $f(x, y)$. An alternative method is to partially differentiate this version of $f(x, y)$ with respect to y and compare the result with $N(x, y)$. In other words,

$$f_y(x, y) = \frac{\partial}{\partial y}[x^2y - x^3 + g(y)] = x^2 + g'(y) = \overbrace{x^2 - 2y}^{N(x,\,y)}.$$

$$\boxed{g'(y) = -2y}$$

So, $g'(y) = -2y$, and it follows that $g(y) = -y^2 + C_1$. Therefore,

$$f(x, y) = x^2y - x^3 - y^2 + C_1$$

and the general solution is $x^2y - x^3 - y^2 = C$. Figure C.1 shows the solution curves that correspond to $C = 1$, 10, 100, and 1000.

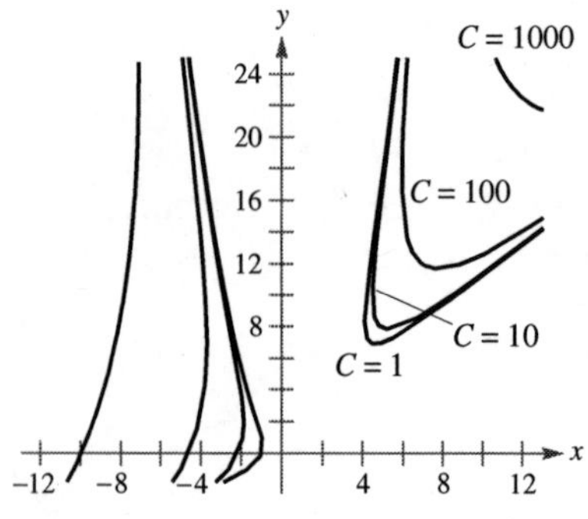

Figure C.1

EXAMPLE 3 Solving an Exact Differential Equation

Find the particular solution of

$$(\cos x - x \sin x + y^2)\,dx + 2xy\,dy = 0$$

that satisfies the initial condition $y = 1$ when $x = \pi$.

Solution The differential equation is exact because

$$\overbrace{\frac{\partial}{\partial y}[\cos x - x \sin x + y^2]}^{\frac{\partial M}{\partial y}} = 2y = \overbrace{\frac{\partial}{\partial x}[2xy]}^{\frac{\partial N}{\partial x}}.$$

Because $N(x, y)$ is simpler than $M(x, y)$, it is better to begin by integrating $N(x, y)$.

$$f(x, y) = \int N(x, y)\,dy = \int 2xy\,dy = xy^2 + g(x)$$

$$f_x(x, y) = \frac{\partial}{\partial x}[xy^2 + g(x)] = y^2 + g'(x) = \overbrace{\cos x - x \sin x + y^2}^{M(x, y)}$$

$$\boxed{g'(x) = \cos x - x \sin x}$$

So, $g'(x) = \cos x - x \sin x$ and

$$\begin{aligned} g(x) &= \int (\cos x - x \sin x)\,dx \\ &= x \cos x + C_1 \end{aligned}$$

which implies that $f(x, y) = xy^2 + x \cos x + C_1$, and the general solution is

$$xy^2 + x \cos x = C. \qquad \text{General solution}$$

Applying the given initial condition produces

$$\pi(1)^2 + \pi \cos \pi = C$$

which implies that $C = 0$. So, the particular solution is

$$xy^2 + x \cos x = 0. \qquad \text{Particular solution}$$

The graph of the particular solution is shown in Figure C.2. Notice that the graph consists of two parts: the ovals are given by $y^2 + \cos x = 0$, and the y-axis is given by $x = 0$.

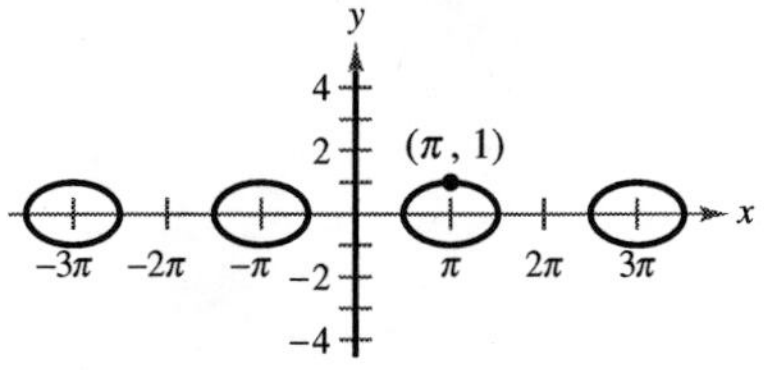

Figure C.2

In Example 3, note that if $z = f(x, y) = xy^2 + x \cos x$, the total differential of z is given by

$$\begin{aligned} dz &= f_x(x, y)\,dx + f_y(x, y)\,dy \\ &= (\cos x - x \sin x + y^2)\,dx + 2xy\,dy \\ &= M(x, y)\,dx + N(x, y)\,dy. \end{aligned}$$

In other words, $M\,dx + N\,dy = 0$ is called an *exact* differential equation because $M\,dx + N\,dy$ is exactly the differential of $f(x, y)$.

Integrating Factors

If the differential equation $M(x, y)\,dx + N(x, y)\,dy = 0$ is not exact, it may be possible to make it exact by multiplying by an appropriate factor $u(x, y)$, which is called an **integrating factor** for the differential equation.

EXAMPLE 4 Multiplying by an Integrating Factor

a. If the differential equation

$$2y\,dx + x\,dy = 0 \qquad \text{Not an exact equation}$$

is multiplied by the integrating factor $u(x, y) = x$, the resulting equation

$$2xy\,dx + x^2\,dy = 0 \qquad \text{Exact equation}$$

is exact—the left side is the total differential of x^2y.

b. If the equation

$$y\,dx - x\,dy = 0 \qquad \text{Not an exact equation}$$

is multiplied by the integrating factor $u(x, y) = 1/y^2$, the resulting equation

$$\frac{1}{y}\,dx - \frac{x}{y^2}\,dy = 0 \qquad \text{Exact equation}$$

is exact—the left side is the total differential of x/y.

Finding an integrating factor can be difficult. However, there are two classes of differential equations whose integrating factors can be found routinely—namely, those that possess integrating factors that are functions of either x alone or y alone. The following theorem, which is presented without proof, outlines a procedure for finding these two special categories of integrating factors.

THEOREM C.2 Integrating Factors

Consider the differential equation $M(x, y)\,dx + N(x, y)\,dy = 0$.

1. If

$$\frac{1}{N(x, y)}[M_y(x, y) - N_x(x, y)] = h(x)$$

is a function of x alone, then $e^{\int h(x)\,dx}$ is an integrating factor.

2. If

$$\frac{1}{M(x, y)}[N_x(x, y) - M_y(x, y)] = k(y)$$

is a function of y alone, then $e^{\int k(y)\,dy}$ is an integrating factor.

STUDY TIP If either $h(x)$ or $k(y)$ is constant, Theorem C.2 still applies. As an aid to remembering these formulas, note that the subtracted partial derivative identifies both the denominator and the variable for the integrating factor.

EXAMPLE 5 Finding an Integrating Factor

Solve the differential equation $(y^2 - x)\,dx + 2y\,dy = 0$.

Solution The given equation is not exact because $M_y(x, y) = 2y$ and $N_x(x, y) = 0$. However, because

$$\frac{M_y(x, y) - N_x(x, y)}{N(x, y)} = \frac{2y - 0}{2y} = 1 = h(x)$$

it follows that $e^{\int h(x)\,dx} = e^{\int dx} = e^x$ is an integrating factor. Multiplying the given differential equation by e^x produces the exact differential equation

$$(y^2e^x - xe^x)\,dx + 2ye^x\,dy = 0$$

whose solution is obtained as follows.

$$f(x, y) = \int N(x, y)\,dy = \int 2ye^x\,dy = y^2e^x + g(x)$$

$$f_x(x, y) = y^2e^x + g'(x) = \overbrace{y^2e^x - xe^x}^{M(x, y)}$$

$$\boxed{g'(x) = -xe^x}$$

Therefore, $g'(x) = -xe^x$ and $g(x) = -xe^x + e^x + C_1$, which implies that

$$f(x, y) = y^2e^x - xe^x + e^x + C_1.$$

The general solution is $y^2e^x - xe^x + e^x = C$, or $y^2 - x + 1 = Ce^{-x}$.

The next example shows how a differential equation can help in sketching a force field given by $\mathbf{F}(x, y) = M(x, y)\mathbf{i} + N(x, y)\mathbf{j}$.

EXAMPLE 6 An Application to Force Fields

Sketch the force field given by

$$\mathbf{F}(x, y) = \frac{2y}{\sqrt{x^2 + y^2}}\mathbf{i} - \frac{y^2 - x}{\sqrt{x^2 + y^2}}\mathbf{j}$$

by finding and sketching the family of curves tangent to $\mathbf{F}$.

Solution At the point (x, y) in the plane, the vector $\mathbf{F}(x, y)$ has a slope of

$$\frac{dy}{dx} = \frac{-(y^2 - x)/\sqrt{x^2 + y^2}}{2y/\sqrt{x^2 + y^2}} = \frac{-(y^2 - x)}{2y}$$

which, in differential form, is

$$2y\,dy = -(y^2 - x)\,dx$$

$$(y^2 - x)\,dx + 2y\,dy = 0.$$

From Example 5, you know that the general solution of this differential equation is $y^2 - x + 1 = Ce^{-x}$, or $y^2 = x - 1 + Ce^{-x}$. Figure C.3 shows several representative curves from this family. Note that the force vector at (x, y) is tangent to the curve passing through (x, y).

Force field:
$\mathbf{F}(x, y) = \dfrac{2y}{\sqrt{x^2 + y^2}}\mathbf{i} - \dfrac{y^2 - x}{\sqrt{x^2 + y^2}}\mathbf{j}$
Family of tangent curves to $\mathbf{F}$:
$y^2 = x - 1 + Ce^{-x}$

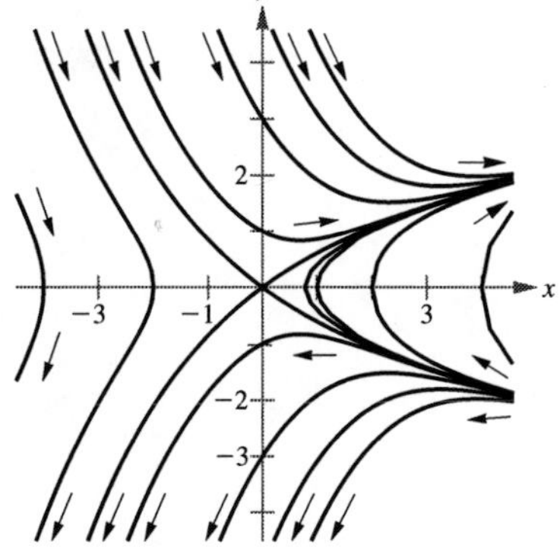

Figure C.3

EXERCISES FOR APPENDIX C.1

In Exercises 1–10, determine whether the differential equation is exact. If it is, find the general solution.

1. $(2x - 3y)\,dx + (2y - 3x)\,dy = 0$
2. $ye^x\,dx + e^x\,dy = 0$
3. $(3y^2 + 10xy^2)\,dx + (6xy - 2 + 10x^2y)\,dy = 0$
4. $2\cos(2x - y)\,dx - \cos(2x - y)\,dy = 0$
5. $(4x^3 - 6xy^2)\,dx + (4y^3 - 6xy)\,dy = 0$
6. $2y^2e^{xy^2}\,dx + 2xye^{xy^2}\,dy = 0$
7. $\dfrac{1}{x^2 + y^2}(x\,dy - y\,dx) = 0$
8. $e^{-(x^2+y^2)}(x\,dx + y\,dy) = 0$
9. $\dfrac{1}{(x - y)^2}(y^2\,dx + x^2\,dy) = 0$
10. $e^y\cos xy\,[y\,dx + (x + \tan xy)\,dy] = 0$

In Exercises 11 and 12, (a) sketch an approximate solution of the differential equation satisfying the initial condition on the slope field, (b) find the particular solution that satisfies the initial condition, and (c) use a graphing utility to graph the particular solution. Compare the graph with the sketch in part (a).

Differential Equation	*Initial Condition*
11. $(2x\tan y + 5)\,dx + (x^2\sec^2 y)\,dy = 0$	$\left(\frac{1}{2}, \frac{\pi}{4}\right)$
12. $\dfrac{1}{\sqrt{x^2 + y^2}}(x\,dx + y\,dy) = 0$	$(4, 3)$

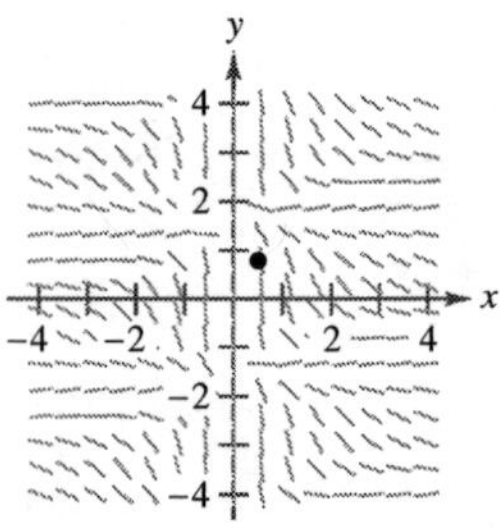

Figure for 11

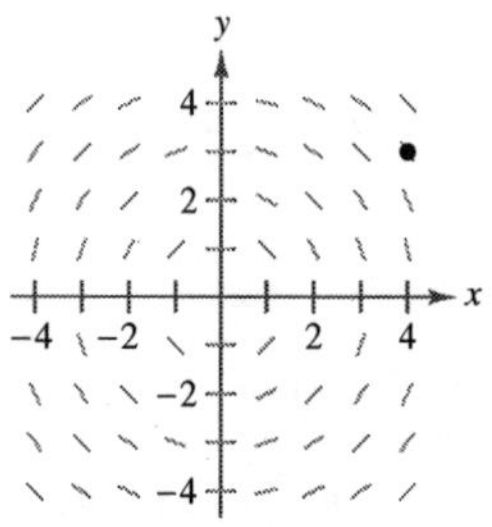

Figure for 12

In Exercises 13–16, find the particular solution that satisfies the initial condition.

Differential Equation	*Initial Condition*
13. $\dfrac{y}{x - 1}\,dx + [\ln(x - 1) + 2y]\,dy = 0$	$y(2) = 4$
14. $\dfrac{1}{x^2 + y^2}(x\,dx + y\,dy) = 0$	$y(0) = 4$
15. $e^{3x}(\sin 3y\,dx + \cos 3y\,dy) = 0$	$y(0) = \pi$
16. $(x^2 + y^2)\,dx + 2xy\,dy = 0$	$y(3) = 1$

In Exercises 17–26, find the integrating factor that is a function of x or y alone and use it to find the general solution of the differential equation.

17. $y\,dx - (x + 6y^2)\,dy = 0$
18. $(2x^3 + y)\,dx - x\,dy = 0$
19. $(5x^2 - y)\,dx + x\,dy = 0$
20. $(5x^2 - y^2)\,dx + 2y\,dy = 0$
21. $(x + y)\,dx + \tan x\,dy = 0$
22. $(2x^2y - 1)\,dx + x^3\,dy = 0$
23. $y^2\,dx + (xy - 1)\,dy = 0$
24. $(x^2 + 2x + y)\,dx + 2\,dy = 0$
25. $2y\,dx + \left(x - \sin\sqrt{y}\right)dy = 0$
26. $(-2y^3 + 1)\,dx + (3xy^2 + x^3)\,dy = 0$

In Exercises 27–30, use the integrating factor to find the general solution of the differential equation.

Integrating Factor	*Differential Equation*
27. $u(x, y) = xy^2$	$(4x^2y + 2y^2)\,dx + (3x^3 + 4xy)\,dy = 0$
28. $u(x, y) = x^2y$	$(3y^2 + 5x^2y)\,dx + (3xy + 2x^3)\,dy = 0$
29. $u(x, y) = x^{-2}y^{-3}$	$(-y^5 + x^2y)\,dx + (2xy^4 - 2x^3)\,dy = 0$
30. $u(x, y) = x^{-2}y^{-2}$	$-y^3\,dx + (xy^2 - x^2)\,dy = 0$

31. Show that each expression is an integrating factor for the differential equation

$y\,dx - x\,dy = 0.$

(a) $\dfrac{1}{x^2}$ (b) $\dfrac{1}{y^2}$ (c) $\dfrac{1}{xy}$ (d) $\dfrac{1}{x^2 + y^2}$

32. Show that the differential equation

$(axy^2 + by)\,dx + (bx^2y + ax)\,dy = 0$

is exact only if $a = b$. If $a \neq b$, show that x^my^n is an integrating factor, where

$$m = -\frac{2b + a}{a + b}, \qquad n = -\frac{2a + b}{a + b}.$$

In Exercises 33–36, use a graphing utility to graph the family of curves tangent to the given force field.

33. $\mathbf{F}(x, y) = \dfrac{y}{\sqrt{x^2 + y^2}}\mathbf{i} - \dfrac{x}{\sqrt{x^2 + y^2}}\mathbf{j}$
34. $\mathbf{F}(x, y) = \dfrac{x}{\sqrt{x^2 + y^2}}\mathbf{i} - \dfrac{y}{\sqrt{x^2 + y^2}}\mathbf{j}$
35. $\mathbf{F}(x, y) = 4x^2y\mathbf{i} - \left(2xy^2 + \dfrac{x}{y^2}\right)\mathbf{j}$
36. $\mathbf{F}(x, y) = (1 + x^2)\mathbf{i} - 2xy\mathbf{j}$

In Exercises 37 and 38, find an equation for the curve with the specified slope passing through the given point.

Slope	*Point*
37. $\dfrac{dy}{dx} = \dfrac{y - x}{3y - x}$	$(2, 1)$
38. $\dfrac{dy}{dx} = \dfrac{-2xy}{x^2 + y^2}$	$(0, 2)$

39. ***Cost*** If $y = C(x)$ represents the cost of producing x units in a manufacturing process, the **elasticity of cost** is defined as

$$E(x) = \frac{\text{marginal cost}}{\text{average cost}} = \frac{C'(x)}{C(x)/x} = \frac{x}{y}\frac{dy}{dx}.$$

Find the cost function if the elasticity function is

$$E(x) = \frac{20x - y}{2y - 10x}$$

where $C(100) = 500$ and $x \geq 100$.

40. ***Euler's Method*** Consider the differential equation $y' = F(x, y)$ with the initial condition $y(x_0) = y_0$. At any point (x_k, y_k) in the domain of F, $F(x_k, y_k)$ yields the slope of the solution at that point. Euler's Method gives a discrete set of estimates of the y values of a solution of the differential equation using the iterative formula

$$y_{k+1} = y_k + F(x_k, y_k)\,\Delta x$$

where $\Delta x = x_{k+1} - x_k$.

(a) Write a short paragraph describing the general idea of how Euler's Method works.

(b) How will decreasing the magnitude of Δx affect the accuracy of Euler's Method?

41. ***Euler's Method*** Use Euler's Method (see Exercise 40) to approximate $y(1)$ for the values of Δx given in the table if $y' = x + \sqrt{y}$ and $y(0) = 2$. (Note that the number of iterations increases as Δx decreases.) Sketch a graph of the approximate solution on the direction field in the figure.

Δx	0.50	0.25	0.10
Estimate of $y(1)$			

The value of $y(1)$, accurate to three decimal places, is 4.213.

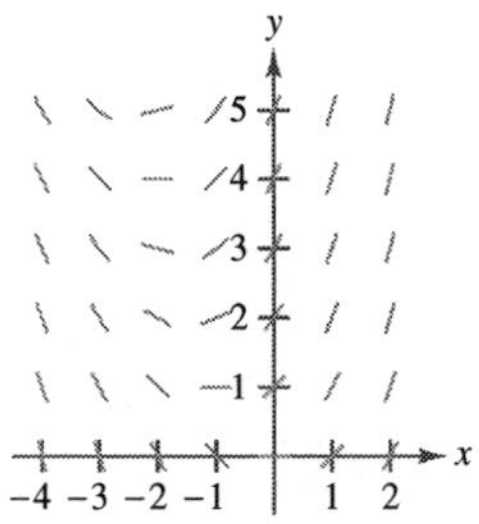

42. ***Programming*** Write a program for a graphing utility or computer that will perform the calculations of Euler's Method for a specified differential equation, interval, Δx, and initial condition. The output should be a graph of the discrete points approximating the solution.

Euler's Method **In Exercises 43–46, (a) use the program you wrote in Exercise 42 to approximate the solution of the differential equation over the indicated interval with the specified value of Δx and the initial condition, (b) solve the differential equation analytically, and (c) use a graphing utility to graph the particular solution and compare the result with the graph of part (a).**

Differential Equation	*Interval*	Δx	*Initial Condition*
43. $y' = x\sqrt[3]{y}$	$[1, 2]$	0.01	$y(1) = 1$
44. $y' = \dfrac{\pi}{4}(y^2 + 1)$	$[-1, 1]$	0.1	$y(-1) = -1$
45. $y' = \dfrac{-xy}{x^2 + y^2}$	$[2, 4]$	0.05	$y(2) = 1$
46. $y' = \dfrac{6x + y^2}{y(3y - 2x)}$	$[0, 5]$	0.2	$y(0) = 1$

47. ***Euler's Method*** Repeat Exercise 45 for $\Delta x = 1$ and discuss how the accuracy of the result changes.

48. ***Euler's Method*** Repeat Exercise 46 for $\Delta x = 0.5$ and discuss how the accuracy of the result changes.

True or False? **In Exercises 49–52, determine whether the statement is true or false. If it is false, explain why or give an example that shows it is false.**

49. The differential equation $2xy\,dx + (y^2 - x^2)\,dy = 0$ is exact.

50. If $M\,dx + N\,dy = 0$ is exact, then $xM\,dx + xN\,dy = 0$ is also exact.

51. If $M\,dx + N\,dy = 0$ is exact, then $[f(x) + M]\,dx + [g(y) + N]\,dy = 0$ is also exact.

52. The differential equation $f(x)\,dx + g(y)\,dy = 0$ is exact.

APPENDIX C.2 Second-Order Homogeneous Linear Equations

Second-Order Linear Differential Equations •
Higher-Order Linear Differential Equations • Application

Second-Order Linear Differential Equations

In this section and the following section, you will learn methods for solving higher-order linear differential equations.

Definition of Linear Differential Equation of Order n

Let $g_1, g_2, \ldots, g_n$ and f be functions of x with a common (interval) domain. An equation of the form

$$y^{(n)} + g_1(x)y^{(n-1)} + g_2(x)y^{(n-2)} + \cdots + g_{n-1}(x)y' + g_n(x)y = f(x)$$

is called a **linear differential equation of order n**. If $f(x) = 0$, the equation is **homogeneous**; otherwise, it is **nonhomogeneous**.

Homogeneous equations are discussed in this section, and the nonhomogeneous case is discussed in the next section.

The functions $y_1, y_2, \ldots, y_n$ are **linearly independent** if the *only* solution of the equation

$$C_1y_1 + C_2y_2 + \cdots + C_ny_n = 0$$

is the trivial one, $C_1 = C_2 = \cdots = C_n = 0$. Otherwise, this set of functions is **linearly dependent**.

EXAMPLE 1 Linearly Independent and Dependent Functions

a. The functions

$$y_1(x) = \sin x \quad \text{and} \quad y_2 = x$$

are linearly independent because the only values of C_1 and C_2 for which

$$C_1 \sin x + C_2x = 0$$

for all x are $C_1 = 0$ and $C_2 = 0$.

b. It can be shown that two functions form a linearly dependent set if and only if one is a constant multiple of the other. For example,

$$y_1(x) = x \quad \text{and} \quad y_2(x) = 3x$$

are linearly dependent because

$$C_1x + C_2(3x) = 0$$

has the nonzero solutions $C_1 = -3$ and $C_2 = 1$.

The following theorem points out the importance of linear independence in constructing the general solution of a second-order linear homogeneous differential equation with constant coefficients.

THEOREM C.3 Linear Combinations of Solutions

If y_1 and y_2 are linearly independent solutions of the differential equation $y'' + ay' + by = 0$, then the general solution is

$$y = C_1y_1 + C_2y_2$$

where C_1 and C_2 are constants.

Proof This theorem is proved in only one direction. If y_1 and y_2 are solutions, you obtain the following system of equations.

$$y_1''(x) + ay_1'(x) + by_1(x) = 0$$
$$y_2''(x) + ay_2'(x) + by_2(x) = 0$$

Multiplying the first equation by C_1, multiplying the second by C_2, and adding the resulting equations together produces

$$[C_1y_1''(x) + C_2y_2''(x)] + a[C_1y_1'(x) + C_2y_2'(x)] + b[C_1y_1(x) + C_2y_2(x)] = 0$$

which means that

$$y = C_1y_1 + C_2y_2$$

is a solution, as desired. The proof that all solutions are of this form is best left to a full course on differential equations.

Theorem C.3 states that if you can find two linearly independent solutions, you can obtain the general solution by forming a **linear combination** of the two solutions.

To find two linearly independent solutions, note that the nature of the equation $y'' + ay' + by = 0$ suggests that it may have solutions of the form $y = e^{mx}$. If so, then $y' = me^{mx}$ and $y'' = m^2e^{mx}$. So, by substitution, $y = e^{mx}$ is a solution if and only if

$$\begin{aligned} y'' + ay' + by &= 0 \\ m^2e^{mx} + ame^{mx} + be^{mx} &= 0 \\ e^{mx}(m^2 + am + b) &= 0. \end{aligned}$$

Because e^{mx} is never 0, $y = e^{mx}$ is a solution if and only if

$$m^2 + am + b = 0.$$ Characteristic equation

This is the **characteristic equation** of the differential equation

$$y'' + ay' + by = 0.$$

Note that the characteristic equation can be determined from its differential equation simply by replacing y'' with m^2, y' with m, and y with 1.

EXAMPLE 2 Characteristic Equation with Distinct Real Zeros

Solve the differential equation

$$y'' - 4y = 0.$$

Solution In this case, the characteristic equation is

$$m^2 - 4 = 0 \qquad \text{Characteristic equation}$$

so, $m = \pm 2$. Therefore, $y_1 = e^{m_1x} = e^{2x}$ and $y_2 = e^{m_2x} = e^{-2x}$ are particular solutions of the given differential equation. Furthermore, because these two solutions are linearly independent, you can apply Theorem C.3 to conclude that the general solution is

$$y = C_1e^{2x} + C_2e^{-2x}. \qquad \text{General solution}$$

The characteristic equation in Example 2 has two distinct real zeros. From algebra, you know that this is only one of *three* possibilities for quadratic equations. In general, the quadratic equation $m^2 + am + b = 0$ has zeros

$$m_1 = \frac{-a + \sqrt{a^2 - 4b}}{2} \quad \text{and} \quad m_2 = \frac{-a - \sqrt{a^2 - 4b}}{2}$$

which fall into one of three cases.

1. Two distinct real zeros, $m_1 \neq m_2$
2. Two equal real zeros, $m_1 = m_2$
3. Two complex conjugate zeros, $m_1 = \alpha + \beta i$ and $m_2 = \alpha - \beta i$

In terms of the differential equation $y'' + ay' + by = 0$, these three cases correspond to three different types of general solutions.

THEOREM C.4 Solutions of $y'' + ay' + by = 0$

The solutions of

$$y'' + ay' + by = 0$$

fall into one of the following three cases, depending on the solutions of the characteristic equation, $m^2 + am + b = 0$.

1. *Distinct Real Zeros* If $m_1 \neq m_2$ are distinct real zeros of the characteristic equation, then the general solution is

$$y = C_1e^{m_1x} + C_2e^{m_2x}.$$

2. *Equal Real Zeros* If $m_1 = m_2$ are equal real zeros of the characteristic equation, then the general solution is

$$y = C_1e^{m_1x} + C_2xe^{m_1x} = (C_1 + C_2x)e^{m_1x}.$$

3. *Complex Zeros* If $m_1 = \alpha + \beta i$ and $m_2 = \alpha - \beta i$ are complex zeros of the characteristic equation, then the general solution is

$$y = C_1e^{\alpha x}\cos\beta x + C_2e^{\alpha x}\sin\beta x.$$

EXAMPLE 3 Characteristic Equation with Complex Zeros

Find the general solution of the differential equation

$$y'' + 6y' + 12y = 0.$$

Solution The characteristic equation

$$m^2 + 6m + 12 = 0$$

has two complex zeros, as follows.

$$\begin{aligned} m &= \frac{-6 \pm \sqrt{36 - 48}}{2} \\ &= \frac{-6 \pm \sqrt{-12}}{2} \\ &= \frac{-6 \pm 2\sqrt{-3}}{2} \\ &= -3 \pm \sqrt{-3} \\ &= -3 \pm \sqrt{3}i \end{aligned}$$

So, $\alpha = -3$ and $\beta = \sqrt{3}$, and the general solution is

$$y = C_1 e^{-3x} \cos \sqrt{3}x + C_2 e^{-3x} \sin \sqrt{3}x.$$

NOTE In Example 3, note that although the characteristic equation has two *complex* zeros, the solution of the differential equation is *real.*

EXAMPLE 4 Characteristic Equation with Repeated Zeros

Solve the differential equation

$$y'' + 4y' + 4y = 0$$

subject to the initial conditions $y(0) = 2$ and $y'(0) = 1$.

Solution The characteristic equation

$$m^2 + 4m + 4 = (m + 2)^2 = 0$$

has two equal zeros given by $m = -2$. So, the general solution is

$$y = C_1 e^{-2x} + C_2 x e^{-2x}. \qquad \text{General solution}$$

Now, because $y = 2$ when $x = 0$, you have

$$2 = C_1(1) + C_2(0)(1) = C_1.$$

Furthermore, because $y' = 1$ when $x = 0$, you have

$$\begin{aligned} y' &= -2C_1 e^{-2x} + C_2(-2xe^{-2x} + e^{-2x}) \\ 1 &= -2(2)(1) + C_2[-2(0)(1) + 1] \\ 5 &= C_2. \end{aligned}$$

Therefore, the solution is

$$y = 2e^{-2x} + 5xe^{-2x}. \qquad \text{Particular solution}$$

Try checking this solution in the original differential equation.

Higher-Order Linear Differential Equations

For higher-order homogeneous linear differential equations, you can find the general solution in much the same way as you do for second-order equations. That is, you begin by determining the n zeros of the characteristic equation. Then, based on these n zeros, you form a linearly independent collection of n solutions. The major difference is that with equations of third or higher order, zeros of the characteristic equation may occur more than twice. When this happens, the linearly independent solutions are formed by multiplying by increasing powers of x, as demonstrated in Examples 6 and 7.

EXAMPLE 5 Solving a Third-Order Equation

Find the general solution of $y''' - y' = 0$.

Solution The characteristic equation is

$$m^3 - m = 0$$
$$m(m-1)(m+1) = 0$$
$$m = 0, 1, -1.$$

Because the characteristic equation has three distinct zeros, the general solution is

$$y = C_1 + C_2 e^{-x} + C_3 e^{x}.$$ General solution

EXAMPLE 6 Solving a Third-Order Equation

Find the general solution of $y''' + 3y'' + 3y' + y = 0$.

Solution The characteristic equation is

$$m^3 + 3m^2 + 3m + 1 = 0$$
$$(m+1)^3 = 0$$
$$m = -1.$$

Because the zero $m = -1$ occurs three times, the general solution is

$$y = C_1 e^{-x} + C_2 x e^{-x} + C_3 x^2 e^{-x}.$$ General solution

EXAMPLE 7 Solving a Fourth-Order Equation

Find the general solution of $y^{(4)} + 2y'' + y = 0$.

Solution The characteristic equation is as follows.

$$m^4 + 2m^2 + 1 = 0$$
$$(m^2 + 1)^2 = 0$$
$$m = \pm i$$

Because each of the zeros $m_1 = \alpha + \beta i = 0 + i$ and $m_2 = \alpha - \beta i = 0 - i$ occurs twice, the general solution is

$$y = C_1 \cos x + C_2 \sin x + C_3 x \cos x + C_4 x \sin x.$$ General solution

Application

One of the many applications of linear differential equations is describing the motion of an oscillating spring. According to Hooke's Law, a spring that is stretched (or compressed) y units from its natural length l tends to *restore* itself to its natural length by a force F that is proportional to y. That is, $F(y) = -ky$, where k is the **spring constant** and indicates the stiffness of the given spring.

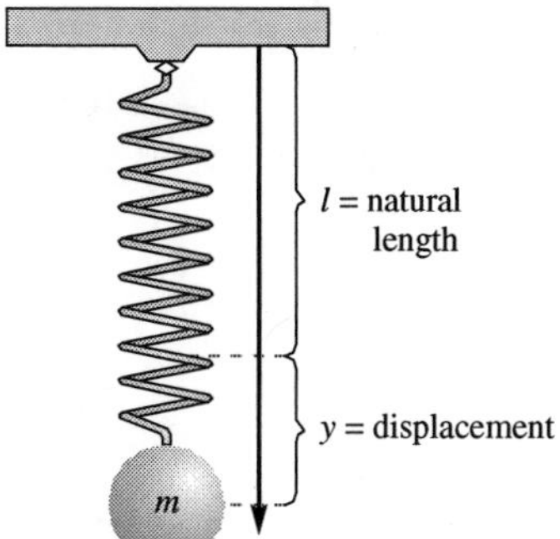

A rigid object of mass m attached to the end of the spring causes a displacement of y.
Figure C.4

Suppose a rigid object of mass m is attached to the end of a spring and causes a displacement, as shown in Figure C.4. Assume that the mass of the spring is negligible compared with m. If the object is pulled downward and released, the resulting oscillations are a product of two opposing forces—the spring force $F(y) = -ky$ and the weight mg of the object. Under such conditions, you can use a differential equation to find the position y of the object as a function of time t. According to Newton's Second Law of Motion, the force acting on the weight is $F = ma$, where $a = d^2y/dt^2$ is the acceleration. Assuming that the motion is **undamped**—that is, there are no other external forces acting on the object—it follows that $m(d^2y/dt^2) = -ky$, and you have

$$\frac{d^2y}{dt^2} + \left(\frac{k}{m}\right)y = 0.$$ Undamped motion of a spring

EXAMPLE 8 Undamped Motion of a Spring

A four-pound weight stretches a spring 8 inches from its natural length. The weight is pulled downward an additional 6 inches and released with an initial upward velocity of 8 feet per second. Find a formula for the position of the weight as a function of time t.

Solution By Hooke's Law, $4 = k\left(\frac{2}{3}\right)$, so $k = 6$. Moreover, because the weight w is given by mg, it follows that $m = w/g = \frac{4}{32} = \frac{1}{8}$. So, the resulting differential equation for this undamped motion is

$$\frac{d^2y}{dt^2} + 48y = 0.$$

Because the characteristic equation $m^2 + 48 = 0$ has complex zeros $m = 0 \pm 4\sqrt{3}i$, the general solution is

$$\begin{aligned} y &= C_1e^0 \cos 4\sqrt{3}\,t + C_2e^0 \sin 4\sqrt{3}\,t \\ &= C_1 \cos 4\sqrt{3}\,t + C_2 \sin 4\sqrt{3}\,t. \end{aligned}$$

Using the initial conditions, you have

$$\frac{1}{2} = C_1(1) + C_2(0) \quad \Rightarrow \quad C_1 = \frac{1}{2}$$ $y(0) = \frac{1}{2}$

$$y'(t) = -4\sqrt{3}\,C_1 \sin 4\sqrt{3}\,t + 4\sqrt{3}\,C_2 \cos 4\sqrt{3}\,t$$

$$8 = -4\sqrt{3}\left(\frac{1}{2}\right)(0) + 4\sqrt{3}\,C_2(1) \quad \Rightarrow \quad C_2 = \frac{2\sqrt{3}}{3}.$$ $y'(0) = 8$

Consequently, the position at time t is given by

$$y = \frac{1}{2}\cos 4\sqrt{3}\,t + \frac{2\sqrt{3}}{3}\sin 4\sqrt{3}\,t.$$

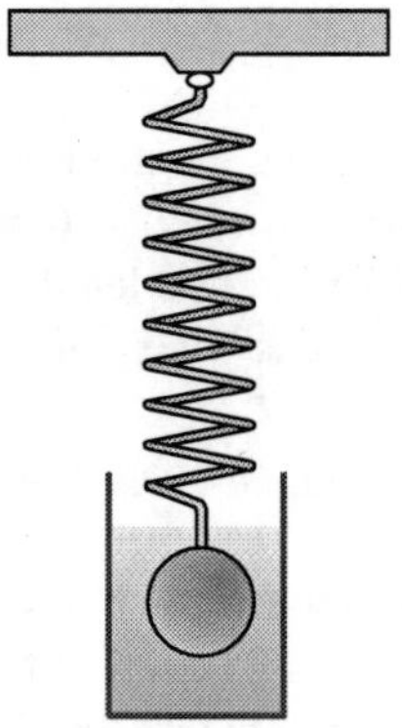

A damped vibration could be caused by friction and movement through a liquid.
Figure C.5

Suppose the object in Figure C.5 undergoes an additional damping or frictional force that is proportional to its velocity. A case in point would be the damping force resulting from friction and movement through a fluid. Considering this damping force, $-p(dy/dt)$, the differential equation for the oscillation is

$$m\frac{d^2y}{dt^2} = -ky - p\frac{dy}{dt}$$

or, in standard linear form,

$$\frac{d^2y}{dt^2} + \frac{p}{m}\left(\frac{dy}{dt}\right) + \frac{k}{m}y = 0.$$

Damped motion of a spring

EXERCISES FOR APPENDIX C.2

In Exercises 1–4, verify the solution of the differential equation.

Solution	*Differential Equation*
1. $y = (C_1 + C_2x)e^{-3x}$	$y'' + 6y' + 9y = 0$
2. $y = C_1e^{2x} + C_2e^{-2x}$	$y'' - 4y = 0$
3. $y = C_1 \cos 2x + C_2 \sin 2x$	$y'' + 4y = 0$
4. $y = e^{-x} \sin 3x$	$y'' + 2y' + 10y = 0$

In Exercises 5–30, find the general solution of the linear differential equation.

5. $y'' - y' = 0$

6. $y'' + 2y' = 0$

7. $y'' - y' - 6y = 0$

8. $y'' + 6y' + 5y = 0$

9. $2y'' + 3y' - 2y = 0$

10. $16y'' - 16y' + 3y = 0$

11. $y'' + 6y' + 9y = 0$

12. $y'' - 10y' + 25y = 0$

13. $16y'' - 8y' + y = 0$

14. $9y'' - 12y' + 4y = 0$

15. $y'' + y = 0$

16. $y'' + 4y = 0$

17. $y'' - 9y = 0$

18. $y'' - 2y = 0$

19. $y'' - 2y' + 4y = 0$

20. $y'' - 4y' + 21y = 0$

21. $y'' - 3y' + y = 0$

22. $3y'' + 4y' - y = 0$

23. $9y'' - 12y' + 11y = 0$

24. $2y'' - 6y' + 7y = 0$

25. $y^{(4)} - y = 0$

26. $y^{(4)} - y'' = 0$

27. $y''' - 6y'' + 11y' - 6y = 0$

28. $y''' - y'' - y' + y = 0$

29. $y''' - 3y'' + 7y' - 5y = 0$

30. $y''' - 3y'' + 3y' - y = 0$

31. Consider the differential equation $y'' + 100y = 0$ and the solution $y = C_1 \cos 10x + C_2 \sin 10x$. Find the particular solution satisfying each of the following initial conditions.

(a) $y(0) = 2, \quad y'(0) = 0$

(b) $y(0) = 0, \quad y'(0) = 2$

(c) $y(0) = -1, \quad y'(0) = 3$

32. Determine C and ω such that $y = C \sin\sqrt{3}\, t$ is a particular solution of the differential equation $y'' + \omega y = 0$, where $y'(0) = -5$.

In Exercises 33–36, find the particular solution of the linear differential equation.

33. $y'' - y' - 30y = 0$
$y(0) = 1, \quad y'(0) = -4$

34. $y'' + 2y' + 3y = 0$
$y(0) = 2, \quad y'(0) = 1$

35. $y'' + 16y = 0$
$y(0) = 0, \quad y'(0) = 2$

36. $y'' + 2y' + 3y = 0$
$y(0) = 2, \quad y'(0) = 1$

Think About It **In Exercises 37 and 38, give a geometric argument to explain why the graph cannot be a solution of the differential equation. It is not necessary to solve the differential equation.**

37. $y'' = y'$

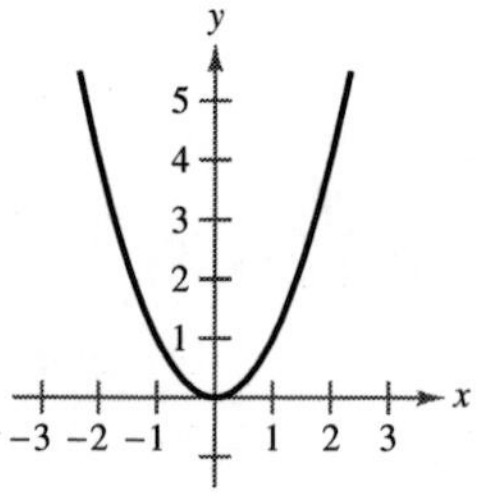

38. $y'' = -\frac{1}{2}y'$

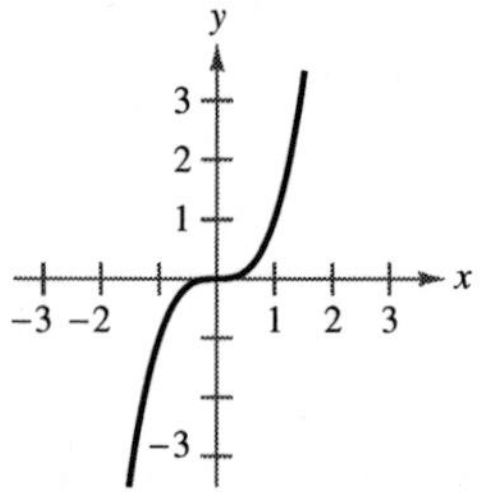

Vibrating Spring **In Exercises 39–44, describe the motion of a 32-pound weight suspended on a spring. Assume that the weight stretches the spring $\frac{2}{3}$ foot from its natural position.**

39. The weight is pulled $\frac{1}{2}$ foot below the equilibrium position and released.

40. The weight is raised $\frac{2}{3}$ foot above the equilibrium position and released.

41. The weight is raised $\frac{2}{3}$ foot above the equilibrium position and started off with a downward velocity of $\frac{1}{2}$ foot per second.

42. The weight is pulled $\frac{1}{2}$ foot below the equilibrium position and started off with an upward velocity of $\frac{1}{2}$ foot per second.

43. The weight is pulled $\frac{1}{2}$ foot below the equilibrium position and released. The motion takes place in a medium that furnishes a damping force of magnitude $\frac{1}{8}$ speed at all times.

44. The weight is pulled $\frac{1}{2}$ foot below the equilibrium position and released. The motion takes place in a medium that furnishes a damping force of magnitude $\frac{1}{4}|v|$ at all times.

Vibrating Spring **In Exercises 45–48, match the differential equation with the graph of a particular solution. [The graphs are labeled (a), (b), (c), and (d).] The correct match can be made by comparing the frequency of the oscillations or the rate at which the oscillations are being damped with the appropriate coefficient in the differential equation.**

(a)

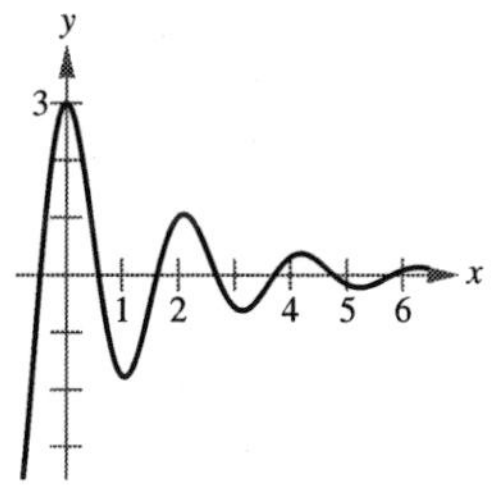

(b)

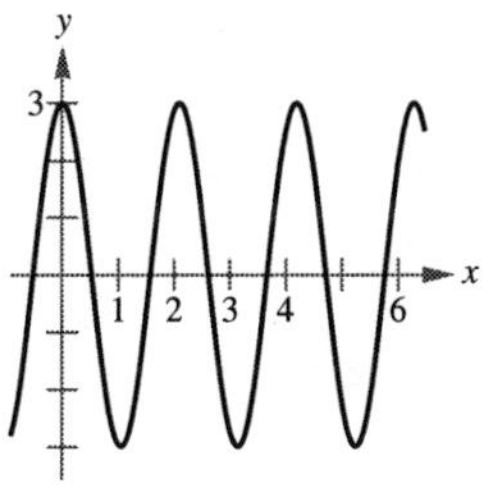

(c)

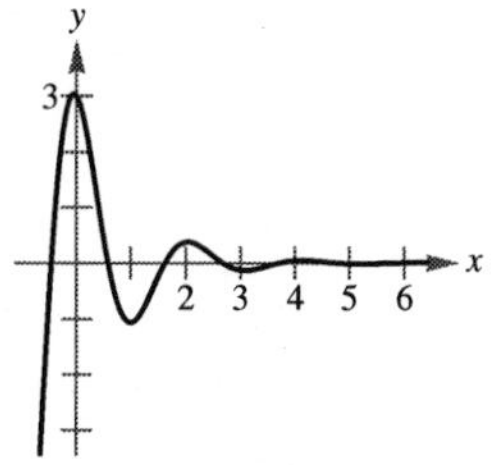

(d)

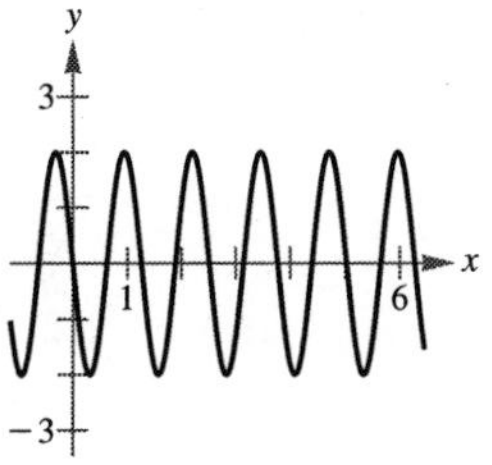

45. $y'' + 9y = 0$

46. $y'' + 25y = 0$

47. $y'' + 2y' + 10y = 0$

48. $y'' + y' + \frac{37}{4}y = 0$

49. If the characteristic equation of the differential equation

$$y'' + ay' + by = 0$$

has two equal real zeros given by $m = r$, show that

$$y = C_1e^{rx} + C_2xe^{rx}$$

is a solution.

50. If the characteristic equation of the differential equation

$$y'' + ay' + by = 0$$

has complex zeros given by $m_1 = \alpha + \beta i$ and $m_2 = \alpha - \beta i$, show that

$$y = C_1e^{\alpha x}\cos\beta x + C_2e^{\alpha x}\sin\beta x$$

is a solution.

True or False? **In Exercises 51–54, determine whether the statement is true or false. If it is false, explain why or give an example that shows it is false.**

51. $y = C_1e^{3x} + C_2e^{-3x}$ is the general solution of $y'' - 6y' + 9 = 0$.

52. $y = (C_1 + C_2x)\sin x + (C_3 + C_4x)\cos x$ is the general solution of $y^{(4)} + 2y'' + y = 0$.

53. $y = x$ is a solution of $a_ny^{(n)} + a_{n-1}y^{(n-1)} + \cdots + a_1y' + a_0y = 0$ if and only if $a_1 = a_0 = 0$.

54. It is possible to choose a and b such that $y = x^2e^x$ is a solution of $y'' + ay' + by = 0$.

The *Wronskian* of two differentiable functions f and g, denoted by $W(f, g)$, is defined as the function given by the determinant

$$W(f, g) = \begin{vmatrix} f & g \\ f' & g' \end{vmatrix}.$$

The functions f and g are linearly independent if there exists at least one value of x for which $W(f, g) \neq 0$. In Exercises 55–58, use the Wronskian to verify the linear independence of the two functions.

55. $y_1 = e^{ax}$
$y_2 = e^{bx},\ a \neq b$

56. $y_1 = e^{ax}$
$y_2 = xe^{ax}$

57. $y_1 = e^{ax}\sin bx$
$y_2 = e^{ax}\cos bx,\ b \neq 0$

58. $y_1 = x$
$y_2 = x^2$

59. Euler's differential equation is of the form

$$x^2y'' + axy' + by = 0, \qquad x > 0$$

where a and b are constants.

(a) Show that this equation can be transformed into a second-order linear equation with constant coefficients by using the substitution $x = e^t$.

(b) Solve $x^2y'' + 6xy' + 6y = 0$.

60. Solve

$$y'' + Ay = 0$$

where A is constant, subject to the conditions $y(0) = 0$ and $y(\pi) = 0$.

APPENDIX C.3 Second-Order Nonhomogeneous Linear Equations

Nonhomogeneous Equations • Method of Undetermined Coefficients • Variation of Parameters

Nonhomogeneous Equations

In the preceding section, damped oscillations of a spring were represented by the *homogeneous* second-order linear equation

$$\frac{d^2y}{dt^2} + \frac{p}{m}\left(\frac{dy}{dt}\right) + \frac{k}{m}y = 0. \qquad \text{Free motion}$$

This type of oscillation is called **free** because it is determined solely by the spring and gravity and is free of the action of other external forces. If such a system is also subject to an external periodic force such as $a \sin bt$, caused by vibrations at the opposite end of the spring, the motion is called **forced,** and it is characterized by the *nonhomogeneous* equation

$$\frac{d^2y}{dt^2} + \frac{p}{m}\left(\frac{dy}{dt}\right) + \frac{k}{m}y = a \sin bt. \qquad \text{Forced motion}$$

In this section, you will study two methods for finding the general solution of a nonhomogeneous linear differential equation. In both methods, the first step is to find the general solution of the corresponding homogeneous equation.

$$y = y_h \qquad \text{General solution of homogeneous equation}$$

Having done this, you try to find a particular solution of the nonhomogeneous equation.

$$y = y_p \qquad \text{Particular solution of nonhomogeneous equation}$$

By combining these two results, you can conclude that the general solution of the nonhomogeneous equation is $y = y_h + y_p$, as stated in the following theorem.

THEOREM C.5 Solution of Nonhomogeneous Linear Equation

Let

$$y'' + ay' + by = F(x)$$

be a second-order nonhomogeneous linear differential equation. If y_p is a particular solution of this equation and y_h is the general solution of the corresponding homogeneous equation, then

$$y = y_h + y_p$$

is the general solution of the nonhomogeneous equation.

Method of Undetermined Coefficients

You already know how to find the solution y_h of a linear *homogeneous* differential equation. The remainder of this section looks at ways to find the particular solution y_p. If $F(x)$ in

$$y'' + ay' + by = F(x)$$

consists of sums or products of x^n, e^{mx}, $\cos \beta x$, or $\sin \beta x$, you can find a particular solution y_p by the method of **undetermined coefficients.** The object of this method is to guess that the solution y_p is a generalized form of $F(x)$. Here are some examples.

1. If $F(x) = 3x^2$, choose $y_p = Ax^2 + Bx + C$.

2. If $F(x) = 4xe^x$, choose $y_p = Axe^x + Be^x$.

3. If $F(x) = x + \sin 2x$, choose $y_p = (Ax + B) + C \sin 2x + D \cos 2x$.

Then, by substitution, determine the coefficients for the generalized solution.

EXAMPLE 1 Method of Undetermined Coefficients

Find the general solution of the equation

$$y'' - 2y' - 3y = 2 \sin x.$$

Solution To find y_h, solve the characteristic equation.

$$\begin{aligned} m^2 - 2m - 3 &= 0 \\ (m + 1)(m - 3) &= 0 \\ m = -1 \quad \text{or} \quad m &= 3 \end{aligned}$$

So, $y_h = C_1e^{-x} + C_2e^{3x}$. Next, let y_p be a generalized form of $2 \sin x$.

$$\begin{aligned} y_p &= A \cos x + B \sin x \\ y_p' &= -A \sin x + B \cos x \\ y_p'' &= -A \cos x - B \sin x \end{aligned}$$

Substitution into the original differential equation yields

$$\begin{aligned} y'' - 2y' - 3y &= 2 \sin x \\ -A \cos x - B \sin x + 2A \sin x - 2B \cos x - 3A \cos x - 3B \sin x &= 2 \sin x \\ (-4A - 2B)\cos x + (2A - 4B)\sin x &= 2 \sin x. \end{aligned}$$

By equating coefficients of like terms, you obtain

$$-4A - 2B = 0 \quad \text{and} \quad 2A - 4B = 2$$

with solutions $A = \frac{1}{5}$ and $B = -\frac{2}{5}$. Therefore,

$$y_p = \frac{1}{5} \cos x - \frac{2}{5} \sin x$$

and the general solution is

$$\begin{aligned} y &= y_h + y_p \\ &= C_1e^{-x} + C_2e^{3x} + \frac{1}{5} \cos x - \frac{2}{5} \sin x. \end{aligned}$$

In Example 1, the form of the homogeneous solution

$$y_h = C_1e^{-x} + C_2e^{3x}$$

has no overlap with the function $F(x)$ in the equation

$$y'' + ay' + by = F(x)$$

However, suppose the given differential equation in Example 1 were of the form

$$y'' - 2y' - 3y = e^{-x}.$$

Now, it would make no sense to guess that the particular solution was $y = Ae^{-x}$, because you know that this solution would yield 0. In such cases, you should alter your guess by multiplying by the lowest power of x that removes the duplication. For this particular problem, you would guess

$$y_p = Axe^{-x}.$$

EXAMPLE 2 Method of Undetermined Coefficients

Find the general solution of

$$y'' - 2y' = x + 2e^x.$$

Solution The characteristic equation $m^2 - 2m = 0$ has solutions $m = 0$ and $m = 2$. So,

$$y_h = C_1 + C_2e^{2x}.$$

Because $F(x) = x + 2e^x$, your first choice for y_p would be $(A + Bx) + Ce^x$. However, because y_h *already* contains a constant term C_1, you should multiply the *polynomial part* by x and use

$$\begin{aligned} y_p &= Ax + Bx^2 + Ce^x \\ y_p' &= A + 2Bx + Ce^x \\ y_p'' &= 2B + Ce^x. \end{aligned}$$

Substitution into the differential equation produces

$$\begin{aligned} y'' - 2y' &= x + 2e^x \\ 2B + Ce^x - 2(A + 2Bx + Ce^x) &= x + 2e^x \\ (2B - 2A) - 4Bx - Ce^x &= x + 2e^x. \end{aligned}$$

Equating coefficients of like terms yields the system

$$2B - 2A = 0, \qquad -4B = 1, \qquad -C = 2$$

with solutions $A = B = -\frac{1}{4}$ and $C = -2$. Therefore,

$$y_p = -\frac{1}{4}x - \frac{1}{4}x^2 - 2e^x$$

and the general solution is

$$\begin{aligned} y &= y_h + y_p \\ &= C_1 + C_2e^{2x} - \frac{1}{4}x - \frac{1}{4}x^2 - 2e^x. \end{aligned}$$

In Example 2, the polynomial part of the initial guess

$$(A + Bx) + Ce^x$$

for y_p overlapped by a constant term with $y_h = C_1 + C_2e^{2x}$, and it was necessary to multiply the polynomial part by a power of x that removed the overlap. The next example further illustrates some choices for y_p that eliminate overlap with y_h. Remember that in all cases the first guess for y_p should match the types of functions occurring in $F(x)$.

EXAMPLE 3 Choosing the Form of the Particular Solution

Determine a suitable choice for y_p for each differential equation, given its general solution of the homogeneous equation.

	$y'' + ay' + by = F(x)$	y_h
a.	$y'' = x^2$	$C_1 + C_2x$
b.	$y'' + 2y' + 10y = 4 \sin 3x$	$C_1e^{-x} \cos 3x + C_2e^{-x} \sin 3x$
c.	$y'' - 4y' + 4 = e^{2x}$	$C_1e^{2x} + C_2xe^{2x}$

Solution

a. Because $F(x) = x^2$, the normal choice for y_p would be $A + Bx + Cx^2$. However, because $y_h = C_1 + C_2x$ already contains a linear term, you should multiply by x^2 to obtain

$$y_p = Ax^2 + Bx^3 + Cx^4.$$

b. Because $F(x) = 4 \sin 3x$ and each term in y_h contain a factor of e^{-x}, you can simply let

$$y_p = A \cos 3x + B \sin 3x.$$

c. Because $F(x) = e^{2x}$, the normal choice for y_p would be Ae^{2x}. However, because $y_h = C_1e^{2x} + C_2xe^{2x}$ already contains an xe^{2x} term, you should multiply by x^2 to get

$$y_p = Ax^2e^{2x}.$$

EXAMPLE 4 Solving a Third-Order Equation

Find the general solution of

$$y''' + 3y'' + 3y' + y = x.$$

Solution From Example 6 in the preceding section, you know that the homogeneous solution is

$$y_h = C_1e^{-x} + C_2xe^{-x} + C_3x^2e^{-x}.$$

Because $F(x) = x$, let $y_p = A + Bx$ and obtain $y_p' = B$ and $y_p'' = 0$. So, by substitution, you have

$$0 + 3(0) + 3(B) + A + Bx = (3B + A) + Bx = x.$$

So, $B = 1$ and $A = -3$, which implies that $y_p = -3 + x$. Therefore, the general solution is

$$\begin{aligned} y &= y_h + y_p \\ &= C_1e^{-x} + C_2xe^{-x} + C_3x^2e^{-x} - 3 + x. \end{aligned}$$

Variation of Parameters

The method of undetermined coefficients works well if $F(x)$ is made up of polynomials or functions whose successive derivatives have a cyclical pattern. For functions such as $1/x$ and $\tan x$, which do not have such characteristics, it is better to use a more general method called **variation of parameters.** In this method, you assume that y_p has the same *form* as y_h, except that the constants in y_h are replaced by variables.

Variation of Parameters

To find the general solution to the equation $y'' + ay' + by = F(x)$, use the following steps.

1. Find $y_h = C_1y_1 + C_2y_2$.
2. Replace the constants by variables to form $y_p = u_1y_1 + u_2y_2$.
3. Solve the following system for u_1' and u_2'.

$$u_1'y_1 + u_2'y_2 = 0$$
$$u_1'y_1' + u_2'y_2' = F(x)$$

4. Integrate to find u_1 and u_2. The general solution is $y = y_h + y_p$.

EXAMPLE 5 Variation of Parameters

Solve the differential equation

$$y'' - 2y' + y = \frac{e^x}{2x}, \quad x > 0.$$

Solution The characteristic equation $m^2 - 2m + 1 = (m - 1)^2 = 0$ has one repeated solution, $m = 1$. So, the homogeneous solution is

$$y_h = C_1y_1 + C_2y_2 = C_1e^x + C_2xe^x.$$

Replacing C_1 and C_2 by u_1 and u_2 produces

$$y_p = u_1y_1 + u_2y_2 = u_1e^x + u_2xe^x.$$

The resulting system of equations is

$$u_1'e^x + u_2'xe^x = 0$$
$$u_1'e^x + u_2'(xe^x + e^x) = \frac{e^x}{2x}.$$

Subtracting the second equation from the first produces $u_2' = 1/(2x)$. Then, by substitution in the first equation, you have $u_1' = -\frac{1}{2}$. Finally, integration yields

$$u_1 = -\int \frac{1}{2}\,dx = -\frac{x}{2} \quad \text{and} \quad u_2 = \frac{1}{2}\int \frac{1}{x}\,dx = \frac{1}{2}\ln x = \ln\sqrt{x}.$$

From this result it follows that a particular solution is

$$y_p = -\frac{1}{2}xe^x + \left(\ln\sqrt{x}\right)xe^x$$

and the general solution is

$$y = C_1e^x + C_2xe^x - \frac{1}{2}xe^x + xe^x\ln\sqrt{x}.$$

EXAMPLE 6 Variation of Parameters

Solve the differential equation

$$y'' + y = \tan x.$$

Solution Because the characteristic equation $m^2 + 1 = 0$ has solutions $m = \pm i$, the homogeneous solution is

$$y_h = C_1 \cos x + C_2 \sin x.$$

Replacing C_1 and C_2 by u_1 and u_2 produces

$$y_p = u_1 \cos x + u_2 \sin x.$$

The resulting system of equations is

$$\begin{aligned} u_1' \cos x + u_2' \sin x &= 0 \\ -u_1' \sin x + u_2' \cos x &= \tan x. \end{aligned}$$

Multiplying the first equation by $\sin x$ and the second by $\cos x$ produces

$$\begin{aligned} u_1' \sin x \cos x + u_2' \sin^2 x &= 0 \\ -u_1' \sin x \cos x + u_2' \cos^2 x &= \sin x. \end{aligned}$$

Adding these two equations produces $u_2' = \sin x$, which implies that

$$\begin{aligned} u_1' &= -\frac{\sin^2 x}{\cos x} \\ &= \frac{\cos^2 x - 1}{\cos x} \\ &= \cos x - \sec x. \end{aligned}$$

Integration yields

$$\begin{aligned} u_1 &= \int (\cos x - \sec x)\, dx \\ &= \sin x - \ln|\sec x + \tan x| \end{aligned}$$

and

$$\begin{aligned} u_2 &= \int \sin x\, dx \\ &= -\cos x \end{aligned}$$

so that

$$\begin{aligned} y_p &= \sin x \cos x - \cos x \ln|\sec x + \tan x| - \sin x \cos x \\ &= -\cos x \ln|\sec x + \tan x| \end{aligned}$$

and the general solution is

$$\begin{aligned} y &= y_h + y_p \\ &= C_1 \cos x + C_2 \sin x - \cos x \ln|\sec x + \tan x|. \end{aligned}$$

EXERCISES FOR APPENDIX C.3

In Exercises 1–4, verify the solution of the differential equation.

Solution	*Differential Equation*
1. $y = 2(e^{2x} - \cos x)$	$y'' + y = 10e^{2x}$
2. $y = \left(2 + \frac{1}{2}x\right)\sin x$	$y'' + y = \cos x$
3. $y = 3\sin x - \cos x \ln\lvert\sec x + \tan x\rvert$	$y'' + y = \tan x$
4. $y = \left(5 - \ln\lvert\sin x\rvert\right)\cos x - x\sin x$	$y'' + y = \csc x \cot x$

In Exercises 5–20, solve the differential equation by the method of undetermined coefficients.

5. $y'' - 3y' + 2y = 2x$

6. $y'' - 2y' - 3y = x^2 - 1$

7. $y'' + y = x^3$
$y(0) = 1, y'(0) = 0$

8. $y'' + 4y = 4$
$y(0) = 1, y'(0) = 6$

9. $y'' + 2y' = 2e^x$

10. $y'' - 9y = 5e^{3x}$

11. $y'' - 10y' + 25y = 5 + 6e^x$

12. $16y'' - 8y' + y = 4(x + e^x)$

13. $y'' + y' = 2\sin x$
$y(0) = 0, y'(0) = -3$

14. $y'' + y' - 2y = 3\cos 2x$
$y(0) = -1, y'(0) = 2$

15. $y'' + 9y = \sin 3x$

16. $y'' + 4y' + 5y = \sin x + \cos x$

17. $y''' - 3y' + 2y = 2e^{-2x}$

18. $y''' - y'' = 4x^2$
$y(0) = 1, y'(0) = 1, y''(0) = 1$

19. $y' - 4y = xe^x - xe^{4x}$
$y(0) = \frac{1}{3}$

20. $y' + 2y = \sin x$
$y\left(\frac{\pi}{2}\right) = \frac{2}{5}$

21. *Think About It*

(a) Explain how, by observation, you know that a particular solution of the differential equation $y'' + 3y = 12$ is $y_p = 4$.

(b) Use your explanation in part (a) to give a particular solution of the differential equation $y'' + 5y = 10$.

(c) Use your explanation in part (a) to give a particular solution of the differential equation $y'' + 2y' + 2y = 8$.

22. *Think About It*

(a) Explain how, by observation, you know that a form of a particular solution of the differential equation $y'' + 3y = 12\sin x$ is $y_p = A\sin x$.

(b) Use your explanation in part (a) to find a particular solution of the differential equation $y'' + 5y = 10\cos x$.

(c) Compare the algebra required to find particular solutions in parts (a) and (b) with that required if the form of the particular solution were $y_p = A\cos x + B\sin x$.

In Exercises 23–28, solve the differential equation by the method of variation of parameters.

23. $y'' + y = \sec x$

24. $y'' + y = \sec x \tan x$

25. $y'' + 4y = \csc 2x$

26. $y'' - 4y' + 4y = x^2 e^{2x}$

27. $y'' - 2y' + y = e^x \ln x$

28. $y'' - 4y' + 4y = \dfrac{e^{2x}}{x}$

Electrical Circuits **In Exercises 29 and 30, use the electrical circuit differential equation**

$$\frac{d^2q}{dt^2} + \left(\frac{R}{L}\right)\frac{dq}{dt} + \left(\frac{1}{LC}\right)q = \left(\frac{1}{L}\right)E(t)$$

where R is the resistance (in ohms), C is the capacitance (in farads), L is the inductance (in henrys), $E(t)$ is the electromotive force (in volts), and q is the charge on the capacitor (in coulombs). Find the charge q as a function of time for the electrical circuit described. Assume that $q(0) = 0$ and $q'(0) = 0$.

29. $R = 20, C = 0.02, L = 2$
$E(t) = 12\sin 5t$

30. $R = 20, C = 0.02, L = 1$
$E(t) = 10\sin 5t$

Vibrating Spring **In Exercises 31–34, find the particular solution of the differential equation**

$$\frac{w}{g}y''(t) + by'(t) + ky(t) = \frac{w}{g}F(t)$$

for the oscillating motion of an object on the end of a spring. Use a graphing utility to graph the solution. In the equation, y is the displacement from equilibrium (positive direction is downward) measured in feet, and t is time in seconds (see figure). The constant w is the weight of the object, g is the acceleration due to gravity, b is the magnitude of the resistance to the motion, k is the spring constant from Hooke's Law, and $F(t)$ is the acceleration imposed on the system.

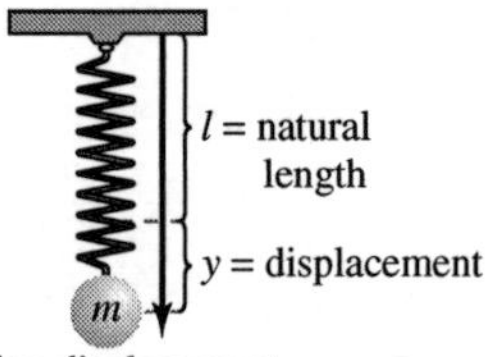

Spring displacement

31. $\frac{24}{32}y'' + 48y = \frac{24}{32}(48\sin 4t)$
$y(0) = \frac{1}{4}, y'(0) = 0$

32. $\frac{2}{32}y'' + 4y = \frac{2}{32}(4\sin 8t)$
$y(0) = \frac{1}{4}, y'(0) = 0$

33. $\frac{2}{32}y'' + y' + 4y = \frac{2}{32}(4\sin 8t)$
$y(0) = \frac{1}{4}, y'(0) = -3$

34. $\frac{4}{32}y'' + \frac{1}{2}y' + \frac{25}{2}y = 0$
$y(0) = \frac{1}{2}, y'(0) = -4$

35. ***Vibrating Spring*** Rewrite y_h in the solution for Exercise 31 by using the identity

$$a \cos \omega t + b \sin \omega t = \sqrt{a^2 + b^2} \sin(\omega t + \phi)$$

where $\phi = \arctan a/b$.

36. ***Vibrating Spring*** The figure shows the particular solution of the differential equation

$$\frac{4}{32} y'' + by' + \frac{25}{2} y = 0$$

$$y(0) = \frac{1}{2}, \; y'(0) = -4$$

for values of the resistance component b in the interval $[0, 1]$. (Note that when $b = \frac{1}{2}$, the problem is identical to that of Exercise 34.)

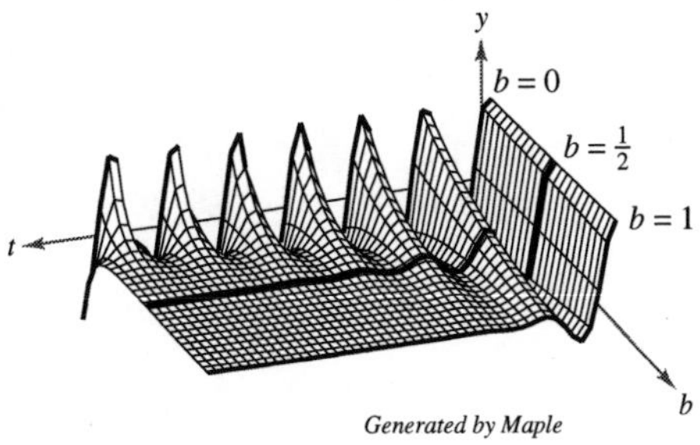

Generated by Maple

(a) If there is no resistance to the motion ($b = 0$), describe the motion.

(b) If $b > 0$, what is the ultimate effect of the retarding force?

(c) Is there a real number M such that there will be no oscillations of the spring if $b > M$? Explain your answer.

37. ***Parachute Jump*** The fall of a parachutist is described by the second-order linear differential equation

$$\frac{w}{g}\frac{d^2y}{dt^2} - k\frac{dy}{dt} = w$$

where w is the weight of the parachutist, y is the height at time t, g is the acceleration due to gravity, and k is the drag factor of the parachute. If the parachute is opened at 2000 feet, $y(0) = 2000$, and at that time the velocity is $y'(0) = -100$ feet per second, then for a 160-pound parachutist, using $k = 8$, the differential equation is

$$-5y'' - 8y' = 160.$$

Using the given initial conditions, verify that the solution of the differential equation is

$$y = 1950 + 50e^{-1.6t} - 20t.$$

38. ***Parachute Jump*** Repeat Exercise 37 for a parachutist who weighs 192 pounds and has a parachute with a drag factor of $k = 9$.

39. Solve the differential equation

$$x^2y'' - xy' + y = 4x \ln x$$

given that $y_1 = x$ and $y_2 = x \ln x$ are solutions of the corresponding homogeneous equation.

40. ***True or False?*** $y_p = -e^{2x} \cos e^{-x}$ is a particular solution of the differential equation

$$y'' - 3y' + 2y = \cos e^{-x}.$$

APPENDIX C.4 Series Solutions of Differential Equations

Power Series Solution of a Differential Equation • Approximation by Taylor Series

Power Series Solution of a Differential Equation

Power series can be used to solve certain types of differential equations. This section begins with the general **power series solution** method.

A power series represents a function f on an interval of convergence, and you can successively differentiate the power series to obtain a series for f', f'', and so on. These properties are used in the power series solution method demonstrated in the first two examples.

EXAMPLE 1 Power Series Solution

Use a power series to solve the differential equation $y' - 2y = 0$.

Solution Assume that $y = \Sigma a_n x^n$ is a solution. Then, $y' = \Sigma n a_n x^{n-1}$. Substituting for y' and $-2y$, you obtain the following series form of the differential equation. (Note that, from the third step to the fourth, the index of summation is changed to ensure that x^n occurs in both sums.)

$$y' - 2y = 0$$

$$\sum_{n=1}^{\infty} n a_n x^{n-1} - 2 \sum_{n=0}^{\infty} a_n x^n = 0$$

$$\sum_{n=1}^{\infty} n a_n x^{n-1} = \sum_{n=0}^{\infty} 2a_n x^n$$

$$\sum_{n=0}^{\infty} (n+1) a_{n+1} x^n = \sum_{n=0}^{\infty} 2a_n x^n$$

Now, by equating coefficients of like terms, you obtain the **recursion formula** $(n + 1)a_{n+1} = 2a_n$, which implies that

$$a_{n+1} = \frac{2a_n}{n+1}, \qquad n \ge 0.$$

This formula generates the following results.

a_0	a_1	a_2	a_3	a_4	a_5	. . .
a_0	$2a_0$	$\frac{2^2 a_0}{2}$	$\frac{2^3 a_0}{3!}$	$\frac{2^4 a_0}{4!}$	$\frac{2^5 a_0}{5!}$	. . .

Using these values as the coefficients for the *solution* series, you have

$$\begin{aligned} y &= \sum_{n=0}^{\infty} \frac{2^n a_0}{n!} x^n \\ &= a_0 \sum_{n=0}^{\infty} \frac{(2x)^n}{n!} \\ &= a_0 e^{2x}. \end{aligned}$$

In Example 1, the differential equation could be solved easily without using a series. The differential equation in Example 2 cannot be solved by any of the methods discussed in previous sections.

EXAMPLE 2 Power Series Solution

Use a power series to solve the differential equation $y'' + xy' + y = 0$.

Solution Assume that $\sum_{n=0}^{\infty} a_n x^n$ is a solution. Then you have

$$y' = \sum_{n=1}^{\infty} na_n x^{n-1}, \qquad xy' = \sum_{n=1}^{\infty} na_n x^n, \qquad y'' = \sum_{n=2}^{\infty} n(n-1)a_n x^{n-2}.$$

Substituting for y'', xy', and y in the given differential equation, you obtain the following series.

$$\sum_{n=2}^{\infty} n(n-1)a_n x^{n-2} + \sum_{n=0}^{\infty} na_n x^n + \sum_{n=0}^{\infty} a_n x^n = 0$$

$$\sum_{n=2}^{\infty} n(n-1)a_n x^{n-2} = -\sum_{n=0}^{\infty} (n+1)a_n x^n$$

To obtain equal powers of x, adjust the summation indices by replacing n by $n + 2$ in the left-hand sum, to obtain

$$\sum_{n=0}^{\infty} (n+2)(n+1)a_{n+2} x^n = -\sum_{n=0}^{\infty} (n+1)a_n x^n.$$

By equating coefficients, you have $(n+2)(n+1)a_{n+2} = -(n+1)a_n$, from which you obtain the recursion formula

$$a_{n+2} = -\frac{(n+1)}{(n+2)(n+1)} a_n = -\frac{a_n}{n+2}, \qquad n \geq 0,$$

and the coefficients of the solution series are as follows.

$$a_2 = -\frac{a_0}{2} \qquad\qquad a_3 = -\frac{a_1}{3}$$

$$a_4 = -\frac{a_2}{4} = \frac{a_0}{2 \cdot 4} \qquad\qquad a_5 = -\frac{a_3}{5} = \frac{a_1}{3 \cdot 5}$$

$$a_6 = -\frac{a_4}{6} = -\frac{a_0}{2 \cdot 4 \cdot 6} \qquad\qquad a_7 = -\frac{a_5}{7} = -\frac{a_1}{3 \cdot 5 \cdot 7}$$

$$\vdots \qquad\qquad \vdots$$

$$a_{2k} = \frac{(-1)^k a_0}{2 \cdot 4 \cdot 6 \cdots (2k)} = \frac{(-1)^k a_0}{2^k (k!)} \qquad\qquad a_{2k+1} = \frac{(-1)^k a_1}{3 \cdot 5 \cdot 7 \cdots (2k+1)}$$

So, you can represent the general solution as the sum of two series—one for the even-powered terms with coefficients in terms of a_0 and one for the odd-powered terms with coefficients in terms of a_1.

$$y = a_0\left(1 - \frac{x^2}{2} + \frac{x^4}{2 \cdot 4} - \cdots\right) + a_1\left(x - \frac{x^3}{3} + \frac{x^5}{3 \cdot 5} - \cdots\right)$$

$$= a_0 \sum_{k=0}^{\infty} \frac{(-1)^k x^{2k}}{2^k (k!)} + a_1 \sum_{k=0}^{\infty} \frac{(-1)^k x^{2k+1}}{3 \cdot 5 \cdot 7 \cdots (2k+1)}$$

The solution has two arbitrary constants, a_0 and a_1, as you would expect in the general solution of a second-order differential equation.

Approximation by Taylor Series

A second type of series solution method involves a differential equation *with initial conditions* and makes use of Taylor series.

EXAMPLE 3 Approximation by Taylor Series

Use a Taylor series to find the series solution of

$$y' = y^2 - x$$

given the initial condition $y = 1$ when $x = 0$. Then, use the first six terms of this series solution to approximate values of y for $0 \le x \le 1$.

Solution Recall that, for $c = 0$,

$$y = y(0) + y'(0)x + \frac{y''(0)}{2!}x^2 + \frac{y'''(0)}{3!}x^3 + \cdots.$$

Because $y(0) = 1$ and $y' = y^2 - x$, you obtain the following.

$$\begin{aligned} & & y(0) &= 1 \\ y' &= y^2 - x & y'(0) &= 1 \\ y'' &= 2yy' - 1 & y''(0) &= 2 - 1 = 1 \\ y''' &= 2yy'' + 2(y')^2 & y'''(0) &= 2 + 2 = 4 \\ y^{(4)} &= 2yy''' + 6y'y'' & y^{(4)}(0) &= 8 + 6 = 14 \\ y^{(5)} &= 2yy^{(4)} + 8y'y''' + 6(y'')^2 & y^{(5)}(0) &= 28 + 32 + 6 = 66 \end{aligned}$$

Therefore, you can approximate the values of the solution from the series

$$\begin{aligned} y &= y(0) + y'(0)x + \frac{y''(0)}{2!}x^2 + \frac{y'''(0)}{3!}x^3 + \frac{y^{(4)}(0)}{4!}x^4 + \frac{y^{(5)}(0)}{5!}x^5 + \cdots \\ &= 1 + x + \frac{1}{2}x^2 + \frac{4}{3!}x^3 + \frac{14}{4!}x^4 + \frac{66}{5!}x^5 + \cdots. \end{aligned}$$

Using the first six terms of this series, you can compute values for y in the interval $0 \le x \le 1$, as shown in the below.

x	0.0	0.1	0.2	0.3	0.4	0.5	0.6	0.7	0.8	0.9	1.0
y	1.0000	1.1057	1.2264	1.3691	1.5432	1.7620	2.0424	2.4062	2.8805	3.4985	4.3000

EXERCISES FOR APPENDIX C.4

In Exercises 1–6, verify that the power series solution of the differential equation is equivalent to the solution found using perviously learned solution techniques.

1. $y' - y = 0$
2. $y' - ky = 0$
3. $y'' - 9y = 0$
4. $y'' - k^2y = 0$
5. $y'' + 4y = 0$
6. $y'' + k^2y = 0$

In Exercises 7–10, use power series to solve the differential equation and find the interval of convergence of the series.

7. $y' + 3xy = 0$
8. $y' - 2xy = 0$
9. $y'' - xy' = 0$
10. $y'' - xy' - y = 0$

In Exercises 11 and 12, find the first three terms of each of the power series representing independent solutions of the differential equation.

11. $(x^2 + 4)y'' + y = 0$
12. $y'' + x^2y = 0$

In Exercises 13 and 14, use Taylor's Theorem to find the series solution of the differential equation under the specified initial conditions. Use *n* terms of the series to approximate *y* for the given value of *x* and compare the result with the approximation given by Euler's Method for $\Delta x = 0.1$.

13. $y' + (2x - 1)y = 0, y(0) = 2, n = 5, x = \frac{1}{2}$
14. $y' - 2xy = 0, y(0) = 1, n = 4, x = 1$

15. ***Investigation*** Consider the differential equation

$$y'' + 9y = 0$$

with initial conditions $y(0) = 2$ and $y'(0) = 6$.

(a) Find the solution of the differential equation using the techniques of Section C.2.

(b) Find the series solution of the differential equation.

(c) The figure shows the graph of the solution of the differential equation and the third-degree and fifth-degree polynomial approximations of the solution. Identify each.

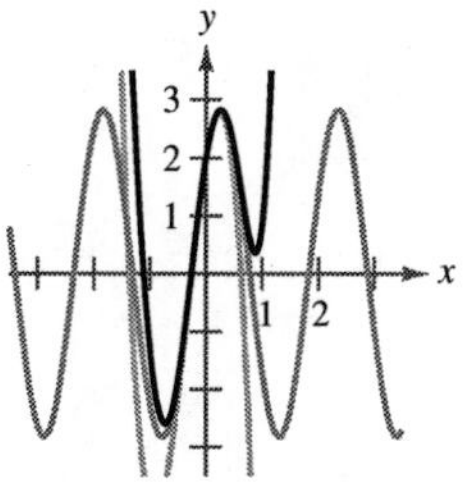

16. Consider the differential equation $y'' - xy' = 0$ with the initial conditions $y(0) = 0$ and $y'(0) = 2$. (See Exercise 9.)

(a) Find the series solution satisfying the initial conditions.

(b) Use a graphing utility to graph the third-degree and fifth-degree series approximations of the solution. Identify the approximations.

(c) Identify the symmetry of the solution.

In Exercises 17 and 18, use Taylor's Theorem to find the series solution of the differential equation under the specified initial conditions. Use *n* terms of the series to approximate *y* for the given value of *x*.

17. $y'' - 2xy = 0, y(0) = 1, y'(0) = -3, n = 6, x = \frac{1}{4}$
18. $y'' - 2xy' + y = 0, y(0) = 1, y'(0) = 2, n = 8, x = \frac{1}{2}$

In Exercises 19–22, verify that the series converges to the given function on the indicated interval. (*Hint:* Use the given differential equation.)

19. $\displaystyle\sum_{n=0}^{\infty} \frac{x^n}{n!} = e^x, (-\infty, \infty)$

Differential equation: $y' - y = 0$

20. $\displaystyle\sum_{n=0}^{\infty} \frac{(-1)^n x^{2n}}{(2n)!} = \cos x, (-\infty, \infty)$

Differential equation: $y'' + y = 0$

21. $\displaystyle\sum_{n=0}^{\infty} \frac{(-1)^n x^{2n+1}}{2n + 1} = \arctan x, (-1, 1)$

Differential equation: $(x^2 + 1)y'' + 2xy' = 0$

22. $\displaystyle\sum_{n=0}^{\infty} \frac{(2n)!x^{2n+1}}{(2^n n!)^2(2n + 1)} = \arcsin x, (-1, 1)$

Differential equation: $(1 - x^2)y'' - xy' = 0$

23. Find the first six terms in the series solution of Airy's equation $y'' - xy = 0$.

APPENDIX D

Precalculus Review

APPENDIX D.1 Real Numbers and the Real Number Line

Real Numbers and the Real Number Line • Order and Inequalities • Absolute Value and Distance

Real Numbers and the Real Number Line

Real numbers can be represented by a coordinate system called the **real number line** or x-axis (see Figure D.1). The real number corresponding to a point on the real number line is the **coordinate** of the point. As Figure D.1 shows, it is customary to identify those points whose coordinates are integers.

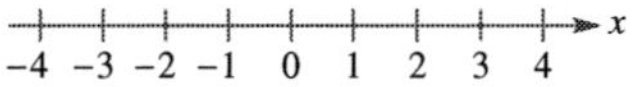

The real number line
Figure D.1

The point on the real number line corresponding to zero is the **origin** and is denoted by 0. The **positive direction** (to the right) is denoted by an arrowhead and is the direction of increasing values of x. Numbers to the right of the origin are **positive.** Numbers to the left of the origin are **negative.** The term **nonnegative** describes a number that is either positive or zero. The term **nonpositive** describes a number that is either negative or zero.

Each point on the real number line corresponds to one and only one real number, and each real number corresponds to one and only one point on the real number line. This type of relationship is called a **one-to-one-correspondence.**

Each of the four points in Figure D.2 corresponds to a **rational number**—one that can be written as the ratio of two integers. (Note that $4.5 = \frac{9}{2}$ and $-2.6 = -\frac{13}{5}$.) Rational numbers can be represented either by *terminating decimals* such as $\frac{2}{5} = 0.4$, or by *repeating decimals* such as $\frac{1}{3} = 0.333\ldots = 0.\overline{3}$.

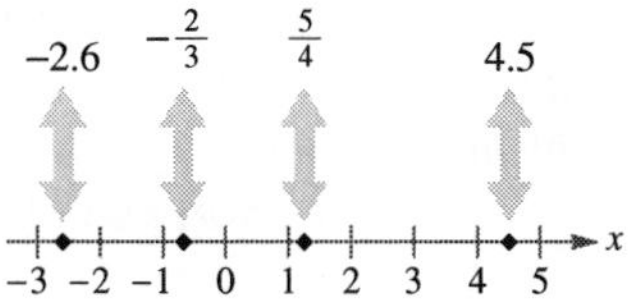

Rational numbers
Figure D.2

Real numbers that are not rational are **irrational.** Irrational numbers cannot be represented as terminating or repeating decimals. In computations, irrational numbers are represented by decimal approximations. Here are three familiar examples.

$$\sqrt{2} \approx 1.414213562$$
$$\pi \approx 3.141592654$$
$$e \approx 2.718281828$$

(See Figure D.3.)

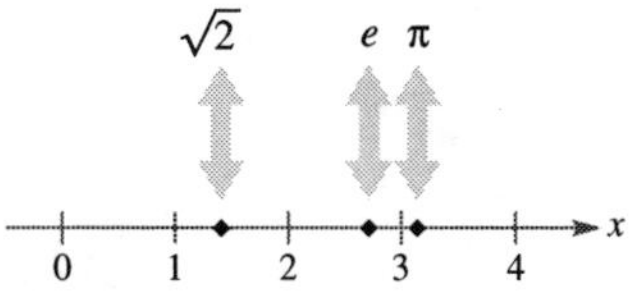

Irrational numbers
Figure D.3

Order and Inequalities

One important property of real numbers is that they are **ordered.** If a and b are real numbers, a is **less than** b if $b - a$ is positive. This order is denoted by the **inequality**

$$a < b.$$

This relationship can also be described by saying that b is **greater than** a and writing $b > a$. When three real numbers a, b, and c are ordered such that $a < b$ and $b < c$, you say that b is **between** a and c and $a < b < c$.

Geometrically, $a < b$ if and only if a lies to the *left* of b on the real number line (see Figure D.4). For example, $1 < 2$ because 1 lies to the left of 2 on the real number line.

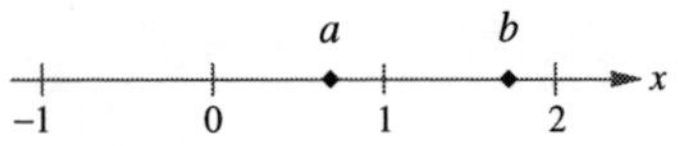

$a < b$ if and only if a lies to the left of b.
Figure D.4

The following properties are used in working with inequalities. Similar properties are obtained if $<$ is replaced by $\leq$ and $>$ is replaced by $\geq$. (The symbols $\leq$ and $\geq$ mean **less than or equal to** and **greater than or equal to,** respectively.)

Properties of Inequalities

Let a, b, c, d, and k be real numbers.

1. If $a < b$ and $b < c$, then $a < c$. — Transitive Property
2. If $a < b$ and $c < d$, then $a + c < b + d$. — Add inequalities.
3. If $a < b$, then $a + k < b + k$. — Add a constant.
4. If $a < b$ and $k > 0$, then $ak < bk$. — Multiply by a positive constant.
5. If $a < b$ and $k < 0$, then $ak > bk$. — Multiply by a negative constant.

NOTE Note that you *reverse the inequality* when you multiply the inequality by a negative number. For example, if $x < 3$, then $-4x > -12$. This also applies to division by a negative number. So, if $-2x > 4$, then $x < -2$.

A **set** is a collection of elements. Two common sets are the set of real numbers and the set of points on the real number line. Many problems in calculus involve **subsets** of one of these two sets. In such cases, it is convenient to use **set notation** of the form $\{x\text{: condition on } x\}$, which is read as follows.

The set of all x such that a certain condition is true.

$$\{ \quad x \quad : \quad \text{condition on } x\}$$

For example, you can describe the set of positive real numbers as

$$\{x\text{: } x > 0\}. \qquad \text{Set of positive real numbers}$$

Similarly, you can describe the set of nonnegative real numbers as

$$\{x\text{: } x \geq 0\}. \qquad \text{Set of nonnegative real numbers}$$

The **union** of two sets A and B, denoted by $A \cup B$, is the set of elements that are members of A *or* B *or both*. The **intersection** of two sets A and B, denoted by $A \cap B$, is the set of elements that are members of A *and* B. Two sets are **disjoint** if they have no elements in common.

The most commonly used subsets are **intervals** on the real number line. For example, the **open** interval

$(a, b) = \{x: a < x < b\}$ Open interval

is the set of all real numbers greater than a and less than b, where a and b are the **endpoints** of the interval. Note that the endpoints are not included in an open interval. Intervals that include their endpoints are **closed** and are denoted by

$[a, b] = \{x: a \leq x \leq b\}.$ Closed interval

The nine basic types of intervals on the real number line are shown in the table below. The first four are **bounded intervals** and the remaining five are **unbounded intervals.** Unbounded intervals are also classified as open or closed. The intervals $(-\infty, b)$ and (a, ∞) are open, the intervals $(-\infty, b]$ and $[a, \infty)$ are closed, and the interval $(-\infty, \infty)$ is considered to be both open *and* closed.

Intervals on the Real Number Line

	Interval Notation	Set Notation	Graph
Bounded open interval	(a, b)	$\{x: a < x < b\}$	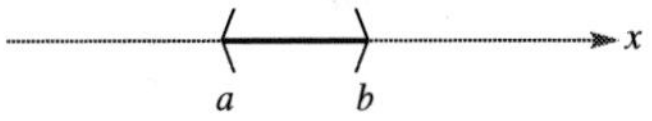
Bounded closed interval	$[a, b]$	$\{x: a \leq x \leq b\}$	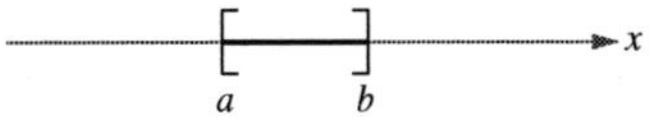
Bounded intervals (neither open nor closed)	$[a, b)$	$\{x: a \leq x < b\}$	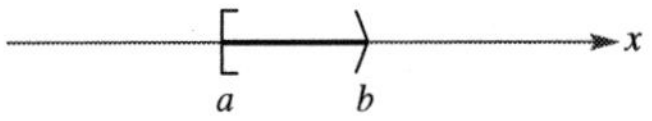
	$(a, b]$	$\{x: a < x \leq b\}$	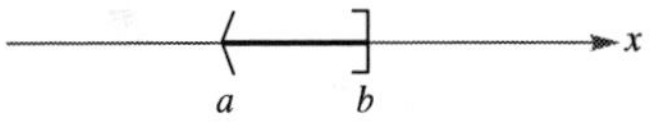
Unbounded open intervals	$(-\infty, b)$	$\{x: x < b\}$	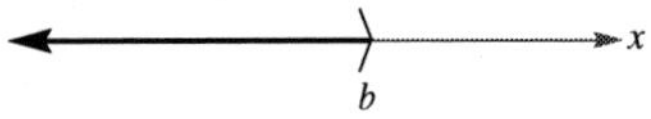
	(a, ∞)	$\{x: x > a\}$	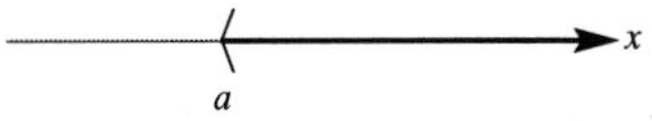
Unbounded closed intervals	$(-\infty, b]$	$\{x: x \leq b\}$	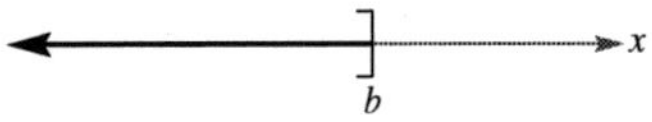
	$[a, \infty)$	$\{x: x \geq a\}$	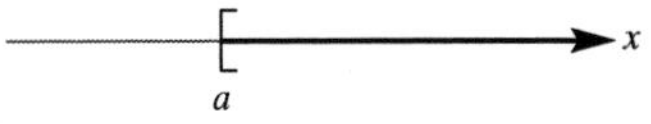
Entire real line	$(-\infty, \infty)$	$\{x: x \text{ is a real number}\}$	

NOTE The symbols ∞ and $-\infty$ refer to positive and negative infinity, respectively. These symbols do not denote real numbers. They simply enable you to describe unbounded conditions more concisely. For instance, the interval $[a, \infty)$ is unbounded to the right because it includes *all* real numbers that are greater than or equal to a.

EXAMPLE 1 Liquid and Gaseous States of Water

Describe the intervals on the real number line that correspond to the temperature x (in degrees Celsius) for water in

a. a liquid state. **b.** a gaseous state.

Solution

a. Water is in a liquid state at temperatures greater than 0°C and less than 100°C, as shown in Figure D.5(a).

$$(0, 100) = \{x:\ 0 < x < 100\}$$

b. Water is in a gaseous state (steam) at temperatures greater than or equal to 100°C, as shown in Figure D.5(b).

$$[100, \infty) = \{x:\ x \geq 100\}$$

(a) Temperature range of water (in degrees Celsius)

(b) Temperature range of steam (in degrees Celsius)

Figure D.5

A real number a is a **solution** of an inequality if the inequality is **satisfied** (is true) when a is substituted for x. The set of all solutions is the **solution set** of the inequality.

EXAMPLE 2 Solving an Inequality

Solve $2x - 5 < 7$.

Solution

$2x - 5 < 7$ — Write original inequality.

$2x - 5 + 5 < 7 + 5$ — Add 5 to each side.

$2x < 12$ — Simplify.

$\dfrac{2x}{2} < \dfrac{12}{2}$ — Divide each side by 2.

$x < 6$ — Simplify.

The solution set is $(-\infty, 6)$.

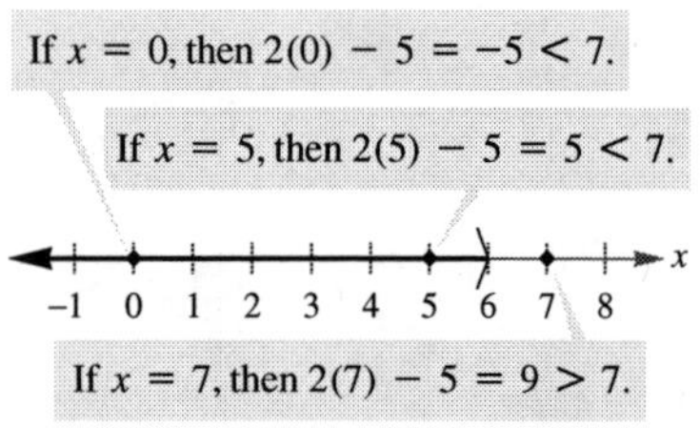

Checking solutions of $2x - 5 < 7$
Figure D.6

NOTE In Example 2, all five inequalities listed as steps in the solution are called **equivalent** because they have the same solution set.

Once you have solved an inequality, check some x-values in your solution set to verify that they satisfy the original inequality. You should also check some values outside your solution set to verify that they *do not* satisfy the inequality. For example, Figure D.6 shows that when $x = 0$ or $x = 5$ the inequality $2x - 5 < 7$ is satisfied, but when $x = 7$ the inequality $2x - 5 < 7$ is not satisfied.

EXAMPLE 3 Solving a Double Inequality

Solve $-3 \le 2 - 5x \le 12$.

Solution

$$-3 \le 2 - 5x \le 12 \qquad \text{Write original inequality.}$$

$$-3 - 2 \le 2 - 5x - 2 \le 12 - 2 \qquad \text{Subtract 2 from each part.}$$

$$-5 \le -5x \le 10 \qquad \text{Simplify.}$$

$$\frac{-5}{-5} \ge \frac{-5x}{-5} \ge \frac{10}{-5} \qquad \text{Divide each part by } -5 \text{ and reverse both inequalities.}$$

$$1 \ge x \ge -2 \qquad \text{Simplify.}$$

The solution set is $[-2, 1]$, as shown in Figure D.7.

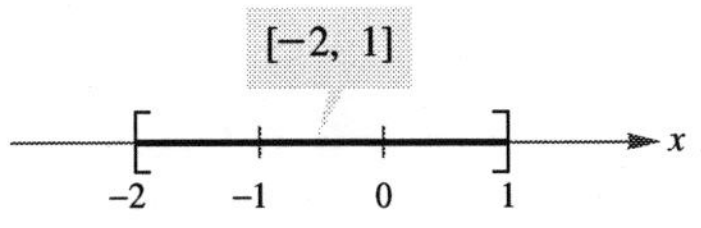

Solution set of $-3 \le 2 - 5x \le 12$
Figure D.7

The inequalities in Examples 2 and 3 are **linear inequalities**—that is, they involve first-degree polynomials. To solve inequalities involving polynomials of higher degree, use the fact that a polynomial can change signs *only* at its real **zeros** (the x-values that make the polynomial equal to zero). Between two consecutive real zeros, a polynomial must be either entirely positive or entirely negative. This means that when the real zeros of a polynomial are put in order, they divide the real number line into **test intervals** in which the polynomial has no sign changes. So, if a polynomial has the factored form

$$(x - r_1)(x - r_2) \cdot \cdot \cdot (x - r_n), \qquad r_1 < r_2 < r_3 < \cdot \cdot \cdot < r_n$$

the test intervals are

$$(-\infty, r_1), \quad (r_1, r_2), \ldots, \quad (r_{n-1}, r_n), \quad \text{and} \quad (r_n, \infty).$$

To determine the sign of the polynomial in each test interval, you need to test only *one value* from the interval.

EXAMPLE 4 Solving a Quadratic Inequality

Solve $x^2 < x + 6$.

Solution

$$x^2 < x + 6 \qquad \text{Write original inequality.}$$

$$x^2 - x - 6 < 0 \qquad \text{Write in general form.}$$

$$(x - 3)(x + 2) < 0 \qquad \text{Factor.}$$

The polynomial $x^2 - x - 6$ has $x = -2$ and $x = 3$ as its zeros. So, you can solve the inequality by testing the sign of $x^2 - x - 6$ in each of the test intervals $(-\infty, -2)$, $(-2, 3)$, and $(3, \infty)$. To test an interval, choose any number in the interval and compute the sign of $x^2 - x - 6$. After doing this, you will find that the polynomial is positive for all real numbers in the first and third intervals and negative for all real numbers in the second interval. The solution of the original inequality is therefore $(-2, 3)$, as shown in Figure D.8.

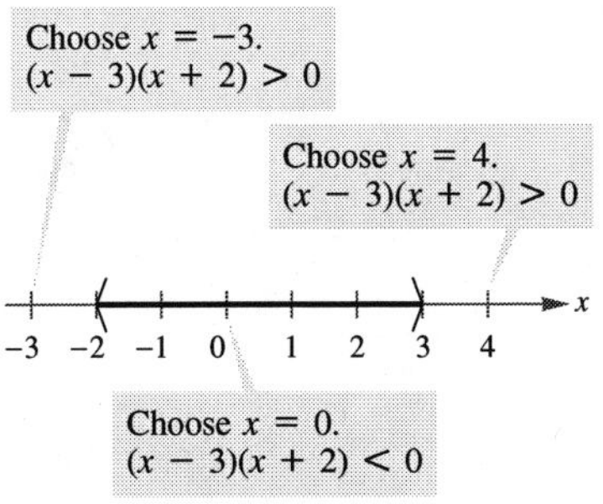

Testing an interval
Figure D.8

Absolute Value and Distance

If a is a real number, the **absolute value** of a is

$$|a| = \begin{cases} a, & \text{if } a \geq 0 \\ -a, & \text{if } a < 0. \end{cases}$$

The absolute value of a number cannot be negative. For example, let $a = -4$. Then, because $-4 < 0$, you have

$$|a| = |-4| = -(-4) = 4.$$

Remember that the symbol $-a$ does not necessarily mean that $-a$ is negative.

Operations with Absolute Value

Let a and b be real numbers and let n be a positive integer.

1. $|ab| = |a|\,|b|$ **2.** $\left|\frac{a}{b}\right| = \frac{|a|}{|b|}$, $b \neq 0$

3. $|a| = \sqrt{a^2}$ **4.** $|a^n| = |a|^n$

NOTE You are asked to prove these properties in Exercises 73, 75, 76, and 77.

Properties of Inequalities and Absolute Value

Let a and b be real numbers and let k be a positive real number.

1. $-|a| \leq a \leq |a|$

2. $|a| \leq k$ if and only if $-k \leq a \leq k$.

3. $|a| \geq k$ if and only if $a \leq -k$ or $a \geq k$.

4. *Triangle Inequality:* $|a + b| \leq |a| + |b|$

Properties 2 and 3 are also true if ≤ is replaced by <.

EXAMPLE 5 Solving an Absolute Value Inequality

Solve $|x - 3| \leq 2$.

Solution Using the second property of inequalities and absolute value, you can rewrite the original inequality as a double inequality.

$$\begin{aligned} -2 \leq \quad x - 3 \quad &\leq 2 && \text{Write as double inequality.} \\ -2 + 3 \leq x - 3 + 3 &\leq 2 + 3 && \text{Add 3 to each part.} \\ 1 \leq \quad x \quad &\leq 5 && \text{Simplify.} \end{aligned}$$

The solution set is $[1, 5]$, as shown in Figure D.9.

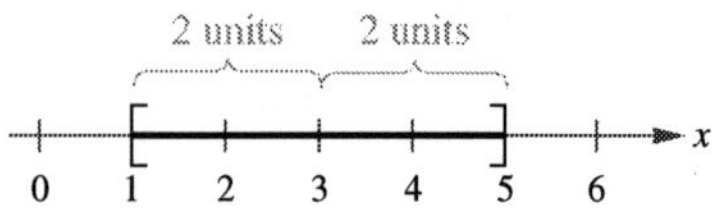

Solution set of $|x - 3| \leq 2$
Figure D.9

EXAMPLE 6 A Two-Interval Solution Set

Solve $|x + 2| > 3$.

Solution Using the third property of inequalities and absolute value, you can rewrite the original inequality as two linear inequalities.

$$x + 2 < -3 \quad \text{or} \quad x + 2 > 3$$
$$x < -5 \qquad\qquad x > 1$$

The solution set is the union of the disjoint intervals $(-\infty, -5)$ and $(1, \infty)$, as shown in Figure D.10.

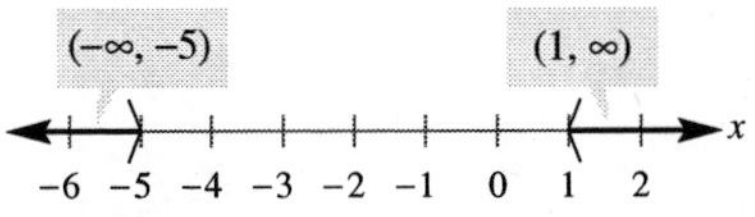

Solution set of $|x + 2| > 3$
Figure D.10

Examples 5 and 6 illustrate the general results shown in Figure D.11. Note that if $d > 0$, the solution set for the inequality $|x - a| \leq d$ is a *single* interval, whereas the solution set for the inequality $|x - a| \geq d$ is the union of *two* disjoint intervals.

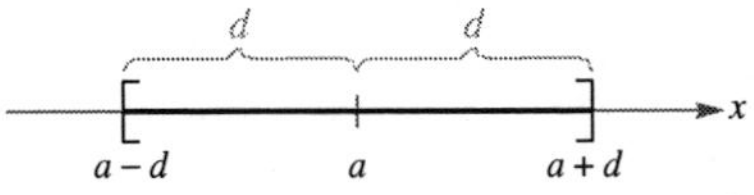

Solution set of $|x - a| \leq d$

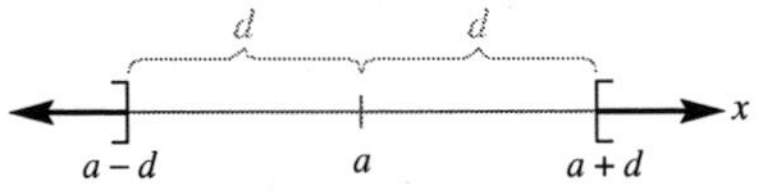

Solution set of $|x - a| \geq d$
Figure D.11

The **distance between two points** a and b on the real number line is given by

$$d = |a - b| = |b - a|.$$

The **directed distance from *a* to *b*** is $b - a$ and the **directed distance from *b* to *a*** is $a - b$, as shown in Figure D.12.

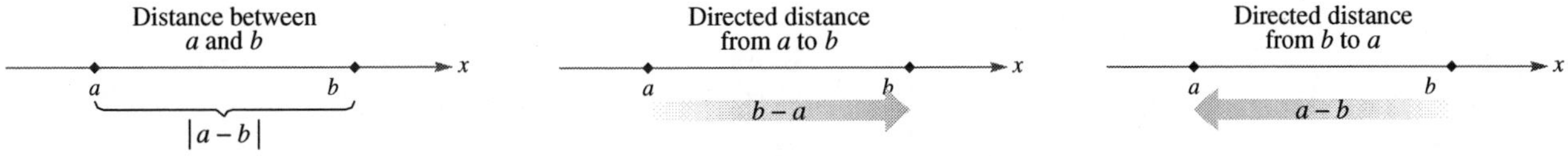

Figure D.12

EXAMPLE 7 Distance on the Real Number Line

a. The distance between -3 and 4 is

$$|4 - (-3)| = |7| = 7 \quad \text{or} \quad |-3 - 4| = |-7| = 7.$$

(See Figure D.13.)

b. The directed distance from -3 to 4 is $4 - (-3) = 7$.

c. The directed distance from 4 to -3 is $-3 - 4 = -7$.

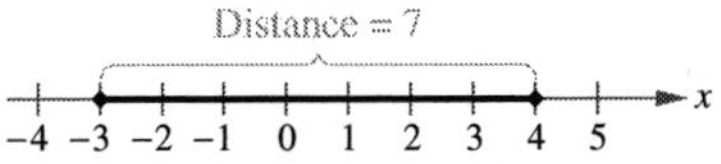

Figure D.13

The **midpoint** of an interval with endpoints a and b is the average value of a and b. That is,

$$\text{Midpoint of interval } (a, b) = \frac{a + b}{2}.$$

To show that this is the midpoint, you need only show that $(a + b)/2$ is equidistant from a and b.

EXERCISES FOR APPENDIX D.1

In Exercises 1–10, determine whether the real number is rational or irrational.

1. 0.7
2. -3678
3. $\dfrac{3\pi}{2}$
4. $3\sqrt{2} - 1$
5. $4.3\overline{451}$
6. $\frac{22}{7}$
7. $\sqrt[3]{64}$
8. $0.\overline{8177}$
9. $4\frac{5}{8}$
10. $(\sqrt{2})^3$

In Exercises 11–14, write the repeating decimal as a ratio of two integers using the following procedure. If $x = 0.6363\ldots$, then $100x = 63.6363\ldots$. Subtracting the first equation from the second produces $99x = 63$ or $x = \frac{63}{99} = \frac{7}{11}$.

11. $0.\overline{36}$
12. $0.3\overline{18}$
13. $0.\overline{297}$
14. $0.\overline{9900}$

15. Given $a < b$, determine which of the following are true.

(a) $a + 2 < b + 2$
(b) $5b < 5a$
(c) $5 - a > 5 - b$
(d) $\dfrac{1}{a} < \dfrac{1}{b}$
(e) $(a - b)(b - a) > 0$
(f) $a^2 < b^2$

16. Complete the table with the appropriate interval notation, set notation, and graph on the real number line.

Interval Notation	Set Notation	Graph
		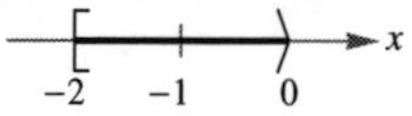
$(-\infty, -4]$		
	$\{x: 3 \le x \le \frac{11}{2}\}$	
$(-1, 7)$		

In Exercises 17–20, verbally describe the subset of real numbers represented by the inequality. Sketch the subset on the real number line, and state whether the interval is bounded or unbounded.

17. $-3 < x < 3$
18. $x \ge 4$
19. $x \le 5$
20. $0 \le x < 8$

In Exercises 21–24, use inequality and interval notation to describe the set.

21. y is at least 4.
22. q is nonnegative.
23. The interest rate r on loans is expected to be greater than 3% and no more than 7%.
24. The temperature T is forecast to be above 90°F today.

In Exercises 25–44, solve the inequality and graph the solution on the real number line.

25. $2x - 1 \ge 0$
26. $3x + 1 \ge 2x + 2$
27. $-4 < 2x - 3 < 4$
28. $0 \le x + 3 < 5$
29. $\dfrac{x}{2} + \dfrac{x}{3} > 5$
30. $x > \dfrac{1}{x}$
31. $|x| < 1$
32. $\dfrac{x}{2} - \dfrac{x}{3} > 5$
33. $\left|\dfrac{x - 3}{2}\right| \ge 5$
34. $\left|\dfrac{x}{2}\right| > 3$
35. $|x - a| < b,\ b > 0$
36. $|x + 2| < 5$
37. $|2x + 1| < 5$
38. $|3x + 1| \ge 4$
39. $\left|1 - \dfrac{2}{3}x\right| < 1$
40. $|9 - 2x| < 1$
41. $x^2 \le 3 - 2x$
42. $x^4 - x \le 0$
43. $x^2 + x - 1 \le 5$
44. $2x^2 + 1 < 9x - 3$

In Exercises 45–48, find the directed distance from a to b, the directed distance from b to a, and the distance between a and b.

45.
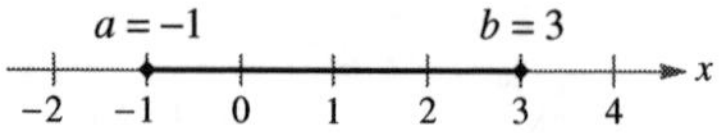

46.
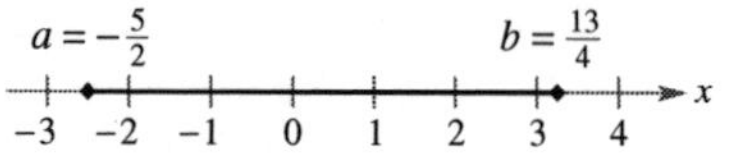

47. (a) $a = 126,\ b = 75$
(b) $a = -126,\ b = -75$

48. (a) $a = 9.34,\ b = -5.65$
(b) $a = \frac{16}{5},\ b = \frac{112}{75}$

In Exercises 49–54, use absolute value notation to define the interval or pair of intervals on the real number line.

49.
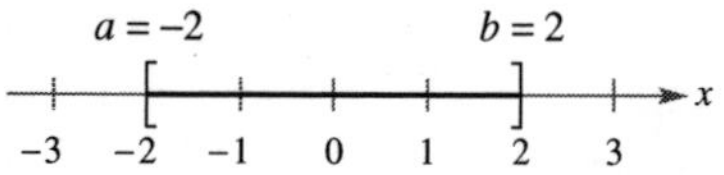

50.
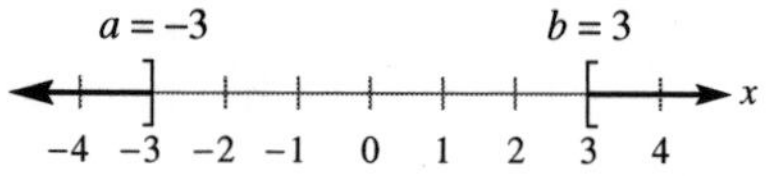

51.
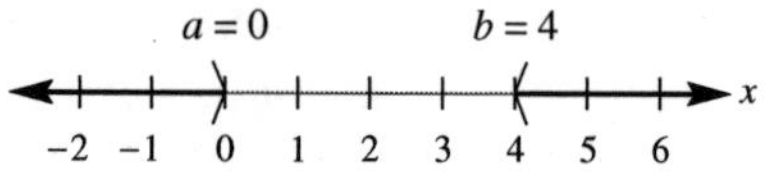

52.
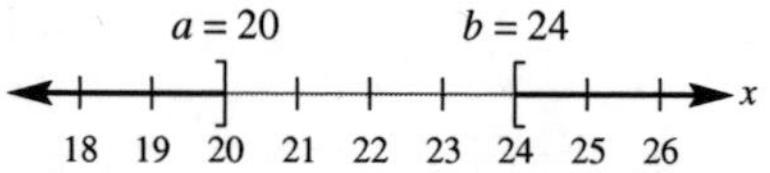

53. (a) All numbers that are at most 10 units from 12.

(b) All numbers that are at least 10 units from 12.

54. (a) y is at most two units from a.

(b) y is less than δ units from c.

In Exercises 55–58, find the midpoint of the interval.

55.

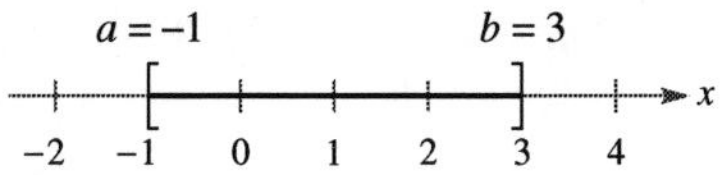

56.

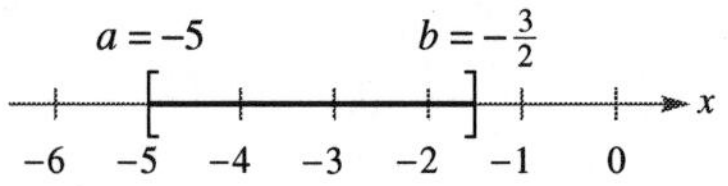

57. (a) $[7, 21]$

(b) $[8.6, 11.4]$

58. (a) $[-6.85, 9.35]$

(b) $[-4.6, -1.3]$

59. ***Profit*** The revenue R from selling x units of a product is

$$R = 115.95x$$

and the cost C of producing x units is

$$C = 95x + 750.$$

To make a (positive) profit, R must be greater than C. For what values of x will the product return a profit?

60. ***Fleet Costs*** A utility company has a fleet of vans. The annual operating cost C (in dollars) of each van is estimated to be

$$C = 0.32m + 2300$$

where m is measured in miles. The company wants the annual operating cost of each van to be less than \$10,000. To do this, m must be less than what value?

61. ***Fair Coin*** To determine whether a coin is fair (has an equal probability of landing tails up or heads up), you toss the coin 100 times and record the number of heads x. The coin is declared unfair if

$$\left|\frac{x - 50}{5}\right| \geq 1.645.$$

For what values of x will the coin be declared unfair?

62. ***Daily Production*** The estimated daily oil production p at a refinery is

$$|p - 2{,}250{,}000| < 125{,}000$$

where p is measured in barrels. Determine the high and low production levels.

In Exercises 63 and 64, determine which of the two real numbers is greater.

63. (a) π or $\frac{355}{113}$

(b) π or $\frac{22}{7}$

64. (a) $\frac{224}{151}$ or $\frac{144}{97}$

(b) $\frac{73}{81}$ or $\frac{6427}{7132}$

65. ***Approximation—Powers of 10*** Light travels at the speed of 2.998×10^8 meters per second. Which best estimates the distance in meters that light travels in a year?

(a) 9.5×10^5 (b) 9.5×10^{15}

(c) 9.5×10^{12} (d) 9.6×10^{16}

66. ***Writing*** The accuracy of an approximation to a number is related to how many significant digits there are in the approximation. Write a definition for significant digits and illustrate the concept with examples.

True or False? **In Exercises 67–72, determine whether the statement is true or false. If it is false, explain why or give an example that shows it is false.**

67. The reciprocal of a nonzero integer is an integer.

68. The reciprocal of a nonzero rational number is a rational number.

69. Each real number is either rational or irrational.

70. The absolute value of each real number is positive.

71. If $x < 0$, then $\sqrt{x^2} = -x$.

72. If a and b are any two distinct real numbers, then $a < b$ or $a > b$.

In Exercises 73–80, prove the property.

73. $|ab| = |a||b|$

74. $|a - b| = |b - a|$

$\left[Hint\text{: } (a - b) = (-1)(b - a)\right]$

75. $\left|\frac{a}{b}\right| = \frac{|a|}{|b|}, \quad b \neq 0$

76. $|a| = \sqrt{a^2}$

77. $|a^n| = |a|^n, \quad n = 1, 2, 3, . . .$

78. $-|a| \leq a \leq |a|$

79. $|a| \leq k$, if and only if $-k \leq a \leq k, \quad k > 0.$

80. $|a| \geq k$ if and only if $a \leq -k$ or $a \geq k, \quad k > 0.$

81. Find an example for which $|a - b| > |a| - |b|$, and an example for which $|a - b| = |a| - |b|$. Then prove that $|a - b| \geq |a| - |b|$ for all a, b.

82. Show that the maximum of two numbers a and b is given by the formula

$$\max(a, b) = \tfrac{1}{2}\left(a + b + |a - b|\right).$$

Derive a similar formula for $\min(a, b)$.

APPENDIX D.2 The Cartesian Plane

The Cartesian Plane • The Distance and Midpoint Formulas • Equations of Circles

The Cartesian Plane

Just as you can represent real numbers by points on a real number line, you can represent ordered pairs of real numbers by points in a plane called the **rectangular coordinate system,** or the **Cartesian plane,** after the French mathematician René Descartes.

The Cartesian plane is formed by using two real number lines intersecting at right angles, as shown in Figure D.14. The horizontal real number line is usually called the ***x*-axis,** and the vertical real number line is usually called the ***y*-axis.** The point of intersection of these two axes is the **origin.** The two axes divide the plane into four parts called **quadrants.**

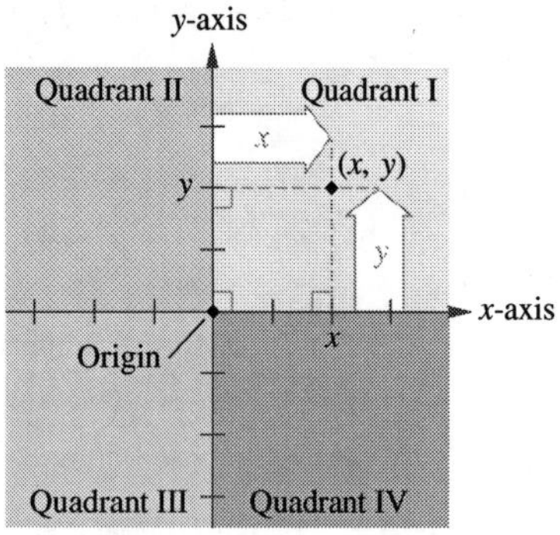

The Cartesian plane
Figure D.14

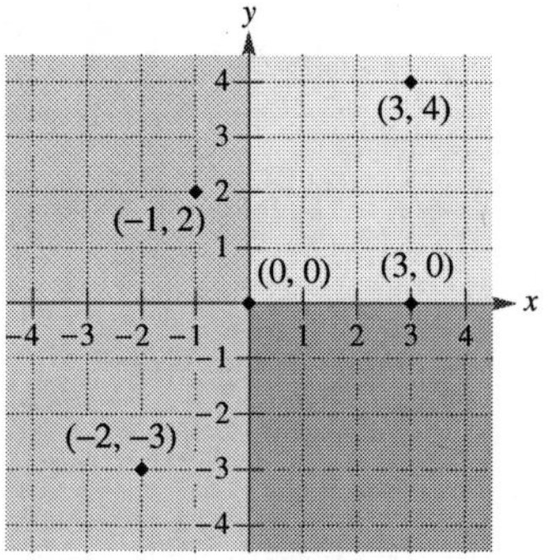

Points represented by ordered pairs
Figure D.15

Each point in the plane is identified by an **ordered pair** (x, y) of real numbers x and y, called **coordinates** of the point. The number x represents the directed distance from the y-axis to the point, and the number y represents the directed distance from the x-axis to the point (see Figure D.14). For the point (x, y), the first coordinate is the ***x*-coordinate** or **abscissa,** and the second coordinate is the ***y*-coordinate** or **ordinate.** For example, Figure D.15 shows the locations of the points $(-1, 2)$, $(3, 4)$, $(0, 0)$, $(3, 0)$, and $(-2, -3)$ in the Cartesian plane.

NOTE The signs of the coordinates of a point determine the quadrant in which the point lies. For instance, if $x > 0$ and $y < 0$, then the point (x, y) lies in Quadrant IV.

Note that an ordered pair (a, b) is used to denote either a point in the plane *or* an open interval on the real number line. This, however, should not be confusing—the nature of the problem should clarify whether a point in the plane or an open interval is being discussed.

The Distance and Midpoint Formulas

Recall from the Pythagorean Theorem that, in a right triangle, the hypotenuse c and sides a and b are related by $a^2 + b^2 = c^2$. Conversely, if $a^2 + b^2 = c^2$, then the triangle is a right triangle (see Figure D.16).

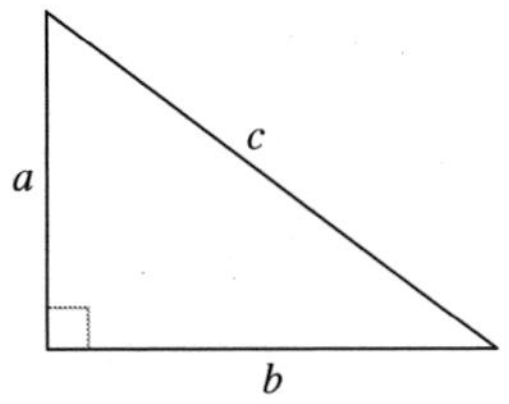

The Pythagorean Theorem:
$a^2 + b^2 = c^2$
Figure D.16

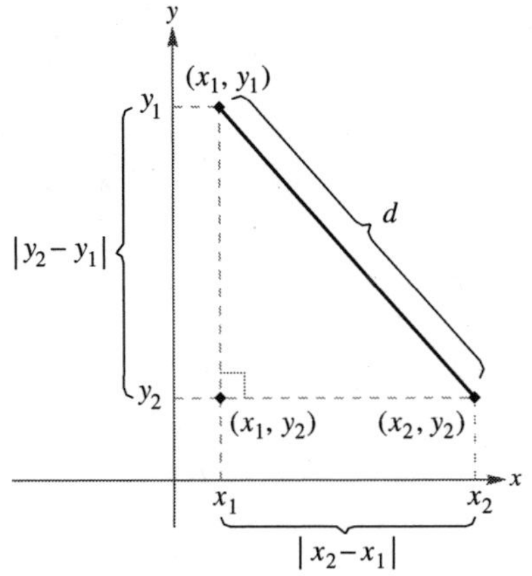

The distance between two points
Figure D.17

Suppose you want to determine the distance d between the two points (x_1, y_1) and (x_2, y_2) in the plane. If the points lie on a horizontal line, then $y_1 = y_2$ and the distance between the points is $|x_2 - x_1|$. If the points lie on a vertical line, then $x_1 = x_2$ and the distance between the points is $|y_2 - y_1|$. If the two points do not lie on a horizontal or vertical line, they can be used to form a right triangle, as shown in Figure D.17. The length of the vertical side of the triangle is $|y_2 - y_1|$, and the length of the horizontal side is $|x_2 - x_1|$. By the Pythagorean Theorem, it follows that

$$d^2 = |x_2 - x_1|^2 + |y_2 - y_1|^2$$
$$d = \sqrt{|x_2 - x_1|^2 + |y_2 - y_1|^2}.$$

Replacing $|x_2 - x_1|^2$ and $|y_2 - y_1|^2$ by the equivalent expressions $(x_2 - x_1)^2$ and $(y_2 - y_1)^2$ produces the following result.

Distance Formula

The distance d between the points (x_1, y_1) and (x_2, y_2) in the plane is given by

$$d = \sqrt{(x_2 - x_1)^2 + (y_2 - y_1)^2}.$$

EXAMPLE 1 Finding the Distance Between Two Points

Find the distance between the points $(-2, 1)$ and $(3, 4)$.

Solution

$$d = \sqrt{(x_2 - x_1)^2 + (y_2 - y_1)^2} \qquad \text{Distance Formula}$$
$$= \sqrt{[3 - (-2)]^2 + (4 - 1)^2} \qquad \text{Substitute for } x_1, y_1, x_2, \text{ and } y_2.$$
$$= \sqrt{5^2 + 3^2}$$
$$= \sqrt{25 + 9}$$
$$= \sqrt{34}$$
$$\approx 5.83$$

EXAMPLE 2 Verifying a Right Triangle

Verify that the points (2, 1), (4, 0), and (5, 7) form the vertices of a right triangle.

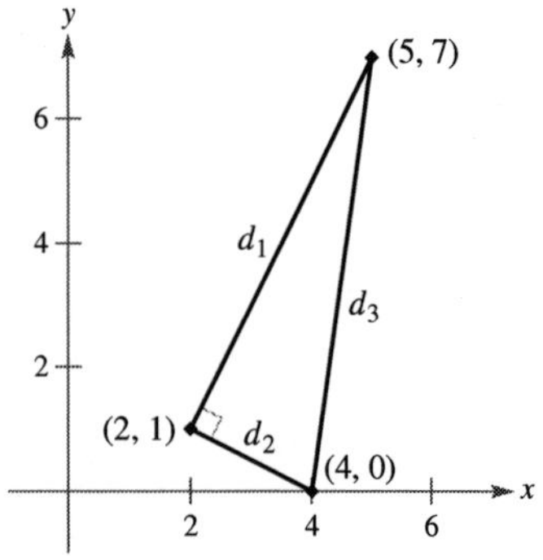

Verifying a right triangle
Figure D.18

Solution Figure D.18 shows the triangle formed by the three points. The lengths of the three sides are as follows.

$$d_1 = \sqrt{(5-2)^2 + (7-1)^2} = \sqrt{9+36} = \sqrt{45}$$
$$d_2 = \sqrt{(4-2)^2 + (0-1)^2} = \sqrt{4+1} = \sqrt{5}$$
$$d_3 = \sqrt{(5-4)^2 + (7-0)^2} = \sqrt{1+49} = \sqrt{50}$$

Because

$$d_1^2 + d_2^2 = 45 + 5 = 50 \qquad \text{Sum of squares of sides}$$

and

$$d_3^2 = 50 \qquad \text{Square of hypotenuse}$$

you can apply the Pythagorean Theorem to conclude that the triangle is a right triangle.

EXAMPLE 3 Using the Distance Formula

Find x such that the distance between $(x, 3)$ and $(2, -1)$ is 5.

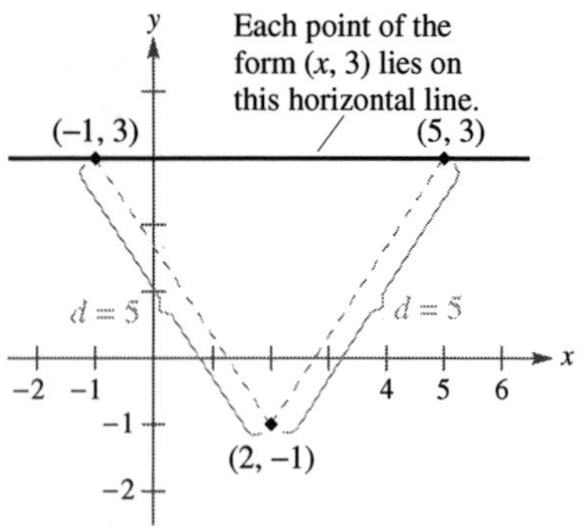

Given a distance, find a point.
Figure D.19

Solution Using the Distance Formula, you can write the following.

$$5 = \sqrt{(x-2)^2 + [3-(-1)]^2} \qquad \text{Distance Formula}$$
$$25 = (x^2 - 4x + 4) + 16 \qquad \text{Square each side.}$$
$$0 = x^2 - 4x - 5 \qquad \text{Write in general form.}$$
$$0 = (x-5)(x+1) \qquad \text{Factor.}$$

Therefore, $x = 5$ or $x = -1$, and you can conclude that there are two solutions. That is, each of the points $(5, 3)$ and $(-1, 3)$ lies five units from the point $(2, -1)$, as shown in Figure D.19.

The coordinates of the **midpoint** of the line segment joining two points can be found by "averaging" the x-coordinates of the two points and "averaging" the y-coordinates of the two points. That is, the midpoint of the line segment joining the points (x_1, y_1) and (x_2, y_2) in the plane is

$$\left(\frac{x_1 + x_2}{2}, \frac{y_1 + y_2}{2}\right). \qquad \text{Midpoint Formula}$$

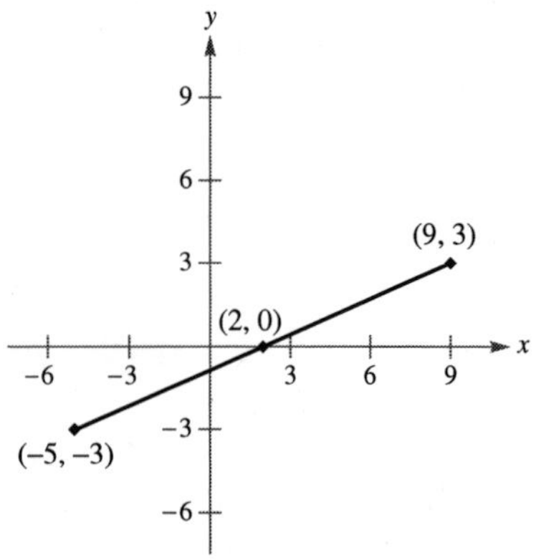

Midpoint of a line segment
Figure D.20

For instance, the midpoint of the line segment joining the points $(-5, -3)$ and $(9, 3)$ is

$$\left(\frac{-5+9}{2}, \frac{-3+3}{2}\right) = (2, 0)$$

as shown in Figure D.20.

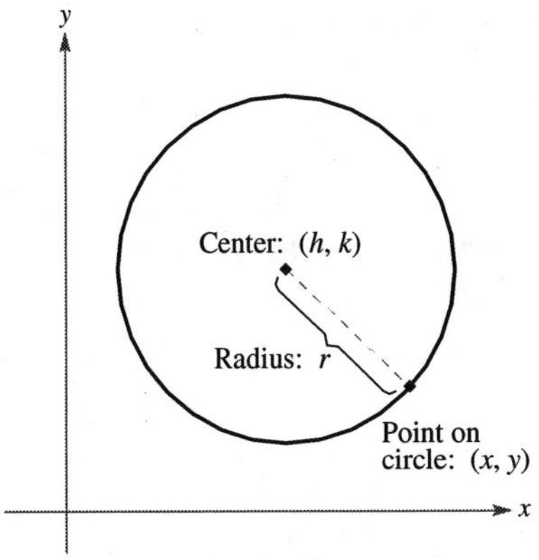

Definition of a circle
Figure D.21

Equations of Circles

A **circle** can be defined as the set of all points in a plane that are equidistant from a fixed point. The fixed point is the **center** of the circle, and the distance between the center and a point on the circle is the **radius** (see Figure D.21).

You can use the Distance Formula to write an equation for the circle with center (h, k) and radius r. Let (x, y) be any point on the circle. Then the distance between (x, y) and the center (h, k) is given by

$$\sqrt{(x-h)^2 + (y-k)^2} = r.$$

By squaring each side of this equation, you obtain the **standard form of the equation of a circle.**

Standard Form of the Equation of a Circle

The point (x, y) lies on the circle of radius r and center (h, k) if and only if

$$(x-h)^2 + (y-k)^2 = r^2.$$

The standard form of the equation of a circle with center at the origin, $(h, k) = (0, 0)$, is

$$x^2 + y^2 = r^2.$$

If $r = 1$, the circle is called the **unit circle.**

EXAMPLE 4 Writing the Equation of a Circle

The point $(3, 4)$ lies on a circle whose center is at $(-1, 2)$, as shown in Figure D.22. Write the standard form of the equation of this circle.

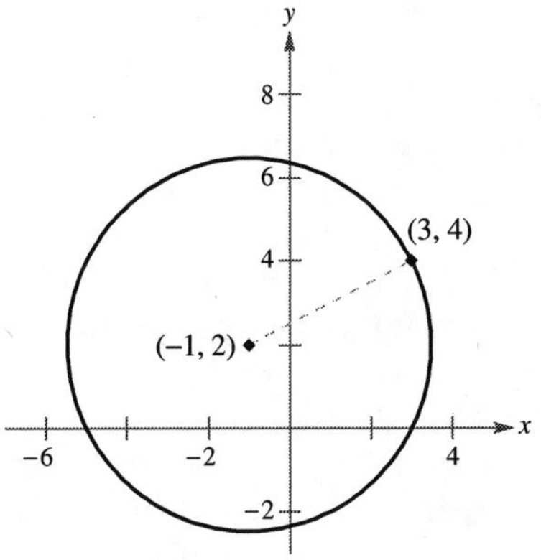

Figure D.22

Solution The radius of the circle is the distance between $(-1, 2)$ and $(3, 4)$.

$$r = \sqrt{[3-(-1)]^2 + (4-2)^2} = \sqrt{16+4} = \sqrt{20}$$

You can write the standard form of the equation of this circle as

$$[x-(-1)]^2 + (y-2)^2 = (\sqrt{20})^2$$

$$(x+1)^2 + (y-2)^2 = 20. \quad \text{Write in standard form.}$$

By squaring and simplifying, the equation $(x-h)^2 + (y-k)^2 = r^2$ can be written in the following **general form of the equation of a circle.**

$$Ax^2 + Ay^2 + Dx + Ey + F = 0, \quad A \neq 0$$

To convert such an equation to the standard form

$$(x-h)^2 + (y-k)^2 = p$$

you can use a process called **completing the square.** If $p > 0$, the graph of the equation is a circle. If $p = 0$, the graph is the single point (h, k). If $p < 0$, the equation has no graph.

EXAMPLE 5 Completing the Square

Sketch the graph of the circle whose general equation is

$$4x^2 + 4y^2 + 20x - 16y + 37 = 0.$$

Solution To complete the square, first divide by 4 so that the coefficients of x^2 and y^2 are both 1.

$$4x^2 + 4y^2 + 20x - 16y + 37 = 0$$ Write original equation.

$$x^2 + y^2 + 5x - 4y + \frac{37}{4} = 0$$ Divide by 4.

$$(x^2 + 5x + \quad) + (y^2 - 4y + \quad) = -\frac{37}{4}$$ Group terms.

$$\left(x^2 + 5x + \frac{25}{4}\right) + (y^2 - 4y + 4) = -\frac{37}{4} + \frac{25}{4} + 4$$ Complete the square by adding $\left(\frac{5}{2}\right)^2 = \frac{25}{4}$ and $\left(\frac{4}{2}\right)^2 = 4$ to each side.

(half)^2 (half)^2

$$\left(x + \frac{5}{2}\right)^2 + (y - 2)^2 = 1$$ Write in standard form.

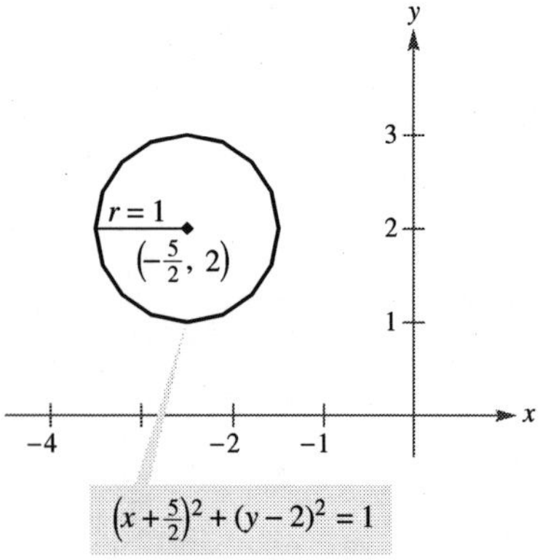

A circle with a radius of 1 and center at $\left(-\frac{5}{2}, 2\right)$
Figure D.23

Note that you complete the square by adding the square of half the coefficient of x *and* the square of half the coefficient of y to each side of the equation. The circle is centered at $\left(-\frac{5}{2}, 2\right)$ and its radius is 1, as shown in Figure D.23.

You have now reviewed some fundamental concepts of *analytic geometry*. Because these concepts are in common use today, it is easy to overlook their revolutionary nature. At the time analytic geometry was being developed by Pierre de Fermat and René Descartes, the two major branches of mathematics—geometry and algebra—were largely independent of each other. Circles belonged to geometry and equations belonged to algebra. The coordination of the points on a circle and the solutions of an equation belongs to what is now called analytic geometry.

It is important to become skilled in analytic geometry so that you can move easily between geometry and algebra. For instance, in Example 4, you were given a geometric description of a circle and were asked to find an algebraic equation for the circle. So, you were moving from geometry to algebra. Similarly, in Example 5 you were given an algebraic equation and asked to sketch a geometric picture. In this case, you were moving from algebra to geometry. These two examples illustrate the two most common problems in analytic geometry.

1. Given a graph, find its equation.

   Geometry ⇨ Algebra

2. Given an equation, find its graph.

   Algebra ⇨ Geometry

In the next section, you will review other examples of these two types of problems.

EXERCISES FOR APPENDIX D.2

In Exercises 1–6, (a) plot the points, (b) find the distance between the points, and (c) find the midpoint of the line segment joining the points.

1. $(2, 1), (4, 5)$ **2.** $(-3, 2), (3, -2)$

3. $\left(\frac{1}{2}, 1\right), \left(-\frac{3}{2}, -5\right)$ **4.** $\left(\frac{2}{3}, -\frac{1}{3}\right), \left(\frac{5}{6}, 1\right)$

5. $\left(1, \sqrt{3}\right), (-1, 1)$ **6.** $(-2, 0), \left(0, \sqrt{2}\right)$

In Exercises 7–10, determine the quadrant(s) in which (x, y) is located so that the condition(s) is (are) satisfied.

7. $x = -2$ and $y > 0$

8. $y < -2$

9. $xy > 0$

10. $(x, -y)$ is in Quadrant II.

In Exercises 11–14, show that the points are the vertices of the polygon. (A rhombus is a quadrilateral whose sides are all of the same length.)

	Vertices	*Polygon*
11.	$(4, 0), (2, 1), (-1, -5)$	Right triangle
12.	$(1, -3), (3, 2), (-2, 4)$	Isosceles triangle
13.	$(0, 0), (1, 2), (2, 1), (3, 3)$	Rhombus
14.	$(0, 1), (3, 7), (4, 4), (1, -2)$	Parallelogram

15. ***Number of Stores*** The table shows the number y of Target stores for each year x from 1994 through 2003. *(Source: Target Corp.)*

Year, x	1994	1995	1996	1997	1998
Number, y	960	1029	1101	1130	1182

Year, x	1999	2000	2001	2002	2003
Number, y	1243	1307	1381	1475	1553

Select reasonable scales on the coordinate axes and plot the points (x, y).

16. ***Conjecture*** Plot the points $(2, 1)$, $(-3, 5)$, and $(7, -3)$ on a rectangular coordinate system. Then change the sign of the x-coordinate of each point and plot the three new points on the same rectangular coordinate system. What conjecture can you make about the location of a point when the sign of the x-coordinate is changed? Repeat the exercise for the case in which the sign of the y-coordinate is changed.

In Exercises 17–20, use the Distance Formula to determine whether the points lie on the same line.

17. $(0, -4), (2, 0), (3, 2)$

18. $(0, 4), (7, -6), (-5, 11)$

19. $(-2, 1), (-1, 0), (2, -2)$

20. $(-1, 1), (3, 3), (5, 5)$

In Exercises 21 and 22, find x such that the distance between the points is 5.

21. $(0, 0), (x, -4)$ **22.** $(2, -1), (x, 2)$

In Exercises 23 and 24, find y such that the distance between the points is 8.

23. $(0, 0), (3, y)$ **24.** $(5, 1), (5, y)$

25. Use the Midpoint Formula to find the three points that divide the line segment joining (x_1, y_1) and (x_2, y_2) into four equal parts.

26. Use the result of Exercise 25 to find the points that divide the line segment joining the given points into four equal parts.

(a) $(1, -2), (4, -1)$ (b) $(-2, -3), (0, 0)$

In Exercises 27–30, match the equation with its graph. [The graphs are labeled (a), (b), (c), and (d).]

(a)

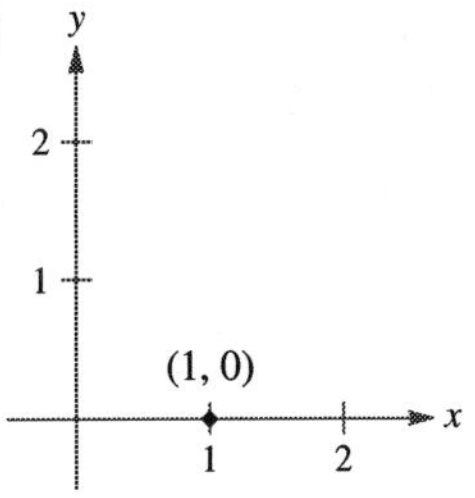

(b)

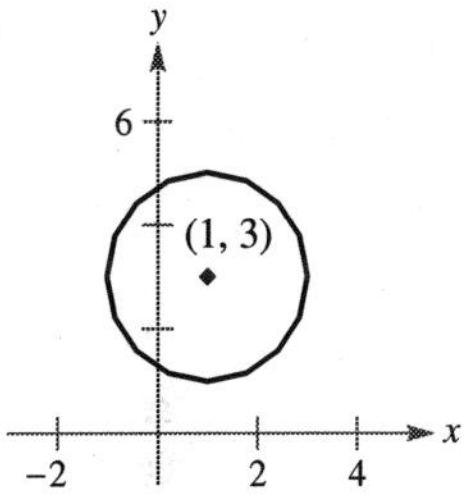

(c)

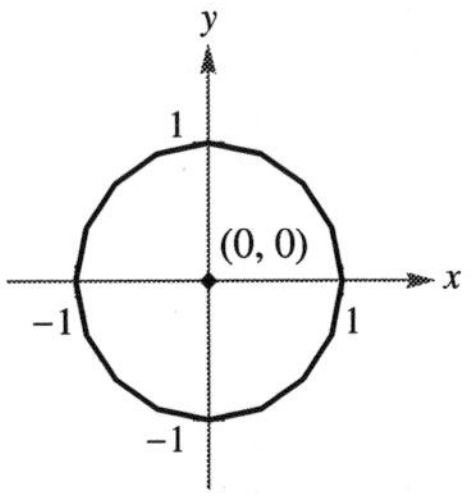

(d)

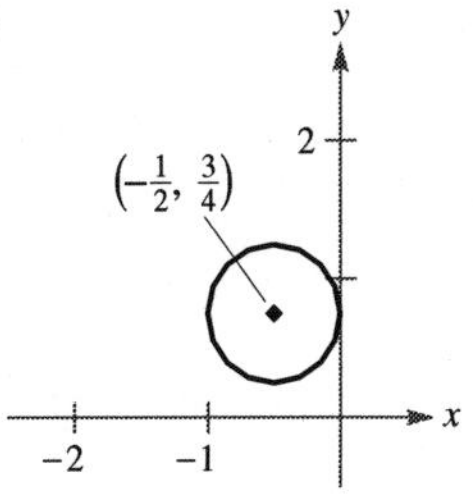

27. $x^2 + y^2 = 1$

28. $(x - 1)^2 + (y - 3)^2 = 4$

29. $(x - 1)^2 + y^2 = 0$

30. $\left(x + \frac{1}{2}\right)^2 + \left(y - \frac{3}{4}\right)^2 = \frac{1}{4}$

In Exercises 31–38, write the general form of the equation of the circle.

31. Center: $(0, 0)$
Radius: 3

32. Center: $(0, 0)$
Radius: 5

33. Center: $(2, -1)$
Radius: 4

34. Center: $(-4, 3)$
Radius: $\frac{5}{8}$

35. Center: $(-1, 2)$
Point on circle: $(0, 0)$

36. Center: $(3, -2)$
Point on circle: $(-1, 1)$

37. Endpoints of diameter: $(2, 5), (4, -1)$

38. Endpoints of diameter: $(1, 1), (-1, -1)$

39. ***Satellite Communication*** Write the standard form of the equation for the path of a communications satellite in a circular orbit 22,000 miles above Earth. (Assume that the radius of Earth is 4000 miles.)

40. ***Building Design*** A circular air duct of diameter D is fit firmly into the right-angle corner where a basement wall meets the floor (see figure). Find the diameter of the largest water pipe that can be run in the right-angle corner behind the air duct.

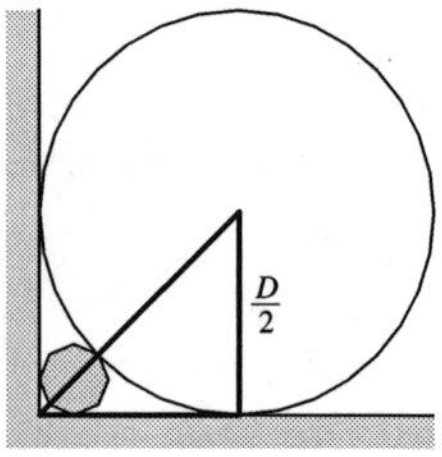

In Exercises 41–48, write the standard form of the equation of the circle and sketch its graph.

41. $x^2 + y^2 - 2x + 6y + 6 = 0$

42. $x^2 + y^2 - 2x + 6y - 15 = 0$

43. $x^2 + y^2 - 2x + 6y + 10 = 0$

44. $3x^2 + 3y^2 - 6y - 1 = 0$

45. $2x^2 + 2y^2 - 2x - 2y - 3 = 0$

46. $4x^2 + 4y^2 - 4x + 2y - 1 = 0$

47. $16x^2 + 16y^2 + 16x + 40y - 7 = 0$

48. $x^2 + y^2 - 4x + 2y + 3 = 0$

In Exercises 49 and 50, use a graphing utility to graph the equation. Use a square setting. (*Hint:* It may be necessary to solve the equation for *y* and graph the resulting two equations.)

49. $4x^2 + 4y^2 - 4x + 24y - 63 = 0$

50. $x^2 + y^2 - 8x - 6y - 11 = 0$

In Exercises 51 and 52, sketch the set of all points satisfying the inequality. Use a graphing utility to verify your result.

51. $x^2 + y^2 - 4x + 2y + 1 \le 0$

52. $(x - 1)^2 + \left(y - \frac{1}{2}\right)^2 > 1$

53. Prove that

$$\left(\frac{2x_1 + x_2}{3}, \frac{2y_1 + y_2}{3}\right)$$

is one of the points of trisection of the line segment joining (x_1, y_1) and (x_2, y_2). Find the midpoint of the line segment joining

$$\left(\frac{2x_1 + x_2}{3}, \frac{2y_1 + y_2}{3}\right)$$

and (x_2, y_2) to find the second point of trisection.

54. Use the results of Exercise 53 to find the points of trisection of the line segment joining the following points.

(a) $(1, -2), (4, 1)$ (b) $(-2, -3), (0, 0)$

True or False? **In Exercises 55–58, determine whether the statement is true or false. If it is false, explain why or give an example that shows it is false.**

55. If $ab < 0$, the point (a, b) lies in either Quadrant II or Quadrant IV.

56. The distance between the points $(a + b, a)$ and $(a - b, a)$ is $2b$.

57. If the distance between two points is zero, the two points must coincide.

58. If $ab = 0$, the point (a, b) lies on the x-axis or on the y-axis.

In Exercises 59–62, prove the statement.

59. The line segments joining the midpoints of the opposite sides of a quadrilateral bisect each other.

60. The perpendicular bisector of a chord of a circle passes through the center of the circle.

61. An angle inscribed in a semicircle is a right angle.

62. The midpoint of the line segment joining the points (x_1, y_1) and (x_2, y_2) is

$$\left(\frac{x_1 + x_2}{2}, \frac{y_1 + y_2}{2}\right).$$

APPENDIX D.3 Review of Trigonometric Functions

Angles and Degree Measure • Radian Measure • The Trigonometric Functions • Evaluating Trigonometric Functions • Solving Trigonometric Equations • Graphs of Trigonometric Functions

Angles and Degree Measure

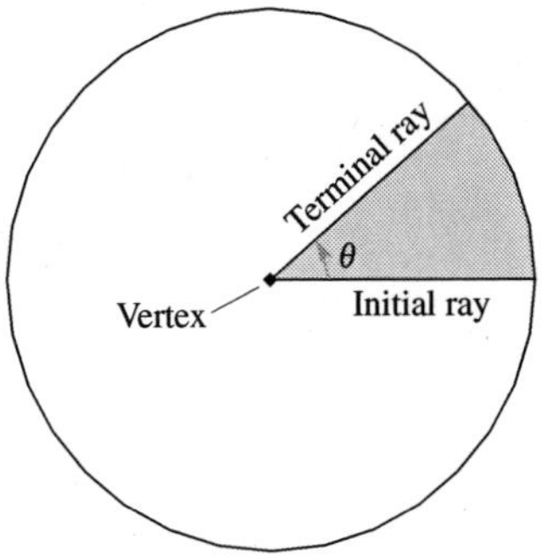

Standard position of an angle
Figure D.24

An **angle** has three parts: an **initial ray,** a **terminal ray,** and a **vertex** (the point of intersection of the two rays), as shown in Figure D.24. An angle is in **standard position** if its initial ray coincides with the positive x-axis and its vertex is at the origin. It is assumed that you are familiar with the degree measure of an angle.* It is common practice to use θ (the Greek lowercase *theta*) to represent both an angle and its measure. Angles between 0° and 90° are **acute,** and angles between 90° and 180° are **obtuse.**

Positive angles are measured *counterclockwise*, and negative angles are measured *clockwise*. For instance, Figure D.25 shows an angle whose measure is $-45°$. You cannot assign a measure to an angle by simply knowing where its initial and terminal rays are located. To measure an angle, you must also know how the terminal ray was revolved. For example, Figure D.25 shows that the angle measuring $-45°$ has the same terminal ray as the angle measuring 315°. Such angles are **coterminal.** In general, if θ is any angle, then

$$\theta + n(360), \qquad n \text{ is a nonzero integer}$$

is coterminal with θ.

An angle that is larger than 360° is one whose terminal ray has been revolved more than one full revolution counterclockwise, as shown in Figure D.26. You can form an angle whose measure is less than $-360°$ by revolving a terminal ray more than one full revolution clockwise.

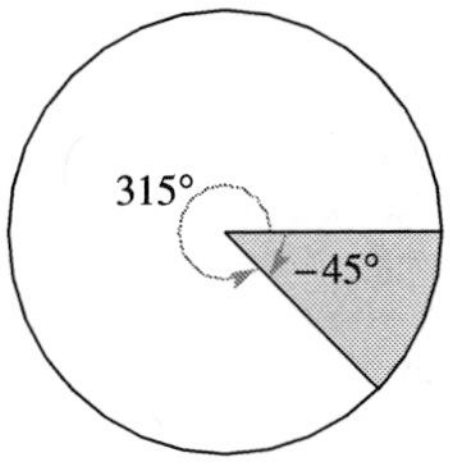

Coterminal angles
Figure D.25

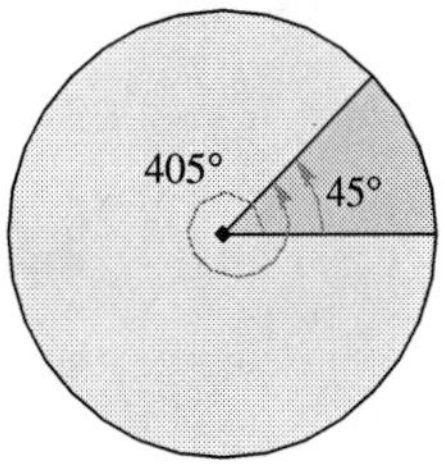

Coterminal angles
Figure D.26

NOTE It is common to use the symbol θ to refer to both an *angle* and its *measure*. For instance, in Figure D.26, you can write the measure of the smaller angle as $\theta = 45°$.

**For a more complete review of trigonometry, see Precalculus, 6th edition, by Larson and Hostetler (Boston, Massachusetts: Houghton Mifflin, 2004).*

Radian Measure

To assign a radian measure to an angle θ, consider θ to be a central angle of a circle of radius 1, as shown in Figure D.27. The **radian measure** of θ is then defined to be the length of the arc of the sector. Because the circumference of a circle is $2\pi r$, the circumference of a **unit circle** (of radius 1) is 2π. This implies that the radian measure of an angle measuring 360° is 2π. In other words, $360° = 2\pi$ radians.

Using radian measure for θ, the length s of a circular arc of radius r is $s = r\theta$, as shown in Figure D.28.

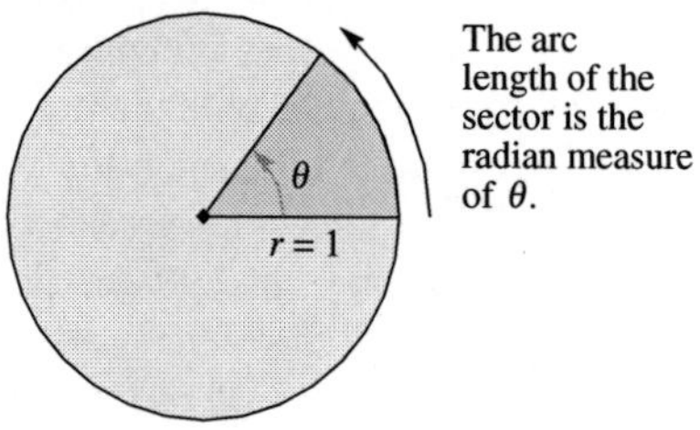

Unit circle
Figure D.27

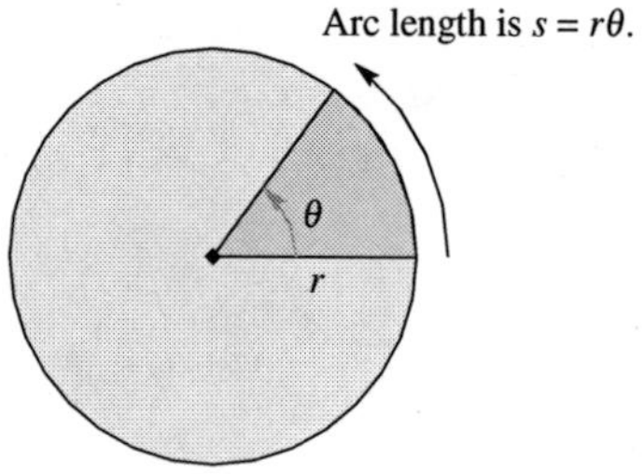

Circle of radius r
Figure D.28

You should know the conversions of the common angles shown in Figure D.29. For other angles, use the fact that 180° is equal to π radians.

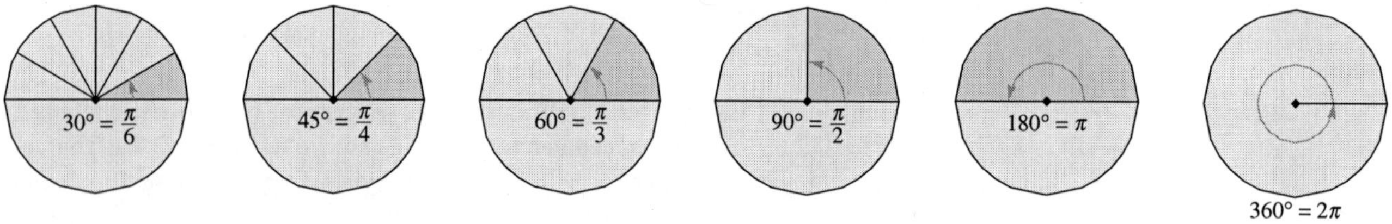

Radian and degree measure for several common angles
Figure D.29

EXAMPLE 1 Conversions Between Degrees and Radians

a. $40° = (40 \text{ deg})\left(\dfrac{\pi \text{ rad}}{180 \text{ deg}}\right) = \dfrac{2\pi}{9}$ radian

b. $540° = (540 \text{ deg})\left(\dfrac{\pi \text{ rad}}{180 \text{ deg}}\right) = 3\pi$ radians

c. $-270° = (-270 \text{ deg})\left(\dfrac{\pi \text{ rad}}{180 \text{ deg}}\right) = -\dfrac{3\pi}{2}$ radians

d. $-\dfrac{\pi}{2}$ radians $= \left(-\dfrac{\pi}{2} \text{ rad}\right)\left(\dfrac{180 \text{ deg}}{\pi \text{ rad}}\right) = -90°$

e. 2 radians $= (2 \text{ rad})\left(\dfrac{180 \text{ deg}}{\pi \text{ rad}}\right) = \dfrac{360}{\pi} \approx 114.59°$

f. $\dfrac{9\pi}{2}$ radians $= \left(\dfrac{9\pi}{2} \text{ rad}\right)\left(\dfrac{180 \text{ deg}}{\pi \text{ rad}}\right) = 810°$

The Trigonometric Functions

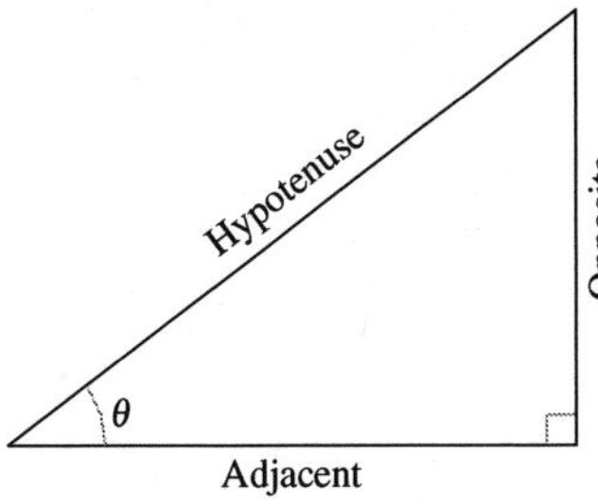

Sides of a right triangle
Figure D.30

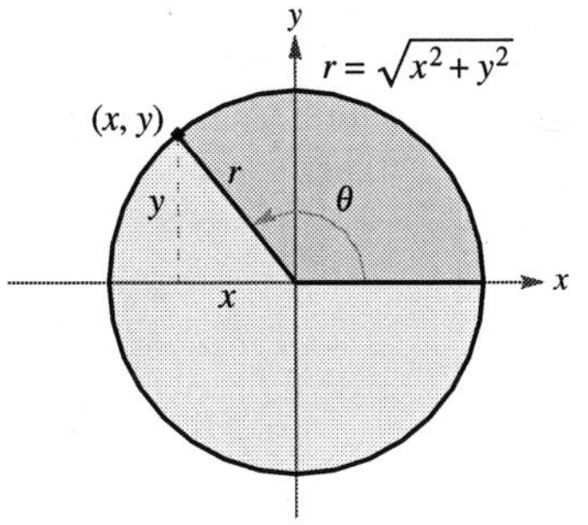

An angle in standard position
Figure D.31

There are two common approaches to the study of trigonometry. In one, the trigonometric functions are defined as ratios of two sides of a right triangle. In the other, these functions are defined in terms of a point on the terminal side of an angle in standard position. The six trigonometric functions, **sine, cosine, tangent, cotangent, secant,** and **cosecant** (abbreviated as sin, cos, etc.), are defined below from both viewpoints.

Definition of the Six Trigonometric Functions

Right triangle definitions, where $0 < \theta < \frac{\pi}{2}$ (see Figure D.30).

$$\sin\theta = \frac{\text{opposite}}{\text{hypotenuse}} \qquad \cos\theta = \frac{\text{adjacent}}{\text{hypotenuse}} \qquad \tan\theta = \frac{\text{opposite}}{\text{adjacent}}$$

$$\csc\theta = \frac{\text{hypotenuse}}{\text{opposite}} \qquad \sec\theta = \frac{\text{hypotenuse}}{\text{adjacent}} \qquad \cot\theta = \frac{\text{adjacent}}{\text{opposite}}$$

Circular function definitions, where θ *is any angle* (see Figure D.31).

$$\sin\theta = \frac{y}{r} \qquad \cos\theta = \frac{x}{r} \qquad \tan\theta = \frac{y}{x},\ x \neq 0$$

$$\csc\theta = \frac{r}{y},\ y \neq 0 \qquad \sec\theta = \frac{r}{x},\ x \neq 0 \qquad \cot\theta = \frac{x}{y},\ y \neq 0$$

The following trigonometric identities are direct consequences of the definitions. (ϕ is the Greek letter *phi.*)

Trigonometric Identities [*Note that* $\sin^2\theta$ *is used to represent* $(\sin\theta)^2$.]

Pythagorean Identities:

$$\sin^2\theta + \cos^2\theta = 1$$
$$1 + \tan^2\theta = \sec^2\theta$$
$$1 + \cot^2\theta = \csc^2\theta$$

Reduction Formulas:

$$\sin(-\theta) = -\sin\theta \qquad \sin\theta = -\sin(\theta - \pi)$$
$$\cos(-\theta) = \cos\theta \qquad \cos\theta = -\cos(\theta - \pi)$$
$$\tan(-\theta) = -\tan\theta \qquad \tan\theta = \tan(\theta - \pi)$$

Sum or Difference of Two Angles:

$$\sin(\theta \pm \phi) = \sin\theta\cos\phi \pm \cos\theta\sin\phi$$
$$\cos(\theta \pm \phi) = \cos\theta\cos\phi \mp \sin\theta\sin\phi$$
$$\tan(\theta \pm \phi) = \frac{\tan\theta \pm \tan\phi}{1 \mp \tan\theta\tan\phi}$$

Half-Angle Formulas:

$$\sin^2\theta = \frac{1 - \cos 2\theta}{2}$$
$$\cos^2\theta = \frac{1 + \cos 2\theta}{2}$$

Double-Angle Formulas:

$$\sin 2\theta = 2\sin\theta\cos\theta$$
$$\cos 2\theta = 2\cos^2\theta - 1 = 1 - 2\sin^2\theta = \cos^2\theta - \sin^2\theta$$

Law of Cosines:

$$a^2 = b^2 + c^2 - 2bc\cos A$$

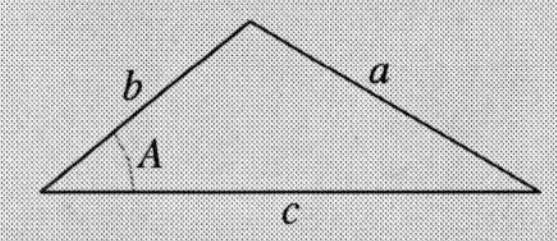

Reciprocal Identities:

$$\csc\theta = \frac{1}{\sin\theta}$$
$$\sec\theta = \frac{1}{\cos\theta}$$
$$\cot\theta = \frac{1}{\tan\theta}$$

Quotient Identities:

$$\tan\theta = \frac{\sin\theta}{\cos\theta}$$
$$\cot\theta = \frac{\cos\theta}{\sin\theta}$$

Evaluating Trigonometric Functions

There are two ways to evaluate trigonometric functions: (1) decimal approximations with a calculator and (2) exact evaluations using trigonometric identities and formulas from geometry. When using a calculator toevaluate a trigonometric function, remember to set the calculator to the appropriate mode—*degree* mode or *radian* mode.

EXAMPLE 2 Exact Evaluation of Trigonometric Functions

Evaluate the sine, cosine, and tangent of $\dfrac{\pi}{3}$.

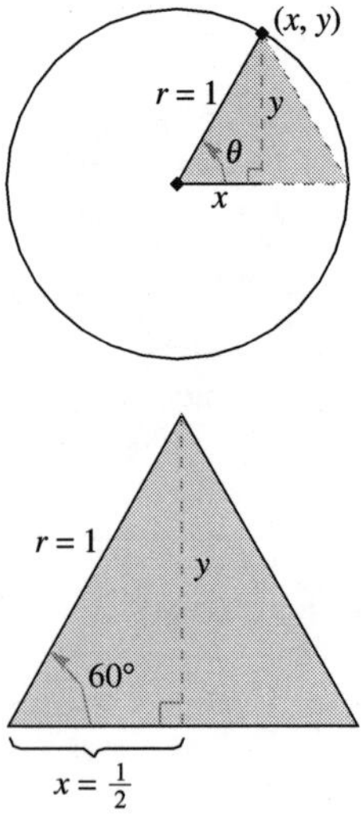

The angle $\pi/3$ in standard position
Figure D.32

Solution Begin by drawing the angle $\theta = \pi/3$ in standard position, as shown in Figure D.32. Then, because $60° = \pi/3$ radians, you can draw an equilateral triangle with sides of length 1 and θ as one of its angles. Because the altitude of this triangle bisects its base, you know that $x = \frac{1}{2}$. Using the Pythagorean Theorem, you obtain

$$y = \sqrt{r^2 - x^2} = \sqrt{1 - \left(\frac{1}{2}\right)^2} = \sqrt{\frac{3}{4}} = \frac{\sqrt{3}}{2}.$$

Now, knowing the values of x, y, and r, you can write the following.

$$\sin\frac{\pi}{3} = \frac{y}{r} = \frac{\sqrt{3}/2}{1} = \frac{\sqrt{3}}{2}$$

$$\cos\frac{\pi}{3} = \frac{x}{r} = \frac{1/2}{1} = \frac{1}{2}$$

$$\tan\frac{\pi}{3} = \frac{y}{x} = \frac{\sqrt{3}/2}{1/2} = \sqrt{3}$$

NOTE All angles in this text are measured in radians unless stated otherwise. For example, when sin 3 is written, the sine of 3 radians is meant, and when sin 3° is written, the sine of 3 degrees is meant.

The degree and radian measures of several common angles are shown in the table below, along with the corresponding values of the sine, cosine, and tangent (see Figure D.33).

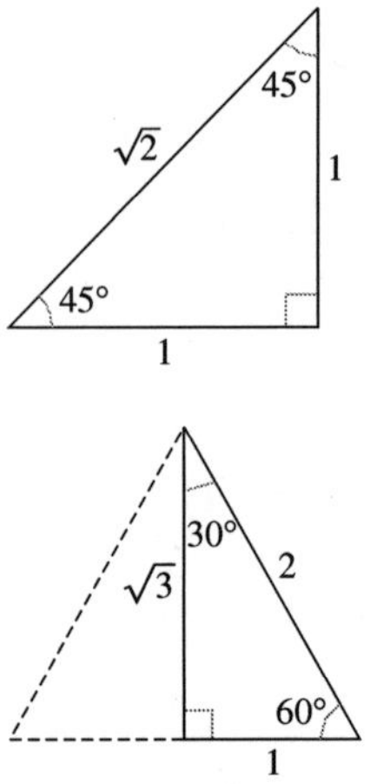

Common angles
Figure D.33

Common First Quadrant Angles

Degrees	0	30°	45°	60°	90°
Radians	0	$\frac{\pi}{6}$	$\frac{\pi}{4}$	$\frac{\pi}{3}$	$\frac{\pi}{2}$
sin θ	0	$\frac{1}{2}$	$\frac{\sqrt{2}}{2}$	$\frac{\sqrt{3}}{2}$	1
cos θ	1	$\frac{\sqrt{3}}{2}$	$\frac{\sqrt{2}}{2}$	$\frac{1}{2}$	0
tan θ	0	$\frac{\sqrt{3}}{3}$	1	$\sqrt{3}$	Undefined

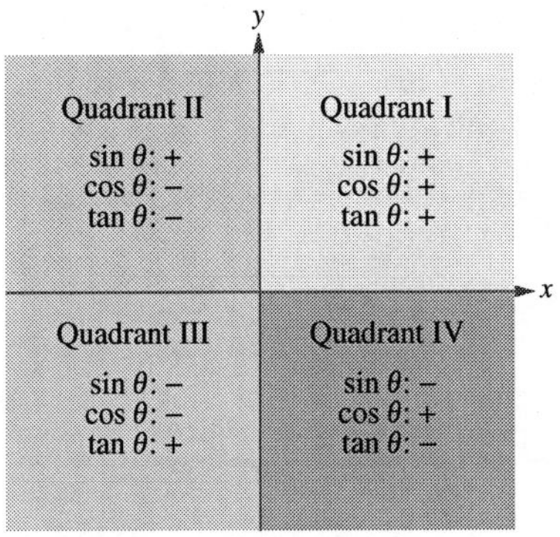

Quadrant signs for trigonometric functions
Figure D.34

The quadrant signs for the sine, cosine, and tangent functions are shown in Figure D.34. To extend the use of the table on the preceding page to angles in quadrants other than the first quadrant, you can use the concept of a **reference angle** (see Figure D.35), with the appropriate quadrant sign. For instance, the reference angle for $3\pi/4$ is $\pi/4$, and because the sine is positive in Quadrant II, you can write

$$\sin\frac{3\pi}{4} = +\sin\frac{\pi}{4} = \frac{\sqrt{2}}{2}.$$

Similarly, because the reference angle for 330° is 30°, and the tangent is negative in Quadrant IV, you can write

$$\tan 330° = -\tan 30° = -\frac{\sqrt{3}}{3}.$$

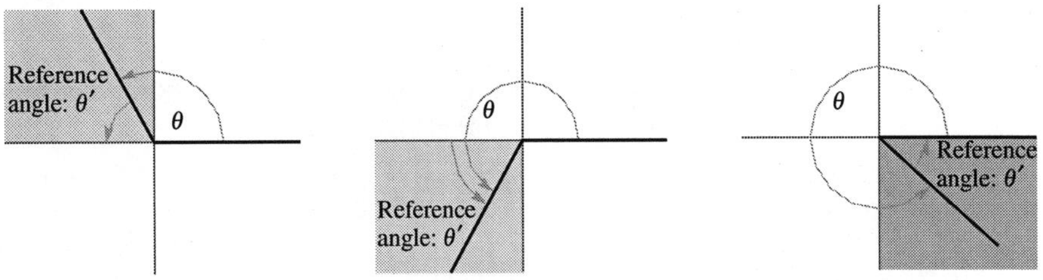

Quadrant II
$\theta' = \pi - \theta$ (radians)
$\theta' = 180° - \theta$ (degrees)

Quadrant III
$\theta' = \theta - \pi$ (radians)
$\theta' = \theta - 180°$ (degrees)

Quadrant IV
$\theta' = 2\pi - \theta$ (radians)
$\theta' = 360° - \theta$ (degrees)

Figure D.35

EXAMPLE 3 Trigonometric Identities and Calculators

Evaluate each trigonometric expression.

a. $\sin\left(-\frac{\pi}{3}\right)$ **b.** $\sec 60°$ **c.** $\cos(1.2)$

Solution

a. Using the reduction formula $\sin(-\theta) = -\sin\theta$, you can write

$$\sin\left(-\frac{\pi}{3}\right) = -\sin\frac{\pi}{3} = -\frac{\sqrt{3}}{2}.$$

b. Using the reciprocal identity $\sec\theta = 1/\cos\theta$, you can write

$$\sec 60° = \frac{1}{\cos 60°} = \frac{1}{1/2} = 2.$$

c. Using a calculator, you obtain

$$\cos(1.2) \approx 0.3624.$$

Remember that 1.2 is given in *radian* measure. Consequently, your calculator must be set in *radian* mode.

Solving Trigonometric Equations

How would you solve the equation $\sin \theta = 0$? You know that $\theta = 0$ is one solution, but this is not the only solution. Any one of the following values of θ is also a solution.

$$\ldots, -3\pi, -2\pi, -\pi, 0, \pi, 2\pi, 3\pi, \ldots$$

You can write this infinite solution set as $\{n\pi\colon n \text{ is an integer}\}$.

EXAMPLE 4 Solving a Trigonometric Equation

Solve the equation

$$\sin \theta = -\frac{\sqrt{3}}{2}.$$

Solution To solve the equation, you should consider that the sine is negative in Quadrants III and IV and that

$$\sin \frac{\pi}{3} = \frac{\sqrt{3}}{2}.$$

So, you are seeking values of θ in the third and fourth quadrants that have a reference angle of $\pi/3$. In the interval $[0, 2\pi]$, the two angles fitting these criteria are

$$\theta = \pi + \frac{\pi}{3} = \frac{4\pi}{3} \quad \text{and} \quad \theta = 2\pi - \frac{\pi}{3} = \frac{5\pi}{3}.$$

By adding integer multiples of 2π to each of these solutions, you obtain the following general solution.

$$\theta = \frac{4\pi}{3} + 2n\pi \quad \text{or} \quad \theta = \frac{5\pi}{3} + 2n\pi, \quad \text{where } n \text{ is an integer.}$$

See Figure D.36.

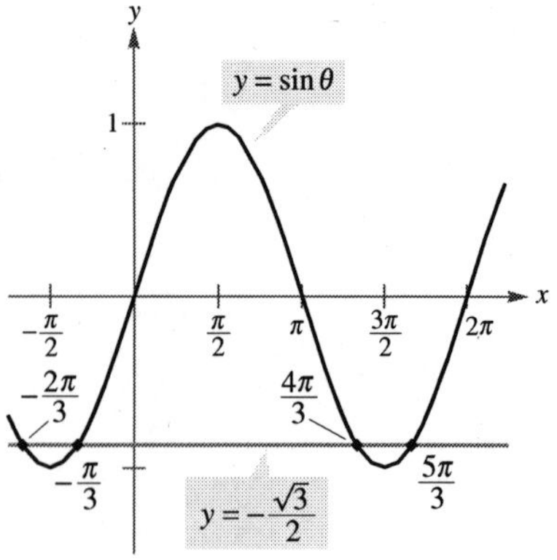

Solution points of $\sin u = -\frac{\sqrt{3}}{2}$
Figure D.36

EXAMPLE 5 Solving a Trigonometric Equation

Solve $\cos 2\theta = 2 - 3 \sin \theta$, where $0 \le \theta \le 2\pi$.

Solution Using the double-angle identity $\cos 2\theta = 1 - 2 \sin^2 \theta$, you can rewrite the equation as follows.

$$\cos 2\theta = 2 - 3 \sin \theta \qquad \text{Write original equation.}$$

$$1 - 2\sin^2 \theta = 2 - 3 \sin \theta \qquad \text{Trigonometric identity}$$

$$0 = 2 \sin^2 \theta - 3 \sin \theta + 1 \qquad \text{Quadratic form}$$

$$0 = (2 \sin \theta - 1)(\sin \theta - 1) \qquad \text{Factor.}$$

If $2 \sin \theta - 1 = 0$, then $\sin \theta = 1/2$ and $\theta = \pi/6$ or $\theta = 5\pi/6$. If $\sin \theta - 1 = 0$, then $\sin \theta = 1$ and $\theta = \pi/2$. So, for $0 \le \theta \le 2\pi$, there are three solutions.

$$\theta = \frac{\pi}{6}, \ \frac{5\pi}{6}, \ \text{or} \ \frac{\pi}{2}$$

Graphs of Trigonometric Functions

A function f is **periodic** if there exists a nonzero number p such that $f(x + p) = f(x)$ for all x in the domain of f. The smallest such positive value of p (if it exists) is the **period** of f. The sine, cosine, secant, and cosecant functions each have a period of 2π, and the other two trigonometric functions, tangent and cotangent, have a period of π, as shown in Figure D.37.

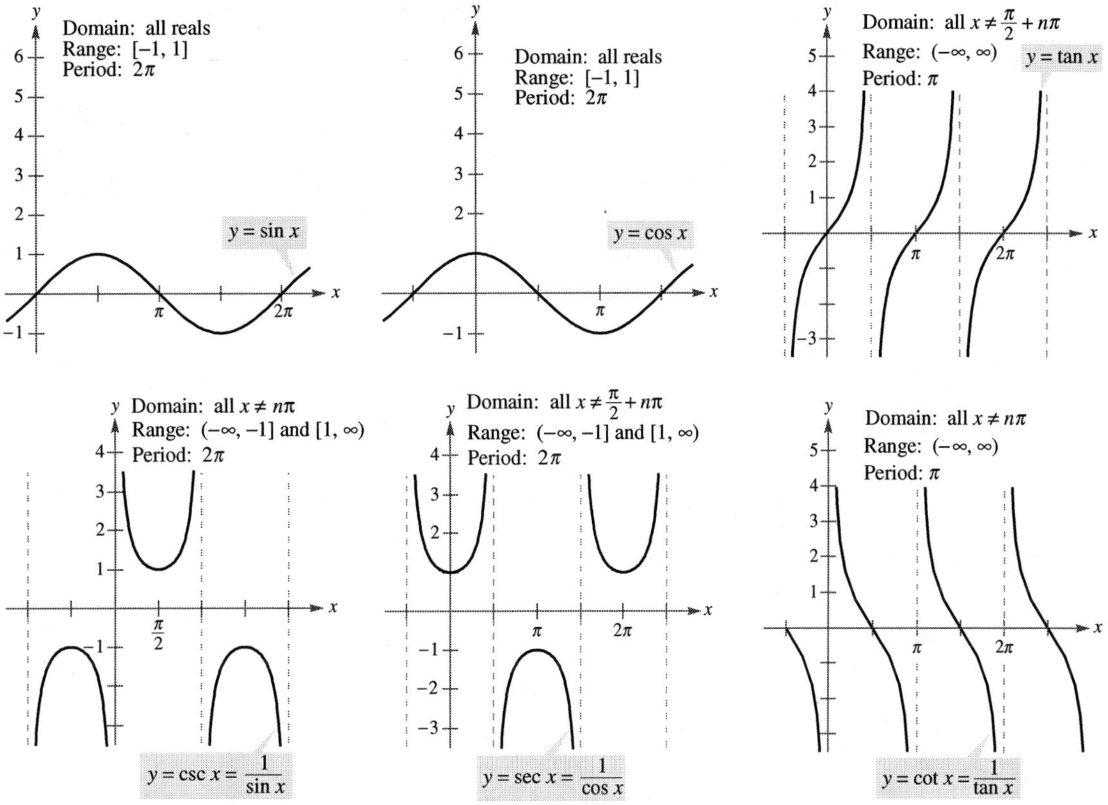

The graphs of the six trigonometric functions
Figure D.37

Note in Figure D.37 that the maximum value of $\sin x$ and $\cos x$ is 1 and the minimum value is -1. The graphs of the functions $y = a \sin bx$ and $y = a \cos bx$ oscillate between $-a$ and a, and so have an **amplitude** of $|a|$. Furthermore, because $bx = 0$ when $x = 0$ and $bx = 2\pi$ when $x = 2\pi/b$, it follows that the functions $y = a \sin bx$ and $y = a \cos bx$ each have a period of $2\pi/|b|$. The table below summarizes the amplitudes and periods for some types of trigonometric functions.

Function	Period	Amplitude
$y = a \sin bx$ or $y = a \cos bx$	$\dfrac{2\pi}{\lvert b\rvert}$	$\lvert a\rvert$
$y = a \tan bx$ or $y = a \cot bx$	$\dfrac{\pi}{\lvert b\rvert}$	Not applicable
$y = a \sec bx$ or $y = a \csc bx$	$\dfrac{2\pi}{\lvert b\rvert}$	Not applicable

EXAMPLE 6 Sketching the Graph of a Trigonometric Function

Sketch the graph of $f(x) = 3 \cos 2x$.

Solution The graph of $f(x) = 3 \cos 2x$ has an amplitude of 3 and a period of $2\pi/2 = \pi$. Using the basic shape of the graph of the cosine function, sketch one period of the function on the interval $[0, \pi]$, using the following pattern.

Maximum: $(0, 3)$

Minimum: $\left(\frac{\pi}{2}, -3\right)$

Maximum: $(\pi, 3)$

By continuing this pattern, you can sketch several cycles of the graph, as shown in Figure D.38.

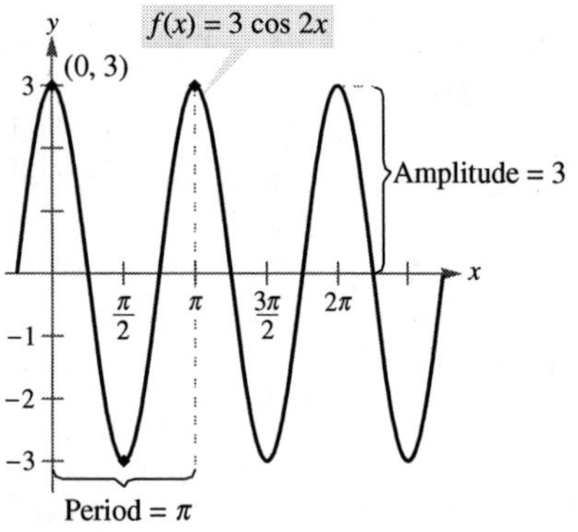

Figure D.38

Horizontal shifts, vertical shifts, and reflections can be applied to the graphs of trigonometric functions, as illustrated in Example 7.

EXAMPLE 7 Shifts of Graphs of Trigonometric Functions

Sketch the graph of each function.

a. $f(x) = \sin\left(x + \frac{\pi}{2}\right)$ **b.** $f(x) = 2 + \sin x$ **c.** $f(x) = 2 + \sin\left(x - \frac{\pi}{4}\right)$

Solution

a. To sketch the graph of $f(x) = \sin(x + \pi/2)$, shift the graph of $y = \sin x$ to the left $\pi/2$ units, as shown in Figure D.39(a).

b. To sketch the graph of $f(x) = 2 + \sin x$, shift the graph of $y = \sin x$ upward two units, as shown in Figure D.39(b).

c. To sketch the graph of $f(x) = 2 + \sin(x - \pi/4)$, shift the graph of $y = \sin x$ upward two units and to the right $\pi/4$ units, as shown in Figure D.39(c).

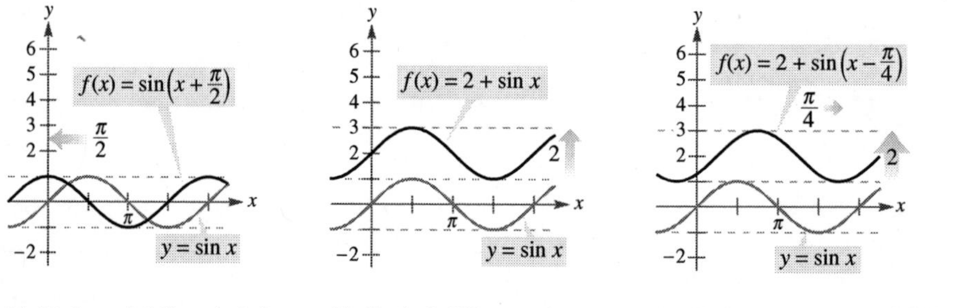

(a) Horizontal shift to the left (b) Vertical shift upward (c) Horizontal and vertical shifts

Transformations of the graph of $y = \sin x$
Figure D.39

EXERCISES FOR APPENDIX D.3

In Exercises 1 and 2, determine two coterminal angles in degree measure (one positive and one negative) for each angle.

1. (a)

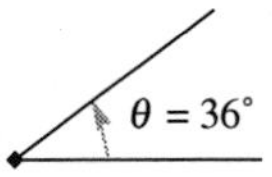

(b)

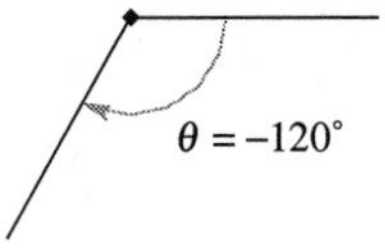

2. (a)

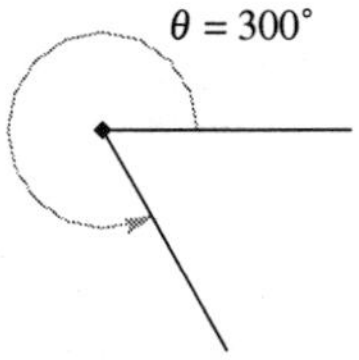

(b)

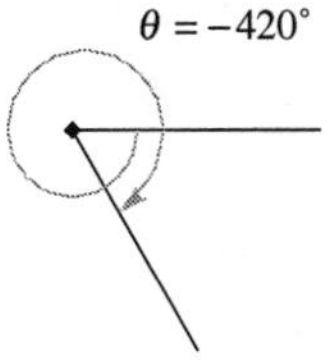

In Exercises 3 and 4, determine two coterminal angles in radian measure (one positive and one negative) for each angle.

3. (a)

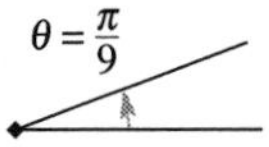

(b)

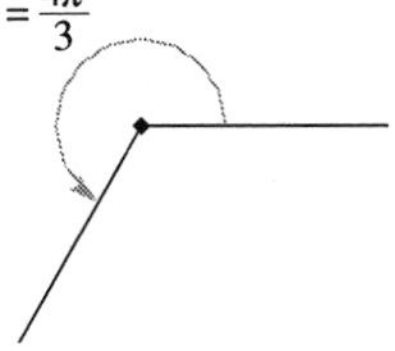

4. (a)

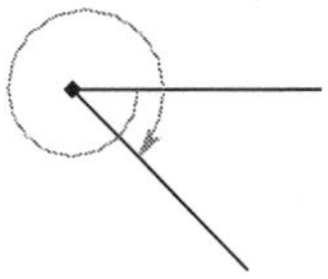

(b)

In Exercises 5 and 6, rewrite each angle in radian measure as a multiple of π and as a decimal accurate to three decimal places.

5. (a) 30° (b) 150° (c) 315° (d) 120°

6. (a) −20° (b) −240° (c) −270° (d) 144°

In Exercises 7 and 8, rewrite each angle in degree measure.

7. (a) $\frac{3\pi}{2}$ (b) $\frac{7\pi}{6}$ (c) $-\frac{7\pi}{12}$ (d) −2.367

8. (a) $\frac{7\pi}{3}$ (b) $-\frac{11\pi}{30}$ (c) $\frac{11\pi}{6}$ (d) 0.438

9. Let r represent the radius of a circle, θ the central angle (measured in radians), and s the length of the arc subtended by the angle. Use the relationship $s = r\theta$ to complete the table.

r	8 ft	15 in.	85 cm		
s	12 ft			96 in.	8642 mi
θ		1.6	$\frac{3\pi}{4}$	4	$\frac{2\pi}{3}$

10. ***Angular Speed*** A car is moving at the rate of 50 miles per hour, and the diameter of its wheels is 2.5 feet.

(a) Find the number of revolutions per minute that the wheels are rotating.

(b) Find the angular speed of the wheels in radians per minute.

In Exercises 11 and 12, determine all six trigonometric functions for the angle θ.

11. (a)

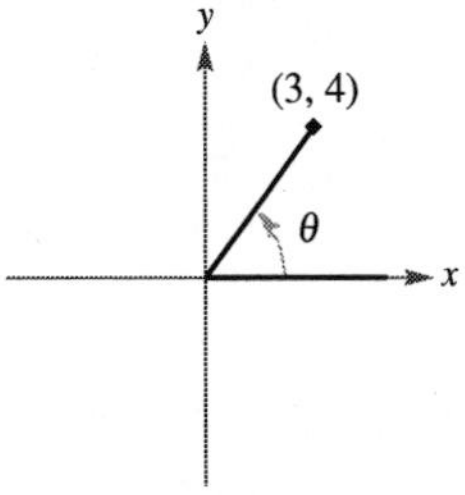

(b)

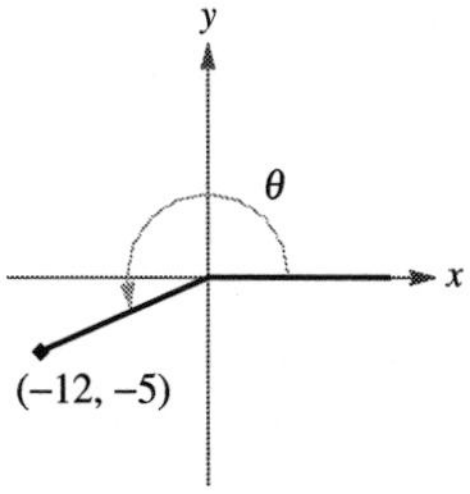

12. (a)

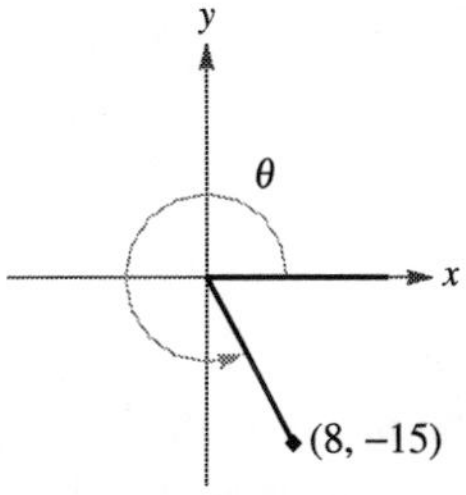

(b)

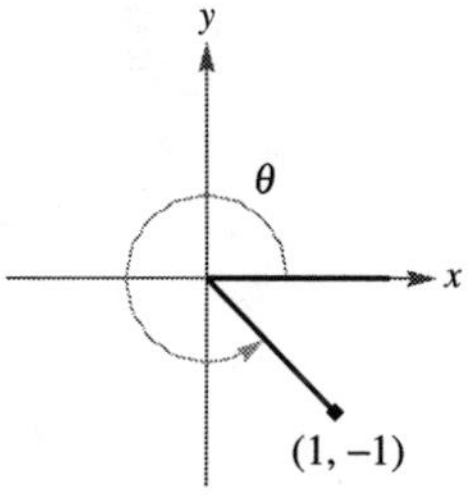

In Exercises 13 and 14, determine the quadrant in which θ lies.

13. (a) $\sin\theta < 0$ and $\cos\theta < 0$

(b) $\sec\theta > 0$ and $\cot\theta < 0$

14. (a) $\sin\theta > 0$ and $\cos\theta < 0$

(b) $\csc\theta < 0$ and $\tan\theta > 0$

In Exercises 15–18, evaluate the trigonometric function.

15. $\sin\theta = \frac{1}{2}$

$\cos\theta =$ ▒

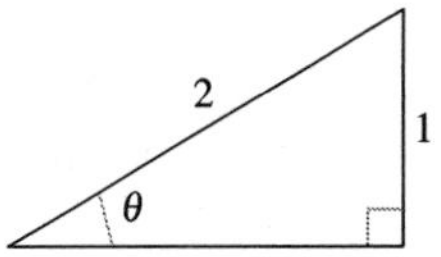

16. $\sin\theta = \frac{1}{3}$

$\tan\theta =$ ▒

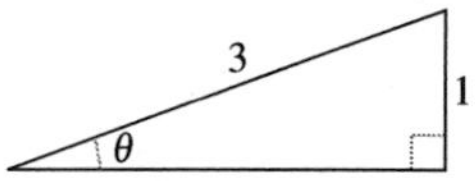

17. $\cos\theta = \frac{4}{5}$

$\cot\theta =$ ▒

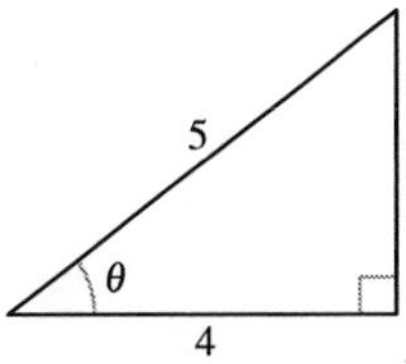

18. $\sec\theta = \frac{13}{5}$

$\csc\theta =$ ▒

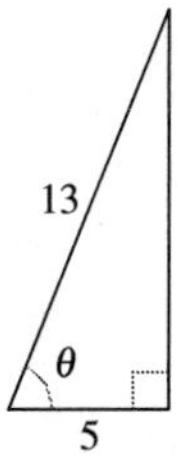

In Exercises 19–22, evaluate the sine, cosine, and tangent of each angle *without* using a calculator.

19. (a) $60°$
(b) $120°$
(c) $\frac{\pi}{4}$
(d) $\frac{5\pi}{4}$

20. (a) $-30°$
(b) $150°$
(c) $-\frac{\pi}{6}$
(d) $\frac{\pi}{2}$

21. (a) $225°$
(b) $-225°$
(c) $\frac{5\pi}{3}$
(d) $\frac{11\pi}{6}$

22. (a) $750°$
(b) $510°$
(c) $\frac{10\pi}{3}$
(d) $\frac{17\pi}{3}$

In Exercises 23–26, use a calculator to evaluate each trigonometric function. Round your answers to four decimal places.

23. (a) $\sin 10°$
(b) $\csc 10°$

24. (a) $\sec 225°$
(b) $\sec 135°$

25. (a) $\tan \frac{\pi}{9}$
(b) $\tan \frac{10\pi}{9}$

26. (a) $\cot(1.35)$
(b) $\tan(1.35)$

In Exercises 27–30, find two solutions of each equation. Give your answers in radians ($0 \le \theta < 2\pi$). Do not use a calculator.

27. (a) $\cos \theta = \frac{\sqrt{2}}{2}$
(b) $\cos \theta = -\frac{\sqrt{2}}{2}$

28. (a) $\sec \theta = 2$
(b) $\sec \theta = -2$

29. (a) $\tan \theta = 1$
(b) $\cot \theta = -\sqrt{3}$

30. (a) $\sin \theta = \frac{\sqrt{3}}{2}$
(b) $\sin \theta = -\frac{\sqrt{3}}{2}$

In Exercises 31–38, solve the equation for θ ($0 \le \theta < 2\pi$).

31. $2 \sin^2 \theta = 1$

32. $\tan^2 \theta = 3$

33. $\tan^2 \theta - \tan \theta = 0$

34. $2 \cos^2 \theta - \cos \theta = 1$

35. $\sec \theta \csc \theta = 2 \csc \theta$

36. $\sin \theta = \cos \theta$

37. $\cos^2 \theta + \sin \theta = 1$

38. $\cos \frac{\theta}{2} - \cos \theta = 1$

39. *Airplane Ascent* An airplane leaves the runway climbing at an angle of 18° with a speed of 275 feet per second (see figure). Find the altitude a of the plane after 1 minute.

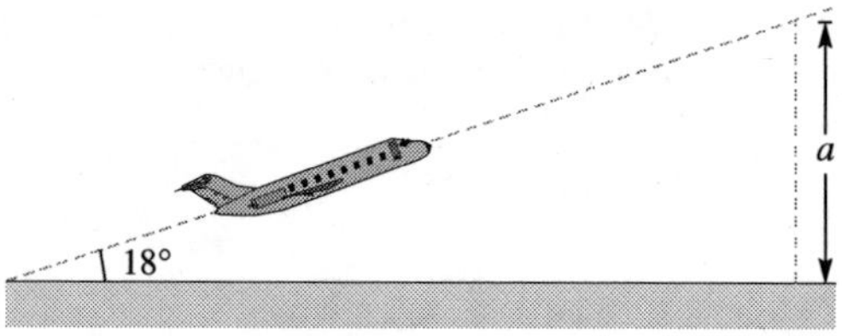

40. *Height of a Mountain* In traveling across flat land, you notice a mountain directly in front of you. Its angle of elevation (to the peak) is 3.5°. After you drive 13 miles closer to the mountain, the angle of elevation is 9°. Approximate the height of the mountain.

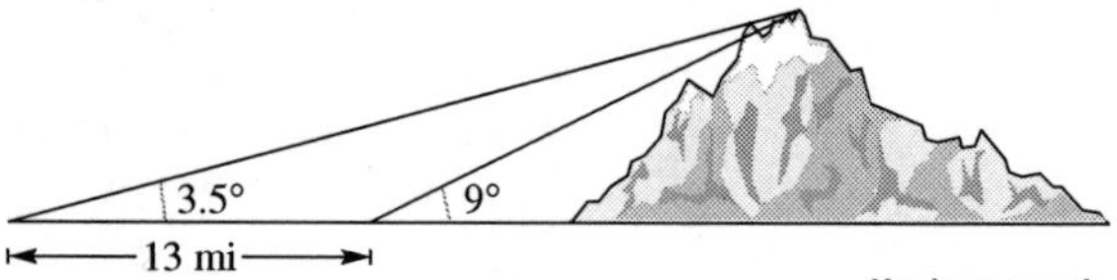

Not drawn to scale

In Exercises 41–44, determine the period and amplitude of each function.

41. (a) $y = 2 \sin 2x$

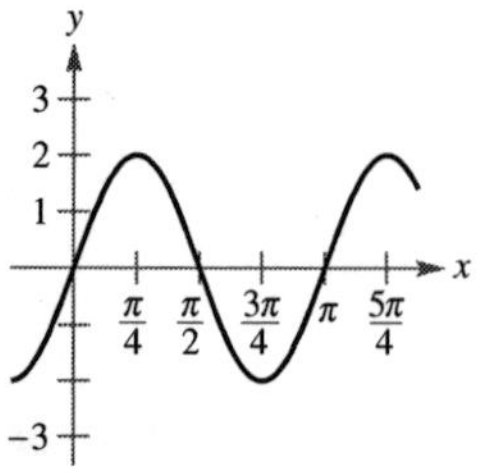

(b) $y = \frac{1}{2} \sin \pi x$

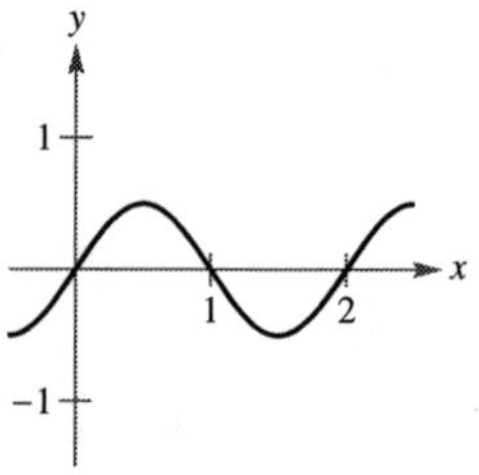

42. (a) $y = \frac{3}{2} \cos \frac{x}{2}$

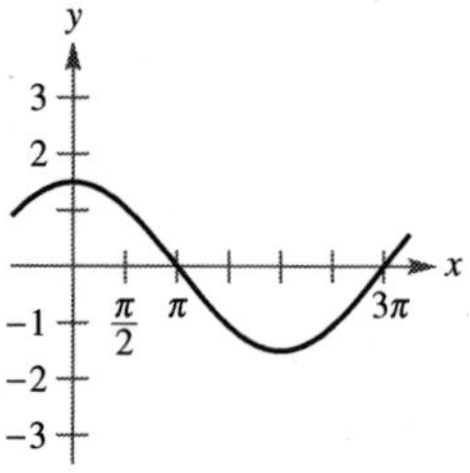

(b) $y = -2 \sin \frac{x}{3}$

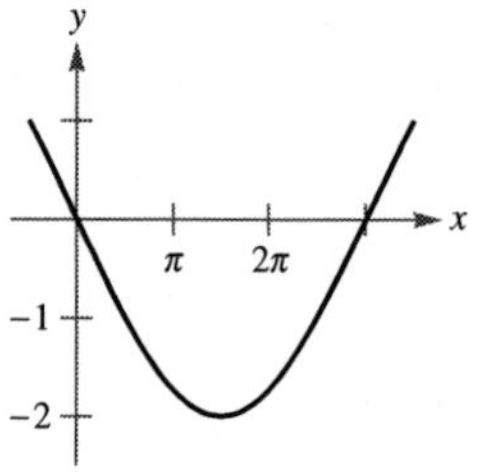

43. $y = 3 \sin 4\pi x$

44. $y = \frac{2}{3} \cos \frac{\pi x}{10}$

In Exercises 45–48, find the period of the function.

45. $y = 5 \tan 2x$

46. $y = 7 \tan 2\pi x$

47. $y = \sec 5x$

48. $y = \csc 4x$

Writing **In Exercises 49 and 50, use a graphing utility to graph each function f in the same viewing window for $c = -2$, $c = -1$, $c = 1$, and $c = 2$. Give a written description of the change in the graph caused by changing c.**

49. (a) $f(x) = c \sin x$
(b) $f(x) = \cos(cx)$
(c) $f(x) = \cos(\pi x - c)$

50. (a) $f(x) = \sin x + c$
(b) $f(x) = -\sin(2\pi x - c)$
(c) $f(x) = c \cos x$

In Exercises 51–62, sketch the graph of the function.

51. $y = \sin \dfrac{x}{2}$

52. $y = 2 \cos 2x$

53. $y = -\sin \dfrac{2\pi x}{3}$

54. $y = 2 \tan x$

55. $y = \csc \dfrac{x}{2}$

56. $y = \tan 2x$

57. $y = 2 \sec 2x$

58. $y = \csc 2\pi x$

59. $y = \sin(x + \pi)$

60. $y = \cos\left(x - \dfrac{\pi}{3}\right)$

61. $y = 1 + \cos\left(x - \dfrac{\pi}{2}\right)$

62. $y = 1 + \sin\left(x + \dfrac{\pi}{2}\right)$

Graphical Reasoning **In Exercises 63 and 64, find a, b, and c such that the graph of the function matches the graph in the figure.**

63. $y = a \cos(bx - c)$

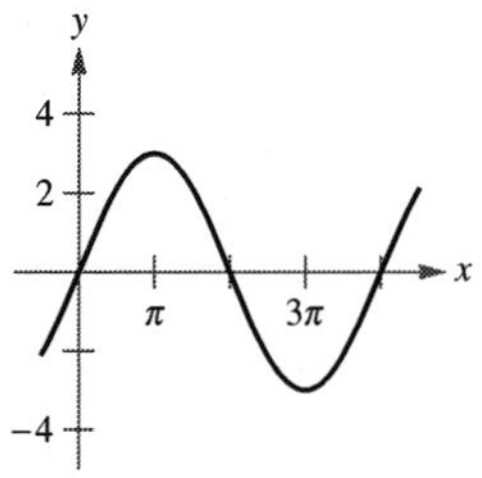

64. $y = a \sin(bx - c)$

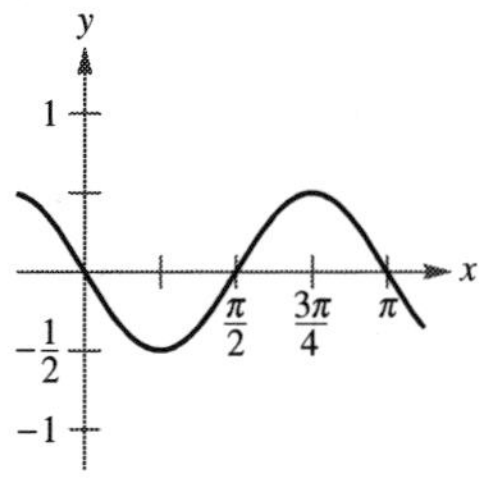

65. ***Think About It*** Sketch the graphs of $f(x) = \sin x$, $g(x) = |\sin x|$, and $h(x) = \sin(|x|)$. In general, how are the graphs of $|f(x)|$ and $f(|x|)$ related to the graph of f?

66. ***Think About It*** The model for the height h of a Ferris wheel car is

$$h = 51 + 50 \sin 8\pi t$$

where t is measured in minutes. (The Ferris wheel has a radius of 50 feet.) This model yields a height of 51 feet when $t = 0$. Alter the model so that the height of the car is 1 foot when $t = 0$.

67. ***Sales*** The monthly sales S (in thousands of units) of a seasonal product are modeled by

$$S = 58.3 + 32.5 \cos \frac{\pi t}{6}$$

where t is the time (in months) with $t = 1$ corresponding to January. Use a graphing utility to graph the model for S and determine the months when sales exceed 75,000 units.

68. ***Investigation*** Two trigonometric functions f and g have a period of 2, and their graphs intersect at $x = 5.35$.

(a) Give one smaller and one larger positive value of x where the functions have the same value.

(b) Determine one negative value of x where the graphs intersect.

(c) Is it true that $f(13.35) = g(-4.65)$? Give a reason for your answer.

Pattern Recognition **In Exercises 69 and 70, use a graphing utility to compare the graph of f with the given graph. Try to improve the approximation by adding a term to $f(x)$. Use a graphing utility to verify that your new approximation is better than the original. Can you find other terms to add to make the approximation even better? What is the pattern? (In Exercise 69, sine terms can be used to improve the approximation and in Exercise 70, cosine terms can be used.)**

69. $f(x) = \dfrac{4}{\pi}\left(\sin \pi x + \dfrac{1}{3} \sin 3\pi x\right)$

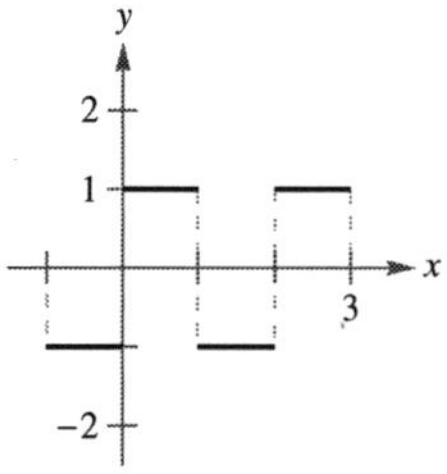

70. $f(x) = \dfrac{1}{2} - \dfrac{4}{\pi^2}\left(\cos \pi x + \dfrac{1}{9} \cos 3\pi x\right)$

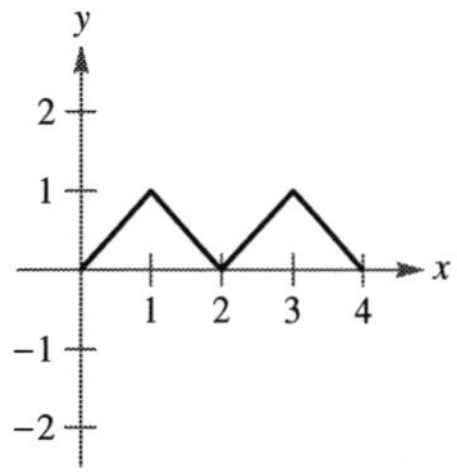

APPENDIX E

Rotation and the General Second-Degree Equation

Rotation of Axes • Invariants Under Rotation

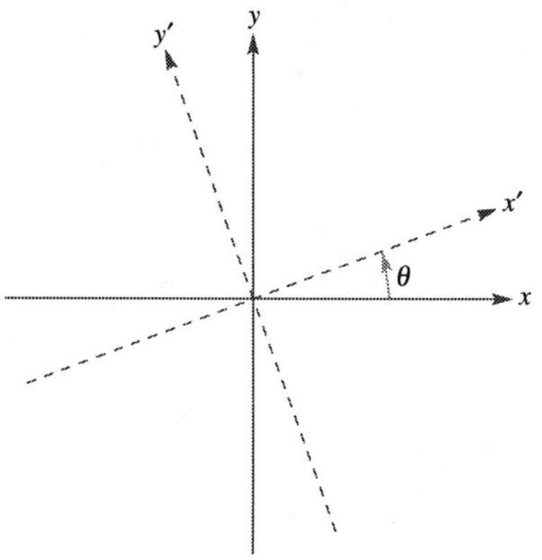

After rotation of the x- and y-axes counterclockwise through an angle θ, the rotated axes are denoted as the x'-axis and y'-axis.
Figure E.1

Rotation of Axes

Previously, you learned that equations of conics with axes parallel to one of the coordinate axes can be written in the general form

$$Ax^2 + Cy^2 + Dx + Ey + F = 0. \qquad \text{Horizontal or vertical axes}$$

Here you will study the equations of conics whose axes are rotated so that they are *not* parallel to either the x-axis or the y-axis. The general equation for such conics contains an xy-term.

$$Ax^2 + Bxy + Cy^2 + Dx + Ey + F = 0 \qquad \text{Equation in } xy\text{-plane}$$

To eliminate this xy-term, you can use a procedure called **rotation of axes.** You want to rotate the x- and y-axes until they are parallel to the axes of the conic. (The rotated axes are denoted as the x'-axis and the y'-axis, as shown in Figure E.1.) After the rotation has been accomplished, the equation of the conic in the new $x'y'$-plane will have the form

$$A'(x')^2 + C'(y')^2 + D'x' + E'y' + F' = 0. \qquad \text{Equation in } x'y'\text{-plane}$$

Because this equation has no $x'y'$-term, you can obtain a standard form by completing the square.

The following theorem identifies how much to rotate the axes to eliminate an xy-term and also the equations for determining the new coefficients A', C', D', E', and F'.

THEOREM E.1 Rotation of Axes

The general second-degree equation of

$$Ax^2 + Bxy + Cy^2 + Dx + Ey + F = 0,$$

where $B \neq 0$, can be rewritten as

$$A'(x')^2 + C'(y')^2 + D'x' + E'y' + F' = 0$$

by rotating the coordinate axes through an angle θ, where

$$\cot 2\theta = \frac{A - C}{B}.$$

The coefficients of the new equation are obtained by making the substitutions

$$x = x'\cos\theta - y'\sin\theta$$

$$y = x'\sin\theta + y'\cos\theta.$$

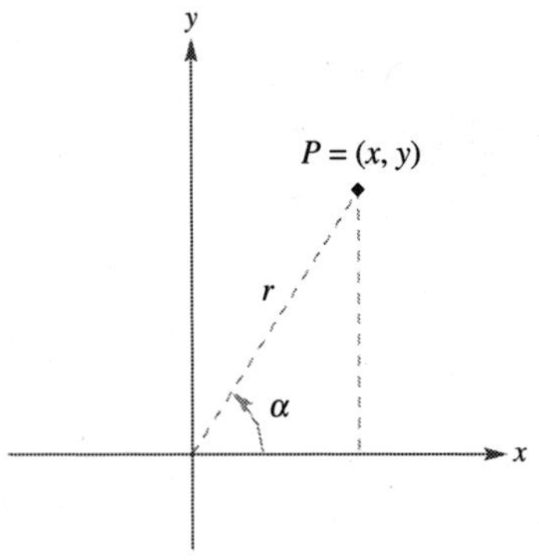

Original: $x = r\cos\alpha$
$y = r\sin\alpha$

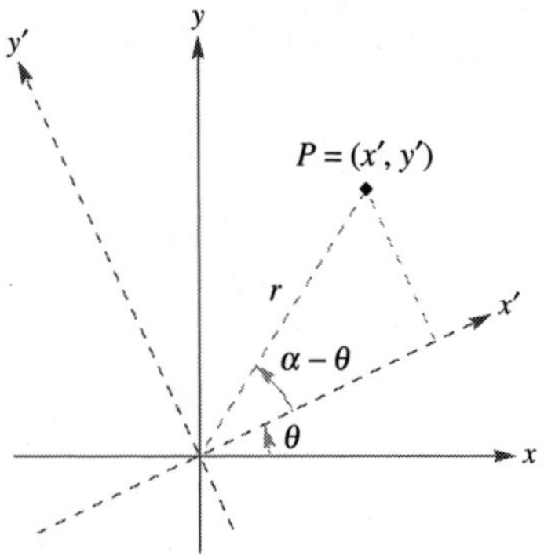

Rotated: $x' = r\cos(\alpha - \theta)$
$y' = r\sin(\alpha - \theta)$

Figure E.2

Proof To discover how the coordinates in the xy-system are related to the coordinates in the $x'y'$-system, choose a point $P = (x, y)$ in the original system and attempt to find its coordinates (x', y') in the rotated system. In either system, the distance r between the point P and the origin is the same, and so the equations for x, y, x', and y' are those given in Figure E.2. Using the formulas for the sine and cosine of the difference of two angles, you obtain

$$\begin{aligned}
x' &= r\cos(\alpha - \theta)\\
&= r(\cos\alpha\cos\theta + \sin\alpha\sin\theta)\\
&= r\cos\alpha\cos\theta + r\sin\alpha\sin\theta\\
&= x\cos\theta + y\sin\theta\\
y' &= r\sin(\alpha - \theta)\\
&= r(\sin\alpha\cos\theta - \cos\alpha\sin\theta)\\
&= r\sin\alpha\cos\theta - r\cos\alpha\sin\theta\\
&= y\cos\theta - x\sin\theta.
\end{aligned}$$

Solving this system for x and y yields

$$x = x'\cos\theta - y'\sin\theta \qquad \text{and} \qquad y = x'\sin\theta + y'\cos\theta.$$

Finally, by substituting these values for x and y into the original equation and collecting terms, you obtain the following.

$$\begin{aligned}
A' &= A\cos^2\theta + B\cos\theta\sin\theta + C\sin^2\theta\\
C' &= A\sin^2\theta - B\cos\theta\sin\theta + C\cos^2\theta\\
D' &= D\cos\theta + E\sin\theta\\
E' &= -D\sin\theta + E\cos\theta\\
F' &= F
\end{aligned}$$

Now, in order to eliminate the $x'y'$-term, you must select θ such that $B' = 0$, as follows.

$$\begin{aligned}
B' &= 2(C - A)\sin\theta\cos\theta + B(\cos^2\theta - \sin^2\theta)\\
&= (C - A)\sin 2\theta + B\cos 2\theta\\
&= B(\sin 2\theta)\left(\frac{C - A}{B} + \cot 2\theta\right)\\
&= 0, \quad \sin 2\theta \neq 0
\end{aligned}$$

If $B = 0$, no rotation is necessary, because the xy-term is not present in the original equation. If $B \neq 0$, the only way to make $B' = 0$ is to let

$$\cot 2\theta = \frac{A - C}{B}, \quad B \neq 0.$$

So, you have established the desired results.

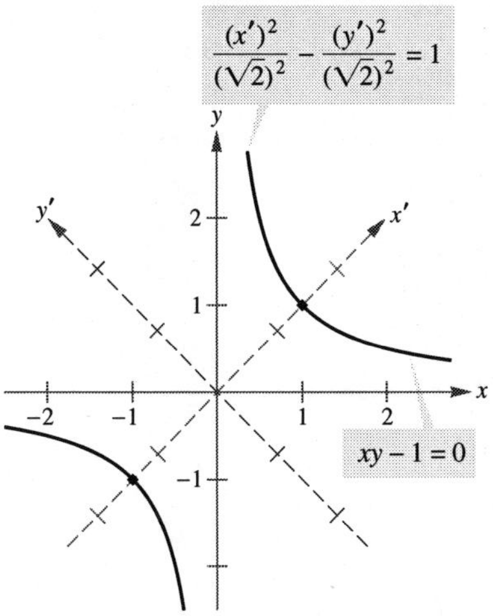

Vertices:
$(\sqrt{2}, 0), (-\sqrt{2}, 0)$ in $x'y'$-system
$(1, 1), (-1, -1)$ in xy-system
Figure E.3

EXAMPLE 1 Rotation of a Hyperbola

Write the equation $xy - 1 = 0$ in standard form.

Solution Because $A = 0, B = 1$, and $C = 0$, you have (for $0 < \theta < \pi/2$)

$$\cot 2\theta = \frac{A - C}{B} = 0 \quad \Longrightarrow \quad 2\theta = \frac{\pi}{2} \quad \Longrightarrow \quad \theta = \frac{\pi}{4}.$$

The equation in the $x'y'$-system is obtained by making the following substitutions.

$$x = x'\cos\frac{\pi}{4} - y'\sin\frac{\pi}{4} = x'\left(\frac{\sqrt{2}}{2}\right) - y'\left(\frac{\sqrt{2}}{2}\right) = \frac{x' - y'}{\sqrt{2}}$$

$$y = x'\sin\frac{\pi}{4} + y'\cos\frac{\pi}{4} = x'\left(\frac{\sqrt{2}}{2}\right) + y'\left(\frac{\sqrt{2}}{2}\right) = \frac{x' + y'}{\sqrt{2}}$$

Substituting these expressions into the equation $xy - 1 = 0$ produces

$$\left(\frac{x' - y'}{\sqrt{2}}\right)\left(\frac{x' + y'}{\sqrt{2}}\right) - 1 = 0$$

$$\frac{(x')^2 - (y')^2}{2} - 1 = 0$$

$$\frac{(x')^2}{(\sqrt{2})^2} - \frac{(y')^2}{(\sqrt{2})^2} = 1. \qquad \text{Write in standard form.}$$

This is the equation of a hyperbola centered at the origin with vertices at $(\pm\sqrt{2}, 0)$ in the $x'y'$-system, as shown in Figure E.3.

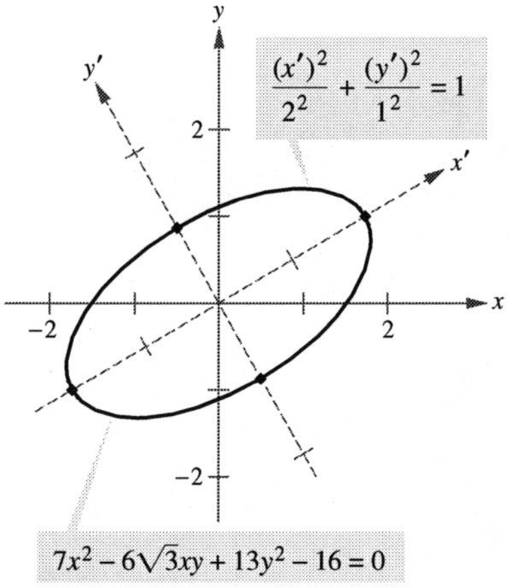

Vertices:
$(\pm 2, 0), (0, \pm 1)$ in $x'y'$-system
$(\pm\sqrt{3}, \pm 1), \left(\pm\frac{1}{2}, \pm\frac{\sqrt{3}}{2}\right)$ in xy-system

Figure E.4

EXAMPLE 2 Rotation of an Ellipse

Sketch the graph of $7x^2 - 6\sqrt{3}xy + 13y^2 - 16 = 0$.

Solution Because $A = 7, B = -6\sqrt{3}$, and $C = 13$, you have (for $0 < \theta < \pi/2$)

$$\cot 2\theta = \frac{A - C}{B} = \frac{7 - 13}{-6\sqrt{3}} = \frac{1}{\sqrt{3}} \quad \Longrightarrow \quad \theta = \frac{\pi}{6}.$$

Therefore, the equation in the $x'y'$-system is derived by making the following substitutions.

$$x = x'\cos\frac{\pi}{6} - y'\sin\frac{\pi}{6} = x'\left(\frac{\sqrt{3}}{2}\right) - y'\left(\frac{1}{2}\right) = \frac{\sqrt{3}x' - y'}{2}$$

$$y = x'\sin\frac{\pi}{6} + y'\cos\frac{\pi}{6} = x'\left(\frac{1}{2}\right) + y'\left(\frac{\sqrt{3}}{2}\right) = \frac{x' + \sqrt{3}y'}{2}$$

Substituting these expressions into the original equation eventually simplifies (after considerable algebra) to

$$4(x')^2 + 16(y')^2 = 16$$

$$\frac{(x')^2}{2^2} + \frac{(y')^2}{1^2} = 1. \qquad \text{Write in standard form.}$$

This is the equation of an ellipse centered at the origin with vertices at $(\pm 2, 0)$ and $(0, \pm 1)$ in the $x'y'$-system, as shown in Figure E.4.

In Examples 1 and 2, the values of θ were the common angles 45° and 30°, respectively. Of course, many second-degree equations do not yield such common solutions to the equation

$$\cot 2\theta = \frac{A - C}{B}.$$

Example 3 illustrates such a case.

EXAMPLE 3 Rotation of a Parabola

Sketch the graph of $x^2 - 4xy + 4y^2 + 5\sqrt{5}y + 1 = 0$.

Solution Because $A = 1, B = -4$, and $C = 4$, you have

$$\cot 2\theta = \frac{A - C}{B} = \frac{1 - 4}{-4} = \frac{3}{4}.$$

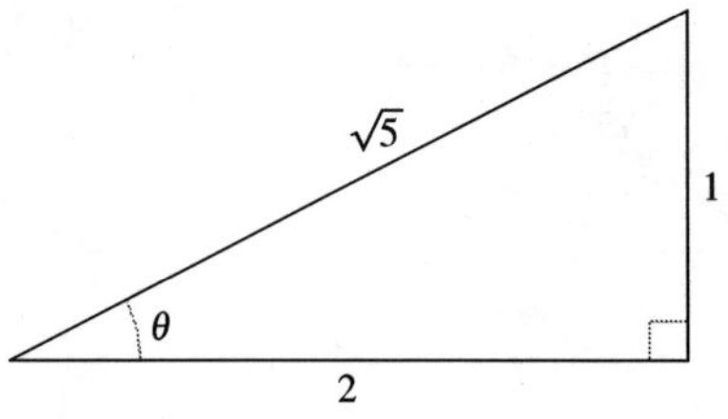

Figure E.5

The trigonometric identity $\cot 2\theta = (\cot^2 \theta - 1)/(2 \cot \theta)$ produces

$$\cot 2\theta = \frac{3}{4} = \frac{\cot^2 \theta - 1}{2 \cot \theta}$$

from which you obtain the equation

$$6 \cot \theta = 4 \cot^2 \theta - 4 \quad \Longrightarrow \quad 4 \cot^2 \theta - 6 \cot \theta - 4 = 0$$

$$(2 \cot \theta - 4)(2 \cot \theta + 1) = 0.$$

Considering $0 < \theta < \pi/2$, it follows that $2 \cot \theta = 4$. So,

$$\cot \theta = 2 \quad \Longrightarrow \quad \theta \approx 26.6°.$$

From the triangle in Figure E.5, you obtain $\sin \theta = 1/\sqrt{5}$ and $\cos \theta = 2/\sqrt{5}$. Consequently, you can write the following.

$$x = x' \cos \theta - y' \sin \theta = x'\left(\frac{2}{\sqrt{5}}\right) - y'\left(\frac{1}{\sqrt{5}}\right) = \frac{2x' - y'}{\sqrt{5}}$$

$$y = x' \sin \theta + y' \cos \theta = x'\left(\frac{1}{\sqrt{5}}\right) + y'\left(\frac{2}{\sqrt{5}}\right) = \frac{x' + 2y'}{\sqrt{5}}$$

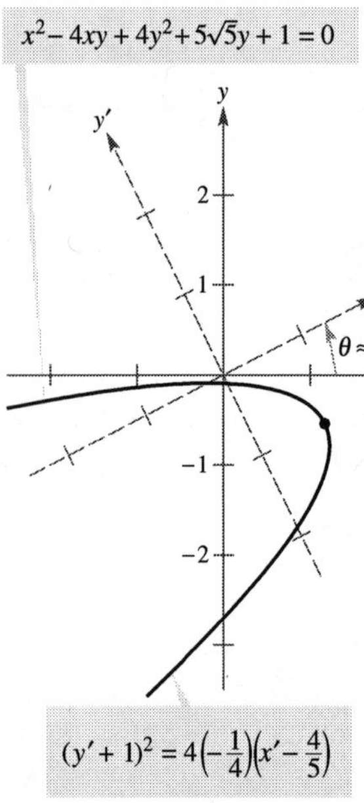

Vertex:
$\left(\frac{4}{5}, -1\right)$ in $x'y'$-system
$\left(\frac{13}{5\sqrt{5}}, -\frac{6}{5\sqrt{5}}\right)$ in xy-system

Figure E.6

Substituting these expressions into the original equation produces

$$\left(\frac{2x' - y'}{\sqrt{5}}\right)^2 - 4\left(\frac{2x' - y'}{\sqrt{5}}\right)\left(\frac{x' + 2y'}{\sqrt{5}}\right) + 4\left(\frac{x' + 2y'}{\sqrt{5}}\right)^2 +$$

$$5\sqrt{5}\left(\frac{x' + 2y'}{\sqrt{5}}\right) + 1 = 0$$

which simplifies to

$$5(y')^2 + 5x' + 10y' + 1 = 0.$$

By completing the square, you obtain the standard form

$$5(y' + 1)^2 = -5x' + 4$$

$$(y' + 1)^2 = 4\left(-\frac{1}{4}\right)\left(x' - \frac{4}{5}\right). \quad \text{Write in standard form.}$$

The graph of the equation is a parabola with its vertex at $\left(\frac{4}{5}, -1\right)$ and its axis parallel to the x'-axis in the $x'y'$-system, as shown in Figure E.6.

Invariants Under Rotation

In Theorem E.1, note that the constant term is the same in both equations—that is, $F' = F$. Because of this, F is said to be **invariant under rotation.** Theorem E.2 lists some other rotation invariants. The proof of this theorem is left as an exercise (see Exercise 34).

THEOREM E.2 Rotation Invariants

The rotation of coordinate axes through an angle θ that transforms the equation $Ax^2 + Bxy + Cy^2 + Dx + Ey + F = 0$ into the form

$$A'(x')^2 + C'(y')^2 + D'x' + E'y' + F' = 0$$

has the following rotation invariants.

1. $F = F'$
2. $A + C = A' + C'$
3. $B^2 - 4AC = (B')^2 - 4A'C'$

You can use this theorem to classify the graph of a second-degree equation *with* an xy-term in much the same way you do for a second-degree equation *without* an xy-term. Note that because $B' = 0$, the invariant $B^2 - 4AC$ reduces to

$$B^2 - 4AC = -4A'C' \qquad \text{Discriminant}$$

which is called the **discriminant** of the equation

$$Ax^2 + Bxy + Cy^2 + Dx + Ey + F = 0.$$

Because the sign of $A'C'$ determines the type of graph for the equation

$$A'(x')^2 + C'(y')^2 + D'x' + E'y' + F' = 0$$

the sign of $B^2 - 4AC$ must determine the type of graph for the original equation. This result is stated in Theorem E.3.

THEOREM E.3 Classification of Conics by the Discriminant

The graph of the equation

$$Ax^2 + Bxy + Cy^2 + Dx + Ey + F = 0$$

is, except in degenerate cases, determined by its discriminant as follows.

1. *Ellipse or circle* $\quad B^2 - 4AC < 0$
2. *Parabola* $\quad B^2 - 4AC = 0$
3. *Hyperbola* $\quad B^2 - 4AC > 0$

EXAMPLE 4 Using the Discriminant

Classify the graph of each equation.

a. $4xy - 9 = 0$

b. $2x^2 - 3xy + 2y^2 - 2x = 0$

c. $x^2 - 6xy + 9y^2 - 2y + 1 = 0$

d. $3x^2 + 8xy + 4y^2 - 7 = 0$

Solution

a. The graph is a hyperbola because

$$B^2 - 4AC = 16 - 0 > 0.$$

b. The graph is a circle or an ellipse because

$$B^2 - 4AC = 9 - 16 < 0.$$

c. The graph is a parabola because

$$B^2 - 4AC = 36 - 36 = 0.$$

d. The graph is a hyperbola because

$$B^2 - 4AC = 64 - 48 > 0.$$

EXERCISES FOR APPENDIX E

In Exercises 1–12, rotate the axes to eliminate the xy-term in the equation. Write the resulting equation in standard form and sketch its graph showing both sets of axes.

1. $xy + 1 = 0$
2. $xy - 4 = 0$
3. $x^2 - 10xy + y^2 + 1 = 0$
4. $xy + x - 2y + 3 = 0$
5. $xy - 2y - 4x = 0$
6. $13x^2 + 6\sqrt{3}xy + 7y^2 - 16 = 0$
7. $5x^2 - 2xy + 5y^2 - 12 = 0$
8. $2x^2 - 3xy - 2y^2 + 10 = 0$
9. $3x^2 - 2\sqrt{3}xy + y^2 + 2x + 2\sqrt{3}y = 0$
10. $16x^2 - 24xy + 9y^2 - 60x - 80y + 100 = 0$
11. $9x^2 + 24xy + 16y^2 + 90x - 130y = 0$
12. $9x^2 + 24xy + 16y^2 + 80x - 60y = 0$

In Exercises 13–18, use a graphing utility to graph the conic. Determine the angle θ through which the axes are rotated. Explain how you used the graphing utility to obtain the graph.

13. $x^2 + xy + y^2 = 10$
14. $x^2 - 4xy + 2y^2 = 6$
15. $17x^2 + 32xy - 7y^2 = 75$
16. $40x^2 + 36xy + 25y^2 = 52$
17. $32x^2 + 50xy + 7y^2 = 52$
18. $4x^2 - 12xy + 9y^2 + (4\sqrt{13} + 12)x - (6\sqrt{13} + 8)y = 91$

In Exercises 19–26, use the discriminant to determine whether the graph of the equation is a parabola, an ellipse, or a hyperbola.

19. $16x^2 - 24xy + 9y^2 - 30x - 40y = 0$
20. $x^2 - 4xy - 2y^2 - 6 = 0$
21. $13x^2 - 8xy + 7y^2 - 45 = 0$
22. $2x^2 + 4xy + 5y^2 + 3x - 4y - 20 = 0$
23. $x^2 - 6xy - 5y^2 + 4x - 22 = 0$
24. $36x^2 - 60xy + 25y^2 + 9y = 0$
25. $x^2 + 4xy + 4y^2 - 5x - y - 3 = 0$
26. $x^2 + xy + 4y^2 + x + y - 4 = 0$

In Exercises 27–32, sketch the graph (if possible) of the degenerate conic.

27. $y^2 - 4x^2 = 0$
28. $x^2 + y^2 - 2x + 6y + 10 = 0$
29. $x^2 + 2xy + y^2 - 1 = 0$
30. $x^2 - 10xy + y^2 = 0$
31. $(x - 2y + 1)(x + 2y - 3) = 0$
32. $(2x + y - 3)^2 = 0$

33. Show that the equation $x^2 + y^2 = r^2$ is invariant under rotation of axes.
34. Prove Theorem E.2.

APPENDIX F

Complex Numbers

Operations with Complex Numbers • Complex Solutions of Quadratic Equations • Polar Form of a Complex Number • Powers and Roots of Complex Numbers

Operations with Complex Numbers

Some equations have no real solutions. For instance, the quadratic equation

$$x^2 + 1 = 0 \qquad \text{Equation with no real solution}$$

has no real solution because there is no real number x that can be squared to produce -1. To overcome this deficiency, mathematicians created an expanded system of numbers using the **imaginary unit *i*,** defined as

$$i = \sqrt{-1} \qquad \text{Imaginary unit}$$

where $i^2 = -1$. By adding real numbers to real multiples of this imaginary unit, you obtain the set of **complex numbers.** Each complex number can be written in the **standard form** $a + bi$.

Definition of a Complex Number

For real numbers a and b, the number

$$a + bi$$

is a **complex number.** If $b \neq 0$, $a + bi$ is called an **imaginary number,** and bi is called a **pure imaginary number.**

To add (or subtract) two complex numbers, you add (or subtract) the real and imaginary parts of the numbers separately.

Addition and Subtraction of Complex Numbers

If $a + bi$ and $c + di$ are two complex numbers written in standard form, their sum and difference are defined as follows.

Sum: $(a + bi) + (c + di) = (a + c) + (b + d)i$

Difference: $(a + bi) - (c + di) = (a - c) + (b - d)i$

The **additive identity** in the complex number system is zero (the same as in the real number system). Furthermore, the **additive inverse** of the complex number $a + bi$ is

$$-(a + bi) = -a - bi \qquad \text{Additive inverse}$$

So, you have

$$(a + bi) + (-a - bi) = 0 + 0i = 0.$$

EXAMPLE 1 Adding and Subtracting Complex Numbers

a. $(3 - i) + (2 + 3i) = 3 - i + 2 + 3i$ — Remove parentheses.

$= 3 + 2 - i + 3i$ — Group like terms.

$= (3 + 2) + (-1 + 3)i$

$= 5 + 2i$ — Write in standard form.

b. $2i + (-4 - 2i) = 2i - 4 - 2i$ — Remove parentheses.

$= -4 + 2i - 2i$ — Group like terms.

$= -4$ — Write in standard form.

c. $3 - (-2 + 3i) + (-5 + i) = 3 + 2 - 3i - 5 + i$

$= 3 + 2 - 5 - 3i + i$

$= 0 - 2i$

$= -2i$

Notice in Example 1(b) that the sum of two complex numbers can be a real number.

Many of the properties of real numbers are valid for complex numbers as well. Here are some examples.

Associative Properties of Addition and Multiplication
Commutative Properties of Addition and Multiplication
Distributive Property of Multiplication over Addition

Notice below how these properties are used when two complex numbers are multiplied.

$(a + bi)(c + di) = a(c + di) + bi(c + di)$ — Distributive Property

$= ac + (ad)i + (bc)i + (bd)i^2$ — Distributive Property

$= ac + (ad)i + (bc)i + (bd)(-1)$ — Definition of i

$= ac - bd + (ad)i + (bc)i$ — Commutative Property

$= (ac - bd) + (ad + bc)i$ — Associative Property

STUDY TIP Rather than trying to memorize the multiplication rule at the right, you can simply remember how the Distributive Property is used to multiply two complex numbers. The procedure is similar to multiplying two polynomials and combining like terms.

EXAMPLE 2 Multiplying Complex Numbers

a. $(3 + 2i)(3 - 2i) = 9 - 6i + 6i - 4i^2$ Product of binomials

$= 9 - 4(-1)$ $i^2 = -1$

$= 9 + 4$ Simplify.

$= 13$ Write in standard form.

b. $(3 + 2i)^2 = 9 + 6i + 6i + 4i^2$ Product of binomials

$= 9 + 12i + 4(-1)$ $i^2 = -1$

$= 9 - 4 + 12i$ Group like terms.

$= 5 + 12i$ Write in standard form.

Notice in Example 2(a) that the product of two complex numbers can be a real number. This occurs with pairs of complex numbers of the form $a + bi$ and $a - bi$, called **complex conjugates.**

$$\begin{aligned}(a + bi)(a - bi) &= a^2 - abi + abi - b^2i^2 \\ &= a^2 - b^2(-1) \\ &= a^2 + b^2\end{aligned}$$

To write the quotient of $a + bi$ and $c + di$ in standard form, where c and d are not both zero, multiply the numerator and denominator by the complex conjugate of the denominator to obtain

$$\begin{aligned}\frac{a + bi}{c + di} &= \frac{a + bi}{c + di}\left(\frac{c - di}{c - di}\right) \\ &= \frac{(ac + bd) + (bc - ad)i}{c^2 + d^2}.\end{aligned}$$

EXAMPLE 3 Writing a Quotient of Complex Numbers in Standard Form

$\frac{2 + 3i}{4 - 2i} = \frac{2 + 3i}{4 - 2i}\left(\frac{4 + 2i}{4 + 2i}\right)$ Multiply numerator and denominator by complex conjugate of denominator.

$= \frac{8 + 4i + 12i + 6i^2}{16 - 4i^2}$ Expand.

$= \frac{8 - 6 + 16i}{16 + 4}$ $i^2 = -1$

$= \frac{2 + 16i}{20}$ Simplify.

$= \frac{1}{10} + \frac{4}{5}i$ Write in standard form.

Complex Solutions of Quadratic Equations

When using the Quadratic Formula to solve a quadratic equation, you often obtain a result such as $\sqrt{-3}$, which you know is not a real number. By factoring out $i = \sqrt{-1}$, you can write this number in standard form.

$$\sqrt{-3} = \sqrt{3(-1)} = \sqrt{3}\sqrt{-1} = \sqrt{3}i$$

The number $\sqrt{3}i$ is called the *principal square root* of -3.

STUDY TIP The definition of principal square root uses the rule

$$\sqrt{ab} = \sqrt{a}\sqrt{b}$$

for $a > 0$ and $b < 0$. This rule is not valid if *both a* and *b* are negative. For example,

$$\begin{aligned}\sqrt{-5}\sqrt{-5} &= \sqrt{5(-1)}\sqrt{5(-1)}\\ &= \sqrt{5}i\sqrt{5}i\\ &= \sqrt{25i^2}\\ &= 5i^2 = -5\end{aligned}$$

whereas

$$\sqrt{(-5)(-5)} = \sqrt{25} = 5.$$

To avoid problems with multiplying square roots of negative numbers, be sure to convert to standard form *before* multiplying.

Principal Square Root of a Negative Number

If a is a positive number, the **principal square root** of the negative number $-a$ is defined as

$$\sqrt{-a} = \sqrt{a}i.$$

EXAMPLE 4 Writing Complex Numbers in Standard Form

a. $$\begin{aligned}\sqrt{-3}\sqrt{-12} &= \sqrt{3}i\sqrt{12}i\\ &= \sqrt{36}i^2\\ &= 6(-1)\\ &= -6\end{aligned}$$

b. $$\begin{aligned}\sqrt{-48} - \sqrt{-27} &= \sqrt{48}i - \sqrt{27}i\\ &= 4\sqrt{3}i - 3\sqrt{3}i\\ &= \sqrt{3}i\end{aligned}$$

c. $$\begin{aligned}\left(-1 + \sqrt{-3}\right)^2 &= \left(-1 + \sqrt{3}i\right)^2\\ &= (-1)^2 - 2\sqrt{3}i + \left(\sqrt{3}\right)^2(i^2)\\ &= 1 - 2\sqrt{3}i + 3(-1)\\ &= -2 - 2\sqrt{3}i\end{aligned}$$

EXAMPLE 5 Complex Solutions of a Quadratic Equation

Solve $3x^2 - 2x + 5 = 0$.

Solution

$$x = \frac{-(-2) \pm \sqrt{(-2)^2 - 4(3)(5)}}{2(3)} \qquad \text{Quadratic Formula}$$

$$= \frac{2 \pm \sqrt{-56}}{6} \qquad \text{Simplify.}$$

$$= \frac{2 \pm 2\sqrt{14}i}{6} \qquad \text{Write in } i\text{-form.}$$

$$= \frac{1}{3} \pm \frac{\sqrt{14}}{3}i \qquad \text{Write in standard form.}$$

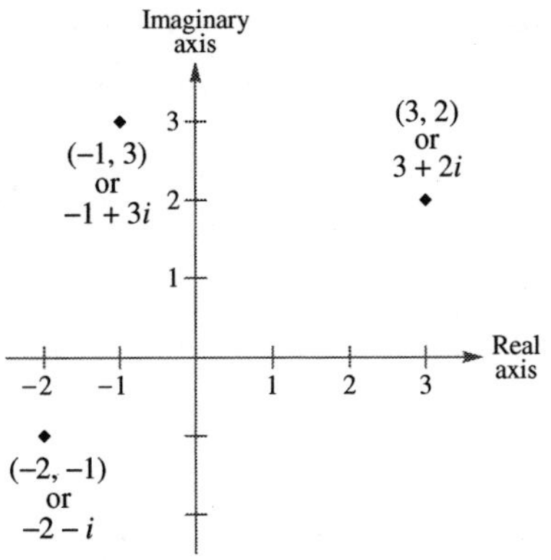

Figure F.1

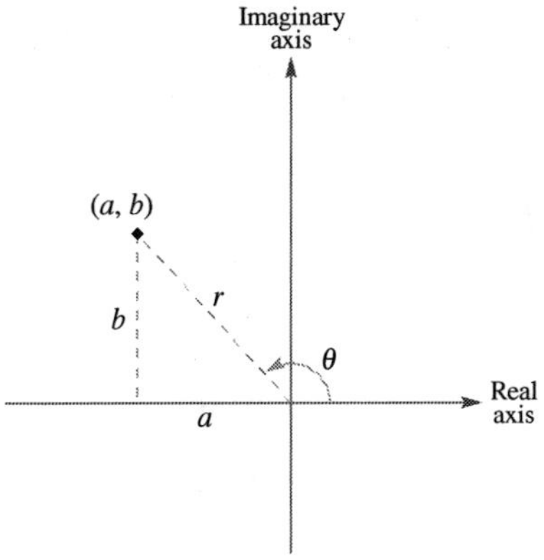

Figure F.2

Polar Form of a Complex Number

Just as real numbers can be represented by points on the real number line, you can represent a complex number

$$z = a + bi$$

as the point (a, b) in a coordinate plane (the **complex plane**). The horizontal axis is called the **real axis** and the vertical axis is called the **imaginary axis,** as shown in Figure F.1.

The **absolute value** of a complex number $a + bi$ is defined as the distance between the origin $(0, 0)$ and the point (a, b).

The Absolute Value of a Complex Number

The **absolute value** of the complex number $z = a + bi$ is given by

$$|a + bi| = \sqrt{a^2 + b^2}.$$

If the complex number $a + bi$ is a real number (that is, if $b = 0$), then this definition agrees with that given for the absolute value of a real number.

$$|a + 0i| = \sqrt{a^2 + 0^2} = |a|.$$

To work effectively with *powers* and *roots* of complex numbers, it is helpful to write complex numbers in **polar form.** In Figure F.2, consider the nonzero complex number $a + bi$. By letting θ be the angle from the positive real axis (measured counterclockwise) to the line segment connecting the origin and the point (a, b), you can write

$$a = r\cos\theta \qquad \text{and} \qquad b = r\sin\theta$$

where $r = \sqrt{a^2 + b^2}$. Consequently, you have

$$a + bi = (r\cos\theta) + (r\sin\theta)i$$

from which you can obtain the **polar form of a complex number.**

Polar Form of a Complex Number

The **polar form** of the complex number $z = a + bi$ is given by

$$z = r(\cos\theta + i\sin\theta)$$

where $a = r\cos\theta$, $b = r\sin\theta$, $r = \sqrt{a^2 + b^2}$, and $\tan\theta = b/a$. The number r is the **modulus** of z, and θ is called an **argument** of z.

NOTE The polar form of a complex number is also called the **trigonometric form.** Because there are infinitely many choices for θ, the polar form of a complex number is not unique. Normally, θ is restricted to the interval $0 \le \theta < 2\pi$, although on occasion it is convenient to use $\theta < 0$.

EXAMPLE 6 Writing a Complex Number in Polar Form

Write the complex number $z = -2 - 2\sqrt{3}i$ in polar form.

Solution The absolute value of z is

$$r = |-2 - 2\sqrt{3}i| = \sqrt{(-2)^2 + (-2\sqrt{3})^2} = \sqrt{16} = 4$$

and the angle θ is given by

$$\tan \theta = \frac{b}{a}$$

$$= \frac{-2\sqrt{3}}{-2}$$

$$= \sqrt{3}.$$

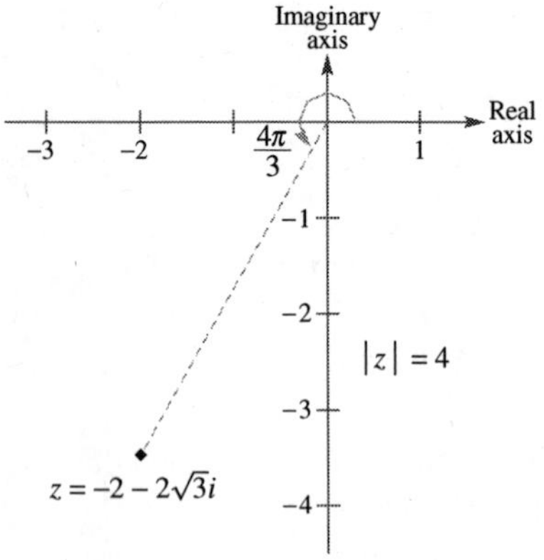

Figure F.3

Because $\tan(\pi/3) = \sqrt{3}$ and because $z = -2 - 2\sqrt{3}i$ lies in Quadrant III, choose θ to be $\theta = \pi + \pi/3 = 4\pi/3$. So, the polar form is

$$z = r(\cos \theta + i \sin \theta)$$

$$= 4\left(\cos \frac{4\pi}{3} + i \sin \frac{4\pi}{3}\right).$$

See Figure F.3.

The polar form adapts nicely to multiplication and division of complex numbers. Suppose you are given two complex numbers

$$z_1 = r_1(\cos \theta_1 + i \sin \theta_1) \quad \text{and} \quad z_2 = r_2(\cos \theta_2 + i \sin \theta_2).$$

The product of z_1 and z_2 is

$$z_1 z_2 = r_1 r_2(\cos \theta_1 + i \sin \theta_1)(\cos \theta_2 + i \sin \theta_2)$$

$$= r_1 r_2[(\cos \theta_1 \cos \theta_2 - \sin \theta_1 \sin \theta_2) + i(\sin \theta_1 \cos \theta_2 + \cos \theta_1 \sin \theta_2)].$$

Using the sum and difference formulas for cosine and sine, you can rewrite this equation as

$$z_1 z_2 = r_1 r_2[\cos(\theta_1 + \theta_2) + i \sin(\theta_1 + \theta_2)].$$

This establishes the first part of the following rule. Try to establish the second part on your own.

Product and Quotient of Two Complex Numbers

Let $z_1 = r_1(\cos \theta_1 + i \sin \theta_1)$ and $z_2 = r_2(\cos \theta_2 + i \sin \theta_2)$ be complex numbers.

$$z_1 z_2 = r_1 r_2[\cos(\theta_1 + \theta_2) + i \sin(\theta_1 + \theta_2)] \qquad \text{Product}$$

$$\frac{z_1}{z_2} = \frac{r_1}{r_2}[\cos(\theta_1 - \theta_2) + i \sin(\theta_1 - \theta_2)], \quad z_2 \neq 0 \qquad \text{Quotient}$$

Note that this rule says that to multiply two complex numbers you multiply moduli and add arguments, whereas to divide two complex numbers you divide moduli and subtract arguments.

EXAMPLE 7 Multiplying Complex Numbers in Polar Form

Find the product z_1z_2 of the complex numbers.

$$z_1 = 2\left(\cos\frac{2\pi}{3} + i\sin\frac{2\pi}{3}\right), \qquad z_2 = 8\left(\cos\frac{11\pi}{6} + i\sin\frac{11\pi}{6}\right)$$

Solution

$$\begin{aligned}
z_1z_2 &= 2\left(\cos\frac{2\pi}{3} + i\sin\frac{2\pi}{3}\right) \cdot 8\left(\cos\frac{11\pi}{6} + i\sin\frac{11\pi}{6}\right)\\
&= 16\left[\cos\left(\frac{2\pi}{3} + \frac{11\pi}{6}\right) + i\sin\left(\frac{2\pi}{3} + \frac{11\pi}{6}\right)\right]\\
&= 16\left[\cos\frac{5\pi}{2} + i\sin\frac{5\pi}{2}\right]\\
&= 16\left[\cos\frac{\pi}{2} + i\sin\frac{\pi}{2}\right]\\
&= 16[0 + i(1)]\\
&= 16i
\end{aligned}$$

Check this result by first converting to the standard forms $z_1 = -1 + \sqrt{3}i$ and $z_2 = 4\sqrt{3} - 4i$ and then multiplying algebraically.

EXAMPLE 8 Dividing Complex Numbers in Polar Form

Find the quotient z_1/z_2 of the complex numbers.

$$z_1 = 24(\cos 300° + i\sin 300°), \qquad z_2 = 8(\cos 75° + i\sin 75°)$$

Solution

$$\begin{aligned}
\frac{z_1}{z_2} &= \frac{24(\cos 300° + i\sin 300°)}{8(\cos 75° + i\sin 75°)}\\
&= \frac{24}{8}[\cos(300° - 75°) + i\sin(300° - 75°)]\\
&= 3[\cos 225° + i\sin 225°]\\
&= 3\left[\left(-\frac{\sqrt{2}}{2}\right) + i\left(-\frac{\sqrt{2}}{2}\right)\right]\\
&= -\frac{3\sqrt{2}}{2} - \frac{3\sqrt{2}}{2}i
\end{aligned}$$

Powers and Roots of Complex Numbers

To raise a complex number to a power, consider repeated use of the multiplication rule.

$$\begin{aligned} z &= r(\cos\theta + i\sin\theta) \\ z^2 &= r^2(\cos 2\theta + i\sin 2\theta) \\ z^3 &= r^3(\cos 3\theta + i\sin 3\theta) \\ &\vdots \end{aligned}$$

This pattern leads to the following important theorem, which is named after the French mathematician Abraham DeMoivre (1667–1754).

THEOREM F.1 DeMoivre's Theorem

If $z = r(\cos\theta + i\sin\theta)$ is a complex number and n is a positive integer, then

$$z^n = [r(\cos\theta + i\sin\theta)]^n = r^n(\cos n\theta + i\sin n\theta).$$

EXAMPLE 9 Finding Powers of a Complex Number

Use DeMoivre's Theorem to find $\left(-1 + \sqrt{3}i\right)^{12}$.

Solution First convert to polar form.

$$-1 + \sqrt{3}i = 2\left(\cos\frac{2\pi}{3} + i\sin\frac{2\pi}{3}\right)$$

Then, by DeMoivre's Theorem, you have

$$\begin{aligned} \left(-1 + \sqrt{3}i\right)^{12} &= \left[2\left(\cos\frac{2\pi}{3} + i\sin\frac{2\pi}{3}\right)\right]^{12} \\ &= 2^{12}\left[\cos\left(12\cdot\frac{2\pi}{3}\right) + i\sin\left(12\cdot\frac{2\pi}{3}\right)\right] \\ &= 4096(\cos 8\pi + i\sin 8\pi) \\ &= 4096. \end{aligned}$$

NOTE Notice in Example 9 that the answer is a real number.

Recall that a consequence of the Fundamental Theorem of Algebra is that a polynomial equation of degree n has n solutions in the complex number system. Each solution is an nth root of the equation. The *nth root* of a complex number is defined as follows.

Definition of nth Root of a Complex Number

The complex number $u = a + bi$ is an **nth root** of the complex number z if

$$z = u^n = (a + bi)^n.$$

STUDY TIP The nth roots of a complex number are useful for solving some polynomial equations. For instance, you can use DeMoivre's Theorem to solve the polynomial equation

$$x^4 + 16 = 0$$

by writing -16 as $16(\cos \pi + i \sin \pi)$.

To find a formula for an nth root of a complex number, let u be an nth root of z, where

$$u = s(\cos \beta + i \sin \beta) \qquad \text{and} \qquad z = r(\cos \theta + i \sin \theta).$$

By DeMoivre's Theorem and the fact that $u^n = z$, you have

$$s^n(\cos n\beta + i \sin n\beta) = r(\cos \theta + i \sin \theta).$$

Taking the absolute value of each side of this equation, it follows that $s^n = r$. Substituting back into the previous equation and dividing by r, you get

$$\cos n\beta + i \sin n\beta = \cos \theta + i \sin \theta.$$

So, it follows that

$$\cos n\beta = \cos \theta \qquad \text{and} \qquad \sin n\beta = \sin \theta.$$

Because both sine and cosine have a period of 2π, these last two equations have solutions if and only if the angles differ by a multiple of 2π. Consequently, there must exist an integer k such that

$$n\beta = \theta + 2\pi k$$

$$\beta = \frac{\theta + 2\pi k}{n}.$$

By substituting this value for β into the polar form of u, you get the following result.

THEOREM F.2 *n*th Roots of a Complex Number

For a positive integer n, the complex number $z = r(\cos\theta + i \sin\theta)$ has exactly n distinct nth roots given by

$$\sqrt[n]{r}\left(\cos \frac{\theta + 2\pi k}{n} + i \sin \frac{\theta + 2\pi k}{n}\right)$$

where $k = 0, 1, 2, \ldots, n - 1$.

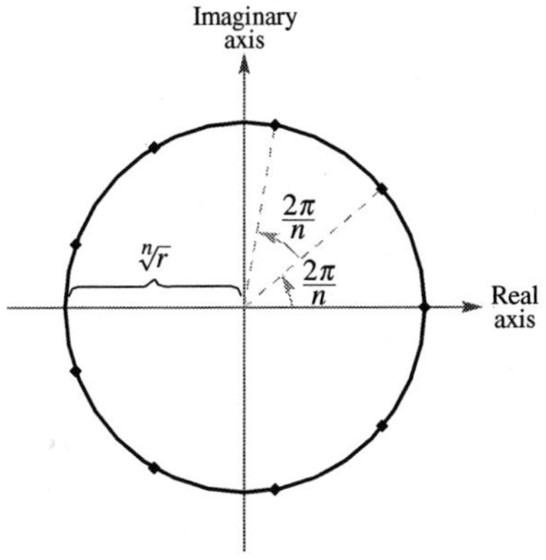

Figure F.4

When k exceeds $n - 1$, the roots begin to repeat. For instance, if $k = n$, the angle

$$\frac{\theta + 2\pi n}{n} = \frac{\theta}{n} + 2\pi$$

is coterminal with θ/n, which is also obtained when $k = 0$.

This formula for the nth roots of a complex number z has a nice geometric interpretation, as shown in Figure F.4. Note that because the nth roots of z all have the same magnitude $\sqrt[n]{r}$, they all lie on a circle of radius $\sqrt[n]{r}$ with center at the origin. Furthermore, because successive nth roots have arguments that differ by $2\pi/n$, the n roots are equally spaced along the circle.

EXAMPLE 10 Finding the nth Roots of a Complex Number

Find the three cube roots of $z = -2 + 2i$.

Solution

Because z lies in Quadrant II, the polar form for z is

$$z = -2 + 2i = \sqrt{8}\,(\cos 135^\circ + i \sin 135^\circ).$$

By the formula for nth roots, the cube roots have the form

$$\sqrt[6]{8}\left(\cos\frac{135^\circ + 360^\circ k}{3} + i\sin\frac{135^\circ + 360^\circ k}{3}\right).$$

Finally, for $k = 0, 1,$ and 2, you obtain the roots

$$\sqrt{2}\,(\cos 45^\circ + i \sin 45^\circ) = 1 + i$$

$$\sqrt{2}\,(\cos 165^\circ + i \sin 165^\circ) \approx -1.3660 + 0.3660i$$

$$\sqrt{2}\,(\cos 285^\circ + i \sin 285^\circ) \approx 0.3660 - 1.3660i.$$

EXERCISES FOR APPENDIX F

In Exercises 1–24, perform the operation and write the result in standard form.

1. $(5 + i) + (6 - 2i)$
2. $(13 - 2i) + (-5 + 6i)$
3. $(8 - i) - (4 - i)$
4. $(3 + 2i) - (6 + 13i)$
5. $\left(-2 + \sqrt{-8}\right) + \left(5 - \sqrt{-50}\right)$
6. $\left(8 + \sqrt{-18}\right) - \left(4 + 3\sqrt{2}i\right)$
7. $13i - (14 - 7i)$
8. $22 + (-5 + 8i) + 10i$
9. $-\left(\frac{3}{2} + \frac{5}{2}i\right) + \left(\frac{5}{3} + \frac{11}{3}i\right)$
10. $(1.6 + 3.2i) + (-5.8 + 4.3i)$
11. $\sqrt{-6} \cdot \sqrt{-2}$
12. $\sqrt{-5} \cdot \sqrt{-10}$
13. $\left(\sqrt{-10}\right)^2$
14. $\left(\sqrt{-75}\right)^2$
15. $(1 + i)(3 - 2i)$
16. $(6 - 2i)(2 - 3i)$
17. $6i(5 - 2i)$
18. $-8i(9 + 4i)$
19. $\left(\sqrt{14} + \sqrt{10}i\right)\left(\sqrt{14} - \sqrt{10}i\right)$
20. $\left(3 + \sqrt{-5}\right)\left(7 - \sqrt{-10}\right)$
21. $(4 + 5i)^2$
22. $(2 - 3i)^2$
23. $(2 + 3i)^2 + (2 - 3i)^2$
24. $(1 - 2i)^2 - (1 + 2i)^2$

In Exercises 25–32, write the complex conjugate of the complex number. Then multiply the number by its complex conjugate.

25. $5 + 3i$
26. $9 - 12i$
27. $-2 - \sqrt{5}i$
28. $-4 + \sqrt{2}i$
29. $20i$
30. $\sqrt{-15}$
31. $\sqrt{8}$
32. $1 + \sqrt{8}$

In Exercises 33–42, write the quotient in standard form.

33. $\dfrac{6}{i}$
34. $-\dfrac{10}{2i}$
35. $\dfrac{4}{4 - 5i}$
36. $\dfrac{3}{1 - i}$
37. $\dfrac{2 + i}{2 - i}$
38. $\dfrac{8 - 7i}{1 - 2i}$
39. $\dfrac{6 - 7i}{i}$
40. $\dfrac{8 + 20i}{2i}$
41. $\dfrac{1}{(4 - 5i)^2}$
42. $\dfrac{(2 - 3i)(5i)}{2 + 3i}$

In Exercises 43–46, perform the operation and write the result in standard form.

43. $\dfrac{2}{1 + i} - \dfrac{3}{1 - i}$
44. $\dfrac{2i}{2 + i} + \dfrac{5}{2 - i}$
45. $\dfrac{i}{3 - 2i} + \dfrac{2i}{3 + 8i}$
46. $\dfrac{1 + i}{i} - \dfrac{3}{4 - i}$

In Exercises 47–54, use the Quadratic Formula to solve the quadratic equation.

47. $x^2 - 2x + 2 = 0$

48. $x^2 + 6x + 10 = 0$

49. $4x^2 + 16x + 17 = 0$

50. $9x^2 - 6x + 37 = 0$

51. $4x^2 + 16x + 15 = 0$

52. $9x^2 - 6x - 35 = 0$

53. $16t^2 - 4t + 3 = 0$

54. $5s^2 + 6s + 3 = 0$

In Exercises 55–62, simplify the complex number and write it in standard form.

55. $-6i^3 + i^2$

56. $4i^2 - 2i^3$

57. $-5i^5$

58. $(-i)^3$

59. $(\sqrt{-75})^3$

60. $(\sqrt{-2})^6$

61. $\dfrac{1}{i^3}$

62. $\dfrac{1}{(2i)^3}$

In Exercises 63–68, plot the complex number and find its absolute value.

63. $-5i$

64. -5

65. $-4 + 4i$

66. $5 - 12i$

67. $6 - 7i$

68. $-8 + 3i$

In Exercises 69–76, represent the complex number graphically, and find the polar form of the number.

69. $3 - 3i$

70. $2 + 2i$

71. $\sqrt{3} + i$

72. $-1 + \sqrt{3}i$

73. $-2(1 + \sqrt{3}i)$

74. $\frac{5}{2}(\sqrt{3} - i)$

75. $6i$

76. 4

In Exercises 77–82, represent the complex number graphically, and find the standard form of the number.

77. $2(\cos 150° + i \sin 150°)$

78. $5(\cos 135° + i \sin 135°)$

79. $\frac{3}{2}(\cos 300° + i \sin 300°)$

80. $\frac{3}{4}(\cos 315° + i \sin 315°)$

81. $3.75\left(\cos \dfrac{3\pi}{4} + i \sin \dfrac{3\pi}{4}\right)$

82. $8\left(\cos \dfrac{\pi}{12} + i \sin \dfrac{\pi}{12}\right)$

In Exercises 83–86, perform the operation and leave the result in polar form.

83. $\left[3\left(\cos \dfrac{\pi}{3} + i \sin \dfrac{\pi}{3}\right)\right]\left[4\left(\cos \dfrac{\pi}{6} + i \sin \dfrac{\pi}{6}\right)\right]$

84. $\left[\dfrac{3}{2}\left(\cos \dfrac{\pi}{2} + i \sin \dfrac{\pi}{2}\right)\right]\left[6\left(\cos \dfrac{\pi}{4} + i \sin \dfrac{\pi}{4}\right)\right]$

85. $\left[\frac{5}{3}(\cos 140° + i \sin 140°)\right]\left[\frac{2}{3}(\cos 60° + i \sin 60°)\right]$

86. $\dfrac{\cos(5\pi/3) + i \sin(5\pi/3)}{\cos \pi + i \sin \pi}$

In Exercises 87–94, use DeMoivre's Theorem to find the indicated power of the complex number. Write the result in standard form.

87. $(1 + i)^5$

88. $(2 + 2i)^6$

89. $(-1 + i)^{10}$

90. $(1 - i)^{12}$

91. $2(\sqrt{3} + i)^7$

92. $4(1 - \sqrt{3}i)^3$

93. $\left(\cos \dfrac{5\pi}{4} + i \sin \dfrac{5\pi}{4}\right)^{10}$

94. $\left[2\left(\cos \dfrac{\pi}{2} + i \sin \dfrac{\pi}{2}\right)\right]^8$

In Exercises 95–100, (a) use Theorem F.2 on page F9 to find the indicated roots of the complex number, (b) represent each of the roots graphically, and (c) write each of the roots in standard form.

95. Square roots of $5(\cos 120° + i \sin 120°)$

96. Square roots of $16(\cos 60° + i \sin 60°)$

97. Fourth roots of $16\left(\cos \dfrac{4\pi}{3} + i \sin \dfrac{4\pi}{3}\right)$

98. Fifth roots of $32\left(\cos \dfrac{5\pi}{6} + i \sin \dfrac{5\pi}{6}\right)$

99. Cube roots of $-\dfrac{125}{2}(1 + \sqrt{3}i)$

100. Cube roots of $-4\sqrt{2}(1 - i)$

In Exercises 101–108, use Theorem F.2 on page F9 to find all the solutions of the equation and represent the solutions graphically.

101. $x^4 - i = 0$

102. $x^3 + 1 = 0$

103. $x^5 + 243 = 0$

104. $x^4 - 81 = 0$

105. $x^3 + 64i = 0$

106. $x^6 - 64i = 0$

107. $x^3 - (1 - i) = 0$

108. $x^4 + (1 + i) = 0$

APPENDIX G

Business and Economic Applications

Business and Economics Applications

Previously, you learned that one of the most common ways to measure change is with respect to time. In this section, you will study some important rates of change in economics that are not measured with respect to time. For example, economists refer to **marginal profit, marginal revenue,** and **marginal cost** as the rates of change of the profit, revenue, and cost with respect to the number of units produced or sold.

Summary of Business Terms and Formulas

Basic Terms	*Basic Formulas*
x is the number of units produced (or sold).	
p is the price per unit.	
R is the total revenue from selling x units.	$R = xp$
C is the total cost of producing x units.	
$\overline{C}$ is the average cost per unit.	$\overline{C} = \dfrac{C}{x}$
P is the total profit from selling x units.	$P = R - C$

The **break-even point** is the number of units for which $R = C$.

Marginals

$\dfrac{dR}{dx}$ = Marginal revenue ≈ *extra* revenue from selling one additional unit

$\dfrac{dC}{dx}$ = Marginal cost ≈ *extra* cost of producing one additional unit

$\dfrac{dP}{dx}$ = Marginal profit ≈ *extra* profit from selling one additional unit

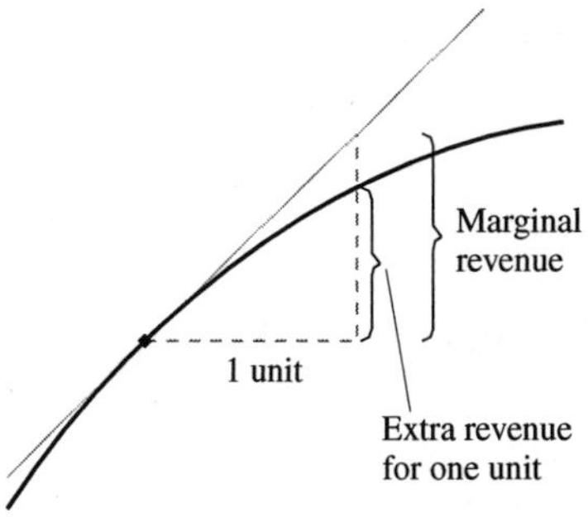

A revenue function
Figure G.1

In this summary, note that marginals can be used to approximate the *extra* revenue, cost, or profit associated with selling or producing one additional unit. This is illustrated graphically for marginal revenue in Figure G.1.

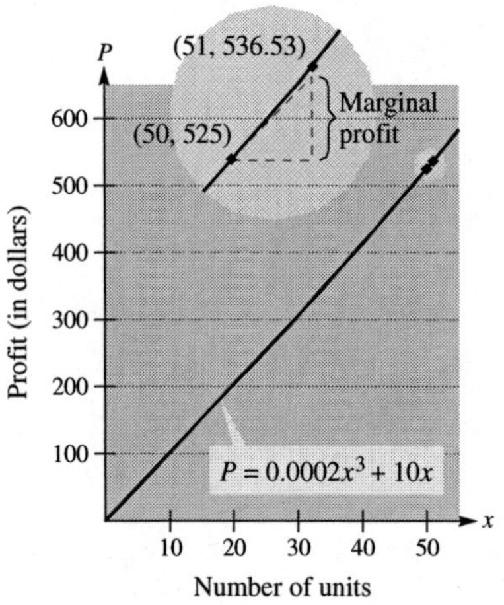

Marginal profit is the extra profit from selling one additional unit.
Figure G.2

EXAMPLE 1 Using Marginals as Approximations

A manufacturer determines that the profit P (in dollars) derived from selling x units of an item is given by

$$P = 0.0002x^3 + 10x.$$

a. Find the marginal profit for a production level of 50 units.

b. Compare this with the actual gain in profit obtained by increasing production from 50 to 51 units. (See Figure G.2.)

Solution

a. Because the profit is $P = 0.0002x^3 + 10x$, the marginal profit is given by the derivative

$$\frac{dP}{dx} = 0.0006x^2 + 10.$$

When $x = 50$, the marginal profit is

$$\frac{dP}{dx} = (0.0006)(50)^2 + 10 \qquad \text{Marginal profit for } x = 50$$
$$= \$11.50.$$

b. For $x = 50$ and 51, the actual profits are

$$P = (0.0002)(50)^3 + 10(50)$$
$$= 25 + 50$$
$$= \$525.00$$
$$P = (0.0002)(51)^3 + 10(51)$$
$$= 26.53 + 510$$
$$= \$536.53.$$

So, the additional profit obtained by increasing the production level from 50 to 51 units is

$$536.53 - 525.00 = \$11.53. \qquad \text{Extra profit for one unit}$$

The profit function in Example 1 is unusual in that the profit continues to increase as long as the number of units sold increases. In practice, it is more common to encounter situations in which sales can be increased only by lowering the price per item. Such reductions in price ultimately cause the profit to decline.

The number of units x that consumers are willing to purchase at a given price p per unit is defined as the **demand function**

$$p = f(x). \qquad \text{Demand function}$$

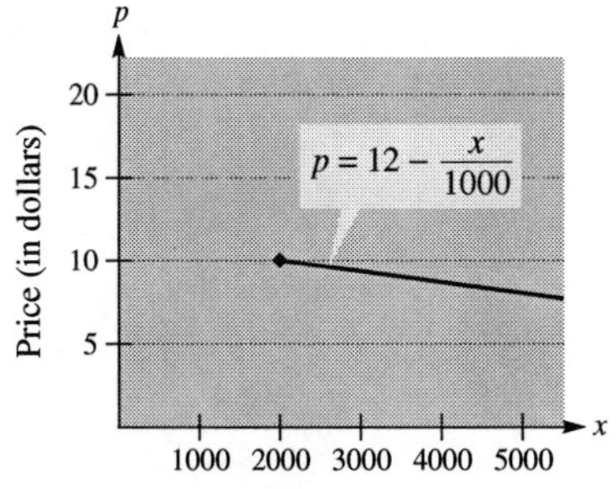

A demand function p
Figure G.3

EXAMPLE 2 Finding a Demand Function

A business sells 2000 items per month at a price of $10 each. It is estimated that monthly sales will increase by 250 items for each $0.25 reduction in price. Find the demand function corresponding to this estimate.

Solution From the given estimate, x increases 250 units each time p drops $0.25 from the original cost of $10. This is described by the equation

$$x = 2000 + 250\left(\frac{10 - p}{0.25}\right)$$
$$= 12{,}000 - 1000p$$

or

$$p = 12 - \frac{x}{1000}, \quad x \geq 2000. \qquad \text{Demand function}$$

The graph of the demand function is shown in Figure G.3.

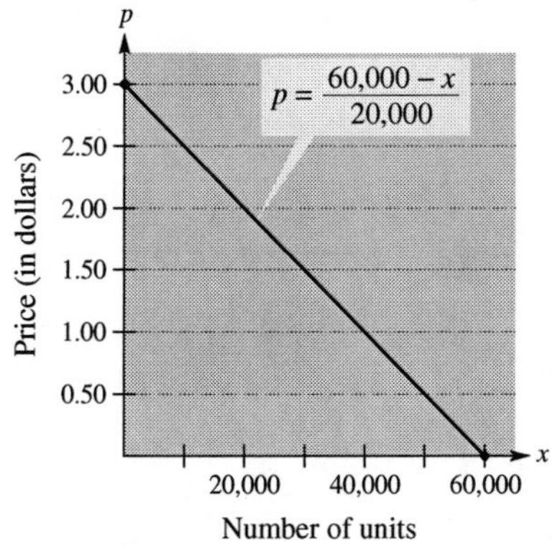

As the price decreases, more hamburgers are sold.
Figure G.4

EXAMPLE 3 Finding the Marginal Revenue

A fast-food restaurant has determined that the monthly demand for its hamburgers is

$$p = \frac{60{,}000 - x}{20{,}000}.$$

Find the increase in revenue per hamburger (marginal revenue) for monthly sales of 20,000 hamburgers. (See Figure G.4.)

Solution Because the total revenue is given by $R = xp$, you have

$$R = xp = x\left(\frac{60{,}000 - x}{20{,}000}\right) = \frac{1}{20{,}000}(60{,}000x - x^2).$$

By differentiating, you can find the marginal revenue to be

$$\frac{dR}{dx} = \frac{1}{20{,}000}(60{,}000 - 2x).$$

When $x = 20{,}000$, the marginal revenue is

$$\frac{dR}{dx} = \frac{1}{20{,}000}[60{,}000 - 2(20{,}000)]$$
$$= \frac{20{,}000}{20{,}000}$$
$$= \$1 \text{ per unit.}$$

NOTE The demand function in Example 3 is typical in that a high demand corresponds to a low price, as shown in Figure G.4.

EXAMPLE 4 Finding the Marginal Profit

Suppose that in Example 3 the cost C (in dollars) of producing x hamburgers is

$$C = 5000 + 0.56x, \quad 0 \le x \le 50{,}000.$$

Find the total profit and the marginal profit for 20,000, 24,400, and 30,000 units.

Solution Because $P = R - C$, you can use the revenue function in Example 3 to obtain

$$P = \frac{1}{20{,}000}(60{,}000x - x^2) - 5000 - 0.56x$$
$$= 2.44x - \frac{x^2}{20{,}000} - 5000.$$

So, the marginal profit is

$$\frac{dP}{dx} = 2.44 - \frac{x}{10{,}000}.$$

The table shows the total profit and the marginal profit for each of the three indicated demands. Figure G.5 shows the graph of the profit function.

Demand	20,000	24,400	30,000
Profit	\$23,800	\$24,768	\$23,200
Marginal profit	\$0.44	\$0.00	−\$0.56

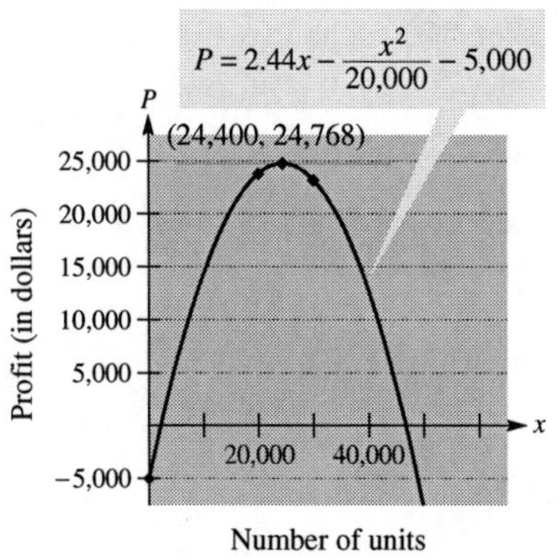

The maximum profit corresponds to the point where the marginal profit is 0. When more than 24,400 hamburgers are sold, the marginal profit is negative—increasing production beyond this point will *reduce* rather than increase profit.
Figure G.5

EXAMPLE 5 Finding the Maximum Profit

In marketing an item, a business has discovered that the demand for the item is represented by

$$p = \frac{50}{\sqrt{x}}. \qquad \text{Demand function}$$

The cost C (in dollars) of producing x items is given by $C = 0.5x + 500$. Find the price per unit that yields a maximum profit (see Figure G.6).

Solution From the given cost function, you obtain

$$P = R - C = xp - (0.5x + 500). \qquad \text{Primary equation}$$

Substituting for p (from the demand function) produces

$$P = x\left(\frac{50}{\sqrt{x}}\right) - (0.5x + 500) = 50\sqrt{x} - 0.5x - 500.$$

Setting the marginal profit equal to 0

$$\frac{dP}{dx} = \frac{25}{\sqrt{x}} - 0.5 = 0$$

yields $x = 2500$. From this, you can conclude that the maximum profit occurs when the price is

$$p = \frac{50}{\sqrt{2500}} = \frac{50}{50} = \$1.00.$$

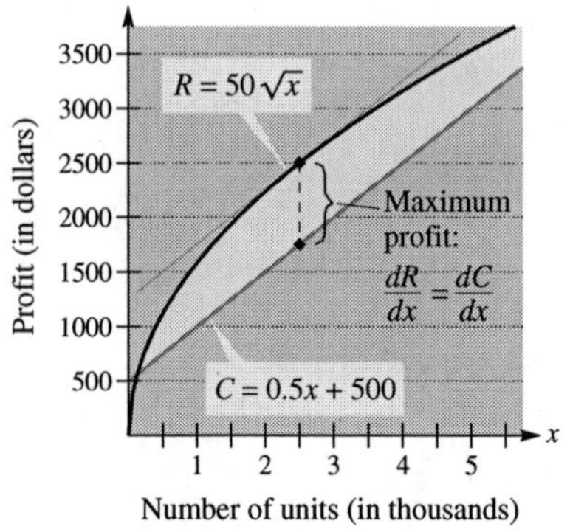

Maximum profit occurs when $\frac{dR}{dx} = \frac{dC}{dx}$.
Figure G.6

NOTE To find the maximum profit in Example 5, the profit function, $P = R - C$, was differentiated and set equal to 0. From the equation

$$\frac{dP}{dx} = \frac{dR}{dx} - \frac{dC}{dx} = 0$$

it follows that the maximum profit occurs when the marginal revenue is equal to the marginal cost, as shown in Figure G.6.

EXAMPLE 6 Minimizing the Average Cost

A company estimates that the cost C (in dollars) of producing x units of a product is given by $C = 800 + 0.04x + 0.0002x^2$. Find the production level that minimizes the average cost per unit.

Solution Substituting from the given equation for C produces

$$\overline{C} = \frac{C}{x} = \frac{800 + 0.04x + 0.0002x^2}{x} = \frac{800}{x} + 0.04 + 0.0002x.$$

Setting the derivative $d\overline{C}/dx$ equal to 0 yields

$$\frac{d\overline{C}}{dx} = -\frac{800}{x^2} + 0.0002 = 0$$

$$x^2 = \frac{800}{0.0002} = 4{,}000{,}000 \Rightarrow x = 2000 \text{ units.}$$

See Figure G.7.

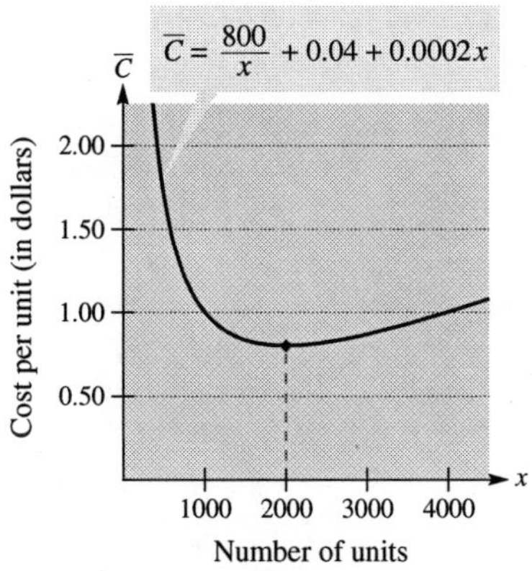

Minimum average cost occurs when $\frac{d\overline{C}}{dx} = 0$.

Figure G.7

EXERCISES FOR APPENDIX G

1. ***Think About It*** The figure shows the cost C of producing x units of a product.
   (a) What is $C(0)$ called?
   (b) Sketch a graph of the marginal cost function.
   (c) Does the marginal cost function have an extremum? If so, describe what it means in economic terms.

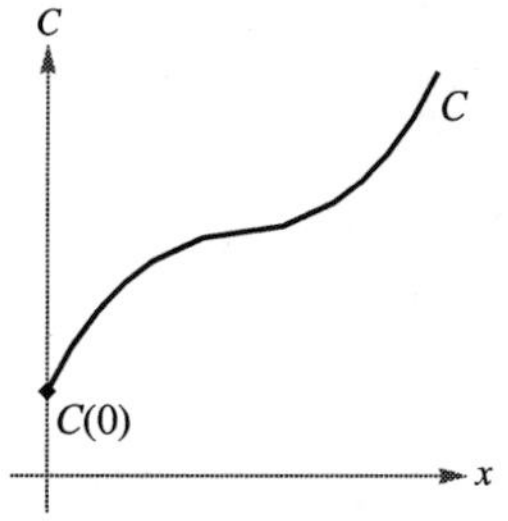

Figure for 1

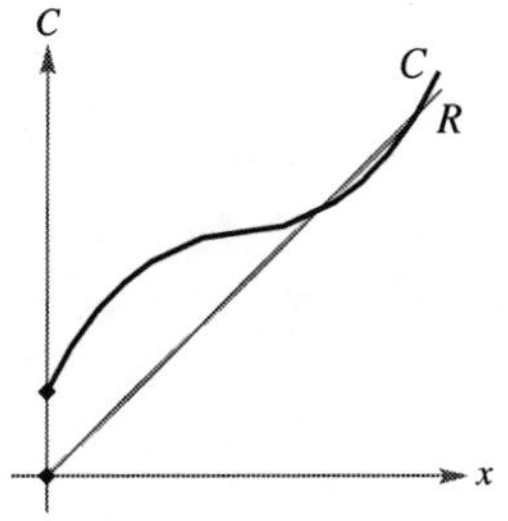

Figure for 2

2. ***Think About It*** The figure shows the cost C and revenue R for producing and selling x units of a product.
   (a) Sketch a graph of the marginal revenue function.
   (b) Sketch a graph of the profit function. Approximate the position of the value of x for which profit is maximum.

In Exercises 3–6, find the number of units x that produces a maximum revenue R.

3. $R = 900x - 0.1x^2$
4. $R = 600x^2 - 0.02x^3$
5. $R = \dfrac{1{,}000{,}000x}{0.02x^2 + 1800}$
6. $R = 30x^{2/3} - 2x$

In Exercises 7–10, find the number of units x that produces the minimum average cost per unit $\overline{C}$.

7. $C = 0.125x^2 + 20x + 5000$
8. $C = 0.001x^3 - 5x + 250$
9. $C = 3000x - x^2\sqrt{300 - x}$
10. $C = \dfrac{2x^3 - x^2 + 5000x}{x^2 + 2500}$

In Exercises 11–14, find the price per unit p (in dollars) that produces the maximum profit P.

	Cost Function	*Demand Function*
11.	$C = 100 + 30x$	$p = 90 - x$
12.	$C = 2400x + 5200$	$p = 6000 - 0.4x^2$
13.	$C = 4000 - 40x + 0.02x^2$	$p = 50 - \dfrac{x}{100}$
14.	$C = 35x + 2\sqrt{x - 1}$	$p = 40 - \sqrt{x - 1}$

Average Cost **In Exercises 15 and 16, use the cost function to find the value of x at which the average cost is a minimum. For that value of x, show that the marginal cost and average cost are equal.**

15. $C = 2x^2 + 5x + 18$ **16.** $C = x^3 - 6x^2 + 13x$

17. Prove that the average cost is a minimum at the value of x where the average cost equals the marginal cost.

18. ***Maximum Profit*** The profit P for a company is

$$P = 230 + 20s - \tfrac{1}{2}s^2$$

where s is the amount (in hundreds of dollars) spent on advertising. What amount of advertising produces a maximum profit?

19. ***Numerical, Graphical, and Analytic Analysis*** The cost per unit for the production of a radio is \$60. The manufacturer charges \$90 per unit for orders of 100 or less. To encourage large orders, the manufacturer reduces the charge by \$0.15 per radio for each unit ordered in excess of 100 (for example, there would be a charge of \$87 per radio for an order size of 120).

(a) Analytically complete six rows of a table such as the one below. (The first two rows are shown.)

x	*Price*	*Profit*
102	$90 - 2(0.15)$	$102[90 - 2(0.15)] - 102(60) = 3029.40$
104	$90 - 4(0.15)$	$104[90 - 4(0.15)] - 104(60) = 3057.60$

(b) Use a graphing utility to generate additional rows of the table. Use the table to estimate the maximum profit. (*Hint:* Use the *table* feature of the graphing utility.)

(c) Write the profit P as a function of x.

(d) Use calculus to find the critical number of the function in part (c) and find the required order size.

(e) Use a graphing utility to graph the function in part (c) and verify the maximum profit from the graph.

20. ***Maximum Profit*** A real estate office handles 50 apartment units. When the rent is \$720 per month, all units are occupied. However, on the average, for each \$40 increase in rent, one unit becomes vacant. Each occupied unit requires an average of \$48 per month for service and repairs. What rent should be charged to obtain a maximum profit?

21. ***Minimum Cost*** A power station is on one side of a river that is $\frac{1}{2}$-mile wide, and a factory is 6 miles downstream on the other side. It costs \$12 per foot to run power lines over land and \$16 per foot to run them underwater. Find the most economical path for the transmission line from the power station to the factory.

22. ***Maximum Revenue*** When a wholesaler sold a product at \$25 per unit, sales were 800 units per week. After a price increase of \$5, the average number of units sold dropped to 775 per week. Assume that the demand function is linear, and find the price that will maximize the total revenue.

23. ***Minimum Cost*** The ordering and transportation cost C (in thousands of dollars) of the components used in manufacturing a product is

$$C = 100\left(\frac{200}{x^2} + \frac{x}{x + 30}\right), \qquad 1 \le x$$

where x is the order size (in hundreds). Find the order size that minimizes the cost. (*Hint:* Use Newton's Method or the *zero* feature of a graphing utility.)

24. ***Average Cost*** A company estimates that the cost C (in dollars) of producing x units of a product is

$$C = 800 + 0.4x + 0.02x^2 + 0.0001x^3.$$

Find the production level that minimizes the average cost per unit. (*Hint:* Use Newton's Method or the *zero* feature of a graphing utility.)

25. ***Revenue*** The revenue R for a company selling x units is

$$R = 900x - 0.1x^2.$$

Use differentials to approximate the change in revenue if sales increase from $x = 3000$ to $x = 3100$ units.

26. ***Analytic and Graphical Analysis*** A manufacturer of fertilizer finds that the national sales of fertilizer roughly follow the seasonal pattern

$$F = 100{,}000\left\{1 + \sin\left[\frac{2\pi(t - 60)}{365}\right]\right\}$$

where F is measured in pounds. Time t is measured in days, with $t = 1$ corresponding to January 1.

(a) Use calculus to determine the day of the year when the maximum amount of fertilizer is sold.

(b) Use a graphing utility to graph the function and approximate the day of the year when sales are minimum.

27. *Modeling Data* The table shows the monthly sales G (in thousands of gallons) of gasoline at a gas station in 2004. The time in months is represented by t, with $t = 1$ corresponding to January.

t	1	2	3	4	5	6	7	8	9	10	11	12
G	8.91	9.18	9.79	9.83	10.37	10.16	10.37	10.81	10.03	9.97	9.85	9.51

A model for these data is

$$G = 9.90 - 0.64 \cos\left(\frac{\pi t}{6} - 0.62\right).$$

(a) Use a graphing utility to plot the data and graph the model.

(b) Use the model to approximate the month when gasoline sales were greatest.

(c) What factor in the model causes the seasonal variation in sales of gasoline? What part of the model gives the average monthly sales of gasoline?

(d) Suppose the gas station added the term $0.02t$ to the model. What does the inclusion of this term mean? Use this model to estimate the maximum monthly sales in the year 2008.

28. *Airline Revenues* The annual revenue R (in millions of dollars) for an airline for the years 1995–2004 can be modeled by

$$R = 4.7t^4 - 193.5t^3 + 2941.7t^2 - 19{,}294.7t + 52{,}012$$

where $t = 5$ corresponds to 1995.

(a) During which year (between 1995 and 2004) was the airline's revenue the least?

(b) During which year was the revenue the greatest?

(c) Find the revenues for the years in which the revenue was the least and greatest.

(d) Use a graphing utility to confirm the results in parts (a) and (b).

29. *Modeling Data* The manager of a department store recorded the quarterly sales S (in thousands of dollars) of a new seasonal product over a period of 2 years, as shown in the table, where t is the time in quarters, with $t = 1$ corresponding to the winter quarter of 2002.

t	1	2	3	4	5	6	7	8
S	7.5	6.2	5.3	7.0	9.1	7.8	6.9	8.6

(a) Use a graphing utility to plot the data.

(b) Find a model of the form $S = a + bt + c \sin \beta t$ for the data. (*Hint:* Start by finding β. Next, use a graphing utility to find $a + bt$. Finally, approximate c.)

(c) Use a graphing utility to graph the model with the data and make any adjustments necessary to obtain a better fit.

(d) Use the model to predict the maximum quarterly sales in the year 2006.

30. *Think About It* Match each graph with the function it best represents—a demand function, a revenue function, a cost function, or a profit function. Explain your reasoning. [The graphs are labeled (a), (b), (c), and (d).]

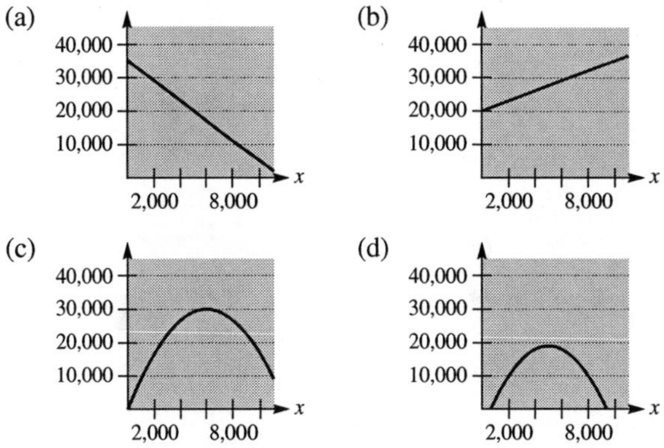

Elasticity **The relative responsiveness of consumers to a change in the price of an item is called the *price elasticity of demand*. If $p = f(x)$ is a differentiable demand function, the price elasticity of demand is**

$$\eta = \frac{p/x}{dp/dx}.$$

For a given price, if $|\eta| < 1$, the demand is *inelastic*, and if $|\eta| > 1$, the demand is *elastic*. In Exercises 31–34, find η for the demand function at the indicated x-value. Is the demand elastic, inelastic, or neither at the indicated x-value?

31. $p = 400 - 3x$, $x = 20$

32. $p = 5 - 0.03x$, $x = 100$

33. $p = 400 - 0.5x^2$, $x = 20$

34. $p = \dfrac{500}{x + 2}$, $x = 23$

APPENDIX C

Appendix C.1 *(page C6)*

1. $x^2 - 3xy + y^2 = C$ **3.** $3xy^2 + 5x^2y^2 - 2y = C$

5. Not exact **7.** $\arctan \dfrac{x}{y} = C$ **9.** Not exact

11. (a) Answers will vary.

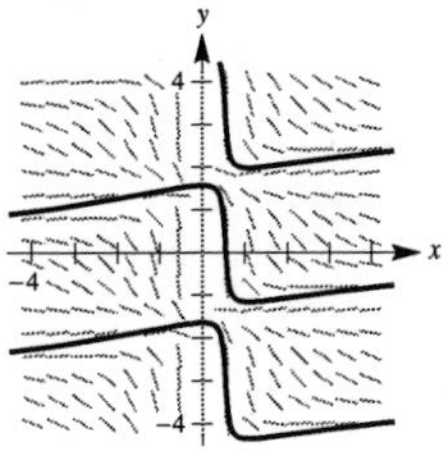

(b) $x^2 \tan y + 5x = \dfrac{11}{4}$

(c)

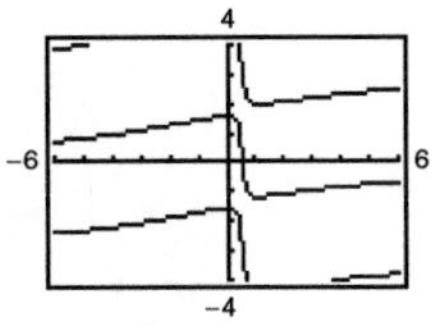

13. $y \ln(x - 1) + y^2 = 16$ **15.** $e^{3x} \sin 3y = 0$

17. Integrating factor: $\dfrac{1}{y^2}$

$\dfrac{x}{y} - 6y = C$

19. Integrating factor: $\dfrac{1}{x^2}$

$\dfrac{y}{x} + 5x = C$

21. Integrating factor: $\cos x$

$y \sin x + x \sin x + \cos x = C$

23. Integrating factor: $\dfrac{1}{y}$

$xy - \ln y = C$

25. Integrating factor: $\dfrac{1}{\sqrt{y}}$

$x\sqrt{y} + \cos\sqrt{y} = C$

27. $x^4y^3 + x^2y^4 = C$

29. $\dfrac{y^2}{x} + \dfrac{x}{y^2} + C$ **31.** Proof

33. $x^2 + y^2 = C$

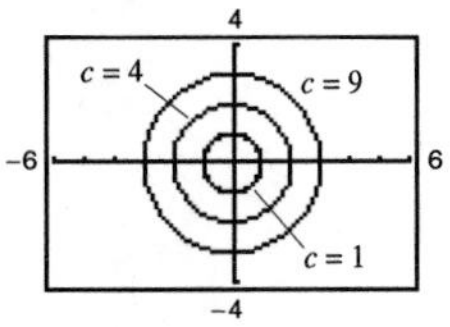

35. $2x^2y^4 + x^2 = C$

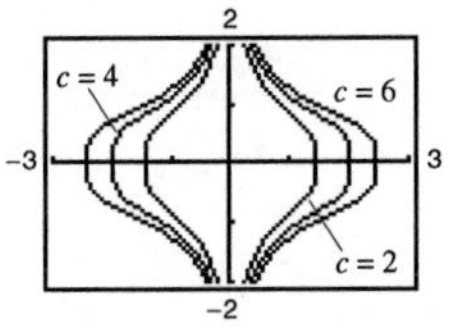

37. $x^2 - 2xy + 3y^2 = 3$ **39.** $C = \dfrac{5\left(x^2 + \sqrt{x^4 - 1{,}000{,}000x}\right)}{x}$

41.

Δx	0.50	0.25	0.10
Estimate	3.7798	3.9875	4.1207

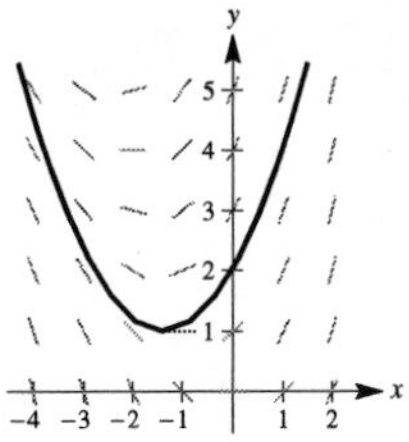

43. (a)

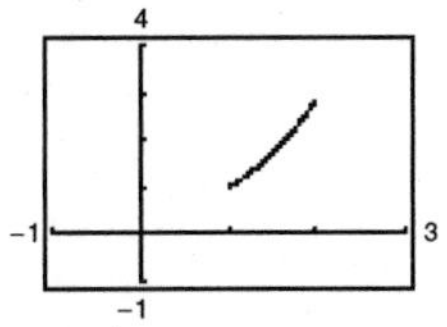

(b) $3y^{2/3} - x^2 = 2$

(c)

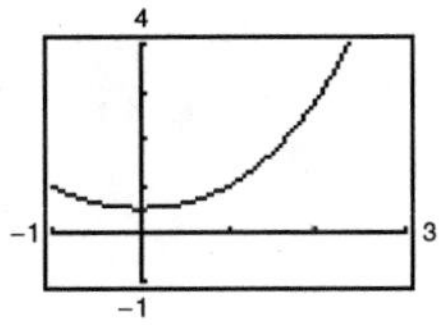

45. (a)

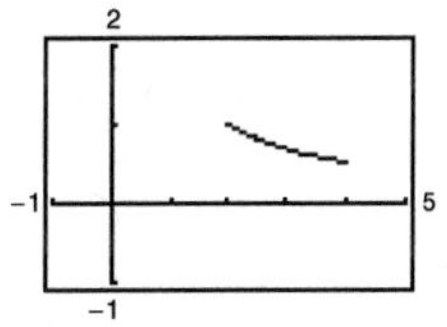

(b) $y^2(2x^2 + y^2) = 9$

(c)

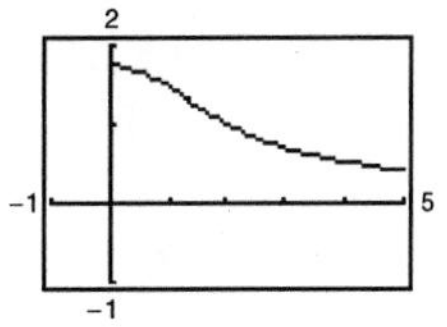

47. (a)

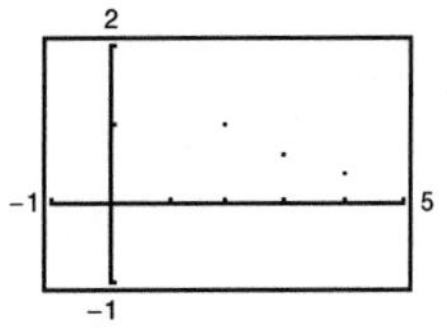

(b) $y^2(2x^2 + y^2) = 9$

(c)

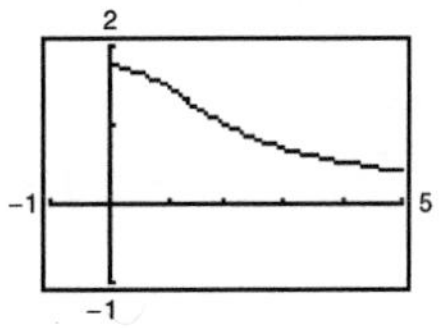

Less accurate

49. False; $\dfrac{\partial M}{\partial y} = 2x, \quad \dfrac{\partial N}{\partial x} = -2x.$

50. False; $ydx + xdy = 0$ is exact, but $xydx + x^2dy = 0$ is not exact.

51. True **52.** True

Appendix C.2 *(page C14)*

1. Proof **3.** Proof **5.** $y = C_1 + C_2e^x$

7. $y = C_1e^{3x} + C_2e^{-2x}$ **9.** $y = C_1e^{x/2} + C_2e^{-2x}$

11. $y = C_1e^{-3x} + C_2xe^{-3x}$ **13.** $y = C_1e^{x/4} + C_2xe^{x/4}$

15. $y = C_1 \sin x + C_2 \cos x$ **17.** $y = C_1e^{3x} + C_2e^{-3x}$

19. $y = e^x\left(C_1 \sin\sqrt{3}x + C_2 \cos\sqrt{3}x\right)$

21. $y = C_1e^{(3+\sqrt{5})x/2} + C_2e^{(3-\sqrt{5})x/2}$

23. $y = e^{2x/3}\left(C_1 \sin \dfrac{\sqrt{7}x}{3} + C_2 \cos \dfrac{\sqrt{7}x}{3}\right)$

25. $y = C_1e^x + C_2e^{-x} + C_3 \sin x + C_4 \cos x$

27. $y = C_1e^x + C_2e^{2x} + C_3e^{3x}$

29. $y = C_1e^x + e^x(C_2 \sin 2x + C_3 \cos 2x)$

31. (a) $y = 2 \cos 10x$ (b) $y = \frac{1}{5} \sin 10x$
(c) $y = -\cos 10x + \frac{3}{10} \sin 10x$

33. $y = \frac{1}{11}(e^{6x} + 10e^{-5x})$

35. $y = \dfrac{1}{2} \sin 4x$

37. y'' and y' are not equal for $x < 0$. $y'' > 0$ for all x, but $y' < 0$ for $x < 0$.

39. $y = \frac{1}{2} \cos 4\sqrt{3}t$

41. $y = \dfrac{2}{3} \cos 4\sqrt{3}t - \dfrac{\sqrt{3}}{24} \sin 4\sqrt{3}t$

43. $y = \dfrac{e^{-t/16}}{2}\left(\cos \dfrac{\sqrt{12{,}287}\,t}{16} + \dfrac{\sqrt{12{,}287}}{12{,}287} \sin \dfrac{\sqrt{12{,}287}\,t}{16}\right)$

45. b **46.** d **47.** c **48.** a **49.** Proof

51. False; the general solution is $y = C_1e^{3x} + C_2xe^{3x}$.

52. True **53.** True

54. False; the solution $y = x^2e^x$ requires that $m = 1$ is a triple zero of the characteristic equation. Because the characteristic equation is quadratic, $m = 1$ can be at most a double zero.

55. Proof **57.** Proof

59. (a) Proof (b) $y = \dfrac{C_1}{x^3} + \dfrac{C_2}{x^2}$

Appendix C.3 *(page C22)*

1. Proof **3.** Proof **5.** $y = C_1e^x + C_2e^{2x} + x + \frac{3}{2}$

7. $y = \cos x + 6 \sin x + x^3 - 6x$

9. $y = C_1 + C_2e^{-2x} + \frac{2}{3}e^x$

11. $y = (C_1 + C_2x)e^{5x} + \frac{3}{8}e^x + \frac{1}{5}$

13. $y = -1 + 2e^{-x} - \cos x - \sin x$

15. $y = \left(C_1 - \dfrac{x}{6}\right)\cos 3x + C_2 \sin 3x$

17. $y = C_1e^x + C_2xe^x + \left(C_3 + \dfrac{2x}{9}\right)e^{-2x}$

19. $y = \left(\dfrac{4}{9} - \dfrac{1}{2}x^2\right)e^{4x} - \dfrac{1}{9}(1 + 3x)e^x$

21. (a) $y''_p = 0$ and $3y_p = 12$ (b) $y_p = 2$ (c) $y_p = 4$

23. $y = (C_1 + \ln|\cos x|)\cos x + (C_2 + x)\sin x$

25. $y = \left(C_1 - \dfrac{x}{2}\right)\cos 2x + \left(C_2 + \dfrac{1}{4}\ln|\sin 2x|\right)\sin 2x$

27. $y = (C_1 + C_2x)e^x + \dfrac{x^2e^x}{4}(\ln x^2 - 3)$

29. $q = \frac{3}{25}(e^{-5t} + 5te^{-5t} - \cos 5t)$

31. $y = \frac{1}{4} \cos 8t - \frac{1}{2} \sin 8t + \sin 4t$

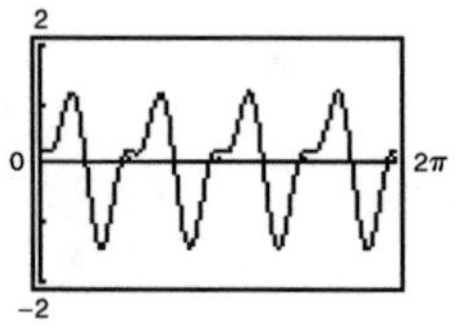

33. $y = \left(\frac{9}{32} - \frac{3}{4}t\right)e^{-8t} - \frac{1}{32} \cos 8t$

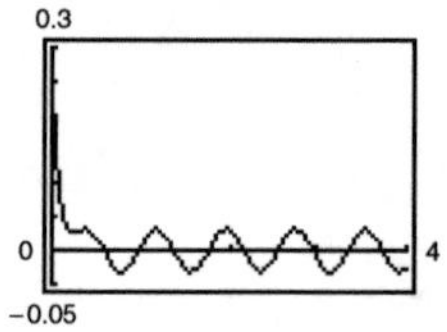

35. $y = \dfrac{\sqrt{5}}{4} \sin\left(8t - \arctan \dfrac{1}{2}\right)$
$= \dfrac{\sqrt{5}}{4} \sin(8t - 0.4636)$

37. Proof

39. $y = C_1x + C_2x \ln x + \frac{2}{3}x(\ln x)^3$

Appendix C.4 *(page C27)*

1. Proof **3.** Proof **5.** Proof

7. $y = a_0 \displaystyle\sum_{k=0}^{\infty} \frac{(-3)^k}{2^k k!} x^{2k}$

Interval of convergence: $(-\infty, \infty)$

9. $y = a_0 + a_1 \displaystyle\sum_{k=0}^{\infty} \frac{x^{2x+1}}{2^k(k!)(2k + 1)}$

Interval of convergence: $(-\infty, \infty)$

11. $y = a_0\left(1 - \dfrac{x^2}{8} + \dfrac{x^4}{128} - \cdots\right) +$
$a_1\left(x - \dfrac{x^3}{24} + \dfrac{7x^5}{1920} - \cdots\right)$

13. Taylor's Theorem: $y = 2 + \frac{2x}{1!} - \frac{2x^2}{2!} - \frac{10x^3}{3!} + \frac{2x^4}{4!} + \cdots$

$y\left(\frac{1}{2}\right) \approx 2.547$

Euler's Method: $y\left(\frac{1}{2}\right) \approx 2.672$

15. (a) $y = 2(\cos 3x + \sin 3x)$

(b) $y = 2\left[\sum_{n=0}^{\infty} \frac{(-1)^n(3x)^{2n}}{(2n)!} + \sum_{n=0}^{\infty} \frac{(-1)^n(3x)^{2n+1}}{(2n+1)!}\right]$

(c)

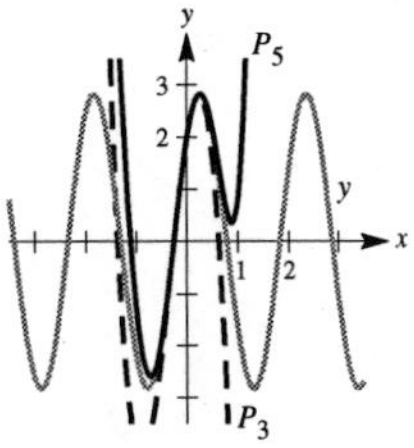

17. $y = 1 - \frac{3x}{1!} + \frac{2x^3}{3!} - \frac{12x^4}{4!} + \frac{16x^6}{6!} - \frac{120x^7}{7!} + \cdots$

$y\left(\frac{1}{4}\right) \approx 0.253$

19. Proof **21.** Proof

23. $y = a_0 + a_1x + \frac{a_0}{6}x^3 + \frac{a_1}{12}x^4 + \frac{a_0}{180}x^6 + \frac{a_1}{504}x^7$

APPENDIX D

Appendix D.1 *(page D8)*

1. Rational **3.** Irrational **5.** Rational **7.** Rational

9. Rational **11.** $\frac{4}{11}$ **13.** $\frac{11}{37}$

15. (a) True (b) False (c) True (d) False
(e) False (f) False

17. x is greater than -3 and less than 3.

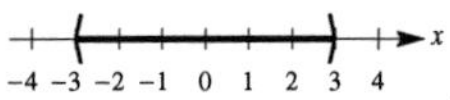

The interval is bounded.

19. x is no more than 5.

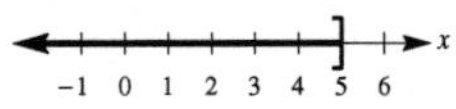

The interval is unbounded.

21. $y \ge 4, [4, \infty)$ **23.** $0.03 < r \le 0.07, (0.03, 0.07]$

25. $x \ge \frac{1}{2}$

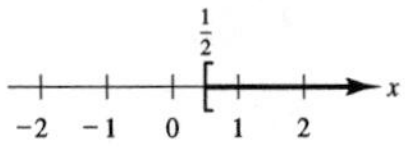

27. $-\frac{1}{2} < x < \frac{7}{2}$

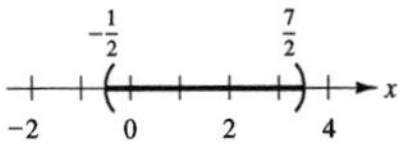

29. $x > 6$

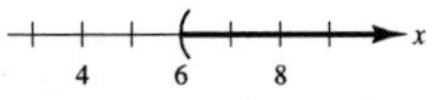

31. $-1 < x < 1$

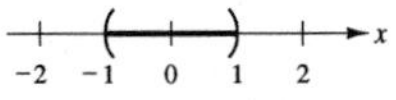

33. $x \ge 13, x \le -7$

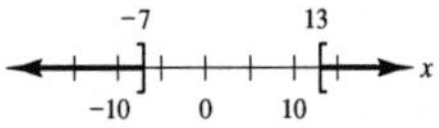

35. $a - b < x < a + b$

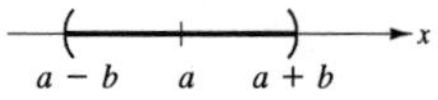

37. $-3 < x < 2$

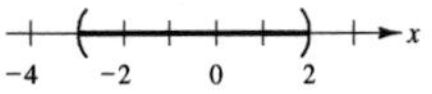

39. $0 < x < 3$

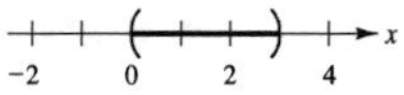

41. $-3 \le x \le 1$

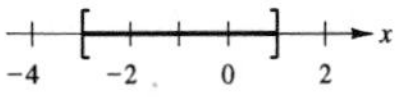

43. $-3 \le x \le 2$

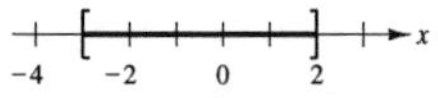

45. 4, −4, 4 **47.** (a) −51, 51, 51 (b) 51, −51, 51

49. $|x| \le 2$ **51.** $|x - 2| > 2$

53. (a) $|x - 12| \le 10$ (b) $|x - 12| \ge 10$

55. 1 **57.** (a) 14 (b) 10

59. $x \ge 36$ units **61.** $x \le 41$ or $x \ge 59$

63. (a) $\frac{355}{112} > \pi$ (b) $\frac{22}{7} > \pi$ **65.** b

67. False; the reciprocal of 2 is $\frac{1}{2}$, which is not an integer.

68. True **69.** True **70.** False; $|0| = 0$. **71.** True

72. True **73.** Proof **75.** Proof **77.** Proof

79. Proof

81. $|-3 - 1| > |-3| - |1|$
$|3 - 1| = |3| - |1|$

Appendix D.2 *(page D15)*

1. (a)

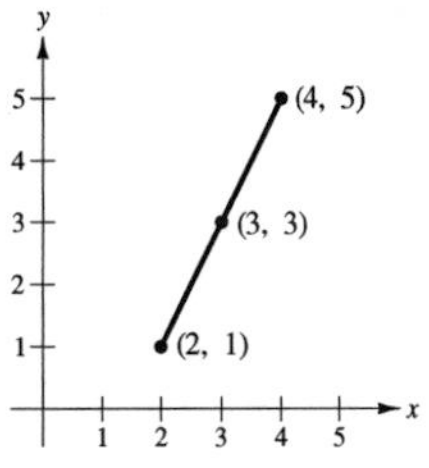

(b) $2\sqrt{5}$

(c) (3, 3)

3. (a)

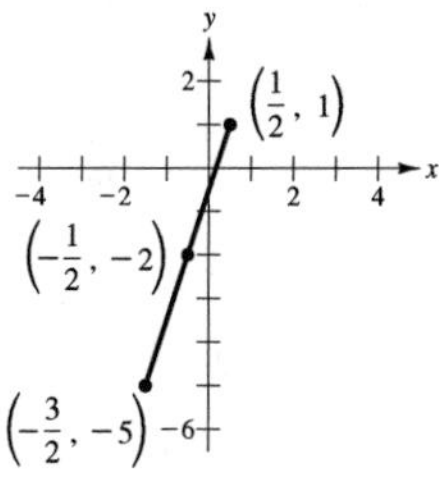

(b) $2\sqrt{10}$

(c) $\left(-\frac{1}{2}, -2\right)$

5. (a)

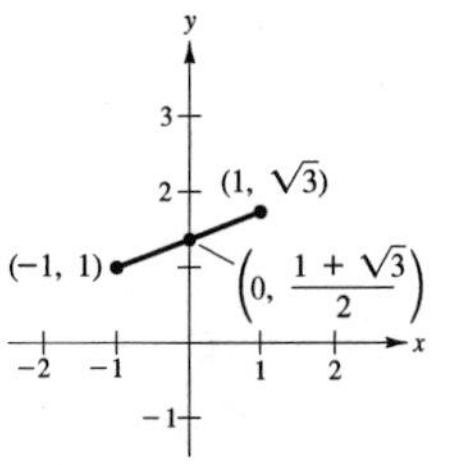

(b) $8\sqrt{8 - 2\sqrt{3}}$

(c) $\left(0, \frac{1 \pm \sqrt{3}}{2}\right)$

7. Quadrant II **9.** Quadrants I and III

11. Right triangle:
$d_1 = \sqrt{45}, d_2 = \sqrt{5}$
$d_3 = \sqrt{50}$
$(d_1)^2 + (d_2)^2 = (d_3)^2$

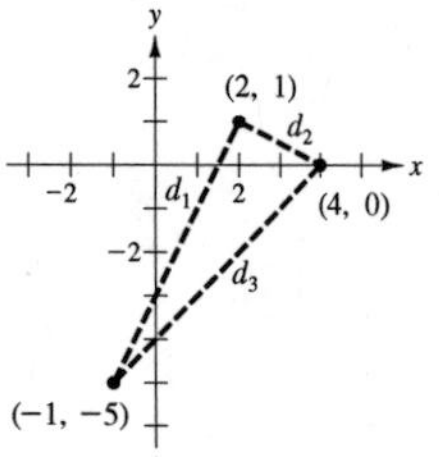

13. Rhombus: the length of each side is $\sqrt{5}$.

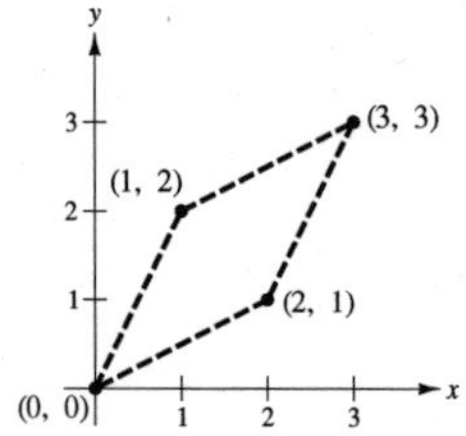

15.

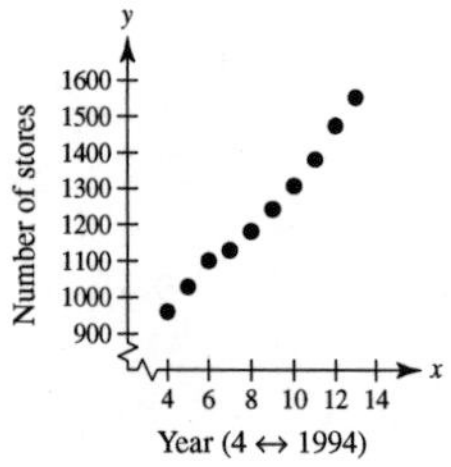

17. $d_1 = 2\sqrt{5}, d_2 = \sqrt{5}, d_3 = 3\sqrt{5}$
Collinear, because $d_1 + d_2 = d_3$.

19. $d_1 = \sqrt{2}, d_2 = \sqrt{13}, d_3 = 5$
Not collinear, because $d_1 + d_2 > d_3$.

21. $x = \pm 3$ **23.** $y = \pm\sqrt{55}$

25. $\left(\frac{3x_1 + x_2}{4}, \frac{3y_1 + y_2}{4}\right)$ $\left(\frac{x_1 + x_2}{2}, \frac{y_1 + y_2}{2}\right)$
$\left(\frac{x_1 + 3x_2}{4}, \frac{y_1 + 3y_2}{4}\right)$

27. c **28.** b **29.** a **30.** d **31.** $x^2 + y^2 - 9 = 0$

33. $x^2 + y^2 - 4x + 2y - 11 = 0$

35. $x^2 + y^2 + 2x - 4y = 0$

37. $x^2 + y^2 - 6x - 4y + 3 = 0$ **39.** $x^2 + y^2 = 26{,}000^2$

41. $(x - 1)^2 + (y + 3)^2 = 4$ **43.** $(x - 1)^2 + (y + 3)^2 = 0$

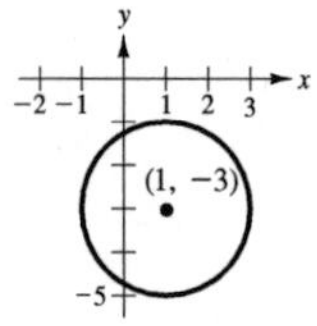

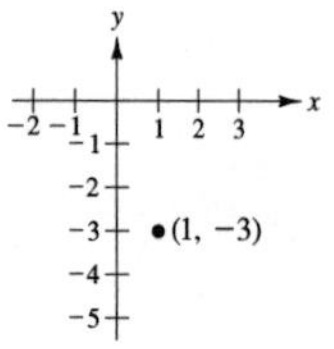

45. $\left(x - \frac{1}{2}\right)^2 + \left(y - \frac{1}{2}\right)^2 = 2$ **47.** $\left(x + \frac{1}{2}\right)^2 + \left(y + \frac{5}{4}\right)^2 = \frac{9}{4}$

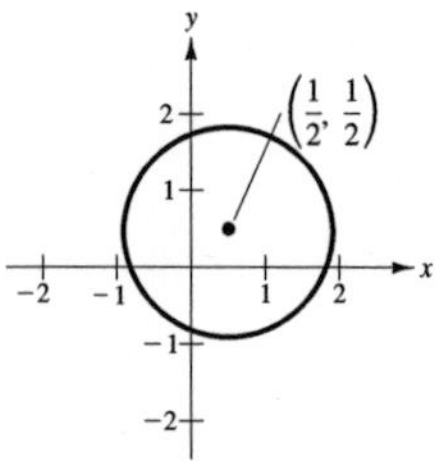

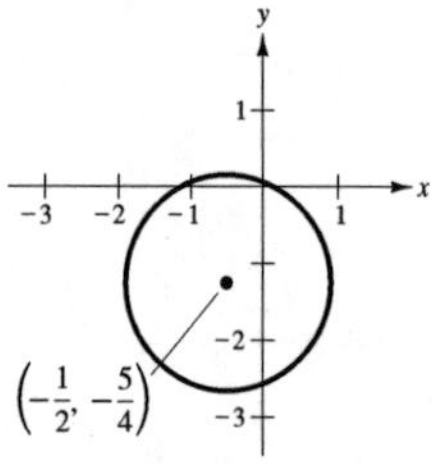

49.

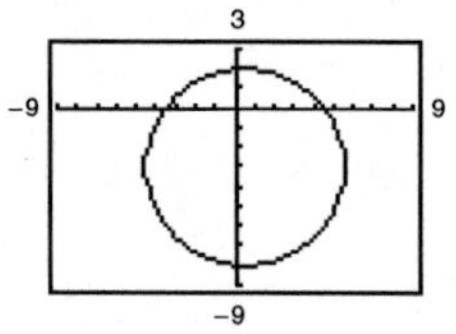

51.

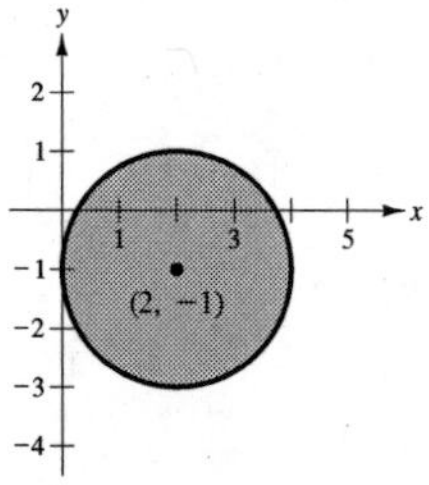

53. Proof **55.** True **56.** False; the distance is $|2b|$.
57. True **58.** True **59.** Proof **61.** Proof

Appendix D.3 *(page D25)*

1. (a) 396°, −324° (b) 240°, −480°

3. (a) $\frac{19\pi}{9}, -\frac{17\pi}{9}$ (b) $\frac{10\pi}{3}, -\frac{2\pi}{3}$

5. (a) $\frac{\pi}{6}$, 0.524 (b) $\frac{5\pi}{6}$, 2.618
(c) $\frac{7\pi}{4}$, 5.498 (d) $\frac{2\pi}{3}$, 2.094

7. (a) 270° (b) 210° (c) −105° (d) −135.6°

9.

r	8 ft	15 in.	85 cm	24 in.	$\frac{12{,}963}{\pi}$ mi
s	12 ft	24 in.	63.75π cm	96 in.	8642 mi
θ	1.5	1.6	$\frac{3\pi}{4}$	4	$\frac{2\pi}{3}$

11. (a) $\sin\theta = \frac{4}{5}$ $\csc\theta = \frac{5}{4}$
$\cos\theta = \frac{3}{5}$ $\sec\theta = \frac{5}{3}$
$\tan\theta = \frac{4}{3}$ $\cot\theta = \frac{3}{4}$
(b) $\sin\theta = -\frac{5}{13}$ $\csc\theta = -\frac{13}{5}$
$\cos\theta = -\frac{12}{13}$ $\sec\theta = -\frac{13}{12}$
$\tan\theta = \frac{5}{12}$ $\cot\theta = \frac{12}{5}$

13. (a) Quadrant III (b) Quadrant IV

15. $\frac{\sqrt{3}}{2}$ **17.** $\frac{4}{3}$

19. (a) $\sin 60° = \frac{\sqrt{3}}{2}$
$\cos 60° = \frac{1}{2}$
$\tan 60° = \sqrt{3}$
(b) $\sin 120° = \frac{\sqrt{3}}{2}$
$\cos 120° = -\frac{1}{2}$
$\tan 120° = -\sqrt{3}$
(c) $\sin\frac{\pi}{4} = \frac{\sqrt{2}}{2}$
$\cos\frac{\pi}{4} = \frac{\sqrt{2}}{2}$
$\tan\frac{\pi}{4} = 1$
(d) $\sin\frac{5\pi}{4} = -\frac{\sqrt{2}}{2}$
$\cos\frac{5\pi}{4} = -\frac{\sqrt{2}}{2}$
$\tan\frac{5\pi}{4} = 1$

21. (a) $\sin 225° = -\dfrac{\sqrt{2}}{2}$ (b) $\sin(-225°) = \dfrac{\sqrt{2}}{2}$

$\cos 225° = -\dfrac{\sqrt{2}}{2}$ $\cos(-225°) = -\dfrac{\sqrt{2}}{2}$

$\tan 225° = 1$ $\tan(-225°) = -1$

(c) $\sin\dfrac{5\pi}{3} = -\dfrac{\sqrt{3}}{2}$ (d) $\sin\dfrac{11\pi}{6} = -\dfrac{1}{2}$

$\cos\dfrac{5\pi}{3} = \dfrac{1}{2}$ $\cos\dfrac{11\pi}{6} = \dfrac{\sqrt{3}}{2}$

$\tan\dfrac{5\pi}{3} = -\sqrt{3}$ $\tan\dfrac{11\pi}{6} = -\dfrac{\sqrt{3}}{3}$

23. (a) 0.1736 (b) 5.759 **25.** (a) 0.3640 (b) 0.3640

27. (a) $\theta = \dfrac{\pi}{4}, \dfrac{7\pi}{4}$ (b) $\theta = \dfrac{3\pi}{4}, \dfrac{5\pi}{4}$

29. (a) $\theta = \dfrac{\pi}{4}, \dfrac{5\pi}{4}$ (b) $\theta = \dfrac{5\pi}{6}, \dfrac{11\pi}{6}$

31. $\theta = \dfrac{\pi}{4}, \dfrac{3\pi}{4}, \dfrac{5\pi}{4}, \dfrac{7\pi}{4}$ **33.** $\theta = 0, \dfrac{\pi}{4}, \pi, \dfrac{5\pi}{4}$

35. $\theta = \dfrac{\pi}{3}, \dfrac{5\pi}{3}$ **37.** $\theta = 0, \dfrac{\pi}{2}, \pi$ **39.** 5099 feet

41. (a) Period: π (b) Period: 2 **43.** Period: $\frac{1}{2}$

Amplitude: 2 Amplitude: $\frac{1}{2}$ Amplitude: 3

45. Period: $\dfrac{\pi}{2}$ **47.** Period: $\dfrac{2\pi}{5}$

49. (a)

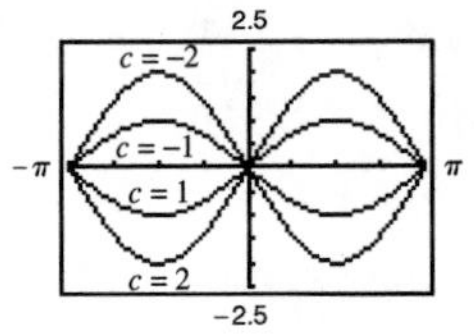

Change in amplitude

(b)

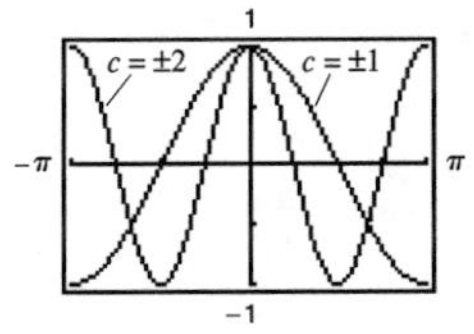

Change in period

(c)

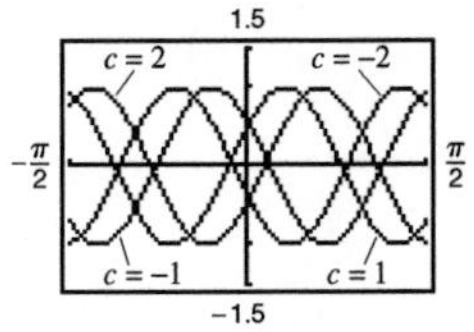

Horizontal translation

51.

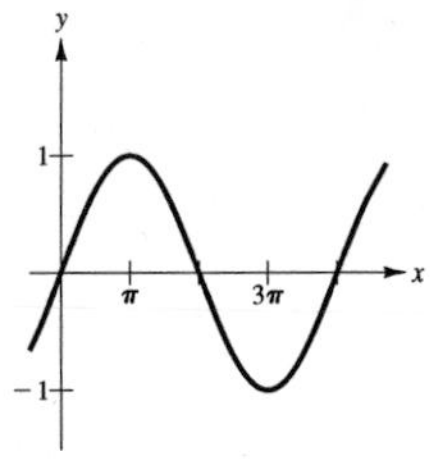

53.

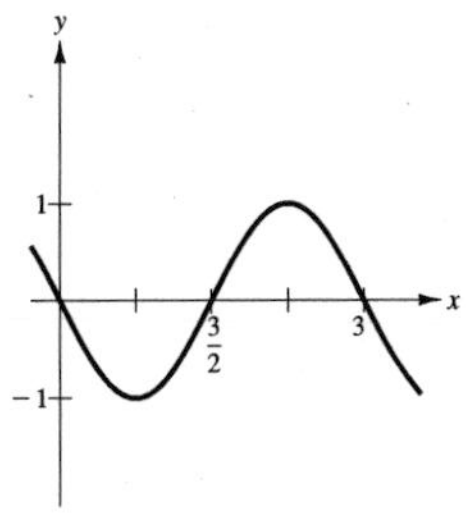

55.

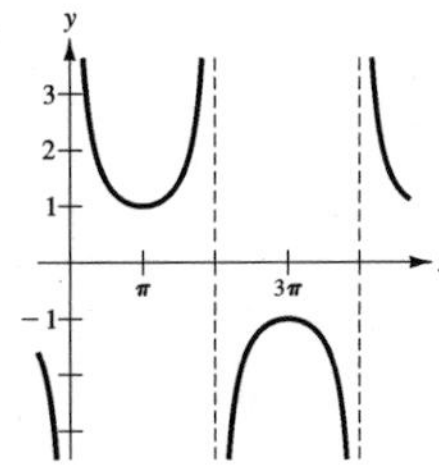

57.

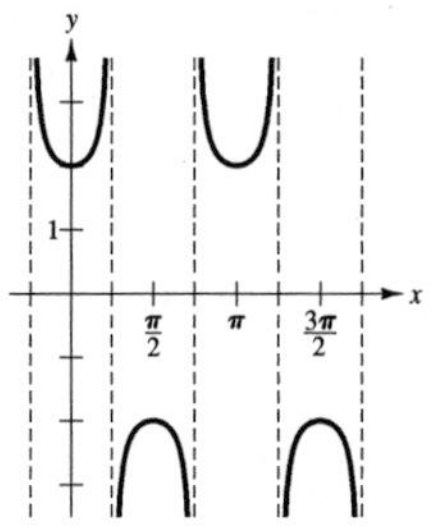

59.

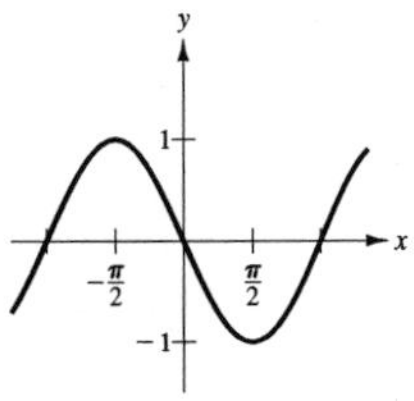

61.

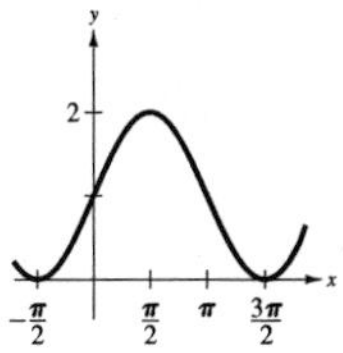

63. $a = 3, b = \dfrac{1}{2}, c = \dfrac{\pi}{2}$

65.

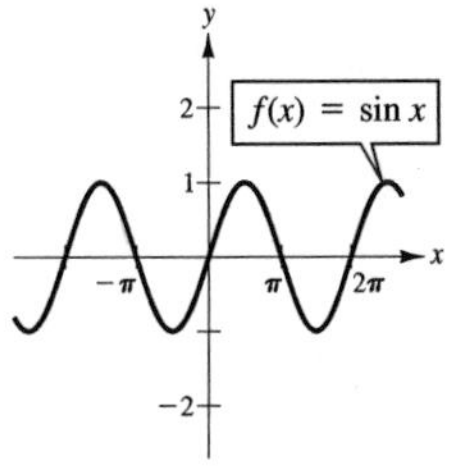

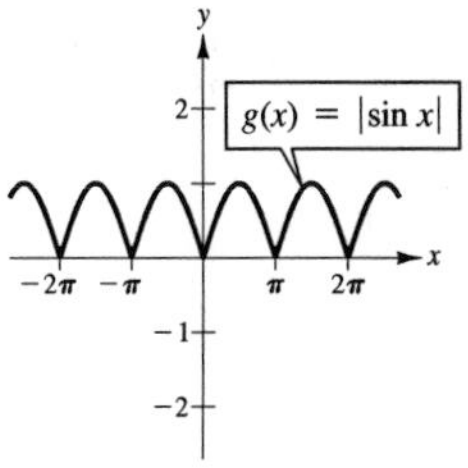

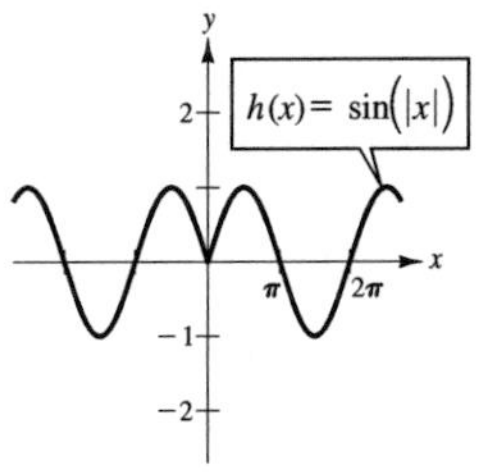

The graph of $|f(x)|$ will reflect any parts of the graph of $f(x)$ below the x-axis about the x-axis. The graph of $f(|x|)$ will reflect the part of the graph of $f(x)$ left of the y-axis about the x-axis.

67.

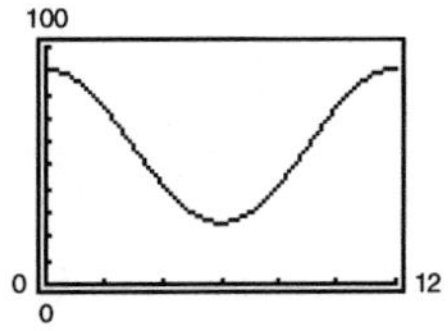

January, November, December

69. $f(x) = \dfrac{4}{\pi}\left(\sin \pi x + \dfrac{1}{3}\sin 3\pi x + \dfrac{1}{5}\sin 5\pi x + \cdots\right)$

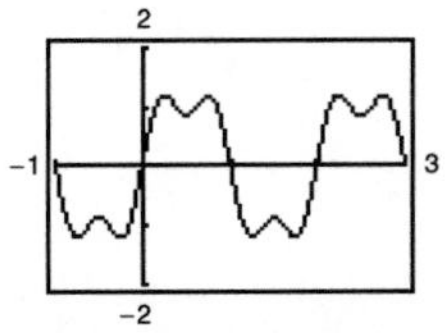

APPENDIX E *(page E6)*

1. $\frac{(y')^2}{2} - \frac{(x')^2}{2} = 1$

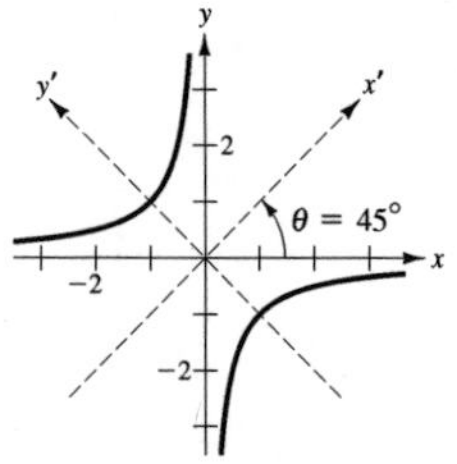

3. $\frac{(x')^2}{1/4} - \frac{(y')^2}{1/6} = 1$

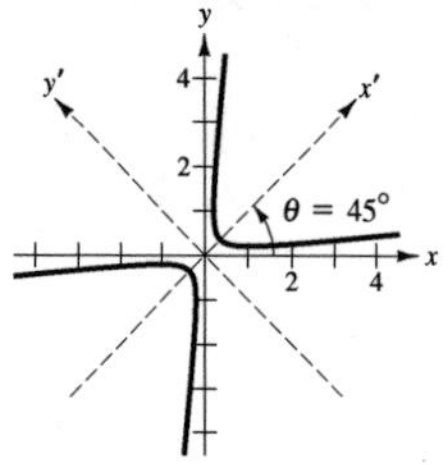

5. $\frac{(x' - 3\sqrt{2})^2}{16} - \frac{(y' - \sqrt{2})^2}{16} = 1$

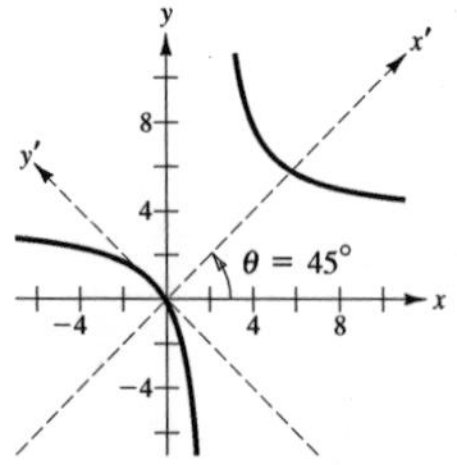

7. $\frac{(x')^2}{3} + \frac{(y')^2}{2} = 1$

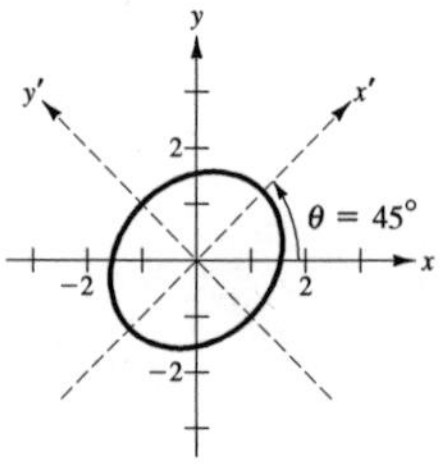

9. $x' = -(y')^2$

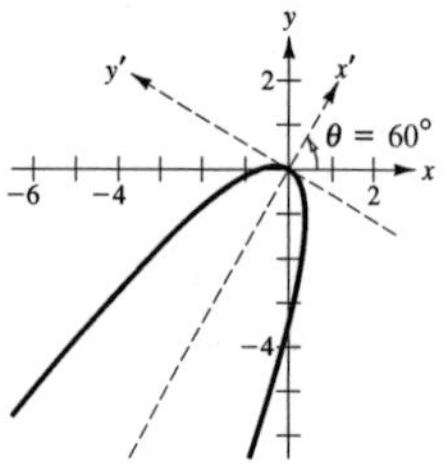

11. $y' = \frac{(x')^2}{6} - \frac{x'}{3}$

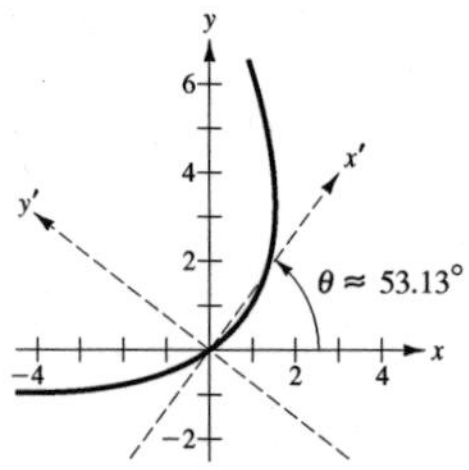

13. $\theta = 45°$

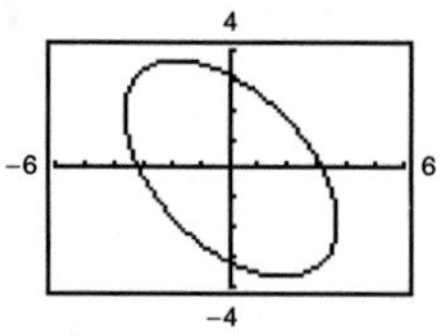

15. $\theta \approx 26.57°$

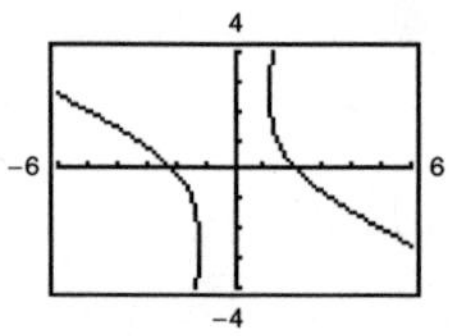

17. $\theta \approx 31.72°$

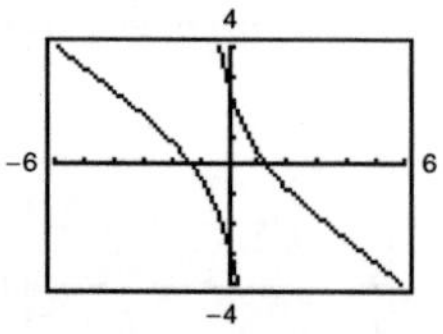

19. Parabola **21.** Ellipse **23.** Hyperbola **25.** Parabola

27. Two lines

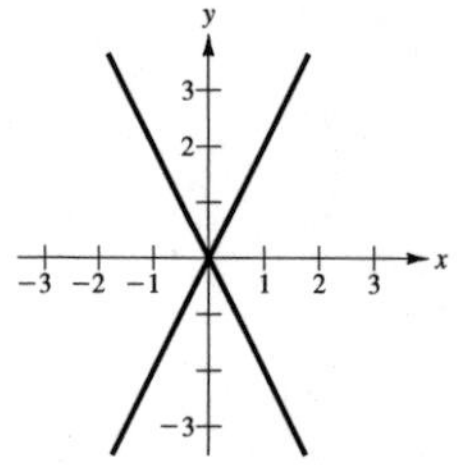

29. Two parallel lines

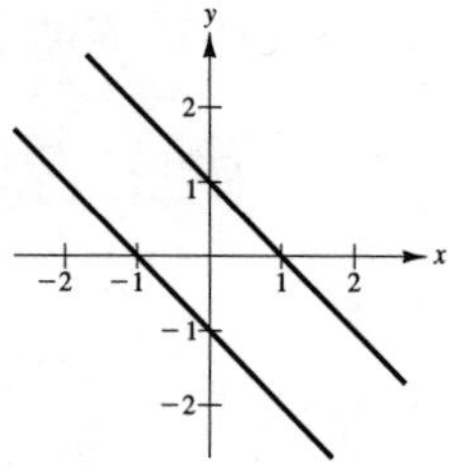

31. Two lines

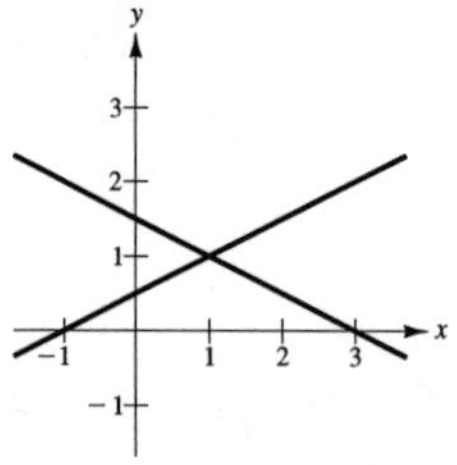

33. Proof

APPENDIX F *(page F10)*

1. $11 - i$ **3.** 4 **5.** $3 - 3\sqrt{2}i$ **7.** $-14 + 20i$

9. $\frac{1}{6} + \frac{7}{6}i$ **11.** $-2\sqrt{3}$ **13.** -10 **15.** $5 + i$

17. $12 + 30i$ **19.** 24 **21.** $-9 + 40i$ **23.** -10

25. $5 - 3i$; 34 **27.** $-2 + \sqrt{5}i$; 9 **29.** $-20i$; 400

31. $\sqrt{8}$; 8 **33.** $-6i$ **35.** $\frac{16}{41} + \frac{20}{41}i$ **37.** $\frac{3}{5} + \frac{4}{5}i$

39. $-7 - 6i$ **41.** $-\frac{9}{1681} + \frac{40}{1681}i$ **43.** $-\frac{1}{2} - \frac{5}{2}i$

45. $\frac{62}{949} + \frac{297}{949}i$ **47.** $1 \pm i$ **49.** $-2 \pm \frac{1}{2}i$ **51.** $-\frac{5}{2}, -\frac{3}{2}$

53. $\frac{1}{8} \pm \frac{\sqrt{11}}{8}i$ **55.** $-1 + 6i$ **57.** $-5i$

59. $-375\sqrt{3}i$ **61.** i

63.

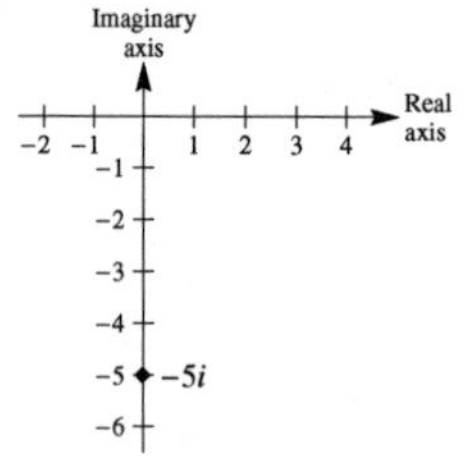

5

65.

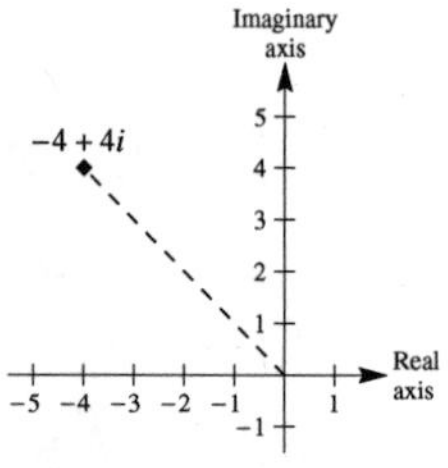

$4\sqrt{2}$

67.

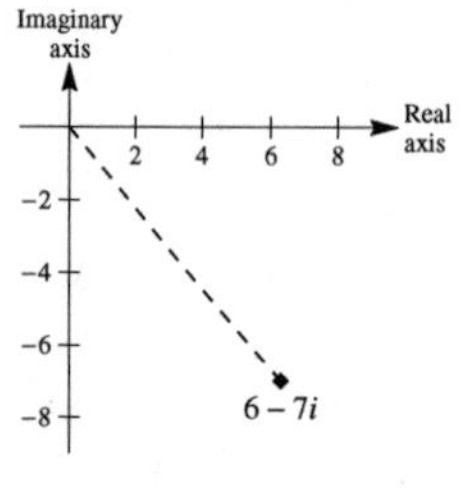

$\sqrt{85}$

69.

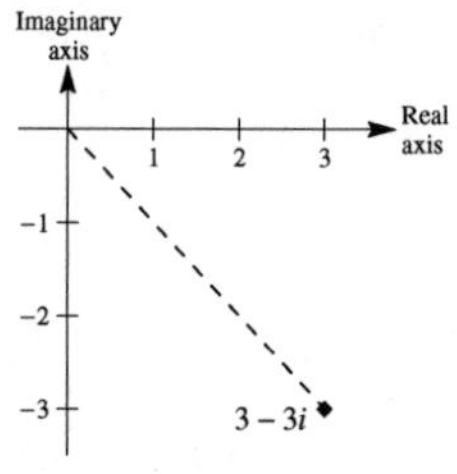

$3\sqrt{2}\left(\cos\frac{7\pi}{4} + i\sin\frac{7\pi}{4}\right)$

71.

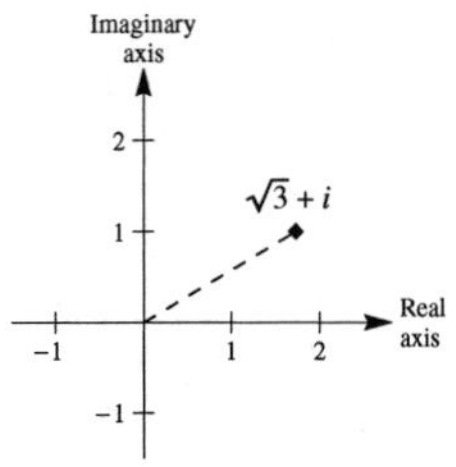

$2\left(\cos\frac{\pi}{6} + i\sin\frac{\pi}{6}\right)$

73.

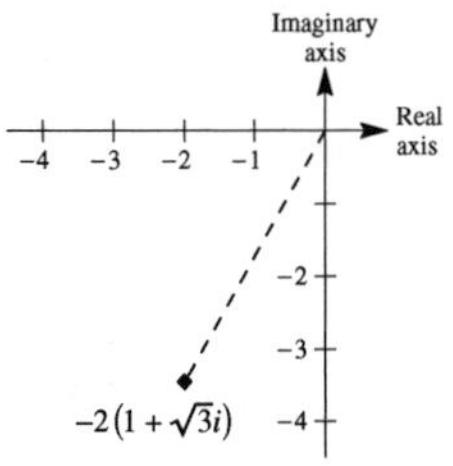

$4\left(\cos\frac{4\pi}{3} + i\sin\frac{4\pi}{3}\right)$

75.

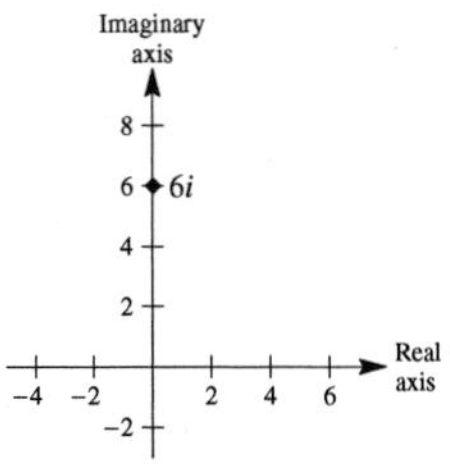

$6\left(\cos\frac{\pi}{2} + i\sin\frac{\pi}{2}\right)$

77.

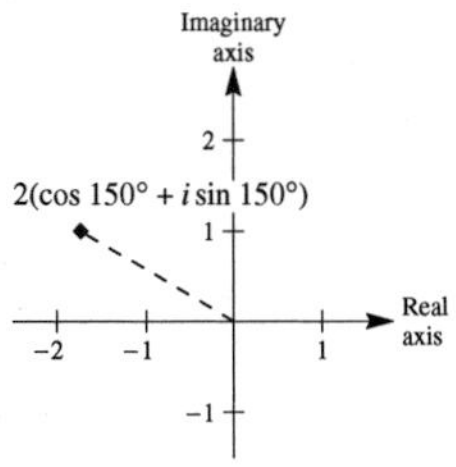

$-\sqrt{3} + i$

79.

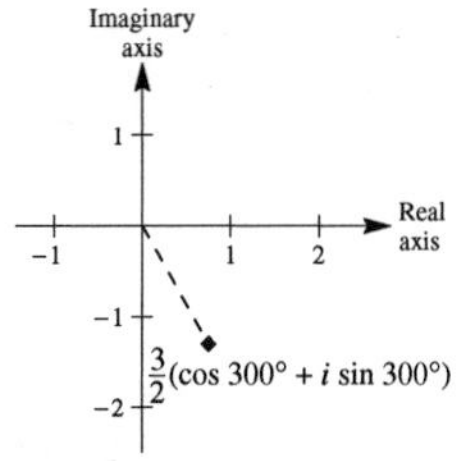

$\frac{3}{4} - \frac{3\sqrt{3}}{4}i$

81.

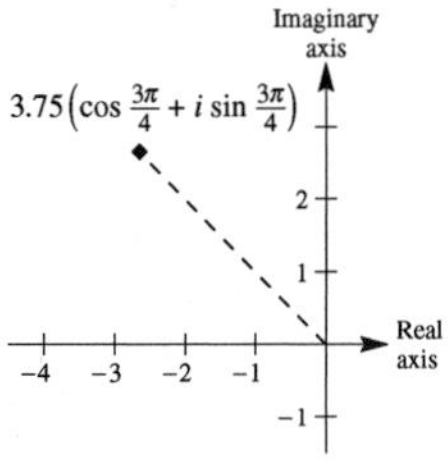

$\frac{-15\sqrt{2}}{8} + \frac{15\sqrt{2}}{8}i$

83. $12\left(\cos\frac{\pi}{2} + i\sin\frac{\pi}{2}\right)$ **85.** $\frac{10}{9}(\cos 200° + i\sin 200°)$

87. $-4 - 4i$ **89.** $-32i$ **91.** $-128\sqrt{3} - 128i$ **93.** i

95. (a) $\sqrt{5}(\cos 60° + i\sin 60°)$
$\sqrt{5}(\cos 240° + i\sin 240°)$

(b)

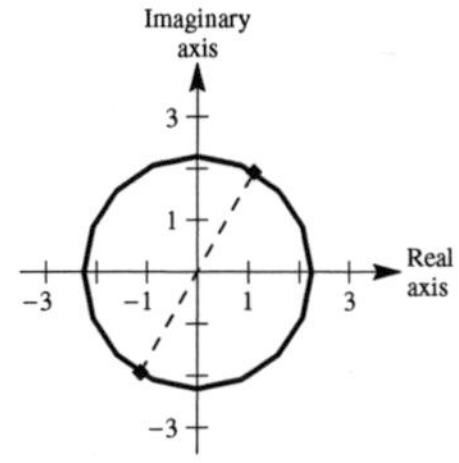

(c) $\frac{\sqrt{5}}{2} + \frac{\sqrt{15}}{2}i, -\frac{\sqrt{5}}{2} - \frac{\sqrt{15}}{2}i$

97. (a) $2\left(\cos\frac{\pi}{3} + i\sin\frac{\pi}{3}\right)$

$2\left(\cos\frac{5\pi}{6} + i\sin\frac{5\pi}{6}\right)$

$2\left(\cos\frac{4\pi}{3} + i\sin\frac{4\pi}{3}\right)$

$2\left(\cos\frac{11\pi}{6} + i\sin\frac{11\pi}{6}\right)$

(b)

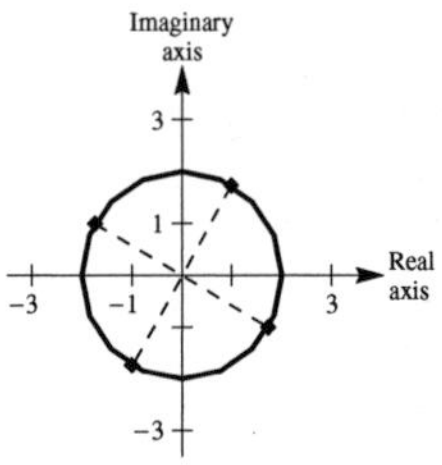

(c) $1 + \sqrt{3}i, -\sqrt{3} + i, -1 - \sqrt{3}i, \sqrt{3} - i$

99. (a) $5\left(\cos\frac{4\pi}{9} + i\sin\frac{4\pi}{9}\right)$
$5\left(\cos\frac{10\pi}{9} + i\sin\frac{10\pi}{9}\right)$
$5\left(\cos\frac{16\pi}{9} + i\sin\frac{16\pi}{9}\right)$

(b)

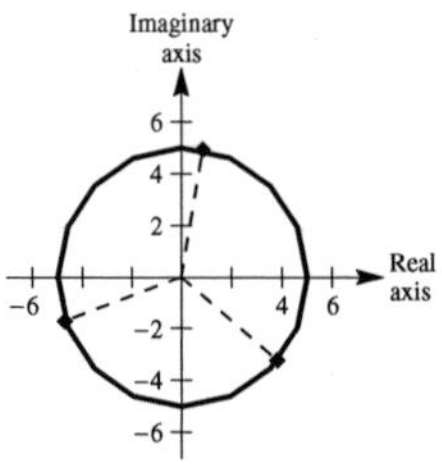

101. $\cos\frac{\pi}{8} + i\sin\frac{\pi}{8}$
$\cos\frac{5\pi}{8} + i\sin\frac{5\pi}{8}$
$\cos\frac{9\pi}{8} + i\sin\frac{9\pi}{8}$
$\cos\frac{13\pi}{8} + i\sin\frac{13\pi}{8}$

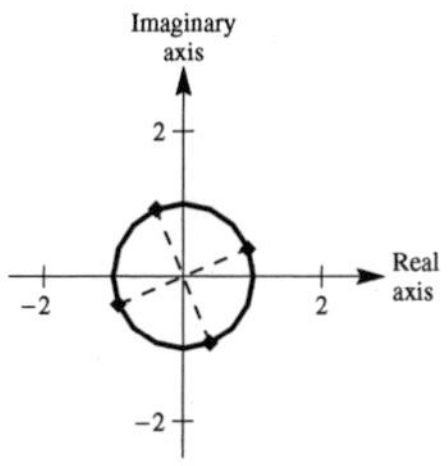

103. $3\left(\cos\frac{\pi}{5} + i\sin\frac{\pi}{5}\right)$
$3\left(\cos\frac{3\pi}{5} + i\sin\frac{3\pi}{5}\right)$
$3(\cos\pi + i\sin\pi)$
$3\left(\cos\frac{7\pi}{5} + i\sin\frac{7\pi}{5}\right)$
$3\left(\cos\frac{9\pi}{5} + i\sin\frac{9\pi}{5}\right)$

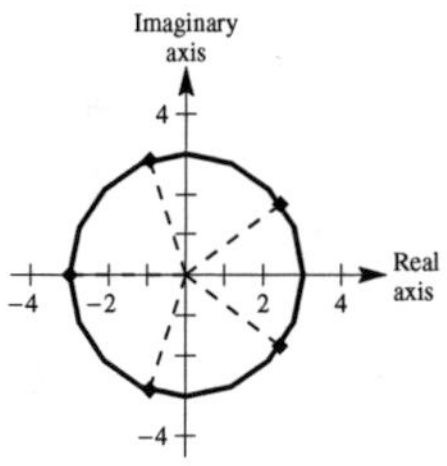

105. $4\left(\cos\frac{\pi}{2} + i\sin\frac{\pi}{2}\right)$
$4\left(\cos\frac{7\pi}{6} + i\sin\frac{7\pi}{6}\right)$
$4\left(\cos\frac{11\pi}{6} + i\sin\frac{11\pi}{6}\right)$

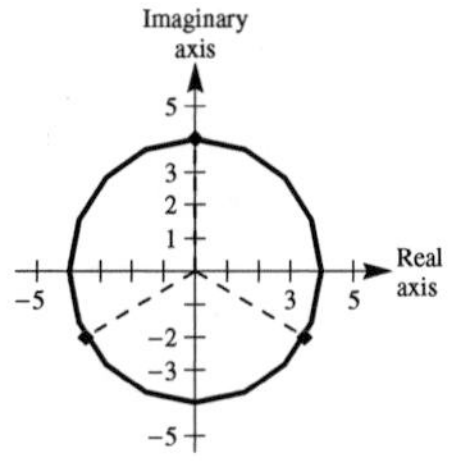

107. $\sqrt[6]{2}(\cos 105° + i\sin 105°)$
$\sqrt[6]{2}(\cos 225° + i\sin 225°)$
$\sqrt[6]{2}(\cos 345° + i\sin 345°)$

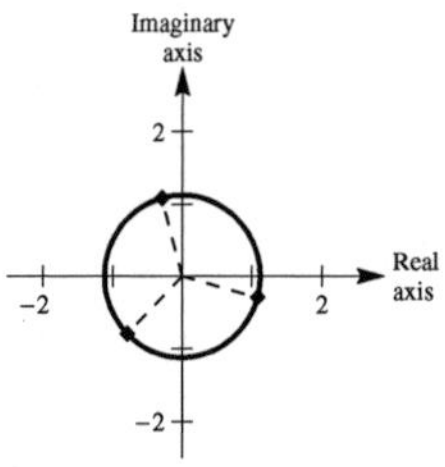

APPENDIX G *(page G5)*

1. (a) Fixed cost

(b)

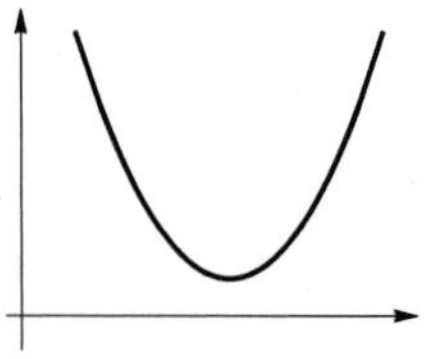

(c) Yes, it occurs when production costs are increasing at their slowest rate.

3. 4500 **5.** 300 **7.** 200 **9.** 200

11. \$60 **13.** \$35 **15.** $x = 3$ **17.** Proof

19. (a)

Order size, x	*Price*	*Profit, P*
102	$90 - 2(0.15)$	$102[90 - 2(0.15)] - 102(60) = 3029.40$
104	$90 - 4(0.15)$	$104[90 - 4(0.15)] - 104(60) = 3057.60$
106	$90 - 6(0.15)$	$106[90 - 6(0.15)] - 106(60) = 3084.60$
108	$90 - 8(0.15)$	$108[90 - 8(0.15)] - 108(60) = 3110.40$
110	$90 - 10(0.15)$	$110[90 - 10(0.15)] - 110(60) = 3135.00$
112	$90 - 12(0.15)$	$112[90 - 12(0.15)] - 112(60) = 3158.40$

(b)

Order size, x	*Price*	*Profit, P*
⋮	⋮	⋮
146	$90 - 46(0.15)$	$146[90 - 46(0.15)] - 146(60) = 3372.60$
148	$90 - 48(0.15)$	$148[90 - 48(0.15)] - 148(60) = 3374.40$
150	$90 - 50(0.15)$	$150[90 - 50(0.15)] - 150(60) = 3375.00$
152	$90 - 52(0.15)$	$152[90 - 52(0.15)] - 152(60) = 3374.40$
154	$90 - 54(0.15)$	$154[90 - 54(0.15)] - 154(60) = 3372.60$
⋮	⋮	⋮

Maximum profit: \$3375.00

(c) $P = x[90 - (x - 100)(0.15)] - x(60) = 45x - 0.15x^2,$

$x \geq 100$

(d) 150 units

(e)

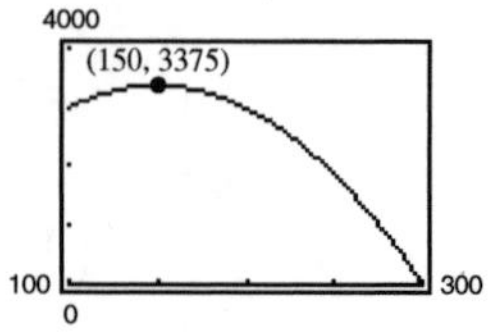

21. Line should run from the power station to a point across the river $3/(2\sqrt{7})$ mile downstream.

23. $x \approx 40$ units **25.** \$30,000

27. (a)

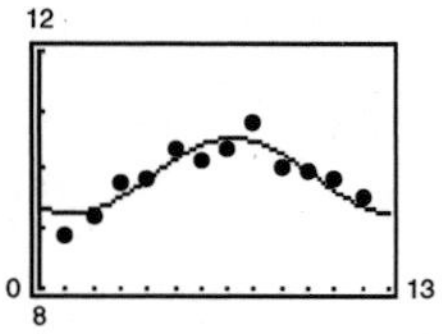

(b) July

(c) The cosine factor; 9.90

(d) $0.02t$ would mean a steady growth of sales over time. In this case, the maximum sales in 2008 (that is, on $49 \leq t \leq 60$) would be about 11.6 thousand gallons.

29. (a)

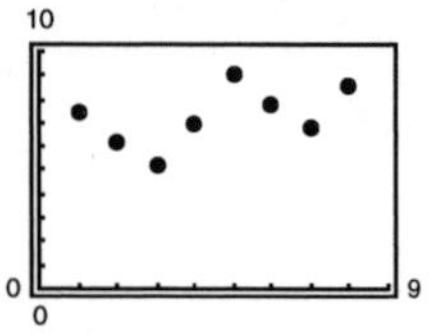

(b) $y = 6.2 + 0.25x + 1.5 \sin\left(\frac{\pi}{2}x\right)$

(c)

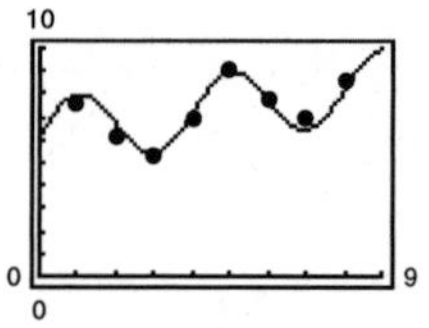

(d) \$12,000

31. $\eta = -\frac{17}{3}$, elastic **33.** $\eta = -\frac{1}{2}$, inelastic

ALGEBRA

Factors and Zeros of Polynomials

Let $p(x) = a_n x^n + a_{n-1}x^{n-1} + \cdot \cdot \cdot + a_1 x + a_0$ be a polynomial. If $p(a) = 0$, then a is a *zero* of the polynomial and a solution of the equation $p(x) = 0$. Furthermore, $(x - a)$ is a *factor* of the polynomial.

Fundamental Theorem of Algebra

An nth degree polynomial has n (not necessarily distinct) zeros. Although all of these zeros may be imaginary, a real polynomial of odd degree must have at least one real zero.

Quadratic Formula

If $p(x) = ax^2 + bx + c$, and $0 \le b^2 - 4ac$, then the real zeros of p are $x = \left(-b \pm \sqrt{b^2 - 4ac}\right)/2a$.

Special Factors

$x^2 - a^2 = (x - a)(x + a)$

$x^3 - a^3 = (x - a)(x^2 + ax + a^2)$

$x^3 + a^3 = (x + a)(x^2 - ax + a^2)$

$x^4 - a^4 = (x^2 - a^2)(x^2 + a^2)$

Binomial Theorem

$(x + y)^2 = x^2 + 2xy + y^2$

$(x - y)^2 = x^2 - 2xy + y^2$

$(x + y)^3 = x^3 + 3x^2y + 3xy^2 + y^3$

$(x - y)^3 = x^3 - 3x^2y + 3xy^2 - y^3$

$(x + y)^4 = x^4 + 4x^3y + 6x^2y^2 + 4xy^3 + y^4$

$(x - y)^4 = x^4 - 4x^3y + 6x^2y^2 - 4xy^3 + y^4$

$$(x + y)^n = x^n + nx^{n-1}y + \frac{n(n-1)}{2!}x^{n-2}y^2 + \cdot \cdot \cdot + nxy^{n-1} + y^n$$

$$(x - y)^n = x^n - nx^{n-1}y + \frac{n(n-1)}{2!}x^{n-2}y^2 - \cdot \cdot \cdot \pm nxy^{n-1} \mp y^n$$

Rational Zero Theorem

If $p(x) = a_n x^n + a_{n-1}x^{n-1} + \cdot \cdot \cdot + a_1 x + a_0$ has integer coefficients, then every *rational zero* of p is of the form $x = r/s$, where r is a factor of a_0 and s is a factor of a_n.

Factoring by Grouping

$acx^3 + adx^2 + bcx + bd = ax^2(cx + d) + b(cx + d) = (ax^2 + b)(cx + d)$

Arithmetic Operations

$ab + ac = a(b + c)$

$\dfrac{a}{b} + \dfrac{c}{d} = \dfrac{ad + bc}{bd}$

$\dfrac{a + b}{c} = \dfrac{a}{c} + \dfrac{b}{c}$

$\dfrac{\left(\frac{a}{b}\right)}{\left(\frac{c}{d}\right)} = \left(\dfrac{a}{b}\right)\left(\dfrac{d}{c}\right) = \dfrac{ad}{bc}$

$\dfrac{\left(\frac{a}{b}\right)}{c} = \dfrac{a}{bc}$

$\dfrac{a}{\left(\frac{b}{c}\right)} = \dfrac{ac}{b}$

$a\left(\dfrac{b}{c}\right) = \dfrac{ab}{c}$

$\dfrac{a - b}{c - d} = \dfrac{b - a}{d - c}$

$\dfrac{ab + ac}{a} = b + c$

Exponents and Radicals

$a^0 = 1, \quad a \neq 0$

$(ab)^x = a^x b^x$

$a^x a^y = a^{x+y}$

$\sqrt{a} = a^{1/2}$

$\dfrac{a^x}{a^y} = a^{x-y}$

$\sqrt[n]{a} = a^{1/n}$

$\left(\dfrac{a}{b}\right)^x = \dfrac{a^x}{b^x}$

$\sqrt[n]{a^m} = a^{m/n}$

$a^{-x} = \dfrac{1}{a^x}$

$\sqrt[n]{ab} = \sqrt[n]{a}\sqrt[n]{b}$

$(a^x)^y = a^{xy}$

$\sqrt[n]{\dfrac{a}{b}} = \dfrac{\sqrt[n]{a}}{\sqrt[n]{b}}$

FORMULAS FROM GEOMETRY

Triangle

$h = a \sin \theta$

$\text{Area} = \frac{1}{2}bh$

(Law of Cosines)

$c^2 = a^2 + b^2 - 2ab \cos \theta$

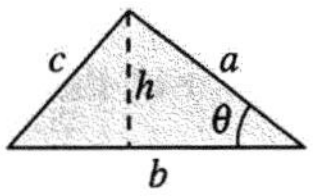

Right Triangle

(Pythagorean Theorem)

$c^2 = a^2 + b^2$

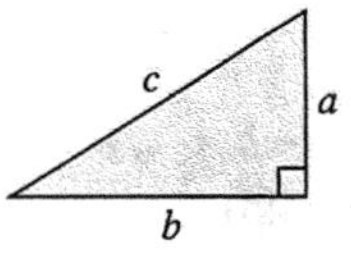

Equilateral Triangle

$h = \frac{\sqrt{3}s}{2}$

$\text{Area} = \frac{\sqrt{3}s^2}{4}$

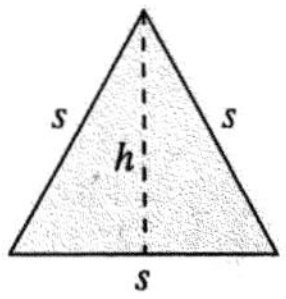

Parallelogram

$\text{Area} = bh$

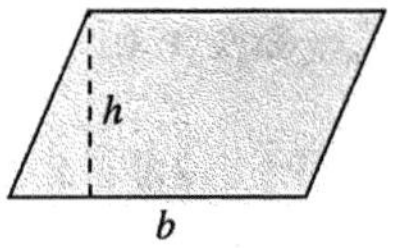

Trapezoid

$\text{Area} = \frac{h}{2}(a + b)$

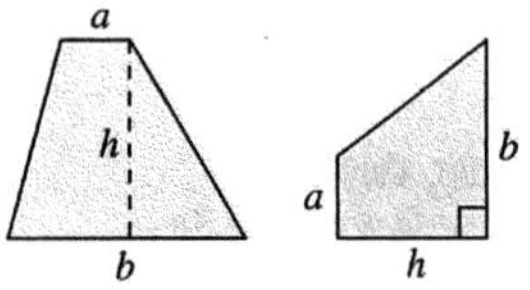

Circle

$\text{Area} = \pi r^2$

$\text{Circumference} = 2\pi r$

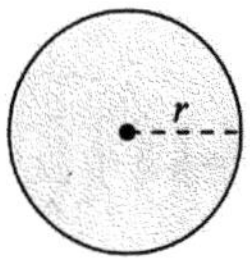

Sector of Circle

(θ in radians)

$\text{Area} = \frac{\theta r^2}{2}$

$s = r\theta$

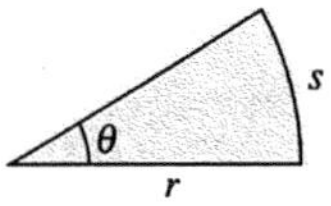

Circular Ring

(p = average radius,
w = width of ring)

$\text{Area} = \pi(R^2 - r^2)$
$= 2\pi pw$

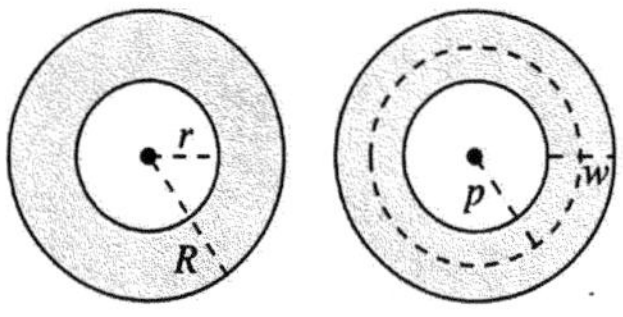

Sector of Circular Ring

(p = average radius,
w = width of ring,
θ in radians)

$\text{Area} = \theta pw$

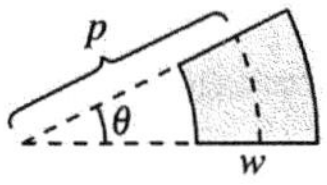

Ellipse

$\text{Area} = \pi ab$

$\text{Circumference} \approx 2\pi\sqrt{\frac{a^2 + b^2}{2}}$

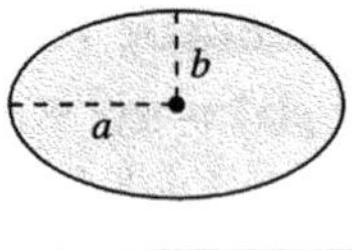

Cone

(A = area of base)

$\text{Volume} = \frac{Ah}{3}$

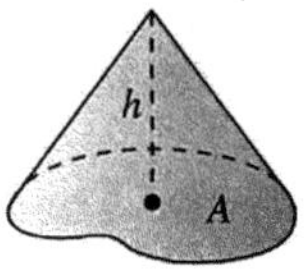

Right Circular Cone

$\text{Volume} = \frac{\pi r^2 h}{3}$

$\text{Lateral Surface Area} = \pi r\sqrt{r^2 + h^2}$

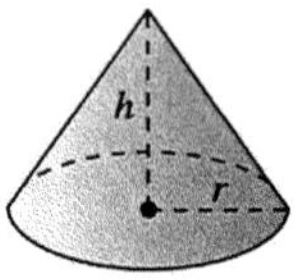

Frustum of Right Circular Cone

$\text{Volume} = \frac{\pi(r^2 + rR + R^2)h}{3}$

$\text{Lateral Surface Area} = \pi s(R + r)$

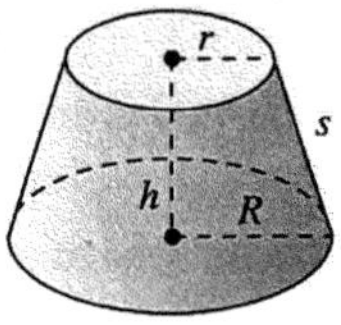

Right Circular Cylinder

$\text{Volume} = \pi r^2 h$

$\text{Lateral Surface Area} = 2\pi rh$

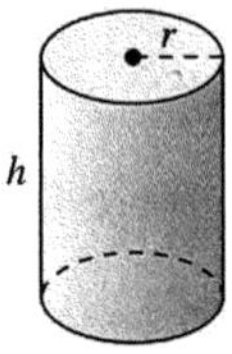

Sphere

$\text{Volume} = \frac{4}{3}\pi r^3$

$\text{Surface Area} = 4\pi r^2$

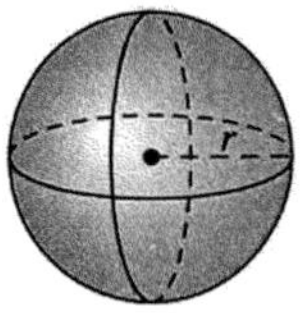

Wedge

(A = area of upper face,
B = area of base)

$A = B \sec \theta$

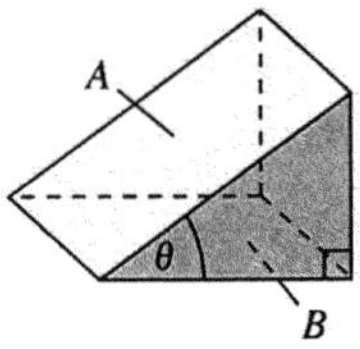